Geographic Subdivisions of California

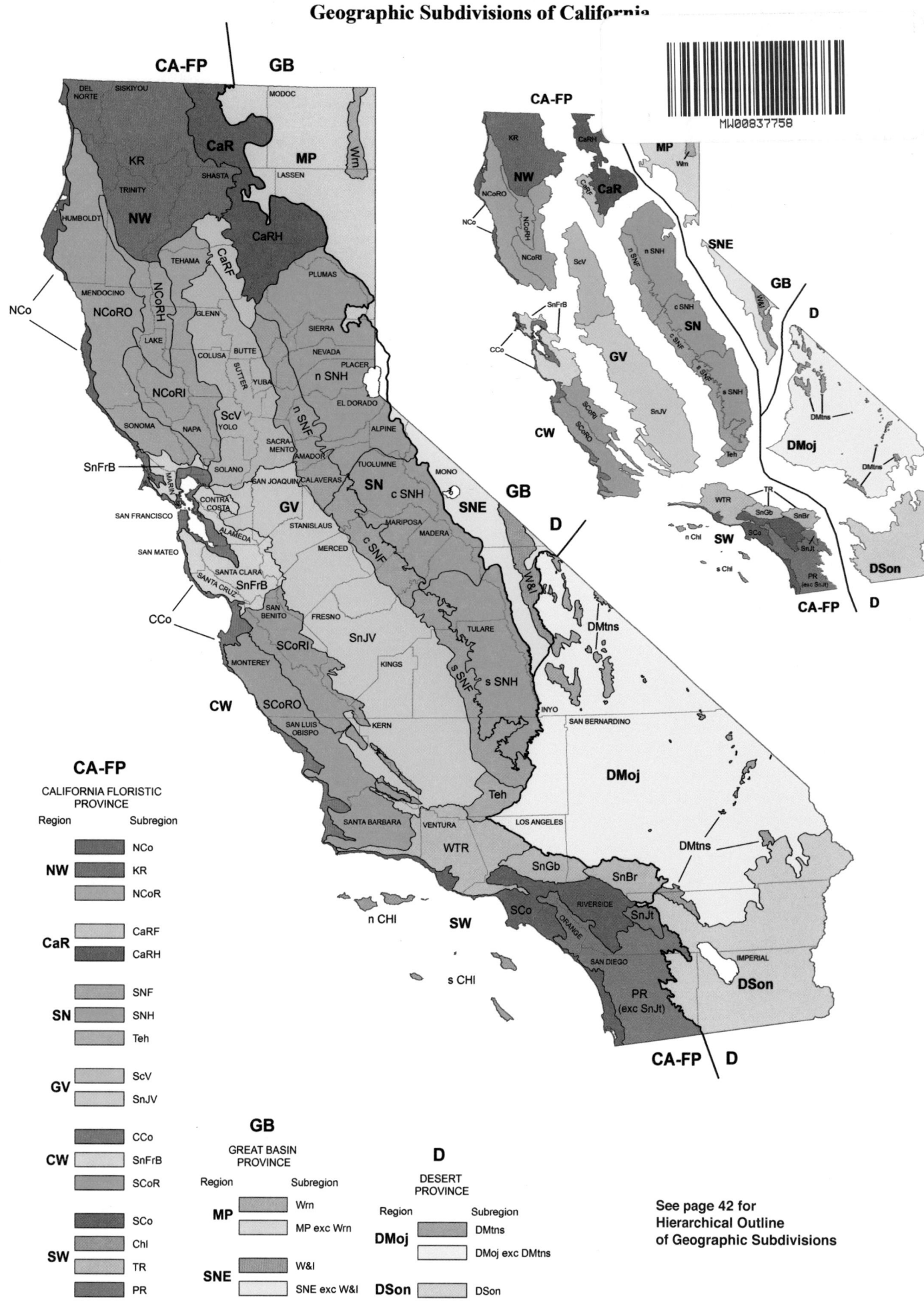

See page 42 for
Hierarchical Outline
of Geographic Subdivisions

MW00837758

CA-FP
CALIFORNIA FLORISTIC
PROVINCE

Region	Subregion
NW	NCo
	KR
	NCoR
CaR	CaRF
	CaRH
SN	SNF
	SNH
	Teh
GV	ScV
	SnJV
CW	CCo
	SnFrB
	SCoR
SW	SCo
	ChI
	TR
	PR

GB
GREAT BASIN
PROVINCE

Region	Subregion
MP	Wrn
	MP exc Wrn
SNE	W&I
	SNE exc W&I

D
DESERT
PROVINCE

Region	Subregion
DMoj	DMtns
	DMoj exc DMtns
DSon	DSon

The Jepson Manual

SECOND EDITION

SUPPORT FOR THE JEPSON FLORA PROJECT

2003–2010

*We are grateful to all listed here for their support of the Jepson Flora Project;
without them this book would not have been possible.*

*We also gratefully acknowledge contributions from the following endowment funds of
the University and Jepson Herbaria: Rimo Bacigalupi, Mary L. Bowerman,
Lawrence R. Heckard, Willis Linn Jepson, and Robert Ornduff.*

$100,000 and above

California Digital Library

University of California Press

$50,000 to $99,999

Bureau of Land Management

Global Biodiversity Information Facility

Institute of Museum and Library Services

Resource Legacy Fund Foundation, Preserving
Wild California

$25,000 to $49,999

Lowell Ahart, in memory of Ida Geary, John
Thomas Howell, and Vernon H. Oswald

National Fish and Wildlife Foundation

Roderic and Catherine Park

USDA Forest Service, Pacific Southwest Region

$10,000 to $24,999

Anonymous

California Department of Food and Agriculture

William and Wilma Follette

JiJi Foundation

Giles W. and Elise G. Mead Foundation

Elvenia J. Slosson Endowment Fund

USDA Forest Service, Shasta-Trinity National
Forest, sponsor of *Campanula* and *Sedum*

Sally Jane Weed, sponsor of *Aesculus* in memory of
David Gideon Weed

$5,000 to $9,999

Heath Bartosh, Nomad Ecology, sponsor of
Ericaceae and the San Francisco Bay Area subregion

Elizabeth Crispin

Frank W. Ellis, sponsor of *Fritillaria biflora* and
Brodiaea santarosae

Claire Englander
  sponsor of *Sequoiadendron* and *Viola* in honor of
    Ralph Marco Boemio and the Boemio family,
    Frances and Herman Englander, and Margaret
    Condliffe Kessler;
  sponsor of *Sequoia* in honor of David Durbin, Ed
    Fonseca, Phil Hoehn, Jim Houillion, Margaret
    Condliffe Kessler, Jim Stenquist, the Gay Burris
    family, and the Rabinovitz Roller family;
  sponsor of *Linum* in honor of the family of Miriam
    and Robbie Rabinovitz Roller;
  sponsor of Cupressaceae in honor of David
    Durbin, Ed Fonseca, Phil Hoehn, Jim Houillion,
    Margaret Condliffe Kessler, Jim Stenquist, the
    Boemio family, the Gay Burris family, and the
    Rabinovitz Roller family

ICF International

Georgie Robinett, sponsor of Inner North Coast
Ranges District, in memory of Jim Robinett

James P. Smith, Jr.

$2,500 to $4,999

Family and Friends of Larry Abers, sponsors of
Liliaceae in memory of Larry Abers

Anonymous, sponsor of Compositae in memory of
H. Cassini

California Native Plant Society, Marin Chapter
  sponsor of *Alnus* in memory of Ida Geary;
  sponsor of *Triantha* in memory of Robert Soost;
  sponsor of *Horkelia* in memory of Joe Kohn

Beth Lowe Corbin, sponsor of *Sidalcea*, in memory of Vernon H. Oswald

Patrick Creehan, in memory of Yvonne Doreen Creehan

Hazel and Wilson Flick

Kenneth Fuller, in memory of Thomas C. Fuller

Peter Garcia, sponsor of *Gilia,* in memory of Annetta Carter

Lawrence Janeway, sponsor of *Eriophorum, Rhynchospora,* and *Trichophorum*

Dwight L. Johnson

R. John Little, sponsor of Violaceae

S. B. Meyer, sponsor of *Platanus*

Richard Rawson, sponsor of Poaceae

The Reeve Family, sponsors of Ranunculaceae in loving memory of Marian E. Reeve and Roger M. Reeve

Thomas J. Zavortink, in memory of David M. Zavortink

$1,000 to $2,499

Anonymous, in honor of John L. Strother

Anonymous, sponsor of *Corethrogyne* in memory of A. de Candolle

Anonymous, sponsor of *Eriastrum* in memory of Herbert Louis Mason

Anonymous, sponsor of *Galium* in memory of Lauramay Tinsley Dempster

Anonymous, sponsor of *Jepsonia* in memory of Robert Ornduff

Anonymous, sponsor of *Pogogyne* in memory of G. Bentham and W. Kelly

Anonymous, sponsor of *Swallenia* in memory of Misses Annie M. Alexander and Louise Kellogg

Susan J. Bainbridge, sponsor of *Fragaria* in honor of Janet Bainbridge

Bruce G. Baldwin, in memory of June McCaskill

Bob Battagin, sponsor of *Deschampsia* in honor of Jim Battagin

Ann and David Bauer, sponsors of *Castilleja*

Kathleen A. Becker

Richard and Linda Beidleman, sponsors of *Cerastium* in memory of Lincoln Constance

Robert J. Berman

F. Thomas Biglione, sponsor of *Populus*

June M. Bilisoly

Linda Brodman

Richard and Trisha Burgess, sponsors of *Imperata*

Beth Burnside

California Native Plant Society, East Bay Chapter, sponsor of *Umbellularia*

California Native Plant Society, State Office

California Natural Diversity Database, Department of Fish and Game, sponsor of *Hesperolinon*

Margaret Colbert, sponsor of *Dodecatheon* in memory of Richard and Sue Colbert

Alison E.L. Colwell, sponsor of *Orobanche*

Toni and Richard Corelli

Gerald and Buff Corsi, sponsors of *Calochortus*

Ellen A. Crumb, sponsor of *Ceanothus* in memory of and thankfulness for Howard E. McMinn and his student Marie Louise Elliott Locke and their teaching

Katherine L.C. Cuneo

Christopher Davidson, sponsor of *Mentzelia* and *Paeonia*

Sally Davis, in memory of Julie Davis Leap

Paula Dawson

Dudek & Associates, Inc., sponsors of *Lathyrus*

Jim Duncan and Elaine Plaisance, sponsors of *Cirsium*

Brian Elliott and Samantha Mackey Hillaire, sponsors of *Cryptantha*

Phyllis M. Faber, sponsor of *Fremontodendron* in memory of Robert Ornduff, Larry Heckard, and Herbert Baker

Gordon W. and Jutta Frankie, in honorable memory of Herbert and Irene Baker

John Game, sponsor of *Erythronium*

Verne Garcia, sponsor of *Arctostaphylos*

Garcia and Associates (GANDA), sponsors of *Lasthenia*

Robert Garner, sponsor of *Pinus*

John R. Gibson, sponsor of *Epilobium*, in honor of Mary Ann Matthews

(continued)

Norman F. and Catherine R. Weeden, sponsors of *Rubus*

M.H. Wolfe and Associates, sponsors of *Atriplex*, in memory of Dr. James Hickman

Vern Yadon, sponsor of *Xerophyllum*

Stella Yang and Stephen Buckhout, sponsors of *Trifolium* in memory of Ssu-Sung Yang

Desi and Karen Zamudio, sponsors of *Salix* in honor of Elly Platou

The Jepson Manual
Vascular Plants of California

SECOND EDITION

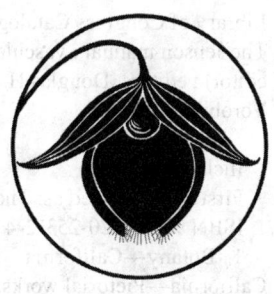

Bruce G. Baldwin
(Convening Editor)

Douglas H. Goldman

David J. Keil — Editors

Robert Patterson

Thomas J. Rosatti

Dieter H. Wilken

Jeffrey Greenhouse

Staci Markos

Richard L. Moe — Staff

Scott Simono

Margriet Wetherwax

Linda Ann Vorobik — Principal Illustrator

UNIVERSITY OF CALIFORNIA PRESS

University of California Press, one of the most distinguished university presses in the United States, enriches lives around the world by advancing scholarship in the humanities, social sciences, and natural sciences. Its activities are supported by the UC Press Foundation and by philanthropic contributions from individuals and institutions. For more information, visit www.ucpress.edu.

University of California Press
Oakland, California

© 2012 by The Regents of the University of California

Third printing with corrections, 2016

Library of Congress Cataloging-in-Publication Data
The Jepson manual : vascular plants of California / Bruce G. Baldwin (convening editor) ; editors, Douglas H. Goldman ... [et al.] ; principal illustrator, Linda Ann Vorobik. — 2nd ed.
    p.   cm.
  Includes index.
  First ed. published as: The Jepson manual: higher plants of California, c1993.
  ISBN 978-0-520-25312-4 (cloth : alk. paper)
  1. Botany—California.   2. Plants—California—Identification.   3. Plants—California—Pictorial works.   I. Baldwin, Bruce G., 1957–   II. Goldman, Douglas H., 1969–
  QK149.J56   2012
  581.9794—dc23   2011027207

19
10 9 8 7 6 5

Some families and genera have been taxonomically revised in the Jepson eFlora. Details are described online at http://ucjeps.berkeley.edu/eflora/supplement_summary.html

The paper used in this publication meets the minimum requirements of ANSI/NISO Z39.48-1992 (R 1997) (*Permanence of Paper*). ⊗

Preferred citation: Baldwin, B. G., D. H. Goldman, D. J. Keil, R. Patterson, T. J. Rosatti, and D. H. Wilken, editors. 2012. The Jepson manual: vascular plants of California, second edition. University of California Press, Berkeley.

Front-cover image: Spring bloom at Carrizo Plain National Monument: California poppy (*Eschscholzia californica*), sky lupine (*Lupinus nanus*), purple owl's clover (*Castilleja exserta*), yellow pincushion (*Chaenactis glabriuscula*). Purestock/Getty Images.

Photo credits for Additional Resources page (inside front cover): *Calochortus venustus,* John C. Game; *Arctostaphylos ohloneana,* Neal Kramer; *Harmonia guggolziorum,* John C. Game; *Caulanthus inflatus,* John C. Game; Willis Linn Jepson in the Vaca Hills, 1942, Neat M. Tate.

In accordance with federal law and U.S. Department of Agriculture policy, this institution is prohibited from discriminating on the basis of race, color, national origin, sex, age, or disability. (Not all prohibited bases apply to all programs.)

To file a complaint of discrimination, write USDA, Director, Office of Civil Rights, Room 326-W, Whitten Building, 1400 Independence Avenue, SW, Washington, DC 20250-9410 or call (202) 720-5964 (voice and TDD). USDA is an equal opportunity employer.

The findings and conclusions in this article/publication are those of the authors and do not necessarily represent the views of the U.S. Fish and Wildlife Service.

The Jepson Manual, Second Edition is dedicated to

JAMES C. HICKMAN

(1941–1993)

Jim Hickman could not have imagined the far-reaching impact *The Jepson Manual* (1993) would have on California botany and on botanists, ecologists, conservationists, and others. As a teacher and ecologist, Jim brought fresh ideas to the Jepson Herbarium when he arrived in 1977. With the leadership of Jepson Curator Larry Heckard, Jim built an amazing team of authors, artists, editors, and volunteers who worked together for about a decade to produce the *Manual*.

The Jepson Manual (1993) was a revolutionary book developed to reach both experienced and inexperienced botanists, the casual observer, and the professional. *The Jepson Manual* (1993) successfully combined, in a single volume, the expertise of hundreds of contributors. The editorship that Jim provided made the *Manual* a comprehensible, complete, and authoritative treatment of the California flora, and we are honored to acknowledge Jim's life and spirit that continue on through *The Jepson Manual, Second Edition*.

The Jepson Manual, Second Edition, is dedicated to

JAMES C. HICKMAN
(1941–1993)

CONTENTS

Willis Linn Jepson, Yolla Bolly Mountains, July 1897. Image courtesy of the University and Jepson Herbaria Archives. University of California, Berkeley.

PREFACE

WILLIS LINN JEPSON (1867–1946) laid much of the foundation of California floristics through his life-long efforts to document and describe vascular plant diversity in the state. His extensive plant collections and library formed the nucleus of the Jepson Herbarium, established at the University of California, Berkeley in 1950; the herbarium continues Jepson's work to understand and conserve the state's flora and to communicate that knowledge to others. One of Jepson's major achievements as a botanist was his publication in 1925 of the first comprehensive statewide guide to vascular plant diversity, *A Manual of the Flowering Plants of California*. That book inspired and informed generations of botanists during a period when the California flora became famous as the subject of pioneering investigations in plant evolution and ecology. Jepson emphasized that a floristic work is never truly complete and final, and that further study must continue to advance knowledge. The productive years since publication of *The Jepson Manual* (1993) are testimony to such change and were the impetus for the international collaboration that produced this book.

ACKNOWLEDGMENTS

*T*HE JEPSON MANUAL, Second Edition, is the direct result of the dedication of expertise from authors, editors, and reviewers and financial support from private donors, organizations, and agencies. We are grateful for the service and contribution of each and every person involved and recognize that this work could not have been completed without a collaborative effort. In that spirit, we thank everyone who contributed comments, distribution records, specimens, and improved systematic understanding of the plants of California. Of special note are those who were unable to participate as authors but who served as authors in *The Jepson Manual: Higher Plants of California* (1993) and whose treatments provided valuable information that was utilized here.

Herbaria played an especially important role in the development of the taxonomic treatments, and the use of specimen data was facilitated greatly by the Consortium of California Herbaria. We thank the collections management staff at herbaria around the world for the assistance they provided to authors.

We thank all of the illustrators, some of whom contributed to *The Jepson Manual* (1993) and some of whom contributed solely to this work. The illustrations for over 4,800 taxa make the technical descriptions accessible in a way that only a drawing can. We are grateful for the work of Linda Ann Vorobik, Principal Illustrator, Bobbie Angell, Maggie Day, Annette Filice, Karen Klitz, Valerie Layne, Lesley Randall, Emily E. Reid, Susan Stanley, Sarah A. Young, and Genevieve Walden.

The Trustees of the Jepson Herbarium, Roderic C. Park, Beth Burnside, Paul Licht, and John W. Taylor, provided guidance throughout the project. Brent D. Mishler, Director of the University and Jepson Herbaria, championed the project from day one by making it an institutional priority and allocating staff time and resources. His ideas helped define the philosophy and design of the book and his ability to negotiate difficult situations helped keep the project focused and cohesive. The University Herbarium at UC Berkeley also supported this project by providing collections management and other staff funding.

We relied on the careful work of the California Native Plant Society, the State of California Department of Food and Agriculture (Division of Plant Health and Pest Prevention Services), the California Invasive Plant Council,

and the Bay Area Early Detection Network for identification of native taxa of special concern and invasive non-native taxa.

Heath Bartosh, Senior Botanist of Nomad Ecology, carefully revised the boundaries of the Geographic Subdivisions of California, produced a new image for use in this book, and developed a GIS layer that represents the modified regions.

Many individuals contributed in a variety of ways and we thank each of them:

Jeanne Marie Acceturo	Wayne R. Ferren, Jr.	L. Maynard Moe
David D. Ackerly	Sean Flynn	Gary A. Monroe
Kelly Agnew	Holly Forbes	James D. Morefield
Lowell Ahart	Peter W. Fritsch	Nancy R. Morin
Robert Lee Allen	John C. Game	Julie Kierstead Nelson
James M. André	Arthur C. Gibson	Elizabeth L. Painter
Susan J. Bainbridge	John R. Gibson	Ana Penny
Michael G. Barbour	Judy Gibson	Dan Post
Roxanne Bittman	David Gowen	Robert E. Preston
Peter A. Bowler	Tonya Haff	Barry Prigge
Steve Boyd	Wes Hildreth	Betsy Ringrose
Barry and Judy Breckling	David S. Hollombe	Fred Marke Roberts, Jr.
	G. Frederic Hrusa	Georgie Robinett
Elizabeth Brusati	Diane Ikeda	Kristina Schierenbeck
Carolyn Chainey-Davis	Nick Jensen	Steve Schoenig
Kenton L. Chambers	Doug Johnson	Aaron Schusteff
Michael Charters	Steve Junak	James R. Shevock
Tom Chester	Amy Kasameyer	Michael G. Simpson
Joanna Clines	Tim Kask	Aaron Sims
Toni Corelli	Jon E. Keeley	Doreen Smith
Patricia Cruse	Dean Kelch	John L. Strother
Susan D'Alcamo	Kim Kersh	Edith Summers
Diane L. Delany	Neal Kramer	Dean Wm. Taylor
Joseph M. DiTomaso	Tasha La Doux	David Tibor
Andrew S. Doran	Kristi Lazar	Michael C. Vasey
Stephen W. Edwards	Aaron Liston	J. Giles Waines
Mark A. Elvin	Steve Matson	Peter J. Warner
Barbara Ertter	Rodney G. Myatt	John W. Willoughby
Diane M. Erwin	Lucinda A. McDade	Michael D. Windham
Phyllis Faber	Misa Milliron	Rick York

AUTHORS CONTRIBUTING TO
THE JEPSON MANUAL, SECOND EDITION

James D. Ackerman, University of Puerto Rico, Río Piedras

Robert P. Adams, Baylor University, Waco, Texas

Carlos Aedo, Real Jardín Botánico, Madrid, Spain

Susan G. Aiken, Canadian Museum of Nature, Ottawa, Ontario

Marisa Alarcón, Real Jardín Botánico, Madrid, Spain

Lawrence A. Alice, Western Kentucky University, Bowling Green

Charles Allen, Colorado State University, Fort Collins

Geraldine A. Allen, University of Victoria, British Columbia, Canada

Robert Lee Allen, Rancho Santa Ana Botanic Garden, Claremont, California

Kelly W. Allred, New Mexico State University, Las Cruces

Ihsan A. Al-Shehbaz, Missouri Botanical Garden, St. Louis

Edward Alverson, Oregon State University, Corvallis

Edward F. Anderson†, Whitman College, Walla Walla, Washington

Katarina Andreasen, Uppsala University, Sweden

George W. Argus, Canadian Museum of Nature, Ottawa, Ontario, Canada

Wayne P. Armstrong, Palomar College, San Marcos, California

T.A. Atkinson, Burlington, North Carolina

Daniel F. Austin, Arizona-Sonora Desert Museum, Tucson

Deborah Engle Averett, Austin, Texas

Tina Ayers, Northern Arizona University, Flagstaff

Susan J. Bainbridge, University of California, Berkeley

Gary I. Baird, Brigham Young University-Idaho, Rexburg

† Deceased

John R. Baird, Ohio Department of Transportation, Columbus

Marc Baker, Arizona State University, Tempe

Riccardo M. Baldini, Università degli Studi, Firenze, Italy

Bruce G. Baldwin, University of California, Berkeley

Peter W. Ball, University of Toronto, Mississauga, Ontario, Canada

Theodore M. Barkley†, Botanical Research Institute of Texas, Fort Worth

Kerry Barringer, Brooklyn Botanic Garden, New York

Jim A. Bartel, United States Fish and Wildlife Service, Carlsbad, California

Robyn Battaglia, Pleasanton, California

Bernard R. Baum, Agriculture and Agri-Food Canada, Ottawa, Ontario, Canada

Randall J. Bayer, University of Memphis, Tennessee

Charles D. Bell, University of New Orleans, Louisiana

Hester L. Bell, Rancho Santa Ana Botanic Garden, Claremont, California

Robin Bencie, Humboldt State University, Arcata, California

Nuri Benet-Pierce, San Diego State University, California

Mark W. Bierner, The University of Texas at Austin; The University of Arizona, Tucson

John Bleck, University of California, Santa Barbara

David Bogler, Missouri Botanical Garden, St. Louis

Allan J. Bornstein, Southeast Missouri State University, Cape Girardeau

Steve Boyd, Rancho Santa Ana Botanic Garden, Claremont, California

David M. Brandenburg, The Dawes Arboretum, Newark, Ohio

Joshua M. Brokaw, Washington State University, Pullman

Luc Brouillet, Université de Montréal, Québec, Canada

Gregory K. Brown, University of Wyoming, Laramie

R.K. Brummitt, Royal Botanic Gardens, Kew, Richmond, United Kingdom

Mark S. Brunell, University of the Pacific, Stockton, California

Christopher S. Campbell, University of Maine, Orono

Gerald D. Carr, Oregon State University, Corvallis

Robert L. Carr, Eastern Washington University, Cheney

Rosa Cerros-Tlatilpa, Universidad Autónoma del Estado de Morelos, Cuernavaca, Mexico

Adolf Ceska, Victoria, British Columbia, Canada

Oldriska Ceska, Victoria, British Columbia, Canada

Henrietta L. Chambers, Oregon State University, Corvallis

Kenton L. Chambers, Oregon State University, Corvallis

Raymund Chan, University of California, Berkeley

Anita F. Cholewa, University of Minnesota, St. Paul

T.I. Chuang†, Illinois State University, Normal

Timothy Chumley, The University of Texas at Austin

Curtis Clark, California State Polytechnic University, Pomona

Lynn G. Clark, Iowa State University, Ames

W.D. Clark, Arizona State University, Tempe

Steven E. Clemants†, Brooklyn Botanic Garden, New York

Ronald A. Coleman, The University of Arizona, Tucson

J. Travis Columbus, Rancho Santa Ana Botanic Garden and Claremont Graduate University, California

Alison E.L. Colwell, National Park Service, Yosemite National Park, El Portal, California

Steven A. Conley, Redding, California

H.E. Connor, University of Canterbury, Christchurch, New Zealand

Lincoln Constance†, University of California, Berkeley

Mihai Costea, Wilfrid Laurier University, Waterloo, Ontario, Canada

Lyn A. Craven, Australian National Herbarium, Canberra, Australia

William J. Crins, Ontario Ministry of Natural Resources, Peterborough, Ontario, Canada

Michael Curto, California Polytechnic State University, San Luis Obispo

Thomas F. Daniel, California Academy of Sciences, San Francisco

S.J. Darbyshire, Agriculture and Agri-Food Canada, Ottawa, Ontario, Canada

Jerrold I. Davis, Cornell University, Ithaca, New York

W.S. Davis, Louisville, Kentucky

Alva G. Day, California Academy of Sciences, San Francisco

Sarah De Groot, Rancho Santa Ana Botanic Garden and Claremont Graduate University, California

Alfonso Delgado-Salinas, Universidad Nacional Autónoma de México, México City

Lauramay T. Dempster†, University of California at Berkeley

Melinda F. Denton†, University of Washington, Seattle

James C. Dice, Natural History Museum, San Diego, California

John E. Ebinger, Eastern Illinois University, Charleston

Patrick E. Elvander†, University of California at Santa Cruz

Mark A. Elvin, United States Fish and Wildlife Service, Ventura, California

Barbara Ertter, University of California, Berkeley

Frederick Essig, University of South Florida, Tampa

Dwayne Estes, Austin Peay State University, Clarksville, Tennessee

Donald R. Farrar, Iowa State University, Ames

Carolyn J. Ferguson, Kansas State University, Manhattan

Wayne R. Ferren, Jr., Rutgers, The State University of New Jersey, New Brunswick

Peggy L. Fiedler, University of California Natural Reserve System, Oakland

Petra Foerster, Fredericton, New Brunswick, Canada

Bruce A. Ford, University of Manitoba, Winnipeg, Canada

Robert W. Freckmann, University of Wisconsin-Stevens Point

Peter W. Fritsch, California Academy of Sciences, San Francisco

Paul A. Fryxell†, The University of Texas at Austin

John C. Game, University of California, Berkeley

Fred R. Ganders, University of British Columbia, Vancouver, Canada

Laura M. Garrison, Brown University, Providence, Rhode Island

John F. Gaskin, United States Department of Agriculture, Agricultural Research Service, Sidney, Montana

Peter Goldblatt, Missouri Botanical Garden, St. Louis

Douglas H. Goldman, Harvard University, Cambridge, Massachusetts

L.D. Gottlieb, University of California, Davis

David Gowen, University of California, Berkeley

Shirley A. Graham, Missouri Botanical Garden, St. Louis

Craig W. Greene†, College of the Atlantic, Bar Harbor, Maine

Brenda J. Grewell, United States Department of Agriculture, Agricultural Research Service, Davis, California; University of California, Davis

James R. Griffin†, University of California at Berkeley

James W. Grimes, New York Botanical Garden, Bronx

Evanielis U. Grissom, University of Illinois, Urbana-Champaign

C. Matt Guilliams, University of California, Berkeley

Erich Haber, Canadian Museum of Nature, Ottawa, Ontario, Canada

J. Robert Haller, Santa Barbara Botanic Garden; University of California, Santa Barbara

Richard R. Halse, Oregon State University, Corvallis

Gary L. Hannan, Eastern Michigan University, Ypsilanti

Debra R. Hansen, The University of Texas at Austin

Danica T. Harbaugh, University of California, Berkeley

Raymond M. Harley, Royal Botanic Gardens, Kew, Richmond, United Kingdom

Ronald L. Hartman, University of Wyoming, Laramie

Michael J. Harvey, Dalhousie University, Halifax, Canada; Victoria, British Columbia, Canada

Kristen E. Hasenstab-Lehman, Rancho Santa Ana Botanic Garden and Claremont Graduate University, California

Richard L. Hauke, University of Rhode Island, Kingston and Atlanta, Georgia

Robert R. Haynes, The University of Alabama, Tuscaloosa

Lawrence R. Heckard†, University of California, Berkeley

Alice Long Heikens, Franklin College, Indiana

C. Barre Hellquist, Massachusetts College of Liberal Arts, North Adams

James Henrickson, The University of Texas at Austin

William J. Hess, The Morton Arboretum, Lisle, Illinois

Steven R. Hill, Illinois Natural History Survey, University of Illinois, Urbana-Champaign

Matthew H. Hils, Hiram College, Ohio

Andrew L. Hipp, The Morton Arboretum, Lisle, Illinois; The Field Museum, Chicago, Illinois; University of Wisconsin-Madison

Peter C. Hoch, Missouri Botanical Garden, St. Louis

Carol A. Hoffman, University of Georgia, Athens

Scott A. Hodges, University of California, Santa Barbara

Noel H. Holmgren, New York Botanical Garden, Bronx

Charles N. Horn, Newberry College, South Carolina

G. Frederic Hrusa, California Department of Food and Agriculture, Sacramento

Larry Hufford, Washington State University, Pullman

Stefanie M. Ickert-Bond, University of Alaska, Fairbanks

Duane Isely†, Iowa State University, Ames

Judith A. Jernstedt, University of California, Davis

Dale E. Johnson, Seattle, Washington

George P. Johnson, Arkansas Tech University, Russellville

Leigh A. Johnson, Brigham Young University, Provo, Utah

James D. Jokerst†, Jones and Stokes Associates, Sacramento, California

William Jones, California State University, Los Angeles

Walter S. Judd, University of Florida, Gainesville

Steve Junak, Santa Barbara Botanic Garden, California

Glenn Keator, Sebastopol, California

Jon E. Keeley, Rancho Santa Ana Botanic Garden, Claremont, California

David J. Keil, California Polytechnic State University, San Luis Obispo

Ronald B. Kelley, Eastern Oregon University, La Grande

Sylvia Kelso, Colorado College, Colorado Springs

Kim R. Kersh, University of California, Berkeley

Carrie Kiel, Rancho Santa Ana Botanic Garden and Claremont Graduate University, California

Ruth E.B. Kirkpatrick, University of California, Berkeley

Jan Kirschner, Institute of Botany, Academy of Sciences, Průhonice, Czech

Genevieve J. Kline, Northern Illinois University, DeKalb

Jason A. Koontz, Augustana College, Rock Island, Illinois

Daryl Koutnik, Impact Sciences, Inc., Camarillo, California

Kathleen A. Kron, Wake Forest University, Winston-Salem, North Carolina

Job Kuijt, University of Victoria, British Columbia, Canada

John C. La Duke, University of North Dakota, Grand Forks

Leslie R. Landrum, Arizona State University, Tempe

Meredith A. Lane, Reston, Virginia

Richard V. Lansdown, Ardeola Environmental Services, Stroud, United Kingdom

Robert Lauri, Rancho Santa Ana Botanic Garden, Claremont, California

Matt Lavin, Montana State University, Bozeman

Thomas Lemieux, University of Colorado, Boulder

Gordon Leppig, California Department of Fish and Game, Eureka

Donald H. Les, University of Connecticut, Storrs

Deborah Q. Lewis, Iowa State University, Ames

Harlan Lewis†, University of California, Los Angeles

Richard A. Lis, University of California at Berkeley

R. John Little, Sycamore Environmental Consultants, Inc., Sacramento, California

Lúcia G. Lohmann, Universidade de São Paulo, Brazil

Timothy K. Lowrey, University of New Mexico, Albuquerque

Karen Lu, Humboldt State University, Arcata, California

B.H. Macmillan, Landcare Research, Lincoln, New Zealand

L.K. Magrath†, University of Science and Arts of Oklahoma, Chickasha

Alison M. Mahoney, Minnesota State University, Mankato

Staci Markos, University of California, Berkeley

Joy Mastrogiuseppe, Washington State University, Pullman

Mark H. Mayfield, Kansas State University, Manhattan

Stephen Ward McCabe, University of California, Santa Cruz

Elizabeth McClintock†, University of California, Berkeley

Joshua R. McDill, The University of Texas at Austin

Robert J. McKenzie, Rhodes University, Grahamstown, South Africa

Michelle M. McMahon, The University of Arizona, Tucson

Dale W. McNeal, University of the Pacific, Stockton, California

T. Lawrence Mellichamp, University of North Carolina, Charlotte

Michael R. Mesler, Humboldt State University, Arcata, California

Timothy C. Messick, Jones and Stokes Associates, Sacramento, California

Constance I. Millar, United States Department of Agriculture, Forest Service, Pacific Southwest Research Station, Albany, California

John M. Miller, University of California, Berkeley

Abigail J. Moore, University of California, Berkeley

Michael O. Moore, University of Georgia, Athens

John S. Mooring, Santa Clara University, California

James D. Morefield, Nevada Natural Heritage Program, Carson City

David R. Morgan, University of West Georgia, Carrollton

Nancy R. Morin, Flora of North America, Point Arena, California

Andy Murdock, University of California, Berkeley

Carmen Navarro, Universidad Complutense, Madrid, Spain

Michael H. Nee, New York Botanical Garden, Bronx

Elizabeth Chase Neese†, University of California, Berkeley

John B. Nelson, University of South Carolina, Columbia

Guy L. Nesom, Fort Worth, Texas

Bryan D. Ness, Pacific Union College, Angwin, California

Lorin I. Nevling, Jr., Illinois Natural History Survey, University of Illinois, Urbana-Champaign

Gilberto Ocampo, California Academy of Sciences, San Francisco

Sang-Hun Oh, Cornell University, Ithaca, New York

Richard G. Olmstead, University of Washington, Seattle

Robert Ornduff†, University of California, Berkeley

Elizabeth L. Painter, Santa Barbara Botanic Garden, California

Bruce D. Parfitt†, The University of Michigan-Flint

Michael S. Park, University of California, Berkeley

V. Thomas Parker, San Francisco State University, California

Robert Patterson, San Francisco State University, California

Paul M. Peterson, Smithsonian Institution, Washington, District of Columbia

James B. Phipps, The University of Western Ontario, London, Canada

Michael B. Piep, Utah State University, Logan

J. Chris Pires, University of Missouri, Columbia

Duncan M. Porter, Virginia Polytechnic Institute and State University, Blacksburg

J. Mark Porter, Rancho Santa Ana Botanic Garden and Claremont Graduate University, California

Daniel Potter, University of California, Davis

A. Michael Powell, Sul Ross State University, Alpine, Texas

Robert E. Preston, ICF International, Sacramento, California

Barry Prigge, University of California, Los Angeles

James S. Pringle, Royal Botanical Gardens, Hamilton, Ontario, Canada

Charles F. Quibell, Sonoma State University, Rohnert Park, California

Richard K. Rabeler, The University of Michigan-Ann Arbor

Jon Rebman, San Diego Natural History Museum, California

John R. Reeder†, The University of Arizona, Tucson

Nancy F. Refulio-Rodriguez, Rancho Santa Ana Botanic Garden, Claremont, California

James L. Reveal, Cornell University, Ithaca, New York

Barry A. Rice, University of California, Davis

Matt Ritter, California Polytechnic State University, San Luis Obispo

Joseph R. Rohrer, University of Wisconsin-Eau Claire

Thomas J. Rosatti, University of California, Berkeley

Jeffery M. Saarela, Canadian Museum of Nature, Ottawa, Ontario, Canada

Andrew C. Sanders, University of California, Riverside

John O. Sawyer, Jr., Humboldt State University, Arcata, California

H. Jochen Schenk, California State University Fullerton

John J. Schenk, Washington State University, Pullman

Robert L. Schlising, California State University, Chico

Lisa M. Schultheis, Foothill College, Los Altos Hills, California

Randall W. Scott, Northern Arizona University, Flagstaff

David Seigler, University of Illinois, Urbana-Champaign

John C. Semple, University of Waterloo, Ontario, Canada

James R. Shevock, California Academy of Sciences, San Francisco

Teresa Sholars, College of the Redwoods, Fort Bragg, California

Leila M. Shultz, Utah State University, Logan

Michael Silveira, San Diego State University, California

Scott Simono, University of California, Berkeley

Beryl B. Simpson, The University of Texas at Austin

Michael G. Simpson, San Diego State University, California

Elizabeth M. Skendzic, Pacific Lutheran University, Tacoma, Washington

Mark W. Skinner, United States Department of Agriculture, Forest Service, Pacific Northwest Region, Portland, Oregon

Tracey Slotta, Montgomery College, Rockville, Maryland

Alan R. Smith, University of California, Berkeley

James P. Smith, Jr., Humboldt State University, Arcata, California

S. Galen Smith, University of Wisconsin-Whitewater; University of Wisconsin-Madison

Douglas E. Soltis, University of Florida, Gainesville

Pamela S. Soltis, University of Florida, Gainesville

Robert J. Soreng, Smithsonian Institution, Washington, District of Columbia

Valerie Soza, University of Washington, Seattle

Richard Spellenberg, New Mexico State University, Las Cruces

G. Ledyard Stebbins†, University of California, Davis

Kelly Steele, Arizona State University, Mesa

Saša Stefanović, University of Toronto, Mississauga, Ontario, Canada

William J. Stone, University of California, Berkeley

Suzanne C. Strakosh, Kansas State University, Manhattan

John L. Strother, University of California, Berkeley

Fosiée Tahbaz, University of California, Berkeley

Jennifer Talbot, Washington University, St. Louis, Missouri

David C. Tank, University of Idaho, Moscow

Sarah Taylor, The University of Texas at Austin

W. Carl Taylor, National Science Foundation, Arlington, Virginia

John W. Thieret†, Northern Kentucky University, Highland Heights

David M. Thompson, Rancho Santa Ana Botanic Garden, Claremont, California

Robert F. Thorne, Rancho Santa Ana Botanic Garden, Claremont, California

Steven L. Timbrook, Ganna Walska Lotusland Foundation, Santa Barbara

Ruth E. Timme, The University of Texas at Austin

Debra K. Trock, California Academy of Sciences, San Francisco

Arthur O. Tucker, Delaware State University, Dover

Gordon C. Tucker, Eastern Illinois University, Charleston

John M. Tucker†, University of California, Davis

Charles E. Turner†, United States Department of Agriculture, Agricultural Research Service, Albany, California

Lowell E. Urbatsch, Louisiana State University, Baton Rouge

Jesús Valdés-Reyna, Universidad Autónoma Agraria Antonio Narro, Saltillo, Coahuila, México

Henk van der Werff, Missouri Botanical Garden, St. Louis

Brian Vanden Heuvel, Colorado State University, Pueblo

Staria S. Vanderpool, Arkansas State University, Jonesboro

Michael C. Vasey, San Francisco State University, California

Michael A. Vincent, Miami University, Oxford, Ohio

Nancy J. Vivrette, Santa Barbara Botanic Garden, California

Linda Ann Vorobik, University of California, Berkeley

Piet Vorster, University of Stellenbosch, South Africa

Eric B. Wada, Folsom Lake College, California

Warren L. Wagner, Smithsonian Institution, Washington, District of Columbia

M. Andrew Walker, University of California, Davis

Gary D. Wallace, Rancho Santa Ana Botanic Garden, Claremont, California

Livia Wanntorp, Stockholm University, Sweden

Michael J. Warnock, University of Missouri, Columbia

Linda E. Watson, Miami University, Oxford, Ohio

Grady L. Webster†, University of California, Davis

Robert Webster, United States Department of Agriculture, United States National Arboretum, Washington, District of Columbia

Thomas L. Wendt, Colegio de Postgraduados, Chapingo, Mexico

Margriet Wetherwax, University of California, Berkeley

R. David Whetstone, Jacksonville State University, Alabama

Justen Whittall, Santa Clara University, California

Alan T. Whittemore, United States Department of Agriculture, United States National Arboretum, Washington, District of Columbia

Jeannette Whitton, University of British Columbia, Vancouver, Canada

John H. Wiersema, United States Department of Agriculture, Agricultural Research Service, Beltsville, Maryland

Dieter H. Wilken, Santa Barbara Botanic Garden, California

Michael P. Williams, Butte College, Oroville, California

Carol A. Wilson, Rancho Santa Ana Botanic Garden and Claremont Graduate University, California

Paul Wilson, California State University, Northridge

Michael D. Windham, Duke University, Durham, North Carolina

Dennis W. Woodland, Andrews University, Berrien Springs, Michigan

Martin F. Wojciechowski, Arizona State University, Tempe

Steven J. Wolf, California State University Stanislaus, Turlock

Lindsay P. Woodruff, The University of Texas at Austin

Thomas Worley, Eureka, California

George Yatskievych, Missouri Botanical Garden, St. Louis

Elizabeth H. Zacharias, Harvard University, Cambridge, Massachusetts

Peter F. Zika, University of Washington, Seattle

Wendy B. Zomlefer, University of Georgia, Athens

PHILOSOPHY

In the years since publication of *The Jepson Manual: Higher Plants of California*, or *TJM* (1993), the study of evolutionary relationships and classification (systematics) has undergone revolutionary change, with rapid innovations in molecular biology, evolutionary methods and philosophy, and computer technology. Those changes have allowed for rapid progress in resolving evolutionary relationships (phylogeny) at deep- and fine-scale levels of divergence, with much focus on California's vascular plants. At the same time, botanists have been targeting under-collected areas of California with high potential for harboring new diversity (e.g., on unusual soils, such as serpentine), aided by the rise of electronic databases of herbarium (botanical museum) specimens and innovative search capabilities across those collections (i.e., the Consortium of California Herbaria). Other botanists, with close attention to alien taxa, have documented the recent rise (and occasional extirpation) of introduced plants, which continue to become naturalized in the state.

All of the above advances have translated to accelerated botanical discovery for California, including about 150 minimum-rank taxa (species, subspecies, and varieties) that have been described as new-to-science since publication of *TJM* (1993). In addition, higher-level classification of Californian plants (e.g., genera and families) has been revised to reflect monophyletic groups (clades, i.e., containing all and only descendants of a common ancestor) more precisely. All of these improvements to California plant classification, evident throughout *The Jepson Manual, Second Edition* (*TJM 2*), are essential to making taxonomy reflect evolutionary groups that are of primary importance to conservation and comparative biology.

TJM 2 is organized to follow the new understanding of vascular plant phylogeny. Families are grouped together into eight major clades of vascular plants. The relationships of all families included in *TJM 2,* as currently understood, are shown on the back endpaper, along with the page numbers where they are treated (also see the Angiosperm Phylogeny Website: http://www.mobot.org/MOBOT/Research/APweb/welcome.html). Within each major clade, families continue to be presented alphabetically (for convenience), but the relationships indicated on the back endpaper should aid identification of plants to family. Taxonomies presented here include some that depart from those in other floras; authors were encouraged to produce such treatments if adherence to the book philosophy (e.g., the criterion of monophyly) warranted it.

TJM 2 has been guided by the importance of the above developments and the conviction that all evolutionary groups warranting taxonomic recognition should be included, so that *TJM 2* can serve well as a primary reference on California's native and naturalized vascular plant diversity. In some instances, the need to be comprehensive has required that authors recognize taxa that are cryptic; that is, difficult or impossible to recognize reliably based on morphology alone, but distinct evolutionarily (and often ecologically) based on multiple lines of evidence. Such examples force us to confront the relative importance in taxonomy of practicality, on the one hand (e.g., identification difficulty), and evolutionary real-

ity, on the other hand. For reasons of scientific accuracy, the latter approach was adopted here. Plants do not necessarily change in ways that are readily evident to human senses, although they may still be divergent in ways that are important to their survival or even to humans (e.g., in producing different chemical compounds). Sometimes the decision as to whether or not to recognize cryptic taxa is forced on botanists by the finding that such entities are not even each other's closest relatives (e.g., *Lasthenia californica* and *L. gracilis*, in Asteraceae; *Downingia pulcherrima* and *D. willamettensis*, in Campanulaceae).

From the standpoint of practicality, production of *TJM 2* was marked by a strong commitment to improving the ability of botanists to identify California vascular plants. Even without considering cryptic diversity, the difficulty of distinguishing closely related plants in the California flora is often challenging, in part because of the youth of much of the plant diversity in the state. Major effort went into ensuring that keys and descriptions are internally consistent and functional. Another strategy to aid plant identification in *TJM 2* was to refine, augment, and add illustrations, with 272 full plates compared to 242 in *TJM* (1993), including a substantial increase in the number of taxa illustrated and in the amount of illustrative detail provided for taxa included in plates in *TJM* (1993). Illustrations, like written treatments, underwent rigorous editing to find and correct problems. Posting of treatments upon completion of editing (but well prior to publication) on the Jepson Herbarium website also helped to allow input from botanists worldwide, who provided helpful reviews.

Accuracy of other components of treatments beyond keys and morphological descriptions also were editorial priorities. For example, geographic and elevational ranges of taxa were improved by drawing attention to outlying records in the Consortium of California Herbaria and aiding authors in resolving such discrepancies, where practicable. In addition, some of the changes in plant names evident in *TJM 2* are not the result of taxonomic changes (i.e., changes that reflect revised understanding of plant relationships) but instead of nomenclatural matters, where a name previously applied to a particular plant group was found to be incorrectly used or to be superseded by another name. In other cases, the authorship assigned to scientific names in *TJM* (1993) and elsewhere was found to be in error and was corrected.

The magnitude of change in *TJM 2* relative to *TJM* (1993) is difficult to express quantitatively. For example, inclusion of 185 plant families in *TJM 2* and 172 in *TJM* (1993) fails to capture the magnitude of revision involved, with 13 families from *TJM* (1993) now treated within other families, 30 families within *TJM 2* that were not recognized in *TJM* (1993) — some (22) segregated from families recognized in *TJM* (1993) and some (8) comprised of taxa that are new to California (all naturalized) — and 1 family extirpated from the state (Cymodoceaceae; an alien here). Similar considerations mask much of the revision that is represented in considering that more than 6500 native minimum-rank taxa (species, subspecies, and varieties) are recognized in *TJM 2*, about 310 more than in *TJM* (1993). In addition, the relatively modest increase in number of naturalized minimum-rank taxa in *TJM 2* (now numbering about 1100 total), about 30 more than in *TJM* (1993), is explained in part by downgrading of some taxa to waif status that were treated as naturalized in *TJM* (1993).

Important elements of the philosophy of *TJM* (1993) and the philosophy of Willis Linn Jepson are still embraced in *TJM 2*, including a commitment to producing a field portable volume that will serve botanists of diverse backgrounds. Like *TJM* (1993), *TJM 2* owes much to the collaborative spirit and international scope of this effort, with hundreds of authors and a sizable group of editors, staff, and illustrators. Most of these individuals worked without compensation, apart from the satisfaction of contributing their knowledge, time, and effort to further floristic knowledge and conservation. Jepson's philosophy that floristics is a never-ending study also aligns with the *TJM 2* effort, which helped to refocus attention on Californian plants that led to discoveries made too late for inclusion here. The growing urgency of rapidly communicating new discoveries to the broader community requires a new approach, one of continual revision and immediate distribution to the botanical community, as is possible now with the new *Jepson eFlora* (see inside front cover).

CONVENTIONS USED IN *THE JEPSON MANUAL,* *SECOND EDITION*

Producing a field-portable manual on the California flora that is accessible to a wide audience requires balancing somewhat competing interests. For example, if technical terms are reduced in the interest of user-friendliness, more words are generally necessary to convey descriptive information and the book must be larger. In part to balance the increase in new plant taxa for California, *The Jepson Manual, Second Edition* (*TJM 2*) includes more technical terms — about 100 additional glossary entries — than were included in *The Jepson Manual,* or *TJM* (1993), and has continued other space-saving conventions (also see Glossary introduction). These other conventions include a concise, abbreviation-rich format that is described here (also see Abbreviations and Symbols). Those familiar with *TJM* (1993) will find many of the same conventions in *TJM 2* but also some that might not be understood without the explanation provided below.

General Conventions

Comprehensiveness. A primary goal of *TJM 2* was to include all native and naturalized vascular plant taxa in California that are accepted by *TJM 2* authors as scientifically sound (also see Philosophy). Natives include both endemic and non-endemic, indigenous taxa. Naturalized plants are defined here as aliens growing in wild or approximately wild conditions and reproducing either sexually or asexually. Aliens that occur in such conditions but are not reproducing and therefore not persisting and becoming established parts of the flora are considered here to be waifs. Recently documented waifs are included in the keys (with names in square brackets) but their descriptions appear only online, in the *Jepson eFlora* (see Associated Electronic Resources, inside front cover). Waifs that have not been collected recently (in the last half-century or so) are not treated at all in *TJM 2*, under the assumption that such "historical waifs" have not persisted to become established parts of the flora. Also excluded here, in general, are taxa represented in California by non-reproducing but long-persisting individuals (e.g., planted fruit trees) or clones. Alien taxa occurring outside cultivation but only in highly modified environments, such as urban, suburban, or agricultural lands, are not included in *TJM 2*.

The establishment status of alien plant taxa in California is dynamic and sometimes questionable based on the best available evidence. Extensive consultation with weed scientists and other botanists focused on naturalized taxa in California helped greatly to improve treatment here of aliens. Nonetheless, some fully naturalized taxa as well as waifs probably have not been included in *TJM 2*; documentation was lacking to substantiate some reports that were considered during treatment preparation. In addition, alien plants in general are often given low priority by collectors, with the result that there are gaps in our knowledge of alien plants occurring outside of cultivation in California. Because of the potential harm that alien taxa can inflict on populations of native plants and animals and their habitats,

3

there is an ever-growing need to monitor their occurrences, as well as to document them with the addition of properly accessioned and curated specimens to herbaria.

Uniformity. A consequence of treating an exceptionally diverse and complex flora is that many specialists are needed to describe it well. Differences among the hundreds of contributing authors in philosophy and style were constrained by editorial conventions on terminology, structure of treatments, and taxonomic philosophy. Nonetheless, differences remain that are unavoidable in a work with such a high level of taxonomic coverage and scientific participation.

Organization. The book is organized into eight major monophyletic groups or clades that reflect the most recent classification systems of vascular plants (also see Philosophy): Lycophytes, Ferns, Gymnosperms, Nymphaeales, Magnoliids, Ceratophyllales, Eudicots, and Monocots. Families within these groups, genera within families, species within genera, and infraspecific taxa within species are arranged alphabetically for convenience, but relationships among the families are depicted on the back inside cover. Difficulty in identifying the family of an unknown plant may be aided by considering families that are most closely related to the one in consideration.

Measurements. All linear measurements are given in metric units. Conversion scales (centimeters to inches, meters to feet) are provided near the end of the book (after the index).

Illustrations. Illustrations are provided here to convey descriptive information that the written word cannot adequately express. All native genera are at least partially illustrated. Because all taxa could not be illustrated, priority was given to native taxa considered to be sufficiently uncommon or threatened to warrant monitoring or, conversely, to aliens that are very likely to be encountered, as well as aliens that have become increasingly problematic because of their growing numbers. Unusual or difficult diagnostic features were also given priority in illustrations. Illustration plates of taxa appear on right-handed pages and, with few exceptions, appear after the descriptions of the taxa they represent; order within a plate is primarily alphabetical.

Index. All family names, genus names, common names, names in notes, and names here considered to be synonyms (or to have been misapplied to plants in California) are listed alphabetically in the index. Accepted specific and infraspecific names are excluded from the index because they are ordered alphabetically under the appropriate genus and therefore are easily found.

Appendices from _TJM_ (1993). Appendix I (floristic summary) has been revised. Appendix II (classification of California plant families) has been replaced in part by online resources, and in part by the phylogenetic tree of California vascular-plant groups and families on the back inside cover. Appendix III (name changes) has been replaced by various resources on our website, including primarily the Index to California Plant Names (ICPN), which is intended to account for names that have been applied to plants in California, as accepted names, synonyms, or misapplied names.

Conventions Applying to Keys

In addition to comprehensively treating California's native and naturalized vascular plants, a central goal of _TJM 2_ is to facilitate plant identification. Dichotomous keys are the primary means to identify plants treated here (descriptions are primarily useful for confirming identifications). Although so-called natural keys also serve to convey relationships among the included taxa, this book primarily uses artificial keys, which are organized principally for ease of usage rather than for expressing relationships (natural keys can be difficult to use if trait evolution within groups has been dynamic). Where characters used in keys here correspond with characters that are useful for recognizing natural groups or clades of taxa, then those components of the keys can be considered natural, and artificial

keys often contain a mix of natural and artificial constructs. Artificiality of keys should not be confused with artificiality of taxonomy; classification of plants recognized in *TJM 2* is guided by a commitment to recognizing natural or monophyletic taxa (see Philosophy).

A dichotomous key is a series of paired, mutually exclusive statements that divides a group of unknowns into progressively smaller subsets until all possibilities but one have been eliminated. Keys are used here to identify plants to family, genus, species, and, if pertinent, subspecies and/or variety. To enhance efficiency, for each pair of statements (couplet), the choice (lead) under which fewer taxa appear is given first. In this way, for example, a very unusual species or group of species within a large genus may be identified without having to read through the entire key; this convention also tends to bring the two leads of a couplet closer together physically, thereby enhancing the ability to compare and contrast and then choose between them. Within this arrangement, taxa appear alphabetically.

Ideally, key leads are mutually exclusive and allow for a straightforward solution, but complex patterns of variation in some groups (e.g., as a result of recent evolutionary divergence of taxa or a history of hybridization) or absence in a partial specimen of a key character for a particular couplet may necessitate trying both leads, and reading through subsequent couplets, in order to make the correct choice. Some taxa are so complex or variable that an easily usable key cannot be constructed in which each taxon keys out in only one place. The names of such taxa are preceded in the key by a superscript indicating the number of times that taxon occurs in that key (or, for subdivided keys in some large families or genera, in a particular key "group").

Keys are generally written so that the more effective and more easily determined traits appear earlier in the lead. Sometimes, unfortunately, only a single feature allows for a choice between leads. To facilitate the choice between two leads in a couplet, traits are presented in the same order and described in the same form, to the extent possible, in both leads. Unilateral statements are included in some couplets, but in each case they are set off, after an em-dash, at the end of the lead to alert users that the trait or traits involved are not addressed in the other lead. As in descriptions (see below), character states indicated in parentheses generally are rare or exceptional conditions, although sometimes parentheses are used to provide additional context or explanatory information. Square brackets are sometimes used for the last-mentioned purpose in keys (e.g., in the key to families), unlike in descriptions, where square brackets are used to indicate conditions that apply only to plants outside California (see below).

The sequence of keys that are used to obtain an increasingly refined identification of an unknown vascular plant follows from the key to groups of families, to the key to families within each group, to the key to genera within each family, and finally to the key to species, subspecies, and varieties within each genus. In some cases there are keys to groups of genera within a family or to groups of species and infraspecific taxa within a genus. One or more of the early keys in the above sequence can be bypassed, of course, if the family, genus, or species of the plant in hand is already known. In some keys, e.g., the key to families, accommodation was made for common mistakes and misinterpretations, so that the correct answer may be obtained even if a technical error is made.

A key to the genera of a family follows each family description, if there is more than one genus in California. Normally, a key to species as well as any infraspecific taxa (subspecies or varieties) follows each genus description, if there is more than one taxon at the level of species or below in California. For a few very large, complex genera (e.g., *Astragalus*), keys to some infraspecific taxa occur after the description of their respective species.

Conventions Applying to Descriptions

All Descriptions. Descriptions address variation among Californian members of a taxon, with information pertaining only to (some or all) members of that taxon occurring outside California

enclosed in square brackets. Space constraints did not allow for addressing all characters for all taxa or even for addressing in descriptions all characters addressed in the associated keys, except in taxa of low diversity. Characters judged by authors to be most important in identification of component taxa were given priority for inclusion. For a given taxonomic level (e.g., species within a genus), characters are addressed in the same sequence from description to description, primarily to facilitate comparisons between taxa.

Characteristics are addressed at the highest rank at which they apply universally and are not repeated in lower-level descriptions, both to save space and to aid in learning diagnostic features of taxa at different ranks. In the description of a taxon, if a character state is indicated to be generally true, that means it is present in over half of the included taxa; in the descriptions of the included taxa, the character is then addressed only for those differing from the general state. This is the so-called gen rule. Also, at the species level and below, if a state exists in over half — but not all — of the included individual plants, the abbreviation "gen" is used.

Exceptional or rare character states (both quantitative and qualitative) of a taxon are often set off by parentheses, without mention of rarity of the condition if the context is unambiguous. Parentheses also are used for explanatory purposes or to aid in interpretation of complex statements (see below).

Descriptive information is organized into fields, most of which are headed and highlighted by all-capital, bold-face abbreviations of a major organ or structure (most descriptions include ST, stem; LF, leaf; INFL, inflorescence; FL, flower; FR, fruit; SEED, seed). In some cases, such headings are given to specialized, complex, or compound structures as well (e.g., SPORANGIA, SEED CONE, STAMINATE INFL, DISK FL, PISTILLATE FL).

Each statement within a field is a noun (plant part) followed by its modifiers; a given noun is being described up to the place where another noun appears, at which point it becomes the noun being modified. Adjectives follow the nouns they modify and different kinds of punctuation are used to make descriptions clear. For example, in the field for flowers (a primary noun), statements about sepals, petals, stamens, and carpels (secondary nouns) are separated by semicolons, leaving only commas and word order as ways to ensure clarity between the semicolons. Where a complex description might be ambiguous, parentheses or other clarifying punctuation may be present: e.g., "FR 3–4 mm, ovoid; segments 10–14, margins winged, tips bristly (or segments 8–9, margins rounded, tips glabrous)." Note that the tertiary noun "tips" refers to the secondary noun "segments," not to the previous tertiary noun "margins" or the even more removed primary noun "FR."

Within descriptions, plant parts are addressed from lower to upper on the plant, and from proximal to distal on a plant part.

Articles and conjunctions are generally used only where their absence would create ambiguity. For example, the statement, "LF: blade lanceolate, margins ciliate, bases, tips tapered . . ." means that the leaf blade bases (as well as tips) are tapered and should not cause the user to wonder if an adjective modifying "bases" has been dropped inadvertently.

Descriptions of Families. Family names are based on the name of an included genus (though not necessarily a genus represented in California), to which is added the termination "-aceae." Some that do not conform to this format were in wide use from (or before) the time of Linnaeus and appear in *TJM 2* treatments only in parentheses after the family name of conventional form, i.e., Apiaceae (Umbelliferae), Arecaceae (Palmae), Asteraceae (Compositae), Brassicaceae (Cruciferae), Fabaceae (Leguminosae), Lamiaceae (Labiatae), and Poaceae (Gramineae). A common or colloquial name is given for each family, based on what is used primarily in California; many families (and other taxa) have more than one common name and in some cases more than one is given here, if more than one is likely to be familiar to users of this book.

For each family, approximate numbers of genera and species worldwide are indicated. Overall geographic range also is summarized. Where appropriate, notes on significance to humans also are included (nutritional, medicinal, agricultural, horticultural value; toxicity). A reference (or two) often is given in brackets, primarily as an entry into the literature, so that a more recent, less relevant citation may be indicated rather than an older, more relevant citation, if the former includes a reference to the latter. Sometimes additional notes are given, including summaries of recent changes in classifications (e.g., lumping or splitting of families, genera), as well as special information to keep in mind in using the treatment that follows.

After each family description, scientific editors are indicated in order of handling of the manuscript and/or relative contribution.

The general form of the descriptions of families is the same as that for species (see below).

Descriptions of Genera. If a family comprises only one genus, the family description serves as the genus description. In such cases, as well as for genera in families with more than one genus, the number of species and the overall geographic range of the genus are given, as well as any importance to humans, followed by the derivation of the genus name (in parentheses), an appropriate reference (or two) in brackets, and sometimes additional notes, such as important information to know in using the key and included descriptions.

A common name (printed in Times-Roman SMALL CAPITALS) is indicated if one exists. As with families, the general form of the description is the same as that for species (see below).

Descriptions of Species, Subspecies, and Varieties. Descriptive conventions not already covered above are addressed here; except as specified, they generally apply to family and genus descriptions as well.

Scientific Names. All plant names below the level of genus, collectively termed "scientific names," are in *italics*, to indicate they are latinized; they comprise a genus name followed by one epithet (for a species), forming a "binomial," or two epithets (for a subspecies or variety, that is, infraspecific taxa), yielding a "trinomial." Trinomials appear intact for species represented in California by only one of its included infraspecific taxa; for species with more than one infraspecific taxon in the flora, only the rank of such infraspecific taxa and the associated epithets are indicated, at the beginning of each description. Genus names begin with a capital letter, whereas epithets for taxa at the level of species and below do not.

Names of taxa considered native to California are in bold-face italic *Times-Roman*; names of alien taxa (naturalized plants as well as waifs) are in italic *Helvetica*, without bolding; names of the few taxa of uncertain status (native or naturalized) are in bold-face italic *Helvetica*.

The etymology of genus names is given after the genus description; for the meanings of epithets there are a number of references online.

Authors of Plant Names. Epithets in scientific names appearing at the beginning of descriptions are followed immediately by an abbreviation of the name of the person(s) who validly published the name, in standardized form. Otherwise, author citations are given only for names that do not appear in such positions. Author citations serve in part to distinguish between two or more scientific names of exactly the same form (genus name and epithets the same), in cases where they refer to different plants, although this kind of problem does not occur very frequently. Author citations also help to locate the original description or other supporting material. *Authors of Plant Names* (1992), compiled at The Herbarium, Royal Botanic Gardens, Kew, serves as the standard for author abbreviations here, as updated online in The International Plant Names Index (IPNI).

Common Names. Long before scientific names were used, people used names in their native languages to refer to plants that were important to them. These have been called colloquial or common names, and those in English are sometimes called English names. Such common names are included here, if their usage has continued. In addition, common names that have been used in the California Native Plant Society's *Inventory of Rare and Endangered Plants of California* are generally included here. These taxa often are referenced elsewhere in the literature only by their common names, which may have been invented recently. Although common names in plants are not subject to regulation and may be used for more than one taxon, they do not need to change when the scientific name of a taxon changes, as happens for example when a species is transferred to another genus. All common names for taxa at the level of species and below are printed here in Times-Roman SMALL CAPITAL letters.

Chromosome Numbers. Chromosome numbers have been considered to be essential components of floristic works since their use in taxonomy first rose to prominence, in about the 1940s. However, those included here should not be taken as definitive. Because this book is intended primarily for use in the field, priority was given to verifying information about the plants that could be determined with the naked eye, aided or not by a hand lens. In other words, whereas morphological data were scrutinized by both authors and editors in preparing treatments for inclusion here, cytological data were mostly determined by workers other than our authors, and simply are being reported here from primary sources. A common problem with these reports is the failure to indicate the specimens used to obtain counts; that is, the frequency with which a voucher specimen is not cited (a problem that has lessened through time). This can be a serious problem, for example, when taxonomic concepts change, one species is recognized as two, and it is not known, without a specimen to consult, to which of the resulting species the previously reported count applies. Another problem is the small number of plants often sampled to determine chromosome numbers, which therefore may vary more than is indicated.

Given the value of chromosome numbers for understanding evolutionary patterns and processes in plants, they are included here where reported, for over 50% of the taxa, as an aid to researchers, with the above cautions. Absence of indicated chromosome numbers also should help to focus attention on which taxa remain uncounted, so that botanists can prioritize their efforts.

Habitats, Elevations, Geography. Accounts of habitats, elevations, and geographic areas where a taxon is expected to occur are included primarily to help predict where it may be found. Secondarily, such information may help to corroborate the identity of a plant thought to belong to a particular taxon based on other descriptive information. Many plant taxa occur in a wide range of habitats, so distributional information may be of limited value in predicting locations of populations and confirming identifications, while for plants of narrow distribution the information may be highly useful in both regards.

A limited number of terms used to characterize habitats, including vegetation, are included in treatments and in the glossary. These terms are defined generally and are not intended to represent a rigorous system of habitat or vegetation classification, which is beyond the scope of this book. For more detailed and/or alternate treatments of vegetation and habitats of Californian plants, and for discussion of environmental factors that influence the occurrence of plants in California, the following two works are recommended: *A Manual of California Vegetation* (Sawyer et al., 2009; California Native Plant Society, Sacramento) and *Terrestrial Vegetation of California* (Barbour et al., 2007; University of California Press, Berkeley).

Elevational ranges of taxa (in meters; a conversion scale to feet is near the back of the book, after the index) are approximations, mostly based on data from herbarium specimens. As with habitats, the information is perhaps most useful for predictive or identification purposes if taxa occur within a nar-

row range of values. Elevational ranges, even if highly accurate, may not reflect potential ranges under current climatic conditions.

Geographic Range Statements. A four-tiered hierarchical geographic system was developed for *TJM* (1993), primarily to allow for the presentation of geographic ranges in a way that is more biologically meaningful than doing so by county (the more traditional way), and to save space as well. The four-tiered system also is used here, with significant refinements (see Geographic Subdivisions and associated maps, also inside the front cover). The Consortium of California Herbaria aided description and refinement of geographic ranges for many taxa, in part by facilitating discovery of outlying records and pinpointing specimens worthy of close scrutiny by authors.

Occurrences of a given taxon within its geographic range are further restricted by elevation and habitat, as noted above. At the most general, "CA" means that a taxon occurs in all three floristic provinces in California (though not necessarily all counties in California), but within that area the taxon may be especially rare or common, depending on the elevations and habitats indicated for it. So too, "CA-FP" (California Floristic Province) as a sole area of occurrence means that the taxon should not be expected in "GB" (Great Basin Province) or "D" (Desert Province), but might be found anywhere within each of the geographic units that constitute the CA-FP where the appropriate elevations and habitats occur.

In general, all areas indicated for a taxon in California are documented by herbarium specimens, with notable exceptions. Alien taxa, for example, are of limited interest to many collectors and therefore tend to be undercollected (except by people studying them and concerned about their spread), so observational data as well as reports of occurrence in the literature were sometimes accepted in lieu of voucher specimens; additionally, alien taxa with currently expanding ranges are generally indicated to be expected elsewhere. Geographic subunits from which native taxa have been reported or are expected but for which no specimens have been seen are followed by a question mark (e.g., "s SNH, SNE, w DMoj?"), and should be given high priority by collectors.

Synonyms, Misapplied Names, Unresolved Variants. Including references to all names that have ever been applied to vascular plants in California (i.e., to provide a complete synonymy for each of over 7,600 terminal taxa) is impractical in a single, field-portable floristic manual. Instead, all names used for accepted, recognized taxa in *TJM* (1993) are accounted for in some way here, in most cases by using them as the accepted name for a taxon or by listing them as synonyms of such names, but in some cases by indicating they have been used by mistake, either by misinterpretation of the rules of nomenclature or by misidentification of plant material (misapplied names), or by indicating them to represent unclearly or questionably differentiated variants (unresolved variants). The Index to California Plant Names (an online resource of the Jepson Flora Project) includes much more extensive coverage of names previously applied to California vascular plant taxa and the status of such names.

Synonyms (based on either the same or a different type specimen; that is, nomenclatural or taxonomic synonyms, respectively), followed by misapplied names, appear in brackets after the statement of geographic distribution.

Unresolved variants are those about which an author wishes to remain noncommittal with respect to taxonomic recognition. They are included after the brackets containing synonyms and misapplied names, along with a very brief diagnosis and geographic range (if different from the species). This account is noncommittal in form (e.g., "If recognized taxonomically, smaller, denser plants from higher elevations assignable to *Planta pumila*"). The wording used in *TJM* (1993) (e.g., "Smaller, denser plants from higher elevations have been called *Planta pumila*") often has been avoided because of the common, incorrect practices of treating such names as accepted names or as synonyms; they are neither.

The number of unresolved variants treated here is considerably smaller than the number treated in *TJM* (1993), primarily because of changes in taxonomic concepts (e.g., the recognition and inclusion of cryptic taxa here) and the fact that research conducted since *TJM* (1993) has resolved many of these issues, elevating some unresolved variants to full taxonomic treatment (as varieties, subspecies, or species) while reducing others to synonymy. Authors were strongly encouraged to keep the number of unresolved variants here to a minimum, by giving a high priority to resolving them one way or another, but in some cases it was not possible to complete the research necessary in time for inclusion here.

Notes. Notes that address unresolved taxonomic problems, hybridization and intergrading variation in general, difficulties or clues in identification, threats from human activity as well as from other plants and animals, and other topics appear near the end of descriptions, after any synonyms and before the following information.

Toxicity. Some California plants (e.g., poison hemlock, poison oak, Klamath weed) are seriously toxic, causing deaths or illness in humans or livestock. Fuller & McClintock's (1986) *Poisonous Plants of California* (University of California Press) overviews plants that are both major and minor sources of poisoning. Plants that have been toxic to animals or people in California (or are expected to be) are noted with an all-capital "TOXIC." Usually, more specific information is included as well, e.g., "TOXIC to livestock from concentrated oxalates"; "TOXIC: resin on lvs, sts, frs causes severe contact dermatitis; one of the most hazardous pls in CA"; or "TOXIC: ingested seeds, lvs, bark may be fatal to humans, livestock."

Commonness and Rarity. Formal designations of invasive, non-native taxa, and native taxa of special concern, including those that are rare, threatened, or endangered, are not provided in *TJM 2* because such rankings are potentially dynamic. Instead, symbols are applied in *TJM 2* treatments to call attention to particular invasive taxa or native taxa of special concern. For more information on the status of the indicated taxa (and perhaps others that are not indicated with a symbol), the reader should contact the agencies and organizations mentioned in the following paragraphs.

Native Taxa of Special Concern. A ★ symbol is applied to taxa as recognition of their inclusion in the California Native Plant Society's (CNPS's) *Inventory of Rare and Endangered Plants of California* (8th Edition), lists 1–4. That set of lists includes all vascular plant taxa that are legally listed as rare, threatened, or endangered by the State of California or the U.S. Fish and Wildlife Service. A limited number of taxa, proposed for inclusion in the CNPS Inventory, also received the ★ symbol.

Invasive Non-native Taxa. The ◆ symbol is applied to taxa as recognition of their inclusion in (1) the Pest Ratings of Noxious Weed Species and Noxious Weed Seed, developed by the State of California, Department of Food and Agriculture, Division of Plant Health and Pest Prevention Services (July, 2010) and/or (2) the 4500 Noxious Weed Species list from Section 5004 of the Food and Agricultural Code. All names with an internal rating of A, B, C, or Q were considered. The ◆ symbol is applied to taxa that occur in California or are considered to be of probable risk of establishment in the state. Some taxa included in the above-mentioned lists are not treated here because they either do not occur in California or their arrival and/or naturalization in the state is unlikely. The ❖ symbol is applied to taxa not included in the above-mentioned state or federal lists, but included in (1) the California Invasive Plant Inventory Database (December 2010) developed by the California Invasive Plant Council (Cal-IPC) and/or (2) the Priority Species List of the Bay Area Early Detection Network (December 2010). Neither symbol ◆ nor symbol ❖ is applied to native taxa.

Flowering/Coning Time. For angiosperm species, subspecies, or varieties, the time period, usually in months (abbreviated to three letters) but sometimes in seasons, when flowers are known to be

present on the plants are presented at the end of the description (the words "flowering time" do not appear). For *Ephedra,* the time periods refer to the presence of cones.

Horticultural Value. Information about horticultural value and growth requirements, included for many native taxa in *TJM* (1993), is not included here, but the Jepson Flora Project's database of such information may be accessed online.

Things to Remember When Using This Book

Coverage. Make sure the organism you wish to identify is included in *TJM 2*. Only vascular plants that are native, naturalized, or active waifs in California may be identified using this book; excluded are all bryophytes, lichens, algae, fungi, animals, and non-living things (see introduction to Key to California Vascular Plant Families for information on distinguishing vascular plants from other organisms).

Reproductive Condition. With rare exceptions, reproductive parts are needed for accurate identification of California vascular plants, whether the unknown is a pteridophyte (indusia, sporangia, spores), gymnosperm (cones, seeds), or angiosperm (flowers, fruit). Aquatic plants in vegetative condition may be identified to family using *TJM 2* but most keys herein require access to reproductive states.

Glossary. Sometimes difficulty in using the keys and descriptions is attributable to the fact that a term has been defined in a way that might seem unusual; check the glossary as a possible solution. Unfamiliar terms that are not included in the glossary may be defined only in descriptions of the relevant family or genus, if those terms do not apply to other taxa in California.

Notes. If an element of a key or description is not making sense, check the notes under the descriptions of relevant families, genera, or species. For example, the number of flowers indicated for *Persicaria*, 1–14, may appear much too low, given that the inflorescence in the genus generally has hundreds of flowers. In a note under the family description, it is stated that "fl number is per fl cluster or involucre, unless otherwise stated."

Gen Rule. If a plant characteristic is not mentioned in a taxon description, information on that state should be sought in descriptions of higher-level taxa to which the plant belongs. As noted above, in the description of a taxon, if a character state is indicated to be generally true, it means it is present in over half of the included taxa; in the descriptions of the included taxa, the character is then addressed only for those taxa differing from the general state. For example, if "stamens gen 5" is stated in the genus description, in the descriptions of species and below the character is not addressed if the state is 5, but it is addressed for those taxa for which the number of stamens is other than 5.

Assumptions. Unless stated otherwise, number of flowers is per inflorescence, of petals is per flower, of seeds is per fruit (if per chamber, that is specified), etc.

Parentheses. In morphological descriptions and keys, parentheses enclose character states that are rare to uncommon, without qualification if the context is unambiguous, as well as explanations of foregoing statements and exceptions to those statements.

Square Brackets. In morphological descriptions, square brackets enclose information pertaining only to members of a taxon (but not necessarily to all members of that taxon) occurring outside of California. If used in keys, square brackets enclose explanatory information, as opposed to rare or exceptional states (in parentheses).

Keys with only 2 Taxa. Character states appearing in keys with only 2 taxa are not repeated in the descriptions of those taxa, as a space-saving measure.

ABBREVIATIONS AND SYMBOLS

Abbreviations

The abbreviations below were selected because they save considerable space, are relatively unambiguous, and are easily remembered. They are used throughout this book, with the exception of introductory material. Words not appearing below are not abbreviated, except that the official, two-letter, postal abbreviations for states in the United States are used. Abbreviations that appear in both lowercase and capital letters are indicated. Periods are used only where their absence could cause confusion. Entries referring to parts of California are marked with asterisks and discussed more fully under Geographic Subdivisions of California (p. 35).

AB = Alberta, Canada
Afr = Africa
Am = Americas (w hemisphere), America(n)
ambig. (or nom. ambig.) = nomen ambiguum, ambiguous name; name commonly used by mistake for more than one taxon
ann = annual

b = born
Baja CA = Baja California
BC = British Columbia, Canada
bien = biennial

c = central
CA-FP = California Floristic Province*
C.Am = Central America(n)
Can = Canada
CaR = Cascade Range*
CaRF = Cascade Range Foothills*
CaRH = High Cascade Range*
CCo = Central Coast*
ChI = Channel Islands*
cm = centimeter, 0.01 meter
Co. = County
cos. = counties
cult = cultivated, cultivation
cv. = cultivar, cultivated variety
CW = Central Western California*

D = Desert Province*
diam = diameter
dm = decimeter, 0.1 meter
DMoj = Mojave Desert*
DMtns = Desert Mountains*
DSon = Sonoran (Colorado) Desert*

e = east(ern)
e-c = east-central
esp = especially
Eur = Europe
exc = except, excluded, excludes, excluding

f. = form, forma; son of (L.f. means son of Linnaeus, see Authors of Plant Names)
fl, fls (FL, FLS) = flower(s), floral, flowering
fld = flowered
FNANM = Flora of North America North of Mexico
fr (FR) = fruit

GB = Great Basin Province*
gen = generally, mostly, usually, over half (e.g., petals gen red in genus description means over half of subordinate taxa have red corollas, with the rest requiring that petal color be addressed)
geog = geographic(al, ally), geography
GV = Great Central Valley*

illeg. (or nom. illeg.) = nomen illegitimum, illegitimate name; name validly published but otherwise not conforming to the rules
incl = included, includes, including
ined. (or nom. ined.) = nomen ineditum, unpublished name; name not published or not validly published
infl, infls (INFL, INFLS) = inflorescence(s)
inval. (or nom. inval.) = nomen invalidum, invalid name; name not validly published according to the rules

KR = Klamath Ranges*

lf (LF) = leaf
lfless = leafless

lflet = leaflet
lfy = leafy
lvs (LVS) = leaves
lvd = leaved

m = meter
MB = Manitoba, Canada
MP = Modoc Plateau*
Medit = Mediterranean
Mex = Mexico
misappl. = misapplied; name used incorrectly for a CA
        plant, through misidentification and other means
mm = millimeter, 0.001 meter (μm, a micrometer, is
        0.0001 meter, previously also called a micron)
(M)mtn(s) = (M)mountain(s)

n = north(ern)
n-c = north-central
N.Am = North America(n)
NB = New Brunswick, Canada
NCo = North Coast*
NCoR = North Coast Ranges*
NCoRH = High North Coast Ranges*
NCoRI = Inner North Coast Ranges*
NCoRO = Outer North Coast Ranges*
ne = northeast(ern)
NL = Newfoundland and Labrador, Canada
notho- = prefix indicating that a taxon is the result of
        hybridization, when at least one of the parental taxa
        involved is known or can be postulated, affixed to
        the term denoting the rank at which that taxon is
        recognized; e.g., nothosubspecies
NS = Nova Scotia, Canada
NT = Northwest Territories, Canada
nud. (or nom. nud.) = nomen nudum, naked name; name
        naked usually in the sense of lacking a description with
        its publication

NW = Northwestern California*
nw = northwest(ern)

occ = occasionally
ON = Ontario, Canada
orn = ornamental
orth. var. = orthographic variant; variant spelling of a name

PE = Prince Edward Island, Canada
per = perennial herb
pl(s) (PL) = plant(s)
PR = Peninsular Ranges*

QC = Quebec, Canada

rej. (or nom. rej.) = nomen rejiciendum, rejected name;
        name prohibited by legislation

s = south(ern)
s-c = south-central

S.Am = South America(n)
SCo = South Coast*
SCoR = South Coast Ranges*
SCoRI = Inner South Coast Ranges*
SCoRO = Outer South Coast Ranges*
ScV = Sacramento Valley*
se = southeast(ern)
sect(s). = section(s) (abbreviated only as taxonomic rank)
SK = Saskatchewan, Canada
s.l. = sensu lato, in the broad sense; broad circumscription
        of a taxon
SON = Sonora, Mexico
SN = Sierra Nevada*
SNE = East of Sierra Nevada*
SNF = Sierra Nevada Foothills*
SNH = High Sierra Nevada*
SnBr = San Bernardino Mountains*
SnFrB = San Francisco Bay Area*
SnGb = San Gabriel Mountains*
SnJt = San Jacinto Mountains*
SnJV = San Joaquin Valley*
sp. = species (singular)
spp. = species (plural)
s.s. = sensu stricto, in the narrow sense; narrow
        circumscription of a taxon
st(s) (ST(S)) = stem(s)
subg. = subgenus, subgenera
subsect(s). = subsection(s)
subsp. = subspecies (singular)
subspp. = subspecies (plural)
superfl. (or nom. superfl.) = nomen superfluum, superfluous
        name; name for a taxon that has an earlier, legitimate
        name
SW = Southwestern California*
sw = southwest(ern)

Teh = Tehachapi Mountain Area*
temp = temperate(s), temperate zone(s)
TJM (1993) = *The Jepson Manual,* 1993 edition
TR = Transverse Ranges*
trop = tropical, tropic(s), tropical zone(s)

US = United States

var. = variety
vars. = varieties
vs = versus

w = west(ern)
w-c = west-central
W&I = White and Inyo Mountains*
WTR = Western Transverse Ranges*
Wrn = Warner Mountains*

yr(s) = year(s)
YT = Yukon, Canada (Yukon Territory)

Symbols

The following symbols are used whenever possible. Most are quantitative, referring to number, height, length, width, etc., while "±" may be qualitative as well, referring to color, fusion, symmetry, etc. Note that "<<," "<," "=," ">," and ">>" do not include the concepts of "greatly exceeded by," "exceeded by," "held at the same level as," "exceeding," and "greatly exceeding," respectively, as defined in *TJM* (1993). Here, these concepts are expressed in words. The symbols "<<," "<," "=," ">," and ">>" are restricted in meaning to "much less than," "less than," "equal to," "greater than," and "much greater than," respectively, in length, or height (or width or diameter if qualified as such). Those symbols also may be used in the sense of "fewer than" (rather than "less than") or "more than" (rather than "greater than") if qualified as referring to counts rather than dimensions. Use of these symbols to include the ideas of "exceeding" or "exceeded by" confuses the concepts of absolute length and what it means for one structure to exceed another or not.

<< much less than (or much fewer than, if qualified as numeric)

<   less than (or fewer than, if qualified as numeric)

≤   less than or equal to (or fewer than or equal to, if qualified as numeric)

=   equal to, equal, equals (e.g., "sepals = petals" or "blade = petiole," but not "sepals equal" in the sense of sepals all equal to each other)

≥   greater than or equal to (or more than or equal to, if qualified as numeric)

>   greater than (or more than, if qualified as numeric)

>> much greater than (or much more than, if qualified as numeric)

0   none, absent

1   solitary, in context of arrangement of sporangia, cones, flowers, or inflorescences

+   1) after a number or range of numbers, it means "or more" (e.g., "6–10+ mm" means "6 to 10 or more mm"); 2) between words (e.g., plant structures, word elements in etymologies) it means "plus" (e.g., "tube + throat 12–16 mm" means that the tube and throat taken together are 12–16 mm)

±   more or less, approximately, nearly, rather, slightly, somewhat (e.g., ± sessile may include sessile)

°   degree of angle, compoundedness, or branching

×   multiplication sign, meaning "times" or indicating hybridity when used with taxon names; meaning "cross" when used in plant descriptions (e.g., ×-section)

-   hyphen, for: compound adjectives (e.g., 5-lobed, saucer-shaped, needle-like, red-brown, glandular-hairy, ovate-elliptic); in common names that are inconsistent with current taxonomy (e.g., Douglas-fir because *Pseudotsuga* is not currently included in *Abies*, fir), or in common names that are used as adjectives (e.g.,

lodgepole-pine forest, but not forest of lodgepole pine); to indicate (as a double hyphen or en-dash) quantitative ranges (e.g., "lvs 5–8 mm"); and to indicate intermediacy in condition (e.g., "lvs ovate-elliptic" means the leaves are intermediate between ovate and elliptic). Qualitative (non-quantitative) ranges are expressed with the word "to": "lvs ovate to elliptic" means the leaves range in shape from ovate to elliptic, possibly including ovate-elliptic

[ ]   1) square brackets enclose information in descriptions pertaining only to members of a taxon (but not necessarily to all members of that taxon) occurring outside of California, in addition to delimiting pertinent literature and nomenclatural/taxonomic synonyms; 2) in descriptive parts of keys, square brackets are sometimes used to delimit information that provides additional clarification or explanation, where use of parentheses might be confusing; 3) square brackets enclosing names in keys refer to taxa (generally waifs) with descriptions that appear only online

( )   parentheses enclose information in keys and descriptions that is only rarely to uncommonly true [e.g., tree (2)4–10 m]; less frequently, they enclose explanations of, elaborations of, or exceptions to foregoing descriptive information; if the word "rarely" appears within parentheses, it is because without it the enclosed information could be interpreted in one of these other ways

µm   micrometer, 0.0001 meter (previously also called a micron)

★   rare and/or endangered native taxon; see Conventions (p. 10) for further explanation

◆   weedy, alien taxon; see Conventions (p. 10) for further explanation

❖   weedy, alien taxon; see Conventions (p. 10) for further explanation

GLOSSARY

One of the main goals of *The Jepson Manual, Second Edition* (*TJM 2*) is to facilitate plant identification, which often relies on extremely subtle differences in plant characters. As in *The Jepson Manual*, or *TJM* (1993), the terminology used in *TJM 2* is constrained to make the work accessible to a broad audience. About 100 new terms not used in *TJM* (1993) were judged to be of high value for improving precision and brevity of descriptions and are included here. All users are encouraged to routinely consult the glossary, and beginners are encouraged to study the glossary, especially the illustrations, as a way to acquire the basic botanical knowledge needed to identify plants.

abaxial. Side or surface of a structure away from the axis distal to the point at which the structure is borne (e.g., lower surface of a leaf, outer surface of a petal). (see adaxial)

abundant. Very likely present in appropriate habitats, sometimes forming dense stands (see common, rare, uncommon)

achene. (pp. 27, 28, 31) Dry, indehiscent, 1-seeded fruit from a 1-chambered ovary in which the fruit wall is free from the seed, sometimes winged; often appearing to be a naked seed. 1-seeded dry fruit derived from an inferior ovary of > 1 carpel (e.g., Asteraceae) is sometimes called a cypsela.

acid (acidic). Soil or water with a low pH, often found in habitats such as coniferous forests and bogs where decomposition of plant remains liberates an excess of hydrogen ions.

acroscopic. In ferns, facing or directed toward the tip of the leaf (e.g., on any pinna, acroscopic pinnules are those on the side closest to the leaf tip). (see basiscopic, distal)

acuminate. (p. 22) Having a long-tapered, sharp tip, the sides concave. (see acute, attenuate, awl-like)

acute. (p. 22) Having a short-tapered, sharp tip, the sides convex or straight and converging at less than a right angle. (see acuminate, attenuate, awl-like, obtuse)

adaxial. Side or surface of a structure toward the axis distal to the point at which the structure is borne (e.g., upper surface of a leaf, inner surface of a petal). (see abaxial)

adherent. Sticking to and sometimes appearing fused to another part of like or unlike kind, but separable from it, such as "perianth adherent to fruit." (see appressed, fused)

adventitious. Arising at unusual times or places, such as roots on aerial stems.

aggressive. Growing or spreading rapidly or invasively, outcompeting other plants, difficult to control.

alien. Not native; introduced purposely or accidentally into an area. (see native, naturalized, ruderal, waif)

alkali, alkaline. Soil or water with a high pH (i.e., basic), often found in areas where evaporation concentrates dissolved solutes.

alkali sink. Basin area in region of interior drainage characterized by soils with high salinity and high pH.

alluvial. Pertaining to sediments deposited by flowing water.

alluvial fan. Fan-shaped deposit of rocks, gravel, and finer sediments, in California generally on lower slopes of mountains.

alpine. Pertaining to the vegetational/altitudinal zone above timberline; zone above the subalpine.

alternate. 1. (p. 21) Arranged singly, often spirally, along an axis (e.g., one leaf per node). (see opposite, whorled) 2. Occurring in different ranks, appearing to be between, not directly above or below, as "stamens alternate petals." (see rank)

angiosperm. Plant that bears flowers (hence, "flowering plant"), in which "vesseled seeds" (hence, angio-sperm) are enclosed in an ovary; woody to herbaceous.

annual. (p. 18) Completing life cycle (germination through death) in one year or growing season, generally non-woody. (see biennial, herb, perennial)

anther. (pp. 26, 30, 31) Pollen-bearing portion of a stamen, including one, two, or four pollen sacs. (see filament)

apogamous. Forming a sporophyte (the generation in a vascular plant life cycle that is conspicuous, 2*n*, and produces spores, which give rise to gametophytes) from a gametophyte (the generation in a vascular plant life cycle that is inconspicuous, *n*, and produces sperms and eggs, which usually unite in fertilization to give rise to sporophytes) by direct, asexual development, rather than by fertilization of eggs by sperms.

appressed. (pp. 19, 28) Parallel or nearly parallel to and often in contact with surface of origin; used to describe

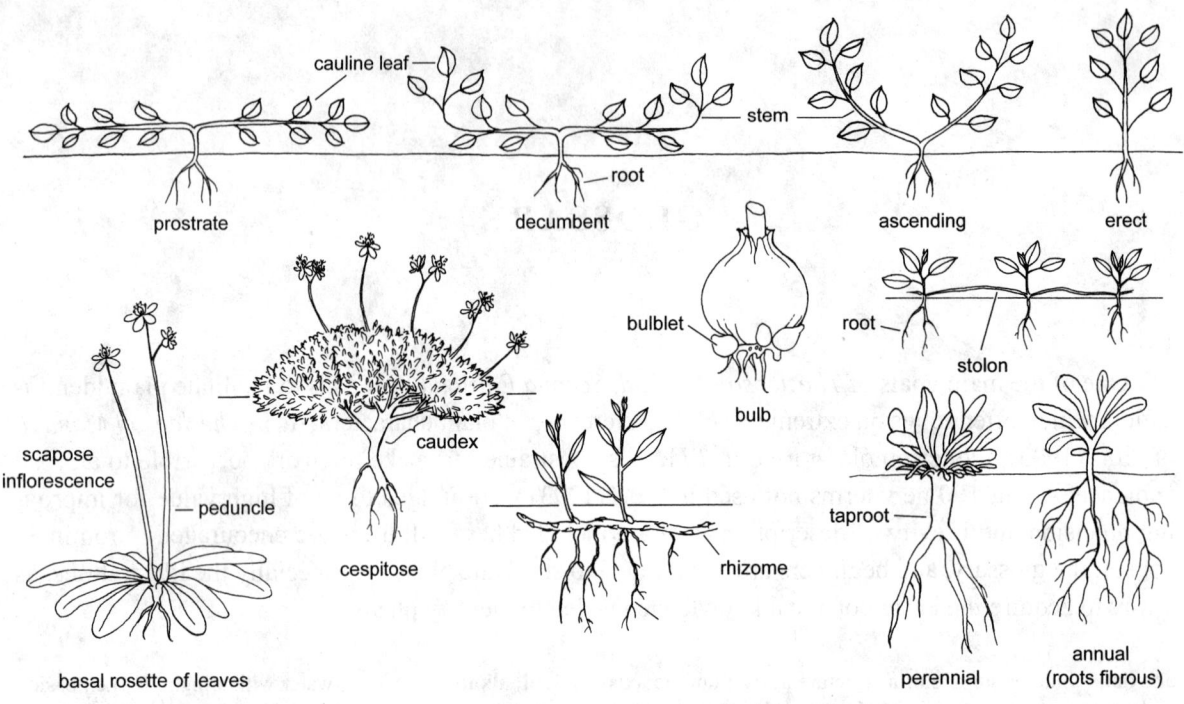

the disposition of hairs, leaves, pedicels, etc. (see adherent, fused)

aquatic.   Growing under, in, or on water (generally fresh; if brackish, saline, or marine, so indicated), whether rooted in bottom or floating, and including plants with parts of shoots submersed but with other parts above water (e.g., *Potamogeton gramineus*); excluding plants of seeps or wet rocks. (see emergent, submersed)

areole.   1. In Cactaceae, a well-defined, axillary area (short shoot) generally bearing one to many spines and other, shorter structures (e.g., *Ferocactus cylindraceus* glands, *Opuntia* glochids). 2. In general, each of many areas defined by smallest veins on a leaf.

aril.   (p. 149) Fleshy, corky, or bony appendage arising at or near the point of seed attachment, sometimes completely covering the seed.

armed.   Bearing prickles, spines, or thorns.

ascending.   (pp. 18, 19) Curving or angling upward from base, or about 30–60° less than vertical or away from axis of attachment. (see decumbent, erect)

asymmetric.   (p. 25) Not divisible into identical or mirror-image halves. (see bilateral, biradial, radial)

attenuate.   Having a long-tapered, sharp tip. (see acuminate, acute, awl-like)

auricle.   (p. 30) In Poaceae, a structure, often lanceolate, projecting from both margins of the leaf lateral to the ligule. (see ligule)

awl-like.   (p. 24) Narrow throughout, but broader at the base and tapered to a sharp tip. (see acuminate, acute, attenuate)

awn.   (pp. 22, 30) 1. Bristle-like appendage or elongation, generally terminal. 2. Stiff, needle-like pappus element in Asteraceae.

axil.   (p. 21) Distal, adaxial angle between an appendage or branch and a main axis (e.g., between leaf and stem, or between lateral vein and midrib on a leaf).

axile.   (p. 26) Pertaining to an axis, as of a placenta along the central axis in a compound ovary with more than one chamber.

axillary.   (pp. 20, 29) Pertaining to or within an axil, especially a leaf axil.

axis (axes).   (p. 30) Line of direction, growth, or extension; structure occupying such a position (e.g., the main stem of a plant or inflorescence, the midrib of a leaf).

banner.   (p. 25) Uppermost, often largest petal of many members of Fabaceae.

bar.   Mound-like temporary deposit of sand or gravel in the channel or mouth of a waterway.

barbed.   (p. 28) Having sharp, normally downward- or backward-pointing projections. Said of an awn, bristle, or other structure.

bark.   Tough tissue (including phloem) covering the wood (hardened xylem) of subshrubs, shrubs, trees, and some vines. (see wood)

barren.   Area in which vegetation is sparse due to harsh or limiting growing conditions, such as those associated with shallow, infertile, rocky soil.

basal.   (p. 18) At or near the base of a plant or plant part. Especially said of leaves clustered near the ground or of a placenta confined to the base of an ovary.

basiscopic.   In ferns, facing or directed toward the base of the leaf (e.g., on any pinna, basiscopic pinnules are those on the side closest to the leaf base). (see acroscopic, proximal)

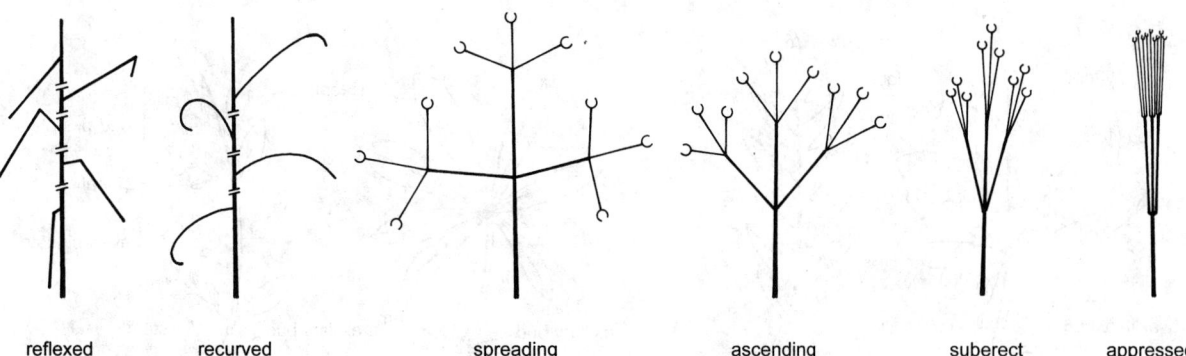

reflexed   recurved   spreading   ascending   suberect   appressed

beak. (p. 287) Generally distal or terminal, narrowed, generally sterile part of fruit or surrounding structures (e.g., *Ambrosia* bur, *Carex* perigynium).

bell-shaped. (p. 25) Widening more or less abruptly at the base and then generally more gradually toward the tip. (see funnel-shaped, rotate, salverform, urn-shaped)

berry. Fleshy, indehiscent fruit in which the seeds are generally more than 1 and are not encased in a stone (e.g., *Solanum americanum*). (see drupe, pome)

biennial. Completing life cycle (germination through death) in two years or growing seasons (generally flowering only in the second), non-woody (at least above ground), often with a rosette the first growing season. (see annual, herb, perennial)

bilateral. (p. 25) Divisible into mirror-image halves in only one way. (see asymmetric, biradial, radial)

biradial. (p. 25) Divisible into mirror-image halves in two ways; isobilateral. (see asymmetric, bilateral, radial)

bisexual. (p. 26) Both male and female reproductive parts occurring and functional in the same plant or structure (e.g., flower, spikelet, inflorescence). (see unisexual, pistillate, staminate, dioecious, monoecious)

blade. (pp. 21, 26, 29, 30) Expanded portion of a leaf, petal, or other structure, generally flat but sometimes rolled, cylindric, wavy, or cupped.

brackish. Somewhat salty, generally a mixture of saline and fresh water.

bract. 1. (pp. 24, 25) Generally reduced, leaf- or scale-like structure subtending a branch, cone scale, inflorescence (infl bract), or sessile flower or pedicel (fl bract). 2. Generally reduced, leaf- or scale-like structure on a peduncle or scape that may or may not subtend another structure. (see bractlet)

bractlet. (p. 25) 1. Relatively small, generally secondary bract within an inflorescence. 2. Bract-like structure on a pedicel that may or may not subtend another structure. (see bract)

bristle. 1. Relatively large, generally stiff, more or less straight hair (e.g., *Navarretia breweri*). 2. In Asteraceae, fine, cylindric or minutely flattened pappus element (e.g., *Calycoseris parryi*) or epidermal outgrowth on the receptacle (e.g., *Centaurea solstitialis*).

bud. (p. 21) 1. Incompletely developed, more or less embryonic shoot, usually covered with bud scales. 2. Unopened flower, often protected by sepals.

bulb. (p. 18) Short underground stem and the fleshy leaves or leaf bases attached to and surrounding it (e.g., an onion). (see stem, corm, caudex, tuber)

bulblet. (p. 18) 1. Small bulb generally produced at the base of a bulb. 2. Any small, bulb-like structure that propagates a plant, often in a leaf or bract axil.

bur. Fruit or fruiting inflorescence with awns or bristles, often barbed (e.g., *Xanthium strumarium*).

callus. 1. (p. 30) In some Poaceae, enlarged or projected hard base of floret; sometimes hairy or sharp-pointed. (see floret) 2. Firm protuberance.

calyx (calyces). (p. 26) Collective term for sepals; outermost or lowermost whorl of flower parts, generally green and enclosing remainder of flower in bud. Sometimes indistinguishable from corolla.

canescent. Covered with dense, fine, generally grayish white hairs (e.g., *Phoenicaulis cheiranthoides* leaf).

capsule. (p. 27) Dry fruit from compound pistil, nearly always dehiscent (irregularly or by pores, slits, or lines of separation). (see circumscissile, loculicidal, septicidal)

carpel. (p. 26) Basic female structure of a flower, evolved from a fertile leaf. Carpels are free or variously fused into a compound pistil, the number of carpels then often equal to the number of stigmas, styles, or chambers of the ovary. (see pistil)

cartilaginous. Thickened, usually whitish, sometimes flexible; in ferns, applied especially to margins of blades.

catkin. (p. 25) Spike or spike-like (e.g., *Alnus*) inflorescence of unisexual flowers with inconspicuous perianths (generally wind-pollinated), usually pendent and often with conspicuous bracts.

caudex (caudices). (p. 18) Generally short, sometimes woody, more or less vertical stem of a perennial, at or beneath ground level. (see stem, bulb, corm, tuber)

cauline. (p. 18) Pertaining to structures, especially leaves, borne along (i.e., not confined to the base of) an elongate, above-ground stem; not basal.

centimeter. One-hundredth of a meter; 10 millimeters (abbreviation: cm).

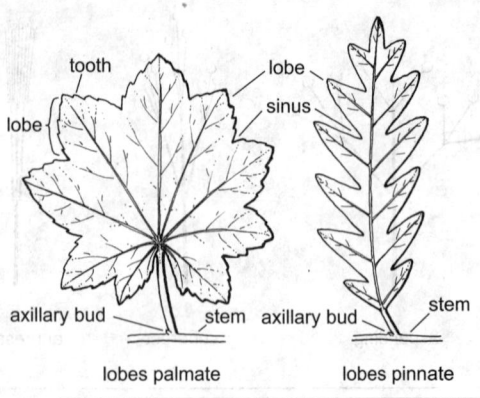

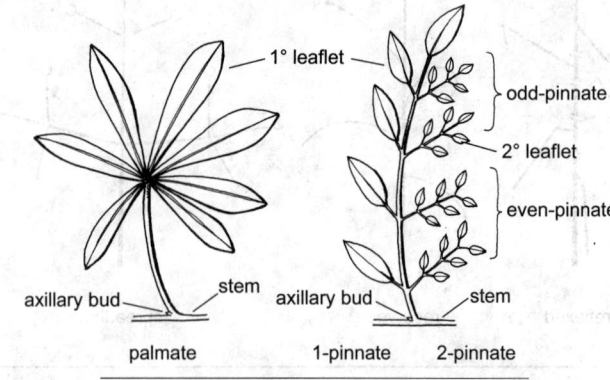

simple leaf   compound leaf

cespitose. (p. 18) Having a densely clumped, tufted, matted, or cushion-like growth form.

chamber. (p. 26) Compartment or cavity within an ovary, capsule, or other hollow structure.

chaparral. Vegetation characterized by mostly evergreen shrubs with thick, leathery leaves and stiff branches.

ciliate. (p. 28) Having generally straight, conspicuous hairs (cilia) along margins or edges.

circumboreal. Occurring around the world at northern latitudes.

circumscissile. (p. 27) Dehiscence, usually of a fruit (capsule), by a transverse line, the top coming off as a lid. (see loculicidal, septicidal)

claw. (p. 26) Stalk-like base of some free or nearly free sepals or petals. (see limb)

cleistogamous. Bud-like, unopening flowers that are generally self-fertilized.

clone. Genetically identical individuals resulting from asexual reproduction (fragmentation of rhizomes or stolons, budding, etc.); often used for an apparent population, the members of which are or were connected (e.g., aspens, cattails, duckweeds, sumacs).

closed-cone conifer forest or woodland. Vegetation characterized by species of *Pinus* or *Hesperocyparis* in which the seed cones persist unopened on the branches for extended periods of time.

coastal scrub. Coastal vegetation characterized by shrubs with flexible branches (e.g., *Baccharis pilularis*, *Artemisia californica*).

coastal strand. Beach and foredune habitat, characterized by sandy soils, strong winds, salt spray, and wave action.

collar. 1. In Poaceae, the abaxial junction of leaf sheath and blade. 2. Raised, inflated, or wing-like, encircling projection (e.g., seeds of *Delphinium luteum, D. nudicaule*).

column. (p. 26) Structure at the center of an orchid flower formed by fusion of stamen(s) and style.

common. Likely present in appropriate habitats. (see abundant, rare, uncommon)

compound. 1. Composed of two or more parts, as a compound leaf composed of leaflets (see compound leaf) or a compound pistil composed of fused or partly fused carpels. 2. Repeating a structural pattern (a compound umbel is an umbel of umbels). (see simple)

compound leaf. (p. 20) Leaf divided into distinct parts. In a 1-compound leaf, the blade is divided into primary leaflets connected by an axis but no blade material, in a 2-compound leaf, the primary leaflets are so divided into secondary leaflets, etc. (see palmate, pinnate, lobed, dissected)

compressed. (p. 28) Flattened side-to-side (laterally compressed) or front-to-back. (see depressed)

concave. Hollowed or indented, as the interior of a curved surface. (see convex)

cone. Reproductive structure composed of an axis, scales, and sometimes bracts. 1. Non-woody structure producing spores (e.g., clubmosses, horsetails) or pollen (e.g., male cone of conifers). 2. Generally woody structure producing seeds (e.g., female cones of most conifers).

conic. Three-dimensional, defined by a wide, more or less round base, the sides evenly tapered to a narrow tip.

conifer forest. Vegetation characterized by trees belonging to various species of conifers (e.g., firs, pines, redwoods).

continuous. Having parts spaced evenly and without interruption, not clumped; pertaining especially to inflorescences in which the flowers are evenly spaced. (see interrupted)

convex. Rounded outward, as the exterior of a curved surface. (see concave)

cordate. (p. 23) Heart-shaped; often pertaining to a leaf in which the blade base on both sides of the petiole is rounded and convex. (see reniform)

corm. Short, thick, unbranched, underground stem often surrounded by dry (not fleshy) leaves or leaf bases (e.g., *Muilla maritima*). (see bulb, stem)

corolla. (pp. 26, 31) Collective term for petals; whorl of flower parts immediately inside or above calyx, often large and brightly colored. Sometimes indistinguishable from calyx.

costa (costae). (p. 29) In ferns, primary axis of a pinna.

cotyledon. Seed-leaf; a modified leaf present in the seed, often functioning for food storage. Persistent in some annuals and of aid in their identification (e.g., *Lupinus microcarpus*).

crenate. (p. 22) Pertaining to margins with shallow, rounded teeth, between which are usually acute sinuses (i.e., scalloped).

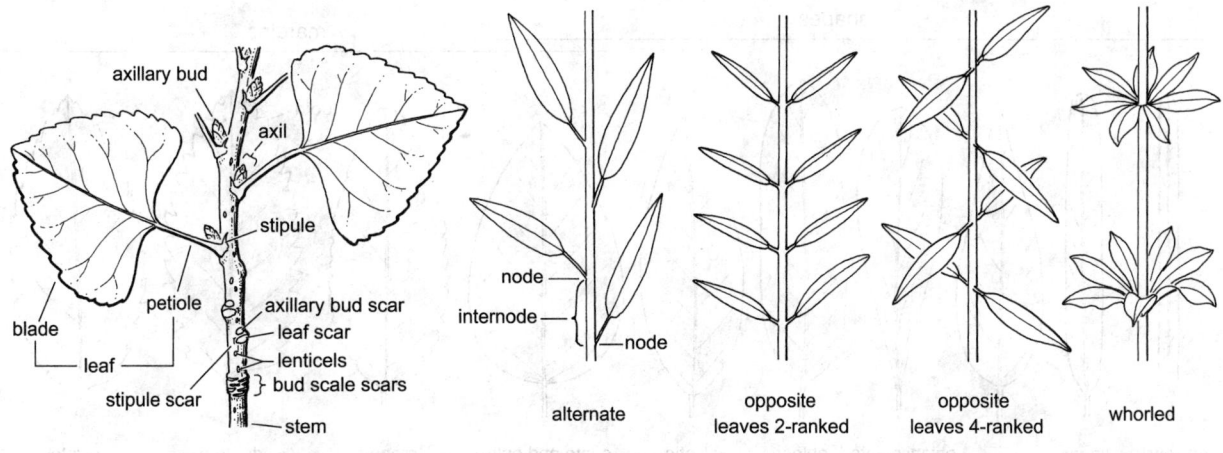

cylindric.   (p. 22) Elongate, with parallel sides and, at any point, round in transverse section.

cyme.   1. (p. 25) In flowering plants excluding Asteraceae and some other groups, a branched inflorescence in which the central or uppermost flower opens before the peripheral or lowermost flowers on any axis. (see panicle) 2. In Asteraceae and some other groups, a cyme-like inflorescence is one in which the central or uppermost inflorescence units (e.g., heads in Asteraceae, umbels enclosed by involucres in *Eriogonum*), instead of individual flowers, develop and mature before the peripheral or lowermost inflorescence units on any axis.

cypsela.   (p. 31) (see achene)

deciduous.   Falling off naturally. 1. Pertaining to leaves that all fall seasonally, or to plants that are seasonally leafless. (see evergreen) 2. Pertaining to structures, such as hairs or flower parts, that fall early or readily.

decimeter.   One-tenth of a meter; 10 centimeters (abbreviation: dm).

decumbent.   (p. 18) Lying mostly flat on the ground but with tips curving up. (see ascending, prostrate)

decurrent.   (p. 23) Pertaining to a wing-like or ridge-like extension basal to the apparent or actual point of attachment, particularly a leaf base that appears to continue onto the stem.

dehiscent.   (p. 27) Opening at maturity to release contents; usually pertaining to anthers or fruits. (see indehiscent)

deltate.   (p. 22) More or less equilaterally triangular, with the corners rounded or not.

dense.   Congested or compact; especially pertaining to the disposition of flowers in an inflorescence. (see open)

dentate.   (p. 22) Having margins with sharp, relatively coarse teeth pointing outward, not tipward. (see serrate)

depressed.   (p. 28) Flattened from above and below, or with the center lower than the margins. (see compressed)

desert.   Region and associated communities characterized by low and irregular precipitation and prolonged periods of drought.

desert woodland.   Vegetation in desert region or on slopes of adjacent mountains characterized by small, drought-tolerant trees; may be classified by characteristic species (e.g., Joshua tree woodland characterized by *Yucca brevifolia*, pinyon/juniper woodland characterized by *Pinus monophylla* and *Juniperus*).

digitate.   In Poaceae, pertaining to an inflorescence of two or more spike-like branches attached at the same point at the apex of the inflorescence stalk.

dioecious.   Pertaining to a taxon in which individuals produce either male or female reproductive structures, and do not produce bisexual reproductive structures. (see monoecious) (e.g., *Salix laevigata*)

diploid.   Having two sets of chromosomes (maternal and paternal); 2*n*. (see haploid, n, polyploid)

disciform head.   In Asteraceae, a head composed of disk flowers and marginal pistillate (or sterile) flowers with minute or missing rays, superficially similar to discoid head. (see discoid head, liguliflorous head, radiant head, radiate head)

discoid head.   (p. 31) In Asteraceae, a head composed entirely of disk flowers. (see disciform head, liguliflorous head, radiant head, radiate head)

disk.   1. (p. 26) Fleshy, often nectar-secreting structure near (often surrounding) an ovary or style base. 2. In Asteraceae, the aggregation of disk flowers in the center of a discoid or radiate head.

disk flower.   (p. 31) In Asteraceae, a generally bisexual (occasionally staminate or sterile, never pistillate), generally radial flower with a 5- (rarely 4-) lobed corolla; appearing without other flower types (in discoid head), or with marginal flowers of a different type (in radiate, radiant, or disciform heads). (see ligulate flower, ray flower)

dissected.   Deeply, often sharply cut but not compound; usually pertaining to leaves (e.g., *Cymopterus deserticola*). (see compound leaf, leaflet, lobe, segment)

distal.   Farther away from the base, origin, or point of attachment, or closer to the edge or tip. (see proximal)

drupe.   Fleshy or pulpy, indehiscent, superficially berry-like fruit in which 1 seed is encased in a stone (as in cherries; e.g., *Prunus emarginata* fruit), or more than 1 seed is encased in an equal number of free or variously fused stones (as in manzanitas). (see berry, nut, pome, stone)

shapes | margins

narrowly linear | linear | lanceolate | oblanceolate | oblong | elliptic | ovate and entire | crenate | tooth | dentate | serrate

cylindric | deltate | awned | acuminate | acute | obtuse | truncate

dune.   Hill or ridge of sand formed by the wind.

e-.   Prefix meaning without, lacking (e.g., in Asteraceae, an epaleate receptacle is one that lacks paleae).

ellipsoid.   In the shape of a flattened or elongated sphere, widest at the middle and tapered equally to both ends, as a fruit; wider than linear. (see elliptic, linear, oblong)

elliptic.   (p. 22) In the shape of a flattened or elongated circle, widest at the middle and tapered equally to both ends, as a leaf; wider than linear. (see ellipsoid, linear, oblong)

emergent.   Pertaining to a plant normally rooted underwater and extending above the water surface, or to a part of such a plant normally held above the water surface (e.g., *Persicaria amphibia*). (see aquatic, submersed)

endemic.   Native to and restricted to a defined geographic area.

entire.   (p. 22) Having margins that are continuous and smooth (i.e., without teeth, lobes, etc.).

ephemeral.   Lasting a short time. 1. Pertaining to individual plants, completing the life cycle (germination through death) or growth cycle in much less than one year. 2. Pertaining to plant parts, falling early or remaining functional for a relatively short time (e.g., less than a day for flower parts).

epidermis.   Outermost cell layer (or layers) of non-woody plant parts.

epipetalous.   (p. 26) Pertaining to stamens that are fused to the petals to various extents and therefore appear to arise from them.

erect.   (p. 18) Upright; vertically oriented. (see ascending)

estuarine.   Pertaining to aquatic habitats where freshwater from streams mixes with sea water in a protected area, resulting in a gradation of brackish waters with varying degrees of salinity.

evergreen.   Never leafless; usually pertaining to leaves that remain green and on the plant for more than one season, and that do not all fall seasonally, or to plants that are never leafless. (see deciduous)

exceeding.   Surpassing tipward, due to relative orientation or length of the structures involved (e.g., lateral branches exceeding inflorescences; hoods exceeded by anther head in *Asclepias californica*; hoods slightly exceeding anther head in *Asclepias erosa*). (see exserted)

exserted.   (p. 26) Protruding out of surrounding structure(s) (e.g., stamens exserted from corolla). (see exceeding, included)

extant.   Currently existing or surviving somewhere. (see extinct, extirpated)

extinct.   No longer existing or surviving anywhere. (see extant, extirpated)

extirpated.   No longer existing or surviving in a defined geographic area, as either a direct or indirect result of human activity. (see extant, extinct)

exudate.   Material discharged (exuded) from a plant, often with characteristic odor, color, or texture (e.g., sticky, gummy, slippery).

false indusium (false indusia).   (p. 29) In many ferns, but especially Pteridaceae and Dennstaedtiaceae, a reflexed or rolled under, often modified leaf blade margin that covers a sorus and protects the young sporangia; also called a marginal indusium. (see indusium)

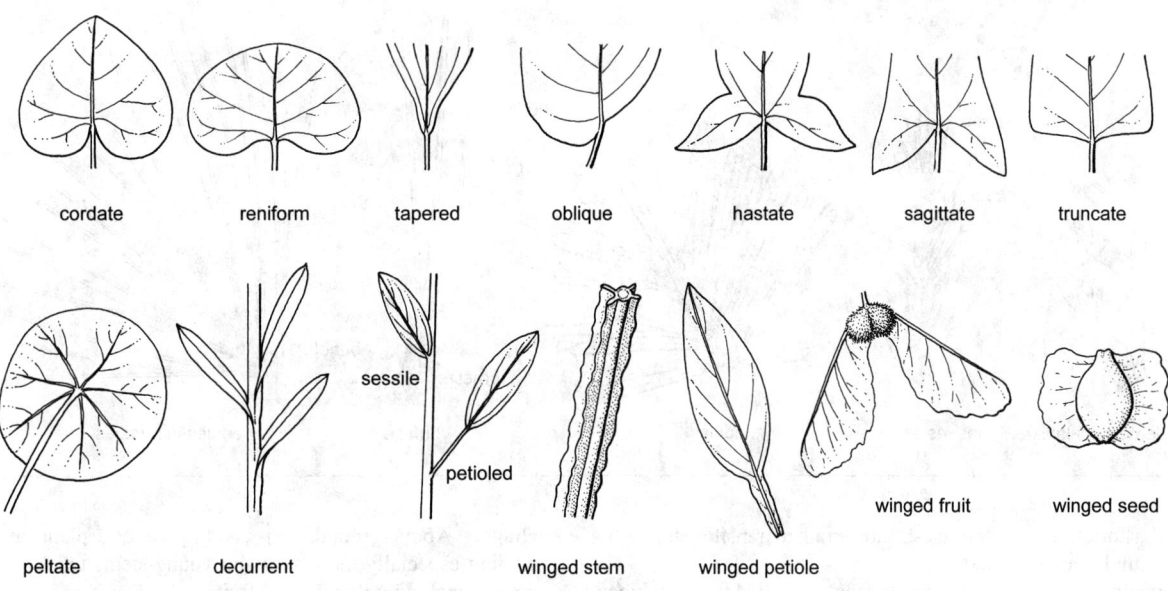

cordate   reniform   tapered   oblique   hastate   sagittate   truncate

sessile

petioled

winged fruit   winged seed

peltate   decurrent   winged stem   winged petiole

fertile. (p. 29) Reproductively functional; pertaining to a plant or plant part that produces or is associated with the production of functional spores, pollen, ovules, or seeds (e.g., fertile leaf, fertile stamen, fertile flower, fertile floret). (see sterile)

fibrous. 1. Pertaining to structures that are composed at least in part of more or less thread-like but usually tough elements (e.g., *Yucca* leaves). 2. (p. 18) Pertaining to a root system composed of many roots similar in length and thickness (e.g., grass roots). (see taproot)

filament. (p. 26) Anther-stalk portion of a stamen, often thread-like.

fleshy. Thick and juicy; succulent (e.g., *Sesuvium verrucosum*).

floret. (p. 30) In Poaceae, a single flower and its immediately subtending bracts (lemma and generally palea, the lemma subtending the palea when the latter is present); in a sterile floret, the flower and sometimes palea are rudimentary or absent. (see glume, lemma, palea, spikelet)

flower. (p. 26) Primary reproductive structure of angiosperms, with stamens and/or carpels and usually a perianth of sepals and/or petals. (see stamen, carpel, perianth, sepal, petal)

follicle. (p. 27) Dry fruit from a simple pistil, dehiscent on generally only one side, along a single suture. A single flower may develop into a simple fruit of 1 follicle or an aggregate fruit of several follicles. (see capsule, legume)

foothill. Slope at the base of a mountain; especially applied to such features in CA-FP.

foothill woodland. Vegetation in foothills characterized by small- to medium-sized trees, composed of one or more species of *Quercus*, often mixed with *Pinus sabiniana* and/or *Aesculus californica*.

forest. Vegetation characterized by closely spaced ± tall trees; with more canopy cover than a woodland (canopies often overlap).

forked. (p. 28) Pertaining to a hair or other structure that branches into two parts. (see stellate)

free. Neither fused to nor adherent to other parts; distinct, separate.

free-central. (p. 26) Pertaining to a placenta along the central axis in a compound ovary with only one chamber. (see axile, basal, parietal)

fringed. (p. 28) Having ragged or finely cut margins.

fruit. (p. 23) Ovary or ovaries and sometimes associated structures after ovule fertilization (i.e., seed initiation). A simple fruit develops from one ovary (e.g., cherry, apple, the latter derived largely from the hypanthium); aggregate and multiple fruits develop from ovaries of one and more than one flower, respectively, that remain distinct yet held together as a unit (e.g., a strawberry is an aggregate fruit of achenes held together by a juicy, red flower receptacle; a fig is a multiple fruit of achenes surrounded by a fleshy inflorescence receptacle).

funnel-shaped. (p. 25) Widening from the base more or less gradually through the throat into an ascending, spreading, or recurved limb; often applied to a fused calyx or corolla. (see bell-shaped, rotate, salverform, urn-shaped)

fused. United, as the petals together into a corolla tube or stamens onto petals; neither free nor adherent.

fusiform. (p. 28) Elongate, widest at the middle, tapered to both ends.

glabrous. (p. 28) Without hairs.

gland (glandular). Small, often spheric body, on or embedded in the epidermis or at the tip of a hair, that exudes a generally sticky substance (e.g., *Psorothamnus arborescens*).

glaucous. Covered with a generally whitish or bluish, waxy or powdery film that is sometimes easily rubbed off.

glochid. In Cactaceae, reduced, barbed, deciduous, bristle-like spine (e.g., *Opuntia*).

glume. (p. 30) In Poaceae, each of generally two sheathing bracts that are the lowermost parts of a spikelet, subtending one or more florets. (see floret, lemma, palea, spikelet)

graduated. In Asteraceae, pertaining to an involucre in which the phyllaries are of unequal length, with the outer

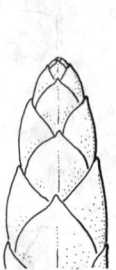  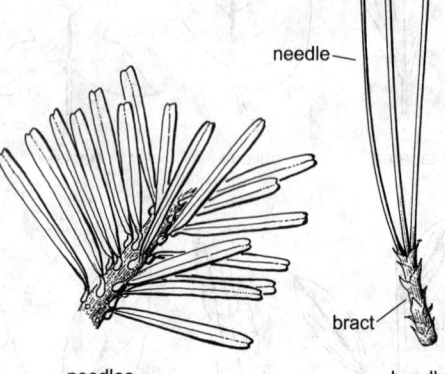 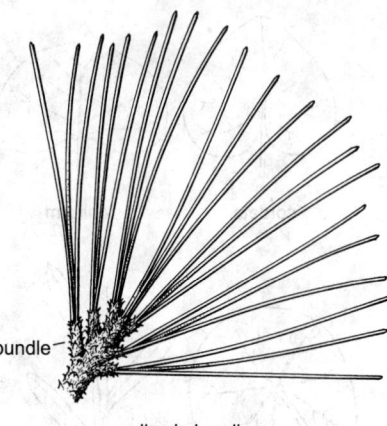

leaves scale-like   leaves awl-like   needles   bundle   needles in bundles

shortest, the inner longest, and a gradual transition through multiple series between.

grain. In Poaceae, dry, indehiscent, 1-seeded fruit in which the fruit wall is fused to the seed; often appearing to be a naked seed.

granular. (p. 28) Covered with minute bumps. (see papillate, tubercle)

grassland. Vegetation characterized by various species of grasses, often mixed with various other kinds of herbs (not grasses) and sometimes scattered, low-growing shrubs.

gymnosperm. Plant that bears woody or fleshy cones, not flowers, in which "naked seeds" (hence, gymno-sperm) are not enclosed in an ovary; woody (e.g., pine, sequoia, ephedra, yew).

habit. Characteristic mode of growth, general form, or shape of a plant (e.g., cespitose, herb, scapose, shrub).

habitat. Natural setting or conditions under which a plant lives (e.g., saltbush scrub, vernal pool, granitic soil among pines, montane forest).

hair. (p. 28) Thread-like epidermal outgrowth. (see glabrous, canescent, ciliate, prickle, puberulent, scabrous, scale, strigose, tomentose)

haploid. Having one set of chromosomes (maternal or paternal); *n*. (see diploid, *n*, polyploid)

hastate. (p. 23) Arrowhead-shaped, with two basal lobes oriented more or less perpendicularly to the long axis. (see sagittate)

head. 1. (p. 25) In flowering plants excluding Asteraceae and some other groups, a dense, often spheric inflorescence of sessile or subsessile flowers. 2. In Asteraceae and some other groups, a head-like inflorescence is one in which sessile or subsessile inflorescence units (e.g., heads in Asteraceae, fl clusters enclosed by involucres in *Eriogonum*), instead of individual flowers, are attached in a short dense cluster without an evident axis or branches.

hemispheric. Shaped like a dome or half sphere.

herb. Plant that, at least above ground, is generally nonwoody and of less than one year or growing season in duration. (see annual, biennial, perennial, subshrub)

herbaceous. Lacking wood; having the characteristics of an herb.

herbage. Above-ground, non-woody parts of a plant, including especially the leaves and young stems taken together, excluding flowers and fruits.

heterostylous. Pertaining to a taxon in which individual plants produce only one of two or more flower types, each type differing in the lengths of styles relative to stamens; a rare condition in angiosperms. (see homostylous)

homostylous. Pertaining to a taxon in which only one type of flower is produced, in which there is no significant variation in the lengths of styles relative to stamens; the usual condition in angiosperms. (see heterostylous)

hypanthium (hypanthia). (p. 26) Structure generally in the shape of a tube, cup, or bowl, derived from the fused lower portions of the perianth and stamens, from which these parts seem to arise, and to which the ovary wall is fused in an inferior ovary (to which the ovary wall is partially fused in a half-inferior ovary; from which the ovary is free in a superior ovary).

included. (p. 26) Not protruding out of surrounding structure(s) (e.g., stamens included in corolla). (see exserted)

indehiscent. (p. 27) Not opening inherently to release contents; usually pertaining to fruits. (see dehiscent)

indusium (indusia). (p. 29) In many ferns, a usually thin, often scale-like outgrowth of the leaf blade surface that covers a sorus and protects the young sporangia; also called a true indusium. (see false indusium)

inferior ovary. (pp. 26, 31) Ovary that is fused to the fused lower portions of the perianth and stamens (i.e., to the hypanthium), to the extent that these structures appear to arise at or above its summit. (see superior ovary)

inflorescence. (p. 18) Entire aggregation of flowers or flower clusters and associated structures (e.g., axes, bracts, bractlets, pedicels); often difficult to determine as to type and boundaries but generally excluding full-sized foliage leaves.

infraspecific. Below the species level or within a species; pertaining to variation within a species, whether taxonomically significant (e.g., characterizing subspecies or varieties) or not (e.g., characterizing forms or minor variants).

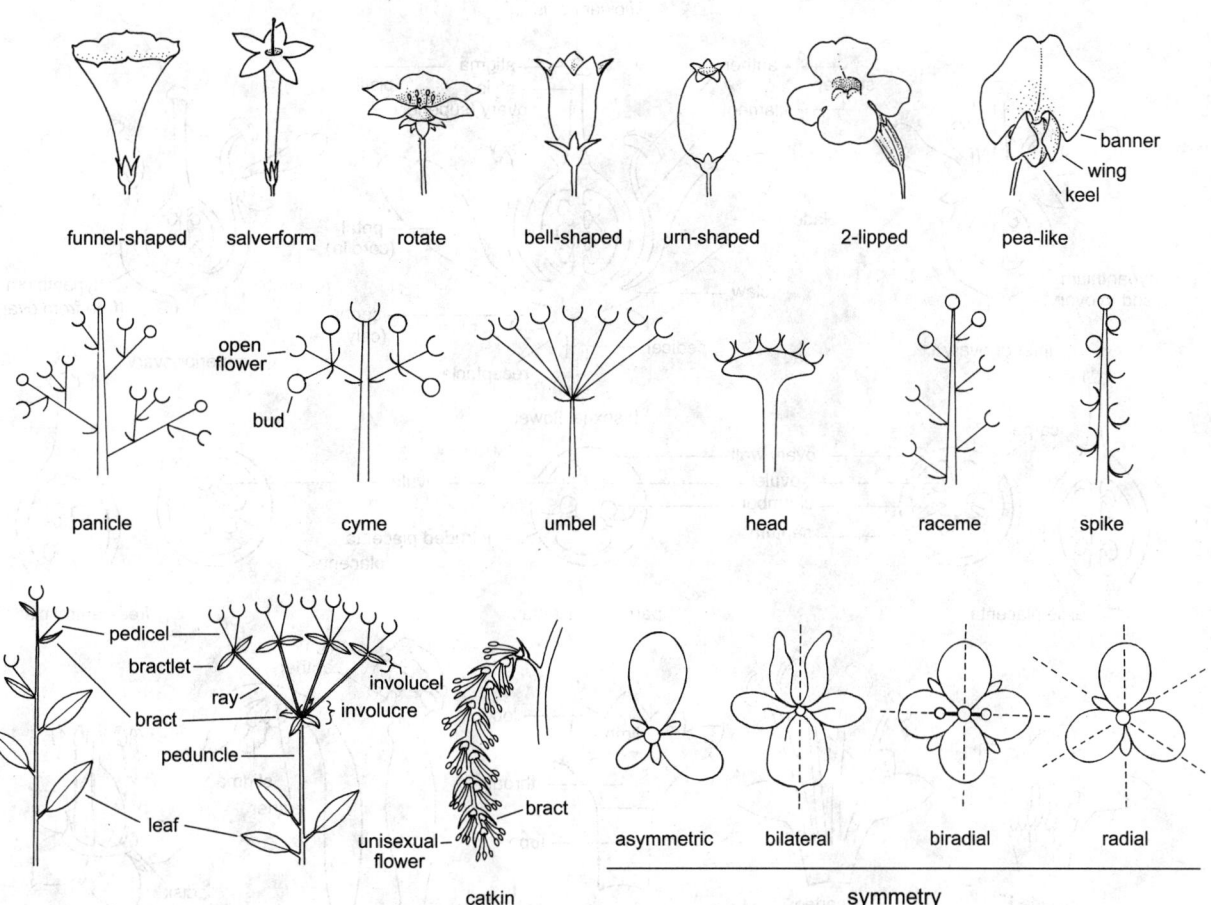

intergrade. To merge gradually from one extreme to another through a more or less continuous series of intermediates.

intermediate. Between extremes or parental taxa in size, shape, color, flowering time, habitat preferences, geographic ranges, or other ways.

internode. (p. 21) Segment of an axis (generally a stem) between successive positions (nodes) from which one or more structures (especially leaves, buds, branches, or flowers) arise.

interrupted. Having parts spaced unevenly, clustered; pertaining especially to inflorescences in which the flowers are clustered. (see continuous)

intertidal. Pertaining to marine habitats that are submerged at high tide and exposed at low tide.

involucel. (p. 25) Secondary involucre (group of bracts) within an inflorescence (e.g., those subtending the secondary umbels in members of Apiaceae).

involucre. (pp. 25, 31) Group of bracts more or less held together as a unit, subtending a flower, fruit (acorn cup), or inflorescence.

keel. 1. Ridge or crease more or less centrally located on the long axis of a structure, generally on the abaxial side.

2. (p. 25) Two lowermost, fused petals of many members of Fabaceae.

lanceolate (lance- in combining forms). (p. 22) Narrowly elongate, widest in the basal half, often tapered to an acute tip.

lateral. Pertaining to the side(s) of a structure (e.g., laterally compressed, meaning flattened side-to-side; lateral branch; lateral appendage). (see terminal)

leaf. (pp. 21, 29) Organ arising from a stem, generally composed of a stalk (petiole) and a flat, expanded, green, photosynthetic area (blade); distinguished from a leaflet by the presence in its axil of a bud, branch, thorn, or flower; sometimes with lateral, basal appendages (stipules); either simple (toothed, lobed, or dissected but not divided into leaflets) or compound (divided into leaflets).

leaflet. (p. 20) Leaf-like unit of a compound leaf; distinguished from a leaf by the absence in its axil of a bud, branch, thorn, or flower; either simple (leaf 1-compound, with 1° leaflets) or compound (leaf 2-compound, with 1° and 2° leaflets; 3-compound, with 1°, 2°, and 3° leaflets; etc.).

legume. (p. 27) In Fabaceae (legume family), a dry or somewhat fleshy, 1- to many-seeded fruit from a simple

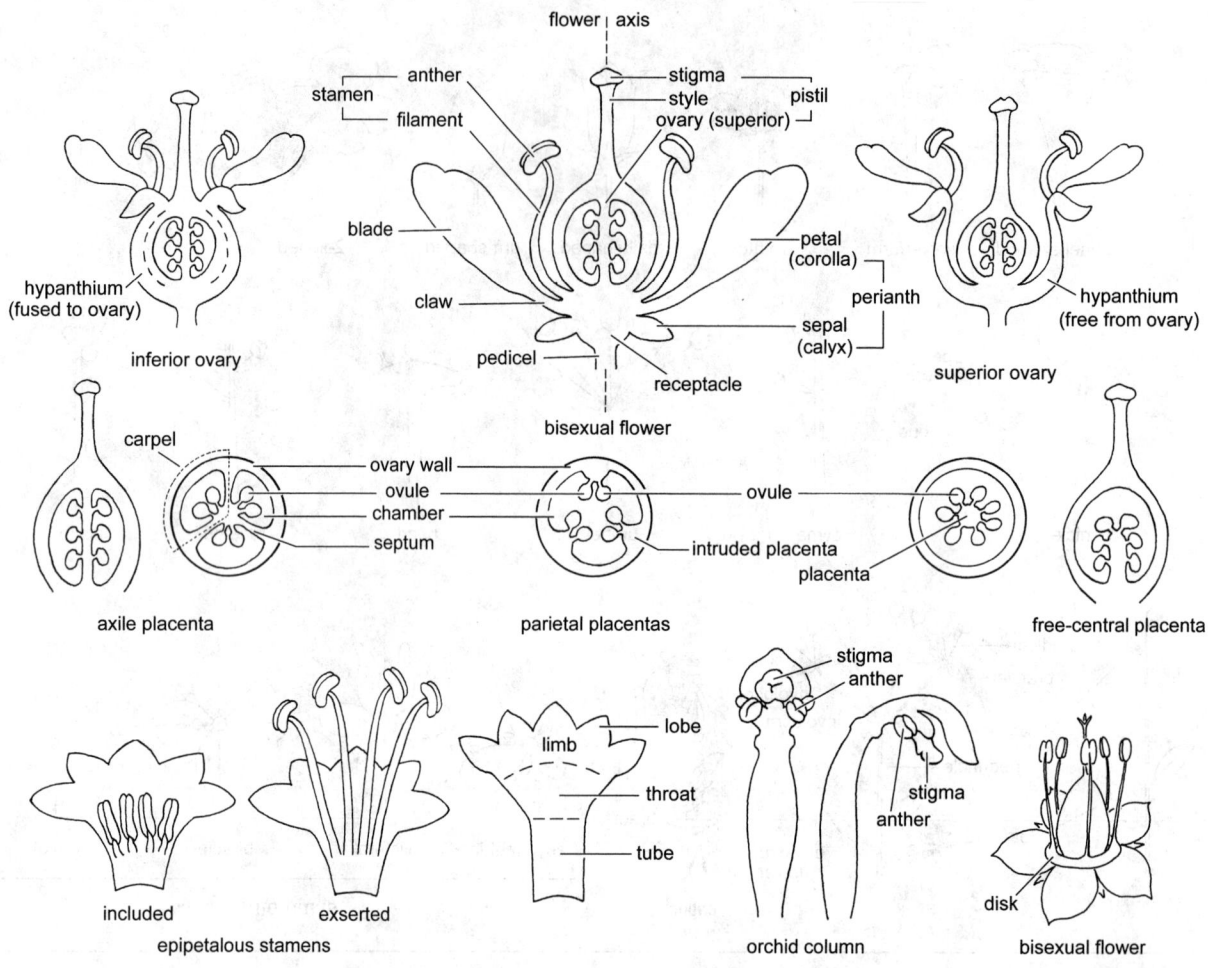

pistil, typically dehiscent longitudinally along two sutures and splitting into halves that remain joined at the base, sometimes indehiscent or breaking crosswise into 1-seeded segments; a plant with such a fruit.

lemma.   (p. 30) In Poaceae, the lower, generally larger of two sheathing bracts subtending a flower, generally ensheathing the palea (in a sterile lemma, the associated flower and sometimes palea are rudimentary or absent); with the palea and flower, comprising the floret. (see floret, glume, palea, spikelet)

lenticel.   (p. 21) Each of many spongy or calloused areas of various shapes, sizes, and colors, most commonly on surfaces of young stems (including twigs) or fruits.

lenticular.   Lens- or discus-shaped, with both major sides convex.

ligulate flower.   (p. 31) In Asteraceae, a bisexual, bilateral flower with the outer portion of the corolla (the ligule) strap- or fan-shaped, 5-lobed; appearing only with other ligulate flowers in a liguliflorous head. (see disk flower, ray flower)

ligule.   1. (p. 31) In Asteraceae, the 5-lobed, strap- or fan-shaped outer portion of the corolla of a ligulate flower. 2. (p. 30) In most Poaceae and some other grass-like plants, an appendage at the adaxial junction of leaf sheath and blade, generally membranous, sometimes formed of hairs. 3. In *Isoetes* and, more obscurely, in *Selaginella*, a membrane that wholly or partially covers a sporangium.

liguliflorous head.   (p. 31) In Asteraceae, a head composed entirely of ligulate flowers. (see disciform head, discoid head, radiant head, radiate head)

limb.   (p. 26) In calyces or corollas with fused sepals or petals, the expanded, often lobed portion distal to the tube or throat; in some free or nearly free sepals and petals, the expanded portion distal to the stalk-like base (claw).

linear.   (p. 22) Elongate, with nearly parallel sides; narrower than elliptic or oblong.

lip.   1. (p. 25) Upper or lower of two parts in a bilateral, unequally divided calyx or corolla. 2. In Orchidaceae, generally the largest, lowest, most highly modified perianth part.

lobe (lobed).   1. (pp. 20, 29, 31) Major expansion or bulge, such as on the margin of a leaf, sepal, or petal, or on the surface of an ovary. 2. (p. 26) Free tips of otherwise fused structures, such as sepals or petals; larger than teeth.

loculicidal.   (p. 27) Pertaining to dehiscence of a fruit (capsule) by a longitudinal line through the wall at or near the

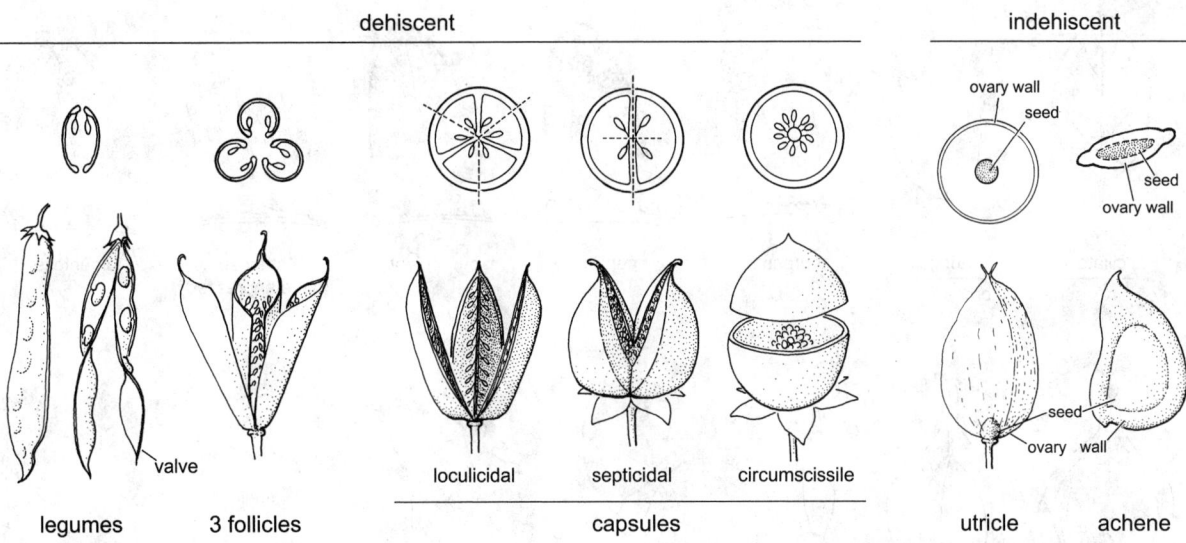

dehiscent

indehiscent

ovary wall

seed

seed

ovary wall

valve

loculicidal   septicidal   circumscissile

seed

ovary wall

legumes   3 follicles   capsules   utricle   achene

center of each chamber, such that each resulting segment corresponds to the two adjacent halves of two adjacent chambers, usually with a placenta-bearing septum centrally. (see circumscissile, septicidal)

longitudinal.   Pertaining to length or the lengthwise dimension; parallel to the axis. (see transverse)

margin.   (p. 22) Edge, generally of a leaf or perianth part.

marsh.   Permanently or periodically inundated, mostly or completely treeless vegetation characterized by semi-aquatic herbs or subshrubs.

meadow.   Open grass- or sedge-characterized vegetation more or less surrounded by woodland or forest; meadow soils are generally seasonally moist and frequently are composed of fine-grained sediments.

membranous.   Thin, dry or moist, pliable, often more or less translucent or variously colored, sometimes green (e.g., *Elymus cinereus* ligule). (see scarious)

mericarp.   One of the generally dry, generally indehiscent, 1–few-seeded, 1-carpelled segments into which certain fruits separate at maturity (e.g., those of Apiaceae).

meter.   Basic unit of length in the metric system, equal to 39.4 inches, slightly more than a yard (abbreviation: m).

millimeter.   One-thousandth of a meter; one-tenth of a centimeter (abbreviation: mm).

mixed-evergreen forest.   Vegetation characterized by a variable mixture of mostly or only hardwood tree species, most of which retain their leaves throughout the year.

monoecious.   Pertaining to a taxon in which individuals produce both male and female reproductive structures and do not produce bisexual reproductive structures (e.g., *Alnus rhombifolia*). (see dioecious)

montane.   Pertaining to mountains; vegetational/altitudinal zone between the foothill and subalpine zones.

mucro (mucronate).   Abrupt, short, sharp, narrow, terminal point, tip, or projection (e.g., *Isolepis* fruit).

n.   Number of chromosomes in haploid cells. (see diploid, polyploid)

native.   Occurring naturally in an area, as neither a direct nor indirect consequence of human activity; indigenous; not alien. (see naturalized, waif)

naturalized.   Alien (not native) and reproducing either sexually (e.g., by spores, seeds) or vegetatively (e.g., by sprouts, suckers) in the absence of any benefit, intentional or not, direct or indirect, of human activity, and thereby persisting beyond initial generation or establishment. (see native, waif)

nectar.   Sugary solution, produced in nectaries, consumed primarily as an energy source by animal visitors, usually pollinators.

nectary.   Variously shaped, nectar-producing structure(s) usually at or near the base of the inside of a flower (or sometimes elsewhere, such as in a perianth spur or on a petiole) (e.g., *Symphoricarpos rotundifolius*).

needle.   (p. 24) Narrowly linear, often waxy, generally evergreen leaf, especially of conifers.

nodding.   Pertaining to a structure (e.g., flower, fruit) borne on a stalk that is curved downward (e.g., *Aquilegia formosa* flower).

node.   (pp. 21, 30) Position on a stem from which one or more structures (especially leaves, buds, branches, or flowers) arise. (see internode)

nut.   Mostly dry, sometimes fleshy or pulpy, usually indehiscent fruit in which a single seed is encased in a hard shell (e.g., *Quercus palmeri*). (see drupe)

nutlet.   Small, dry nut or nut-like fruit, usually several of which are produced by a single flower (e.g., Boraginaceae, Lamiaceae). (see nut, drupe)

ob-.   (p. 22) Prefix indicating inversion of shape (e.g., lanceolate and oblanceolate leaf blades are widest below and above the middle, respectively).

oblique.   (p. 23) Having unequal sides or an asymmetric base.

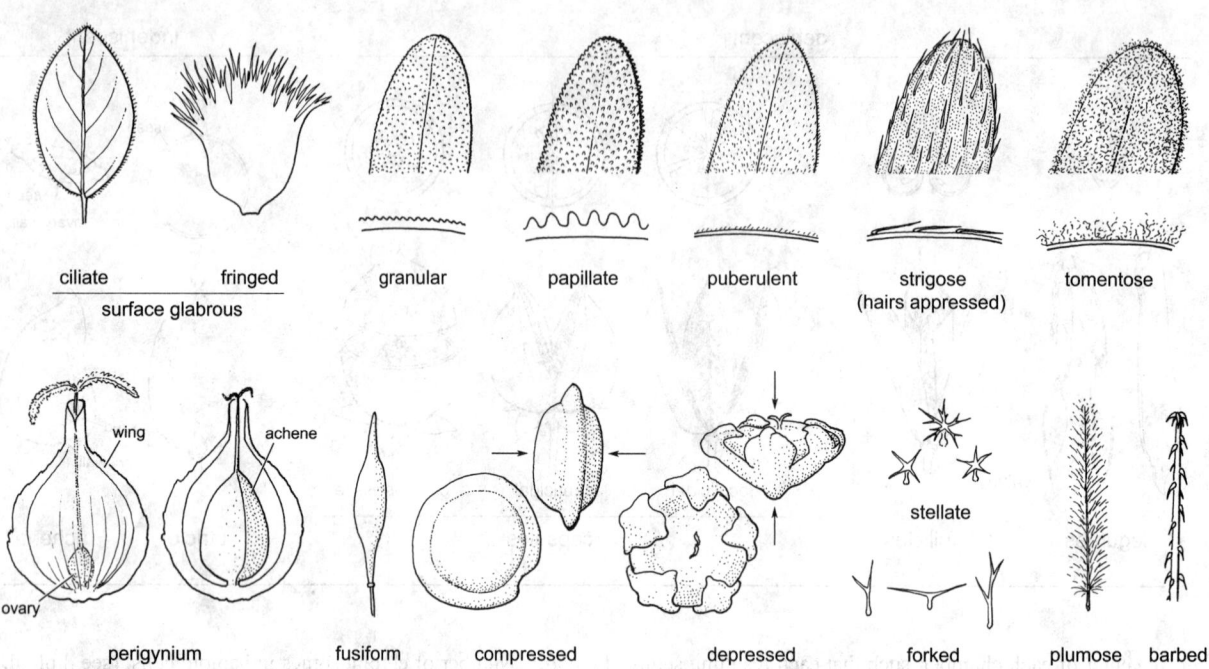

ciliate   fringed   granular   papillate   puberulent   strigose (hairs appressed)   tomentose

surface glabrous

wing   achene

ovary

perigynium   fusiform   compressed   depressed   stellate   forked   plumose   barbed

oblong. (p. 22) Longer than wide, with nearly parallel sides; wider than linear. (see elliptic)

obtuse. (p. 22) Having a short-tapered, blunt tip or base, the sides convex or straight and converging at more than a right angle. (see acute)

ocrea, ochrea (ocreae, ochreae). In Polygonaceae, generally scarious sheath around the stem formed by the fusion of stipules.

open. Uncongested or diffuse; especially pertaining to the disposition of flowers in an inflorescence. (see dense)

opposite. 1. (p. 21) Arranged in pairs along an axis (e.g., two leaves per node). (see alternate, whorled) 2. Occurring in the same rank, directly above or below, as "stamens opposite petals." 3. Located directly across from.

ovary. (pp. 26, 28) Ovule-bearing, usually wider, basal portion of pistil, normally developing into a fruit as ovules become seeds; may be simple (one carpel, one chamber) or compound (two or more carpels, one or more chambers).

ovary stalk (fruit stalk). Pedestal-like, apical prolongation of a floral receptacle (often termed elsewhere a carpophore) or basal constriction of an ovary (often termed elsewhere a gynophore), above the level of perianth insertion, each with the result that the ovary or fruit appears to be stalked over and above the pedicel (whereas the demarcation between pedicel and ovary- or fruit-stalk is observable as the point of perianth insertion, carpophores generally are distinguished from gynophores only by anatomical study).

ovate. (p. 22) Egg-shaped (i.e., widest below the middle) in two dimensions (i.e., in one plane), as a leaf. (see ovoid)

ovoid. Egg-shaped (i.e., widest below the middle) in three dimensions, as a fruit. (see ovate)

ovule. (p. 26) In gymnosperms and angiosperms, structure containing an egg, and normally developing into a seed after fertilization.

palea (paleae, paleate). 1. (p. 31) In Asteraceae, a scale-like bract that subtends an individual flower on the receptacle (equal to "chaff scale" in *TJM* [1993]), absent in some genera, restricted to a ring separating ray and disk flowers in most tarweed species. 2. (p. 30) In Poaceae, the distal, generally smaller of two sheathing bracts subtending a flower, generally 2-veined and -keeled and ensheathed by the lemma; with the lemma and flower, comprising the floret. (see floret, glume, lemma, spikelet)

palmate. (p. 20) More than two structures or parts (e.g., veins, lobes, or leaflets) radiating from a common point in two dimensions (i.e., in one plane). (see pinnate, ternate)

panicle. 1. (p. 25) In flowering plants excluding Asteraceae, Cyperaceae, Poaceae, and some other groups, a branched inflorescence in which the basal or lateral flowers (or some of them) open before the terminal or central flowers on any axis. (see cyme) 2. In Asteraceae, Cyperaceae, Poaceae, and some other groups, a panicle-like inflorescence is one in which at least some of the inflorescence units (e.g., heads in Asteraceae; spikelets in Cyperaceae and Poaceae), instead of individual flowers, are attached (stalked or unstalked) to branches and not directly to the main axis of the inflorescence and in which floral development may or may not proceed as in 1.

papillate. (p. 28) Pertaining to a surface (e.g., of a leaf, stigma, fruit) bearing small, rounded or conic protuberances (papillae; singular papilla).

pappus. (p. 31) In Asteraceae, the aggregate of structures such as awns, bristles, or scales arising from the top of the inferior ovary, in place of the calyx.

parasite. Plant that benefits by taking resources from a physical connection to a host plant of another species; green parasites (hemiparasites) derive water and dissolved inorganic substances (e.g., mineral nutrients) from

FERNS

fern with leaves all alike

4 sorus examples

fern with fertile and sterile leaves

the connection and often are able to survive without it, while non-green parasites (holoparasites) obtain in addition energy-rich, organic compounds (products of photosynthesis) from the connection and cannot survive without it; the connection may or may not involve a fungal intermediate, and may or may not be detrimental to the host.

parietal. (p. 26) Pertaining to placentas on the inside surface of the ovary wall in a compound ovary with one or more chambers.

peat. Material formed by the partial decomposition in water of plant tissues, especially mosses (*Sphagnum*) or sedges.

peatland. Moss- or herb-characterized freshwater wetland with nutrient-deficient substrate and accumulated peat; often said elsewhere to be bogs if acidic, fens if basic.

pedicel. (pp. 25, 26) Stalk of an individual flower in an inflorescence, or the corresponding structure in fruit. (see peduncle, ray)

peduncle. (pp. 18, 25, 31) Stalk of an individual flower borne singly, not in an inflorescence, or of an entire inflorescence, or the corresponding structure in fruit; the stalk subtending an involucre (e.g., in Asteraceae, Polygonaceae). (see pedicel, ray)

peltate. (p. 23) With the stalk attached toward the middle, not at a margin, of a flat structure such as an indusium, scale, or leaf.

pendent. Drooping, hanging, or suspended from a point of attachment above (e.g., *Amelanchier utahensis* fruit).

perennial. (p. 18) Completing life cycle (germination through death) in more than two years or growing seasons, generally non-woody (at least above ground) to woody; includes perennial herbs as well as subshrubs to trees; the abbreviation "per" only refers to perennial herb, not to the word "perennial" alone. (see annual, biennial)

perianth. (p. 26) Calyx and corolla collectively, whether or not they are distinguishable.

perianth part. Individual member of a perianth; used whether or not calyx and corolla are distinguishable, but usually when they are not.

perigynium. (p. 28) Variously shaped, sac-like structure enclosing the ovary and achene in *Carex* and *Kobresia*.

persistent. Not falling off; remaining attached. (see deciduous, ephemeral)

petal. (p. 26) Individual member of the corolla, whether fused or not; if fused, often equal in number to the number of corolla lobes; often conspicuously colored. (see sepal)

petiole. (p. 21) Leaf stalk, connecting leaf blade to stem; sometimes more or less indistinct.

phyllary. (p. 31) In Asteraceae, a bract of the involucre.

pinna (pinnae). (p. 29) In ferns, primary division of a compound or dissected leaf blade.

pinnate. (p. 20) Feather-like; pertaining to veins, lobes, leaflets, or other structures arranged in two dimensions (i.e., in one plane) along either side of an axis; a leaf is odd-pinnate if there is a terminal leaflet, even-pinnate if there is not, and either odd- or even-pinnate may be 1-pinnate (blade divided into primary leaflets), 2-pinnate (primary leaflets divided into secondary leaflets), 3-pinnate (secondary leaflets divided into tertiary leaflets), etc. (see compound leaf, palmate, ternate, plumose)

pinnule. (p. 29) In ferns, secondary division of a compound or dissected leaf blade, primary division of a pinna.

pistil. (p. 26) Female reproductive structure of a flower, composed of an ovule-containing ovary at the base, one or more pollen-receiving stigmas at the tip, and generally one or more styles between ovary and stigma. A flower may have one or more simple pistils (each a single, free carpel with a single ovary chamber, placenta, and stigma) or one compound pistil (two or more fused or partially fused carpels, the exact number often equaling the number of ovary lobes, ovary chambers, placentas, styles, or stigmas).

GRASS FAMILY

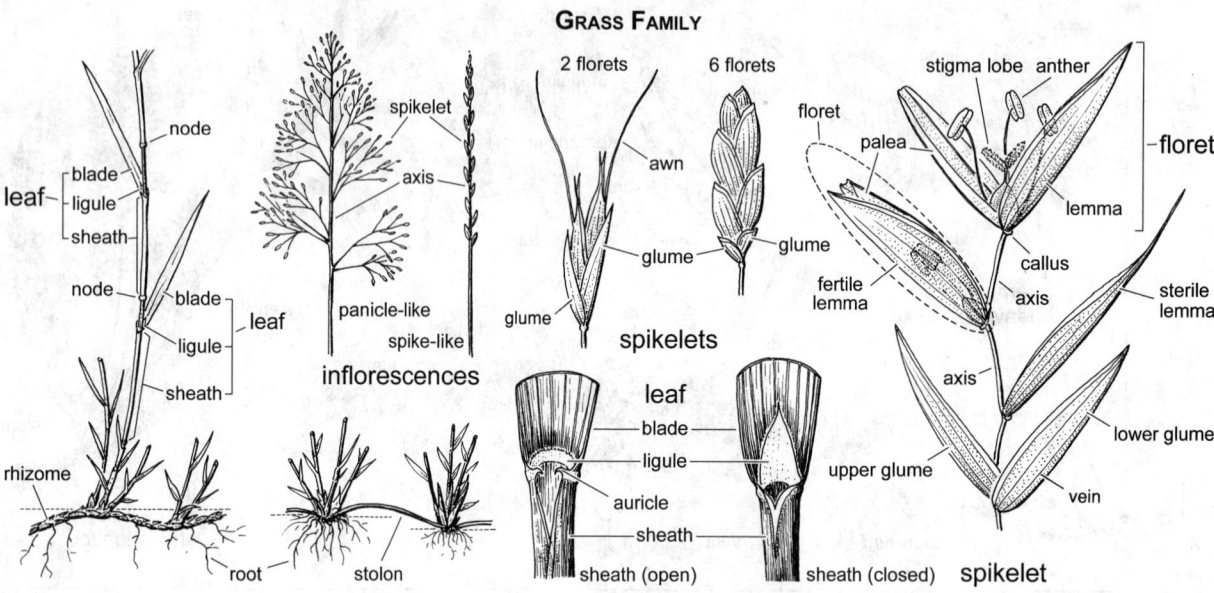

pistillate. Pertaining to flowers, inflorescences, or plants with fertile pistils but sterile or missing stamens (e.g., *Salix laevigata* pistillate flower). (see staminate)

placenta. (p. 26) Structure or area to which ovules are attached in an ovary; variously shaped and positioned.

planoconvex. Flat or nearly so on one side, rounded on the other (e.g., *Carex harfordii* perigynium).

pleated. Having accordion-like folds.

plumose. (p. 28) Plume-like, usually with the parts arrayed in three dimensions around an axis, or in tufts held together at the base; usually pertaining to small, finely divided structures, such as certain stigmas and pappus elements. (see pinnate)

pollen (pollen grain). In gymnosperms and angiosperms, structure containing the sperm; when sperm fertilizes an egg, the egg and surrounding ovule normally develop into a seed.

pollen sac. Each of the one, two, or four pollen-bearing portion(s) of an anther.

pollination. Placement of pollen, by an insect, the wind, or other vector, on a stigmatic or ovular surface, through which pollen tube growth and fertilization may occur; self-pollination involves only one plant, cross-pollination occurs between plants.

pollinium (pollinia). Especially in *Asclepias* and related genera, and in Orchidaceae, a mass of adherent pollen grains disseminated as a unit.

polyploid. Having three or more sets of chromosomes; $3n$, $4n$, etc. (see diploid, haploid, *n*)

pome. In Rosaceae, a fleshy, indehiscent fruit, such as an apple or pear; derived from a hypanthium (represented as outer fleshy material and skin) surrounding and ± fused to a compound ovary (represented as (1)2–5 papery-walled, radiating segments alternating with fleshy material, as in, e.g., *Amelanchier utahensis* fruit) or to (1)2–5 free ovaries, each with a ± stony outer layer. (see berry, drupe)

prickle. Sharp-pointed, stiff or somewhat flexible projection, originating at or near the surface and neither subtending an axillary bud or branch nor subtended by a leaf or leaf scar, without leaves, leaf scars, buds, or branches (e.g., *Rosa woodsii* stem); loosely used for any sharp projection. (see armed, spine, thorn)

prostrate. (p. 18) Lying flat on the ground. (see ascending, decumbent)

protandrous. Pertaining to a bisexual flower in which pollen release precedes stigma receptivity, or to a plant with staminate and pistillate flowers in which this is true, with the result that cross-pollination is favored.

protogynous. Pertaining to a bisexual flower in which stigma receptivity precedes pollen release, or to a plant with staminate and pistillate flowers in which this is true, with the result that cross-pollination is favored.

proximal. Closer to the base, origin, or point of attachment, or farther away from the edge or tip. (see distal)

puberulent. (p. 28) Minutely hairy.

raceme. 1. (p. 25) In flowering plants excluding Asteraceae, Cyperaceae, Poaceae, and some other groups, an unbranched inflorescence in which the flowers are borne on pedicels and nearly always open from the bottom to the top of the inflorescence. (see panicle, spike) 2. In Asteraceae, Cyperaceae, Poaceae, and some other groups, a raceme-like inflorescence is one in which the inflorescence units (e.g., heads in Asteraceae; spikelets in Cyperaceae and Poaceae), instead of individual flowers, are stalked and attached directly to the main axis of the inflorescence, not to branches, and in which floral development may or may not proceed as in 1.

rachis. (p. 29) In ferns, primary axis of a compound or dissected leaf blade.

radial. (p. 25) Divisible into mirror-image halves in three or more ways. (see asymmetric, bilateral, biradial)

COMPOSITE FAMILY

liguliflorous head   ligulate flower   discoid head   disk flower   ray flower   radiate head

radiant head. In Asteraceae, a discoid head with a peripheral ring of flowers having much enlarged, often bilateral corollas.

radiate head. (p. 31) In Asteraceae, a head composed of central disk flowers and marginal ray flowers. (see disciform head, discoid head, liguliflorous head, radiant head)

rank. 1. (p. 21) Row or column of parts along an axis (e.g., leaves on an erect stem arranged in four vertical rows are 4-ranked). (see alternate, opposite) 2. In classification, a taxonomic level (e.g., family, genus, species, subspecies, variety). (see taxon)

rare. Extremely unlikely to be present in appropriate habitats, often restricted to a small number of sites. (see uncommon)

ray. 1. (p. 25) Each of a number of radiating axes, as a primary branch in a compound umbel. (see pedicel, peduncle) 2. (p. 31) In Asteraceae, the flat, strap- or fan-shaped, often 3-lobed outer portion of the corolla of a ray flower.

ray flower. (p. 31) In Asteraceae, a generally pistillate or sterile, bilateral flower with a flat, strap- or fan-shaped, often 3-lobed outer portion of the corolla (ray); appearing in a ring around a central cluster of disk flowers. (see ligulate flower, disk flower)

receptacle. 1. (p. 26) In individual flowers, the structure to which flower parts are attached. 2. (p. 31) In heads or head-like inflorescences, especially in Asteraceae, the structure to which flowers or sometimes heads are attached.

recurved. (p. 19) Gradually curved downward or backward.

reduced. Smaller, less lobed, simpler, etc.

redwood forest. Vegetation characterized by *Sequoia sempervirens*, occurring on slopes and canyons of coastal mountain ranges.

reflexed. (p. 19) Abruptly bent or curved downward or backward.

reniform. (p. 23) Kidney-shaped; often pertaining to a leaf in which the blade base on both sides of the petiole is rounded and concave. (see cordate)

rhizome. (pp. 18, 29, 30) 1. In seed plants, stem that is often elongate, generally more or less horizontal and underground; distinguished from roots by bearing of leaves, leaf scars, axillary buds, etc. (see stolon) 2. In ferns, stem that is located underground, embedded in leaf litter, on rocks or in rock crevices, or on trees or tree branches, often scaly or hairy; distinguished from roots by bearing of leaves (roots rarely bear leaves) and their greater diameter.

rib. 1. Ridge, as on a fruit. 2. Raised vein, as on a leaf or other part (e.g., *Carex hendersonii* perigynium).

riparian. Pertaining to communities that occupy the banks, channels, and flood plains of waterways.

root. (pp. 18, 29, 30) Generally underground axis or axes of a plant; distinguished from stems by not bearing leaves, leaf scars, axillary buds, flowers, etc.; generally growing into the ground from the base of a stem, its functions include anchorage, absorption of water and nutrients, and food storage. (see bulb, corm, rhizome, caudex, tuber, stolon)

rosette. (p. 18) Radiating cluster of leaves generally at or near ground level.

rotate. (p. 25) Wheel-shaped, spreading, or saucer-shaped; often applied to a fused corolla with a short or nonexistent tube and a spreading limb. (see bell-shaped, funnel-shaped, salverform, urn-shaped)

ruderal. Plant, usually alien, occurring in waste areas, along roadsides, and in other places disturbed by humans; pertaining to such a plant.

sagittate. (p. 23) Arrowhead-shaped, with two basal lobes oriented nearly parallel to the long axis. (see hastate)

salverform. (p. 25) Having a slender tube and an abruptly spreading, flat limb; often applied to a fused corolla. (see bell-shaped, funnel-shaped, rotate, urn-shaped)

savanna. Vegetation characterized by various species of grasses with scattered individual trees; with less canopy cover than a woodland (canopies do not touch).

scabrous. Rough to the touch, generally owing to short stiff hairs (e.g., *Brickellia* pappus).

scale. 1. (p. 29) Wide, appressed, membranous, epidermal outgrowth (e.g., *Cheilanthes covillei*). (see hair) 2. Structure (bud scale) partially or entirely covering an overwintering bud (e.g., *Salix gooddingii* bud). 3. In gymnosperms, a woody, seed-bearing structure (cone scale) attached to the cone axis (e.g., *Abies magnifica*). 4. In Asteraceae, flat, membranous pappus element (e.g., *Hymenoxys hoopesii*). Leaves or bracts may be scale-like in one or more of the preceding ways.

scapose. (p. 18) Pertaining to a plant or an inflorescence in which a relatively long peduncle (scape) arises, sometimes with leaf- or scale-like bracts but without true foliage leaves, from a rosette or other arrangement of leaves at ground level.

scar. (p. 21) Mark left by the natural separation of two structures, as a leaf scar on a stem.

scarious. Thin, dry, pliable, translucent or variously colored but not green (e.g., *Carex incurviformis* pistillate flower bract). (see membranous)

scree. Relatively unstable, sloping accumulation of small rock fragments, often at a cliff base. (see talus)

scrub. Vegetation characterized by shrubs; may be classified by habitat type or by characteristic species; shrubland.

sculpture. Surface ornamentation or topography, often visible only when magnified, as on a seed or pollen grain (e.g., *Plagiobothrys nothofulvus* nutlet).

seed. (pp. 23, 26) Any fertilized ovule, but in descriptions pertaining to the fully mature condition (i.e., at full cone or fruit maturation), unless noted otherwise.

segment. 1. Ultimate or smallest division of a compound leaf (then the segment is also a leaflet) or dissected leaf — not a marginal lobe, tooth, bristle, etc. 2. Part into which an organ is naturally or apparently divisible, such as of a calyx, corolla, fruit, etc. 3. Specified length, such as of a stem, root, style, etc.

sepal. (p. 26) Individual member of the calyx, whether fused or not; if fused, often equal in number to the number of calyx lobes; generally green. (see petal)

septicidal. (p. 27) Pertaining to dehiscence of a fruit (capsule) by a longitudinal line through the wall at or near the center of each septum, such that each resulting segment corresponds to a single chamber, with placentas placed variously. (see circumscissile, loculicidal)

septum (septa, septate). (p. 26) Wall between chambers in a hair, root, compound ovary, or other structure.

series. Group of structures of one kind (e.g., involucral bracts, sepals, petals, stamens) of similar size or shape, usually more or less in a row or whorl.

serpentine. Pertaining to rocks, or soils derived from them, with generally low levels of calcium and other nutrients, and high levels of magnesium, iron, and certain toxic metals; many plant taxa are restricted to or excluded from serpentine.

serrate. (p. 22) Having margins with sharp, fine to coarse teeth generally pointing tipward, not outward; margins with such teeth on such primary teeth are doubly serrate. (see dentate)

sessile. (p. 23) Without a petiole, peduncle, pedicel, or other kind of stalk.

sheath. (p. 30) Surrounding or partially surrounding, often tubular structure or part of a structure, such as a leaf base in Apiaceae or Poaceae.

shoot. Pertaining collectively to a young stem or twig and its appendages (e.g., new growth in the spring), or to all above-ground parts of a plant (i.e., shoot system). A shortshoot (i.e., spur) is a shoot on some woody plants that undergoes only limited elongation, so that the internodes between appendages (leaves, flowers, etc.) are very short.

shrub. Woody plant of relatively short maximum height, with generally many branches from the base. (see tree, subshrub, perennial)

simple. (p. 20) Composed of a single part; undivided; unbranched. (see compound)

sinus. (p. 20) Usually pertaining to margins of leaves, sepals, petals, or other parts, an indentation between adjacent lobes or teeth.

sorus (sori). (p. 29) In many ferns, a distinct cluster of sporangia.

spheric. Globe- or ball-shaped; circular in three dimensions.

spike. 1. (p. 25) In flowering plants excluding Asteraceae, Cyperaceae, Poaceae, and some other groups, an unbranched inflorescence in which the flowers are sessile and nearly always open from the bottom to the top of the inflorescence. (see panicle, raceme, head) 2. In Asteraceae, Cyperaceae, Poaceae, and some other groups, a spike-like inflorescence is one in which the inflorescence units (e.g., heads in Asteraceae; spikelets in Poaceae and most Cyperaceae), instead of individual flowers, are sessile and attached directly to the main axis of the inflorescence, not to branches, and in which floral development may or may not proceed as in 1. (see spikelet)

spikelet. (p. 30) 1. In Poaceae, one or more florets (each a flower with subtending lemma and generally palea) and generally two subtending glumes; in a sterile spikelet, the flower(s) and sometimes palea(e) are rudimentary or absent. (see floret, glume, lemma, palea) 2. For definitions in Cyperaceae, see note after family description.

spine. Sharp-pointed, usually stiff projection, derived from a leaf or leaf part (e.g., stipule, vein tip), and therefore often subtending an axillary bud or branch, without buds of its own (e.g., marginal spine of *Cirsium arvense* leaf); loosely used for any sharp projection. (see armed, prickle, thorn)

sporangium (sporangia). (p. 29) In non-seed plants, a case or sac in which spores are produced, and from which they are released.

spore. (p. 29) In non-seed plants, one of very many minute, haploid cells (in mass, often appearing dust-like) dispersed from sporangia on a diploid parent plant (sporophyte), normally developing into a small haploid plant (gametophyte) that produces eggs, sperm, or both, the fusion of which results in new diploid offspring.

spreading. (p. 19) Oriented more or less perpendicularly to the axis of attachment; often, more or less horizontal.

spur. 1. Hollow projection or expansion, generally of a perianth part and containing nectar (e.g., *Aquilegia formosa* petals). 2. Shoot on some woody plants that undergoes only limited elongation, so that the internodes between appendages (leaves, flowers, etc.) are very short. (see shoot)

stamen.   (p. 26) Male reproductive structure of a flower, typically composed of a stalk-like filament and a terminal, pollen-producing anther. Filaments sometimes partly fuse to the corolla, or to other filaments to form a tube. (see anther, filament, pistil)

staminate.   Pertaining to flowers, inflorescences, or plants with fertile stamens but sterile or missing pistils (e.g., *Salix laevigata* staminate flower). (see pistillate)

staminode.   Sterile stamen, usually modified in appearance, sometimes petal-like or elaborate in structure (e.g., *Penstemon palmeri* var. *palmeri*).

stellate.   (p. 28) Pertaining to a hair or other structure with three or more branches radiating in two or three dimensions from a common point. (see forked)

stem.   (pp. 18, 20) Generally above-ground but sometimes below-ground axis or axes of a plant; distinguished from roots by bearing leaves, leaf scars, axillary buds, flowers, nodes, etc. (see bulb, caudex, corm, rhizome, root, stolon, tuber)

sterile.   (p. 29) Not reproductively functional; pertaining to a plant or plant part that does not produce or is not associated with the production of functional spores, pollen, ovules, or seeds (e.g., sterile leaf, sterile stamen, sterile flower, sterile floret). (see fertile)

stigma.   (pp. 26, 30) Part of a pistil on which pollen germinates, generally terminal and elevated above the ovary on a style, usually sticky or hairy, sometimes lobed.

stipe.   (p. 29) In ferns, a leaf stalk (analogous to a petiole of a leaf), connecting blade to rhizome.

stipule.   (p. 21) Appendage at base of a petiole, generally paired, variable in form but often leaf- or scale-like, sometimes a spine.

stolon.   (pp. 18, 30) Normally thin, elongate stem lying more or less flat on the ground and forming roots as well as erect stems or shoots (which become new, clonal plants) at generally widely spaced nodes; runner. (see rhizome)

stomate (stoma, stomata).   Minute pore on a leaf (less often, stem or other structure) through which gases pass into or out of a plant (generally, carbon dioxide in, oxygen and water vapor out); generally closed during times of water stress; sometimes used in identification.

stone.   In a drupe, the very hard inner ovary wall and the generally single seed it surrounds; occurring one or more per flower, free or variously fused (e.g., *Prunus emarginata* fruit).

stout.   Thick, sturdy, not slender.

striate.   With fine, longitudinal channels, lines, or ridges.

strigose.   (p. 28) With stiff, straight, sharp, appressed hairs.

style.   (pp. 26, 31) In many but not all pistils, the stalk-like part that connects ovary to stigma.

sub-.   Prefix meaning almost, just below, or somewhat imperfectly.

subalpine.   Pertaining to the vegetational/altitudinal zone just below timberline, between the montane and the alpine.

submersed.   Pertaining to a plant normally rooted and remaining underwater, or to a part of such a plant normally held underwater. (see aquatic, emergent)

subshrub.   Plant with the proximal above-ground stems woody, the distal stems and twigs not woody (or less so) and dying back seasonally. (see perennial, shrub)

subtend.   Occurring immediately below, as sepals subtending petals or leaves subtending axillary buds.

subtidal.   Pertaining to marine aquatic habitats that are continuously submerged, even at low tide.

superior ovary.   (p. 26) Ovary that is free from the perianth and stamens, or free from the fused lower portions of these structures (i.e., free from the hypanthium), to the extent that these structures appear to arise at its base, and it appears to arise from the top of the receptacle. (see hypanthium, inferior ovary)

suture.   Groove or line of dehiscence or fusion.

swamp.   Shrub- or tree-characterized vegetation that occurs in permanently wet soils with standing water.

talus.   Relatively stable, sloping accumulation of large rock fragments, often at a cliff base. (see scree)

tapered.   (p. 23) Gradually (not abruptly) narrower or smaller at base or tip. (see truncate)

taproot.   (p. 18) Primary root that grows more or less straight down into soil, is tapered to the end, and has smaller, lateral branches (e.g., carrot). (see fibrous)

taxon (taxa).   In classification, a group of organisms (such as plants) at any rank (e.g., species, genus, family); taxonomy is the science of classifying organisms. (see rank)

tendril.   Slender, generally coiling structure (generally stem, stipule, or leaf tip) by which a climbing plant becomes attached to its support (e.g., *Lathyrus lanszwertii* leaf).

terminal.   Pertaining to the tip of a structure (e.g., terminal bud). (see lateral)

ternate.   Lobed, dissected, or compound into three parts, once, as a clover leaf (e.g., *Trifolium wormskioldii*), or more than once; in a ternate-pinnate leaf, the leaf is divided into three leaflets, each of which is pinnately compound (e.g., *Aralia californica*); three structures or parts (e.g., veins, lobes, or leaflets) radiating from a common point. (see palmate, pinnate)

thorn.   Sharp-pointed, usually stiff projection, derived from a branch and therefore often subtended by a leaf or leaf scar, sometimes with leaf scars and buds of its own (e.g., *Castela emoryi*); loosely used for any sharp projection. (see armed, prickle, spine)

throat.   (p. 26) In calyces or corollas with fused sepals or petals, the expanded, fused portion distal to the tube and proximal to the limb.

timberline.   Region in high mountains where subalpine forests give way to treeless alpine vegetation.

tomentose.   (p. 28) Covered with densely interwoven, generally matted hairs.

tooth (teeth).   1. (p. 20) Small, pointed projection, such as on the margin of a leaf, sepal, or petal. 2. Free tips of otherwise fused structures, such as sepals or petals (somewhat archaic usage); smaller than lobes. (see dentate, serrate)

transverse.   Pertaining to width or the widthwise dimension; perpendicular to the axis. (see longitudinal)

tree.   Woody plant of medium to tall maximum height, with generally one trunk from the base (e.g., *Sequoia sempervirens*). (see shrub, perennial)

truncate.   (pp. 22, 23) Abruptly (not gradually) narrower or smaller at base or tip, as if cut straight across or nearly so. (see tapered)

tube.   (p. 26) In calyces or corollas with fused sepals or petals, the often more or less cylindric, fused portion at the base, proximal to the throat and limb.

tuber.   Short, thickened, fleshy, usually starchy, underground stem for storage (of water, food, or both) and sometimes propagation (e.g., potato). (see stem)

tubercle.   Small, wart-like projection (e.g., *Cryptantha muricata* nutlet).

twig.   In woody plants, a terminal stem segment, produced during the current or most recent growth period.

twining.   Climbing by the twisting or coiling of stems, tendrils, or other structures (e.g., *Antirrhinum filipes*).

ultimate.   Last, most distal, or smallest, as all the tips of a branching stem or the smallest divisions (segments) of a compound leaf or dissected leaf.

umbel.   1. (p. 25) In flowering plants excluding Asteraceae and some other groups, an inflorescence in which three to many pedicels and, if compound, branches (rays) radiate from a common point; characteristic of but not confined to Apiaceae. (see ray) 2. In Asteraceae and some other groups, an umbel-like inflorescence is one in which three to many stalked inflorescence units (e.g., heads in Asteraceae, umbels enclosed by involucres in *Eriogonum*), instead of individual flowers, radiate from a common point of attachment without an evident axis or branches.

uncommon.   Unlikely to be encountered; sometimes not present in appropriate habitats. (see abundant, common, rare)

understory.   Layer of vegetation growing beneath a canopy of taller plants.

unisexual.   (p. 25) Either male or female reproductive parts occurring and functional in the same plant or structure (e.g., flower, spikelet, inflorescence). (see bisexual, pistillate, staminate, dioecious, monoecious)

urn-shaped.   (p. 25) Widening more or less abruptly at the base and then gradually or abruptly narrowed toward the tip. (see bell-shaped, funnel-shaped, rotate, salverform)

utricle.   (p. 27) Mostly dry, dehiscent or indehiscent fruit from a generally compound pistil in which a balloon- or bladder-like ovary wall loosely encloses (or, in some Chenopodiaceae, is adherent to) a single seed.

valve.   1. (p. 27) In legumes or capsules opening by longitudinal lines, one of the parts into which the fruit is dehiscent (such fruits valvate). 2. In anthers, the flap of tissue resulting from dehiscence by a curved line (e.g., Lauraceae, some Berberidaceae). 3. In Ophioglossaceae, one of the parts into which the sporangium is dehiscent.

vascular.   Pertaining to plant veins or to plants with veins.

vein.   (pp. 29, 30) 1. Tissue specialized for transport of substances within a plant: water and dissolved inorganic substances (e.g., mineral nutrients) through the xylem; energy-rich, organic compounds (products of photosynthesis) through the phloem. 2. Strand of such tissue, often seen as a line in surface view and as a bundle in transverse section.

vernal.   Pertaining to the spring season.

vernal pool.   Shallow, ephemeral body of water (i.e., one that becomes dry by spring or early summer) that occupies a depression, with underlying hardpan, claypan, or bedrock, in a grassland, foothill woodland, or chaparral.

vestigial.   Rudimentary; pertaining to a structure that is undeveloped, poorly developed, or degenerate and therefore non-functional.

vine.   Trailing, twining, or climbing plant, usually attached to its support by the twisting or coiling of stems, tendrils, or other structures (e.g., *Phaseolus filiformis*).

waif.   Alien, adventive; reproducing neither sexually (e.g., by spores, seeds) nor vegetatively (e.g., by sprouts, suckers) in the absence of any benefit, intentional or not, direct or indirect, of human activity, and therefore not persisting beyond initial generation or establishment, or reproducing to some extent but not persisting for more than a few generations or well beyond initial establishment and therefore not completely naturalized; generally not considered to be part of the flora, but of interest because of their potential to become naturalized, and thereby to have become so. (see alien, naturalized)

wash.   Normally dry drainage channel with only occasional surface flow (e.g., flash floods), in some cases with water movement and availability below in times of no surface flow.

whorled (whorl).   (p. 21) Arranged in groups of three or more at nodes or positions along an axis (e.g., three leaves per node). (see alternate, opposite)

wing.   1. (pp. 23, 28) Thin, flat extension or appendage of a surface or margin. 2. (p. 25) In many members of Fabaceae and in some other groups, each of the two lateral petals.

wiry.   Pertaining to roots, stems, hairs, and other structures that are slender, stiff, and tough.

wood.   Hardened, thickened, vascular tissue (xylem) under the bark of subshrubs, shrubs, trees, and some vines; number of concentric rings in wood often corresponds to years or growing seasons. (see bark)

woodland.   Vegetation characterized by small- to medium-sized trees, often with less continuous canopy cover than a forest and more than a savanna (canopies do not always touch).

GEOGRAPHIC SUBDIVISIONS OF CALIFORNIA

Inclusion of geographic ranges of taxa, in addition to their habitats and elevational ranges, in *The Jepson Manual, Second Edition* (*TJM 2*) provides an eco-geographic context for plant diversity that can aid in locating known populations of particular taxa and predicting where unknown populations may occur. Formulating such predictions can be a challenge, especially in a large state with the topographic complexity and climatic and habitat diversity of California, where sizable areas remain insufficiently explored botanically.

To enhance the effectiveness of geographic data in predicting plant occurrences, a system was developed for *The Jepson Manual*, or *TJM* (1993), that departed from the widespread practice of simply listing the counties in which a taxon is known to occur or indicating those counties on a map. The geographic system used in *TJM* (1993), slightly modified here, combines features of natural landscapes and biota to delimit the units, as opposed to using the often arbitrary and unnatural boundaries of counties for that purpose. The Jepson geographic system most importantly reflects broad patterns of natural vegetation (and, at a finer scale, more specific plant assemblages), geology, topography, and climate.

Patterns of vegetation and flora that influenced the Jepson geographic system were drawn largely from A. W. Kuchler's (1977) "The Map of the Natural Vegetation of California" (pp. 909–938 in M. G. Barbour and J. Major, eds, *Terrestrial Vegetation of California*, J. Wiley & Sons, New York; reprinted in 1988 by the California Native Plant Society, Sacramento) and P. H. Raven and D. I. Axelrod's (1978) *Origin and Relationships of the California Flora* (Univ Calif Publ Bot 72). Minor refinements of the geographic system in *TJM 2* were based on improved resolution of boundaries as described in *TJM* (1993) or on adjustment of those boundaries in light of additional geographic and vegetation data from satellite photographs and other sources, and from finer-scale adherence to elevational criteria, where appropriate. For further, detailed information on Californian vegetation, see J. Sawyer, T. Keeler-Wolf, and J. Evens's (2009) *A Manual of California Vegetation, Second Edition* (California Native Plant Society, Sacramento) and M. G. Barbour, T. Keeler-Wolf, and A. A. Schoenherr's (2007) *Terrestrial Vegetation of California, Third Edition* (University of California Press, Berkeley).

The Jepson geographic system is organized hierarchically, starting with broadly defined provinces and ending with districts (third-order subdivisions of provinces). Directional modifiers on the geographic units, e.g., "sw NCoRI," are used to increase precision. Combining geographic range statements with habitat descriptions and elevation ranges increases the predictiveness of overall range statements.

There are 50 geographic units in this system. Each has a unique abbreviation, used in descriptions and keys. If a user is already familiar with California geography, the units and their abbreviations may be readily understood. For those less familiar with the state's landforms and vegetation, the system (with abbreviations) should be of help in learning California geography.

Each of the 50 units is defined and described below. These descriptions will be easier to follow if studied in conjunction with the map and hierarchical outline of subdivisions (inside front cover and pp. 42–43). In some cases, counties are indicated parenthetically after a geographic unit in the range state-

ment to provide more distributional detail (for units that span multiple counties). This was frequently done for rare plants because county-by-county information is often sought for them.

The system of geographic units is four-tiered: provinces, regions, subregions, and districts (see outline, p. 42). There are three provinces at the most inclusive level. All three extend outside of California but the California Floristic Province includes most of the state and only small parts of adjacent Oregon, Nevada, and Baja California, Mexico. The other two provinces are the Great Basin and Desert. Each province is subdivided into regions. The California Floristic Province is made up of six regions; in California, each of the other provinces has two regions. Together, these three provinces and ten regions delineate the broad physiographic and biologic geography of California. Provinces and regions are shown most clearly on the "exploded" inset map (inside front cover and p. 43). Like California as a whole, most of the units are elongate in a more or less north-south direction. Nine of the ten regions are further divided, into a total of 20 subregions (1–4 subregions per region). Subregions are based on topographic, climatic, and vegetation variation within the region. Seven of the subregions are further divided into districts, based on more localized environmental variation.

In contrast to the use of arbitrary, often politically determined delimiters, such as county lines, the use of biologically meaningful criteria to delimit geographic units results in sometimes frustratingly indefinite or fuzzy boundaries. Wherever possible, subdivisions are defined on the basis of all three of the main biologically relevant variables: topography, climate, and vegetation. These three variables do not always shift in concert, and in such cases vegetation differences generally take priority. For example, the grassland-covered, treeless hills that qualify on geologic and topographic criteria as lower foothills of the Sierra Nevada instead support vegetation like that of the adjacent Great Central Valley and are therefore considered part of that region, up to the elevation where oak/pine woodlands begin. In other situations, transitions in vegetation occur gradually and there is no apparent botanical basis for drawing a sharp line. Where this occurs, boundaries were established primarily using a combination of geological and topographic criteria (e.g., the North Fork of the Feather River divides the Cascade Ranges from the Sierra Nevada, with volcanic substrates predominant on the Cascade side) or easy-to-follow, man-made corridors (e.g., State Highway 58 through Tehachapi Pass divides the Tehachapi Mountains from the southern Sierra Nevada Foothills).

Presence of a taxon in broad or imprecise transitional areas between two geographic subunits is indicated by use of a "/" between the adjacent subunits involved. For example, the distribution of a plant that occurs in the Sacramento Valley (ScV) and also in an area that is in a transition zone between the ScV and the northern Sierra Nevada Foothills (n SNF) is indicated as "ScV, ScV/n SNF." If more than two adjacent subunits are involved, they may all be listed, each pair separated with a "/."

California Floristic Province (CA-FP)

The largest and most botanically diverse geographic unit in California is the California Floristic Province (CA-FP). It comprises all of the state west of the two other provinces, the Great Basin Province (GB), in the north, and the Desert Province (D), in the south. The CA-FP includes all of the "cismontane" region, as used by Jepson, Munz, and others, in addition to the adjacent, leeward, high montane slopes of the westernmost "transmontane" region of those authors; the GB and D together contain the remainder of the transmontane region. The border between the CA-FP, on the one hand, and the GB and D, on the other, is the main phytogeographic boundary in the state.

North of Lake Tahoe, the boundary between the CA-FP and GB lies between the Cascade Ranges and the Sierra Nevada to the west, with their montane conifer forests, and the Modoc Plateau of the GB to the east, with its juniper woodland and sagebrush steppe. Vegetational, topographic, and geologic boundaries are all indistinct in the north; there are inclusions of sagebrush steppe in the Cascade Ranges (an especially large one in Shasta Valley in north-central Siskiyou County) and of montane forest at

higher elevations in the GB. In the Cascade Ranges, volcanic cones and mountains are more numerous, while in the GB the terrain is generally flatter, with a greater predominance of lava flows that have been faulted into small mountains with intervening basins.

The boundary between the CA-FP and GB runs south from the Oregon border at US Highway 97, along the south side of Lava Beds National Monument, and around Glass Mountain and Black Mountain (barely in Modoc County); it curves west again around the Burnt Lava Flow area, and (from the Shasta County border) approximately follows Highways 89, 44, 36 (through Susanville), and 395 south, along the northeastern base of the Diamond Mountains. There is a floristically interesting indentation of the boundary at Sierra Valley (Plumas and Sierra counties), which is included in the Modoc Plateau. The CA-FP extends slightly into Nevada east of Lake Tahoe (e.g., in the Mount Rose area), with the boundary between the CA-FP and GB nearly following Highways 395 and 88 through Nevada.

South of Lake Tahoe, the boundary between the CA-FP and GB follows the east slope of the Sierra Nevada, generally defined by the indefinite break between either upper montane (red-fir/lodgepole-pine) forest or Jeffrey-pine forest on the CA-FP side and either pinyon/juniper woodland or sagebrush steppe on the GB side; there also is Jeffrey-pine forest in the GB (e.g., in the Mono Craters area). In some places, the boundary between the CA-FP and GB is approximated by Highway 395, but south of Bishop it lies to the west of Highway 395, farther up the east slope of the Sierra Nevada.

South of Owens Valley, the provincial boundary lies between chaparral or pinyon/juniper woodland on the CA-FP side, and vegetation including Joshua tree or creosote bush and white bur-sage on the D side. Montane vegetation in the southeastern Sierra Nevada, northeastern Transverse Ranges, and eastern Peninsular Ranges — all in the CA-FP — tends to grade into desert vegetation on the lower slopes of these mountains in the transition to the D. In Riverside County, San Diego County, and southwesternmost Imperial County, the San Jacinto, Santa Rosa, Volcan, Laguna, and Jacumba mountains make up the eastern edge of the CA-FP, and are included within it.

Outside of California, the CA-FP extends north into southwestern Oregon, south into northwestern Baja California, and east into the Lake Tahoe region of Nevada, as described above. Those out-of-state parts of the CA-FP are not covered by *TJM 2*. In California, the CA-FP is divided into six regions, 17 subregions, and 17 districts, as described below.

NORTHWESTERN CALIFORNIA REGION (NW).

This region has the wettest and most predictable climate in California. The boundary between the NW and the Cascade Ranges Region (CaR) is approximated by Interstate 5 and the Sacramento River south to the Great Central Valley Region (GV, which reaches its northern limit near Red Bluff). Substrates derived from metamorphic rock support oak woodland or montane fir/pine forest with hemlock on the NW side; those developed from volcanic material support sagebrush scrub or montane conifer forest much like that of the High Sierra Nevada Subregion (SNH), with sugar pine but without hemlock, on the CaR side.

From near Red Bluff south to southwestern Solano County, the NW meets the GV and the boundary is defined primarily by blue oak/foothill-pine woodland on the NW side, and grassland (or agricultural land) on the GV side. From southwestern Solano County, the southern boundary of the NW turns westward along a vegetational boundary that excludes salt marsh, coastal prairie, and other maritime communities of the Central Western California Region (CW) to the south, and then proceeds through southern Sonoma County to the Pacific Ocean near Bodega Bay. The NW is divided into three subregions.

North Coast Subregion (NCo). This subregion extends along the Pacific Coast the full length of the NW, from the Oregon border south to Bodega Bay. It is a strip of land of variable width that supports truly coastal vegetation, including predominantly coastal prairie, along with coastal marsh, coastal scrub, closed-cone-pine/cypress forest, and grand-fir/Sitka-spruce forest. In some places (e.g., the northern Mendocino coast), the NCo is reduced to coastal bluffs.

Klamath Ranges Subregion (KR). The California portion of this geologically old and distinct, serpentine-rich subregion is bounded to the north by Oregon and in the northwest by the coastal vegetation of the NCo. Its southwestern and southeastern boundaries abut the North Coast Ranges Subregion (NCoR). In the southwestern KR, the boundary with the NCoR has a geological basis, with the mostly sedimentary Franciscan Complex of the NCoR faulted against the older, plutonic and metamorphic rocks of the KR. This fault boundary generally coincides with the northwest-flowing Klamath and South Fork of the Trinity rivers. The transition in forest types across the boundary between the KR and NCoR is gradual, with the KR containing forests of globally exceptional conifer diversity.

In the east, the boundary between the predominantly metamorphic KR and the volcanic CaR lies east and north of Shasta Lake, incorporating the McCloud and Hosselkus limestone formations. In the southeast, the boundary generally excludes the chaparral and foothill-pine/blue-oak woodland vegetation of the Inner North Coast Ranges District (NCoRI) in southwestern Shasta and northwestern Tehama counties.

The KR includes the Marble, Salmon, Scott, Scott Bar, Siskiyou, and Trinity mountains, the Trinity Alps, and Mount Eddy. Red Mountain, near the point where Trinity, Shasta, and Tehama counties meet, is one of the southernmost peaks in the KR that exceeds 1500 m.

North Coast Ranges Subregion (NCoR). This subregion is the largest in the NW and includes widespread serpentine. It is divided into three districts:

Outer North Coast Ranges District **(NCoRO).** This district, the largest in the NCoR, is characterized by very high rainfall, as well as by redwood, mixed-evergreen, and mixed-hardwood forests. Notable mountain peaks include Mount Lassic, Grouse Mountain, and Horse Mountain, all of which are exceeded in elevation by peaks to the east in the High North Coast Ranges District (NCoRH).

High North Coast Ranges District **(NCoRH).** This district is characterized by heavy snow cover, as well as by montane and subalpine conifer forests, treeless high peaks, and floristic similarities to the SNH. Major peaks of the NCoRH all rise above 1500 m (most are above 2000 m), and extend from South Fork Mountain in Humboldt County southeast to the Yolla Bolly Mountains, and from there south to Pine Mountain in Lake County. Somewhat lower, more western, and more isolated peaks similar in vegetation to South Fork Mountain (e.g., Mount Lassic, Grouse Mountain, Horse Mountain) are included instead in the NCoRO. Snow Mountain and Mount Sanhedrin are in the NCoRH.

Inner North Coast Ranges District **(NCoRI).** This district is characterized by low rainfall and hot, dry summers, as well as by chaparral and pine/oak woodland. It extends from the Anderson area in southwestern Shasta County, southward along the east slope of the North Coast Ranges, with a conspicuous westward bulge near the southern end of the NCoRH, to an area west of the Russian River (from north of Ukiah south to Mount St. Helena). Serpentine is widespread in the NCoR, but especially common in this district.

CASCADE RANGES REGION (CaR).

This region, characterized by volcanics, is bounded to the north by Oregon, to the west by the predominantly metamorphic KR, to the southwest by agricultural land or grassland of the GV, to the southeast by the Sierra Nevada Region (SN), and to the east by the juniper woodland of the GB.

Differences in vegetation between the CaR and the Modoc Plateau Region (MP) of the GB are especially unclear. For example, a major island of GB vegetation (sagebrush steppe and juniper woodland) occurs well within the CaR, in Shasta Valley (east of Yreka, near the boundary between the CaR and KR).

The interface between the CaR and SN is defined geologically by the contact between the rela-

tively recent volcanics of the CaR and the predominant metamorphics (with both granitic intrusions and volcanics) of the northern SN (n SN). This contact, located slightly northwest of the canyon of the North Fork of the Feather River, serves as a reasonably distinct topographic marker. The geologic and topographic aspects to the interface between the CaR and SN are not reflected in any vegetational change; rather, the forests of these regions change gradually with latitude. The CaR is divided into two subregions.

Cascade Range Foothills Subregion (CaRF). This subregion, in the southwestern part of CaR, is characterized by chaparral and blue-oak/foothill-pine woodland at about 100–500 m in elevation. The northern CaRF and northern NCoRI combined with the southern KR form a continuous horseshoe-shaped band of similar foothill vegetation along the northern margin of the GV.

High Cascade Range Subregion (CaRH). This subregion (generally above 500 m) comprises the remainder of the CaR and is characterized by ponderosa-pine, montane fir/pine, and lodgepole-pine forests, with treeless alpine vegetation on Mount Shasta and Lassen Peak.

SIERRA NEVADA REGION (SN).

This primarily igneous region meets the volcanic CaR to the north. To the west it shares a long north-south border with the GV (grassland on the GV side versus foothill vegetation on the SN side), and meets the Southwestern California Region (SW) at Tejon Pass (on Interstate 5). On the east, the SN contacts the provincial boundaries of the GB and D.

The SN is divided into three subregions, the two larger of which (SNF, SNH) comprise all but the southernmost extremity of the region, the Tehachapi Mountains Subregion (Teh). Each of the two larger subregions is divided into three districts (northern, central, southern) along contiguous, more or less east-west lines. Although vegetation changes more or less gradually with latitude in the SN, the lines between the northern, central, and southern districts were chosen, somewhat arbitrarily, to coincide with areas of more or less abrupt floristic transition and with major rivers or drainage systems.

Sierra Nevada Foothills Subregion (SNF). This subregion comprises a lower, mostly narrow, north-south strip in the westernmost one-third to one-fifth of the SN, with the GV to the west, the SNH or the Mojave Desert Subregion (DMoj) to the east, and the Teh to the south. The upper elevational limit of the SNF is approximately 1500 m (in the north) to 1000 m (in the south) except near Lake Isabella, where the upper limit is approximately 1500 m.

Throughout most of its area, the SNF is characterized by blue-oak/foothill-pine woodlands (versus ponderosa-pine forest of higher elevations in the SNH) and chaparral, with some serpentine. It is best differentiated from the SNH and GV by vegetation, as opposed to climatic, topographic, geologic, or other considerations. The SNF is divided into northern, central, and southern districts, as discussed under the SN, and as defined under each.

***Northern Sierra Nevada Foothills District* (n SNF).** This district meets the CaRF to the north (northwest of Oroville) and is bounded more or less arbitrarily in the south, where it meets the c SNF, by the Stanislaus River, which corresponds to the Calaveras-Tuolumne county line. Oroville, Auburn, and Placerville are all well within the n SNF, whereas Grass Valley, at about 800 m, is near the border with the n SNH.

***Central Sierra Nevada Foothills District* (c SNF).** This district meets the n SNF to the north and is bounded in the south by the divide (in Fresno County) between the San Joaquin and Kings river drainages, which is approximated by Highway 168. Sonora, Incline, and Mariposa are all within the c SNF.

Southern Sierra Nevada Foothills District (s SNF). This district meets the c SNF to the northwest and the Teh to the south, at Highway 58 through Tehachapi Pass, which approximates the division between the Tehachapi Creek and Cache Creek drainages. The district runs the width of the SN at its southern end (i.e., the SNH does not extend all the way to the southern end of the SN). Like the Teh, the s SNF is complex, with gradual transitions into surrounding areas of the GV, s SNH, and DMoj.

High Sierra Nevada Subregion (SNH). This large subregion is elongate in a north-south direction, extending from Lassen and Plumas counties in the north to Kern County in the south, and is bounded by the SNF to the west and the Great Basin Province (GB) and Desert Province (D), including parts of Nevada, to the east. It is vegetationally complex, with forests of ponderosa pine, white fir, and giant sequoia in lower montane areas, forests of red fir, Jeffrey pine, and lodgepole pine in upper montane areas, forests of mountain hemlock and whitebark pine in subalpine areas, and treeless alpine areas at the highest elevations (about 3000–4400+ m).

The long border between the SNH to the west and the GB and D to the east, extending more than half the length of California, is in places difficult to define (see the CA-FP, above). The SNH is divided (as is the SNF) into northern, central, and southern districts, as discussed under the SN.

Northern High Sierra Nevada District (n SNH). This district in the north meets the CaRH of the CA-FP and the MP of the GB; the boundary with the CaRH more or less coincides with the North Fork of the Feather River, from northeastern Butte County to southwestern Lassen County. In the south, the border with the c SNH approximately follows the Calaveras-Tuolumne, Alpine-Tuolumne, and Alpine-Mono county lines to the border with the GB. Quincy, Downieville, Truckee, and Markleeville are within the n SNH.

Central High Sierra Nevada District (c SNH). This district meets the n SNH to the north, as defined above. The southern boundary, west of the Sierran crest, is the divide between the San Joaquin and Kings river drainages (as it is in the c SNF). This divide winds to the south in eastern Fresno County, reaching the Sierran crest along the Goddard Divide, near Mount Darwin (4200 m). East of the Sierran crest, the boundary with the s SNH follows Bishop Creek, down to the border with the GB at about 2000 m. Yosemite National Park and Mammoth Lakes are within the c SNH.

Southern High Sierra Nevada District (s SNH). This district meets the c SNH to the north-northwest and the s SNF to the west and south. All but the northern tip of Kings Canyon National Park and all of Sequoia National Park are included in the s SNH. In the northern part of this district are the highest mountains in California, including Mount Whitney at 4000+ m. In this area, peaks average about 3000 m, while in the southernmost part of the district this figure is 2000–2500 m. The boundary with the s SNF in the south, defined by vegetation, is convoluted and relatively indistinct. To the east, the s SNH meets the DMoj, in the south, and the GB, in the north, at the transition between montane and desert vegetation (as discussed under the CA-FP). The higher mountains of the southern part of this district (e.g., Piute Mountains, Scodie Mountains, Breckenridge Mountain) support yellow or pinyon pines, but not the oak/pine woodland, chaparral, or desert scrub of neighboring geographic units.

Tehachapi Mountain Area Subregion (Teh). This small foothill and montane subregion, in which elevations rarely exceed 2000 m, has floristic elements of all surrounding geographic units. Highway 58 through Tehachapi Pass constitutes the boundary between this subregion and the s SNF. In the west, the subregion is bounded by the GV, where included foothill and mixed-woodland vegetation meets grassland and agricultural land. To the southwest, the subregion ends at Tejon Pass on Interstate 5, where it meets the northern part of the Western Transverse Range District (WTR). The eastern-southeastern boundary with the D is indistinct, as discussed under the CA-FP, with chaparral or pinyon/juniper woodland on the Teh side and creosote-bush scrub on the D side.

GREAT CENTRAL VALLEY REGION (GV).

This region is an elongate, north-south oriented lowland surrounded by all other regions of the CA-FP but bordered mostly by coast ranges to the west and the SN to the east. On all borders (i.e., those with the NW, CW, SW, SN, and CaR) it ends where oak-pine woodlands or mixed hardwood forests begin. Although now predominantly agricultural, the GV still supports some grasslands, marshes, vernal pools, riparian woodlands, alkali sink vegetation, and stands of valley oak. Toward the southern end of the GV, some desert elements occur. The region is divided into two subregions.

Sacramento Valley Subregion (ScV). This subregion comprises the northern, smaller, wetter, cooler area of the GV, extending from near Red Bluff in Tehama County to the salt marshes of Suisun Slough in southwestern Solano County. The boundary between the ScV and the San Joaquin Valley Subregion (SnJV) follows the northern borders of Contra Costa and San Joaquin counties, which approximately bisect "the delta" area of the Sacramento and San Joaquin rivers.

San Joaquin Valley Subregion (SnJV). This subregion comprises the southern, larger, drier, hotter area of the GV; its northern limits are defined under the ScV, while its other boundaries equal those of the GV. Islands of higher (± 800 m), moister habitats in the Temblor Range and on associated ridges, located geographically in the sw SnJV, are included instead in the Inner South Coast Ranges District (SCoRI) of CW. The Caliente Range is also in the SCoRI based on floristics and topography. These and other eastern ranges of the SCoR flank western extensions of the SnJV, such as the Carrizo Plain and San Juan Valley in eastern San Luis Obispo County, and Cuyama Valley in southernmost San Luis Obispo and northernmost Santa Barbara counties. Further north, the Livermore Valley (Alameda County) is another western extension of SnJV.

CENTRAL WESTERN CALIFORNIA REGION (CW).

This north-south oriented region is bounded by the NW to the north, the Pacific Ocean to the west, SW to the south, and GV to the east. The boundary between the CW and SW follows the crest of the Santa Ynez Mountains from Point Conception to just north of Santa Barbara, where it turns approximately north along Mono Creek and beyond; the region thus includes most of the San Rafael Mountains but excludes Mount Pinos, which is in the SW. Many, often small outcrops of serpentine are scattered throughout the region. The CW is divided into three subregions, one of which comprises two districts.

Central Coast Subregion (CCo). This subregion extends along the Pacific Coast (and San Francisco Bay) the full length of the CW, from near Bodega Bay in the north to Point Conception in the south. Like the NCo in the NW, the CCo is variable in width and coastal vegetation predominates. In places (e.g., the southern Monterey coast), the CCo is reduced to coastal bluffs. Salt marshes and coastal prairie occur around the San Francisco Bay; coastal-sage scrub is prevalent in the south. In the southern part of the CCo, from Morro Bay to San Luis Obispo, the Seven Sisters support chaparral and other non-coastal vegetation.

San Francisco Bay Area Subregion (SnFrB). This subregion occupies the northern one-third of the CW, east of the CCo. It is reasonably well defined physiographically, by features such as Mount Tamalpais, the Santa Cruz Mountains, and the northern Diablo Range, including Mount Diablo and Mount Hamilton. The southern boundary is somewhat arbitrary, following Highways 156 and 152 from the CCo east of Castroville, through Hollister and Pacheco Pass, to the GV near San Luis Reservoir. The subregion is less well defined vegetationally, encompassing a diversity of vegetation types, from very wet redwood forest to dry oak/pine woodland and chaparral.

Hierarchical Outline of Geographic Subdivisions

CA-FP
California
Floristic Province

NW Northwestern California
  NCo North Coast
  KR Klamath Ranges
  NCoR North Coast Ranges
    NCoRO Outer North Coast Ranges
    NCoRH High North Coast Ranges
    NCoRI Inner North Coast Ranges

CaR Cascade Ranges
  CaRF Cascade Range Foothills
  CaRH High Cascade Range

SN Sierra Nevada
  SNF Sierra Nevada Foothills
    n SNF northern Sierra Nevada Foothills
    c SNF central Sierra Nevada Foothills
    s SNF southern Sierra Nevada Foothills
  SNH High Sierra Nevada
    n SNH northern High Sierra Nevada
    c SNH central High Sierra Nevada
    s SNH southern High Sierra Nevada
  Teh Tehachapi Mountains Area

GV Great Central Valley
  ScV Sacramento Valley
  SnJV San Joaquin Valley

CW Central Western California
  CCo Central Coast
  SnFrB San Francisco Bay Area
  SCoR South Coast Ranges
    SCoRO Outer South Coast Ranges
    SCoRI Inner South Coast Ranges

SW Southwestern California
  SCo South Coast
  ChI Channel Islands
    n ChI northern Channel Islands
    s ChI southern Channel Islands
  TR Transverse Ranges
    WTR Western Transverse Ranges
    SnGb San Gabriel Mountains
    SnBr San Bernardino Mountains
  PR Peninsular Ranges
    SnJt San Jacinto Mountains
    PR exc SnJt Peninsular Ranges except San Jacinto Mountains

GB
Great Basin Province

MP Modoc Plateau
  Wrn Warner Mountains
  MP exc Wrn Modoc Plateau except Warner Mountains

SNE East of Sierra Nevada
  W&I White and Inyo Mountains
  SNE exc W&I East of Sierra Nevada except White and Inyo Mountains

D
Desert Province

DMoj Mojave Desert
  DMtns Desert Mountains
  DMoj exc DMtns Mojave Desert except Desert Mountains

DSon Sonoran Desert (also known as Colorado Desert)

Geographic Subdivisions of California

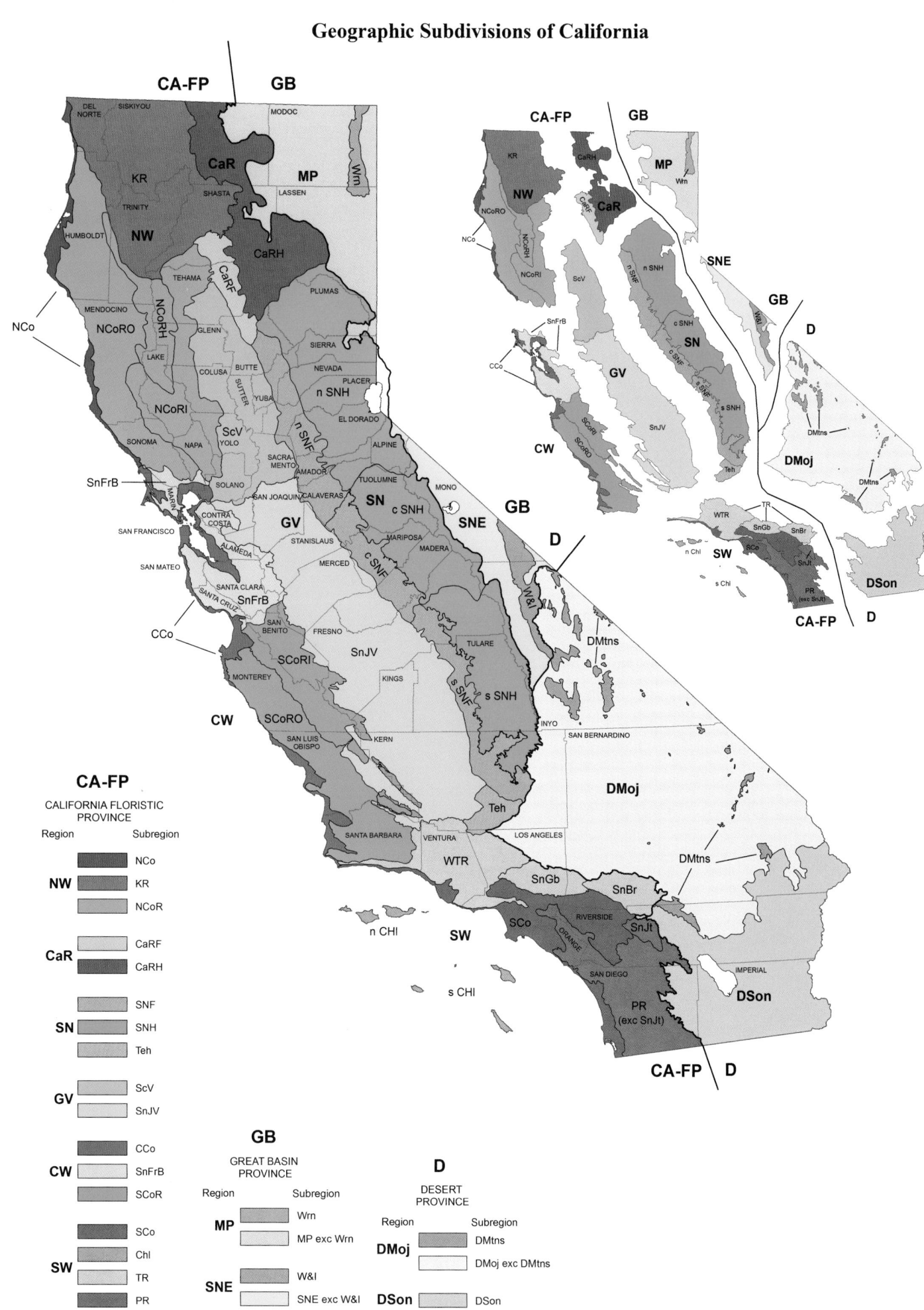

CA-FP
CALIFORNIA FLORISTIC PROVINCE

Region	Subregion
NW	NCo
	KR
	NCoR
CaR	CaRF
	CaRH
SN	SNF
	SNH
	Teh
GV	ScV
	SnJV
CW	CCo
	SnFrB
	SCoR
SW	SCo
	Chl
	TR
	PR

GB
GREAT BASIN PROVINCE

Region	Subregion
MP	Wrn
	MP exc Wrn
SNE	W&I
	SNE exc W&I

D
DESERT PROVINCE

Region	Subregion
DMoj	DMtns
	DMoj exc DMtns
DSon	DSon

South Coast Ranges Subregion (SCoR). This subregion is bounded by the SnFrB to the north (boundary defined under the SnFrB), CCo to the west, SW to the south, and SnJV to the east. It is divided into two districts.

Outer South Coast Ranges District (SCoRO). The boundary between this district and the Inner South Coast Ranges District (SCoRI) to the east runs along the Salinas River (approximated by Highway 101), from near Salinas south to about San Miguel in northern San Luis Obispo County, and from there up the Estrella River to the western edge of the SnJV near Shandon. The SCoRO includes the Sierra de Salinas, Santa Lucia Range, and San Rafael Mountains, and extends to as far south as the boundary between the CW and SW, which corresponds to the crest of the Santa Ynez Mountains and Mono Creek. Near the coast, there are small stands of redwood and mixed-evergreen forests in the north, and oak forests in the south, with pockets of montane conifer forest at the highest elevations. Hotter, drier, more inland slopes support primarily blue-oak/foothill-pine woodland and chaparral.

Inner South Coast Ranges District (SCoRI). Located east of the SCoRO, this district includes the southern Diablo Range from Hollister and Pacheco Pass south to (and including) San Benito Mountain, the Gabilan Range, Cholame Hills, and the higher elevations of the Temblor Range, Caliente Range, and associated ridges (isolated within the southern part of the SnJV). The SCoRI supports a mosaic of blue-oak/foothill-pine woodland, juniper woodland, chaparral, and elements of desert scrub.

SOUTHWESTERN CALIFORNIA REGION (SW).

The mainland part of this region is a wide band, oriented northwest to southeast, that is bounded by the Pacific Ocean to the west and Mexico to the south. The SW also includes the Channel Islands. It is separated from the CW to the northwest by the crest of the Santa Ynez Mountains, Mono Creek, and most of the San Rafael Mountains, from the GV to the north at the woodland/grassland interface, from the Teh to the northeast at Tejon Pass along Interstate 5, and from the D to the northeast and east where chaparral or pinyon/juniper woodland on the CA-FP side meets desert vegetation including Joshua tree or creosote bush on the D side (and otherwise as described under the CA-FP). The SW is divided into four subregions and six districts.

South Coast Subregion (SCo). This subregion extends along the Pacific Coast, from Point Conception of the CCo (CW) to Mexico. It is comparable to the NCo and CCo of the NW and CW regions, respectively, but is hotter and drier and extends much farther inland — to San Gorgonio Pass at Banning, which marks the boundary between the CA-FP and D. Coastal-sage scrub and chaparral vegetation predominated in the SCo before urbanization.

Channel Islands Subregion (ChI). The eight major islands in the Pacific Ocean off the coast of southern California are floristically similar to the SCo, but include enough endemics to justify recognition of the ChI as a separate geographic unit. The subregion is divided into two districts. Counties are indicated below (but not on the map) for each of the eight major islands because information on this subject is commonly incorrect and/or not readily verified. Santa Barbara Island was originally in Santa Barbara County, placed in Ventura County for a period, and is presently in Santa Barbara County.

Northern Channel Islands District (n ChI). This district includes the islands of San Miguel (Santa Barbara County), Santa Rosa (Santa Barbara County), Santa Cruz (Santa Barbara County), and Anacapa (Ventura County), which are separated from the mainland by the Santa Barbara Channel. These islands are geologically related to (and probably represent the westernmost peaks of) the Santa Monica Mountains, located in the southern part of the Western Transverse Ranges District (WTR).

Southern Channel Islands District (s ChI). This district includes the islands of Santa Barbara (Santa Barbara County), Santa Catalina (Los Angeles County), San Clemente (Los Angeles County), and San Nicolas (Ventura County). These islands are geologically and floristically more isolated and more diverse among themselves than those of the northern group, probably in part because they were not as readily colonized from the mainland during periods of lowered sea levels that accompanied various glaciations.

Transverse Ranges Subregion (TR). This subregion, the northernmost in the SW, includes mountain ranges that are oriented in an east-west direction. The TR shares nearly all of its southern boundary with the SCo; in the easternmost extreme of this boundary the TR is separated from the Peninsular Ranges Subregion (PR) by San Gorgonio Pass (Interstate 10), which lies between the San Bernardino Mountains (TR) to the north and the San Jacinto Mountains (in PR) to the south. San Gorgonio Pass, near Banning, also marks the division between the SCo (CA-FP) to the west and the D to the east.

The TR is characterized at lower elevations by chaparral and at higher elevations by oak forest and dry montane forests of white fir, incense cedar, or Jeffrey, sugar, or lodgepole pines. The boundary between the TR and D lies between these communities and desert vegetation that includes Joshua tree or creosote bush on the D side. Some high peaks in the TR extend above treeline. The TR is divided into three districts that are progressively higher, hotter, and drier eastward.

Western Transverse Ranges District (WTR). This district meets the SN, GV, and CW to the north, the SCo to the south (a narrow strip of which separates the WTR from the Pacific Ocean), and the D and the San Gabriel Mountains District (SnGb) to the east. It includes Mount Pinos (at 2700 m, the highest point in the WTR), the Santa Ynez Mountains (south of its crest west of Mono Creek), Sierra Pelona, and the Topatopa, Santa Susana, Santa Monica, and Liebre mountains. At the north end of the San Fernando Valley, a topographic boundary with the SnGb follows Interstate 5 north to the Santa Clara River, and from there northeast through Soledad Canyon and Soledad Pass to the boundary between the WTR and D south of Palmdale.

San Gabriel Mountains District (SnGb). This district is a topographically well-defined mountain range situated northeast of Los Angeles. It is bounded by the D to the north and northeast, the WTR to the northwest and west, the SCo to the south, and the San Bernardino Mountains District (SnBr) to the east. The SnGb is separated from the SnBr by the northwest-southeast oriented Cajon Canyon, which is occupied by Highway 138 and Interstate 15. Mount San Antonio ("Old Baldy"), straddling the Los Angeles-San Bernardino county line at 3070 m, is the highest point in the SnGb. It supports alpine taxa near its summit.

San Bernardino Mountains District (SnBr). This is a topographically well-defined mountain range, east of the SnGb. This district is adjacent to the D on its north, east, and southeast boundaries, the SCo to the southwest, and the San Jacinto Mountains District (SnJt) of the PR to the south, from which the SnBr is separated by San Gorgonio Pass (D). The highest point in the SnBr is San Gorgonio Mountain (3500 m), which has the most well-developed alpine vegetation in California south of the SN. The Little San Bernardino Mountains to the southeast of the SnBr are here considered part of the Desert Mountains Subregion (DMtns) because the vegetation is more similar to the D than to the SnBr.

Peninsular Ranges Subregion (PR). This subregion occupies approximately the southeastern one-third of the SW. It includes Mount Palomar, as well as the Santa Ana, Cuyamaca, Santa Rosa, Laguna, Jacumba, Volcan, and San Jacinto mountains. The last range comprises its own district within the PR.

San Jacinto Mountains District (SnJt). This district in the PR is an area with a high level of local endemism. The San Jacinto Mountains include the highest elevations in the

PR, with San Jacinto Peak at about 3300 m. The Santa Rosa Mountains to the southeast, with elevations to 2650 m, are the only other range in the PR that supports well-developed montane to subalpine forests.

Great Basin Province (GB)

The Great Basin Province (GB) lies to the east of the CA-FP in the northern two-thirds of California and meets the D at its southern margin. The boundary with the CA-FP is described above; it follows the high eastern margin of the CaR and SN. The boundary with the D at its northern extent is the transition from sagebrush steppe or pinyon/juniper woodland on the GB side to creosote-bush scrub on the D side. Deep Springs and Fish Lake valleys are in the GB, Eureka and Saline valleys are in the D. Southward, the mixed vegetation of the Owens Valley is included in the GB. This province is characterized by low rainfall, hot to very hot summers, and relatively cold winters compared to much of the D. It is divided into two regions and two subregions.

MODOC PLATEAU REGION (MP).

This region, entirely north of Lake Tahoe, is a high plateau (mostly about 1300–1800 m) in the northeastern corner of California, occupying most of Modoc and Lassen counties and parts of Plumas, Shasta, Sierra, and Siskiyou counties. The MP is characterized primarily by juniper woodland and sagebrush steppe, but also has extensive areas of ponderosa-pine and Jeffrey-pine forests, and lesser areas of montane pine/fir forest. Substrates are volcanic, with faulted lava flows predominating over cones (see the CaR, above).

Warner Mountains Subregion (Wrn). The Warner Mountains, a faulted volcanic range situated mostly in eastern Modoc County, is the most outstanding topographic feature of the MP. Its highest point is Eagle Peak, which exceeds 3000 m. The Wrn is recognized as a distinct subregion because it supports a unique flora that includes an alpine component at the higher elevations.

EAST OF THE SIERRA NEVADA REGION (SNE).

This region, entirely south of Lake Tahoe, has a wide elevational range, from Owens Lake at 1100 m to White Mountain Peak at 4330 m. The part of the SNE excluding the White-Inyo Mountains Subregion (W&I) supports primarily a mosaic of sagebrush steppe, pinyon/juniper woodland, and cottonwood-dominated riparian vegetation. There are also extensive areas of Jeffrey-pine forest in the Mono Craters area, subalpine fir/pine forest on Glass Mountain (3400 m), and alpine vegetation at the top of the Sweetwater Mountains (3550 m). The SNE extends along the eastern edge of the SN to the southern limit of Owens Valley and the W&I, where there is a gradual transition to the DMoj, with creosote bush and white bur-sage dominated scrub vegetation. To the east of the junction of the W&I at Westgard Pass lies a low (1500–2000 m) outlier of the SNE that includes the Deep Springs and Fish Lake valleys.

The boundary between the SNE and CA-FP along the eastern edge of the SN is generally defined by an indefinite break between either upper montane (red-fir/lodgepole-pine) forest or Jeffrey-pine forest on the CA-FP side and either pinyon/juniper woodland or sagebrush scrub on the SNE side. As noted above, there is also Jeffrey-pine forest in the SNE (e.g., Mono Craters area). The boundary between the SNE and CA-FP is west of Highway 395 from 1800 m (south of Bishop) to 2000 m (north of Bishop).

White and Inyo Mountains Subregion (W&I). The White-Inyo Range (W&I) is considered a separate subregion because it supports subalpine bristlecone-pine and limber-pine woodlands as well as unique, treeless, alpine vegetation (White Mountain Peak 4330 m; Inyo and Waucoba peaks both ± 3400 m).

DESERT PROVINCE (D)

The Desert Province (D) of southeastern California encompasses the Mojave Desert Region (DMoj) and Sonoran Desert Region (DSon). This province lies east of the CA-FP and south of the GB. A matrix of scrub vegetation dominated by creosote bush and white bur-sage occurs throughout much of the lowlands, with saltbush scrub characteristic of alkaline basins. The boundary of the D with the SNE is described above under the GB.

South of Owens Valley, the provincial boundary with the CA-FP lies between chaparral or pinyon/juniper woodland on the CA-FP side, and vegetation dominated by Joshua tree or creosote bush and white bur-sage on the D side. Montane vegetation of adjacent areas in the southeastern SN, northeastern TR, and eastern PR tends to grade into desert vegetation on the lower slopes of these mountains. Some taxa are limited to this interface, which may be specified as "w edge D," "w edge DMoj," or "w edge DSon," as appropriate.

MOJAVE DESERT REGION (DMoj).

This region, occupying the northern two-thirds of the D, exhibits greater temperature ranges and more extreme elevational relief than the DSon to the south. Joshua tree and Mojave yucca are conspicuous, widespread members of DMoj vegetation that are absent from the DSon.

Desert Mountains Subregion (DMtns). Although the entire DMoj is a series of mountains and intervening (often wide) valleys, some ranges reach sufficient elevation (generally above 1700 m) to support pinyon/juniper woodland vegetation and are therefore recognized as a distinct subregion, the DMtns. These high ranges include, but are not limited to, the Last Chance Range, Grapevine Mountains, Panamint Range, Coso Range, Argus Range, Kingston Range, Clark Mountain Range, Ivanpah Mountains, New York Mountains, Providence Mountains, Granite Mountains, Old Woman Mountains, and Little San Bernardino Mountains (discussed below). The Panamint, Kingston, and Clark Mountain ranges and the New York Mountains also support white fir or limber pine at their highest elevations. The DMtns have unique elements but also overlap floristically with pinyon/juniper woodland vegetation of the adjacent CA-FP. Some of the eastern DMtns support taxa that occur more widely, in D or GB outside of the state, but are otherwise unknown in California.

The Little San Bernardino Mountains, across Morongo and Yucca valleys from the SnBr (of TR, CA-FP) and mostly included in Joshua Tree National Park, are included as part of the DMtns because the vegetation in this range is more similar to the D than to the SnBr, as noted above.

SONORAN DESERT REGION (DSon).

This region, the California portion of which is also known as the Colorado Desert, occupies the southern one-third of the D, south of the DMoj. The physiographic line separating the two desert regions is not always clear, but overall the DSon is lower, warmer, and somewhat distinct floristically. Conspicuous members of the flora in the DSon that are absent from the DMoj or confined to the southeastern limits of the DMoj include blue palo verde, ocotillo, chuparosa, and ironwood.

The approximate boundary between the DMoj and DSon, from west to east, is along the south edge of the Little San Bernardino, Cottonwood, and Eagle mountains (all in the DMoj), then north along the eastern edge of the Coxcomb Mountains (DMoj) and around the Old Woman, Turtle, and Chemehuevi mountains (all in the DMoj) to the Colorado River. The Chuckwalla and Whipple mountains are in the DSon.

GEOLOGIC, CLIMATIC, AND VEGETATION HISTORY OF CALIFORNIA

Constance I. Millar

Introduction

The dawning of the "Anthropocene," the era of human-induced climate change, exposes what paleoscientists have documented for decades: earth's environment—land, sea, air, and the organisms that inhabit these—is in a state of continual flux. Change is part of global reality, as is the relatively new and disruptive role humans superimpose on environmental and climatic flux. Historic dynamism is central to understanding how plant lineages exist in the present—their journey through time illuminates plant ecology and diversity, niche preferences, range distributions, and life-history characteristics, and is essential grounding for successful conservation planning.

The editors of the current *Manual* recognize that the geologic, climatic, and vegetation history of California belong together as a single story, reflecting their interweaving nature. Advances in the sciences of geology, climatology, and paleobotany have shaken earlier interpretations of earth's history and promoted integrated understanding of the origins of land, climate, and biota of western North America. In unraveling mysteries about the "what, where, and when" of California history, the respective sciences have also clarified the "how" of processes responsible for geologic, climatic, and vegetation change.

This narrative of California's prehistory emphasizes process and scale while also portraying pictures of the past. The goal is to foster a deeper understanding of landscape dynamics of California that will help toward preparing for changes coming in the future. This in turn will inform meaningful and effective conservation decisions to protect the remarkable diversity of rock, sky, and life that is our California heritage.

California's Prehistory: A Tale of Time and Space

The concept of scale is central to understanding history. Time scale can be especially difficult to untangle in resolving past landscapes because there are many histories depending on context. These range from details of the last 200 years in the Lake Tahoe Basin, for example, to the grand sweep of time since the origin of North America. When millions of years are swept into a single phrase ("the Sierra Nevada was uplifted"), it becomes easy to forget that shorter processes also ensued in the distant past and were as important in shaping the landscape and biota as they are at present. In a similar manner, spatial complexities challenge interpretation of landscape-defining events. The land that is now California has been fragmented, stretched, rearranged, uplifted, and submerged in many ways, shapes, and forms. In the past as in the future, there is no California distinct from its continental and global context. This perspective is adopted in outlining the history that follows.

Forces That Shape Change

Whereas time, like a river, flows one way, many processes that affect landscapes, climate, and vegetation recur. Knowing something about these forces helps to make sense of the big and small pictures of the past. Our ability to understand history in turn relies on methods for resolving conditions now long gone. While these scientific techniques have improved dramatically, bias always remains, and in historic vision we continue to "see through a glass darkly."

At the longest historic scale, center stage is taken by geologic drama. Plate tectonics demonstrate that continents are land masses riding on buoyant lithospheric plates, which move over the earth's viscous upper mantle (asthenosphere) powered by convection currents created by the immense heat generated from the hot molten core. Over hundreds of millions of years, earth's crust oscillated through phases of aggregation and dispersal. When continents collided, supercontinents formed. In contrast, breaking-up (rifting) of supercontinents led to dispersal and fragmentary landmasses. These super-continental cycles take about 300–500 million years to complete. Earth is currently in a dispersed-continent phase.

Plate tectonic processes and super-continental cycling affect landscape-building forces. When plates move toward each other, they collide in a boundary that is active or convergent (which may or may not result in subduction, where one plate passes under the other); when plates move away from each other, the boundaries are passive or divergent (plates spread apart along a rift zone). Active boundaries are associated with volcanism, mountain building, faulting, and earthquakes in the adjacent regions; passive boundaries are associated with quiescent continental margins, and erosion dominates. Over time, boundaries can change from active to passive. To the extent that we can trace the land we call California through billions of years, the region has drifted through many degrees of longitude and latitude, switched from active to passive boundaries multiple times, witnessed mountain ranges rise and erode, harbored inland seas, and at times in part been submerged beneath the ocean. The California margin has changed from passive to convergent to the present situation of a combination of transform (side-by-side movement) and convergent.

Climatic changes at this scale were similarly enormous, involving evolution of the atmosphere as well as responses that reflect movement of the continents. Tectonics of super-continental cycles influence an analogous icehouse-greenhouse climate cycle, whereby global climate regimes alternated over hundreds of millions of years between end states. Icehouse conditions tend to (but do not always) occur when global continents accrete and supercontinents form, sea levels are low, polar and continental ice caps are extensive, and global climates are cold-arid. Contrasting greenhouse periods have high sea levels, little or no land ice, and warm, humid climates. Earth at present is in a warm interglacial interval of a longer icehouse phase.

Geologic and climatic cycling strongly influence organic evolution. Dominant at the longest time scale are processes that led to the origin and diversification of life and the rise of the first land plants. Much of our knowledge of the earliest living forms derives from fossils exposed in eastern California. Nested within these long cycles are mid- and short-term processes. Plate tectonics influences geologic processes not only at continental scales but at regional and local scales as well, affecting locations and magnitudes of earthquakes, volcanism and mountain-building, sea level and tides, and the erosion and exposure of underlying rocks. Similarly, superimposed within the current icehouse climate phase are shorter glacial-interglacial oscillations of tens of thousands of years in duration. Modern orbital theory explains these as paced by the oscillating pattern of earth's relationship to the sun. At successively shorter times, diverse climate cycles come into focus, driven by fluxes in solar variability, atmospheric dust concentrations, and ocean circulation. Interannual modes such as the El Niño/La Niña cycle, for example, are paced by changes in ocean patterns.

Plants and animals respond to geologic and climatic processes at each of these scales via changes in

distribution as well as through evolution. As the earth below and the atmosphere above changes, plants migrate, expand and contract in range, and die out locally and recolonize. Vegetation composition is scrambled as new patterns emerge, often in quasi-cyclic manner. In so doing, these changes result in differential birth and death of lineages, subjecting plant populations to natural selection as well as random forces of genetic change. Subspecies evolve, hybridization and gene flow dissolve taxonomic boundaries, and species go extinct as new biodiversity flourishes.

Finally, anomalous events have created many of the defining trajectories on earth. From asteroid impacts to methane hydrate releases, volcanic eruptions to the rise of *Homo sapiens*, surprise events have changed the history of earth — and the California landscape — in unparalleled ways many times.

Late Precambrian through Paleozoic Eras: 1.2 Billion Years (Ga) to 250 Million Years (Ma) Ago

(Refer to the stratigraphic chart at the end of the chapter for reference to time periods)

The early phases of earth's history involved major geologic construction as continents developed, rifted, and re-formed. The Precambrian era includes the earliest period of history, starting with the origin of the earth about 4.6 billion years ago (Ga). By 1.2 Ga, land was beginning to emerge in western North America, and California history comes into focus. Climates of early California were influenced by varying paleolatitudes as the land masses drifted north and south. This in turn affected the course of biotic—and eventually plant—evolution. California was submerged below shallow (east) to deep (west) seas during much of the early period. Starting about 400–350 million years ago (Ma), major mountain building occurred offshore (Antler and Sonoma episodes). As the continent drifted westward, the continental margin repeatedly collided with these offshore island mountain chains, which were added successively to the continent, contributing land mass to California and extending the shoreline from western Nevada into present-day California.

This period of earth's history also included dramatic climatic change. From the earliest time, the sun was young and faint, atmospheric methane and carbon dioxide concentrations were much higher than at present, and atmospheric oxygen evolved only in association with the origin and expansion of photosynthetic life. This period experienced one of the most severe icehouse climates in earth's history, known as Snowball Earth, when ice sheets covered the continents and extended to equatorial latitudes. Glacial till in the Kingston Range near Death Valley documents that California was locked in ice, as was much of the rest of the earth. With the end of the Snowball Earth interval, climate entered a multi-million-year greenhouse phase, beginning about 600 Ma. California lay at low latitudes then, and climates were correspondingly warm.

The first life on earth evolved during this period, with radiation into the major clades or kingdoms. Whereas most life forms were long thought to have evolved after 542 Ma, fossils have been found in recent decades that incontrovertibly document an earlier origin. Although these bear little resemblance to modern plants and animals, the eukaryotic plan is clearly recognizable by 1.7 Ga. In California, photosynthetic cyanobacteria dating to about 650 Ma are recorded in stomatolites (fossil mats) found in the Kingston Range near Death Valley and outcrops of the Nopah Range near Tecopa. Vascular plants first appeared about 420 Ma; seed plants appeared in the fossil record abruptly, at about 360 Ma. Gymnosperms evolved in this interval and are represented by radiations of *Ginkgo,* now with only one (E. Asian) species, *G. biloba,* and now-extinct conifer forms. Extant conifer lineages did not appear for another 100 million years.

Mesozoic Era: 250 Million Years to 65 Million Years Ago

Geology. The direction of subduction along the western margin of North America reversed ~215 Ma, when sea floor (the Farallon plate) began subducting under the North American continent in the

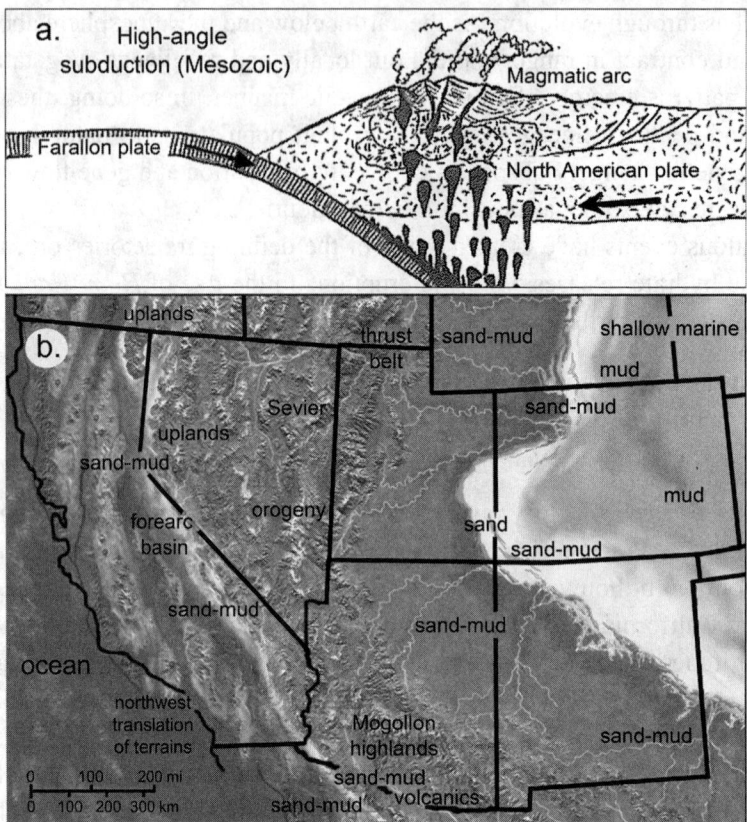

Figure 1. Southwestern North America, ~75 million years ago. (a) The western margin of the continent was an active subduction zone, catalyzing volcanism and mountain building inland of the Sevier and Nevadan ranges and intrusion below-ground of magmatic batholiths that would later be exposed as granitic rocks in mountains of California. The volcanic uplands of present-day Nevada were important for the development of California flora. (b) Much of California was submerged as the so-called fore arc basin west of the continental margin. To the east, the Western Interior Seaway divided the continent nearly in half. Reconstructions (a) modified from F.L. DeCourten, 2003 *The Broken Land,* University of Utah Press; (b) modified from Ron Blakey, Colorado Plateau Geosystems and Northern Arizona University, http://jan.ucc.nau.edu/~rcb7/RCB.html; cross-sections.

vicinity of the present-day Sierra Nevada (Fig. 1a). This resulted in part as the supercontinent Pangea split and cores of Africa and South America rifted off the eastern and southern parts of Pangea, triggering an increase in the rate at which North America moved west over the adjacent sea floor. Subduction under North America led to a dramatically altered geologic history in California, and marked the beginning of several long, complex, and significant mountain-building episodes inland. Between 200 Ma and 70 Ma, two major episodes of thrust faulting led to the Nevadan and Sevier orogenies (Sierran arc volcanoes), which resulted in extensive north-south volcanic mountain chains. Rocks of this age are exposed in the Klamath Ranges, Sierra Nevada, Basin and Range Province, Mojave Desert, and Peninsular Ranges. Much of the interior of California and the Great Basin became elevated plateaus (steppe) built by this volcanic activity.

Subduction during this period also resulted in placement of intruded (unerupted) magma, mostly granitic, far underground. The above-ground mountain ranges and below-ground plutons extended from the latitude of Baja California to Canada. Plutons from this age are found in the Klamath Ranges, Sierra Nevada, Basin and Range Province, Mojave Desert, and Peninsular Ranges. The greatest volume

of magma was intruded ~100 Ma, which resulted from subduction of the Farallon plate under North America. By ~85 Ma emplacement of the granitic batholiths ended.

East of the large Nevadan and Sevier mountain chains lay a large inland sea, the Western Interior Seaway, which divided the continent in half. West of the volcanic arc and east of the subduction trench, a large shallow marine (forearc) basin extended the length of California in what is now the Central Valley (Fig. 1b). Over subsequent millions of years, erosion from the volcanic mountains of the Nevadan and Sevier orogenies deposited material into this basin, sediments of which today represent the primary evidence for these immense mountain chains. By the end of the interval (~65 Ma), the Interior Seaway had disappeared and the continental halves united as land.

Offshore older terranes (fragments of crustal material, in this case from offshore volcanism) continued to be added onto the North American continent by accretion. These appear as NE-SW trending discontinuous belts of rocks of different age and composition in the Klamath Ranges and northern Sierra Nevada. The subduction zone moved westward after each accretion event and triggered successive cycles of accretion of increasingly younger terranes onto California. Accretion by subduction along the northern California margin in the Coast Ranges from Sonoma County to Oregon continued well into the Cenozoic. California drifted on the continental plate northwest during this period from sub-tropical and tropical latitudes at 250 Ma to middle latitudes by ~65 Ma.

Climate.   The climate of this interval continued to be highly variable, alternating from extremely cold glacial periods to the highest global temperatures documented in the last 545 million years. California was at tropical latitudes at the beginning of the Mesozoic, about 250 Ma. Annual temperatures were 10°C warmer and winter temperatures 15–20°C warmer than present. The extensive Western Interior Seaway mitigated climate extremes and enforced warm conditions throughout the American Southwest. Rainshadows developing over the Nevadan and Sevier volcanic ranges began to wring humidity from the air, creating drier climates. By 140 Ma, mild icehouse conditions developed following the breakup of Pangea, which nonetheless left the California region on average warmer than present. The latest part of this interval was characterized by multiple abrupt climate events.

Vegetation.   Some ancient gymnosperm lineages, such as cycads, Taxodiaceae (in the old sense), and Ginkgoaceae, had their heyday during the Mesozoic. Many forms of the last two families extended across the Northern Hemisphere, including western North America. Taxodiaceous taxa appeared in the Mesozoic, with *Sequoia* and relatives dating to ~200 Ma. *Sequoiadendron* is not known from the Mesozoic although the diversity of forms appearing after 65 Ma suggests that it had earlier origins. By ~200 Ma modern conifer families are recognizable. Early forms of Pinaceae appeared by 150 Ma, although their radiation lagged those of other conifers. *Pityostrobus,* a group of Pinaceae taxa that disappeared by 33–30 Ma, and *Pinus* are among the oldest records for this family. In addition to gymnosperms, ferns (including *Equisetum*) radiated and expanded worldwide starting ~150 Ma.

Rapid diversification of angiosperm taxa began ~110 Ma with almost exponential increase in taxonomic diversity. By this time, angiosperms were abundant on a worldwide basis, and by 65 Ma, they had become the most diverse and floristically dominant group of plants, as evidenced by the composition of numerous macrofossil and pollen floras. In North America, the Cretaceous Western Interior Seaway separated two principal floristic provinces. The western province is distinguished by the abundance of *Aquilapollenites*, an early angiosperm pollen taxon resembling grains of modern Santalales (e.g., Comandraceae, Viscaceae) but likely representing a broad polyphyletic clade. Closed-canopy forests of broad-leaved evergreen angiosperms and conifer forests dominated in the warm humid environments, suggesting little seasonality and annual mean temperatures of 20–25°C. Middle latitude west-coast forests contained araucarian, rosid, platanoid, and hamamelid elements including species of Betulaceae, Ulmaceae, Tiliaceae, Juglandaceae, and Santalales. There is also evidence for a continental

margin floristic province based in part on pollen samples from California. This province is recognized by absence or low abundance of *Aquilapollenites*. Indications are that angiosperms first spread to California between 120 Ma and 100 Ma. During this and subsequent tens of millions of years, angiosperms in California appear to have been most extensive and abundant in coastal and fluvial environments, while conifers remained dominant in well-drained and upland areas.

The end of the Cretaceous period was marked by earth's second largest global extinction event, the Cretaceous-Paleogene extinction at 65.5 Ma. This event is attributed to collision of an asteroid or comet with the earth. The impactor probably measured more than 10 km wide, and it left an impact crater 180 km in diameter in the Gulf of Mexico near the Yucatan Peninsula. In addition to non-avian dinosaurs and many other animal lineages, many plant genera went extinct in this event, especially at locations near the impact site. Broad-leaved evergreen trees were at higher risk of extinction whereas taxa with dormancy adaptations (e.g., deciduous leaves) fared better during the "impact winter" that followed. Although California was relatively near the impact site and would thus have been severely affected, no records from the time are firmly documented in our region. The best records are in western interior North America, in a zone from New Mexico north into Canada. Sites in this belt clearly indicate mass plant kills, with estimates of 50–75% extinction of earlier taxa.

Cenozoic Era: 65 Million Years to Present

1. Tertiary Period: 65 Million Years to 2.6 Million Years Ago

Geology.   Before the asteroid impact, North America had begun to rift away from Europe, increasing the speed at which it moved westward. This increase in rate of movement is thought to have lessened the angle of subduction of the Farallon (oceanic) plate under the western margin of North America, transferring volcanic activity from the Pacific west (California-Nevada) into the interior (Colorado-Montana). This catalyzed initial uplift of the Rocky Mountains and began to shut off arc volcanism in much of California and the Great Basin. As a result, the early Tertiary was a period of relative volcanic quiescence in this region.

In the early Tertiary, the region at the eastern margin of California (now the Great Basin) was an elevated upland that drained to the west via rivers that flowed through California to the Pacific Ocean (Fig. 2). A steep gradient existed along what is now the west slope of the Sierra Nevada, but the Sierra Nevada was not the major hydrologic divide that it is now. Rather, to the east lay mountains of significant and apparently greater elevation (likely > 2750 m), with the hydrologic crest in the vicinity of central Nevada. The uplands of what is now the Sierra Nevada were the western edge of a generally mountainous region that extended eastward. This region has been called the Nevadaplano as it reflects similar character to South America's Altiplano.

About 40 Ma, for still poorly understood reasons, the angle of subduction of the Farallon plate re-steepened again (Fig. 3a). As the steeply diving portion of the Farallon plate migrated eastward, it fell away from the bottom of the overriding continent like a trap door slowly opening from east to west (Fig. 3b). This had many consequences, one of which was to relieve compressive forces that had existed in the region of the Rocky Mountains and eastern Great Basin. The change in plate angle also exposed the bottom of the continent to the underlying mantle's heat. Partial melting of the deep crust in response to upwelling hot mantle led to massive volcanic eruptions regionally. Exposure to deep heat also caused the continent to become less rigid, and to thin and stretch. As this occurred, the entire Nevadaplano region subsided, like the domed top of a cake sinking as it comes from the oven (Fig. 3c). This subsidence marked the beginning of the evolution of internal drainage and the birth of the hydrologic Great Basin.

These events also set the stage for a new era of mountain building. Regional subsidence, release of

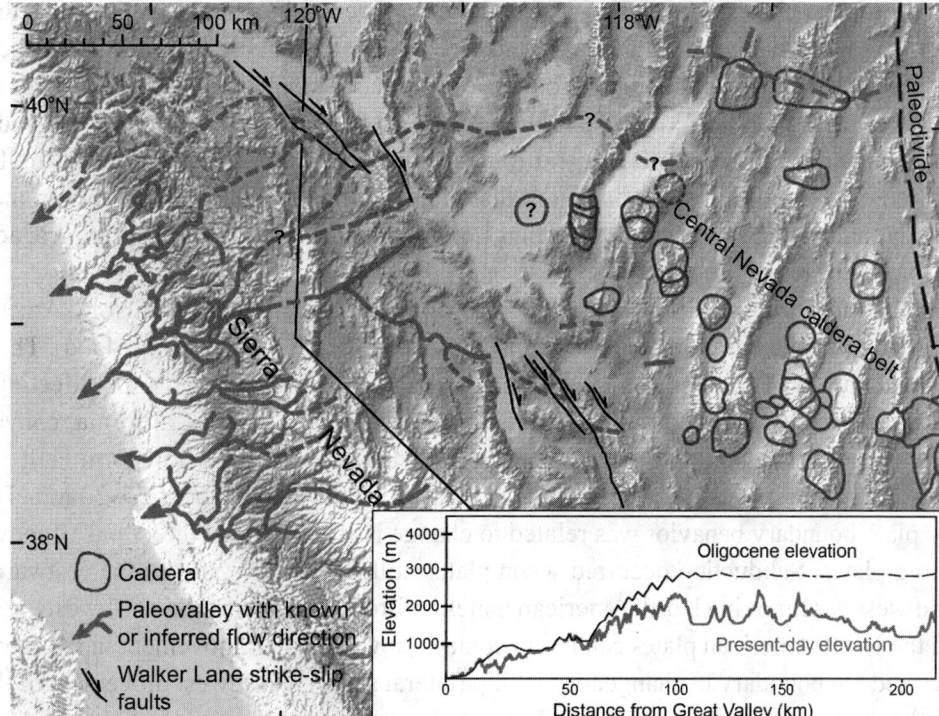

Figure 2. Reconstruction of the Nevadaplano, ~30 million years ago. The Sierra Nevada formed the western flank of a large volcanic upland that extended through the present Great Basin region. Elevation was ~2800 m at the latitude of Lake Tahoe and summit elevations increased eastward to a paleo-divide in central Nevada. Streams flowed westward from the divide, crossing through the Sierra Nevada to the Pacific Ocean, which filled the current Central Valley. Inset shows the greater elevation and different topographic profile of the Nevadaplano relative to the modern mountain crests. Much of the evidence for this reconstruction comes from analyzing tuff deposits from volcanoes and calderas of the central Nevadaplano. Modified from C. Henry, Uplift of the Sierra Nevada, California, 2009, *Geology* 37:575–576.

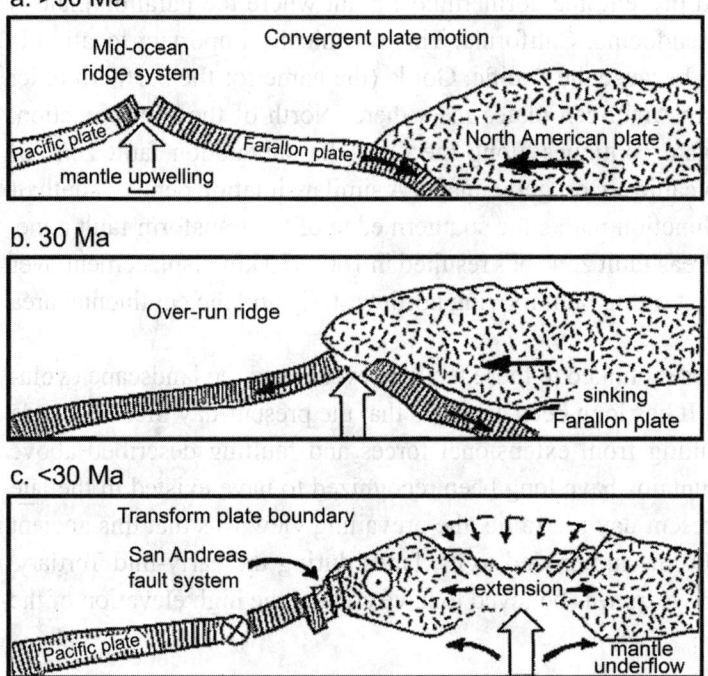

Figure 3. Plate tectonics and the relationship of Great Basin extension and origin of the San Andreas fault system over the past 30 million years. (a) Subduction prior to 30 Ma was active as the Farallon plate dove under the North American plate. (b) Extension of the Great Basin region began when the North American plate overrode the Farallon ridge system in the Pacific Basin about 30 Ma. (c) Contact of the Pacific plate with the North American plate changed boundary dynamics from subduction to lateral shear in northwest-southeast directions. This marked the beginning of the San Andreas fault system. Lateral motion is shown by symbols: the circled X indicates motion away from the viewer and the circled dot indicates motion toward the viewer. Modified from F.L. DeCourten, 2003, *The Broken Land,* University of Utah Press.

compression forces, and crustal extension affected the upper crust in a new way. Because the crust is brittle, it responded to changes deep below the surface by breaking along increasingly numerous normal (extensional) faults to accommodate stretching. Continuous stretching caused blocks of the continental crust to tilt along faults, giving rise to more than 300 fault-block mountain ranges and adjacent basins that characterize the present Basin and Range province. Continued extension of the Basin and Range over the past 30 million years has more than doubled the amount of land between the western (Sierra Nevada) and eastern (Wasatch Mountains-Colorado Plateau) edges of the province, adding 400 km of new landscape in the process.

At about the same time, a major change occurred along the western margin of North America as the eastward-moving oceanic Farallon plate was consumed under North America (Fig. 4a–d). This allowed the North American and Pacific plates to come into direct contact for the first time. Meeting of these two plates fundamentally changed the nature of the contact along western California, converting the boundary from one of subduction to lateral shear along what is known as a transform fault zone. This shear zone was the ancestral San Andreas fault system, which developed about 25–20 Ma. The reason for the new plate boundary behavior was related to change in the dominant directions of movement of the contacting plates. Subduction occurred when plates collided directly, as did the eastward-moving Farallon and westward-moving North American plates before 30 Ma. When the northwestward-moving Pacific and the North American plates came into contact, however, their movement in the same general direction caused the boundary to change to side-slip (lateral shear). Northwest movement of the Pacific plate exerted a drag effect on the continent, adding to the crustal extension of the Great Basin. Subduction continued north and south of the contact between the Pacific and North American plates, where the Farallon plate was disappearing under the continent. The meeting points of the three plates are known as triple junctions. As the Farallon plate was consumed increasingly under North America, the San Andreas fault boundary lengthened accordingly, and the two triple junctions became more widely separated (Fig. 4b–d).

This change from subduction to shear marked a significant transition in the geologic processes that influence California to this day. The transform zone has continued to grow over the past 25 million years as more of the Farallon plate is consumed, leading to an ever-larger area of contact between the North American and Pacific plates. At present, the northernmost point where the Farallon plate is passing under the continent is at Cape Mendocino, California. This seismically important location is known as the Mendocino Triple Junction, because the Pacific, Gorda (the name for the northern relict fragment of the Farallon plate), and North American plates meet there. North of the triple junction, subduction under the North American plate continues along the Cascadia subduction fault zone of Oregon and Washington, with ongoing volcanic arc orogeny inland. A similar situation persists south of Baja California where the Rivera Triple Junction marks the southern edge of the transform fault zone. Cumulative movement along the San Andreas fault zone has resulted in 160–370 km displacement over the past 25 million years, with the area west of the fault moving northwesterly and the continental area now moving southeasterly.

The history of the Sierra Nevada is closely linked to these tectonic events and the landscape evolution of the Nevadaplano and Great Basin. It has long been assumed that the present-day Sierra Nevada is a young uplifted mountain range resulting from extensional forces and faulting described above for the Great Basin ranges. Although mountains have long been recognized to have existed in the late Mesozoic and early Tertiary where the present day Sierra lie, the prevailing view was that this ancient range had never gained elevation > ~2000 m, and eroded to lowlands during the early-mid Tertiary. Fault-block tilting in the past 10–5 million years was believed to have created the high elevation of the modern Sierra Nevada.

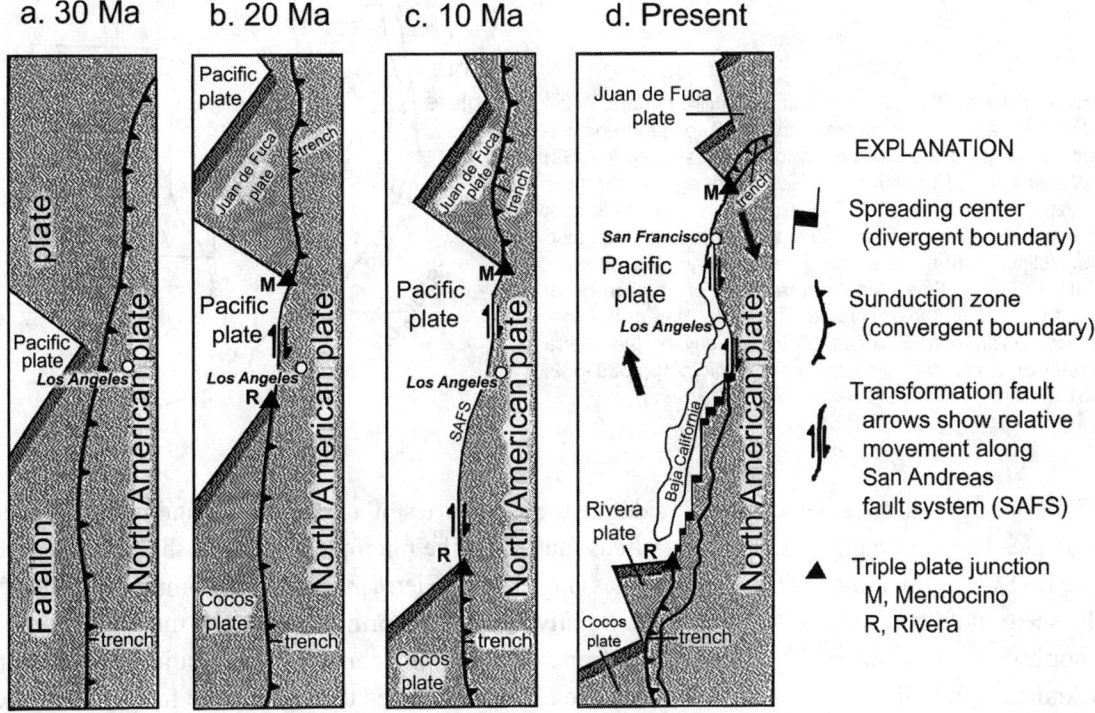

Figure 4. Development of the San Andreas fault system from 30 Ma to present. (a) As the Farallon plate was consumed under the North American plate, the Pacific plate was brought into contact with the North American plate and the San Andreas fault system was initiated. (b and c) As the San Andreas system expanded over time, the two triple junctions (Mendocino in the north, Rivera in the south) migrated farther from each other. (d) Cumulative movement along the San Andreas fault system has resulted in about 300 km displacement over 30 million years. Modified from F.L. DeCourten, 2003, *The Broken Land,* University of Utah Press.

Although some lines of evidence still support this view, an increasing body of research, including paleobotanic records, suggests that the Sierra Nevada achieved heights > 2800 m in the early Tertiary and remained high through subsequent millennia. The mountains of the Nevadaplano to the east of the ancient Sierra Nevada were even higher during this interval (Fig. 2).

This new evidence about elevation of the Sierra Nevada does not suggest that the range was exempt from effects of the extensional and faulting processes that were occurring. The form, topography, and elevation of the modern Sierra Nevada were strongly influenced by those events. For one, the processes that led to general subsidence of the domed Nevadaplano and created the internal drainage of the Great Basin appear similarly to have lowered rather than elevated (as was earlier interpreted) the Sierra Nevada relative to its early Tertiary heights (Fig. 3c).

Further, by the middle to late Tertiary, new patterns of tectonism in the California and western Nevada region strongly influenced the form of the present Sierra Nevada (Fig. 5). At about 10 Ma, shear stress of the Pacific and North American plates along the southern San Andreas began to be displaced inland. This is recognized by a series of fault complexes that describe the chronological development of the displacement. These faults extend eastward along what is known as the Eastern California Shear Zone (ECSZ) in the region of the present Transverse Ranges, continue eastward around the base of the Sierra Nevada and White Mountains, then turn abruptly northward along overlapping sets of echelon faults to the south end of the Tahoe Basin. There the fault zone forks, with one branch extending through Lake Tahoe and the other along the eastern foot of the Carson Range. These zones then become diffuse but eventually strike westward through northern California, converging at the Mendocino Triple Junction.

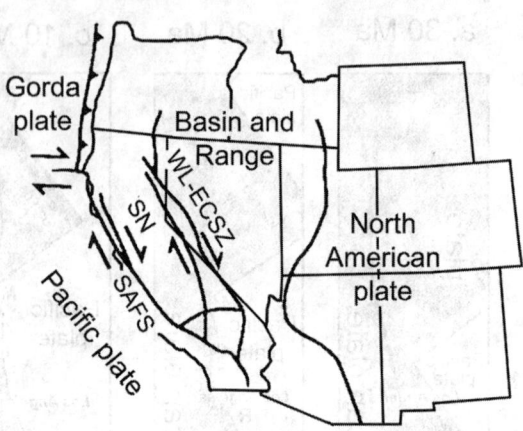

Figure 5. Major fault zones of the past 10 million years. Activity along the San Andreas fault system (SAFS) began to be displaced inland along the Eastern California Shear Zone (ECSZ) and Walker Lane (WL) about 10 million years ago, forming the new Sierran microplate (SN). Tectonic activity along this fault zone is shaping Sierra Nevada topography by processes related to (but distinct from) those affecting the Basin and Range province. [Note: Gorda plate = northern fragment of Farallon plate] Modified from J. Wakabayashi and T. Sawyer, Stream Incision, Tectonics, Uplift, and Evolution of Topography of the Sierra Nevada, California, 2001, *The Journal of Geology* 109:539–562. © 2001 by The University of Chicago.

This arc of faults is now recognized to define the boundaries of a new contintental plate, the Sierra microplate. The west margin is the San Andreas fault, thus the microplate contains the Sierra Nevada, the Central Valley, and some of the Coast Ranges (Fig. 5). The Sierra microplate accommodates 15–25% of the shear motion of the San Andreas zone. Relative to stable North America, the microplate is moving northwestward at about 12 mm/yr. This compares to the movement of the Pacific plate along the San Andreas Fault of about 50 mm/yr. Because the faults that carved this microplate from the continent are relatively young, they exist as a zone of multiple short faults, rather than coalesced into a single prominent fault such as the San Andreas, which by comparison is considered mature. The eastern portion of the microplate fault zone is an obvious topographic belt of low relief that extends northwesterly through eastern California and western Nevada, known as the Walker Lane. West of the Walker Lane are mountains of the Sierra microplate; east are mountains of Basin and Range origin.

Despite being a separate plate, the Sierra microplate remains coupled to the Pacific plate along the San Andreas, and tectonic activity of the San Andreas Fault translates to the Walker Lane and ECSZ. This plate-edge tectonic action, rather than extensional faulting related to the passage of the Farallon plate under the continent, appears responsible for having given shape (tilting and fault boundaries) to the present Sierra Nevada during the last 10 million years. Landmarks such as the Tahoe Basin, Carson Valley, and Owens Valley owe their origins to these forces, all within the past 5 million years.

Uplifting and tilting of the Sierra Nevada and down-dropping of basins had an important effect on the nature of exposures in this region. As slopes were tilted, fault-bordered surfaces were exposed and erosion accelerated, exposing underlying rocks. Included in the exposures are the granitic batholiths emplaced during subduction between 200 Ma and 70 Ma, as well as rocks from far older eras when California was submerged below seas. The latter are exposed as so-called roof-pendants in places such as the steep escarpment faces above Convict Lake in the eastern Sierra Nevada and throughout the southern part of the range.

Extensional forces that thinned the crust and generated basin and range topography also influenced topography in interior California. Especially in the Mojave Desert, faults formed as the crust stretched starting ~35 Ma. In many desert locations, so much crust was displaced that much older rocks below were exposed. The creation of faults via extension triggered volcanic activity in the Basin and Range and Mojave provinces, and many eruption centers arose around fault zones starting ~20 Ma. The Transverse Ranges derive their origin and orientation from lateral shear action along the Pacific and North American plates, but with a unique twist. As the Pacific plate moved northwest relative to the continent, a piece of the North American plate broke off in southern California but remained attached at the eastern margin. Detachment and continued shearing transferred this portion to the Pacific plate and in

the process rotated the mountain axis clockwise, creating the east-west oriented Transverse Range. As this rotation proceeded, it created extension forces to the south that led to the development of the Los Angeles Basin and offshore islands.

The current Coast Ranges are geologically young and owe their origin to diverse and still poorly understood activities of plate contact as the lateral shear zone has increased. Extension, fault-block tilting, and uplift contributed relief to this region as well as volcanic activity along the newly propagating fault areas of the San Andreas. Elsewhere forces remained that derived from subduction and included compression forces, bends in regional faults, and thrust uplifting. Between the ancient Sierra Nevada and Coast Ranges lay the San Pablo Sea, a shallow inland water body, which dried at its north end ~9 Ma. A shallow sea persisted in the San Joaquin Valley to ~2 Ma.

Climate.   New analytic methods have prompted re-interpretation of Tertiary climate processes. Early views regarded the climate since 65 Ma to have begun in warm greenhouse conditions followed by gradual cooling to the current icehouse regime beginning ~2 Ma. The picture now emerging is of much greater complexity and variability (Fig. 6). The time interval began with a greenhouse climate regime, with peak warmth at about 52–50 Ma (Fig. 6a). A slight cooling trend followed that terminated with an abrupt and defining global cooling at 33.5 Ma, the Eocene-Oligocene event. Temperatures at California latitudes during this event dropped by 6–8°C.

The Eocene-Oligocene event marked the return of a global icehouse regime that continues to present (Fig. 6a). Ice-cap development began in Antarctica, and only much later extended into the Northern Hemisphere. Global sea levels dropped by 70 m, reflecting the buildup of polar ice. Two global warming periods interrupted the background icehouse conditions. Peaking at 15–17 Ma was the middle Miocene climatic optimum, after which global temperatures gradually declined and Northern Hemisphere glaciations began. Another brief warming period, the early Pliocene climatic optimum, occurred from 4.5 to 3.5 Ma, when Northern Hemisphere ice melted and temperatures were much warmer than present (as much as 19°C in the Arctic). This was followed by climatic deterioration into fluctuations of the ensuing ice ages, which started about 2.6 Ma.

Most of the warming and cooling trends from 65 to 2.6 Ma are explained by the pacing of tectonic and orbital cycles. Superimposed on these trends, however, were four major climatic aberrations or anomalous periods with highly non-linear response. Two warm events are the hot spike at 65.5 Ma, attributed to the asteroid impact, and a short hot pulse centered at 55.8 Ma, the Paleocene-Eocene Thermal Maximum (PETM), which lasted 170,000 yrs (Fig. 6a). During the PETM, global temperatures increased by 5–10°C in less than 20,000 years. The cause of the PETM is still being debated but is widely attributed to spontaneous release of massive stores of methane hydrates from the ocean floor.

Two other anomalous pulses were global cooling events. These resulted from unusual coincidences in earth's orbital and tectonic cycles. The switch to an icehouse regime at 33.5 Ma appears related to the tectonic opening of the deep-water passage between South American and Antarctica. Superimposed on this were peaks in several orbital cycles of the earth's orientation toward the sun. The cumulative effect of these conditions turned what would have been a gradual trend into an abrupt temperature decline and catalyzed a deep 400,000 year glaciation.

By all indications, Early Tertiary warm humid lowlands and the cooler uplands alike in the California region were characterized by precipitation that was distributed throughout the year; persistent drought was uncommon. Truly arid climates and dry environments did not develop until middle-late Tertiary, and seasonality increased only after the Eocene-Oligocene event. The California Current, an ocean circulation pattern that exists at present, began to evolve about 15 Ma. This current is a primary driver of Mediterranean climates in the California region, and also regulates the steep summer thermal gradient from the coast to the interior. Loss of summer rain as a result and extension of a long sum-

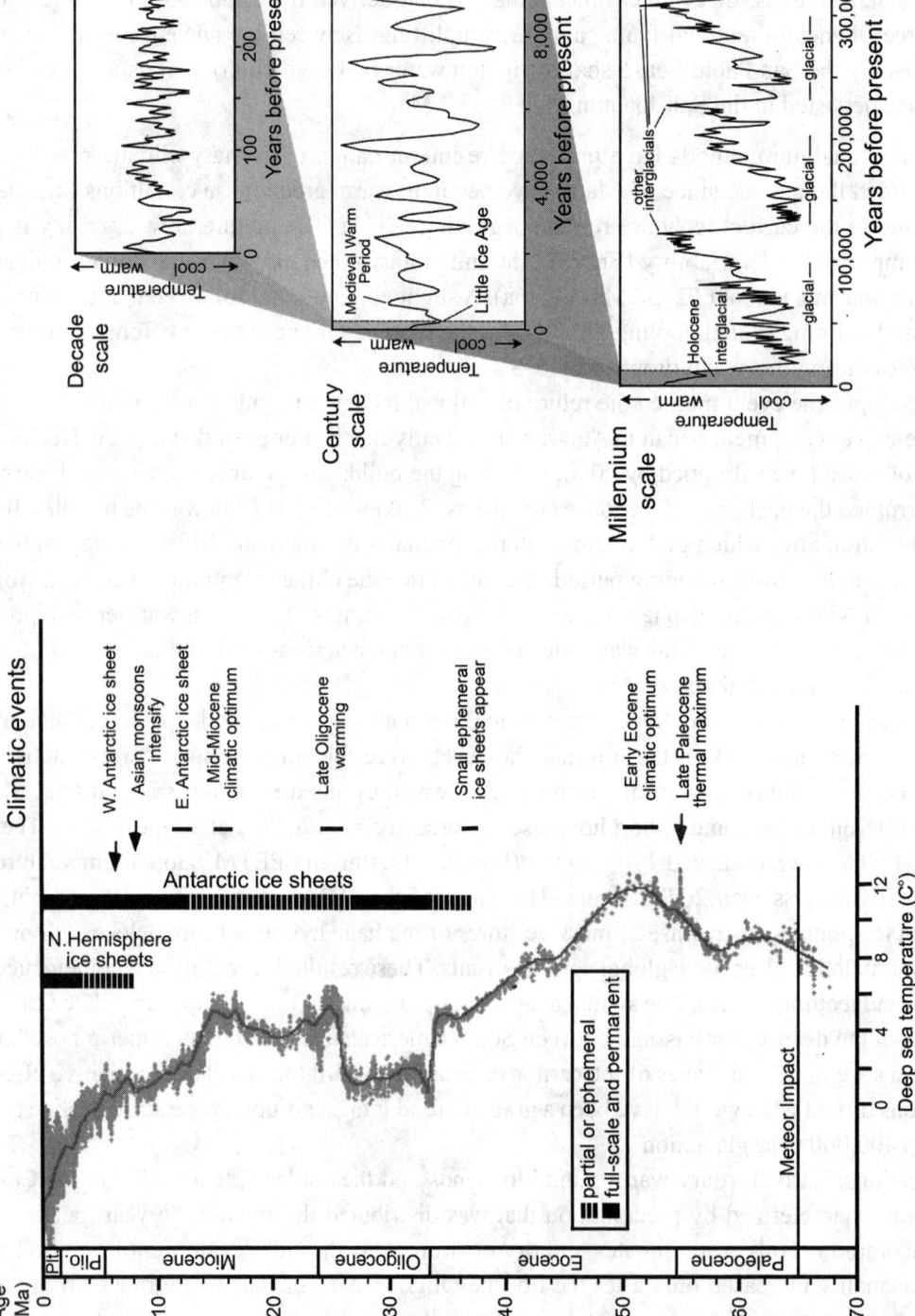

Figure 6. Major trends in global temperature at multiple scales. (a) Climate, tectonics, and biota over the past 65 million years. Modified from J. Zachos et al., 2001. Trends, Rhyths, and Aberrations in Global Climate 65 Ma to Present, *Science* 292:686–693. Reprinted with permission from AAAS. (b) Nested temperature cycles of the past 400,000 years, showing major glacial cycles (bottom), driven by variations in the orbit of the earth around the sun; ~1470 yr Bond cycles, related to variation in solar activity (middle); and cycles of the Pacific Decadal Oscillation, forced by varying patterns of ocean circulation (top). Modified from C. Millar, 2003, USFS, *Science Perspectives*, and sources therein.

mer drought became important influences on the evolution of the modern California flora. Significant regional rain shadows developed with evolution of Sierran and Basin and Range topography, marking initiation of summer-dry climates and the first appearance of desert environments. A Mediterranean climate pattern appears to have evolved in California by 7–4 Ma as the California Current strengthened, although some regions retained a pattern of summer precipitation.

Vegetation.   During the Tertiary we can trace the roots of California's modern flora with satisfying detail. The story of this development mirrors events in the geologic and climatic history of the interval. Early in the period, angiosperm and gymnosperm taxa and community assemblages reflected adaptations to warm-temperate conditions similar to those of the late Mesozoic, i.e., to conditions warmer than present and with precipitation distributed year-round. Records from Wyoming during the PETM indicate that this hot episode created a global floristic upheaval likely experienced in California as well. Evidence points to massive plant species range shifts of 1500 km that occurred in less than 10,000 years in response to rapid warming. These dynamics were highly individualistic: some taxa persisted in place while others underwent significant displacement.

In California, angiosperm diversity appears to have been relatively low before 55 Ma. Fossil taxa bear scant affinity to modern lineages, but show warm-temperate and some subtropical adaptations (Fig. 7a, b). Increasing temperatures and humidity ~50–52 Ma triggered significant floristic shifts toward species adapted to tropical conditions and having affinities to taxa now in rain forests of eastern Asia, southern Mexico, and Amazonia. The Chalk Bluffs fossil flora near Colfax in Nevada County contains one of the richest floras in the West from this climatic period (Fig. 7b). Many species belong to families long extinct in California. More than 71 taxa are identified, including many evergreen angiosperms, as well as deciduous species. Few taxa overlap current native species. Five genera of laurels (including *Persea*), a palm, and *Viburnum* are included, as well as now-exotic genera such as *Perminalia, Phytocrene, Magnolia, Cedrela, Hyperbaena, Artocarpus, Ficus*, and *Meliosma*. Only one gymnosperm, a cycad, is present as a leaf fossil, although temperate conifers including *Pinus, Abies*, and *Picea* are represented by pollen. Such conifers are not recorded in other floras of this age in California. These and other taxa recorded only as pollen in these fossil beds, such as *Platycarya, Juglans, Carya*, and *Liquidambar*, have pollen widely dispersed by wind that may have drifted from uplands either in the Klamath or proto-Sierra ranges to the east. Warm-humid tropical adaptations are reflected in floras elsewhere in California, which had multi-storied rain forests containing, for example, *Cinnamomum, Laurus, Juglans, Magnolia,* and *Zamia*, and rich understories and diverse ground layers.

Whereas western California was blanketed by rich subtropical plant communities prior to about 33.5 Ma, upland regions to the east in the Great Basin high plateau harbored refugial populations of temperate-adapted species, including many conifers and associates now present in the modern flora. These upland populations were important not only as sources for colonizing California following climatic cooling but also as biogeographic crucibles for significant conifer evolution.

Following the climatic deterioration at about 33.5 Ma, abrupt changes in floristic composition and structure took place in California. Tropical-adapted woody angiosperm species disappeared within two million years. Throughout California, temperate-adapted species re-appeared, especially cool-adapted broad-leaved deciduous species and conifers, although the taxa differed from those previously present in California. These new plant communities had affinities to modern communities and high diversity, reflecting the heterogeneous climate and environmental conditions at that time. Taxa such as *Metasequoia, Sequoia, Pinus, Liquidambar, Carya, Juglans, Sorbus, Platanus, Acer, Crataegus, Ulmus, Zelkova, Rhus,* and *Tilia* appear. Notable for the first time in western records are terrestrial herb groups. Pollen records in particular document the expansion and widespread diversification of Asteraceae in the Oligocene. Increasing winter cold was likely a trigger for herb expansion.

Figure 7. Eocene fossils of the northern Sierra Nevada, California. (a) At the late Eocene LaPorte Flora south of Quincy, CA, sediments are exposed of a paleochannel that was part of the ancestral Yuba River system (see Fig. 2). These preserve a diversity of fossils of warm-humid tropical and subtropical plant species that were characteristic of California in the early Tertiary. (b) Impression leaf fossil from another Eocene site, the Chalk Bluffs Flora east of Nevada City, CA, of the extinct genus *Macginitiea* (Platanaceae). Photos courtesy of Diane M. Erwin, UC Museum of Paleontology.

Floras younger than 23 Ma include highly diverse assemblages with taxa present in California today as well as many native to climates warmer and milder than California and having year-round rainfall. They indicate distinctions between upland and coastal communities, and reveal earliest adaptations to summer drying. During the warm climatic optimum at 17–15 Ma global temperatures rose to the highest levels reached during the past 23 million years. Floras throughout the West from this period reflect adaptations and range shifts in response to these conditions, with increasing latitudinal gradients from coastal environments to inland mountains. Associations of taxa unknown at present persisted in many locations, however, such as in the Tehachapi Mountains, where dry-adapted species occurred together, including *Hesperocyparis arizonica, Pinus cembroides, Arctostaphylos, Arbutus, Umbellularia, Cercocarpus,* several shrubby *Quercus, Viburnum,* and *Fremontodendron* alongside now-exotic taxa such as *Ficus, Persea, Dodonea viscosa,* and *Cedrela.*

Inland, in the higher ranges of western Nevada and northeast California, fossil assemblages contained diverse conifers, including *Chamaecyparis, Ginkgo, Abies, Pinus,* and *Torreya,* and hardwoods such as *Castanea, Cedrella, Fagus, Quercus, Carya, Umbellularia, Cercis, Fraxinus, Platanus, Prunus, Sorbus, Tilia,* and *Ulmus* in the Upper Cedarville Flora (16–15.5 Ma); the conifers *Abies, Pinus, Chamaecyparis,* and *Picea breweriana,* and hardwoods *Carya, Quercus, Alnus, Betula, Acer, Platanus,*

Ulmus, and *Zelkova* in the Fingerrock Flora (15.5 Ma); and *Thuja, Abies, Picea, Pinus, Sequoiadendron,* along with *Acer, Berberis, Arbutus, Quercus, Persea, Robinia, Platanus, Cercocarpus,* and *Styrax* in the Middlegate Flora (15.5 Ma).

Intensification of the Mediterranean climate with decreased summer rainfall is reflected in younger floras (< 7 Ma) of the California region, which have increasing representation of taxa adapted to mild and cooling temperatures with dry summers. These floras also show spatial partitioning. Increasing abundance of hypsodont fossil horses corroborate the evolution and spread of California grasslands. A flora of this age in Contra Costa County, for example, is dominated by evergreen *Quercus,* with abundant *Platanus, Populus, Salix,* and understory taxa allied to *Dendromecon, Celtis, Berberis, Cercocarpus, Prunus, Lyonothamnus, Arctostaphylos, Ceanothus, Fremontodendron, Rhus,* and many grasses.

By the end of the Tertiary (2.6 Ma), many species and vegetation elements of modern California and recognizable species affinities were in place. Here and there remained species that are exotic to the modern flora, and many locations of native species were different than at present.

2. Quaternary: 2.6 Million Years Ago to Near-Present

Zooming the focus to the Quaternary introduces processes that occur at increasingly shorter time scales, dynamics that are blurred in discussion of deep time. In this section geology and climate are discussed together, as this time interval reveals how processes are interwined when resolution is high.

Geology and Climate.   As the earth cooled over the last 4 million years, and earth's orbital relationships intensified icehouse conditions, a discernible transition in climate variability began, ~2.6 Ma. Early understanding of Quaternary climate relied on geomorphic evidence that painted the time interval in broad strokes. Collectively these data led to interpretations of the Pleistocene (2.6 to 0.01 Ma) as a long cold interval, or the "great ice period" of Agassiz. By the late 19th century, evidence for multiple glaciations accumulated and led to description of four major glacial periods. The ice ages were regarded as ending about 10,000 years ago (10 ka) with the arrival of our recent warm epoch, which was called the Holocene to signify its novel character.

New high-resolution methods that analyze stratified ice from polar ice caps and deep ocean sediments, however, revealed surprising variability. Rather than one or a few long-persistent ice ages, ice-core records show a pattern of over 40 cycles of glacial (cold) and interglacial (warm) intervals, each lasting from 40,000 to 100,000 years. The ice-core data reveal further variability nested within major glacial and interglacial phases. During a glacial episode extensive cold-glacial periods (stadials) were regularly interrupted by shorter warmer periods (interstadials) as well as by very short (~1,000 year) flip-flops between extreme cold and relatively warm conditions (Fig. 6b). Interglacials, by contrast, began abruptly, although not without a series of short (~1,000 year) reversals, peaked in temperature during early to middle cycle (the middle 4,000–5,000 years), and ended (last 4,000–5,000 years) in a series of steps of decreasing temperature, each with rather abrupt transitions, into the cold of another glacial period. The cumulative effect is a sawtooth pattern typical of Quaternary climate records from around the world. A startling insight from this revised view of the Quaternary is the overall similarity of the Holocene (now formally starting 11,700 years ago) to interglacial periods throughout the Pleistocene. The Holocene is, in fact, not novel.

During glacial periods of the Quaternary, polar ice sheets expanded in Greenland and Antarctica. Continental ice sheets developed across northern North America and parts of northern Eurasia, and glaciers formed on continental mountains to the south of the ice sheets. In California, glaciers formed in the Trinity Alps, Salmon Mountains, Cascade Ranges (Mt. Shasta, Mt. Lassen, Medicine Lake), Warner Mountains, Sweetwater Range, White Mountains, Sierra Nevada, and San Bernardino Mountains. By far the most extensive glaciations occurred in the Sierra Nevada, where, during the coldest parts of gla-

cial periods, an ice cap extended over most of high parts of the range. During the last glacial maximum (~20 ka), the Sierran ice cap was 125 km long, 65 km wide, and extended downslope to about 2600 m in elevation. Valley glaciers, fed by the ice cap, extended 65 km down Sierra Nevadan west slope canyons, and at most 30 km down the shorter but steeper eastern escarpment canyons. Glacial meltwater flowed into valleys below.

As a result of ice buildup on land during glacial periods, global ocean levels fluctuated greatly throughout the Quaternary, declining about 150 m relative to present during the last and penultimate glacial maxima (20 ka and 140 ka). Declines of 60 m in sea level characterize less severe stadial periods during the last two glacial cycles. Along the Pacific margin, for example, the California coastline retreated about 80 km to a position west of the Farallon Islands, rendering San Francisco Bay, Eureka Bay, and other low basins as dry land. Increased precipitation and decreased evaporation during glacial periods led to formation of inland lakes and waterways. These included large lakes in the Central Valley (e.g., Lake Clyde, which filled the San Joaquin Valley 700–600 ka) and Great Basin (e.g., Pleistocene versions of Mono and Owens lakes), and higher river levels throughout California.

The present-day California Current system, which is responsible for maintaining the dry Mediterranean climate of our region as well as the cool coastal fog belt, wavered in its intensity through the Quaternary. When strong, as now, the California Current brings cool, relatively fresh water from the Oregon coast equator-ward along the California margin to just south of the U.S.-Mexico border. This current promotes favorable conditions for upwelling of cold water throughout much of the year, particularly in the summer months. During the peaks of glacial periods, however, continental ice sheets reached a large enough size to reorganize the wind systems over the North Pacific Ocean. These perturbations to wind fields caused the California Current to weaken, triggering large differences in ocean-surface temperatures relative to those of interglacial times. Collapse of the California Current during these millennia translated to weakening of the Mediterranean climate regime over California, reducing thermal gradients from coast to inland, and diminishing fog belts along the California coastal zone as warmer waters came near the coast.

During interglacial periods, most of these patterns reversed. As global ice melted, ocean levels rose, coastlines moved eastward forming bays and inlets, and inland water levels lowered or dried. The oldest evidence for the San Francisco Bay estuary system is about 600 ka; at 10 ka rising water began to fill the San Francisco Bay, which retreated partially during the middle Holocene dry and warm period, and then reached a maximum extent about 4 ka. The California Current, with correlated summer coastal fog belt, thermal gradients, and long summer droughts, developed most strongly during peak interglacial times and the modern pattern of the current evolved about 3 ka.

Over the last 2.6 million years, ongoing tectonic changes resulting from the Great Basin expansion and processes along the California Shear Zone contributed to the increasing development of the southern Sierra Nevada, White Mountains, and Carson Range escarpments, producing, for example, the deep and sharp-bordered Owens and Carson valleys, as well as deepening of the Lake Tahoe Basin. As the mountain ranges acquired their modern geometry, glacial action in turn carved the landscape in new ways. The Quaternary glaciers of California deposited prominent moraines, etched glacial cirques and valleys, and sculpted arêtes and matterhorn topography.

As the Sierra Nevada continued to be influenced by extensional and micro-plate tectonic processes of tilting and subsidence, the rivers running off both slopes eroded deep incisions and charted new courses. An example of the combined effect of river and glacier forces is Yosemite Valley. *Deepening* of the valley is attributed equally to the forces of glacial and river erosion. *Widening* of the valley, by contrast, is considered primarily the work of glaciers.

Major volcanic events continued in California throughout the Quaternary, centered along extensive fault zones of the Sierra Nevada and Coast Ranges. A globally significant example is the Long Valley

eruption of eastern California. Basaltic eruptions began around Long Valley about 4 Ma, coinciding with fault subsidence of Panamint Valley, Death Valley, Owens Valley, Saline Valley, and many other valleys in southeastern California. Volcanism began in the Glass Mountains about 2 Ma, and peaked in a cataclysmic eruption of 600 km³ of high-silica rhyolite at 760 ka. This massive eruption resulted in ash clouds extending as far as Nebraska, and widespread deposition in California of the Bishop Tuff. Simultaneous 2–3 km subsidence of the magma chamber roof formed the present Long Valley Caldera, the western-most portion of which approaches the modern Sierra Nevada crest near Mammoth Lakes. Subsequent volcanism in this region shaped much of the current landscape, including, for example, Mammoth Mountain, which erupted as a series of small extrusions over a period from 110 ka to 50 ka. Volcanism shifted north, first forming the Mono-Inyo Craters chain (50 ka to 650 years before present) and then farther northward to form the islands of Mono Lake, where volcanism continued to just before the historic period (~200 years ago) and is still active, as attested by hot springs on the islands.

Vegetation.   California plant species and communities were significantly influenced by climatic and geologic events of the Quaternary and responded to both major and minor climate cycles. Relatively few plant speciation or extinction events (contrasting with abundant animal extinctions) are documented at this time in California, although evidence points to significant genetic adaptation at population levels. A significant new factor influencing vegetation patterns in California during the past 10,000 years is the presence of humans. In the California region, Native American activity likely had its greatest effect on vegetation in the past 6,000 to 4,000 years, as migrations of people into California took place, populations grew, and sophisticated methods of plant use and vegetation control developed. Invasion of modern Eurasians starting in the 1700s and increasing greatly from the middle 1800s vastly altered the scope, rate, and nature of vegetation change in the region.

Several categories of vegetation response to Quaternary glacial-interglacial climatic change (and later human manipulation) occurred in the greater California region. These include:

1. *North-south shifts of distribution ranges, primarily in low-relief areas.* An example is single-leaf pinyon pine (*Pinus monophylla*) during the last glacial cycle. Pollen and woodrat-midden records document that singleleaf pinyon pine distribution was widespread in the late Pleistocene south of its current range and in the current range of the Mojave Desert. As climates warmed during the early Holocene, singleleaf pinyon pine moved gradually northward, reaching central Nevada about 5 ka, near Monitor Pass in eastern California 1.4 ka, the Reno area 400 years ago, and its current northern limit on the west side of the Great Basin near Pyramid Lake in western Nevada 200 years ago. Similar shifts appear for species of California's Great Central Valley.

2. *Vertical shifts in elevation in mountainous areas.* Elevational shifts that correspond to glacial-interglacial climatic phases are documented for many California species. In the Sierra Nevada, for example, during coldest glacial periods when an ice cap covered the range, montane conifer ranges shifted downslope by as much as 1000 m relative to their present elevations.

3. *Population contractions (refugia and extirpations) and expansions (colonizations).* Contractions and expansions were common for many California plant species in response to glacial-interglacial climate dynamics. These shifts occurred sometimes with little significant effect on elevational limits of species ranges. For example, coast redwood, the California closed-cone pines, coastal cypresses, and many of California's oak species followed this pattern, contracting to fewer populations of smaller size during unfavorable periods. Such contractions rarely amounted to significant directional shift in the overall species range; rather to loss of connectivity and small population sizes. For example, California's oak populations expanded during

interglacials, becoming more connected and covering large areas of the California landscape. During unfavorable climate periods (glacial), populations contracted into disjunct, isolated locations, with many population extirpations. Similarly, during interglacials when the California Current was strongest and coastal fog belts extensive, coast redwood expanded; the converse occurred during glacial periods. The scattered distributions resulting from such contractions included important refugial populations for many species during unfavorable climatic periods. These were not only sources for rapid recolonization following return to favorable conditions but critical for conservation of population-level genetic diversity. Such refugial populations also existed throughout mountainous areas, where habitat heterogeneity afforded considerable opportunity for maintenance of small populations.

4. *Changes in community composition, including development of non-analog assemblages.* Plant communities at any time and place on the California landscape reflect to some degree the interaction of climate with individual species' ecologies. In some situations, especially for broadly adapted taxa, species responded synchronously to Quaternary climatic changes, and community compositions remained relatively similar as species shifted. In other situations, species responded individualistically, and community compositions changed over time. Unusual assemblages, such as combinations of species not found at present, resulted from unique combinations of climate, other environmental conditions, species adaptation, differential migration, and chance events.

Past, Present, and Future: 1,200 Years Ago and Forward

The last 1,200 years in California warrant special attention. Not only is this interval our immediate heritage where we have sufficient detail to illustrate key processes, but the period also provides context for the future. Further, this interval marks the beginning of a transition to human dominance and influence. The background for this 1,200 year period is a general cooling trend that began ~4 ka, which varied in timing regionally and distinguished the late Holocene from the warm middle Holocene (6–4 ka). Climate cycles paced by fluctuations in solar activity and in ocean cycling rendered significant climate variability at century and shorter scales within this time. The Medieval climatic anomaly, about 700 to 1,100 years ago (900–1350 CE), was a worldwide interval of temperature and precipitation divergence, varying in expression regionally. In California, abundant evidence documents two major droughts each lasting more than a century. These caused many currently large lakes and rivers to dry, and salinities to increase in those that remained. In mountain regions, evidence exists for increased warmth relative to present (as much as 3°C). Shifts occurred in response to this interval, with upslope movements of mountain taxa and considerable rearrangement of vegetation communities.

About 600 years ago (1400 CE), shifts in the solar cycle resulted in return to cooler conditions, with the advent of the global Little Ice Age. Cool temperatures for this period were intensified by the coincidence of several significant volcanic eruptions that injected abundant ash into the atmosphere and by the persistence of several anomalous sunspot events. In California, the Little Ice Age triggered the largest glacial advance in over 11,000 years, and cirque glaciers in the Sierra Nevada, Cascades, and Klamath Ranges formed. The coldest part of the Little Ice Age in California was during the late 1800s and into the early decades of the 20th century. The primary effect of this period on vegetation was to dampen productivity. Forests were sparse and growth rates low, whereas high water tables and cold soils maintained extensive mountain meadows. Relative to the Medieval climatic anomaly, forest fires were generally of low severity, long duration, and broad extent across the landscape, rather than the high-intensity crown fires of earlier. Fire patterns and their consequences to vegetation were influenced in many regions by activities of Native Californians as well as climate. Upper treeline in California's

mountains was distinct, and persistent snowpacks maintained significant gaps in mountain vegetation. Lake and river waters were cool and riparian corridors extensive. Species such as aspen (*Populus tremuloides*) that thrive under high soil-water tables expanded on slopes as well as along rivers and streams (riparian habitats) and in meadows.

The Little Ice Age ended about 1925 CE in California as solar cycles shifted once again, triggering warming trends and drought that are known from the 1930s and 1940s. Temperatures plateaued in the mid 20th century and began to climb sharply in the early 1970s. The impact of anthropogenic greenhouse gases became significant during these decades, compounding climatic processes and forcing conditions beyond natural variability. Vegetation dynamics of recent decades are further influenced by short natural climate cycles such as the Pacific Decadal Oscillation (~40–60 years) and the El Niño/La Niña (~2–8 year) modes. These ocean circulation oscillations stimulate alternating periods of warm-wet years with cool-dry conditions, modulating forest and grassland cycles of fuel buildup followed by vegetation dry-down and heightened fire impact. Trends that began in response to natural warming after the Little Ice Age have accelerated in recent decades, including species range and elevation shifts, forest densification, forest mortality by insects and disease, and altered fire regimes.

The present condition of California's environment and the future that is unfolding are complex expressions of natural forces intertwined with increasing anthropogenic influences. A key lesson from history is to work with change and harness inherent capacities for adaptation. That plant species have been subject to continuous change over time highlights the value of understanding natural processes as we develop conservation strategies for an uncertain future.

Acknowledgments

I thank Bruce G. Baldwin, Diane M. Erwin (UC Berkeley), Stephen W. Edwards (East Bay Regional Park District), Wes Hildreth (U.S. Geological Survey), and David J. Keil for critical reviews of the manuscript, and Diane L. Delany (USDA Forest Service) for rendering the figures.

Additional Reading

Edwards, S.W. 2004. Paleobotany of California. *The Four Seasons.* Vol 4(2):3–75.

Harden, D.R. 2004. *California geology. 2nd Edition.* Pearson-Prentice Hall, NJ.

Sierra Nevada Ecosystem Project, Final report to Congress, Vol II, Assessments and scientific basis for management options. 1996. Centers for Water and Wildland Resources, Report 37, University of California, Davis, California: Section 1 (chapters 1–5): Past Sierra Nevada landscapes. http://ceres.ca.gov/snep/pubs/v2.html

Jones, T.L., and K.A. Klar (eds.). 2007. *California prehistory: Colonization, culture, and complexity* (chapters 2 and 3). Alta Mira Press.

International Stratigraphic Chart

International Commission on Stratigraphy

Eonothem Eon	Erathem Era	System Period	Series Epoch	Age Ma
Phanerozoic	Cenozoic	Quaternary	Holocene	0.0117
			Pleistocene	2.588
		Neogene	Pliocene	5.332
			Miocene	23.03
	Paleogene		Oligocene	33.9 ±0.1
			Eocene	55.8 ±0.2
			Paleocene	65.5 ±0.3
	Mesozoic	Cretaceous	Upper	99.6 ±0.9
			Lower	145.5 ±4.0

Eonothem Eon	Erathem Era	System Period	Series Epoch	Age Ma
				145.5 ±4.0
Phanerozoic	Mesozoic	Jurassic	Upper	161.2 ±4.0
			Middle	175.6 ±2.0
			Lower	199.6 ±0.6
		Triassic	Upper	~ 228.7
			Middle	~ 245.9
			Lower	251.0 ±0.4
	Paleozoic	Permian	Lopingian	260.4 ±0.7
			Guadalupian	270.6 ±0.7
			Cisuralian	299.0 ±0.8
		Carboniferous — Pennsylvanian	Upper	307.2 ±1.0
			Middle	311.7 ±1.1
			Lower	318.1 ±1.3
		Carboniferous — Mississippian	Upper	328.3 ±1.6
			Middle	345.3 ±2.1
			Lower	359.2 ±2.5

Eonothem Eon	Erathem Era	System Period	Series Epoch	Age Ma
				359.2 ±2.5
Phanerozoic	Paleozoic	Devonian	Upper	385.3 ±2.6
			Middle	397.5 ±2.7
			Lower	416.0 ±2.8
		Silurian	Pridoli	418.7 ±2.7
			Ludlow	422.9 ±2.5
			Wenlock	428.2 ±2.3
			Llandovery	443.7 ±1.5
		Ordovician	Upper	460.9 ±1.6
			Middle	471.8 ±1.6
			Lower	488.3 ±1.7
		Cambrian	Furongian	~ 499
			Series 3	~ 510
			Series 2	~ 521
			Terreneuvian	542.0 ±1.0

Eonothem Eon	Erathem Era	Age Ma
		542
Precambrian	Proterozoic	Neoproterozoic — 1000
		Mesoproterozoic — 1600
		Paleoproterozoic — 2500
	Archean	Neoarchean — 2800
		Mesoarchean — 3200
		Paleoarchean — 3600
		Eoarchean — 4000
		Hadean — ~4600

Stratigraphic Time Chart, from the International Commission on Stratigraphy (IUGS), 2009 revision

KEY TO CALIFORNIA VASCULAR PLANT FAMILIES

David J. Keil

As in *The Jepson Manual: Higher Plants of California* (1993), only vascular plants are treated in *The Jepson Manual, Second Edition*. Vascular plants are a monophyletic group or clade that includes (1) Lycophytes (often superficially moss-like but with thicker scale-like leaves and axillary sporangia that are often borne in cones), (2) Ferns (most with sporangia borne on abaxial leaf surfaces, a few with sporangia on modified leaf segments or in hard sporangium cases), and *Equisetum*, with whorled scale leaves and sporangia grouped in terminal cones, (3) Gymnosperms (with seeds generally produced in cones of various forms), and (4) flowering plants (with seeds developing in ovaries of flowers). Flowering plants in California include five clades: Nymphaeales, Magnoliids, Ceratophyllales, Eudicots, and Monocots (see back inside cover). Unlike all other organisms, vascular plants have true roots, stems, and leaves, with water- and nutrient-conducting (vascular) tissues that also provide structural support.

In terrestrial (and aquatic) environments, true mosses and leafy liverworts may appear superficially similar to vascular plants, but their minute leaves are filmy (generally only 1 cell layer in thickness), and their delicate, usually short stems are attached to soil or other substrate by fragile filaments of 1 or more cells (rhizoids) that lack conducting tissue. Mosses, the most diverse land plants in California other than vascular plants, have alternate, closely overlapping, scale-like or narrowly linear leaves and produce solitary sporangia (generally finely stalked) at the tips of leafy shoots. Some freshwater liverworts can be confused with free-floating vascular plants that lack clearly differentiated stems and leaves, but the liverworts can be recognized by their forking body plan or forked grooves on their upper surface.

Although in aquatic environments an emergent habit distinguishes vascular plants from other aquatics, potential for confusion of vascular plants with organisms other than mosses and liverworts in fully submersed settings warrants attention. In particular, some marine algae, such as kelps and rockweeds, appear superficially similar to vascular plants because of leaf-like, stem-like, and root-like (holdfast) structures. These algae, however, have accessory pigments ranging in color from olive-green to reddish or brownish that distinguish them from marine vascular plants, which are grass-green in color. Some of these algae have gas-filled floats, which are not found in marine vascular plants, and all of them are somewhat rubbery in texture, in contrast to the papery texture of marine vascular plants. Green algae are similar in color to vascular plants; of these, charophytes, with a central axis and whorled, cylindric branches (but no leaves) are most readily confused with vascular plants.

Keys in *The Jepson Manual, Second Edition*, as in other field manuals, are based primarily on structures associated with sexual reproduction (e.g., sporangia, cones, flowers, fruits). The manual does not provide a means for keying terrestrial vascular plants that are in strictly vegetative condition. For aquatic vascular plants, some of which reproduce mostly by vegetative means, a key is provided that allows identification using only vegetative characteristics. A key is also provided for identification of the few species in which bulblets (small bulbs or corms) take the places of flowers in modified inflorescences or inflorescence-like clusters.

Scientific Editor: Bruce G. Baldwin

Key to Groups

1. Specimens available for examination without fls, these either not present on the specimen or not produced at all
  2. Bulblets [small bulbs or corms that are dispersed in place of seeds or frs] formed in place of fls in infl or
    infl-like cluster . . . . **Group 1** (p. 70)
  2′ Bulblets 0
    3. Sporangia, sporangium cases, seeds, cones, or cone-like structures 0 or not readily apparent [pls in
      strictly vegetative condition]
      4. Pl aquatic . . . . **Group 2** (p. 71)
      4′ Pl terrestrial [Note: terrestrial pls in strictly vegetative condition cannot be identified with this key]
    3′ Sporangia, sporangium cases, seeds, cones, or cone-like structures present
      5. Herbs reproducing by spores released directly from sporangia, the sporangia variously located [on
        abaxial lf face, in stalked cluster arising from petiole or blade base, in hardened case at petiole base,
        in axils of linear or scale-like lvs, or in terminal cone-like structure], seeds and pollen never formed
        [LYCOPHYTES and FERNS] . . . . **Group 3** (p. 74)
      5′ Herbs to trees reproducing by seeds
        6. Ovules exposed to air at time of pollination, not enclosed in ovary; pollen grains deposited directly
          on ovule, stigma 0; seeds, if present, borne between scales of cone or naked on branches, fr not
          produced; shrubs or trees [GYMNOSPERMS] . . . . **Group 4** (p. 75)

6′ Ovules enclosed in ovaries that on available specimens have matured as dehiscent or indehiscent frs; seeds enclosed within ripened pericarp of fr [if dry remains of fls are present it may be possible to determine the family of a fruiting specimen under 1′ in this key, but not all fruiting specimens may be identifiable here]
1′ Specimens available for examination with fls, often also producing seeds within frs [ANGIOSPERMS]
7. Specimens with only cleistogamous fls — frs developing without fls opening . . . . **Group 5** (p. 75)
7′ Specimens with open fls
  8. Specimens with unisexual fls of only 1 kind [staminate or pistillate, but not both]; pls dioecious or monoecious
    9. Specimen with only staminate fls
      10. Herb or subshrub, woody only at base . . . . **Group 6** (p. 76)
      10′ Tree or shrub, conspicuously woody . . . . **Group 7** (p. 78)
    9′ Specimen with only pistillate fls
      11. Herb or subshrub, woody only at base . . . . **Group 8** (p. 81)
      11′ Tree or shrub, conspicuously woody . . . . **Group 9** (p. 83)
  8′ Specimens with bisexual fls or with both staminate and pistillate fls [on same or different individuals]; pls bisexual, dioecious, or monoecious (occ with mixture of bisexual and unisexual fls)
    12. Pistils 2 or more per fl [carpels > 1, free to the base] . . . . **Group 10** (p. 85)
    12′ Pistil 1 per fl [carpel 1 or carpels > 1 fused at least proximally]
      13. Perianth 0 or in a single whorl [appearing to be either sepals or petals but not both], sometimes reduced to scale-like or bristle-like structures
        14. Tree or shrub, conspicuously woody
          15. Infl, at least the staminate, a catkin or catkin-like spike; fls unisexual . . . . **Group 11** (p. 86)
          15′ Infl various but not catkin-like; fls bisexual or unisexual . . . . **Group 12** (p. 87)
        14′ Herb or subshrub, woody proximally if at all
          16. Lf venation parallel or lvs and st not differentiated . . . . **Group 13** (p. 89)
          16′ Lf venation gen pinnate or palmate (sometimes only midvein evident) or lvs reduced to bladeless scales
            17. Ovary inferior . . . . **Group 14** (p. 90)
            17′ Ovary superior . . . . **Group 15** (p. 91)
      13′ Perianth in 2 or more whorls (gen both sepals and petals) or perianth parts spiraled 2 or more times around floral axis
        18. Perianth parts 2 or 3 per whorl . . . . **Group 16** (p. 93)
        18′ Perianth parts gen 4 or 5 per whorl (rarely some other number or whorls differing in number of parts) or in a spiral with the number of parts indefinite
          19. Petals fused into a ring or a tube, the corolla gen falling as a unit [corolla of free petals fused to a hypanthium should be keyed under 19′]
            20. Ovary inferior or partly so . . . . **Group 17** (p. 95)
            20′ Ovary superior
              21. Corolla bilateral . . . . **Group 18** (p. 96)
              21′ Corolla radial . . . . **Group 19** (p. 97)
          19′ Petals free, at least at base, attached and gen falling singly (sometimes individually joined to a hypanthium; in a few families ± joined and falling in groups, but not forming a ring or a tube, or petal 1)
          22. Lf compound or so deeply divided as to appear compound . . . . **Group 20** (p. 100)
          22′ Lf simple, sometimes much reduced
            23. Ovary inferior or partly so . . . . **Group 21** (p. 102)
            23′ Ovary superior
              24. Stamens > 2 × as many as petals . . . . **Group 22** (p. 103)
              24′ Stamens ≤ 2 × as many as petals
                25. Tree or shrub, conspicuously woody . . . . **Group 23** (p. 104)
                25′ Herb or subshrub, woody only at base . . . . **Group 24** (p. 105)

Group 1

Herbs; bulblets dispersed in place of seeds or frs; available specimens without fls

1. Lvs linear, ± entire, sheathing at base
  2. Pl with strong odor of onion or garlic; bulblets in head or umbel
  . . . . **ALLIACEAE** ([*Allium sativum*], *Allium vineale*)
  2′ Pl unscented; bulblets in spike- or panicle-like cluster
    3. Lvs stiff, sword-like, vertically folded and attached "edge-on" to the st, <60 mm wide
    . . . . **IRIDACEAE** (*Watsonia meriana*)
    3′ Lvs soft, flat or folded, not attached "edge-on" to the st, 1–2 mm wide . . . **POACEAE** (*Poa bulbosa* subsp. *vivipara*)
1′ Lvs linear-elliptic to round, entire, toothed, or palmately lobed, not sheathing at base
  4. Lvs opposite, ± fleshy . . . . **MONTIACEAE** (*Montia chamissoi*)
  4′ Lvs alternate or all basal, not fleshy
    5. Proximal lvs sessile or tapered to short petioles; lf blade entire or sharply toothed
    . . . . **SAXIFRAGACEAE** (*Micranthes*)
    5 Proximal lvs abruptly slender-petioled; lf blade palmately lobed or divided
      6. Proximal lvs 5–15 cm wide; pl stout, 30–150(200) cm . . . . **RANUNCULACEAE** (*Aconitum columbianum*)
      6′ Proximal lvs 1.5–4 cm wide; pl slender, 8–50 cm . . . . **SAXIFRAGACEAE** (*Lithophragma*)

Group 2

Aquatic pls in vegetative condition; sporangia, sporangium cases, seeds, cones, or
cone-like structures 0 or not readily apparent

1. Pl body consisting of a jointed, hollow-stemmed central axis and whorled branches; lvs reduced to whorled
 scales; pl emergent. . . . . **EQUISETACEAE** (G3)
1′ Pl body various, but not consisting of a jointed, hollow-stemmed central axis and whorled branches; lvs
 various, but not whorled scales; pl free-floating, emergent, or submersed
  2. Pl raft-like, free-floating on water surface or stranded along shore or in drying bottom sediments,
 breaking apart into individuals or small clumps
    3. Individual pl gen 0.4–30 mm in length or diam, sometimes not differentiated into lvs and sts; roots unbranched
      4. Pl body spheric to disk-shaped or oblong, not differentiated into sts and lvs. . . . . [3]**ARACEAE** (G6,8,13,15)
      4′ Pl body differentiated into a short, often branched st with small, minutely velvety-papillate, scale-like
 lvs . . . . **AZOLLACEAE** (G3)
    3′ Individual pl often >> 30 mm, differentiated into lvs and sts; roots gen branched (0 in Salviniaceae,
 with finely dissected, root-like submersed lvs)
      5. Lf blades glossy green, glabrous, finely parallel-veined; petioles often enlarged, forming gas-filled
 floats . . . . **PONTEDERIACEAE** (*Eichhornia*)
      5′ Lf blades light to dark green or ± purple, velvety-hairy, ± palmately or obscurely veined, sessile, lacking floats
        6. Lvs in floating rosettes, evidently ± palmately veined . . . . **ARACEAE** ([*Pistia*]) (G15)
        6′ Lvs in whorls of 3, 2 floating, entire, obscurely veined, the 3rd submersed, finely dissected, root-like
 . . . . **SALVINIACEAE** (G3)
  2′ Pl gen anchored in bottom sediments or, if free-floating, most of pl body below water surface or not
 readily breaking apart — rhizomes or stolons present or 0, sometimes elongated
    7. Lf peltate, cordate, sagittate, or deeply notched
      8. Lf peltate
        9. Submersed parts not gelatinous; lf blades gen emergent. . . . . [2]**ARALIACEAE** (*Hydrocotyle*) (G14,21)
        9′ Submersed parts gelatinous; lf blades floating . . . . **CABOMBACEAE** (*Brasenia*)
      8′ Lf sagittate or cordate to ovate or ± round, the base shallowly to deeply notched
        10. Lf blades pinnately veined; petioles long, stout; rhizome stout
          11. Lf sagittate; rhizome ± vertical. . . . . [2]**ARACEAE** ([*Peltandra*])
          11′ Lf ovate to round; rhizome horizontal . . . . **NYMPHAEACEAE**
        10′ Lf blades ± palmately veined; petioles long or short, slender or stout; rhizome, if present, slender
          12. Lf blades palmately lobed, never sagittate, the lobes often toothed — veinlets forming an irregular
 network . . . . [2]**ARALIACEAE** (*Hydrocotyle*) (G14,21)
          12′ Lf blades ovate to cordate or sagittate, entire
            13. Palmate veins 3 — smaller veins parallel, extending from main veins toward lf margin
 . . . . [2]**ARACEAE** ([*Peltandra*])
            13′ Palmate veins gen 5–13+
              14. Stipules 1–6 cm. . . . . [3]**PONTEDERIACEAE** (*Heteranthera limosa*)
              14′ Stipules 0
                15. Smaller veins passing like ladder-rungs between prominent main veins
 . . . . **ALISMATACEAE** or **HYDROCHARITACEAE**
                15′ Smaller veins forming network. . . . . **MENYANTHACEAE** (*Nymphoides*) (G19)
    7′ Lf not peltate, cordate, sagittate, or deeply notched
      16. Lvs all basal or basal and cauline on a gen erect, gen emergent st, sometimes from rhizome or
 corm-like base anchored in bottom sediments
        17. Lvs petioled, with expanded blades
          18. Lf serrate or dentate to lobed or compound
            19. Lf palmate; lflets 4 . . . . [2]**MARSILEACEAE** (*Marsilea*) (G3)
            19′ Lf toothed to pinnately lobed or 1–3-pinnate
              20. Lf bases sheathing; lvs, lobes, or lflets serrate to sharply dissected . . . . [2]**APIACEAE** (G6,8,14,20,21)
              20′ Lf bases not sheathing; lobes or lflets entire or dentate with ± rounded teeth
 . . . . **BRASSICACEAE** (*Nasturtium, Rorippa*)
          18′ Lf entire, simple
            21. Lvs gen << 1 cm wide. . . . . [2]**SCROPHULARIACEAE** (*Limosella*) (G18,19)
            21′ Lvs gen ≥ 1 cm wide, often much larger
              22. Wider lvs floating . . . . **APONOGETONACEAE** (G10)
              22′ Wider lvs gen emergent
                23. Stipules 0 . . . . [2]**ALISMATACEAE** (G10)
                23′ Stipules present, 1–6 cm. . . . . [3]**PONTEDERIACEAE** (*Heteranthera limosa*)
      17′ Lvs sessile, ± linear or reduced to bladeless sheaths
        24. Lvs vertically folded and attached "edge-on" to the st, obviously 2-ranked . . . . [4]**JUNCACEAE** (*Juncus*)
        24′ Lvs cylindric, ± angled, flat, or reduced to bladeless sheaths, but not folded, not obviously 2-ranked
          25. Lvs with flat blades or reduced to bladeless sheaths
            26. Lvs bladeless or nearly so, composed mainly of a tubular sheath

27. Sts ± triangular .... [4]**CYPERACEAE** (G8,13)
27′ Sts round, flattened, or several-angled
  28. Lf sheaths closed, forming a continuous cylinder around st .... [4]**CYPERACEAE** (G8,13)
  28′ Lf sheaths open on one side, with overlapping margins .... [4]**JUNCACEAE** (*Juncus*)
26′ Lvs with ordinary flat blades and gen also ± tubular sheaths
  29. Lf blade with a well-developed midvein or keel
    30. Lf sheaths closed, forming a continuous cylinder around st; st gen triangular; lvs 3-ranked
    .... [4]**CYPERACEAE** (G8,13)
    30′ Lf sheaths open on one side, with overlapping margins; st round; lvs 2-ranked
    .... [3]**TYPHACEAE** (*Sparganium*) (G6,8,13)
  29′ Lf blade flat or rounded on back, lacking a prominent midvein or keel
    31. Ann; lvs all basal .... **RANUNCULACEAE** (*Myosurus*)
    31′ Per, or if ann, some lvs cauline
      32. Pl submersed; seawater or brackish water of bays and estuaries .... [2]**ZOSTERACEAE** (G6,8,13)
      32′ Pl emergent; fresh or brackish water of coastal or inland areas
        33. Nodes swollen; st gen hollow................................. [2]**POACEAE** (G5,6,8,13)
        33′ Nodes not swollen; st gen filled to center with spongy tissue
          34. Lf gen > 5 mm wide; remnants of old infl a dense spike; inner surface of lf sheaths
            mucilaginous .... **TYPHACEAE** (*Typha*)
          34′ Lf ≤ 5 mm wide; remnants of old infl not evident or a raceme; inner surface
            of sheaths not mucilaginous
            35. Lvs ± circular to semicircular in ×-section .... [3]**JUNCAGINACEAE** (G10,13,16)
            35′ Lvs flat .... [3]**PONTEDERIACEAE** (*Heteranthera limosa*)
25′ Lvs cylindric or angled
  36. Rhizomes and stolons 0; lvs densely tufted
    37. Lf divided lengthwise into 4 hollow chambers; a sporangium gen embedded in adaxial base of
      each lf, containing either powdery or larger spores .... **ISOETACEAE** (G3)
    37′ Lf solid or hollow, but not regularly divided into 4 chambers; a sporangium never
      embedded in lf base
      38. Pl per; lf often with internal cross-partitions .... [4]**JUNCACEAE** (*Juncus*)
      38′ Pl ann; lf without internal cross-partitions or partitions obscure
        39. Lf bases not sheathing .... **BRASSICACEAE** (*Subularia*)
        39′ Lf bases sheathing .... [3]**JUNCAGINACEAE** (G10,13,16)
  36′ Rhizomes or stolons present; lvs tufted or borne along rhizome or stolon
    40. Lf sharply angled
      41. St triangular; lvs 3-ranked................................... [4]**CYPERACEAE** (G8,13)
      41′ St ± round; lvs 2-ranked .... [3]**TYPHACEAE** (*Sparganium*) (G6,8,13)
    40′ Lf cylindric
      42. Lf with internal cross-partitions
        43. Lf soft, blunt, gen ≤ 15 cm, loosely sheathing; cross-partitions often externally prominent
        .... **APIACEAE** (*Lilaeopsis*)
        43′ Lf gen stiff, acute, often > 15 cm, tightly sheathing; cross-partitions readily apparent by feel
        but often externally obscure .... [4]**JUNCACEAE** (*Juncus*)
      42′ Lf without internal cross-partitions
        44. Lf gen >> 6 cm
          45. St ± straight; lf blade tip without evident terminal pore; ligule 0.5–5 mm
          .... [3]**JUNCAGINACEAE** (G10,13,16)
          45′ St zigzag; lf blade tip with evident terminal pore; ligule 2–12 mm ..... **SCHEUCHZERIACEAE** (G10)
        44′ Lf 1–7(11) cm
          46. Pl spreading by rhizomes, producing ± spheric, ball-like structures [sporangium cases] ± 2–3
            mm diam at bases of some lvs; new lvs uncoiling at tips .... **MARSILEACEAE** (*Pilularia*) (G3)
          46′ Pl spreading by stolons, not producing ball-like structures at bases of lvs; new lvs not
            uncoiling at tips .... [2]**SCROPHULARIACEAE** (*Limosella*) (G18,19)
16′ Lvs gen all cauline; sts emergent or taking the form of underwater or floating rhizomes or
  stolons that are unable to support the pl body outside of water
  47. Lvs (or lf-like branches) compound or very deeply divided
    48. Lf palmately compound — lflets 4, floating or emergent .... [2]**MARSILEACEAE** (*Marsilea*) (G3)
    48′ Lf pinnately compound or variously dissected
      49. Lflets lanceolate or wider; lvs mostly aerial
        50. Lf bases sheathing; lobes or lflets serrate to sharply dissected .... [2]**APIACEAE** (G6,8,14,20,21)
        50′ Lf bases not sheathing; lobes or lflets entire or dentate with ± rounded teeth
        .... **BRASSICACEAE** (*Nasturtium, Rorippa*)
      49′ Lflets or lobes linear, at least on submersed lvs; all or many lvs submersed

51. Underwater parts bearing small, hollow bladders that suck in and trap small organisms when hairs at opening are triggered; pl carnivorous — highly dissected underwater parts are actually modified sts and the bladders are the lvs . . . . **LENTIBULARIACEAE** (*Utricularia*) (G18)

51′ Underwater parts lacking bladder-like traps; pl not carnivorous

  52. Submersed lvs alternate . . . . [3]**RANUNCULACEAE** (*Ranunculus*)

  52′ Submersed lvs opposite or whorled

    53. Submersed lvs pinnately divided, central lf axis evident . . . . **HALORAGACEAE** (G6,8,14,20)

    53′ Submersed lvs palmately dissected or repeatedly forked, central lf axis 0

      54. Submersed lvs palmately dissected; floating lvs 0–few, linear-elliptic, ± entire . . . . **CABOMBACEAE** (*Cabomba*)

      54′ Submersed lvs repeatedly forked; floating lvs 0 . . . . **CERATOPHYLLACEAE** (G6,8,14,15)

47′ Lvs simple, the margins entire, toothed, or shallowly lobed

  55. Lvs long, narrow, strap-shaped or ribbon-like, often >> 20 cm, gen at least some basal

    56. Pl of freshwater habitat . . . . [3]**TYPHACEAE** (*Sparganium*) (G6,8,13)

    56′ Pl of seawater or brackish water of bays and estuaries . . . . [2]**ZOSTERACEAE** (G6,8,13)

  55′ Lvs linear to round, gen < 20 cm, gen all cauline

    57. Lf with stipules or sheathing base

      58. Principal veins of lf pinnate

        59. Lvs opposite or whorled, 2–15 mm . . . . **ELATINACEAE**

        59′ Lvs alternate, >> 15 mm . . . . **POLYGONACEAE** (G6,8,15,16,19)

      58′ Principal veins of lf parallel or pinnate-parallel or only 1 lf vein apparent

        60. Lvs gen ± opposite or appearing whorled

          61. Lvs emergent or floating; nodes swollen, internodes gen hollow . . . . [2]**POACEAE** (G5,6,8,13)

          61′ Lvs submersed or some floating; nodes gen not swollen, internodes gen solid

            62. Lvs minutely toothed (teeth sometimes appearing ± 0 to naked eye); lf base abruptly expanded, forming sheath, stipules 0 . . . . **HYDROCHARITACEAE** (*Najas*) (G8,13)

            62′ Lvs entire; lf base not abruptly expanded, stipules present, free or fused to lf base forming a sheath . . . . **ZANNICHELLIACEAE** (G8,10,13)

        60′ Lvs gen alternate

          63. Stipules ± completely fused to lf blade base . . . . **RUPPIACEAE** (G8,10,13)

          63′ Stipules fused to lf blade base ≤ 2/3 their length

            64. Roots branched; lvs linear; midvein of lf not differentiated from other veins; floating lvs 0 . . . . **PONTEDERIACEAE** (*Heteranthera dubia*)

            64′ Roots gen unbranched; lvs linear to ovate; midvein of lf more prominent than other veins; floating lvs often present . . . . **POTAMOGETONACEAE** (G10,13)

    57′ Lf without stipules or sheathing base

      65. Lvs alternate

        66. Lvs sessile; ann . . . . **CAMPANULACEAE** (G5,14,17)

        66′ Lvs petioled; per

          67. Lvs toothed or lobed . . . . [3]**RANUNCULACEAE** (*Ranunculus*)

          67′ Lvs entire

            68. Lateral lf veins strongly diverging from midvein . . . . [2]**ONAGRACEAE** (*Ludwigia*)

            68′ Lateral lf veins strongly ascending to nearly parallel to midvein . . . . [3]**RANUNCULACEAE** (*Ranunculus*)

      65′ Lvs opposite or whorled

        69. Lvs whorled

          70. Lf minutely toothed (teeth sometimes appearing ± 0 to naked eye) . . . . [2]**HYDROCHARITACEAE** (G6,8,16)

          70′ Lf entire . . . . **PLANTAGINACEAE** (*Hippuris*) (G14,15)

        69′ Lvs opposite

          71. Lvs fleshy

            72. Lvs 2–6 mm, linear or oblanceolate; ann; pl gen ± red . . . . **CRASSULACEAE** (*Crassula aquatica*)

            72′ Lvs 3–50 mm linear to obovate; ann or per; pl gen not red . . . . **MONTIACEAE** (*Montia*) (G5)

          71′ Lvs not or scarcely fleshy

            73. Lvs 0.5–5 mm wide

              74. Lvs minutely toothed (teeth sometimes appearing ± 0 to naked eye) to coarsely dentate, gen enlarged at base, all similar in form and submersed . . . . [2]**HYDROCHARITACEAE** (G6,8,16)

              74′ Lvs entire, linear to ovate, not enlarged at base, sometimes with floating distal lvs different from submersed lvs . . . . **PLANTAGINACEAE** (*Callitriche*) (G8,15)

            73′ Lvs 6–15+ mm wide

              75. Lf veins palmate . . . . **PHRYMACEAE** or **PLANTAGINACEAE**

              75′ Lf veins pinnate

                76. Lvs toothed . . . . **PHRYMACEAE** or **PLANTAGINACEAE**

                76′ Lvs entire

                  77. Lvs (3)5–13 cm; sts often swollen, hollow . . . . **AMARANTHACEAE** (*Alternanthera*)

                  77′ Lvs < 5 cm; sts not swollen . . . . . . [2]**ONAGRACEAE** (*Ludwigia*) or **PLANTAGINACEAE** (*Veronica*)

Group 3

Pls reproducing by spores released directly from sporangia;
fls and seeds never formed (Lycophytes and Ferns)

1. Pl free-floating, raft-like, or stranded along shoreline or on drying mud
   2. Lvs 0.5–1.5 mm, alternate [each with a thick green or ± red-tinged floating lobe and a thin, colorless
      submersed lobe]; upper surface of floating lobe papillate or with short, inconspicuous hairs.... **AZOLLACEAE** (G2)
   2′ Lvs 5–25 mm, appearing opposite [actually 2 floating lvs plus a 3rd submersed lf dissected into root-like
      lobes]; upper surface of floating lvs conspicuously hairy .... **SALVINIACEAE** (G2)
1′ Pl anchored in soil or growing on another pl [if pl aquatic, rooted in bottom sediments]
   3. Lvs linear or scale- or needle-like, entire or toothed (sometimes divided into linear lobes in *Asplenium*
      septentrionale); veins gen 0 or 1 (–few in *Asplenium septentrionale*)
      4. Lvs basal, linear
         5. Sporangia in 1 or more linear sori on abaxial lf face .... **ASPLENIACEAE** (*Asplenium septentrionale*)
         5′ Sporangia underground, each embedded in adaxial face of widened lf base or several enclosed in hard,
            spheric sporangium case near lf base
            6. Pl tufted, base corm-like; new lvs erect, never coiled; sporangium solitary, embedded in adaxial face
               of widened lf base, ± covered by a translucent membrane .... **ISOETACEAE** (G2)
            6′ Pl creeping by rhizomes; new lvs ± coiled, unrolling as they develop; sporangia several, enclosed in
               short-stalked, hard case attached near lf base........................ **MARSILEACEAE** (*Pilularia*) (G2)
      4′ Lvs cauline, scale- or needle-like; sporangia in cone at tip of aerial shoot or in axils of distal lvs
         7. St hollow, jointed at nodes; lvs whorled, fused at base; sporangia several on undersides of peltate
            structures that are clustered in terminal cone .... **EQUISETACEAE** (G2)
         7′ St solid, not jointed; lvs alternate or opposite; sporangia in axils of distal scale-like lvs, the aggregation
            of fertile lvs sometimes cone-like
            8. Fertile lvs not in well defined ranks, the aggregation appearing round when viewed along axis from
               tip; sporangia all of 1 kind, producing only 1 kind of spore.......................... **LYCOPODIACEAE**
            8′ Fertile lvs 4-ranked, the aggregation appearing square when viewed along axis from tip; sporangia of
               2 kinds, some with 4 large spores and others with many small spores .... **SELAGINELLACEAE**
   3′ Lvs with well developed blades, not linear or scale- or needle-like, often lobed or compound; veins many
      9. Pl aquatic or in dried pools or creek beds; lf palmately compound, lflets 4; sporangia borne underground
         or underwater, enclosed in short-stalked, hard case attached near petiole base .. **MARSILEACEAE** (*Marsilea*) (G2)
      9′ Pl terrestrial or strongly emergent if in wet soil, or growing on another pl; lf simple or 1–4-pinnate,
         if compound, lflets gen >> 4; sporangia borne on aerial portion of lf
         10. Lf divided into dissimilar sterile (trophophore) and fertile (sporophore) parts; sterile part with an
            entire or 1–3-pinnate blade; fertile part bladeless, sporangia borne in stalked cluster arising from
            petiole or blade base; new lvs never coiled .... **OPHIOGLOSSACEAE**
         10′ Lf not divided into dissimilar sterile and fertile parts; sporangia borne on underside of lf blade,
            sometimes near margin; new lvs gen coiled, unrolling as they develop
            11. Lvs of 2 dissimilar kinds, the fertile (sporangium-bearing) blades with greatly narrowed segments
               12. Blade deeply pinnately lobed or 1-pinnate; lf gen > 30 cm................... **BLECHNACEAE** (*Blechnum*)
               12′ Blade 2 or more × pinnate or divided; lf ≤ 30 cm .... ⁴**PTERIDACEAE**
            11′ Lvs all alike or nearly so, the fertile [sporangium-bearing] blades very similar in size and shape to
               sterile blades
               13. Sori borne along or near margin of lf; indusia 0 or sori covered by reflexed or recurved lf margin or
                  lf margin segment
                  14. Pinnae all entire or minutely toothed .... ⁴**PTERIDACEAE**
                  14′ Pinnae [some or all of them] conspicuously toothed, lobed, or compound
                     15. Rhizome and base of petiole covered by hairs; rhizome long-creeping; lf often > 80 cm,
                        sometimes ≥ 2 m — blade widely triangular, 2–4-pinnate .... **DENNSTAEDTIACEAE**
                     15′ Rhizome and base of petiole bearing linear to ovate scales; rhizome short and compact to
                        long-creeping; lf gen < 80 cm, often much shorter .... ⁴**PTERIDACEAE**
               13′ Sori borne away from margin on underside of lf or lflet, or sporangia scattered along veins and not
                  clustered into sori; indusia present or 0
                  16. Sporangia scattered along veins, not clustered into distinct sori; indusia 0 .... ⁴**PTERIDACEAE**
                  16′ Sporangia clustered in distinct sori; indusia sometimes present
                     17. Indusia 0
                        18. Blade deeply pinnately lobed .... **POLYPODIACEAE**
                        18′ Blade 2–3 pinnate .... **WOODSIACEAE** (*Athyrium distentifolium* var. *americanum*)
                     17′ Indusia present
                        19. Sori linear to oblong or J-shaped
                           20. Veins of lf segment forming a network; sori gen 2–4 mm, linear-oblong, end-to-end in rows
                              parallel to midvein of lf segment — lvs 1 pinnate with deeply pinnately lobed pinnae
                              .... **BLECHNACEAE** (*Woodwardia*)

20′ Veins of lf segment free, minor veins extending to segment margin without rejoining; sori
    0.5–15 mm, linear, oblong, J-shaped, or reniform, not end-to-end in rows parallel to midvein of lf segment
    21. Lf 3–30 cm, 1-pinnate with pinnae < 1.5 cm or lf simple, linear, undivided or forked
        . . . . **ASPLENIACEAE**
    21′ Lf 20–200 cm, 1–2 pinnate with pinnae gen 5–25 cm, toothed or pinnately lobed
        . . . . **WOODSIACEAE** (*Athyrium filix-femina* var. *cyclosorum*)
  19′ Sori ± round
    22. Blade with 1-celled, clear, needle-like hairs. . . . . **THELYPTERIDACEAE**
    22′ Blade without needle-like hairs
      23. Indusium peltate or round-reniform, attached ± in center of sorus, gen present and readily
        observable in late-season specimens. . . . . **DRYOPTERIDACEAE**
      23′ Indusium attached to one side of sorus or cup-like, attached beneath sorus and often of
        hair- or scale-like fragments or lobes, gen delicate, often difficult to observe in late-season
        specimens. . . . . [2]**WOODSIACEAE**

Group 4

Woody pls; pollen grains borne in 2–10(15) sacs on each scale or filament-like stalk of pollen cone;
seeds borne between scales of seed cone or naked on branches; fls never formed [Gymnosperms]

1. Specimens with pollen-producing cones only; ovule/seed-producing structures unavailable
  2. Lvs opposite or whorled
    3. Sts jointed at nodes, the internodes [main sites of photosynthesis] long, ± green; lvs dry, scale-like, often
      early deciduous, not overlapping; pollen cones with stamen-like structures, the pollen sacs borne on
      stout, filament-like stalks. . . . . [2]**EPHEDRACEAE** (G7,9)
    3′ Sts not jointed at nodes, the internodes gen very short; lvs green, scale-, awl-, or needle-like, persistent,
      crowded, gen closely overlapping; pollen cones with sessile pollen sacs . . . . [3]**CUPRESSACEAE** (G7)
  2′ Lvs alternate
    4. Pollen cones in lf axils on 1-year old twigs; pollen sacs in groups of 4–16 on stout, filament-like stalks
      . . . . [2]**TAXACEAE** (G7)
    4′ Pollen cones terminal or in axils of newly formed lvs of current season; pollen sacs 2–6, sessile on
      abaxial faces of cone scales
      5. Lvs needle-like; pollen sacs 2 per cone scale . . . . [2]**PINACEAE** (G7)
      5′ Lvs needle-, awl-, or scale-like; pollen sacs 2–6 per cone scale. . . . . [3]**CUPRESSACEAE** (G7)
1′ Specimens with ovule/seed-producing structures; pollen-producing cones present or 0
  6. Ovules/seeds borne in a cone that becomes ± woody at maturity
    7. Cone bracts present, individually subtending and ± free from cone scales, exserted beyond cone scales
      or ± < scales and then ± hidden by them; lvs needle-like; ovules/seeds 2 per cone scale; seeds distally
      1-winged or wingless. . . . . [2]**PINACEAE** (G7)
    7′ Cone bracts not evident, ± wholly fused to cone scales; lvs scale-like, awl-like, or needle-like; ovules/
      seeds 1–many per cone scale; seeds laterally 2-winged (wing sometimes very narrow) or distally
      1-winged in *Calocedrus*. . . . . [3]**CUPRESSACEAE** (G7)
  6′ Ovules/seeds exposed on branches *or* surrounded by a ± perianth-like cluster of membranous scales, *or*
    enclosed in a cone that becomes fleshy and berry-like at maturity
    8. Sts jointed at nodes, the internodes [main sites of photosynthesis] long, green; lvs dry, scale-like, often
      early deciduous, not overlapping; ovules/seed(s) surrounded by a ± perianth-like cluster of membranous
      scales; pollen cones with stamen-like structures, the pollen sacs borne on stout, filament-like stalks
      . . . . [2]**EPHEDRACEAE** (G7,9)
    8′ Sts not jointed at nodes, the internodes gen very short; lvs green, scale-like, awl-like, or needle-like,
      persistent, crowded, sometimes overlapping; ovules/seeds enclosed in fleshy cones *or* exposed on
      branches and seeds individually subtended or surrounded by fleshy aril; pollen cones with sessile or
      stalked pollen sacs
      9. Lvs opposite or whorled, 1–20+ mm, scale-, awl-, or needle-like; seeds 1–3, enclosed in a fleshy,
        berry-like cone; pollen sacs 2–6, sessile on abaxial faces of cone scales. . . . . **CUPRESSACEAE** (*Juniperus*)
      9′ Lvs alternate, 12–70 mm, needle-like; seed 1, ± enclosed by a fleshy, gen colored aril; pollen cones
        with stamen-like structures, pollen sacs in groups of 4–16 on stout filament-like stalks . . . . [2]**TAXACEAE** (G7)

Group 5

Frs developing from cleistogamous fls; pls sometimes also producing open fls

1. Cleistogamous fls formed at or below ground or water surface
  2. Pl ± aquatic, submersed or emergent
    3. Lvs opposite, palmately veined. . . . . [2]**PLANTAGINACEAE** (*Bacopa*)
    3′ Lvs alternate, parallel-veined . . . . **PONTEDERIACEAE** (*Heteranthera*)

2′ Pl terrestrial
  4. Lvs petioled
    5. Fr 3-angled achene, tightly enclosed by hardened perianth; cleistogamous fls at st base
    . . . . **POLYGONACEAE** (*Emex spinosa*)
    5′ Fr ovoid to oblong capsule, not enclosed by hardened perianth; cleistogamous fls in proximal lf axils
    . . . . [2]**VIOLACEAE** (G15,20,24)
  4′ Lvs sessile
    6. Lvs 0.5–4 cm, ± linear, ± 1-veined, not sheathing; cleistogamous fls at st base. . . . **BORAGINACEAE** (*Pectocarya*)
    6′ Lvs 2–9 cm, ovate to lance-elliptic, parallel-veined, sheathing; cleistogamous fls underground
    . . . . **COMMELINACEAE** (*Commelina*)
1′ Cleistogamous fls formed above ground or water surface
  7. Lvs sheathing, linear, parallel-veined; spikelets of cleistogamous fls enclosed within lf sheaths or borne in
    panicle-like clusters . . . . **POACEAE** (G2,6,8,13)
  7′ Lvs not sheathing, linear to ovate, not parallel-veined; spikelets of cleistogamous fls 0
    8. Lvs all basal, infl or solitary fl scapose
      9. Pl aquatic; lvs on rhizome or stolon, often bearing small, hollow bladders that suck in and trap small
        organisms when hairs at opening are triggered; fls 1–few . . . . **LENTIBULARIACEAE** (*Utricularia subulata*)
      9′ Pl terrestrial; lvs in basal rosette, not bearing traps; infl a spike (fl 1)
      . . . . **PLANTAGINACEAE** (*Plantago*) (G8,18,19)
    8′ Lvs all or mostly cauline
      10. Ovary inferior
        11. Infl a closed head enclosed by spine-tipped involucre bracts . . . . **ASTERACEAE** (*Centaurea melitensis*)
        11′ Infl a spike or spike-like raceme or fls sessile in lf axils
          12. Ovary/fr not narrowed distally; sepals ascending or ± spreading
          . . . . **CAMPANULACEAE** (*Githopsis, Heterocodon, Triodanis*)
          12′ Ovary/fr distally narrowed; sepals not separating. . . . . **ONAGRACEAE** (*Epilobium, Neoholmgrenia*)
      10′ Ovary superior
        13. Lvs ± fleshy
          14. Pl 50–230 cm; lvs entire or toothed . . . . [2]**BALSAMINACEAE** ([*Impatiens capensis*])
          14′ Pl 1–30 cm; lvs entire . . . . **MONTIACEAE** (*Montia*) (G2)
        13′ Lvs not fleshy
          15. Lvs opposite
            16. Lvs 2–7 mm, awl-shaped to oblong, bristle-tipped . . . . **CARYOPHYLLACEAE** (*Loeflingia*)
            16′ Lvs linear, oblong, or lanceolate to ovate or round, not bristle-tipped
              17. St square in ×-section; fls in head-like axillary and terminal clusters — fr 4 nutlets
              . . . . **LAMIACEAE** (*Lamium amplexicaule*)
              17′ St round in ×-section; fls axillary or in terminal raceme
                18. Lvs of a pair unequal; ovule 1; fr achene; perianth base hardened or 5-winged or -ribbed around fr
                . . . . **NYCTAGINACEAE** (*Acleisanthes, Mirabilis albida*)
                18′ Lvs of a pair equal; ovules several–many; fr capsule; perianth base not hardened or 5-winged around fr
                  19. Fr valves 3, ovary chambers 3 . . . . **CISTACEAE** (*Tuberaria*)
                  19′ Fr valves 2, ovary chambers 1–2
                    20. Sepals strongly fused. . . . . **PHRYMACEAE** (*Mimulus*)
                    20′ Sepals ± free . . . . [2]**PLANTAGINACEAE** (*Bacopa*)
          15′ Lvs alternate
           21. Fr 1–4 hard, shiny gray nutlets. . . . . **BORAGINACEAE** (*Lithospermum incisum*)
           21′ Fr a capsule
             22. Per
              23. Lvs ± sessile; capsule flat . . . . **POLYGALACEAE** (*Polygala californica*)
              23′ Lvs petioled; capsule ovoid to oblong. . . . . [2]**VIOLACEAE** (G15,20,24)
             22′ Ann
              24. Infl or solitary fls axillary; ovary chambers 5 . . . . [2]**BALSAMINACEAE** ([*Impatiens capensis*])
              24′ Infl terminal (axillary); ovary chambers 2–3
                25. Infl a raceme . . . . **PLANTAGINACEAE** (*Antirrhinum, Dopatrium, Nuttallanthus*)
                25′ Infl a head . . . . **POLEMONIACEAE** (*Collomia grandiflora*)

Group 6

Herbs, subshrubs, or herbaceous vines; only staminate fls present;
pistillate or bisexual fls unavailable for examination

1. Pl fully aquatic, submersed, floating in water, or stranded on mud
  2. Pl gen free-floating, at or just below water surface, raft-like, pl body 0.4–10 mm; lvs and sts not
    differentiated; roots 0–few, unbranched — stamens emerging from tiny lateral, membrane-covered
    pouch or from minute cavity on top of pl body . . . . **ARACEAE** (G2,8,13,15)

2′ Pl rooted in bottom sediments or free-floating, >> 10 mm; lvs and sts clearly differentiated; roots often branched
  3. Lvs alternate
    4. Fls in dense, spheric heads, these solitary or in axillary or terminal clusters, not enclosed in lf sheaths;
      freshwater habitats . . . . **TYPHACEAE** (*Sparganium*) (G2,8,13)
    4′ Fls in axillary spikes, these gen enclosed and concealed in sheaths of subtending lvs;
      marine habitats. . . . . **ZOSTERACEAE** (G2,8,13)
  3′ Lvs opposite or whorled
    5. Lf blades entire, toothed, or shallowly lobed. . . . . **HYDROCHARITACEAE** (G2,8,16)
    5′ Lf blades [at least those of submersed lvs] divided into linear or thread-like lobes
      6. Blades of lvs repeatedly forked . . . . **CERATOPHYLLACEAE** (G2,8,14,15)
      6′ Blades of lvs pinnately divided . . . . **HALORAGACEAE** (G2,8,14,20)
1′ Pl terrestrial or parasitic on st of woody host plant, or if growing in wet habitat, rooted in place and
  emerging well above water surface
  7. Pl parasitic on st of woody host pl
    8. Fls of parasite borne directly on sts of host; remainder of parasite internal within tissues of host; sts and
      lvs not differentiated; on *Psorothamnus* . . . . **APODANTHACEAE** (G8)
    8′ Fls borne on lfy branches of parasite; shoots of parasite external; lvs differentiated though sometimes
      reduced to scales; on various woody hosts [not *Psorothamnus*]. . . . . **VISCACEAE** (G7,8,9,11,12,14)
  7′ Pl free-living
    9. Lvs opposite or whorled, not all basal
      10. Sts or lvs thick and fleshy; infl a terminal spike
        11. Lvs ± cylindric or 3-angled in ×-section; petal-like perianth parts 4 . . . . **BATACEAE** (G8,15,24)
        11′ Lvs fleshy, scale-like; petal-like perianth parts 0. . . . . [2]**CHENOPODIACEAE** (G7,8,9,11,15)
      10′ Sts and lvs of normal texture, not thick and fleshy; infls various
        12. Lvs palmately compound or simple and deeply lobed. . . . . **CANNABACEAE** (G8,15)
        12′ Lvs simple, entire or toothed
          13. Lvs whorled. . . . . **RUBIACEAE** (*Galium*) (G7,8,9,12)
          13′ Lvs opposite
            14. Lvs sessile or nearly so, entire . . . . **CARYOPHYLLACEAE** (G8,15,24)
           14′ Lvs conspicuously petioled, toothed
             15. Stamens 8–15(20); stinging hairs 0. . . . . **EUPHORBIACEAE** (*Mercurialis*) (G8)
             15′ Stamens 4; stinging hairs present or 0 . . . . **URTICACEAE** (G8,15)
    9′ Lvs alternate or all basal
      16. Lvs stiff and sword-like, 5–15 dm; infl a large panicle; perianth parts 6 . . . . **RUSCACEAE** (*Nolina*) (G7,8,9,16)
      16′ Lvs not sword-like, often smaller; infls various; perianth parts often other than 6
        17. Lf blades linear or narrowly lanceolate, simple and entire, veins parallel; lf bases sheathing
          18. Infl a panicle or raceme of showy fls; perianth parts all petal-like, white to ± yellow, each with a
            glistening, ± green-yellow gland in proximal half. . . . . **MELANTHIACEAE** (*Toxicoscordion*)
          18′ Infl a cluster of spikelets (sometimes panicle- or raceme-like); perianth parts 0 or minute scales
            19. St triangular, nodes not swollen; lf blades channeled . . . . **CYPERACEAE** (*Carex*)
            19′ St round; nodes gen swollen and knot-like; lf blades gen flat . . . . **POACEAE** (G2,5,8,13)
        17′ Lf blades variously shaped, sometimes toothed, lobed, or compound, veins mostly pinnate or
          palmate; lf bases gen not sheathing
          20. Ocreae present, persistent or not; nodes gen swollen . . . . **POLYGONACEAE** (G2,8,15,16,19)
          20′ Ocreae 0; nodes gen not swollen
            21. Lvs all basal
              22. Lvs compound, lflets 3; perianth parts in 5s; stolons present . . . . **ROSACEAE** (*Fragaria*) (G8)
             22′ Lvs simple, entire; perianth parts in 3s; stolons 0
                23. Perianth clearly differentiated into sepals and petals; pl ± emergent aquatic, ann or per
                . . . . **ALISMATACEAE** (*Sagittaria*) (G8)
                23′ Perianth whorls not strongly differentiated, outer and inner similar; pl terrestrial, matted per
                . . . . **POLYGONACEAE** (*Eriogonum*) (G8,12)
            21′ At least some lvs cauline
             24. Pls vines
               25. Tendrils 0; perianth parts not at all petal-like; stamens free . . . . **CANNABACEAE** (*Humulus*) (G8,15)
               25′ Tendrils present; perianth parts petal-like
                 26. Perianth parts fused, at least at base; stamens fused, number difficult to determine
                 . . . . **CUCURBITACEAE** (G8,14,17,21)
                 26′ Perianth parts free; stamens 6, free . . . . **SMILACACEAE** (G7,8,9,16)
            24′ Pls prostrate to erect herbs
             27. Fls in heads or umbels
              28. Infl a simple or compound umbel . . . . [2]**APIACEAE** (G2,8,14,20,21)
              28′ Infl of 1 or more heads
               29. Heads without involucres. . . . . [2]**CHENOPODIACEAE** (G7,8,9,11,15)

29′ Heads each subtended by an involucre
 30. Stamens free; petals free or 0 . . . . [2]**APIACEAE** (G2,8,14,20,21)
 30′ Stamens fused by their anthers; petals fused . . . . **ASTERACEAE** (G7,8,9,12,14,17)
27′ Fls in axillary clusters, racemes, or panicles
 31. Lvs deeply lobed or compound
  32. Lvs palmately compound, the lflets serrate. . . . . **CANNABACEAE** (*Cannabis*) (G8,15)
  32′ Lvs pinnately dissected or compound, the lobes or lflets entire to coarsely toothed
   33. Lvs simple or 1-pinnate; anthers subsessile, 3–6 mm — sepals 4–9, infl a lfy raceme or
    panicle. . . . . **DATISCACEAE** (G14)
   33′ Lvs 2–4-pinnate; anthers exserted on slender filaments
    34. Lflets coarsely crenate or shallowly lobed; sepals 4–5; petals 0; anthers linear, 1.5–5 mm;
     infl an open panicle . . . . **RANUNCULACEAE** (*Thalictrum*) (G8)
    34′ Lflets 2-serrate; sepals 5; petals 5; anthers oval, < 1 mm; infl a panicle of dense spikes
    . . . . **ROSACEAE** (*Aruncus*) (G8)
 31′ Lvs entire or toothed
  35. Blades of lvs with several main veins; perianth parts ≥ 6 mm . . . . . **MELANTHIACEAE** (*Veratrum*) (G16)
  35′ Blades of lvs with 1–3 main veins; perianth parts gen ≤ 5 mm
   36. Herbage armed with stinging, nettle-like hairs. . . . . **EUPHORBIACEAE** (*Tragia*) (G15)
   36′ Herbage glabrous or with non-stinging hairs
    37. Lvs covered with stellate scales or hairs . . . . **EUPHORBIACEAE** (*Croton*) (G8,12)
    37′ Lvs without stellate scales or hairs
     38. Lvs gen bearing bead-like, sessile hairs or powdery scales
     . . . . **CHENOPODIACEAE** (*Atriplex*) (G8,12)
     38′ Lvs glabrous or glaucous
      39. Lvs fleshy; sepals, petals each 4(5). **CRASSULACEAE** (*Rhodiola*) (G8)
      39′ Lvs thin or fleshy; sepals 2–6, petals 0
       40. Milky latex present; lf blade base gen with 2 glands. . . . . **EUPHORBIACEAE** (*Stillingia*)
       40′ Milky latex 0; lf blade base without glands
        41. Lf ± round, blade with translucent dots; pl mat-forming . . . . **URTICACEAE** (*Soleirolia*) (G8)
        41′ Lf linear to (ob)ovate, blade without translucent dots; pl mat-forming to erect
        . . . . **AMARANTHACEAE** (*Amaranthus*) (G8)

Group 7

Trees, shrubs, or woody vines; only staminate fls present;
pistillate or bisexual fls unavailable for examination

1. Pl parasitic on st of woody host pl. . . . . **VISCACEAE** (G6,8,9,11,12,14)
1′ Pl free-living, rooted in ground
 2. Trunk unbranched, gen stout, covered with persistent woody lf bases; lvs 4–7 m, pinnately compound
 . . . . **ARECACEAE** (*Phoenix*) (G9,10)
 2′ Trunk(s) gen branched, stout or slender, gen not covered with persistent lf bases; lvs gen smaller, simple
  or compound
  3. Lvs stiff and sword-like, 0.6–2 m, base much expanded, white, fleshy; infl a large panicle
  . . . . **RUSCACEAE** (*Nolina*) (G6,8,9,16)
  3′ Lvs not sword-like, mostly smaller, base not much expanded, white, and fleshy; infls various
   4. Lvs opposite or whorled
    5. Lvs compound
     6. Lvs palmate. . . . . **SAPINDACEAE**
     6′ Lvs pinnate
      7. Woody vine; sepals gen 4, conspicuous, petal-like . . . . **RANUNCULACEAE** (*Clematis*) (G9)
      7′ Tree or large shrub; sepals gen 5 or number difficult to determine, very small
       8. Fls sessile or borne on short, stout pedicels. **OLEACEAE** (*Fraxinus*) (G9,12,16)
       8′ Fls borne on elongated, thread-like pedicels . . . . **SAPINDACEAE** (*Acer negundo*) (G9)
    5′ Lvs simple
     9. Lf toothed or lobed
      10. Lf palmately veined and lobed . . . . **SAPINDACEAE** (*Acer*) (G12,23)
      10′ Lf pinnately veined and toothed
       11. Petals 0; stamens 4–10. . . . . [4]**PICRODENDRACEAE** (G9,12)
       11′ Petals 5, united in salverform corolla; stamens 4, epipetalous
       . . . . [2]**SCROPHULARIACEAE** (*Buddleja*) (G9,19)
     9′ Lf entire (strongly wavy in some *Garrya* spp.)
      12. Infl a catkin
       13. Lf blades well developed; catkins elongate, pendent . . . . **GARRYACEAE** (G9,11)

13′ Lvs scale-, awl-, or needle-like; "catkins" short, not pendent — actually catkin-like pollen cone of gymnosperm

    14. Internodes very short; st concealed by persistent, overlapping, green, scale-, awl-, or needle-like lvs; pollen sacs sessile beneath scales of small cone . . . . [2]**CUPRESSACEAE** (G4)

    14′ Internodes long; st green, bearing dry, scale-like, often early-deciduous lvs; pollen sacs borne on filament-like stalk . . . . **EPHEDRACEAE** (G4,9)

12′ Infl not a catkin

  15. Lvs whorled

    16. Lf with abaxial groove . . . . [2]**ERICACEAE** (*Empetrum*) (G9,12,16)

    16′ Lf flat or with midvein raised on abaxial face

      17. Stamens 4; weak-stemmed shrub; sts 4-angled when young . . . . **RUBIACEAE** (*Galium*) (G6,8,9,12)

      17′ Stamens 5–10; stiff shrub; sts round . . . . [4]**PICRODENDRACEAE** (G9,12)

  15′ Lvs opposite

    18. Herbage covered with silvery scales . . . . **ELAEAGNACEAE** (*Shepherdia*) (G9,12)

    18′ Herbage glabrous or variously hairy

      19. Petals fused proximally; stamens fused to corolla

        20. Lvs glossy . . . . **RUBIACEAE** ([*Coprosma*]) (G9)

        20′ Lvs densely hairy . . . . [2]**SCROPHULARIACEAE** (*Buddleja*) (G9,19)

      19′ Petals free or 0; stamens free from corolla

        21. Lvs thick, leathery, evergreen; fls in peduncled, head-like clusters; sepals well developed

        . . . . **SIMMONDSIACEAE** (G9,12)

        21′ Lvs thin, deciduous; fls in sessile axillary clusters or short-peduncled racemes; sepals very small

          22. Bark of young sts ± red; fls pedicelled in short axillary racemes; stamens 5–10

          . . . . [4]**PICRODENDRACEAE** (G9,12)

          22′ Bark of young sts gray; fls ± sessile in axillary clusters; stamens 1–4

          . . . . **OLEACEAE** (*Forestiera*) (G9,12)

4′ Lvs alternate (lf often opposite a tendril or infl in Vitaceae)

  23. Tendrils present; pl a vine; infl of umbels or a panicle of head- or umbel-like clusters

    24. Tendrils each opposite a lf; calyx 0 or 5-lobed; petals 5, adherent at tips, ± yellow

    . . . . **VITACEAE** (*Vitis*) (G9,12,23)

    24′ Tendrils 2+ near petiole base; perianth parts 6 in 2 whorls of 3, petal-like, free, white to ± green or ± yellow . . . . **SMILACACEAE** (G6,8,9,16)

  23′ Tendrils 0; pl a shrub or tree, rarely a vine; infl various

    25. Infl a catkin or spike

      26. Lvs compound

        27. Lvs 1- and 2-pinnate, lflets 20–28 on 1-pinnate lvs . . . . **FABACEAE** (*Gleditsia*) (G9)

        27′ Lvs 1-pinnate, lflets 5–21

          28. Pl evergreen — infl resembles catkin when young, but is actually a branched panicle

          . . . . **ANACARDIACEAE** (*Pistacia*) (G11,12)

          28′ Pl deciduous . . . . **JUGLANDACEAE** (G9,11)

      26′ Lvs simple

        29. Shrubs of ± saline desert habitats; lvs sometimes thick, ± fleshy; spikes erect or very short and axillary

          30. Lf blades gen flat; perianth present though gen small and inconspicuous; bracts 0 or not peltate

          . . . . [2]**CHENOPODIACEAE** (G6,8,9,11,15)

          30′ Lf blades gen ± cylindric; perianth 0; bracts peltate . . . . **SARCOBATACEAE** (G9,11)

        29′ Shrubs or trees, gen of non-saline habitats; lvs mostly thinner, not fleshy; spikes or catkins erect to pendent

          31. Lvs scale-, awl- or needle-like; pls evergreen — catkin-like structure actually pollen cone of a gymnosperm

            32. Pollen sacs borne on filament-like stalks . . . . **TAXACEAE** (G4)

            32′ Pollen sacs sessile on abaxial side of scale-like structures

              33. Scales with 2 pollen sacs; needle-like lvs 1–30 cm; bark gen separating in flakes or plates

              . . . . **PINACEAE** (G4)

              33′ Scales with several pollen sacs; needle-like lvs (if present) 1–2.5 cm; bark thick, fibrous, persistent . . . . [2]**CUPRESSACEAE** (G4)

          31′ Lvs with well developed blades; pls evergreen or deciduous

            34. Bractlets of infl well developed and readily visible at time of fl

              35. Catkins pendent

                36. Catkin-bractlets fringe-margined, early-deciduous; stamens 8–60; trees . . . . . . **SALICACEAE** (*Populus*)

                36′ Catkin-bractlets entire or appearing 3-lobed, persistent; stamens 1–10; shrubs or trees . . . . **BETULACEAE** (G9,11)

              35′ Catkins ± stiff, ± erect to spreading but gen not pendent

                37. Low, brittle shrubs; lvs thin, crenate . . . . **EUPHORBIACEAE** (*Acalypha*) (G9,11)

                37′ Shrubs or trees, not brittle; lvs entire or toothed — catkin bracts sometimes early-deciduous . . . . **SALICACEAE** (*Salix*)

            34′ Bractlets of infl 0 or very inconspicuous at time of fl

              38. Evergreen shrubs or trees

39. Main veins of lf parallel or nearly so — 2 perianth whorls actually present but sepals small
   and easily overlooked . . . . [2]**FABACEAE** (*Acacia*)
39′ Main veins of lf clearly pinnate
   40. Lvs glabrous or with stellate hairs, not gland-dotted; sepals present; gen of dry-land
      habitats; catkins slender, pendent, gen ≥ 2 cm . . . . [2]**FAGACEAE** (G9,11)
   40′ Lvs glabrous or with minute, straight hairs, dotted with tiny resin glands; sepals 0; gen of
      moist, often wetland habitats; catkins short, dense, gen < 2 cm. . . . . [2]**MYRICACEAE** (G9,11)
38′ Deciduous shrubs or trees
   41. Sepals small but evident
      42. Sepals 5–7; stamens 5–20; veins pinnate; lf bases gen symmetric. . . . . [2]**FAGACEAE** (G9,11)
      42′ Sepals 4; stamens 4; veins palmate; lf bases oblique . . . . **MORACEAE** (G11,12)
   41′ Sepals 0 or calyx modified into cup- or saucer-shaped structure or reduced to a nectary
      43. Lvs dotted with tiny resin glands; individual fls difficult to distinguish . . . . . . [2]**MYRICACEAE** (G9,11)
      43′ Lvs not gland-dotted; individual fls clearly distinguishable on close examination
         . . . . **SALICACEAE** (G9,11)
25′ Infl not a catkin or spike
   44. Lvs and sts armed with prickles
      45. Petals free; stamens many. . . . . **ROSACEAE** (*Rubus ursinus*) (G9)
      45′ Petals fused; stamens 5. . . . . **SOLANACEAE** (*Solanum marginatum*)
   44′ Foliage not prickly, sometimes sts thorny or lvs spiny-toothed
      46. Lvs palmately lobed
         47. Lvs peltate; infl a panicle. . . . . **EUPHORBIACEAE** (*Ricinus*) (G9,12)
         47′ Lvs not peltate; infl a head . . . . **PLATANACEAE** (G9,10,12)
      46′ Lvs not palmately lobed
         48. Sts and lvs densely covered with stellate hairs. . . . . **EUPHORBIACEAE** (G9,15)
         48′ Sts and lvs glabrous or hairy but hairs not stellate
            49. Stamens many
               50. Lvs strongly aromatic; sepals 4; petals 0; anthers dehiscing by uplifted lids
                  . . . . [2]**LAURACEAE** ([*Laurus*]) (G9)
               50′ Lvs not aromatic; sepals 5; petals 5; anthers dehiscent by slits
                  51. Fls in spheric heads; petals yellow, gen concealed by stamens; lvs simple or 2-pinnate
                     . . . . [2]**FABACEAE** (*Acacia*)
                  51′ Fls in racemes; petals white, conspicuous; lvs simple . . . . **ROSACEAE** (*Oemleria*) (G9)
            49′ Stamens 12 or fewer
               52. Lvs compound
                  53. Lflets with a few, gland-tipped teeth near base. . . . . **SIMAROUBACEAE** (*Ailanthus*) (G9,10,20)
                  53′ Lflets entire or teeth not gland-tipped
                     54. Lflets 3–20, lance-linear to ovate, 10–130 mm . . . . [2]**ANACARDIACEAE** (G9,20,23)
                     54′ Lflets 7–33, linear-oblong or narrowly elliptic, 2–10 mm. . . . . **BURSERACEAE** (G9,20)
               52′ Lvs simple (occ reduced to scales)
                  55. Shrub with thorny, gen lfless green sts; lvs scale-like to narrowly oblong, early deciduous
                     . . . . **SIMAROUBACEAE** (*Castela*) (G9,10,23)
                  55′ Tree, shrub, or woody vine, thorns occ present but sts gen lfy; lvs various, sometimes
                     deciduous in winter or dry season
                     56. Fl parts in 3s — lvs with abaxial groove. . . . . [2]**ERICACEAE** (*Empetrum*) (G9,12,16)
                     56′ Fl parts gen in 4s, 5s, or 6s
                        57. Fls in involucred heads, these gen in few- to many-headed clusters
                           . . . . **ASTERACEAE** (G6,8,9,12,14,17)
                        57′ Fls solitary or variously clustered, not in involucred heads
                           58. Lvs gen ± gray, either softly hairy or covered with powdery scales
                              . . . . [2]**CHENOPODIACEAE** (G6,8,9,11,15)
                           58′ Lvs green
                              59. Petals present
                                 60. Lvs entire to lobed, not spiny-toothed; infl a terminal or axillary panicle or umbel-like
                                    cluster, or fls 1; fl parts gen in 5s
                                    61. Petals 1–5 mm . . . . [2]**ANACARDIACEAE** (G9,20,23)
                                    61′ Petals 10–15 mm . . . . **PITTOSPORACEAE** (*Pittosporum*)
                                 60′ Lvs spiny-toothed or entire; infl an axillary cyme; fl parts in 4s or 5s
                                    62. Stamens gen 4, alternate petals. . . . . **AQUIFOLIACEAE** (G9)
                                    62′ Stamens 4–5, opposite petals . . . . [2]**RHAMNACEAE** (*Rhamnus*) (G9)
                              59′ Petals 0
                                 63. Lvs strongly aromatic; anthers dehiscing by uplifted lids . . . . [2]**LAURACEAE** ([*Laurus*]) (G9)
                                 63′ Lvs not aromatic; anthers dehiscent by slits
                                    64. Stamens = in number to and alternate sepals. . . . . [2]**RHAMNACEAE** (*Rhamnus*) (G9)
                                    64′ Stamens = in number to and opposite sepals or > in number than sepals

65. Lf blade 2–12 mm; shrubs . . . . [4]**PICRODENDRACEAE** (G9,12)
65′ Lf blade 30–140 mm; trees or large shrubs
  66. Lf 3-veined from base; infl a sessile axillary cluster . . . . **CANNABACEAE** (*Celtis*) (G12)
  66′ Lf 1-veined from base; infl a stalked umbel or umbel-like raceme
  . . . . **MORACEAE** (*Maclura*) (G9)

Group 8

Herbs, subshrubs, or herbaceous vines; only pistillate fls present;
staminate or bisexual fls unavailable for examination

1. Pl fully aquatic, submersed, floating in water, or stranded on mud
 2. Pl gen free-floating, at or just below water surface, raft-like, the pl body 0.4–10 mm; lvs and sts not
   differentiated; roots 0–few, unbranched — pistil located in tiny lateral, membrane-covered pouch
   or in minute cavity on top of pl body . . . . **ARACEAE** (G2,6,13,15)
2′ Pl rooted in bottom sediments or free-floating, gen >> 15 mm; lvs and sts clearly differentiated; roots
   often branched
 3. Lvs alternate
  4. Frs borne in umbel-like clusters — fls actually bisexual but stamens easily overlooked and ephemeral
  . . . . **RUPPIACEAE** (G2,10,13)
  4′ Frs or fls borne in spikes or heads
   5. Fls in dense, spheric heads, these solitary or in axillary or terminal clusters, not enclosed in lf sheaths;
    freshwater habitats . . . . **TYPHACEAE** (*Sparganium*) (G2,6,13)
   5′ Fls in axillary spikes, these gen enclosed and concealed in sheaths of subtending lvs; marine habitats
   . . . . **ZOSTERACEAE** (G2,6,13)
 3′ Lvs opposite or whorled or all basal
  6. Lvs all basal; fls solitary in lf-axils; style very elongated, gen 6–20 cm
  . . . . [2]**JUNCAGINACEAE** (*Triglochin scilloides*) (G13)
  6′ Lvs cauline; fls axillary or variously clustered; styles gen << 6 cm (exc in some Hydrocharitaceae)
   7. Lf blades [at least those of submersed lvs] divided into linear or thread-like lobes
    8. Blades of lvs repeatedly forked . . . . **CERATOPHYLLACEAE** (G2,6,14,15)
    8′ Blades of lvs pinnately forked . . . . **HALORAGACEAE** (G2,6,14,20)
   7′ Lf blades entire, toothed, or shallowly lobed
    9. Petals and sepals both 3(4), evident; perianth and stigmas borne at water surface at end of long,
     tubular hypanthium; ovary inferior, sessile in lf axil . . . . **HYDROCHARITACEAE** (G2,6,16)
    9′ Petals and sepals both 0; stigmas submersed, borne at end of short styles; ovary or ovaries superior,
     sessile or short-stalked in lf axil
     10. Pistils 2–10, simple, each with a cup-like stigma; lvs entire . . . . **ZANNICHELLIACEAE** (G2,10,13)
     10′ Pistil 1, compound, bearing 2–4 slender stigmas
      11. Lvs subentire to finely or coarsely toothed; ovary chamber 1, ovule 1; fr achene-like
      . . . . **HYDROCHARITACEAE** (*Najas*) (G2,13)
      11′ Lvs entire; ovary chambers 4, ovules 4; fr splitting into 4 nutlets
      . . . . **PLANTAGINACEAE** (*Callitriche*) (G2,15)
1′ Pl terrestrial or parasitic on st of woody host, or if growing in wet habitat, rooted in place and emerging
  well above water surface
 12. Pls parasitic on sts of woody host pls
  13. Fls of parasite borne directly on sts of host; remainder of parasite internal within tissues of host; sts
   and lvs not differentiated; on *Psorothamnus* . . . . **APODANTHACEAE** (G6)
  13′ Fls borne on lfy branches of parasite; shoots of parasite external; lvs differentiated, though sometimes
   reduced to scales; on various woody hosts [not *Psorothamnus*] . . . . **VISCACEAE** (G6,7,9,11,12,14)
 12′ Pls free-living
 14. St fleshy, bearing spiny tubercles; lvs 0 . . . . **CACTACEAE** (*Mammillaria dioica*)
 14′ St gen not fleshy, without spiny tubercles; lvs gen present
  15. Lvs opposite or whorled, not all basal
  16. Sts or lvs thick and fleshy; infl a terminal spike
   17. Lvs ± cylindric or 3-angled in x-section; stigma 1, head-like, 2-lobed; ovary chambers 4; ovaries
    joining into a fleshy multiple fr; perianth 0 . . . . **BATACEAE** (G6,15,24)
   17′ Lvs fleshy, scale-like; stigmas 2, linear; ovary chamber 1; ovaries maturing as utricles, sometimes
    surrounded by fleshy bracts and perianth elements . . . . [2]**CHENOPODIACEAE** (G6,7,9,11,15)
  16′ Sts and lvs of normal texture, not thick and fleshy; infl various
   18. Lvs compound or deeply lobed, or also some simple
    19. Compound lvs palmate . . . . **CANNABACEAE** (G6,15)
    19′ Compound or deeply lobed lvs pinnate . . . . **VALERIANACEAE** (*Valeriana occidentalis*)
   18′ Lvs simple, entire or toothed
    20. Lvs whorled; ovary inferior . . . . **RUBIACEAE** (*Galium*) (G6,7,9,12)
    20′ Lvs opposite; ovary superior
     21. Lvs sessile or nearly so, entire; petals free . . . . **CARYOPHYLLACEAE** (G6,15,24)

21′ Lvs conspicuously petioled, toothed; petals fused or 0

  22. Corolla showy, bilateral . . . . **LAMIACEAE** (G18,19)

  22′ Corolla 0

    23. Styles 2, ovary 2-chambered; stinging hairs 0 . . . . **EUPHORBIACEAE** (*Mercurialis*) (G6)

    23′ Style 0 or 1, ovary 1-chambered; stinging hairs present or 0. . . . . **URTICACEAE** (G6,15)

15′ Lvs alternate or all basal

  24. Lvs stiff and sword-like, 0.5–1.5 m; infl a large panicle; perianth parts 6 . . . . . **RUSCACEAE** (*Nolina*) (G6,7,9,16)

  24′ Lvs not sword-like, often smaller; infls various; perianth parts mostly other than 6

    25. Blades of lvs linear or narrowly lanceolate, simple and entire; veins parallel; lf bases sheathing st

      26. Lvs all basal; fls solitary in lf axils; styles very elongated, gen 6–20 cm

      . . . . [2]**JUNCAGINACEAE** (*Triglochin scilloides*) (G13)

      26′ Lvs basal and cauline or all cauline; fls in spikes or spikelets, these gen in 2° clusters; styles gen << 5 cm

        27. St triangular; nodes not swollen; lf blades channeled . . . . **CYPERACEAE** (G2,13)

        27′ St round; nodes gen swollen and knot-like; lf blades gen flat . . . . **POACEAE** (G2,5,6,13)

    25′ Blades of lvs variously shaped, sometimes toothed, lobed, or compound; veins mostly pinnate or palmate; lf bases often not sheathing sts — fls not in spikelets

      28. Ocreae present, persistent or not; nodes gen ± swollen . . . . **POLYGONACEAE** (G2,6,15,16,19)

      28′ Ocreae 0; nodes gen not swollen

        29. Lvs all basal

          30. Lvs compound, lflets 3; perianth parts in 5s; stolons present. . . . . **ROSACEAE** (*Fragaria*) (G6)

          30′ Lvs simple, entire; perianth parts in 3s or 4s; stolons 0

            31. Perianth parts in 4s; fr a circumscissile capsule . . . . **PLANTAGINACEAE** (*Plantago*) (G5,18,19)

            31′ Perianth parts in 3s; fr 1–many achenes

              32. Perianth clearly differentiated into sepals and petals; pistils many; pl ± emergent aquatic, ann or per . . . . **ALISMATACEAE** (*Sagittaria*) (G6)

              32′ Perianth whorls not strongly differentiated, outer and inner similar; pistil 1; pl terrestrial matted per . . . . **POLYGONACEAE** (*Eriogonum*) (G6,12)

        29′ At least some lvs cauline

          33. Pls vines

            34. Tendrils 0; perianth not at all corolla-like; ovary superior. . . . . **CANNABACEAE** (*Humulus*) (G6,15)

            34′ Tendrils present; perianth corolla-like

              35. Ovary inferior . . . . **CUCURBITACEAE** (G6,14,17,21)

              35′ Ovary superior. . . . . **SMILACACEAE** (G6,7,9,16)

          33′ Pls prostrate to erect herbs

           36. Pistils 2–22, simple; fr a cluster of achenes or follicles

            37. Lvs simple, fleshy. . . . . **CRASSULACEAE** (*Rhodiola*) (G6)

            37′ Lvs compound, lflets thin

              38. Lflets coarsely crenate or shallowly lobed; sepals gen 4; petals 0; infl an open panicle

              . . . . **RANUNCULACEAE** (*Thalictrum*) (G6)

              38′ Lflets 2-serrate; sepals 5; petals 5; infl a panicle of dense spikes. . . . . **ROSACEAE** (*Aruncus*) (G6)

          36′ Pistil 1, simple or compound; fr achene, utricle, capsule, or splitting into 2–13 mericarps

           39. Fls in umbels or heads

            40. Infl a simple or compound umbel . . . . **APIACEAE** (G2,6,14,20,21)

            40′ Infl of 1 or more heads

              41. Head maturing as bur enclosing 1–5 frs . . . . [2]**ASTERACEAE** (G6,7,9,12,14,17)

              41′ Head not maturing as bur

                42. Corolla 0; head without involucre; ovary superior . . . . [2]**CHENOPODIACEAE** (G6,7,9,11,15)

                42′ Corolla tubular; head subtended by an involucre; ovary inferior . . . . [2]**ASTERACEAE** (G6,7,9,12,14,17)

           39′ Fls solitary or in axillary clusters or in racemes or panicles

           43. Style branches 5–19; fr splitting into 5–19 wedge-shaped mericarps at maturity

           . . . . **MALVACEAE** (G19,20,22)

           43′ Style unbranched or branches 2 or 3 (dissected into linear lobes in *Croton*); fr unsegmented or splitting into 2–3 segments

            44. Lvs palmately compound, lflets serrate. . . . . **CANNABACEAE** (*Cannabis*) (G6,15)

            44′ Lvs entire or toothed

              45. Lvs densely stellate-scaly or -hairy; ovary chambers gen 3; fr a capsule

              . . . . **EUPHORBIACEAE** (*Croton*) (G6,12)

              45′ Lvs not stellate-hairy; ovary chamber 1; fr an achene or utricle

                46. Bractlets subtending fl 3, winged, abaxially bearing hooked hairs; stigma 1, sessile, hair-tufted; sepals 4, fused. . . . . **URTICACEAE** (*Soleirolia*) (G6)

              46′ Bractlets subtending fl 1–2, without hooked hairs; stigmas 2, linear; sepals 0 or 3–5, free

                47. Fl enclosed between a pair of bractlets; sepals gen 0; lvs gen bearing bead-like, sessile hairs or powdery scales. . . . . **CHENOPODIACEAE** (*Atriplex*) (G6,12)

                47′ Fl not enclosed by a pair of bractlets; sepals 3–5; lvs glabrous or glaucous

              . . . . **AMARANTHACEAE** (*Amaranthus*) (G6)

Group 9

Trees, shrubs, or woody vines; only pistillate fls present;
staminate or bisexual fls unavailable for examination

1. Pl parasitic on st of woody host pl. . . . . **VISCACEAE** (G6,7,8,11,12,14)
1′ Pl free-living, rooted in ground
  2. Trunk unbranched, gen stout, often covered with persistent woody lf bases; lvs 4–7 m, pinnately
    compound . . . . **ARECACEAE** (*Phoenix*) (G7,10)
  2′ Trunk(s) gen branched, slender to stout, gen not covered with persistent lf bases; lvs gen much smaller,
    simple or compound
    3. Lvs stiff and sword-like, 0.6–2 m, base much expanded, white, fleshy; infl a large panicle
     . . . . **RUSCACEAE** (*Nolina*) (G6,7,8,16)
    3′ Lvs not sword-like, mostly smaller, base not much expanded, white, and fleshy; infls various
      4. Lvs opposite or whorled
        5. Lvs compound
          6. Lvs palmate; lflets 3 . . . . **SAPINDACEAE** (*Acer glabrum*) (G20)
          6′ Lvs pinnate; lflets often > 3
            7. Woody vine; sepals gen 4, conspicuous, petal-like; pistils many; styles elongated and plumose in fr
             . . . . **RANUNCULACEAE** (*Clematis*) (G7)
            7′ Tree or large shrub; sepals gen 5 or number difficult to determine, very small; pistil 1; styles never plumose
              8. Stigmas short, on slender style; ovary not winged in fl; fr a 1-winged achene, not splitting
                . . . . **OLEACEAE** (*Fraxinus*) (G7,12,16)
              8′ Stigmas linear, ± sessile; ovary with 2 short wings in fl; fr gen splitting into winged mericarps
                . . . . **SAPINDACEAE** (*Acer negundo*) (G7)
        5′ Lvs simple
          9. Margins of lvs toothed . . . . [3]**PICRODENDRACEAE** (G7,12)
          9′ Margins of lvs entire
            10. Weak-stemmed shrub
              11. Perianth parts in 3s, free; ovary superior, unlobed, smooth; NCo . . . . . . [2]**ERICACEAE** (*Empetrum*) (G7,12,16)
              11′ Perianth parts in 4s, fused; ovary inferior, ± bilobed, smooth to covered with short bristles;
                widespread. . . . . **RUBIACEAE** (*Galium*) (G6,7,8,12)
          10′ Stouter shrub or tree
            12. Sts and lvs covered with silvery scales. **ELAEAGNACEAE** (*Shepherdia*) (G7,12)
            12′ Sts and lvs glabrous or variously hairy
              13. Petals fused proximally
                14. Lvs glossy. . . . . **RUBIACEAE** ([*Coprosma*]) (G7)
                14′ Lvs densely hairy. . . . . **SCROPHULARIACEAE** (*Buddleja*) (G7,19)
             13′ Petals free or 0
               15. Fls in pendent catkins . . . . **GARRYACEAE** (G7,11)
              15′ Fls not in catkins
                16. Sts green, jointed; lvs thin, dry, scarious, scale-like, often early-deciduous — structures
                appearing to be fls actually ovulate cones of a gymnosperm. . . . . **EPHEDRACEAE** (G4,7)
                16′ Sts gray or brown or only young sts green, not jointed; lvs thin or thick, green, not dry and
                scarious, not scale-like, persistent or deciduous
                  17. Lvs thick, leathery, evergreen; fls solitary; sepals well developed . . . . **SIMMONDSIACEAE** (G7,12)
                  17′ Lvs thin, deciduous; fls in sessile, axillary clusters or short-peduncled racemes; sepals very small
                    18. Fls subsessile, solitary or few in axils; styles 3–4; ovary 3–4-lobed; fr a capsule
                     . . . . [3]**PICRODENDRACEAE** (G7,12)
                  18′ Fls pedicelled, in axillary clusters; style 1; ovary unlobed; fr a drupe
                   . . . . **OLEACEAE** (*Forestiera*) (G7,12)
  4′ Lvs alternate (lf often opposite a tendril or an infl in Vitaceae)
    19. Tendrils present; pl a vine; infl an umbel or panicle; fr a berry
      20. Infl an umbel. . . . . **SMILACACEAE** (G6,7,8,16)
      20′ Infl a panicle. . . . . **VITACEAE** (*Vitis*) (G7,12,23)
    19′ Tendrils 0; pl a shrub or tree, rarely a vine; infl, fr various
      21. Fls in catkins or catkin-like spikes
        22. Lvs compound
          23. Lvs 1–2-pinnate, lflets entire; fr a long, flat ± indehiscent legume . . . . **FABACEAE** (*Gleditsia*) (G7)
          23′ Lvs 1-pinnate, lflets serrate; fr a small 2-winged nut. . . . . **JUGLANDACEAE** ([*Pterocarya*])
        22′ Lvs simple
          24. Shrubs of ± saline desert habitats; lvs ± thick and fleshy or very narrow and densely hairy; spikes
          erect or very short and axillary. . . . . [3]**CHENOPODIACEAE** (G6,7,8,11,15)
          24′ Shrubs or trees, gen of moister habitats; lvs thin and wide; spikes or catkins erect or pendent
            25. Stipules 0

    26. Lvs sessile; blades ± gray, entire, not resin-dotted; fls enclosed by a pair of bractlets; fr an achene or utricle, enclosed between and falling together with papery or hardened bractlets . . . . [3]**CHENOPODIACEAE** (G6,7,8,11,15)

    26′ Lvs petioled; blades green, crenate or serrate, resin-dotted; fls subtended by 2–4 bractlets; fr a papillate, wax-coated drupe, free of subtending bractlets, or an achene enclosed by and falling together with a pair of spongy bractlets . . . . **MYRICACEAE** (G7,11)

  25′ Stipules present, evident at least on new growth, sometimes deciduous but leaving evident scars

    27. Styles or stigmas repeatedly branched into >> 4 fine, thread-like divisions; ovary 3-lobed, maturing as a capsule; sts brittle; shrub . . . . **EUPHORBIACEAE** (*Acalypha*) (G7,11)

    27′ Styles, style branches, or stigmas 2–4, gen not further divided; ovary gen unlobed, the frs various, incl capsules; sts mostly not brittle; shrub or tree

      28. Catkins very dense at time of fl; only stigmas exserted beyond tips of tightly appressed bractlets; at maturity either catkin ± cone-like with ovaries ripening as winged achenes or bractlets expanding into 1 or more tubular involucres, each surrounding a nut . . . **BETULACEAE** (G9,11)

      28′ Catkins looser at time of fl; bractlets, if present, not tightly appressed, the ovaries or calyces gen visible on close inspection; at maturity either catkin becoming a ± open cluster of capsules or a fleshy multiple fr

        29. Ovary tightly enwrapped by 4 appressed sepals; catkin bractlets 0; sepals fleshy at maturity, together with ovaries maturing as a multiple fr . . . . **MORACEAE** (*Morus*)

        29′ Ovary not enclosed by sepals, calyx 0 or represented by an entire or slightly lobed rim; fls subtended by bractlets, these sometimes deciduous; ovary ripening as a capsule containing several–many hair-tufted seeds. . . . . **SALICACEAE** (G7,11)

21′ Fls not in catkins or spikes — ovules never produced in cone-like catkins

  30. Lvs and sts gen with many straight, weak, slender prickles . . . . **ROSACEAE** (*Rubus ursinus*) (G7)

  30′ Lvs and sts without many straight, weak prickles (rarely with stout prickles), sometimes sts thorny or lvs spiny-margined

    31. Lvs palmately lobed

      32. Infl an erect terminal panicle; shrubs; bark not flaking in large plates; lvs peltate, glabrous . . . . **EUPHORBIACEAE** (*Ricinus*) (G7,12)

      32′ Infl axillary, gen drooping, of 1–7 spheric heads along an axis; trees; bark flaking from trunks in large plates; lvs obtuse to cordate at base, densely tomentose when young . . . . . **PLATANACEAE** (G7,10,12)

    31′ Lvs not palmately lobed

      33. Sts and lvs stellate-hairy

        34. Ovary clearly superior, 2–4-lobed; fr a capsule; involucre 0; (sub)shrub with entire lvs or desert shrub with crenate lvs . . . . **EUPHORBIACEAE** (G7,15)

        34′ Ovary inferior but sepals very small and often concealed by bractlets of involucre; fr either an ovoid nut, subtended by a cup-like involucre, or a triangular nut, surrounded in groups of 1–3 by a bur-like involucre; shrub or tree with lvs entire, toothed, or pinnately lobed but not crenate . . . . **FAGACEAE** (G7,11)

    33′ Sts and lvs glabrous or hairy but not stellate

      35. Lvs compound

        36. Petals 0; ovary inferior — fr a nut with a ± dehiscent husk . . . . **JUGLANDACEAE** (G7,11)

        36′ Petals present; ovary or ovaries superior

          37. Lflets with a few gland-tipped teeth near base; ovaries 5–6, joined only at style; fr a cluster of winged achenes . . . . **SIMAROUBACEAE** (*Ailanthus*) (G7,10,20)

          37′ Lflets entire or teeth not gland-tipped; ovary 1; fr of 1 drupe, capsule, or winged achene

            38. Lvs strongly aromatic, dotted with minute, translucent glands; fr a winged achene . . . . **RUTACEAE** (*Ptelea*) (G20)

            38′ Lvs aromatic or not, not dotted with translucent glands; fr a drupe or capsule

              39. Lflets 3–20, lance-linear to ovate, 10–130 mm; fr a drupe . . . . [2]**ANACARDIACEAE** (G7,20,23)

              39′ Lflets 7–33, linear-oblong or narrowly elliptic, 2–10 mm; fr drupe-like when young, capsule-like in age . . . . **BURSERACEAE** (G7,20)

    35′ Lvs simple

      40. Corolla 0; perianth parts 0 or in 1 series

        41. Infl a dense, spheric head; styles long, spreading; frs and knobby-tipped, fleshy sepals ripening as a large multiple fr . . . . **MORACEAE** (*Maclura*) (G7)

        41′ Infl not a spheric head; styles not long, spreading; frs and sepals not forming multiple fr

          42. Fls subtended or surrounded at base by an involucre of tiny scale-like bractlets; fr an ovoid nut, subtended by a cup-like involucre; ovary inferior . . . . **FAGACEAE** (*Quercus*)

          42′ Fls without an involucre; fr an achene, utricle, capsule, or drupe; ovary inferior or superior

            43. Calyx 0; fls or frs each enclosed by a pair of tightly appressed bracts . . . . [3]**CHENOPODIACEAE** (G6,7,8,11,15)

            43′ Calyx present; fls or frs not enclosed by a pair of tightly appressed bracts

              44. Lvs strongly aromatic. . . . . **LAURACEAE** ([*Laurus*]) (G7)

              44′ Lvs not aromatic

45. Lvs fleshy, linear, ± cylindric; calyx fused to ovary, winged in fr . . . . . **SARCOBATACEAE** (G7,11)
45′ Lvs not fleshy, linear to round, flat; calyx free from ovary, not winged in fr
    46. Sepals free; fr a capsule . . . . [3]**PICRODENDRACEAE** (G7,12)
    46′ Sepals fused; fr a drupe . . . . [2]**RHAMNACEAE** (*Rhamnus*) (G7)
40′ Corolla present; perianth parts in 2 series
  47. Shrub with thorny green sts; lvs early deciduous, scale-like to narrowly oblong
    . . . . **SIMAROUBACEAE** (*Castela*) (G7,10,23)
  47′ Tree, shrub, or woody vine, gen not thorny; lvs gen present, sometimes deciduous in winter
    or dry season
    48. Fl parts in 3s . . . . [2]**ERICACEAE** (*Empetrum*) (G7,12,16)
    48′ Fl parts gen in 4s or 5s
      49. Fls in involucred heads, these gen in open to dense clusters; ovary 1 per fl, inferior; calyx
        modified as a white to ± brown pappus . . . . **ASTERACEAE** (G6,7,8,12,14,17)
      49′ Fls not in heads; ovaries 1–5, superior; calyx green, not modified
      50. Hypanthium present
        51. Ovary 1; petals ≤ 1 mm, ± yellow; infl an axillary cyme . . . . [2]**RHAMNACEAE** (*Rhamnus*) (G7)
        51′ Ovaries gen 5; petals 3–6 mm, white; infl a raceme . . . . **ROSACEAE** (*Oemleria*) (G7)
      50′ Hypanthium 0
        52. Fl parts in 4s; lvs spiny-toothed or entire . . . . **AQUIFOLIACEAE** (G7)
        52′ Fl parts gen in 5s; lvs entire to lobed, not spiny-toothed
          53. Petals 1–5 mm; fr a drupe . . . . [2]**ANACARDIACEAE** (G7,20,23)
          53′ Petals 10–15 mm; fr a capsule — seeds sticky . . . . **PITTOSPORACEAE** (G23)

Group 10

Fls with 2 or more free (or apparently free) pistils

1. Perianth 0 or of 1 whorl, gen called sepals even when petal-like
  2. Hypanthium present . . . . [4]**ROSACEAE** (G15,21,22,23)
  2′ Hypanthium 0
   3. Trees; lvs palmately lobed . . . . **PLATANACEAE** (G7,9,12)
   3′ Herbs or woody vines; lvs various
    4. Pl terrestrial or if aquatic with emergent lvs . . . . [2]**RANUNCULACEAE**
    4′ Pl aquatic with submersed or floating lvs
     5. Perianth parts 1–4
      6. Perianth part 1, white, petal-like, enlarging and turning green in age; stamens 6–18; fr follicle,
        4-seeded . . . . **APONOGETONACEAE** (G2)
      6′ Perianth parts 4, green or ± brown, sepal-like; stamens 4; fr achene-like drupe, 1-seeded
        . . . . **POTAMOGETONACEAE** (G2,13)
     5′ Perianth ± 0
      7. Fls bisexual, borne along axis of initially short-peduncled spike, peduncle elongating in fr; stamens
        2; pistils (2)4(8); ovary stalk in fr 12–35 mm; fr ± ovoid, smooth, beak 0.5–1 mm . . . . . **RUPPIACEAE** (G2,8,13)
      7′ Fls unisexual, staminate and pistillate together in sessile, 2-fld axillary cluster; stamen 1; pistils
        (1)4–5(9); ovary stalk in fr 0.1–1.5 mm; fr abaxially curved, minutely roughened, beak 0.7–2 mm
        . . . . **ZANNICHELLIACEAE** (G2,8,13)
1′ Perianth of 2 or more whorls or spirals, the outer gen called sepals, the inner gen called petals
  8. Lvs and often sts thick and fleshy; pistils gen as many as petals or sepals — stamens 1–2 × as many
    . . . . **CRASSULACEAE** (G19)
  8′ Lvs and sts not very thick and fleshy; number of pistils same as or different from number of sepals or petals
   9. Perianth parts in 3s
    10. Lvs compound
     11. Tree with unbranched trunk; lvs 4–7 m, lflets many . . . . **ARECACEAE** (*Phoenix*) (G7,9)
     11′ Ann; lvs ≤ 6 cm, lflets 3–5 — carpels actually united by common style
      . . . . **LIMNANTHACEAE** (*Floerkea*) (G16)
    10′ Lvs simple
     12. Fls solitary — sepals and petals all similar, ± white or ± red-purple; lvs floating and peltate or some
      or all submersed and dissected . . . . **CABOMBACEAE**
     12′ Fls several–many in raceme- to panicle-like infl
      13. Sepals green, petals white (pink) . . . . **ALISMATACEAE** (G2)
      13′ Sepals and petals the same color, ± white to green or ± red
       14. Infl bractless — carpels actually ± fused, separating in fr . . . . **JUNCAGINACEAE** (G2,13,16)
       14′ Infl bracted
        15. Ovules many per carpel; frs erect — carpels actually proximally fused [may appear ± free]
        . . . . **MELANTHIACEAE** (G16)
        15′ Ovules 1–3 per carpel; frs spreading . . . . **SCHEUCHZERIACEAE** (G2)

9′ Perianth parts gen in 4s or 5s or indefinite in number
  16. Stamens > 2 × as many as petal-like perianth segments or > 15
    17. Pls obviously woody; shrubs or trees
      18. Crushed lvs strongly aromatic; lvs opposite . . . . **CALYCANTHACEAE**
      18′ Crushed lvs not aromatic; lvs alternate
        19. Fls solitary; seeds with arils . . . . [2]**CROSSOSOMATACEAE**
        19′ Fls clustered; seeds without arils . . . . [4]**ROSACEAE** (G15,21,22,23)
    17′ Pls herbaceous or slightly woody at base
      20. Hypanthium present . . . . [4]**ROSACEAE** (G15,21,22,23)
      20′ Hypanthium 0
        21. Sepals enlarging and persisting around fr; stamens maturing from fl center toward edge . . . . **PAEONIACEAE**
        21′ Sepals deciduous as fl opens or withering as fr matures; stamens maturing from fl edge toward center
          22. Sepals 2–3(4), falling as fl opens, not petal-like; carpels actually weakly fused in fl, ± separating
            in age . . . . **PAPAVERACEAE** (*Platystemon*)
          22′ Sepals 4–many, present in open fls, sometimes petal-like; carpels gen wholly free . . . [2]**RANUNCULACEAE**
  16′ Stamens 2 × as many as petals or fewer
    23. Petals fused, at least at base
      24. Pl erect, prostrate, or twining; fls gen conspicuous, not concealed by lvs; ovules many per ovary;
        milky latex gen present — carpels proximally free but actually united by common stigma
        . . . . **APOCYNACEAE** (G19)
      24′ Pl prostrate, mat-forming; fls minute, subsessile, axillary, gen concealed by lvs; ovules 1–2 per
        ovary; milky latex 0 . . . . **CONVOLVULACEAE** (*Dichondra*) (G19)
    23′ Petals free but sometimes attached to a disk-like to tubular hypanthium
      25. Hypanthium well developed
        26. Lvs simple, basal or cauline; stipules often 0; veins gen palmate; ovules gen several–many per
          carpel — carpels slightly fused proximally but free most of their length . . . . [2]**SAXIFRAGACEAE** (G20,21,24)
        26′ Lvs simple or compound, gen cauline; stipules gen present; veins often pinnate; ovules gen 1–2 per carpel
          27. Shrub . . . . **CROSSOSOMATACEAE** (*Glossopetalon spinescens* var. *aridum*)
          27′ Herb . . . . [4]**ROSACEAE** (G15,21,22,23)
      25′ Hypanthium 0 or inconspicuous
        28. Pl herbaceous
          29. Ann; ovule 1 per ovary; lvs pinnately divided or compound — carpels actually united by
            common style . . . . **LIMNANTHACEAE** (*Limnanthes*) (G20)
          29′ Per; ovules several–many per ovary; lvs simple and entire or variously toothed to palmately lobed
            or compound . . . . [2]**SAXIFRAGACEAE** (G20,21,24)
        28′ Pl woody
          30. Lvs pinnately compound . . . . **SIMAROUBACEAE** (*Ailanthus*) (G7,9,20)
          30′ Lvs simple, entire, sometimes scale-like
            31. Fls solitary in lf axils; fr a cluster of follicles; lvs elliptic, persistent or drought-deciduous
            . . . . [2]**CROSSOSOMATACEAE**
            31′ Fls in dense panicles; fr cluster of ± dry drupes; lvs scale-like to narrowly oblong,
              early-deciduous — shrub with thorny green sts . . . . **SIMAROUBACEAE** (*Castela*) (G7,9,23)

Group 11

Shrubs or trees; perianth in one whorl or absent; fls unisexual; staminate fls in catkins or catkin-like spikes;
pistillate fls with 1 pistil

1. Pl parasitic on st of woody host pl . . . . **VISCACEAE** (G6,7,8,9,12,14)
1′ Pl free-living, rooted in ground
  2. Lvs fleshy, scale-like or linear to oblanceolate; catkins erect
    3. Lvs scale-like; distal sts ± fleshy, jointed . . . . [2]**CHENOPODIACEAE** (G6,7,8,9,15)
    3′ Lvs linear to oblanceolate, ± cylindric; distal sts not jointed . . . . **SARCOBATACEAE** (G7,9)
  2′ Lvs non-fleshy, ± thin to leathery, with well-developed flat blades, gen wider; catkins erect to drooping
    4. Lvs opposite . . . . **GARRYACEAE** (G7,9)
    4′ Lvs alternate
      5. Lf compound
        6. Pls dioecious, evergreen; staminate catkin-like infl branched — infl resembles catkin when young,
          but is actually a branched panicle . . . . **ANACARDIACEAE** (*Pistacia*) (G7,12)
        6′ Pls monoecious, deciduous; staminate catkin unbranched . . . . **JUGLANDACEAE** (G7,9)
      5′ Lf simple
        7. Fr capsule; seeds 3–many
          8. Ovary 3-lobed; seeds 3, without a tuft of hairs; sts brittle . . . . **EUPHORBIACEAE** (*Acalypha*) (G7,9)
          8′ Ovary unlobed; seeds several–many, each with a tuft of hairs; sts not brittle . . . . **SALICACEAE** (G7,9)
        7′ Fr dry or fleshy, indehiscent; seed gen 1

9. Shrubs of ± saline desert habitats; lvs densely stellate-hairy or scaly-puberulent and ± glabrous in age. . . . . ²**CHENOPODIACEAE** (G6,7,8,9,15)

9′ Shrubs or trees, gen of moister habitats; lvs glabrous or variously hairy

  10. Milky latex present; sepals of pistillate infl enlarged and fleshy in fr, ripening together with the ovaries as a spheric to ellipsoid or ovoid multiple fr . . . . **MORACEAE** (G7,12)

  10′ Milky latex 0; sepals 0 or if present, not becoming fleshy; ovaries ripening as individual drupes, nuts, or winged achenes

    11. Lf blade dotted with minute, bead-like, peltate resin glands, aromatic; stipules 0; staminate catkins short, densely fld; sepals 0; fr a papillate, wax-coated drupe, free of subtending bracts *or* an achene enclosed between a pair of spongy bractlets . . . . **MYRICACEAE** (G7,9)

    11′ Lf blade without peltate glandular hairs, not dotted with minute, bead-like resin glands (exc in some Betulaceae), gen not aromatic; stipules formed but gen early-deciduous; staminate catkins elongated, either ± flexible and drooping or stiff and spike-like; sepals present in staminate catkins (exc *Corylus* in Betulaceae); fr a nut or winged achene

      12. Frs many, small, flattened, winged achenes, clustered in a cone-like fr catkin; pl deciduous; pistillate fl calyx 0. . . . . **BETULACEAE** (G7,9)

      12′ Frs 1–3, each a well-developed nut, subtended or surrounded by a husk-, cup-, or spiny, bur-like involucre; pl deciduous or evergreen; pistillate fl calyx fused to ovary, sepal tips surrounding style base

        13. Fr a nut enclosed in a tubular, husk-like involucre; involucral bractlets 2, papery; staminate catkins drooping, densely fld, bractlets ± concealing fls, sepals 0; deciduous shrub . . . . **BETULACEAE** (*Corylus*)

        13′ Fr a nut in a scaly cup-like involucre or 1–3 nuts enclosed in a spiny, bur-like involucre; involucral bractlets many, scaly or spiny; staminate catkins very slender and drooping or stiff and spike-like, bractlets inconspicuous, sepals present; deciduous or evergreen tree or shrub . . . . **FAGACEAE** (G7,9)

Group 12

Shrubs, trees, or woody vines; perianth in one whorl or absent; fls bisexual or unisexual;
infl various but not catkin or catkin-like spike; pistil 1

1. Pl parasitic on st of woody host pl. . . . . **VISCACEAE** (G6,7,8,9,11,14)

1′ Pl rooted in soil, not parasitic

  2. Infl a dense spike or a spheric head, or fls enclosed within a fleshy, hollow receptacle

    3. Lvs pinnately compound or modified into linear to oblong or narrowly elliptic, entire, flattened axes without lflets; fr a legume — perianth actually of 2 whorls, the calyx easily overlooked. **FABACEAE** (Mimosoideae)

    3′ Lvs simple, ovate or wider in overall outline, entire, toothed, or palmately lobed; fr not a legume

      4. Infl a hollow receptacle enclosing many tiny fls; ripened infl an externally smooth, fleshy multiple fr [a fig]; milky latex present . . . . **MORACEAE** (*Ficus*)

      4′ Infl a short, dense spike or a head; ripened pistillate infl an externally roughened spike or spheric head; milky latex present or 0

        5. Perianth parts fused into a tube; fls bisexual. . . . . ²**ASTERACEAE** (G6,7,8,9,14,17)

        5′ Perianth parts free; fls unisexual

          6. Lf-blade entire to toothed, sometimes irregularly few-lobed, faces glabrous or sparsely short-hairy; ripened pistillate infl a fleshy spike or head; milky latex present; petiole base not expanded, not covering axillary bud; stipules membranous, free, gen early-deciduous . . . . **MORACEAE** (G7,11)

          6′ Lf-blade ± deeply palmately lobed, the lobes entire or ± toothed, faces densely stellate-tomentose when young; ripened pistillate infl a dry head of long-stiff-hairy achenes; milky latex 0; petiole base expanded, covering axillary bud; stipules large and green, fused around st, persistent or deciduous — pistils actually > 1 . . . . **PLATANACEAE** (G7,9,10)

2′ Infl not a dense spike or a spheric head, fls not enclosed in a fleshy, hollow receptacle

  7. Fls unisexual, pl monoecious; pistillate infl 1–5-fld, maturing as a bur with spiny, knob-like, or membranous bract tips; staminate infl a head. . . . . **ASTERACEAE** (*Ambrosia*)

  7′ Fls bisexual, or if unisexual, pistillate infl not 1–5-fld and maturing as a bur with spiny, knob-like, or membranous bract tips; staminate infl various

    8. Ovary inferior, partly so, or appearing so

      9. Herbage covered with silvery scales; fr drupe-like — ovary actually superior, ripening as an achene, tightly enclosed by hypanthium base that becomes fleshy at maturity. . . . . **ELAEAGNACEAE**

      9′ Herbage without silvery scales; fr various

        10. Infl unit an involucred head, superficially resembling a single fl . . . . ²**ASTERACEAE** (G6,7,8,9,14,17)

        10′ Infl open or fls solitary or paired

          11. Trees, evergreen; stamens many; lvs aromatic — perianth early-deciduous as cap . . . . **MYRTACEAE** (*Eucalyptus*)

          11′ Shrubs or woody vines, evergreen or deciduous; stamens 4–12; lvs gen not aromatic

            12. Perianth rotate, lobes >> tube; lvs whorled; fr 2-lobed . . . . **RUBIACEAE** (*Galium*) (G6,7,8,9)

12′ Perianth tubular, lobes << tube; lvs alternate or opposite; fr not lobed

   13. Lvs alternate; perianth tube U-shaped; twining vine . . . . **ARISTOLOCHIACEAE** (*Aristolochia*)

   13′ Lvs opposite; perianth tube straight; erect shrub . . . . **CAPRIFOLIACEAE** (*Lonicera involucrata*)

8′ Ovary superior

  14. Lvs opposite or whorled

   15. Fr winged, indehiscent; lvs gen lobed or compound

    16. Style evident, stigma 1, entire or slightly 2-lobed; fr 1-winged, not splitting; lvs pinnately

     compound (lflet gen 1 in *Fraxinus anomala*) . . . . **OLEACEAE** (*Fraxinus*) (G7,9,16)

    16′ Style very short, stigmas 2(3), elongate; fr 2(3)-winged, gen splitting into mericarps; lvs

     palmately lobed or compound or pinnately compound . . . . **SAPINDACEAE** (*Acer*) (G7,23)

   15′ Fr not winged; lvs simple, entire or toothed

    17. Fls bisexual; stamens 20–25(30) . . . . **ROSACEAE** (*Coleogyne*)

    17′ Fls unisexual; stamens gen 3–12

     18. Lf covered with silvery scales . . . . **ELAEAGNACEAE** (*Shepherdia*) (G7,9)

     18′ Lf glabrous or ± hairy

      19. Lvs needle-like, abaxially grooved . . . . [2]**ERICACEAE** (*Empetrum*) (G7,9,16)

      19′ Lvs not needle-like, not grooved

       20. Staminate and pistillate fls in sessile umbels; style 1, stigma entire or 2-lobed

        . . . . **OLEACEAE** (*Forestiera*) (G7,9)

       20′ Staminate fls clustered, pistillate fls gen solitary; styles or stigmas (2)3–5

        21. Staminate infl a raceme or panicle, fls clearly stalked; fr 3–4-lobed . . . . . **PICRODENDRACEAE** (G7,9)

        21′ Staminate infl head-like, fls subsessile; fr unlobed . . . . **SIMMONDSIACEAE** (G7,9)

14′ Lvs alternate

  22. Lvs pinnately compound or very deeply pinnately divided

   23. Perianth parts ± brown-green, inconspicuous, bract-like; fls unisexual, pls dioecious

    . . . . **ANACARDIACEAE** (*Pistacia*) (G7,11)

   23′ Perianth parts petal-like, conspicuous; fls bisexual

    24. Shrub; fls solitary; perianth radial, parts 6, white — sepals actually present in bud, but

     early-deciduous, 0 in open fl . . . . **PAPAVERACEAE** (*Romneya*) (G22)

    24′ Tree; fls in raceme; perianth bilateral, parts 4, yellow or orange, proximally ± red . . . . **PROTEACEAE**

22′ Lvs simple, sometimes lobed

  25. Infl fl-like with a cup-like involucre bearing colored, often petal-like nectaries; each stamen

   [actually a staminate fl] with a jointed stalk [composed of a filament and a pedicel]; pistil

   [actually pistillate fl] 3-lobed, gen exserted from involucre; milky latex present

   . . . . **EUPHORBIACEAE** (*Euphorbia*)

  25′ Infls various, but not fl-like; stamens gen > 1 per fl, the filaments not jointed; pistils various;

   milky latex gen 0

   26. Anthers opening by pores with hinged lids; lf strongly aromatic — 2-whorled nature of perianth

    may not be evident in open fls. . . . . **LAURACEAE** (G16)

   26′ Anthers opening by slits; lf gen not aromatic

    27. Lvs fleshy, cylindric to ± flat . . . . **CHENOPODIACEAE** (*Suaeda*)

    27′ Lvs not fleshy, gen flat

     28. Lf palmately veined and lobed

      29. Woody vine; tendrils present — calyx 0 or reduced to an entire or slightly lobed rim, easily

       overlooked . . . . **VITACEAE** (*Vitis*) (G7,9,23)

      29′ Erect or spreading shrub; tendrils 0

       30. Lf peltate, glabrous; fls unisexual, perianth parts ± green, free; stamens very many, the

        filaments irregularly joined in clusters . . . . **EUPHORBIACEAE** (*Ricinus*) (G7,9)

       30′ Lf not peltate, stellate-hairy; fls bisexual, perianth parts yellow or orange to ± red,

        petal-like; stamens 5, filaments fused in 1 group . . . . **MALVACEAE** (*Fremontodendron*) (G19)

     28′ Lf pinnately veined or only midvein evident, gen unlobed

      31. Perianth parts 2–10 cm; ovules many — sepals actually present in bud, but early-deciduous,

       0 in open fl . . . . **PAPAVERACEAE** (Papaveroideae) (G15,20,22,24)

      31′ Perianth parts gen < 1.5 cm; ovules 1–2 per chamber

       32. Pl covered with ± gray or silvery scales

        33. Fls tubular in proximal half, some or all bisexual; tree or large shrub; lvs linear-lanceolate

         to oblong . . . . **ELAEAGNACEAE** (*Elaeagnus*)

        33′ Fls not tubular, all unisexual, pl dioecious; shrub; lvs various

         34. Ovary of pistillate fl enclosed by 2 appressed bracts; lf surface scales powder-like, not

          stellate . . . . **CHENOPODIACEAE** (*Atriplex*) (G6,8)

         34′ Ovary of pistillate fl subtended by 5 sepals; lf surface scales stellate

          . . . . **EUPHORBIACEAE** (*Croton*) (G6,8)

       32′ Pl without scales

        35. Perianth parts 6— 2-whorled nature of perianth may not be evident in open fls

36. Fls solitary, axillary, unisexual; fr a berry . . . . . : . . . . [2]**ERICACEAE** (*Empetrum*) (G7,9,16)

36′ Fls in involucred umbels, bisexual; fr an achene . . . . **POLYGONACEAE** (*Eriogonum*) (G6,8)

35′ Perianth parts gen 4 or 5

  37. Stamens = in number to and alternate sepals . . . . **RHAMNACEAE** (G21,23)

  37′ Stamens = in number to and opposite sepals or > in number than sepals

    38. Style and stigma 1

      39. Style feathery with long, stiff hairs; fr an achene surrounded by a tubular hypanthium; branches stiff . . . . **ROSACEAE** (*Cercocarpus*)

      39′ Style glabrous; fr an exposed drupe; branches pliable — corolla 0 or reduced to minute scales, easily overlooked . . . . **THYMELAEACEAE**

    38′ Styles or stigmas 2–4

      40. Fr a drupe; fls staminate and bisexual . . . . **CANNABACEAE** (*Celtis*) (G7)

      40′ Fr a winged nutlet or splitting into 1–2-seeded segments that release seeds; fls unisexual or bisexual

        41. Tree; fr a flattened, round to ovate, 2-winged nutlet . . . . **ULMACEAE**

        41′ Shrub; fr splitting into segments

          42. Lvs elliptic, crenate; fr segments 1-seeded . . . . **EUPHORBIACEAE** (*Bernardia*)

          42′ Lvs oblanceolate to obovate, entire; fr segments 1–2-seeded

          . . . . **PICRODENDRACEAE** (*Tetracoccus hallii*)

Group 13

Herbs; lf venation gen parallel or lvs and st not differentiated; perianth in 1 whorl or absent, sometimes reduced to bristles or minute scales; pistil 1

1. Pl terrestrial, or if aquatic, strongly emergent

  2. Fls all unisexual; infl either a dense erect spike of staminate fls directly distal to a spike of pistillate fls, or a group of several spheric unisexual heads attached to a zigzag main axis, spikes or heads 1–4 cm diam . . . . **TYPHACEAE**

  2′ Fls bisexual or some or all unisexual; infls of various types, gen < 1 cm diam if spike or head

    3. Fr a berry or capsule with 3–many seeds or breaking into 1-seeded mericarps; perianth segments 4–6, well developed — actually in two whorls, but this may not be evident in open fls

      4. Lvs with expanded blades; perianth parts petal-like, white or cream . . . . **RUSCACEAE** (*Maianthemum*)

      4′ Lvs narrowly linear; perianth parts scale-like, green to brown or ± purple-black or tinged ± red-purple

        5. Infl of 1–many fls, variously clustered, subtended individually and/or in small groups by bracts; style 1, stigmas (2)3, linear, spreading; fr a capsule . . . . **JUNCACEAE** (G16)

        5′ Infl a bractless raceme; stigmas 3 or 6, ± sessile, papillate or plumose; fr breaking into mericarps . . . . **JUNCAGINACEAE** (G2,10,16)

    3′ Fr an achene or grain, 1-seeded; perianth inconspicuous or 0 or modified as ± conspicuous bristles

      6. Infl a bractless emergent spike of bisexual and sometimes staminate fls, plus 2 submerged, long-styled, sessile pistillate fls enclosed by lf sheath at base of peduncle; stamen 1; stigma 1 . . . . [2]**JUNCAGINACEAE** (*Triglochin scilloides*) (G8)

      6′ Infl 1–many spikelets with fls individually subtended or enclosed by scale-like bracts; stamens (1)3(6); stigmas 2–3

        7. St lfless, the lvs all basal, sometimes reduced to bladeless sheaths . . . . [2]**CYPERACEAE** (G2,8)

        7′ St at least proximally lfy

          8. Sts gen 3-angled (cylindric), gen solid; nodes not swollen; lvs gen 3-ranked; spikelet gen without 2 bractlets at base that do not directly subtend fl parts; each fl subtended by 1 bract (in *Carex* and *Kobresia* a second, gen hollow, ± flask-shaped bract [perigynium] surrounds or encloses each pistillate fl); perianth of gen inconspicuous to long-exserted bristles or 0; fr an achene . [2]**CYPERACEAE** (G2,8)

          8′ Sts cylindric, the internodes gen hollow; nodes swollen, knot-like; lvs 2-ranked or not in obvious ranks; each spikelet gen with 2 bractlets [glumes] at base that do not directly subtend fl parts; each fl gen enclosed by 2 additional bractlets [lemma and palea]; perianth of 2 tiny scales or 0; fr a grain . . . . **POACEAE** (G2,5,6,8)

1′ Pl aquatic, submersed or floating

9. Pl raft-like, free-floating or stranded on shore or drying mud, either pl body 0.4–10 mm, not differentiated into sts and lvs with roots unbranched or 0, or pl with floating rosette of widely wedge-shaped, velvety-hairy lvs and many fine, branched roots . . . . **ARACEAE** (G2,6,8,15)

9′ Pl not raft-like, gen rooted, >> 1 cm, differentiated into sts and lvs, not forming floating rosettes; roots gen present, often branched; lvs linear, glabrous

  10. Lvs opposite or whorled (occ alternate at some nodes), all cauline

    11. Lf coarsely to finely toothed (teeth sometimes appearing ± 0 to naked eye); stipules 0 (lf base may have ear-like lobes); fr without a stiff beak . . . . **HYDROCHARITACEAE** (*Najas*) (G2,8)

    11′ Lf entire; stipules fused to blade or free, membranous, entire; fr with a stiff beak . . . . **ZANNICHELLIACEAE** (G2,8,10)

10′ Lvs alternate, basal and/or cauline
 12. Lvs all ± basal, sheaths overlapping and blades emerging at different levels, but st internodes very
    short, not evident — pistillate fls paired, axillary, submersed, style 6–20 cm, thread-like, stigma
    floating, head-like; staminate and bisexual fls in peduncled spike
    . . . . [2]**JUNCAGINACEAE** (*Triglochin scilloides*) (G8)
 12′ Lvs basal and cauline or all cauline, st internodes evident
  13. Infl a dense, spheric head. . . . . **TYPHACEAE** (*Sparganium*) (G2,6,8)
  13′ Infl a spike or fls solitary
   14. Infl a flattened spike enwrapped by a sheathing bract; subtidal or intertidal seawater or brackish
     water of bays and estuaries. . . . . **ZOSTERACEAE** (G2,6,8)
   14′ Infl a cylindric spike, short- to long-stalked; freshwater, brackish, or inland saltwater habitat —
     carpels actually free but easily misinterpreted
    15. Sepals 4; stamens 4; frs sessile; fresh water in lakes, streams, ponds. . . . . . **POTAMOGETONACEAE** (G2,10)
    15′ Sepals 0; stamens 2; frs on thread-like stalks; fresh to brackish or salt water in coastal to inland
     areas . . . . **RUPPIACEAE** (G2,8,10)

Group 14

Herbs or subshrubs; lf venation gen pinnate or palmate; perianth 0 or in 1 whorl; ovary 1, inferior

1. Pl parasite on st of woody host
 2. Fls of parasite emerging directly from st of host, parasite otherwise growing within tissues of host; lvs 0;
   on *Psorothamnus* . . . . **APODANTHACEAE**
 2′ Fls of parasite borne in infls on branches of parasite; lvs scale-like or with expanded blades; on various
   woody hosts [not *Psorothamnus*]. . . . . **VISCACEAE** (G6,7,8,9,11,12)
1′ Pl free-living and rooted in ground or root-parasite
 3. Lf blade 1–2+ m, covered with prickles — perianth actually of 2 whorls but petals very small, easily
   overlooked. . . . . **GUNNERACEAE** (G16)
 3′ Lf blade << 1 m, not prickly
  4. Lvs [at least proximal] whorled, all cauline
   5. Pl terrestrial . . . . [2]**RUBIACEAE** (G17)
   5′ Pl ± aquatic
    6. Lvs entire . . . . **PLANTAGINACEAE** (*Hippuris*) (G2,15)
    6′ Lvs toothed to deeply divided
     7. Lvs repeatedly forked. . . . . **CERATOPHYLLACEAE** (G2,6,8,15)
     7′ Lvs [at least submersed] pinnately divided . . . . **HALORAGACEAE** (G2,6,8,20)
  4′ Lvs alternate or opposite (or sometimes some or all basal)
   8. Infl bracts large, petal-like; perianth 0 — infl a fl-like spike . . . . **SAURURACEAE**
   8′ Infl bracts not at all petal-like; perianth present
    9. Perianth parts free or nearly so
     10. Lvs glabrous, round-peltate or -reniform; pl aquatic with stolons or rhizomes floating in shallow
       water or rooted in wet soil . . . . **ARALIACEAE** (*Hydrocotyle*) (G2,21)
     10′ Lvs glabrous or hairy, lf shape various; pl terrestrial if lvs round-reniform
      11. Infl a simple or compound umbel. . . . . **APIACEAE** (G2,6,8,20,21)
      11′ Infl various or fls solitary
       12. Pl aquatic, floating, stranded on mud, or rooted in bottom sediments
        13. Sepals, stamens 5; pedicel 1–4(8) mm; lvs alternate (sometimes ± opposite or whorled), blade
          linear . . . . **CAMPANULACEAE** (*Howellia*)
        13′ Sepals, stamens 4; pedicel 0–0.5 mm; lvs opposite, blade narrowly elliptic to ± ovate
          . . . . **ONAGRACEAE** (*Ludwigia palustris*)
       12′ Pl terrestrial or sometimes growing in damp soil
        14. Lvs all basal. . . . . **SAXIFRAGACEAE** (*Heuchera cylindrica*)
        14′ Lvs cauline, at least in part
         15. Fr an achene; sepals 3–5; lvs alternate, ± entire, petiole ± = blade . . . . **CHENOPODIACEAE** (*Beta*)
         15′ Fr a capsule or ± nut-like [hard, indehiscent, 3–9-seeded with 2–5-horns]; sepals 3–9; lvs
           alternate or opposite, entire to pinnately lobed, petiole often < blade
          16. Per; sts erect; lvs gen unequally pinnately lobed; styles long, forked . . . . **DATISCACEAE** (G6)
          16′ Ann; sts prostrate to erect; lvs entire or toothed; styles short, not forked
           17. Styles 2–9; lvs ± fleshy, often ± papillate. . . . . **AIZOACEAE** (G15,17,21)
           17′ Style 1; lvs not fleshy, not papillate . . . . **CAMPANULACEAE** (G2,5,17)
    9′ Perianth parts fused
     18. Infl of involucred heads or umbels
      19. Ovary truly inferior; anthers fused around style or, if free, the fls unisexual; stigma lobes 2
        . . . . **ASTERACEAE** (G6,7,8,9,12,17)
      19′ Ovary actually superior, but appearing inferior [tightly enclosed by perianth base]; anthers free;
        stigma unlobed . . . . [2]**NYCTAGINACEAE** (G5,15,17,19)

18′ Infl not of involucred heads or umbels, fls solitary or variously clustered
  20. Fls large, solitary, red to maroon or brown; perianth lobes 3 — lvs cordate to reniform,
     long-petioled . . . . **ARISTOLOCHIACEAE** (*Asarum*)
  20′ Fls gen ± small, gen clustered, not red to maroon or brown; perianth lobes (3)4–5
    21. Lvs alternate
      22. Tendrils 0; lvs entire; fls bisexual . . . . **COMANDRACEAE**
      22′ Tendrils present; lvs palmately lobed; fls unisexual . . . . **CUCURBITACEAE** (G6,8,17,21)
    21′ Lvs opposite
      23. Fls bilateral . . . . **VALERIANACEAE** (G17)
      23′ Fls ± radial
        24. Perianth lobes 5; ovary actually superior [tightly enwrapped by hardened or winged calyx
          base]; lvs of a pair equal or unequal
          25. Calyx green, not at all corolla-like, very inconspicuous; sepals erect, free or nearly so above
            the level of the ovary, thickened and linear, erect; lvs narrowly linear, needle-like, gen 1–2 cm
            or shorter, gen those of a pair equal . . . . **CARYOPHYLLACEAE** (*Scleranthus*) (G15)
          25′ Calyx corolla-like in color and texture, often very showy; sepals fused well above level of
            ovary, calyx lobes wide and thin, ± spreading; lvs linear to ovate, often > 2 cm, those of a pair
            often unequal . . . . ²**NYCTAGINACEAE** (G5,15,17,19)
        24′ Perianth lobes 3–4; ovary truly inferior; lvs of a pair equal
          26. Stamens 3–4; lvs entire or teeth minute, prickle-like. . . . . ²**RUBIACEAE** (G17)
          26′ Stamens 8; lvs with rounded teeth . . . . **SAXIFRAGACEAE** (*Chrysosplenium*)

Group 15

Herbs, subshrubs, or herbaceous vines; lf venation gen pinnate or palmate;
perianth in 1 whorl or 0; pistil 1, ovary superior

1. Lvs opposite or whorled
  2. Pl aquatic, weak-stemmed; submersed, floating, or stranded on mud
    3. Lvs opposite; ovary lobed or unlobed; seeds 4–many
      4. Ovary unlobed, many-seeded . . . . **LYTHRACEAE** (*Lythrum portula*)
      4′ Ovary 2- or 4-lobed, 4-seeded. . . . . **PLANTAGINACEAE** (*Callitriche*) (G2,8)
    3′ Lvs whorled; ovary unlobed; seed 1
      5. Lvs dissected into narrow lobes. . . . . **CERATOPHYLLACEAE** (G2,6,8,14)
      5′ Lvs entire . . . . **PLANTAGINACEAE** (*Hippuris*) (G2,14)
2′ Pl terrestrial, sometimes in damp soil, or in areas flooded at high tide
  6. Ovary chambers 2 or more (doubtful cases should be keyed both ways)
    7. Fls unisexual
      8. Style 0, stigma sessile; lvs ± cylindric or 3-angled in ×-section, fleshy; milky latex 0; pls of coastal
        salt-marshes . . . . **BATACEAE** (G6,8,24)
      8′ Styles or stigmas 2–3, free; lvs thin; milky latex present or 0; pls of various habitats but not salt marsh
        . . . . ³**EUPHORBIACEAE** (G7,9)
    7′ Fls bisexual
      9. Lvs opposite, equal; fr circumscissile — 1/2-inferior ovary may appear superior. . . . . . . **AIZOACEAE** (*Sesuvium*)
      9′ Lvs whorled, often unequal; fr loculicidal . . . . **MOLLUGINACEAE**
  6′ Ovary chamber 1
    10. Ovules 2–many; fr a capsule
      11. Perianth parts 6(8) — sepals actually present in bud, but early-deciduous, 0 in open fl
      . . . . ²**PAPAVERACEAE** (Papaveroideae) (G12,20,22,24)
    11′ Perianth parts (4)5
      12. Sepals free; fls gen in cymes . . . . ²**CARYOPHYLLACEAE** (G6,8,24)
      12′ Sepals fused; fls solitary in lf axils
        13. Lvs of a pair unequal; stipules present; fr circumscissile . . . . **AIZOACEAE** (G14,17,21)
        13′ Lvs of a pair equal; stipules 0; fr 5-valved. . . . . **MYRSINACEAE** (*Glaux*)
    10′ Ovule 1; fr an achene, 1-seeded capsule, or utricle
      14. Style 1, undivided or stigma 1, sessile
        15. Sepals petal-like, fused into a tube; lvs entire or slightly lobed, stinging hairs 0
        . . . . **NYCTAGINACEAE** (G5,14,17,19)
        15′ Sepals green, small, inconspicuous; lvs toothed or entire, sometimes with stinging hairs
          16. Herbage densely covered with branched hairs. ²**AMARANTHACEAE** (*Tidestromia*)
          16′ Herbage glabrous or ± hairy, hairs not branched (stinging hairs present in *Hesperocnide*, *Urtica*)
          . . . . ²**URTICACEAE** (G6,8)
    14′ Styles or style branches 2 or more
      17. Lvs palmately compound or deeply lobed . . . . **CANNABACEAE** (G6,8)
      17′ Lvs simple, entire or toothed

18. Fr a 3-angled achene; perianth parts 6 [actually in 2 whorls of 3, but this may not be evident in
   open fls] . . . . **³POLYGONACEAE** (G2,6,8,16,19)
18′ Fr not 3-angled; perianth parts ≤ 5
 19. Lvs with stipules
  20. Lvs petioled, fleshy; fr a circumscissile capsule. . . . . **AIZOACEAE** (*Trianthema*)
  20′ Lvs sessile, not fleshy; fr 1-seeded, indehiscent. . . . . **²CARYOPHYLLACEAE** (G6,8,24)
 19′ Lvs without stipules
  21. Lvs reduced to fleshy scales; infl a fleshy spike . . . . **³CHENOPODIACEAE** (G6,7,8,9,11)
  21′ Lvs linear to ovate; infls various or fls solitary
   22. Sepals erect, linear, fused below and hardened around ovary; stamens fused to calyx; slender
      ann, gen ≤ 15 cm; lvs narrowly linear, ≤ 15 mm . . . . **CARYOPHYLLACEAE** (*Scleranthus*) (G14)
   22′ Sepals erect or spreading, free, not hardened around ovary (sepals gen 0 in pistillate fls of
      Atriplex, which has a pair of ± hardened bractlets around fr); stamens free from calyx; pls often
      larger, ann or per; lvs various
    23. Bracts subtending fls dry, scarious; pls neither fleshy nor with powdery or beaded surface;
       habitats gen not saline . . . . **²AMARANTHACEAE**
    23′ Bracts subtending fls lf-like or fleshy; pls often fleshy or with powdery or beaded surface;
       habitats often ± saline
     24. Sepals pink or white, papery; pl from rhizome. . . . . **AMARANTHACEAE** (*Nitrophila*)
     24′ Sepals green or ± red-tinged, herbaceous; rhizome 0. . . . . **³CHENOPODIACEAE** (G6,7,8,9,11)
1′ Lvs alternate or all basal
25. Pl aquatic, raft-like, floating on water surface or stranded on shoreline — lvs in rosettes, wedge-shaped,
   sessile, velvety; fls enclosed by small, sheathing bracts . . . . **ARACEAE** ([*Pistia*]) (G2)
25′ Pl terrestrial or rooted in shallow water
 26. Infl a ± fleshy spike enclosed by a large, basally sheathing, often petal-like bract; petiole stout
    . . . . **ARACEAE** (G2,6,8,13)
 26′ Infl various but not a fleshy spike subtended by a sheathing bract; lvs sessile or ± slender-petioled
  27. Lvs with stipules, gen well developed
   28. Style 1, unbranched, or stigma sessile
    29. Ovule 1; fr an achene . . . . **ROSACEAE** (G10,20,21,22,23)
    29′ Ovules several to many; fr a capsule — late season cleistogamous fls formed in most *Viola* spp.
       . . . . **VIOLACEAE** (G5,20,24)
   28′ Styles or stigmas 2 or more
    30. Lvs palmately compound . . . . **CANNABACEAE** (*Cannabis*) (G6,8)
    30′ Lvs simple, entire to lobed
     31. Pl a vine, sts twining
      32. Lvs toothed, at least the proximal also palmately lobed . . . . **CANNABACEAE** (*Humulus*) (G6,8)
      32′ Lvs entire. . . . . **POLYGONACEAE** (*Fallopia*)
     31′ Pl prostrate to erect, sts not twining
      33. Stipules fused into a membranous sheath [ocrea] around st, the sheath sometimes torn or
         shredded as branches develop. . . . . **³POLYGONACEAE** (G2,6,8,16,19)
      33′ Stipules free, not fused around st, not torn or shredded as branches develop
       34. Ann; sts prostrate; stinging hairs 0; fls bisexual . . . . **CARYOPHYLLACEAE** (*Herniaria*)
       34′ Per; sts spreading to erect; stinging hairs present; fls unisexual. . . . . **EUPHORBIACEAE** (*Tragia*) (G6)
  27′ Lvs without stipules, sometimes reduced to scale-like bracts
   35. Infl fl-like with a cup-like involucre bearing 1–4(5) nectary glands that sometimes have flattened,
      white to ± pink, petal-like, marginal appendages — stamens [actually staminate fls] each with
      a jointed stalk [composed of a filament and a pedicel], pistillate fl 1, gen exserted from
      involucre, ovary 3-lobed . . . . **³EUPHORBIACEAE** (G7,9)
   35′ Infl not fl-like or with a cup-like, nectary-gland-bearing involucre
    36. Perianth parts 4 or 6, ± petal-like
     37. Perianth parts 4 or 6, free; stigmas 1–many, gen sessile; fr a capsule — sepals actually present in
        bud, but early-deciduous, 0 in open fl. . . . . **²PAPAVERACEAE** (*Papaveroideae*) (G12,20,22,24)
     37′ Perianth parts 6, proximally fused; stigmas 3, at tips of slender styles; fr a 3-angled achene
        . . . . **³POLYGONACEAE** (G2,6,8,16,19)
    36′ Perianth parts ≤ 5, sometimes 0, gen not petal-like
     38. Lvs all basal, palmately compound; lflets 3; perianth 0. . . . . **BERBERIDACEAE** (*Achlys*)
     38′ Lvs cauline, simple; perianth gen present
      39. Ovule 1; infls various
       40. Herbage ± densely covered with branched hairs
        41. Fls bisexual; hairs soft, irregularly branched, branches all short; fr indehiscent — lf arrange-
           ment variable, often difficult to interpret . . . . **²AMARANTHACEAE** (*Tidestromia*)
        41′ Fls unisexual; hairs harshly stellate, central branches often long, spreading, bristle-like; fr
           dehiscent . . . . **EUPHORBIACEAE** (*Croton setiger*)

40′ Herbage glabrous or ± hairy, hairs not branched
  42. Stigma 1 . . . . [2]**URTICACEAE** (G6,8)
  42′ Stigmas 2–4
    43. Bracts subtending fls scarious-margined or dry and scarious throughout . . . . [2]**AMARANTHACEAE**
    43′ Bracts subtending fls 0 or not scarious or scarious-margined . . . . . . [3]**CHENOPODIACEAE** (G6,7,8,9,11)
39′ Ovules 2–many; infl gen a raceme (spike, axillary cluster)
  44. Fls unisexual . . . . [3]**EUPHORBIACEAE** (G7,9)
  44′ Fls bisexual
    45. Pl red- and white-striped, non-photosynthetic, green pigment 0 . . . . **ERICACEAE** (*Allotropa*)
    45′ Pl green and photosynthetic
      46. Sepals 4 . . . . **BRASSICACEAE** (G2,20,24)
      46′ Sepals 5
        47. Sts ± erect, ≤ 3(7) m; lf 6–36 cm; fls many in spike or raceme; fr a berry, chambers 5–12;
             perianth radial . . . . **PHYTOLACCACEAE**
        47′ Sts ± decumbent, 0.5–3.5 dm; lf ≤ 6 cm; fls 1–few in raceme; fr a capsule; chambers 2;
             perianth bilateral — some fls often ± cleistogamous . . . . **POLYGALACEAE** (G18,22,23,24)

Group 16

Perianth parts in 2 or more whorls, parts 2 or 3 per whorl; pistil 1

1. Ovary fully inferior
 2. Lf parallel veined
  3. Lvs fleshy, toothed, spine-tipped . . . . **AGAVACEAE** (*Agave*)
  3′ Lvs not fleshy, entire, not spine-tipped
   4. Fertile stamens 1–2, filament(s) and anther(s) wholly fused above ovary to stigma and style, forming a
       column; ovary with a half twist; ovules many, microscopic; seeds dust-like . . . . **ORCHIDACEAE**
   4′ Fertile stamens 3 or more, not fused to stigma and style; ovary straight; ovules not microscopic; seeds
       well developed
    5. Stamens 6; lvs not 2-ranked and folded (0 at fl in *Amaryllis*); adaxial and abaxial faces both visible
        . . . . **AMARYLLIDACEAE**
    5′ Stamens 3; lvs 2-ranked, gen folded lengthwise and apparently attached "edge-on", only abaxial face
        of lf visible . . . . **IRIDACEAE**
 2′ Lf pinnately or palmately veined
  6. Lvs basal, palmately veined and lobed, prickly — petiole 1–1.5 m, blade 1–2+ m . . . . **GUNNERACEAE** (G14)
  6′ Lvs cauline, pinnately veined, entire or toothed, smooth
   7. Lvs opposite; petals 2, 1–1.5 mm, white; fr indehiscent, with hooked hairs . . . . **ONAGRACEAE** (*Circaea*)
   7′ Lvs alternate; petals 3, yellow; fr dehiscent, without hooked hairs
    8. Petals 0.8–2.3 mm; lvs narrowly oblanceolate, 1–3 cm. . . . . **ONAGRACEAE** (*Neoholmgrenia*)
    8′ Petals ≥ 30 mm; lvs ± elliptic, 20–55 cm. . . . . **[ZINGIBERACEAE]**
1′ Ovary superior or partially inferior
 9. St nodes with small, dry, sometimes spiny-spurred, scale-like lvs that subtend 1–20 thread-like, needle-
     like, or ovate, lf-like branchlets — erect to sprawling herb, sub-shrub, or twining vine . . . . **ASPARAGACEAE**
 9′ St nodes without small, dry, scale-like lvs that subtend lf-like branches
  10. Lvs needle-like, abaxially grooved; low shrub . . . . **ERICACEAE** (*Empetrum*) (G7,9,12)
  10′ Lvs gen not needle-like, gen not abaxially grooved; herb to tree
   11. Pl submersed aquatic; lvs very thin, 1-veined, opposite or whorled; fls gen unisexual
        . . . . [2]**HYDROCHARITACEAE** (G2,6,8)
   11′ Pl terrestrial, or if aquatic, lvs not very thin, 1-veined, and opposite or whorled
    12. Lf venation pinnate or palmate, smaller veins gen forming a network (occ only midvein evident)
     13. Tendrils present; vine; lvs ovate — fls unisexual; pls dioecious. . . . . **SMILACACEAE** (G6,7,8,9)
     13′ Tendrils 0; herb, shrub, or tree (occ vine with twining st); lvs various
      14. Lvs compound or deeply lobed
       15. Herb
        16. Per; lvs basal, 2–3-ternate; petals in 2 tight whorls of 3; sepals in 4–5 whorls of 3
             . . . . **BERBERIDACEAE** (*Vancouveria*) (G20)
        16′ Ann; lvs cauline, 1-ternate or 1-odd-pinnate; petals 3; sepals 3 . . . . . **LIMNANTHACEAE** (*Floerkea*) (G10)
       15′ Shrub or tree
        17. Trunk stout, unbranched, crowned by a tight terminal rosette of very large lvs . . . . **ARECACEAE**
        17′ Trunk(s) branched, with lvs < 50 cm, distributed on twigs
         18. Lvs alternate; lflets spiny-toothed; fr berry . . . . **BERBERIDACEAE** (*Berberis*) (G20)
         18′ Lvs opposite; lflets serrate; fr winged achene. . . . . **OLEACEAE** (*Fraxinus*) (G7,9,12)
      14′ Lvs simple
       19. Pl woody
        20. Stamens many . . . . **HYDRANGEACEAE** (*Carpenteria*) (G21,22)
        20′ Stamens 2–9
         21. Lf blade gland-dotted, aromatic; anthers dehiscing by uplifted lids; fr fleshy . . . . **LAURACEAE** (G12)

  21′ Lf blade not gland-dotted, not aromatic; anthers dehiscing by slits; fr dry

    22. Inner perianth parts 2, petal-like, outer minutely 4-lobed, ± green; fr winged, 22–30 mm
. . . . **OLEACEAE** (*Fraxinus parryi*)

    22′ Inner and outer perianth parts in 3s, all petal-like; fr 3-angled achene, < 5 mm
. . . . ²**POLYGONACEAE** (G2,6,8,15,19)

  19′ Pl herbaceous

    23. Fl bilateral, proximal portion of calyx with a protruding, hollow spur . . . . **BALSAMINACEAE** (G24)

    23′ Fl radial, calyx without a spur

      24. Cauline lvs in a whorl of 3 subtending 1 large terminal fl . . . . ²**MELANTHIACEAE** (G10)

      24′ Cauline lvs alternate, opposite, or whorled, sometimes reduced to bracts; fls >> 1, small

        25. Fr achene, (2)3-angled — inner and outer perianth gen similar, both whorls petal-like or both
sepal-like (inner perianth parts much enlarged in fr in *Rumex*) . . . . . . ²**POLYGONACEAE** (G2,6,8,15,19)

        25′ Fr capsule

          26. Lvs not fleshy; pl ± aquatic; fls axillary; outer perianth parts 2–3, ± = inner; stigmas 3; seeds
many, surface net-like, brown to yellow-brown . . . . **ELATINACEAE** (*Elatine*)

          26′ Lvs fleshy; pl terrestrial; fls in ± 1-sided raceme-, panicle-, or umbel-like cluster;
outer perianth parts 2, > inner; stigmas 2; seeds 1–many, shiny, black
. . . . **MONTIACEAE** (*Calyptridium*) (G24)

  12′ Lf venation parallel, sometimes in wider lvs several larger veins originating from lf base but smaller
veins parallel to them and not forming a network

  27. Shrub, tree, or densely lfy rosette pl; lvs ± sword-like or ± grass-like, fibrous and/or fleshy, persisting > 1 yr

    28′ Ovules 2 per chamber

      29. Fls bisexual; fr a berry . . . . **LAXMANNIACEAE**

      29′ Fls gen unisexual (occ some bisexual); fr a papery capsule . . . . **RUSCACEAE** (*Nolina*) (G6,7,8,9)

    28. Ovules 3–many per chamber

      30. Perianth orange or red . . . . **ASPHODELACEAE**

      30′ Perianth ± white

        31. Leaves sword-like; style 1 or stigmas sessile . . . . **AGAVACEAE** (2)

        31′ Leaves grass-like; styles 3 . . . . **MELANTHIACEAE** (*Xerophyllum*)

  27′ Herb; lvs gen thin, gen not persisting (sometimes ± cylindric)

  32. Sepals and petals clearly different

    33. Pl aquatic, pl body submersed or lvs floating to emergent . . . . ²**HYDROCHARITACEAE** (G2,6,8)

    33′ Pl terrestrial

      34. Above-ground st 0, lvs 2–3(4), all basal; sepals much wider than narrowly linear petals; stamens
3; pedicels twisted, recurved in age; fr touching soil . . . . **LILIACEAE** (*Scoliopus*)

      34′ Above-ground st well developed, ± lfy; sepals much narrower than reniform to gen widely ovate
petals; stamens 6, sometimes 3 sterile; pedicels not recurved in age; fr not touching soil

        35. Petals glabrous, withering after 1 day; ovary ± spheric; seeds few . . . . **COMMELINACEAE**

        35′ Petals hairy, gen persistent several days; ovary ± cylindric; seeds many . . . . . . **LILIACEAE** (*Calochortus*)

  32′ Sepals and petals very similar

    36. Perianth parts scale-like, green to brown or ± purple-black or tinged red-purple

      37. Infl of 1–many fls, variously clustered, subtended individually and/or in small groups by bracts;
style 1, stigmas 2–3, slender, spreading; fr a capsule . . . . **JUNCACEAE** (G13)

      37′ Infl a bractless raceme; stigmas 3 or 6, ± sessile; fr breaking into mericarps
. . . . **JUNCAGINACEAE** (G2,10,13)

    36′ Perianth parts petal-like, gen not green to brown or ± purple-black or tinged red-purple

      38. Infl an umbel (rarely fl solitary or in short, dense raceme); lvs all basal, arising from bulb or corm

        39. Fr a berry . . . . **LILIACEAE** (*Clintonia*)

        39′ Fr a capsule

          40. Pl with onion odor; infl bracts 2 (sometimes fused), wholly enclosing immature infl, but gen
not formed among pedicels; sts and lvs arising from bulb . . . . **ALLIACEAE** (G1)

          40′ Pl without onion odor; infl bracts 3 or more, not wholly enclosing immature infl, and some
formed among pedicels; sts and lvs arising from corm . . . . **THEMIDACEAE**

      38′ Infl various, often a raceme or panicle, if fls in umbels st ± lfy or with scale-like bracts; lvs basal
and/or cauline, underground parts various

      41. Perianth parts proximally ± fused into tube

        42. Pl aquatic, floating on water surface, stranded on mud, or rooting in bottom sediments; fls gen
open only 1 day . . . . **PONTEDERIACEAE**

        42′ Pl gen terrestrial, sometimes in damp soil; fls gen open > 1 day

          43. Ovary underground; perianth tube long, stalk-like . . . . **AGAVACEAE** (*Leucocrinum*)

          43′ Ovary above-ground; perianth tube not stalk-like

            44. Perianth red (yellow) . . . . **ASPHODELACEAE** (*Kniphofia*)

            44′ Perianth white to ± yellow

45. Perianth 45–60 mm; lvs strongly wavy-margined. . . . . **AGAVACEAE** (*Hesperocallis*)
45′ Perianth 8–12 mm; lvs not wavy-margined. . . . . **TECOPHILAEACEAE**
41′ Perianth parts free or nearly so
  46. Lvs mostly or all cauline
    47. Styles 3; fls bisexual or staminate . . . . **MELANTHIACEAE** (*Veratrum*) (G6)
    47′ Style 1 or stigmas sessile; fls bisexual
      48. Perianth 8–110+ mm . . . . **[3]LILIACEAE**
      48′ Perianth 1–7 mm . . . . **RUSCACEAE** (*Maianthemum*) (G15)
  46′ Lvs all basal or basal and cauline
    49. All or some filaments densely hairy
      50. Inner 3 filaments hairy. . . . . **ASPHODELACEAE** ([*Bulbine*])
      50′ All filaments hairy. . . . . **NARTHECIACEAE**
    49′ Filaments glabrous (or expanded filament base puberulent)
      51. Styles 3
        52. St glabrous . . . . **[2]MELANTHIACEAE** (G10)
        52′ St glandular. . . . . **TOFIELDIACEAE**
      51′ Style 1 or 0 and stigmas sessile
        53. Cauline lvs present, ± well developed . . . . **[3]LILIACEAE**
        53′ Cauline lvs 0 or reduced to linear or scale-like bracts
          54. Lf lanceolate or ovate to elliptic or obovate. . . . . **[3]LILIACEAE**
          54′ Lf linear
            55. Lvs ± flat, often keeled or wavy-margined; st gen 1 from bulb. . . . . **[2]AGAVACEAE**
            55′ Lvs ± cylindric, hollow; sts several–many from tuber-like rhizome
. . . . **ASPHODELACEAE** (*Asphodelus*)

Group 17

Perianth in 2 or more whorls or spirals; petals 4 or more, fused; pistil 1, ovary ± inferior

1. Stamens > in number than corolla lobes or > 15
  2. Pl spiny or bristly
    3. St very fleshy, gen spiny (in some *Opuntia* spp. only with glochids); lvs 0 or conic to cylindric, fleshy;
      petal-like parts many, overlapping in several series . . . . **CACTACEAE** (G21)
    3′ St not or slightly fleshy, bristly with stinging hairs; lvs well developed, with expanded blade; petals 5, in
      1 series . . . . **LOASACEAE** (*Eucnide*)
  2′ Pl glabrous or ± hairy but not spiny or bristly
    4. Petals or petal-like stamens many; lvs gen fleshy. . . . . **AIZOACEAE** (G14,15,21)
    4′ Petals 4 or 5, petal-like stamens 0; lvs gen thin
      5. Herbage glabrous to soft-hairy, sometimes glandular. **ERICACEAE** (G18,19,24)
      5′ Herbage stellate-hairy. . . . . **STYRACACEAE** (G19)
1′ Stamens = in number to corolla lobes or fewer, gen ≤ 5
  6. Ovary actually superior, surrounded by but not fused to hardened or winged base of corolla-like calyx
    that simulates an inferior ovary; style passing through constriction separating calyx base from petal-like
    portion and joining to top of ovary . . . . **NYCTAGINACEAE** (G5,14,15,19)
  6′ Ovary truly inferior, fused to bases of surrounding fl parts; style not passing through a calyx constriction
    7. Lvs alternate or all basal
      8. Tendrils gen present; fls unisexual — pl gen monoecious, lvs palmately veined, gen palmately lobed
. . . . **CUCURBITACEAE** (G6,8,14,21)
      8′ Tendrils 0; fls bisexual or unisexual
        9. Infl a head subtended or ± enclosed by involucre; calyx modified as pappus of bristles, scales, and/or
          awns, not green and lf-like. . . . . **[2]ASTERACEAE** (G6,7,8,9,12,14)
        9′ Infl various but not a head; calyx of unmodified sepals, gen green
          10. Stamens alternate corolla lobes; staminodes 0; fls radial or bilateral; placentas axile or parietal
. . . . **CAMPANULACEAE** (G2,5,14)
          10′ Stamens opposite corolla lobes; staminodes alternate fertile stamens; fls radial; placentas
            free-central. . . . . **THEOPHRASTACEAE** (G21)
    7′ Lvs opposite or whorled
      11. Infl a peduncled pair of fls — creeping subshrub, twining vine, or erect shrub
        12. Fls sessile; frs berries, sometimes fused . . . . **CAPRIFOLIACEAE** (*Lonicera*)
        12′ Fls on slender pedicels; frs capsules, not fused. . . . . **LINNAEACEAE**
      11′ Infl a head, spike, or cyme, or fls solitary
        13. Infl a head or dense, head-like spike, subtended or ± enclosed by ± calyx-like involucre
          14. Anthers fused into tube around style. . . . . **[2]ASTERACEAE** (G6,7,8,9,12,14)
          14′ Anthers free
            15. Lvs opposite. . . . . **DIPSACACEAE**

15′ Lvs whorled. . . . . **RUBIACEAE** (*Sherardia*)
13′ Infl various, not subtended or enclosed by ± calyx-like involucre
  16. Herb or subshrub
    17. Stamens 4 or 5; corolla radial, not spurred or with proximal bulge. . . . . **²RUBIACEAE** (G14)
    17′ Stamens 1 or 3; corolla often bilateral, often spurred or with proximal bulge . . . . . **VALERIANACEAE** (G14)
  16′ Shrub, tree, or woody vine
    18. Infl a raceme or ± interrupted spike or fls 1–2 in axils . . . . **CAPRIFOLIACEAE**
    18′ Infl a head or compound cyme
      19. Lvs compound . . . . **ADOXACEAE** (*Sambucus*)
      19′ Lvs simple
      20. Lvs 3–5-veined from base. . . . . **ADOXACEAE** (*Viburnum*)
      20′ Lvs 1-veined from base. . . . . **²RUBIACEAE** (G14)

Group 18

Perianth in 2 whorls; petals 4 or more, all fused, corolla bilateral; pistil 1, ovary superior

1. Stamens > in number than corolla lobes
  2. Lf compound — petals proximally free but strongly adherent and may appear fused . . . **FABACEAE** (Papilionoideae)
  2′ Lf simple, sometimes reduced to linear or scale-like bracts
    3. Filaments fused, forming a U-shaped tube around the style — petals actually free but may appear fused
    . . . . **POLYGALACEAE** (G15,22,23,24)
    3′ Filaments free
      4. Shrub; lvs alternate; anthers dehiscing by pores; ovules many; fr a capsule
      . . . . **ERICACEAE** (*Rhododendron macrophyllum*)
      4′ Herb or weak-stemmed subshrub; lvs opposite; anthers dehiscing by lateral slits; ovule 1; fr an achene
      tightly enwrapped by hardened base of corolla-like calyx — petals actually 0 but sepals petal-like and
      involucre bracts sepal-like. . . . . **NYCTAGINACEAE** (*Allionia*)
1′ Stamens = in number to corolla lobes or fewer
5. Corolla scarious, persistent in fr. . . . . **PLANTAGINACEAE** (*Plantago*) (G5,8,19)
5′ Corolla not scarious, gen deciduous after fl
  6. Fertile stamens = in number to corolla lobes
    7. Ovary deeply 4-lobed, esp in fr; style attachment surrounded by lobes of ovary (style readily detaches
    in Lamiaceae as corolla falls); fr breaking apart into 1-seeded nutlets (sometimes only 1–3 nutlets maturing)
      8. St round in ×-section; lvs alternate (proximal opposite in some *Plagiobothrys*); herbage not aromatic;
      infl or infl branches coiled when young, unrolling in fl or fr . . . . **BORAGINACEAE** (G19)
      8′ St 4-angled, at least when young; lvs opposite throughout; herbage gen strongly aromatic, esp if
      crushed; infl or infl branches not coiled . . . . **³LAMIACEAE** (G8,19)
    7′ Ovary entire or shallowly lobed; style clearly attached to tip of ovary; fr a capsule or splitting into nutlets
      9. Shrub; filaments attached directly to receptacle, falling free from corolla or only weakly adhering;
      anthers dehiscent by pores. . . . . **ERICACEAE** (G17,19,24)
      9′ Herb; filaments fused to and falling with corolla; anthers dehiscent by slits
      10. Calyx lobes bordered by or ± fused by a much thinner, transparent or translucent membrane; corolla
      gen funnel-shaped to salverform; stigmas 3; ovary chambers 3. . . . . **POLEMONIACEAE** (G19)
      10′ Calyx lobes not bordered by or fused by transparent or translucent membrane; corolla ± rotate;
      stigmas 1–2; ovary chambers 1–4
        11. Corolla white to blue or purple, 2–8 mm diam; infl cyme-like, or fls solitary; fr splitting into gen 4
        one-seeded nutlets. . . . . **BORAGINACEAE** (*Tiquilia*) (G19)
        11′ Corolla yellow (white to ± purple), 15–40 mm diam; infl a narrow raceme or panicle; fr a
        many-seeded capsule. . . . . **SCROPHULARIACEAE** (*Verbascum*) (G19)
  6′ Fertile stamens fewer than corolla lobes — 1 or more sterile stamens sometimes present
    12. Pl entirely non-green, ± fleshy. . . . . **²OROBANCHACEAE**
    12′ Pl green and photosynthetic
      13. Proximal portion of lower corolla lip with backward-projecting, conic or tubular spur
        14. Pl aquatic; lf-like underwater sts gen dissected, gen submersed, gen bearing small, hollow bladders
        that suck in and trap small organisms when hairs at opening are triggered
        . . . . **LENTIBULARIACEAE** (*Utricularia*) (G2)
        14′ Pl terrestrial; lvs entire to shallowly lobed, without hollow bladders
          15. Lvs all basal; adaxial lf face sticky or slimy with insect-capturing glands
          . . . . **LENTIBULARIACEAE** (*Pinguicula*)
          15′ Lvs basal and cauline or all cauline; adaxial lf surface not sticky or slimy with insect-capturing
          glands . . . . **⁴PLANTAGINACEAE** (G5,19)
      13′ Proximal portion of lower corolla lip without backward-projecting, conic or tubular spur, occ with
      expanded pouch
        16. Ovary gen deeply 4-lobed; style attachment surrounded by lobes of ovary [style readily detaches as
        corolla falls] — fr breaking into 4 one-seeded nutlets. . . . . **³LAMIACEAE** (G8,19)

16′ Ovary entire or ± shallowly lobed; style arising from tip of ovary
  17. Ovules 1–4 per ovary chamber
    18. Fr a capsule
      19. Shrub . . . . **ACANTHACEAE**
      ⸳19′ Ann . . . . [4]**PLANTAGINACEAE** (G5,19)
    18′ Fr 2–4 nutlets
      20. Stamens exserted. . . . . [3]**LAMIACEAE** (G8,19)
      20′ Stamens included . . . . **VERBENACEAE** (G19)
  17′ Ovules several to many per ovary chamber
    21. Lvs compound . . . . [2]**BIGNONIACEAE**
    21′ Lvs simple
      22. Capsule 5–35 cm
        23. Shrub or tree; capsule body 15–35 cm, straight, linear . . . . [2]**BIGNONIACEAE**
        23′ Ann; capsule body 5–10 cm, curved, tipped with a hooked beak several cm long. . . . . **MARTYNIACEAE**
      22′ Capsule < 2 cm
        24. Corolla upper lip forming a concave beak or hood, ± enclosing or covering stamens and style
          25. Sterile stamen 0; lobes of upper corolla lip obscure or 0 . . . . [2]**OROBANCHACEAE**
          25′ Sterile stamen present, inserted on corolla tube between lobes of upper corolla lip; lobes of
            upper lip evident . . . . [4]**PLANTAGINACEAE** (G5,19)
        24′ Corolla upper lip not forming a beak or hood, not enclosing or covering stamens and style
          [though these may be included within corolla tube or throat]
          26. Stigma lobes flat, disk-like, moving together when touched; cauline lvs opposite
            . . . . **PHRYMACEAE** (G2,19)
          26′ Stigma dot-like or head-like, unlobed, or lobes not moving together when touched; cauline lvs
            opposite or alternate (whorled) or lvs all basal
            27. Lvs all basal
              28. Fls solitary; stamens 4; lvs linear or awl-like to spoon-shaped or ovate, entire
                . . . . **SCROPHULARIACEAE** (*Limosella*) (G2,19)
              28′ Fls in raceme; stamens 2; lvs ovate to round, toothed or shallowly lobed
                . . . . **PLANTAGINACEAE** (*Synthyris*)
            27′ Lvs basal and cauline or all cauline
              29. St 4-angled; corolla tube urn-shaped to spheric . . . . **SCROPHULARIACEAE** (*Scrophularia*)
              29′ St gen round in ×-section; corolla tube cylindric to bell-shaped, sometimes very short
                . . . . [4]**PLANTAGINACEAE** (G5,19)

Group 19

Perianth in 2 whorls; petals 4 or more, all fused, corolla radial; pistil 1, ovary superior

1. Stamens > in number than corolla lobes
  2. Filaments fused proximally, forming a cup or tube
    3. Lvs compound
      4. Shrub or tree; lvs 2-pinnate; fr a legume; stamens many, much more conspicuous than corolla
        . . . . **FABACEAE** (Mimosoideae)
      4′ Herb; lvs palmate with 3 lflets; fr a capsule; stamens 10, less conspicuous than corolla. . . . **OXALIDACEAE** (G20)
    3′ Lvs simple
      5. Stamens many; filaments strongly fused, forming tube around pistil; anthers crescent-shaped, chamber
        1; style 1, distally branched or styles 5–many, stigmas linear to head-like — petals gen free from each
        other but fused to and falling with filament tube . . . . **MALVACEAE** (G8,20,22)
      5′ Stamens 10(16); filaments fused at base; anthers oblong, 2-chambered; style 1, stigma dot-like or
        minutely lobed . . . . **STYRACACEAE** (G17)
  2′ Filaments free
    6. Fl parts in 6s; fr a triangular achene — 6-parted perianth easily mistaken for corolla, esp in pls with 1–
      few-fld, calyx-like involucre . . . . [3]**POLYGONACEAE** (G2,6,8,15,16)
    6′ Fl parts in 4s or 5s; fr a capsule, berry, drupe, or cluster of follicles
      7. Pl fleshy
        8. Pl photosynthetic; lvs well developed, green; carpels fused only at base, styles 5 . . . . . . **CRASSULACEAE** (G10)
        8′ Pl non-photosynthetic; lvs scale-like, white to cream or red; carpels wholly fused, style 1
          . . . . [2]**ERICACEAE** (G17,18,24)
      7′ Pl not fleshy
        9. St spineless; corolla urn-shaped to rotate or widely funnel-shaped, variously colored . . [2]**ERICACEAE** (G17,18,24)
        9′ St bearing a stout, petiolar spine at each node; corolla tubular, bright red. . . . . **FOUQUIERIACEAE**
1′ Stamens = in number to corolla lobes or fewer
  10. Fertile stamens fewer than corolla lobes — 1 or more sterile stamens sometimes present
    11. Perianth parts in 6s; stigmas and style branches 3 — 6-parted perianth easily mistaken for corolla in pl
      with 1-fld, calyx-like involucre . . . . [3]**POLYGONACEAE** (G2,6,8,15,16)

11′ Perianth parts in 4s or 5s; stigmas 1 or 2, style unbranched or branches 2
  12. Ovary deeply 4-lobed; style attachment surrounded by lobes of ovary [style readily detaches as
     corolla falls] — fr 4 nutlets . . . . [2]**LAMIACEAE** (G8,18)
  12′ Ovary entire or ± shallowly lobed; style arising from tip of ovary
    13. Ovules several to many per ovary chamber
      14. Lvs all basal; infl scapose, fls solitary — lvs linear or awl-shaped to spoon-shaped or ovate, entire
         . . . . **SCROPHULARIACEAE** (*Limosella*) (G2,18)
      14′ Lvs basal and cauline or all cauline; infl not scapose, fls 1–many
        15. Stigma lobes flat, disk-like, moving together when touched . . . . **PHRYMACEAE** (G2,18)
        15′ Stigma dot-like or head-like, unlobed, or lobes not moving together when touched
           . . . . **PLANTAGINACEAE** (G5,18)
    13′ Ovules 1–4 per ovary chamber
      16. Corolla-like calyx constricted above ovary, base persistent around ovary, becoming hardened or
         winged at maturity; stigma, ovary chamber, and ovule 1; fr an achene — corolla-like calyx easily
         mistaken for corolla, esp in pl with 1-fld, calyx-like involucre . . . . [2]**NYCTAGINACEAE** (G5,14,15,17)
      16′ True corolla present, not constricted above ovary, base not persistent as hardened or winged
         structure around ovary; stigmas, ovary chambers, and ovules > 1; fr various
        17. Stamens 2 . . . . **OLEACEAE**
        17′ Stamens 4
          18. Lvs gland-dotted; shrub or small tree . . . . **SCROPHULARIACEAE** (*Myoporum*)
          18′ Lvs not gland-dotted though sometimes glandular-hairy . . . . **VERBENACEAE** (G18)
10′ Fertile stamens = in number to corolla lobes
  19. Pls entirely non-green, parasitic
    20. Pl an erect or mound-shaped, fleshy root-parasite with many crowded, scale-like lvs
       . . . . **BORAGINACEAE** (*Pholisma*)
    20′ Pl a thread-like, lfless, yellow to orange, twining vine that wraps around sts and lvs of host pl
       . . . . **CONVOLVULACEAE** (*Cuscuta*)
  19′ Pls green and photosynthetic
    21. Ovaries 2, ± pressed together in fl but not fused, carpels distally united by a shared style or by a
       complex stigmatic structure; fr of 1–2 follicles; milky latex gen present . . . . **APOCYNACEAE** (G10)
    21′ Ovary 1, sometimes deeply lobed; frs various; milky latex 0
      22. Lvs whorled
        23. Shrub — lvs evergreen, leathery; corolla throat with fringed appendages . . . . **APOCYNACEAE** (*Nerium*)
        23′ Herb
          24. Per, sometimes dying after fl; perianth parts in 4s; fr a capsule . . . . **GENTIANACEAE** (*Frasera*)
          24′ Ann; perianth parts in 6s; fr an achene — 6-parted perianth easily mistaken for corolla, involucre
            1-fld, calyx-like . . . . **POLYGONACEAE** (*Chorizanthe*)
      22′ Lvs alternate, opposite, or all basal
        25. Corolla dry, scarious, 4-lobed; infl a dense spike (rarely fl only 1); lvs gen all basal (opposite,
           appearing clustered at nodes in *Plantago arenaria*) . . . . **PLANTAGINACEAE** (*Plantago*) (G5,8,18)
        25′ Corolla with normal, petal-like texture, 4–7-lobed; infl various or fls solitary; lvs basal and/or cauline
          26. Ovary deeply 2–4-lobed, esp in fr; style attachment surrounded by lobes of ovary (style readily
             detaches in Lamiaceae as corolla falls); fr breaking apart into 1-seeded nutlets or segments
             (sometimes only 1–3 nutlets maturing)
            27. Lvs opposite, at least proximally; st 4-angled or round in ×-section; st and lvs sometimes aromatic
              28. St round; lvs gen alternate distally or lvs of a pair unequal; herbage not aromatic; infl or
                 branches of infl gen ± coiled or 1-sided when young, unrolling during development, at maturity
                 ± open, gen resembling a 1-sided spike or raceme or fls sometimes solitary . . . . . [6]**BORAGINACEAE** (G18)
              28′ St 4-angled; lvs opposite throughout; herbage gen strongly aromatic; infl not coiled
                 . . . . [2]**LAMIACEAE** (G8,18)
            27′ Lvs alternate; st ± round; st and lvs gen not aromatic
              29. Lvs linear to ovate; sts mostly not rooting; infl or branches of infl gen ± coiled or 1-sided when
                 young, unrolling during development, at maturity ± open, gen resembling a 1-sided spike or
                 raceme, fls seldom concealed by lvs; ovary 2–4-lobed . . . . [6]**BORAGINACEAE** (G18)
              29′ Lvs reniform; sts matted, rooting at nodes; fls solitary, ± concealed by lvs; ovary 2-lobed
                 . . . . **CONVOLVULACEAE** (*Dichondra*) (G10)
          26′ Ovary entire or shallowly lobed; style clearly attached to tip of ovary; fr gen a capsule, berry, drupe, or achene
            30. Filaments fused, forming a tube around style
              31. Shrub; fls solitary, yellow or orange to ± red; sts and lvs densely stellate — corolla-like whorl
                 actually a calyx, subtended by sepal-like bracts . . . . **MALVACEAE** (*Fremontodendron*) (G12)
              31′ Herb; fls in scapose umbels, white to pink or purple; sts and lvs glabrous or glandular-hairy
                 . . . . **PRIMULACEAE** (*Dodecatheon*)
            30′ Filaments gen free
              32. Sepals 2; sts and lvs ± fleshy

33. Vine with twining sts; fls in many-fld raceme or panicle; fr fleshy, indehiscent, 1-seeded
. . . . **[BASELLACEAE]**
33′ Herb, sts prostrate to erect, not twining; fls solitary or in few-fld clusters; fr dry, dehiscent,
2–many-seeded . . . . **MONTIACEAE** (G22,24)
32′ Sepals 4 or more; sts and lvs gen not fleshy
34. Perianth parts in 6s or 7s
35. Style and stigma 1; placentas free-central; ovules several–many; fr a capsule. . . ²**MYRSINACEAE** (G24)
35′ Styles and stigmas 3; placentas basal; ovule 1; fr a 3-angled achene — 6-parted perianth
easily mistaken for corolla, esp in pl with 1-fld, calyx-like involucre . . ³**POLYGONACEAE** (G2,6,8,15,16)
34′ Perianth parts in 4s or 5s
36. Styles or style-branches 5; sepal tips and subtending bracts widely obtuse, scarious
(sometimes brightly pigmented) *or* calyx stalked-glandular . . . . **PLUMBAGINACEAE**
36′ Styles, style-branches, or stigmas gen 1–3; sepal tips and bracts gen not widely obtuse and
scarious, calyx glandular or not
37. Corolla-like calyx constricted above the ovary, persistent base becoming hardened or winged
at maturity, enclosing fr; ovary chamber 1, ovule 1; fr an achene — corolla-like calyx easily
mistaken for corolla, esp in pls with 1-fld, calyx-like involucre . . . . . ²**NYCTAGINACEAE** (G5,14,15,17)
37′ True corolla present, gen not constricted above ovary, base not persistent as hardened or
winged structure enclosing fr; ovary chambers gen 1–4, ovules gen > 1; frs various
38. Stamens opposite corolla lobes; placentas free-central
39. Fls axillary or in racemes. . . . . ²**MYRSINACEAE** (G24)
39′ Fls in scapose umbels. . . . . **PRIMULACEAE**
38′ Stamens alternate corolla lobes; placentas parietal or axile (rarely basal)
40. Filaments of stamens not fused to corolla, sometimes adhering weakly to base of corolla
but readily separable
41. Ann; corolla 5-lobed; lvs pinnately lobed . . . . **BORAGINACEAE** (*Emmenanthe*)
41′ Shrub; corolla 4-lobed; lvs entire. . . . . **ERICACEAE** (*Menziesia*)
40′ Filaments of stamens definitely fused to corolla tube or throat, not separable without
tearing stamen or corolla tissue
42. Infl or branches of infl gen ± coiled or 1-sided when young, unrolling during development,
at maturity ± open, gen resembling a 1-sided spike or raceme. . . . . ⁶**BORAGINACEAE** (G18)
42′ Infl not coiled and unrolling, gen not 1-sided
43. Calyx lobes gen bordered by or ± fused by a much thinner, transparent or translucent
membrane; stigmas and ovary chambers gen 3 (1 or 2). . . . . ²**POLEMONIACEAE** (G18)
43′ Calyx lobes gen without a transparent or translucent marginal membrane; stigmas gen 2
(1 or 3), ovary chambers gen 1 or 2
44. St twining . . . . ²**CONVOLVULACEAE**
44′ St not twining
45. Lvs palmately compound — rhizomed per of bogs, marshes, lake margins
. . . . **MENYANTHACEAE** (*Menyanthes*)
45′ Lvs simple to pinnately compound
46. Pl aquatic; lvs floating on water surface, deeply cordate; infl an umbel or fl 1
. . . . **MENYANTHACEAE** (*Nymphoides*) (G2)
46′ Pl gen terrestrial, sometimes in damp soil; lvs not floating on water surface, shapes
various; infl not an umbel
47. Stigmas 3 . . . . ²**POLEMONIACEAE** (G18)
47′ Stigmas 1–2
48. Lvs alternate or of unusual arrangement [in some *Solanum* spp. 2 lvs that are not
opposite may be attached at the same node, and infls may arise from internodes],
not opposite or all basal
49. Stigma 1, gen undivided
50. Some or all stamens densely hairy; infl a narrow raceme or panicle
. . . . **SCROPHULARIACEAE** (*Verbascum*) (G18)
50′ Stamens glabrous or nearly so; infl variously cyme-like, or fls solitary. . . . . . **SOLANACEAE**
49′ Stigmas or stigma lobes 2
51. Style undivided
52. Corolla < 15 mm; fls in axillary or terminal clusters, several to many per
peduncle, not subtended by paired bracts; lvs lobed or compound, not cordate,
sagittate, or hastate. . . . . ⁶**BORAGINACEAE** (G18)
52′ Corolla 20–70 mm; fls gen 1(5) in lf axils, often subtended by a pair of opposite
or subopposite bracts appressed against the calyx or borne below the fl(s) on the
peduncle; lvs often cordate, sagittate, or hastate. . . . . ²**CONVOLVULACEAE**
51′ Styles or style branches 2
53. Herbage glabrous or variously hairy and/or glandular; fls solitary or clustered;
lvs entire to lobed or compound. . . . . ⁶**BORAGINACEAE** (G18)

53′ Herbage densely silky-canescent; fls solitary in distal lf axils, ± on 1 side of st;
    lvs entire . . . . **CONVOLVULACEAE** (*Cressa*)
48′ Lvs opposite or all basal
  54. Corolla lobes 4
    55. Herb. . . . . [2]**GENTIANACEAE** (G24)
    55′ Shrub
      56. Corolla ± rotate, lobes ± leathery, ± green-white; coastal lagoons or salt marshes
      . . . . **ACANTHACEAE** (*Avicennia*)
      56′ Corolla ± salverform, lobes thin, ± yellow or white to purple; gen dry places
      . . . . **SCROPHULARIACEAE** (*Buddleja*) (G7,9)
  54′ Corolla lobes 5
    57. Fls solitary, infl scapose . . . . **BORAGINACEAE** (*Hesperochiron*)
    57′ Fls 1–many, borne on a ± lfy st
      58. Lvs toothed, lobed, or compound. [6]**BORAGINACEAE** (G18)
      58′ Lvs entire
        59. Pls glandular-hairy . . . . **SOLANACEAE** (*Petunia*)
        59′ Pls glabrous or hairy, but not glandular
          60 Sts spreading or mat-forming; sts and lvs densely strigose; fls sessile or
          subsessile in forks of st; lvs of a pair often unequal . . . **BORAGINACEAE** (*Tiquilia*) (G18)
          60′ Sts mostly erect or ascending; sts and lvs glabrous or short-hairy; fls axillary
          or in terminal infls; lvs of a pair equal. . . . . [2]**GENTIANACEAE** (G24)

Group 20

Lvs compound or nearly so; perianth in 2 or more whorls or spirals;
petals 4 or more (rarely fewer), free; pistil 1

1. Lvs 2 or more × compound or divided
 2. Ovary inferior
  3. Fr dry, splitting into 2 mericarps; infl gen a compound umbel (occ simple umbel, head)
  . . . . [2]**APIACEAE** (G2,6,8,14,21)
  3′ Fr a berry; infl a simple or branched cluster of umbels . . . . [2]**ARALIACEAE**
 2′ Ovary superior
  4. Shrub or tree
   5. Main axis of lf bearing many small lflets or lobes in addition to main lflets. . . . . **ROSACEAE** (*Chamaebatia*)
   5′ Main axis of lf without small lflets or lobes in addition to main lflets
    6. Lflets entire; fls cream to yellow, orange, red, or pink . . . . [2]**FABACEAE** (G12,18,19,22,23,24)
    6′ Lflets serrate; fls white to dark purple. . . . . **MELIACEAE**
  4′ Herb or subshrub
   7. Ovary very deeply 4–6-lobed; style arising from a deep indentation in middle of ovary
    8. Fls white or pink; herbage not scented . . . . [3]**LIMNANTHACEAE** (*Limnanthes*) (G10)
    8′ Fls yellow; herbage strongly scented. . . . . **RUTACEAE** (*Ruta*)
   7′ Ovary unlobed or only shallowly lobed; style or sessile stigma(s) terminal or from shallow depression
   at tip of ovary
    9. Stamens 12–50
     10. Infl a raceme . . . . **RANUNCULACEAE** (*Actaea, Consolida*)
     10′ Infl a cyme or fls solitary
      11. Petals 5; sepals present in open fl . . . . **NITRARIACEAE** (G22)
      11′ Petals 4 or 6 (very rarely 5); sepals deciduous as fl opens, 0 in
      open fl . . . . **PAPAVERACEAE** (Papaveroideae) (G12,15,22,24)
    9′ Stamens 2–10
     12. Petals 5
      13. Fls solitary on peduncles ≥ 2 cm . . . . [2]**VIOLACEAE** (G5,15,24)
      13′ Fls several to many in umbels or racemes
        14. Petals yellow to orange or ± red; lflets entire; infl a raceme . . . . **FABACEAE** (*Hoffmannseggia*)
        14′ Petals pink to purple; lflets toothed; infl an umbel . . . . [2]**GERANIACEAE** (*Erodium*)
     12′ Petals 4 or 6
      15. Petals 6; lvs all basal. . . . . **BERBERIDACEAE** (*Vancouveria*) (G16)
      15′ Petals 4; lvs basal and cauline
        16. Petals all equal; fls radial; infl a bractless raceme . . . . [3]**BRASSICACEAE** (G2,15,24)
        16′ Petals unequal; fls bilateral or biradial; infl a bracted raceme or panicle-like cluster
        . . . . **PAPAVERACEAE** (Fumarioideae)

1′ Lvs once compound or divided
  17. Petal 1 . . . . **FABACEAE** (*Amorpha*)
  17′ Petals 4–6
    18. Ovary > 1/2 inferior
      19. Shrub or tree; stamens gen 20; fr a pome . . . . **ROSACEAE** (*Sorbus*)
      19′ Herb; stamens 5–10; fr a capsule, splitting into mericarps, or a berry
        20. Infl a raceme or spike (sometimes branched)
          21. Pl a freshwater aquatic; lvs all cauline, submersed lvs whorled, pinnately dissected into thread-like
            segments . . . . **HALORAGACEAE** (G2,6,8,14)
          21′ Pl terrestrial; lvs basal and cauline, alternate, ± palmately compound
            . . . . **SAXIFRAGACEAE** (*Lithophragma parviflorum*)
        20′ Infl a simple or compound umbel, a cluster of simple umbels, or 1–several heads
          22. Styles 2; fr splitting into 2 mericarps; infl gen a compound umbel (simple umbel, 1 or more heads)
            . . . . [2]**APIACEAE** (G2,6,8,14,21)
          22′ Styles 5; fr a berry; infl of 2–many umbels in simple or branched cluster . . . . [2]**ARALIACEAE**
    18′ Ovary superior
      23. Sepals deciduous as fl opens, 0 in open fl . . . . [2]**PAPAVERACEAE** (Papaveroideae) (G12,15,22,24)
      23′ Sepals present in open fl
        24. Stamens many, filaments fused in tube around style; styles 5+ . . . . **MALVACEAE** (G8,19,22)
        24′ Stamens gen 4–15, or if many, free; styles gen 1–3
          25. Fls strongly bilateral, 1 petal markedly different from others
            26. Odd petal uppermost in fl; stamens 10(5), gen all filaments or 9 of them fused, forming a tube
              around ovary (filaments free); carpel 1; fr a legume or indehiscent pod . . . . **FABACEAE** (Papilionoideae)
            26′ Odd petal lowermost in fl; stamens 5, free but tightly appressed against ovary; carpels 3;
              fr a capsule . . . . [2]**VIOLACEAE** (G5,15,24)
          25′ Fls ± radial or if bilateral, petals 4 in dissimilar pairs
            27. Lflets or lf segments 2 or 3
              28. Herb or subshrub
                29. Petals 4; stamens 6
                  30. Ovary sessile on receptacle; fls of raceme not subtended by bracts . . . . [3]**BRASSICACEAE** (G2,15,24)
                  30′ Ovary separated from receptacle by a stalk; fls of raceme gen subtended by bracts or fls solitary,
                    axillary . . . . **CLEOMACEAE**
                29′ Petals 5; stamens 5 or 10
                  31. Lflet margins toothed or lobed . . . . **SAXIFRAGACEAE** (G10,21,24)
                  31′ Lflet margins entire (sometimes apically notched)
                    32. Lvs alternate or all basal; stamens ± fused, at least at base; ovules or seeds many
                      . . . . **OXALIDACEAE** (G19)
                    32′ Lvs opposite; stamens free; ovules or seeds 1 per chamber. . . . . [2]**ZYGOPHYLLACEAE**
              28′ Tree or well-developed shrub
                33. Lvs opposite
                  34. Lflets 2, fused at base [lvs deeply 2-lobed], entire, resinous . . . . **ZYGOPHYLLACEAE** (*Larrea*)
                  34′ Lflets 3, occ ± fused, toothed, not resinous
                    35. Lflets coarsely serrate or crenate; stigmas 2, ± sessile; ovary 2-winged; fr gen splitting at
                      maturity into winged mericarps . . . . **SAPINDACEAE** (*Acer glabrum*) (G9)
                    35′ Lflets finely serrate; stigmas and styles 3; ovary wingless; fr a bladdery, 3-chambered capsule
                      . . . . **STAPHYLEACEAE**
                33′ Lvs alternate
                  36. Lflets gland-dotted; fr a round, winged achene . . . . **RUTACEAE** (*Ptelea*) (G9)
                  36′ Lflets not gland-dotted; fr a capsule or berry
                    37. Petals 4; ovary separated from receptacle by a stalk . . . . **CLEOMACEAE** (*Peritoma arborea*)
                    37′ Petals 5–6; ovary sessile on receptacle
                      38. Petals 5; lflets entire to lobed, but not spiny-toothed . . . . [2]**ANACARDIACEAE** (G7,9,23)
                      38′ Petals 6; lflets spiny-toothed . . . . [2]**BERBERIDACEAE** (*Berberis*) (G16)
            27′ Lflets or lf segments 4 or more
              39. Lvs palmately compound or divided
                40. Pl woody
                  41. Lvs opposite; erect shrub or tree . . . . **SAPINDACEAE** (*Aesculus*)
                  41′ Lvs alternate; woody vine with tendrils . . . . **VITACEAE** (*Parthenocissus*)
                40′ Pl herbaceous
                  42. Lflets entire; infl a raceme; ovary separated from receptacle by a stalk . . . . **CLEOMACEAE** (*Peritoma*)
                  42′ Lflets or lf segments toothed; infl (1)2-fld; ovary sessile on receptacle
                    . . . . **GERANIACEAE** (*Geranium*) (G24)
              39′ Lvs pinnately compound or divided
                43. Petals 4
                  44. Ovary entire or ± 2-lobed; infl a gen bractless raceme . . . . [3]**BRASSICACEAE** (G2,15,24)

    44′ Ovary very deeply 4-lobed; fls solitary and axillary or infl a lfy-bracted raceme
    . . . . [3]**LIMNANTHACEAE** (*Limnanthes*) (G10)
   43′ Petals 5 or more
    45. Shrub or tree (occ low-creeping)
     46. Petals 6; lflets spiny-toothed; sts and lvs non-scented . . . . [2]**BERBERIDACEAE** (*Berberis*) (G16)
     46′ Petals 5(6); lflets entire or toothed but not spiny; sts and lvs often strongly scented
      47. Ovary divided into 2–5 flattened lobes; fr a cluster of elongated, winged achenes; lflets with
        gland-tipped teeth near base . . . . **SIMAROUBACEAE** (*Ailanthus*) (G7,9,10)
      47′ Ovary entire; fr a drupe or legume; lflets without gland-tipped teeth
       48. Fls solitary or few; lflets 2–10 mm, ± entire — small tree of DSon. . . . . **BURSERACEAE** (G7,9)
       48′ Fls in racemes or panicles, often many; lflets mostly > 7 mm, often much longer, toothed or entire
        49. Corolla << 1 cm diam, white to ± green or pale yellow; fr a drupe; lflets often toothed
         . . . . [2]**ANACARDIACEAE** (G7,9,23)
        49′ Corolla > 1 cm diam, bright yellow; fr a legume; lflets entire . . . . [2]**FABACEAE** (*Senna*)
   45′ Herb or subshrub
    50. Lvs ± cylindric [lflets and lobes gen closely overlapping around axis], gen basal; fr an achene;
      hypanthium present . . . . **ROSACEAE** (*Ivesia*)
    50′ Lvs ± flat, basal and/or cauline; fr nutlets, mericarps, a capsule, or a legume; hypanthium 0
     51. Anther-bearing stamens as many as petals — fr with long beak, splitting into 5 mericarps
      . . . . [2]**GERANIACEAE** (*Erodium*)
     51′ Anther-bearing stamens 2 × as many as petals (occ some anthers much smaller than others)
      52. Fls in racemes or panicles; ovary chamber 1; fr dehiscent
      53. Terminal lflet 0; fr a legume. . . . . [2]**FABACEAE** (*Senna*)
      53′ Terminal lflet present; fr a capsule . . . . **RESEDACEAE**
      52′ Fls solitary, in pairs, or in a lfy-bracted raceme; ovary chambers gen 5; fr splitting into mericarps
       54. Stipules 0; ovary very deeply 5-lobed; style base surrounded by ovary lobes
        . . . . [3]**LIMNANTHACEAE** (*Limnanthes*) (G10)
       54′ Stipules present; ovary entire or shallowly lobed at time of pollination; style at tip of ovary
        . . . . [2]**ZYGOPHYLLACEAE**

Group 21

Lvs simple; perianth in 2 or more whorls or spirals; petals 4 or more, free; pistil 1, ovary inferior

1. Stamens > 2 × as many as petals or > 15
 2. Style 1
  3. St fleshy, spiny (in some *Opuntia* spp. only with glochids); sepals and petals indefinite in number and
    not sharply differentiated, in spirals . . . . **CACTACEAE** (G17)
  3′ St not fleshy, gen not spiny; sepals and petals each 3–7, clearly differentiated, in whorls
   4. Herb or subshrub with rough, barbed, or stinging hairs. . . . . **LOASACEAE**
   4′ Tree or shrub lacking such hairs
    5. Ovary only 1/2 inferior, the free portion conic, esp in fr . . . . **HYDRANGEACEAE** (*Carpenteria*) (G16,22)
    5′ Ovary ± wholly inferior, the free portion, if any, rounded
     6. Lf gen dotted with embedded oil glands, often strongly scented when crushed; ovary chambers gen
      2–5, all at 1 level; fls variously colored . . . . **MYRTACEAE**
     6′ Lf not gland-dotted, not strongly scented; ovary chambers gen 5 or more, some at different levels in
      the ovary; fls bright red-orange to pale yellow . . . . **LYTHRACEAE** (*Punica*)
 2′ Styles or sessile stigmas > 1, sometimes partly fused at base
  7. Shrub or tree, woody throughout, not fleshy; petals gen 4–5
   8. Lvs opposite; fr a capsule. . . . . **HYDRANGEACEAE** (*Philadelphus*)
   8′ Lvs alternate; fr a pome. . . . . [2]**ROSACEAE** (G10,15,22,23)
  7′ Herb or pl woody only at base, gen ± fleshy; petals 4–many
   9. Pl aquatic; lf blades and fls floating at water surface . . . . **NYMPHAEACEAE** (*Nymphaea*)
   9′ Pl terrestrial; lvs and fls not floating
    10. Sepals 4–8; petals many. . . . . **AIZOACEAE** (G14,15,17)
    10′ Sepals 2; petals 5 . . . . [2]**PORTULACACEAE**
1′ Stamens 2 × as many as petals or fewer
 11. Stamens opposite and = in number to petals
  12. Shrub or tree. . . . . **RHAMNACEAE** (G12,23)
  12′ Herb — petals actually slightly fused at base, easily misinterpreted as free. . . . . . . **THEOPHRASTACEAE** (G17)
 11′ Stamens alternate petals or different in number
  13. Style 1, sometimes distally branched
   14. Pl a tendril-bearing vine; fls unisexual, pl monoecious — corolla sometimes so deeply divided that
    petals appear to be free . . . . **CUCURBITACEAE** (G6,8,14,17)
   14′ Pl not tendril-bearing; fls gen bisexual

15. Ovules > 1 per chamber
  16. Herb (subshrub). . . . . **ONAGRACEAE** (G5)
  16′ Shrub
    17. Stamens (4)5, free; lvs ovate or obovate to round, gen palmately 3–5-lobed, gen toothed; glands
       stalked or 0 . . . . [2]**GROSSULARIACEAE**
    17′ Stamens 10, fused in ring with staminodes; lvs needle-like, entire, opposite; glands embedded
       in lvs . . . . **MYRTACEAE** ([*Chamelaucium*])
15′ Ovule 1 per chamber or per ovary
  18. Lvs alternate
    19. Pl a non-desert, sprawling, woody vine (fl sts ± erect); lvs smooth, dark green, often palmately
       lobed; infl a cluster of umbels; fr a berry . . . . **ARALIACEAE** (*Hedera*)
    19′ Pl a low-desert per; lvs gen scabrous, pale green, unlobed; infl ± a raceme; fr an achene
       . . . . **LOASACEAE** (*Petalonyx*)
  18′ Lvs opposite; sepals and petals 4
    20. Per to tree; infl an umbel, head, or cyme; fr a drupe . . . . **CORNACEAE**
    20′ Ann; infl a spike or spike-like raceme; fr indehiscent, nut-like . . . . **ONAGRACEAE** (*Clarkia heterandra*)
13′ Styles > 1
  21. Seeds 1–2 per chamber
    22. Infls of umbels or heads (spike in *Hydrocotyle verticillata*)
      23. Lvs various [if round-reniform, pl terrestrial], glabrous or hairy; infl gen compound umbel (occ
         simple umbel, head) . . . . **APIACEAE** (G2,6,8,14,20)
      23′ Lvs round-peltate or -reniform, glabrous; pl aquatic with stolons or rhizomes, floating or rooted in
         shallow water or on mud; infl simple umbel or spike . . . . **ARALIACEAE** (*Hydrocotyle*) (G2,14)
    22′ Infls various but not of umbels, heads, or spikes
      24. Lvs alternate; hypanthium well developed; fr a pome. . . . . [2]**ROSACEAE** (G10,15,22,23)
      24′ Lvs opposite; hypanthium very short or 0; fr a capsule . . . . **HYDRANGEACEAE** (G24)
  21′ Seeds gen several–many per chamber
    25. Sepals 2; lvs fleshy . . . . [2]**PORTULACACEAE**
    25′ Sepals 4–5; lvs gen not fleshy
      26. Herb. . . . . **SAXIFRAGACEAE** (G10,20,24)
      26′ Shrub
        27. Stamens 4–5 . . . . [2]**GROSSULARIACEAE**
        27′ Stamens 10 . . . . **HYDRANGEACEAE** (*Jamesia*)

Group 22

Lvs simple; perianth in 2 or more whorls or spirals; petals 4 or more (rarely fewer), free;
stamens > 2 × as many as petals; pistil 1, ovary superior

1. Ovary chamber 1
  2. Hypanthium present
    3. Sepals fused into conical cap that falls away as fl opens, 0 in open fl; petals 4 — funnel-shaped
       receptacle readily misinterpreted as a hypanthium . . . . **PAPAVERACEAE** (*Eschscholzia*)
    3′ Sepals present in open fl; petals (2)5
      4. Fls ± radial; petals gen 5, oblanceolate to ± round, not recurved . . . . [2]**ROSACEAE** (G10,15,21,23)
      4′ Fls strongly bilateral; petals 2, narrowly awl-like, recurved — sepals obovate; fr opening before seeds
         mature, becoming widely dehiscent . . . . **SAXIFRAGACEAE** (*Bensoniella*)
  2′ Hypanthium 0
    5. Infl a spike or 1–many heads
      6. Shrub or tree; fr a legume . . . . **FABACEAE** (Mimosoideae)
      6′ Herb; fr a capsule, dehiscent at tip. . . . . **RESEDACEAE** (*Reseda*)
    5′ Infl a cyme, raceme, panicle, or umbel, or fl solitary
      7. Sepals 2–3 (fused into conical cap in *Eschscholzia*), early-deciduous, 0 in open fl; latex gen present [at
         least in roots], gen milky or colored. . . . . **PAPAVERACEAE** (Papaveroideae) (G12,15,20,24)
      7′ Sepals 2–5(8), persistent; milky or colored latex 0
        8. Filaments fused at base into 3–5 bunches . . . . [2]**HYPERICACEAE** (G24)
        8′ Filaments free to base
          9. Sepals 3 or 5; placentas parietal, ovary chamber 1 (or ± 3–12 from intruded placentas); ann to shrub;
             pl hairy (± glabrous), not at all fleshy; capsule splitting by valves . . . . **CISTACEAE**
          9′ Sepals 2–8; placentas free-central, ovary chamber 1; ann or taprooted per; pl ± glabrous, gen ±
             fleshy; capsule circumscissile or splitting by valves . . . . **MONTIACEAE** (G19,24)
1′ Ovary chambers 2–many
  10. Lvs hollow, tubular, often holding watery fluid within . . . . **SARRACENIACEAE**
  10′ Lvs with ordinary flat blades

11. Pl aquatic; lvs 1–4 dm, oblong or ovate to round, floating or rising a short distance above the water
  surface . . . . **NYMPHAEACEAE** (*Nuphar*)
11′ Pl terrestrial, sometimes in wet soil; lvs often < 1 dm
 12. Fls unisexual — ovary 3-lobed . . . . **EUPHORBIACEAE** (*Ditaxis*) (G24)
 12′ Fls bisexual
  13. Filaments fused into tube around style; anther chamber 1 or appearing so
   14. Fls radial; stamens many; filament tube cylindric . . . . **MALVACEAE** (G8,19,20)
   14′ Fls bilateral; stamens 6–8; filament tube open on one side . . . . **POLYGALACEAE** (G15,18,23,24)
  13′ Filaments free or fused at base into groups; anther chambers 2
   15. Lvs opposite or whorled (distal occ alternate), entire
    16. Petals yellow (salmon) . . . . [2]**HYPERICACEAE** (G24)
    16′ Petals white to purple
     17. Shrub — ovary actually 1/2 inferior but easily interpreted as superior
      . . . . **HYDRANGEACEAE** (*Carpenteria*) (G16,21)
     17′ Herb . . . . **LYTHRACEAE** (G24)
   15′ Lvs alternate throughout, entire to lobed
    18. Hypanthium present . . . . [2]**ROSACEAE** (G10,15,21,23)
    18′ Hypanthium 0
     19. Petals 4–5; sepals 4–5, present in open fl, persistent in fr; stamens 12–15 . . . . **NITRARIACEAE** (G20)
     19′ Petals 6; sepals 3, early-deciduous, 0 in open fl; stamens many . . . . **PAPAVERACEAE** (*Romneya*) (G12)

Group 23

Shrubs, trees, and woody vines; lvs simple; perianth in 2 whorls; petals 4 or more, free;
stamens 2 × as many as petals or fewer; pistil 1, ovary superior

1. Lvs opposite or whorled
 2. Lvs 2-lobed, resin-coated; fls bright yellow; fr covered with white, spreading hairs — desert shrub;
  lf sometimes treated as compound with 2 lflets fused at base . . . . **ZYGOPHYLLACEAE** (*Larrea*)
 2′ Lvs entire or toothed to palmately lobed or divided; fl ± green or white to creamy yellow, blue, or purple;
  fr not covered with white, spreading hairs
  3. Lvs lobed or divided; fr splitting into mericarps
   4. Shrub ≤ 1 m; stigmas 5, on long, stiff style; ovary not winged; mericarps wingless, each tipped by a
    spirally coiled beak . . . . [2]**GERANIACEAE** (*Pelargonium*) (G24)
   4′ Tree or large shrub; stigmas 2(3), ± sessile; ovary 2(3)-winged; mericarps winged, beaks 0
    . . . . **SAPINDACEAE** (*Acer*) (G7,12)
  3′ Lvs entire or toothed; fr a capsule or drupe
   5. Lvs ± fleshy, salt encrusted, gen < 1 mm wide, united across node by a sheath
    . . . . **FRANKENIACEAE** (*Frankenia palmeri*)
   5′ Lvs gen thin or leathery, not salt encrusted, gen ≥ 2 mm wide, not united by a sheath
    6. Stamens > in number than petals . . . . **RUTACEAE** (*Cneoridium*)
    6′. Stamens = in number to petals
     7. Petals opposite stamens, gen ± cupped . . . . **RHAMNACEAE** (G12,21)
     7′ Petals alternate stamens, gen ± flat
      8. Lvs opposite . . . . **CELASTRACEAE**
      8′ Lvs ± whorled at some nodes, alternate at others . . . . **PITTOSPORACEAE**
1′ Lvs alternate, sometimes reduced to minute scales or so quickly deciduous that the pls are gen lfless
 9. Lvs lobed
  10. Tendrils present; pl a woody vine, gen trailing or climbing on other pls
   11. Fls solitary, large and showy . . . . **PASSIFLORACEAE**
   11′ Fls in panicles, small and inconspicuous . . . . **VITACEAE** (*Vitis*) (G7,9,12)
  10′ Tendrils 0; pl ± erect, not climbing
   12. Infl a panicle; fls radial; 2–4 mm diam; fr a drupe . . . . **ANACARDIACEAE** (*Rhus aromatica*)
   12′ Infl an umbel; fls ± bilateral, 15–40 mm wide; fr splitting into 5 mericarps
    . . . . [2]**GERANIACEAE** (*Pelargonium*) (G24)
 9′ Lvs entire or toothed, unlobed
  13. Lf palmately veined, the blade well developed
   14. Fls strongly bilateral . . . . **FABACEAE** (*Cercis*)
   14′ Fls radial . . . . **RHAMNACEAE** (*Ceanothus*)
  13′ Lf pinnately veined, 1-veined, or reduced to a bladeless scale
   15. Fls strongly bilateral
    16. Stamens 4, all inserted on 1 side of ovary, free or nearly so; sepals rose-purple, petal-like, widely
     spreading; upper 3 petals similar, stalked, free or fused, lower 2 petals reduced to fleshy scales or
     glands; fr an indehiscent pod bearing slender, barb-tipped prickles . . . . **KRAMERIACEAE**

16′ Stamens 6–8 or 10, all filaments or 9 of 10 fused into a tube around the ovary [tube often open on 1 side]; sepals variously colored, sometimes petal-like; no petals reduced to fleshy scales or glands; fr a capsule, legume, or unarmed, indehiscent pod

  17. Sepals all fused, at least at base, forming a cup-shaped to cylindric calyx tube, gen not petal-like; the odd petal [banner] uppermost, overlapping the margins of the 2 upper lateral petals; lower 2 lateral petals gen free at base but fused toward the tip, forming a keel that encloses stamens and ovary; fr a legume or indehiscent pod . . . . **FABACEAE** (Papilionoideae)

  17′ Sepals free, petal-like, 2 of them spreading and very different from the other 3; the odd petal lowermost, often appendaged, folded and forming a keel enclosing stamens and ovary, a banner petal never present; fr a flattened capsule . . . . **POLYGALACEAE** (G15,18,22,24)

15′ Fls radial or nearly so

  18. Stamens > in number than petals (sterile stamens as many as petals in pistillate fls of *Castela* (Simaroubaceae))

    19. Petals dark blue-purple; sts and lvs gland-dotted, very strongly scented . . . . **RUTACEAE** (*Thamnosma*)

    19′ Petals white to yellow or green; sts and lvs not scented

      20. Hypanthium present, well developed, tubular or cup-shaped . . . . **ROSACEAE** (G10,15,21,22)

      20′ Hypanthium 0 or small and disk-shaped

        21. Anthers opening by small round pores; petals 5; st not thorn-tipped; moist sites . . . . **ERICACEAE** (*Rhododendron columbianum*)

        21′ Anthers opening along sides by long slits; petals 4–6; sts (exc *Glossopetalon pungens*) gen thorn-tipped; arid sites

          22. Fls unisexual; ovary deeply 4–8-lobed . . . . **SIMAROUBACEAE** (*Castela*) (G7,9,10)

          22′ Fls bisexual; ovary unlobed

            23. Petals gen 5; ovary sessile on receptacle, chamber 1; st angled; lvs oblong or narrowly elliptic to obovate, deciduous or ± persistent; fr a follicle . . . . ²**CROSSOSOMATACEAE** (*Glossopetalon*)

            23′ Petals 4; ovary raised ± 1 mm above receptacle on short stalk, chambers 2; st smooth; lvs scale-like, soon deciduous; fr a berry . . . . **KOEBERLINIACEAE**

18′ Stamens = in number to petals

  24. Lvs all reduced to bladeless scales; twigs very slender, jointed, green . . . . **TAMARICACEAE**

  24′ Lvs with expanded blades, linear to (ob)ovate, sometimes early deciduous; twigs not jointed, green or brown

    25. Stamens opposite petals; petals gen with cupped blade . . . . ²**RHAMNACEAE** (G12,21)

    25′ Stamens alternate petals; petals gen flat

      26. Lvs 6–17 mm

        27. Fls many in tight, lfy-bracted panicle; ovary chambers 5 . . . . **CELASTRACEAE** (*Mortonia*)

        27′ Fls solitary, axillary; ovary chamber 1 . . . . ²**CROSSOSOMATACEAE** (*Glossopetalon*)

      26′ Lvs 15–150 mm

        28. Petals 8–15 mm . . . . ²**PITTOSPORACEAE** (G9)

        28′ Petals 1.5–3 mm

          29. Lvs 20–80 mm wide, elliptic to ovate; fls many in terminal panicle . . . . **ANACARDIACEAE** (G7,9,20)

          29′ Lvs 5–10 mm wide, ± lanceolate; fls 1–few, axillary . . . . **CELASTRACEAE** (*Maytenus*)

Group 24

Herbs or subshrubs; lvs simple; perianth in 2 whorls; petals 4 or more, free;
stamens 2 × as many as petals or fewer; pistil 1, ovary superior

1. Fl bilateral

  2. Proximal part of calyx with a protruding, hollow spur

    3. Lf not peltate; ovary chambers 5, several-seeded; fr an exploding capsule . . . . **BALSAMINACEAE** (G16)

    3′ Lf peltate; ovary chambers 3, 1-seeded; fr breaking into 3 1-seeded mericarps . . . . **TROPAEOLACEAE**

  2′ Proximal part of calyx without a protruding spur, spur 0 or fused to pedicel

    4. Filaments of all or all but 1 stamen fused, at least proximally

      5. Petals not strongly overlapping; ovary chambers and style branches 5; style base elongating, forming a stiff beak, beak segments coiled on segments of dry fr . . . . **GERANIACEAE** (*Pelargonium*) (G23)

      5′ Petals strongly overlapping; ovary chambers 1 or 2, style unbranched, not forming beak on fr

        6. Sepals all fused, at least at base, forming a cup-shaped to cylindric calyx tube, gen not petal-like; the odd petal [banner] uppermost, overlapping the margins of the 2 upper lateral petals; lower 2 lateral petals gen free at base but fused toward the tip, forming a keel that encloses stamens and ovary; fr a legume or indehiscent pod . . . . **FABACEAE** (Papilionoideae)

        6′ Sepals free, 2 of them petal-like, spreading and very different from the other 3; the odd petal lowermost, often appendaged, folded and forming a keel enclosing stamens and ovary, a banner petal never present; fr a flattened capsule . . . . **POLYGALACEAE** (G15,18,22,23)

  4′ Filaments free (0 or fused in pairs)

    7. Fls solitary; petals 5 . . . . **VIOLACEAE** (G5,15,20)

    7′ Fls in raceme or spike; petals 2 or 4

      8. Infl a spike; ovary 4-lobed, stigmas 4; developing fr open at tip; stamens 3–4 . . . . . . **RESEDACEAE** (*Oligomeris*)

8′ Infl a raceme or panicle; ovary unlobed or 2-lobed, stigma 1–2; developing fr closed at tip
 9. Sepals free to base; stamens 6; fls not subtended by bracts. . . . . [2]**BRASSICACEAE** (G2,15,20)
 9′ Sepals proximally fused to hypanthium; stamens 3; each fl subtended by a bract
 . . . . **SAXIFRAGACEAE** (*Tolmiea*)
1′ Fl radial
 10. Pls without green pigmentation . . . . [2]**ERICACEAE** (G17,18,19)
 10′ Pls green and photosynthetic
  11. Petals 4
   12. Lvs alternate, sometimes all basal
    13. Lvs ± fleshy; sepals 2 . . . . **MONTIACEAE** (*Calyptridium*) (G16)
    13′ Lvs not fleshy; sepals 4. . . . . [2]**BRASSICACEAE** (G2,15,20)
   12′ Lvs opposite or whorled
    14. Lvs ± cylindric or 3-angled in ×-section, fleshy; fls unisexual, dioecious per or subshrub; petals
     present only in staminate fls. . . . . **BATACEAE** (G6,8,15)
    14′ Lvs not fleshy; fls bisexual, ann to subshrub; petals present in all fls
     15. Sepals free or nearly so, hypanthium 0
      16. Petals 6–20 mm, each bearing a prominent, fringed appendage — petals actually fused but easily
       misinterpreted as free . . . . **GENTIANACEAE**
      16′ Petals ≤ 6 mm, not appendaged
       17. Fls pedicelled, terminal or from distal axils, sometimes in cymes; lf without stipules; stamens 4
        or 8. . . . . [4]**CARYOPHYLLACEAE** (G6,8,15)
       17′ Fls ± sessile in lf axils (pedicel elongating in fr); lf with stipules; stamens 8
       . . . . **ELATINACEAE** (*Elatine californica*)
     15′ Sepals fused, at least toward base, or hypanthium present
      18. Lvs gen shallowly few-toothed; infl a dense cyme or raceme borne on a terminal peduncle —
       ovary actually 1/2 inferior but easily misinterpreted as superior . . . . **HYDRANGEACEAE** (G21)
      18′ Lvs entire; infl of axillary or terminal cymes or fls solitary, axillary
       19. Petals borne on receptacle; style 2–3(4)-branched, rarely unbranched . . . . [2]**FRANKENIACEAE**
       19′ Petals borne on inner face of hypanthium; style unbranched . . . . **LYTHRACEAE** (G22)
  11′ Petals 5 or more
   20. Fl unisexual . . . . **EUPHORBIACEAE** (*Ditaxis*) (G22)
   20′ Fl bisexual
    21. Sepals 2 or 3
     22. Lvs ± fleshy; sepals persistent on opened fl and fr . . . . **MONTIACEAE** (G19,22)
     22′ Lvs not fleshy; sepals early-deciduous, 0 in open fl and fr
     . . . . **PAPAVERACEAE** (Papaveroideae) (G12,15,20,22)
    21′ Sepals 5 or more
     23. Stamens > in number than petals
      24. Lvs alternate, sometimes all basal
       25. Style branches and stigmas 5; style elongating, forming a beak; ovules 1 per ovary chamber; fr
        breaking into 5 one-seeded mericarps, each tipped with a coiled beak segment
        . . . . **GERANIACEAE** (*Geranium*) (G20)
       25′ Styles or style branches 1–3, stigmas 1–3; style not elongating or forming a beak; ovules several–
        many per ovary chamber; fr a capsule
        26. Stigmas 2–3, sessile or borne on 2–3 styles or style branches. . . . . [2]**SAXIFRAGACEAE** (G10,20,21)
        26′ Stigma 1, sessile or borne on an unbranched style
         27. Fl disk-shaped, without a hypanthium; stamens dehiscent by pores . . . . [2]**ERICACEAE** (G17,18,19)
         27′ Fl with a tubular hypanthium; stamens dehiscent along the sides by slits . . . . . [2]**LYTHRACEAE** (*Lythrum*)
      24′ Lvs opposite or whorled
       28. Stigma and style 1
        29. Fl without a hypanthium; lf toothed . . . . **ERICACEAE** (*Chimaphila*)
        29′ Fl with a tubular hypanthium; lf entire . . . . [2]**LYTHRACEAE** (*Lythrum*)
       28′ Stigmas 2–6
        30. Lf lobed; opposite lvs a single pair, the remainder all basal; stigmas 3 — ovary chamber 1,
         placentas parietal — ovary actually ± 1/2 inferior. . . . . **SAXIFRAGACEAE** (*Lithophragma cymbalaria*)
        30′ Lf entire or shallowly toothed; opposite lvs gen several–many pairs; stigmas 2–6
         31. Lf entire; ovary chamber 1 (rarely 2–4)
          32. Lf not gland-dotted; petals white to pink, purple, or red; placentas free-central (axile)
          . . . . [4]**CARYOPHYLLACEAE** (G6,8,15)
          32′ Lf gland-dotted; petals yellow; placentas parietal . . . . **HYPERICACEAE** (G22)
         31′ Lf crenate-serrate to serrate; ovary chambers 5–6
          33. Fls 1–few, sessile or short-pedicelled in lf axils; ann; lf teeth gland-tipped . . . . . **ELATINACEAE** (*Bergia*)
          33′ Fls in a dense cyme or raceme borne on a terminal peduncle; per to subshrub; lf teeth not
           gland-tipped — ovary actually 1/2 inferior but easily misinterpreted as superior
           . . . . **HYDRANGEACEAE** (*Whipplea*)

23′ Stamens = in number to petals or fewer

  34. Lvs opposite or whorled

    35. Lvs, or at least distal-most, toothed or lobed (proximally gen entire in *Sclerolinon*)

      36. Petals white to pink or purple; infl an umbel . . . . [2]**GERANIACEAE**

      36′ Petals yellow; infl a cyme . . . . **LINACEAE** (*Sclerolinon*)

    35′ Lvs entire

      37. Stamens opposite petals — petals actually fused but easily misinterpreted as free

        . . . . **MYRSINACEAE** (G19)

      37′ Stamens alternate petals or of different numbers

        38. Sepals evidently fused above base

          39. Lf ± linear, acute . . . . [4]**CARYOPHYLLACEAE** (G6,8,15)

          39′ Lf oblong to obovate, obtuse, margin often curled under and lf appearing narrower

            . . . . [2]**FRANKENIACEAE**

        38′ Sepals free or nearly so

          40. Petals without fringed nectary pits . . . . [4]**CARYOPHYLLACEAE** (G6,8,15)

          40′ Base or center of each petal with 2 fringed nectary pits (sometimes appearing as 1 due to overlapping fringes) — petals actually fused but easily misinterpreted as free

            . . . . **GENTIANACEAE** (*Swertia*)

  34′ Lvs alternate, sometimes all basal

    41. Lf linear, entire

      42. Lvs cauline; fls solitary or in racemes, panicles, or cymes . . . . [2]**LINACEAE**

      42′ Lvs all basal; fls in scapose heads . . . . **PLUMBAGINACEAE** (*Armeria*)

    41′ Lf wider, often toothed or lobed

      43. Lf covered with long, gland-tipped, insect-trapping hairs . . . . **DROSERACEAE**

      43′ Lf without insect-trapping hairs

        44. Styles, style branches, or stigmas 2–4

          45. Fls solitary; stamens alternating with toothed or fringed staminodes; stigmas 4 . . . . . **PARNASSIACEAE**

          45′ Fls 2–many; staminodes 0; stigmas 2–3

            46. Ann; lf veins pinnate or lateral veins obscure; hypanthium 0; ovary unlobed . . . . [2]**LINACEAE**

            46′ Per; lf veins ± palmate; hypanthium present; ovary gen 2–3-lobed distally

              . . . . [2]**SAXIFRAGACEAE** (G10,20,21)

        44′ Styles or style branches 5 or more

          47. Filaments fused in tube around style; petals thread-like . . . . **MALVACEAE** (*Ayenia*)

          47′ Filaments free or fused only at base; petals wider

            48. Style elongating, forming a beak; fr breaking into 5 one-seeded mericarps, each tipped with a coiled beak segment; fls in umbels . . . . [2]**GERANIACEAE**

            48′ Styles not elongating and forming a beak; fr not breaking into mericarps; fls in a raceme, cyme, or panicle-like infl

              49. Sepals free, ± green; fr a capsule, 5- or 10-seeded; fls in a raceme or cyme . . . . . . **LINACEAE** (*Linum*)

              49′ Sepals fused, gen pink to blue; fr indehiscent, 1-seeded; fls in a panicle-like infl

                . . . . **PLUMBAGINACEAE** (*Limonium*)

LYCOPHYTES

Free-sporing pls; sporangia solitary in lf axils

ISOETACEAE QUILLWORT FAMILY

W. Carl Taylor & Jon E. Keeley

Per, aquatic to terrestrial. **ST**: buried, corm-like, 2–3-lobed, corky, brown. **LF**: simple, in grass-like tufts, spirally arranged on st top, erect to spreading, < 30 cm, linear above base. **SPORANGIUM**: solitary, embedded in wide lf base, < 1 cm, ± covered by a translucent membrane, male or female; male spores > 10000, < 0.045 mm, ± bean-shaped, gray or brown in mass; female spores 20–200, 0.2–0.7 mm, spheric, white, ± smooth, ridged, tubercled, or prickly. 1 genus, 200+ spp.: worldwide. [Taylor et al. 1993 FNANM 2:64–75] Scientific Editors: Alan R. Smith, Thomas J. Rosatti.

ISOETES QUILLWORT

(Greek: evergreen, from habit of some spp.) Perhaps most poorly known lycophyte genus. Mature female spores, found in decaying lf bases or soil, critical for identification, as are hand lens for texture when dry, microscope with micrometer for size. Hybrids (spores of variable size, shape) common between aquatic spp., making them less distinct.

1. Pl terrestrial (or becoming so), of seasonally wet soil, lake margins, temporary streams, vernal pools; gen < 1500 m
  2. Translucent membrane covering < 75% of sporangium . . . . *I. howellii*
  2′ Translucent membrane covering > 75% of sporangium
    3. Pl of wet soil; lf gen > 8 cm, > 1 mm wide at middle, rigid, ± brittle. . . . . *I. nuttallii*
    3′ Pl of vernal pools; lf gen < 8 cm, < 1 mm wide at middle, soft, flexible . . . . *I. orcuttii*
1′ Pl underwater in persistent lakes or pools; gen > 1500 m
  4. Female spores prickly. . . . . *I. echinospora*
  4′ Female spores ridged or tubercled
    5. Lf abruptly tapered to tip; female spore 0.3–0.5 mm diam . . . . *I. bolanderi*
    5′ Lf gradually tapered to tip; female spore 0.5–0.7 mm diam . . . . *I. occidentalis*

I. bolanderi Engelm. (p. 115) Pl underwater. **LF**: deciduous, < 20 cm, rigid, not brittle, abruptly narrowed to tip, bright green; base white to ± brown. **SPORANGIUM**: membrane covering < 30%; male spores 0.02–0.03 mm, brown in mass; female spores 0.3–0.5 mm, ridged, tubercled. 2*n*=22. Persistent lakes, pools; > 1300 m (250 m in Marin Co.). KR, NCoRH, CaRH, SNH, SnFrB, SnBr, Wrn; to BC, WY, NM. [*I. b.* var. *pygmaea* (Engelm.) Clute] Hybridizes with *I. echinospora, I. occidentalis*. Spores mature late summer.

I. echinospora Durieu (p. 115) Pl underwater. **LF**: ± evergreen, < 20 cm, soft, flexible, tapered to tip, bright green; base white to ± brown. **SPORANGIUM**: membrane covering < 50%; male spores 0.02–0.03 mm, gray in mass; female spores 0.4–0.5 mm, prickly. 2*n*=22. Persistent lakes, pools; > 1500 m. KR, n SNH; to AK, e N.Am; Eurasia. Hybridizes with *I. bolanderi, I. occidentalis*. Spores mature late summer.

I. howellii Engelm. (p. 115) Pl becoming terrestrial. **LF**: deciduous, < 30 cm, rigid, not brittle, tapered to tip, bright green; base ± brown to black. **SPORANGIUM**: membrane covering < 75%; male spores 0.025–0.035 mm, brown in mass; female spores 0.3–0.5 mm, ridged. 2*n*=22. Vernal pools, lake margins; gen < 1500(2900) m. KR, NCoR, CaR, SNF, n SNH, GV, CCo, SnFrB, SCoR, SCo, PR; to WA, MT, UT. Small pls of SCo, WA, Baja CA (lf < 10 cm, female spore

< 0.42 mm) are assignable to *I. howellii* var. *minima* (A.A. Eaton) N. Pfeiff., recognition of which remains questionable pending further research, although it is ± clear that it should not be treated as an infraspecific taxon of *I. howellii*. Spores mature late spring, summer.

I. nuttallii Engelm. (p. 115) Pl terrestrial. **LF**: deciduous, gen > 8 cm, > 1 mm wide at middle, rigid, ± brittle, tapered to tip, light green to gray-green; base white to ± brown, outermost sterile, often surrounded by several black scales. **SPORANGIUM**: membrane covering > 75%; male spores 0.02–0.03 mm, brown in mass; female spores 0.35–0.6 mm, ± shiny, ± tubercled. 2*n*=22. Seasonally wet soil, temporary streams; gen < 1500(2760) m. NCoR, CaR, SN, ScV, n SnJV, CW (exc SCoRI), SCo, PR, w MP; to BC. Spores mature late spring, summer.

I. occidentalis L.F. Hend. (p. 115) Pl underwater. **LF**: evergreen, < 20 cm, rigid, brittle, tapered to tip, dark green; base brown-white. **SPORANGIUM**: membrane covering < 50%; male spores 0.035–0.045 mm, gray in mass; female spores 0.5–0.7 mm, ridged, tubercled. 2*n*=66. Persistent lakes, pools; > 1500 m. KR, SNH; to BC, CO. Hybridizes with *I. bolanderi, I. echinospora*. Spores mature late summer.

I. orcuttii A.A. Eaton Pl becoming terrestrial. **LF**: deciduous, gen < 8 cm, < 1 mm wide at middle, soft, flexible, tapered to tip, bright

green; base white to ± brown, outermost fertile, often surrounded by several black scales. **SPORANGIUM**: membrane covering > 75%; male spores 0.02–0.025 mm, brown in mass; female spores 0.2–0.4 mm, ± shiny, ± smooth. $2n=22$. Vernal pools; < 1700 m. NCoRI (Howell Mtn, Napa Co.), CaR, n&c SNF, SNH, GV, CCo, SnFrB, SCo, PR; sw OR, Baja CA. Spores mature spring.

LYCOPODIACEAE CLUB-MOSS FAMILY

Andy Murdock, Alan R. Smith & Thomas Lemieux

Pl on ground (or other pls), creeping (to ± vine-like). **ST**: branches few to many, forked often unequally, also branched laterally or not. **LF**: many, simple, ± alternate, spirally arranged, small, needle- or scale-like, 1-veined, few-toothed to gen entire, those subtending sporangia gen unlike others [or not], in distinct cones [or not]. **CONE**: terminal on erect st, erect [or 0]. **SPORANGIA**: 1 in lf axils; spores of 1 kind. 4 genera, ± 400 spp.: worldwide, gen trop. [Wagner & Beitel 1993 FNANM 2:18–37] Scientific Editors: Alan R. Smith, Thomas J. Rosatti.

1. Fertile st unbranched, with 1 cone at tip . . . . **LYCOPODIELLA**
1′ Fertile st branched, with few to many cones at tip . . . . **LYCOPODIUM**

LYCOPODIELLA BOG CLUB-MOSS

ST: sterile creeping, branched in 1 horizontal plane; fertile erect. ± 40 spp.: gen Am. (Diminutive of *Lycopodium*)

L. inundata (L.) Holub (p. 115) INUNDATED BOG CLUB-MOSS **ST**: sterile 5–15(25) cm, 0.5–1 cm wide, incl lvs; fertile gen 1–2, ± 4–6(9) cm. **LF**: ± 3–8 mm, 1 mm wide, those subtending sporangia ± wider at base, ± not bristle-tipped. **CONE**: 1.5–3(6) cm, ± 8–10 mm wide. $2n=156$. Peat bogs, muddy depressions, pond margins; < 50 m (NCo), ± 1000 m (n SNH). NCo (Humboldt Co.), n SNH (Nevada Co.); to AK, e N.Am, Eur, Asia. ★

LYCOPODIUM RUNNING-PINE, CLUB-MOSS

ST: sterile branched in > 1 plane; fertile erect. ± 40 spp.: worldwide. (Greek: wolf foot, from branch tips)

L. clavatum L. (p. 115) RUNNING-PINE **ST**: sterile wide-creeping to ± vine-like, < 0.5 m, 0.5–1.5 cm wide, incl lvs; fertile ± 10–20 cm terminated by 2–5 cones. **LF**: ± 4–7 mm, 0.5–1 mm wide, with narrow bristle tip, those subtending sporangia ovate-triangular, abruptly bristle-tipped. **CONE**: 1–3(5) cm, ± 4–6 mm wide. $2n=68$. Moist ground, swamps (on trees); < 200 m. NCo, NCoRO; to AK, MT, NM, e N.Am; Caribbean, S.Am, Eur, Afr, Asia. ★

SELAGINELLACEAE SPIKE-MOSS FAMILY

Paul Wilson & Thomas J. Rosatti

ST: wiry, gen rooting adventitiously [or not], branching variable, gen not fragile when dry. **LF**: many, simple, overlapped, appressed, small, ± scale-like, 1-veined, gen grooved abaxially [or not] nearly to tip, gen ± of 2 kinds ("under-lvs" under main st, "over-lvs" over it). **CONE**: paired or 1, terminal, gen 4-sided, fertile lvs not like sterile, gen strongly keeled. **SPORANGIA**: 1 per lf axil, 2 kinds, male (gen more distal in cones, spores many, small), female (spores (1)4, large, gen orange-yellow). 1 genus, ± 700 spp.: worldwide, gen trop, warm temp. [Valdespino 1993 FNANM 2:38–63] Scientific Editors: Alan R. Smith, Thomas J. Rosatti.

SELAGINELLA SPIKE-MOSS

(Latin: small *Selago*, ancient name for some *Lycopodium*) Some cult as groundcover, curiosity: *S. kraussiana* (Kunze) A. Braun; *S. lepidophylla* (Hook. & Grev.) Spring, resurrection plant. Hand lens, gen at ± 20×, required for lvs (shape, margin, awn at tip), cones. *S. kraussiana* may be naturalized in CA, differs from native CA taxa in lvs lacking abaxial groove.

1. Lf tip awn 0 or inconspicuous
  2. Lvs of main st not pointed up, under-lvs ± like over-lvs; lvs near tip of sterile shoots gen < 0.4 mm wide; pl pale green aging tan . . . . ***S. cinerascens***
  2′ Lvs of main st with tips pointed up, under-lvs longer, more curved around st than over-lvs; lvs near tip of sterile shoots often > 0.4 mm wide; pl green aging orange-brown . . . . ***S. eremophila***
1′ Lf tip awn >> lf margin teeth
  3. Pl in festoons, curly when dry; on, under trees; humid NCo, KR, NCoRO . . . . ***S. oregana***
  3′ Pl erect, creeping, or mat-forming, not curly when dry; not on, under trees; humid NCo, KR, NCoRO or not
    4. Pl with ± erect rootless sts arising from ground; lf base hairs on margins, abaxial surface . . . . ***S. bigelovii***
    4′ Pl without ± erect rootless sts arising from ground; lf base hairs gen on margins only (also on surfaces in *Selaginella asprella*)
      5. For most sterile lvs of main st, distance from end of abaxial groove to base of awn gen > 1/3 maximum lf width; awns gen entire; 1350–4100 m . . . . ***S. watsonii***

5′ For most sterile lvs of main st, distance from end of abaxial groove to base of awn < 1/3 maximum lf
width; awns toothed; < 2700 m
6. Pl with some lvs red, pink-red, or red-streaked; pl not dense cushions, at margins loosely spreading,
branching ± pinnate; s NCoRI, CaR, SN, ScV (Sutter Buttes), n SCoR; 250–1700 m . . . . *S. hansenii*
6′ Pl without any lvs red, pink-red, or red-streaked; pls dense cushions or not, at margins not loosely
spreading, branching not pinnate; NW, n SNF (Butte Co.), n&c SNH, s SN, n SnFrB (Marin Co.), TR,
PR, DMtns; 600–2700 m
7. Awn length often > lf width, margins with cilia >> 3 × longer than wide; lf hairs on surface at base
on some lvs . . . . *S. asprella*
7′ Awn length gen < lf width, margins with teeth gen < 3 × longer than wide; lf hairs on margins only
8. Branches not elongate, not spreading, 2° branches rarely above substrate; st fragile when dry;
DMtns . . . . *S. leucobryoides*
8′ Branches ± elongate, often spreading, 2° branches frequently above substrate; st ± not fragile when
dry; n CA
9. Tip of lf near sterile st tip with end of abaxial groove not surrounded by thickening; lvs of main st
gen swept upward, under-lvs curved, ± > over-lvs, decurrent; sterile shoot internodes gen < 7 mm,
tip with distal-most awns more exserted than those just below . . . . *S. scopulorum*
9′ Tip of lf near sterile st tip with end of abaxial groove surrounded by thickening; lvs of main st not
swept upward, ± alike, under-lvs rarely unlike over-lvs, not decurrent; sterile shoot internodes
often > 7 mm, tip with distal-most awns not more exserted than those just below . . . . *S. wallacei*

S. asprella Maxon (p. 115) BLUISH SPIKE-MOSS Pl in loose mat;
green with white cast aging tan then gray. **ST**: fragile when dry. **LF**:
of main sts 1.5–2.3 mm, 0.4–0.6 mm wide, lanceolate to narrowly
so, decurrent, base often sparse-hairy on surface; awns in tufts at st
tips, 0.5–1.4 mm, often > lf width, wavy, ciliate. **CONE**: 0.4–2.5 cm.
Exposed rocky spots, montane conifer woodland; 1080–2700 m. c
SNH, s SN, TR, PR; Baja CA. Report in FNANM of nw CA based
on mislabeled specimen. ★

S. bigelovii Underw. (p. 115) Pl a series of ± erect, rootless shoots
of ± determinate growth from rhizome; green aging tan then gray.
LF: of main sts 1.5–2.7 mm, 0.4–0.6 mm wide, linear to narrowly
lanceolate, not decurrent, base hairy on surface; awn 0.3–0.7 mm, ±
rigid, toothed. **CONE**: 0.4–1.5 cm. Open sites, mineral soil on rock
outcrops, amid shrubs; < 2000 m. s NW, s SN, SnJV, CW, SW, sw
edge DMoj, DSon; Baja CA.

S. cinerascens A.A. Eaton (p. 115) ASHY SPIKE-MOSS Pl a
loose tangle of prostrate runners; pale green aging tan. **LF**: of main
sts 1.1–2 mm, 0.2–0.5 mm wide, lanceolate, not decurrent, base glabrous, tips acute near st tip, with small marginal teeth; awn 0. **CONE**:
gen 1, 2–5 mm, ± like sterile branches. Sunny spots or under shrubs,
often "red clay"; < 550 m. SCo, PR; Baja CA. ★

S. eremophila Maxon (p. 115) DESERT SPIKE-MOSS Pl a mat,
loosely spreading at margins where branching sub-pinnate; green
aging orange-brown. **LF**: of main st 0.4–0.6 mm wide, tip pointed up,
acute; under-lf of main st 1.5–3 mm, ± lanceolate or sickle-shaped,
decurrent, over-lf 1.3–1.5 mm, lanceolate, not decurrent, base hairy
or not; awn 0–0.03 mm. **CONE**: sparse, 3–10 mm. Shaded sites,
sandy or gravelly soils, at base of rocks, in cracks; < 1100 m. e PR,
DSon; to AZ, n Mex. ★

S. hansenii Hieron. (p. 115) Pl a sprawling mat, loosely spreading at margins where branching ± pinnate; green aging tan then gray,
some lvs red, pink-red, or red-streaked. **LF**: 1–4.5 mm, 0.4–0.7 mm
wide, not decurrent; under-lvs lance-linear, grading to linear-deltate
over-lvs; base glabrous; awn 0.4–1.4 mm, in tufts at st tips. **CONE**:
Sparse, 4–8 mm. Rocks of outcrops, river gorges, foothill woodland;
250–1700 m. s NCoRI, CaR, SN, ScV (Sutter Buttes), SnJV, n SCoR.

S. leucobryoides Maxon MOJAVE SPIKE-MOSS Pl cushion-like,
in little tufts; green with white cast aging to tan. **ST**: fragile when dry.

LF: of main sts 1.5–4.5 mm, 0.4–0.6 mm wide, lanceolate, decurrent,
base glabrous; awn 0.2–0.6 mm, rigid. **CONE**: 4–15 mm. Base of
rocks, cracks, gen limestone, amid scrub, pinyon/juniper woodland;
600–2300 m. DMtns; sw NV, n AZ. ★

S. oregana D.C. Eaton (p. 115) Pl in festoons, curly when dry,
internodes > 1 cm; green or ± yellow aging to tan, then gray. **LF**: of
main sts 1.4–2.4 mm, 0.4–0.06 mm wide, lanceolate, ± decurrent,
base glabrous, margin ± toothed; awn 0.1–0.4 mm, ± rigid, entire.
CONE: often paired, 10–60 mm. On, under trees on rocks, shaded
to open sites, streambanks, humid conifer forest; < 800 m. NCo, KR,
NCoRO; w of Cascades to WA.

S. scopulorum Maxon (p. 115) ROCKY MOUNTAIN SPIKE-MOSS
Pl a mat (often with dense shag of cones), sterile shoot internodes gen
< 7 mm; green aging orange-tan. **LF**: of main sts 1.6–4 mm, 0.3–0.5
mm wide; under-lf of main st ± lance-linear or sickle-shaped, decurrent, over-lf lanceolate, not decurrent; base glabrous; awn 0.3–0.8
mm, ± toothed. **CONE**: (5)10–30(45) mm. Open, rocky spots, conifer
forest; 1400–2200 m. n KR; to BC, MT, CO, NM. [*S. densa* Rydb.
var. *s.* (Maxon) R.M. Tryon] ★

S. wallacei Hieron. (p. 115) Pl mat-forming in sun, to elongate-
creeping with sterile shoot internodes often > 7 mm in shade; green
aging tan then gray. **LF**: of main sts 1.4–2.5 mm, 0.3–0.7 mm wide,
lanceolate, gen not decurrent, color unlike brown of st, base glabrous,
tips of lvs of sterile sts thickened around end of abaxial groove; awn
0.2–0.5(0.9) mm, rigid, ± toothed. **CONE**: often paired, 10–45(90)
mm. Rocky outcrops, open to shady, dry to moist sites, chaparral to
mixed-evergreen forest; < 1820 m. NW, CaRF, n SNF (Butte Co.), n
SNH, n SnFrB (Marin Co.); to BC, MT.

S. watsonii Underw. (p. 115) Pl in shaggy cushions, dense to ±
open but branches gen close enough to ± obscure internodes, shoots
with a robust, braided appearance; green aging ± orange-tan then
gray. **LF**: of main sts 1.5–4 mm, 0.4–0.7 mm wide, lance-linear to
oblong, decurrent, base glabrous; distance from end of abaxial groove
to base of awn gen > 1/3 maximum lf width, tips of lvs of sterile sts ±
thickened around end of abaxial groove; awn 0.2–0.5 mm, gen entire,
rigid. **CONE**: 5–35 mm. Open rocky sites, conifer forest, alpine;
1350–4100 m. KR (Trinity Alps), SNH, TR, PR, n SNE, W&I, n
DMtns (Telescope Peak); to OR, MT, UT.

FERNS

Free-sporing pls; sporangia in cones in *Equisetum* or on lvs or modified
lf segments, along veins or in clusters, bands, or cases

ASPLENIACEAE SPLEENWORT FAMILY

Alan R. Smith

Rhizome-scale cells with lateral walls dark brown to ± black, surficial walls clear. **LF:** stipe ×-section with 1 X-shaped or 2 back-to-back C-shaped vascular strands; segment veins gen free. **SPORANGIA:** in linear [to oblong] sori along veins; indusia linear, opening away from veins; stalk cells in 1 row; spores elliptic, winged. 1 or 2 genera (most segregates now subsumed in *Asplenium*), 700 spp.: worldwide, esp trop. Scientific Editor: Thomas J. Rosatti.

ASPLENIUM SPLEENWORT

Pl in soil or on rocks; rhizome gen short-creeping to erect. **LF:** often tufted, gen glabrous; rachis often ± winged; blade simple or 1[many]-pinnate or forked; pinnae often more developed acroscopically, often without obvious midrib. **SPORANGIA:** indusia persistent, covering sori when young, later reflexed. (Greek: spleen)

1. Lvs simple or 1–2-forked, blades narrowly linear . . . . *A. septentrionale*
1′ Lvs 1-pinnate, blades linear, pinnae many, blades wider than linear
  2. Stipe ± green or straw-colored above red-brown base . . . . *A. viride*
  2′ Stipe dark red- to purple-brown ± throughout
    3. Stipe 0.4–0.6 mm wide, narrowly winged adaxially; pinnae 3–6(7) mm, gen shallowly crenate on
      acroscopic and distal margins . . . . *A. trichomanes* subsp. *trichomanes*
    3′ Stipe gen 0.5–1 mm wide, unwinged; pinnae gen 5–12 mm, margin shallowly lobed or toothed ±
      throughout . . . . *A. vespertinum*

A. septentrionale (L.) Hoffm. (p. 115) NORTHERN SPLEENWORT
LF: simple or 1–2-forked, many, densely clustered, 5–15 cm, grass-like; stipe 2.5–12 cm, ± 0.5 mm wide, unwinged, red-brown at base, ± green above, dull; blade gen 1–2 mm wide, narrowly linear, sometimes with small, ± linear, gen sterile teeth near or at tip. **SPORANGIA:** sori 5–15 mm, gen 1 per segment but sometimes appearing to have as many as 4, corresponding to gen the same number of teeth per segment. 2*n*=144. Crevices of granite rocks; 2500–3350 m. CaRH, s SNH; to OR, SD, TX, n Baja CA; also WV, Eur, Asia. ★

A. trichomanes L. subsp. ***trichomanes*** (p. 115) MAIDENHAIR SPLEENWORT **LF:** 1-pinnate, many, clustered, 8–25 cm; stipe 1–3(5+) cm, 0.4–0.6 mm wide, narrowly winged adaxially, dark purple-brown, shiny; blade gen 0.5–1.5 cm wide, linear; pinnae 15–30(37) pairs, 3–6(7) mm, 2–3 mm wide, oblong, gen shallowly crenate on acroscopic and distal margins. **SPORANGIA:** sori gen 1–1.5 mm, gen 2–4 pairs per pinna. 2*n*=72. On rocks; 200 m. nw KR (Del Norte Co.); widespread in N.Am, Eur, Asia, s temp. ★

A. vespertinum Maxon (p. 115) WESTERN SPLEENWORT **LF:** 1-pinnate, many, clustered, 8–30 cm; stipe 1–5(8) cm, gen 0.5–1 mm wide, unwinged, dark red- to purple-brown, shiny; blade gen 1–2.5 cm wide, linear; pinnae 20–30 pairs, gen 5–12 mm, 2–8 mm wide, oblong, margin shallowly lobed or toothed ± throughout. **SPORANGIA:** sori gen 0.5–1.5 mm, gen 2–6 pairs per pinna. Base of overhanging boulders; 200–1000 m. SCo, SnGb, PR (exc SnJt); n Baja CA. ★

A. viride Huds. GREEN SPLEENWORT **LF:** 1-pinnate, gen few, clustered, 3–20 cm; stipe 2–5 cm, 0.5–1 mm wide, unwinged, red-brown at base, ± green or straw-colored above, dull; blade gen 8–12 mm wide, linear; pinnae 8–15 pairs, 3–7 mm, 3–6 mm wide, ± rhombic or broadly ovate, gen shallowly crenate on acroscopic and distal margins. **SPORANGIA:** sori gen 1–1.5 mm, gen 1–3 pairs per pinna. 2*n*=72. On rocks in limestone seams in metamorphic rocks; 2050 m. n SNH (e side of Sierra Buttes, Sierra Co.); to AK, e N.Am, Eur, Asia. [*A. trichomanes-ramosum* L., nom. rej.] ★

AZOLLACEAE MOSQUITO FERN FAMILY

Alan R. Smith & Andy Murdock

Pl free-floating or stranded on mud, gen 1–5 cm, often fan-shaped; roots pendent from st forks, unbranched. **ST:** forked repeatedly or pinnate, thread-like, easily fragmented at joints. **LF:** alternate, in 2 rows, sessile, often overlapped, 0.5–1.5 mm,

seemingly paired but actually of 2 ± round to ovate lobes; upper lobe floating or emergent, thick, ± green or ± red, margin ± white, adaxial surface smooth or gen with papillae; lower lobe submersed, gen ± larger, thinner, ± white. **SPORANGIA**: in seemingly axillary cases of 2 kinds, cases gen in pairs of 1 kind. **MALE SPORANGIUM CASE**: 1.2–2 mm diam, spheric; tip dark-pointed; wall transparent; sporangia gen 20–100+, long-stalked; spores 32 or 64, spheric, in gen 3–6 barbed masses. **FEMALE SPORANGIUM CASE**: 0.2–0.4 mm diam, hemispheric or spheric; tip obtuse, covered by dark, conic, spongy structures that aid in flotation; wall ± opaque; sporangium 1, sessile; spore 1, spheric. 1 genus, ± 6 spp.: ± worldwide. When *Salvinia* and *Azolla* in same family, the name is Salviniaceae. Scientific Editor: Thomas J. Rosatti.

AZOLLA

(Greek: dry kill, from pl death in dried habitats) [Reid et al. 2006 Int J Pl Sci 167:529–538] Used as green manure in rice paddies because of nitrogen-fixing algae in upper lf lobe; spp. identification difficult, depends in part on fertile material (gen 0 on herbarium specimens). Spp. hybridize in culture.

1. Sts equally to often unequally dichotomously branched or pinnately branched, forming round to elongate
   pls to 3 cm, 2 cm wide; lvs ovate or oblong-ovate, gen 1.2–2 mm, 1 mm wide; lf margins of upper (± green
   or ± red) lf lobes with broad ± white band of cells ≥ 4 cells wide; male spore mass barbs lacking partitions
   . . . . *A. filiculoides*
1′ Sts ± equally dichotomously branched, forming ± round pls gen 1–1.3 cm wide; lvs ovate to ± round,
   0.5–0.8 mm, 0.5–0.7 mm wide; lf margins of upper (green or ± red) lf lobes with narrow ± white band of
   cells 1–2 cells wide; male spore mass barbs with 1–3 partitions . . . . *A. microphylla*

A. filiculoides Lam. (p. 115) Pl green to ± red. **ST**: gen 1–3 cm: immature prostrate, internodes < 5 mm; mature ascending, internodes < 1 mm. **LF**: smooth or with gen inconspicuous papillae on upper lf lobe. **SPORANGIUM CASES**: male and female, often 0, female with distinct equatorial girdle, wall tubercled and pitted. Common. Ponds, slow streams; < 1300 m. CA-FP, GB, DMoj(?); to WA, AZ, S.Am; also e US, Eurasia, Afr.

A. microphylla Kaulf. MEXICAN MOSQUITO FERN Pl green or blue-green to dark red. **ST**: prostrate, gen 1–1.3 cm; internodes < 1 mm. **LF**: with gen conspicuous papillae on upper lf lobe. **SPORANGIUM CASES**: male and female, often 0, female with equatorial girdle, wall ± smooth to pitted. Ponds, slow streams; < 1200 m. n&s SNH, GV, CCo, SnFrB, SnBr, SNE; to BC, c US, S.Am. [*A. mexicana* C. Presl] Geog ± uncertain due to confusion with *A. filiculoides*. ★

BLECHNACEAE DEER FERN FAMILY

John C. Game, Alan R. Smith & Thomas Lemieux

Pls in soil [climbing]; rhizome short- to long-creeping to erect, scaly; new growth often ± red. **LF**: of 1 or 2 kinds, fertile, sterile; stipe in ×-section with vascular strands in circle; blade [(simple or 2-pinnate)] deeply pinnately lobed to 1-pinnate, pinnae deeply pinnately lobed or not, hairs gen 0; veins free or netted. **SPORANGIA**: sori linear to oblong, along veins parallel to nearest midrib; indusium shaped ± like sorus, opening towards nearest midrib; stalk cells in 2–3 rows; spores elliptic, scar linear. ± 9 genera, ± 250 spp.: worldwide, esp trop; several spp. cult. New classification badly needed, in which all genera but *Woodwardia* are placed under *Blechnum* or, preferably, ± 25 genera, incl ± 10 new ones, are recognized. Scientific Editors: Alan R. Smith, Thomas J. Rosatti.

1. Blade deeply pinnately lobed to 1-pinnate, pinnae unlobed; lvs of 2 kinds, fertile, sterile; sori 2 per pinna,
   linear, > 10 × longer than wide, 1 along each side of costa. . . . . **BLECHNUM**
1′ Blade 1-pinnate, pinnae deeply pinnately lobed; lvs of ± 1 kind; sori many per lobe, oblong, 2–4 × longer
   than wide, end-to-end along each side of lobe midrib, some oblong to linear along costa also . . . . **WOODWARDIA**

BLECHNUM

Rhizome suberect [or long-creeping, climbing, or erect]. **LF**: of [± 1] or 2 kinds, fertile > sterile, with much narrower pinnae or lobes; lower pinnae reduced or not; veins of sterile lvs free. > 200 spp.: worldwide, esp trop, subtrop. (Greek: ancient name for ferns)

B. spicant (L.) Roth (p. 115) DEER FERN **LF**: stipe short, base ± black, with persistent ± brown scales; sterile lvs forming rosette, spreading to arching, firm, deeply pinnately lobed to 1-pinnate, gen < 1 m, 2–10 cm wide; pinnae gen 20–80 pairs, 5–8 mm wide, entire to shallowly crenate, lower gradually reduced to semicircular lobes < 5 mm; fertile lvs appearing after sterile, in center of rosette, ± erect, 1-pinnate, > sterile, pinnae ± 2 mm wide, unlobed. 2*n*=68. Shaded, neutral to acid moist areas; ± 0–1500 m. NCo, NCoRO, n SNH, CCo, SnFrB; to AK, ID; Eur, sw Asia, n Afr. [*Struthiopteris s.* (L.) Weiss] Locally common.

WOODWARDIA CHAIN FERN

Rhizome prostrate to ascending; scales dense, orange-brown. **LF**: all alike, lower pinnae gen slightly reduced or not; veins of sterile lvs netted but free at margin. **SPORANGIA**: sori oblong. 14 spp.: gen temp, subtrop, N.Am, Eur, e Asia. (T.J. Woodward, British phycologist, 1745–1820)

W. fimbriata Sm. (p. 121) GIANT CHAIN FERN Rhizome prostrate, short, stout. **LF**: evergreen, gen 1–3 m, coarse; stipe gen 5–15 mm wide at base, scales large, orange-brown to straw-colored; pinnae gen 15–30 cm, often glandular, lobed ± to midrib, lower ± reduced. **SPORANGIA**: sori gen ± 2–4 mm. 2*n*=68. Near streams, springs, seeps; ± 0–2300 m. CA-FP (rare GV); to BC, NV, AZ, nw Mex. [*W. chamissoi* Brack.; *W. radicans* (L.) Sm., misappl.]

115

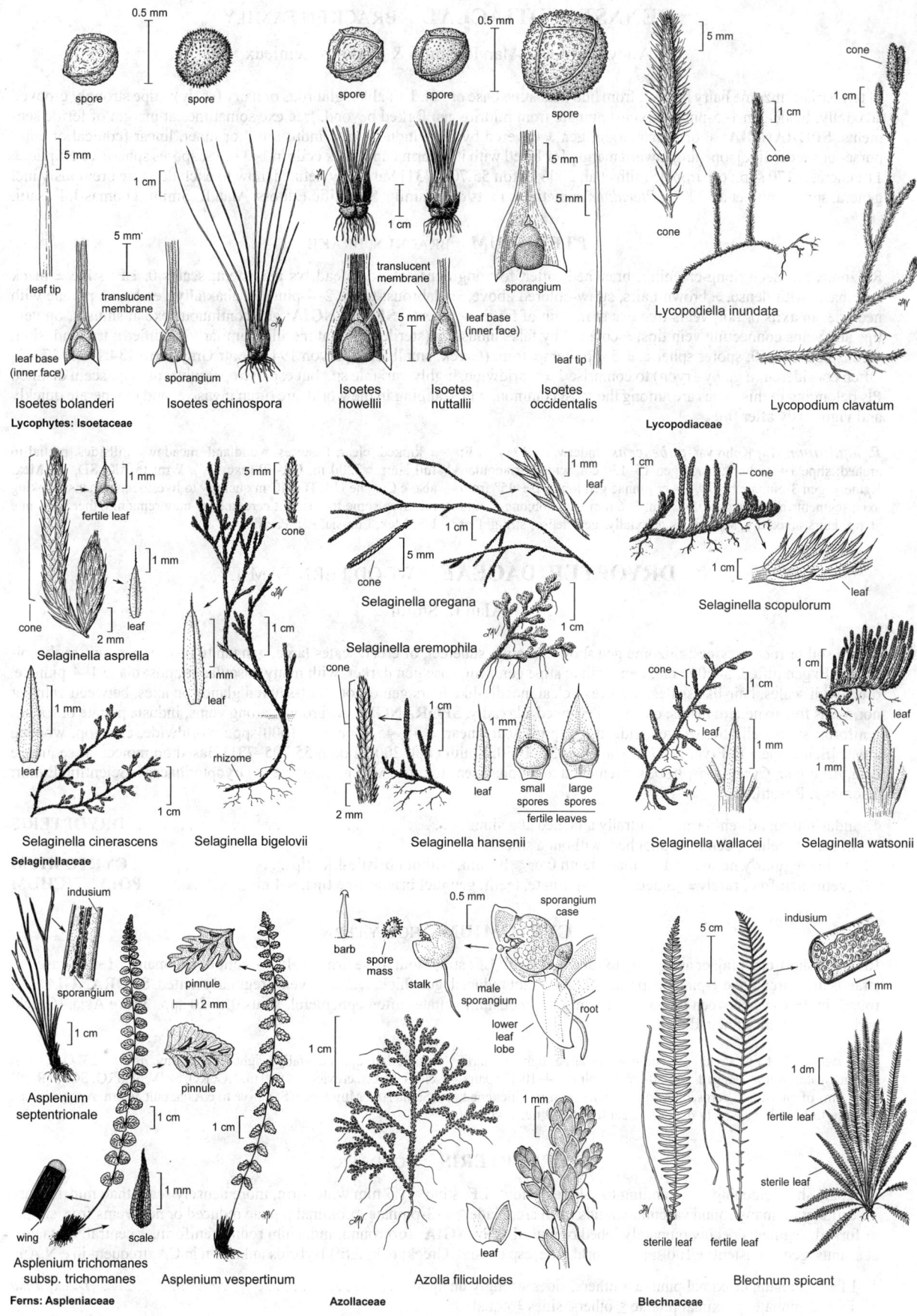

spore

spore

spore

spore

spore

leaf tip

translucent membrane

leaf base (inner face)

sporangium

Isoetes bolanderi

Lycophytes: Isoetaceae

Isoetes echinospora

translucent membrane

sporangium

Isoetes howellii

Isoetes nuttallii

leaf base (inner face)

Isoetes occidentalis

leaf tip

Lycopodiella inundata

cone

cone

cone

cone

Lycopodium clavatum

Lycopodiaceae

fertile leaf

cone

leaf

cone

Selaginella asprella

cone

leaf

rhizome

Selaginella bigelovii

leaf

Selaginella cinerascens

Selaginellaceae

cone

Selaginella oregana

leaf

Selaginella eremophila

cone

Selaginella hansenii

leaf

small spores

large spores

fertile leaves

cone

Selaginella scopulorum

leaf

cone

leaf

Selaginella wallacei

leaf

Selaginella watsonii

indusium

sporangium

Asplenium septentrionale

pinnule

pinnule

Asplenium vespertinum

wing

scale

Asplenium trichomanes subsp. trichomanes

Ferns: Aspleniaceae

barb

spore mass

stalk

male sporangium

sporangium case

lower leaf lobe

root

leaf

Azolla filiculoides

Azollaceae

indusium

sterile leaf

fertile leaf

fertile leaf

sterile leaf

Blechnum spicant

Blechnaceae

DENNSTAEDTIACEAE   BRACKEN FAMILY

Andy Murdock, Alan R. Smith & Thomas Lemieux

Pl terrestrial; rhizome hairy [scaly], from bud near stipe base or not. **LF**: alike, glabrous or hairy (scaly); stipe strongly grooved adaxially; blade gen 1–5-pinnate; veins pinnate from midrib, gen forked beyond, free exc sometimes at margin of fertile segments. **SPORANGIA**: at or near margin, gen ± covered by false indusia; true indusium 0 or inner, linear [conical or cup-, purse- or saucer-like], opening toward margin [or fused with it to form cup]; stalk cells in 1–3 rows; spores spheric or elliptic. ± 11 genera, ± 170 spp.: esp trop. [Smith et al. 2006 Taxon 55:705–731] Variously defined, now to exclude some previously incl genera, spp. (Smith et al. 2006); *Pteridium* sometimes in its own family. Scientific Editors: Alan R. Smith, Thomas J. Rosatti.

PTERIDIUM   BRACKEN, BRAKE

Rhizome gen deep, long-creeping, branched; often forming dense stands; dead lvs persistent; scales 0. **LF**: stipe ± black near base, with dense, ± brown hairs, straw-colored above, ± glabrous; blade 2–4-pinnate, abaxially gen hairy; pinnae with nectaries in axils or not; veins free exc at margin of fertile segments. **SPORANGIA**: gen continuous exc at sinuses, on vein tips and veins connecting vein tips, ± covered by false indusium (sterile segment margins similarly modified); true indusium inconspicuous or 0; spores spheric. ± 5 spp.: temp, trop. (Greek: small fern) [Tryon 1941 Contr Gray Herb 134:1–31, 37–67] Often considered (e.g., by Tryon) to comprise 1 ± worldwide, highly variable sp., but esp in trops, subtrops, spp. seem distinct. Pls belonging to this genus are among the most common, wide-ranging in the world, are often invasive, and regenerate quickly and vigorously after fires.

P. aquilinum (L.) Kuhn var. ***pubescens*** Underw. (p. 121)   **LF**: ± arched; stipe 10–100 cm; blade gen 15–150 cm, widely-triangular, leathery, gen 3-pinnate below, lower pinnae gen longest, ± 45° from axis; segments or lobes gen 0.5–2 cm, 3–6 mm wide, oblong, round at tip, hairs abaxially, sometimes adaxially, gen dense, straight or ± kinked, clear. Pastures, woodland, meadows, hillsides, partial to full sun; < 3200 m. CA-FP (exc GV), Wrn; to AK, SD, nw Mex; also e Can, ne US. TOXIC in quantity to livestock, humans; cooking removes some toxins, but carcinogens may remain. Other vars. in e US, Mex, Eurasia, Afr, Pacific.

DRYOPTERIDACEAE   WOOD FERN FAMILY

Alan R. Smith

Per, in soil or rock crevices; rhizome gen short-creeping, suberect, or erect, scales large, gen tan to brown, gen uniformly colored. **LF**: gen tufted, 5–200+ cm, gen ± alike; stipe gen firm, base gen darker, with many vascular strands; blade 1–4-pinnate, often with scales, hair-like scales, hairs (exc clear, needle-like hairs gen 0), or short-stalked glands on axes, between veins or not, veins free to netted; rachis, costa gen grooved adaxially. **SPORANGIA**: sori round, along veins; indusia peltate or round-reniform; spores elliptic, winged, ridged, or spiny, scar linear. ± 40–45 genera, > 1600 spp.: worldwide, esp trop, wooded areas. [Schuettpelz & Pryer 2007 Taxon 56:1037–1050; Smith et al. 2006 Taxon 55:705–731] Based on molecular sequence data, *Athyrium*, *Cystopteris*, *Woodsia* removed to Woodsiaceae to preserve a monophyletic Dryopteridaceae. Scientific Editor: Thomas J. Rosatti.

1. Indusium round-reniform, ± centrally attached at a sinus . . . . **DRYOPTERIS**
1′ Indusium peltate, centrally attached, without a sinus
  2. Veins regularly netted; lf 1-pinnate, teeth 0 or < 10 mm, without bristle-like tips . . . . **CYRTOMIUM**
  2′ Veins gen free, rarely ± joined; lf 1–3-pinnate, teeth, gen incl bristle-like tips, < 4 mm . . . . **POLYSTICHUM**

CYRTOMIUM   HOLLY FERN

Rhizome short-creeping or ascending to suberect, stout. **LF**: stipe stout, base firm, scaly, ×-section with many ± round vascular strands in an arc; blade 1-pinnate, proximal pinnae not reduced, gen thick, leathery, veins regularly netted. **SPORANGIA**: sori round, in 2+ rows between pinna midrib, margin; indusium peltate, often ephemeral, sinus 0. ± 20 spp.: gen e Asia. (Greek: arch, from pattern of netted veins)

C. falcatum (L. f.) C. Presl   Rhizome scales large, light- to dark-brown, ovate, entire to jagged. **LF**: 30–80 cm; pinnae 4–10(12) pairs, 8–12 cm, often with 1 acroscopic lobe basally, margin thickened, ± entire to wavy or coarsely dentate, teeth 0 or < 10 mm, without bristle-like tips, adaxially bright green, shiny. *n*=2*n*=123. Gen moist cliffs, banks, crevices; < 900 m. NCoRO, SnJV, SCoRO, SCo, TR, PR (Santa Ana Mtns); se US; native to e Asia; cult as orn. Apogamous.

DRYOPTERIS   WOOD FERN

Rhizome short-creeping or ascending to suberect, stout. **LF**: stipe > 1.5 mm wide, firm, more densely scaly than midrib, base ×-section with many round vascular strands in an arc; blade ≥ 1–3-pinnate, proximal pinnae reduced or not, veins free, simple or forked; segments deeply pinnately lobed or not. **SPORANGIA**: sori round; indusium round-reniform, ± centrally attached at a sinus, gen persistent. ± 100 spp.: ± worldwide, esp e Asia. (Greek: oak, fern) Hybrids unknown in CA, frequent in e N.Am.

1. Lf ± 3-pinnate, proximal pinnae > others, sides strongly unequal . . . . ***D. expansa***
1′ Lf 1–2-pinnate, proximal pinnae ≤ others, sides ± equal

2. Stipe gen > 1/3 lf length; longest pinnae gen near blade base; veins into segment teeth, adaxially obscure; scales of pinna midribs lance-ovate to lanceolate; blade, indusial glands sparse to ± dense; blades thick; NW, SN, ScV, CW, SW, MP, s DMoj . . . . ***D. arguta***

2′ Stipe gen ≤ 1/3 lf length; longest pinnae gen near blade middle; veins gen not into segment teeth, adaxially ± widened, prominent at tip; scales of pinna midribs hair-like, linear; blade, indusial glands gen 0 or very sparse; blades thin; SnBr, W&I . . . . ***D. filix-mas***

D. arguta (Kaulf.) Maxon (p. 121) **LF**: 30–60(100+) cm, 12–18(30) cm wide; stipe, midrib minutely glandular; blade lanceolate, 1–2-pinnate, proximal pinnae ≤ others, longest gen near blade base, sides ± equal, basiscopic pinnules 1–1.3 × acroscopic on same pinna, segments deeply pinnately lobed or not, teeth with bristle-like tips or not, veins into teeth; scales of pinna midribs lance-ovate to ± lanceolate. 2*n*=82. Locally common. Open, wooded slopes, caves; < 2500 m. NW, SN, ScV (Sutter Buttes), CW, SW, MP (caves in Lava Beds National Monument), s DMoj; to BC, AZ.

D. expansa (C. Presl) Fraser-Jenk. & Jermy (p. 121) **LF**: 30–80(100+) cm, 10–30(40) cm wide; stipe gen darker abaxially, scales gen with a dark central stripe; blade broadly deltate, ± 3-pinnate, proximal pinnae > others, sides strongly unequal, basiscopic pinnules > 2 × acroscopic on same pinna, segments deeply pinnately lobed, teeth with bristle-like tips or not, veins gen not into teeth; longest pinnae near base. 2*n*=82. Caves, shaded, wooded areas, esp banks of streams; < 500 m (1300 m in MP). NCo, NCoRO, CCo, MP (caves in Lava Beds National Monument); to AK, Can, Rocky Mtns.

D. filix-mas (L.) Schott (p. 121) MALE FERN **LF**: ± 40–70(100+) cm, 15–25(30+) cm wide; stipe, midrib nonglandular; blade elliptic, 2-pinnate, proximal pinnae ≤ others, longest gen near blade middle, sides ± equal, basiscopic pinnules 1 × acroscopic on same pinna, segments deeply pinnately lobed or not, teeth ± without bristle-like tips, veins gen not into teeth; scales of pinna midribs ± linear or hair-like. 2*n*=164. Granitic cliffs; 2400–3100 m. SnBr, W&I; to BC, ne N.Am, Eur, Afr. ★

POLYSTICHUM SWORD FERN

Rhizome gen suberect to erect, often stout. **LF**: stipe stout, firm, gen densely scaly, ×-section with many round vascular strands in an arc; blade 1–3-[> 3–]pinnate, proximal pinnae reduced or not, thin to leathery, scaly, veins gen free, rarely ± jointed; pinna bases often wider acroscopically; teeth, gen incl bristle-like tips, < 4 mm [or teeth 0]. **SPORANGIA**: sori round; indusium peltate [0 or reniform], sinus 0. ± 175+ spp.: ± worldwide. (Greek: many rows, from rows of sori on type sp.)

1. Lf gen 1-pinnate, rarely to partly 2-pinnate; pinnae gen simple, ± entire to serrate, in *Polystichum kruckebergii* sometimes 1-lobed
  2. Lf 10–15(25) cm; pinnae simple or gen lobed at bases of lowest . . . . ²***P. kruckebergii***
  2′ Lf 10–120(200) cm; pinnae simple
    3. Proximal pinnae ± deltate, ± 1/2 longest; stipe 1/10–1/6 blade. . . . . ***P. lonchitis***
    3′ Proximal pinnae ovate to lanceolate, ± = to ± 2/3 longest; stipe gen 1/5–1/2 blade
      4. Stipe base scales ovate, ± 3(6) mm wide, those above proximal pinnae gen > 1 mm wide, persistent; pinnae gen in 1 plane; indusium ciliate. . . . . ***P. munitum***
      4′ Stipe base scales lanceolate, ± 2–3 mm wide, those above proximal pinnae gen < 1 mm wide, falling early; pinnae ± in 1 plane or not; indusium ± entire to toothed . . . . ***P. imbricans***
        5. Lf 15–80 cm; pinnae ± in 1 plane, longest 3–6(10) cm; sori submarginal . . . . subsp. ***curtum***
        5′ Lf 15–50 cm; pinnae often not ± in 1 plane, longest 2–5 cm; sori ± near midvein . . . . subsp. ***imbricans***
1′ Lf 1–2-pinnate, rarely to partly 3-pinnate; pinnae deeply pinnately lobed to 1-pinnate, at least at bases of lowest
  6. Segment teeth without bristle-like tips; serpentine . . . . ***P. lemmonii***
  6′ Segment teeth with bristle-like tips; serpentine or not
    7. Lf gen 2- to rarely partly 3-pinnate; pinnules abruptly narrowed to attachment < 0.5 mm wide, often with a lobe on distal side of base; rachis with persistent scales . . . . ***P. dudleyi***
    7′ Lf gen 1- to partly 2-pinnate; pinnules gradually tapered to attachment < 3 mm wide, lobes 0; rachis with deciduous scales
      8. Smallest scales on abaxial blade surface hair-like; blade gen > 5 cm wide . . . . ***P. californicum***
      8′ Smallest scales on abaxial blade surface linear, few hair-like; blade gen < 5 cm wide
        9. Proximal pinnae ± deltate, longest < 1(1.5) cm . . . . ²***P. kruckebergii***
        9′ Proximal pinnae ± lanceolate, longest 1.5–3 cm . . . . ***P. scopulinum***

P. californicum (D.C. Eaton) Diels (p. 121) **LF**: 40–100 cm; stipe < 1/2 blade, base scales 3–4 mm wide, lance-ovate; blade lanceolate, 1- to partly 2-pinnate; pinnae gen 4–7 cm. **SPORANGIA**: indusium ciliate. 2*n*=82,164. Woodland, streambanks, to rocky open slopes; < 1100 m. NCoRI, NCoRO, s CaRF, SN (3 sites), CW (exc SCoRI), SnBr (1 site); to BC. Probably sterile hybrid between *P. munitum* or *P. imbricans* and *P. dudleyi*. Chromosome doubling (2*n*=164) restores fertility, allows backcrossing to parents.

P. dudleyi Maxon (p. 121) **LF**: 50–100 cm; stipe ± 1/4–1/2 blade, base scales 3–7(9) mm wide, ovate; blade lanceolate, gen 2- to rarely partly 3-pinnate; pinnae ± 5–12 cm. **SPORANGIA**: indusium ciliate. 2*n*=82. Moist forests; < 500 m. NCoRO, CW (exc SCoRI). Hybridizes with *P. californicum*, *P. munitum*.

P. imbricans (D.C. Eaton) D.H. Wagner (p. 121) **LF**: stipe gen 1/5–1/2 blade, base scales ± 2–3 mm wide, lanceolate, those above proximal pinnae gen < 1 mm wide, falling early; blade narrow-lanceolate to -elliptic, 1-pinnate. **SPORANGIA**: indusium ± entire to toothed. 2*n*=82. Some pls from Butte, Tehama cos. difficult to assign to subsp. Hybridizes with *P. dudleyi* (called *P. californicum*), *P. lemmonii* (called *P. scopulinum*), *P. munitum*.

  subsp. ***curtum*** (Ewan) D.H. Wagner **LF**: 15–80 cm; pinnae ± in 1 plane, longest 3–6(10) cm. **SPORANGIA**: sori submarginal; indusium gen toothed. Rocky slopes, crevices, woodland; 400–1600 m. NCoRI, CaRH, n&c SNH, SnFrB, SCoRO, SCo, TR, PR.

  subsp. ***imbricans*** **LF**: 15–50 cm; pinnae often not ± in 1 plane, longest 2–5 cm. **SPORANGIA**: sori ± near midvein; indusium ± entire to toothed. Shaded or exposed outcrops, banks, slopes, rocky areas; 300–2500 m. KR, NCoR, CaRH, n SNF, SNH, SnFrB, SCoRO, TR, PR, MP; to BC. Some specimens from n SNH (Ahart 7561, UC) and KR (Oswald & Ahart 8756, UC), from ± 1100 m, show significant spore malformation and may be hybrids between *P. imbricans*, *P. munitum*. Differences between subsp. not always clear in n&c SN.

P. kruckebergii W.H. Wagner (p. 121) KRUCKEBERG'S SWORD FERN **LF**: 10–15(25) cm; stipe gen ± 1/10 blade, base scales 2–3 mm wide, ovate to lanceolate; blade linear to narrow-elliptic, 1- to partly 2-pinnate; pinnae gen 6–15 mm, proximal ± deltate, longest < 1(1.5) cm. **SPORANGIA**: indusium entire or minutely toothed. $2n$=164. Rocky slopes, crevices; 2100–3200 m. KR, CaRH, n&c SNH; to BC, MT, UT. Probably fertile hybrid between *P. lonchitis*, *P. lemmonii*. ★

P. lemmonii Underw. (p. 121) **LF**: 15–40 cm; stipe 1/4–1/3 blade, base scales ± 2 mm wide, gen lance-linear; blade narrow-lanceolate, 2-pinnate; pinnae gen 1–3 cm; teeth without bristle-like tips. **SPORANGIA**: indusium entire or minutely toothed. $2n$=82. Serpentine, rocks, ledges, rocky streambeds; 1200–2600 m. KR, NCoRH, CaRH, n SNH; to BC.

P. lonchitis (L.) Roth (p. 121) HOLLY FERN **LF**: gen 10–60 cm; stipe gen 1/6–1/10 blade, base scales 2–4(5) mm wide, lance-ovate; blade linear, 1-pinnate; pinnae gen 1–3(4) cm, proximal ± deltate, ± 1/2 longest. **SPORANGIA**: indusium entire or minutely toothed. $2n$=82. Gen shaded, moist or wet, granite or limestone crevices or bluffs; 1700–2600 m. KR, NCoRH (Yolla Bolly Mtns), n SNH; to AK, e Can, Rocky Mtns, Eurasia. ★

P. munitum (Kaulf.) C. Presl (p. 121) WESTERN SWORD FERN **LF**: gen 50–120(200) cm; stipe gen 1/5–1/2 blade, base scales ± 3(6) mm wide, ovate, those above proximal pinnae gen > 1 mm wide, persistent; blade lanceolate to narrow-elliptic, 1-pinnate; pinnae gen in 1 plane, 2–8(14) cm. **SPORANGIA**: indusium ciliate. $2n$=82. Common. Wooded hillsides, shaded slopes, rarely cliffs, outcrops; < 1600 m. NW, n SNF, n&s SNH, CW, n ChI, TR, MP (caves in Lava Beds National Monument); to AK, MT, SD; also Guadalupe Island (Mex). Hybrids with *P. dudleyi* are called *P. californicum*.

P. scopulinum (D.C. Eaton) Maxon (p. 121) **LF**: 10–50 cm; stipe 1/4–1/2 blade, base scales 1.5–2(3) mm wide, lanceolate to elliptic; blade narrow-lanceolate 1- to partly 2-pinnate; pinnae gen 1–3 cm, lance-oblong, proximal ± lanceolate, longest 1.5–3 cm. **SPORANGIA**: indusium entire. $2n$=164. Serpentine to acidic soils, gen full sun, rock crevices, boulder bases; 400–3200 m. KR, NCoRO, NCoRH, CaRH, SNH, SnGb, SnBr, SnJt, MP, DMtns (Surprise Canyon, Panamint Range); to BC, Rocky Mtns, AZ. Probably fertile hybrid between *P. imbricans*, *P. lemmonii*. DMtns distribution based on Rompert 229 (RSA), 1977, which differs from others in having pinna lobes 0 or shallow.

EQUISETACEAE HORSETAIL FAMILY

Robert E. Preston & Richard L. Hauke

Per from rhizome. **ST**: ann to per, gen erect, of 2 kinds (sterile, fertile) or not; internodes with lengthwise alternating ridges and grooves, hollow exc at nodes, branches 0 or whorled, alternate lvs. **LF**: scale-like, whorled, fused into nodal sheath with as many teeth as lvs, gen not green. **SPORANGIA**: several on inner surface of peltate scales that are clustered into a terminal cone; spores of 1 kind, spheric, green, unmarked, with 4 strap-like appendages. 1 genus, 15 spp.: worldwide exc Australia, New Zealand. [Hauke 1993 FNANM 2:76–84] Scientific Editors: Alan R. Smith, Thomas J. Rosatti.

EQUISETUM HORSETAIL, SCOURING RUSH

(Latin: horse, bristle, from roots of *Equisetum fluviatile* L.)

1. St branches regular
    2. Basal internode of branch > subtending nodal sheath . . . . ²***E. arvense***
    2′ Basal internode of branch < subtending nodal sheath
        3. Sheath teeth 5–10 . . . . ²***E. palustre***
        3′ Sheath teeth 14–28 on sterile st, 20–30 on fertile st . . . . ²***E. telmateia*** subsp. ***braunii***
1′ St branches 0 or irregular
    4. St brown, fleshy, lasting < 1 month irregularly in spring
        5. Sheath teeth 6–14 on sterile st, 6–10 on fertile st . . . . ²***E. arvense***
        5′ Sheath teeth 14–28 on sterile st, 20–30 on fertile st . . . . ²***E. telmateia*** subsp. ***braunii***
    4′ St green, firm, lasting > 1 month
        6. Spores white, misshapen . . . . **E. ×*ferrissii***
        6′ Spores green, spheric
            7. Sheath teeth 5–10, persistent . . . . ²***E. palustre***
            7′ Sheath teeth 10–50, gen deciduous
                8. Cone tip pointed; sheath gen with 2 dark bands, 1 at base, 1 at tip; st per, gen scabrous
                . . . . ***E. hyemale*** subsp. ***affine***
                8′ Cone tip rounded; sheath gen with 1 dark band, at tip; st ann (per in s), gen not scabrous . . . . ***E. laevigatum***

E. arvense L. (p. 121) COMMON HORSETAIL **ST**: ann, of 2 kinds. **STERILE ST**: 10–60 cm, green; basal internode of branch > subtending sheath; sheath 3–8.5 mm, ± as long as wide, teeth 6–14, 1.5–3.5 mm, dark, often joined but not fused; branch with 3–4 rounded ridges, solid. **FERTILE ST**: 11–32 cm, unbranched, fleshy, brown, ephemeral; sheath 5–11 mm, > that of sterile st, teeth 6–10, 3–7.5 mm. Streambanks, wet meadows, springs, other wet, shaded places; < 3050 m. NW (exc NCoRH), CaR, SN (exc Teh), GV, CCo, SnFrB, SCoRO, SW (exc ChI), MP, SNE, DMoj (exc DMtns); N.Am, Eur, Asia.

E. ×ferrissii Clute FERRISS' HORSETAIL **ST**: ann to per, of 1 kind, 20–180 cm, green; sheath 7–17 mm, 3–12 mm wide, often with dark bands; teeth 14–32, deciduous or persistent; branches 0. **SPORANGIA**: cone tip pointed; spores white, misshapen. Moist,

sandy or gravelly areas; < 3000 m. NCo, KR, NCoRO, CaRF, SNH, GV, CW (exc SCoRI), SCo, SnGb, PR (exc SnJt), DMtns; to BC, e US. Hybrids between *E. hyemale*, *E. laevigatum*, reproducing vegetatively from fragmented sts, sometimes forming large populations, sometimes at great distance from parents.

E. hyemale L. subsp. ***affine*** (Engelm.) Calder & Roy L. Taylor (p. 121) COMMON SCOURING RUSH **ST**: per, of 1 kind, 60–210 cm, green, scabrous; sheath 7–17 mm, ± as long as wide, gen with 2 dark bands, 1 at base, 1 at tip, teeth 22–50, gen deciduous; branches 0. **SPORANGIA**: cone tip pointed; spores green, spheric. Streams, moist, sandy, gravelly areas; < 2500 m. NCo, KR, NCoRH, NCoRO, CaR, SN (exc Teh), GV, CW (exc SCoRI), SW (exc s ChI), Wrn, SNE, D; N.Am, Mex, Guatemala. 1 other subsp., in Eur, Asia.

E. laevigatum A. Braun (p. 121) SMOOTH SCOURING RUSH **ST**: ann (per in s), of 1 kind, 30–180 cm, green; sheath 6–15 mm, longer than wide, gen with 1 dark band at tip, teeth 10–26, gen deciduous; branches 0. **SPORANGIA**: cone tip rounded. Moist, sandy or gravelly areas; < 3000 m. NW, CaR, SN (exc Teh), GV, CW (exc SCoRI), SW, GB (exc Wrn), D; to BC, e US. [*E. funstonii* A.A. Eaton]

E. palustre L. (p. 121) MARSH HORSETAIL **ST**: ann, of 1 kind, 20–80 cm, green; basal internode of branch < subtending sheath; sheath 4–9 mm, longer than wide, teeth 5–10, 2–5 mm, free; branches with 4–6 rounded ridges, hollow. Marshes; < 300 m. NCoRI, CCo (San Francisco); OR to AK, e N.Am. Presence in CA questionable;

hybrids with *E. telmateia* subsp. *braunii* reported questionably for CA, from San Mateo, Marin cos. ★

E. telmateia Ehrh. subsp. ***braunii*** (J. Milde) Hauke (p. 121) GIANT HORSETAIL **ST**: ann, of 2 kinds. **STERILE ST**: 30–100 cm, light green; basal internode of branch < subtending sheath; sheath 7–18 mm, ± as long as wide, teeth 14–28, 4–10 mm; branches with 4–5 grooved ridges, solid. **FERTILE ST**: 17–45 cm, fleshy, brown, ephemeral; sheath 1.5–4 cm, > that of sterile st, teeth 20–30, 5–16 mm; branches 0. Streambanks, roadside ditches, seepage areas; < 1000 m. NW (exc NCoRH), CaRH, n SNF, n&c SNH, CW (exc SCoRI), SW (exc SnJt); along coast to BC. 1 other subsp., in Eur, Asia. Records from higher elevations are likely *E. arvense*.

MARSILEACEAE MARSILEA FAMILY

Andy Murdock, Alan R. Smith & Thomas Lemieux

Pl gen aquatic or in perennially wet areas, gen rooted in mud; rhizome creeping, slender, branched. **LF**: floating, emergent, or out of water, ± alike; blade 1-palmate or 0, << stipe; veins not or repeatedly forked, free or netted. **SPORANGIA**: in stalked, spheric or ± flat-ovoid, hard cases of 1 kind, attached near stipe base. **SPORES**: large megaspores (female) and small microspores (male), in separate sporangia. 3 genera, ± 70 spp.: esp temp. Scientific Editors: Alan R. Smith, Bruce G. Baldwin.

1. Lf blade 1-palmate; lflets 4, wedge-shaped, hairy; lf like that of clover or wood sorrel . . . . **MARSILEA**
1′ Lf blade 0; lf grass-like . . . . **PILULARIA**

MARSILEA WATER-CLOVER

LF: blades floating and emergent; stipe of floating lvs weak. **SPORANGIUM CASES**: fused to stalk 0.8–1.7 mm, ± flat-ovoid; hairs long, dense, deciduous or not; distal tooth present or 0, tip of stalk often appearing tooth-like. ± 60 spp. (L.F. Marsigli, Italian botanist, 1656–1730) [Johnson 1986 Syst Bot Monogr 11:1–87] *M. mutica* Mett. (native to Australia, New Caledonia), *M. drummondii* A. Braun (native to Australia) cult in US in aquatic gardens, occ escape in urban areas, may persist in CA; weed of concern in se US [Knepper et al. 2002 Amer Fern J 92:243–244].

1. Distal tooth of sporangium case obtuse, < 0.4 mm or 0; sporangial stalk nodding, fused to case 0.8–1 mm
. . . . ***M. oligospora***
1′ Distal tooth of sporangium case acute, (0.2)0.4–1.2 mm; sporangial stalk ± erect, fused to case 1.1–1.7 mm
. . . . ***M. vestita*** subsp. ***vestita***

M. oligospora Goodd. (p. 127) **LF**: hairy; stipe of floating lvs ± 15 cm, others 3–6 cm; lflet ± symmetrical, sides gen straight to slightly concave, distal margin truncate or convex, faintly fine-crenate. **SPORANGIUM CASES**: 5–6 mm, 3–4 mm wide; stalk unbranched. Creek beds, flood basins, vernal pools; 1400–2000 m. KR, CaRH, n SNH, MP; to WA, MT, UT. [*M. vestita* var. *o.* (Goodd.) Dorn]

M. vestita Hook. & Grev. subsp. ***vestita*** (p. 127) **LF**: hairy; stipe of floating lvs 6–35 cm, others 3–8 cm; lflet sides gen unequally concave causing lflet asymmetry, distal margin convex, ± entire. **SPORANGIUM CASES**: 3–8 mm, 3–7 mm wide; stalk unbranched. Creek beds, flood basins, vernal pools; < 2200 m. KR, NCoRI, CaR, n&s SNF, SNH, GV, CW (exc SCoRI), SCo, WTR, SnBr, PR, MP, DSon; to w&c Can, Mex; also Peru.

PILULARIA

Aquatic, submerged or in mud, forming dense clumps. **SPORANGIUM CASES**: subterranean, fused to stalk tip, spheric, hairy; teeth 0. ± 6 spp.: gen temp. (Latin: little ball, from sporangium case)

P. americana A. Braun (p. 127) **LF**: gen 2–6(11) cm. **SPORANGIUM CASES**: ± 2–3 mm diam; stalk 1–3 mm. $2n=20$. Vernal pools, mud flats, lake margins; < 2000 m. NCoRI, CaR, SNF, n SNH, GV, CCo, SnFrB, SCoR, SCo, WTR, PR, MP; OR, Baja CA; also scattered c&se US, S.Am. Poorly collected, often overlooked due to its small, grass-like appearance.

OPHIOGLOSSACEAE ADDER'S-TONGUE FAMILY

Donald R. Farrar, except as noted

Per, small, fleshy, gen glabrous; caudex gen underground, unbranched; roots glabrous with bulblets or plantlets or not. **LF**: gen 1 per caudex per yr, divided into 2 facing parts with a common stalk,(0)1 sterile and 1(2) fertile (fertile occ aborted); sterile photosynthetic part (trophophore) separated from spore-bearing part (sporophore) at to well above ground level; trophophore simple to compound, veins free and forked or netted with incl veinlets; sporophore simple to compound, or 0 in young pls. **SPORANGIA**: dehiscent into 2 valves, ± 1 mm wide, thick-walled. 10 genera, 80–100 spp.: ± worldwide, gen rare or overlooked. [Hauk et al. 2003 Molec Phylogen Evol 28:131–151; Kato 1987 Gard Bull Straits Settlem 40:1–14] Scientific Editors: Alan R. Smith, Bruce G. Baldwin, Thomas J. Rosatti.

1. Trophophore simple, entire, not midribbed, veins netted with incl veinlets; sporangia sunken in simple axis
   of sporophore . . . . **OPHIOGLOSSUM**
1´ Trophophore gen compound (small, simple, entire or 0), gen midribbed, veins free, forked; sporangia sessile
   or short-stalked, not sunken, in gen pinnately branched sporophore
  2. Lf bud glabrous, trophophore gen < 10 cm wide, gen 1–2-pinnate (0) . . . . **BOTRYCHIUM**
  2´ Lf bud hairy, trophophore gen > 10 cm wide, gen 2–4-pinnate
    3. Lf deciduous, sporophore and trophophore joined well above ground level; trophophore sessile, blade
       thin, membranaceous . . . . **BOTRYPUS**
    3´ Lf evergreen for 1 yr, sporophore and trophophore joined at to slightly below ground level; trophophore
       stalked, blade thick, leathery . . . . **SCEPTRIDIUM**

BOTRYCHIUM MOONWORT

Roots smooth, pale yellow, without bulblets or plantlets. **LF**: deciduous; bud glabrous; sporophore and trophophore (or 2 sporophores) joined at or well above ground level; trophophore gen 1–2-pinnate (simple or entire or 0), oblong to deltate to ternately triangular, thin to fleshy, pinnae ovate to oblong and midribbed or wedge- to fan-shaped and not midribbed, veins free, forked; sporophore 1–2-pinnate, rarely absent. **SPORANGIA**: not sunken in axis; stalk 0 or short. 25–35 spp.: gen temp to arctic or alpine. (Greek: bunch of grapes, from clusters of sporangia) [Stensvold 2007 Ph.D. Dissertation, Iowa State Univ; Wagner & Wagner 1993 FNANM 2:85–106] Difficult, needs study; most spp. uncommon, sporadic; good sampling of populations highly desirable in specimens, which must be carefully spread and pressed for identification. *B. multifidum* moved to *Sceptridium*. *B. pedunculosum* W.H. Wagner, differing from *B. pinnatum* in having trophophore stalk ± = trophophore rachis (vs trophophore stalk 0 to 1/10 trophophore rachis), recently confirmed for CA, based on discovery in summer of 2010 near Reynolds Creek, w of Yosemite National Park, Calaveras Co.

1. Lf divided into 2 nearly identical sporophores and no trophophore . . . . **B. paradoxum**
1´ Lf divided into trophophore and sporophore (latter occ aborted)
  2. Trophophore ternate, appearing to be divided into 3 ± equal segments due to great enlargement and
     dissection of basal pinna pair
    3. Trophophore sessile or nearly so, stalk of sporophore 0.3–0.7 × trophophore . . . . **B. pumicola**
    3´ Trophophore clearly stalked, sporophore stalk gen ≥ trophophore . . . . **B. simplex**
      4. Middle pinnae narrowly attached to rachis, slightly or not decurrent; pls yellow-green; gen in
         seasonally moist meadows . . . . var. **compositum**
      4´ Middle pinnae broadly attached to rachis, strongly decurrent; pls blue-green; gen in hard-water-
         saturated substrate . . . . var. **simplex**
  2´ Trophophore pinnate; basal pinnae not disproportionately enlarged
    5. Basal pinna pair pinnately dissected, midribbed; pinnae ovate to elliptic, broadest near base or middle
       . . . . **B. pinnatum**
    5´ Basal pinna pair entire or palmately dissected, fan-shaped, broadest at outer margin
      6. Trophophore and sporophore joined well below mid-lf, gen at ground level . . . .
         . . . . **B. simplex** or **B. pumicola** (small or shade pls, see couplet 3 to separate the spp.)
      6´ Trophophore and sporophore joined gen near or distal to mid-lf
        7. Side margins of simple pinnae ± parallel or, if cleft into segments, side margins of segments ± parallel
          8. Trophophore ± sessile; sporophore stalk very short, pinnae elongate, longer than wide, not decurrent
             to rachis . . . . **B. lineare**
          8´ Trophophore and sporophore stalk long, pinnae rhomboid, as long as wide, strongly decurrent . . . . **B. montanum**
        7´ Side margins of simple pinnae converging at > 30°, or, if cleft into segments, segments wedge-shaped
          9. Side margins of basal pinnae converging at (90)150–180°; pinnae touching to overlapping,
             especially toward tip
            10. Middle pinnae spreading, ± perpendicular to rachis; pinna texture delicate, outer margins gen
                finely crenate to dentate; basal sporophore branches often downturned, occurring in or near
                saturated substrate . . . . [2]**B. crenulatum**
            10´ Middle pinnae ascending; pinna texture firm, outer margins entire to wavy to coarsely toothed;
                basal sporophore branches gen not downturned
              11. Spores 33–39 μm in longest diam . . . . **B. lunaria**
              11´ Spores 46–57 μm in longest diam . . . . **B. yaaxudakeit**
          9´ Side margins of basal pinnae converging at 30–120(160)°; pinnae gen well-spaced to touching
            12. Pinnae spreading, ± perpendicular to rachis; fertile portion of sporophore triangular to deltate;
                basal sporophore branches often downturned
              13. Pinnae fan-shaped, texture delicate, deep green, outer margins gen finely crenate to dentate;
                  occurring in or near saturated substrate . . . . [2]**B. crenulatum** (sun pls)
              13´ Pinnae oblong or rounded without a sharp angle between side and outer margin, texture thick,
                  yellow-green, outer margins entire or shallowly cleft; occurring in well-drained sites . . . . **B. tunux**
            12´ Pinnae ascending; fertile portion of sporophore oblong to deltate; basal sporophore branches not downturned
              14. Pinnae entire to symmetrically cleft into 2–4 segments, outer margins dentate; sporophore
                  branches short, stiffly erect . . . . **B. ascendens**
              14´ Pinnae entire to irregularly cleft or shallowly lobed, outer margins entire to coarsely toothed or
                  lobed; sporophore branches long, spreading . . . . **B. minganense**

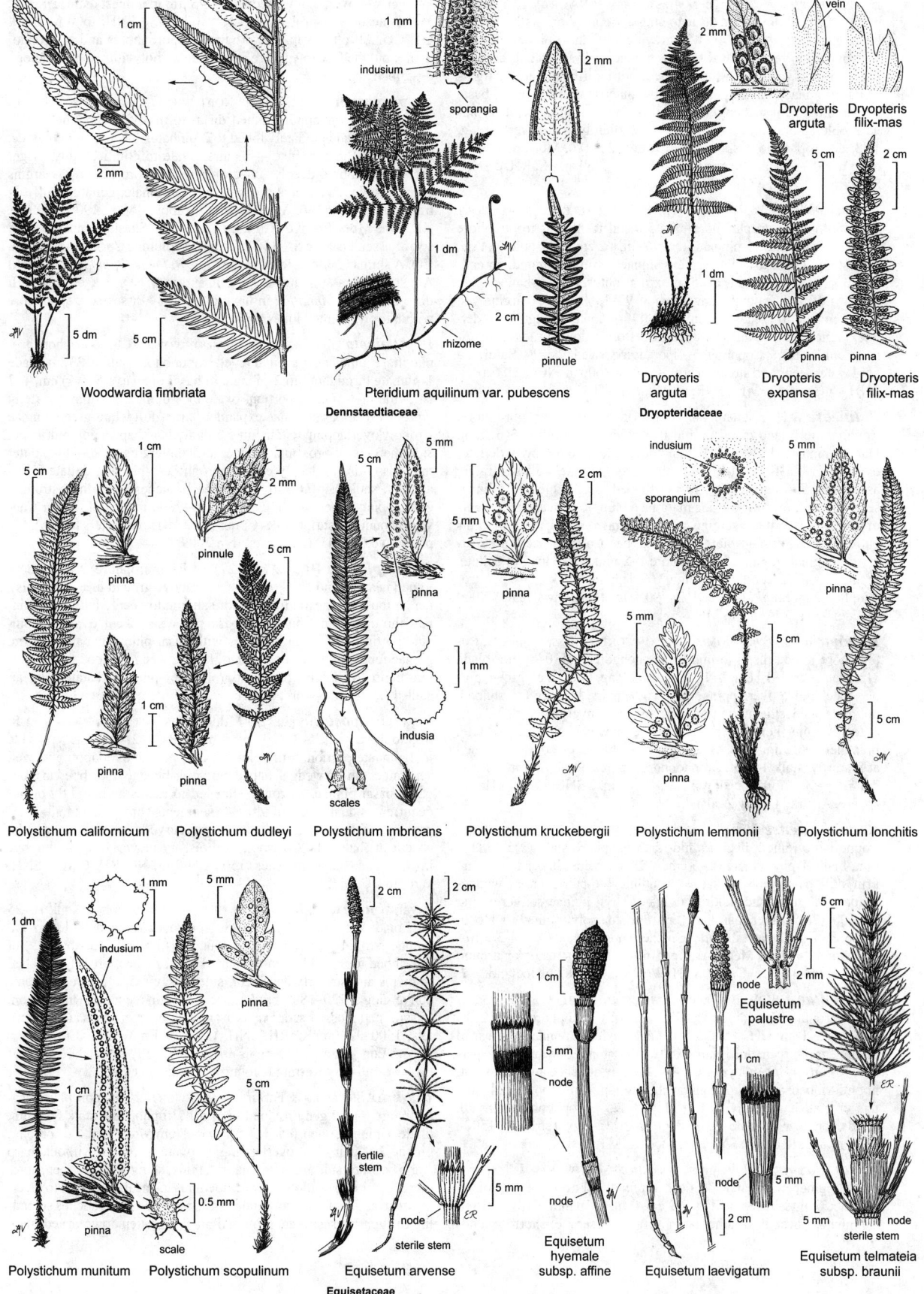

Woodwardia fimbriata

Pteridium aquilinum var. pubescens
Dennstaedtiaceae

Dryopteris arguta
Dryopteris filix-mas

Dryopteris arguta
Dryopteris expansa
Dryopteris filix-mas
Dryopteridaceae

Polystichum californicum
Polystichum dudleyi
Polystichum imbricans
Polystichum kruckebergii
Polystichum lemmonii
Polystichum lonchitis

Polystichum munitum
Polystichum scopulinum
Equisetum arvense
Equisetaceae
Equisetum hyemale subsp. affine
Equisetum palustre
Equisetum laevigatum
Equisetum telmateia subsp. braunii

B. ascendens W.H. Wagner UPSWEPT MOONWORT **LF:** sporophore, trophophore joined distal to mid-lf; trophophore stalk < 1 mm, blade 1-pinnate, < 6 cm, < 1.5 cm wide, oblong to oblong-deltate, thin but firm, veiny, yellow-green, pinnae well separated, ascending, < 5 pairs, acutely wedge-shaped, not midribbed, side margins of basal pinnae converging at 30–90°, outer margins dentate, basal pinnae often with scattered marginal sporangia; sporophore 1–2-pinnate, oblong, branches short, stiffly erect, slightly to not overlapping, stalk ± = trophophore, sporangia crowded. 2*n*=180. Moist meadows, open woodland near streams or seeps; 1500–3200 m. s CaRH, c SNH, SNE; to AK, MN, e Can. ★

B. crenulatum W.H. Wagner (p. 127) SCALLOPED MOONWORT **LF:** sporophore, trophophore joined distal to mid-lf; trophophore stalk 2–8 mm, blade 1-pinnate, gen < 6 cm, < 2 cm wide, oblong, thin, soft, shiny, green to yellow-green, pinnae ± well separated, spreading, 3–5 pairs, not midribbed, proximal obtusely fan-shaped, side margins of basal pinnae converging at 90–160°, outer margins gen finely crenate to dentate; sporophore 1–2-pinnate, triangular to deltate, branches spreading, basal branches often downturned at maturity, stalk 1–1.3 × trophophore, sporangia spaced. 2*n*=90. Saturated hard water seeps and stream margins; 1500–3600 m. NCoRH, CaRH, SNH, SnGb, SnBr, Wrn, SNE; to WA, MT, UT. ★

B. lineare W.H. Wagner SLENDER MOONWORT **LF:** sporophore, trophophore joined distal to mid-lf; trophophore stalk 0–3(6) mm, blade 1-pinnate, 1–4 cm, < 1.5 cm wide, oblong to narrowly deltate-ovate, thin but firm, green to white-green, pinnae well separated, ascending, 3–6 pairs, linear or fan-shaped, deeply cleft into linear segments, not midribbed, side margins of simple pinnae or segments of cleft pinnae not converging or converging at < 20°, outer margins entire to coarsely and bluntly toothed, basal pinnae occ with scattered marginal sporangia; sporophore 1–2-pinnate, oblong to deltate, branches stiff, ascending, strongly overlapping, stalk 0.2–0.5 × trophophore, sporangia crowded. 2*n*=90. Moist meadows; 2500–4000 m. c&s SNH; to AK, e to MN. ★

B. lunaria (L.) Sw. COMMON MOONWORT **LF:** sporophore, trophophore joined distal to mid-lf; trophophore stalk 0–8 mm, blade 1-pinnate, gen 6–10 cm, 2–4 cm wide, oblong, thick, dark green, pinnae touching to overlapping, gen 4–6(9) pairs, broadly fan-shaped, not midribbed, side margins of basal pinnae converging at 120–180°, outer margins gen entire; sporophore mostly 1-pinnate, deltate, branches spreading to ± ascending, basal branches not downturned at maturity, stalk 1–1.5 × trophophore; spores 33–39 μm in longest diam. 2*n*=90. Moist meadows; 2300–3400 m. c SNH, Wrn; N Hemisphere, Australia, New Zealand. ★

B. minganense Vict. MINGAN MOONWORT **LF:** sporophore, trophophore joined distal to mid-lf; trophophore stalk (2)5–10(15) mm, blade 1-pinnate, gen 2–5 cm, < 1.5 cm wide, linear to oblong, firm, dull, pale green, pinnae ± ascending, 3–6 pairs, fan- or wedge-shaped, not midribbed, side margins of basal pinnae converging at 50–100(120)°, outer margins ± entire to coarsely toothed or lobed; sporophore 1-pinnate, deltate, branches spreading, stalk 1–1.5 × trophophore. 2*n*=180. Meadows, open forest along streams or around seeps; 1500–3100 m. CaRH, SNH, Wrn; to AK, e N.Am, Iceland. ★

B. montanum W.H. Wagner WESTERN GOBLIN **LF:** sporophore, trophophore joined distal to mid-lf; trophophore stalk 2–15 mm, blade 1-pinnate, 0.5–2.5 cm, < 1 cm wide, irregularly linear to oblong, gray-green, dull, pinnae square or oblong, side margins ± parallel to converging at an angle < 45°, broadly attached and decurrent, not midribbed; sporophore unbranched to 1-pinnate, sporangial clusters or branches widely spaced, stalk 1–2.5 × trophophore. 2*n*=90. Shady conifer woodland, esp under *Calocedrus* along streams; 1500–2100 m. s CaRH, n&s SNH, Wrn; to AK, MT. ★

B. paradoxum W.H. Wagner PARADOX MOONWORT **LF:** sporophore segments 2, nearly identical, 1 a modified trophophore, joined at or above mid-lf; segments 1.5–3 cm, stalk 2–10 mm, fertile portion 1-pinnate, linear in outline, lateral branches sessile, short, ascend-ing, gen not overlapping; spores 36–43 μm in longest diam. 2*n*=180. Moist meadows, shrubby slopes; ±4000 m. c SNH; to WA, s AB, UT, CO. All other spp. occ produce 2 sporophores and no trophophore, but the 2 segments are not identical, not with the morphology described above. ★

B. pinnatum H. St. John NORTHWESTERN MOONWORT **LF:** sporophore, trophophore joined distal to mid-lf; trophophore stalk 0–0.2 mm, blade pinnately lobed to 2-pinnate, 3–6 cm, 2–4 cm wide, oblong to deltate, pinnae 4–7 pairs, ovate to elliptic, shiny, bright green, midribbed, deeply lobed at base to entire near tip, margins entire to minutely crenate; sporophore 2-pinnate, deltate, stalk ± = trophophore. 2*n*=180. Moist fields, shrubby slopes; 1900–2800 m. KR (Etna Mills, Siskiyou Co.), CaRH (Mount Shasta, Domingo Lake se of Lassen Peak), c SNH (Bond Pass, Tuolumne Co.); to AK, c Can, CO. A similar 2 ×-dissected moonwort, *B. lanceolatum* (S.G. Gmel.) Angstr., occurs at Crater Lake, OR and may be expected in CA. It differs from *B. pinnatum* in having ternately dissected trophophore with narrow, pointed lobes. ★

B. pumicola Underw. PUMICE MOONWORT **LF:** sporophore, trophophore joined near ground level; trophophore stalk 0–3 mm, blade 1-pinnate throughout to 2-pinnate in basal pinnae, 1.5–5(7) cm, 1–3 cm wide in 1-pinnate portion, ovate in 1-pinnate portion, ternate in large pls with basal pinnae expanded, firm, dull white-green, pinnae strongly overlapping to touching, broadly fan-shaped, not midribbed, side margins of proximal pinnae gen converging at 90–180°, outer margins entire to deeply cleft; sporophore 1-pinnate, deltate to tri-angular, stalk 0.4–0.6 × trophophore, branches ascending, strongly overlapping. Open volcanic soil; 2700–2800 m. CaRH (Diller Can-yon, Mount Shasta); to OR (Crater Lake to Three Sisters region). CA pls, found in 1941, rediscovered in 2009. ★

B. simplex E. Hitchc. (p. 127) **LF:** sporophore, trophophore joined near ground level (may be well above ground in ± young pls), gen at top of lf sheath, trophophore simple to deeply lobed to 1-pin-nate throughout to 2-pinnate in basal pinnae, < 12 cm, ovate in 1-pin-nate portion, ternate in large pls with basal pinnae expanded, firm, pinnae touching to well separated, fan- to wedge shaped, outer mar-gins entire to slightly crenate; sporophore 1-pinnate, deltate to linear, stalk 1–2 × trophophore, branches ascending, well spaced.

var. ***compositum*** (Lasch) Milde YOSEMITE MOONWORT **LF:** trophophore stalk 5–15(20) mm, blade 3–9(12) cm, 1–3 cm wide in 1-pinnate portion, dull yellow-green, pinnae overlapping to well separated, fan- to wedge-shaped, not midribbed to midribbed in elon-gated basal pinnae, narrowly attached to rachis and slightly or not decurrent, side margins of ultimate segments converging at 90–150°, proximal side margins not strongly recurved; sporophore may be absent in shade pls. Common in moist meadows over granite, occ in soft water seeps, marshes; 1500–3800 m. NCoRH, CaRH, SNH, Wrn; to BC, MT, CO.

var. ***simplex*** LEAST MOONWORT **LF:** trophophore stalk 10–25 mm, blade 2–5 cm, 1–3 cm wide in 1-pinnate portion, dull blue-green, pinnae strongly overlapping to touching, broadly fan-shaped, not midribbed to midribbed in elongated basal pinnae, broadly attached to rachis and strongly decurrent, side margins of ultimate segments converging at 120–180°, proximal side margin recurved. Uncommon. In saturated moss or sedge mats around hard water seeps and stream-lets; 1500–3200 m. CaRH, SNH, WTR, SnBr, Wrn, SNE; to OR, e N.Am, Eur, Japan. W N.Am pls differ genetically from e N.Am pls of this var. and may warrant recognition as distinct var.

B. tunux Stensvold & Farrar MOOSEWORT **LF:** sporophore, tro-phophore joined gen proximal to mid-lf; trophophore stalk 0–3 mm, blade 1-pinnate, gen 0.8–1.5 cm, 0.6–1 cm wide, ovate to oblong, pinnae 2–4 pairs, not overlapping, obovate to round, without sharp angle between side and outer margin, thick, shiny, yellow green, outer margin entire to shallowly cleft, side margins nearly parallel to meet-ing at an angle of 90° sporophore 1-pinnate, deltate, branches spread-ing to slightly ascending, proximal branches often downturned, stalk

0.5 to 1 × trophophore; spores 38–42 µm in longest diam. $2n=90$. Well-drained, rocky meadows; 3600 m. c SNH (Mount Hoffman, Mariposa Co.); to AK, CO, s NV. S US pls small, not developing morphology typical of sp. in n habitats. ★

B. yaaxudakeit Stensvold & Farrar GIANT MOONWORT **LF:** sporophore, trophophore joined distal to mid-lf; trophophore stalk 0–2 mm, blade 1-pinnate, gen 1.5–2.5 cm, 1.3–1.5 cm wide, ovate to oblong, thick, dark green, pinnae touching to overlapping, gen 4–5 pairs, broadly fan-shaped, not midribbed, side margins of basal pinnae converging at 90–160°, outer margins gen entire; sporophore mostly 1-pinnate, deltate, branches spreading, basal branches gen not downturned at maturity, stalk 1–1.5 × trophophore; spores 46–57 µm in longest diam. $2n=180$. Moist alpine meadows; 3200 m. c SNH (Virginia Canyon); ne OR to AK. CA pls small, not developing morphology typical of the sp. in n habitats. ★

BOTRYPUS RATTLESNAKE FERN

Roots smooth, without bulblets or plantlets. **LF:** deciduous; bud hairy; sporophore, trophophore joined well above ground level; trophophore blade gen 3–4-pinnate, deltate to ternately triangular, thin, membranous, ultimate segments midribbed; sporophore 1–3-pinnate, absent in young pls. **SPORANGIA:** not sunken in axis; stalk 0 or short. 2–3 spp.: worldwide. (Greek: bunch of grapes, from clusters of sporangia)

B. virginianus (L.) Michx. (p. 127) Pl often robust, herbaceous, deciduous; roots 2 mm thick (1 cm from base), smooth, yellow to brown. **LF:** bud hairy; trophophore sessile, < 20 cm wide, ultimate segments linear to ovate, veins free, forked, margins entire to coarsely serrate to deeply lobed; sporophore stalk long, 2–3-pinnate. $2n=184$. Moist shaded valleys along small streams; 700–1200 m. KR, CaR; throughout Am, Eur, Asia. [*Botrychium v.* (L.) Sw.] ★

OPHIOGLOSSUM ADDER'S-TONGUE

Alan R. Smith

Roots smooth, pale, gen with bulblets or plantlets. **LF:** trophophore simple, linear to lanceolate or cordate, not midribbed, entire, firm, herbaceous, tip rounded, acuminate, or often mucronate, veins netted with incl free branched or unbranched veinlets; sporophore gen > sterile, unbranched, slender. **SPORANGIA:** in 2 rows, sunken in a linear, long-stalked axis. 20–25 spp.: gen warm temp, trop. (Greek: snake's tongue, from extended sporophore of lf) Incl highest chromosome numbers known in vascular pls.

1. Trophophore 1–2(3) per caudex, separated from sporophore ± at ground, blade lanceolate, < 3(4) cm, < 1 cm wide, tip acuminate or mucronate . . . . ***O. californicum***
1′ Trophophore 1 per caudex, separated from sporophore well above ground, blade oblanceolate to obovate, gen 4–10 cm, 1.8–4 cm wide, tip rounded. . . . . ***O. pusillum***

O. californicum Prantl (p. 127) CALIFORNIA ADDER'S-TONGUE Caudex < 20 mm, 4 mm wide. **LF:** blade gen ± folded, thick, green, sporophore 1–2.5 × sterile. **SPORANGIA:** 8–15 pairs; rows of sporangia 8–15 mm; sterile tip of fertile stalk 0.3–1 mm. Uncommon but sometimes locally abundant, often overlooked. Grassy pastures, chaparral, vernal pool margins; 60–450 m. n&c SNF, GV, CCo, SCo, sw PR; Mex. Emerges in early spring during rainy periods; probably under-collected. ★

O. pusillum Raf. (p. 127) NORTHERN ADDER'S-TONGUE Caudex gen < 10 mm, 3 mm wide. **LF:** blade flat, pale green, sporophore 1.3–3 × sterile. **SPORANGIA:** 10–30 pairs; rows of sporangia 10–30 mm; sterile tip of fertile stalk 1–3 mm. $2n=\pm960$. Marsh edges, low pastures, grassy roadside ditches, vernal pool margins; 1100–2000 m. e KR (Siskiyou Co.), NCoRH (Lake, Mendocino cos.), n SNH (El Dorado Co.); to AK, ne&n-c N.Am. True *O. vulgatum* L., adder's-tongue fern, unknown in N.Am. ★

SCEPTRIDIUM GRAPE-FERN

Roots smooth or cork-ridged, dark gray, without bulblets or plantlets. **LF:** evergreen for 1 yr; bud hairy; sporophore and trophophore joined near or at ground level; trophophore gen 2–3-pinnate, deltate to ternately triangular, fleshy, leathery, ultimate segments midribbed, veins free, forked; sporophore 1–3-pinnate, aborted in young pls. **SPORANGIA:** not sunken in axis; stalk 0 or short. 15–20 spp.: gen temp to arctic or alpine. (Greek: scepter, staff, from tall upright sporophore)

S. multifidum (S.G. Gmel.) Tagawa (p. 127) LEATHER GRAPE-FERN Pl often robust, fleshy; roots 5 mm thick (1 cm from base), encircled by coarse, ± black, corky ridges. **LF:** bud densely hairy; trophophore stalk gen < blade, blade thick, leathery, gen ± 2–3-pinnate, < 35 cm wide, ultimate segments ovate, margins entire to shallowly crenate; sporophore stalk long, 2–3-pinnate. $2n=90$. Common. Wet meadows, edges of lakes and streams, among willows; < 2900 m. NW, CaRH, SNH, CCo, SnFrB, Wrn; to AK, e N.Am, Eur. [*Botrychium m.* (S.G. Gmel.) Rupr.]

POLYPODIACEAE POLYPODY FAMILY

Alan R. Smith

Per, on pls, rocks, in rock crevices, or in soil, humus, or on dunes; rhizome short- to long-creeping, branched, glaucous to not, scaly. **LF:** ± alike or of 2 kinds, fertile and sterile; stipe thin to thick, gen straw-colored or green to brown, base persistent on rhizome; blade gen simple to 1-pinnate, membranous to fleshy or leathery, veins free to gen fused, often netted. **SPORANGIA:** sori round to elongate (linear), gen 1 per areole, in 1–several rows on each side of segment midrib; indusium 0; spores elliptic, ± smooth to coarse-tubercled or -ridged, scar linear. ± 40 genera, ± 650 spp.: worldwide, esp trop; many spp. cult. Scientific Editor: Thomas J. Rosatti.

POLYPODIUM POLYPODY

Rhizome long-creeping; scales lanceolate, gen ± brown, 1-colored or often with darker central area or midstripe. **LF**: 0.2–10(20) dm, ± alike or fertile > sterile; stipe glabrous to scaly; blade 1-pinnate to gen deeply pinnately lobed (or simple, unlobed), hairy to not, glandular or not, scales on abaxial midrib near base gen lanceolate or lance-linear, gen ± brown; veins free to fused. **SPORANGIA**: sori in 1 row on each side of segment midrib, gen raised, sometimes incl sporangium-like structures, shriveled sporangia, or branched or unbranched glandular hairs; spores yellow. ± 40 spp.: gen New World, temp, trop, few boreal. (Latin: many feet, from persistent petiole bases) [Hildebrand et al. 2002 Amer Fern J 92:214–228] Identification complicated in CA by fact that 2 or more co-occurring spp. often hybridize (often indicated by malformed spores), esp in CCo (esp Point Reyes), NCo, where the sterile hybrids may outnumber the parental spp., and because coastal ecotypes of several spp. often have thicker, more succulent blades than inland forms. *P. australe* Fée exc (dubiously reported from but not persisting on San Clemente Island).

1. Lf blade midrib adaxially glabrous, blade ± membranous to leathery, not fleshy
  2. Lf segment gen < 2.5 cm, gen < 1 cm wide; lf blade ± membranous to ± thick, not leathery, ± firm, not brittle; sori 1–2.5 mm . . . . ***P. hesperium***
  2′ Lf segment 2.5–7(10) cm, 0.9–1.8(2.5) cm wide; lf blade thick, leathery, firm, brittle; sori 2–6 mm . . . . . . ***P. scouleri***
1′ Lf blade midrib adaxially hairy, if glabrous, blade membranous to fleshy, not leathery, often firm
  3. Veins free; sori gen round . . . . ***P. glycyrrhiza***
  3′ Veins free and fused; sori round to gen ovate or oblong
    4. Lf blade deltate to ovate, often ± irregular in outline, lower 1–3 segment pairs often ≥ those above; sori gen ± sunken, round to gen ovate; CCo, SCoRO, SW . . . . ***P. californicum***
    4′ Lf blade oblong-ovate, ± regular in outline, lower 1–3 segment pairs gen < those above; sori not sunken, ovate to oblong; NW (exc NCoRH), CaRF, SN (exc s SNF), GV (rare), CW . . . . ***P. calirhiza***

P. californicum Kaulf. CALIFORNIA POLYPODY Rhizome (3)5–10 mm diam, ± glaucous or not, taste bland or acrid; scales ± 1-colored. **LF**: summer-deciduous; blade (5)10–25(35) cm, deltate to ovate, membranous to fleshy, often firm, midrib adaxially hairy to ± glabrous, segments serrate, tips obtuse to acute, veins gen 10–50% fused. **SPORANGIA**: sori 1.5–3.5 mm, round to gen ovate, gen ± sunken, flat, with short, branched, glandular hairs or not. 2*n*=74. Shaded canyons, streambanks, n-facing slopes, roadcuts, cliffs, coastal bluffs, rocks, often granitic or volcanic, humus, not on pls; < 1520 m. CCo, SCoRO, SW; to Baja CA, s Mex. Hybrids with *P. hesperium* (SnBr) uncommon, sterile, 2*n*=111.

P. calirhiza S.A. Whitmore & A.R. Sm. (p. 127) Rhizome 5–10 mm diam, ± glaucous or not, taste acrid, ± sweet; scales 1-colored. **LF**: ± summer-deciduous if conditions dry; blade (5)10–20(40) cm, oblong-ovate, ± membranous to ± firm, midrib adaxially hairy, segments serrate, tips acute to obtuse, veins free to 35(50)% fused. **SPORANGIA**: sori 1.5–4 mm, ovate to oblong, with short, many-branched, glandular hairs or not. 2*n*=148. On pls, rocky cliffs or outcrops, roadcuts, often granitic or volcanic, rarely dunes; < 1400 m. NW (exc NCoRH), CaRF, SN (exc s SNF), GV (rare), CW; to OR. Hybrids with *P. glycyrrhiza* sterile, 2*n*=111, abundant (outnumbering parental spp. or not), ± throughout range of geog overlap; also hybridizes with *P. scouleri*, forming probable triploids (3*n*).

P. glycyrrhiza D.C. Eaton (p. 127) LICORICE FERN Rhizome 3–6 mm diam, ± glaucous or not, taste sweet-licorice, aftertaste ± bitter; scales 1-colored. **LF**: summer-deciduous, or alive until new lvs formed; blade (5)8–23(38) cm, lanceolate to lance-ovate, membranous to ± firm, midrib adaxially hairy, segments serrate, tips gen acute to acuminate, veins free. **SPORANGIA**: sori 1–2.5 mm, gen round, with short, branched, glandular hairs or not. 2*n*=74. Gen near coast, on pls, rocks, moist rocky banks, mossy logs; < 600(1200) m. NCo, KR, NCoRO, CCo, SnFrB; to AK (incl Aleutian Islands), reported from Asia (Kamchatka Peninsula). Hybrids with *P. scouleri* possible; see *P. calirhiza, P. hesperium*.

P. hesperium Maxon (p. 127) WESTERN POLYPODY Rhizome 3–6 mm diam, white-glaucous or not, taste acrid to sweet; scales 1-colored or gen with ± darker central area. **LF**: alive until new lvs formed; blade 2–25 cm, oblong to oblong-ovate, ± membranous to ± thick, ± firm, midrib adaxially glabrous, segments entire to minutely serrate, tips gen obtuse to acute, veins free. **SPORANGIA**: sori 1–2.5 mm, round to ovate, each with 0–5(10) dark brown or red-black, shriveled, glandular sporangia. 2*n*=148. Rock crevices, talus slopes, under rock ledges; 1400–2980 m. KR, n&c SNH, SnBr, SnJt, W&I, e DMtns (New York Mtns); to BC, Rocky Mtns, n Mex. Hybrids with *P. glycyrrhiza* (SNH) uncommon, sterile, 2*n*=111; see *P. californicum*.

P. scouleri Hook. & Grev. (p. 127) LEATHER-LEAF FERN Rhizome 3–12 mm diam, conspicuously white-glaucous, taste bland; scales with darker midstripe, often partly deciduous in age. **LF**: evergreen 1–several seasons; blade 6–18(50) cm, deltate to oblong-ovate, thick, leathery, firm, brittle, midrib adaxially glabrous, abaxially with deltate, dark brown, shiny scales near base, segments crenate, margins thickened, tips gen round or round-obtuse, veins ± often fused. **SPORANGIA**: sori 2–6 mm, ovate to round, often ± merged, with short, branched, glandular hairs or not. 2*n*=74. Coast, gen in heavy fog-drip or salt-spray zones, on pls (esp Douglas fir, *Eucalyptus*), dunes, rocky cliffs, bluffs, mtn ridges, granitic or volcanic rocks, mossy logs, or in soil; < 600 m. NCo, NCoRO, CCo, SnFrB, SCo; to BC, Baja CA (Guadalupe Island). See *P. calirhiza*. Pls called *P. scouleri* from n ChI are fertile or sterile hybrids (2*n*=111,148) involving *P. californicum*.

PTERIDACEAE BRAKE FAMILY

Ruth E.B. Kirkpatrick, Alan R. Smith & Thomas Lemieux, except as noted

Per, in soil or on or among rocks; rhizome creeping to erect, scaly. **LF**: gen all ± alike (or of 2 kinds, fertile, sterile), gen < 50 cm, often < 25 cm; stipe gen thin, wiry, often dark, ×-section with vascular strands gen 1–3, less often many in circle; blade gen pinnate or ± palmate-pinnate (see *Adiantum*), often ≥ 2-compound, abaxially often with glands, ± powdery exudate, hairs, or scales; segments round, oblong, fan-shaped, or other, veins gen free. **SPORANGIA**: in sori or not, marginal, submarginal, or along veins, covered by recurved, often modified segment margins (false indusia) or not; true indusia 0; spores spheric, sides flat or not, scar with 3 radiating branches. ± 40 genera, 500 spp.: worldwide, esp dry areas. [Windham 1993 FNANM 2:122–186] Definition of *Cheilanthes*, related genera problematic; traditional limits often untenable. Scientific Editors: Alan R. Smith, Thomas J. Rosatti.

1. Lvs of 2 kinds, fertile more erect, with longer stipes, longer, narrower segments than sterile . . . . . . **CRYPTOGRAMMA**
1′ Lvs of ± 1 kind (in *Aspidotis densa* fertile, sterile ± dissimilar)
  2. Lf segment margin gen not recurved, ± unmodified, not covering sporangia; sporangia along veins
    3. Lf 1-pinnate; lf adaxially glabrous or with scales
      4. Pinna adaxially with stellate scales, margins shallowly (not deeply) pinnately lobed or dissected or not
      . . . . **ASTROLEPIS**
      4′ Pinna adaxially glabrous, margins ± wavy, not lobed or dissected . . . . *Pellaea bridgesii*
    3′ Lf either 1-pinnate with pinnae pinnately dissected or lf gen more divided; lf adaxially glabrous,
      glandular, or ± covered with exudate, scales 0
      5. Sporangia along veins for outer 1/3–2/3; lf segments narrowed at base. . . . . ²**ARGYROCHOSMA**
      5′ Sporangia along veins ± throughout (best seen on immature, fertile lf); lf segments not narrowed at
      base. . . . . **PENTAGRAMMA**
  2′ Lf segment margin gen recurved at least partly, often modified, gen covering sporangia at least partly;
    sporangia at or near vein tips, so appearing marginal
    6. Sporangia borne on and covered by highly modified, recurved part of segment margin (false indusium);
      segments fan-shaped or oblong, thin-textured . . . . **ADIANTUM**
    6′ Sporangia borne on unmodified segment surface, gen covered at least partly by modified or unmodified,
      recurved part of segment margin (false indusium); segments lanceolate, round, or other, gen thick-textured
      7. Lf gen > 40 cm, stipe green to brown, ± thick. . . . . **PTERIS**
      7′ Lf gen < 40 cm, if larger, stipe ± black and wiry (exc in *Pellaea andromedifolia*)
      8. Lf abaxially with scales, hairs, or glands. . . . . **CHEILANTHES**
      8′ Lf abaxially glabrous or covered with colored exudate, scales 0
      9. Lf abaxially densely covered with white or yellow exudate, adaxially sparsely dotted with same
      . . . . **NOTHOLAENA**
      9′ Lf without exudate (exc some spp. of *Pellaea* and abaxially in *Argyrochosma limitanea*)
      10. Sterile lf segments ± sessile, connected by blade tissue or not, toothed or not; false indusium wide,
      scarious . . . . **ASPIDOTIS**
      10′ Sterile lf segments (and fertile) stalked (exc *Pellaea breweri*), not connected by blade tissue, not
      toothed; false indusium 0 or narrow or wide, scarious, not scarious, or ± scarious at margin
      11. Sporangia along veins for outer 1/3–2/3; rhizome scales without dark mid-stripe; false indusium 0
      . . . . ²**ARGYROCHOSMA**
      11′ Sporangia along veins only at tips; rhizome scales often with dark mid-stripe; false indusium
      present (exc *Pellaea bridgesii*) . . . . **PELLAEA**

ADIANTUM

Pl in soil or rock crevices; rhizome short-creeping, scales variously colored. **LF:** < ± 1 m; stipe cylindric, gen dark red-brown to ± black, shiny, ± scaly at base; blade 2–3-pinnate or ± palmate-pinnate (1st division ± palmate, subsequent ones pinnate), pinnae stalked, fan-shaped or oblong, gen lobed, toothed, or both; axes, blades lacking colored exudate. **SPORANGIA:** borne along veins on and covered by highly modified, recurved part of segment margin, appearing to run together at maturity; false indusia ± semicircular to linear; spores gen smooth, tan. ± 200 spp.: trop, temp. (Greek: unwettable) Widely cult.

1. Lf ± palmate-pinnate . . . . *A. aleuticum*
1′ Lf 2–3-pinnate
  2. Pinnules cut or lobed often > 1/4 way to base, margins at base converging at 45–90°, sori (and false
    indusia) (2)3–11, gen < 5 mm . . . . *A. capillus-veneris*
  2′ Pinnules cut or lobed often < 1/4 way to base, margins at base converging at 90–180(240)°, sori (and false
    indusia) 1–3(5), gen > 5 mm. . . . . *A. jordanii*

A. aleuticum (Rupr.) C.A. Paris (p. 127) FIVE-FINGER FERN
LF: 20–75(100+) cm; stipe red-brown to ± black; blade ± palmate-pinnate; pinnules cut or lobed gen < 1/2 way to midrib, often with > 4 regular lobes, margins at base converging at 45–90°, stipe color extending gradually into base, midvein gen part way along 1 margin. **SPORANGIA:** sori (and false indusia) gen 4–6 per pinnule, gen < 3 mm. 2*n*=58. Shady, moist banks, streamsides, serpentine; < 3400 m. NW, CaR, SNH, CCo, SnFrB, SCoRO, n ChI, TR, SnJt; to AK, w Can, MT, CO, NM; e US, adjacent Can, nw Mex. Sterile hybrids with *A. jordanii* (*A.* ×*tracyi* W.H. Wagner).

A. capillus-veneris L. (p. 127) SOUTHERN MAIDENHAIR **LF:** gen (7)20–40(50+) cm; stipe dark brown to ± black; blade 2–3-pinnate; pinnules cut or lobed often > 1/4 way to base, often with < 4 ± irregular lobes, margins at base converging at 45–90°, stipe color often extending gradually into base, midvein often part way along 1 margin. **SPORANGIA:** sori (and false indusia) (2)3–11 per pinnule, gen < 5 mm. 2*n*=60. Uncommon (or locally common). Shaded, rocky or moist banks, exposed sites or not; < 2000 m. KR, NCoR, CaRF, n SNF, s SNH, SnJV, CCo, SCoRO, SW, GB, D; gen s US; worldwide, esp temp. Widely cult.

A. jordanii Müll. Hal. (p. 127) CALIFORNIA MAIDENHAIR **LF:** 20–50(70+) cm; stipe red-brown to ± black; blade 2–3-pinnate; pinnules cut or lobed often < 1/4 way to base, gen with < 4 ± irregular lobes, margins at base converging at 90–180(240)°, stipe color often ending ± abruptly at base, midvein forked into ± equal branches, not along margin. **SPORANGIA:** sori (and false indusia) 1–3(5) per pinnule, gen > 5 mm. 2*n*=60. Shaded hillsides, moist woodland; < 1200 m. CA-FP (exc uncommon or absent > 1200 m CaR, SNH); OR, Baja CA. Cult; sterile hybrids with *A. aleuticum* (*A.* ×*tracyi* W.H. Wagner).

ARGYROCHOSMA

Pl in soil or rock crevices; rhizome short-creeping, scales lance-linear, tan to ± red. **LF**: < 40 cm; hairs 0; scales 0; stipe cylindric, dark, glabrous or ± scaly at base; blade 2–5-pinnate, segments stalked, gen < 5 mm, round to oblong, blue- to gray-green, gen thick, veins obscure; axes, blades covered with ± white exudate abaxially or not. **SPORANGIA**: along veins for outer 1/3–2/3 of segments; segment margin unmodified, often only ± recurved; spores tan, coarsely ridged. ± 20 spp.: Am. (Greek: silver ornament) Closer to *Pellaea* than to *Notholaena*.

1. Lf blade without ± white exudate abaxially, 2–3-pinnate; basal pinnae spreading to ± ascending . . . . *A. jonesii*
1′ Lf blade covered with ± white exudate abaxially, 3–5-pinnate; basal pinnae ± strongly ascending
. . . . *A. limitanea* subsp. *limitanea*

A. jonesii (Maxon) Windham   JONES' FALSE CLOAK-FERN   **LF**: 5–15 cm; stipe dark brown; basal pinna stalk < 5 mm; pinnule stalk 0–1.5 mm; segments gen 2–5 mm. **SPORANGIA**: 64-spored. $2n$=54,108. Gen calcareous rock crevices, cliff bases; 400–1800 m. s SNH, SCoRO, SnGb, W&I, DMtns; to sw UT, AZ.

A. limitanea (Maxon) Windham subsp. *limitanea* (p. 131) SOUTHWESTERN FALSE CLOAK-FERN   **LF**: 10–25 cm; stipe dark brown to black; blade ovate to triangular; basal pinna stalk 5–10 mm; pinnule stalk 3–6 mm; segments gen 1.5–3 mm. **SPORANGIA**: 32-spored. n=$2n$=81. In crevices, esp bases of calcareous rocks; 1800 m. e DMtns (New York Mtns); to UT, NM, nw Mex. [*A. l.* var. *l.*, ined.] Apogamous. ★

ASPIDOTIS

Pl in soil or rock crevices; rhizome short-creeping-decumbent, scales lance-elongate, gen dark, with narrow, lighter margin or not. **LF**: ± alike or fertile, sterile ± dissimilar; axes grooved, light to dark brown; blade 3–4-pinnate, ovate-triangular to 5-sided, glabrous, adaxially ± glossy; pinnae long, narrow, tip pointed, sterile. **SPORANGIA**: in small clusters or continuous along margin; false indusia scarious, irregularly toothed. 4 spp.: gen w N.Am, also e Can, ± dry, mtn areas. (Greek: shield-bearer, from shield-like false indusia in *Aspidotis californica*)

1. Sporangia continuous along both sides of segment midvein; false indusia with many shallow, regular teeth;
   lf segments linear, ± entire . . . . *A. densa*
1′ Sporangia not continuous along both sides of segment midvein; false indusia ± entire or with few to many
   coarse or deep, irregular teeth or lobes; lf segments lanceolate to triangular, teeth few
   2. False indusia ± as wide as long, free, 1–2(5) per segment, ± entire or with few coarse or deep, irregular
      teeth . . . . *A. californica*
   2′ False indusia ± as wide as or wider than long, gen fused at base, 3–5(7) per segment, with many coarse or
      deep, irregular teeth or lobes . . . . *A. carlotta-halliae*

A. californica (Hook.) Copel. (p. 131)   CALIFORNIA LACE FERN   **LF**: 3(4)-pinnate, 10–20(40) cm, thin; pinnae pinnately dissected, segments lanceolate to triangular, teeth few. **SPORANGIA**: false indusia ± as wide as long, free, 1–2(5) per segment, ± entire or with few coarse or deep, irregular teeth. $2n$=60,120. Rock outcrops, crevices; 20–1300 m. NCoR, CaRF, SN, GV (Sutter Buttes), CW, SW; Baja CA.

A. carlotta-halliae (W.H. Wagner & E.F. Gilbert) Lellinger (p. 131)   CARLOTTA HALL'S LACE FERN   **LF**: 3-pinnate, 8–15(25) cm, leathery; pinnae pinnately dissected, segments narrowly lanceolate to triangular, teeth few. **SPORANGIA**: false indusia ± as wide as or

wider than long, gen fused at bases, 3–5(7) per segment, with many coarse or deep, irregular teeth or lobes. $2n$=120. Gen serpentine slopes, crevices, outcrops; 100–1400 m. CW. Fertile hybrid between *A. californica* and *A. densa*; sometimes backcrosses. ★

A. densa (Brack.) Lellinger (p. 131)   DENSE LACE FERN   **LF**: (2)3-pinnate, 15–20(30+) cm, leathery, fertile, sterile ± dissimilar; pinnae pinnately dissected or not, segments linear, ± entire. **SPORANGIA**: false indusia continuous along both sides of segment midvein, with many shallow, regular teeth. $2n$=60. Slopes, crevices, outcrops, esp serpentine; 100–3400 m. NW, CaR, SN, CW, PR; to sw Can, MT, WY, UT; also se Can (Gaspé Peninsula).

ASTROLEPIS

Pl in soil or rock crevices; rhizome ± short-creeping-decumbent, scales gen linear to lance-linear, toothed, pale to red-brown, older with a dark, irregular central area or not. **LF**: axes gen orange to red-brown, scaly; blade 1-pinnate, linear, pinnae shallowly pinnately lobed or dissected or not, adaxially with stellate scales. **SPORANGIA**: along veins, obscured by dense scales; segment margin unmodified, not recurved. ± 6 spp.: sw US through S.Am. (Greek: star scale)

A. cochisensis (Goodd.) D.M. Benham & Windham (p. 131) SCALY CLOAK-FERN   Rhizome short; scales ± 10 mm, 0.1–0.5 mm wide, linear; teeth ± sparse, more pronounced in distal 1/2 or not. **LF**: stipe 3–6(10) cm, 1–1.5 mm wide, scales appressed, 0.5 mm, ± white; blade 1-pinnate, 8–15(20) cm, tapered to tip; pinnae < 0.5 cm, stalk jointed, lobes 0–3 pairs, gen obtuse at tip, adaxially with persis-

tent, stellate scales, abaxially with lanceolate, ± white to tan, densely toothed and finely dissected scales covering small (< 0.1 mm) glandular hairs. **SPORANGIA**: 32-spored, in submarginal band when mature, ± visible, erupting through scales. n=$2n$=87. Limestone slopes, crevices; 900–1800 m. DMtns; to TX, Mex. Apogamous. ★

CHEILANTHES

Pl in soil or rock crevices; rhizome short- to long-creeping[-decumbent], gen many-branched, scales gen lance-linear, pale to dark, mid-stripe dark or not. **LF**: < 75 cm, young lf tip hooked or coiled; stipe cylindric, red-brown to ± black; blade gen 2–3-pinnate, gen oblong to narrowly triangular; segments gen small, ± flat or abaxially concave (from recurved margins).

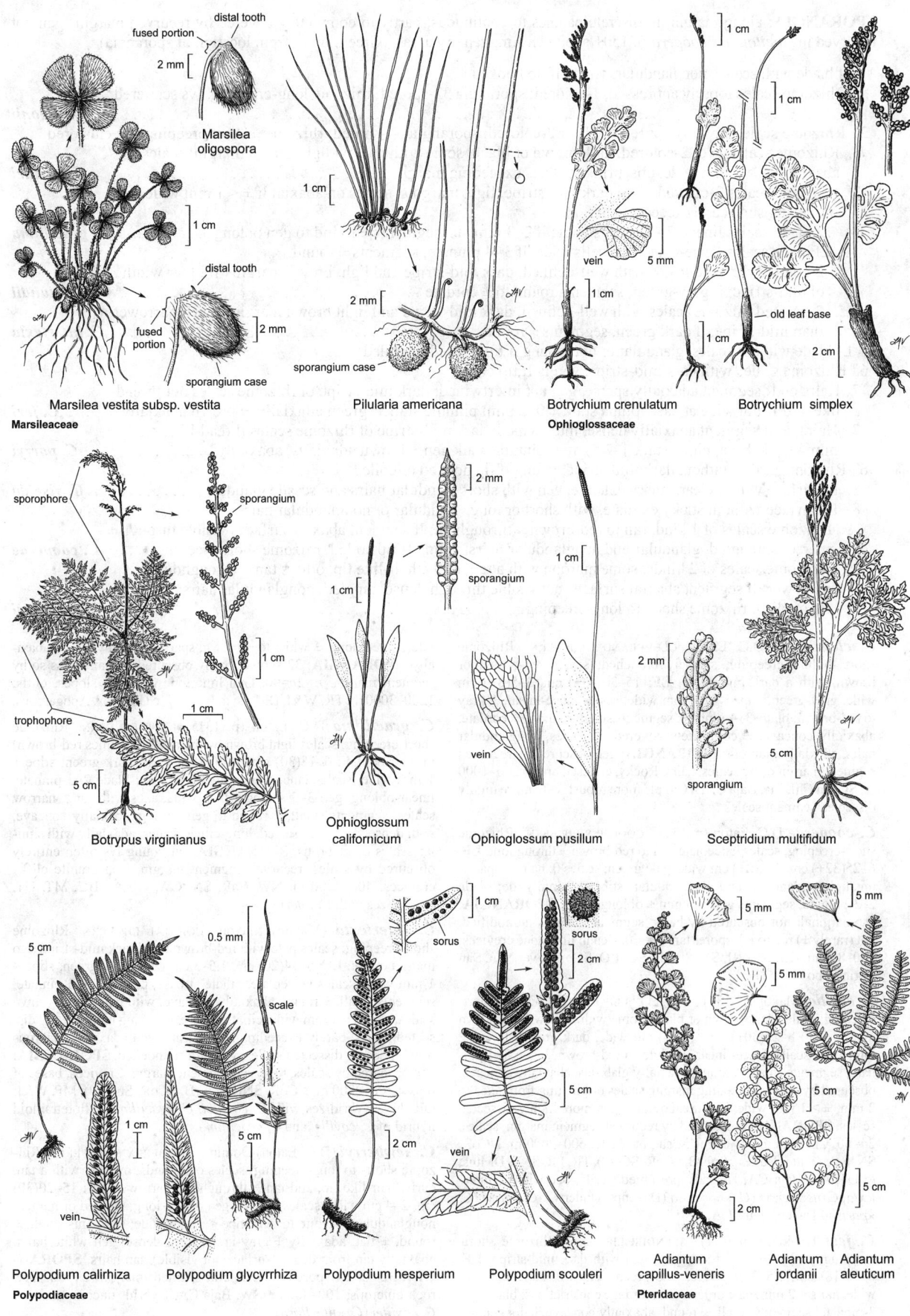

Marsilea oligospora

distal tooth
fused portion
2 mm

Marsilea vestita subsp. vestita
1 cm
distal tooth
fused portion
sporangium case
2 mm
Marsileaceae

Pilularia americana
1 cm
2 mm
sporangium case

Botrychium crenulatum
vein
5 mm
1 cm

Botrychium simplex
1 cm
1 cm
old leaf base
1 cm
2 cm
Ophioglossaceae

Botrypus virginianus
sporophore
sporangium
1 cm
trophophore
1 cm
5 cm

Ophioglossum californicum
1 cm
sporangium
vein

Ophioglossum pusillum
1 cm
2 mm
sporangium
vein

Sceptridium multifidum
2 mm
5 cm

Polypodium calirhiza
5 cm
0.5 mm
scale
1 cm
vein
Polypodiaceae

Polypodium glycyrrhiza
5 cm

Polypodium hesperium
sorus
1 cm
2 cm

Polypodium scouleri
sorus
2 cm
5 cm
vein

Adiantum capillus-veneris
5 mm
5 mm
5 cm
2 cm

Adiantum jordanii

Adiantum aleuticum
5 mm
5 cm
Pteridaceae

SPORANGIA: along margin, in discrete patches to continuous, partly to completely covered by recurved margin (gen not recurved in *Cheilanthes cooperae*). 150+ spp.: gen Am, gen dry areas. (Greek: lip fl, from location of sporangia)

1. Lf blade with scales, nonglandular; young lf tip hooked
   2. Rhizome scales loosely appressed, 1-colored; sporangia 32-spored; rhizome long-creeping, lvs scattered
      . . . . ***C. wootonii***
   2′ Rhizome scales strongly appressed, 1- or 2-colored; sporangia 64-spored; rhizome ± short-creeping, lvs clustered
      3. Rhizome scales 1- or 2-colored, dark brown or black, some with narrow light brown margin; scales on
         abaxial lf ± > 1 mm wide, obscuring surface, exceeding margin. . . . . ***C. covillei***
      3′ Rhizome scales 2-colored, with dark mid-stripe, light margins; scales on abaxial lf ± < 1 mm wide, not
         obscuring surface, ± exceeding margin
         4. Lf blade scales linear, 2–3(5) cells wide; lf 2–3 pinnate, segments ± round to gen oblong . . . . ***C. gracillima***
         4′ Lf blade scales lance-ovate, > 5 cells wide; lf 3–4 pinnate, segments ± round
            5. 2-colored rhizome scales with well-defined, dark mid-stripe and light brown margins 1/2 to = width
               of mid-stripe; lf gray-green, segments round to ± cordate . . . . ***C. clevelandii***
            5′ 2-colored rhizome scales with well-defined, dark mid-stripe and light brown margins much narrower
               than mid-stripe; lf dark green, segments ± round . . . . ***C. intertexta***
1′ Lf blade without scales, glandular or not; young lf tip hooked or coiled
   6. Rhizome scales with dark mid-stripe; young lf tip coiled
      7. Hairs on lf segment adaxially sparse, gen not intertwined; dark mid-stripe of rhizome scales not thread-
         like, > 0.1 mm wide at base; pinna stalk ± 0–1 mm; pinnule stalk ± green adaxially, ± brown abaxially . . . . . . ***C. feei***
      7′ Hairs on lf segment adaxially dense, intertwined; dark mid-stripe of rhizome scales thread-like, < 0.1
         mm wide at base; pinna stalk 1–2(5) mm; pinnule stalk gen ± brown adaxially, abaxially . . . . ***C. parryi***
   6′ Rhizome scales without dark mid-stripe; young lf tip hooked or coiled
      8. Lf surface with ± clear, sticky exudate, gen with short glandular hairs and sessile glands. . . . . ***C. viscida***
      8′ Lf surface without sticky exudate, with short or long, glandular or nonglandular hairs
         9. Rhizome scales of 1 kind, tan to red-brown ± throughout; lf segment abaxial surfaces visible through ±
            sparse, untangled, glandular and nonglandular hairs; young lf tip coiled; rhizome short-creeping . . . . . ***C. cooperae***
         9′ Rhizome scales of 2 kinds, some maroon with a tan, curly, hair-like tip, others tan ± throughout,
            narrower; lf segment abaxial surfaces not visible through dense, tangled, nonglandular hairs; young lf
            tip hooked; rhizome short- to long-creeping. . . . . ***C. newberryi***

C. clevelandii D.C. Eaton CLEVELAND'S LIP FERN Rhizome short to long-creeping, sparsely branched; scales red-brown or brown, with a dark mid-stripe. **LF**: 15–30(40+) cm, 3–5(8+) cm wide, gray-green; stipe < 2(3) mm wide, scales lance-linear, ± gray to red-brown; blade 3–4-pinnate; segments small, round to ± cordate, abaxially concave, ± completely covered by scales, nonglandular hairs, adaxially glabrous. **SPORANGIA**: gen obscured by recurved segment margin or by scales, hairs. Rocky, exposed areas; 200–1000 m. SCo, n ChI, PR; Baja CA. n ChI pls more robust, with more highly dissected segment scales.

C. cooperae D.C. Eaton (p. 131) COOPER'S LIP FERN Rhizome short-creeping; scales lanceolate, tan to red-brown ± throughout. **LF**: 6–25(32+) cm, 3–6(8+) cm wide, pale green; scales 0; hairs ± sparse, untangled, glandular and nonglandular; stipe < 2 mm wide; blade 2–3-pinnate; segments gen 1–3 mm ± oblong, ± flat. **SPORANGIA**: submarginal, not obscured by hairs; segment margins unmodified, not (rarely ±) recurved; spores tan. $2n$=60. Gen in limestone crevices; 100–800 m. KR, NCoRI, SN, SnFrB, SCoRO, SCo (Slover Mtn, San Bernardino Co.), TR.

C. covillei Maxon (p. 131) COVILLE'S LIP FERN Rhizome short-creeping; scales dark brown or black, some with narrow light brown margin. **LF**: 8–22(30+) cm, 2–4(6) cm wide, dark green; stipe < 2 mm wide, scales lance-linear, ± white to red-brown; blade 3–4-pinnate; segments small, ± round, adaxially glabrous, abaxially concave, obscured by scales exceeding margin, scales originating from axes, > 2 mm, ± > 1 mm wide, ± entire, covering gen more highly dissected scales. **SPORANGIA**: obscured by recurved segment margin, scales. $2n$=60. Crevices, bases of rocks, sun or shade; 600–2400 m. NCoR, SN, ScV (Sutter Buttes), SnFrB, SCoR, SCo(?), TR, PR, SNE, DMtns; to UT, AZ, Baja CA. Hybrids, presumed sterile, occ with *C. intertexta, C. newberryi* (*C.* ×*fibrillosa* (Davenp.) Underw.), *C. parryi* (*C.* ×*parishii* Davenp.) in s CA.

C. feei T. Moore (p. 131) SLENDER LIP FERN Rhizome short-creeping; scales light to red-brown, gen with dark mid-stripe. **LF**: 6–15(18) cm, 1.5–3 cm wide, pale green, scales 0; stipe ± 1 mm wide, hairs < 2 mm, pale or tan with ± orange constrictions; blade gen 3-pinnate; segments small, ± round, abaxially concave, hairs gen not

intertwined, long, ± white to ± brown, sparse adaxially, dense abaxially. **SPORANGIA**: 32-spored, partly obscured by hairs, less so by segment margin. n=2n=90. Gen limestone crevices, slopes, cliffs; 1200–3000 m. TR, W&I, DMtns; to BC, MT, c US, Mex. Apogamous.

C. gracillima D.C. Eaton (p. 131) LACE LIP FERN Rhizome short-creeping; scales light brown, with dark (sometimes red-brown) mid-stripe. **LF**: 6–18(30) cm, 1–2(3) cm wide, dark green; stipe ± 1 mm wide, scales lance-linear, ciliate at base; blade 2–3-pinnate, linear-oblong, gen 3–5 × longer than wide, axes with long, narrow scales; segments small, ± round to gen oblong, abaxially concave, with dense, deeply dissected, long-ciliate scales, adaxially with similar scales or glabrous. **SPORANGIA**: on young lvs often entirely obscured by scales, recurved segment margin. Gen granite cliffs, crevices; 400–3200 m. NW, CaR, SN, CW, GB; to BC, MT, UT. Hybridizes with *C. intertexta*.

C. intertexta (Maxon) Maxon COASTAL LIP FERN Rhizome short-creeping; scales pale with red-brown to ± black mid-stripe ± to margin or not. **LF**: 6–14(20) cm, 1.5–3 cm wide, dark green; stipe < 1 mm wide, scales lanceolate, ciliate at base, pale; blade 3-pinnate; segments small, ± round, abaxially concave, with pale to red-brown scales, scales < 1 mm wide, ciliate at base, covering more deeply dissected scales, barely exceeding segment margin, adaxially glabrous or with deeply dissected scales; young lf tip hooked. **SPORANGIA**: gen obscured by scales, recurved segment margin. Crevices, bases of rocks; 300–2800 m. NCoR, SNH, SnFrB, SCoR, SnBr(?), MP, W&I; OR, NV. Hybridizes with *C. covillei, C. gracillima*. Allotetraploid hybrid of *C. covillei* and *C. gracillima*.

C. newberryi (D.C. Eaton) Domin NEWBERRY'S LIP FERN Rhizome short- to long-creeping; scales of 2 kinds, maroon with a tan, curly, hair-like tip, and tan ± throughout, narrower. **LF**: 15–20(30) cm, 2–4 cm wide; scales 0; hairs gen dense, long, tangled or matted, nonglandular, ± white to tan; stipe < 1 mm wide; segments small, ± round, ± flat, adaxially ± gray-green from dense, gen white hairs, abaxially tan from dense (surface not visible), tan hairs. **SPORANGIA**: 32- or 64-spored, ± visible at segment margin. $2n$=60. Dry, rock outcrops; 100–800 m. SW; Baja CA. Hybrids uncommon with *C. covillei* (*C.* ×*fibrillosa*).

C. parryi (D.C. Eaton) Domin (p. 131) PARRY'S LIP FERN Rhizome short-creeping, > 6 cm; scales medium brown, most with dark, thread-like mid-stripe. **LF:** 6–15(25) cm, 1–2(3) cm wide; scales 0; stipe < 1 mm wide, hairs short to long, bent, appressed to ± spreading, glandular and not, pale; segments small, ± round, ± flat, hairs 4+ mm, tangled, gen nonglandular, dense both surfaces, adaxially silver-white, abaxially tan to brown or golden. **SPORANGIA:** ± visible through hairs at segment margin; spores ± black. $2n=60$. Limestone, granite crevices, rocks; 100–1500 m. PR, SNE, D; to UT, AZ, Mex.

C. viscida Davenp. (p. 131) VISCID LACE FERN Rhizome short-creeping; scales red-brown, dark mid-stripe 0. **LF:** 10–15(25) cm, 2(3) cm wide, pale to dull green; stipe < 1 mm wide, scales sparse, at base, gland stalks 0–0.3 mm; segments small, ± oblong, ± flat, both surfaces covered with ± clear, sticky exudate from glands; young lf tip coiled. **SPORANGIA:** visible at recurved segment margins; spores ± brown. Limestone, granite crevices, rocks; 100–1600 m. e edge SnBr, e PR (incl SnJt), DMoj, w edge DSon; Baja CA.

C. wootonii Maxon (p. 131) WOOTON'S LIP FERN Rhizome long-creeping; scales tan to brown, dark mid-stripe 0. **LF:** 10–20 cm, 2–3 cm wide; stipe 1–2 mm wide; blade 3–4-pinnate; segments small, ± round, abaxially concave, densely covered with ciliate, lance-linear scales, adaxially glabrous. **SPORANGIA:** 32-spored, gen obscured by dense, overlapped scales. $n=2n=90$. Rocky outcrops; 1600–1800 m. DMtns (Providence, New York mtns); to w OK, n Mex. Apogamous. ★

CRYPTOGRAMMA PARSLEY FERN, ROCK-BRAKE

Ruth E.B. Kirkpatrick, Alan R. Smith, Thomas Lemieux & Edward Alverson

Pl in rocky places; rhizome creeping-decumbent, scales brown. **LF:** tufted, deciduous or evergreen, of 2 kinds, fertile more erect, with longer stipes, longer, narrower segments than sterile; stipes dark, scaly at base, tan to ± green, glabrous above base; blades 2–4-pinnate, triangular, lanceolate, or elliptic; veins free. **SPORANGIA:** along veins, submarginal, appearing to cover surface at maturity; false indusia linear, from segment base to tip. 11 spp.: temp N.Am, S.Am, Eur, Asia. (Greek: hidden line, from protected sori)

1. Sterile lvs leathery, opaque when held up to light, hairs small, 0.1 mm, appressed, esp in grooves of axes adaxially, stipe bases 1–2 mm wide, persistent, blades persistent or not . . . . ***C. acrostichoides***
1′ Sterile lvs thin, translucent when held up to light, glabrous, stipe bases < 1 mm wide, shriveling or deciduous, blades deciduous . . . . ***C. cascadensis***

C. acrostichoides R. Br. (p. 131) AMERICAN PARSLEY FERN Rhizome forming small clumps. **LF:** fertile 10–30 cm; sterile 6–22 cm, blade lance-ovate, ± dark green. $2n=60$. Moist to ± dry rocky slopes, crevices; 1400–3400 m. KR, NCoRH, CaRH, SNH, SnBr, SnJt; to AK, c Can, MI, CO, NM.

C. cascadensis E.R. Alverson CASCADE PARSLEY FERN Rhizome forming small to large clumps. **LF:** fertile 5–25 cm; sterile 3–20 cm, blade lance-ovate to deltate, green. $2n=60$. Gen ± moist talus slopes, crevices, often granitic or volcanic rock; 1800–3650 m. KR, CaRH, SNH, Wrn; to BC, MT.

NOTHOLAENA

Pl in soil or granite rock crevices; rhizome short-creeping to ± erect, scales lance-linear. **LF:** stipe gen cylindric, dark brown to black, glabrous to ± scaly; blade 2–4-pinnate, segments gen sessile, ± narrower at base or not. **SPORANGIA:** in ± continuous, marginal bands; segment margin recurved, partly covering sporangia, unmodified; spores finely ridged or granular, often ± black. 34 spp.: gen Mex, sw US, few in Caribbean, S.Am. (Greek: false cloak, from lf blade margin not reflexed as it is in *Cheilanthes*)

N. californica D.C. Eaton (p. 131) CALIFORNIA CLOAK-FERN Rhizome scales rigid, with ± black midrib ± to margins, finely ciliate. **LF:** 3-pinnate, ± 3–13 cm; blade axes brown to black, glabrous or with white to yellow exudate; lowermost pinnae each more developed on basal side; segment abaxially covered with white to yellow exudate, hairs 0, scales 0, adaxially sparsely dotted with white to yellow exudate. **SPORANGIA:** 32-spored. $n=2n=150$. Dry rocky slopes, rock crevices, under rock ledges; 200–1300 m. SCo, s ChI, TR, PR, DMtns, DSon; AZ, nw Mex. Apogamous. At least 2 entities in CA chemically distinct: 1 with pale to bright yellow exudate on lf abaxially (*N. californica* subsp. *c.*), 1 with white exudate on lf abaxially (*N. californica* subsp. *leucophylla* Windham). Gene flow between them where they overlap geog (s CA) 0 due to apogamy, and pentaploids in AZ suggest further study needed to decide if taxonomic recognition warranted.

PELLAEA CLIFF-BRAKE

Pl in soil or rock crevices; rhizome short- to long-creeping, scales overlapped, narrowly linear, light- to red- or medium-brown, often with dark mid-stripe. **LF:** erect, persistent, < 1 m; stipes ± cylindric, gen dark or red-brown to ± black, ± shiny, glabrous; blade 1–4-pinnate; segments gen stalked, gen free, linear to rounded, lobed or not, often folded lengthwise when dried; veins gen free. **SPORANGIA:** in ± continuous, submarginal bands, among a ± white to ± yellow exudate or not; segment margin gen recurved, gen modified; spores tan to light yellow. ± 35 spp.: trop, temp, few in Eur, 0 in Asia. (Greek: dusky, from blue-gray lvs) [Kirkpatrick 2007 Syst Bot 32:504–518] Occ cult; as defined by Tryon (1957), polyphyletic (Kirkpatrick, 2007).

1. Lf 1-pinnate, pinnae unlobed or deeply 2(3)-lobed
  2. Pinnae lance-ovate, deeply 2(3)-lobed; fracture lines at base of stipe many; rhizome scales without dark mid-stripe . . . . ***P. breweri***
  2′ Pinnae rounded, unlobed; fracture lines at base of stipe 0; rhizome scales with dark mid-stripe . . . . ***P. bridgesii***
1′ Lf 2–4-pinnate, pinnae compound.
  3. Lf 2–4-pinnate
    4. Segment without mucro, gen 6–15 mm, 3–10 mm wide; stipe ± light brown . . . . ***P. andromedifolia***
    4′ Segment with mucro gen 2–6(8) mm, 0.5–2(4) mm wide; stipe dark brown to ± black
    . . . . ***P. mucronata*** var. ***mucronata***

3′ Lf 2-pinnate
5. Segment 5–15 × longer than wide, > costa . . . . ***P. brachyptera***
5′ Segment 1–4 × longer than wide, < costa
  6. Fertile segment gen appearing folded in half, recurved margins ± meeting abaxially; c&s SNH, TR, PR, SNE, DMtns (exc Providence, New York mtns). . . . . ***P. mucronata*** var. ***californica***
  6′ Fertile segment not appearing folded in half, recurved margins not meeting abaxially; DMtns (Providence, New York mtns) . . . . ***P. truncata***

P. andromedifolia (Kaulf.) Fée (p. 137) COFFEE FERN Rhizome long-creeping, branched, > 20 cm, 0.5 cm wide; scales 2–3 mm, tan to orange-brown, mid-stripe dark or not. **LF:** ± unclustered, 20–60(80) cm, 10–20(30) cm wide, green to ± purple; stipe < ± 3 mm wide, ± light brown; blade (2–4)3-pinnate, elongate-triangular; segments gen 6–15 mm, 3–10 mm wide, tip ± rounded to obtuse, notched or not. **SPORANGIA:** 32- or 64-spored. $2n=58$, $n=2n=87,116$. Gen rocky or dry areas; 30–1800 m. NCoR, CaRF, SN, CW, SW; Baja CA. Apogamous or gen sexual. CA pls diploid (for which the name *P. a.* var. *rubens* D.C. Eaton has been published), triploid, or tetraploid; *P. a.* var. *pubescens* D.C. Eaton (Baker instead might be correct) has been published for hairy pls near coast in s CA, ChI; further study needed to determine whether or not taxonomic recognition of either entity is warranted.

P. brachyptera (T. Moore) Baker SIERRA CLIFF-BRAKE Rhizome short-creeping, many-branched, compact, > 8 cm, 1 cm wide; scales red-brown, mid-stripe dark. **LF:** clustered, 15–30(40) cm, 2–3 cm wide, gray-green; stipe 1–2 mm wide; blade 2-pinnate, oblong; segments 5–11(13), < 1(2) cm, linear, 5–15 × longer than wide, with mucro, > costa. **SPORANGIA:** 64-spored. Rocky crevices, slopes, serpentine or not; 700–2200 m. KR, NCoR, CaR, n SNH; to WA. Hybridizes with *P. mucronata* var. *m.*

P. breweri D.C. Eaton (p. 137) BREWER'S CLIFF-BRAKE Rhizome short-creeping, branched, > 10 cm, 5(7) mm wide; scales ± thread-like, red-brown, dark mid-stripe 0. **LF:** clustered, 8–20(25) cm, 2–3(4) cm wide, pale greenish; stipe < 2 mm wide, fracture lines at base many; blade 1-pinnate, oblong, rachis green at tip; pinnae < 2 cm, < 1.5 cm wide, lance-ovate, deeply 2(3)-lobed. **SPORANGIA:** 64-spored; spores dark to light brown. $2n=58$. Gen n-facing granite rock crevices, slopes; 1500–3700 m. KR, CaRH, SNH, SnBr, GB, DMtns; to WA, ID, CO.

P. bridgesii Hook. (p. 137) BRIDGES' CLIFF-BRAKE Rhizome short-creeping, many-branched, > 15 cm, 0.5 cm wide; scales light to medium brown, mid-stripe dark. **LF:** clustered, 12–25(35) cm, 1.5–2(3.5) cm wide, blue-green; stipe < 1.5 mm wide; blade 1-pinnate, oblong; pinnae < 2 cm, < 1.5 cm wide, rounded, unlobed, often folded lengthwise. **SPORANGIA:** in marginal bands mixed with yellow exudate; segment margin modified, not recurved. $2n=58$. Gen granite rock crevices, slopes; 1200–3200 m. SNH; to OR, ID. Hybrids with *P. mucronata* (*P.* ×*glaciogena* W.H. Wagner et al.) sterile, ± common, c&s SNH, 1500–2400 m, intermediate between parents (Wagner et al. 1983 Madroño 30:69–83).

P. mucronata (D.C. Eaton) D.C. Eaton BIRD'S-FOOT FERN Rhizome short-creeping, branched, > 8 cm, 0.5–1 cm wide; scales ± brown, mid-stripe dark. **LF:** ± clustered, ± green to ± purple; stipe < 2(3) mm wide; blade 2–3(4)-pinnate, narrowly triangular to oblong; segments 2–6(8) mm, 0.5–2(4) mm wide, linear to oblong, with mucro. **SPORANGIA:** 64-spored. $2n=58$.

var. ***californica*** (Lemmon) Munz & I.M. Johnst. (p. 137) **LF:** 15–25(33) cm, 2–4(8) cm wide; blade 2-pinnate; pinnae often overlapped, ascending; fertile segments gen appearing folded in half, recurved margins ± meeting abaxially. Rocky or dry areas; 1800–3000 m. c&s SNH, TR, PR, SNE, DMtns (exc Providence, New York mtns).

var. ***mucronata*** **LF:** 20–40(60) cm, 5–15 cm wide; blade 2–3(4)-pinnate; pinnae not overlapped, gen ± spreading to widely ascending; fertile segments not appearing folded in half, recurved margins not meeting abaxially. Rocky or dry areas; 20–2400 m. KR, NCoR, CaR, SN, GV (Sutter Buttes), CW, SW, SNE, DMoj (desert mtns and lower montane slopes, canyons); Baja CA. Hybrids with *P. brachyptera* (pinnae linear, well-spaced on elongate 2 ×-pinnate axes), *P. truncata* (some pinnae irregularly lobed, costae ± 90° from rachis) ± common.

P. truncata Goodd. (p. 137) SPINY CLIFF-BRAKE Rhizome short-creeping, branched, to 8+ cm, 2–3 mm wide; scales ± brown, mid-stripe dark. **LF:** clustered, 15–30(36) cm, 4–11 cm wide, olive-green; stipe 1(2) mm wide, ± flat or adaxially grooved; blade 2-pinnate, narrowly triangular to oblong-triangular; pinnae gen not overlapped, ± spreading; segments 4–8 mm, 2–4 mm wide, linear to oblong, with mucro, margins wavy-crenate, often (esp sterile) ± white, fertile not appearing folded in half, recurved margins not meeting abaxially. **SPORANGIA:** 64-spored. $2n=58$. Gen in crevices of or at bases of granite (in CA) or igneous rock; 1200–1900 m. e DMtns (Providence, New York mtns); to CO, TX, Baja CA. Hybrids with *P. mucronata* var. *m.* uncommon. ★

PENTAGRAMMA GOLDBACK OR SILVERBACK FERN

Pl in soil or rock crevices; rhizome short-creeping-decumbent, gen 3–5(8) mm wide, scales lance-linear, mid-stripe dark. **LF:** stipe 5–20(32) cm, 0.5–2(3) mm wide; blade gen 2–3-pinnate, 2–8(15) cm, triangular or gen 5-sided, with white or yellow exudate abaxially, with exudate or not adaxially, main axis shallowly to deeply grooved adaxially; lowermost pinnae more strongly developed on basal side; veins free. **SPORANGIA:** along veins ± throughout; segment margins unmodified, recurved or not. 2 spp.: w N.Am. (Greek: 5 lines, for lf blades) A puzzling complex of intergrading chemical, chromosomal, and morphological variants (see Yatskievych et al. 1990 Amer Fern J 80:9–17).

1. Stipe dark brown to gen ± black; lf blade adaxially with white exudate; SN, SCoRI. . . . . ***P. pallida***
1′ Stipe brown to red-brown; lf blade adaxially gen without exudate; CA-FP, GB, DMtns. . . . . ***P. triangularis***
  2. Lf blade adaxially not sticky, gen glabrous . . . . subsp. ***triangularis***
  2′ Lf blade adaxially sticky or ± not, ± densely or sparsely glandular
    3. Lf blade adaxially ± not sticky, gen sparsely glandular; pinnules on basal side of lowermost pinnae deeply pinnately lobed to ± 1-pinnate . . . . subsp. ***maxonii***
    3′ Lf blade adaxially gen sticky, ± densely glandular; pinnules on basal side of lowermost pinnae ± entire to shallowly pinnately lobed . . . . subsp. ***viscosa***

P. pallida (Weath.) Yatsk. et al. (p. 137) SILVERBACK FERN Rhizome tip, incl scales, covered with white exudate. **LF:** stipe dark brown to gen ± black, with white exudate; blade 2–8(12) cm, pale to olive green, both surfaces with white exudate; lower pinnae 2–6(8) cm. $2n=60$. Gen in or near granitic soil, rock, boulders; 100–1200 m. SN, SCoRI. [*Pityrogramma triangularis* (Kaulf.) Maxon var. *p.* Weath.]

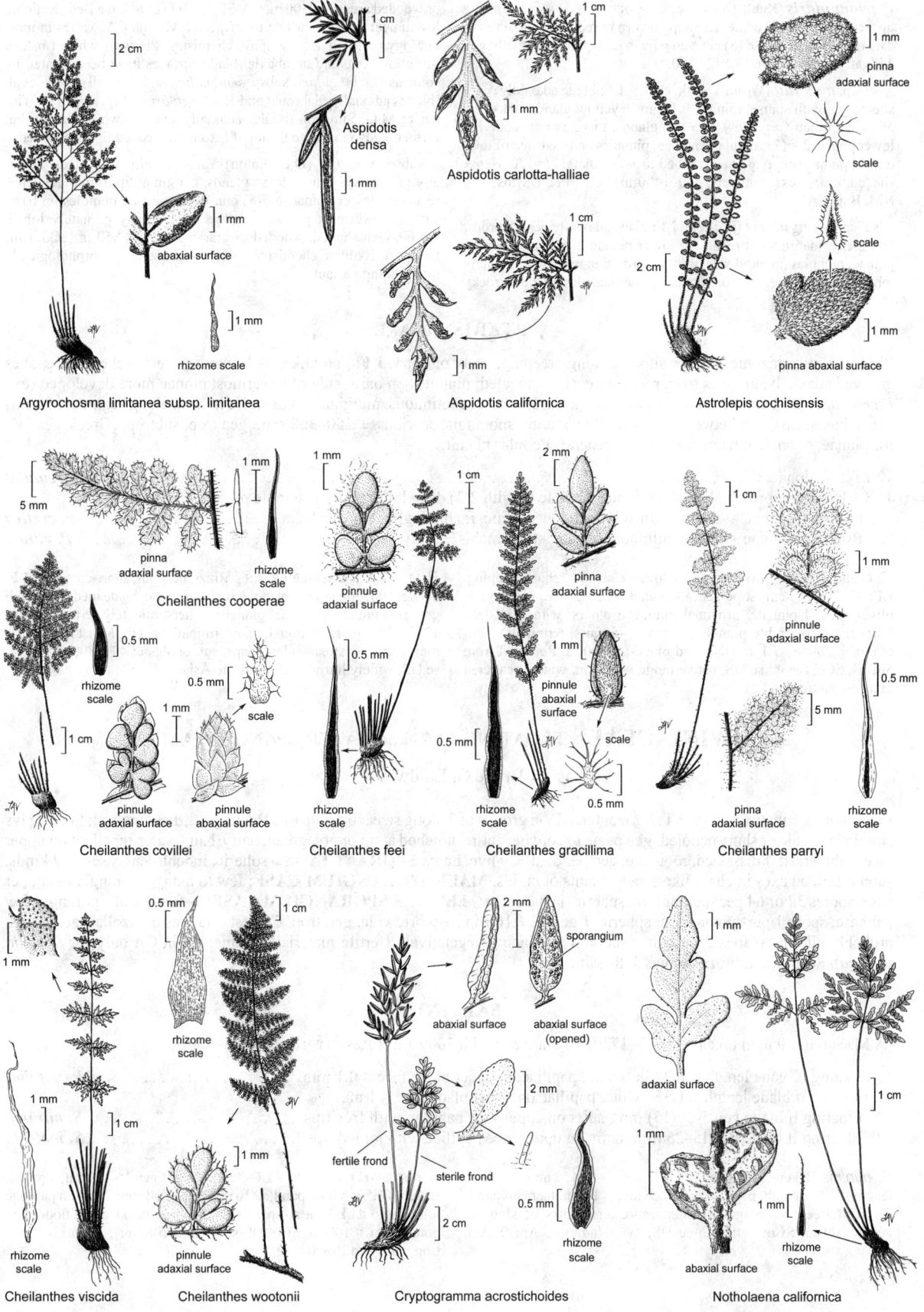

Argyrochosma limitanea subsp. limitanea

Aspidotis densa

Aspidotis carlotta-halliae

Aspidotis californica

Astrolepis cochisensis

Cheilanthes cooperae

Cheilanthes covillei

Cheilanthes feei

Cheilanthes gracillima

Cheilanthes parryi

Cheilanthes viscida

Cheilanthes wootonii

Cryptogramma acrostichoides

Notholaena californica

P. triangularis (Kaulf.) Yatsk. et al.   GOLDBACK FERN   Rhizome tip, scales without exudate. **LF**: stipe brown to red-brown, with exudate or not; blade 3–10(18) cm, gen pale to dark green, adaxially gen without exudate. 2*n*=60,90,120,150.

subsp. ***maxonii*** (Weath.) Yatsk. et al.   **LF**: blade adaxially ± not sticky, gen with sparse, minute, 0.1 mm, ± yellow glands, abaxially often with many ± yellow or ± red glands, margins not recurved; lower pinnae 2–5(7) cm; upper pinnae, pinnules on basal side of lowermost pinnae deeply pinnately lobed to ± 1-pinnate. 2*n*=120. Gen ± shaded, near rocks, boulders; 300–1400 m. SnBr, PR, DMtns; AZ, NM, Baja CA.

subsp. ***triangularis*** (p. 137)   **LF**: blade adaxially gen glabrous, not sticky, margins not recurved; lower pinnae 2–6(11) cm; upper pinnae, pinnules on basal side of lowermost pinnae deeply pinnately lobed to ± 1-pinnate. 2*n*=60. Common. Gen shaded, sometimes rocky or wooded areas; < 2300 m. CA-FP, MP (caves in Lava Beds National Monument), SNE, DMtns; to BC, ID, NV, Baja CA. Varies in morphology, cytology, geography, chemistry. Pls with white (instead of yellow) exudate on abaxial blade surfaces have been treated by some as *P. triangularis* subsp. *semipallida* (J.T. Howell) Yatsk. et al. (blades adaxially glabrous) and *P. triangularis* subsp. *rebmanii* Winner & M.G. Simpson (blades adaxially sparsely white-mealy), but further study needed to decide if taxonomic recognition warranted.

subsp. ***viscosa*** (D.C. Eaton) Yatsk. et al.   **LF**: blade adaxially gen sticky, with ± dense glands, margin at tip of pinnule gen ± recurved; lower pinnae 2–5(8) cm; upper pinnae, pinnules on basal side of lowermost pinnae ± entire to shallowly pinnately lobed. 2*n*=60. Gen shaded, wooded or grassy slopes; < 850 m. SCo, ChI; Baja CA. Exudate chemistry unique in genus; incl morphologically intergrading variants.

PTERIS   BRAKE

Pl gen in soil; rhizome erect or short- to long-creeping, scaly or hairy. **LF**: gen alike, 1–4-pinnate, erect-arched; stipe, axes grooved adaxially, grooves from rachis to costa connected; pinnules on basal side of lowermost pinnae more developed (exc *Pteris vittata*). **SPORANGIA**: among hair-like structures in continuous, marginal bands; false indusia along segment margins exc at bases, tips, and between lobes, partly covering sporangia, scarious. ± 250–300 spp.: gen trop, subtrop. (Greek: feather, for pinnae, or ancient name for ferns in general) Popular in cult.

1. Lf 2–4-pinnate . . . . **[*P. tremula*]**
1′ Lf 1-pinnate, proximal pinnae minutely serrate or with 2(3) deep lobes that ± resemble pinnules
 2. Proximal pinnae > others, with 1(3) deep lobes; stipe, rachis ± glabrous exc base . . . . ***P. cretica***
 2′ Proximal pinnae < others, minutely serrate; stipe, rachis ± scaly. . . . . ***P. vittata***

P. cretica L.   CRETAN BRAKE   Rhizome slender, short-creeping. **LF**: 15–70(100) cm; stipe gen > blade, ± glabrous exc base; blade olive-green, 1-pinnate, proximal pinnae > others, with 1(3) deep lobes that ± resemble pinnules; pinnae 1–5 pairs, terminal > subterminal. 2*n*=58,87,116. Disturbed places; < 500 m. Reported from SnFrB, SCo, SnGb; se US; native range uncertain, widely scattered in trops, subtrops.

P. vittata L.   LADDER BRAKE   Rhizome stout, short-creeping. **LF**: 30–75(100) cm; stipe < blade, ± scaly, esp base; blade medium to dull green, 1-pinnate, proximal pinnae < others, minutely serrate; pinnae 12–20(30+) pairs, terminal > subterminal. 2*n*=116. Moist rock walls, rock crevices, streamsides, seeps, sun or shade; 400–800 m. SnGb; se US; widely introduced; native to Asia.

SALVINIACEAE   FLOATING-FERN or WATER-SPANGLE FAMILY

Bruce G. Baldwin

Free-floating aquatic; roots 0. **ST**: horizontal. **LF**: in groups of 3 along st, each group of 2 floating lvs, 1 submersed; floating lvs simple, ± sessile or short-petioled, gen ovate to cordate, entire, notched at tip or not, green, hairy, hairs water-repellent on upper face; submersed lf dissected, root-like, gen petioled, ± white, hairy. **SPORANGIA**: in ± spheric, indehiscent cases of 2 kinds, submersed, on axes in chain-like arrangements or on lvs. **MALE SPORANGIUM CASE**: few to many; sporangia many per case; spores 32 or 64 per sporangium, spheric, in hard mass. **FEMALE SPORANGIUM CASE**: 1–few, basal; sporangia few per case; spore 1 per sporangium, spheric. 1 genus, ± 10 spp.: ± worldwide, gen trop. Closest relatives of Azollaceae. In CA, probably all are escapes from cult, aquaria, reproducing vegetatively (fertile material documented in CA only for *Salvinia minima*). Scientific Editor: Thomas J. Rosatti.

SALVINIA

(A.M. Salvini, Italian botanist, 1633–1729) [Riefner & Smith 2009 J Bot Res Inst Texas 3:855–866]

1. Floating lf blade length gen 2–3 × width, papillae on upper blade face < 0.1 mm . . . . **[*S. oblongifolia*]**
1′ Floating lf blade length < 1.5 × width, papillae on upper blade face ≤ 1 mm
 2. Floating lf blade gen 5–10(15) mm, hairs on upper face papillae with free tips. ***S. minima***
 2′ Floating lf blade gen 15–25 mm, hairs on upper face papillae with joined tips. ***S. molesta***

S. minima Baker   **LF**: floating blade gen 5–10(15) mm, ± round, length < 1.5 × width, papillae on upper face ≤ 1 mm, hairs on papillae with free tips. Stagnant water, mud, algal mats of sloughs; gen < 100 m. SCo; native to se US, West Indies, C.Am, S.Am.

S. molesta D.S. Mitch.   **LF**: floating blade gen 15–25 mm, ± round, length < 1.5 × width, papillae on upper face ≤ 1 mm, hairs on papillae joined into dark knot at tips. River channels, backwaters, floodplain ponds; gen < 100 m. SCo, DSon (Colorado River); ± worldwide in trop; native to Brazil. ◆

THELYPTERIDACEAE   THELYPTERIS FAMILY

John C. Game, Alan R. Smith & Thomas Lemieux

Pls gen in soil; rhizome short- to long-creeping, prostrate to erect, [trunk-like], scales gen hairy, ± brown. **LF:** stipe ×-section at base with 2 crescent-shaped vascular strands fusing into 1 U-shaped strand distally; blade 1[2]-pinnate, gen with needle-like, clear hairs on axes and/or between veins; rachis, costae gen grooved adaxially exc between axes; veins free or in regular, net-like pattern. **SPORANGIA:** sori on veins, gen round; indusia reniform or round-reniform [0]; stalk cells in 3 rows; spores gen elliptic, scar gen linear. 1 genus (5–32 genera), ± 1000 spp.: worldwide, esp trops; several cult. Scientific Editors: Alan R. Smith, Thomas J. Rosatti.

THELYPTERIS

LF: pinnae entire to gen deeply lobed, veins gen not forked. (Greek: female fern; ancient name for delicate fern) [Smith & Cranfill 2002 Amer Fern J 92:131–149]

1. Lower pinnae << upper, blade widest above base; lowermost veins to margin above deepest point of sinus; KR, NCoRO, n SNH . . . . ***T. nevadensis***
1′ Lower pinnae ± ≥ upper, blade widest at or near base; lowermost veins to margin at deepest point of sinus; SW . . . . ***T. puberula*** var. ***sonorensis***

T. nevadensis (Baker) C.V. Morton (p. 137) NEVADA MARSH FERN Rhizome creeping, 1.5–3 mm wide; gen dormant in winter. **LF:** densely clustered, gen 40–100 cm, 8–15 cm wide; stipe scales ovate, tan, persistent; blade thin, abaxially with many short-stalked or sessile, resinous glands between and on veins, nonglandular hairs sparse on axes, veins, 0 between veins; pinnae deeply lobed, lobes ± oblong, entire to shallowly crenate. **SPORANGIA:** sori small, round; indusia hairs 0 or sparse. Springy hillsides, seepage areas; 365–1700 m. KR, NCoRO, n SNH; to BC, ID. Locally forming large colonies. Differs from *Athyrium filix-femina* in having longer, thinner rhizome, thinner lvs, more numerous reduced proximal pinnae, more entire lobes, rounder sori.

T. puberula (Baker) C.V. Morton var. ***sonorensis*** A.R. Sm. (p. 137) SONORAN MAIDEN FERN Rhizome creeping, 3–8 mm wide; evergreen. **LF:** ± regularly spaced, gen 50–120 cm, 15–30 cm wide; stipe scales lanceolate, brown, not persistent; blade thick, abaxially ± nonglandular, ± densely hairy on axes, veins, and between veins; pinnae deeply lobed, lobes narrowly oblong, entire. **SPORANGIA:** sori small, round; indusia densely hairy. Along streams, seepage areas; 50–800 m. SCo, WTR, SnGb, SnJt; to AZ, s Mex. Rarely cult in s CA. 1 other var., *T. puberula* var. *p.*, Mex to Costa Rica. ★

WOODSIACEAE   CLIFF FERN FAMILY

Alan R. Smith, except as noted

Per in soil or rock crevices; rhizome gen short-creeping, ascending, or erect, scales small to large, gen tan to brown, gen uniformly colored. **LF:** gen tufted or short-spaced, 5–200+ cm, gen ± alike; stipe firm or fleshy (easily crushed), base darker or not, with 2 vascular strands; blade gen 1–3-pinnate, ± glabrous or with hairs, hair-like scales, or gland-tipped hairs on axes, veins gen free (or netted); rachis, costa gen grooved adaxially. **SPORANGIA:** sori round, oblong, J-shaped, or linear along veins; indusia 0 or oblong, J-shaped, reniform, or linear, or of many segmented hair- or scale-like fragments or lobes encircling sorus from below; spores elliptic, winged, ridged, or spiny, scar linear. ± 15 genera, 700 spp.: worldwide, esp trop, wooded areas, but some genera (e.g., *Cystopteris*, *Woodsia*) gen temperate. See note, reference (Smith et al. 2006 Taxon 55:705–731) under Dryopteridaceae for removal of genera from that family to this. Scientific Editor: Thomas J. Rosatti.

1. Lvs gen 30–150+ cm; stipe gen > 1.5 mm diam, if < then indusia 0; blade gen 25–130 cm, (5)10–60 cm wide; indusia 0, oblong, J-shaped, or reniform . . . . **ATHYRIUM**
1′ Lvs gen < 30 cm; stipe gen < 1.5 mm diam; blade gen 10–25 cm, 1–10 cm wide; indusia hood- or cup-like and of many segmented hair- or scale-like fragments or lobes
  2. Indusium hood-like; pinnae sides unequal, acroscopic pinnules more spreading, larger, more incised; lf blade axes abaxially glabrous . . . . **CYSTOPTERIS**
  2′ Indusium cup-like, of many segmented hair- or scale-like fragments or lobes encircling sorus from below, often obscure in age; pinnae sides ± equal; lf, blade axes abaxially with (without) short-stalked glands or gland-tipped hairs 0.1–1 mm . . . . **WOODSIA**

ATHYRIUM   LADY FERN

Rhizome short-creeping to suberect, stout. **LF:** stipe stout, fleshy, easily crushed, straw-colored exc base gen blackened, base scaly, ×-section with 2 crescent-shaped vascular strands; blade gen ≥ 2-pinnate, pinnae of equal sides, ± glabrous or minutely hairy, veins free. **SPORANGIA:** sori ± round, ± oblong, or J-shaped; indusia 0, oblong, J-shaped, or reniform, laterally attached. ± 100 spp.: gen n temp, esp e Asia. (Greek: doorless, from enclosed sori)

1. Sori gen round, ± near margin; indusia 0, marginal tooth often reflexed over sorus. . . ***A. distentifolium*** var. ***americanum***
1′ Sori oblong or J-shaped, ± not near margin; indusia oblong, J-shaped, or reniform, margin ± reflexed or not, but not over sorus . . . . ***A. filix-femina*** var. ***cyclosorum***

A. distentifolium Opiz var. **americanum** (Butters) Cronquist (p. 137) **LF**: blade narrow-elliptic or lanceolate, 2–3-pinnate, lower 1 or 2 pinna pairs gen ± < those above, ultimate segments shallowly to deeply pinnately lobed, midribs below glabrous or with scattered scales. 2*n*=80. Moist or wet rock crevices, talus, cliffs, boulder bases, streamsides; 1700–3700 m. KR, CaRH, SNH, Wrn, W&I; to AK, w Can, MT, CO; also e Can, Greenland, e Asia. [*A. alpestre* Milde var. *a.* Butters, illeg.] *A. distentifolium* var. *d.* in Eurasia.

A. filix-femina (L.) Roth var. **cyclosorum** Rupr. (p. 137) **LF**: blade elliptic to lanceolate, 1–2-pinnate-pinnatifid, lower 2–4 pinna pairs gen << those above, ultimate segments pinnately lobed to ± toothed, midribs below near base often with minute, branched hairs 0.1–0.2 mm. 2*n*=80. Woodland, along streams, seepage areas; < 3200 m. CA-FP (exc Teh, ScV, SCoRI, SCo, WTR, PR exc SnJt), MP; to AK, w Can, ID, CO, n C.Am. [*A. f.* var. *californicum* Butters] Highly variable, but named vars. in w N.Am seem indistinct; other vars. worldwide.

CYSTOPTERIS FRAGILE FERN

Rhizome gen short-creeping. **LF**: stipe ± fleshy, often with few scales, base ×-section with 2 vascular strands; blade 2–4-pinnate, pinnae sides unequal, acroscopic pinnules more spreading, larger, more incised; veins free. **SPORANGIA**: sori round; indusia hood-like, arched over sorus, attached on side away from margin, often obscure in mature sori. ± 20 spp.: gen temp, a few at high elevations in tropics. (Greek: bladder fern, from indusia) Often confused with *Woodsia* (pinnae sides equal; more fragmented indusia encircling sorus base). Sole sp. of *Cystopteris* in CA highly polymorphic, may represent a sp. complex. Malformed spores produced by some pls, possibly indicative of hybridization, existence of different ploidal levels (numbers of chromosome sets).

C. fragilis (L.) Bernh. (p. 137) Rhizome 2–4 mm diam; scales at tip, lanceolate, ± brown, shining, glabrous, entire. **LF**: 8–30(37) cm; stipe gen < blade, < 1.5 mm wide, glabrous, base straw-colored to red-brown; blade gen 10–24 cm, 3–9 cm wide, lance-ovate, lowest 2–4 pinnae ± < others. **SPORANGIA**: indusia gen ± white. 2*n*=168. Shady, moist rock crevices, meadows, streamsides; 50–4100 m. KR, NCoR, CaR, n&c SNF, SNH, SnFrB, SCoR, n ChI (Santa Cruz Island), TR, PR, GB, DMtns; worldwide.

WOODSIA CLIFF FERN

John C. Game, Alan R. Smith & Thomas Lemieux

Rhizome gen ascending to suberect, short, old stipe bases many. **LF**: often glandular or hairy; stipe base ×-section with 2 vascular strands; blade 1–2-pinnate, segments ± toothed to pinnately lobed, veins free, ending just short of margin. **SPORANGIA**: sori round, gen not at margins; indusium cup-like, often of many segmented hair- or scale-like fragments or lobes encircling sorus from below, often of crusty, ± white beads, often obscure in age. ± 30 spp.: gen n temp. (J. Woods, Britain, b. 1776) [Windham 1993 FNANM 2:270–280]

1. Hairs on abaxial lf axes ± 0.5–1 mm, ± flat, segmented, nonglandular, and ± 0.1 mm, cylindric,
    non-segmented, glandular . . . . ***W. scopulina***
1′ Hairs on abaxial lf axes 0 or ± 0.1 mm, cylindric, non-segmented, glandular
  2. Indusium of segmented hairs . . . . ***W. oregana***
  2′ Indusium of scale-like fragments or lobes . . . . ***W. plummerae***

W. oregana D.C. Eaton (p. 137) **LF**: 5–25 cm, 1–3.5 cm wide, tip ± acute, unforked; hairs on abaxial lf axes 0 or ± 0.1 mm, cylindric, non-segmented, glandular; pinnae 0.5–2.5 cm, 0.3–1.3 cm wide, pinnately lobed to 1-pinnate, margin fine-toothed. **SPORANGIA**: indusium of segmented hairs. 2*n*=76,152. Crevices, rock bases; 900–2800 m. KR, CaRH, n&s SNH, SnBr, PR, MP, W&I, DMtns; to BC, e Can, n US, OK, AZ. Variable. Sterile, nonglandular pls distinguished from *Cystopteris fragilis* by lf veins that do not quite reach the margin in *Woodsia* but do in *Cystopteris*. If recognized taxonomically, pls differing in chromosome number, spore size [Windham 1993], perhaps other ways, assignable to *W. o.* subsp. *cathcartiana* (B.L. Rob.) Windham (2*n* = 152), *W. o.* subsp. *o.* (2*n* = 76); study needed.

W. plummerae Lemmon (p. 137) PLUMMER'S WOODSIA **LF**: < 25 cm, < 4 cm wide, tip often blunt, sometimes forked; hairs on abaxial lf axes ± 0.1 mm, cylindric, non-segmented, glandular; pinnae < 3 cm, < 1.5 cm wide, pinnately lobed to 1-pinnate, margin toothed to shallowly lobed. **SPORANGIA**: indusium of scale-like fragments or lobes ending in hairs or not. 2*n*=152. Crevices, rock bases; 1600–2000 m. DMtns; to TX, n Mex. ★

W. scopulina D.C. Eaton (p. 137) **LF**: < 32 cm, 1.5–2 cm wide, tip ± acute, unforked; hairs on abaxial lf axes ± 0.5–1 mm, ± flat, segmented, nonglandular, and ± 0.1 mm, cylindric, non-segmented, glandular; pinnae < 12–27 mm, 5–12 mm wide, pinnately lobed to 1-pinnate, margin toothed to shallowly lobed. **SPORANGIA**: indusium of narrow scale-like lobes. 2*n*=76,152. Crevices, rock bases; 1300–3500 m. KR, CaRH, SNH, SnBr, MP, W&I; to AK, e Can, w US. If recognized taxonomically, pls differing in chromosome number, spore size [Windham 1993], perhaps other ways assignable to *W. s.* subsp. *laurentiana* Windham (2*n* = 152), *W. s.* subsp. *s.* (2*n* = 76); study needed.

GYMNOSPERMS

Seed pls; seeds naked in cones or on stalks

CUPRESSACEAE   CYPRESS FAMILY

Jim A. Bartel, except as noted

Shrub, tree, gen evergreen; monoecious or dioecious. **LF**: simple, cauline, alternate or opposite (either ± 4-ranked) or whorled in 3s (6-ranked), linear or scale-, awl- or needle-like (sometimes linear and awl-like on 1 pl, or on juvenile or injured pls), gen decurrent, covering young sts. **POLLEN CONE**: axillary or terminal. **SEED CONE**: ± fleshy to gen woody, gen hard at maturity; scales opposite or whorled, peltate or not. **SEED**: 1–many per scale, angled or lateral winged, gen wind-dispersed. *n*=11. 30 genera, 130+ spp.: ± worldwide, esp N.Am, Eurasia. [Farjon 2005 Monogr Cupressaceae *Sciadopitys*. RBG, Kew] Incl (paraphyletic) Taxodiaceae. Taxa of (polyphyletic) *Cupressus* in TJM (1993) now in *Callitropsis, Chamaecyparis, Hesperocyparis*. Scientific Editor: Thomas J. Rosatti.

1. Lvs alternate, linear or awl-like
  2. Lvs on sprouting, rapidly growing, or fertile sts awl-like, appressed, not ranked, < 8 mm, on other sts linear (to lance-linear), spreading, ± 2-ranked, 5–25 mm; seed cone 13–35 mm . . . . **SEQUOIA**
  2′ Lvs awl-like, appressed, ± 4-ranked, < 15 mm; seed cone 40–90 mm . . . . **SEQUOIADENDRON**
1′ Lvs opposite or whorled, scale-like to less often awl- or needle-like
  3. Seed cone ± fleshy, berry-like, scales fused; seed wing 0; lvs whorled in 3s or not . . . . **JUNIPERUS**
  3′ Seed cone woody, not berry-like, scales ± free; seed wing present, vestigial or not; lvs not whorled in 3s
    4. Seed cone ± oblong, scales not peltate, ± overlapping; seeds (1)2 per fertile scale
      5. Seed cone ± pendent, scales in 3 pairs, middle pair fertile; lvs opposite, appearing whorled in 4s, exposed part of each lf longer than wide; seed wings unequal . . . . **CALOCEDRUS**
      5′ Seed cone erect to reflexed, scales in 4–6 pairs, middle 2–3 pairs fertile; lvs opposite, ± not appearing whorled in 4s; exposed part of each lf ± as long as wide; seed wings ± equal . . . . **THUJA**
    4′ Seed cone ± spheric to widely cylindric, scales peltate, abutting; seeds gen many per fertile scale
      6. Seed cone maturing 1st yr, scale projection gen present, ≤ 1 mm, base depressed from edge; lvs of 2 kinds . . . . **CHAMAECYPARIS**
      6′ Seed cone maturing 2nd yr, scale projection gen > 1 mm, esp 1st yr, base level with or rising from edge; lvs of 1 kind (exc *Callitropsis*)
        7. Ultimate 3 orders of shoots in 2-dimensional, pendent clusters; seed cone opening, falling at maturity, scales gen 4–6 . . . . **CALLITROPSIS**
        7′ Ultimate 3 orders of shoots in 3-dimensional, spreading clusters (exc sometimes *Hesperocyparis macnabiana*); seed cone gen closed, falling many yrs after maturity, scales gen 6–12 . . . . . . **HESPEROCYPARIS**

CALLITROPSIS   NOOTKA CYPRESS

1 sp. (Greek: resembling *Callitris*) [Debreczy et al. 2009 Phytologia 91:140–159; Little 2006 Syst Bot 31:461–480] *Callitropsis* Oerst. probably will not be available for use in plant names after 2012, when *Xanthocyparis* or another genus name probably will be needed for *C. nootkatensis*; all other CA taxa formerly in *Callitropsis* (and before that, in TJM (1993), in *Cupressus*) are here incl in *Hesperocyparis* Bartel & R.A. Price.

C. nootkatensis (D. Don) D.P. Little (p. 137)   ALASKA CEDAR   Tree 20–30 m, pyramidal in youth. **ST**: ultimate 3 orders of shoots in 2-dimensional, pendent clusters; bark fibrous, fissured, peeling, orange-red to gray-brown to purple-brown; ultimate branches 1.1–1.8 mm diam, ± 4-sided to flat. **LF**: opposite, 4-ranked, of 2 kinds, both scale-like, closely appressed, overlapping, green, not glaucous; glands gen obscure; lateral lf tips gen parallel to axis of attachment or diverging. **POLLEN CONE**: terminal, 2.4–4 mm, 1.3–2.6 mm diam, gen yellow. **SEED CONE**: 6–12 mm, spheric, ash-gray; scales gen 4–6; maturing 2nd yr, opening, falling at maturity. **SEED**: < 15 per cone, 3–4.6 mm, brown to red-brown, not glaucous, flat, wings gen 2, vestigial, attachment inconspicuous; cotyledons 2. Cool, moist, forested, well-drained mtn slopes; 20–2500 m. KR (Del Norte, Siskiyou cos.); to AK. [*Cupressus n.* D. Don; *Chamaecyparis n.* (D. Don) Spach; *Xanthocyparis n.* (D. Don) Farjon & D.K. Harder, nom. superfl.] ★

CALOCEDRUS INCENSE CEDAR

John W. Thieret, final revision by Jim A. Bartel

Tree, gen widely conic; monoecious. **ST**: young shoots flat, in flat clusters. **LF**: opposite, 4-ranked, appearing whorled in 4s, scale-like, closely appressed, overlapping, exposed part longer than wide; lf bases elongated on smallest shoots. **SEED CONE**: ± pendent, woody, ± oblong, maturing 1st yr; scales in 3 partially overlapping pairs, middle pair fertile, free, ± spreading, distal pair united, sterile. **SEED**: 2 per fertile scale, flat, wings 2, unequal; cotyledons 2. 3 spp.: w N.Am, e Asia. (Greek: beautiful cedar)

C. decurrens (Torr.) Florin (p. 137)   Tree 20–69 m. **ST**: trunk < 4 m diam, tapered from wide base; bark 1–2.5+ cm thick, fibrous, cinnamon-red; lower branches down-curved. **LF**: 3–10 mm. **POLLEN CONE**: 5–7 mm, oblong, light yellow. **SEED CONE**: 17–35 mm, light red-brown. **SEED**: 14–25 mm, incl wing, ± yellow to red-brown, maturing fall. Common. Mixed-evergreen, yellow-pine forests; 350–2500 m. KR, NCoR, CaR, SN, SCoR, TR, PR, MP; OR, w NV, n Baja CA. Frequently planted as an orn.

CHAMAECYPARIS FALSE CYPRESS

Single-trunked tree, pyramidal in youth; monoecious. **ST**: young shoots flat, in drooping or pendent clusters, in 1 plane. **LF**: opposite, 4-ranked, scale-like, overlapping, of 2 kinds (facials smaller, appressed, laterals with tips spreading above facials). **POLLEN CONE**: terminal. **SEED CONE**: 5–14 mm, woody, ± spheric to widely cylindric, maturing 1st yr, open at maturity; scales 8–12, peltate, abutting, shield- or wedge-shaped, projection gen present, ≤ 1 mm, pointed, base depressed from edge. **SEED**: (1)2–5(7) per scale, flat, wings 2, narrow; cotyledons 2. *n*=11. 5 spp.: w&e N.Am, e Asia. (Greek: dwarf or on-the-ground cypress) [Farjon 2005 Monograph Cupressaceae *Sciadopitys*. RBG, Kew]

C. lawsoniana (A. Murray bis) Parl. (p. 141)   PORT ORFORD CEDAR, LAWSON CYPRESS   Tree 20–65 m. **ST**: trunk < 6 m diam; bark 15–25 cm thick, red-brown to tan, fibrous, fire-resistant; young shoots flat, in flat clusters, held ± horizontally, lower surfaces often paler. **LF**: green, gen glaucous; glands gen visible; lateral lf tips gen curved toward st axis. **POLLEN CONE**: yellow-green, becoming dark red to purple black. **SEED CONE**: 6–11 mm, spheric, red-brown; scales 7–10, projection ± 0. **SEED**: 2–5 per scale, 2–4 mm, gen with wart-like pitch pockets, light chestnut-brown, gen glandular. Coastal conifer, mixed-evergreen, yellow-pine forests, often on serpentine; < 1700 m. KR; sw OR. [*Cupressus l.* A. Murray bis]

HESPEROCYPARIS WESTERN CYPRESS

Large shrub, 1- or multi-trunked tree, often pyramidal in youth; monoecious. **ST**: young shoots or branches gen cylindric (sometimes ± 4-angled or flat), gen in 3-dimensional clusters. **LF**: opposite, 4-ranked, of 1 kind, scale-like, closely appressed, overlapping. **POLLEN CONE**: terminal, 2–6.5 mm, 1.3–3 mm diam, gen yellow. **SEED CONE**: 10–50 mm, woody, ± spheric to widely cylindric, maturing 2nd yr, gen closed and attached beyond maturity (> 2 yrs); scales (4)6–12, peltate, abutting, shield- or wedge-shaped, projection gen > 1 mm esp 1st yr, pointed, base level with or rising from edge. **SEED**: 60–150 per cone, flat, wings gen 2, vestigial; cotyledons gen 3–5. 2*n*=22(23,24). 16 spp.: w&c N.Am, C.Am. (Greek: western cypress) [Adams & Bartel 2009 Phytologia 91:287–299; Adams et al. 2009 Phytologia 91:160–185; Little 2006 Syst Bot 31:461–480]

1. Ultimate 2 orders of shoots or branches in 1 plane; large shrub; resin copious, sticky when fresh . . . . ***H. macnabiana***
1′ Ultimate 2 orders of shoots or branches in 3-dimensions; gen tree (exc gen *Hesperocyparis forbesii*); resin not copious, not sticky
  2. Lf glands or pits on ultimate branches, resin on lf glands 0 or not
   3. Resin ± 0 on lf glands on ultimate branches; PR
    4. Multi-trunked shrub to small tree; lf not glaucous, foliage not gray-green . . . . ***H. forbesii***
    4′ Single-trunked tree; lf glaucous, foliage gray-green . . . . ***H. stephensonii***
   3′ Resin gen on lf glands on ultimate branches; n of SW
    5. Ultimate branches < 1 mm diam; seed cones gen < 20 mm diam; KR, CaRH, n SNH, MP . . . . ***H. bakeri***
    5′ Ultimate branches > 1 mm diam; seed cones gen > 20 mm diam; s SN . . . . ***H. nevadensis***
  2′ Lf glands or pits ± 0 on ultimate branches, resin on lf glands 0 (exc *Hesperocyparis sargentii* in Santa Barbara Co.)
   6. Crown asymmetric, often open, flat-topped to widely conic in age; seed cones 20–32(38) mm; outside native range widely planted, naturalized. . . . . ***H. macrocarpa***
   6′ Crown symmetric, often dense, gen pyramidal in age; seed cones (10)12–27(35) mm; outside native range not widely planted, not naturalized
    7. Foliage dull, dusty to gray-green, glaucous or not; on serpentine; NCoR, SnFrB, SCoR . . . . ***H. sargentii***
    7′ Foliage bright to dark green, not glaucous; not on serpentine; NCo, CCo, SnFrB
     8. Pl 1–2 m on sterile soil, 10–20(50) m on rich soil; foliage dark dull green; seed not glaucous; s NCo (Mendocino, Sonoma cos.). . . . . ***H. pygmaea***
     8′ Pl 6–10(15) m; foliage yellow-green or light bright to deep green; seed glaucous or not; CCo, SnFrB
      9. Pl 5–7 m; seed cones (10)14–20 mm; ultimate branches 0.8–1.1 mm diam; CCo (Monterey Peninsula) . . . . ***H. goveniana***
      9′ Pl 6–10(15) m; seed cones 16–30(35) mm; ultimate branches 0.9–1.2 mm diam; SnFrB (Santa Cruz Mtns). . . . . ***H. abramsiana***
       10. Seed cones (14)16–25 mm, 14–22 mm diam; Santa Cruz Co. . . . . var. ***abramsiana***
       10′ Seed cones 22–32(35) mm, 22–31 mm diam; San Mateo Co. . . . . var. ***butanoensis***

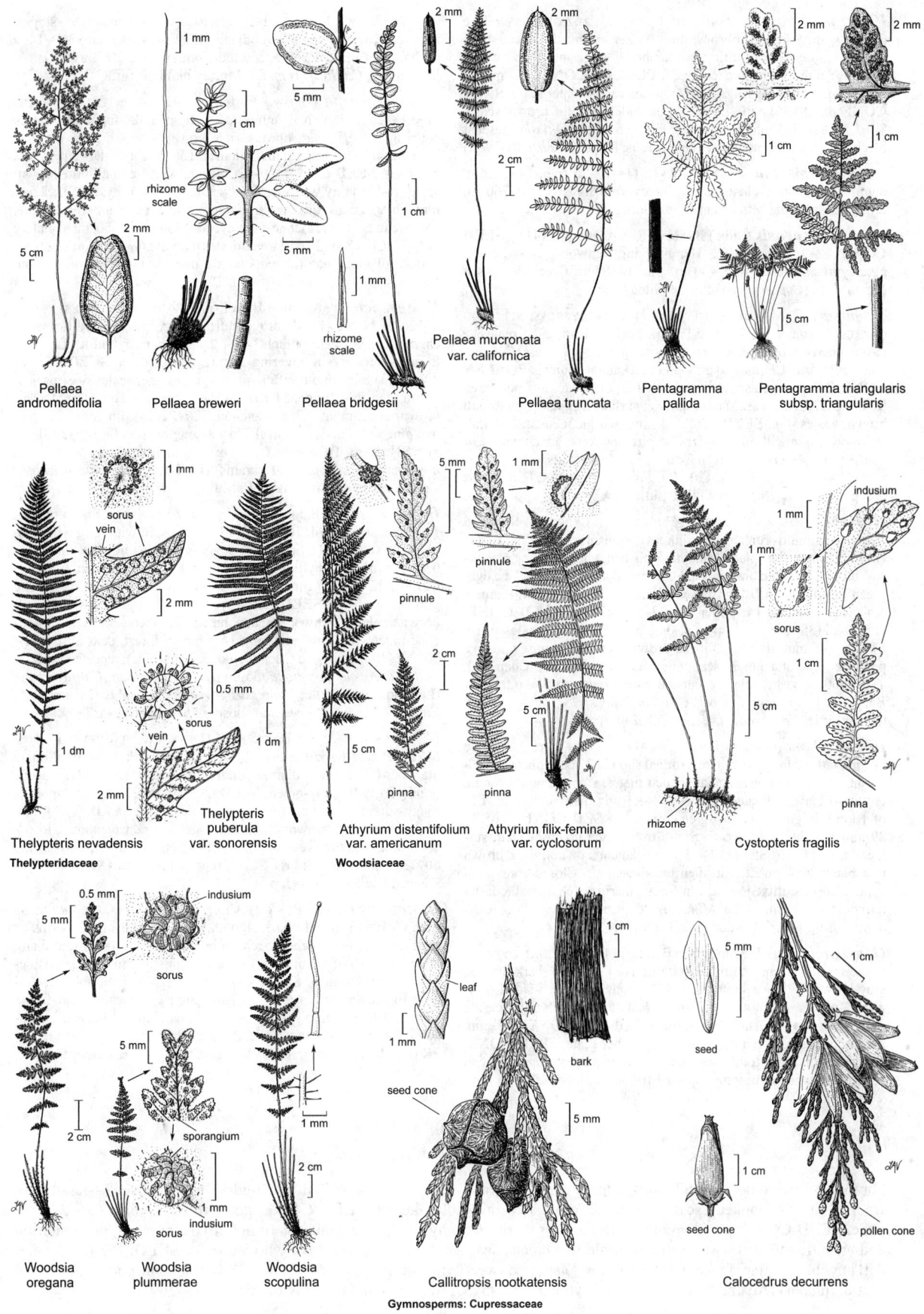

Pellaea
andromedifolia

Pellaea breweri

Pellaea bridgesii

Pellaea mucronata
var. californica

Pellaea truncata

Pentagramma
pallida

Pentagramma triangularis
subsp. triangularis

Thelypteris nevadensis

Thelypteridaceae

Thelypteris
puberula
var. sonorensis

Athyrium distentifolium
var. americanum

Athyrium filix-femina
var. cyclosorum

Woodsiaceae

Cystopteris fragilis

Woodsia
oregana

Woodsia
plummerae

Woodsia
scopulina

Callitropsis nootkatensis

Gymnosperms: Cupressaceae

Calocedrus decurrens

H. abramsiana (C.B. Wolf) Bartel SANTA CRUZ CYPRESS Tree 6–10(15) m. **ST:** bark fibrous, thin, broken in thick vertical strips or plates, gray-brown; ultimate branches 0.9–1.2 mm diam, cylindric. **LF:** light bright to deep green. **POLLEN CONE:** 3–4 mm, 2 mm diam, ± 4-sided; scales 10–16; pollen sacs 4–6 per scale. **SEED CONE:** 16–30(35) mm, spheric to gen widely elliptic, brown; scales 8–10. **SEED:** 3–5 mm, glaucous or not, gen dull brown to iridescent black, rough; attachment scar conspicuous. [*Cupressus a.* C.B. Wolf]

var. ***abramsiana*** **SEED CONE:** (14)16–25 mm, 14–22 diam. Yellow-pine, closed-cone-pine/cypress forests; 370–760 m. SnFrB (Santa Cruz Mtns, Santa Cruz Co.). ★

var. ***butanoensis*** (Silba) Bartel & R.P. Adams (p. 141) **SEED CONE:** 22–32(35) mm, 22–31 mm diam. Redwood, closed-cone-pine/cypress forests; 400–490 m. SnFrB (Santa Cruz Mtns, San Mateo Co.). [*Cupressus a.* subsp. *b.* Silba] ★

H. bakeri (Jeps.) Bartel (p. 141) BAKER CYPRESS, SISKIYOU CYPRESS Tree 7–30(40) m. **ST:** bark peeling in irregular plates, ± red- or cherry-brown, aging gray-brown; ultimate branches 0.7–1 mm diam, cylindric. **LF:** dark to gray-green, resin not copious. **POLLEN CONE:** 2–3 mm, gen spheric; scales 6–10; pollen sacs 3–5 per scale. **SEED CONE:** 10–18(25) mm, spheric; surface warty, silvery to dull brown; scales 4–8. **SEED:** (2.4)3–4 mm, gen glaucous, tan to dull red-brown, gen with many wart-like pitch pockets; attachment scar conspicuous. Mixed-evergreen forest, open slopes, flats, often on serpentine; 1100–1800 m. KR, CaRH, n SNH, MP; sw OR. [*Cupressus b.* Jeps.; *Callitropsis b.* (Jeps.) D.P. Little] ★

H. forbesii (Jeps.) Bartel (p. 141) TECATE CYPRESS Shrub to tree < 10 m, multi-trunked, gen without dominant terminal shoot. **ST:** bark peeling in thin plates, smooth, polished, cherry-red or mahogany-brown; ultimate branches 0.9–1.1 mm diam, cylindric. **LF:** light green to dull green. **POLLEN CONE:** 3–4 mm, ± 2 mm diam, ± 4-sided; scales 12–14; pollen sacs 3–5 per scale. **SEED CONE:** (17)20–32(35) mm, gen spheric, dull brown or gray; scales 6–10. **SEED:** 3–6 mm, dark to dull red-brown, gen with many wart-like pitch pockets; attachment scar conspicuous. 2n=22(23). Chaparral; 450–1500 m. w PR, planted outside native range in SW mtns; nw Baja CA. [*Cupressus f.* Jeps.; *C. guadalupensis* S. Watson var. *f.* (Jeps.) Little; *Callitropsis f.* (Jeps.) D.P. Little] ★

H. goveniana (Gordon) Bartel (p. 141) GOWEN CYPRESS Tree 5–7 m, without long, whip-like terminal shoot. **ST:** bark fibrous, peeling in linear strips or irregular-shaped plates, smooth, aging rough, brown to gray; ultimate branches 0.8–1.1 mm diam, cylindric. **LF:** of 1 kind, bright- to yellow-green, resin 0. **SEED CONE:** (10)14–20 mm, ± spheric, brown to gray-brown, often green beneath surface; scales 6–10. **SEED:** 2–4.5 mm, glaucous or not, dark brown to ± black; attachment scar often inconspicuous. Closed-cone-pine/cypress forests, mixed-evergreen forest, maritime chaparral, coastal terraces; 50–160 m. c CCo (Monterey Peninsula). [*Cupressus g.* Gordon; *Callitropsis g.* (Gordon) D.P. Little] ★

H. macnabiana (A. Murray bis) Bartel (p. 141) MCNAB CYPRESS Large shrub 3–10 m (multi-trunked tree 10–18 m). **ST:** bark fibrous, gray-brown; ultimate branches 0.8–1.1 mm diam, ± 4-sided, ultimate 2 orders of shoots in 1 plane. **LF:** blue- to dull gray-green to gray; resin copious, sticky when fresh. **POLLEN CONE:** 2–3 mm, 2 mm diam; scales gen 8; pollen sacs 3–5 per scale. **SEED CONE:** 15–21(25) mm, gen longer than wide, gray to red-brown; scales gen 6, projection conic, pointing upward to incurved. **SEED:** 3–5 mm,

± glaucous, dull brown; attachment scar conspicuous. Dry slopes, flats, chaparral, pine/oak woodland, often on serpentine; 300–1460 m. NCoR, CaRF, n SNF; reported from sw OR. [*Cupressus m.* A. Murray bis; *Callitropsis m.* (A. Murray bis) D.P. Little]

H. macrocarpa (Hartw.) Bartel (p. 141) MONTEREY CYPRESS Tree 18–25 m; crown asymmetric, often open, flat-topped to widely conic in age. **ST:** bark fibrous, rich brown aging ash-gray; ultimate branches 0.9–1.2 mm diam, cylindric. **LF:** bright to dark green, not glaucous. **SEED CONE:** 20–32(38) mm, spheric to elliptic, brown; scales 8–12, shiny when fresh; ± opening at maturity. **SEED:** 2.5–5 mm, not glaucous, dull red-brown to black; attachment scar ± white, conspicuous. Closed-cone-pine/cypress forests; < 50 m in native range. NCo, CCo, SCo (native to Monterey Peninsula, Point Lobos), widely planted, naturalized outside native range. [*Cupressus m.* Hartw.; *Callitropsis m.* (Hartw.) D.P. Little] ★

H. nevadensis (Abrams) Bartel (p. 141) PIUTE CYPRESS Tree 5–12(25) m. **ST:** bark fibrous, not peeling, gray- to red-brown to cherry-red; ultimate branches 1–1.2 mm diam, cylindric. **LF:** dull to gray-green; resin covering glands, conspicuous. **SEED CONE:** 20–30(35) mm, ovoid, often silver-gray in age; scales 6–8, projection less visible in age. **SEED:** 3–6 mm, gen glaucous, light to red-brown, attachment ± inconspicuous. 2n=22(23). Pinyon/juniper or oak/pine woodland, chaparral, closed-cone-cypress forest; 750–1800 m. s SN (Kern, Tulare cos.). [*Cupressus macnabiana* A. Murray var. *n.* Abrams; *Callitropsis n.* (Abrams) D.P. Little; *Cupressus arizonica* Greene subsp. *n.* (Abrams) A.E. Murray] ★

H. pygmaea (Lemmon) Bartel PYGMY CYPRESS Shrub or tree, 1–2 m on sterile soil, 10–20(50) m on rich soil; with long whip-like ultimate shoot. **ST:** bark fibrous, gray-brown; ultimate branches 0.9–1.1 mm diam, cylindric. **LF:** gen dark dull green, resin 0. **SEED CONE:** 12–27(35) mm, spheric to gen widely elliptic, tan aging gray; scales 6–10. **SEED:** 2.5–4.7(5.5) mm, not glaucous, dark red-brown to black, shiny or not; attachment scar inconspicuous. Closed-cone-pine/cypress forests, mixed-evergreen forest, coastal terraces; 50–200(300) m. s NCo (Mendocino, Sonoma cos.); reported from sw OR. [*Cupressus goveniana* Gordon var. *p.* Lemmon; *Callitropsis p.* (Lemmon) D.P. Little; *Cupressus goveniana* subsp. *p.* (Lemmon) A. Camus; *Cupressus goveniana* subsp. *pigmaea*, orth. var.] ★

H. sargentii (Jeps.) Bartel (p. 141) SARGENT CYPRESS Tree 6–20 m. **ST:** bark fibrous, thick, gray or dark brown; ultimate branches 0.9–1.2 mm diam, cylindric to 4-sided. **LF:** dull, glaucous or not, dusty- to gray-green. **SEED CONE:** 15–25(30) mm, spheric, rough-surfaced, dull brown to gray, scales 6–10. **SEED:** 3–5.3 mm, ± glaucous, dark brown; attachment scar gen conspicuous. Closed-cone-pine/cypress, yellow-pine forests, chaparral, on serpentine; 60–1370 m. NCoR, SnFrB, SCoR; reported from sw OR. [*Cupressus s.* Jeps.; *Callitropsis s.* (Jeps.) D.P. Little]

H. stephensonii (C.B. Wolf) Bartel (p. 141) CUYAMACA CYPRESS Tree (5)10–16 m. **ST:** bark smooth, thin, peeling in thin strips or plates, red-brown to cherry-red; ultimate branches 1–1.3 mm diam, cylindric. **LF:** of 1 kind, gray-green; glands gen inactive, conspicuous or not, resin 0. **SEED CONE:** 17–27(30) mm, spheric, dull gray to brown; scales 6–8(10), projection conspicuous and conic. **SEED:** 3.6–6(8) mm, glaucous or not, dark to red-brown; attachment scar conspicuous. 2n=22(23,24). Chaparral; 910–1800 m. sw PR (Cuyamaca Mtns); nw Baja CA. [*Cupressus arizonica* subsp. *a.*, misappl.] ★

JUNIPERUS JUNIPER

Robert P. Adams & Jim A. Bartel

Shrub, tree; gen dioecious. **ST:** bark thin, peeling in strips; young shoots 4-angled to cylindric. **LF:** opposite (4-ranked) or whorled in 3s (6-ranked), scale-like to less often awl- or needle-like. **POLLEN CONE:** gen terminal; pollen sacs 2–6 per scale. **SEED CONE:** gen terminal, 5–18 mm, ± spheric, ± fleshy, berry-like, glaucous or not, dry or resinous, gen maturing 2nd yr, surrounded at base by minute scale-like bracts; scales 3–8, fused, opposite or whorled in 3s. **SEED:** 1–3 per cone, ± flat, unwinged, often not angled, gen animal-dispersed over 2 yrs; cotyledons 2–6. 67 spp., 28 var.: n hemisphere exc ne Afr. (Latin: juniper) [Adams & Nguyen 2007 Phytologia 89:43–57; Adams et al. 2006 Phytologia 88:299–309]

1. Lf awl- or needle-like, suberect to spreading but not appressed; pollen, seed cones axillary; shrub < 1 m
   (sect. *Juniperus*)... ***J. communis***
     2. Lf 10–20+ mm; glaucous stomatal band ± 1–1.5 × as wide as green lf margin .... var. ***depressa***
     2′ Lf 5–10(12) mm; glaucous stomatal band 2–4 × as wide as green lf margin
        3. Mature seed cone elongate to ± spheric; glaucous stomatal band 3–4 × as wide as green lf margin .... var. ***jackii***
        3′ Mature seed cone spheric; glaucous stomatal band 2 × as wide as green lf margin .... var. ***saxatilis***
1′ Lf gen scale-like, closely appressed; pollen, seed cones terminal; tree, shrub > 1 m (sect. *Sabina*)
     4. Seed cone maturing blue-black; tree 5–15(30) m
        5. Bark red-brown; lvs gen whorled in 3s; gen dioecious; NCoRH, SNH, SnGb, SnBr, SNE, DMtns .... ***J. grandis***
        5′ Bark brown; lvs opposite or whorled in 3s; gen ± monoecious; CaRH, MP .... ***J. occidentalis***
     4′ Seed cone maturing red-brown; shrub, tree 1–8(10) m
        6. Lf gland obvious; trunks several at base; gen dioecious................................... ***J. californica***
        6′ Lf gland obscure; trunk gen 1 at base; monoecious .... ***J. osteosperma***

J. californica Carrière (p. 141) CALIFORNIA JUNIPER Shrub, tree 1–4(10) m; gen dioecious. **ST:** trunks several at base; bark gray, thin, outer layers persistent. **LF:** gen whorled in 3s, 6-ranked; scale-like, closely appressed; gland obvious. **POLLEN CONE:** 2–3 mm, oblong. **SEED CONE:** 7–12 mm, spheric to ovoid, ± blue maturing red-brown, dry. **SEED:** 5–7 mm, pointed, angled, brown. Dry slopes, flats, pinyon/juniper woodland; 50–1500 m. NCoRI, SNF, ScV (Sutter Buttes), SnFrB, SCoR, TR, PR, D; s NV, nw AZ, Baja CA (Cedros, Guadalupe islands).

J. communis L. COMMON JUNIPER Shrub < 1 m, gen ± prostrate; gen dioecious. **ST:** bark brown, peeling in papery sheets. **LF:** gen whorled in 3s, 6-ranked; suberect to spreading, awl- or needle-like, jointed, not decurrent. **POLLEN CONE:** axillary, 4–5 mm. **SEED CONE:** axillary, 5–9 mm, green maturing bright blue to black, glaucous, resinous. **SEED:** 2–5 mm, ovoid, acute, gen 3-angled. 3 other vars. extend sp. range to incl Eur, Asia.

 var. ***depressa*** Pursh DEPRESSED JUNIPER Low, occ spreading. **LF:** 10–20+ mm; glaucous stomatal band ± 1–1.5 × as wide as green lf margin. **SEED CONE:** 6–9 mm. **SEED:** 3 per cone. Rocky soil, slopes, summits; < 2800 m. SNH; ± to AK, ne N.Am.

 var. ***jackii*** Rehder (p. 141) JACK'S JUNIPER Spreading, mat-like, occ erect. **LF:** 7–9(10) mm; glaucous stomatal band 3–4 × as wide as green lf margin. **SEED CONE:** mature elongate to ± spheric. **SEED:** 1(2) per cone. Rocky or wooded slopes, high-elevation forests, often on serpentine or lava; 1900–3400 m. KR, CaRH; OR. [*J. c.* var. *montana* Aiton, in part]

 var. ***saxatilis*** Pall. MOUNTAIN JUNIPER Spreading, mat-like, occ erect. **LF:** 5–10(12) mm; glaucous stomatal band 2 × as wide as green lf margin. **SEED CONE:** mature spheric. **SEED:** 1–3 per cone. Rocky areas; 600–2400 m. KR, SNH, Wrn; to YT, Eur, Asia. [*J. c.* var. *montana* Aiton, in part]

J. grandis R.P. Adams (p. 141) SIERRA JUNIPER, GRAND JUNIPER Tree 8–20(30) m; gen dioecious. **ST:** bark red-brown. **LF:** gen whorled in 3s, 6-ranked; closely appressed, scale-like. **SEED CONE:** 5–9 mm, blue-green maturing blue-black, resinous. **SEED:** cotyledons 2–4. 2*n*=22. Exposed, dry, rocky slopes, flats, forest, pinyon/juniper woodland; 100–3100 m. NCoRH, SNH, SnGb, SnBr, SNE, DMtns; w NV. [*J. occidentalis* var. *australis* (Vasek) A.H. Holmgren & N.H. Holmgren]

J. occidentalis Hook. WESTERN JUNIPER Tree 5–15(20) m; gen ± monoecious. **ST:** bark brown. **LF:** opposite, 4-ranked or whorled in 3s, 6-ranked. **POLLEN CONE:** 2–3 mm, oblong. **SEED CONE:** 7–12 mm, blue-green maturing blue-black, resinous. **SEED:** cotyledons gen 2. 2*n*=22. Dry slopes, flats, sagebrush, juniper woodland; 700–2300 m. CaRH, MP; to WA, ID, w NV.

J. osteosperma (Torr.) Little (p. 141) UTAH JUNIPER Tree 1–8 m; monoecious. **ST:** trunk gen 1; bark thin, gray-brown aging ash-white. **LF:** gen opposite, 4-ranked; closely appressed, scale-like; gland obscure. **POLLEN CONE:** 2–3 mm, cylindric. **SEED CONE:** 5–13 mm, spheric, brown maturing red-brown, dry. **SEED:** 1(2), 3–4 mm, ovoid, strongly angled. Pinyon/juniper woodland; 1300–2600 m. SnGb, SnBr, GB, DMtns; to MT, NM.

SEQUOIA REDWOOD

Steve Boyd & James R. Griffin

1 sp.: w N.Am. (Sequoyah, Cherokee chief, 1776?–1843)

S. sempervirens (D. Don) Endl. (p. 141) Tree, pl gen sprouting vigorously from base if cut, from entire crown if burned. **ST:** trunk < 110 m, to 9 m diam; old crown narrowly conic to ± cylindric, gen unbranched in lower 1/2; bark < 30 cm thick near base, fibrous, ridged, red-brown; branches downswept to ± ascending; twigs persistent < 4 yrs. **LF:** alternate, green < 3 yrs, persistent < 4; of 2 kinds, those on sprouting, rapidly growing, or fertile sts appressed, not ranked, < 8 mm, awl-like, others spreading, ± 2-ranked, 5–25 mm, linear (to lance-linear), gen flat. **POLLEN CONE:** 2–5 mm, ± spheric to ovoid. **SEED CONE:** 13–35 mm, ± spheric, woody, maturing in 1 yr, persistent < 2; scales peltate, fused to bracts. **SEED:** 2–7 per scale, 3–6 mm, wings 2, narrow, lateral. *n*=33. Redwood forest; < 1100 m. w KR, NCoRO, w NCoRI, n&c CCo, SnFrB, n SCoRO; sw OR. Tallest trees in N.Am.

SEQUOIADENDRON GIANT SEQUOIA

Steve Boyd & James R. Griffin

1 sp.: CA. (Greek: sequoia tree)

S. giganteum (Lindl.) J. Buchholz (p. 141) Tree, gen not sprouting. **ST:** trunk < 90 m, to 11 m diam; old crown irregular, with large branches throughout; bark to ± 60 cm thick near base, fibrous, ridged, red-brown; branches spreading to downswept, ends upturned; twigs persistent < 20 yrs. **LF:** alternate, green < 4 yrs, persistent < 20; of 1 kind, appressed, ± 4-ranked, < 15 mm, awl-like. **POLLEN CONE:** 4–8 mm, ± spheric to ovoid. **SEED CONE:** 40–90 mm, oblong, woody, maturing in 2 yrs, persistent < 20; scales peltate, fused to bracts. **SEED:** 3–9 per scale, 3–6 mm, wings 2, unequal, lateral. *n*=11. Uncommon. Mixed-conifer forest, esp with favorable soil moisture; 825–2700 m. SNH, probably naturalizing in nw SnJt (nw Black Mtn). Most massive trunks in N.Am.

THUJA ARBORVITAE

ST: trunk gen flared at base; branches spreading to ± pendent, tips often upturned; young shoots 4-sided, in flat clusters, held horizontally. **LF**: opposite, 2-ranked, ± not appearing whorled in 4s, closely appressed, scale-like, exposed part ± as long as wide. **POLLEN CONE**: terminal. **SEED CONE**: erect to reflexed, woody, ovoid-oblong, tapered to point, maturing 1st yr; scales in 4–6 pairs, ± overlapping, sharp-pointed near tip, thin, oblong, acute, leathery, middle 2–3 pairs fertile. **SEED**: (1)2 per fertile scale, wings 2, ± equal; cotyledons 2. 5 spp.: n N.Am, e Asia. (Greek: resinous tree)

T. plicata D. Don (p. 147) WESTERN RED CEDAR, CANOE CEDAR
Tree 30–70 m. **ST**: bark 1–2 cm thick, cinnamon-red, fibrous; young shoot upper surface glossy dark green, lower surface gen faintly white-streaked. **POLLEN CONE**: red-brown turning dark gray.

SEED CONE: 10–19 mm, light brown. **SEED**: 4–6 mm, narrow, elliptic, light brown. n=11. Coastal conifer forest; < 1800 m. NCo, KR, NCoRO (Del Norte, Humboldt cos.); to AK, MT.

EPHEDRACEAE EPHEDRA FAMILY

Stefanie M. Ickert-Bond

Shrub; gen dioecious; taproot deep. **ST**: erect, prostrate, occ climbing, jointed, much branched, green in youth; branchlets opposite or whorled, cylindric, longitudinally grooved; pith brown or white. **LF**: simple, 2 (each pair at right angles to adjacent pairs) or 3 per node, scale-like, ± fused at base into membranous sheath, gen ephemeral; resin canals 0. **CONE**: axillary (terminal), sessile or not, 1–many per node, ovoid or ellipsoid. **POLLEN CONE**: bracts 2–3 per node, proximal sterile, distal subtending 2 basally fused bractlets subtending a stalk with 2–10(15) sessile to stalked, anther-like parts with 2 apical pores. **SEED CONE**: stalk 1–11(50) mm, gen straight, naked or with reduced basal bracts; distal bracts in (2)4–10 pairs or whorls, free or fused, brown or green, scarious or fleshy, subtending 2 fused leathery bractlets enclosing 1 ovule. **SEED**: 1–3(4) per cone (in ×-section, ± round if 1, ± semicircular if 2, ± triangular if 3, etc.), ± spheric to ellipsoid or lance-ovoid, beaked or not. 1 genus, ± 50 spp.: arid, semi-arid (to trop) regions of n Afr, Asia, Eur, N.Am, S.Am. [Ickert-Bond 2003 Ph.D. Dissertation Arizona State Univ] Scientific Editor: Thomas J. Rosatti.

EPHEDRA EPHEDRA, MORMON TEA

ST: spreading to gen erect, 0.3–2(4.5) m; uppermost bud gen conic, gen not thorn-like. (Greek: sitting upon, from jointed sts and/or ancient name for *Equisetum*, also with jointed sts)

1. Lvs, cone bracts gen 3 per node; seed cones sessile
  2. Uppermost bud thorn-like; lf sheath persistent, fibrous . . . . ***E. trifurca***
  2′ Uppermost bud narrowed but not thorn-like; lf sheath deciduous, not fibrous
    3. Seed not beaked; twig yellow-green in youth; lf base in age recurved, swollen . . . . ***E. californica***
    3′ Seed beaked; twig gray-green in youth; lf base in age not recurved, not swollen. . . . . ***E. funerea***
1′ Lvs, cone bracts gen 2 per node; seed cones sessile or stalked
  4. Distal bracts of seed cone fused ≥ 2/3, fleshy at maturity. . . . . ***E. foeminea***
  4′ Distal bracts of seed cone free or fused < 1/3, scarious at maturity
    5. Seed gen 1; seed cone stalk ≤ 2 mm . . . . ***E. aspera***
    5′ Seeds gen 2; seed cone stalk ≤ 11(50) mm
      6. Twig pale green, glaucous in youth; lf base gray, deciduous; bracts of seed cone broadly ovate . . . . . . ***E. nevadensis***
      6′ Twig bright- to yellow-green in youth; lf base brown, persistent; bracts of seed cone elliptic . . . . ***E. viridis***

E. aspera S. Watson (p. 147) BOUNDARY EPHEDRA **ST**: 0.3–2 m; twig ± green, aging yellow, glaucous or not; pith dark brown. **LF**: 2 per node, 1–3.5 mm, gen persistent; sheath splitting, base in age swollen. **POLLEN CONE**: 1–many per node, 2.5–7 mm; bracts 2 per node. **SEED CONE**: 1–few per node, 4.5–8(10) mm, ovoid, stalk ≤ 2 mm, basal bracts reduced; distal bracts 2 per node, free, ovate, red-brown, margins membranous, stalk-like base 0. **SEED**: gen 1, 4–7(8) mm, gen < 2 × as long as wide, smooth to ± roughened. $2n$=14,28. Creosote-bush scrub, Joshua-tree woodland; 50–2133 m. SnBr, SnJt, SNE, D; to TX, Mex. [*E. fasciculata* A. Nelson var. *clokeyi* (H.C. Cutler) Clokey; *E. f.* var. *f.*] Cones present Mar–Apr

E. californica S. Watson (p. 147) DESERT TEA **ST**: 0.3–1.5 m; twig yellow-green, aging gray-brown; uppermost bud narrowed but not thorn-like; pith dark brown. **LF**: 3 per node, 2–6 mm; ± white margins wearing off; base in age recurved, swollen. **POLLEN CONE**: 1–many per node, 6–7.5 mm; bracts 3 per node. **SEED CONE**: 1–few per node, 4.5–8(10) mm, ovoid, stalk 1–2 mm, straight, basal bracts reduced; distal bracts 3 per node, ovate, free to base, scarious, center orange- or green-yellow, ± thickened, margins entire, translucent, stalk-like base short. **SEED**: gen 1, < 10 mm, gen > 2 × as long as wide, not beaked, smooth. Scattered in arid grassland, chaparral, creosote-bush scrub; 70–1300 m. s SNF, Teh, w SnJV, SCoR, SW, D;

Baja CA. Pls of coast ranges, relative to pls of desert, taller, may have different ecological tolerances. Cones present Mar–Apr

E. foeminea Forssk. Climbing shrub, ≤ 5 m, or hanging (prostrate). **ST**: twigs glabrous, gray-green, aging red-brown at first, later gray; uppermost bud elliptic; pith white. **LF**: 2 per node, ≤ 2.5 mm, sheath base not swollen. **POLLEN CONE**: 1–few per node, ≤ 5 mm; bracts 2 per node. **SEED CONE**: 1–several per node, 8–9 mm, narrow-cylindric before full maturity, stalk 2–10(20) mm, gen curved, naked; distal bracts 2 per node, round, fused ≥ 2/3, fleshy, red. **SEED**: (1)2, 6–9 mm, gen > 2 × as long as wide, smooth. $2n$=14 (acc. to older records also 12 and 28). Oak woodland; 600–750 m. SCo; e Medit, w Saudi Arabia, Ethiopia, Somalia. [*E. distachya* L., misappl.] Cones present Mar–Apr, Jul–Aug

E. funerea Coville & C.V. Morton DEATH VALLEY EPHEDRA **ST**: < 1.5 m; twig gray-green, aging gray; uppermost bud narrowed but not thorn-like; pith brown. **LF**: gen 3 per node, < 5 mm; base persistent, not recurved, not swollen. **POLLEN CONE**: 1–3 per node, 5–8 mm; bracts 3 per node. **SEED CONE**: 1–3 per node, 9–13 mm, elliptic, stalk 0 or < 1 mm, straight, basal bracts reduced; distal bracts 3 per node, obovate, membranous, ± green, ± thickened, margin wide, yellow-translucent, stalk-like base wide. **SEED**: 1(3), 6–9 mm,

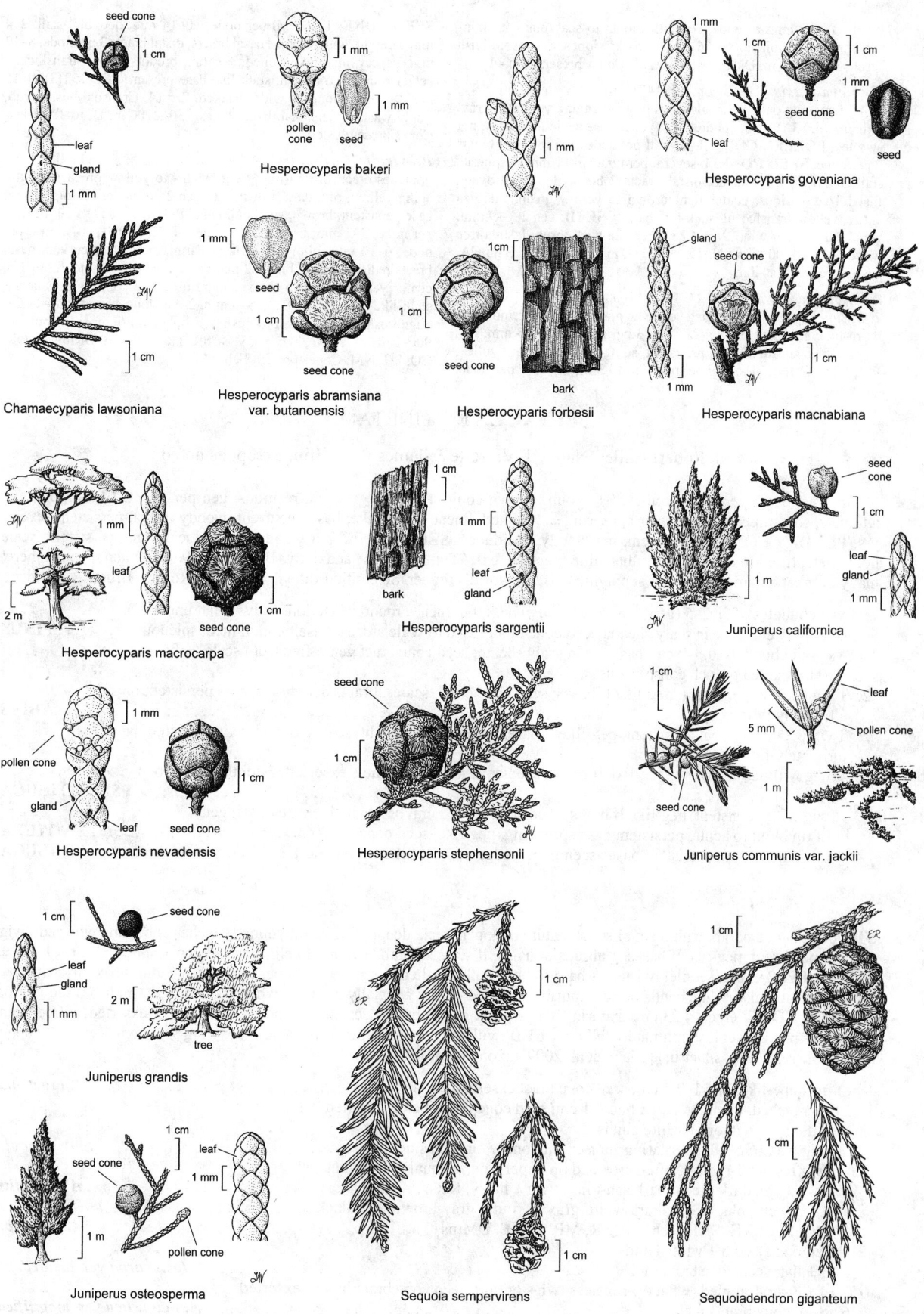

seed cone

1 cm

leaf

gland

1 mm

1 mm

pollen cone

seed

Hesperocyparis bakeri

1 mm

1 mm

1 cm

leaf

1 cm

seed cone

1 mm

seed

Hesperocyparis goveniana

Chamaecyparis lawsoniana

1 cm

1 mm

seed

1 cm

seed cone

Hesperocyparis abramsiana var. butanoensis

1cm

1 cm

seed cone

bark

Hesperocyparis forbesii

gland

seed cone

1 cm

1 mm

Hesperocyparis macnabiana

1 mm

leaf

2 m

1 cm

seed cone

Hesperocyparis macrocarpa

1 cm

1 mm

bark

leaf

gland

Hesperocyparis sargentii

seed cone

1 cm

leaf

gland

1 mm

1 m

Juniperus californica

1 mm

pollen cone

gland

leaf

1 cm

seed cone

Hesperocyparis nevadensis

seed cone

1 cm

Hesperocyparis stephensonii

1 cm

leaf

5 mm

pollen cone

1 m

seed cone

Juniperus communis var. jackii

1 cm

seed cone

leaf

gland

2 m

1 mm

tree

Juniperus grandis

1 cm

seed cone

leaf

1 mm

pollen cone

1 m

Juniperus osteosperma

1 cm

1 cm

Sequoia sempervirens

1 cm

1 cm

Sequoiadendron giganteum

barely 3 × as long as wide, beaked, smooth to scabrous with minute, spreading papillae. 2*n*=56+B. Dry rocky slopes in creosote-bush scrub; 360–1700 m. DMoj; w NV, w AZ. Cones present Mar–Apr

E. nevadensis S. Watson (p. 147) NEVADA EPHEDRA **ST**: 0.3–1.5 m; twig pale green, glaucous in youth, aging yellow to gray; pith brown. **LF**: 2–3 per node, 2–6(8) mm; base deciduous, gray, not swollen. **POLLEN CONE**: 1–several per node, 3–5 (8) mm; bracts 2 per node. **SEED CONE**: 1–several per node, 5–11 mm, ± spheric, stalk 1–11(50) mm, naked; distal bracts 2 per node, broad-ovate, fused 1/5, scarious, center light brown to yellow-green, margins entire, yellow-translucent, stalk-like base 0. **SEED**: gen 2, 6–9 mm, gen 2 × as long as wide. 2*n*=14,28. Creosote-bush scrub, Joshua-tree woodland; < 1100 m. s SN, TR, SNE, D; to OR, UT, AZ, Baja CA. Cones present Mar–Jun

E. trifurca S. Watson (p. 147) LONGLEAF EPHEDRA **ST**: 0.4–2(4.5) m; twig pale green, aging yellow to gray-green; uppermost bud thorn-like; pith dark brown. **LF**: gen 3 per node, (5)7–15 mm; tips ± spine-like; sheath fibrous, persistent, aging gray; base not swollen. **POLLEN CONE**: 1–many per node, 6–10 mm; bracts 3 per node.

SEED CONE: 1–several per node, 10–14 mm, obovoid, stalk 2–4 mm, straight, with reduced basal bracts; distal bracts 3 per node, 8–12 mm, papery, margins inrolled, ± entire, broad-scarious, translucent, center round, red-brown, stalk-like base present. **SEED**: 1(3), 9–15 mm, > 3 × as long as wide, beaked. 2*n*=14. Creosote-bush scrub, sandy washes, flats, stabilized dunes; -30–2100 m. D; to TX, Mex. Cones present Dec–Apr

E. viridis Coville (p. 147) GREEN EPHEDRA **ST**: 0.5–1 m, branches erect, broom-like; twig bright- to yellow-green in youth, aging yellow; pith dark brown. **LF**: gen 2 per node, 1.5–6(13) mm; base persistent, brown, swollen in age. **POLLEN CONE**: 1–several per node, 5–7 mm; bracts 2 per node. **SEED CONE**: 1–several per node, 6–10 mm, obovoid, stalk ≤ 4 mm, straight or curved, basal bracts reduced; distal bracts 2 per node, scarious, elliptic, fused 1/5, center yellow-green to brown, margins entire, yellow-translucent, stalk-like base 0. **SEED**: 2, 5–8 mm, 2 × as long as wide. 2*n*=14,28. Sagebrush scrub, creosote-bush scrub, pinyon/juniper woodland; 900–2300 m. s SNF, Teh, SnJV, SCoR, TR, SnJt, GB, DMtns; to OR, CO, UT, NM. Cones present Feb–Jun

PINACEAE PINE FAMILY

J. Robert Haller, Nancy J. Vivrette & James R. Griffin, except as noted

Shrub, tree, evergreen; monoecious. **ST**: young crown conic; twig not grooved, resinous, gen persistent. **LF**: simple, gen alternate, sometimes in bundles or appearing ± 2-ranked, linear or awl-like; base decurrent, woody or not, persistent several yrs. **POLLEN CONE**: gen < 6 cm, not woody, deciduous. **SEED CONE**: gen woody; bracts, scales gen persistent; scale not peltate, fused to or free from subtending bract. **SEED**: 2, on scale base adaxially. 10 genera, 193 spp.: gen n hemisphere; many of great commercial value, supplying > 1/2 of world's timber. Scientific Editors: Thomas J. Rosatti, Bruce G. Baldwin.

1. Lvs in bundles of (1)2–5 (gen 1 in *Pinus monophylla*, but then ± round in section, vs lvs ± flat under 1′), gen 2.5–35 cm, base in scaly sheath; seed cone bract fused to scale at least basally, incl, inconspicuous . . . . **PINUS**
1′ Lvs not in bundles, 0.5–9 cm, base not in scaly sheath; seed cone bract gen ± free from scale (± fused in *Abies*), exserted or incl, conspicuous
  2. Twig without persistent, peg-like lf bases; seed cone erect, scales, bracts deciduous, axis persistent on st, ultimately falling . . . . **ABIES**
  2′ Twig with or without persistent, peg-like lf bases; seed cone pendent, scales, bracts, axis persistent on st, ultimately falling as unit
    3. Twig without persistent, peg-like lf bases; seed cone 4–20 cm, bracts exserted, 3-toothed or -lobed
. . . . **PSEUDOTSUGA**
    3′ Twig with persistent, peg-like lf bases; seed cone 1.2–12 cm, bracts incl, entire or fringed
      4. Lf tip blunt to acute, persistent base spreading, peg-like; seed cone 5–12 cm . . . . **PICEA**
      4′ Lf tip gen blunt, persistent base ascending, wedge or scale-like; seed cone 1.2–7.5 cm . . . . **TSUGA**

ABIES FIR

ST: young bark smooth, with resin blisters, mature bark gen thick, deeply furrowed; young branches appearing whorled; twig without persistent, peg-like lf bases, glabrous or hairy; lf scars smooth, ± round to elliptic; bud gen ± spheric, gen < 1 cm, ± resinous. **LF**: 2–9 cm, sessile, twisted at base to be 2-ranked, often upcurved on upper twigs, gen ± flat; adaxially with 2 ± faint, longitudinal, ± white bands or not, midrib depressed or not; abaxially with 2 ± white bands or not, midrib ridge-like or not **SEED CONE**: erect, < 23 cm, maturing 1st yr; stalk gen 0; bracts, scales deciduous; bract incl or exserted, ± free from scale; axis persistent on st, ultimately falling. **SEED**: with obvious resin deposits on surface; wing < 2.5 cm. 2*n*=24. 39 spp.: n hemisphere. (Latin: silver fir) [Xiang et al. 2009 Taxon 58:141–152]

1. Lf tip spine-like; bud 1–2.5 cm; seed cone bract exserted 1.5–4.5 cm . . . . ***A. bracteata***
1′ Lf tip notched, blunt, or acute; bud < 1 cm; seed cone bract incl or exserted < 1 cm
  2. Lf adaxially without ± white bands
    3. Twig lvs 2-ranked, alternating shorter and longer on each side . . . . ***A. grandis***
    3′ Twig lvs not 2-ranked, often crowded on upper side, ± equal throughout
      4. Lf < 3 cm, dark green; bark ash-gray; twig ± hairy; KR . . . . ***A. amabilis***
      4′ Lf 3–9 cm, blue-green; bark white-gray in youth, gray-brown to ± black in age; twig glabrous; KR, NCoR, CaRH, SNH, Teh, TR, PR, MP, n SNE, DMtns . . . . ***A. concolor***
  2′ Lf adaxially with ± white bands
    5. Lf ± flat; seed cone bract incl . . . . ***A. lasiocarpa*** var. ***lasiocarpa***
    5′ Lf ± 3- or 4-angled (± flat on coneless twigs or not); seed cone bract incl or exserted
      6. Seed cone bracts incl . . . . ***A. magnifica*** var. ***magnifica***

6′ Seed cone bracts exserted
  7. Seed cone bract awns < bract body . . . . *A. magnifica* var. *shastensis*
  7′ Seed cone bract awns > bract body . . . . *A. procera*

A. amabilis J. Forbes PACIFIC SILVER FIR **ST:** trunk < 75 m, < 2.6 m wide; mature crown steeple-like; bark ± scaly, ash-gray; twig ± hairy; bud resinous. **LF:** not 2-ranked, spreading, < 3 cm, ± equal throughout, dark green; adaxially grooved, ± white bands 0; tip notched or blunt. **SEED CONE:** < 15 cm; bract incl. Subalpine forest; 1700–2140 m. KR (w Siskiyou Co., 2 populations); to BC. ★

A. bracteata (D. Don) A. Poit. (p. 147) BRISTLECONE FIR **ST:** trunk < 55 m, < 1.3 m wide; mature crown steeple-like; bark thin; branches ± drooping, to ground or not; twig glabrous; bud 1–2.5 cm, sharp-pointed, not resinous. **LF:** < 6 cm, dark green, faintly grooved adaxially; tip spine-like. **SEED CONE:** < 9 cm; stalk < 15 mm; bract exserted 1.5–4.5 cm, spreading, tip with slender spine. Steep, rocky, fire-resistant slopes, gen in canyon-live-oak phase of mixed-evergreen forest; 210–1600 m. n SCoRO (Santa Lucia Range). Bark not fire-resistant. ★

A. concolor (Gordon & Glend.) Hildebr. (p. 147) WHITE FIR **ST:** trunk < 61 m, < 2.7 m wide; mature crown rounded; bark white-gray in youth, gray-brown to ± black, thick, deep-furrowed, with alternate dark, light layers in age; twig glabrous; bud resinous. **LF:** ± 2-ranked on lower branches, twisted upward on higher, 3–9 cm, ± flat; adaxial ± white bands 0; tip gen blunt or acute. **SEED CONE:** 7–13 cm; stalk < 5 mm; bract incl. Mixed-conifer to lower red-fir forest; 900–3100 m. KR, NCoR, CaRH, SNH, Teh, TR, PR, MP, n SNE, DMtns; to c OR, ID, CO, n Baja CA. [*A. lowiana* (Gordon) A. Murray bis] Further study needed: most CA, some Rocky Mtns pls would be called *A. concolor* var. *lowiana* (Gordon) Lemmon were that taxon recognized; relationship of s CA pls (esp from DMtns) to gen Rocky Mtns *A. concolor* var. *c.*

A. grandis (D. Don) Lindl. GRAND FIR **ST:** trunk < 73 m, < 1.6 m wide; mature crown rounded, branched large; bark white-gray in youth, red-brown, thin in age; twig hairy; bud resinous. **LF:** 2-ranked, < 5 cm, on each side, shorter toward twig tip; adaxially ± flat, ± white bands 0; tip notched or blunt. **SEED CONE:** 8–15 cm; stalk < 5 mm;

bract incl. Redwood, Douglas-fir, mixed-evergreen forests; < 700 m. NW; to BC, MT.

A. lasiocarpa (Hook.) Nutt. var. *lasiocarpa* (p. 147) SUBALPINE FIR **ST:** trunk < 30 m, < 1.3 m wide; mature crown narrow, steeple-like; bark gray, < 5 cm thick; twig hairy several yrs; bud resinous. **LF:** < 3 cm, ± flat; adaxially with ± white bands; tip notched or blunt. **SEED CONE:** < 10 cm; bract incl. Subalpine forests, meadows; 1700–2100 m. KR (w Siskiyou Co., 6 populations); to nw Can, in Rocky Mtns to NM. ★

A. magnifica A. Murray bis CALIFORNIA RED FIR **ST:** trunk < 57 m, < 2.5 m wide; mature crown ± cylindric, top rounded; bark gray, with resin blisters in youth, deeply furrowed, with dark ± red ridges in age; twig hairy 1st yr; bud ± resinous. **LF:** < 3.5 cm, ± 3–4-angled; adaxially with ± white bands; tip notched or blunt. **SEED CONE:** 12–23 cm; bract incl or ± exserted.

  var. ***magnifica*** Cotyledons 6–13. **SEED CONE:** bract incl. Mixed-conifer to subalpine forests; 1200–2600 m. KR, NCoRH, CaRH, SNH, SNE; w-c NV.

  var. ***shastensis*** Lemmon (p. 147) SHASTA RED FIR Cotyledons 5–8. **SEED CONE:** bract exserted, covering < 25% cone surface. Upper mixed-conifer to subalpine forests; 1350–2800 m. KR, CaRH, s SNH; s OR (Cascade Range). Pls in s SNH recently described as *A. magnifica* var. *critchfieldii* Lanner (Madroño 57:141–144, 2010).

A. procera Rehder (p. 147) NOBLE FIR **ST:** trunk < 85 m, < 2.8 m wide; mature crown cylindric, top rounded; bark smooth, gray, with resin blisters in youth, ± red-brown in age; twig hairy; bud ± resinous. **LF:** < 4 cm, ± 3–4-angled, ± flat on coneless twigs or not; adaxially grooved, with ± white bands; tip notched or blunt. **SEED CONE:** < 15 cm; bracts ± tapered, exserted, strong-reflexed, covering > 90% cone surface. Upper mixed-conifer to subalpine forests; > 1500 m. n KR, CaRH; to BC. Intergrades with *A. magnifica* var. *shastensis*.

PICEA SPRUCE

ST: crown conic; young branches appearing whorled, drooping or not; bark thin, scaly; twig with persistent, peg-like, spreading, < 2 mm lf bases, glabrous or hairy; bud conic to ovoid, resinous. **LF:** often crowded toward upper side of twigs, gen < 3 cm, sessile, ± 4-angled or flat, stiff, often strong-smelling when crushed. **SEED CONE:** conspicuously terminal or not, pendent, 5–12 cm, maturing 1st yr; stalk 0–1 cm; bract incl, ± free from scale. **SEED:** wing terminal. $2n=24$. 34 spp.: n hemisphere. (Latin: pitch) Trunks, esp of *P. sitchensis*, more flared or buttressed at base than other CA conifers.

1. Lf ± 4-angled, with 2 obvious, ± white bands adaxially, abaxially . . . . *P. engelmannii*
1′ Lf ± flat, gen with 2 faint, ± white bands adaxially, also abaxially or not
  2. Lf tip blunt; seed cone scale margin ± entire, bract < 1/4 scale . . . . *P. breweriana*
  2′ Lf tip acute; seed cone scale margin ± jagged, bract > 1/2 scale . . . . *P. sitchensis*

P. breweriana S. Watson (p. 147) BREWER SPRUCE **ST:** trunk < 53 m, < 1.3 m wide; branches drooping, to ground or not; twig hairy. **LF:** < 3 cm, flexible; adaxially ± flat, gen with 2 faint, ± white bands; tip blunt. **SEED CONE:** < 12 cm, oblong; scale woody, margin ± entire; bract < 1/4 scale. Uncommon. Cool, moist, forested slopes; 560–2300 m. KR; sw OR.

P. engelmannii Engelm. (p. 147) ENGELMANN SPRUCE **ST:** trunk < 55 m, < 2.4 m wide; branches not drooping; young twigs ± hairy. **LF:** < 3 cm, rigid, 4-angled; adaxially, abaxially with 2 obvious, ± white bands; tip flat, acute, not sharp to touch. **SEED CONE:**

< 7 cm, ovoid-oblong; scale papery, margin ± jagged. Cool, moist, mixed-conifer, subalpine forests; 1200–2100 m. KR, CaRH; to BC, in Rocky Mtns to NM. ± 3 populations known in CA. ★

P. sitchensis (Bong.) Carrière (p. 147) SITKA SPRUCE **ST:** trunk < 66 m, < 5.1 m wide; branches ± drooping; twig glabrous. **LF:** 1–3 cm, rigid; adaxially ± flat, with 2 faint, ± white bands, abaxially rounded, darker green, with 2 faint, ± white bands or not; tip acute, sharp to touch. **SEED CONE:** < 10 cm, oblong; scale ± papery, margin ± jagged; bract > 1/2 scale. Moist soils, esp near coastal river mouths; < 450 m. NCo; to AK.

PINUS PINE

J. Robert Haller & Nancy J. Vivrette

ST: young crown conic, mature often rounded or flat; branches ± whorled in young pls; young bark smooth, mature furrowed; bud ± conic, gen resinous. **LF:** gen 2.5–35 cm, gen sessile, in bundles of (1)2–5; bundles 1 in axils of alternate, awl-like bracts, base in a sometimes deciduous, scaly sheath of bracts, gen persistent several yrs. **SEED CONE:** often whorled, gen maturing, opening 2nd yr, persistent on st or not; stalk 0 or < 16 cm; bract incl, fused to scale at least basally, minute; scale tip reflexed,

elongated 3–7 cm or often with a rounded or angled, often prickled knob < 3 cm. **SEED**: coat hard, woody or not. 2*n*=24. 94 spp.: n hemisphere. (Latin: pine) *P. pinea* L., stone pine (lvs 2 per bundle, 10–30 cm; seed cone 8–15 cm, maturing in 3 yrs) cult in Eur for over 6000 yrs for edible seeds (pine nuts), reportedly naturalized in SnFrB, n ChI.

1. Lf-bundle sheath deciduous or mostly so; seed cone scale tip gen without prickle
 2. Lvs gen < 5 per bundle
  3. Lvs (3)4(5) per bundle . . . . *P. quadrifolia*
  3′ Lvs gen 1–2 per bundle
   4. Lvs gen 2 per bundle . . . . *P. edulis*
   4′ Lvs gen 1 per bundle . . . . *P. monophylla*
 2′ Lvs 5 per bundle
  5. Seed cone gen closed, gen torn apart by animals, gen not on ground intact; seed wings persistent on scale . . . . *P. albicaulis*
  5′ Seed cone opening and releasing seeds, not torn apart by animals, gen on ground intact (exc for released seeds); seed wings persistent on scale or not
   6. Seed cone scale thinnest at tip, without prickle
    7. Seed cone stalk < 2 cm; seed wings narrow, gen persistent on scale . . . . *P. flexilis*
    7′ Seed cone stalk > 2 cm; seed wings wide, deciduous from scale
     8. Seed cone 20–60 cm; mature bark dark purple-brown. . . . . *P. lambertiana*
     8′ Seed cone 9–25 cm; mature bark dark gray to red-brown . . . . *P. monticola*
   6′ Seed cone scale thickest at tip, with prickle
    9. Scale knob tip prickle 1–6 mm; trunks often many . . . . *P. longaeva*
    9′ Scale knob tip prickle ≤ 1 mm; trunks gen 1 . . . . *P. balfouriana*
     10. Lf ± yellow-green; bark red-brown, in ± square plates; s SNH. . . . . subsp. *austrina*
     10′ Lf ± blue-green; bark gray-brown, in ± narrow ridges; KR, NCoRH. . . . . subsp. *balfouriana*
1′ Lf-bundle sheath persistent; seed cone scale tip with prickle, at least when immature (sometimes 0 at maturity)
 11. Lvs 2 per bundle
  12. Seed cone 5–9.7 cm; lf gen 5–15 cm . . . . *P. muricata*
  12′ Seed cone 2–6 cm; lf 2.5–8.6 cm . . . . *P. contorta*
   13. Seed cone gen not opening, asymmetric. . . . . subsp. *bolanderi*
   13′ Seed cone opening, asymmetric or ± symmetric
    14. Seed cone on st many yrs after opening . . . . subsp. *contorta*
    14′ Seed cone deciduous soon after opening. . . . . subsp. *murrayana*
 11′ Lvs 3 or 5 per bundle (2–5 in *Pinus ponderosa*)
  15. Lvs 5 per bundle, often 3 in young trees . . . . *P. torreyana*
   16. Crown ± closed, gen wider than long, most open seed cones with combined length and width > 27 cm, width > length; light colored scale tips gen prominent; n ChI (Santa Rosa Island) . . . . subsp. *insularis*
   16′ Crown ± open, gen longer than wide, most open seed cones with combined length and width < 27 cm, width ≤ length; light colored scale tips gen not prominent; s SCo (mostly Torrey Pines State Reserve) . . . . subsp. *torreyana*
  15′ Lvs 3 per bundle (rarely 2 in *Pinus ponderosa*)
   17. Proximal seed cone scale tips recurved, elongated 1.5–7 cm
    18. Seed cone ± yellow; seed < wing; lf rigid, not drooping . . . . *P. coulteri*
    18′ Seed cone ± brown; seed > wing; lf flexible, gen drooping . . . . *P. sabiniana*
   17′ Proximal seed cone scale tips rarely recurved, then elongated < 3 cm
    19. Seed cones on sts many yrs
     20. Seed cone scale tips, exc distal, with knob angled, > 2 cm . . . . *P. attenuata*
     20′ Seed cone scale tips, exc distal, with knob rounded, < 2 cm . . . . *P. radiata*
    19′ Seed cones deciduous exc stalks, proximal scales persistent
     21. Bark crevices with definite odor of vanilla/banana, esp on warm days; scales of open seed cone ± crowded, gen not darker abaxially than adaxially; lf gray-blue-green, glaucous; bark outer scales with ± pink inner surfaces; vegetative buds not resinous; seed cone gen 13–25 cm, scales each with a gen incurved prickle, bracts with white fringing hairs . . . . *P. jeffreyi*
     21′ Bark crevices gen lacking odor of vanilla/banana; scales of open seed cone well separated to very crowded, gen darker abaxially than adaxially; lf gray-green, shiny deep green, or light green; bark outer scales with ± yellow inner surfaces; vegetative buds resinous; seed cone 7–15(18) cm, scales each with a straight or outcurved prickle, bracts with light brown fringing hairs . . . . *P. ponderosa*
      22. Mature bark light to medium yellow-brown; maturing closed cones green; open mature cones gen 8–18 cm, scales abaxially black. . . . . var. *pacifica*
      22′ Mature bark gen medium to dark red- or yellow-brown; maturing closed cones ± green to dark purple; open mature cones 6–12(15) cm, scales abaxially brown with black striations or black
       23. Immature seed cones green-brown to dark purple; mature open cones ovate to ± conic, 7–12(15) cm, scales abaxially gen black, occ brown with black striations; CaRH, n SNH, MP. . . . . . . var. *ponderosa*
       23′ Immature seed cones dark red-purple; mature open cones ovate or gen conic, gen 6–11 cm, scales abaxially gen brown with black striations, occ black; CaRH, n SNH, Wrn . . . . var. *washoensis*

P. albicaulis Engelm. (p. 147) WHITEBARK PINE **ST:** gen prostrate to shrubby when exposed; trunks 1–many, < 26 m, < 1.5 m wide, much wider at base; mature bark gray-white, smooth, thin; mature crown often deformed by wind. **LF:** 5 per bundle, 3–7 cm, ± curved, dark green, stiff; sheath deciduous. **SEED CONE:** sessile, erect, 3.5–9 cm, ovate, purple-brown, gen torn apart, seeds dispersed by animals; scale tip knobs angled, prickled. **SEED:** wing persistent on scale. Upper red-fir forest to timberline, esp subalpine forest; 2000–3700 m. KR, CaRH, SNH, Wrn, SNE; to BC, WY.

P. attenuata Lemmon (p. 147) KNOBCONE PINE **ST:** trunk < 36 m, < 1.1 m wide; bark gray-brown; mature crown with several tops or not, branches many. **LF:** 3 per bundle, 6–16 cm, yellow-green; sheath persistent. **SEED CONE:** recurved to reflexed, 6–18 cm, asymmetric, yellow-brown, often closed unless burned, on st many yrs; stalk < 2 cm, ± 0 with st growth; proximal scale tip knobs > 2 cm, angled, prickled. **SEED:** < wing. Closed-cone-pine forest, chaparral; < 2000 m. NW, CaR, SN, e SnFrB, SCoR, SnBr, PR, MP; sw OR, Baja CA.

P. balfouriana Grev. & Balf. FOXTAIL PINE **ST:** trunk gen 1, < 22 m, < 2.6 m wide; mature crown branches short, thick. **LF:** 5 per bundle, 1.5–4 cm, curved, stiff, ± blue- or yellow-green, adaxially ± white, tip acute, sharp to touch; sheath deciduous. **SEED CONE:** pendent, 6–19 cm, ovoid; stalk 7–16 mm; scale tip knob angled, prickle ≤ 1 mm. **SEED:** < wing.

subsp. ***austrina*** R.J. Mastrog. & J.D. Mastrog. **ST:** mature bark thick, red-brown, in ± squared plates. **LF:** ± yellow-green, persisting < 30 yrs. Subalpine forest; 2700–3700 m. s SNH. In some ways more like *P. longaeva* than *P. balfouriana* subsp. *b*.

subsp. ***balfouriana*** (p. 147) **ST:** mature bark thin, gray-brown, in ± narrow ridges. **LF:** ± blue-green, persisting < 15 yrs. Subalpine forest; 2100–2500 m. KR, NCoRH.

P. contorta Loudon **ST:** mature bark scaly, thin; trunk < 34 m (extremely variable at maturity). **LF:** 2 per bundle, 2.5–8.6 cm; sheath persistent. **SEED CONE:** pendent, 2–6 cm, brown; stalk ± 0; scale tip knobs angled, prickle < 6 mm.

subsp. ***bolanderi*** (Parl.) Critchf. BOLANDER'S BEACH PINE **ST:** trunk gen < 2 m. **SEED CONE:** asymmetric, closed, on st many yrs. Pygmy forest on coastal terrace soils with clay- or hardpan; < 250 m. NCo (Mendocino Co.). Cone woodiest in sp. Threatened by development, off-road vehicles. ★

subsp. ***contorta*** SHORE PINE **ST:** trunk gen < 15 m. **SEED CONE:** asymmetric, on st many yrs after opening. Coastal dunes, bluffs lacking clay-, hardpan; < 150 m. NCo; to AK.

subsp. ***murrayana*** (Grev. & Balf.) Critchf. (p. 147) LODGEPOLE PINE **ST:** trunk < 34 m. **SEED CONE:** ± symmetric, deciduous soon after opening. Lodgepole forest, wet meadows, cold places in mixed-conifer forest; 1000–3500 m. KR, CaRH, SNH, SnGb, SnBr, SnJt, GB; OR, n Baja CA.

P. coulteri D. Don (p. 147) COULTER PINE **ST:** trunk < 42 m, < 1.5 m wide; young bark dark brown to ± black, with irregular furrows, mature with ± yellow plates; mature crown branches thick, lower reaching ground or not. **LF:** 3 per bundle, 15–30 cm, stiff; sheath persistent. **SEED CONE:** pendent, 19–35 cm, ± ovoid-oblong, ± yellow, opening slowly in 2nd or 3rd yr, deciduous over many yrs; stalk < 10 cm, persistent with proximal scales; scale tips reflexed, elongated 3–5 cm, angled. **SEED:** < wing. Chaparral, lower mixed-conifer, mixed-hardwood forests; < 3000 m. CW, SW; n Baja CA.

P. edulis Engelm. (p. 147) COLORADO PINYON **ST:** trunk < 15 m, < 1.1 m wide; bark shallow-furrowed, red-brown; mature crown ± rounded. **LF:** gen 2 per bundle, 2–6 cm, dark green; sheath deciduous. **SEED CONE:** spreading, 3–5 cm, ovoid, yellow-brown; stalk ± 0; scale tip knobs < 1 cm, angled, truncate. **SEED:** wing persistent on scale. Pinyon/juniper woodland; 1300–2700 m. DMtns (New York Mtns); to s WY, w TX, n Mex. Extent of hybridization with *P. monophylla* under study. ★

P. flexilis E. James (p. 147) LIMBER PINE **ST:** trunk < 20 m, < 2.9 m wide; mature bark dark brown, deep-furrowed, in rectangular plates; mature crown branches reaching ground or not. **LF:** 5 per

bundle, 2–9 cm, stiff, gen curved; bundles in dense tufts at branch ends; sheath deciduous. **SEED CONE:** spreading, 7–15 cm, oblong, yellow-brown, opening late 2nd yr; stalk < 2 cm; scales thinnest at tips, angled, prickle 0. **SEED:** wing gen persistent on scale. Lodgepole, subalpine, bristlecone forests; 1830–3700 m. SNH, TR, PR, SNE, n DMtns; to w Can, SD, NM.

P. jeffreyi Grev. & Balf. (p. 149) JEFFREY PINE **ST:** trunk < 53 m, < 2.3 m wide; mature bark gen red-brown, furrows close-spaced, deep, outer scales with ± pink inner surfaces, crevices with definite odor of vanilla/banana, esp on warm days; mature crown rounded; buds not resinous, scales light brown, white-hairy. **LF:** 3 per bundle, 12–27 cm, thick, gray-blue-green, glaucous; sheath persistent. **SEED CONE:** spreading or recurved, gen 13–25 cm, ovate to ± oblong, brown, when immature light green to red-purple; stalk < 3 cm, persistent with proximal scales; scales gen not darker abaxially than adaxially, in open cone ± crowded, tips recurved, elongated < 3 cm; knob prickles incurved; bracts with white fringing hairs. Upper mixed-conifer, red-fir forests, elsewhere on serpentine; 450–3100 m. KR, NCoR, CaR, SN, SCoRI, TR, PR, GB; sw OR, w NV, n Baja CA.

P. lambertiana Douglas (p. 149) SUGAR PINE **ST:** trunk < 70 m, < 3.3 m wide; mature bark thick, dark purple-brown, irregularly furrowed into plate-like ridges; mature crown ± flat, with large, ± horizontal branches. **LF:** 5 per bundle, 5–11 cm, gen stiff, twisted or not; sheath deciduous. **SEED CONE:** pendent, 20–60 cm, cylindric, yellow-brown; stalk 5–16 cm; scales thinnest at tips, angled, prickle 0. **SEED:** < wing. Mixed-conifer, mixed-evergreen forests; < 3200 m. NW, CaR, SN, SCoRO (Santa Lucia Range), SW, w GB; OR, n Baja CA.

P. longaeva D.K. Bailey (p. 149) BRISTLECONE PINE **ST:** trunks often many, twisted, strong-tapered, < 16 m, < 2 m wide; bark red-brown; mature crown bushy, irregular. **LF:** 5 per bundle, 1–4 cm, often with white resin spots; sheath deciduous. **POLLEN CONE:** deep red when mature. **SEED CONE:** ± spreading to pendent, < 5–14 cm, ovoid-oblong, dark red-brown; stalk 0–2 cm; scale tip knob prickle 1–6 mm, slender, stiff. **SEED:** < wing. Subalpine forest; 2200–3700 m. W&I, n DMtns; to e UT. ★

P. monophylla Torr. & Frém. (p. 149) SINGLELEAF PINYON PINE **ST:** trunk < 15 m, < 40 cm wide; mature crown much-branched, rounded. **LF:** gen 1 per bundle, 2–7 cm, often curved, gray or blue-green; sheath deciduous. **SEED CONE:** spreading, 3–12 cm, spheric-ovoid, light- or red-brown; stalk < 1 cm; scale tip knobs < 1 cm, angled, truncate. **SEED:** wing persistent on scale. Pinyon/juniper woodland; < 2800 m. c&s SNH, Teh, se SCoRI, TR, PR, SNE, DMtns; to se ID, n Baja CA.

P. monticola D. Don (p. 149) WESTERN WHITE PINE **ST:** trunk < 73 m, < 2 m wide; mature bark dark gray to red-brown, in ± square blocks, ± thin; mature crown narrowly conic. **LF:** 5 per bundle, 3–10 cm, gen persistent < 4 yrs, gen straight, flexible, blue-green, glaucous; sheath deciduous. **SEED CONE:** pendent, 9–25 cm, cylindric, yellow-brown; stalk 2–5 cm; scales thinnest at tips, angled, prickle 0. **SEED:** < wing. Upper mixed-conifer to subalpine forests; 150–3400 m. KR, NCoRH, CaRH, SNH, GB; to BC, MT.

P. muricata D. Don (p. 149) BISHOP PINE **ST:** trunk < 51 m, < 1.2 m wide; bark brown, ridges rough; mature crown variable, branches often large. **LF:** 2 per bundle, gen 5–15 cm, twisted or not, gen green; sheath persistent. **SEED CONE:** gen whorled, 5–9.7 cm, ovoid, brown, weathering gray, gen closed, persistent many yrs, either ± spreading, symmetric, with scale tip knobs < 3 mm, prickled, or reflexed, asymmetric, with proximal, middle scale tip knobs < 15 mm, angled, prickled; stalk 0–2 cm. Redwood forest, n coastal conifer forest, closed-cone-pine forest, chaparral; < 300 m. NCo, CCo, n SnFrB, n ChI; n Baja CA, Cedros Island.

P. ponderosa Lawson & C. Lawson (p. 149) PONDEROSA PINE, WESTERN YELLOW PINE **ST:** trunk gen < 68 m, gen < 2.2 m wide; branched in lower 1/2 when mature or not; mature bark furrows shallow, well spaced, forming plates, outer scales with ± yellow inner surfaces; mature crown short, conic or flat-topped; buds resinous, scales red-brown, dark-hairy. **LF:** (2)3[5] per bundle, 12–26 cm, < 2 mm thick, ± or not glaucous, deep yellow-green; sheath persis-

tent. **SEED CONE:** ± spreading or recurved, 7–15(18) cm, ovate to ± conic, when immature green-brown to dark purple; stalk < 2 cm, persistent with proximal scales; scales gen darker abaxially than adaxially, in open cone well separated to very crowded; knob prickles < 3 mm, straight or outcurved; bracts with light brown fringing hairs. **SEED:** < wing.

var. *pacifica* J.R. Haller & Vivrette   PACIFIC PONDEROSA PINE **ST:** mature bark light to medium yellow-brown. **LF:** 15–28 cm, thin, not glaucous, deep green, shiny, not glaucous; sheath persistent?. **SEED CONE:** gen 8–18 cm, gen ovate, when immature light yellow-green, maturing closed cones green; scales black abaxially, brown adaxially, in open cone well separated; distal knob prickles gen outcurved. **SEED:** ± 1/3–1/5 wing. Coastal-draining slopes of major mtn ranges, streambanks; 100–2700 m. NCoR, CaR, SN, SnFrB (Santa Cruz Mtns), TR, PR; sw OR, WA. Some very large-coned pls in Santa Cruz Mtns, unassigned taxonomically in this treatment, may be indistinct from var. *pacifica* (then the earlier var. *benthamiana* (Hartw.) Vasey would be correct for the unassigned pls, with var. *pacifica* as a synonym), or may be a distinct var. within *P. ponderosa* (then var. *benthamiana* would be correct for the unassigned pls), or may be a distinct species (then *P. benthamiana* Hartw. would be correct for the unassigned pls); study needed.

var. *ponderosa*   NORTH PLATEAU PONDEROSA PINE **ST:** mature bark gen medium to dark red- or yellow-brown. **LF:** 14–26 cm, thick, gray-green, ± glaucous. **SEED CONE:** 7–12(15) cm, ovate to ± conic, when immature green-brown to dark purple, maturing closed cones ± green to dark purple; scales abaxially gen black, occ brown with black striations, in open cone ± separated. Gen semi-arid plateaus, low mtns; 1200–1900 m. CaRH, n SNH, MP; to BC, MT, NE; n Mex.

var. *washoensis* (H. Mason & Stockw.) J.R. Haller & Vivrette WASHOE PINE **ST:** trunk < 35 m, < 1.5 m wide; mature bark gen medium to dark red- or yellow-brown, gen shallow-furrowed; mature crown short, conic or flat-topped. **LF:** 3 per bundle, 12–17 cm, 1.8–2.4 mm wide, very thick, light green, ± glaucous; sheath persistent. **SEED CONE:** spreading, gen 5–11 cm, ovate or gen conic, when immature dark red-purple, maturing closed cones ± green to dark purple; stalk < 2 cm; scales adaxially gen brown, abaxially gen brown with black striations, occ black, in open cone very crowded; knob prickles varied, gen straight, parallel to cone edge. Upper mixed-conifer to lower subalpine; (1400)2000–3000 m (gen hybridizes with *Pinus ponderosa* at 1700–2000 m; growing near but not hybridizing with *Pinus jeffreyi* at 1800–2100 m). CaRH, n SNH, Wrn; to BC, w NV. [*P. w.* H. Mason & Stockw.]

P. quadrifolia Sudw. (p. 149)   PARRY PINYON PINE **ST:** trunk < 16 m, < 70 cm wide; mature bark red-brown. **LF:** (3)4(5) per bundle, 2–5 cm, curved, blue-green, adaxially ± white; sheath deciduous. **SEED CONE:** spreading, 4–8 cm, ± spheric, light brown; stalk < 1 cm; scale tip knobs < 1 mm, angled, truncate. **SEED:** wing persistent on scale. Pine forest, pinyon/juniper woodland, chaparral; 1100–1800 m. s PR, SnJt; n Baja CA.

P. radiata D. Don (p. 149)   MONTEREY PINE **ST:** trunk < 38 m, < 2.1 m wide, in youth < 2 m growth per yr; mature bark black, deep-grooved; mature crown irregular, round-topped. **LF:** (2)3 per bundle, 6–15 cm, dark green; sheath persistent. **SEED CONE:** recurved, 6–15 cm, asymmetric, light brown, opening slowly 2nd yr, persistent < 25 yrs; stalk < 15 mm; proximal scale tip knobs < 2 cm, rounded, minute-prickled. **SEED:** < wing. Closed-cone-pine forest, oak woodland; < 1300 m. CCo (near Point Año Nuevo; vicinity of Monterey Peninsula; Cambria-San Simeon area; all < 300 m) (naturalized NCo, CCo, SCo, PR (Santa Ana Mtns)); islands off Baja CA; naturalized? where escaped or introduced s OR, Eur, Afr, Australia, New Zealand. Lvs gen 2 per bundle on Cedros, Guadalupe islands.   ★

P. sabiniana D. Don (p. 149)   GRAY, GHOST, OR FOOTHILL PINE **ST:** trunk < 38 m, < 2 m wide, often leaning, with several major branches after 20–30 yrs; bark dark gray, furrows irregular, forming yellow plates in age. **LF:** 3 per bundle, 9–38 cm, gray-green, gen drooping; sheath persistent. **SEED CONE:** pendent, 10–28 cm, ovate-oblong, ± brown, opening slowly 2nd yr; stalk < 7 cm, persistent with proximal cone scales several yrs; scale tip recurved, elongated 3–7 cm, angled. **SEED:** > wing. Foothill woodland, n oak woodland, chaparral, infertile soils in mixed-conifer and hardwood forests; 150–1500 m. CA-FP (exc n NW, n CaR, SnJV), w GB, w D; s OR.

P. torreyana Carrière (p. 149)   **ST:** trunk < 23 (< 33 in cult) m, < 1 m wide; mature crown open, rounded, branches many large; mature bark with red-brown plates between irregular furrows. **LF:** 5 per bundle, often 3 in young trees, 15–26 cm, stiff, gray-yellow- to gray-blue-green; sheath persistent. **SEED CONE:** spreading to recurved, < 16 cm, ± symmetric, ± longer than wide to much wider, scales dark brown, opening slowly 3rd yr, persistent < 15 yrs; stalk < 4 mm; scale tip knobs pyramidal, light-colored, < 2 cm, minute-prickly. **SEED:** 16–24 mm, > wing, light brown to ± black, mottled or not, in cone < 13 yrs.

subsp. *insularis* J.R. Haller   SANTA ROSA ISLAND TORREY PINE **ST:** trunk < 15 m; mature crowns of sheltered pls gen wider than pl height, overall gen compact, with crowded branches. **LF:** gen gray-blue-green. **SEED CONE:** width gen > 13.5 cm, > length; scale tip knobs gen > 6 mm, > 1/3 scale thickness, entire knob curved outward. **SEED:** gen > 11 mm wide, medium brown to ± black. Forming densely forested strip on ± n-facing slope parallel to coast, with scattered chaparral, island woodland components; < 150 m. n ChI (Santa Rosa Island). < 1000 trees. Threatened by browsing of young pls, small population size. Not well established when planted at a few localities in coastal Santa Barbara Co.   ★

subsp. *torreyana*   TORREY PINE **ST:** trunk < 23 m; mature crown ± open, gen taller than wide on sheltered pls, branches well-spaced. **LF:** gen gray-yellow-green. **SEED CONE:** width gen < 13.5 cm, ≤ length; scale tip knobs < 6 mm. **SEED:** gen < 11 mm wide, light to dark brown. Ocean bluffs, rapidly eroding Eocene sandstone; coastal scrub, chaparral; < 200 m. s SCo. ± 7000 native trees, most in Torrey Pines State Reserve, others on private property.   ★

PSEUDOTSUGA

ST: young crown conic; branches ± whorled in young pls, smaller ± drooping; young bark smooth, with resin blisters, mature thick, deep-furrowed, dark brown; twig without persistent, peg-like lf bases; lf scars ± smooth, ± elliptic; bud < 15 mm, fusiform. **LF:** persistent < 8 yrs, 2–4.5 cm, tapered to a short petiole above persistent base, ± spreading, ± flat, with 2 ± white bands abaxially. **SEED CONE:** pendent, 4–20 cm, maturing 1st yr, persistent or not; stalk < 2 cm; bract exserted, 3-toothed or -lobed, free from scale. **SEED:** plus wing < 25 mm. 4 spp.: N.Am, Asia. (Latin, Japanese: false hemlock)

1. Seed cone 9–20 cm; lf tip gen acute .... *P. macrocarpa*
1′ Seed cone 5–9 cm; lf tip gen blunt .... *P. menziesii* var. *menziesii*

P. macrocarpa (Vasey) Mayr   BIGCONE DOUGLAS-FIR, BIGCONE SPRUCE **ST:** trunk < 44 m, < 2.1 m wide, strong-tapered; mature crown rounded to flat; lower branches often many, large. **LF:** 2–4.5 cm. **SEED CONE:** persistent several yrs or not; bracts little exserted exc for central tooth or lobe. Scattered on fire-resistant slopes, mixed-evergreen forest; 200–2400 m. s SCoRO (Sierra Madre Mtns), TR, PR.

P. menziesii (Mirb.) Franco var. *menziesii* (p. 149)   DOUGLAS-FIR **ST:** trunk < 67 m, < 4.4 m wide; mature crown rounded; upper branches large. **LF:** 2–4 cm. **SEED CONE:** deciduous; bracts well exserted. $2n=26$. Widespread in mixed-evergreen, mixed-conifer forests; < 2200 m. KR, NCoR, CaRH, n&c SNH, CCo, SnFrB, SCoRO; to sw BC.

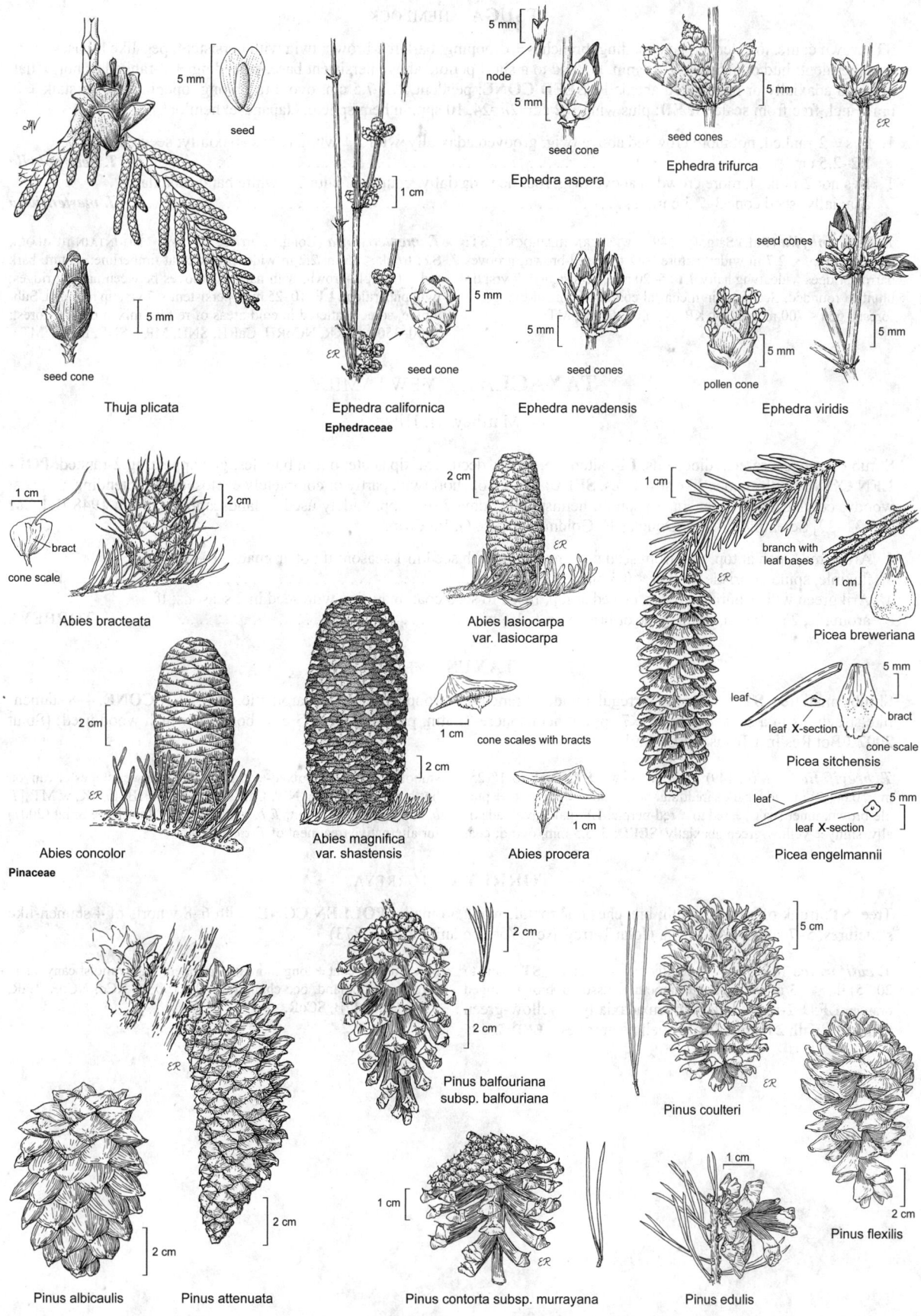

5 mm

seed

seed cone

Thuja plicata

5 mm

node

5 mm

seed cone

Ephedra aspera

seed cones

Ephedra trifurca

1 cm

5 mm

seed cone

Ephedra californica
Ephedraceae

5 mm

seed cones

Ephedra nevadensis

seed cones

pollen cone

5 mm

5 mm

Ephedra viridis

1 cm

bract

cone scale

Abies bracteata

2 cm

2 cm

Abies concolor
Pinaceae

2 cm

2 cm

Abies magnifica
var. shastensis

2 cm

cone scales with bracts

1 cm

1 cm

Abies procera

2 cm

Abies lasiocarpa
var. lasiocarpa

1 cm

branch with
leaf bases

1 cm

Picea breweriana

leaf

5 mm

bract

cone scale

leaf **X**-section

Picea sitchensis

leaf

leaf **X**-section

5 mm

Picea engelmannii

2 cm

2 cm

2 cm

Pinus balfouriana
subsp. balfouriana

2 cm

Pinus albicaulis

Pinus attenuata

1 cm

Pinus contorta subsp. murrayana

5 cm

Pinus coulteri

1 cm

Pinus edulis

2 cm

Pinus flexilis

TSUGA HEMLOCK

ST: crown conic, top slender, gen nodding; branches ± drooping; bark red-brown; twig with persistent, peg-like lf bases, hairy or puberulent; bud ovoid. **LF**: 5–25 mm, tapered to a short petiole above persistent base, spreading, ± 2-ranked or not, ± flat, grooved adaxially or not, ridged abaxially. **SEED CONE**: pendent, 1.2–7.5 cm, ovoid to oblong, opening 1st yr; stalk ± 0; bract incl, free from scale. **SEED**: plus wing < 2 cm. 2*n*=24. 10 spp.: n hemisphere. (Japanese: hemlock)

1. Lvs ± 2-ranked, not more crowded above twig, grooved adaxially, with 2 ± white bands abaxially; seed cone
   1.2–2.5 cm . . . . ***T. heterophylla***
1′ Lvs not 2-ranked, more crowded above twig, rounded adaxially, glaucous, with 2 ± white bands abaxially,
   adaxially; seed cone 3–7.5 cm . . . . ***T. mertensiana***

T. heterophylla (Raf.) Sarg. (p. 149) WESTERN HEMLOCK **ST**: trunk < 50 m, < 2.7 m wide; mature bark thin, red-brown, grooves narrow, ridges wide; twig hairy. **LF**: 5–20 mm, persistent 4–7 yrs; tip blunt to rounded. Scattered in n coastal conifer to mixed-evergreen forests; gen < 700 m. NCo, w KR, NCoRO; to AK, MT.

T. mertensiana (Bong.) Carrière (p. 149) MOUNTAIN HEMLOCK **ST**: trunk < 35 m, 2.2 m wide, prostrate at timberline; mature bark red- or purple-brown, with narrow grooves between narrow ridges; twig puberulent. **LF**: 10–25 mm, persistent < 7 yrs, tip rounded. Subalpine, some scattered in cold areas of red-fir, mixed-conifer forest; 1200–3500 m. KR, NCoRH, CaRH, SNH, MP, n SNE; to AK, MT.

TAXACEAE YEW FAMILY

Matthew H. Hils

Shrub or tree, evergreen; dioecious. **LF**: alternate, linear, decurrent, tip acute; not in bundles, gen appearing 2-ranked. **POLLEN CONE**: with stamen-like structures. **SEED**: 1 at tip of short twig, partly or completely enclosed by subtending aril; coat woody; cotyledons 2. 5 genera, 16 spp.: n hemisphere; some *Taxus* spp. widely used in landscaping. [Florin 1948 Bot Gaz 110:31–39] Scientific Editors: Douglas H. Goldman, Bruce G. Baldwin.

1. Aril ± red, open at top, free from seed coat, maturing with seed in 1 season; lf not aromatic, 10–29 mm, ±
   flexible, spine or bristle tip 0 or < 0.5 mm . . . . **TAXUS**
1′ Aril green with ± purple streaks, closed at top, fused to seed coat, maturing with seed in 2 seasons; lf
   aromatic, 25–80 mm, rigid, spine or bristle tip 1–1.5 mm . . . . **TORREYA**

TAXUS YEW

Shrub, small tree. **ST**: trunk gen of irregular width; branches ± drooping; wood not aromatic. **POLLEN CONE**: 4–8 stamen-like structures in a stalked cluster. ± 7 spp.: n hemisphere. (Latin, probably from Greek: bow, for which wood used) [Spjut 2007 J Bot Res Inst Texas 1:203–289]

T. brevifolia Nutt. (p. 149) PACIFIC YEW **ST**: trunk to 18(25) m, < 0.6(1.2) m wide; bark shredding, outer scales ± purple to ± purple-brown, inner scales ± red to ± red-purple. **LF**: pale green adaxially, shiny ± yellow-green abaxially. **SEED**: 5–6.5 mm, ovoid; coat smooth. Gen dense, mixed-evergreen forest, lower slopes or canyon bottoms; 10–2150 m. NW, CaR, n&c SN, SnFrB; to AK, w MT. [*T. b.* var. *polychaeta* Spjut; *T. b.* var. *reptaneta* Spjut] See Spjut (2007) for alternative treatment of *Taxus* in CA.

TORREYA TORREYA

Tree. **ST**: trunk of regular width; branches horizontal; wood aromatic. **POLLEN CONE**: with 6–8 whorls of 4 stamen-like structures. 5–7 spp.: N.Am, Asia. (John Torrey, New York botanist, 1796–1873)

T. californica Torr. (p. 149) CALIFORNIA-NUTMEG **ST**: trunk to 20(25) m, < 1.5 m wide; bark ± smooth or fissured, brown, tinged orange. **LF**: ± 2-ranked, dark green adaxially, ± yellow-green abaxially, with 2 longitudinal, ± yellow grooves. **SEED**: 25–45 mm, oblong; coat ± longitudinally grooved. Shady moist canyons in forest or woodland, occ chaparral; 10–2100 m. NCo, NCoR, CaR, SN, CCo, SnFrB, SCoR.

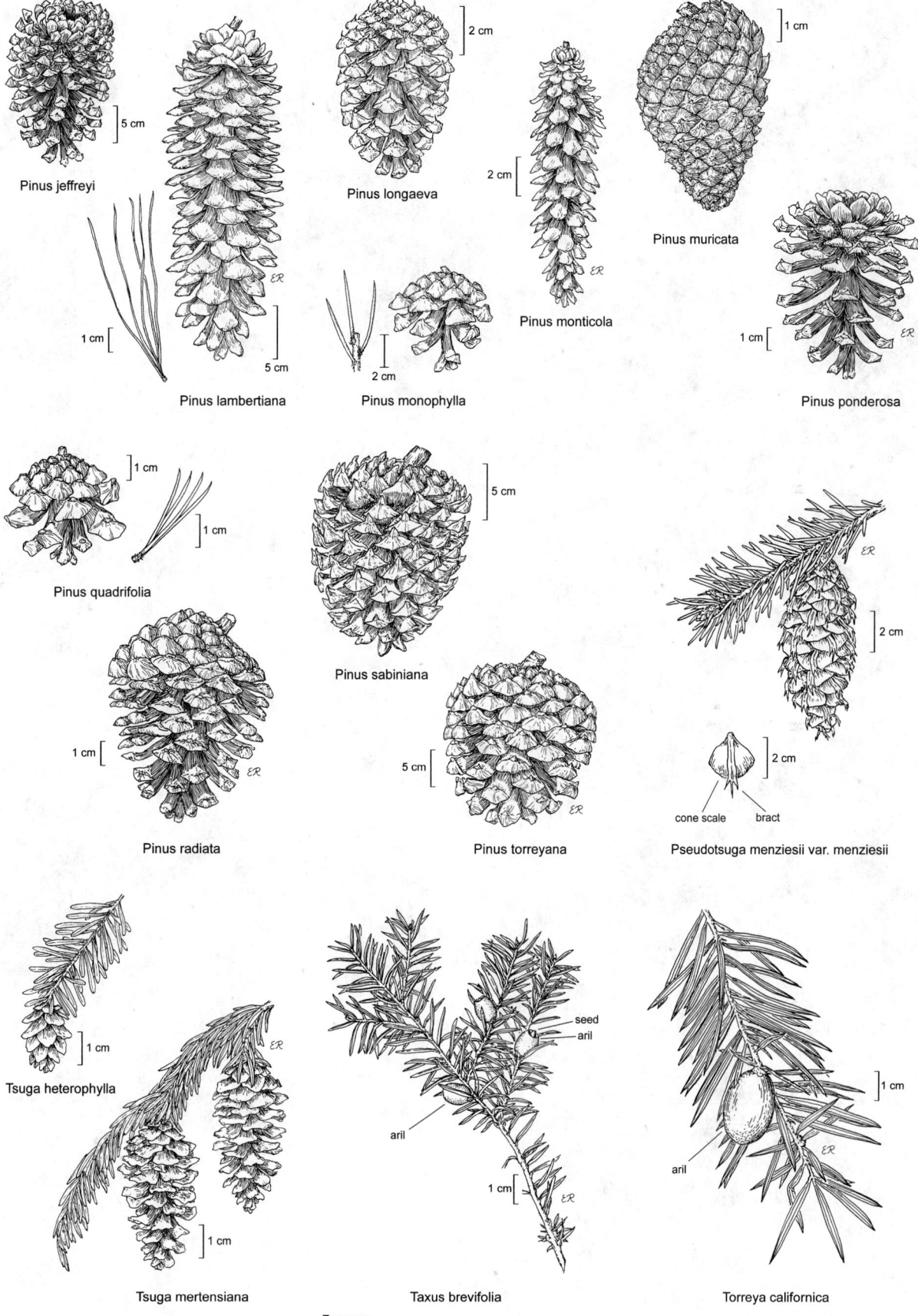

Pinus jeffreyi

Pinus lambertiana

Pinus longaeva

Pinus monophylla

Pinus monticola

Pinus muricata

Pinus ponderosa

Pinus quadrifolia

Pinus sabiniana

Pinus radiata

Pinus torreyana

Pseudotsuga menziesii var. menziesii

cone scale   bract

Tsuga heterophylla

Tsuga mertensiana

Taxus brevifolia

seed
aril

aril

Torreya californica

aril

Taxaceae

NYMPHAEALES

Aquatic fl pls, rhizomed; cotyledons 2; some or all lf blades floating; fl 1,
bisexual, parts spirally arranged or in 3s; pollen aperture 1

CABOMBACEAE WATERSHIELD FAMILY

Thomas J. Rosatti

Per, aquatic, rhizomes in mud. **ST**: elongate, lfy, ends floating. **LF**: of 2 kinds: submersed, palmately dissected, opposite
[whorled], short-petioled, not peltate; floating, simple, alternate, petioled, peltate, entire; stipules 0. **INFL**: fls 1 on long axillary
peduncles, gen on or above water surface. **FL**: bisexual, parts ± free; sepals, petals each 3[4], alternate, persistent; stamens 3–6
or 18–36(51), filaments slender, anthers opening lengthwise; carpels [1]2–18, ovaries superior, 1-chambered, ovules [1]2–3[5].
FR: achene- or follicle-like, indehiscent. **SEED**: 1–3, aril 0. 2 genera, 6 spp.: temp, trop Am, Afr, e Asia, Australia; some
Cabomba spp. cult for aquaria. [Taylor & Osborn 2006 Amer J Bot 93:344–356] Separation from Nymphaeaceae (carpels
fused, frs berry-like) confirmed by molecular data (Les et al. 1999 Syst Bot 24:28–46). Scientific Editors: Bruce G. Baldwin,
Robert E. Preston.

1. Lvs floating; stamens 18–36(51); submersed pl parts covered with thick mucilage.................... **BRASENIA**
1′ Lvs submersed, in fl gen also a few floating; stamens 3–6; submersed pl parts not or ± covered with thin
 mucilage .... **CABOMBA**

BRASENIA WATERSHIELD

1 sp. (C. Brasen, Danish surgeon, pl collector, 1738–1774)

B. schreberi J.F. Gmel. (p. 157) Submersed parts covered with
thick mucilage. **ST**: slender, 3–20 dm, branching, red-brown. **LF**:
centrally peltate; blade << petiole, 3.5–12(13.5) cm, 2–8 cm wide,
elliptic to ovate, ± entire. **INFL**: fls 1 in axils. **FL**: perianth 10–20
mm, ± red-purple or purple-brown; petals ± > sepals, linear-oblong;
filaments thread-like; carpels (2)4–18, stigmas linear-decurrent. **FR**:
6–8(10) mm, oblong, leathery. **SEED**: 1–2. 2n=80. Ponds, slow
streams; < 2200 m. KR, NCoR, CaRH, SNH, ScV, MP (exc Wrn); to
AK, MT; e N.Am, C.Am, S.Am, Afr, e Asia, e Australia. Apr–Oct ★

CABOMBA FANWORT

LF: blade ± ≥ petiole; submersed lvs dissected into many ± thread-like to linear segments. **FL**: perianth white to ± yellow
[purple]; carpels [1]3[4], stigmas head-like. **SEED**: 1–3. 5 spp.: gen trop Am. (Probably aboriginal)

C. caroliniana A. Gray (p. 157) **LF**: submersed petiole 7–30(40)
mm, blade 10–40 mm, (15)30–60 mm wide, ± deltate to reniform;
floating 0–few, 6–20(30) mm, 1–4 mm wide, linear-elliptic, ± entire,
base ± notched. **FL**: sepals (5)7–12 mm, white or cream; petals ± >
sepals, colored as sepals exc base with 2 ± yellow nectaries. **FR**: 4–7
mm. 2n=±78,±104. Ponds, slow streams; 0 m. SnJV; OR; native to
e US (exc n part of range), s S.Am. Reportedly spreading rapidly in
Sacramento Delta, expected widely. May–Sep ◆

NYMPHAEACEAE WATERLILY FAMILY

John H. Wiersema

Per, aquatic, rhizomed, stoloned or not; herbage with air chambers. **LF**: alternate, from rhizome; blades floating, submersed,
or ± emergent, << petiole, ± deeply notched at base into 2 lobes [peltate]. **INFL**: 1-fld, axillary; peduncle long. **FL**: bisexual,
gen on or above water; sepals 4–14, free, petal-like or not; petals many [(0)], stamen-like or not; stamens many, attached to
receptacle or ovary side, filaments gen wide; ovary compound, superior to inferior, chambers 3–many, many-ovuled, stigmas
in radiating lines on stigmatic disk. **FR**: berry-like, ± dehiscent or not, spongy. 5 genera, ± 70 spp.: ± worldwide. [Wiersema
& Hellquist 1997 FNANM 3:66–77] Scientific Editor: Thomas J. Rosatti.

1. Lf veins ± pinnate; sepals 5–14 (inner increasingly petal-like; petals inconspicuous), round to broad-obovate, ± erect; ovary superior; stigmatic disk margin entire to crenate . . . . **NUPHAR**
1′ Lf veins ± palmate; sepals gen 4 (petals conspicuous), gen longer than wide, ± spreading; ovary partly inferior; stigmatic disk margin with ± erect-incurved appendages . . . . **NYMPHAEA**

NUPHAR COW-LILY, YELLOW POND-LILY

Rhizomes prostrate, branched, not stoloned. **LF**: blade gen ovate to round. **FL**: sepals persistent; petals 10–20, stamen-like, thick, oblong to thin, spoon-shaped; stamens many, gen yellow, attached to receptacle, erect to incurved at dehiscence; ovary > stamens. **SEED**: ovoid, not arilled. ± 8 spp.: n temp. (Arabic name) [Padgett 2007 Rhodora 109:1–95]

N. polysepala Engelm. (p. 157) **LF**: blade gen floating, 1–4 dm, oblong to ovate, basal lobes gen ± rounded. **FL**: 5–10 cm wide; sepals 7–9(12), < 5 cm, ovate to wide-obovate, green or yellow-green, red-tinged or not; petals, stamens yellow, red-tinged or not; stigmatic disk 2–3.5 cm wide. **SEED**: 3–5 mm. 2*n*=34. Ponds, slow streams; < 2500 m. NW, CaRH, n&c SN, GV, n&c CW, MP; to AK, Rocky Mtns. [*N. lutea* (L.) Sm. subsp. *p.* (Engelm.) E.O. Beal] Apr–Sep

NYMPHAEA WATERLILY, WATER-NYMPH

Rhizomes prostrate to erect, branched or not, stoloned or not. **LF**: blade gen floating, elliptic to round, basal lobes gen ± acute. **FL**: sepals < petals, ± green; petals 8–many, white, ± red, [blue], or yellow; stamens many, attached to ovary side, erect to ascending at dehiscence, outer filaments flat, petal-like or not, inner linear; ovary < stamens. **SEED**: ± spheric to elliptic, arilled. ± 50 spp.: ± worldwide. (Greek: water nymph) [Woods et al. 2005 Syst Bot 30:471–480] Pls of both CA taxa problematic weeds in waterways.

1. Corolla bright yellow; pl stoloned . . . . ***N. mexicana***
1′ Corolla white or ± pink; pl not stoloned . . . . ***N. odorata***

N. mexicana Zucc. YELLOW OR BANANA WATERLILY Rhizomes erect, unbranched. **LF**: blade 7–14(18) cm wide, ovate to ± round. **FL**: 6–11 cm wide, floating or emergent; sepals, petals lanceolate to narrow-elliptic; petals 12–30; outer stamens gen 2–2.5 cm, inner anthers 4–6 mm; styles 7–9. **FR**: 2–2.5 cm, ovoid. **SEED**: 4–5 mm. 2*n*=56. Lakes, ponds, slow streams; < 100 m. SnJV; native to se US, Mex. CA pls referred to this sp. may be instead belong to *Nymphaea* ×*thiona* D.B. Ward, a natural hybrid between *N. mexicana* and *N. odorata*. Apr–Jul ◆

N. odorata Aiton FRAGRANT OR WHITE WATERLILY Rhizome prostrate, branched or not. **LF**: blade 5–25 cm wide, ± round. **FL**: 6–19 cm wide, floating; sepals, petals lanceolate to ovate; petals 17–43; outer stamens gen 3–4 mm, inner anthers 7–12 mm; styles gen 20. **FR**: 2.5–3 cm, depressed-spheric. **SEED**: 1.5–2.5 mm. 2*n*=56,84. Quiet waters, ponds, edges of lakes; gen < 2700 m. CaRF, n SNF, SNH (Lake Tahoe), ScV (Butte Co.), SnGb, expected elsewhere; native to e N.Am. Cult widely as orn. Pls in w US introduced, but available material ± unassignable to subsp. (2 recognized in FNANM; CA pls mentioned under *N. odorata* subsp. *o.*). Apr–Aug ◆

MAGNOLIIDS

Terrestrial fl pls, often scented from ethereal oils; cotyledons 2;
fl parts gen spirally arranged or in 3s; pollen aperture 1

ARISTOLOCHIACEAE PIPEVINE FAMILY

Michael R. Mesler & Karen Lu

Per, woody vine, [shrub], rhizomed, aromatic. **ST**: branched, occ ± underground. **LF**: simple, basal, cauline, or arising from rhizome, alternate; blade gen cordate, entire. **INFL**: fl gen 1, axillary or terminal. **FL**: bisexual, radial or bilateral; sepals 3, free or fused; petals gen 0; stamens gen 6 or 12, free or fused to style; pistil gen 1, ovary ± or partly inferior, chambers gen 6. **FR**: gen capsule. **SEED**: many. 5–8 genera, ± 500 spp.: mainly trop, warm temp; some cult (*Aristolochia, Asarum, Saruma*). [Neinhaus et al. 2005 Pl Syst Evol 250:7–26] Scientific Editors: Douglas H. Goldman, Bruce G. Baldwin.

1. Woody vine; fl bilateral . . . . **ARISTOLOCHIA**
1′ Per from rhizome; fl radial . . . . **ASARUM**

ARISTOLOCHIA PIPEVINE, BIRTHWORT

ST: gen climbing. **LF**: cauline. **INFL**: fl axillary. **FL**: [often foul smelling] sepals fused into a gen curved tube, deciduous, lobes 1–3; stamens 6, fused to style. **SEED**: flat. ± 400 spp.: gen trop, warm temp. (Greek: best birth, from use as medication in childbirth) [Ohi-Toma et al. 2006 Syst Bot 31:481–492]

A. californica Torr. (p. 157) Pl soft-hairy. **ST**: gen < 5 m, twining. **LF**: deciduous; blade 3–15 cm, ovate-cordate to sagittate. **FL**: fragrance musty; sepals 3, tube 2–4 cm, U-shaped, ± green to pale brown, with thick ± yellow to red lining, veins purple. **FR**: winged capsule. $2n=28$. Streamsides, forest, chaparral; < 700 m. KR, NCoR, CaRF, n&c SNF, ScV, n SnJV, CCo, SnFrB, n SCoRO. [*Isotrema c.* (Torr.) H. Huber] Pollinated by fungus gnats. Jan–Apr

ASARUM WILD-GINGER

Rhizome shallowly horizontal or deeply ± vertical, pls spreading or clumped; roots gingery-aromatic. **LF**: from rhizome, gen evergreen; blade cordate to reniform. **INFL**: fl terminal, at ground level. **FL**: gen dark colored; sepals forming tube, persistent; stamens 12, free from style, tips gen appendaged. **FR**: fleshy capsule. **SEED**: with fleshy appendage, ant-dispersed. 90 spp.: n temp. (Greek: derivation unknown) [Kelly 2001 Syst Bot 26:17–53]

1. Lf blade gen ± white along major veins; rhizomes deep, ± vertical; pls clumped; stamen tip > pollen sacs
  2. Calyx tube white inside with red stripes covered with white hairs; lf margin hairs curved toward lf tip;
    stamen tips pale . . . . *A. hartwegii*
  2′ Calyx tube dark red-maroon inside with scattered dark hairs; lf margin hairs ± erect-spreading; stamen
    tips dark . . . . *A. marmoratum*
1′ Lf blade all green; rhizomes near soil surface, horizontal; pls gen spreading; stamen tip < pollen sacs
  3. Calyx lobes spreading in fl, long-tapered. . . . . *A. caudatum*
  3′ Calyx lobes strongly reflexed in fl, obtuse to acute or abruptly pointed. . . . . *A. lemmonii*

A. caudatum Lindl. (p. 157) Pls gen spreading, forming loose mat; rhizomes near soil surface, horizontal. **LF**: blade all green. **INFL**: peduncle ± horizontal. **FL**: calyx lobes 2–9 cm, spreading in fl, long-tapered, maroon (rarely ± green), tube white inside with 1 median red stripe; stamen tip < pollen sacs, dark. $2n=26$. Moist places in forest; < 2200 m. NW, CaRH, CCo, SnFrB; to BC, MT. Pls with short, reflexed, tapered calyx lobes from CaRH (McCloud River) resemble *A. lemmonii* but have horizontal infl and longer calyx lobe taper. Mar–Aug

A. hartwegii S. Watson (p. 157) Pls densely clumped; rhizomes deep, ± vertical. **LF**: blade ± white along major veins, margin hairs curved toward tip. **INFL**: peduncle ± erect. **FL**: calyx lobes gen spreading from tube in fl, tapered, maroon, tube white inside with red stripes covered with white hairs; stamen tip > pollen sacs, pale. $2n=26$. Dry rocky slopes in open forest; 150–2200 m. KR, CaRH, SNH; sw OR. Apr–Jul

A. lemmonii S. Watson (p. 157) Pls gen spreading, forming loose mat; rhizomes near soil surface, horizontal. **LF:** blade all green. **INFL:** peduncle reflexed in fl. **FL:** calyx lobes 0.8–1.5 cm, strongly reflexed in fl, obtuse to acute or abruptly pointed, red, tube white inside, gen not striped (or partial median stripe); stamen tip < pollen sacs. 2*n*=26. Shady wet places; 1000–1900 m. s CaRH (Powellton, Butte Co.), SNH. May–Sep

A. marmoratum Piper (p. 157) MARBLED WILD-GINGER Pls clumped; rhizomes deep, ± vertical. **LF:** blade ± white along major veins (green), margin hairs ± erect-spreading. **INFL:** peduncle ± erect. **FL:** calyx lobes ± erect in fl, olive-brown, tube dark red-maroon inside with scattered dark hairs; stamen tip > pollen sacs, dark. Moist forest, exposed rocky slopes; 200–1800 m. KR (n Del Norte, Siskiyou cos.); sw OR. Apr–Jun ★

CALYCANTHACEAE SWEET-SHRUB or CALYCANTHUS FAMILY

George P. Johnson & Fosiée Tahbaz

Shrub, deciduous [evergreen], aromatic. **LF:** opposite, simple, entire; stipules 0; petiole short. **INFL:** fls 1, terminal on short branches. **FL:** bisexual, radial; receptacle hollow; parts gen many, spirally arranged; perianth parts ± petal-like, grading into bracts below; stamens on top of receptacle, smaller, sterile inward, filaments < anthers [0]; pistils simple, on inner face of receptacle, 2-ovuled, style thread-like, exserted from receptacle. **FR:** achenes many, incl in ± leathery receptacle. 3 genera, ± 10 spp.: N.Am, China, Australia; *Calycanthus*, *Chimonanthus* cult as orn, for fragrance. [Zhou et al. 2006 Molec Phylogen Evol 39:1–15] Notable for disjunctions. Scientific Editor: Thomas J. Rosatti.

CALYCANTHUS

FL: perianth maroon or red-brown, inner parts shorter. 2 spp.: CA, e US. (Greek: cup fl, from hollow receptacle)

C. occidentalis Hook. & Arn. (p. 157) SWEET-SHRUB, SPICEBUSH Erect, 1–3+ m, ± rounded. **LF:** petiole 3–10 mm; blade 5–15 cm, ovate- to lance-oblong or ovate-elliptic, firm, base gen rounded (cordate), tip acute (obtuse), abaxially ± glabrous to ± hairy, adaxially ± scabrous. **FL:** ± 5 cm diam; perianth ± fleshy, hairy; filaments < 1 mm, anthers 4–5 mm. **FR:** receptacle ovoid (bell-shaped), with scars of perianth parts; achenes 7–10 mm, ± velvety, margins granular. *n*=11. Moist, shady places, canyons, streamsides; gen < 1500 m. se KR, s NCoRO, NCoRI, s CaR, SNF, c&s SNH, SnJV (Kings River floodplain, e of Fresno), apparently naturalized n PR (Palomar Mtns). Mar–Aug

LAURACEAE LAUREL FAMILY

Henk van der Werff

[Shrub], tree, [parasitic vine], gen evergreen, aromatic; [dioecious or ± so]. **LF:** gen alternate, simple, unlobed [(lobed)], entire, gen thick; stipules 0. **INFL:** [(fls 1, head), raceme, panicle], umbel-like, enclosed by bracts [or not]. **FL:** gen bisexual, gen ± yellow to ± green; hypanthium often calyx-tube-like, perianth parts in 2(3) whorls of 3, ± sepal-like; stamens [(3)]9[(12)], in whorls of 3, inner often with 2 stalked orange glands at base, 1 [or more] whorls often staminodes [or not], anthers [2] 4-celled, opening by uplifting valves; pistil 1, simple, ovary gen superior, chamber 1, ovule 1, style 1, very short. **FR:** ± berry, often with swollen hypanthium, sepals. ± 54 genera, ± 3500 spp.: widespread in trop, less so in temp; some cult (*Laurus*, laurel, bay; *Persea*, avocado; *Cinnamomum*, cinnamon, camphor). [Buzgo et al. 2007 Int J Pl Sci 168:261–284; Carpenter et al. 2007 Int J Pl Sci 168:1191–1198] Scientific Editor: Thomas J. Rosatti.

1. Fls unisexual; anthers 2-celled . . . . [*Laurus nobilis*]
1′ Fls bisexual; anthers 4-celled
  2. Infl umbel-like . . . . *Umbellularia californica*
  2′ Infl a panicle or panicle-like cyme
    3. Basal pair of lateral veins stronger than distal, with axillary pit visible adaxially . . . . *Cinnamomum camphora*
    3′ Basal pair of lateral veins as strong as distal, without axillary pit . . . . [*Persea americana*]

CINNAMOMUM

INFL: in lf axils, panicle. **FL:** perianth parts 6; stamens 9, anthers 4-celled. ± 350 spp.: trop, subtrop Am, Asia, Pacific Islands, Australia. (Greek: cinnamon)

C. camphora (L.) J. Presl CAMPHOR TREE Pl < 30 m. **LF:** 6–12 cm, 2.5–5.5 cm wide, ovate-elliptic, hairs 0 or sparse abaxially, 0 adaxially; basal pair of lateral veins stronger than distal, with axillary pit visible adaxially; petiole < blade. **INFL:** 3–7 cm. **FL:** perianth parts ± 2 mm, hairs 0 abaxially, present adaxially. **FR:** ovoid to spheric, 6–9 mm diam, purple. 2*n*=24. Moist places; < 150 m. nw SnJV (Antioch Dunes), SCo, sw PR; native to trop, subtrop e Asia. Camphor distilled from wood; fr eaten, seeds dispersed by birds; reported pest in parts of Australia. Apr–May

UMBELLULARIA   CALIFORNIA BAY, CALIFORNIA LAUREL, PEPPERWOOD

1 sp.: w N.Am. (Latin: partial umbel, from infl)

U. californica (Hook. & Arn.) Nutt. (p. 157)   **ST:** < 45 m, bark ± green to red-brown. **LF:** 3–10 cm, 1.5–3 cm wide, narrowly ovate to oblong, shiny, gen deep yellow-green, minute-gland-dotted, abaxially glabrous, sparse-appressed-hairy, or minute-gray-tomentose, adaxially glabrous; petiole < blade. **INFL:** in upper axils, umbel-like, peduncled, 5–10-fld, subtending bracts ≤ 7 mm. **FL:** perianth parts 6, 3–4.5 mm, oblong-ovate; stamens 9, staminodes 3, < glands, anthers 4-celled. **FR:** gen 1, 2–2.5 cm, round-ovoid, ± green, dark purple when dry, olive-like. 2*n*=24. Common. Canyons, valleys, chaparral; < 1600 m. NW, CaRF, SNF, SNH, ScV (Sutter Buttes), deltaic SnJV, CCo, SnFrB, SCoRO, SCoRI, SCo, scattered TR, PR; s OR. [*U. c.* var. *fresnensis* Eastw.] Used in cooking, woodworking. Nov–May

SAURURACEAE   LIZARD'S-TAIL FAMILY

John W. Thieret & Elizabeth McClintock, final revision by Thomas J. Rosatti

Per, gen rhizomed, of ± wet places. **LF:** simple, alternate; stipules gen fused to petiole. **INFL:** spike [raceme], dense, many-fld, gen terminal, subtended by involucre of petal-like bracts [or not], each fl gen subtended by 1 bract. **FL:** small, bisexual; perianth 0; stamens [3]6(8), appearing to arise from infl axis [or not]; ovary inferior [or superior], sunken into infl axis [or not], compound, chamber gen 1 [or carpels fused only at base]; styles 3–4[5(7)], distinct. **FR:** capsule, ± fleshy, dehiscent near top [or mericarps]. **SEED:** [1 or] many, spheric or ovoid. 5 genera, 7 spp.: e Asia, N.Am. [Meng 2003 Ann Missouri Bot Gard 90:592–602] Scientific Editor: Thomas J. Rosatti.

ANEMOPSIS

1 sp. (Greek: anemone-like, from infl) [Howell 1971 Wasmann J Biol 29:97–100]

A. californica (Nutt.) Hook. & Arn. (p. 157)   YERBA MANSA Rhizome thick, woody. **ST:** 8–80 cm, hollow, glabrous or hairy. **LF:** basal several, blade 3–20 cm, elliptic to oblong, base cordate or not, petiole 2–40 cm; cauline < basal, 1–few, 1 ovate, gen subsessile to clasping, subtending 1–3 with short petioles or not. **INFL:** 1–4 cm, conic; involucre bracts 4–9, 0.3–3.5 cm, petal-like, white, often tinged ± red; fl bracts 3.5–6 mm, ± spoon-shaped, white. Common. Saline or alkaline soil, wet or moist areas, seeps, springs; < 2000 m. CaRH, s SN, Teh(?), sw ScV, SnJV, CW, SCo, ChI, WTR, SnGb(?), SnBr(?), PR, SNE, DMoj, DSon(?); to OR, KS, TX, nw Mex. Pls aromatic, once used to treat diseases of skin, blood. Mar–Sep

CERATOPHYLLALES

1 family

CERATOPHYLLACEAE HORNWORT FAMILY

Donald H. Les

Per, submersed aquatic; overwintering on bottom as detached, dense shoot tips (winter buds); roots 0; monoecious; water-pollinated. **ST**: flexible with water currents, internodes clustered near tip. **LF**: whorled, gen compound, forked; segments linear, toothed; stipules 0; petiole 1–2 mm, translucent. **INFL**: reduced axillary cymes gen near shoot tips, ± sessile, 1–several per node, gen staminate below, pistillate above, or sometimes mixed at a node; fls 1–3 per cyme but appearing to be 1 subtended by calyx-like whorl of 8–15 bracts, bracts linear, lf-like, fused at base, toothed at tip, persistent in fr. **FL**: unisexual; perianth 0; staminate round-topped, stamens 3–50, spirally arranged, anthers ± sessile, exserted from bracts, pollen shed underwater; pistillate of 1 simple, ovoid pistil, ovary superior, chamber 1; placenta pendulous; style elongate, spine-like, with decurrent groove; stigma pocket-like, at style base; ovule 1. **FR**: achene, tubercled or not, spines 0 or elongate, basal, facial, marginal or stylar; lateral margin winged or not. 1 genus, 6 spp.: ± worldwide. [Les 1997 FNANM 3:81–84] Scientific Editor: Thomas J. Rosatti.

CERATOPHYLLUM HORNWORT

(Greek: horn lf, for antler-like appearance) *Ceratophyllum* grown as orn or for oxygen generation in aquaria, garden ponds. Lvs, fr eaten by migrating waterfowl. Vigorous growth (even of native spp.) can result in weedy infestations.

C. demersum L. (p. 157) **ST**: 1–3+ m, glabrous, limp, suspended by water, brittle, branches 0–3 per node. **LF**: 3–11 per node, 0.8–3 cm; segments 2–4, with 2 rows small teeth distally, 1 multicellular, glandular appendage at tip. **INFL**: uncommon. **FL**: staminate 2–5 mm, stamens pink to ± red; pistillate 1–2 mm, pistil gen ± yellow, margins red-tinged or not, style gen 1 mm. **FR**: 3.5–6 mm (exc spines), dark green to red-brown, gen tubercled, base with 2 tubercles or 2 spines, margin spineless, wingless; spines 0.1–12 mm, stiff, bases gen webbed; style persistent on fr, stiff, erect, 0.5–14 mm. **SEED**: cotyledons large, fleshy; terminal bud highly developed; first lvs simple, opposite, awl-like. $2n=24,38,40,48$. Common. Ditches, lakes, ponds, pools, slow watercourses; water 0.1–4 m deep, fresh to ± brackish, medium to high nutrient levels, acidic to alkaline (pH 5.9–9.4) but gen alkaline (pH > 7); gen < 1700 m. CA (exc CaRF, Teh, ChI, W&I, DMtns); common worldwide. Pls self-fertile; populations gen > 50% staminate. Jun–Aug

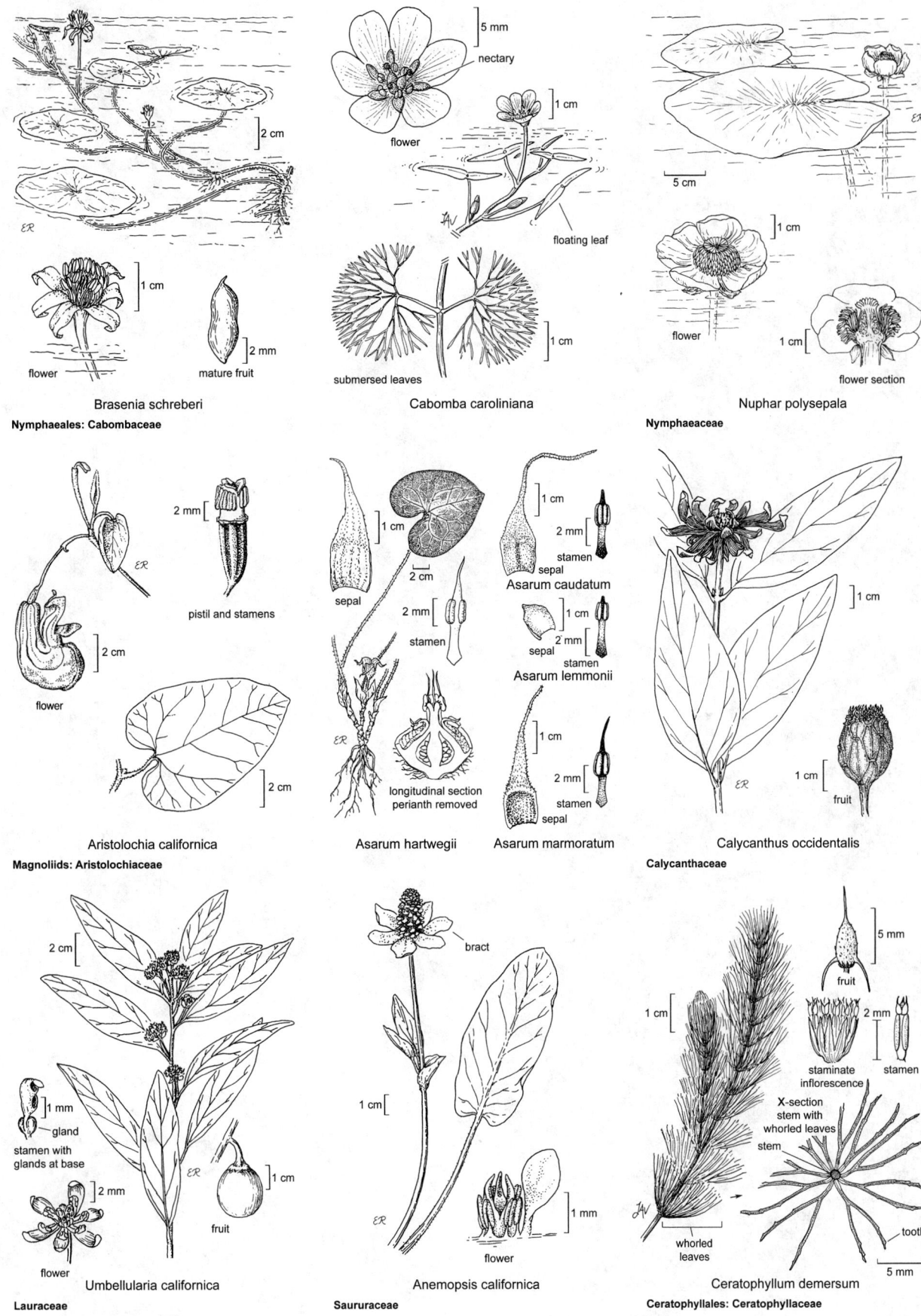

157

Brasenia schreberi
Nymphaeales: Cabombaceae

Cabomba caroliniana

Nuphar polysepala
Nymphaeaceae

Aristolochia californica
Magnoliids: Aristolochiaceae

Asarum hartwegii

Asarum caudatum

Asarum lemmonii

Asarum marmoratum

Calycanthus occidentalis
Calycanthaceae

Umbellularia californica
Lauraceae

Anemopsis californica
Saururaceae

Ceratophyllum demersum
Ceratophyllales: Ceratophyllaceae

EUDICOTS

Fl pls; cotyledons 2; fl parts gen in 4s or 5s; pollen apertures 3+ [*Diospyros virginiana* L. var. *virginiana*, American persimmon (Ebenaceae), documented from two clonal colonies in SnBr (Hrusa et al. 2002 Madroño 49:61–98.]

ACANTHACEAE   ACANTHUS FAMILY

Thomas F. Daniel, Margriet Wetherwax & Lawrence R. Heckard, except as noted

[Ann, per to] shrub, [tree], nodes gen swollen. **LF:** simple, gen opposite, entire (toothed [lobed]); stipules 0. **INFL:** variable, with bracts, gen also bractlets. **FL:** bisexual; calyx deeply (3)4–5 lobed [(sepals free)]; corolla 4–5 lobed, radial to 2-lipped; stamens 2 or 4, epipetalous, anther sacs sometimes dissimilar in size or placement; ovary superior, 4–many-ovuled, chambers 1–2, placentas axile (free central), stigmas 1–2. **FR:** capsule, loculicidal, gen dehiscing explosively, valves 2. **SEED:** 1–4[+] each gen subtended by hook-like outgrowth that remains in fr. 220 genera, 4000 spp.: esp trop; some orn: *Justicia* (Beloperone, shrimp-plant), *Acanthus*, *Thunbergia*. [Daniel 1998 Proc Calif Acad Sci 50:217–256; Hilu et al. 2003 Amer J Bot 90:1758–1766; Schwarzbach & McDade 2002 Syst Bot 27:84–98] Scientific Editor: Thomas J. Rosatti.

1. Shrub; roots exposed at low tide; hook-like outgrowth subtending seed 0; fls white; coastal salt marshes, SCo. . . . . **AVICENNIA**
1′ Subshrub, shrub; roots not exposed; hook-like outgrowth subtending seed present; fls white with yellow or ± red; dry sandy or rocky soils, esp washes, rocky slopes, w PR, DSon
　2. Corolla 1–1.8 cm, white with yellow, maroon-streaked spot on upper lip, tube < lips; anther sacs ± equally placed on filament, not spurred; fr glabrous. . . . . **CARLOWRIGHTIA**
　2′ Corolla 2–4 cm, dull scarlet (yellow), tube gen ≥ lips; anther sacs unequally placed on filament, lower sac spurred; fr canescent . . . . **JUSTICIA**

AVICENNIA

Thomas F. Daniel, Margriet Wetherwax & Dieter H. Wilken

Shrub [to tree]; roots spreading, exposed at low tide. **LF:** persistent, petioled; blade thick. **INFL:** axillary, terminal, head-like, < peduncle; bractlets 3, gen < calyx, outer 1 > inner 2. **FL:** calyx deeply 5-lobed; corolla 4-lobed, radial, rotate; stamens 4, attached at 1 level. **FR:** ellipsoid to ovoid, often asymmetric; 1-chambered, gen indehiscent; stalk not flat. ± 8 spp.: temp, subtrop worldwide, gen coastal marshes, lagoons. (Avicenna, Persian physician, philosopher, 980–1037) [Duke et al. 1998 Evolution 52:1612–1626] Placement in Acanthaceae s.l. fairly well established; sister group relationship with Thunbergioideae weakly supported (Schwarzbach & McDade 2002; Hilu et al. 2003).

A. marina (Forssk.) Vierh. var. *australasica* (Walp.) Moldenke GRAY MANGROVE **ST:** < 2.5 m; trunk < 10 cm diam. **LF:** 4–11 cm, elliptic to ovate, leathery; adaxially ± shiny, abaxially felt-like. **INFL:** 4–9-fld; peduncles 2–4 cm; bractlets ± ovate. **FL:** corolla ± yellow-white, lobes ± ovate, leathery, puberulent abaxially. **FR:** ± 2 cm diam, ovoid. Coastal salt marshes; ± 0 m. SCo (Mission Bay, San Diego Co.); native to New Zealand, Australasia [*A. m.* var. *resinifera* (G. Forst.) Bakh.]. First planted in 1960s, then weedy, extirpation failed in 1990. Jun–Aug

CARLOWRIGHTIA

Subshrub, shrub. **INFL:** gen spike- [or panicle-]like. **FL:** corolla 4-lobed (but petals 5), tube < lobes, slender, barely wider at throat, uppermost lobe gen notched (2 petals ± entirely fused), 2 lateral ascending or reflexed, lowest ± keeled, incl stamens, style; stamens 2, anther sacs ± equally placed on filament, opening toward uppermost lobe, not spurred. **FR:** compressed-ovoid; stalk ± flat. **SEED:** 4, disk-like, notched at base. 24 spp.: esp sw US, Mex. (Charles Wright, Am botanical collector, 1811–1885) [Daniel et al. 2008 Syst Bot 33:416–436]

C. arizonica A. Gray (p. 165) ARIZONA CARLOWRIGHTIA, ARIZONA WRIGHTWORT **ST:** multi-branched, < 1 m; hairs erect or recurved, minute. **LF:** variable in size, shape; blade gen < 5 cm, lanceolate to elliptic or cordate, puberulent; petiole 0–8 mm. **INFL:** fls 1–several per bract. **FL:** calyx 1.5–5 mm, lobes 5, narrow-triangular; corolla 1–1.8 cm, white with yellow, maroon-streaked spot on upper lobe. **FR:** 7–11 mm, glabrous. **SEED:** 3–4 mm, tubercled; margin finely dentate. 2*n*=36. Rocky slopes; ± 300 m. w DSon (Flat Cat, Palm canyons, Anza Borrego); to TX, C.Am. Apr–Jun ★

JUSTICIA

[Ann, per] shrub. **INFL**: gen raceme [variable]. **FL**: corolla 2-lipped, tube gen ≥ lips, wider upward, upper lip notched or 2-lobed, lower 3-lobed; stamens 2, gen appressed to upper lip, anther sacs unequally placed on filament, opening toward lower lip, lower sac spurred. **FR**: club-shaped; stalk ± flat. **SEED**: 4, outlined on fr surface [or not]. 400 spp.: trop, subtrop. (James Justice, Scottish horticulturist, 1698–1763) [Graham 1988 Kew Bull 43:551–624]

J. californica (Benth.) D.N. Gibson (p. 165) BELOPERONE, CHUPAROSA **ST**: < 2 m, gen lfless in fl, canescent. **LF**: blade 1–6 cm, ovate, triangular, or ± round, puberulent; petiole < 20 mm. **INFL**: bracts < 1 cm, lance-elliptic, falling early; bractlets 2–5 mm, narrow-tapered. **FL**: calyx 5–8 mm, lobes 5, lanceolate; corolla 2–4 cm, puberulent, dull scarlet (yellow), lips 1–2 cm, < tube, lobes of lower 1–4 mm. **FR**: 1.5–2 cm, canescent. **SEED**: 2.5–3.5 mm, ± round, mottled. 2*n*=28. Dry sandy or rocky soils, esp washes; < 950 m. e PR, D; AZ, nw Mex. Mar–Jun

ADOXACEAE MUSKROOT FAMILY

Charles D. Bell

[Per], subshrub to tree; hairs often stellate or glandular. **LF**: gen opposite, simple or compound, gen toothed; stipules gen 0. **FL**: gen bisexual; calyx teeth or lobes [2]5; corolla small, radial, rotate, lobes [3–4]5; stamens [4]5, epipetalous; ovary ± inferior, chambers 1 or 3–5, 1-ovuled; styles ± 0 or 3–5. **FR**: drupe. 5 genera, 200 spp.: esp n temp, also S.Am, se Asia, trop Afr. [Backlund & Bremer 1997 Pl Syst Evol 207:225–254] Incl in Caprifoliaceae in TJM (1993), and possibly in future. Scientific Editor: Thomas J. Rosatti.

1. Lf compound; fr berry-like, 3–5-seeded . . . . **SAMBUCUS**
1′ Lf simple; fr drupe-like, 1-seeded . . . . **VIBURNUM**

SAMBUCUS ELDERBERRY

Gen shrub to small tree, deciduous; main trunk gen 0. **ST**: pith large, spongy. **LF**: 1(2)-odd-pinnately compound; lflets serrate. **INFL**: panicle of cymes, terminal, gen ± dome-shaped. **FL**: ovary chambers 3–5, ovules pendent; style ± 0, stigma lobes 3–5. **FR**: drupe, berry-like. **SEED**: 3–5. 20 spp.: temp, subtrop; some cult as orn. (Greek: for stringed instrument made from wood of genus) Toxic in quantity (exc cooked frs). [Bolli 1994 Diss Bot 223:1–256]

1. Fr ± black, densely white-glaucous (appearing ± blue); infl ± flat-topped, central axis gen not dominant; petals ± spreading . . . . *S. nigra* subsp. *caerulea*
1′ Fr red or purple-black, not glaucous; infl ± dome-shaped, central axis gen dominant; petals often reflexed . . . . *S. racemosa*
  2. Fr purple-black; infl gen 4–7 cm diam. . . . . var. *melanocarpa*
  2′ Fr red; infl gen 3–12 cm diam . . . . var. *racemosa*

S. nigra L. subsp. *caerulea* (Raf.) Bolli (p. 165) BLUE ELDERBERRY Pl 2–8 m, gen as wide as tall. **LF**: lflets 3–9, 3–20 cm, elliptic to ovate, glabrous to hairy, axis often bowed, base often asymmetric, tip acute to acuminate. **INFL**: 4–33 cm diam, ± flat-topped; central axis gen not dominant. **FL**: petals ± spreading. **FR**: ± black, densely white glaucous (appearing ± blue). 2*n*=36. Common. Streambanks, open places in forest; < 3000 m. CA-FP, GB, DMtns; to BC, UT, NM. [*S. mexicana* DC., misappl.] Variable, in need of study. Mar–Sep

S. racemosa L. Pl 1–6 m. **LF**: lflets 5–7, 4–16 cm, lanceolate to oblong-ovate, base gen asymmetric, tip ± acuminate. **INFL**: 4–12 cm diam, ± dome-shaped; central axis gen dominant. **FL**: petals often reflexed. **FR**: red or purple-black, not glaucous.

var. *melanocarpa* (A. Gray) McMinn BLACK ELDERBERRY **LF**: abaxially glabrous or ± stiff-hairy on veins. **INFL**: gen 4–7 cm diam. **FR**: purple-black. Streamsides, edges of meadows or conifer forest; 1800–3600 m. CaR, SNH; to w Can, UT, NM. [*S. m.* A. Gray] Apr–Jul

var. *racemosa* (p. 165) RED ELDERBERRY **LF**: abaxially glabrous or stiff-hairy, esp on veins. **INFL**: gen 6–12 cm diam. **FR**: red. Moist places; < 3300 m. NW, CaR, SN, n CCo, SnFrB, SnBr; N.Am, Eur, Asia. [*S. r.* var. *microbotrys* (Rydb.) Kearney & Peebles] May–Jul

VIBURNUM

Shrub, slender, gen hairy, also ± glandular, esp in infl, gen deciduous. **LF**: simple. **INFL**: compound cyme, gen terminal, rounded or ± flat-topped, gen with oblanceolate bracts; peduncles 1.5–4 cm; rays gen 7. **FL**: ovary chambers 1 (2 abort), ovules pendent; style short, stigma lobes 3. **FR**: drupe, drupe-like. **SEED**: 1. ± 250 spp.: n temp, subtrop. (Latin: for pliable branches used in binding) *V. rigidum* Vent. possibly naturalized in SnFrB (Tilden Park).

1. Lvs (exc upper) ± widely elliptic, 3-lobed, dentate . . . . *V. edule*
1′ Lvs elliptic to round or cordate, unlobed, coarsely dentate above middle . . . . *V. ellipticum*

V. edule (Michx.) Raf. SQUASHBERRY **LF**: petiole 6–8 mm; blade 4–12 cm, palmately 3–5-veined. **INFL**: 1.5–2.5 cm. **FL**: 6–8 mm diam. **FR**: 6–12 mm, ovoid, yellow to orange-red. **SEED**: ovoid, 3–5 grooved. Spring-fed montane riparian/montane meadow habitats; 300–900 m. CaRH; to AK, ne US. May–Jul ★

V. ellipticum Hook. (p. 165) OVAL-LEAVED VIBURNUM **LF**: petiole 6–12 mm; blade 2–6 cm, gen 3-veined from base. **INFL**: 1–3 cm. **FL**: 6–8 mm diam. **FR**: 10–12 mm, elliptic, ± red, in age black. **SEED**: oblong, 5-grooved. Chaparral, yellow-pine forest, gen n-facing slopes; 300–1400 m. NW, n&c SNF, SnFrB; to WA. Jun–Aug ★

AIZOACEAE FIG-MARIGOLD or ICEPLANT FAMILY

Nancy J. Vivrette, John Bleck & Wayne R. Ferren, Jr., family description, key to genera

Ann, per, shrub, gen fleshy. **ST**: underground to erect; root fibrous or tuberous. **LF**: gen simple, entire, flat, cylindric, 3-angled, or scale-like, gen cauline, gen opposite; stipules gen 0; blade papillate, pubescent, or gen glabrous, often glaucous. **INFL**: cyme or 1-fld, gen terminal. **FL**: gen bisexual, radial; hypanthium present; sepals (3)5(8), often unequal; petals 0 or many in several whorls, free or fused at base, linear; stamens 1–many, free or fused in groups, outer often petal-like; nectary a ring or separate glands; pistil 1, ovary superior to inferior, chambers 1–20, placentas gen parietal, styles 0–20. **FR**: berry, nut, or gen capsule, opening by flaps or circumscissile. **SEED**: 1–many per chamber, gen ovoid, arilled or not. 130 genera, 2500 spp.: gen subtrop, esp s Afr; many cult. [Hartmann 2002 Illus Handbook Succulent Pls Aizoaceae A-E (Vol 1) and F-Z (Vol 2). Springer; Vivrette et al. 2003 FNANM 4:75–91] *Galenia pubescens* (Eckl. & Zeyh.) Druce, a waif, may be naturalizing in s CA. Scientific Editors: Bruce G. Baldwin, Thomas J. Rosatti.

1. Petals, petal-like stamens 0
   2. Ovary inferior (subfam. Tetragonioideae) . . . . **TETRAGONIA**
   2′ Ovary ± 1/2-inferior to superior (subfam. Aizooideae)
      3. Lvs alternate . . . . **[GALENIA]**
      3′ Lvs ± opposite
         4. Stamens 30; styles (2)3–5; lvs of opposite pair equal . . . . **SESUVIUM**
         4′ Stamens ≤ 10; styles 1–2; lvs of opposite pair unequal
            5. Stamens gen 3; stipules fringed . . . . **CYPSELEA**
            5′ Stamens 5–10; stipules of 2 teeth . . . . **TRIANTHEMA**
1′ Petals, petal-like stamens many
   6. Placentas axile (subfam. Aptenioideae)
      7. Sepals, ovary chambers 4 . . . . **APTENIA**
      7′ Sepals, ovary chambers (4)5 . . . . **MESEMBRYANTHEMUM**
   6′ Placentas parietal (subfam. Ruschioideae)
      8. Fr fleshy or a capsule, valves separating; stigmas, ovary chambers 8–20
         9. Basal rosette 0, cauline lvs opposite; fr fleshy, indehiscent (tribe Carpobroteae) . . . . **CARPOBROTUS**
         9′ Basal rosette present, cauline lvs gen alternate; fr capsule, tardily dehiscent (tribe Apatesieae) . . . . . . **CONICOSIA**
      8′ Fr a capsule, valves not separating; stigmas, ovary chambers 4–8(12) (tribe Ruschieae)
         10. Ovary chambers 8(12) . . . . **MALEPHORA**
         10′ Ovary chambers 4–6
            11. Sepals unequal; stigmas awl-shaped . . . . **DELOSPERMA**
            11′ Sepals equal; stigmas thread-like . . . . **DROSANTHEMUM**

APTENIA

Nancy J. Vivrette

Per, subshrub. **ST**: prostrate or climbing, base woody; green, papillate. **LF**: opposite, near infl alternate; petioled; blade flat, lanceolate or cordate, gen glabrous, often glaucous. **INFL**: fls 1 [or whorled], axillary [terminal], peduncled [not]. **FL**: ± 1 cm diam; hypanthium obconic; sepals 4, 2 larger, lf-like, 2 smaller, awl-shaped; petals fused at base; stamens many, outer sterile, petal-like, inner incurved, white or yellow; ovary inferior, chambers 4, placentas axile, style 0, stigmas 4. **FR**: capsule; valves 4, lids 0, wings 0. **SEED**: flat, tubercled, black-brown. 4 spp.: s Afr. (Greek: wingless, for fr)

A. cordifolia (L. f.) Schwantes BABY SUN-ROSE **ST**: 3–6 dm; nodes wide-spaced. **LF**: 1–3 cm, cordate, minutely papillate. **INFL**: peduncle 8–15 mm. **FL**: hypanthium 6–7 mm; sepals ± 5 mm; petals ± 3 mm, purple (red in cv. 'Red Apple'). **FR**: 13–15 mm. *n*=6. Uncommon. Disturbed places, margins of coastal wetlands; < 100 m. CCo, SCo, s ChI; native to s Afr. Apr–May

CARPOBROTUS

Nancy J. Vivrette

Shrub. **ST**: trailing, forming mats, < 20 dm, nodes rooting; fl-branches ascending. **LF**: opposite, ± linear-elliptic, fleshy, glabrous, ± triangular in ×-section. **INFL**: 1-fld, terminal; bracts 2. **FL**: 3–15 cm diam; sepals 5, unequal, margins of smaller 2–3 expanded, papery; petals free, magenta, pink, yellow, or white; stamens many, erect, white or yellow; ovary inferior, chambers 6–14[20], styles 0, stigmas 6–14[20], sessile, linear, hairy. **FR**: berry-like, fleshy, indehiscent. **SEED**: many, glossy brown, in fleshy pulp. 14 spp.: s Afr, Chile, Australia. (Greek: fr edible)

1. Fl 3–5 cm diam, petals rose-magenta; ovary chambers 6–8; lf 4–7 cm, rounded-triangular in ×-section, abaxial angle smooth . . . . *C. chilensis*
1′ Fl 8–10 cm diam, petals yellow (± white) aging pink; ovary chambers 7–14; lf 6–10 cm, sharp-triangular in ×-section, abaxial angle serrate near tip . . . . *C. edulis*

C. chilensis (Molina) N.E. Br. (p. 165) SEA FIG **ST:** < 2 m. **LF:** widest above middle, glaucous. **FL:** sessile; sepals 1–2 cm; petals 1–2.5 cm. **FR:** soft when ripe. Common on coastal sandy shores; < 100 m. NCo, CCo, SCo, n ChI; to OR, Mex; probably native to s Afr. ± all year. ❖

C. edulis (L.) N.E. Br. (p. 165) FREEWAY ICEPLANT **ST:** < 3 m. **LF:** widest below middle, not glaucous. **FL:** pedicelled; sepals 3–4 cm, sharp-triangular in ×-section, outer angle serrate near tip; petals 3–4 cm. **FR:** triangular, ± flat toward pedicel. n=18. Common. Many coastal habitats, esp sand; < 100 m. NCo, CCo, SCo, ChI; to Mex; native to s Afr. Hybridizes with *C. chilensis*; extensively planted along highways and for dune stabilization; invasive. ± all year. ❖

CONICOSIA

John Bleck

Per, gen short-lived; roots diffuse or tuberous. **ST:** caudex underground; ann shoots prostrate to ascending. **LF:** basal rosetted; cauline gen alternate, linear, ± triangular in ×-section. **INFL:** 1-fld, axillary; bracts 0. **FL:** 5–13 cm diam; sepals 5, ± unequal, wider at base, tip cylindric, basal margins of inner 3 papery; petals free, linear, yellow, ciliate in basal 1/2; stamens many, base hairy; ovary ± inferior, chambers [8]10–20[22], thread-like, sessile. **FR:** capsule, opening by 10–20 flaps at conic top that do not spread when moistened, valves 10–20, separating, dehiscing tardily. **SEED:** spheric, smooth. 2 spp.: s Afr. (Greek: cone-shaped)

C. pugioniformis (L.) N.E. Br. (p. 165) NARROWLEAF ICEPLANT **ST:** caudex 1, < 30 cm, 1–2 cm diam. **LF:** cauline crowded toward st tip, 15–20 cm, 10–12 mm wide, gray-green, glabrous, adaxially grooved or flat. **INFL:** pedicel < 12 cm. **FL:** 5–8 cm diam, shiny yellow, lasting several days; odor unpleasant. **SEED:** 1.5–1.8 mm. $2n$=18. Uncommon. Sandy places, esp coastal dunes; < 100 m. CCo; native to s Afr. May–Oct ❖

CYPSELEA

Wayne R. Ferren, Jr.

Ann [per], mat-forming, glabrous. **ST:** prostrate, branched. **LF:** opposite, petioled; stipule sheathing st, scarious, fringed, attached to lower petiole margins. **INFL:** 1-fld, axillary; bracts fringed. **FL:** calyx bell-shaped, sepals 4–5, unequal, scarious; petals 0; stamens gen 3, alternate sepals; ovary superior, ovoid to spheric, chamber 1, placenta free-central, styles, stigmas (1)2. **FR:** capsule, circumscissile. **SEED:** many; aril slender, persistent on placenta. 3 spp.: Caribbean, S.Am. (Greek: beehive)

C. humifusa Turpin (p. 165) Mats < 2 cm wide. **ST:** slender, diffuse-branched from base. **LF:** pair unequal, axil of smaller with st, fl, or gen both, with fl between main and axillary st; blade ± = petiole, larger 5–10 mm, elliptic, obtuse. **FL:** ± 2 mm diam; sepals ± 1.5 mm, erect, ovate, ± green, margin scarious, white. **FR:** 1.5 mm, ± spheric, thin-walled. **SEED:** ± 0.3 mm, round-reniform, smooth, brown. Uncommon. Seasonally dry margins of wetlands; < 1500 m. NCoR, GV, SnFrB, SCoR, PR; to NV, se US; native to Caribbean, S.Am. Jul–Oct

DELOSPERMA

John Bleck

Per, often taprooted. **ST:** prostrate to erect. **LF:** opposite, sessile, < 7 cm, linear, triangular to round in ×-section. **INFL:** 1-fld or cyme; bracts present. **FL:** 1.5–8 cm diam; sepals 4–8, unequal; petals free; outer stamens sterile, petal-like, inner erect, ± white; nectary glands separate; ovary 1/2-inferior, chambers 5, placentas parietal, stigmas 5, awl-shaped. **FR:** capsule, valves not separating; lid gen 0. **SEED:** 0.5–1.5 mm, ± spheric, pale brown, gen arilled. ± 160 spp.: Afr. (Greek: visible seed)

D. litorale (Kensit) L. Bolus **ST:** prostrate, thin, mat-forming, nodes widely spaced, rooting. **LF:** tip often hooked. **INFL:** few-fld cyme; pedicel to 2 cm. **FL:** 1.5–2.2 cm diam; sepals 5–6, ± unequal; petals linear, white; inner stamens free, forming cone around stigmas. **FR:** not seen. $2n$=36. Uncommon. Margins of coastal wetlands, bluffs, stabilized dunes; < 35 m. SCo, s ChI; native to s Afr. May–Jul

DROSANTHEMUM

John Bleck

Per, shrub, papillate. **ST:** prostrate [to erect], gen rough. **LF:** opposite, 4-ranked, [1]1.2–1.4[2.5] cm, 2.5 mm wide, linear, triangular to round in ×-section, often papillate. **INFL:** fls 1 [few in cyme]. **FL:** 1.8[3.5] cm diam; sepals gen 5, equal; petals in 1–2 whorls, free; outer stamens sterile, inner erect; nectaries separate; ovary ± inferior, top flat or convex, chambers 4–6, placentas parietal, stigmas 4–6, thread-like. **FR:** capsule, valves not separating; flaps present. **SEED:** light brown, rough. ± 105 spp.: s Afr. (Greek: dew fl) *D. hispidum* (L.) Schwantes (indicated as a waif in CA in TJM (1993)) occ collected in CA but only from cult.

D. floribundum (Haw.) Schwantes **ST:** mat-forming, thin; older nodes rooting. **LF:** cylindric, ± curved, wider toward tip, light green. **INFL:** 1-fld; pedicels 2–3 cm. **FL:** sepals ± 3–4 mm, linear, obtuse; petals 8–9 mm, gen pink or pale purple. **SEED:** 0.5 mm. Uncommon. Coastal habitats; < 35 m. NCo, CCo, SCo, ChI; n Mex; native to s Afr. Apr–Jul

MALEPHORA

John Bleck

Per, shrub. **ST**: prostrate [to erect]. **LF**: opposite, ± fused at base, ± triangular [to ± round] in ×-section, smooth, often glaucous. **INFL**: 1-fld (or cyme). **FL**: < 5 cm diam; sepals 4–6, unequal; petals free; stamens many in several whorls; nectary a ring; ovary ± inferior, top flat, placentas parietal, styles 0, stigmas wide, feathery. **FR**: capsule, valves not separating; valve winged and with lids. **SEED**: flat, rough, with tubercles in rows. 15 spp.: s Afr. (Greek: armhole, to bear)

M. crocea (Jacq.) Schwantes **ST**: stout, pale, corky; nodes often rooting. **LF**: on short shoots, 2.5–6 cm, 6 mm wide, ± linear-elliptic or -oblanceolate, pale glaucous-green, sometimes ± red. **INFL**: pedicel 1–6 cm. **FL**: calyx 0.8–1.5 cm diam, at least 2 sepals short, acuminate, with translucent margins; petals adaxially orange, abaxially purple, or entirely yellow or orange; outer stamens sterile, yellow, petal-like, inner erect, often hairy at base; ovary chambers 8(12), stigmas 8(12), feathery. **FR**: placental tubercles at outer margin of chamber, in adaxial seed pockets, 2-lobed. **SEED**: many, 1 mm, 0.8 mm wide, lenticular. $2n=27,36$. Common. Margins of wetlands, coastal bluffs; < 50 m. CCo, SCo, s ChI; native to s Afr. [*M. purpureo-c.* (Haw.) Schwantes] ± invasive. Mar–Dec

MESEMBRYANTHEMUM ICEPLANT

Nancy J. Vivrette

Ann, bien, glabrous. **ST**: prostrate to ascending, cylindric, angled or winged, papillate. **LF**: alternate or opposite, petioled or not, cylindric or flat, ± red in age or stress, papillae prominent to flat, inconspicuous; bases of pair fused. **INFL**: 1-fld or cyme. **FL**: sepals, petals free or fused at base; sepals (4)5, 2 often lf-like; petals white; stamens many; nectar gland ± grooved; ovary 1/2-inferior, chambers (4)5(20), styles (4)5(20). **FR**: capsule, valves 5, dehiscing when moist. **SEED**: many, round, compressed, often D-shaped, minutely tubercled, light or dark red-brown. 15 spp.: sw Afr, Medit, w Asia. (Greek: midday-blooming)

1. Lf flat; fl 10–30 mm diam . . . . *M. crystallinum*
1′ Lf ± cylindric; fl 4–12 mm diam . . . . *M. nodiflorum*

M. crystallinum L. (p. 165) CRYSTALLINE ICEPLANT **ST**: trailing, forked, < 1 m. **LF**: blade 2–20 cm, ovate to spoon-shaped, lower ± cordate. **INFL**: cyme, ± sessile. **FL**: hypanthium spheric, aging red; sepals 5, equal; petals aging pink. **FR**: course-papillate. **SEED**: brown, rough. Common, coastal bluffs, cliffs, disturbed ground; < 100 m. NCo, CCo, SCo, ChI; to AZ, Mex; also Medit, S.Am, Australia, Afr; native to s Afr. Reported from roadside in DSon but not persisting. Mar–Oct ❖

M. nodiflorum L. (p. 165) SLENDER-LEAVED ICEPLANT **ST**: prostrate to ascending, branched from base, 15–20 cm. **LF**: blade 1–2 cm, linear. **INFL**: 1-fld; pedicel short. **FL**: hypanthium obconic; sepals 5, equal; petals aging white to yellow. **FR**: fine-papillate. **SEED**: ± white to light brown, smooth. $2n=36$. Uncommon. Coastal bluffs, margins of saline wetlands; < 100 m. SnFrB, SCo, ChI; to AZ, Mex; also Australia; native to s Afr. Apr–Nov

SESUVIUM SEA-PURSLANE

Wayne R. Ferren, Jr.

[Ann] per [shrub], glabrous, gen papillate. **ST**: prostrate to erect, forming mats < 2 m diam; nodes rooting or not. **LF**: opposite, gen < 6 cm; stipule 0; petiole base gen wide with scarious margins; blade linear to ovate, entire. **INFL**: 1-fld [cyme]; bracts 0 or 2. **FL**: hypanthium obconic; sepals gen hooded near tip, ± red adaxially; petals 0; stamens [1]30, often fused at base; ovary 1/2-inferior, chambers 2–5, placentas axile, styles (2)3–5, papillate. **FR**: capsule, circumscissile, ovoid to conic, thin-walled. **SEED**: many, ± reniform, gen smooth, shiny, black or brown, arilled. 8 spp.: gen trop, subtrop coasts, deserts.

S. verrucosum Raf. (p. 169) WESTERN SEA-PURSLANE Branched from base, minute-papillate. **ST**: many, < 9 dm; nodes not rooting. **LF**: 0.5–4 cm, linear to widely spoon-shaped; base clasping st. **INFL**: axillary; pedicel 0 or short. **FL**: sepals 4–10 mm, hooded, beaked, or both near or at tip, abaxially papillate, margins scarious; filaments fused to midlength, ± red. **FR**: 4–5 mm. **SEED**: 0.8–1 mm, smooth. Uncommon. Moist or seasonally dry flats, margins of gen saline wetlands; < 1400 m. NCo, GV, SCoRO, SCo, WTR, PR, GB, D; to OR, KS, S.Am. Apr–Nov

TETRAGONIA

Nancy J. Vivrette

Ann [per, subshrub]; ± papillate. **ST**: prostrate. **LF**: alternate or sometimes basal opposite. **INFL**: fls 1 or 2–3 in cluster, axillary, ± sessile. **FL**: sepals 3–5; petals 0; stamens 1–many; ovary inferior, chambers 3–9, ovule 1, styles 3–9. **FR**: nut-like, 8–10 mm diam, angled, with 2–5 horns. **SEED**: ± pear-shaped [reniform]. 50 spp.: s hemisphere. (Greek: four-angled)

T. tetragonoides (Pall.) Kuntze (p. 169) NEW ZEALAND SPINACH Papillae widely spaced, large, crystalline. **ST**: many, spreading, 3+ dm. **LF**: 2–5(10) cm; petiole winged, 1–3 cm; blade triangular-ovate, base truncate, margin entire, ± wavy. **FL**: gen bisexual, self-fertile, 5 mm; sepals 4–5, spreading, ± yellow. **SEED**: pear-shaped, brown. $2n=32$. Common. Sand dunes, bluffs, margins of coastal wetlands; < 100 m. NCo, CCo, SCo; native to s hemisphere. Apr–Sep

TRIANTHEMA HORSE-PURSLANE

Wayne R. Ferren, Jr.

Ann [per, shrub], branched from base, glabrous, hairy, or papillate, ± fleshy. **ST:** gen prostrate. **LF:** ± opposite, 2 of pair unequal; stipules papery; blade linear to round, base tapered, margin entire. **INFL:** gen 1-fld; bracts 2. **FL:** sepals 5, with abaxial mucro just below tip; petals 0; stamens 5–10[20]; ovary superior, chambers 1–2, ovules 1–few, placentas basal, styles 1–2. **FR:** capsule, papery or leathery, circumscissile; lid winged. **SEED:** reniform, rough, arilled. 20 spp.: trop, subtrop, esp Australia. (Greek: 3-fld)

T. portulacastrum L. (p. 169) **ST:** < 10 dm, in youth with lines of hairs below petioles. **LF:** stipules widened, occ ± toothed at base; blade < 4 cm, smaller on twigs, gen = petiole, elliptic to ± round, base tapered, tip often notched. **INFL:** fl sessile, ± covered by stipule. **FL:** sepals 3–5, ± 2.5 mm, lanceolate, ± purple adaxially; ovary chamber 1, stigmas 2. **FR:** 4–5 mm, cylindric, ± curved; wings of lid 2, prominent, erect. **SEED:** 1.5–2 mm, ridged, red-brown to black. Uncommon. Moist or seasonally dry wetlands, disturbed areas; < 1000 m. SnJV, D; to TX, e N.Am, S.Am; trop Old World. Jun–Nov

AMARANTHACEAE AMARANTH FAMILY

Mihai Costea, except as noted

Ann to subshrub; monoecious and/or dioecious; occ spiny; hairs simple (branched). **LF:** blade simple, alternate or opposite, margins entire or serrate; veins pinnate; stipules 0. **INFL:** axillary or terminal; 3-fld cymes in dense spikes, heads or panicles; bracts 0 or 1–5, persistent; bractlets 0–2. **FL:** bisexual or unisexual, small, green (± white), yellow or purple; perianth parts 0 or (1)3–5, free or fused basally, scarious or hardened, persistent; stamens 1–5, opposite perianth parts, free or basally fused as a tube, gen unequal, occ alternate with appendages on stamen tubes (pseudostaminodes), anthers 2- or 4-chambered; ovary superior, chamber 1; ovule 1 (2–many); style (0)1–3, stigmas 1–3(5). **FR:** utricle; gen with persistent perianth or bracts. **SEED:** 1 [2+], small, lenticular to spheric, smooth or dotted to striate or tubercled. ± 75 genera, 900 spp.: cosmopolitan, esp disturbed, arid, saline or alkaline soils; some cult for food, orn; many naturalized, ruderal or agricultural weeds. [Müller & Borsch 2005 Ann Missouri Bot Gard 92:66–102] Amaranthaceae incl Chenopodiaceae by some. Polycnemoideae, represented in CA by *Nitrophila*, formerly considered subfamily of Chenopodiaceae, but needs further research. *Guilleminea densa* (Willd.) Moq. var. *aggregata* Uline & W.L. Bray is a waif. *Froelichia gracilis* (Hook.) Moq. is an historical waif. Scientific Editors: Douglas H. Goldman, Bruce G. Baldwin.

1. Cauline lvs alternate; fl unisexual, pls monoecious or dioecious . . . . **AMARANTHUS**
1′ Cauline lvs gen opposite; fl bisexual
  2. Hairs forked . . . . **TIDESTROMIA**
  2′ Hairs simple or 0
    3. Lvs fleshy; infl axillary, fls 1–3 . . . . **NITROPHILA**
    3′ Lvs not fleshy; infl axillary, terminal, or both, fls > 3
      4. Perianth parts free . . . . **ALTERNANTHERA**
      4′ Perianth parts fused ≥ 1/2 . . . . **[GUILLEMINEA]**

ALTERNANTHERA

[Ann] per, aquatic to terrestrial. **ST:** prostrate or ascending; hairs simple. **LF:** opposite, subsessile to short-petioled, blade margins entire to serrate. **INFL:** axillary or axillary and terminal, sessile or peduncled, many-fld cylindric spikes or spheric heads; bracts and bractlets persistent, scarious. **FL:** bisexual; perianth parts 5, free, ± equal or not, scarious, persistent, white, glabrous or woolly; stamens 3–5, fused at base into short tube or cup, anthers 2-chambered; pseudostaminodes 5, alternate with stamens; ovary obovoid to ± spheric; style short, ± 0.2 mm; stigma spheric (2-lobed), ovule 1, pendent. **FR:** indehiscent; fr wall membranous. **SEED:** 1, lenticular or ovoid-oblong, ± red brown, smooth, shiny. ± 170 spp.: gen trop Am. (Latin: alternating pseudostaminodes and stamens) *A. sessilis* (L.) DC. occ a greenhouse weed ◆; *A. ficoidea* (L.) P. Beauv. var. *bettzickiana* (Regel) Backer widely cult, CA reports unconfirmed.

1. St not hollow; infl axillary, sessile; perianth parts woolly . . . . ***A. caracasana***
1′ St hollow; infl axillary and terminal, peduncled; perianth parts glabrous . . . . ***A. philoxeroides***

A. caracasana Kunth Terrestrial. **ST:** prostrate, mat-forming, 10–50 cm, woolly, ± glabrous in age. **LF:** petiole 2–10 mm; blade 5–25 mm, 3–15 mm wide, rhombic-ovate to obovate, tip rounded, short-pointed, sparsely woolly, ± glabrous in age. **INFL:** 5–8 mm, 4–6 mm wide, ovoid, white to straw-colored; bracts 3–3.5 mm, ovate to lanceolate, keeled, midrib hairy, acuminate to awned. **FL:** perianth parts unequal, spine-tipped, outer 3.5–5 mm, ± lanceolate, not keeled, 3-veined abaxially, inner 2.5 mm, ± linear, keeled, gen 1-veined; stamens 5, filaments 0.2–0.4 mm, pseudostaminodes < stamens, triangular or awl-like, entire to dentate. **FR:** ± 1.5 mm, ovoid-spheric, brown, tip ± truncate. **SEED:** 1–1.5 mm, ovate-spheric. Gen disturbed areas; < 150 m. SCo; to se US, c Mex, also Afr; native to C.Am, S.Am. Aug–Oct

A. philoxeroides (Mart.) Griseb. (p. 169) ALLIGATOR WEED Aquatic to semi-terrestrial, stoloniferous. **ST:** prostrate, mat-forming, 50–120 cm, young sts and lf axils white-hairy, in age ± glabrous or hairs in 2 lines. **LF:** petiole 2–10 mm; blade 30–130 mm, 5–35 mm wide, elliptic to obovate, glabrous or ± hairy near wedge-shaped base, tip obtuse or acute. **INFL:** 1.2–1.7 cm diam, spheric; peduncle 1–9 cm; bract 2–2.5 mm, ovate to lanceolate, acute to acuminate, not keeled, glabrous. **FL:** perianth parts 5–7 mm, ± equal, lanceolate or oblong, 1-veined, tip acute; stamens 5, filaments 2.5–3 mm, pseudostaminodes ± = stamens, oblong-linear; ovary obovoid. **FR:** unknown in N.Am. 2*n*=100. Shallow water (< 2 m) or wet soils, marshes, edges of ponds; < 200 m. SnJV, SCo; to se US, C.Am; native to S.Am. Only vegetative reproduction in N.Am. May–Oct ◆

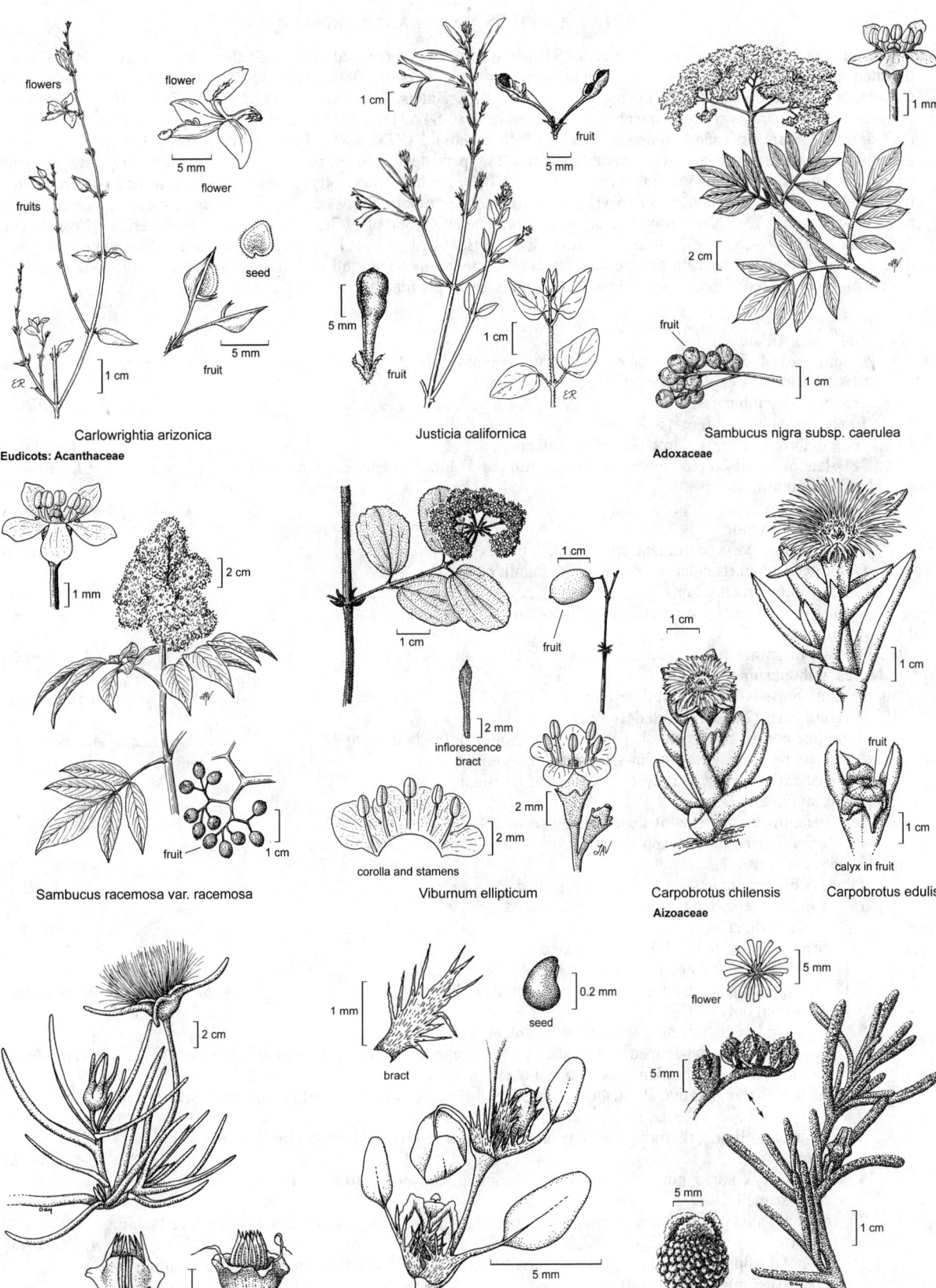

Carlowrightia arizonica
Eudicots: Acanthaceae
flowers
flower
5 mm
flower
fruits
seed
5 mm
fruit
1 cm

Justicia californica
1 cm
fruit
5 mm
5 mm
fruit
1 cm

Sambucus nigra subsp. caerulea
Adoxaceae
1 mm
2 cm
fruit
1 cm

Sambucus racemosa var. racemosa
1 mm
2 cm
fruit
1 cm

Viburnum ellipticum
1 cm
fruit
2 mm
inflorescence bract
2 mm
2 mm
corolla and stamens

Carpobrotus chilensis
1 cm
1 cm

Carpobrotus edulis
Aizoaceae
fruit
1 cm
calyx in fruit

Conicosia pugioniformis
2 cm
fruit long-section
fruit
1 cm

Cypselea humifusa
1 mm
0.2 mm
seed
bract
5 mm

Mesembryanthemum crystallinum
flower
5 mm
5 mm
5 mm
calyx in fruit

Mesembryanthemum nodiflorum
1 cm

AMARANTHUS AMARANTH, PIGWEED

Ann (short-lived per); monoecious or dioecious. **ST**: prostrate to erect, branched or not. **LF**: alternate, petioled, ovate to linear, tip often ± notched (2-lobed), midvein ending in sharp point, margin entire, flat or wavy. **INFL**: 3-fld cymes, in dense axillary clusters, or large, terminal, panicle- or spike-like infl; bract 1, bractlets 0–2, alike, persistent, spine-like to ± lf-like or ± membranous, at least margins scarious-membranous. **FL**: unisexual. **STAMINATE FL**: perianth parts (2)3–5, scarious; stamens (1)2–5, filaments free; pseudostaminodes 0; anthers 4-chambered. **PISTILLATE FL**: perianth parts (1)3–5, membranous or scarious, free, persistent; ovary ovoid, style 0, stigmas 2–3, persistent, slender, papillate; ovule 1, erect. **FR**: circumscissile or indehiscent, ovoid to obovoid, smooth or ± wrinkled, tip ± gradually (abruptly) narrowed to stigmas or beak; walls thin, membranous. **SEED**: 1, lenticular to ± spheric, round to obovate, smooth, shiny, occ obscure-dotted or -net-like, ± white-ivory to brown-red or black. ± 70 spp.: worldwide; weeds, ornamentals, food pls. (Greek: unfading, non-withering) [Costea et al. 2001 Sida 19:931–974, 975–992; Sauer 1967 Ann Missouri Bot Gard 54:103–137]. Hybrids common, F1 gen with numerous, densely packed bractlets beneath gen sterile pistillate fls, abnormal-shaped infl with dense, twisted or fan-shaped branches. Unless otherwise noted, descriptions of bracts and fl parts are of pistillate fls.

1. Dioecious
  2. Pl in hand pistillate
    3. Perianth parts 1–2, 1 narrowly lanceolate, 1 rudimentary or 0 . . . . ²*A. tuberculatus* var. *rudis*
    3′ Perianth parts 5, well developed, fan- or spoon-shaped
      4. Bracts ≤ perianth parts. . . . . ²*A. arenicola*
      4′ Bracts > than perianth parts
        5. Pl ± glabrous; outer perianth parts ± acuminate . . . . ²*A. palmeri*
        5′ Pl glandular-hairy esp on bracts; outer perianth parts obtuse to ± notched, mucronate . . . . ²*A. watsonii*
  2′ Pl in hand staminate
    6. Bracts ≤ 2 mm . . . . ²*A. tuberculatus* var. *rudis*
    6′ Bracts (1.4)2–6 mm
      7. Outer perianth parts acuminate, spine-tipped; bracts ≥ perianth parts . . . . ²*A. palmeri*
      7′ Outer perianth parts mucronate; bracts ≤ perianth parts
        8. Infl axis and bracts glabrous . . . . ²*A. arenicola*
        8′ Infl axis and bracts gen sparsely glandular-hairy . . . . ²*A. watsonii*
1′ Monoecious
  9. Nodes with 2 spines . . . . [*A. spinosus*]
  9′ Nodes without spines
    10. Perianth parts 1–3, only 1 well developed. . . . . *A. californicus*
    10′ Perianth parts (2)3–5, all well developed.
      11. Pistillate perianth parts fan-shaped, margins fringed or finely dentate . . . . *A. fimbriatus*
      11′ Pistillate perianth parts not fan-shaped, margins entire
        12. Infl of axillary clusters only, or axillary and terminal
          13. Perianth parts (3)4–5
            14. Perianth parts narrow-ovate to oblong; seeds 1.3–1.7 mm. . . . . *A. blitoides*
            14′ Perianth parts narrow-spoon-shaped; seeds 0.9–1.1 mm. . . . . *A. torreyi*
          13′ Perianth parts (2)3
            15. Fr dehiscent; bracts 1.5–4 mm, tip ± spined . . . . *A. albus*
            15′ Fr indehiscent; bracts ≤ 1 mm, tip not spined
              16. St hairy distally; fr 2–3 mm, smooth . . . . *A. deflexus*
              16′ St ± glabrous; fr 1.2–1.5 mm, wrinkled
                17. Lf tip entire or ± notched; seeds dull, dotted . . . . [*A. viridis*]
                17′ Lf tip deeply notched to 2-lobed; seeds shiny, only margins dotted . . . . [*A. blitum* subsp. *emarginatus*]
        12′ Infl terminal only
          18. Infl green; bracts > stigma; seeds brown to black; often weedy
            19. Perianth parts spoon-shaped to obovate, overlapping, reflexed in age, tip rounded to ± notched . . . *A. retroflexus*
            19′ Perianth parts linear to lance-oblong, not overlapping, erect, tip acute to awned
              20. Lf blades rhombic-ovate or widely lanceolate; infl of many thin, flexible branches; bracts 2–4 mm
                . . . . *A. hybridus*
              20′ Lf blades elliptic, rhombic to widely lanceolate; infl of few rigid, erect branches; bracts 4.5–6(8) mm. . . . . *A. powellii*
          18′ Infl brightly colored, not green; bracts gen ≤ stigma; seeds white to ± red-brown or black; cult, rarely escaped
            21. Infl main axis gen drooping or nodding; perianth parts widely spoon-shaped, margins overlapping, tip obtuse to rounded. . . . . [*A. caudatus*]
            21′ Infl not drooping or nodding; perianth parts linear to elliptic (obovate), margins not overlapping, acute to acuminate (rounded)
              22. Infl erect to flexible, lateral axes thin, flexible, 3–8(12) cm, 0.5–1 cm wide; bracts 1.8–2.5 mm; fr tip abruptly narrowed to a beak . . . . [*A. cruentus*]
              22′ Infl erect, lateral axes thick, stiff, 9–20(30) cm, 1.2–2 cm wide; bracts 3–5(6) mm; fr tip gradually narrowed. . . . . [*A. hypochondriacus*]

A. albus L. (p. 169) TUMBLEWEED Monoecious. **ST:** prostrate to erect, 10–100 cm, gen many-branched, ± green, white when dry, moderately to densely woolly distally. **LF:** petiole 5–40 mm; blade elliptic to obovate or narrow-spoon-shaped, base wedge-shaped, tip obtuse to acute, margins flat or ± wavy; st lvs gen deciduous, 40–80 mm, 15–30 mm wide, replaced by axillary lvs, 7–20 mm, 3–10 mm wide. **INFL:** axillary clusters only, green; bracts 1.5–4 mm, 1.5–2 × perianth, lance-linear to awl-like, tip ± spined. **FL:** staminate fls intermixed with pistillate, perianth parts 3, stamens 3; pistillate perianth parts 3, 0.7–2 mm, ± equal, lance-oblong to linear, tip acute; stigmas 3, erect. **FR:** circumscissile; 1.5–2 mm, ≥ perianth, ellipsoid-ovoid, green-white to brown, wrinkled esp near tip. **SEED:** 0.8–1.1 mm, lenticular, red-brown, smooth, shiny. 2*n*=32. Disturbed areas, roadsides, riverbanks, sandy places, agricultural fields; < 2500 m. CA; N.Am; cosmopolitan; native to c N.Am. Jun–Oct

A. arenicola I.M. Johnst. SAND AMARANTH Dioecious. **ST:** erect, 4–100(200) cm, gen many-branched, green-white, ± glabrous. **LF:** petiole 10–70 mm; blade 15–75 mm, 5–30 mm wide, narrow-ovate to elliptic or lanceolate, base wedge-shaped to ± round, tip ± acute to obtuse, margins flat to irregular-wavy. **INFL:** terminal (and axillary), erect to nodding, spike- or panicle-like, main axes 15–60 cm, 1.1–2.2 cm wide (incl fls), green; staminate bracts 2–3.5 mm, < perianth parts, tip acute to acuminate; pistillate bracts (1.4)2–2.5 mm, ≤ perianth, ovate to lance-ovate, tip acute to acuminate with midrib short-elongated, not spiny. **FL:** staminate perianth parts 5, 2–3.5 mm, ± equal, tip obtuse to ± acute, inner perianth parts ± mucronate, stamens 5; pistillate perianth parts 5, 1.6–2.5 mm, spoon-shaped, tip obtuse, mucronate; stigmas 2–3, ± erect. **FR:** circumscissile; ± 2 mm, < perianth, ± spheric, smooth, ± brown. **SEED:** 0.9–1.2 mm, lenticular, dark red-brown, smooth, shiny. Sandy soils; < 200 m. ScV, CCo, SCo; native to c&e US. Aug–Sep

A. blitoides S. Watson PROCUMBENT PIGWEED Monoecious. **ST:** prostrate, mat-forming, 30–120 cm, green or occ purple, glabrous. **LF:** petiole 3–15 mm; blade (5)10–30(40) mm, 4–10(15) mm wide, obovate to spoon-shaped, base wedge-shaped, tip obtuse, rounded, margins flat (± wavy). **INFL:** axillary clusters, green; bracts ± lf-like, 1.5–3 mm, ± ≥ perianth, narrowly ovate, tip acute. **FL:** staminate perianth parts (3)4, 1.5–2.5 mm; stamens (3)4–5; pistillate perianth parts 4–5, 1.3–3.5 mm, narrowly ovate to oblong, unequal, outer 2 > others, 2–3 mm; stigmas gen 3, spreading. **FR:** circumscissile; 2–2.5 mm, ± spheric to obovoid, green to occ purple, ± wrinkled near tip. **SEED:** 1.3–1.7 mm, lenticular, black. 2*n*=32. Disturbed areas, roadsides, agricultural fields, sandy soils; < 2500 m. CA-FP, W&I, D; to WA, e US, c&s Eur. Jul–Nov

A. californicus (Moq.) S. Watson (p. 169) CALIFORNIAN AMARANTH Monoecious. **ST:** prostrate, mat-forming, 10–50 cm, green, glabrous. **LF:** petiole 3–8 mm; blade 3–20(30) mm, 2–1.5 mm wide, obovate to oblanceolate, base wedge-shaped, tip obtuse to ± acute, margins flat or ± wavy. **INFL:** axillary clusters, green, bracts 0.7–1 mm, ± = perianth parts, linear, ± spine- or awl-like. **FL:** staminate perianth parts (2)3, 1–1.5 mm, stamens 1–2; pistillate perianth parts 1–3, only 1 well developed, 1–1.2 mm, lance-linear, acute to acuminate; stigmas 3, erect. **FR:** circumscissile; 1–1.2 mm, ± spheric to ovoid, ± smooth or ± wrinkled near tip. **SEED:** 0.8–1.1 mm, lenticular, obovate, dark red-brown, shiny, smooth. Seasonally moist flats, lake margins, disturbed areas; < 2800 m. CA; to s Can, w TX. Jul–Oct

A. deflexus L. Short-lived per; monoecious. **ST:** prostrate to ascending, 10–50 cm, branched at base, ± brown or occ green, hairy distally. **LF:** petiole 12–30 mm; blade (5)10–30(50) mm, 5–15(25) mm wide, rhombic-ovate to lanceolate, base tapered, wedge-shaped, tip ± acute or obtuse, margins flat or ± wavy. **INFL:** axillary and terminal, spike-like or with few axes, erect to flexible, main axis 2–8 cm, 1.5–2.5 cm wide (incl fls), green or ± brown; bracts 0.5–1 mm, < perianth, membranous, triangular-ovate. **FL:** staminate perianth parts 2–3, 1.2–1.4 mm, stamens 2–3; pistillate perianth parts 2–3, 1–2 mm, < fr, narrow-oblong to linear; stigmas 3, erect. **FR:** indehiscent; 2–3 mm, ovoid, inflated, brown, smooth. **SEED:** 1–1.1 mm, elliptic to obovate, dark-brown to black, smooth, shiny. 2*n*=34. Railroad right-of-ways, disturbed areas; < 650 m. ScV, CW, SW; to e US, native to S.Am, Eur, Asia. May–Nov

A. fimbriatus (Torr.) S. Watson (p. 169) FRINGED AMARANTH Monoecious. **ST:** erect or ascending, 30–100 cm, gen branched, green or purple, glabrous. **LF:** petiole 8–40 mm; blade (10)20–60(100) mm, 1–5(10) mm wide, linear to narrow-lanceolate, base narrowly wedge-shaped, tip acute, margins flat. **INFL:** axillary and terminal, spike-like, ± flexible to nodding, main axis 10–30 cm, 1–2 cm wide (incl fls), green-silver to pink-purple; bracts 1–1.8 mm, < perianth, membranous, ovate to lance-ovate. **FL:** staminate fl gen at infl tips, perianth parts 5, stamens (2)3; pistillate perianth parts 5, ± equal, 1.5–3.3 mm, reflexed, fan-shaped, tip obtuse, margins finely dentate to minutely toothed; stigmas 3(4) erect to ± spreading. **FR:** circumscissile; 1.3–2 mm, ± spheric to ovoid, green to tan, wrinkled esp near tip. **SEED:** 0.9–1 mm, lenticular, round, dark red-brown to black, smooth, shiny. 2*n*=34. Sandy, gravelly slopes, washes, esp after summer rains; 600–1700 m. D; to s Can, MT, w TX, Baja CA. Aug–Nov

A. hybridus L. SMOOTH PIGWEED, GREEN AMARANTH Monoecious. **ST:** erect, 30–120(200) cm, branched or not, green, hairy distally. **LF:** petiole 30–70 mm; blade (30)50–120 mm, (10)20–60 mm wide, rhombic-ovate or wide-lanceolate, base wedge-shaped, tip acute, margins flat. **INFL:** terminal, panicle-like, erect or nodding, branches many, thin, flexible, 2.5–7(12) cm, 0.5–1 cm wide (incl fls), green or olive green; bracts 2–4 mm, 1.2–2 × perianth, spine-like, lance-linear. **FL:** staminate fls gen at infl tips, perianth parts 5, 2–3 mm, stamens 5; pistillate perianth parts 5, 1.5–2.5(3) mm, ± = or 1 longer, erect, lance-oblong to linear, tip acute or acuminate; stigma branches 3, erect. **FR:** circumscissile; 1.5–2.2 mm, ovoid to obovoid, irregular-dehiscent or indehiscent, occ on same pl, lid strongly wrinkled at base. **SEED:** 0.9–1.4 mm, lenticular, widely elliptic to round, brown to black, smooth to finely dotted on margin, shiny. 2*n*=32. Disturbed areas, agricultural fields; < 300 m. GV, CW, SW; US, s Can; native to e N.Am. Naturalized in trop, subtrop, and temp regions worldwide. Jun–Nov

A. palmeri S. Watson (p. 169) PALMER'S AMARANTH Dioecious. **ST:** erect, 20–200(250) cm, branched or not, green, ± glabrous. **LF:** petiole 60–90 mm; blade 15–150 mm, 10–35 mm wide, ovate, rhombic-ovate to lanceolate, base ± wedge-shaped, tip ± obtuse to acute, margins flat. **INFL:** terminal, simple or panicle-like, ± flexible to drooping, main axis 10–35(50) cm, 1–1.8(2.5) cm wide (incl fls), green; staminate bracts 4–6 mm, ≥ outer perianth parts, spine-like; pistillate bracts 4–6 mm, 1.3–2 × perianth, spine-like, lance-ovate. **FL:** staminate perianth parts 5, unequal, outer (2.5)3–6 mm, acuminate, spine-tipped, inner 2.5–3 mm, obtuse, notched, stamens 5; pistillate perianth parts 5, unequal, outer 3–4 mm, lanceolate, ± acuminate, inner 1.7–2.5 mm, obovate to spoon-shaped, rounded to ± notched and mucronate; stigmas 2(3), spreading. **FR:** circumscissile; 1.5–2 mm, < perianth, ± spheric to obovoid, brown to occ red-brown, smooth or ± rough near tip. **SEED:** 1–1.4 mm wide, lenticular, elliptic to obovate, dark red-brown to black, smooth, shiny. 2*n*=32, 34. Disturbed areas, agricultural fields; < 1200 m. GV, CW, SCo, WTR, D; to c US, c Mex; introduced to Eur, Asia, Australia. Aug–Nov

A. powellii S. Watson (p. 169) POWELL'S AMARANTH Monoecious. **ST:** erect, 25–150(200) cm, branched or not, green with purple stripes, ± glabrous. **LF:** petiole 30–60 mm; blade 30–80 mm, 20–50 mm wide, elliptic to rhombic or widely lanceolate, base wedge-shaped, margins flat, tip ± obtuse. **INFL:** terminal, erect, gen with a few rigid-erect branches (well-branched, dense panicle), (3)4–12 cm, 1.2–1.6 cm wide (incl fls); green to pale green; bracts 4.5–6(8) mm, 2–3(4) × > perianth, spine-like, narrowly triangular to lanceolate. **FL:** staminate gen at infl tips, perianth parts 3–5, ± equal, lanceolate, acute, stamens 3–5; pistillate perianth parts 3–5, unequal, longest 2.3–3.5 mm, others 1.2–2 mm, oblong to lance-linear, tip acute to awned; stigmas 2–3, spreading. **FR:** circumscissile or indehiscent; 2–3 mm, ovoid, gen > perianth parts, ± wrinkled near tip, ± smooth when indehiscent, gray-cream to light brown. **SEED:** 0.9–1.3 mm, lenticular, elliptic, ovate, dark brown to black, smooth, shiny. 2*n*=32,34. Agricultural fields, disturbed areas; < 1500 m. NCoRO, CaRH, SNF, n&c SNH, GV, CW, SW, W&I, MP; native to sw US, adjacent Mex; ± cosmopolitan weed. Jun–Oct

A. retroflexus L. REDROOT PIGWEED Monoecious. **ST:** erect, 15–150(200) cm, branched or not, green or red, densely to moderately woolly. **LF:** petiole 25–80 mm; blade 20–150 cm, 10–70 cm

wide, ovate to rhombic-ovate, base wedge-shaped, tip obtuse to acute. **INFL:** terminal, panicle-like, gen erect, many-branched, lateral axes short, thick, 2–5(10) cm, 1.2–2.5 cm wide (incl fls), green or silver-green; bracts (2.5)4–6(8) mm, 1.5–2(3) × > perianth, spine-like, lanceolate to awl-shaped. **FL:** staminate perianth parts 4–5, 2–3 mm, ± equal, stamens (3)4–5; pistillate perianth parts 5, 2–3.2 mm, ± equal, spoon-shaped to obovate, overlapping, reflexed in age, tip rounded to ± notched, mucronate; stigmas 3, reflexed. **FR:** circumscissile; 1.7–2.5 mm, < perianth, ovoid to obovoid, lid ± wrinkled near base, gray-cream to ± brown. **SEED:** 1–1.3 mm, lenticular, obovate, dark-brown to black, smooth, shiny. 2*n*=34. Agricultural fields, disturbed areas; < 2400 m. NCo, NCoRI, GV, CW, SW, SNE; native to c&e N.Am; cosmopolitan weed. Jun–Nov

A. torreyi (A. Gray) S. Watson (p. 169) TORREY'S AMARANTH Monoecious. **ST:** erect or ascending, 10–70 cm, many-branched, green. **LF:** petiole 10–30 mm; blade (12)15–50(70) mm, 3–20 mm wide, oblanceolate to lanceolate or ovate, base narrowly wedge-shaped, tip acute to ± obtuse, mucronate, margins flat or ± wavy. **INFL:** axillary, clusters distally grouped in a lfy spike (tip occ ± lfless), green; bracts 1.3–3.5 mm, ± > perianth parts, lanceolate to awl-like. **FL:** staminate perianth parts 5, 2.5–4.4 mm, ± equal, stamens 3(5); pistillate perianth parts 5, 1.4–2.5 mm, narrowly spoon-shaped, tips obtuse, rounded to ± notched; stigmas 3, erect. **FR:** circumscissile; 1.4–2 mm, ≤ perianth parts, ± spheric to obovoid; lid coarse-wrinkled near base. **SEED:** 0.9–1.1 mm, lenticular, dark brown to black, smooth, shiny. Sandy flats, arroyos, after late-summer rain; 1200–1700 m. DMoj; w AZ. Aug–Oct

A. tuberculatus (Moq.) J.D. Sauer var. *rudis* (J.D. Sauer) Costea & Tardif COMMON WATERHEMP Dioecious. **ST:** erect or ascending, 50–150(200) cm, branched or occ not, green or purple, ± glabrous. **LF:**

petiole 20–60 mm; blade 15–150 mm, 5–30 mm wide, ovate, oblong, elliptic to narrowly lanceolate distally, base narrowly wedge-shaped, margins flat, tip obtuse to acute. **INFL:** terminal, ± erect, linear spikes to dense panicles, main axis occ interrupted, lfy or not, 3–15 cm, 0.5–1.2 cm wide, green to purple; staminate bracts 1.5–2 mm, < perianth, triangular-ovate, midrib less prominent; pistillate bracts 1.5–2.2 mm, = perianth, spine-like, lance-ovate. **FL:** staminate perianth parts 5, ± equal, inner 2–2.5 mm, elliptic-obovate, obtuse or notched, outer 2.5–3 mm, lance-elliptic, tip acuminate, mucronate, stamens 5; developed pistillate perianth parts 1–2, others reduced; stigmas 3, erect. **FR:** circumscissile; 1.5–2 mm, obovoid, wrinkled, green-brown or ± red. **SEED:** 0.8–1 mm, lenticular, elliptic to obovate, dark red-brown, smooth, shiny. 2*n*=34. Disturbed areas, agricultural fields; < 250 m. ScV, SCo; US, Can, Eur; native to c US. Jul–Oct

A. watsonii Standl. (p. 169) WATSON'S AMARANTH Dioecious. **ST:** erect, 10–100 cm, many-branched, green, glandular-hairy esp on bracts. **LF:** petiole 15–60 mm; blade 10–80 mm, 5–40 mm wide, ovate to obovate or oblong-elliptic, base wide-wedge shaped to ± round, tip obtuse, margins flat or ± wavy. **INFL:** terminal with some proximal axillary clusters, erect, spike- or panicle-like, main axis gen thick, uninterrupted, 3–15 cm, 1–2 cm wide (incl fls); staminate bracts 2.2–4 mm, ± = outer perianth parts, spine-like; pistillate bracts 3–4.2 mm, 1.3–2 × > perianth, spine-like, ovate, glandular. **FL:** staminate perianth parts 5, 1.5–2(3) mm, ± equal, tip acute to ± obtuse, stamens 3–5; pistillate perianth parts 5, 1.7–2.2 mm, ± equal, spoon- to fan-shaped, margin jagged, tip obtuse to ± notched, mucronate; stigmas 2–3, spreading. **FR:** circumscissile; 1.5–2 mm, < perianth, ± spheric to obovoid, ± smooth, light-brown to brown. **SEED:** 0.9–1.2 mm, lenticular, ovate to round, red-black, smooth, shiny. Coastal dunes, beaches, sandy inland areas, ± saline flats, disturbed areas after winter rain; < 100 m. SnGb, s DSon (Imperial Co.); AZ, nw Mex. Aug–Sep ★

NITROPHILA

Margriet Wetherwax, Dieter H. Wilken & Noel H. Holmgren

Per, glabrous, rhizomed. **ST:** decumbent to erect, ± ribbed. **LF:** opposite, sessile to clasping, linear to ovate, fleshy. **INFL:** axillary, fls 1–3; bractlets 1 or 2 per fl. **FL:** calyx 5(7)-parted, enclosing fr, papery, lobes erect, overlapping, ovate, back ribbed, persistent in fr; stamens 5, incl; style slender, = ovary, stigmas 2. **FR:** indehiscent, ± 2 mm. **SEED:** vertical, lenticular, black or brown. x=9. ± 5 spp.: w US, C.Am, S.Am. (Greek: soda loving) [Müller & Borsch 2005 Ann Missouri Bot Gard 92:66–102] In Chenopodiaceae in TJM (1993).

1. Internodes < lvs, lvs (2)3–4.5 mm, widely ovate; pl 3–10 cm . . . . ***N. mohavensis***
1′ Internodes > lvs, lvs 5–30 mm, ± linear; pl 7–30 cm . . . . ***N. occidentalis***

N. mohavensis Munz & J.C. Roos (p. 169) AMARGOSA NITROPHILA **ST:** many, gen erect. **LF:** ± clasping. **INFL:** bracts 2; fls 1 in axils, sessile. **FL:** calyx 2–3.5 mm, ± pink. **SEED:** ± 1 mm, shiny. Alkaline flats; 300–750 m. n DMoj (Amargosa Desert); w NV. May–Nov ★

N. occidentalis (Moq.) S. Watson (p. 169) **ST:** decumbent to erect. **LF:** sessile, occ scale-like on decumbent sts. **INFL:** bracts 2; fls 1, sessile, or 2–3, short-pedicelled. **FL:** calyx 2–3.3 mm, pink, white in age. **SEED:** ± 1 mm, black or brown, dull. Moist, alkaline soils; < 2100 m. KR/CaRH (Shasta Valley), GV, SCo, w PR, GB, DMoj; to e OR, UT, n Mex. May–Oct

TIDESTROMIA

Ann or subshrub, canescent to densely woolly, or glabrous in age; hairs forked. **LF:** gen opposite, occ alternate at base, sessile or short-petioled, petiole ≤ 2.5 mm; blade round to lanceolate, cordate, base wedge-shaped or oblique, tip acute to obtuse, margins entire, papery to fleshy. **INFL:** axillary head-like cymes, fls 1–3(5), sessile, ± enclosed by 2 ± opposite lvs hardened and fused in age; bracts, bractlets persistent, ovate, scarious, woolly. **FL:** bisexual; perianth parts 5, free, keeled, outer 3 > inner 2, scarious or leathery, glabrous or woolly, tip acute or obtuse; stamens 5, filaments fused at base into short tube, anthers 2-chambered, pseudostaminodes 0 or short, triangular, ± 0.2 mm; ovary ± spheric, ovule 1, style 0 or short, ± 0.1 mm, stigmas 2-lobed, deltoid (irregularly 3-lobed). **FR:** ± spheric, wall membranous. **SEED:** obovoid, brown-red. 7 spp.: w N.Am, West Indies, deserts. (Ivar T. Tidestrom, Swedish-born botanist of sw USA, 1864–1956)

1. Ann; buds at st base 0 . . . . ***T. lanuginosa***
1′ Subshrub; buds gen present at st base . . . . ***T. suffruticosa*** var. ***oblongifolia***

T. lanuginosa (Nutt.) Standl. WOOLLY TIDESTROMIA **ST:** 10–50 cm, prostrate to ascending; canescent to woolly, occ glabrous in age. **LF:** 6–32 mm, 9–30 mm wide, gray-green. **INFL:** 1–3-fld; involucre lf blades 3–40 mm, 2–31 mm wide, ovate, obovate, lanceolate, papery to fleshy; bracts 1–1.5 mm, 0.8–1 mm wide; bractlets

1–1.5 mm, 0.6–0.8 mm wide, tip obtuse. **FL:** perianth 1.5–2.8 mm, ± yellow; staminal tube (0.3)0.5–1 mm, filaments 0.5–1.5 mm. **FR:** 1.3–1.5 mm. **SEED:** 1–1.4 mm. Slopes, gravelly to sandy soils; ± 1200 m. e DMtns (Granite Mtns); to CO, TX, Mex. Jul–Oct

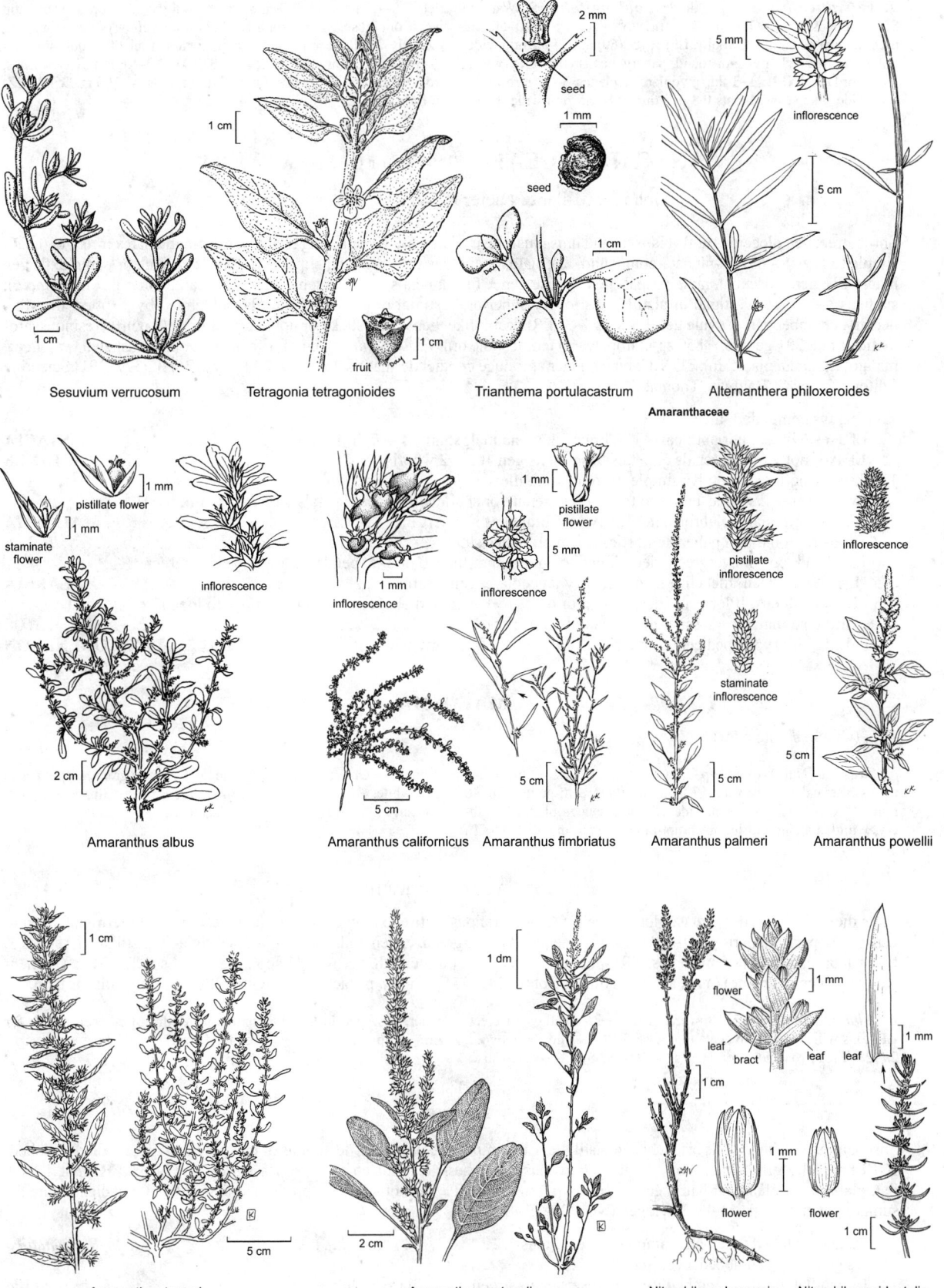

Sesuvium verrucosum

Tetragonia tetragonioides

Trianthema portulacastrum

seed

2 mm

seed

1 mm

1 cm

1 cm

fruit

1 cm

Alternanthera philoxeroides

inflorescence

5 mm

5 cm

Amaranthaceae

pistillate flower

1 mm

staminate flower

1 mm

inflorescence

inflorescence

1 mm

inflorescence

5 cm

Amaranthus albus

2 cm

5 cm

Amaranthus californicus

pistillate flower

1 mm

inflorescence

5 mm

Amaranthus fimbriatus

5 cm

pistillate inflorescence

staminate inflorescence

5 cm

Amaranthus palmeri

inflorescence

5 cm

Amaranthus powellii

Amaranthus torreyi

1 cm

5 cm

Amaranthus watsonii

2 cm

1 dm

Nitrophila mohavensis

flower

leaf

bract

leaf

1 mm

1 cm

flower

1 mm

flower

1 mm

Nitrophila occidentalis

leaf

1 mm

1 cm

T. suffruticosa (Torr.) Standl. var. ***oblongifolia*** (S. Watson) Sánch.Pino & Flores Olv. (p. 177) HONEYSWEET **ST:** 10–60 cm, ascending or decumbent, woolly. **LF:** 10–50(60) mm, 4–22 mm wide, oblong to narrowly ovate or round, papery, persistently gray-white, base cordate. **INFL:** 1–3-fld; involucre lf blades 3–25 mm, 2–20 mm wide, like st lvs; bracts 0.8–1.5 mm, 0.8–1.5 mm wide, bract-lets 0.8–1.5 mm, 0.6–1.5 mm wide, woolly, occ glabrous in age, tip acute or obtuse. **FL:** perianth 1.7–2.7 mm, ± yellow or yellow-brown, densely woolly, occ glabrous in age; stamen tube 0.3–0.8 mm; filaments 0.4–0.8 mm. **FR:** 1–2 mm. **SEED:** 0.9–1.6 mm. Dry washes, rocky hillsides, sandy or slightly alkaline soils; < 1200 m. D; NV, AZ, n Baja CA. [*T. o.* (S. Watson) Standl.] Apr–Dec

ANACARDIACEAE SUMAC or CASHEW FAMILY

John M. Miller & Dieter H. Wilken, except as noted

Shrub, tree; gen dioecious or fls bisexual and unisexual; resin clear, often weathering black, gen aromatic, latex milky or 0. **LF:** simple to ternate- or odd-pinnate-compound, alternate, deciduous or evergreen; stipules 0. **INFL:** raceme or panicle; fls gen many. **FL:** gen unisexual, radial; sepals gen 5, base gen ± fused; petals 5, gen > sepals, free; (perianth parts 1–7 in *Pistacia*); stamens 4–7 or 10, vestigial in pistillate fls; ovary superior, vestigial or 0 in staminate fls, subtended by ± lobed, disk-like nectary, chamber gen 1, ovule gen 1, styles 1–3. **FR:** drupe-like, gen ± flat, sticky or not, hairs short or 0; pulp ± resinous, aromatic or not. 70+ genera, ± 850 spp.: trop, warm temp; some orn (*Rhus, Schinus*), cult for fr (*Anacardium*, cashew; *Mangifera*, mango; *Pistacia*, pistachio). TOXIC: many genera produce contact dermatitis. [Yi et al. 2007 Syst Bot 32:379–391] Scientific Editors: Bruce G. Baldwin, Thomas J. Rosatti.

1. Tree; lvs compound, lflets 7–16+
  2. Lf axes winged; perianth parts 1–7, bract-like, unequal; stamens 4–7, in 1 whorl . . . . **PISTACIA**
  2′ Lf axes not winged; sepals 5, petals 5; stamens gen 10, in 2 whorls . . . . **SCHINUS**
1′ Shrub, vine-like or not; lvs simple to compound, lflets 3–5
  3. Lvs evergreen, simple, entire to toothed or crenate or shallowly 3-lobed; petals white to ± pink
    4. Fr 2–3 mm diam, glabrous, not sticky; infl branches slender . . . . **MALOSMA**
    4′ Fr 6–10 mm diam, puberulent, sticky; infl branches stout . . . . ²**RHUS**
  3′ Lvs deciduous or evergreen, deep-lobed to compound, lflets 0 or 3–5; petals yellow to yellow-green
    5. Lvs ± evergreen, lflets linear to narrowly lanceolate, gen entire . . . . **SEARSIA**
    5′ Lvs deciduous, lf lobes or lflets oblong or ovate to ± diamond-shaped or ± round, entire to lobed
      6. Infl terminal; fls ± sessile; lflets adaxially dull; fr red . . . . ²**RHUS**
      6′ Infl axillary; fls pedicelled; lflets adaxially shiny; fr creamy white . . . . **TOXICODENDRON**

MALOSMA

1 sp. (Greek: strong odor)

M. laurina (Nutt.) Abrams (p. 177) LAUREL SUMAC Shrub, 2–6 m; fls bisexual or unisexual. **LF:** simple, evergreen; petiole 10–40 mm; blade 3–10 cm, 2–4.5 cm wide, elliptic to lance-oblong, ± leathery, ± folded along midrib, tip abrupt-pointed, margin entire. **INFL:** branches slender; bractlets < 1.5 mm. **FL:** sepals green, entire; petals gen white. **FR:** 2–3 mm diam, glabrous, ± white. Slopes, canyons, chaparral; < 1000 m. SW; Baja CA. Jun–Jul

PISTACIA PISTACHIO

Tree; dioecious. **LF:** deciduous; lflets [3]7–9[16], membranous, entire to toothed. **INFL:** panicle, axillary or terminal, open to dense. **FL:** perianth parts gen 1–7, bract-like, unequal, brown-green, ephemeral; stamens 4–7, vestigial in pistillate fls; styles 3, fused at base, gen 0 in staminate fls. **FR:** spheric to obovoid, ± purple; pulp fleshy. ± 11 spp.: Medit, e Asia, TX, Mex. (Ancient Arabic or Persian name) [Yi et al. 2008 Amer J Bot 95:241–251] *P. vera*, pistachio, gen with 3 lflets, widely cult for food.

P. atlantica Desf. Pl 3–10 m. **ST:** branches spreading to erect. **LF:** axis winged; lflet tip acute to obtuse. **FR:** 6–8 mm, ± obovoid. 2*n*=28. Flats, roadsides, drainages; < 100 m. ScV, expected else-where; escaped in UT, TX; native to Medit, Middle East. Cult for orn, escaping; used as rootstock for *P. vera* L. Feb–Apr

RHUS

Shrub, tree; dioecious or fls bisexual and pistillate. **LF:** simple or compound, deciduous or evergreen, entire, toothed, or lobed. **INFL:** panicle, terminal on short twigs, open to dense; fls ± sessile. **FL:** stamens 5; styles 3, free or ± fused. **FR:** spheric or ± flat, glabrous or glandular-hairy, gen ± red; pulp thin or thick, ± resinous. ± 150 spp.: warm temp. (Greek: ancient name for sumac) [Yi et al. 2004 Molec Phylogen Evol 33:861–879]

1. Lvs deeply lobed to ternate-compound, deciduous . . . . ***R. aromatica***
1′ Lvs gen simple, unlobed, evergreen
  2. Lf gen flat, entire to toothed, tip ± obtuse; sepals green, glandular-ciliate . . . . ***R. integrifolia***
  2′ Lf gen folded along midrib, entire, tip acute to acuminate; sepals red, ciliate . . . . ***R. ovata***

R. aromatica Aiton (p. 177) SKUNK BUSH Pl 0.5–2.5 m. **LF**: deciduous, thin, flat; petiole 5–15 mm; lobes or lflets gen 3, crenate to ± lobed, abaxially tomentose to ± glabrous; terminal lobe or lflet 10–35 mm, ± diamond-shaped, lateral 5–18 mm, gen ovate. **INFL**: appearing before lvs; branches short, stiff. **FL**: sepals yellow-green to ± red; petals gen yellow. **FR**: 5–8 mm diam, sparsely hairy, sticky, gen bright red-orange. Slopes, washes, scrub; < 2200 m. CA-FP, DMoj (exc c), n DSon; to s Can, c US, n Mex. [*R. trilobata* Nutt.] Mar–May

R. integrifolia (Nutt.) Rothr. (p. 177) LEMONADE BERRY Pl 1–8 m. **LF**: simple, evergreen; petiole 2–7 mm; blade 2.5–6 cm, 2–4 cm wide, wide-elliptic to lance-elliptic, entire to toothed, ± leathery, gen flat, tip ± obtuse. **INFL**: branches stout; bractlets 2–4 mm. **FL**:

sepals green, glandular-ciliate; petals white to ± pink. **FR**: 7–10 mm diam, glandular-hairy, ± red. Canyons, gen n-facing slopes, chaparral; < 900 m. SW; Baja CA. Cult elsewhere. Hybridizes with *R. ovata*. Feb–May

R. ovata S. Watson (p. 177) SUGAR BUSH Pl 2–10 m. **LF**: simple, evergreen; petiole 10–30 mm; blade 3–8 cm, 3–8 cm wide, wide-ovate to -elliptic, entire, ± leathery, gen folded along midrib, tip acute to acuminate. **INFL**: branches stout; bractlets < 2 mm. **FL**: sepals red, ciliate; petals white to ± pink. **FR**: 6–8 mm diam, glandular-hairy, ± red. Canyons, gen s-facing slopes, chaparral; < 1300 m. SW; AZ, Baja CA. Cult elsewhere. Hybridizes with *R. integrifolia*. Mar–May

SCHINUS

Tree, branches ending in thorn or not; dioecious. **LF**: simple or compound; lflets 5–20, ± leathery, ± resinous, entire to toothed. **INFL**: panicle, axillary or terminal, open to ± dense; pedicels short. **FL**: sepals, petals ± white to ± yellow; stamens 10, 2 whorls of 5, vestigial in pistillate fls; styles 3, fused at base. **FR**: drupe, spheric, leathery, shiny, gen pink to red; pulp resinous to oily, aromatic. ± 25 spp.: trop, warm temp S.Am. (Greek: ancient name) [Carmello-Guerreiro & Paoli 2002 Brazil Arch Biol Technol 45:73–79]

1. Pl < 5 m, branches ending in thorn; lvs simple . . . . ***S. polygamus***
1′ Pl ≥ 5 m, branches not ending in thorn; lvs compound
  2. Distal branches, twigs flexible, drooping; lflets gen > 15, gen < 1 cm wide. ***S. molle***
  2′ Distal branches, twigs stiff, spreading to erect; lflets 7(13), gen > 1 cm wide . . . . ***S. terebinthifolius***

S. molle L. PEPPER TREE Pl 5–18 m, root-sprouting or not. **LF**: 10–30 cm; lflets sessile, 1–6 cm, lanceolate to lance-linear, gen entire. **INFL**: pedicel 2–5 mm in fr. **FL**: < 3 mm. **FR**: 5–8 mm diam. 2*n*=28. Washes, slopes, abandoned fields; < 700 m. SNF, Teh, GV, CW, SW; to TX, Mex; native to S.Am. Jun–Aug ❖

S. polygamus (Cav.) Cabrera CHILEAN PEPPER TREE Pl < 5 m. **LF**: simple, blade 1.5–3 cm, 0.5–2 cm wide, oblong to oblanceolate,

entire to wavy. **INFL**: pedicel 2–5 mm in fr. **FL**: 4–5 mm. **FR**: 5 mm diam. 2*n*=28. Abandoned fields, slopes, scrub; < 1000 m. SCo, PR; weed in se US, native to S.Am. May–Sep

S. terebinthifolius Raddi BRAZILIAN PEPPER TREE Pl 5–10 m. **LF**: 8–15 cm; lflets sessile, 2.5–7 cm, elliptic to oblong, entire to toothed. **INFL**: pedicel 2–4 mm in fr. **FR**: 4–7 mm diam. Washes, canyons; < 200 m. s SCo; FL; native to S.Am. May–Sep ❖

SEARSIA
John M. Miller & Bruce G. Baldwin

Shrubs or trees; gen monoecious. **LF**: ternate-pinnate-compound, ± evergreen [deciduous], gen entire [to lobed]. **INFL**: panicle [raceme], terminal and/or axillary; fls ± sessile to pedicelled. **FL**: stamens 5; styles 3, fused at base. **FR**: ± spheric, glabrous [hairy]; pulp thin, resinous. 120 spp.: Medit, Asia, Afr. (P.B. Sears, Am ecologist, 1891–1990) [Pell et al. 2008 Syst Bot 33:375–383]

S. lancea (L. f.) F.A. Barkley AFRICAN SUMAC Pl 1–8 m. **LF**: petiole 2–5 mm; lflets 2.5–12 cm, 0.5–2 cm wide, linear to narrowly lanceolate, gen entire, ± leathery, gen flat, tip acute to ± obtuse. **INFL**: pendant; bractlets 2–4 mm. **FL**: sepals green, glandular; petals white

to ± yellow. **FR**: 5–8 mm diam, smooth, yellow or yellow-brown, glabrous. Canyons, alluvial fans in desert, coastal scrub, riparian woodland, disturbed places; < 500 m. SCo, DSon; native to s Afr. [*Rhus l.* L. f.] Cult elsewhere. Mar–Jun

TOXICODENDRON POISON OAK, POISON IVY

Shrub, vine-like or not; gen dioecious. **LF**: ± resinous; lflets 3–9, thin to ± leathery, entire, toothed, or lobed. **INFL**: raceme or panicle, axillary, ± open; fls pedicelled. **FL**: stamens 5, vestigial in pistillate fls; styles ± fused, stigmas 3. **FR**: gen spheric, papery or leathery in age, cream to brown; pulp resinous. 15 spp.: Am, e Asia. (Latin: poisonous tree) TOXIC: resin on lvs, sts, frs causes severe contact dermatitis; one of the most hazardous pls in CA.

T. diversilobum (Torr. & A. Gray) Greene (p. 177) WESTERN POISON OAK Shrub, 0.5–4 m, or vine-like, < 25 m. **ST**: twigs gray- to red-brown, tapered, hairs 0 to sparse. **LF**: petiole 1–10 cm; lflets 3(5), ± round to oblong, entire, wavy, or ± lobed, thin to ± leathery, bright red in fall, adaxially glabrous, shiny, abaxially sparse-short-hairy, base truncate to rounded, tip obtuse to rounded; terminal lflet 1–13 cm, 1–8 cm wide, lateral 1–7 cm, 1–6 cm wide. **INFL**: branches

loose, gen arched, slender; pedicels 2–8 mm; bractlets < 1 mm. **FL**: sepals green; petals > sepals, gen ovate, yellow- to white-green. **FR**: 1.5–6 mm diam, spheric to ± compressed, glabrous to fine-bristly, creamy white, in age leathery; pulp white, black-striate. 2*n*=30. Canyons, slopes, chaparral, coastal scrub, oak woodland; < 1650 m. CA-FP, sw edge DMoj; to BC, w NV, n Baja CA. Apr–Jun

APIACEAE (Umbelliferae) CARROT FAMILY
Lincoln Constance & Margriet Wetherwax, except as noted

Ann to per [shrub, tree], gen from taproot. **ST**: gen ± scapose, gen ribbed, hollow. **LF**: basal and gen cauline, gen alternate; stipules gen 0; petiole base gen sheathing st; blade gen much dissected, occ compound. **INFL**: umbel or head, simple or com-

pound, gen peduncled; bracts present in involucres or 0; bractlets gen present in "involucels". **FL**: many, small, gen bisexual (or some staminate), gen radial (or outer bilateral); calyx 0 or lobes 5, small; petals 5, free, gen ovate or spoon-shaped, gen incurved at tips, gen ± ephemeral; stamens 5; pistil 1, ovary inferior, 2-chambered, gen with a ± conic, persistent projection or platform at tip subtending 2 free styles. **FR**: 2 dry, 1-seeded halves (= mericarps), separating from each other but gen ± persistent to central axis; ribs on halves 5, 2 marginal, 3 to back; oil tubes 1–several per interval between ribs. 300 genera, 3000 spp.: ± worldwide, esp temp; many cult for food or spice (e.g., *Carum*, caraway; *Daucus*; *Petroselinum*); *Bupleurum lancifolium* Hornem. is historical garden weed; some toxic (e.g., *Conium*). Mature fr gen critical in identification, shape given in outline. *Hydrocotyle* moved to Araliaceae. *Petroselinum crispum* (Mill.) A.W. Hill is a waif. Scientific Editors: Douglas H. Goldman, Bruce G. Baldwin.

1. Fls in simple umbels or heads (spikes or racemes); fr central axis not obvious
  2. Lf base not sheathing but stipules present, occ ± fused distal to node when lvs opposite
    3. Herbage hairs stellate; fr not or ± compressed front-to-back, inflated . . . . **BOWLESIA**
    3′ Herbage hairs simple or 0; fr very compressed side-to-side. . . . . **HYDROCOTYLE** (treated in Araliaceae)
  2′ Lf conspicuously sheathing at base, stipules 0
    4. Fls in open umbels; fr glabrous, with well-defined, thickened ribs or not . . . . **LILAEOPSIS**
    4′ Fls in dense umbels or heads; fr with scales, prickles, or tubercles, without distinct ribs
      5. Fls bisexual, sessile, each with 1 bract . . . . **ERYNGIUM**
      5′ Fls bisexual and staminate in same umbel or head (staminate only), staminate gen long-pedicelled; fls not each with 1 bract . . . . **SANICULA**
1′ Fls in compound or occ simple umbels, occ compound and head-like; fr central axis gen obvious, separate (exc *Podistera*)
  6. Ann or bien, slender-taprooted or roots fibrous, reproducing only from seed
    7. Corolla yellow . . . . **ANETHUM**
    7′ Corolla gen white
      8. Fr gen elongate, much longer than wide, beaked or long-tapered to tip
        9. Bractlets entire; fr ribs 0 or obscure, sterile beak << fertile body . . . . **ANTHRISCUS**
        9′ Bractlets entire to lobed or dissected; fr ribs prominent, sterile beak > fertile body. . . . . **SCANDIX**
      8′ Fr ± round to oblong or elliptic, not or only ± longer than wide, not beaked, not long-tapered to tip
        10. Fr ± round, not compressed, halves not separating readily . . . . **CORIANDRUM**
        10′ Fr ovate or cordate to oblong or elliptic, compressed either side-to-side or front-to-back, halves separating readily
          11. Bracts lf-like, gen dissected
            12. Herbage, fr glabrous . . . . **AMMI**
            12′ Herbage hairy; fr conspicuously bristly
              13. Fr very compressed front-to-back; rays gen many; umbel dense, nest-like in fr . . . . **DAUCUS**
              13′ Fr compressed side-to-side; rays few (1–9); umbel open, not nest-like in fr. **YABEA**
          11′ Bracts 0 or inconspicuous, not lf-like
            14. Lvs gen opposite; rays 2–3; fr elliptic-cordate . . . . **APIASTRUM**
            14′ Lvs gen alternate; rays 0–14; fr elliptic to narrowly ovate
              15. Petal tips narrowed, incurved; fr prickly throughout or only on outer half; herbage hairy . . . . **TORILIS**
              15′ Petal tips not narrowed, not incurved; fr prickly throughout, tubercled, sharply scabrous, or glabrous; herbage glabrous or ± scabrous, not hairy
                16. Involucel 0; fr glabrous . . . . **CYCLOSPERMUM**
                16′ Involucel present; fr conspicuously bristly or sharply scabrous, at least on ribs
                  17. Infl exc frs puberulent; fr oblong-ovate. . . . . **AMMOSELINUM**
                  17′ Infl exc frs glabrous; fr wide-ovate. . . . . **SPERMOLEPIS**
  6′ Per or bien, from taproot or from persistent underground structures, not reproducing only from seed
    18. Pl glabrous; gen in flooded to wet or moist (dry) soil
      19. St gen conspicuously purple-spotted or -streaked; herbage musty-scented; fr oil tubes not apparent . . . ²**CONIUM**
      19′ St green or occ ± purple, unspotted; herbage not musty-scented; fr oil tubes apparent
        20. St from well-developed single or clustered tubers, or clustered, fibrous roots, gen not rooting at proximal nodes
          21. Pl rhizomatous; sap from rhizome becoming ± red-brown in air . . . . **CICUTA**
          21′ Pl not rhizomatous; sap from rhizome not changing color in air
            22. Calyx lobes a persistent crown on fr; fr corky; fr central axis not obvious . . . . ²**OENANTHE**
            22′ Calyx lobes evident but not a persistent crown on fr; fr not evidently corky; fr central axis an apparent, separate structure
              23. Fr very compressed front-to-back, marginal ribs wide, thin-winged . . . . **OXYPOLIS**
              23′ Fr ± or not compressed side-to-side, ribs ± equal, thread-like to prominent but not clearly winged . . . . **PERIDERIDIA**
        20′ St gen from stout taproot, or gen rooting at proximal nodes or along rhizomes, stolons
          24. Calyx lobes a persistent crown on fr; fr central axis not obvious. . . . . ²**OENANTHE**
          24′ Calyx lobes 0 or minute, gen not a persistent crown on fr; fr central axis an apparent, separate structure
            25. Fr halves not adhering to central axis; bracts, bractlets conspicuous to 0 . . . . **APIUM**

25′ Fr halves gen adhering to central axis; bracts, bractlets conspicuous
26. Fr ribs thread-like, inconspicuous in corky fr wall . . . . **BERULA**
26′ Fr ribs prominent, corky. . . . . **SIUM**
18′ Pl glabrous or variously hairy; wet to dry soil
27. Herbage anise- or licorice-scented; lf segments thread-like. . . . . **FOENICULUM**
27′ Herbage scented or not, but not anise-scented; lf segments wider than thread-like
28. Fr ribs ± equal, none expanded as wings
29. Bien of disturbed or cult places
30. Sts gen purple-spotted; cauline lvs like basal; fr ribs prominent, wavy. . . . . ²**CONIUM**
30′ Sts unspotted; cauline lvs with more, narrower segments than basal; fr ribs thread-like, not wavy
. . . . **[PETROSELINUM]**
29′ Per of gen undisturbed places
31. Pl dwarfed, 2–15 cm; lf segments not sharply serrate
32. Umbels not head-like; involucel inconspicuous, not 1-sided, bractlets 0 or free; calyx lobes 0;
styles cylindric; pl erect or spreading, not cushion-forming; lvs loosely recurved or occ spreading
at soil surface . . . . ²**OROGENIA**
32′ Involucel conspicuous, 1-sided, bractlets gen partly fused; calyx lobes conspicuous or not; styles
compressed, flattened; pls cushion-forming or lvs spreading at soil surface; umbels head-like
33. Herbage coarsely hairy to tomentose; sterile pedicels many, > fr . . . . **OREONANA**
33′ Herbage thinly to densely puberulent; sterile pedicels 0–few, << fr. . . . . **PODISTERA**
31′ Pl not dwarfed, 10–120 cm, if < 20 cm, lf segments sharply serrate
34. Fr linear to oblong, 8–22 mm; fr oil tubes obscure; roots licorice-scented. . . . . **OSMORHIZA**
34′ Fr oblong to round, 2–9 mm; fr oil tubes evident, 1–5 per rib-interval; roots not licorice-scented
35. Corolla white; involucel gen 0, or < pedicels . . . . ²**LIGUSTICUM**
35′ Corolla yellow; involucel present, gen > pedicels . . . . **TAUSCHIA**
28′ Fr ribs equal or not, some expanded as definite wings
36. Ribs winged at margins of fr halves, others not or more narrowly so; fr compressed front-to-back
37. Pl robust, 1–3 m; outer petals of marginal fls > others; oil tubes extending part-way to base of fr
. . . . **HERACLEUM**
37′ Pl not robust, gen < 1 m; outer petals of marginal fls = others; oil tubes extending to base of fr
38. Projection at ovary tip small, conic, persistent on fr
39. Lf 1–4-pinnately or -ternate-pinnately dissected or compound; corolla white; per . . . . . . **CONIOSELINUM**
39′ Lf 1-pinnate; corolla yellow or orange; gen bien, or per . . . . **PASTINACA**
38′ Projection at ovary tip 0
40. Marginal fr wings thin or corky, not incurved; fr axis divided to base. . . . . **LOMATIUM**
40′ Marginal fr wings very corky, incurved; fr axis a corky ridge along middle of each fr half . . . . ²**OROGENIA**
36′ Ribs winged at margins of fr halves, some others ± winged; fr cylindric to ± compressed front-to-
back or side-to-side (strongly compressed)
41. Pl low, gen spreading, st gen 0; rays well developed or not; ovary tip projection 0
42. Pl erect or ± spreading; lf adaxial surface gen glaucous, glabrous to scabrous or finely hairy,
abaxial surface not tomentose; inland, gen dry places . . . . **CYMOPTERUS**
42′ Pl prostrate or ± spreading; lf adaxial surface green, abaxial tomentose; seashore . . . . **GLEHNIA**
41′ Pl erect, st gen present; rays well developed; ovary tip projection conic
43. Pedicels reduced to a disk; 2° umbels head-like. . . . . **SPHENOSCIADIUM**
43′ Pedicels distinct or ± webbed at base, not reduced to a disk; 2° umbels open, not head-like
44. St base not fibrous; lf segments gen coarse; fr gen compressed front-to-back (± compressed to
cylindric) . . . . **ANGELICA**
44′ St base conspicuously fibrous; lf segments gen fine; fr compressed side-to-side . . . . ²**LIGUSTICUM**

AMMI

Ann, bien, taprooted, glabrous. **ST:** erect, branched. **LF:** blade oblong to ovate, pinnately or ternately dissected, segments lanceolate to thread-like. **INFL:** umbels compound; bracts, bractlets many, lf-like; rays, pedicels many, spreading-ascending to incurved in fr. **FL:** calyx lobes minute; petals wide, white, tips 2-lobed. **FR:** oblong to ovate, ± cylindric, compressed side-to-side; ribs ± equal, thread-like; oil tube 1 per rib-interval; fr axis entire or divided. **SEED:** face flat. ± 6 spp.: Eurasia, Medit. (Ancient name used by Dioscorides) TOXIC: rarely eaten by livestock.

1. Infl disk-like receptacle 0; rays spreading in fr; cauline lf segments lanceolate to round, finely serrate . . . . *A. majus*
1′ Infl with disk-like receptacle; rays incurving in fr to form a nest-like umbel; cauline lf segments linear . . . . . . *A. visnaga*

A. majus L. Pl 2–8 dm. **LF:** petiole 1–5 cm; blade 6–20 cm, oblong, segments 10–15 mm; cauline lvs 2-pinnately dissected. **INFL:** scabrous; peduncle 8–14 cm; rays 20–60, 2–7 cm, slender; pedicels 1–10 mm. **FR:** 1.5–2 mm, oblong. 2*n*=22. Fields, roadsides, disturbed areas; gen < 1000 m. NCo, NCoRO, ScV; native to Eurasia. May–Jul

A. visnaga (L.) Lam. (p. 177) BISNAGA Pl 2–8 dm. **LF:** petiole ± 1 cm; blade 5–20 cm, triangular-ovate, segments 5–35 mm; cauline lvs 1- or 2-pinnately or -ternately dissected. **INFL:** glabrous; peduncle 8–14 cm; rays 60–100, 2–5 cm, slender; pedicels 3–13 mm. **FR:** 2–2.5 mm, oblong-ovate to ovate. 2*n*=20. Roadsides, railroad tracks, disturbed areas; gen < 1000 m. CCo, SCo; native to Eurasia. Jun–Jul

AMMOSELINUM

Ann, taprooted. **ST**: erect or gen loosely branched, glabrous or roughened. **LF**: petiole entirely sheathing; blade oblong to obovate, ternately or ternate-pinnately dissected, segments linear to spoon-shaped. **INFL**: umbels compound, peduncled or occ sessile, puberulent; bracts gen 0; bractlets several, narrow; rays, pedicels few, spreading or spreading-ascending, unequal. **FL**: calyx 0; petals ovate, white, tips obtuse, not narrowed, not incurved. **FR**: oblong-ovate, compressed side-to-side; ribs ± equal, prominent, conspicuously bristly or sharply scabrous; oil tubes 1–3 per rib-interval; fr axis notched at tip. **SEED**: face flat to concave. 4 spp.: 3 N.Am, 1 S.Am. (Greek: sand-parsley)

A. giganteum J.M. Coult. & Rose (p. 177)   DESERT SAND-PARSLEY
Pl 1–2 dm. **LF**: petiole 3–8 mm; blade 1.5–2.5 cm, obovate, segments 4–13 mm, linear, glabrous or roughened. **INFL**: peduncles 0–4 cm; rays 4–8, 0–2 cm; pedicels 1–10, 0–8 mm. **FR**: 3–5 mm, oblong-ovate; ribs corky, sharply scabrous. $2n=38$. Heavy soil under shrubs; 152 m. DSon (Hayfield Lake, Riverside Co., 1922); AZ, n Mex. Possibly introduced in CA. Mar–Apr ★

ANETHUM

1 sp. (Latin: dill)

A. graveolens L.   DILL   Ann, 0.5–2 m, taprooted; herbage glabrous, glaucous, anise-scented. **ST**: erect, branched, hollow. **LF**: gen cauline; petiole 5–6 cm; blade 12–35 cm, oblong to obovate, pinnately dissected, segments 4–20 mm, thread-like. **INFL**: umbels compound, peduncled; bracts, bractlets 0; rays, pedicels many, slender, ± equal, spreading or spreading-ascending. **FL**: calyx lobes 0; petals wide, yellow, tips narrowed. **FR**: 2–4.5 mm, ovate, compressed front-to-back, glabrous; ribs unequal, marginal narrowly winged, others thread-like; oil tubes 1 per rib-interval; fr axis divided to base. **SEED**: face ± flat. $2n=22$. Disturbed places; gen < 1000 m. SW; native to Medit. Widely but sporadically escaped from cult. Jun–Aug

ANGELICA

Per, taprooted. **ST**: erect, lfy, hollow. **LF**: petioles sheathing, cauline sheaths often inflated, bladeless; blades compound (dissected), lflets gen wide, distinct. **INFL**: umbels compound, peduncled; bracts 0; bractlets 0 or many and conspicuous; rays, pedicels many, spreading-ascending to ascending. **FL**: calyx lobes 0 or minute; petals wide, white, pink, red, or purple. **FR**: oblong to round, gen compressed front-to-back (± compressed or cylindric), glabrous to hairy; ribs unequal, winged but marginal gen wider than others; oil tubes 1–several per rib-interval, adhering to fr wall (to seed); fr axis divided to base. **SEED**: face flat. 50–60 spp.: temp N.Am, Asia. (Latin: angelic, for cordial and medicinal properties) [DiTomaso 1984 Madroño 31:69–79]

1. Lf 2–3-ternate-pinnately dissected, segments linear or linear-oblong, entire, 2–10 cm . . . . ***A. lineariloba***
1′ Lf 1–3-pinnate or -ternate-pinnate, lflets lanceolate to widely ovate, gen serrate to dentate or irregularly cut (entire), 3–15 cm
  2. Bractlets conspicuous, 5–15 mm; rays, pedicels not webbed at base
    3. 1° lflets gen sharply reflexed; fr compressed, ovate to round, 3–4 mm, ribs on back much narrower than marginal, thin. . . . . ***A. genuflexa***
    3′ 1° lflets not reflexed; fr cylindric, ± oblong, 4–9 mm, ribs ± equal, widely, thickly corky-winged . . . . ***A. lucida***
  2′ Bractlets gen 0 or inconspicuous; rays, pedicels webbed at base
    4. Basal lvs 1-ternate-pinnate
      5. Rays ± equal, spreading-ascending to reflexed . . . . ***A. callii***
      5′ Rays unequal, ascending
        6. Lf oblong; lflets entire to sparsely serrate; rays 7–14; ovary minutely bristly; fr oblong, 4–5 mm . . . . ***A. kingii***
        6′ Lf triangular-ovate; lflets sharply serrate; rays 15–50; ovary glabrous to minutely hairy; fr oblong to ovate, 6–7 mm . . . . ***A. californica***
    4′ Basal lvs 2–3-ternate-pinnate
      7. Pl glaucous . . . . ***A. tomentosa***
      7′ Pl bright green, at least adaxially on lvs
        8. Pl coastal; lf thick, abaxially gen white-tomentose. . . . . ***A. hendersonii***
        8′ Pl not coastal; lf thin, green, both sides gen glabrous
          9. Petals, ovary glabrous. . . . . ***A. arguta***
          9′ Petals, ovary hairy . . . . ***A. breweri***

A. arguta Nutt.   ANGELICA   Pl 1–2 m, gen glabrous. **LF**: < 1 m diam, ovate-triangular, 2–3-ternate-pinnate; lflets 6–9 cm, lanceolate to elliptic, acute, sharply serrate. **INFL**: gen glabrous; bracts, bractlets gen 0; rays 20–60, 2–10 cm; rays, pedicels webbed at base. **FL**: petals, ovary glabrous. **FR**: 8–9 mm, oblong to ovate. $2n=22$. Uncommon. Conifer forest; 60–2300 m. KR, CaRH; to BC, MT, UT. Jul–Aug

A. breweri A. Gray   Pl 1–2 m, gen hairy. **LF**: < 1 m diam, triangular-ovate, 2–3-ternate-pinnate; both surfaces gen glabrous; lflets 6–10 cm, lanceolate, acuminate, serrate. **INFL**: hairy; bracts, bractlets gen 0; rays 20–50, 2–10 cm; rays, pedicels webbed at base. **FL**: petals, ovary hairy. **FR**: 8–12 mm, oblong to ovate. $2n=66$. Conifer forest; 800–3000 m. CaRH, n&c SNH; NV. Jun–Aug

A. californica Jeps.   Pl 1–2.5 m, glabrous to sparsely hairy. **LF**: 1–12 dm, triangular-ovate, 1-ternate-pinnate; lflets 4–8 cm, lanceolate to oblong, sharply serrate. **INFL**: gen glabrous; bracts, bractlets gen 0; rays 15–50, 2–13 cm, unequal, ascending; rays, pedicels webbed at base. **FL**: petals, ovary glabrous to minutely hairy. **FR**: 6–7 mm, oblong to ovate. $2n=22$. Dry slopes; 15–1500 m. NCoR, CaR, n SNF, SnFrB. Jun–Aug

A. callii Mathias & Constance (p. 177) CALL'S ANGELICA Pl 1–2 m, gen roughened. **LF:** 1–4 dm, ovate, 1-ternate-pinnate; lflets 3–13 cm, ovate, acute to obtuse, sharply serrate. **INFL:** roughened; bracts, bractlets gen 0; rays 25–50, 2.5–7 cm, ± equal, spreading-ascending to reflexed; rays, pedicels webbed at base. **FL:** petals, ovary gen hairy. **FR:** 3.5–5 mm, oblong to obovate, moderately compressed front-to-back; marginal ribs thick. 2*n*=22. Streambanks in conifer forest; 1000–2000 m. s SNH (Tulare, Kern cos.). Jun–Jul ★

A. genuflexa Nutt. Pl 1–2 m, glabrous to hairy. **LF:** 1–8 dm, ovate to triangular-ovate, 2–3-ternate-pinnate; 1° lflets gen sharply reflexed, ultimate 4–10 cm, lanceolate to ovate, acute to acuminate, coarsely serrate to irregularly cut. **INFL:** hairy; bracts 0; bractlets several, 5–10 mm, conspicuous; rays 25–50, 2–7 cm; rays, pedicels not webbed at base. **FL:** petals glabrous, ovary gen glabrous. **FR:** 3–4 mm diam, ovate to round. 2*n*=22. Streambanks, wet areas in conifer forest; < 1500 m. KR, NCoRO; to AK; e Asia. Jul–Aug

A. hendersonii J.M. Coult. & Rose Pl 8–20 dm, gen sprawling, tomentose (± glabrous). **LF:** < 6 dm, triangular-ovate, 2–3-ternate-pinnate; lflets 5–10 cm, lanceolate to oblong, obtuse to acute, double-serrate to crenate, adaxially green, glabrous, abaxially gen white-tomentose. **INFL:** tomentose; bracts, bractlets 0; rays 20–65, 2–8 cm; rays, pedicels webbed at base. **FL:** petals, ovary tomentose. **FR:** 6–9 mm, oblong to ovate. 2*n*=22. Coastal bluffs, scrub; < 150 m. NCo, CCo; to WA. Jun–Jul

A. kingii (S. Watson) J.M. Coult. & Rose (p. 177) KING'S ANGELICA Pl 3–20 dm, glabrous to roughened. **LF:** 1.5–4 dm, oblong, 1-ternate-pinnate (± 2-pinnate); lflets 3–12 cm, ± lanceolate, acute to acuminate, entire to sparsely serrate. **INFL:** roughened;

bracts, bractlets 0; rays 7–14, 0.5–10 cm, unequal, ascending; rays, pedicels webbed at base. **FL:** petals hairy; ovary minutely bristly. **FR:** 4–5 mm, oblong. 2*n*=44. Subalpine streambanks; 2100–3100 m. W&I (White Mtns); to UT. Jun–Aug ★

A. lineariloba A. Gray (p. 177) Pl 5–15 dm, ± glabrous to scabrous. **LF:** 1–3.5 dm, triangular-ovate, 2–3-ternate-pinnately dissected; segments 2–10 cm, linear to linear-oblong, acute, entire. **INFL:** scabrous; bracts, bractlets 0; rays 20–40, 3–7 cm, ± equal; rays, pedicels not webbed at base. **FL:** petals, ovary roughened to glabrous in age. **FR:** 10–13 mm, oblong to wedge-shaped. Rocky open slopes; 1700–3300 m. c&s SNH, SNE; NV. Jun–Aug

A. lucida L. (p. 179) SEA-WATCH Pl 1–1.5 m, glabrous. **LF:** 1–3 dm, ovate to triangular-ovate, 2–3-ternate-pinnate; lflets 4–10 cm, widely lanceolate to ovate, acute, serrate to crenate-dentate; both surfaces green, glabrous. **INFL:** hairy; bracts 0; bractlets 5–12, 5–15 mm, conspicuous; rays 20–45, 3–8 cm; rays, pedicels not webbed at base. **FL:** petals, ovary glabrous. **FR:** 4–9 mm, ± oblong, cylindric; ribs ± equal, widely, thickly corky-winged. *n*=28. Coastal bluffs, beaches; < 50 m. NCo; to AK; coastal e N.Am, e Asia. May–Aug ★

A. tomentosa S. Watson Pl 7–20 dm, glaucous, glabrous to hairy. **LF:** < 1 m diam, triangular-ovate, 2–3-ternate-pinnate; lflets 2–12 cm, lanceolate to ovate, acute to acuminate, serrate to entire. **INFL:** glabrous to hairy; bracts, bractlets 0; rays 20–60, 2–11 cm; rays, pedicels webbed at base. **FL:** petals, ovary glabrous to hairy. **FR:** 6–10 mm, oblong to ovate. 2*n*=22. Gen wooded areas; 30–2400 m. KR, NCoR, CW (exc SCoRI), SnGb, SnBr, PR; s OR. ± intergrading with *A. hendersonii* in NCoRO. Jun–Aug

ANTHRISCUS

Ann, bien, taprooted, glabrous or bristly. **ST:** erect, branched. **LF:** gen cauline; blade oblong to ovate, pinnately or ternately dissected or compound, segments or lflets linear-oblong to ovate. **INFL:** umbels compound, gen peduncled; bracts gen 0; bractlets several, reflexed, entire; rays, pedicels few, spreading. **FL:** outer occ ± bilateral; calyx lobes 0; petals narrow, white. **FR:** narrowly elongate to ovoid, beaked, smooth or bristly; each half ± cylindric; ribs, oil tubes 0 or obscure; fr axis entire or notched at tip. **SEED:** face grooved. ± 15 spp.: Eurasia, Afr. (Ancient Greek name)

A. caucalis M. Bieb. (p. 179) BUR-CHERVIL Ann, 4.5–10 dm. **LF:** petiole 3–8 cm; blade 5–15 cm, oblong- to triangular-ovate, pinnately dissected, segments 1–5 mm, linear-oblong, obtuse. **INFL:** peduncle 0–2 cm; rays 3–6, 1–2.5 cm. **FR:** ± 4 mm, ovoid; beak << body. 2*n*=14. Gen shady places; < 1500 m. CA-FP, MP; native to Eurasia. Apr–Jun

APIASTRUM

1 sp. (Latin: wild celery)

A. angustifolium Nutt. (p. 179) Ann, 5–50 cm, taprooted. **ST:** many-branched throughout. **LF:** mostly cauline, gen opposite; petiole 2–4 cm; blade 1–5 cm, round, finely ternately dissected, segments 5–25 mm, linear to oblong. **INFL:** umbels compound, peduncled or not, from axils or opposite distal lvs; bracts, bractlets 0; rays 2–3, pedicels few, unequal, spreading. **FL:** calyx lobes 0; petals wide, white, tip acute, not incurved; ovary tip projection flat to ± conic. **FR:** 1–1.5 mm, elliptic-cordate, compressed side-to-side, papillate-roughened, glabrous in age; ribs thread-like; oil tube 1 per rib-interval; fr axis divided to base. **SEED:** face concave. 2*n*=22. Chaparral, coastal scrub; < 1500 m. CA-FP; Baja CA. Mar–Apr

APIUM

Bien, per, taprooted or fibrous-rooted from rhizome. **ST:** prostrate to erect, hollow, gen rooting from proximal nodes, glabrous. **LF:** blade oblong to obovate, 1-pinnate, lflets paired, lanceolate to ± round. **INFL:** umbels compound, peduncled or not; bracts, bractlets conspicuous to 0; rays, pedicels few, spreading-ascending. **FL:** calyx lobes 0 or minute; petals wide, white to ± green-white; ovary tip projection occ flat. **FR:** ovate-oblong to round, compressed side-to-side; ribs ± equal, thread-like to obtuse and ± corky; oil tube 1 per rib-interval; fr axis entire or notched at tip. **SEED:** face flat. 2*n*=22. ± 20 spp.: gen s hemisphere, also Eurasia. (Classical name for celery)

1. Pl ± erect, terrestrial; bracts, bractlets 0 . . . . ***A. graveolens***
1′ Pl prostrate to ascending, gen aquatic; bracts few or 0, bractlets ≥ pedicels . . . . [***A. nodiflorum***]

A. graveolens L. CELERY Pl 5–15 dm. **ST:** not rooting at nodes. **LF:** petiole 0.3–2.5 dm; blade 7–18 cm, oblong to obovate, lflets 2–4.5 cm, ovate to ± round, gen lobed. **INFL:** rays 7–16, 0.7–2.5 cm, ± equal; calyx lobes minute. **FR:** 1.5–2 mm diam, elliptic to nearly round; fr axis tip notched. Wet places; gen < 1000 m. CA; temp zones worldwide; native to Eurasia. Cult and naturalized widely. May–Jul

BERULA CUTLEAF WATER-PARSNIP

1 sp. (Latin: water-cress)

B. erecta (Huds.) Coville (p. 179) Per, < 15 dm, stoloned, glabrous. **ST**: ascending or erect, branched, hollow, rooting at proximal nodes. **LF**: petiole 4–12 cm, narrowly sheathing; blade 1.5 dm, oblong, 1-pinnate, much dissected if submerged, lflets 7–12 pairs, 1–8 cm, oblong to ovate, sessile, lobed or serrate. **INFL**: umbels compound, peduncled, terminal or opposite lvs; bracts, bractlets lf-like, lanceolate; rays, pedicels ± unequal, spreading. **FL**: calyx lobes minute, persistent; petals wide, white. **FR**: 1.5–2 mm, ± round, compressed side-to-side, glabrous; ribs thread-like, inconspicuous in corky fr wall; oil tubes apparent, deeply embedded in fr wall; fr axis divided to base, branches adhering to and falling with fr-halves. 2*n*=18. Marshy areas, streams; < 2000 m. CA; to BC, NY, TX, Mex; Eurasia, Afr. Possibly TOXIC: possibly poisonous to livestock. Jul–Oct

BOWLESIA

Ann, per, taprooted, stellate-hairy. **ST**: decumbent to erect, branched. **LF**: opposite to alternate; stipules occ ± fused distal to node when lvs opposite; blade ovate to round, palmately lobed to compound (± entire). **INFL**: umbels simple, few-fld, peduncled or sessile; bracts gen present; pedicels 0 to short. **FL**: calyx lobes gen 0; petals gen wide, obtuse to ± acute, tip not narrowed, not incurved. **FR**: ovate to ovoid-oblong, round, or spheric, cylindric to ± 4-angled or ± compressed front-to-back, stellate-hairy to prickly or ± glabrous; ribs obscure; fr central axis not obvious. **SEED**: face ± flat. 16 spp.: temp S.Am, 1 in CA. (Wm. Bowles, Irish writer on Spanish natural history, 1705–1780) [Mathias & Constance 1965 Univ Calif Publ Bot 38:1–73]

B. incana Ruiz & Pav. (p. 179) Pl 0.5–6 dm. **LF**: gen opposite; petiole 1.5–12 cm; blade 0.5–3 cm, round-reniform, lobes 5–9, ± 1/2 to base, widely lanceolate to round, gen obtuse. **INFL**: axillary, ± sessile; bracts lanceolate. **FL**: calyx lobes minute; petals oblong-ovate, ± yellow-green. **FR**: 1.5–2 mm, ovoid to spheric, inflated. 2*n*=16. Shade of trees, rocks, shrubs; < 1400 m. NCoRI, c&s SNF, SnJV, CW, SW, DMoj; to TX, S.Am. Mar–Apr

CICUTA WATER-HEMLOCK

Per, glabrous; rhizome internally chambered, sap becoming ± red-brown in air, fibrous- or tuberous-rooted. **ST**: erect, hollow. **LF**: blade oblong to triangular-ovate, 1–3-pinnate or ternate-pinnate, lflets linear to lance-ovate, serrate or irregularly cut. **INFL**: umbels compound; bracts gen 0; bractlets gen inconspicuous; rays, pedicels many, spreading. **FL**: calyx lobes minute; petals wide, white, tips narrowed. **FR**: ovoid to spheric, ± compressed side-to-side; ribs low, corky, occ unequally spaced; oil tube 1 per rib-interval; fr axis divided to base. **SEED**: face flat or concave. ± 4 spp.: Eurasia, N.Am. (Ancient Latin name) TOXIC: the most lethally toxic native plant spp. [Lee & Downie 2006 Canad J Bot 84:453–468]

1. Fr gen round, rib width >> intervals between; areas defined by major veinlets on abaxial lf surface large, gen elongate . . . . ***C. douglasii***
1′ Fr gen ovate, rib width gen ≤ intervals between; areas defined by major veinlets on abaxial lf surface small, gen ± circular or ± square, less often ± elongate . . . . ***C. maculata***
   2. Lf gen 1–2-pinnate; styles < 1 mm; inland . . . . var. ***angustifolia***
   2′ Lf gen 2-pinnate; styles gen > 1 mm; coastal . . . . var. ***bolanderi***

C. douglasii (DC.) J.M. Coult. & Rose (p. 179) Pl 15–30 dm. **LF**: 1.5–4.5 dm, narrowly ovate to triangular-ovate, 1–2(3)-pinnate; lflets 1–10(15) cm, linear to widely lanceolate, acute or acuminate, ± entire to coarsely serrate, areas defined by major veinlets on abaxial surface large, gen elongate. **INFL**: umbels compound, terminal and lateral; peduncles 2–18 cm; rays 15–30(35), 2–8 cm; pedicels 20–30, 2–10 mm. **FR**: 2–4 mm, gen round; rib width >> intervals between. 2*n*=44. Wet places, gen aquatic; < 2800 m. KR, NCo, NCoRH, NCoRI, CaRH, s SNF, SNH, CCo, SCo, GB; to BC, MT. Jun–Sep

C. maculata L. Pl 10–15 dm. **LF**: 1–4 dm, ovate to triangular-ovate, 1–2-pinnate; lflets 2–10 cm, lanceolate, acute or acuminate, coarsely to sparsely serrate, areas defined by major veinlets on abaxial surface small, gen ± circular or ± square, less often ± elongate. **INFL**: umbels compound, terminal and lateral; peduncles 2.5–12 cm; rays 15–30, 2–4.5 cm; pedicels 15–30, 2–10 mm. **FR**: 3–4 mm, gen ovate; rib width gen ≤ intervals between. 2*n*=22. Vars. best distinguishable in CA by habitat, geog. Other vars. in e N.Am.

  var. ***angustifolia*** Hook. **LF**: gen 1–2-pinnate. **FL**: styles < 1 mm. Wet meadows; 900–2100 m. KR, SnBr, PR, GB. Jun–Sep

  var. ***bolanderi*** (S. Watson) G.A. Mulligan BOLANDER'S WATER-HEMLOCK **LF**: gen 2-pinnate. **FL**: styles gen > 1 mm. Coastal wetlands; < 200 m. s ScV (Suisun marshes), CCo, SCo. Jul–Sep ★

CONIOSELINUM

Per, taprooted or short-rhizomed, glabrous or ± scabrous in infl. **ST**: erect, branched. **LF**: blade ovate to widely triangular, 1–4-pinnately or -ternate-pinnately dissected or compound; segments or lflets toothed or lobed to deeply cut; sheaths conspicuously dilated. **INFL**: umbels compound; bracts gen 0; bractlets gen present, ± scarious; rays, pedicels many, spreading-ascending. **FL**: calyx lobes 0 or minute; petals wide, white, tip notched. **FR**: oblong to ovate, compressed front-to-back; marginal ribs thin-winged, wide, abaxial low, corky, or all ribs winged; oil tubes 1–4 per rib-interval; fr axis divided to base. **SEED**: face flat or ± concave. ± 10 spp.: N.Am, Eurasia. (Combined generic names *Conium* + *Selinum*)

C. pacificum (S. Watson) J.M. Coult. & Rose (p. 179) Pl 10–40 dm. **LF**: triangular-ovate; petiole 3–15 cm; blade 0.5–2 dm, 2–3-pinnate, lflets 1–5 cm, oblong to ovate, coarsely serrate, pinnately cut. **INFL**: terminal and gen some lateral; peduncles 6–20 cm; rays 15–30, 2–4 cm, spreading-ascending; bractlets narrow, scarious; pedicels 5–10 mm. **FR**: 5–8 mm, oblong to oblong-ovate; ribs winged, marginal widest. 2*n*=44. Ocean bluffs, gen in coastal scrub; < 400 m. NCo, CCo; to AK; n Japan. Jun–Aug

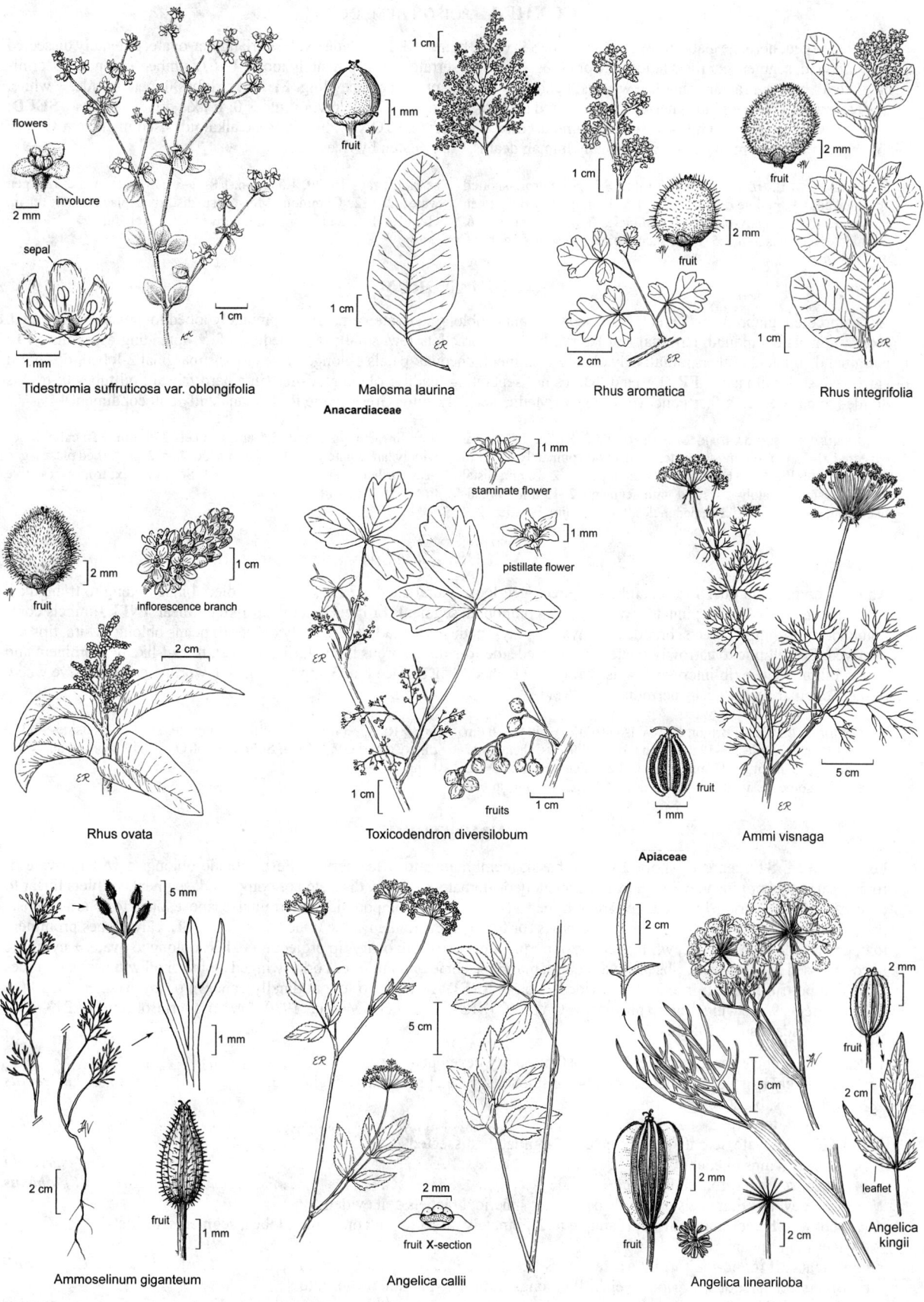

177

Tidestromia suffruticosa var. oblongifolia

flowers
involucre
2 mm
sepal
1 mm
1 cm

Malosma laurina
Anacardiaceae
fruit
1 cm
1 mm
1 cm

Rhus aromatica
1 cm
fruit
2 mm
2 cm

Rhus integrifolia
fruit
2 mm
1 cm

Rhus ovata
fruit
2 mm
inflorescence branch
1 cm
2 cm

Toxicodendron diversilobum
staminate flower
1 mm
pistillate flower
1 mm
1 cm
fruits
1 cm

Ammi visnaga
Apiaceae
fruit
1 mm
5 cm

Ammoselinum giganteum
5 mm
1 mm
fruit
2 cm
1 mm

Angelica callii
5 cm
fruit X-section
2 mm

Angelica lineariloba
2 cm
5 cm
fruit
2 mm
2 cm

Angelica kingii
2 mm
fruit
2 cm
leaflet

CONIUM POISON HEMLOCK

Bien, taprooted; herbage glabrous, musty-scented. **ST**: erect, branched. **LF**: blade ovate to triangular-ovate, pinnately dissected or compound, segments or lflets lanceolate or oblong to ovate, serrate to 1–2-pinnately lobed. **INFL**: umbels compound, terminal, lateral; bracts, bractlets small, few; rays, pedicels ± many, spreading-ascending. **FL**: calyx lobes 0; petals wide, ± white, tips narrowed. **FR**: ovoid to spheric, ± compressed side-to-side; ribs ± equal, low; oil tubes 0; fr axis divided to base. **SEED**: face grooved. ± 6 spp.: Eur, s Afr. (Greek name used by Dioscorides) TOXIC: highly toxic alkaloids used in ancient Greece for capital punishment (e.g., Socrates); many human deaths, rarely eaten by livestock.

C. maculatum L. (p. 179) Pl 5–30 dm. **ST**: gen purple-spotted or -streaked. **LF**: petiole dilated; blade 1.5–3 dm, widely ovate, gen 2-pinnate. **INFL**: many-branched; peduncles 2–8 cm; bracts 4–6, acuminate; bractlets 5–6, 1.5–2 mm, acuminate, gen ± fused at base, scarious; rays 10–20, 1.5–5 cm. **FR**: 2–3 mm wide, ovate; ribs gen wavy. 2*n*=22. Common. Moist, esp disturbed places; < 1500 m. CA-FP, GB (uncommon); Am; native to Eur. Apr–Jul ❖

CORIANDRUM

Ann, taprooted, glabrous. **ST**: erect, branched. **LF**: blade oblong to ovate, ternately or pinnately lobed to pinnately dissected. **INFL**: umbels compound, terminal and lateral; bracts 0; bractlets few, small; rays, pedicels few, spreading-ascending. **FL**: marginal fls bilateral, others radial; calyx lobes prominent, unequal; petals oblong, white or rosy, marginal 2-lobed, tips of all narrowed; styles elongate. **FR**: ± round; halves not separating readily; ribs thread-like, in a hard fr wall; oil tubes 0; fr axis divided to base. **SEED**: face concave. 1–2 spp.: Medit, 1 widely cult. (Greek name for this anciently cult condiment)

C. sativum L. CORIANDER, CILANTRO Pl 2–8 dm. **LF**: basal clustered, 3–15 cm, oblong to ovate, ternately or pinnately lobed to 1-pinnate with lflets 1–2 cm, ovate to round, petiole 2–10 cm; distal cauline ovate, pinnately dissected with segments 2–15 mm, thread-like to linear, petiole 0. **INFL**: peduncles 0 or 3–10 cm; bractlets 2–4 mm, linear; rays 2–8, 1–2.5 cm; pedicels 2–5 mm. **FL**: calyx lobes widely lanceolate. **FR**: 2.5–5 mm wide. 2*n*=22. Disturbed places, gen near gardens; gen < 1000 m. NCoRI, ScV; to Mex, trop Am; native to Medit. May–Jul

CYCLOSPERMUM

Ann, taprooted, glabrous. **ST**: decumbent to erect, gen loosely branched. **LF**: sessile or petioled; blade oblong to triangular-ovate, pinnately or ternate-pinnately dissected or compound, segments or lflets gen thread-like to linear. **INFL**: umbels compound or some simple; bracts, bractlets 0; rays few; rays, pedicels spreading. **FL**: calyx lobes 0; petals oblong, white, tips not incurved. **FR**: elliptic to narrowly ovate, compressed side-to-side, glabrous [hairy]; ribs ± equal, thread-like to prominent and corky; oil tube 1 per rib-interval; fr axis shallowly notched. **SEED**: face flat. 3 spp.: S.Am, 1 ± worldwide aggressive weed. (Greek: circular seed) "Ciclospermum" in TJM (1993).

C. leptophyllum (Pers.) Britton & P. Wilson (p. 179) Pl 0.5–6 dm. **LF**: 3.5–10 cm; petiole 2.5–12 mm, gen 0 on cauline lvs, sheath margin scarious; segments 3–15 mm, thread-like to linear, entire. **INFL**: < 2 cm, gen some sessile; rays 1–3; pedicels 6–20, 2–16 mm, spread-ing. **FR**: 1.2–3 mm wide, elliptic to ovate. 2*n*=14. Roadsides; < 350 m. NCoRO, GV, c CCo, SnFrB, SCoRO, SCo; ± worldwide, warm temp. Apr–Aug

CYMOPTERUS

Per, taprooted. **ST**: gen 0 or short. **LF**: gen basal, membranous to ± leathery or fleshy; blade oblong to widely ovate or round, palmately or pinnately lobed to 1–2-pinnately or -ternate-pinnately dissected or compound, segments or lflets linear to obovate, entire to variously lobed, gen spine-tipped. **INFL**: umbels compound, gen terminal, scapose, open to spheric, dense, peduncled; bracts, bractlets conspicuous, scarious (or 0); rays few to many, occ 0, pedicels occ 0. **FL**: calyx lobes prominent to 0; petals oblong to obovate, white, yellow, or purple, tips narrowed; ovary tip projection 0. **FR**: oblong to ovate, ± cylindric to compressed front-to-back; ribs unequal[equal], marginal, some or all thin- or corky-winged (some or all wingless); oil tubes 1–several per rib-interval; fr axis 0 or divided to base. **SEED**: face flat to longitudinally concave or grooved. ± 50 spp.: w N.Am. (Greek: wave wing) Some spp. outside CA are TOXIC to livestock. [Mathias 1930 Ann Missouri Bot Gard 17:213–476] Generic boundaries fluctuating.

1. Rays ± 0, umbels dense, spheric; bractlets 0 or poorly developed
  2. Lf minutely hairy or roughened; bracts conspicuous, fused proximally; fr glabrous . . . . *C. cinerarius*
  2′ Lf glabrous; bracts 0; fr hairy
    3. Lf round, ternate . . . . *C. ripleyi*
    3′ Lf oblong-ovate, ternate to 2-pinnate or 2-pinnately dissected
      4. St 0; fr wings unequal . . . . *C. deserticola*
      4′ St short, underground; fr wings ± equal . . . . *C. globosus*
1′ Rays ± developed, umbels open to ± dense, not spheric; bractlets gen evident
  5. Pl not woody; corolla ± purple; lf simple to 2-pinnately dissected or compound, lobes, segments, or lflets wider than linear
    6. Bracts 0; lf round-reniform, ternate . . . . *C. gilmanii*
    6′ Bracts conspicuous, scarious, veiny; lf oblong-ovate, 1–2-pinnate or -ternate
      7. Bracts and bractlets ± green or ± purple, with many green or purple veins; pedicels < 1 mm . . . . . *C. multinervatus*
      7′ Bracts and bractlets white, with 1–5 green or white veins; pedicels 3–8 mm . . . . *C. purpurascens*

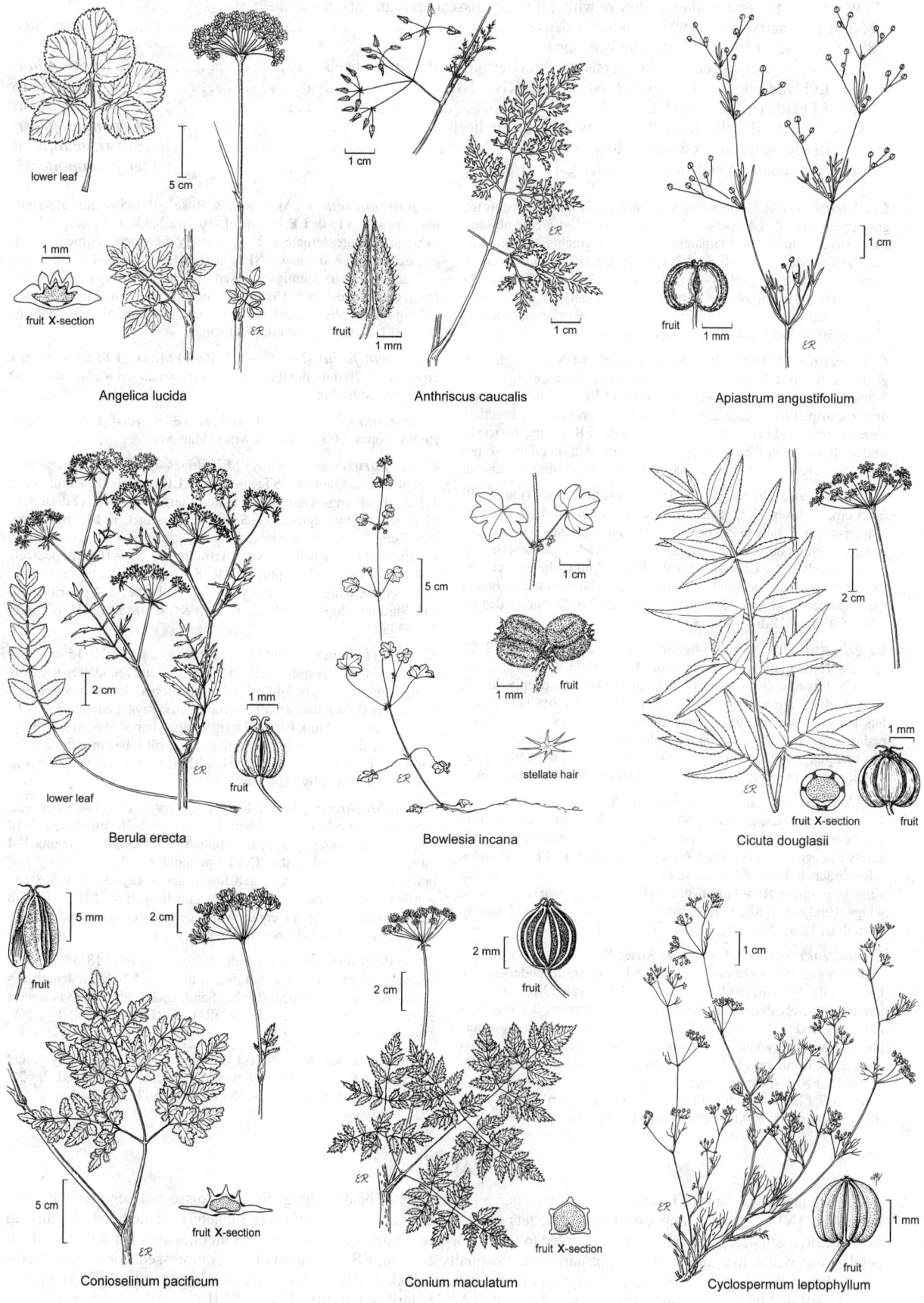

lower leaf

5 cm

1 mm

fruit **X-section**

Angelica lucida

1 cm

fruit

1 mm

1 cm

ER

Anthriscus caucalis

1 cm

fruit

1 mm

ER

Apiastrum angustifolium

2 cm

1 mm

ER

lower leaf

fruit

Berula erecta

5 cm

1 cm

fruit

1 mm

stellate hair

ER

Bowlesia incana

2 cm

1 mm

ER

fruit **X-section**

fruit

Cicuta douglasii

5 mm

2 cm

fruit

5 cm

ER

fruit **X-section**

Conioselinum pacificum

2 mm

fruit

2 cm

2 cm

ER

fruit **X-section**

Conium maculatum

1 cm

1 mm

ER

fruit

Cyclospermum leptophyllum

5′ Pl woody at base; corolla ± yellow or white; lf finely dissected, segments gen ± linear
  8. Lf finely hairy; corolla white; umbels ± dense . . . . ***C. aboriginum***
  8′ Lf glabrous; corolla ± yellow; umbels open
    9. Rays gen unequal; calyx lobes persistent; bractlets gen < mature pedicels . . . . ***C. terebinthinus***
    10. Lf blade length ± = width; KR, NCoRO, CaRH, SNH, n SNE. . . . . var. ***californicus***
    10′ Lf blade length >> width; c SNE . . . . var. ***petraeus***
    9′ Rays ± equal; calyx lobes 0; bractlets > mature pedicels . . . . ***C. panamintensis***
    11. Lf segments not crowded, 3–20 mm, flexible . . . . var. ***acutifolius***
    11′ Lf segments ± crowded, 1–5 mm, rigid . . . . var. ***panamintensis***

C. aboriginum M.E. Jones   Pl 1–3.5 cm, finely hairy or scabrous, gray-green. **ST**: 0. **LF**: petiole 2–13 cm; blade 1–4.5 cm, oblong, ternate to 2-pinnately or 3-ternately dissected, segments 2–8 mm, linear. **INFL**: peduncles 8–30 cm, ≥ lvs; bracts gen 0; bractlets linear, ± scarious, acute; rays 3–10, 4–20 mm; pedicels 3–7 mm. **FL**: corolla white. **FR**: 6–11 mm, oblong to ovate; ribs ± equal, wings 2 × body in width; oil tubes 2–8 per rib-interval. $2n=22$. Rocky mtn slopes; 1500–2850 m. W&I, DMtns; NV. Apr–Jun

C. cinerarius A. Gray   Pl 7–8 cm, minutely hairy or roughened, glaucous. **ST**: 0. **LF**: petiole 3–5 cm; blade 1–2.5 cm, oblong-ovate, 2-pinnately dissected, segments 1–3 mm. **INFL**: peduncles > lvs; bracts conspicuous, scarious-margined, fused proximally; bractlets obscure; rays, pedicels ± 0. **FL**: corolla white. **FR**: 6 mm, narrowly wedge-shaped; ribs ± equal, wings < body in width; oil tubes 5–8 per rib-interval. Rocky mtn slopes; 2100–3500 m. SNH, SNE; NV. Jun–Jul

C. deserticola Brandegee (p. 185)   DESERT CYMOPTERUS   Pl ± 15 cm, glabrous. **ST**: 0. **LF**: petiole 4–10 cm; blade 2–6.5 cm, oblong-ovate, ternate to 2-pinnately dissected, segments 1–4 cm, lanceolate, acuminate. **INFL**: peduncles ≥ lvs; bracts gen bractlets 0; rays, pedicels ± 0. **FL**: corolla purple. **FR**: 5–7 mm, oblong-ovate to wedge-shaped; ribs unequal, marginal wings < body in width, others inconspicuous; oil tubes 3–5 per rib-interval. $2n=22$. Sandy desert; 700–1500 m. w DMoj. Apr ★

C. gilmanii C.V. Morton   GILMAN'S CYMOPTERUS   **ST**: 12–23 cm, glabrous, glaucous; base fibrous. **LF**: petiole 8–18 cm; blade 2.5–4.5 cm, round-reniform, ternate, lflets obovate, deeply lobed, lobes toothed, spine-tipped. **INFL**: peduncles > lvs; bracts 0; bractlets inconspicuous, lf-like, linear to lanceolate; rays ± 8, 1–2 cm; pedicels 2–5 mm. **FL**: corolla ± purple. **FR**: 7–8 mm, widely ovate; ribs unequal, marginal and 1 or 2 other wings > body in width. Limestone, gypsum slopes; 900–2000 m. ne DMtns; NV. Apr–May ★

C. globosus (S. Watson) S. Watson   GLOBOSE CYMOPTERUS   Pl 3–20 cm, glabrous, glaucous. **ST**: short, underground. **LF**: petiole 1–10 cm; blade 1–7 cm, oblong-ovate, ternate to 2-pinnate or 2-pinnately dissected, segments 0.5–6 mm, ± indistinct. **INFL**: peduncles ≥ lvs; bracts 0; bractlets linear, small; rays, pedicels ± 0. **FL**: corolla white or purple. **FR**: 6–11 mm, narrowly wedge-shaped; ribs ± equal, wings < body in width; oil tubes gen 1 per rib-interval. $2n=22$. Sandy open flats; 1200–2100 m. SNE; to UT. Mar–May ★

C. multinervatus (J.M. Coult. & Rose) Tidestr.   PURPLE-NERVE CYMOPTERUS   Pl 4–20 cm, glabrous. **ST**: 0 or short, underground. **LF**: petiole 2–7 cm; blade 1–8.5 cm, oblong-ovate, 2-pinnately or ternate-pinnately dissected, glaucous, fleshy, segments 0.5–6 mm, ± indistinct. **INFL**: peduncles ≥ lvs, 2–14 cm; bracts ± green or ± purple, scarious, many-veined, forming a shallow sheath or cup; bractlets like bracts; fertile rays 1–5, 0.5–2.5 cm; pedicels < 1 mm. **FL**: corolla ± purple. **FR**: 8–17 mm wide, oblong-ovate to ovate; ribs ± equal, wings 2–3 × body in width; oil tubes 3–9 per rib-interval. Sandy and rocky slopes; 630–1800 m. DMoj; to UT, NM. Mar–Apr ★

C. panamintensis J.M. Coult. & Rose   Pl 0.5–4 dm, glabrous; base woody. **ST**: 0. **LF**: petiole 1–10 cm; blade 1–14 cm, oblong-ovate to obovate, ternate to 2–3-pinnately dissected, segments ± linear, acute, gen ± distinct. **INFL**: peduncles > lvs, 3–25 cm; bracts 0; bractlets linear, acuminate, fused at base; fertile rays 5–15, 1–6.5 cm, ± equal; pedicels 4–13 mm. **FL**: corolla ± yellow. **FR**: 6–10 mm, oblong-ovate; ribs ± equal, wings ≥ body in width; oil tubes per rib-interval 1–5. Vars. questionably distinct.

var. ***acutifolius*** (J.M. Coult. & Rose) Munz   **LF**: segments not crowded, 3–20 mm, flexible. $2n=22$. Rocky canyon walls; 700–1000 m. DMoj. Mar–Apr

var. ***panamintensis***   **LF**: segments ± crowded, 1–5 mm, rigid. Rocky slopes; 800–2500 m. DMtns. Mar–May

C. purpurascens (A. Gray) M.E. Jones   Pl 3–15 cm, with persistent fibers, glabrous. **ST**: 0–15 cm. **LF**: petiole 1–4 cm; blade 1.2–5 cm, oblong-ovate, 1–2-pinnately (ternate-pinnately) dissected, glaucous, fleshy, segments 1–8 mm, ± indistinct. **INFL**: peduncles 1.5–7 cm, ≥ lvs; bracts white, with 1–5 green or white veins, fused proximally; bractlets like bracts; fertile rays 3–5, 4–10 mm; pedicels 3–8 mm. **FL**: corolla ± purple. **FR**: 8–18 mm wide, widely ovate; ribs ± equal, wings 2–3 × body in width; oil tubes 3–4 per rib-interval. Shrubby slopes; 1300–2200 m. W&I, e DMoj; to ID, NV, AZ. Mar–May

C. ripleyi Barneby   RIPLEY'S CYMOPTERUS   Pl 10–15 cm, glabrous. **ST**: 0. **LF**: petiole 3–10 cm; blade 2–5 cm, round, ternate, lflets wedge-shaped, deeply 3-lobed, lobes also lobed. **INFL**: peduncles > lvs; bracts 0; bractlets small, membranous; rays, pedicels ± 0. **FL**: corolla purple (white). **FR**: 6–7 mm, wedge-shaped to obovate, hairy; ribs unequal, marginal winged, others not; oil tubes minute. $2n=22$. Sandy soil; 1000–1600 m. s SNH, s SNE, n DMtns; NV. [*C. r.* var. *saniculoides* Barneby] Apr–Jun ★

C. terebinthinus (Hook.) Torr. & A. Gray   Pl 1.5–4.5 dm, glabrous; base woody. **ST**: 0 to short. **LF**: petiole 2–16 cm; blade 1.5–18 cm, ± ovate, pinnately or ternate-pinnately dissected, segments 1–4 mm, linear, ± rigid, acute. **INFL**: peduncles 1–3.5 cm, gen < lvs; bracts 0; bractlets 2–6 mm, gen linear, acute; rays 3–24, 0.5–8 cm, gen unequal; pedicels 1–8 mm. **FL**: corolla yellow. **FR**: 5–10 mm wide, ± ovate; ribs gen ± equal, wings gen irregularly curled, ≥ body in width; oil tubes 3–12 per rib-interval.

var. ***californicus*** (J.M. Coult. & Rose) Jeps. (p. 185)   Pl herbage gray-green to bright green. **ST**: 0 to short. **LF**: blade length ± = width, segments crowded. $2n=22$. Sand, rocks, serpentine at lowest elevations; 150–3500 m. KR, NCoRO, CaRH, SNH, n SNE; n NV. May–Jun

var. ***petraeus*** (M.E. Jones) Goodrich   Pl herbage gray-green. **ST**: short. **LF**: blade length >> width, segments not crowded. $2n=22$. Rocky alpine slopes; 1800–3400 m. c SNE; to UT. May–Jun

DAUCUS

Ann, bien, taprooted, hairy. **ST**: decumbent or erect, gen ± branched. **LF**: blade oblong, pinnately dissected, segments linear to lanceolate. **INFL**: umbels compound; bracts, bractlets gen present; bracts conspicuous, gen pinnately lobed; bractlets entire to toothed; rays gen many, spreading, in fr incurved to form nest-like umbel. **FL**: outer occ ± bilateral; calyx lobes 0 or evident; petals wide, white, margins occ ± red, tips narrowed, unequally 2-lobed. **FR**: oblong to ovate, compressed front-to-back; ribs 10, 1° thread-like, bristly, 2° winged, prickly; oil tubes 1 beneath each 2° rib; fr axis entire or notched at tip. ± 20 spp.: Am, Eurasia, n Afr, Australia. (Greek: carrot) [Sáenz Laín 1980 Anales Jard Bot Madrid 37:481–533]

1. Bien; bract segments elongate, narrowly linear to ± thread-like; umbel central fl gen purple; fr widest at middle
. . . . ***D. carota***
1′ Ann; bract segments short, linear to lanceolate; umbel central fl white; fr widest proximal to middle . . . . ***D. pusillus***

D. carota L. CARROT, QUEEN ANNE'S LACE Pl 1.5–12 dm, gen branched. **LF**: petiole 3–10 cm; blade 5–15 cm, segments 2–12 mm, linear to lanceolate, acute, entire or with a few irregular cuts, bristly to glabrous. **INFL**: peduncles 2.5–6 cm, bristles reflexed to spreading; rays 3–7.5 cm; pedicels 3–10 mm. **FR**: 3–4 mm. 2*n*=18. Roadsides, disturbed places; < 1650 m. CA-FP, SNE; to e N.Am; native to Eur. Sporadic in CA; widely naturalized, occ hybridizing with cultivars. May–Sep

D. pusillus Michx. (p. 185) Pl 0.3–9 dm, gen simple or few-branched. **LF**: petiole 4–15 cm; blade 3–10.5 cm, segments 1–5 mm, linear, acute, entire, ± bristly. **INFL**: peduncles 1–4.5 cm, bristles reflexed to spreading; rays 0.4–4 cm; pedicels 2–9 mm. **FR**: 3–5 mm. *n*=22. Rocky or sandy places; < 1650 m. CA-FP (esp coastal), DMtns; to BC, se US, S.Am. Apr–Jun

ERYNGIUM

Robert E. Preston, Michael S. Park & Lincoln Constance

Bien, per; taprooted, roots clustered, or rhizomes; gen glabrous, ± spiny. **ST**: decumbent to erect, gen branched. **LF**: basal rosette, cauline; petioles present or 0; blades linear to triangular-ovate or round, gen pinnately or palmately lobed or dissected (entire), gen sharp-toothed or spine-tipped, net-veined; juvenile lvs linear, segmented. **INFL**: heads 1–many in cymes, racemes [panicles]; bracts each gen subtend 1 fl, with scarious membrane enclosing ovary, outer > to >> inner, spiny or not on margins and abaxially; rays, pedicels 0. **FL**: sepals spine-tipped, gen persistent; petals oblong to ovate or oblanceolate, white to blue or purple, tip long; anthers, styles gen green, occ blue; ovary tip projection 0. **FR**: obconic to obovate or narrowly elliptic [round], compressed or not, densely scaly; scales at fr tip and along juncture of carpels gen larger, longer than on face or base; ribs 0; oil tubes obscure; fr central axis not obvious. **SEED**: face gen flat. ± 230 spp.: Am, Eurasia, Australia, New Zealand. (Ancient Greek name used by Theophrastus) [Marsden & Simpson 1999 Madroño 46:61–64] CA spp. variable, intergrading, need study.

1. Pl stout, erect; bracts and sepals blue-tinged; inner bracts lobed at or distal to middle . . . . ***E. articulatum***
1′ Pl slender to stout, prostrate to erect; bracts and sepals green, styles or anthers occ blue; inner bracts not lobed
  2. Bract margins thickened, marginal spines 0
    3. Lf gen unlobed, margin sharply serrate to irregularly cut . . . . ***E. armatum***
    3′ Lf pinnately to bipinnately lobed
      4. Pl decumbent, ± green; main st branched 1–6 cm distal to rosette; SCo . . . . ***E. pendletonense***
      4′ Pl erect, silvery; main st branched 7–32 cm distal to rosette; n&c SNF . . . . ***E. pinnatisectum***
  2′ Bract margins not thickened, at least outer gen with spines
    5. Lf blade >> petiole, deeply pinnately or bipinnately lobed, lobes broad to narrow
      6. Sepals pinnately lobed or toothed to entire, lanceolate, tapered to tip-spine
        7. Bracts spiny on margin and abaxial surface; fr scales lanceolate, acuminate . . . . ***E. castrense***
        7′ Bracts spiny on margins only, or 0–few abaxially; fr scales oblong to ovate, acute . . . . ***E. spinosepalum***
      6′ Sepals gen entire, ovate, abruptly narrowed to tip-spine . . . . ***E. vaseyi***
        8. Main st branching 1–5 cm distal to rosette; sepal tips erect; n SNF, se ScV, ne SnJV . . . . var. ***vallicola***
        8′ Main st 0, pl branched within rosette; sepal tips gen reflexed; SCoRO . . . . var. ***vaseyi***
    5′ Lf blade < petiole or petiole ± 0, lf spiny-margined to dentate, serrate or irregularly cut, lobes 0 or
      shallow, broad
      9. Pl prostrate or decumbent, with roots and juvenile lvs at nodes; infl raceme-like . . . . ***E. racemosum***
      9′ Pl decumbent to erect, not rooting nor with juvenile lvs at nodes; infl cyme-like
        10. Bracts with many spines on margins and abaxial surface. . . . . ***E. mathiasiae***
        10′ Bracts with 0 or few spines on abaxial surface
          11. Main st gen 0, pl branched within rosette; basal lvs gen > branches. . . . . ***E. alismifolium***
          11′ Main st present, branched distal to base; basal lvs < branches
            12. Pl densely puberulent throughout; heads 5–7-fld. . . . . ***E. constancei***
            12′ Pl glabrous, or occ puberulent only on lvs or bracts; heads > 10-fld
              13. Pl stout, erect; lf blade tapering to base, petiole obscure; sepal tip-spine gen 1–2 mm; outer bracts
                > 2 × inner bracts . . . . ***E. jepsonii***
              13′ Pl slender to stout, decumbent to erect; lf blade abrupt-narrowed at base, < petiole; sepal tip-spine
                < 1 mm; outer bracts < 2 × inner bracts. . . . . ***E. aristulatum***
                14. Styles in fr >> calyx; NCo, NCoR, ScV, SnFrB . . . . var. ***aristulatum***
                14′ Styles in fr ± = calyx; SnFrB, SCoR, SW
                  15. Bract margins spiny. . . . . var. ***hooveri***
                  15′ Outer bract margin spines 0–3 pairs, inner bract spines 0 . . . . var. ***parishii***

E. alismifolium Greene ALISMA-LEAVED BUTTON-CELERY **ST**: Pl 0.5–3 dm, gen branched within rosette, decumbent to erect with downward-arched branches. **LF**: basal gen > branches; blade 6–15 cm, < petiole, lanceolate to narrowly obovate, sharply serrate or irregularly cut (pinnately lobed). **INFL**: heads 5–12 mm, spheric, in cymes; peduncle 0.5–1.5 cm; bracts lance-linear, outer 7–16 mm, > heads, with 0–many marginal spines, inner 4–8 mm, spines gen 0. **FL**: sepals 1.5–3 mm, lanceolate, entire; petals oblong, white; styles 3–3.5 mm, 2 × sepals. **FR**: 2–2.5 mm, obovate; scales dense, gen ± equal, lanceolate, acuminate. 2*n*=32. Vernal pools, wet areas, dry lakebeds; 785–2020 m. CaRH, n SNH (Donner Lake), MP (exc Wrn); to OR, ID, NV. [*Eryngium alismaefolium* Greene, orth. var.] Jul–Aug

E. aristulatum Jeps.   Gen glabrous, occ puberulent only on lvs or bracts. **ST:** decumbent to erect, 1–9 dm, slender to stout, main st branching 2–5 cm distal to rosette. **LF:** basal < branches; petiole 5–27 cm; blade 3–10 cm, < petiole, lanceolate to oblanceolate, coarsely sharp-serrate, irregularly cut, or lobed. **INFL:** heads 5–12 mm, ± spheric, in cymes, glabrous, puberulent or rough; peduncle 0.5–1.5 cm; bracts 6–27 mm, linear to lance-linear, margins with 0–few spines proximal to middle, 0–few spines adaxially. **FL:** sepals 1.7–2.8 mm, lanceolate to ovate, entire, tip-spine < 1 mm; petals oblanceolate, white; styles 1.5–3.5 mm, occ ± purple. **FR:** 1.5–2.5 mm, narrowly elliptic; scales dense, unequal, lanceolate to ovate, acuminate, minutely bristled. Highly variable.

var. ***aristulatum***   Glabrous (puberulent). **ST:** slender, sprawling, to stout, erect. **INFL:** outer bract margins spiny, occ few spines adaxially, inner bracts with 0–few marginal spines. **FR:** styles >> calyx; scales densely to sparsely bristly. 2*n*=32,64. Vernal pools, lakeshores, drying lakes, wet depressions; < 1070 m. NCo, NCoR, ScV, SnFrB. May–Aug

var. ***hooveri*** M.Y. Sheikh   HOOVER'S BUTTON-CELERY   Pl stout, ascending to erect. **INFL:** bract margins spiny. **FR:** styles ± = calyx. 2*n*=32. Vernal pools, seasonal wetlands, occ alkaline; < 50 m. SnFrB, SCoRO (San Luis Obispo Co.). Relationship between SnFrB and SCoRO populations needs study. Jul   ★

var. ***parishii*** (J.M. Coult. & Rose) Mathias & Constance   SAN DIEGO BUTTON-CELERY   **ST:** erect or spreading. **LF:** petiole 8–10 cm; blade 3–5 cm, < petiole, lanceolate to oblanceolate, gen pinnately lobed, occ coarsely sharp-serrate. **INFL:** heads 5–9 mm; bracts 1–3 cm, outer bract margin spines 0–3 pairs, inner bract spines 0. **FL:** sepals 1.5–2.5 mm, glabrous or puberulent. **FR:** 2 mm, obovate; persistent styles ± = calyx; scales awl-like, or ovate and acuminate. 2*n*=32. Vernal pools, marshes; < 705 m. s SCo, PR; Baja CA. Now only on mesas near San Diego, Santa Rosa Plateau. May–Jun   ★

E. armatum (S. Watson) J.M. Coult. & Rose   COASTAL BUTTON-CELERY   Gen glabrous, bracts occ puberulent. **ST:** decumbent to erect, 1–5 dm, main st gen branched 2–5 cm distal to rosette. **LF:** thick; petiole < blade to 0; blade 10–30 cm, oblanceolate, sparsely sharp-serrate to irregularly cut (pinnate). **INFL:** heads 5–15 mm, ± spheric, in cymes; peduncle 0–1 cm; bracts 7.5–23 mm, 1.5–2 × heads, lanceolate, margins thickened, spines 0. **FL:** sepals 3–4.5 mm, lanceolate to lance-ovate, gen entire; petals oblanceolate, white (± purple); styles 3.5–4 mm, > to < sepals. **FR:** 1.5–2.5 mm, narrowly-elliptic; scales dense, unequal, ovoid, acute. 2*n*=32. Coastal prairie, bluffs, grassland, marsh margins, often on clay soils; < 200 m. NCo, s NCoRO, s ScV (Suisun Marsh, Montezuma Hills), CCo, SnFrB, n SCo. May–Aug

E. articulatum Hook.   BEETHISTLE   Glabrous. **ST:** stout, erect, 0.3–1.2 m, main st branching 2–5 cm distal to rosette. **LF:** petiole >> blade; blade 5–10 cm, lanceolate to narrowly ovate, coarsely sharp-serrate or irregularly cut. **INFL:** heads 1–2.5 cm, ovoid to ± spheric, in cymes; peduncle 0.5–2.5 cm; bracts 8–25 mm, outer 2–3 mm wide, inner 1–1.2 mm wide, gen 3-lobed at or distal to middle, ± blue-tinged. **FL:** sepals 3–3.5 mm, lance-linear, gen entire, blue-tinged; petals oblanceolate, bright blue or purple; styles 2.5–3.5 mm, ± ≥ sepals. **FR:** 2.5–4 mm, narrowly elliptic; scales dense, lanceolate, acuminate, with club-shaped papillae. 2*n*=32. Seasonally wet areas on lake and stream margins, marshes; < 1800 m. KR (Trinity Co.), NCoRO (Napa Co.), CaR, ScV, MP; to WA, ID. Jun–Aug

E. castrense Jeps.   GREAT VALLEY COYOTE-THISTLE   Glabrous. **ST:** ascending to erect, 2–6 dm, many-branched 2–5 cm distal to rosette, spiny, glabrous. **LF:** petiole short to 0, << blade; blade 10–30 cm, oblong to lanceolate, deeply pinnately or bipinnately lobed. **INFL:** heads 8–15 mm, ± spheric, in cymes; peduncle 0.5–2.5 cm; bracts 15–30 mm, linear, densely spiny on margins and abaxially. **FL:** sepals 3–3.5 mm, lanceolate, margin scarious, entire (pinnately lobed or toothed in CaRF & c SNF), tapered to tip-spine; petals oblanceolate, white or faintly purple; styles 3.5–4.5 mm, gen > sepals. **FR:** 3–3.5 mm, narrowly elliptic; scales dense, ± unequal, lanceolate, acuminate, papillate. 2*n*=32,64. Vernal pools, wet swales, intermittent streambeds; < 900 m. CaRF, n&c SNF, e GV, e SnFrB (Diablo

Range). Intergrades with *E. spinosepalum* in s SNF and *E. vaseyi* var. *vallicola* in SnJV. Apr–Jul

E. constancei M.Y. Sheikh (p. 185)   LOCH LOMOND BUTTON-CELERY   Densely puberulent. **ST:** branches 2–3 dm, from 1–2 cm distal to rosette, decumbent or ascending, slender. **LF:** petiole 8–12 cm; blade 3–4 cm, << petiole, ± lanceolate, spiny-margined to sharp-serrate or -lobed. **INFL:** heads 5–7-fld, 3–5 mm, spheric, in cymes; peduncle 5–8 cm; bracts 5–7 mm, lance-linear, margin spiny. **FL:** sepals ± 2 mm, lanceolate, entire; petals oblanceolate, gen white; styles 3–3.5 mm, >> sepals. **FR:** 1.6–2.2 mm, obovate; scales dense, unequal, ovate, acuminate, scabrous. 2*n*=32. Vernal pools; ± 800 m. NCoR (Lake Co.). Closely related to *E. aristulatum* var. *a.* Apr–Jun   ★

E. jepsonii J.M. Coult. & Rose   Glabrous. **ST:** erect, 2–8 dm, few–many-branched. **LF:** blade 10–30 cm, narrowly oblanceolate, tapering to obscure petiole, margins serrate, spiny, occ crisped or curled. **INFL:** heads in cymes, 5–15 mm, spheric; outer bracts 10–30 mm, inner 4–10 mm. **FL:** sepal body 2–2.5 mm, lanceolate, entire, tip-spine gen 1–2 mm. **FR:** 2 mm, obovate; scales ovate to lanceolate, tapered to spiny tip, glabrous. Moist clay soil; < 500 m. s NCoRI, deltaic GV, SnFrB. [*E. aristulatum* var. *a.*, in part, misappl.] If recognized taxonomically, pls with reflexed bracts that key here assignable to *E. elongatum* J.M. Coult. & Rose. Apr–Aug

E. mathiasiae M.Y. Sheikh   MATHIAS' BUTTON-CELERY   Glabrous. **ST:** stout, ascending to erect, 3–4 dm, main st branched 1.5–2 cm distal to rosette. **LF:** petiole 6–10 cm; blade 3–5.5 cm, < petiole, lanceolate to narrowly obovate, sharp-serrate to -lobed. **INFL:** heads 8–12 mm, ± spheric, in cymes; peduncle 0.8–1.8 cm; bracts linear, 9–23 mm, 1–1.3 mm wide, densely spiny abaxially and on margins. **FL:** sepals 3–3.5 mm, lanceolate, gen entire; petals oblong, white; styles 2.5–3 mm, ± = sepals. **FR:** 2.5–3 mm, elliptic; scales dense, unequal, lanceolate, acuminate, smooth or scabrous. 2*n*=64. Vernal pools, wet areas, dry lakebeds; 970–1510 m. CaRH, MP. Jun–Aug

E. pendletonense K.L. Marsden & M.G. Simpson   PENDLETON BUTTON-CELERY   Glabrous. **ST:** sprawling, 0–2 dm, main st branched 1–6 cm distal to rosette. **LF:** 8–25 cm, oblanceolate, pinnately to bipinnately lobed, lobes gen lanceolate to narrowly elliptic. **INFL:** heads in cymes, 9–19-fld; peduncle 2–3 cm; bracts 5–21 mm, narrow-triangular to lanceolate, margins thickened, entire. **FL:** sepals 2 mm, oblong to ovate, with wide, scarious margins; petals 1 mm, white. **FR:** obovoid, scales lanceolate to lance-ovate, acuminate. 2*n*=32. Clay soil, coastal bluffs, coastal-sage scrub; < 50 m. SCo. Apr–Jun   ★

E. pinnatisectum Jeps. (p. 185)   TUOLUMNE BUTTON-CELERY   Glabrous, or infl short-stiff-hairy. **ST:** erect, 1.5–5 dm, stout, main st branched 7–32 cm distal to rosette. **LF:** petiole << blade, blade 10–30 cm, lanceolate, sharply pinnately lobed, lobes gen opposite, white, margins thickened. **INFL:** heads 1.5–2 cm, spheric, in cymes; peduncle 0.2–1 cm; bracts 9–17 mm, lance-linear, margins, veins thickened, marginal spines 0. **FL:** sepals 3.5–4 mm, entire, lanceolate; petals oblanceolate, white; styles 3.5–4 mm, gen = sepals. **FR:** 3–3.5 mm, narrow-elliptic; scales dense, unequal, lanceolate, acuminate. 2*n*=32. Vernal pools, swales, intermittent streams; 70–950 m. n&c SNF (Sacramento, Amador, Calaveras, Tuolumne cos.). Jun–Aug   ★

E. racemosum Jeps.   DELTA BUTTON-CELERY   Glabrous. **ST:** prostrate or decumbent, 1–5 dm, slender, main st branched 1.5–3 cm distal to rosette, roots and juvenile lvs at nodes. **LF:** blade 3–5 cm, < petiole, lanceolate to ± oblong, spiny-margined to sharp-serrate or -lobed. **INFL:** heads 5–8 mm, spheric-ovoid, raceme-like; peduncle 0.5–1 cm; bracts 6–12 mm, linear to narrow-lanceolate, margin spiny. **FL:** sepals 1–1.5 mm, ovate, entire; petals oblong to oblanceolate, white to faint-purple; styles 2.5 mm, > sepals. **FR:** 1.5 mm, obovate; scales dense, unequal, lanceolate, scabrous. 2*n*=32. Seasonally flooded clay depressions in floodplains; 3–30 m. n SNF (Calaveras Co.), n SnJV. Jun–Aug   ★

E. spinosepalum Mathias   SPINY-SEPALED BUTTON-CELERY   Glabrous. **ST:** erect, 3–7.5 dm, stout, main st branched 2–5 cm distal to rosette. **LF:** petiole 0.5–2 cm; blade 9–35 cm, >> petiole, oblong to oblanceolate, pinnately lobed. **INFL:** heads 0.8–2 cm, ovoid to

spheric, in cymes; peduncle 0–2 cm; bracts 7–30 mm, lance-linear, spines 3–4 pairs on margins, gen 0 abaxial spines. **FL:** sepals 3.5–4.5 mm, lanceolate, pinnately lobed or toothed (entire in SnJV), tapered to tip-spine; petals oblong, white; styles 3–4 mm, < sepals. **FR:** 2.5–3 mm, narrowly elliptic; scales dense, unequal, oblong to ovate, acute, acute, glabrous. $2n=32$. Vernal pools, swales, roadside ditches; 100–1270 m. s SNF, SnJV. Intergrades with *E. castrense* in c&s SNF, *E. vaseyi* in sw SnJV, *E. jepsonii* in nw SnJV. Apr–Jul ★

E. vaseyi J.M. Coult. & Rose COYOTE-THISTLE Glabrous. **ST:** decumbent to ascending or erect, 1.5–5 dm, main st branched 0–5 cm distal to rosette. **LF:** petiole 1–4 cm; blade 8–24 cm, >> petiole, lanceolate to oblong, deeply pinnately and sharply lobed. **INFL:** heads 8–13 mm, ± spheric, in cymes; peduncle gen 1–4 cm; bracts 8–25 mm, linear, margins spiny. **FL:** sepals 2–3 mm, ovate, gen entire, abruptly narrowed to tip-spine; petals oblanceolate to oblong, white;

styles 2–2.5 mm, ≥ sepals. **FR:** 2–2.5 mm, obovate; scales dense, ± equal to unequal, ovate or awl-like, acute. $2n=32$. Occ misapplied to *E. aristulatum* and *E. castrense*, now with persistent nomenclatural confusion.

var. ***vallicola*** (Jeps.) Munz **ST:** erect, branched 1–5 cm distal to rosette. **FL:** sepal tips erect. **FR:** scales unequal, ovate or awl-like, acute. Vernal pools; 10–335 m. n SNF (Amador, Calaveras cos.), se ScV, ne SnJV. Pls keying here from c SNH (Tuolumne Co.) may be an undescribed taxon. May–Jul

var. ***vaseyi*** (p. 185) **ST:** 0, branched inside rosette, branches gen decumbent. **FL:** sepal tips gen reflexed. **FR:** scales ± equal, ovate, acute. Vernal pools; 220–320 m. SCoRO. Pls keying here from SCo (Santa Barbara Co.) possibly related to *E. aristulatum* var. *parishii*. May–Jul

FOENICULUM

1 sp. (Latin: fennel)

F. vulgare Mill. (p. 185) FENNEL Per, taprooted, 0.9–2 m, glabrous, glaucous, anise- or licorice-scented. **ST:** erect, branched, solid. **LF:** petiole 7–14 cm, conspicuously sheathing; blade 3–4 dm wide, triangular-ovate, finely pinnately dissected, segments 4–40 mm, thread-like. **INFL:** umbels compound, peduncled; bracts, bractlets 0; rays 15–40, unequal, 1–4 cm, spreading-ascending to ascending; pedicels 18–25, 1–10 mm, ± equal. **FL:** calyx lobes 0; petals wide, yellow, tips narrowed. **FR:** 3.5–4 mm, oblong-ovate, compressed side-to-side, glabrous; ribs ± equal, prominent, acute; oil tube 1 per rib-interval; fr axis divided to base. **SEED:** face gen flat. $2n=22$. Roadsides, disturbed sites; < 1600 m. CA-FP, W&I; native to s Eur; widely escaped from cult in w hemisphere. Locally abundant, invasive. May–Sep ❖

GLEHNIA

1 sp. (P. von Glehn, Russian botanist, 19th century)

G. littoralis Miq. subsp. ***leiocarpa*** (Mathias) Hultén (p. 185) AMERICAN GLEHNIA Per, taprooted, low, prostrate or ± spreading, sparsely puberulent to tomentose. **ST:** ± 0. **LF:** fleshy; petiole 2.5–15 cm; blade 2.5–15 cm wide, widely ovate, 1–2-ternate or ternate-pinnate, lflets 0.5–5 cm wide, ovate, often 3-lobed, serrate, adaxial surface green, glabrous or glabrous in age, abaxial white-tomentose. **INFL:** umbels compound, peduncled; bracts 0–few; bractlets several, lanceolate, long-acuminate; rays 5–15, 0.5–4.5 cm, stout, hairy; pedicels 0. **FL:** calyx lobes minute; petals widely lanceolate, white, tips narrowed; ovary tip projection 0. **FR:** 4–12 mm, oblong to round, compressed front-to-back, glabrous to sparsely tomentose; ribs ± equal, conspicuously corky-winged; oil tubes several per rib-interval, large; fr axis divided to base. **SEED:** face concave. $2n=22$. Ocean beaches; ± 0 m. NCo; to AK. Other subsp. coastal e Asia. May–Jun ★

HERACLEUM HOGWEED

Per from taproot or clustered roots. **ST:** stout, erect, gen branched, hollow. **LF:** blades oblong to round, ternately, pinnately, or palmately compound (simple), lflets large, lobed or serrate; distal-most cauline gen reduced to large sheaths. **INFL:** umbels compound, large, margins gen sterile; bracts 0–few, gen deciduous; bractlets gen present, persistent; rays, pedicels many, spreading-ascending. **FL:** marginal bilateral, outer petals > others, 2-lobed; calyx lobes gen 0; petals wide, white, ± yellow. **FR:** oblong-ovate to round or obcordate, compressed front-to-back; ribs unequal, marginal thin-winged, veined near outer margin, others thread-like; oil tubes 1–2 per rib-interval, unequal in length; fr axis divided to base. **SEED:** face flat. ± 80 spp.: Eurasia, e Afr, 1 in N.Am. (Hercules, presumably from large stature of some spp.)

H. maximum W. Bartram (p. 185) COW PARSNIP, GIANT HOGWEED Pl 1–3 m, stout, tomentose, strong-scented. **LF:** round to reniform; petioles 1–4 dm, widely sheathing, distal sheaths enlarged, bladeless; blades 2–5 dm wide, ternate, lflets 1–4 dm wide, ovate to round, cordate, lobed, coarsely serrate. **INFL:** tomentose or long-hairy; peduncle 5–20 cm; rays 15–30, 5–10 cm, unequal; pedicels 8–20 mm. **FL:** petals obovate, white. **FR:** 8–12 mm, obovate to obcordate, ± hairy. $2n=22$. Moist places, wooded or open; < 2900 m. CA-FP, GB; to AK, e US, AZ. [*H. lanatum* Michx.] Only native Apiaceae sp. found from e to w N.Am. Apr–Jul

LIGUSTICUM

Per, taprooted, glabrous to minutely scabrous. **ST:** erect, lfy, conspicuously fibrous at base, gen branched. **LF:** blade oblong to round, ternately or pinnately compound or dissected, lflets oblong to obovate, entire to deeply pinnately lobed, segments linear to oblong. **INFL:** compound umbels; bracts gen 0; bractlets 0 or inconspicuous; rays, pedicels few to many, spreading-ascending. **FL:** calyx lobes minute; petals wide, white [± pink]. **FR:** oblong to elliptic, ± compressed side-to-side; ribs ± equal, thread-like to narrowly winged; oil tubes gen several per rib-interval; fr axis divided to base. **SEED:** face flat to concave. ± 25 spp.: Eurasia, N.Am. (Liguria, Italy, home of related *Levisticum*, lovage) [Leute 1970 Ann Naturhist Mus Wien 74:457–519] Genus and spp. poorly defined.

1. Herbage, infl minutely scabrous or puberulent; fr ribs thread-like, ± unwinged . . . . *L. apiifolium*
1′ Herbage, infl glabrous; fr ribs distinctly but narrowly thin-winged
  2. Cauline lvs (1)2–3, ± = basal; rays gen 16–30; lf segments gen wide, obtuse . . . . *L. californicum*
  2′ Cauline lvs (0)1(2), << basal; rays gen 5–18; lf segments gen narrow, acute . . . . *L. grayi*

L. apiifolium (Nutt.) A. Gray   Pl 3–15 dm. **LF**: petiole gen 1–3 dm; blade 0.8–2.5 dm wide, triangular-ovate, ternate-pinnate, lflets 1.5–4.5 cm, ovate, segments obtuse or acute, deeply pinnately lobed, margins minutely scabrous; cauline lvs ± = basal, gen 2–3, distal ± sessile, occ paired. **INFL**: gen ± puberulent or scabrous; peduncles alternate or whorled, 1–3.5 dm; rays 12–23, 2–6 cm, unequal; pedicels 5–10 mm, unequal. **FR**: 3–5 mm, oblong; ribs thread-like, ± unwinged; oil tubes 3–6 per rib-interval. **SEED**: face concave. 2*n*=22,44. Coastal meadows, scrub or woodland; < 1850 m. NW (exc NCoRI), CCo; to WA. Jun–Jul

L. californicum J.M. Coult. & Rose   Pl 6–12 dm, glabrous. **LF**: petiole 0.5–4 dm; blade 1–3 dm wide, triangular-ovate, ternate-pinnate, lflets 1.5–4 cm, oblong to ovate, segments gen wide, obtuse, shallowly to deeply pinnately lobed; cauline lvs ± = basal, gen 2–3, distal ± sessile, occ paired. **INFL**: peduncles gen whorled, terminal 1.5–4 dm, lateral gen staminate; rays gen 16–30, 2–6 cm, unequal; pedicels 5–10 mm, unequal. **FR**: 4–6 mm, oblong-ovate; ribs narrowly winged; oil tubes several per rib-interval. **SEED**: face concave. Chaparral, woodland, often on serpentine; 15–1500 m. NCo, KR, CaRH, n SN; sw OR. Jun–Aug

L. grayi J.M. Coult. & Rose (p. 191)   Pl 2–8 dm, glabrous. **LF**: petiole 0.2–3 dm; blade 1–2.5 dm, oblong to triangular-ovate, 2-pinnate or ternate-pinnate, lflets 1–4 cm, oblong to ovate, segments narrow, acute, deeply pinnately lobed; cauline lvs << basal, gen 1. **INFL**: glabrous; peduncles gen whorled, terminal 1–3 dm, lateral < terminal, gen staminate; rays 5–18, 1.5–5 cm, unequal; pedicels 5–10 mm, unequal. **FR**: 4–5 mm, oblong-ovate; ribs narrowly winged; oil tubes 3–5 per rib-interval. **SEED**: face concave. 2*n*=22,44. Wet soil of subalpine meadows, conifer forest; 1000–3300 m. KR, CaRH, SNH, MP; to WA, MT, NV. Jun–Sep

LILAEOPSIS

Per, glabrous; rhizomes fibrous-rooted. **ST**: prostrate, creeping. **LF**: 1 or tufted, linear to spoon-shaped, cylindric or ± flattened, segmented, entire, without definite blade or petiole, scarious-sheathing at base. **INFL**: umbels simple, open, gen peduncled; bracts several, inconspicuous; pedicels few, spreading to recurved. **FL**: calyx lobes minute; petals wide, white or maroon, short-acuminate, tip not incurved. **FR**: ovate to obovate, ± compressed side-to-side, glabrous; ribs equal or not, 0–all conspicuously spongy-thickened; oil tubes several to many per rib-interval; fr central axis not obvious. **SEED**: face rounded or flat. 13 spp.: Am, Australia, New Zealand, Afr. (Greek: like *Lilaea*) [Affolter 1985 Syst Bot Monogr 6:1–140]

1. Lf cylindric, 0.4–1.2 mm wide, lf internal cross-walls 3–8, obscure . . . . *L. masonii*
1′ Lf cylindric to flattened, 0.7–4.5 mm wide, lf internal cross-walls 4–13, evident . . . . *L. occidentalis*

L. masonii Mathias & Constance (p. 191)   MASON'S LILAEOPSIS **LF**: gen tufted at vertical branch tips, 1.5–7.5 cm, linear or thread-like. **INFL**: peduncles 2–20 mm; bracts 0.5–1 mm; pedicels 3–8, 1–6 mm. **FR**: 1.2–1.6 mm, elliptic or ovate, only marginal ribs rounded, thickened; oil tubes 5–6 per rib-interval. 2*n*=44. Intertidal marshes, streambanks; < 36 m. s ScV, n SnJV, CCo (Inverness, Point Reyes), ne SnFrB. Locally abundant; threatened by development, flood control, agriculture. Jun–Aug ★

L. occidentalis J.M. Coult. & Rose   **LF**: 1 or tufted, 2.5–30 cm, linear or tapered. **INFL**: peduncles 5–65 mm; bracts 0.5–2 mm; pedicels 5–12, 1–12 mm. **FR**: 1.3–2.4 mm, elliptic or round, only marginal ribs wide, thickened; oil tubes gen 6–9 per rib-interval. 2*n*=44. Salty or brackish soil, esp coastal; < 765 m. NCo, CaRH (Shasta Valley), CCo; to BC. Jun–Aug

LOMATIUM

Per, from taproot or gen deep-seated tuber, glabrous to tomentose. **ST**: 0 or erect, simple or branched; base fibrous or not. **LF**: blade oblong to triangular-ovate or obovate, ternately, pinnately, or ternate-pinnately dissected or compound, segments or lflets thread-like to wide; old basal lf sheaths fibrous-persistent or not. **INFL**: umbels compound, peduncled; bracts gen 0; bractlets gen present, 0 to conspicuous; rays, pedicels spreading to erect, gen webbed at base. **FL**: calyx lobes gen 0; petals wide, yellow, white, or purple, tips narrowed; ovary tip projection 0. **FR**: linear to obovate, compressed front-to-back; marginal ribs widely to narrowly thin or thick-winged, others thread-like; oil tubes 1–several per rib-interval; fr axis divided to base. **SEED**: face flat to concave. ± 75 spp.: c&s N.Am. (Greek: bordered, from prominent marginal fr wing) [Constance & Ertter 1996 Madroño 43:515–521] Fr wing width given as width of 1 wing, not both together. *L. roseanum* Cronquist is reported from CaRH, n SNH, MP.

1. At base and tip fr evidently notched and wings narrower; lflets or segments gen large
  2. Lflets coarsely dentate or some 3-lobed but not pinnately so
    3. Lf 1–2-ternate; fr wings thickened, >> body in width; s CA . . . . *L. lucidum*
    3′ Lf ternate-pinnate or 1–2-pinnate; fr wings thin, ± ≥ body in width; n CA
      4. Pl < 8 dm; fr wide-elliptic to round; KR . . . . *L. howellii*
      4′ Pl < 5 dm; fr wide-oblong; NCoR . . . . *L. repostum*
  2′ Lflets pinnately lobed, gen also toothed or irregularly cut
    5. Lf blade > petiole; fr 15–18 mm — s ChI . . . . *L. insulare*
    5′ Lf blade < to ± = petiole; fr 6–14 mm
      6. Lflet spines weak; fr wings ≥ body in width; CCo, SCoR . . . . *L. parvifolium*
      6′ Lflet spines strong; fr wings < body in width; SNH, SNE
        7. Mature rays 25–50 mm; fertile pedicels 5–10 mm . . . . *L. rigidum*
        7′ Mature rays 1–11 mm; fertile pedicels 0.1–1 mm . . . . *L. shevockii*

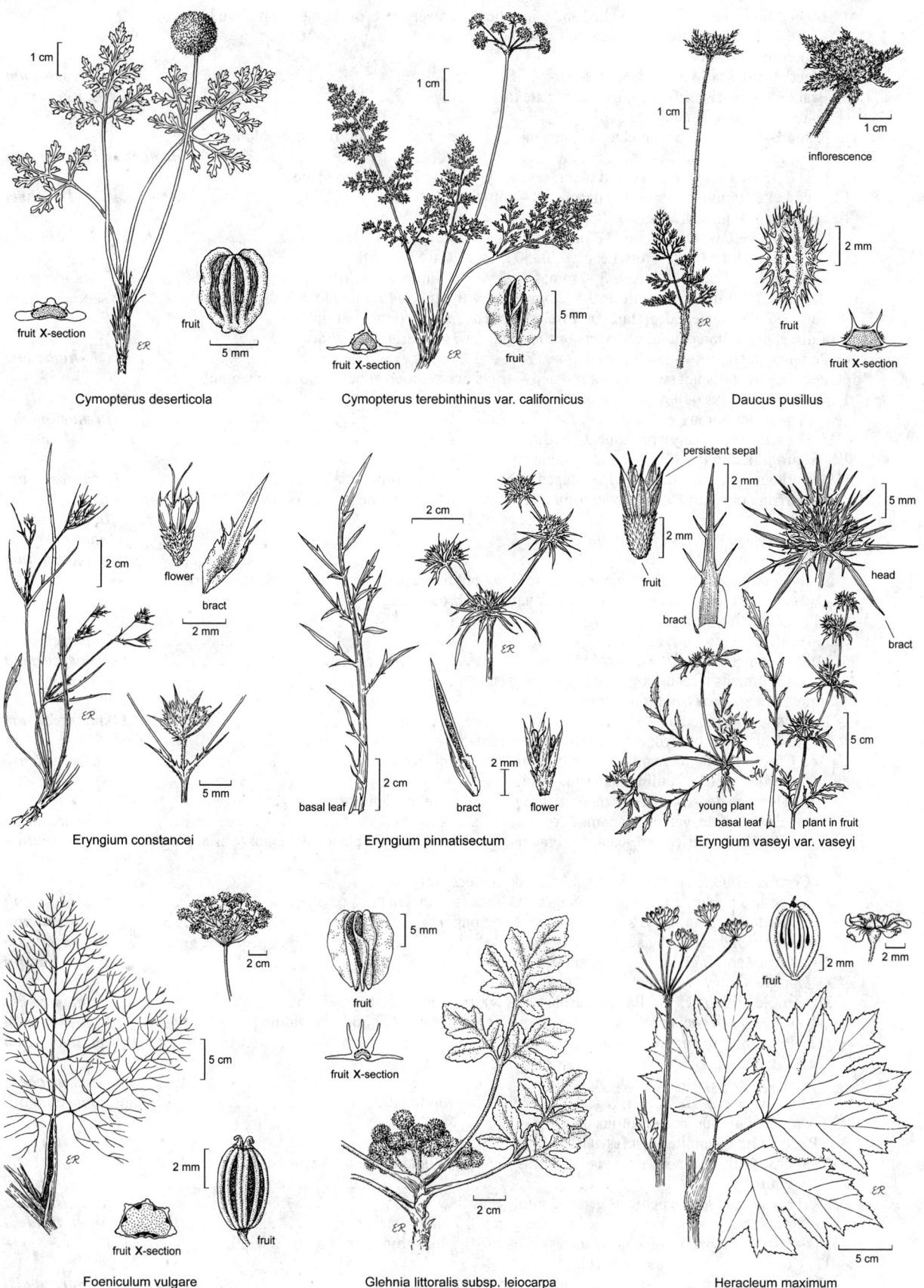

1 cm

fruit X-section
fruit
5 mm

Cymopterus deserticola

1 cm

fruit X-section
fruit
5 mm

Cymopterus terebinthinus var. californicus

1 cm
1 cm

inflorescence

2 mm

fruit

fruit X-section

Daucus pusillus

2 cm

flower
bract
2 mm

5 mm

Eryngium constancei

2 cm

2 cm

basal leaf
bract
flower
2 mm

Eryngium pinnatisectum

persistent sepal
2 mm

2 mm

fruit
bract

5 mm

head

bract

5 cm

young plant
basal leaf
plant in fruit

Eryngium vaseyi var. vaseyi

2 cm

5 cm

2 mm

fruit X-section
fruit

Foeniculum vulgare

5 mm

fruit

fruit X-section

2 cm

Glehnia littoralis subsp. leiocarpa

fruit
2 mm
2 mm

5 cm

Heracleum maximum

1´ At base and tip fr not evidently notched and wings not narrower; lflets or segments small to large
  8. Pl from a definite, spheric to elongate or irregularly thickened, gen ± shallow tuber
    9. Pl puberulent to hairy
      10. Bractlets not scarious, ± fused, reflexing. . . . .[3]***L. macrocarpum***
      10´ Bractlets ± scarious, free to fused, not reflexing. . . . .[5]***L. nevadense***
    9´ Pl glabrous to ± scabrous but not hairy
      11. Pl short-stemmed, from an elongate, gen tapered tuber; rays ± erect; fr linear-oblong
        . . . .[3]***L. bicolor*** var. ***leptocarpum***
      11´ Pl st 0, from a ± spheric to ovoid tuber; rays ± spreading; fr oblong to ovate
        12. Pedicel 5–16 mm; bractlet ± = fl; tuber < 4 cm wide. . . . . ***L. canbyi***
        12´ Pedicel 0–4 mm; bractlet 0 or < fl; tuber ≤ 2.5 cm wide
          13. Corolla yellow; lvs ternate-1–2-pinnate; rays 1–12 mm, n&c SNH . . . . ***L. stebbinsii***
          13´ Corolla white; lvs 2–3-ternate; rays 10–50 mm; KR, n SNH, MP
            14. Rays 2–5, 1–2 cm; pedicels 3–4 mm; fr ± 5 mm, ± 4 mm wide; MP . . . . ***L. hendersonii***
            14´ Rays 3–20, 1–6 cm; pedicels 0.5–2.5 mm; fr 5–9 mm, 3–4 mm wide; KR, n SNH, MP. . . . . ***L. piperi***
  8´ Pl from a gen elongated, deep taproot (rarely with a deep-seated tuber at tip)
    15. Lf dissected or compound, segments or lflets gen few, gen large, gen wide
      16. Peduncle distally swollen, inflated. . . . . ***L. nudicaule***
      16´ Peduncle distally not swollen, not inflated — rays occ webbed at base into prominent disk
        17. St 0 or short; lvs 0–few. 1–3-pinnate
          18. Pl base not fibrous; calyx 0. . . . .[2]***L. martindalei***
          18´ Pl base fibrous; calyx prominent . . . . ***L. parryi***
        17´ St prominently lfy; lf 1–2-ternate-pinnate
          19. Herbage glaucous; lflets wedge-shaped to obovate, 10–50 mm wide . . . . ***L. californicum***
          19´ Herbage green to ± gray, not conspicuously glaucous; lflets linear to lance-ovate, 0.5–10 mm wide
            . . . . ***L. triternatum***
            20. Ovary, gen fr densely puberulent. . . . . var. ***macrocarpum***
            20´ Ovary, fr glabrous. . . . . var. ***triternatum***
    15´ Lf dissected, segments gen many, small, occ long but then narrow (see also *Lomatium parryi*)
      21. Ovary puberulent to scabrous or hairy, fr hairy (glabrous in age)
        22. Pls gen with definite st — st lvs ≥ 1
          23. Petals tomentose. . . . . ***L. dasycarpum***
            24. Lf segments linear; pedicels gen > fr. . . . . subsp. ***dasycarpum***
            24´ Lf segments thread-like; pedicels gen < fr. . . . . subsp. ***tomentosum***
          23´ Petals ± glabrous, not tomentose
            25. Involucel 1-sided, reflexing, not scarious. . . . .[3]***L. macrocarpum***
            25´ Involucel not 1-sided, not reflexing, gen ± scarious
              26. Corolla white to cream; cauline lf sheaths inconspicuous. . . . .[5]***L. nevadense***
              26´ Corolla yellow; cauline lf sheaths swollen
                27. Herbage densely puberulent to glabrous; ovaries puberulent; bractlets gen obovate to
                  oblanceolate, gen overlapping. . . . .[2]***L. utriculatum***
                27´ Herbage, ovaries gen scabrous; bractlets gen lanceolate to oblanceolate, not overlapping . . . . .[2]***L. vaginatum***
        22´ St 0
          28. Corolla white to cream, occ purple-tinged; anthers gen purple
            29. Bractlets puberulent to hairy; lf segments linear to lanceolate. . . . . ***L. ravenii***
            29´ Bractlets ± glabrous; lf segments linear to oblong . . . .[5]***L. nevadense***
            30. Ovary and gen fr densely fine-hairy. . . . . var. ***nevadense***
            30´ Ovary and fr glabrous to ± scabrous . . . . var. ***parishii***
          28´ Corolla yellow or purple; anthers gen yellow
           31. Petiole > lf blade; corolla gen purple (yellow); fr wings ≥ body in width . . . . ***L. mohavense***
           31´ Petiole < lf blade; corolla yellow (± purple); fr wings gen 1/2 body in width . . . . ***L. foeniculaceum***
            32. Petal margin minutely ciliate . . . . subsp. ***fimbriatum***
            32´ Petal margin glabrous
              33. Rays gen 1, < 1.5 cm; subalpine scrub . . . . subsp. ***inyoense***
              33´ Rays gen 2–17, < 6 cm; sagebrush scrub, pine woodland . . . . subsp. ***macdougalii***
      21´ Ovary, fr glabrous, occ scabrous
        34. Pl gen with a definite st, st lvs gen ≥ 2
          35. St base with 1 or more scarious, bladeless sheaths; fr oblong-ovate to elliptic, wings thick, << body
            in width . . . . ***L. dissectum***
            36. Fr ± sessile; fertile pedicels gen 1–3 mm, < sterile . . . . var. ***dissectum***
            36´ Fr distinctly pedicelled; fertile pedicels gen 5–15 mm, > sterile . . . . var. ***multifidum***
          35´ St base without conspicuous bladeless sheaths; fr wings thin, < to > body in width
           37. Corolla white, cream or ± purple
            38. Pl glabrous to ± scabrous; bractlets 0–few, inconspicuous . . . .[2]***L. martindalei***

38′ Pl densely puberulent to tomentose; bractlets gen conspicuous, occ fused into a sheath
   39. Involucel 1-sided, reflexing, not scarious . . . . ³*L. macrocarpum*
   39′ Involucel gen radial, spreading, scarious . . . . ⁵*L. nevadense*
37′ Corolla yellow, occ white in age
   40. Lvs gen not crowded at st base; fr wings ≥ body in width
      41. Herbage puberulent to glabrous; ovaries gen puberulent — fr gen glabrous . . . . ²*L. utriculatum*
      41′ Herbage, gen ovaries minutely scabrous. . . . . ²*L. vaginatum*
   40′ Lvs crowded at st base; fr wings < body in width
      42. Fr linear-oblong; lf segments linear, 10–30 mm, not crowded. . . . . ³*L. bicolor* var. *leptocarpum*
      42′ Fr oblong to ovate; lf segments linear to oblong, 2–7 mm, crowded
         43. Lf green, shiny, petiole sheathing to ± middle; bractlets free, inconspicuous; KR . . . . *L. hallii*
         43′ Lf ± gray, dull, petiole ± sheathing throughout; bractlets fused at base into 1-sided scarious
            cup; CaRH, n SNH, GB. . . . . *L. plummerae*
34′ Pl st ± 0, cauline lvs 0, occ 1 or more at base of short st
  44. Fr linear-oblong to oblong
     45. Pl base fibrous; bractlets 0; fr wings narrow, < 1/2 body; SNH. . . . . *L. torreyi*
     45′ Pl base not fibrous; bractlets several, linear; fr wings narrow to broad, < or > 1/2 body; CaRH, MP
       46. Fr narrowly oblong, 10–12 mm, 2–5 mm wide, wings narrow, < 1/2 body; CaRH, MP
          . . . . ³*L. bicolor* var. *leptocarpum*
       46′ Fr oblong, 8–16 mm, 5–9 mm wide, wings broad, > 1/2 body; MP . . . . *L. grayi*
44′ Fr elliptic or ovate to ± round
  47. Lf segments thread-like to linear, long, not crowded; herbage green, gen shiny
     48. Bractlets gen > 5, lanceolate to obovate, scarious, occ 0; pedicels stout — gen < fr . . . . *L. caruifolium*
       49. Bractlets wide; fr wings ± thin, wide . . . . var. *caruifolium*
       49′ Bractlets narrow, occ 0; fr wings thick, narrow. . . . . var. *denticulatum*
     48′ Bractlets 0–5, thread-like to lance-linear, ± scarious-margined; pedicels slender. . . . . *L. marginatum*
       50. Corolla yellow; KR, CaRF, NCoRI, n&c SNF, n SNH, ScV . . . . var. *marginatum*
       50′ Corolla ± red-purple; NCoR . . . . var. *purpureum*
  47′ Lf segments lanceolate or oblong to ovate (linear to thread-like), short, gen crowded; herbage gen
     glaucous or ± gray
     51. Herbage clearly puberulent to tomentose
       52. Lf 2–3-pinnately dissected or compound; petiole sheathing in proximal 1/2; corolla white; GB
          . . . . ⁵*L. nevadense*
       52′ Lf ternate-pinnately dissected; petiole sheathing throughout; corolla yellow or purple; coast ranges
         53. Corolla purple; bractlets scarious ± throughout . . . . *L. hooveri*
         53′ Corolla gen yellow; bractlets narrowly scarious-margined
            54. Ultimate lf sections overlapping, broad, obtuse; fr ± elliptic, wings thick; serpentine
               substrates, NCoRH, NCoRI . . . . *L. ciliolatum*
            54′ Ultimate lf sections less crowded, narrow, gen acute; fr ovoid, wings thin; volcanic substrate,
               SnFrB. . . . . *L. observatorium*
     51′ Herbage glabrous to minutely scabrous or sparsely puberulent
       55. Pedicels gen > fr; bractlets 0
         56. Pl base fibrous; corolla pale yellow. . . . . *L. congdonii*
         56′ Pl base not fibrous; corolla ± purple . . . . *L. engelmannii*
       55′ Pedicels < fr; bractlets present
         57. Lf blade obovate, basal sheaths straw-colored; volcanic soils . . . . *L. peckianum*
         57′ Lf blade oblong to ovate, basal sheaths ± purple; serpentine substrates . . . . *L. tracyi*

L. bicolor (S. Watson) J.M. Coult. & Rose var. ***leptocarpum*** (Torr. & A. Gray) Schlessman Pl 2–5 dm, glabrous to minutely scabrous; taproot 0 or elongate; tubers 0 or rarely deep-seated. **ST:** 0 or short. **LF:** petiole 9–14 cm; blade 5–18 cm, widely obovate, ternate-pinnately dissected, segments gen 1–3 cm, linear; cauline lvs 0 or like basal. **INFL:** glabrous to minutely scabrous; peduncle 8–25 cm; bractlets several, linear; rays 4–15, 2–13 cm, ± erect; pedicels 1–2(5) mm. **FL:** corolla yellow. **FR:** 8–17 mm, linear to oblong, glabrous; wings narrow, < 1/2 body in width; oil tubes several per rib-interval. $2n=22$. Drying adobe, sagebrush scrub, slopes; 1000–2250 m. KR, CaRH, MP; to WA, CO, AZ. *L. bicolor* var. *b.* in ID, WY, MT, CO, UT. Apr–May

L. californicum (Nutt.) Mathias & Constance (p. 191) Pl 3–12 dm, glabrous, glaucous; taproot stout, thickened. **ST:** base fibrous. **LF:** petiole 5–25 cm; blade 1–3 dm wide, triangular-ovate, 1–2-ternate-pinnate, lflets 2–5 cm, wedge-shaped to obovate, gen with 3 major lobes, each coarsely toothed to lobed; cauline lvs few, like basal. **INFL:** glabrous; peduncle 1.5–3 dm; bractlets 0 or inconspicuous; rays 8–20, 3–8(15) cm, ± equal, gen forming disk at base; pedicels 4–12 mm, gen forming disk at base. **FL:** corolla yellow. **FR:** 10–15 mm, oblong-ovate to elliptic, glabrous; wings thickened, < body in width; oil tubes 3–4 per rib-interval. $2n=22$. Woodland, brushy slopes; 150–1800 m. KR, NCoR, s SNH, Teh, SnFrB, SCoR, WTR; s OR. Apr–Jun

L. canbyi (J.M. Coult. & Rose) J.M. Coult. & Rose CANBY'S LOMATIUM Pl 7–25 cm, glabrous, glaucous; tuber < 4 cm wide, spheric. **ST:** 0. **LF:** petiole 4–6 cm, scarious sheath conspicuous; blade 1–9 cm, oblong to ovate, ternate-2-pinnately or ternate-pinnately dissected, segments 1–5 mm, linear, obtuse. **INFL:** glabrous; peduncle 5–10 cm; bractlets ± = fls, linear; rays 5–17, 1–6 cm, ± spreading; pedicels 5–16 mm. **FL:** corolla white. **FR:** 6–13 mm, widely oblong; wings thin, narrow; oil tubes 1–3 per rib-interval. $2n=22$. Barren or rocky places, sagebrush steppe; 2000 m. CaRH, MP; to WA, ID, NV. Apr–May ★

L. caruifolium (Hook. & Arn.) J.M. Coult. & Rose Pl 1.5–4.5 dm, glabrous to finely scabrous or hairy; taproot gen slender. **ST:** ± 0.

LF: petiole 4–7 cm; blade 5–30 cm wide, triangular-ovate to obovate, 1–3-ternately or ternate-2-pinnately dissected, segments 2–60 mm, linear, pointed. **INFL**: peduncle 1–4 dm, erect or spreading; bractlets sessile or not, gen = fls, lanceolate to obovate, entire or toothed, scarious-margined, veiny, green to ± purple; rays 6–15, 1–12 cm, gen spreading-ascending; pedicels 2–8 mm. **FL**: corolla yellow (± purple). **FR**: 6–13 mm, ovate to obovate, glabrous; wings ± thick, < body in width. Vars. poorly defined.

var. *caruifolium* **INFL**: bractlets obovate, gen overlapping side-to-side. **FR**: wings ± thin, wide. 2*n*=22,44. Wet, clay depressions, open grassland; 60–600 m. NCoR (Mendocino Co.), c&s SNF, CCo, SnFrB, SCoR, ChI. Mar–May

var. *denticulatum* Jeps. **INFL**: bractlets lanceolate to oblanceolate (0), not overlapping. **FR**: wings thick, narrow. 2*n*=44. Vernal pools, open grassland; 60–500 m. KR, NCoRH, CaRF, ScV. Apr–May

L. ciliolatum Jeps. (p. 191) Pl low, stout, 9–15 cm, densely puberulent to tomentose; taproot slender. **ST**: 0. **LF**: petiole 2–7 cm, sheathing throughout; blade 3–13 cm, oblong-ovate to triangular-ovate, ternate-pinnately dissected, margins ciliate, segments 1–2 cm, lanceolate to ovate, ultimate overlapping. **INFL**: peduncle 7–25 cm; bractlets sessile or not, gen = fls, lanceolate to obovate, scarious-margined, prominently veined; rays 3–5, 0.8–3 cm, unequal; pedicels 2–4 mm. **FL**: corolla yellow. **FR**: 7–8 mm, ± elliptic, glabrous; wings narrow, ± 0.5 mm. 2*n*=22. Serpentine ridges; 1200–2100 m. NCoRH, NCoRI. Jun–Jul

L. congdonii J.M. Coult. & Rose (p. 191) CONGDON'S LOMATIUM Pl 18–36 cm, taprooted, glaucous, glabrous to minutely scabrous. **ST**: short; base fibrous. **LF**: petiole 2–6 cm, scarious-sheathing throughout, persistent; blade 6.5–15 cm, wide-oblong, ternate to pinnately dissected, segments 3–10 mm, linear, tips sharp-pointed. **INFL**: glabrous to scabrous; peduncle 12–30 cm; bractlets 0; rays 6–16, 3–13.5 cm, ascending; pedicels 6–15 mm. **FL**: corolla pale yellow. **FR**: 7–10 mm, ovate to obovate, glabrous; wings ± 1/2 body in width; oil tubes gen 1 per rib-interval. 2*n*=22. Serpentine, woodland; 300–1200 m. c SNF (Tuolumne, Mariposa cos.). Mar–Jun ★

L. dasycarpum (Torr. & A. Gray) J.M. Coult. & Rose Pl 1–5 dm, taprooted, gen densely short-hairy to tomentose. **ST**: ascending to erect (0). **LF**: petiole 2.5–12 cm; blade 2–12 cm, oblong to obovate, pinnately or ternate-pinnately dissected, segments 2–6 mm, thread-like to linear; cauline lvs 0 or like basal. **INFL**: peduncle 1–3.5 dm; bractlets linear to narrowly ovate, acute, fused or not; rays 10–21, 1–8.5 cm, spreading; pedicels 5–20 mm. **FL**: corolla ± green-white or ± purple, tomentose. **FR**: 8–22 mm, oblong-ovate to round, ± hairy; wings ≥ body in width; oil tubes 1–4 per rib-interval.

subsp. *dasycarpum* **LF**: petiole gen sheathing to middle. **FR**: body sparsely hairy, gen < pedicel, gen < wings in width. 2*n*=22. Rocky (gen serpentine), chaparral, woodland; < 1600 m. NCoR, SnFrB, SCoR, PR; Baja CA. Mar–Jun

subsp. *tomentosum* (Benth.) W.L. Theob. **LF**: petiole sheathing at base. **FR**: body tomentose, gen > pedicel, gen = wings in width. 2*n*=22. Stony flats, grassland, oak woodland; 25–1500 m. CaRF, SNF, GV. Mar–May

L. dissectum (Nutt.) Mathias & Constance Pl 3–14 dm, glabrous to puberulent or minutely scabrous, ± glaucous; taproot stout, thickened. **ST**: base with ≥ 1 scarious sheaths. **LF**: petiole 3–30 cm; blade 15–35 cm wide, triangular-ovate to round, ternate-pinnately dissected, segments 2–22 mm, linear-oblong; cauline lvs gen few, like basal. **INFL**: glabrous; peduncle 1.5–6 dm; bractlets several, > to < fls, linear; rays 10–30, 3–10 cm, spreading; pedicels 1–20 mm. **FL**: corolla maroon-red or yellow. **FR**: 12–16 mm, oblong-ovate to elliptic, glabrous; wings thick, << body in width; oil tubes obscure. 2*n*=22.

var. *dissectum* **LF**: segments 2–8 mm, gen 1.5–3 (>> 3) mm wide. **FL**: corolla gen maroon-red, occ yellow. **FR**: pedicels gen 1–3 mm. Wooded or brushy slopes; 150–2000 m. NCo, KR, c SNF; to WA, ID. May–Jul

var. *multifidum* (Nutt.) Mathias & Constance **LF**: segments 2–22 mm, 0.5–2 mm wide. **FL**: corolla yellow. **FR**: pedicels gen

5–15 mm. Wooded or brushy slopes, gen conifer forest; 600–3000 m. KR, CaRH, SNH, Teh, SCo, SnGb, SnBr, GB; to w Can, Baja CA. Apr–Jul

L. engelmannii Mathias ENGELMANN'S LOMATIUM Pl 1–3 dm, glabrous to sparsely puberulent; taproot slender. **ST**: 0. **LF**: petiole 2–10 cm, sheathing throughout; blade 2.5–20 cm, oblong to ovate, ternately to pinnately dissected, segments 1–15 mm, lance-linear to lanceolate. **INFL**: gen glabrous; peduncle 12–20 cm; bractlets 0; rays 2–12 (1–8 fertile), 1–13 cm, spreading to ascending; pedicels 2–12 mm. **FL**: corolla ± purple. **FR**: 7–14 mm, oblong-ovate, glabrous; wings 1/2 body in width; oil tubes 1–2 per rib-interval. 2*n*=22. Serpentine slopes in conifer forest; 1150–2300 m. KR (Siskiyou, Trinity cos.); sw OR. Jun–Aug ★

L. foeniculaceum (Nutt.) J.M. Coult. & Rose Pl 3–40 cm, taprooted, densely puberulent to soft-hairy or tomentose. **ST**: 0. **LF**: petiole 1–15 cm, < blade, sheathing throughout; blade gen 2.5–18 cm wide, oblong to obovate, pinnately to ternate-pinnately dissected, segments 1–6 mm, linear to obovate, pointed. **INFL**: hairy to ± glabrous; peduncle 3–40 cm; bractlets fused or not, ≤ fls, linear to lance-linear, entire to lobed, acute; rays 1–17, gen < 6 cm, spreading-ascending or spreading; pedicels 1–15 mm. **FL**: corolla yellow (± purple). **FR**: 4–12 mm, oblong-ovate, gen hairy; wings gen 1/2 body in width; oil tubes 1–7 per rib-interval. Other subspp. to BC, TX.

subsp. *fimbriatum* W.L. Theob. Pl 7–30 cm. **LF**: segments 1–5 mm, linear. **INFL**: peduncle 10–30 cm; rays 2–14, 0.5–6 cm. **FL**: petals yellow or ± purple, minutely ciliate. 2*n*=22. Sagebrush scrub, pine woodland; 1600–3300 m. SNE, DMtns (Last Chance Range); to UT. Apr–Jun

subsp. *inyoense* (Munz & J.C. Roos) W.L. Theob. INYO LOMATIUM Pl 3–12 cm. **LF**: segments 1–2 mm, linear to obovate. **INFL**: peduncle 3–12 cm; rays gen 1, < 5 cm. **FL**: petals pale yellow, glabrous. Open summits, subalpine scrub; 2195–3200 m. W&I; possibly to ID, NV. May be a form induced by high-elevation conditions. Jun–Jul ★

subsp. *macdougalii* (J.M. Coult. & Rose) W.L. Theob. MACDOUGAL'S LOMATIUM Pl 7–40 cm. **LF**: segments 1–6 mm, linear to obovate. **INFL**: peduncle 10–40 cm; rays 2–17, 0.5–6 cm; pedicels 2–13 mm. **FL**: petals yellow (± purple), glabrous. Sagebrush scrub, pine woodland; 1390–1800 m. n SNH/MP, MP, DMtns (Cottonwood Mtns); e OR, to UT, MT, AZ. May–Jun ★

L. grayi (J.M. Coult. & Rose) J.M. Coult. & Rose GRAY'S LOMATIUM Pl 1.5–5 dm, ± glabrous; taproot long, thickened. **ST**: ± 0. **LF**: petiole 3.5–10 cm, sheathing throughout; blade 10–16 cm, ± ovate, finely ternate-pinnately dissected, segments 1–11 mm, thread-like to linear, acute. **INFL**: peduncle 7.5–22 cm; bractlets < fls; rays 7–22, 2–15 cm, ascending to spreading-ascending; pedicels 6–16 mm. **FL**: corolla yellow. **FR**: 8–16 mm, oblong; wings > 1/2 body in width; oil tubes 1 per rib-interval. Rocky banks, slopes; 1300–1400 m. MP; to WA, WY, UT. May–Jun ★

L. hallii (S. Watson) J.M. Coult. & Rose Pl 2–5 dm; taproot stout; herbage glabrous to minutely scabrous. **ST**: several; base branched, not fibrous. **LF**: petiole 3–6 cm, sheathing ± to middle; blade 5–15 cm, oblong-ovate, gen 3-pinnately dissected, segments 2–4 mm, linear to oblong, pointed. **INFL**: scabrous or glabrous; peduncle 1.2–3 dm; bractlets several, free, inconspicuous, linear; rays 8–18, 1.3–5 cm, spreading-ascending; pedicels 4–7 mm. **FL**: corolla yellow. **FR**: 5–9 mm, oblong-ovate, glabrous; wings gen 1/2 body in width; oil tubes 2–3 per rib-interval. 2*n*=22. Rocky bluffs or slopes; ± 150 m. nw KR (Del Norte Co.); OR. May–Aug

L. hendersonii (J.M. Coult. & Rose) J.M. Coult. & Rose HENDERSON'S LOMATIUM Pl 10–30 cm; herbage green, shiny, glabrous; tuber 1–2.5 cm diam, ± ovoid. **ST**: 0. **LF**: petiole 2–3 cm, sheathing to middle or throughout; blade 5–9 cm, triangular-ovate, ternate-1–2-pinnate, lflets 2–5 mm, linear to oblong, obtuse. **INFL**: peduncle 5–15 cm, spreading; bractlets scarious; rays 2–5, 1–2 cm; pedicels 3–4 mm. **FL**: corolla white. **FR**: ± 5 mm, ± 4 mm wide, elliptic to ovate; wings < 1/2 body in width; oil tubes 1–4 per rib-interval. Gravelly or rocky soil, sagebrush flats; 1400–2300 m. MP;

OR, ID, NV. Yellow-fld pls that key here could be *L. roseanum* Cronquist. Mar–Jun ★

L. hooveri (Mathias & Constance) Constance & Ertter (p. 191) HOOVER'S LOMATIUM Pl ± erect to spreading, slender, 1.5–3 dm; densely puberulent. **ST:** 0. **LF:** petiole 3–7 cm, sheathing throughout; blade 6–13 cm, segments linear, acute, ultimate not crowded. **INFL:** peduncle 8–25 cm; fertile rays 3–14, 3–10 cm; bractlets scarious ± throughout; pedicels 3–8 mm. **FL:** corolla purple. **FR:** wings wide. 2*n*=22. Serpentine, woodland, chaparral; 300–600 m. NCoRI. [*L. ciliolatum* var. *h.* Mathias & Constance] Apr–May ★

L. howellii (S. Watson) Jeps. HOWELL'S LOMATIUM Pl 1.2–8 dm, glabrous, glaucous; taproot stout, branched. **ST:** short to ± 0; base not fibrous. **LF:** petiole 2–20 cm, sheath short; blade 5–15 dm, triangular-ovate, ternate-pinnate to 2-pinnate, lflets 1–3 cm wide, gen obovate, sharply dentate, gen irregularly cut, occ 3-lobed; cauline lvs like basal. **INFL:** peduncle 1–4 dm; bractlets several, 1–3 mm, thread-like to lanceolate; rays 8–20, 2.5–6.5 cm, spreading; pedicels 5–10 mm. **FL:** corolla yellow or ± purple. **FR:** 6–14 mm, widely elliptic to round, glabrous; wings = body in width; oil tubes 2–3 per rib-interval. 2*n*=22. Serpentine in chaparral, conifer forest; 150–1000 m. KR (Del Norte, Siskiyou cos.); sw OR. May–Jun ★

L. insulare (Eastw.) Munz (p. 191) SAN NICOLAS ISLAND LOMATIUM Pl 1–3.5 dm, glabrous, glaucous; taproot stout, fleshy-thickened. **ST:** short; base fibrous. **LF:** petiole 2–15 cm, < blade, sheath wide; blade 6–24 cm wide, triangular-ovate to round, ternate-pinnate, lflets 2–6 cm wide, gen ovate, pinnately lobed, sharply serrate; cauline lvs like basal. **INFL:** glabrous; peduncle 1–3 dm; bractlets several, thread-like, gen > fls; rays gen 12–20, 4–10 fertile, 3–12 cm, spreading; pedicels 5–15 mm. **FL:** corolla yellow. **FR:** 15–18 mm, widely elliptic to obovate, glabrous; wings = body in width; oil tubes 2 per rib-interval. 2*n*=22. Sandy soil among rocks; < 800 m. s ChI (San Nicolas Island); Baja CA (Guadalupe Island). Feb–Apr ★

L. lucidum (Torr. & A. Gray) Jeps. Pl 1.5–12 dm; taproot slender; herbage glabrous, green, ± fleshy. **ST:** short; base not fibrous. **LF:** petiole 2–12 cm, sheath short; blade 4–12 cm wide, triangular-ovate, 1–2-ternate, lflets 1.5–5 cm, oblong to obovate, coarsely sharp-dentate, gen 3-lobed; cauline lvs like basal. **INFL:** glabrous; peduncle gen 1.5–5 dm; bractlets several, 3–7 mm, lance-linear; rays gen 10–20, 4–18 fertile, 3–8 cm, spreading or spreading-ascending; pedicels 5–12 mm. **FL:** corolla yellow. **FR:** gen 10–15 mm, wide-elliptic to round, glabrous; wings thickened, >> body in width; oil tubes 1 per rib-interval. 2*n*=22. Chaparral, esp on burns; 450–1500 m. SCo, SnGb, SnBr, SnJt; Baja CA. Apr–May

L. macrocarpum (Torr. & A. Gray) J.M. Coult. & Rose Pl 1–5 dm; taproot or occ tubers slender or basally swollen; herbage gray, gen tomentose to densely short-hairy. **ST:** short; base not fibrous. **LF:** petiole 1.5–7 cm; blade 2.5–15 cm, oblong to obovate, pinnately or ternate-pinnately dissected, segments 1–7 mm, linear to oblong, entire; cauline lvs like basal. **INFL:** gen tomentose; peduncle 0.5–3 dm; involucel 1-sided; bractlets several, ≥ fls, lance-linear to ovate, acute, ± fused, reflexing, not scarious; rays 5–25, 1–8.5 cm, spreading-ascending; pedicels 1–14 mm. **FL:** corolla white, cream, or ± purple; ovary gen hairy. **FR:** 9–20 mm, lanceolate or oblong to narrowly elliptic, minutely hairy to ± glabrous; wings gen < body in width; oil tubes 1–3 per rib-interval. 2*n*=22. Gen serpentine rocky slopes in chaparral or woodland; 150–3000 m. NCo, KR, CaR, SNF, Teh, SnFrB, SCo; to BC, c Can, ND, UT. Apr–Jun

L. marginatum (Benth.) J.M. Coult. & Rose Pl 1.5–5 dm, glabrous to minutely scabrous; taproot slender. **ST:** ± 0. **LF:** petiole gen 3–8 cm, narrowly sheathing throughout, gen ± purple; blade 5–20 cm wide, triangular-ovate to obovate, 1–3-ternately or ternate-pinnately dissected, segments 0.5–8 cm, thread-like or linear, pointed. **INFL:** peduncle 1–4 cm; bractlets 0–5, 1–5 mm, thread-like to lance-linear, entire, ± scarious-margined; rays 6–12, spreading to ascending. **FR:** 8–12 mm, narrowly ovate to obovate, glabrous; wings thin, < body in width; oil tubes obscure.

var. ***marginatum*** **INFL:** rays 1.5–10 cm; pedicels 2–6 mm. **FL:** corolla yellow. 2*n*=22. Serpentine slopes, gen chaparral; 250–1000 m. KR, CaRF, NCoRI, n&c SNF, n SNH, ScV. Mar–May

var. ***purpureum*** (Jeps.) Jeps. **INFL:** rays 2–15 cm; pedicels 5–12 mm. **FL:** corolla ± red-purple. 2*n*=22,44. Serpentine slopes, chaparral, woodland; 300–800 m. NCoR. Mar–May

L. martindalei (J.M. Coult. & Rose) J.M. Coult. & Rose COAST RANGE LOMATIUM Pl 1.5–4 dm; taproot stout, gen swollen proximally, carrot-like; herbage glabrous to ± scabrous, ± glaucous, ± fleshy. **ST:** 0 or short; base not fibrous. **LF:** petiole 1.5–5 cm, sheath wide, scarious; blade 2.5–15 cm, oblong to obovate; 1–2-pinnate to ternate-pinnately dissected, lflets or segments 0.8–3 cm, oblong to ovate, gen crowded, gen obtuse, dentate to irregularly cut or pinnately lobed; cauline lvs like basal, petioles gen sheathing throughout. **INFL:** gen glabrous; peduncle 0.5–3 dm; bractlets 0 or inconspicuous; rays 3–15, 1–8 cm, ± unequal, spreading to ascending; pedicels 3–10 mm. **FL:** corolla yellow, fading in age. **FR:** 8–15 mm, oblong to ± diamond-shaped, glabrous; wings ≤ body in width; oil tubes gen 1 per rib-interval. 2*n*=22. Conifer forest, rocks, meadows, talus, pumice, coastal bluffs; 240–3000 m. KR (Del Norte, Siskiyou cos.); to BC. May–Jun ★

L. mohavense (J.M. Coult. & Rose) J.M. Coult. & Rose (p. 191) Pl 1–4 dm, ± gray, densely short-hairy, base fibrous; taproot long, gen thickened. **ST:** 0. **LF:** petiole 2–12 cm, > blade; blade 2–10 cm, oblong to ovate, 3–4-pinnately dissected, segments 2–5 mm, linear-oblong to obovate, ± crowded. **INFL:** finely hairy; peduncle 8–22 cm; bractlets 8–12, 2–4 mm, linear to lance-linear, acute, free or fused basally, scarious-margined, gen obscured by hairs; rays 8–18, 1–5 cm, spreading-ascending, unequal; pedicels 1–10 mm. **FL:** petals yellow or purple, glabrous. **FR:** 4.5–11 mm, ovate to round, ± glabrous to densely short-hairy; wings ≥ body in width; oil tubes 1–4 per rib-interval. 2*n*=22. Desert flats, slopes, scrub, or woodland; 1000–2000 m. SCoR, WTR, D; Baja CA. Apr–May

L. nevadense (S. Watson) J.M. Coult. & Rose Pl 1–4.5 dm; taproot slender, occ swollen proximally; herbage ± gray, gen finely hairy. **ST:** 0 or short. **LF:** petiole 4–6 cm; blade 3.5–10 cm, oblong to obovate, 2–3-pinnately dissected, segments gen < 3 mm, linear or oblong, pointed, gen crowded; cauline lvs 0 or like basal. **INFL:** peduncle 0.5–3 dm; bractlets 1–10, linear and free to obovate and ± fused, conspicuously scarious or scarious-margined, ± glabrous; rays 8–22, 1–2.5 cm, unequal, spreading; pedicels 3–10 mm. **FL:** corolla white to cream. **FR:** 6–11 mm, oblong to round or obovate, densely hairy to glabrous; wings < to > body in width; oil tubes 1–9 per rib-interval. 2*n*=22. Vars. poorly defined.

var. ***nevadense*** **LF:** 2–3-pinnately dissected. **FL:** ovary, gen fr densely fine-hairy. Sagebrush, woodland; 1500–3000 m. CaRH, n&c SNH, GB, DMtns; to OR, UT, AZ. Apr–Jul

var. ***parishii*** (J.M. Coult. & Rose) Jeps. **LF:** gen 1–2-pinnately dissected. **FL:** ovary, fr glabrous to ± scabrous. Sagebrush, desert scrub, pine woodland; 1000–3000 m. c&s SNH, SnGb, SnBr, GB, DMtns; to NV, NM, SON. Apr–Jul

L. nudicaule (Pursh) J.M. Coult. & Rose Pl 2.5–7 dm, glabrous, glaucous; taproot long, thickened. **ST:** ± 0. **LF:** petiole 0.4–2.5 dm, wide-sheathing to middle; blade 9–20 cm, wide-ovate, 1–2-ternate-pinnate, lflets 15–90 mm, lanceolate to wide-ovate, entire or toothed or lobed, gen near tip. **INFL:** peduncle 1–4 dm, conspicuously swollen distally, gen translucent; involucel 0; rays 10–20, 1–20 cm, ascending, ± swollen distally; pedicels 3–15 mm. **FL:** corolla yellow (purple). **FR:** 10–14 mm, oblong to elliptic; wings < body in width; oil tubes 1–several per rib-interval. 2*n*=22. Rocky slopes, flats, gen pine woodland; 180–2000 m. KR, NCoR, CaRH, n SNF, SnFrB, MP; to BC, ID, UT. Apr–Jun

L. observatorium Constance & Ertter (p. 191) MOUNT HAMILTON LOMATIUM Pl 1–3 dm, puberulent to short-hairy; taproot to 15 cm. **ST:** ± 0. **LF:** petiole 3–6 cm, sheathing throughout; blade 4–12 cm, ternate-pinnate, ultimate segments lance-linear, acute, 1–8 mm, ± 1 mm wide, ciliate. **INFL:** peduncle 0.8–20 cm, exceeding lvs; bractlets 5–10, 3–4 mm, lanceolate to obovate, narrowly scarious, margins ciliate; rays 1–3(7), 1–8 cm; pedicels 1–5 mm. **FL:** ovary, fr glabrous. **FR:** 7–10 mm, ovate; wings ± 0.6 mm wide. Volcanic soil, rocky openings in pine/oak woodland; 1280–1330 m. se SnFrB (Mount Hamilton Range). Mar–May ★

L. parryi (S. Watson) J.F. Macbr.   Pl 2–4 dm, taprooted, glabrous, glaucous. **ST:** ± 0; base fibrous. **LF:** petiole 6–10.5 cm, wide-sheathing basally; blade 5–20 cm, narrowly oblong, 2–3-pinnate, lflets 2–9 mm, linear, entire, sharp-pointed. **INFL:** peduncle 1–2.5 dm, ± swollen distally; bractlets 3–8, 3–6 mm, linear, acute, scarious, entire to lobed; rays 8–15, ascending to ± erect, 2–4.5 cm, ± equal; pedicels 10–17 mm. **FL:** calyx conspicuous; corolla yellow. **FR:** 9–15 mm, oblong to ± diamond-shaped; wings ≥ body in width; oil tubes 2–3 per rib-interval. Rocky slopes, gen in pinyon woodland; 1500–2500 m. DMtns; to UT, AZ. May–Jun

L. parvifolium (Hook. & Arn.) Jeps.   SMALL-LEAVED LOMATIUM   Pl 1.5–4 dm; taproot slender; herbage glabrous, ± glaucous, ± fleshy. **ST:** short. **LF:** petiole 3–15 cm, sheathing basally; blade 3–15 cm wide, oblong- to triangular-ovate, ternate-pinnate, lflets 1–4 cm wide, ovate to obovate, weakly spine-toothed and -tipped or irregularly cut. **INFL:** peduncle 1–5 dm; bractlets gen 2–5, 2–8 mm, thread-like to lance-linear, ± fused basally; rays 8–15, spreading, 0.5–6.5 cm, ± equal; pedicels gen 5–12 mm. **FL:** corolla bright yellow. **FR:** 8–14 mm, elliptic to round; wing ≥ body in width; oil tubes gen 1 per rib-interval. 2*n*=44. Pine woodland, serpentine outcrops; 70–150 m. CCo, SCoR. Feb–May ★

L. peckianum Mathias & Constance (p. 195)   PECK'S LOMATIUM   Pl 1–4.5 dm, glabrous or minutely scabrous, glaucous; taproot slender, scabrous. **ST:** 0. **LF:** petiole < 5 cm, gen scarious-sheathing throughout, straw-colored; blade 5–10 cm wide, obovate, ternately or ternate-pinnately dissected, segments 1–15 mm, linear to oblong, ± acute. **INFL:** peduncle 1.2–4 dm, spreading; bractlets gen 5–10, 2–7 mm, linear, ± scarious; rays gen 8–12, 1–6.5 cm, spreading; pedicels 3–6 mm. **FL:** corolla cream to lemon-yellow. **FR:** 8–14 mm, oblong-elliptic to elliptic, glabrous; wings < body in width; oil tubes several per rib-interval, obscure. 2*n*=22. Volcanics, pine/oak woodland; 800–1800 m. KR (Siskiyou Co.); s OR. Locally abundant. May ★

L. piperi J.M. Coult. & Rose   Pl 0.5–2.5 dm, glabrous to minutely scabrous; tuber 0.5–1 cm wide, spheric. **ST:** 0. **LF:** petiole 3.5–10 cm, sheathing basally; blade 1.5–5.5 cm wide, triangular-ovate, 2-ternate-pinnately or 2-ternately dissected, segments 2–40 mm, linear, acute to obtuse; cauline lvs 1–2, gen linear, entire. **INFL:** gen glabrous; terminal umbel 1, lateral gen 1–2; peduncle 2–8 cm; bractlets 2–6, 0.5–2.2 mm, < fls, narrowly elliptic, acute; rays 3–20, gen 1–6 cm, spreading-ascending; pedicels 0.5–2.5 mm. **FL:** corolla white; anthers purple. **FR:** 5–9 mm, 3–4 mm wide, ovate to oblong; wings ± 1/2 body in width; oil tubes 1–8 per rib-interval. 2*n*=22. Rocky slopes, sagebrush, pine woodland; 1000–1500 m. KR, n SNH, MP; to WA. Mar–May

L. plummerae (J.M. Coult. & Rose) J.M. Coult. & Rose   Pl 1.2–3.5 dm; taproot slender; herbage ± gray, dull, ± fleshy; hairs ± 0 to dense, fine, soft. **ST:** short, lvs crowded at base. **LF:** petiole 3–6 cm, sheathing throughout, scarious-margined; blade 5–10 cm, oblong to ovate, ternate-pinnately dissected, segments 3–7 mm, linear to oblong, obtuse to ± acute; cauline lvs like basal. **INFL:** glabrous to finely soft-hairy; peduncle 0.7–3 dm, spreading-ascending; bractlets 5–10, lance-linear to obovate, fused into 1-sided, scarious, veiny, irregularly cut cup ≥ fls; rays 10–25, 0.5–7.5 cm, unequal, spreading-ascending, ± webbed; pedicels 3–8 mm. **FL:** corolla light yellow. **FR:** 9–13 mm, oblong to oblong-ovate, glabrous; wings < body in width; oil tubes 1–several per rib-interval. Rocky places, sagebrush, pine woodland; 1500–2300 m. CaRH, n SNH, GB; w NV. Extremely variable in hairiness. May–Jun

L. ravenii Mathias & Constance   RAVEN'S LOMATIUM   Pl 0.5–4 dm; base fibrous; taproot ending in deep-seated tuber; herbage ± gray; hairs dense, fine, soft. **ST:** 0. **LF:** petiole 2–10 cm, sheathing to middle; blade 4–10 cm, oblong to widely ovate, ternate-pinnately to pinnately dissected, segments 1–4 mm, linear to lanceolate, acute. **INFL:** densely puberulent; peduncle 0.5–3 dm; bractlets 5–10, 1–5 mm, linear, puberulent to hairy, ± fused basally, involucel 1-sided; rays 3–18, 1–7 cm, unequal, spreading-ascending; pedicels 2–10 mm. **FL:** corolla white, occ purple-tinged; anthers purple; ovary hairs dense, fine. **FR:** 6–8 mm, ovate to round, soft-hairy to ± glabrous; wings < body in width; oil tubes 1–2 per rib-interval. 2*n*=22. Flats,

slopes, ridges, gen ± alkaline soils, sagebrush, pinyon/juniper woodland; 1000–3000 m. MP (e Lassen Co.); to ID, w UT. Abundant in se OR. Apr–Jun ★

L. repostum (Jeps.) Mathias   NAPA LOMATIUM   Pl low, 1.2–5 dm, glabrous, glaucous; taproot slender. **ST:** ± 0; lvs clustered at base. **LF:** petiole 1.5–15 cm, sheathing basally; blade 5–15 cm wide, ovate to triangular-ovate, 1–2-ternate or ternate-pinnate, lflets 1–6 cm wide, ovate to widely wedge-shaped, sharply serrate and occ sharply lobed; cauline lvs 0 or like basal. **INFL:** peduncle 5–35 cm; bractlets 5–10, 2–8 mm, lanceolate, acuminate, ± fused basally, reflexed; rays 8–20, 3–8 cm, spreading-ascending, webbed; pedicels 3–12 mm. **FL:** corolla ± green-yellow or ± purple. **FR:** 8–12 mm, widely oblong, glabrous; wings ≥ body in width; oil tubes 1–3 per rib-interval. 2*n*=44. Pine/oak woodland, chaparral, gen serpentine; 100–800 m. s NCoR. Apr–May ★

L. rigidum (M.E. Jones) Jeps.   STIFF LOMATIUM   Pl 1.5–6 dm; taproot massive; herbage glabrous, green. **ST:** ± 0; lvs clustered at base. **LF:** petiole 5–15 cm; blade 7–15 cm, oblong to ovate, ternate-pinnate or 2-pinnate, lflets 1–2 cm, ovate, pinnately sharply lobed; cauline lvs 0 or like basal. **INFL:** peduncle 1–5 dm; bractlets 5–8, 3–8 mm, lanceolate, acuminate, ± fused basally, reflexed; rays 10–16, 2.5–5 cm, spreading-ascending; pedicels 5–10 mm. **FL:** corolla yellow; calyx lobes < 1 mm, evident. **FR:** 6–12 mm, oblong-ovate, glabrous; wings < body in width; oil tubes 3 per rib-interval. Rocky slopes near streams, among sagebrush, pinyon/juniper woodland; 1200–2200 m. SNE (Big Pine, Bishop creeks, Inyo Co.). Vegetatively like *Tauschia parishii*. Apr–May ★

L. shevockii R.L. Hartm. & Constance (p. 195)   OWENS PEAK LOMATIUM   Pl 4–12 cm, glabrous, glaucous; taproot elongated. **ST:** ± 0; base fibrous. **LF:** petiole 1.5–5 cm; blade 1.5–4 cm, ovate to triangular-ovate, 1-pinnate, lflets gen 3–5, ovate to elliptic, pinnately sharp-lobed. **INFL:** peduncle 4–12 cm, spreading; bractlets 3–6, 1–3.5 mm, lanceolate to ovate, scarious-margined, acute to acuminate, gen distinct; rays 5–9, 1–11 mm, spreading; pedicels 0.1–1 mm, webbed. **FL:** calyx lobes < 0.6 mm, evident; corolla purple. **FR:** 8–10 mm, widely elliptic to round, glabrous; wings < body in width; oil tubes 3–5 per rib-interval. Rocky slopes, talus, conifer forest, pine/oak woodland; 2200–2500 m. s SNH (Kern Co.). Apr–May ★

L. stebbinsii Schlessman & Constance (p. 195)   STEBBINS' LOMATIUM   Pl 5–15 cm; tuber < 2 cm wide, ovoid to spheric; herbage green, shiny, glabrous. **ST:** 0. **LF:** petiole 2–3 cm, sheathing to middle or throughout; blade 2–5 cm, triangular-ovate, ternate-1–2-pinnate, lflets 2–12 mm, linear, acute. **INFL:** peduncle 5–15 cm, spreading; bractlets 0; rays 2–7, 1–12 mm, ascending; pedicels 1–2 mm. **FL:** corolla yellow. **FR:** 6–9 mm, narrowly elliptic; wings << body in width; oil tubes 1–4 per rib-interval. 2*n*=22. Yellow-pine forest on gravelly volcanic soil; 1250–1700 m. n SNH/c SNH (Calaveras, Tuolumne cos.). Mar–May ★

L. torreyi (J.M. Coult. & Rose) J.M. Coult. & Rose   Pl 1–3 dm, glabrous to minutely scabrous; base fibrous, lvs clustered; taproot long. **ST:** ± 0. **LF:** petiole 2–5 cm, sheathing throughout; blade 2–15 cm, oblong to ovate, ternate-pinnately dissected, segments 3–10 mm, thread-like to linear, acute. **INFL:** peduncle 0.8–2 dm; bractlets 0; rays 3–9, 1–4 cm, ascending to spreading-ascending; pedicels 1–10 mm. **FL:** corolla yellow. **FR:** 10–16 mm, narrowly oblong, glabrous; wings < 1/2 body in width; oil tubes 1 per rib-interval. 2*n*=22. Granite crevices in pine forests; 1100–3300 m. SNH. May–Aug

L. tracyi Mathias & Constance (p. 195)   TRACY'S LOMATIUM   Pl 1–3.5 dm, glabrous to minutely scabrous; taproot slender. **ST:** ± 0. **LF:** petiole 2.5–8 cm, scarious, gen sheathing throughout; blade 4.5–10 cm, oblong to ovate, ternate-pinnately or 2-pinnately dissected, segments 1–7 mm, thread-like to oblong, acute or obtuse, gen overlapping. **INFL:** peduncle 1–2 dm, spreading; bractlets 2–6, 1–4 mm, linear to oblanceolate, scarious; rays 6–12, 1–6 fertile, 0.5–8 cm, ascending, unequal; pedicels 1–5 mm. **FL:** corolla yellow. **FR:** 6–10 mm, ± oblong-ovate, glabrous; wings << body in width; oil tubes obscure. 2*n*=22. Open pine forest, on serpentine; 500–1500 m. KR, n NCoR, CaRH. May–Jun ★

191

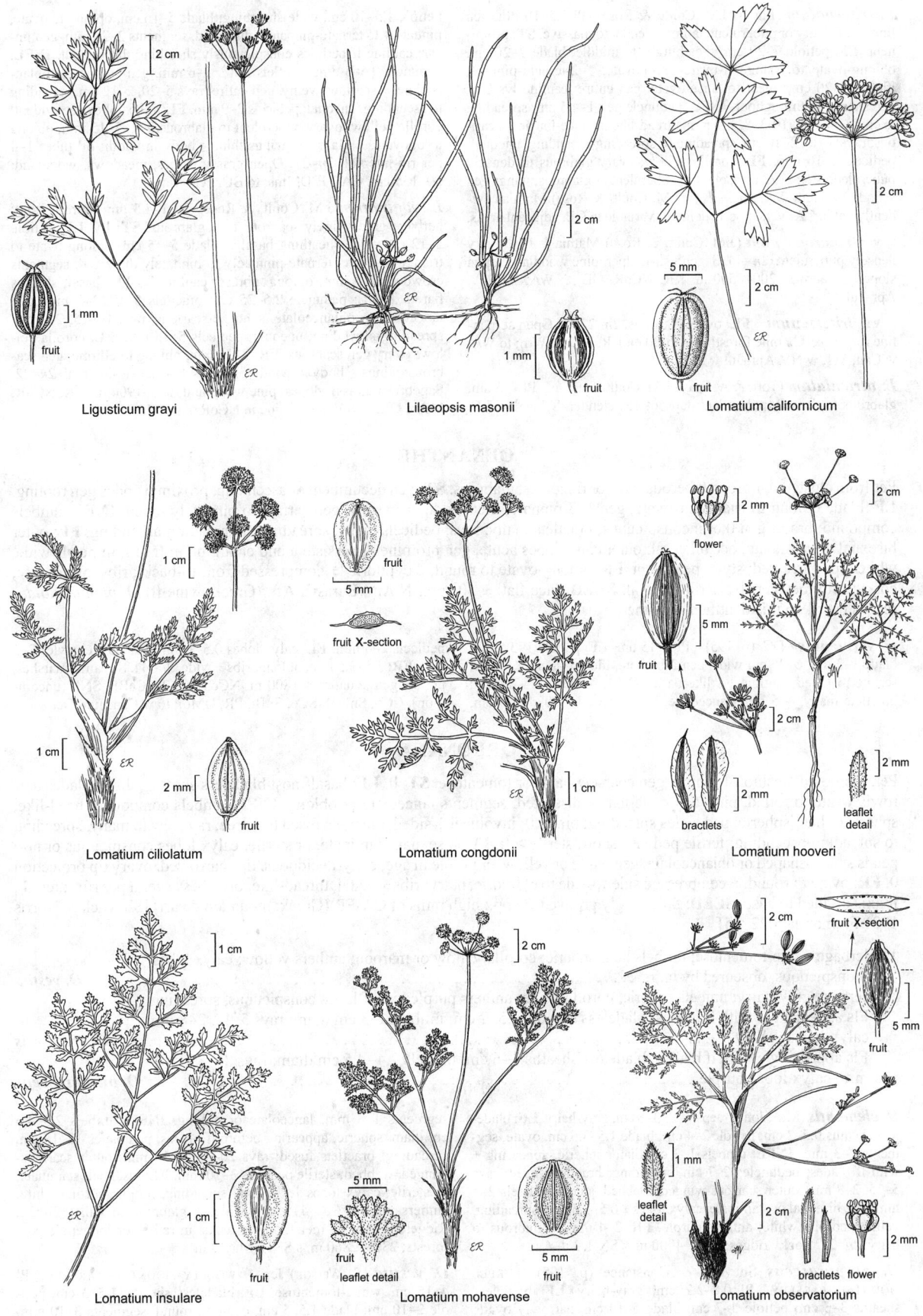

Ligusticum grayi

Lilaeopsis masonii

Lomatium californicum

Lomatium ciliolatum

Lomatium congdonii

Lomatium hooveri

Lomatium insulare

Lomatium mohavense

Lomatium observatorium

L. triternatum (Pursh) J.M. Coult. & Rose   Pl 1.5–10 dm, gen finely soft-hairy or puberulent; taproot slender to massive. **ST**: prominent. **LF**: petiole 7–20 cm, sheathing ± to middle; blade 7–20 cm, oblong-ovate to triangular-ovate or obovate, 1–2-ternate-pinnate, lflets 1.5–20 cm, linear to lance-ovate, gen entire; cauline lvs 0 or gen sheathing throughout. **INFL**: peduncle gen 1–4.5 dm, spreading to erect; bractlets (0)3–8, 1–5 mm, thread-like to lance-linear, ± scarious; rays 5–20, 2–10 cm, spreading or spreading-ascending, unequal; pedicels 1–10 mm. **FL**: corolla yellow; ovary glabrous to densely puberulent. **FR**: 6–22 mm, oblong, puberulent or glabrous; wings gen < body in width. Var. *anomalum* (J.M. Coult. & Rose) Mathias evidently not in CA. Vars. poorly defined. Variable in GB, adjacent areas.

   var. ***macrocarpum*** (J.M. Coult. & Rose) Mathias   **FL**: ovary densely puberulent. 2*n*=44. Sagebrush-juniper, pine woodland, open slopes, meadows; 200–1500 m. KR, NCoR, GB; to WA, ID, NV. Apr–Jul

   var. ***triternatum***   **FL**: ovary glabrous. 2*n*=22,44. Open serpentine ridges, scrub, pine forest; 200–2000 m. KR, NCoR, Wrn; to WA, w Can, MT, w NV. Apr–Jul

L. utriculatum (Torr. & A. Gray) J.M. Coult. & Rose   Pl 1–5 dm, glabrous to densely puberulent; taproot gen slender. **ST**: ± lfy. **LF**: petiole 1.5–10 cm, wide-sheathing; blade 5–16 cm, oblong to ovate, pinnately to ternate-pinnately dissected, segments 2–25 mm, gen linear; cauline lf petioles conspicuously sheathing throughout. **INFL**: peduncle 0.5–3 dm; bractlets 3–12, 3–6 mm, gen obovate to oblanceolate, ± scarious, veiny, gen entire; rays 5–20, 2–12 cm, spreading to ascending, unequal; pedicels 2–9 mm. **FL**: calyx lobes occ evident; corolla yellow; ovary puberulent to glabrous. **FR**: 5–15 mm, oblong to obovate, ± glabrous; wings thin, ≥ body in width; oil tubes 1–4 per rib-interval. 2*n*=22. Open grassy slopes, meadows, woodland; 50–1550 m. CA-FP, DMtns; to BC. Feb–May

L. vaginatum J.M. Coult. & Rose   Pl 1–4.5 dm; taproot stout; herbage green, finely scabrous to ± glabrous. **ST**: lfy. **LF**: petiole 2–12 cm, wide-sheathing basally; blade 5–15 cm, oblong-ovate to triangular-ovate, ternate-pinnately to pinnately dissected, segments crowded, 1–5 mm, oblong, obtuse; cauline lvs like basal, sheaths flared. **INFL**: peduncle 2.5–25 cm; bractlets 5–10, 3–7 mm, gen lanceolate to oblanceolate, acute, ± scarious; rays 10–15, 2–7 cm, spreading-ascending, unequal; pedicels 5–15 mm. **FL**: corolla yellow; ovary gen scabrous. **FR**: 8–15 mm, oblong to elliptic, gen scabrous; wings ≥ body in width; oil tubes 1–4 per rib-interval. 2*n*=22. Sagebrush, grassy slopes, pine woodland; 600–1900 m. KR, NCoR, MP; c OR, w NV. Serpentine in NCoR. Apr–May

OENANTHE

Per from clustered, fibrous-tuberous roots or rhizomes, glabrous. **ST**: gen decumbent or ascending, proximal nodes gen rooting. **LF**: blade oblong to triangular-ovate, gen 1–3-pinnate, lflets wide or narrow, gen serrate to pinnately lobed. **INFL**: umbels compound; bracts gen 0 or inconspicuous; bractlets many; rays, pedicels many, spreading or spreading-ascending. **FL**: outer bisexual or staminate, occ bilateral; outer calyx lobes acute, gen prominent, persistent and enlarging in fr or not; petals wide, white, tips narrowed; styles persistent. **FR**: oblong-ovate to round, ± cylindric [± compressed front-to-back]; ribs low, obtuse, corky; oil tubes gen 1 per rib-interval. **SEED**: face flat. ± 30 spp.: N.Am, Eurasia, Afr. (Greek: wine fl) *O. pimpinelloides* record based on misidentified specimen.

O. sarmentosa DC. (p. 195)   Pl 5–15 dm. **LF**: petiole 1–3.5 dm; blade 1–3 dm, 6–25 cm wide, gen 2-pinnate, lflets 1–6 cm, ± ovate, serrate to lobed; cauline lvs like basal. **INFL**: peduncle 5–13 cm; bractlets many, 4–5 mm, lanceolate, acute; rays 10–20, 1.5–3 cm; pedicels 2–6 mm. **FL**: calyx lobes 0.5–1 mm, lanceolate; styles 2–3 mm. **FR**: 2.5–3.5 mm, oblong; ribs ± wide. 2*n*=44. Streams, marshes, ponds, gen aquatic; < 1800 m. NCo, NCoRI, CaRF, SNF (uncommon), CCo, SnFrB, SCo, SnBr, PR, DMoj; to w Can. Jun–Oct

OREONANA

Per, taprooted, cushion-forming, gen coarsely hairy or tomentose. **ST**: 0. **LF**: basal-most bladeless sheaths; distal blades narrowly ovate to round, pinnately or ternately dissected, segments lanceolate or oblong. **INFL**: umbels compound, head-like, spheric or hemispheric; peduncles spreading; bracts 0; involucel 1-sided, bractlets fused to ± free; rays few to many, spreading to spreading-ascending; fertile pedicels short, sterile > fr. **FL**: bisexual, staminate, or sterile; calyx lobes conspicuous or not; petals spoon-shaped or oblanceolate, gen white or yellow, or purple in age, early-deciduous, tips narrowed; ovary tip projection 0. **FR**: ovate to round, ± compressed side-to-side to cylindric, hairy; ribs ± equal, thread-like; oil tubes several per rib-interval; fr axis divided to base. **SEED**: face deeply grooved. 3 spp.: high mtns of CA-FP. (Greek: mountain dwarf) [Shevock & Norris 1981 Fremontia 9:22–25]

1. Herbage white-tomentose; umbels hemispheric; corolla yellow or maroon; anthers yellow; calyx lobes inconspicuous, obscured by hairs . . . . ***O. vestita***
1′ Herbage gray-hairy; umbels spheric; corolla white; anthers purple; calyx lobes conspicuous, spreading
  2. Fls appearing ± with lf blades; bladeless sheaths 0.5–2 cm; umbels 1–3 cm diam; rays 5–15, 2–8 mm; calyx lobes yellow. . . . . ***O. clementis***
  2′ Fls appearing before lf blades; bladeless sheaths 3–6 cm; umbels 2.4–4.5 cm diam; rays 20–35, 5–15 mm; calyx lobes purple . . . . ***O. purpurascens***

O. clementis (M.E. Jones) Jeps.   Pl 3–8 cm, gray-hairy. **LF**: bladeless sheaths 0.5–2 cm; petiole 2–4 cm; blade 1.5–3.5 cm, ovate, segments 1–3 mm. **INFL**: umbels 1–3 cm diam, spheric, appearing ± with lf blades; peduncles 2–7 cm, spreading; bractlets fused; rays 5–15, 2–8 mm, outer scarious-winged, webbed; sterile pedicels 2–5 mm. **FL**: bisexual or staminate; calyx lobes ± 0.5–1.5 mm, spreading, yellow; corolla white; anthers purple. **FR**: 3–4 mm, ± glabrous to hairy. 2*n*=22. Rocky ridges; 1500–4000 m. s SNH. Jul–Aug

O. purpurascens Shevock & Constance (p. 195)   PURPLE MOUNTAIN-PARSLEY   Pl 0.8–2.2 cm, gray-hairy. **LF**: bladeless sheaths 3–6 cm; petiole 4–7 cm; blade 5–10 cm, narrowly ovate, segments 1–3 mm, lanceolate or oblong. **INFL**: umbels 2.5–4.5 cm diam, spheric, appearing before lf blades; peduncles 12–18 cm, spreading; bractlets fused; rays 20–35, 5–15 mm, outer scarious-winged, webbed; sterile pedicels 3–10 mm. **FL**: bisexual, staminate, or sterile; calyx lobes 1.5–3 mm, spreading, purple; corolla white; anthers purple. **FR**: 4–5 mm wide, ± glabrous to hairy. 2*n*=22. Ridgetops, gen on metamorphic rocks, in red-fir or lodgepole-pine forests; 2375–2860 m. s SNH. May–Jun   ★

O. vestita (S. Watson) Jeps.   WOOLLY MOUNTAIN-PARSLEY   Pl 4–15 cm, white-tomentose. **LF**: bladeless sheaths 1.5–3 cm; petiole 2–10 cm; blade 1.5–5 cm, ovate to round, segments 3–10 mm,

oblong. **INFL:** umbels 2–5 cm, hemispheric, appearing ± with lf blades; peduncles 4–12 cm, spreading; bractlets ± free; rays 10–25, 1–2 cm, not scarious-winged, not webbed; sterile pedicels 10–15 mm. **FL:** bisexual or staminate; calyx lobes inconspicuous, obscured by hairs; corolla yellow or maroon; anthers yellow. **FR:** 5–6 mm, 3–4 mm wide, tomentose. 2*n*=22. Ridge tops; 1670–3500 m. SnGb, SnBr. Mar–Jul ★

OROGENIA

Per, glabrous, ± glaucous; root tuberous. **ST:** 0 or short. **LF:** basal, gen recurved or spreading; lowest bladeless, scarious sheaths; blades ovate to triangular-ovate, 1–3-ternately dissected, segments gen linear to lanceolate, elongate. **INFL:** umbels compound, peduncled; bracts 0; bractlets 0–few, minute; rays few, spreading; pedicels few, 0 to short. **FL:** calyx 0; petals obovate, white, tips narrowed; ovary tip projection 0. **FR:** ± ovate, compressed front-to-back; marginal ribs corky-winged, incurved, not wing-like, others thread-like; oil tubes per rib-interval several; fr axis a corky ridge along middle of each fr half. **SEED:** face ± concave. 2 spp.: w N.Am. (Greek: mountain race)

O. fusiformis S. Watson (p. 195) Pl 5–15 cm; tuber 3–10 mm wide, carrot-like. **LF:** bladeless sheath 2–7 cm, oblong; petiole 3–5 cm; blade 2–8 cm wide, segments 5–60 mm, linear to lance-linear, acute or obtuse. **INFL:** peduncles 1–3, 2–15 cm, recurving; fertile rays 1–8, 5–60 mm; fertile pedicels 2–15, < 1 mm. **FR:** 3–4 mm, ovate; ribs on back obscure. Gravelly flats, fl near melting snow; 1000–2500 m. KR, CaRH, n SNH; to c OR. May–Jul

OSMORHIZA SWEET-CICELY

Per, ± glabrous to hairy; roots thick, clustered, licorice-scented. **ST:** branched, lfy. **LF:** blade oblong to triangular-ovate, 2-pinnate or ternate-pinnate or 2–3-ternate, lflets lanceolate to round. **INFL:** umbels compound; bracts 0; bractlets 0–several, conspicuous; rays, pedicels few, spreading-ascending to spreading. **FL:** calyx lobes 0; petals obovate, white, purple, or ± green-yellow (± green-white), tips narrowed; disk occ present. **FR:** linear to oblong, cylindric to club-shaped, ± compressed side-to-side, bristly to glabrous; base obtuse or long-tapered into tail, tip tapered into beak or obtuse; ribs thread-like; oil tubes obscure; fr axis divided in distal 1/2. **SEED:** face concave or grooved. ± 10 spp.: Am, e&s Asia. (Greek: sweet root) [Lowry & Jones 1985 Ann Missouri Bot Gard 71:1128–1171]

1. Fr not long-tapered at base, glabrous; lf 2-pinnate . . . . ***O. occidentalis***
1′ Fr long-tapered into tail at base, bristly; lf 2–3-ternate
  2. Involucel conspicuous; corolla ± green-yellow . . . . ***O. brachypoda***
  2′ Involucel 0 or vestigial; corolla white or purple (± green-white)
    3. Fr club-like, beakless; rays, pedicels spreading . . . . ***O. depauperata***
    3′ Fr linear-fusiform or -oblong, beaked; rays, pedicels spreading-ascending
      4. Ovary tip projection conic; disk 0; fr 12–25 mm; corolla white . . . . ***O. berteroi***
      4′ Ovary tip projection depressed-conic; disk conspicuous; fr 8–15 mm; corolla purple (± green-white) . . . ***O. purpurea***

O. berteroi DC. Pl 3–12 dm, ± glabrous to finely hairy. **LF:** petiole 5–16 cm; blade 4–20 cm wide, widely ovate to obovate, 2-ternate, lflets 2–8 cm, widely lanceolate to round, serrate to irregularly cut or lobed. **INFL:** peduncle 5–25 cm; bractlets gen 0; rays 3–8, 2–12 cm, spreading-ascending; pedicels 4–20 mm. **FL:** corolla white; styles gen < 1 mm; ovary tip projection conic; disk 0. **FR:** 12–25 mm, linear-fusiform or -oblong; tail 2–8.5 mm; beak slender; ribs bristly. 2*n*=22. Conifer forest, woodland, disturbed areas; < 2800 m. CA-FP, MP; to AK, AB, CO; also c&e N.Am, s S.Am. [*O. chilensis* Hook. & Arn.] Apr–Jul

O. brachypoda Torr. (p. 195) Pl 3–8 dm, finely hairy. **LF:** petiole 5–20 cm; blade 1–2 dm, ovate to triangular-ovate, 2–3-ternate, lflets 2–8 cm, ovate, serrate to irregularly cut or pinnately lobed. **INFL:** peduncle 9–20 cm; bractlets 2–6, 2–10 mm, linear to lanceolate; rays 2–5, 3.5–12 cm, spreading-ascending; pedicels 1–5 mm. **FL:** corolla ± green-yellow; styles gen < 1 mm; disk gen conspicuous. **FR:** 12–20 mm, oblong-fusiform; tail 1–4 mm; beak slender; ribs bristly. 2*n*=22. Moist ravines, conifer forest, woodland; 200–2000 m. c&s SNF, n SNH, Teh, SnFrB, SCoR, TR, PR; c AZ. Mar–May

O. depauperata Phil. BLUNT-FRUITED SWEET-CICELY Pl 1.5–8 dm, gen sparsely short-hairy to glabrous. **LF:** petiole 3–20 cm; blade 4–12 cm wide, widely ovate to round, 2–3-ternate, lflets 1.5–5 cm, widely lanceolate to ovate, coarsely serrate to deeply cut. **INFL:** peduncle 5–15 cm; bractlets gen 0; rays 2–5, 3–9 cm, spreading; pedicels 8–15 mm. **FL:** corolla white; styles < 0.5 mm; ovary tip projection low- or depressed-conic; disk 0 or vestigial. **FR:** 10–18 mm, club-like; tail 3–8.5 mm; tip obtuse, beakless; ribs ± bristly. 2*n*=22. Conifer forest, aspen woodland; 500–3300 m. MP; to AK, MT, CO, NM; also in c&e N.Am, s S.Am. May–Jul ★

O. occidentalis (Nutt.) Torr. (p. 195) Pl 4–12 dm, glabrous to sparsely fine-hairy. **LF:** petiole 5–25 cm; blade 1–2 dm, oblong to ovate, 2-pinnate, lflets 2–10 cm, lance-oblong to ovate, serrate and gen irregularly cut or lobed. **INFL:** peduncle 6–20 cm; bractlets gen 0; rays 5–12, gen 3–8 cm, ascending to spreading-ascending; pedicels 3–8 mm. **FL:** corolla yellow; styles 0.8–1.4 mm; disk conspicuous. **FR:** 12–22 mm, linear-fusiform, not long-tapered at base; tail 0; tip narrowed proximal to beak; ribs (and intervals) glabrous. 2*n*=22. Conifer forest, oak woodland; 200–3200 m. KR, NCoR, CaRH, n&c SNH, MP, n SNE; to w Can, CO. May–Jul

O. purpurea (J.M. Coult. & Rose) Suksd. Pl 2–6 dm, glabrous to sparsely fine-hairy. **LF:** petiole 5–12 cm; blade 3–10 cm wide, triangular-ovate to round, 2–3-ternate, lflets 1.5–7 cm, lanceolate to ovate, coarsely serrate to irregularly cut or lobed. **INFL:** peduncle 3–10 cm; bractlets 0; rays 2–6, 3–9.5 cm, spreading-ascending; pedicels 5–25 mm. **FL:** corolla purple (± green-white); styles 0.4–0.8 mm; ovary tip projection depressed-conic; disk conspicuous. **FR:** 8–15 mm, linear-fusiform; tail 1–5 mm; tip narrowed proximal to beak; ribs bristly at base. 2*n*=22. Damp conifer forest; 150–2200 m. NCo (Del Norte Co.); to AK, MT. Jun–Jul

OXYPOLIS

Per, glabrous; tubers clustered. **ST:** erect, gen branched. **LF:** blade oblong [linear, triangular-ovate], pinnate, [simple, ternate]; main axis segmented, hollow, occ bladeless. **INFL:** umbels compound; involucre, involucel variable; bracts, bractlets 0 or

1–several, ± inconspicuous; rays, pedicels few to many, spreading-ascending. **FL:** calyx lobes evident [minute] but not a persistent crown on fr; petals wide, white, tips narrowed. **FR:** oblong to obovate, compressed front-to-back; ribs unequal, marginal widely thin-winged, wings veined on back at inner margin, others thread-like; oil tube 1 per rib-interval; fr axis divided to base. ± 6 spp.: w&e US, Caribbean. (Greek: sharp white)

O. occidentalis J.M. Coult. & Rose (p. 201)   Pl 6–15 dm. **LF:** petiole 1–5 dm; blade 12–30 cm, oblong, 1-pinnate, lflets 5–13, 3.5–9.5 cm, lanceolate to widely ovate, crenate to serrate or irregularly cut; cauline lvs with enlarged petiole, smaller lflets than basal. **INFL:** peduncle 6–30 cm; bracts 0–1(8), 5–20 mm, linear, scarious; bractlets like bracts, < 10 mm; rays 12–24, 2–8.5 cm; pedicels 3–15 mm. **FL:** calyx lobes conspicuous. **FR:** 5–6 mm, oblong or ovate. 2*n*=36. Bogs, wet meadows, stream sides, gen in conifer forest; < 2700 m. CaRF, SN, SnGb, SnBr, W&I; OR, BC. Jul–Aug

PASTINACA   PARSNIP

Bien, per, taprooted, ± glabrous to hairy. **ST:** erect, branched. **LF:** blade oblong to triangular-ovate, 1–2-pinnate, lflets oblong to ovate. **INFL:** umbels compound; bracts gen 0; bractlets gen 0; rays 5–20, pedicels many, rays and pedicels spreading-ascending. **FL:** calyx lobes minute; petals wide, yellow [orange], tips narrowed. **FR:** oblong to obovate, compressed front-to-back; ribs unequal, marginal narrowly winged, others thread-like; oil tube 1 per rib-interval, all equal in length; fr axis divided to base. **SEED:** face flat. ± 10 spp.: Eurasia. (Ancient name for parsnip)

P. sativa L.   Bien, 0.5–2 m, ± glabrous to puberulent. **ST:** conspicuously angled, grooved. **LF:** petiole 1–1.5 cm; blade 1.5–3 dm, oblong to ovate, 1-pinnate, lflets 5–11, 5–10 cm, oblong to ovate, coarsely serrate and lobed or divided. **INFL:** peduncle 7–15 cm. **FR:** 4–6 mm wide, oblong to round. 2*n*=22. Roadsides, disturbed areas; < 1600 m. KR, NCoRO, NCoRI, CaRH, ScV, CCo, TR, MP; to e US; native to Eurasia. Sporadic. Jul–Aug

PERIDERIDIA   YAMPAH

Per, glabrous, gen glaucous; roots tuberous, single or clustered, or clustered-fibrous. **ST:** erect, branched. **LF:** blade lanceolate to triangular-ovate, gen 1–2-ternate-pinnate or 1–2-pinnately or ternate-pinnately dissected, lflets or segments gen linear to lance-linear. **INFL:** umbels compound; bracts 0–many, conspicuous and reflexed or not; bractlets several to many, narrow, ± scarious; rays, pedicels few to many, gen spreading-ascending; 2° umbels gen convex distally. **FL:** calyx lobes evident; petals gen obovate, white, tips narrowed. **FR:** linear-oblong, ± compressed side-to-side or not at all, glabrous; ribs ± equal, thread-like to prominent, not winged; oil tubes 1–several per rib-interval; fr axis divided to base. **SEED:** face flat to grooved. ± 12 spp.: gen w Am. (Greek: around the neck, from involucre) [Chuang & Constance 1969 Univ Calif Publ Bot 55:1–74] Roots, basal lvs needed for identification.

1. Roots fibrous or ± thickened, gen 5–20-clustered; styles 0.8–1 mm
  2. Basal blade lflets widely lanceolate to ovate, 10–40 mm wide; petals 5–7-veined; fr ribs prominent . . . . . . ***P. howellii***
  2′ Basal blade lflets linear to lanceolate, 2–8 mm wide; petals 1-veined; fr ribs thread-like. . . . . ***P. kelloggii***
1′ Roots tuberous, 2–6-clustered or 1; styles gen 1–2.5 mm
  3. Basal lvs 1–2-ternate or 1–2-pinnate, with 1–3 pairs of 1° lflets
    4. Rays in fr unequal, spreading-ascending; oil tubes per rib-interval 1. . . . . ***P. lemmonii***
    4′ Rays in fr ± equal and ascending, or unequal and spreading-ascending; oil tubes per rib-interval 3–4 . . . . ***P. parishii***
      5. Rays gen 12–14, ± equal in fr; bractlets ± = pedicels; fr ovate to ± round, 2.5–3.5 mm . . . . subsp. ***latifolia***
      5′ Rays gen 6–11, unequal in fr; bractlets gen < pedicels; fr oblong to ± ovate, 3–5 mm . . . . subsp. ***parishii***
  3′ Basal lvs gen 1-pinnate with 3–5 pairs of 1° lflets, or 2–3-pinnately or ternate-pinnately dissected or compound
    6. Basal lvs gen 1-pinnate with 3–5 pairs of 1° lflets — proximal 1° lflets occ lobed or ternately dissected
    . . . . ***P. gairdneri***
      7. Tuberous roots 2–3-clustered; petals 5- or 7-veined . . . . subsp. ***borealis***
      7′ Tuberous roots 1; petals 1-veined . . . . subsp. ***gairdneri***
    6′ Basal lvs 2–3-pinnately or ternate-pinnately dissected or compound
      8. Terminal lf segments unlike lateral in size, form
        9. Fr 4–6 mm; bractlets scarious throughout, 3–9 mm, acuminate . . . . ***P. bolanderi*** subsp. ***bolanderi***
        9′ Fr 5–8 mm; bractlets scarious-margined, 1–4 mm, acute
          10. Pl 5–15 dm; tuberous roots cylindric, 5–12 cm; lf segments flattened, 5–10 mm wide; oil tubes per
            rib-interval 1 . . . . ***P. californica***
          10′ Pl 3.5–7.5 dm; tuberous roots oblong to ovoid, 1–6 cm; lflets cylindric, < 1 mm wide; oil tubes per
            rib-interval 3–4. . . . . [2]***P. pringlei***
      8′ All lf segments or lflets similar in size, form
        11. Longest rays in fr 5–9 cm
          12. 2° umbels gen spheric, gen 40–50(60)-fld; 1° umbels ± concave distally, rays unequal; oil tube 1 per
            rib-interval . . . . ***P. bacigalupii***
          12′ 2° umbels convex distally, gen 20–25-fld; 1° umbels convex distally, rays ± equal to unequal; oil
            tubes 3–4 per rib-interval. . . . . [2]***P. pringlei***
        11′ Longest rays in fr 1–4.5 cm
          13. Bracts 0–2, inconspicuous
            14. Umbels strongly concave distally; rays in fr unequal; fr linear-club-shaped, 5–7 mm . . . . ***P. leptocarpa***
            14′ Umbels convex or flat distally; rays in fr ± equal; fr oblong, 3–6 mm . . . . [2]***P. oregana***

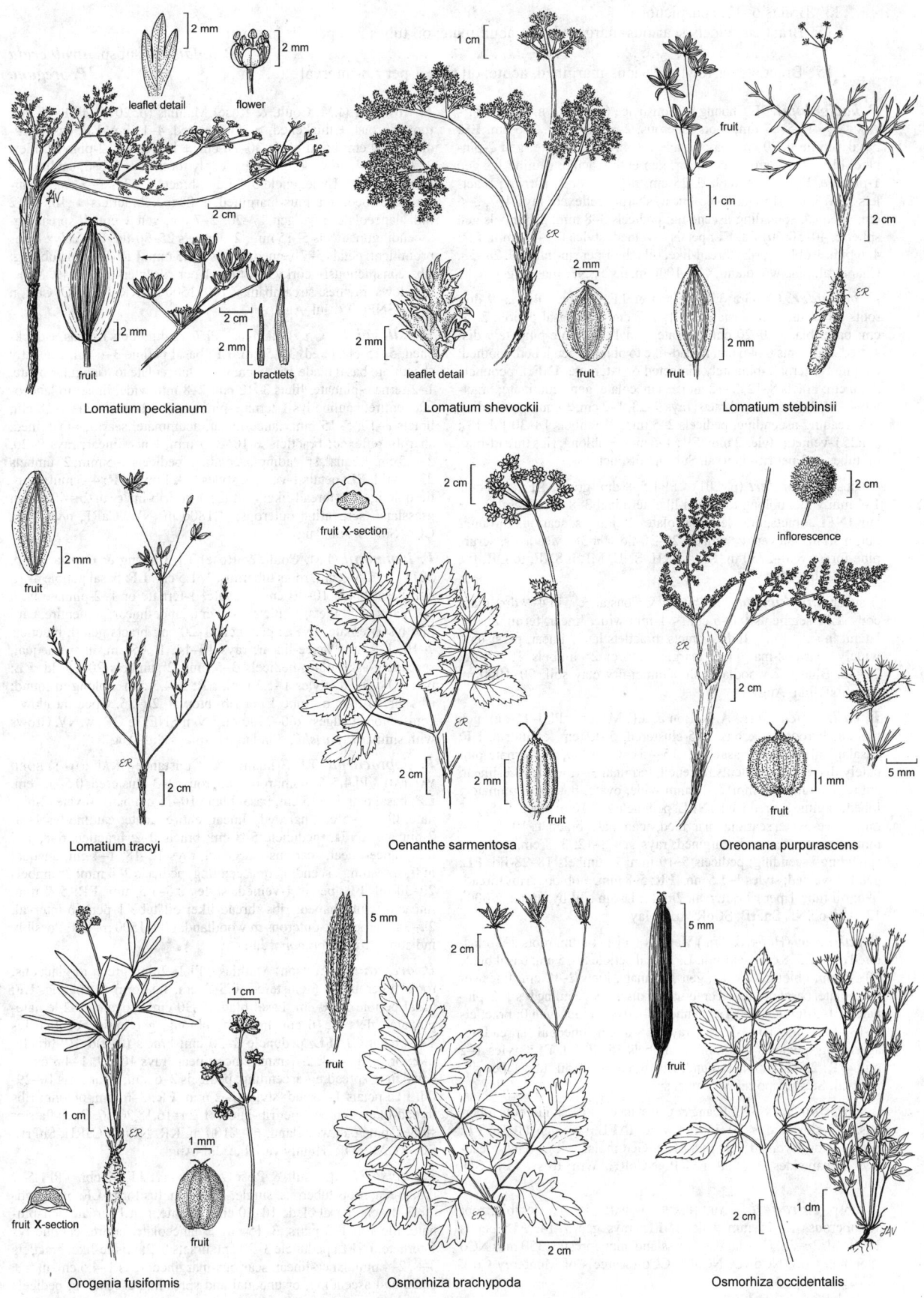

leaflet detail

flower

2 mm

2 mm

2 cm

fruit

2 mm

bractlets

2 mm

Lomatium peckianum

1 cm

leaflet detail

2 mm

fruit

2 mm

Lomatium shevockii

fruit

1 cm

2 cm

fruit

2 mm

Lomatium stebbinsii

fruit

2 mm

Lomatium tracyi

2 cm

fruit X-section

2 cm

2 cm

2 mm

fruit

Oenanthe sarmentosa

inflorescence

2 cm

2 cm

fruit

1 mm

5 mm

Oreonana purpurascens

fruit X-section

1 cm

fruit

1 mm

Orogenia fusiformis

fruit

5 mm

2 cm

Osmorhiza brachypoda

2 cm

fruit

5 mm

2 cm

1 dm

Osmorhiza occidentalis

13′ Bracts 6–12, conspicuous
    15. Bractlets widely scarious-margined, gen acuminate; oil tubes 2–3 per rib-interval
      .... ***P. bolanderi*** subsp. ***involucrata***
    15′ Bractlets narrowly scarious-margined, acute; oil tube 1 per rib-interval....................... ²***P. oregana***

P. bacigalupii T.I. Chuang & Constance (p. 201) BACIGALUPI'S
YAMPAH Pl 5–17 dm; roots tuberous, 2–6-clustered, 2–6 cm. **LF:**
basal petiole 6–20 cm; basal blade 20–40 cm, ± ovate, gen 2-pin-
nate, lflets 2–15 cm, lance-linear, gen entire, acute; cauline lvs gen
1-pinnate. **INFL:** peduncle 3–25 cm; bracts 1 or 2, narrow; bractlets 12–16, 0.5–15 cm, lance-linear, sharply reflexed; rays 4–7, 3–6
cm, unequal, spreading-ascending; pedicels 3–8 mm; 2° umbels gen
spheric, 40–50(60)-fld. **FL:** petals 1-veined; styles 0.7–1.2 mm. **FR:**
4–6 mm, ± oblong; ribs thread-like; oil tube 1 per rib-interval. 2*n*=38.
Chaparral, pine woodland; 450–1000 m. n&c SNF. Jun–Aug ★

P. bolanderi (A. Gray) A. Nelson & J.F. Macbr. Pl 1.5–9 dm;
roots tuberous, 2–3-clustered or 1, 1–7 cm. **LF:** basal petiole 2–15
cm; basal blade 10–20 cm, ± ovate, gen 1–2-ternate-pinnately dissected, segments 0.5–6 cm, thread-like to oblong, gen lobed, toothed;
cauline lvs ternate-pinnately dissected or 1-ternate. **INFL:** peduncle
2–20 cm; bracts 8–12, 3–12 mm, ± lanceolate, gen acuminate; bractlets 4–10, 3–9 mm, like bracts; rays 9–23, 1–2 cm, ± equal, ascending
or spreading-ascending; pedicels 2–5 mm; 2° umbels 18–30-fld. **FL:**
petals 1-veined; styles 2 mm. **FR:** 4–6 mm, ± oblong; ribs thread-like;
oil tubes 2–3 per rib-interval. Subspp. distinct.

    subsp. ***bolanderi*** (p. 201) Pl 1.5–8 dm, green. **LF:** segments
1–4 mm wide, oblong to thread-like, terminal 3–8 cm, lateral 0.5–3
cm. **INFL:** bracts, bractlets lanceolate, deciduous, scarious throughout, margins uneven; 2° umbels 25–30-fld. 2*n*=38. Meadows, scrub,
pine forest; 850–2750 m. KR, CaRH, SNH, MP, n SNE; to OR, ID,
WY, NV. Jun–Aug

    subsp. ***involucrata*** T.I. Chuang & Constance Pl 4–9 dm, glau-
cous. **LF:** segments 0.6–6 cm, < 1 mm wide, linear, terminal like
lateral in size, form. **INFL:** bracts, bractlets lance-linear, persistent,
widely scarious-margined, margins entire; 2° umbels 18–25-fld.
2*n*=38. Blue-oak woodland, in summer-dry clay soil; 90–1000 m.
n&c SNF. Jun–Aug

P. californica (Torr.) A. Nelson & J.F. Macbr. Pl 5–15 dm, gen
branched; roots tuberous, 2–5-clustered, 5–12 cm, cylindric. **LF:**
basal petiole 2–6 cm; basal blade 15–45 cm, ± ovate, 1–2-ternate-pin-
nately dissected, segments flattened, terminal 3–15 cm, lance-linear,
entire, lateral 3–30 mm, 5–10 mm wide, ovate, flattened, pinnately
lobed; cauline lvs reduced. **INFL:** peduncle 2–20 cm: bracts 5–8, 2–5
mm, lance-ovate, scarious-margined, acuminate; bractlets 10–12, 1–4
mm, ± ovate, scarious-margined; rays gen 8–12, 3–8 cm, ± equal,
spreading-ascending; pedicels 5–10 mm; 2° umbels 18–28-fld. **FL:**
petals 1-veined; styles 1–1.5 mm. **FR:** 5–8 mm, ± oblong; ribs thread-
like; oil tube 1 per rib-interval. 2*n*=44. Damp soil by streams; 300–
1250 m. c SNF, SnFrB, SCoR. Apr–May

P. gairdneri (Hook. & Arn.) Mathias Pl 3–14 dm; roots 2–3-clus-
tered or 1, 1.5–8 cm, fusiform. **LF:** basal petiole 8–15 cm; basal blade
20–35 cm, oblong to ovate, gen 1-pinnate, lflets 2–12 cm, linear or
lanceolate, basal occ lobed or ternately dissected; cauline lvs 1–2-pin-
nate or 1–2-ternate. **INFL:** peduncle 3–20 cm; bracts gen 0; bractlets
8–13, 1–4 mm, lance-linear; rays 1.5–7 cm, unequal, spreading-
ascending; pedicels 4–8 mm; 2° umbels 15–40-fld. **FL:** styles 1–2
mm. **FR:** 2.5–3.5 mm, ± round; ribs thread-like; oil tube 1 per rib-
interval. Subspp. possibly distinct spp.

    subsp. ***borealis*** T.I. Chuang & Constance Pl ± rigid; roots tuber-
ous, 2–3-clustered, < 10–15 mm wide. **INFL:** rays gen 11–16. **FL:**
petals 5- or 7-veined. 2*n*=40,80,120. Gen moist soil of flats, meadows, streamsides; < 3000 m. KR, NCoRH, Wrn; to sw Can, SD.
Jun–Aug

    subsp. ***gairdneri*** GAIRDNER'S YAMPAH Pl gen flexible; root
1, tuberous, 10–15 mm wide. **INFL:** rays gen 7–14. **FL:** petals
1-veined. 2*n*=38. Coastal flats, grassland, pine forest; < 350 m. s NCo
(Sonoma Co.), NCoRO, NCoRI, CCo (scarce s of Monterey Co.),
SCo. Jun–Jul ★

P. howellii (J.M. Coult. & Rose) Mathias (p. 201) Pl 8–16 dm;
roots fibrous, ± thickened, > 15-clustered, 4–15 cm. **LF:** basal petiole 10–20 cm; basal blade 30–50 cm, ± lanceolate, 2-pinnate, lflets
2–7 cm, 1–4 cm wide, ± ovate, deeply toothed or lobed; cauline lvs
like basal. **INFL:** peduncle 5–25 cm; bracts 8–12, 0.5–2 cm, linear-
spoon-shaped, scarious-margined, reflexed; bractlets ± 10, 5–12
mm, lanceolate; rays gen 17–20, 2–7 cm, gen ± equal, spreading-
ascending; pedicels 5–15 mm; 2° umbels 25–30-fld. **FL:** calyx lobes
prominent; petals 5–7-veined; styles < 0.5 mm. **FR:** 4–6 mm, oblong;
ribs conspicuously corky; oil tube 1 per rib-interval. 2*n*=40. Moist
meadows, ravines, streambanks; 100–1500 m. KR, NCoRO, CaR, n
SN, c SNF; OR. Jul–Aug

P. kelloggii (A. Gray) Mathias Pl 7–15 dm; roots fibrous, ± thickened, 5–15-clustered, 3–15 cm. **LF:** basal petiole 3–6 cm, gen fully
sheathing; basal blade 15–45 cm wide, lanceolate to triangular-ovate,
1–2-ternate-pinnate, lflets 3–12 cm, 2–8 mm wide, linear to lanceolate, entire; cauline lvs 1-ternate-pinnate. **INFL:** peduncle 5–18 cm;
bracts 8–10, 5–15 mm, lance-linear, acuminate, scarious-margined,
sharply reflexed; bractlets ± 10, 3–5 mm, lance-linear; rays 9–20,
2–6.5 cm, ± equal, spreading-ascending; pedicels 2–5 mm; 2° umbels
23–27-fld. **FL:** petals 1-veined; styles 0.8–1 mm. **FR:** 4–5 mm, elliptic-oblong; ribs thread-like; oil tube 1 per rib-interval. 2*n*=40. Open
grassland, serpentine outcrops; < 1800 m. NW, CaRF, n&c SNF,
CCo, SnFrB. Jul–Aug

P. lemmonii (J.M. Coult. & Rose) T.I. Chuang & Constance (p.
201) Pl 2.5–9 dm; roots tuberous, 1, 1.5 cm. **LF:** basal petiole 4–15
cm; basal blade 10–30 cm, ± ovate, 1-ternate or 1–2-pinnate with
1–2 pairs of 1° lflets, 1° lflets 3–10 cm, lance-linear, gen entire; cauline lvs 1-ternate. **INFL:** peduncle 3–20 cm; bracts gen 0; bractlets
7–10, 1–2 mm, lance-linear; rays 10–14, 1–3.5 cm, in fr unequal,
spreading-ascending; pedicels 3–5 mm; 2° umbels 20–35-fld. **FL:**
petals 1-veined; styles 1–1.5 mm. **FR:** 3–4.5 mm, oblong to round;
ribs thread-like; oil tube 1 per rib-interval. 2*n*=36. Open meadows,
conifer forest edges; 760–2500 m. SN, n SNE; se OR, w NV. Grows
with similar *P. parishii*, fl in large displays. Jul–Aug

P. leptocarpa T.I. Chuang & Constance NARROW-SEEDED
YAMPAH Pl 4.5–7.5 dm; roots tuberous, 2–3-clustered, 0.5–3.5 cm.
LF: basal petiole 1–5 cm; basal blade 10–15 cm, lance-ovate, 2-pin-
nate, lflets 1–5 cm, narrowly linear, entire, acute; cauline lvs 1- or
2-pinnate. **INFL:** peduncle 5–8 cm; bracts 0–2; bractlets 6–8, 1–4
mm, lance-linear, scarious-margined; rays 12–18, 1–4 cm, unequal
in fr, spreading-ascending or ascending; pedicels 2–4 mm; 2° umbels
20–30-fld. **FL:** petals 1-veined; styles 0.5–1.5 mm. **FR:** 5–7 mm,
linear to club-shaped; ribs thread-like; oil tube 1 per rib-interval.
2*n*=34. Serpentine outcrops in woodland; 600–1500 m. KR. Possibly
indistinct from *P. oregana*. Jun–Aug ★

P. oregana (S. Watson) Mathias Pl 1–9 dm, green to glaucous;
roots tuberous, 2–6-clustered, 0.5–3 cm, fusiform to spheric. **LF:**
basal petiole 2–10 cm; basal blade 3–30 cm, ± ovate, 1–2-ternate-
pinnate, lflets 0.5–6 cm, linear or oblong, gen entire; cauline lvs
1–2-pinnate. **INFL:** peduncle 3–20 cm; bracts (0–2)6–10, bristle-
like; bractlets 4–8, 2–7 mm, lance-linear; rays 10–29, 1–4.5 cm, ±
equal in fr, spreading-ascending; pedicels 2–6 mm; 2° umbels 10–29-
fld. **FL:** petals 1-veined; styles 1–2 mm. **FR:** 3–6 mm, oblong; ribs
thread-like; oil tube 1 per rib-interval. 2*n*=16,18,20,26. Open flats or
slopes, pine/oak woodland; 60–2100 m. KR, NCoR, CaRH, SnFrB,
SCoRI, MP; OR. Highly variable. Jul–Aug

P. parishii (J.M. Coult. & Rose) A. Nelson & J.F. Macbr. Pl 1.5–9
dm, green; roots tuberous, single, 1–2.5 cm, fusiform. **LF:** basal petiole 3–10 cm; basal blade 10–20 cm, ± ovate, gen 1-ternate or 1-pin-
nate, lflets in 1–3 pairs, 3–15 cm, ± lanceolate, entire; cauline lvs
1-ternate. **INFL:** peduncle 3–20 cm; bracts 0–2, bristle-like; bractlets
3–8, 2–4 mm, lance-linear, scarious-margined; rays 1–4.5 cm, in fr ±
equal and ascending, or unequal and spreading-ascending; pedicels

3–4 mm; 2° umbels 13–27-fld. **FL:** petals 1-veined. **FR:** 2.5–5 mm wide; ribs thread-like; oil tubes 3–4 per rib-interval.

subsp. *latifolia* (A. Gray) T.I. Chuang & Constance **INFL:** umbels flat to convex distally; rays gen 12–14, ± equal in fr; bractlets 6–8, ± = pedicels. **FL:** styles 1.5 mm. **FR:** 1.8–3.5 mm, ovate to ± round. 2*n*=38. Wet meadows, open conifer forests; 2000–3400 m. KR, CaR, SNH, TR, PR, SNE; NV. Jun–Aug

subsp. *parishii* PARISH'S YAMPAH **INFL:** umbels concave distally; rays gen 6–11, unequal in fr; bractlets 3–5, gen < pedicels. **FL:** styles 1 mm. **FR:** 3–5 mm, oblong to ± ovate. 2*n*=38. Damp meadows; 2000–3000 m. SnBr; AZ, NM. Jun–Aug ★

P. pringlei (J.M. Coult. & Rose) A. Nelson & J.F. Macbr. ADOBE YAMPAH Pl 3.5–7.5 dm; roots tuberous, 2–4-clustered, 1–6 cm. **LF:** basal petiole 4–15 cm; basal blade 10–30 cm, ± ovate, 2-pinnate, lflets 0.5–8 cm, < 1 mm wide, cylindric; cauline lvs 1-pinnate. **INFL:** peduncle 3–12 cm; bracts 0–5, bristle-like; bractlets 8–12, 2–4 mm, lance-linear to ovate; rays gen 5–7, 2–8 cm, ± equal to unequal, spreading-ascending; pedicels 5–7 mm; 2° umbels (18)20–25(29)-fld. **FL:** petals 1-veined; styles 1 mm. **FR:** 5–8 mm, oblong; ribs thread-like; oil tubes 3–4 per rib-interval. 2*n*=40. Grassy slopes, serpentine outcrops; 300–1800 m. Teh, SCoR, WTR. Apr–Jun ★

PODISTERA

Per, taprooted, low, fibrous at base, ± puberulent. **ST:** 0. **LF:** basal; blades oblong to widely ovate, 1–2-pinnate, lflets linear to round. **INFL:** umbels compound, dense, head-like; bracts 0 or linear; involucel 1-sided; bractlets several, narrow or wide, gen partly fused; rays few, cylindric to flattened or winged; pedicels like rays. **FL:** calyx lobes conspicuous; petals wide, yellow, ± purple, or white, tips narrowed. **FR:** oblong-ovate to ovate, ± compressed side-to-side, glabrous; ribs ± equal, thread-like to prominently corky, obtuse; oil tubes 2–several per rib-interval; fr axis not obvious. **SEED:** compressed front-to-back; face flat to ± concave. 4 spp.: mtns of w N.Am. (Greek: solid foot, from compact habit)

P. nevadensis (A. Gray) S. Watson (p. 201) SIERRA PODISTERA Compact cushion 2–5 cm, 2–5 dm diam. **LF:** petiole 3–15 mm, white scarious-sheathing; blade 3–10 mm, oblong to ovate, 1-pinnate, lflets 1–6 mm, linear to lanceolate, entire, pointed. **INFL:** peduncle 5–30 mm; bracts 0; bractlets 2–4 mm, ± = fls and frs, ovate, strongly fused into a cup; rays winged, short; pedicels 0–few, < fr. **FL:** corolla yellow. **FR:** 4–4.5 mm, oblong-ovate; ribs thread-like; oil tubes 3–4 per rib-interval. 2*n*=22. Unglaciated granitic gravel, scree, crevices; 3000–4000 m. n&c SNH, SnBr, W&I. Jul–Sep ★

SANICULA

Bien, per, rhizomed or tap- or tuberous-rooted, glabrous or minutely scabrous. **ST:** gen spreading to erect. **LF:** blade oblong-ovate to obovate, entire to ternately, palmately, or pinnately lobed, dissected, or compound. **INFL:** heads simple, in cymes or racemes, dense, of bisexual and staminate (staminate only) fls; bracts entire or lobed, < to > heads; bisexual fls pedicelled or not, staminate gen long-pedicelled. **FL:** calyx lobes prominent, persistent, occ fused; petals wide, yellow, purple, or ± white (pale red-orange), tips narrowed, gen lobed; styles long or short; ovary tip projection 0. **FR:** oblong-ovate to round, ± compressed side-to-side; fr-halves ± cylindric, prickly to scaly or tubercled; ribs 0; oil tubes evident or obscure, regularly or irregularly arranged; fr central axis not obvious. **SEED:** face flat or grooved. ± 40 spp.: temp, ± worldwide. (Latin: to heal) [Bell 1954 Univ Calif Publ Bot 27:133–230]

1. Lvs compound, base of main division petiole-like and unwinged
  2. Lvs palmate, occ palmate-ternate, distal segment of main axis 4–30 mm wide
    3. Main lf divisions gen ± 5, central division gen 2–4 cm; styles 2 × calyx lobes; bracts 1–2 mm; fr occ stalked distal to receptacle; common, widespread . . . . [2]*S. crassicaulis*
    3′ Main lf divisions ± 3, central division gen 4–7 cm; styles ± = calyx lobes; bracts 3–5 mm; fr ± sessile distal to receptacle; uncommon, coastal . . . . [2]*S. hoffmannii*
  2′ Lvs ± pinnate, first division gen ternate, others pinnate, distal segment of main axis 1–5 mm wide
    4. Fr with many hooked prickles
      5. Staminate fls 4–6 per umbel, inconspicuous; st lfy, ± erect . . . . *S. bipinnata*
      5′ Staminate fls 7–12 per umbel, conspicuous; st ± lfless, branches gen spreading widely from near base
        . . . . *S. graveolens*
    4′ Fr with ± rounded tubercles, hooked prickles 0 or few
      6. Pl from taproot (tuber-like); staminate fls ± > fr, inconspicuous — NCo, NCoR, CaRF . . . . *S. tracyi*
      6′ Pl from a distinct, spheric or irregular tuber; staminate fls gen >> fr, conspicuous
        7. Corolla ± yellow or pale red-orange; fr 2.5–3 mm, most tubercles nipple-like or with bristle; SnFrB . . . . *S. saxatilis*
        7′ Corolla yellow; fr 1.5–2 mm, tubercles unarmed; widespread . . . . *S. tuberosa*
1′ Lvs simple, entire to deeply lobed or cut, axis of main division with at least a narrow, toothed wing throughout
  8. Bracts yellow, conspicuous, >> heads; pl prostrate; lvs bright ± yellow-green at fl . . . . *S. arctopoides*
  8′ Bracts ± green, inconspicuous, ≤ heads; pl spreading to erect; lvs green at fl
    9. Lf margins not or ± toothed, some basal lvs entire . . . . *S. maritima*
    9′ Lf margins sharply toothed, all lvs ± lobed
      10. Lvs palmately or ± ternately lobed or dissected; main divisions 3–5, gen not deeply cut
        11. Main lf division narrowed to ± petiole-like base . . . . [2]*S. hoffmannii*
        11′ Main lf division gen ≥ 10 mm wide at base
          12. Outline of lf margin ± rounded, teeth gen ± 1 mm; pl stout, 24–120 cm, gen branched well distal to base; widespread . . . . [2]*S. crassicaulis*
          12′ Outline of lf margin sharply angled, teeth gen ± 2 mm; pl slender, 5–30 cm, gen branched near base; gen coastal . . . . *S. laciniata*

. 10′ Lvs ± pinnately lobed or dissected; main divisions gen ≥ 7, gen deeply cut
  13. Lvs appearing ± palmate, with some pinnate divisions — corolla yellow; CCo, SCoRO, SW . . . . ***S. arguta***
  13′ Lvs clearly pinnate, with some ternate divisions
    14. Corolla purple or yellow; frs 3–8 per head; staminate fls inconspicuous, < fr; widespread . . . . . ***S. bipinnatifida***
    14′ Corolla yellow; frs 1–5 per head; staminate fls conspicuous, > fr; KR (Del Norte Co.) . . . . ***S. peckiana***

S. arctopoides Hook. & Arn. FOOTSTEPS OF SPRING, YELLOW MATS Pl prostrate, 10–30 cm wide, taprooted. **LF:** simple, ± palmately or ternately dissected, bright ± yellow-green at fl; blade 2–6.5 cm, triangular-ovate to round, lobes coarsely toothed to lobed. **INFL:** peduncle 3–21 cm; bracts fused at base, 8–17, 5–18 mm, >> heads, oblanceolate, entire or 3-lobed, yellow; pedicel of bisexual fl ± 2 mm, staminate 3–4 mm. **FL:** bisexual 10–12, staminate 10–13; calyx lobes fused proximal to middle, 1–2 mm, ovate, acute; corolla yellow; styles 2–3 × calyx lobes. **FR:** 2–5 mm, obovate to ± round, smooth or with stout, curved, inflated, bulbous-based prickles distally, unarmed tubercles proximally. **SEED:** face concave. 2n=16. Open coastal bluffs, headlands, dunes; < 250 m. NCo, CCo; to BC. Feb–May

S. arguta J.M. Coult. & Rose Pl 10–50 cm, minutely scabrous; taproot turnip-shaped. **LF:** simple, ± pinnately dissected but appearing ± palmately dissected, green; blade 3–10 cm, triangular-ovate, main divisions gen ≥ 7, some divisions pinnate, margins sharply toothed. **INFL:** peduncle 2.5–11 cm; bracts ± 10, 2–4 mm, < heads, lanceolate, entire or 3-lobed; pedicel of bisexual fl < 2 mm, staminate 2–4 mm. **FL:** bisexual 5–10, staminate ± 10; calyx lobes fused proximal to middle, ovate, obtuse; corolla yellow; styles 2–3 × calyx lobes. **FR:** 4–6 mm, obovate; prickles stout, curved, inflated, bulbous-based distally, little developed proximally. **SEED:** face concave to flat. 2n=16. Open grassy or rocky slopes; < 1400 m. CCo, SCoRO, SW; Baja CA. [*S. simulans* Hoover] Mar–Apr

S. bipinnata Hook. & Arn. POISON SANICLE Pl 12–60 cm; taproot ± swollen. **ST:** lfy. **LF:** compound, ternate then 1–2-pinnate, green; blade 3–10 cm, ± ovate, lflets well separated, stalked, margins entire to lobed, distal segment of main axis 1–5 mm wide. **INFL:** peduncle 1.5–9 cm; bracts 6–8, 1–2 mm, < heads, linear to lanceolate, entire; bisexual fl 3–10 per umbel, pedicel 0, staminate 4–6 per umbel, 1–1.5 mm. **FL:** inconspicuous; calyx lobes fused at base, ovate, acute; corolla yellow; styles 2–3 × calyx lobes. **FR:** 2–3 mm wide, ovate to ± round, bulbous-tubercled, each gen with short, hooked prickles. **SEED:** face grooved. 2n=16. Open grassland or pine/oak woodland; 20–1000 m. NCoR, s CaR, SNF, ScV, SnFrB, SCoR, n&c SW. Possibly TOXIC, but no record of poisonings. Apr–May

S. bipinnatifida Hook. PURPLE SANICLE, SHOE BUTTONS Pl 12–60 cm, taprooted. **LF:** simple, 1–2-pinnately dissected, green, glaucous, or ± purple; blade 4–19 cm, oblong-ovate to ± round, main divisions gen 7 or more, gen narrow, sharply toothed. **INFL:** peduncle 0.5–16 cm; bracts ± fused at base, 6–8, 2.5 mm, < heads, lanceolate; pedicel of bisexual fl 0, of staminate 2 mm, < fr. **FL:** bisexual 8–10, staminate 10–12; calyx lobes ± fused at base, 0.8–1 mm, widely lanceolate, acute; corolla purple or yellow; styles 2 × calyx lobes. **FR:** 3–8 per head, 3–6 mm, ovate to round; prickles stout, curved, inflated, bulbous-based. **SEED:** face concave. 2n=16. Open grassland, gen on serpentine, or pine/oak woodland; 20–1850 m. NW, CaRH, SNF, n&c SNH, ScV (Sutter Buttes), CW, SW, MP; to BC, Baja CA. Mar–May

S. crassicaulis DC. Pl 24–120 cm, stout, taprooted. **LF:** gen palmate or palmate-ternate, green; blade 3–12 cm, gen ± rounded-cordate, lobes 3–5, obovate, gen ± deeply cut, margins finely sharp-serrate, teeth gen ± 1 mm, central lobe gen ≥ 10 mm wide at base, distal segment 4–20 mm wide. **INFL:** peduncle 0.7–8 cm; bracts ± 5, 1–2 mm, < heads, narrowly lanceolate; pedicels short. **FL:** bisexual 3–8, staminate 3–5; calyx lobes fused at base, 0.5–0.7 mm, ± lanceolate, acute; corolla yellow; styles 2 × calyx lobes. **FR:** 2–5 mm, ± round, occ stalked, with stout, bulbous-based prickles, occ little-developed proximally. **SEED:** face grooved. 2n=32,48,64. Open slopes, ravines, woodland; < 1500 m. NW, SNF, s SNH, ScV (Sutter Buttes), CW, SW; to BC, Baja CA; s S.Am. Highly variable. Mar–May

S. graveolens DC. Pl low-spreading to erect, 5–45 cm, slender, taprooted (roots tuberous). **LF:** compound, ternate then 1–2-pinnate,

green or ± purple; blade 1.5–4 cm, ± ovate, lflets stalked, margins crenate to lobed, distal segment of main axis 1–5 mm wide. **INFL:** peduncle 0.5–8 cm; bracts fused at base, 6–10, ± 1 mm, < heads, linear to ovate; pedicel of bisexual fl 0, staminate 2–2.5 mm. **FL:** bisexual 3–5, staminate 7–12; calyx lobes ± fused, 0.5–1 mm, ovate, acute; corolla yellow; styles 3 × calyx lobes. **FR:** 3–5 mm, ovate to ± round, prickles stout, hooked, inflated, bulbous-based, proximal occ poorly developed. **SEED:** face concave. 2n=16. Open forest or rocky slopes, occ serpentine; 600–2600 m. NW, CaRH, SNH, SCoRO, WTR, SnBr, PR, MP; to BC, MT; s S.Am. Mar–May

S. hoffmannii (Munz) R.H. Shan & Constance HOFFMANN'S SANICLE Pl 30–90 cm, stout, taprooted. **LF:** gen compound, palmate, ± blue-green; blade 4.5–13.5 cm, triangular, lflets or deepest lobes 3, obovate, ± cut, margins irregularly serrate, central lflet gen 4–7 cm, 4–30 mm wide, narrowed to a petiole-like base. **INFL:** peduncle 1.5–12 cm; bracts 5–7, 3–5 mm, < heads, lanceolate; pedicels 0 or < 1 mm. **FL:** bisexual 4–10, staminate 3–5; calyx lobes fused at base, 1–2.3 mm, widely lanceolate, acute; corolla ± green-yellow; styles ± = calyx lobes. **FR:** 3–5 mm, ovate to obovate; prickles stout, curved, poorly developed proximally. **SEED:** face flat. 2n=16. Shrubby coastal hills, pine woodland; < 500 m. CCo (Santa Cruz Co.), SCo, n ChI. Mar–May ★

S. laciniata Hook. & Arn. Pl 5–30 cm, slender, taprooted. **ST:** gen branched near base. **LF:** simple, palmately or ± ternately lobed, green; blade 1.5–4 cm, ovate to round, lobes 0–5, margins sharply angled, teeth gen ± 1 mm, central lobe gen > 10 mm wide at base. **INFL:** peduncle 1.5–8 cm; bracts ± 10, < heads, lanceolate to ovate; pedicel of bisexual fl 0, staminate 1.5 mm. **FL:** bisexual 4–8, staminate 10–12; calyx lobes fused proximal to middle, 0.5–0.8 mm, ovate, acute; corolla yellow; styles 2 × calyx lobes. **FR:** 2–4 mm wide, ovate to ± round; prickles slender, curved, ± bulbous-based distally, poorly developed proximally. **SEED:** face grooved. 2n=16. Coastal, open or shrubby slopes, woodland; 30–900 m. NCo, NCoR, CCo, SnFrB, SCoRO; s OR. Mar–May

S. maritima S. Watson (p. 201) ADOBE SANICLE Pl 8–40 cm, taprooted. **LF:** simple, entire to ± pinnately lobed or dissected, green at fl; blade 3–8 cm, ovate-cordate to obovate, lobes or segments 0, 3, or 5, obtuse, entire to ± toothed. **INFL:** peduncle 1.5–12 cm; bracts ± 10, 1–4 mm, < heads, lanceolate, acute; pedicel of bisexual fl 0, staminate 2.5–3 mm. **FL:** bisexual 3–8, staminate 10–12; calyx lobes fused at base, 1.5–2 mm, lance-ovate, obtuse; corolla yellow; styles 3 × calyx lobes. **FR:** ± 5 mm, obovate; prickles stout, curved, inflated, bulbous-based distally, ± 0 proximally. **SEED:** face concave. 2n=16. Coastal, grassy, open wet meadows, ravines; ± 150 m. CCo (Monterey, San Luis Obispo cos.), SnFrB (apparently extirpated). Apr–May ★

S. peckiana J.F. Macbr. PECK'S SANICLE Pl 16–38 cm, glabrous. **LF:** simple, 1–2-pinnately dissected, green, ± glaucous; blade 5–10 cm, oblong-ovate, main divisions gen 7 or more, margins sharply toothed. **INFL:** peduncle 2–75 mm; bracts fused at base, ± 6, ± 1 mm, < heads, lance-ovate, acute; pedicel of bisexual fl 0, staminate 2.5–5 mm, > fr. **FL:** bisexual 1–5, staminate 5–18; calyx lobes fused proximal to middle, 0.5–0.8 mm, ± lanceolate, acute; corolla yellow; styles 3–4 × calyx lobes. **FR:** 1–5 per head, 1.5–3 mm wide, ovate to round, with unarmed, bulbous tubercles exc distally with curved prickles. **SEED:** face concave. 2n=16. Serpentine, chaparral, woodland; 150–800 m. nw KR (Del Norte Co.); sw OR. May ★

S. saxatilis Greene ROCK SANICLE Pl 10–25 cm, stout; tuber 2–3.5 cm wide, ± spheric, deep-seated. **LF:** compound, ternate then 1–2-pinnate, ± purple or glaucous; blade 3–9 cm, triangular, lflets stalked, margins finely serrate. **INFL:** peduncle 0.5–4 cm; bracts fused at base, 3–5, < heads, widely lanceolate, entire or 3-lobed; pedicel of bisexual fl 0, staminate 3–8 mm, >> fr. **FL:** bisexual 3–5, sta-

minate 6–12; calyx lobes ± 0.5 mm, triangular-ovate, acute; corolla pale red-orange or ± yellow; styles ± 3 × calyx lobes. **FR**: 2.5–3 mm wide, ovate to ± round, tubercles ± rounded, most nipple-like or with a bristle. **SEED**: face ± plane. $2n=16$. Rocky ridges or talus, chaparral, woodland; 900–1100 m. e&s SnFrB (Mount Diablo, Contra Costa Co.; Mount Hamilton, Santa Clara Co.). May–Jun ★

S. tracyi R.H. Shan & Constance (p. 201) TRACY'S SANICLE Pl 35–60 cm, slender, taprooted. **LF**: compound, ternate then 1–2-pinnate, green or ± purple; blade 2.5–3.5 cm, ± ovate, lflets irregularly toothed or lobed, margins serrate. **INFL**: peduncle 0.5–4 cm; bracts fused proximal to middle, 5–8, < heads, lanceolate or ovate; pedicel of bisexual fl 0, staminate 2–4 mm, ± > fr. **FL**: bisexual 2–3, staminate 4–7; calyx lobes fused proximal to middle, 0.3–0.5 mm, ovate, acute; corolla yellow; styles 3 × calyx lobes. **FR**: 2–3 mm wide, obovate or ± round; tubercles inflated, unarmed or few-prickly

distally. **SEED**: face concave. $2n=16$. Openings in conifer forest, woodland; 40–1550 m. NCo (Humboldt Co.), NCoRO, NCoRH, CaRF (Butte Co.). Mar–May ★

S. tuberosa Torr. Pl 5–80 cm, slender; tuber 5–15 mm wide, spheric. **LF**: compound, ternate then 1–2-pinnate, green, glaucous, or ± purple; blade 2–13 cm, triangular to ovate, lflets entire to deeply pinnately lobed. **INFL**: peduncles 0.5–8.5 cm; bracts fused, 6–10, 1–3 mm, < heads, oblong to lanceolate; pedicel of bisexual fl 0, of staminate 2–7 mm, gen >> fr. **FL**: bisexual 3–5, staminate 4–10; calyx lobes fused proximal to middle, 0.3–0.5 mm, ovate, acute; corolla yellow; styles ± 3 × calyx lobes. **FR**: 1.5–2 mm, ovate to ± round, with rounded, unarmed tubercles. **SEED**: face flat. $2n=16$. Open gravelly meadows, chaparral, woodland, pine forest; 30–2700 m. NW, CaR, SN, ScV (Sutter Buttes), SnFrB, SCoR, TR, PR; Baja CA. Variable. Mar–Jul

SCANDIX

Ann, taprooted, ± hairy. **ST**: spreading or erect, branched. **LF**: blade oblong to ovate, pinnately dissected, segments thread-like to linear. **INFL**: umbels compound, gen opposite a lf; bracts gen 0; bractlets few, entire to lobed or dissected, conspicuous; rays, pedicels few. **FL**: ± bilateral; calyx lobes gen 0; petals white, ± notched, narrowed at tips, outer > others. **FR**: gen ± bristly; body linear to oblong, ± compressed side-to-side; beak > body, sterile, compressed front-to-back or cylindric; ribs thread-like; oil tubes inconspicuous; fr axis entire or tip divided. **SEED**: face grooved. 15–20 spp.: Medit. (Greek: chervil)

S. pecten-veneris L. (p. 201) VENUS' NEEDLE Pl 1.5–5 dm. **LF**: petiole gen 2–10 cm; blade 2–10 cm, 1–5 cm wide, segments 1.5 mm, linear. **INFL**: peduncle 1–6 cm; bractlets 2–5, 5–10 mm, lance-linear to obovate, ciliate; rays 1–3, gen 1–2(5) cm, ascending; pedicels 2–3 mm, stout. **FR**: body 6–15 mm, 1–2 mm wide; beak 2–7 cm, compressed front-to-back, bristly-ciliate; ribs low, wide, rounded. $2n=16,26$. Grassy slopes, roadsides; 15–1000 m. CA-FP; native to Medit. Apr–Jun

SIUM

Per, glabrous; roots clustered, fibrous or ± tuberous. **ST**: erect or ascending, branched. **LF**: blade oblong to ovate, 1-pinnate, occ 2-pinnate when submerged, lflets distinct, serrate, irregularly cut to pinnately lobed. **INFL**: umbels compound, gen opposite a lf; bracts, bractlets lf-like, gen reflexed, conspicuous; rays, pedicels many, spreading-ascending. **FL**: occ ± bilateral; calyx lobes 0 or minute; petals wide, white, narrowed at tips, outer ± > others. **FR**: ovate to round, ± compressed side-to-side; ribs prominent, ± equal, corky; oil tubes 1–3 per rib-interval; fr axis entire or divided to base, adhering to fr halves or not. **SEED**: face flat. $2n=12$. ± 10 spp.: N.Am, Eurasia, Afr. (Greek: for an aquatic member of family)

S. suave Walter (p. 205) Pl 6–12 dm, stout. **LF**: petiole 1–8 dm, segmented; blade 6–25 cm, 7–18 cm wide, lflets 1–4 cm, linear or lanceolate, serrate, irregularly cut. **INFL**: peduncle 4–10 cm; bracts 6–10, 3–15 mm, linear or lanceolate, acute, entire to irregularly cut; bractlets 4–8, 1–3 mm, lance-linear; rays 10–20, 1.5–3 cm, ± equal, slender; pedicels 3–5 mm. **FR**: 2–3 mm wide. $2n=12$. Swamps, marshes, streambanks; < 2350 m. CaRH, c SNH, deltaic GV, GB; to BC, e N.Am; e Asia. Jul–Aug

SPERMOLEPIS

Ann, taprooted, glabrous. **ST**: gen spreading, branched. **LF**: blade [oblong] ovate, ± ternate-pinnately dissected, segments thread-like to linear. **INFL**: umbels compound, terminal and lateral, peduncled or not; bracts 0; bractlets few, narrow; rays, pedicels 0–few, ± erect, gen ± spreading. **FL**: calyx lobes 0; petals oblong to ovate, white, tips not narrowed, not incurved. **FR**: widely ovate, ± compressed side-to-side; ribs low, thread-like; oil tubes 1–3 per rib-interval; fr axis divided at tip. **SEED**: face grooved. 5 spp.: s US, HI, s S.Am. (Greek: seed scale, from tubercled or bristly fr)

S. echinata (DC.) A. Heller (p. 205) BRISTLY SCALESEED Pl low, spreading, 5–40 cm. **LF**: ovate; petiole 3–20 mm; blade 7–25 mm wide, segments 2–18 mm, thread-like. **INFL**: peduncles 1–5 cm; bractlets few, thread-like to linear, entire or toothed; rays 5–14, 1–15 mm, gen ± ascending, unequal; pedicels gen < 7 mm, those of central fl of 2° umbels gen 0. **FR**: 1.5–2 mm wide; ribs prominent, short-bristly. $2n=16,20$. Rocky slopes, sandy flats; 60–1500 m. DSon (Borrego Valley); to se US, n Mex. Mar–Apr ★

SPHENOSCIADIUM

1 sp. (Greek: wedge umbrella, from umbel)

S. capitellatum A. Gray (p. 205) SWAMP WHITE HEADS, RANGER'S BUTTONS Per, ± scabrous; root tuberous. **ST**: erect, 5–18 dm, gen branched, lfy. **LF**: petiole 1–4 dm; blade 1–4 dm, oblong to ovate, 1–2-pinnate or ternate-pinnate, lflets 1–12 cm, gen ± lanceolate, acute, sparsely toothed to irregularly cut or pinnately lobed; cauline lf sheaths enlarged. **INFL**: umbels compound, tomentose; peduncle 7–40 cm; bracts 0; bractlets many, linear, bristle-like; rays 4–18, 1.5–10 cm, ascending to reflexed; pedicels reduced to a disk; 2° umbels head-like, spheric. **FL**: calyx lobes 0; petals obovate, white or ± purple, tips narrowed; styles slender. **FR**: 5–8 mm, wedge-shaped-obovate, compressed front-to-back, tomentose; ribs unequally winged, marginal wider than others; oil tubes per rib-interval 1; fr axis divided to base. **SEED**: face ± flat. $2n=22$. Wet meadows, streamsides, lakeshores; gen higher elevations; < 3500 m. NCoRI, CaR, SNH, TR, PR, GB; to OR, ID, NV, Baja CA. TOXIC to livestock, but rarely eaten. Jul–Aug

TAUSCHIA

Per, taprooted or roots tuberous; glabrous to hairy. **ST:** 0 or short. **LF:** blade oblong to obovate, 1–2-pinnate or -ternate, lflets wide, margins entire to pinnately lobed. **INFL:** umbels compound, terminal; peduncle gen > lf; bracts gen 0; involucel 1-sided; bractlets inconspicuous to lf-like; rays, pedicels few to many, ascending to reflexed, gen few fertile. **FL:** calyx lobes occ 0; petals wide, yellow, tips narrowed; styles slender; ovary tip projection inconspicuous. **FR:** oblong to round, ± compressed side-to-side, glabrous; ribs prominent to thread-like, ± equal, unwinged; oil tubes several per rib-interval; fr axis gen divided ± to base. **SEED:** face gen grooved or concave. ± 35 spp.: w N.Am to n S.Am. (I.F. Tausch, Czech botanist, 1793–1848)

1. Peduncle gen < lf; bractlets large, >> umbels, wide, sharply pinnately lobed, lf-like. . . . . ***T. howellii***
1′ Peduncle > lf; bractlets small, < to ± > umbels, narrow, ± entire, not lf-like
  2. Fr oblong to narrowly elliptic, ribs prominent, acute, ± wing-like; lf ± leathery
    3. Lf 1-pinnate; rays unequal, spreading-ascending . . . . ***T. arguta***
    3′ Lf 2-pinnate; rays ± equal, spreading, reflexed . . . . ***T. parishii***
  2′ Fr ± round, ribs low, obtuse, thread-like; lf not leathery
    4. Pl glabrous, smooth; rays 5–12; pedicel 1–3 mm . . . . ***T. glauca***
    4′ Pl minutely scabrous; rays 10–30; pedicel 2–15 mm
      5. Gen several bractlets > fls, frs; fr 4–7 mm, 4–5 mm wide . . . . ***T. hartwegii***
      5′ Gen all bractlets < fls, frs; fr 3–5 mm, 4–6 mm wide. . . . . ***T. kelloggii***

T. arguta (Torr. & A. Gray) J.F. Macbr. (p. 205) Pl 3–7 dm, glabrous. **ST:** gen > lvs. **LF:** petiole 6–20 cm; blade 8–16 cm, oblong to ovate, 1-pinnate, lflets 3–8 cm, oblong to ovate, sharply dentate. **INFL:** peduncle 17–45 cm; bractlets several, 2–10 mm, linear to lanceolate, entire or few-lobed; rays 12–25, 2–12 cm, unequal, spreading-ascending; pedicels 3–9 mm. **FL:** calyx lobes evident; corolla yellow; styles slender. **FR:** 6–9 mm, oblong; ribs acute, prominent; oil tubes 3–5 per rib-interval; fr axis divided to base. 2*n*=22. Chaparral, woodland; 100–1500 m. SCoRO, SW; Baja CA. Apr–Jun

T. glauca (J.M. Coult. & Rose) Mathias & Constance GLAUCOUS TAUSCHIA Pl 2–4 dm, glabrous. **ST:** 0 or gen < lvs. **LF:** petiole 2–12 cm; blade 6–12 cm, ovate to round, 2-ternate or ternate-pinnate, lflets 10–17 mm, ovate to round, coarsely serrate or lobed. **INFL:** peduncle 2–4 dm; bractlets several, 1–7 mm, lanceolate; rays 5–12, 1–6 cm, unequal; pedicel 1–3 mm. **FL:** calyx lobes minute; corolla yellow; styles slender. **FR:** 2–3 mm, ± round; ribs thread-like; oil tubes 2–3 per rib-interval; fr axis divided to base or nearly so. 2*n*=22. Gravelly, gen serpentine flats in conifer forest; 80–1700 m. KR, NCoRO; s OR. Apr–Jun ★

T. hartwegii (A. Gray) J.F. Macbr. Pl 3–10 dm, minutely scabrous. **ST:** 0. **LF:** petiole 5–25 cm; blade 12–24 cm, oblong to widely ovate, 1–2-ternate-pinnate, lflets 25–60 mm, oblong to ovate, coarsely serrate, gen lobed. **INFL:** peduncle 2.5–8 dm; bractlets several, 5–12 mm, ± lanceolate, entire, reflexed, some ± > fls, frs; rays 10–30, 2–12 cm, unequal; pedicel 2–7 mm. **FL:** calyx lobes minute; corolla yellow; styles slender. **FR:** 4–7 mm, ± round; ribs thread-like; oil tubes 3–5 per rib-interval; fr axis divided to base. 2*n*=22. Chaparral, pine/oak woodland; < 1800 m. NCoRI, CaRF, SNF, c&s SNH, ScV (Sutter Buttes), SnFrB, SCoR, SCo, WTR. Mar–May

T. howellii (J.M. Coult. & Rose) J.F. Macbr. (p. 205) HOWELL'S TAUSCHIA Pl 5–8 cm, glabrous. **ST:** short, < lvs. **LF:** petiole 2–3 cm; blade 1.5–3 cm, ovate, 1-pinnate or -ternate, lflets 5–15 mm, oblong to ovate, irregularly sharply toothed. **INFL:** peduncle < 2 cm, gen < lf; bractlets several, 1–2 mm, divided, sharply toothed, >> umbels; rays 3–5, 8–16 mm, unequal, or some 2° umbels sessile; pedicels < 5 mm. **FL:** calyx lobes prominent; corolla yellow; styles slender. **FR:** 2–4 mm, oblong; ribs thread-like; oil tubes several per rib-interval; fr axis divided ± to base. 2*n*=22. Granitic gravel, ridge tops, *Abies* forest; 2000–2500 m. KR (Salmon Mtns), n SNH; s OR. Jun–Jul ★

T. kelloggii (A. Gray) J.F. Macbr. Pl 2–7 dm, minutely scabrous. **ST:** gen 0. **LF:** petiole 5–15 cm; blade 8–20 cm, ovate to round, 1–3-ternate or -ternate-pinnate, lflets 15–35 mm, oblong to ovate, coarsely serrate, gen irregularly cut or lobed. **INFL:** peduncle 2–5 dm; bractlets few, 3–8 mm, linear, entire, < fls, frs; rays 10–20, 2–12 cm, unequal; pedicel 3–15 mm. **FL:** calyx lobes minute; corolla yellow; styles slender. **FR:** 3–5 mm, ± round; ribs thread-like; oil tubes 2–3 per rib-interval; fr axis divided ± to base. 2*n*=22. Scrub, chaparral, woodland, conifer forest; 50–1700 m. NW, s CaR, n SN, ScV (Sutter Buttes), n CCo, SnFrB; s OR. Apr–Jun

T. parishii (J.M. Coult. & Rose) J.F. Macbr. (p. 205) Pl 1–4 dm, glabrous, ± glaucous. **ST:** 0. **LF:** petiole 5–15 cm; blade 8–15 cm, oblong to ovate, 2-pinnate, lflets 15–40 mm, oblong to ovate, sharply serrate to pinnately lobed. **INFL:** peduncle 10–30 cm; bractlets few, 5–12 mm, linear, entire; rays 12–18, 3–6 cm, ± equal, spreading, reflexed; pedicels 2–7 mm. **FL:** calyx lobes evident; styles slender. **FR:** 5–8 mm, oblong to narrowly elliptic; ribs narrow, prominent, acute; oil tubes 4–5 per rib-interval; fr axis divided in distal 2/3. 2*n*=22. Rocky or sandy soil, pine woodland; 1200–2600 m. s SNH, Teh, TR, SnJt, SNE. May–Jul

TORILIS

Ann, taprooted, hairy or bristly. **ST:** spreading or erect, branched. **LF:** blade lanceolate to triangular, 1-pinnately dissected. **INFL:** umbels compound, terminal or opposite lvs, peduncled or sessile; bracts gen 0 or small; bractlets several, thread-like to linear; rays 0–few, spreading-ascending; pedicels 0 or short. **FL:** gen ± bilateral; calyx lobes 0 or evident; petals obcordate, white or ± red, tips narrowed, outer petals ± > others; styles short. **FR:** oblong to ovate, ± compressed side-to-side; 1° ribs thread-like, prickly, 2° ribs densely prickly or tubercled; oil tubes 1 per rib interval; fr axis divided in distal 1/2. **SEED:** face grooved. 10–15 spp.: Eurasia. (Name used by Adanson in 1763, meaning obscure)

1. Pl erect; umbel open, not head-like; outer, inner fr-halves equally prickly. . . . . ***T. arvensis***
1′ Pl spreading; umbel dense, head-like; outer fr-half prickly, inner tubercled only . . . . ***T. nodosa***

T. arvensis (Huds.) Link (p. 205) TALL SOCK-DESTROYER Pl 3–10 dm, slender. **LF:** petiole 2–8 cm; blade 5–12 cm, ± ovate, 2–3-pinnate, lflets 5–60 mm, lanceolate to ovate, regularly pinnately cut; distal cauline lvs gen 1-pinnate. **INFL:** peduncle 2–12 cm; bracts 0–2; bractlets several, 2–4 mm, linear; rays 2–10, ± equal; pedicels 1–4 mm. **FR:** 3–5 mm, oblong-ovate; prickles spreading, length gen = fr width. 2*n*=12. Disturbed places; < 1600 m. CA-FP (esp NW, n SNF, CW); widely introduced, native to c&s Eur. Apr–Jul ❖

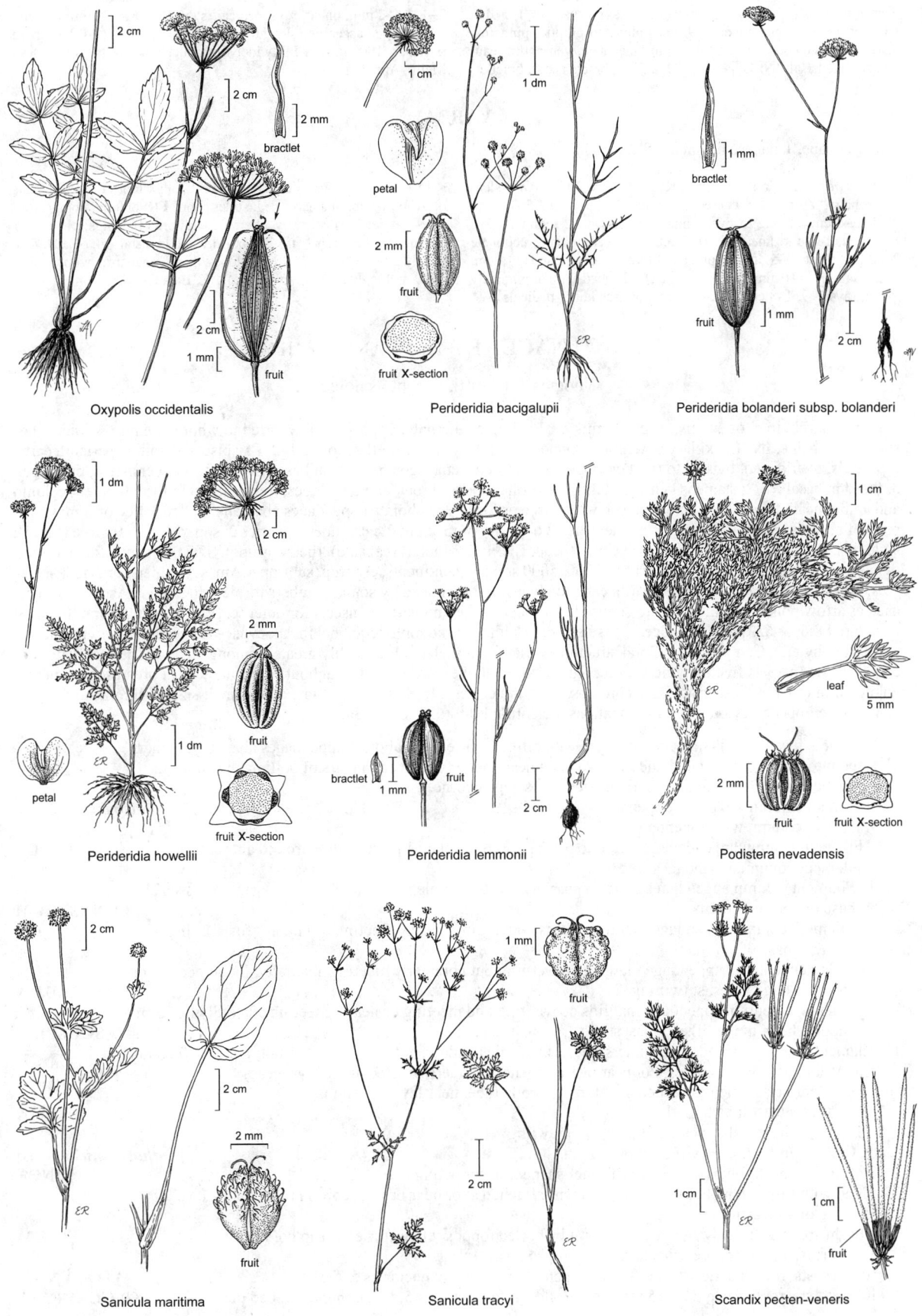

Oxypolis occidentalis

Perideridia bacigalupii

Perideridia bolanderi subsp. bolanderi

Perideridia howellii

Perideridia lemmonii

Podistera nevadensis

Sanicula maritima

Sanicula tracyi

Scandix pecten-veneris

T. nodosa (L.) Gaertn. SHORT SOCK-DESTROYER Pl 1–5 dm. **LF**: petiole 3–10 cm; blade 3–9 cm, oblong to elliptic, pinnately dissected, segments 3–8 mm, linear to lanceolate, gen entire; cauline lvs like basal. **INFL**: peduncle 0–2.5 cm; bracts gen 0; bractlets 2–4, 1.5–2 mm, linear; rays, pedicels short to 0. **FR**: 2–3 mm, ovate; prickles gen ascending, length < to > fr width. $2n$=22,24. Disturbed places; < 1800 m. CA-FP; widely introduced, native to Eurasia, esp Medit. Apr–Jun

YABEA

1 sp. (Y. Yabe, Japanese botanist, 1876–1931)

Y. microcarpa (Hook. & Arn.) Koso-Pol. (p. 205) Ann, slender, ± hairy, taprooted. **ST**: erect, 3–40 cm. **LF**: petiole 2.5–3.5 cm; blade 2–6 cm, oblong to ovate, pinnately dissected, segments 2–8 mm, thread-like to linear. **INFL**: umbels gen compound; peduncle 2–10 cm; bracts 2–5, lf-like, pinnately lobed to compound, 1–5 cm; bractlets 1–5, 1–10 mm, entire or some gen 3-lobed at tip or pinnately lobed; rays 1–9, 1–8 cm, unequal, erect or ascending; pedicels 2–9, < 15 mm, erect. **FL**: minute, ± bilateral; calyx lobes evident; petals obovate, white, tips narrowed; styles short. **FR**: oblong, compressed side-to-side; 1° ribs bristly, alternate with prickly wings; oil tubes 1 per interval between 1° ribs; fr axis divided in distal 1/4. **SEED**: face deeply grooved. $2n$=12. Grassy slopes, dunes, chaparral, woodland; < 1500 m. CA-FP, DMoj; to WA, ID, AZ, Baja CA. Apr–Jun

APOCYNACEAE DOGBANE FAMILY

Thomas J. Rosatti, except as noted

Ann, per, shrub, tree, often vine; sap gen milky. **LF**: simple, alternate, opposite, subwhorled to whorled, entire; stipules 0 or small, finger-like. **INFL**: axillary or terminal, cyme, gen umbel- or raceme-like, or fls 1–2. **FL**: bisexual, radial; perianth parts, esp petals, overlapped, twisted to right or left, at least in bud; sepals gen 5, fused at base, often reflexed, persistent; petals gen 5, fused in basal ± 1/2; stamens gen 5, attached to corolla tube or throat, alternate lobes, free or fused to form filament column and anther head, filament column then gen with 5 free or fused, ± elaborate appendages abaxially, pollen ± free or removed in pairs of pollinia; nectaries 0 or near ovaries, then 2 or 5[10], or in stigmatic chambers; ovaries 2, superior or ± so, free [fused]; style tips, stigmas gen fused into massive pistil head. **FR**: 1–2 follicles, (capsule), [berry, drupe]. **SEED**: many, often with tuft of hairs at 1 or both ends. 200–450 genera, 3000–5000 spp.: all continents, esp trop, subtrop S.Am, s Afr; many orn (incl *Asclepias, Hoya, Nerium, Plumeria, Stapelia*); cardiac glycosides, produced by some members formerly treated in Asclepiadaceae, used as arrow poisons, in medicine to control heart function, and by various insects for defense. [Fishbein 2001 Ann Missouri Bot Gard 88:603–623] Asclepiadaceae ("asclepiads"), although monophyletic, incl in Apocynaceae because otherwise the latter is paraphyletic. Complexity of floral structure, variation in asclepiads arguably greatest among all angiosperms. Pattern of carpel fusion (carpels free in ovule-bearing region, fused above), present ± throughout Apocynaceae (in broad sense), nearly unknown in other angiosperms. Base chromosome number gen 11; abundance of latex, gen small size of chromosomes evidently have impeded cytological investigations. Scientific Editor: Bruce G. Baldwin.

1. Filaments fused into filament column, gen with 5 free or fused, ± elaborate appendages abaxially; anthers fused into anther head around and fused to pistil head; pollen removed in pairs of pollinia; nectaries in stigmatic chambers (genera formerly treated as Asclepiadaceae)
  2. Filament column ± without appendages . . . . ***Funastrum utahense***
  2′ Filament column with appendages
   3. Filament column appendages fused into 5-lobed, cup- or plate-like structure around anther head . . . . . . **MATELEA**
   3′ Filament column appendages free
    4. Filament column appendages hollow (possibly due to complete fusion of margins); ring of tissue at base of corolla present . . . . **FUNASTRUM**
    4′ Filament column appendages solid (margins converging, nearly meeting or not, not fused); ring of tissue at base of corolla 0
     5. Filament column appendage margins converging but not nearly meeting abaxially; fls in raceme- or panicle-like cymes; st twining . . . . **ARAUJIA**
     5′ Filament column appendage margins converging and meeting or nearly meeting adaxially (to form hood); fls in umbel-like cymes; st prostrate to erect . . . . **ASCLEPIAS**
1′ Filaments free, unappendaged; anthers free, lying against, adherent to pistil head or not; pollen ± free, not removed in pollinia; nectaries 0 or near ovaries (genera formerly treated as Apocynaceae)
  6. Seed glabrous; stamens attached near top of corolla tube, near level of stigma
   7. Lvs alternate to subwhorled; pl erect . . . . **AMSONIA**
   7′ Lvs opposite to subopposite; pl erect or sprawling
    8. Corolla pink, tube ± cylindric; pl erect . . . . [***Catharanthus roseus***]
    8′ Corolla purple-blue (white), tube funnel-shaped; pl sprawling . . . . **VINCA**
  6′ Seed with tuft of long hairs at 1 end; stamens attached at or near base of corolla tube, below level of stigma, or appearing so
   9. Pl shrub, small tree; lvs gen whorled or subwhorled (opposite); nectaries 0; sap not milky . . . . **NERIUM**
   9′ Pl per; lvs opposite; nectaries 5, free or fused; sap milky
    10. Lvs, sts, roots not fleshy; fl < 8 mm, in gen >> 6-fld cyme; nectaries 5, free . . . . **APOCYNUM**
    10′ Lvs, sts, roots fleshy; fl > 15 mm, in 2–6-fld cyme; nectaries 5, fused into a 5-lobed disk . . . . **CYCLADENIA**

AMSONIA

Thomas J. Rosatti & Lauramay T. Dempster

Per, erect, semi-woody. **LF**: alternate to subwhorled. **INFL**: ± terminal, compound cyme. **FL**: corolla salverform; filaments free, attached near top of corolla tube, unappendaged, anthers free from each other and stigma, pollen ± free; nectary 0 or a shallow ring around ovaries; style ± thread-like, stigma skirted at base. **SEED**: glabrous. 5–25 spp.: N.Am, Japan. (John Amson, VA physician, 18th century)

A. tomentosa Torr. & Frém. (p. 205) Pl glabrous or gray-tomentose. **ST**: several to many from woody crown, 16–36 cm, branches few to many. **LF**: 2–4 cm; petiole short or 0; blade ovate-lanceolate, acute at both ends. **FL**: calyx lobes erect, thread-like above base; corolla ± white, blue, or ± green, tube ± 15 mm, inflated above middle, narrowed just below spreading lobes; style with spheric thickening just below stigma. **FR**: 3–8 cm, constricted between seeds, often breaking into 1-seeded segments. $2n=22$. Desert plains, canyons; 300–1800 m. SnBr (n slope), D; to UT. Tomentose and glabrous pls (latter assignable to *A. brevifolia* A. Gray) have identical ranges, do not intergrade, and show no other differences, suggesting hairiness is governed by a single gene. Mar–May

APOCYNUM DOGBANE, INDIAN HEMP

Thomas J. Rosatti & Lauramay T. Dempster

Per, ascending or erect. **LF**: opposite. **INFL**: gen cyme, >> 6-fld. **FL**: < 8 mm; corolla bell- or urn-shaped to cylindric, 5-lobed, with 5 triangular appendages alternate stamens; filaments free, attached at base of corolla tube, unappendaged, short, broad, anthers forming cone around and adherent to stigma, each partly sterile, sharply sagittate, pollen ± free; nectaries 5, free, around but not exceeding ovaries; ovaries free, not adherent, style ± 0, stigma massive, ovoid, obscurely 2-lobed. **FR**: slender, cylindric, pointed. **SEED**: with tuft of long hairs at 1 end. 7 spp.: N.Am. (Greek: away from, dog, from ancient use as dog poison) 2 geog overlapped but ecologically different spp. in CA, many hybrids between them, many of these named.

1. Pl 1.6–3 dm; lf drooping to spreading, ovate to round, dark green adaxially, often pale abaxially; corolla red-purple or pink to white, pink-striped or not, lobes spreading to recurved; calyx << corolla tube
. . . . ***A. androsaemifolium***
1′ Pl 3–12 dm; lf ascending, lanceolate to narrow-ovate, yellow-green adaxially, abaxially; corolla ± green or white, lobes ± erect; calyx ± = corolla tube. . . . . ***A. cannabinum***

A. androsaemifolium L. (p. 209) BITTER DOGBANE **ST**: diffuse-branched. **LF**: blade 4–6 cm, >> petiole, base gen round or cordate, tip round or obtuse to ± acute. **INFL**: terminal, also abortive fls in upper lf axils. **FL**: corolla 4–8 mm, ± bell-shaped. **FR**: 7–11 cm, pendent or erect. $2n=16,22$. Open slopes, rocky places, with conifers, chaparral; 200–2500 m. NW, CaRH, n SNF, SNH, CW (exc SCoRO), SnGb, SnBr, PR, MP, W&I; to Can, e N.Am. May–Oct

A. cannabinum L. (p. 209) INDIAN HEMP **ST**: stout, ± stiff-erect, branched near top. **LF**: blade 5–8 cm, >> petiole, base tapered to cordate, clasping st or not, tip obtuse to acute. **INFL**: terminal, fls in upper lf axils 0. **FL**: corolla 2.5–5 mm, cylindric to urn-shaped. **FR**: 6–9 cm, ± pendent. $2n=16,22$. Moist places, near streams, springs, or as weed in orchards; 150–2000 m. CA (exc evidently NCo, CaRF, SNF, CCo, SCo, ChI); to BC, e N.Am. Apr–Oct

ARAUJIA BLADDER-FLOWER

Thomas J. Rosatti & Carol A. Hoffman

Per. **ST**: twining. **LF**: opposite; blade cordate, hastate, or ovate. **INFL**: at nodes, raceme- or panicle-like cyme. **FL**: sepals large, lf-like, ± erect; corolla ± erect (exceeding stamens, pistils), ring of tissue at base 0; filament column appendages free, attached to base of filament column and base of corolla, without projections, margins converging but not nearly meeting abaxially, anthers fused into anther head around and fused to pistil head, pollen in pollinia; nectaries in stigmatic chambers. **FR**: pendent, gen ovoid, with coarse longitudinal grooves. 5 spp.: S.Am. (António de Araújo de Azevedo, Portuguese statesman, 1752–1817) [Forster & Bruyns 2001 Taxon 41:746–749]

A. sericifera Brot. (p. 209) **ST**: < 12 m, soft-tomentose when young. **LF**: petiole > 1 cm; blade 5–12 cm, adaxially glabrous, abaxially gen dense-puberulent. **FL**: corolla 2–3 cm, bell- or funnel-shaped, white; pistil head with 2 erect, elongate lobes on top. **FR**: 10–12 cm. $2n=22$. Chaparral, woodland, esp citrus groves; 100–400 m. CCo, s SCoRO, SCo, WTR, PR (exc SnJt); GA; native to S.Am. Aug–Oct ◆

ASCLEPIAS MILKWEED

Thomas J. Rosatti & Carol A. Hoffman

Ann, per, shrub. **ST**: prostrate to erect. **LF**: gen opposite (alternate, whorled), each pair at right angles to those below, above, gen persistent; blade narrow-linear to ovate or cordate. **INFL**: terminal or at gen upper nodes, umbel-like cyme. **FL**: ring of tissue at base of corolla 0; filament column appendages (hoods) free, elevated above corolla base or not, each often with an elongate projection (horn) attached to inside, margins converging and meeting or nearly meeting adaxially but not fused; anthers fused into anther head around and fused to pistil head, pollen in pollinia; pistil head flat or conic on top; nectaries in stigmatic chambers. **FR**: erect (but gen on pendent pedicel) or pendent, lance-ovoid to ovoid, smooth or with tubercles. In narrow sense of genus, 100 spp.: N.Am, C.Am, perhaps S.Am. (Greek physician Aesculapius) Fresh fls gen better for determining relative positions of parts; hoods may have near anther head 2 ± sickle shaped lobes each that may ± resemble horns. *A. linaria* not outside cult in CA.

1. Lvs ephemeral (st gen lfless), narrow-linear
2. Hoods exceeded by anther head. . . . . ***A. albicans***
2′ Hoods much exceeding anther head. . . . . ***A. subulata***

1′ Lvs persistent, narrow-linear to wide-ovate or -elliptic
  3. Lvs 1 per node (alternate or subwhorled)
    4. Lvs alternate to subwhorled, gen linear-lanceolate; hoods not elevated above corolla base, ± reflexed to
      spreading at base, at ± same level as anther head . . . . *A. asperula* subsp. *asperula*
    4′ Lvs alternate, linear; hoods gen elevated above corolla base, ± erect at base, exceeding anther head . . . . . . *A. linaria*
  3′ Lvs > 1 per node (opposite or whorled)
    5. Horn 0 or incl in hood
      6. Hoods not or ± exceeded by anther head
        7. Lf sessile, blade base clasping st; corolla dark red-purple . . . . *A. cordifolia*
        7′ Lf short-petioled or sessile, blade base rarely clasping st; corolla green-yellow or white
          8. Lf ± cordate or ovate to ± round; corolla green-yellow, lobes 10–15 mm; fr glabrous, prickles 0 . . . *A. cryptoceras*
          8′ Lf linear or narrow- to oblong-lanceolate, acuminate; corolla white, lobes 7–8 mm; fr puberulent,
            with thread-like prickles . . . . [*A. fruticosa*]
      6′ Hoods exceeded by anther head
        9. St decumbent to ± ascending; pl dense-hairy, ± white (appearing woolly at arm's length); tops of
          hoods exceeding base of anther head (hoods, anther head overlapped) . . . . *A. californica*
        9′ St prostrate (and ± flat); pl ± hairy, green (appearing glabrous at arm's length); tops of hoods exceeded
          by base of anther head (hoods, anther head not overlapped) . . . . *A. solanoana*
    5′ Horn exserted from hood
      10. Horn gen exceeding hood. . . . . *A. fascicularis*
      10′ Horn at ± same level as to much exceeded by hood
        11. Hoods exceeding anther head by at least 1/2 their lengths
          12. Hoods at ± same level as exserted horns, ± erect, tips rounded . . . . *A. nyctaginifolia*
          12′ Hoods much exceeding exserted horns, ± ascending, tips acute . . . . *A. speciosa*
        11′ Hoods not exceeding anther head by at least 1/2 their lengths
          13. Infls at nodes sessile or nearly so . . . . *A. vestita*
          13′ Infls at nodes peduncled
            14. Lvs gen broad-elliptic, rarely lanceolate to ovate, tip obtuse to truncate or notched. (Extirpated) . . . *A. latifolia*
            14′ Lvs elliptic, lanceolate, or ovate, tip obtuse to acuminate
              15. Lvs opposite or whorled, short-petioled to sessile, blade bases tapered to obtuse, rarely cordate,
                margins gen not wavy, gen entire; hoods exceeded by to at ± same level as anther head; seed 8–9
                mm . . . . *A. eriocarpa*
             15′ Lvs opposite, gen sessile, blade bases tapered to cordate, margins gen ± wavy and/or ± jagged;
                hoods ± exceeding anther head; seed 10–13 mm . . . . *A. erosa*

A. albicans S. Watson WHITE-STEMMED MILKWEED Shrub, ± hairy, waxy. **ST**: ± erect. **LF**: opposite or whorled in 3s, ephemeral; blade narrow-linear. **INFL**: gen terminal; peduncle 1–3 cm. **FL**: corolla reflexed, green-white, occ with brown or pink tinge; hoods elevated above corolla base, exceeded by anther head, ± yellow; horns exceeding hoods. **FR**: pendent on spreading to ± reflexed pedicels. **SEED**: ± 6 mm. Dry washes, gravelly hillsides; 200–1100 m. D; AZ, Baja CA. Mar–Jun

A. asperula (Decne.) Woodson subsp. ***asperula*** (p. 209) ANTELOPE-HORNS Per, ± hairy. **ST**: decumbent to ascending. **LF**: alternate to subwhorled in 3s; petiole short; blade gen linear-lanceolate. **INFL**: terminal; peduncle gen 3–10 cm. **FL**: corolla spreading to ascending, green-white; hoods not elevated above corolla base, ± reflexed to spreading at base, at ± same level as anther head, dark purple; horns incl in hoods. **FR**: erect on ± reflexed pedicels. **SEED**: 6–8 mm. Dry, open, rocky places; 1500–2000 m. DMoj; sw US, Mex. Mar–Sep ★

A. californica Greene (p. 209) CALIFORNIA MILKWEED Per, dense-hairy. **ST**: decumbent to ± ascending. **LF**: opposite; petiole short to 0; blade ovate. **INFL**: terminal and at nodes; peduncle 0–3 cm. **FL**: corolla reflexed, ± purple; hoods elevated above corolla base (gen n CA) or not (gen s CA), exceeded by anther head (but tops of hoods exceeding base of anther head), dark purple; horns 0 or minute, incl in hood. **FR**: erect on ± reflexed pedicels. **SEED**: ± 1 cm. Flats, grassy or brushy hillsides; 200–2100 m. c&s SNF, CW, SW, D; n Baja CA. [*A. c.* subsp. *greenei* Woodson] Apr–Jul

A. cordifolia (Benth.) Jeps. PURPLE MILKWEED Per, ± glabrous. **ST**: ascending. **LF**: opposite; petiole 0; blade cordate, base clasping st. **INFL**: terminal and at upper nodes; peduncle 1–4(7) cm. **FL**: corolla spreading to reflexed, dark red-purple; hoods ± elevated above corolla base, not exceeded by anther head; horns 0. **FR**: erect on ± reflexed pedicels. **SEED**: 7–8 mm. 2*n*=22. Rocky slopes, talus, woodland, chaparral, lava flows; 50–2000 m. NW, CaR, SN, ScV (Sutter Buttes), CCo, n SnFrB, MP; OR, NV. May–Jul

A. cryptoceras S. Watson HUMBOLDT MOUNTAINS MILKWEED Per, ± glabrous. **ST**: prostrate to decumbent. **LF**: opposite; petiole 0 to short; blade ± cordate or ovate to ± round, base rarely clasping st. **INFL**: gen terminal; peduncle 0–6 cm. **FL**: corolla reflexed, green-yellow; hoods not elevated above corolla base, ± exceeded by anther head, pink-tan; horns incl in hoods or reportedly 0. **FR**: erect on erect pedicels. **SEED**: 6–8 mm. Sand, gravel, clay, or shale of slopes, canyon bottoms, washes, arid plains; 1400–2500 m. SNE (Mono Co.); to WA, WY, CO, AZ. [*A. c.* subsp. *davisii* (Woodson) Woodson] Apr–Jun

A. eriocarpa Benth. KOTOLO Per, very hairy (less so in age or not). **ST**: erect. **LF**: opposite or whorled in 3s or 4s; petiole short to 0; blade elliptic, lanceolate, or ovate, base tapered to obtuse (to cordate), tip obtuse to short-acuminate, margins gen not wavy, gen entire. **INFL**: terminal and at upper nodes; peduncle 1–10 cm. **FL**: corolla reflexed to ascending, cream, occ pink-tinged; hoods ± elevated above corolla base, exceeded by to at ± same level as anther head, cream, purple-tinged or not; horns exserted, at ± same level as hoods, not converging over anther head. **FR**: erect on ± reflexed pedicels. **SEED**: 8–9 mm. Dry, barren areas; 200–1900 m. CA (exc possibly GB, D, elsewhere); NV, n Baja CA. May–Oct

A. erosa Torr. (p. 209) DESERT MILKWEED Per, very hairy to glabrous. **ST**: ascending to erect. **LF**: opposite; petiole gen 0; blade elliptic, lanceolate, or ovate, base tapered to cordate, clasping st or not, margins gen ± wavy and/or ± jagged. **INFL**: terminal and at nodes; peduncle 2–12 cm. **FL**: corolla reflexed to spreading, pale cream or green-white; hoods elevated above corolla base, ± exceeding anther head, gen cream, occ yellow or ± red; horns exserted, exceeded by to at ± same level as hoods, converging over anther head or not. **FR**: erect on ± reflexed pedicels. **SEED**: 10–13 mm. 2*n*=22. Dry slopes, washes; 150–1900 m. s SN, SnJV, SW, W&I, D; to UT, AZ, Baja CA. Apr–Oct

A. fascicularis Decne. NARROW-LEAF MILKWEED Per, gen glabrous. **ST**: ascending to erect. **LF**: opposite (and reduced, on sterile branches) or whorled gen in 3s to 5s, often with clusters of smaller lvs

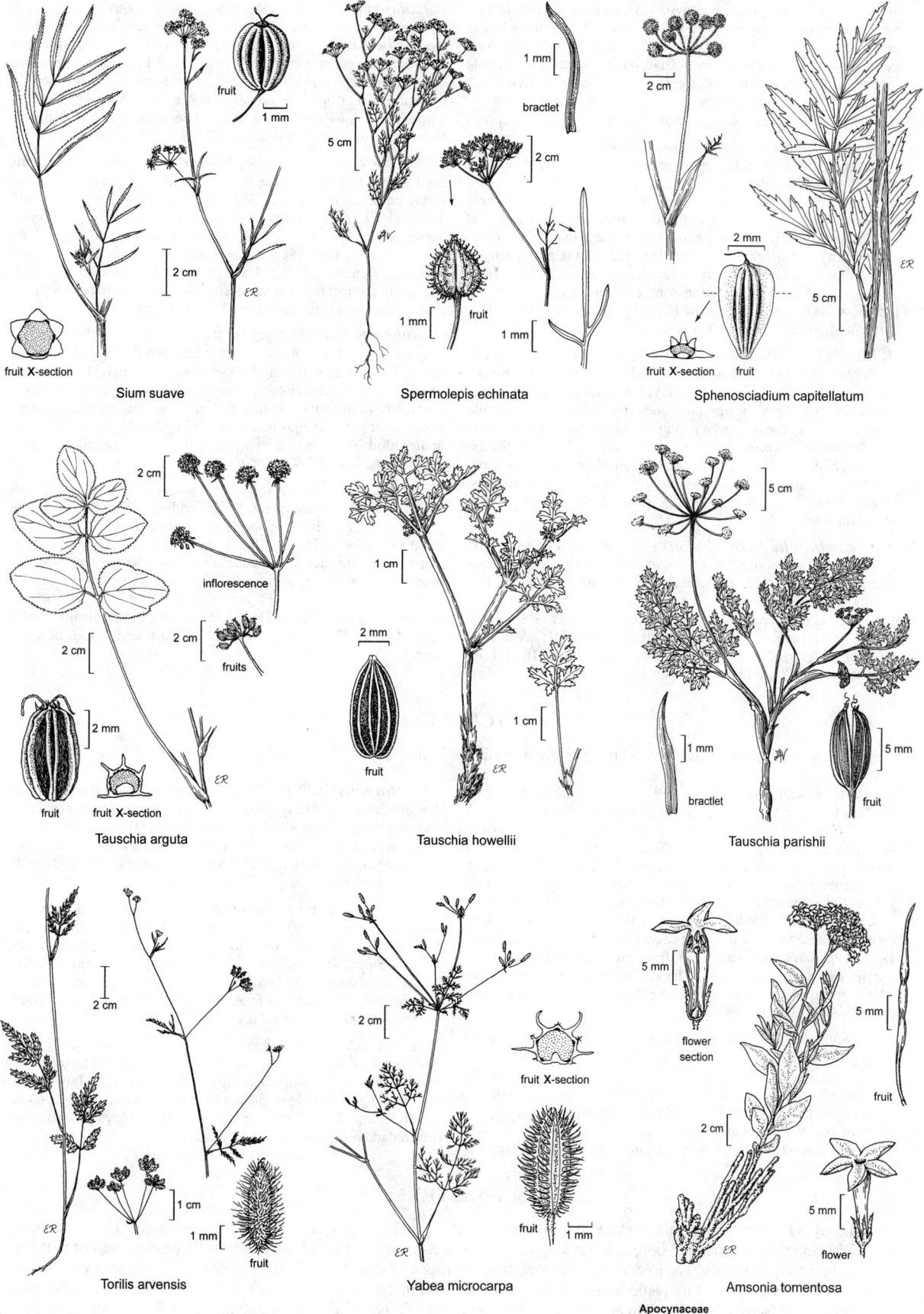

Sium suave

fruit

1 mm

2 cm

ER

fruit X-section

Spermolepis echinata

1 mm
bractlet

2 cm

5 cm

1 mm   fruit

1 mm

Sphenosciadium capitellatum

2 cm

2 mm

5 cm

ER

fruit X-section   fruit

Tauschia arguta

2 cm

inflorescence

2 cm

fruits

2 mm

2 mm

fruit   fruit X-section

ER

Tauschia howellii

1 cm

2 mm

fruit

1 cm

ER

Tauschia parishii

5 cm

1 mm

bractlet

5 mm

fruit

Torilis arvensis

2 cm

1 cm

1 mm

fruit

ER

Yabea microcarpa

2 cm

fruit X-section

fruit

1 mm

ER

Amsonia tomentosa

5 mm

flower
section

2 cm

5 mm

fruit

5 mm

flower

ER

Apocynaceae

in axils; petiole short; blade narrow-lanceolate, base tapered. **INFL:** terminal and at upper nodes; peduncle 1–3(5) cm. **FL:** corolla reflexed, green-white, purple-tinged or not; hoods elevated above corolla base, gen exceeded by anther head, green-white; horns exserted, gen exceeding hoods (and anther head). **FR:** erect on erect pedicels. **SEED:** 5.5–7 mm. $2n$=22. Dry ground, valleys, foothills; 50–2200 m. CA (exc NCo, n CCo); to WA, UT, Baja CA. Gen poisonous to stock. May–Oct

A. latifolia (Torr.) Raf.   BROADLEAF MILKWEED   Per, very hairy, less so in age. **ST:** erect. **LF:** opposite; petiole short to 0; blade very wide-elliptic (lanceolate, ovate), base tapered, obtuse, or cordate, tip obtuse to truncate or notched, gen mucronate. **INFL:** peduncle [0–]1.5–2.7 cm. **FL:** corolla reflexed to spreading, cream-green, purple-tinged abaxially or not; hoods ± elevated above corolla base, exceeded by to at ± same level as anther head, cream, purple-tinged or not; horns exserted, at ± same level as hoods, converging over anther head or not. **FR:** erect on ± reflexed pedicels. **SEED:** 7–8 mm. $2n$=22. Dry washes; ± 150 m. NCoRI (Rumsey, Yolo Co., 1912, extirpated); to SD, TX, Mex. May–Jun

A. linaria Cav.   Per, shrub, ± hairy. **ST:** erect. **LF:** alternate; petiole 0; blade linear (± like pine needle). **INFL:** terminal and at nodes; peduncle 0.5–2 cm. **FL:** corolla reflexed, adaxially white or green-white, abaxially ± pink or ± purple; hoods gen elevated above corolla base, ± erect at base, exceeding anther head, white or green-white; horns exserted, exceeded by hoods. **FR:** ± erect on gen ± reflexed pedicels. **SEED:** ± 6 mm. $2n$=22. Uncommon. Open woodland, limestone ridges, rocky hills, canyons, arroyos, dry abandoned pastures; 1000–1400 m. D (excluded from CA in note under genus); to NM, Mex. Apr–Nov

A. nyctaginifolia A. Gray   MOJAVE MILKWEED   Per, hairy. **ST:** decumbent to ascending. **LF:** opposite; petioled; blade elliptic, lanceolate, or ovate. **INFL:** terminal and at nodes; peduncle ± 0. **FL:** corolla reflexed, green-white, abaxially purple or not; hoods not to ± elevated above corolla base, much exceeding anther head, green-white; horns exserted, at ± same level as hoods. **FR:** erect on ± reflexed pedicels. **SEED:** 7–8 mm. Arroyos, dry slopes; 1000–1700 m. DMoj; to NV, NM. May–Aug ★

A. solanoana Woodson   SERPENTINE MILKWEED   Per, ± hairy. **ST:** prostrate. **LF:** opposite; petioled; blade reniform, cordate, or ovate, base clasping st or not. **INFL:** terminal (and at nodes), an esp well-defined sphere; peduncle 1.5–4.5 cm. **FL:** corolla reflexed to ascending, purple; hoods ± elevated above corolla base, exceeded by anther head (tops of hoods exceeded by base of anther head), ± white; horns 0. **FR:** erect on ± reflexed pedicels. **SEED:** 5.5–8 mm. Serpentine outcrops; 700–1600 m. KR, NCoR. Jun ★

A. speciosa Torr.   SHOWY MILKWEED   Per, hairy. **ST:** ascending to erect. **LF:** opposite; petiole short; blade elliptic to ovate, base rarely cordate, clasping st. **INFL:** terminal and at nodes; peduncle 1–10 cm. **FL:** corolla reflexed, rose-purple; hoods not to ± elevated above corolla base, much exceeding anther head, pink aging yellow; horns exserted, much exceeded by hoods, at ± same level as anther head. **FR:** ± erect on gen ± reflexed pedicels. **SEED:** 6–9 mm. $2n$=22. Many habitats incl fields, roadsides; < 1900 m. CA (exc possibly CW, SW, DSon, elsewhere); to w Can, TX. May–Sep

A. subulata Decne.   RUSH MILKWEED   Per, gen glabrous, exc infl. **ST:** erect. **LF:** opposite, ephemeral; petiole 0; blade narrow-linear. **INFL:** terminal and at upper nodes; peduncle 1–2 cm. **FL:** corolla reflexed, yellow- to green-white; hoods ± elevated above corolla base, much exceeding anther head, yellow-white; horns exserted, at same level as or ± exceeded by hoods, much exceeding anther head. **FR:** pendent on ± reflexed pedicels. **SEED:** ± 6 mm. Arroyos, washes; < 700 m. D; to s NV, AZ, nw Mex. ± all yr

A. vestita Hook. & Arn.   WOOLLY MILKWEED   Per, gen dense-hairy, ± glabrous in age or not. **ST:** ascending. **LF:** opposite; petiole gen short; blade elliptic, lanceolate, or ovate. **INFL:** terminal and at upper nodes; peduncle ± 0. **FL:** corolla reflexed, cream to purple; hoods ± elevated above corolla base, at ± same level as anther head, yellow-white, occ with vertical brown stripe; horns exserted, at ± same level as hoods. **FR:** erect on ± reflexed pedicels. **SEED:** ± 10 mm (largest in CA). Dry plains, brushy flats, hillsides, desert canyons; 50–1350 m. GV, CW, TR, DMoj. [*A. v.* var. *parishii* Jeps.] Apr–Jul

CYCLADENIA

1 sp.: CA. (Greek: ring gland, from nectary) [Sipes & Wolf 1997 Amer J Bot 84:401–409]

C. humilis Benth.   Per, ± erect, 6–12 cm, fleshy (incl large root), herbage tomentose to gen glabrous, glaucous. **LF:** opposite, 2–5 pairs, < 9 cm; petiole < to > blade; blade ovate or ± round, base ± truncate to tapered. **INFL:** cyme, 2–6-fld. **FL:** > 15 mm; calyx lobes narrow-triangular; corolla 15–20 mm, funnel-shaped, with 5 ± round appendages behind anthers, rose-purple, lobes obovate or round, margins wavy; filaments free, appearing to be attached at base of corolla tube but fused to it up to level of stigma, unappendaged, hairy, anthers forming cone around but free from stigma, each partly sterile, sharply sagittate, pollen ± free; nectaries 5, fused into a 5-lobed disk around but not exceeding ovaries; style thread-like; stigma skirted at base. **FR:** 3–5 cm. **SEED:** with tuft of long hairs at 1 end. $2n$=14. Vars. possibly untenable, merit study.

1. Perianth glabrous (tomentose) abaxially; corolla glabrous or papillate adaxially; KR, NCoRH, CaRH, n SNH, SCoRO .... var. ***humilis***
1' Perianth with ± not interwoven hairs abaxially or at least on margins; corolla sparse-hairy adaxially; SCoRO, SnGb, SNE
   2. Corolla lobes 4–5 mm; W&I................. var. ***jonesii***
   2' Corolla lobes 8–12 mm; SCoRO, SnGb, SNE ... var. ***venusta***

var. ***humilis*** (p. 209)   Pl gen glabrous, glaucous, less often dense-tomentose. **FL:** perianth glabrous (tomentose) abaxially; corolla glabrous or papillate adaxially, lobes (5)7–9(11) mm. Loose gravel or sand, talus slopes, often with *Pinus ponderosa*; 1200–2800 m. KR, NCoRH, CaRH, n SNH (Plumas, Sierra cos.), SCoRO (Santa Lucia Range). Apr–Aug

var. ***jonesii*** (Eastw.) S.L. Welsh & N.D. Atwood   Pl gen glabrous, glaucous exc infl hairy. **FL:** perianth with ± not interwoven hairs abaxially or at least on margins, corolla sparse-hairy adaxially, lobes 4–5 mm. Shale slides; 2530 m. W&I (Inyo Mtns); UT, AZ. Unpublished molecular evidence (Slakey, et al., 2011, pers. comm.) suggests that CA members of var. *jonesii* belong instead either to var. *venusta* or to an undescribed taxon. Jul

var. ***venusta*** (Eastw.) Munz   Pl glabrous, glaucous exc infl hairy. **FL:** perianth with ± not interwoven hairs abaxially or at least on margins; corolla sparse-hairy adaxially, lobes 8–12 mm. Talus, loose gravel, dry ground in light shade of pines, chaparral; 1550–2500 m. SCoRO (Santa Lucia Range), SnGb, SNE. Pls of Inyo Co. have narrower, darker red corolla lobes. May–Jul

FUNASTRUM

Per [shrub]. **ST:** gen twining or trailing. **LF:** opposite, gen ± persistent; blade thread-like to narrow-lanceolate. **INFL:** at nodes, umbel-[raceme-]like cyme. **FL:** corolla lobes ± spreading to erect-incurved, ring of tissue at corolla base present or not; filament column appendages ± 0 or free from each other, fused to ring of tissue at corolla base or not, ± spheric, attached to base of filament column, without projections, hollow (possibly due to complete fusion of margins), anthers fused into anther head around and fused to pistil head, pollen in pollinia; pistil head flat or, if ± conic, 2-lobed or not; nectaries in stigmatic chambers. **FR:** gen 1, erect or pendent, narrow-fusiform to lance-ovoid, with fine longitudinal grooves [or smooth]. $2n$=20,22,40,44

(reports not incl CA pls). ± 40 spp.: N.Am, Afr to Australia. (Greek: fleshy crown or wreath, from sac-like filament column appendages of some spp.) [Liede & Täuber 2002 Syst Bot 27:789–800] Our spp. treated as *Cynanchum, Sarcostemma* in TJM (1993), both shown to be polyphyletic in previous, broader circumscriptions (Liede & Täuber 2000, 2002).

1. Corolla lobes erect-incurved, meeting above anther head, giving appearance of closed fl bud; filament
   column appendages ± 0; ring of tissue at corolla base 0 . . . . *F. utahense*
1′ Corolla lobes ± spreading to ± erect, not meeting, not giving appearance of closed fl bud; filament column
   appendages free from each other; ring of tissue at corolla base present
   2. Lf narrow- to wide-lanceolate, margins gen wavy; calyx lobes 3–6 mm; corolla 8–12 mm; filament
      column 1–2 mm, appendage bases exceeding ring of tissue at corolla base; fr gen 1, 8.5–12.5 cm . . . . *F. crispum*
   2′ Lf linear to lanceolate, margins not wavy; calyx lobes 2–3 mm; corolla 3–8 mm; filament column 0.5–1
      mm, appendage bases not exceeding ring of tissue at corolla base; fr 1–2, 3.6–4.8 or 7.2–11.2 cm
      3. Corolla 4–8 mm, ring of tissue at base free from filament column appendages; pl green, gen with sparse,
         ± appressed hairs; fr gen 1, 7.2–11.2 cm . . . . *F. cynanchoides* var. *hartwegii*
      3′ Corolla 3–6 mm, ring of tissue at base fused to filament column appendages; pl gray-green, gen with
         dense, short, erect hairs; fr gen 2, 3.6–4.8 cm . . . . *F. hirtellum*

F. crispum (Benth.) Schltr. WAVYLEAF TWINEVINE Pl gray-green, gen ± white-puberulent. **LF**: petiole 4–8 mm, blade 20–90 mm, narrow- to wide-lanceolate, base hastate, cordate, or truncate, margins gen wavy. **FL**: corolla 8–12 mm, ± green-purple, esp abaxially, ring of tissue at base free from filament column appendages, lobes ± spreading to ± erect. **FR**: gen 1, 8.5–12.5 cm. Uncommon. Open, dry, stony or rocky ground; ± 1200[2200] m. PR (Pinyon Flat, Santa Rosa Mtns); to s CO, TX, c Mex. May–Aug

F. cynanchoides (Decne.) Schltr. var. ***hartwegii*** (Vail) Krings (p. 209) CLIMBING MILKWEED Pl green, gen with sparse, ± appressed hairs. **LF**: petiole 7–15 mm, blade 30–60 mm, narrow-lanceolate, base hastate (or cordate to wedge-shaped). **FL**: corolla 4–8 mm, pink to purple or lobes white, each with purple streak centrally, ring of tissue at base free from filament column appendages, lobes ± spreading to ± erect. **FR**: gen 1, 7.2–11.2 cm. Dry, sandy, rocky arroyos or plains; 30–1600 m. SCo, PR, D; to UT, AR, TX, Mex. [*Sarcostemma c.* subsp. *h.* (Vail) R.W. Holm] Apr–Jul

F. hirtellum (A. Gray) Schltr. TRAILING TOWNULA Pl gray-green, gen with dense, short, erect hairs. **LF**: petiole 1–15 mm, blade 10–36 mm, linear to narrow-lanceolate, base obtuse (truncate). **FL**: corolla 3–6 mm, white to green-white, ring of tissue at base fused to filament column appendages, lobes ± spreading to ± erect. **FR**: gen 2 (often ± 180° divergent), 3.6–4.8 cm. Hard desert pavement, washes; 150–1200 m. D; NV, AZ. [*Sarcostemma h.* (A. Gray) R.W. Holm] Mar–May

F. utahense (Engelm.) Liede & Meve (p. 209) UTAH VINE MILKWEED Pl ± green, minute-hairy esp at nodes, dense-white-hairy at base. **LF**: petiole ± 0, blade 15–40 mm, thread-like, in age reflexed, persistent or not. **FL**: corolla 1.5–3 mm, yellow, in age orange, ring of tissue at base 0, lobes erect-incurved, meeting above anther head, adaxially concave (hood-like); filament column appendages ± 0. **FR**: gen 1, 4–6 cm. Open, dry, sandy or gravelly areas; < 1000 m. D; UT, AZ. [*Cynanchum u.* (Engelm.) Woodson] Apr–Sep ★

MATELEA

Per [shrub]. **ST**: twining [prostrate to erect]. **LF**: opposite; blade ovate, cordate, hastate [round]. **INFL**: at nodes, fls 1–2, peduncles < pedicels or 0 [raceme-, umbel-, or panicle-like cyme]. **FL**: corolla spreading to ± erect, ring of tissue at base 0; filament column appendages fused into 5-lobed, cup- or plate-like structure around anther head [free], attached to base of filament column, each with a vertical, flap-like projection fused to filament column, forming compartments within cup [projections otherwise or 0], solid (margins fused to those of adjacent filament column appendages), anthers fused into anther head around and fused to pistil head, pollen in pollinia; pistil head flat; nectaries in stigmatic chambers. **FR**: erect or pendent, fusiform to lance-ovoid or ovoid, smooth, with tubercles [longitudinal wings]. $2n=22$ (reports not incl CA pls). In broad sense of genus, 200 spp.: trop, warm temp Am. [Ezcurra & Belgrano 2007 Syst Bot 32:856–861]

M. parvifolia (Torr.) Woodson (p. 209) SPEARLEAF **ST**: slender, much branched, to 0.5 m. **LF**: blade gen << 1 cm. **FL**: corolla with acute, turned out tooth in each sinus between lobes, ± green or purple. **FR**: ± 7 cm, with fine longitudinal grooves. Dry rocky areas; 700–1000 m. D; to NV, TX, Baja CA. Mar–May ★

NERIUM OLEANDER

Shrub, small tree; sap not milky. **LF**: gen whorled or subwhorled (opposite). **INFL**: cyme, terminal, branched. **FL**: corolla funnel-shaped, 5-lobed, with 5 petal-like appendages alternate stamens; filaments free, attached at base of corolla tube, short, broad, unappendaged, anthers free, forming cone around and adherent to stigma, each partly sterile, sharp-sagittate, with a long, twisting, hairy appendage, pollen ± free; nectaries 0; ovaries initially free, ± adherent, becoming fused, style ± 0, stigma massive, ovoid, obscurely 2-lobed. **FR**: septicidal capsule, slender, cylindric, pointed. **SEED**: with tuft of long hairs at 1 end. 1–3 spp.: Medit, subtrop Asia, Japan. (Greek for Oleander, from resemblance of lvs to those of olive, *Olea*)

N. oleander L. COMMON OLEANDER **LF**: 6–20 cm, linear- to oblong-lanceolate, evergreen. **INFL**: gen exceeding lvs. **FL**: often double, sterile; calyx ± lf-like, lobes 4–6 mm; corolla 2–6 cm wide, white to ± yellow to red-purple. **FR**: 8–20 cm. $2n=22$. Highway medians, roadsides, streamsides; < 700 m. KR, n SNF, ScV, CCo, SCo, SnGb, SnBr, PR, DSon; UT, TX to NC, FL. Widely planted as orn, esp along highways; pls often persist long after cult, may reproduce vegetatively but evidently not sexually without human fostering; latex (± colorless) of all parts lethally poisonous, even in small quantities. Jun–Sep

VINCA PERIWINKLE

Thomas J. Rosatti & Lauramay T. Dempster

Per, ± glabrous (exc ciliate lf, sepal margins). **LF:** opposite to subopposite. **INFL:** fls gen 1 in lf axils. **FL:** calyx lobes long, slender; corolla tube funnel-shaped, lobes asymmetric; filaments free, attached near top of corolla tube, sharply bent near base, unappendaged, anthers held around top of but free from stigma, each partly sterile, pollen ± free; nectaries 2, alternate ovaries, wide-spaced, gen not exceeding ovaries; style cylindric, widened distally, stigma skirted at base. **SEED:** glabrous. 6–7 spp.: s Eur, n Afr, to Afghanistan. (Latin: possibly, to bind or wind about)

V. major L. (p. 233) GREATER PERIWINKLE Pl sprawling. **ST:** arching, rooting at tips. **LF:** petiole gen < 1 cm; blade ± 7 cm, ovate, base ± truncate, tip acute. **FL:** corolla 3–5 cm wide at top, purple-blue (white). **FR:** curved, rare. 2*n*=90,92. Coastal bluffs, sheltered places, esp along stream beds; 2–200 m. NCo, NCoRO, n SN, ScV, CCo, SnFrB, s SCoRO, SCo, SnGb, SnBr, PR; s US; native probably to Eur. Cult. Mar–Jun(Jan) ❖

APODANTHACEAE STEMSUCKER FAMILY

George Yatskievych

Per? (duration of pls uncertain), st parasites, non-green (chlorophyll 0), dioecious? (number of pls contributing to fls observed on given host uncertain). **ST:** reduced, thread-like, inside host st. **LF:** simple, bract- and/or scale-like, 2–6, subtending fl, gen in 2 series. **INFL:** fl 1. **FL:** unisexual, radial, fleshy; sepals 4–5[7], often fused at base; petals 0; stamens many, gen fused to style to form column that is expanded at tip into disk or knob (column, disk completely of carpels in female fls according to Blarer et al.); anthers or stigmatic hairs on column, below disk margin; ovary ± inferior, chamber 1, placentas parietal. **FR:** berry-like, ± fleshy, gen dehiscing irregularly in age. **SEED:** many, minute. ± 2–3 genera, ± 12–20 spp.: sw US to S.Am, e Afr, sw Asia, w Australia. [Barkman et al. 2004 Proc Natl Acad Sci USA 101:787–792; Blarer et al. 2004 Pl Syst Evol 245:119–142] Poorly known taxonomically, ecologically, reproductively. Molecular evidence (Blarer et al., and online references) indicates Rafflesiaceae, in which our pls were incl in TJM (1993), are polyphyletic, consisting of 3 main lineages belonging to different orders, evidently Malvales or Cucurbitales for our pls. Scientific Editor: Thomas J. Rosatti.

PILOSTYLES

Per. ± 12 spp.: sw US to trop Am, Afr, Australia, sw Asia (esp Iran). (Latin: hair pillar, from hairs on fl column) Only lvs, fls, frs visible on host.

P. thurberi A. Gray (p. 233) THURBER'S PILOSTYLES Lvs, sepals brown to maroon. **LF:** overlapping, 4–7, 1–1.5 mm, round or ovate. **FL:** < 2 mm; sepals ± like lvs, ± yellow to cream or not adaxially and/or on margins; column ± dark green, disk < 1 mm diam, yellow; anthers in ring of ± 3 rows. Open desert scrub; < 300 m. s DSon (Riverside, San Diego, Imperial cos.); to NV, TX, Mex. Parasitic on *Psorothamnus*, esp *Psorothamnus emoryi*. Jan ★

AQUIFOLIACEAE HOLLY FAMILY

Elizabeth McClintock

Shrub, tree, evergreen [deciduous]; dioecious or ± so. **LF:** gen alternate, simple, often spiny, toothed, or lobed. **INFL:** gen cyme, gen axillary. **FL:** gen unisexual, radial; sepals gen 4, gen fused at base; petals 4, fused at base [free], white [green]; stamens gen 4, gen alternate petals; ovary superior, chambers often 4, gen 1-ovuled, style 0 or terminal, short, stigma head-like or lobed. **FR:** drupe, gen red [purple to black or yellow to ± white]; stones gen 4. 4 genera, 300–400 spp.: trop, temp. Scientific Editor: Thomas J. Rosatti.

ILEX HOLLY

LF: petioled; adaxially gen shiny. **FL:** sepals persistent in fr; petals oblong or obovate. **FR:** pulpy. 300–400 spp.: esp trop. (Latin: name for Medit holly-oak, *Quercus ilex* L.) [Zika 2010 Madroño 57:1–10] *I. cornuta* Lindl. & Paxton occ persisting from cult.

I. aquifolium L. ENGLISH HOLLY Shrub, tree, < 20 m. **ST:** erect; branches many, branchlets minutely hairy. **LF:** 2.5–6 cm, ovate or oblong-ovate, entire to lobed, even on 1 pl, teeth 0 or widely spaced, stiff, spine-like. **FL:** petals 2–3 mm, dull white. **FR:** 7–8 mm wide, smooth. Cool, wooded areas; < 200 m. NCo, KR?, NCoRO, CCo, SnFrB, expected elsewhere; native to Eur, w Asia. Cult since ancient time for showy fr. May–Jun ❖

ARALIACEAE GINSENG FAMILY

Robert E. Preston & Elizabeth McClintock, except as noted

Per, shrub, tree, woody vine; juvenile, fl pls gen unlike. **ST:** gen branched. **LF:** simple or compound, gen alternate; stipules ± fused to ± sheathing, petiole base or 0. **INFL:** umbels 1 to panicled; bracts deciduous or not. **FL:** gen bisexual, gen radial,

209

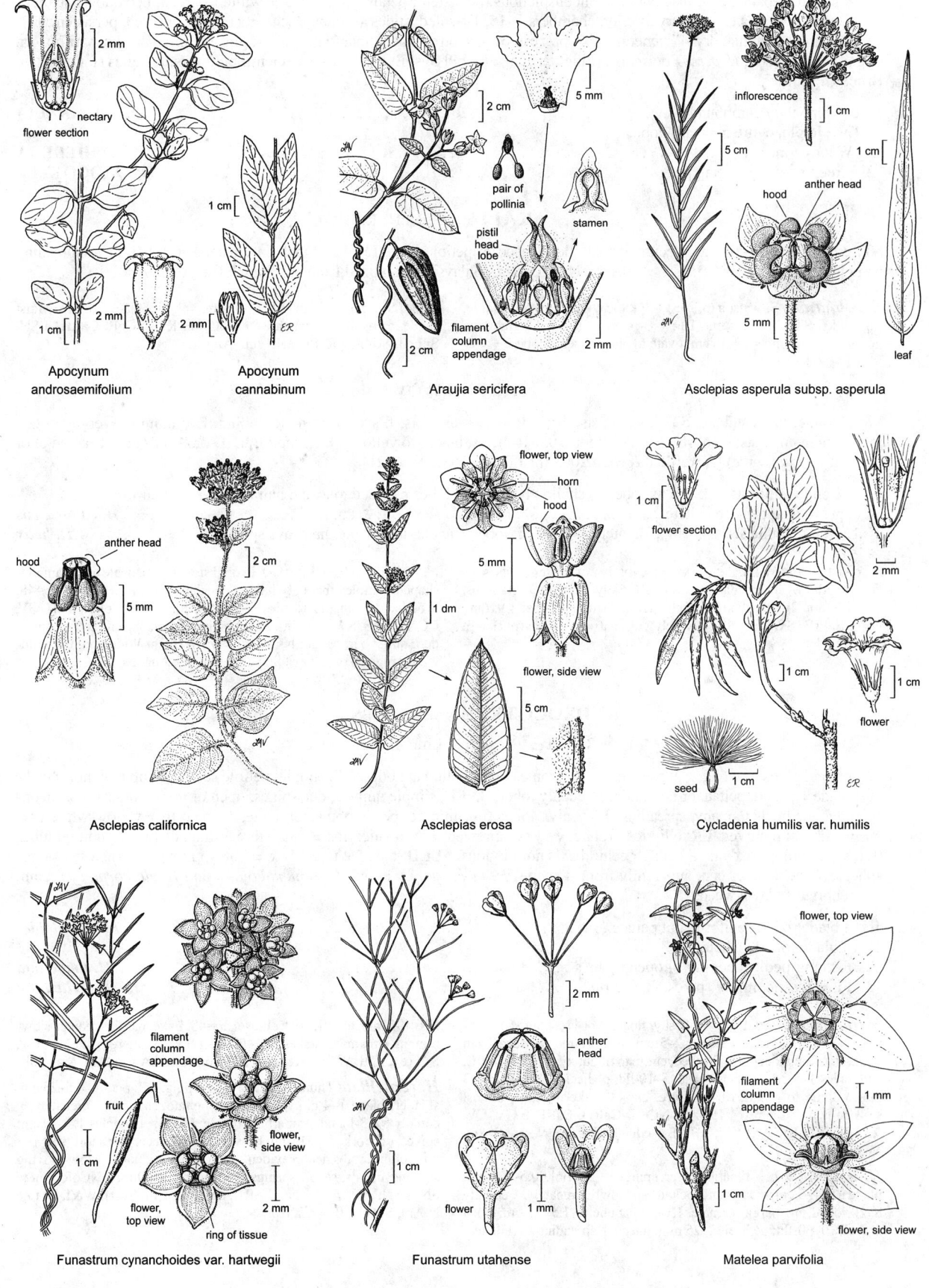

Apocynum
androsaemifolium

Apocynum
cannabinum

Araujia sericifera

Asclepias asperula subsp. asperula

Asclepias californica

Asclepias erosa

Cycladenia humilis var. humilis

Funastrum cynanchoides var. hartwegii

Funastrum utahense

Matelea parvifolia

gen < 5 mm; sepals gen 5, fused at base, inconspicuous, persistent; petals gen 5, free, ± white to green, deciduous; stamens gen 5, gen alternate petals; ovary inferior, chambers 1–15, 1-ovuled, styles as many as chambers, free or fused, persistent. **FR:** berry or drupe, occ flat, dry. 47 genera, 1350 spp.: esp trop, subtrop; medicinal (e.g., *Panax*, ginseng; *Aralia*, sarsaparilla), orn (e.g., *Aralia*, *Fatsia*, *Hedera*, *Polyscias*). [Plunkett et al. 2004 Pl Syst Evol 245:1–39] Scientific Editors: Douglas H. Goldman, Bruce G. Baldwin.

1. Lvs pinnately compound . . . . **ARALIA**
1′ Lvs simple, entire to deeply lobed
  2. Woody vine . . . . **HEDERA**
  2′ Creeping or sprawling per . . . . **HYDROCOTYLE**

ARALIA SPIKENARD

ST: erect. **LF:** 1–3-pinnate, deciduous; stipule ± fused to petiole base. **INFL:** umbels in clusters of 2–few or in spreading panicles. **FR:** berry. **SEED:** 3–5. ± 30 spp.: N.Am, Asia, Malaysia. (Latinized from old French Canadian, Aralie)

A. californica S. Watson (p. 233) ELK CLOVER Per; roots large; sap milky. **ST:** 2–3 m, stout. **LF:** 1–3-pinnate, 1–2 m, ± glabrous; petiole < 3 dm; lflets 15–30 cm, ovate to oblong, serrate, base ± cordate. **INFL:** 30–45 cm. **FR:** ± 5 mm, spheric, black. 2*n*=48. Moist shade, canyons, stream sides; < 2500 m. KR, NCoR, CaRH, n SN, SnFrB, SCoR, TR, PR; OR. Jun–Aug

HEDERA IVY

Woody vine; hairs stellate. **ST:** juvenile sts climbing by aerial roots; fl sts fewer, nonclimbing. **LF:** simple, evergreen; gen lobed on juvenile sts; ± entire on fl sts; stipules 0. **FR:** berry, black [to yellow]. ± 13 spp.: Eur, n Afr, Asia. (Latin: sacred pl of Bacchus, god of wine) [Ackerfield & Wen 2003 Int J Pl Sci 163:593–602]

1. Lvs on juvenile sts ≤ 25 cm, unlobed to shallowly 3-lobed, lobes rounded to acute; hair red-orange, hair rays
    appressed, appearing peltate . . . . ***H. canariensis***
1′ Lvs on juvenile sts ≤ 35 cm, palmately 3–5-lobed, lobes ± acute; hair white, hair rays spreading . . . . ***H. helix***

H. canariensis Willd. CANARY ISLANDS IVY **LF:** on fl sts < 15 cm, broadly ovate, cordate; petiole, abaxially hairy to ± glabrous. **FR:** ± 5 mm. 2*n*=48. Woodland, chaparral, disturbed areas; < 916 m. NCo, SnFrB, SW (exc SnJt); widely cult, native to Macaronesia, n Afr. Aug–Nov ❖

H. helix L. ENGLISH IVY **LF:** on fl sts < 15 cm, ovate to ± diamond-shaped; petiole, abaxially hairy to ± glabrous. **FR:** ± 5 mm. 2*n*=48. Woodland, open, disturbed areas; < 1280 m. NCo, NCoR, n SNH, GV, CCo, SnFrB, SW (exc ChI, SnJt); native to Eur. Sap can cause contact dermatitis; berries and lvs TOXIC when eaten. Widely cult in mild-winter regions, gen spreading aggressively. Aug–Nov ❖

HYDROCOTYLE MARSH PENNYWORT
Robert E. Preston & Lincoln Constance

Per, creeping or sprawling, glabrous [hairy]; rhizomes or st rooting at nodes. **LF:** simple; petiole scarious-stipuled, not sheathing; blade ± round, peltate or not, entire to deeply lobed. **INFL:** simple umbels, occ spikes, open or dense; bracts 0 or inconspicuous; pedicels 0–many, spreading. **FL:** calyx lobes 0 or minute; petals obtuse or acute, ± green to ± yellow-white or ± purple, tip not incurved. **FR:** elliptic to round, very compressed side-to-side; ribs ± equal, thread-like, distinct or not; oil tubes 0, fr wall with individual oil cells; fr central axis not obvious. **SEED:** face flat to convex. ± 100 spp.: worldwide, esp s hemisphere. (Greek: water cup, apparently from lf shape) *H. moschata* G. Forst., *H. ranunculoides*, and *H. sibthorpioides* Lam. occ reported as lawn weeds.

1. Lf blade round-reniform, not peltate . . . . ***H. ranunculoides***
1′ Lf blade round, peltate
  2. Fls long-pedicelled; infl of open umbels . . . . ***H. umbellata***
  2′ Fls ± sessile; infl of spikes, fls whorled . . . . ***H. verticillata***

H. ranunculoides L. f. Pl fleshy, floating or creeping. **LF:** petiole 5–35 cm, stout; blade gen 2–5 cm wide, round-reniform, gen wider than long, deeply 3–7-lobed, entire to minutely crenate. **INFL:** peduncles 1–5 cm; umbels dense, 5–10-fld; pedicels short. **FR:** 1–3 mm, elliptic to round; ribs obscure. 2*n*=48. Lake margins, ponds, slow-moving streams; < 1500 m. NCo, NCoRO, CaRF, c GV, CW (exc SCoRI), SW (exc ChI, WTR), w edge DMoj; to e N.Am, S.Am. Mar–Aug

H. umbellata L. Floating or creeping. **LF:** petiole 0.5–4.5 cm; blade gen 1–5 cm wide, round, peltate, shallowly, crenately, ± equally 8–20-lobed, or margin crenate. **INFL:** peduncles 1.5–35 cm; open umbel, 10–60-fld; pedicels 2–25 mm, unequal, spreading to reflexed. **FR:** 1–2 mm, elliptic; ribs obtuse. 2*n*=48. Lake margins, ponds, slow-moving streams, marshes; < 1400 m. n SNF (scattered), c SNH, GV, SW (exc ChI, WTR), s&w edge DMoj; to e N.Am, S.Am. Mar–Jul

H. verticillata Thunb. (p. 233) Creeping. **LF:** petiole 0.5–25 cm, slender; blade 1–4 cm wide, round, peltate, shallowly, crenately, ± equally 8–13-lobed, or margin crenate. **INFL:** peduncles 1.5–20 cm; spikes, occ forked, to 15-fld, fls whorled; pedicels 0 or short. **FR:** 1–3 mm, elliptic; ribs acute, evident. Lake margins, ponds, slow-moving streams, canals, seeps, springs, marshes; < 1400 m. s NCoRO, n&c SN, s SNF, GV, CW (exc SCoRI), SW (exc ChI, SnJt), W&I, D; to e N.Am, S.Am, HI, s Afr. Apr–Sep

ASTERACEAE (Compositae) SUNFLOWER FAMILY

David J. Keil, except as noted

Ann to tree. **LF**: basal and/or cauline, alternate, opposite, rarely whorled, simple to 2+ × compound. **INFL**: 1° infl a head, resembling a fl, of several types (see below), 1–many in gen ± cyme-like cluster; each head gen with ± calyx-like involucre of 1–many series of phyllaries (involucral bracts); receptacle of head flat to conic or columnar, paleate (bearing paleae = receptacle bracts) or epaleate; fls 1–many per head. **FL**: bisexual, unisexual, or sterile, ± small, of several types (see below); calyx 0 or modified into ± persistent pappus of bristles, scales, and/or awns; corolla radial or bilateral (0), lobes gen (0)3–5; stamens 4–5, filaments gen free, gen fused to corolla at tube/throat junction, anthers gen fused into cylinder around style, anther base gen rounded or cordate (deeply sagittate or with tail-like appendages), tip (= flattened appendage) gen projecting beyond pollen sac; pistil 1, 2-carpeled, ovary inferior, 1-chambered, 1-seeded, placenta basal, style 1, tip gen ± 2-branched (exc in some staminate disk fls), branch tips truncate or gen bearing ± brush-like appendages; stigmas 2, gen on adaxial faces of style branches. **FR**: achene (also called a cypsela) (drupe in *Chrysanthemoides*), cylindric to ovoid, sometimes compressed, gen deciduous with pappus attached. ± 1500 genera, 23000 spp.: worldwide, many habitats. Fl and head types differ in form and sexual condition. A disk fl has a gen radial corolla, with a cylindric tube, expanded throat, and gen 5 lobes. Disk fls are gen bisexual and fertile but occ staminate with reduced ovaries. Discoid heads comprise only disk fls. A radiant head is a variant of a discoid head, with peripheral disk fl corollas expanded, often bilateral. A ray fl corolla is bilateral, gen with a slender tube and flattened petal-like ray (single lip composed of gen 3 lobes). Ray fls are gen pistillate or sterile (occ lacking styles). Radiate heads have peripheral ray fls and central disk fls. Disciform heads superficially resemble discoid heads, with pistillate or sterile fls that lack rays, together with or separate from disk fls. A ligulate fl is bisexual, with a bilateral, gen ephemeral corolla and 5-lobed ligule. Liguliflorous heads comprise only ligulate fls. See glossary p. 31 for illustrations of family characteristics. *Echinops sphaero-cephalus* L., *Gaillardia aristata* Pursh, *G. pulchella* Foug., *Hymenothrix loomisii* S.F. Blake, *Tagetes erecta* L., *Thelesperma megapotamicum* (Spreng.) Kuntze are waifs. *Melampodium perfoliatum* Kunth, historic urban waif. *Ageratum conyzoides* L., *Guizotia abyssinica* (L. f.) Cass., *Santolina chamaecyparisus* L., orth. var. are rare or uncommon escapes from cult. *Dyssodia papposa*, *Ismelia carinata* (Schousb.) Sch. Bip. [*Chrysanthemum carinatum* Schousb.], *Mantisalca salmantica* (L.) Briq. & Cavill. are historical or extirpated waifs in CA. *Inula helenium* L. not documented in CA. Taxa of *Aster* in TJM (1993) treated here in *Almutaster*, *Eucephalus*, *Eurybia*, *Ionactis*, *Oreostemma*, *Sericocarpus*, *Symphyotrichum*; *Chamomilla* in *Matricaria*; *Cnicus* in *Centaurea*; *Conyza* in *Erigeron* and *Laennecia*; *Dugaldia* in *Hymenoxys*; *Erechtites* in *Senecio*; *Hymenoclea* in *Ambrosia*; *Lembertia* in *Monolopia*; *Osteospermum ecklonis* in *Dimorphotheca*; *Picris echioides* in *Helminthotheca*; *Prionopsis* in *Grindelia*; *Raillardiopsis* in *Anisocarpus* and *Carlquistia*; *Schkuhria multiflora* in *Bahia*; *Trimorpha* in *Erigeron*; *Venidium* in *Arctotis*; *Whitneya* in *Arnica*. Scientific Editors: David J. Keil, Bruce G. Baldwin.

Key to Groups

1. None of fls of head with strap-shaped corollas that resemble spreading petals [ligulate fls 0, ray fls 0 or rays very short and easily overlooked]; heads discoid, disciform, or radiant
  2. Fls of 2 kinds, in same or different heads
    3. Heads radiant; corollas of outer fls [bisexual or sterile] conspicuously expanded, often ± bilateral. . . . . **Group 1**
    3′ Heads disciform [pistillate or sterile fls and bisexual or staminate fls] or unisexual, corollas of pistillate or sterile fls inconspicuous or 0. . . . . **Group 2**
  2′ Fls of 1 kind [disk fls] — heads discoid
    4. Receptacle bearing scale-like bracts [paleae] that gen individually subtend disk fls or bearing hair-like bristles or tooth-like to membranous or bristle-like scales among fls
      5. Receptacle paleate or bearing tooth-like to membranous scales . . . . **Group 3**
      5′ Receptacle bristly or bearing long, bristle-like scales (free or fused at base) . . . . **Group 4**
    4′ Receptacle epaleate or paleae only marginal and simulating phyllaries (sometimes bearing minute scales or short hairs among fls)
      6. Pappus 0 or only a low crown . . . . **Group 5**
      6′ Pappus well developed
        7. Pappus of bristles (sometimes with an additional series of shorter bristles or scales) . . . . **Group 6**
        7′ Pappus of flat, ± membranous scales or stiff, ± needle-like awns . . . . **Group 7**
1′ Some or all fls of head with strap-shaped corollas or corolla lips [ligules or rays] that resemble spreading petals; heads liguliflorous, composed of 2-lipped fls, or radiate
  8. Head composed of 1 kind of fl; central fls sometimes smaller or less mature but all bisexual and with same kind of corolla
    9. Head liguliflorous; corolla 1-lipped, the spreading, petal-like ligule tipped by 5 short lobes [ligulate fls]; corolla readily withering; sap milky. . . . . **Group 8**
    9′ Head composed of 2-lipped fls; outer corolla lip shallowly 3-lobed, spreading, resembling ray of ray fl, inner lip deeply 2-lobed, recurved or coiled; corolla not readily withering; sap clear . . . . **Group 9**
  8′ Head radiate, composed of 2 kinds of fls; outer fls [ray fls] pistillate or sterile, corolla bilateral with gen spreading, petal-like lip [ray]; inner fls [disk fls] bisexual (staminate) with gen radial, (4)5-lobed tubular corolla
    10. Receptacle bearing scale-like bracts [paleae] that gen individually subtend disk fls (sometimes only a single ring of paleae separate ray and disk fls)
      11. Phyllaries in 1(2) series, each subtending a ray fl . . . . **Group 10**

11′ Phyllaries in 2+ series, not all subtending ray fls . . . . **Group 11**
10′ Receptacle epaleate or bearing minute scales or hairs (rarely long bristles or awn-like projections) among fls
 12. Pappus 0 or only a low crown . . . . **Group 12**
 12′ Pappus well developed on ray or disk frs (or both)
  13. Pappus of flat, ± membranous scales or stiff, ± needle-like awns. . . . . **Group 13**
  13′ Pappus of bristles (sometimes with an additional series of shorter bristles or scales)
   14. Rays white to pink, blue, or purple . . . . **Group 14**
   14′ Rays yellow to orange or red . . . . **Group 15**

Group 1

Heads radiant; corollas of outer fls enlarged, often ± bilateral

1. Phyllaries graduated in 4–8 series
 2. Outer fls fruiting
  3. Pappus of bristles; ann, taprooted . . . . **LESSINGIA** (G6,7)
  3′ Pappus a crown of short scales; per with rhizome . . . . *Tanacetum bipinnatum*
 2′ Outer fls sterile
  4. Distal phyllary margin expanded as a spiny-margined, fringed, or irregularly toothed appendage
   . . . . **CENTAUREA** (G2,4)
  4′ Distal phyllary margin entire, tapered smoothly to tip-spine . . . . **VOLUTARIA** (G4)
1′ Phyllaries ± equal or weakly graduated in 1–3 series
 5. Fr 10–15 mm . . . . **PALAFOXIA** (G7)
 5′ Fr 2–9 mm
  6. Phyllary margin thin, ± scarious, brown to purple . . . . **HYMENOTHRIX** (G7)
  6′ Phyllary margin not or scarcely scarious, variously colored
   7. Lvs opposite . . . . *Arnica discoidea*
   7′ Lvs alternate
    8. Pappus scales entire or fringed; lf entire to ± deeply 1–2-pinnately lobed. . . . . **CHAENACTIS** (G7)
    8′ Pappus scales dissected into bristles; lf sharply dentate or with a few short, sharp lobes
     . . . . **TRICHOPTILIUM** (G6,7)

Group 2

Heads disciform or unisexual; fls of 2 kinds or of 1 unisexual kind; corollas of pistillate or sterile fls inconspicuous

1. Pistillate and staminate fls in different heads
 2. Pistillate and staminate heads on same pl [monoecious]
  3. Subshrub or shrub (to small tree) . . . . [2]**AMBROSIA**
  3′ Herb
   4. St armed with 3-branched spines. . . . . *Xanthium spinosum*
   4′ St unarmed
    5. Staminate heads well-spaced in long terminal infl; bur 2–10 mm . . . . [2]**AMBROSIA**
    5′ Staminate heads congested; bur 10–30+ mm . . . . *Xanthium strumarium*
 2′ Pistillate and staminate heads on different pls [dioecious]
  6. Shrub or erect herb with sticky lvs, not woolly. . . . . **BACCHARIS** (G6)
  6′ Herb, sometimes prostrate, not sticky; pl gen ± woolly
   7. St from rhizome; lf rosette withered at fl; cauline lvs long, all exc distal-most ± equal . . . . [2]**ANAPHALIS** (G6)
   7′ St from persistent lf rosette or thick, often branched caudex; cauline lvs gen ± reduced, at least
    distal-most short, 1–3 cm . . . . **ANTENNARIA** (G6)
1′ Pistillate or sterile fls in same head as staminate or bisexual fls
 8. Outer fls without corollas; heads not embedded in woolly hairs
  9. Heads solitary; pl low, sts often spreading or prostrate
   10. Head on slender ± lfless peduncle; outer frs stalked, not spine-tipped. . . . . **COTULA**
   10′ Head sessile near st base or at branch tips, ± enveloped by sheathing bases of opposite or whorled
    subtending lvs, overtopped by branches; fr sessile, spine-tipped. **SOLIVA**
  9′ Heads in raceme- or panicle-like clusters
   11. Ann; fr smooth or warty, gen winged, glabrous or hairy . . . . **DICORIA**
   11′ Subshrub or shrub; fr long-soft-hairy, not winged . . . . *Euphrosyne acerosa*
 8′ Outer fls with (sometimes narrowly cylindric) corollas; heads sometimes embedded in woolly hairs
  12. Outer fls sterile; head not very small, not embedded in woolly hairs
   13. Cauline lvs gen bladeless, sheathing; basal lvs from creeping rhizome, often appearing after fl st,
    long-petioled, palmately lobed. . . . . **PETASITES** (G14)
   13′ Cauline lvs with blades, sometimes reduced distally on st, not sheathing; basal lvs 0 or from taproot,
    appearing before fl st, sessile or petioled, entire to 1–2-pinnately lobed
    14. Fertile fls 1–2 per head . . . . **CRUPINA** (G3)
    14′ Fertile fls many per head
     15. Lf spiny . . . . *Centaurea benedicta* (G4)

15′ Lf not spiny
16. Pappus of awns with reflexed barbs; phyllaries in 2 series, unarmed . . . . **BIDENS** (G3,11)
16′ Pappus 0 or of smooth or minutely rough bristles or narrow scales; phyllaries graduated in 6+
series, often fringed or spine-tipped (see also *Volutaria canariensis*) . . . . **CENTAUREA** (G1,4)
12′ Outer fls pistillate [corolla often narrowly cylindric]; heads sometimes very small, embedded in woolly hairs
17. Phyllaries (or outermost bracts of head) papery, membranous, or scarious, sometimes green only in
proximal 1/2 or in narrow, central band; heads often embedded in woolly hairs
18. Phyllaries 0 or 1–6, not overlapping, < fls; outer receptacle paleate; some or all pistillate fls
individually subtended by paleae (which often resemble phyllaries)
19. Disk fls bisexual, pappus gen of 13–28+ bristles, ± exserted; inner pistillate fls with pappus
20. Receptacle ± cylindric, height 5–15 × diam; most pistillate paleae open to folded around pistillate
fls, but not enclosing them, acuminate or bristle-tipped; innermost paleae erect or ascending in fr;
outer fr ± = inner. . . . . **FILAGO**
20′ Receptacle mushroom-shaped to obovoid, length 0.4–15 × width; most pistillate paleae ±
enclosing pistillate fls, obtuse to acute; innermost paleae spreading in fr; outer fr > inner. . . . . **LOGFIA**
19′ Disk fls staminate, pappus 0 or 1–12 bristles, incl; pistillate fl pappus 0
21. Paleae subtending disk fls (disk paleae) enlarged, rigid, ± spreading, very different from scales
subtending pistillate fls (pistillate paleae)
22. Disk palea tips strongly incurved to hooked inward, acuminate or abruptly pointed, often
spine-like; pistillate paleae closed, strongly 3-veined or middle vein obscure, tips gen strongly
scarious-winged; receptacle not bristly, gen widest at tip . . . . **ANCISTROCARPHUS**
22′ Disk palea tips erect or spreading, ± flat to folded, not spine-like, obtuse; pistillate paleae open,
concave, veins obscure, tips not or barely scarious-winged; receptacle bristly, widest at base
. . . . **HESPEREVAX**
21′ Disk paleae 0 or scales ± gradually reduced, scarious, erect
23. Lvs gen opposite; paleae net-veined, inner edge with scarious wing that is hidden in head;
pappus 0. . . . . **PSILOCARPHUS**
23′ Lvs alternate or seeming whorled; paleae parallel-veined, scarious wing visible in head; pappus 0
or gen of 1–12 bristles
24. Phyllaries 4–6, equal, scarious, obovate, rounded, abruptly different from paleae; style on inner
edge of fr; receptacle length ± 1–2 × width . . . . **MICROPUS**
24′ Phyllaries 0 or 1–4, unequal, vestigial or ± like paleae; style ± at fr tip; receptacle length ± 3–8 ×
width . . . . **STYLOCLINE**
18′ Phyllaries gen many, overlapping, inner often ≥ fls; receptacle epaleate; pistillate fls not subtended by bracts
25. Pappus 0 or a minute crown . . . . ³**ARTEMISIA** (G5)
25′ Pappus of bristles
26. Herbage short-appressed-hairy, glandular, or silky; distal 1/2 of outer phyllaries leathery, scarious
margin 0 or narrow. . . . . ²**PLUCHEA**
26′ Herbage ± long-hairy (lf gen densely woolly abaxially); phyllaries thin, membranous, margin
scarious distal to middle or throughout, gen wide
27. Shrub or subshrub; st slender, much-branched. . . . . **PLECOSTACHYS**
27′ Herb (if subshrub, st erect or ascending)
28. Pl with creeping rhizome . . . . ²**ANAPHALIS** (G6)
28′ Pl taprooted, fibrous-rooted (sometimes from lfy stolons), or with caudex
29. Heads gen in tight groups in gen spike- or narrow, panicle-like cluster (reduced to terminal
cluster in small pls); pappus bristles fused at base, falling in ring. . . . . **GAMOCHAETA**
29′ Heads in axillary or terminal head-like clusters or in ± open, flat-topped to panicle-like
clusters; pappus bristles free or ± fused at base and falling in groups
30. Heads in dense, hemispheric to ± spheric terminal cluster closely subtended by involucre-like
ring of lfy bracts, sometimes also in head-like axillary clusters . . . . **EUCHITON**
30′ Heads in axillary clusters or terminal clusters, not closely subtended by involucre-like
cluster of lfy bracts
31. Ann, 1–30 cm, low-growing; sts gen prostrate or spreading; head clusters ± sessile, axillary
or terminal; drying mud, shorelines, other moist habitats . . . . **GNAPHALIUM**
31′ Ann to per, 15–200 cm; sts gen erect or ascending, head clusters gen terminal, sessile to
stalked, often branched; gen dry habitats . . . . **PSEUDOGNAPHALIUM**
17′ Phyllaries gen green (± purple) throughout, sometimes scarious-margined or inner scarious
throughout; heads gen not embedded in woolly hairs
32. Pappus well developed
33. Pappus of flattened scales or barbed awns
34. Lvs gen opposite throughout (occ distal-most alternate). . . . . **LASTHENIA** (G12,13)
34′ Lvs mostly alternate, sometimes very crowded
35. Pl dotted with sessile resin glands; fr 3–4-angled . . . . ²**AMBLYOPAPPUS** (G12,13)
35′ Pl densely woolly; fr compressed
36. Lvs entire; pappus scales 2; pl densely tufted; GB. . . . . **EATONELLA** (G13)
36′ Lvs gen wavy-dentate or shallowly lobed; pappus scales 2–7; pl not densely tufted;
s SnJV . . . . *Monolopia congdonii* (G12,13)

33′ Pappus of bristles
  37. Main phyllaries in 1 series, equal . . . . **SENECIO** (G6,15)
  37′ Main phyllaries in 2+ series
    38. Subshrub or shrub
      39. Sts ± lfless; receptacle paleate; pappus bristles plumose . . . . **BEBBIA**
      39′ Sts evidently lfy; receptacle epaleate; pappus bristles ± smooth to barbed
        40. Corolla yellow or becoming ± red-purple; lvs gen ± toothed; pistillate fls 5–18 . . . . **HAZARDIA**
        40′ Corolla pink to deep rose; lvs entire; pistillate fls many . . . . *Pluchea sericea*
    38′ Ann or per
      41. St thread-like; pistillate fls 1–5 . . . . [3]**PENTACHAETA** (G12,14,15)
      41′ St stouter; pistillate fls several to many
        42. Pappus bristles 1–2; lf palmately lobed or toothed . . . . [2]*Perityle emoryi* (G12,13,14)
        42′ Pappus bristles gen many; lf entire to pinnately lobed
          43. Head ± flat, button-like . . . . [2]**ERIGERON** (G6,14,15)
          43′ Head gen cylindric, bell-shaped, or ± obconic
            44. Disk fl corolla purple; herbage strongly scented; anther bases short-tailed . . . . [2]**PLUCHEA**
            44′ Disk fl corolla cream or yellow; herbage not strongly scented; anther bases not tailed
              45. Disk corolla 9–13 mm . . . . *Pyrrocoma carthamoides* var. *cusickii*
              45′ Disk corolla 3–5 mm
                46. Lf not clasping, entire, toothed, or shallowly few-lobed . . . . [2]**ERIGERON** (G6,14,15)
                46′ Lf clasping, regularly dentate or spreading-lobed . . . . **LAENNECIA**
32′ Pappus 0 or reduced to a minute crown
  47. Phyllary margins widely scarious or transparent
    48. Delicate ann; pistillate corollas thread-like . . . . [3]**PENTACHAETA** (G12,14,15)
    48′ Per or stout ann; pistillate corollas short, wider than thread-like
      49. Heads in spikes, racemes, or panicles . . . . [3]**ARTEMISIA** (G5)
      49′ Heads 1 or few to many in ± flat-topped clusters
        50. Involucre gen 3–5 mm diam, phyllaries gen 12–20; pappus 0 . . . . **SPHAEROMERIA**
        50′ Involucre 5–22 mm diam, phyllaries gen 30–60; pappus crown-like, 0.1–0.5+ mm . . . **TANACETUM** (G12)
  47′ Phyllary margins gen not scarious or transparent
    51. Ovary of disk fl much reduced; style tip truncate or tack-shaped
      52. Lf ovate to widely triangular, gen ± basal or proximal cauline, green adaxially, white abaxially;
          infl open; corolla white; fr bearing stalked glands . . . . **ADENOCAULON**
      52′ Lf linear to (ob)ovate or ± round, basal and cauline or all cauline, faces ± alike, gen green to gray
          on 1 or both faces; infl ± dense; corolla ± pale yellow, yellow-white, or ± pink; fr not glandular
        53. Anthers fused . . . . [3]**ARTEMISIA** (G5)
        53′ Anthers ± free or only weakly fused
          54. Ann; lf 1–2-pinnately divided . . . . *Euphrosyne nevadensis*
          54′ Per or subshrub; lf gen entire . . . . **IVA**
    51′ Ovary of disk fls well developed; style tip ± branched
      55. Lf wide, palmately lobed or toothed . . . . [2]*Perityle emoryi* (G12,13,14)
      55′ Lf linear to narrowly oblanceolate, entire or pinnately toothed to lobed
        56. Pl not glandular, not strongly scented
          57. Lvs opposite . . . . *Lasthenia microglossa*
          57′ Lvs alternate . . . . [3]**PENTACHAETA** (G12,14,15)
        56′ Pl glandular, ± strongly scented
          58. Pappus present . . . . [2]**AMBLYOPAPPUS** (G12,13)
          58′ Pappus 0
            59. Fr of pistillate fl compressed front-to-back . . . . **HEMIZONELLA** (G10)
            59′ Fr of pistillate fl compressed side-to-side or ± 3-angled in ×-section . . . . **MADIA** (G10)

Group 3

Heads discoid; receptacle paleate or bearing tooth-like to membranous scales among fls

1. Lf margins spiny
  2. Heads many-fld . . . . **CARTHAMUS** (G4,5)
  2′ Heads 1-fld, grouped in spheric 2° head . . . . [**ECHINOPS**] (G5,7)
1′ Lf margins not spiny
  3. Pappus of bristles, bristles and small scales, or stout needle-like awns
    4. Pappus of awns
      5. Lf ± lanceolate, simple, or pinnately compound, lflets lanceolate to ovate . . . . **BIDENS** (G2,11)
      5′ Lf linear or dissected into linear lobes . . . . [**THELESPERMA**]
    4′ Pappus of bristles or of bristles and small scales
      6. Corolla purple; fr 1–2 per head; herb . . . . **CRUPINA** (G2)
      6′ Corolla yellow; fr 18–50 per head; shrub or subshrub
        7. Lf linear to narrowly elliptic or triangular, rough-hairy or becoming glabrous; corolla 6–10 mm;
          SW, D . . . . **BEBBIA** (G2)

      7′ Lf ovate to ± round, tomentose; corolla ± 2.5 mm; CCo, SCo . . . . **HELICHRYSUM** (G6)

3′ Pappus 0 or of thin, membranous scales or narrow, scale-like awns

  8. Ovary and fr strongly flattened, long-ciliate

    9. Shrub, strigose . . . . *Encelia frutescens*

    9′ Per, sticky-glandular . . . . *Geraea viscida*

  8′ Ovary and fr ± thick, not ciliate

    10. Ann; all or inner phyllaries tapered to long, needle-like tips . . . . *Chaenactis carphoclinia*

    10′ Bien to shrub; phyllaries obtuse to acute or acuminate

      11. Lf ± cylindric, lobes many, small . . . . [**SANTOLINA**]

      11′ Lf flat, entire to few-lobed

        12. Lf deeply and coarsely pinnately 3–7-lobed, abaxially gray-green-canescent . . . . *Artemisia palmeri*

        12′ Lf entire or toothed, abaxially glabrous to short-hairy or ± glandular, green

          13. Shrub; lf ± linear or narrowly oblanceolate . . . . **EASTWOODIA**

          13′ Per; lf elliptic to widely ovate or triangular

            14. Receptacle strongly conic or columnar; corolla dark brown-purple . . . . *Rudbeckia occidentalis*

            14′ Receptacle flat to convex; corolla yellow or orange . . . . *Wyethia invenusta*

Group 4

       Heads discoid; receptacle bristly or bearing long, bristle-like scales (free or ± fused at base)

1. Lvs not spiny

  2. Phyllaries spine-tipped

    3. Phyllary tip-spines hooked . . . . **ARCTIUM**

    3′ Phyllary tip-spines straight

      4. Phyllary tip fringed with spines or slender teeth . . . . **[2]CENTAUREA** (G1,2)

      4′ Phyllary tip entire, tapered to spine . . . . **VOLUTARIA** (G1)

  2′ Phyllaries not spine-tipped

    5. Phyllary tips (at least inner) ± prominently expanded, ± fringed with short spines or irregular teeth

    . . . . **[2]CENTAUREA** (G1,2)

    5′ Phyllary tips not expanded or fringed

      6. Proximal lvs linear to oblong or oblanceolate, often lobed or divided; attachment scar of fr lateral

      . . . . **ACROPTILON**

      6′ Proximal lvs lanceolate to triangular, sharply dentate; attachment scar of fr basal . . . . **SAUSSUREA** (G6)

1′ Lvs spiny or spiny-toothed

  7. Corolla yellow to orange-red

    8. Pappus of many, unequal, narrow scales in several series (sometimes 0), gen 0 on outer frs; fls all fertile;

    fr 4-angled . . . . **CARTHAMUS** (G3,5)

    8′ Pappus of 20 stiff bristles or awns in 2 series, on all frs; outer fls sterile [corolla 3-lobed; ovary

    vestigial]; fr cylindric, 20-ribbed . . . . *Centaurea benedicta* (G2)

  7′ Corolla white to blue, red, or purple

  9. Pappus of long-plumose bristles

    10. Largest lvs toothed to deeply lobed; involucre body 1–6 cm diam; receptacle gen not fleshy; phyllaries

    linear to ovate . . . . **CIRSIUM**

    10′ Largest lvs often ± compound; involucre body 4–15 cm diam; receptacle fleshy; phyllaries ovate . . . . . . **CYNARA**

  9′ Pappus of ± rough or barbed bristles or slender scales

    11. St spiny-winged . . . . **CARDUUS**

    11′ St not spiny-winged

      12. Lf not blotched along veins; heads not long-peduncled; pappus of many, unequal, narrow scales in

      several series . . . . [*Carthamus leucocaulos*]

      12′ Lf white-blotched along veins; heads long-peduncled; pappus of many long bristles, deciduous in a

      ring . . . . **SILYBUM**

Group 5

     Heads discoid; receptacle epaleate (or paleae only marginal and simulating phyllaries), not bristly;

                        pappus 0 or reduced to crown

1. Fls 1–7 per head

  2. Lvs coarsely 1–2-pinnately lobed, spiny-margined—heads 1-fld, grouped in spheric 2° head . . . . . [**ECHINOPS**] (G3,7)

  2′ Lvs entire or toothed

    3. Shrub; lvs sometimes forming spines

      4. Heads in raceme- or panicle-like cluster; 1° lvs soft, entire or distally (2)3(5)-toothed . . . . **[3]ARTEMISIA** (G2)

      4′ Heads in ± tight head-like cluster; 1° lvs ± persistent as spines

        5. Heads 1-fld, clustered in 2° heads; phyllaries ± 15 in several overlapping series . . . . . . **HECASTOCLEIS** (G7)

        5′ Heads 5–6-fld, short-peduncled in terminal clusters; phyllaries 5–6 in 1 series . . . . *Tetradymia comosa*

    3′ Herb; lvs never forming spines

      6. Herbage glabrous . . . . **FLAVERIA** (G12)

6′ Herbage hairy or glandular
  7. Pl densely glandular, soft-hairy to bristly, 5–120 cm . . . . *Madia glomerata*
  7′ Pl not glandular, soft-spreading-hairy to woolly, 1–15 cm
    8. Heads ± sessile; pl 1–2.5 cm, tufted, ± woolly . . . . *Eriophyllum mohavense*
    8′ Heads slender-peduncled; pl 2–15 cm, soft-spreading-hairy . . . . [3]*Pentachaeta exilis* subsp. *exilis* (G6)
1′ Fls gen 8–many per head — heads not in 2° heads
9. Lvs with spine-tipped lobes or teeth. . . . . **CARTHAMUS** (G3,4)
9′ Lvs not spiny
  10. Fr compressed
    11. Ann
      12. Proximal lvs 2–4 × dissected. . . . . *Chaenactis artemisiifolia*
      12′ Proximal lvs entire . . . . [3]*Pentachaeta exilis* subsp. *exilis* (G6)
    11′ Per or subshrub
      13. Phyllaries in 1 series, weakly fused; lvs long-acuminate . . . . **PERICOME** (G6)
      13′ Phyllaries in 2–3 series, ± equal, free; lvs not long-acuminate. . . . . **PERITYLE** (G6)
  10′ Fr not compressed
    14. Lvs 1–2 × pinnately dissected
      15. Involucre ovoid to hemispheric, ± enclosing fls; head ± hemispheric; heads many in spike- to
         panicle-like cluster . . . . [3]**ARTEMISIA** (G2)
      15′ Involucre shallowly cup-like or disk-like; head conic to spheric; heads 1–4 in cyme-like cluster
        16. Mature head spheric; herbage strongly pungent-scented; proximal portion of corolla tube sac-like,
           enclosing top of ovary. . . . . **ONCOSIPHON**
        16′ Mature head ovoid; herbage sweetly scented; proximal portion of corolla tube not enclosing top
           of ovary . . . . **MATRICARIA**
    14′ Lvs entire or toothed to shallowly few-lobed
      17. Shrub, strongly scented. . . . . [3]**ARTEMISIA** (G2)
      17′ Herb, not or weakly scented
        18. Corolla pink to lavender . . . . *Lessingia arachnoidea*
        18′ Corolla yellow to ± red, red-brown or purple
          19. Pl 50–160 cm — heads spheric, many-fld, long-peduncled, solitary or in open clusters
             . . . . *Helenium puberulum*
          19′ Pl 1–15 cm
            20. Pl densely woolly; heads ± sessile in lfy clusters at branch tips . . . . *Eriophyllum pringlei*
            20′ Pl soft-spreading-hairy; heads slender-peduncled. . . . . [3]*Pentachaeta exilis* subsp. *exilis* (G6)

Group 6

Heads discoid; receptacle epaleate (or paleae only marginal and simulating phyllaries), not bristly;
pappus of bristles

1. Lvs and phyllaries dotted or streaked with embedded, translucent oil glands, otherwise glabrous; pl strongly
   pungent-scented . . . . **POROPHYLLUM**
1′ Lvs and phyllaries without embedded, translucent oil glands (sometimes stalked- or sessile-glandular, or
   resin-gland-dotted); pl scented or not
  2. Shrub or subshrub (if per, pappus of 1–2 bristles + crown of scales)
    3. Corolla white or cream to dull purple or ± green
      4. Ovaries very reduced, sterile; style incl . . . . **BACCHARIS** (G2)
      4′ Ovaries forming frs; style often exserted
        5. Fr 10-ribbed . . . . [2]**BRICKELLIA**
        5′ Fr 5-ribbed
          6. Lf sessile, narrowly linear, entire. . . . . *Ericameria albida*
          6′ Lf sessile or petioled, blade diamond-shaped to lanceolate, elliptic, ovate or triangular, gen toothed
            7. Petiole >> lf blade; pappus of bristles and short scales . . . . **PLEUROCORONIS** (G7)
            7′ Petiole 0 or gen ≤ blade; pappus of bristles only
              8. Per or subshrub; lvs serrate, opposite throughout or distal alternate . . . . [2]**AGERATINA**
              8′ Shrub — rigidly branched; lvs spiny-toothed, alternate throughout. . . . . *Hazardia brickellioides*
    3′ Corolla yellow to orange
      9. Pappus bristles 1–2
        10. Lf blade 3–12 cm, petiole 1.5–5 cm; phyllaries in 1 series, weakly fused . . . . [2]**PERICOME** (G5)
        10′ Lf blade gen < 2.5 cm, petiole ≤ 0.6 cm; phyllaries in 2–3 equal series, free . . . . **PERITYLE** (G5)
      9′ Pappus bristles 18–many
        11. Involucre hemispheric to ± spheric; phyllaries elliptic or widely ovate . . . *Acamptopappus sphaerocephalus* (G7)
        11′ Involucre gen cylindric to bell-shaped or obconic (hemispheric); phyllaries linear to narrowly ovate
          12. Phyllaries in 1–2 series, ± equal
            13. Phyllaries 9–18; pl glabrous, resinous-glandular; fls 12–21. . . . . **PEUCEPHYLLUM**
            13′ Phyllaries 4–6; pl ± woolly; fls 4–8 . . . . **TETRADYMIA**

12′ Phyllaries graduated in several series, unequal
  14. Lvs (at least distal) scale-like
    15. Heads solitary; fr appressed-hairy . . . . *Arida carnosa*
    15′ Heads in panicle-like clusters; fr ± glabrous . . . . *Lepidospartum squamatum*
  14′ Lvs linear to oblong, (ob)ovate, or ± round
    16. Lvs toothed
      17. Corolla throat expanded gradually above tube . . . . **HAZARDIA** (G15)
      17′ Corolla throat expanded abruptly above tube. . . . . [2]**ISOCOMA**
    16′ Lvs entire
      18. Phyllaries in 5 ± distinct vertical ranks
        19. Young st densely tomentose, hairs often tightly matted. . . . . [4]**ERICAMERIA** (G15)
        19′ Young st glabrous or puberulent
          20. Lvs glandless or glands, if present, not in deep, round pits . . . . [2]**CHRYSOTHAMNUS**
          20′ Lvs gland-dotted, glands in deep, round pits . . . . [4]**ERICAMERIA** (G15)
      18′ Phyllaries not in distinct vertical ranks
        21. St ± tomentose
          22. Phyllary tips thin, papery, ascending to ± spreading, creamy white; corolla ± 2.5 mm
          . . . . **HELICHRYSUM** (G3)
          22′ Phyllary margins thick, gen appressed or stiffly ascending, gen green; corolla 5–12.5 mm
            23. Phyllaries acuminate . . . . [4]**ERICAMERIA** (G15)
            23′ Phyllaries obtuse to acute
              24. St loosely tomentose; pappus 3–5 mm. . . . . *Isocoma menziesii*
              24′ St closely tomentose exc glabrous ribs; pappus 8–11 mm . . . . *Lepidospartum latisquamum*
        21′ St glabrous or ± hairy but not tomentose
          25. Corolla throat expanded abruptly above tube. . . . . [2]**ISOCOMA**
          25′ Corolla throat expanded gradually above tube
            26. Lvs gland-dotted in deep, round pits. . . . . [4]**ERICAMERIA** (G15)
            26′ Lvs glandless or glands, if present, not in deep, round pits
              27. Involucre 5–10 mm. . . . . [2]**CHRYSOTHAMNUS**
              27′ Involucre 10–17.5 mm
                28. Lvs 30–85 mm, ± lanceolate; fr glabrous; s W&I, n DMtns . . . . **CUNICULOTINUS**
                28′ Lvs 5–15 mm, linear; fr appressed-hairy; s KR, n NCoRH . . . . *Ericameria ophitidis*
2′ Herb
29. Corollas white to maroon or purple
  30. Lvs and phyllaries spiny. . . . . **ONOPORDUM**
  30′ Lvs and phyllaries not spiny
    31. Pappus bristles 2–6
      32. Lvs oblong to ovate, toothed; involucre widely bell-shaped to hemispheric; wet places, GV, SCo
      . . . . **TRICHOCORONIS**
      32′ Lvs linear, gen entire; involucre cylindric to narrowly bell-shaped
        33. Corolla 5–6 mm; DSon. . . . . **MALPERIA** (G7)
        33′ Corolla 2–3 mm; NCoR, SNF, GV, CW. . . . . [2]*Pentachaeta exilis* subsp. *exilis* (G5)
    31′ Pappus bristles gen >> 6
      34. Phyllaries ± equal, in 1–2 series
        35. Fls 2(3); ann; pappus bristles plumose . . . . **DIMERESIA**
        35′ Fls 10–many; per; pappus bristles minutely barbed
          36 Abaxial lf face glabrous or glandular to short-bristly. . . . . **AGERATINA** (2)
          36′ Abaxial lf face white-tomentose. . . . . **LUINA**
      34′ Phyllaries unequal, graduated in several series
        37. Ann; outer corollas often bilateral, ± ray-like . . . . [2]**LESSINGIA** (G1)
        37′ Per; corollas all radial
          38. Herbage glabrous, sticky. . . . . *Baccharis glutinosa*
          38′ Herbage puberulent to tomentose
            39. Lvs entire; involucre 3–9 mm; fls unisexual — pls dioecious . . . . [2]**ANTENNARIA**
            39′ Lvs sharply toothed; involucre 10–15 mm; fls bisexual
              40. Corollas cream, lobes ± deltate; style branches long, club-shaped . . . . [2]**BRICKELLIA**
              40′ Corollas white to gen ± purple, lobes linear; style branches short, oblong. . . . . **SAUSSUREA** (G4)
29′ Corollas yellow to orange or red
  41. Phyllaries in 2–several series, unequal, often strongly graduated
    42. Phyllaries thin-membranous, white to pink or red
      43. St from rhizome; lf rosette withered at fl; cauline lvs long, all exc distal-most ± equal. . . . . . **ANAPHALIS** (G2)
      43′ St from persistent lf rosette or thick, often branched caudex; cauline lvs gen ± reduced, at least
      distal-most short, 1–3 cm. . . . . **ANTENNARIA** (G2)
    42′ Phyllaries not thin-membranous, not white
      44. Pl low, mounded, densely gray-woolly, strongly scented; lvs often wider than long; phyllaries
      in 2 series, outer phyllaries wide, tips spreading. . . . . [2]**PSATHYROTES**

44′ Pl spreading to erect, gen not densely woolly, scented or not; lvs gen longer than wide; phyllaries in 3–several series
  45. Pappus bristles flat, readily deciduous; involucre gummy . . . . **GRINDELIA** (G7,13,15)
  45′ Pappus bristles ± cylindric, ± persistent; involucre gen not gummy
    46. Ann
      47. Outer-fl corollas ± bilateral, often enlarged, ± ray-like . . . . [2]**LESSINGIA** (G1)
      47′ Outer-fl corollas radial
        48. Pl 10–50 cm; phyllaries gen in 3–5 series . . . . [2]***Dieteria canescens*** var. ***shastensis***
        48′ Pl 2–6 cm; phyllaries in 2–3 series . . . . ***Pentachaeta exilis*** (G5)
    46′ Per — outer-fl corollas radial, not enlarged
      49. Phyllary tips spreading
        50. Lvs lanceolate to triangular; corollas creamy yellow . . . . ***Brickellia grandiflora***
        50′ Lvs linear to ± obovate; corollas bright yellow . . . . [2]***Dieteria canescens*** var. ***shastensis***
      49′ Phyllary tips gen appressed (lowest sometimes spreading)
        51. Lvs linear to narrowly oblong; disk wide, button-like . . . . [2]**ERIGERON** (G2,14,15)
        51′ Lvs linear to ovate; disk ± narrow
          52. Hairs smooth; various habitats . . . . **EUCEPHALUS** (G14)
          52′ Hairs minutely knobby (at 20×); gen dry streambanks, gravel bars . . . . ***Heterotheca oregona***
41′ Phyllaries in 1–2 series, ± equal
  53. Pappus bristles ± plumose
    54. Lvs all or mostly basal . . . . **RAILLARDELLA** (G7,10)
    54′ Some or all lvs cauline
      55. Phyllaries in 2 series; pappus bristles ± plumose . . . . [2]**ARNICA** (G15)
      55′ Phyllaries or phyllary-like paleae in 1 series; pappus bristles plumose
        56. Glandless hairs restricted to margins of phyllary-like paleae . . . . ***Anisocarpus scabridus*** (G7)
        56′ Glandless hairs not restricted to margins of phyllary-like paleae . . . . **CARLQUISTIA** (G7)
  53′ Pappus bristles smooth or barbed
    57. Cauline lvs opposite (distal-most sometimes alternate)
      58. Pappus bristles many; phyllaries in 2 series, free . . . . [2]**ARNICA** (G15)
      58′ Pappus bristles 1–2; phyllaries in 1 series, fused at base . . . . [2]**PERICOME** (G5)
    57′ Cauline lvs alternate
      59. Ann or short-lived per with slender taproot
        60. Main phyllaries in 1 series, outer, if present, << inner . . . . [2]**SENECIO** (G2,15)
        60′ Main phyllaries in 1–2 series, ± equal
          61. Lf thread-like to linear . . . . [2]***Pentachaeta exilis*** subsp. ***exilis*** (G5)
          61′ Lf oblanceolate to ovate or wider
            62. Pappus bristles ± free; heads short-peduncled; lvs entire or blunt-toothed . . . . [2]**PSATHYROTES**
            62′ Pappus bristles fused at base in 5 groups; heads long-peduncled; lvs sharply toothed or lobed
            . . . . **TRICHOPTILIUM** (G1)
      59′ Per, often with caudex or rhizome
        63. Lvs palmately veined and lobed
          64. Pl ± erect . . . . **CACALIOPSIS**
          64′ Pl a twining vine . . . . **DELAIREA**
        63′ Lvs pinnately veined (or lateral veins obscure), entire to pinnately lobed
          65. Pappus double, outer pappus of minute bristles or scales . . . . [2]**ERIGERON** (G2,14,15)
          65′ Pappus single (double, bristles ± equal)
            66. Distal lf blades 1.5–4 cm; phyllary tips green or red . . . . **PACKERA** (G15)
            66′ Distal lf blades 6–25 cm; phyllary tips black or green . . . . [2]**SENECIO** (G2,15)

Group 7

Heads discoid; receptacle epaleate (or paleae only marginal and simulating phyllaries), not bristly;
pappus of scales or awns

1. Heads ± sessile, sometimes in dense 2° clusters
  2. Heads 1-fld, in dense 2° heads; pl ± spiny
    3. Herb, 10–20 dm, stout, erect; lvs large, 1–2-pinnately lobed, spiny-margined; 2° heads spheric
    . . . . **[ECHINOPS]** (G3,5)
    3′ Shrub, 4–7 dm; lvs small, spine-tipped, margins weakly spiny; 2° heads not spheric, ± concealed by
    veiny bracts . . . . **HECASTOCLEIS** (G5)
2′ Heads 3–20-fld, solitary or clustered; pl not spiny
  4. Lf tips gen 3-lobed; pappus not falling in a ring . . . . **ERIOPHYLLUM** (G12,13)
  4′ Lf tips entire; pappus falling in a ring . . . . **OROCHAENACTIS**

1′ Heads peduncled
  5. Phyllaries unequal, graduated in 2–many series, outer << inner
    6. Pappus of short scales alternating with longer awns or bristles
      7. Ann; lvs linear, gen entire . . . . **MALPERIA** (G6)
      7′ Per or subshrub; lvs lanceolate to ± diamond-shaped, gen few-toothed . . . . **PLEUROCORONIS** (G6)
    6′ Pappus of readily deciduous awns or narrow, bristle-like scales, or of scales that are ± dissected into bristles
      8. Shrub; phyllary tips appressed. . . . . *Acamptopappus sphaerocephalus* (G6)
      8′ Herb; phyllary tips often spreading to recurved
        9. Involucre gummy; pappus awns readily deciduous. . . . . **GRINDELIA** (G6,13,15)
        9′ Involucre hairy or glandular; pappus scales often ± dissected into bristles . . . . **LESSINGIA** (G1,6)
  5′ Phyllaries or phyllary-like paleae ± equal (or outer > inner), in 1–3 series
    10. Head ± spheric; lvs decurrent . . . . **HELENIUM** (G12,13)
    10′ Head cylindric to hemispheric; lvs not decurrent
      11. Phyllary margins ± scarious, white to yellow, red, or purple
        12. Pappus scales bristle-tipped; corollas cream to white or purple-tinged . . . . **HYMENOTHRIX** (G1)
        12′ Pappus scales obtuse to acute; corollas gen yellow
          13. Ann; pappus scales gen 8 . . . . **BAHIA**
          13′ Per; pappus scales 12–22 . . . . **HYMENOPAPPUS**
      11′ Phyllary or phyllary-like palea margins not evidently scarious, gen ± green
        14. Corollas white to purple
          15. Fr 3–9 mm; pappus scale midrib 0; lvs often toothed or lobed . . . . **CHAENACTIS** (G1)
          15′ Fr 10–15 mm; pappus scale midrib well developed; lvs entire . . . . **PALAFOXIA** (G1)
        14′ Corollas yellow
          16. Outer corollas ± bilateral, much enlarged, ray-like; pappus scales entire or toothed. . . . *Chaenactis glabriuscula*
          16′ Outer corollas gen radial, not or only slightly enlarged; pappus scales plumose or dissected into bristles
            17. Ann or short-lived per; rhizome 0
              18. Lf teeth and lobes blunt; involucre densely glandular; pappus scales plumose. . . . . *Layia discoidea*
              18′ Lf teeth and lobes sharp; involucre not glandular; pappus scales dissected into bristles
                . . . . **TRICHOPTILIUM** (G1,6)
            17′ Per with rhizome
              19. Lvs mostly or all basal. . . . . **RAILLARDELLA** (G6,10)
              19′ Lvs basal and cauline, well distributed on st
                20. Glandless hairs restricted to margins of phyllary-like paleae . . . . *Anisocarpus scabridus* (G6)
                20′ Glandless hairs not restricted to margins of phyllary-like paleae . . . . **CARLQUISTIA** (G6)

Group 8

    Heads liguliflorous; all fls of head with petal-like, 5-lobed ligules; corollas gen readily withering; sap gen milky

1. Receptacle with long paleae among fls
  2. Lvs mostly cauline, spiny, thistle-like; paleae wide, individually enclosing and falling with frs . . . . **SCOLYMUS**
  2′ Lvs mostly or all basal, not spiny; paleae narrow, membranous, not enclosing frs
    3. Peduncle bracted; scape branched or unbranched. . . . . **HYPOCHAERIS**
    3′ Peduncle bractless; scape unbranched
      4. Pappus bristles not longer on 1 side of frs, ± minutely barbed; phyllary margins not wide and papery
      . . . . [2]**AGOSERIS**
      4′ Pappus bristles much longer on 1 side of all frs, plumose; phyllary margins papery, much wider than
        midrib. . . . . [2]**ANISOCOMA**
1′ Receptacle epaleate (sometimes bristly or minutely scaly)
  5. Pappus 0 (or occ low crown in *Phalacroseris*)
    6. Heads solitary; st lfless . . . . **PHALACROSERIS**
    6′ Heads few to many in cyme-like or panicle-like infl; st often lfy
      7. Outer frs 10–15 mm, rigidly spreading, individually enfolded by inner phyllaries. . . . . **RHAGADIOLUS**
      7′ Outer fr 3–5 mm, gen straight (curved), not individually enfolded by phyllaries
        8. Lvs mostly basal; corolla white; D. . . . . **ATRICHOSERIS**
        8′ Lvs mostly cauline; corolla yellow; NW, SN, GV, CCo, SnFrB, SW . . . . **LAPSANA**
  5′ Pappus present
    9. Corolla blue
      10. Pappus of short scales. . . . . **CICHORIUM**
      10′ Pappus of long bristles . . . . [2]**LACTUCA**
    9′ Corolla yellow or white to purple
      11. Pappus of scales, stiff awns (sometimes with shorter bristles), or awn-tipped scales, not plumose (exc
        Microseris nutans, *Microseris sylvatica*, inner fr pappus of *Leontodon*)
        12. Outer frs often enfolded by phyllaries; pappus of outer and inner frs dissimilar
          13. Lvs basal and cauline; peduncle thickened below fr head; inner fr pappus bristles minutely barbed
          . . . . **HEDYPNOIS**

13′ Lvs basal; peduncle not or scarcely thickened; inner fr pappus bristles plumose . . . . [3]**LEONTODON**
12′ Outer frs gen not enfolded by phyllaries; pappus of outer and inner frs ± alike
  14. Corolla pink to ± white; st much-branched — cauline lvs short, linear, entire . . . . **CHAETADELPHA**
  14′ Corolla white to yellow or orange; st branches 0–few
    15. Outer phyllaries long, linear; pappus scales stiff, bristle-like . . . . **TOLPIS**
    15′ Outer phyllaries ≤ inner, gen wider than linear (if linear, short); pappus scales few to many, bristle-tipped
      16. Per, scapose; rosette lvs entire, narrow, long-tapered; heads erect; pappus scales 10–30, narrowly lanceolate, silvery, bristle-tip not plumose . . . . *Nothocalais troximoides*
      16′ Ann or per, scapose or not; lvs entire to lobed; pappus scales 1–many, silvery to ± black, bristle-tip smooth, barbed, or plumose
        17. Ann; head erect; involucre glabrous, outer phyllaries not < 1/4 inner; pappus scales 5, lanceolate, silvery, bristle short, smooth, from evenly notched scale tip . . . . **UROPAPPUS**
        17′ Ann or per; head ± nodding in bud; involucre glabrous to hairy, outer phyllaries often < 1/4 inner; pappus scales 1–30, variously colored, tip entire to unevenly cut, bristle from tip smooth to plumose
          18. Per with fleshy taproot(s); ligules >> involucre; pappus scales 5–30 . . . . [3]**MICROSERIS**
          18′ Ann; ligules ± = involucre; pappus scales 5 (often < 5 in *Microseris douglasii*)
            19. Head gen strongly nodding in bud; pappus scales gen deltate to ovate (if lanceolate, tapered to bristle); lvs basal. . . . . [3]**MICROSERIS**
            19′ Head gen not strongly nodding in bud; pappus scales narrowly lanceolate, tip notched or irregularly cut; cauline lvs sometimes present . . . . **STEBBINSOSERIS**
11′ Pappus of simple bristles or of ± plumose bristles or awns
20. Pappus bristles or awns plumose
  21. Pappus bristles wide at base, scale-like . . . . [3]**MICROSERIS**
  21′ Pappus of bristle or awns, slender or thick at base, not scale-like
    22. Pappus bristles very unequal
      23. Pappus bristles much longer on 1 side of all frs; phyllary margins papery, much wider than midrib . . . . [2]**ANISOCOMA**
      23′ Outer fr pappus of short scales, inner fr pappus of long and short bristles; phyllary margins not wide and papery . . . . [3]**LEONTODON**
    22′ Pappus bristles ± equal
      24. Lvs linear to lanceolate or ovate-elliptic, grass-like, entire or few-toothed; involucre 2.5–5 cm
        25. Per; phyllary series several; fr beak 0. . . . . **SCORZONERA**
        25′ Ann or bien; phyllary series 1; fr beak > body. . . . . **TRAGOPOGON**
      24′ Lvs lanceolate to elliptic (or scale-like), larger lvs ± toothed or lobed; involucre 0.5–2 cm
        26. Corollas white to pink or lavender
          27. Fr beaked; corollas white or cream, sometimes red-veined . . . . **RAFINESQUIA**
          27′ Fr beak 0; corollas pink to pale lavender . . . . **STEPHANOMERIA**
        26′ Corollas yellow
          28. Lvs in basal rosette; heads gen 1 on scapose peduncle — outer phyllaries inconspicuous. . . [3]**LEONTODON**
          28′ Lvs basal and cauline, or basal 0 at fl; heads 1–many, not on scapose peduncle
            29. Phyllaries in 2 series, outer wide, free, inner narrow . . . . **HELMINTHOTHECA**
            29′ Phyllaries in 1 series, wide, bases slightly fused . . . . **UROSPERMUM**
20′ Pappus bristles smooth or barbed, stiff or soft
30. Fr ± compressed
  31. Fr beaked (beak sometimes short, thick) . . . . [2]**LACTUCA**
  31′ Fr beak 0 . . . . **SONCHUS**
30′ Fr not compressed
  32. St much-branched
    33. St thorny; corollas pink or lavender to red-purple (white). . . . . **PLEIACANTHUS**
    33′ St not thorny; corollas yellow or white to light pink
      34. Fls 3–4 per head, corollas white or light pink; ann, 5–30 cm . . . . **PRENANTHELLA**
      34′ Fls 6–60 per head, corollas white or yellow; bien or per, 10–150 cm
        35. Fr beaked . . . . **CHONDRILLA**
        35′ Fr beak 0 . . . . [2]**HIERACIUM**
  32′ St little-branched, not thorny; corolla yellow, white, lavender, or rosy to ± purple
    36. Shrub; s ChI (San Clemente Island) . . . . **MUNZOTHAMNUS**
    36′ Herb
      37. Fr widely cylindric or base ± narrowed, not beaked
        38. Some or all pappus bristles falling from fr — long glandless hairs 0 . . . . [2]**MALACOTHRIX**
        38′ Pappus bristles persistent
          39. Pl long-hairy, esp below; heads few to many; st gen branched. . . . . [2]**HIERACIUM**
          39′ Pl glabrous or minutely puberulent; heads solitary on scapose peduncle
            40. Ann; pappus bristle 5 or fewer . . . . *Microseris douglasii*

40′ Per; pappus bristles > 20
  41. Pappus bristles ± brown, barbed . . . . ***Microseris borealis***
  41′ Pappus bristles white, smooth . . . . [2]***Nothocalais alpestris***
37′ Fr tapered at both ends (or distally), often beaked
  42. Pappus bristles deciduous, sometimes a few persistent
    43. Fr not beaked . . . . [2]**MALACOTHRIX**
    43′ Fr short-beaked
      44. Fr tapered to beak; lf margins not hard . . . . **CALYCOSERIS**
      44′ Fr abruptly beaked; lf margins white, hard . . . . **GLYPTOPLEURA**
  42′ Pappus bristles persistent
    45. St branched, lfy (lvs often reduced); beak present or 0 . . . . **CREPIS**
    45′ St unbranched, lfless (exc some ann *Agoseris*)
      46. Phyllaries unequal, outer << inner, inner equal; fr beak long, slender . . . . **TARAXACUM**
      46′ Phyllaries equal or ± graduated in several series; fr beak 0 or short to long, slender to stout
        47. Pl gen ± hairy; outer phyllaries gen < inner, often striped ± purple, speckled or not . . . . . . [2]**AGOSERIS**
        47′ Pl ± glabrous (or petioles ciliate); outer phyllaries ± = inner, finely speckled purple —
        KR, n&c SNH . . . . [2]***Nothocalais alpestris***

Group 9

All fls of head with 2-lipped corollas

1. Per; corolla white to pink-purple; SCoR, SW . . . . **ACOURTIA**
1′ Shrub; corolla yellow; D . . . . **TRIXIS**

Group 10

Heads radiate; phyllaries in 1(2) series, individually subtending ray fls;
receptacle paleate (at least between ray and disk fls)

1. Phyllary margins ± flat, not clasping or enclosing ray ovaries
  2. Pappus 2–4 mm, of plumose bristles . . . . ***Layia hieracioides***
  2′ Pappus 0 or < 1 mm, of scales or awns
    3. Lvs woolly; ray corollas thin, deciduous; receptacle center with 1–6 palea-like scales
    . . . . ***Eriophyllum ambiguum*** var. ***paleaceum***
    3′ Lvs scabrous; ray corollas leathery, persistent; receptacle paleate throughout, paleae awn-tipped . . . . **SANVITALIA**
1′ Phyllary margins folded, ± clasping or enclosing ray ovaries
  4. Mature ray fr compressed front-to-back, adaxial face gen ± covered by phyllary margins
    5. Disk fls 1(2); ann 1–20 cm . . . . **HEMIZONELLA** (G2)
    5′ Disk fls 3–120+; ann or per 2–150 cm
      6. Per, rhizomed; disk corolla white . . . . **HOLOZONIA**
      6′ Ann; disk corolla yellow (sometimes aging ± red)
        7. Pappus of 10 obtuse scales, 3–13 mm; fr 10-ribbed . . . . **ACHYRACHAENA**
        7′ Pappus 0 or of awns, bristles, or acute scales, 0.5–7 mm; fr not ribbed
          8. Ray fls 5; disk fls 6, staminate; pappus 0 . . . . **LAGOPHYLLA**
          8′ Ray fls 3–27; disk fls 5–125, bisexual; pappus of (0)1–32 awns, bristles, or scales . . . . . **LAYIA**
  4′ Mature ray fr compressed side-to-side or ± 3-angled or round in ×-section, adaxial face covered by
    phyllary margins or not
    9. Disk fl style bristly-puberulent proximal to minutely notched tip . . . . **BLEPHARIPAPPUS**
    9′ Disk fl style glabrous proximal to tapered branches
      10. Per; lvs all or mostly basal, cauline 0 or proximal, few; disk pappus of narrowly awl-shaped or ±
        bristle-like, ciliate-plumose scales . . . . **RAILLARDELLA** (G6,7)
      10′ Ann, lfy-stemmed per, or subshrub to shrub; disk pappus 0 or of scales, often not awl-shaped or
        bristle-like and ciliate-plumose
        11. Distal lvs and peduncle bracts gen spine-tipped; disk fls staminate, most or all subtended by paleae
        . . . . **CENTROMADIA**
        11′ Distal lvs and peduncle bracts not spine-tipped; disk fls bisexual or staminate, outermost to all
        subtended by paleae
          12. Rays yellow, sometimes proximally ± red; ray fr gen compressed side-to-side or ± 3-angled in
          ×-section, abaxially rounded, adaxially ± 2-faced, angle between adaxial faces 15–70°; peduncle
          bracts not tipped by pit- or tack-gland
            13. Disk pappus 0 . . . . **MADIA** (G2)
            13′ Disk pappus of 5–21 wide to bristle-like scales
              14. Per
                15. Anthers yellow to ± brown; ray fr short-beaked; involucre bell-shaped, ellipsoid, or spheric
                . . . . **ANISOCARPUS**
                15′ Anthers ± dark purple; ray fr beak 0; involucre bell-shaped to hemispheric . . . . **KYHOSIA**

    14′ Ann
      16. Anthers ± yellow to ± brown . . . . **²HARMONIA**
      16′ Anthers ± dark purple . . . . **JENSIA**
  12′ Rays yellow, white, or rose; ray fr round in ×-section or nearly so (exc ± flattened adaxially) or ±
      3-angled [abaxially gen ± widely 2-faced, angle between those faces gen 90+°, adaxially gen ±
      flattened to slightly bulging]; peduncle bracts sometimes tipped by pit- or tack-gland
    17. Ray fr ± completely enfolded by a phyllary; rays yellow, sometimes ± red at base
      18. Disk pappus of fringed or plumose scales . . . . **²HARMONIA**
      18′ Disk pappus 0 . . . . *Madia anomala*
    17′ Ray fr gen partly enfolded by a phyllary, adaxial face exposed; rays yellow, white, or rose
      19. Ann; lvs ± linear, gen narrowly so; ray corolla lobed (1/4)1/3–3/4+ length, lobes often spreading;
        peduncle bracts gen tack-gland-tipped (exc in *Osmadenia*, ray lobes widely spreading); disk fr not
        strongly ribbed
      20. Ray fr beak 0; tack glands present . . . . **CALYCADENIA**
      20′ Ray fr short-beaked; tack glands 0 . . . . **OSMADENIA**
      19′ Ann or subshrub to shrub; lvs linear or wider; ray corolla lobed gen 1/10–1/2 length, lobes ±
        parallel; peduncle bracts not tack-gland-tipped unless frs 10-ribbed
      21. Ray gen white or yellow, often abaxially red- or purple-veined; ray fr beak 0 or inconspicuous, straight
        22. Paleae restricted to outermost disk fls; disk fls bisexual; ray fr hairy . . . . **BLEPHARIZONIA**
        22′ Paleae throughout receptacle, subtending all disk fls; disk fls staminate; ray fr glabrous . . . . **HEMIZONIA**
      21′ Ray yellow, not abaxially purple-veined; ray fr beak adaxial, ascending
        23. Ann or subshrub to shrub; peduncle bracts not pit-gland-tipped; if ann, paleae subtending only
          outermost disk fls. . . . . **DEINANDRA**
        23′ Ann; peduncle bracts pit-gland-tipped; paleae throughout receptacle, subtending all or most
          disk fls . . . . **HOLOCARPHA**

Group 11

Heads radiate; phyllary series 2 or more, none or only some individually subtending ray fls;
receptacle paleate, gen ± throughout (only between ray and disk fls in *Rigiopappus*)

1. Rays white to pink or purple
  2. Disk corollas white or pink
    3. Lvs finely dissected, aromatic; heads many in flat-topped clusters; rays ovate to round. . . . . **ACHILLEA**
    3′ Lvs entire or toothed, not aromatic; heads solitary or in small cymes; rays narrowly linear . . . . **ECLIPTA**
  2′ Disk corollas yellow
    4. Lvs entire or toothed
      5. Lvs opposite, ± ovate . . . . **GALINSOGA**
      5′ Lvs alternate, linear or lance-linear to narrowly oblanceolate . . . . **²RIGIOPAPPUS** (G12,13)
    4′ Lvs pinnately lobed or dissected to compound
      6. Phyllaries in 2 very different series [inner ± membranous]
        7. Lvs pinnate; lflets flat, serrate; rays short, inconspicuous. . . . . *Bidens pilosa*
        7′ Lvs 1–3-pinnately dissected, segments narrowly linear, entire; rays long, conspicuous. . . . . **COSMOS**
      6′ Phyllaries graduated in 2–several series, unequal, margins scarious or transparent
        8. Fr initially hairy, densely tomentose in age. . . . . **LASIOSPERMUM**
        8′ Fr glabrous
          9. Base of corolla narrow . . . . **ANTHEMIS**
          9′ Base of corolla wide, enclosing fr tip. . . . . **CHAMAEMELUM**
1′ Rays yellow to orange (sometimes multi-colored)
  10. Paleae flat, linear to ovate, not folded around disk ovaries
    11. Phyllaries ± alike, in 2+ series
      12. Phyllaries graduated in several series, unequal, margins scarious or transparent — pappus crown-like,
        gen 0.2–0.3 mm . . . . **[COTA]**
      12′ Phyllaries not strongly graduated, ± equal, margins not scarious or transparent (exc inner phyllaries
        of *Rigiopappus*)
        13. Lvs oblong to ovate, entire to pinnately lobed; pappus scales awn-tipped . . . . **[GAILLARDIA]** (G13)
        13′ Lvs linear, entire; pappus scales stiff, tapered . . . . **²RIGIOPAPPUS** (G12,13)
    11′ Phyllaries in 2 very different series, outer with lf-like texture, inner ± membranous
      14. Pappus of awns with reflexed barbs. . . . . **BIDENS** (G2,3)
      14′ Pappus 0 or of flat scales
        15. Lvs alternate or all basal; ray fls fertile . . . . **LEPTOSYNE**
        15′ Lvs opposite; ray fls sterile. . . . . **COREOPSIS**

10′ Paleae each folded around a disk fl ovary
  16. Ray fls fruiting [style present, ovary well developed]
    17. Ann; disk fr compressed, winged . . . . *Verbesina encelioides* subsp. ***exauriculata***
    17′ Per; disk fr gen 4-angled
      18. Lvs mostly ± basal, cauline 0 or few; pappus 0 . . . . **BALSAMORHIZA**
      18′ Lvs basal and cauline; pappus of scales or 0 . . . . **WYETHIA**
  16′ Ray fls sterile [ovary vestigial, style gen 0]
    19. Fr strongly compressed, margin ± thin
      20. Fr margin ciliate
        21. Subshrub or shrub . . . . **ENCELIA**
        21′ Ann or per (sometimes with woody caudex)
          22. Lvs all ± basal, gray or silvery, entire . . . . [2]**ENCELIOPSIS**
          22′ Lvs mostly cauline, green, entire or toothed . . . . *Geraea canescens*
      20′ Fr margin not ciliate
        23. Lvs all ± basal, silvery or gray; fr margin corky . . . . [2]**ENCELIOPSIS**
        23′ Lvs basal and cauline or all cauline, green; fr margin thin or winged
          24. Per; lvs basal and cauline, entire . . . . **HELIANTHELLA**
          24′ Subshrub; lvs cauline, some often toothed . . . . *Verbesina dissita*
    19′ Fr weakly compressed or not, margin gen not thin
      25. Receptacle widely cylindric to conic gen ± spheric; disk corollas yellow-green or dark brown-purple
        26. Involucre at early fl 8–16 mm diam; fr compressed, ciliate abaxially — phyllary series unequal,
          outer >> inner . . . . **RATIBIDA**
        26′ Involucre at early fl 15–30+ mm diam; fr not compressed, not ciliate abaxially . . . . **RUDBECKIA**
      25′ Receptacle gen flat to convex, if conic, disk corollas yellow or yellow-orange
        27. Pappus 0 . . . . **HELIOMERIS**
        27′ Pappus present
          28. Herb (subshrub); pappus scales gen readily deciduous . . . . **HELIANTHUS**
          28′ Subshrub or shrub; pappus scales persistent or readily deciduous (if subshrub, pappus persistent)
            29. Lf oblong or narrowly lanceolate to ovate or triangular, base obtuse or rounded to truncate or
              cordate . . . . **BAHIOPSIS**
            29′ Lf elliptic to widely obovate, base tapered, wedge-shaped . . . . **VIGUIERA**

Group 12

Heads radiate; receptacle epaleate; pappus 0 or a low crown

1. Lvs opposite, at least proximally
  2. Distal lvs opposite
    3. Phyllaries in 3+ series, widely overlapping; per, fleshy, coastal . . . . **JAUMEA**
    3′ Phyllaries in 1–2 series, barely overlapping; ann or per, gen not fleshy, coastal or interior
      4. Pl white-hairy and sessile-glandular; old ray corolla papery, persistent on fr when dry; disk fls
        staminate . . . . *Arnica dealbata*
      4′ Pl glabrous or ± hairy; old ray corolla deciduous from fr; disk fls gen fruiting
        5. Ray fl gen 1 per head; ray < 1 mm . . . . **FLAVERIA** (G5)
        5′ Ray fls 4–many; ray gen >> 1 mm
          6. Lf margin without embedded oil glands, base not bristly-ciliate; herbage gen not scented;
          spring-fl . . . . [2]**LASTHENIA** (G2,13)
          6′ Lf margin with embedded oil glands, base bristly-ciliate; herbage strongly scented;
          summer/fall-fl . . . . **PECTIS** (G13,15)
  2′ Distal lvs alternate
    7. Ray white
      8. Ray not red-veined abaxially; e SnFrB, SnGb, SnBr, PR, SNE, D . . . . [2]*Eriophyllum wallacei*
      8′ Ray red-veined abaxially; SCoRI, TR, SnJt . . . . [2]*Syntrichopappus lemmonii*
    7′ Ray yellow
      9. Ray corolla with small lobe opposite ray . . . . [2]**MONOLOPIA**
      9′ Ray corolla without small lobe opposite ray
        10. Lvs mostly or all opposite . . . . [2]**LASTHENIA** (G2,13)
        10′ Lvs mostly alternate
          11. Lvs glabrous . . . . [2]**AMBLYOPAPPUS** (G2,13)
          11′ Lvs ± tomentose . . . . [2]**ERIOPHYLLUM** (G7,13)
1′ Lvs alternate throughout or in rosettes
  12. Ray ≤ 1 mm
    13. Shrub . . . . *Artemisia bigelovii*
    13′ Herb
      14. Lf finely dissected; phyllaries unequal, overlapping in several series, scarious-margined . . . . [2]**TANACETUM** (G2)

14′ Lf entire to ± lobed; phyllaries ± equal, in 1–3 series, margins barely scarious or not (widely scarious in *Pentachaeta*)

  15. Phyllaries in 2–3 series; lvs linear, entire . . . . **²PENTACHAETA** (G2,14,15)

  15′ Phyllaries in 1 series; proximal lvs often toothed or lobed

    16. Pl glabrous or sticky-glandular; fr 1.5–2 mm . . . . **²AMBLYOPAPPUS** (G2,13)

    16′ Pl ± woolly; fr 2.5–3 mm . . . . *Monolopia congdonii* (G2,13)

12′ Ray > 1 mm

  18. Ray white to purple (sometimes multi-colored)

    19. Phyllaries unequal, in 2+ series, gen conspicuously overlapping, gen scarious-margined

      20. Subshrub — ray fr 1–3-winged . . . . **ARGYRANTHEMUM**

      20′ Herb

        21. Lf 1–3-pinnately dissected; heads few to many in rounded to ± flat-topped clusters

          22. Ultimate lf segments flat. . . . . *Tanacetum parthenium* (G13)

          22′ Ultimate lf segments thread-like

            23. Herbage strongly scented; fr obconic, weakly compressed or ± round in ×-section, faintly 5-ribbed, glandular between ribs, ribs without embedded resin glands . . . . *Matricaria chamomilla*

            23′ Herbage not scented; fr 3-angled, ± compressed, 3–5-ribbed, minutely roughened between ribs, 2–3 ribs distally with embedded resin glands . . . . **TRIPLEUROSPERMUM**

        21′ Lf entire, serrate, or pinnately lobed; heads solitary or few in open, cyme-like cluster

          24. Per; heads 6–9 cm diam . . . . **LEUCANTHEMUM**

          24′ Ann; heads 2.5–4(5.5) cm diam

            25. Outer phyllary margins and tips not scarious; ray adaxially uniformly pale yellow or with darker yellow proximal band, abaxially steely blue; fr densely woolly . . . . *Arctotheca calendula*

            25′ Outer phyllary margins and tips scarious; ray adaxially and abaxially white (aging pink), proximally yellow; fr faces glabrous . . . . **[MAURANTHEMUM]**

    19′ Phyllaries ± equal, in 1–2 series, gen little overlapping, gen not scarious-margined

      26. Ray 25–40 mm; coarse per or subshrub . . . . **²DIMORPHOTHECA**

      26′ Ray < 10 mm; ann or slender per

        27. Lf widely ovate, coarsely dentate or palmately lobed, abruptly petioled . . . . *Perityle emoryi* (G2,13,14)

        27′ Lf linear to obovate, entire to pinnately lobed (tip sometimes 3-lobed), sessile or tapered to petiole

          28. Proximal lvs pinnately lobed . . . . **²BLENNOSPERMA**

          28′ Lvs entire, toothed, or 3-lobed

            29. Per; moist habitats . . . . **BELLIS**

            29′ Ann; dry habitats

              30. Pl tomentose or becoming glabrous

                31. Ray not red-veined abaxially; e SnFrB, SnGb, SnBr, PR, SNE, D. . . . . **²*Eriophyllum wallacei***

                31′ Ray red-veined abaxially; SCoRI, TR, SnJt . . . . **²*Syntrichopappus lemmonii***

              30′ Pl short-hairy

                32. Pl tufted or gen prostrate to ascending; D . . . . **MONOPTILON** (G13,14)

                32′ Pl ± erect; CW . . . . **²PENTACHAETA** (G2,14,15)

18′ Ray yellow to orange (occ multi-colored)

  33. Fr drupe; shrub. . . . . **CHRYSANTHEMOIDES**

  33′ Fr achene; gen herb or subshrub (exc *Venegasia*)

    34. Phyllaries unequal, overlapping in several series

      35. Phyllary margins conspicuously scarious

        36. Pl sticky-glandular . . . . **HETERANTHEMIS** (G13)

        36′ Pl not glandular

          37. Heads solitary or in open, lfy cymes; outer frs 3-angled. . . . . **GLEBIONIS**

          37′ Heads in dense, flat-topped clusters; frs cylindric to obconic. . . . . **²TANACETUM** (G2)

      35′ Phyllary margins not or barely scarious

        38. Lf oblong to obovate, pinnately lobed, tomentose at least below; phyllaries narrow, appressed or tips spreading or reflexed . . . . **ARCTOTHECA**

        38′ Lf widely ovate, entire to coarsely toothed, puberulent; phyllaries wide, outer spreading or reflexed . . . . **VENEGASIA**

    34′ Phyllaries ± equal (or outer longer), in 1–3 series, slightly overlapping or not

      39. Ray fr prickly, knobby, papillate, or ridged on back; phyllary margins gen narrowly scarious

        40. Leaves glabrous or sparsely tomentose; involucre gen 5–9 mm . . . . **²BLENNOSPERMA**

        40′ Leaves puberulent and/or glandular; involucre 7–20 mm

          41. Ray fr strongly curved [crescent-shaped or almost loop-like]; disk fls staminate. . . . . **CALENDULA**

          41′ Ray fr 3-angled, not curved; disk fls fruiting, fr flat . . . . **²DIMORPHOTHECA**

      39′ Ray fr gen smooth on back; phyllary margins gen not scarious

        42. Head hemispheric to ovoid or ± spheric; phyllaries spreading or reflexed . . . . **HELENIUM** (G7,13)

42′ Head cylindric to hemispheric; phyllaries erect or ascending
  43. Pl green, glabrous to strigose, glandular or puberulent
    44. Phyllaries in 2–3 series; pl glandular . . . . **AMAURIOPSIS** (G13)
    44′ Phyllaries in 1 series; pl not glandular, ± glabrous . . . . *Blennosperma nanum*
  43′ Pl ± gray-tomentose
    45. Rays when dry persistent, papery, reflexed . . . . **BAILEYA**
    45′ Rays when dry withering, deciduous
      46. Ray corolla with small lobe opposite ray . . . . ²**MONOLOPIA**
      46′ Ray corolla without small lobe opposite ray
        47. Ann, per, or subshrub; fr 4(5)-angled or outer frs compressed. . . . . ²**ERIOPHYLLUM** (G7,13)
        47′ Ann; fr ± compressed . . . . **PSEUDOBAHIA**

Group 13

Heads radiate; receptacle epaleate; pappus of scales or awns

1. Lvs and phyllaries dotted or streaked with embedded, translucent oil glands — herbage strongly scented
  2. Ray pale pink to bright pink-purple. . . . . **NICOLLETIA** (G14)
  2′ Ray yellow to red
    3. Phyllaries in 1 series
      4. Phyllaries free, falling with ray frs; disk pappus gen of bristles. . . . . **PECTIS** (G12,15)
      4′ Phyllaries fused, persistent; disk pappus of scales . . . . **TAGETES**
    3′ Phyllaries in ± 2 series — pappus scales ± divided into 5–10 bristles
      5. Ann; peduncles 1–5(10) mm . . . . [**DYSSODIA**]
      5′ Subshrubs; peduncles 2–15 cm
        6. Lf toothed or divided into ± flat segments; involucre 10–18 mm . . . . **ADENOPHYLLUM** (G15)
        6′ Lf divided into stiff, needle-like segments; involucre 5–6 mm . . . . **THYMOPHYLLA**
1′ Lvs and phyllaries without embedded oil glands (sometimes glandular-hairy or gland-dotted)
  7. Ray white to pink or purple (occ multi-colored)
    8. Bien or per
      9. Lf compound; pappus scales short. . . . . *Tanacetum parthenium* (G12)
      9′ Lf simple; pappus scales long, bristle-like . . . . **TOWNSENDIA** (G14)
    8′ Ann
      10. Lf wide, palmately lobed or toothed . . . . *Perityle emoryi* (G2,12,14)
      10′ Lf ± narrow, entire to fiddle-shaped or pinnately lobed
        11. Pl puberulent or short-stiff-hairy
          12. Pl gen prostrate; ray corollas 5–12 mm . . . . **MONOPTILON** (G12,14)
          12′ Pl erect; ray corollas < 2 mm . . . . ²**RIGIOPAPPUS** (G11,12)
        11′ Pl tomentose
          13. Ray 1.5–2.5 mm; pappus scales 2. . . . . ²**EATONELLA** (G2)
          13′ Ray 3–30 mm; pappus scales 6–16
            14. Phyllaries in 3–6+ series; ray (17)20–30 mm. **ARCTOTIS**
            14′ Phyllaries in 1 series; ray 3–7 mm . . . . *Eriophyllum lanosum*
7′ Ray yellow to red, sometimes multi-colored or fading to cream
  15. Receptacle with stiff awn-like projections . . . . [**GAILLARDIA**] (G11)
  15′ Receptacle smooth or pitted and/or minutely hairy, without stiff awn-like projections
    16. Phyllaries in several series, strongly graduated
      17. Lf abaxially tomentose; st creeping or decumbent to erect . . . . **ARCTOTHECA**
      17′ Lf glabrous to short-hairy; st erect or ascending (rarely decumbent)
        18. Phyllary tips spreading to recurved; involucre gummy — pappus of 2–6 slender deciduous awns
        . . . . **GRINDELIA** (G6,7,15)
        18′ Phyllary tips ± appressed; involucre not or barely gummy
          19. Ann; fr winged; lvs toothed or lobed . . . . **HETERANTHEMIS** (G12)
          19′ Per to shrub; fr not winged; lvs entire
            20. Body of head ± spheric; disk fls 30–80 . . . . *Acamptopappus shockleyi* (G15)
            20′ Body of head narrower than length; disk fls 1–13
              21. Lf elliptic to obovate; disk pappus scales twisted, bristle-like. . . . . **AMPHIPAPPUS** (G15)
             21′ Lf linear; disk pappus scales straight, wide, flat. **GUTIERREZIA**
  16′ Phyllaries in 1–3(4) series, gen ± equal (sometimes outer longer), not strongly graduated
    22. Lvs basal
      23. Head 3.5–8 cm diam; ray base gen dark-spotted; fr hairs concealing pappus; sap milky . . . . **GAZANIA**
      23′ Head gen < 3 cm diam; ray base not dark-spotted; fr hairs not concealing pappus; sap clear
        24. Pl ± densely glandular-hairy. . . . . *Hulsea vestita*
        24′ Pl not glandular-hairy, often dotted with sessile resin glands. . . . . **TETRANEURIS**

   22′ Lvs basal and cauline or all cauline
     25. Lvs gen opposite throughout (occ distal-most alternate)......................**LASTHENIA** (G2,12)
     25′ Lvs gen alternate (sometimes proximally opposite)
       26. Old ray corolla papery, persistent on fr when dry.........................**PSILOSTROPHE**
       26′ Old ray corolla withering, deciduous from fr
         27. Pappus of rigid, tapered awns; lvs linear, entire.......................[2]**RIGIOPAPPUS** (G11,12)
         27′ Pappus of ± thin scales; lvs various, entire to lobed
           28. Phyllaries in 2–3 gen ± equal series, sometimes outer phyllaries longer
             29. Lvs mostly or all deeply linear-lobed or 1–3 × dissected
               30. Distal st and involucres stalked-glandular; lvs 1–3 × pinnately dissected.........**AMAURIOPSIS** (G12)
               30′ Distal st and involucres not stalked-glandular, glands sessile or sunken; lvs entire or gen 1–2 ×
                  dissected into linear lobes...**HYMENOXYS**
             29′ Lvs entire to shallowly lobed
               31. Pappus scales 4, in 2 unequal pairs; pl glandular-hairy ....**HULSEA**
               31′ Pappus scales 5–8, equal or not; pl gen not glandular
                 32. Lf narrowly linear, or if blade wider, base long-decurrent on st; disk ± spheric; phyllaries
                  gen ± reflexed..**HELENIUM** (G7,12)
                 32′ Lf not narrowly linear or decurrent; disk hemispheric; phyllaries gen spreading (or reflexed
                  in age)...*Hymenoxys hoopesii*
           28′ Phyllaries in 1(2) series
             33. Ray < 1 mm
               34. Pl dotted with sessile resin glands; fr 3–4-angled ....**AMBLYOPAPPUS** (G2,12)
               34′ Pl densely woolly; fr compressed ....*Monolopia congdonii* (G2,12)
             33′ Ray 1.5–10+ mm
               35. Disk fr ± compressed; ann.................................[2]**EATONELLA** (G2)
               35′ Disk fr (exc sometimes outermost) ± cylindric, club-shaped, or 3–4-angled; ann to shrub
                 36. Subshrub; lvs 8–25 cm, distinctly petioled; pappus scales 2–6+, unequal or, if equal, scales 2,
                  opposite ....**CONSTANCEA**
                 36′ Ann to shrub; lvs ≤ 8 cm, sessile or ± winged-petioled; pappus scales 0 or 6–12+, in 1–2
                  equal or unequal series ....**ERIOPHYLLUM** (G7,12)

Group 14

Heads radiate; rays white to pink, blue, or purple; receptacle epaleate; pappus of bristles

1. Lvs palmately lobed or dentate
  2. Ann; lvs cauline; pappus bristle 1; SCo, ChI, PR, D....................*Perityle emoryi* (G2,12,13)
  2′ Per; lvs basal, cauline lvs reduced to sheathing scales; pappus bristles many; moist forests, NW, n SNH,
    nw&n-c CW ....**PETASITES** (G2)
1′ Lvs entire to pinnately lobed
  3. Rays inconspicuous, barely surpassing disk or not
    4. Phyllaries with 1–3 resin-filled veins that become orange when dry
      5. Resin-filled phyllary vein 1 ....[5]**ERIGERON** (G2,6,15)
      5′ Resin-filled phyllary veins gen 3
        6. Subshrub; branches often thorny...[3]**CHLORACANTHA**
        6′ Herb; branches not thorny ....[5]**ERIGERON** (G2,6,15)
    4′ Phyllaries without resin-filled veins that become orange when dry
      7. Disk wide, flat; disk fls many; pappus gen double [outer, inner elements distinct], outer of short bristles
        or scales ....[5]**ERIGERON** (G2,6,15)
      7′ Disk longer than wide; disk fls 4–25+; pappus gen single
        8. Fr 1–3 mm, not distally long-tapered; st branches gen not overtopping main st ....[5]**SYMPHYOTRICHUM**
        8′ Fr 5–6 mm, distally long-tapered; st branches stiffly ascending, often overtopping main st ....**TRACYINA** (G15)
  3′ Rays conspicuous, gen much exceeding disk
    9. Pappus bristles alternating with well-developed scales — pl with strong, pungent odor ....**NICOLLETIA** (G13)
    9′ Pappus of bristles only or scales very short
      10. Pappus bristles ≤ 15
        11. Pappus double, of 1–15 bristles and 5 narrow scales or of 1 plumose-tipped bristle and a low crown
          ....**MONOPTILON** (G12,13)
        11′ Pappus single, of 3–5 bristles..**PENTACHAETA** (G2,12,15)
      10′ Pappus bristles gen > 20
        12. Bristles distinctly flattened, barbed to ± plumose.................................**TOWNSENDIA** (G13)
        12′ Bristles ± cylindric or weakly flattened, smooth to barbed
          13. Main phyllaries ± equal in 1–3(4) series
            14. Pappus ± 1 mm, bristles fused at base; pl woolly.........................**SYNTRICHOPAPPUS** (G15)

14′ Pappus gen >> 1 mm, bristles free; pl gen not woolly

  15. Pappus gen double, outer of short bristles or scales (0), inner of soft bristles; rays gen
     < 2 mm wide . . . . **⁵ERIGERON** (G2,6,15)

  15′ Pappus single; rays often > 2 mm wide

    16. Ann; main phyllaries equal, in 1 series [often with very short outer phyllaries]. . . . . ***Senecio elegans***

    16′ Per; main phyllaries ± equal, in 2–3(4) series

      17. Heads solitary; lvs all or mostly basal, cauline reduced or 0 . . . . **OREOSTEMMA**

      17′ Heads clustered; lvs basal and cauline or all cauline . . . . **⁵SYMPHYOTRICHUM**

13′ Main phyllaries unequal, graduated in 2–several series

  18. Shrub or subshrub

    19. Phyllary veins resin-filled, orange when dry; pl often thorny . . . . **³CHLORACANTHA**

    19′ Phyllary veins not resin-filled or orange; pl unarmed

      20. Ray fls sterile [style 0, ovary vestigial]; pappus brown or ± red . . . . **²CORETHROGYNE**

      20′ Ray fls fruiting [style present]; pappus gen ± white

        21. Ray fls 15–40 mm; lvs often sharply toothed

          22. Outer phyllaries ± oblanceolate, tips acute to obtuse, gen reflexed in age; e PR
            . . . . ***Dieteria canescens*** var. ***ziegleri***

          22′ Outer phyllaries linear to lance-linear, tips sharply acute, not reflexed in age; SNE, D. . . . **²XYLORHIZA**

        21′ Ray fls 5–10 mm; lvs entire

          23. Slender subshrub, minutely stalked-glandular, or nonglandular and strigose . . . . **²CHAETOPAPPA**

          23′ Stiff shrub, glabrous, sticky-resinous . . . . ***Ericameria gilmanii***

  18′ Herb

    24. Ray fls sterile (style 0, ovary ± vestigial)

      25. Disk style branch tips densely tufted with stiff yellow hairs; pappus brown or ± red
        . . . . **²CORETHROGYNE**

      25′ Disk style branches without stiff yellow hairs; pappus ± white

        26. Involucre 3–6 mm . . . . ***Arida arizonica***

        26′ Involucre 6–9 mm . . . . ***Dieteria canescens*** var. ***shastensis***

    24′ Ray fls fruiting, style present

      27. Pappus double, outer of short bristles or scales, inner of long bristles

        28. Resin-filled phyllary veins gen 3; st often thorny. . . . . **³CHLORACANTHA**

        28′ Resin-filled phyllary veins 0–1; st unarmed

          29. Rays narrowly linear, often very many . . . . **⁵ERIGERON** (G2,6,15)

          29′ Rays oblong, 1–16

            30. Disk corolla white — heads in tight clusters at tips of infl branches . . . . **²SERICOCARPUS**

            30′ Disk corolla yellow

              31. Heads in cyme-like cluster. . . . . **EUCEPHALUS** (G14)

              31′ Heads solitary. . . . . **IONACTIS**

      27′ Pappus single, or bristles ± equal if more than 1 series

        32. Phyllaries ± herbaceous throughout (or margin scarious to ± white)

          33. Lvs 4–12 mm; pl 5–15(20) cm — sts slender, gen branched, very lfy . . . . **²CHAETOPAPPA**

          33′ Lvs gen 30–200 mm; pl often > 20 cm . . . . **⁵SYMPHYOTRICHUM**

        32′ Phyllary tips ± herbaceous, bases white to straw-colored

          34. Ann to per with taproot; lvs ± entire to toothed or pinnately dissected, teeth sometimes
            bristle-tipped; phyllary tips often spreading

            35. Heads 3.5–6 cm diam, solitary . . . . **²XYLORHIZA**

            35′ Heads gen < 3 cm diam, often in cyme-like clusters

              36. Rays 1.3–4 mm. . . . . **⁵SYMPHYOTRICHUM**

              36′ Rays 6–20 mm

                37. Lvs entire to toothed. . . . . **DIETERIA**

                37′ Lvs 1–2-pinnately dissected. . . . . **MACHAERANTHERA**

        34′ Per with rhizome or caudex; lvs entire to toothed; phyllary tips gen appressed

          38. Heads in tight clusters at tips of infl branches; disk corolla white . . . . **²SERICOCARPUS**

          38′ Heads in raceme-like or flat-topped cluster or 1 at tips of long infl branches; disk corolla
            yellow or tinged pink or purple

            39. Phyllaries glabrous or hairy

              40. Disk corolla tube ≥ throat; fr ribs 7–10 . . . . **EURYBIA**

              40′ Disk corolla tube < throat; fr ribs (1)2–6 (5–8 in *Symphyotrichum defoliatum*)
                . . . . **⁵SYMPHYOTRICHUM**

            39′ Phyllaries densely glandular

              41. Head solitary at tips of long infl branches; cauline lvs ± linear, grass-like . . . . **ALMUTASTER**

              41′ Heads in raceme-like, panicle-like, or ± flat-topped cluster; cauline lvs linear to
                (ob)lanceolate or obovate, grass-like or not

                42. Involucre 8–14 mm; disk corolla 6–7.8 mm, pale yellow, aging ± pink or purple . . . . ***Eurybia integrifolia***

                42′ Involucre 5.5–8 mm; disk corolla 4.5–6 mm, yellow . . . . ***Symphyotrichum campestre***

Group 15

Heads radiate; rays yellow to red; receptacle epaleate; pappus of bristles

1. Lvs opposite at least proximally
  2. Lvs and phyllaries dotted or streaked with embedded, translucent oil glands, strongly scented
    3. Lvs sharply toothed to pinnately dissected; phyllaries in 2 series . . . . **³ADENOPHYLLUM** (G13)
    3′ Lvs linear, entire, base bristly-ciliate; phyllaries 8, in 1 series . . . . **PECTIS** (G12,13)
  2′ Lvs and phyllaries without embedded oil glands, not or faintly scented
    4. Trailing fleshy per; pappus bristles 1–5; coastal saline habitats . . . . **JAUMEA**
    4′ Erect ann or per, not fleshy; pappus bristles many; inland
      5. Per, gen from rhizome; involucre gen 8–20 mm; pappus bristles ≫ 2 mm, free . . . . **ARNICA** (G6)
      5′ Ann; involucre 5–7 mm; pappus bristles 1–2 mm, fused at base . . . . **²SYNTRICHOPAPPUS** (G14)
1′ Lvs alternate throughout
  6. Ann or bien
    7. Phyllary tip with a prominent tack-shaped gland; disk fls staminate . . . . **BENITOA**
    7′ Phyllary tip without a prominent tack-shaped gland; disk fls bisexual
      8. Ray fr pappus 0 [disk fr pappus present]
        9. Phyllaries in 1 series, fused at base . . . . **²CROCIDIUM**
        9′ Phyllaries in 2–several series, free . . . . **²HETEROTHECA**
      8′ Ray [and disk] frs with pappus
        10. Main phyllaries unequal, in 2–several series
          11. Pappus bristles flat, deciduous; involucre gummy-resinous . . . . **³GRINDELIA** (G6,7,13)
          11′ Pappus bristles gen ± cylindric or flattened only at base, gen persistent; involucre not gummy-resinous
            12. Pappus double, outer of short scales, inner of 8–16+ bristles . . . . **³PULICARIA**
            12′ Pappus single or, if double, of bristles throughout
              13. Fr distally long-tapered; rays 1–1.5+ mm . . . . **TRACYINA** (G14)
              13′ Fr not distally long-tapered; rays ≤ 12 mm
                14. Lvs serrate to 1–2-pinnately lobed . . . . *Xanthisma gracile*
                14′ Lvs entire to finely toothed
                  15. Pl sticky-glandular, strongly aromatic; heads many in dense raceme- or ± panicle-like clusters
                    . . . . **DITTRICHIA**
                  15′ Pl not glandular, not aromatic; heads solitary or few in open, ± flat-topped cluster
                    . . . . **²PENTACHAETA** (G2,12,14)
        10′ Main phyllaries ± equal, in 1–3 series
          16. Phyllary margins evidently scarious
            17. Rays 3–12 mm; pappus single, of 5–20 bristles . . . . **²PENTACHAETA** (G2,12,14)
            17′ Rays 1.5–2 mm; pappus double, outer of short scales, inner of 8–16+ bristles . . . . **³PULICARIA**
         16′ Phyllary margins not or only slightly scarious (exc alternate phyllaries in *Syntrichopappus*)
            18. Receptacle conic; phyllaries fused at base . . . . **²CROCIDIUM**
            18′ Receptacle ± flat; phyllaries free
              19. Phyllaries gen 5, ± enfolding ray frs . . . . **²SYNTRICHOPAPPUS** (G14)
              19′ Phyllaries gen 8, 13, or 21, not enfolding ray frs
                20. Pl with short, button-like caudex, with many fleshy-fibrous, unbranched roots . . . . **⁷SENECIO** (G2,6)
                20′ Pl tap- or fibrous rooted, without fleshy-fibrous, unbranched roots
                  21. Phyllary tips black . . . . **⁷SENECIO** (G2,6)
                  21′ Phyllary tips green
                    22. Rays 7–15+ mm; bien . . . . **²PACKERA** (G6)
                    22′ Rays ≤ 3 mm; slender ann . . . . **⁷SENECIO** (G2,6)
6′ Per to shrub
  23. Shrub or subshrub
    24. Outer phyllaries with an embedded, translucent oil gland near tip; pl strongly pungent-scented
    . . . . **³ADENOPHYLLUM** (G13)
    24′ Outer phyllaries without oil glands; pl scented or not
      25. Main phyllaries in 1 series, equal (often a few, gen much shorter, outer phyllaries present) . . . . **⁷SENECIO** (G2,6)
      25′ Main phyllaries in 2–7 series, often graduated, often unequal
        26. Pappus bristles flattened; phyllaries ± ovate
          27. Phyllary tips recurved to coiled; involucre gummy; ray fls 20–60 . . . . **³GRINDELIA** (G6,7,13)
          27′ Phyllary tips ± appressed; involucre ± resinous but not gummy; ray fls 1–14
            28. Rays 5–14, ≫ involucre; disk fls 30–80, fruiting . . . . *Acamptopappus shockleyi* (G13)
            28′ Rays 1–2, barely > involucre; disk fls 3–7, staminate . . . . **AMPHIPAPPUS** (G13)
      26′ Pappus bristles ± cylindric; phyllaries gen linear to narrowly lanceolate
        29. Woody sts ± prostrate, pl cushion-forming
          30. Lvs 1-veined, linear, of uniform size throughout . . . . **²NESTOTUS**
          30′ Lvs 3-veined, narrowly elliptic to oblanceolate, distal reduced . . . . **²STENOTUS**

29′ Woody sts gen erect or ascending
  31. Phyllaries ± graduated in 2–4 series, bases loose to tightly appressed, tips weakly thickened, or
    not, green or straw-colored, never tomentose; lvs entire . . . . **ERICAMERIA** (G6)
  31′ Phyllaries strongly graduated in 5–7 series, bases tightly appressed, straw-colored, tips clearly
    thickened, green (sometimes phyllaries densely tomentose); lvs often toothed
    32. Distal lvs gen not much reduced . . . . **HAZARDIA** (G6)
    32′ Distal lvs reduced to scale-like bracts . . . . ²**XANTHISMA**
23′ Herb
  33. Outer phyllaries with an embedded, translucent oil gland near tip; pl strongly pungent-scented
    . . . . ³**ADENOPHYLLUM** (G13)
  33′ Outer phyllaries without oil glands; pl scented or not
    34. Main phyllaries in 1 series, equal [often with a few, very short, outer phyllaries]
      35. St from an abruptly shortened, button-like caudex, with many fleshy-fibrous, unbranched roots
      . . . . ⁷**SENECIO** (G2,6)
      35′ St from a taproot, caudex, or rhizome, never with an abruptly shortened, button-like caudex
      36. Lvs ± equal, ± evenly distributed along st or crowded distally at fl . . . . ⁷**SENECIO** (G2,6)
      36′ Lvs reduced distally; proximal lvs prominent, gen persistent
        37. St from thin taproot, caudex, or rhizome; roots fibrous, branched; lf ± entire to pinnately lobed,
        marginal teeth not hard, not translucent . . . . ²**PACKERA** (G6)
        37′ St from thick, creeping or erect rhizome; roots fleshy, unbranched; lf dentate to minutely dentate,
        marginal teeth hard, translucent — lf tapered to winged petiole . . . . ⁷**SENECIO** (G2,6)
  34′ Phyllaries in 2+ series, ± equal to strongly graduated
    38. Disk pappus double, outer of very short bristles or scales
      39. St simple, erect; phyllaries ± equal — fr 2-ribbed . . . . **ERIGERON** (G2,6,14)
      39′ St gen branched, prostrate to erect; phyllaries gen ± unequal
        40. Rays 3–12 mm; disk fr 1.5–4 mm, ± flat; pappus 3–7 mm . . . . ²**HETEROTHECA**
        40′ Rays 1.5–2.5 mm; disk fr ± 1 mm, ± cylindric or fusiform; pappus 2–3 mm
          41. Outer pappus of bristles, 1.5 mm; anther base appendages 0 . . . . ²**EUTHAMIA**
          41′ Outer pappus a crown of ± fused scales, < 0.4 mm; anther base appendages bristle-like. . . . . . ³**PULICARIA**
    38′ Disk pappus single
      42. Pappus bristles flat, deciduous; involucre strongly gummy-resinous, esp in bud . . . . . . ³**GRINDELIA** (G6,7,13)
      42′ Pappus bristles ± cylindric, gen persistent; involucre not strongly gummy-resinous
        43. Lf toothed or lobed
        44. Heads gen small, in racemes or panicles, often clustered on 1 side of branches. . . . . ²**SOLIDAGO**
        44′ Heads gen not small, not in 1-sided clusters
          45. Basal rosette well developed
          46. Pl with stout taproot; sts, lvs glabrous or tomentose to woolly and/or stalked-glandular;
          fr 3–4 angled . . . . ²**PYRROCOMA**
          46′ Pl with branched caudex, taprooted; sts, lvs stalked-glandular, sticky, otherwise glabrous;
          fr not 3–4 angled . . . . ²**TONESTUS**
        45′ Rosette 0 or poorly developed
        47. Fr 2–3 mm, obconic; lf teeth or lobes bristle- or minutely spine-tipped . . . . ²**XANTHISMA**
        47′ Fr 3–10 mm, ± cylindric or compressed; lf margin not bristle- or minutely spine-tipped
          48. Fr 5–10 mm, glabrous; phyllaries in 4–5 series, strongly graduated . . . . *Hazardia whitneyi*
          48′ Fr 2.5–6 mm, hairy; phyllaries inn 3–4 series, ± equal or weakly graduated. . . . . ²**TONESTUS**
      43′ Lf entire
        49. Phyllaries in vertical ranks; disk fls staminate . . . . **PETRADORIA**
        49′ Phyllaries not in vertical ranks; disk fls fruiting
        50. Heads small, in clusters at branch tips; infl lfy, panicle-like or ± flat-topped; st erect from
        rhizome; fr ± 1 mm . . . . ²**EUTHAMIA**
        50′ Heads small to large, 1–several; infl various; st from taproot, caudex, or rhizome;
        fr gen >> 1 mm
          51. Pl with stout taproot — basal rosette well developed . . . . ²**PYRROCOMA**
          51′ Pl with ± branched caudex or rhizome
          52. Pl densely long-stalked-glandular . . . . *Tonestus lyallii*
          52′ Pl glandless or short-stalked-glandular
            53. Rays 1–6 mm; cauline lvs distributed well up st. . . . . ²**SOLIDAGO**
            53′ Rays 7–12 mm; cauline lvs gen only at st base
            54. Lvs 1-veined, linear to oblanceolate, of uniform size throughout. . . . . ²**NESTOTUS**
            54′ Lvs 3-veined, narrowly elliptic to oblanceolate, distal reduced . . . . ²**STENOTUS**

ACAMPTOPAPPUS GOLDENHEAD

Guy L. Nesom & Meredith A. Lane

Subshrub, appearing glabrous. **ST**: decumbent to erect, striate; old sts gray, bark sometimes shreddy with age; young st proximally white, distally green. **LF**: simple, proximally sometimes in axillary clusters, spreading-ascending to appressed-erect, linear to oblong or narrowly oblanceolate, gen minutely stiff-pointed, pale green to light gray-green. **INFL**: heads radiate or discoid, 1 or in rounded to ± flat-topped clusters, small in bud, expanding rapidly in fr; involucre widely bell-shaped to ± spheric; phyllaries 20–24, graduated in 2–3 series, ovate to ovate-elliptic, bases cream-yellow, tips green, margins scarious; receptacle deeply pitted, with projections between fls, epaleate. **RAY FL**: present or 0; corolla yellow. **DISK FL**: 14–80; corolla funnel-shaped, yellow, sinuses deep, lobes spreading to reflexed; anther tip acute; style tips lanceolate. **FR**: densely long-hairy, hairs white, bronze, or ± brown; pappus of 20–25 wide, stiff, wide-spreading bristle-like scales, slightly > fr. 2 spp.: sw US. (Greek: unbending pappus) [Nesom 2006 FNANM 20:184–185]

1. Heads radiate, gen solitary; involucre bell-shaped to hemispheric, 7–10 mm diam . . . . *A. shockleyi*
1′ Heads discoid, gen in rounded to flat-topped clusters; involucre hemispheric to spheric,
   8–16 mm diam. . . . . *A. sphaerocephalus*
2. Lf faces densely rough-puberulent. . . . . var. *hirtellus*
2′ Lf faces glabrous (margins sometimes sparsely rough-puberulent) . . . . var. *sphaerocephalus*

A. shockleyi A. Gray (p. 233) SHOCKLEY'S GOLDENHEAD **ST**: decumbent to ascending, gen ≤ 4 dm, gen minutely hairy to scabrous. **LF**: ≤ 2.5 cm, ≤ 4 mm wide, gen oblanceolate, densely minutely hairy to scabrous. **INFL**: heads gen 1; involucre bell-shaped to hemispheric, 7–10 mm diam. **RAY FL**: 5–14; corolla ≤ 2 cm, ray ≤ 6 mm wide. **DISK FL**: 30–80; corolla 2–5 mm. **FR**: ≤ 5 mm. 2*n*=18. Mesas, slopes, ridges, ravines, washes; 500–2000 m. SNE, ne DMoj; s NV. Apr–Jun

A. sphaerocephalus (Harv. & A. Gray) A. Gray (p. 233) RAYLESS GOLDENHEAD **ST**: much-branched, ascending to erect. **LF**: ≤ 1.5 cm, ≤ 3 mm wide, gen linear to oblanceolate. **INFL**: heads gen in loose, rounded to flat-topped clusters; involucre hemispheric to spheric, 8–16 mm diam. **RAY FL**: 0. **DISK FL**: 13–27; corolla 2–3.5 mm. **FR**: ≤ 3 mm. 2*n*=18. Vars. sometimes occur together and intergrade.

   var. ***hirtellus*** S.F. Blake **ST**: gen ≤ 6 dm. **LF**: Lf faces densely rough-puberulent. Gravelly or rocky soils, flats or slopes, juniper woodland; < 1750 m. Teh, SnGb, SnBr, PR, SNE, DMoj; nw AZ, s NV. Apr–Jun

   var. ***sphaerocephalus*** **ST**: gen ≤ 1 m. **LF**: Lf faces glabrous, margins sometimes sparsely rough-puberulent. Gravelly or rocky soils, flats or slopes, juniper woodland; < 2200 m. Teh, SnGb, SnBr, PR, D; to UT, AZ. Mar–Jun

ACHILLEA YARROW

Per, strongly scented. **LF**: [simple to] 3-pinnately dissected, basal and cauline, alternate, ± reduced distally on st, ± hairy. **INFL**: heads radiate, many, small, in gen flat-topped clusters; involucre bell-shaped or ovoid; phyllaries graduated in 3–4 series, ovate, obtuse; margins membranous; receptacle flat to rounded, paleate; paleae narrow, transparent, ± folded. **RAY FL**: (3)5–8; ray short, gen round, white or pink [yellow]. **DISK FL**: ± many; corolla short, white to purple [yellow]. **FR**: oblong to obovate, compressed, thick-margined, glabrous; pappus 0. ± 115 spp.: N.Am, Eurasia, n Afr. (Greek: Achilles of ancient mythology) [Trock 2006 FNANM 19:492–494]

A. millefolium L. (p. 233) Pl 10–200 cm. **LF**: finely 3-pinnately divided; cauline lvs ± clasping. **INFL**: phyllaries 4–9 mm. **RAY FL**: ray 2.5–4 mm, ovate to round, white to pink. **DISK FL**: 15–40; corolla 2–3 mm, white to pink. **FR**: ± 2 mm. 2*n*=[18,27],36,45,54,[63,72]. Many habitats; < 3650 m. CA-FP, GB; circumboreal. Highly variable polyploid complex; lf size and hairiness esp variable. Apr–Sep

ACHYRACHAENA BLOW-WIVES

Bruce G. Baldwin

1 sp. (Greek, Latin: scaly achene, from fr) [Baldwin & Strother 2006 FNANM 21:258–259]

A. mollis Schauer (p. 233) Ann 4–62 cm. **ST**: erect, branched distally. **LF**: simple, gen cauline, proximal opposite, distal alternate, ± sessile, 2–15 cm, 1–7 mm wide, linear, entire or toothed, hairy, distal glandular-hairy. **INFL**: heads radiate, 1–10+ in loose, rounded or ± flat-topped cluster; involucre 3–12 mm diam; phyllaries 3–8 in 1 series, each fully enveloping a ray ovary, falling with fr, 10–20 mm, ± lance-linear, abaxially hairy and sparsely glandular or glands 0; receptacle flat to convex, short-bristly; paleae in 1 series between ray and disk fls, free, deciduous. **RAY FL**: 3–8; corolla yellow, turning ± red, ray 4–7 mm, erect or spreading, inconspicuous, lobes ± parallel. **DISK FL**: 4–35; corolla 6–10 mm, yellow to ± red, abaxially hairy, tube ± = throat, lobes lance-deltate; anthers ± dark purple, tips ovate; style glabrous proximal to branches, branches free, tips lance-linear, short-bristly. **FR**: 4.5–9 mm; ray fr ± club-shaped, compressed front-to-back, 10-ribbed, glabrous or ± scabrous, black, pappus 0; disk fr similar to ray frs, ± scabrous, brown to black, pappus of 10, oblong, shiny-white, obtuse scales in 2 series, 5 outer scales 3–9 mm, 5 inner scales 6–13 mm, spreading in fr. 2*n*=16. Grassy sites, often clay soils; < 1220 m. CA-FP (exc NCo, NCoRH, SNH); s OR, n Baja CA. Self-fertile. Mar–Jun

ACOURTIA

Per. **ST**: erect or spreading. **LF**: simple, cauline, alternate, sessile, pinnately veined. **INFL**: heads appearing ± radiate, 1–many in panicle-like clusters; involucre cylindric to bell-shaped; phyllaries graduated in several series; receptacle epaleate, pitted. **FL**: few to many; corolla white to pink or purple, strongly 2-lipped, inner lip deeply 2-lobed, recurved or coiled, outer lip entire to shallowly 3-lobed, spreading, resembling a ray; anther tips lance-oblong, basal appendages stiff, tail-like; style branches

short, tips rounded to truncate. **FR**: cylindric, ribbed, glandular; pappus of many finely barbed bristles. ± 40 spp.: N.Am, C.Am. (Mrs. A'Court, English amateur botanist) [Simpson 2006 FNANM 19:72–74]

A. microcephala DC. (p. 233) **ST**: several from woody caudex, 6–16+ dm, very lfy, gen branched only in distal 1/2. **LF**: 2.5–15 cm, widely ovate to elliptic or oblong; base truncate to widely clasping; tip acute to widely obtuse, finely dentate; faces densely glandular. **INFL**: heads many; peduncles 1–10 mm; involucre 5–10 mm diam; phyllaries 7–10 mm, linear to lanceolate, acute or obtuse, glandular. **FL**: 2–20; corolla 8–11 mm, white to pink-purple. **FR**: 1.5–4 mm; pappus 7–10 mm. 2*n*=54. Shrubby and wooded slopes, esp prominent after fires; < 1550 m. SCoR, SW; nw Baja CA. May–Aug

ACROPTILON

1 sp.: Eurasia. (Greek: feather-tipped, from pappus bristles) [Keil 2006 FNANM 19:171–172]

A. repens (L.) DC. (p. 233) RUSSIAN KNAPWEED Per 3–10 dm, from dark rhizome. **ST**: erect; branches ascending, ± cobwebby-tomentose. **LF**: proximal 4–10 cm, oblong, 1–2-pinnately lobed, distal 1–3 cm, linear to narrowly lanceolate, entire to toothed; blades glabrous to ± tomentose. **INFL**: heads discoid in ± flat-topped or panicle-like, lfy-bracted clusters; involucre 10–14 mm, ± ovoid; phyllaries graduated in several series, entire, ± soft-hairy, inner narrower, tips widely scarious; receptacle flat, epaleate, bristly, bristles flattened. **FL**: 15–36; corolla ± 15 mm, white to pink or purple, tube very slender, throat abruptly wider, lobes linear; anther base short-tailed, tip oblong; style with minutely hairy and swollen distal node, papillate above node, branches ± linear, tips triangular. **FR**: 3–4 mm, obovoid, slightly compressed, glabrous; pappus bristles many, 6–8 mm, ± deciduous, barbed proximally, short-plumose distally. 2*n*=26. Fields, roadsides; < 2250 m. CA (exc wettest NW, driest GB); N.Am; native to c Asia. May–Sep ◆

ADENOCAULON

Per. **ST**: slender; base ± lfy. **LF**: basal and cauline, alternate, reduced distally on st; proximal lvs long-petioled; blades wide, thin. **INFL**: heads small, disciform, in open panicle-like clusters; involucre inconspicuous; phyllaries in 1–2 series, ± equal, reflexed in fr, deciduous; receptacle convex, epaleate. **PISTILLATE FL**: few in 1 series; corolla deeply 4–5-lobed, white, early-deciduous. **DISK FL**: few, staminate, ± persistent; corolla cylindric, white; anther bases sagittate, tips narrowly triangular; style undivided. **FR**: club-shaped to obovoid, weakly veined, prominently stalked-glandular; pappus 0. 6 spp.: w(c) N.Am, C.Am, s S.Am, e Asia. (Greek: gland st) [Keil 2006 FNANM 19:77–78]

A. bicolor Hook. (p. 237) TRAIL PLANT Per 3–10 dm, gen ± erect, openly branched, proximally ± tomentose, distally stalked-glandular. **LF**: petiole ± winged; blade 3–25 cm, ± triangular to ovate, ± entire to shallowly lobed, abaxially white-tomentose, adaxially glabrous, base truncate to cordate or hastate. **INFL**: phyllaries 1–2.5 mm, ovate. **PISTILLATE FL**: 3–7; corolla 0.5–1 mm. **DISK FL**: 2–10; corolla 1.5–2 mm. **FR**: 5–9 mm, club-shaped. 2*n*=46. Gen in shade, woodland, forest; < 2000 m. NW, CaR, SN, n CCo, SnFrB; to BC, n-c US. Jun–Oct

ADENOPHYLLUM

Bruce G. Baldwin, adapted from Strother (2006)

[Ann] per, subshrub [shrub], 20–50+ cm, pungently scented. **ST**: erect, branched from base or throughout. **LF**: alternate to ± opposite [opposite], petioled or sessile; blades or lobes linear to oblanceolate or ovate, simple or 1-pinnately lobed, with ± bristle-tipped teeth, glabrous or ± minutely coarse-hairy to minutely scabrous, embedded oil-glands gen at bases of lobes or blade and near lf tip. **INFL**: heads gen radiate, 1, peduncled; bracts closely subtending head 12–22, linear to awl-shaped, sometimes 1–2-lobed, often bristle-tipped, gen with oil-glands; involucre bell-shaped to obconic [hemispheric], [5]7–20+ mm diam; phyllaries [8]12–20+ in ± 2 series, weakly fused 1/3–3/4 lengths, free in age, gen lanceolate to linear, persistent, margins of outer ± free to base, oil-glands on faces and near margins; receptacle convex, epaleate, ± pitted, margins of pits minutely fringed to bristly. **RAY FL**: (0)7–14[16]; corolla yellow to orange or red-orange. **DISK FL**: 25–80[100+]; corolla yellow to yellow-orange, tube < cylindric throat, lobes lance-ovate to -linear; anther tip elliptic or ovate, acute; style tips linear, tapered, densely bristly. **FR**: obpyramidal [obconic], glabrous or ± strigose or silky-hairy, pappus of 8–20 scales in 1[2] series, each scale composed of 5–11 basally fused bristles. 10 spp.: sw US, Mex, C.Am. (Greek: gland-lf) [Strother 1986 Sida 11:371–378; Strother 2006 FNANM 21:237–239]

1. Lvs gen not lobed, blades ovate to oblanceolate, toothed; disk fls 50–80+; pappus of 15–20 scales . . . . ***A. cooperi***
1′ Lvs 3–5 lobed, lobes linear to wedge-shaped or oblanceolate, ultimate margin entire or toothed; disk fls 25–40; pappus of 8–12 scales . . . . ***A. porophylloides***

A. cooperi (A. Gray) Strother (p. 237) Pl 25–50+ cm; herbage glaucous. **LF**: blade 8–25 mm, 4–15 mm wide, sometimes ± lobed at base, ± glabrous to densely puberulent, oil-glands 1–2 pairs near base, 1 near tip of each lf. **INFL**: peduncle 6–15 cm; bracts closely subtending head 12–22, 5–8 mm, linear, long-tapered, glandular; involucre 15–18 mm, bell-shaped to obconic; phyllaries ± 20, lanceolate to linear. **RAY FL**: (0)7–14; corolla yellow-orange, becoming red-orange, tube 5 mm, ray 8–10 mm, 2.5–4 mm wide. **DISK FL**: corolla 8–10 mm, yellow. **FR**: 5–7 mm, 3-angled; pappus scales 7–10 mm, each composed of 5–9 basally fused bristles. 2*n*=26. Washes, alluvial fans, sometimes rocky slopes, in open scrub, woodland; 450–1600 m. n edge SnBr, DMoj; to sw UT, nw AZ. Apr–Jun, Sep–Nov

A. porophylloides (A. Gray) Strother (p. 237) Pl 20–50+ cm. **LF**: blade 15–40 mm, pinnately lobed, glabrous, oil-glands at lobe bases and near lf tip. **INFL**: peduncle 2–8 cm; bracts closely subtending head 12–16, 3–8 mm, awl-shaped, glandular; involucre 10–15 mm, obconic; phyllaries 12–20, lanceolate. **RAY FL**: 10–14; corolla yellow, becoming red-orange, tube 2–5 mm, ray 5–6 mm, 1–2 mm wide. **DISK FL**: corolla 7–11 mm, yellow-orange or distally red. **FR**: 5 mm; pappus scales 7–12 mm, each composed of 7–11 basally fused bristles. 2*n*=26. Washes, alluvial fans, rocky slopes, in open scrub, woodland; 50–1300 m. e edge PR, s DMoj, DSon; to NV, AZ, n Mex. Mar–Jun, Oct–Dec

AGERATINA SNAKEROOT

Per, subshrub [shrub]. **LF:** cauline, opposite or distal alternate, petioled; blade elliptic to triangular, entire or toothed, gen 3-veined from base. **INFL:** heads discoid, 1 or in ± flat-topped cyme-like clusters; involucre bell-shaped, phyllaries ± equal, in 1–2(3) series, linear to narrowly lanceolate; receptacle flat to conic, epaleate. **FL:** 10–60; corolla ± white to pink-tinged or lavender, cylindric (or throat wider); style branches linear. **FR:** 5-angled, gen 5-ribbed; pappus of 5–40 slender scabrous bristles, often easily detached. ± 250 spp.: N.Am to S.Am. (Latin: resembling *Ageratum*) [King & Robinson 1987 Monogr Syst Bot Missouri Bot Gard 22:428–436; Nesom 2006 FNANM 21:547–553]

1. Heads gen solitary (rarely 2–3) . . . . *A. shastensis*
1′ Heads clustered
  2. Lvs alternate distally on st; heads 8–10 mm . . . . *A. occidentalis*
  2′ Lvs opposite throughout; heads 6–8 mm
    3. Lf blade gen 40–100 mm; petiole gen 20–40 mm . . . . *A. adenophora*
    3′ Lf blade gen 15–55 mm; petiole gen 5–10 mm . . . . *A. herbacea*

A. adenophora (Spreng.) R.M. King & H. Rob. CROFTON WEED Per or subshrub; base woody. **ST:** gen 5–15 dm, erect, ± purple, glandular-hairy. **LF:** opposite; blade gen 40–100 mm, deltate-ovate, serrate, glandular-puberulent, esp adaxially, purple adaxially. **INFL:** heads ± 6.5 mm, clustered; phyllaries glandular-puberulent. **FL:** corolla white or pink-tinged. **FR:** 1.7–2 mm. $2n=3x=51$. Disturbed places, streambanks, canyons, hillslopes; < 1000 m. CCo, SnFrB, SCoRO, SW; native of Mex, widely naturalized. Reproduces by asexual seed; cult as orn; may be seriously invasive in mild coastal situations. All yr ◆

A. herbacea (A. Gray) R.M. King & H. Rob. (p. 237) DESERT AGERATINA Per; caudex woody. **ST:** 4.5–7 dm, erect or spreading, green, puberulent. **LF:** gen opposite throughout; blade gen 15–55 mm, gen triangular to ± cordate, glabrous to puberulent, ± yellow to light or ± gray-green. **INFL:** heads 6–8 mm, in dense clusters; phyllaries puberulent. **FL:** corolla white. **FR:** 2–3 mm. $2n=34$. Locally common. Rocky pinyon/juniper woodland; 1500–2200 m. e DMtns (Clark, New York, Providence mtns); to CO, w TX, n Mex. May–Jun, Oct–Nov ★

A. occidentalis (Hook.) R.M. King & H. Rob. (p. 237) WESTERN SNAKEROOT Per, subshrub; caudex woody, ± rhizomatous. **ST:** 1.5–7 dm, erect or ascending, ± green or purple, puberulent. **LF:** proximal opposite, distal alternate; blade 25–50 mm, ± triangular, ± serrate, abaxially glandular. **INFL:** heads 8–10 mm, clustered; phyllaries puberulent and/or glandular. **FL:** corolla ± white to purple. **FR:** 3–3.5 mm. $2n=34$. Rocky sites, montane chaparral, conifer forest, alpine; < 3700 m. NW, CaRH, SN, Wrn, n SNE, W&I; to WA, ID, UT. Pls from SN are gen smaller than those from NCoR. Jun–Oct

A. shastensis (D.W. Taylor & Stebbins) R.M. King & H. Rob. (p. 237) SHASTA AGERATINA Per; caudex woody. **ST:** 1.3–4.5 dm, erect or ascending, glandular-hairy. **LF:** proximal opposite, distal alternate; blade 15–30 mm, ± deltate-ovate, serrate, ± hairy, abaxially glandular. **INFL:** heads 1(3), ± 1.2 cm; phyllaries glabrous to glandular-hairy. **FL:** corolla white. **FR:** 3–5.5 mm. $2n=34$. Limestone or metavolcanic cliffs, chaparral, conifer forest; 400–1800 m. KR. Jun–Oct ★

AGOSERIS

Gary I. Baird

Ann or per from taproot, occ with branched caudex, gen scapose; sap milky. **LF:** gen all basal, gen lanceolate to oblanceolate, long-tapered to obtuse, entire, toothed, or 1–2-pinnately lobed, 1° lobes gen ± opposite, gen on proximal 2/3; 2° lobes 0–1; faces glabrous to densely hairy, hairs gen white-opaque and glandless. **INFL:** heads liguliflorous, 1, erect, long-peduncled; involucre cylindric to ovoid or bell-shaped in fl, ovoid in fr, glabrous to densely hairy, hairs white-opaque and glandless or colorless- or yellow-translucent (occ with purple cross-walls) and glandular, glands gen yellow, occ purple; phyllaries in 2–5(+) series, ± equal in fl, often strongly graduated in fr, gen entire, green to rosy-purple, often with darker spots or midstripe, outer erect or tips spreading to recurved, inner erect, elongating with fr or not, ± reflexed when dry; receptacle flat, epaleate (paleate), gen pitted. **FL:** 5–500; ligules ± equaling to much exceeding involucre, yellow, or orange, pink, red, or purple, readily withering. **FR:** cylindric to fusiform or obconic (inflated), ± 10-ribbed, gen ± white to brown or purple-black (gray), glabrous to minutely coarse-hairy or puberulent; beak << to >> body (0); pappus of many fine, simple, white bristles. 11 spp.: Am. (Greek: chief or goat + chicory, meaning is obscure) [Baird 2006 FNANM 19:323–335] Closely related to *Nothocalais*. Self-pollination complicates variation in some spp.; polyploidy and hybridization blur distinctions between some taxa.

1. Ann from slender taproot; st often lfy above base; involucre 10–25 mm in fr; outer frs sometimes
   wavy-ribbed (or inflated and ribless) . . . . *A. heterophylla*
  2. Ligule 10–15 mm, much exceeding involucre; anthers 2–4 mm; lf lobes gen 3–5 pairs . . . . var. *cryptopleura*
  2′ Ligule 2–6 mm, ± equaling involucre; anthers 1–1.5 mm; lf lobes gen 2–3 pairs . . . . var. *heterophylla*
1′ Per from stout taproot, gen with caudex; st gen not or obscurely lfy above base; involucre 10–60 mm in fr;
  outer frs gen straight-ribbed
  3. Corolla orange, pink, or ± red, gen drying ± purple
    4. Involucre not glandular-hairy; petiole gen ± purple . . . . [2]*A. aurantiaca* var. *aurantiaca*
    4′ Involucre ± glandular-hairy; petiole gen not purple . . . . [2]*A. ×elata*
  3′ Corolla yellow, outer gen ± purple abaxially, drying ± white but purple still evident
    5. Fr beak stout, << body; inner phyllaries gen not elongating in fr
      6. Peduncle gen 5–20 cm, ± glandular-hairy, base gen ± glandular-woolly; lf blade gen 2–10 cm, gen
        toothed or lobed (entire), lobes gen 2(+) pairs, ± divergent, proximal lobes angled towards lf base, distal
        angled towards lf tip or spreading; pl gen decumbent or prostrate . . . . *A. monticola*

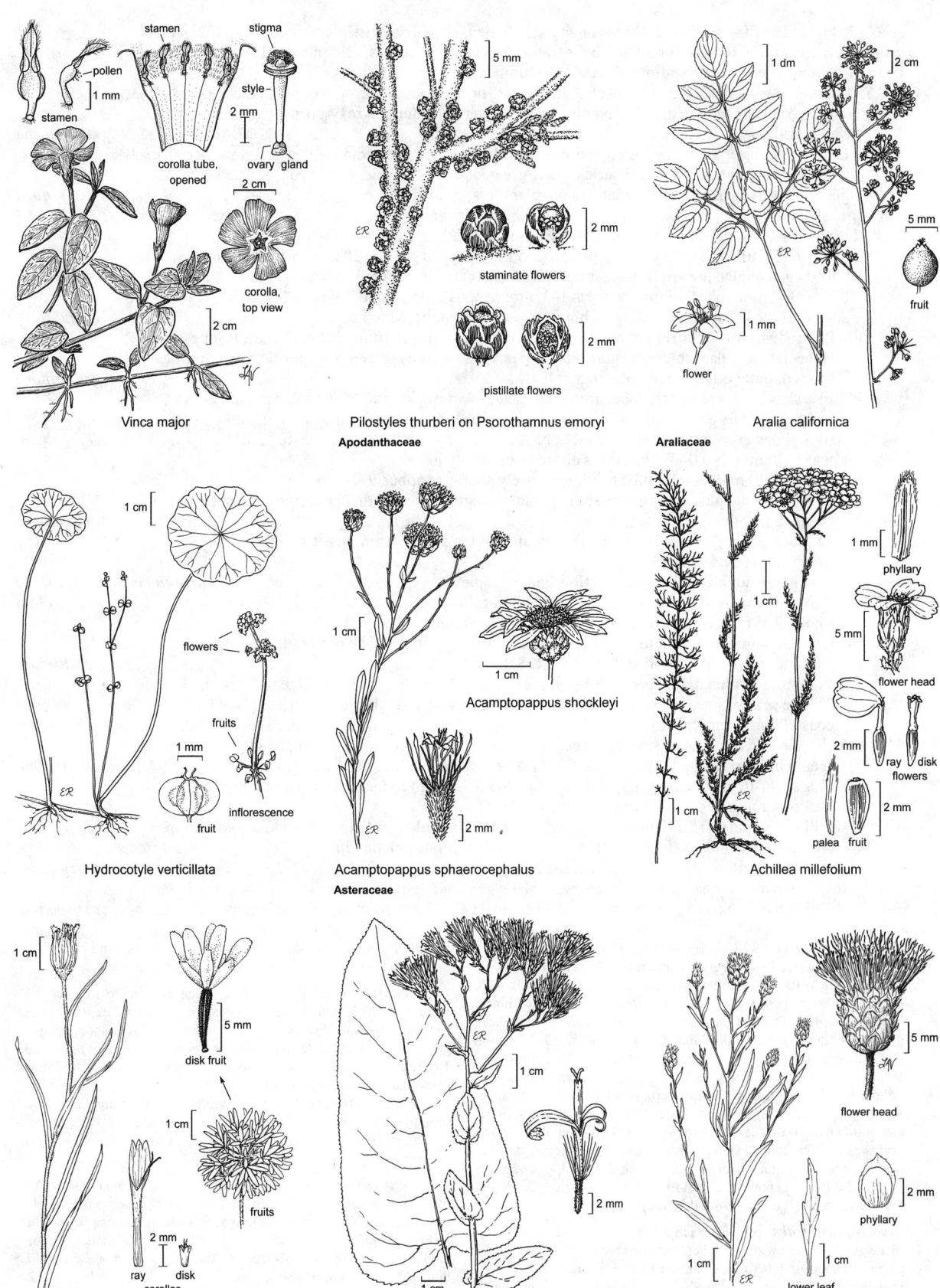

Vinca major

Pilostyles thurberi on Psorothamnus emoryi
Apodanthaceae

Aralia californica
Araliaceae

Hydrocotyle verticillata

Acamptopappus sphaerocephalus
Asteraceae

Acamptopappus shockleyi

Achillea millefolium

Achyrachaena mollis

Acourtia microcephala

Acroptilon repens

6′ Peduncle gen 20–60 cm, ± glabrous or sparsely to densely hairy, hairs glandular or not; lf blade gen 3–30 cm, entire, toothed, or lobed, lobes (if present) gen > 2 pairs, spreading or all angled in same direction; pl erect to spreading or petioles decumbent

    7. Lvs glabrous and glaucous to densely hairy, entire or few-toothed or -lobed; peduncle gen hairy or ± glabrous in age exc distally; phyllaries in 3–4 series, outer gen rosy-purple exc margin, gen ± glandular-hairy . . . . ***A. ×dasycarpa***

    7′ Lvs ± glabrous, ± glaucous, gen ± entire or weakly few-toothed; peduncle gen glabrous, occ sparsely hairy distally; phyllaries in 2–3 series, outer green or ± rosy-purple, gen purple-spotted, gen ± glabrous proximally (weakly ciliate), hairs glandless . . . . ***A. glauca*** var. ***glauca***

5′ Fr beak ± slender, (< 1)1–4 × body; inner phyllaries ± elongating with fr

    8. Fr beak 10–25 mm, (< 3)3–4 × body; lvs gen pinnately lobed or toothed (entire)

        9. Ligule 6–15 mm, tube 10–20 mm; anthers 2–5 mm; pappus gen 15–20 mm, bristles in 4–6 series; lf lobes gen angled toward lf base; fr body tip ± truncate, not tapered to beak . . . . ***A. retrorsa***

        9′ Ligule 3–7 mm, tube 4–7 mm; anthers 1–3 mm; pappus 7–15 mm, bristles in 2–3 series; lf lobes gen angled toward lf tip or spreading; fr body tip ± abruptly tapered to beak . . . . ***A. grandiflora***

           10. Pl ± robust; lvs (± entire) toothed to pinnately lobed, gen ≥ 10 mm wide exc lobes; lf lobes ± lanceolate to oblanceolate; fls gen 150–500; phyllaries green or gen rosy-purple, rarely purple-spotted, outer often exceeding inner at fl . . . . var. ***grandiflora***

           10′ Pl ± slender; lvs pinnately lobed (toothed), gen 2–4 mm wide exc lobes; lf lobes thread-like to lance-linear; fls gen < 100; phyllaries gen rosy-purple and often purple-spotted, outer ± = or sometimes exceeding inner at fl . . . . var. ***leptophylla***

    8′ Fr beak ≤ 10 mm, (< 1)1–3 × body; lvs entire to pinnately lobed

        11. Fr body 5–10 mm; corolla tube 6–15 mm, rarely shorter; pappus 9–20 mm

           12. Lf lobes gen angled toward lf base or spreading; ligule 10–20 mm, exceeding involucre; fr beak (< 1)1–2 × body . . . . ***A. parviflora***

           12′ Lf lobes gen angled toward lf tip or spreading; ligule 5–10(+) mm, gen ± equaling (exceeding) involucre; fr beak gen = body

               13. Involucre not glandular-hairy; petiole gen ± purple. [2]***A. aurantiaca*** var. ***aurantiaca***

               13′ Involucre ± glandular-hairy; petiole gen not purple . . . . [2]***A. ×elata***

        11′ Fr body 3–5 mm; corolla tube 2–6 mm; pappus 4–10 mm

           14. Lvs gen pinnately lobed (toothed); lf lobes gen 5–7(+) pairs, 2° lobe occ 1; pl gen erect, lvs all basal; ± inland in ± fine-textured soils — NCoR, ScV, CW . . . . ***A. hirsuta***

           14′ Lvs entire to pinnately lobed; lf lobes if present gen 3–5(+) pairs, thread-like to ± spoon-shaped, 2° lobe gen 0; pl gen decumbent to ± erect, lfy st present and ± buried by drifting sand, or lfy st 0; coastal dunes and sand hills . . . . ***A. apargioides***

               15. Ligule 3–6 mm, not or scarcely exceeding phyllaries; anthers 1.5–2.5 mm; head glabrous or hairs gen white-opaque, glandless — n NCo (Humboldt Bay n) . . . . var. ***maritima***

               15′ Ligule 8–16 mm, exceeding phyllaries; anthers 3.5–4.5 mm; head glabrous or hairs gen white- or yellow-translucent, glandular

                  16. Pl gen glabrous to ± densely hairy, hairs ± white; lf blade gen 1–15 mm wide exc lobes, linear to oblanceolate, lobes if present thread-like to ± lanceolate (spoon-shaped) — CCo (San Francisco s to Cambria) . . . . var. ***apargioides***

                  16′ Pl gen densely hairy, hairs ± yellow; lf blade gen 15–30 mm wide exc lobes, blade and lobes oblanceolate to ± spoon-shaped — c&s NCo, n CCo (Point Reyes n to Mendocino) . . . . var. ***eastwoodiae***

A. apargioides (Less.) Greene   Per 10–45 cm, decumbent to ± erect, sts occ buried by drifting sand and rhizome-like. **LF:** petiole ± purple; blade 3–10(15) cm, rounded to short-tapered, entire to lobed; lobes, if present, gen 3–5(+) pairs, ± spreading. **INFL:** peduncle becoming ± glabrous, base gen remaining densely hairy, base of head glabrous to densely hairy, glandular or not; involucre 10–15 mm in fl, 15–25 mm in fr; phyllaries green to rosy-purple, sometimes spotted, outer gen obtuse, glabrous to tomentose, hairs gen colorless- or yellow-translucent (occ with purple cross-walls) and glandular, or white-opaque and glandless, margin ciliate to densely hairy; inner elongating with fr. **FL:** 25–200; tube 2–5 mm, ligule yellow. **FR:** outermost gen different or all alike; body 3–5 mm, tip ± abruptly tapered; ribs glabrous to minutely rough-hairy; beak 1–8 mm, gen slender, (< 1)1–2 × body; pappus bristles 4–9 mm, in 2–3 series. 2*n*=20,36. Hybridizes with *A. grandiflora*, *A. heterophylla*, *A. hirsuta*.

var. ***apargioides*** (p. 237)   Gen ± prostrate to decumbent. **LF:** blade gen 1–15 mm wide exc lobes, gen linear to oblanceolate, entire to gen toothed or lobed; lobes thread-like to ± lanceolate (spoon-shaped); gen glabrous to ± densely hairy, hairs ± white-translucent. **INFL:** involucre glabrous to densely hairy, hairs gen ± colorless-translucent, glandular; outer phyllaries lanceolate to oblanceolate, margin ± hairy. **FL:** ligules 8–16 mm, much exceeding phyllaries; anthers 3.5–4.5 mm. **FR:** body gen fusiform to obconic; ribs straight or

slightly wavy, ± reduced proximally. Coastal dunes, sand hills; < 100 m. c CCo (San Francisco s to Cambria). Apr–May

var. ***eastwoodiae*** (Fedde) Munz   Gen reclining to prostrate. **LF:** blade gen 15–30 mm wide exc lobes, gen oblanceolate to ± spoon-shaped, gen toothed to lobed; lobes oblanceolate to ± spoon-shaped; gen densely hairy to tomentose, hairs mostly ± yellow-translucent. **INFL:** involucre hairy to tomentose, hairs gen yellow-translucent and glandular; outer phyllaries ovate to obovate, margin gen tomentose. **FL:** ligules 8–16 mm, much exceeding phyllaries; anthers 3.5–4.5 mm. **FR:** body gen fusiform to obconic; ribs straight, ± reduced proximally. Coastal dunes, sand hills; < 50 m. c&s NCo, n CCo (Point Reyes n to Mendocino). Gen Apr–May

var. ***maritima*** (E. Sheld.) Cronquist   Gen erect to spreading. **LF:** blade ± oblanceolate, 4–20 mm wide exc lobes, entire to lobed; lobes lanceolate to oblanceolate; glabrous to hairy, hairs gen ± opaque-white. **INFL:** involucre glabrous or hairs gen white-opaque, glandless; outer phyllaries lanceolate to oblanceolate, margin glabrous or ± hairy. **FL:** ligules 3–6 mm, not or scarcely exceeding phyllaries; anthers 1.5–2.5 mm. **FR:** all ± alike; body ± fusiform; ribs straight, ± uniform. Coastal dunes, sand hills; < 50 m. n NCo (Humboldt Bay n); to WA. Jul–Aug

A. aurantiaca (Hook.) Greene var. ***aurantiaca*** (p. 237)   Per 10–50 cm, gen erect. **LF:** petiole gen ± purple; blade gen 10–30

cm, narrowly to broadly oblanceolate, acute to long-tapered, entire or toothed to irregularly lobed; lobes gen 2–4 pairs, ± lanceolate, ± spreading; faces ± glabrous, ± glaucous. **INFL:** peduncle ± hairy or glabrous in age, base of head densely hairy to tomentose (± glabrous); involucre 15–20 mm in fl, 25–30 mm in fr; phyllaries (exc margin) green or ± rosy-purple, often darker speckled or spotted (and/or midstripe darker); outer lance-linear to ovate, long-tapered to acute, glabrous or hairs ± white-opaque, glandless, margin ciliate esp towards base; inner elongating with fr. **FL:** 15–100; tube gen 7–9 mm, ligule gen 5–10 mm, equaling or exceeding involucre, gen orange (rarely yellow, pink, red, or purple), gen drying ± purple; anthers 2–5 mm. **FR:** gen all similar; body 6–9 mm, cylindric to obconic, tip abruptly to gradually tapered; ribs straight, often thickened upward, glabrous to minutely rough-hairy (esp upward), sometimes minutely puberulent; beak 5–10 mm, gen slender, ± = body; pappus 9–15 mm, in 2–4 series. 2*n*=18,36. Meadows, scrub, forest, streamsides; 1500–3500 m. KR, NCoRH, CaRH, SNH, MP (mostly Wrn), SNE (Sweetwater Mtns); to AK, w Can, QC, SD, NM. Hybridizes with *A. glauca, A. grandiflora, A. monticola, A. parviflora.* A 2nd var. occurs from NV to WY, CO, NM. Jun–Sep

A. ×*dasycarpa* Greene   Per 15–45 cm, erect or petiole decumbent. **LF:** petiole not purple; blade (6)10–30 cm, ± oblanceolate, obtuse to long-tapered, entire or few-toothed or -lobed; lobes 1–4 pairs, (ob)lanceolate, gen angled towards lf tip or spreading; glabrous and glaucous to densely hairy. **INFL:** peduncle gen hairy or ± glabrous in age exc distally, base of head ± hairy (± glabrous), ± glandless; involucre 10–30 mm in fl, ± same in fr; phyllaries in 3–4 series, gen rosy-purple exc margin; outer ± lanceolate, tapered to acute, ± hairy (± glabrous), hairs gen ± yellow-translucent, glandular, margin ± ciliate (glabrous); inner not elongating in fr. **FL:** gen 10–50; tube 4–10 mm, ligule 5–15 mm, exceeding involucre, yellow; anthers 3–5 mm. **FR:** outer ± different; body 6–9 mm, ± fusiform, tip gen ± narrowed; ribs straight, uniform, glabrous to ± minutely rough-hairy; beak 1–2 mm, gen << body; pappus 8–10 mm, in gen 2 series. Moist meadows, marshes, dry grassland, open forest; 1100–2500 m. n SNH, MP; to OR. Localized hybrid populations (*A. glauca* var. *g.* × *A. monticola*) gen appearing as hairy *A. glauca* var. *g.* or misplaced *A. monticola*, and gen misidentified (often as *A. glauca* var. *agrestis*, which is restricted to CO). May–Aug

A. ×*elata* (Nutt.) Greene Per 15–65(+) cm, gen erect. **LF:** petiole gen not purple; blade 15–45 cm, ± oblanceolate, acute (to obtuse), entire to lobed; lobes gen 2–4 pairs, ± lanceolate, ± spreading or angled toward lf tip; ± glabrous, glaucous or not, hairs, if present, white-opaque, glandless. **INFL:** peduncle becoming ± glabrous, base gen remaining hairy, base of head ± becoming glabrous (hairy), ± glandless; involucre 15–20 mm in fl, 20–40 mm in fr; phyllaries gen rosy-purple exc margin, ± darker spotted or tipped; outer lance-ovate to obovate, tapered to acute, ± (sparsely) hairy to tomentose, hairs gen ± yellow-translucent, glandular (occ mixed with white-opaque, glandless hairs), margin ± ciliate; inner elongating with fr. **FL:** gen 50–150; tube 8–10 mm, ligule gen 6–8 mm, ± equaling (exceeding) involucre, orange or yellow, gen drying ± purple; anthers 2–4 mm. **FR:** outermost gen different; body 8–10 mm, ± fusiform, tip gen gradually tapered; ribs straight, uniform, gen minutely coarse-hairy; beak 5–10 mm, gen = body; pappus 10–14 mm, in 2–3 series. Montane meadows, lake margins, streamsides; gen 1400–2800 m. SNH; OR, WA. [*A. e.* (Nutt.) Greene] Uncommon and confusing sp. of hybrid origin, the CA populations involving *A. aurantiaca* × *A. grandiflora* and/or *A. monticola.* Jun–Sep

A. glauca (Pursh) Raf. var. *glauca* (p. 237)   Per 20–60 cm, erect or petioles decumbent. **LF:** petiole not purple; blade gen 10–30 cm, lance-linear to oblanceolate, long-tapered (rarely obtuse), gen ± entire or weakly few-toothed; lobes 0; ± glabrous, ± glaucous. **INFL:** peduncle gen 20–60 cm, gen glabrous, occ sparsely hairy distally, base of head ± glabrous; involucre 10–20 mm in fl, ± same in fr; phyllaries in 2–3 series, green or ± rosy-purple, gen darker spotted or tipped; outer ± lanceolate, long-tapered, gen ± glabrous proximally, hairs, if present, white-opaque, glandless, margin glabrous (weakly ciliate); inner not elongating in fr. **FL:** 15–150; tube 9–14 mm, ligule 10–18 mm, much exceeding involucre, yellow; anthers 4–6.5 mm. **FR:** outermost gen different; body 6–9 mm, fusiform to narrowly

conic, tip ± tapered (or merely narrowed); ribs straight, uniform, glabrous or minutely rough-hairy (minutely puberulent); beak 1–4 mm, gen < 1/2 body; pappus gen 10–15 mm, in gen 2 series. 2*n*=18. Gen wet, often alkaline or saline meadows, stream and lake margins, seeps, gen silty or clay soils; 1400–2500 m. SN (rare c&s SNH), GB; to BC, ON, MI, CO, NM. Hybridizes with *A. monticola, A. parviflora.* A 2nd var. occurs from n Can to OR, SD, CO. Jun–Aug

A. grandiflora (Nutt.) Greene (p. 237)   Per 25–100 cm, ± erect. **LF:** petiole gen purple; lobes gen 3–5 pairs, angled toward lf tip or spreading; sparsely to densely hairy (± glabrous), hairs gen white-opaque, glandless. **INFL:** peduncle becoming ± glabrous, base of head ± tomentose; outer phyllaries ± glabrous to hairy, hairs white-opaque, glandless, margin ciliate to woolly; inner much elongating in fr. **FL:** tube 4–7 mm, ligule 3–7 mm, ± equaling involucre, yellow; anthers 1–3 mm. **FR:** ± all alike, body 3–7 mm, fusiform, tip ± abruptly tapered; ribs straight, uniform, glabrous to minutely rough-hairy (or minutely puberulent); beak gen 10–20 mm, thread-like, flexuous, gen (1)3–4 × body; pappus 7–15 mm, in 2–3 series. 2*n*=18.

var. *grandiflora*   Pl ± robust. **LF:** gen 10–35 mm wide exc lobes, narrowly to broadly oblanceolate, obtuse to long-tapered, (± entire) toothed to lobed; lobes ± lanceolate to oblanceolate. **INFL:** peduncle gen 5–6 mm wide, infrequently bracted; involucre gen 3–5 cm; phyllaries gen rosy-purple, rarely purple-spotted; outer often exceeding inner in fl, lanceolate to (ob)ovate, gen obtuse to acute, entire to toothed. **FL:** gen 150–500. Grassland, scrub, woodland; gen 300–2000 m. CA-FP (exc GV, ChI), MP; to WA, ID, NV, rare in MT and UT. Hybridizes with *A. apargioides, A. aurantiaca.* Apr–Jul

var. *leptophylla* G.I. Baird   Pl ± slender. **LF:** gen 2–4 mm wide exc lobes, ± linear (or thread-like) to linear-oblanceolate, long-tapered, lobed (toothed); lobes thread-like to lance-linear. **INFL:** peduncle gen 3–4 mm wide, not bracted; involucre gen 2–4 cm; phyllaries gen rosy-purple and often purple-spotted; outer ± = or sometimes exceeding inner in fl, ± lanceolate, gen acute to long-tapered, gen entire. **FL:** gen < 100. Coastal prairie, grassland, meadows, forest; gen 100–1500 m. NW (mostly KR); to BC, also in ID. Possibly hybridizes with *A. aurantiaca.* Ranges of vars. completely overlap in n CA and intermediate pls are not uncommon. Gen May–Jun

A. heterophylla (Nutt.) Greene (p. 237)   Ann 5–60 cm, gen erect. **LF:** basal and often cauline; petiole not purple; blade gen 1–25 cm, ± oblanceolate, obtuse to long-tapered, entire to pinnately lobed; lobes linear to ± spoon-shaped, angled toward lf tip or spreading; glabrous to densely hairy, hairs white-opaque, glandless. **INFL:** peduncle ± glabrous, base of head ± tomentose (glabrous); involucre 5–15 mm in fl, 10–25 mm in fr; phyllaries green or rosy-purple (exc margin), sometimes darker spotted and/or tipped; outer phyllaries lanceolate to obovate, obtuse to long-tapered, glabrous or hairs gen colorless- or yellow-translucent (often with purple cross-walls), glandular, occ mixed with white-opaque, glandless hairs, margin glabrous or ciliate (tomentose); inner elongating in fr. **FL:** tube 1–5 mm, ligule yellow. **FR:** outermost gen different (all alike); body 2–5 mm, gen fusiform but highly variable, sometimes inflated, tip gradually tapered to truncate; ribs straight to wavy (ribless), uniform or smaller downward, glabrous to minutely coarse-hairy; beak 5–11 mm, slender, gen 2–3 × body; pappus 4–9 mm, in 2–3 series.

var. *cryptopleura* Greene   **LF:** lobes gen 3–5 pairs; 2° lobes gen 1; ± hairy (glabrous). **INFL:** outer phyllaries lanceolate to (ob)ovate, glabrous to densely long-soft hairy. **FL:** gen 20–100+, gen open well into day; ligule 10–15 mm, much exceeding involucre; anthers 2–4 mm. **FR:** highly variable but ribs ± reduced proximally. 2*n*=18. Many open habitats; gen 150–1000(2000) m. NCoRI, SNF, Teh, GV (uncommon, gen near edges), SnFrB, SCoR, WTR. *A. heterophylla* var. *cryptopleura* only outcrosses; almost impossible to separate from *A. heterophylla* var. *h.* when out of fl. Hybridizes (uncommonly) with *A. hirsuta.* May–Jun

var. *heterophylla*   **LF:** lobes gen 2–3 pairs; 2° lobes gen 0; glabrous to densely hairy. **INFL:** outer phyllaries ± lanceolate, glabrous to long-soft-hairy. **FL:** gen 20–50, open ± briefly in morning; ligule 2–6 mm, ± equaling involucre; anthers 1–1.5 mm. **FR:** highly variable but ribs gen not reduced proximally. 2*n*=18,36. Many open habitats; gen < 2000 m. CA-FP (exc s ChI, uncommon GV), MP; to BC,

MT, UT, NM, Baja CA; adventive Eur (Sweden). Var. *heterophylla* self-pollinates; almost impossible to separate from *A. heterophylla* var. *cryptopleura* when out of fl. Hybridizes (uncommonly) with *A. apargioides*, *A. hirsuta*. May–Jun

A. hirsuta (Hook.) Greene (p. 237) Per 10–45 cm, gen erect. **LF:** petiole rarely purple; blade gen 10–30 cm, narrowly to broadly oblanceolate, acute to long-tapered, gen pinnately lobed (toothed); lobes gen 5–7(+) pairs, lanceolate to oblanceolate, angled toward lf tip or spreading; ± hairy, hairs white-opaque, glandless. **INFL:** peduncle becoming ± glabrous, base remaining gen ± densely hairy, base of head densely hairy to ± glabrous; involucre 10–15 mm in fl, 15–25 mm in fr; phyllaries green or rosy-purple exc margin, occ spotted; outer lance-linear to ovate, gen obtuse to bluntly tapered, gen ± densely hairy, hairs gen ± yellow-translucent (occ with purple cross-walls) and glandular, margin ciliate; inner elongating in fr. **FL:** 50–250; tube 3–6 mm, ligule 6–16 mm, much exceeding involucre, yellow; anthers 3–5 mm. **FR:** outermost gen different; body 3–5 mm, fusiform to narrowly conical, tip gradually to (occ) abruptly tapered; ribs straight or slightly wavy, uniform, glabrous to minutely coarse-hairy; beak gen 6–10 mm, slender, flexuous, gen 1–3 × body; pappus gen 6–10 mm, in 3–4 series. 2*n*=18. Grassland, scrub, marshes (rare); gen < 500 m. NCoR, ScV, CW; OR (type collection only). Hybridizes with *A. apargioides*, *A. grandiflora* var. *g.*, *A. heterophylla* var. *cryptopleura*, *A. retrorsa*. Apr–Jun

A. monticola Greene (p. 237) Per 2–25 cm, gen decumbent to prostrate. **LF:** petiole gen not purple; blade gen 2–10 cm, narrowly to broadly oblanceolate, obtuse to long-tapered, gen toothed or lobed (entire); lobes gen 2(+) pairs, linear to oblanceolate, ± divergent, proximal angled toward lf base, distal angled toward lf tip, or spreading; ± densely hairy (± glabrous, glaucous), hairs ± white- or yellow-translucent. **INFL:** peduncle gen 5–20 cm, ± glandular-hairy, base gen ± glandular-woolly, base of head glandular-hairy; involucre 10–20 mm in fl, ± same in fr; phyllaries gen rosy-purple exc thin margin (rarely all green), gen with darker midstripe; outer lance-linear to deltate, obtuse to long-tapered, ± hairy, hairs gen colorless- or yellow-translucent (rarely with purple cross walls) and glandular, margin ± ciliate; inner not elongating in fr. **FL:** gen 10–40; tube 4–10 mm, ligule 5–10 mm, exceeding involucre, yellow; anthers gen 3–5 mm. **FR:** outermost gen different; body 6–9 mm, narrowly fusiform, tip ± narrowed (or not); ribs straight, uniform, glabrous to minutely rough-hairy; beak (0)1–3 mm, stout, << body; pappus 8–11 mm, in ±

2 series. 2*n*=18,36. Subalpine meadows, forest to alpine tundra, rocky slopes; 1500–3800 m. KR, CaRH, SNH, Wrn, W&I; to WA, NV. [*A. glauca* var. *m.* (Greene) Q. Jones] Hybridizes with *A. aurantiaca, A. glauca, A. parviflora*. Gen Jul–Aug

A. parviflora (Nutt.) D. Dietr. (p. 237) Per gen 5–25(45) cm, erect to decumbent. **LF:** petiole ± purple; blade gen 10–20 cm, ± lanceolate, long-tapered, toothed to pinnately lobed (entire); lobes gen 5–8 pairs, lanceolate to linear, angled toward lf base or spreading; glabrous and glaucous or hairs white-opaque, glandless. **INFL:** peduncle ± glabrous in age, base of head ± hairy; involucre 10–30 mm in fl, 20–35 mm in fr; phyllaries gen rosy-purple exc margin (all green or spotted), with ± darker midstripe; outer lanceolate to (ob) ovate, gen long-tapered, glabrous or hairs white-opaque, glandless, margin gen ciliate to woolly (esp proximally); inner gen ± elongating in fr. **FL:** 30–100; tube 6–15 mm, ligule 10–20 mm, exceeding involucre, yellow; anthers 3–5 mm. **FR:** outermost ± different; body 5–9 mm, fusiform to narrowly obconic, tip gradually to abruptly tapered; ribs straight, ± uniform, glabrous to minutely coarse-hairy; beak 3–10 mm, gen slender, (< 1)1–2 × body; pappus 10–20 mm, in ± 3 series. 2*n*=18. Gen dry, open sagebrush scrub, open pine woodland, meadows; gen 1400–2400(3400) m. CaRH/GB, n&c SNH, GB; to OR, MT, SD, NM. [*A. glauca* var. *laciniata* (D.C. Eaton) Kuntze] Hybridizes with *A. aurantiaca, A. glauca, A. monticola, A. retrorsa*. Apr–Aug

A. retrorsa (Benth.) Greene (p. 237) Per 10–65(+) cm, ± erect. **LF:** petiole ± purple; blade gen 10–30 cm, linear to lance-elliptic, long-tapered, gen pinnately lobed (toothed or entire); lobes gen 7–9 pairs, gen lance-linear, gen angled toward lf base (spreading); ± hairy, hairs white-opaque, glandless. **INFL:** peduncles becoming ± glabrous; base of head hairy; involucre 20–30 mm in fl, 40–60 mm in fr; phyllaries gen rosy-purple (esp toward base) exc margin (rarely all green), not spotted; outer phyllaries lanceolate to deltate (or oblong), obtuse to long-tapered, ± glabrous or hairs white-opaque, glandless, margin ciliate to tomentose; inner phyllaries gen early-elongating in fr. **FL:** 10–100; tube 10–20 mm, ligule 6–15 mm, exceeding (equaling) involucre, yellow; anthers 2–5 mm. **FR:** ± all alike; body 5–7 mm, ± obconic, tip ± truncate; ribs straight, ± uniform, glabrous or minutely puberulent; beak 15–25 mm, slender, (< 3)3–4 × body; pappus gen 15–20 mm, in 4–6 series. 2*n*=18. Scrub, oak woodland, conifer forest; gen 900–1800 m. CA-FP (exc coast, GV, ChI); to WA, UT, AZ, Baja CA. Apr–Aug

ALMUTASTER MARSH ALKALI ASTER

John C. Semple

1 sp.: N.Am. (Almut G. Jones, Am botanist, b. 1923) [Brouillet 2006 FNANM 20:461]

A. pauciflorus (Nutt.) Á. Löve & D. Löve (p. 237) Per, rhizome long. **ST:** 3–12 dm, ascending to erect, distally branched, proximally glabrous, glandular distally. **LF:** basal and cauline, alternate, 6–10 cm; basal petioled, petiole base widened, sheathing; cauline sessile, smaller distally (bract-like in infl), ± linear, acute to acuminate, ± glabrous. **INFL:** heads radiate, in open panicle-like clusters; involucre 4–7 mm, ± bell-shaped, phyllaries in 3–4 graduated series, ± oblong to narrowly oblanceolate, acute, green, ± glandular, midrib gen ± swollen, translucent; receptacle convex, pitted, epaleate. **RAY FL:** 15–30(45); corolla 5–8 mm, ± white to pale purple. **DISK FL:** corolla 4–6 mm, yellow, ± glabrous; anther tip ovate; style branches finely papillate, tips narrowly triangular. **FR:** fusiform, hairy, finely 7–10-ribbed; pappus of 25–45 bristles in 2–3 series, innermost longest, tip ± wider, flattened. 2*n*=18. Damp alkaline places; 200–700 m. s SNF, n DMoj (Inyo Co.); to c Can, TX, Mex. [*Aster p.* Nutt.] Jun–Oct

AMAURIOPSIS YELLOW RAGLEAF

1 sp. [Strother 2006 FNANM 21:392–393] (Greek: like *Amauria*)

A. dissecta (A. Gray) Rydb. (p. 237) Gen bien (ann or short-lived per). **ST:** gen 1, erect, 2–12 dm, openly branched distally, glandular. **LF:** basal and cauline, alternate, petioled, 1–several × ternately lobed or divided into linear or oblong segments 1–5 mm wide, strigose or short-spreading-hairy, gen gland-dotted. **INFL:** heads radiate, gen in ± flat-topped to panicle-like clusters; peduncle 0.7–7 cm; involucre 7–10 mm diam, ± bell-shaped to hemispheric; phyllaries 12–24, ± equal in 2–3 series, 4.5–6 mm, oblanceolate to lanceolate, acuminate, soft-hairy, glandular; receptacle rounded, epaleate, pitted. **RAY FL:** 10–21+; corolla yellow, ray 5–10+ mm. **DISK FL:** 30–80+; corolla radial or bilateral, 2.5–4.5 mm, yellow, glandular; anther tips triangular; style tips flat, widely triangular. **FR:** 3–4.5 mm, obpyramidal, 4-angled, glabrous or short-rough-hairy; pappus gen 0 (scales 1–13, 1.5–3 mm, lanceolate). 2*n*=3*x*=36. Dry, open forest slopes; 1800–2650 m. SnBr, e PR (Santa Rosa Mtns), DMtns (New York Mtns); to WY, CO, TX, n Mex. [*Bahia d.* (A. Gray) Britton] Triploid, reproducing asexually. Aug–Oct

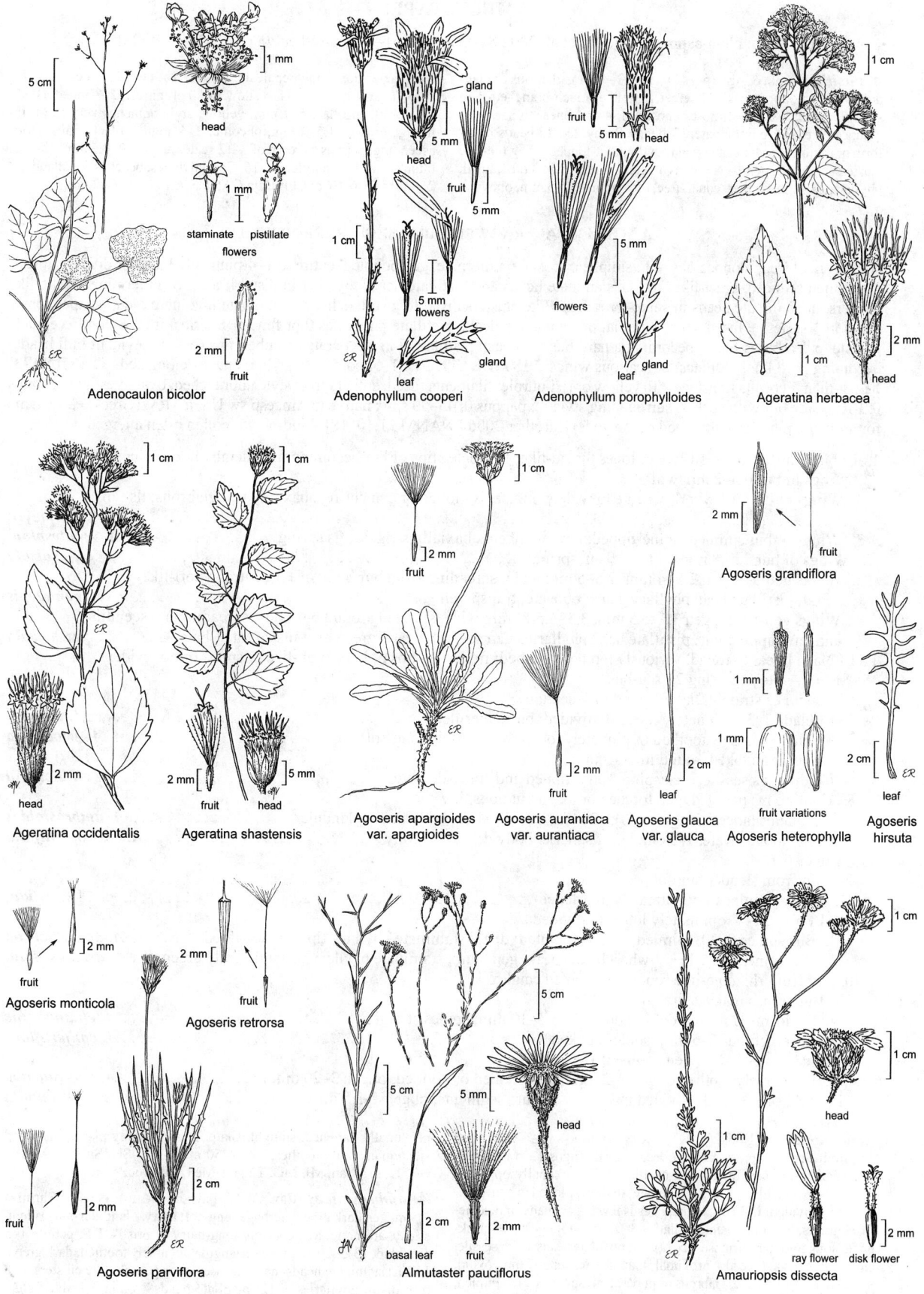

Adenocaulon bicolor

Adenophyllum cooperi

Adenophyllum porophylloides

Ageratina herbacea

Ageratina occidentalis

Ageratina shastensis

Agoseris apargioides var. apargioides

Agoseris aurantiaca var. aurantiaca

Agoseris glauca var. glauca

Agoseris heterophylla

Agoseris hirsuta

Agoseris grandiflora

Agoseris monticola

Agoseris retrorsa

Agoseris parviflora

Almutaster pauciflorus

Amauriopsis dissecta

AMBLYOPAPPUS

1 sp. (Greek, Latin: blunt pappus) [Baldwin et al. 2002 Syst Bot 27:161–198; Strother 2006 FNANM 21:348]

A. pusillus Hook. & Arn. (p. 241)   Ann 3–40 cm, glabrous or glandular, sticky; odor strong. **ST**: erect, simple or much-branched. **LF**: simple, alternate or lowermost opposite, 1–4 cm, linear to narrowly oblanceolate, entire or pinnately lobed, ± fleshy. **INFL**: heads disciform or minutely radiate, many in round- or flat-topped, lfy cyme-like clusters; peduncle gen 1–6 mm; involucre 3–6 mm diam, bell-shaped; phyllaries 3–8, in 1 series, equal, free, overlapping, 2–5 mm, obovate; receptacle short-conic, epaleate. **PISTILLATE FL**: 1 per phyllary; corolla 0.5–1 mm, tubular, yellow, ray 0 or minute, 2–3-lobed. **DISK FL**: 2–30; corolla 0.5–1 mm, yellow; style branches very short, tip triangular. **FR**: 1.5–2 mm, obconic, 3–4 angled, black; hairs short, ascending; pappus a crown of 7–12 scales, each 0.2–0.5 mm, oblong, blunt, white or ± purple. 2*n*=16. Coastal dunes, beaches, headlands; < 250 m. s CCo, SCo, ChI; Baja CA; w S.Am. Gen Mar–Jun

AMBROSIA   RAGWEED, BUR-SAGE, BURROBRUSH

Ann to small tree; monoecious. **LF**: simple, opposite or alternate, gen petioled, entire to 1–4-pinnately lobed or divided, distalmost often reduced, bract-like. **INFL**: staminate heads and pistillate heads together in distal lf axils or in terminal, spike-like clusters, or staminate heads in spike- or raceme-like clusters distal to pistillate heads; staminate involucre ± cup-shaped, phyllaries in 1 series, ± fused, receptacle flat or rounded, paleate; pistillate phyllaries 0 or few, in 1 series, free, ± thin, receptacle paleate, palea bases fused, becoming a hard bur with each pistillate fl in a separate chamber, tips (occ not evident in fl heads) becoming hard knobs, spines, or scarious wings. **PISTILLATE FL**: 1–5; corolla 0; style branches elongated. **STAMINATE FL**: 3–many; corolla translucent to yellow or red-purple; filaments fused, anthers free; style unbranched, ovary vestigial, pappus 0. **FR**: inside bur with beak(s) surrounding style(s); pappus 0. 45–50 spp.: native to Am, esp sw US, n Mex. (Greek: early name for aromatic plants; mythic food of the gods) [Strother 2006 FNANM 21:10–18] Wind-blown pollen often allergenic.

1. Lf blade linear to thread-like or lobes thread-like; free palea tips of bur becoming flat, membranous, wings
   2. Wings of bur 0.5–2 mm wide
      3. Wings of bur in 1 whorl, ± acute to widely obtuse; lvs lobed or gen entire, abaxially gen glabrous; fls fall
         . . . . ***A. monogyra***
      3′ Wings of bur spiraled, spine-tipped; lvs gen lobed, abaxially strigose; fls spring . . . . [***A. ×platyspina***]
   2′ Wings of bur 2.5–8 mm wide — fls in spring . . . . — ***A. salsola***
      4. Wings gen 5–9, gen 2.5–4 mm, 2.5–5 mm wide, spreading from bur body in ± 1 central whorl-like
         spiral; pistillate head phyllary gen 1, obovate, transparent . . . . var. ***pentalepis***
      4′ Wings gen 10–13, gen 3.5–5.5 mm, 3.5–6.5(8) mm wide, spiraled around bur body, spreading or ascending
         and enwrapping bur; pistillate head phyllaries gen several, ovate, green or transparent, ± hairy . . . . var. ***salsola***
1′ Lf blade linear to round, variously toothed or lobed; free palea tips of burs knob-like or spiny, or not evident
   5. Shrub — sts persisting 2+ seasons
      6. Bur spines straight, flat or round in ×-section
         7. Lf blade 1–3 × pinnately lobed or divided; bur puberulent . . . . ***A. dumosa***
         7′ Lf blade coarsely toothed or pinnately lobed; bur densely long-soft-hairy . . . . ***A. eriocentra***
      6′ Bur spines hooked, round in ×-section
         8. Cauline lvs sessile, ± clasping, spine-tipped and spiny-dentate . . . . ***A. ilicifolia***
         8′ Cauline lvs petioled, gen toothed or lobed but not spiny
            9. Lf blades lance-triangular, (3)5–20 cm; bur ellipsoid, body finely glandular . . . . ***A. ambrosioides***
            9′ Lf blades ± ovate, 1–3 cm; bur ± spheric, body densely woolly . . . . ***A. chenopodiifolia***
   5′ Ann or per
      10. Ann from slender taproot
         11. Lf blade unlobed or palmately 3–5-lobed . . . . [***A. trifida***]
         11′ Lf blade 1–4 × pinnately lobed or divided
            12. Bur spines 0–30, spiraled, sharply pointed; distal staminate phyllary tips ± black-lined . . . . ***A. acanthicarpa***
            12′ Bur spines 4–12, in ± 1 whorl below beak, gen blunt; staminate phyllary tips uniformly green . . . ***A. artemisiifolia***
      10′ Per from rhizome-like roots or taprooted caudex
         13. Bur spines many, spiraled
            14. Sts forming mat on coastal dunes; bur 5–10 mm; spines straight . . . . ***A. chamissonis***
            14′ Sts erect; bur 2–5 mm; spines hooked . . . . ***A. confertiflora***
         13′ Bur spines 0–few, gen on distal 1/2
            15. Lf coarsely toothed or gen 1–2 × pinnately lobed or divided; pl gen 3–20 dm . . . . ***A. psilostachya***
            15′ Lf 2–4-pinnately divided into many minute segments; pl gen 1–3.5 dm . . . . ***A. pumila***

A. acanthicarpa Hook. (p. 241)   ANNUAL BUR-SAGE   Ann 4–15 dm, from slender taproot; herbage ± canescent-strigose, ± bristly. **ST**: 1–several from base, branches ascending. **LF**: proximally opposite, petiole 1–6+ cm; blade 1–8 cm, ± deltate, shallowly lobed to gen 1–3 × pinnately divided. **INFL**: staminate heads few to gen many in raceme-like clusters, 2–5 mm diam, phyllaries 3–9, tips of longest 3 ± black-lined along midvein, strigose, occ bristly; pistillate heads 1-fld, sessile, clustered in distal (occ also proximal) lf axils. **FR**: bur 5–7 mm, ovoid, ± golden or dark purple, glabrous or puberulent; spines 0–30, spiraled, proximally flattened, straight, sharp. 2*n*=36. Sandy plains, disturbed sites, many communities; -30–2450 m. NCoR, SN, SnJV, CW, SW, GB, D; to s Can, MI, MO, TX, nw Mex. Jul–Dec

A. ambrosioides (Cav.) W.W. Payne   BIG BUR-SAGE   Shrub < 2.5 m, gen dark green; herbage long-soft hairy, glandular-puberulent, sticky, aromatic. **ST**: coarsely long-hairy or bristly. **LF**: petiole 1–5 cm; blade (3)5–20 cm, lance-triangular, coarsely toothed, dark green. **INFL**: staminate heads many in raceme- or panicle-like clusters, 6–8 mm diam, phyllaries 7–12; pistillate heads 1 or in ± stalked clus-

ters proximal to staminate, 3–5-fld. **FR:** bur 10–15 mm, ellipsoid, ± brown, finely glandular; spines gen > 50, spiraled, 4–6 mm, ± cylindric, hooked. 2*n*=36. Coastal scrub, disturbed sites; < 150 m. s SCo (near San Diego); AZ, nw Mex. Mar–May

A. artemisiifolia L. (p. 241) COMMON RAGWEED Ann < 7 dm, much-branched; herbage long-soft hairy and/or ± bristly. **ST:** occ red- or black-marked. **LF:** proximally opposite, petiole 1–4 cm; blade 3–12 cm, widely ovate, gen 2–3 × pinnately divided, strigose and ± long-soft hairy. **INFL:** staminate heads many in raceme-like clusters, 2–5 mm diam, involucre asymmetric, lobes 5–10, green; pistillate heads 1-fld, sessile, single or clustered in distal lf axils. **FR:** bur 2–4 mm, widely obconic, green to brown, ± puberulent; spines 4–12, gen blunt, ± vestigial, ± in 1 whorl below beak. 2*n*=36. Uncommon. Disturbed sites; < 1050 m. KR, CaRF, n&c SN, ScV, SCoRO, SCo, PR; to AK, Can, ± worldwide; native to e US. Jul–Oct

A. chamissonis (Less.) Greene (p. 241) BEACH BUR-SAGE Per from caudex or taproot, mat-forming, herbage gen densely canescent-strigose, occ also long-soft hairy. **ST:** < 4 m, decumbent or prostrate. **LF:** petiole 1–5 cm; blade 2–5+ cm, oblanceolate to elliptic or widely triangular, crenate to deeply toothed or 1–3 × pinnately lobed or divided. **INFL:** staminate heads gen many in raceme-like clusters, 4–8 mm diam, phyllaries ± 10, ± black-lined above, strigose (glabrous); pistillate heads 1-fld, proximal on infl axis to staminate or clustered in distal lf axils. **FR:** bur 5–10 mm, ovoid, ± brown, minutely glandular, puberulent to tomentose; spines 10–20+, spiraled, 0.5–1.5 mm; ± cylindric, straight, sharp, bases wide. 2*n*=32,36. Beaches, dunes; < 25 m. NCo, CCo, SCo, ChI; to BC, Baja CA, also w S.Am. May–Oct

A. chenopodiifolia (Benth.) W.W. Payne (p. 241) SAN DIEGO BUR-SAGE Shrub < 3.5 m, rounded, much-branched. **ST:** slender, tough, tomentose or becoming glabrous. **LF:** petiole 5–20 mm; blade 1–3 cm, ± ovate, entire to palmately 3-lobed, abaxially densely tomentose, adaxially ± green, thinly tomentose or becoming ± glabrous. **INFL:** staminate heads few, in raceme-like clusters, 4–8 mm diam, phyllaries 5–8, ± veiny, tomentose; pistillate heads 2–3-fld, on infl axis or clustered on short infl branches proximal to staminate heads. **FR:** bur 5–7 mm, ± spheric, body densely woolly; spines 12–25, spiraled, 2–3 mm, slender, gen hooked. 2*n*=36. Coastal scrub; < 250 m. s SCo (sw San Diego Co.); Baja CA. Feb–May ★

A. confertiflora DC. Per 3–18 dm, from rhizome-like roots; herbage strigose, minutely resin-gland-dotted. **ST:** strigose and ± bristly. **LF:** sessile or petiole to 4 cm, with lobed wing; blade < 16 cm, ± ovate, 1–4 × pinnately divided. **INFL:** staminate heads many in raceme-like clusters, 1.5–3+ mm diam, phyllaries 5–9, green, minutely strigose; pistillate heads 1(2)-fld, clustered in distal lf axils. **FR:** bur 2–5 mm, 1.5–5 mm diam, ovoid, ± tapered at base, distally pitted at spine bases; spines 5–12, spiraled, 0.5–1 mm (or vestigial), cylindric, hooked, puberulent and minutely gland-dotted. 2*n*=72,108. Disturbed areas; < 1250 m. SnJV, CCo, SCo, e DMoj; to KS, OK, TX, n Mex. Often forms large colonies. Jul–Oct

A. dumosa (A. Gray) W.W. Payne (p. 241) WHITE BUR-SAGE Shrub 2–9 dm, rounded, much-branched; herbage softly canescent-strigose. **ST:** stiff. **LF:** sessile or petiole to ± 1 cm; blade 0.5–4 cm, ± ovate, 1–3 × pinnately lobed or divided. **INFL:** staminate and pistillate heads mixed along axis of spike- or raceme-like cluster; staminate heads 3–5 mm diam, phyllaries 5–8, canescent; pistillate heads 2-fld. **FR:** bur 4–9 mm, ± spheric, golden to purple or brown, puberulent; spines 12–35, 2–4 mm, spiraled, flat, straight, sharp. 2*n*=36,72,108,126,144. Creosote-bush scrub; -80–1700 m. s SNE, W&I, D; to sw UT, AZ, nw Mex. Hybridizes with *A. salsola.* Dec–Jun

A. eriocentra (A. Gray) W.W. Payne (p. 241) WOOLLY BUR-SAGE Shrub 3–18 dm, ± spheric. **ST:** gray-brown, ± woolly, becoming glabrous. **LF:** sessile, 1–9 cm, ± lanceolate, base tapered, coarse-toothed or pinnately lobed, ± rolled under, abaxially gray-tomentose, minutely resin-gland-dotted, adaxially ± green, thinly tomentose, gland-dotted. **INFL:** staminate heads few, in raceme-like clusters, 5–7 mm diam, phyllaries 5–8, long-soft hairy; pistillate heads gen 1-fld, clustered in distal lf axils. **FR:** bur 8–11 mm, green-brown, densely long-soft-hairy; spines 12–20, gen near middle, 3–4 mm, straight, flat, stout, sharp, tips ± hair-tufted. 2*n*=36. Dry washes and slopes; 450–1750 m. e DMoj, DMtns; to nw AZ, sw UT. Apr–Jun

A. ilicifolia (A. Gray) W.W. Payne (p. 241) HOLLY-LEAVED BUR-SAGE Shrub < ± 1 m, ± matted; herbage densely glandular, short-stiff hairy, sticky. **ST:** erect, few-branched. **LF:** sessile, ± clasping, 2–10 cm, ovate to round, spine-tipped and spiny dentate, leathery, brittle, dark green, prominently veiny. **INFL:** staminate heads few, in raceme-like clusters, 10–15 mm diam, phyllaries 10–15, lanceolate, spine-tipped, glandular, sparsely stiff-hairy; pistillate heads gen 2-fld, sessile or short-peduncled in axils of distal, bract-like lvs. **FR:** bur 10–20 mm, spheric, ± brown, sticky; spines 20–70, spiraled, < 6 mm, ± cylindric, strongly hooked. 2*n*=36. Sandy washes, rocky canyons, creosote-bush scrub; < 700 m. DSon; w AZ; Baja CA. Jan–Apr

A. monogyra (Torr. & A. Gray) Strother & B.G. Baldwin (p. 241) SINGLEWHORL BURROBRUSH Shrub or small tree 1–4 m, much-branched above. **ST:** glabrous, resinous. **LF:** 2–5 cm, thread-like or with few thread-like lobes, abaxially gen glabrous, resinous, adaxially grooved and minutely puberulent in the groove. **INFL:** staminate and pistillate heads mixed, ± sessile in tight groups on short, spike-like branches of open, lfy, panicle-like cluster; staminate heads 2–4 mm diam, phyllaries 5–7, glabrous; pistillate heads 1-fld. **FR:** bur 2–5 mm, body fusiform; wings 7–12 in 1 central whorl, 1.5–2 mm, 0.5–2 mm wide, oblanceolate to obovate, ± acute to widely obtuse. Washes, dry riverbeds; < 500 m. SW, SNE, DSon; to w NV, TX, n Mex. [*Hymenoclea m.* Torr. & A. Gray] Sep–Nov ★

A. psilostachya DC. (p. 241) WESTERN RAGWEED Per 3–20 dm, colonial from rhizome-like roots; herbage strigose and/or ± bristly. **ST:** 1–few from common base, often little-branched. **LF:** proximally opposite, petiole ≤ 2.5 cm, winged, distal lvs gen sessile; blade 2–12 cm, lanceolate to ovate, coarsely toothed or gen 1–2 × pinnately lobed or divided, resin-gland-dotted. **INFL:** staminate heads gen many in raceme-like clusters, 2–5 mm diam, phyllaries 3–5, strigose or short-rough-hairy; pistillate heads 1-fld, clustered in distal lf axils. **FR:** bur 3–4.5 mm, obovoid, brown, puberulent; spines 0–7, (0.1)0.5–1 mm, conic, straight. 2*n*=36,72,108,144. Common. Roadsides, dry fields; < 1000 m. CA-FP; to ID, e N.Am, TX, n Mex. Jun–Nov

A. pumila (Nutt.) A. Gray (p. 241) SAN DIEGO AMBROSIA Per 1–3.5(5) dm, colonial from rhizome-like roots. **ST:** erect, gen few-branched, densely short-hairy. **LF:** petiole 1–4 cm, with lobed wings; blade 3–13 cm, lanceolate to widely ovate, 2–4 × pinnately divided into minute lobes, ± densely canescent-strigose. **INFL:** staminate heads few to many in raceme-like clusters, 3–5 mm diam, phyllaries 5–8, widely obtuse, puberulent; pistillate heads 1-fld. **FR:** burs 2–2.5 mm, obovoid, ± brown, puberulent; spines (0)1–5, vestigial. 2*n*=36. Disturbed sites; 50–600 m. s SCo, PR (Riverside, San Diego cos.); Baja CA. Apr–Jul ★

A. salsola (Torr. & A. Gray) Strother & B.G. Baldwin COMMON BURROBRUSH, CHEESEBUSH Subshrub ≤ 2 m. **ST:** branched throughout. **LF:** 2–5 cm, linear to thread-like or with few thread-like lobes, abaxially glabrous or puberulent, resinous, adaxially grooved, densely minutely puberulent in groove. **INFL:** staminate and pistillate heads ± sessile, mixed, or staminate ± distal to pistillate, 1 or in tight groups on short, spike-like branches; staminate head 2–4 mm diam, phyllaries 4–8, ± glabrous or puberulent; pistillate head 1-fld. **FR:** bur 3–6 mm, body widely fusiform; wings 5–18, whorled or spiraled, 2–7 mm, often reniform. Hybridizes with *A. dumosa.* Vars. intergrade.

var. ***pentalepis*** (Rydb.) Strother & B.G. Baldwin **INFL:** pistillate head phyllary gen 1, obovate, transparent. **FR:** bur wings 5–9(13), in 1 central whorl-like spiral, gen 2.5–4 mm, 2.5–5 mm wide, spreading from bur body. Dry flats, washes, fans; < 550 m. D; s NV, sw AZ, nw Mex. [*Hymenoclea s.* Torr. & A. Gray var. *p.* (Rydb.) L.D. Benson] Mar–Apr

var. ***salsola*** (p. 241) **INFL:** pistillate head phyllaries gen several, ovate, green or transparent, ± hairy. **FR:** bur wings (6)10–13(18), spiraled, gen 3.5–5.5 mm, 3.5–6.5(8) mm wide, spreading from bur body or ascending and enwrapping bur. Dry flats, washes, fans; -70–1850 m. s SnJV, s SCoRI, SW, s GB, D; to sw UT, w AZ, Baja CA. [*Hymenoclea s.* Torr. & A. Gray var. *s.*; *A. s.* var. *fasciculata* (A. Nelson) Strother & B.G. Baldwin; *Hymenoclea s.* var. *f.* (A. Nelson) K.M. Peterson & W.W. Payne; *H. s.* var. *patula* (A. Nelson) K.M. Peterson & W.W. Payne, illeg.] Feb–Jun

AMPHIPAPPUS CHAFF-BUSH

David J. Keil & Meredith A. Lane

1 sp. (Greek: double pappus) [Nesom 2006 FNANM 20:186]

A. fremontii Torr. & A. Gray Shrub, gen < 6 dm, glabrous or herbage minutely stiff-hairy. **ST:** much-branched, widely spreading to ascending, striate, smooth, white distally, ± gray proximally; lfless sts thorny. **LF:** alternate, short-petioled, gen < 2 cm, obovate or elliptic, entire, sometimes ± thick, light yellow- or gray-green. **INFL:** heads radiate, in groups of 2–4, crowded in flat-topped cyme-like clusters gen 3–5 cm wide; involucres obconic-cylindric, ± 5 mm, < 3 mm wide; phyllaries 7–12, ovate, ± white or pale green; receptacle flat, pitted, epaleate. **RAY FL:** 1–2; ray barely exceeding involucre, 2- or 3-toothed, yellow. **DISK FL:** 3–7, staminate; corolla narrowly funnel-shaped, sinuses deep, lobes reflexed, style-branch appendages lanceolate to rounded. **FR:** ray fr ≤ 3 mm, hairy; pappus of 15–20 stout bristles fused at base, gen 1 mm; ovary of disk fl ≤ 1 mm, glabrous; pappus of 25 flattened, twisted bristle-like scales gen ≤ 3 mm. Vars. not known to intergrade. Phylogenetic studies by Roberts & Urbatsch 2004 Syst Bot 29:199–215, suggest that *Amphipappus* and some other small genera should be merged into *Chrysothamnus*.

1. Lvs glabrous, sometimes gummy . . . . var. *fremontii*
1′ Lvs minutely scabrous on faces and margins. . . . var. *spinosus*

var. *fremontii* (p. 241) 2*n*=18. Rocky or gravelly flats, slopes, canyons; < 1100 m. ne DMoj; NV. Mar–Jun

var. *spinosus* (Nelson) Ced. Porter Rocky or gravelly flats, slopes, canyons; < 1600 m. e-c DMoj (San Bernardino Co.); to UT, NV, AZ. Mar–Jun

ANAPHALIS PEARLY EVERLASTING

Guy L. Nesom

Per [subshrub], fibrous-rooted, from rhizome; ± dioecious. **ST:** gen erect. **LF:** basal and cauline, alternate, sessile [petioled], linear to lanceolate or oblanceolate, narrowed at base, entire, adaxially green and ± glabrous or ± gray-tomentose, abaxially gen white to gray, tomentose or becoming glabrous, glandular or not. **INFL:** heads discoid or disciform, ± unisexual (sometimes a few pistillate fls peripheral in gen staminate heads or 1–9 staminate fls central in gen pistillate heads), in tight groups in ± flat-topped or panicle-like cluster; involucre ± spheric; receptacle flat, epaleate, glabrous; phyllaries ± graduated in 8–12 series, bright white, opaque, often proximally woolly. **PISTILLATE FL:** many; corolla ± yellow, narrowly tubular, minutely lobed; anthers 0. **STAMINATE FL:** corolla ± yellow; anther tip ovate; style tips truncate. **FR:** oblong to obconic, 2-veined, papillate with club-shaped hairs; pappus bristles gen deciduous, free or basally fused, tips club-shaped in staminate fls. ± 100 spp.: mostly Eurasia, 1 N.Am. (Ancient name or perhaps from generic name *Gnaphalium*) [Nesom 2006 FNANM 19:426–427]

A. margaritacea (L.) Benth. (p. 241) Rhizome ± slender. **ST:** white, densely, closely tomentose. **LF:** blade 1–3-veined, 3–10(15) cm, base ± clasping, decurrent, margins rolled under. **INFL:** involucre 5–7 mm, 6–8(10) mm diam; outer phyllaries ovate, inner to nearly linear, ± equal to unequal; tips ± spreading. **FR:** 0.5–1 mm, base narrowed, forming stalk. 2*n*=28. Woodland, disturbed places; < 3200 m. NW, CaRH, SNH, CW, SnBr, MP; to AK, n Can, e US, Baja CA; Eurasia. Gen Jul–Oct

ANCISTROCARPHUS GROUNDSTAR, WOOLLY FISHHOOKS

James D. Morefield

Ann 0.5–14 cm, ± green to ± gray, cobwebby or tomentose. **ST:** 0, or 1, erect, or 2–10, ascending to prostrate, ± forked, ± lfless between distal forks. **LF:** basal and/or cauline, ± alternate, ± sessile or petioled, linear-oblanceolate to spoon-shaped, entire; distal lvs subtending heads, crowded, largest > proximal lvs (sometimes ± grading into paleae in *Ancistrocarphus keilii*). **INFL:** heads disciform, sessile, 1 or 2–5 per group; involucre 0, simulated by paleae or lvs, or involucre ± cup-like, then phyllaries 3–6, ± equal, round, << paleae, membranous, persistent, sometimes vestigial; receptacle length 1–2 × width, mushroom- or hourglass-shaped, glabrous; paleae of pistillate fls each enclosing a fl, ± persistent with fr, ± boat-shaped, ± acute, woolly abaxially, prominently 3-veined or middle vein obscure, margin reflexed distally as a scarious wing, wing terminal, erect to curved inward, hidden or visible in head; paleae of disk fls gen 5, ± enlarged, spreading proximally, hooked or curved inward distally, open, concave, ± lanceolate or ± obovate, acuminate or abruptly papery-pointed, cartilaginous, ± green, tomentose. **PISTILLATE FL:** 5–11, each subtended by a palea; corolla obscure, narrowly cylindric. **DISK FL:** 3–6, staminate; pappus 0; corolla 4–5(6)-lobed; anther base tailed, tip ± triangular; style tips ± linear-oblong. **FR:** each enclosed by a palea, obovoid, compressed front-to-back, smooth, dull, corolla scar terminal; pappus 0. 2 spp.: w N.Am. (Greek: fishhook chaff) [Morefield 2006 FNANM 19:465–466] Closely related to *Hesperevax*.

1. St (1)2–10(14) cm, ± forked; lvs basal and cauline; paleae of disk fls becoming 2.7–4.1 mm, ± lanceolate; palea tip exserted from head, hooked inward, acuminate, stiff, forming a spine; fr 1.4–2 mm, 0.6–0.9 mm wide, proximally dark-banded . . . . *A. filagineus*
1′ St 0; lvs basal; paleae of disk fls becoming 1.8–2.8 mm, ± obovate; palea tip incl in head, curved inward, rounded, abruptly papery-pointed, not forming a spine; fr 1–1.4 mm, 0.5–0.6 mm wide, not dark-banded . . . . . . *A. keilii*

A. filagineus A. Gray (p. 241) WOOLLY FISHHOOKS **LF:** ± sessile, largest 8–16(28) mm, 2–3(4) mm wide, linear-oblanceolate to elliptic; base narrowed, ± acuminate, 1-veined, ± green, scarcely involucre-like, unlike pistillate paleae. **INFL:** heads 2–5 per group, 3.5–8 mm; involucre ± cup-like; receptacle mushroom-shaped; wings of paleae of pistillate fl visible in head, ovate. **DISK FL:** 1–1.5 mm; corolla lobes (4)5(6), ± equal, radial. Bare or grassy, often serpentine or clay, drainages, road beds, burns, vernally moist sites; 60–1900(2100) m. CA-FP (exc NCo, CCo, ChI), MP (exc Wrn), w DMoj (exc DMtns); to OR, sw ID, n NV, nw Baja CA. Hooked paleae likely aid dispersal. Mar–Jun

A. keilii Morefield (p. 241) SANTA YNEZ GROUNDSTAR **LF:** petioled, largest 8–10(18) mm, 1(2) mm wide, ± spoon-shaped; base

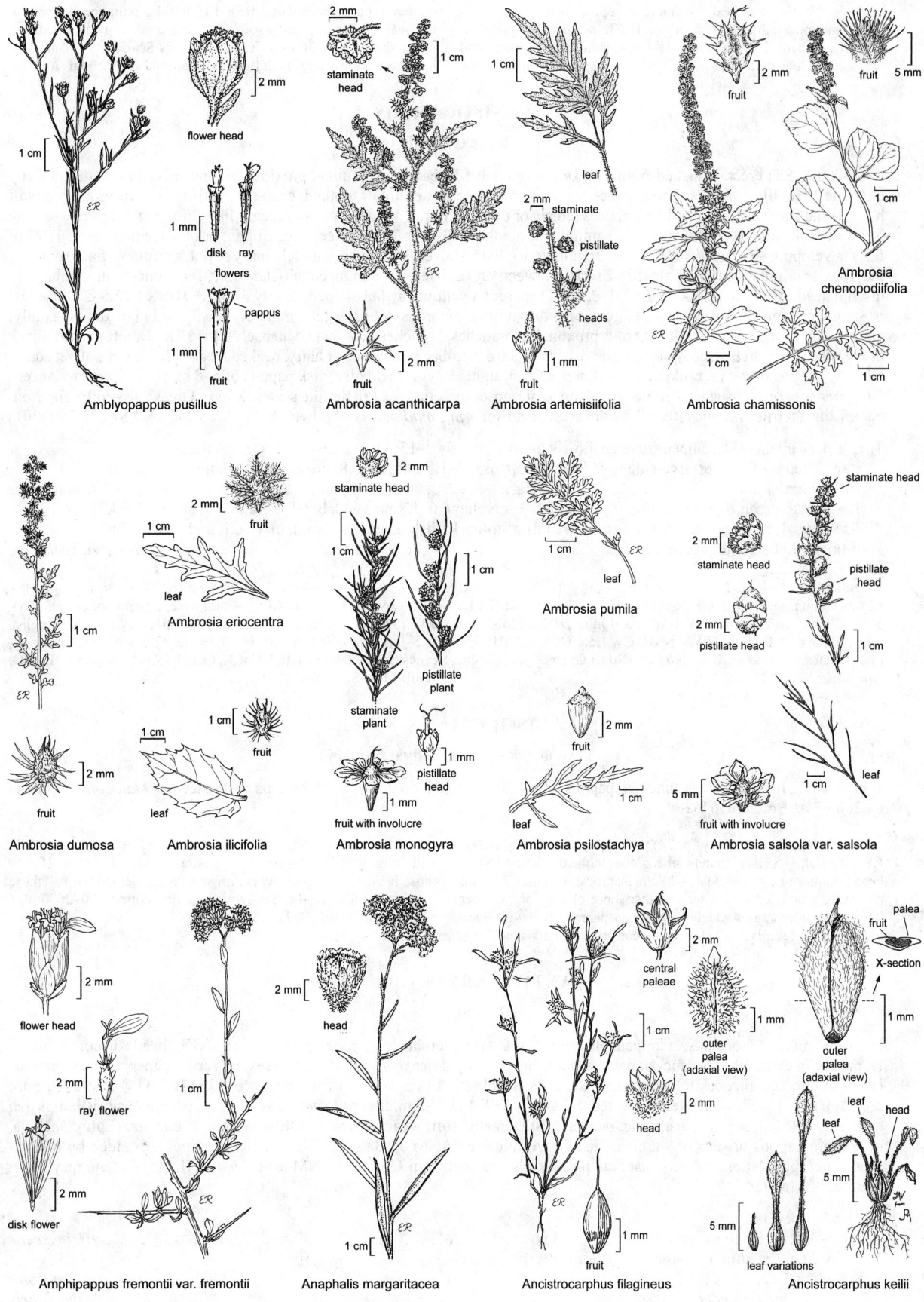

2 mm
flower head
1 cm
ER
1 mm
disk   ray
flowers
pappus
1 mm
fruit
Amblyopappus pusillus

2 mm
staminate head
1 cm
ER
fruit
2 mm
Ambrosia acanthicarpa

1 cm
leaf
2 mm   staminate
pistillate
heads
1 mm
fruit
Ambrosia artemisiifolia

2 mm
fruit
5 mm
fruit
1 cm
1 cm
ER
1 cm
1 cm
Ambrosia chenopodiifolia

1 cm
ER
fruit
2 mm
Ambrosia dumosa

2 mm
fruit
1 cm
leaf
Ambrosia eriocentra

1 cm
fruit
leaf
Ambrosia ilicifolia

staminate head
2 mm
1 cm
1 cm
pistillate plant
staminate plant
pistillate head
1 mm
fruit with involucre
1 mm
Ambrosia monogyra

1 cm
leaf
ER
Ambrosia pumila

fruit
2 mm
leaf
1 cm
Ambrosia psilostachya

staminate head
2 mm
staminate head
pistillate head
2 mm
pistillate head
1 cm
leaf
1 cm
5 mm
fruit with involucre
Ambrosia salsola var. salsola

flower head
2 mm
ray flower
2 mm
1 cm
disk flower
2 mm
ER
Amphipappus fremontii var. fremontii

2 mm
head
1 cm
ER
1 cm
Anaphalis margaritacea

2 mm
central paleae
1 cm
outer palea (adaxial view)
1 mm
2 mm
head
ER
fruit
1 mm
Ancistrocarphus filagineus

palea
fruit
X-section
1 mm
outer palea (adaxial view)
leaf
leaf
head
5 mm
5 mm
leaf variations
Ancistrocarphus keilii

expanded, ± spoon-shaped, 3-veined, ± yellow to tan, involucre-like, sometimes ± grading into paleae. **INFL:** heads 1 (but pls often densely grouped), 2.5–3.5 mm; involucre 0, simulated by paleae and lvs; receptacle ± hourglass-shaped; wings of paleae of pistillate fls hidden in head, lanceolate. **DISK FL:** 0.9–1.3 mm; corolla lobes 4, unequal, ± bilateral. Sandy soils, chaparral bordering oak woodland, under shrubs; 40–130 m. s CCo/SCoRO, sw SCoRO (sw Santa Barbara Co.). Rarity may result from low dispersal ability. Mar–Apr ★

ANISOCARPUS

Bruce G. Baldwin

Per 1–8 dm. **ST:** erect, branched from base or throughout. **LF:** basal and cauline, proximally opposite or in rosettes, distally alternate, ± sessile, oblong to linear, lance-linear, or oblanceolate, entire or toothed, coarsely to stiffly appressed-hairy or soft hairy, distal stalked-glandular. **INFL:** heads radiate or discoid, 1 or in ± flat-topped or raceme-like clusters; involucre ± spheric or widely ellipsoid to bell-shaped, 4–6+ mm diam; phyllaries 0–3 or 7–15, lanceolate, elliptic, or oblanceolate, each ± 1/2 or fully enveloping a ray ovary, falling with fr, minutely ciliate, stalked-glandular, minutely hairy or not; receptacle flat to convex, glabrous or minutely bristly; paleae in 1 series between ray and disk fls, gen fused, phyllary-like, functionally an involucre in discoid heads, deciduous. **RAY FL:** 0, 1–3, or 7–15; corolla yellow, ray fan-shaped, deeply 3-lobed. **DISK FL:** 5–30, bisexual or staminate; corolla yellow, tube < throat, tube/throat often glabrous, lobes deltate, minutely bristly and glandular abaxially; anthers yellow, tips ovate; style glabrous proximal to branches, branches awl-shaped, densely hairy. **FR:** ± club-shaped, black or ± gray, ray fr ± arched, compressed side-to-side or ± front-to-back, glabrous or hairy, beaked, beak 0.2–0.3 mm, offset adaxially, ray pappus 0 or crown-like; disk fr if present ± straight, ± cylindric, hairy, disk pappus of 5–8 or 11–21 linear to lanceolate, ± square, or awl-shaped, fringed or ciliate to plumose, sometimes ± bristle-like scales. 2 spp. (Greek: dissimilar frs, from contrasting (fertile) ray and (sterile) disk ovaries of *Anisocarpus madioides*) [Baldwin & Strother 2006 FNANM 21:299–301]

1. Lvs dark green, 4–13 cm, entire or shallowly toothed; ray fls 7–15; involucre 4–6 mm, ± spheric; disk fls staminate; ray fr compressed side-to-side; disk pappus 0.2–1.5 mm, of 5–8, linear to lanceolate or ± square, fringed scales . . . . ***A. madioides***
1′ Lvs ± blue-green, 1–3 cm, entire; ray fls 0 or 1–3; involucre 6–12 mm, widely ellipsoid to bell-shaped; disk fls bisexual, forming frs; ray fr ± compressed front-to-back; disk pappus 4–7 mm, of 11–21, awl-shaped, or ± bristle-like, ciliate-plumose scales . . . . ***A. scabridus***

A. madioides Nutt. (p. 247) WOODLAND TARWEED Pl 1.5–8 dm. **LF:** 5–15 mm wide, tip acute. **RAY FL:** ray 4–10 mm. **DISK FL:** 5–30. **FR:** ray fr 3–5 mm; disk ovaries 3–4 mm. 2*n*=14. Moist forest, woodland; < 1300 m. NW, n SNH, CW (rare CCo, SCoRI), PR (n Palomar Mtns); to s BC. [*Madia m.* (Nutt.) Greene] Self-fertile. Apr–Sep

A. scabridus (Eastw.) B.G. Baldwin (p. 247) SCABRID ALPINE TARPLANT Pl 1–5 dm. **LF:** 1–4 mm wide, tip acute or obtuse. **RAY FL:** ray 3.5–11.5 mm. **DISK FL:** 5–23. **FR:** ray fr 6–8 mm; disk fr 5–9 mm, ± cylindric. 2*n*=14. Open ridges or slopes on metamorphics; 1600–2400 m. KR, NCoRH, CaRH. [*Raillardiopsis s.* (Eastw.) Rydb.] Jun–Sep ★

ANISOCOMA SCALEBUD

David J. Keil & G. Ledyard Stebbins

1 sp.: sw US, n Mex. (Greek: unequal pappus) [Keil 2006 FNANM 19:309–310] Perhaps best placed in *Malacothrix* (Lee et al. 2003 Syst Bot 28:616–626).

A. acaulis Torr. & A. Gray (p. 247) Ann, ± scapose; sap milky. **LF:** all basal, 3–5 cm, pinnately lobed; lobes toothed, ± hairy. **INFL:** head liguliflorous, 1; peduncle 5–20 cm, glabrous; involucre 2–3 cm; phyllaries graduated in 4–5 series, outer short, oblong, blunt, inner linear, pointed, margins widely papery-transparent, often with ± red tips and dots; receptacle flat to weakly convex, paleate, paleae long, narrow, bristle-like. **FL:** ± 40; ligule cream to bright yellow, readily withering. **FR:** 10–15-veined; pappus readily deciduous, of 10 plumose bristles in 1 series, white, bristles on adaxial side of fr < those on abaxial side. 2*n*=14. Sandy washes, dry slopes; 100–2600 m. s SN, Teh, SnJV, SCoR, TR, PR, GB, D; NV, AZ, Baja CA. Mar–Jun

ANTENNARIA PUSSY-TOES

Randall J. Bayer

Per, often matted; dioecious; staminate pls present or 0. **LF:** alternate, entire, gen ± tomentose. **INFL:** heads discoid or disciform, 1 or in cyme-like (raceme- or panicle-like) clusters; phyllaries many, graduated in several series, papery or membranous (wider, more conspicuous in staminate heads); receptacle flat to convex or ovoid, epaleate. **PISTILLATE FL:** 2–10 mm; corolla barely lobed, white, yellow, or red. **STAMINATE FL:** 2–5 mm; corolla white, yellow, or red; pappus bristle tips gen enlarged. **FR:** 0.5–3.5 mm, ± elliptic; pappus bristles many, soft, weakly barbed. ± 40 spp.: Am, n Eurasia. (Latin: antenna, describing pappus bristles of staminate fls) [Bayer 2006 FNANM 19:388–415] Races of some spp. reproduce by asexual seeds, their populations entirely pistillate pls. KR pls reported from CA in FNANM as *A. lanata* (Hook.) Greene may be an undescribed taxon.

1. Heads 1 per fl st
  2. St 5–12 cm; lvs thick, green adaxially, tips notched . . . . ***A. suffrutescens***
  2′ St < 4 cm; lvs thin, gray-tomentose adaxially, tips acute
    3. Stolons 0 . . . . ***A. dimorpha***
    3′ Stolons long, slender, ± lfless . . . . ***A. flagellaris***

1′ Heads 2–many per fl st
  4. Stolons 0, pls not forming mats
    5. Involucre proximally densely tomentose; phyllary tips gen pink to red. . . . . *A. geyeri*
    5′ Involucre ± glabrous
      6. Proximal lvs 7–20 mm wide; st 18–40 cm; phyllaries white or cream, unequal . . . . *A. argentea*
      6′ Proximal lvs 1–10 mm wide; sts 7–25 cm; phyllaries straw-colored or pale, ± equal . . . . *A. luzuloides*
        7. Heads 10–30 in raceme- or panicle-like cluster. . . . . subsp. *aberrans*
        7′ Heads 10–110 in ± flat-topped, cyme-like cluster . . . . subsp. *luzuloides*
4′ Stolons present, pls forming mats
    8. Rosette lvs elliptic, green adaxially; infl raceme- or panicle-like; peduncles 10–30 mm; infl finely
      glandular . . . . *A. racemosa*
    8′ Rosette lvs linear or oblanceolate to spoon- or wedge-shaped; infl cyme-like; peduncles < 5 mm; infl
      glandular or not
      9. Rosette lvs green or lightly tomentose adaxially, 1–3-veined
        10. Rosette lvs > 6 mm wide, 1–3-veined; cauline lvs 20–40 mm; infl not glandular; KR
          . . . . *A. howellii* subsp. *howellii*
        10′ Rosette lvs ≤ 6 mm wide, 1-veined; cauline lvs 7–16 mm; infl gen glandular; SnBr. . . . . *A. marginata*
      9′ Rosette lvs gray- to silvery-tomentose adaxially, 1-veined
        11. Phyllaries dark brown to black; st 3–13 cm
          12. Most proximal cauline lf 11–20 mm; herbage gen not glandular; pistillate fl corolla 3–4.5 mm in fr;
            staminate corolla gen 2.8–4.5 mm in fl . . . . *A. media*
          12′ Most proximal cauline lf ≤ 11 mm; herbage gen glandular; pistillate corolla 2–3 mm in fr;
            staminate corolla gen ≤ 2.8 mm in fl . . . . *A. pulchella*
        11′ Phyllaries white to pale yellow, rose or pale brown; st 6–40 cm
          13. Phyllaries white-tipped with a prominent ± black-brown spot near base of scarious part . . . . *A. corymbosa*
          13′ Phyllaries white or white-tipped, rose, straw-colored or (paler) brown, not dark-spotted near base
            14. St distally stalked-glandular. . . . . *A. microphylla*
            14′ St not glandular
              15. Stolons slightly woody, ascending; phyllaries pale yellow to pale brown distally, narrow and acute
                (pistillate) or wide and blunt (staminate); staminate pls present — n SNH, W&I. . . . . *A. umbrinella*
              15′ Stolons not woody, horizontal to ascending; phyllaries various colors, tips acute; staminate pls ± 0. . . . *A. rosea*
                16. Longest lf of fl rosettes 8–20 mm; phyllaries often brown . . . . subsp. *confinis*
                16′ Longest lf of fl rosettes 20–40 mm; phyllaries gen not brown . . . . subsp. *rosea*

A. argentea Benth. (p. 247) **ST:** 18–40 cm, from branched caudex; stolons 0. **LF:** basal 20–50 mm, 7–20 mm wide, oblanceolate to elliptic, 1–3-veined; cauline 15–45 mm, lanceolate. **INFL:** heads 10–75; involucre 4–5 mm, ± glabrous; phyllaries unequal, wide, acute, white or cream. **FL:** pistillate corolla 3–4 mm, staminate 2.5–3.5 mm. **FR:** 1–1.5 mm, glandular; pappus 3–4 mm. $2n=28$. Dry conifer forest; 600–2000 m. KR, NCoRH, CaRH, SNH, MP; to WA, NV. May–Jul

A. corymbosa E.E. Nelson (p. 247) **ST:** 6–15 cm, slender; stolons horizontal, 1–10 cm. **LF:** basal 18–22 mm, spoon-shaped, 1-veined, thin, ± gray-tomentose; cauline 8–13 mm. **INFL:** heads 3–7; infl cyme-like; involucre 4–5.3 mm, densely hairy proximally; phyllaries wide, tips acute (pistillate) or blunt (staminate), white, distinctly spotted ± black-brown near base. **FL:** pistillate corolla 2.5–3.5 mm, staminate 2–3.2 mm. **FR:** 0.5–1 mm, slightly papillate; pappus 3.5–4.5 mm. $2n=28$. Moist meadows, streamsides; 1900–3200 m. SNH; to WA, MT, NM. Jun–Aug

A. dimorpha (Nutt.) Torr. & A. Gray (p. 247) **ST:** < 4 cm, cespitose from much-branched caudex; stolons 0. **LF:** basal, 8–11 mm, linear to narrowly spoon-shaped, acute, 1-veined, thin, ± gray-tomentose. **INFL:** head 1; involucre 10–11 mm (pistillate) or 6–8 mm (staminate), base hairy; phyllaries narrow, acute-acuminate, dingy brown. **FL:** pistillate corolla 8–10 mm; staminate corolla 3–5 mm, pappus bristle tips slender. **FR:** 2–3.5 mm, hairy; pappus 10–12 mm. $2n=28,56$. Dry places; 800–2400 m. KR, NCoRH, SNH, TR, GB; to sw Can, MT, NE, NM. May–Jul

A. flagellaris (A. Gray) A. Gray STOLONIFEROUS PUSSY-TOES **ST:** < 1.5 cm, from slender caudex; stolons 3–10 cm, slender, lfless exc tip. **LF:** basal, 16–18 mm, linear-oblanceolate, acute, 1-veined, ± gray-tomentose. **INFL:** head 1; involucre 7–9 mm (pistillate) or 6–7 mm (staminate), base hairy; phyllaries wide, tips acute, brown to ± black or ± white. **FL:** pistillate corolla 5–7 mm; staminate corolla 3–4.5 mm, pappus bristle tips slender. **FR:** 2–3 mm, papillate; pappus

6–8 mm. $2n=28$. Seasonally moist sagebrush scrub; 1750 m. MP; to BC, ID, SD, NV. May–Jun ★

A. geyeri A. Gray (p. 247) **ST:** 3–14 cm; base woody, branched; stolons 0. **LF:** cauline, 11–35 mm, lance-linear, 1-veined, ± gray-tomentose. **INFL:** heads 3–25; involucre 6–8 mm, proximally densely tomentose; phyllaries wide, tips gen pink to red (light brown or white), tips acute (pistillate) or blunt (staminate). **FL:** pistillate corolla 5–6 mm; staminate corolla 3–4.5 mm, pappus bristle tips slender. **FR:** 2–2.5 mm, hairy, papillate; pappus 6–7 mm. $2n=28$. Dry, open woodland, scrub; 900–2400 m. KR, NCoRH, n SNH, MP; to WA, NV. Jun–Aug

A. howellii Greene subsp. ***howellii*** Pls pistillate. **ST:** 15–30 cm; stolons 1–4 cm. **LF:** basal 25–40 mm, > 6 mm wide, oblanceolate to spoon-shaped, 1–3-veined, green adaxially, tomentose abaxially; cauline 20–40 mm, linear. **INFL:** heads 5–12; infl cyme-like; peduncles < 5 mm; involucre 6–7.5 mm, base hairy; phyllaries narrow, acute, distally light brown to white, not glandular. **FL:** corolla 5–6 mm. **FR:** 1.5–2 mm, papillate; pappus 6–8 mm. $2n=56,84$. Open, pine woodland, rocky slopes; 1300–2000 m. KR; to w Can, WI, CO. 3 other vars. in US, Can. Jul–Aug

A. luzuloides Torr. & A. Gray **ST:** 7–25 cm, from partly woody caudex; stolons 0. **LF:** basal 18–45 mm, 1–3-veined, ± gray-tomentose; cauline 12–45 mm. **INFL:** involucres narrow, glabrous; phyllaries narrow, acute, ± equal, phyllaries straw-colored or pale. **FR:** 1–2 mm, papillate; pappus 3–4.5 mm. $2n=28$.

subsp. ***aberrans*** (E.E. Nelson) R.J. Bayer & Stebbins **LF:** basal 1–5 mm wide, linear. **INFL:** heads 10–30 in raceme- or panicle-like cluster; pistillate involucre 3.5–4.5 mm, staminate 3.5–4 mm. **FL:** pistillate corolla 2–2.5 mm, staminate 2.5–3.2 mm. **FR:** 1–1.5 mm, papillate-strigose; pistillate pappus 2.5–3 mm, staminate 3–3.5 mm. Moist open areas, meadows in montane forest; ± 1500 m. CaRH, MP; to OR, NV. May–Jul

subsp. ***luzuloides*** **LF:** basal 1–10 mm wide, linear or narrowly oblanceolate. **INFL:** heads 10–110 in ± flat-topped, cyme-like cluster; pistillate involucre 5–6.5 mm, staminate 4–5.5 mm. **FL:** pistillate corolla 2.5–4 mm, staminate 3–4 mm. **FR:** 1–2 mm, papillate; pistillate pappus 3–4 mm, staminate 3–4.5 mm. ± dry meadows, drainages in lower montane forest; 1200–2000 m. NCoRH, n SNH, MP; to sw Can, MT, SD, CO. May–Jul

A. marginata Greene   WHITE-MARGINED EVERLASTING   Pls pistillate (in CA). **ST:** 15–18 cm; stolons horizontal, 2–7 cm. **LF:** basal 15–20 mm, ≤ 6 mm wide, spoon-shaped, 1-veined, green adaxially with white-woolly margins, gray-tomentose abaxially; cauline 7–16 mm, linear. **INFL:** heads 5–8; infl cyme-like; peduncles < 5 mm; involucre 5–7 mm, sparsely hairy at base; phyllaries wide, tips acuminate, distally white, gen glandular. **FL:** corolla 4.5–6.5 mm. **FR:** 0.8–2 mm, papillate; pappus 5.5–8.5 mm. 2*n*=28,56,84,112,140. Dry woodland; 1950–3350 m. SnBr (San Gorgonio Mtn, s Fork Santa Ana River); to CO, TX. Jun–Jul ★

A. media Greene (p. 247)   **ST:** 5–13 cm; stolons many, matted. **LF:** basal 6–19 mm, ± linear to spoon-shaped, 1-veined, densely gray-tomentose; cauline 5–20 mm (most proximal 11–20 mm), linear. **INFL:** heads 2–7; infl cyme-like; peduncles < 5 mm; involucre 4–8 mm, base woolly; phyllaries narrow and acute (pistillate) or wide and blunt (staminate), distally dark brown or black. **FL:** pistillate corolla 3–4.5 in fr, staminate corolla gen 2.8–4.5 mm. **FR:** 0.6–1.6 mm, papillate or smooth; pappus 4–5.5 mm. 2*n*=56,98,112. Meadows, snow basins, ridges; 1800–3900 m. KR, NCoRH, CaRH, SNH, SnBr, W&I; to AK, nw Can, MT, NM. Jul–Aug

A. microphylla Rydb.   **ST:** 9–30 cm, distally stalked-glandular; stolons 1–5 cm. **LF:** basal 6–16 mm, spoon-shaped, 1-veined, silvery-hairy on faces; cauline 5–25 mm, linear. **INFL:** heads 6–13; infl cyme-like; pistillate involucre 5.5–7 mm, staminate 5–6.5 mm; phyllaries green, distally white to pale yellow. **FL:** pistillate corolla 3–4.3 mm, staminate 2.5–3 mm. **FR:** 0.7–1.2 mm, glabrous or papillate; pistillate pappus 3–5 mm, staminate 3.5–4 mm. 2*n*=28. Dry meadows; 2500–3250 m. SNE; to AK, n Can, MN, NB, NM. Jun–Jul

A. pulchella Greene   BEAUTIFUL PUSSY-TOES   Herbage gen glandular. **ST:** 3–12 cm; stolons many, matted. **LF:** basal 6–12 mm, ± linear to spoon-shaped, 1-veined, densely woolly, often purple-glandular; cauline 3–11 mm, linear. **INFL:** heads 4–6; infl cyme-like; peduncles < 5 mm; involucre 3.5–5 mm, base woolly; phyllaries narrow and acute (pistillate) or wide and blunt (staminate), distally gen dark brown-black (to ± white at tip). **FL:** pistillate corolla 2–3 mm, staminate 1.9–2.8 mm. **FR:** 0.7–1.3 mm, papillate or smooth; pappus 2.5–3.5. 2*n*=28. Meadows, snow basins, ridges; 2800–3700 m. SNH, n SNE (Sweetwater Mtns); NV. Jul–Aug ★

A. racemosa Hook.   **ST:** 12–50 cm, glandular distally; stolons horizontal, 3–8 cm. **LF:** basal 30–100 mm, elliptic, 3-veined, green adaxially, tomentose abaxially; cauline 10–30 mm. **INFL:** heads 3–12; cluster raceme- or panicle-like; peduncles 10–30 mm; involucre 4–9 mm (pistillate > staminate), wide, blunt or acute, finely glandular, distally white to light brown. **FL:** pistillate corolla 3.5–5.5 mm; staminate corolla 3–4 mm, pappus bristle tips slender. **FR:** 1–1.5 mm, barely papillate; pappus 4.5–7 mm. 2*n*=28. Montane forest; 1200–2000 m. KR; to sw Can, SD, WY. Jun–Sep

A. rosea Greene   Pls gen all pistillate. **ST:** not glandular; stolons short. **LF:** basal 8–40 mm, spoon- to wedge-shaped, 1-veined, ± gray-tomentose. **INFL:** heads 3–16; infl cyme-like; involucre base hairy; phyllaries wide, acute, tips white, rose, ± yellow, or ± brown. **FR:** 0.7–1.8 mm, papillate or smooth. 2*n*=42,56,70. Subsp. *arida* (E.E. Nelson) R.J. Bayer, subsp. *pulvinata* (Greene) R.J. Bayer in adjacent states, expected in CA.

subsp. ***confinis*** (Greene) R.J. Bayer (p. 247)   **ST:** 9–25 cm; stolons horizontal to ascending, 15–45 mm, not woody. **LF:** longest lf of fl rosettes 8–20 mm; cauline 6–20 mm. **INFL:** heads 4–11; involucre 4–6.5 mm; phyllaries often brown. **FL:** corolla 2.5–4 mm. **FR:** pappus 3.5–5 mm. Woodland, meadow edges, rock barrens, dry ridges; 1200–3700 m. KR, CaRH, SNH, SnGb, SnBr, SnJt, GB; to AK, Greenland, MT, WY, CO, NM. Jun–Aug

subsp. ***rosea***   **ST:** 10–40 cm; stolons decumbent, 20–70 mm. **LF:** longest lf of fl rosettes 20–40 mm; cauline 8–36 mm. **INFL:** heads 6–20; involucre 5–8 mm; phyllaries gen not brown. **FL:** corolla 3–4.5 mm. **FR:** pappus 4–6 mm. Woodland, meadow edges, rock barrens, dry ridges; 1200–3700 m. KR, CaRH, SNH, SnGb, SnBr, SnJt, GB; to AK, n Can, MI, CO. Jun–Aug

A. suffrutescens Greene   EVERGREEN EVERLASTING   **ST:** many, 5–12 cm, densely tufted, woody at base; stolons 0. **LF:** dense, 5–12 mm, spoon-shaped, evergreen, 1-veined, thick, green adaxially, tomentose abaxially; tip notched. **INFL:** head 1; involucre 10–15 mm (pistillate) or 5–9 mm (staminate), woolly and finely glandular at base; phyllaries wide, blunt or acute, green-yellow proximally, distally white. **FL:** pistillate corolla 5–8 mm, staminate 4–5 mm. **FR:** 1–2 mm, papillate; pappus 7–9 mm. 2*n*=28. Dry, open conifer woodland, serpentine barrens; 500–1600 m. KR, NCoRO; OR. Jun–Jul ★

A. umbrinella Rydb.   **ST:** 7–16 cm; base ± woody, not glandular; stolons or sterile shoots ascending, slightly woody. **LF:** basal 10–17 mm, spoon- to wedge-shaped, 1-veined, ± gray-tomentose; cauline 8–18 mm. **INFL:** heads 3–8; infl cyme-like; involucre 3–6.5 mm, base woolly; phyllaries narrow and acute (pistillate) or wide and blunt (staminate), distally pale yellow to pale brown. **FL:** corolla 2.5–3.5 mm. **FR:** 0.5–1.2 mm, glabrous; pappus 3–5 mm. 2*n*=28,56. Uncommon. Dry sagebrush scrub, open yellow-pine forest; 1800–3900 m. SNH, W&I; to sw Can, MT, CO, AZ. May–Aug

ANTHEMIS   DOG-FENNEL, CHAMOMILE

Linda E. Watson

Ann, per, often aromatic; herbage strigose to long-soft-hairy or ± glabrous. **ST:** 1–5+, decumbent to erect, gen branched. **LF:** most cauline, alternate, petioled or sessile, ± obovate to spoon-shaped, 1–3-pinnately divided. **INFL:** heads radiate [discoid], 1 or in rounded to flat-topped clusters; involucre obconic to hemispheric or wider, phyllaries persistent, gen 21–35+, ± graduated in 3–5 series, free, gen lanceolate, oblong, or elliptic, margin and tip scarious, receptacle hemispheric to narrowly conic, paleate throughout or distally; paleae awl-shaped or elliptic to obovate. **RAY FL:** (0)5–20+, pistillate or sterile (style present); corolla white [yellow or pink]. **DISK FL:** gen 100–300+; corolla gen yellow, tube << funnel-shaped throat, swollen, lobes ± triangular; anther tip ± ovate; style tips truncate. **FR:** obovoid to obconic or top-shaped, round or 4-angled, gen 10-ribbed, smooth or tubercled, glabrous; pappus 0 or crown-like. ± 175 spp.: Eur, w Asia, n Afr. (Greek: fl) [Watson 2006 FNANM 19:537–538] *Cota tinctoria* (L.) Guss. [*A. tinctoria* L. in TJM (1993)] occ escape from cult.

1. Ray fl pistillate; receptacle paleate throughout, paleae lanceolate to oblanceolate, weakly folded, ± keeled, tip spiny-acuminate; fr ribs smooth or weakly tubercled; st branched mostly proximally . . . . ***A. arvensis***
1′ Ray fl sterile; receptacle paleate distally, paleae needle-like to awl-shaped; fr ribs ± tubercled; st branched distally or ± throughout . . . . ***A. cotula***

A. arvensis L. CORN CHAMOMILE Ann (occ persisting), ≤ 3(8) dm, not notably scented. **ST**: decumbent (sometimes rooting at nodes) or ascending to erect. **LF**: 15–35 mm, 8–16 mm wide, 1–2-pinnately lobed (ultimate lobes linear or narrowly elliptic to triangular). **INFL**: involucre 6–13 mm diam. **RAY FL**: 5–20. **DISK FL**: corolla 2–3(4) mm. **FR**: 1.7–2+ mm; pappus 0 or a minute crown. $2n=18$. Uncommon. Escape from cult in disturbed areas, fields; > 1000 m. KR, NCoRO, c SNH, expected elsewhere; native to Eur. Mar–Apr

A. cotula L. (p. 247) MAYWEED Ann; ≤ 6(9) dm, strongly scented. **ST**: gen erect, ± glabrous, puberulent, or ± minutely strigose, resin-gland-dotted. **LF**: 10–55 mm, 15–30 mm wide, 1–3-pinnately divided into thread-like lobes. **INFL**: involucres 5–11 mm diam. **RAY FL**: (0)10–15. **DISK FL**: corolla 2–3 mm. **FR**: 1.3–2 mm; pappus 0. $2n=18$. Common. Disturbed areas, fields, coastal dunes, chaparral, oak woodland; < 2600 m. NW, SN, GV, CW, SCo, TR, PR, MP; native to Eur. Apr–Aug

ARCTIUM BURDOCK

Bien. **ST**: 1–several, branched distally. **LF**: basal and cauline, alternate, long-petioled, widely ovate; base deeply cordate; margin entire or toothed; gradually reduced distally on st. **INFL**: heads discoid, in lfy-bracted clusters; involucre ± spheric; phyllaries graduated in many series, ± linear, bases appressed, tips stiffly radiating, hooked-spiny; receptacle ± flat, epaleate, bristly. **FL**: many; corolla pink to ± purple, lobes narrowly triangular; anther base tailed, tips ovate, acute to obtuse; style branched just above distal hairy ring, branches oblong, obtuse. **FR**: ± compressed, rough or ribbed, glabrous, attachment basal; pappus several series of rough bristles, readily deciduous. 10 spp.: Eurasia, n Afr. (Greek: bear) [Keil 2006 FNANM 19:168–171]

1. Heads gen 25–40 mm diam, gen long-peduncled in ± rounded or flat-topped clusters; inner phyllaries gen green, margins minutely hairy; pappus 2–6 mm . . . . ***A. lappa***
1′ Heads gen 10–25 mm diam, sessile to short-peduncled in raceme- or panicle-like clusters; inner phyllaries gen purple-tinged, margins minutely serrate; pappus ± 2 mm . . . . ***A. minus***

A. lappa L. Pl ≤ 25 dm. **ST**: branches ascending. **LF**: blade ≤ 80 cm, abaxially thinly gray-tomentose, adaxially green, nearly glabrous. **FL**: corolla 9–12 mm. **FR**: 6–7 mm, ± brown, sometimes dark-spotted. $2n=36$. Uncommon. Disturbed places; < 500 m. s NCoRI, SnFrB; widely naturalized in N.Am; native to Eur. Cult for edible roots. Jun–Oct

A. minus (Hill) Bernh. (p. 247) Pl ≤ 15 dm. **ST**: branches stiffly ascending. **LF**: blade ≤ 80 cm, green, abaxially ± gray-tomentose, adaxially puberulent or soft-hairy, becoming glabrous. **FL**: corolla 8–10 mm. **FR**: 4–7.5 mm, dark brown or dark-spotted. $2n=36$. Disturbed places; < 1500 m. NW, n ChI, MP; widely naturalized in N.Am; native to Eur. Jul–Oct

ARCTOTHECA

Alison M. Mahoney & Robert J. McKenzie

Ann or per, ± glandular and/or with segmented hairs. **ST**: 0, or prostrate and creeping with adventitious roots at nodes, or shortly decumbent to erect, ribbed, white-cobwebby. **LF**: alternate, basal in rosettes and/or cauline, petioled, (2)5–20(30+) cm, (1)2–5(7) cm wide, [ovate] obovate to spoon-shaped, ± deeply pinnately lobed, ultimate margin ± dentate-prickly [entire or shallowly toothed] or distal ± entire, adaxially finely cobwebby [densely woolly], abaxially densely white-woolly. **INFL**: heads radiate, 1, peduncled; phyllaries graduated in 3–6 series, proximal narrowed to spreading or reflexed, awn-like, tips fringed, distal with expanded membranous tips; receptacle flat, ± pitted, pit margins with short, membranous scales, epaleate. **RAY FL**: sterile; ray adaxially yellow throughout or with ± deeper yellow proximal band (dried rays yellow or proximally ± white with distal blue margin and distally blue to ± brown-purple), abaxially yellow to ± blue, red, or purple; style 0. **DISK FL**: corolla yellow throughout, or distally ± green to ± black; anther tip ovate-triangular; style tip proximally slender, thickened distal to a minutely hairy node, branches minute. **FR**: obovoid-elliptic, compressed, 4–5-ribbed, faces roughened, woolly to ± glabrous, occ with proximal cluster of segmented hairs; pappus of [2–3] 6–10 scales < fr, crown-like, or 0. 5 spp.: native s Afr. (Greek: bear-capsule, probably for woolly fr) [Mahoney 2006 FNANM 19:197]

1. Ann from rosette, initially stless, later sts short-decumbent to ± erect, branched in age, rarely rooting at st bases; ray 6–15(22) mm; disk corolla proximally yellow, distally ± green to black . . . . ***A. calendula***
1′ Per from rosette, initially stless, later spreading by creeping sts, rooting at nodes, forming new rosettes; ray (20)25–40 mm; disk corolla yellow throughout . . . . ***A. prostrata***

A. calendula (L.) Levyns (p. 247) CAPEWEED **LF**: rosette and proximal cauline lvs ± 0 on older pls. **INFL**: heads 2–4(5.5) cm diam. **RAY FL**: 6–13(17); ray 2–3(4) mm wide, adaxially uniformly pale yellow or with darker yellow proximal band, abaxially steely blue. **FR**: ± 3 mm, densely ± purple- to brown-woolly; pappus scales 6–9 or 0, gen obscured by hairs. $2n=18$. Disturbed coastal areas; < 100 m. NCo, CCo, SCoRO; ± suitable habitats worldwide. Highly invasive, still uncommon in CA. Mar–Aug ◆

A. prostrata (Salisb.) Britten PROSTRATE CAPEWEED **LF**: rosette lvs persistent. **INFL**: heads 4–7 cm diam. **RAY FL**: (13)16–25; ray (2)3–5.5 mm wide, adaxially uniformly yellow, abaxially yellow or distally ± red-, purple-, or brown-tinged. **FR**: gen not maturing, ovary white-woolly to ± glabrous; pappus scales 6–10, or 0. $2n=18$. Escape from cult, disturbed areas; < 400 m. NCo, CCo, SCo; ± suitable habitats worldwide. [*A. calendula* (L.) Levyns, in part, misappl.] All yr

ARCTOTIS

Alison M. Mahoney & Robert J. McKenzie

Ann [per, subshrub]. **ST**: erect [decumbent], branched. **LF**: simple, basal and cauline, alternate; petioled or not; blades entire to pinnately lobed, white-woolly or sparsely cobwebby. **INFL**: heads radiate, 1, long-peduncled; phyllaries graduated in ± 3–6 series, cobwebby to bristly; receptacle flat, pitted (pit margins often ciliate), epaleate. **RAY FL**: corolla white, yellow, orange, pink or purple, often with proximal markings, abaxially gen red-purple or violet. **DISK FL**: gen bisexual; anther base minutely

sagittate, tip ovate-triangular; style slender proximally, thickened distally to a minutely hairy node, branches minute. **FR**: ovoid to oblong; abaxially with 2–3 wings ± bent inward forming 1–2 cavities or furrows; adaxially ± longitudinally ridged, glabrous to hairy; pappus 0, or of ± overlapping scales in 1–2 series (occ minute). ± 60 spp.: s Afr. (Greek: bear's ear, probably for woolly fr) [Mahoney 2006 FNANM 19:198–199; Mahoney & McKenzie 2008 Madroño 55:244–257] *A. fastuosa* Jacq. [*Venidium fastuosum* (Jacq.) Stapf], a rare escape from cult, has bright orange to yellow rays, obconic fr 1–1.5 mm, glabrous, wrinkled, with 1 abaxial furrow and no basal cluster of hairs, and pappus 0.

A. venusta Norl. BLUE-EYED AFRICAN DAISY Pl taprooted. **ST**: stout, (2)4–7(10) dm, cobwebby to woolly. **LF**: blade 5–20 × 1–4.5 cm, oblong to obovate, entire to wavy, occ fiddle-shaped to pinnately lobed, ultimate margin entire or irregularly toothed, faces evenly gray-tomentose. **INFL**: heads < 8 cm diam. **RAY FL**: 22–30; ray (17)20–30 × 2–4 mm, adaxially white with narrow yellow proximal band, abaxially violet. **DISK FL**: corolla gray-violet. **FR**: 2–3 mm, oblong, wings 3, cavities 2, lateral wings ± toothed and incurved into cavities; adaxially short-hairy; basal ring of many straight, forked hairs > fr; pappus scales 16 in 2 unequal series, translucent, outer << than inner, inner > fr. 2*n*=18. Uncommon. Escape from cult in coastal areas, roadsides; < 300 m. SCo; native to s Afr. [*A. stoechadifolia* P.J. Bergius, misappl.] Apr–Nov

ARGYRANTHEMUM

Bruce G. Baldwin, adapted from Strother (2006)

Subshrub or shrub, 10–80+[150] cm; herbage scented. **ST**: gen 1, prostrate to erect, gen branched, glabrous [hairy]. **LF**: alternate, petioled or sessile, blades ± obovate [oblong to lanceolate or linear] (bases sometimes ± clasping st), glabrous [hairy], [0](1)2–3-pinnately lobed, lobes wedge-shaped to linear, ultimate margins dentate [entire]. **INFL**: heads radiate [discoid], 1 or in open, flat-topped arrays; involucres [6]10–18[22+] mm diam, hemispheric or broader, phyllaries 28–45+, graduated in 3–4 series, free, persistent in fr, oblancolate or ovate to lance-deltate or lanceolate, not ridged abaxially, margins and tips ± yellow to brown, scarious, tips of inner often ± expanded; receptacle convex to conic, epaleate. **RAY FL**: 12–35+; corolla gen white, sometimes yellow or pink, ray ± ovate to linear. **DISK FL**: [50]80–150+; corolla yellow [red, purple], tube ± cylindric, throat bell-shaped, lobes deltate; anther tip ± ovate; style tips truncate, papillate. **FR**: ray fr 3-angled, each angle gen ± winged; all frs ± ribbed, gen glabrous, sometimes gland-dotted between ribs; pappus 0, fr wall sometimes forming a crown, teeth, or tubes at fr tip. 24 spp.: Canary Islands, Madeira. (Greek: silver-fl) [Strother 2006 FNANM 19:552] Cultivars of *A. foeniculaceum* and/or *A. frutescens* (L.) Sch. Bip., sometimes persisting near abandoned nurseries, sold as florists' daisies, marguerites, or Paris daisies.

A. foeniculaceum (Willd.) Sch. Bip. **LF**: scattered or crowded at base of peduncle, ± fleshy, blades 20–45(100) mm, 10–30(65) mm wide, ± glaucous. **INFL**: peduncle 5–10 cm. **FR**: 2–4(6) mm. 2*n*=18. Escape from cult in disturbed coastal areas; < 100 m. n CCo, SCo; native to Canary Islands. Mar–Aug

ARIDA

David R. Morgan

Ann [per] from taproot, or subshrub. **LF**: simple, alternate, entire to deeply pinnately lobed; teeth or lobes often ± bristle-tipped. **INFL**: heads radiate or discoid, 1 or in cyme-like clusters; involucre obconic, bell-shaped, or hemispheric; phyllaries graduated in 2–5[8] series, persistent, spreading or reflexed in fr, proximally gen straw-colored, distally green; receptacle convex, shallowly pitted, glabrous, epaleate. **RAY FL**: 0 or 25–35; corolla white to lavender [blue]. **DISK FL**: 12–45; corolla yellow; anther tip lanceolate; style tips triangular to linear. **FR**: narrowly oblong, several- to many-ribbed, moderately to densely appressed-hairy; pappus of many unequal bristles (ray pappus occ 0), gen ± white. 9 spp.: sw N.Am. (Latin: dry, from habitat) [Hartman & Bogler 2006 FNANM 20:401–405]

1. Ray fls present . . . . *A. arizonica*
1′ Ray fls 0 . . . . *A. carnosa*

A. arizonica (R.C. Jacks. & R.R. Johnson) D.R. Morgan & R.L. Hartm. (p. 247) SILVER LAKE DAISY Ann 5–30 cm. **ST**: 1–several from base, branched ± throughout, glandular-puberulent and non-glandular-hairy. **LF**: gen sessile, 1–30 mm, 1–10 mm wide, proximal oblong, toothed to pinnately lobed, lobes and teeth bristle-tipped; distal reduced, appressed. **INFL**: heads radiate; involucre 3–6 mm, 5–10 mm wide, hemispheric; phyllaries in 2–3 series, oblong to oblanceolate, acute, glandular. **RAY FL**: 25–35; corolla ≤ 7 mm, white to lavender. **DISK FL**: 28–45; corolla 2.5–3 mm. **FR**: 1.4–1.9 mm, disk fr weakly compressed; ray pappus 0 or obscure; disk pappus 2–3 mm. 2*n*=10. Uncommon. Riverbanks, sandy alkaline flats, roadsides; 30–1100 m. DMoj; s NV, s AZ, n Mex. [*Machaeranthera arida* B.L. Turner & D.B. Horne] Mar–Aug

A. carnosa (A. Gray) D.R. Morgan & R.L. Hartm. (p. 247) SHRUBBY ALKALI ASTER Subshrub 5–9 dm, from long rhizome. **ST**: much-branched, ± glabrous, glaucous. **LF**: sessile, 1–2 cm, 2–3 mm wide, proximal linear, entire, fleshy, glabrous; distal reduced to appressed, awl-shaped scales 1–4 mm. **INFL**: heads discoid, 1 at branch tips; involucre 5.5–7 mm, 4–7 mm wide; phyllaries gen in 4–5 series, lanceolate, acute or acuminate, glabrous, appressed or outer spreading. **DISK FL**: 12–18; corolla ± 6 mm. **FR**: 2–3 mm, ± cylindric; pappus ± 6 mm. 2*n*=10. Alkaline soils; 100–1600 m. SnJV, e SCo, SNE, DMoj; s NV, w&s AZ. [*Machaeranthera c.* (A. Gray) G.L. Nesom] Jun–Oct

ARNICA

Steven J. Wolf & Theodore M. Barkley

Per, gen from long, naked rhizome. **LF**: basal 0 or gen withered by fl, or present as sterile rosettes; cauline opposite. **INFL**: heads radiate or discoid (radiant), 1–many in ± flat-topped clusters; gen erect in bud (nodding); involucre hemispheric to

247

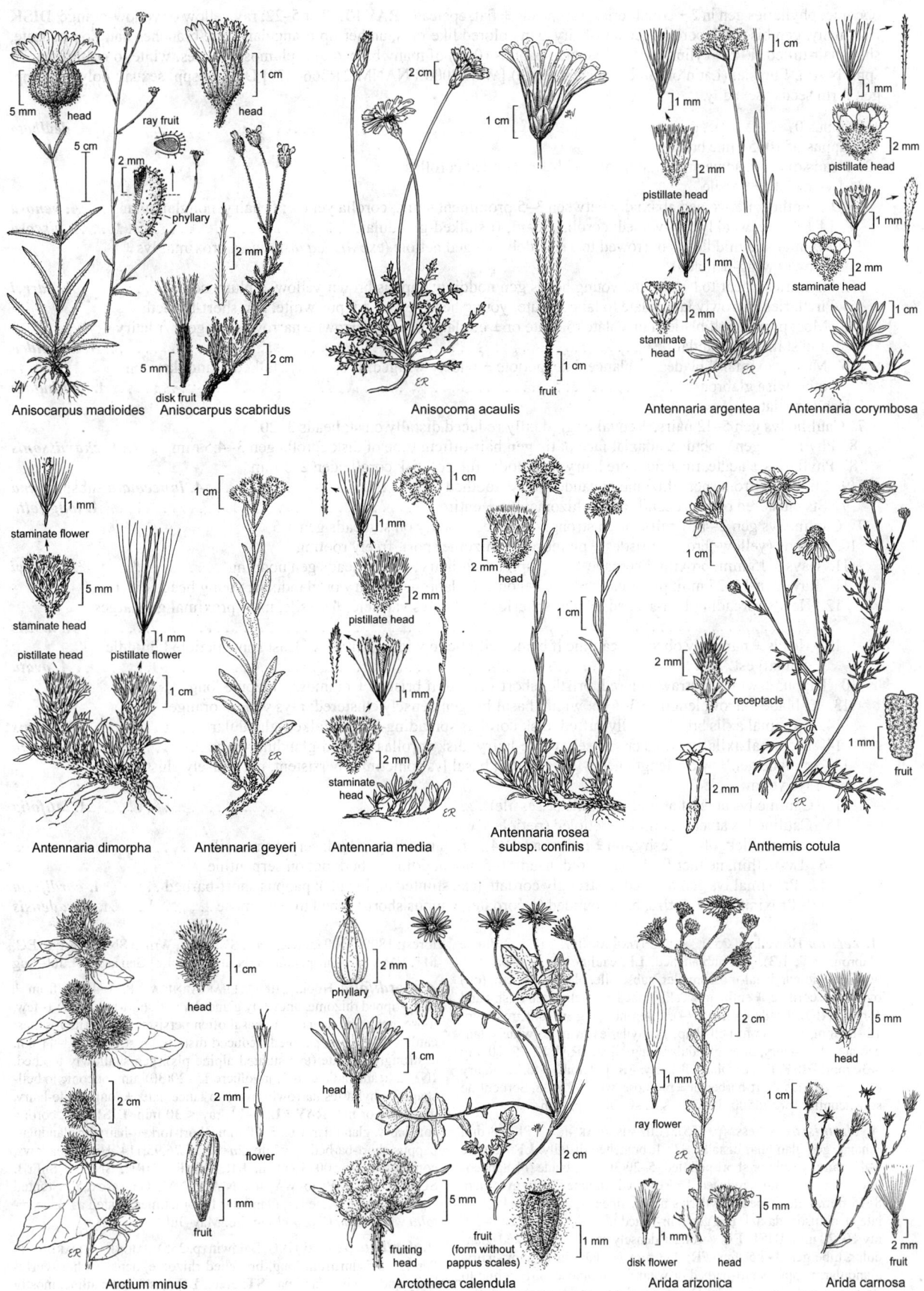

Anisocarpus madioides Anisocarpus scabridus Anisocoma acaulis Antennaria argentea Antennaria corymbosa

Antennaria dimorpha Antennaria geyeri Antennaria media Antennaria rosea subsp. confinis Anthemis cotula

Arctium minus Arctotheca calendula Arida arizonica Arida carnosa

obconic; phyllaries gen in 2 ± equal series; receptacle ± flat, epaleate. **RAY FL**: 0 or 5–22; ray yellow or yellow-orange. **DISK FL**: many, gen bisexual; corolla gen soft-hairy, gen colored like rays; anther tip triangular; style branches flat, tips truncate, short-hair-tufted. **FR**: ± cylindric, 5–10-veined; pappus (0 or) of many barbed to ± plumose bristles, white to red-brown. 29 spp.: N.Am, Eurasia. (Latin or Greek: ancient name) [Wolf 2006 FNANM 21:366–377] Diploid spp. sexual; polyploid spp. gen form seeds asexually.

1. Pappus 0. . . . . *A. dealbata*
1′ Pappus of 10–50 fine bristles
  2. Heads discoid, some marginal fls occ with ± expanded corollas
    3. Lvs widely ± sessile
      4. Lf toothed, net-veined abaxially between 3–5 prominent veins; corolla yellow; fr hairy, nonglandular . . . *A. venosa*
      4′ Lf ± entire, weakly 3–5-veined; corolla cream; fr stalked-glandular. . . . . *A. viscosa*
    3′ Proximal and middle lvs narrowed to a ± widely winged petiole (exc *Arnica discoidea*, proximal lvs ± narrowly wing-petioled)
      5. Phyllaries linear to lanceolate; young heads gen nodding; pappus brown-yellow, ± plumose . . . . ²*A. parryi*
      5′ Phyllaries narrowly lanceolate to lance-ovate; young heads erect; pappus white, gen short-barbed
        6. Most proximal lf blades lanceolate to ovate or ± cordate; petiole narrow, ± narrowly winged; fr hairy and stalked-glandular. . . . . *A. discoidea*
        6′ Most proximal lf blades ± oblanceolate; petiole ± widely winged; fr sparsely stalked-glandular, gen otherwise glabrous. . . . . *A. spathulata*
  2′ Heads radiate
    7. Cauline lvs gen 5–12 pairs, ± equal or gradually reduced distally on st; heads 3–20
      8. Phyllaries gen ± obtuse, adaxial face of tip gen hair-tufted; tube of disk corolla gen 3–4.5 mm . . . . . *A. chamissonis*
      8′ Phyllaries ± acute, tip ± no more hairy than body; tube of disk corolla gen 2–3 mm
        9. Sts 1–few from short rhizome or caudex; lvs ± toothed . . . . *A. lanceolata* subsp. *prima*
        9′ Sts clustered on short caudex-like rhizome; lvs ± entire. . . . . *A. longifolia*
    7′ Cauline lvs gen 2–4(6) pairs, often strongly reduced distally on st; heads gen < 5
      10. Pappus (yellow-)brown, bristles ± plumose — rhizome short, freely rooting
        11. Rays < 15 mm; proximal herbage ± long-spreading-hairy; young heads gen nodding . . . . ²*A. parryi*
        11′ Rays gen 15–25 mm; proximal herbage becoming glabrous to hairy or glandular; young heads erect
          12. Heads spreading-bell-shaped to hemispheric; cauline lvs variable, ± sessile, most proximal gen largest . . . . *A. mollis*
          12′ Heads ± narrowly obconic; cauline lf blades elliptic to widely deltate, at least some petioled, middle gen largest. . . . . *A. ovata*
      10′ Pappus ± white to straw-colored, bristles short-barbed, if bristles ± plumose, rhizome long
        13. Lf blade narrow, length ± 3–6+ × width; basal lvs gen densely clustered; rays yellow-orange
          14. Proximal axils brown-woolly-tufted; disk corollas spreading-hairy, stalked-glandular . . . . *A. fulgens*
          14′ Proximal axils glabrous or sparsely white-hairy; disk corollas stalked-glandular only. . . . . *A. sororia*
        13′ Lf blade gen ± wide, length gen < 2.5 × width; basal lvs sometimes persistent, not densely clustered; rays yellow
          15. Cauline lvs at least at mid-st sessile (± sessile) . . . . *A. latifolia*
          15′ Cauline lvs at least at mid-st petioled (petiole ± wide)
            16. Lvs ± thick, often fleshy and ± red; heads 1(4), often nodding in bud; serpentine soils. . . . . *A. cernua*
            16′ Lvs ± thin, neither fleshy nor ± red; heads 1–5, not nodding in bud; not on serpentine
              17. Proximal lvs gen toothed, ± strongly cordate (exc stunted alpine pls); pappus short-barbed . . . . . . *A. cordifolia*
              17′ Proximal lvs ± entire, base rounded (± cordate); pappus short-barbed to ± plumose . . . . *A. nevadensis*

A. cernua Howell (p. 255) SERPENTINE ARNICA Pl 1–3 dm, ± glabrous. **ST**: 1(3), gen unbranched. **LF**: cauline gen 3–4 pairs, ± widely petioled, distal-most smaller, subsessile, blade 1–5+ cm, (ob)ovate to ± cordate, ± entire to weakly lobed, ± thick, often fleshy and ± red. **INFL**: heads radiate, 1(4), often nodding in bud; involucre 12–18 mm, obconic to bell-shaped; phyllaries ovate to widely lanceolate, short-hairy, often glandular, long-ciliate. **RAY FL**: 5–10; ray < 30 mm. **DISK FL**: corolla soft-hairy. **FR**: 6–8 mm, ± forked-hairy distally; pappus short-barbed or ± plumose, white. 2*n*=38. Serpentine soils, conifer forest; 500–1500 m. KR; sw OR. Apr–Jun ★

A. chamissonis Less. (p. 255) CHAMISSO ARNICA Pl 2–8 dm, ± hairy, gen glandular distally. **ST**: 1, branched distally. **LF**: cauline 4–10 pairs, sessile or short-petioled, 5–20(30) cm, blade (ob)lanceolate, ± entire, distal ± reduced. **INFL**: heads radiate, 3–10; involucre 8–15 mm, bell-shaped; phyllaries nearly linear to narrowly lanceolate, ± obtuse, adaxial face gen hair-tufted near tip. **RAY FL**: 8–20; ray 10–20 mm. **DISK FL**: corolla ± densely soft-hairy, stalked-glandular, tube gen 3–4.5 mm. **FR**: 3–8 mm, ± glabrous to short-hairy, glandular; pappus short-barbed or weakly ± plumose, dirty white to straw-colored. 2*n*=38,57,76. Damp meadows, rocky places, conifer

forest; 1800–3500 m. KR, CaR, SN, SnBr, Wrn, n SNE, W&I; to BC, MT, NM. [*A. c.* subsp. *foliosa* (Nutt.) Maguire] Gen asexual. Jul–Aug

A. cordifolia Hook. HEARTLEAF ARNICA Pl 1–4 dm from ± scaly-tipped rhizome, unevenly glandular and short-hairy. **ST**: 1–few, loose, gen unbranched. **LF**: basal often persistent on sterile rosettes; cauline 2–4 pairs, petioled, reduced distally on st, blades 3–11 cm, ± strongly cordate (exc stunted alpine pls), gen shallowly toothed. **INFL**: heads radiate, 1–5; involucre 15–20(30) mm, obconic to bell-shaped; phyllaries narrowly ovate to lanceolate, ± long-white-hairy, glandular or not. **RAY FL**: 6–13; ray < 30 mm. **DISK FL**: corolla soft-hairy, glandular. **FR**: 5–10 mm, short-forked-hairy or glandular; pappus short-barbed, ± white. 2*n*=38,57,76,95,114. High meadows, conifer forest; 500–3000 m. KR, NCoR, CaRH, SNH, e SnFrB, SCoRO, PR, MP; to WY, n-c N.Am, NM. Gen asexual. Stunted alpine pls with ovate lvs, glandular fr sometimes treated as *A. cordifolia* var. *pumila* (Rydb.) Maguire. May–Jul

A. dealbata (A. Gray) B.G. Baldwin (p. 255) MOCK LEOPARDBANE Per 1.5–3.5 dm from long, branched rhizome, densely short-curly-hairy and sessile-glandular. **ST**: erect. **LF**: basal and cauline, mostly

crowded toward st base; cauline 3–8 pairs, proximal petioled, blades 5–10 cm, elliptic to oblanceolate or obovate; distal sessile, smaller, narrower, entire. **INFL:** heads radiate, 1–6; involucre 1–2 cm diam, bell-shaped or hemispheric; phyllaries oblong or elliptic. **RAY FL:** 5–12 (1 per phyllary); ray 10–25 mm, ± persistent on fr when dry. **DISK FL:** many, staminate; style incl. **FR:** 4–10 mm, oblong, ± flattened, brown or black, short-hairy; pappus 0. 2*n*=38,76. Open forest, meadows, slopes; 1200–2400 m. CaR, n&c SN. [*Whitneya d.* A. Gray] Jun–Aug

A. discoidea Benth. RAYLESS ARNICA Pl 1.5–6 dm, gen ± long-hairy, short-glandular. **ST:** gen 1, branched or not. **LF:** basal often in prominent sterile rosettes; cauline 3–7 pairs, proximal ± narrowly wing-petioled, blade 2–12 cm, lanceolate to ovate or ± cordate, ± toothed, reduced distally on st. **INFL:** heads discoid (radiant), 3–10(30); involucre 10–17 mm, obconic to ± hemispheric; phyllaries narrowly lanceolate to lance-ovate, long-hairy, stalked-glandular. **RAY FL:** 0 (corolla of peripheral disk florets rarely dilated, resembling rays). **DISK FL:** corolla ± soft-hairy, glandular. **FR:** 6–8 mm, forked-hairy and stalked-glandular distally; pappus short-barbed (± plumose), white. 2*n*=38,57,76. Chaparral, foothill woodland; 100–1500 m. NW, CaR, n&c SN, s SNH, CW, WTR, SnBr, w PR, MP; to WA, w NV. Sexual or not. Highly variable; pls with expanded outer disk corollas resemble *A. cordifolia.* May–Jul

A. fulgens Pursh HILLSIDE ARNICA Pl 1–7.5 dm from short, densely scaly, branched rhizome, stalked-glandular, gen hairy, esp distally; proximal axils brown-woolly-tufted. **ST:** 1–several, unbranched. **LF:** basal densely clustered, persistent, petioled, blade 4.5–20 cm, narrowly elliptic to oblanceolate, ± entire; cauline 3–5 pairs, reduced distally on st. **INFL:** heads radiate, 1(3); involucre 10–15 mm, widely hemispheric; phyllaries elliptic-oblong or narrowly to widely lanceolate, glandular, hairy on faces, ± ciliate. **RAY FL:** 8–16; ray 10–30 mm, yellow-orange. **DISK FL:** corolla spreading-hairy, stalked-glandular. **FR:** 3.5–7 mm, forked-hairy; pappus short-barbed, white or pale straw-colored. 2*n*=38, 57. Open, damp depressions in sagebrush scrub or grassland; 1800–2700 m. n SNH (e slope), MP; to BC, AB, CO. May–Jul ★

A. lanceolata Nutt. subsp. **prima** (Maguire) Strother & S.J. Wolf (p. 255) CLASPING ARNICA Pl 5–8 dm from short rhizome or caudex, gen hairy and glandular, esp distally. **ST:** 1–few, sometimes branched distal to middle. **LF:** basal petioled; cauline 4–10 pairs, ± sessile, 4–12 cm, narrowly lanceolate or oblanceolate, ± toothed. **INFL:** heads radiate, 3–10; involucre 8–15 mm, bell-shaped or widely obconic; phyllaries narrowly lanceolate, ± acute, ± equally hairy throughout. **RAY FL:** 7–17; ray 10–20 mm. **DISK FL:** corolla sparsely soft-hairy, tube gen 2–3 mm. **FR:** 4–8 mm, sparsely hairy, sometimes glandular; pappus ± plumose, ± brown. 2*n*=38,57,76. Moist areas, along stream banks, snow-melt areas, montane to alpine meadows; 2200–3500 m. KR, CaRH, SNH, SNE; to s AK, WY. [*A. amplexicaulis* Nutt.] Gen asexual. Other var. in ne N.Am. Jul–Aug

A. latifolia Bong. BROADLEAF ARNICA Pl 1–5 dm from ± scaly caudex and long rhizomes, becoming glabrous to unevenly soft-hairy and glandular. **ST:** 1–several, gen unbranched. **LF:** basal sometimes in sterile rosettes, petioled; cauline gen 2–4 pairs, middle gen largest and sessile (± subsessile), blade 2–10(15) cm, lance-elliptic to ovate, gen toothed. **INFL:** heads radiate, 1–5; involucre 7–18 mm, obconic to ± hemispheric; phyllaries lanceolate to oblanceolate, gen glandular, sometimes sparsely long-hairy. **RAY FL:** 8–15; ray 10–30 mm. **DISK FL:** corolla sparsely soft-hairy. **FR:** 5–9 mm, short- ± forked-hairy or short-glandular distally; pappus short-barbed, white. 2*n*=38,76. Meadows, open conifer forest to subalpine meadows; 1800–2400 m. KR, c SNH; to AK, CO. Sexual or not. Jul–Aug

A. longifolia D.C. Eaton SPEARLEAF ARNICA Pl 3–6 dm from caudex-like rhizome, scabrous, short-hairy distally, often ± sticky. **ST:** clustered, often forming large patches. **LF:** basal few, < cauline; cauline 5–7 pairs, most proximal pairs often fused around st, blade 5–12 cm, ± lanceolate, ± entire, distal ± reduced. **INFL:** heads radiate, 3–20; involucre 7–10 mm, bell-shaped or widely obconic; phyllaries narrow to widely lanceolate, ± acute, glandular, ± long-hairy, esp tip. **RAY FL:** 6–15; ray 10–20 mm. **DISK FL:** corolla sparsely soft-hairy, ± stalked-glandular. **FR:** 3–7 mm, ± glabrous to glandu-

lar and hairy; pappus short-barbed or ± plumose, yellow-brown to brown. 2*n*=57,76. Gen wet meadows, late snow-melt areas, or open conifer forest; 1300–3500 m. KR, s NCoRH, CaRH, SNH, GB; to w Can, MT, CO. Gen asexual in CA. Jul–Aug

A. mollis Hook. HAIRY ARNICA Pl 1.5–7 dm from short rhizome or loose caudex, proximally ± hairy, glandular, or becoming glabrous. **ST:** 1–several, simple or few-branched. **LF:** cauline 3–5 pairs, ± sessile, most proximal gen largest, blade 4–20 cm, (ob)lanceolate to (ob)ovate, entire to unevenly dentate. **INFL:** heads radiate, 1–3(7); involucre 10–16 mm, spreading-bell-shaped to hemispheric; phyllaries widely lanceolate, rarely lanceolate or oblanceolate, sparsely soft-hairy. **RAY FL:** 10–22; ray (10)15–30 mm. **DISK FL:** corolla ± densely soft-hairy, ± stalked-glandular. **FR:** 4–8 mm, forked-hairy, gen glandular; pappus plumose, (yellow-)brown. 2*n*=38,57,76,95,114,133,152. Meadows, streambanks in subalpine zone; 2500–3500 m. KR, NCoRH, CaRH, SNH, Wrn, n SNE, W&I; to WA, w Can, MT, CO. Sexual or not. Jul–Sep

A. nevadensis A. Gray SIERRA ARNICA Pl 1–5 dm, short- (or long-)hairy, glandular; rhizome long, tip scaly. **ST:** 1–several, loose, gen unbranched. **LF:** basal sometimes in sterile rosettes; cauline 2–3 pairs, short-petioled, proximal largest, blade 3–8 cm, elliptic to widely ovate, ± entire, base rounded (± cordate). **INFL:** heads radiate, 1–3; involucre 10–18 mm, ± obconic; phyllaries oblanceolate, stalked-glandular, base often ± long-hairy. **RAY FL:** 6–14; ray gen 10–20 mm. **DISK FL:** corolla glandular. **FR:** 6–9 mm, short-forked-hairy or glandular; pappus short-barbed to ± plumose, white to straw-colored. 2*n*=76. Conifer forest, meadows; 1500–3000 m. KR, NCoRH, CaRH, SNH; to WA, NV. [*A. tomentella* Greene] Sexual or not. Alpine forms intergrade with *A. cordifolia.* Jul–Aug

A. ovata Greene STICKY LEAF ARNICA Pl 1–5 dm from short, freely rooting rhizome, ± glandular-hairy, or becoming glabrous. **ST:** 1–several, unbranched. **LF:** cauline 2–3(4) pairs, short-petioled, middle gen largest, blade 4–8 cm, elliptic or ovate to widely deltate, irregularly toothed. **INFL:** heads radiate, 1–3(5); involucre 10–14 mm, ± narrowly obconic; phyllaries linear to narrowly lanceolate, soft-hairy or short-glandular. **RAY FL:** 8–16; ray (10)15–20 mm. **DISK FL:** corolla ± soft-hairy. **FR:** 5–7 mm, sparsely to moderately long-soft-hairy, short-stalked-glandular; pappus ± plumose, straw-colored to ± brown. 2*n*=57,76. Uncommon. Grassy or wet, rocky sites, conifer forest; 1800–3600 m. KR, CaRH, SNH; to AK, MT, CO. [*A. diversifolia* Greene] Gen asexual. Jul–Sep

A. parryi A. Gray NODDING ARNICA Pl 1–6 dm from short, freely rooting rhizome; proximally ± long-spreading-hairy. **ST:** 1, gen unbranched. **LF:** basal ± sessile, ± ovate, < cauline; cauline 3–4 pairs, proximal pairs ± crowded, ± widely wing-petioled, blade 5–20 cm, lanceolate to ± ovate, ± entire; distal pairs sessile, reduced. **INFL:** heads gen radiate, (1)3–9, gen nodding in bud; involucre 15–20 mm, ± obconic; phyllaries linear to lanceolate, sparsely soft-hairy. **RAY FL:** (0)5–10; ray < 15 mm. **DISK FL:** corolla ± densely soft-hairy, ± stalked-glandular. **FR:** 4–6 mm, glabrous to unevenly hairy and glandular; pappus ± plumose, brown-yellow. 2*n*=38,57,76. Open conifer forest to alpine meadows; 500–3630 m. KR, s NCoRH, SNH, W&I; to nw Can, MT, CO. Jul–Aug

A. sororia Greene (p. 255) TWIN ARNICA Pl 1–5 dm from short, sparsely scaly rhizome, stalked-glandular, unevenly hairy distally; proximal axils glabrous or sparsely white-hairy. **ST:** 1–several, branched or not. **LF:** basal densely clustered, cauline 3–6 pairs, crowded toward base, proximal petioled, blades 3.5–14.5 cm, ± oblanceolate, ± entire; distal lvs narrower, reduced. **INFL:** heads radiate, 1–5; involucre 10–15 mm, widely hemispheric; phyllaries narrowly to ± widely lanceolate, hairy, stalked-glandular, ciliate, adaxial face glabrous. **RAY FL:** 9–17; ray 10–30 mm, yellow-orange. **DISK FL:** corolla stalked-glandular. **FR:** 3–5.5 mm, hairy, sometimes sparsely glandular; pappus short-barbed, white. 2*n*=38. Uncommon. Open sagebrush scrub; 1400–1800 m. GB; to w Can, WY. Similar to *A. fulgens.* May–Jul

A. spathulata Greene (p. 255) KLAMATH ARNICA Pl 1.5–4.5 dm, gen spreading-hairy and ± stalked-glandular. **ST:** 1–few, loose, often branched distally. **LF:** basal gen prominent, < 15 cm, ± oblan-

ceolate, ± widely wing-petioled, ± toothed; cauline 3–5 pairs, most proximal like basal, reduced distally on st. **INFL:** heads discoid, 1 or 3–9(25); involucre obconic to narrowly bell-shaped; phyllaries narrowly to widely lanceolate, glandular, coarsely spreading-hairy. **RAY FL:** 0. **DISK FL:** corolla sparsely soft-hairy and glandular proximally. **FR:** 5–10 mm, sparsely stalked-glandular, gen otherwise glabrous; pappus short-barbed, white. 2*n*=38,76. Open, dry, disturbed oak/conifer woodland, gen on serpentine; 200–1500 m. KR; sw OR. Sexual or not. Apr–Jul ★

A. venosa H.M. Hall (p. 255) SHASTA COUNTY ARNICA Pl 2–6 dm from branched, scaly caudex, unevenly hairy, stalked-glandular. **ST:** 0–few-branched. **LF:** basal 0; cauline 6–10 pairs, widely ± sessile, middle largest, blade 3–7 cm, lanceolate to ± ovate, unevenly toothed, strongly 3–5-veined, strongly net-veined abaxially between main veins; distal often bract-like. **INFL:** head discoid, 1; involucre 15–20(+) mm, ± obconic; phyllaries widely lanceolate to ovate,

hairy, stalked-glandular. **RAY FL:** 0. **DISK FL:** corolla densely soft-hairy proximally. **FR:** 6–8 mm, densely forked-hairy; pappus short-barbed, white. 2*n*=38. Open, often disturbed oak/pine woodland; 400–1400 m. KR. Some pls seem intermediate to *A. discoidea.* May–Jun ★

A. viscosa A. Gray (p. 255) MOUNT SHASTA ARNICA Pl 2–5 dm from woody, scaly caudex, densely stalked-glandular, sparsely hairy distally. **ST:** 1–several, much-branched. **LF:** basal 0; cauline 5–10 pairs, fewer on branches, widely sessile, blade 2–4(5+) cm, ovate-oblong, ± entire, weakly 3–5-veined, not otherwise strongly veined. **INFL:** heads discoid, gen 10–20; involucre 8–15 mm, ± obconic; phyllaries widely lanceolate, short-hairy, stalked-glandular. **RAY FL:** 0. **DISK FL:** corolla cream, glandular. **FR:** 4.5–6.5 mm, stalked-glandular; pappus short-barbed to ± plumose, gen white (± brown). 2*n*=38. Open, rocky, subalpine to alpine sites; 2000–2500 m. KR, nw CaR; s OR. Aug–Sep ★

ARTEMISIA MUGWORT, SAGEBRUSH, SAGEWORT

Leila M. Shultz

Ann to shrub, gen aromatic. **LF:** entire to ± lobed, glabrous to densely hairy; hairs glandular (resin-filled) or T-shaped, hollow. **INFL:** heads discoid or disciform, in spike-, raceme-, or gen panicle-like clusters; involucre ovoid to hemispheric, gen concealing fls; phyllaries in 1–several series, persistent, margins scarious; receptacle hemispheric to conic, gen epaleate, gen glabrous. **PISTILLATE FL:** 0–many; corolla gen ≤ 2 mm, tubular, rarely short ray-like extension present. **DISK FL:** 3–many, gen bisexual, sometimes staminate (ovary vestigial, style branches remaining erect or style unbranched, tip expanded, tack-shaped); corolla ≤ 2.5 mm, ± pale yellow, tube < throat, lobes short-triangular; anther tip acute-ovate to awl-shaped; style tips flat, truncate, hair-fringed. **FR:** < 2 mm, obovoid or fusiform, sometimes ribbed, glabrous, hairy, or resin-gland-dotted; pappus gen 0 or minute crown. ± 400 spp.: esp n hemisphere. (Greek: Artemis, goddess of the hunt, noted herbalist, Queen of Anatolia) [Shultz 2006 FNANM 19:503–534; Shultz 2009 Syst Bot Monogr 89:1–131] Reports of *A. campestris* L. subsp. *pacifica* (Nutt.) H.M. Hall & Clem. for CA unconfirmed. *A. arbuscula* var. *thermopola* Beetle not in CA. *A. absinthium* L. often cult, may persist, perhaps naturalizing in ne CA. Recent studies place *Sphaeromeria* in the *Artemisia* clade (Garcia et al. 2011 Amer J Bot).

1. Ann to per, base sometimes woody; axillary lf clusters gen 0
 2. Rhizome present
  3. Lvs gen 1–8 cm wide, ± glabrous to sparsely tomentose adaxially, glabrous or hairy abaxially; sts 3–25 dm
   4. Proximal-most lvs 2-pinnately divided, lateral lobes acute; lvs gen glabrous adaxially, tomentose (or becoming glabrous) abaxially
    5. Pl sage-scented (not lemony); phyllaries gray-green, densely tomentose . . . . *A. ludoviciana* subsp. *incompta*
    5′ Pl strongly lemon-scented; phyllaries often ± purple and dotted with yellow glands, gen ± glabrous
    . . . . *A. michauxiana*
   4′ Lvs entire to pinnately lobed, lateral lobes obtuse to acute, glabrous to ± sparsely tomentose adaxially, densely tomentose abaxially
    6. Involucre bell-shaped, 2–4 mm diam; phyllaries gray-hairy, ± widely (ob)ovate; fls > 15; lvs green to gray-green adaxially; widespread . . . . *A. douglasiana*
    6′ Involucre narrow, ≤ 2 mm diam; phyllaries ± yellow, glabrous to sparsely hairy, oblong to ± ovate; fls < 15; lvs dark green adaxially; NCo, s ChI . . . . *A. suksdorfii*
  3′ Lvs 0.5–1(4) cm wide, densely gray-hairy (or becoming ± glabrous adaxially); sts 2–8(12) dm . . . . . . *A. ludoviciana*
   7. Infl gen open, occ dense and narrow; lvs gen 1–2 cm . . . . subsp. *albula*
   7′ Infl dense, narrow; lvs gen 2–11 cm
    8. Lvs entire to toothed or shallowly lobed . . . . subsp. *ludoviciana*
    8′ Lvs deeply 1–2-divided into narrow lobes
     9. Involucre 4–5 mm, 4–7 mm diam; lvs gray-white-tomentose. . . . . subsp. *candicans*
     9′ Involucre 2.5–4 mm, 3–5 mm diam; adaxial lf face often green . . . . subsp. *incompta*
 2′ Rhizome 0
  10. St hairy, branched from base, from basal lf cluster
   11. Lvs densely hairy; phyllaries densely gray-hairy, margins widely scarious; coastal . . . . *A. pycnocephala*
   11′ Lvs glabrous to densely silky-hairy; phyllaries sparsely hairy, margins dark or scarious; montane
    12. Phyllary margins scarious; involucre 3.5–4 mm diam; bien or per from taproot; heads gen erect in spike- or narrowly panicle-like clusters. . . . . *A. borealis* subsp. *borealis*
    12′ Phyllary margins ± black; involucre (4)7–11 mm diam; per from branched caudex; heads nodding, slender-peduncled in raceme-like or few-branched, panicle-like cluster . . . . *A. norvegica* subsp. *saxatilis*
  10′ St gen glabrous, simple, basal lvs 0
   13. Bien or per; lvs entire to 1-pinnately lobed
    14. Lvs linear, entire or with few linear lobes; heads disciform, disk fls staminate; receptacle epaleate; widespread . . . . *A. dracunculus*
    14′ Lvs deeply and coarsely pinnately 3–7-lobed; heads discoid; receptacle paleate; SCo, PR . . . . *A. palmeri*

13′ Ann or bien; st erect; lvs 2–3-pinnately divided

15. Heads nodding; ann, aromatic; lvs 2–3-pinnately divided . . . . **[*A. annua*]**

15′ Heads erect; ann or bien, unscented; lvs 2-pinnately divided. . . . . **A. biennis**

1′ Subshrub or shrub; axillary lf clusters present

16. Pl spiny, compact; infl short, heads ± hidden by lvs; lvs palmately 2–5-divided; fl spring/early summer

. . . . **A. spinescens**

16′ Pl unarmed, compact to tall; infl long, heads not hidden by lvs; lf lobes 0–5; fl late summer/fall

17. Lvs (1)2–10 cm, linear, often ± thread-like or lobes ≤ 3 mm wide, gen > 1 cm

18. Lf segments gen < 1 mm wide, margins curled under; CW, SW . . . . **A. californica**

18′ Lf segments 1–3 mm wide, flat; s ChI . . . . **A. nesiotica**

17′ Lvs 0.3–2(6) cm, gen oblong or wedge-shaped (to linear), gen > 3 mm wide, entire to shallowly 3-lobed at tip, lobes < 1 cm

19. Heads nodding; lvs not clustered, entire or teeth or lobes 3, sharply pointed — W&I, DMoj . . . . **A. bigelovii**

19′ Heads erect; lvs often in axillary clusters, entire or teeth 3(6), rounded to acute

20. Lvs gen linear, entire, winter-deciduous; heads gen > 5 mm wide . . . . **A. cana** subsp. **bolanderi**

20′ Lvs gen wedge-shaped, gen 3-toothed or -lobed (entire or 6-toothed in *Artemisia spiciformis*), persistent in winter; heads gen < 5 mm wide (4.5–6 mm in *Artemisia spiciformis*)

21. Pl gen < 4 dm; infl narrow, gen < 3 cm wide

22. Heads short-peduncled; involucre gen 1–2 mm diam

23. Phyllaries shiny-resinous, straw-colored; lvs of fl-sts gen entire . . . . **A. nova**

23′ Phyllaries densely gray-tomentose; lvs of fl-sts (at least proximal) lobed . . . . **A. tridentata** subsp. **wyomingensis**

22′ Heads ± sessile; involucre 2–4.5 mm diam, gray-hairy . . . . **A. arbuscula**

24. Fls 3–6 per head

25. Lvs of fl-st shallowly lobed or distal entire; involucre 3–4.5 mm diam. . . . . subsp. **arbuscula**

25′ Lvs of fl-st deeply lobed; involucre 2–3 mm diam . . . . subsp. **longiloba**

24′ Fls 8–23 per head

26. Pl sticky-resinous; lvs evergreen . . . . [2]**A. rothrockii**

26′ Pl not sticky; lvs ± winter-deciduous . . . . [2]**A. spiciformis**

21′ Pl gen > 4 dm; infl gen > 3 cm wide

27. Heads ≥ 3 mm diam; pl sticky-resinous; meadows, c&s SNH, SnGb, SnBr, n SNE, W&I . . . . [2]**A. rothrockii**

27′ Heads gen < 3 mm diam (gen 4.5–6 mm in *Artemisia spiciformis*); pl not sticky; mtns, dry valleys, widespread

28. Lvs entire to irregularly 3–6-toothed, partly deciduous; pl sprouting from roots; moist slopes, rocky meadows, SNH, W&I > 2000 m. . . . . [2]**A. spiciformis**

28′ Lvs gen regularly 3-lobed or toothed at tip, wedge-shaped, evergreen; pl not root-sprouting; widespread . . . . **A. tridentata** (in part)

29. Pls with a flat-topped crown; infl rising above truncate vegetative portion, narrow, branches gen erect. . . . . subsp. **vaseyana**

29′ Pls with a rounded crown; infl amongst as well as surpassing distal-most vegetative branches, widely branching, branches erect to spreading or drooping

30. Infl branches gen drooping; fr hairy. . . . . subsp. **parishii**

30′ Infl branches erect to gen spreading; fr glandular . . . . subsp. **tridentata**

A. arbuscula Nutt. LOW SAGEBRUSH Shrub < 3 dm, mounded, gray, evergreen. **ST:** much-branched. **LF:** 3–9 mm, wedge-shaped, 3-lobed (or distal entire), hairy, gray-green. **INFL:** heads discoid, 2–4.5 mm diam, ± sessile, gen 1 (in groups of 2–5) in narrow, panicle-like cluster; phyllaries ovate, densely hairy, margins transparent. **PISTILLATE FL:** 0. **DISK FL:** 4–6. **FR:** 0.7–0.8 mm, light brown, finely resin-gland-dotted. 2*n*=18,36. Other subspp. in ID, WY, UT.

subsp. **arbuscula** **LF:** of fl-sts shallowly 3-lobed or distal entire. **INFL:** involucre 3–4.5 mm diam. 2*n*=18, 36. Clay soils, valleys, slopes; 1500–3800 m. KR, NCoRH, SNH, GB. Jul–Aug

subsp. **longiloba** (Osterh.) L.M. Shultz **LF:** of fl-sts deeply 3-lobed. **INFL:** involucre 2–3 mm diam. Uncommon. Igneous rock; 2200–2500 m. SNE. Jun–Jul

A. biennis Willd. (p. 255) BIENNIAL WORMWOOD Ann or bien, gen 3–20 dm, glabrous, green, unscented. **ST:** 1, erect, finely striate, often ± red. **LF:** 4–13 cm, widely lanceolate, 2-pinnately divided ± to midrib; lobes sharply toothed. **INFL:** heads disciform, 2–4 mm diam, erect, sessile or short-peduncled on branches of ± dense, lfy-bracted, panicle-like cluster; phyllaries widely elliptic to obovate, green, margins widely scarious. **PISTILLATE FL:** 6–25. **DISK FL:** 15–40. **FR:** 0.7–1.3 mm, 4–5-angled, glabrous. 2*n*=18. Disturbed moist sites; < 2200 m. NCoRO, GV, CW, SCo, WTR, SnBr, GB, DMtns; N.Am; native to Eur. Jul–Nov

A. bigelovii A. Gray BIGELOW SAGEBRUSH Subshrub 4–6 dm, rounded, branched from base. **ST:** many, slender, curved, silvery-canescent. **LF:** 0.5–3 cm, linear to narrowly wedge-shaped, entire or sharply 3-toothed at tip, densely hairy. **INFL:** heads discoid or disciform, 2–2.5 mm diam, nodding, in tight groups on branches of narrow, panicle-like cluster; phyllaries ovate, obtuse, densely hairy, margins narrowly scarious. **PISTILLATE FL:** 0–2, ray-like extension ≤ 1 mm. **DISK FL:** 1–3. **FR:** elliptic, 5-ribbed, glabrous. 2*n*=18,36. Sandy, often limestone soils; 1300–1900 m. W&I, DMtns; to CO, TX. Aug–Oct

A. borealis Pall. subsp. **borealis** BOREAL SAGEWORT Bien or per from taproot, 1–4 dm, mounded, light green, unscented. **ST:** erect, ± silky, ± red. **LF:** mostly basal, 2–7 cm, 2–3 cm wide, ± ovate, 2–3-pinnately or -ternately divided into linear lobes, ± glabrous to densely silky; cauline widely spaced, 3–5 cm, entire to pinnately lobed. **INFL:** heads disciform, < 5 mm diam, gen erect in spike- or narrowly panicle-like clusters; phyllaries ovate to widely elliptic, glabrous to loosely tomentose, margins widely scarious, midrib prominent. **PISTILLATE FL:** 10–25. **DISK FL:** 15–30, staminate. **FR:** ≤ 1 mm, glabrous. 2*n*=18,36. Meadows; ± 2200 m. KR; to AK, Can, n Eurasia. [*A. campestris* subsp. *b.* (Pall.) H.M. Hall & Clem.] Other subsp. in AK, n Can. Jul–Aug

A. californica Less. (p. 255) CALIFORNIA SAGEBRUSH Shrub (2)6–25 dm, rounded, branched from base. **ST:** slender, flexible, wand-like, glabrous to canescent. **LF:** 1–10 cm, thread-like and entire or 1–2-pinnately divided into thread-like lobes, ± hairy, light green to gray; margins curled under. **INFL:** heads disciform, < 5 mm diam, nodding, short-peduncled along branches of lfy, ± narrow, panicle-like cluster; phyllaries widely (ob)ovate, sparsely canescent, margins wide, scarious. **PISTILLATE FL:** 6–10. **DISK FL:** 15–30. **FR:** 0.8–1.5 mm, resin-gland-dotted; pappus ± 0. 2*n*=18. Coastal scrub, chaparral, open woodland; < 800 m. CW, SW; n Baja CA. Aug–Nov

A. cana Pursh subsp. ***bolanderi*** (A. Gray) G.H. Ward (p. 255) SILVER SAGEBRUSH Shrub < 9 dm, from woody trunk. **ST:** white-felty. **LF:** (1.5)3–4 cm, linear to narrowly lanceolate, gen entire, winter-deciduous. **INFL:** heads discoid, gen > 5 mm diam, gen sessile in groups of 2–3 along branches of lfy, erect, panicle-like cluster; phyllaries elliptic to ovate, outer acute, densely hairy, inner obtuse, membranous, margins scarious. **PISTILLATE FL:** 0. **DISK FL:** 8–16. **FR:** < 1.2 mm, resin-gland-dotted. 2*n*=18,36. Gravelly soils, meadows, streambanks; 1200–3300 m. CaRH, s SNF, SNH, GB; to OR, NV. Sep–Oct

A. douglasiana Besser (p. 255) MUGWORT Per 5–25 dm, from rhizome. **ST:** many, erect, brown to gray-green, ± tomentose. **LF:** evenly spaced, 1–11(15) cm, narrowly elliptic to widely oblanceolate, entire or coarsely 3–5(+)-lobed near tip, ± sparsely tomentose adaxially, densely white-tomentose abaxially. **INFL:** heads disciform, 2–4 mm diam, bell-shaped, erect to ± nodding along branches of lfy-bracted panicle-like clusters; phyllaries ± widely (ob)ovate, gray-tomentose, margins wide, transparent. **PISTILLATE FL:** 5–9. **DISK FL:** 6–25. **FR:** 0.8–1 mm, glabrous. 2*n*=18,36,54. Common. Open to shady areas, often in drainages; < 2200 m. CA-FP, MP, n SNE; to WA, ID, Baja CA. May–Nov

A. dracunculus L. TARRAGON Per 5–15 dm, from woody caudex, gen glabrous (exc D), odorless or tarragon-scented. **ST:** many, stiff, erect, brown. **LF:** basal and cauline, 1–7 cm, linear, entire or with few linear lobes, bright green, glabrous. **INFL:** heads disciform, 2–3.5 mm diam, gen nodding, short-peduncled along branches of lfy, panicle-like cluster; phyllaries widely ovate, glabrous, light brown, membranous, margins widely transparent. **PISTILLATE FL:** 14–25. **DISK FL:** 8–20, staminate; style tip scarcely divided, ± tack-shaped. **FR:** 0.5–0.8 mm, glabrous. 2*n*=18. Common. Many habitats, esp disturbed sites; < 3400 m. CA (exc NCo, NCoRH, NCoRI, CaR); to AK, n Can, n-c US, n Mex, Eurasia. Gen Aug–Oct

A. ludoviciana Nutt. SILVER WORMWOOD Per 2–8(12) dm, from rhizome. **ST:** many, simple, gray- to white-tomentose. **LF:** 1–11 cm, linear to elliptic or (ob)ovate, entire to deeply pinnately lobed, densely tomentose. **INFL:** heads disciform, < 7 mm diam, gen nodding, sessile to short-peduncled along branches of narrow, ± dense to open, diffuse panicle-like cluster; phyllaries lanceolate to (ob)ovate, densely gray tomentose, margins narrowly transparent. **PISTILLATE FL:** 5–12. **DISK FL:** 6–30. **FR:** < 0.5–1 mm, glabrous.

subsp. ***albula*** (Wooton) D.D. Keck **LF:** gen 1–2(5) cm, linear to obovate, entire to ± serrate or lobed, ± persistently white-tomentose on faces, often ± curled under. **INFL:** gen open, occ dense, narrow; involucre ± 3 mm. **DISK FL:** 8–13. Dry sandy soils, scrub, forest; < 1400 m. PR, D; to CO, w TX, nw Mex. May–Oct

subsp. ***candicans*** (Rydb.) D.D. Keck **LF:** proximal gen 5–10 cm, deeply divided into narrow, entire to toothed lobes; distal entire to divided, gray-white-tomentose on faces, margins flat. **INFL:** narrow; involucre 4–5 mm. **DISK FL:** 17–42. Dry woodland, scrub; < 2400 m. n SNH, MP; to WA, MT, UT. Jul–Sep

subsp. ***incompta*** (Nutt.) D.D. Keck **LF:** proximal gen 2–8 cm, 1–2-divided into lance-linear lobes; distal entire to divided, white-tomentose abaxially, becoming ± glabrous adaxially, ± flat. **INFL:** narrow; involucre 2.5–4 mm. **DISK FL:** 15–30. 2*n*=36. Shrubland, woodland, conifer forest; < 3500 m. CaRH, SN, TR, PR, GB, DMtns; to OR, MT, CO. Often confused with *A. michauxiana*. Jul–Sep

subsp. ***ludoviciana*** **LF:** proximal gen 3–11 cm, entire to few-toothed or shallowly lobed ± near tip, white-tomentose on faces or

becoming ± glabrous adaxially, flat. **INFL:** narrow; involucre 3–4 mm. **DISK FL:** 6–20. 2*n*=36. Rocky soils, scrub, conifer forest; < 2600 m. KR, NCoRO, SN, SnJV, SW (exc ChI), GB, DMoj; to BC, e Can, e US, n Mex. Jul–Sep

A. michauxiana Besser LEMON SAGEWORT Per 3–10 dm, from rhizome, lemon-scented. **ST:** many, unbranched, green. **LF:** 1.5–11 cm, ovate to obovate, 1–2-pinnately divided, adaxially green, glabrous and dotted with yellow glands, abaxially thinly white-tomentose or becoming ± glabrous. **INFL:** heads disciform, < 5.5 mm diam, nodding, sessile or short-peduncled in raceme-like or narrow, panicle-like cluster; phyllaries elliptic to widely (ob)ovate, often ± purple and dotted with yellow glands, gen ± glabrous, margins widely scarious. **PISTILLATE FL:** 9–12. **DISK FL:** 15–35. **FR:** 0.5–1 mm, glabrous. 2*n*=18,36. Subalpine to alpine scree, talus, drainages; 3000–3500 m. W&I; to BC, MT, CO. Often confused with *A. ludoviciana* subsp. *incompta*. May–Aug

A. nesiotica P.H. Raven ISLAND SAGEBRUSH Subshrub < 1.5 dm, rounded. **ST:** slender, soft, wand-like from brittle woody base, ± canescent. **LF:** 1–10 cm, blunt, pinnately divided; lobes few, 1–3 mm wide, flat, ± canescent. **INFL:** heads disciform, ± 5 mm diam, ± nodding, subsessile on branches of narrow, sparse, lfy, panicle-like clusters; phyllaries widely ovate, canescent, margins widely scarious. **PISTILLATE FL:** 10–15. **DISK FL:** 20–40. **FR:** 1–1.5 mm, resinous; pappus present. Rocky slopes; gen < 320 m. s ChI. Jul–Oct ★

A. norvegica Fr. subsp. ***saxatilis*** (Besser) H.M. Hall & Clem. ALPINE SAGEWORT Per 3–6 dm, from branched caudex, mildly fragrant. **ST:** erect, loosely tomentose. **LF:** basal petioled, 6–15 cm; cauline < 10 cm; 1–2-pinnately divided; lobes ± linear, 1–2 mm wide. **INFL:** heads disciform, (4)7–11 mm diam, nodding, slender-peduncled in raceme-like or few-branched, panicle-like cluster; phyllaries widely (ob)ovate, sparsely hairy, margins ± black, scarious. **PISTILLATE FL:** 6–12. **DISK FL:** 30–80. **FR:** 2–2.5 mm, glabrous. Rocky slopes; 2300–3800 m. KR, SNH, SnJt, W&I; to AK, n Can, MT, CO. Jul–Sep

A. nova A. Nelson BLACK SAGEBRUSH Shrub 1–3 dm, loosely branched from short trunk, evergreen. **ST:** canescent, becoming ± glabrous. **LF:** 0.5–2 cm, wedge-shaped, gen 3-toothed at tip, gen entire on fl-sts. **INFL:** heads discoid, 1–2 mm diam, erect, short-peduncled on erect branches or axis of slender panicle- or spike-like clusters; phyllaries elliptic to ovate, inner gen glabrous, shiny-resinous, straw-colored, margins transparent. **PISTILLATE FL:** 0. **DISK FL:** 3–6. **FR:** 0.8–1 mm, ribbed, glabrous or resin-gland-dotted. 2*n*=18,36. Shallow rocky soils in desert valleys, dry slopes; < 2300 m. CaRH, SnJt, GB, DMtns; to OR, MT, NM. Often confused with *A. arbuscula*. Sep–Nov

A. palmeri A. Gray SAN DIEGO SAGEWORT Bien or per 8–30 dm, from ± woody base, strongly scented. **ST:** brittle, wand-like, glabrous. **LF:** cauline, 3.5–12(15) cm, deeply and coarsely pinnately 3–7-lobed, glabrous to sparsely hairy adaxially, gray-green-canescent abaxially. **INFL:** heads discoid, 2.5–3.5 mm diam, gen nodding, subsessile or short-peduncled on branches of open, panicle-like cluster; phyllaries ± widely ovate, glabrous to sparsely hairy, membranous, margins ± transparent; receptacle paleate. **PISTILLATE FL:** 0. **DISK FL:** 8–30. **FR:** 1–1.2 mm, smooth to shiny-glandular. 2*n*=18. Moist drainages, sandy soil; < 600 m. SCo, PR; Baja CA. Threatened by development. Gen Jun–Oct ★

A. pycnocephala (Less.) DC. (p. 255) COASTAL SAGEWORT Per from taproot, forming mounds 3–7 dm, densely white-gray tomentose throughout, faintly aromatic. **ST:** ≤ 1 m, decumbent, densely lfy. **LF:** proximal cauline petioled, 3–8+ cm, finely 2–3-pinnately divided, lobes 1–2 mm wide; distal 1–2-divided. **INFL:** heads disciform, 3–4.5 mm diam, erect, sessile, in narrow, lfy panicle-like cluster; phyllaries ovate, densely soft-hairy, margins widely scarious. **PISTILLATE FL:** 8–15. **DISK FL:** 8–25, staminate; style tip unbranched, tack-shaped. **FR:** 1–1.5 mm, glabrous. 2*n*=18. Rocky or sandy soils, coastal strand; < 200 m. NCo, CCo; OR. Jun–Sep

A. rothrockii A. Gray ROTHROCK SAGEBRUSH Shrub 2–5 dm from narrow trunk, sticky-resinous, dark gray-green throughout (sts sometimes ± white-hairy), pungently aromatic. **LF:** 1–1.5(2) cm,

(0)3-lobed (gen entire on fl-sts), evergreen, canescent, sometimes becoming ± glabrous, sticky-resinous (exc White Mtns). **INFL**: heads discoid, 3–5 mm diam, erect, sessile to short-peduncled on branches of narrow, panicle-like clusters; phyllaries ± ovate, sparsely hairy, straw-colored, shiny, margins wide, scarious. **PISTILLATE FL**: 0. **DISK FL**: gen 10–16. **FR**: 0.8–2 mm, resin-gland-dotted; pappus present on outer frs. 2*n*=18,36,54. Clay soils, meadows; 2000–3100 m. c&s SNH, SnGb, SnBr, n SNE, W&I. Aug–Sep

A. spiciformis Osterh. SNOWFIELD SAGEBRUSH Shrub 3–8 dm, gray-tomentose, widely branched, root-sprouting. **LF**: (1)2.5–5.5 cm, entire to irregularly 3–6-toothed (gen entire on fl-sts), turning yellow, ± winter-deciduous. **INFL**: heads discoid, (3)4.5–6 mm diam, erect, in tight groups on axis and short branches of narrow, lfy spike- or panicle-like cluster; phyllaries oblanceolate to obovate, ± hairy, margins ± brown-scarious. **PISTILLATE FL**: 0. **DISK FL**: 8–23. **FR**: ± 1 mm, glabrous. 2*n*=18,36,54. Common. Moist open slopes to rocky meadows; 2100–3700 m. SNH, W&I; to WY, CO. Often confused with *A. rothrockii*; perhaps hybrid of *A. cana* and *A. tridentata*. Jul–Sep

A. spinescens D.C. Eaton (p. 255) BUDSAGE Shrub, ≤ 3 dm, stout, mound-like, pungently aromatic. **ST**: canescent; old branches forming thorns. **LF**: petiole ≤ 2 cm, base expanded; blade ± round in outline, palmately 2–5-divided into linear to spoon-shaped segments, densely soft-hairy. **INFL**: heads disciform, 2–3(5) mm diam, in ± flat-topped or ball-like clusters; phyllaries 5–8, in 1–2 series, widely obovate, densely soft-hairy, margins obscurely scarious; receptacle conic. **PISTILLATE FL**: 2–8. **DISK FL**: 4–13, staminate; corolla widely bell-shaped, densely soft-hairy; style unbranched. **FR**: 1–1.5 mm, obovoid, obscurely ribbed, lightly veined, long-hairy. 2*n*=18,36. Clay or gravelly, often saline soils, saltbush scrub; 650–2050 m. SnBr, GB, DMoj; to OR, MT, NM. [*Picrothamnus desertorum* Nutt.] Apr–Jun

A. suksdorfii Piper COASTAL MUGWORT Per 5–20 dm from rhizome. **ST**: many, erect, brown. **LF**: 5–10(15) cm, sessile, lanceolate, coarsely and irregularly lobed, adaxially dark green, abaxially white-tomentose. **INFL**: heads disciform, 1.5–2 mm diam, gen erect, (±)

sessile along branches of erect, dense, panicle-like cluster; phyllaries oblong to ± ovate, shiny, glabrous to sparsely hairy, straw-colored to yellow-green, margins scarious. **PISTILLATE FL**: 3–7. **DISK FL**: 2–8. **FR**: < 0.8 mm, glabrous. 2*n*=18. Coastal drainages, roadsides; < 300 m. NCo, s ChI; to BC. Hybridizes with *A. douglasiana*. Gen Jun–Sep

A. tridentata Nutt. (p. 255) Shrub < 30 dm, from thick trunk, gray-hairy. **ST**: gen glabrous. **LF**: 1–3(6) cm, gen wedge-shaped, gen (0)3(5)-toothed or lobed at tip, often in axillary clusters, persistent, gray-green, densely hairy. **INFL**: heads discoid, 1–3 mm diam, gen erect; phyllaries oblanceolate to widely obovate, densely tomentose, margins ± transparent. **PISTILLATE FL**: 0. **DISK FL**: 3–6. **FR**: 1–2 mm, glandular or hairy. 2*n*=18,36.

subsp. ***parishii*** (A. Gray) H.M. Hall & Clem. Pl 10–30 dm. **LF**: gen 3–6 cm, linear to narrowly wedge-shaped. **INFL**: < 30 cm, wide, often surrounded by lvs; branches gen drooping. **FR**: hairy. Uncommon. Dry sandy soils; 300–2000 m. SCoRI, SCo, WTR, W&I, DMoj. Sep–Nov

subsp. ***tridentata*** BIG SAGEBRUSH Pl < 20 dm. **LF**: gen 1.2–4 cm, narrowly wedge-shaped. **INFL**: < 30 cm, 5–15 cm wide, often surrounded by lvs; branches erect to gen spreading. **FR**: glandular. 2*n*=18,36. Valleys, benches, sandy to coarse gravelly soils; 800–1900 m. CaRH, s SnJV, SCoRI, TR, GB, DMtns; to WA, MT, NM. Aug–Nov

subsp. ***vaseyana*** (Rydb.) Beetle MOUNTAIN SAGEBRUSH Pl gen 5–10 dm. **LF**: gen 2–3 cm, wedge-shaped. **INFL**: < 30 cm, narrow, exposed; branches erect, exceeding vegetative branches. **FR**: glandular. 2*n*=18,36. Common. Dry slopes; 1800–3000+ m. CaRH, SNH, TR, Wrn, n SNE, W&I; to ID, ND, CO, NM. Aug–Oct

subsp. ***wyomingensis*** Beetle & A.L. Young WYOMING SAGEBRUSH Pl gen < 4 dm. **LF**: gen < 1.2 cm, wedge- to fan-shaped. **INFL**: gen < 6 cm, narrow, often surrounded by vegetative branches; branches short, stiffly spreading, persistent, giving pl a twiggy appearance. **FR**: glandular. 2*n*=36. Uncommon. Valleys, slopes; < 2200 m. GB; to ID, WY, NM. Aug–Nov

ATRICHOSERIS GRAVEL-GHOST

David J. Keil & G. Ledyard Stebbins

1 sp.: sw N.Am. (Greek: chicory-like pl without pappus) [Keil 2006 FNANM 19:309] Perhaps best placed in *Malacothrix* (Lee et al. 2003 Syst Bot 28:616–626).

A. platyphylla (A. Gray) A. Gray (p. 259) Ann, 2–18 dm, glabrous; sap milky. **ST**: erect. **LF**: basal and proximal cauline gen flat against soil, sessile or tapering to a short, winged petiole, 3–10 cm, obovate, obtuse, finely dentate, gray-green, often purple-tinged, esp abaxially; cauline few, reduced distally on st to inconspicuous, triangular scales. **INFL**: heads liguliflorous, 1.5–2.5 cm diam, few to many in an open, ± scapose cyme-like cluster; involucre 6–10 mm; phyllaries in 2–4 series, outer short, triangular to lanceolate, inner 10–15, ± equal; receptacle ± rounded, epaleate. **FL**: 30–60, fragrant; ligule white, readily withering. **FR**: 4–4.5 mm, ± cylindric or broadly club-shaped with 4–5 blunt angles, ± white; pappus 0. 2*n*=18. Desert valleys, washes; < 1600 m. D; to UT, AZ, ne Baja CA. Feb–May

BACCHARIS

David Bogler

Per from woody base, shrub, dioecious, glabrous or finely hairy, often glandular and sticky or shiny. **ST**: gen erect or ascending, round or often striate-angled, gen green. **LF**: cauline, alternate, sessile to short-petioled, linear to (ob)ovate, entire or toothed, distally occ reduced to bracts. **INFL**: heads of 2 kinds, unisexual; discoid (staminate) and disciform (pistillate), 1 or in terminal or lateral raceme- or panicle-like or ± flat-topped clusters; phyllaries lanceolate to ovate, graduated in several series; receptacle flat to conic, epaleate, smooth or pitted. **PISTILLATE FL**: 8–150; corolla narrowly tubular, ± white, lobes vestigial; style branches linear, ± glabrous. **STAMINATE FL**: 8–48; corolla white to pale yellow, 5-lobed; anther tip oblong-triangular; style tips erect, club-shaped or oblong to linear, densely papillate to minutely bristly, ovary much reduced; pappus of many equal bristles. **FR**: ± cylindric, 5–10-ribbed; pappus of many bristles, gen elongating and exceeding phyllaries in fr, ± white to tawny. 350–450 spp.: Am. (Latin: Bacchus, god of wine) [Sundberg & Bogler 2006 FNANM 20:23–34]

1. St gen short-hairy, at least distally among heads
  2. Lvs regularly minutely dentate with closely spaced sharp teeth, oblong or oblanceolate, (3)5–13 mm
    wide. . . . . ***B. plummerae*** subsp. ***plummerae***
  2´ Lvs entire or weakly toothed, linear to linear-(ob)lanceolate, 1–4(8) mm wide
    3. Proximal lvs present at fl, blades (15)20–45(68) mm, 1–4(8) mm wide, linear to linear-oblanceolate,
      resin-gland-dotted adaxially; distal lvs gen well developed. . . . . ***B. malibuensis***

3′ Proximal lvs deciduous and 0 at fl, blades 5–17 mm, 1–2 mm wide, lance-linear, not resin-gland-dotted or sticky; distal lvs reduced, bract-like . . . . *B. brachyphylla*
1′ St gen glabrous throughout
  4. Pl broom-like, sts dense, with numerous parallel branches; lvs sparse or 0 at fl; heads single at branch tips or ± sessile on lateral branches, these in raceme- or panicle-like clusters
    5. Lvs 5–15 mm wide, widely oblanceolate to obovate; heads ± sessile on lateral branches in crowded panicle-like clusters; staminate corolla 2–3.5 mm . . . . *B. sergiloides*
    5′ Lvs 1–3 mm wide, thread-like to narrowly oblanceolate; heads single, stalked, or in small groups in raceme- or panicle-like clusters; staminate corolla > 3.5 mm
      6. St striate, sharply angled; involucre of pistillate head ± cylindric; phyllaries ovate to lanceolate, tips acute to rounded, glabrous; gravelly or sandy disturbed areas. . . . . *B. sarothroides*
      6′ St smooth, rounded; involucre of pistillate head ± funnel-shaped; phyllaries lanceolate, narrowly tapered, usually glandular-scurfy; dense chaparral . . . . *B. vanessae*
4′ Pl bushy or sparingly branched, not broom-like; lvs present at fl; heads in flat-topped or panicle-like clusters (or smaller, lateral clusters in spring-fl pls).
  7. Lf blades obovate or oblanceolate . . . . *B. pilularis*
    8. Sts gen erect, brittle, forming erect or rounded shrub; lvs gen 15–40 mm . . . . subsp. *consanguinea*
    8′ Sts prostrate, flexible, forming mat; lvs gen 5–15 mm; exposed sandy beaches and sea bluffs . . . . . . subsp. *pilularis*
  7′ Lf blades linear, lanceolate, elliptic, oblong to narrowly oblanceolate
    9. Lf blades 1–2(5) mm wide, linear to narrowly oblanceolate, margin entire to minutely dentate with closely spaced sharp teeth; faces not sticky. . . . . *B. plummerae* subsp. *glabrata*
    9′ Lf blades 3–30 mm wide, lanceolate, lance-elliptic or oblong to narrowly oblanceolate, faces gen sticky
      10. Lf blades linear-oblong to narrowly oblanceolate, entire or distally coarsely and irregularly toothed; fr 8–10-ribbed, pappus 8–12 mm. . . . . *B. salicina*
      10′ Lf blades lanceolate or lance-elliptic, entire or finely serrate distally or throughout; fr 5-ribbed, pappus 2.5–6(7) mm
        11. Rhizomatous per, forming large colonies; heads in dense, compact, terminal rounded to ± flat-topped clusters; fr minutely hairy . . . . *B. glutinosa*
        11′ Shrub, not rhizomatous, not forming colonies; heads in terminal flat-topped or pyramidal panicle-like clusters in summer–fall-fl pls, in smaller lateral clusters in spring-fl pls; fr glabrous . . . . *B. salicifolia* subsp. *salicifolia*

B. brachyphylla A. Gray Shrub < 1 m, densely branched, glandular. **ST:** erect, finely branched, slender and wand-like, short-hairy. **LF:** sessile, < 17 mm, 1–2 mm wide, lance-linear, entire, distal reduced, proximal 0 at fl; main vein 1. **INFL:** heads in raceme-like and open panicle-like clusters; involucre funnel- or bell-shaped; phyllaries 1–4 mm, lanceolate, finely hairy, tips acute to acuminate. **PISTILLATE FL:** 8–18; corolla 2–2.8 mm. **STAMINATE FL:** (8)12–18(29); corolla 3.3–4.2 mm; pappus 3–4.4 mm. **FR:** 1.5–2.5 mm, finely hairy; ribs 5; pappus 4.5–7 mm. $2n=18$. Canyons, dry washes, sandy deserts; 180–1200 m. SnBr, SnJt, D; to TX, n Mex. Jul–Aug

B. glutinosa Pers. (p. 259) MARSH BACCHARIS Per < 2 m, from rhizome, glabrous, sticky. **ST:** gen 1 from base; simple, mostly unbranched, erect or ascending, resin-gland-dotted, sticky. **LF:** present at fl, short-petioled; blade < 130 mm, 8–30 mm wide, lanceolate, entire or finely serrate, gland-dotted; main veins 3. **INFL:** heads in rounded to ± flat-topped cluster; involucre hemispheric; phyllaries linear to lanceolate, 2–4 mm, narrow, acuminate, glabrous, gen resin-gland-dotted and sticky. **PISTILLATE FL:** 80–150; corolla 1.7–3 mm. **STAMINATE FL:** 26–40; corolla 3.5–4 mm; pappus 3–4(5) mm. **FR:** 0.6–1.5 mm, ± glandular, tip hairy; ribs 5; pappus 2.5–4(7) mm. $2n=18$. Coastal freshwater and saltwater marshes, streambanks; < 1200 m. NCo, w KR, NCoRI, NCoRO, CaRF, SNF, GV, CW, SW; OR, Baja CA, S.Am. [*B. douglasii* DC.] [Müller 2006 Syst Bot Monogr 76:1–341] The name *B. glutinosa* was mistakenly treated as a synonym of, rather than as a name that had been misapplied to, *B. salicifolia* in TJM (1993) and elsewhere previously. Jul–Oct

B. malibuensis R.M. Beauch. & Henrickson MALIBU BACCHARIS Shrub < 2 m, branched from base, woody, rounded. **ST:** erect to arching, striate-angled, glabrous to sparsely short-hairy. **LF:** short-petioled; blade (15)20–45(68) mm, 1–4(8) mm wide, linear to linear-oblanceolate, acute or acuminate, entire or weakly toothed, glabrous or occ hairy, resin-gland-dotted adaxially, proximal present at fl, distal gen well developed; main veins 1–3. **INFL:** heads in cylindrical panicle-like clusters; phyllaries lance-linear, 2–5 mm, tips becoming brown. **PISTILLATE FL:** 35–38; corolla 2.2–4.2 mm. **STAMINATE FL:** 23–36; corolla 3.7–4.5 mm. **FR:** 2.4–3 mm, ribs 5, face

with thick glandular hairs; pappus 6.5–7.5 mm. $2n=18$. Chaparral, grassy openings; 50–300 m. WTR, PR. Aug–Sep ★

B. pilularis DC. COYOTE BRUSH Shrub < 4.5 m, prostrate and mat-forming to erect and rounded, glabrous, gen sticky. **ST:** branches many, spreading, erect, or ascending, striate, dark brown, sticky. **LF:** sessile or short-petioled; blade 5–40 mm, 2–15 mm wide, oblanceolate to obovate, entire to toothed, glabrous, gland-dotted; main veins 3. **INFL:** heads in a lfy panicle-like cluster; involucre hemispheric to bell-shaped; phyllaries 1–3 mm, lanceolate to ovate, glabrous, tip rounded to acute. **PISTILLATE FL:** 19–43; corolla 2.5–3.5 mm. **STAMINATE FL:** 20–34; corolla 3–4 mm; pappus 3–4 mm. **FR:** 1–2 mm, glabrous; ribs 8–10; pappus 5.5–9 mm. $2n=18$.

subsp. *consanguinea* (DC.) C.B. Wolf (p. 259) **ST:** erect (prostrate), brittle, forming erect or rounded shrub, branchlets evenly distributed around branches. **LF:** gen 15–40 mm. Coastal bluffs, woodland, grassland, disturbed sites, occ on serpentine; < 750(1500) m. NW, CaRF, SNF, ScV, w CW, SCo, ChI, WTR, PR; OR, n Mex. Jul–Dec

subsp. *pilularis* **ST:** prostrate, flexible, forming mat, branchlets mostly turned to one side. **LF:** gen 5–15 mm. Sandy beaches, exposed coastal bluffs; < 100 m. NCo, n&c CCo; OR. Jul–Dec

B. plummerae A. Gray Subshrub or shrub, < 2 m, loosely branched, rounded and bushy, hairy or glandular (at least infl). **ST:** erect, simple, slender, wand-like, short-hairy to ± glabrous, sometimes glandular. **LF:** sessile, 1–13 mm wide, densely hairy to ± glabrous, base wedge-shaped, tip obtuse, entire to regularly minutely dentate, teeth closely-spaced, sharp. **INFL:** heads in compact panicle-like or flat-topped clusters; involucre bell-shaped; phyllaries 2–6 mm, lance-linear, hairy, glandular, tips acute to acuminate. **PISTILLATE FL:** 20–30; corolla 3.5–5 mm. **STAMINATE FL:** 19–26; corolla 4–7 mm; pappus 3.5–4.5 mm. **FR:** 2.5–3.5 mm, hairy, ribs 5; pappus 7–8.5 mm. $2n=18$.

subsp. *glabrata* Hoover (p. 259) SAN SIMEON BACCHARIS < 80 cm. **ST:** ± glabrous, glandular. **LF:** 8–35 mm, linear to narrowly

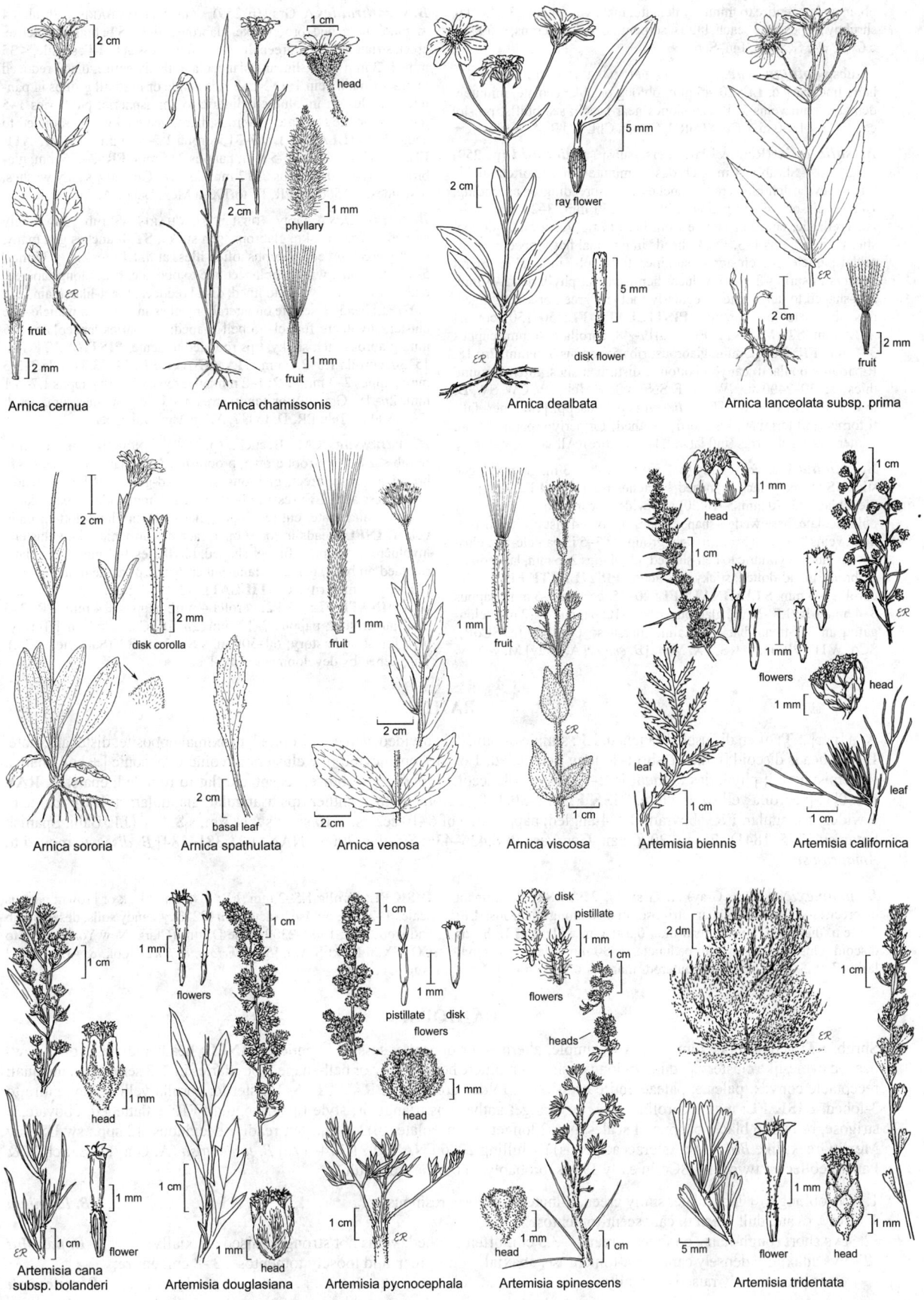

Arnica cernua

Arnica chamissonis

Arnica dealbata

Arnica lanceolata subsp. prima

Arnica sororia

Arnica spathulata

Arnica venosa

Arnica viscosa

Artemisia biennis

Artemisia californica

Artemisia cana subsp. bolanderi

Artemisia douglasiana

Artemisia pycnocephala

Artemisia spinescens

Artemisia tridentata

oblanceolate, entire to minutely dentate; main veins gen 1(3). Rocky, shrubby slopes near beach, bluffs, serpentine rock outcrops; < 500 m. c CCo, w-c SCoRO. Jun–Sep ★

subsp. *plummerae* (p. 259) PLUMMER'S BACCHARIS **ST**: hairs fine, curled. **LF**: 20–45 mm, oblong or oblanceolate, minutely dentate; main veins 3. Rocky slopes near beach, sea bluffs, brushy canyons; < 1850 m. CCo, SCoRO, SCo, n ChI, WTR. Aug–Nov ★

B. salicifolia (Ruiz & Pav.) Pers. subsp. *salicifolia* (p. 259) MULE FAT Shrub < 4 m, glabrous to minutely hairy, often sticky. **ST**: 1–few, often clustered; branches few, spreading or ascending. **LF**: sessile or weakly petioled; blade < 150 mm, 3–20 mm wide, lance-elliptic, entire or serrate from base to tip, resin gland-dotted, sticky; main veins 1–3. **INFL**: heads in terminal flat-topped or pyramidal panicle-like cluster in summer–fall-fl pls, in smaller lateral clusters in spring-fl pls; involucre hemispheric; phyllaries 2–4 mm, awl-shaped to lanceolate, irregularly toothed, green or ± red-tinged, tips obtuse to long-tapered. **PISTILLATE FL**: 50–150; corolla 2–3.5 mm. **STAMINATE FL**: (10)17–48; corolla 4–6 mm, pappus 3–4 mm. **FR**: 0.8–1.3 mm, glabrous; ribs 5; pappus 3–6 mm. 2*n*=18. Riparian woodland, canyon bottoms, disturbed sites, often forming thickets; -30–2400 m. NW, CaRF, SNF, s SNH, Teh, GV, CW, SW, D; to CO, TX, Mex, S.Am. [*B. glutinosa* Pers., misappl.] Summer–fall-fl forms (infl terminal, lvs mostly toothed) formerly separated from winter–spring-fl forms (infl lateral, lvs ± entire). All yr

B. salicina Torr. & A. Gray (p. 259) Shrub 1–3 m, glabrous, gen sticky. **ST**: erect, much-branched, branches ascending. **LF**: short-petioled; blade 25–70 mm, 5–10(20) mm wide, linear-oblong to narrowly oblanceolate, base wedge-shaped, margin toothed distally, teeth 1–3; main veins 3. **INFL**: heads in tight groups of 3–5 in panicle-like cluster; involucre cylindric to bell-shaped; phyllaries 2–6 mm, lanceolate, glabrous, gland-dotted, sticky, tips acute. **PISTILLATE FL**: 25–30; corolla 3–4 mm. **STAMINATE FL**: 20–25; corolla 3–5 mm; pappus 3–4 mm. **FR**: 1.2–2 mm, glabrous; ribs 8–10; pappus 8–12 mm, elongating in fr. Stream banks, alkaline marshes; -60–1600 m. SCoRI, SCo, WTR, PR, D; to KS, TX, Mex. [*B. emoryi* A. Gray] May–Nov

B. sarothroides A. Gray (p. 259) BROOM BACCHARIS Shrub < 4 m, much-branched, broom-like, glabrous, sticky. **ST**: branches dense, erect, striate-angular, green distally, often lfless at fl. **LF**: sessile, < 35 mm, 1–2 mm wide, linear-oblanceolate, thick, entire, distal reduced to bracts; main vein 1. **INFL**: heads single or in small groups in panicle-like cluster; involucre cylindric to hemispheric; phyllaries 1–5 mm, lanceolate to ovate, glabrous, hard, gen sticky, tips rounded to acute. **PISTILLATE FL**: 19–31; corolla 2.5–3.5 mm. **STAMINATE FL**: 18–35; corolla 4.2–5 mm; pappus 3–5 mm. **FR**: 2–2.6 mm, glabrous; ribs 8–10; pappus 7–12 mm. 2*n*=18. Gravelly, sandy washes, roadsides; < 1500 m. PR, D; to TX, n Mex, Baja CA. Aug–Nov

B. sergiloides A. Gray DESERT BACCHARIS Shrub < 3 m, freely branched, broom-like, glabrous, gen sticky. **ST**: branches gen many, strictly erect, green, glabrous, often lfless at fl. **LF**: sessile, < 35 mm, 5–15 mm wide, widely oblanceolate to obovate, bases long-tapered, entire or irregularly 1–4 toothed, distal reduced, bract-like; main vein 1. **INFL**: heads ± sessile on lateral branches in crowded panicle-like cluster; involucre funnel- to bell-shaped; phyllaries lanceolate, 1–5 mm, glabrous, gen sticky, tips rounded to acute. **PISTILLATE FL**: 15–30; corolla 1.6–2.7 mm. **STAMINATE FL**: 24–33; corolla 2–3.5 mm; pappus 2–3 mm. **FR**: 1–2 mm, glabrous; ribs 10; pappus 1.7–3.1 mm. 2*n*=18. Gravelly or sandy streambeds, dry washes, grassland; -70–1800 m. Teh, PR, D; to UT, AZ, n Mex. Jul–Oct

B. vanessae R.M. Beauch. (p. 259) ENCINITAS BACCHARIS Shrub < 2 m from root crown, broom-like, gen glabrous, sticky. **ST**: branches dense, erect, glabrous or stalked-glandular below heads, often lfless at fl. **LF**: sessile, 1–45 mm, 1–3 mm wide, thread-like to linear-oblanceolate, entire, slightly fleshy, resin-gland-dotted; main vein 1. **INFL**: heads in loose open raceme- or panicle-like clusters; involucre 3–5 mm, ± funnel-shaped, phyllaries 1–4 mm, lanceolate, rounded on back, gen glandular-puberulent, tips acute to acuminate, reflexed at maturity. **PISTILLATE FL**: ± 25; corolla ± 2.5 mm. **STAMINATE FL**: 15–22; corolla 4 mm; pappus ± 4 mm. **FR**: 2–3 mm, puberulent; pappus 7–10 mm. 2*n*=18. Chaparral, incl Torrey-pine forest understory; 60–300 m. s SCo, nw PR (San Diego Co.). Threatened by development. Aug–Dec ★

BAHIA

Ann [per]. **ST**: often diffusely branched. **LF**: simple or pinnately divided, gen gland-dotted; proximal opposite; distal alternate. **INFL**: heads discoid [radiate], short- to long-peduncled, 1 or in ± open cyme-like clusters; involucre obconic [± bell-shaped to hemispheric]; phyllaries ± equal in 2–3 series, oblanceolate [obovate], obtuse; receptacle flat to rounded, epaleate. **RAY FL**: 0[15]; corolla yellow or white. **DISK FL**: 15–30[120]; corolla yellow; anther tips triangular, glandular; style tips acute to ± widely triangular. **FR**: obpyramidal, 4-angled; pappus gen of 6–10 scales. ± 10 spp.: sw N.Am, s S.Am. (J.F. Bahi, Spanish botanist, 1775–1841) [Panero 2006 Fam Gen Vasc Pl 8:435–436; Strother 2006 FNANM 21:381–384] *B. dissecta* moved to *Amauriopsis*.

B. neomexicana (A. Gray) A. Gray (p. 259) **ST**: decumbent or erect, 3–25 cm, glandular, strigose, or becoming glabrous. **LF**: 2–4 cm, dissected, lobes thread-like, 0.5–1 mm wide. **INFL**: heads discoid; clusters few-headed; peduncles 5–30 mm, glandular; phyllaries 7–9, 5–6 mm, , green-centered, margins often red or yellow. **DISK FL**: corolla 1.5–2 mm. **FR**: 3–4 mm, black or brown; pappus scales 1–2 mm, obtuse to acute. 2*n*=22. Dry, sandy soils, desert scrub and woodland; 1500–2300 m. se DMtns (Clark, New York mtns); to CO, TX, n Mex; S.Am. [*Schkuhria multiflora* Hook. & Arn.var. *m.*] Aug–Oct ★

BAHIOPSIS

Shrub. **ST**: gen several from base. **LF**: simple, alternate or opposite, sessile or petioled. **INFL**: heads radiate, 1 or in few-headed cyme-like clusters; peduncles long or short; involucre hemispheric or bell-shaped; phyllaries in 2–3 series, gen unequal; receptacle convex, paleate; paleae entire or 3-lobed, folded around frs. **RAY FL**: 8–15, sterile; corolla yellow; ray entire to 3-lobed. **DISK FL**: many; corolla yellow or orange; anther tips triangular; style tips triangular. **FR**: ± flattened, obovate, ± strigose, brown to black; pappus of scales, gen 2 longer, ± lanceolate, (0)2–6 shorter, readily-deciduous. 12 spp.: sw US, nw Mex. (Greek: like *Bahia*, an Asteraceae genus) [Schilling 2006 FNANM 21:174–176] *B. tomentosa* (A. Gray) E.E. Schill. & Panero collected twice in s SCo in early 1900s, probably an escape from cult.

1. Lvs oblong to ± lanceolate, shiny green, minutely strigose, resinous . . . . *B. laciniata*
1′ Lvs ± ovate, dull green to canescent-tomentose
2. Lvs short rough-hairy on faces, green, 1–3.5 cm, often toothed; veins not strongly raised abaxially . . . . *B. parishii*
2′ Lvs adaxially densely canescent-tomentose, abaxially dull green and loosely tomentose, 3–8 cm, entire; veins prominently raised abaxially . . . . *B. reticulata*

B. laciniata (A. Gray) E.E. Schill. & Panero (p. 259) SAN DIEGO VIGUIERA Pl ≤ 2 m diam, short-rough-hairy, covered with varnish-like resin throughout. **ST:** 6–13 dm, slender. **LF:** gen alternate; petiole 0–10 mm; blade 1–5 cm, oblong to narrowly lanceolate, base rounded to truncate or hastate, tip obtuse to acute, margin entire to coarsely round-toothed or shallowly lobed, often ± rolled under, faces shiny green, minutely strigose, resinous, gen 3-veined, veins ± prominently raised abaxially. **INFL:** heads 1 or in ± flat-topped cyme-like clusters; peduncle 0.5–5 cm with 1–several, scattered, lfy or scale-like bracts; involucre 8–10 mm diam, hemispheric; phyllaries 3–7 mm, ovate to lanceolate; paleae 5.5–7.3 mm. **RAY FL:** 5–13; ray 6–12 mm. **DISK FL:** corolla 3.5–4.5 mm. **FR:** 2–4 mm, obovate; long pappus scales 1.7–3 mm, short scales 0.4–1 mm. 2*n*=36. Coastal scrub, chaparral slopes; 90–750 m. s SCo, PR (probably introduced n SCo, s WTR); Baja CA, w SON. [*Viguiera l.* A. Gray] Feb–Aug(Oct–Dec) ★

B. parishii (Greene) E.E. Schill. & Panero (p. 259) PARISH'S GOLDENEYE Pl ≤ 2 m diam, short-rough-hairy throughout. **ST:** 6–13 dm, much-branched. **LF:** proximal opposite, distal alternate; petiole 2–8 mm; blade gen 1–3.5 cm, triangular-ovate, 3-veined from obtuse to truncate or ± cordate base, tip obtuse to acute, margin entire or often few-toothed, faces green to lightly canescent. **INFL:** heads 1 or in open, few-headed cyme-like clusters; peduncle 3–15 cm, slender, bracts 0 or few and lf-like; involucre 10–13 mm diam, hemispheric or appearing disk-like when pressed; phyllaries equal or unequal, 3–9 mm, lance-oblong, green to canescent, tips abruptly narrowed; paleae 5–6 mm. **RAY FL:** 8–15; ray 1–1.5 cm. **DISK FL:** corolla 3.5–5 mm. **FR:** 2.7–3.8 mm; long pappus scales 2–3 mm, short scales 0.5–1 mm. 2*n*=36. Common. Washes, dry, rocky slopes; < 1500 m. e PR, D; NV, AZ, nw Mex. [*Viguiera p.* Greene] Feb–Jun, Sep–Oct

B. reticulata (S. Watson) E.E. Schill. & Panero (p. 259) DEATH VALLEY GOLDENEYE Pl ≤ 1.5 m diam. **ST:** many, 5–15 dm, soft-hairy; bark peeling in age. **LF:** proximal opposite, distal alternate; petiole 3.5–30 mm; blade 2–9 cm, ovate, 3-veined from truncate to cordate base, tip acute, margin entire, adaxially densely canescent-tomentose, abaxially gray-green, loosely tomentose, with veins prominently raised. **INFL:** heads few in cyme-like clusters, on long, ± naked branch; bracts reduced to scales 3–10 mm; peduncle 5–50 mm; involucre hemispheric or appearing disk-like when pressed; phyllaries 2–5 mm, oblong to ovate, obtuse, short-white-hairy; paleae 4–5.5 mm. **RAY FL:** 10–15; ray 7–15 mm. **DISK FL:** corolla 3–4 mm. **FR:** 2.5–4 mm, obovate; long pappus scales 1.5–2.8 mm, short scales 0.5–1 mm. Arid slopes; < 1500 m. DMoj; w NV. [*Viguiera r.* S. Watson] Feb–Jun, Sep–Oct

BAILEYA DESERT-MARIGOLD

Ann, per, ± tomentose throughout. **ST:** 1–many from base. **LF:** simple, basal and alternate, entire to deeply lobed, ± reduced upward, petioled or distal sessile, tomentose and glandular. **INFL:** heads radiate, 1 or in loose cyme-like clusters; peduncles short to very long; involucre cylindric to bell-shaped or hemispheric; phyllaries in 1–2 ± equal series, lance-linear; receptacle flat to slightly rounded, pitted, epaleate or with scattered, narrow paleae. **RAY FL:** 4–many; ray ovate, ± 3-lobed, sessile on ovary, yellow, drying cream, papery, reflexed and persistent on fr when dry. **DISK FL:** 8–many; corolla yellow, gland-dotted, lobes triangular, long-hairy; anther tips triangular; style tips truncate or short-triangular. **FR:** linear to club-shaped, cylindric or ± angled, short-rough-hairy and gland-dotted to glabrous; pappus 0. 3 spp.: sw US, Mex. (J.W. Bailey, Am microscopist, 1811–1857) [Turner 2006 FNANM 21:444–447]

1. Ray fls 4–8, pale yellow; involucre gen 4–6 mm diam; heads 2–3 . . . . ***B. pauciradiata***
1′ Ray fls 15–many, bright yellow; involucre 7–25 mm diam; heads solitary
  2. Rays 10–20 mm, widely linear or oblong, prominently 3-lobed; fr cylindric or barely angled; ribs ± equal; peduncles gen 10–30 cm, ± naked, scape-like . . . . ***B. multiradiata***
  2′ Rays 6–10 mm, widely elliptic to obovate, shallowly 3-lobed to truncate; fr ribbed and angled; ribs of angles most prominent; peduncles gen 2–10 cm, gen lfy-bracted. . . . . ***B. pleniradiata***

B. multiradiata Torr. (p. 263) Pl canescent-tomentose. **ST:** 2–5 dm, gen branched only at base. **LF:** mostly basal and proximal cauline, these 2–10 cm; petioles winged; blades 1–3-pinnately divided, lobes linear to ovate; distal cauline 0 or reduced to linear, entire bracts. **INFL:** heads 1, showy; peduncle gen 1–3 dm, ± naked, scape-like; involucre 10–25 mm diam, hemispheric; phyllaries 5–8 mm. **RAY FL:** 34–60 in > 1 series; ray 10–20 mm, widely linear or oblong, bright yellow; lobes prominent, lanceolate to ovate. **DISK FL:** many; corolla 3–4 mm. **FR:** 2.5–4 mm, cylindric or slightly angled, ± equally ribbed. 2*n*=32. Desert roadsides, flats, washes, hillsides; < 1600 m. SCoRO (waif), D; to UT, TX, n Mex. Pls from ne DSon apparently native; sw DSon records probably naturalized. Apr–Jul, Oct

B. pauciradiata Harv. & A. Gray (p. 263) Ann (rarely per), ± loosely tomentose. **ST:** 1–7.5 dm, gen branched above base. **LF:** mostly cauline; basal soon withering; cauline lvs 4–14 cm, not markedly reduced distally on st, linear-oblong to lanceolate or oblanceolate, entire or divided into linear lobes. **INFL:** heads 2–3 in cymes; peduncles 2–5 cm; involucre 4–6 mm diam, cylindric to bell-shaped; phyllaries 8–13, 4–8 mm. **RAY FL:** 4–8; ray 4–10 mm, obovate, shallowly 3-lobed, pale yellow. **DISK FL:** 8–20; corolla 2.5–3 mm. **FR:** 4–5 mm, narrowly club-shaped, evenly ribbed. 2*n*=32. Sandy desert soils, esp dunes; < 1700 m. D; w AZ, n Mex. Dec–Jun, Oct

B. pleniradiata Harv. & A. Gray (p. 263) Ann (rarely per), canescent-tomentose. **ST:** 1–5 dm, branched mostly in proximal 1/2. **LF:** mostly cauline; basal gen withering by fl time; basal and proximal cauline similar, 2–8 cm, petioled; blade 1–3 × divided into oblong to ovate lobes; distal cauline simple, sessile, linear to narrowly oblanceolate, gen entire. **INFL:** heads 1; peduncle gen 2–10 cm; involucre 7–12 mm diam, hemispheric; phyllaries 20–30, 4–6 mm. **RAY FL:** 20–60, in 2 or more series; ray 6–10 mm, widely elliptic to obovate, ± entire to shallowly 3-lobed, bright yellow. **DISK FL:** many; corolla 3 mm. **FR:** 3–4 mm, cylindric, distinctly angled; ribs of angles most prominent. 2*n*=32. Desert roadsides, sandy soils; < 1900 m. SCo (waif), SNE, D; to UT, AZ, n Mex. Mar–Jun, Oct–Nov

BALSAMORHIZA BALSAMROOT

Per from fleshy taproot; caudices 1–many. **ST:** erect. **LF:** basal and few cauline, alternate or opposite, long-petioled; blade entire to 1–3-pinnately lobed. **INFL:** heads 1–few, radiate; peduncles long, bracts 0–few; involucre hemispheric to bell-shaped; phyllaries in 2–4 series; receptacle flat, paleate; paleae folded around frs. **RAY FL:** showy; ray yellow. **DISK FL:** many; corolla yellow, tube short, throat cylindric to narrowly bell-shaped; style branches tapered. **FR:** oblong, 3–4-angled; pappus 0. 12 spp.: w N.Am. (Greek: balsam root, from sticky sap of taproot) [Weber 2006 FNANM 21:93–99] Hybrids common.

1. Lf blade widely triangular, base cordate and ± hastate, margin entire to coarsely dentate
  2. Lf gen scabrous to thinly stiff-hairy; phyllaries glandular and loosely long-soft-hairy . . . . ***B. deltoidea***
  2′ Lf densely canescent-tomentose, at least when young; phyllary bases permanently tomentose . . . . ***B. sagittata***
1′ Lf blade lanceolate to elliptic or ovate, base gen not cordate or hastate, margin coarsely toothed to
     deeply divided
  3. Lf blade coarsely serrate but gen not deeply lobed . . . . ***B. serrata***
  3′ Lf blade gen deeply pinnately lobed to divided
    4. St densely glandular-puberulent (and gen also sparsely long-hairy) . . . . ***B. hirsuta***
    4′ St strigose to densely hairy (sometimes also gland-dotted), sometimes becoming ± glabrous
      5. Outer phyllaries lanceolate, 2–4 mm wide, acuminate; herbage densely ± white-tomentose . . . . ***B. lanata***
      5′ Outer phyllaries lance-ovate to widely ovate, 4–10 mm wide, obtuse to acute; herbage finely strigose to
          silky or bristly, green to densely silvery-canescent
        6. Outer phyllaries 20–40 mm, >> disk; lvs thinly strigose, green or lightly silvery-canescent . . . . ***B. macrolepis***
        6′ Outer phyllaries 12–24 mm, gen = or slightly > disk; lvs gen densely strigose, lightly canescent
            to silvery
          7. Lf ± gray-green (gen not silvery), lobes gen again divided; e CaR, n SNH, MP . . . . ***B. hookeri***
          7′ Lf silvery, lobes gen entire; e KR . . . . ***B. sericea***

B. deltoidea Nutt.   **ST**: 2–9 dm, densely glandular, sparsely long-hairy. **LF**: basal 20–60 cm, blade widely triangular, entire to coarsely dentate, gen scabrous to thinly stiff-hairy, base cordate and ± hastate; cauline 2–several, alternate or opposite, oblanceolate, entire, gen short-stiff-hairy, ± gland-dotted. **INFL**: heads 1(2); outer phyllaries 10–40+ mm, 3–9 mm wide, lance-oblong, obtuse to acute, glandular and loosely long-soft-hairy, often ciliate. **RAY FL**: ray 2–5 cm. **DISK FL**: corolla 5–7 mm. **FR**: 7–8 mm. 2*n*=38. Common. Grassy slopes, open forests, shrubby areas; < 2400 m. NW, SN, CW, WTR, MP; to BC. Apr–Jul

B. hirsuta Nutt.   **ST**: 1–3 dm, densely glandular-puberulent, sparsely long-hairy. **LF**: basal 15–30 cm, blade lanceolate to elliptic or ovate, pinnately lobed, further divided into linear-oblong segments 2–3 mm wide, faces green, strigose to silky or bristly, glandular-puberulent; cauline gen 2 near base, opposite. **INFL**: head 1; outer phyllaries 15–22 mm, 4–5 mm wide, lance-ovate, long-acuminate, ciliate, minutely glandular, tips often spreading or reflexed. **RAY FL**: ray 2–4 cm. **DISK FL**: corolla 7–8 mm. **FR**: 6–7 mm. Sagebrush scrub, open forest, dry meadows; 1000–1800 m. MP; to WA, nw NV. May–Jul

B. hookeri Nutt.   **ST**: 0.5–4 dm, ± densely hairy, becoming ± glabrous. **LF**: basal 20–40 cm, blade lanceolate to elliptic or ovate, pinnately divided into linear to oblong or lanceolate segments, these gen also divided, finely strigose, to silky or bristly, ± gray-green, glandular; cauline often 2–3 near base. **INFL**: head 1; outer phyllaries 10–24 mm, 4–8 mm wide, tip wide, flat, obtuse to acute, appressed to spreading, densely strigose-canescent. **RAY FL**: ray 1.5–3 cm. **DISK FL**: corolla 7–10 mm. **FR**: 5–7 mm. 2*n*=38. Dry slopes, valleys; 1300–1800 m. e CaRH, n SNH, MP; to e WA, nw NV. [*B. macrolepis* var. *platylepis* (W.M. Sharp) Ferris] May–Jul

B. lanata (W.M. Sharp) W.A. Weber (p. 263)   WOOLLY BALSAMROOT   **ST**: 1–3 dm, densely tomentose. **LF**: basal 10–30 cm, blade lanceolate to widely ovate, pinnately lobed, further divided into linear-oblong segments 2–3 mm wide, faces densely ± white tomentose; cauline gen 2 near base, opposite. **INFL**: head 1; outer phyllaries 11–20 mm, 2–4 mm wide, lance-oblong, acuminate, tomentose. **RAY FL**: ray 1.5–2.5 cm. **DISK FL**: corolla 7–8 mm. **FR**: 5–6 mm. Open woodland, grassy slopes; 800–1050 m. KR (Scott Mtns), n CaRH (Shasta Valley, Siskiyou Co.). [*B. hookeri* var. *l.* W.M. Sharp] Apr–Jun ★

B. macrolepis W.M. Sharp (p. 263)   **ST**: 2–6 dm, finely hairy, gen also gland-dotted. **LF**: basal 15–40 cm, blade lanceolate to elliptic, pinnately divided into linear to ovate segments, finely strigose, green to lightly silvery canescent, sometimes gland-dotted; cauline often 2–3 near base. **INFL**: head gen 1; outer phyllaries 20–40 mm, gen 6–10 mm wide, lance-ovate, tip wide, flat, obtuse to acute, appressed to spreading, finely strigose, minutely glandular. **RAY FL**: ray 2–3 cm. **DISK FL**: corolla 7–10 mm. **FR**: 7–8 mm. Open grassy or rocky slopes, valleys; gen ≤ 1400 m. SNF, c SNH, ScV, e SnFrB. Mar–Jul ★

B. sagittata (Pursh) Nutt. (p. 263)   **ST**: 2–6 dm, ± short-tomentose, minutely glandular. **LF**: basal 20–50 cm, blade widely triangular, entire, acute or obtuse, base cordate and ± hastate, adaxial face soft-hairy, abaxial face short-tomentose to finely strigose, densely canescent-tomentose, at least when young; cauline gen several, linear to oblanceolate. **INFL**: heads 1–3; outer phyllaries 10–25 mm, 4–9 mm wide, lance-oblong to ovate, obtuse to acute, ± tomentose. **RAY FL**: ray 2.5–4 cm. **DISK FL**: corolla 6–8 mm. **FR**: 7–9 mm. 2*n*=38. Open forest, scrub; 1400–2600 m. CaRH, SNH, GB; to BC, AB, SD, CO. May–Aug

B. sericea W.A. Weber   SILKY BALSAMROOT   **ST**: 1–4 dm. **LF**: basal 10–30 cm, blade lanceolate to elliptic, pinnately lobed (lobes gen entire), finely silvery canescent; cauline gen 2–3 near base. **INFL**: head 1; outer phyllaries 12–20 mm, gen 7–8 mm wide, widely ovate, tip wide, flat, obtuse to acute, appressed to spreading, finely strigose, minutely glandular. **RAY FL**: ray 1.5–3 cm. **DISK FL**: corolla 7–10 mm. **FR**: 5–8 mm. Serpentine outcrops, rocky slopes; 400–1800 m. e KR (Siskiyou, Trinity cos.); sw OR. May–Jun ★

B. serrata A. Nelson & J.F. Macbr. (p. 263)   SERRATED BALSAMROOT   **ST**: 1–3 dm, glandular-puberulent, thinly tomentose to soft-hairy. **LF**: basal 8–20 cm, blade lanceolate to widely ovate, gen serrate, rarely irregularly pinnately lobed, lobes wide, serrate, green, strigose, bristly, or scabrous; cauline gen 2 near base, opposite. **INFL**: head 1; outer phyllaries 10–22 mm, 2–3 mm wide, lance-linear, long-acuminate, tomentose, tips often spreading or reflexed. **RAY FL**: ray 2–4 cm. **DISK FL**: corolla 7–8 mm. **FR**: 5–8 mm. 2*n*=38. Sagebrush scrub, forest openings, meadow margins; 1400–1500 m. n MP; to WA, nw NV. Apr–Jun ★

BEBBIA   SWEETBUSH

Subshrub, shrub, ± strongly scented. **ST**: many, slender, from thick, woody root-crown, short-lived, brittle, often lfless. **LF**: simple, opposite (or distal alternate), sessile or petioled; blades linear [triangular], entire to dentate or irregularly lobed. **INFL**: heads discoid or disciform, 1 or in open, rounded cyme-like clusters; peduncle slender; involucre cylindric to bell-shaped; phyllaries graduated in 3–5 series; receptacle rounded, paleate, paleae folded around frs. **PISTILLATE FL**: 0–8; corolla yellow to orange, radial. **DISK FL**: 20–50; corolla yellow to orange; anther tips ovate, acute; style tips tapered, acute. **FR**: club-shaped, compressed, 3-angled, brown to black; hairs ascending, white; pappus of 15–30 ± plumose bristles. 2 spp.: sw US, nw Mex. (M.S. Bebb, Am botanist, 1833–1895) [Whalen 2006 FNANM 21:177]

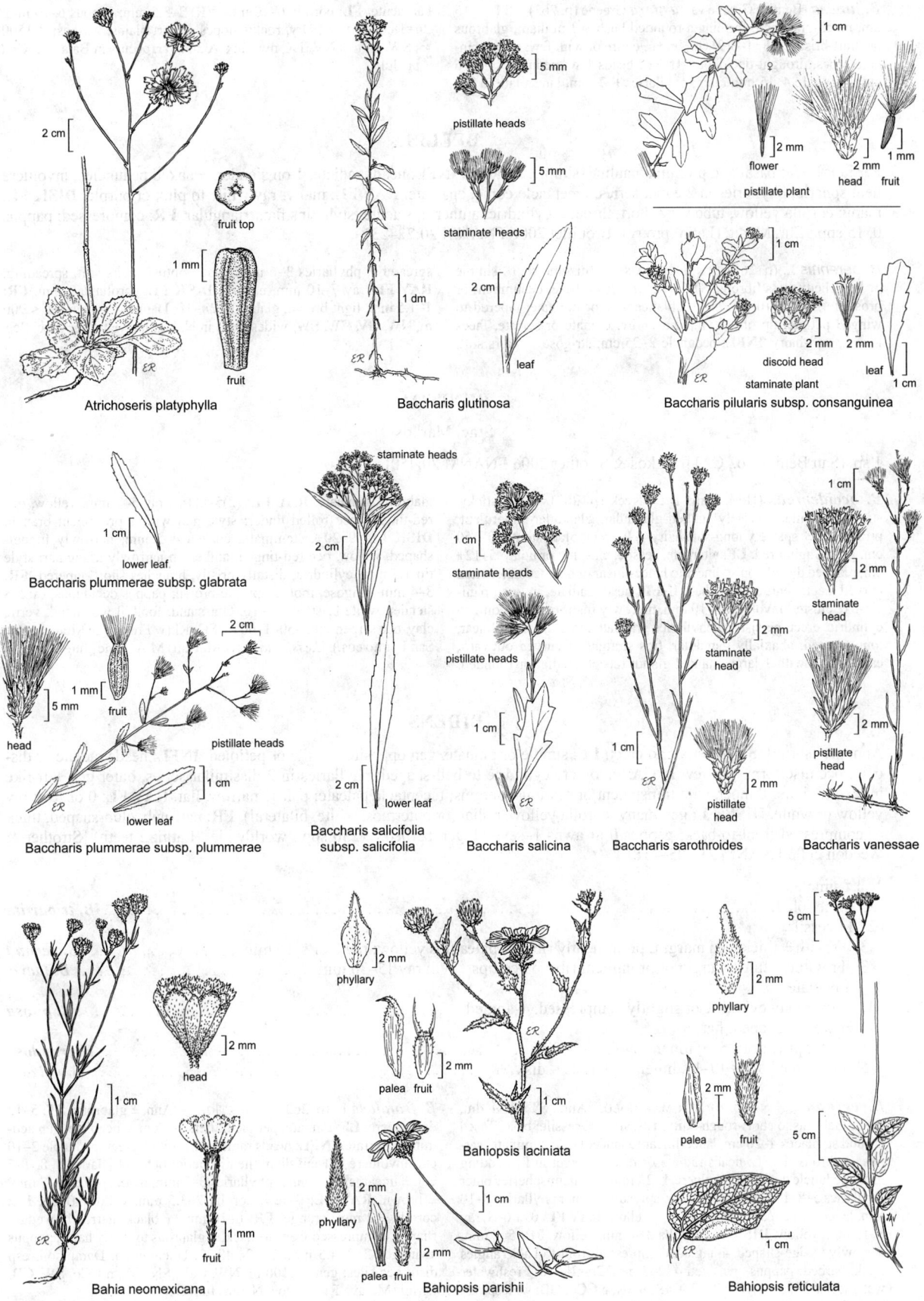

Atrichoseris platyphylla

Baccharis glutinosa

Baccharis pilularis subsp. consanguinea

Baccharis plummerae subsp. glabrata

Baccharis plummerae subsp. plummerae

Baccharis salicifolia subsp. salicifolia

Baccharis salicina

Baccharis sarothroides

Baccharis vanessae

Bahia neomexicana

Bahiopsis laciniata

Bahiopsis parishii

Bahiopsis reticulata

B. juncea (Benth.) Greene var. ***aspera*** Greene (p. 263) **ST**: 5–15 dm, much-branched, forming a rounded bush ≤ 3 m diam, glabrous or short-bristly. **LF**: 1–3(9) cm, linear, entire or with few, sharp, pinnate lobes, drought-deciduous. **INFL**: heads few; peduncles 1.5–6 cm; involucre 4–15 mm diam; phyllaries 1–7 mm, lanceolate to linear, acute. **FL**: corolla 6–10 mm. **FR**: 2–3.5 mm; pappus 6–10 mm. $2n=18$. Common. Dry, rocky slopes, desert plains, washes; < 1500 m. SW, D; to s NV, TX, nw Mex. A 2nd var. occurs in Baja CA. Gen Apr–Jul

BELLIS .

Per [ann]. **LF**: basal and proximal cauline, simple, petioled. **INFL**: heads radiate, 1 on slender, ± naked peduncles; involucre hemispheric; phyllaries in 2 equal series; receptacle conic, epaleate. **RAY FL**: many; ray white to pink or purple. **DISK FL**: many; corolla yellow, tube very short, throat ± cylindric; anther tips acute; style tips flat, triangular. **FR**: compressed; pappus 0. 15 spp.: Eur, Medit. (Latin: pretty) [Brouillet 2006 FNANM 20:22–23]

B. perennis L. (p. 263) ENGLISH DAISY Rhizome short, simple or branched; roots fibrous. **ST**: 1–several from base, decumbent or prostrate. **LF**: 2–10 cm; blade oblanceolate or obovate, tapered to winged petiole, tip obtuse, margin entire, crenate or serrate, faces loosely soft-hairy. **INFL**: peduncle 2–25 cm, strigose or hairs soft, spreading; phyllaries 3–6 mm, ovate, obtuse, hairs soft, spreading. **RAY FL**: ray 7–10 mm, narrow. **DISK FL**: corolla ± 2 mm. **FR**: 1–1.5 mm, light brown, glabrous. $2n=18$. Damp, grassy areas; < 900 m. NW, GV, CW, SW; widespread in N.Am; native to Eur. Dec–Sep

BENITOA

Staci Markos

1 sp. (San Benito Co., CA) [Markos & Strother 2006 FNANM 20:450]

B. occidentalis (H.M. Hall) D.D. Keck (p. 263) Ann; sticky, strongly scented, densely stalked-glandular, glandular-puberulent, proximally ± sparsely long-soft-hairy. **ST**: erect, often stout, 10–150 cm, green, aging red. **LF**: alternate, sessile, clasping or not, 0.5–12+ cm, reduced distally on st, linear to lance-linear or oblanceolate, often curved, acute, entire or toothed. **INFL**: heads radiate, in open panicle-like cluster; involucre 8–10 mm, narrowly fusiform or obconic to cylindric, erect, persistent; phyllaries graduated in 5–8 series, linear, long-tapered, abaxially glandular, tips straight to curved outward, each tipped with 1 large, stalked gland; receptacle flat or rounded, glabrous, epaleate. **RAY FL**: (2)5–8(14); ray 5–7 mm, yellow, occ red-tinged, occ rolled under; style gen with 1 prominent branch. **DISK FL**: 9–20+, staminate; corolla 4–5 mm, narrowly funnel-shaped, yellow, occ red-tinged; anther tip narrowly triangular; style tip 1.5 mm, cylindric, distally bristly, branches short, tapered. **FR**: 3–4 mm, strigose, mottled purple-brown; pappus deciduous, of 2–8 slender, white bristles. $2n=10$. Grassland, foothill woodland, vertic clay, occ serpentine; 350–1100 m. SCoRI (w Fresno, se Monterey, se San Benito cos.). [*Lessingia o.* (H.M. Hall) M.A. Lane] Jun–Nov ★

BIDENS

Ann, per [shrub]. **ST**: prostrate to erect. **LF**: simple or pinnate, gen opposite, sessile or petioled. **INFL**: heads radiate or discoid, occ disciform, gen few in CA; involucre cylindric to bell-shaped; phyllaries in 2 dissimilar series, outer gen ± lf-like in texture, inner thinner, with transparent or scarious margins; receptacle paleate; paleae narrow, flat. **RAY FL**: 0 or few; ray yellow or white. **DISK FL**: gen many; corolla yellow, radial (or outermost white, bilateral). **FR**: narrowly club-shaped, thick or compressed front-to-back; pappus 0 or awns 1–several, gen barbed. ± 230 spp.: worldwide. (Latin: 2 teeth) [Strother & Weedon 2006 FNANM 21:205–218]

1. Lf simple
  2. Lf petioled . . . . ***B. tripartita***
  2′ Lf sessile
    3. Fr with ± thickened margin, prominently ribbed; palea tips yellow; ray 0 or < 15 mm . . . . ***B. cernua***
    3′ Fr without thick margin or prominent ribs; palea tips ± red; ray 15–30 mm . . . . ***B. laevis***
1′ Lf pinnate
  4. Fr narrowly cylindric or slightly compressed, 4-angled . . . . ***B. pilosa***
  4′ Fr wedge-shaped, flat
    5. Outer phyllaries 5–8; inner phyllaries = disk . . . . ***B. frondosa***
    5′ Outer phyllaries 10–16; inner phyllaries < disk . . . . ***B. vulgata***

B. cernua L. NODDING BUR-MARIGOLD Ann. **ST**: 1–9 dm, erect, glabrous to short-rough-hairy. **LF**: simple, sessile; bases fused around st; blades 4–20 cm, lance-linear to lanceolate, acuminate, serrate, glabrous. **INFL**: heads radiate or discoid, erect in fl, nodding in fr; peduncle 1–7 cm; involucre 1–2 cm diam, hemispheric; outer phyllaries 5–8, 1–3 cm, lance-linear, spreading; inner phyllaries 8–10 mm, lance-ovate; paleae 6–8 mm, tips yellow. **RAY FL**: 0 or 6–8; ray 8–15 mm, yellow. **DISK FL**: corolla 3–4 mm, yellow. **FR**: 5–7 mm, narrowly wedge-shaped, 4-angled, compressed front-to-back; angles thick, barbed; pappus awns gen 4, 2–3 mm. $2n=24, 48$. Freshwater wetlands; < 2000 m. KR, NCoR, CaR, n SN, n CCo, GB; widespread n hemisphere. Much like *B. laevis*. Jul–Oct

B frondosa L. (p. 263) STICKTIGHT Ann, ± glabrous. **ST**: 5–12 dm, square. **LF**: pinnate, petioled; lflets 2–8 cm, lanceolate, gen acuminate, serrate. **INFL**: heads radiate or discoid, erect; peduncle 2–10 cm; involucre ± 1 cm diam, hemispheric; outer phyllaries 5–8, 1–5 cm, ± linear, ciliate; inner phyllaries 5–7 mm, ovate; paleae 5–7 mm, ± brown. **RAY FL**: 0–few; corolla 2–3.5 mm, yellow. **DISK FL**: corolla ± 2 mm, orange. **FR**: 6–10 mm, ± black, narrowly wedge-shaped, compressed front-to-back, ± glabrous to stiffly hairy; pappus awns gen 2, 3–4.5 mm. $2n=24,48,72$. Uncommon. Damp soil, esp disturbed sites; gen ≤ 2100 m. NW, CaR, SN, GV, n CCo, SW, GB, DMoj (Mojave River); to e N.Am. Jun–Oct

B. laevis (L.) Britton et al. (p. 263) BUR-MARIGOLD Ann, per, gen ± glabrous. **ST**: 2–25 dm, ± decumbent to erect, ± cylindric. **LF**: simple, sessile; blades 5–15 cm, ± lanceolate, acute to acuminate, serrate, bases sometimes fused around st. **INFL**: heads radiate, erect in fl, often nodding in fr; peduncle 2–10 cm; involucre 1–2 cm diam, bell-shaped; outer phyllaries 6–8, 1–2 cm, lance-linear, sparsely ciliate; inner phyllaries 8–16 mm, obovate; paleae 6–8 mm, tips ± red. **RAY FL**: 7–8; ray 1.5–3 cm, yellow. **DISK FL**: corolla 4–6 mm, yellow. **FR**: narrowly wedge-shaped, compressed front-to-back or 3–4-angled; angles thin, barbed; pappus awns 2–4, 3–5 mm. 2*n*=22, 24. Freshwater wetlands; < 2000 m. GV, CW, SW, SNE, DMoj (Mojave River); to e N.Am. Much like *B. cernua*. Aug–Nov

B. pilosa L. (p. 263) COMMON BEGGAR-TICKS Ann, glabrous or ± soft-hairy. **ST**: erect, 3–18 dm, square. **LF**: pinnate, petioled; axis sometimes winged; lflets gen 3–5, 2–6 cm, lanceolate to ovate, acute, serrate, base often asymmetric. **INFL**: heads discoid or radiate, few to many, erect; peduncle 1–9 cm; involucre 7–8 mm diam in fl, hemispheric; outer phyllaries 7–9, 4–5 mm, linear; inner phyllaries 4–7 mm, lanceolate; paleae 6–7 mm, linear, acuminate, ± brown. **RAY FL**: 0 or vestigial (2–3 mm); corolla white. **DISK FL**: alike, corolla ± 2 mm, radial, yellow (or not alike, outer corollas 2–3 mm, white, ± bilateral). **FR**: 4–16 mm, narrowly cylindric or slightly compressed, 4-angled, black, short-rough-hairy; pappus awns 2–4, 2–4 mm, ± yellow. 2*n*=72. Disturbed sites; gen ≤ 750 m. CW, SW; subtrop, trop worldwide. All yr

B. tripartita L. Ann, glabrous. **ST**: erect, 4–15 dm. **LF**: simple, petioled; blades ≤ 20 cm, lanceolate, acuminate, serrate. **INFL**: heads gen discoid, erect; peduncle 1–6 cm; involucre 7–20 mm diam (exc lfy tips of outer phyllaries), hemispheric; outer phyllaries 4–5, gen 1–3 cm, ± linear, spreading; inner phyllaries 7–8 mm, ovate; paleae 8–10 mm, black or ± green with fine black lines. **RAY FL**: gen 0. **DISK FL**: corolla ± 3 mm. **FR**: wedge-shaped, ± compressed; outer fr 3.5–4 mm, 2–3-angled; inner fr 5–8 mm, 4-angled; barbs of angles ascending below, reflexed above; pappus awns gen 4, 1–3.6 mm. 2*n*=48. Freshwater wetlands; 50–1600 m. c SNH, GV; native to e N.Am, Eur. Jul–Aug

B. vulgata Greene Ann, ± puberulent. **ST**: erect, 3–15 dm, 4-angled. **LF**: pinnate, petioled; lflets 2–8 cm, lanceolate, acuminate, serrate. **INFL**: heads radiate; peduncle 2–23 mm; involucre bell-shaped; outer phyllaries 10–16, 1–2 cm, ± linear, ciliate; inner phyllaries 7–9 mm, lance-ovate; paleae 6–9 mm, tips ± brown. **RAY FL**: 0–few; ray 2.5–3.5 mm, yellow. **DISK FL**: corolla ± 3 mm, yellow. **FR**: 6–12 mm, wedge-shaped, ± yellow or olive-brown, compressed; margin with stiff, ascending hairs or reflexed barbs; faces glabrous or short-rough-hairy; pappus awns 2, 3–4 mm. 2*n*=24,48. Uncommon. Freshwater wetlands; < 300 m. NCoR, ScV; to WA; native to e N.Am. Sep–Oct

BLENNOSPERMA
Bruce G. Baldwin

Ann 3–12(30) cm, taprooted. **ST**: gen 1, erect, gen branched ± throughout. **LF**: basal and cauline, alternate, mostly sessile; blade linear or pinnately divided into 2–15 linear lobes, ultimate margins entire; faces glabrous or sparsely woolly-tomentose. **INFL**: heads radiate, 1; involucre ± hemispheric, 3–6 mm diam; phyllaries 5–13+ in ± 2 series, basally fused, ± erect, curved inward in late fl, reflexed in fr, elliptic to ovate, ± equal, ± membranous, veiny, tips gen purple; receptacle flat to convex, smooth or pitted, epaleate. **RAY FL**: 5–13+; ray sessile, linear to ovate, gen yellow (white), often ± purple abaxially; some pistillate fls sometimes lack corollas. **DISK FL**: 20–60(100+), staminate; corolla ± 2–3 mm, gen not exceeding involucre, yellow, tube < to > bell-shaped throat, lobes ± erect, deltate to lanceolate; anther tip concave; style undivided, tip head-like, white. **FR**: ± ellipsoid, gen 5–6(10)-ribbed or -angled, gen papillate and, if so, sticky when wet; pappus 0. 3 spp.: CA, Chile. (Greek: mucus-seed, the fr becoming slimy when wetted) [Strother 2006 FNANM 20:640–641]

1. Stigma of ray fl red; proximal lvs unlobed or 2–3(5)-lobed . . . . ***B. bakeri***
1′ Stigma of ray fl yellow; proximal lvs gen 5–15-lobed . . . . ***B. nanum***
  2. Fr gen 2.5–3 mm; disk fls gen 12–60(70+); pollen gen white; widespread, gen more inland in n CA . . . . . . var. ***nanum***
  2′ Fr gen 3–4.5 mm; disk fls gen 60–100+; pollen gen yellow; c NCo (Mendocino Co.), n CCo . . . . var. ***robustum***

B. bakeri Heiser (p. 269) SONOMA SUNSHINE **LF**: proximal 5–15+ cm, lobes 0 or 2–3(5). **INFL**: involucre 7–9+ mm. **RAY FL**: ray 5–7 mm, yellow; stigma red. **DISK FL**: 30–70. **FR**: 3–4 mm, papillate. 2*n*=18. Grassy margins of swales, vernal pools; 20–40 m. NCoRO, ne SnFrB (s Sonoma Co.). Feb–Apr ★

B. nanum (Hook.) S.F. Blake **LF**: proximal mostly 3–6+ cm, gen 5–15-lobed. **INFL**: involucre 5–6+ mm. **RAY FL**: ray (0)3–7 mm, gen yellow (white), often purple abaxially; stigma yellow. **DISK FL**: 20–100+. **FR**: 2.5–4.5 mm, sometimes papillate. 2*n*=14.

var. ***nanum*** (p. 269) **DISK FL**: mostly 20–60(70+); pollen gen white. **FR**: gen 2.5–3 mm. Open, grassy areas, often margins of seeps or vernal pools; < 1500 m. NCoR, CaRF, SNF, GV, CW (exc SCoRI), SCo, n ChI, PR. Jan–May

var. ***robustum*** J.T. Howell POINT REYES BLENNOSPERMA **DISK FL**: mostly 60–100+; pollen gen yellow. **FR**: gen 3–4.5 mm. Sandy bluffs, grassy places among shrubs; 10–120 m. c NCo (Fort Bragg, Mendocino Co.), n CCo (Point Reyes Peninsula, Marin Co.). Some populations intermediate to *B. nanum* var. *n.* in fr length, pollen color. Jan–Apr ★

BLEPHARIPAPPUS EYELASH TARWEED
Bruce G. Baldwin

1 sp. (Greek: eyelash pappus) [Baldwin & Strother 2006 FNANM 21:259]

B. scaber Hook. (p. 269) Ann 5–20(40+) cm, slender-branched, scabrous to variously hairy, distally glandular. **LF**: simple, mostly cauline, proximal opposite, most alternate, sessile, 6–25+ mm, 0.5–1.5 mm wide, narrowly spoon-shaped to linear, entire, gen ascending, margins gen rolled under. **INFL**: heads 1 or in open, ± flat-topped cluster, radiate; involucre 3–6+ mm diam, obconic to hemispheric; phyllaries 2–8, each partly surrounding ray fr, falling with fr, 3–8 mm, lanceolate or oblanceolate, ± hairy and/or glandular; receptacle convex, glabrous, paleate throughout; paleae deciduous, outer phyllary-like, tips of inner green, hairy. **RAY FL**: 2–8; corolla white, ray 2–11 mm, widely fan-shaped, abaxially purple-veined. **DISK FL**: 6–25(60+); corolla 2–3.5 mm, white, tube < or ± = throat, tube/throat glabrous, lobes deltate, minutely bristly; anthers ± dark purple, tips hemispheric or conic; style barely branched, bristly-puberulent proximal to branches. **FR**: 2–3.5 mm, ± obconic, ± round in x-section, black, gen ± white-hairy; pappus of gen (0)12–18(26), awl-like, fringed to plumose scales. 2*n*=16. Openings in sagebrush scrub, yellow-pine forest, juniper woodland; 300–2300 m. KR, CaRH, n SNH, MP; to WA, ID, NV. Self-sterile. Apr–Sep

BLEPHARIZONIA BIG TARWEED

Bruce G. Baldwin

Ann 1–18+ dm, strongly aromatic. **ST**: erect, branches arched-ascending or ± wand-like, long- and short-bristly, ± glandular, at least distally. **LF**: basal and cauline, proximal opposite, often withered before fl, 4–15+ cm, most alternate, ± sessile, narrowly spatula-shaped to linear or distal lanceolate, entire or serrate, soft-hairy to bristly, distal glandular-hairy and/or with marginal tack-shaped glands. **INFL**: heads radiate, 1 or in ± raceme- to panicle- or spike-like clusters; peduncle bracts with 1+ tack-shaped glands on tip or margins; involucre bell-shaped or obconic to spheric, 3–9+ mm diam; phyllaries 5–13, elliptic to lanceolate or oblanceolate, each ± half enclosing a ray ovary, falling with fr, 4.5–7.5 mm, hairy or ± glabrous, gen with tack-shaped glands; receptacle flat, minutely bristly; paleae in 1 series between ray and disk fls, free, deciduous. **RAY FL**: 5–13; corolla ± white, ray 5–10+ mm, gen red- or purple-veined abaxially, lobes ± parallel, 1/4–1/2 length of ray. **DISK FL**: 5–35(60+); corolla 4–5.5 mm, ± white to ± purple, tube < throat, lobes lanceolate; anthers ± dark purple, tips ovate to ovate-deltate; style glabrous proximal to branches, style tips awl-shaped, densely short-bristly. **FR**: 3.5–4 mm, club-shaped, ± round in ×-section or compressed front-to-back, 10-ribbed, densely ascending-hairy, ray pappus 0 or crown-like; disk pappus similar or of 12–20+, awl-shaped, ± red-brown, ciliate to plumose scales. 2 spp. (For resemblance to *Blepharipappus* and position in *Hemizonia*, where originally named as a subgenus) [Baldwin & Strother 2006 FNANM 21:289–291] Self-sterile.

1. Disk pappus 0 or 0.1–1 mm; lvs ± yellow-green, distal gen densely stalked-glandular and with scattered to abundant tack-shaped glands; heads 1 or in ± open, raceme- to panicle-like clusters; branches ± wand-like . . . . . . ***B. laxa***
1′ Disk pappus 1.5–3 mm; lvs gray-green, sparsely or not stalked-glandular exc for scattered tack-shaped glands; heads gen in ± spike- to panicle-like clusters; branches often arched-ascending . . . . ***B. plumosa***

B. laxa Greene (p. 269) **INFL**: involucre ± coarse- to stiff-hairy or becoming glabrous, gen also with scattered to abundant tack-shaped glands. 2*n*=28. Openings in woodland, chaparral, grassland; < 1500 m. w SnJV, e SnFrB, SCoR. [*B. plumosa* subsp. *viscida* D.D. Keck] Jul–Nov

B. plumosa (Kellogg) Greene (p. 269) BIG TARPLANT **INFL**: involucre gen ± canescent, gen also with scattered tack-shaped glands. 2*n*=28. Dry slopes in grassland; < 500 m. nw SnJV, e SnFrB. Jul–Nov ★

BRICKELLIA BRICKELLBUSH

Randall W. Scott

[Ann] per to shrub. **LF**: simple, alternate or opposite, gen resinous-dotted, veiny, main veins gen 3. **INFL**: heads discoid, gen clustered; involucre cylindric to bell-shaped; phyllaries gen graduated, ± green, veiny-striate, spreading in age; receptacle gen flat, epaleate. **FL**: corolla cylindric, ± white to pale yellow-green, occ tinged red or purple; anther tip ovate; style branches long, club-shaped, tips rounded. **FR**: 10-ribbed, gen cylindric, gen hairy; pappus of 10–40+ gen minutely barbed bristles, gen white. 110 spp.: w US, Mex, C.Am. (John Brickell, early botanist in Georgia) [Scott 2006 FNANM 21:491–507]

1. Head 3–7-fld
  2. Lf blade lanceolate to lance-ovate, dentate to serrate; faces gland-dotted and sparsely puberulent or hairy, not appearing glossy or varnished . . . . ***B. knappiana***
  2′ Lf blade linear to lance-elliptic or -ovate, entire or nearly so; faces gland-dotted, appearing varnished or glossy, often also puberulent . . . . ***B. longifolia***
   3. Lf blade linear or lance-linear to lance-elliptic, flat, not folded or sickle-shaped, 2–9 mm wide, length 8–25 × width, base tapered . . . . var. ***longifolia***
   3′ Lf blade lance-ovate to narrowly lanceolate, folded along midvein and/or sickle-shaped, 2–15 mm wide, length 4–8 × width, base rounded . . . . var. ***multiflora***
1′ Head 8–90-fld
 4. Per or subshrub from woody caudex
  5. Petiole 10–70 mm . . . . ***B. grandiflora***
  5′ Petiole 0–2 mm
   6. Lf blade ovate, serrate . . . . ***B. greenei***
   6′ Lf blade elliptic, lance-linear or oblong, entire . . . . ***B. oblongifolia*** var. ***linifolia***
 4′ Shrub
  7. Head 40–90-fld
   8. St, lvs densely white-tomentose . . . . ***B. incana***
   8′ St minutely glandular; lvs glabrous or puberulent, often strigose, or with mix of glandular and glandless hairs . . . . ***B. atractyloides***
    9. Outer phyllaries widely ovate . . . . var. ***atractyloides***
    9′ Outer phyllaries lance-linear to -ovate
     10. Margin of outer phyllaries ± entire . . . . var. ***arguta***
     10′ Margin of outer phyllaries dentate . . . . var. ***odontolepis***
  7′ Head 8–35-fld
   11. Lf blade oblong to spoon-shaped, entire . . . . ***B. frutescens***
   11′ Lf blade deltate, ovate, to nearly round or cordate, gen toothed (entire)
    12. Phyllaries in 4–7 series, erect
     13. Petiole 5–60+ mm; lf blade 10–100 mm . . . . ***B. californica***

Baileya multiradiata

1 cm · head · 1 cm · involucre · 2 mm · ray flower · 2 mm · fruit

Baileya pauciradiata

1 cm · 2 mm · ray flower · disk flower

Baileya pleniradiata

2 mm · ray flower · 2 mm · disk flower · 2 cm

Balsamorhiza lanata

2 mm · 1 cm · head · fruit · 5 cm

Balsamorhiza macrolepis

2 cm · 2 mm · palea · fruit

Balsamorhiza sagittata

5 cm · 2 cm

Balsamorhiza serrata

Bebbia juncea var. aspera

1 cm · 5 mm · 2 mm · fruit · palea

Bellis perennis

1 cm · 1 mm · disk flower

Benitoa occidentalis

gland · phyllary tip · 2 mm · head · 1 cm · 2 cm

Bidens frondosa

flower · head · 1 cm · 1 cm · fruit · head · 5 mm · palea · fruit

Bidens pilosa

1 cm · head · 5 mm · 5 mm · palea · fruit

Bidens laevis

2 mm · fruit · palea · 1 cm

13' Petiole 1–2.5 mm; lf blade 3–15 mm . . . . *B. desertorum*
12' Phyllaries in 5–9 series, tips recurved or spreading
  14. Pl densely tomentose and glandular; head 16–24-fld. . . . . *B. nevinii*
  14' Pl glandular-puberulent or glands mixed with long, crooked hairs, occ short-tomentose; head
     8–24-fld. . . . . *B. microphylla*
    15. Head 15–28(34)-fld; pl with glands mixed with long, crooked hairs, occ short-tomentose . . . . . var. *microphylla*
    15' Head gen 8–15-fld; pl glandular-puberulent. . . . . var. *scabra*

B. atractyloides A. Gray Shrub 20–50 cm. **ST:** densely branched, minutely glandular. **LF:** alternate or opposite, petiole 0–3 mm, blade deltate, lanceolate, or ovate, gen sharply dentate or dentate-serrate (entire), base acute to truncate or cordate, tip acute to acuminate, faces glabrous or puberulent, often strigose, or with mix of erect glandular and glandless hairs. **INFL:** heads 1 on short lfy branches; peduncle 10–70 mm, with small, sharp bristles and gland-tipped hairs; involucre 10–15 mm cylindric to widely bell-shaped; phyllaries 24–33, ± equal or graduated in 3–4 series, striate, acute to acuminate, margin narrowly scarious, outer often bright green, 4–16-striate, lance-linear to widely ovate, inner pale green, 3–4-striate, linear or narrowly lanceolate, minutely scabrous, often glandular. **FL:** 40–90; corolla 6–8 mm, pale yellow-green or cream, often purple-tinged. **FR:** 3–5.5 mm, minutely scabrous; pappus bristles 18–25, smooth or short-barbed. Vars. intergrade.

var. ***arguta*** (B.L. Rob.) Jeps. (p. 269) PUNGENT BRICKELLBUSH **LF:** blade 10–20 mm, 5–15 mm wide, ovate, dentate to serrate, strigose, or with mix of erect glandular and glandless hairs. **INFL:** outer phyllaries lance-ovate to -linear, ± entire, 6–10-striate, inner linear, ± papery. **FL:** 40–55; corolla cream, often purple-tinged. 2*n*=18. Mtn slopes, washes; 100–1500 m. D; to NV, AZ, Baja CA. [*B. arguta* B.L. Rob. var. *arguta*] Apr–Jun

var. ***atractyloides*** SPEARLEAF BRICKELLBUSH **LF:** blade 10–50 mm, 5–25 mm wide, deltate, lanceolate, or ovate, sharply dentate or entire, glabrous or minutely glandular. **INFL:** outer phyllaries widely ovate, entire, 7–16-striate, inner narrowly lanceolate. **FL:** 40–90; corolla pale yellow-green. 2*n*=18. Rock crevices, cliff faces, talus slopes; 600–1600 m. DMoj; to CO, AZ. Mar–Sep

var. ***odontolepis*** (B.L. Rob.) Jeps. **LF:** blade 10–20 mm, 5–15 mm wide, ovate, dentate to serrate, strigose, often glandular. **INFL:** outer phyllaries lance-ovate to -linear, dentate, 5–12-striate, inner linear, ± papery. **FL:** 40–55; corolla cream, often purple-tinged. Mtn slopes, washes; 100–1500 m. s DMoj, DSon; Baja CA. [*B. arguta* var. *o.* B.L. Rob.] Apr–Jun

B. californica (Torr. & A. Gray) A. Gray (p. 269) CALIFORNIA BRICKELLBUSH Shrub 50–200 cm. **ST:** branched from near base, short-hairy, glandular. **LF:** alternate or proximal opposite, petiole 5–60+ mm, blade 10–100 mm, 10–90 mm wide, ovate to deltate, crenate to serrate, base cordate to truncate, tip acute to rounded, faces puberulent to becoming glabrous. **INFL:** heads in lfy panicle-like cluster; peduncle 1–5 mm, puberulent, glandular; involucre 7–12 mm, cylindric to obconic; phyllaries 21–35, in 5–6 series, often purple-tinged, 3–6-striate, margin scarious, outer ovate to lance-ovate, glabrous to sparsely short-hairy, glandular, tip obtuse to acuminate, inner linear-oblong to lanceolate, glabrous, tip obtuse to acuminate. **FL:** 8–18; corolla 5.5–8 mm, pale yellow-green. **FR:** 2.5–3.5 mm, puberulent; pappus bristles 24–30. 2*n*=18. Dry, rocky hillsides, canyons, sea bluffs; < 2700 m. NW, CaRF, SN, SnJV, CW, SW, D; to ID, WY, CO, TX. Jul–Dec

B. desertorum Coville Shrub 80–200 cm. **ST:** intricately branched, densely puberulent, often gland-dotted. **LF:** opposite or alternate, petiole 1–2.5 mm, blade 3–15 mm, 4–14 mm wide, ovate, crenate to serrate, base obtuse to truncate, tip obtuse, faces slightly tomentose to densely puberulent. **INFL:** heads in panicle-like clusters; peduncle 0–3 mm, gland-dotted; involucre 8–10 mm cylindric; phyllaries 20–24, in 4–7 series, ± green or ± brown, 3–4-striate, tip acute to short-pointed, margin scarious; outer widely ovate, granular-puberulent, inner lance-linear to lanceolate, glabrous or slightly tomentose, often gland-dotted. **FL:** 8–12; corolla 4.5–5 mm, white. **FR:** 2–3 mm, scabrous; pappus bristles 12–15, smooth or minutely barbed. Dry hillsides, canyons; 200–1400 m. D; to NV, AZ; also West Indies. Sep–May

B. frutescens A. Gray (p. 269) Shrub 30–60 cm. **ST:** intricately branched, short-rough-hairy, sparsely gland-dotted. **LF:** alternate, petiole 1–2 mm, blade 3–12 mm, 1–4 mm wide, oblong to spoon-shaped, entire, base acute to acuminate, tip obtuse to acute, faces tomentose, veins obscure. **INFL:** heads 1–3; peduncle 10–30 mm, long-soft-wavy-hairy, gland-dotted; involucre 9–11 mm, cylindric; phyllaries 20–25, in 5–6 series, often purple-tinged, 4–5-striate, outer lanceolate to ovate, ciliate, inner linear-oblong, tip rounded, mucronate, margin narrowly scarious, faces sparsely long-wavy-hairy, gland-dotted. **FL:** 20–30; corolla 6.5–7.5 mm, ± purple. **FR:** 3.5–4.8 mm, minutely bristly; pappus bristles 26–30. Granitic desert slopes, sand; 600–1200 m. DSon; NV, Baja CA. Mar–Jun, Oct–Nov

B. grandiflora (Hook.) Nutt. (p. 269) Per from thick taprooted caudex, 30–95 cm. **ST:** branched, puberulent. **LF:** alternate or opposite, petiole 10–70 mm, blade 15–120 mm, 20–70 mm wide, lance- to deltate-ovate or ± cordate, crenate, dentate, or serrate, base acute, truncate, or ± cordate, tip long-tapered, faces puberulent. **INFL:** heads in loose, cyme- or panicle-like clusters, nodding; peduncle 4–30 mm, puberulent to hairy; involucre 7–12 mm, cylindric or obconic; phyllaries 30–40, in 5–7 series, 4–5-striate, margin scarious; outer lanceolate to lance-ovate, puberulent, tip long-acuminate, ciliate, inner lance-linear to lanceolate, glabrous, tip acute to acuminate. **FL:** gen 20–40(70); corolla 6.5–7.5 mm, pale yellow-green. **FR:** 4–5 mm, minutely rough-hairy or bristly; pappus bristles 20–30. 2*n*=18. Rocky hillsides, shaded forest, dry slopes, canyons; 1200–3000 m. KR, NCoRH, CaRH, c SNF, SNH; to WA, MT, WY, NE, MO, TX, Mex. Jul–Oct

B. greenei A. Gray (p. 269) Per from woody caudex, 20–50 cm. **ST:** branched, glandular- and long-soft-wavy-hairy. **LF:** alternate, petiole 0–2 mm, blade 15–30 mm, 5–25 mm wide, ovate, serrate, base rounded to truncate, tip acute, faces sticky-glandular. **INFL:** heads 1 or in open cyme-like clusters, each head subtended by lf-like bracts; peduncle 0–2 mm; involucre 14–18 mm, widely cylindric to bell-shaped; phyllaries (24)36–42, in (3)5–7 series, ± green or straw-colored, often purple-tinged, 3–5-striate, tip acute to long-acuminate, margin narrowly scarious; outer lanceolate to lance-ovate, sparsely gland-dotted, inner narrowly lanceolate, glabrous. **FL:** 44–60; corolla 8–10 mm, yellow-green, often purple-tinged. **FR:** 5.5–7 mm, minutely bristly or ± glabrous; pappus bristles 20–30. 2*n*=18. Open rocky slopes, canyon bottoms, riparian areas, serpentine soils; 800–2700 m. NW (exc NCoRO), CaRH, n SNH; OR. Often sweet-scented. Jul–Sep

B. incana A. Gray (p. 269) Shrub 40–130 cm. **ST:** branched from base, densely white-tomentose. **LF:** alternate, petiole 0–2 mm, blade 10–30 mm, 5–20 mm wide, ovate, entire or minutely serrate, base obtuse to truncate or cordate, tip acute, faces densely white-tomentose. **INFL:** heads 1 or in open, few-headed cyme-like clusters; peduncle 30–60 mm, canescent; involucre 15–20 mm, bell-shaped; phyllaries 38–44, graduated in 6–8 series, ± green to purple, 6–9 striate, tip obtuse to acuminate, margin narrowly scarious-ciliate, faces white-tomentose and gland-dotted; outer ovate, inner lanceolate. **FL:** 45–60; corolla 8.5–14 mm, pale yellow or cream, often red or purple-tinged. **FR:** 8.5–10 mm, finely silky-hairy, gland-dotted; pappus bristles 42–44. 2*n*=18. Sandy and gravelly washes, flats; 300–1600 m. D; NV, AZ. Apr–Dec

B. knappiana Drew (p. 269) Shrub 100–200 cm. **ST:** arching, branched from near base, gland-dotted. **LF:** alternate, petiole 4–5 mm, blade 10–35 mm, 1–16 mm wide, lanceolate to lance-ovate, dentate to serrate, base acute to attenuate, tip acute to acuminate, faces densely gland-dotted and sparsely puberulent or hairy, not appearing glossy or varnished. **INFL:** heads in panicle-like clusters,

peduncle 2–5 mm, densely short-hairy and glandular; involucre 6–9 mm, cylindric to narrowly bell-shaped; phyllaries 18–22, in 5–7 series, 3–4-striate, tip acute to acuminate, margin scarious; outer lance-ovate, sparsely puberulent, gland-dotted, inner lanceolate, glabrous. **FL**: 5–7; corolla 5–6.5 mm, white. **FR**: 2.5–3, puberulent to sparsely hairy; pappus bristles 28–32. Gravelly washes; 700–1700 m. n&e DMtns (Panamint, n Kingston, Funeral, sw Argus ranges), DSon. Possibly a hybrid between *B. californica* and *B. longifolia* var. *multiflora* or between *B. desertorum* and *B. longifolia*. Sep–Oct

B. longifolia S. Watson   Shrub 20–200 cm. **ST**: branched from above base, ± glabrous. **LF**: alternate, petiole 0–5 mm, blade entire or nearly so, base rounded or tapered, tip acuminate, faces often puberulent, appearing varnished or glossy. **INFL**: heads in lfy panicle-like clusters; peduncle 0–3 mm, glabrous or puberulent, sticky; involucre 5–7 mm, cylindric; phyllaries 10–24, in 6–8 series, pale green to straw-colored, 3–5-striate, sticky, margin scarious; outer ovate, tip acute, inner lanceolate, tip obtuse. **FL**: 3–7; corolla 3.5–4.5 mm, cream. **FR**: 1.8–2.5 mm, scabrous; pappus bristles 30–40. Vars. often difficult to distinguish.

var. ***longifolia*** (p. 269)   **LF**: blade 10–130 mm, 2–9 mm wide, length 8–25 × width, linear or lance-linear to lance-elliptic, flat, not folded or sickle-shaped, base tapered. Uncommon. Washes, springs; 900–1700 m. s SNF, n&w DMoj; to UT, AZ. Sep–Nov

var. ***multiflora*** (Kellogg) Cronquist   **LF**: blade 10–130 mm, 2–15 mm wide, length 4–8 × width, narrowly lanceolate to lance-ovate, folded along midveins and/or sickle-shaped, base rounded. Uncommon. Washes, springs; 900–1700 m. W&I, DMoj; NV. [*B. m.* Kellogg] Sep–Nov

B. microphylla (Nutt.) A. Gray   Shrub 30–70 cm. **ST**: much-branched, puberulent or with long, crooked hairs mixed with glands or gland-tipped hairs, occ tomentose. **LF**: alternate, petiole 0–3 mm, blade 3–20 mm, 1–15 mm wide, ovate to ± round, entire, coarsely dentate, or serrate, base acute to obtuse or rounded, tip rounded to acute, faces glandular, and with long, crooked hairs or small, sharp, bristle-like hairs. **INFL**: heads in loose, panicle-like clusters, often grouped at branch ends; peduncle bracted, 2–10 mm, sticky glandular or glands mixed with long, crooked hairs; involucre 7–12 mm, cylindric to narrowly bell-shaped; phyllaries 30–48, in 6–9 series, often purple-tinged, 3–5-striate, recurved or spreading, tip acute to acuminate, margin scarious; outer obovate to ± round, puberulent and glandular, middle occ 3-toothed with center tooth elongated, inner linear-oblong, glabrous or sparsely gland-dotted. **FL**: 8–28; corolla 5.5–7, mm pale yellow, often purple-tinged. **FR**: 3.5–4.7 mm, glabrous or

minutely stiff-spreading-hairy; pappus bristles 18–24. 2*n*=18. Vars. intergrade.

var. ***microphylla***   **ST**: glands dense to intermixed with long, crooked hairs, occ tomentose. **INFL**: peduncle glands dense to intermixed with long, crooked hairs; involucre 8–12 mm. **FL**: 15–28(34). Dry rocky places, canyon walls, sand dunes, washes; 1200–2400 m. s SNF, SNH, SnGb, Wrn, SNE, DMoj; to WA, WY, UT, n&w AZ. [*B. watsonii* B.L. Rob.] Jul–Oct

var. ***scabra*** A. Gray   **ST**: sparsely to densely glandular. **INFL**: peduncle densely glandular; involucre 7–11 mm. **FL**: gen 8–15. 2*n*=18. Dry rocky places, canyons, granitics or limestone; 1200–2400 m. DMoj; to WY, CO, AZ, Mex. Jul–Oct

B. nevinii A. Gray (p. 269)   Shrub 30–80 cm. **ST**: much-branched near base, tomentose, gland-dotted. **LF**: alternate, petiole 0–1 mm, blade 7–11 mm, 3–8 mm wide, ovate to cordate, dentate-serrate, base rounded to ± cordate, tip obtuse, faces tomentose, main veins 3–5, obscure. **INFL**: heads at long-branch tips, in panicle-like clusters; peduncle 1–5 mm, tomentose; involucre 10–12 mm, cylindric to top-shaped; phyllaries 30–38, in 7–9 series, often red- or ± purple-tinged, 3–6 striate, recurved, tomentose, gland-dotted, tip acute to acuminate, margin scarious, ciliate; outer lance-ovate, inner lanceolate. **FL**: 16–24; corolla 6.2–7.5 mm, pale yellow or cream, often purple-tinged. **FR**: 3.5–5 mm, minutely bristly; pappus bristles 18–24, white or tawny. 2*n*=18. Rock crevices, rocky slopes, desert scrub; 300–1900 m. SCoRI, WTR, SnGb, DMoj; NV. Aug–Nov

B. oblongifolia Nutt. var. ***linifolia*** (D.C. Eaton) B.L. Rob. (p. 269)   Per or subshrub from woody caudex, 10–60 cm. **ST**: branched, dotted with stalked glands to short-hairy and glandular. **LF**: gen alternate (± opposite), sessile, 9–40 mm, 1–15 mm wide, elliptic, lance-linear, or oblong, entire, base acute to attenuate, tip acute or obtuse, faces puberulent to ± densely covered with long, crooked hairs, these occ mixed with gland-tipped hairs, veins obscure. **INFL**: heads 1 or in ± flat-topped clusters; peduncle 2–50 mm, short-hairy and glandular; involucre 10–20 mm, cylindric to bell-shaped; phyllaries 25–35 in 4–6 series, 4–5-striate, unequal, tip acute to acuminate, margin scarious, occ ciliate; outer lanceolate to ovate, gland-dotted to glandular-puberulent, inner linear to lance-linear, glabrous. **FL**: 25–50; corolla 5–10 mm, pale yellow-green or cream, often purple-tinged. **FR**: 3–7 mm, minute-bristly or -hairy (glandular); pappus of 18–25, minutely barbed to ± plumose bristles. 2*n*=18. Desert, grassland, dry rocky hillsides; 1200–2800 m. SnJt, SNE, D; to BC, MT, CO, NM. Other var. in w US, BC. Apr–Oct

CACALIOPSIS

1 sp. (Greek: like *Cacalia*, a relative of *Senecio*) [Strother 2006 FNANM 20:627]

C. nardosmia (A. Gray) A. Gray (p. 269)   Per 1.5–9 dm, from rhizome. **ST**: gen erect. **LF**: simple, basal and cauline, alternate, distal reduced; petiole 7–30 cm; blade 7–35 cm, widely reniform, palmately lobed and further divided, abaxially ± tomentose, adaxially ± glabrous. **INFL**: heads gen 3–11 in raceme-like to ± flat-topped cluster, discoid; involucre gen 12–16 mm, ± bell-shaped; phyllaries in 1–2 equal series, linear to lanceolate; receptacle flat, epaleate. **FL**: 20–50; corolla gen 5–7 mm, ± yellow, lobes lanceolate, spreading; anther base entire, tip narrowly lanceolate; style branches ± short-bristly, tips flattened, conic. **FR**: 6–8.5 mm, cylindric, 12–15-veined, glabrous, brown; pappus 10–18 mm, of many fine, minutely barbed bristles, white. 2*n*=60. Meadows, open forest, sometimes serpentine; 200–2150 m. KR, NCoR; to s BC. Apr–Jul

CALENDULA

Ann [per], gen glandular-hairy. **ST**: erect or ascending, branched. **LF**: simple, alternate. **INFL**: heads radiate, 1, peduncled; phyllaries ± equal in 2–3 series, linear or ± lanceolate, margins narrowly scarious; receptacle flat, epaleate. **RAY FL**: 5–50+; ray yellow or orange. **DISK FL**: 20–150+, staminate; corolla yellow to brown; anther base sagittate, short-tailed, tip ovate or triangular-ovate; style with ring of hairs just below tip, branches very short. **FR**: gen of 3–4 kinds per head, winged or not, beaked or not, straight to strongly incurved and almost forming a circle, smooth, wrinkled, or with many prickles or bristles on back; pappus 0. ± 15 spp.: Eur, n Afr, w Asia. (Latin: 1st day of the month, perhaps for fl all yr) [Strother 2006 FNANM 19:381–382]

1. Head, incl rays, gen 1–4 cm diam; ray fls 5–20, ray gen 5–8 mm; lf ± lanceolate . . . . ***C. arvensis***
1′ Head, incl rays, gen > 5 cm diam; ray fls gen 30–50+, ray gen 12–20 mm; lf oblong to obovate . . . . ***C. officinalis***

C. arvensis L. FIELD-MARIGOLD Finely glandular-hairy. **ST**: slender, ≤ 60 cm. **LF**: petioled; blade ≤ 7 cm, ± thin, becoming sessile distally. **INFL**: head nodding at maturity. **RAY FL**: ray yellow to orange. **DISK FL**: corolla gen 2.5–4 mm. **FR**: 3–12 mm. $2n$=18,36,44. Uncommon. Escape from cult in disturbed areas, sometimes established from seed mixes; < 200 m. c SNF, CW (exc SCoRI), SCo, expected more widely; native to c Eur, Medit. Mar–Apr

C. officinalis L. POT-MARIGOLD Finely hairy. **ST**: slender to ± coarse, sparingly branched. **LF**: sessile, ≤ 15 cm, ± thick; base gen clasping. **INFL**: heads erect at maturity, fls closing at night. **RAY FL**: ray pale yellow to orange. **DISK FL**: corolla gen 5–6 mm. **FR**: 10–20+ mm. $2n$=14,32. Uncommon. Escape from cult in disturbed areas, occ from seed mixes; < 500 m. CCo, SnFrB, SCoRO, SCo, SnBr; native origin unknown. Mar–May

CALYCADENIA
Robert L. Carr & Gerald D. Carr

Ann, erect to spreading, with coarse tack-shaped glands, and often simple glandular hairs, aromatic; gen self-sterile. **LF**: basal ± many, mostly deciduous by fl, occ persistent, margins often curled under, and cauline, gen alternate and reduced distally on st, ± sessile, ± linear, entire or sparsely toothed, ± scabrous, gen with longer stiff hairs, esp on proximal margins. **INFL**: heads radiate, axillary, often ± sessile, and terminal, 1 or clustered; peduncle bracts linear or narrowly lanceolate to widely oblanceolate, margins thickened, rounded, defining a ± prominent central abaxial channel, esp proximally, ± bristly-ciliate, esp proximally, some or all with 1+ tack-like glands; phyllaries ± equal to shorter than paleae, each partly enfolding a ray fr; receptacle ± flat, hairy; paleae in one series between ray and disk fls, ± fused, forming a receptacle cup, phyllaries, paleae gen with abaxial simple and/or tack-like glands and appressed and/or spreading hairs. **RAY FL**: 1–6; ray 2–12 mm, 3-lobed, white to rose or yellow, occ with red spot near base. **DISK FL**: 1–25; corolla 2.5–10 mm, colored ± like rays; anthers gen purple, tips ovate to triangular; style branches long, bristly. **FR**: ray fr 1.5–4 mm, ± 3-angled, adaxial face ± flat, abaxial face rounded to ± angled, pappus 0; disk fr 2–4 mm, angled, base tapered, ± appressed-hairy to glabrous, pappus 0 or gen of 6–16 scales, all ± lanceolate or some blunt and shorter. ± 10 spp.: esp CA (OR). (Greek: cup gland) [Carr & Carr 2006 FNANM 21:270–276]

1. Tack-like glands of peduncle bracts (0)1, terminal; central ray lobe gen << laterals, widest at base
  2. Corolla yellow; ray fr gen rough-wrinkled
    3. Pl 1–5 dm, ± slender; ray 2–2.5(3.5) mm; disk fl 1–3 . . . . *C. micrantha*
    3′ Pl 2–12 dm, occ robust; ray (4)5–12 mm; disk fl 3–25 . . . . *C. truncata*
  2′ Corolla gen white, occ aging ± red; ray fr gen smooth
    4. Branchlets gen many, thread-like, very flexible, ray fl (0)1(2); disk fl 1–2; ray fr ± glabrous . . . . *C. hooveri*
    4′ Branchlets, if present, coarser, ± rigid, ray fl 1–5; disk fl 4–15; ray fr ± densely appressed-hairy
      5. Tack-like glands of phyllaries and paleae gen many, variable in size; disk fl corolla 7–10 mm . . . . *C. spicata*
      5′ Tack-like glands gen 0 on phyllaries and paleae; disk fl corolla 5–6 mm . . . . *C. villosa*
1′ Tack-like glands of peduncle bracts gen 2–6+, terminal and/or scattered; central ray lobe ≤ laterals, gen widest near or beyond middle
  6. Peduncle bracts ± widely oblanceolate, central channel ± not narrowed at tip, tip ± widely rounded, tack-like glands marginal, ± uniform in size; phyllaries and paleae gen without tack-like glands . . . . *C. mollis*
  6′ Peduncle bracts ± linear or elliptic to lanceolate or club-like, central channel narrowed or 0 distally, tip acute, rounded, or truncate, tack-like glands variable in size and position; phyllaries and paleae often with tack-like glands
    7. Central lobe of ray widest at base, << and narrower than laterals . . . . *C. multiglandulosa*
    7′ Central lobe of ray widest near or beyond middle, ± similar to laterals
      8. Lvs all opposite; heads in congested, axillary and terminal clusters, appearing whorled . . . . *C. oppositifolia*
      8′ Lvs gen alternate; heads 1 or clustered in axils, in spike-like or diffusely branched clusters, not appearing whorled
        9. St gen simple or few-branched, ± rigid, main axis gen obvious; paleae (3)4–6+, receptacle cup ± bell-shaped, length ± = diam; ray fl (1)2–6, ray yellow, cream, or white, aging ± pink; disk fl (4)6–21 . . . . *C. fremontii*
        9′ St branched, main st axis gen 0, branches ± many, slender, flexible, gen divergent, often zigzag; paleae 3(4), receptacle cup ± fusiform to club-shaped, length ± 2 × diam; ray fl 1(2), ray white, aging ± pink, never yellow; disk fl 2–5 . . . . *C. pauciflora*

C. fremontii A. Gray **ST**: 1–10 dm, gen simple or few-branched, main axis gen obvious, ± rigid, ± finely strigose (some hairs longer). **LF**: proximal 2–8 cm. **INFL**: heads 1–3+ per node; peduncle bracts 2–10 mm, ± narrowly linear-elliptic to club-shaped, rigid, tack-like glands gen 2–5+; phyllaries 3–7 mm, abaxially finely scabrous, sparsely long-hairy, gen glandular, tack-like glands (0)1–5+; paleae (3)4–6+, 3.5–8 mm, receptacle cup ± bell-shaped, length ± = diam, with face features of phyllaries. **RAY FL**: (1)2–6; ray 4.5–7 mm, ray yellow, cream, or white, aging ± pink, lobes similar, ± obovate, middle lobe = or narrower than laterals, ± symmetric, widest beyond middle, laterals ± symmetric, sinuses ± = ray. **DISK FL**: (4)6–21, corolla 4–5 mm. **FR**: ray fr smooth, gen ± glabrous; disk pappus scales gen 8–14, alternately long and short. $2n$=12. Common. Open, dry meadows, hillsides, gravelly washes; 50–1400 m. NW, CaR, n SNF, ScV; OR. Highly variable but not easily divisible. Intergrades with *C. pauciflora*. Apr–Oct

C. hooveri G.D. Carr (p. 269) HOOVER'S CALYCADENIA Self-fertile. **ST**: 1–6 dm; branchlets many distally, thread-like, flexible, minutely scabrous, glandular. **LF**: proximal many, 1–8 cm. **INFL**: heads 1–4 per node; peduncle bracts few, 1–5 mm, ± narrowly club-shaped, tack-like gland gen 1, terminal; phyllaries 2.5–3.5 mm, abaxially minutely glandular, tack-like glands (0)1, terminal; paleae 3–5 mm, abaxially glandular, tack-like gland 0. **RAY FL**: (0)1(2); ray 2–3.5 mm, white, middle lobe << laterals, symmetric, broadest at base, laterals ± asymmetric, sinuses ± 2/3 ray. **DISK FL**: 1–2, corolla 2.5–3.5 mm. **FR**: ray fr smooth or occ rough-wrinkled, ± glabrous; disk pappus scales 6–13, lanceolate, acuminate. $2n$=14. Rocky, exposed places, grassland, oak savanna; 100–400 m. n&c SNF, SnJV. Jun–Sep ★

C. micrantha R.L. Carr & G.D. Carr SMALL-FLOWERED CALYCADENIA Self-fertile. **ST:** 1–5 dm, ± slender, gen < 2(3) mm diam at base, ± purple distally, gen much-branched from near mid-st, glabrous. **LF:** proximal 2–5 cm. **INFL:** heads 1(3) per node, peduncle bracts 2–4 mm, lance-oblong to linear-oblanceolate, gen thickened to ± cylindric toward tip, tack-like glands (0)1, terminal; phyllaries 3.5–5 mm, paleae 4–6 mm, both abaxially glabrous to sparsely bristly, tack-like gland 0(1) (terminal). **RAY FL:** 1–3(6); ray 2–2.5(3.5) mm, yellow, central lobe smaller than laterals, widest at base, symmetric, laterals asymmetric, sinuses ± 1/4 ray. **DISK FL:** 1–3, corolla 3–4 mm. **FR:** ray fr rough-wrinkled, glabrous; disk pappus 0. 2*n*=14. Dry, open rocky ridges, hillsides, talus; openings in scrub, woodland; 500–1500 m. NCoR, SCoRO. [*C. truncata* subsp. *microcephala* D.D. Keck, in part] Jun–Oct ★

C. mollis A. Gray (p. 269) **ST:** 3–9 dm, gen simple, often zigzag-curved, long-soft-hairy, ± glandular. **LF:** 2–8 cm, often longest at mid-st. **INFL:** heads 3–10+ in dense cyme-like clusters; peduncle bracts 3–9 mm, flat, ± flexible, ± widely oblanceolate, ± widely rounded at tip, glandular throughout, channel ± uniformly broad, tack-like glands gen 3–6+, ± uniform in size, confined to narrow, rolled or thickened margins; phyllaries 4–6 mm, paleae 4.5–7.5 mm, both abaxially soft-spreading-hairy, densely glandular, tack-like gland gen 0. **RAY FL:** 1–4; ray 5–6 mm, yellow, white, or rose, central lobe slightly narrower than laterals, nearly symmetric, widest beyond middle, sinuses gen 3/4+ ray. **DISK FL:** 2–10, corolla ± 5 mm. **FR:** ray fr gen rough-wrinkled, ± glabrous; disk pappus scales gen 6–8, gen lanceolate, acuminate, 1–3 often shorter, ± rounded. 2*n*=14. Common. Open, dry meadows, fields; 100–1500 m. c&s SNF, c SNH, SnJV. May–Sep

C. multiglandulosa DC. (p. 273) **ST:** 1–7 dm, ± simple or branched, often with diffuse, relatively short, many-bracted branchlets, gen hairy, glandular, esp distally. **LF:** gen alternate, occ opposite to mid-st or beyond, proximal often persistent, 3–8 cm. **INFL:** spike-like to highly congested, esp distally; peduncle bracts 4–20 mm, often ± red, linear to narrowly lanceolate, occ narrowed and ± cylindric toward tip, tack-like glands gen 2–6+; phyllaries 4–8.5 mm, paleae 4.5–10 mm, both abaxially often long-hairy, glandular, ± red, tack-like glands (0)1–15+. **RAY FL:** 2–6; ray 5–10 mm, white or cream to rose or yellow; central lobe << and narrower than lateral lobes, widest at base, symmetric, laterals strongly asymmetric, sinuses gen 1/3–1/2 ray. **DISK FL:** 4–20, corolla 6–10 mm. **FR:** ray fr smooth, glabrous to ± appressed-hairy; disk pappus scales ± 11, lanceolate, acuminate, 2–3 shorter, blunt. 2*n*=12. Common. Gen dry, open valleys, hillsides, rocky ridges; 50–1100 m. NCoR, SN, GV, SnFrB, SCoRI. Variable. May–Oct

C. oppositifolia (Greene) Greene (p. 273) BUTTE COUNTY CALYCADENIA **ST:** gen 1–3 dm, slender, simple or sparingly branched, strigose and sparsely ± spreading-hairy. **LF:** opposite, 1–5 cm, little reduced distally on st. **INFL:** heads in congested axillary and terminal clusters, appearing whorled; peduncle bracts 4–10 mm, lanceolate, flat, occ ± cylindric towards tip, tack-like glands (0)1–5+; phyllaries and paleae 4–7 mm, abaxially minutely scabrous, often sparsely long-hairy, tack-like gland 0(rarely 1, terminal). **RAY FL:** 2–4; ray 6–9 mm, white to ± red, central lobe ± equaling or narrower than laterals, elliptic, widest at middle, ± symmetric, ray lobed to base. **DISK FL:** 4–20, corolla ± 6 mm. **FR:** ray fr gen smooth, glabrous; disk pappus scales gen 8–10, ± lanceolate, acuminate, often 2–4 shorter, blunt. 2*n*=14. Locally common; open, dry meadows, hillsides, openings in foothill woodland; 50–900 m. CaRF, n SNF. Apr–Jul ★

C. pauciflora A. Gray **ST:** 1–4 dm, gen much-branched, main st axis gen 0, branchlets ± many, slender, flexible, ± divergent, often zigzag, ± fan-shaped, finely strigose (some hairs longer). **LF:** proximal 1–5 cm. **INFL:** heads 1–4 per node; peduncle bracts 2–3 mm, ± narrowly linear-elliptic to club-shaped, tack-like glands gen 1–3+; phyllaries 3.5–5.5 mm, abaxially finely scabrous, sparsely long-hairy, tack-like glands (0)1–5+; paleae 3(4), 4–6.5 mm, receptacle cup ± fusiform to club-shaped, length ± 2 × diam, with face features of phyllaries. **RAY FL:** 1(2); ray 5–7 mm, white, aging ± pink, lobes ± equal, ± elliptic, middle lobe widest near or beyond middle, sinuses nearly equaling ray. **DISK FL:** 2–5, corolla 5–6 mm. **FR:** ray fr smooth, glabrous; disk pappus scales gen 8–10, alternately long and short. 2*n*=10,12. Common. Open, dry, gen rocky meadows, hillsides, openings in chaparral, foothill woodland; 50–1100 m. NCoRI, ScV. Intergrades with *C. fremontii*. Apr–Sep

C. spicata (Greene) Greene (p. 273) **ST:** 2–6 dm, simple or few-branched beyond middle, densely hairy and glandular distally. **LF:** 2–5 cm, often longest at mid-st. **INFL:** heads 1–3+ per node; peduncle bracts 15–20+, ± concealing head prior to fl, 3–7(8) mm, ± narrowly lance-linear, distal 1–3+ mm cylindric, tip truncate to concave, tack-like gland 1, terminal; phyllaries 5–8 mm, paleae 5.5–9 mm, both abaxially often ± red, often long-hairy, tack-like glands gen many, variable in size, terminal large, others variable, often much smaller. **RAY FL:** 1–5; ray 6–11 mm, white, aging ± red, central lobe smaller than laterals, widest at base, symmetric, laterals asymmetric, sinuses 1/3–2/3 ray. **DISK FL:** 4–11, corolla 7–10 mm. **FR:** ray fr smooth, ± densely appressed-long-hairy; disk pappus scales 9–16, lanceolate, acuminate. 2*n*=8. Common. Dry, open meadows, hillsides, grassland, openings in foothill woodland; 50–1400 m. SNF, GV. May–Sep

C. truncata DC. (p. 273) ROSIN WEED **ST:** 2–12 dm, gen ± red, branched beyond middle, glabrous to distally ± scabrous. **LF:** proximal 2–10 cm. **INFL:** heads gen 1(3) per node; peduncle bracts 2–12 mm, ± narrowly linear-elliptic to linear-oblanceolate, gen distally thickened to ± cylindric, tack-like gland (0)1, terminal; phyllaries 5–9 mm, paleae 5.5–10 mm, both abaxially glabrous or sparsely stiff-appressed- or long-straight-hairy, esp distally, tack-like glands gen 0. **RAY FL:** 3–6; ray (4)5–12 mm, yellow, central lobe smaller than laterals, symmetric, widest at base, laterals asymmetric, sinuses < 1/3 ray. **DISK FL:** 3–25, corolla 4–6 mm. **FR:** ray fr rough-wrinkled (smooth), glabrous; disk pappus scales 0 or 7–12, short, often blunt, toothed. 2*n*=14. Common. Dry, open hillsides, rocky ridges, talus, grassland, openings in foothill woodland, scrub; 50–1600 m. NW, CaR, SNF, n&c SNH, GV, CW; OR. May–Oct

C. villosa DC. (p. 273) DWARF CALYCADENIA **ST:** 1–4.5 dm, ± red distally, simple or branches few and ascending or many and spreading, rigid, scabrous and ± densely long-hairy. **LF:** proximal gen many, often persistent, 2–5 cm. **INFL:** heads 1–3 per node; peduncle bracts 3–12 mm, crowded, narrow, linear to lance-elliptic, gen thickened distally, occ tip cylindric ≤ 1 mm, tack-like gland (0)1; phyllaries 4.5–6.5 mm, paleae 5–7.5 mm, both abaxially ± glabrous to short-rough-hairy or bristly, tack-like glands gen 0. **RAY FL:** 1–4; ray ± 5 mm, white to ± pink, central lobe much narrower than laterals, widest at base, symmetric, laterals asymmetric, sinuses 1/3–1/2 ray. **DISK FL:** 5–15, corolla 5–6 mm. **FR:** ray fr smooth, ± densely appressed long-hairy; disk pappus scales ± 10, lanceolate, acuminate. 2*n*=14. Dry, rocky hills, ridges, grassland, openings in foothill woodland; 250–850 m. c&s SCoRO. May–Sep ★

CALYCOSERIS TACK-STEM

L.D. Gottlieb

Ann; taprooted; sap milky. **ST:** 1–3, 5–30 cm, erect or ascending, branched from near base; proximally glabrous, distally conspicuously dotted with tack-shaped glands. **LF:** alternate, sessile; basal blades pinnately lobed, lobes narrow, linear, spreading, ultimate margins entire, smooth; distal cauline reduced to linear bracts. **INFL:** heads liguliflorous, 1 or in open, few-headed clusters, peduncles 1–3 cm; phyllaries in 2 series, outer unequal, < 1/2 × inner, reflexed, inner lance-linear, equal, scarious-margined, acute, reflexed in fr; receptacle flat, smooth, each fl subtended by 1 fine, smooth bristle. **FL:** many; ligules yellow or white, readily withering. **FR:** fusiform, body tapered to beak, ribs 5, separated by grooves, faces smooth or roughened; pappus of 50+ smooth, white, basally fused bristles, borne on finely toothed cup at beak tip, deciduous. 2 spp.: sw US, nw

Mex. (Greek: alluding to shallow cup on fr tip and to chicory) [Gottlieb 2006 FNANM 19:307–308] Perhaps best placed in *Malacothrix* (Lee et al. 2003 Syst Bot 28:616–626).

1. Ligules yellow; tack-shaped glands red- or purple-tinged; fr deeply grooved between ribs. . . . . *C. parryi*
1′ Ligules white; tack-shaped glands straw-colored; fr shallowly grooved between ribs. *C. wrightii*

C. parryi A. Gray (p. 273) YELLOW TACK-STEM **LF**: basal 5–12 cm. **INFL**: involucre 13–17 mm, inner phyllaries 10–13 mm. **FL**: ligules 1.5–2.5 cm. **FR**: light brown or gray; ribs ± smooth; pappus 6–8 mm. 2*n*=14. Sandy to gravelly soils, washes, slopes; 200–1800 m. Teh, e SCo, SnGb, e SnBr, PR, SNE, D; to UT, AZ, Baja CA. Apr–May

C. wrightii A. Gray WHITE TACK-STEM **LF**: basal 6–15 cm. **INFL**: involucre 10–15 mm, inner phyllaries 10–12 mm. **FL**: ligules 2–3 cm, abaxially often with fine ± red veins. **FR**: dark brown; roughened, ribs tubercled; pappus 5–6.5 mm. 2*n*=14. Desert plains, pavement, washes, gravelly slopes; 100–1600 m. W&I, D; to UT, TX; nw Mex. Mar–May

CARDUUS

Ann, bien [per]. **ST**: erect. **LF**: simple, basal and cauline, alternate, reduced upward, decurrent as spiny wings, spiny-dentate and pinnately lobed, glabrous to tomentose; basal tapered to winged petiole; cauline sessile. **INFL**: heads discoid, 1–20 at branch tips; involucre cylindric to spheric; phyllaries graduated in several series, spine-tipped; receptacle flat, epaleate, ± white bristly. **FL**: corolla white to pink or purple, tube long, slender, throat abruptly expanded, short, lobes linear; anther base short-sagittate, tip oblong; style tip cylindric above slightly swollen distal node, branches very short. **FR**: ovoid, slightly compressed, glabrous; base slightly angled; pappus of many flat, minutely barbed bristles. ± 90 spp.: Eurasia, e Afr. (Latin: ancient name for a kind of thistle) [Keil 2006 FNANM 19:91–94]

1. Phyllaries gen > 2 mm wide; heads gen solitary, conspicuously peduncled; involucre 2–7 cm diam . . . . *C. nutans*
1′ Phyllaries gen ≤ 2 mm wide; heads 1–several, often clustered at branch tips, short-peduncled or sessile;
  involucre 1–3 cm diam
  2. Involucre ± spheric or hemispheric . . . . *C. acanthoides* subsp. *acanthoides*
  2′ Involucre cylindric or narrowly ellipsoid
    3. Heads gen 2–5 per cluster; phyllaries not scarious-margined, bases ± persistently loosely tomentose,
      tips scabrous . . . . *C. pycnocephalus* subsp. *pycnocephalus*
    3′ Heads 5–20 per cluster; phyllaries scarious-margined, bases glabrous to sparsely tomentose,
      tips glabrous or barely scabrous, sometimes minutely scabrous-ciliate . . . . *C. tenuiflorus*

C. acanthoides L. subsp. *acanthoides* (p. 273) PLUMELESS THISTLE Bien. **ST**: 3–15 dm, ± glabrous to loosely woolly, strongly spiny-winged. **LF**: basal 10–30 cm, spiny-toothed to deeply 1–2-pinnately lobed, gen sparsely hairy. **INFL**: heads 1–5 in clusters; involucre 1–2.5 cm diam, ± spheric or hemispheric; phyllary base glabrous or thinly tomentose, margin not scarious, tip linear or narrowly lanceolate, appressed to spreading, minutely scabrous-ciliate. **FL**: corolla 13–20 mm, purple; tube 5–10 mm, throat ± 2 mm, lobes 6–8 mm. **FR**: 2.5–3 mm, golden to brown; pappus 11–13 mm. 2*n*=22. Roadsides, pastures, disturbed areas; < 1300 m. NCoRI, n SN, n CCo, n SCoRO, MP; N.Am; native to Eur. Jul–Nov ◆

C. nutans L. (p. 273) MUSK THISTLE Bien. **ST**: 4–15 dm, glabrous to woolly, narrowly spiny-winged. **LF**: basal 10–40 cm, 1–2-pinnately lobed; cauline glabrous or sparsely hairy. **INFL**: heads gen 1, often nodding; involucre 2–7 cm diam, ± spheric; phyllary base glabrous, margin not scarious, tip lanceolate to ovate, spreading, minutely scabrous-ciliate. **FL**: corolla 20–25 mm, purple; tube 12–14 mm, throat 4–5 mm, lobes 4–6 mm. **FR**: 4–5 mm, golden to brown; pappus 13–25 mm. 2*n*=16. Roadsides, pastures, disturbed areas; 100–2100 m. KR, CaR, n SNH, SCo, SnBr, MP, DMoj; N.Am; native to Eur. Jun–Jul ◆

C. pycnocephalus L. subsp. *pycnocephalus* (p. 273) ITALIAN THISTLE Ann. **ST**: 2–20 dm, glabrous or slightly woolly, narrowly spiny-winged. **LF**: basal 10–15 cm, 4–10-lobed; cauline ± tomentose. **INFL**: heads gen 2–5 per cluster, sessile or short-peduncled; involucre 1–2 cm diam, cylindric to ellipsoid; phyllary base ± persistently loosely tomentose, margin not scarious, tip ascending, lance-linear, scabrous. **FL**: corolla 10–14 mm, pink to purple; tube 5–8 mm, throat 2–3 mm, lobes 4–5 mm. **FR**: 4–6 mm, golden to brown; veins 20; pappus 10–15 mm. 2*n*=60–62. Roadsides, pastures, disturbed areas; < 1200 m. KR, NCoR, CaRF, SNF, ScV, CW, SCo, TR, PR; N.Am; native to Medit. Mar–Jul ◆

C. tenuiflorus Curtis Ann. **ST**: 2–20 dm, glabrous or slightly woolly, widely spiny-winged. **LF**: basal 10–15 cm, 12–20-lobed; cauline ± tomentose. **INFL**: heads 5–20 per cluster, sessile or short-peduncled; involucre 1–2 cm diam, cylindric to ellipsoid; phyllary bases glabrous or thinly tomentose, margins scarious, tips ascending, lance-linear, glabrous or barely scabrous, sometimes minutely scabrous-ciliate. **FL**: corolla 10–14 mm, pink to purple; tube 4–6 mm, throat 2–3 mm, lobes 4–5 mm. **FR**: 4–5 mm, brown; veins 10–13; pappus 10–15 mm. 2*n*=54. Roadsides, pastures, disturbed areas; < 1000 m. NCo, NCoR, SNF, CW, SW; OR, PA; native to c Eur. Apr–Jun ◆

CARLQUISTIA

Bruce G. Baldwin

1 sp. (Sherwin Carlquist, CA botanist, b. 1930) [Baldwin & Strother 2006 FNANM 21:302–303]

C. muirii (A. Gray) B.G. Baldwin (p. 273) MUIR'S TARPLANT Per 7–54 cm, gen matted, from woody rhizome. **ST**: erect. **LF**: simple, proximally opposite, distally alternate, 9–42 cm, 2–4 mm wide, linear to lance-linear, entire, coarse- to soft-hairy and glandular-hairy. **INFL**: heads discoid, 1 or in loose, ± flat-topped clusters, involucre 8–13 mm, 5–10+ mm diam; phyllaries 0; receptacle flat, short-bristly; paleae (5)7–16 in 1 involucre-like series, ± fused, lance-linear to lanceolate, herbaceous, coarsely hairy and glandular-hairy, deciduous. **RAY FL**: 0. **DISK FL**: 7–29; corolla 6.5–10 mm, yellow, tube ≤ throat, lobes deltate, throat/lobes minutely hairy abaxially; anthers yellow to ± brown, tips widely ovate-deltate; style glabrous proximal to branches, tips awl-shaped, densely bristly. **FR**: 4–7.5 mm, 0.5–1.2

269

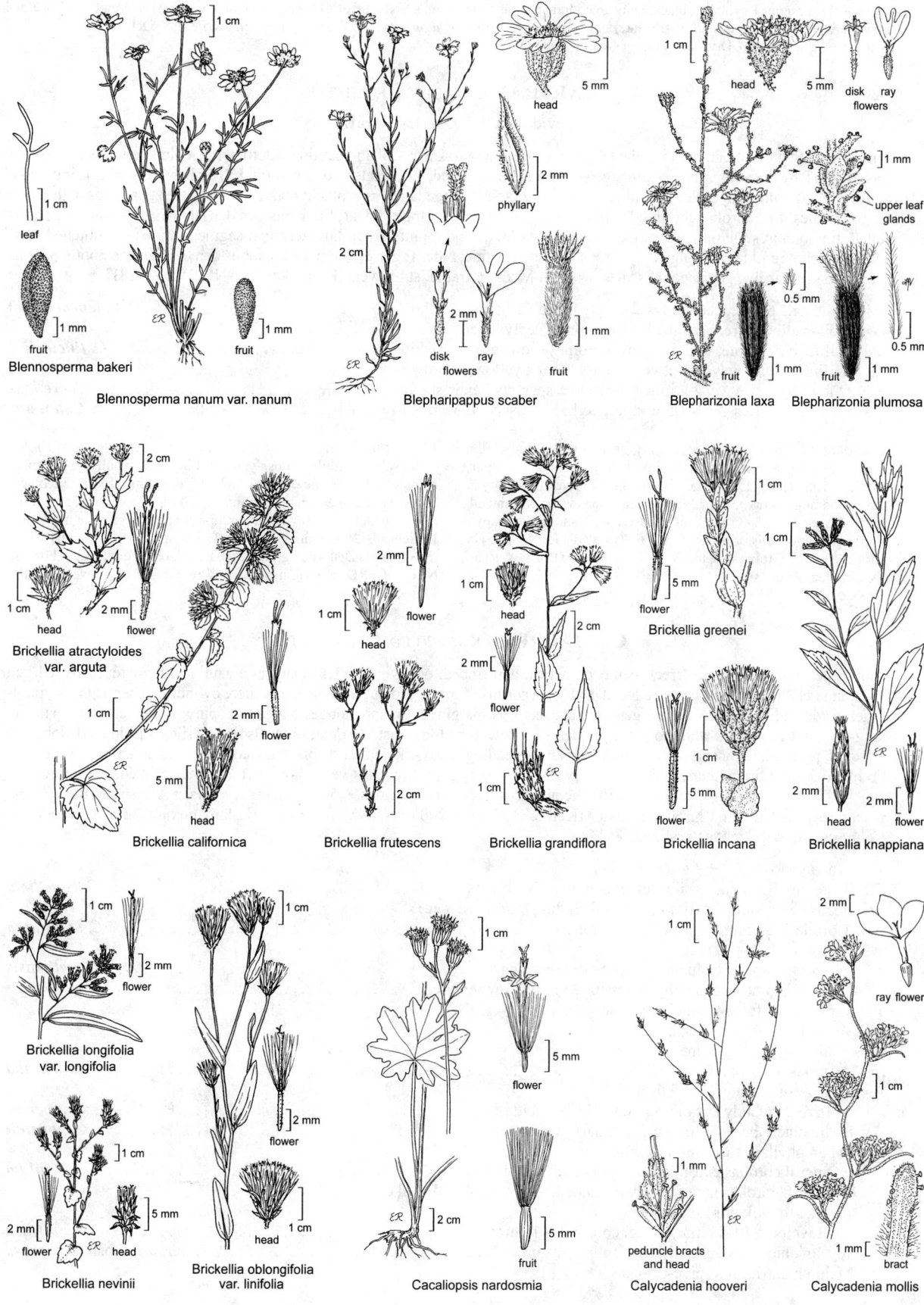

Blennosperma bakeri

Blennosperma nanum var. nanum

Blepharipappus scaber

Blepharizonia laxa   Blepharizonia plumosa

Brickellia atractyloides
var. arguta

Brickellia californica

Brickellia frutescens

Brickellia grandiflora

Brickellia greenei

Brickellia incana

Brickellia knappiana

Brickellia longifolia
var. longifolia

Brickellia nevinii

Brickellia oblongifolia
var. linifolia

Cacaliopsis nardosmia

Calycadenia hooveri

Calycadenia mollis

mm wide, ± straight, ± cylindric, black, hairs ascending; pappus 5–11 mm, of 9–17 awl-shaped or ± bristle-like, ciliate-plumose scales, white to ± brown. 2*n*=16. Dry, open sites on granitic soils; 1100–2500 m. s SNH, SCoRO (Ventana Double Cone, Monterey Co.). [*Raillardiopsis m.* (A. Gray) Rydb.] Self-sterile. Jun–Oct ★

CARTHAMUS DISTAFF THISTLE

David J. Keil & Charles E. Turner

Ann in CA. **ST**: gen erect, lfy, branched distally or throughout. **LF**: basal and cauline, alternate, gen pinnately lobed, ± spiny; basal often 0 by fl; cauline gen clasping, gen spreading to recurved, lanceolate to ovate, rigid. **INFL**: heads discoid, 1; involucre ± urn-shaped; outer phyllaries ± lf-like, inner with ± spiny appendages; receptacle convex to conic, epaleate but with many narrow scales. **FL**: corolla tube slender, throat abruptly expanded, lobes linear; filaments gen densely hairy, anther base short-tailed, tip oblong; style tip with minutely hairy distal node and minutely papillate terminal segment, tip barely notched. **FR**: oblong to obpyramidal, ± 4-angled, glabrous, attached at side; outer fr gen ± roughened, pappus gen 0; inner fr smooth, pappus 0 or of many, narrow, gen unequal scales. 14 spp.: Medit. (Arabic: safflower) [Keil 2006 FNANM 19:178–181]

1. Corolla pale purple; cauline lvs ± deeply lobed . . . . **[*C. leucocaulos*]**
1′ Corolla yellow to red; cauline lvs toothed to deeply lobed
  2. Cauline lvs dentate, weakly spiny; corolla yellow to red; fr white . . . . **[*C. tinctorius*]**
  2′ Cauline lvs ± deeply lobed, very spiny; corolla yellow; fr brown
    3. Outer phyllaries gen 2 × inner; herbage ± sparsely hairy; st white . . . . *C. creticus*
    3′ Outer phyllaries gen ≤ 1.5 × inner; herbage loosely cobwebby to ± woolly; st straw-colored . . . . *C. lanatus*

C. creticus L. (p. 273) SMOOTH DISTAFF THISTLE Pl 4–10 dm, ± sparsely hairy. **ST**: white. **LF**: cauline ± deeply lobed, very spiny. **INFL**: involucre 20–25 mm, gen becoming glabrous; outer phyllaries 35–55 mm, recurved, gen 2 × inner, tip appendages prominently spiny. **FL**: corolla 25–35 mm, yellow. **FR**: 4–6 mm, brown; pappus 8–10 mm. 2*n*=64. Disturbed ground; < 650 m. s NCoR, c SNF, c GV, SnFrB, SW, w DMoj; native to Medit. [*C. baeticus* (Boiss. & Reut.) Nyman] Jun–Aug ◆

C. lanatus L. (p. 273) WOOLLY DISTAFF THISTLE Pl 4–10 dm, ± densely glandular, loosely cobwebby to ± woolly. **ST**: straw-colored. **LF**: cauline ± deeply lobed, very spiny. **INFL**: involucre 25–35 mm, gen ± tomentose; outer phyllaries 35–50 mm, ascending or ± spreading, gen ≤ 1.5 × inner, tip appendages prominently spiny. **FL**: corolla 25–35 mm, yellow. **FR**: 4–6 mm, brown; pappus 10–13 mm. 2*n*=44. Disturbed ground, grassland, oak woodland; < 1100 m. NCo, NCoRO, c SNF, n SnJV, CW (exc SCoRI); native to Medit. May–Sep ◆

CENTAUREA KNAPWEED, STAR-THISTLE

Ann to per. **ST**: prostrate to erect, gen ± branched, gen ribbed, occ winged. **LF**: basal and cauline, alternate; proximal gen 1–2-pinnately lobed; distal gen ± reduced. **INFL**: heads disciform or radiant (discoid); involucre cylindric to hemispheric; phyllaries graduated in 6–many series, gen ± ovate, scarious-margined, tip appendages fringed to spiny; receptacle flat, epaleate, long-bristly. **FL**: corolla white to pink, purple, or yellow, tube long, distally bent; outer fls gen sterile, corolla 3–10-lobed, ± bilateral, reduced, inconspicuous or expanded and spreading, ± ray-like; inner fls bisexual, corolla ± radial; anther base tailed, tip oblong; style tip cylindric, minutely hairy distal to hairy ring, branches very short. **FR**: ± barrel-shaped, ± compressed, attached ± at side; pappus 0 or gen of stiff, unequal bristles or narrow scales. ± 500 spp.: esp Eurasia, n Afr; some cult. (Greek: pl name associated with Chiron, a centaur) [Keil & Ochsmann 2006 FNANM 19:181–194] Many noxious or invasive weeds. *C. nigrescens* Willd. not naturalized.

1. Corolla yellow
  2. Heads sessile or subsessile, subtended by lf-like bracts . . . . *C. benedicta*
  2′ Heads gen peduncled (if not, then subtending bracts reduced)
    3. Corolla 25–30 mm . . . . *C. sulphurea*
    3′ Corolla gen 10–20 mm
      4. Central spine of main phyllaries 5–10 mm, ± purple . . . . *C. melitensis*
      4′ Central spine of main phyllaries 10–25 mm, straw-colored . . . . *C. solstitialis*
1′ Corolla white to pink, blue, or purple
  5. Main phyllaries ± spine-tipped
    6. Main phyllary tip-spine 6–25 mm
      7. Pappus 0 . . . . *C. calcitrapa*
      7′ Pappus present, 0.3–2.5 mm
        8. Involucre body 8–14 mm diam, widely ovoid . . . . *C. iberica*
        8′ Involucre body 5–10 mm diam, narrowly ovoid . . . . *C. pouzinii*
    6′ Main phyllary tip-spine 1–5 mm
      9. Outer fl corollas expanded, ± ray-like, >> disk fl corollas, 25–35 mm, pink-purple . . . . [2]*C. diluta*
      9′ Outer fl corollas not or barely expanded, = or slightly > disk fl corollas, 7–13 mm, white to pink or pale purple
        10. Involucre 10–13 mm; disk fl corolla 12–13 mm . . . . *C. diffusa*
        10′ Involucre 7–8 mm; disk fl corolla 7–9 mm . . . . *C. virgata* subsp. *squarrosa*
  5′ Main phyllaries not spine-tipped

11. Ann
  12. Herbage ± tomentose; disk fl corolla 10–15 mm . . . . *C. cyanus*
  12′ Herbage puberulent or ± glabrous, sometimes thinly cobwebby; disk fl corolla ± 20 mm . . . . ^{2}*C. diluta*
11′ Bien or per
  13. Outer and inner fls bisexual; corollas all ± equal . . . . *C. jacea* subsp. *nigra*
  13′ Outer fls sterile, inner fls bisexual; sterile fl corollas > bisexual fl corollas
    14. Herbage densely gray-tomentose . . . . *C. cineraria*
    14′ Herbage ± green, sometimes thinly cobwebby
      15. Involucre 10–13 mm . . . . *C. stoebe* subsp. *micranthos*
      15′ Involucre 15–18 mm
        16. Phyllary appendages scarious, light brown, concave, entire to coarsely dentate . . . . *C. jacea* subsp. *jacea*
        16′ Phyllary appendages light to dark brown, flat to ± concave, coarsely dentate to dissected into fine,
          wire-like lobes. . . . . *C. jacea* nothosubsp. *pratensis*

C. benedicta (L.) L. BLESSED THISTLE Ann to 10 dm. **ST:** erect to wide-spreading, gen branched throughout, not winged, loosely to densely tomentose, with long, crinkled hairs and finer, cobwebby hairs. **LF:** becoming ± glabrous, resin-dotted; basal and proximal cauline 10–20 cm, wing-petioled, oblanceolate to elliptic, coarsely lobed, dentate, teeth and lobes weakly spine-tipped; distal cauline sessile, not much reduced. **INFL:** heads disciform, 1–few, sessile or subsessile, often closely subtended by involucre-like set of wide, lf-like bracts; involucre 20–40 mm, ovoid or spheric; main phyllaries ± green or straw-colored, outer tightly appressed with spreading spine tips, inner with pinnately divided, gen purple or straw-colored spine tips. **FL:** many, corolla yellow; sterile fl corolla << disk; disk fl corolla 19–24 mm. **FR:** 8–11 mm, straw-colored, 20-ribbed, tipped with a 10-toothed rim, glabrous; pappus of 2 series of awns, outer 9–10 mm, inner 2–5 mm. $2n=22$. Roadsides, disturbed places; < 1700 m. NW, SNF, s SNH, GV, CW, SW, w DMoj; to e US; native to Eur. [*Cnicus b.* L.] Apr–Jun

C. calcitrapa L. (p. 273) PURPLE STAR-THISTLE Ann to per, 2–10+ dm, bushy, often ± mounded, puberulent to ± tomentose. **ST:** often wide-spreading, gen branched throughout, not winged. **LF:** ± glabrous in age, resin-dotted; basal petioled, 10–20(35) cm, ± deeply 1–2 × divided into narrow lobes, cauline gen sessile, linear or oblong, entire, toothed, or divided into narrow lobes. **INFL:** heads disciform, sessile or short-peduncled, often closely subtended by involucre-like set of spreading, lance-linear or oblong bracts; involucre 13–20 mm, body 6–8 mm diam, ovoid; main phyllaries ± green or straw-colored, appendages spiny-fringed at base, tip-spine 10–25 mm, stout. **FL:** 25–40, corolla purple; sterile fl corolla slender, ≤ disk; disk fl corolla 15–24 mm. **FR:** 2.5–3.5 mm, white or brown-streaked, glabrous; pappus 0. $2n=20$. Pastures, disturbed places; gen < 1000 m. NW, s CaRF, SNF, n SNH, GV, CW, SW; native to s Eur. Hybridizes with *C. iberica* in s NCoRI. Apr–Nov ◆

C. cineraria L. DUSTY MILLER Per 3–9 dm from woody base, densely gray-tomentose. **ST:** 1–many, branched distally or throughout, not winged. **LF:** proximal 5–25 cm, shallowly lobed or 1–2 × dissected, distal gen narrower, less divided. **INFL:** heads radiant, 1–few, ± peduncled; involucre 10–30 mm, ovoid; main phyllaries ± green to brown, striate, tip appendages straw-colored to black, obtuse to acuminate, fringed with tapered teeth, decurrent on distal margins of phyllaries. **FL:** many; corolla purple; sterile fl corolla > disk; disk fl corolla 12–20 mm. **FR:** ± 3.5 mm, tan, finely short-hairy; pappus bristles, 0.5–2.5 mm, white. $2n=18$. Disturbed areas, coastal scrub; < 100 m. CCo, SCo, ChI; native to Italy. Escape from cult. May–Sep

C. cyanus L. (p. 273) BACHELOR'S BUTTON, CORNFLOWER Ann 2–10 dm, ± tomentose. **ST:** gen 1, erect, often branched distally, not winged. **LF:** thinly to densely tomentose, green to gray; proximal 3–10+ cm, entire or few-lobed, gen 0 or withered at fl; distal cauline little reduced, linear, entire or minutely few-toothed. **INFL:** heads radiant, few to many in open, ± panicle-like cluster, slender-peduncled; involucre 10–16 mm, ovoid or bell-shaped; main phyllaries green or purple, tip appendages fringed with slender, white to black teeth. **FL:** 25–35, corolla gen blue (white to purple); sterile fl corolla 20–25 mm, 5–8-lobed, >> disk; disk fl corolla 10–15 mm. **FR:** 4–5 mm, straw-colored (± blue), finely short-hairy; pappus bristles 1–4

mm. $2n=24$. Grassland, open woodland, disturbed areas; gen ≤ 1500 m. CA-FP, MP; native to s Eur. Cult, often escaped; often in wildflower mixes. Apr–Sep

C. diffusa Lam. DIFFUSE KNAPWEED Bien or per, puberulent, ± gray-tomentose. **ST:** 2–8+ dm, much-branched, not winged. **LF:** resin-dotted; basal and proximal cauline 10–20 cm, ± deeply 2-pinnately divided, lobes linear to oblong, mid-cauline sessile, deeply lobed, distal cauline ± undivided. **INFL:** heads disciform or ± radiant, many in open panicle-like clusters, gen short-peduncled; involucre 10–13 mm, cylindric to narrowly ovoid; main phyllaries pale green, prominently parallel-veined, tip appendages fringed with slender, straw-colored spines, tip-spine 1–3 mm. **FL:** 12–13, corolla creamy white (lavender or pale purple); sterile fl corolla slender, ± = disk; disk fl corolla 12–13 mm. **FR:** ± 2.5 mm, dark brown; pappus 0 or scales ≤ 1 mm, white. $2n=18$. Fields, roadsides, open woodland; < 2300 m. NW, CaR, n&c SN, ScV, SnFrB, SCoR, SnBr, MP, W&I; N.Am; native to se Eur. [Blair & Huffbauer 2009 Inv Pl Sci Managem 2:55–69] May–Oct ◆

C. diluta Aiton NORTH AFRICAN KNAPWEED Ann ≤ 13 dm, puberulent or ± glabrous, resin-dotted, sometimes thinly cobwebby. **ST:** distally openly branched. **LF:** proximal 10–15 cm, coarsely lobed, distal lobe gen largest, distal lvs smaller, lobed or toothed to entire. **INFL:** heads radiant, peduncled, in open, panicle-like clusters; involucre 15–18 mm, ovoid; main phyllaries ± green or straw-colored, tip appendages fringed with slender teeth, tip-spine 0–5 mm, slender. **FL:** many, corolla pink-purple; sterile fl corolla 25–35 mm, >> disk; disk fl corolla ± 20 mm. **FR:** ± 3.5 mm; pappus bristles ± 4.5 mm, white. $2n=20$. Disturbed places; < 170 m. w CW, SCo; native to sw Eur. Apr–Dec

C. iberica Spreng. IBERIAN STAR-THISTLE Per 5–10 dm, ± loosely tomentose, becoming ± glabrous, resin-dotted. **ST:** openly branched, not winged, puberulent to thinly tomentose. **LF:** short-rough hairy when young; proximal 10–20 cm, coarsely lobed; distal ± oblong, entire or toothed. **INFL:** heads disciform, peduncled, few to many in lfy clusters, involucre 15–18 mm, body 8–14 mm diam, widely ovoid; main phyllaries ± green or straw-colored, tip appendage short-spined at base, tip-spine 10–25 mm, stout. **FL:** 40–many; corolla rose-pink or ± white; sterile fl corolla ± = disk; disk fl corolla 15–20 mm. **FR:** ± 3 mm, white or straw-colored, or brown-marked, glabrous; pappus bristles 1–2.5 mm, white. $2n=16,20$. Disturbed places; < 1000 m. s NCoRI, n&c SNF, c GV, CW, s SCo, PR; native to se Eur. Hybridizes with *C. calcitrapa* in s NCoRI. Jun–Aug ◆

C. jacea L. Per 3–12 dm, short-stiff hairy or ± soft-crinkly-hairy, thinly cobwebby-tomentose when young, green. **ST:** 1–many, branched distally or throughout, not winged. **LF:** proximal 5–25 cm, entire to irregularly lobed, distal gen narrower, entire. **INFL:** heads discoid or radiant, few to many, on lfy-bracted peduncles in open, cyme-like clusters; involucre 15–18 mm, hemispheric; main phyllaries proximally ± green, bodies gen concealed by overlapping appendages of adjacent phyllaries, appendages scarious, light to dark brown or black, flat or concave, entire to coarsely dentate or variably dissected into fine, wire-like lobes. **FL:** many; corolla purple; sterile fl (if present) corolla gen >> disk. **FR:** 2.5–3 mm, tan, finely hairy; pappus 0 or scales 0.5–1 mm.

subsp. *jacea* BROWN KNAPWEED **INFL**: heads radiant; peduncle not swollen below head; phyllary tip appendages scarious, light brown, concave, entire to coarsely dentate. **FL**: disk fl corolla 15–22 mm. **FR**: pappus bristles ± 0.5 mm, white or tan. 2*n*=22, 44. Grassland, disturbed places, montane forest; 150–1100 m. NCoRO, CaR; to s Can, e US; native to Eur. Aug–Oct

subsp. *nigra* (L.) Bonnier & Layens BLACK KNAPWEED **INFL**: heads discoid; peduncle swollen below head; phyllary tip appendages dark brown to black, flat, conspicuously fringed with long, wire-like, minutely bristly lobes. **FL**: disk fl corolla 15–18 mm. **FR**: pappus bristles (0) many, gen ≤ 1 mm, ± black. 2*n*=22, 44. Disturbed places; < 600 m. NCo, KR, NCoRO, CaRF; to s Can, e US; native to Eur. [*C. n.* L.] Aug–Nov

nothosubsp. *pratensis* (W.D.J. Koch) Celak. **INFL**: heads radiant (discoid); peduncle swollen below head or not; phyllary tip appendages ± scarious, light to dark brown, flat to ± concave, coarsely dentate to dissected into fine, wire-like lobes. **FL**: disk fl corolla 15–22 mm. **FR**: pappus scales ≤ 1 mm, white or tan. 2*n*=22, 44. Grassland, disturbed places, montane forest; 150–1100 m. NCoRO, CaR; to s Can, e US; native to Eur or locally derived from hybridization of parental taxa native to Eur. [*C.* ×*moncktonii* C.E. Britton; *C.* ×*p.* Thuill., illeg.] Variable, outcrossing derivative of fertile hybrid (*C. jacea* subsp. *j.* × *C. jacea* subsp. *nigra*). Populations show much pl-to-pl variation, often with some individuals approaching one or the other parental subsp. Aug–Oct ◆

C. melitensis L. TOCALOTE Ann 1–10 dm, ± gray-hairy, resin-dotted. **ST**: gen 1, distally ± branched, winged. **LF**: ± thinly tomentose, ± scabrous, and puberulent with short, crinkled hairs; proximal 2–15 cm, entire to lobed, often 0 at fl; distal cauline entire or toothed, long-decurrent. **INFL**: heads disciform, 1–many, peduncled (or sessile in axils), sometimes of cleistogamous fls; involucre 10–15 mm, ovoid, ± cobwebby or becoming glabrous; main phyllaries ± straw-colored; tip appendage ± purple, base spine-fringed, central spine 5–10 mm, slender. **FL**: many; corolla yellow; sterile fl corolla ± = disk; disk fl corolla 10–12 mm. **FR**: ± 2.5 mm, ± light brown, finely hairy; pappus bristles 2.5–3 mm, white. 2*n*=18,24,36. Disturbed fields, open woodland; < 2200 m. CA-FP, D (uncommon); native to s Eur. Apr–Jul ◆

C. pouzinii DC. POUZIN'S STAR-THISTLE Ann to per, 2–10+ dm, bushy, often ± mounded, ± tomentose. **ST**: erect or often widespreading, branched distally or throughout, angled, not winged. **LF**: finely cobwebby and/or with thicker, jointed hairs, resin-dotted; distal sometimes short-rough-hairy; basal petioled, 10–20(35) cm, ± deeply 1–2 × divided into narrow lobes, cauline gen sessile, distal linear or divided into linear lobes. **INFL**: heads disciform, sessile or short-peduncled, often closely subtended by involucre-like set of spreading, lance-linear or oblong bracts; involucre 14–17 mm, body 5–10 mm diam, narrowly ovoid; main phyllaries ± green or straw-colored, tip appendages stiffly spreading, proximally fringed with 2–3 pairs of ascending to spreading spines, tip-spine 6–20 mm, stout. **FL**: 25–50+, corolla pink-purple; sterile fl corolla slender to slightly expanded, ≤ disk; disk fl corolla 15–20 mm. **FR**: 2.5–4 mm, white to dark brown or brown-streaked, glabrous or thinly fine-hairy; pap-

pus of scales and/or bristles 0.3–2 mm. 2*n*=42. Pastures, grassland, disturbed places; < 120 m. NCoR (Humboldt Co.), SCo (near Corona, Riverside Co.); native to sw Eur. Jun–Dec

C. solstitialis L. (p. 273) YELLOW STAR-THISTLE Ann 1–10 dm, often rounded, ± bushy, ± gray-tomentose. **ST**: branched distally or throughout, winged. **LF**: ± scabrous-bristly beneath tomentose hairs; proximal 5–15 cm, 1–2 × lobed or dissected, gen 0 or withered at fl; distal cauline linear, entire, long-decurrent. **INFL**: heads disciform, 1–many, peduncled, in open, cyme-like clusters; involucre 13–17 mm, ovoid; main phyllaries pale green to straw-colored or brown; appendages straw-colored, palmately spiny, central spine 10–25 mm, stout. **FL**: many; corolla yellow; sterile fl corolla ± = disk; disk fl corolla 13–20 mm. **FR**: 2–3 mm, glabrous; outer fr dark brown, pappus 0; inner fr ± mottled light brown, pappus bristles 2–4 mm, fine, white. 2*n*=16. Invasive, roadsides, disturbed grassland or woodland; < 1300 m. CA-FP, MP, w DMoj; native to s Eur. Cumulatively TOXIC to horses. May–Oct ◆

C. stoebe L. subsp. *micranthos* (Gugler) Hayek (p. 273) SPOTTED KNAPWEED Bien, per, 3–10 dm, ± thinly tomentose, ± green. **ST**: 1–many from base, distally branched, not winged. **LF**: short-stiff hairy and resin-dotted; proximal 10–15 cm, ± deeply 1–2 × lobed, distal entire or lobed. **INFL**: heads radiant, gen many in open, panicle-like clusters; involucre 10–13 mm, ovoid; phyllaries pale green or pink-tinged, prominently parallel-veined; tip appendages dark brown-purple to black, fringed with slender, dark brown or straw-colored teeth or lobes. **FL**: 30–40; corolla gen pink or purple (white); sterile fl corolla >> disk; disk fl corolla 12–25 mm. **FR**: 3–3.5 mm, ± pale brown, finely hairy; pappus bristles 1–2 mm, white. 2*n*=18,36. Disturbed areas; < 2600 m. NW, CaR, SN, n ScV, n CW, TR, s PR, MP, n SNE; native to Eur. [*C. maculosa* Lam.] Jul–Sep ◆

C. sulphurea Willd. SICILIAN STAR-THISTLE Ann 1–10 dm, ± soft-crinkly-hairy. **ST**: simple or openly branched, winged. **LF**: puberulent, resin-dotted; proximal 10–15 cm, lobed; distal entire or ± spiny-serrate, long-decurrent. **INFL**: heads disciform, 1–few, peduncled, bracts reduced; involucre 2–3 cm, ovoid; main phyllaries ± green or straw-colored; tip appendages straw-colored or purple-black, sometimes reflexed, palmately spiny, tip-spine 10–25 mm, stout. **FL**: many; corolla yellow; sterile fl corolla ± = disk; disk fl corolla 25–30 mm. **FR**: ± 5 mm; pappus bristles 6–7 mm, dark. 2*n*=24. Disturbed places; < 300 m. c SNF, c GV, s SnFrB; native to sw Eur. May–Jun ◆

C. virgata Lam. subsp. *squarrosa* (Boiss.) Gugler (p. 273) Per 2–5 dm, scabrous, ± loosely tomentose, resin-dotted. **ST**: 1–several from base, much-branched distally, not winged. **LF**: proximal 10–15 cm, deeply 1–2 × lobed, gen 0 at fl, distal entire to deeply lobed. **INFL**: heads disciform or ± radiant, ± many, in panicle-like clusters, gen short-peduncled; involucre 7–8 mm, ± cylindric; main phyllaries pale green to straw-colored (or tinged purple), appendages fringed with slender, straw-colored spines, tip-spine 1–3 mm. **FL**: few; corolla pink or pale purple; sterile fl corolla = or slightly > disk; disk fl corolla 7–9 mm. **FR**: 2.5–3.5 mm, ± light brown; pappus bristles (0)2–2.5 mm, white. 2*n*=18. Disturbed places; 900–1400 m. KR, CaR, n SNH, MP; native to Asia. [*C. s.* Willd., illeg.] Jun–Sep ◆

CENTROMADIA SPIKEWEED

Bruce G. Baldwin

Ann [per] 1–12 dm. **ST**: prostrate to ± erect. **LF**: simple, basal and cauline, opposite proximally, often withered before fl, most alternate, ± sessile, oblanceolate to linear or lance-linear, proximal gen 1–2 pinnately lobed, lobes toothed or entire, sometimes bristly-ciliate, distal entire, gen spine-tipped, faces glabrous or scabrous and/or coarse- to soft-hairy or puberulent, often also glandular. **INFL**: heads radiate, congested or in spike- to panicle- or umbel-like clusters; peduncle bracts gen spine-tipped; involucre ± obconic or urn-shaped, 3–8+ mm diam; phyllaries 5–75+, lanceolate or oblanceolate, each ± 1/2-enclosing ray ovary, falling with fr or both persistent and head or fl branch dispersing as unit, scabrous and/or coarse- to soft-hairy or puberulent, often also glandular; receptacle flat to convex, minutely bristly; paleae subtending all or most disk fls, free. **RAY FL**: 5–75+; corolla yellow, ray 2–6 mm. **DISK FL**: 6–200+, gen staminate; corolla 2–5 mm, yellow, tube < throat, tube/throat glabrous, lobes deltate, abaxially often minutely bristly; anthers ± red to dark purple or yellow to ± brown, tips ovate; style glabrous proximal to branches, tips ovate to awl-shaped, densely bristly. **FR**: ray fr 2–3 mm, ± compressed, bulging abaxially, glabrous, beaked or elevated adaxially; pappus 0; disk pappus 0 or of 3–12 linear, awl-shaped or oblanceolate scales. 4 spp.:

273

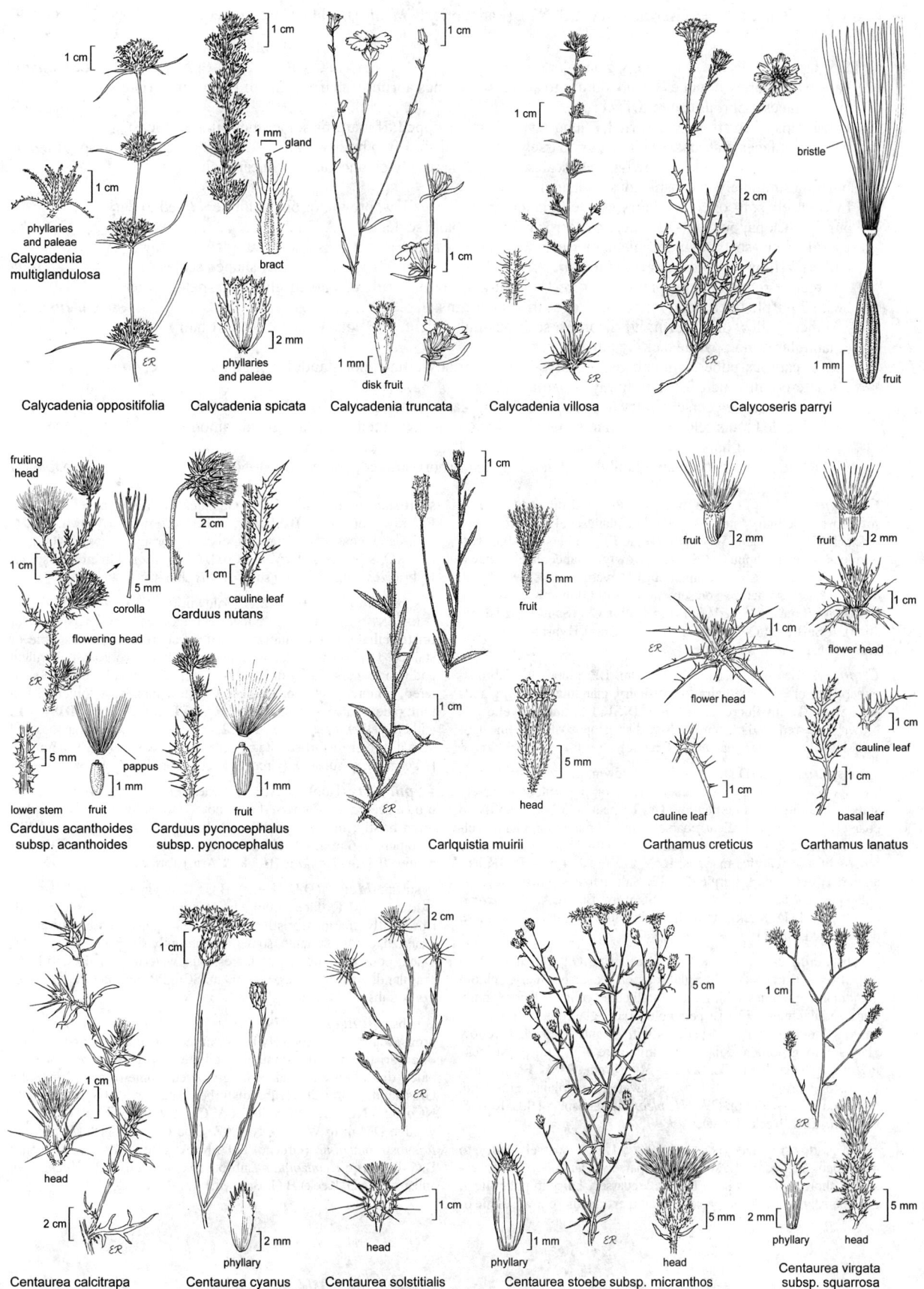

Calycadenia oppositifolia

Calycadenia multiglandulosa

phyllaries and paleae

Calycadenia spicata

gland

bract

phyllaries and paleae

Calycadenia truncata

disk fruit

Calycadenia villosa

Calycoseris parryi

bristle

fruit

Carduus acanthoides subsp. acanthoides

fruiting head

flowering head

corolla

lower stem

fruit

pappus

Carduus pycnocephalus subsp. pycnocephalus

fruit

Carduus nutans

cauline leaf

Carlquistia muirii

fruit

head

Carthamus creticus

fruit

flower head

cauline leaf

Carthamus lanatus

fruit

flower head

cauline leaf

basal leaf

Centaurea calcitrapa

head

Centaurea cyanus

phyllary

Centaurea solstitialis

head

Centaurea stoebe subsp. micranthos

phyllary

head

Centaurea virgata subsp. squarrosa

phyllary

head

CA, s OR, n Baja CA (alien elsewhere in w US, NY). (Latin: prickly *Madia*) [Baldwin & Strother 2006 FNANM 21:276–279] Self-sterile.

1. Disk pappus 0; lvs not or minutely glandular . . . . *C. pungens*
  2. Paleae tips barely or not exserted, obtuse to acute, sometimes abruptly sharp-pointed; lf margins bristly-
     ciliate, faces glabrous exc midribs bristly . . . . subsp. *laevis*
  2′ Paleae tips often strongly exserted, ± long-tapered, spine-tipped; lf margins sometimes ciliate, at least
     near base, faces glabrous, scabrous, or coarsely hairy, midribs often bristly. . . . . subsp. *pungens*
1′ Disk pappus of 3–5, linear to awl-shaped scales, or 8–12, narrowly oblanceolate to linear scales; lvs
  sometimes minutely to conspicuously glandular
  3. Lvs densely soft- or coarse-hairy, stalked-glandular, glands yellow, brown, or black; anthers ± red to dark
    purple; disk pappus of 8–12, linear or narrowly oblanceolate scales . . . . *C. fitchii*
  3′ Lvs glabrous, scabrous-puberulent, or ± coarsely to softly hairy, sometimes glandular, glands yellow;
    anthers yellow to ± brown or ± red to dark purple; disk pappus of 3–5, linear to awl-shaped scales . . . . *C. parryi*
    4. Anthers ± red to dark purple; lvs soft- or ± coarse-hairy or puberulent, stalked-glandular; paleae often
      with 2 purple lines along inner edges of scarious margins. . . . . subsp. *australis*
    4′ Anthers yellow or ± brown; lvs glabrous, scabrous-puberulent, or ± coarse-hairy, not soft-hairy,
      glandular or not; paleae lacking purple lines.
      5. Lvs scabrous-puberulent and gen ± coarse-hairy or bristly-ciliate, not glandular or glands minute,
        scattered; involucre 3.5–5 mm; ray 2–3 mm . . . . subsp. *rudis*
      5′ Lvs glabrous or coarsely hairy to puberulent, glandular or not; involucre 4–10 mm; ray 2.5–5(6) mm.
        6. Peduncle bracts seldom glandular, sometimes with minute, stalked, ± yellow glands among
          nonglandular hairs; ray 2.5–3(4.5) mm . . . . subsp. *congdonii*
        6′ Peduncle bracts coarsely glandular, glands sessile or short-stalked, yellow; ray 3–5(6) mm . . . . subsp. *parryi*

C. fitchii (A. Gray) Greene (p. 279) **Pl** 0.5–5 dm. **LF:** densely soft- to coarse-hairy, stalked-glandular, glands yellow, brown, or black. **INFL:** involucre 5–10 mm. **DISK FL:** anthers ± red to dark purple. **FR:** disk pappus of 8–12, narrowly oblanceolate to linear scales. 2*n*=26. Grassland, ± alkaline flats, vernal pools, woodland, disturbed sites, sometimes on serpentine; < 1000 m. e KR, NCoRO, NCoRI, CaR, n&c SNF, GV, SnFrB, SCoR, n ChI (Santa Cruz Island, alien), n SnBr; sw OR. [*Hemizonia f.* A. Gray] Hybridizes with *C. parryi* subsp. *rudis*. May–Nov

C. parryi (Greene) Greene **Pl** 1–7 dm. **LF:** glabrous, scabrous-puberulent, or ± coarse-hairy to soft-hairy, glandular or not, glands yellow. **INFL:** involucre 2.5–10 mm. **DISK FL:** anthers yellow, ± brown, or ± red to dark purple. **FR:** disk pappus of 3–5, linear or awl-shaped scales. [*Hemizonia p.* Greene]

  subsp. *australis* (D.D. Keck) B.G. Baldwin (p. 279) SOUTHERN TARPLANT **LF:** softly to coarsely hairy or puberulent, coarsely stalked-glandular, glands yellow. **INFL:** peduncle bracts softly to coarsely hairy or puberulent, coarsely stalked-glandular, glands yellow; involucre 2.5–6(7) mm; paleae often with 2 purple lines along inner edges of scarious margins. **RAY FL:** ray 2–4 mm. **DISK FL:** anthers ± red to dark purple. 2*n*=22. Salt marshes, grassland, vernal pools, coastal scrub; < 200 m. SCo; nw Baja CA. [*Hemizonia p.* subsp. *a.* D.D. Keck] Sterile hybrids with *Deinandra fasciculata* documented. Jun–Oct ★

  subsp. *congdonii* (B.L. Rob. & Greenm.) B.G. Baldwin (p. 279) CONGDON'S TARPLANT **LF:** glabrous or ± coarsely hairy, seldom glandular, sometimes with minute, stalked, ± yellow glands among nonglandular hairs. **INFL:** peduncle bracts glabrous or ± coarse-hairy, seldom glandular, sometimes with minute, stalked, ± yellow glands among nonglandular hairs; involucre 4–9 mm; palea lacking purple lines. **RAY FL:** ray 2.5–3(4.5) mm. **DISK FL:** anthers yellow or ± brown. 2*n*=24. Terraces, swales, floodplains, grassland, disturbed sites; < 300 m. CW. [*Hemizonia p.* subsp. *c.* (B.L. Rob. & Greenm.) D.D. Keck] Jun–Oct ★

  subsp. *parryi* PAPPOSE TARPLANT **LF:** ± coarsely hairy to puberulent, gen coarsely, sometimes sparsely, glandular, glands sessile or short-stalked, yellow. **INFL:** ± coarsely hairy to puberulent, gen coarsely, sometimes sparsely, glandular, glands yellow, sessile or short-stalked; involucre 5–10 mm; paleae lacking purple lines. **RAY FL:** ray 3–5(6) mm. **DISK FL:** anthers yellow or ± brown. 2*n*=24. Grassland, coastal salt marshes, alkaline springs, seeps; < 400 m. s NCoRO, s NCoRI, s ScV, n CCo. [*Hemizonia p.* Greene subsp. *p.*] Hybridizes with *C. parryi* subsp. *rudis*. Jun–Oct ★

  subsp. *rudis* (Greene) B.G. Baldwin (p. 279) PARRY'S ROUGH TARPLANT **LF:** scabrous-puberulent and gen ± coarsely hairy or bristly-ciliate, not glandular or with scattered, minute, sessile or stalked, ± yellow glands. **INFL:** peduncle bracts scabrous-puberulent and gen ± coarse-hairy or bristly-ciliate, not glandular or with scattered, minute, sessile or stalked, ± yellow glands; involucre 3.5–5 mm; paleae lacking purple lines. **RAY FL:** ray 2–3 mm. **DISK FL:** anthers yellow or ± brown. 2*n*=22. Grassland, edges of marshes and vernal pools, disturbed sites; < 500 m. s NCoRI (rare), n&c GV. [*Hemizonia p.* subsp. *r.* (Greene) D.D. Keck] Jun–Oct ★

C. pungens (Hook. & Arn.) Greene **Pl** 1–12 dm. **LF:** faces glabrous, scabrous, or coarse-hairy, not or minutely glandular, midrib often bristly; margin gen bristly-ciliate, at least near base. **INFL:** involucre 3–6 mm. **DISK FL:** anthers yellow to ± brown. **FR:** disk pappus 0. [*Hemizonia p.* (Hook. & Arn.) Torr. & A. Gray]

  subsp. *laevis* (D.D. Keck) B.G. Baldwin (p. 279) SMOOTH TARPLANT **LF:** faces minutely glandular or glabrous exc midribs bristly, margins bristly-ciliate. **INFL:** palea tips barely or not exserted, obtuse to acute, sometimes abruptly sharp-pointed. 2*n*=18. Open, poorly drained flats, depressions, waterway banks and beds, grassland, disturbed sites; 90–500 m. SCo, PR; n Baja CA. [*Hemizonia p.* subsp. *l.* D.D. Keck] Apr–Sep ★

  subsp. *pungens* (p. 279) COMMON SPIKEWEED **LF:** faces glabrous, scabrous, or coarsely hairy, minutely glandular or not, midrib often bristly, margins sometimes ciliate, at least near base. **INFL:** palea tips often exserted, ± long-tapered, spine-tipped. 2*n*=18,20. Grassland, saltbush scrub, disturbed sites; < 1200(1800) m. s NCoRO, s NCoRI, CaR, s SNF, GV, CW, SW (exc ChI, WTR), s MP, w edge DMoj; to WA, ID, NV, AZ, Baja CA, also NY. [*Hemizonia p.* subsp. *maritima* (Greene) D.D. Keck; *C. p.* subsp. *m.* (Greene) B.G. Baldwin; *Hemizonia p.* subsp. *septentrionalis* D.D. Keck; *C. p.* subsp. *s.* (D.D. Keck) B.G. Baldwin] Apr–Nov

CHAENACTIS PINCUSHION, DUSTY-MAIDENS

James D. Morefield

Ann to subshrub (2)5–70(200) cm; proximal st and lvs glabrous to variously hairy. **ST**: 1, erect, or 2–25+, prostrate to ascending, simple or branched. **LF**: simple, alternate, often crowded proximally, petioled, gen ± elliptic to ovate or obovate, gen deeply 1–4-pinnately lobed, occ entire (then linear), lobes gen not or scarcely overlapping; distal gen ± reduced. **INFL**: heads discoid or radiant, 1–25+ per st, gen in terminal cyme-like cluster; peduncle gen erect, hairs gen as on phyllary bases; involucre gen ≤ 15[25] mm diam, cylindric to obconic or hemispheric; phyllaries in 1–2 ± = series, gen linear to lanceolate, persistent, tips gen ± flat, gen ± green; receptacle flat to rounded, glabrous, gen epaleate. **FL**: 8–70+; corolla white, ± pink, or yellow, gen open during day; anthers gen exserted, tips lanceolate to ovate; style tips linear, minutely bristly. **FR**: ± club-shaped, gen not compressed, ± hairy; pappus 0, crown-like, or gen of (1)4–20 persistent, ± fringed scales in 1–4 series, scales often fewer and/or shorter on outer fr. 18 spp.: w N.Am. (Greek: gaping ray, for enlarged outer corollas of type sp.) [Morefield 2006 FNANM 21:400–414] Spp. of sect. *Chaenactis* hybridize, esp in SCoRI/s SnJV, where identification can be difficult.

1. Per (bien) or subshrub, rarely fl 1st yr; inner fr pappus gen of 10–20 scales in 3–4 ± equal series; lf blade gland-pitted under hairs; heads discoid (sect. *Macrocarphus*)
  2. Lvs gen cauline (also basal in *Chaenactis douglasii*); heads not scapose, (1)2–25+ per st; pls not or scarcely cespitose or matted
    3. Per or bien, rarely fl 1st yr or slightly woody; proximal hairs cobwebby to tomentose, ± gray, thinning with age; basal lvs gen persistent, largest lf blades ± elliptic, not flat . . . . *C. douglasii* var. *douglasii*
    3′ Gen subshrub; proximal hairs (esp sts) ± felty, ± white, persistent; basal lvs gen 0, largest lf blades ± deltate, ± flat
      4. Longest phyllaries 10–13 mm, outer ± tomentose, not or scarcely glandular; e PR. . . . . *C. parishii*
      4′ Longest phyllaries 14–18 mm, outer gen glandular-hairy, other hairs 0 or scarce; e KR
      . . . . ²*C. suffrutescens*
  2′ Lvs ± basal; heads ± scapose, 1(3) per st; pls cespitose to ± matted
    5. Outer phyllaries silky-tomentose, nonglandular. . . . . *C. alpigena*
    5′ Outer phyllaries gen glandular-hairy, often also cobwebby to woolly or shaggy-hairy
      6. Largest lf blades ± deltate to ovate, ± flat; 1° lobes gen 2–4(5) pairs, tips ± flat
        7. Pl 2–10(12) cm; lvs 2.5–5 cm; longest phyllaries 9–12(14) mm; corolla 5.5–8 mm; longest pappus scales 3–5 mm. . . . . *C. nevadensis*
        7′ Pl (10)25–45(60) cm; lvs 5–10 cm; longest phyllaries 14–18 mm; corolla 8.5–10 mm; longest pappus scales 7–9 mm . . . . ²*C. suffrutescens*
      6′ Largest lf blades ± elliptic or cylindric to ± fusiform, not flat; 1° lobes (4)5–18+ pairs, tips curled and/or twisted
        8. Largest lf blades ± elliptic; 1° lobes (4)5–9(12) pairs, not or scarcely overlapping . . . . *C. douglasii* var. *alpina*
        8′ Largest lf blades cylindric to ± fusiform; 1° lobes (7)10–18+ pairs, ± overlapping. . . . . *C. santolinoides*
1′ Ann; inner fr pappus 0, crown-like, or of 4(5) or (7)8 scales in 1 or 2 unequal series, outer << inner; lf blades (exc *Chaenactis macrantha*) not gland-pitted; heads discoid or radiant
  9. Proximal hairs mealy, scaly, or powdery; largest lf blades gen ± deltate, (2)3–4-pinnately lobed (sect. *Acarphaea*)
    10. Phyllary tips acute or scarcely acuminate, ± flat, ± green; lf lobe tips ± flat; receptacle epaleate; fr compressed; pappus 0 or longest scales ≤ 0.5 mm . . . . *C. artemisiifolia*
    10′ All or inner phyllary tips long-acuminate, needle-like, ± red; lf lobe tips ± cylindric; receptacle paleate; fr not compressed; longest inner fr pappus scales gen 3–5 mm . . . . *C. carphoclinia*
      11. Lvs basal and cauline, longest 1–6(7) cm; petiole slender, flexible, base scarcely enlarged . . . . . . var. *carphoclinia*
      11′ Lvs gen basal, longest 7–10 cm; petiole thick, stiff, base strongly enlarged . . . . var. *peirsonii*
  9′ Proximal hairs cobwebby to woolly or 0; largest lf blades linear, entire, or ± elliptic (to ovate), 1–2-pinnately lobed (sect. *Chaenactis*)
    12. Corolla white to ± pink or rarely pale yellow; gen in or near GB, D
      13. Heads radiant; peduncle distally gen glandular-hairy, or ± glabrous in fr; outer corollas bilateral, spreading, > inner; inner fr pappus gen of 4 scales in 1 series
        14. Longest pappus 6–8.5 mm, exceeding corolla, tips visible among buds; proximal hairs 0 by fl time; largest lf blades entire or 1-pinnately lobed, lobes gen 1–2(5) pairs; outer phyllaries ± glabrous in fr, tip acute. . . . . *C. fremontii*
        14′ Longest pappus 1.5–6 mm, not exceeding corolla, tips hidden; proximal hairs persisting at least at nodes by fl time; largest lf blades gen 1–2-pinnately lobed, 1° lobes 4–8 pairs; outer phyllaries glandular-hairy and/or ± cobwebby in fr, tip ± obtuse . . . . *C. stevioides*
      13′ Heads discoid; peduncle nonglandular; outer corollas radial, spreading or not, ± = inner; pappus of (6)8 scales in 2 unequal series, outer << inner
        15. Corolla open at night, anthers ± incl; inner corollas 9–12(15) mm, 1.8–2.2 × fr; lf blades ± gland-pitted under hairs . . . . *C. macrantha*
        15′ Corolla open during day, anthers exserted; inner corolla 4–10 mm, ± = fr; lf blades not gland-pitted
          16. Longest phyllaries 10–18 mm, outer nonglandular, glabrous proximally, densely puberulent distally, tip gen recurved, flexible; pappus scales 8. . . . . *C. xantiana*

16′ Longest phyllaries 4.5–10(12) mm, outer gen glandular-hairy, hairs otherwise various, uniform, tip
erect, stiff; pappus scales gen < 8 (hybrids; see genus note)
12′ Corolla bright to dark yellow; gen CA-FP . . . . *C. glabriuscula*
17. Inner fr pappus gen of 7–8 scales in 2 series, outer << inner . . . . var. *heterocarpha*
17′ Inner fr pappus gen of 4 scales in 1 series
18. Proximal hairs tomentose to woolly, ± white; lvs ± basal, persistent . . . . var. *lanosa*
18′ Proximal hairs 0 or ± cobwebby, ± gray; lvs basal (withering) and cauline
19. Largest lf blades gen 2-pinnately lobed, fleshy; st ± spreading; SCo, coastal dunes, bluffs . . . . . . var. *orcuttiana*
19′ Largest lf blades 1(2)-pinnately lobed, scarcely fleshy; st erect to ascending; inland from coast
20. Longest phyllaries 5–7 mm, 1–2 mm wide, outer ± silky-cobwebby at least medially in fr, often
also ± glandular-hairy; longest pappus 2–4 mm, 0.4–0.7 × corolla . . . . var. *glabriuscula*
20′ Longest phyllaries 7–9 mm, 2–3 mm wide, outer becoming glabrous in fr; longest pappus gen 5–8
mm, ± 0.9(1) × corolla . . . . var. *megacephala*

C. alpigena Sharsm. (p. 279) SHARSMITH PINCUSHION Per, cespitose to ± matted, 2–7 cm; proximal hairs woolly, ± gray to ± yellow, sometimes becoming nearly 0. **ST:** gen 5–15+; branches 0, basal, or below ground. **LF:** basal, persistent, 1–2.5(3.5) cm; largest blades widely elliptic to ± obovate or linear, ± flat and 1-± palmately lobed to entire (gen n pls) to 3-dimensional and 1–2-pinnately lobed (gen s pls), scarcely fleshy, gland-pitted under hairs; 1° lobes 2–5 pairs, tips ± flat (gen n pls) to 4–7 pairs, tips curled and/or twisted (gen s pls). **INFL:** heads discoid, 1 per st; involucre ± cylindric to obconic; longest phyllaries 9–14 mm, outer silky-tomentose, non-glandular, tip erect, ± stiff, obtuse. **FL:** corolla white to ± pink, 5.5–8 mm, outer radial, ± erect, = inner. **FR:** 5–8 mm; inner fr pappus of 10–20 scales in 3–4 ± equal series, longest gen 5–8 mm. Open, loose, gen granitic sand, gravel, scree; 2200–3900 m. SNH, W&I (n White Mtns); to w NV. Jul–Sep

C. artemisiifolia (Harv. & A. Gray) A. Gray (p. 279) WHITE PINCUSHION Ann (15)25–90(200) cm; proximal hairs mealy, scaly, or powdery, gen ± white. **ST:** gen 1; branches gen distal. **LF:** cauline in fl, 3–15(20) cm; largest blades ± deltate, ± flat, (2)3–4-pinnately lobed, not fleshy, nonglandular; 1° lobes gen 5–10 pairs, tips ± flat. **INFL:** heads discoid, 3–20+ per st; involucre ± hemispheric; longest phyllaries 7–10(12) mm, outer ± shaggy-hairy, tip erect, stiff, acute or scarcely acuminate. **FL:** corolla white to ± pink, 5–7 mm, outer radial, ± erect, = inner. **FR:** 4–7 mm, compressed; pappus 0 or crown-like, scales ± 10, ≤ 0.5 mm. 2n=16. Dry canyons, open slopes, often granitic, esp chaparral burns; 80–1600 m. s SCoRO (in 1889), SW (exc ChI); n Baja CA. Apr–Jul

C. carphoclinia A. Gray PEBBLE PINCUSHION Ann; proximal hairs powdery or scaly, gen ± white. **ST:** gen 1; branches gen distal. **LF:** basal, withering or not, and gen cauline; largest blades ± deltate, not flat, (2)3–4-pinnately lobed, ± fleshy or not, nonglandular; 1° lobes gen 2–7(10) pairs, tips gen curved, ± cylindric. **INFL:** heads discoid, 3–20+ per st; involucre ± cylindric to obconic or hemispheric; longest phyllaries 7–10 mm, outer ± granular-glandular and shaggy-hairy, tip (all or inner) erect or curved inward, stiff, long-acuminate, needle-like, ± red; paleae (0)3–10+, like phyllaries, length > buds. **FL:** corolla white to ± pink, 4–6 mm, outer radial, ± spreading, = inner. **FR:** 3–4.5 mm; inner fr pappus gen of 4 scales in 1 series, longest 3–5 mm. Often mistaken for *C. stevioides* but closer to *C. artemisiifolia*. Paleae unique in genus.

var. **carphoclinia** (p. 279) Pl (5)10–30(40) cm. **LF:** basal ± withering, and cauline, longest 1–6(7) cm; petiole slender, flexible, base scarcely enlarged. 2n=16. Common. Open, gen rocky or gravelly slopes, flats; -90–1900 m. s SNE, D; to UT, sw NM, nw Mex. Mar–Jun

var. **peirsonii** (Jeps.) Munz (p. 279) PEIRSON'S PINCUSHION Pl gen 40–60 cm. **LF:** gen basal, persistent, longest 7–10 cm; petiole thick, stiff, base strongly enlarged. Open, gen rocky or gravelly slopes, flats; < 200 m. e PR (e Santa Rosa Mtns), adjacent w DSon. Mar–Apr ★

C. douglasii (Hook.) Hook. & Arn. Gen per or bien; proximal hairs cobwebby to tomentose, ± gray, thinning with age. **ST:** branches proximal and/or distal. **LF:** basal gen persistent, and gen cauline, (1)2–12(15) cm; largest blades ± elliptic, not flat, gen 2-pinnately lobed,

not fleshy, gland-pitted under hairs; 1° lobes (4)5–9(12) pairs, tips ± curled and/or twisted. **INFL:** heads discoid; involucre obconic to ± hemispheric; outer phyllaries gen glandular-hairy, often also cobwebby to woolly or shaggy, tip ± recurved, ± flexible, obtuse. **FL:** corolla gen ± pink to white, 5–8 mm, outer radial, ± erect, = inner. **FR:** 5–8 mm; inner fr pappus of 10–20 scales in 3–4 ± equal series, longest 3–6 mm.

var. **alpina** A. Gray ALPINE DUSTY-MAIDENS Per, cespitose to ± matted, gen (2)5–10 cm. **ST:** gen (1)10–25+. **LF:** basal, persistent, (1)2–6 cm, sometimes becoming glabrous. **INFL:** heads ± scapose, 1(2) per st; longest phyllaries 9–12 mm. 2n=12. Rocky or gravelly ridges, talus, fell-fields, crevices; 3000–3400 m. n SNH, n DMtns (Panamint Range); to OR, MT, CO (exc NV). Intergrades downslope with *C. douglasii* var. *d.*; populations often more uniform outside CA (CA pls ± atypical). Jul–Sep ★

var. **douglasii** (p. 279) DUSTY-MAIDENS Per or bien, rarely fl 1st yr or slightly woody, not or scarcely cespitose or matted, (3)8–50[60] cm. **ST:** gen 1–5(12). **LF:** basal (sometimes withering) and ± cauline, 1.5–12(15) cm, hairs gen persistent. **INFL:** heads (1)2–25+ per st; longest phyllaries 9–15(17) mm. 2n=12,24,36 (plus triploids, dysploids). Common. Rocky or gravelly ridges, talus, fell-fields, crevices; (400)900–3500 m. KR, NCoRI, NCoRH, CaRH, SNH, GB, n DMtns; to s BC, s AB, w ND, CO, nw NM. More distinctive diploid variants occur outside CA. May–Sep

C. fremontii A. Gray (p. 279) FREMONT PINCUSHION Ann 10–30(40) cm; proximal hairs ± cobwebby early, ± gray, 0 by fl time. **ST:** gen 1–12; branches gen proximal. **LF:** basal, withering, and ± cauline, 1–7(10) cm; largest blades linear, cylindric, entire, or ± elliptic, flat, 1-pinnately lobed, ± fleshy, nonglandular; lobes gen 1–2(5) pairs, tips ± cylindric. **INFL:** heads radiant, gen 1–5 per st; peduncle distally gen glandular-hairy, or ± glabrous in fr; involucre ± obconic to hemispheric; longest phyllaries 8–10(12) mm, outer ± glabrous in fr, tip erect, stiff, acute. **FL:** corolla white to ± pink, inner 5–8 mm, outer bilateral, spreading, > inner. **FR:** (3)6–8 mm; inner fr pappus of 4 scales in 1 series, longest 6–8.5 mm, tips visible among buds. 2n=10. Loose sand or gravel, often growing through shrubs; -10–1600(2200) m. s SNF, Teh, s SnJV, SCoRI, n&e edges SW (exc ChI), s SNE (exc W&I), D; to sw UT, w AZ, n Baja CA. Feb–May

C. glabriuscula DC. YELLOW PINCUSHION Ann; proximal hairs cobwebby to woolly, ± gray to white, sometimes 0 by fl time. **ST:** branches 0 or proximal, often also distal. **LF:** basal (often withering) and gen cauline, 1–10 cm; largest blades gen ± elliptic, flat or not, 1–2-pinnately lobed (sometimes linear, cylindric, entire), fleshy or not, nonglandular; 1° lobes 1–7 pairs, tips various. **INFL:** heads radiant, 1–20+ per st; involucre obconic or widely cylindric to ± hemispheric; longest phyllaries 4.5–10 mm, outer variously hairy to glabrous in fr, tip erect, stiff, ± obtuse. **FL:** corolla bright to dark yellow, inner 4–8 mm, outer bilateral, spreading, > inner. **FR:** 3–9 mm; inner fr pappus gen of 4 scales in 1 series, or (7)8 scales in 2 unequal series, outer << inner, longest (1)2–8 mm. 2n=12. Highly variable; some forms differ from *C. stevioides* [*C. glabriuscula* var. *g.*] or *C. fremontii* [*C. glabriuscula* var. *megacephala*] only by yellow fls and 2n=12.

var. **glabriuscula** Pl 10–60 cm; proximal hairs ± cobwebby, ± gray. **ST:** gen 1–5, erect to ascending; branches proximal, often also

distal. **LF**: basal, withering, and cauline, 3–8 cm; largest blades ± flat or not, 1(2)-pinnately lobed, scarcely fleshy; 1° lobes 2–7 pairs, tips flat, curled, twisted, or cylindric. **INFL**: heads 2–20+ per st; peduncle 1–4(10) cm; involucre obconic to ± hemispheric; longest phyllaries 5–7 mm, 1–2 mm wide, outer ± silky-cobwebby at least medially in fr, often also ± glandular-hairy. **FL**: inner corollas 4–6 mm. **FR**: 3–5.5 mm; inner fr pappus gen of 4 scales in 1 series, longest 2–4 mm, 0.4–0.7 × corolla. 2*n*=12. Open sandy slopes, openings in chaparral, woodland; 100–2300 m. NCoRI, SN (exc e SNH), GV, CW (exc CCo), SW (exc s ChI), sw edge D (exc DMtns); to n Baja CA. Feb–Jul

var. **heterocarpha** (Torr. & A. Gray) H.M. Hall INNER COAST RANGE PINCUSHION Pl 6–40 cm; proximal hairs cobwebby to tomentose, ± gray. **ST**: gen 1–5, erect to spreading; branches proximal, often also distal. **LF**: basal (often withering) and ± cauline, 1–6(10) cm; largest blades not flat, gen 2-pinnately lobed, scarcely fleshy; 1° lobes 2–7 pairs, tips ± curled and/or twisted. **INFL**: heads gen (1)2–5 per st; peduncle 2–7(9) cm; involucre widely cylindric to ± hemispheric; longest phyllaries 6–10 mm, 1.5–2.5 mm wide, outer ± silky, often becoming glabrous or ± glandular in fr. **FL**: inner corollas 4.5–7 mm. **FR**: 4–9 mm; inner fr pappus gen of 7–8 scales in 2 unequal series, longest gen 2–7 mm, 0.4–0.9(1) × corolla. 2*n*=12. Slopes, ridges, openings in chaparral, woodland, gen serpentine or shale; 100–1500(1900) m. s KR, NCoRH, NCoRI, CaR, n&c SNF, c SNH, ScV edges, CW (exc CCo), rare w WTR. Possibly a separate sp. Mar–Jun

var. **lanosa** (DC.) H.M. Hall SAND BUTTONS Pl 8–15(35) cm; proximal hairs tomentose to woolly, ± white. **ST**: gen 1–12, erect to decumbent; branches 0 or proximal. **LF**: ± basal, persistent, 2–10 cm; largest blades cylindric, entire, or ± flat, 1-pinnately lobed, scarcely fleshy; lobes 1–2(5) pairs, tips flat to cylindric. **INFL**: heads scapose, 1(3) per st; peduncle 8–20(30) cm; involucre obconic to ± hemispheric; longest phyllaries 6–8 mm, 1–2 mm wide, outer tomentose to woolly in fr. **FL**: inner corollas 5–6.5 mm. **FR**: 4–6 mm; pappus of 4 scales in 1 series, longest gen 4–6 mm, 0.8–0.9(1) × corolla. 2*n*=12. Open loose sand, gravel, often coastal dunes; 10–700(2100) m. Typical forms gen CW (exc SnFrB); intermediates gen n SW (exc s ChI), nw edge DSon. Forms called *C. glabriuscula* var. *denudata* (Nutt.) Munz intermediate toward *C. glabriuscula* var. *g.*, *C. glabriuscula* var. *megacephala*. Mar–Jul

var. **megacephala** A. Gray (p. 279) Pl 15–40 cm; proximal hairs cobwebby early, ± gray, gen becoming 0. **ST**: gen 1–5, erect to ascending; branches gen proximal. **LF**: basal, withering, and cauline, 1–8 cm; largest blades ± flat, gen 1-pinnately lobed, scarcely fleshy; lobes 2–7 pairs, tips flat to cylindric. **INFL**: heads gen 1–3 per st; peduncle 5–20 cm; involucre widely cylindric to ± hemispheric; longest phyllaries 7–9 mm, 2–3 mm wide, outer becoming glabrous in fr. **FL**: inner corollas 5–8 mm. **FR**: 5–8.5 mm; pappus of 4(5) scales in 1 series, longest gen 5–8 mm, ± 0.9(1) × corolla. Dry, often sandy slopes, openings in chaparral, woodland; 300–1500 m. s SNF, SnJV, SCoR, WTR. ± intermediate between *C. glabriuscula* var. *g.*, *C. glabriuscula* var. *heterocarpha*. Mar–Jun

var. **orcuttiana** (Greene) H.M. Hall ORCUTT'S PINCUSHION Pl 10–30 cm; proximal hairs cobwebby, ± gray. **ST**: gen 3–7(12), ± spreading; branches proximal and distal. **LF**: basal, withering, and cauline, gen 2–5 cm; largest blades ± flat or not, gen 2-pinnately lobed, fleshy; 1° lobes 2–7 pairs, tips ± flat to curled. **INFL**: heads gen 2–5 per st; peduncle 1–4 cm; involucre ± hemispheric; longest phyllaries 4.5–6.5(9) mm, 1–2 mm wide, outer gen shaggy- and glandular-hairy. **FL**: inner corollas 4.5–5.5 mm. **FR**: 4–5 mm; pappus of 4 scales in 1 series, longest gen 1–2.5 mm, 0.2–0.5 × corolla. 2*n*=12. Coastal dunes, bluffs; < 100 m. SCo; to nw Baja CA. Apr–Jul ★

C. macrantha D.C. Eaton (p. 279) MOJAVE PINCUSHION Ann 5–25(35) cm; proximal hairs cobwebby to tomentose, ± gray. **ST**: gen 1–5; branches gen proximal. **LF**: basal, withering, and cauline, 1.5–7 cm; largest blades ± elliptic to ovate, ± flat, 1(2)-pinnately lobed, not fleshy, ± gland-pitted under hairs; 1° lobes gen 2–5 pairs, tips ± flat. **INFL**: heads discoid (± radiant at night), gen 1–5(7) per st; peduncle nodding in bud, nonglandular; involucre widely cylindric to ± obconic; longest phyllaries 12–18 mm, outer ± tomentose, nonglandular, tip ± recurved, flexible, ± obtuse. **FL**: corolla white to ± pink

or cream, open at night, anthers ± incl; inner corollas 9–12(15) mm, 1.8–2.2 × fr; outer corollas radial, spreading at night, ± = inner. **FR**: 5–6(7) mm; pappus of 8 scales in 2 unequal series, outer << inner, longest 5–7 mm. 2*n*=12. Open, silty, often gravel-covered, calcareous or alkaline soils; 600–2200 m. n edge TR, s SNE, DMoj; to se OR, sw ID, se UT, c AZ. Nocturnal fls unique in genus. Mar–Jul

C. nevadensis (Kellogg) A. Gray (p. 279) SIERRA PINCUSHION Per, cespitose to ± matted, 2–10(12) cm; proximal hairs woolly, ± white. **ST**: gen 10–20+; branches 0, basal, or below ground. **LF**: ± basal, persistent, 2.5–5 cm; largest blades ovate to deltate, ± flat, (1)2-pinnately lobed, not fleshy, gland-pitted under hairs; 1° lobes 2–4 pairs, tips ± flat. **INFL**: heads scapose, discoid, 1(2) per st; involucre ± cylindric to obconic; longest phyllaries 9–12(14) mm, outer glandular-hairy, tip erect, ± stiff, obtuse. **FL**: corolla white to ± pink, 5.5–8 mm, outer radial, ± erect, = inner. **FR**: 5.5–7.5 mm; pappus of 10–16 scales in 3–4 ± equal series, longest 3–5 mm. 2*n*=12. Loose, gen volcanic or ultramafic sand, gravel, scree; 1900–3200 m. s KR (Bully Choop Mtn), s CaRH (Lassen Park area), n SNH; to w NV. Habit of s KR pls approaches *C. suffrutescens*. Jul–Sep

C. parishii A. Gray (p. 279) PARISH'S CHAENACTIS Subshrub (10)20–40(60) cm; proximal hairs (esp sts) ± felty, ± white, persistent. **ST**: gen 5–15+; branches proximal. **LF**: gen cauline, (1)2–5 cm; largest blades ± deltate, ± flat, 1-pinnately lobed, not fleshy, gland-pitted under hairs; lobes gen 2–5 pairs, tips ± flat. **INFL**: heads discoid, gen 1–3 per st; involucre ± obconic; longest phyllaries 10–13 mm, outer ± tomentose, not or scarcely glandular, tip recurved, flexible, ± obtuse. **FL**: corolla white to ± pink, 7–8.5 mm, outer radial, ± erect, = inner. **FR**: 6–8 mm; inner fr pappus of 10–16 scales in 3–4 ± equal series, longest 6–8 mm. 2*n*=12. Rocky to sandy openings in chaparral, woodland; 1300–2500 m. e PR (gen SnJt, Cuyamaca Peak area); to n Baja CA. May–Jul ★

C. santolinoides Greene (p. 279) SANTOLINA PINCUSHION Per, cespitose to ± matted, 10–25(35) cm; proximal hairs woolly, ± white to ± gray. **ST**: gen 5–15+; branches 0, basal, or below ground. **LF**: basal, persistent, (1)3–11 cm; largest blades cylindric to ± fusiform, not flat, 1–2-pinnately lobed, not fleshy, gland-pitted under hairs; 1° lobes (7)10–18+ pairs, ± overlapping, tips curled, twisted. **INFL**: heads scapose, discoid, 1(3) per st; involucre ± cylindric to obconic; longest phyllaries 8–13 mm, outer glandular-hairy, sometimes also ± cobwebby, tip erect, ± stiff, ± obtuse. **FL**: corolla white to ± pink, inner 5–7 mm, outer radial, ± erect, = inner. **FR**: 4–6 mm; pappus of 10–16 scales in 3–4 ± equal series, longest 3–4.5 mm. 2*n*=12. Open sandy to rocky ridges, talus, scree, road cuts; (1100)1500–2900 m. s SNH, Teh, s SCoRO, TR. Related to *C. douglasii* var. *alpina*. May–Aug

C. stevioides Hook. & Arn. (p. 279) DESERT PINCUSHION Ann 5–30(45) cm; proximal hairs ± cobwebby-silky, ± gray, persisting at least by fl time. **ST**: 1–12; branches proximal and/or distal. **LF**: basal (gen withering) and ± cauline, 1–8(10) cm; largest blades ± elliptic, not flat, gen 1–2-pinnately lobed, rarely fleshy, nonglandular; 1° lobes 4–8 pairs, tips curled and/or twisted. **INFL**: heads radiant, gen 3–20+ per st, peduncle distally gen glandular-hairy, often also ± cobwebby; involucre obconic to ± hemispheric; longest phyllaries 5.5–8(10) mm, outer glandular-hairy and/or ± cobwebby in fr, tip erect, stiff, ± obtuse. **FL**: corolla white to ± pink or pale yellow, inner 4.5–6.5 mm, outer bilateral, spreading, > inner. **FR**: (3)4–6.5 mm; inner fr pappus of 4 scales in 1 series (sometimes a 5th outer scale << inner), longest 1.5–6 mm, tips hidden among buds. 2*n*=10. Abundant in higher DMoj, SNE. Open sandy or gravelly flats, slopes; -30–2100(2400) m. s SNH, Teh, s SnJV, SCoRI, n edge TR, GB, D; to se OR, sw WY, w CO, sw NM, nw Mex. Feb–Jun

C. suffrutescens A. Gray (p. 279) SHASTA CHAENACTIS Gen subshrub, sometimes cespitose, (10)25–45(60) cm; proximal hairs (esp sts) ± felty, ± white, persistent. **ST**: gen 5–15+; branches proximal. **LF**: gen ± cauline, sometimes ± basal, 5–10 cm; largest blades ± deltate, ± flat, 1–2-pinnately lobed, not fleshy, gland-pitted under hairs; 1° lobes gen 2–5 pairs, tips ± flat. **INFL**: heads discoid, gen 1–3 per st; involucre ± cylindric; longest phyllaries 14–18 mm, outer gen glandular-hairy, other hairs 0 or scarce, tip recurved, flexible, ± obtuse. **FL**: corolla white to ± pink, 8.5–10 mm, outer radial, ±

erect, = inner. **FR**: 7–9 mm; inner fr pappus of 10–16 scales in 3–4 ± equal series, longest 7–9 mm. 2*n*=12. Unstable, sandy to rocky, gen serpentine soils, scree, drainages; 700–2300 m. e KR. Habit of small pls approaches *C. nevadensis*. May–Aug ★

C. xantiana A. Gray (p. 279) FLESHY PINCUSHION Ann 10–40 cm; proximal hairs ± cobwebby early, ± gray, gen becoming 0 by fl time. **ST**: gen 1–5(12); branches proximal and/or distal. **LF**: basal (withering) and cauline, (1)2–6 cm; largest blades linear, cylindric, entire, or ± elliptic, ± flat, 1-pinnately lobed, ± fleshy, nonglandular;

lobes gen 1–2(5) pairs, tips ± cylindric. **INFL**: heads discoid, gen 1–5(7) per st, peduncle erect, distally ± expanded, glabrous; involucre widely obconic to bell-shaped; longest phyllaries 10–18 mm, outer nonglandular, proximally glabrous, distally densely puberulent, tip gen recurved, flexible, ± obtuse. **FL**: corolla dirty white to ± pink; inner corolla 6–10 mm, ± = fr; outer corolla radial, scarcely spreading, ± = inner. **FR**: 5–9 mm; pappus of 8 scales in 2 unequal series, outer << inner, longest 5–9 mm. 2*n*=14. Gen open loose sand, often in burns; (100)300–2600 m. s SN, Teh, s SnJV, se SCoRO, SCoRI, n TR, GB, w DMoj; to se OR, w NV, nw AZ. Mar–Jul

CHAETADELPHA

L.D. Gottlieb

1 sp.: w US. (Greek: bristles, sister, alluding to fusion of pappus parts) [Gottlieb 2006 FNANM 19:368]

C. wheeleri S. Watson (p. 287) WHEELER'S DUNE-BROOM Per from stout rhizome, glabrous; sap milky. **ST**: 2–many, ascending to ± erect, much-branched, 1.5–4 dm. **LF**: alternate, basal 1–5 cm, linear to lance-linear, entire, withered at fl; cauline linear or reduced to scales. **INFL**: heads liguliflorous, gen 1 at st and branch tips; involucre 11–14 mm, narrowly cylindric; phyllaries in 2 series, outer scale-like, inner 5, equal, linear, margin membranous; receptacle ±

flat, epaleate. **FL**: 5; ligules pale lavender to white, gen not readily withering. **FR**: 8–12 mm, cylindric, 5 faces separated by ridges, smooth, glabrous, light tan; pappus persistent, of 5 stiff awns and 35–50+ shorter bristles in 1–2 series, all ± fused basally in groups that alternate with awns. 2*n*=18. Sand dunes, alkali flats, creosote-bush and sagebrush scrub; 800–1800 m. SNE, n DMoj; se OR, w NV. May–Sep ★

CHAETOPAPPA

David J. Keil & Geraldine A. Allen

Per to subshrub [ann]. **LF**: basal and cauline, alternate, linear to oblanceolate, entire. **INFL**: heads radiate, 1; phyllaries graduated in 2–6 series, margins white-scarious; receptacle ± flat, epaleate. **RAY FL**: 5–24; ray coiled at maturity, white to pink-purple or pale blue. **DISK FL**: 5–many; corolla yellow; anther tip triangular; style tips triangular. **FR**: gen linear-fusiform, ± compressed or not; pappus 0 or of scales, bristles, or both. ± 10 spp.: sw US, n Mex. (Greek: bristle-pappus) [Nesom 2006 FNANM 20:206–209]

C. ericoides (Torr.) G.L. Nesom (p. 287) ROSE-HEATH Bushy per to subshrub from rhizome and ± woody caudex; herbage minutely stalked-glandular, or nonglandular and strigose. **ST**: gen several from base, 5–15(20) cm, slender, ascending to erect, gen branched, lfy throughout. **LF**: 4–12 mm, linear to obovate, obtuse to abruptly

pointed. **INFL**: phyllaries lanceolate to oblong, acute to acuminate, green, ± purple-tipped. **RAY FL**: gen 12–21; corolla 5–10 mm, white to ± pink. **FR**: brown, hairy; pappus of ± 25 minutely barbed bristles in 1 series, ± white. 2*n*=16,32. Dry slopes, desert woodland, blackbush scrub; 1150–2900 m. W&I, DMtns; to WY, NE, TX, n Mex. Apr–Sep

CHAMAEMELUM

Linda E. Watson

Ann [per], aromatic; herbage ± glabrous, puberulent, or long-soft-hairy to silvery-strigose. **ST**: gen 1, erect or ascending [prostrate], gen branched. **LF**: mostly cauline at fl, alternate, petioled or sessile, elliptic to ovate, [oblong or spoon-shaped], 1–3-pinnately lobed, ultimate lobes thread-like to linear, or narrowly spoon-shaped, entire, abruptly short-pointed. **INFL**: heads radiate [discoid], 1 or in loose ± flat-topped clusters; involucre hemispheric to saucer-shaped; phyllaries persistent, 22–45+ in 3–4+ series, sometimes reflexed in fr, margins and tips scarious; receptacle hemispheric to conic, paleate; paleae weakly folded to ± flat. **RAY FL**: [0]12–15+, pistillate [sterile]; corolla white, ray oblong, often persistent, reflexed in fr. **DISK FL**: 100–200+; corolla yellow, tube proximally sac-like, weakly clasping ovary top, > bell-shaped throat, lobes deltate; anther tip ovate; style tips truncate, brush-like. **FR**: ± obovoid, weakly compressed front-to-back, with 2 weak lateral veins and 1 adaxial vein, faces finely striate, glabrous; pappus 0. 6 spp.: Eur, n Afr. (Greek: ground + melon) [Watson 2006 FNANM 19:496] *C. nobile* (L.) All., historic waif, apparently not naturalized in CA.

C. fuscatum (Brot.) Vasc. CHAMOMILE Pl 5–20(35+) cm. **ST**: glabrous or puberulent. **LF**: proximal petioled, 1–4 cm, ± ovate to elliptic, gen 2-pinnately divided; distal sessile, 1–2 cm, ± elliptic, entire to pinnately divided. **INFL**: involucre 6–10 mm diam, phylla-

ries 3–4 mm, ± glabrous, tip and margins brown; paleae 2–3 mm, tips brown. **RAY FL**: ray 8–15 mm. **DISK FL**: corolla 2.5–3 mm. **FR**: ray and disk fr equal, 1–1.3 mm. 2*n*=18. Disturbed sites, grassland; < 100 m. s NCoRO (Sonoma Co.), ScV; native to Medit. Mar–Apr

CHLORACANTHA SPINY ASTER

David J. Keil & Guy L. Nesom

1 sp.: s US, Mex, C.Am. (Latin: green thorns) [Sundberg & Nesom 2006 FNANM 20:358]

C. spinosa (Benth.) G.L. Nesom var. ***spinosa*** (p. 287) Per or subshrub from stout rhizome, ± glabrous. **ST**: erect, 5–15(25) dm, sometimes ± fleshy, armed with thorns or unarmed. **LF**: alternate, 10–50 mm, oblanceolate, gen entire, 1-veined, early-deciduous.

INFL: heads radiate, 1 or in loose raceme-like clusters; phyllaries graduated in 4–5 series, inner 4.5–7.5 mm, ± lanceolate, veins (1)3(5), parallel, tips gen rounded, margins transparent; receptacle flat or slightly convex, epaleate. **RAY FL**: 20–33; ray 3.5–5(7) mm,

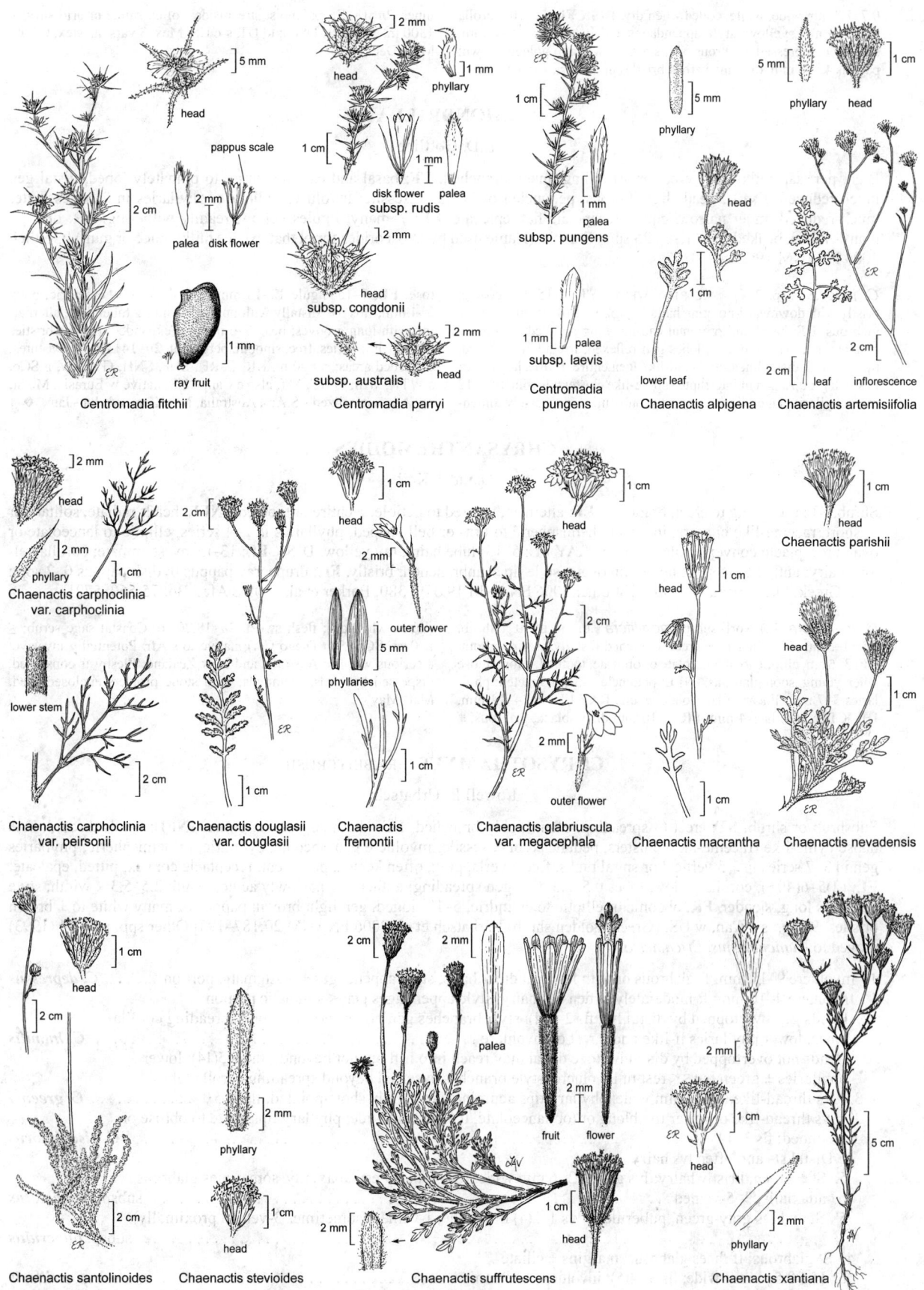

Centromadia fitchii

head

pappus scale

palea   disk flower

ray fruit

2 cm

5 mm

2 mm

1 mm

Centromadia parryi

head
subsp. rudis

disk flower   palea

head
subsp. congdonii

head
subsp. australis

2 mm

1 mm

2 mm

2 mm

phyllary

subsp. pungens

palea

palea
subsp. laevis

Centromadia
pungens

1 cm

1 mm

1 mm

5 mm

phyllary   head

head

lower leaf

leaf   inflorescence

Chaenactis alpigena

Chaenactis artemisiifolia

5 mm

1 cm

1 cm

2 cm

Chaenactis carphoclinia
var. carphoclinia

head

phyllary

lower stem

Chaenactis carphoclinia
var. peirsonii

2 mm

2 mm

1 cm

2 cm

Chaenactis douglasii
var. douglasii

2 cm

1 cm

Chaenactis
fremontii

head

phyllaries

outer flower

1 cm

2 mm

5 mm

1 cm

Chaenactis glabriuscula
var. megacephala

head

outer flower

1 cm

2 cm

2 mm

Chaenactis macrantha

head

1 cm

1 cm

Chaenactis parishii

head

head

Chaenactis nevadensis

1 cm

1 cm

1 cm

head

Chaenactis santolinoides

2 cm

1 cm

2 cm

Chaenactis stevioides

phyllary

head

2 mm

1 cm

Chaenactis suffrutescens

palea

fruit   flower

head

2 cm

2 mm

2 mm

1 cm

2 mm

Chaenactis xantiana

head

phyllary

2 mm

1 cm

5 cm

2 mm

0.7–1.3 mm wide, white, coiled when dry. **DISK FL:** 25–70; corolla gen 4–5 mm, yellow; style appendages ± deltate. **FR:** 1.5–2 mm, slightly compressed, glabrous; veins 5(6), ± white to golden-brown; pappus 4.5–6 mm, of many barbed bristles in ± 2 series, outer gen << inner. $2n$=18. Seeps, moist streamsides, often saline or arid sites; < 1300 m. SCo, PR, DSon; to UT, s-c US, Mex. 3 vars. in Mex, C.Am. May–Dec

CHONDRILLA

L.D. Gottlieb

Bien, per; sap milky. **ST:** erect or ascending, much-branched. **LF:** basal and cauline, entire to pinnately lobed, distal gen much-reduced. **INFL:** heads liguliflorous, peduncled or lateral ± sessile; involucre cylindric; phyllaries in 2 series, outer much-reduced, inner narrow, equal; receptacle flat, epaleate. **FL:** 7–many; ligules yellow, readily withering. **FR:** oblong, many-ribbed, beaked [beakless]. 25 spp.: Eurasia. (Name used by Dioscorides for pl that exudes milky juice or gum) [Gottlieb 2006 FNANM 19:252–253]

C. juncea L. (p. 287) SKELETON WEED   **ST:** 4–15 dm; coarse, bristly with downward-pointing hairs in proximal 10–15 cm; distally glabrous. **LF:** basal and proximal cauline wing-petioled, oblong to obovate, occ withered at fl, lobes gen reflexed, irregularly toothed, tips pointed; distal linear to thread-like, gen entire. **INFL:** heads 1 or in small groups, gen in interrupted spike-like clusters; involucre 9–12 mm; phyllaries lance-linear, glabrous, puberulent, or sparsely tomen- tose. **FL:** 7–15; ligule 12–18 mm. **FR:** 8–10 mm, cylindric, body 3–4 mm, ribs 5+ (distally with upward-pointing tubercles), alternat- ing with long grooves; beak 5–6 mm, tip expanded; pappus bristles 40–50+ in 1 series, free, smooth, persistent. $2n$=14+1B, 30. Pastures, disturbed areas; < 600 m. KR, CaRH, SNF, SNH, GV, CW, n SCo, s WTR, Wrn; to BC, MT, also e Can, e US; native w Eurasia, Medit, nw Afr; naturalized s S.Am, Australia, New Zealand. Jun–Jan ◆

CHRYSANTHEMOIDES

David J. Keil

Shrub. **ST:** ascending to erect, branched. **LF:** alternate, tapered to petiole, ± entire or toothed. **INFL:** heads radiate, solitary or in short, raceme-like clusters; involucre hemispheric to urn- or bell-shaped; phyllaries in 2–4 series, elliptic to lanceolate or ovate; receptacle convex, epaleate, hairy. **RAY FL:** 5–13; tube hairy, ray yellow. **DISK FL:** 13–many, staminate; corolla yel- low, hairy; anther base short-tailed, tip ovate; style tip ± unbranched, bristly. **FR:** drupe, ray pappus 0; disk pappus 0. 2 spp.: Afr. (Greek: like *Chrysanthemum*) [Strother 2006 FNANM 19:379–380; Barker et al. 2009 S Afr J Bot 75:560–572]

C. monilifera (L.) Norl. subsp. *monilifera* BITOU BUSH   Shrub. **ST:** 1–3 m, ascending to ± erect, branched distally or throughout. **LF:** 3–6 cm, elliptic to oblanceolate or obovate, toothed, ± tomentose when young, soon glabrous. **INFL:** peduncle 1–3 cm, bracted; phyl- laries 3–7 mm, linear to lanceolate, acute. **RAY FL:** ray 4–14 mm. **DISK FL:** corolla ± 4 mm. **FR:** 6–10 mm diam, black, glabrous, ± ribbed when dry; flesh sweet. $2n$=18,20,36. Coastal sage scrub; ≤ 200 m. SCo (San Diego Co.); native to s Afr. Potentially invasive; a serious weed in Australia and New Zealand. Fleshy fr consumed, dispersed by birds, mammals; hard stone protects enclosed seed. Mar–May

CHRYSOTHAMNUS RABBITBRUSH

Lowell E. Urbatsch

Subshrub or shrub. **ST:** erect to spreading, often highly branched. **LF:** alternate, sessile, entire. **INFL:** heads discoid, in ± dense cyme-like (raceme-like) clusters, peduncled or ± sessile; involucre gen obconic, cylindric, or hemispheric; phyllaries gen in 3–7 series in ± 5 vertical or spiral ranks, free, overlapping, often keeled, persistent; receptacle convex, pitted, epaleate. **FL:** (2)5–6(40+); corolla yellow, lobes 0.5–2.3 mm, gen spreading; anther tips narrowly acute, length 2.5–5.3 × width; style branches long, slender. **FR:** obconic or elliptic to cylindric, 5–10 ridged, gen light brown; pappus of many white to ± brown bristles. 9 spp.: sw Can, w US. (Greek: golden shrub) [Urbatsch et al. 2006 FNANM 20:187–193] Other spp. in TJM (1993) moved to *Cuniculotinus, Ericameria*.

1. Involucre 9–15 mm; fr glabrous or with few glandular hairs; style appendages ± = stigmatic portion . . . . . . *C. depressus*
1′ Involucre 4–10 mm; fr moderately to densely hairy; style appendages gen < stigmatic portion
  2. Heads gen overtopped by distal lvs; fls 2–3(4); style branches gen not exserted beyond spreading corolla lobes; lower phyllaries lf-like and ≥ rest of involucre . . . . *C. humilis*
  2′ Heads not overtopped by distal lvs (occ distal may reach into infl but not beyond); fls 3–5(14); lower phyllaries ± green but << rest of involucre; style branches exserted beyond spreading corolla lobes.
    3. Lvs thread-like, 0.5–2 mm wide; phyllary tips acuminate to rigidly short-pointed; fls 4–5 . . . . *C. greenei*
    3′ Lvs thread-like or linear to oblong or (ob)lanceolate, 0.5–10 mm wide; phyllary tips acute to obtuse or rounded; fls 3–14 . . . . *C. viscidiflorus*
      4. Distal sts and often lvs hairy
        5. St ± green, bristly hairy; lvs green, 2–6 mm wide, rough-hairy esp abaxially, sometimes glabrous adaxially, 3–5-veined . . . . subsp. *lanceolatus*
        5′ St and lvs gray-green, puberulent; lvs 1–2(4) mm wide; 1-veined (sometimes 3-veined proximally) . . . . subsp. *puberulus*
      4′ St glabrous; lf faces glabrous, margins ± ciliate
        6. Lvs 0.5–1 mm wide; fls 3–4(5); involucre ± obconic . . . . subsp. *axillaris*
        6′ Lvs 1–10 mm wide; fls 4–14; involucre narrowly cylindric . . . . subsp. *viscidiflorus*

C. depressus Nutt. (p. 287)   Subshrub 1–5 dm; caudex much-branched. **ST:** unbranched proximal to infl, brittle, rough, ± gray-hairy. **LF:** 1–3 cm, oblanceolate, 1-veined, flat, rough-hairy. **INFL:** heads in small clusters; involucre 9–15 mm, ± cylindric; phyllaries gen in 5–6 well-defined vertical ranks, strongly keeled, acute to acuminate, straw-colored or ± purple. **FL:** 5–6; corolla 7–11 mm; style branches exserted, appendage gen ± = stigmatic portion. **FR:** 5–6.5 mm, ± cylindric, glabrous or with few glandular hairs; pappus ± = corolla. 2*n*=18. Rocky crevices, sagebrush, pinyon/juniper woodland; 1000–2100 m. DMtns; to CO, NM. Aug–Oct

C. greenei (A. Gray) Greene (p. 287)   GREENE'S RABBITBRUSH   Shrub 1–5 dm; caudex moderately branched. **ST:** gen glabrous. **LF:** 1–4 cm, 0.5–2 mm wide, thread-like, not twisted, sometimes reflexed, veins not evident, faces glabrous, margins often minutely ciliate. **INFL:** involucre 5–8 mm, cylindric to obconic; phyllaries elliptic to ovate, vertically ranked, weakly keeled, tips (at least outer) gen ascending, long-tapered, acuminate to stiffly short-pointed, sticky. **FL:** 4–5; corolla 4–5.5 mm; style branches exserted, appendage < stigma. **FR:** 3–4 mm, top-shaped, hairy; pappus < corolla. 2*n*=18. Sandy washes, sagebrush, saltbush scrub; 1340–1830 m. SNE (Fish Lake Valley), n DMtns (Cottonwood Mtns); to WY, CO, NM. Oct ★

C. humilis Greene (p. 287)   Subshrub 1–3 dm; caudex branched. **ST:** lfy, gen covered with spreading conic hairs. **LF:** 2–3 cm, ± oblanceolate, 1-veined, flat, often covered with spreading conic hairs. **INFL:** heads in small clusters gen overtopped by lvs; involucre 6–10 mm, cylindric to obconic; phyllaries weakly keeled and ranked, outermost often green, covered with spreading conic hairs, lf-like, gen ≥ length of rest of involucre, inner ovate-oblong, straw-colored, not very sticky, tips acute or gen obtuse to rounded, margins irregularly cut or toothed. **FL:** 2–3(4); corolla (5)7–8 mm, pale yellow, lobes erect; style branches ± incl, appendage < stigmatic portion. **FR:** 4.5–7 mm, top-shaped, moderately strigose; pappus ± = corolla. 2*n*=18. Uncommon. Sagebrush grassland; 1400–3040 m. ne SNH, MP; to WA, ID, NV. [*C. viscidiflorus* subsp. *h.* (Greene) H.M. Hall & Clem.] Jul–Aug

C. viscidiflorus (Hook.) Nutt.   YELLOW RABBITBRUSH   Shrub 1–15 dm; caudex branched. **ST:** gen erect to spreading, brittle, gla-

brous to pubescent, green when young, becoming ± white. **LF:** 1–7.5 cm, 0.5–10 mm wide, thread-like or linear to oblong or (ob)lanceolate, 1–5-veined, flat or twisted, often wavy-margined, (± gray-) green, ± sticky. **INFL:** heads in flat-topped or rounded clusters; involucre 5–10 mm, gen cylindric to obconic; phyllaries gen ± lanceolate, in ± 5 vertical ranks, keeled, yellow-green, ± sticky, tips acute to obtuse or rounded. **FL:** 3–5(14); corolla 3.5–7.5 mm; style branches exserted, appendage gen < stigma. **FR:** 3–5 mm, obconic, hairy; pappus ± = corolla. 2*n*=18,36,54. Highly variable; 5 subspp. in w N.Am.

subsp. ***axillaris*** (D.D. Keck) L.C. Anderson   **ST:** glabrous. **LF:** 1–3 cm, 0.5–1 mm wide, thread-like, glabrous, margins ± ciliate. **INFL:** involucre ± obconic. **FL:** 3–4(5); corolla 3.5–4.5 mm. 2*n*=18. Gravelly washes, sagebrush, pinyon/juniper woodland; 1300–2000 m. s W&I, ne DMtns; to w CO, n AZ. Jul–Oct

subsp. ***lanceolatus*** (Nutt.) H.M. Hall & Clem.   **ST:** bristly near infl with ± spreading, conic hairs. **LF:** 1.5–4.5 cm, 2–6 mm wide, ± lanceolate, rough-hairy esp abaxially, 3–5-veined. **FL:** gen 5; corolla 5–6 mm. 2*n*=18,36. Uncommon. Juniper/sagebrush scrub; 1200–2500 m. c SNH, MP; to BC, MT, SD, CO, NM. Intergrades with *C. viscidiflorus* subsp. *v.* esp in n GB. Jul–Oct

subsp. ***puberulus*** (D.C. Eaton) H.M. Hall & Clem. (p. 287)   Pl gray-green, densely puberulent with spreading, conic hairs. **LF:** 1–4 cm, gen 1–2(4) mm wide, thread-like to ± oblanceolate, puberulent, 1(3)-veined. **INFL:** involucre narrowly cylindric. **FL:** gen 5 (4–7); corolla 4.5–6 mm. 2*n*=18,36 (tetraploids larger, lower elevations). Sagebrush scrub, pinyon/juniper woodland, subalpine slopes; 1500–3000 m. SN (e slope), SnBr, GB, DMtns; to OR, s ID, UT, n AZ. Jul–Oct

subsp. ***viscidiflorus*** (p. 287)   **ST:** glabrous. **LF:** 0.5–7.5 cm, 1–10 mm wide, linear to ± lanceolate, gen twisted, glabrous, margins ± ciliate. **INFL:** involucre narrowly cylindric. **FL:** 4–14; corolla 4–7.5 mm. 2*n*=18,36,54. Common. Sagebrush, pinyon/juniper, alpine talus; 900–4000 m. KR, CaR, SN, n WTR, SnBr, GB, s DMtns; to WA, MT, SD, NB, CO, NM. [*C. v.* subsp. *latifolius* (D.C. Eaton) H.M. Hall & Clem.; *C. v.* subsp. *pumilus* (Nutt.) H.M. Hall & Clem.; *C. v.* subsp. *stenophyllus* (A. Gray) H.M. Hall & Clem.] Jul–Sep

CICHORIUM   CHICORY

David J. Keil & G. Ledyard Stebbins

Per [ann, bien]; sap milky. **ST:** 3–10+ dm, branched. **LF:** basal and cauline; proximal 1–2 dm, toothed or pinnately lobed, wing-petioled; middle sessile, sometimes clasping; distal greatly reduced. **INFL:** heads liguliflorous, showy, terminal and axillary, lateral sessile; involucre ± cylindric; phyllaries in 2 series, hardened in basal 1/2, outer short, spreading, inner elongate, erect; receptacle ± flat, epaleate, minutely scaly. **FL:** 10–25; ligules blue to purple, readily withering. **FR:** oblong, glabrous, 5-angled; pappus of minute blunt scales. 6 spp.: Eur, Medit, Afr. (Old Arabic name) [Strother 2006 FNANM 19:221–222] *C. endivia* L. a waif in CA.

C. intybus L. (p. 287)   Per 4–20 dm, from deep, woody taproot; herbage glabrous to short-bristly, esp near base. **ST:** erect; branches stout, ascending. **LF:** proximal oblong to elliptic in outline, ± entire to coarsely pinnately lobed. **INFL:** peduncle of terminal heads not thickened. **FL:** ligule blue (pink, white); anthers gen dark blue. **FR:** 1.5–2.5 mm; pappus scales < 0.5 mm. 2*n*=18. Common. Roadsides, disturbed places; < 1500 m. CA-FP, n DMtns (Panamint Range); widespread in N.Am; native to Eur. Roasted roots used as coffee flavoring or substitute. Apr–Oct

CIRSIUM   THISTLE

Taprooted ann, bien, or short-lived per that fls once, or multi-fl per with taprooted rosettes arising from runner roots or from simple to branched caudex; glabrous to cobwebby or ± densely tomentose with long, fine, slender hairs, sometimes with thicker multicellular, jointed hairs that often appear crinkled, shining, iridescent when dry. **ST:** gen erect. **LF:** basal and proximal cauline gen tapered or ± wing-petioled, gen wavy-margined, dentate to gen pinnately lobed and ± dentate, lobes and teeth spine-tipped, gen spiny-ciliate, faces glabrous to tomentose, esp abaxially; distal gen sessile, ± reduced. **INFL:** heads discoid, 1–many, center head of cluster gen larger, gen erect; involucre ± cylindric to ovoid, spheric, or bell-shaped, persistent when dry; phyllaries many, graduated in 5–20 series, gen entire (spiny-ciliate with irregularly toothed or cut scarious margin or distal appendage), outer and middle gen spine-tipped, in some spp. midrib with sticky-resinous ridge (milky when fresh, dark when dry, occ very narrow); inner phyllaries gen narrow, flat, tips straight or twisted; receptacle flat, long-bristly, epaleate. **FL:** ± many, gen bisexual (unisexual in *Cirsium arvense*); corolla ± radial, white to red or purple, tube long, narrowly cylindric, throat cylindric, lobes linear; anther tube colored same as corolla or not, anther base sharply sagittate, tip linear or oblong;

style gen exserted, tip cylindric, branches very short. **FR**: ovoid, thick or ± compressed, straw-colored or tan to dark brown, glabrous; attachment scar slightly angled; pappus bristles many, ± flattened proximally, plumose, weakly fused at base, often deciduous in ring, white to brown. ± 200 spp.: N.Am, Eurasia. (Greek: thistle) [Keil 2006 FNANM 19:95–164] Taxa difficult, variable, incompletely differentiated, hybridize. Exceptional white-fld pls occur in most taxa with pigmented corollas; these gen not treated in key.

1. Adaxial lf face with slender appressed bristle-like prickles . . . . *C. vulgare*
1′ Adaxial lf face without prickles
  2. Pls dioecious or nearly so . . . . *C. arvense*
  2′ Pls with bisexual fls
   3. Corolla lobes 1–2+ × tube
    4. Corolla pink-purple to red-purple, red-pink, or red; heads gen erect; lf adaxially tomentose (occ ±
      glabrous in age) . . . . *C. arizonicum*
     5. Corolla bright red or ± red-pink; corolla throat 2–5 mm, lobes 13–18 mm; SNH, W&I . . . . var. *arizonicum*
     5′ Corolla pink- to ± red-purple; corolla throat 5–8 mm, lobes 10–14 mm; e DMtns . . . . var. *tenuisectum*
    4′ Corolla white to pink or lavender; heads gen nodding; lf adaxially densely velvety, with short, erect,
      jointed hairs, also cobwebby — serpentine wetlands; SnFrB, SCoRO . . . . *C. fontinale*
     6. Fr 4.1–5 mm; phyllary spines 1–2 mm . . . . var. *fontinale*
     6′ Fr 3.4–4.1 mm; phyllary spines gen 2–6 mm
      7. Longer spines of cauline lvs 10–18 mm; most phyllaries without marginal spines . . . . var. *campylon*
      7′ Longer spines of cauline lvs 4–7 mm; many outer phyllaries with marginal spines . . . . var. *obispoense*
   3′ Corolla lobes < tube
    8. Abaxial face of outer and middle phyllaries with long, sticky-resinous ridge (milky when fresh, dark
      when dry, occ very narrow), occ poorly developed or 0 in *Cirsium mohavense*
     9. Margins of phyllaries minutely spiny-fringed . . . . *C. hydrophilum*
      10. Heads gen 2.5–3 cm; fr ± 5 mm, oblong, tapered near base; tidal marshes; deltaic GV (Suisun
        Marsh, Solano Co.) . . . . var. *hydrophilum*
      10′ Heads gen 3–3.5 cm; fr 4–5 mm, oblong or elliptic; serpentine springs; n SnFrB (Mount Tamalpais,
        Marin Co.) . . . . var. *vaseyi*
     9′ Margins of phyllaries entire
      11. St evidently coarse-hairy with multicellular hairs, also cobwebby to densely tomentose with long,
        fine hairs; cauline lvs gen ± clasping, with expanded basal lobes; lf veins gen prominently raised
        abaxially . . . . *C. cymosum*
       12. Larger heads 15–25 mm diam; outer phyllaries gen << inner, sticky-resinous ridge prominent, well
         developed, appearing dark brown on dry specimens . . . . var. *canovirens*
       12′ Larger heads 20–35 mm diam; outer phyllaries long, often nearly = inner, sticky-resinous ridge
         weakly developed . . . . [2]var. *cymosum*
      11′ St with few or 0 coarse, multicellular hairs, thinly to densely cobwebby or tomentose with long
        fine hairs; cauline lvs gen not clasping with expanded basal lobes; lf veins gen not prominently
        raised abaxially
       13. Involucre 2.5–4.5 cm
        14. Mid and distal lvs not or scarcely decurrent . . . . *C. undulatum*
        14′ Mid and distal lvs evidently decurrent as spiny wings to 5 cm
         15. Phyllary spines gen 2–4 mm; corolla ± white . . . . *C. canescens*
         15′ Phyllary spines gen 5–12 mm; corolla white to pale lavender or pink . . . *C. ochrocentrum* var. *ochrocentrum*
       13′ Involucre 1.5–2.5(3) cm
        16. Rosette arising from runner roots; lf adaxially green, thinly gray-tomentose — n KR . . . . *C. ciliolatum*
        16′ Rosette arising from taproot or woody caudex; lf adaxially gray-tomentose or becoming ± green
         17. Phyllaries ± loosely appressed, gen ascending to spreading above appressed base
          18. Phyllary tip-spine gen 1–3 mm; SCoR and c SNF to mtns of SW . . . . [4]*C. occidentale* var. *californicum*
          18′ Phyllary tip-spine gen 3–10 mm
           19. Involucre, incl phyllary tips gen ovoid to ± spheric; tip-spines of phyllaries gen 3–7 mm; outer
            phyllaries gen spreading or ascending — SNH, GB . . . . [2]*C. inamoenum* var. *inamoenum*
           19′ Involucre, incl phyllary tips often wider than long; tip-spines of phyllaries gen 5–10 mm;
            outer phyllaries often reflexed . . . . [2]*C. neomexicanum*
        17′ Phyllaries tightly appressed, only tip-spines spreading
         20. Corolla ± white to pink or pale lavender; involucre pale green or straw-colored
          21. Distal lvs gen conspicuously decurrent as spiny wings; wet soils, streams, springs; SNE, DMoj
           . . . . *C. mohavense*
          21′ Distal lvs not or only weakly decurrent; dry sites; SCoR and c SNF to mtns of SW
           . . . . [4]*C. occidentale* var. *californicum*
         20′ Corolla gen ± dark rose-purple (white); involucre gen dark green or ± purple — n CA-FP, MP
          . . . . *C. douglasii*
          22. Larger spines of distal cauline lvs gen ≤ 7 mm . . . . var. *breweri*
          22′ Larger spines of distal cauline lvs gen 7–20 mm . . . . var. *douglasii*

8′ Abaxial face of outer and middle phyllaries without sticky-resinous ridge
  23. Margins of outer phyllaries irregularly toothed to spiny-ciliate
    24. Lvs densely gray- or white-felty-tomentose on faces — coastal dunes, headlands; s CCo . . . . . ***C. rhothophilum***
    24′ Lvs adaxially green and glabrous to ± gray-hairy with multicellular hairs and/or finely cobwebby, abaxially sometimes densely gray- or white-tomentose
      25. Corolla 30–45 mm, ± bright rose-purple . . . . [2]***C. andersonii***
      25′ Corolla 17–28 mm, white or cream to purple
        26. Heads gen not closely subtended by well-developed distal lvs or clustered lfy bracts; phyllary tips flattened, gen ascending, connected side-to-side by cobwebby hairs or not
          27. Pl stout, gen 10–30 dm; st 2–10 cm diam near base, evidently hollow; middle and inner phyllaries entire — SnJV . . . . ***C. crassicaule***
          27′ Pl slender, gen 1–11 dm; sts gen < 2 cm diam near base; middle and inner phyllaries irregularly toothed or spiny-ciliate . . . . ***C. remotifolium***
            28. Phyllary margins irregularly toothed or cut . . . . var. ***odontolepis***
            28′ Phyllary margins ciliate with minute spreading to recurved spines . . . . var. ***rivulare***
        26′ Heads gen closely subtended by ± well-developed distal lvs or clustered bract-like lvs; phyllary tips needle-like, ascending to radially spreading, connected side-to-side by long, cobwebby, or stouter, multicellular hairs
          29. Corolla lobes linear, tips not knob-like; style conspicuously exserted; st cobwebby, ± glabrous in age, coarser, multicellular hairs 0 . . . . ***C. andrewsii***
          29′ Corolla lobes thread-like, tips knob-like; style incl or exserted 1–2 mm; st evidently hairy with ± coarse multicellular hairs, often cobwebby to densely tomentose with long, fine hairs . . . . . [2]***C. brevistylum***
  23′ Margins of outer phyllaries ± entire
    30. Outer and middle phyllaries lance-linear to ovate, appressed; tip-spines 1–12 mm, gen ascending to erect
      31. Pl erect; st openly branched distally — bien; dry sites; SCoR and c SNF to mtns of SW . . . . [4]***C. occidentale*** var. ***californicum***
      31′ Pl gen ± compact, stless or low-growing and branched ± throughout (if erect, gen distally unbranched)
        32. Per from runner roots; gen from dry sites; NCo, NCoRO, CW . . . . ***C. quercetorum***
        32′ Bien or per that fls once, then dies, taprooted; gen from moist sites; widespread . . . . ***C. scariosum***
          33. Corolla pink to purple; SNH, SnBr, SNE . . . . var. ***congdonii***
          33′ Corolla white or faintly pink- or lilac-tinged
            34. Abaxial lf face gen gray-tomentose; CaRH, SNH, Wrn . . . . var. ***americanum***
            34′ Abaxial lf face gen ± glabrous (occ tomentose in *Cirsium scariosum* var. *citrinum*)
              35. Pl gen stless, or stless and occ short-stemmed individuals in population
                36. Main lf spines gen 2–7(10) mm; Teh, WTR, PR . . . . [2]var. ***citrinum***
                36′ Main lf spines gen (3)7–10(12) mm; s CCo (sw San Luis Obispo, nw Santa Barbara cos.) . . . . [2]var. ***loncholepis***
              35′ Pl gen with evident branched or unbranched st, or occ stless individuals present in population
                37. St gen erect, proximally unbranched; heads gen ± tightly clustered at st tip; corolla 20–28 mm . . . . var. ***scariosum***
                37′ St gen branched proximally or throughout, pl often growing as low, rounded mound; most heads on short to ± long lateral branches; corolla 26–36 mm
                  38. Inner phyllary tips gen expanded as scarious, irregularly toothed appendage — KR, nw MP . . . . var. ***robustum***
                  38′ Inner phyllary tips acuminate, entire or rarely toothed
                    39. Main lf spines gen 2–7(10) mm; Teh, WTR, PR . . . . [2]var. ***citrinum***
                    39′ Main lf spines gen (3)7–10(12) mm; s CCo (sw San Luis Obispo, nw Santa Barbara cos.) . . . . [2]var. ***loncholepis***
    30′ Outer and middle phyllaries with short-appressed bases and lanceolate to linear or needle-like, ascending to stiffly spreading tips; tip-spines gen 1–35 mm
      40. Outer phyllaries ± equal, narrowly lanceolate, rigidly spreading, tip-spines 5–10 mm; middle and inner phyllaries ± graduated, bodies appressed, tips spreading, spineless — s SnFrB (Palo Alto, Santa Clara Co.) . . . . ***C. praeteriens***
      40′ Outer and middle phyllaries ± equal or graduated, tips ascending to spreading, needle-like to linear or lanceolate; middle phyllaries spine-tipped
        41. Phyllary tips conspicuously connected side-to-side by network of cobwebby or multicellular hairs
          42. Corolla lobes thread-like, tips knob-like; style incl or exserted 1–2 mm; st ± hairy, hairs coarse, jointed, gen also cobwebby to densely tomentose with long, fine hairs . . . . [2]***C. brevistylum***
          42′ Corolla lobes linear, tips not knob-like; style conspicuously exserted; st cobwebby to densely tomentose, hairs long, fine . . . . [2]***C. occidentale***
            43. Pl compact, low-growing, stless or proximally branched, rounded, mound-like; heads gen not much elevated above lvs — CCo . . . . var. ***compactum***
            43′ Pl erect, main st gen 1, proximally unbranched; heads gen elevated above lvs
              44. Phyllary tips ± equal, all needle-like, 13–30 mm; corolla light purple to rich red-purple or red . . . var. ***coulteri***

44′ Phyllary tips ± graduated, outer gen < middle and inner, ± needle-like to lanceolate, 5–15 mm; corolla lavender to bright purple . . . . **var. *occidentale***
41′ Phyllary tips glabrous or hairy to densely tomentose but gen not conspicuously connected side-to-side by network of cobwebby or multicellular hairs
  45. Corolla red to ± red-pink or bright rose-purple
    46. Per, rosettes from runner roots; corolla throat ± abruptly narrowed to tube; corolla ± bright rose-purple. . . . . ²***C. andersonii***
    46′ Bien; corolla throat gradually narrowed to tube, without a clear external separation between tube and throat. . . . . ²***C. occidentale***
      47. Pl densely white-tomentose; phyllaries persistently white-tomentose (exc tip-spines); outer phyllaries gen very long, spreading or reflexed. . . . . **var. *candidissimum***
      47′ Pl variably tomentose, occ ± glabrous in age; phyllaries glabrous to ± loosely tomentose, occ some loosely cobwebby hairs extending between adjacent phyllary tips; outer phyllaries short to long, erect to spreading or reflexed
        48. Corolla 20–24 mm, red or dark red-purple; s SCoRO (s Santa Lucia Range, San Luis Obispo Co.). . . . . **var. *lucianum***
        48′ Corolla 23–35 mm, gen bright red-pink to red; widespread in CA-FP, W&I, w DMoj. . . . . . **var. *venustum***
  45′ Corolla white or cream to lavender, purple, or pale rose
    49. At least some inner phyllaries ± fringed. . . . . **C. remotifolium** var. *remotifolium*
    49′ Inner phyllaries entire
      50. Base of cauline lvs gen clasping with expanded basal lobes; lf veins gen prominently raised abaxially . . . . ²***C. cymosum** var. **cymosum***
      50′ Base of cauline lvs gen not clasping with expanded basal lobes; lf veins not prominently raised abaxially
        51. Phyllary tip-spine gen 1–3 mm; SCoR and c SNF to mtns of SW . . . . . . ⁴***C. occidentale** var. **californicum***
        51′ Phyllary tip-spine gen 3–10 mm
          52. Involucre, incl phyllary tips gen ovoid to ± spheric; tip-spines of phyllaries (2)3–6 mm; outer phyllaries gen spreading or ascending — SNH, GB . . . . ²***C. inamoenum** var. **inamoenum***
          52′ Involucre, incl phyllary tips often wider than long; tip-spines of phyllaries gen 5–10 mm; outer phyllaries often reflexed . . . . ²***C. neomexicanum***

C. andersonii (A. Gray) Petr. (p. 287)   Per (1.5)4–7(10) dm, from runner roots. **ST:** 1–several, simple or branches ascending, ± glabrous to minutely jointed-hairy and/or ± cobwebby-tomentose. **LF:** gen green adaxially, ± persistently gray-tomentose abaxially; proximal 8–35 cm, petiole spiny-winged, blade coarsely dentate or shallowly to deeply 1–2 × lobed, main spines 1–5 mm; mid-cauline ± clasping; distal gen much reduced, linear-oblong, often spinier than proximal. **INFL:** heads 1–few; peduncle ≤ 20 cm (lateral heads occ ± sessile); involucre 3–5 cm, 2–4 cm diam, widely cylindric to narrowly bell-shaped, closely subtended by 1–several bracts, loosely tomentose or glabrous; phyllaries strongly graduated, bases appressed, tips ascending, ± lanceolate, long-acuminate, outer and middle entire or spiny-ciliate, tip-spine 1–3 mm, inner tips entire, long, flat, purple. **FL:** corolla 30–45 mm, ± bright rose-purple, tube 10–20 mm, throat 10–16 mm, ± abruptly narrowed to tube, lobes 9–11 mm; style tip 3.5–5 mm. **FR:** 6–7 mm; pappus 25–40 mm. 2*n*=32,64. Open places, woodland, forest; (250)1100–3150 m. KR, CaR, SN, MP; to se ID, NV. Jul–Sep

C. andrewsii (A. Gray) Jeps. (p. 287) FRANCISCAN THISTLE Bien (or short-lived per) 6–20 dm. **ST:** gen much-branched proximally, ± fleshy, thinly cobwebby, soon becoming glabrous. **LF:** thinly cobwebby adaxially, becoming glabrous, gray-cobwebby-tomentose abaxially or becoming glabrous; proximal 3–7.5 dm, gen 1–2 × lobed, main spines 2–9 mm; mid and distal smaller, clasping, bases wide, spiny-margined, ear-like; distal-most much reduced, gen very spiny, spines ≤ 15 mm. **INFL:** heads 1–few in loose cymes, closely subtended by distal lvs, sessile or peduncle ≤ 7 cm; involucre 1.5–3 cm, 1.5–5 cm diam, ovoid to hemispheric or bell-shaped, densely cobwebby; phyllaries ± linear, tips widely spreading, spine 5–15 mm, outer and middle phyllaries spiny-ciliate. **FL:** corolla 17–24 mm, dark red-purple (cream), tube 8–11 mm, throat 3.5–6 mm, lobes 5–7 mm, linear, tips not knob-like; style tip 3–4 mm. **FR:** 4–5 mm; pappus ± 15 mm. 2*n*=32. Bluffs, ravines, seeps, occ on serpentine; < 100 m. NCo, n CCo. Reports from interior CA are misidentifications. May–Sep ★

C. arizonicum (A. Gray) Petr.   Per 2–15 dm, from ± woody caudex. **ST:** often > 1 from base, thinly tomentose or ± glabrous. **LF:** ± tomentose (esp abaxially), occ ± glabrous in age; proximal 1–2 dm, tapered to spiny-winged petiole, oblong-obovate, gen ± lobed to 2 × divided, main spines 2–30 mm; mid and distal not strongly reduced, clasping or short-decurrent. **INFL:** heads 1–few in loose cymes, ± closely subtended by distal lvs, peduncle ≤ 7(15) cm; involucre 3–4 cm, 1.5–2.5 cm diam when fresh, cylindric or ± narrowly ovoid to bell-shaped, sparsely tomentose to ± glabrous; phyllaries linear to lanceolate, entire, gen erect or ascending, often with narrow, inconspicuous sticky-resinous ridge, outer and middle tipped with spine (1)5–25 mm, inner with tips flat or short-spined, straight, often red or purple, puberulent. **FL:** exserted; corolla 24–35 mm, lavender to pink, pink-purple, or red, tube 7–13 mm, throat 2–8 mm, lobes 10–18 mm; style tip 1–2.5[4] mm. **FR:** 3.5–7 mm; pappus 17–28 mm. Variable; 3 other vars. in sw US.

var. **arizonicum** ARIZONA THISTLE **LF:** unlobed to deeply divided; main spines 2–15 mm. **INFL:** phyllary spines 1–20 mm. **FL:** corolla 24–34 mm, bright red or ± red-pink, tube 7–12 mm, throat 2–5 mm, lobes 13–18 mm; style tips ± 2.5 mm. 2*n*=32. Open forest, sagebrush scrub; 2300–3500 m. s SNH, W&I; to UT, NM. Jul–Aug

var. **tenuisectum** D.J. Keil DESERT MOUNTAIN THISTLE **LF:** deeply divided, abaxially cobwebby-tomentose to ± glabrous in age, adaxially thinly cobwebby or ± glabrous; main spines 5–30 mm. **INFL:** cylindric to bell-shaped; phyllary spine 5–25 mm. **FL:** corolla 25–35 mm, pink- to ± red-purple, tube 10–13 mm, throat 5–8 mm, lobes 10–14 mm; style tips 1–2 mm. 2*n*=34. Washes, rocky slopes, scrubland, woodland; 1600–2300 m. e DMtns (Clark, New York mtns); to s NV. [*C. nidulum* (M.E. Jones) Petr., misappl.] Jul–Nov ★

C. arvense (L.) Scop. (p. 287) CANADA THISTLE Per 5–10 dm, from runner roots; herbage green to ± appressed gray-tomentose; dioecious or nearly so. **ST:** colonial, lfy. **LF:** 3–30 cm, mostly cauline, gradually reduced distally on st, sessile, tapered at base, occ decurrent as spiny wings, ± entire to coarsely dentate or 1–2 × lobed, main spines 1–7 mm. **INFL:** heads several to many, in tight to ± open, rounded or flat-topped cluster; peduncle 0–7 cm; involucre hemispheric to ovoid in fl, 1–2 cm, 1–2 cm diam, ± tomentose when young, gen ± purple; outer phyllaries ovate, tip-spine 0.1–1 mm, inner lanceolate, tips flat, membranous. **FL:** corolla gen purple (white or pink). **PISTILLATE FL:** corolla 13–20 mm, gen < pappus in fr, tube 10–16 mm, throat ± 1 mm, lobes 2–3 mm; style tip 1–2 mm. **STAMINATE FL:** corolla 12–18 mm, > pappus at maturity, tube

8–11 mm, throat 1–1.5 mm, lobes 3–5 mm. **FR**: 2–4 mm; pappus 13–32 mm. 2*n*=34. Disturbed areas; < 2450 m. NW, CaR, SN, SnFrB, n SCoRI, SCo, GB; N.Am; native to Eur. Jun–Sep ◆

C. brevistylum Cronquist (p. 287) Ann or bien (short-lived per), 3–35 dm. **ST**: gen 1, gen branched distally, ± jointed-hairy, also gen cobwebby to densely tomentose, esp proximal to heads. **LF**: abaxially gray-tomentose, occ becoming ± glabrous, adaxially sparsely jointed-hairy; proximal 15–25 cm, tapered to spiny-winged petiole, gen shallowly lobed and dentate, main spines 3–7 mm; cauline gradually reduced, occ ± clasping or short-decurrent. **INFL**: heads 1–several at st tips, occ also in distal axils, gen ± sessile, closely subtended by bract-like lvs; involucre 2.5–3.5 cm, 2.5–4 cm diam, hemispheric to ± bell-shaped, loosely cobwebby; phyllaries ± equal, not strongly graduated, bases short, appressed, tips needle-like, ascending to spreading, connected side-to-side by network of cobwebby or jointed hairs, outermost occ spiny-ciliate near base, tip-spine 3–5+ mm. **FL**: corolla 20–25 mm, white to purple, tube 10–17 mm, throat 4–5 mm, lobes 3–5 mm, thread-like exc tips knob-like; style tip 2–4 mm, incl or exserted 1–2 mm. **FR**: 3–4.5 mm; pappus 10–22 mm. 2*n*=34. Moist places; < 500 m. NW, CW, SCo, n ChI; to BC, MT. Mar–Aug

C. canescens Nutt. PRAIRIE THISTLE Bien or short-lived per, 2–10 dm. **ST**: gen 1, erect, densely gray-tomentose, hairs long, fine; branches gen few, ascending. **LF**: ± gray-tomentose, esp abaxially, occ ± glabrous adaxially; proximal 1–2.5 dm, wing-petioled, coarsely dentate to deeply lobed, lobes well-separated, main spines 2–3(10) mm; cauline well developed, gradually reduced distally on st, decurrent 1–5 cm as spiny wings, shallowly to deeply lobed, distal-most bract-like. **INFL**: heads 1–10+ in ± flat-topped to raceme-like cluster; peduncle 0–10 cm; involucre 3–4 cm, 2.5–4 cm diam, hemispheric to widely bell-shaped, ± truncate at base, loosely cobwebby at phyllary margins or ± glabrous; phyllaries strongly graduated, lance-ovate (outer) to ± lance-linear, tips ascending to spreading, midribs of outer and middle with long, sticky-resinous ridge, tip-spine 2–4(8) mm, ascending to spreading. **FL**: corolla 20–35 mm, dull white to lavender-tinged, tube 10–17 mm, throat 6–11 mm, lobes 4–9 mm; style tip 5–8 mm. **FR**: 5–7 mm; pappus 18–30 mm. 2*n*=34,36. Grassland, arid scrubland, disturbed areas; 1500–1600 m. MP; native to n-c US. [*C. undulatum* (Nutt.) Spreng., in part, misappl.] May–Aug

C. ciliolatum (L.F. Hend.) J.T. Howell (p. 291) ASHLAND THISTLE Per 6–20 dm, from runner roots. **ST**: gen simple proximally, ± cobwebby to densely white-tomentose, hairs long, fine; branches few, ascending. **LF**: abaxially ± white, adaxially green, thinly gray-tomentose; proximal 1–2.5 dm, wing-petioled, entire to shallowly lobed, spines 0 or 1–2 mm; cauline well developed, gradually reduced distally on st, clasping or short-decurrent as spiny wings, shallowly to deeply lobed, distal-most bract-like; main spines 1–6 mm. **INFL**: heads 1–few in ± flat-topped cluster; peduncle 1–15 cm; involucre 1.5–2.3 cm, 1.5–3 cm diam, ovoid to hemispheric, thinly cobwebby or ± glabrous; phyllaries strongly graduated, lanceolate (outer) to ± linear, tips ascending to spreading, midribs of outer and middle with long, sticky-resinous ridge, tip-spine 1–3 mm, ascending. **FL**: corolla 15–25 mm, dull white to lavender, tube 7–11 mm, throat 5–7 mm, lobes 5–7 mm; style tip 5–7 mm. **FR**: 3.5–7 mm; pappus 15–20 mm. Grassy areas, open woodland; ± 800 m. n KR; s OR. Closely related to *C. undulatum*; perhaps should be treated as *C. undulatum* var. *ciliolatum* L.F. Hend. Jun–Jul ★

C. crassicaule (Greene) Jeps. (p. 291) SLOUGH THISTLE Ann or bien, (6)10–30 dm. **ST**: gen 1, gen openly branched adaxially, thinly cobwebby, also at least proximally jointed-hairy; base gen 2–10 cm diam, evidently hollow. **LF**: abaxially gray-tomentose, midvein occ jointed-hairy, adaxially thinly cobwebby-tomentose; proximal 1.5–7 dm, petiole spiny-lobed, winged, blade elliptic to widely oblanceolate, 1–2 × lobed, lobes occ dentate, main spines 3–8 mm; mid and distal smaller, narrower, sessile, spiny-margined, bases wide-clasping or short-decurrent, distal much reduced, gen spiny, spines ≤ 10 mm. **INFL**: heads 1–several at branch tips, forming loose to crowded flat-topped to panicle-like clusters (occ on short axillary branches), ± closely subtended by ± reduced distal-most lvs; peduncle 0–15 cm; involucre 1.5–3 cm, 1.5–3 cm diam, ovoid to ± bell-shaped, ± glabrous; outer phyllaries lanceolate, irregularly spiny-fringed, middle

and inner entire, outer and middle ascending or ± spreading, tip-spine 3–5 mm. **FL**: corolla 19–26 mm, pale rose-purple (white), tube 9–12 mm, throat 4–6 mm, lobes 5–9 mm; style tip 3.5–4.5 mm. **FR**: 5–5.5 mm; pappus 1.5–2 cm. 2*n*=32. Freshwater marshes; < 100 m. SnJV. Mar–Jun ★

C. cymosum (Greene) J.T. Howell Bien or per, 2.5–12 dm. **ST**: gen distally branched, ± cobwebby and soft-jointed-hairy. **LF**: rosettes from runner roots or taproot; faces ± cobwebby or becoming glabrous adaxially, soft-jointed-hairy, esp on abaxial midrib, veins abaxially gen prominently raised; proximal 1–2(5) dm, petioled, gen 1–2 × lobed, main spines 2–7 mm; middle and distal gen clasping with expanded basal lobes (short-decurrent), distal much reduced, spines many. **INFL**: heads 1–few in ± flat-topped (occ raceme-like) cluster, peduncle (0)2–15 cm; involucre 2–3 cm, 1.5–5 cm diam, ovoid to hemispheric, ± tomentose; outer and middle phyllaries linear to lanceolate, gen with narrow, ± conspicuous sticky-resinous ridge, tips ascending to spreading, spine 2–4 mm. **FL**: corolla 20–33 mm, dull white, tube 8–15 mm, throat 5.5–12 mm, lobes 5.5–7 mm; style tip 4–6 mm. **FR**: 5–8 mm; pappus 15–25 mm. Some specimens may not be readily assignable to var.

var. *canovirens* (Rydb.) D.J. Keil GRAY-GREEN THISTLE Bien or per from taproot. **INFL**: larger heads 15–25 mm diam; outer phyllaries gen << inner, sticky-resinous ridge prominent, appearing dark brown on dry pl. 2*n*=34. Sagebrush scrubland, grassland, woodland, open forest; 1350–2300 m. MP; OR, MT, WY. [*C. canovirens* (Rydb.) Petr., in part] Jun–Jul

var. *cymosum* PEREGRINE THISTLE Per from creeping rootstock. **INFL**: larger heads 20–35 mm diam; outer phyllaries long, often nearly = inner, sticky-resinous ridge weakly developed. 2*n*=30,34. Scrubland, woodland, open forest, meadows, occ on serpentine; 100–2100 m. KR, NCoR, CaR, n&s SN, SnFrB, n SCoRI; to s OR, n NV. Variable. Apr–Jul

C. douglasii DC. SWAMP THISTLE Bien or per from taproot or woody caudex, 6–25 dm, gray-tomentose, hairs appressed, felt-like. **ST**: 1–several, gen much-branched. **LF**: faces gray tomentose or becoming ± green adaxially; basal 3–11 dm, petiole spiny-toothed, blade coarsely dentate to deeply 1–2 × lobed and dentate, main spines 3–10 mm; cauline smaller, bases expanded to ear-like, spiny-winged, decurrent 1–3 cm, blade shallowly to deeply lobed, spines often 7–20 mm. **INFL**: heads 10–many in ± lfy-bracted groups at branch tips, forming ± panicle-like cluster; peduncle (0)1–4(8) cm; involucre 1.5–2.5(3) cm, 2–4.5 cm diam, ovoid to hemispheric, gen dark green or ± purple; phyllaries graduated, ovate (outer) to lance-ovate (middle), tightly appressed, often with dark purple patch near tip, midrib with sticky-resinous ridge, outer and middle tipped by spreading spine 1–9 mm. **FL**: corolla 18–21 mm, ± dark rose-purple (white), tube 8–9 mm, throat 5–6 mm, lobes 5–6 mm; style tips 3–4.5 mm. **FR**: 5–6 mm; pappus 15–20 mm. Some pls may not be readily assignable to var. Closely related to *C. mohavense* and *C. hydrophilum*.

var. *breweri* (A. Gray) D.J. Keil & C.E. Turner (p. 291) **ST**: ≤ 2.5 m. **LF**: cauline gen shallowly lobed or not, larger spines of distal lvs gen ≤ 7 mm. **INFL**: heads gen 2–3 cm. 2*n*=30,34. Wet places, occ on serpentine; 1000–2200 m. KR, NCoR, CaR, n SNH, MP; s OR. Jun–Sep

var. *douglasii* **ST**: ≤ 1.5 m. **LF**: cauline gen deeply lobed, larger spines of distal lvs gen 7–20 mm. **INFL**: heads gen 2.5–3.5 cm. Wet places, occ on serpentine; < 1400 m. NCo, NCoR, n CCo, SnFrB, n SCoRI. Jun–Aug

C. fontinale (Greene) Jeps. FOUNTAIN THISTLE Bien or short-lived per, 5–20 dm. **ST**: 1–several. **LF**: gen strongly wavy, velvety with short, erect, jointed hairs, also cobwebby, esp abaxially; proximal 1–7 dm, petiole spiny-lobed or toothed, blade shallowly to deeply 1–2 × lobed and dentate, main spines 1–10 mm, abrupt; cauline gradually reduced, well distributed, ± lobed, distal clasping (bases expanded, ear-like), occ short-decurrent, longer spines 4–18 mm. **INFL**: heads in ± panicle-like cluster, gen nodding, closely subtended by distal-most lvs; peduncle 0–7 cm; involucre 1.5–3 cm, 2–5 cm diam, hemispheric or bell-shaped, green to purple; phyllaries lanceolate to ovate, spreading to reflexed, spine 1–6 mm. **FL**: corolla 18–19 mm, white to

pink or lavender, tube 6–7 mm, throat 6–8 mm, lobes 5–8 mm; style tip 2.5–4.5 mm. **FR:** 3.4–5 mm; pappus 12–15 mm.

var. *campylon* (H. Sharsm.) D.J. Keil & C.E. Turner (p. 291) MOUNT HAMILTON FOUNTAIN THISTLE **ST:** ≤ 2 m, green. **LF:** densely cobwebby, velvety hairs ± concealed; longer spines of cauline lvs 10–18 mm. **INFL:** heads strongly, permanently nodding; phyllaries 65–85, outer 25–35 strongly recurved, 20–30 mm, green, channeled, widest in proximal 1/2, distal 1/2 tapered to spine 3–6 mm, marginal spines gen 0. **FL:** style tip 4–6 mm. **FR:** 3.4–4.1 mm. Serpentine seeps and streams; < 750 m. e SnFrB. Mar–Oct ★

var. *fontinale* (p. 291) CRYSTAL SPRINGS FOUNTAIN THISTLE **ST:** ≤ 1.3 m, ± red. **LF:** thinly cobwebby, velvety hairs ± evident. **INFL:** heads ± nodding in fl, gen erect in fr; phyllaries 100–120, outer 35–45 moderately recurved, 15–20 mm, ± red, flat to ± channeled, widest in distal 1/2, abruptly tipped by 1–2 mm spine. **FL:** style tip 3–4 mm. **FR:** 4.1–5 mm. 2*n*=34+1. Serpentine seeps and streams; 120–150 m. sw SnFrB (Crystal Springs Reservoir, San Mateo Co.). Hybridizes with *C. quercetorum.* May–Aug ★

var. *obispoense* J.T. Howell SAN LUIS OBISPO FOUNTAIN THISTLE **ST:** ≤ 2 m, green to ± purple. **LF:** densely cobwebby, esp abaxially, velvety hairs ± concealed; longer spines of cauline lvs 4–7. **INFL:** heads ± nodding in fl, gen erect in fr; phyllaries 70–120, outer 28–38 strongly recurved, 15–20 mm, green to dark purple, ± channeled, widest in proximal 1/2, distal 1/2 tapered to spine (1)2–5 mm, marginal spines often present. **FL:** style tip 4–6 mm. **FR:** 3.8–4 mm. Serpentine seeps and streams; < 350 m. c SCoRO (San Luis Obispo Co.). Apr–Oct ★

C. hydrophilum (Greene) Jeps. Bien (or short-lived per) 1–2.2 m. **ST:** ± cobwebby, becoming glabrous. **LF:** faces thinly cobwebby-tomentose or becoming glabrous adaxially; proximal 3–9 dm, petiole spine-margined, blade gen ± lobed, lobes gen with 2–4(6) wide 2° lobes or coarse teeth, main spines 2–9 mm; mid and distal smaller, clasping or short-decurrent with ear-like bases; distal-most much reduced, gen very spiny, lobes narrower. **INFL:** heads 1–few in ± open flat-topped or panicle-like clusters; peduncle 0–10 cm; involucre 1.5–2.5 cm, 1.5–3 cm diam, ovoid to bell-shaped, thinly cobwebby, becoming glabrous; phyllaries graduated, ovate (outer) to lance-ovate (middle), minutely spiny-fringed, appressed, midribs with narrow sticky-resinous ridge, tip-spine 1–3 mm, spreading. **FL:** corolla 18–23 mm, pale rose-purple, tube 8–10 mm, throat 5–6 mm, lobes 5–7 mm; style tip 3.5–4.5 mm. **FR:** 4–5 mm; pappus ± 15 mm. Related to *C. douglasii* and *C. mohavense.*

var. *hydrophilum* (p. 291) SUISUN THISTLE **ST:** gen branched in distal 1/2. **INFL:** head gen 2.5–3 cm. **FR:** ± 5 mm, oblong, tapered near base. 2*n*=32. Tidal marsh; ± 0 m. Deltaic GV (Suisun Marsh, Solano Co.). Jun–Sep ★

var. *vaseyi* (A. Gray) J.T. Howell MOUNT TAMALPAIS THISTLE **ST:** gen branched in proximal 1/2. **INFL:** head gen 3–3.5 cm. **FR:** 4–5 mm, oblong or elliptic. 2*n*=32. Serpentine seeps; 300–450 m. n SnFrB (Mount Tamalpais, Marin Co.). Jun–Sep ★

C. inamoenum (Greene) D.J. Keil var. *inamoenum* GREENE'S THISTLE Bien or per, 2–10 dm. **ST:** 1–several, erect, distally branched. **LF:** adaxially thinly gray-tomentose or ± glabrous, abaxially gen densely gray-tomentose; basal often 0 at fl, 10–35 cm; proximal linear or oblanceolate, unlobed, dentate or ± deeply lobed; distal wing-petioled, gradually reduced, decurrent 1–3 cm as spiny wings, often spinier than proximal, distal-most ± bract-like, main spines 2–7 mm. **INFL:** heads 1–many, in open ± flat-topped cluster or crowded near st tips; peduncle 0–25 cm; involucre gen ovoid to ± spheric (bell-shaped), 2–3 cm, 1.5–3 cm diam, glabrous or ± cobwebby-tomentose; phyllaries gen strongly graduated, ± loosely appressed, with indistinct to ± prominent sticky-resinous ridge, tip-spine (2)3–6 mm, ascending to abruptly spreading. **FL:** corolla 19–31 mm, white or pale lavender, tube 7–13 mm, throat 6.5–9.5 mm, lobes 4–8 mm; style tip 3.5–7 mm. **FR:** 5–8 mm; pappus 12–25 mm. 2*n*=34. Dry slopes, grassland, scrubland, woodland, forest; 1700–3000 m. SNH, GB; to WA, WY, UT. [*C. canovirens,* in part, misappl.; *C. subniveum* Rydb.; *C. utahense* Petr., misappl.] Jun–Aug

C. mohavense (Greene) Petr. MOJAVE THISTLE Bien or per, from taproot or woody caudex. **ST:** 1–several, ± branched distally, ± gray-tomentose. **LF:** ± densely tomentose, esp abaxially; proximal 1.5–6 dm, spiny-petioled, elliptic to oblanceolate, toothed to deeply lobed, lobes gen rigidly spreading, simple or with 1–2 pairs of coarse teeth or 2° lobes, main spines 3–20 mm; middle and distal smaller, narrower, conspicuously decurrent as spiny wings, distal-most much reduced, often long-acuminate, gen very spiny, spines 4–30 mm. **INFL:** heads in loose to crowded flat-topped or panicle-like clusters (occ on short axillary branches); peduncle lfy, 0–10 cm; involucre 1.5–2.5 cm, 1.5–2 cm diam, ± ovoid, ± loosely tomentose, becoming glabrous; phyllaries graduated, ovate (outer) to lance-ovate (middle), appressed, midribs often with sticky-resinous ridge, tip-spines 3–7 mm, ascending or spreading. **FL:** corolla 16–25 mm, white to pink or lavender, tube 7–12 mm, throat 4–7 mm, lobes 4–8 mm; style tip 3–4 mm. **FR:** 3–6 mm; pappus 14–16 mm. 2*n*=30,32. Damp soil around springs, canyons, streams; 250–2800 m. SNE, DMoj; to UT, AZ. [*C. virginense* S.L. Welsh] Related to *C. douglasii* and *C. hydrophilum.* Jul–Oct

C. neomexicanum A. Gray DESERT THISTLE Bien 4–29 dm. **ST:** gen 1, distally with a few, ascending branches, ± white-cobwebby-tomentose and puberulent. **LF:** ± persistently gray-tomentose, esp abaxially; proximal 6–35 cm, petioled or tapered to spiny-winged base, oblong-elliptic to oblanceolate, ± lobed, lobes gen rigidly spreading, simple or with 2–4 coarse teeth or 2° lobes, main spines 5–15 mm; middle and distal gen smaller, narrower, decurrent as spiny wings, distal-most well separated, ± bract-like, occ reduced to a cluster of long spines. **INFL:** heads 1–few in open, ± flat-topped cluster (occ on short axillary branches); peduncle 2.5–30 cm; involucre 2–2.5 cm, 2.5–5 cm diam, width gen > length, hemispheric or bell-shaped, ± loosely tomentose or ± glabrous; phyllaries lance-linear, outer and middle linear, spreading to reflexed, occ with sticky-resinous ridge, spine (4)5–10(15) mm. **FL:** corolla 18–27 mm, white to pale lavender or pink, tube 8–14 mm, throat 4–7 mm, lobes 5–9 mm; style tip 4–5 mm. **FR:** 5–6 mm; pappus 15–20 mm. 2*n*=30,32. Canyons, slopes, roadsides; 800–2100 m. e DMoj, nw DSon; to CO, NM. Closely related to *C. occidentale.* Apr–May

C. occidentale (Nutt.) Jeps. Bien 0.3–30 dm. **ST:** gen 1, gen erect, branched distally (st ± 0 or branched ± throughout in dwarf pls), ± tomentose. **LF:** ± densely gray- or ± white-tomentose, esp abaxially, occ ± glabrous in age; proximal 1–4 dm, petiole spiny-winged, blade oblanceolate, lobed 1/2+ to midvein, lobes widely triangular, dentate or further lobed, main spines 1–10 mm; distal gradually reduced, ± clasping or short-decurrent, linear or oblong, often entire, often spinier than proximal, distal-most often bract-like. **INFL:** heads 1–several in loose to tight cluster (barely raised above rosette in dwarf pls); peduncle 1–30 cm; involucre 1.5–5 cm, 1.5–8 cm diam, ovoid to spheric or wider; phyllaries ± equal to strongly graduated, needle-like to lance-linear, straight or twisted or bent, appressed to widely radiating, ± glabrous to densely tomentose, sometimes connected side-to-side by network of conspicuous cobwebby hairs, tip-spine 1–10+ mm. **FL:** corolla 18–35 mm, white to lavender, purple, or red, tube 8–18 mm, throat 5–7 mm, lobes 5–10 mm; style tip 4–5 mm. **FR:** 5–6 mm; pappus 15–30 mm. Variable complex; some pls may not be assignable to var. Needs study. Related to *C. neomexicanum.*

var. *californicum* (A. Gray) D.J. Keil & C.E. Turner CALIFORNIA THISTLE Pl gen 5–20 dm, erect. **LF:** green to gray adaxially, gray abaxially. **INFL:** heads short- to long-peduncled in ± open cluster, well elevated above proximal lvs; involucre 1.5–5 cm diam, ± glabrous to densely cobwebby; phyllaries occ with sticky-resinous ridge, middle phyllary tips gen ≤ 1 cm, occ much longer, 1–3 mm wide, appressed to loosely spreading, occ twisted or bent, tip-spine gen 1–3. **FL:** corolla 18–35 mm, white to purple or rose. 2*n*=28,30. Woodland, open forest, disturbed sites; < 2300 m. SN, c&s SCoR, SW. Intergrades with *C. occidentale* var. *venustum* in s SN, s SCoR; pls intermediate to *C. occidentale* var. *candidissimum* in n SN. Apr–Jul

var. *candidissimum* (Greene) J.F. Macbr. (p. 291) SNOWY THISTLE Pl gen 4–20 dm, erect or ± bushy, densely white-tomentose. **INFL:** heads gen long-peduncled, occ in tight clusters at

287

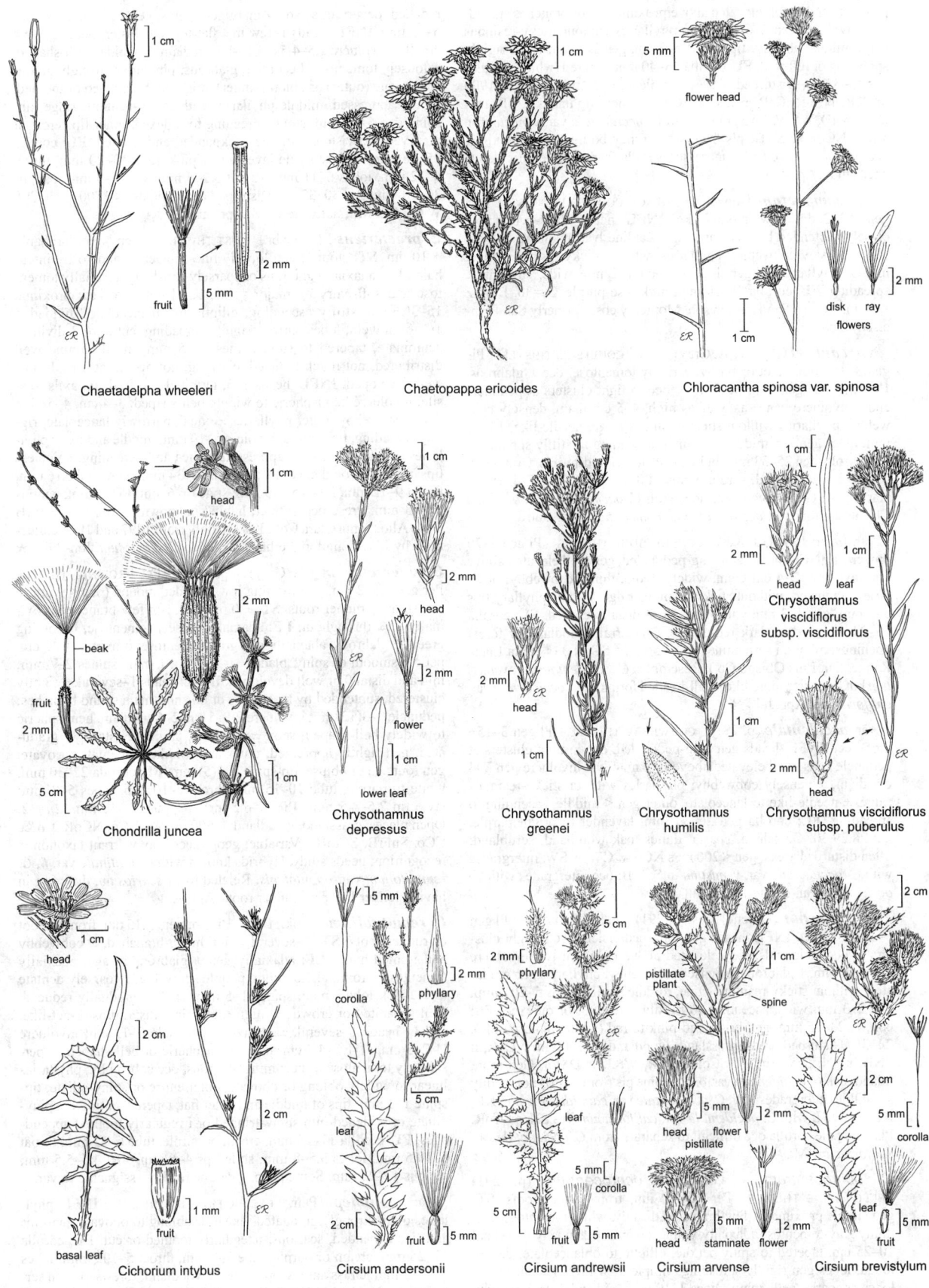

Chaetadelpha wheeleri

Chaetopappa ericoides

Chloracantha spinosa var. spinosa

Chondrilla juncea

Chrysothamnus depressus

Chrysothamnus greenei

Chrysothamnus humilis

Chrysothamnus viscidiflorus subsp. viscidiflorus

Chrysothamnus viscidiflorus subsp. puberulus

Cichorium intybus

Cirsium andersonii

Cirsium andrewsii

Cirsium arvense

Cirsium brevistylum

peduncle ends, well elevated above proximal lvs or branches spreading; involucre gen 2–6 cm diam; phyllaries without sticky-resinous ridge, middle phyllary tips gen 1.5–3 cm, gen 2–3 mm wide, rigidly spreading or reflexed. **FL**: corolla 26–40 mm, gen red (white or pink). $2n$=30–33,60. Disturbed places, scrubland, open woodland; < 1900 m. KR, NCoR, CaR, n SN, c CCo (Carmel Highlands), MP; to s OR, sw ID, w NV. Intergrades with *C. occidentale* var. *venustum* in s KR, NCoR, c SNH; pls from s NCo may be intergrades with *C. occidentale* var. *o.* (need more study). Pls from Carmel Highlands need study. Apr–Sep

var. ***compactum*** Hoover   COMPACT COBWEBBY THISTLE   Pl gen 0.3–3 dm, low, mound-like. **INFL**: heads short-peduncled, closely subtended by basal and large cauline lvs; involucre 5–8 cm diam, densely cobwebby; phyllaries without sticky-resinous ridge, middle phyllary tips gen 1–2 cm, gen 1–2 mm wide, straight, ± spreading. **FL**: corolla 25–30 mm, dark rose-purple. $2n$=30. Bluffs; < 50 m. CCo (n San Luis Obispo, Monterey cos., formerly San Francisco). Feb–Jul ★

var. ***coulteri*** (Harv. & A. Gray) Jeps.   COULTER'S THISTLE   Pls gen 3–15+ dm, erect or bushy, variably tomentose, occ ± glabrous. **INFL**: heads gen long-peduncled, occ in tight clusters at peduncle ends; involucre gen ± as wide as high, 4–5 cm diam, densely cobwebby; phyllaries without sticky-resinous ridge, needle-like, 13–30 mm, outer gen ± = middle and inner, ascending to stiffly spreading. **FL**: corolla gen 25–33 mm, light purple to rich red-purple. Grassland, dunes, oak woodland, scrub; gen < 700 m. CCo, SCo, ChI. [*C. c.* Harv. & A. Gray] Occ intergrades with *C. occidentale* var. *o.*; some ChI pls approach *C. occidentale* var. *compactum*. Mar–Jun

var. ***lucianum*** D.J. Keil   CUESTA RIDGE THISTLE   Pl gen 3–20 dm, erect. **INFL**: heads gen long-peduncled, gen well elevated above lvs; involucre 2–4 cm diam, widely ovoid, thinly cobwebby; phyllaries graduated, without sticky-resinous ridge, middle phyllary tips 5–8 mm, gen 1–3 mm wide, straight or distally curved. **FL**: corolla 20–24 mm, red or dark red-purple. Chaparral, woodland or forest openings, often on serpentine; 500–750 m. s SCoRO (s Santa Lucia Range, San Luis Obispo Co.). Resembles *C. occidentale* var. *venustum* but head size more like small-headed forms of *C. occidentale* var. *californicum*. Apr–Jul ★

var. ***occidentale*** (p. 291)   COBWEBBY THISTLE   Pl gen 3–15+ dm, erect. **INFL**: heads gen long-peduncled, occ in tight clusters at peduncle ends, well elevated above proximal lvs; involucre gen 3–4 cm diam, ± densely cobwebby; phyllaries without sticky-resinous ridge, ± needle-like to lanceolate, outer gen < middle, ascending to spreading. **FL**: corolla gen 25–35 mm, lavender to bright purple. $2n$=28,29,30. Grassland, coastal dunes, oak woodland, scrubland, often disturbed areas; gen < 200 m. s NCo, w CW, w SW. Intergrades with *C. occidentale* var. *venustum* in SnFrB; occ intergrades with *C. occidentale* var. *coulteri*. Mar–Jul

var. ***venustum*** (Greene) Jeps. (p. 291)   VENUS THISTLE   Pl gen 5–30 dm, erect. **INFL**: heads gen long-peduncled, occ in tight clusters at peduncle ends, well elevated above proximal lvs; involucre 2–6 cm diam, ± glabrous to densely cobwebby; phyllaries gen graduated, without sticky-resinous ridge, middle phyllary tips 5–20+ mm, gen 2–3 mm wide, ascending to rigidly spreading or reflexed. **FL**: corolla 23–35 mm, gen bright red-pink to red (white, pink, purple). $2n$=30. Disturbed areas, grassland, woodland; < 3600 m. NCoR, n SNH, s SN, ScV, SnFrB, SCoR, WTR, W&I, w DMoj. Gen more inland than *C. occidentale* var. *o.* but some pls from SnFrB not readily separable. Intergrades with *C. occidentale* var. *candidissimum* in KR, NCoR, SNH, with *C. occidentale* var. *californicum* in s SN, s SCoR. Pls with pale corolla occ difficult to separate from *C. occidentale* var. *californicum*. May–Jul

C. ochrocentrum A. Gray var. ***ochrocentrum*** (p. 291)   YELLOWSPINE THISTLE   Per 2.5–10 dm, from runner roots. **ST**: gen simple proximally, few-branched distally, white-tomentose. **LF**: thinly gray-tomentose adaxially, white-tomentose abaxially; proximal 10–25 cm, tapered to spiny petiole, elliptic to oblanceolate, deeply lobed, lobes gen rigidly spreading, simple or with 2–4 ± narrow 2° lobes or coarse teeth, main spines 3–10 mm; mid and distal gradually

reduced, decurrent as spiny-margined wings, gen very spiny, spines 5–15 mm. **INFL**: heads 1–few in ± flat-topped cluster; peduncle 0–4 cm, lfy; involucre 2.5–4.5 cm, 2–4.5 cm diam, ± ovoid to bell-shaped, ± loosely tomentose, becoming glabrous; phyllaries strongly graduated, ovate (outer) or oblong (inner), minutely roughened or toothed, tightly appressed, middle phyllaries with sticky-resinous ridge, tip-spine gen 5–12 mm, stout, spreading to reflexed, inner tips erect or recurved, ± twisted, flat, occ ± expanded and fringed. **FL**: corolla 25–38 mm, white to pale lavender or pink, tube 13–20 mm, throat 6–12 mm, lobes 6–11 mm; style tip 2–8 mm. **FR**: 6–9 mm; pappus 20–30 mm. $2n$=30–32,34. Disturbed areas, fields; < 1700 m. n ChI, WTR, PR, MP; native to c US. Apr–Jul ◆

C. praeteriens J.F. Macbr.   LOST THISTLE   Bien or per, probably > 10 dm. **ST**: stout, erect, loosely fine-cobwebby and soft-jointed-hairy. **LF**: adaxially glabrous or sparsely soft-hairy, abaxially tomentose and soft-hairy on major veins; basal not observed; proximal 15–30+ cm, stiffly ascending, elliptic to oblanceolate, divided ≥ 1/2 to midvein, lobes narrow, rigidly spreading, entire or 3-divided, acuminate, tapered to stout spines 5–15 mm; distal cauline well distributed, not much reduced, clasping, not decurrent, distal-most well developed. **INFL**: heads 1–5, terminal and in distal axils, sessile; involucre hemispheric to widely bell-shaped, 3–4 cm, 4–5+ cm diam, cobwebby; outer phyllaries ± equal, narrowly lanceolate, rigidly spreading, tip-spine spreading, 5–10 mm; middle and inner phyllaries ± graduated, with appressed bodies and spreading, spineless tips, minutely roughened. **FL**: corolla 30–34 mm, white, tube 16 mm, throat 9–12 mm, lobes 5.5–9 mm; style tip 6 mm. **FR**: 6 mm; pappus 25–33 mm. Presumed extinct; habitat unknown; < 100 m. s SnFrB (Palo Alto, Santa Clara Co.). Known only from 1897 and 1901 collections by J.W. Congdon. Perhaps related to *C. scariosum*. Jun–Jul ★

C. quercetorum (A. Gray) Jeps. (p. 291)   BROWNIE THISTLE   Per gen 0.5–2 dm, gen forming low, rounded mound (occ ≤ 9 dm, ± erect), from runner roots. **ST**: ± 0 or short, gen few-branched proximally or ± throughout. **LF**: adaxially loosely tomentose, becoming green, ± glabrous, abaxially gen gray-tomentose; proximal ≤ 35 cm, petiole smooth or spiny, blade ± 1–3 × lobed, main spines 2–7 mm; mid and distal 0 or well developed. **INFL**: heads 1–several, ± tightly clustered, subtended by rosette lvs or on short main st and branches; peduncle 0–3(12) cm; involucre 2.5–5 cm, 2–6 cm diam, hemispheric to widely bell-shaped, soon ± glabrous; phyllaries strongly graduated, gen tightly appressed, outer and middle lanceolate to ovate, gen rounded or obtuse, tip-spine 1–2(5) mm. **FL**: corolla 22–30 mm, white to purple, tube 10–15 mm, throat 7–10 mm, lobes 5–8 mm; style tip 2.5–4.5 mm. **FR**: 5–6.5 mm; pappus 20–40 mm. $2n$=32. Open places, grassland, woodland; < 500 m. c&s NCo, NCoRO, n&c CCo, SnFrB, SCoRO. Variable; geog races may warrant taxonomic recognition; needs study. Hybrids known with *C. fontinale* var. *f.*, *C. remotifolium* var. *odontolepis*. Related to *C. scariosum*, differing in having longer lifespan, runner roots. Apr–Aug

C. remotifolium (Hook.) DC.   Bien or per, 3–15 dm, from taproot or runner roots. **ST**: 1–several, often much-branched, ± cobwebby and jointed-hairy. **LF**: adaxially gen ± glabrous in age, abaxially tomentose; proximal 1.5–5 dm, petiole spiny, blade coarsely dentate to ± 1–2 × lobed, main spines 1–5 mm; cauline gradually reduced, well separated or crowded, often ± clasping, distal-most bract-like. **INFL**: heads 1–several, ± clustered; peduncle 0–12 cm; involucre (1)2–3 cm, (1)1.5–4.5 cm diam, hemispheric or bell-shaped, ± persistently tomentose or becoming glabrous, occ cobwebby; phyllaries linear to widely oblong or obovate, outer entire or spiny-ciliate, tip-spine 1–3 mm; tips of middle and inner flat, tapered, entire, or spiny-ciliate or widened into spiny-fringed or irregularly toothed appendages. **FL**: corolla 17–25 mm, cream or purple, tube 6–12 mm, throat 5–10.5 mm, lobes 3.5–8 mm; style tips 4–6 mm. **FR**: 4.5–5.5 mm; pappus 13–23 mm. Some pls may not be readily assignable to var.

var. ***odontolepis*** Petr.   PACIFIC FRINGED THISTLE   **INFL**: phyllaries often strongly graduated, linear or oblong to obovate, margins and tips expanded, scarious, irregularly toothed or cut. **FL**: corolla 17–25 mm, cream or purple, tube 7–9 mm, throat 5–10.5 mm, lobes 3.5–6.5 mm. Grassy areas, openings in woodland, forests, occ on serpentine; < 1850 m. NW, n SnFrB; OR. [*C. amblylepis* Petr.; *C. callil-*

epis (Greene) Jeps.; *C. c.* var. *pseudocarlinoides* (Petr.) J.T. Howell] Hybridizes with *C. quercetorum.* Jun–Sep

var. ***remotifolium*** (p. 291) REMOTE-LEAVED THISTLE **INFL:** phyllaries graduated or ± equal, linear or narrowly oblong, entire or gen some inner tips weakly expanded, ± fringed. **FL:** corolla 18–25 mm, cream, tube 6–12 mm, throat 8–10 mm, lobes 4.5–6.5 mm. Grassy areas, openings in woodland, forests; < 1000 m. n NCo, KR; to WA. May–Aug

var. ***rivulare*** Jeps. KLAMATH THISTLE **INFL:** phyllaries graduated or ± equal, linear, ciliate with minute spreading to recurved spines, occ tips weakly expanded, scarious. **FL:** corolla 19–25 mm, purple (cream), tube 8–12 mm, throat 5–6 mm, lobes 4.5–8 mm. 2*n*=32. Grassy areas, openings in woodland, forest; < 750 m. n NW; OR. [*C. acanthodontum* S.F. Blake] May–Aug

C. rhothophilum S.F. Blake (p. 291) SURF THISTLE Bien or short-lived per, 1–10 dm, ± fleshy, bush-like or low, mounded, gray-tomentose throughout; hairs appressed, felt-like. **ST:** 1–several. **LF:** gen strongly wavy, densely gray- or white-felty-tomentose; proximal 10–25 cm, wing-petioled, entire or ± widely lobed, lobes entire or few-toothed, main spines 1–4 mm; cauline gradually reduced, well distributed, distal sessile, clasping, ± lobed, often spinier than proximal, spines ≤ 8 mm. **INFL:** heads 1–several, ± clustered, closely subtended by distal-most lvs; peduncle 0–7 cm; involucre 3–4 cm, 4–6 cm diam, hemispheric or bell-shaped; phyllaries lance-linear, persistently tomentose, outer and middle spiny-ciliate, tip-spine 2–5 mm. **FL:** corolla 21–25 mm, white to pale yellow, tube 11–15 mm, throat 5–8 mm, lobes 5–8 mm; style tip 2.5–4.5 mm. **FR:** 5–7 mm; pappus 15–20 mm. 2*n*=34. Dunes, bluffs; < 20 m. s CCo (s San Luis Obispo, n Santa Barbara cos.). Hybrids known with *C. occidentale* var. *o.*, *C. scariosum* var. *loncholepis.* Apr–Aug ★

C. scariosum Nutt. MEADOW THISTLE Bien or short-lived per, 0.5–10+ dm. **ST:** often 0 or short, occ erect or bushy-branched, often ± fleshy, glabrous to loosely tomentose, occ coarse-hairy. **LF:** glabrous to loosely tomentose adaxially, glabrous to densely tomentose abaxially; often all basal, ± petioled, cauline 0 or well distributed, proximal 1–4 dm, tapered or spiny-petioled, oblong to oblanceolate, unlobed to deeply lobed, occ with 2° lobes or teeth, main spines 2–10+ mm. **INFL:** heads gen crowded, ± sessile, closely subtended by basal rosette or in ± tight, gen lfy clusters at st tips; involucre 2–4.5 cm, 2–5 cm diam, ovoid to bell-shaped; outer phyllaries lance-linear to ovate, ± entire, tip-spine 1–12 mm, ascending, inner tips narrow and entire or expanded and irregularly toothed or cut, flat or crinkled. **FL:** corolla 20–40 mm, ± white to purple, tube 7–24 mm, throat 4–12 mm, lobes 4–10 mm; style tip 3.5–8 mm. **FR:** 3–6.5 mm; pappus 15–35 mm. Variable complex of intergrading races. Extreme forms very different. Stemmed or stless forms may occur in same population or not. Some pls not readily assignable to variety. Needs study.

var. ***americanum*** (A. Gray) D.J. Keil DINNERPLATE THISTLE **ST:** gen 0. **LF:** adaxially glabrous or jointed-hairy, abaxially gray-tomentose, hairs jointed or not; main spines gen 2–7 mm. **INFL:** heads 1–many, sessile or subsessile, closely subtended by basal rosette (occ stemmed pls in same population); involucre 1.5–3 cm; outer phyllary tip-spine 1–12 mm, inner tips entire or abruptly expanded to scarious, irregularly toothed appendage. **FL:** corolla white or pink-tinged. 2*n*=34. Wet to dry meadows, grassy forest openings, open sites; 1600–3500 m. CaRH, SNH, Wrn; to OR, ID, WY, CO, n Baja CA. Jun–Aug

var. ***citrinum*** (Petr.) D.J. Keil (p. 291) SOUTHERN MEADOW THISTLE **ST:** 0 or 1–5 dm, gen ± branched throughout. **LF:** adaxially glabrous (occ cobwebby or jointed-hairy); abaxially glabrous or thinly (densely) tomentose; main lf spines gen 2–7(10) mm. **INFL:** heads 1–many, sessile or short-peduncled, closely subtended by basal rosette or clustered at branch tips; involucre 2.5–4+ cm; outer phyllary tip-spine 1–6 mm, inner tips gen entire (expanded and toothed). **FL:** corolla white to faintly purple-tinged. Meadows, damp soil, openings in forest; 400–2000 m. Teh, WTR, PR. May–Sep

var. ***congdonii*** (R.J. Moore & Frankton) D.J. Keil (p. 291) ROSETTE THISTLE **ST:** 0. **LF:** adaxially glabrous to cobwebby-tomentose and/or jointed-hairy, abaxially glabrous to white-tomen-

tose. **INFL:** heads 1–many, sessile, closely subtended by basal rosette; involucre 2–3 cm; outer phyllary tip-spine 1–4 mm, inner tips gen entire. **FL:** corolla lavender to purple. 2*n*=34. Meadows, springs, streambanks; 1900–3100 m. SNH, SnBr, SNE; NV. [*C. tioganum* (Congdon) Petr.] Jun–Aug

var. ***loncholepis*** (Petr.) D.J. Keil LA GRACIOSA THISTLE **ST:** 0 or 1–10 dm, gen ± branched throughout. **LF:** abaxially gen glabrous; main lf spines (3)7–10(12) mm. **INFL:** heads 1–many, sessile or short-peduncled, closely subtended by basal rosette or clustered at branch tips; involucre 2–3.5 cm; outer phyllary tip-spine 1–6 mm, inner tips gen entire (expanded and toothed). **FL:** corolla white to faintly purple-tinged. Marshes, dune wetlands; < 50 m. s CCo (sw San Luis Obispo, nw Santa Barbara cos.). [*C. l.* Petr.; *C. s.* Nutt. var. *citrinum* (Petr.) D.J. Keil, in part] Localized in lower valley of Santa Maria River; probably derived from similar pls of *C. scariosum* var. *citrinum* near headwaters of Cuyama River (tributary of Santa Maria River ± 150 km to e). Apr–Sep ★

var. ***robustum*** D.J. Keil SHASTA VALLEY THISTLE **ST:** 25–70 cm, gen branched proximally or throughout. **LF:** adaxially jointed-hairy and/or cobwebby-tomentose; abaxially thinly to densely tomentose. **INFL:** heads 3–many, peduncled in ± head-like to flat-topped clusters; involucre 2.5–4 cm; outer phyllary tip-spine 1–6 mm, inner tips gen expanded, as scarious, irregularly toothed appendages. **FL:** corolla white. Wet ground, meadows, pastures, marshes; 900–1900 m. KR, nw MP; OR. Jun–Jul

var. ***scariosum*** ELK THISTLE **ST:** 15–200 cm, often unbranched. **LF:** adaxially glabrous or jointed-hairy; abaxially glabrous to thinly or densely tomentose. **INFL:** heads gen several to many, ± sessile in head- to raceme- or spike-like cluster, gen overtopped by crowded distal cauline lvs; involucre 2–3.5 cm; outer phyllary tip-spine 1–8 mm, inner tips entire or expanded, as scarious, irregularly toothed appendages. **FL:** corolla white. 2*n*=34,36. Meadows, wet soil; ± 1000 m. KR; to BC, AB, MT, WY, CO. May–Jul

C. undulatum (Nutt.) Spreng. (p. 291) WAVYLEAF THISTLE Per 2–23 dm, from runner roots. **ST:** gen simple proximally, few-branched distally, white-tomentose. **LF:** adaxially gray-tomentose, abaxially white-gray-tomentose; proximal 15–30 cm, tapered to spiny petioles, elliptic to oblanceolate, ± shallowly lobed, lobes simple or with 2–4 narrow to wide 2° lobes or coarse teeth, main spines 2–4 mm, yellow; middle and distal gradually reduced, clasping or short-decurrent as spiny-margined wings, often spinier than proximal, main spines 2–10 mm. **INFL:** heads 1–few in ± flat-topped cluster; peduncle 0–25 cm, ± lfy; involucre 2.5–4.5 cm, 1.5–4.5 cm diam, ± ovoid to bell-shaped, ± loosely tomentose on phyllary margins, becoming glabrous; phyllaries strongly graduated, outer ovate, entire or minutely scabrous, tightly appressed, midrib with sticky-resinous ridge, tip-spines 2–5 mm, spreading to reflexed. **FL:** corolla 24–50 mm, white to pale lavender or pink, tube 11–28 mm, throat 6–14 mm, lobes 7–15 mm; style tip 5–7.5 mm. **FR:** 6–7 mm; pappus 20–40 mm. 2*n*=26. Disturbed areas, dry grassland, open scrubland; < 1600 m. MP; to s Can, c US. Waif (probably extirpated) in SnFrB, SCo. Considered noxious weed outside native (ne CA) range. Closely related to *C. ciliolatum.* May–Oct

C. vulgare (Savi) Ten. (p. 297) BULL THISTLE Bien 3–20 dm. **ST:** gen 1, ± openly branched in distal 1/2, loosely tomentose, often glandular-hairy. **LF:** adaxially with slender appressed bristle-like prickles, sometimes ± tomentose when young, abaxially ± tomentose; main veins prominently raised abaxially, ± glandular; proximal 10–40 cm, sessile or wing-petioled, shallowly to deeply 1–2 × lobed, main lobes gen rigidly spreading, spine-margined, otherwise entire, tip prolonged, cauline gradually reduced, long-decurrent as spiny wings, gen spinier than basal, main spines ≤ 15 mm. **INFL:** heads 1–several, ± clustered, closely subtended by bract-like distal-most lvs; peduncle 1–6 cm; involucre 3–4 cm, 2–4 cm diam, hemispheric or bell-shaped; phyllaries graduated, tips linear to lance-linear, spreading to reflexed, spines 1–5 mm. **FL:** corolla 25–35 mm, purple, tube 18–25 mm, throat 5–6 mm, lobes 5–7 mm; style tip 3.5–6 mm. **FR:** 3–4.5 mm; pappus 20–30 mm. 2*n*=68. Common. Disturbed areas; < 2350 m. CA-FP, GB; N.Am; native to Eur. May–Oct ◆

CONSTANCEA

Bruce G. Baldwin

1 sp. (Lincoln Constance, CA botanist, 1909–2001) [Baldwin & Strother 2006 FNANM 21:362–363]

C. nevinii (A. Gray) B.G. Baldwin (p. 297) NEVIN'S WOOLLY SUNFLOWER Subshrub 50–150(200) cm. **ST:** decumbent to ± erect, branched at base or throughout. **LF:** cauline, alternate, petioled, 8–25 cm; blade widely ovate, 1–2-pinnately divided into linear to oblong or oblanceolate lobes ± rounded at tips, margins slightly rolled under, faces white-tomentose, often ± glabrous adaxially in age. **INFL:** heads radiate, 50–100+ in flat-topped or panicle-like clusters; peduncle 1–5 mm; involucre 3–5 mm diam, cylindric to bell-shaped; phyllaries 8–16 in ± 2 series, free, ± keeled, 4–6 mm, oblong to linear, remaining erect in fr, persistent; receptacle flat to convex, epaleate, shallowly pitted, glabrous. **RAY FL:** 4–9; corolla yellow, ray 2–3 mm. **DISK FL:** 10–25+; corolla 2–4 mm, yellow, lobes deltate; anther tips ovate; style-branch tips obtuse to ± deltate, papillate adaxially. **FR:** 2–3 mm, obpyramidal to club-shaped, minutely scabrous or ± glabrous in age, ± black, dull; pappus scales 2–6+, 0.5–2.5 mm, basally fused, unequal (or 2 opposite, ± equal), oblong to awl-shaped, tip acute to fringed. 2*n*=38. Coastal bluffs, cliff faces; < 300 m. s ChI (exc San Nicolas Island). [*Eriophyllum n.* A. Gray] Apr–Sep ★

COREOPSIS

Ann, per. **ST:** erect. **LF:** simple or 1–2 × pinnately lobed or compound, basal or cauline, opposite [alternate], sessile or petioled. **INFL:** heads radiate, 1 or in cyme-like clusters; peduncles short to long; involucre hemispheric or bell-shaped; phyllaries in 2 series, outer ± spreading, thick, green, inner thin, membranous; receptacle flat to rounded, paleate; paleae flat, scarious. **RAY FL:** sterile; ray showy, yellow or proximally red-brown and distally yellow. **DISK FL:** many; corolla 4–5-lobed, yellow to red-brown; style tips truncate to short triangular. **FR:** gen compressed front-to-back, often winged; pappus 0 or of 2 awns or scales. 14–21 spp.: e N.Am. (Greek: bedbug-like, from fr) [Kimball & Crawford 2004 Molec Phylogen Evol 33:127–139; Strother 2006 FNANM 21:185–198] *C. wrightii* (A. Gray) E.B. Sm. not naturalized in CA. Native spp. moved to *Leptosyne*.

1. Per; ray yellow throughout; st nodes gen 2–5 . . . . ***C. lanceolata***
1′ Ann; ray yellow or proximally red-brown; st nodes gen > 5 . . . . ***C. tinctoria***

C. lanceolata L. (p. 297) GARDEN COREOPSIS Pl 3–6 dm from branched rootstock, glabrous to ± hairy. **ST:** 1–few, erect, simple or few-branched. **LF:** basal and cauline, opposite, proximal petioled, distal sessile; blade 5–15 cm, simple and oblanceolate or pinnate with 3–7 linear to oblanceolate lflets. **INFL:** heads in lfy-bracted cymes; peduncle 15–30 cm; involucre hemispheric; outer phyllaries 8–10, 5–10 mm, narrowly lanceolate; inner phyllaries 8–12 mm, lanceolate to ovate, obtuse or acute, margin scarious; palea 4–6 mm, lanceolate or ovate. **RAY FL:** gen 8; ray 1.5–3 cm, oblanceolate to obovate, gen 4-lobed. **DISK FL:** corolla ± 4 mm, 5-lobed, yellow. **FR:** 2.5–3 mm, round in ×-section or compressed front-to-back; faces black, rough; wing wide, thin; pappus scales 2, ≤ 1 mm. 2*n*=26. Disturbed places, escaped from cult; < 500 m. SnJV, CCo, SCo; native to e US. May–Jun

C. tinctoria Nutt. (p. 297) CALLIOPSIS Pl 3–12 dm, glabrous. **ST:** gen 1, erect, much-branched above. **LF:** cauline, opposite, petioled; blade 3–12 cm, proximal 1–2 × divided into linear to narrowly lanceolate, entire lobes, distal simple, linear. **INFL:** heads many in open, lfy-bracted cyme-like clusters; peduncle 4–10 cm; involucre bell-shaped; outer phyllaries 7–8, 1–3 mm, triangular; inner phyllaries 4.5–8 mm, ovate, obtuse or acute; palea 4–5 mm, linear, red-orange. **RAY FL:** gen 8; ray 6–20 mm. **DISK FL:** corolla ± 3 mm, 4-lobed, ± red. **FR:** 1.5–2.8 mm, linear-oblong to widely elliptic, black, smooth or minutely rough; wing 0 to wide; pappus 0. 2*n*=24,26. Disturbed places, escaped from cult; < 1000 m. GV, CW, SW; native nw US to e US. Gen Jun–Sep

CORETHROGYNE CALIFORNIA-ASTER, SAND-ASTER

Staci Markos & John L. Strother

1 sp. (Greek: female broom, for style-branch appendages) [Markos & Strother 2006 FNANM 20:450–452]

C. filaginifolia (Hook. & Arn.) Nutt. (p. 297) Per, subshrub, gen 10–100 cm. **ST:** 1–many from base, decumbent to ascending or erect, simple or distally branched, gen densely white-tomentose, sometimes ± glabrous and/or glandular, esp distally. **LF:** cauline at fl, often crowded proximally, alternate; sessile or wing-petioled; blade 10–70+ × 3–19 mm, linear to oblanceolate, spoon-shaped, or ovate, entire or toothed, hairy, sometimes with sunken glands and/or with sessile bead-like and/or stalked glands; distal smaller, sessile, bract-like. **INFL:** heads radiate, 1 or in cyme-like cluster; involucre hemispheric to bell-shaped, top-shaped, or cylindric, 6–14 × 3–10 mm; phyllaries 30–90+, graduated in 3–9 series, often spreading or with spreading tips, reflexed in age, linear to narrowly lanceolate, stiff-papery to scarious, flat, tips lf-like in texture, variously hairy and/or glandular; receptacle ± flat or rounded, pitted, with small scale-like projections, glabrous, epaleate. **RAY FL:** 10–43, sterile; ray white or pink to purple; style 0. **DISK FL:** 12–120+; corolla 4–8 mm, yellow, tube 0.6–1.4 mm, < narrowly cylindric throat, lobes erect, equal, narrowly lanceolate; anther tip awn-like; style branches linear, ± yellow-bristly; appendages blunt to awl-shaped, length 1/3–1/2 stigmatic bands. **FR:** cylindric to obconic, 2–5 mm, 5–7-ribbed, faces puberulent to long-soft-hairy; pappus 3–8 mm, persistent, of 35–65 free, unequal, coarse, minutely barbed bristles in 1–2 series, ± brown to ± red. 2*n*=10. Coastal scrub, chaparral, grassland, foothill woodland, forest; < 2600 m. NCo, KR, NCoRO, c&s SN, Teh, CW, SW, w DMoj; sw OR, n Baja CA. [*Lessingia f.* (Hook. & Arn.) M.A. Lane var. *californica* (DC.) M.A. Lane; *L. f.* var. *f.*] Jul–Nov

COSMOS

Ann [per]. **ST:** gen erect. **LF:** 1–3 × pinnately divided [simple], opposite. **INFL:** heads radiate, 1–few in cyme-like clusters; peduncle long; involucre hemispheric; phyllaries in 2 series, fused at base, outer gen ± lf-like in texture, inner thinner, wider; receptacle ± flat, paleate, paleae ± flat, entire. **RAY FL:** sterile; ray yellow, orange, white, pink, or purple. **DISK FL:** many; corolla yellow or orange; style tips thickened, acute. **FR:** ± cylindric, gen 4-angled, often beaked or much narrowed toward tip; pappus 0 or of stiff, barbed awns. ± 26 spp.: trop, subtrop Am, esp Mex. (Greek: ornament) [Kiger 2006 FNANM 21:203–205] In CA, *C. sulphureus* Cav. is a garden pl only.

291

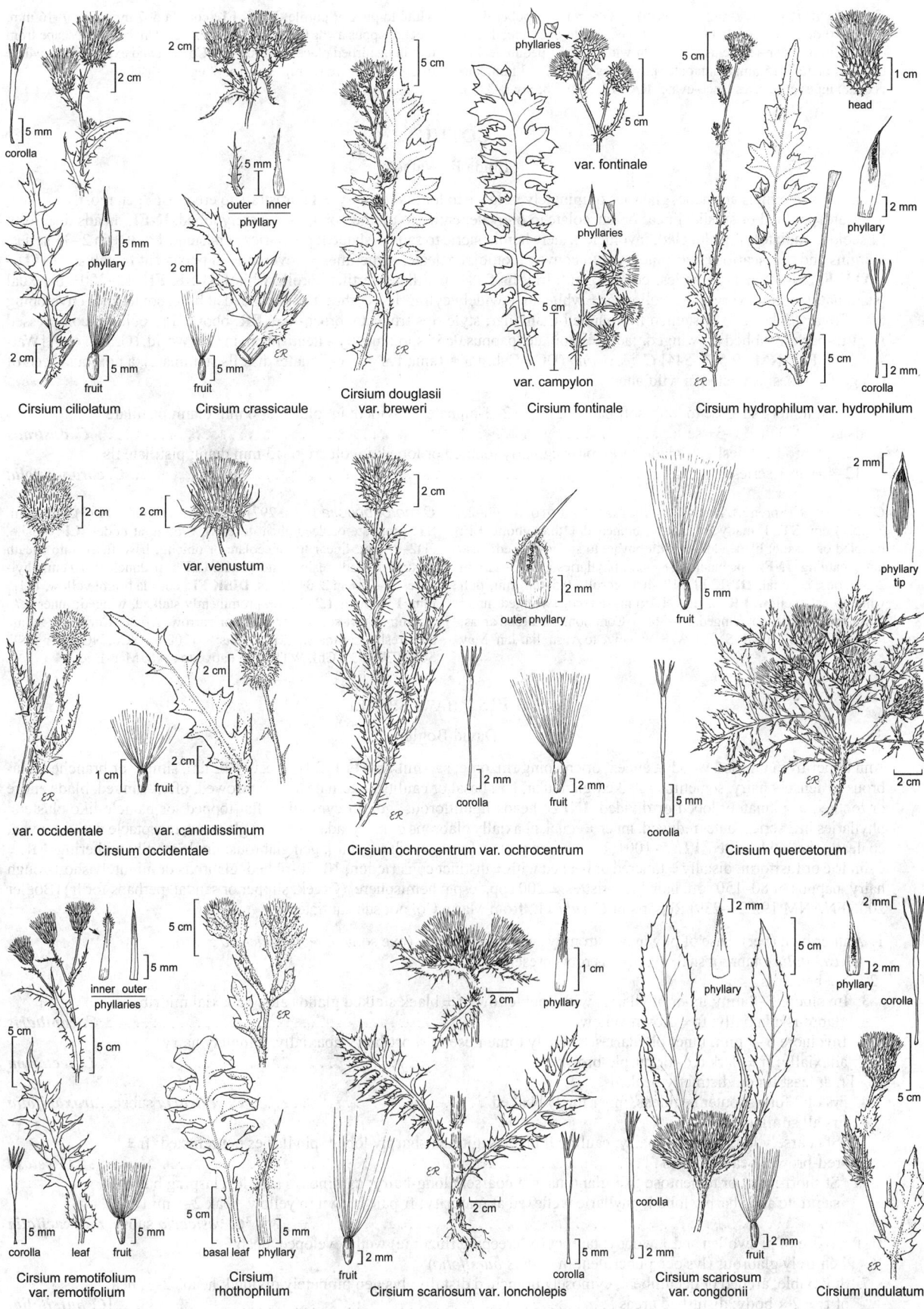

Cirsium ciliolatum

Cirsium crassicaule

Cirsium douglasii var. breweri

Cirsium fontinale

Cirsium hydrophilum var. hydrophilum

var. occidentale   var. candidissimum
Cirsium occidentale

Cirsium ochrocentrum var. ochrocentrum

Cirsium quercetorum

Cirsium remotifolium var. remotifolium

Cirsium rhothophilum

Cirsium scariosum var. loncholepis

Cirsium scariosum var. congdonii

Cirsium undulatum

C. bipinnatus Cav. GARDEN COSMOS Glabrous or puberulent. **ST**: 1–20 dm. **LF**: sessile or short-petioled; blade 6–11 cm, 1–2 × divided into linear segments ≤ 1.5 mm wide. **INFL**: peduncle 1–2 dm; involucre 7–15 mm diam; outer phyllaries gen 8, 9–13 mm, lanceolate; inner phyllaries lance-ovate. **RAY FL**: gen ± 8; ray 1–3 cm, white to pink or purple. **DISK FL**: corolla 5–7 mm. **FR**: 7–16 mm, black; pappus awns 0–3, 1–1.5 mm. 2*n*=24. Uncommon escape from cult in disturbed places; < 1000 m. SCo, cult and expected elsewhere in CA-FP; native to trop Am. Apr–Aug

COTULA

Linda E. Watson

Ann, per, sometimes aromatic, glabrous or minutely strigose to long-soft-hairy. **ST**: prostrate to erect. **LF**: gen mostly cauline, alternate, petioled or sessile, linear or lanceolate to obovate, entire to toothed or 1–3-pinnately lobed. **INFL**: heads disciform [discoid or radiate], 1, peduncled; involucre widely hemispheric to saucer-shaped; phyllaries persistent, 13–30+ in 2–3+ series, margins and tips scarious; receptacle flat to convex [conic], epaleate, sometimes ± covered with persistent fr stalks. **PISTILLATE FL**: 8–80 in 1–3+ series; corolla gen 0 [ray fls 5–8+, pistillate, fertile; corolla white]. **DISK FL**: 12–200+ bisexual [staminate]; corolla very short, yellow or ± white, tube widely cylindric, << throat, ± expanded at base, sometimes enveloping top of ovary, lobes gen 4; stamen tip rounded-triangular; style tips truncate, brush-like. **FR**: obovoid to oblong, compressed front-to-back, 2-ribbed or -winged, faces ± papillate; pappus 0. 55 spp.: mostly s hemisphere in Old World. (Greek: cup) [Watson 2006 FNANM 19:543–544] *C. mexicana* (DC.) Cabrera ◆ (ann, lvs gen 1-pinnate, disk fls staminate), a noxious weed of CA golf courses, expected in wildlands.

1. Ann, minutely strigose to long-soft-hairy; lf blade 2–3-pinnately divided; involucre 3–6 mm diam; pistillate fls gen 8–40 in ± 1–3+ series .... ***C. australis***
1′ Per, glabrous, ± fleshy; lf blade entire or irregularly toothed or lobed; involucre 6–15 mm diam; pistillate fls 12–40+ in 1 series .... ***C. coronopifolia***

C. australis (Spreng.) Hook. f. AUSTRALIAN COTULA Pl 2–20(25+) cm. **ST**: 1–many from base, branched ± throughout. **LF**: petioled or sessile; blade (1)2–6 cm, obovate to spoon-shaped, base not sheathing. **INFL**: peduncle 1–8+ cm; phyllaries gen 13–22+ in 2–3 series, ± equal. **DISK FL**: 12–40+; corolla 0.5–0.8 mm, dull white to pale yellow. **FR**: outer 1–1.2 mm, stalked, ± winged; inner 0.8–1 mm, sessile, not winged. 2*n*=36,40. Common. Disturbed areas; < 1650 m. NCo, NCoR, SNF, CW, SW; native to Australia. Jan–May

C. coronopifolia L. (p. 297) BRASS-BUTTONS Pl (3)5–40+ cm. **ST**: prostrate or decumbent to erect, rooting at nodes. **LF**: sessile, (1)2–7+ cm, linear to lanceolate or oblong, base fused into sheath around st, blade resin-gland-dotted. **INFL**: peduncle 2–10 cm; phyllaries 21–30+ in 2–3+ series. **DISK FL**: corolla bright yellow, 1–1.5 mm. **FR**: outer 1.2–2 mm, prominently stalked, winged; inner 0.7–1.5 mm, short-stalked, wings 0 or narrow. 2*n*=20. Common. Saline and freshwater marshes, mud flats; < 1200 m. NCo, NCoRI, n SNF, ScV, CW, SCo, ChI, WTR, PR; native to s Afr. Mar–Dec ❖

CREPIS HAWKSBEARD

David Bogler

Ann to per from taproot, woody caudex, or creeping rhizome; sap milky. **ST**: 1–20+, erect, ≤ 12 dm, simple or branched, glabrous to densely hairy, sometimes stalked-glandular. **LF**: basal or cauline, alternate, base narrowed, often winged, blade entire or toothed to pinnately lobed or divided. **INFL**: heads liguliflorous, gen in cyme-like, flat-topped, or panicle-like clusters; phyllaries in 2 series, outer reduced, inner ± equal, abaxially glabrous or hairy, adaxially gen glabrous; receptacle flat to convex, epaleate, shallowly pitted. **FL**: 5–100+; corolla gen yellow [orange, white, pink], gen glabrous; ligule readily withering. **FR**: ± cylindric or fusiform, distally ± tapered or beaked with ± distinct constriction, 10–20-ribbed, glabrous or minutely short-rough hairy; pappus of 80–150 soft, hair-like bristles. ± 200 spp.: esp n hemisphere. (Greek: slipper or sandal, perhaps for fr) [Bogler 2006 FNANM 19:222–239] Reports of *C. rubra* L. from Marin Co. not substantiated.

1. Ann or bien (per); base of pl gen not strongly swollen or woody (exc some *Crepis vesicaria* subsp. *taraxacifolia*); taproot shallow, fibrous roots present
  2. Fr ± beakless
    3. Involucre 5–8 mm; inner phyllaries with double row of ± black stalked glands along abaxial midrib, glabrous adaxially; fr ± brown-yellow .... ***C. capillaris***
    3′ Involucre 6–9 mm; inner phyllaries minutely tomentose to short-hairy abaxially, minutely hairy adaxially; fr ± dark red- or purple-brown... ***C. tectorum***
  2′ Fr, at least inner, distinctly beaked
    4. Frs of 2 forms, outer beakless, inner finely beaked .... ²***C. vesicaria*** subsp. ***taraxacifolia***
    4′ Frs all similar, beaked
      5. St coarsely long- or short-hairy; cauline lf blade sagittate, sharply lobed; phyllaries not reflexed; fr ± red-brown, beak 1–2 mm... ***C. setosa***
      5′ St short-hairy or tomentose to ± glabrous, not coarsely long-hairy; cauline lvs sessile, clasping but not sagittate and sharply lobed; phyllaries reflexed at maturity; fr pale brown to yellow, beak 2–5 mm
        .... ²***C. vesicaria*** subsp. ***taraxacifolia***
1′ Per; base of pl swollen and ± woody; taproot (or creeping rhizome) well developed, deep
  6. Pl entirely glabrous (lvs occ puberulent in *Crepis bursifolia*)
    7. St flexible, arched or decumbent, cymosely branched distally; lvs gen pinnately divided; heads 2–3; beak of fr ± 2 × body; disturbed areas .... [***C. bursifolia***]

7′ St ± erect or ascending, gen branched; lvs gen round to spoon- or fiddle-shaped, entire or few-toothed to
    shallowly lobed; heads 5–10(100); fr tapered but not beaked; alpine. . . . . ***C. nana***
6′ Pl gen ± hairy (*Crepis runcinata* sometimes glabrous)
  8. Lvs gen dentate, sometimes pinnately lobed. . . . . ***C. runcinata***
    9. Lf green, teeth prominently white-tipped; inner phyllaries widely lanceolate or elliptic to oblong
    . . . . subsp. ***andersonii***
    9′ Lf glaucous, teeth not prominently white-tipped; inner phyllaries linear to narrowly lanceolate . . . . . . subsp. ***hallii***
  8′ Lvs gen pinnately lobed or divided or sharply serrate
    10. Herbage and involucres coarsely hairy, hairs 1–3 mm; phyllaries acute or long-acuminate
      11. Herbage and involucres densely coarse-hairy and stalked-glandular; inner phyllaries long-acuminate
      . . . . ***C. monticola***
      11′ Herbage and involucres bristly and ± short-tomentose, gen not stalked-glandular; inner phyllaries
      acute (see 13. for vars.). . . . . ²***C. modocensis***
    10′ Herbage and involucres ± short-tomentose, hairs or bristles gen < 1 mm; phyllaries acute
      12. Inner phyllaries with ± black, ± white, or green bristles; fr ± dark green to deep red, dark brown or ±
      black, weakly ribbed or striate. . . . . ²***C. modocensis***
        13. St branched near or distal to middle; involucre 11–16 mm; pappus 5–10 mm . . . . subsp. ***modocensis***
        13′ St low, branched near base; involucre 13–21 mm; pappus 9–13 mm . . . . subsp. ***subacaulis***
      12′ Inner phyllaries gen glabrous, tomentose, or less often short-stalked-glandular or with few black
      bristles; fr ± yellow, light to dark brown, red-brown, or deep green, distinctly ribbed
        14. Largest heads with 5–8 inner phyllaries; fls gen 5–10
          15. Cauline lvs well developed; heads 30–70(100+) in compound flat-topped cluster; inner phyllaries
          gen glabrous or occ evenly tomentose, not strongly keeled; fr ± yellow or yellow-brown. . . . . ***C. acuminata***
          15′ Cauline lvs much reduced; heads 7–10(30) in small cyme-like cluster; inner phyllaries
          conspicuously white-tomentose near margins, bases becoming strongly keeled, midrib gen
          glabrous; fr red-brown . . . . ***C. pleurocarpa***
        14′ Largest heads with 7–14 inner phyllaries; fls 6–40
          16. Pl 25–60 cm; heads gen > 20, in large ± flat-topped or panicle-like cluster; involucre 3–5 mm
          diam; fls 7–12 . . . . ***C. intermedia***
          16′ Pl 5–40 cm; heads gen < 20 in cyme-like or panicle-like cluster; involucre 5–15 mm diam; fls 10–40
            17. Herbage green, stalked-glandular; lf midribs often conspicuously red-purple when fresh; inner
            phyllaries conspicuously stalked-glandular (glabrous) . . . . ***C. bakeri***
              18. Involucre narrowly cylindric to ± obconic, 18–21 mm in fr; outer phyllaries deltate, longest <<
              inner; pappus > fr . . . . subsp. ***idahoensis***
              18′ Involucre widely cylindric, 13–20 mm in fr; outer phyllaries lanceolate, longest ± 1/2 inner;
              pappus ± ≤ fr
                19. Involucre 16–20 mm in fr; fr 8–10.5 mm, tip slightly tapered; pappus 9–10.5 mm . . . . subsp. ***bakeri***
                19′ Involucre 13–17 mm in fr; fr 6–9 mm, tip strongly tapered; pappus 6–9 mm . . . . subsp. ***cusickii***
           17′ Herbage ± gray-tomentose, stalked-glandular or not; lf midribs not red-purple, inconspicuous;
           phyllaries stalked-glandular or not. . . . . ***C. occidentalis***
            20. Inner phyllaries with at least some stalked glands
              21. Inner phyllaries, peduncles, and distal cauline lvs stalked-glandular and with large dark or
              black bristles; largest heads with 8 inner phyllaries and 12–14 fls. . . . . subsp. ***costata***
              21′ Inner phyllaries, peduncles and distal cauline lvs stalked-glandular, but lacking large dark or
              black glandular bristles; largest heads with 10–13 inner phyllaries and 18–30 fls . . . . subsp. ***occidentalis***
            20′ Inner phyllaries glandless, or if with a few stalked glands then involucre with 8 phyllaries
              22. St gen low, 5–20 cm, branched near base; inner phyllaries 8–12; lvs deeply pinnately lobed,
              lobes narrow, remote, lanceolate, few-toothed. . . . . subsp. ***conjuncta***
              22′ St well developed, 10–40 cm, with definite primary axis below heads; inner phyllaries gen 8;
              lvs strongly dentate or pinnately lobed, lobes closely spaced . . . . subsp. ***pumila***

C. acuminata Nutt. (p. 297) Per from deep taproot; caudex swollen, woody. **ST:** 1–5, erect, 2–7 dm, branched near or distal to middle, tomentose. **LF:** basal and cauline, 12–40 cm, elliptic to lanceolate, long-acuminate, deeply pinnately lobed, lobes narrowly acute-triangular, sometimes with short 2° lobes, faces gray-green, minutely tomentose; cauline similar, smaller. **INFL:** heads 30–70(100+) in compound flat-topped cluster; involucre 8–16 mm, cylindric to bell-shaped; outer phyllaries 1–2 mm, narrowly triangular; inner 5–8, 8–12 mm, lanceolate, gen glabrous, smooth and shining or lightly tomentose. **FL:** 5–10. **FR:** 6–9 mm, ± cylindric, pale yellow-brown, tip tapered but not beaked, ribs 12; pappus white. 2*n*=22,33,44,55,88. Open rocky hillsides, ridges, grassy flats; 1000–3300 m. KR, NCoRH, CaRH, SNH, Teh, TR, GB; to WA, MT, NE, NM. May–Aug

C. bakeri Greene Per, taproot thick, caudex woody. **ST:** 1–3, erect, 1–3 dm, mostly simple, dark green or ± red, ± tomentose, often short-stalked-glandular. **LF:** basal and cauline; basal ± elliptic, pinnately divided, lobes widely lanceolate, recurved, coarsely dentate, midribs often red-purple, faces ± tomentose, stalked-glandular; cauline similar, smaller. **INFL:** heads in cyme-like cluster; outer phyllaries 3–8 mm, deltate to lanceolate; inner phyllaries 10–14, 10–14 mm, lanceolate, glabrous or ± tomentose and stalked-glandular, occ coarsely long-hairy. **FL:** 11–40. **FR:** fusiform, dark brown to ± yellow, ribs 10–13; pappus dusky or ± yellow-white. 2*n*=22,33,44,55.

subsp. ***bakeri*** Pl 10–30 cm. **LF:** 8–12 cm, deeply lobed, lobes lanceolate to elliptic, faces minutely tomentose. **INFL:** heads 2–13; involucre 16–20 mm in fr, widely cylindric; outer phyllaries lanceolate, longest ± 1/2 inner. **FR:** 8–10.5 mm, tip slightly tapered; pappus 9–10.5 mm. 2*n*=44. Dry open slopes; 550–1900 m. KR, NCoRH, CaR, SNH, MP; to OR, NV. May–Jul

subsp. *cusickii* (Eastw.) Babc. & Stebbins Pl 8–16 cm. **LF**: 8–12 cm, deeply lobed, lobes triangular, dentate, faces tomentose. **INFL**: heads 1–4(10); involucre 13–17 mm in fr, widely cylindric; outer phyllaries lanceolate, longest ± 1/2 inner. **FR**: 6–9 mm, tip strongly tapered; pappus 6–9 mm. $2n=22,33$. Sagebrush scrub; 1200–2200 m. CaRH; to OR, UT. Jun–Jul

subsp. *idahoensis* Babc. & Stebbins Pl 25–30 cm. **LF**: 15–18 cm, shallowly lobed, lobes triangular, dentate, faces ± glabrous. **INFL**: heads 7–22; involucre 18–21 mm in fr, narrowly cylindric to ± obconic; outer phyllaries deltate, longest << inner. **FR**: 8 mm, narrow, tip not strongly tapered; pappus 12–13 mm. $2n=55$. Dry open places; 400–2200 m. MP; to ID. May–Jul

C. capillaris (L.) Wallr. (p. 297) Ann, bien with shallow taproot, many fibrous lateral roots. **ST**: gen 1(many), 0.1–9 dm, decumbent to erect, short-hairy proximally or throughout, not glandular. **LF**: basal and cauline, 5–30 cm, ± oblanceolate, clasping, pinnately divided or sharply toothed, lobes lanceolate, remote, unequal, recurved, faces glabrous to sparsely hairy. **INFL**: heads 10–15(30+) in flat-topped cluster; involucre 5–8 mm, cylindric; outer phyllaries 2–4 mm, linear, minutely tomentose or stalked-glandular; inner 8–16, 6–7 mm, lanceolate, with prominent double row of black glandular bristles along midrib. **FL**: 20–60; corolla hairy. **FR**: 1.5–2.5 mm, fusiform, ± brown-yellow, tip slightly tapered, ribs 10; pappus white, fluffy. $2n=6$. Weed in pastures, disturbed places; < 1300 m. NW, CaR, n&c SN, CCo, SnFrB, SCo; much of US and Can; native to Eur. May–Nov

C. intermedia A. Gray Per from stout taproot; caudex swollen, woody. **ST**: 1–2, erect, 2.5–6 dm, stout, branched near middle or distally, ± tomentose. **LF**: basal and cauline; basal 15–40 cm, elliptic to lanceolate, pinnately lobed, lobing variable but often cleft ± 1/2 to midrib, lobes narrowly triangular, often with 2° lobes, faces gray-tomentose; cauline similar, smaller. **INFL**: heads (10)20–60, in flat-topped or panicle-like cluster; involucre 10–16 mm, narrowly cylindric, 3–5 mm diam; outer phyllaries 2–4 mm, lance-deltate, minutely tomentose; inner phyllaries 7–10, 10–13 mm, lanceolate, ± tomentose, occ with a few ± green bristles, not stalked-glandular. **FL**: 7–12. **FR**: 6–9 mm, ± cylindric, ± yellow-brown, tip tapered, ribs 10–12; pappus dusky white. $2n=33,44,55,88$. Dry slopes, ridges, open forest; 800–3300 m. KR, NCoRH, CaRH, n&c SN, GB; to w Can, WY, CO. Complex series of asexually reproducing polyploid forms, probably of hybrid origin, combining characters of *C. acuminata, C. modocensis, C. occidentalis, C. pleurocarpa*. May–Jul

C. modocensis Greene Per from slender taproot; caudex branched. **ST**: 1–4, erect, 5–35 cm, slender to stout, simple or sparsely branched, tomentose or coarsely long-hairy, hairs ± yellow (white). **LF**: basal and cauline, 7–25 cm, lanceolate, deeply pinnately lobed, lobes lanceolate, dentate and secondarily lobed, teeth mucronate, faces minutely tomentose and coarsely hairy with ± black, ± white, or green bristles. **INFL**: heads 1–9, in cyme-like cluster; involucre 11–21 mm, cylindric; outer phyllaries 2–4 mm, lanceolate, acute; inner phyllaries 8–18, 10–16 mm, lanceolate, acute, often densely ± black- or ± white-tomentose or bristly. **FL**: 10–60. **FR**: 7–12 mm, ± cylindric to fusiform, ± dark green to deep red, dark brown or ± black, tip tapered or beaked, weakly ribbed or striate; pappus dusky white. $2n=22,33,44,55,66,88$.

subsp. *modocensis* Pl 10–35 cm. **ST**: slender, branched near or distal to middle; sparsely bristly. **LF**: 10–20 cm. **INFL**: involucre 11–16 mm in fr; phyllaries with ± black bristles. **FR**: ± deep green or ± black to ± deep red; pappus 5–10 mm. Rocky slopes; 900–2500 m. CaRH, MP; to BC, MT, WY, CO. Incl the diploid form and several apomictic polyploid forms. May–Jul

subsp. *subacaulis* (Kellogg) Babc. & Stebbins Pl 6–30 cm. **ST**: stout, branched near base; bristly. **LF**: 10–15 cm. **INFL**: heads 1–6; involucre 13–21 mm; phyllaries sparsely coarse-hairy with black bristles, or glabrous. **FR**: ± black to ± brown or ± dark red; pappus 9–13 mm. Wet meadows, steep slopes; 1800–2100 m. n SNH; to OR, MT, NV. May–Jul

C. monticola Coville (p. 297) Per from deep taproot. **ST**: 1–3.5 dm, lfy, simple or branched, densely long-coarse-hairy and stalked-glandular. **LF**: basal and cauline 10–25 cm, elliptic or oblanceolate,

pinnately lobed or sharply serrate, lobes lanceolate, toothed, faces long-soft- or long-coarse-hairy, glandular-bristly, hairs 1–3 mm. **INFL**: heads 4–20 in loose cyme-like cluster; involucre 18–24 mm, cylindric to bell-shaped; outer phyllaries 3–5 mm, lance-linear; inner phyllaries 7–12, 14–20 mm, linear to narrowly lanceolate, long acuminate, densely coarse-hairy with long, gland-tipped hairs. **FL**: 16–20. **FR**: 5.5–9 mm, fusiform, ± red-brown, distally narrowed but not beaked, strongly ribbed; pappus creamy white. $2n=22,33,44,55,77,88$. Conifer forest, open woodland; 700–2400 m. KR, NCoRH, CaRH, n SNH, se SnFrB (Mount Hamilton Range), MP; to OR. May–Jul

C. nana Richardson (p. 297) Per from taproot or creeping rhizome, ± purple-green, glabrous. **ST**: 1–10+, 2–10 cm, erect or ascending, slender, simple or gen much-branched, in dense clumps. **LF**: mostly basal, round to spoon- or fiddle-shaped, entire or few-toothed to shallowly lobed. **INFL**: heads 5–10(100) in cyme-like clusters borne among or distal to lvs; involucre 10–13 mm, cylindric; outer phyllaries 2–3 mm, lanceolate or ovate, dark green or ± black; inner phyllaries 10–11 mm, 10, oblong, dark green or purple. **FL**: 9–12. **FR**: 4–6 mm, ± cylindric to fusiform, golden-brown, tip tapered but not beaked, ribs 10–13; pappus bright white, deciduous. $2n=14$. Talus slopes, gravel bars; 2000–4000 m. c&s SNH, SnGb, n SNE (Sweetwater Mtns), W&I, n DMtns (Panamint Range); to AK, MT, WY, CO, ne N.Am; Asia. May–Sep

C. occidentalis Nutt. Per from deep taproot; caudex woody. **ST**: erect, stout, branched from base or middle, ± gray-tomentose and/or stalked-glandular. **LF**: basal and cauline; basal and proximal ± elliptic, pinnately lobed or dentate, lobes widely lanceolate, recurved, often dentate; faces gray-tomentose, ± stalked-glandular; cauline reduced distally on st. **INFL**: heads in loose flat-topped cluster; involucre 11–19 mm, cylindric; outer phyllaries 2–6 mm, linear to deltate; inner phyllaries 7–13 mm, lanceolate, acute, tomentose and occ with short, green or black gland-tipped hairs. **FR**: 6–10 mm, fusiform, light to dark brown, tip beakless, ribs 10–18; pappus dusky white. $2n=22,33,44,55,66,77,88$. Subspp. intergrade extensively. Highly variable; incl genes from most or all dry-land montane spp.

subsp. *conjuncta* Babc. & Stebbins **ST**: 5–20 cm, tomentose but not stalked-glandular. **LF**: 10–18 cm, deeply pinnately lobed, lobes narrow, remote. **INFL**: heads 2–9, peduncle tomentose, not stalked-glandular; inner phyllaries 8–12, minutely tomentose, not stalked-glandular. **FL**: 12–40. **FR**: dark brown. Ridgetops, gravelly soils; 1400–2100 m. KR, NCoRH, CaRH, n&c SNH, Wrn; to WA, MT, CO. Jun–Jul

subsp. *costata* (A. Gray) Babc. & Stebbins **ST**: 8–40 cm, tomentose, ± stalked-glandular. **LF**: 5–20 cm, pinnately lobed, lobes dentate. **INFL**: heads 15–30, peduncle stalked-glandular; inner phyllaries 7–8, stalked-glandular with at least a few large dark or black bristles. **FL**: (10)12–14. **FR**: golden brown. $2n=44$. Rocky hillsides, oak/juniper woodland; 1200–2500 m. CaRH, n SNH, MP; to Can, MT, CO, WY. Jun–Jul

subsp. *occidentalis* (p. 297) **ST**: 10–40 cm, stalked-glandular. **LF**: 10–20 cm, sharply dentate to pinnately lobed, lobes dentate, stalked-glandular. **INFL**: heads 10–30, peduncle minutely tomentose, stalked-glandular; inner phyllaries 10–13, stalked-glandular. **FL**: 18–30. **FR**: golden brown. $2n=22,33$. Dry rocky hillsides, sagebrush scrub; 1000–2200 m. CaRH, SNH, SnBr, MP; to Can, c US, n Mex. May–Jun

subsp. *pumila* (Rydb.) Babc. & Stebbins **ST**: 10–40 cm, minutely tomentose, not glandular. **LF**: 10–20 cm, strongly dentate or pinnately lobed, lobes closely spaced, coarse, dentate. **INFL**: heads 5–15, peduncle minutely tomentose, stalked-glandular; inner phyllaries gen 8, ± tomentose, stalked-glandular or not. **FL**: 12–20. **FR**: brown. Open pine woodland, serpentine slopes; 800–1800 m. NCoRH, CaRH, n SNH, SnFrB, WTR, GB; to Can, MT, UT. Jun–Jul

C. pleurocarpa A. Gray (p. 297) Per from slender woody taproot. **ST**: 1.5–6 dm, gen branched near base, ± glabrous to ± tomentose. **LF**: basal and cauline; basal 7–28 mm, elliptic or ± oblanceolate, margins pinnately lobed to dentate, lobes remote, often recurved; cauline much reduced, gen not lobed. **INFL**: heads 7–10(30) in small

cyme-like cluster; involucre 8.5–17 mm, cylindric to bell-shaped; outer phyllaries 1.5–4 mm, lanceolate to deltate; inner phyllaries 5–8, 10–16 mm, lanceolate to deltate, bases becoming strongly keeled and swollen, margins densely tomentose. **FL:** 5–8. **FR:** 5–8 mm, ± cylindric, ± deep red-brown, tip beakless, ribs 10, prominent; pappus dusky to ± yellow-white. 2*n*=22,33,44, 55,77,88. Along streams in mixed-conifer forest, steep rocky slopes, often on serpentine; 400–2200 m. KR, NCoRO, NCoRH, CaRH, n&c SNH; to WA. Jun–Aug

C. runcinata (E. James) Torr. & A. Gray Per from long woody taproot and swollen caudex. **ST:** 2.5–8 dm, branched near middle, glabrous to short-hairy, sometimes stalked-glandular. **LF:** mostly basal, oblanceolate or elliptic, 7–27 cm, gen dentate, sometimes pinnately lobed, glabrous or short-hairy, sometimes glaucous; cauline inconspicuous. **INFL:** heads 1–5(30) in compound flat-topped clusters; involucre 7–21 mm, ± bell-shaped; outer phyllaries 1–3 mm, narrowly triangular; inner phyllaries 8–10 mm, narrowly to widely lanceolate, glabrous to minutely tomentose, sometimes stalked-glandular. **FL:** 20–50. **FR:** 3.5–7.5 mm, fusiform, pale yellow to brown, tip tapered or shortly beaked, ribs 10–13, strong; pappus white. 2*n*=22. Other subspp. to WA, s-c Can, n-c US, NV. ★

subsp. ***andersonii*** (A. Gray) Babc. & Stebbins Pl 25–50 cm. **LF:** 2–3 cm wide, oblanceolate, strongly, closely dentate, teeth conspicuously white-tipped, faces glabrous or short-rough-hairy, green. **INFL:** inner phyllaries widely lanceolate or elliptic to oblong. **FR:** ± distinctly beaked. 2*n*=22. Alkaline seeps, grassland, valleys; 1200–1500 m. s MP (Sierra Valley, se Plumas Co.); NV. [*C. r.* var. *a.* (A. Gray) Cronquist] May–Jul

subsp. ***hallii*** Babc. & Stebbins (p. 297) HALL'S MEADOW HAWKSBEARD Pl 20–60 cm. **LF:** 1.5–3 cm wide, oblanceolate or narrowly obovate, coarsely dentate or pinnately lobed, teeth not prominently white-tipped, faces glabrous, glaucous. **INFL:** inner phyllaries linear to narrowly lanceolate. **FR:** tapered but not distinctly beaked. 2*n*=22. Moist, alkaline valleys; 1250–1450 m. SNE; NV. Declining due to grazing, habitat drainage. Jun–Jul

C. setosa Haller f. Ann with shallow taproot and fibrous roots. **ST:** 1–many, 2–8 dm, stout, hollow, often ± red, coarsely long- or short-hairy, hairs spreading, not glandular. **LF:** basal and proximal cauline 5–30 cm, basal ± oblanceolate, blade sagittate, proximally dentate or pinnately lobed, terminal lobe often large, lateral lobes ± recurved, faces finely hairy; cauline sagittate, sharply lobed, distal cauline sessile, narrower. **INFL:** heads 5–30 in cyme- or panicle-like cluster; involucre 8–11 mm, narrowly bell-shaped; outer phyllaries 2–4 mm, linear, coarsely long-hairy; inner phyllaries 12–16, 6–7 mm, lanceolate, coarsely long-hairy, not glandular. **FL:** 20–25; corolla yellow, sometimes proximally ± red. **FR:** 3–5 mm, fusiform, ± red-brown, tapered to narrow, beak 1–2 mm, ribs 10; pappus white, fine and soft, deciduous. 2*n*=8. Openings in mixed-conifer forest, disturbed areas; 50–500 m. NCoRO; native to Eur. May–Nov

C. tectorum L. Ann with shallow taproot and fibrous roots. **ST:** 1, 1–10 dm, hollow, branched, minutely tomentose to short-hairy. **LF:** basal and cauline; basal 5–15 cm, lanceolate to oblanceolate, entire, dentate, or pinnately lobed, lobes remote, coarse, unequal, ± recurved, faces glabrous or tomentose; cauline reduced, with ear-like basal lobes. **INFL:** heads 5–20, in panicle-like or flat-topped clusters; involucre 6–9 mm, narrowly bell-shaped; outer phyllaries 2–5 mm, awl-like, hairy; inner phyllaries 12–15, 5–9 mm, lanceolate, abaxially tomentose to short-hairy, adaxially minutely hairy. **FL:** 30–70. **FR:** 3–4 mm, fusiform, ± dark red- or purple-brown, constricted towards tip but not beaked, ribs 10; pappus white, fine, soft. 2*n*=8. Dry sandy pine woodland, forest clearings; 2100–2500 m. SnBr, SNE; widespread US; native to Eur. May–Sep

C. vesicaria L. subsp. ***taraxacifolia*** (Thuill.) Thell. Ann to per, from slender taproot; caudex swollen, woody. **ST:** 1, decumbent to erect, 0.3–12 dm, much-branched, short-hairy or tomentose to ± glabrous. **LF:** basal and cauline 10–35 cm, oblanceolate to ovate, pinnately lobed to dentate, terminal lobe large, toothed, lateral lobes ± recurved, faces short-hairy to ± glabrous; cauline sessile, clasping. **INFL:** heads 10–20, in loose flat-topped cluster; involucre 8–12 mm, cylindric to bell-shaped or ± obconic; outer phyllaries 3–4 mm, ovate to lance-linear, glabrous; inner phyllaries 9–13, lanceolate, reflexed at maturity, often stalked-glandular. **FL:** 50–70; corolla yellow, ± red abaxially. **FR:** all similar or of 2 forms, 4–9 mm, fusiform, pale brown to yellow, outer distally narrowed but not beaked, inner with beak 2–5 mm, ± 1/3 body, ribs 10; pappus white, fine, soft. 2*n*=8,16. Sandy clearings, hillsides, disturbed places; < 300 m. NCoRO, SnFrB, SCo; native to Eur. Feb–Oct

CROCIDIUM SPRING GOLD

1 sp. (Greek: little loose tufts, from hair in lf axils) [Strother 2006 FNANM 20:641]

C. multicaule Hook. (p. 297) Ann 5–30 cm. **ST:** 1–few from base, gen unbranched, glabrous to ± tomentose (esp lower axils). **LF:** basal and cauline, alternate, 10–25 mm, linear to obovate, ± fleshy, entire or few-toothed, cauline reduced. **INFL:** heads 1, radiate, long-peduncled; involucre ± bell-shaped; phyllaries 5–13 in 1 series, fused at base, ovate or elliptic, thin, veiny; receptacle conic, epaleate. **RAY FL:** 5–13; ray 4–10 mm, yellow. **DISK FL:** 12–100+; corolla 2–3 mm, yellow; anther bases rounded, tips narrowly triangular; style branches flattened, tips triangular, minutely papillate. **FR:** 1.5–2.5 mm, elliptic, ribbed, brown, puberulent, gelatinous when wet; pappus 1–3 mm, of barbed bristles, early-deciduous (often 0 on ray fr). 2*n*=18. Sandy soils, grassland, open woodland; gen ≤ 1600 m. KR, NCoR, CaRF, SNF, e SnFrB, SCoR (uncommon), MP; to BC, ID. Feb–Jun

CRUPINA

David J. Keil & Charles E. Turner

Ann. **ST:** erect, openly branched above. **LF:** basal entire or toothed; cauline alternate, pinnately lobed. **INFL:** heads disciform; involucre cylindric to ovoid; phyllaries graduated in several series, lance-oblong, acute; receptacle epaleate, scaly. **FL:** 3–15; outer sterile, central 1–2 fertile; corolla purple, tube slender, gradually wider, lobes linear; anther base short-tailed, tip narrowly triangular; style with minutely hairy cylindric distal node and papillate terminal segment, branches short, triangular. **FR:** cylindric or ± compressed; base puberulent, soft-hairy upward; pappus (on fertile fr only) of small scales and stiff bristles. 2 spp.: Eur. (Latin: ancient name) [Keil 2006 FNANM 19:177–178]

C. vulgaris Cass. (p. 301) BEARDED CREEPER **ST:** gen 2–10 dm, lfy up to branches. **LF:** basal sessile or petioled, oblong to obovate, scabrous, gen 0 at fl; cauline sessile, 1–3.5 cm, lobes linear, minutely toothed. **INFL:** peduncle 1–3 cm; involucre 8–20 mm. **FL:** 3–5; corolla ± 14 mm. **FR:** gen 1 per head, 3–4 mm, barrel-shaped, attached at base; pappus scales ≤ 1 mm, bristles 5–7 mm, black-brown. 2*n*=28, 30. Grassy places; < 1000 m. s NCoR (Sonoma Co.), MP; to WA, ID; native to s Eur. May–Aug ◆

CUNICULOTINUS PANAMINT ROCK-GOLDENROD

Lowell E. Urbatsch

1 sp. (Latin: rabbit shrub) [Urbatsch et al. 2006 FNANM 20:100–101]

C. gramineus (H.M. Hall) Urbatsch et al. (p. 301)   Subshrub 2–6 dm; woody base branched, ≤ 1 dm. **ST**: ± glabrous, ridged from bases of major lf veins. **LF**: basal and cauline, alternate, sessile, ascending, 30–85 mm, 3–9 mm wide, ± lanceolate, firm, glabrous exc margins finely rough-hairy, prominently 3–5-veined. **INFL**: heads discoid, few; involucre 11–17.5 mm, cylindric; phyllaries graduated, oblong to ovate, firm, not keeled, tips irregularly toothed or notched, abruptly pointed; receptacle flat, finely pitted, epaleate. **FL**: 4–7; corolla 9.5–11.5 mm, yellow; anther tip narrowly triangular; style tips linear, appendage > stigmatic portion. **FR**: 7–9 mm, glabrous; pappus ± = corolla. 2*n*=18. Pinyon/juniper woodland, bristlecone-pine forest; 2200–2900 m. s W&I, n DMtns (Panamint Range); s NV. [*Chrysothamnus g.* H.M. Hall] Jul–Aug ★

CYNARA

David J. Keil & Charles E. Turner

Per 5–25 dm, ± tomentose in CA. **ST**: gen erect, lfy, branched, stout. **LF**: basal and cauline, alternate, 1–3-pinnately lobed, divided, or compound, ± spiny. **INFL**: heads discoid, large, 1–few in cyme-like clusters; involucre ovoid or hemispheric; phyllaries graduated in 5–8+ series, gen ovate, leathery, entire, glabrous, tips gen triangular; receptacle flat, fleshy, epaleate, bristly. **FL**: many; corolla white to blue-purple, tube very slender, throat widened abruptly, lobes linear; anther base long-sagittate, tip oblong; style tip with terminal segment long, cylindric, minutely papillate, tip barely notched. **FR**: cylindric to obconic, ± 4-angled or ± compressed, glabrous, attached at base; pappus of many stiff bristles in several series, white or ± brown, plumose below, fused and falling together. 8 spp.: Medit, Canary Islands, w Asia. (Greek: artichoke) [Keil 2006 FNANM 19:89–90]

C. cardunculus L.   Pls 5–25 dm. **LF**: 3–20 dm, abaxially densely tomentose, gray or white, adaxially thinly cobwebby, green; lobes oblong to lanceolate, entire or coarsely toothed; spine tips of teeth and lobes small and weak to stiff and needle-like. **INFL**: involucre 3–15 cm, 4–15 cm diam (not incl phyllary tips); phyllaries weakly to strongly spine-tipped. **FL**: corolla 3–5 cm, gen blue or purple; styles long exserted. **FR**: 4–8 mm; pappus 2–4 cm. 2*n*=34. ❖

1. Middle phyllaries widely obtuse, truncate, or notched, distally with or without spine tip 1–2 mm or acuminate with pointed tip 22–38 mm and slender spine 6–9 mm; phyllaries indistinctly ± yellow-margined or not. . . . . subsp. *cardunculus*

1′ Middle phyllaries acute to short-acuminate, with pointed tip 10–21 mm and spine 2–5(6) mm; phyllaries prominently ± yellow-margined . . . . subsp. *flavescens*

subsp. *cardunculus* (p. 301) ARTICHOKE **LF**: lobes unarmed or minutely spiny or tipped with needle-like spines 1–3 cm. Disturbed places; < 500 m. s SNF, GV, CW, SW; native to Medit. [*C. scolymus* L.] Nearly spineless forms cult for edible phyllary bases and receptacles. Some globe artichoke escapes apparently have reverted to much spinier wild forms. Apr–Jul

subsp. *flavescens* Wiklund (p. 301)   Pl 8–20 dm, bushy. **LF**: lobes tipped with needle-like spines 1–3 cm margin. Disturbed places; < 500 m. NCoR, SNF, GV, CW, SW; native to s Eur. Apr–Jul

DEINANDRA TARWEED, TARPLANT

Bruce G. Baldwin

Ann [per] or subshrub to shrub, 4–120(150) cm, gen aromatic. **ST**: gen ± erect, gen ± solid. **LF**: proximal in basal rosette or opposite (gen not persistent), most alternate, sessile, linear to lanceolate or oblanceolate, proximal gen pinnately lobed to serrate, distal gen entire, hairy, often glandular. **INFL**: heads radiate, gen in flat-topped or ± panicle-like clusters or in tight groups; involucre bell- or urn-shaped, hemispheric, or ± obconic, 2–13+ mm diam; phyllaries 3–35 in 1 series, ± linear to lanceolate or oblanceolate, each gen 1/2 enclosing a ray ovary, falling with fr, ± hairy and sessile- or stalked-glandular; receptacle flat to convex, glabrous or minutely bristly; paleae in 1 series between ray and disk fls in ann or in 2–3+ series or subtending all or most disk fls in shrubs, fused or free, phyllary-like, more scarious. **RAY FL**: 3–35; corolla deep or pale yellow. **DISK FL**: 3–70, gen staminate, sometimes bisexual; corolla yellow, tube ≤ throat, lobes deltate; anthers gen ± red to dark purple or yellow to brown, tips lance-ovate to deltate; style glabrous proximal to branches, tips awl-shaped, densely bristly. **FR**: ray nearly round in ×-section (exc ± flattened adaxially) or ± 3-angled (abaxially gen ± widely 2-faced, adaxially ± flattened to slightly bulging), gen ± arched, glabrous, tip ± beaked, beak offset adaxially, ascending, pappus 0; disk fr gen 0, disk pappus gen of 1–15 gen linear to lanceolate, entire or fringed to deeply cut scales, sometimes 0 or crown-like. 21 spp.: CA, w AZ, n Baja CA. (Greek: fierce man, probably for name it replaced, *Hartmannia* DC., meaning "stag man," stags being fiercely territorial) [Baldwin & Strother 2006 FNANM 21:280–286; Baldwin 2007 Amer J Bot 94:237–248] Self-sterile exc *D. arida* and *D. mohavensis*.

1. Subshrub or shrub; paleae in 2 or 3+ series or subtending all or most disk fls
   2. Ray fls (11)13(20); anthers ± red to dark purple; paleae in 2 series (1 between ray and disk fls, 1 between outermost and adjacent disk fls). . . . . ***D. clementina***
   2′ Ray fls (4)8; anthers yellow or ± brown; paleae in 3+ series or throughout receptacle . . . . ***D. minthornii***
1′ Ann; paleae in 1 series, between ray and disk fls
   3. Ray fls 3–5; disk fls 3–6; pappus of (4)5–12 scales
      4. Ray fls 3(4); disk fls 3(4) . . . . ***D. lobbii***
   4′ Ray fls 5; disk fls 6
      5. Phyllaries gen sessile-glandular, at least near margins, rarely stalked-glandular, stalks < glands; anthers ± red to dark purple
         6. Phyllaries glandular near margins, sometimes also with slender-based glandless hairs; bracts subtending head gen overlapping at least proximal 1/2 of involucre . . . . ***D. fasciculata***

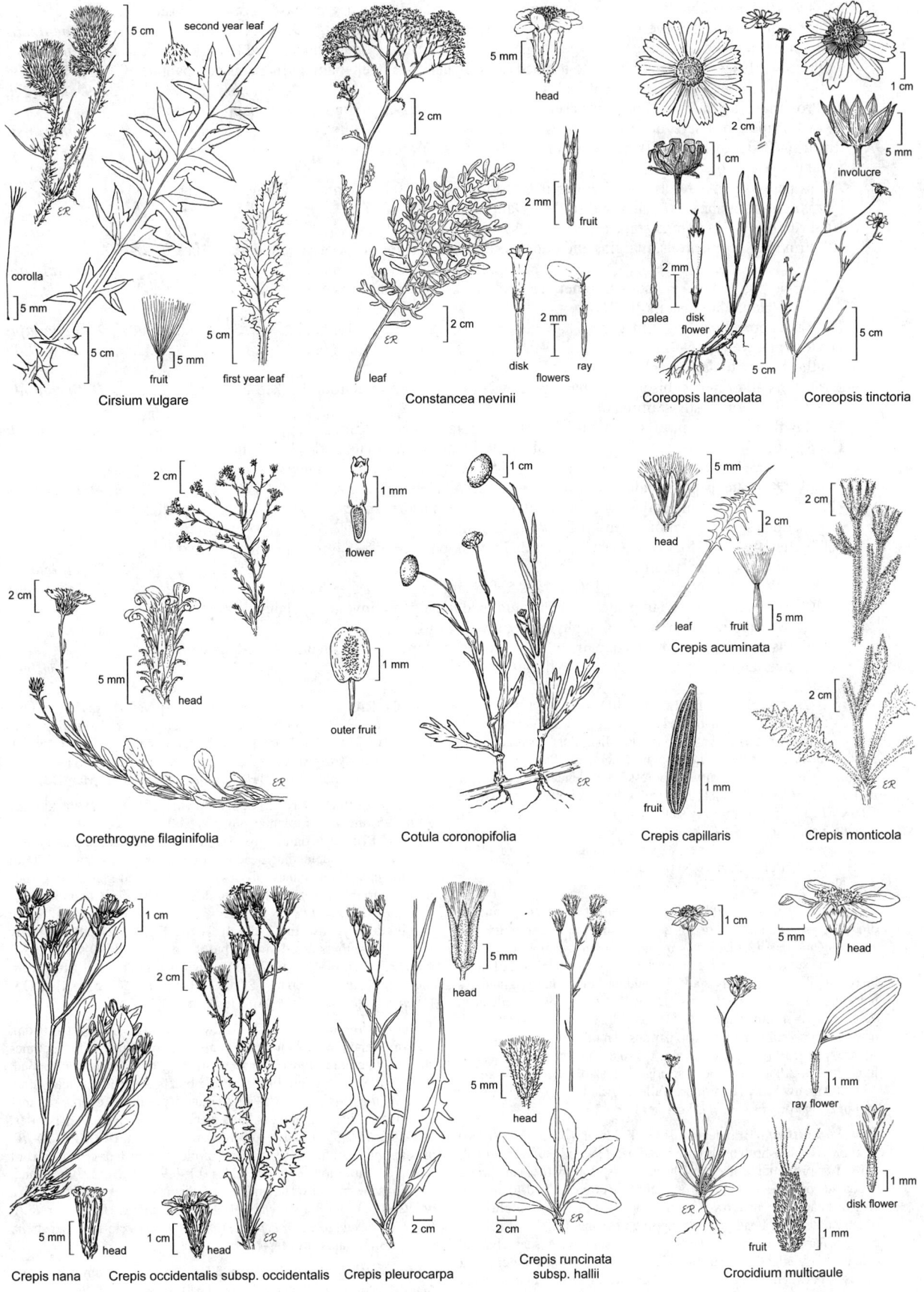

Cirsium vulgare

Constancea nevinii

Coreopsis lanceolata

Coreopsis tinctoria

Corethrogyne filaginifolia

Cotula coronopifolia

Crepis acuminata

Crepis capillaris

Crepis monticola

Crepis nana

Crepis occidentalis subsp. occidentalis

Crepis pleurocarpa

Crepis runcinata subsp. hallii

Crocidium multicaule

6′ Phyllaries ± evenly glandular and with swollen-based glandless hairs, at least on midribs; bracts
  subtending head gen overlapping proximal 1/2 or less of involucre. . . . . ***D. pentactis***
5′ Phyllaries stalked-glandular, stalks gen ≥ glands; anthers yellow or ± brown (± red to dark purple or maroon).
  7. Proximal lvs gen pinnately lobed to toothed (entire); heads in open, panicle-like clusters; pappus of
    linear to oblong, entire or fringed scales . . . . ***D. kelloggii***
  7′ Proximal lvs gen entire, sometimes serrate; heads gen ± sessile in tight groups; pappus of irregular,
    awl-shaped to square, deeply cut scales . . . . ***D. mohavensis***
3′ Ray fls (4)8–35; disk fls 8–70; pappus 0, or of 1–14 scales, or crown-like
  8. Anthers yellow or ± brown
    9. Disk pappus gen 0, rarely of 1–5, linear to bristle-like scales 0.1–0.6 mm or rudimentary; faces of
      proximal lvs glabrous or hairy and stalked-glandular; disk fls 17–60
      10. Proximal lvs hairy and stalked-glandular; st solid; disk fls 17–25 . . . . ***D. arida***
      10′ Proximal lvs glabrous, margins and midribs sometimes scabrous or bristly; st hollow; disk fls 28–60
        . . . . ***D. halliana***
    9′ Disk pappus gen of 4–13 scales, sometimes crown-like, rarely 0; faces of proximal lvs ± hairy and
      stalked-glandular; disk fls 10–21
      11. Ray deep yellow, 2–4 mm . . . . ***D. bacigalupii***
      11′ Ray pale yellow, 6–12 mm . . . . ***D. pallida***
  8′ Anthers ± red to dark purple
    12. Disk fls all or mostly bisexual — pappus scales often maroon or maroon-flecked . . . . ***D. floribunda***
    12′ Disk fls all or mostly staminate
      13. Ray fls 15–35; pappus 0 or crown-like, with irregular scales < 1 mm . . . . ***D. corymbosa***
      13′ Ray fls (7)8–13; pappus of 4–14, gen oblong or linear to lanceolate scales 0.5–2 mm.
        14. Phyllaries unevenly sessile- and stalked-glandular, glands highly variable in size; phyllary body
          3–7+ striate; pappus scales 0.5–1 mm, << 1/2 corolla length . . . . ***D. conjugens***
        14′ Phyllaries ± evenly stalked-glandular, glands gen ± uniform in size; phyllary body not striate;
          pappus scales 1–2 mm, often ≥ 1/2 corolla length
          15. Disk fls 8–14(15); ray fls (7)8(10), rays 3–7.5 mm; phyllaries 5–6 mm, < peduncle — basal lvs
            sometimes persistent in SCoR . . . . ***D. paniculata***
          15′ Disk fls (11)13–32; ray fls 8–13(15), rays 5–9.5 mm; phyllaries 5–8.5 mm, < or > peduncle . . . . . . ***D. increscens***
            16. Pls ≤ 10 dm; heads in panicle-like clusters; peduncle gen > involucre, peduncle bracts gen
              overlapping < proximal 3/4 of phyllaries; ray fls 8–13. . . . . subsp. ***increscens***
            16′ Pls gen ≤ 4.5 dm; heads in pairs or tight groups; peduncle gen < involucre, peduncle bracts gen
              overlapping ≥ proximal 3/4 of phyllaries; ray fls (8)13(15) . . . . subsp. ***villosa***

D. arida (D.D. Keck) B.G. Baldwin (p. 301) RED ROCK TARPLANT
Ann 2–8 dm. **LF**: proximal toothed to entire, bristly and coarse-hairy,
stalked-glandular. **INFL**: heads in panicle-like clusters; bracts subtending head gen not overlapping involucre; phyllaries ± evenly
stalked-glandular and ± hairy, glandless hairs slender-based; paleae
in 1 series. **RAY FL**: (4)8(10); corolla deep yellow, ray 5–7 mm.
DISK FL: 17–25, all or mostly staminate; anthers yellow or ± brown;
pappus gen 0, rarely of 1–5, linear to bristle-like scales 0.1–0.6 mm.
2*n*=24. Washes, canyon slopes, edges of springs, seeps; 600–1000
m. w DMoj (El Paso Mtns, e Kern Co.). [*Hemizonia a.* D.D. Keck]
Apr–Nov ★

D. bacigalupii B.G. Baldwin LIVERMORE TARPLANT Ann 1–4
dm. **LF**: proximal entire or irregularly lobed, ± coarse-hairy and
stalked-glandular. **INFL**: heads in flat-topped or panicle-like clusters; bracts subtending head gen overlapping proximal 0–1/2 of involucre; phyllaries ± evenly stalked-glandular and ± hairy, glandless
hairs slender-based; paleae in 1 series. **RAY FL**: (6)8(9); corolla deep
yellow, ray 2–4 mm. **DISK FL**: (10)15–18(21), all or mostly staminate; anthers yellow or ± brown; pappus gen of 8–13, awl-shaped to
square, fringed to deeply cut scales 0.1–0.8 mm, sometimes crown-
like. 2*n*=24. Alkaline meadows, edges of alkali barrens or sinks;
100–200 m. nw SnJV (Livermore Valley, e Alameda Co.). Previously
incl in *D. increscens* subsp. *i.* Jun–Oct ★

D. clementina (Brandegee) B.G. Baldwin (p. 301) ISLAND
TARPLANT Subshrub or shrub 1.5–8 dm. **ST**: woolly tufts in some lf
axils. **LF**: proximal toothed or entire, long-soft hairy to ± canescent,
± coarse- or appressed-hairy, or scabrous, often stalked-glandular.
INFL: heads gen in crowded, flat-topped or panicle-like clusters;
bracts subtending head gen overlapping proximal 0–1/2+ of involucre; phyllaries ± evenly stalked-glandular, sometimes sparsely, and
often ± hairy, glandless hairs swollen- or slender-based; paleae in 2
series–1 between ray and disk fls, 1 between outermost and adjacent

disk fls. **RAY FL**: (11)13(20); corolla deep yellow, ray 4.5–7 mm.
DISK FL: 18–30, all or most staminate; anthers ± red to dark purple;
pappus of 7–10(15), lance-linear, fringed scales 1–3 mm. 2*n*=24.
Coastal scrub, open sites, salt marsh edges; < 200 m. n ChI (Anacapa,
Santa Cruz islands), s ChI. [*Hemizonia c.* Brandegee] Mar–Dec ★

D. conjugens (D.D. Keck) B.G. Baldwin (p. 301) OTAY TARPLANT
Ann 1–5 dm. **LF**: proximal pinnately lobed to toothed, ± coarse-
hairy. **INFL**: heads tightly grouped or well separated, in panicle-like
clusters; bracts subtending head gen overlapping proximal 0–1/2+ of
involucre; phyllaries unevenly sessile- and stalked-glandular, glands
highly variable in size, sometimes ± hairy, glandless hairs slender-
based; phyllary body 3–7+ striate; paleae in 1 series. **RAY FL**: 7–10;
corolla deep yellow, ray 3–6 mm. **DISK FL**: 13–21, all or mostly
staminate; anthers ± red to dark purple; pappus of 6–9 scales 0.5–1
mm, << 1/2 corolla length. 2*n*=24. Clayey soils of coastal scrub openings, grassland; 20–300 m. s SCo, w PR (San Diego Co.); n Baja CA.
[*Hemizonia c.* D.D. Keck] Apr–Jun ★

D. corymbosa (DC.) B.G. Baldwin (p. 301) Ann 0.6–10 dm.
LF: proximal pinnately lobed, ± coarse- to soft-hairy and sometimes
stalked-glandular. **INFL**: heads in tight groups or in flat-topped,
raceme-like, or panicle-like clusters; bracts subtending head often
overlapping proximal 0–1/2+ of involucre; phyllaries evenly stalked-
glandular, ± hairy, glandless hairs slender-based; paleae in 1 series.
RAY FL: 15–35; corolla deep yellow, ray 4–8 mm. **DISK FL**: 24–70,
all staminate; anthers ± red to dark purple; pappus 0 or crown-like, of
irregular, entire to deeply cut scales 0.1–0.9 mm. 2*n*=20. Grassland,
openings in scrub or woodland, dunes, disturbed sites; < 600 m. NCo,
w NCoRO, CCo, SnFrB, w SCoRO. [*Hemizonia c.* (DC.) Torr. & A.
Gray; *H. c.* subsp. *macrocephala* (Nutt.) D.D. Keck; *D. c.* subsp. *m.*
(Nutt.) B.G. Baldwin] Mar–Nov

D. fasciculata (DC.) Greene (p. 301) Ann 0.4–10 dm. **LF**: proximal toothed, ± coarse-hairy. **INFL**: heads often in tight groups or in

pairs, sometimes well separated, in raceme-like or panicle-like clusters; bracts subtending head gen overlapping at least proximal 1/2 of involucre; phyllaries sessile-glandular near margins, sometimes ± hairy, glandless hairs slender-based; paleae in 1 series. **RAY FL:** 5; corolla deep yellow, ray 6–14 mm. **DISK FL:** 6, all or mostly staminate; anthers ± red to dark purple; pappus of 5–12, lanceolate to oblong or linear, entire or fringed scales 1–1.5 mm. 2*n*=24. Open or disturbed sites, grassland, scrub, woodland, vernal pools, sometimes on serpentine; < 1200 m. s CCo, w SCoRO, SW; to c Baja CA. [*Hemizonia f.* (DC.) Torr. & A. Gray] Apr–Sep

D. floribunda (A. Gray) Davidson & Moxley (p. 301) TECATE TARPLANT Ann 3–10 dm. **LF:** proximal entire or toothed, soft-hairy and stalked-glandular. **INFL:** heads in often narrow raceme- to panicle-like clusters, side branches short; bracts subtending head overlapping proximal 0–1/2+ of involucre or not; phyllaries ± evenly stalked-glandular, gen ± hairy, glandless hairs slender-based; paleae in 1 series. **RAY FL:** 13–20; corolla deep yellow, ray 4–7 mm. **DISK FL:** 24–31, all or mostly bisexual; anthers ± red to dark purple; pappus of 6–10, oblong to elliptic, fringed scales 0.7–1.5 mm, often maroon or maroon-flecked. 2*n*=26. Moist openings in chaparral, streambeds, disturbed sites; 70–1200 m. s PR (s San Diego Co.); n Baja CA. [*Hemizonia f.* A. Gray] Aug–Nov ★

D. halliana (D.D. Keck) B.G. Baldwin (p. 301) HALL'S TARPLANT Ann 1.5–12 dm. **ST:** hollow. **LF:** proximal entire to serrate, faces glabrous, margins and midribs sometimes scabrous or bristly. **INFL:** heads in open, flat-topped, raceme-like, or panicle-like clusters; bracts subtending head gen not overlapping involucre; phyllaries evenly stalked-glandular, gen sparsely hairy, glandless hairs slender-based; paleae in 1 series. **RAY FL:** (8)10–14; corolla deep yellow, ray 5–10 mm. **DISK FL:** 28–60, all staminate; anthers yellow or ± brown; pappus 0 or rudimentary. 2*n*=20. Grasslands, open slopes, sink edges, vertic clay, rarely serpentine; 300–1000 m. w SnJV, SCoRI. [*Hemizonia h.* D.D. Keck] Apr–May ★

D. increscens (D.D. Keck) B.G. Baldwin Ann. **LF:** proximal pinnately lobed, ± coarse- or soft-hairy and sometimes stalked-glandular. **INFL:** heads tightly grouped or paired or in panicle-like clusters; bracts subtending head gen overlapping proximal 0–3/4+ of involucre; phyllaries 5–8.5 mm, < or > peduncle, evenly stalked-glandular, gen ± hairy, glandless hairs slender-based; paleae in 1 series. **RAY FL:** 8–13(15); corolla pale to deep yellow, ray 5–9.5 mm. **DISK FL:** (11)13–32, all or mostly staminate; anthers ± red to dark purple; pappus of 5(14), linear, lance-linear, or oblong, fringed to deeply cut scales 1–2 mm, often ≥ 1/2 corolla length. 2*n*=24.

   subsp. **increscens** (p. 301) Pl ≤ 10 dm. **INFL:** heads in panicle-like clusters; peduncles gen > involucres, peduncle bracts gen overlapping < proximal 3/4 of phyllaries. **RAY FL:** 8–13. **DISK FL:** (11)13–29. Sandy or clayey soils (sometimes serpentine), grassland, openings in scrub or woodland, disturbed sites; < 300 m. CCo, w SCoRO, nw edge SCoRI, n SCo, n ChI. [*Hemizonia i.* (D.D. Keck) Tanowitz subsp. *i.*] Apr–Nov

   subsp. **villosa** (Tanowitz) B.G. Baldwin GAVIOTA TARPLANT Pl 0.6–4.5 dm. **INFL:** heads in tight groups or paired; peduncles gen < involucres, peduncle bracts gen overlapping ≥ proximal 3/4 of phyllaries. **RAY FL:** (8)13(15). **DISK FL:** (12)16–32. Coastal bluffs, fields; 30–50 m. n SCo (Gaviota to Point Conception, Santa Barbara Co.), w WTR (w Santa Ynez Mtns). [*Hemizonia i.* subsp. *v.* Tanowitz] Jun–Sep ★

D. kelloggii (Greene) Greene (p. 301) Ann 1–10(15) dm. **LF:** proximal gen pinnately lobed to toothed (entire), bristly to coarse-hairy and sometimes stalked-glandular. **INFL:** heads in open, panicle-like clusters; bracts subtending heads gen not overlapping involucres; phyllaries ± evenly stalked-glandular, ± hairy or not, glandless hairs slender-based, if present; paleae in 1 series. **RAY FL:** 5; corolla deep yellow, ray 4–8 mm. **DISK FL:** 6, all or mostly staminate; anthers gen yellow or ± brown, ± red to dark purple or maroon in some, mostly SW pls; pappus of 6–12, linear to oblong, entire or fringed scales 1–2 mm. 2*n*=18. Sandy or clayey soils, grassland, openings in scrub or woodland, disturbed sites; < 1200(2100) m. c&s SNF, SnJV, e SnFrB, SCoRI, SW (exc ChI), w edge DMoj; n Baja CA. [*Hemizonia k.* Greene] Intergrades with *D. pallida* in s SNF. Mar–Nov

D. lobbii (Greene) Greene (p. 301) Ann 0.5–7 dm. **LF:** proximal pinnately lobed to toothed, bristly and coarse-hairy, sometimes sessile- or short-stalked-glandular. **INFL:** heads in panicle-like clusters, bracts subtending head overlapping proximal 0–1/2 of involucre or not; phyllaries ± evenly sessile- or short-stalked-glandular, ± hairy, glandless hairs swollen-based, at least on midribs; paleae in 1 series. **RAY FL:** 3(4); corolla deep yellow, ray 3–5 mm. **DISK FL:** 3(4), all or mostly staminate; anthers ± red to dark purple; pappus of (4)6–8(12), square or oblong to lance-linear, fringed scales 0.5–1 mm. 2*n*=22. Grassland, open woodland, sagebrush scrub, disturbed areas; < 700(1800) m. CaRH, nw SnJV, SnFrB, n SCoR, MP (exc Wrn). [*Hemizonia l.* Greene] May–Dec

D. minthornii (Jeps.) B.G. Baldwin (p. 301) SANTA SUSANA TARPLANT Subshrub or shrub 1.5–10 dm. **LF:** proximal pinnately lobed to toothed, ± short-coarse-hairy, sometimes stalked-glandular. **INFL:** heads 1 or in loose, raceme- to panicle-like clusters; bracts subtending head gen overlapping proximal 0–1/2 of involucre; phyllaries ± evenly stalked-glandular, often ± hairy, glandless hairs slender-based; paleae in 3+ series or subtending most or all disk fls. **RAY FL:** (4)8; corolla deep yellow, ray 5.5–6.5 mm. **DISK FL:** 18–23, all or mostly staminate; anthers yellow or ± brown; pappus of 8–12, linear to lance-linear, entire or fringed scales 1–3 mm. 2*n*=24. Chaparral, coastal scrub, often on sandstone; 200–800 m. s WTR (Santa Monica, Santa Susana mtns). [*Hemizonia m.* Jeps.] Jun–Nov ★

D. mohavensis (D.D. Keck) B.G. Baldwin MOJAVE TARPLANT Ann 1–10(15) dm. **LF:** proximal gen entire, sometimes serrate, hairy and stalked-glandular. **INFL:** heads gen in tight groups or ± crowded, sometimes in open, panicle-like clusters; bracts subtending head gen overlapping proximal 0–1/2+ of involucre; phyllaries ± evenly stalked-glandular, gen ± hairy, glandless hairs slender-based; paleae in 1 series. **RAY FL:** 5; corolla deep yellow, ray 4–7 mm. **DISK FL:** 6, staminate or bisexual; anthers yellow or ± brown; pappus of 5–9, often ± fused, irregular, awl-shaped to square, deeply cut scales 0.1–0.6 mm. 2*n*=22. Moist sites, openings in chaparral, desert scrub, woodland; 460–1600 m. s SNH, n SnBr (extirpated), PR, w edge DMoj. [*Hemizonia m.* D.D. Keck] May–Jan ★

D. pallida (D.D. Keck) B.G. Baldwin (p. 301) Ann 0.9–10 dm. **ST:** ± solid or hollow. **LF:** proximal pinnately lobed to toothed, ± coarse-hairy, sometimes also sparsely stalked-glandular. **INFL:** heads in crowded to open, flat-topped or panicle-like clusters; bracts subtending head sometimes overlapping proximal 0–1/2 of involucre; phyllaries ± evenly and minutely stalked-glandular, ± hairy, glandless hairs slender-based; paleae in 1 series. **RAY FL:** (7)8–12(13); corolla pale yellow, ray 6–12 mm. **DISK FL:** 10–21, all or mostly staminate; anthers yellow or ± brown; pappus gen of 4–9, linear or oblong scales 0.8–1.1 mm, or of 1–5, awl-shaped to bristle-like scales 0.1–0.9 mm, rarely 0. 2*n*=18. Grassland, open scrub and woodland, disturbed sites, often in ± alkaline soils; 70–1100 m. s SNF, Teh, s SnJV, s SCoRI. [*Hemizonia p.* D.D. Keck] Mar–Jul

D. paniculata (A. Gray) Davidson & Moxley (p. 301) PANICULATE TARPLANT Ann 1–8(15) dm. **LF:** proximal pinnately lobed to toothed, bristly and coarse-hairy and sometimes stalked-glandular; basal lvs sometimes persistent in SCoR. **INFL:** heads in gen open, panicle-like clusters; bracts subtending head gen overlapping proximal 0–1/2 of involucre; phyllaries 5–6 mm, < peduncles, ± evenly stalked-glandular, often ± hairy, glandless hairs slender-based; paleae in 1 series. **RAY FL:** (7)8(10); corolla deep yellow, ray 3–7.5 mm. **DISK FL:** 8–14(15), all or mostly staminate; anthers ± red to dark purple; pappus of 6–12, oblong, fringed scales 1–2 mm, often ≥ 1/2 corolla length. 2*n*=24. Grassland, open chaparral and woodland, disturbed areas, often in sandy soils; < 1320 m. s CCo/SCoRO, s SCoRO, SCo, WTR (e Santa Ynez Mtns), PR; n Baja CA. [*Hemizonia p.* A. Gray; *H. increscens* subsp. *foliosa* (Hoover) Tanowitz; *D. i.* subsp. *f.* (Hoover) B.G. Baldwin] May–Nov ★

D. pentactis (D.D. Keck) B.G. Baldwin Ann 0.4–7.5 dm. **LF:** proximal pinnately lobed to toothed, sharp- and coarse-hairy and sometimes sessile- or short-stalked-glandular, rarely glabrous. **INFL:** heads in open, panicle-like clusters; bracts subtending head gen overlapping proximal 0–1/2 of involucre; phyllaries ± evenly sessile- or stalked-glandular, gen ± hairy, glandless hairs swollen-based, at least on midribs; paleae in 1 series. **RAY FL:** 5; corolla deep yellow, ray

3–5 mm. **DISK FL**: 6, all or mostly staminate; anthers ± red to dark purple; pappus of (5)6–8(11), linear or oblong, fringed scales 0.8–1 mm. 2*n*=22. Grassland, open woodland, disturbed areas; (0)200–900 m. w edge SnJV, sw SnFrB (alien), c SCoR. [*Hemizonia p.* (D.D. Keck) D.D. Keck] Apr–Oct

DELAIREA CAPE IVY, GERMAN IVY

Debra K. Trock

1 sp.: S.Afr, Australia (introduced). (Eugene Delaire, French gardener, 1810–1856) [Barkley 2006 FNANM 20:608–609]

D. odorata Lem. (p. 305)   Per vine, glabrous; new growth fleshy. **ST**: 1–6 m, much-branched, twining, green or purple, proximally becoming woody, creeping, rooting at nodes. **LF**: alternate, evenly spaced; petiole base often with 2 stipule-like lobes; blade 3–8(12) cm, widely deltate, palmately 5–9-lobed, lobes acute to acuminate, faces green to yellow-green, shiny. **INFL**: heads discoid, ± 20 in terminal or axillary cyme-like clusters; involucre cylindric to obconic; phyllaries ± 8, equal in 1–2 series, erect, narrowly linear, green, glabrous, persistent. **DISK FL**: 8–45; corolla narrowly bell- to funnel-shaped, bright yellow; anther tip oblong; style tips truncate. **FR**: cylindric, 5-angled, glabrous; pappus of white, minutely barbed bristles, exceeding phyllaries. 2*n*=40. Shady, ± disturbed places, riparian woodland, coastal scrub; < 1500 m. NCo, s NCoR, w-c GV, CCo, SnFrB, w SCoRO, w SW; native to s Afr. [*Senecio mikanioides* Walp.] Highly invasive, difficult to eradicate. Nov–Mar ◆

DICORIA

Bruce G. Baldwin, adapted from Strother (2006)

Ann [per], 10–90+ cm, gen canescent. **ST**: erect, branches wand-like to widely divergent. **LF**: proximally opposite, 2–10+ pairs in Feb–Apr, otherwise alternate, petioled, blades proximally lance-linear to lanceolate, distally elliptic or lanceolate to ± deltate or ovate, entire or toothed, minutely strigose to silky-hairy, sometimes coarse-erect-hairy, gen gland-dotted, 3-veined. **INFL**: heads disciform or discoid and staminate, 1 or loosely in 2s or 3s in gen bractless, raceme- to panicle-like clusters; involucre 3–5 mm diam, ± cup- to saucer-shaped; outer phyllaries (4)5(7) in 1 series, free, ± green, inner 0–4, each subtending a pistillate fl, scarious to membranous, enlarging with age, ovate to elliptic in fr; receptacle convex, gen paleate, paleae wedge-shaped to linear, membranous, ± soft-hairy to bristly distally. **PISTILLATE FL**: (0)1–4; corolla 0. **STAMINATE FL**: 5–15+; corolla ± white to maroon, funnel-shaped, lobes 5, erect, reflexed, or curved inward; filaments fused, anthers free or gen fused at tips. **FR**: strongly compressed front-to-back, ± obovoid to ellipsoid, smooth or warty, sometimes gland-dotted, margins corky-winged, toothed, fr tip often white-tufted; pappus 0. 2 spp.: sw US, nw Mex. (Greek: 2 bugs, from 2-fr heads) [Strother 2006 FNANM 21:24–25]

D. canescens A. Gray (p. 305)   **LF**: petiole 5–20+ mm; blade gen 1–3(12) cm, 3–20(30) mm wide. **INFL**: peduncle 1–3+ mm; involucre 2–3+ mm at fl; outer phyllaries ± lanceolate to ovate, minutely strigose to ± silky-hairy, inner becoming obovate to ± round, gen ± hood-shaped, gen glandular-hairy, each ± surrounding a fr; paleae 1.8–2+ mm. **STAMINATE FL**: corolla 2.5–3 mm. **FR**: 3–8+ mm. 2*n*=36. Alkaline or sandy soils, dunes, washes, flats; < 1300 m. SCo (Colton Dunes, San Bernardino Co.), SNE (exc W&I), D; to CO, NM, nw Mex. Highly variable; features vary ± independently. Sep–Jan

DIETERIA

David R. Morgan

Ann to subshrub. **ST**: 1–several from taproot or ± branched caudex, gen branched distally and ± bushy. **LF**: alternate, entire to irregularly serrate or dentate, proximal tapered at base; teeth gen bristle-tipped. **INFL**: heads gen radiate, in cyme-like clusters; involucre bell-shaped, hemispheric, or obconic; phyllaries in 3–12 series of unequal length, proximally straw-colored, distally green; receptacle convex, with short, triangular scales, glabrous, epaleate. **RAY FL**: (0)7–many; corolla white, blue, or purple. **DISK FL**: 15–gen many; corolla 5.5–8 mm, yellow; anther tip lanceolate; style tips lanceolate. **FR**: 2.5–3.5 mm, gen narrowly obovate, weakly curved and ± flattened, smooth or gen with 5–7 ribs on each face, glabrous or ± silky; pappus 6–8 mm, of many unequal bristles. 3 spp.: temp w N.Am. (Greek: biennial, from duration of type sp.) [Morgan 2006 FNANM 20:395–401]

1. Ray fls 0 . . . . [2]*D. canescens* var. ***shastensis***
1′ Ray fls present
  2. Phyllary tips acute to long-acuminate, 1–6 mm; phyllaries gen hairy and/or glandular throughout . . . . . . ***D. asteroides***
    3. Involucre hemispheric; phyllary tips long-acuminate, 3–6 mm; lvs of mid-st serrate or minutely serrate
    . . . . var. ***asteroides***
    3′ Involucre widely obconic to ± hemispheric; phyllary tips acute to short-acuminate, 1–3 mm; lvs of
      mid-st entire or nearly so . . . . var. ***lagunensis***
  2′ Phyllary tips acute to short-acuminate, 1–3 mm; phyllaries gen hairy and/or glandular only on green tips
  . . . . ***D. canescens***
    4. Heads gen 14–20 mm; subshrub or long-lived per; e PR (Santa Rosa Mtns) . . . . var. ***ziegleri***
    4′ Heads gen 6–12 mm; ann to short-lived per; widespread
      5. Sts with stalked glands, occ also canescent-puberulent . . . . var. ***leucanthemifolia***
      5′ Sts canescent-puberulent, stalked glands sparse or lacking
        6. Sts stiffly erect, ≤ 60 cm; branches many, stiff, spreading; heads many . . . . var. ***incana***

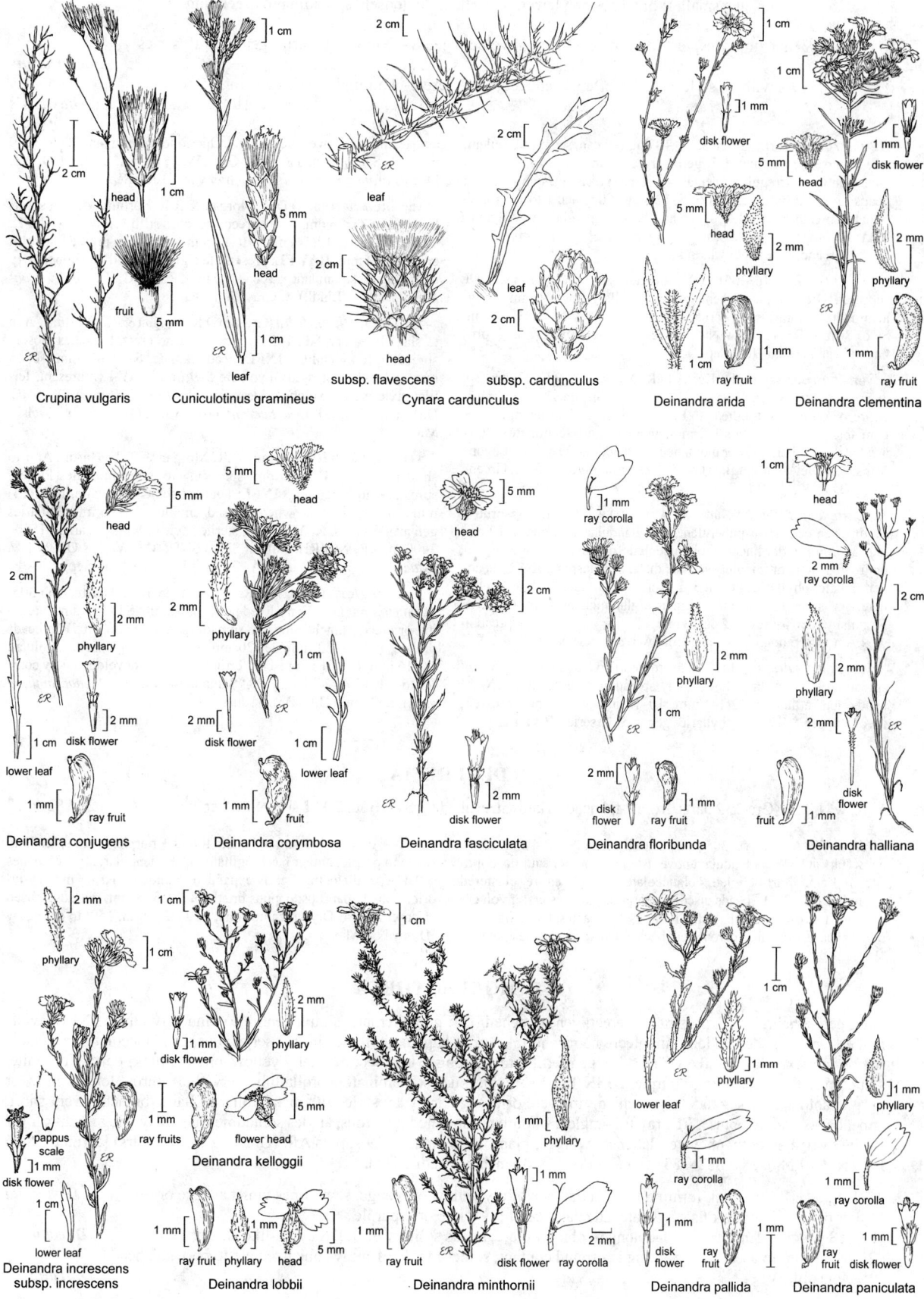

Crupina vulgaris

Cuniculotinus gramineus

Cynara cardunculus
subsp. flavescens   subsp. cardunculus

Deinandra arida

Deinandra clementina

Deinandra conjugens

Deinandra corymbosa

Deinandra fasciculata

Deinandra floribunda

Deinandra halliana

Deinandra increscens
subsp. increscens

Deinandra lobbii

Deinandra minthornii

Deinandra pallida

Deinandra paniculata

6′ Sts various, gen smaller; branches gen fewer, more flexible, loosely spreading to ascending; heads few to many

7. Ray fls with styles, fertile, well developed; involucre gen 8–12 mm; phyllaries gen in 5–10 series
. . . . var. ***canescens***

7′ Ray fls without styles, sterile, often ± reduced; involucre 6–9 mm; phyllaries gen in 3–5 series
. . . . [2]var. ***shastensis***

D. asteroides Torr.   Bien, per, ≤ 10 dm, gen canescent-puberulent, often sparsely glandular. **LF:** gen 3–10 cm, lanceolate to oblanceolate, ± entire to irregularly dentate or serrate; distal clasping. **INFL:** heads radiate; phyllaries gen in 5–12 series, tips acute to long-acuminate, spreading to reflexed, gen hairy and/or glandular throughout. **RAY FL:** many; ray 1–2 cm, blue or purple. [*Machaeranthera a.* (Torr.) Greene] One other var. in sw US.

var. ***asteroides*** (p. 305)   **LF:** mid-cauline 6–15(25) mm wide, ± serrate. **INFL:** involucre hemispheric; phyllary tips 3–6 mm, long-acuminate. 2*n*=8. Desert scrub, coastal scrub, dry fields; ± 100 m. SCo, SnJt, e DSon; to sw NM, n Mex. [*Machaeranthera a.* (Torr.) Greene var. *a.*] Mar–Jun, Sep–Oct

var. ***lagunensis*** (D.D. Keck) D.R. Morgan & R.L. Hartm. (p. 305)   MOUNT LAGUNA ASTER   **LF:** mid-cauline gen 2–5 mm wide, entire or obscurely toothed. **INFL:** involucre widely obconic to ± hemispheric; phyllary tips 1–3 mm, acute to short-acuminate. Chaparral, oak woodland, lower montane forest; 800–2400 m. PR (Laguna Mtns, San Diego Co.); n Baja CA. [*Machaeranthera a.* (Torr.) Greene var. *l.* (D.D. Keck) B.L. Turner] Jul–Oct ★

D. canescens (Pursh) Nutt.   HOARY-ASTER   Ann to subshrub ≤ 12 dm, gen canescent-puberulent, occ glandular. **LF:** gen 3–10 cm, gen 2–6 mm wide, linear to oblanceolate (occ ovate to obovate), ± entire to dentate or minutely serrate; distal occ clasping. **INFL:** heads gen radiate; phyllaries gen in 3–10 series, tips acute to short-acuminate, gen spreading to reflexed, gen ± glandular or glabrous. **RAY FL:** many (occ 0); ray 1–2 cm, white, blue, or purple. Widespread in w N.Am with 5 additional vars. from w US to n Mex.

var. ***canescens*** (p. 305)   Ann to short-lived per. **ST:** 1–5 dm, spreading to erect; branches loosely spreading to ascending. **INFL:** heads gen radiate, 6–12(14) mm, 10–15 mm wide when pressed; involucre gen 8–12 mm; phyllaries gen in 5–10 series. **RAY FL:** pres-

ent, fertile; style well developed. Open montane areas; 2000–3000 m. CaR, SN, e SCo, TR, SnJt, GB, DMoj; to WA, c Can, CO, AZ. [*Machaeranthera c.* (Pursh) A. Gray var. *c.*] Jul–Sep

var. ***incana*** (Lindl.) D.R. Morgan & R.L. Hartm.   Ann to short-lived per. **ST:** ≤ 6 dm, stiffly erect; branches stiffly spreading. **INFL:** heads radiate, 6–12(14) mm, 10–15 mm wide when pressed; involucre 7–9(12) mm. **RAY FL:** present, fertile; style well developed. Dry fields, sandy streamsides; 300–1100 m. KR; to BC, ID. [*Machaeranthera c.* var. *i.* (Lindl.) A. Gray] Jun–Oct

var. ***leucanthemifolia*** (Greene) D.R. Morgan & R.L. Hartm.   Ann to short-lived per. **ST:** 1–5 dm, spreading to erect; branches loosely spreading to ascending. **INFL:** heads radiate, 6–12(14) mm, 10–15 mm wide when pressed; involucre 5–9 mm. **RAY FL:** present, fertile; style well developed. Desert scrub; 750–2000 m. e PR, SNE, DMoj; to e OR, UT. [*Machaeranthera c.* var. *l.* (Greene) S.L. Welsh] May–Jun

var. ***shastensis*** (A. Gray) D.R. Morgan & R.L. Hartm. Ann to short-lived per. **ST:** 1–5 dm, spreading to erect; branches loosely spreading to ascending. **INFL:** heads discoid or radiate, 6–12(14) mm, 10–15 mm wide when pressed; involucre 6–9 mm; phyllaries gen in 3–5 series. **RAY FL:** 0 or sterile; style 0. Montane areas; 1500–3400 m. KR, NCoRH, CaR, n SN, c SNH, MP, W&I; s OR, w NV. [*Machaeranthera c.* var. *s.* (A. Gray) B.L. Turner] Jul–Sep

var. ***ziegleri*** (Munz) D.R. Morgan & R.L. Hartm. (p. 305) ZIEGLER'S ASTER   Long-lived per or subshrub. **ST:** 1–5 dm, spreading or erect; branches loosely spreading to ascending. **INFL:** heads radiate, (12)14–20 mm, 15–20 mm wide when pressed; involucre 14–15 mm. **RAY FL:** present, fertile; style well developed. Dry conifer forest; 1400–2470 m. e PR (Santa Rosa Mtns). [*Machaeranthera c.* var. *z.* (Munz) B.L. Turner] Jul–Oct ★

DIMERESIA

1 sp.: w N.Am. (Greek: 2 parts, from 2-fld head) [Baldwin et al. 2002 Syst Bot 27:161–198; Strother 2006 FNANM 21:182–183]

D. howellii A. Gray (p. 305)   DOUBLET   Ann gen 1–4 cm, tufted, cobwebby at base, ± glandular above. **LF:** basal and/or cauline, opposite, 1–3 cm, short-petioled, oblanceolate to ovate, entire, clustered around heads. **INFL:** heads discoid, gen in sessile clusters; involucre 4–6 mm, narrowly bell-shaped; phyllaries 2–3, each subtending a fl, ± fused at base, obovate, tip rounded, back convex; receptacle epaleate. **FL:** 2–3; corolla 4–6 mm, tube and throat ± purple, lobes spreading, white to purple; anther base sagittate, tip ± triangular; style branches ± flat, ± papillate, tip slightly expanded, rounded. **FR:** ± 3 mm, cylindric, ribbed, glabrous; pappus bristles ± 20, 3–5 mm, plumose, fused at base. 2*n*=14. Dry volcanic soils; 1500–2400 m. MP; to s OR, sw ID, nw NV. May–Aug ★

DIMORPHOTHECA

Ann, per, subshrub. **ST:** prostrate to erect, gen branched. **LF:** gen alternate, simple, entire to pinnately lobed. **INFL:** heads radiate, gen 1; peduncle long; involucre hemispheric or bell-shaped; phyllaries in 1–2 series, linear to lanceolate; receptacle flat or convex, epaleate. **RAY FL:** ≤ 20, gen fertile; ray oblong, 3-toothed, adaxially yellow, orange, pink, purple, or white, sometimes abaxially blue or purple. **DISK FL:** many, bisexual or staminate; corolla tube very short, throat long, yellow or purple; anther base sagittate, short-tailed, tip ovate or triangular-ovate; style with distal collar of hairs, branches very short, appendages widely obtuse. **FR:** ray fr 3-angled, smooth or tubercled, glabrous, angles sometimes narrowly winged; disk fr (or sterile ovary) smooth, obovate, flattened, winged, glabrous; pappus 0. 19 spp.: s Afr. (Greek: 2 forms of frs) [Strother 2006 FNANM 19:380–381, 382–383] *D. pluvialis* (L.) Moench not naturalized.

1. Ann; disk fls bisexual, forming frs; adaxial face of ray yellow or orange, sometimes violet at base or tip . . . . . ***D. sinuata***
1′ Per or subshrub; disk fls staminate; adaxial face of ray white or ± purple
    2. St ± erect, ≤ 1 m; lvs linear-oblong to oblanceolate; heads solitary or in lfy, cyme-like cluster . . . . ***D. ecklonis***
    2′ St trailing to ascending, rooting at ground contact, sometimes >> 1 m; lvs oblanceolate to obovate; heads gen solitary . . . . ***D. fruticosa***

D. ecklonis DC. (p. 305) BLUE & WHITE DAISYBUSH Subshrub, ± puberulent and minutely stalked-glandular. **ST:** ≤ 1 m, erect or ascending, branched distally or throughout. **LF:** 3–6 cm, slightly fleshy; proximal wing-petioled, distal ± sessile; blade linear-oblong to obovate, entire or sparingly dentate. **INFL:** heads 1 or in lfy cyme-like clusters, 4–7 cm diam; involucre ± bell-shaped; phyllaries 12–15 mm, lance-linear, acuminate, narrowly scarious-margined. **RAY FL:** ray 2.5–3.5 cm, adaxially white and abaxially blue-purple, sometimes deep violet at base or adaxially pink-purple to rose-purple and abaxially purple or deep rose-purple. **DISK FL:** staminate (but ovary enlarging); corolla 5–6 mm, blue-purple. **FR:** ray fr 5–8 mm, faces ribbed; sterile disk ovary 5–8 mm. Uncommon. Disturbed places, beaches; escape from cult; < 200 m. CCo, SCo, PR; native to s Afr. [*Osteospermum e.* (DC.) Norl.] Many cult forms are hybrids with *D. fruticosa* and combine spp. features; these may escape or persist from cult. Gen Mar–Jul(± all yr in cult)

D. fruticosa (L.) DC. (p. 305) TRAILING AFRICAN DAISY Per, subshrub, ± puberulent and minutely stalked-glandular. **ST:** sometimes >> 1 m, trailing to ascending, sparingly branched, rooting at ground contact. **LF:** 4–8 cm, ± fleshy; proximal wing-petioled, distal ± sessile; blade oblanceolate to obovate, entire or sparingly dentate. **INFL:** heads gen 1, 4–7 cm diam; involucre ± bell-shaped; phyllaries 12–15 mm, lance-linear, acuminate, narrowly scarious-margined.

RAY FL: ray 2.5–3.5 cm, adaxially pink-purple to rose-purple and abaxially purple or deep rose-purple or adaxially white and abaxially blue-purple, sometimes deep violet at base. **DISK FL:** staminate (but ovary enlarging); corolla 5–6 mm, blue-purple. **FR:** ray fr 5–7 mm, faces ribbed; sterile disk ovary 5–7 mm. 2*n*=20. Uncommon. Disturbed places, beaches, coastal areas; < 200 m. CCo, SCo, PR; native to s Afr. [*Osteospermum f.* (L.) Norl.] Many cult forms are hybrids with *D. ecklonis* and combine spp. features; these may escape or persist from cult. Gen Mar–Jul(± all yr in cult)

D. sinuata DC. (p. 305) NAMAQUALAND DAISY Ann, glandular-hairy. **ST:** 10–30 cm, erect, simple or sparingly branched from base. **LF:** < 10 cm; proximal tapered to petiole-like base, distal sessile; blade oblong to oblanceolate, entire to coarsely dentate, distal smaller, sometimes linear. **INFL:** heads 3–7 cm diam; involucre ± bell-shaped; phyllaries 10–15 mm, lance-linear, acuminate, narrowly scarious-margined. **RAY FL:** ray 2–2.5 cm, adaxially orange to yellow, sometimes violet at base or tip, abaxially yellow to orange, often purple-tinged. **DISK FL:** bisexual; corolla 4.5–5.5 mm, proximally yellow or orange, distally often purple. **FR:** ray fr 4–5 mm, faces knobby; disk fr 6–7 mm. 2*n*=18. Escape from cult, roadsides, disturbed places, occ sown in wildflower mixes; < 1000 m. SnJV, CW, SW, D; native to s Afr. Feb–Jun

DITTRICHIA

Thomas J. Rosatti

Ann [to subshrub], sticky-glandular, strongly aromatic. **LF:** basal and cauline, alternate, sessile. **INFL:** heads radiate, many, in dense raceme- or ± panicle-like clusters; involucre ± ovoid, ± bell-shaped when pressed; phyllaries graduated in 3–4 series, reflexed in fr; receptacle flat, pitted, epaleate. **RAY FL:** corolla yellow, ray 3-lobed, ± not [to clearly] exceeding phyllaries. **DISK FL:** corolla 4–5-lobed, yellow; anther base tailed, tip acute; style tips linear. **FR:** ellipsoid to ± cylindric, abruptly narrowed below pappus, glandular-hairy; pappus of barbed bristles, in 1 row, fused at base. 5 spp.: w Eur, Medit, sw Asia; introduced, naturalized elsewhere. (M. Dittrich, German student of Asteraceae, b. 1934) [Preston 2006 FNANM 19:473]

D. graveolens (L.) Greuter (p. 305) STINKWORT Ann, erect, 2–6 dm, camphor-scented. **ST:** gen 1, much-branched, proximally ± red. **LF:** entire to finely toothed; basal 2–7 cm, lanceolate to oblanceolate, soon withering; cauline 1–4 cm, linear to linear-oblanceolate. **INFL:** peduncle 0–5 mm; involucre 4–7 mm; outer phyllaries wider, with narrower scarious margins than inner. **RAY FL:** 10–12; ray 2–5(–7) mm, erect or spreading. **DISK FL:** 9–14; corolla ± 4.5 mm. **FR:** ± 2 mm; pappus bristles 25–30, ± 4 mm. 2*n*=18, 20+. Disturbed areas; < 700 m. KR, s NCoRI, s NCoRO, CaRF, n SN, GV, CCo, SnFrB, n SCoRI, SCo, w PR, expected elsewhere; native to w Eur, Medit, sw Asia, introduced, possibly naturalized in NY, CT, NJ, Eur, Asia, Afr, S.Am, Australia. Increasingly problematic, potentially threatening to agriculture, livestock, wildlands; possibly causing contact dermatitis. Sep–Nov ❖

EASTWOODIA YELLOW MOCK ASTER

David J. Keil & Meredith A. Lane

1 sp. (Alice Eastwood, w Am botanist, 1859–1953) [Nesom 2006 FNANM 20:169]

E. elegans Brandegee (p. 305) Rounded shrub, minutely scabrous but appearing glabrous, resinous. **ST:** ≤ 1 m, ± striate; new growth yellow to tan; bark of older growth gray-white, shredding with age. **LF:** alternate, proximal sometimes in axillary clusters, ≤ 4 cm, ≤ 4 mm wide, ± linear or narrowly oblanceolate, entire. **INFL:** heads discoid, ± spheric, 1 or in open cyme-like clusters; involucre bell-shaped, < 2 cm, ≤ 2 cm wide; phyllaries graduated in 3–5 series, narrow, green or ± white, tips acuminate, green; receptacle hemispheric, paleate, paleae oblanceolate, readily deciduous. **FL:** 30–180+; corolla ≤ 6 mm, cylindric to funnel-shaped, yellow; style branch appendages lanceolate. **FR:** 1.7–2 mm, narrowly obconic, obscurely 3–4-angled, hairy, esp on angles; pappus of 5–8 lance-linear scales, > fr. 2*n*=18. Banks, arid hillsides, brushy slopes, juniper woodland; 60–1300 m. s SNF, Teh, s SnJV, se SnFrB, SCoR, WTR. Apr–Jul

EATONELLA

1 sp.: w US. (Daniel C. Eaton, Am botanist, 1834–1895) [Baldwin et al. 2002 Syst Bot 27:161–198; Strother 2006 FNANM 21:348–349]

E. nivea (D.C. Eaton) A. Gray (p. 305) Ann 1–4 cm, densely tomentose. **ST:** congested. **LF:** basal and alternate, crowded, gen ≤ 1 cm, linear to obovate, entire. **INFL:** heads radiate (rays gen inconspicuous), 1, terminal, peduncle 1–35 mm, in fl ± concealed by lvs, in fr sometimes elongating; involucre 4–5 mm diam, bell-shaped; phyllaries 8–12 in 1 series, 4–5 mm, linear-oblong, free, in fr reflexed; receptacle flat, epaleate. **RAY FL:** 1 per phyllary; ray 1.5–2.5 mm, light yellow, often drying ± purple. **DISK FL:** many; corolla 2–2.5 mm, yellow. **FR:** ± 3 mm, oblanceolate, flattened; faces glabrous, black and shiny; margins ciliate with long, white hairs; pappus of 2 fringed scales 1–2.5 mm. 2*n*=38. Sandy or volcanic soil, sagebrush scrub; 1250–2900 m. GB; se OR, sw ID, w NV. May–Jul

ECLIPTA

Ann. ST: prostrate to erect, often rooting proximally, simple to much-branched. **LF:** simple, opposite, sessile. **INFL:** heads radiate, small, 1 or few in cyme-like clusters; peduncle short or 0; involucre hemispheric; phyllaries in 1–2 ± equal series, free, ovate; receptacle rounded, paleate; paleae narrowly linear, bristle-like. **RAY FL:** 20–40; corolla white, ray short, narrowly linear. **DISK FL:** 15–30+; corolla white; style branches short, incl; anthers brown, incl. **FR:** 4-angled, ± flat, brown, glabrous; pappus a crown of minute bristles or 0. 1–4 spp.: Am, esp trop. (Greek: deficient, from lack of pappus) [Strother 2006 FNANM 21:128–129]

E. prostrata (L.) L. (p. 311) FALSE DAISY Pl ± strigose throughout. **ST:** 1–10 dm. **LF:** blade 2–10 cm, linear, lanceolate, or narrowly elliptic, entire or short-toothed, tip acute. **INFL:** peduncle 0–15+ mm; involucre 4–10 mm diam; phyllaries 4–5 mm, acute. **RAY FL:** corolla 1.5–3 mm. **DISK FL:** corolla 1.5–2 mm. **FR:** 1.7–2.2 mm, obovate, smooth or ± tubercled; pappus ≤ 0.2 mm. 2*n*=22. Damp places; < 1350 m. GV, SCoR, SW, DSon; a weed on all continents. Source of dark dye, medicine against roundworm parasites. All yr

ENCELIA

David J. Keil & Curtis Clark

[Subshrub], shrub. **ST:** gen many from base. **LF:** alternate, gen drought-deciduous, simple, petioled, entire or rarely toothed. **INFL:** heads radiate or discoid, 1 or in panicle-like cluster; peduncle gen long; involucre hemispheric; phyllaries graduated in 2–4+ series, free; receptacle paleate, palea folded around and falling with fr. **RAY FL:** sterile; style 0; ray yellow. **DISK FL:** many; corolla yellow or brown-purple, tube slender, throat abruptly expanded, lobes triangular; anther tip ovate, ± acute; style tips triangular. **FR:** strongly compressed, obovate or wedge-shaped; edges long-ciliate; faces glabrous or short-hairy; pappus of 2 narrow scales or 0. 11–12 spp.: w N.Am, w S.Am. (Christoph Entzelt, German naturalist, 1517–1583) [Clark 2006 FNANM 21:118–122; Fehlberg & Ranker 2007 Syst Bot 32:692–699] Commonly hybridizing, esp in disturbed areas; *E. farinosa* × *E. frutescens* is common; *E. farinosa* × *E. californica*, *E. farinosa* × *E. actoni*, *E. actoni* × *E. frutescens*, *E. frutescens* × *E. virginensis*, *E. farinosa* × *Geraea canescens* have been reported.

1. Heads in panicle-like cluster, always radiate; lvs ± hairy, hairs curled
  2. Heads 3–9 in tight panicle-like cluster; rays well developed; lvs densely silver- or gray-tomentose, not at all strigose. . . . . ***E. farinosa***
  2′ Heads 2–5 in loose panicle-like cluster; rays often short, few, deeply lobed; lvs moderately hairy, often partly strigose. . . . . [***E. farinosa* × *E. frutescens***]
1′ Heads gen solitary, radiate or discoid; lvs glabrous or hairy
  3. Disk corolla brown-purple; lvs green, glabrous to sparsely and minutely strigose or bristly . . . . ***E. californica***
  3′ Disk corolla yellow or orange; lvs green to gray, glabrous or hairy
    4. Ray fls 0; lvs green, strigose but not canescent. . . . . ***E. frutescens***
    4′ Ray fls present; lvs gray-green to green, canescent
      5. Lvs silvery green, canescent, not at all strigose; ray fls 14–25, ray 10–25 mm, shallowly toothed; sw SnJV and adjacent WTR (Cuyama Valley), W&I, w D and adjacent CA-FP, DMtns . . . . ***E. actoni***
      5′ Lvs sparsely strigose and lightly soft-canescent; ray fls 11–21, ray 8–15 mm, deeply toothed; e DMoj, DMtns. . . . . ***E. virginensis***

E. actoni Elmer (p. 311) Shrub 5–15 dm, with many slender branches from base. **ST:** branched below; young st hairy; older st with fissured bark. **LF:** scattered along st; petiole 6–12 mm; blade 2.5–4 cm, ovate to deltate, acute, silvery green, canescent. **INFL:** heads radiate, 1; peduncle canescent; involucre 8–14 mm; phyllaries ovate. **RAY FL:** 14–25; ray 10–25 mm. **DISK FL:** corolla 5–6 mm, yellow. **FR:** 5–7 mm; pappus gen 0. 2*n*=36. Open areas, rocky slopes, roadsides; 200–2100 m. s SNF, s SnJV, n TR, e PR, W&I, s SNE, w D; sw NV, n Baja CA. Feb–Jul

E. californica Nutt. (p. 311) Shrub 5–15 dm, with many slender branches from base. **ST:** branched proximally; young st glabrous; older st with smooth or roughened bark. **LF:** scattered along st; petiole 5–25 mm; blade 3–6 cm, diamond-shaped or narrowly ovate, acute, green, glabrous to sparsely and minutely strigose or bristly. **INFL:** heads radiate, gen 1; peduncle hairy; involucre 10–12 mm; phyllaries lanceolate. **RAY FL:** 15–25; ray 15–35 mm. **DISK FL:** corolla 5–6 mm, brown-purple. **FR:** 5–7 mm; pappus 0. 2*n*=36. Coastal scrub; < 600 m. s CCo, SCo, WTR, w PR; n Baja CA. Feb–Jun

E. farinosa Torr. (p. 311) BRITTLEBUSH Shrub 3–15 dm, from 1 or several trunks; sap fragrant. **ST:** much-branched above; young st tomentose; older st with smooth bark. **LF:** clustered near st tip; petiole 10–20 mm; blade 2–7 cm, ovate to lanceolate, obtuse or acute, silver- or gray-tomentose. **INFL:** heads radiate, 3–9 in tight panicle-like cluster; peduncle ± yellow, glabrous exc just below heads; involucre 4–10 mm; phyllaries lanceolate. **RAY FL:** 11–21; ray 8–12 mm. **DISK FL:** corolla 5–6 mm, yellow or brown-purple. **FR:** 3–6 mm; pappus 0. 2*n*=36. Coastal scrub, stony desert hillsides; < 1500 m. e SCo, adjacent PR, D; to sw UT, nw Mex. Widely planted, esp along highways; may persist or escape outside native range. Dried resin used as incense. Jan–Jun, Aug–Sep

E. frutescens (A. Gray) A. Gray (p. 311) Shrub 5–15 dm, with many slender branches from 1–several short trunks. **ST:** branched below; young st glabrous; older st with fissured bark. **LF:** scattered along st; petiole 2–7 mm; blade 1–2.5 cm, elliptic or narrowly ovate, obtuse, green, strigose. **INFL:** heads discoid, 1; peduncle strigose; involucre 6–12 mm; phyllaries lanceolate. **RAY FL:** 0. **DISK FL:** corolla 5–6 mm, yellow. **FR:** 6–9 mm; pappus of 2 slender scales or 0. 2*n*=36. Desert washes, flats, slopes, roadsides; < 800 m. D; s NV, w AZ, Baja CA. Feb–May

E. virginensis A. Nelson Shrub 5–15 dm, with many slender branches from base. **ST:** proximally branched; young st hairy; older st with fissured bark. **LF:** scattered along st; petiole 2–7 mm; blade 1.2–2.5 cm, narrowly ovate to deltate, acute or obtuse, gray-green, sparsely strigose and lightly soft-canescent. **INFL:** heads radiate, 1; peduncle canescent; involucre 9–13 mm; phyllaries narrow ovate. **RAY FL:** 11–21; ray 8–15 mm, yellow. **FR:** 5–8 mm; pappus gen 0. 2*n*=36. Desert flats, rocky slopes, roadsides; 500–1500 m. e DMoj, DMtns; to sw UT, NM. Mar–Jun, Dec

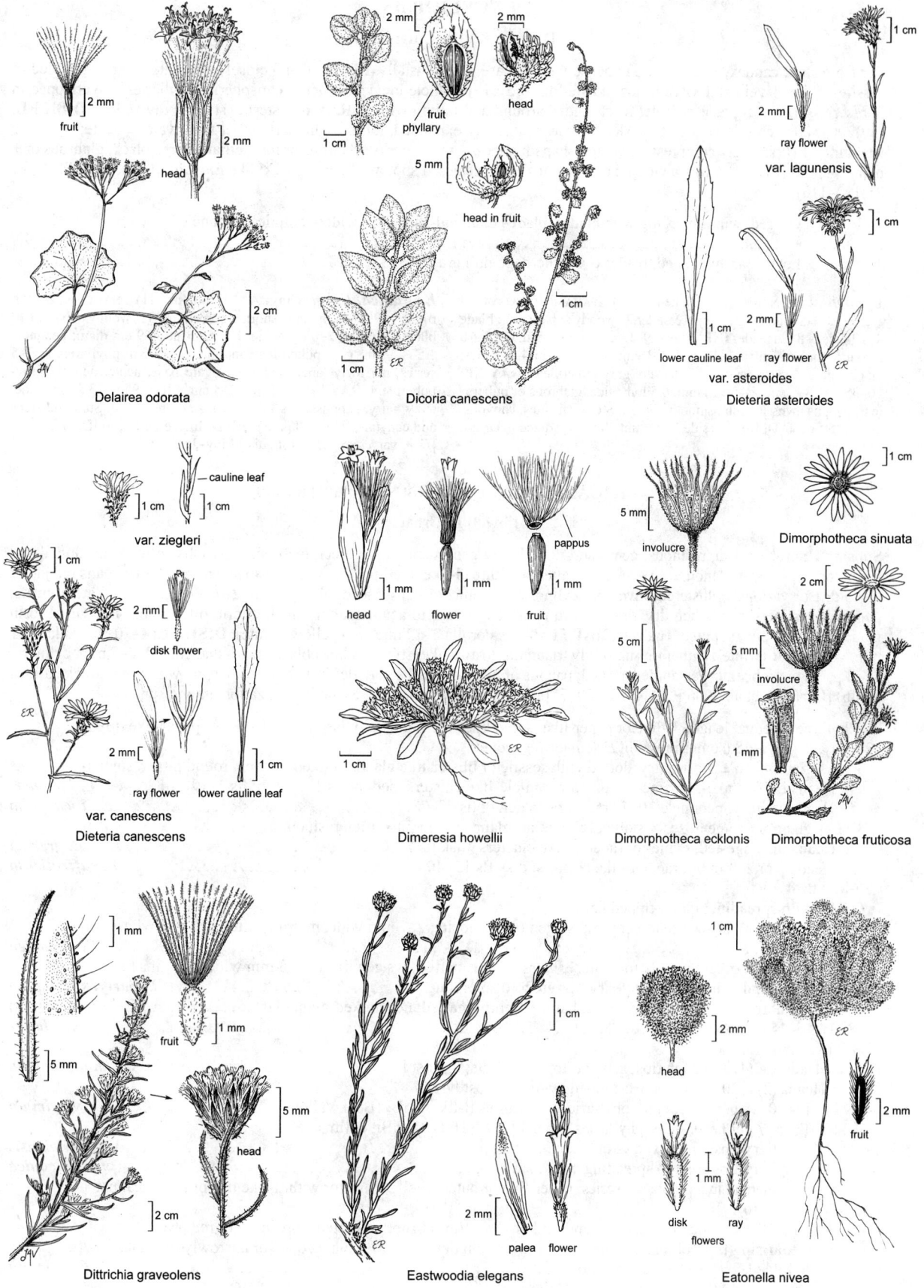

Delairea odorata

Dicoria canescens

var. lagunensis

var. asteroides
Dieteria asteroides

var. ziegleri

var. canescens
Dieteria canescens

Dimeresia howellii

Dimorphotheca sinuata

Dimorphotheca ecklonis   Dimorphotheca fruticosa

Dittrichia graveolens

Eastwoodia elegans

Eatonella nivea

ENCELIOPSIS

David J. Keil & Curtis Clark

Per from stout caudex, ± scapose. **ST**: densely lfy at base, lfless distally. **LF**: basal and closely alternate, simple, petioled or sessile, entire, 3-veined. **INFL**: heads radiate [discoid], 1; peduncle long; involucre hemispheric; phyllaries ± overlapped in 3–6 series, free; receptacle paleate, palea folded around and falling with fr. **RAY FL**: sterile; style 0; ray yellow. **DISK FL**: (50)200–500+; corolla yellow, tube slender, throat abruptly expanded, lobes triangular; anther tips ovate, ± acute; style tips triangular. **FR**: strongly compressed, wedge-shaped; edges ± white, corky, glabrous or long-ciliate; faces black, glabrous or ± hairy; pappus [0] of 2 narrow awns and a crown of shorter scales. 4 spp.: w N.Am. (Greek: like *Encelia*) [Clark 2006 FNANM 21:112–113]

1. Petioles winged, wings merging with blades, blades diamond-shaped or widely elliptic; herbage silvery-canescent . . . . *E. covillei*
1′ Petioles not or barely winged, blades ovate; herbage dull gray . . . . *E. nudicaulis*

E. covillei (A. Nelson) S.F. Blake (p. 311) PANAMINT DAISY Pl 1.5–8(10+) dm; hairs fine, ± appressed. **ST**: woody at base. **LF**: blade 4–10 cm, 2–8 cm wide. **INFL**: head 9–13 cm diam; peduncle 3–10 dm, gray-puberulent; involucre 1.8–3 cm; phyllaries in 4–6 series, lanceolate to ovate, acuminate, densely gray-puberulent. **RAY FL**: 20–35; ray 3–5 cm. **FR**: ± 10 mm, 6.5 mm wide, glabrous or puberulent; pappus awns ± 1 mm, smooth. 2*n*=36. Stony hillsides, canyons; 400–1250 m. n DMtns (w side Panamint Range), adjacent DMoj. Mar–Jun ★

E. nudicaulis (A. Gray) A. Nelson (p. 311) NAKED-STEMMED DAISY Pl 1–4 dm; hairs short, ± spreading. **ST**: woody at base. **LF**: blades 2–6 cm, 2–6 cm wide. **INFL**: head 4–9 cm diam; peduncle 1.5–4.5 dm, gray-puberulent; involucre 1–2 cm; phyllaries in 3–5 series, narrowly lanceolate from ovate base, acute, densely gray-puberulent. **RAY FL**: ± 21; ray 2–4 cm. **FR**: ± 9 mm, 3.5 mm wide, silky-hairy; pappus awns 1–1.5 mm, smooth. 2*n*=36. Stony hillsides and canyons; 950–2000 m. W&I, DMtns, e DMoj; to ID, UT, n AZ. [*E. n.* var. *corrugata* Cronquist] May–Jun ★

ERICAMERIA   GOLDENBUSH, RABBITBRUSH

Lowell E. Urbatsch

Subshrub, shrub, ≤ 5 m, resinous, gen gland-dotted. **ST**: gen ascending to erect, glabrous to woolly; often resin-dotted and resin-coated, stalked-glandular in some. **LF**: alternate, thread-like to elliptic or wedge-shaped, entire. **INFL**: heads radiate or discoid, 1 or variously clustered; involucre cylindric, obconic, or bell-shaped; phyllaries in 2–7 series, ± lanceolate to ovate, gen resinous, persistent when dry, tips erect to recurved, obtuse to acuminate or tailed, midrib often thickened with a resin gland; ± flat to convex, pitted, epaleate. **RAY FL**: 0–30; corolla 2–12 mm, gen yellow (white). **DISK FL**: 4–70+; corolla 3–11 mm, gen yellow (white); anther tips narrowly triangular to awl-like; style tips lanceolate to awl-shaped. **FR**: 2–8 mm, cylindric to ellipsoid or ± obconic, gen angled, ribbed; pappus of minutely barbed bristles in 1 series. ± 36 spp.: w N.Am. (Greek: golden shrub) [Urbatsch et al. 2006 FNANM 20:50–77] Most spp. fl summer/fall. Some hybridization among spp.

1. Phyllaries ± equal length, lf-like or paper-like; sometimes each outer phyllary with a lf-like or paper-like tip
  2. Ray fls gen 1–18 (sometimes 0 in *Ericameria greenei*)
    3. Lvs glabrous to ± short-hairy, dotted with sessile or blister-like glands sunken in deep, round pits, resinous
      4. Heads in lfy spike-, raceme-, or narrow, panicle-like clusters; peduncles 1–15 mm; fls 5–19 . . . . *E. bloomeri*
      4′ Heads solitary; peduncle 20–70 mm, bracts gen 0; fls 28–78. . . . . *E. linearifolia*
    3′ Lvs glabrous, tomentose, or stalked-glandular, glands not in pits, often resinous
      5. Heads in congested, cyme- or raceme-like clusters; disk fls 7–20 . . . . *E. greenei*
      5′ Heads 1 or 2–3 in lfy, raceme-like clusters; disk fls 15–40 . . . . *E. suffruticosa*
  2′ Ray fls 0
    6. Lvs with spreading, gland-tipped hairs
      7. St densely, silvery-white tomentose; lvs gen oblanceolate, 2–7 mm wide, margins often wavy; phyllary
        tips erect . . . . *E. discoidea*
      7′ St white to ± green, closely tomentose, hairs tightly matted; lvs gen linear, 1–3 mm wide, margins flat
        or rolled under, not wavy; phyllary tips erect to spreading . . . . *E. parryi* var. *aspera*
    6′ Lvs ± glabrous or ± white tomentose, hairs ± tightly-matted, not stalked-glandular . . . . *E. parryi* (in part)
      8. Lvs 5–14 mm wide. . . . . var. *latior*
      8′ Lvs 0.5–3 mm wide
        9. Heads 1–2(4) at branch tips, gen overtopped by distal-most lvs. . . . . var. *monocephala*
        9′ Heads (2)4–20+, gen not overtopped by distal-most lvs
          10. Fls 10–18 per head; outer phyllaries sometimes lf-like; SnBr (Bear Valley) . . . . var. *imula*
          10′ Fls 4–7 per head; outer phyllaries paper-like; SNH, GB, SnBr, DMtns
            11. Phyllary tips ± recurved . . . . var. *nevadensis*
            11′ Phyllary tips erect to spreading . . . . var. *vulcanica*
1′ Phyllaries unequal in a graduated series, paper-like or outer wholly lf-like or with lf-like tips or subapical patches
  12. Ray fls gen (0)1–10
    13. Ray and disk corollas white; outer and mid phyllary tips abruptly pointed, spreading or recurved,
      lf-like, arising from obtuse to notched summit of phyllary body; lf oblanceolate or narrowly obovate,
      often folded, 2–4 mm wide; Inyo Co. . . . . *E. gilmanii*

13′ Ray and disk corollas yellow; phyllary tips acuminate, acute to rounded or abruptly pointed, erect, sometimes lf-like

  14. Mid phyllaries < 1 mm wide

    15. Style appendage ± = stigma; rays 0–3; fl Mar–Jun . . . . *[2]E. cooperi* var. *cooperi*

    15′ Style appendage > stigma; rays 1–20; fl (Jul)Aug–Dec

      16. Mid phyllary midvein scarcely evident to slightly raised and slightly widened toward tip; disk fls 4–8; fr glabrous to densely hairy . . . . *[2]E. nana*

      16′ Mid phyllary midvein conspicuously raised throughout its length or on the distal 1/2 and slightly to greatly widened toward tip; disk fls 5–20; fr hairy

        17. Mid phyllary tips acute to long-acuminate; midvein gen prominently raised along phyllary length, slightly widened toward tip; heads in cyme-like clusters; DMtns . . . . *[2]E. laricifolia*

        17′ Mid phyllary tips rounded to acute; midvein gen distally prominently raised as a much darker, abaxial, narrowly obovate structure, proximal portion of midvein not noticeably raised or not at all evident; heads in raceme- to panicle-like clusters; SCo, DSon. . . . . *E. palmeri*

          18. Lvs 5–20(30) mm; axillary lvs 2–10 per node; outer and mid phyllaries strongly thickened near tip; midveins often not evident on proximal 1/2 of each phyllary; fls 6–12(14); Riverside to Ventura cos. . . . . var. *pachylepis*

          18′ Lvs 20–40 mm; axillary lvs 1–2 per node; phyllary tips not strongly thickened; midveins gen evident on proximal 1/2 of each phyllary; fls 12–30; San Diego Co. var. *palmeri*

  14′ Mid phyllaries > 1.0 mm wide

    19. Style appendage ± = stigma; rays 0–3; fl Mar–Jun . . . . *[2]E. cooperi* var. *cooperi*

    19′ Style appendage ≥ stigma; rays 1–30; fl (Jul)Aug–Dec (*Ericameria pinifolia* occ fl in spring)

      20. Mid phyllary midvein scarcely evident to slightly raised and slightly widened distally; phyllary margins entire, glabrous; disk flrs 4–8; lvs without regularly spaced deep resin pits; DMtns. . . . . *[2]E. nana*

      20′ Mid phyllary midvein prominently raised on distal 1/2 or throughout length of phyllary; phyllary margins entire, glabrous or ciliate to fringed; disk fls 6–25; lvs gen with regularly spaced deep resin pits

        21. Mid phyllary tips acute to long-acuminate; midvein gen prominently raised throughout phyllary length, slightly widened distally; margins glabrous or ciliate; DMtns . . . . *[2]E. laricifolia*

        21′ Mid phyllary tips mucronate to long-acuminate; margins ciliate, fringed, or irregularly toothed; widespread

          22. Tips of outermost phyllaries rounded, acute, acuminate, or abruptly pointed; fr gen silky to densely hairy. . . . . *E. fasciculata*

          22′ Tips of outermost phyllaries gen abruptly pointed, lf-like; fr glabrous or hairy, more densely distally

            23. Lvs 3–18(23) mm; ray fls 2–6; disk fls 5–14; gen dunes or sand hills along or near coast (occ inland), Sonoma to Los Angeles cos. . . . . *E. ericoides*

            23′ Lvs 10–40 mm; ray fls 3–10 in fall, 15–30 in spring; disk fls 11–25; sandy to stony, often disturbed soils away from immediate coast, Ventura Co. and s . . . . *E. pinifolia*

12′ Rays 0

  24. Lvs wedge-shaped, obovate, or spoon-shaped, 2–16 mm wide, tips gen obtuse, rounded, or notched. . . . *E. cuneata*

    25. Disk fls 36–70 — San Diego Co. . . . . var. *macrocephala*

    25′ Disk fls 7–33

      26. Lvs sessile, blades wedge-shaped, largest 3–14(18) mm, 2–9(12) mm wide . . . . var. *cuneata*

      26′ Lvs stalked, blades spoon-shaped, largest (9)12–25 mm, 4–16 mm wide . . . . var. *spathulata*

  24′ Lvs thread-like, linear, elliptic, oblanceolate, or oblong, 0.3–12 mm wide, tips acute

    27. Lvs and often young st dotted with round, ± evenly spaced, resin-filled pits

      28. Outer and mid phyllaries with lf-like, recurved tips; corolla white . . . . *E. albida*

      28′ Outer and mid-level phyllary tips rounded to acute or acuminate; corolla yellow

        29. Lvs oblong to oblanceolate, flat to slightly concave, 3–12 mm wide. . . . . *E. parishii* var. *parishii*

        29′ Lvs thread-like to narrowly oblong or oblanceolate, grooved, 0.5–3 mm wide

          30. Phyllaries thin, midvein often evident but not raised or enlarged at tip . . . . *E. paniculata*

          30′ Phyllary midvein obscure or very prominently raised and expanded at tip

            31. Phyllary midvein gen obscure proximal 1/2, tip bearing a prominent, ± spheric resin gland. . . . . . *E. teretifolia*

            31′ Phyllary midvein prominently raised and expanded at tip

              32. Lvs 25–90 mm; heads in flat-topped or convex cyme-like clusters, peduncle 1–15 mm, bracts 0–7, scale-like; fls 10–25. . . . . *E. arborescens*

              32′ Lvs 10–25 mm; heads 1 or in raceme- to panicle-like clusters, peduncle 3–20 mm, lfy or bracted; disk fls 6–16(22) . . . . *E. brachylepis*

    27′ Lvs and young st lacking round resin pits

      33. Phyllaries spiraled, outermost lf-like, inner parchment-like, bodies of outer and mid phyllaries with tooth-like, spreading to recurved tips; Shasta, Tehama, and Trinity cos., serpentine soils . . . . *E. ophitidis*

      33′ Phyllaries gen aligned vertically, dry, paper-like (if spiraled, outer and mid phyllary tips not tooth-like); tips long-tapered, erect or recurved; widespread, often abundant on various soils . . . . . . *E. nauseosa*

        34. Fr glabrous . . . . var. *leiosperma*

        34′ Fr hairy, ± long-hairy or silky (only distally in *Ericameria nauseosa* var. *washoensis*)

          35. Style appendages gen < stigmas (± =)

            36. Corolla lobes 0.5–1 mm . . . . var. *hololeuca*

36′ Corolla lobes 1.5–2.5 mm
   37. Phyllary tips recurved — s CA . . . . var. *ceruminosa*
   37′ Phyllary tips erect . . . . var. *oreophila*
35′ Style appendages > stigmas
   38. Corolla lobes sparsely long-hairy; fr distally hairy . . . . var. *washoensis*
   38′ Corolla lobes ± glabrous; fr hairy throughout
      39. St loosely tomentose, gray-white to dark green; lvs present at fl . . . . var. *speciosa*
      39′ St closely tomentose with tightly matted hairs, gen ± yellow-green, sometimes becoming ±
         white; lvs present or 0 at fl
         40. St gen lfy at fl; lvs 20–60 mm . . . . var. *bernardina*
         40′ St nearly lfless at fl; lvs 10–30 mm . . . . var. *mohavensis*

E. albida (A. Gray) L.C. Anderson (p. 311) WHITE-FLOWERED RABBITBRUSH Pl 3–15 dm, herbage resinous. **ST:** glabrous. **LF:** 2–3.5 cm, linear, ± cylindric, gland-dotted in pits, glabrous. **INFL:** heads discoid, in cyme-like clusters; peduncle gen < 10 mm, bractless; involucre 6–10 mm, 2–4 mm wide, ± cylindric; phyllaries 15–20, graduated in 3–4 series, in poorly defined ranks, ± lanceolate, barely keeled, straw-colored, tips abruptly pointed, green, outer and mid phyllaries with lf-like, recurved tips. **RAY FL:** 0. **DISK FL:** 5–7; corolla 5–8 mm, white, lobes gen recurved; style appendage > stigma. **FR:** 4–5 mm, ± cylindric to narrowly ellipsoid or top-shaped, hairy; pappus slightly < corolla, ± white. 2*n*=18. Saline or alkaline soils; 300–1300 m. SNE, n DMoj; to w UT. [*Chrysothamnus a.* (A. Gray) Greene] Aug–Nov ★

E. arborescens (A. Gray) Greene (p. 311) GOLDEN-FLEECE Pl ≤ 5 m. **ST:** distally densely branched forming a dense, rounded, lfy crown, glabrous or sparsely hairy, gland-dotted, resinous. **LF:** 25–90 mm, thread-like to linear, acute, sparsely gland-dotted in pits. **INFL:** heads discoid, many, in flat-topped or convex cyme-like clusters at branch tips; peduncle 1–15 mm, bracts 0–7, scale-like; involucre ≤ 4.5 mm, 3.5–4.5 mm wide, obconic; phyllaries 20–25, graduated in 3–4 series, spiraled, lanceolate, acute, (±) glabrous, midrib raised, brown, expanded at tip. **RAY FL:** 0. **DISK FL:** 10–25; corolla 4.7–5.5 mm; style appendage > stigma. **FR:** 2–3 mm, narrowly top-shaped, 5-angled, densely appressed-hairy; pappus > disk corolla, dull white. 2*n*=18. Woodland, open forest, chaparral, esp after fire; gen < 1700 m (< 2900 m in SNH). KR, NCoR, SNF, s SNH, SnFrB, SCoR, WTR. Closely related to *E. parishii.* Aug–Nov

E. bloomeri (A. Gray) J.F. Macbr. (p. 311) BLOOMER'S GOLDENBUSH Pl 2–6 dm. **ST:** glabrous to barely woolly, often glandular. **LF:** 20–70 mm, thread-like to ± oblanceolate, acute or acuminate, glabrous or sparsely short hairy, dotted with ± blister-like glands. **INFL:** heads radiate, in lfy spike-, raceme-, or narrow panicle-like clusters; peduncle 1–15 mm, gen bractless; involucre 8–11 mm, 8–12 mm wide, narrowly bell-shaped; phyllaries ± equal, ± lf-like, loosely overlapping in 3–6 series, rigid, linear to lanceolate, resinous, tips recurved, margin scarious, woolly-ciliate, all but innermost abruptly narrowed to linear green tips. **RAY FL:** 1–5; ray 8–12 mm. **DISK FL:** 4–14; corolla 7–11 mm, yellow; style appendage > stigma. **FR:** 5.5–8 mm, 5-angled, brown, ± glabrous to hairy; pappus 6–9 mm, tan to ± red. 2*n*=18. Open, gravelly, conifer forest; 900–4000+ m. KR, CaR, SN, MP; to WA, w NV. Jul–Oct

E. brachylepis (A. Gray) H.M. Hall BOUNDARY GOLDENBUSH Pl 10–20 dm. **ST:** ± glabrous, gland-dotted in pits, resinous. **LF:** crowded, 10–25 mm, 0.8–1.5 mm wide, thread-like to narrowly oblong, obtuse to ± acute, curled under in age, glabrous or sparsely hairy, gland-dotted, resinous. **INFL:** heads discoid, 1 or in raceme- or narrow panicle-like clusters at ends of lfy branchlets; peduncle 3–20 mm, ± bracted; involucre 4.5–6 mm, 4–6 mm wide, obconic; phyllaries 16–22, graduated in 3–4 series, spiraled, linear to ovate, obtuse to acute, midrib brown, raised, expanded at tip, resinous, margin white-scarious, minutely ciliate. **RAY FL:** 0(2). **DISK FL:** 6–16(22); corolla 5–7 mm; style appendage < stigma. **FR:** ≤ 5 mm, densely soft-hairy; pappus 6–7 mm, ± white to tan. 2*n*=36. Chaparral; 300–1400 m. s SCo, s PR; AZ, n Baja CA. Sep–Dec

E. cooperi (A. Gray) H.M. Hall var. *cooperi* (p. 311) COOPER'S GOLDENBUSH Pl 3–10 dm. **ST:** puberulent, gland-dotted, resinous. **LF:** 3–15 mm, ± linear, acute, puberulent, gland-dotted in pits. **INFL:**

heads gen radiate or some discoid, in rounded cyme-like clusters; peduncle 5–15 mm, bracts 0–2; involucre 4–5 mm, 3–4 mm wide, narrowly bell-shaped; phyllaries 9–15, graduated in 3–4 series, spiraled, oblong to ovate, obtuse or outer acute; mid phyllaries ± 1 mm wide. **RAY FL:** 0–3; corolla 4–9 mm. **DISK FL:** 4–12; corolla 3–5 mm; style appendage ± = stigma. **FR:** 3–3.5 mm, obconic to ± cylindric, softly silky-hairy, veins 10–12, thin; pappus 3–5 mm, white. 2*n*=18. Rocky slopes, valleys, in creosote-bush scrub, Joshua-tree woodland; 300–2000 m. WTR, SnGb, SNE, DMoj; s NV. Hybridizes with *E. linearifolia.* Other var. occurs in Baja CA. Mar–Jun

E. cuneata (A. Gray) McClatchie WEDGELEAF GOLDENBUSH Pl 1–10 dm. **ST:** glabrous, ± gland-dotted in pits, resinous. **LF:** 3–25 mm, 2–16 mm wide, wedge-shaped, obovate, or spoon-shaped, gen obtuse, rounded, or notched, glabrous. **INFL:** heads radiate or discoid, 1 or in small compact, rounded, cyme-like clusters; peduncle 2–10 mm, bracts 0–10, scale-like; involucre 5–12 mm, 4–14 mm wide, obconic; phyllaries 20–60, graduated in 4–7 series, lanceolate to obovate, obtuse to gen acuminate, glabrous, sometimes resinous. **RAY FL:** 0(7); corolla < 5 mm. **DISK FL:** 7–70; corolla ± 5.5 mm; style appendage ± = stigma. **FR:** 2.5–3 mm, 5-ribbed, silky-hairy; pappus < corolla, sparse, brown.

var. *cuneata* CLIFF GOLDENBUSH **LF:** largest 3–14(18) mm, 2–9(12) mm wide, wedge-shaped, sessile. **INFL:** heads radiate or discoid; involucre 5–9 mm, 4–7 mm wide. **DISK FL:** 12–33. 2*n*=18. Granite outcrops; 600–2800 m. SN, WTR, SnGb, PR, SNE. Sep–Nov

var. *macrocephala* Urbatsch LAGUNA MOUNTAINS GOLDENBUSH **LF:** largest gen 12–20 mm, 6–14 mm wide, ± obovate; base petiole-like. **INFL:** heads discoid; involucre 9–12 mm, 6–14 mm wide. **DISK FL:** 36–70. 2*n*=18. Rock crevices; 1200–1830 m. c PR (Laguna Mtns, San Diego Co.). Sep–Nov ★

var. *spathulata* (A. Gray) H.M. Hall (p. 311) **LF:** largest (9)12–25 mm, 4–16 mm wide, ± spoon-shaped; base petiole-like; tip gen widely obtuse or notched. **INFL:** heads gen discoid; involucre 5–8 mm, 4–7 mm diam. **DISK FL:** 7–15. 2*n*=18. Rock outcrops; 100–1900 m. Teh, SCoR, D; s NV, AZ, nw Mex. Intergrades with *E. cuneata* var. *c.* in Kern Co. Sep–Nov

E. discoidea (Nutt.) G.L. Nesom WHITESTEM GOLDENBUSH Pl 1–4 dm. **ST:** densely silvery-white tomentose. **LF:** 10–35 mm, oblong to gen oblanceolate, sessile, obtuse to acute, often wavy-margined, stalked-glandular. **INFL:** heads discoid, 1–few in cyme- or raceme-like clusters; peduncle 3–15, bracts 0–3, lf-like; involucre 9–13 mm, 8–12 mm diam, obconic to bell-shaped; phyllaries ± equal, in 2–3 series, lanceolate, acuminate, tips erect, outer grading into distal lvs, inner scarious. **RAY FL:** 0. **DISK FL:** 10–26; corolla 9–11 mm; style appendage >> stigma. **FR:** 5–6 mm, narrowly obconic, hairy; pappus 8.5–11 mm, ± brown. 2*n*=18. Rocky slopes; 2300–3800 m. SNH, Wrn, SNE (Sweetwater Mtns), W&I; to OR, MT, CO. If recognized taxonomically, stabilized hybrids with *E. nauseosa* var. *speciosa* [*Chrysothamnus nauseosus* subsp. *albicaulis*] assignable to *C. parryi* (A. Gray) Greene subsp. *bolanderi* (A. Gray) H.M. Hall & Clem. Jul–Sep

E. ericoides (Less.) Jeps. (p. 311) MOCK HEATHER, CALIFORNIA GOLDENBUSH Pl ≤ 2 m. **ST:** ± lightly puberulent. **LF:** crowded, 3–18(23) mm, ± cylindric, adaxial face grooved, ± obtuse, glabrous or puberulent, gland-dotted in pits; fan-shaped axillary lf clusters gen present. **INFL:** heads radiate, many in panicle-like clusters; pedun-

cle 1–30 mm, bracts 4+, lf-like; involucre 5–8 mm, 4–6 mm wide, obconic; phyllaries 16–24, graduated in 3–4 series, outer lanceolate, acuminate, outermost with lf-like tips, middle > 1 mm wide, mucronate to long-acuminate, midvein prominently raised, inner narrowly oblong, acute, tomentose-ciliate, midrib a thread-like gland. **RAY FL**: 2–6; corolla 6–8 mm. **DISK FL**: 5–14; corolla 5–8 mm; style appendage ≥ stigma. **FR**: ≤ 4 mm, cylindric, 8–10-ribbed, striate; glabrous to ± silvery hairy, more densely distally; pappus 5–7 mm, white to tan. 2*n*=18. Dunes, inland sandy soils; < 600 m. s NCo, s SnJV (Cuyama Valley), CW (exc SCoRI), c SCo. Perhaps 1 variable sp. with *E. pinifolia*. Hybridizes with *E. nauseosa*. Sep–Nov

E. fasciculata (Eastw.) J.F. Macbr. (p. 311) EASTWOOD'S GOLDENBUSH Pl ≤ 15 dm. **ST**: densely lfy, glabrous to sparsely puberulent, resinous. **LF**: 5–25 mm, ± cylindric, reflexed with age, ± obtuse, glabrous or sparsely hairy, gland-dotted in pits; axillary clusters of short lvs gen present. **INFL**: heads gen radiate, 1–few in cyme-like clusters; peduncle 1–15 mm, bracts 3+, lf-like; involucre 6–8 mm, 5–7 mm wide, obconic; phyllaries 22–26, graduated in 3–5 series, scarious, yellow-tan, resinous, outer lanceolate or ovate, rounded, acute, acuminate, or abruptly pointed, without lf-like tips, midvein ± thickened; mid phyllaries > 1 mm wide, mucronate to long-acuminate; inner oblong, acute. **RAY FL**: 4–6; corolla 5–6.5 mm. **DISK FL**: 18–25; corolla 6–9 mm; style appendage > stigma. **FR**: < 4 mm, ± cylindric, gen silky to densely hairy; pappus > corolla, dull white to ± red-brown. 2*n*=18. Sandy soils; < 100 m. c CCo (n Monterey Co.). Jul–Oct ★

E. gilmanii (S.F. Blake) G.L. Nesom (p. 311) GILMAN'S GOLDENBUSH Pl 2–5 dm, aromatic. **ST**: divergent to ascending or erect, glabrous to sparsely hairy, gland-dotted to resinous. **LF**: 6–12 mm, oblanceolate or narrowly obovate, obtuse, often folded, glabrous, gland-dotted to resinous. **INFL**: heads radiate, 1–few in cyme-like clusters; peduncle 1–15 mm, bracts 0–4, scale-like; involucre 7–12 mm, ± 5 mm wide, narrowly bell-shaped; phyllaries ± 25, graduated in 4–6 series, outer and mid lf-like, widely lanceolate, tip elongated, often thickened, green, spreading or recurved, midvein darker, inner phyllaries oblong, parchment-like, resinous. **RAY FL**: 4–7; corolla 8–10 mm, white. **DISK FL**: 10–18; corolla 5.5–7.5 mm; appendage < stigma. **FR**: 3–4 mm, ± cylindric, 5-ribbed, silky-hairy; pappus ≤ 6.5 mm, ± white. 2*n*=18. Open conifer forest, gen on limestone; 2100–3400 m. W&I, n DMtns (Panamint Range). Aug–Sep ★

E. greenei (A. Gray) G.L. Nesom GREENE'S GOLDENBUSH Pl 1–3 dm. **ST**: glabrous to ± sticky-glandular, distally puberulent or white-tomentose, gen with stalked glands. **LF**: 15–30 mm, ± oblanceolate, obtuse, glabrous, tomentose, or with stalked glands. **INFL**: heads radiate, gen in congested, cyme- or raceme-like clusters; peduncle gen 5–20(100) mm, lfy-bracted; involucre 8–12 mm, 12–15 mm wide, obconic; phyllaries 18–28, ± equal in 2–3 series, linear to lanceolate, acute, glandular and sticky to minutely tomentose, all but innermost with lf-like tips. **RAY FL**: (0)1–7; corolla 7–10 mm. **DISK FL**: 7–20; corolla 8–9.5 mm; style appendage >> stigma. **FR**: 5–7 mm, narrowly oblong, ± glabrous to densely appressed-soft-hairy; pappus 7–9 mm, white to light brown. 2*n*=18. Rocky areas in open conifer forest; 1500–2200 m. KR, NCoRH, CaRH, n SNH, Wrn; to WA, ID. Jul–Sep

E. laricifolia (A. Gray) Shinners TURPENTINE-BRUSH Pl 3–10 dm, glabrous, aromatic. **ST**: glabrous, gland-dotted, resinous. **LF**: 10–30 mm, gen ± cylindric, ± acute, glabrous, gland-dotted in pits; axillary lf clusters sometimes present. **INFL**: heads radiate, in cyme-like clusters; peduncle 3–15 mm, bracts 3–20+, proximal lf-like, distal scale-like; involucre 3–5 mm, 3–5 mm wide, obconic; phyllaries 12–20, graduated in 3–4 series, spiraled, ± linear, acute, glabrous, midrib a ± brown to ± yellow gland; mid phyllaries ± 1 mm wide, tips acute to long-acuminate, midvein gen prominently raised throughout phyllary length, slightly widened toward tip, margin glabrous or ciliate. **RAY FL**: 3–6; corolla 8–11 mm. **DISK FL**: 6–18; corolla 5–6.5 mm; style appendage > stigma. **FR**: 3.5–4 mm, narrowly obconic, obscurely 4-ribbed, densely white-soft-hairy; pappus 4–6 mm, tan. 2*n*=18. Rocky canyons, pinyon/juniper woodland, creosote-bush scrub; 1000–2000 m. DMtns; to TX, n Mex. Sep–Oct

E. linearifolia (DC.) Urbatsch & Wussow (p. 311) INTERIOR GOLDENBUSH Pl 4–15 dm. **ST**: glabrous to ± puberulent, resinous.

LF: 10–55 mm, linear, acute; base narrowed, glabrous to ± hairy, gland-dotted in pits. **INFL**: heads radiate, 1; peduncle 20–70 mm, gen bractless; involucre 8–14 mm, 10–18 mm wide, hemispheric; phyllaries ± lf-like, ± equal in 2–3 series, linear to lanceolate, acuminate, stalked-glandular, center green, margin cut-ciliate, scabrous. **RAY FL**: 12–18; corolla 9–20 mm. **DISK FL**: 16–60; corolla 6–10 mm; style appendage ± = stigma. **FR**: 4–5 mm, compressed, 6–8-veined, densely silky-hairy; pappus 5.5–7 mm, white. 2*n*=18. Dry slopes, valleys, foothill and desert woodland, saltbush and creosote-bush scrub; < 2000 m. NCoRI, s SNF, Teh, ScV (Sutter Buttes), s SnJV, e CW, TR, PR, DMoj, w DSon; to sw UT, w AZ, n Baja CA; sw CO?. Hybridizes with *E. cooperi*. Mar–May

E. nana Nutt. DWARF GOLDENBUSH Pl 1–5 dm. **ST**: recurved to spreading or erect, glabrous, gen resinous. **LF**: 10–15 mm, linear to narrowly oblanceolate, gen curved, acute, gen glandular, sometimes irregularly gland-dotted in shallow pits, ± sticky. **INFL**: heads radiate, in dense lfy cyme-like clusters; peduncle 0–5 mm, gen bractless; involucre 5.5–7.5 mm, 3–4 mm wide, obconic; phyllaries 20–30, graduated in 4–5 series, spiraled, lanceolate, acute to acuminate, glabrous; mid phyllaries ± 1 mm wide, midvein scarcely evident to slightly raised and slightly widened toward tip. **RAY FL**: 1–7; corolla 2–3 mm. **DISK FL**: 4–8; corolla 4.5–6.5 mm; style appendage > stigma. **FR**: 4–5.5 mm, cylindric, faintly 5-angled, glabrous to densely hairy; pappus 4–6 mm, light brown. 2*n*=18. Rocky soils, cliffs; 2100–2800 m. c SNE, DMtns; to WA, MT, UT. Jul–Nov ★

E. nauseosa (Pursh) G.L. Nesom & G.I. Baird RUBBER RABBITBRUSH Pl 2–28 dm; strongly scented. **ST**: spreading or ascending to erect, ± flexible, very lfy or lfless at fl, ± white to green, loosely to densely tomentose. **LF**: 10–70 mm, thread-like to narrowly (ob)lanceolate, glabrous to tomentose, gland-dotted or not. **INFL**: heads discoid, many in dense, flat-topped or rounded to ± panicle-like clusters; peduncle 1–20 mm, bracts gen 0(1–5, scale-like); involucre 6–14 mm, 2–4 mm wide, cylindric; phyllaries 10–31, graduated in 3–5 series, gen in vertical ranks (appearing spiraled), ± lanceolate to ovate, gen ± strongly keeled, firm, dry, paper-like, obtuse to acute, tips long-tapered, erect to recurved. **RAY FL**: 0. **DISK FL**: gen 5; corolla 6–13 mm, lobes gen glabrous (long-soft-hairy); style appendage < to > stigma. **FR**: 3–10 mm, gen hairy; pappus 3–13 mm, ± white. 2*n*=18. Highly variable; 13 more vars. in w N.Am, nw Mex. [*Chrysothamnus n.* (Pursh) Britton]

var. ***bernardina*** (H.M. Hall) G.L. Nesom & G.I. Baird BERNARDINA RABBITBRUSH Pl 5–15 dm, white-tomentose to green. **ST**: lfy at fl, closely tomentose, ± yellow-green, becoming ± white. **LF**: 25–60 mm, ± narrowly lanceolate. **INFL**: involucre 10–14 mm, glabrous to hairy. **FL**: corolla 10–12 mm, lobes 1.7–2.3 mm, ascending to spreading; style appendage > stigma. Open yellow-pine forest; 1200–3000 m. s SN, TR, n PR (incl SnJt), W&I; n Baja CA. [*Chrysothamnus n.* subsp. *b.* (H.M. Hall) H.M. Hall & Clem.] Much like robust *E. nauseosa* var. *speciosa*. Sometimes also confused with *E. nauseosa* var. *oreophila*, *E. nauseosa* var. *mohavensis*. Aug–Oct

var. ***ceruminosa*** (Durand & Hilg.) G.L. Nesom & G.I. Baird (p. 311) DESERT RABBITBRUSH Pl 6–15 dm. **ST**: ± lfless at fl, closely tomentose, yellow-green. **LF**: (0)1–3 cm, thread-like, curved. **INFL**: involucre 7–8.5 mm, glabrous, sticky; phyllary tips abruptly narrowed, recurved. **FL**: corolla 6–7.5 mm, glabrous, lobes 1.5–2.5 mm, ascending to recurved; style appendage < stigma. Gravelly arroyos; 700–1700 m. DMoj. [*Chrysothamnus n.* subsp. *c.* (Durand & Hilg.) H.M. Hall & Clem.] Locally common. Aug–Oct

var. ***hololeuca*** (A. Gray) G.L. Nesom & G.I. Baird WHITE RABBITBRUSH Pl 3–25 dm, fragrant. **ST**: lfy at fl, closely to loosely tomentose, white to gray- or yellow-green. **LF**: 30–70 mm, ± narrowly lanceolate, sometimes reflexed. **INFL**: involucre gen 7–9 mm, straw-colored, less hairy inward; phyllary midribs ± brown. **FL**: corolla 8–9.5 mm, lobes 0.5–1 mm, erect or bent inward; style appendage < stigma. Common. Well-drained granitic or limestone soils in scrub or woodland; 150–2500 m. SNH, Teh, SCoRO, WTR, SNE, w DMoj, DMtns; to s OR, ID, UT, n AZ. [*Chrysothamnus n.* subsp. *h.* (A. Gray) H.M. Hall & Clem.] *Ericameria* ×*viscosa* (D.D. Keck) G.L. Nesom & G.I. Baird is hybrid of *E. nauseosa* var. *hololeuca* × *E. cuneata*. Aug–Oct

var. *leiosperma* (A. Gray) G.L. Nesom & G.I. Baird (p. 311) SMOOTH-FRUIT RABBITBRUSH Pl gen 3–6 dm. **ST**: often lfless at fl, closely tomentose, ± yellow. **LF**: (0)10–30 mm, thread-like. **INFL**: involucre 8–11.5 mm, gen straw-colored (± purple), glabrous. **FL**: corolla 6–8.5 mm, lobes 0.5–1.1 mm, erect or bent inward; style appendage > stigma. **FR**: glabrous. Common. Dry sand, gravel, rocky crevices; 700–2400 m. SNE, DMtns; to UT, CO, n AZ. [*Chrysothamnus n.* subsp. *l.* (A. Gray) H.M. Hall & Clem.] Aug–Oct

var. *mohavensis* (Greene) G.L. Nesom & G.I. Baird MOJAVE RABBITBRUSH Pl 5–28 dm. **ST**: nearly lfless at fl, densely tomentose, ± yellow-green becoming white. **LF**: 15–30 mm, thread-like. **INFL**: involucre 8.5–12 mm, straw-colored, glabrous. **FL**: corolla 7–10.5 mm, lobes 0.9–2 mm, spreading; style appendage > stigma. Common. Dry scrub; 400–2400 m. SnFrB (Mount Hamilton Range), SCoRO, s SCoRI, TR, DMoj; n NV. [*Chrysothamnus n.* subsp. *m.* (Greene) H.M. Hall & Clem.] Sometimes intergrades with *E. nauseosa* var. *oreophila*, *E. nauseosa* var. *hololeuca*. Aug–Oct

var. *oreophila* (A. Nelson) G.L. Nesom & G.I. Baird GREAT BASIN RABBITBRUSH Pl 5–25 dm. **ST**: gen lfy at fl, closely tomentose, ± yellow-green. **LF**: 20–60 mm, ± thread-like. **INFL**: ± narrow; involucre 6–10 mm, glabrous; phyllaries spiraled, weakly keeled or rounded, midribs ± brown. **FL**: corolla 6–9 mm, lobes 1.5–2.5 mm, spreading to recurved; style appendage ≤ stigma. Common. Gen alkaline soils; 1000–3000 m. SN (e slope), Teh, s SCoRI, SnBr, PR, GB; to OR, MT, WY, NM, Baja CA. [*Chrysothamnus n.* subsp. *consimilis* (Greene) H.M. Hall & Clem.] May intergrade with *E. nauseosa* var. *speciosa* (gen higher elevation), *E. nauseosa* var. *mohavensis*. Aug–Oct

var. *speciosa* (Nutt.) G.L. Nesom & G.I. Baird (p. 311) SHOWY RABBITBRUSH Pl 2–10 dm. **ST**: lfy at fl, loosely tomentose, gray-white to dark green. **LF**: 30–70 mm, ± narrowly lanceolate. **INFL**: involucre 7–13.5 mm, ± tomentose. **FL**: corolla gen 9–13 mm, lobes 1.1–2.1 mm, ascending to spreading; style appendage > stigma. Common. Many dry habitats; 50–3500 m. NW, CaR, SN, W&I; to BC, MT, CO. [*Chrysothamnus n.* subsp. *albicaulis* (Nutt.) H.M. Hall & Clem.] Intergrades with *E. nauseosa* var. *oreophila*, *E. nauseosa* var. *hololeuca*. The name *E.* ×*bolanderi* (A. Gray) G.L. Nesom & G.I. Baird in c SNH (sw Mono Co.) applies to hybrids between *E. nauseosa* var. *speciosa* and *E. discoidea*. Aug–Oct

var. *washoensis* (L.C. Anderson) G.L. Nesom & G.I. Baird (p. 311) WASHOE RABBITBRUSH Pl 3–10 dm. **ST**: gen ± white, sparsely lfy, tomentose. **LF**: ± 30 mm, ± narrowly oblanceolate. **INFL**: involucre 10–12 mm, sparsely long-hairy and short-tomentose. **FL**: corolla 7.5–9 mm, lobes 1.3–1.6 mm, nearly erect, sparsely long-hairy; style appendage > stigma. **FR**: glabrous exc distal white hair-tuft. Uncommon. Dry juniper or pinyon grassland; 1500–1700 m. MP; nw NV. [*Chrysothamnus n.* subsp. *w.* L.C. Anderson] Aug–Oct

E. ophitidis (J.T. Howell) G.L. Nesom SERPENTINE GOLDENBUSH Pl ≤ 3 dm. **ST**: (±) glabrous, ± resinous. **LF**: 5–15 mm, linear, recurved, abruptly pointed, glabrous or sparsely hairy, gland-dotted. **INFL**: heads discoid, 1 or sometimes few per cluster; peduncle 1–10 mm, bracts 0–5, lf-like; involucre 10–15 mm, 4–8 mm wide, ± cylindric to obconic; phyllaries ± 25, graduated in 5–7 series, spiraled, ovate to elliptic, outermost lf-like, inner parchment-like, bodies of outer and mid phyllaries abruptly narrowed or truncate (sometimes notched) proximal to herbaceous, tooth-like, spreading to recurved tips. **RAY FL**: 0. **DISK FL**: 5–6; corolla ± 10 mm; style appendage > stigma. **FR**: ± 7 mm, 5-angled, appressed-hairy; pappus ± 10 mm, ± brown. Open conifer forest, gen on serpentine; ± 1600 m. s KR? NCoRH. Jul–Aug ★

E. palmeri (A. Gray) H.M. Hall Pl 5–40 dm. **ST**: erect, glabrous to puberulent, often gland-dotted in pits when young. **LF**: 5–40 mm, linear to oblanceolate, acute, glabrous to ± hairy, gland-dotted in pits; axillary clusters of short lvs gen present. **INFL**: heads radiate, many in raceme- to panicle-like clusters; peduncle 2–15 mm, bracts 5–10+, lf-like; involucre 5–8.5 mm, 3.5–4.5 mm wide, obconic to ± cylindric; phyllaries 16–24, graduated in 3–5 series, spiraled < disk fls, oblong, tan, margins narrow, white, ciliate; mid phyllaries < 1 mm wide, tips

rounded to acute, ± green; midvein gen distally prominently raised as a much darker, abaxial, narrowly obovate structure, proximal portion of midvein not noticeably raised or not evident. **RAY FL**: 1–8; corolla 4–6 mm. **DISK FL**: 5–20; corolla 5–8 mm; style appendage > stigma. **FR**: 3–4 mm, ± cylindric, 4–7-angled, hairy; pappus of disk fls > corolla, brown.

var. *pachylepis* (H.M. Hall) G.L. Nesom THICKBRACTED GOLDENBUSH Pl 5–15 dm, very lfy, puberulent. **LF**: 5–20(30) mm, thread-like; axillary lvs 2–10 per node. **INFL**: involucre 6–7 mm; outer and mid phyllaries strongly thickened near tip; midveins often not evident on proximal 1/2 of each phyllary. **FL**: 6–12(14); ray fl 1–6. Coastal scrub, disturbed chaparral; < 800 m. n SCo, ChI, WTR, w DSon. Aug–Dec

var. *palmeri* (p. 319) PALMER'S GOLDENBUSH Pl 10–40 dm, stout. **LF**: 20–40 mm, often curved; axillary lvs 1–2 per node. **INFL**: involucre 5–6.5 mm; phyllary tips not strongly thickened; midveins gen evident on proximal 1/2 of each phyllary. **FL**: 12–30; ray fl 4–8. 2*n*=18. Coastal scrub; < 600 m. s SCo (s San Diego Co.); n Baja CA. Threatened by urban development. Sep–Nov ★

E. paniculata (A. Gray) Rydb. (p. 319) BLACK-BANDED RABBITBRUSH Pl 5–18 dm. **ST**: erect, clustered, glabrous, gland-dotted, resinous, gen ± black-banded (from fungal or insect attack). **LF**: 10–35 mm, 0.5–1.5 mm wide, thread-like, grooved, light green, glabrous, gland-dotted in pits, resinous. **INFL**: heads discoid, many in panicle-like clusters; peduncle 2–12 mm, bracts 0–10, lf-like; involucre 6–9 mm, 3.5–7 mm wide, obconic; phyllaries 13–20, graduated in 3–5 series, weakly 5-ranked, oblong to ovate, thin, weakly keeled, obtuse, resinous, midvein often evident but not raised or enlarged at tip. **RAY FL**: 0. **DISK FL**: 5–8; corolla 5–7 mm; style appendage ± = stigma. **FR**: 3.5–4.5 mm, hairy; pappus 5.5–6.5 mm, ± brown. 2*n*=18. Common. Gravelly washes; 400–1600 m. W&I, D; to sw UT, nw AZ. [*Chrysothamnus p.* (A. Gray) H.M. Hall] Jun–Dec

E. parishii (Greene) H.M. Hall var. *parishii* PARISH'S GOLDENBUSH Pl 15–50 dm, very lfy. **ST**: lfy, glabrous to ± hairy, gland-dotted, resinous. **LF**: 20–70 mm, 3–12 mm wide, oblong to oblanceolate, flat to slightly concave, acute, leathery, glabrous or sparsely hairy, gland-dotted in pits. **INFL**: heads discoid, in dense, terminal cyme-like clusters; peduncle 1–5 mm, bracts 0–5, scale-like; involucre 4–6.5 mm, 3–4 mm wide, obconic; phyllaries 18–24, graduated in 3–4 series, spiraled, ± lanceolate, acute, ± white, midrib thick, brown. **RAY FL**: 0. **DISK FL**: 8–18; corolla 5–6 mm; style appendage ± = stigma. **FR**: 2–2.5 mm, obconic, minutely hairy, veins 12 or fewer; pappus 4–5 mm, copious, dull white to brown. 2*n*=18. Dry slopes, chaparral, open forest, esp after fires; 400–2200 m. SnGb, SnBr, PR. *E. parishii* var. *peninsularis* (Moran) G.L. Nesom in n Baja CA. Jul–Oct

E. parryi (A. Gray) G.L. Nesom & G.I. Baird PARRY'S GOLDENBUSH Pl 1–10 dm. **ST**: prostrate to erect, white to ± green, closely tomentose with tightly matted hairs. **LF**: 10–80 mm, 0.5–14 mm wide, thread-like to ± oblanceolate, ± glabrous or ± white tomentose with ± tightly matted hairs, gen not stalked-glandular, gland-dotted or not, margins flat or rolled under. **INFL**: heads discoid, 1–gen ± many, in long or rounded cyme-, raceme-, or spike-like clusters; peduncle 1–10+ mm, 0–3, ± lf-like; involucre 10–18 mm, 4–8 mm wide, widely cylindric; phyllaries ± equal in 3–6 series, ± lanceolate, weakly 5-ranked, keeled, ± membranous, tips erect to recurved, green. **RAY FL**: 0. **DISK FL**: 5–18; corolla 8–12.5 mm; style appendage > stigma. **FR**: 4–7 mm, hairy; pappus gen < corolla, tan. 2*n*=18. [Chrysothamnus p. (A. Gray) E. Greene]

var. *aspera* (Greene) G.L. Nesom & G.I. Baird ROUGH RABBITBRUSH Pl 1.5–7 dm. **ST**: erect to spreading, white to ± green. **LF**: 15–50 mm, straight or curved, gray or green; distal-most often overtopping heads; glands short-stalked. **INFL**: spike-like or branched, heads 4–10, ± dense; phyllaries acute, tips erect to spreading, ± white, straw-colored, or ± purple. **FL**: 5–10. Dry forest to alpine barrens, often in pumice or gravel; 1900–3300 m. c&s SNH, SnBr, SNE, DMtns; NV. [*Chrysothamnus p.* subsp. *a.* (Greene) H.M. Hall & Clem.] Jul–Sep

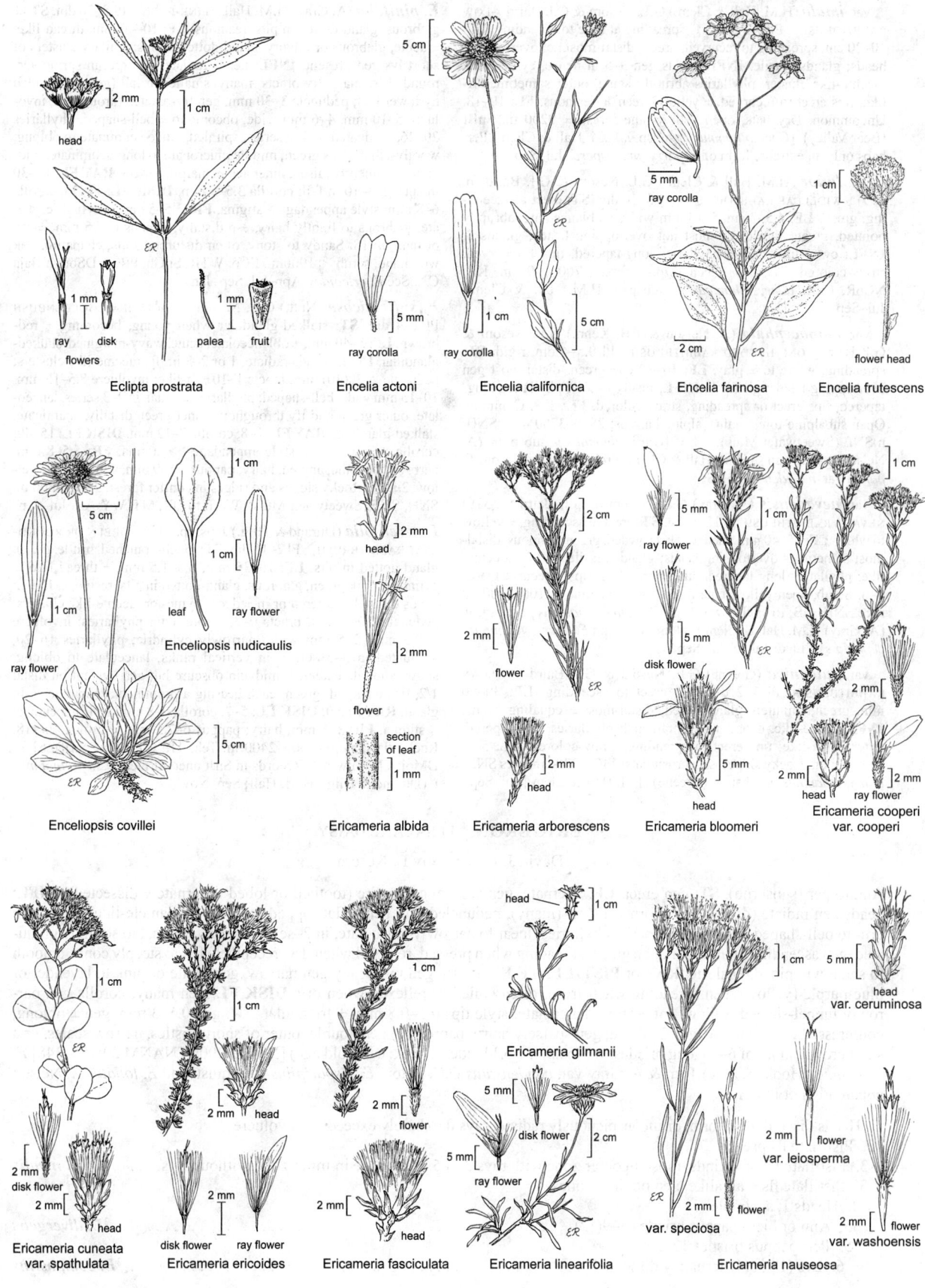

Eclipta prostrata

Encelia actoni

Encelia californica

Encelia farinosa

Encelia frutescens

Enceliopsis nudicaulis

Enceliopsis covillei

Ericameria albida

Ericameria arborescens

Ericameria bloomeri

Ericameria cooperi
var. cooperi

Ericameria cuneata
var. spathulata

Ericameria ericoides

Ericameria fasciculata

Ericameria gilmanii

Ericameria linearifolia

Ericameria nauseosa

var. ceruminosa

var. leiosperma

var. speciosa

var. washoensis

var. *imula* (H.M. Hall & Clem.) G.L. Nesom & G.I. Baird   LOW RABBITBRUSH   Pl 1–2 dm. **ST:** spreading at base to ascending. **LF:** 10–20 cm, spreading to recurved, green; distal-most not overtopping heads; glands sessile. **INFL:** heads gen 3–5 in dense, cyme- or ± raceme-like cluster; phyllaries abruptly acute, outer sometimes lf-like, tips erect to recurved, ± yellow-green, ± resinous. **FL:** 10–18. Uncommon. Dry flats, open yellow-pine forest; ± 2200 m. SnBr (Bear Valley). [*Chrysothamnus p.* subsp. *i.* H.M. Hall & Clem.] Perhaps only an extreme form of *E. parryi* var. *aspera.* Jul–Sep

var. *latior* (H.M. Hall & Clem.) G.L. Nesom & G.I. Baird (p. 319)   WIDELEAF RABBITBRUSH   Pl 3–10 dm. **ST:** erect to ascending, green. **LF:** 4.5–7 cm, 5–14 mm wide, ± oblanceolate, abruptly pointed, green, flat; distal-most not overtopping heads; glands 0. **INFL:** cyme-like, dense; phyllaries long-tapered, tips ± recurved, straw-colored. **FL:** 5–7. Open fir/pine forest; 700–1900 m. KR, NCoR, CaR. [*Chrysothamnus p.* subsp. *l.* H.M. Hall & Clem.] Jul–Sep

var. *monocephala* (A. Nelson & P.B. Kenn.) G.L. Nesom & G.I. Baird   ONE-HEADED RABBITBRUSH   Pl 0.5–4 dm, rigid. **ST:** spreading, white to ± gray. **LF:** 10–30 cm, green, distal-most gen overtopping heads; glands 0. **INFL:** heads 1–2(4); phyllaries long-tapered, tips erect or spreading, straw-colored. **FL:** 5–8. Common. Open subalpine forest, talus, alpine barrens; 2800–3700 m. c SNH, n SNE (Sweetwater Mtns); w NV. [*Chrysothamnus p.* subsp. *m.* (A. Nelson & P.B. Kenn.) H.M. Hall & Clem.] Probably derived from *E. parryi* var. *nevadensis.* Jul–Sep

var. *nevadensis* (A. Gray) G.L. Nesom & G.I. Baird (p. 319)   NEVADA RABBITBRUSH   Pl 2–6 dm. **ST:** erect to ascending, ± yellow to white. **LF:** 15–40 mm, moderately crowded, green, resinous, distal-most sometimes overtopping heads; glands sessile. **INFL:** raceme-like; phyllaries long-tapered, outer paper-like, tips ± recurved. **FL:** 4–6. Scrub, open yellow-pine forest (on serpentine); 1100–2700 m. n&c SNH, GB; to s OR, sw UT, n AZ. [*Chrysothamnus p.* subsp. *n.* (A. Gray) H.M. Hall & Clem.] Intergrades with *E. parryi* var. *monocephala* s of Lake Tahoe. Jul–Sep

var. *vulcanica* (Greene) G.L. Nesom & G.I. Baird   VULCAN RABBITBRUSH   Pl 1–2 dm. **ST:** erect to ascending. **LF:** 30–50 mm, green; minutely gland-dotted; distal-most ± equaling heads. **INFL:** cyme-like, dense, much-branched; phyllaries long-tapered, outer paper-like, tips erect to spreading, straw-colored. **FL:** 5–7. Uncommon. Rocky slopes, sagebrush flats; 1400–3200 m. c&s SNH. [*Chrysothamnus p.* subsp. *v.* (Greene) H.M. Hall & Clem.] Jul–Sep

E. pinifolia (A. Gray) H.M. Hall   PINE-BUSH   Pl 5–30 dm. **ST:** ± glabrous, gland-dotted in pits, resinous. **LF:** 10–40 mm, thread-like, ± acute, glabrous or ± hairy, gland-dotted in pits; axillary clusters of short lvs gen present. **INFL:** heads radiate, 1, large, in spring surrounded by many lfy bracts, many, smaller in fall (gen surrounded by fewer lvs); peduncle 3–30 mm, gen lfy-bracted throughout; involucre 5–10 mm, 4–6 mm wide, obconic to ± bell-shaped; phyllaries 20–26, graduated in 4–6 series, spiraled, lance-acuminate to oblong, woolly-ciliate, tips green, middle mucronate to long-acuminate, midvein prominently raised, inner acute, margin narrow. **RAY FL:** 15–30 in spring, 3–10 in fall; corolla 3.5–5 mm. **DISK FL:** 11–25, corolla 6–8 mm; style appendage > stigma. **FR:** 3.5–5 mm, ± cylindric, striate, glabrous to lightly hairy, esp distally; pappus 6–7.5 mm, ± red or tan. 2*n*=18. Sandy to stony, often disturbed soils; chaparral, oak woodland, scrub; < 1900 m. SCo, WTR, SnGb, PR, w DSon; n Baja CA. See *E. ericoides.* Apr–Jul, Sep–Jan

E. suffruticosa (Nutt.) G.L. Nesom   SINGLEHEAD GOLDENBUSH   Pl 1–4 dm. **ST:** stalked-glandular when young, becoming ± red-brown. **LF:** 5–40 mm, ± oblanceolate, acute, wavy-margined, stalked-glandular. **INFL:** heads radiate, 1 or 2–3 in lfy, raceme-like clusters; peduncle 2–20(40) mm, bracts 1–10+, lf-like; involucre 8.5–15 mm, 10–15 mm wide, bell-shaped; phyllaries ± equal, in 2–3 series, lanceolate, outer green and lfy throughout, inner green distally, acuminate, stalked-glandular. **RAY FL:** 1–8; corolla 7–12 mm. **DISK FL:** 15–40; corolla 8.5–10.5 mm; style appendage >> stigma. **FR:** 5.5–8 mm, narrowly obconic, angled, hairy; pappus 7.5–9 mm, white to pale yellow. 2*n*=18. Rocky slopes and ridges in conifer forest; 2100–3800 m. SNH, SNE (Sweetwater Mtns), W&I; to OR, MT, WY, AZ. Jul–Sep

E. teretifolia (Durand & Hilg.) Jeps. (p. 319)   GREEN OR ROUND-LEAF RABBITBRUSH   Pl 2–15 dm. **ST:** much-branched, brittle, green, gland-dotted in pits. **LF:** 10–35 mm, 0.5–1.5 mm, ± thread-like, ± cylindric, dark green, glabrous, gland-dotted in pits, resinous. **INFL:** heads discoid, scattered or in ± dense cyme- or raceme-like clusters; peduncle 0.5–2 mm, bracts 0–3, ± like outer phyllaries; involucre 5–9.5 mm, 2–5 mm wide, narrowly cylindric; phyllaries 16–20, graduated in 3–5 series, in vertical ranks, lanceolate to oblong, straw-colored, ± keeled, midvein obscure but gen evident on distal 1/2, tips enlarged, green, each bearing a prominent, ± spheric resin gland. **RAY FL:** 0. **DISK FL:** 5–7; corolla 6–8 mm; style appendage < stigma. **FR:** 3–5 mm, hairy; pappus 6–7.5 mm, ± brown. 2*n*=18. Rocky flats, slopes; 600–2400 m. Teh, TR, e PR (exc SnJt), SNE, DMoj; s NV, nw AZ. Records in SnJt unconfirmed. [*Chrysothamnus t.* (Durand & Hilg.) H.M. Hall] Sep–Nov

ERIGERON   FLEABANE DAISY

David J. Keil & Guy L. Nesom

Ann to per (subshrub). **ST:** gen erect. **LF:** alternate, gen sessile, gen entire (toothed or lobed to ternately dissected). **INFL:** heads gen radiate (discoid, disciform), 1–few (many), peduncled; infl gen ± flat-topped (raceme- to panicle-like); involucre urn- to bell-shaped or gen hemispheric; phyllaries linear to narrowly lanceolate, in 2–several series, ± equal to strongly graduated, gen ascending or erect in fl, gen green, spreading when pressed, reflexed when dry; receptacle flat to steeply conic, smooth to shallowly pitted, epaleate. **RAY** or **PISTILLATE FL:** (0)10–gen many; ray gen narrow, gen white or pink to lavender or blue-purple (yellow), gen spreading when fresh, often coiled or reflexed when dry. **DISK FL:** gen many; corolla gen narrowly funnel-shaped, yellow; anther tip ± lanceolate; style tips 0.1–0.8 mm, ± triangular. **FR:** gen 0.5–3 mm, gen ± oblong, compressed to ± cylindric, gen 2-ribbed, gen sparsely hairy; pappus (0) gen double, outer of short bristles, narrow scales, or a short crown, inner of 6–50 long bristles. ± 375 spp.: worldwide. (Greek: early old age) [Nesom 2006 FNANM 20:256–348] *E. concinnus* (Hook. & Arn.) Torr. & A. Gray var. *condensatus* D.C. Eaton, *E. disparipilus* Cronquist, and *E. lobatus* A. Nelson apparently not in CA.

1. Heads discoid, disciform, or inconspicuously radiate, rays 0 to barely exceeding involucre
 2. Pistillate fls present
  3. Pistillate fls of 2 kinds, those in outer zone with rays 3–4.5 mm, those in inner zone without rays . . . . ²*E. nivalis*
  3′ Pistillate fls ± all alike, gen not in 2 zones
   4. Heads 1; st simple
    5. Ann or bien; pappus bristles 6–9(12) . . . . ³*E. divergens*
    5′ Per; pappus bristles 12–25
     6. Lvs (1)2–3(4)-ternately dissected . . . . ²*E. compositus*

6′ Lvs entire

  7. St minutely glandular; disk corolla gen abruptly inflated to throat; fr base ringed ± white; outer pappus of narrow scales or obvious bristles . . . . **E. aphanactis** var. **congestus**

  7′ St glandless; disk corolla narrowly funnel-shaped; fr base ringed ± yellow; outer pappus of minute bristles . . . . **E. chrysopsidis** var. **austiniae**

4′ Heads 1–many; st often branched

  8. Rays 1–3 mm

    9. Infl panicle-like, heads many; rays ≤ 1 mm; involucre glabrous or sparsely strigose to minutely bristly; central colored part of middle phyllaries resin-filled, ± red-brown when dry . . . . ³**E. canadensis**

    9′ Infl gen raceme-like, heads 1–12; rays 2–3 mm; involucre ± rough-hairy; central colored part of middle phyllaries not resin-filled, not red-brown when dry . . . . ²**E. lonchophyllus**

  8′ Rays 0–1 mm

    10. Central st and axes of stiffly ascending branches gen strongly dominant; infl gen raceme- or panicle-like, often many-headed; disk fls 6–25

      11. Central st often overtopped by distal branches in age; pl ± gray-hairy; heads ± long-peduncled, gen in raceme-like clusters; receptacle (2.5)3–5 mm diam in fr; pistillate fls 60–150+; phyllaries gen minutely bristly and/or densely strigose; pappus 3–4+ mm . . . . **E. bonariensis**

      11′ Central st gen exceeding distal branches in age; pl green, glabrous to hairy; heads ± short-peduncled in panicle-like clusters; receptacle 1–3 mm diam in fr; pistillate fls 20–90+; phyllaries glabrous to densely strigose or densely bristly; pappus 2–3(4) mm

        12. Lf ± long-bristly-ciliate; phyllaries gen glabrous, occ sparsely strigose; central colored part of middle phyllaries narrower than light-colored margin, resin-filled, ± red-brown when dry; ray 0.3–1 mm; disk corolla gen 4-lobed; fr uniformly pale tan to light gray-brown . . . . ³**E. canadensis**

        12′ Lf margin gen ± fine-strigose; phyllaries glabrous to densely strigose or spreading-hairy; central part of middle phyllaries gen wider than lighter margin, not red-brown or resin-filled when dry; ray 0–0.3 mm; disk corolla gen 5-lobed; fr pale tan, gen some in head with ± red veins . . . . . . **E. sumatrensis**

    10′ Central st and axes of ascending branches not strongly dominant, pl ± openly branched; infl ± cyme-like, often 1–few-headed; disk fls gen ≥ 25

      13. Per; pappus bristles 12–17 . . . . **E. aphanactis** var. **aphanactis**

      13′ Ann, bien or short-lived per; pappus bristles 6–9(12) or 15–20

        14. Pappus bristles 15–20; bien or short-lived per . . . . **E. calvus**

        14′ Pappus bristles 6–9(12); ann or bien . . . . ³**E. divergens**

2′ Pistillate fls 0

  15. Basal lvs present at fl; cauline lvs gen much-reduced or 0 distally

    16. Caudex simple . . . . **E. lassenianus** var. **deficiens**

    16′ Caudex branched

      17. St ascending, 15–40 cm, occ sparsely spreading-hairy; caudex branches slender, rhizome-like; basal lvs elliptic to oblanceolate to spoon-shaped . . . . **E. supplex**

      17′ St erect, 4–15 cm, ± glabrous to densely strigose; caudex branches short, stout; basal lvs ± linear . . . . **E. bloomeri**

        18. Herbage and phyllaries strigose . . . . var. **bloomeri**

        18′ Herbage and phyllaries ± glabrous . . . . var. **nudatus**

  15′ Basal lvs 0 at fl; cauline lvs gen evenly sized and spaced

    19. Glandless st hairs ± spreading

      20. St hairs dense, short, stiff, slightly reflexed, ± gray . . . . **E. inornatus** var. **keilii**

      20′ St hairs sparse to dense, long, ± crinkly-spreading, white

        21. St 30–90 cm, ascending to erect; lvs 20–40 mm; involucre 12–15 mm diam . . . . ²**E. biolettii**

        21′ St 5–30 cm, decumbent to ascending; lvs 7–25 mm; involucre 7–15 mm diam

          22. Inner phyllaries 3.5–5 mm . . . . **E. miser**

          22′ Inner phyllaries 5.5–7 mm . . . . **E. petrophilus**

            23. Herbage ± glandless; hairs stiff, straight or curved . . . . var. **viscidulus**

            23′ Herbage densely glandular; glandless hairs gen loose, often crinkly

              24. Phyllary tips obviously widened and ± purple . . . . var. **petrophilus**

              24′ Phyllary tips not widened or purple . . . . var. **sierrensis**

    19′ Glandless st hairs 0–sparse, appressed

      25. Phyllaries gen glabrous . . . . **E. inornatus** (in part)

        26. St 10–20 cm, decumbent to ascending; proximally ± reflexed-hairy; proximal and middle lf blades and margins stiffly spreading-hairy and -ciliate; inner phyllaries 5–6.5 mm . . . . var. **calidipetris**

        26′ St 30–90 cm, ascending; proximally ± glabrous or appressed-hairy; proximal and middle lf blades and margins gen short-strigose or ± glabrous; inner phyllaries 4.5–5.5 mm . . . . var. **inornatus**

      25′ Phyllaries glandular (sometimes also with sparse, glandless hairs)

        27. Herbage glandular; lf narrowly oblanceolate . . . . ²**E. biolettii**

        27′ Herbage glandless; lf thread-like to oblanceolate

          28. Lf ± oblanceolate; st long-ascending from woody taproot . . . . **E. inornatus** var. **inornatus**

28′ Lf thread-like to linear; st from stout woody caudex or thin, rhizome-like caudex branches
  29. St 30–90 cm, many, from stout woody caudex . . . . ***E. greenei***
  29′ St 8–20(30) cm, 1, from slender, rhizome-like caudex branches . . . . ***E. reductus***
    30. Phyllary tips gen ± purple (variable n SnFrB); pappus bristles 38–61 . . . . var. ***angustatus***
    30′ Phyllary tips ± green; pappus bristles 20–30 . . . . var. ***reductus***
1′ Heads conspicuously radiate, rays exceeding to much-exceeding involucre
  31. Pistillate fls of 2 kinds, those in outer zone with rays 3–4.5 mm, those in inner zone without rays . . . . [2]***E. nivalis***
  31′ Pistillate fls all alike, not in 2 zones
    32. Rays 1–3 mm
      33. Infl panicle-like, heads many; rays ≤ 1 mm; involucre glabrous or sparsely strigose to minutely
        bristly; central colored part of middle phyllaries resin-filled, ± red-brown when dry . . . . [3]***E. canadensis***
      33′ Infl gen raceme-like, heads 1–12; rays 2–3 mm; involucre ± rough-hairy; central colored part of
        middle phyllaries not resin-filled, not red-brown when dry . . . . [2]***E. lonchophyllus***
    32′ Rays 4–25 mm
      34. Ann (bien to short-lived per) from slender taproot or shallow fibrous roots (see also *Erigeron
        philadelphicus* var. *philadelphicus*)
        35. Ray fl pappus 0; pl fibrous-rooted
          36. Lvs gen coarsely toothed; st hairs sparsely spreading or appressed . . . . ***E. annuus***
          36′ Lvs gen entire; st hairs appressed . . . . ***E. strigosus*** var. ***strigosus***
        35′ Ray fl pappus present; pl taprooted (*Erigeron flagellaris* gen fibrous-rooted)
          37. St hairs appressed; pl with herbaceous, lfy runners . . . . ***E. flagellaris***
          37′ St hairs spreading; pl without runners . . . . [3]***E. divergens***
    34′ Per or subshrub from woody taproot, branched caudex, or rhizome
      38. Lvs 1–4-ternately dissected
        39. Lvs gen (1)2–3(4)-ternately dissected; ray < 1 mm wide; caudex branches short, thick, ascending
          . . . . [2]***E. compositus***
        39′ Lvs deeply 3-lobed at tip; ray 1–2 mm wide; caudex branches thin, rhizome-like . . . . ***E. vagus***
      38′ Lvs entire to shallowly toothed
        40. Lf bases hardened, white-shiny, expanded, enclosing proximal st
          41. Caudex branches thick, lf bases persistent; rays yellow . . . . ***E. linearis***
          41′ Caudex branches slender, rhizome-like, lf bases not persistent; rays white, pink to blue or lavender
            42. Involucre 13–18 mm diam; basal lvs gen 2–5 mm wide; pappus bristles 25–40; phyllary hairs
              spreading; ≥ 2100 m; CaRH, n&c SNH . . . . ***E. barbellulatus***
            42′ Involucre 6–11 mm diam; basal lvs gen 0.5–1 mm wide; pappus bristles 20–30; phyllary hairs
              appressed; 1000–2200 m; MP . . . . ***E. elegantulus***
        40′ Lf bases ± thin, not hardened, barely or not expanded
          43. Fr ribs 4–8; herbage gen ± silvery, hairs dense, appressed
            44. Fr ribs 6–8; basal lvs tufted, persistent . . . . ***E. argentatus***
            44′ Fr ribs 4(6); basal lvs gen 0 at fl
              45. Outer pappus conspicuous, of bristles or narrow scales; st silvery-white . . . . ***E. parishii***
              45′ Outer pappus inconspicuous; st gray-green, densely strigose . . . . ***E. utahensis***
        43′ Fr ribs gen 2 (if more, herbage not silvery)
          46. Phyllaries strongly graduated; lvs entire, not clasping, basal 0, cauline gen evenly sized and spaced
            47. St minutely glandular . . . . ***E. aequifolius***
            47′ St gen not glandular
             48. St hairs dense, minutely curled- or bent; fr glabrous . . . . ***E. blochmaniae***
             48′ St hairs 0 to dense, straight, stiffly spreading to ± reflexed or appressed; fr sparsely strigose
              49. St hairs dense, stiffly spreading, gen slightly reflexed
                50. Herbage hairs 0.5–1 mm; phyllaries densely minutely glandular, inner phyllary tips widely
                  white-thick-margined; st often from woody basal offsets and roots . . . . ***E. klamathensis***
                50′ Herbage hairs ≤ 0.1–0.4 mm; phyllaries glandular or glandless, inner phyllary tips green; st
                  from ± slender, rhizome-like bases . . . . ***E. breweri***
                  51. Phyllaries densely glandular, glandless hairs 0 or very sparse
                    52. St gen ≥ 20 cm, ascending to erect; lvs gen ≥ 15 mm . . . . var. ***breweri***
                    52′ St 7–15 cm, prostrate to decumbent; lvs 5–12 mm . . . . var. ***jacinteus***
                  51′ Phyllaries not or barely glandular, glandless hairs prominent
                    53. St 20–30 cm, wiry, brittle; phyllary hairs long, thick-based, stiffly spreading, translucent
                      (also some glandular) . . . . var. ***porphyreticus***
                  53′ St 30–75 cm, not wiry or brittle; phyllary hairs short, white, ± appressed, glandless
                    54. Phyllary hairs thin-based, ± appressed, slightly less dense on inner phyllaries . . . . var. ***bisanctus***
                    54′ Phyllary hairs thick-based, ascending, much less dense on middle and inner phyllaries . . . var. ***covillei***
              49′ St hairs ± 0 or sparse, appressed
                55. St prostrate to ± erect, from slender rhizome . . . . ***E. elmeri***
                55′ St decumbent to erect, from woody taproot
                  56. Ray fls 9–13; lf ciliate, hairs thin-based; phyllaries with minute glandless hairs and densely
                  minutely glandular . . . . ***E. serpentinus***

56′ Ray fls 15–60; lf ciliate, hairs thick-based; phyllaries glandular or glandless
  57. Phyllaries ± glandless, 0.5–0.8 mm wide, margins gen thick to ± scarious
    58. Lvs 2–4(5) cm, oriented variously; inner phyllaries 3.2–4.5 mm . . . . ***E. foliosus*** var. ***foliosus***
    58′ Lvs (2.5)4–6(8) cm, often ± oriented to 1 side of st; inner phyllaries (4)5–6 mm
      59. St 30–85 cm; axillary lf tufts 0; lvs 35–80 mm, 1–2(4) mm wide; inner phyllaries not
        strongly scarious; ray fls 26–50 . . . . ***E. foliosus*** var. ***hartwegii***
      59′ St 15–28 cm; axillary lf tufts gen present; lvs 25–45 mm, (2)5–8 mm wide; inner
        phyllary margins strongly scarious; ray fls 18–22 . . . . ***E. mariposanus***
  57′ Phyllaries minutely but prominently glandular, 0.8–1 mm wide, margins gen widely
    scarious. . . . . ***E. foliosus*** (in part)
    60. Phyllary hairs all glandular; lvs gen 1–2 mm wide . . . . var. ***confinis***
    60′ Phyllary hairs glandular and glandless, ± densely bristly-strigose; lvs gen 2–4 mm wide
      61. Lf blades ± strigose; phyllary midvein raised, orange-resinous; ray corolla 7–10 mm
        . . . . var. ***franciscensis***
      61′ Lf blades ± glabrous (exc ciliate hairs); phyllary midvein gen not raised or orange-
        resinous; ray corolla 10–15 mm . . . . var. ***mendocinus***
46′ Phyllaries gen ± equal; lvs entire to lobed, sometimes clasping, basal sometimes present, cauline
  often strongly reduced distally on st, if present
  62. Pl fibrous-rooted, gen from obvious rhizome or ± decumbent caudex; cauline lvs often ± clasping;
    63. Glandless phyllary hairs with black cross-walls (sometimes only on basal cells of hairs)
      64. Rays white . . . . ***E. coulteri***
      64′ Rays blue to purple . . . . [2]***E. sanctarum***
    63′ Glandless phyllary hairs 0 or with clear cross-walls
      65. Ray fls 150–400 . . . . ***E. philadelphicus*** var. ***philadelphicus***
      65′ Ray fls 20–165 (80–300+ in *E. glaucus*)
        66. Lvs thick, ± fleshy, entire or toothed; coastal . . . . ***E. glaucus***
        66′ Lvs ± thin, entire; coastal or inland
          67. Glandless phyllary hairs prominent, glandular hairs sometimes present
            68. St from thin, often stolon-like caudex branches with few fibrous roots . . . . [2]***E. sanctarum***
            68′ St from short, thick, strongly fibrous-rooted rhizome or caudex
              69. St 2–20(30) cm, hairs spreading; phyllaries ± black-purple; pappus bristles 12–16(20). . . . . [2]***E. algidus***
              69′ St 30–100 cm, hairs appressed; phyllaries ± green; pappus bristles 20–30 . . . . [2]***E. aliceae***
          67′ Glandless phyllary hairs 0, glandular hairs dense
            70. St glabrous or sparsely minutely glandular; ray fls 20–30(45) . . . . ***E. cervinus***
            70′ St ± appressed-hairy, sometimes also ± bristly; ray fls 30–105
              71. Fr 2(4)-ribbed, moderately to densely strigose; proximal lvs gen coarsely dentate,
                sometimes entire. . . . . [2]***E. aliceae***
              71′ Fr 4–7-ribbed, sparsely strigose; proximal lvs entire . . . . ***E. glacialis***
                72. Peduncles densely minutely strigose with loosely appressed, slightly crinkled hairs . . . . var. ***glacialis***
                72′ Peduncles bristly or ± stiffly long-soft-hairy. . . . . var. ***hirsutus***
  62′ Pl taprooted or with taprooted caudex; cauline lvs not at all clasping
    73. St spreading- or reflexed-hairy
      74. Lf clearly petioled, blade oblanceolate, elliptic, obovate to spoon-shaped
        75. Basal lvs 20–70 mm; cauline lvs much reduced; st 2–30 cm; phyllaries ± black-purple; ray fls
          (30)50–125, ray 7–13 mm. . . . . [2]***E. algidus***
        75′ Basal lvs 5–25 mm; cauline lvs 0; st 1–4 cm; phyllaries green; ray fls 15–40, ray 4–6 mm
          . . . . ***E. uncialis*** var. ***uncialis***
      74′ Petiole indistinct, blade linear or oblanceolate, gradually tapered to base
        76. Margins of proximal lvs gen hairy but not stiffly spreading-ciliate; gen > 2200 m
          77. Basal lvs 20–80 mm; st sparsely minutely glandular at least distally, and moderately hairy
            with short, stiffly spreading to reflexed, glandless hairs; phyllaries green; ray fls 25–55,
            ray 6–11 mm. . . . . ***E. clokeyi*** var. ***pinzliae***
          77′ Basal lvs 6–35 mm; st densely minutely glandular, also with short-stiff and ± spreading
            glandless hairs; phyllaries ± black-purple; ray fls 20–37, ray 4–7(10) mm . . . . ***E. pygmaeus***
        76′ Margins of proximal lvs gen stiffly long-spreading-ciliate; 1200–1800 m
          78. Disk corolla (±) glabrous; pappus bristles 12–22; outer pappus of inconspicuous bristles or
            narrow scales; rays (40)55–115, 8–12 mm. ***E. pumilus*** var. ***intermedius***
          78′ Disk corolla sharply scabrous; pappus bristles 5–15; outer pappus of prominent wide scales;
            rays 40–60, 7–9 mm . . . . ***E. concinnus*** var. ***concinnus***
    73′ St glabrous or ascending- to appressed-hairy (to ± spreading-hairy in *Erigeron lassenianus*)
      79. Ray fls 75–125; disk corolla throat ± widened . . . . ***E. multiceps***
      79′ Ray fls gen 14–80; disk corolla ± narrowly cylindric
        80. Basal lvs 0 at fl
          81. Lvs sharply reduced distally on st; axillary lf tufts 0; ray fls 12–27(45). . . . . ***E. oxyphyllus***
          81′ Only most distal cauline lvs reduced; lf tufts in axils; ray fls 45–80 . . . . ***E. karvinskianus***

80′ Basal lvs gen persistent
  82. Basal lvs thread-like to linear
    83. St 2–8 cm, lfless; st hairs straight, appressed; basal lvs 5–25 mm; fr ribs densely ciliate, faces glabrous . . . . ***E. compactus***
    83′ St 15–30 cm, gen lfy, at least proximally; st hairs curved up, ± loose; basal lvs gen 20–70 mm; fr ribs and faces sparsely hairy . . . . ***E. filifolius***
  82′ Basal lvs narrowly oblanceolate to elliptic or spoon-shaped (or linear in *Erigeron lassenianus* var. *lassenianus* and *Erigeron robustior*)
    84. Pls cespitose; caudex stout, branched; basal lvs oblanceolate to elliptic, petioled; cauline lvs abruptly reduced . . . . ***E. tener***
    84′ Pls not cespitose, caudex gen slender; basal lvs linear to narrowly oblanceolate, petiole indistinct; cauline lvs gradually reduced
      85. Basal lvs gen persistent; cauline lvs gen smaller than basal; heads 1–4(7) . . . . ***E. eatonii***
        86. Involucre (14)17–23 mm diam; longest phyllaries 7–11 mm; disk corolla 4.4–6.8 mm . . . . var. ***nevadincola***
        86′ Involucre 8–16 mm diam; longest phyllaries 5–8 mm; disk corolla 3–5 mm
          87. St 10–23 cm; disk corolla 3–4 mm; pappus bristles 16–20, 3–3.5 mm . . . . var. ***plantagineus***
          87′ St 4–12(21) cm; disk corolla 3.5–5 mm; pappus bristles 18–30, 3.5–5 mm . . . . var. ***sonnei***
      85′ Basal lvs persistent at, or withering before, fl; cauline lvs gradually or little reduced; heads (1)2–7
        88. Phyllaries densely minutely glandular; involucre 4–5.5 mm, 6–10(12) mm wide; lvs gen 1-veined . . . . ***E. lassenianus*** var. ***lassenianus***
        88′ Phyllaries gen glandless; involucre gen 6–8.5 mm, (9)12–20 mm wide; lvs gen 3-veined
          89. Pls 10–22(27) cm; taproots ± thick, (3)5–8 mm wide; basal lvs oblanceolate to ± narrowly spoon-shaped; cauline lvs ± continuing to heads; involucres (5)6–7 mm, 9–12(14) mm wide; phyllaries elliptic-oblanceolate to oblong-oblanceolate, abruptly acuminate . . . . ***E. maniopotamicus***
          89′ Pls (15)25–55 cm; taproots ± thin, 2–3 mm wide; basal lvs linear to very narrowly oblanceolate; cauline lvs ending ± well below heads; involucres 6–8.5 mm, (12)14–20 mm wide; phyllaries narrowly oblanceolate to lanceolate, acute-acuminate . . . . ***E. robustior***

E. aequifolius H.M. Hall (p. 319) HALL'S DAISY Per 10–20 cm, from woody roots and slender-branched caudex. **ST:** ascending to erect, few-branched, short-loose-spreading-hairy, minutely stalked-glandular. **LF:** cauline, 6–20 mm, narrowly elliptic to oblanceolate, evenly sized and spaced, sparsely short-soft-hairy, minutely glandular. **INFL:** heads 1(3), short-peduncled; involucre 3.5–4 mm, 7–10 mm diam; phyllaries strongly graduated in 3–5 series, proximally sparsely strigose or not, minutely glandular. **RAY FL:** 14–30; ray 5–8 mm, white or lavender, drying ± blue, weakly coiled when dry. **FR:** pappus bristles 20–35. 2*n*=18. Rock ledges, crevices; 1500–2100 m. s SNH. Jul–Aug ★

E. algidus Jeps. (p. 319) SIERRA FLEABANE Per 2–20(30) cm, from short, thick, strongly fibrous-rooted caudex or rhizome. **ST:** unbranched, sparsely and loosely spreading-hairy and glandular. **LF:** basal petioled, 2–7 cm, oblanceolate to spoon-shaped; cauline much reduced, strigose or ± bristly, sometimes sparsely glandular. **INFL:** head 1; involucre 5–8 mm, 8–16 mm diam; phyllaries ± equal, ± black-purple, spreading-hairy, minutely glandular, tips spreading to reflexed. **RAY FL:** (30)50–125; ray 7–13 mm, white or pink to lavender, coiled when dry. **FR:** pappus bristles 12–16(20). Alpine meadows, talus; 2600–3700 m. SNH, SNE (exc W&I); w NV. Jul–Aug

E. aliceae Howell (p. 319) ALICE EASTWOOD'S FLEABANE Per 30–100 cm, from short, thick, fibrous-rooted rhizome. **ST:** few-branched near mid-st, sparsely strigose. **LF:** basal petioled, blade oblanceolate, gen coarsely dentate, spreading-hairy; mid-cauline 5–15 cm, lanceolate, gen ± clasping. **INFL:** heads 1–7, long-peduncled; involucre 6–10 mm, (10)12–20 mm diam; phyllaries ± equal, loose-spreading-hairy or not, densely minutely glandular. **RAY FL:** 45–80; ray 10–15 mm, white or lavender-tinged, often drying lavender or ± blue, tips slightly coiled when dry. **FR:** 2(4)-ribbed; pappus bristles 20–30. Meadows, openings in woodland; 1300–2200 m. NW; to WA. Jun–Aug(Sep)

E. annuus (L.) Pers. ANNUAL FLEABANE Ann 50–120 cm, from fibrous roots. **ST:** often branched distally, sparsely spreading- or appressed-hairy. **LF:** basal gen petioled, blades 4–16 cm, elliptic to obovate, gen coarsely serrate; cauline gen slightly reduced distally on st, sparsely strigose or ± bristly. **INFL:** heads 5–50+; involucre 3–5 mm, 5–12 mm diam; phyllaries ± equal, spreading-hairy, sparsely minutely glandular. **RAY FL:** 80–120; ray 5–8 mm, white, not coiled or reflexed when dry. **FR:** ray pappus bristles 0; disk pappus bristles 10–15. 2*n*=27. Disturbed places; < 2000 m. KR, n&c SNH; native to e US; worldwide weed. Asexual triploid. Intergrades with *E. strigosus*. Apr–Jul

E. aphanactis (A. Gray) Greene (p. 319) RAYLESS SHAGGY FLEABANE Per 8–25 cm, from taproot and short-branched caudex, often densely cespitose. **ST:** branched or unbranched, stiffly spreading-hairy, minutely glandular. **LF:** basal sometimes long-petioled, 4–7 cm, linear-oblanceolate, gradually reduced distally on st, stiffly spreading-hairy. **INFL:** heads ± disciform, 1–many; involucre 4–6 mm, 7–15 mm diam; phyllaries ± equal, coarsely long-hairy, minutely glandular. **PISTILLATE FL:** many; ray 0 or < involucre. **DISK FL:** corolla gen abruptly inflated to throat. **FR:** pappus bristles or narrow scales 12–17.

  var. ***aphanactis*** **ST:** gen branched in proximal 1/2. **INFL:** heads gen 2–many. 2*n*=18. Sagebrush or juniper scrub; 1300–2600 m. c SNH, SnBr, GB, DMtns. Apr–Sep

  var. ***congestus*** (Greene) Cronquist **ST:** unbranched. **INFL:** head 1. Sagebrush or juniper scrub; 1800–2600 m. SnBr; to UT. May–Sep

E. argentatus A. Gray (p. 319) SILVER FLEABANE Per 10–40 cm, from woody taproot and short-branched caudex, often cespitose, densely silvery-hairy. **ST:** unbranched. **LF:** basal many, erect, persistent, gen 2–5 cm, narrowly oblanceolate; cauline scattered on proximal 2/3 of st. **INFL:** head 1; involucre 5.5–9 mm, 12–22 mm diam; phyllaries ± equal, silvery strigose, minutely glandular. **RAY FL:** 25–48; ray 10–16 mm, lavender to blue (white or pink), coiled when dry. **FR:** 6–8-ribbed, densely hairy; pappus bristles 25–40. 2*n*=18. Rocky slopes, pinyon/juniper woodland; 2000–2300 m. W&I, ne DMtns (Last Chance Range); to UT, w AZ. May–Jul

E. barbellulatus Greene (p. 319) SHINING FLEABANE Per 5–15 cm, from caudex with slender, rhizome-like branches (or taproot), ± appressed-short-hairy. **ST:** ascending, unbranched, proximal 1/4

gen lfy, white-shiny or ± purple. **LF:** gen basal, 2–5 cm, 2–5 mm wide, narrowly oblanceolate, strigose; bases wider, hard, white-shiny, enclosing proximal st. **INFL:** head 1; involucre 5.5–9 mm, 13–18 mm diam; phyllaries ± equal, sparsely fine-spreading-hairy. **RAY FL:** 15–35; corolla 7–15 mm, ray white to lavender, drying ± blue, not coiled or reflexed when dry. **FR:** pappus bristles 25–40. 2*n*=18. Gravelly or rocky slopes, sagebrush/pine to subalpine forest; 2100–3300 m. CaRH, n&c SNH. Jun–Jul

E. biolettii Greene (p. 319) STREAMSIDE DAISY Per 30–90 cm, from woody caudex. **ST:** ascending to erect, branched distally, sparsely spreading- to appressed-hairy, ± densely glandular. **LF:** cauline, 2–4 cm, narrowly oblanceolate, evenly sized and spaced, sparsely bristly or stiff-hairy, densely glandular. **INFL:** heads discoid, 2–15; involucre 6–8 mm, 12–15 mm diam; phyllaries strongly graduated in 3–5 series, densely glandular, tips of inner gen purple. **RAY FL:** 0. **FR:** pappus bristles 22–38. Dry slopes, rocks, ledges along rivers; < 1100 m. KR, NCoRO. Intergrades with *E. inornatus.* Jun–Sep ★

E. blochmaniae Greene (p. 319) BLOCHMAN'S LEAFY DAISY Per 40–80 cm, from woody caudex (or rhizome). **ST:** gen ascending, branched distally, densely and minutely curled- or bent-hairy. **LF:** cauline, 1–5.5 cm, linear to narrowly oblanceolate, evenly sized and spaced, densely puberulent or short-soft-hairy. **INFL:** heads 1–7; involucre 4.5–6 mm, 9–14 mm diam; phyllaries strongly graduated in 3–5 series, densely short-stiff-spreading-hairy, densely minutely glandular. **RAY FL:** 45–72; ray 8–11 mm, white to pink or lavender, weakly coiled when dry. **FR:** glabrous; pappus bristles 21–36. 2*n*=18. Sand dunes and hills; < 70 m. s CCo. Threatened by coastal development. Jul–Oct ★

E. bloomeri A. Gray BLOOMER'S FLEABANE Per 4–15 cm, from taproot and short-branched caudex, cespitose. **ST:** unbranched, ± glabrous to densely white-strigose. **LF:** basal present at fl, 2–7 cm, ± linear; cauline lvs proximal, glabrous or strigose. **INFL:** heads discoid, 1; involucre 5–10 mm, 7–10 mm diam; phyllaries ± equal, strigose to spreading-hairy, ± minutely glandular. **RAY FL:** 0. **FR:** pappus bristles 25–40.

var. ***bloomeri*** Herbage and phyllaries strigose. 2*n*=18. Rocky slopes, lava beds, meadows; 800–2000 m. CaR, n SNH, GB; to WA, ID, NV. May–Jul

var. ***nudatus*** (A. Gray) Cronquist WALDO DAISY Herbage and phyllaries ± glabrous. Serpentine slopes, rocky ridges; 600–2300 m. KR; sw OR. May–Jul ★

E. bonariensis L. (p. 319) FLAX-LEAVED HORSEWEED Ann 10–100(150+) cm, ± gray-hairy. **ST:** erect or ascending, lfy, branched distally or throughout, densely strigose, bristly; central st often overtopped by distal branches in age. **LF:** ± densely strigose and/or bristly; basal and proximal cauline petioled, 3–8(12+) cm, 10–25+ mm wide, oblanceolate, entire to coarsely toothed or pinnately lobed; distal sessile, 1–5 cm, 2–10 mm wide, linear to narrowly oblanceolate, entire or shallowly toothed. **INFL:** heads disciform, 5–many, ± long-peduncled, gen in raceme-like (panicle-like or ± flat-topped) clusters; involucre gen 3.5–5 mm, widely cylindric or urn-shaped in fl; phyllaries ± equal, densely strigose and/or minutely bristly, green to ± red or purple; central colored part of middle phyllaries gen wider than light-colored margin, not red-brown when dry, not resin-filled; receptacle (2.5)3–5 mm diam in fr. **PISTILLATE FL:** 60–150+; corolla ± = style, green-white to purple, ray 0–0.3 mm. **DISK FL:** 8–25. **FR:** 1–1.5 mm, pale tan, sparsely minutely strigose; pappus bristles 15–25+, 3–4+ mm, cream to ± brown. 2*n*=54. Disturbed sites; < 1300+ m. NCoRI, CaRF, n&c SNF, GV, CW, SW, DMtns; to e N.Am; native to S.Am. [*Conyza b.* (L.) Cronquist] All yr

E. breweri A. Gray BREWER'S FLEABANE Per 7–75 cm, gen from woody roots and slender, rhizome-like branched caudex. **ST:** prostrate to gen ascending or erect, gen not wiry or brittle, often ± purple, gen densely short-bristly, hairs ≤ 0.1–0.4 mm, gen reflexed; glands gen 0. **LF:** cauline, 5–40 mm, linear to oblanceolate, evenly sized and spaced, bristly, gen glandless. **INFL:** heads 1–5(10), gen radiate; involucre 4–6 mm, 8–15 mm diam; phyllaries strongly graduated in 3–5 series, strigose to bristly or glandless hairs 0, densely minutely glandular, tips green, margins straw-colored. **RAY FL:**

12–45; ray 4–7 mm, white to pink or lavender, drying ± blue, weakly coiled when dry. **FR:** pappus bristles 22–46.

var. ***bisanctus*** G.L. Nesom (p. 319) **ST:** 30–75 cm; base ± thick. **LF:** 20–35 mm; hairs < 0.1 mm. **INFL:** phyllary hairs short, white, thin-based, ± appressed, slightly less dense on inner phyllaries, glandless. Open, dry slopes and washes; 300–1600 m. SnGb, SnBr. May–Sep

var. ***breweri*** (p. 319) Strongly developed woody roots gen ± 0. **ST:** (12)20–60 cm. **LF:** (10)15–40 mm; hairs 0.1–0.3 mm. **INFL:** phyllaries narrowly margined, green-tipped, densely minutely glandular, glandless hairs 0 or sparse. 2*n*=18. Open, rocky habitats; 1200–3100 m. SNH, SnBr; c-w NV. Many pls of s SNH are tall, more erect, lack rhizomes, have narrower lvs. Jun–Sep

var. ***covillei*** (Greene) G.L. Nesom **ST:** 30–75 cm; base ± thick. **LF:** 15–30 mm; hairs 0.2–0.4 mm. **INFL:** phyllary hairs glandless, short, white, thick-based, ascending, much less dense on middle and inner phyllaries. Open, rocky sagebrush, chaparral, juniper scrub; ± 1000–1900 m. s SNH (e slope), SnGb, SnBr, SnJt, DMoj. May–Sep

var. ***jacinteus*** (H.M. Hall) Cronquist SAN JACINTO MOUNTAINS DAISY Strongly developed woody roots gen ± 0. **ST:** 7–15 cm, prostrate to decumbent. **LF:** 5–12 mm; hairs 0.2–0.4 mm. **INFL:** phyllaries densely glandular, glandless hairs 0 or sparse. Open, rocky slopes and crests; ± 2700–2900 m. SnGb, SnBr, SnJt. Jun–Sep ★

var. ***porphyreticus*** (M.E. Jones) Cronquist **ST:** 20–30 cm, wiry, brittle. **LF:** 5–30 mm; hairs 0.1–0.2 mm. **INFL:** phyllaries thick-margined, hairs long, thick-based, stiffly spreading, translucent (also some glandular). 2*n*=18. Open, rocky sagebrush to yellow-pine forest; 1200–2600 m. SnBr, SNE, DMoj; sw NV. Some pls in SnBr have hairs like *E. breweri* var. *b.* May–Aug

E. calvus Coville BALD DAISY Bien or short-lived per, 10–14 cm, from taproot. **ST:** much-branched at base, ± long-spreading-hairy. **LF:** basal 3–5 cm, spoon-shaped, proximally ciliate, faces ± long-spreading-hairy, minutely glandular; cauline abruptly reduced. **INFL:** heads disciform or inconspicuously radiate, 1–few; involucre ± 5 mm, 13–14 mm diam; phyllaries ± equal, coarsely hairy, hairs thick-based, minutely glandular. **PISTILLATE FL:** ± 50–100; ray 0 or not exceeding involucre. **FR:** pappus bristles 15–20. Sagebrush and desert scrub; ± 1200 m. s SNE (w base Inyo Mtns). Closely related to *E. divergens*; also confused with *E. aphanactis.* May ★

E. canadensis L. (p. 319) HORSEWEED Ann 20–200(350) cm. **ST:** erect, very lfy, branched distally or throughout, strigose, short-bristly, or ± glabrous; central st gen exceeding distal branches. **LF:** ± long-bristly-ciliate, faces minutely strigose or spreading-bristly on veins, ± glabrous in age; basal and proximal cauline proximally tapered or ± petioled, 2–5(10) cm, 4–10(15+) mm wide, linear to oblanceolate, entire or toothed, gen 0 at fl; distal similar, sessile, smaller, entire. **INFL:** heads inconspicuously radiate, gen many in panicle-like clusters; involucre 3–4 mm, 2–3 mm diam, ± urn-shaped in fl; phyllaries weakly graduated, glabrous or sparsely strigose to minutely bristly, green when fresh; central colored part of middle phyllaries narrower than light-colored margin, resin-filled, ± red-brown when dry; receptacle 1–1.5(3) mm diam in fr. **RAY FL:** 20–45+; ray 0.3–1 mm, white or pink, not coiled. **DISK FL:** 7–30+; corolla gen 4-lobed. **FR:** 1–1.5 mm, tan to light gray-brown, sparsely minutely strigose; pappus bristles 15–25, 2–3 mm, white. 2*n*=18. Disturbed places; gen < 2300 m. CA; to BC, e N.Am, C.Am; introduced ± worldwide. [*Conyza c.* (L.) Cronquist] All yr

E. cervinus Greene (p. 319) SISKIYOU DAISY Per 15–30 cm, from short to long, weakly woody rhizome or caudex. **ST:** 0–3-branched distal to middle, glabrous or sparsely minutely glandular. **LF:** basal 4–12 cm, oblanceolate to narrowly spoon-shaped; cauline slightly clasping, little reduced, glabrous. **INFL:** heads 1–4, short-peduncled; involucre 5–7 mm, 9–14 mm diam; phyllaries ± equal, densely minutely glandular. **RAY FL:** 20–30(45); ray 7–10 mm, white, coiled when dry. **FR:** pappus bristles 12–15. Open, rocky slopes, meadows, forest; 900–1900 m. KR; sw OR. Jun–Aug ★

E. chrysopsidis A. Gray var. ***austiniae*** (Greene) G.L. Nesom DWARF YELLOW FLEABANE Per 3–12 cm, from woody taproot and

short-branched caudex, cespitose. **ST:** unbranched, sparsely short-spreading-hairy. **LF:** ± basal, erect, 2–8 cm, linear to oblanceolate, petiole ciliate, faces short-rough-hairy to loosely strigose. **INFL:** heads disciform or inconspicuously radiate, 1; involucre 4.5–6 mm, 10–15 mm diam; phyllaries ± equal, sparsely rough-hairy, minutely glandular. **PISTILLATE FL:** many; ray 0 or not exceeding involucre, yellow. **FR:** pappus bristles 15–25. 2*n*=18. Crevices, rocky slopes, sagebrush scrub; 1200–1700 m. MP; to se OR, sw ID, n NV. [*E. a.* Greene] 2 other vars. in Pacific NW. May–Jun

E. clokeyi Cronquist var. **pinzliae** G.L. Nesom (p. 319) PINZL'S FLEABANE Per 5–20 cm, from stout taproot and (±) simple caudex. **ST:** ascending to erect, unbranched, sparsely minutely glandular (at least distally), and moderately hairy with short, stiffly spreading to reflexed, glandless hairs. **LF:** basal 2–8 cm, ± oblanceolate, cauline not clasping, gen strongly reduced by mid-st, blade uniformly minutely rough-hairy or minutely bristly, hairs spreading to spreading-arching, stiff. **INFL:** head 1; involucre 4–7 mm, 6–12 mm diam; phyllaries ± equal, hairs short-stiff-spreading- to reflexed-hairy, minutely glandular. **RAY FL:** 25–55; ray 6–11 mm, white to lavender, reflexed when dry. **FR:** 1.8–2 mm; pappus bristles 13–22. 2*n*=18. Sagebrush scrub to alpine talus; 2200–3400 m. s SNH, SNE, DMtns; to UT. Other var. in NV. Jun–Sep

E. compactus S.F. Blake (p. 319) COMPACT DAISY Per 2–8 cm, from taproot and short-branched caudex, cespitose. **ST:** unbranched, short-white-appressed-hairy. **LF:** basal, 5–25 mm, linear, cauline 0, densely fine-strigose. **INFL:** head 1; involucre 5–8 mm, 7–17 mm diam; phyllaries ± equal, fine-strigose, inconspicuously minutely glandular. **RAY FL:** 15–32; ray 7–11 mm, white to lilac, sometimes abaxially rose or lilac, coiled when dry. **FR:** ribs densely ciliate; faces glabrous; pappus bristles 30–40. Rocky slopes, pinyon/juniper woodland; 1800–2300 m. W&I; to UT. May–Jun ★

E. compositus Pursh (p. 319) CUT-LEAF FLEABANE Per 3–15 cm, from stout taproot and caudex with short, thick, ascending branches, cespitose. **ST:** unbranched, densely minutely glandular, sometimes also sparsely soft-bristly. **LF:** gen basal, 1–5 cm, oblanceolate to spoon-shaped, (1)2–3(4)-ternately divided, ± densely puberulent, minutely glandular; cauline lvs (if any) much reduced, gen entire. **INFL:** heads ± radiate or disciform, 1; involucre 5–10 mm, 8–20 mm diam; phyllaries ± equal, rough-spreading-hairy, minutely glandular. **RAY FL:** (0) ± 30–60; ray 0 or 6–12 mm, < 1 mm wide, white to ± pink or lavender, not or weakly coiled when dry. **FR:** pappus bristles 12–15. 2*n*=18,27,36,45,54,63. Rocky slopes, crevices, talus; 2000–4300 m. KR, NCoRH, CaRH, SNH, SnBr, Wrn, SNE; to AK, e Can, CO, AZ. May–Sep

E. concinnus (Hook. & Arn.) Torr. & A. Gray var. **concinnus** (p. 319) NAVAJO FLEABANE Per 6–16 cm, from woody taproot and thick, short-branched caudex, cespitose. **ST:** 1–4-branched proximal to mid-st; hairs spreading or reflexed, soft-bristly and minutely stalked-glandular. **LF:** basal erect, crowded, 2–6 cm, ± linear to oblanceolate, proximal gen long-spreading-ciliate, gen long-rough- or soft-hairy, occ ± strigose or ± glabrous; cauline 0 or gradually reduced distally on st. **INFL:** heads 1–5(50); involucre 4–7 mm, 7–12 mm diam; phyllaries ± equal, ± stiff-spreading-hairy, ± minutely glandular. **RAY FL:** 40–60; ray 7–9 mm, white to pink or lavender, reflexed or weakly coiled when dry. **DISK FL:** corolla ± widely funnel-shaped, sharply scabrous. **FR:** pappus bristles 5–15, outer series of prominent, wide scales 0.2–0.5 mm. 2*n*=18. Sandy to rocky slopes, crevices; 1200–1800 m. DMtns; to ID, WY, CO, NM. 2 other vars. in sw US. Apr–Jun

E. coulteri Porter (p. 323) COULTER'S FLEABANE Per 20–70 cm, from thin rhizome. **ST:** 0–3-branched distal to mid-st, glabrous proximally, sparsely hairy distally. **LF:** basal gen 5–12 cm, ± oblanceolate, entire or with 2–6 pairs of shallow teeth, sparsely strigose to ± appressed-long-hairy; cauline oblanceolate to elliptic-ovate, ± clasping, little reduced. **INFL:** heads 1–4; involucre 7–10 mm, 10–16 mm diam; phyllaries ± equal, spreading-hairy, hairs with black cross-walls and bases, minutely glandular. **RAY FL:** 45–140; ray 9–25 mm, white, coiled when dry. **FR:** pappus bristles 20–25. 2*n*=18. Streambanks, wet meadows, conifer forest; ± 1900–3400 m. SNH, Wrn; to OR, WY, NM. Jul–Sep

E. divergens Torr. & A. Gray SPREADING FLEABANE Ann or bien, 10–45 cm, from slender taproot. **ST:** simple or gen branched near mid-st, evenly puberulent to short-stiff-hairy, hairs gen ≤ 0.5 mm, minutely glandular near and on heads. **LF:** basal and proximal cauline 2–6 cm, ± obovate, entire to lobed, gradually reduced distally on st, not clasping, densely, evenly puberulent. **INFL:** heads gen radiate (disciform), (1)5–many, buds nodding; involucre 3–4 mm, (5)7–11 mm diam; phyllaries ± equal, evenly puberulent to short-stiff-hairy, minutely glandular. **RAY FL:** gen 75–150; ray (0) 5–10 mm, white to purple. **DISK FL:** corolla abruptly wider at throat. **FR:** pappus bristles 6–9(12), outer series short bristles or narrow scales. 2*n*=18,27,36. Desert scrub to yellow-pine forest; 500–2600 m. SN (exc Teh), ScV (Sutter Buttes), SnGb, SnBr, SnJt, GB, D; to BC, TX, nw Mex. [*E. lobatus* A. Nelson, misappl.] Variable. Apr–Aug

E. eatonii A. Gray EATON'S FLEABANE Per 4–33 cm, from taproot and slender, (±) simple caudex; prostrate to erect. **ST:** 0–few-branched proximally, ± ascending-hairy. **LF:** linear to narrowly oblanceolate, 3-veined; basal gen present at fl, sparsely strigose to loosely stiff-hairy; mid-cauline 1–3(5) cm. **INFL:** heads 1–4(7), long-peduncled; involucre 4.5–11 mm, 8–23 mm diam; phyllaries ± equal, short-stiff- to long-soft-hairy, occ minutely glandular. **RAY FL:** 15–39; ray 7–15 mm, adaxially white, gen ± lavender or ± pink abaxially, weakly coiled when dry. **FR:** pappus bristles 16–30. CA vars. intergrade; other vars. in Rocky Mtns.

var. **nevadincola** (S.F. Blake) G.L. Nesom NEVADA DAISY **ST:** 14–33 cm. **INFL:** involucre (14)17–23 mm diam; longest phyllaries 7–11 mm. **DISK FL:** corolla 4.4–6.8 mm. **FR:** pappus bristles 20–26, 4–5 mm. Open grassland, rocky flats, gen in sagebrush or pinyon/juniper scrub; 1400–2900 m. MP; n NV. May–Jul ★

var. **plantagineus** (Greene) Cronquist **ST:** 10–23 cm. **INFL:** involucre 8–16 mm diam; longest phyllaries 5–8 mm. **DISK FL:** corolla 3–4 mm. **FR:** pappus bristles 16–20, 3–3.5 mm. 2*n*=18. Open grassy or sagebrush scrub, gen on volcanic rock; 1000–2500 m. KR, CaR, MP; s-c OR. May–Aug

var. **sonnei** (Greene) G.L. Nesom (p. 323) **ST:** 4–12(21) cm. **INFL:** involucre 8–16 mm diam; longest phyllaries 5–8 mm. **DISK FL:** corolla 3.5–5 mm. **FR:** pappus bristles 18–30, 3.5–5 mm. 2*n*=18. Rocky grassland or sagebrush scrub; 1800–2800 m. n&c SNH, SNE; c-w NV. May–Sep

E. elegantulus Greene (p. 323) VOLCANIC DAISY Per 3–15 cm, from taproot and slender-branched caudex. **ST:** unbranched, sparsely and minutely white-appressed-hairy. **LF:** ± basal, 15–50 mm, 0.5–1 mm wide, linear to narrowly oblanceolate, loosely strigose; base wider, hard, white-shiny, enclosing proximal st. **INFL:** head 1; involucre 3.5–5 mm, 6–11 mm diam; phyllaries ± equal, sparsely appressed-hairy. **RAY FL:** 15–25; ray 6–9 mm, white or pink to blue (lavender), weakly coiled when dry. **FR:** pappus bristles 20–30. 2*n*=27. Open, rocky sites, esp in volcanics, sagebrush scrub, juniper woodland; 1000–2200 m. MP; e OR. Asexual triploid. Jun–Jul ★

E. elmeri (Greene) Greene (p. 323) ELMER'S FLEABANE Per 6–20 cm, from woody roots and slender rhizomes. **ST:** prostrate to ± erect, wiry, gen unbranched, sparsely and minutely appressed-thin-based-hairy. **LF:** cauline, 5–20 mm, ± linear to narrowly oblanceolate, strigose. **INFL:** heads 1(3); involucre 3.5–5 mm, 7–10 mm diam; phyllaries strongly graduated in 3–5 series, minutely glandular, inner often purple-tipped. **RAY FL:** 12–21; ray 6–9 mm, white, drying white to pink or blue, weakly coiled when dry. **FR:** pappus bristles 18–26. Rock ledges, crevices, talus; 1300–3300 m. c SNH. Jun–Sep

E. filifolius Nutt. (p. 323) THREAD-LEAF FLEABANE Per 15–30 cm, from taproot and slender-branched caudex. **ST:** 0–few-branched near mid-st, ± densely, minutely white-upcurved-hairy. **LF:** basal 2–7 cm, thread-like to linear, ± strigose; cauline restricted to proximal st or not. **INFL:** heads 1–5+; involucre 4–7 mm, 5–18 mm diam; phyllaries ± equal, loosely strigose to spreading-hairy, minutely glandular. **RAY FL:** 20–45; ray 4–10 mm, white to pink or gen lavender or blue, weakly coiled when dry. **FR:** pappus bristles 20–30. 2*n*=18,36. Sagebrush scrub, juniper to yellow-pine woodland; 1200–2000 m. CaRH, MP; to BC, MT, UT. [*E. f.* var. *robustior* M. Peck] May–Aug

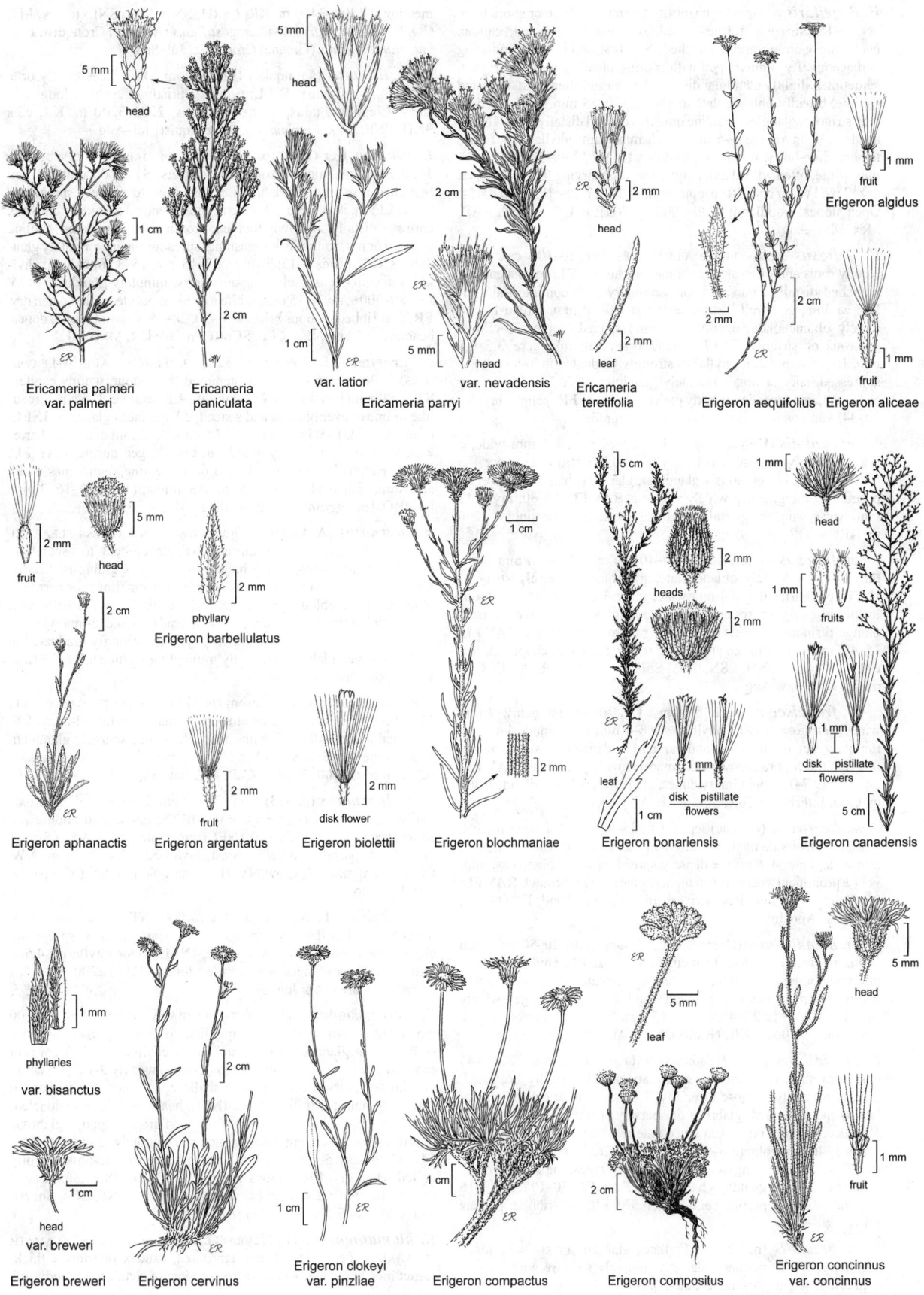

5 mm
head

Ericameria palmeri
var. palmeri

Ericameria
paniculata

2 cm
1 cm
var. latior

5 mm
head

2 cm

var. nevadensis

Ericameria parryi

ER

2 mm
head

2 mm
leaf

Ericameria
teretifolia

ER

2 mm

Erigeron aequifolius

1 mm
fruit

Erigeron algidus

2 cm

1 mm
fruit

Erigeron aliceae

2 mm
fruit

5 mm
head

2 cm

Erigeron aphanactis

2 mm
phyllary

Erigeron barbellulatus

2 mm
fruit

Erigeron argentatus

2 mm
disk flower

Erigeron biolettii

1 cm

ER

2 mm

Erigeron blochmaniae

5 cm

2 mm
heads

2 mm

leaf

1 cm

1 mm
disk   pistillate
flowers

Erigeron bonariensis

1 mm

head

1 mm
fruits

1 mm
disk   pistillate
flowers

5 cm

Erigeron canadensis

1 mm
phyllaries
var. bisanctus

head
1 cm
var. breweri

Erigeron breweri

2 cm

ER

Erigeron cervinus

1 cm
ER

Erigeron clokeyi
var. pinzliae

1 cm
ER

Erigeron compactus

ER
5 mm
leaf

2 cm

Erigeron compositus

5 mm
head

1 mm
fruit

ER

Erigeron concinnus
var. concinnus

E. flagellaris A. Gray   TRAILING FLEABANE   Bien or short-lived per, 3–15 cm; gen fibrous-rooted, sometimes taprooted, caudex becoming woody, rarely branched. **ST:** first erect, then producing herbaceous, lfy runners, gen with rooting plantlets at tips, strigose, sometimes slightly glandular distally. **LF:** basal (often persistent) and cauline, broadly oblanceolate to elliptic, 20–55 mm, entire or dentate, strigose, glandless, cauline quickly reduced distally on st. **INFL:** heads 1(3); involucre 3–5 mm, 6–13 mm diam; phyllaries strigose to spreading-hairy, minutely glandular. **RAY FL:** 40–125; ray 4–10 mm, white, often with abaxial midstripe, often drying lilac, not coiled or reflexed when dry. **FR:** pappus bristles 10–17. 2*n*=18,27,36,45,54. Open slopes, woodland; 2200–3000 m. W&I; w US to s BC, s AB, Mex. May–Aug

E. foliosus Nutt.   LEAFY FLEABANE   Per 20–100 cm, from woody roots and often slender-branched caudex. **ST:** gen ascending, branched distally, sparsely appressed-hairy or becoming glabrous. **LF:** cauline, gen well distributed on st, 10–80 mm, thread-like to widely oblanceolate, evenly sized and spaced, appressed-ciliate, glabrous or strigose. **INFL:** heads 1–several; involucre 3.5–6.5 mm, 10–16 mm diam; phyllaries strongly graduated in 3–5 series, ± appressed hairy or not, ± minutely glandular. **RAY FL:** 15–60; ray 6–15 mm, gen lavender, weakly coiled when dry. **FR:** pappus bristles 20–34. Var. *confinis*, var. *mendocinus* intergrade.

var. ***confinis*** (Howell) Jeps.   **LF:** 20–60 mm, 1–2 mm wide, ± linear, sparsely strigose. **INFL:** phyllaries 0.8–1 mm wide, inner 4–5 mm, minutely but obviously glandular, glandless hairs 0, midvein ± prominent, margins gen widely scarious. **RAY FL:** 19–30; ray 6–12 mm. Rocky sites, chaparral, woodland, often on serpentine; < 2200 m. KR; sw OR. May–Aug

var. ***foliosus*** (p. 323)   **LF:** 20–40(50) mm, 1–5(10) mm wide, thread-like to widely oblanceolate, glabrous to sparsely strigose. **INFL:** phyllaries 0.5–0.8 mm wide, inner 3.2–4.5 mm, ± glabrous to sparsely strigose, sometimes obscurely glandular, midvein raised, orange-resinous, margins gen thick (or narrowly scarious). **RAY FL:** 15–49; ray 6–11 mm. 2*n*=18. Open, rocky grassland, chaparral, forest; < 2900 m. n SNH, s SN, SnJV, SnFrB, SCoRO, TR, PR, DMtns; n Baja CA. May–Aug

var. ***franciscensis*** G.L. Nesom   **LF:** 20–40 mm, gen 2–4 mm wide, ± strigose. **INFL:** phyllaries 0.8–1 mm wide, inner 4–6 mm, minutely but obviously glandular and ± densely bristly-strigose, midvein raised, orange-resinous, margins widely scarious. **RAY FL:** 28–48; ray 7–10 mm. Grassy dunes, chaparral, oak woodland; < 800 m. CCo, SnFrB, n SCoR, SCo/WTR. May–Oct

var. ***hartwegii*** (Greene) Jeps   **LF:** 35–80 mm, 1–2(4) mm wide, ± oriented to 1 side of st, sparsely strigose. **INFL:** phyllaries 0.5–0.8 mm wide, inner 4–6 mm, ± densely spreading-hairy, glandless, midvein ± prominent, margins gen thick (or narrowly scarious). **RAY FL:** 26–50; ray 7–12 mm. Rocky riverbanks, oak woodland; 100–600 m. n&c SNF. Apr–Jul

var. ***mendocinus*** (Greene) G.L. Nesom   **LF:** 20–50 mm, gen 2–4 mm wide, ± glabrous (exc ciliate hairs). **INFL:** phyllaries 0.8–1 mm wide, inner 4–5.5 mm, minutely but prominently glandular and ± densely bristly-strigose, midvein ± indistinct, margins gen widely scarious. **RAY FL:** 25–48; ray 10–15 mm. River bars, banks, ledges, dry slopes; < 800 m. KR, NCoRO. May–Aug

E. glacialis (Nutt.) A. Nelson   SUBALPINE FLEABANE   Per 8–45 cm, from short rhizome and stout caudex. **ST:** distally branched, distally minutely strigose to bristly. **LF:** basal 5–20 cm, oblanceolate to spoon-shaped, glabrous or sparsely spreading-hairy; cauline ± reduced distally on st, gen lanceolate to ovate, ± clasping. **INFL:** heads 1–4(8); involucre 6–9(12) mm, 10–22(25) mm diam; phyllaries ± equal, with long-acuminate, loosely spreading tips, densely stalked-glandular, glandless hairs gen 0. **RAY FL:** 30–105; ray 8–15 mm, ray white to purple, coiled when dry. **FR:** 4–7-ribbed; pappus bristles 20–30.

var. ***glacialis*** (p. 323)   **LF:** faces glabrous or sparsely long-soft-hairy. **INFL:** peduncle densely minutely strigose with loosely appressed, slightly crinkled hairs. 2*n*=18. Clearings, talus, alpine

meadows; 1300–3400 m. KR, CaRH, SNH, Wrn, SNE; to AK, MT, CO, NM. [*E. peregrinus* var. *angustifolius* (A. Gray) Cronquist; *E. p.* var. *callianthemus* (Greene) Cronquist] Jul–Sep

var. ***hirsutus*** (Cronquist) G.L. Nesom   **LF:** faces bristly or ± stiffly long-soft-hairy. **INFL:** peduncle bristly or ± stiffly long-soft-hairy. Clearings, talus, alpine meadows; 2200–3200 m. KR, c&s SNH, SNE. [*E. peregrinus* var. *h.* Cronquist] Jul–Aug

E. glaucus Ker Gawl. (p. 323)   SEASIDE DAISY   Per, subshrub, 5–30 cm, from thick rhizomes and offsets. **ST:** ± decumbent, gen branched near mid-st, glabrous or glandular to densely spreading-hairy. **LF:** thick, ± fleshy, 2–13 cm, spoon-shaped to widely obovate, entire or distally shallowly toothed; proximal wing-petioled, cauline reduced or not, sometimes ± clasping, spreading-hairy, minutely glandular. **INFL:** heads 1–15; involucre 7–15 mm, 15–35 mm diam; phyllaries ± equal, ± densely long-soft-hairy, minutely glandular. **RAY FL:** 80–300+; ray 8–15 mm, white to pink or purple, coiled when dry. **FR:** 2–6-ribbed; pappus bristles 20–30. 2*n*=18. Coastal bluffs, dunes, beaches; < 20 m. NCo, CCo, SCoRO, n ChI; OR. May–Jul

E. greenei G.L. Nesom (p. 323)   GREENE'S NARROW-LEAVED DAISY   Per 30–90 cm, many-stemmed from stout, woody caudex. **ST:** few-branched distally, (±) glabrous. **LF:** cauline, 1–6 cm, thread-like to linear, evenly sized and spaced, ciliate, faces glabrous. **INFL:** heads discoid, 1–5; involucre 5.5–7.5 mm, 8–12 mm diam; phyllaries ± equal, densely minutely glandular, distally gen purple. **RAY FL:** 0. **FR:** pappus bristles 26–38. Gen on serpentine, sometimes rocky alluvium, chaparral, woodland, conifer forest; (100)500–1600 m. s NCoRO. [*E. angustatus* Greene, illeg.] May–Sep ★

E. inornatus (A. Gray) A. Gray   WESTERN RAYLESS FLEABANE   Per 1–9 dm, from woody taproot. **ST:** decumbent to ascending, branched near st-tips, gen ± hairy proximally, ± glabrous distally. **LF:** cauline, gen evenly sized and spaced, 1–6 cm, linear or narrowly oblanceolate to oblong, gen ciliate, glabrous or strigose to stiff-hairy, occ sparsely minutely glandular. **INFL:** heads discoid, 5–many; involucre 4.5–6.5 mm, 7–12 mm diam; phyllaries strongly graduated in 3–5 series, gen glabrous (sparsely minutely glandular). **RAY FL:** 0. **FR:** pappus bristles 28–60.

var. ***calidipetris*** G.L. Nesom (p. 323)   HOT ROCK DAISY   **ST:** 10–20 cm, decumbent to ascending, proximally ± reflexed-hairy. **LF:** proximal and middle stiff-hairy and -ciliate, occ sparsely glandular. **INFL:** inner phyllaries 5–6.5 mm. Loose sand, lava beds, depression edges, forest; 1100–2000 m. CaR, MP. Jun–Aug ★

var. ***inornatus*** (p. 323)   **ST:** 30–90 cm, long-ascending, proximally appressed-hairy or ± glabrous. **LF:** proximal and middle gen short-strigose or ± glabrous. **INFL:** inner phyllaries 4.5–5.5 mm. 2*n*=18. Chaparral to pine/fir forest, lava beds; 400–2300 m. NW, CaR, SN, Wrn; to WA, nw NV. Hairs variable in c SN (e slope), w NV. Jun–Sep

var. ***keilii*** G.L. Nesom   KEIL'S DAISY   **ST:** 40–65 cm, erect, densely short-stiff-hairy throughout, ± gray; hairs ± spreading, slightly reflexed. **LF:** short-strigose. **INFL:** inner phyllaries 4.5–6 mm. Dry slopes, meadows, conifer forest; 1200–2200 m. s SN (Fresno, Tulare cos.). Jun–Sep ★

E. karvinskianus DC.   SANTA BARBARA DAISY   Per 50–100 cm, from woody roots. **ST:** sprawling to erect, sparsely strigose or becoming glabrous. **LF:** basal 0 at fl; cauline gen 1–5 cm (not reduced exc most distal), elliptic to obovate, weakly 3-veined, entire or with 1–2 pairs of acute teeth or shallow lobes near tip, strigose to spreading-hairy; lf tufts in axils. **INFL:** heads 1–5, long-peduncled; involucre 2.5–3.5 mm, 7–10 mm diam; phyllaries ± equal, ± glabrous to strigose or spreading-hairy, sometimes minutely glandular. **RAY FL:** 45–80; ray 5–8 mm, white, drying white to rose-pink, slightly coiled when dry. **FR:** pappus bristles 15–27. 2*n*=18,27,36. Shaded rock walls, moist disturbed habitats; < 1100 m. c SNF, CCo, SnFrB; native Mex to S.Am. Cult. Apr–Aug

E. klamathensis (G.L. Nesom) G.L. Nesom (p. 323)   KLAMATH FLEABANE   Per 6–15(20) cm, taprooted, caudex or roots ± thick, sometimes branched, woody, often producing slender, ± woody off-

sets. **ST**: minutely rough-hairy to soft-bristly, hairs 0.5–1 mm, dense, stiffly spreading, gen slightly reflexed. **LF**: cauline, 20–40 cm, narrowly lance-oblong to narrowly obovate; densely-long-hairy, glandless. **INFL**: heads 1–5; involucre 3–5 mm, 7–10 mm diam; phyllaries sparsely hairy or not, densely minutely glandular, inner phyllary tips widely white-thick-margined. **RAY FL**: 22–40; ray 8–10 mm, lavender to purple, slightly coiled when dry. **FR**: pappus bristles 18–23. Open, rocky slopes, ridges, crevices; (400)700–2200 m. KR, CaRH; s-c OR. [*E. breweri* var. *k.* G.L. Nesom] Jun–Sep

E. lassenianus Greene LASSEN FLEABANE Per 9–35 cm, from taproot and simple caudex. **ST**: decumbent to erect, gen few-branched near mid-st, ± purple, appressed- to ± spreading-hairy. **LF**: basal present at fl, 5–15 cm, linear to spoon-shaped, gen 1-veined; cauline little reduced, much reduced or 0 distally in *Erigeron lassenianus* var. *deficiens*, short-rough- to long-soft hairy, often minutely glandular. **INFL**: heads radiate or discoid, 1–8, long-peduncled; involucre 4–5.5 mm, 6–10(12) mm diam; phyllaries ± equal, spreading-hairy, densely minutely glandular. **RAY FL**: (0)14–36; ray 5–8 mm, white to pink or lavender, slightly coiled when dry. **FR**: pappus bristles 12–24.

var. **deficiens** Cronquist **RAY FL**: 0. Open, rocky sites, barren flats, gravelly soils, sometimes serpentine; 1200–1900 m. n SNH (Plumas Co.). Jun–Sep ★

var. **lassenianus** (p. 323) **RAY FL**: well developed. $2n=18$. Open sites, gravelly and brushy flats, mixed conifer forest, gen in serpentine or glacial moraine; 600–2300 m. KR, CaRH, n SNH. Jun–Sep

E. linearis (Hook.) Piper (p. 323) DESERT YELLOW FLEABANE Per 5–20 cm, from taproot and thick-branched caudex, cespitose. **ST**: unbranched, sparsely short-white-appressed-hairy. **LF**: bases persistent on caudex; basal 2–9 cm, ± linear, loosely strigose, bases wider, hard, white-shiny, enclosing proximal st; cauline 0 on distal 1/3–3/4 of st. **INFL**: head 1; involucre 4–7 mm, 8–13 mm diam; phyllaries ± equal, strigose to soft-spreading-hairy, minutely glandular. **RAY FL**: 25–38; ray 4–8 mm, yellow, weakly coiled when dry. **FR**: pappus bristles 10–20. $2n=18,27,36,45$. Grassland, sagebrush scrub, open, rocky slopes; 1300–3100 m. n SNH, MP; to BC, MT, NV. May–Jul

E. lonchophyllus Hook. (p. 323) SHORT-RAYED FLEABANE Ann to short-lived per, 4–20 cm, fibrous-rooted. **ST**: gen unbranched, sparsely to densely bristly or soft-hairy; glands 0. **LF**: basal 2–8 cm, gradually reduced and becoming linear distally, ciliate, ± glabrous to spreading-hairy. **INFL**: heads inconspicuously radiate or disciform, 1–12 in gen raceme-like cluster; involucre 4–7 mm, 7–10 mm diam; phyllaries ± graduated, ± rough-hairy, glandless, gen purple-tipped. **RAY FL**: in 2–3 series, rays 2–3 mm, narrow, barely extending beyond involucre, not coiled when dry. **FR**: 1.3–1.5 mm; pappus bristles ± 5 mm. $2n=18$. Meadows, creek banks; 1800–3550 m. SN, SnBr, W&I; to AK, e Can, NM. [*Trimorpha l.* (Hook.) G.L. Nesom] Jul–Aug

E. maniopotamicus G.L. Nesom & T.W. Nelson MAD RIVER FLEABANE DAISY Per 10–22(27) cm; taproot ± thick. **ST**: branched at base, strigose, glandless. **LF**: basal and cauline, 3–10 cm, oblanceolate to ± narrowly spoon-shaped, 3-nerved, entire; cauline gradually reduced distally on st or not, spreading-hairy. **INFL**: heads 1(4); involucre (5)6–7 mm, 9–12(14) mm diam; phyllaries ± equal, elliptic-oblanceolate to oblong-oblanceolate, abruptly acuminate, bristly-strigose to ± stiffly long-straight-hairy, densely long-shaggy-hairy at base, glandless. **RAY FL**: (16)21–33; ray 10–12 mm, white to pink or ± purple, not or weakly coiled when dry. **FR**: pappus bristles 16–20. Dry, barren meadows and openings in mixed-conifer woodland; 1300–1500 m. KR. Jun–Aug ★

E. mariposanus Congdon MARIPOSA DAISY Per 15–28 cm, from short-, slender-branched caudex. **ST**: decumbent to ascending, 0–few-branched distally, sparsely appressed-hairy or becoming glabrous. **LF**: gen with axillary lf tufts, cauline 25–45 mm, (2)5–8 mm wide, oblanceolate, reduced distal st, ± loosely strigose. **INFL**: heads gen 1–4; involucre 4–5 mm, 8–12 mm diam; phyllaries strongly graduated in 3–5 series, outer sparsely strigose, midrib orange-resinous, margins of inner widely scarious, ± wing-like. **RAY FL**: 18–22; ray 7–9 mm, ± blue, weakly coiled when dry. **FR**: pappus bristles

28–32. Foothill woodland; 600–800 m. SNF. Similar to *E. aequifolius*. Jun–Aug ★

E. miser A. Gray (p. 323) STARVED DAISY Per 5–25 cm, many-stemmed, from woody caudex. **ST**: decumbent to ascending, gen unbranched, densely long-spreading-hairy. **LF**: cauline, 7–16 mm, evenly sized and spaced, narrowly oblanceolate, white-spreading-hairy, minutely glandular. **INFL**: heads discoid, 1–4, long-peduncled; involucre 4–5, 7–14 mm diam; phyllaries strongly graduated in 3–5 series, densely minutely glandular, inner 3.5–5 mm. **RAY FL**: 0. **FR**: pappus bristles 12–28. $2n=18$. Rocky sites; 1900–2300 m. n SNH. Jul–Oct ★

E. multiceps Greene (p. 323) KERN RIVER DAISY Per 12–20 cm, from taproot and (±) simple caudex. **ST**: decumbent to ascending, few-branched ± mid-st, ± ascending-hairy. **LF**: basal persistent, 2–5 cm, oblanceolate to spoon-shaped, entire or few-toothed; cauline reduced, strigose. **INFL**: heads 1–3; involucre 3.5–4 mm, 7–10 mm diam; phyllaries ± equal, sparsely bristly, minutely glandular. **RAY FL**: 75–125; ray 5–8 mm, white or ± purple, not reflexed or coiled when dry. **DISK FL**: corolla throat ± widened. **FR**: pappus bristles 5–8. Riverbanks, sandy flats, meadows in pine or aspen woodland; 1500–2500 m. s SNH, SnBr. Jun–Aug ★

E. nivalis Nutt. (p. 323) SNOW FLEABANE DAISY Bien, per, 10–35 cm, taprooted or fibrous-rooted. **ST**: sparsely hairy, stalked-glandular. **LF**: basal 3–8 cm, cauline gradually reduced distally on st, sparsely rough-hairy or strigose. **INFL**: heads 1–8, inconspicuously radiate, in ± flat-topped cluster; involucre 5–6 mm, 8–11 mm diam; phyllaries ± equal, sparsely soft-hairy or not, minutely stalked-glandular, ± green. **RAY FL**: in ± 2 series, outer fls with erect rays 3–4.5 mm, white to ± pink, rays not coiled when dry, inner fls rayless. **FR**: 2–2.4 mm. Volcanic rocks, meadows; 2700–2900 m. CaR; to AK, CO. [*Trimorpha acris* (L.) A. Gray var. *debilis* (A. Gray) G.L. Nesom] Jun–Aug ★

E. oxyphyllus Greene WAND-LIKE FLEABANE DAISY Per 5–50 cm, taprooted; caudex branches relatively thick, woody. **ST**: ascending, glabrous, glandless. **LF**: cauline, mid and distal < internodes, all gen withering early in season, oblanceolate, sharply reduced in size and becoming linear to filiform distally, sparsely and minutely strigose to glabrous. **INFL**: heads 1–8; involucre 4–6 mm, 7–12 mm diam; phyllaries ± equal, outer occ sparsely minutely strigose or not, minutely glandular. **RAY FL**: 12–27(45); ray 6–9 mm, white to lavender or ± blue, slightly coiled at tips when dry. **FR**: pappus bristles 17–25. Rocky hillsides around seeps or springs, canyons, cliff bases; 700–1100 m. DSon; AZ, Mex (SON). May–Jun ★

E. parishii A. Gray (p. 323) PARISH'S DAISY Per, subshrub, 10–35 cm, from thick taproot and branched caudex. **ST**: 0–few-branched near mid-st, silvery-hairy, esp distally. **LF**: basal 3–6 cm, ± linear, often 0 by fl; cauline ± reduced, silvery-strigose. **INFL**: heads 1–10; involucre 5–7 mm, 10–15 mm diam; phyllaries ± equal, proximally strigose or not, minutely glandular. **RAY FL**: 30–55; ray 6–13 mm, pink or white, coiled when dry. **FR**: 4-ribbed, ± hairy; pappus bristles 18–26, outer pappus conspicuous, of bristles or narrow scales to 1 mm. Rocky blackbush or creosote-bush scrub to pinyon/juniper woodland, often on limestone; 800–2000 m. n SnBr (Cushenbury Canyon). Threatened by limestone mining. May–Jun ★

E. petrophilus Greene ROCK-LOVING FLEABANE Per 10–30 cm, from woody root and rhizome or caudex. **ST**: many, decumbent to ascending, 0–few-branched distally, ± long-loose-spreading-hairy, densely glandular near heads or not. **LF**: cauline, 10–25 mm, linear or narrowly oblong to oblanceolate, evenly sized and spaced, gen proximally long-ciliate. **INFL**: heads discoid, (1)2–5(10); involucre 5.5–7 mm, 10–15 mm diam; phyllaries strongly graduated in 3–5 series, spreading-hairy or gen not, densely minutely glandular, sometimes distally ± purple. **RAY FL**: 0. **FR**: pappus bristles 22–35.

var. **petrophilus** Herbage gen loose-crinkly-hairy, densely glandular-hairy. **INFL**: phyllary tips obviously widened, ± purple. $2n=18$. Rocky foothills to montane forest, sometimes on serpentine; 500–2100 m. NCoRO, NCoRH, SnFrB, SCoR. May–Sep

var. **sierrensis** G.L. Nesom NORTHERN SIERRA DAISY Herbage gen loose-crinkly-hairy, densely glandular-hairy. **INFL**: phyllary

tips not widened or purple. Rocky foothills to montane forest, sometimes on serpentine; 300–1900 m. n SN. Jul–Oct ★

var. *viscidulus* (A. Gray) G.L. Nesom (p. 323) KLAMATH ROCK DAISY Herbage stiff-straight- or curved-hairy, ± glandless. **INFL:** phyllary tips like body. Rocky foothills to montane forest, sometimes on serpentine; 1500–2700 m. KR, NCoRH, MP; sw OR. Intergrades with *E. reductus*. Jul–Sep ★

E. *philadelphicus* L. var. ***philadelphicus*** (p. 323) PHILADELPHIA FLEABANE Ann to short-lived per, 25–80 cm, from fibrous roots and often short rhizome. **ST:** branched distally, widened below heads, spreading-hairy. **LF:** basal 8–15 cm, oblong-obovate to spoon-shaped, gen coarsely toothed; cauline lanceolate to ovate, clasping, little reduced, ± long-hairy. **INFL:** heads 1–35; involucre 4–6 mm, 6–15 mm diam; phyllaries ± equal, bases often ± fused, spreading-hairy, sometimes minutely glandular. **RAY FL:** ± 150–400; ray 6–9 mm, white or ± pink, coiled when dry. **FR:** pappus bristles 20–30. 2*n*=18. Streamsides, other moist habitats; < 1200 m. CA (exc GV, DMtns); to e US. Scattered. May–Jun

E. *pumilus* Nutt. var. ***intermedius*** (Cronquist) S.L. Welsh (p. 327) SHAGGY FLEABANE Per 8–35 cm, cespitose, from woody taproot and thick-branched caudex. **ST:** 0–4-branched below mid-st, stiffly spreading- or reflexed-hairy, minutely stalked-glandular. **LF:** basal erect, crowded, 2–8 cm, linear to oblanceolate, proximal stiffly long-spreading-ciliate, rough-hairy, ± not glandular; cauline gradually reduced. **INFL:** heads 1–5(50); involucre 4–7 mm, 9–14 mm diam; phyllaries ± equal, rough-hairy, minutely glandular. **RAY FL:** (40)55–115; ray 8–12 mm, white or pink (lavender), reflexed, sometimes weakly coiled when dry. **DISK FL:** corolla abruptly wider at throat, ± glandular. **FR:** pappus bristles 12–22, outer series of inconspicuous bristles or narrow scales. 2*n*=18, 36. Open slopes, meadows; 1200–1800 m. CaR, n SNH, MP, W&I; to BC, MT, WY, UT. [*E. p.* Nutt. var. *gracilior* Cronquist] Other var. on e side Rocky Mtns. May–Aug

E. *pygmaeus* (A. Gray) Greene (p. 327) PYGMY FLEABANE Per 1–6 cm, gen cespitose, from taproot and short-branched caudex. **ST:** unbranched, densely minutely glandular and with short stiff ± spreading glandless hairs. **LF:** gen basal, 6–35 mm, linear to narrowly oblanceolate, lf base ciliate, faces rough-strigose, densely minutely glandular; cauline (if any) proximal to mid-st. **INFL:** head 1; involucre 4–7 mm, 6–15 mm diam; phyllaries ± equal, purple-black, stiff-spreading-hairy, densely minutely glandular. **RAY FL:** 20–37; ray 4–7(10) mm, blue or purple (white), not coiled or reflexed when dry. **FR:** pappus bristles 15–25. 2*n*=18. Rocky sites, subalpine forest to alpine talus; 2900–4100 m. c SNH, SNE; w-c NV. Jul–Aug

E. *reductus* (Cronquist) G.L. Nesom LITTLE RAYLESS FLEABANE Per 8–20(30) cm, from woody roots and slender, rhizome-like branched caudex. **ST:** 1, ascending to erect, 0–few-branched near st tips, (±) glabrous. **LF:** basal 0; cauline gen 1–4 cm, thread-like to linear, evenly sized and spaced, ascending-ciliate, faces glabrous or sparsely strigose. **INFL:** heads discoid, 1–2(5+); involucre 4–6 mm, 8–10 mm diam; phyllaries strongly graduated in 3–5 series, densely minutely glandular, green or distally ± purple. **RAY FL:** 0. **FR:** pappus bristles 20–61.

var. *angustatus* (A. Gray) G.L. Nesom (p. 327) **INFL:** phyllary tips ± purple (variable in n SnFrB). **FR:** pappus bristles 38–61. Rocky sites, sometimes on serpentine, pine/oak woodland; 600–1400 m. NCoRO, NCoRH, SnFrB. Jun–Aug

var. *reductus* CALIFORNIA RAYLESS DAISY **INFL:** phyllary tips ± green. **FR:** pappus bristles 20–30. Crevices, other open, rocky sites; 700–2400 m. KR, n SNH. Jun–Aug

E. *robustior* (Cronquist) G.L. Nesom (p. 327) ROBUST DAISY Per (15)25–55 cm, from slender taproot and simple caudex. **ST:** gen decumbent, few-branched near mid-st, gen ± purple, sparsely ± appressed-hairy. **LF:** basal 8–17 cm, linear to narrowly oblanceolate, 3-veined; cauline gradually reduced distally on st, rough-strigose. **INFL:** heads (1)2–4, long-peduncled; involucre 6–8.5 mm, (12)14–20 mm diam; phyllaries ± equal, phyllaries narrowly oblanceolate to lanceolate, acute-acuminate, ± densely shaggy-hairy, glandless. **RAY FL:** 21–36; ray 9–19 mm, white or pink, not coiled or reflexed

when dry. **FR:** pappus bristles 14–20. Grassy openings, meadows, sometimes on serpentine; 200–500 m. NCoRO. [*E. decumbens* var. *r.* (Cronquist) Cronquist] Jun–Jul ★

E. *sanctarum* S. Watson (p. 327) SAINTS' DAISY Per 5–35 cm, from thin, often stolon-like caudex branches with few fibrous roots. **ST:** ascending, 0–few-branched near mid-st, sparsely short-spreading-hairy. **LF:** basal 2–5 cm, oblanceolate to spoon-shaped; cauline lanceolate, barely < basal, sometimes ± clasping, below mid-st or ± evenly distributed, sparsely spreading-hairy. **INFL:** heads 1(3); involucre 6–9 mm, 12–17 mm diam; phyllaries ± equal, spreading-hairy, hairs glandless, gen with black cross-walls, occ minutely glandular. **RAY FL:** 45–90; ray 7–13 mm, lavender to purple, not coiled or reflexed when dry. **FR:** pappus bristles 18–25. Sandy sites, coastal scrub or woodland; < 500 m. s CCo, s SCoRO, n ChI. Mar–Jun ★

E. *serpentinus* G.L. Nesom (p. 327) SERPENTINE DAISY Per 30–45 cm, from woody taproot. **ST:** many, few-branched near st tips, gen glabrous. **LF:** cauline, mid and distal > internodes, 2–4 cm, linear to thread-like, evenly sized and spaced, ± ciliate, faces glabrous. **INFL:** heads 1–4; involucre 4.5–5 mm, 9–12 mm diam; phyllaries strongly graduated in 3–5 series, minutely puberulent, densely minutely glandular. **RAY FL:** 9–13; ray 7–8 mm, white, drying ± blue, weakly coiled when dry. **FR:** pappus bristles 26–32. Serpentine scrub; 400–600 m. NCoRO (nw Sonoma Co.). Like *E. greenei* G.L. Nesom. May–Aug ★

E. *strigosus* Willd. var. ***strigosus*** (p. 327) PRAIRIE FLEABANE Ann or bien, 30–80 cm, from fibrous roots. **ST:** branched distal to mid-st, ± glabrous to sparsely appressed-hairy. **LF:** basal gen petioled, blades 3–6 cm, elliptic to obovate, strigose to ± rough-hairy; cauline gen reduced distally on st. **INFL:** heads 10–many; involucre 2–4 mm, 5–12 mm diam; phyllaries ± equal, glabrous or strigose to sparsely spreading-hairy. **RAY FL:** ± 60–110; ray 5–6 mm, white, not coiled or reflexed when dry. **FR:** ray pappus 0; disk pappus bristles 10–15. 2*n*=18,27,36. Disturbed sites; < 1100 m. CA-FP; w US; native to e US. [*E. s.* var. *septentrionalis* (Fernald & Wiegand) Fernald] Scattered weed, often producing seeds asexually. Apr–Aug

E. *sumatrensis* Retz. (p. 327) TROPICAL HORSEWEED Ann 30–200+ cm. **ST:** erect, lfy, branched distally or throughout, puberulent or long-soft-hairy to bristly, hairs ± ascending; central st gen exceeding branches. **LF:** margin ± fine-strigose, faces ± glabrous to puberulent, densely strigose, or bristly, hairs ± erect on veins; basal and proximal cauline 5–10 cm, 5–20 mm wide, ± petioled, linear to elliptic or oblanceolate, entire or toothed, gen withered by fl; distal similar, sessile, smaller, gen entire. **INFL:** heads disciform, many in panicle-like to ± flat-topped clusters; involucre 4–5 mm, 3–4 mm diam, ± urn-shaped in fl; phyllaries weakly graduated, ± glabrous to puberulent, densely strigose, or spreading-hairy, green when fresh, central part of middle phyllaries gen wider than lighter margin, not red-brown or resin filled when dry; receptacle 1–2.5 mm diam in fr. **PISTILLATE FL:** 60–90+; ray 0–0.3 mm, white or cream. **DISK FL:** 6–10. **FR:** 1–1.5 mm, pale tan (gen some with ± red ribs), sparsely minutely strigose or ± glabrous; pappus bristles 15–25, 3–4 mm, cream to tan. Disturbed sites; < 600 m. NW (exc NCoRH), CaRF, n SNF, ScV, CW (exc SCoRI), SCo, ChI, PR (exc SnJt); widely naturalized; native to S.Am. [*Conyza s.* (Retz.) E. Walker; *C. floribunda* Kunth; *C. bilbaoana* J. Rémy, misappl.] All yr

E. *supplex* A. Gray (p. 327) SUPPLE DAISY Per 15–40 cm, from taproot and slender-branched, rhizome-like caudex. **ST:** ascending, simple, occ sparsely spreading-hairy. **LF:** basal present at fl, 4–8 cm, elliptic to oblanceolate or spoon-shaped, ± sessile, ciliate, faces ± glabrous; cauline gen gradually reduced, not clasping. **INFL:** heads discoid, 1(2); involucre 7–11 mm, 14–20 mm diam; phyllaries ± equal, spreading-hairy, densely minutely glandular. **RAY FL:** 0. **FR:** pappus bristles 17–30. Coastal areas, bluffs; < 50 m. n&c NCo. Threatened by coastal development. May–Aug ★

E. *tener* (A. Gray) A. Gray THIN DAISY Per 2–15 cm, from taproot and branched caudex, cespitose. **ST:** gen unbranched, short-white-appressed-hairy, glandless. **LF:** basal 1–8 cm, long-petioled, oblanceolate to elliptic; cauline abruptly much reduced, ± strigose. **INFL:** heads 1(2); involucre 3–5 mm, 5–12 mm diam; phyllaries ±

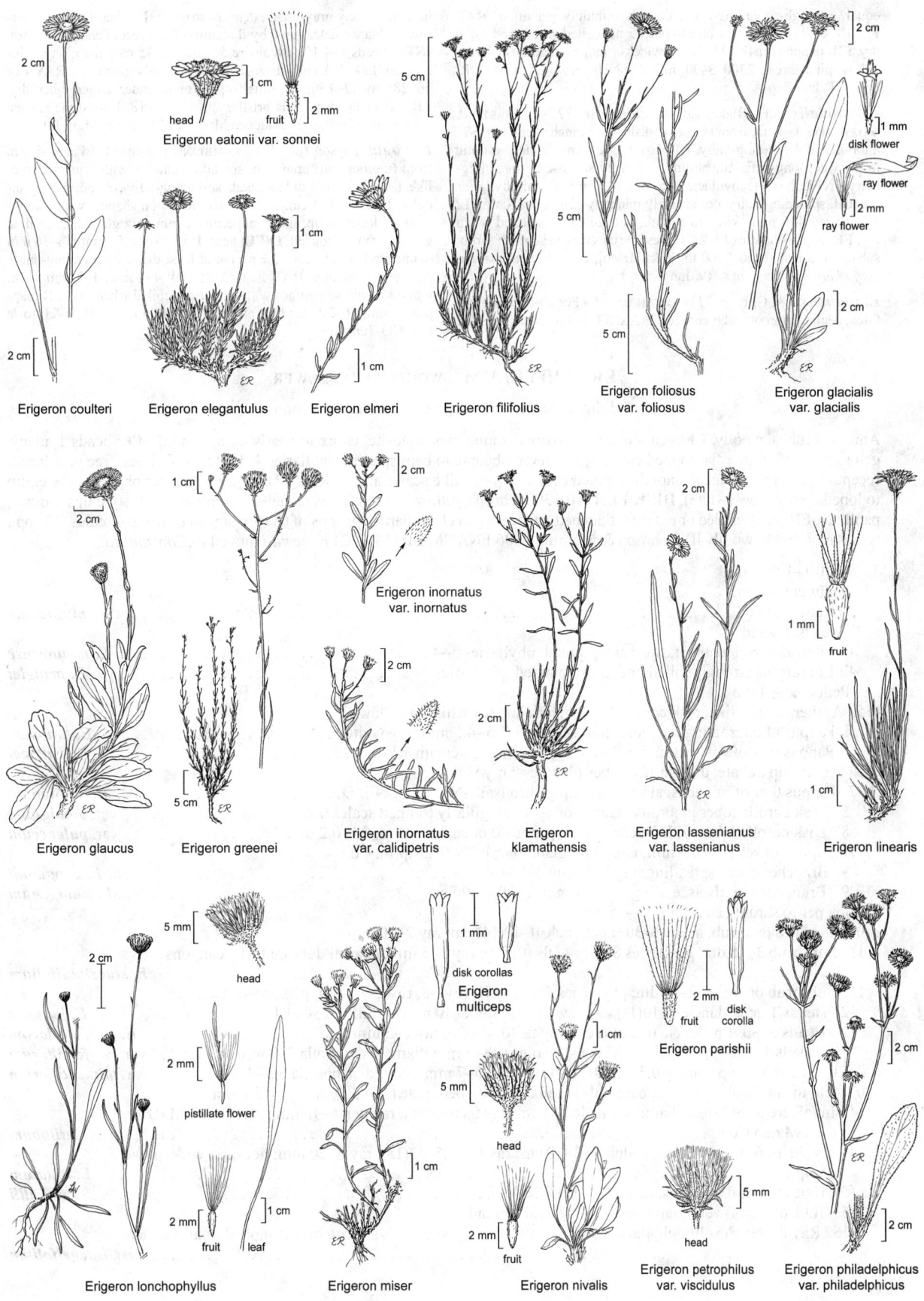

Erigeron eatonii var. sonnei
head   fruit

Erigeron coulteri

Erigeron elegantulus

Erigeron elmeri

Erigeron filifolius

Erigeron foliosus
var. foliosus

Erigeron glacialis
var. glacialis

disk flower
ray flower
ray flower

Erigeron glaucus

Erigeron greenei

Erigeron inornatus
var. inornatus

Erigeron inornatus
var. calidipetris

Erigeron
klamathensis

Erigeron lassenianus
var. lassenianus

Erigeron linearis
fruit

Erigeron lonchophyllus
head
pistillate flower
fruit   leaf

Erigeron miser

Erigeron multiceps
disk corollas

Erigeron nivalis
head
fruit

Erigeron parishii
fruit   disk corolla

Erigeron petrophilus
var. viscidulus
head

Erigeron philadelphicus
var. philadelphicus

equal, ± rough-spreading-hairy, densely minutely glandular. **RAY FL**: 18–40; ray 4–8 mm, ± blue to purple, not coiled or reflexed when dry. **FR**: pappus bristles 15–30. Crevices or ledges, sagebrush scrub, yellow-pine forest; 2300–3400 m. KR, SNH, MP, W&I; to se OR, WY, UT. Jun–Sep

E. uncialis S.F. Blake var. ***uncialis*** (p. 327) LIMESTONE DAISY Per 1–4 cm, from taproot and slender-branched caudex. **ST**: unbranched, spreading-hairy. **LF**: basal, 5–25 mm, widely elliptic to obovate, long-petioled, cauline 0, loosely strigose to spreading-hairy. **INFL**: head 1; involucre 3–5 mm, 6–11 mm diam; phyllaries ± equal, spreading-hairy, occ sparsely minutely glandular. **RAY FL**: 15–40; ray 4–6 mm, white to ± pink, not coiled or reflexed when dry. **FR**: pappus bristles 13–22. Limestone crevices, sagebrush scrub, subalpine forest; 2100–2900 m. W&I, DMoj; c NV. *E. uncialis* var. *conjugans* S.F. Blake in s NV. Jun–Jul ★

E. utahensis A. Gray (p. 327) UTAH DAISY Per 10–50 cm, from thick, peeling taproot and branched caudex. **ST**: 0–few-branched near

mid-st, densely gray-green, densely strigose. **LF**: basal 6–8 cm, narrowly oblanceolate, gen 0 by fl; cauline ± reduced, densely strigose. **INFL**: heads 1–4(10); involucre 5–7 mm, 7–15 mm diam; phyllaries ± equal, loosely rough-strigose, often minutely glandular. **RAY FL**: 16–28; ray 12–15 mm, white to pink or lavender, coiled when dry. **FR**: 4(6)-ribbed; pappus bristles 20–35. $2n=18$. Limestone slopes; ± 1500 m. e DMtns (Providence Mtns); to CO, n AZ. May–Jul ★

E. vagus Payson (p. 327) RAMBLING FLEABANE Per 2–5 cm, from (taproot and) branched, spreading caudex with thin, rhizome-like branches. **ST**: unbranched, spreading-hairy, sometimes glandular. **LF**: gen 1–3 cm, oblanceolate to spoon-shaped, with 3 long-obovate lobes at tip, finely puberulent, minutely glandular; cauline gen 0 or much reduced. **INFL**: head 1; involucre 5–7 mm, 8–16 mm diam; phyllaries ± equal, ± purple at least distally, spreading-hairy, minutely glandular. **RAY FL**: 25–35; ray 4–7 mm, 1–2 mm wide, white to pink, sometimes with midstripe, coiled when dry. **FR**: pappus bristles 16–20. $2n=18$. Talus; 3300–4400 m. c SNH, W&I; to se OR, CO. Jun–Aug

ERIOPHYLLUM WOOLLY SUNFLOWER

John S. Mooring & Dale E. Johnson (ann spp.)

Ann to shrub, ± woolly. **LF**: gen alternate, proximal sometimes opposite, entire to nearly compound. **INFL**: heads 1–many, gen radiate; often in ± flat-topped clusters; involucre obconic to hemispheric; phyllaries 4–13(15) in 1 series, free or ± fused; receptacle flat to columnar, smooth or pitted (occ 1–6 palea-like scales at tip). **RAY FL**: 0 or gen ± 1 per phyllary; ray entire to lobed, gen yellow (white). **DISK FL**: (3)10–300; corolla yellow; anther tip ovate, deltate or awl-shaped; style tips ± ovate, papillate. **FR**: 4(5)-angled or outer frs flattened, inner frs gen club-shaped; pappus of 0–15 ± jagged or fringed scales. 13 spp.: w N.Am. (Greek: woolly lf) [Johnson & Mooring 2006 FNANM 21:353–362] *E. nevinii* moved to *Constancea*.

1. Ann 0.1–1.5(3) dm
  2. Peduncle < 1 cm
    3. Heads radiate . . . . ***E. multicaule***
    3′ Heads discoid
      4. Lf weakly rolled under, tip sharp-pointed; phyllaries 3–4 . . . . ***E. mohavense***
      4′ Lf margins strongly rolled under, tip rounded; phyllaries 6–8 . . . . ***E. pringlei***
  2′ Peduncle ≥ 1 cm
    5. Anther tip awl-like; disk corolla lobes glandular; ray white to yellow
      6. Pappus of alternating long and short scales; fr 2.5–4.5 mm; ray white . . . . ***E. lanosum***
      6′ Pappus 0 or of ± = scales; fr ± 2 mm; ray yellow, occ cream-white . . . . ***E. wallacei***
    5′ Anther tip deltate; disk corolla lobes glandless; ray yellow
      7. Pappus 0 or of minute scales; anther tip glabrous; s SNF, Teh, SNE, D . . . . ***E. ambiguum***
        8. Disk corolla lobes glabrous; pappus of 6–10 irregularly toothed scales 0.2–0.5 mm . . . . var. ***ambiguum***
        8′ Disk corolla lobes with 1-celled hairs; pappus 0 or scales entire, 0.1–0.2 mm . . . . var. ***paleaceum***
      7′ Pappus of scales ≥ 0.5 mm; anther tip glandular; c SN (Mariposa Co.)
        9. Branches open, spreading; ray 3–5 mm; 500–1900 m . . . . ***E. congdonii***
        9′ Branches strictly ascending; ray ± 1 mm; 1800–2500 m . . . . ***E. nubigenum***
1′ Gen per to shrub, occ fl 1st yr, 1–15 dm
  10. Subshrub or shrub; heads 1–30+; peduncle 0–10(14) cm; ray 2–10 mm
    11. Subshrub 3–15 dm; phyllaries 8–11; ray fls (0)6–9; ray 3–5 mm; coastal dunes, bluffs, canyons
      . . . . ***E. staechadifolium***
    11′ Subshrub or shrub, 2–10 dm; phyllaries 4–8; ray fls (0)4–8, ray 2–10 mm; gen non-coastal
      12. Heads 1–5; peduncle 5–10(14) cm; ray fls 5–8, ray 6–10 mm — SnFrB, SCoRI . . . . ***E. jepsonii***
      12′ Heads 3–30+; peduncle 0–2.5(9) cm; ray fls (0)4–6(8), ray 2–5 mm . . . . ***E. confertiflorum***
        13. Heads 10–30+; peduncle 0–1(2) cm; involucre 3–5 mm diam; disk corolla 2–3 mm . . . . var. ***confertiflorum***
        13′ Heads 3–10; peduncle 0.5–2.5(9) cm; involucre 5–7 mm diam; disk corolla 3.5–4 mm . . . . var. ***tanacetiflorum***
  10′ Ann to subshrub, woody at base only; heads 1–5(10); peduncle (1)3–30 cm; ray 6–20 mm
    14. Involucre 5–6(7) mm diam; peduncle 1–8 cm; ray fls 6–8(13), ray gen 6–10 mm; rare; c-w SnFrB
      (San Mateo Co.) . . . . ***E. latilobum***
    14′ Involucre 6–15 mm diam; peduncle 3–30 cm; ray fls (0)5–13(15), ray 6–20 mm; not rare; widespread
      complex . . . . ***E. lanatum***
      15. Tube of disk fl glabrous; rare, s Teh, se SCoRO . . . . var. ***hallii***
      15′ Tube of disk fl very glandular or hairy; widespread
        16. Ray fls gen 8(5–10); phyllaries gen 8(5–10), strongly keeled; lvs entire or 3-(5-)lobed, margins flat
        . . . . var. ***integrifolium***

16′ Ray fls gen 9–13; phyllaries 9–15, keeled to flat; lvs entire to nearly compound, margins flat or rolled under
  17. Lvs densely woolly on faces, entire or coarsely toothed (esp distal to middle), margins flat
    18. Ray 7–10 mm; KR, NCoRH . . . . var. *lanceolatum*
    18′ Ray 6–7 mm; s SNH, SnBr. . . . . var. *obovatum*
  17′ Lvs less hairy, (often green) adaxially, entire to nearly compound, margins rolled under
    19. Lvs linear to ovate, entire to 2-compound, often woolly-tufted adaxially; pappus 0.5–1.5 mm;
      peduncle 3–30 cm; ray yellow
      20. Lvs 1–3 cm, lobed to ± 2-pinnately compound; peduncle 3–10 cm, gen slender; involucre 6–8
        mm; ray 6–9 mm . . . . var. *achilleoides*
      20′ Lvs 3–8 cm, entire to pinnately lobed; peduncle 10–30 cm, gen swollen proximal to heads;
        involucre 8–10 mm; ray 10–20 mm . . . . var. *grandiflorum*
    19′ Lvs narrowly oblanceolate or wider, serrate to sharply lobed, glabrous adaxially; pappus 0 or
      minute; peduncle 3–10 cm; ray golden yellow
      21. Lvs ± diamond-shaped, proximal sharply 3–5 lobed, woolly-tufted abaxially; coastal . . . . var. *arachnoideum*
      21′ Lvs oblanceolate to obovate, coarsely serrate or lobed distally, silky-woolly abaxially; SNH . . . . var. *croceum*

E. ambiguum (A. Gray) A. Gray Ann 5–30 cm, decumbent to ascending. **LF:** < 4 cm, oblong-oblanceolate, entire to shallowly lobed. **INFL:** heads 1; peduncle 1–8 cm; involucre 3–6 mm, obconic to hemispheric; phyllaries 6–10, acuminate; receptacle conic. **RAY FL:** 6–10; ray 2–10 mm. **DISK FL:** many; corolla 1.3–3 mm, tube and throat hairy; anther tip deltate, smooth. **FR:** 2.2–3 mm, ± strigose; pappus 0 or 0.1–0.5 mm. $2n=14$.

var. **ambiguum** (p. 327) BEAUTIFUL WOOLLY SUNFLOWER **LF:** pinnately lobed. **INFL:** involucre 4–5.5 mm; phyllaries occ fused; receptacle tip scales 0. Open chaparral; 200–1900 m. s SNF, Teh. Mar–Jun

var. **paleaceum** (Brandegee) Ferris (p. 327) **LF:** entire or 3-lobed near tip. **INFL:** involucre 5–7 mm; phyllaries free; receptacle tip occ with 1–6 palea-like scales. Desert scrub or woodland; 100–2800 m. s SNF, Teh, SNE, D; s NV. [*E. p.* Brandegee] Jan–Jun

E. confertiflorum (DC.) A. Gray GOLDEN-YARROW, YELLOW-YARROW Subshrub or shrub gen 2–7 dm. **LF:** 1–5 cm, ± obovate, deeply 3–5-lobed to nearly 2-pinnately compound, rolled under, often becoming ± glabrous adaxially. **INFL:** heads 3–30+; peduncle 0–2.5(9) cm; involucre 3–7 mm, bell-shaped; phyllaries 4–7, obtuse, keeled, strongly overlapping, ± free; receptacle ± convex. **RAY FL:** (0)4–6(8); ray 2–5 mm. **DISK FL:** 10–75; corolla 2–4 mm, puberulent to glandular; anther ovate. **FR:** 2–4 mm; pappus gen < 1 mm. Highly variable, intergrading polyploid complex; hybridizes with *E. lanatum.*

var. **confertiflorum** (p. 327) Pl gen persistently tomentose. **INFL:** clusters dense. **RAY FL:** occ 0. **DISK FL:** gen 10–35. **FR:** 2–3 mm; pappus scales 5–14, ± equal. $2n=16,32,48,64$. Many dry habitats; < 3000 m. NCoR, SN, CW, SW, w edge D. Apr–Aug

var. **tanacetiflorum** (Greene) Jeps. TANSY-FLOWERED WOOLLY SUNFLOWER Pl occ ± glabrous. **INFL:** clusters gen loose. **DISK FL:** 35–75. **FR:** 3.5–4 mm; pappus scales ± 8, unequal. $2n=64$. Oak woodland; 600–800 m. n&c SNF. Possible hybrid derivative of *E. confertiflorum* var. *c.* × *E. lanatum.* May–Jul ★

E. congdonii Brandegee (p. 327) CONGDON'S WOOLLY SUNFLOWER Ann 1–3 dm; branches spreading. **LF:** 1–4 cm, oblanceolate, entire or lobed near tip. **INFL:** head 1; peduncle 3–10 cm; involucre 5–8 mm, bell-shaped; phyllaries 8–10, acute, free; receptacle flat to conic. **RAY FL:** 8–10; ray 3–5 mm. **DISK FL:** many; corolla 2–3 mm, glabrous; anther tip deltate, glandular. **FR:** 2.5–3 mm, strigose; pappus scales 1–2 mm, unequal. $2n=14$. Rocky, open, foothill woodland, yellow-pine forest; 500–1900 m. c SNF (Mariposa Co.). Mar–Jun ★

E. jepsonii Greene (p. 327) JEPSON'S WOOLLY SUNFLOWER Subshrub 5–8 dm. **LF:** 3–6 cm, ovate, pinnately 5–7-lobed (lobes linear, obtuse), rolled under, woolly-tufted, becoming glabrous adaxially. **INFL:** heads 1–5; peduncle 4–14 cm; involucre 4–7 mm, widely bell-shaped; phyllaries 5–8, acute, strongly overlapping, keeled, free; receptacle convex to low-conic. **RAY FL:** 5–8; ray 6–10 mm. **DISK FL:** gen 35–50; corolla 3–5 mm, glandular-puberulent or bristly;

anther tip ovate. **FR:** 3–4.5 mm; pappus scales gen 8, ± 1–1.5 mm, unequal. $2n=64$. Dry oak woodland; 200–1000 m. SnFrB, SCoRI. Possible derivative of *E. confertiflorum* × *E. lanatum.* Apr–Jun ★

E. lanatum (Pursh) J. Forbes COMMON WOOLLY SUNFLOWER Ann, bien, or gen per to subshrub, 1–10 dm. **LF:** 1–8 cm, linear to ovate, entire to ± 2-pinnately compound, gen becoming glabrous adaxially. **INFL:** heads 1–5+; peduncle 3–30 cm; involucre 5–12 mm, bell-shaped to hemispheric; phyllaries 5–15; receptacle ± flat to ± conic. **RAY FL:** (0)5–13(15); ray 6–20 mm, oblong to elliptic. **DISK FL:** 20–300; corolla 2.5–5 mm, tube gen glandular; anther tip ovate. **FR:** gen variable, 2–5 mm, glandular or hairy; pappus scales 0 or 6–12, 0–2 mm, translucent. Polyploid pillar complex of intergrading races; key is to modal populations; some vars. hybridize with *E. confertiflorum.*

var. **achilleoides** (DC.) Jeps. Ann to per. **LF:** 1–3 cm, gen ± 1–2-compound. **INFL:** peduncle 3–10 cm, gen slender; involucre 6–8 mm. **RAY FL:** gen 12–13, ray 6–9 mm. **DISK FL:** corolla 2.5–3 mm. **FR:** 2.2–3 mm; pappus gen < 1 mm. $2n=16,32$. Dry, often rocky sites, chaparral, forest; < 1300 m. KR, NCoR, CaRH, n SNF, SnFrB, SCoRO, MP; OR, NV. [*Eriophyllum lanatum* var. *achillaeoides*, orth. var.] Intergrades with *E. lanatum* var. *arachnoideum*, *E. lanatum* var. *grandiflorum* in CA. Rayless NCoRI pls with $2n=32$ may be treated as *E. lanatum* var. *aphanactis* J.T. Howell. May–Jul

var. **arachnoideum** (Fisch. & Avé-Lall.) Jeps. (p. 327) Per, occ stoloned. **LF:** 2–5 cm, ± diamond-shaped, thin; proximal sharply 3–5-lobed, woolly-tufted abaxially. **INFL:** peduncle 3–10 cm; involucre 8–11 mm. **RAY FL:** gen 12–13, ray 8–10 mm, golden-yellow. **DISK FL:** corolla 3–4 mm. **FR:** 2–4 mm, gen glabrous; pappus 0 or minute. $2n=16,32$. Ocean bluffs, ± moist places, often in mixed-evergreen or redwood forest; < 400 m. NCo, NCoRO, CCo, SnFrB. Apr–Aug

var. **croceum** (Greene) Jeps. Per, often stoloned. **LF:** 2–5 cm, oblanceolate to obovate, coarsely serrate or lobed distally, silky-woolly abaxially. **INFL:** peduncle 3–8 cm; involucre 5–8 mm. **RAY FL:** gen 12–13, ray 8–10 mm, golden-yellow. **DISK FL:** corolla 3–4 mm. **FR:** 2–3 mm; pappus 0 or minute. $2n=16,32$. Gen under conifers; 1300–2000 m. SNH. Like *E. lanatum* var. *arachnoideum.* May–Jul

var. **grandiflorum** (A. Gray) Jeps. Bien or short-lived per. **LF:** 3–8 cm, gen entire to pinnately lobed. **INFL:** peduncle 1–3 dm, gen swollen proximal to heads; involucre 8–10 mm. **RAY FL:** gen 12–13, ray 10–20 mm. **DISK FL:** corolla 4 mm. **FR:** 3–4 mm; longest pappus scales gen > 1 mm. $2n=16,32,48,64$. Dry, gen rocky sites, grassland, forest; < 1700 m. KR, NCoR, CaR, n&c SN, ScV; OR, extinct in Mex. May–Jul

var. **hallii** Constance FORT TEJON WOOLLY SUNFLOWER Per. **LF:** 2.5–5 cm, gen opposite proximally, ovate, thin, pinnately lobed. **INFL:** peduncle 5–12 cm; involucre 8–10 mm. **RAY FL:** 8–9; ray 10–13 mm. **DISK FL:** corolla 4–5 mm, tube glabrous. **FR:** 4–5 mm. $2n=16$. Dry sites, woodland; 1200–1500 m. s Teh (near Fort Tejon), se SCoRO (Sierra Madre). Threatened by grazing. Jun–Jul ★

var. ***integrifolium*** (Hook.) Smiley OREGON SUNSHINE Short-lived per to subshrub. **LF**: 1–4 cm, wedge-shaped to obovate, entire to 3-(5-)lobed; margins flat. **INFL**: head 1; peduncle 3–10 cm; involucre 6–8 mm; phyllaries gen 8(5–10), strongly keeled. **RAY FL**: (5)8(10); ray 6–10 mm. **DISK FL**: corolla ± 4 mm. **FR**: 3–4 mm; pappus < 2 mm. $2n=16,32,48,64$. Dry sites, sagebrush, forest, alpine; 1400–3500 m. KR, NCoRH, CaRH, n&c SNH, GB; to WA, WY, UT. Variable and complex. Jul–Aug

var. ***lanceolatum*** (Howell) Jeps. Short-lived per. **LF**: 2–4 cm, lanceolate to ovate, thick, entire to coarsely serrate, ± flat, densely woolly. **INFL**: head 1; peduncle 3–10(15) cm, slender; involucre 8–12 mm; phyllaries 10–15, ± flat. **RAY FL**: 10–15; ray 7–10 mm. **DISK FL**: corolla 3–4 mm. **FR**: 2–3 mm, hairy; pappus < 1 mm. $2n=16,32$. Dry, rocky oak/conifer forest; 200–2200 m. KR, NCoRH; OR. May–Aug

var. ***obovatum*** (Greene) H.M. Hall (p. 327) SOUTHERN SIERRA WOOLLY SUNFLOWER Short-lived per. **LF**: 1–5 cm, entire to few-toothed distally, densely woolly. **INFL**: head 1; peduncle 3–10(15) cm, often ± swollen below head; involucre 7–10 mm. **RAY FL**: gen 12–13, ray 6–7 mm. **DISK FL**: corolla 3–4 mm. **FR**: 2.5–3 mm, ± glabrous; pappus < 1 mm. $2n=16$. Open conifer forest; 1300–2500 m. s SNH, SnBr. Like *E. lanatum* var. *lanceolatum*. Jun–Jul ★

E. lanosum (A. Gray) A. Gray (p. 329) Ann 1–15 cm, decumbent-ascending, often ± red, sparsely woolly. **LF**: 5–20 mm, linear-oblanceolate, entire or lobed at tip. **INFL**: head 1; peduncle 1–5 cm; involucre 5–7 mm, ± cylindric; phyllaries 8–10, acuminate, free; receptacle conic. **RAY FL**: 8–10; ray 3–7 mm, oblong, white, occ red-veined. **DISK FL**: many; corolla 2–3 mm, glabrous; anther tip awl-like, glabrous. **FR**: 2.5–4.5 mm, linear or narrowly club-like, glabrous to minutely strigose; pappus scales 0.5–2.5 mm, gen unequal, longest awned. $2n=14$. Desert scrub; 70–1400 m. D; to sw UT, NM, nw Mex. Feb–May

E. latilobum Rydb. (p. 329) SAN MATEO WOOLLY SUNFLOWER Subshrub 2–5 dm, becoming ± glabrous. **LF**: 2–6 cm, diamond-shaped to ± obovate, thin, deeply triangular-lobed, glabrous adaxially. **INFL**: heads 1–10; peduncle 1–8 cm; involucre 4–7 mm, widely bell-shaped; phyllaries 6–10, acute, barely overlapping, free; receptacle flat (exc conic in center). **RAY FL**: 6–13; ray gen 6–10 mm. **DISK FL**: 40–70; corolla 3–4 mm, glandular; anther tip ovate. **FR**: 3–4 mm; angles gen strigose; pappus 0.3–1 mm, disk scales > ray scales. $2n=32$. Gen oak woodland; 100–150 m. c-w SnFrB (San Mateo Co.). Probable derivative of *E. lanatum* var. *arachnoideum* × *E. confertiflorum*. Threatened by development. May–Jun ★

E. mohavense (I.M. Johnst.) Jeps. (p. 329) BARSTOW WOOLLY SUNFLOWER Ann 1–2.5 cm, tufted, spreading, loosely white-woolly. **LF**: 3–10 mm, spoon- to wedge-shaped, entire or 2–3-lobed (lobes pointed). **INFL**: head 1, discoid, ± sessile; involucre 3–4 mm, ± cylindric; phyllaries 3–4, acute, free; receptacle ± pointed-columnar, flanges 3, protruding between frs, spine-tipped. **RAY FL**: 0. **DISK FL**: ± 3; corolla ± 2 mm, throat minutely puberulent; anther tip narrowly deltate. **FR**: 2–2.5 mm, narrowly obconic, strigose; pappus ± 1.5 mm. Creosote-bush scrub; 500–800 m. DMoj. Apr–May ★

E. multicaule (DC.) A. Gray (p. 329) MANY-STEM WOOLLY SUNFLOWER Ann 2–15 cm, decumbent-ascending, green to purple, ± fleshy, often becoming glabrous. **LF**: ± 1 cm, wedge-shaped, 2–3-lobed, ± woolly. **INFL**: heads ± sessile in lfy clusters at branch tips; involucre 3–4 mm, bell-shaped to hemispheric; phyllaries 5–7, acute, free; receptacle convex. **RAY FL**: 5–7; ray ± 2 mm. **DISK FL**: 13–25; corolla ± 2 mm, throat minutely puberulent; anther tip deltate. **FR**: ± 2 mm, narrowly club-shaped, glabrous to minutely strigose; pappus ± 1 mm. $2n=14$. Sandy soils, open coastal scrub, chaparral; < 1600 m. sw SnJV (e San Luis Obispo, ne Santa Barbara cos.), SCoR, SCo, SnGb, PR. Mar–Jul

E. nubigenum A. Gray (p. 329) YOSEMITE WOOLLY SUNFLOWER Ann 5–15 cm, gray-woolly; branches strictly ascending. **LF**: 1–2 cm, oblanceolate, entire. **INFL**: heads 1 or clustered; peduncle ± 1 cm; involucre 5–6 mm, cylindric; phyllaries 4–6, acute, free; receptacle flat to conic. **RAY FL**: 4–6; ray ± 1 mm. **DISK FL**: 3–16; corolla ± 2 mm; anther tip deltate, glandular. **FR**: 2.5–3 mm, strigose; pappus 0.5–1.5 mm, unequal. $2n=14$. Open, gravelly or rocky forest; 1800–2500 m. c SNH (Mariposa Co.). Jun–Jul ★

E. pringlei A. Gray (p. 329) PRINGLE'S WOOLLY SUNFLOWER Ann 1–5 cm, ± tufted-spreading, white-woolly. **LF**: 3–10 mm, wedge-shaped, gen 3-lobed, very woolly; margins rolled under. **INFL**: heads discoid, ± sessile in lfy clusters at branch tips; involucre 3–6 mm, hemispheric; phyllaries 6–8, acuminate, free; receptacle convex. **RAY FL**: 0. **DISK FL**: 10–20; corolla ± 2 mm, minutely glandular; anther tip deltate. **FR**: 1.5–2 mm, strigose; pappus ± 1 mm. $2n=14+0–1B$ or $0–1I,16$. Chaparral, sagebrush or desert scrub or woodland; 300–2200 m. s SNF, Teh, s SCoRO, SCoRI, TR, SNE, D; to s NV, AZ, n Baja CA. Feb–Jul

E. staechadifolium Lag. (p. 329) SEASIDE WOOLLY SUNFLOWER Subshrub 3–15 dm, much-branched, becoming glabrous. **LF**: 3–7 cm, lanceolate to ovate, entire to ± 1–2-pinnately compound, becoming glabrous adaxially; margins rolled under. **INFL**: heads 5–15+; peduncle < 1 cm; involucre 5–7 mm, bell-shaped; phyllaries 8–11, obtuse or acute, keeled, barely overlapping, free; receptacle convex. **RAY FL**: (0)6–9; ray 3–5 mm. **DISK FL**: corolla 4 mm, glandular; anther tip ovate. **FR**: 3–4 mm, linear-oblong, ± glandular and bristly; pappus often ± 1 mm, unequal, obtuse. $2n=30$. Dunes, sea bluffs, coastal scrub; < 100 m. NCo, CCo, n ChI; s OR. Apr–Sep

E. wallacei (A. Gray) A. Gray (p. 329) WALLACE'S WOOLLY DAISY Ann 1–15 cm, often tufted, woolly. **LF**: 7–20 mm, spoon-shaped to obovate, entire or 3-lobed. **INFL**: head 1; peduncle 1–3 cm; involucre 5–7 mm, bell-shaped; phyllaries 5–10, acute, free; receptacle hemispheric. **RAY FL**: 5–10; ray 3–4 mm, yellow, occ cream-white. **DISK FL**: many; corolla 2–3 mm, throat minutely puberulent; anther tip awl-like, glabrous. **FR**: ± 2 mm, narrowly club-shaped, glabrous or minutely strigose; pappus (0)0.4–0.8 mm. $2n=10+0–1I$ or 0–3B. Chaparral, sagebrush or desert scrub or woodland; 30–2400 m. e SnFrB, SnGb, SnBr, PR, SNE, D; to sw UT, nw AZ, n Baja CA. Dec–Jul

EUCEPHALUS ASTER

Geraldine A. Allen

Per from woody caudex or short rhizome. **ST**: gen erect, 1–16 dm. **LF**: cauline, alternate, proximal gen scale-like, main lvs gen entire. **INFL**: heads radiate or discoid, 1 or in cyme- or panicle-like clusters; involucre obconic to hemispheric; phyllaries ± equal or graduated in 3–6 series, free, ± keeled, margins papery, pale, sometimes ± red-tipped; receptacle ± flat, pitted, epaleate. **RAY FL**: 0–13(21); corolla violet to pink or white. **DISK FL**: many; corolla, anthers gen yellow, tube < throat; anther tip ± triangular; style branches flat on inner face, base ± warty, tip acute, hairy. **FR**: gen ± flattened, ± 1–2-ribbed, ± brown, ± hairy; pappus of bristles, gen in 2 series (outer ± 1 mm, inner gen 5–10 mm), ± white to ± brown. 10 spp.: N.Am. (Greek: good head, describing involucre) [Allen 2006 FNANM 20:39–42]

1. Ray fls 5 or more (gen 5, 8, or 13)
  2. Pl 5–15 dm; lvs 4–10 cm, green on faces, ± glabrous; ray white, ± becoming pink . . . . ***E. engelmannii***
  2′ Pl 2–8 dm; lvs ≤ 6(7) cm, abaxially pale green, ± tomentose, adaxially ± glabrous; ray violet to purple
  . . . . ***E. ledophyllus*** var. ***covillei***

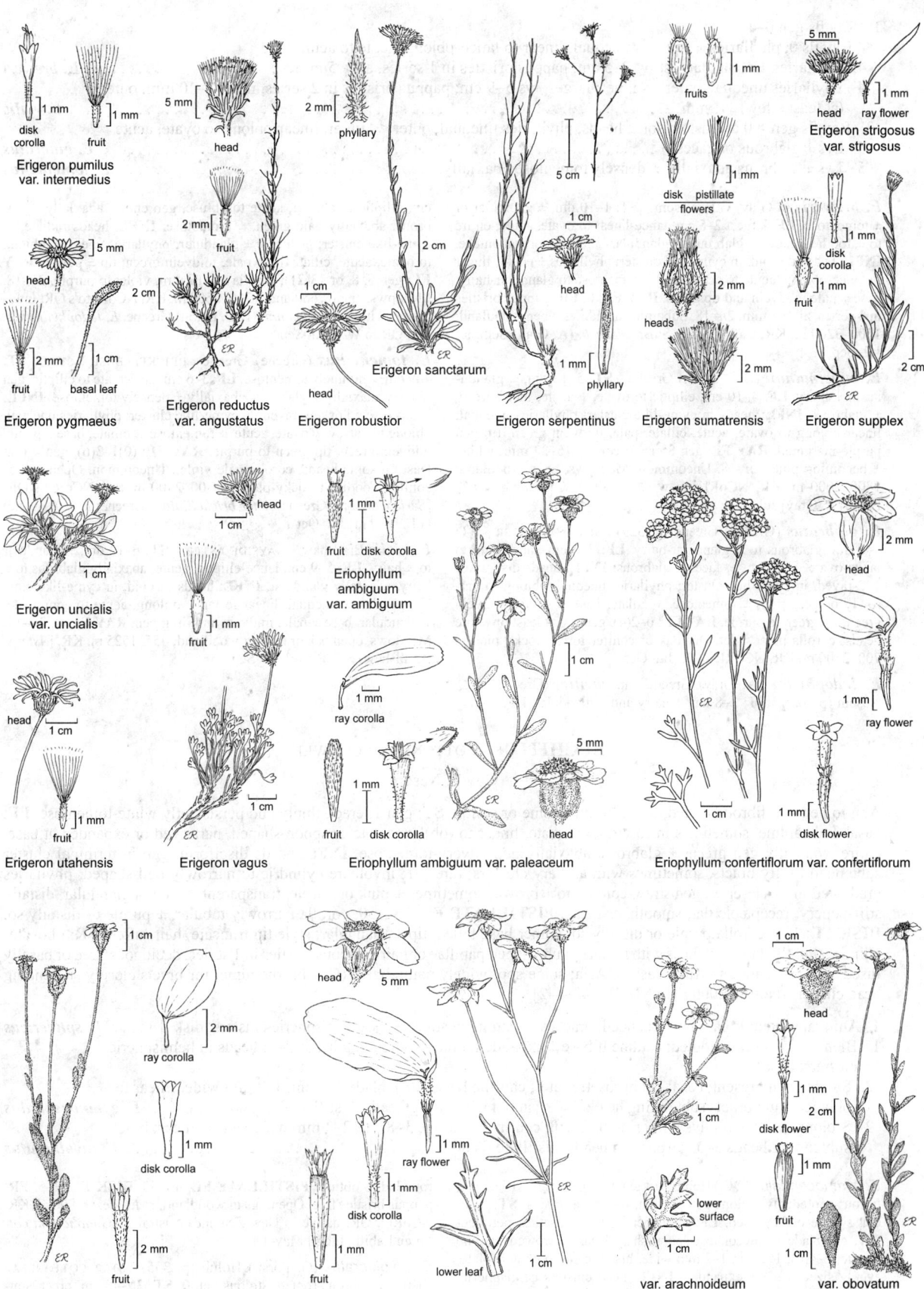

Erigeron pumilus var. intermedius — disk corolla, 1 mm; fruit, 1 mm

head, 5 mm; fruit, 2 mm

phyllary, 2 mm

fruits, 1 mm

head, 5 mm; ray flower, 1 mm
Erigeron strigosus var. strigosus

Erigeron pygmaeus — head, 5 mm; fruit, 2 mm; basal leaf, 1 cm; 2 cm

fruit, 2 mm
Erigeron reductus var. angustatus

Erigeron robustior — head

Erigeron sanctarum — 2 cm; 1 cm

5 cm; head, 1 cm; phyllary, 1 mm
Erigeron serpentinus

disk, pistillate flowers, 1 mm

heads, 2 mm; 2 mm
Erigeron sumatrensis

disk corolla, 1 mm; fruit, 1 mm; 2 cm
Erigeron supplex

Erigeron uncialis var. uncialis — head, 1 cm; 1 cm; fruit, 1 mm

Erigeron utahensis — head, 1 cm; fruit, 1 mm

Erigeron vagus — 1 cm

fruit, 1 mm; disk corolla, 1 mm
Eriophyllum ambiguum var. ambiguum

ray corolla, 1 mm

fruit, 1 mm; disk corolla, 1 mm; head, 5 mm; 1 cm
Eriophyllum ambiguum var. paleaceum

head, 2 mm; ray flower, 1 mm; disk flower, 1 mm; 1 cm
Eriophyllum confertiflorum var. confertiflorum

Eriophyllum congdonii — 1 cm; ray corolla, 2 mm; disk corolla, 1 mm; fruit, 2 mm

head, 5 mm; ray flower, 1 mm; disk corolla, 1 mm; fruit, 1 mm; lower leaf, 1 cm
Eriophyllum jepsonii

var. arachnoideum — 1 cm; disk flower, 1 mm; lower surface, 1 cm

head, 1 cm; fruit, 1 mm; var. obovatum; 2 cm; 1 cm
Eriophyllum lanatum

1′ Ray fls gen 0–4
  3. Ray fls 0; phyllaries ± equal or unequal, linear to lance-oblong, acute to acuminate
    4. Phyllaries ± equal; largest lvs 2–5 cm; pappus bristles in 1 series, all > 5 mm . . . . ***E. breweri***
    4′ Phyllaries unequal, outer << inner; largest lvs 5–9 cm; pappus bristles in 2 series, inner 5–10 mm, outer
      (at least a few) ± 1 mm . . . . ***E. vialis***
  3′ Ray fls gen > 0 at least on some heads; phyllaries unequal, outer << inner, linear-oblong to ovate, acute
    5. Lvs ± glabrous on faces . . . . ***E. glabratus***
    5′ Lvs ± glabrous adaxially, ± densely tomentose abaxially . . . . ***E. tomentellus***

E. breweri (A. Gray) G.L. Nesom **ST**: 1–10 dm, ± glandular or ± tomentose. **LF**: largest 2–5 cm, lance-linear to ovate, acute, entire to ± toothed, faces ± glabrous to glandular-hairy and/or tomentose. **INFL**: heads discoid, in cyme-like cluster; phyllaries ± equal, linear to lance-oblong, acute to acuminate, ± tomentose to glandular-hairy, base ± pale, midvein and tip green. **RAY FL**: 0. **FR**: pappus bristles in 1 series, all > 5 mm. 2*n*=18. Subalpine meadows, open woodland; 1200–3200 m. KR, CaRH, SNH, SnBr. [*Aster b.* (A. Gray) Semple] Jul–Sep

E. engelmannii (D.C. Eaton) Greene **ST**: 5–15 dm, ± glandular to ± hairy. **LF**: 4–10 cm, elliptic to ovate, ± acute, faces green, ± glabrous. **INFL**: heads in cyme-like cluster; phyllaries unequal, linear-oblong to ovate, acute, ciliate, pale, midvein green, tip gen purple-margined. **RAY FL**: gen 8 or 13; corolla 15–22 mm, white, ± becoming pink. 2*n*=18. Uncommon. Meadows, open woodland; 1800–2000 m. KR, NCoRH; to w Can, CO, NV. [*Aster e.* (D.C. Eaton) A. Gray] Jul–Sep

E. glabratus (Greene) Greene (p. 329) SISKIYOU ASTER **ST**: 3–6 dm, glabrous to ± glandular-hairy. **LF**: 3–6 cm, lanceolate to lance-ovate, gen obtuse, faces ± glabrous. **INFL**: heads discoid or 1–2-rayed, in cyme-like cluster; phyllaries unequal, oblong to narrowly ovate, acute, ± tomentose, ± ciliate, base ± white, midvein green, tip green to purple. **RAY FL**: 0–2(4), gen > 0 at least on some heads; corolla pale violet. Dry oak or conifer forest, rocky places; 700–2400 m. KR, NCoRI; s OR. Jul–Oct

E. ledophyllus (A. Gray) Greene var. **covillei** (Greene) G.L. Nesom (p. 329) **ST**: 2–8 dm, ± hairy and ± glandular. **LF**: 2–6(7) cm, elliptic to oblong, acute to obtuse, gen entire, adaxially ± glabrous, abaxially pale green, ± tomentose. **INFL**: heads radiate, in cyme-like cluster; peduncle ± glandular; phyllaries unequal, oblong to ovate, acute, ciliate, base pale, midvein green, tip ± purple. **RAY FL**: gen 5, 8, or 13(21); corolla 10–15 mm, violet to purple. 2*n*=18. Meadows, open woodland; 1300–2500 m. KR, NCoR; to s OR. Intergrades with *E. engelmannii* (D.C. Eaton) Greene. *E. ledophyllus* var. *l.* in OR to WA. Jul–Sep

E. tomentellus (Greene) Greene BRICKELLBUSH ASTER **ST**: 4–9 dm, ± minutely tomentose. **LF**: 3–6 cm, lanceolate to elliptic, gen obtuse, adaxially ± glabrous, abaxially ± densely tomentose. **INFL**: heads discoid or few-rayed, in cyme-like cluster; phyllaries unequal, oblong to narrowly ovate, acute, ± tomentose, ± ciliate, base ± white, midvein green, tip green to purple. **RAY FL**: (0)1–3(6), gen > 0 at least on some heads; corolla pale violet. Uncommon. Open oak or conifer woodland, rocky places; 1300–2400 m. KR, NCoRI; s OR. [*Sericocarpus t.* Greene; *Aster brickellioides* Greene; *E. b.* (Greene) G.L. Nesom] Jul–Oct

E. vialis Bradshaw WAYSIDE ASTER **ST**: 6–12 dm, ± glandular to ± hairy. **LF**: 5–9 cm, lance-elliptic, acute, abaxially glabrous to ± hairy, adaxially glandular. **INFL**: heads discoid, in cyme-like cluster; phyllaries unequal, linear to lance-oblong, acute to acuminate, ± glandular, base ± pale, midvein and tip green. **RAY FL**: 0. 2*n*=18. Meadows, open oak or conifer woodland; 455–1525 m. KR. [*Aster v.* (Bradshaw) S.F. Blake] Jul–Sep ★

EUCHITON COTTONLEAF, CUDWEED

Guy L. Nesom

Ann to per, gen fibrous-rooted, often from rhizome or stolon. **ST**: gen 1, erect, thinly and persistently white-tomentose. **LF**: basal and cauline, sometimes in rosettes, alternate, linear to (ob)lanceolate or spoon-shaped, narrowed or expanded at base, entire, adaxially gen green, ± glabrous, abaxially gen silvery-tomentose. **INFL**: heads disciform, gen in terminal cluster subtended by lfy bracts, sometimes with axillary clusters, rarely 1; involucre cylindric to narrowly bell-shaped; phyllaries graduated in 3–4+ series, gen straw-colored to ± brown, sometimes ± pink or purple, transparent, base not glandular, distally stiff-papery; receptacle flat, smooth, epaleate. **PISTILLATE FL**: 16–150; corolla narrowly tubular, ± purple or distally so. **DISK FL**: 1–7; corolla purple or distally so; anther base tailed, tip ± triangular; style tip truncate, hair-tufted. **FR**: obovate-elliptic, slightly flattened, faces with minute, club-shaped papillae or hairs; pappus bristles in 1 series, deciduous, free or basally fused. 17 spp.: native to Australasia, e Asia; some spp. widely naturalized. (Greek: true tunic, for bracts closely subtending head cluster) [Nesom 2006 FNANM 19:440–442]

1. Ann; taprooted; lf base not expanded; bracts subtending heads 4–8; heads in spheric clusters; disk fl 1 . . . . ***E. sphaericus***
1′ Bien or per; fibrous-rooted; cauline lf base ± expanded; bracts subtending heads 2–5; heads in hemispheric
  clusters; disk fls 3–5
  2. Stolons gen present; basal lvs in rosettes at fl; cauline lvs 2–4(6), blade 1–2 cm, 1–2 mm wide, linear to
    oblanceolate; bracts subtending heads 2–3, gen not surpassing heads; pistillate fls 40–60 . . . . ***E. gymnocephalus***
  2′ Stolons gen 0; basal lvs withering before fl; cauline lvs 6–10, 3–8 cm, 2–3 mm wide, gen ± linear; bracts
    subtending heads 3–5, surpassing heads; pistillate fls 80–150 . . . . ***E. involucratus***

E. gymnocephalus (DC.) Holub CREEPING CUDWEED Bien, per, fibrous-rooted; lfy stolons gen present, rooting at nodes. **ST**: 5–40 cm, simple or branched. **LF**: basal persistent in rosettes, petioled, 2–10 cm, blade oblanceolate to spoon-shaped, base narrowed; cauline sessile, 2–4(6), 1–2 cm, 1–2 mm wide, blade linear to oblanceolate, evenly sized, base ± expanded. **INFL**: head clusters hemispheric, 10–20 mm diam, sometimes axillary; bracts subtending heads 2–3, 6–10 mm, gen not surpassing heads, plus a few shorter ones; involucre 4–4.5 mm; phyllaries tawny or rosy-tinged, oblong, tips rounded to obtuse. **PISTILLATE FL**: 40–60. **DISK FL**: 3–5. **FR**: pappus bristles free. Openings in woodland, roadsides; < 800 m. KR, NCoRO; OR; native to New Zealand, Australia. [*Gnaphalium collinum* Labill., illeg.] May–Oct

E. involucratus (G. Forst.) Holub (p. 335) STAR COTTONLEAF Bien, per, fibrous-rooted, stolons gen 0. **ST**: 30–40 cm, erect, simple. **LF**: basal withering before fl; cauline 6–10, 3–8 cm, 2–3 mm wide, blade linear to linear-(ob)lanceolate, largest at mid-st, base

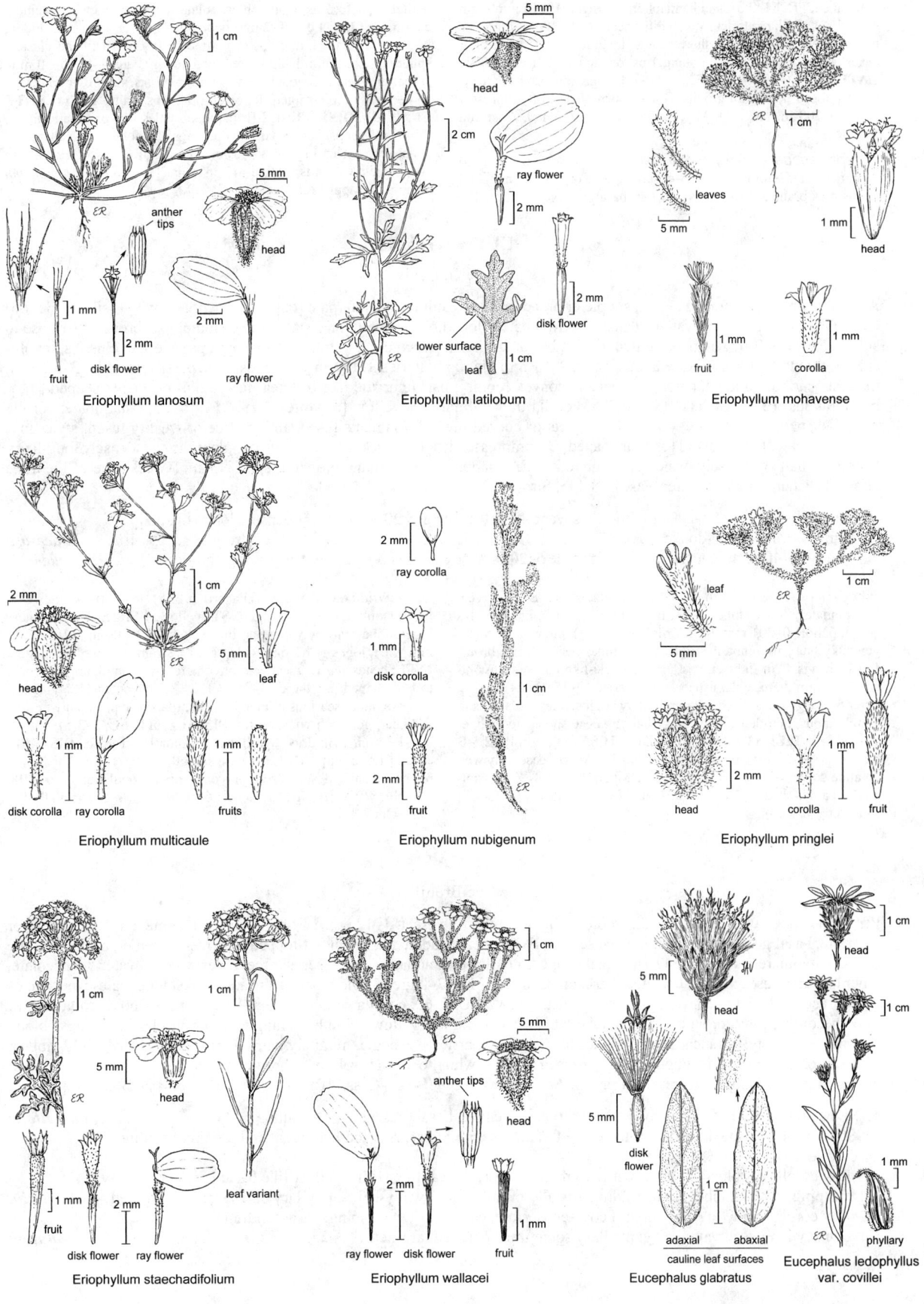

Eriophyllum lanosum

Eriophyllum latilobum

Eriophyllum mohavense

Eriophyllum multicaule

Eriophyllum nubigenum

Eriophyllum pringlei

Eriophyllum staechadifolium

Eriophyllum wallacei

Eucephalus glabratus

Eucephalus ledophyllus
var. covillei

± expanded. **INFL**: heads in hemispheric clusters, 10–15 mm diam, sometimes axillary; bracts subtending heads 3–5, 10–15 mm, surpassing heads, plus a few shorter ones; involucre 4–4.5 mm; phyllaries tawny or rosy-tinged, oblong, tips rounded to obtuse. **PISTILLATE FL**: 80–150. **DISK FL**: 3–5(7). **FR**: pappus bristles free or basally fused, falling in groups. Grassy open places, often moist or wet; 50–700 m. NCoRO; MA; native to New Zealand, Australia. Jul–Oct

E. sphaericus (Willd.) Holub (p. 335) GLOBE COTTONLEAF Ann, taprooted, stolons 0. **ST**: 5–80 cm, simple or basally branched, sometimes branched from lf axils. **LF**: basal and proximal cauline withering before fls, blade oblanceolate to spoon-shaped, 1-veined; cauline 8–12, 2–4 cm, 1–2 mm wide, blade linear, largest at mid-st, base not expanded, margins sometimes wavy. **INFL**: head clusters spheric, 10–20 mm diam; bracts subtending heads 4–8, 10–30 mm, surpassing heads; involucre 3.5–4 mm; phyllaries ± brown to tawny, sometimes purple-tinged, lance-elliptic, tips acute. **PISTILLATE FL**: 16–26. **DISK FL**: 1. **FR**: pappus bristles free or basally fused, falling in groups. 2*n*=28. Grassy openings in wooded areas, disturbed soils, recent clearings; 30–700 m. KR, NCoRO, NCoRH; OR; se Asia, Pacific islands; native to Australia. [*Gnaphalium japonicum* Thunb., misappl.] Jul–Oct

EUPHROSYNE

Bruce G. Baldwin

Ann to shrub, 5–200 cm. **ST**: erect, simple or branched. **LF**: cauline, gen alternate [opposite], petioled or ± sessile, blade (ob)lanceolate or oval to (ob)ovate or deltate, 1–3-pinnately lobed or dissected, or distal entire, minutely scabrous or strigose to silky-hairy [woolly], gen gland-dotted. **INFL**: heads disciform [discoid], 1, in tight groups, or in panicle-like clusters; involucre obconic to bell-shaped or ± hemispheric, 2–5+ mm diam; phyllaries 3–15+ in 1–3 series, free, persistent, outer 3–5 green, the rest scarious to membranous; receptacle convex [conic]; paleae bristle-like or linear to spatula- or wedge-shaped [0], ± membranous. **PISTILLATE FL**: [0]2–8; corolla 0 or cylindric, ± yellow [± white]. **DISK FL**: 5–25+, staminate; corolla funnel-shaped, ± white or ± yellow, lobes erect or curved inward; filaments fused, anthers free or weakly fused; style tip ± flared, truncate. **FR**: obovoid [or pear-shaped], ± compressed front-to-back, smooth or warty, glabrous or ± densely long-hairy [minutely hairy], sparsely or not gland-dotted in CA; pappus 0 or a minute crown. 5 spp.: N.Am. (Greek: one of the three Graces) [Panero 2006 Fam Gen Vasc Pl 8:445; Strother 2006 FNANM 21:29, 31–32]

1. Subshrub or shrub, 50–200 cm; branches wand-like; lf blade gen 20–150 mm; fr densely long-soft-hairy, sparsely or not gland-dotted . . . . *E. acerosa*
1′ Ann 5–25(40) cm; branches diffuse; lf blade 5–20 mm; fr glabrous or sparsely hairy, not gland-dotted. . . . *E. nevadensis*

E. acerosa (Nutt.) Panero (p. 335) COPPERWORT **ST**: ± green. **LF**: pinnately lobed, lobes 3–7+, linear to thread-like, or distal lvs entire; petiole 0–20+ mm; blade or lobes 1–1.5(3) mm wide, ± thick, gen silky-hairy to minutely strigose, sometimes becoming glabrous. **INFL**: heads 1, in tight groups, or in panicle-like clusters; involucre 4–5+ mm diam, ± hemispheric; phyllaries 10–15+ in 2–3 series, outer 5–7 ± green, silky-hairy to minutely strigose, inner scarious to membranous, gen densely long-soft-hairy; paleae spoon- to wedge-shaped. **PISTILLATE FL**: 5; corolla 0. **DISK FL**: corolla 2.5–3 mm. **FR**: 1.5–2.5 mm, plump, smooth. 2*n*=36. Wet or seasonally wet, alkaline soils; < 700 m. n DMoj (exc DMtns); to CO, NM. Poisonous to cattle, sheep. [*Oxytenia a.* Nutt.; *Iva a.* (Nutt.) R.C. Jacks.] Jun–Oct ★

E. nevadensis (M.E. Jones) Panero (p. 335) NEVADA WORMWOOD From stout taproot, ill-scented. **ST**: yellow. **LF**: gen 1–2 pinnately lobed, lobes narrowly ovate to linear; petiole 2–20 mm; blade 5–15 mm wide, lobes 1–1.5 mm wide, faces ± strigose, gen gland-dotted. **INFL**: heads gen 1, ± scattered; involucre 2–3+ mm diam, ± obconic to bell-shaped; phyllaries 3–8+ in 1–2+ series, outer 3+ ± green, ± strigose, inner scarious to membranous, glabrous; paleae bristle-like to linear, gen with widened tip. **PISTILLATE FL**: 2–3(5), corolla 0.5–1+ mm, cylindric, ± yellow, ± minutely strigose. **DISK FL**: corolla 1.5–2 mm. **FR**: 1.5–2 mm, smooth or warty. Alkaline, sandy plains, washes, desert scrub, pinyon/juniper woodland; 1200–2100 m. SNE; w NV. [*Iva n.* M.E. Jones; *Chorisiva n.* (M.E. Jones) Rydb.] May–Oct ★

EURYBIA ASTER

Luc Brouillet

Per from caudex or rhizome, fibrous-rooted. **ST**: decumbent to erect, 1–12 dm. **LF**: basal and/or cauline, alternate, entire or ± serrate, basal gen petioled, cauline gen sessile. **INFL**: heads radiate, (1 or) in flat-topped to cyme-, raceme-, or panicle-like cluster; involucre narrowly to broadly bell-shaped; phyllaries graduated in 3–7 series, free, outer widely ovate to lanceolate, inner linear, obtuse to acute, ± white at base, green in distal 1/3–3/4, green zone basally truncate, at least inner pale-papery-margined; receptacle flat to convex, glabrous, epaleate. **RAY FL**: 5–60; corolla white to violet. **DISK FL**: 8–260; corolla yellow, often becoming pink- or purple-tinged, tube > or < throat; anthers yellow, often becoming pink- or purple-tinged, tips oblong to ± triangular; style branches flat on inner face, base ± warty, tip ± acute, minutely hairy. **FR**: ± compressed, 7–12-ribbed, glabrous or ± hairy, ± brown; pappus of ± barbed bristles, white to ± pink, yellow or brown. 22 spp.: N.Am, Eurasia. (Greek: eurys, wide, and baios, few, for lvs or rays of *Eurybia macrophylla* (L.) Cass.) [Brouillet 2006 FNANM 20:365–382]

1. St distally glandular; disk corolla tube < throat; lvs entire, basal present at fl, > cauline . . . . *E. integrifolia*
1′ St hairy, not glandular; disk corolla tube ≥ throat; lvs entire to sharply toothed, basal and proximal cauline early-deciduous
   2. Ray corolla purple to violet; lf entire to minutely serrate, cauline with small earlike basal lobes or slightly clasping; sts decumbent to ascending, distally minutely soft-wavy-hairy; phyllary dark-purple-margined . . . *E. merita*
   2′ Ray corolla white to pale violet; lf coarsely serrate, cauline often clasping; sts ascending to erect, ± densely long-soft-wavy-hairy; phyllary sometimes purple-margined . . . . *E. radulina*

E. integrifolia (Nutt.) G.L. Nesom (p. 335) Rhizomes short, ± woody. **ST**: erect, 1.5–7 dm, distally glandular. **LF**: basal and cauline, 3–22 cm (distal smaller), entire, glabrous to ± hairy; basal present at fl, long-petioled, lance-ovate to obovate, acute or rounded; cauline lanceolate to oblanceolate, base ± expanded-clasping, tip acute. **INFL**: elongate, raceme- or panicle-like; phyllaries in 3–4 series, outer oblong, with spreading tips, gen acute, densely glandular, outer green ± to base, inner green at tip, ± purple-tinged. **RAY FL**: 8–27; corolla 10–15 mm, violet-purple. **DISK FL**: 20–50; corolla tube < throat. **FR**: ± densely short-hairy. 2*n*=18. Dry meadows, open forest; 1600–3200 m. KR, CaR, SN, Wrn; to WA, MT, WY, UT. [*Aster i.* Nutt.] Jul–Sep

E. merita (A. Nelson) G.L. Nesom SUBALPINE ASTER Rhizomes becoming ± woody. **ST**: decumbent to ascending, (0.2)1–5 dm, distally short-soft-wavy-hairy. **LF**: cauline (proximal early-deciduous) (1)2–8 cm, lanceolate to oblanceolate, base of mid-st lvs with small earlike lobes or slightly clasping, tip obtuse to acute, margin entire to minutely serrate, adaxially ± strigose or glabrous in age, abaxially sparsely long-soft-wavy-hairy on veins or glabrous in age. **INFL**: cyme-like, ± flat-topped; phyllaries in 4–5 series, outer oblong, acute to obtuse, green in distal 1/3–1/2, often ± purple, dark-purple-margined. **RAY FL**: (10)14–32; corolla 7–12(15) mm, purple to violet. **DISK FL**: 30–60; corolla tube ≥ throat. **FR**: finely strigose; pappus white to red-brown or ± yellow, 5–6 mm, > disk corolla. 2*n*=36. Montane forest; 1300–2000 m. KR; to BC, SD. [*Aster sibiricus* L. var. *m.* (A. Nelson) Raup] [Brouillet 2004 Sida 21:459–461] Jul–Aug ★

E. radulina (A. Gray) G.L. Nesom (p. 335) Rhizomes ± woody. **ST**: ascending to erect, 1–7 dm, ± densely long-soft-wavy-hairy. **LF**: cauline (proximal gen early-deciduous), middle 3–8(12) cm, ovate or elliptic to obovate, wedge-shaped or often clasping at base, obtuse- to acute-tipped, coarsely serrate, adaxially scabrous-strigose, abaxially scabrous. **INFL**: cyme-like, ± flat-topped; phyllaries in 4–5 series, outer oblong, ± acute, green in distal 1/3–1/2, sometimes purple-margined. **RAY FL**: 9–15; corolla 8–13 mm, white to pale violet. **DISK FL**: 30–70; corolla tube ≥ throat. **FR**: finely strigose; pappus tawny, 2.7–3 mm, ± = disk corolla. 2*n*=18, 27. Dry forest, oak/pine woodland, brushy slopes; 100–1600 m. NW, CaR, n&c SN, SnFrB, SCoR, n ChI; to BC. [*Aster r.* A. Gray] Jul–Sep

EUTHAMIA GRASS-LEAVED GOLDENROD

John C. Semple

Per from rhizome, ascending to erect, branched distally. **LF**: alternate, sessile, linear to lance-linear, entire, resin-dotted, 3–5-veined, margin finely scabrous. **INFL**: heads radiate, ± sessile in dense, sometimes flat-topped clusters; involucre ± ovoid; phyllaries graduated in 3–5 series, midrib gen ± swollen, translucent; receptacle convex, pitted, epaleate. **RAY FL**: ray yellow. **DISK FL**: corolla yellow, ± glabrous; anther tips narrowly triangular; style branches finely papillate, tips narrowly triangular. **FR**: fusiform; pappus bristles 25–45, in 1–2 series, outer bristles few, 1/2 inner. ± 8 spp.: N.Am. (Greek: well crowded, from dense infl) [Haines 2006 FNANM 20:97–100]

E. occidentalis Nutt. (p. 335) WESTERN GOLDENROD **ST**: < 2 m, smooth, sometimes ± white. **LF**: < 10 cm, ≤ 6(10) mm wide; proximal deciduous; middle largest. **INFL**: often large, panicle-like or ± flat-topped, ± resinous; branches ascending; involucre 3–5 mm, phyllaries in 3–4 series. **RAY FL**: 15–28; ray 1.5–2.5 mm. **DISK FL**: 6–18; corolla 3–4 mm. **FR**: 1 mm, strigose; pappus 2.5–3 mm, few outer-series bristles 1.5 mm. 2*n*=18. Marshes, streambanks, meadows; < 2300 m. CA; to w Can, n-c US, NM, n Baja CA. Jul–Nov

FILAGO

James D. Morefield

Ann (1)5–35[40] cm, ± gray to green, cobwebby to woolly. **ST**: 1, ± erect [0 or 2–10+, ascending to prostrate], forked at least distally, ± evenly lfy [lfless between distal forks]. **LF**: alternate, ± sessile [petioled], oblanceolate to ± spoon-shaped [to lanceolate], entire; distal lvs subtending heads ± crowded, largest > proximal lvs. **INFL**: heads disciform, ± sessile in groups of 8–20; involucre 0 or vestigial (simulated by paleae); receptacle length 5–15 × width, ± cylindric [club-shaped], glabrous; paleae, exc innermost, each ± enfolding pistillate fl, ± persistent [deciduous], lanceolate, acuminate to awned, abaxially cobwebby, obscurely parallel-veined, margin ± reflexed as indistinct, narrow, terminal scarious wing, wing indistinct, narrow, terminal, recurved [erect], visible in head; innermost paleae gen 5, collectively surrounding inner pistillate and disk fls, ≤ outer, ± ascending [spreading], ± concave, lanceolate, gen acuminate, papery, scarious, ± glabrous. **PISTILLATE FL**: [12]20–30[40+], all or [gen only] outer paleate; corolla obscure, narrowly cylindric. **DISK FL**: [1]5–9(11), bisexual; pappus present; corolla 4-lobed; anther base tailed, tip ± triangular; style tips ± linear-oblong. **FR**: ± club-shaped and ± compressed laterally to cylindric, rough or papillate [smooth], dull, corolla scar at [near] tip; outer fr scarcely > inner, falling free of palea; inner fr free of paleae; outer pistillate fl pappus 0, inner and disk fl pappus of [3]13–21 deciduous bristles (visible in head, gen falling together in complete or broken ring). 12 spp.: Eur, w Asia, n Afr, n Atlantic islands, some introduced ± worldwide. (Latin: with threads, for woolly hairs) [Morefield 2006 FNANM 19:447–449] Other taxa in TJM (1993) moved to *Logfia*.

F. pyramidata L. var. ***pyramidata*** BROADLEAF COTTONROSE **LF**: longest (10)15–20 mm, 3–5 mm wide; distal lvs 1.3–2 × heads, obtuse. **INFL**: heads only at st forks and tips, ± ellipsoid, 5-angled, longest 5–7 mm, 2.5–4 mm wide (groups dense, spheric, ± yellow, largest 11–15 mm diam); paleae in vertical ranks, longest 4.5–6 mm, outer keeled. **PISTILLATE FL**: pappus 0 in outer (15)20–25 fls, present in inner 2–7 fls. **DISK FL**: 5–9(11); corolla 2–3 mm, lobes gen ± brown to ± yellow. **FR**: 0.7–1 mm; pappus of inner fr 2–2.8 mm. 2*n*=28. Uncommon. Disturbed, often rocky places; 10–800 m. NCoRI, n CCo, n&c SnFrB; s BC, Australia; native to Medit. Apr–Aug

FLAVERIA

Ann [per, shrub]. **ST**: prostrate to erect. **LF**: simple, opposite, sessile or petioled. **INFL**: heads radiate or discoid, borne in stalked or sessile, open to condensed, head- or cyme-like clusters; peduncles 0 or slender; involucre ± cylindric; phyllaries 2–5; receptacle convex, epaleate. **RAY FL**: 0 or 1; corolla yellow or cream; ray inconspicuous. **DISK FL**: 0 or 1–2 [15]; corolla yellow; style tips flattened, obtuse. **FR**: 10-ribbed, ± flattened, glabrous, shining; pappus 0 [rarely of 2–4 scales]. 21 spp.: N.Am, S.Am, Australia. (Latin: yellow) [Yarborough & Powell 2006 FNANM 21:247–250]

F. trinervia (Spreng.) C. Mohr   Ann, often rounded. **ST**: 15–100 cm. **LF**: 3–15 cm, lanceolate, oblanceolate, elliptic or ovate, petioled or distal sessile; bases often fused around st; tip acute to obtuse; margin dentate or serrate; faces glabrous. **INFL**: heads in dense sessile, head-like clusters at forks of st; involucres ≤ 1 mm diam; phylla-ries gen 2, 4–4.5 mm, oblong, obtuse. **RAY FL**: 0–1; ray 0.5–1 mm, creamy yellow. **DISK FL**: 0–1(2); corolla 2–2.5 mm. **FR**: 2–2.6 mm, black; pappus 0. 2*n*=36. Moist soil in disturbed places, cult areas; < 300 m. SCo; native AZ to se US, S.Am. Sep–Dec

GALINSOGA

Ann. **ST**: gen erect, simple to much-branched. **LF**: simple, opposite, petioled; blade 3-veined. **INFL**: heads radiate [discoid], gen small, in lfy-bracted cyme-like clusters; peduncle slender; involucre bell-shaped; phyllaries in 2 series, free; receptacle conic, paleate, paleae of 2 kinds, outer fused in groups of 2–3 together with a phyllary around a ray fl, inner narrower, each subtending disk fl. **RAY FL**: [0]5–8, ray short, white. **DISK FL**: 8–50; corolla yellow; style tips acute. **FR**: obconic, round to ± angled; pappus of fringed scales [0]. ± 15 spp.: Am trop; some widespread weeds. (D. Mariano Martinez de Galinsoga, Spanish physician, 18th century) [Canne-Hilliker 2006 FNANM 21:180–182] *G. quadriradiata* Ruiz & Pav. not in CA.

G. parviflora Cav. var. *parviflora* (p. 335)   Ann, glabrous or sparsely soft-hairy, sometimes also glandular. **ST**: 10–60 cm, simple or much-branched. **LF**: petiole ≤ 2.5 cm; blade 1–11 cm, ± ovate, acute, finely dentate to coarsely serrate. **INFL**: head 2–6 mm diam; cluster round- to ± flat-topped; peduncle 1–40 mm; phyllaries and paleae persistent; outer phyllaries 2–4, 1.2–2.2 mm, 0.6–1.5 mm wide, margin scarious; inner phyllaries gen 5, 2.5–3.5 mm, 1.3–2.6 mm wide; inner paleae deeply 3-lobed. **RAY FL**: gen 5; ray 1–1.5 mm. **DISK FL**: 8–50; corolla 1.3–1.8 mm. **FR**: 1.2–2.5 mm, glabrous or strigose; pappus of ray fr of 5–8 unequal scales ≤ 1 mm; pappus of disk fr of 15–20 obtuse to acute scales ≤ 2 mm. 2*n*=16. Moist soil, disturbed ground; gen ≤ 1000 m. c SNH, CCo, SW, SNE; worldwide; native to S.Am. Other var. occurs in sw US, n Mex. All yr

GAMOCHAETA   CUDWEED
Guy L. Nesom

Ann to short-lived per; taprooted or fibrous-rooted. **ST**: erect to ± ascending. **LF**: basal and cauline, ± sessile, linear to oblanceolate or spoon-shaped, entire, base tapered to ± cordate, gen ± densely tomentose. **INFL**: heads disciform, gen in tight groups in gen spike- or panicle-like cluster, reduced to terminal cluster in small pls; involucre cylindric to ± urn-shaped, ± bell-shaped when pressed; phyllaries graduated in 3–7 series, bases stiff-papery to scarious, margins and tips transparent-membranous, often shiny, gen ± brown to straw-colored; receptacle flat, concave in fr, smooth, epaleate. **PISTILLATE FL**: pistillate 50–130; corolla narrowly tubular, all yellow or distally ± purple. **DISK FL**: 2–7; corolla yellow or distally ± purple; anther tip triangular; style tips truncate, hair-tufted. **FR**: oblong, slightly flattened, faces papillate; pappus bristles in 1 series, basally fused, deciduous in ring. ± 50 spp.: native to Am, some spp. widely naturalized. (Greek: united hair, for pappus bristles) [Nesom 2006 FNANM 19:431–438]

1. Lf faces contrasting in color, adaxially ± glabrous to ± cobwebby, abaxially densely white-felty tomentose
  2. Adaxial lf face ± glabrous; involucre ± purple, 2.5–3 mm, base glabrous; outer phyllaries elliptic-obovate
   to widely ovate-elliptic, rounded to obtuse; disk fls 2–3; fr 0.5–0.6 mm . . . . **G. coarctata**
  2′ Adaxial lf face ± sparsely cobwebby-tomentose; involucre ± brown, 4.5–5 mm, embedded in woolly
   hairs, base often sparsely cobwebby; outer phyllaries and flattened tips of inner dark or ± green-brown,
   widely ovate-triangular, acute to acuminate; disk fls (3)4–6; fr 0.7–0.8 mm . . . . **G. ustulata**
1′ Lf faces not or weakly contrasting in color, ± green or gray-green, loosely tomentose or cobwebby,
  sometimes ± felty, sometimes adaxially ± glabrous in *Gamochaeta stagnalis*
  3. Basal and proximal cauline lvs 4–16 mm wide; bracts among heads, cauline lvs oblanceolate to
   spoon-shaped, at least proximal bracts surpassing clusters of heads . . . . **G. pensylvanica**
  3′ Basal and proximal cauline lvs 2–9 mm wide; bracts among heads, mid to distal cauline lvs gen linear to
   ± oblanceolate, bracts surpassing heads or not
   4. Mid and distal cauline lvs oblanceolate, ± clasping; bracts among heads gen < clusters; involucre 3.5–4
    mm; phyllaries in 4–5 series . . . . [**G. stachydifolia**]
   4′ Mid and distal cauline lvs linear to lanceolate, oblanceolate, or spoon-shaped, not ± clasping; bracts
    among heads < or > than clusters; involucre 2.5–3.5 mm; phyllaries in 3–4(5) or 5–7 series
    5. Involucre 3–3.5 mm, not purple, base gen ± glabrous to occ sparsely cobwebby; phyllaries in 5–7
     series, outer ovate-triangular, 1/3–1/2 × inner, acute-acuminate; fl Jun–Jul . . . . **G. calviceps**
    5′ Involucre 2.5–3 mm, gen ± purple proximally, base cobwebby; phyllaries in 3–4(5) series, outer
     ovate-triangular, 1/2–2/3 × inner, acute; fl Mar–Apr . . . . **G. stagnalis**

G. calviceps (Fernald) Cabrera   Ann, tap- or fibrous-rooted. **ST**: erect-ascending, 8–45(55) cm, gen branched throughout, felty-tomentose, hairs silver-gray, longitudinally arranged. **LF**: gen cauline, basal gen withering before fl, gen 2–6 cm, 2–9 mm wide, oblanceolate to spoon-shaped, distal becoming linear to linear-oblanceolate, blade gen folded along midvein, faces not or weakly contrasting in color, felty-tomentose, hairs closely appressed. **INFL**: initially 2–4 cm, 8–12 mm wide, continuous or at least distally interrupted, spike-like when pressed, later 4–18 cm, panicle-like, main axes gen visible between heads, peduncles gen evident; involucre 3–3.5 mm, bell-shaped, base sparsely cobwebby or gen ± glabrous; phyllaries in 5–7 series, outer ovate-triangular, 1/3–1/2 × inner, acute-acuminate, inrolled and spreading to recurved, inner oblong, flattened, obtuse, abruptly pointed, slightly brown (not purple). **PISTILLATE FL**: corolla tip purple. **DISK FL**: 2–4; corolla tip purple. **FR**: 0.4–0.5 mm. Disturbed sites; 50–800 m. n SNF, SnJV, SnFrB, SnGb, PR; S.Am, Eur, New Zealand; native to se US. Jun–Jul

G. coarctata (Willd.) Kerguélen   Ann, bien, fibrous-rooted. **ST**: decumbent-ascending, 15–35(50) cm, white, felty-tomentose, mat-

ted hairs gen sheath-like. **LF**: basal, not withered at fl, and cauline, gen 3–8 cm, 6–15 mm wide, spoon-shaped to oblanceolate-obovate, slightly fleshy, often minutely crenate when dry, adaxially ± glabrous, abaxially white, closely felty-tomentose. **INFL**: 2–20 cm, 10–14 mm wide when pressed, gen dense, continuous, spike-like, becoming branched, interrupted; involucre 2.5–3 mm, cylindric or urn-shaped to bell-shaped, ± purple, base glabrous; phyllaries in 4–5 series, outer elliptic-obovate to widely ovate-elliptic, 1/4–1/3 × inner, ± purple or rosy, tip rounded to obtuse, inner oblong, flattened, rounded to obtuse or blunt, abruptly pointed, distally brown-membranous. **PISTILLATE FL**: corolla gen ± purple-tipped. **DISK FL**: 2–3; corolla gen ± purple-tipped. **FR**: 0.5–0.6 mm. 2*n*=28,40. Disturbed areas; < 1050 m. NCoRO, n SNF, SnJV; widely naturalized worldwide; native to S.Am. May–Aug

G. pensylvanica (Willd.) Cabrera   Ann, taprooted. **ST**: decumbent to erect, 10–50 cm, loosely cobwebby-tomentose. **LF**: basal and cauline, proximal gen present at fl, 2–7 cm, 4–16 mm wide, spoon-shaped to oblanceolate-obovate, narrowed to petiole-like base, wavy-margined, often abruptly pointed, becoming spoon-shaped to oblanceolate, faces not or weakly contrasting in color, loosely tomentose; bracts among proximal heads surpassing clusters. **INFL**: 1–12 cm, 10–15 mm wide when pressed, continuous or interrupted, spike-like; involucre 3–3.5 mm, cup-like or bell-shaped, base sparsely cobwebby; phyllaries in 3–4 series, outer ovate-triangular, 1/2–2/3 × inner, abruptly pointed, inner oblong, flattened, acute to obtuse, often proximally purple-tinged, distally transparent, sometimes golden. **PISTILLATE FL**: corolla gen ± purple-tipped. **DISK FL**: 3–4; corolla ± purple-tipped. **FR**: 0.4–0.5 mm. 2*n*=28. Disturbed areas; < 650 m. SnJV, CCo, SCo, TR, PR; widely naturalized worldwide; native(?) to S.Am. Mar–Jul

G. stagnalis (I.M. Johnst.) Anderb.   Ann, gen taprooted, sometimes fibrous-rooted. **ST**: erect to decumbent-ascending, 2.5–20(35) cm; densely, loosely cobwebby-tomentose. **LF**: basal gen withering before fl, gen cauline, 1–2.5(3) cm, 2–6 mm wide, gen oblanceolate to oblong-oblanceolate, faces not or weakly contrasting in color, both loosely tomentose or adaxial greener, becoming ± glabrous. **INFL**: ± 1 cm, head-like in smallest pls or 1–3(12) cm, 8–12 mm wide when pressed, interrupted, spike-like, sometimes branching at proximal nodes, clusters subtended by lf-like bracts; involucre 2.5–3 mm, bell-shaped, base sparsely cobwebby; phyllaries in 3–4(5) series, outer ovate-triangular, 1/2–2/3 × inner, widely acute, inner oblong, flattened, rounded-obtuse, gen purple, distally ± white. **PISTILLATE FL**: corolla ± purple-tipped. **DISK FL**: (2)3(4); corolla ± purple-tipped. **FR**: 0.3–0.5 mm. Disturbed areas, often moist; < 300 m. SCo, PR; AZ, NM; nw&c Mex. Mar–Apr

G. ustulata (Nutt.) Holub (p. 335)   Ann to bien or short-lived per, fibrous-rooted. **ST**: erect to ascending, 10–40 cm, gen decumbent-ascending and stolon-like, white, densely felty-tomentose. **LF**: basal and cauline, basal gen withering before fl, 2–5 cm, 6–12(35) mm wide, spoon-shaped to oblanceolate, faces contrasting in color, adaxially ± sparsely cobwebby-tomentose, abaxially white, felty-tomentose. **INFL**: 1–6(8+) cm, 12–18 mm wide when pressed, gen continuous, ± cylindric; involucre 4.5–5 mm, urn-shaped, base embedded in woolly hairs, often sparsely cobwebby on proximal 1/5–1/2; phyllaries in 4–6 series, outer widely ovate-triangular, ± 1/2 × inner, acute to acuminate, brown or ± green-brown, mid phyllaries ± keeled near tip, inner oblong, flattened, rounded to obtuse, abruptly pointed, gen dark brown, sometimes proximally ± purple. **PISTILLATE FL**: corolla gen ± yellow, sometimes ± purple-tipped. **DISK FL**: (3)4–6; corolla gen ± yellow, sometimes ± purple-tipped. **FR**: 0.7–0.8 mm. Dunes, bluffs, fields, disturbed sites; < 650(1050) m. NW, c SNF, SnJV, w CW, SCoRO, SCo, ChI, PR; to BC. [*Gnaphalium purpureum* L., misappl.] Apr–Jul

GAZANIA   TREASURE FLOWER

Alison M. Mahoney, Robert J. McKenzie & Elizabeth McClintock

Per [ann, shrub], gen cespitose; juice milky. **LF**: basal [cauline], margin ± entire to dentate or pinnately lobed. **INFL**: scapose; heads radiate, showy, 1; phyllaries in 2–4 series, fused proximally; receptacle conic or convex, pitted, epaleate. **RAY FL**: sterile; ray yellow or orange, variously marked, closing in shade and at night; style 0. **DISK FL**: corolla gen yellow or ± orange; anther base minutely sagittate, tip ovate-triangular; style slender proximally, thickened distally to a minutely hairy node, branches minute. **FR**: obovoid, long-hairy; pappus of slender scales ± hidden by hairs. ± 20 spp.: esp s Afr. (Greek: riches or royal treasure, or for Theodorus of Gaza, died 1478, translator of Theophrastus' works) [Mahoney 2006 FNANM 19:196–197] Local, not naturalized populations may be other spp. of *Gazania* or hybrid cultivars involving *G. krebsiana* Less.

G. linearis (Thunb.) Druce   **ST**: short, decumbent. **LF**: in loose rosettes; blades of 2 kinds or not, 10–20(38) × 0.6–1 cm, linear to lanceolate, entire, or 2.5–5 cm wide, oblanceolate to oblong and pinnately-lobed, gradually narrowed to winged petiole, ultimate margin gen entire, occ prickly, rolled under, abaxially white-woolly, adaxially ± dark green, glabrous to cobwebby. **INFL**: heads 3.5–8 cm diam; phyllaries ± pouched at base. **RAY FL**: 13–18; ray 4–5 cm, yellow or orange, gen with dark abaxial stripe and proximal adaxial spot. **DISK FL**: corolla ± 8 mm, ± red-orange. **FR**: 1–2 mm; pappus scales 7–8, 3–4 mm, hidden by hairs. Uncommon. Escape from cult in coastal areas, disturbed areas; < 1350 m. CCo, SnFrB, SCo, PR; NM?; native to s Afr. All yr ❖

GERAEA

David J. Keil & Curtis Clark

Ann, per. **ST**: erect; branches ascending. **LF**: basal and alternate, simple, sessile or petioled, entire or dentate, 3-veined from base. **INFL**: heads radiate or discoid, 1 or in few- to many-headed cyme-like or panicle-like cluster; peduncle ± elongated; involucre hemispheric; phyllaries ± equal or weakly graduated in 2–3 series, free, green; receptacle paleate, palea folded around and falling with fr. **RAY FL**: 0 or 10–21, sterile; style 0; ray yellow. **DISK FL**: many; corolla yellow, tube slender, throat gradually expanded, lobes triangular; anther tips ovate, ± acute; style tips triangular. **FR**: strongly compressed, narrowly wedge-shaped; edges ± white, long-ciliate; faces black, ± hairy; pappus of 2 narrow awns. 2 spp.: sw US, nw Mex. (Greek: old, from white-haired involucre) [Clark 2006 FNANM 21:122]

1. Rays present; ann from taproot; lvs sessile or tapered to winged petiole, ± canescent, bases not ear-like; phyllaries acute, ciliate . . . . ***G. canescens***
1′ Rays 0; per from an underground caudex; lvs sessile, glandular-puberulent, bases ear-like; phyllaries obtuse, glandular . . . . ***G. viscida***

G. canescens Torr. & A. Gray (p. 335) DESERT-SUNFLOWER Herbage bristly or soft-hairy. **ST**: 1–8 dm, simple to openly much-branched. **LF**: 1–10 cm; blade lanceolate or ovate to elliptic or oblanceolate, acute. **INFL**: heads radiate, 1 or few to many in panicle-like cluster; involucre 7–12 mm; phyllaries linear to narrowly lanceolate. **RAY FL**: 10–21; ray 1–2 cm. **DISK FL**: corolla 4–5 mm. **FR**: 6–7 mm; pappus awns 3–4 mm. 2*n*=36. Sandy desert soils; < 1300 m. D; to sw UT, w AZ, n Mex. Occ hybridizes with *Encelia farinosa*. Jan–May, Sep–Nov

G. viscida (A. Gray) S.F. Blake (p. 335) STICKY GERAEA Herbage densely glandular-puberulent and ± bristly. **ST**: several from caudex, 3–10 dm, simple or few-branched. **LF**: 3–9 cm; blade ovate to oblong, obtuse. **INFL**: heads discoid, 1 or several in ± flat-topped cyme-like cluster; involucre 10–15 mm; phyllaries narrowly lance-oblong. **DISK FL**: corolla 6–8 mm. **FR**: 7–10 mm; pappus awns 3–5 mm. 2*n*=36. Openings in chaparral; 450–1700 m. s PR (s San Diego Co.); nw Baja CA. May–Jul ★

GLEBIONIS

Ann. **ST**: erect, gen branched. **LF**: alternate, sessile, toothed or 1–3 pinnately lobed or dissected. **INFL**: heads radiate, peduncled; phyllaries gen in 3 series, overlapping, widely scarious-margined; receptacle convex, epaleate. **RAY FL**: 13–21; ray yellow or cream with yellow base. **DISK FL**: 60–150+; corolla yellow; anther base rounded or ± cordate, tip ovate; style branches truncate with brush-like tips. **FR**: ray fr 3-angled, 2–3-winged; disk fr angled, ribbed, and adaxially winged or ± cylindric and ribbed, pappus 0. 2 spp.: Medit, sw Asia. (Derivation unknown) [Strother 2006 FNANM 19:554–555] Previously treated in *Chrysanthemum*, a name recently conserved for other pls.

1. Lvs deeply 2–3-pinnately dissected; ray fr 3-winged; disk fr 4-angled with adaxial angle winged, faces ribbed . . . . ***G. coronaria***
1′ Lvs simple, toothed or ± shallowly 1(2)-pinnately lobed; ray fr 2-winged; disk fr ± cylindric or obovoid, not winged, 10-ribbed . . . . ***G. segetum***

G. coronaria (L.) Spach GARLAND OR CROWN DAISY **ST**: ≤ 10 dm. **LF**: ≤ 8 cm, obovate. **INFL**: head 2–6 cm diam. **RAY FL**: ray yellow or cream with yellow base. **DISK FL**: corolla 4–5 mm. **FR**: ray fr ≤ 3 mm, wings lateral and adaxial; disk fr ± 2.5 mm. 2*n*=18. Escape in coastal areas, roadsides, disturbed areas; < 1000 m. w CW, SW; native to Medit. [*Chrysanthemum c.* L.] If recognized taxonomically, white-rayed pls assignable to *G. coronaria* var. *discolor* (d'Urv.) Turland. Mar–Jul ❖

G. segetum (L.) Fourr. (p. 335) CORN CHRYSANTHEMUM **ST**: ≤ 8 dm. **LF**: ≤ 6 cm, oblong to obovate. **INFL**: heads 2.5–6 cm diam. **RAY FL**: ray yellow. **DISK FL**: corolla 3–4 mm. **FR**: ray fr ± 2 mm, ± compressed, wings lateral; disk fr ± 2 mm. 2*n*=18. Escape in coastal areas, fields, sea bluffs; < 200 m. NCo, n CCo, SCo (uncommon); native to Eur, sw Asia. [*Chrysanthemum s.* L.] Apr–Oct

GLYPTOPLEURA
David J. Keil & G. Ledyard Stebbins

Ann. **ST**: 1–many, ± prostrate, densely tufted. **LF**: basal and cauline, alternate, crowded, lower ± oblanceolate, lobed, lobes rounded, toothed, margins ± white, crust-like. **INFL**: heads liguliflorous, gen 1(2–3 in cyme-like clusters); involucre cylindric to urn-shaped; phyllaries in 2 series, outer few, linear, with expanded tips with crust-like margins; inner 7–12, equal, narrow; receptacle ± flat, epaleate. **FL**: 7–18; ligule white to pale yellow, becoming pink-purple when dry, readily withering. **FR**: 4–5 mm, ± cylindric, often curved, obtusely 5-angled, abruptly short-beaked; ribs roughened, alternating with 5 rows of pits; pappus of many white bristles, outer falling separately, inner ± persistent. 2 spp.: w N.Am. (Greek: carved rib, from sculptured fr) [Keil 2006 FNANM 19:361–362]

1. Ligules 4–10 mm, equaling involucre or exserted 1–5 mm . . . . ***G. marginata***
1′ Ligules 15–25 mm, exserted 10–20 mm beyond involucre . . . . ***G. setulosa***

G. marginata D.C. Eaton **ST**: 1–6 cm. **LF**: 0.5–5 cm, ± white crust-like margins conspicuous. **INFL**: margins of outer phyllaries crust-like throughout. **FL**: 9–18; ligule white to cream. 2*n*=18. Sandy or rocky deserts, often in alkaline soils; 600–2100 m. s SNH, GB, DMoj; to OR, ID, UT. Apr–Jul

G. setulosa A. Gray (p. 339) **ST**: 2–6 cm. **LF**: 1–6 cm, ± white, crust-like margins narrow. **INFL**: margins of outer phyllaries gen crust-like only at tips. **FL**: 7–14; ligule cream to pale yellow. Local on sandy desert flats, rocky soil, arid grassland, creosote-bush scrub; 400–1400 m. DMoj; to UT, AZ. Incl in *G. marginata* in TJM (1993). Mar–Jun

GNAPHALIUM CUDWEED
Guy L. Nesom

Ann, tap- or fibrous-rooted. **ST**: gen 1, erect, branched from base; ± woolly-tomentose, not glandular. **LF**: most cauline, alternate, oblanceolate to spoon-shaped [linear], entire, base ± wedge-shaped, faces gray-tomentose. **INFL**: heads disciform, gen in head-like axillary groups, sometimes in spike-like clusters; involucre ± bell-shaped; phyllaries in 3–5 series, gen white to brown, opaque or not, often shiny; phyllary base gen glandular distally, ± equal to unequal, stiff-papery toward tips, inner phyllaries protruding distal to outer. **PISTILLATE FL**: 40–130; corolla ± white to ± purple. **DISK FL**: 4–7, corolla ± white to ± purple. **FR**: oblong, gen glabrous, sometimes minutely papillate; pappus bristles in 1 series, free, deciduous. ± 38 spp.: Am, Asia, Afr, Australia. (Greek: downy plant, ancient name for these or similar plants) [Nesom 2006 FNANM 19:428–430] Other spp. in TJM (1993) moved to *Euchiton*, *Gamochaeta*, and *Pseudognaphalium*.

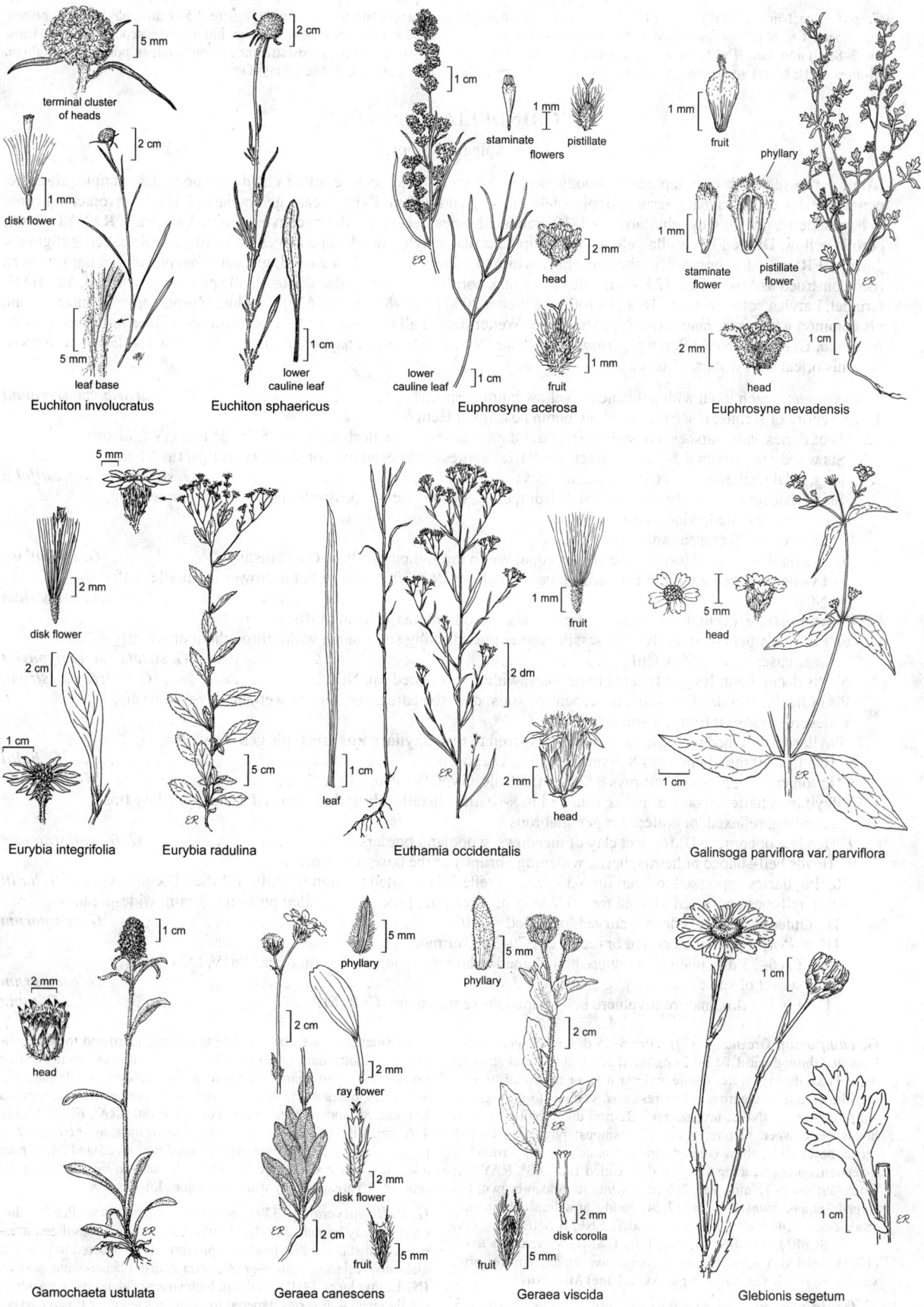

terminal cluster of heads

disk flower

leaf base

Euchiton involucratus

lower cauline leaf

Euchiton sphaericus

staminate flowers

pistillate

head

lower cauline leaf

fruit

Euphrosyne acerosa

fruit

phyllary

staminate flower

pistillate flower

head

Euphrosyne nevadensis

disk flower

Eurybia integrifolia

Eurybia radulina

leaf

fruit

head

Euthamia occidentalis

head

Galinsoga parviflora var. parviflora

head

Gamochaeta ustulata

phyllary

ray flower

disk flower

fruit

Geraea canescens

phyllary

disk corolla

fruit

Geraea viscida

Glebionis segetum

G. palustre Nutt. (p. 339) **ST**: (1)3–15(30) cm, branches gen decumbent. **LF**: blade spoon-shaped to oblanceolate-oblong, 1–3.5 cm, 3–8(10) mm wide. **INFL**: head-like clusters at st tips and in distal-most axils; bracts subtending heads 4–12 mm, 1.5–4 mm wide, oblanceolate to obovate; involucre 2.5–4 mm; phyllaries ± brown, base woolly, inner with opaque, blunt tips; receptacle flat, epaleate. $2n$=14. Arroyos, sandy streambeds, pond edges, potholes; < 3000 m. CA; Can, w US, Mex. May–Oct

GRINDELIA GUMPLANT

Abigail J. Moore

[Ann] per to subshrub from taproot or woody caudex, glabrous or tomentose, often glandular-sticky. **LF**: simple, alternate, gen not fleshy, entire, crenate, serrate, or pinnately lobed, gland-dotted. **INFL**: heads gen radiate (discoid); involucre obconic to hemispheric, gen gummy; phyllaries in 4–10 graduated series; receptacle flat to convex, ± pitted, epaleate. **RAY FL**: 0–60; corolla yellow. **DISK FL**: corolla yellow; anther tip lanceolate; style-branch appendages linear to lanceolate, gen ≥ stigmatic portion. **FR**: cylindric or swollen-obconic, shiny-white to ± brown, smooth or ridged, glabrous; pappus of 1–6 narrow awns (occ construed as bristle-like) [25–40 bristles], ± < disk corolla, gen entire, deciduous. ± 60 spp.: c&w N.Am, S.Am. (D.H. Grindel, Latvian botanist, 1776–1836) [Strother & Wetter 2006 FNANM 20:424–436] Variable. Morphologically intermediate pls common where spp. ranges overlap. Strother & Wetter treated all CA spp. exc *G. squarrosa* and *G. fraxinipratensis* in *G. hirsutula*. *G. ciliata* (Nutt.) Spreng. [*Prionopsis ciliata* (Nutt.) Nutt.], with pappus of many united bristles (falling as a unit), is an historical waif from n CCo, n SCo.

1. Lvs crenate, each tooth with a distinct, ± yellow bump near tip . . . . **G. squarrosa** var. **serrulata**
1′ Lvs entire or serrate; if serrate, ± yellow bump near tip of teeth 0;
  2. Pls of dunes, salt marshes, coastal bluffs, tidal flats, sloughs; lvs ± fleshy; NCo to SCo, deltaic GV (Suisun)
   3. Sts woody in proximal 3–15 dm, erect; phyllaries appressed to head exc for short, erect tips, tip < 3 mm long; tidal wetlands; CCo (San Francisco Bay) . . . . **G. stricta** var. **angustifolia**
   3′ Sts herbaceous or woody in proximal 0–1 dm, erect, decumbent, or prostrate; phyllaries gen spreading, recurved, or coiled; widespread
    4. Sts erect, 6–20 dm; salt marshes, sloughs
     5. Lvs on fl sts gen widest at base or of ± equal width throughout; Deltaic GV (Suisun) . . . . **G. ×paludosa**
     5′ Lvs on fl sts gen widest at rounded tip or ± equally wide at tip and base but narrower in middle of lf; NCo . . . . ²**G. stricta** var. **stricta**
    4′ Sts prostrate, decumbent, or erect, if erect, sts 1–6 dm; dunes, coastal bluffs
     6. Pls decumbent or erect; lvs gen sessile, sometimes clasping sts, ± same width throughout or widest near base; NCo to SCo, ChI . . . . **G. stricta** var. **platyphylla**
     6′ Pls decumbent; lvs gen tapered to petioles, widest at rounded tip; NCo . . . . ²**G. stricta** var. **stricta**
2′ Pls of fields, grassland, woodland, serpentine soils, disturbed areas, or interior wetlands; lvs not fleshy; widespread (absent from Suisun delta)
  7. Phyllaries flattened throughout, gradually tapered to tips; phyllary tips erect; pls gen ± hairy
   8. Heads 7–10 mm diam; rays 8–9 mm; PR (San Diego Co.) . . . . ²**G. hallii**
   8′ Involucres 7–25 mm diam; rays 8–20 mm; NCoR, GV, CW, SCo, TR . . . . **G. hirsutula**
  7′ Phyllaries flattened only at bases, rounded in ×-section distally, abruptly narrowed to tips; phyllary tips spreading, reflexed, or coiled; pls gen glabrous
   9. Heads ± obconic; e DMoj, wet clay of meadows, woodland borders . . . . **G. fraxinipratensis**
   9′ Heads bell-shaped or hemispheric, widening abruptly at the base; widespread
    10. Phyllaries appressed to head for > 3/4 length, reflexed or coiled portion < 3 mm; PR (San Diego Co.) . . . ²**G. hallii**
    10′ Phyllaries appressed to head for < 1/2 length, spreading, reflexed, or coiled portion > 5 mm; widespread
     11. Outer phyllaries reflexed, curved, or coiled < 270° . . . . ²**G. camporum**
     11′ > 75% of phyllaries coiled or recurved 270–360° or more
      12. Pls 6–25 dm; mature involucre bell-shaped to hemispheric, 15–22 mm diam; SnJV, SCo (geog subset of sp.) . . . . ²**G. camporum**
      12′ Pls 1–5 dm; mature involucre bell-shaped, 7–12 mm diam; CaR, MP . . . . **G. nana**

G. camporum Greene (p. 339) Per 6–25 dm, erect, gen much-branched throughout. **LF**: 2–15 cm; basal gen 0 at fl, distal smaller; blade lanceolate to ovate, sessile and clasping or narrowed to petiole-like base, gen glabrous, often resinous, yellow- to gray-green, entire or serrate. **INFL**: involucre 10–22 mm diam, bell-shaped to hemispheric when mature, glabrous, resinous; phyllaries in 5–7 series, bases wide, straw-colored, tips green, acuminate, ± round in ×-section, outer spreading to reflexed, or coiled 180–360°. **RAY FL**: (0)25–39; ray 5–11 mm. **FR**: 2–5 mm, white to golden-brown, top ridged; pappus awns 2–6. $2n$=12,24. Sandy or saline bottomland, roadsides; < 1400 m. KR, NCoR, CaRF, SNF, n SNH, Teh, GV, SnFrB, SCoRO, SW, SNE, DMoj; Baja CA. [*G. c.* var. *bracteosa* (J.T. Howell) M.A. Lane; *Grindelia camporum* var. *bracteosum*, orth. var.; *G. hirsutula* var. *davyi* (Jeps.) M.A. Lane] May–Nov

G. fraxinipratensis Reveal & Beatley ASH MEADOWS GUMPLANT Per 5–12 dm, erect, branched throughout. **LF**: 1–8 cm; basal 0 at fl, distal smaller; blade oblanceolate to oblong, narrowed to base, glabrous, resinous, dark green to yellow-green, entire or serrate. **INFL**: involucre 5–9 mm diam, ± obconic, glabrous, resinous; phyllaries in 4–7 series, bases wide, straw-colored, tips green, acuminate, ± round in ×-section, spreading to reflexed or coiled 180°. **RAY FL**: 8–12; ray 4–6 mm. **FR**: 2.5–4 mm, white to golden-brown, top gen truncate; pappus awns 2. $2n$=24. Wet clay of meadows, woodland edges near alkaline springs; ± 700 m. e DMoj; NV. [*Grindelia fraxino-pratensis*, orth. var.] Threatened by water diversion. Jul–Oct ★

G. hallii Steyerm. (p. 339) SAN DIEGO GUMPLANT Per 2–6 dm, erect, openly branched in distal 50%. **LF**: 1–12 cm; basal gen present at fl, distal smaller; blades of proximal lvs narrowed to petioles, glabrous, resinous, yellow-green, serrate, distal lance-ovate, sessile. **INFL**: involucre 7–10 mm diam, hemispheric, glabrous, ± resinous; phyllaries in 4–6 series, tapered to acute tips, erect or tips recurved. **RAY FL**: 12–20; ray 8–9 mm. **FR**: 4–5 mm, tan to brown, top trun-

cate with triangular projections; pappus awns 2. 2*n*=12. Meadows, dry slopes, open pine/oak woodland; 800–1700 m. PR (San Diego Co.). [*G. hirsutula* var. *hallii* (Steyerm.) M.A. Lane] Jul–Oct ★

G. hirsutula Hook. & Arn. (p. 339) Per 2–15 dm, erect, few-branched, side branches gen not branched. **LF:** 1–10 cm; basal gen present at fl, distal not much smaller; blade oblong to lanceolate, basal sometimes lobed, glabrous or tomentose, gen not resinous, yellow-, red-, or gray-green, base narrowed to ± sessile, margin entire or serrate. **INFL:** involucre 7–25 mm diam, bell-shaped to hemispheric, glabrous or more often tomentose, resinous or not; phyllaries in 4–5 series, gradually tapered to acute tips, tip flat in ×-section, erect, outer gen green throughout. **RAY FL:** 10–60; ray 8–20 mm. **FR:** 2.5–5.5 mm, golden- to red-brown, top truncate to knobby; pappus awns 2–4. 2*n*=24. Sandy, clay, or serpentine slopes or roadsides; < 1700 m. NCoR, GV, CW, SCo, TR. [*G. h.* var. *maritima* (Greene) M.A. Lane] *G. hirsutula* tends to fl earlier than *G. camporum* at a given location. Apr–Jun

G. nana Nutt. (p. 339) Per 1–5 dm, decumbent to erect, branched throughout. **LF:** 3–9 cm; basal gen 0 at fl, distal not much smaller; blade oblanceolate, glabrous, resinous, yellow- to gray-green, base gen tapered, margin entire or serrate. **INFL:** involucre 7–12 mm diam, bell-shaped when mature, glabrous, resinous; phyllaries in 5–7 series, bases wide, straw-colored, tips green, acuminate, ± round in ×-section, coiled 270–360°. **RAY FL:** 11–28; ray 5–11 mm. **FR:** 3.5–4 mm, light brown, top ridged; pappus awns 2. 2*n*=12. Dry, sandy hills, roadsides; 100–1800 m. CaR, MP; to WA, MT. Jun–Sep

G.* ×*paludosa Greene SUISUN GUMPLANT Per 8–20 dm, erect, branched throughout. **LF:** 1–17 cm; basal gen 0 at fl, distal smaller; blade lance-ovate, ± fleshy, sessile or tapered to ± petiole-like base, glabrous, green to red-green, entire or serrate. **INFL:** involucre 10–20 mm diam, hemispheric, glabrous, gen resinous; phyllaries in 4–5 series, bases wide, straw-colored, tips acute to acuminate, flat to ± round in ×-section, outer spreading, reflexed, or coiled 270–360°. **RAY FL:** 20–30; ray 10–17 mm. **FR:** ± 4 mm, tan, top truncate; awns 2–5. Salt marshes, banks of sloughs; < 30 m. Deltaic GV (Suisun). Putative stabilized hybrid between *G. camporum* and *G. stricta* var. *angustifolia*. Jul–Nov

G. squarrosa (Pursh) Dunal var. ***serrulata*** (Rydb.) Steyerm. (p. 339) Bien 1–6 dm, decumbent to erect, much-branched through-out. **LF:** 1.5–7 cm; basal lvs gen 0 at fl, distal not much smaller; blade oblong to ovate, sessile or narrowed at base, glabrous, resinous, gray-green, crenate, each tooth with a ± yellow bump near tip. **INFL:** involucre 10–17 mm diam, bell-shaped, glabrous, resinous; phyllaries in 5–6 series, bases wide, straw-colored, tips green, acuminate, ± round in ×-section, coiled 360°. **RAY FL:** 0 or 24–36; ray 8–10 mm. **FR:** 2.3–3 mm, light brown to ± yellow, top truncate; pappus awns 2–3(6). 2*n*=12. Disturbed roadsides, streamsides; 700–2300 m. CaRH, SNH, TR, GB, DMoj; native WY to NM. TOXIC, concentrates selenium. Jul–Sep

G. stricta DC. Per or subshrub, branched throughout. **LF:** 1–15 cm; basal present at fl or not, distal not much smaller; blade oblong to lanceolate, ± fleshy, sessile or narrowed at base, glabrous or sparsely tomentose, green or red-veined, serrate. **INFL:** involucre 10–45 mm diam, hemispheric, glabrous or tomentose, resinous; phyllaries in 4–6 series, bases wide, straw-colored, tips green, erect, reflexed, spreading, or coiled 270–360°. **RAY FL:** 16–60; ray 12–25 mm. **FR:** 3.5–7 mm, ± white or gray- to red-brown, top knobby; pappus awns 2–6.

var. ***angustifolia*** (A. Gray) M.A. Lane (p. 339) MARSH GUMPLANT Subshrub 10–20 dm, erect; sts woody in proximal 3–15 dm. **LF:** gen tapered to base, glabrous, tip gen acute. **INFL:** heads not subtended by lf-like bracts; phyllary tips acute, flat in ×-section, erect. **RAY FL:** 16–56; ray 12–17 mm. **FR:** 5–7 mm. 2*n*=12,24. Tidal wetlands; < 10 m. CCo (San Francisco Bay). May–Dec

var. ***platyphylla*** (Greene) M.A. Lane (p. 339) Per 1–10 dm, decumbent to erect, herbaceous or sts proximally woody ≤ 1 dm. **LF:** gen sessile, sometimes clasping, glabrous or sparsely tomentose, tip acute or rounded. **INFL:** head often subtended by lf-like bracts; phyllary tips acuminate, ± round in ×-section, spreading, reflexed, or coiled 270–360°. **RAY FL:** gen 20–60; ray 12–20 mm. **FR:** 3.5–5 mm. 2*n*=24. Coastal bluffs, dunes; < 300 m. NCo, CCo, SCo, ChI. All yr

var. ***stricta*** (p. 339) Per 1–10 dm, decumbent to erect, herbaceous or sts proximally woody ≤ 0.5 dm. **LF:** gen long-tapered to base, glabrous or sparsely tomentose, esp near head, tip rounded to acute. **INFL:** heads not subtended by lf-like bracts; phyllary tips acuminate, ± round in ×-section, spreading, reflexed, or coiled 270–360°. **RAY FL:** 30–60; ray 13–25 mm. **FR:** 3.5–7 mm. 2*n*=24. Sloughs, salt marshes, coastal bluffs, dunes; < 60 m. NCo; to AK. Jun–Nov

GUTIERREZIA SNAKEWEED, MATCHWEED

David J. Keil & Meredith A. Lane

Subshrub (in CA), ≤ 1.5 m, appearing glabrous. **ST:** 1–many from base, erect or ascending, branched distally, ± striate, gummy-resinous, minutely scabrous, yellow to tan or gray; older bark gen fibrous. **LF:** alternate, sometimes in axillary clusters, entire, gland-dotted, gummy-resinous, glabrous or minutely scabrous, dark gray-green. **INFL:** heads radiate, gen many in ± open cyme-like clusters, 1 or in short-peduncled clusters at tips of infl branches; involucre ± cylindric or narrowly to widely obconic; phyllaries graduated in 3–4 series, straw-colored, tips green; receptacle flat to conic, epaleate, minutely hairy. **RAY FL:** 1–13; corolla yellow, ray often inconspicuous. **DISK FL:** 1–13 (in CA), bisexual or staminate; corolla yellow, club- or narrowly funnel-shaped, lobes short, recurved; style appendages lanceolate. **FR:** narrowly obconic, 5–8-veined, light tan, with ± white appressed hairs; pappus of 1–2 series of finely toothed, white or ± yellow scales gen 1/2 fr length (in CA) or much reduced. 28 spp.: w N.Am, S.Am. (Gutiérrez, surname of a noble Spanish family) TOXIC to livestock, fresh or dried in hay. [Nesom 2006 FNANM 20:88–94]

1. Involucre cylindric; phyllaries 4–6; ray fls 1–2 . . . . ***G. microcephala***
1' Involucre cylindric to obconic or bell-shaped; phyllaries 8–21; ray fls 2–13
  2. Ray fls 4–13; disk fls 4–13; total fls 8–20; heads gen 1 at 2° infl tips . . . . ***G. californica***
  2' Ray fls 2–8; disk fls 2–9; total fls 6–14; heads gen in clusters of 2–5 at 2° infl tips . . . . ***G. sarothrae***

G. californica (DC.) Torr. & A. Gray (p. 339) CALIFORNIA MATCHWEED Pl 2–10 dm. **ST:** sprawling to erect, sometimes ± red. **LF:** ± linear. **INFL:** heads 8–20-fld, gen 1 (or in groups of 2–3) at 2° infl tips; peduncle gen > 1.5 mm; involucre gen ± bell-shaped, sometimes obconic, gen ≤ 6.5 mm, < 3.5 mm diam; phyllaries gen 9–21 in 3 series. **RAY FL:** 4–13; corolla 2.5–7.2 mm. **DISK FL:** 4–13, bisexual; corolla 2.3–4.2 mm. **FR:** 1–2.8 mm. 2*n*=16,24. Grassland, arid woodland and shrubland, sometimes on serpentine; < 1600 m. NCoRI, SnJV, CW, SCo, WTR, SnGb, PR; nw Baja CA. Variable in

involucre shape, arrangement of heads. Intergrades with *G. sarothrae* in c&s SCo, Baja CA. Jul–Nov

G. microcephala (DC.) A. Gray (p. 339) STICKY SNAKEWEED Pl 2–6 dm, much-branched, often nearly spheric. **ST:** brown below, yellow or green above. **LF:** linear to thread-like. **INFL:** heads 1–3-fld, in groups of 5–6 at 2° infl tips, sessile; involucre gen < 3.2 mm, < 1.2 mm diam, cylindric; phyllaries 4–6 in 2 series. **RAY FL:** 1–2; corolla 2.1–3.5 mm. **DISK FL:** 1–2, staminate; corolla 2.2–3.3 mm.

FR: 1–1.5 mm. $2n$=8,16,24,32. Grassland, sand dunes, desert scrub; 700–2500 m. SCo, SnBr, PR, SNE, D; to CO, TX, c Mex. Jul–Nov

G. sarothrae (Pursh) Britton & Rusby (p. 339) MATCHWEED Pl 1–6 dm. **ST**: sprawling or upright, brown below, green or tan above. **LF**: single lvs lance-linear, clustered lvs thread-like. **INFL**: heads 6–14-fld, in clusters of 2–5 or fewer at 2° infl tips, on peduncles < 1.5

mm or sometimes sessile; involucre gen ≤ 4.5 mm, ≤ 2.5 mm diam, cylindric or narrowly obconic; phyllaries 8–21 in 2–3 series. **RAY FL**: 2–8; corolla 3–5.4 mm. **DISK FL**: 2–9, bisexual; corolla 2.3–3.5 mm. **FR**: 0.9–1.6 mm. $2n$=8,16,32. Grassland, desert, montane areas; 50–3050 m. SCo, WTR, PR, D; to WA, s-c Can, c US, n Mex. Intergrades with *G. californica* in c&s SCo, Baja CA. May–Oct

HARMONIA

Bruce G. Baldwin

Ann 0.5–4 dm. **ST**: erect. **LF**: proximal opposite, distal alternate, sessile, linear, entire or toothed, gen coarsely hairy, sometimes also minutely stalked-glandular, glands gen black, sometimes ± yellow. **INFL**: heads radiate, 1 or in loose, ± umbel-like to flat-topped clusters; involucre obovoid to obconic, 2–5+ mm diam; phyllaries 3–8 in 1 series, lanceolate to oblanceolate, each enclosing a ray ovary, falling with fr; receptacle flat to convex, glabrous or minutely bristly, paleae in 1 series between ray and disk fls, free or fused, phyllary-like (more scarious), deciduous. **RAY FL**: gen 3–8; corolla bright yellow, ray fan-shaped to obovate. **DISK FL**: 7–30, bisexual or staminate, sometimes in same head; corolla bright yellow, tube < throat, lobes deltate; anthers ± yellow to ± brown, tips ovate-dentate to hemispheric; styles glabrous proximal to branches, tips awl-shaped, densely bristly. **FR**: ray fr black, round in ×-section to ± compressed, weakly arched, bowed abaxially or not, beaked or beakless, glabrous, pappus 0 or of 3–12, lanceolate to awl-shaped, fringed to plumose scales; disk fr black, ± club-shaped, ± round, glabrous or hairy, pappus of 7–11, gen linear to lanceolate or awl-shaped, fringed or plumose scales. 5 spp.: nw CA. (Harvey Monroe Hall, CA botanist, 1874–1932) [Baldwin & Strother 2006 FNANM 21:297–299]

1. Heads gen nodding in bud and fr; ray pappus 0; disk pappus of lanceolate, long-tapered, minutely fringed scales 2–3.7 mm. . . . . ***H. nutans***
1′ Heads gen erect in bud and fr; ray pappus 0.2–1.5 mm; disk pappus of linear to lanceolate or awl-shaped, or oblong to square, fringed or plumose scales 0.2–3.5 mm
  2. Lvs ± evenly distributed on st; ray fr strongly bowed abaxially, beaked, beak 0.4–0.5 mm; disk fls staminate. . . . . ***H. doris-nilesiae***
  2′ Lvs unevenly distributed, mostly along main, central st and immediately proximal to branches supporting heads; ray fr slightly or not bowed abaxially, beak 0 or pappus elevated adaxially 0.1–0.2 mm; some or all disk fls bisexual
    3. Phyllaries densely white-soft-hairy near folded edges; ray fr slightly bowed abaxially; disk pappus of linear and long-tapered to awl-shaped, plumose scales 1.2–3.5 mm . . . . ***H. stebbinsii***
    3′ Phyllaries coarsely hairy, or minutely so, near folded edges; ray fr not bowed abaxially; disk pappus of lanceolate to linear or oblong to square, fringed scales 0.2–0.8 mm
      4. Proximal, unbranched part of main, central st gen > branches supporting heads, distal lvs of main st not densely clustered; disk pappus of lanceolate to linear scales 0.6–0.8 mm. . . . . ***H. guggolziorum***
      4′ Proximal, unbranched part of main, central st gen < branches supporting heads; distal lvs of main st densely clustered; disk pappus of oblong or square scales 0.2–0.5 mm. . . . . ***H. hallii***

H. doris-nilesiae (T.W. Nelson & J.P. Nelson) B.G. Baldwin (p. 339) NILES' HARMONIA Pl 9–40 cm. **ST**: proximal, unbranched part of central st < or > branches supporting heads. **LF**: ± evenly distributed on sts. **INFL**: heads gen erect in bud and fr; phyllaries 4–8, densely white-soft-hairy near folded edges. **RAY FL**: 4–8; ray 2.5–7 mm. **DISK FL**: 8–20, staminate. **FR**: ray fr strongly bowed abaxially, beak 0.4–0.5 mm, pappus to 0.9 mm; disk fr 0, pappus of 7–10, lanceolate-long-tapered to awl-shaped, plumose scales 0.2–0.9 mm. $2n$=18. Serpentine slopes; 800–1600 m. s KR. [*Madia d.* T.W. Nelson & J.P. Nelson] May–Jun ★

H. guggolziorum B.G. Baldwin (p. 339) GUGGOLZ'S HARMONIA Pl 10–30 cm. **ST**: proximal, unbranched part of central st gen > branches supporting heads. **LF**: mostly on central st and immediately proximal to branches supporting heads, not densely clustered distally on central st. **INFL**: heads gen erect in bud and fr; phyllaries 3–6, coarsely hairy, or minutely so, near folded edges. **RAY FL**: 3–6; ray 4–5 mm. **DISK FL**: 8–13, some or all bisexual. **FR**: ray fr not bowed abaxially, beak 0, pappus to 0.5 mm; disk fr 3–3.5 mm, pappus of 9–11, lanceolate to linear, fringed scales 0.6–0.8 mm. $2n$=18. Serpentine slopes; 100–200 m. e edge NCoRO (Hopland, Mendocino Co.). Apr–May ★

H. hallii (D.D. Keck) B.G. Baldwin (p. 339) HALL'S HARMONIA Pl 5–18 cm. **ST**: proximal, unbranched part of central st gen < branches supporting heads. **LF**: mostly on central st and immediately proximal to branches supporting heads, densely clustered distally on central st. **INFL**: heads gen erect in bud and fr; phyllaries 3–6, coarsely hairy, or minutely so, near folded edges. **RAY FL**: 3–6;

ray 2–5 mm. **DISK FL**: 8–20, mostly bisexual, sometimes staminate. **FR**: ray fr not bowed abaxially, beak 0, pappus 0.2–0.5 mm; disk fr 2.8–3.2 mm, pappus of 8–10, oblong or square, fringed scales 0.2–0.5 mm. $2n$=18. Open sites, disturbed areas in serpentine chaparral; 500–1000 m. s NCoRI. [*Madia h.* D.D. Keck] Apr–Jun ★

H. nutans (Greene) B.G. Baldwin (p. 339) NODDING HARMONIA Pl 5–25 cm. **ST**: proximal, unbranched part of central st < or > branches supporting heads. **LF**: ± evenly distributed on sts. **INFL**: heads gen nodding in bud and fr; phyllaries 4–8, coarsely hairy, or minutely so, near folded edges. **RAY FL**: 4–8; ray 3–7 mm. **DISK FL**: 7–30, bisexual. **FR**: ray fr not bowed abaxially, beak 0, pappus 0; disk fr 2.3–4.5 mm, pappus of 9–11, lanceolate, long-tapered, minutely fringed scales 2–3.7 mm. $2n$=18. Rocky, gen volcanic, open or disturbed sites in chaparral and woodland; 100–1000 m. s NCoR. [*Madia n.* (Greene) D.D. Keck] Apr–Jun ★

H. stebbinsii (T.W. Nelson & J.P. Nelson) B.G. Baldwin (p. 339) STEBBINS' HARMONIA Pl 5–27 cm. **ST**: proximal, unbranched part of central st gen < branches supporting heads. **LF**: mostly on central st and immediately proximal to branches supporting heads, densely clustered distally on central st. **INFL**: heads gen erect in bud and fr; phyllaries 4–6, densely white-soft-hairy near folded edges. **RAY FL**: 4–6; ray 4–6.5 mm. **DISK FL**: 8–20, outer bisexual, inner staminate. **FR**: ray fr slightly bowed abaxially, elevated 0.1–0.2 mm adaxially, beak 0, pappus 0.5–1.5 mm; disk fr 2–4 mm, pappus of 8–10, linear-long-tapered to awl-shaped, plumose scales 1.2–3.5 mm. $2n$=18. Serpentine slopes; 1100–1600 m. s KR, n NCoRI. [*Madia s.* T.W. Nelson & J.P. Nelson] May–Jul ★

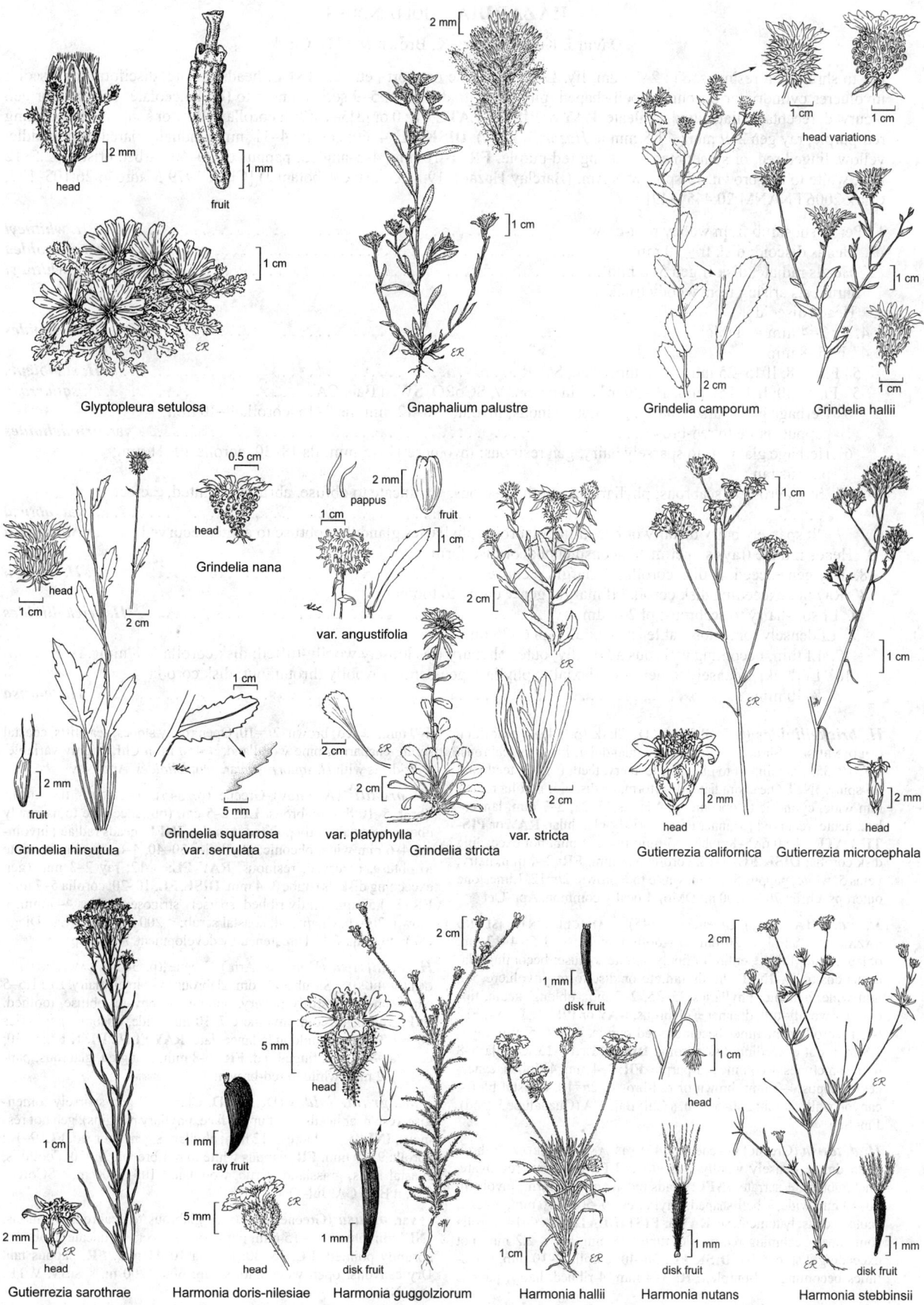

Glyptopleura setulosa

Gnaphalium palustre

head variations

Grindelia camporum

Grindelia hallii

Grindelia hirsutula

Grindelia nana

var. angustifolia

var. platyphylla

var. stricta

Gutierrezia californica

Gutierrezia microcephala

Grindelia squarrosa
var. serrulata

Grindelia stricta

Gutierrezia sarothrae

Harmonia doris-nilesiae

Harmonia guggolziorum

Harmonia hallii

Harmonia nutans

Harmonia stebbinsii

HAZARDIA GOLDENBUSH

David J. Keil, Gregory K. Brown & W.D. Clark

Per to shrub, gen resinous. **ST**: 2–25 dm, lfy. **LF**: gen sessile or short-petioled. **INFL**: heads radiate, disciform, or discoid; involucres cylindric or obconic to bell-shaped; phyllaries graduated in 5–9 series, linear to (ob)lanceolate, tips erect or gen recurved; receptacle flat, pitted, epaleate. **RAY** or **PISTILLATE FL**: 0 or [3]5–18[25]; corolla yellow or sometimes becoming red-purple, ray gen 2–9 mm (≤ 1.8 mm in *Hazardia cana*). **DISK FL**: 4–60; corolla 4–11 mm, gradually flared from middle, yellow, tinged red, or sometimes becoming red-purple. **FR**: 1–10 mm, 4–5-angled; pappus of 20–60 barbed bristles, 2.5–12 mm, white to red-brown. 13 spp.: w N.Am. (Barclay Hazard, 19th century CA botanist) [Clark 1979 Madroño 26:105–127; Clark 2006 FNANM 20:445–449]

1. Per or subshrub from woody root-crown... ***H. whitneyi***
  2. Heads discoid; disk fr ≤ 10 mm .... var. ***discoidea***
  2′ Heads radiate; disk fr gen 5–8 mm .... var. ***whitneyi***
1′ Shrub; sts arising from woody trunk
  3. Heads discoid
    4. Fr 2–4 mm — DMoj .... ²***H. brickellioides***
    4′ Fr 5–8 mm
      5. Fls 4–8; lf 15–25 mm, 7–12 mm wide; SCoR .... ***H. stenolepis***
      5′ Fls 9–30; lf 15–50 mm, 10–20 mm wide; s SnJV, SCoRO, SW, n Baja CA .... ***H. squarrosa***
        6. Herbage gen sparsely hairy, gen not resinous; involucre 8–12 mm; fls 9–16, corolla 9–10 mm; pappus white to red-brown... var. ***grindelioides***
        6′ Herbage glabrous to sparsely hairy, gen resinous; involucre 11–15 mm; fls 18–30, corolla 10–11 mm; pappus tan
          7. St glabrous to scabrous; phyllaries smooth, resinous, ± truncate to obtuse, abruptly pointed, ± erect .... var. ***obtusa***
          7′ St sparsely hairy distally or becoming glabrous; phyllaries glandular, obtuse to acute, recurved ... var. ***squarrosa***
  3′ Heads radiate (rays sometimes inconspicuous) or disciform
    8. Ray gen exceeding disk corolla; lf margin ± entire .... ***H. orcuttii***
    8′ Ray not exceeding disk corolla; lf margin gen ± entire to toothed
      9. Lf soft-hairy to scabrous; pl 2–8 dm .... ²***H. brickellioides***
      9′ Lf densely tomentose, at least abaxially; pl 6–25 dm
        10. Lf thin, becoming glabrous adaxially; outer phyllary tips loosely woolly-tufted; disk corolla 5–8 mm ... ***H. cana***
        10′ Lf thick, ± densely tomentose adaxially; phyllary gen densely woolly throughout; disk corolla 8–10 mm .... ***H. detonsa***

H. brickellioides (S.F. Blake) W.D. Clark (p. 345) BRICKELL GOLDENBUSH Shrub 2–8 dm, yellow-glandular, hairy to scabrous. **LF**: 10–35 mm, elliptic to obovate, leathery, teeth (0)2–8, teeth and tip spiny. **INFL**: heads radiate, disciform, or discoid; involucre 4–5 mm wide, cylindric to obconic; phyllaries 15–25, 3–7 mm, lanceolate, acute, recurved (or inner erect), bristly-glandular. **RAY** or **PISTILLATE FL**: 0 or 5–8; tube 4–5 mm; ray 2–4 mm, not exceeding disk corolla. **DISK FL**: 8–12; corolla 6–8 mm. **FR**: 2–4 mm, hairy; veins 5, white; pappus 5–7 mm, white to ± brown. 2*n*=12. Limestone outcrops, cliffs; 700–2100 m. DMoj. Locally common. Apr–Oct

H. cana (A. Gray) Greene (p. 345) SAN CLEMENTE ISLAND HAZARDIA Shrub 6–20 dm, ± woolly-tomentose. **LF**: 4–12 cm, oblanceolate, thin, ± entire to finely serrate, obtuse; becoming glabrous adaxially. **INFL**: heads radiate or disciform; involucre 5–8 mm wide, obconic; phyllaries 25–35, 2–7 mm, oblong, acute, tips of outer woolly-tufted, inner ± glabrous. **RAY** or **PISTILLATE FL**: 6–14; corolla sometimes becoming red-purple, tube 4–5 mm, ray ≤ 1.8 mm, not exceeding disk corolla. **DISK FL**: 15–25; corolla 5–8 mm, sometimes becoming red-purple. **FR**: 3–4 mm, 4-ribbed, canescent; pappus 4–7 mm, brown or red-brown. 2*n*=10. Coastal bluffs, canyon walls, scrub; 200–500 m. s ChI; Baja CA (Guadalupe Island). Jun–Sep ★

H. detonsa (Greene) Greene (p. 345) ISLAND HAZARDIA Shrub 6–25 dm, ± densely woolly-tomentose. **LF**: 4–14 cm, (ob)ovate, thick, obtuse, ± serrate. **INFL**: heads radiate or disciform; involucre 10–13 mm wide, ± bell-shaped; phyllaries 30–50, 4–10 mm, oblong, acute, ± densely tomentose. **RAY** or **PISTILLATE FL**: 6–14; corolla sometimes becoming red-purple, tube 5–6 mm, ray < 2.2 mm, not exceeding disk corolla. **DISK FL**: 30–40; corolla 8–10 mm, sometimes becoming red-purple. **FR**: 3–4 mm, 4-ribbed, hairy; pappus 6–9 mm, (± red) brown. 2*n*=10. Open rocky slopes, sea cliffs, coastal scrub, chaparral, pine woodland; < 450 m. n ChI. Highly variable; hybridizes with *H. squarrosa* var. *grindelioides*. Apr–Nov ★

H. orcuttii (A. Gray) Greene (p. 345) ORCUTT'S HAZARDIA Shrub 5–10 dm, glabrous. **LF**: 2–5 cm, (ob)lanceolate to narrowly obovate, gen abruptly pointed, ± entire. **INFL**: heads radiate; involucre 4–6 mm wide, obconic; phyllaries 30–40, 4–6 mm, linear, acute to obtuse, recurved, resinous. **RAY FL**: 8–12; ray 2–3 mm (gen exceeding disk fls), tube 3–4 mm. **DISK FL**: 10–20; corolla 5–7 mm. **FR**: 3–4.5 mm, faintly ribbed, sparsely strigose; pappus 4–5 mm, ± brown. 2*n*=10. Chaparral, coastal scrub; < 200 m. s SCo (San Diego Co.); nw Baja CA. Threatened by development. Aug–Sep ★

H. squarrosa (Hook. & Arn.) Greene (p. 345) SAW-TOOTHED GOLDENBUSH Shrub 3–23 dm, glabrous to sparsely hairy. **LF**: 1.5–5 cm, leathery or stiffly papery, oblong to obovate, obtuse, toothed. **INFL**: heads discoid; involucre 7–10 mm wide, obconic; phyllaries 30–60, 3–10 mm, oblong to lanceolate. **RAY FL**: 0. **DISK FL**: 9–30; corolla 9–11 mm, tinged red. **FR**: 5–8 mm, 5-angled, glabrous; pappus 7–12 mm, white to red-brown.

  var. ***grindelioides*** (DC.) W.D. Clark Pl gen sparsely tomentose, (esp near heads, on lf upper face, phyllary margins), gen not resinous. **INFL**: involucre 8–12 mm; phyllaries gen recurved. **FL**: 9–16; corolla 9–10 mm. **FR**: pappus white to red-brown. 2*n*=10. Foothills, coastal mtns, grassland, scrub, woodland; 100–1300 m. s SCoRO, SW; n Baja CA. Jul–Oct

  var. ***obtusa*** (Greene) Jeps. Pl glabrous to scabrous, resinous. **INFL**: involucre 11–15 mm; phyllaries ± erect, ± truncate to obtuse, abruptly pointed. **FL**: 18–30; corolla 10–11 mm. **FR**: pappus tan. Dry canyons, open woodland, scrub; 600–1200 m. s SnJV, WTR. Sep–Nov

var. **squarrosa**   Pl ± glabrous (exc sometimes sts sparsely hairy distally), resinous. **INFL**: involucre 11–15 mm; phyllaries recurved, glandular, obtuse to acute. **FL**: 18–30; corolla 10–11 mm. **FR**: pappus tan. Foothills, coastal mtns, open woodland, scrub; < 700 m. SCoRO. Aug–Oct

H. stenolepis (H.M. Hall) Hoover (p. 345)   Shrub 3–10 dm. **ST**: scabrous. **LF**: 15–25 mm, oblong to obovate, leathery, obtuse, abruptly pointed, toothed, glabrous. **INFL**: heads discoid; involucre 3–6 mm wide, narrowly obconic; phyllaries 20–30, 3–23 mm, linear, acute, glabrous. **RAY FL**: 0. **DISK FL**: 4–8; corolla 9–11 mm. **FR**: 5–8 mm, 5-angled, glabrous; pappus 7–12 mm, red-brown. 2*n*=10. Serpentine or shale, grassland, open woodland, scrub; 150–1200 m. SCoR. Sep–Nov

H. whitneyi (A. Gray) Greene (p. 345)   WHITNEY'S GOLDENBUSH Per or subshrub 2–5 dm, glabrous to scabrous-glandular. **LF**: 25–50 mm, widely oblong to oblanceolate, acute, serrate. **INFL**: heads radiate or discoid; involucre 8–12 mm wide, bell-shaped; phyllaries grading into lvs, 5–12 mm, lance-linear, acute, sometimes recurved. **RAY FL**: 0 or 5–18; corolla 9–13 mm, tube 4–5 mm, ray 5–8 mm, exceeding disk corolla. **DISK FL**: 15–30; corolla 8–10 mm. **FR**: 5–10 mm, 5-angled, glabrous; pappus 7–10 mm, ± brown.

var. **discoidea** (J.T. Howell) W.D. Clark   **RAY FL**: 0. **FR**: ≤ 10 mm. Open montane conifer forest; 1000–2500 m. NW; OR. Jul–Sep

var. **whitneyi**   **RAY FL**: 5–18. **FR**: gen 5–8 mm. 2*n*=8. Open montane conifer forest; 1200–3500 m. SN. Jul–Sep

HECASTOCLEIS

1 sp. (Greek: each enclosed, from 1-fld heads) [Simpson 2006 FNANM 19:71–72]

H. shockleyi A. Gray (p. 345)   Shrub. **ST**: 4–7 dm, stiff, much-branched, glandular-puberulent or becoming glabrous exc for tufts of soft hair in axils of persistent lf bases. **LF**: simple, alternate, some also clustered in axils of older sts; primary lvs 1–3 cm, linear to lance-linear, sessile, spine-tipped, sparsely spiny-dentate, ± persistent as spines when dry; clustered axillary lvs narrower, obtuse to acute, gen not toothed. **INFL**: heads 1-fld, sessile, in dense, head-like clusters surrounded by involucre of persistent, ovate, spiny-toothed, net-veined bracts 1–2 cm; true involucre narrowly cylindric; phyllaries ± 15 in several unequal series, 4–10 mm, linear, acuminate, loosely soft-hairy; receptacle epaleate. **FL**: 1 per head; corolla pink to red-purple in bud, green-white in fl, lobes linear, equal; filaments inserted near base of corolla, anthers purple, exserted, base with stiff, bristle-like tails, tips short-triangular; style tips shallowly lobed. **FR**: cylindric, glabrous; pappus a crown of fringed scales ≤ 1 mm. 2*n*=16. Dry, rocky slopes; 1000–2200 m. SNE, DMtns; w NV. May–Jul   ★

HEDYPNOIS

David J. Keil & G. Ledyard Stebbins

Ann; sap milky. **ST**: branched, spreading. **LF**: basal and cauline, alternate, entire to dentate or pinnately lobed. **INFL**: heads liguliflorous, small; phyllaries in 2 series, outer small, inner equal, becoming hardened, enveloping outer frs; receptacle epaleate. **FL**: 8–40+; ligules yellow, drying ± blue, readily withering. **FR**: cylindric, not beaked, often strongly incurved, inner persistent on receptacle; pappus of long or short scales, sometimes mixed with bristles. 2 spp.: Atlantic islands, Medit, sw Asia. (Greek: name of Pliny for a kind of wild endive) [Strother 2006 FNANM 19:302]

H. rhagadioloides (L.) F. W. Schmidt. (p. 345)   CRETE WEED Herbage finely bristly; hairs minutely forked or barbed at tip. **ST**: 0.5–4 dm. **LF**: 5–18 cm, gen oblong to oblanceolate; proximal tapered to base. **INFL**: heads 1–several in open cyme-like clusters; peduncle gen thickened; involucre 8–10 mm; phyllaries linear, becoming strongly incurved, glabrous to densely bristly. **FR**: 5–7 mm, minutely scabrous; pappus of outer frs a low crown of fused scales; pappus of inner frs short scales ± elongated, bristle-tipped scales. 2*n*=8,11–16,18. Weed of pastures, grassy slopes, roadsides; < 1150 m. NCo, s NCoRO, CaRF, SNF, GV, CW, SW, w DSon; to TX; native to Medit. *H. cretica* (L.) Dum. Cours. Feb–Jun

HELENIUM   SNEEZEWEED

Mark W. Bierner

Ann, per, 10–160 cm. **ST**: erect, unbranched or branched distally, gen ± winged by decurrent lf bases, glabrous or ± hairy. **LF**: basal and cauline; basal, proximal cauline lvs often withered by fl, alternate, simple, entire, toothed or lobed, glabrous or ± hairy, gland-dotted. **INFL**: heads radiate or discoid, hemispheric to ovoid or ± spheric, 1 or gen in panicle-like or ± flat-topped clusters; involucre shallowly cup- or disk-like, phyllaries in 2 series (in CA), free or ± basally fused, ± equal (or outer longer), gen reflexed in fr; receptacle conic, ± spheric, hemispheric, or ovoid, pitted, epaleate. **RAY FL**: 0 or 7–34, pistillate (in CA), corolla yellow (in CA), ray fan-shaped, 3-lobed. **DISK FL**: 75–1000+; corolla gen 5-lobed, yellow, purple, or proximally yellow to yellow-green and distally ± brown or purple; anther tip triangular; style tips truncate. **FR**: obpyramidal, ribbed, glabrous or ± hairy; pappus of 5–12 membranous scales. ± 32 spp.: N.Am, S.Am. (Helen of Troy) [Bierner 2006 FNANM 21:426–435]

1. St not winged; cauline lvs gen narrowly linear; ann . . . . ***H. amarum*** var. ***amarum***
1′ St winged by decurrent lf bases; cauline lvs gen wider; ann or per
  2. Cauline lf margin ± toothed . . . . ***H. autumnale***
  2′ Cauline lf margin entire
    3. Heads (1)4–20(30); ray corolla 3.8–10 mm (occ ray fls 0); disk corolla gen 4-lobed, 1.6–2.7 mm; pappus
       scales 0.4–1 mm . . . . ***H. puberulum***
    3′ Heads 1–20; ray corolla 12–37 mm; disk corolla 5-lobed, 3–5.2 mm; pappus scales 1.3–4.5 mm
      4. Peduncle sparsely to moderately hairy; pappus scales 1.3–2.2(2.7) mm; widespread but not NCo . . . . . . ***H. bigelovii***
      4′ Peduncle gen densely hairy; pappus (2.5)3–4.5 mm; NCo . . . . ***H. bolanderi***

H. amarum (Raf.) H. Rock var. *amarum* Ann (1)2–6(10) dm. **ST**: 1–3(15), gen branched distally, not winged, glabrous or sparsely hairy. **LF**: glabrous or sparsely hairy, proximal gen withered by fl; basal linear to ovate, gen entire or pinnately toothed or lobed; cauline gen narrowly linear, gen entire. **INFL**: heads (1)10–150(250+), 5–9 mm, 6–10 mm diam; peduncle 3–11 cm, sparsely hairy; phyllaries free; receptacle 2–3 mm diam, hemispheric to ± spheric or ovoid. **RAY FL**: 8–10; corolla 6.5–14 mm. **DISK FL**: 75–150+; corolla 2–2.7 mm, proximally yellow, distally yellow to yellow-brown. **FR**: 0.9–1.3 mm, hairy; pappus scales 6–8, 1.2–1.8 mm, awn-tipped. 2*n*=30. Disturbed areas, fields; 20–400 m. SnJV, SnFrB; native to se US. Jul–Oct

H. autumnale L. Per 5–13 dm. **ST**: 1(7), branched distally, strongly winged, ± hairy. **LF**: glabrous or gen ± hairy, proximal gen withered by fl; basal oblanceolate or obovate to spoon-shaped, entire or weakly lobed; cauline lanceolate to oblanceolate to obovate, ± toothed. **INFL**: heads 5–70(100+), 8–20 mm, 8–23 mm diam; peduncle 3–10 cm, ± hairy; phyllaries basally fused; receptacle 3–5 mm diam, ± spheric to ovoid. **RAY FL**: 8–21; corolla 10–23 mm. **DISK FL**: 200–400(800+); corolla 2.4–4 mm, proximally yellow, distally yellow to yellow-brown. **FR**: 1–2 mm, ± hairy; pappus scales 5–7, (0.5)0.9–1.5(1.8) mm, awn-tipped. 2*n*=32,34,36. Roadsides, fields, wet areas; 20–2600 m. NCo, KR, MP; continental US, sw Can. [*H. a.* var. *grandiflorum* A. Gray; *H. a.* var. *montanum* (Nutt.) Fernald] Aug–Oct

H. bigelovii Torr. & A. Gray (p. 345) Per 3–13 dm. **ST**: 1–3(10), unbranched or sparingly branched distally, gen weakly winged, glabrous or sparsely hairy. **LF**: glabrous or sparsely hairy; basal oblanceolate to oblong-elliptic, entire; cauline oblong-elliptic to lanceolate to linear, entire. **INFL**: heads 1–20, 12–20 mm, (14)17–22(25) mm diam; peduncle (6)10–30 cm, sparsely to moderately hairy; phyllaries

basally fused; receptacle 4–8 mm diam, ± spheric. **RAY FL**: 14–20; corolla 13–25 mm. **DISK FL**: 250–500(800+); corolla 3–4.4(4.8) mm, proximally yellow, distally yellow to brown to purple. **FR**: 1.8–2.4 mm, ± hairy; pappus scales 6–8, 1.3–2.2(2.7) mm, awn-tipped. 2*n*=32. Wet meadows, marshes, bogs, fens, streambanks, lake margins; < 3400 m. KR, NCoR, CaR, SN, deltaic GV, CW, TR, PR; OR. Jul–Aug

H. bolanderi A. Gray Per 2.5–14 dm. **ST**: gen 1, gen unbranched, weakly winged, glabrous or ± hairy. **LF**: glabrous or sparsely hairy; basal spoon-shaped, entire; cauline spoon-shaped to ovate to oblanceolate to oblong-elliptic, entire. **INFL**: heads 1(3), 14–24 mm, 18–34 mm diam; peduncle 10–30(50) cm, gen densely hairy, tomentose below involucre; phyllaries basally fused, spreading to reflexed in fr; receptacle 4–8 mm diam, hemispheric to ± spheric or ovoid. **RAY FL**: 15–30; corolla (12)16–28(37) mm. **DISK FL**: 300–750(1000+); corolla 4–5(5.2) mm, proximally yellow, distally yellow to brown to purple. **FR**: 1.5–2.5 mm, hairy; pappus scales 6–8, (2.5)3–4.5 mm, awn-tipped. 2*n*=32. Bogs, seepage areas, wet meadows; 30–200 m. NCo; OR. Jun–Sep

H. puberulum DC. (p. 345) Ann, per, 5–16 dm. **ST**: gen 1, branched distally, strongly winged, glabrous or sparsely hairy. **LF**: glabrous or sparsely hairy; basal oblanceolate to oblong-elliptic, entire; cauline oblong-elliptic to lanceolate to lance-linear, entire. **INFL**: heads (1)4–20(30), 9–15 mm, 9–17(19) mm diam; peduncle (6)9–17(23) cm, ± hairy; phyllaries gen free; receptacle 5–10 mm diam, ± spheric. **RAY FL**: 0 or 13–15; corolla 3.8–10 mm. **DISK FL**: 300–500(1000+); corolla (1.6)1.9–2.7 mm, gen 4-lobed, proximally yellow, distally yellow to ± red-brown or purple. **FR**: 1.2–1.9 mm, ± hairy; pappus scales 5–6, 0.4–1 mm, short-awned. 2*n*=58. Streambanks, seepage areas, lake margins; < 1200 m. KR, NCoR, CaRH, SNF, GV, CW, WTR, w PR; Baja CA. Jun–Aug

HELIANTHELLA

Per from taprooted caudex. **ST**: 1–several, erect. **LF**: basal and cauline, alternate or opposite; petioles short to long; blades gen 3-veined, linear to ± ovate, entire, ± scabrous. **INFL**: heads radiate, 1–few, in cyme-like cluster; peduncles long, bracts 0–few, lf-like; involucre hemispheric; phyllaries in 2–3 series, free; receptacle flat to ± convex, paleate, paleae folded around ovary, entire, obtuse, tip hairy. **RAY FL**: sterile; corolla yellow; style 0. **DISK FL**: many; corolla and anthers yellow or purple; style tips short-triangular. **FR**: thick or compressed; pappus 0 or 2 awns 1–2 mm (sometimes also crown of low scales). 9 spp.: w N.Am. (Latin: diminutive of *Helianthus*) [Weber 2006 FNANM 21:114–117]

1. Outer phyllaries gen enlarged, lf-like, incurved; involucre 2.5–4 cm diam; fr center thick, edge thin . . . . ***H. castanea***
1′ Outer phyllaries seldom enlarged or lf-like, not incurved; involucre gen (1)1.5–2 cm diam; fr compressed, thin . . . . ***H. californica***
  2. Pappus 0; st branched; cauline lvs gen opposite . . . . var. ***californica***
  2′ Pappus present; st gen unbranched; cauline lvs gen alternate
    3. Disk gen 1.5–2 cm diam; pappus gen of 2 awns or narrow scales, 1–2 mm, and crown of short scales; lf blades oblong- or lance-linear, (1.5)2–4(4.5) cm wide, heads gen 2–3 . . . . var. ***nevadensis***
    3′ Disk gen 1–1.5 cm diam; pappus gen of 2 awns, 1 mm; lf blades linear to lance-linear (narrowly ovate), 3–6(18) mm wide; heads gen 1 . . . . var. ***shastensis***

H. californica A. Gray **ST**: 1–6 dm, glabrous to coarsely hairy. **INFL**: peduncle sometimes ± scapose, 10–30 cm, ± short-hairy; involucre gen (1)1.5–2 cm diam; phyllaries seldom lf-like, gen 1–2 cm, short-ciliate. **RAY FL**: 9–21; ray gen 1–2 cm. **DISK FL**: corolla 4–5 mm, yellow; anthers yellow to purple. **FR**: 6–8 mm, obovate, thin, glabrous; pappus 0 or 2 awns or narrow scales, sometimes with crown of short scales.

var. *californica* **ST**: gen 3–6 dm. **LF**: 1–4 cm wide, oblanceolate to narrowly lanceolate. **INFL**: heads 1–few; disk gen 1.5–2 cm diam. **FR**: pappus 0. 2*n*=30. Open, grassy sites; < 1000 m. NCoR, SnFrB. [*H. c.* subsp. *c.*] Apr–Jun

var. *nevadensis* (Greene) Jeps. **ST**: gen 3–6 dm. **LF**: (1.5)2–4(4.5) cm wide, oblong- or lance-linear **INFL**: heads 1–few; disk gen 1.5–2 cm diam. **FR**: pappus of 2 awns or narrow scales 1–2 mm and crown of short scales. Meadows, forest, chaparral; 250–2600 m. NCoRI, CaR, SN, s PR, Wrn; s OR, w NV. [*H. c.* subsp. *n.* (Greene) W.A. Weber] Apr–Aug

var. *shastensis* W.A. Weber **ST**: gen 1–3 dm. **LF**: 3–6(18) mm wide, linear to lance-linear (narrowly ovate); cauline few. **INFL**: head gen 1; disk gen 1–1.5 cm diam. **FR**: pappus of 2 awns or narrow scales 1 mm. Open conifer forest, chaparral; 400–2600 m. KR, CaR. [*H. c.* subsp. *s.* (W.A. Weber) W.A. Weber] Apr–Jul

H. castanea Greene (p. 345) DIABLO HELIANTHELLA **ST**: 1–5 dm, glabrous to coarsely hairy. **LF**: cauline few; petioles long; blades 2–6 cm wide, narrowly to widely elliptic. **INFL**: head gen 1; peduncle 7–20 cm, stout, ± rough-hairy, often with 1–few bracts near tip; involucre 2.5–4 cm diam; outer phyllaries gen lf-like, 3–10 cm, 7–20 mm wide, curving up around head; inner phyllaries 2–2.5 cm, coarsely ciliate. **RAY FL**: 13–21; ray 1–3 cm. **DISK FL**: corolla 6–7 mm, yellow; anthers yellow. **FR**: 8–10 mm, obovate, glabrous; center thick; edges thin; pappus awns 0 or 2, ≤ 1 mm. 2*n*=30. Open, grassy sites; 200–1300 m. n CCo, n SnFrB. Apr–Jun ★

HELIANTHUS SUNFLOWER

Ann or per (subshrub). **ST**: gen erect. **LF**: opposite or alternate, gen reduced distally on st, often 3-veined from near base, gen ± flat, gen green, gen rough-hairy. **INFL**: heads radiate, 1 or in cyme-like clusters; involucre bell-shaped to hemispheric; phyllaries in 1–3 gen ± equal series; receptacle flat to rounded, paleate; paleae 0–3-lobed. **RAY FL**: 10–many, sterile; ray yellow. **DISK FL**: many; corolla yellow to red or purple, tube short, throat base tapered or often swollen, lobes triangular; style appendages triangular. **FR**: oblanceolate to obovate, ± compressed, sides rounded; pappus gen of 2 deciduous, lanceolate to ovate scales (+ 0–several shorter scales). 53 spp.: Am. (Greek: sun fl) [Schilling 2006 FNANM 21:141–169] *H. maximilianii* Schrad. a garden pl only.

1. Ann (or robust subshrub)
  2. Phyllaries gen 5–8 mm wide, abruptly acuminate . . . . *H. annuus*
  2′ Phyllaries gen 1–5 mm wide, acute or gradually acuminate
    3. Central paleae tipped with stiff white hairs . . . . *H. petiolaris*
      4. Lvs and phyllaries densely gray-canescent; corolla throat gradually narrowed distally to slightly hairy
        basal bulge . . . . subsp. *canescens*
      4′ Lvs and phyllaries strigose, green; corolla throat abruptly narrowed distally to densely hairy basal
        bulge . . . . subsp. *petiolaris*
    3′ Central paleae glabrous or finely appressed-hairy
      5. Paleae acute, = or slightly > disk fls, glabrous or finely appressed-hairy; lvs densely white-hairy
        . . . . ²*H. niveus* subsp. *tephrodes*
      5′ Paleae awn-tipped, >> disk fls, tip glabrous; lvs green
        6. Lf blade narrowly lance-ovate to ovate, gen serrate; involucre of larger heads gen 20+ mm diam;
          fr 3.5–4.5 mm . . . . *H. bolanderi*
        6′ Lf blade lance-linear lance-ovate, gen entire or ± entire (shallowly serrate); involucre of larger heads
          15–20 mm diam; fr 2.7–3.5 mm . . . . *H. exilis*
1′ Per (occ fl in 1st yr)
  7. Disk corolla lobes red to purple
    8. Lvs blue-green, glabrous to sparsely short-rough-hairy or bristly, margins often crinkled; roots
      horizontal, rhizome-like; ray gen 8–9 mm . . . . *H. ciliaris*
    8′ Lvs green or densely canescent, margins gen flat; pl taprooted; ray gen >> 1 cm
      9. Lvs green, short-rough-hairy or bristly; phyllaries ≤ disk . . . . ²*H. gracilentus*
      9′ Lvs densely white-hairy; phyllaries = or slightly > disk . . . . ²*H. niveus* subsp. *tephrodes*
  7′ Disk corolla lobes yellow
    10. Phyllaries ≤ disk, obtuse to acute . . . . ²*H. gracilentus*
    10′ Phyllaries ≥ disk, gen acuminate
      11. Phyllaries gen 12–18; most lvs opposite, gen < 12 cm; pl from taproot . . . . *H. cusickii*
      11′ Phyllaries gen > 25; most or all lvs alternate, often > 15 cm; pl from rhizome
        12. Phyllaries gen 3–5 mm wide, tips of outer gen reflexed at maturity . . . . *H. californicus*
        12′ Phyllaries gen < 3 mm wide, tips ± erect to reflexed at maturity
          13. Phyllaries gen 2–3 mm wide, tips gen spreading to reflexed . . . . *H. inexpectatus*
          13′ Phyllaries gen 1–1.5 mm wide, tips of outer ± spreading to erect . . . . *H. nuttallii*
            14. Phyllaries glabrous or strigose; lvs abaxially short-rough-hairy . . . . subsp. *nuttallii*
            14′ Phyllaries densely hairy; lvs abaxially ± finely tomentose . . . . subsp. *parishii*

H. annuus L. Ann ≤ 3 m. **ST**: ± rough-hairy. **LF**: most alternate; petiole 2–20 cm; blade 10–40 cm, ± widely lanceolate to triangular-ovate, base wedge-shaped to truncate or ± cordate, tip obtuse to acute, margin ± entire to serrate, abaxially bristly, sometimes gland-dotted. **INFL**: heads gen 1–9; peduncle 2–20 cm; involucre gen 1.5–4 cm diam; phyllaries 13–25 mm, gen 5–8 mm wide, lanceolate to widely ovate, abruptly acuminate, glabrous to rough-hairy, gen ciliate; paleae deeply 3-lobed, middle lobe long-acuminate. **RAY FL**: (13)15–30+; ray gen 5–50 mm. **DISK FL**: corolla 5–8 mm, throat proximally enlarged, goblet-shaped, lobes red to purple or yellow. **FR**: (3)4–5(15) mm; main pappus scales 2, lanceolate, 2–3.5 mm, smaller scales 0–4, obtuse. 2*n*=34. Disturbed areas, scrub, grassland, many other habitats; < 2000 m. CA; to e N.Am. Highly variable; hybridizes with several other ann spp. Many cult forms may be waifs. Robust subshrub from c SNF may be new sp. Jun–Oct

H. bolanderi A. Gray (p. 345) BOLANDER'S SUNFLOWER Ann ≤ 1.5 m. **ST**: rough-hairy. **LF**: most alternate; petiole 1–4 cm; blade 3–15 cm, narrowly lance-ovate to ovate, base wedge-shaped to truncate, tip gen acute, margin gen serrate, abaxially sparsely bristly, gland-dotted. **INFL**: heads 1–3; peduncle 3–13 cm; involucre 17–25 mm diam; phyllaries 8–27 mm, 3.5–5 mm wide, often >> disk, lanceolate to lance-ovate, long-acuminate, hairs long, soft to stiff; paleae

9.5–10.5 mm, >> disk fls, 3-toothed, middle tooth awn-tipped, tip glabrous. **RAY FL**: 12–17; ray 14–20 mm. **DISK FL**: corolla 5–7 mm, throat proximally enlarged, goblet-shaped, lobes gen red-purple. **FR**: 3.5–4.5 mm; pappus scales 1.7–3 mm. 2*n*=34. Grassy, often disturbed places; < 1800 m. KR, NCoR, CaR, SNF, GV, CW, WTR; s OR. Similar and closely related to *H. exilis*. Jun–Oct

H. californicus DC. (p. 345) CALIFORNIA SUNFLOWER Per 15–35 dm; roots thick, woody; rhizome short. **ST**: glabrous, glaucous, grooved. **LF**: all or most alternate; petiole 0–3 cm; blade 10–20 cm, gen lanceolate, entire or few-toothed, base ± wedge-shaped, tip acute, abaxially rough-hairy to bristly, gland-dotted. **INFL**: heads 3–10; peduncle (1)3–15 cm; involucre 10–25 mm diam; phyllaries (8)10–25 mm, gen 3–5 mm wide, gen >> disk, lanceolate, acuminate, tips of outer gen reflexed at maturity, margin glabrous or rough-ciliate, faces ± glabrous to short-rough-hairy; paleae 10–11 mm, 3-lobed, middle lobe acute, short-rough-hairy. **RAY FL**: 12–21; ray 15–30 mm. **DISK FL**: corolla 6–8 mm, lobes yellow. **FR**: 4.5–5 mm; pappus scales 3–4 mm. 2*n*=102. Springs, marshes, streambanks, canyons; < 1850 m. s NCoR, c SNF, s SNF (Piute Mtns), c SNH, ScV, nw SnJV, n CW, SW; n Baja CA. Jul–Oct

H. ciliaris DC. BLUEWEED Per 4–7 dm, from rhizome-like, horizontal roots, often forming dense colonies. **ST**: ± glabrous, glaucous.

LF: most opposite, sessile, 3–7.5 cm, oblong or lanceolate, base wedge-shaped, tip acute, margin entire to shallowly lobed, often crinkled, glabrous, sometimes ciliate, abaxially glabrous to sparsely short-rough-hairy or bristly, glaucous, blue-green. **INFL:** heads 1–5; peduncle 3–13 cm; involucre 12–25 mm diam; phyllaries gen unequal, 3–8 mm, 2–3 mm wide, < disk, oblong to ovate, tip obtuse or abruptly pointed to acute, ciliate, (±) glabrous; paleae 7–7.5 mm, ± entire or 3-toothed, tip hairy. **RAY FL:** 10–18; ray gen 8–9 mm. **DISK FL:** corolla 4–6 mm, lobes ± red. **FR:** 3–3.5 mm; pappus scales 1.2–1.5 mm. 2*n*=68,102. Roadsides, streambanks, low drainage areas, often in saline soils; < 650 m. GV, CCo, SCo, w DMoj; native to s-c US, n Mex. Jun–Nov ◆

H. cusickii A. Gray CUSICK'S SUNFLOWER Per 6–12 dm, from stout, ± fleshy taproot. **ST:** glabrous to sparsely long-hairy. **LF:** opposite or distal alternate; petiole 0–1 cm; blade 5–15 cm, ± lanceolate, base wedge-shaped, tip acute, margin entire, abaxially glabrous or scabrous to short bristly. **INFL:** heads 1–3; peduncle 2–15 cm; involucre 12–28 mm diam; phyllaries gen 12–18, 11–25 mm, 1.5–3 mm wide, gen > disk, ± lanceolate, gen acuminate, strigose or stiff-spreading-hairy; paleae 9–13 mm, entire or 3-toothed, sparsely hairy. **RAY FL:** 10–21; ray 20–40 mm. **DISK FL:** corolla 6.5–7.5 mm, lobes yellow. **FR:** 4–5 mm; pappus scales 3–4 mm. 2*n*=34. Dry grassy slopes, open woodland; 1050–1300 m. e KR, CaR, MP; to WA, ID, w NV. May–Jul

H. exilis A. Gray SERPENTINE SUNFLOWER Ann 1–10 dm. **ST:** rough-hairy. **LF:** most alternate; petiole 0.7–2.5 cm; blade 3–15 cm, lance-linear to lance-ovate, base wedge-shaped, tip gen acute, margin gen ± entire (shallowly serrate), abaxially sparsely bristly, gland-dotted. **INFL:** heads 1–7; peduncle gen 3–13 cm; involucre (10)15–20 mm diam; phyllaries 8–17 mm, 3–4 mm wide, gen >> disk, lanceolate, long-acuminate, hairs long, soft to stiff; paleae >> disk fls, 3-toothed, middle tooth awn-tipped, glabrous. **RAY FL:** 10–13; ray 1.4–2 cm. **DISK FL:** corolla 4–6 mm, throat proximally enlarged, goblet-shaped, lobes red-purple. **FR:** 2.7–3.5 mm; pappus scales 1.7–2.7 mm. 2*n*=34. Gravelly streamsides, often on serpentine; 300–1300 m. KR, NCoR. Similar and closely related to *H. bolanderi.* Jun–Oct ★

H. gracilentus A. Gray (p. 345) SLENDER SUNFLOWER Per (occ fl in 1st yr), 6–20 dm, from stout taproot. **ST:** rough-hairy. **LF:** opposite or distal alternate; petiole 0–3 cm; blade 5–11 cm, ± lanceolate, acute, entire or serrate, adaxially short-rough-hairy, abaxially bristly, gland-dotted. **INFL:** heads 1–5; peduncle (2.5)4–30 cm; involucre 13–20 mm diam; phyllaries gen unequal, 4–8 mm, 1.5–3 mm wide, ≤ disk, ± lanceolate, obtuse to acute, glabrous to short-rough-hairy, ciliate; paleae 8–9 mm, entire or weakly 3-toothed, tip short-ciliate. **RAY FL:** 13–21; ray 15–25 mm. **DISK FL:** corolla 4–6 mm, lobes yellow or red. **FR:** 3–4 mm; pappus scales 2.5–3 mm. 2*n*=34. Dry slopes, esp after fires, chaparral, woodland; < 2000 m. s NCoR, Teh, e SnFrB, SCoR, SW; n Baja CA. Apr–Oct

H. inexpectatus D.J. Keil & Elvin NEWHALL SUNFLOWER Per 15–50 dm; roots thick, woody; rhizome short. **ST:** ± glabrous. **LF:** alternate or opposite; petiole 0–3 cm; blade 10–20 cm, lanceolate or proximal elliptic, base wedge-shaped, tip acute, margin entire or few-toothed, abaxially strigose to short-rough-hairy. **INFL:** heads 1–7+ in rounded to ± flat-topped clusters; peduncle 3–15 cm; involucre body 1–2.5 cm diam; phyllaries 10–25 mm, 2–3 mm wide, often >> disk, linear or narrowly lanceolate, ascending to reflexed in fr, margin ciliate; paleae 7–8 mm, ± entire or shallowly 3-lobed, middle lobe ovate, obtuse or acute, short-rough-hairy. **RAY FL:** 12–21; ray 2–3 cm. **DISK FL:** corolla 5–6 mm, lobes yellow. **FR:** 3–3.8 mm; pappus scales 3–4 mm. 2*n*=68. Spring-fed marsh in willow woodland; 300 m. WTR (1 site along Santa Clara River near Newhall, Los Angeles Co.). Aug–Oct ★

H. niveus (Benth.) Brandegee subsp. ***tephrodes*** (A. Gray) Heiser (p. 349) ALGODONES DUNES SUNFLOWER Ann or per, ± subshrub, ≤ 15 dm, from taproot. **ST:** soft-white-, ± appressed-hairy. **LF:** most alternate; petiole 1.5–3.5 cm; blade gen 3–7 cm, triangular-ovate, base wedge-shaped, tip obtuse to acute, faces densely white-hairy, abaxially gland-dotted. **INFL:** heads 1–3; peduncle 4–17 cm; involucre 8–28 mm diam; phyllaries 8–10 mm, 1–3.5 mm wide, = or slightly > disk, ± lanceolate, acute, canescent; paleae 8–11 mm, = or slightly > disk fls, ± entire to deeply 3-lobed, acute, central paleae glabrous or finely appressed-hairy. **RAY FL:** 10–13; ray 16–25 mm. **DISK FL:** corolla 4.5–6 mm, throat proximally enlarged, goblet-shaped, lobes ± red. **FR:** 4–8 mm, densely long-hairy; pappus scales 1.5–3 mm, gen also several shorter scales. 2*n*=34. Sand dunes; gen < 100 m. s DSon (Imperial Co.); sw AZ, n Mex. Other subsp. in w Baja CA. Mar–May, Oct–Jan ★

H. nuttallii Torr. & A. Gray Per 5–50 dm, from clustered, tuber-like roots; rhizome short. **ST:** glabrous or hairy. **LF:** all or most alternate; petiole 0.5–1.5 cm; blade 10–20 cm, narrowly lanceolate to ovate, base wedge-shaped, tip acute to acuminate, margin entire or serrate, abaxially glabrous or bristly to short-tomentose, gland-dotted. **INFL:** heads (1) few to many in flat-topped or rounded cluster; peduncle 1–18 cm; involucre 1–2 cm diam; phyllaries 30–38, 8–16 mm, gen 1–1.5 mm wide, ± = or slightly > disk, ± linear, tips of outer spreading to erect at maturity, margins ciliate, faces ± glabrous or strigose to tomentose; paleae 8–12 mm, entire or 3-toothed, acute, short-rough-hairy. **RAY FL:** 12–21; ray 15–25 mm. **DISK FL:** corolla 5–7 mm, lobes yellow. **FR:** 3–4 mm; pappus scales 3–4 mm (sometimes also with shorter scales).

subsp. ***nuttallii*** (p. 349) NUTTALL'S SUNFLOWER **ST:** glabrous to scabrous. **LF:** adaxially scabrous, abaxially short-rough-hairy. **INFL:** peduncle glabrous; phyllaries glabrous or strigose. 2*n*=34. Damp meadows, springs, streams; 1200–1750 m. SnGb, SnBr, GB, w DMoj; to BC, e Can, NM. Jun–Oct

subsp. ***parishii*** (A. Gray) Heiser (p. 349) LOS ANGELES SUNFLOWER **ST:** glabrous to tomentose. **LF:** adaxially rough-hairy to densely tomentose, abaxially ± finely tomentose. **INFL:** peduncle and phyllaries densely hairy. Marshes; < 500 m. c-w SW. Last seen in 1937. Aug–Oct ★

H. petiolaris Nutt. PRAIRIE SUNFLOWER Ann 4–20 dm; strigose or densely gray-canescent. **LF:** most alternate; petiole 2–4 cm; blade 4–15 cm, lanceolate to widely ovate, base wedge-shaped, truncate (± cordate), tip obtuse to acute, margin entire or serrate, abaxially strigose, gland-dotted or not. **INFL:** heads 1–10+; peduncle 4–15+ cm; involucre 15–25 mm diam; phyllaries 9–14 mm, 1–5 mm wide, lanceolate to widely ovate, short acuminate, short rough-hairy to densely gray-canescent; paleae 3-lobed, middle lobe ± ciliate or hair-tufted, central paleae tipped with stiff white hairs. **RAY FL:** 10–30; ray 15–20 mm. **DISK FL:** corolla 4.5–6 mm, throat proximally enlarged, goblet-shaped, lobes red to purple or yellow. **FR:** 3–4.5 mm; pappus scales 1.5–3 mm + 0–2 shorter scales. 2*n*=34.

subsp. ***canescens*** (A. Gray) D.J. Keil **ST:** densely canescent-strigose and stiff-spreading-hairy. **LF:** densely canescent-strigose, abaxially densely glandular. **INFL:** phyllaries 2–3 mm wide. **DISK FL:** corolla throat gradually narrowed distally to slightly hairy basal bulge. Sandy soils; < 300 m. DSon. [*H. niveus* subsp. *c.* A. Gray] Mar–Jun

subsp. *petiolaris* **ST:** short-rough-hairy or strigose. **LF:** strigose, abaxially sparingly or not glandular. **INFL:** phyllaries 3–5 mm wide. **DISK FL:** corolla throat abruptly narrowed distally to densely hairy basal bulge 2–4 mm wide. 2*n*=34. Disturbed areas; gen < 450 m. SnFrB, SW; native to w-c US. May–Oct

HELICHRYSUM

Guy L. Nesom

[Ann to per] subshrub, shrub, often aromatic, taprooted. **ST:** tomentose, gen stalked- or sessile-glandular. **LF:** cauline; alternate, [sessile] petioled, linear to ovate or spoon-shaped, tapered to truncate at base [clasping or decurrent], entire, faces 1-[or 2-] colored, gen gray- to white-tomentose or silvery hairy, sometimes stalked- or sessile-glandular. **INFL:** heads [disciform or]

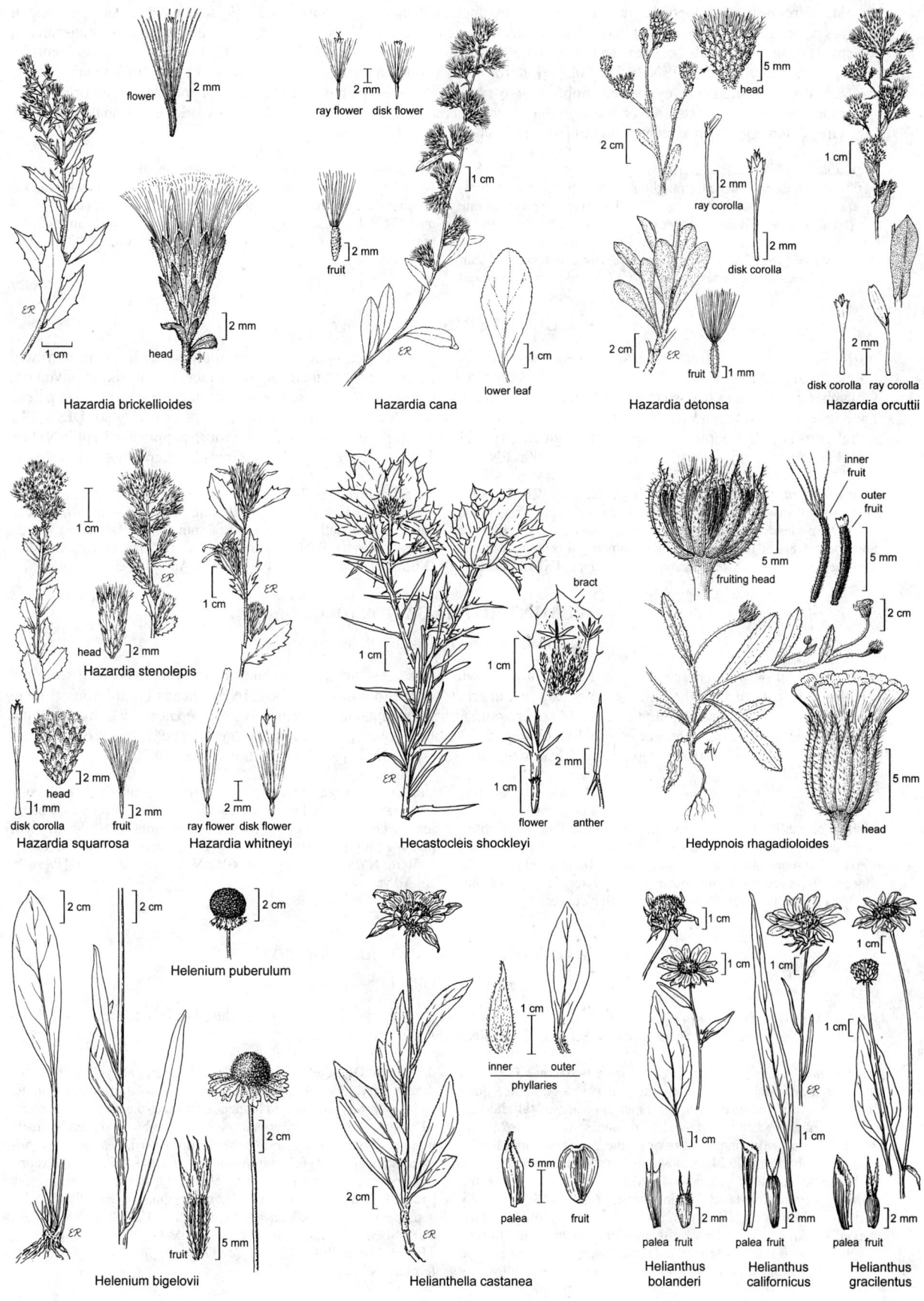

flower

2 mm

1 cm

Hazardia brickellioides

head

2 mm

ray flower   disk flower

2 mm

1 cm

fruit

2 mm

lower leaf

1 cm

Hazardia cana

head

5 mm

2 cm

ray corolla

2 mm

disk corolla

2 mm

2 cm

fruit

1 mm

Hazardia detonsa

1 cm

2 mm

disk corolla   ray corolla

Hazardia orcuttii

1 cm

1 cm

1 cm

head

2 mm

Hazardia stenolepis

head

2 mm

disk corolla

1 mm

fruit

2 mm

Hazardia squarrosa

2 mm

ray flower   disk flower

2 mm

Hazardia whitneyi

bract

1 cm

1 cm

1 cm

2 mm

flower   anther

Hecastocleis shockleyi

inner
fruit

outer
fruit

5 mm

5 mm

fruiting head

2 cm

head

5 mm

Hedypnois rhagadioloides

2 cm

2 cm

2 cm

Helenium puberulum

2 cm

2 cm

fruit

5 mm

Helenium bigelovii

1 cm

inner   outer

phyllaries

5 mm

palea   fruit

2 cm

Helianthella castanea

1 cm

1 cm

1 cm

1 cm

1 cm

1 cm

1 cm

1 cm

palea   fruit

2 mm

Helianthus
bolanderi

palea fruit

2 mm

Helianthus
californicus

palea   fruit

2 mm

Helianthus
gracilentus

discoid, in groups in ± flat-topped cyme-like clusters; involucre bell-shaped; phyllaries in 3–5[7] series, persistent, spreading in fr, bases green, gen sessile-glandular distally, distally ± white [straw-colored, orange, ± red, or ± pink], opaque [or transparent], gen shiny; receptacle [flat to] convex, epaleate, smooth or with short scales. **PISTILLATE FL**: 0 [to few, peripheral; corolla yellow, ± minutely lobed]. **DISK FL**: 3–30[50+]; corolla gen yellow; anther base with bristle-like tail, tip lanceolate; style tips truncate, hair-tufted. **FR**: cylindric, smooth or 4–6-ribbed, glabrous, strigose, or papillate; pappus of barbed or plumose bristles in 1 series, deciduous, free or loosely clinging together at base. ± 600 spp.: mostly Old World, esp Madagascar and s Afr. (Greek: sun + gold, an ancient pl name) [Nesom 2006 FNANM 19:425–426]

H. petiolare Hilliard & B.L. Burtt (p. 349) LICORICE PLANT **ST**: ≤ 100+ cm, loosely branched, straggling or trailing, sometimes rooting at soil contact, loosely gray-tomentose. **LF**: blade 1–3.5 cm, ovate to ± round, base widely tapered to truncate or ± cordate, tip obtuse to ± acute, faces silvery green, tomentose. **INFL**: heads discoid, many, in long-stalked rounded or flat-topped clusters; involucre 3–7 mm diam; phyllaries in ± 5 series, proximally appressed, green-centered with transparent margins, loosely tomentose, distally ascending to ± spreading, creamy white, glabrous, tips rounded; receptacle scales ± 0.8 mm, linear, dark brown in age. **DISK FL**: 18–30, corolla ± 2.5 mm. **FR**: ± 1 mm, widely cylindric, 5-ribbed, glabrous; pappus 3–3.5 mm. Cult as orn, escaped into coastal scrub, forest; < 200 m. CCo, SCo; Eur; native to s Afr. Jul–Aug ❖

HELIOMERIS GOLDEN-EYE

Per [ann]. **ST**: 1–many from base, slender, spreading or erect. **LF**: simple, opposite or distal alternate, linear to narrowly elliptic, entire. **INFL**: heads radiate, 1 or in few-headed cyme-like clusters; peduncle slender, bracts 0–few, linear; involucre hemispheric, appearing disk-shaped when pressed; phyllaries in 2–3 ± equal series, linear; receptacle conic, paleate, paleae lance-linear, folded around frs. **RAY FL**: sterile; corolla yellow-orange; ray showy, oblong, entire or nearly so. **DISK FL**: many; corolla yellow-orange; style tips triangular. **FR**: oblanceolate, ± 4-angled, flat, glabrous; pappus 0. 5 spp.: N.Am. (Greek: sun part, from showy heads) [Schilling 2006 FNANM 21:169–172] *H. hispida* (A. Gray) Cockerell not in CA.

H. multiflora Nutt. var. *nevadensis* (A. Nelson) W.F. Yates (p. 349) NEVADA GOLDENEYE Per from branched, woody rootstock. **ST**: 3–9 dm, slender, erect or spreading, glabrous or finely strigose or puberulent. **LF**: 2–6 cm, linear to oblong or narrowly elliptic, ± stiffly hairy near base; tip obtuse to acute; faces strigose. **INFL**: peduncle 5–15 cm; involucre 7–13 mm diam; phyllaries 4–7 mm, linear, strigose. **RAY FL**: 8–15; ray 15–20 mm, oblong to ovate. **DISK FL**: corolla 3 mm; anthers yellow. **FR**: 2 mm. 2*n*=16. Dry, rocky slopes, upland valleys; (800)1200–2500 m. SNE, DMtns; to UT, NM, n Mex. Other vars. widespread in mtns of w N.Am. May–Sep

HELMINTHOTHECA OX-TONGUE

David J. Keil & G. Ledyard Stebbins

Ann, bien; sap milky; herbage with rigid hairs, most of which have 2–4-hooked branches at tip, often also with scattered prickles. **ST**: 1, branched, lfy. **LF**: basal and cauline, alternate, entire to pinnately lobed. **INFL**: heads liguliflorous, 1 or in cyme- or panicle-like clusters; phyllaries in 2–3 overlapping series; receptacle flat, minutely hairy, epaleate. **FL**: many; ligule yellow, readily withering, outer occ sterile. **FR**: oblong, 5–10-ribbed, covered with minute ridges, beaked; pappus of 1–2 rows of plumose or barbed bristles. 4 spp.: Eur. (Greek: worm case, describing fr) [Strother 2006 FNANM 19:300]

H. echioides (L.) Holub (p. 349) BRISTLY OX-TONGUE **ST**: coarse, stout, 3–20 dm. **LF**: 5–20 cm, oblong, entire, coarsely toothed, or shallowly lobed; proximal tapered to winged petioles; distal sessile, sometimes clasping. **INFL**: heads 2–4 cm diam, terminal and axillary, short- to long-peduncled in lfy-bracted cyme-like clusters; involucre 15–20 mm; outer phyllaries spreading to ascending, lf-like, widely ovate, cordate, >> and often concealing the erect, lanceolate, tapered, inner phyllaries. **FR**: 5–7.5 mm (incl beak); body ± = beak, outer ± white to straw-colored or tan, adaxially ± hairy, inner ± brown, transversely roughened, glabrous; beak slender; pappus 4–7 mm, plumose, white. 2*n*=10. Common. Disturbed areas; < 1050 m. NW, n SNF, GV, CW, SW; N.Am; native to Eur. [*Picris e.* L.] All yr ❖

HEMIZONELLA MINIATURE TARWEED

Bruce G. Baldwin

1 sp. (Diminutive *Hemizonia*) [Carlquist et al. 2003 Tarweeds and silverswords: evolution of the Madiinae (Asteraceae). Missouri Botanical Garden Press; Baldwin & Strother 2006 FNANM 21:296–297]

H. minima (A. Gray) A. Gray (p. 349) Ann 1–20 cm. **ST**: ± erect, branches often widely divergent, minutely coarse-hairy, stalked-glandular. **LF**: basal and/or cauline, proximal opposite, distal alternate or clustered immediately proximal to branches, sessile, 5–25 mm, (0.5)1–2.5 mm wide, linear, entire or toothed, coarse-hairy, distally glandular-puberulent. **INFL**: heads radiate (rays occ obscure), 1 or in flat-topped clusters; peduncle thread-like; involucre (1)2–4 mm diam, ± obovoid; phyllaries 3–5 in 1 series, each mostly or wholly enclosing a ray ovary, falling with fr, (1)2–4 mm, ± oblanceolate, abaxially coarse-hairy and stalked-glandular; receptacle flat to convex, glabrous or sparsely bristly; paleae in 1 series between ray and disk fls, fused, deciduous. **RAY FL**: 3–5; corolla pale yellow, ray 0.5–1 mm. **DISK FL**: 1–2; corolla 1–2.5 mm, pale yellow, tube ± = throat, hairy, lobes deltate; anthers yellow, tips hemispheric, minute; styles glabrous proximal to branches, tips awl-shaped, densely hairy. **FR**: ray fr 1.8–2.8 mm, compressed front-to-back, arched, sparsely minutely bristly or ± glabrous, black, tip beaked, beak straight, oriented adaxially, 0.1–0.2 mm, pappus 0; disk fr 1.8–2.8 mm, ± round in ×-section, club-shaped, ± minutely bristly, black, tip beaked, beak straight, oriented vertically, 0.1–0.2 mm, pappus 0. Gravelly or rocky, gen open sites in scrub, meadows, forest; 300–2900 m. KR, NCoR, CaR, SN, se SnFrB, SCoRO (Hanging Valley, Santa Lucia Mtns), TR, PR, MP; to WA, MT, NV. [*Madia m.* (A. Gray) D.D. Keck] Self-fertile. Apr–Aug

HEMIZONIA   HAYFIELD TARWEED

Bruce G. Baldwin

1 sp. (Greek: 1/2 belt or girdle, fr 1/2 enfolded by phyllary) [Baldwin & Strother 2006 FNANM 21:291–293] Other taxa in TJM (1993) moved to *Centromadia* (spiny-lvd taxa), *Deinandra*.

H. congesta DC.   Ann 5–80 cm, often aromatic. **ST**: ± erect. **LF**: basal and cauline, proximal opposite or in rosette, sometimes persistent, most alternate, sessile, 5–18+ cm, 2–8(12) mm wide, narrowly elliptic to linear or lance-linear, minutely serrate or entire, gen puberulent or coarse- to silky-hairy, distal often also stalked-glandular. **INFL**: heads radiate, 1 or in ± panicle-, raceme-, or spike-like clusters or in tight groups; involucre hemispheric to ± urn-shaped or spheric, 3–8+ mm diam (rarely subtended by calyx-like set of 5–7 bractlets); phyllaries 5–14 in 1 series, linear to lanceolate or oblanceolate, each gen 1/2 enveloping a subtended ray ovary, falling with fr, 3.5–12 mm, coarse- or soft-hairy, stalked-glandular; receptacle flat to conic, glabrous, paleae fused, forming cells around each disk fl, scarious, ± liquifying. **RAY FL**: 5–14; corolla white or yellow, ray 5–12 mm, often purple-veined abaxially. **DISK FL**: 5–60+, staminate; corolla 2.5–3.5 mm, white or yellow, tube < throat, lobes deltate; anthers ± dark purple, tips widely ovate to ovate-deltate; style glabrous proximal to branches, tips lanceolate to awl-shaped, densely hairy. **FR**: ray fr 2–3.5 mm, nearly round in ×-section (exc ± flattened adaxially) or ± 3-angled (abaxially gen ± widely 2-faced, adaxially ± flattened to slightly bulging), glabrous, black, tip sometimes beaked, beak inconspicuous, straight, diam > length, pappus 0; disk fr 0, pappus 0. 2*n*=28. Self-sterile.

1. Heads gen nearly sessile along well-developed side branches of fl sts; ray white, purple-veined abaxially
2. Peduncle bracts and calyx-like bracts subtending head often much exceeding phyllaries; phyllaries 6–12 mm, tips gen > bodies; fr width 0.5–0.6 × length . . . . subsp. *calyculata*
2' Peduncle bracts not or barely exceeding phyllaries; calyx-like bracts subtending head 0; phyllaries 3–7 mm, tips gen < bodies; fr width 0.6–0.75 × length . . . . subsp. *clevelandii*
1' Heads gen terminating elongate side branches of fl sts or ± sessile in tight groups; ray white or yellow, purple-veined abaxially or not
3. Lvs gen puberulent or minutely bristly or strigose and nonglandular, distal rarely long-soft-hairy and glandular; ray white, not purple-veined abaxially exc NCo; heads in panicle-like cluster . . . . subsp. *tracyi*
3' Lvs short-hairy, ± shaggy, or silky-hairy, all or distal glandular; ray white or yellow, purple-veined abaxially; heads in panicle-like cluster or in tight groups
4. Lvs ± shaggy, hairs at margins often longer on distal lvs; some or all heads gen in tight groups, sometimes in flat-topped or panicle-like cluster; phyllaries 6–10 mm, tips gen > bodies; ray white . . . . subsp. *congesta*
4' Lvs short-hairy, ± shaggy, or silky-hairy, hairs at margins not notably longer on distal lvs; heads in panicle-like cluster; phyllaries 3–7(8) mm, tips gen < bodies; ray white or yellow
5. Ray yellow . . . . subsp. *lutescens*
5' Ray white . . . . subsp. *luzulifolia*

subsp. *calyculata* Babc. & H.M. Hall (p. 349)   MENDOCINO TARPLANT   **LF**: minutely bristly or strigose and often coarse-hairy; all or distal densely stalked-glandular. **INFL**: heads gen nearly sessile along well-developed side branches of spike- to raceme- or panicle-like clusters; peduncle 0 or 1–3 mm, bracts > phyllaries; calyx-like bracts subtending head 5–7; phyllaries 6–12 mm, tips gen > bodies. **RAY FL**: 5–8; corolla white, ray purple-veined abaxially. **FR**: width 0.5–0.6 × length. Clay soils, grassland, openings in woodland; 200–1400 m. NCoR. Jul–Nov   ★

subsp. *clevelandii* (Greene) Babc. & H.M. Hall (p. 349)   **LF**: coarse-hairy to ± shaggy and often minutely bristly or strigose; all or distal gen stalked-glandular. **INFL**: heads gen nearly sessile along well-developed side branches of spike- to raceme- or panicle-like clusters; peduncle 0 or 1–4 mm, bracts not or barely > phyllaries; calyx-like bracts subtending head 0; phyllaries 3–7 mm, tips gen < bodies. **RAY FL**: 5–8; corolla white, ray purple-veined abaxially. **FR**: width 0.6–0.75 × length. Dry sites, grassland, opening in chaparral and woodland; 40–1600 m. s KR, NCoR, w edge n SNF, ScV; sw OR. Jun–Nov

subsp. *congesta* (p. 349)   CONGESTED-HEADED HAYFIELD TARPLANT   **LF**: ± shaggy, hairs often longer at margins on distal lvs; all or distal glandular, often sparsely (nonglandular). **INFL**: some or all heads gen in tight groups, sometimes in flat-topped or panicle-like cluster; peduncle 0 or 1–4 mm, bracts not > phyllaries; calyx-like bracts subtending heads 0; phyllaries 6–10 mm, tips gen > bodies. **RAY FL**: 5–13; corolla white, ray purple-veined abaxially. **FR**: width 0.5–0.6 × length. Grassy sites, marsh edges; < 100 m. s NCoRO, sw NCoRI, n CCo, w SnFrB. [*H. c.* subsp. *leucocephala* (Tanowitz) D.J. Keil] Hybridizes with *H. congesta* subsp. *lutescens*. May–Nov   ★

subsp. *lutescens* (Greene) Babc. & H.M. Hall   **LF**: short-hairy, ± shaggy, or silky-hairy, hairs not notably longer at margins on distal lvs; all or distal glandular. **INFL**: heads in panicle-like cluster; peduncle 0 or 1–20 mm, bracts not > phyllaries; calyx-like bracts subtending heads 0; phyllaries 4.5–6.5(8) mm, tips gen < bodies. **RAY FL**: 5–14; corolla yellow, ray purple-veined abaxially. **FR**: width 0.4–0.63 × length. Grassland, barrens, openings in chaparral and woodland, often on serpentine; < 500 m. s NCo, s NCoRO, n CCo, n SnFrB. [*H. c.* subsp. *c.*, misappl. in TJM (1993)] Apr–Dec

subsp. *luzulifolia* (DC.) Babc. & H.M. Hall (p. 349)   **LF**: short-hairy, ± shaggy, or silky-hairy, hairs not notably longer at margins on distal lvs; all or distal lvs glandular. **INFL**: heads in panicle-like cluster; peduncle 0 or 1–30 mm, bracts not > phyllaries; calyx-like bracts subtending heads 0; phyllaries 3.5–6.5 mm, tips gen < bodies. **RAY FL**: 5–11; corolla white, ray purple-veined abaxially. **FR**: width 0.5–0.6 × length. Disturbed, open, or grassy sites, often clayey soils, serpentine; < 1000 m. NCoRO, NCoRI, w edge n SNF, GV, CW (exc SCoRI). Hybridizes with *H. congesta* subsp. *lutescens*. Mar–Dec

subsp. *tracyi* Babc. & H.M. Hall   TRACY'S TARPLANT   **LF**: gen puberulent or minutely bristly or strigose and gen nonglandular throughout; distal long-soft-hairy and glandular. **INFL**: heads in panicle-like cluster; peduncle 0 or 1–18 mm, bracts not > phyllaries; calyx-like bracts subtending heads 0; phyllaries 3–10.5 mm, tips gen slightly < bodies. **RAY FL**: 5–8(13); corolla white, ray not purple-veined abaxially exc NCo. **FR**: width 0.5–0.6 × length. Grassy sites, riverbanks, openings in scrub, woodland, forest; < 1200 m. NCo (Cape Mendocino, Humboldt Co.), n&c NCoRO. May–Nov   ★

HESPEREVAX

James D. Morefield

Ann 0.5–10(20 cm), ± green to gray, woolly to cobwebby. **ST**: 0, or 1, erect, or 2–10, ± decumbent to ascending, gen unbranched or branched proximally, equally or more lfy distally. **LF**: alternate, sessile or petioled, oblanceolate to spoon-shaped, entire, base ± yellow, ± hard; distal lvs subtending heads, ± crowded, largest ≥ proximal lvs. **INFL**: heads disciform, sessile, single or

2–40+ per group; involucre 0, simulated by paleae or lvs; receptacle length 0.8–1.3 or 4–6 × width, ± conic, acuminate, bristly; paleae of pistillate fls open, lanceolate to obovate, acute, flat to concave or ± folded, persistent, ± glabrous, obscurely parallel-veined, veins 5+, margin ± continuous, thinned but scarcely winged; paleae of disk fls gen 5, ± enlarged, erect proximally, erect to spreading distally, open, obovate, obtuse, cartilaginous, proximally ± yellow, distally green and tomentose adaxially. **PISTILLATE FL**: (3)5–25, all subtended by paleae; corolla obscure, narrowly cylindric. **DISK FL**: staminate, 2–6(12); pappus 0; corolla (3)4(5)-lobed, ± bilateral; anther base tailed, tip ± triangular; style tips ± linear-oblong; pappus 0. **FR**: falling free of paleae, ± obovoid, compressed front-to-back, smooth, dull to ± shiny, corolla scar terminal; pappus 0. 3 spp.: w CA, s OR. (Greek: western *Evax*) [Morefield 2006 FNANM 19:467–470] Apparent sister of *Ancistrocarphus*; bristly receptacle unique. Historical CA specimens of *Diaperia* [*Evax* sect. *Diaperia*], native c AZ and eastward, are likely label errors.

1. Distal heads 3–5 per group, ± terminal, ± cylindric, length 1.8–2.5 × width; groups loose, subtended by gen
     1–4 lvs; paleae of disk fls 0.3–0.4 × head, tips incl, ± erect . . . . *H. sparsiflora*
   2. Largest lvs 6–12(14) mm, 3–5(6) mm wide; blade ± round, woolly . . . . var. *brevifolia*
   2´ Largest lvs (10)13–32 mm, 4–8(10) mm wide; blade oblanceolate to obovate, silky-cobwebby . . . . var. *sparsiflora*
1´ Distal heads 1 or (2)10–40+ per group, terminal, bell-shaped or ± obconic, length 1–1.5 × width; groups
     dense, subtended by 6–20 lvs; paleae of disk fls 0.6–0.9 × head, tips exserted, spreading
   3. Longest lvs (25)33–90 mm, 7–20 mm wide; petiole 2–3 × blade, base strongly thickened; heads 10–40+
        per group, largest group 10–25 mm wide, subtended by (not mixed with) lvs . . . . *H. caulescens*
   3´ Longest lvs 4–22(32) mm, < 5 mm wide; petiole 0–1.5 × blade, base scarcely thickened; head gen 1,
        distal heads rarely 2–8 per group, largest group 3–7 mm wide, mixed with lvs . . . . *H. acaulis*
      4. St ± erect; largest lvs (9)12–22(32) mm, 2–4(5) mm wide; longest paleae of disk fls 2.5–3.2 mm . . . . . . var. *robustior*
      4´ St ± prostrate or 0; largest lvs gen 4–12 mm, 0.5–2 mm wide; longest paleae of disk fls 1.6–2.4 mm
         5. Largest lvs gen 7–12 mm, 0.5–2 mm wide; petiole 0–0.8 × blade; blade oblanceolate, acute. . . . . var. *acaulis*
         5´ Largest lvs gen 4–7 mm, 1–2 mm wide; petiole gen 0.5–1.5 × blade; blade obovate to round, obtuse
            . . . . var. *ambusticola*

H. acaulis (Kellogg) Greene DWARF EVAX   **ST**: 0 or 1–10, prostrate to erect, to 7 cm. **LF**: distal > proximal, longest 4–22(32) mm, < 5 mm wide; petiole 0–1.5 × blade, base scarcely thickened. **INFL**: distal heads gen 1, terminal, 2–4 mm, 1.5–3.5 mm wide, ± bell-shaped, length 1–1.5 × width, subtended by gen 6–12 lvs (in vars. *acaulis* and *robustior* distal heads rarely 2–8 per group, largest group dense, 3–7 mm wide, mixed with lvs); proximal heads 0 or axillary, < distal; receptacle length 1–1.3 × width; paleae in spiral ranks; paleae of disk fls 0.6–0.8 × head, tips exserted, spreading. **DISK FL**: 0.6–1 mm. **FR**: 0.6–1.6 mm. Vars. intergrade.

   var. *acaulis* (p. 349) STEMLESS EVAX   Pls 1–2(4) cm. **ST**: gen 0, sometimes 1(7), ± prostrate; branches 0. **LF**: largest gen 7–12 mm, 0.5–2 mm wide, distal ± erect; petiole 0–0.8 × blade; blade oblanceolate, acute. **INFL**: distal heads 2.5–3 mm, 2–2.5 mm wide; receptacle 0.8–1.4 mm; paleae of pistillate fls gen in 3–4 series, 2–3 mm; longest paleae of disk fls 1.9–2.4 mm. **FR**: gen 0.6–0.8 mm. Grassland, swales, moss turf, chaparral; 30–900 m. SNF, n&e GV, e SnFrB, SCoRO; sw OR. Gen at lower elevations when near other vars. Recent OR record may be alien. Mar–May

   var. *ambusticola* Morefield (p. 349) FIRE EVAX   Pls 0.5–2(4) cm. **ST**: (0)4–10, ± prostrate (when depauperate 0 or 1, erect, unbranched); branches proximal and/or distal. **LF**: largest gen 4–7 mm, 1–2 mm wide, distal gen spreading; petiole gen 0.5–1.5 × blade; blade obovate to round, obtuse. **INFL**: distal heads 2–2.5 mm, 1.5–2 mm wide; receptacle 1–1.4 mm; paleae of pistillate fls gen in 2–3 series, 1.7–2.5 mm; longest paleae of disk fls 1.6–2 mm. **FR**: gen 0.8–1 mm. Clearings, barren slopes, burns; 200–1300 m. s NCoRI, n&c SNF, s SNH, sw SnJV, e SnFrB, SCoR, SnGb/SCo, nw PR (exc SnJt). Depauperate pls among smallest Asteraceae. Apr–Jun

   var. *robustior* Morefield (p. 349) Pls gen 2–7 cm. **ST**: 1(7), ± erect; branches gen 0. **LF**: largest (9)12–22(32) mm, 2–4(5) mm wide, distal erect or tips spreading; petiole 0–0.8 × blade; blade oblanceolate to obovate, acute to obtuse. **INFL**: distal heads 3–4 mm, 2.5–3.5 mm wide; receptacle 1.4–1.9 mm; paleae of pistillate fls gen in 3–5 series, 2.5–4 mm; longest paleae of disk fls 2.5–3.2 mm. **FR**: gen 1–1.6 mm. Slopes, flats, swales, canyons, path edges, openings or under shrubs; 60–1100 m. KR, NCoRO, NCoRI, CaRF, SNF, n&e GV, SCoRO; to sw OR. Largest sizes above from cult specimen. Apr–Jun

H. caulescens (Benth.) A. Gray (p. 349) HOGWALLOW STARFISH   **ST**: gen 0, sometimes 1–4, erect to decumbent, to 8(17) cm. **LF**: distal > proximal, longest (25)33–90 mm, 7–20 mm wide; petiole 2–3 × blade, base strongly thickened. **INFL**: distal heads 10–40+ per group, terminal, 3–5 mm, 2.5–4 mm wide, ± obconic, length 1–1.5 × width; largest group dense, 10–25 mm wide, subtended by (not mixed with) (6)10–20 lvs; proximal heads 0; receptacle length 0.8–1 × width; paleae in vertical ranks; paleae of disk fls 0.7–0.9 × head, tips exserted, spreading. **DISK FL**: 1.1–1.6 mm. **FR**: gen 1.5–2 mm. Declining. Drying shrink-swell clay of vernal pools, flats, steep slopes (sometimes serpentine); < 300(500) m. NCoRI, CaRF, n&s SNF, GV, SCoRO, sw PR (reportedly alien, likely extirpated). Scattered tall, gray-tomentose, narrowly lvd pls (*Evax involucrata* Greene) not distinct. Consistently stless, smaller pls in s range may be unnamed var. Mar–Jun ★

H. sparsiflora (A. Gray) Greene ERECT EVAX   **ST**: (0)1–10, ascending to erect, to 17 cm. **LF**: ± equal, longest 6–32 mm, 3–8(10) mm wide; petiole 0.9–1.5 × blade, base thickened. **INFL**: distal heads 3–5 per group,± terminal, 3–4.5 mm, 1.5–2 mm wide, ± cylindric, length 1.8–2.5 × width; largest group loose, 3–4 mm wide, subtended by (and mixed with) gen 1–4 lvs; proximal heads axillary, = distal; receptacle length 4–6 × width; paleae in spiral ranks; paleae of disk fls 0.3–0.4 × head, tips incl, ± erect. **DISK FL**: 0.8–1.1 mm. **FR**: gen 1–1.7 mm. Vars. distinctive; some intermediates may occur in SnFrB.

   var. *brevifolia* (A. Gray) Morefield (p. 349) SHORTLEAVED EVAX   Pl gen 3–9 cm, ± green. **LF**: largest 6–12(14) mm, 3–5(6) mm wide; blade ± round, woolly. **INFL**: longest heads 3–3.7 mm. Sandy, grassy or wooded coastal bluffs, terraces, dunes; < 100(300) m. NCo, n CCo; to sw OR. Mar–Jul ★

   var. *sparsiflora* (p. 349) Pl (2)8–17 cm, ± green to ± gray. **LF**: largest (10)13–32 mm, 4–8(10) mm wide; blade oblanceolate to obovate, silky-cobwebby. **INFL**: longest heads 3.6–4.5 mm. Common. Open, clay and/or rocky, gen serpentine soil; 10–1000 m. s NCoR, c SNF, deltaic GV, CW, ChI, w PR (exc SnJt). Santa Rosa Island pls (uniformly stless) possibly unnamed taxon. Mar–Jun

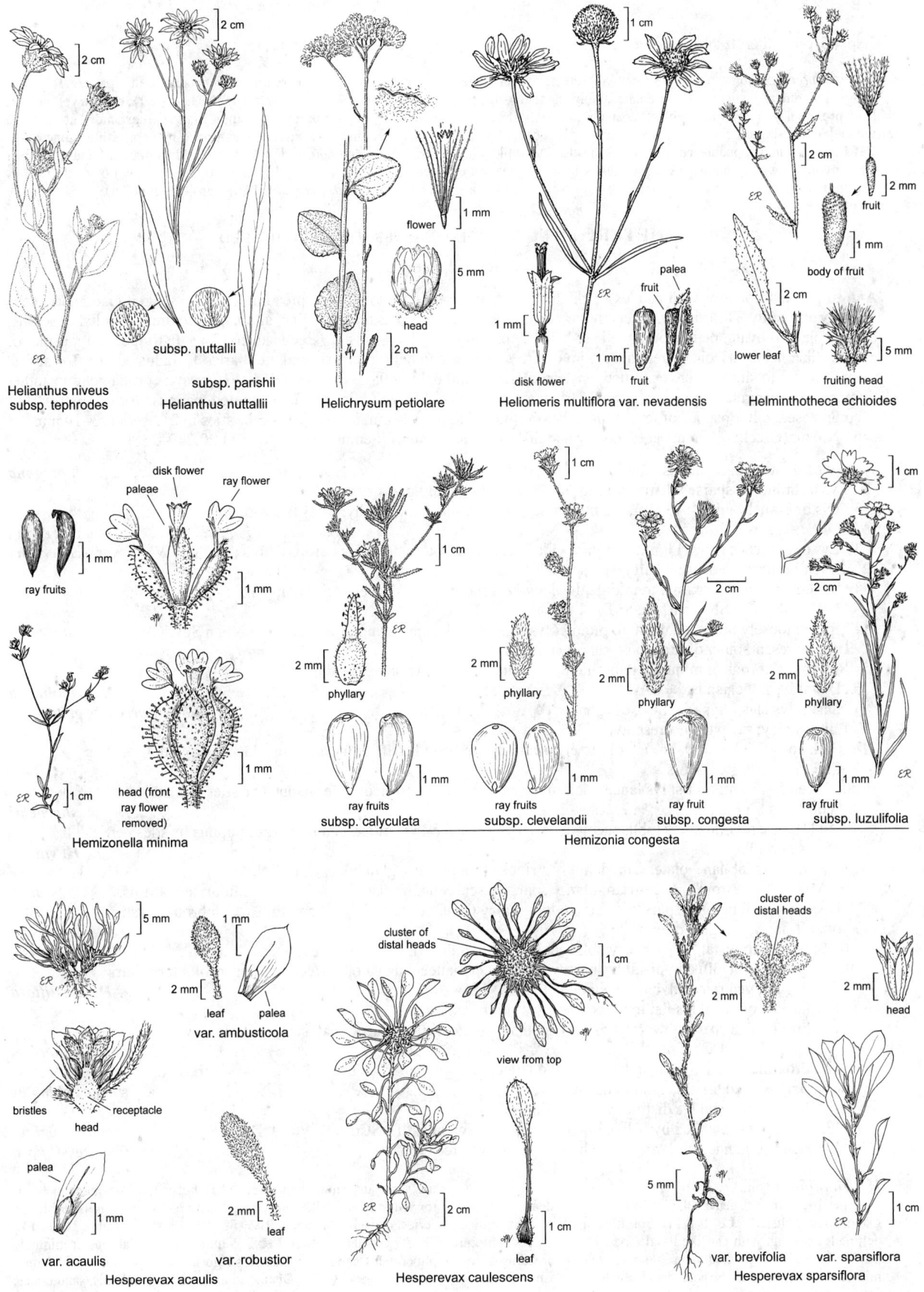

Helianthus niveus subsp. tephrodes

subsp. nuttallii

subsp. parishii
Helianthus nuttallii

Helichrysum petiolare

Heliomeris multiflora var. nevadensis

flower

head

disk flower

palea
fruit
fruit

Helminthotheca echioides

body of fruit

fruit

lower leaf

fruiting head

ray fruits

disk flower
paleae   ray flower

head (front ray flower removed)

Hemizonella minima

phyllary

ray fruits
subsp. calyculata

phyllary

ray fruits
subsp. clevelandii

phyllary

ray fruit
subsp. congesta

phyllary

ray fruit
subsp. luzulifolia

Hemizonia congesta

leaf   palea
var. ambusticola

bristles   receptacle
head

palea

var. acaulis

leaf
var. robustior

Hesperevax acaulis

cluster of distal heads

view from top

leaf

Hesperevax caulescens

cluster of distal heads

head

var. brevifolia   var. sparsiflora
Hesperevax sparsiflora

HETERANTHEMIS

1 sp. (Greek: other *Anthemis*, from similarity to that genus) [Strother 2006 FNANM 19:551]

H. viscidihirta Schott   Ann 20–80 cm. **ST**: erect, few-branched above, glandular, ± sticky. **LF**: alternate, finely glandular-puberulent; proximal 4–6 cm, oblanceolate, coarsely pinnately lobed and toothed; distal shorter, oblong, toothed or shallowly lobed, clasping. **INFL**: heads radiate, peduncled, 1 or in few-headed cyme-like clusters; involucre widely hemispheric; phyllaries in 2–3 series, outer 2–6 mm, ± triangular, inner 6–10 mm, oblong to widely obovate, margins and tip widely scarious; receptacle conic, epaleate. **RAY FL**: 13–21; ray 2–2.5 cm, yellow. **DISK FL**: 40–80+; corolla yellow. **FR**: glabrous; pappus 0; ray fr ± 4 mm, triangular, expanded at tip into a thickened disk, 3-winged, wings tipped by pappus-like spines; disk fr ± 4 mm, compressed, 1–2-winged. Uncommon. Sandy soil near coast, disturbed sites; < 100 m. CCo; native to sw Eur, n Afr. [*Heteranthemis viscidehirta*, orth. var.] (Jan)Mar–Jun(Oct)

HETEROTHECA   GOLDENASTER, TELEGRAPH WEED

John C. Semple

Ann to per, taprooted or with caudex or rhizome; herbage ± strigose-bristly or spreading-hairy, ± stalked-glandular, gen ± strongly aromatic. **ST**: 1–many, ascending or erect, ± branched. **LF**: basal and cauline, alternate; proximal petioled or sessile, oblanceolate to ovate, petiole or blade base gen ± spreading-long-hairy; distal ± reduced. **INFL**: ± flat-topped to raceme- or panicle-like; heads discoid or gen radiate; involucre ± cylindric to bell- or urn-shaped; phyllaries ± graduated in ± 3–7 series; receptacle flat to slightly convex, pitted, epaleate. **RAY FL**: 0 or 10–40; ray yellow. **DISK FL**: many; corolla yellow; anther tip narrowly triangular; style tips narrowly triangular, papillate. **FR**: obconic; ray fr ± 3-angled, pappus 0 or of bristles; disk fr compressed, outer pappus of few to many bristle-like scales 0.2–1 mm, inner of 30–45 bristles in 2–3 series, 3–10 mm. 28 spp.: N.Am. (Greek: different cases, for ray and disk frs of some spp.) [Semple 2006 FNANM 20:230–256]

1. Ray fls 0; outer disk pappus scales few — gen dry streambeds . . . . ***H. oregona***
   2. Lvs of branches sparsely hairy or scabrous, ± densely glandular
      3. Lvs of branches gen > 13 mm, > 3 mm wide, sparsely hairy; outer phyllaries lance-deltate; NW, CW . . . . var. ***oregona***
      3′ Lvs of branches gen < 13 mm, < 3 mm wide, ± scabrous; outer phyllaries ovate-deltate; SnJV, CW . . . var. ***scaberrima***
   2′ Lvs of branches ± densely hairy, sparsely glandular
      4. St gen branched proximal to middle; lvs long-soft-hairy and ± densely bristly, appearing gray-green; NW, CaR, n&c SNF, n SNH, SnJV, CW. . . . . var. ***compacta***
      4′ St gen loosely branched distal to middle; lvs short-bristly, appearing green; NW, n SNF, n ScV, CW . . . . . . var. ***rudis***
1′ Ray fls present; outer disk pappus scales gen many
   5. Ray fr ± glabrous or minutely puberulent; ray pappus 0; ann to short-lived per
      6. Distal lvs not clasping; ± CA-FP . . . . ***H. grandiflora***
      6′ Distal lvs clasping; Teh, SCo, e DSon. . . . . ***H. subaxillaris*** subsp. ***latifolia***
   5′ Ray fr hairy; ray pappus present; per
      7. Disk corolla lobes ± glabrous or sparsely minutely strigose; lf stiff-hairy, margins flat or inrolled — CaR, SN, MP, sw DMtns
         8. St gen 5–13 dm; mid-st lvs lance-triangular, margin gen ± inrolled; inner pappus bristles ≤ disk corolla; s SNF. . . . . ***H. shevockii***
         8′ St (1)2–5 dm; mid-st lvs (ob)lanceolate or narrowly lance-triangular, gen flat; inner pappus bristles gen ≥ disk corolla . . . . ***H. villosa***
            9. Mid-st lvs (ob)lanceolate, ± moderately strigose (not bristly), glandular; CaR, SN, MP . . . . var. ***minor***
            9′ Mid-st lvs narrowly lance-triangular, ± sparsely scabrous to bristly, ± densely glandular; sw DMtns . . . var. ***scabra***
      7′ Disk corolla lobes gen sparsely short- to long-hairy (coastal or SnFrB pls sometimes ± glabrous); lf stiff- or soft-hairy, margins often wavy
         10. St 9–18 cm; distal lvs narrowly oblanceolate, little reduced, bases tapering; SN . . . . ***H. monarchensis***
         10′ St gen 20–70(130) cm; distal lvs narrowly to widely lanceolate (if oblanceolate then corolla lobe hairs to 1 mm), gen reduced, bases rounded; Teh, CW, SW . . . . ***H. sessiliflora***
            11. Margins of distal lvs flat to weakly wavy, lf hairs ± 2 mm
               12. Proximal st sparsely woolly; distal lvs oblanceolate, little reduced, flat, strigose to long-woolly; c&s NCo, n CCo, SnFrB. . . . . subsp. ***bolanderi***
               12′ Proximal st bristly; distal lvs elliptic to lanceolate, gen much reduced, occ ± wavy-margined, strigose and bristly; gen inland c&s CA . . . . subsp. ***echioides***
            11′ Margins of distal lvs distinctly wavy, lf hairs ± 1 mm
               13. Head not subtended by lf-like bracts; distal lvs ± white, stiff; SCo, TR, PR, w D ≥ 100 m. . . . . . , subsp. ***fastigiata***
               13′ Head subtended by large lf-like bracts; distal lvs green, not stiff; s CCo, SCo < 60 m . . . . subsp. ***sessiliflora***

H. grandiflora Nutt. (p. 355)   TELEGRAPH WEED   Ann to short-lived per 1–25 dm. **ST**: gen erect, distally branched, ± densely bristly, glandular, esp distally. **LF**: basal (in rosettes) and proximal cauline petioled, clasping, with ear-like basal lobes, blade ovate to elliptic or oblong, entire or toothed, densely appressed- to spreading-hairy, mid-cauline lanceolate, ± sessile, distal sessile, not clasping, ascending, less hairy, more glandular. **INFL**: heads few–gen many in ± flat-topped to panicle-like cluster; involucre 6–10 mm, phyllaries in 4–6 series, densely glandular. **RAY FL**: 25–40; ray 5–8 mm. **DISK FL**: 30–75; corolla 4–6 mm. **FR**: 2–5 mm; ray fr ± glabrous or minutely puberulent, pappus 0; disk fr strigose, outer pappus 0.2–0.7 mm, inner 3–5 mm. $2n=18$. Disturbed areas, dry streambeds, sand dunes;

< 1100 m. s NW, SNF, s Teh, GV, CW, SW, D (uncommon); nw Mex; introduced in AZ, UT. Gen Jun–Oct(± all yr)

H. monarchensis D.A. York et al. (p. 355) MONARCH GOLDENASTER Per 9–18 cm. **ST:** ascending to erect, not branching, ± densely bristly, sparsely glandular. **LF:** proximal cauline petioled, oblanceolate to obovate, entire, ± wavy, ± densely bristly, appressed-hairy, sparsely glandular, mid to distal narrowly oblanceolate, little reduced. **INFL:** heads 1–5, in ± flat-topped cluster, subtended by few narrowly lanceolate lfy bracts; involucre 6–8 mm, phyllaries in 3–4 series, ± strigose, sparsely glandular. **RAY FL:** 8–19; ray 10–15 mm. **DISK FL:** 30–45; corolla 5–6 mm, lobes sparsely hairy, hairs ascending, gen 0.3–0.5 mm. **FR:** 2–3.5 mm; ray and disk fr similar, ± strigose; outer pappus 0.2–0.5 mm, inner series 5–6.5 mm. 2*n*=18. Cracks, ledges, flats, on limestone; 1000–1900 m. s SN (Kings River canyon). Jun–Oct ★

H. oregona (Nutt.) Shinners (p. 355) RAYLESS GOLDENASTER Per 2–10 dm. **ST:** ascending to erect, branched proximally and/or distally, ± glabrous to densely bristly or glandular. **LF:** basal and proximal cauline sessile, 0 by fl, mid-cauline (ob)lanceolate, entire, ± glabrous to densely bristly or glandular, distal wider. **INFL:** heads discoid, 1–15 per branch in flat-topped, raceme-, or panicle-like clusters; involucre 7.5–14 mm, phyllaries in 5–7 series, ± bristly, glandular. **DISK FL:** 14–60; corolla 7.5–14 mm. **FR:** 2–5 mm, sparsely strigose; outer pappus 0.1–0.5 mm, scales few, inner series 4.5–6.5 mm. 2*n*=18. Vars. intergrade.

var. *compacta* (D.D. Keck) Semple **ST:** ± densely branched proximal to middle. **LF:** branch lvs long-soft-hairy and ± densely bristly, sparsely glandular, appearing gray-green. **INFL:** ± dense; outer phyllaries lance-deltate. Seasonally dry streambeds; 200–1000+ m. NW, CaR, n&c SNF, n SNH, SnJV, CW. Jul–Oct

var. *oregona* **ST:** ± sparsely branched distal to middle. **LF:** of branches gen > 13 mm, > 3 mm wide, sparsely hairy, ± densely glandular. **INFL:** ± open; outer phyllaries lance-deltate. Seasonally dry streambeds; < 500 m. NW, CW; to WA. Jul–Oct

var. *rudis* (Greene) Semple **ST:** ± sparsely gen loosely branched distal to middle. **LF:** of branches gen < 13 mm, < 3 mm wide, ± densely short-bristly, sparsely glandular, appearing green. **INFL:** ± open; outer phyllaries lanceolate. Seasonally dry streambeds; < 500 m. NW, n SNF, n ScV, CW. Jul–Oct

var. *scaberrima* (A. Gray) Semple **ST:** ± sparsely branched distal to middle. **LF:** of branches < 13 mm, < 3 mm wide, ± scabrous, glandular. **INFL:** much-branched; outer phyllaries ovate-deltate. Seasonally dry streambeds; 100–700 m. SnJV, CW. Jul–Oct

H. sessiliflora (Nutt.) Shinners SESSILEFLOWER GOLDENASTER Per gen 2–7(13) dm. **ST:** decumbent to erect, branching in infl, ± bristly, ± glandular, esp distally. **LF:** basal and proximal cauline subsessile, oblanceolate, tapered at base, entire, gen withered at fl, mid-cauline sessile, ± lanceolate, flat or wavy, ± densely bristly-strigose, ± densely glandular, distal gen reduced. **INFL:** heads gen ± many in ± flat-topped or panicle-like cluster, gen not subtended by lf-like bracts; involucre 8–14 mm, phyllaries in 4–6 series, ± sparsely strigose, ± sparsely to densely glandular. **RAY FL:** 4–24; ray gen 7–15 mm. **DISK FL:** 20–50; corolla 3–10 mm; lobes gen sparsely hairy, 0.25–1 mm. **FR:** 1.5–4.5 mm; ray and disk fr similar, ± strigose; outer pappus 0.25–0.5 mm, inner 5–8(10) mm. 2*n*=18. Highly variable, esp in CW; subspp. ± merge where ranges overlap.

subsp. *bolanderi* (A. Gray) Semple (p. 355) Pl sparsely woolly proximally. **LF:** flat; distal little reduced, oblanceolate, not stiff, green, strigose to long-woolly, hairs ± 2 mm. **INFL:** heads gen few, gen not subtended by lf-like bracts. **DISK FL:** 30–50; corolla lobes ± glabrous. 2*n*=18. Dunes, headlands, grassy coastal slopes; < 200 m. c&s NCo, n CCo, SnFrB. [*H. echioides* (Benth.) Shinners var. *b.* (A. Gray) G.L. Nesom] Jun–Sep

subsp. *echioides* (Benth.) Semple (p. 355) Pl bristly proximally. **LF:** flat (occ ± wavy in Teh, TR), distal gen much reduced, elliptic to lanceolate, gen stiff, gray-green, strigose and bristly, hairs ± 2 mm. **INFL:** heads many in tall pls, not subtended by lf-like bracts. **DISK FL:** 30–50; corolla lobes sparsely to moderately hairy. 2*n*=18.

Grassland, scrub, woodland, open forest, disturbed sites; < 2100 m. c&s SNF, Teh, SnJV, CW, SCo, TR, PR. [*H. s.* var. *e.* (Benth.) Semple] Sparsely glandular pls from serpentine in SnFrB with long-hairy disk corolla lobes may be treated as *H. sessiliflora* var. *bolanderioides* Semple [*H. echioides* var. *bolanderioides* (Semple) G.L. Nesom]; densely glandular pls from s SnFrB, n SCo with ± glabrous disk corolla lobes may be treated as *H. sessiliflora* var. *camphorata* (Eastw.) Semple. Jul–Oct

subsp. *fastigiata* (Greene) Semple (p. 355) Pl densely hairy proximally, gen moderately hairy and densely glandular distally. **LF:** wavy-margined, distal reduced, stiff, ± white, hairs gen ± 1 mm. **INFL:** heads few to many, not subtended by lf-like bracts. **DISK FL:** 20–40; corolla lobes sparsely hairy. 2*n*=18. Oak woodland, pine forest, desert washes (rare); 100–2200 m. SCo, TR, PR, w D. [*H. s.* var. *f.* (Greene) Semple] PR pls (more glandular, less hairy) may be treated as *H. sessiliflora* var. *sanjacintensis* Semple. Jul–Oct

subsp. *sessiliflora* (p. 355) BEACH GOLDENASTER Pl gen moderately to densely bristly-strigose proximally, gen sparsely hairy and densely glandular distally. **LF:** wavy-margined, distal reduced, not stiff, green, hairs gen ± 1 mm. **INFL:** heads few, subtended by lf-like bracts. **DISK FL:** 30–50; corolla lobes ± glabrous. Beaches, dunes, mud flats; < 60 m. s CCo, SCo; Baja CA. [*H. s.* var. *s.*] Jun–Sep ★

H. shevockii (Semple) Semple (p. 355) SHEVOCK'S GOLDENASTER Per gen 5–13 dm. **ST:** ascending to erect, branching in infl, proximally sparsely strigose, ± bristly, distally ± bristly-strigose, densely glandular. **LF:** basal and proximal cauline sessile or ± petioled, (ob)lanceolate, entire, mid-cauline lance-triangular, stiff, often reflexed, margin ± inrolled (esp large pls), distal much reduced; blade occ ± clasping, ± sparsely scabrous to bristly, ± densely glandular esp distally. **INFL:** heads 5–20+(70) in panicle-like cluster; involucre 9–13 mm, phyllaries in 5–6 series, sparsely strigose along midvein, densely glandular. **RAY FL:** 9–18; ray 5–12 mm. **DISK FL:** 30–80; corolla 5–8 mm. **FR:** 2.5–4 mm; ray and disk fr similar, ± strigose; outer pappus 0.2–0.5 mm, inner 5–6.5 mm. 2*n*=36. Crevices, shallow sand, grassland, pine/oak woodland; 400–800 m. s SNF (Kern River Canyon). [*H. villosa* var. *s.* Semple] Aug–Sep ★

H. subaxillaris (Lam.) Britton & Rusby subsp. *latifolia* (Buckley) Semple (p. 355) CAMPHOR-WEED Ann 3–15 dm, ± bristly-strigose, ± densely glandular, esp distally. **ST:** ascending to erect, in large pls much-branched. **LF:** basal rosetted, petioled and proximal cauline wing-petioled, clasping, with ear-like basal lobes, lanceolate to elliptic or ovate, toothed or entire, mid-cauline ± sessile, ovate, clasping, distal spreading. **INFL:** heads gen many in ± flat-topped or panicle-like cluster; involucre 6–12 mm, phyllaries in 4–6 series. **RAY FL:** 20–30; ray 4–10 mm. **DISK FL:** 40–75; corolla 4–6 mm. **FR:** 2–4 mm; ray fr ± glabrous, pappus 0; disk fr densely strigose, outer pappus 0.2–0.6 mm, inner 6–9 mm. 2*n*=18. Uncommon. Disturbed sandy soils, roadsides; < 1150 m. Teh, SCo, e DSon; s US, n Mex. [*H. psammophila* B. Wagenkn.] Other subsp. in se US. Probably naturalized in Teh, SCo. Aug–Oct

H. villosa (Pursh) Shinners HAIRY GOLDENASTER Per (1)2–5 dm. **ST:** decumbent to erect, branching in infl, ± bristly-strigose, sometimes densely glandular distally. **LF:** proximal ± (ob)lanceolate, tapered to base, gen entire, gen flat, ± densely bristly-strigose with hairs gen < 1 mm, glandless to densely glandular, distal tapered to ± clasping, not strongly reduced. **INFL:** heads 1–many, gen in open, ± flat-topped cluster; involucre gen 6–9.5 mm, phyllaries in 4–5 series, ± strigose, glandless or glandular. **RAY FL:** 7–21; ray 4–10 mm. **DISK FL:** 15–70; corolla 4–8 mm, lobes glabrous (± glabrous, hairs < 0.2 mm), glands few. **FR:** 2–3 mm; ray and disk fr similar, ± strigose; outer pappus 0.2–1 mm, inner gen 5–6.5 mm. Highly variable; local forms often ± distinct.

var. *minor* (Hook.) Semple Pl 10–35 cm. **LF:** mid-st lvs (ob)lanceolate, ± moderately strigose (not bristly), glandular. **RAY FL:** 7–21; ray 4–10 mm. **DISK FL:** 15–70; corolla 4–8 mm. **FR:** 2–3 mm. 2*n*=18,36. Lava flows, rocky sites; 600–3100 m. CaRH, n&c SNH, MP. [*H. v.* var. *hispida* (Hook.) V.L. Harms; *H. v.* var. *v.*, misappl.] Jun–Aug

var. ***scabra*** (Eastw.) Semple   Pl 20–50 cm. **LF:** mid-st lvs narrowly lance-triangular, ± sparsely scabrous to bristly, ± densely glandular, esp fall growth. **RAY FL:** 9–14; ray 5–6 mm. **DISK FL:** 30–50; corolla 5–7 mm. **FR:** 2–3 mm. 2*n*=18,36. Uncommon. Rock crevices; 1200–1300 m. sw DMtns (Little San Bernardino Mtns); to UT, n AZ. Spring- and fall-fl pls often differ in infl form and hairiness. Apr–May, Oct–Nov

HIERACIUM   HAWKWEED

Per; stolons gen 0 (in CA); sap milky; herbage glabrous or with long simple hairs, shorter, branched hairs, and/or gland-tipped hairs. **ST:** erect, 1–12 dm. **LF:** basal and/or cauline, alternate. **INFL:** heads liguliflorous, few to many in raceme-like, ± flat-topped, or panicle-like clusters; involucre cylindric to ± bell-shaped; phyllaries in 2–4 series of different lengths; receptacle ± flat, epaleate. **FL:** 6–150+; ligule gen yellow (white or orange, occ ± purple-tinged), readily withering. **FR:** gen ± cylindric (± urn-shaped), tan to red-brown or black; pappus of many slender, brittle, minutely barbed bristles, dull white, straw-colored, or tan to ± brown. ± 250 spp.: ± worldwide. (Greek: hawk) [Strother 2006 FNANM 19:278–294]

1. Ligule white or orange
  2. Ligule white; stolons 0; native of forests . . . . ***H. albiflorum***
  2′ Ligule orange; stolons present; weed of disturbed sites . . . . [***H. aurantiacum***]
1′ Ligule yellow
  3. Lf glabrous on 1 or both faces
    4. Lf glabrous abaxially, sparsely to ± densely long-hairy adaxially; phyllaries glabrous or sparsely glandular; fls 6–12 . . . . ²***H. bolanderi***
    4′ Lf glabrous; phyllaries long-hairy, branched-hairy, and black-glandular; fls 20–60 . . . . ²***H. triste***
  3′ Lf ± hairy on faces (glabrous in *Hieracium scouleri*)
    5. Phyllaries gen without long, glandless hairs (occ long-hairy in *Hieracium argutum*), often minutely branched-hairy and/or glandular
      6. Fls 30–60 . . . . ***H. parryi***
      6′ Fls 4–30
        7. Fls 15–30; involucre 7–9(12) mm; gen at least some lvs wavy-toothed (all entire); pappus ± white; c&s CA . . . . ²***H. argutum***
        7′ Fls gen 4–12(15); involucre (7)8–12 mm; lvs gen entire (minutely dentate); pappus tan to ± brown; KR, NCoR, CaRH
          8. Peduncle glabrous or stalked-glandular; fr ± urn-shaped . . . . ²***H. bolanderi***
          8′ Peduncle branched-hairy; fr ± cylindric, not distally narrowed . . . . ***H. greenei***
    5′ Phyllaries with long, glandless hairs, occ also branched-hairy and/or glandular
      9. Lf faces minutely rough-hairy and/or stalked glandular . . . . ²***H. triste***
      9′ Lf faces long-hairy, also sometimes branched-hairy (rarely glabrous in *Hieracium scouleri*)
        10. Fls 6–12(15); involucre ± cylindric to ± bell-shaped . . . . ***H. horridum***
        10′ Fls 15–50; involucre obconic to bell-shaped
          11. Some or all lvs wavy-toothed; c&s CA . . . . ²***H. argutum***
          11′ Lvs all entire or some minutely dentate; n CA
            12. Cauline lvs 0–3(5); pappus 4–6 mm. . . . . ***H. nudicaule***
            12′ Cauline lvs gen (3)5–10+; pappus 6–7 mm . . . . ***H. scouleri***

H. albiflorum Hook. (p. 355)   WHITE HAWKWEED   **ST:** 2–12 dm; proximally densely long-hairy. **LF:** mostly basal, 8–15 cm, oblong to oblanceolate, entire or few-toothed, coarsely long-hairy; cauline gen restricted to proximal st, gen smaller. **INFL:** heads (3)12–50+; peduncle glabrous or occ stalked-glandular; involucre (7)8–10(11) mm, ± bell-shaped, glabrous or glandular, sometimes sparsely coarse-bristly. **FL:** gen 15–30; ligule white. **FR:** 2–4 mm; pappus (4)5–7 mm, dull white or tan. 2*n*=18. Forest; < 3300 m. CA-FP, Wrn; to AK, AB, SD, CO. May–Sep

H. argutum Nutt. (p. 355)   SOUTHERN HAWKWEED   **ST:** 3–10 dm, proximally glabrous to densely long-hairy and/or branched-hairy, distally glabrous or branched-hairy. **LF:** 0.8–1.6 dm, lance-oblong, gen at least some wavy-toothed, teeth widely spaced (all entire), coarsely long- and branched-hairy; cauline gen restricted to proximal st, smaller. **INFL:** heads many in open panicle-like clusters; peduncle branched-hairy, occ stalked-glandular; involucre 7–9(12) mm, ± obconic to bell-shaped, gen branched-hairy and short-glandular, occ long-hairy. **FL:** 15–30. **FR:** 2–3 mm; pappus 3.5–5 mm, ± white. 2*n*=18. Dry slopes, woodland; < 1530 m. c SNF, SCoRO, SCo, n ChI, TR; n Baja CA. Pls of SnBr and n Baja CA resemble *H. bolanderi*. Further study needed. Jun–Oct

H. bolanderi A. Gray (p. 355)   BOLANDER'S HAWKWEED   **ST:** 1–3 dm, simple or few-branched, glabrous or puberulent. **LF:** mostly basal, 2–7 cm, widely oblanceolate, gen entire (minutely dentate); adaxially long-hairy; abaxially gen glabrous (exc long-hairy on mid-

vein). **INFL:** heads few in open or dense clusters; peduncle glabrous or stalked-glandular; involucre (7)9–12 mm, cylindric to bell-shaped, glabrous or sparsely glandular. **FL:** 6–12. **FR:** 4 mm, ± urn-shaped; pappus 5–8 mm, ± brown. Dry forest; 300–2700 m. KR, NCoR; OR. Jun–Aug

H. greenei A. Gray   GREENE'S HAWKWEED   **ST:** 2–4 dm, proximally densely branched-tomentose and coarsely long-hairy. **LF:** basal and proximal cauline 5–10 cm, oblong-oblanceolate, ± entire or blunt-toothed, densely long-hairy and branched-hairy, abaxially occ glandular-hairy; distal lvs much reduced. **INFL:** heads several to many; peduncle branched-hairy; involucre 8–12 mm, cylindric, ± branched-hairy. **FL:** 4–10(15). **FR:** 4–5 mm; pappus 7–9 mm, deep tan. 2*n*=18. Dry slopes in montane forest; 900–2750 m. KR, NCoRH, CaRH; OR. Jul–Sep

H. horridum Fr. (p. 355)   PRICKLY HAWKWEED   Densely long-hairy; hairs ± white or ± brown. **ST:** 1–several from taproot, 1–4.5 dm. **LF:** gen cauline, 3–10 cm, oblong, entire. **INFL:** heads many; peduncle branched-hairy, occ short-straight-hairy; involucre 6–9 mm, ± cylindric to ± bell-shaped, hairs short, branched and long, black-based. **FL:** 6–12(15). **FR:** 3 mm; pappus 4–6 mm, ± brown. 2*n*=18. Rocky places, crevices; 1350–3300 m. KR, CaRH, SNH, SnGb, SnBr, SnJt, GB; OR. Jun–Oct

H. nudicaule (A. Gray) A. Heller   NAKED-STEMMED HAWKWEED   **ST:** 2–5 dm, simple, proximally ± glabrous to bristly and/or branched-

hairy, distally gen ± glabrous. **LF**: gen basal, cauline 0–3(5), 5–12 cm, lanceolate to lance-elliptic, gen entire (minutely dentate), faces gen long- and branched-hairy. **INFL**: heads gen 5–12; peduncle ± branched-hairy, occ glandular; involucre 8–10 mm, ± bell-shaped, short-straight-hairy, branched-hairy, and dark-glandular. **FL**: 20–40. **FR**: 2.5–3 mm; pappus 4–6 mm, white or straw-colored. Openings in chaparral and conifer forest; 1800–3500 m. CaRH, SNH; OR. Incl in *H. scouleri* Hook. in TJM (1993). Jun–Sep

H. parryi Zahn PARRY'S HAWKWEED **ST**: 1.5–5 dm, simple, long-hairy, stalked-glandular throughout. **LF**: basal and cauline, 3–8 cm, lance-elliptic to lanceolate, gen entire, ± long-hairy. **INFL**: densely glandular; heads 3–12+, in flat-topped clusters; peduncle densely glandular; involucre 10–12 mm, glandular. **FL**: 30–60. **FR**: 2.5–3 mm; pappus ± 5 mm, white. Open woodland, shrubby areas, serpentine soils; < 2000 m. KR; OR. Jun–Jul

H. scouleri Hook. WESTERN HAWKWEED **ST**: 3.5–10 dm, simple, ± glabrous to branched-hairy and densely long-hairy. **LF**: basal

and (3)5–10+ cauline, 10–20 cm, lanceolate to oblanceolate, entire, glabrous or sparsely to moderately long-hairy and sparsely branched-hairy. **INFL**: heads many, in flat-topped clusters; peduncle branched-hairy, occ also long-hairy and/or glandular; involucre 8–12 mm, bell-shaped, long-hairy, branched-hairy, and glandular. **FL**: 20–50. **FR**: 3 mm; pappus 6–7 mm, white or dull brown. $2n=18$. Open woodland, brushy areas; 300–2300 m. KR, NCoRH, CaRH, n SNH, MP; to BC, AB, WY, UT. Variable in density of hairs. May–Sep

H. triste Spreng. SLENDER HAWKWEED **ST**: 1–several from basal rosette, gen 1–2 dm, puberulent. **LF**: all or mostly basal, 2–8 cm, lance-oblong, entire or nearly so, gen glabrous (occ short-rough-hairy or glandular). **INFL**: heads few, in raceme-like or flat-topped clusters; peduncle branched-hairy, glandular; involucre 7–8 mm, long-hairy, branched-hairy, and black-glandular. **FL**: 20–60. **FR**: 2 mm; pappus tan. $2n=18$. Moist forest, rocky open areas, meadows; 1650–3550 m. KR, CaRH, SNH; to AK, nw Can, WY, NM. [*H. gracile* Hook.] Jul–Sep

HOLOCARPHA   TARWEED, TARPLANT

Bruce G. Baldwin

Ann 1–12 dm, aromatic. **ST**: ± erect. **LF**: proximal opposite or in rosette, gen withered before fl, most alternate, sessile, linear to oblanceolate, serrate, or minutely so, or entire, coarse-hairy to strigose, silky-, or ± shaggy-hairy, distal sometimes also stalked-glandular and/or gland-dotted, tips of distal lvs gen each with a pit-gland. **INFL**: heads radiate, 1 or in ± flat-topped, panicle-, raceme-, or spike-like clusters or in tight groups; peduncle bracts each with terminal pit-gland; involucre ± obconic or bell-shaped to ± spheric, 4–8+ mm diam; phyllaries 3–16 in 1 series, elliptic, oblanceolate, or obovate, each gen 1/2 enveloping a subtended ray ovary, falling with fr, abaxially with pit-gland-tipped cylindrical outgrowths and glabrous or ± shaggy, bristly, and/or sessile- or stalked-glandular; receptacle flat to convex, glabrous; paleae subtending all or most disk fls. **RAY FL**: 3–16; corolla yellow, ray 4–6.5 mm, lobes 1/8–1/2 length of ray, ± parallel. **DISK FL**: 9–90, some bisexual, most staminate; corolla 2.5–5 mm, yellow, tube ≤ throat, lobes deltate; anthers yellow to ± brown or ± red to dark purple, tips ovate to ovate-deltate; style glabrous proximal to branches, tips awl-shaped to bristle-like, densely hairy. **FR**: ray fr ± 2.5–4 mm, nearly round in ×-section (exc ± flattened adaxially) or ± 3- angled (abaxially gen ± widely 2-faced, adaxially ± flattened to slightly bulging), glabrous, tip beaked, beak adaxial, ascending, pappus 0; disk fr ± club-shaped, glabrous, tip beakless, pappus 0. 4 spp.: CA. (Greek: whole chaff, for bracts throughout receptacle) [Baldwin & Strother 2006 FNANM 21:287–289] ± identical forms often intersterile in *H. heermannii* and *H. virgata*. Self-sterile.

1. Anthers yellow to ± brown
  2. St notably stalked-glandular; phyllaries each with 25–50 pit-gland-tipped outgrowths and minutely
     sessile- or stalked-glandular and puberulent or minutely bristly; involucre bell-shaped to ± spheric; heads
     1 or in panicle- or raceme-like clusters . . . . ***H. heermannii***
  2′ St resinous distally, not notably stalked-glandular; phyllaries each with (0)5–15(20) pit-gland-tipped
     outgrowths and gen glabrous or minutely sessile- or stalked-glandular (bristly); involucre ± obconic to ±
     spheric; heads in flat-topped or panicle-like clusters . . . . ***H. obconica***
1′ Anthers ± red to dark purple
  3. Heads 1, in tight groups, or in spike-like clusters of tight groups; ray fls 8–16; disk fls 40–90 . . . . ***H. macradenia***
  3′ Heads 1, at ends of branches, or in ± raceme- to spike-like clusters; ray fls 3–8; disk fls 9–25+ . . . . ***H. virgata***
    4. St branches ± upcurved, ± pliable; peduncles 0–5(15) cm . . . . subsp. ***elongata***
    4′ St branches ± straight, ± rigid; peduncles 0–5+ cm . . . . subsp. ***virgata***

H. heermannii (Greene) D.D. Keck (p. 355) **ST**: notably stalked-glandular. **INFL**: heads 1, at ends of branches, or in panicle- or raceme-like clusters; involucre bell-shaped to ± spheric; phyllaries each with 25–50 pit-gland-tipped outgrowths and minutely sessile- or stalked-glandular and puberulent or minutely bristly. **RAY FL**: 3–13. **DISK FL**: 9–22; anthers yellow to ± brown. $2n=12$. Grassland; < 1400 m. c&s SNF, Teh, SnJV, e SnFrB, SCoR, n WTR. May–Nov

H. macradenia (DC.) Greene (p. 355) SANTA CRUZ TARPLANT Pl 1–5 dm. **ST**: notably stalked-glandular. **INFL**: heads 1 or in tight groups or spike-like clusters of tight groups; involucres ± spheric; phyllaries each with ± 25 pit-gland-tipped outgrowths and minutely sessile- or stalked-glandular. **RAY FL**: 8–16. **DISK FL**: 40–90; anthers ± red to dark purple. $2n=8$. Grassy areas, clay soil; < 200 m. CCo (n&c Monterey Bay, extirpated elsewhere), sw SnFrB (introduced e SnFrB). Jun–Nov ★

H. obconica (J.C. Clausen & D.D. Keck) D.D. Keck (p. 355) Pl 1–8(12) dm. **ST**: ± resinous distally, not notably stalked-glandular. **INFL**: heads in flat-topped or panicle-like clusters; involucre ±

obconic to ± spheric; phyllaries each with (0)5–15(20) pit-gland-tipped outgrowths and gen glabrous or minutely sessile- or stalked-glandular (± bristly). **RAY FL**: 4–9. **DISK FL**: 11–21; anthers yellow to ± brown. $2n=12$. Grassland; < 500 m. s SNF, nw SnJV, e SnFrB, e SCoRI. Apr–Nov

H. virgata (A. Gray) D.D. Keck Pl 2–12 dm. **ST**: gen notably stalked-glandular, rarely only ± resinous distally. **INFL**: heads 1, at ends of branches, or in ± raceme- to spike-like clusters; involucre ± obconic to bell-shaped (± spheric); phyllaries each with 5–20 gland-tipped outgrowths and glabrous or minutely sessile- or stalked-glandular. **RAY FL**: 3–8. **DISK FL**: 9–25+; anthers ± red to dark purple. $2n=8$.

  subsp. ***elongata*** D.D. Keck GRACEFUL TARPLANT **ST**: branches ± upcurved, pliable. **INFL**: peduncle 0–5(15) cm. Grassland; < 900 m. c&s SCo, PR (exc SnJt). Jul–Nov ★

  subsp. ***virgata*** (p. 355) **ST**: branches ± straight, rigid. **INFL**: peduncle 0–5+ cm. Grassland; < 800(1100) m. NCoRI, s CaRF, n&c SNF, GV, SnFrB (mostly e), c SCoRO, SCoRI, c&s SCo, PR (exc SnJt). May–Nov

HOLOZONIA
Bruce G. Baldwin

1 sp. (Greek: whole girdle, for phyllary fully enclosing fr) [Baldwin & Strother 2006 FNANM 21:294]

H. filipes (Hook. & Arn.) Greene (p. 355) WHITECROWN Per 3–15 dm, from rhizome. **ST:** aerial st ± erect, branches ± stiffly ascending, ± long-soft hairy. **LF:** mostly cauline, proximal-most opposite, bases fused around st, distal alternate, sessile, 3–10 cm, 2–8 mm wide, linear to lanceolate, entire, ± soft-hairy, distal stiff-glandular-hairy, glands cup-shaped. **INFL:** heads radiate, 1 or in loose, ± flat-topped to panicle-like clusters; peduncle thread-like; involucre ± obconic or top-shaped, 2–4+ mm diam; phyllaries 4–10 in 1 series, each mostly or wholly enveloping a ray ovary, 3–5 mm, ± lance-linear, coarse-hairy, with stalked cup-shaped glands or not; receptacle flat to convex, glabrous or minutely bristly; paleae in 1 series between ray and disk fls, fused, deciduous. **RAY FL:** 4–10; corolla ± white, ray fan-shaped, deeply lobed, abaxially ± purple-veined. **DISK FL:** 9–28, staminate; corolla 3–4.5 mm, white, hairy, tube < throat, lobes deltate; anthers ± dark purple, tips ovate to triangular; style glabrous proximal to branches, tips long, awl-shaped, short-hairy. **FR:** 2.5–3.5 mm, compressed front-to-back, ± club-shaped, glabrous, black, tip beakless, broadly cup-shaped; ray pappus 0 or crown-like, 0.1–0.3 mm; disk pappus 0 or of 1–5 awl-shaped, deciduous scales. 2*n*=28. Banks, dry streambeds, pools, rocky or alkaline clay, sometimes serpentine; 30–1000 m. s KR, NCoRO, CaRF, n&c SNF, e edge GV, SnFrB, c SCoRO, sw MP. Self-sterile. Jun–Oct

HULSEA
Dieter H. Wilken

Ann to per. **ST:** 1–5, 1–15 dm, ± hairy, glandular. **LF:** basal and cauline, alternate; petiole gen ciliate; blade gen ± oblanceolate, entire to lobed, ± reduced distally on st. **INFL:** heads radiate, 10–20 mm, 1–15+ in ± flat-topped or raceme-like cluster; bracts ± narrowly lanceolate; involucre hemispheric to obconic, phyllaries many, in 2–3 series, linear to obovate, green, ± glandular, reflexed in age; receptacle flat or slightly convex, shallowly pitted, glabrous, epaleate. **RAY FL:** ray yellow to red. **DISK FL:** many; corolla 5–9 mm, yellow to orange, gen glabrous; anther tip ovate; style tips linear-oblong. **FR:** 4–10 mm, cylindric to club-like, black, ± hairy; pappus scales gen 2 pairs, 4–10 mm, gen deeply cut, gen translucent. 7 spp.: w US. (G.W. Hulse, US Army surgeon, botanist, 1807–1883) [Wilken 1977 Madroño 24:48–55; Wilken 2006 FNANM 21:396–400] Self-sterile.

1. Proximal lvs gen glandular, ± thinly long-hairy (woolly)
  2. St 5–40 cm; heads 1–2; cauline lvs few, abruptly smaller above rosette; gen alpine or subalpine
    3. Phyllaries narrowly oblong, long-tapered; ray fls 25–60 . . . . ***H. algida***
    3′ Phyllaries oblong to obovate, acuminate; ray fls 12–30 . . . . ²***H. nana***
  2′ St 30–150 cm; heads 3–15+; cauline lvs gradually smaller distally; montane
    4. Ray < 2 mm wide; red . . . . ***H. heterochroma***
    4′ Ray ≥ 2 mm wide; yellow
      5. Ray fls 10–23, hairs of ray corolla tube glandular and glandless . . . . ***H. brevifolia***
      5′ Ray fls 20–35, hairs of ray corolla tube all glandular . . . . ***H. mexicana***
1′ Proximal lvs ± woolly or long-soft-wavy-hairy, most hairs glandless
  6. Lvs gradually smaller distally; heads 2–5, distal peduncles < 5 cm . . . . ***H. californica***
  6′ Lvs abruptly smaller above rosette; head gen 1(2), or peduncles gen > 5 cm
    7. Lvs gen oblanceolate, sparsely glandular; n CA . . . . ²***H. nana***
    7′ Lvs gen spoon-shaped, woolly or long-soft-wavy-hairy; s CA . . . . ***H. vestita***
      8. Basal lf blade entire to weakly scalloped, gradually tapered; petiole gen ≤ blade
        9. Bracts densely woolly; st occ ± lfless . . . . subsp. ***vestita***
        9′ Bracts glandular to barely woolly; st ± lfy in proximal 1/3
          10. Petioles green; pl gen > 5 dm; ray gen yellow to orange; n PR . . . . subsp. ***callicarpha***
          10′ Petioles ± red; pl gen < 5 dm; ray ± red-tinged at base; e WTR, SnGb . . . . subsp. ***gabrielensis***
      8′ Basal lf blades scalloped to lobed, abruptly tapered, petiole gen ≥ blade
        11. Ray corolla yellow, 12–18 mm . . . . subsp. ***inyoensis***
        11′ Ray corolla red to orange, 5–10 mm
          12. Hairs of abaxial lf face glandular and glandless; phyllaries gen green, tips tinged red; montane . . . . subsp. ***parryi***
          12′ Hairs of abaxial lf face gen glandular only; phyllaries deep red-purple; alpine . . . . subsp. ***pygmaea***

H. algida A. Gray (p. 359) Per (10)20–40 cm, ± long-soft-hairy, ± glandular. **LF:** basal < 10 cm, ± coarsely toothed, cauline few, abruptly smaller. **INFL:** head 1(2); involucre obconic to hemispheric, 10–25 mm diam, phyllaries 8–15 mm, 1–3 mm wide, narrowly oblong, long-tapered. **RAY FL:** 25–60; corolla 10–15 mm, ray 2–4 mm wide, yellow, puberulent. **FR:** 6–10 mm, sparsely hairy; pappus scales < 1.5 mm, ± equal. 2*n*=38. Subalpine to alpine talus; 3000–4000 m. SNH, n SNE, W&I; to OR, MT, NV. Jul–Aug

H. brevifolia A. Gray (p. 359) SHORTLEAVED HULSEA Per 3–6 dm, ± glandular-hairy. **LF:** basal gen 5–6 cm, coarsely toothed, cauline gradually smaller distally. **INFL:** heads 3–4; involucre 10–16 mm diam, obconic to hemispheric, phyllaries 8–10 mm, 1–3 mm wide, lanceolate, acuminate. **RAY FL:** 10–23; corolla 10–12 mm, ray 2–4 mm wide, yellow, short-hairy. **FR:** 6–8 mm, sparsely hairy;

pappus scales 1–3 mm, unequal, tinged red. 2*n*=38. Gravelly soils, montane forest; 1500–2700 m. c&s SNH. Jun–Aug ★

H. californica Torr. & A. Gray (p. 359) SAN DIEGO SUNFLOWER Bien 4–12 dm, ± woolly, ± glandular. **LF:** basal 6–10 cm, entire, cauline gradually smaller distally. **INFL:** heads 2–5; involucre widely conic to hemispheric, 15–26 mm diam, phyllaries 9–14 mm, lance-linear, long-acuminate. **RAY FL:** 22–40; corolla 15–20 mm, ray 3–5 mm wide, yellow, puberulent. **FR:** 4–6 mm, silky-hairy; pappus scales < 2 mm, ± equal. 2*n*=38. Open sites; 1000–2000 m. s PR (Laguna, Cuyamaca mtns). May–Aug ★

H. heterochroma A. Gray (p. 359) Ann or per, < 15 dm, ± densely glandular-hairy. **LF:** basal 10–20 cm, coarsely toothed, green, cauline gradually smaller distally. **INFL:** heads 3–15+; involucre obconic to

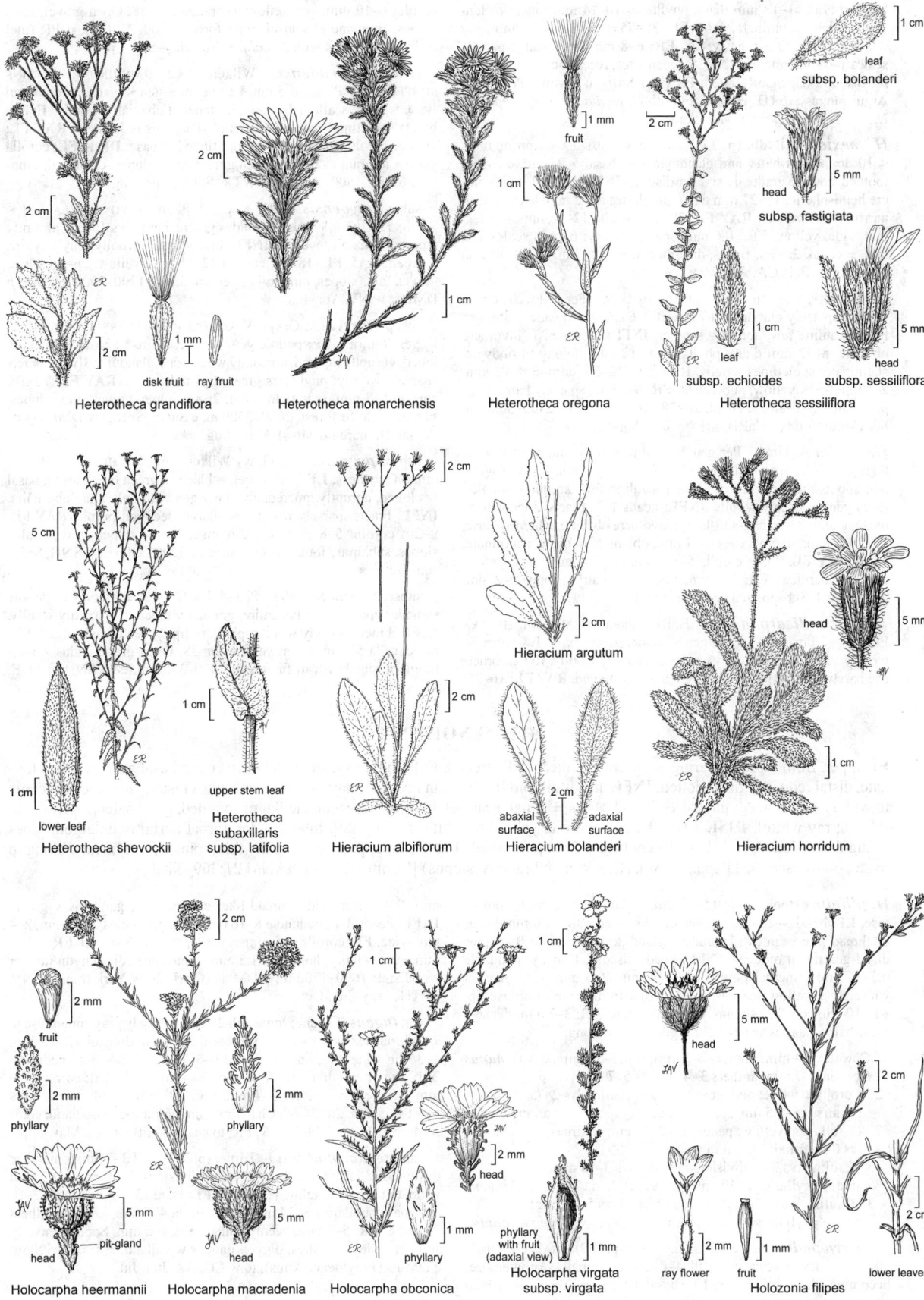

Heterotheca grandiflora

Heterotheca monarchensis

Heterotheca oregona

Heterotheca sessiliflora

leaf
subsp. bolanderi

head
subsp. fastigiata

leaf
subsp. echioides

head
subsp. sessiliflora

fruit

disk fruit   ray fruit

Heterotheca shevockii

lower leaf

upper stem leaf
Heterotheca
subaxillaris
subsp. latifolia

Hieracium albiflorum

Hieracium argutum

Hieracium bolanderi

abaxial
surface

adaxial
surface

Hieracium horridum

head

Holocarpha heermannii

fruit

phyllary

head   pit-gland

Holocarpha macradenia

phyllary

head

Holocarpha obconica

phyllary

head

Holocarpha virgata
subsp. virgata

phyllary
with fruit
(adaxial view)

Holozonia filipes

head

ray flower   fruit

lower leaves

hemispheric, 11–18 mm diam, phyllaries 10–14 mm, linear to lanceolate, long-acuminate. **RAY FL:** 30–75+; corolla 6–10 mm, ray < 2 mm wide, linear, hairy, red. **FR:** 6–8 mm, very hairy; pappus scales 1–3 mm, unequal. $2n$=38. Open sites, recent burns; 300–2700 m. SN, SnFrB, SCoR, TR, n PR (incl SnJt), n DMtns (Panamint, Argus ranges); to UT. Hybridizes with *H. vestita* subsp. *callicarpha*, *H. vestita* subsp. *parryi*. May–Aug

H. mexicana Rydb. (p. 359) MEXICAN HULSEA Ann or bien, < 10 dm, ± soft-hairy and glandular. **LF:** basal 5–15 cm, coarsely toothed, cauline gradually smaller distally. **INFL:** heads 3–5; involucre hemispheric, 15–27 mm diam, phyllaries 8–12 mm, lance-linear, narrowly acuminate. **RAY FL:** 20–35; corolla 12–18 mm, ray 3–5 mm wide, yellow. **FR:** 4–6 mm, sparsely hairy; pappus scales 1–2 mm, ± equal. $2n$=38. Burns, disturbed sites; 1200 m. se PR (se San Diego Co.); Baja CA. Apr–May ★

H. nana A. Gray (p. 359) LITTLE HULSEA Per 5–15(20) cm, ± woolly, sparsely glandular. **LF:** basal 2–6 cm, gen oblanceolate, gen lobed, cauline few, abruptly smaller. **INFL:** head gen 1; involucre obconic, 8–12 mm diam, phyllaries 8–12 mm, oblong to obovate, acuminate, sometimes woolly. **RAY FL:** 12–30; corolla 8–12 mm, 2–5 mm wide, yellow, puberulent. **FR:** 6–8 mm, sparsely hairy; pappus scales 1–2 mm, ± equal. $2n$=38. Volcanic talus; 2400–3000 m. KR (Mount Eddy), CaRH, MP; to WA. Jun–Sep ★

H. vestita A. Gray Per gen 1–10 dm, woolly and/or glandular. **ST:** gen lfy in proximal 1/3–1/2. **LF:** basal < 8 cm, 1–3 cm wide, gen spoon-shaped, entire to lobed, cauline few, abruptly smaller, faces gen ± densely woolly. **INFL:** heads 1–2; bracts lanceolate to ovate, glandular, ± long-soft-hairy; involucre obconic to hemispheric, 10–20 mm diam, phyllaries 8–11 mm, oblong to obovate, acuminate, hairy. **RAY FL:** 9–32; corolla 5–18 mm, ray 2–5 mm wide, yellow to red, puberulent. **FR:** 5–7 mm, moderately hairy; pappus 1–2 mm, gen ± equal. Subspp. gen geog separated.

    subsp. ***callicarpha*** (H.M. Hall) Wilken BEAUTIFUL HULSEA Pl gen > 5 dm. **LF:** petiole gen < blade, green; basal lvs entire to weakly scalloped, gradually tapered, abaxially woolly. **INFL:** bracts occ barely woolly; phyllary tips green or red-tinged. **RAY FL:** 16–25;

corolla 6–10 mm, gen yellow to orange. $2n$=38. Open gravel, talus slopes, montane chaparral, pine forest; 1300–2500 m. n PR (incl SnJt). Intergrades with *H. californica*. May–Sep ★

    subsp. ***gabrielensis*** Wilken SAN GABRIEL MOUNTAINS SUNFLOWER Pl gen < 5 dm. **LF:** petioles gen < blade, ± red; basal lvs ± weakly scalloped, gradually tapered abaxially woolly. **INFL:** bracts sometimes barely woolly; phyllary tips red-tinged. **RAY FL:** 16–23; corolla 6–8 mm, ray ± red-tinged at base. **DISK FL:** corolla yellow to orange. $2n$=38. Open gravel, talus slopes, chaparral, montane forest; 1500–2500 m. e WTR, SnGb. Apr–Aug ★

    subsp. ***inyoensis*** (D.D. Keck) Wilken INYO HULSEA Pl < 7 dm. **LF:** petioles gen > blade, green; basal lvs lobed, abruptly tapered, abaxially woolly. **INFL:** bracts barely woolly; phyllary tips ± green. **RAY FL:** 18–32; corolla 12–18 mm, yellow. $2n$=38. Open gravel, talus slopes, pinyon/juniper woodland; 1700–3000 m. SNE, n DMtns; w NV. Apr–Jun ★

    subsp. ***parryi*** (A. Gray) Wilken (p. 359) PARRY'S SUNFLOWER Pl gen 2–6 dm. **LF:** petiole gen > blade, ± green; basal lvs deeply lobed, abruptly tapered, abaxially woolly and glandular. **INFL:** bracts sparsely woolly; phyllaries green, tips red-tinged. **RAY FL:** 10–16; corolla 5–7 mm, orange to ± red. $2n$=38. Open gravel, talus slopes, sagebrush to fir forest; 2000–2500 m. e SnGb, SnBr, sw DMtns (Little San Bernardino Mtns). May–Aug ★

    subsp. ***pygmaea*** (A. Gray) Wilken PYGMY HULSEA Pl < 1 dm. **ST:** ± lfless. **LF:** petiole gen = blade, ± green or ± purple; basal lvs lobed, abruptly tapered, abaxially gen with only glandular hairs. **INFL:** bracts sparsely woolly; phyllaries deep red-purple. **RAY FL:** 9–20; corolla 5–8 mm, ± red-orange. $2n$=38. Open gravel, talus slopes, subalpine forest, alpine barrens; 3200–3900 m. s SNH, SnBr. Jun–Oct ★

    subsp. ***vestita*** (p. 359) Pl 1–4 dm. **ST:** occ ± lfless. **LF:** petiole = blade, green; basal lvs entire, gradually tapered, abaxially woolly. **INFL:** bracts densely woolly; phyllary tips green or red-tinged. **FL:** ray corolla 5–9 mm, gen yellow. $2n$=38. Open gravel, talus slopes, montane sagebrush to fir forest; 2400–3350 m. c&s SNH, n SNE. May–Sep

HYMENOPAPPUS

Bien, per, from taproot bearing 1–several caudices. **ST:** erect. **LF:** [simple to] 2-pinnately dissected; basal and cauline, alternate, distal reduced, gland-dotted. **INFL:** heads discoid [radiate], in few- to many-headed cyme-like clusters; involucre obconic to widely bell-shaped; phyllaries in 2–3 series, ± equal, scarious-margined; receptacle flat or rounded, gen epaleate. **RAY FL:** 0 [or 8; ray white]. **DISK FL:** 10–many; corolla yellow or white [red-purple], tube slender, throat abruptly enlarged, lobes triangular, reflexed; style branches obtuse. **FR:** obpyramidal, 4-angled; pappus 0 or of many thin, transparent, linear-oblong to ovate, obtuse scales. 11 spp.: N.Am. (Greek: membranous pappus) [Strother 2006 FNANM 21:309–316]

H. filifolius Hook. Per 0.5–10 dm, ± glabrous to densely tomentose. **LF:** basal 3–20 cm, 2-pinnately dissected into 2–50 mm linear or thread-like segments, minutely gland-dotted; cauline 0 or few, distal gen much reduced. **INFL:** heads discoid, 1–many; peduncle 0.5–16 cm; principal phyllaries 3–14 mm, 2–5 mm wide, margin white- or ± yellow-scarious for 1–4 mm from acute to obtuse tip. **FL:** 10–70; corolla 2–7 mm, yellow or white. **FR:** 3–7 mm, densely short-hairy; pappus scales 12–22, gen linear-oblong.

1. Corolla 3–4 mm; anthers 2–3 mm; fr 4.5–5.5 mm. . . var. ***nanus***
1' Corolla 4–7 mm; anthers 3–4 mm; fr 5–7 mm
  2. Corolla ± white; peduncle 8–16 cm; cauline lvs 2–7;
    fr hairs 0.5–1.5 mm . . . . var. ***eriopodus***
  2' Corolla gen yellow; peduncles 2–12 cm; cauline
    lvs 0–7; fr hairs 1–2 mm
    3. Cauline lvs 0–3; divisions of basal lvs 3–15 mm;
      main phyllaries 6–10 mm . . . . var. ***lugens***
    3' Cauline lvs (1)2–7; divisions of basal lvs 5–30 mm;
      main phyllaries 8–12(14) mm . . . . var. ***megacephalus***

    var. ***eriopodus*** (A. Nelson) B.L. Turner (p. 359) HAIRY-PODDED FINE-LEAF HYMENOPAPPUS Pl 4–8 dm, proximally ± tomentose, becoming glabrous distally. **LF:** basal 10–20 cm, divisions 10–20

mm, 0.4–1 mm wide, thread-like; cauline 2–7, gen glossy green. **INFL:** heads 3–8; peduncle 8–16 cm; main phyllaries 7–10 mm, 2–4 mm wide. **FL:** corolla 4–5 mm, ± white; anthers 3–4 mm. **FR:** 5.5–6 mm, evenly hairy; hairs 0.5–1.5 mm. Limestone soil, pinyon/juniper woodland; 1600–1700 m. e DMtns (Clark, New York mtns); s NV, sw UT. May–Jun, Oct ★

    var. ***lugens*** (Greene) Jeps. Pl 2–6 dm, densely gray-tomentose to ± glabrous exc woolly axils. **LF:** basal 5–14 cm; divisions 3–15 mm, 1–3 mm wide, flat, linear; cauline 0–3. **INFL:** heads 3–8; peduncle 2–12 cm; main phyllaries 6–10 mm, 3–4 mm wide. **FL:** corolla 4–6 mm, gen yellow; anthers 3–4 mm. **FR:** 4–6 mm, evenly hairy; hairs gen 1–1.5 mm. $2n$=34,68. Chaparral, pinyon/juniper woodland, conifer forest; 1300–2700 m. TR, PR; to s UT, AZ, Baja CA. May–Sep

    var. ***megacephalus*** B.L. Turner (p. 359) Pl 3–7 dm, tomentose throughout. **LF:** basal 8–20(30) cm, divisions 5–30 mm, 1–2 mm wide, flat, linear; cauline (1)2–7. **INFL:** heads 3–14; peduncle 2–10 cm; main phyllaries 8–14 mm. **FL:** corolla 4–7 mm, yellow; anthers 3–4 mm. **FR:** 5–7 mm, evenly hairy; hairs 1–2 mm. Sandy, gravelly soils in valleys, washes, pinyon/juniper woodland; ± 1000–1500 m. e DMtns (Providence Mtns); to w CO, AZ. Jun–Jul

var. **nanus** (Rydb.) B.L. Turner (p. 359) LITTLE CUTLEAF Pl 0.5–5 dm, ± evenly, sparsely tomentose throughout. **LF**: basal 2–12 cm, divisions 5–15 mm, 0.5–1 mm, thread-like; cauline 0–3. **INFL**: heads 1–6; peduncle 3–15 cm; main phyllaries 6–9 mm. **FL**: corolla 3–4 mm, pale yellow; anthers 2–3 mm. **FR**: 4.5–5.5 mm, evenly hairy; hairs gen 0.2–1 mm. 2*n*=34. Limestone soil, pinyon/juniper woodland, subalpine forest; 1500–3100 m. W&I; also c&e NV to w UT, nw AZ. May–Aug ★

HYMENOTHRIX

Ann, bien, per, taprooted. **ST**: 1–several from base, erect, simple or distally much-branched. **LF**: basal and cauline, alternate; proximal petioled, pinnately or ternately dissected into narrow lobes; distal reduced, sessile. **INFL**: heads radiant or discoid [radiate], in many-headed, ± flat-topped cyme-like clusters; peduncle slender; involucre obconic to hemispheric; phyllaries in 1–2 ± equal series, free; receptacle rounded, epaleate. **RAY FL**: 0 [few; ray yellow]. **DISK FL**: few to many; corolla white or purple-tinged [yellow]; anther tips triangular; style tips narrowly triangular. **FR**: club-shaped, 4–5-angled, ± hairy; pappus of 10–15 bristle-tipped scales. 5 spp.: sw US, Mex. (Greek: membranous bristle, from pappus) [Strother 2006 FNANM 21:387–388]

1. Corolla cream, outer often larger than inner; anthers yellow or light brown . . . . *[H. loomisii]*
1′ Corolla white or purple-tinged, equal; anthers dark purple. . . . . *H. wrightii*

H. wrightii A. Gray (p. 359) WRIGHT'S HYMENOTHRIX Bïen, per ≤ 1.5 m, glabrous to densely glandular or soft-hairy. **LF**: densely crowded on proximal st; lobes gen ≤ 1 mm; margins rolled under. **INFL**: peduncle 1–7 cm; bracts scattered, scale-like; involucre hemispheric; phyllaries 3–8 mm, oblanceolate to obovate, obtuse, tips gen purple, glabrous or glandular. **FL**: corolla 5–8 mm, lobes linear, spreading; anthers ± 3 mm. **FR**: 3–6 mm; pappus scales 3–7 mm. 2*n*=24. Open slopes; 1100–1700 m. s PR (Palomar, Cuyamaca, Laguna mtns); to NM, n Mex. Aug–Oct ★

HYMENOXYS

Mark W. Bierner

Ann to per, 5–150 cm. **ST**: erect, unbranched or branched, green to ± red-purple, glabrous or ± hairy. **LF**: basal and cauline, alternate, simple, entire or 1–2-pinnately lobed or divided, ultimate margins entire or toothed, glabrous or hairy, ± gland-dotted. **INFL**: heads radiate [discoid], 1 or in panicle-like or ± flat-topped clusters; peduncle expanded distally, glabrous or ± hairy; disk ± hemispheric to spheric, bell-shaped or urn-shaped; involucre ± rotate; phyllaries in 2[3] series, gen unequal, outer basally fused, inner free; receptacle conic to ovoid, hemispheric or ± spheric [flat], pitted, epaleate. **RAY FL**: [3]8–26[34+], pistillate; corolla gen yellow or yellow-orange to orange, ray fan-shaped, 3(5)-lobed. **DISK FL**: 25–150(400+), bisexual [6–15, staminate]; corolla yellow; anther tip triangular; style tips truncate. **FR**: ± narrowly obpyramidal [obconic], gen 5-angled, glabrous or hairy; pappus of gen obovate, gen awn-tipped scales [0]. 25 spp.: w N.Am, S.Am. (Greek: sharp membrane, from pappus) [Bierner 2006 FNANM 21:435–443] *H. acaulis* moved to *Tetraneuris*.

1. Lf entire — outer phyllaries ≥ 11 mm, inner ≥ 7.5 mm. . . . . : . . . . . *H. hoopesii*
1′ Lf entire or 1–2-pinnately divided into linear lobes
  2. Ann . . . . *H. odorata*
  2′ Bien or per
    3. Bien or per, (1)2–8(10) dm, fl once; sts 1–3(15); lf ± hairy . . . . *H. cooperi*
    3′ Per 3–5 dm, fl > once; sts 3–10(15); lf glabrous or sparsely hairy . . . . *H. lemmonii*

H. cooperi (A. Gray) Cockerell Bien or per from simple or weakly branched caudex, (1)2–8(10) dm, fl once. **ST**: 1–3(15), branched distally, gen ± red-purple proximally (or throughout), ± hairy. **LF**: 4–9 cm, entire or divided into 3–9 linear lobes, ± hairy; middle blades 3–5-lobed, terminal lobe 1–2(2.5) mm wide. **INFL**: heads (1)7–45(80), in panicle-like to ± flat-topped clusters; peduncle (2)3.5–8(13) cm, ± hairy; disk ± hemispheric, 8–10 mm, 10–17 mm diam; outer phyllaries 4.5–8.9 mm, lanceolate, inner 4.1–6.8 mm, obovate to oblanceolate. **RAY FL**: 9–14; corolla 10.2–17(21.5) mm. **DISK FL**: 30–150+; corolla 2.7–4.8 mm. **FR**: 1.7–3.7 mm; pappus scales 5–6(8), 1.3–3.3 mm. 2*n*=30. Roadsides, open areas, edges of juniper/pine forest; 1000–3500 m. SNE (Sweetwater Mtns), DMtns; OR, ID, NV, UT, AZ. [*H. cooperi* var. *canescens* (Eaton) K.F. Parker] May–Sep

H. hoopesii (A. Gray) Bierner (p. 359) Per from stout, ± black rhizome, 3–10 dm, fl > once. **ST**: 1–4, branched distally, red-purple proximally or green or red-purple throughout, glabrous or ± hairy. **LF**: proximal 5–30 cm, entire, oblong to oblanceolate, distal reduced, linear to lanceolate; glabrous or ± hairy. **INFL**: heads 1–12, in panicle-like to ± flat-topped clusters; peduncle 3–16 cm, white-tomentose below involucre; disk hemispheric to broadly bell-shaped, 12–17 mm, 19–26 mm diam; outer phyllaries 11–16 mm, ovate to lanceolate, inner 7.5–8.5 mm, obovate to elliptic. **RAY FL**: 14–26; corolla 21–45 mm, gen yellow-orange to orange. **DISK FL**: 100–325+; corolla 4.2–5.4 mm. **FR**: 3.5–4.5 mm; pappus scales 5–7,

2.9–4.1 mm, lanceolate to lance-acuminate, not awned. 2*n*=30. Mtn meadows, open forest, streambanks; 1500–3650 m. KR, SN, Wrn, n SNE; to OR, WY, CO, NM. [*Dugaldia h.* (A. Gray) Rydb.] May–Nov

H. lemmonii (Greene) Cockerell (p. 359) ALKALI HYMENOXYS Per from ± branched caudex, fl > once, 3–5 dm. **ST**: 3–10(15), branched distally, purple-red-tinted proximally or green throughout, glabrous or sparsely hairy. **LF**: 2–9 cm, entire or divided into 3–13 linear lobes, glabrous or sparsely hairy; middle blades (3)5(7)-lobed, terminal lobe 1.5–2.3 mm wide. **INFL**: heads 10–85+, in panicle-like to ± flat-topped clusters; peduncle (1)2–4.5 cm, glabrous or ± hairy; disk ± hemispheric to bell-shaped, 8–11 mm, 12–15 mm diam; outer phyllaries 4.5–7 mm, lanceolate to ovate, inner 4–6 mm, obovate. **RAY FL**: 9–12; corolla 10–16 mm. **DISK FL**: 50–125+; corolla 3.5–4.2 mm. **FR**: 2.5–3.5 mm; pappus scales 5(6), (1.5)2.1–2.8 mm. 2*n*=30. Roadsides, open areas, meadows, slopes, drainage areas, stream banks; 800–3200 m. e KR, GB; to se OR, s ID, UT. Jun–Sep ★

H. odorata DC. BITTER HYMENOXYS Ann 1–8 dm. **ST**: 1–25, often branched throughout, gen purple-red-tinted proximally, occ throughout, ± hairy. **LF**: 1–5 cm, entire or divided into 3–19+ linear lobes, ± hairy; middle blades ± 5–11-lobed, terminal lobe 0.3–1 mm wide. **INFL**: heads 15–350+, in panicle-like cluster; peduncle 2–12 cm, sparsely hairy; disk ± hemispheric to bell- or ± urn-shaped, 6–10

mm, 7–12.5 mm diam; outer phyllaries 3.5–5.2 mm, obovate to oblanceolate, inner 3.8–5.9 mm, obovate. **RAY FL**: 8–13; corolla 8.5–11 mm. **DISK FL**: 50–150+; corolla 2.6–4.1 mm. **FR**: 1.7–2.5 mm; pappus scales 5–6, 1.6–2.3 mm. $2n$=22,24,28,30. Roadsides, open flats, drainage areas, streambanks and bottoms; 60–1500 m. e DSon; to KS, OK, TX, n Mex. TOXIC range pl outside CA. Feb–Aug ★

HYPOCHAERIS CAT'S-EAR

Kenton L. Chambers

Ann, per; sap milky. **ST**: erect, 1–8 dm, simple or few-branched. **LF**: basal in rosette, oblanceolate, margin entire to pinnately lobed; cauline alternate, scale-like. **INFL**: heads liguliflorous, 1–few, erect, terminal on st, branches; involucre cylindric to bell-shaped in fl, elongating in fr; phyllaries graduated in 4–5 series, reflexed when dry; receptacle flat to convex, paleate with thin membranous scales, glabrous. **FL**: gen many; ligules yellow, often ± red abaxially, readily withering. **FR**: fusiform, ribbed, long-beaked, or outer cylindric, beakless; pappus of stiff, plumose bristles, or shorter outer bristles merely barbed, tawny or dull white. ± 60 spp.: Eurasia, n Afr, S.Am. (Greek: less than joyous, from weedy habit) [Bogler 2006 FNANM 19:297–299]

1. Ann, ± glabrous; ligules 5–8 mm, barely exceeding involucre; outer fr cylindric, beakless, inner beaked . . . . . . *H. glabra*
1′ Per, rough-hairy; ligules 10–15 mm, much exceeding involucre; fr all beaked . . . . *H. radicata*

H. glabra L. (p. 359) SMOOTH CAT'S-EAR **ST**: 1–many from slender taproot, 1–6 dm. **LF**: 2–10 cm, entire to shallowly lobed, occ with minutely prickle-tipped teeth. **INFL**: involucre 8–16 mm. **FR**: outer fr 3–4 mm, cylindric, tapered to base, pappus bristles densely plumose in proximal 1/3; inner fr longer, fusiform, beak ≤ body, pappus bristles 6–10 mm, lightly plumose throughout. $2n$=10,12. Common. Disturbed areas, grassland, open woodland; < 1630 m. CA-FP; to BC, e US, native to Eur. Mar–Jun ❖

H. radicata L. ROUGH CAT'S-EAR **ST**: 1–several from fleshy per root, 1–8 dm. **LF**: 6–25 cm, toothed or lobed. **INFL**: involucre 12–16 mm. **FR**: fr body fusiform, 3–5 mm, ≤ slender beak; pappus bristles 9–10 mm, inner lightly plumose. $2n$=8. Disturbed areas, grassland, open woodland; < 1500 m. NW, CaR, n SN, ScV, CW, SW; to AK, MT, CO; e N.Am; native to Eur. Apr–Jul ❖

IONACTIS ASTER

Per [subshrub], cespitose, thickly taprooted, sometimes with rhizome and woody caudex. **ST**: erect or ascending. **LF**: simple, cauline, proximally crowded, spreading or ascending, alternate, sessile; blades linear to narrowly lanceolate or oblanceolate, entire, 1-veined. **INFL**: heads radiate, 1 [2–3 in loose cyme-like cluster]; involucre cylindric or bell-shaped; phyllaries 20–60, ± equal or graduated in 2–6 series, lance-linear to oblong, 1-veined, midrib green, thickened, margins scarious or membranous; receptacle flat, pitted, epaleate. **RAY FL**: pistillate, fertile. **DISK FL**: bisexual [or staminate]; corolla yellow, tube < narrowly cylindric throat, lobes 5, triangular; style branch tips lanceolate. **FR**: fusiform, ± compressed, densely strigose, nonglandular (in CA) or not; pappus of 2 series of bristles, outer < inner. 5 spp.: N.Am. (Greek: violet ray) [Nesom & Leary 1992 Brittonia 44: 247–252; Nesom 2006 FNANM 20:82–84]

I. alpina (Nutt.) Greene (p. 359) LAVA-ASTER, CRAG-ASTER Per; caudex fibrous-rooted. **ST**: 4–12 cm, ± hairy. **LF**: 0.3–1.2 cm, narrowly oblanceolate to elliptic, firm, ± abruptly pointed, densely short-hairy. **INFL**: head 1; phyllaries lance-linear to oblong, acute to acuminate, green to ± purple, inner pale-margined below. **RAY FL**: gen 8–16; corolla 7–12 mm, violet to purple. **DISK FL**: corolla 5.5–7.5 mm. **FR**: 5–6 mm; pappus outer bristles ± 1 mm, << inner. $2n$=18,36. Dry, rocky places, often with sagebrush; 1300–3000 m. Wrn, n SNE, W&I; to e OR, WY, NV. [*Aster scopulorum* A. Gray] May–Jul

ISOCOMA GOLDENBUSH

Subshrub; herbage glabrous, minutely scabrous, sessile- or stalked-glandular, long-soft-hairy, or tomentose. **ST**: prostrate to erect, ± striate below, yellow-white or gray to red-brown, glabrous or variously hairy, often dotted with sessile resin glands. **LF**: alternate, sometimes clustered in axils, entire, toothed, or pinnately lobed, gland-dotted, sometimes gummy-resinous, light to dark gray-green. **INFL**: heads discoid, in loose to tight cyme-like clusters, these borne at branch tips or in ± flat-topped or panicle-like 2° clusters; involucre obconic; phyllaries yellow-white proximally, cartilaginous, tips green; receptacle flat, epaleate. **FL**: corolla yellow; tube narrowly cylindric, abruptly expanded into larger cylindric throat; lobes short, erect; style branch appendages triangular. **FR**: narrowly obconic, light tan, silky-hairy; hairs white, yellow, tan, or light red-tan; pappus of 1–2 series of white, ± yellow, or red-tan bristles ± 2 × fr. 16 spp.: sw N.Am, Mex. (Greek: equal hair-tuft, from fls) [Nesom 2006 FNANM 20:439–445] Pls from s San Diego Co. with pinnately lobed lvs resemble *I. tenuisecta* Greene, not in CA, but are more densely hairy.

1. Phyllary tip swollen, with a prominent linear to widely elliptic resin gland (resin glands rarely ≥ 2), often with additional minute dot-like resin glands
   2. Lvs all entire or proximal lvs toothed or lobed . . . . [2]*I. acradenia*
      3. Phyllary tips rounded and not or occ weakly soft-pointed; fls 18–24 . . . . var. *acradenia*
      3′ Phyllary tips blunt to acute (inner ones), abruptly soft-pointed; fls 20–27 . . . . [2]var. *bracteosa*
   2′ Lvs all toothed or pinnately lobed
      4. Fls 10–13; phyllaries minutely awn-tipped . . . . [2]*I. arguta*
      4′ Fls 12–27; phyllaries rounded and not or minutely awn-tipped . . . . [2]*I. acradenia*
         5. Fls 20–27; phyllaries abruptly soft-pointed . . . . [2]var. *bracteosa*
         5′ Fls 12–20; phyllaries rounded and not awn-tipped or very weakly awn-tipped . . . . var. *eremophila*

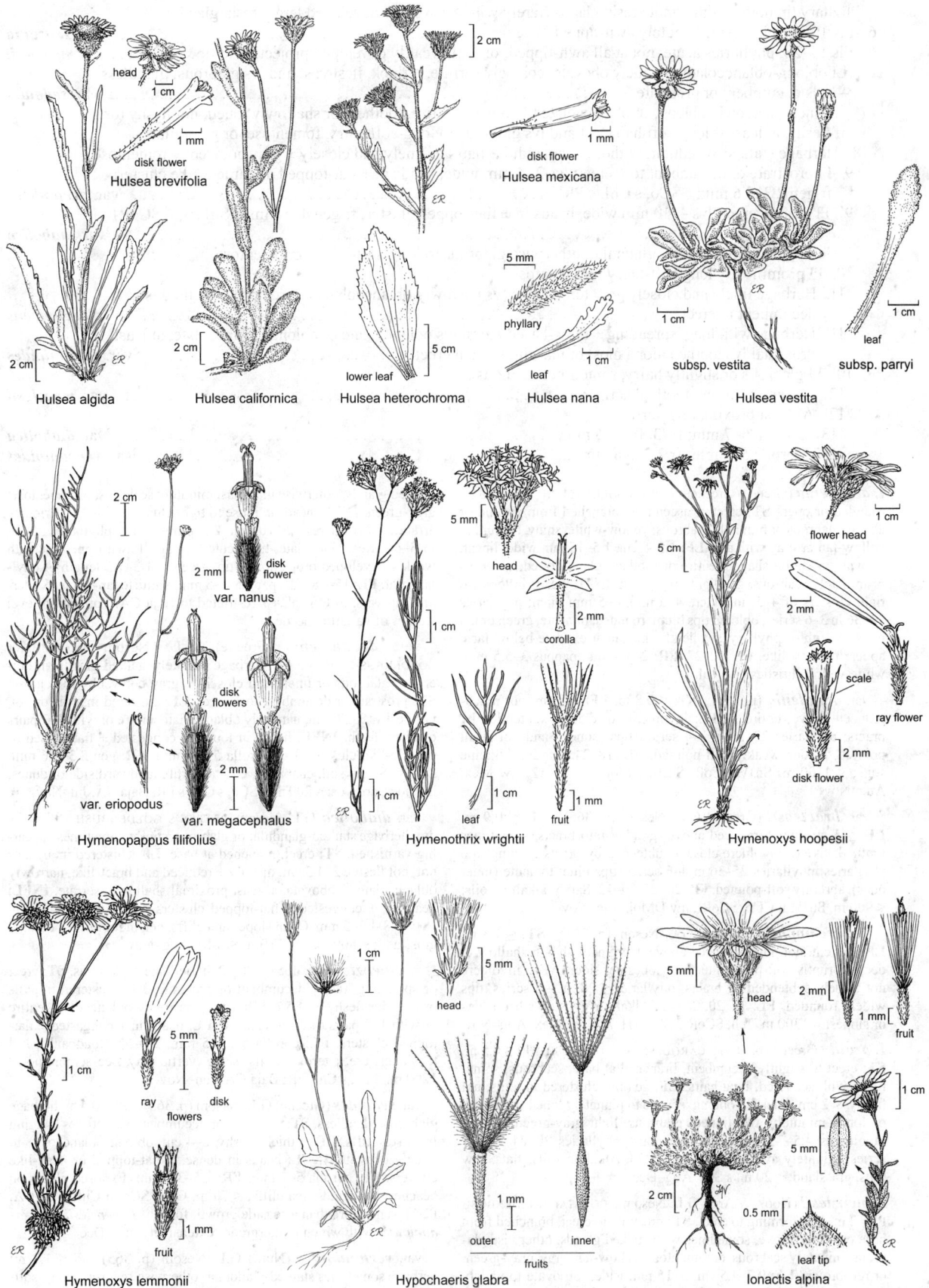

head
1 cm
disk flower
1 mm
Hulsea brevifolia

2 cm
ER

disk flower
1 mm
Hulsea mexicana

5 mm
phyllary
leaf
1 cm

2 cm
ER
subsp. vestita

leaf
1 cm
subsp. parryi

Hulsea vestita

2 cm
ER
Hulsea algida

2 cm
ER
Hulsea californica

lower leaf
5 cm
Hulsea heterochroma

Hulsea nana

2 cm
disk flower
var. nanus

2 mm

2 mm
disk flowers

2 mm

var. eriopodus
var. megacephalus
Hymenopappus filifolius

5 mm
head

corolla
2 mm

1 cm

1 cm
leaf
fruit
1 mm
Hymenothrix wrightii

5 cm

flower head
1 cm

2 mm

scale
ray flower

disk flower
2 mm

ER
Hymenoxys hoopesii

1 cm

ray
flowers
5 mm
disk

fruit
1 mm
ER
Hymenoxys lemmonii

1 cm
head
5 mm

outer
inner
fruits
1 mm
ER
Hypochaeris glabra

5 mm
head
2 mm
1 mm
fruit

2 cm
0.5 mm
leaf tip
5 mm
1 cm
ER
Ionactis alpina

1′ Phyllary tip dotted with minute resin glands (rarely with ≥ 1 weakly developed larger resin glands)
   6. Fls 10–13; phyllaries minutely awn-tipped . . . . ²*I. arguta*
   6′ Fls 15–28; phyllaries acute, not at all awn-tipped, or with weakly developed projection at tip . . . . *I. menziesii*
      7. Lf oblong-oblanceolate to widely obovate, coarsely serrate, ± thick, fleshy; st and lvs glabrous, glandless
         — pls decumbent or prostrate . . . . var. *sedoides*
      7′ Lf linear, narrowly oblong, or oblanceolate to obovate, entire, toothed, or shallowly lobed, not fleshy (or
         if fleshy, at least st long-soft hairy); st and lvs glabrous to long-soft hairy, tomentose, or glandular
        8. Herbage stalked-glandular, without nonglandular hairs or finely and closely ± gray cobwebby-tomentose
          9. Pl prostrate or decumbent to ± erect; lvs 2–4 mm wide; heads in ± flat-topped or panicle-like cluster;
            fr gen 2.3–3.6 mm; s SCo, s ChI, s PR . . . . ²var. *decumbens*
          9′ Pl strictly erect; lvs 4–10 mm wide; heads in ± flat-topped cluster; fr gen 4–5 mm; s SnFrB, n SCoRI
            . . . . ²var. *diabolica*
        8′ Herbage not prominently glandular, otherwise glabrous to long-soft-hairy or tomentose
          10. Pl prominently long-soft-hairy or tomentose
            11. Herbage finely and closely gray tomentose; lvs narrowly oblanceolate, entire or few-toothed; sts
              decumbent to erect . . . . ²var. *decumbens*
            11′ Herbage with long, spreading ± thick-based hairs; lvs oblanceolate or oblong-oblanceolate, at least
              proximal lvs toothed along most of margin; sts gen erect . . . . ²var. *vernonioides*
          10′ Pl glabrous or slightly hairy, sometimes resinous
            12. Lvs entire or few-toothed at tip . . . . var. *menziesii*
            12′ At least proximal lvs serrate
              13. Corolla 6–7 mm; fr (3.8)4.5–5 mm . . . . ²var. *diabolica*
              13′ Corolla 5–7 mm; fr 2.3–3.6 mm . . . . ²var. *vernonioides*

I. acradenia (Greene) Greene ALKALI GOLDENBUSH Pl ≤ 1.3 m, rounded or open. **ST:** erect or ascending, branched from ground or above, glabrous or minutely scabrous, yellow-white, shiny, becoming yellow-tan or gray with age. **LF:** 1.5–6 cm, 1.5–15 mm wide, linear, obovate or spoon-shaped, entire or toothed, gland-dotted, glabrous or minutely scabrous, gen light gray-green. **INFL:** heads in loose to tight clusters of 4–5; involucre 4–5 mm, 4–5 mm diam; phyllaries 22–36 in 3–6 series, oblong, tips blunt, rounded, or acute, green or tan to 1/4 length of phyllary, swollen by glandular exudate below face, appearing wart-like. **FL:** 12–27. **FR:** 2–3.5 mm; pappus 3–5.5 mm, white-yellow, bristles unequal.

   var. ***acradenia*** (p. 365) **ST:** ≤ 0.8 m. **LF:** ≤ 5 cm, not much-reduced above, entire. **INFL:** involucre not closely subtended by bracts; phyllaries 22–28 in 3–4 series, tips rounded and not at all pointed, or occ weakly soft-pointed. **FL:** 18–24. $2n=24$. Alkaline soils; < 1100 m. SnJV, SCoR, SnBr, DMoj; to UT, AZ, nw Mex. Aug–Nov

   var. ***bracteosa*** (Greene) G.L. Nesom (p. 365) **ST:** ≤ 0.9 m. **LF:** ≤ 5 cm, much-reduced above, grading into bracts, sometimes toothed. **INFL:** involucre closely subtended by bracts grading into phyllaries; phyllaries 25–36 in 4–6 series, tips blunt to acute (inner ones), abruptly soft-pointed. **FL:** 20–27. $2n=12$. Sandy, alkaline soils; < 900 m. SnJV, n CCo, SCoRI, nw DMoj. Aug–Nov

   var. ***eremophila*** (Greene) G.L. Nesom (p. 365) **ST:** ≤ 1.3 m. **LF:** ≤ 6 cm, not much-reduced above; margins with 4–6 shallow to deep, abruptly soft-pointed teeth or lobes per side. **INFL:** involucre not closely subtended by bracts; phyllaries ≤ 28 in 3–5 series, tips widely rounded. **FL:** 12–20. $2n=12$. Alkali or gypsum silt on flats or slopes; < 1300 m. Teh, SCoRI, D; to UT, AZ, nw Mex. Aug–Nov

I. arguta Greene (p. 365) CARQUINEZ GOLDENBUSH Pl ≤ 0.5 m. **ST:** erect to slightly decumbent, branched at base, scabrous, sometimes with scattered, long hairs. **LF:** gen not clustered in axils, not fleshy, < 2 cm, linear to oblong, serrate to pinnately lobed with teeth or lobes minutely awn-tipped, glabrous, light gray-green. **INFL:** involucres 4.5–7 mm, 2.5–5 mm diam; phyllaries 20–24 in 3–4 series, minutely awn-tipped. **FL:** 10–13. Alkaline soils, flats, low hills, grassland; < 20 m. s ScV. Aug–Dec ★

I. menziesii (Hook. & Arn.) G.L. Nesom COASTAL GOLDENBUSH Pl ≤ 2 m, mat-forming to erect. **ST:** prostrate to erect, branched from base or rarely above, sometimes with stalked glands, otherwise glabrous, minutely scabrous, or tomentose, yellow-tan, gray, gray-green, or red-brown. **LF:** 0.7–4.5 cm, 5–15 mm wide, (ob)ovate to widely spoon-shaped, entire or toothed, gland-dotted or sometimes with stalked glands, otherwise glabrous, minutely scabrous, or tomentose, gray-green. **INFL:** heads in loose to tight clusters of 4–10, variously arranged; involucres 4.5–10 mm, 2.5–8 mm diam; phyllaries 20–40 in 3–6 series, lanceolate, tips acute, not at all awn-tipped, or with weakly developed projection at tip, green to 1/3–1/2 length of phyllary, flat. **FL:** 15–28. **FR:** pappus 3–5 mm, white to tan-white. 5 intergrading vars. in CA, plus 1 restricted to Baja CA. Population-level studies of variation needed.

   var. ***decumbens*** (Greene) G.L. Nesom DECUMBENT GOLDENBUSH Pl ≤ 5 dm; herbage minutely stalked-glandular, otherwise glabrous or finely and closely ± gray cobwebby-tomentose. **ST:** prostrate or decumbent to ± erect. **LF:** clustered in axils or not, not fleshy, 5–22 mm, narrowly oblanceolate, entire or with 1–2 pairs of distal teeth. **INFL:** heads in loose or congested ± flat-topped or panicle-like clusters. **FL:** corolla 5–6.5 mm. **FR:** gen 2.3–3.6 mm. $2n=24$. Sandy soil, chaparral, coastal scrub, landward side of dunes, hillsides, arroyos; < 200 m. s SCo, s ChI, s PR; Baja CA. Jul–Nov ★

   var. ***diabolica*** G.L. Nesom SATAN'S GOLDENBUSH Pl 4–7+ dm; herbage stalked-glandular or glabrous, sticky, sometimes appearing varnished. **ST:** erect, branched at base. **LF:** clustered in axils or not, not fleshy, 2–4.5 cm, distally ± reduced and bract-like, narrowly oblanceolate to obovate, at least proximal shallowly serrate. **INFL:** heads in ± congested ± flat-topped clusters. **FL:** corolla 6–7 mm. **FR:** (3.8)4.5–5 mm. Open slopes and cliffs, gen in foothill woodland, grassland; < 800 m. s SnFrB, n SCoRI. Aug–Nov ★

   var. ***menziesii*** (p. 365) Pl ≤ 2 m; herbage ± glabrous. **ST:** erect or spreading, rarely decumbent or prostrate. **LF:** clustered in axils or not, not fleshy, < 4.5 cm, linear to narrowly oblanceolate, entire or with 1–2 pairs of distal teeth. **INFL:** heads in ± congested ± flat-topped clusters. **FL:** corolla 5.5–6.5 mm. $2n=24$. Roadbanks, hill toeslopes, creek terrace cliffs, shale, vertic clay, occ serpentine; < 1200 m. SCo, s ChI, PR; Baja CA. Jun–Nov

   var. ***sedoides*** (Greene) G.L. Nesom (p. 365) Pl ≤ 0.5 m; herbage glabrous, glandless. **ST:** prostrate or decumbent, sometimes hanging from sea cliffs. **LF:** ± thick, fleshy, 2–4 cm, oblong-oblanceolate to widely obovate. **INFL:** heads in dense, ± flat-topped or head-like clusters. **FL:** corolla 5–7 mm. **FR:** 2.5–3.5 mm. Exposed places on beaches, headlands, sea cliffs; < 70 m. CCo, SCo, n ChI; Baja CA. Perhaps no more than a seaside growth form of *I. menziesii* var. *vernonioides* with which it intergrades extensively. Jun–Dec

   var. ***vernonioides*** (Nutt.) G.L. Nesom (p. 365) Pl ≤ 1.2 m; herbage sometimes stalked-glandular, otherwise scabrous, long-soft-hairy, gray tomentose, or becoming glabrous. **ST:** erect to decumbent,

branched at base. **LF**: clustered in axils, often so dense as to obscure st, not fleshy, 10–45 mm, linear-oblong to oblanceolate, or obovate. **INFL**: heads in ± dense ± flat-topped clusters. **FL**: corolla 5–7 mm. **FR**: 2.3–3.6 mm. 2*n*=24. Protected sites on dunes, lagoon shores, marshes, dry slopes in grassland and coastal scrub; < 400 m. sw ScV, CW, SCo, ChI; n Baja CA. Intergrading with *I. menziesii* var. *m.*, *I. menziesii* var. *sedoides*. Jun–Dec

IVA MARSH-ELDER

Bruce G. Baldwin, adapted from Strother (2006)

[Ann] per to subshrub [shrub]. **ST**: erect to sprawling, often freely branched. **LF**: gen opposite, distal sometimes alternate, petioled or sessile; blades thread-like to (ob)ovate or deltate, gen entire, rarely toothed (in CA), ± minutely scabrous [glabrous], gland-dotted (in CA). **INFL**: heads disciform [discoid], ± inverted, in ± raceme-like [spike-like], lfy-bracted cluster; involucre ± hemispheric (in CA), 2–10+ mm diam; phyllaries 3–15+ in 1–3+ series, free or ± fused, ± green or inner scarious to membranous; receptacle flat to hemispheric; paleae linear to wedge-shaped, ± membranous, sometimes 0. **PISTILLATE FL**: [0]3–8; corolla ± white, inconspicuous, ± tubular. **DISK FL**: [2]4–20+, staminate; corolla ± white to ± pink, funnel-shaped; anthers free or fused, tip incurved; style tip ± flared, truncate. **FR**: plumply obovate to pear-shaped, often ± compressed, glabrous or minutely scabrous or bristly distally; pappus 0. ± 9 spp.: N.Am. (Latin: from mint *Ajuga iva* (L.) Schreb., with similar odor) [Strother 2006 FNANM 21:25–28] Other taxa in TJM (1993) moved to *Euphrosyne*.

1. Outer phyllaries ± fused; sts gen herbaceous. . . . . ***I. axillaris***
1′ Outer phyllaries free; sts gen woody at base. . . . . ***I. hayesiana***

I. axillaris Pursh (p. 365) POVERTY WEED Per 1–4(6) dm, from rhizome. **ST**: erect. **LF**: petiole 0–3 mm; blade 15–25(45) mm, 3–8(15) mm wide, elliptic to (ob)ovate or spoon-shaped, gen ± strigose to minutely scabrous, gland-dotted. **INFL**: peduncle 1–2 mm; involucre 2–3.5 mm; outer 3–5 phyllaries ± green; paleae 1.5–2 mm, linear. **PISTILLATE FL**: 3–5(8); corolla 0.5–1.5 mm. **DISK FL**: 4–8(20+); corolla 2–2.5 mm. **FR**: 2.5–3 mm. 2*n*=36,54. Seasonally wet, saline habitats, roadsides; < 2500 m. CaRH, SNH, s SNF, GV, CW, SW (exc ChI), GB, DMoj; to BC, MT, c US, TX. [*I. a.* subsp. *robustior* (Hook.) Bassett] Apr–Oct

I. hayesiana A. Gray (p. 365) SAN DIEGO MARSH-ELDER Per to subshrub, < 1 m. **ST**: sprawling to erect. **LF**: petiole 0–3(10) mm; blade 2–5(10) cm, 5–10(18) mm wide, lance-elliptic to oblanceolate, sparsely strigose to minutely scabrous, gland-dotted. **INFL**: peduncle 2–3+ mm; involucre 2.5–3.5 mm; outer 5 phyllaries ± green; paleae 1.5–2.5 mm, linear to wedge-shaped. **PISTILLATE FL**: 5; corolla 0.5–1 mm. **DISK FL**: 5–12(20); corolla 1.5–2.5 mm. **FR**: 2–2.5 mm. Alkaline flats, depressions, streambanks; < 300(900) m. s SCo, sw PR (sw San Diego Co.); to Baja CA. Mar–Sep ★

JAUMEA

Per from stolons and slender rhizomes; herbage glabrous. **ST**: prostrate to ascending, rooting proximally. **LF**: opposite, sessile, narrow, entire, fleshy. **INFL**: heads radiate [discoid], gen 1(3); peduncle bractless; involucre cylindric or ovoid; phyllaries graduated in 3–5 series, tightly appressed; receptacle conic, epaleate. **RAY FL**: [0]3–10; ray yellow. **DISK FL**: 20–50+; corolla yellow; anther tips narrowly triangular; style tips short-triangular. **FR**: ± cylindric, angled, glabrous; pappus 0 or a crown of short bristles. 2 spp.: w coasts of N.Am, s S.Am. (J.H. Jaume St. Hilaire, French botanist, 1772–1845) [Strother 2006 FNANM 21:253–254]

J. carnosa (Less.) A. Gray (p. 365) **ST**: long, weak, trailing. **LF**: gen 1.5–5 cm, linear to narrowly oblong-oblanceolate; bases fused 1–2 mm around st, tip ± obtuse. **INFL**: heads 1.2–2 cm (incl fls); peduncle 1–2 cm, enlarged below head; involucre 5–12 mm diam; phyllaries ovate, gen ± purple, tips rounded. **RAY FL**: ray 1–5 mm, narrow. **DISK FL**: corolla 6–7 mm. **FR**: 2–3 mm, brown; pappus 0 or bristles 1–5, ≤ 0.5 mm. 2*n*=38. Coastal salt marshes, bases of sea cliffs; < 70 m. NCo, CCo, SCo; to BC, n Baja CA. Apr–Dec

JENSIA TARWEED, TARPLANT

Bruce G. Baldwin

Ann 5–60 cm. **ST**: erect. **LF**: basal and cauline, proximal opposite, often crowded, distal alternate, sessile, spoon-shaped to linear, entire or toothed, coarse-hairy to strigose, distal sometimes also stalked-glandular. **INFL**: heads radiate, in ± umbel-like clusters; involucre 3–5 mm diam, ± obconic or urn-shaped to spheric; phyllaries 2–12 in 1 series, lanceolate (often long-tapered), each gen wholly enveloping a ray ovary, falling with fr, coarse-hairy, hair tips ± hooked; receptacle flat to convex, glabrous or minutely bristly; paleae in 1 series between ray and disk fls, fused, phyllary-like to ± scarious, deciduous. **RAY FL**: 2–12; corolla yellow, ray sometimes purple-veined abaxially. **DISK FL**: 1–65, staminate; corolla yellow, tube < throat, lobes deltate; anthers ± dark purple, tips ovate-deltate; style glabrous proximal to branches, tips narrowly lance-awl-shaped, densely hairy. **FR**: ray fr compressed, club-shaped, arched, glabrous, tip beaked, pappus crown-like, scales 0.1–1 mm; disk pappus of 5–7 scales, 2.5–3 mm, white or purple-tipped, awl-shaped, wavy, minutely ciliate. 2 spp.: CA. (Jens C. Clausen, CA botanist, 1891–1969) [Baldwin & Strother 2006 FNANM 21:301–302]

1. Involucre urn-shaped or spheric; ray fls 5–12, ray 4–10 mm; disk fls 16–65; pls 6–60 cm . . . . ***J. rammii***
1′ Involucre broadly obconic; ray fls 2–8, ray 0.5–3 mm; disk fls 1–7; pls 5–15(25) cm. . . . . ***J. yosemitana***

J. rammii (Greene) B.G. Baldwin (p. 365) **LF**: 15–100 mm, 1–3 mm wide. **INFL**: phyllaries (3)4–5 mm. 2*n*=16. Grassy slopes, openings in forest, woodland, often in clayey soils; 400–1100(1600) m. n SN. [*Madia r.* Greene] Self-sterile. Apr–Jul

J. yosemitana (A. Gray) B.G. Baldwin (p. 365) YOSEMITE TARPLANT **LF**: 10–50 mm, 1–2 mm wide. **INFL**: phyllaries 2.5–4 mm. 2*n*=16. Meadows, sandy sites; 1200–2300 m. SNH. [*Madia y.* A. Gray] Self-fertile. May–Aug ★

KYHOSIA

Bruce G. Baldwin

1 sp.: CA, s OR, w NV. (Donald W. Kyhos, CA botanist, b. 1929) [Baldwin & Strother 2006 FNANM 21:295–296]

K. bolanderi (A. Gray) B.G. Baldwin (p. 365) Per 5–12 dm, from rhizome; strongly scented. **ST**: erect. **LF**: basal and cauline, proximal opposite and in rosette, distal alternate, sessile, 5–35 cm, 4–15 wide, lance-linear to linear, entire, coarse-hairy, distal finely stalked-glandular. **INFL**: heads radiate, 1 or in loose, ± flat-topped cluster; involucre ± bell-shaped to hemispheric, 6–12+ mm diam; phyllaries 8–12 in 1 series, each mostly or wholly enveloping a ray ovary, falling with fr, 7–14 mm, lanceolate to lance-linear, coarse-hairy, finely stalked-glandular; receptacle flat to convex, glabrous; paleae in 1 series between ray and disk fls, weakly fused or free, deciduous. **RAY FL**: 8–12; corolla bright yellow, ray 7–13 mm, 3-lobed. **DISK FL**: 28–65; corolla 5–8 mm, bright yellow, hairy, tube < throat, lobes deltate; anthers ± dark purple, tips lance-ovate to ovate; style glabrous proximal to branches, tip awl-shaped, densely hairy. **FR**: ray fr 5–7 mm, compressed, club-shaped, arched, glabrous or minutely bristly, black, tip beakless, pappus 0 or crown-like, ≤ 0.7 mm; disk fr 5–9 mm, ± cylindric, straight or arched, minutely bristly, brown to black, pappus of 5–10 scales, 1–5 mm, straw-colored to ± purple, lanceolate to awl-shaped, ciliate to plumose. 2*n*=12. Meadows, streambanks; 1000–2600 m. KR, CaRH, SNH; s OR. [*Madia b.* (A. Gray) A. Gray] Self-sterile. Jul–Sep

LACTUCA LETTUCE

David J. Keil & G. Ledyard Stebbins

Ann to per; sap milky. **ST**: decumbent to erect. **LF**: basal and cauline, alternate, entire to pinnately lobed. **INFL**: heads liguliflorous, in panicle-like or flat-topped clusters; involucre ± cylindric; phyllaries in 2–several series; receptacle flat or rounded, epaleate. **FL**: 6–50+; ligules yellow or cream to blue, readily withering. **FR**: flattened, short- or long-beaked; pappus of 80–120+ bristles, falling separately. ± 100 spp.: ± worldwide temp. (Latin: milky) [Strother 2006 FNANM 19:258–263] *L. sativa* L. (garden lettuce) occ escapes from cult but does not persist.

1. Fr beak short, thick; bien or per
  2. Bien from short taproot; open heads 1–1.5 cm diam; NW, n CCo . . . . ***L. biennis***
  2′ Per from long, deep rhizome; open heads 2–3 cm diam; CaRH, c SNH, MP . . . . ***L. tatarica*** subsp. ***pulchella***
1′ Fr beak slender, thread-like; ann or bien
  3. Fr with 1 median rib on each face
    4. Open heads 5–10 mm diam; fls 13–25 . . . . ***L. canadensis***
    4′ Open heads 4–5 mm diam; fls 20–56 . . . . ***L. ludoviciana***
  3′ Fr with 5–9 ribs on each face
    5. Bien; basal rosette lvs toothed or shallowly lobed; cauline lvs widely clasping . . . . ***L. virosa***
    5′ Ann; basal lvs gen 0 at fl; cauline lvs with narrow, acute, basal lobes clasping st
      6. Sts proximally decumbent and arching upward or erect, 5–10(20) dm; lvs lance-linear, entire or
        few-lobed, lobes linear; peduncles and infl branches often appressed to axis; fls 5–12 . . . . ***L. saligna***
      6′ Sts erect, 5–30 dm; lvs oblanceolate to oblong-elliptic or obovate in outline, margins prickly-toothed,
        sometimes also coarsely lobed; infl branches often widely spreading; fls 14–20 . . . . ***L. serriola***

L. biennis (Moench) Fernald (p. 365) Bien from taproot. **ST**: erect, 0.5–4 m. **LF**: basal 0 at fl; cauline many, lanceolate to ovate or elliptic in outline, entire to deeply lobed and coarsely toothed, proximal wing-petioled, distal sessile, clasping, reduced to linear bracts in infl; abaxial midribs glabrous or sparsely bristly. **INFL**: heads many in large, panicle-like cluster, open heads 10–15 mm diam; involucre in fr 8–10 mm. **FL**: 15–30+; corolla pale blue to cream or pale yellow. **FR**: 5–7 mm, oblong, ± thick, mottled, 5–6-ribbed on each face, unwinged, tapered to a stout beak 0.1–0.5 mm; pappus 4–6 mm, ± brown. 2*n*=34. Streambanks, conifer forest; < 800 m. NW, n CCo; to AK, e N.Am. Jun–Aug

L. canadensis L. Gen bien from taproot. **ST**: erect, 0.5–2.5 m. **LF**: basal 0 at fl, cauline many, linear to oblong, elliptic, lanceolate, or ovate in outline, toothed or with linear to ovate, curved, gen entire lobes, proximal wing-petioled, distal sessile, ± clasping, reduced to linear bracts in infl; abaxial midribs glabrous or sparsely bristly. **INFL**: heads many in large, panicle-like cluster, open heads 5–10 mm diam; involucre in fr 10–15 mm. **FL**: 13–25; corolla ± blue or pale yellow. **FR**: body 3–4 mm, elliptic, ± black, gen with 1 median rib on each face, unwinged; beak slender, 2–3 mm; pappus 5–7 mm, white. 2*n*=34. Forest, disturbed areas; < 1400 m. KR, n SN; native to c&e N.Am. Jul–Oct

L. ludoviciana (Nutt.) Riddell Bien from taproot. **ST**: erect, 1–2 m. **LF**: basal 0 at fl, cauline oblanceolate to obovate in outline, gen clasping, minutely dentate, unlobed or with widely lanceolate, dentate lobes; abaxial midveins gen hairy or bristly. **INFL**: heads many in panicle-like clusters, open heads 4–5 mm diam; involucre in fr 12–22 mm. **FL**: 20–56; corolla light blue to ± yellow. **FR**: 3–4 mm, elliptic, mottled gray and black, transversely roughened, 1-ribbed on each face, unwinged; beak slender, 2.5–4.5 mm; pappus 7–12 mm, white. 2*n*=34. Disturbed areas, openings in woodland, shaded areas; < 1000 m. NCoR, CaRF, SCo, MP; native to c US. Jun–Sep

L. saligna L. (p. 365) Ann. **ST**: 1–many from base, 0.3–1(2) m, proximally decumbent and arching upward or erect, glabrous or proximally bristly. **LF**: basal gen 0 at fl, cauline few to many, linear to lanceolate, entire or with few, linear lobes, base sagittate or hastate, clasping; abaxial midvein glabrous or prickly-bristly. **INFL**: heads few to many in gen narrow, spike-like or panicle-like clusters; branches and peduncles often appressed; open heads 4–5 mm diam; involucre in fr 10–17 mm. **FL**: 5–12; corolla pale yellow. **FR**: body 2.5–3.5 mm, brown, dark-mottled, rough, glabrous, unwinged, faces 5–7-ribbed; beak thread-like, 5–6 mm; pappus 5–6 mm, white. 2*n*=18. Roadsides, grassland; < 750 m. KR, NCoR, n SNF, ScV, CW, SW; to e US; native to Eur. Jul–Nov

L. serriola L. (p. 365) PRICKLY LETTUCE Ann from taproot. **ST**: 0.5–3 m, erect, prickly-bristly. **LF**: few to many, oblanceolate to oblong-elliptic or obovate, margin prickly-bristly, unlobed or coarsely lobed; base sagittate-clasping; abaxial veins (esp midvein) prickly-bristly. **INFL**: heads gen many, in open, panicle-like clusters, branches often widely spreading; open heads 4–6 mm diam; involucre in fr 10–12 mm. **FL**: 14–20; corolla pale yellow. **FR**: body 2.5–3.5 mm, light to dark brown, rough-hairy, unwinged, faces 5–7-ribbed; beak 2.5–4 mm, thread-like; pappus 3.5–5 mm, white. 2*n*=18. Abundant. Disturbed places; < 2700 m. CA; N.Am; native to Eur. May–Oct

L. tatarica (L.) C.A. Mey. subsp. ***pulchella*** Stebbins (p. 365) Per from extensive rhizomes. **ST**: 3–10 dm erect, glabrous. **LF**: basal 0 at

fl, cauline many, linear to elliptic or lanceolate, entire to lobed, base gen narrowed, not clasping; abaxial veins glabrous. **INFL**: heads few to many in cyme-like or panicle-like cluster, in fl 2–3 cm diam; involucre in fr 1.4–1.8 mm. **FL**: 15–50; corolla bright blue, conspicuous. **FR**: body 4–5 mm, red-brown to black, sometimes mottled, glabrous, faces 4–6-ribbed; beak 0–1 mm; pappus 8–11 mm, white. $2n=18$. Dry to moist alluvial valleys; 1150–2000 m. CaRH, c SNH, MP; to AK, e Can, c US, NM. [*Mulgedium oblongifolium* (Nutt.) Reveal; *M. p.* G. Don, illeg.] Jun–Sep

L. virosa L. (p. 365) Bien from taproot, 0.6–2 m. **ST**: erect, glabrous or proximally bristly. **LF**: basal oblanceolate to obovate, forming persistent rosette, dentate or shallowly lobed; cauline many, proximal wing-petioled, distal sessile, widely clasping, abaxial midvein prickly-bristly. **INFL**: heads many, in cyme-like clusters, in fl 3–4 cm diam; involucre in fr 8–12 mm. **FL**: 15–20; corolla pale yellow. **FR**: 5–9-ribbed on each face, rough; body dark, wing-margined; beak white, $\pm$ = body; pappus 8 mm, white. $2n=18$. Disturbed, shrubby and wooded slopes; < 600 m. SnFrB, n SCoRO; native to Eur. Jun–Aug

LAENNECIA

Bruce G. Baldwin, adapted from Strother (2006)

Ann [per], gen taprooted. **ST**: gen erect, simple or branched. **LF**: mostly cauline at fl, alternate, sessile (in CA), gen 1-veined, linear or elliptic to (ob)ovate or spoon-shaped, entire, toothed, or 1–2-pinnately lobed, minutely coarse-hairy to long-soft-hairy, gen glandular. **INFL**: heads obscurely radiate or disciform, in $\pm$ flat-topped or raceme- or panicle-like clusters; involucres $\pm$ obconic, 2–7+ mm diam; phyllaries 20–40+ in 2–4 series, appressed, gen reflexed in fr, linear to lanceolate, $\pm$ (un)equal, margins $\pm$ membranous, abaxial face $\pm$ green, minutely hairy, gen glandular, midvein not orange to brown; receptacle $\pm$ flat, pitted or smooth, epaleate. **PISTILLATE FL**: [20]60–100[200]; corolla thread-like, $\pm$ yellow, gen distally truncate or 2–5-toothed, ray 0 [0.5–1+ mm]. **DISK FL**: [2]5–20[30+]; corolla $\pm$ yellow, narrowly funnel-shaped, throat > tube, abruptly expanded distally, lobes deltate; anther tip awl-like; style-branch tips $\pm$ deltate, papillate. **FR**: oblong to elliptic or obovate, compressed, each edge 1-veined, faces glabrous or $\pm$ minutely strigose, often stalked- or sessile-glandular; pappus in CA of 9–30+ white to tawny, minutely-barbed, distally long-tapered bristles, in 1 series, $\pm$ equal, persistent. $\pm$ 18 spp.: New World. (R.-T.-H. Laennec, French physician, 1781–1826) [Strother 2006 FNANM 20:35–38]

L. coulteri (A. Gray) G.L. Nesom (p. 371) COULTER'S HORSEWEED Ann 1–15 dm. **ST**: 1–several from base, gen simple proximally, much-branched distally. **LF**: proximal 2–7(10) cm, 1–2.5 cm wide, widely spoon-shaped to oblong, obscurely lobed or coarsely toothed; distal 1–3 cm, 3–10 mm wide, coarsely toothed or entire. **INFL**: heads in raceme- or panicle-like clusters; involucre 2.5–3+ mm; receptacle 2–3 mm diam. **PISTILLATE FL**: 60–100+; corolla 1.5–2 mm, $\pm$ 1/2 style, ray 0. **DISK FL**: 5–20; corolla 3–4 mm. **FR**: 0.5–1 mm, pale tan; pappus (2.5)3.5–4 mm, $\pm$ white. $2n=18$. Disturbed sites, clayey or sandy soils, often seasonally wet, alkaline; < 1300 m. s SNF, SnJV, CW, SW, SNE (exc W&I), D; to CO, TX, Mex. [*Conyza c.* A. Gray] May–Nov

LAGOPHYLLA HARE-LEAF

Bruce G. Baldwin

Ann 1–10(15) dm. **ST**: $\pm$ erect. **LF**: mostly cauline, proximal opposite, most alternate, $\pm$ sessile; blade narrowly elliptic to linear or proximal oblanceolate to spoon-shaped, entire or proximal sometimes toothed, coarse-, soft-, or silky-hairy or strigose, all or distal sometimes also stalked-glandular. **INFL**: heads radiate, in tight groups or $\pm$ panicle-like clusters; involucre 3–6+ mm diam, $\pm$ hemispheric or obovoid to obconic, sometimes subtended by calyx-like set of bracts; phyllaries 5 in 1 series, linear to oblanceolate, each wholly enveloping a subtended ray ovary, spreading and falling with fr, coarsely long-straight-hairy to minutely coarse-hairy or scabrous on angles; receptacle flat to convex, densely bristly; paleae in ring between ray and disk fls, fused or free, scarious. **RAY FL**: 5; corolla yellow, ray fan-shaped, deeply lobed, abaxially often red- to purple-veined. **DISK FL**: 6, staminate; corolla yellow, tube < throat, lobes deltate; anthers $\pm$ dark purple, tips triangular-ovate to $\pm$ rounded; style glabrous proximal to undivided, awl-shaped, short-hairy tip. **FR**: ray fr $\pm$ compressed front-to-back, glabrous, black, tip beakless, pappus 0; disk fr 0, pappus 0. 4 spp.: CA, to WA, MT, NV. (Greek: hare lf, for soft-hairy lvs) [Baldwin & Strother 2006 FNANM 21:260–261] Easily overlooked; lvs wither early, heads close at mid-day.

1. Ray (4)3–6 mm; lvs $\pm$ gray, distal stalked-glandular abaxially, glands gen $\pm$ white or yellow, sometimes golden-brown; st glandless; heads in tight groups or open, panicle-like clusters; calyx-like bracts subtending involucre 2–5; phyllaries coarsely long-straight-hairy on angles . . . . ***L. ramosissima***
1′ Ray (4)7–13 mm; lvs green or gray-green, distal glandless or stalked-glandular, glands yellow, golden, $\pm$ white, or purple; st glandless or distally stalked-glandular; heads in panicle-like clusters; calyx-like bracts subtending involucre 0 or 2–5; phyllaries scabrous or minutely coarsely hairy to coarsely long-straight-hairy on angles
   2. Main st axis obvious; distal lvs stalked-glandular, glands yellow or golden; calyx-like bracts subtending involucre 3–5; involucre obconic; phyllaries minutely coarsely hairy to coarsely long-straight-hairy on angles, hairs 0.3–1+ mm, $\pm$ widely spreading or curved toward phyllary tip . . . . ***L. glandulosa***
   2′ Main st axis often not obvious, $\pm$ zigzag; distal lvs glandless or stalked-glandular, most glands purple, some yellow; calyx-like bracts subtending involucre 0 or 2–3; involucre $\pm$ hemispheric to obovoid; phyllaries $\pm$ minutely coarsely hairy or scabrous or coarsely long-straight-hairy on angles, hairs 0.1–1+ mm, $\pm$ curved toward phyllary tip
      3. Phyllaries 4–6.5 mm, $\pm$ minutely coarsely hairy to scabrous on angles, hairs 0.1–0.6 mm; fr dull . . . . . ***L. dichotoma***
      3′ Phyllaries 4–5 mm, coarsely long-straight-hairy on angles, hairs 0.5–1+ mm; fr shiny . . . . ***L. minor***

L. dichotoma Benth. (p. 371) Pl 1–10 dm; self-sterile. **ST**: distally glandless or sparsely stalked-glandular; main axis often not obvious, $\pm$ zigzag. **LF**: green, proximal gen glandless, distal glandless or stalked-glandular, most glands purple, some yellow or $\pm$ white. **INFL**: heads in panicle-like clusters; calyx-like bracts subtending involucre 0 or 2–3; involucre $\pm$ hemispheric to obovoid; phyllaries

4–6.5 mm, ± minutely coarse-hairy to scabrous on angles, hairs 0.1–0.6 mm, ± curved toward phyllary tip. **RAY FL**: ray (4)7–13 mm. **FR**: dull, striate. $2n=14$. Grassland, openings in woodland; 50–900 m. c&s SNF, e SnJV, SCoRI. Apr–Jun ★

L. glandulosa A. Gray (p. 371) Pl 1–10(15) dm; ± self-incompatible. **ST**: distally gen sparsely to densely stalked-glandular (nonglandular); main axis obvious. **LF**: green or gray-green, all or distal stalked-glandular, glands yellow or golden. **INFL**: heads in panicle-like clusters; calyx-like bracts subtending involucre 3–5; involucre obconic; phyllaries 5–7 mm, minutely coarsely hairy to coarsely long-straight-hairy on angles, hairs ± widely spreading or curved toward phyllary tip, 0.3–1+ mm. **RAY FL**: ray 7–13 mm. **FR**: glossy, not striate. $2n=14$. Grassland, openings in chaparral, woodland; 10–900 m. c NCoRO (e edge), c NCoRI, CaRF, SNF (rare s), n GV. May–Nov

L. minor (D.D. Keck) D.D. Keck (p. 371) Pl 8–30+ cm; self-incompatible. **ST**: glandless or distally sparsely glandular; main axis not obvious, ± zigzag. **LF**: green, proximal gen glandless, dis-tal glandless or stalked-glandular, most glands purple, some yellow. **INFL**: heads in panicle-like clusters; calyx-like bracts subtending involucre 0; involucre ± hemispheric to obovoid; phyllaries 4–5 mm, coarsely long-straight-hairy on angles, hairs 0.5–1+ mm, ± curved toward phyllary tip. **RAY FL**: ray 7–13 mm. **FR**: shiny, not striate. $2n=14$. Openings in chaparral, woodland, on serpentine; 70–900 m. NCoRI, n SNF (El Dorado Co.). Apr–Jun

L. ramosissima Nutt. (p. 371) Pl 1–10(15) dm; self-compatible. **ST**: glandless; main axis obvious or not (± zigzag). **LF**: ± gray, most glandless, distal stalked-glandular abaxially, glands gen ± white or yellow, sometimes golden-brown. **INFL**: heads tightly grouped or in open, panicle-like clusters; calyx-like bracts subtending involu-cre 2–5; involucre obconic to obovoid; phyllaries 4–7 mm, coarsely long-spreading-hairy on angles, hairs ± widely spreading to curved toward phyllary tip, 0.5–1+ mm. **RAY FL**: ray 3–6 mm. **FR**: dull to ± shiny, weakly striate. $2n=14$. Grassland, openings in scrub, wood-land, forest; < 1800 m. CA-FP (exc NCo, ChI, SnJt), MP; to WA, MT, NV. [*L. r.* subsp. *congesta* (Greene) D.D. Keck] Apr–Oct

LAPSANA NIPPLEWORT

Kenton L. Chambers

1 sp. (Greek: name used by Dioscorides) [Bogler 2006 FNANM 19:257]

L. communis L. Ann, bien, glabrous to ± soft-hairy; sap milky. **ST**: gen 1 from base, erect, 2–15 dm. **LF**: gen cauline, alternate, 2–20 cm, petioled or distal ± sessile, ovate, entire to toothed, often few-lobed proximally on petiole. **INFL**: panicle-like; heads liguliflorous, erect; peduncles slender, ascending, subtended by scale-like bracts; involucre ovoid to ± cylindric, glabrous; phyllaries in 2 series, outer few, 0.5–1 mm, inner 8–10, equal, 3–9 mm, linear-oblong, distally keeled, erect in fr; receptacle flat, epaleate, smooth. **FL**: 6–15; lig-ules 4–10 mm, exceeding involucre, yellow, readily withering. **FR**: 3–5 mm, curved, fusiform, ± compressed, ± 20-veined, pale brown, glabrous; pappus 0. $2n=12,14,16$. Shady places, disturbed areas; < 1600 m. NW, SN, GV, CCo, SnFrB, SW; widely adventive in N.Am; native to Eur. Jun–Sep

LASIOSPERMUM COCOONHEAD

Thomas J. Rosatti

[Ann] per. **LF**: alternate, deeply 1–2-pinnately lobed. **INFL**: head radiate [disciform, discoid], 1, long-peduncled; phyllaries unequal, weakly graduated in 2–4 series, reflexed in fr, margins, tips scarious; receptacle flat or convex; paleae scarious, each with conspicuous resin canal. **RAY FL**: ray linear, entire, short-erect or long-spreading, white [± pink]. **DISK FL**: corolla tubular, ± yellow; anther tip ± ovate; style tips truncate. **FR**: oblong, 8–10-ribbed, initially hairy, densely tomentose in age; pappus 0. 4 spp.: s Afr, Egypt. (Greek: hairy, seed) [Ross & Boyd 1996 Madroño 43:433–434]

L. bipinnatum (Thunb.) Druce Pl glabrous. **ST**: gen ± decum-bent, with adventitious roots, < 60 cm. **LF**: 2–6 cm. **INFL**: peduncle 10–20 cm, bracts 5–10, not overlapping, pinnately lobed or toothed to entire, 2–6 mm; involucre 4–5 mm. **RAY FL**: ± 30, ray 5–10 mm. $2n=18,20+$. Disturbed areas; < 500 m. SCo; native to s Afr, adventive in adjacent areas, naturalized in Tasmania. Collection by Ross and Boyd in CA evidently the only record for N.Am. Apr–May

LASTHENIA GOLDFIELDS

Raymund Chan & Robert Ornduff

Ann, per, glabrous or hairy. **ST**: gen branched, gen erect, < 60 cm. **LF**: opposite (or distal-most alternate), < 20 cm, entire to pinnately cut. **INFL**: heads radiate (disciform), 1 or in cyme-like cluster; involucre cylindric to obconic, bell-shaped, or hemispheric; phyllaries 4–18 in 1(2) series, free or ± fused; receptacle narrowly conic to hemispheric, smooth, pitted, or rough, epaleate. **RAY FL**: 4–16; ray gen yellow (short or 0). **DISK FL**: gen many; corolla gen (4)5-lobed, gen yellow to ± orange; anther tip awl-shaped to triangular; style tips triangular or dome-shaped, gen hair-tufted. **FR**: < 5 mm, cylindric to obovoid, black to gray; pappus of awns, scales, or 0. 18 spp.: w N.Am, Chile. (Greek: female student of Plato who dressed as a man) [Chan & Ornduff 2006 FNANM 21:336–347] Gen self-sterile.

1. Phyllaries ± fused
  2. Phyllaries fused < 1/2; lvs entire to pinnately lobed . . . . ***L. conjugens***
  2′ Phyllaries fused > 2/3; lvs entire
    3. Pappus present . . . . ***L. glaberrima***
    3′ Pappus 0
      4. Fr strongly flattened, margin fringed with blunt hairs . . . . ***L. chrysantha***
      4′ Fr not strongly flattened, glabrous or face papillate or hairy
        5. Fr hairs short, curved . . . . ***L. ferrisiae***
        5′ Fr glabrous or papillate . . . . ***L. glabrata***
          6. Fr papillate . . . . subsp. ***coulteri***
          6′ Fr glabrous . . . . subsp. ***glabrata***

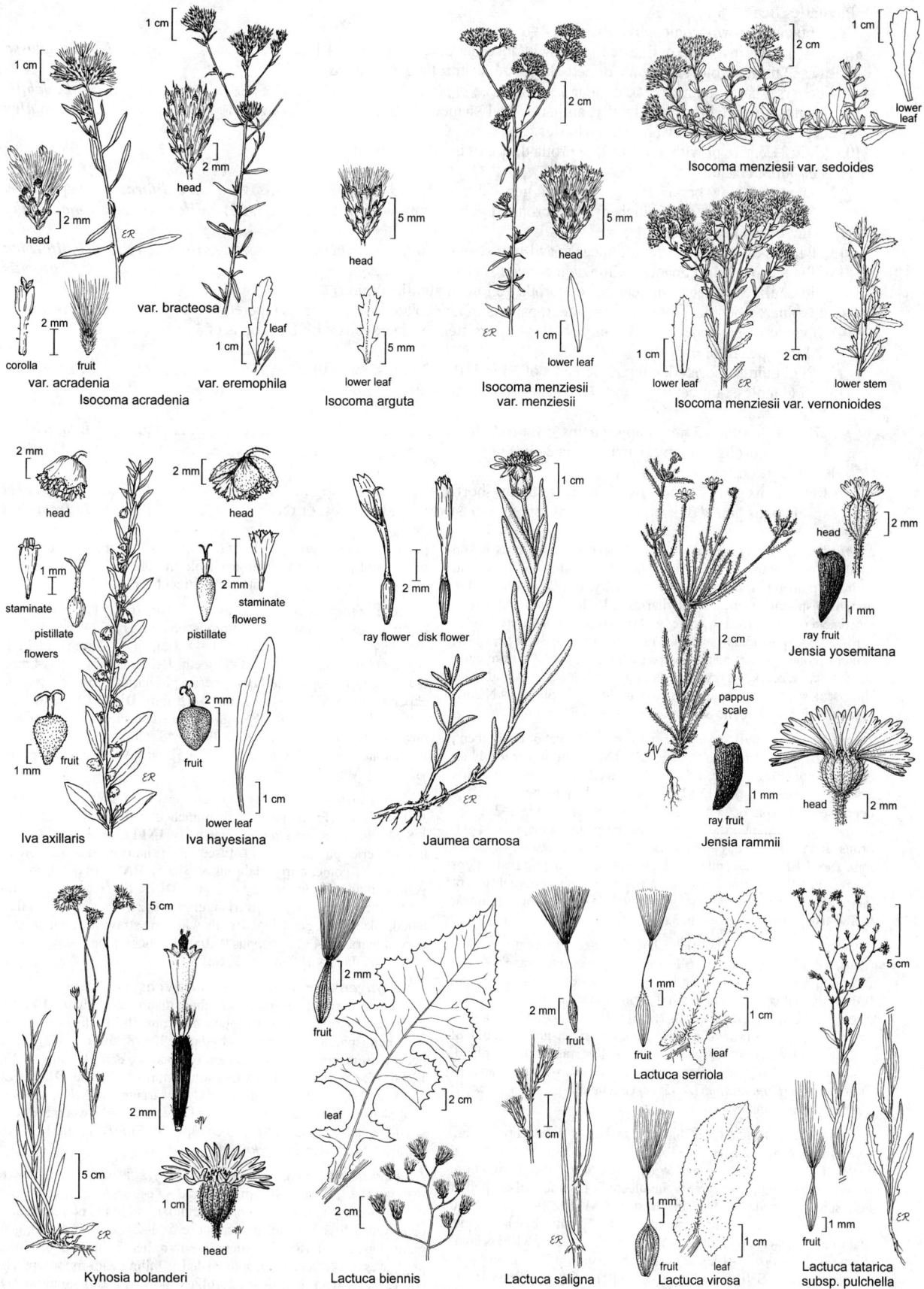

1 cm

1 cm

head

head

2 mm

var. bracteosa

2 mm

head

corolla   fruit

2 mm

var. acradenia   var. eremophila

leaf

1 cm

Isocoma acradenia

5 mm

head

5 mm

lower leaf

Isocoma arguta

2 cm

2 cm

head

5 mm

5 mm

1 cm

lower leaf

Isocoma menziesii
var. menziesii

2 cm

lower
leaf

Isocoma menziesii var. sedoides

1 cm

lower leaf

2 cm

lower leaf   lower stem

Isocoma menziesii var. vernonioides

2 mm

head

staminate
flowers

pistillate
flowers

1 mm

fruit

1 mm

Iva axillaris

2 mm

head

2 mm

staminate
flowers

pistillate

2 mm

fruit

1 cm

lower leaf

Iva hayesiana

2 mm

ray flower   disk flower

1 cm

Jaumea carnosa

2 cm

head

2 mm

ray fruit

1 mm

Jensia yosemitana

pappus
scale

ray fruit

1 mm

head

2 mm

Jensia rammii

5 cm

2 mm

5 cm

head

1 cm

Kyhosia bolanderi

fruit

2 mm

leaf

2 cm

2 cm

Lactuca biennis

2 mm

fruit

leaf

1 cm

Lactuca saligna

1 mm

fruit   leaf

Lactuca serriola

1 cm

fruit   leaf

1 mm

Lactuca virosa

1 cm

5 cm

fruit

1 mm

Lactuca tatarica
subsp. pulchella

1′ Phyllaries free
  7. Receptacle narrowly conic; phyllaries gen 4–6
    8. Ray ≤ 1 mm; involucre cylindric to narrowly obconic; disk corolla gen 4-lobed . . . . *L. microglossa*
    8′ Ray > 2 mm; involucre obconic or bell-shaped; disk corolla gen 5-lobed
      9. St coarse, hairy; anther tip triangular . . . . *L. debilis*
      9′ St fine, wiry, glabrous proximally; anther tip awl-shaped . . . . *L. leptalea*
  7′ Receptacle conic or hemispheric; phyllaries gen > 6
    10. All lvs ± entire or with 3–5+ teeth; corolla dark red in alkali solution
      11. Per (ann); coastal
        12. St erect, 0–few-branched; root fleshy; lf gen < 2 mm wide . . . . *L. californica* subsp. *bakeri*
        12′ St ± decumbent, gen branched; root fibrous; lf gen > 2 mm wide . . . . *L. californica* subsp. *macrantha*
      11′ Ann; coastal or inland
        13. Pappus of 1–7 clear, brown, linear to awl-like scales or 0 . . . . *L. californica* subsp. *californica*
        13′ Pappus of (2)4(6) opaque, white, lance-ovate scales or 0 . . . . *L. gracilis*
    10′ Mid-cauline lvs gen pinnately lobed; corolla yellow in alkali solution
      14. Involucre obconic; phyllaries persistent; pappus of lanceolate to ovate scales tapered to tip . . . . *L. platycarpha*
      14′ Involucre hemispheric to obconic; phyllaries persistent or deciduous with fr; pappus of 1 or
          2 kinds per fr, or 0
        15. Pl glandular, scented; pappus elements of 1 kind (if of 2 kinds, awns 0) or 0 . . . . *L. coronaria*
        15′ Pl glandless, unscented; pappus elements of 2 kinds or 0
          16. Fr ≥ 2 mm
            17. Ray corolla ≤ 3 mm; pappus awns ≥ 4, rarely 0 . . . . *L. maritima*
            17′ Ray corolla > 4 mm; pappus awns 2–3(4) or 0 . . . . *L. minor*
          16′ Fr < 1.5 mm
            18. Pappus gen of 1 long awn and few to many short scales . . . . *L. burkei*
            18′ Pappus gen of 3 or more long awns mixed with several short scales, or rarely 0 . . . . *L. fremontii*

L. burkei (Greene) Greene (p. 371) BURKE'S GOLDFIELDS Ann < 30 cm. **ST:** simple or freely branched, hairy. **LF:** < 5 cm, linear, entire or pinnately lobed, glabrous or ± hairy. **INFL:** involucre 4–6 mm, hemispheric or obconic; phyllaries 7–16, free, hairy, persistent; receptacle dome-shaped or conic, glabrous or hairy. **RAY FL:** 8–13; corolla yellow in alkali solution; ray < 6 mm. **DISK FL:** many; anther tip linear to ± ovate; style tips triangular. **FR:** < 1.5 mm, club-shaped, hairy, black to gray; pappus of gen 1(2) long awns and 3–6+ short scales. 2n=12. Vernal pools, wet meadows; < 500 m. s NCoRI (s Mendocino, s Lake, ne Sonoma cos.). Apr–Jun ★

L. californica Lindl. Ann, per < 40 cm. **ST:** erect or decumbent, simple or freely branched, ± hairy. **LF:** 0.8–21 cm, linear to oblanceolate or oblong, ± entire or with 3–5+ teeth, glabrous or ± hairy, ± fleshy in coastal forms. **INFL:** involucre 5–14 mm, bell-shaped to depressed-hemispheric or hemispheric; phyllaries 4–16 in 1–2 series, free, hairy, persistent or falling with frs; receptacle conic, rough, glabrous. **RAY FL:** 6–16; corolla dark red in alkali solution; ray 5–18 mm. **DISK FL:** many; anther tip ± lanceolate to triangular; style tips triangular. **FR:** < 4 mm, linear to narrowly club-shaped, glabrous or hairy, black to gray or silver-gray; pappus of 1–7 clear, brown, linear to awl-like scales, or 0. 2n=16, 32, 48.

  subsp. ***bakeri*** (J.T. Howell) R. Chan BAKER'S GOLDFIELDS Per (ann), roots fleshy, clustered. **ST:** erect, simple or few-branched. **LF:** 2–21 cm, gen < 2 mm wide, entire, basal clustered. **INFL:** involucre 9–14 mm, bell-shaped to depressed-hemispheric; phyllaries 13–16 in 2 series, ± persistent. **RAY FL:** 8–16; ray 5–16 mm. **DISK FL:** anther tip ± lanceolate to triangular. **FR:** glabrous, silver-gray; pappus of 1–4 clear, brown, awl-like scales, or 0. 2n=48. Grassland, woodland; < 500 m. c&s NCo (Mendocino, Sonoma cos.), CCo (Marin, San Luis Obispo cos.). [*L. macrantha* (A. Gray) Greene subsp. *b.* (J.T. Howell) Ornduff] May–Jun ★

  subsp. ***californica*** (p. 371) CALIFORNIA GOLDFIELDS Ann, roots fibrous, not clustered. **ST:** erect, occ decumbent in coastal forms. **LF:** 0.8–7 cm, 1–6 mm wide, linear to oblanceolate, entire or with 3–5+ teeth, hairy. **INFL:** involucre 5–10 mm, bell-shaped or hemispheric; phyllaries 4–13 in 1 series. **RAY FL:** 6–13; ray 5–10 mm. **DISK FL:** anther tip triangular. **FR:** < 3 mm, black to gray, glabrous or hairy; pappus of 1–7 clear, brown, linear to awl-like, awn-tipped scales, or 0. 2n=16,32,48. Many habitats; < 1500 m. NW (exc NCoRH), CaRF, SNF, GV, CW, SCo, WTR; sw OR. [*L. chrysostoma* (Fisch. & C.A. Mey.) Greene; *L. hirsutula* Greene] Circumscription

previously incl *L. gracilis*. Pls of *L. californica* subsp. *c.* and *L. gracilis* without pappus not distinguishable morphologically; molecular studies show them as separate and distinct taxa. Feb–Jun

  subsp. ***macrantha*** (A. Gray) R. Chan (p. 371) PERENNIAL GOLDFIELDS Per (ann), roots fibrous, not clustered. **ST:** gen decumbent, gen branched at base. **LF:** 2.8–8.8 cm, 1.5–15 mm wide, linear to oblong, entire or with 3–5+ teeth. **INFL:** involucre 9–14 mm, bell-shaped to depressed-hemispheric; phyllaries 9–16 in 2 series, ± persistent. **RAY FL:** 8–16; ray 6–18 mm. **DISK FL:** anther tip ± lanceolate to triangular. **FR:** glabrous, silver-gray; pappus of 1–4 clear, brown, awl-like scales, or 0. 2n=48. Grassland, dunes along immediate coast; < 500 m. NCo, CCo. [*L. m.* (A. Gray) Greene] All yr, mostly May–Aug ★

L. chrysantha (A. Gray) Greene ALKALI-SINK GOLDFIELDS Ann < 28 cm. **ST:** simple or freely branched, glabrous or ± hairy. **LF:** < 8 cm, linear, entire, glabrous or ± hairy. **INFL:** involucre 5–7 mm, hemispheric; phyllaries 8–14, fused > 2/3, hairy at tips, persistent; receptacle ± conic, warty, glabrous or ± hairy. **RAY FL:** 6–10; corolla yellow in alkali solution; ray 6–7 mm. **DISK FL:** many; anther tip ovate or triangular; style tips triangular. **FR:** 2–3 mm, obovoid, flattened, black; margin fringed with white or straw-like, blunt, gen curved hairs < 0.4 mm; pappus 0. 2n=14. Uncommon. Vernal pools, wet saline flats; < 100 m. s ScV, SnJV. Feb–Apr

L. conjugens Greene (p. 371) CONTRA COSTA GOLDFIELDS Ann < 40 cm. **ST:** simple or freely branched, glabrous or ± hairy. **LF:** < 8 cm, linear, entire or pinnately lobed, glabrous. **INFL:** involucre 6–10 mm, hemispheric or obconic; phyllaries 12–18, fused < 1/2, hairy, persistent; receptacle dome-shaped or obconic, densely hairy. **RAY FL:** 6–13; corolla yellow in alkali solution; ray 5–10 mm. **DISK FL:** many; anther tip linear to ± ovate; style tips triangular. **FR:** < 1.5 mm, club-shaped, glabrous; pappus 0. 2n=12. Vernal pools, wet meadows; < 100 m. s ScV (Napa, Solano cos.), CCo, SnFrB, formerly NCo, NCoRO, SCo. Mar–Jun ★

L. coronaria (Nutt.) Ornduff (p. 371) CROWNED OR ROYAL GOLDFIELDS Ann < 40 cm; herbage sweet-scented. **ST:** simple or much-branched; hairs short, glandular, or long, nonglandular, or mixed. **LF:** 1.5–6 cm, linear, entire or 1–2-pinnately lobed, gen glandular-hairy. **INFL:** involucre 4–7 mm, hemispheric to obconic; phyllaries 6–14, free, gen glandular-hairy, falling with frs; receptacle conic, hairy. **RAY FL:** 6–15; corolla yellow in alkali solution; ray

3–10 mm. **DISK FL**: many; anther tip elliptic; style tips triangular or dome-shaped. **FR**: < 2.5 mm, linear to narrowly club-shaped, hairy, black; pappus gen of 5–6+ lanceolate to ovate scales or 4–5 oblong, truncate scales or 0, gen different in ray and disk frs. 2*n*=8,10. Sunny, open grassy places, uncommon; < 700 m. SCo, TR, PR, w D; nw Baja CA. Only *Lasthenia* sp. with glands that produce a characteristic scent. Mar–May

L. debilis (A. Gray) Ornduff (p. 371) GREENE'S GOLDFIELDS Ann < 30 cm. **ST**: simple or branched, ascending, hairy. **LF**: 1–8 cm, ± entire (1–2-toothed), linear to linear-oblong, hairy. **INFL**: involucre 5–7 mm, bell-shaped to obconic; phyllaries 5, free, slightly hairy, falling with frs; receptacle narrowly conic, with small ridges and grooves, glabrous. **RAY FL**: 5–10; corolla dark red in alkali solution; ray 3–5 mm, yellow to white. **DISK FL**: many; corolla yellow to white; anther tip triangular with 1–4 wart-like glands on face; style tips triangular, with wide tuft of hairs. **FR**: < 3 mm, ± linear, hairy, black; pappus 0 or scales 2–4, lanceolate, white or brown. 2*n*=8. ± shaded, moist woodland slopes; < 500 m. c&s SNF, SCoRI (Temblor Range), w DMoj. Mar–May

L. ferrisiae Ornduff (p. 371) FERRIS' GOLDFIELDS Ann < 40 cm. **ST**: erect, simple or branched, glabrous or slightly hairy. **LF**: 1–8 cm, linear, entire, glabrous, fleshy. **INFL**: involucre 5–10 mm, hemispheric; phyllaries 6–14, fused > 2/3, hairy at tips, persistent; receptacle conic, papillate, glabrous. **RAY FL**: 6–13; corolla yellow in alkali solution; ray 6–10 mm. **DISK FL**: many; anther tip ovate or triangular; style tips triangular. **FR**: 2–2.5 mm, ± club-shaped, barely flattened, sparsely to densely short-hairy and papillate, black; pappus 0. 2*n*=14. Vernal pools or wet saline flats; < 700 m. ScV (2 stations), SnJV. Variable; putatively derived from hybridization between *L. chrysantha* and *L. glabrata* subsp. *coulteri*. Feb–May ★

L. fremontii (A. Gray) Greene FREMONT'S GOLDFIELDS Ann < 35 cm. **ST**: erect, simple or branched from base, ± hairy. **LF**: 1–6 cm, linear and entire, or with 1–3 pairs of linear lobes, glabrous to sparsely hairy. **INFL**: involucre 4–7.5 mm, hemispheric or obconic; phyllaries 8–16, free, hairy, persistent; receptacle hemispheric, hairy. **RAY FL**: 6–13; corolla yellow in alkali solution; ray 5–7 mm. **DISK FL**: many; anther tip linear to narrowly ovate; style tips triangular, short hair-tufted. **FR**: < 1.5 mm, club-shaped, gen hairy; pappus of 3–5 narrow awns intermixed with short scales, (narrow awns only, or 0). 2*n*=12. Vernal pools, wet meadows; < 700 m. CaRF, n SNF, GV. Mar–May

L. glaberrima DC. SMOOTH GOLDFIELDS Ann < 35 cm. **ST**: sprawling or erect, simple or freely branched, glabrous. **LF**: 3–10 cm, linear, entire, glabrous. **INFL**: heads radiate or disciform; involucre 5–7 mm, hemispheric or bell-shaped; phyllaries 5–10, fused > 2/3, hairy, persistent; receptacle conic, papillate, glabrous. **PISTILLATE FL**: 6–13; corolla yellow in alkali solution; ray 0–2 mm, pale yellow. **DISK FL**: many; corolla gen 4-lobed, pale yellow or ± green; anther tip obovate or oblong, blunt; style tips triangular or dome-shaped, short, glabrous. **FR**: < 4 mm, ± linear, ± flattened, hairy, ± gray; pappus of 5–10 narrowly tapered or elliptic scales. 2*n*=10. Vernal pools, wet areas; 1300 m. NCoR, ScV, n SnJV, CW, MP; to sw BC. Gen self-pollinated. Mar–Jul

L. glabrata Lindl. Ann < 60 cm. **ST**: erect, simple or branched, glabrous or slightly hairy. **LF**: 4–15 cm, linear or awl-shaped, entire, glabrous. **INFL**: involucre 5–10 mm, hemispheric; phyllaries 10–14, fused > 2/3, glabrous, persistent; receptacle conic, papillate, glabrous or sparsely hairy. **RAY FL**: 7–15; corolla yellow in alkali solution; ray 4–14 mm. **DISK FL**: many; anther tips ovate or triangular; style tips triangular. **FR**: 2–3.5 mm, club-shaped or obovoid, glabrous or papillate, gray; pappus 0. 2*n*=14. Subspp. almost identical exc frs. The only *Lasthenia* sp. known to have been used for food by aboriginal Californians.

  subsp. ***coulteri*** (A. Gray) Ornduff (p. 371) COULTER'S GOLDFIELDS **FR**: covered with rusty or ± yellow, wart-like papillae. Saline places, vernal pools; < 1000 m. NCoRI, s SNF, Teh (1 station), GV, CW, SCo, n ChI (Santa Rosa Island), PR, w DMoj. Apr–May ★

  subsp. ***glabrata*** (p. 371) YELLOW-RAY GOLDFIELDS **FR**: glabrous, not papillate. Saline places, vernal pools; < 550 m. NCoRI, s SNF, GV, CW. Mar–May

L. gracilis (DC.) Greene (p. 371) COMMON GOLDFIELDS Ann < 40 cm. **ST**: simple or freely branched, branching occ basal in desert forms, ± hairy. **LF**: 0.8–7 cm, linear to oblanceolate, entire, ± hairy, ± fleshy in coastal forms. **INFL**: involucre 5–10 mm, bell-shaped or hemispheric; phyllaries 4–13, free, hairy, persistent or falling with frs; receptacle conic, rough, glabrous. **RAY FL**: 6–13; corolla dark red in alkali solution; ray 5–10 mm. **DISK FL**: gen many; anther tip triangular; style tips ± triangular. **FR**: < 3 mm, linear to ± club-shaped, glabrous or hairy; pappus of (2)4(6) opaque, white, lance-ovate scales, or 0. 2*n*=16,32. Abundant, many habitats; < 1500 m. CA-FP (exc NCoRH, CaRH, SNH), w DMoj; c AZ, nw Baja CA. Most commonly found *Lasthenia*. Distinguished from *L. californica* subsp. *c.* by its pappus (when present) and wider range, extending into s CA. Highly variable; needs further study. Feb–Jun

L. leptalea (A. Gray) Ornduff (p. 371) SALINAS VALLEY GOLDFIELDS Ann < 15 cm, erect. **ST**: simple or branched, peduncle occ sinuous, glabrous proximally. **LF**: 3–20 mm, linear, entire, sparsely hairy. **INFL**: involucre 4–6 mm, obconic to bell-shaped; phyllaries 4–6, hairy only at tips, ± persistent; receptacle narrowly conic, glabrous. **RAY FL**: 6–9; corolla dark red in alkali solution; ray 2.5–5 mm. **DISK FL**: many; anther tip awl-shaped; style tips ± triangular, long hair-tufted. **FR**: < 2 mm, narrowly club-shaped, sparsely hairy, gray; pappus of < 4 awns, each narrowly tapered, white to ± yellow (0). 2*n*=16. Openings in woodland; < 500 m. SCoRO (Monterey, San Luis Obispo cos.). Feb–May ★

L. maritima (A. Gray) M.C. Vasey MARITIME OR SEASIDE GOLDFIELDS Ann < 25 cm. **ST**: prostrate or decumbent, branched; nodes gen hairy. **LF**: 1–9 cm, narrowly to widely strap-shaped, blunt, fleshy, entire or variously lobed, glabrous. **INFL**: involucre 4–7 mm, hemispheric; phyllaries 6–14, free, hairy on margins and midribs, falling with frs; receptacle conic, rough, glabrous. **RAY FL**: 7–12; corolla yellow in alkali solution; ray 1–3 mm. **DISK FL**: many; anther tip ± oblong, obtuse; style tips triangular or dome-shaped, often not hair-tufted. **FR**: 2–3.2 mm, linear to narrowly club-shaped, hairy, gray; pappus of 4–12 ± brown awns and 4–5+ narrow scales, or rarely 0. 2*n*=8. Uncommon. Seabird nesting, roosting areas, gen offshore rocks and islands; < 100 m. NCo, CCo; to sw BC. Guano endemic. Mostly self-pollinated. May–Jul

L. microglossa (DC.) Greene (p. 371) SMALL-RAY GOLDFIELDS Ann < 25 cm. **ST**: sprawling or erect, simple or much-branched, hairy. **LF**: 1.5–8 cm, linear or awl-shaped, ± entire, hairy. **INFL**: involucre 6–8.5 mm, cylindric to narrowly obconic; phyllaries ± 4, hairy, falling with frs; receptacle narrowly conic, glabrous. **RAY FL**: ± 4; corolla dark red in alkali solution; ray ≤ 1 mm or occ 0. **DISK FL**: few; corolla gen 4-lobed; anther tip narrowly tapered; style tips lanceolate, glabrous. **FR**: < 5 mm, ± linear, hairy, black; pappus of 1–4 scales, each lanceolate, ± yellow or white, awn-tipped, or 0. 2*n*=24. Shaded slopes of woodland, chaparral, desert scrub; < 1000 m. NCoRI, s SNF, ScV (1 collection), SnFrB, SCoR, TR, PR, DMoj. Gen self-pollinated. Mar–May

L. minor (DC.) Ornduff COASTAL GOLDFIELDS Ann < 35 cm. **ST**: erect, simple or much-branched, sparsely to densely woolly. **LF**: 2–12 cm, linear, glabrous to soft-hairy, entire or irregularly toothed or lobed, lobes < 1.5 cm. **INFL**: involucre 4–6 mm, hemispheric, phyllaries 7–14, free, hairy at margins, falling with frs; receptacle conic, rough, glabrous. **RAY FL**: < 13; corolla yellow in alkali solution; ray 4–8 mm. **DISK FL**: many; anther tip ovate or elliptic; style tips triangular or dome-shaped. **FR**: < 2.6 mm, narrowly club-shaped, glabrous or hairy, black; pappus of 2–3(4) narrowly tapered to lanceolate, brown or white awns intermixed with 4–5+ ± truncate, fringed shorter scales, or 0. 2*n*=8. Grassland; < 700 m. NCo, SNF, s ScV, SnJV, CCo, SCoR. Mar–Jun

L. platycarpha (A. Gray) Greene (p. 371) ALKALI GOLDFIELDS Ann < 30 cm. **ST**: erect, simple or branched from base, glabrous or ± hairy. **LF**: 1–6 cm, gen pinnately lobed, sometimes entire, gla-

brous or hairy. **INFL**: involucre 6–8 mm, obconic; phyllaries 6–9, free, glabrous or hairy, persistent; receptacle conic, rough, glabrous or sparsely hairy. **RAY FL**: 6–13; corolla yellow in alkali solution; ray 7–8 mm. **DISK FL**: many; anther tip triangular; style tips trian-gular. **FR**: < 3.5 mm, narrowly club-shaped, hairy; pappus of 4–6 scales, each lanceolate to ovate, tapered to tip, white or ± yellow. $2n=8$. Alkali flats; < 120 m. GV, SnFrB (uncommon). Mar–Apr

LAYIA

Bruce G. Baldwin & Susan J. Bainbridge

Ann 2–6(13) dm. **ST**: gen ascending to erect, often glandular, gen ± purple or brown. **LF**: basal in rosette or opposite, cauline gen alternate, sessile, gen linear to lanceolate, or oblanceolate, minutely dentate to (2-)pinnately lobed, glabrous or hairy, distal often stalked-glandular. **INFL**: heads gen radiate, 1 or in ± open clusters; involucre ± hemispheric to bell-shaped, obconic, or urn-shaped, 2–15+ mm diam; phyllaries 1 per ray fl, in 1(2) series, lanceolate to oblanceolate, gen folded completely around ray ovary, falling with fr, gen ± hairy or scabrous, often glandular; receptacle flat to slightly convex, minutely bristly, paleae free, gen in 1 series between ray and disk fls, or subtending ± each disk fl, phyllary-like, more scarious. **RAY FL**: (0)3–27; corolla white, often aging ± pink, to yellow or proximally yellow and distally pale yellow or ± white. **DISK FL**: 5–125; corolla yellow, puberulent, sometimes glandular, tube < throat, lobes deltate; anthers ± dark purple or yellow to ± brown, tips narrowly triangular or ± lanceolate to ovate; style branches awl-shaped, bristly. **FR**: gen 2–5 mm, gen club-shaped, black; ray fr compressed front-to-back, ± curved, beakless, glabrous or sparsely hairy, pappus 0; disk fr ± straight, gen ± hairy, pappus 0 or of 1–32 awns, scales, or bristles. 14 spp.: w N.Am. (George T. Lay, early 19th century English pl collector) [Baldwin et al. 2006 FNANM 21:262–269] Gen self-sterile (exc *L. carnosa*, *L. hieracioides*, sometimes *L. chrysanthemoides*).

1. Disk pappus 0
  2. Pl glandless; paleae subtending ± all disk fls. . . . . ^{2}***L. chrysanthemoides***
  2′ Pl glandular; paleae in 1 involucre-like series between ray and disk fls
   3. Pls apple- or banana-scented; basal rosette lvs gen minutely dentate to minutely serrate (coarsely toothed); ray corolla white or cream . . . . ^{2}***L. heterotricha***
   3′ Pls unscented or not apple- or banana-scented; basal rosette lvs lobed; ray corolla white, yellow, or 2-colored, yellow proximally, distally ± white or pale yellow
    4. St gen purple-streaked; involucre bell-shaped to hemispheric, ± spheric, or ± urn-shaped
     5. Involucre ± spheric to bell-shaped or hemispheric; ray fls 6–18 in 1 series, ray yellow throughout or distally white or pale-yellow . . . . ^{2}***L. gaillardioides***
     5′ Involucre ± widely urn-shaped; ray fls 13–27 in 2 series, ray 2-colored . . . . ^{2}***L. jonesii***
    4′ St gen not purple-streaked; involucre ± hemispheric
     6. Pl not strongly scented; ray yellow throughout or distally white; anthers gen ± dark purple, sometimes yellow to ± brown in SW . . . . ^{2}***L. platyglossa***
     6′ Pl strongly lemon- or acrid-scented; ray white or yellow; anthers yellow to ± brown . . . . ^{2}***L. pentachaeta***
      7. Ray white . . . . subsp. ***albida***
      7′ Ray yellow . . . . subsp. ***pentachaeta***
1′ Disk pappus of 1–32 awns, bristles, or scales
  8. Heads discoid . . . . **L. discoidea**
  8′ Heads radiate, ray sometimes inconspicuous
   9. Paleae subtending ± all disk fls; pl glandless
    10. Disk pappus of gen very unequal, awl-shaped or bristle-like awns and scales . . . . ^{2}***L. chrysanthemoides***
    10′ Disk pappus of ± equal, lanceolate scales . . . . **L. fremontii**
   9′ Paleae in 1 involucre-like series between ray and disk fls; pl glandular
    11. Disk pappus of elliptic or lance-linear to ovate, non-plumose, non-woolly scales, 0.5–3.5 mm, bases sparsely bristly
     12. Ray white; anthers yellow to ± brown; ray fr sparsely hairy . . . . **L. leucopappa**
     12′ Ray proximally yellow, distally white; anthers ± dark purple; ray fr glabrous or sparsely hairy
      13. Ray fls 13–27 in 2 series; st gen purple-streaked; pappus 0.5–2 mm; ray fr shiny, glabrous; s CCo, c SCoRO . . . . ^{2}***L. jonesii***
      13′ Ray fls 6–15 in 1 series; st not purple-streaked; pappus 2–3.5 mm; ray fr ± dull, glabrous or sparsely hairy; s SnJV . . . . **L. munzii**
    11′ Disk pappus gen of bristles or bristle-like scales, 1–7 mm, if scales linear and long-tapering or awl-shaped, then proximally plumose and often proximally woolly adaxially
     14. Disk pappus bristles or scales proximally plumose and adaxially woolly, or ± scabrous throughout, if proximally plumose and not woolly, then scales linear and long-tapering or awl-shaped; pl gen not strongly scented if not touched
      15. Disk pappus of 10–15 linear and long-tapering or awl-shaped scales; ray gen ± white, sometimes pale or golden yellow . . . . **L. glandulosa**
      15′ Disk pappus of (11)14–32 bristles or bristle-like scales; ray yellow throughout or distally white
       16. Disk pappus gen scabrous throughout, sometimes proximally plumose and adaxially woolly in SW; ray yellow throughout or distally white; anthers gen ± dark purple, sometimes yellow to ± brown in SW; involucre ± hemispheric; phyllary tips often > folded bases . . . . ^{2}***L. platyglossa***
       16′ Disk pappus densely plumose proximally and adaxially woolly; ray yellow; anthers yellow to ± brown; involucre narrower, bell-shaped to ± ellipsoid; phyllary tips gen < folded bases . . . . . . **L. septentrionalis**

14′ Disk pappus bristles or bristle-like scales proximally plumose, gen not woolly adaxially; pl gen
  strongly scented, even when not touched (exc fleshy pls of coastal dunes)

  17. Disk pappus readily falling as a unit; basal rosette lvs gen minutely dentate or minutely serrate
  (coarsely toothed); ray white or cream . . . . *²L. heterotricha*

  17′ Disk pappus persistent; basal rosette lvs lobed; ray white or yellow throughout, or proximally
  yellow and distally white or pale yellow

    18. Anthers yellow to ± brown . . . . *²L. pentachaeta*

    19. Ray white . . . . subsp. *albida*

    19′ Ray yellow . . . . subsp. *pentachaeta*

    18′ Anthers ± dark purple

      20. St not purple-streaked; ray white, 1.5–3.5 mm; ray fr sparsely hairy . . . . **L. carnosa**

      20′ St purple-streaked; ray yellow throughout or distally ± white or pale yellow, 1–18 mm;
  ray fr glabrous

        21. Ray yellow throughout or distally ± white or pale-yellow, 3.5–18 mm; disk pappus of 15–24
  bristles or bristle-like scales, main st ascending, not strictly erect . . . . *²L. gaillardioides*

        21′ Ray yellow, 1–4 mm; disk pappus of 10–16 bristles or bristle-like scales; main st strictly erect

        . . . . **L. hieracioides**

L. carnosa (Nutt.) Torr. & A. Gray (p. 371) BEACH LAYIA Pl 2–18 cm, glandular, not strongly scented. **ST:** prostrate to erect, not purple-streaked. **LF:** 3–45 mm, oblong to ovate, fleshy; proximal lvs gen lobed < 1/2 to midvein. **INFL:** peduncle < 3 cm; involucre 3–7+ mm diam, ± bell-shaped; phyllaries 4–8 mm, tip < folded base, basal margins strongly overlapping. **RAY FL:** 4–10; ray 1.5–3.5 mm, white. **DISK FL:** 5–45; corolla 2–4 mm; anthers ± dark purple. **FR:** ray fr sparsely hairy; disk pappus of 24–32 bristles or bristle-like scales, 2.5–3.5 mm, ± equal, white to ± brown (esp at base), ± long-plumose proximally, scabrous above, not adaxially woolly. 2*n*=16. Coastal dunes; < 70 m. n NCo, CCo; sw OR. Apr–Jul ★

L. chrysanthemoides (DC.) A. Gray (p. 371) Pl 4–53 cm, glandless, not strongly scented. **ST:** not purple-streaked. **LF:** < 12 cm, linear to lanceolate or oblanceolate, gen scabrous-ciliate; proximal lvs often lobed ± to midvein. **INFL:** peduncle < 10 cm; involucre 4–14 mm diam, ± hemispheric; phyllaries 4–12 mm, tip often > folded base, folded basal edge and tip margin gen papillate-scabrous, basal margins interlocked by cottony hairs; receptacle paleate ± throughout. **RAY FL:** 6–16; ray 3–18(24) mm, gen proximally yellow, distally white or light yellow (yellow throughout). **DISK FL:** 28–100+; corolla 3–5 mm; anthers ± dark purple. **FR:** ray fr glabrous; disk pappus of (0)2–18 awl-shaped or bristle-like awns and scales, 1–4 mm, very unequal, ± white to ± brown, ± scabrous, not adaxially woolly. 2*n*=14. Grassy or open heavy soil, sometimes ± alkaline; < 800 m. NCo, e&s edge NCoRO, NCoRI, GV, CW; sw OR. Mar–Jun

L. discoidea D.D. Keck (p. 371) RAYLESS LAYIA Pl 3–20 cm, glandular, not strongly scented. **ST:** not purple-streaked. **LF:** 2–35 mm, linear to lanceolate or oblanceolate; basal rosette lvs lobed < 1/2 to midvein. **INFL:** heads discoid; peduncle < 4 cm; false-involucre 2–6+ mm diam, cylindric to narrowly obconic or bell-shaped; phyllaries 0; paleae 4–7 mm, phyllary-like, in 1 series. **RAY FL:** 0. **DISK FL:** 5–35+; corolla 2.5–4 mm; anthers yellow to ± brown. **FR:** disk pappus of 8–15 lanceolate to awl-shaped, irregular scales, gen < 1.5 mm, ± equal, often irregularly notched or cut, ± white to ± brown, ± plumose or long-soft-hairy, not adaxially woolly. 2*n*=16. Open serpentine soil or talus; 800–1600 m. SCoRI (Fresno, San Benito cos.). Apr–Jun ★

L. fremontii (Torr. & A. Gray) A. Gray (p. 375) Pl < 40 cm, glandless, not strongly scented. **ST:** not purple-streaked. **LF:** < 7(9) cm, linear to lanceolate or oblanceolate; proximal lvs < 30-lobed, often ± to midvein. **INFL:** peduncle < 9 cm; involucre 3–11+ mm diam, ± hemispheric; phyllaries 4–11 mm, ± papillate-scabrous, tip often > folded base, basal margins interlocked by cottony hairs; receptacle paleate ± throughout. **RAY FL:** 3–15; ray 5–18(23) mm, proximally yellow, distally white- or light-yellow. **DISK FL:** 4–100+, 3.5–5.5 mm; anthers ± dark purple. **FR:** ray fr glabrous; disk pappus of 9–12 scales, 2–5 mm, ± equal, lanceolate, white to ± brown, glabrous, tip long-tapered. 2*n*=14. Grassy or open, heavy or shallow soil, incl serpentine; < 800 m. c NCoRO (rare, Mendocino Co.), CaRF, SNF, GV. Feb–May

L. gaillardioides (Hook. & Arn.) DC. (p. 375) Pl 6–60 cm, glandular, often strongly scented. **ST:** purple-streaked, ascending, not strictly erect. **LF:** < 12 cm, linear to lanceolate or oblanceolate; proximal lvs serrate or lobed < or > 1/2 to midvein. **INFL:** peduncle < 7 cm; involucre 4–12+ mm diam, ± spheric to bell-shaped or hemispheric; phyllaries 4–9+ mm, tips < folded bases, basal margins interlocked by cottony hairs. **RAY FL:** 6–18; ray 3.5–18 mm, yellow throughout or distally ± white or pale-yellow. **DISK FL:** 14–100+; corolla 3–5 mm; anthers ± dark purple. **FR:** ray fr glabrous; disk pappus of (0)15–24 bristles or bristle-like scales, slightly expanded at base, 1–4 mm, ± equal, ± white to red-brown, plumose proximally or throughout, not adaxially woolly. 2*n*=16. Open or semi-shaded slopes, in sandy or clayey soil (incl serpentine); < 1300 m. NCo, NCoRO, NCoRI, CCo (Marin Co.), SnFrB, SCoRO (Old Creek, San Luis Obispo Co.), SCoRI. Mar–Aug

L. glandulosa (Hook.) Hook. & Arn. (p. 375) WHITE LAYIA Pl 3–60 cm, glandular, often not strongly scented, sometimes spicy-scented. **ST:** not purple-streaked, often uniformly dark purple. **LF:** < 10 cm, linear to obovate, proximal lvs gen ± irregularly toothed to lobed. **INFL:** peduncle 0–7 cm; involucre 3–11 mm diam, ± bell-shaped to hemispheric; phyllaries 4–11 mm, tip often < (sometimes >) folded base, basal margins interlocked by cobwebby hairs. **RAY FL:** 3–14; ray 3–22 mm, gen ± white, sometimes pale yellow, or golden yellow. **DISK FL:** 17–105; corolla 3.5–6.5 mm; anthers yellow to ± brown. **FR:** ray fr glabrous; disk pappus of 10–15 scales, 2–5 mm, ± equal, linear and long-tapering or awl-shaped, white, scabrous and proximally plumose, often woolly on adaxial face. 2*n*=16. Open gravelly or sandy soil, dunes; < 2700 m. CaRH, s SNF, SNH, Teh, c&s SnJV, CW (extirpated SnFrB), SW (exc ChI), GB, D; to WA, ID, UT, NM, Baja CA. Feb–Jul

L. heterotricha (DC.) Hook. & Arn. (p. 375) PALE-YELLOW LAYIA Pl 13–90 cm, glandular, apple- or banana-scented. **ST:** not purple-streaked, often stout and hollow. **LF:** < 12 cm, elliptic to ovate, often clasping, proximal lvs entire or minutely dentate to minutely serrate (coarsely toothed). **INFL:** peduncle 0–7 cm; involucre 6–13+ mm diam, hemispheric; phyllaries 7–12 mm, tip gen < folded base, basal margins overlapping. **RAY FL:** 7–13; ray 5–24 mm, white or cream. **DISK FL:** 15–90+; corolla 4–7 mm; anthers yellow to ± brown. **FR:** ray fr gen glabrous, sometimes sparsely hairy; disk pappus of (0)14–20 bristles or bristle-like scales, readily falling as a unit, 3–6 mm, ± equal, long-plumose proximally, scabrous distally, not adaxially woolly. 2*n*=16. Open clayey or sandy soil, sometimes ± alkaline; 200–1800 m. s SNF, Teh, e&w edges SnJV, SCoR, n WTR. Apr–Jun ★

L. hieracioides (DC.) Hook. & Arn. (p. 375) Pl 5–130 cm, glandular, sweetly or pungently scented. **ST:** purple-streaked, strictly erect. **LF:** < 15 cm, elliptic or linear to lanceolate or oblanceolate, sessile to ± clasping, proximal toothed or ± irregularly lobed. **INFL:** peduncle < 6 cm; involucre 4–9+ mm diam, ± ellipsoid to ± obconic; phyllaries 4–9 mm, tip < folded base, bases sometimes not enfolding

ray frs. **RAY FL**: 6–16; ray 1–4 mm, yellow. **DISK FL**: 9–80; corolla 2.5–4.5 mm; anthers ± dark purple. **FR**: ray fr glabrous; disk pappus of 10–16 bristles or bristle-like scales, 2–4 mm, ± equal, ± white to red-brown, plumose proximally, scabrous distally, not adaxially woolly. 2*n*=16,32. Open, semi-shady, or disturbed sites, in light soil; < 1200 m. CW, w WTR. Sometimes hybridizes with *L. glandulosa* in s CCo. Apr–Jul

L. jonesii A. Gray (p. 375)   JONES' LAYIA   Pl 7–55 cm, glandular, not strongly scented. **ST**: gen purple-streaked. **LF**: < 7(9) cm, linear to lanceolate or oblanceolate, proximal lobed > 1/2 to midvein. **INFL**: peduncle 1–8 cm; involucre 4–8(12) mm diam, ± widely urn-shaped; phyllaries 4–8 mm, tip gen > (sometimes <) folded base, interlocked by cobwebby hairs. **RAY FL**: 13–27, in 2 series; ray 5–10(14) mm, proximally yellow, distally white. **DISK FL**: 35–100+; corolla 3–5 mm; anthers ± dark purple. **FR**: ray fr glabrous, shiny; disk pappus of (0)8–14 scales, 0.5–2 mm, ± equal, ± ovate or elliptic, ± white, ciliate, not adaxially woolly, base sparsely bristly. 2*n*=14. Open serpentine or clayey slopes; < 300 m. s CCo, c SCoRO (San Luis Obispo Co.). Mar–May ★

L. leucopappa D.D. Keck (p. 375)   COMANCHE POINT LAYIA   Pl 8–60 cm, glandular, not strongly scented. **ST**: straw-colored, not purple-streaked. **LF**: < 5(8) cm, oblong to oblanceolate, ± glaucous, gen scabrous-ciliate, proximal ± lobed. **INFL**: peduncle 1–12 cm; involucre 4–10(13) mm, ± hemispheric; phyllaries 3.5–8(11) mm, tip < or > folded base, interlocked by cottony hairs. **RAY FL**: 6–15; ray 3–12(19) mm, white. **DISK FL**: 20–100+; corolla 2.5–5 mm; anthers yellow to ± brown. **FR**: ray fr sparsely hairy; disk pappus of 10–13 scales, 2–3.5 mm, ± equal, lanceolate, acuminate, white, ciliate, not adaxially woolly, base sparsely bristly. 2*n*=14. Grassy or open heavy soil; 100–350 m. w edge Teh (Comanche Point, Tejon Hills), s SnJV (extirpated). Threatened by development. Mar–Apr ★

L. munzii D.D. Keck (p. 375)   MUNZ'S TIDY-TIPS   Pl 6–50 cm, decumbent to erect, glandular, not strongly scented. **ST**: not purple-streaked. **LF**: < 6 cm, linear to oblanceolate; proximal lvs lobed > 1/2 to midvein. **INFL**: peduncle 1–10 cm; involucre 5–10+ mm diam, hemispheric to ± urn-shaped; phyllaries 5–9 mm, tip < or > folded base, basal margins interlocked by cobwebby hairs. **RAY FL**: 6–15; ray 3–14 mm, proximally yellow, distally white. **DISK FL**: 16–100+; corolla 3.5–5 mm; anthers ± dark purple. **FR**: ray fr glabrous or sparsely hairy, ± dull; disk pappus of 9–12 scales, 2–3.5 mm, ± equal, lance-linear, distally long-tapered, ± white, ± scabrous, not adaxially woolly, base sparsely bristly. 2*n*=14. Alkaline clay soils; 50–800 m. s SnJV. Mar–Apr ★

L. pentachaeta A. Gray   Pl 5–100 cm, glandular, strongly acrid- or lemon-scented. **ST**: not purple-streaked. **LF**: < 11 cm, linear to lanceolate or oblanceolate; proximal lvs ± irregularly 1–2 × lobed. **INFL**: peduncle < 6 cm; involucre 2–12+ mm diam, ± hemispheric; phyllaries 5–12 mm, tip < or > folded base, basal margins interlocked by hairs. **RAY FL**: 4–14; ray 3–26 mm. **DISK FL**: 7–125; corolla 3–6 mm; anthers yellow to ± brown. **FR**: ray fr glabrous; disk pappus of (0)1–22 bristles or bristle-like scales, 1.5–3.5 mm, ± equal, ± white, gen plumose proximally, gen not adaxially woolly. 2*n*=16. Subspp. hybridize in s SNF.

subsp. ***albida*** D.D. Keck (p. 375)   **RAY FL**: ray white. Grassy or open, clayey or sandy soils; 70–900 m. s SNF, Teh, SnJV, SCoRI. Mar–May

subsp. ***pentachaeta***   **RAY FL**: ray yellow. Grassy or open, clayey or sandy soils, sometimes serpentine; 100–1200 m. SNF, e edge SnJV, SCoRI (rare), n WTR. Mar–Jun

L. platyglossa (Fisch. & C.A. Mey.) A. Gray (p. 375)   TIDY-TIPS   Pl 3–70 cm, decumbent to erect, glandular, not strongly scented. **ST**: gen not purple-streaked. **LF**: 4–100(120) mm, linear to lanceolate or oblanceolate, proximal lobed ± 1/2 to midvein. **INFL**: peduncle < 13 cm; involucre 4–15+ mm diam, hemispheric; phyllaries 4–18 mm, tip often > (sometimes <) folded base, basal margins interlocked by cottony hairs. **RAY FL**: 5–18; ray 3–21 mm, yellow throughout or distally white. **DISK FL**: 6–120+; corolla 3.5–6 mm; anthers gen ± dark purple, sometimes yellow to ± brown in SW. **FR**: ray fr glabrous or sparsely hairy, dull; disk pappus of (0 or 11)14–32 bristles or bristle-like scales, 1–6 mm, ± equal, white to ± brown, gen scabrous throughout, gen not plumose and not adaxially woolly, sometimes short-plumose proximally and woolly adaxially in SW. 2*n*=14. Common. Many habitats; < 2000 m. s NCo, s NCoRI, GV, CW, SW, w edge DMoj; sw OR, Baja CA. Feb–Jul

L. septentrionalis D.D. Keck (p. 375)   COLUSA LAYIA   Pl 6–35 cm, glandular, not strongly scented. **ST**: purple-streaked or not. **LF**: 4–70 mm, linear to lanceolate or oblanceolate, proximal gen lobed > 1/2 to midvein. **INFL**: peduncle < 8 cm; involucre 3–12+ mm diam, ± ellipsoid to bell-shaped; phyllaries 5–12 mm, tip gen < folded base; basal margins interlocked by cottony hairs. **RAY FL**: 5–9; ray 4–15 mm, yellow. **DISK FL**: 10–65+; corolla 5–8 mm; anthers yellow to ± brown. **FR**: ray fr glabrous or sparsely hairy; disk pappus of 16–22 bristles or bristle-like scales, 4–7 mm, ± equal, white, densely plumose proximally, densely woolly adaxially. 2*n*=16. Serpentine or sandy soils; 100–900 m. c&s NCoRI, ScV (Sutter Buttes) sw OR. Apr–Jun ★

LEONTODON   HAWKBIT

Kenton L. Chambers

Ann to per, scapose; sap milky. **ST**: 1–many from base; branches 0 or few. **LF**: basal entire or toothed to deeply pinnately lobed; cauline 0 or reduced to scale-like bracts. **INFL**: heads liguliflorous, terminal on st, branches; involucre bell-shaped in fl, ovoid in fr, phyllaries in 2–several series, outer << inner, inner ± equal, reflexed when dry; receptacle ± flat, epaleate, shallowly pitted, minutely roughened. **FL**: 25–many; ligules exceeding involucre, yellow, often abaxially ± purple, readily withering. **FR**: fusiform or cylindric, ribbed, narrowed distally or beaked; pappus of short scales or stiff, smooth to plumose bristles. ± 45 spp.: Old World. (Greek: lion tooth) [Bogler 2006 FNANM 19:294–296]

L. saxatilis Lam.   HAIRY HAWKBIT   **ST**: many, 1–3 dm, unbranched, curved-ascending. **LF**: oblanceolate, 2–25 cm, lightly to densely stiff-bristly-hairy. **INFL**: heads gen 1, nodding in bud; involucre 6–12 mm, glabrous to bristly; outer phyllaries short, ± linear, main phyllaries ± equal, linear to narrowly lanceolate. **FL**: corolla 8–15 mm. **FR**: outer fr smooth, pappus a crown of short, fringed scales; inner fr roughened, pappus of short and long plumose bristles with expanded bases. 2*n*=8. [*L. taraxacoides* (Vill.) Mérat, illeg.]

1. Ann (bien); beak of inner fr 2–3 mm . . . . . . subsp. ***longirostris***
1′ Per (bien); beak of inner fr ± 1 mm . . . . subsp. ***saxatilis***

subsp. ***longirostris*** (Finch & P.D. Sell) P. Silva   Disturbed areas; < 1000 m. NCo, CaRF, n SNF, ScV, n SnJV, CCo, SnFrB; to OR, native to Eur. [*L. taraxacoides* subsp. *l.* Finch & P.D. Sell] Jun–Oct

subsp. ***saxatilis*** (p. 375)   Disturbed areas; < 1000 m. NW, n SNF, CCo, SnFrB; to BC, e US, native to Eur. Jun–Oct

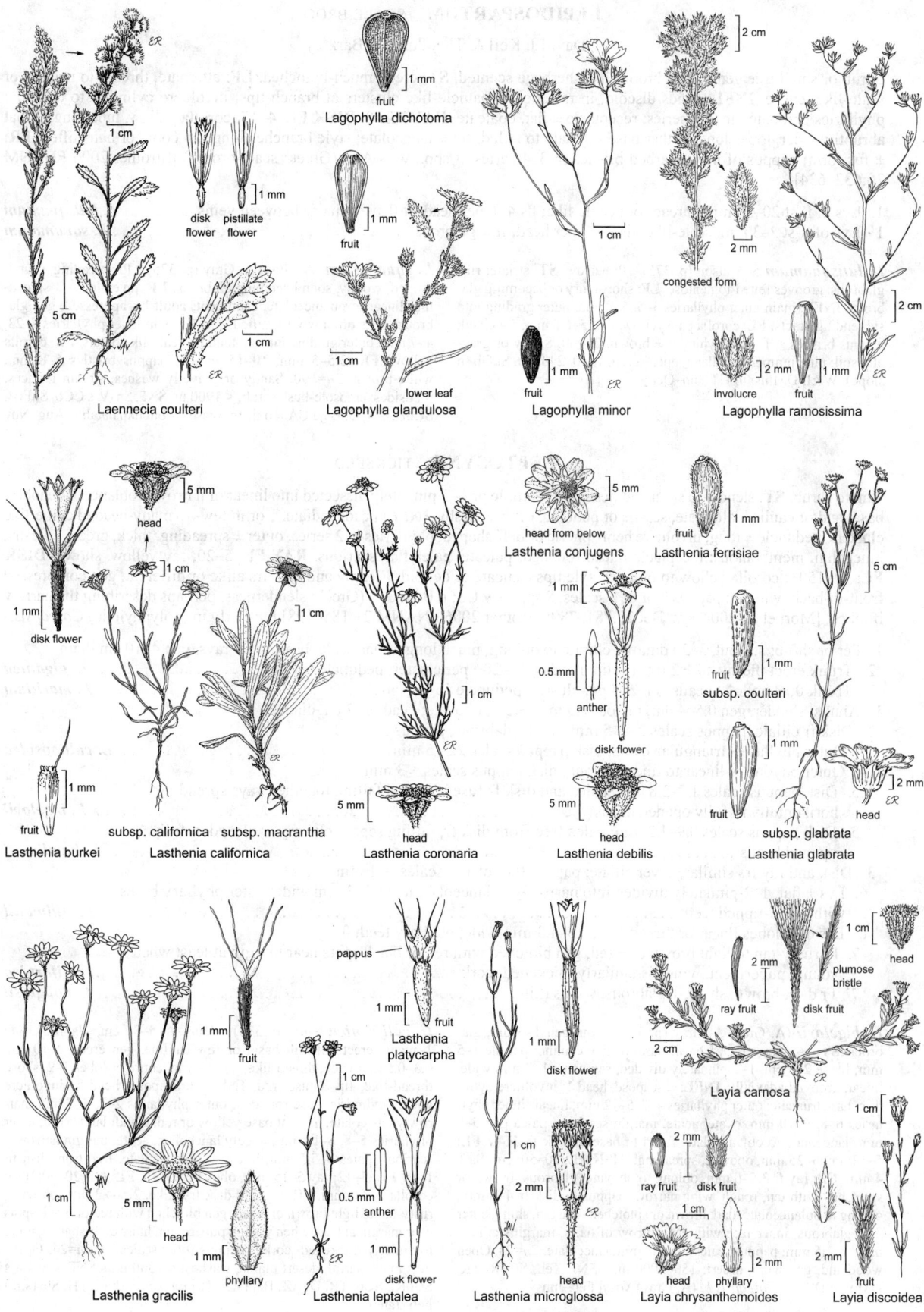

Laennecia coulteri

disk flower
pistillate flower

fruit
Lagophylla dichotoma

fruit
lower leaf
Lagophylla glandulosa

fruit
Lagophylla minor

congested form

involucre
fruit
Lagophylla ramosissima

disk flower
head

fruit
Lasthenia burkei
subsp. californica
subsp. macrantha
Lasthenia californica

head
Lasthenia coronaria

head from below
Lasthenia conjugens

fruit
Lasthenia ferrisiae

anther
disk flower
head
Lasthenia debilis

fruit
subsp. coulteri

fruit
head
subsp. glabrata
Lasthenia glabrata

fruit
head
phyllary
Lasthenia gracilis

pappus
fruit
Lasthenia platycarpha

anther
disk flower
Lasthenia leptalea

disk flower
head
Lasthenia microglossa

ray fruit
plumose bristle
disk fruit
head
Layia carnosa

ray fruit
disk fruit
head
phyllary
Layia chrysanthemoides

fruit
Layia discoidea

LEPIDOSPARTUM SCALE-BROOM

David J. Keil & Theodore M. Barkley

Shrub or small tree gen ≤ 3 m, broom-like; herbage scented. **ST**: erect, much-branched. **LF**: alternate, thread- to needle- or scale-like, entire. **INFL**: heads discoid, in raceme- or panicle-like clusters at branch tips; involucre cylindric to obconic; phyllaries graduated in 2–4 series; receptacle ± flat, epaleate. **RAY FL**: 0. **DISK FL**: 4–17; corolla yellow, tube long, throat abruptly wider, lobes long; anther base sagittate to tailed, tip ± lanceolate; style branches long, tip conic or hair-tufted. **FR**: ± fusiform; pappus of many barbed bristles in 3–4 series. 3 spp.: w N.Am. (Greek: scale-broom) [Strother 2006 FNANM 20:632–634]

1. Lvs of fl-st 20–30 mm, thread- or needle-like; fls 4–6 per head; fr densely hairy between veins . . . . *L. latisquamum*
1′ Lvs of fl-st 2–3 mm, scale-like; fls 9–17 per head; fr ± glabrous . . . . *L. squamatum*

L. latisquamum S. Watson (p. 375)   Pl narrow. **ST**: striate; ribs glabrous, grooves felted-tomentose. **LF**: short-hairy or becoming glabrous. **INFL**: main inner phyllaries 3–5, 6–8 mm, outer grading into subtending bracts. **FL**: corolla pale yellow. **FR**: 5–6.5 mm, 5-veined; pappus bristles ≤ 11 mm, white to ± brown. 2*n*=60. Sandy or gravelly soil; pine/juniper woodland, open scrub; 1400–2400 m. SnGb (n slope), W&I, DMtns; to UT. Jun–Oct

L. squamatum (A. Gray) A. Gray (p. 375)   Pl spreading, round-topped, woolly, soon becoming glabrous. **LF**: juvenile lvs ± spreading, linear to oblanceolate, canescent; adult lvs appressed, gen glabrous; axils often woolly-tufted. **INFL**: main inner phyllaries 7–23, 4–7 mm, outer grading into subtending scale-like bracts. **FL**: corolla yellow. **FR**: 3.5–5 mm, 10–15-veined; pappus bristles 5–8 mm, white-brown. 2*n*=±90. Sandy or gravelly washes, stream terraces, roadsides; creosote-bush scrub; < 1900 m. SNF, SnJV, s CCo, SnFrB, SCoR, SW, D; Baja CA. Toxic to livestock but unpalatable. Aug–Nov

LEPTOSYNE TICKSEED

Ann to shrub. **ST**: slender to stout and fleshy. **LF**: simple or 1–3-pinnately dissected into linear or narrowly oblanceolate lobes, basal and/or cauline, alternate, sessile or petioled, often ± fleshy. **INFL**: heads radiate, 1 or in few- to many-headed cyme-like clusters; peduncle ± long; involucre hemispheric or bell-shaped; phyllaries in 2 series, outer ± spreading, thick, green, ± fleshy, inner thin, membranous; receptacle flat to rounded, paleate; palea flat, scarious. **RAY FL**: 5–20; ray yellow, showy. **DISK FL**: 10–150+; corolla yellow to orange; style tips truncate or deltoid. **FR**: ray and disk frs alike or different, gen compressed front-to-back, winged; pappus 0 or of 2 scales. 8 spp.: sw US; n Baja CA. (Greek: slenderness, perhaps describing the narrow lf lobes) [Mort et al. 2004 Syst Bot 29:781–789; Strother 2006 FNANM 21:185–198] Formerly in (polyphyletic) *Coreopsis*.

1. Per or shrub; st stout, 3–20 dm; ray elliptic to oblong, much longer than wide; heads, incl rays, gen 4–10 cm diam
  2. Trunk erect, fleshy, to ± 2 m; st solid; heads (1)8–20+ per cluster; peduncle 6–20 cm . . . . *L. gigantea*
  2′ Trunk 0; st hollow; heads gen 2–4 per cluster; peduncle 15–50 cm . . . . *L. maritima*
1′ Ann; st slender, gen 0.5–4 dm; ray oblong to ovate or obovate; heads 1–7 cm diam
  3. Disk fr ciliate, pappus scales 2, 1–5 mm; ray fr glabrous, pappus scales 0
    4. Outer phyllaries triangular-ovate; disk pappus scales 2.5–5 mm . . . . *L. calliopsidea*
    4′ Outer phyllaries linear to linear-oblong; disk pappus scales ≤ 3 mm
      5. Disk pappus scales 1.7–2.8 mm; palea and disk fr fused at base, falling together; rays spreading
        horizontally in fully opened heads . . . . *L. bigelovii*
      5′ Disk pappus scales 0.9–1.3 mm; palea free from disk fr, falling separately; rays reflexed in fully
        opened heads . . . . *L. hamiltonii*
  3′ Disk and ray frs similar, never ciliate; pappus 0 or of 1–2 scales, < 1 mm
    6. Lvs ± flat, 1–2-pinnately divided into narrowly oblanceolate lobes 1–3 mm wide; outer phyllary bases
    with gland-tipped teeth . . . . *L. stillmanii*
    6′ Lvs or lf lobes linear or thread-like, 0.3–1.3 mm wide; phyllary teeth 0
      7. Fr rusty-tan to light brown or ± red, gen blotched with red or black spots near margin at least when
      young, puberulent, wings irregularly thickened, corky . . . . *L. californica*
      7′ Fr dark brown, shiny, ± glabrous, wings thin . . . . *L. douglasii*

L. bigelovii (A. Gray) A. Gray (p. 375)   Ann gen 1–3 dm, glabrous. **ST**: 1–many, erect. **LF**: all basal or few cauline; petiole 1–5 mm; blade 2–8 cm, 1–2-pinnately divided, segments 1–2 mm wide, linear, grooved adaxially. **INFL**: ± scapose, head 1; involucre cylindric, base truncate; outer phyllaries 4–7, 5–12 mm, linear; inner phyllaries 6–8, 6–10 mm, ovate, acute, margin scarious; palea gen 5–8 mm, lanceolate to oblanceolate, fused to base of disk fr. **RAY FL**: 5–13; ray 5–25 mm, obovate, spreading. **DISK FL**: 20–50; corolla ± 4 mm. **FR**: ray fr 3.5–6 mm, oblong to obovate, glabrous, brown or splotched with tan, rough, wing narrow, pappus 0; disk fr 4–6 mm, oblong to oblanceolate, dark brown or splotched with tan, shiny, outer face glabrous, inner face with central row of hairs, margins ciliate, hairs 1–1.5 mm; pappus scales 1.7–2.8 mm, lanceolate. 2*n*=24. Open woodland, grassland, desert; 150–2000 m. s SNF, Teh, SCoR, TR, DMoj, n DSon. [*Coreopsis b.* (A. Gray) Voss] Feb–Jun

L. californica Nutt. (p. 375)   Ann gen 5–30 cm, glabrous. **ST**: 1–many, erect. **LF**: all basal or few cauline, gen erect, 2–10 cm, 0.3–0.5 mm wide, thread-like, ± cylindric, entire or lobes 1–2, short, thread-like, tip obtuse, red. **INFL**: ± scapose, head 1; involucre widely cylindric, base rounded; outer phyllaries 2–7, 4–7 mm, narrowly lanceolate, hairs at base yellow or red, glandular, tip red; inner phyllaries 5–8, 6–10 mm, widely lanceolate, acute, margin narrowly scarious; palea 4–5.5 mm, linear to oblanceolate, free from disk fr. **RAY FL**: 5–12; ray 5–15 mm, obovate. **DISK FL**: (10)20–60(100); corolla 2–3.6 mm. **FR**: ray and disk frs alike, 2.5–4.3 mm, obovate, rusty-tan to light brown or ± red, gen blotched with red or black spots near margin at least when young, puberulent, hairs club-shaped, wing irregularly thickened, corky; pappus 0 (or scales 2). 2*n*=24. Openings in chaparral, desert plains, washes; < 1300 m. s SNF, s SnJV, s SCoRI, SCo, TR, D; AZ, Baja CA. [*Coreopsis c.* (Nutt.) H. Sharsm.] Feb–Jun

L. calliopsidea (DC.) A. Gray (p. 375) Ann gen 1–4(6+) dm, glabrous. **ST**: 1–many, erect, simple or few-branched. **LF**: basal and alternate; petiole 1–5 mm; blade 1–5 cm, 1–2-pinnately divided, segments 0.5–2 mm wide, grooved adaxially; distal lvs sometimes simple. **INFL**: head 1; involucre bell-shaped; outer phyllaries 4–6, 3–8 mm, triangular-ovate, fused at base; inner phyllaries gen 8, 8–10 mm, ovate, acute, margin narrowly scarious; palea 6–7 mm, lanceolate to oblanceolate, fused to base of disk fr. **RAY FL**: 8(10); ray 10–35 mm, obovate. **DISK FL**: 15–50+; corolla ± 5 mm. **FR**: ray fr 5–6 mm, ovate, glabrous, tan or brown, wing smooth, flat, pappus 0; disk fr 4–6 mm, linear to oblanceolate, outer face glabrous, dark brown, shiny, inner face white-hairy, margin ciliate, hairs 2–3 mm; pappus scales 2.5–5 mm, lanceolate. 2*n*=24. Desert, dry grassy areas; 100–1100 m. sw SnJV, SnFrB, SCoRI, TR, w DMoj. [*Coreopsis c.* (DC.) A. Gray] Feb–Jun

L. douglasii DC. (p. 375) Ann gen 5–25 cm, glabrous, glaucous. **ST**: 1–few, erect. **LF**: basal and alternate on proximal st, 2–8 cm, 1 mm wide, linear, entire (or lobes 1–2, short, linear), ± fleshy, grooved adaxially. **INFL**: ± scapose, head 1; involucre widely cylindric, base rounded; outer phyllaries 2–7, 4–7 mm, narrowly lanceolate; inner phyllaries 5–8, 6–10 mm, obovate, acute; palea 4–5 mm, linear, free from disk fr. **RAY FL**: 5–8; ray 5–12(16+) mm, ovate. **DISK FL**: 10–60(100+); corolla 4–4.5 mm. **FR**: ray and disk frs alike, 2.5–5 mm, obovate, ± glabrous, dark brown, shiny; wing ± yellow; pappus 0. 2*n*=24. Dry, rocky slopes; 150–600 m. e SnFrB, SCoR, n WTR. [*Coreopsis d.* (DC.) H.M. Hall] Mar–May

L. gigantea Kellogg (p. 375) GIANT TICKSEED Shrub, glabrous. **ST**: trunk gen 1–2 m, 4–10 cm diam, erect, fleshy, few-branched. **LF**: alternate, tightly clustered at st tips; petiole 3–7 cm; blade 3–25 cm, 3–4-pinnate, segments 1–5 cm, 0.5–1.5 mm wide, linear, fleshy; mop-like tangle of withered lvs persistent in dry season. **INFL**: heads (1)8–20+ in cyme-like clusters; peduncle 6–20 cm, lfy-bracted; involucre bell-shaped; outer phyllaries 5–12, 5–20 mm, lanceolate to oblong; inner phyllaries 10–15, 10–15 mm, oblong-ovate, obtuse to acute; palea 8–10 mm, linear, free from disk fr. **RAY FL**: 10–16; ray 20–30 mm, elliptic to oblong. **DISK FL**: many; corolla 6–6.5 mm. **FR**: ray and disk frs alike, 5–6 mm, oblong to obovate, ± glabrous, dark brown; wing narrow, thin; pappus 0. 2*n*=24. Shrubby hillsides, coastal dunes, sea bluffs; < 500 m. s CCo, n&c SCo, ChI. [*Coreopsis g.* (Kellogg) H.M. Hall] Jan–May

L. hamiltonii Elmer (p. 375) MOUNT HAMILTON TICKSEED Ann 5–25 cm, glabrous. **ST**: 1–many, erect. **LF**: basal (or few cauline, alternate); petiole 1–3 cm; blade 5–20 mm, 1–2-pinnately divided, segments 1 mm wide, linear, grooved adaxially. **INFL**: ± scapose, head 1; involucre cylindric, base truncate; outer phyllaries 4–7, 3–6 mm, linear-oblong; inner phyllaries 6–8, 5–8 mm, ovate, acute, margin scarious; palea gen 5–6 mm, linear, free from disk fr. **RAY FL**: 5–8; ray 3–8 mm, oblong to obovate, reflexed. **DISK FL**: 20–30; corolla ± 3–4.5 mm. **FR**: ray fr 5 mm, obovate, glabrous, smooth, brown or tan splotched with brown, shiny; wing narrow; pappus 0; disk fr 5–6 mm, obovate, tan splotched with brown, shiny, faces ascending-hairy, margins ciliate, hairs 1–1.5 mm; pappus scales ± 1 mm, obovate. 2*n*=24. Dry exposed slopes; 600–1300 m. e SnFrB (Diablo Range). [*Coreopsis h.* (Elmer) H. Sharsm.] Mar–May ★

L. maritima (Nutt.) A. Gray SEA DAHLIA Per 3–8 dm, from fleshy taproot, glabrous. **ST**: few to many, stout, hollow, much-branched. **LF**: alternate, fleshy; petiole 2–15 cm; blade 2–3-pinnately divided, lobes 1.5–50 mm, 2–3 mm wide, linear. **INFL**: heads gen 2–4 in cyme-like clusters; peduncle 1.5–5 dm; involucre bell-shaped; outer phyllaries 6–10, 10–25 mm, oblong to obovate; inner phyllaries 10–15, 12–15 mm, widely lanceolate to ovate, acute; palea 8–12 mm, linear to oblanceolate, free from disk fr. **RAY FL**: 15–20; ray 25–40 mm, oblong. **DISK FL**: many; corolla 6–6.5 mm. **FR**: ray and disk frs alike, 6–7 mm, oblong to obovate, smooth or minutely rough, dark brown; wing narrow, thin; pappus 0. 2*n*=24. Seabluffs; < 20 m. s SCo (San Diego Co.); Baja CA. [*Coreopsis m.* (Nutt.) Hook. f.] Feb–Jun ★

L. stillmanii A. Gray Ann 5–30 cm, glabrous. **ST**: 1–few, erect. **LF**: basal and alternate on proximal st; petiole 1–5 cm; blade 0.5–5 cm, 1–2-pinnate, lobes 1–3 mm wide, narrowly oblanceolate, ± flat, terminal often widest. **INFL**: ± scapose, head 1; involucre widely cylindric, base rounded; outer phyllaries 4–8, 3–10 mm, linear or narrowly oblanceolate, green or ± red; inner phyllaries 5–10, 5–10 mm, ovate, acute, narrowly scarious-margined; palea 5–6 mm, lanceolate, free from disk fr. **RAY FL**: 5–8; ray 5–8 mm, ovate. **DISK FL**: 10–40; corolla 2.5–4 mm. **FR**: ray and disk frs alike, 2.5–5 mm, obovate, flat, dark brown, smooth, not shiny, glabrous or hairs few, short; wing ± yellow; pappus 0 or awns 1–2, < 1 mm. 2*n*=24. Grassy slopes; < 900 m. NCoRI, SNF, GV, SnFrB, n SCoRI. [*Coreopsis s.* (A. Gray) S.F. Blake] Mar–May

LESSINGIA

Staci Markos

Ann, taprooted; often strongly scented. **ST**: decumbent to erect, 1–several from base, 2–90 cm, simple or gen ± openly branched, distally glabrous or sparsely hairy to woolly, often glandular. **LF**: simple, alternate, entire to pinnately lobed; basal gen withered at fl (persistent), petioled or sessile, 4–11 cm, linear to oblanceolate or spoon-shaped, lobes (if present) toothed; cauline sessile (occ clasping), reduced distally on st, 1-veined, linear to ovate or obovate, glabrous or sparsely hairy to woolly, occ with stalked or bead-like sessile glands. **INFL**: heads radiant, gen slender-peduncled in open cyme- or panicle-like clusters (± sessile in spike- or head-like clusters); involucre 4–13 mm diam, narrowly cylindric or fusiform to obconic or hemispheric; phyllaries 10–55, graduated in 4–8 series, erect or recurved, persistent, spreading or reflexed in age, texture lf-like or scarious (papery), tips green or ± purple-tinged, glabrous or sparsely hairy to woolly, often glandular; receptacle slightly convex, shallowly pitted, epaleate. **DISK FL**: 3–40; corolla funnel-shaped to tubular, white, pink, lavender, or yellow, limb ± palmately expanded in peripheral fls, lobes erect or spreading; anther tip awn-like; style branches 0.8–2.5 mm, appendage 0.3–1.3 mm. **FR**: 1–5 mm, cylindric to obconic, not compressed, smooth or obscurely 5–10 nerved, faces tan or mottled purple-brown, densely puberulent to long-soft-hairy; pappus of 3–55 persistent bristles, free, fused at base, or fused throughout into awns, occ reduced to crown, white, tan, or ± red. 2*n*=10, exc *Lessingia nemaclada*, 2*n*=10,12. 12 spp.: CA; w NV, nw AZ, n Baja CA. (C.F. Lessing, German-born botanist, 1809–1862) [Markos 2006 FNANM 20:452–458] Other spp. now treated in *Benitoa* and *Corethrogyne.*

1. Corolla yellow (externally occ pink or ± purple in peripheral fls of *Lessingia tenuis*).
   2. Style-branch appendage lanceolate, 0.3–1.3 mm; corolla tube not purple-brown banded inside ...... ***L. glandulifera***
     3. Phyllaries persistently tomentose .... var. ***peirsonii***
     3′ Phyllaries sparsely hairy or glandless hairs 0
      4. St and lvs glabrous or slightly hairy to tomentose .... var. ***glandulifera***
      4′ St and lvs densely, persistently tomentose — e PR (San Diego Co.) .... var. ***tomentosa***
   2′ Style-branch appendage short-triangular, 0.1–0.4 mm; corolla tube purple-brown banded inside
     5. Pl glandless or phyllaries rarely with sparse bead-like, sessile glands......................... ***L. germanorum***

5′ Pl ± glandular

  6. Fls 3–25; cauline lvs gen glandless, occ with bead-like sessile glands; phyllary tips ± purple; corolla of peripheral fls yellow or occ pink or ± purple externally . . . . **L. tenuis**

  6′ Fls 15–30; cauline lvs with stalked and bead-like sessile glands; phyllary tips gen green (occ purple); corolla of peripheral fls yellow . . . . **L. pectinata**

    7. St ± red to dark brown; lf margin dentate to pinnately lobed, segments with short stiff points . . . . . . var. **pectinata**

    7′ St green or tan; lvs entire or dentate to pinnately lobed, segments without short stiff points . . . . var. **tenuipes**

1′ Corolla white, pink, or lavender

  8. Phyllaries gen glabrous, rarely sparsely hairy, occ with stalked and/or bead-like sessile glands

    9. Basal lvs gen persistent at fl; cauline lvs with stalked glands; involucre hemispheric to bell-shaped; corolla lavender (never white) . . . . **L. ramulosa**

    9′ Basal lvs gen withering before fl; cauline lvs occ with stalked glands; involucre cylindric to broadly obconic; corolla white (occ pale lavender)

      10. Cauline lvs with sunken glands and occ with bead-like sessile and/or stalked glands; style-branch appendage gen lanceolate, occ short-triangular . . . . **L. nemaclada**

      10′ Cauline lvs gen glandless, occ with sparse stalked glands; style-branch appendage short-triangular . . . . **L. micradenia**

        11. Phyllary glands 0 or bead-like, sessile . . . . var. **glabrata**

        11′ Phyllaries with stalked glands . . . . var. **micradenia**

  8′ Phyllaries sparsely hairy, tomentose, or woolly, occ with bead-like sessile glands

    12. St decumbent; outer phyllary tip green; inner phyllaries stiff, white; pappus pink to red . . . . **L. nana**

    12′ St decumbent to erect; outer phyllary tip green or purple; inner phyllaries scarious; pappus white to tan

      13. Cauline lvs glandless

        14. Basal lvs withered at fl; involucre 4–8 mm; style-branch appendage short-triangular; pappus gen < fr (exc in Sonoma Co.) . . . . **L. arachnoidea**

        14′ Basal lvs gen persistent at fl; involucre (5)8–13 mm; style-branch appendage lanceolate; pappus ≥ fr . . . . **L. hololeuca**

      13′ Cauline lvs with sunken glands (rarely also with stalked or bead-like sessile glands in *Lessingia virgata*), glands occ obscured by hairs

        15. Cauline lvs glabrous or sparsely hairy to tomentose; infl open, cyme- or panicle-like, heads gen at ends of branchlets; fls 6–25; corolla lavender (never white) . . . . **L. leptoclada**

        15′ Cauline lvs gen tomentose (sparsely hairy); infl spike-like, heads gen in lf axils; fls 3–6; corolla gen white, occ pale lavender . . . . **L. virgata**

L. arachnoidea Greene (p. 379) CRYSTAL SPRINGS LESSINGIA **ST**: erect, 15–80 cm, branches ascending, tan, distally glabrous or thinly hairy. **LF**: basal petioled; cauline 0.2–3.5 cm, lanceolate, entire, reduced distally on st to awl-shaped bracts, glandless, glabrous or sparsely hairy. **INFL**: involucre 4–8 mm, narrowly obconic or fusiform; phyllaries lanceolate, cobwebby tomentose; tips erect or slightly spreading, ± purple tinged. **FL**: (3)8–18; corolla funnel-shaped, pink to lavender, tube darker; style branch appendage 0.1–0.3 mm, short-triangular, abruptly pointed. **FR**: 2–3 mm; pappus gen < fr, occ a crown (≥ fr, Sonoma Co.), white or ± tan-white. Serpentinite soil in grassland, coastal scrub, chaparral, woodland; 40–300 m. SnFrB (near Crystal Springs Reservoir, San Mateo Co.; near Camp Meeker, Sonoma Co.). Jul–Oct ★

L. germanorum Cham. (p. 379) SAN FRANCISCO LESSINGIA **ST**: erect, 5–30 cm, tan to ± red-brown, distally glabrous or sparsely hairy. **LF**: basal petioled; cauline 0.5–3 cm, oblong to oblanceolate, entire to pinnately lobed, glandless, glabrous or long-soft-hairy. **INFL**: involucre 4–8 mm, obconic to bell-shaped; phyllaries lanceolate, glabrous, puberulent, and/or thinly tomentose, rarely with sparse bead-like sessile glands, tips recurved, gen purple. **FL**: 20–40; corolla funnel-shaped to tubular, yellow, tube purple-brown banded inside; style branch appendage 0.1–0.4 mm, short-triangular, abruptly pointed. **FR**: 1–3 mm; pappus ≥ fr, tan. Sandy soil; < 100 m. CCo (Presidio, San Francisco Co.), SnFrB (Hillside Park, San Mateo Co.). Jun–Nov ★

L. glandulifera A. Gray **ST**: ± erect, 5–40 cm, tan, distally puberulent and/or sparsely hairy to tomentose (glabrous). **LF**: basal petioled, sometimes not withered at fl; cauline oblong to oblanceolate, gen entire, occ toothed to pinnately lobed, reduced distally on st to awl-shaped bracts, 0.2–2.5 cm, puberulent and/or sparsely hairy to tomentose (glabrous), margins gen with bead-like sessile glands. **INFL**: involucre 4–7 mm, obconic to cylindric; phyllaries oblong, puberulent and/or sparsely hairy to tomentose (glabrous), tips erect,

green, gen with bead-like sessile glands. **FL**: 10–30; corolla funnel-shaped, yellow, tube not purple-brown banded inside; style branch appendage 0.3–1.3 mm, lanceolate, tapered. **FR**: 1.5–3.5 mm; pappus ≥ fr, white to tan.

var. **glandulifera** (p. 379) Herbage glabrous or slightly hairy to tomentose. **LF**: 0.2–2 cm. **INFL**: phyllaries sparsely hairy. **FL**: 12–30; style branch appendage 0.3–1.3 mm. Chaparral, pine forest, desert, gen sandy soil; 50–2200 m. s SNH, s SnJV, SnFrB, SW, SNE, w DMoj; w NV, nw AZ, n Baja CA. [*L. lemmonii* var. *l.*; *L. l.* var. *ramulosissima* (A. Nelson) Ferris] Some pls incl here, which form tumbleweeds and have puberulent phyllaries and overlapping cauline lvs that clasp bases of involucre, are sometimes segregated as *L. lemmonii* var. *ramulosissima*. May–Oct

var. **peirsonii** (J.T. Howell) Markos (p. 379) **ST**: sparsely hairy to tomentose. **LF**: 0.2–2 cm, persistently tomentose. **INFL**: phyllaries persistently tomentose. **FL**: 10–25, style branch appendages 0.6–1.3 mm. Dry foothills, desert washes, roadsides, gen in sandy soil; 300–1850 m. s SNF, e WTR. [*L. lemmonii* var. *p.* (J.T. Howell) Ferris] May–Sep

var. **tomentosa** (Greene) Ferris WARNER SPRINGS LESSINGIA Herbage densely, persistently tomentose. **LF**: 0.2–2 cm. **INFL**: phyllaries ± glabrous or sparsely hairy. **FL**: 15–30, style branch appendages 0.3–0.6 mm. Grassland, hillsides, roadsides, gen sandy soil; 900 m. e PR (Warner Springs, San Diego Co.). Aug–Nov ★

L. hololeuca Greene (p. 379) WOOLLY-HEADED LESSINGIA **ST**: decumbent to erect, 5–60 cm, tan, distally long-shaggy-hairy to tomentose. **LF**: basal petioled, gen persistent; cauline 0.2–3 cm, lanceolate, entire, reduced distally on st to awl-shaped bracts, glandless, tomentose. **INFL**: sometimes congested (heads rarely axillary); involucre (5)8–13 mm, widely (narrowly) obconic; phyllaries lanceolate, cobwebby tomentose, tips erect, green or ± purple. **FL**: 10–20; corolla funnel-shaped, lavender, tube darker; style branch appendage

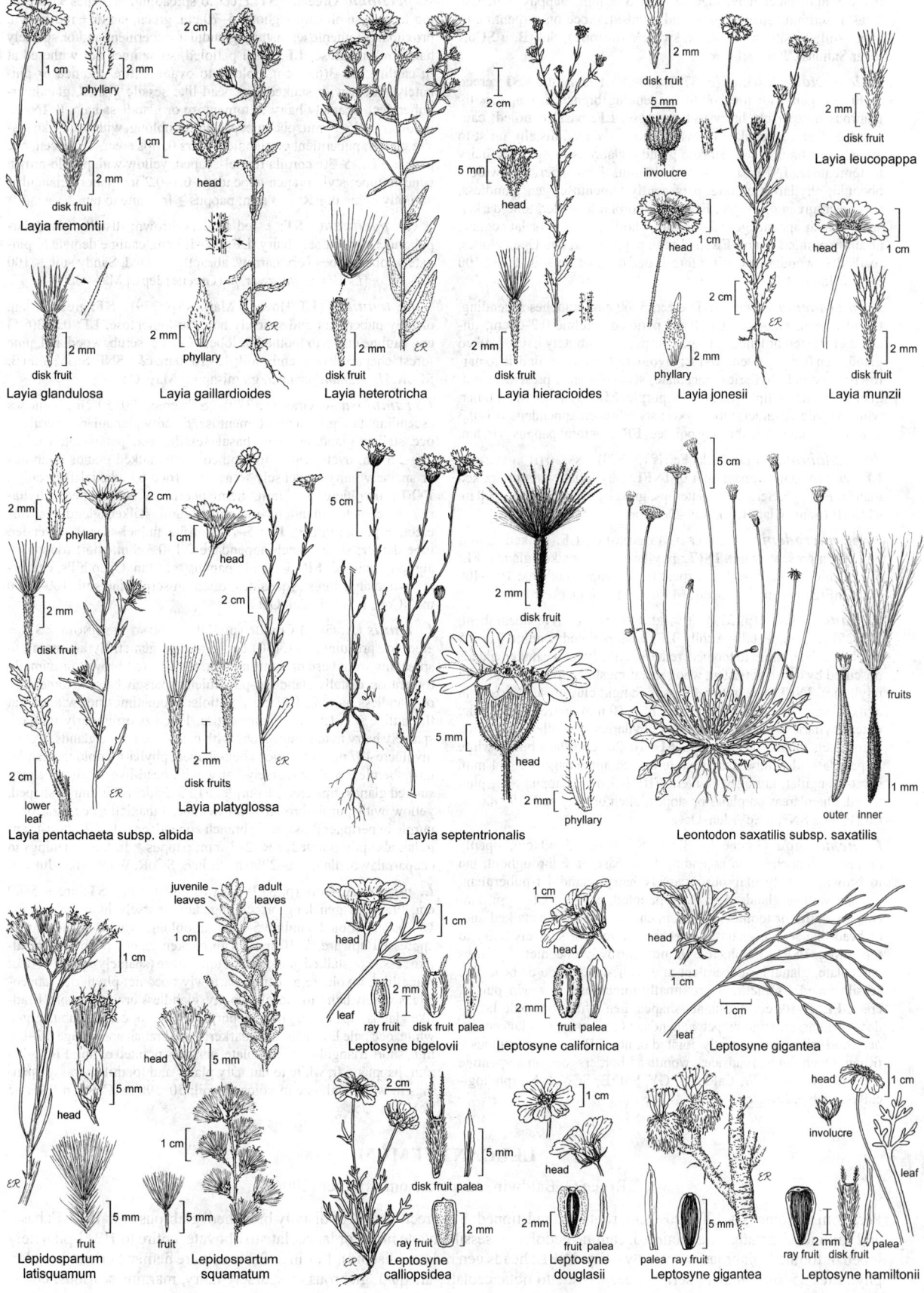

Layia fremontii

Layia glandulosa

Layia gaillardioides

Layia heterotricha

Layia hieracioides

Layia jonesii

Layia munzii

Layia leucopappa

Layia pentachaeta subsp. albida

Layia platyglossa

Layia septentrionalis

Leontodon saxatilis subsp. saxatilis

Lepidospartum latisquamum

Lepidospartum squamatum

Leptosyne bigelovii

Leptosyne californica

Leptosyne gigantea

Leptosyne calliopsidea

Leptosyne douglasii

Leptosyne gigantea

Leptosyne hamiltonii

0.4–0.8 mm, lanceolate, tapered. **FR**: 3–5 mm; pappus ≥ fr, tan. Coastal scrub, chaparral, grassland, roadsides, occ on serpentine or alkali soil; 10–600 m. s NCoR, sw ScV (historic), SnFrB, n SCoR (near Salinas). Jun–Oct ★

L. leptoclada A. Gray (p. 379) SIERRA LESSINGIA **ST**: erect, 5–90 cm, gen with long, stiffly ascending branches, tan, distally glabrous or sparsely hairy to tomentose. **LF**: basal petioled; cauline 0.5–5 cm, lanceolate or ovate, entire, reduced distally on st to awl-shaped bracts, with sunken glands, glabrous or sparsely hairy to tomentose. **INFL**: involucre 5–10 mm, hemispheric or widely obconic; phyllaries oblong, persistently tomentose, gen glandless, tips erect, green or purple. **FL**: 6–25; corolla lavender, tube darker, style branch appendage 0.3–0.6 mm, short-triangular or lanceolate, abruptly pointed or not. **FR**: 3–4 mm; pappus > fr, tan. Open slopes, roadsides, woodland, conifer forest, occ on granitic soil; 150–2100 m. SN. Jul–Oct

L. micradenia Greene **ST**: erect, 5–60 cm, branches ascending, tan to brown, glabrous. **LF**: basal petioled; cauline 0.2–2 cm, linear, awl-shaped or lanceolate, gen entire, occ minutely hairy, stalked glands gen 0, occ present. **INFL**: involucre 4–6 mm, cylindric to narrowly obconic; phyllaries lanceolate, stalked glands present or not, glandless hairs 0, tips appressed, ± purple. **FL**: 3–10; corolla tubular, white to pale lavender, tube darker; style branch appendage 0.1–0.3 mm, short-triangular, abruptly pointed. **FR**: 2–4 mm; pappus ≤ fr, tan.

var. ***glabrata*** (D.D. Keck) Ferris (p. 379) SMOOTH LESSINGIA **LF**: stalked glands on margin 0. **INFL**: phyllaries without stalked glands. **FL**: 3–5. Serpentine outcrops, gravelly roadcuts; 100–500 m. s SnFrB (Santa Clara Co.). Aug–Oct ★

var. ***micradenia*** TAMALPAIS LESSINGIA **LF**: stalked glands on margin present or not. **INFL**: phyllaries with stalked glands. **FL**: 5–10. Thin, gravelly soil of serpentine outcrops, roadcuts; 100–400 m. n SnFrB (Mount Tamalpais, Marin Co.). Jul–Oct ★

L. nana A. Gray (p. 379) DWARF LESSINGIA **ST**: decumbent, 2–5(25) cm, tan, distally woolly. **LF**: basal petioled; cauline 0.5–2.5 cm, lanceolate, entire to toothed, reduced distally on st, sunken glands obscured by dense persistent wool; distal-most gen intergrading with phyllaries. **INFL**: heads 1 or ± sessile in tight clusters at st tips or in axils of basal or cauline lvs; involucre 7–10 mm, narrowly obconic; outer phyllaries green, woolly; inner phyllaries > corolla, stiff, white, with erect, tough, abrupt point. **FL**: 10–20; corolla tubular, white to pale lavender, tube darker; style branch appendage 0.2–0.4 mm, short-triangular, abruptly pointed. **FR**: 2–3 mm; pappus > fr, pink to red. Open areas on plains or slopes, often on clay soil; 50–900 m. CaRF, n&c SNF, n ScV. Jun–Oct

L. nemaclada Greene (p. 379) **ST**: erect, 5–60 cm, openly branched, branches gen ascending from base or ± throughout, tan to brown, distally glabrous, sparsely hairy, glandular-puberulent, or with stalked glands. **LF**: basal petioled; cauline 0.2–3 cm, lanceolate, entire or toothed, with sunken glands and occ stalked and/or bead-like sessile glands, otherwise glabrous or sparsely hairy to tomentose. **INFL**: involucre 5–6 mm, narrowly obconic; phyllaries lanceolate, glandular-puberulent and with stalked and/or bead-like sessile glands, sometimes proximally tomentose, tips gen purple, erect. **FL**: 3–10; corolla funnel-shaped, gen white, occ pale lavender, tube darker; style branch appendage 0.4–1 mm, gen lanceolate, occ short-triangular, abruptly pointed or not. **FR**: 3–4 mm; pappus ≥ fr, tan. Open fields, roadsides, woodland borders, occ on serpentine soil; 50–2100 m. NW, CaR, SN, GV, SnFrB, SCoRI. Morphologically variable. Jul–Oct

L. pectinata Greene **ST**: erect to spreading, branches ± spreading from base or ± throughout, 5–70 cm, green, tan or ± red-brown, proximally tomentose, distally glandular-puberulent and/or sparsely hairy to tomentose. **LF**: basal petioled, sometimes not withered at fl; cauline 0.2–3(6.5) cm, oblong to ovate, entire, toothed or pinnately lobed, with stalked and bead-like sessile glands, glandular-puberulent, sparsely hairy to tomentose or glandless hairs 0. **INFL**: involucre (4)5–8 mm, obconic; phyllaries oblong, with bead-like sessile glands, puberulent or glandless hairs 0, tips erect, gen green, occ purple. **FL**: 15–30; corolla funnel-shaped, yellow with purple-brown band in tube; style branch appendage 0.1–0.2 mm, short-triangular, abruptly pointed. **FR**: 2–3 mm; pappus ≥ fr, white to tan.

var. ***pectinata*** **ST**: ± red to dark brown; distally glandular-puberulent or sparsely hairy. **LF**: 0.2–1.5 cm, cauline dentate to pinnately lobed, lobes very narrow, abruptly pointed. Sandy soil; < 100 m. s CCo. [*L. glandulifera* var. *p.* (Greene) Jeps.] May–Oct

var. ***tenuipes*** (J.T. Howell) Markos (p. 379) **ST**: green or tan, distally puberulent and sparsely hairy to tomentose. **LF**: 0.2–3(6.5) cm, cauline entire to toothed or lobed. Coastal scrub, woodland, pine forest, chaparral, occ sandy soil; 15–1600 m. c&s SNF, SnJV, SnFrB, SCoR. [*L. glandulifera* var. *g.*, misappl.] May–Oct

L. ramulosa A. Gray (p. 379) **ST**: erect, 20–50 cm, branches ascending, tan, proximally tomentose, distally glandular-puberulent, occ stalked-glandular. **LF**: basal sessile, gen persistent; cauline 0.5–2.5 cm, ovate, entire to toothed, with stalked glands, glabrous or sparsely hairy to densely tomentose (occ on adaxial face only). **INFL**: involucre 5–7 mm, hemispheric to bell-shaped; phyllaries lanceolate, glandular-puberulent and stalked-glandular, tips erect, ± purple-tinged. **FL**: 5–15; corolla funnel-shaped, lavender, tube darker; style branch appendage 0.1–0.3 mm, short-triangular, abruptly pointed. **FR**: 3–4 mm; pappus > fr, tan. Open hills, chaparral, woodland, forest, roadsides, often on serpentine soil; 100–1000 m. NCoR, n SnFrB. Jul–Oct

L. tenuis (A. Gray) Coville (p. 379) SPRING LESSINGIA **ST**: ± erect to spreading, 2–15(30) cm, branches gen stiffly ascending to spreading from base or ± throughout, tan to ± red-brown, proximally tomentose, distally glandular-puberulent, sparsely hairy to tomentose or glandless hairs 0. **LF**: basal petioled, sometimes not withered at fl; cauline 0.3–1.5 cm, ovate, entire, lobed or irregularly toothed, sparsely hairy to tomentose, occ with bead-like sessile glands. **INFL**: involucre 4–7 mm, obconic to bell-shaped; phyllaries oblong, glandular-puberulent and/or sparsely hairy, with bead-like sessile or short-stalked glands, tips erect, ± purple. **FL**: 3–25; corolla funnel-shaped, yellow with purple-brown band in tube, (abaxially occ pink or ± purple in peripheral fls); style branch appendage 0.1–0.2, short-triangular, abruptly pointed. **FR**: 2–3 mm; pappus ≥ fr, tan. Openings in chaparral, woodland; 50–2200 m. SnFrB, SCoR, WTR. May–Jul ★

L. virgata A. Gray (p. 379) WAND LESSINGIA **ST**: erect, 5–60 cm, branches gen long, ascending, tan, sparsely hairy to woolly. **LF**: basal petioled; cauline 5–10 mm, oblong-ovate to ovate, entire, appressed upward, 5–10 mm, with sunken glands (also with bead-like sessile or stalked glands), gen tomentose (sparsely-hairy). **INFL**: spike-like; involucre 5–7 mm, narrowly obconic; phyllaries lanceolate, sparsely hairy to loosely woolly, glandless or with sessile, bead-like glands, tips erect, green or purple. **FL**: 3–6; corolla tubular, gen white, occ pale lavender, tube darker; style branch appendage 0.3–0.9 mm, short-triangular or lanceolate, abruptly pointed or not. **FR**: 2–2.5 mm; pappus > fr, white to tan. Dry plains and foothills, grassy openings in woodland, occ in volcanic soil; 50–500 m. CaRF, n SNF, ne ScV. Jun–Oct

LEUCANTHEMUM

Bruce G. Baldwin, adapted from Strother (2006)

Per, from rhizome, 10–120(200+) cm; roots gen red-tipped. **ST**: erect, simple or distally branched, glabrous or hairy. **LF**: basal and cauline, alternate; basal petioled, cauline petioled or sessile; blade linear or lanceolate to obovate, entire to 1[2+]-pinnately lobed or toothed, glabrous or sparsely hairy. **INFL**: heads gen radiate (discoid), 1 or in 2–3s; involucre hemispheric or broader; phyllaries 35–60+ in 3–4+ series, free, ± ovate to oblanceolate, unequal, glabrous or sparsely hairy, margins scarious, color-

less to ± brown; receptacle convex, epaleate, glabrous. **RAY FL**: gen 13–34+ (0); corolla white, drying ± pink, ray linear to ovate. **DISK FL**: 120–200+; corolla yellow, tube ± cylindric, proximally swollen, throat bell-shaped, lobes deltate, resin sacs 0; anther tip triangular-ovate; style tips truncate, papillate. **FR**: ± columnar to obovoid, gen 10-ribbed, glabrous; pappus 0 or fr wall crown-like at fr tip. 20–40+ spp.: Eur, n Afr. (Greek: white fl) [Strother 2006 FNANM 19: 557–559]

1. Basal lvs gen pinnately 3–7+ lobed and/or irregularly toothed; mid-st lvs gen irregularly toothed throughout
. . . . ***L. vulgare***
1′ Basal lvs not lobed, gen toothed (entire); mid-st lvs gen proximally entire, distally regularly serrate
  2. Larger phyllaries 3–5 mm wide; ray fr 3–4 mm, tip adaxially gen with ear-like projections; cauline lf
    blades 3–12+ cm, 12–25(35+) mm wide, elliptic to oblanceolate . . . . ***L. lacustre***
  2′ Larger phyllaries 2–3 mm wide; ray fr 2–3(4) mm, tip gen entire (adaxially with ear-like projections);
    cauline lf blades 5–12+ cm, 8–22+ mm wide, oblanceolate to lanceolate or linear . . . . ***L. maximum***

L. lacustre (Brot.) Samp. PORTUGUESE DAISY Pl 30–120(200+) cm. **LF**: basal lf petiole 30–60+ mm, blade 50–100+ mm, 15–30+ mm wide, lanceolate, not lobed, gen toothed (entire); cauline ± petioled or sessile, blade 30–120+ mm, 12–25(35+) mm, elliptic to oblanceolate, mid-st lvs gen proximally entire, distally regularly serrate. **INFL**: involucre (18)25–35 mm diam; larger phyllaries 3–5 mm wide. **RAY FL**: 21–34+; ray 20–25(35) mm. **FR**: ray fr 3–4 mm, tip adaxially gen with ear-like projections. 2*n*=198. Uncommon. Disturbed sites, meadows, seeps, sea-cliffs; < 50 m. NCo, CCo; native to Portugal. Jul–Aug

L. maximum (Ramond) DC. SHASTA DAISY Pl 20–60(80+) cm. **LF**: basal lf petiole 50–80(200+) mm, blade 50–80(120+) mm, 15–25(35+) mm wide, obovate to spoon-shaped, not lobed, gen toothed (entire); cauline petioled or sessile, blade 50–120+ mm, 8–22+ mm wide, oblanceolate to lanceolate or linear, mid-st lvs gen proximally entire, distally regularly serrate. **INFL**: involucre 18–28+ mm diam; larger phyllaries 2–3 mm wide. **RAY FL**: 21–34+; ray 20–30(40+) mm. **FR**: ray fr 2–3(4) mm, tip gen entire (adaxially with ear-like projections). 2*n*=90,108. Uncommon. Disturbed areas, forest, streambanks; 600–1500 m. CaRH, n SNH; native to Eur. Jun–Aug

L. vulgare Lam. (p. 379) OX-EYE DAISY Pls 10–30(100+) cm. **LF**: basal lf petiole 10–30(120) mm, blade 12–35(50+) mm, 8–20(30) mm wide, obovate to spoon-shaped, pinnately 3–7+ lobed or irregularly toothed; cauline petioled or sessile, blade 30–80+ mm, 2–15+ mm wide, mid-st lvs gen irregularly toothed throughout. **INFL**: involucre 12–20+ cm diam; larger phyllaries 2–3 mm wide. **RAY FL**: (0)13–34+; ray 12–20(35+) mm. **FR**: ray fr 1.5–2.5 mm, tip gen crown-like or with ear-like projections. 2*n*=18,36,54,72,90. Common. Disturbed areas, meadows, seeps; < 2600 m. NCo, KR, NCoRO, CaR, n&c SNH, ScV, CCo, SnFrB, PR, MP; native to Eur; widely naturalized. Jun–Aug ❖

LOGFIA

James D. Morefield

Ann 1–50[70] cm, ± gray, cobwebby to tomentose. **ST**: 1, erect, or 2–10+, gen ascending to prostrate, ± forked at least distally, gen evenly lfy proximally, ± lfless between distal forks. **LF**: alternate, ± sessile, linear to obovate or awl-shaped, entire; distal lvs subtending heads, crowded, largest > proximal lvs. **INFL**: heads disciform, ± sessile, gen 2–10(14) per group or some [all] single; involucre 0 or vestigial, simulated by paleae, or ± cup-like, then phyllaries various, << paleae; receptacle length [0.4]0.7–1.6 × width, obovoid to mushroom-shaped, glabrous; paleae, exc innermost, each ± enclosing pappus-lacking pistillate fl, deciduous with fr (sometimes tardily), ± boat-shaped, obtuse to acute, ± tomentose abaxially, obscurely parallel-veined, veins 5+, margin reflexed distally as scarious wing, wing prominent, terminal, erect to curved inward, visible in head; innermost paleae ± 5[8], collectively surrounding pappus-bearing disk + inner pistillate fls, > outer, spreading at maturity, concave, lance-ovate, acute, cartilaginous, ± brown, ± glabrous. **PISTILLATE FL**: (11)14–45+, gen only outer subtended by paleae; corolla obscure, narrowly cylindric. **DISK FL**: bisexual, 2–10; pappus present; corolla 4–5-lobed; anther base tailed, tip ± triangular; style tips ± linear-oblong. **FR**: obovoid to oblong; outer pistillate-fl fr > inner, ± enclosed by palea, gen compressed laterally, smooth, shiny, corolla scar ± terminal, pappus 0; inner pistillate-fl fr and disk fr free of paleae, ± cylindric, gen ± papillate, dull, corolla scar terminal, pappus of (11)13–28+ deciduous bristles, visible in head, gen falling together in intact or broken ring. 12 spp.: sw N.Am, Eur, sw Asia, n Afr, some alien ± worldwide. (Anagram of *Filago*) [Morefield 2006 FNANM 19:443–447] Characters may be unreliable in young or dwarf pls. Formerly incl in *Filago* in N.Am.

1. Lf gen awl-shaped, stiff; phyllaries gen 5, equal, unlike paleae; outer paleae bent inward 70–90°, longest
  3.3–4.1 mm, body hard . . . . ***L. gallica***
1′ Lf gen elliptic to obovate or ± linear, flexible; phyllaries 0 or 1–4, unequal, like vestigial paleae; outer
  paleae curved inward 20–60°, longest 2.1–3.3 mm, body membranous to cartilaginous
  2. St gen lfless between proximal forks, becoming ± purple to black, ± glabrous; heads 4–10 per group, only
    at st forks and tips; distal lvs ± linear, gen 2–5 × heads; pappus-bearing inner fls 4–12, (0)1–2 pistillate . . . ***L. arizonica***
  2′ St ± evenly lfy proximally, green to ± gray or ± white, cobwebby to tomentose; heads (1)2–5 per group,
    not only at st forks and tips; distal lvs elliptic to obovate, 0.8–2(3) × heads; pappus-bearing inner fls
    (12)17–40+, (4)10–35 pistillate
    3. St (1)3–10+, forked ± throughout; largest lvs 6–8(10) mm, 1–2 mm wide; distal lvs obtuse; outer palea
      body, exc midvein, membranous; disk corolla lobes gen 5, ± yellow to ± brown; outer fr 0.7–0.9 mm;
      inner fr gen smooth, pappus bristles (11)13–15, falling separately or in 2s . . . . ***L. depressa***
    3′ St 1(7), gen forked distally, central axis dominant; largest lvs 10–15(20) mm, 2–3(4) mm wide; distal
      lvs gen acute; outer palea body ± cartilaginous; disk corolla lobes gen 4, ± bright red to purple; outer fr
      0.9–1 mm; inner fr ± papillate, pappus bristles 17–23+, falling together in intact or broken ring . . . . . ***L. filaginoides***

L. arizonica (A. Gray) Holub (p. 379) ARIZONA COTTONROSE Pl 2–10(20) cm. **ST**: (1)3–10, forked ± throughout, becoming ± purple to black, ± glabrous, gen lfless between proximal forks. **LF**: linear to narrowly oblanceolate, flexible, largest 15–20(25) mm, 1–1.5 mm wide; distal lvs ± linear, gen 2–5 × heads, acute. **INFL**: heads 4–10 per group, only at st forks and tips, ± pyramidal, largest ± 4 mm, ± 3 mm wide; phyllaries 0 or vestigial; outer paleae in 5 vertical ranks, curved inward 20–60°, longest 2.2–2.7 mm, body cartilaginous. **PISTILLATE FL**: outer 9–13(17); inner (with pappus) (0)1–2. **DISK FL**: 4–10; corolla 1.2–1.7 mm, lobes gen 5, ± brown to ± yellow. **FR**: outer curved inward, 0.9–1 mm, compressed front-to-back; pappus bristles of inner 17–23, 1.3–2 mm. 2*n*=28. Locally or seasonally moist, gen clay soils; < 800 m. s SCo, s ChI, PR, w DSon; s-c AZ, to nw Mex (incl islands). [*Filago a.* A. Gray] Feb–May

L. depressa (A. Gray) Holub (p. 379) HIERBA LIMPIA Pls 1–5(10) cm. **ST**: (1)3–10+, forked ± throughout, ± gray to ± white, gen tomentose. **LF**: elliptic to obovate, flexible, largest 6–8(10) mm, 1–2 mm wide; distal lvs gen 0.8–1.5 × heads, obtuse. **INFL**: heads 2–5 per group, some between st forks and tips, ± pear-shaped, largest 3–4 mm, 2–2.5 mm wide; phyllaries 0, or 1–4, unequal, like vestigial paleae; outer paleae in spiral ranks, curved inward 20–60°, longest 2.1–3.1 mm, body, exc midvein, membranous. **PISTILLATE FL**: outer 7–13; inner (with pappus) (4)10–21. **DISK FL**: 2–5; corolla 1.3–2 mm, lobes gen 5, ± yellow to ± brown. **FR**: outer 0.7–0.9 mm, ± straight; inner gen smooth, pappus bristles (11)13–15, 1.3–2.4 mm, falling separately or in 2s. Dry, open, gen sandy flats, slopes, drainages; < 1500 m. s SNF, SnJV/SCoRO/WTR, SCo, PR, s SNE, D; to s NV, se AZ, nw Mex. [*Filago d.* A. Gray] See *L. filaginoides.* Common only outside CA-FP. Feb–May

L. filaginoides (Hook. & Arn.) Morefield (p. 379) CALIFORNIA COTTONROSE Pls 1–30(55) cm. **ST**: 1(7), gen forked distally, central axis dominant, ± gray to green, gen cobwebby. **LF**: gen oblanceolate, flexible, largest 10–15(20) mm, 2–3(4) mm wide; distal lvs 1–2(3) × heads, gen acute. **INFL**: heads (1)2–4 per group, at and between st forks and tips, ± pear-shaped, largest 3.5–4.5 mm, 2.5–3 mm wide; phyllaries 0, or 1–4, unequal, like vestigial paleae; outer paleae in spiral ranks, curved inward 20–60°, longest 2.7–3.3 mm, body ± cartilaginous. **PISTILLATE FL**: outer 7–13; inner (with pappus) 14–35. **DISK FL**: 4–7; corolla 1.9–2.8 mm, lobes gen 4, ± bright red to purple. **FR**: outer 0.9–1 mm, ± straight; pappus bristles of inner 17–23+, 1.9–3 mm. 2*n*=28. Common, ± weedy. Bare, rocky, or grassy sites, drainages; < 1800 m. CA-FP (uncommon NW, CaR), SNE, D (esp DMtns); to sw UT, w TX, nw Mex. [*Filago californica* Nutt.] Dwarf pls like *L. depressa* exc pappus bristles adherent, disk corollas 4-lobed. Feb–May

L. gallica (L.) Coss. & Germ. (p. 379) DAGGERLEAF COTTONROSE Pls 2–50 cm. **ST**: 1–5, forked ± throughout, central axis rarely dominant, ± gray to green, gen cobwebby. **LF**: gen awl-shaped, stiff, largest 20–30(40) mm, 1–1.5(2) mm wide; distal lvs gen 2–5 × heads, acute to ± spiny. **INFL**: heads (2)3–10(14) per group, only at st forks and tips, ± flask-shaped, largest (3)3.5–4.5 mm, 2–3 mm wide; phyllaries gen 5, equal, obovate, translucent, unlike paleae; outer paleae in vertical ranks, bent inward 70–90°, longest 3.3–4.1 mm, body hard. **PISTILLATE FL**: outer 9–12; inner (with pappus) 8–14(30). **DISK FL**: 3–5; corolla 2.2–3 mm, lobes gen 4, ± brown to ± yellow. **FR**: outer bent inward, 0.9–1 mm; pappus bristles of inner 18–28+, 2.2–3 mm. 2*n*=28. Bare or grassy openings, burns; < 1100 m. NW (exc NCoRH), CaRF, SNF, w SNH, GV, CW, SW; to sw OR, nw Baja CA; native to Medit, introduced ± worldwide. [*Filago g.* L.] Mar–Jul

LUINA

Per from branched caudex. **ST**: slender, gray-tomentose. **LF**: cauline, alternate, lanceolate to ovate, entire or toothed. **INFL**: heads discoid, gen in ± flat-topped clusters; involucre obconic; phyllaries in 1–2 equal series; receptacle flat or convex, epaleate. **DISK FL**: gen 11–23; corolla white to cream [yellow], tube > throat; anther base short-tailed, tip narrowly lanceolate; style branches rounded-truncate. **FR**: ± cylindric; pappus of many bristles in 1–2 series. 2 spp.: nw N.Am. (Anagram of *Inula*) [Pelser et al. 2007 Taxon 56:1077–1104; Strother 2006 FNANM 20:627–628]

L. hypoleuca Benth. (p. 379) Pl 1.5–6 dm. **LF**: ± sessile, 2–6.5 cm, glabrous adaxially, white-tomentose abaxially. **INFL**: heads gen 4–12; peduncle 1–5 cm; phyllaries 5–9 mm, lanceolate to narrowly ovate, gen tomentose, sometimes glandular-puberulent. **FL**: 11–15(23); corolla gen 8–10 mm, tube 3–4 mm, throat slightly expanded, lobes 1–2 mm, spreading to recurved. **FR**: 3–4 mm, ± 9-veined, glabrous or strigose; pappus 8–10 mm. Rocky places, cliffs, sometimes on serpentine; mixed-evergreen forest, riparian; < 2100 m. KR, NCoR, sw SnFrB; to BC. Jun–Sep

MACHAERANTHERA TANSY-ASTER

David R. Morgan

Ann, bien from taproot. **ST**: gen erect or ascending, branched distally or throughout. **LF**: simple, alternate, pinnately dissected; lobes gen bristle-tipped. **INFL**: heads radiate, in cyme-like clusters; involucre bell-shaped or hemispheric; phyllaries graduated in 3–6 series, proximally straw-colored to ± purple, distally green; receptacle flat or convex, pitted, epaleate. **RAY FL**: 8–many; corolla blue or purple. **DISK FL**: 14–many; corolla yellow, lobes gen 0.3–0.7 mm, gen ± glabrous [0.7–1 mm, hairy]; anther tip lanceolate; style tips lanceolate to linear, distally bristly. **FR**: narrowly to broadly obovoid, ± flattened, 4–9-ribbed on each face, sparsely to densely hairy; pappus of many unequal, minutely barbed bristles. 2 spp.: temp w N.Am. (Greek: sword-like anthers) [Morgan & Hartman 2006 FNANM 20:394–395] Other taxa in TJM (1993) moved to *Arida, Dieteria, Xanthisma.*

M. tanacetifolia (Kunth) Nees (p. 385) Pl 1–10 dm, ± bushy, puberulent to densely glandular. **ST**: 1–several from base, gen branched distally. **LF**: gen 3–12 cm, 1–2-pinnately dissected. **INFL**: phyllaries gen in 3–5 series, tips linear, long-acuminate, spreading to reflexed. **RAY FL**: many; corolla blue-purple; ray 1–2 cm. **DISK FL**: many; corolla 5–7 mm. **FR**: 3–4 mm, narrowly obovate, silky; pappus 4–6 mm. 2*n*=8. Uncommon. Desert scrub, pinyon/juniper woodland; ± 1700 m. e DMtns (New York Mtns); to MT, SD, TX, Mex. May–Jul

MADIA TARWEED, TARPLANT

Bruce G. Baldwin

Ann, 0.5–25 dm. **ST**: erect. **LF**: proximal opposite, often in rosettes, distal alternate, sessile; blades lanceolate or oblong-linear to linear, gen entire, seldom toothed, coarse- to soft-hairy, gen also glandular. **INFL**: heads gen radiate, occ obscurely so (discoid), in flat-topped or panicle-, raceme-, or spike-like clusters; involucre 1–10+ mm diam, gen ± spheric to ovoid or

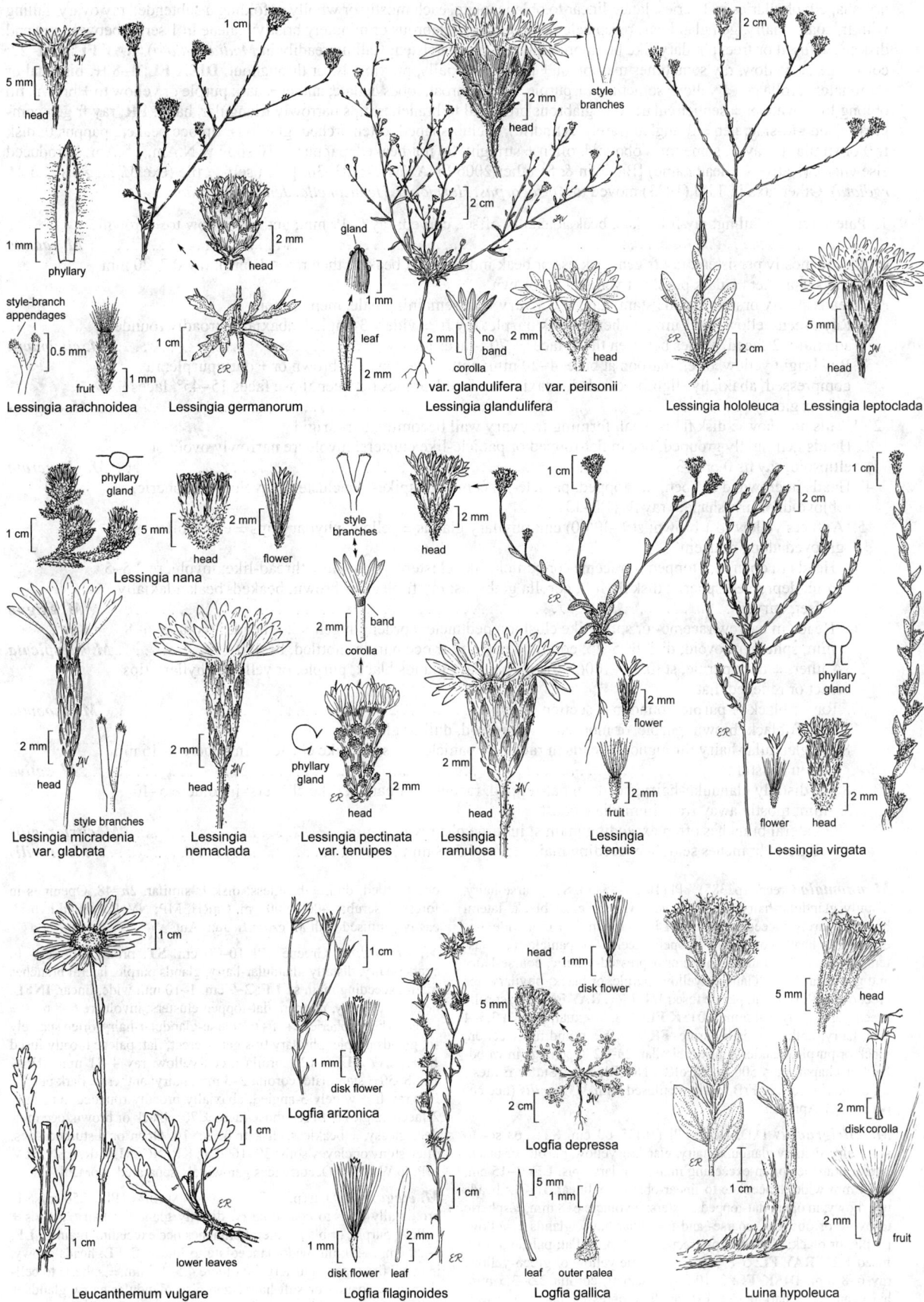

Lessingia arachnoidea

Lessingia germanorum

var. glandulifera

var. peirsonii

Lessingia glandulifera

Lessingia hololeuca

Lessingia leptoclada

Lessingia nana

Lessingia micradenia
var. glabrata

Lessingia
nemaclada

Lessingia pectinata
var. tenuipes

Lessingia
ramulosa

Lessingia
tenuis

Lessingia virgata

Leucanthemum vulgare

Logfia arizonica

Logfia depressa

Luina hypoleuca

Logfia filaginoides

Logfia gallica

urn-shaped; phyllaries in 1 series, lance-linear to oblanceolate, each mostly or wholly enfolding a subtended ray ovary, falling with fr, coarse-hairy, gen glandular; receptacle flat to convex, glabrous or minutely bristly; paleae in 1 series between ray and disk fls, ± fused or free, phyllary-like but more scarious, gen persistent (falling readily in *Madia radiata*). **RAY FL**: (0)1–22; corolla gen ± yellow, ray sometimes maroon or ± purple adaxially, proximally, or throughout. **DISK FL**: 1–80+, bisexual or staminate; corolla gen ± yellow, sometimes ± purple, tube ≤ throat, lobes deltate; anther ± dark purple or yellow to ± brown, tip oblong to ± ovate or ± semicircular; style glabrous proximal to branches, tips narrowly triangular, hairy. **FR**: ray fr gen compressed side-to-side, gen ± 3-angled (rarely cylindric), ± club-shaped, often arched, glabrous, tip occ beaked, pappus 0; disk fr 0 or similar to ray fr, sometimes obovoid, often ± straight, tip not beaked, pappus 0. 10 spp.: w N.Am, s S.Am; introduced elsewhere. (Native Chilean name) [Baldwin & Strother 2006 FNANM 21:303–308] Gen self-fertile (exc *M. elegans* and *M. radiata*). Other taxa in TJM (1993) moved to *Anisocarpus, Harmonia, Hemizonella, Jensia, Kyhosia.*

1. Paleae readily falling; ray fr beaked, beak adaxially offset, curved; ray 6–19 mm; anthers yellow to ± brown
. . . . ***M. radiata***
1′ Paleae mostly persistent; ray fr gen beakless or beak indistinct (if beaked, then ray ≤ 1 mm); ray 0.7–20 mm or rays 0; anthers ± dark purple or yellow to ± brown
  2. Heads showy or not; disk fls staminate, fr 0, ovary wall remaining pale, membranous
    3. Ray green-yellow, 4–11 mm; anthers ± dark purple; ray fr ± widely 3-angled, abaxially broadly rounded, adaxially 2-faced, angles between those faces ± 70°, faces glossy . . . . ***M. citriodora***
    3′ Ray bright yellow, often maroon at base, 4–20 mm; anthers yellow to ± brown or ± dark purple; ray fr compressed, abaxially slightly rounded, adaxially 2-faced, angles between those faces 15–45°, faces dull or glossy . . . . ***M. elegans***
  2′ Heads not showy; disk fl bisexual, forming fr, ovary wall becoming dark, rigid
    4. Heads gen tightly grouped, occ in flat-topped or panicle-like clusters; involucre narrowly ovoid or ellipsoid; ray fls 0 or 1–3 . . . . ***M. glomerata***
    4′ Heads in crowded or open, flat-topped, panicle-, raceme-, or spike-like clusters; involucre ± spheric, (ob)ovoid, or urn-shaped; ray fls (1)3–13
      5. Anthers yellow to ± brown; st 1–30(60) cm; phyllary glands ± yellow; phyllary tips ± erect, often grooved along midvein
        6. Heads in open, flat-topped or raceme- or panicle-like clusters; peduncle ± thread-like; involucre 2.5–5 mm, depressed-spheric; disk fls 1(2), corolla glabrous; ray fr black or brown, beaked, beak adaxially offset, curved . . . . ***M. exigua***
        6′ Heads in narrow raceme- or spike-like clusters; peduncle 0 or length gen < 2 × head; involucre 6–8 mm, spheric or ovoid; disk fls 5–15, corolla hairy; ray fr occ purple-mottled, beak 0 . . . . ***M. subspicata***
      5′ Anthers ± dark purple; st (0.3)6–100(240) cm; phyllary glands black, purple, or yellow; phyllary tips erect or reflexed, flat
        7. Ray fr black or purple, round in ×-section, glossy. . . . . ***M. anomala***
        7′ Ray fr black, brown, purple, or mottled, compressed, dull or glossy
          8. St glandular-hairy throughout; heads in raceme-, panicle-, or spike-like clusters; involucre 6–16 mm; often coastal . . . . ***M. sativa***
          8′ St distally glandular-hairy; heads in flat-topped, raceme-, or panicle-like clusters; involucre 5–10 mm; mostly away from immediate coast
            9. Lateral branches often exceeding main st in large pls; ray 6–8 mm . . . . ***M. citrigracilis***
            9′ Lateral branches seldom exceeding main st; ray 1.5–8 mm . . . . ***M. gracilis***

M. anomala Greene (p. 385) Pl (10)20–55 cm. **ST**: coarse-hairy, distally glandular-hairy, glands ± yellow, purple, or black, lateral branches rarely exceeding main st. **LF**: 2–10 cm, 2–7 mm wide, linear. **INFL**: heads not showy, in open, raceme- or panicle-like clusters; involucre 6–10 mm, spheric or depressed-spheric, coarse-hairy and glandular-hairy, glands ± yellow, purple, or black; phyllary tips erect or ± reflexed, flat; paleae fused 1/4–1/2+. **RAY FL**: 3–8; corolla green-yellow, ray 3–4.5 mm. **DISK FL**: 3–8, bisexual; corolla 3.5–4 mm, hairy; anthers ± dark purple. **FR**: ray fr ± round in ×-section, black or purple, beakless; disk fr similar. 2*n*=32. Openings in woodland or chaparral; < 500 m. NCoRO, NCoRI, ScV (Sutter Buttes), CCo (Marin Co.), SnFrB. Easily confused with *M. gracilis* (occ co-occurring). Apr–Jun

M. citrigracilis D.D. Keck Pl (10)25–60 cm. **ST**: coarse- to soft-hairy, distally glandular-hairy, glands ± yellow, purple, or black, lateral branches often exceeding main st in large pls. **LF**: 3–15 cm, 2–14 mm wide, lanceolate to linear-oblong or linear. **INFL**: heads not showy, in open, flat-topped clusters; involucre 6–8 mm, ± spheric to ovoid or obovoid, coarse- and glandular-hairy, glands ± yellow, purple, or black; phyllary tips erect or ± reflexed, flat; paleae mostly fused 1/2+. **RAY FL**: 5–8(14); corolla pale yellow or green-yellow, ray 6–8 mm. **DISK FL**: 3–10(30), bisexual; corolla 2.5–3.5 mm, hairy; anthers ± dark purple. **FR**: ray fr compressed, black or brown,

occ mottled, dull, ± beakless; disk fr similar. 2*n*=48. Openings in forest or scrub; 1400–1900+ m. CaRH, MP; NV. Polyploid hybrid; easily confused with *M. gracilis.* Jun–Aug

M. citriodora Greene Pl 10–70 cm. **ST**: proximally soft- to coarse-hairy, distally glandular-hairy, glands purple, lateral branches often exceeding main st. **LF**: 2–9 cm, 1–10 mm wide, linear. **INFL**: heads not showy, in open, flat-topped clusters; involucre 6–8 mm, ± ovoid to hemispheric, ± soft- to coarse-glandular-hairy, often sparsely so, glands purple; phyllary tips gen ± erect, flat; paleae mostly fused 1/2+. **RAY FL**: 5–12; corolla green-yellow, ray 4–11 mm. **DISK FL**: 8–50+, staminate; corolla 2–3 mm, hairy; anthers ± dark purple. **FR**: ray fr ± widely 3-angled, abaxially broadly rounded, adaxially 2-faced, angles between those faces ± 70°, black or brown, occ mottled, glossy, ± beakless; disk fr 0. 2*n*=16. Open or disturbed sites, often stony or clayey soils; 30–1600 m. KR, NCoRI, CaR, n SN, ScV, MP; to WA, NV. Occurrences gen widely separated. Apr–Jul

M. elegans D. Don (p. 385) COMMON MADIA Pl 6–250 cm. **ST**: proximally soft- to coarse-hairy, distally glandular-hairy, glands ± yellow, purple, or black, lateral branches occ exceeding main st. **LF**: 3–20 cm, 2–20 mm wide, lanceolate to linear. **INFL**: heads showy, in open, flat-topped clusters; involucre 4.5–12 mm, ± spheric to bell-shaped, ± coarse- or soft-hairy, gen also glandular-hairy, glands ± yellow, purple, or black; phyllary tips erect or reflexed, flat; paleae

mostly fused 1/2+. **RAY FL**: (2)5–22; corolla bright yellow, often maroon at base, ray 4–20 mm. **DISK FL**: 25–80+, staminate; corolla 2.5–5 mm, hairy; anthers yellow to ± brown or ± dark purple. **FR**: ray fr compressed or ± 3-angled, slightly rounded abaxially, angled 15–45° adaxially, black or brown, occ mottled, dull or glossy, ± beakless; disk fr 0. 2*n*=16. Grassy, open, or disturbed sites, in coarse or clayey soils, incl serpentine; < 3400 m. CA-FP (exc ChI), GB (exc Wrn, W&I); to WA, NV, Baja CA. [*M. e.* subsp. *densifolia* (Greene) D.D. Keck; *M. e.* subsp. *vernalis* D.D. Keck; *M. e.* subsp. *wheeleri* (A. Gray) D.D. Keck] Highly variable; forms sterile hybrids with *M. sativa*. Apr–Nov

M. exigua (Sm.) A. Gray (p. 385) Pl 1–30(50) cm. **ST**: coarse- and glandular-hairy, glands ± yellow or purple, lateral branches occ exceeding main st. **LF**: 0.2–4 cm, 0.5–2 mm wide, linear. **INFL**: heads not showy, in open, flat-topped or raceme- or panicle-like clusters; peduncle ± thread-like; involucre 2.5–5 mm, depressed-spheric, ± coarse- and glandular-hairy, glands golden yellow; phyllary tips ± erect, grooved along midvein; paleae fused 1/2+. **RAY FL**: 1–8; corolla pale yellow, ray 0.7–1 mm. **DISK FL**: 1(2), bisexual; corolla 1–1.8 mm, glabrous; anthers yellow to ± brown. **FR**: ray fr compressed, strongly arched, black or brown, dull, with adaxially offset, curved beak; disk fr obovoid, slightly compressed. 2*n*=32. Grassy, open, or disturbed sites, in sandy or clayey soils, incl serpentine; 30–2500 m. CA-FP (exc s SNF, Teh, SnJV, SCo, SnJt), MP; to BC, MT, NV, Baja CA. Often confused with *Hemizonella minima*. Apr–Jul

M. glomerata Hook. (p. 385) MOUNTAIN TARWEED Pl 0.5–12 dm. **ST**: proximally soft-hairy to bristly, glandular-hairy distally, glands ± yellow or black, lateral branches sometimes exceeding main st. **LF**: 2–10 cm, 2–7 mm wide, linear to lance-linear. **INFL**: heads not showy, gen tightly grouped, occ in flat-topped or panicle-like clusters; involucre 0 or 5.5–9 mm, narrowly ovoid or ellipsoid, ± soft- and glandular-hairy, glands ± yellow or black; phyllary tips erect or reflexed, flat; paleae free. **RAY FL**: 0 or 1–3; corolla green-yellow to ± purple, ray 1–3 mm. **DISK FL**: 1–5(12), bisexual; corolla 3–4.5 mm, hairy; anthers ± dark purple. **FR**: ray fr black, dull, compressed, beakless; disk fr similar. 2*n*=28. Meadows, open or disturbed sites, often sandy or gravelly soils; (0)1050–2700 m. KR, CaRH, n&c SNH, SnBr, GB (exc W&I); introduced NCo, sw SnFrB; to BC, SK, MT, SD, WY, CO, NM, scattered to AK, ME. Jun–Sep

M. gracilis (Sm.) Applegate (p. 385) GUMWEED Pl 6–100 cm. **ST**: proximally soft- to coarse-hairy, distally glandular-hairy, glands ± yellow, purple, or black, lateral branches seldom exceeding main st. **LF**: 1–10(15) cm, 1–8(10) mm wide, oblong to linear. **INFL**: heads not showy, in ± open, panicle- or raceme-like clusters; involucre 5–10 mm, depressed-spheric to urn-shaped, occ coarse-hairy, always finely or coarsely glandular-hairy, glands ± yellow, purple, or black; phyllary tips erect or ± reflexed, flat; paleae fused 1/2+. **RAY FL**: 3–10; corolla lemon-yellow or green-yellow, ray 1.5–8 mm. **DISK FL**:

2–16+, bisexual; corolla 2.5–5 mm, hairy; anthers ± dark purple. **FR**: ray fr compressed, black, purple, or mottled, dull, ± beakless; disk fr similar. 2*n*=32,48. Open, semi-shaded, or disturbed sites, many habitats, incl serpentine; < 2500 m. CA-FP (exc SnJV, SCo), GB (exc W&I); to BC, MT, UT, Baja CA. Apr–Aug

M. radiata Kellogg (p. 385) SHOWY GOLDEN MADIA Pl 1–9 dm. **ST**: glandular-hairy, glands ± yellow or purple, lateral branches often exceeding main st. **LF**: 2–10 cm, 4–15 mm wide, lanceolate to linear. **INFL**: heads showy, in open, ± flat-topped clusters; involucre 4–7 mm, depressed-spheric, soft-hairy to bristly, hairs hooked at tip, and glandular-hairy, glands ± yellow or purple; phyllary tips ± erect or reflexed, flat; paleae free, readily falling. **RAY FL**: 8–16; corolla golden yellow, ray 6–19 mm. **DISK FL**: 18–65, bisexual; corolla 3.5–5.5 mm, hairy; anthers yellow to ± brown. **FR**: ray fr compressed, strongly arched, black, purple, or mottled, dull or glossy, beaked, beaks adaxially offset, curved; disk fr similar, not beaked. 2*n*=16. Grassy or open slopes, vertic clay, rarely serpentine; 20–1200 m. SnJV/SnFrB, SCoRI. Often confused with *Deinandra halliana*. Mar–May ★

M. sativa Molina (p. 385) COAST TARWEED Pl (0.3)35–100(240) cm. **ST**: coarse- and glandular-hairy throughout, glands ± yellow, purple, or black, lateral branches rarely exceeding main st. **LF**: 2–18 cm, 3–18(29) mm wide, broadly lanceolate to linear-oblong or linear. **INFL**: heads not showy, in gen crowded, panicle-, raceme-, or spike-like clusters; involucre 6–16 mm, ovoid to urn-shaped, coarse- and glandular-hairy, glands ± yellow, purple, or black; phyllary tips erect or ± reflexed, flat; paleae fused 1/2+. **RAY FL**: (5)8–13; corolla green-yellow or occ purple-red abaxially or throughout, ray 1.5–4 mm. **DISK FL**: 11–45+, bisexual; corolla 2–5 mm, hairy; anthers ± dark purple. **FR**: ray fr compressed, black or brown, sometimes mottled, dull, beakless; disk fr similar. 2*n*=32. Grassy, open, or disturbed sites; < 1000 m. NW (exc NCoRH), n SNF, n SNH (uncommon), ScV (rare), CW, SCo, ChI, WTR; to BC; also s S.Am; introduced elsewhere. Occ occurs with (and tends to fl later than) *M. gracilis*. May–Oct

M. subspicata D.D. Keck (p. 385) Pl 5–60 cm. **ST**: proximally ± soft-hairy, distally glandular-hairy, glands ± yellow, lateral branches not exceeding main st. **LF**: 2–7 cm, 1–5 mm wide, linear to lance-linear. **INFL**: heads not showy, in narrow raceme- or spike-like clusters, peduncle 0 or lengths gen < 2 × heads; involucre 6–8 mm, spheric or ovoid, ± coarse-hairy and thick-stalked-glandular, glands golden yellow; phyllary tips ± erect, grooved along midvein or flat; paleae free or fused < 1/2. **RAY FL**: 5–8; corolla pale yellow, ray 1–2.5 mm. **DISK FL**: 5–15, bisexual; corolla 3–3.5 mm, hairy; anthers yellow to ± brown. **FR**: ray fr compressed, ± club-shaped, black or brown, occ purple-mottled, dull, beakless; disk fr similar. 2*n*=16. Grassland, open woodland, often in shade; 50–800 m. CaRF, n&c SNF, ScV (Sutter Buttes). Apr–Jun

MALACOTHRIX

W.S. Davis

Ann, per, < 70 (200) cm; sap milky. **ST**: gen ± branched, gen erect, gen glabrous. **LF**: gen basal and cauline, alternate, sessile, sometimes reduced distally on st, entire, toothed, or ± deeply pinnately lobed or divided. **INFL**: heads liguliflorous; involucre gen bell-shaped; phyllaries in 3–6 series, mid-stripe green or ± red; receptacle flat to convex, shallowly pitted, glabrous or with fragile, smooth bristles < 5 mm, epaleate. **FL**: corolla yellow or white, readily withering; ligules of outermost fls exserted 1–15 mm, gen ± red- or purple-striped abaxially. **FR**: 0.9–4 mm, ± fusiform, tip truncate, gen smooth; ribs 15 (5 gen prominent, 10 ± obscure), gen extending to tip; outer pappus 0 or of teeth < 0.5 mm, or rounded-toothed crown, and 0–6 smooth, persistent bristles, inner pappus of 12–32 bristles fused at base, readily deciduous, minutely barbed proximally. 20 spp.: w N.Am, s S.Am. (Greek: soft hair) [Davis 2006 FNANM 19:310–321] *Anisocoma*, *Atrichoseris*, and *Calycoseris* perhaps best placed here (Lee et al. 2003 Syst Bot 28:616–626).

1. Per (occ fl first yr)
2. Cauline lvs ± fleshy; ligules ± yellow; fr smooth, outer pappus 0; dunes . . . . ***M. incana***
2′ Cauline lvs gen not fleshy; ligules white; fr minutely spiny, outer pappus of irregular teeth . . . . ***M. saxatilis***
  3. Distal lvs gen 1–2-pinnately lobed
    4. St densely lfy; distal lvs (ob)ovate, 2-pinnately divided into linear segments . . . . var. ***implicata***
    4′ St ± sparsely lfy; distal lvs linear to lanceolate, sharply and narrowly lobed . . . . var. ***tenuifolia***

3′ Distal lvs gen entire (toothed)
  5. Distal lvs gen linear to ovate, tip occ obtuse; coastal bluffs . . . . var. *saxatilis*
  5′ Distal lvs lance-linear to elliptic, gen acute
    6. Herbage tomentose . . . . var. *arachnoidea*
    6′ Herbage glabrous to ± hairy . . . . var. *commutata*
1′ Ann
  7. Pl scapose, st unbranched, gen not visible above ground; heads borne singly on long, ± naked peduncles;
     outer phyllaries with ± tangled long hairs . . . . *M. californica*
  7′ Pl with ± lfy st, or if ± scapose, st gen evident above ground, branched or not; heads gen borne in
     clusters; outer phyllaries glabrous, or ± short white-hairy
    8. Cauline lvs gen with ear-like basal lobes; involucre gen spheric; outermost phyllaries widely ovate to
       round, margin widely scarious . . . . *M. coulteri*
    8′ Cauline lvs gen without ear-like basal lobes; involucre gen bell-shaped; outermost phyllaries lanceolate
       to widely ovate
      9. Cauline lvs not or seldom notably reduced distally on st; ChI
        10. Fr with outer pappus
          11. Distal lvs gen oblanceolate, with few, short lobes; involucre 7–8.5 mm; fr ribs ± equal, extending to
             tip; outer pappus of minute teeth and 1–2 bristles; Anacapa Island . . . . *M. junakii*
          11′ Distal lvs narrowly ovate, with 5–10 narrow, ± sharp teeth or lobes; involucre 9–12 mm; 5 fr ribs
             prominent, extending ± beyond tip; outer pappus of minute teeth, bristles 0(1); Anacapa, Santa Cruz
             islands . . . . *M. squalida*
        10′ Fr gen without outer pappus
          12. Cauline lvs gen fleshy, margins ± equally lobed, tips obtuse; San Miguel, Santa Cruz, Santa Rosa
             islands . . . . *M. indecora*
          12′ Cauline lvs gen not fleshy, lobes ± equal, gen sharp; Anacapa, San Clemente, San Nicolas, Santa
             Barbara islands . . . . *M. foliosa*
            13. Ligules of outermost fls exserted < 5 mm
              14. St 1–6, gen branched at base and distally; distal-most lvs distally sharply 4–10-lobed, ultimate
                 margins curled; outer phyllaries ± overlapping; Anacapa Island . . . . subsp. *crispifolia*
              14′ St gen 1, branched above base; distal-most lvs gen 2-lobed at base, ultimate margins gen not
                 curled; outer phyllaries ± 1/2 inner phyllaries; San Nicolas Island . . . . subsp. *polycephala*
            13′ Ligules of outermost fls exserted gen > 5 mm
              15. St gen 1, gen erect, branched distally; distal-most lvs gen 2-lobed at base; San Clemente Island
                 . . . . subsp. *foliosa*
              15′ Sts (1)3–5(10), erect, or decumbent to ascending; distal-most lvs gen pinnately lobed from base to
                 near tip; Santa Barbara Island . . . . subsp. *philbrickii*
      9′ Cauline lvs gen much reduced distally on st; widespread, gen mainland habitats (*Malacothrix*
         clevelandii, Malacothrix similis also ChI)
        16. Ligules of outermost fls exserted < 5 mm
          17. Proximal cauline lvs ± fleshy, lobes in 3–8 pairs, short, ± equal, gen toothed; fr outer pappus a
             rounded-toothed crown, bristles 0 . . . . *M. phaeocarpa*
          17′ Proximal cauline lvs gen not fleshy, unequally toothed or narrowly lobed; fr outer pappus of
             needle-like teeth, bristle gen 1
            18. Basal lvs gen obovate in outline; corolla gen 6–8 mm, white or pale yellow; fr gen 1.7–2.3 mm,
               ribs 0 near tip . . . . *M. stebbinsii*
            18′ Basal lvs gen lance-linear to (ob)lanceolate in outline; corolla 4–9 mm, pale yellow to yellow; fr
               1.2–1.8 mm, ribs extending to tip
              19. St gen branched mostly distally; cauline lvs gen toothed; pollen 70–100% 3-pored; common,
                 widespread . . . . *M. clevelandii*
              19′ St gen branched mostly from base; cauline lvs ± entire; pollen 70–100% 4-pored; uncommon,
                 SCo, n ChI . . . . *M. similis*
        16′ Ligules of outermost fls exserted ≥ 5 mm
          20. Ligules gen white (± yellow); fr 1.2–2 mm; outer pappus 0; not of desert habitats . . . . *M. floccifera*
          20′ Ligules yellow (white); fr 1.8–3.3 mm; outer pappus present; often desert-like habitats
            21. Lf lobes ± equal, gen thread-like . . . . *M. glabrata*
            21′ Lf lobes short, ± equal, oblong, or triangular to linear
              22. Distal-most lvs gen widest at base; outer pappus a rounded-toothed crown, bristles 0 . . . . *M. sonchoides*
              22′ Distal-most lvs gen narrowed to strap-like base; outer pappus of irregular teeth, bristles (0)1–6 . . . . *M. torreyi*

M. californica DC. (p. 385) Ann ≤ 45 cm, scapose. **ST**: 1, short, unbranched, gen not visible. **LF**: all basal, not fleshy, gen linear to oblanceolate, conspicuously long-hairy at base, lobes gen linear to thread-like, well-spaced. **INFL**: heads 1 on long, ± naked peduncle; involucre gen 10–15 mm; outer phyllaries ± 1/2 inner, lanceolate, bases with long tangled hairs, mid-stripe ± red; receptacle glabrous or sparsely bristly. **FL**: corolla 17–22 mm, gen yellow (white), ligules of outermost fls exserted 11–13 mm. **FR**: 2–3.4 mm; outer pappus teeth irregular, bristles 2. 2*n*=14. Open, sandy soil in coastal dunes, grassland, oak woodland, chaparral, desert margins; < 1700 m. s SNF, SnJV, CW, SW, D; Mex. Mar–May

M. clevelandii A. Gray (p. 385) Ann ≤ 40+ cm. **ST:** 1–5, erect or ascending, gen branched distally, glabrous. **LF:** basal not fleshy, gen (ob)lanceolate, teeth or short lobes ± equal, well spaced; distal much reduced, 2–4-toothed at base. **INFL:** heads gen many in open cyme- or panicle-like clusters; involucre 4–8 mm; outer phyllaries ± 1/2 inner, gen lanceolate, glabrous; receptacle glabrous. **FL:** corolla 4.6–7.7 mm, pale yellow, ligules of outer fls exserted 1–3 mm; 70–100% of pollen 3-pored. **FR:** 1.2–1.8 mm, ribs ± equal; outer pappus teeth needle-like, bristle 1. 2*n*=14. Cleared areas (burns, slides), gen chaparral, ± desert margins; < 1500 m. NW, SNF, s SNH, SnJV, CW, SW (exc s ChI); to UT, S.Am. Mar–Jun

M. coulteri Harv. & A. Gray (p. 385) SNAKE'S-HEAD Ann ≤ 75 cm. **ST:** simple or ± branched, ascending or erect, glabrous, ± glaucous. **LF:** basal ± fleshy, lance-linear to ovate, wavy-lobed, base slightly narrowed; proximal gen (ob)lanceolate, toothed, short lobed, or parted to midrib, distal reduced, gen ear-like at base. **INFL:** heads gen few in cyme-like clusters; involucre 10–22 mm, gen spheric; outer phyllaries widely overlapping, ± equal, widely ovate to round, glabrous, midstripe gen ± red, margin widely scarious; receptacle densely bristly. **FL:** corolla 8–20 mm, gen pale yellow (white), ligules of outermost fls exserted 2–12 mm. **FR:** 1.6–3.2 mm, prominent ribs, ± winged, extending beyond tip; outer pappus teeth uneven, bristles 2–6. 2*n*=14. Sandy, open areas, in coastal-sage scrub, grassland, desert; < 1500 m. SnJV, CW, n ChI (extirpated), WTR, PR, SNE, DMoj, n DSon; to UT, AZ, S.Am. Mar–May

M. floccifera (DC.) S.F. Blake (p. 385) Ann ≤ 45 cm. **ST:** 1–many, branched at and above base, erect or ascending, glabrous. **LF:** basal ± fleshy, oblong or oblong-spoon-shaped, with wide, toothed lobes; proximal gen oblanceolate, lobes gen 4–6 pairs, ± equal, obtuse, ± fleshy, lobe base white hair-tufted; distal reduced, entire or basally 2–4-toothed (or distal lobed ± throughout). **INFL:** heads few to many in cyme- or panicle-like clusters; involucre 5–9 mm; outer phyllaries ± 1/2 inner, lanceolate to ovate, glabrous; receptacle densely bristly. **FL:** corolla 7–15 mm, gen white (± yellow), ligules of outermost fls exserted 5–9 mm. **FR:** 1.2–2 mm; outer pappus 0. 2*n*=14. In loose soil of open burns, slides, in chaparral, pinyon/juniper woodland, yellow-pine forest; < 2000 m. NW, CaR, n&c SNH, ScV, CW, WTR, PR; NV. Mar–Nov

M. foliosa A. Gray Ann ≤ 45 cm. **ST:** decumbent to ascending or erect, branched variously, gen red-tinged, gen lfy near tip, glabrous. **LF:** basal not fleshy, gen (ob)lanceolate to (ob)ovate, toothed, lobed, or divided, lobe tips wide-acute to acuminate; cauline gen lobed, tip gen acute; distal-most seldom notably reduced, toothed or lobed. **INFL:** heads gen many in open cyme- or panicle-like clusters; involucre 5–12 mm, gen red-tinged; outer phyllaries ± 1/2 inner or ± overlapping, red-tinged, glabrous; receptacle glabrous. **FL:** corolla 10–17 mm, pale to medium yellow; ligules of outermost fls exserted 1.5–10 mm. **FR:** 0.9–1.7 mm; ribs ± equal or 5 prominent; outer pappus gen 0(1). 2*n*=14.

subsp. ***crispifolia*** W.S. Davis WAVY-LEAVED MALACOTHRIX **ST:** 1–6, gen branched at base and distally. **LF:** distal oblanceolate or obovate, distally with 2–5 pairs of narrow lobes, ultimate margins curled, tip acute. **INFL:** involucre 7–9.2 mm; outer phyllaries in 4–5 series. **FL:** corolla 6–10 mm, medium yellow; ligules of outermost fls exserted 2.5–4 mm. **FR:** 1.3–1.6 mm, ± 5-angled, 5 ribs prominent. Openings in island chaparral; 20–100 m. n ChI (Anacapa Island). Mar–Jul ★

subsp. ***foliosa*** (p. 385) LEAFY MALACOTHRIX **ST:** gen 1, gen erect, branched distally. **LF:** distal ovate to lanceolate, with 1–2 pairs of narrow lobes at base. **INFL:** involucre 7–11 mm; outer phyllaries in 3 series. **FL:** corolla 10–17 mm; ligules of outermost fls exserted 6–10 mm. **FR:** 0.9–1.5 mm, ribs ± equal. Sandy, open areas or among shrubs in island chaparral; < 150 m. s ChI (San Clemente Island). Mar–Jul ★

subsp. ***philbrickii*** W.S. Davis PHILBRICK'S MALACOTHRIX **ST:** 1–10, erect when single, or decumbent to ascending, branched at or above base. **LF:** distal oblanceolate to lanceolate, gen lobed from base to near tip. **INFL:** involucre 6–9 (11) mm; outer phyllaries in 2–3 series. **FL:** corolla 8–16 mm, medium yellow; ligules

of outermost fls exserted 4–8 mm. **FR:** 1.3–1.7 mm, 5-angled, 5 ribs prominent. Windward side of island on gravelly ridges, rocky canyon slopes, bluffs, compacted soils; < 80 m. s ChI (Santa Barbara Island). Mar–Jul ★

subsp. ***polycephala*** W.S. Davis MANY-HEADED MALACOTHRIX **ST:** gen 1, erect, branched distally (or ± throughout). **LF:** distal narrowly triangular, gen with 1–2 pairs of narrow lobes at base. **INFL:** involucre 5–7 mm; outer phyllaries in 2–3 series. **FL:** corolla 5–9 mm; ligules of outermost fls exserted 1.5–3.5 mm. **FR:** 0.9–1.5 mm, ribs ± equal. Canyons, bluffs, open sandy loam on flats near but not on dunes, near ocean; 2–200 m. s ChI (San Nicolas Island). Mar–Jul ★

M. glabrata (D.C. Eaton) A. Gray (p. 385) DESERT DANDELION Ann ≤ 50+ cm. **ST:** (1)3–10+, gen branched at and above base, erect to ascending, glabrous, ± glaucous. **LF:** basal not fleshy, gen oblong-oblanceolate, remotely deeply lobed with narrow, wide-spreading segments to 3(4) cm, glabrous; proximal oblanceolate to obovate, with 3–6 pairs of well-spaced teeth or long narrow lobes; distal reduced, ± lobed. **INFL:** heads 1–few in cyme-like clusters, gen ± long-peduncled; involucre 9–17 mm; outer phyllaries gen 1/2 inner, lanceolate, glabrous or short white-hairy; receptacle densely bristly. **FL:** corolla 15–23 mm, gen pale yellow (white), ligules of outermost fls gen exserted 9–13 mm. **FR:** 2–3.3 mm; outer pappus of irregular teeth, bristles 1–5. 2*n*=14. Coarse soils in open areas or among shrubs in desert areas, foothill woodland; < 2000 m. SnJV, SCoRO, WTR, SNE, D; to OR, ID, UT; Mex. Mar–Jun

M. incana (Nutt.) Torr. & A. Gray (p. 385) DUNEDELION Per ≤ 70 cm, gen mounded. **ST:** branched at and above base, glabrous to densely hairy, buried sts ± woody, much-branched. **LF:** basal ± fleshy, spoon-shaped, entire or with few wide lobes or teeth; cauline ± fleshy, obovate to narrowly spoon-shaped, entire or with 1–2(3) pairs of ± equal blunt lobes. **INFL:** heads 1–few in cyme-like clusters; involucre 10–14 mm; outer phyllaries < inner, gen ± ovate, glabrous; receptacle glabrous. **FL:** corolla 11–20 mm, medium yellow, ligules of outermost fls exserted 5–10 mm. **FR:** 1.5–2.2 mm, smooth; ribs ± equal; outer pappus 0. 2*n*=14. Dunes; < 300 m. CCo, SCo, n ChI, s ChI (San Nicolas Island). All yr ★

M. indecora Greene (p. 385) SANTA CRUZ ISLAND MALACOTHRIX Ann ≤ 12(45) cm. **ST:** gen 1, erect or ± prostrate, gen branched at and above base, lfy throughout, gen glabrous. **LF:** basal ± fleshy, oblong or oblanceolate, with wide, obtuse lobes; cauline gen fleshy, obovate to spoon-shaped, gen with 1–3+ pairs of distal obtuse lobes. **INFL:** heads few in cyme-like clusters; involucre 6–8 mm; outer phyllaries ± 1/2 inner, gen ovate, glabrous; receptacle glabrous. **FL:** corolla 4–8 mm, medium yellow, ligules of outermost fls exserted 1–4 mm. **FR:** 1.2–1.5 mm, ribs ± equal; outer pappus 0. 2*n*=14. Shallow soils of ocean bluffs, or open rocky areas; < 30 m. n ChI (exc Anacapa Island). Apr–Sep ★

M. junakii W.S. Davis JUNAK'S MALACOTHRIX Ann ≤ 30 cm. **ST:** ± branched at and above base, ascending to erect, glabrous, lfy to tip. **LF:** basal not fleshy, oblanceolate, distal 1/2 with 2–3 teeth or with 2–7 short, acute lobes; proximal oblanceolate, with 2–4 pairs of lobes; distal gen oblanceolate, not much reduced, proximally with 1–2 pairs of ± linear, acute-tipped lobes. **INFL:** heads gen many in open cyme- or panicle-like clusters; involucre 7–8.5 mm; outer phyllaries ± 1/2 inner, gen lanceolate, glabrous; receptacle glabrous. **FL:** corolla 7–11 mm, medium yellow, ligules of outermost fls exserted 3.5–5.5 mm. **FR:** 1.6–2 mm, ribs ± equal; outer pappus minute irregular teeth, bristles 1–2. 2*n*=28. Shallow soils of ocean bluffs, open rocky areas; < 30 m. n ChI (Middle Anacapa Island). May–Jun ★

M. phaeocarpa W.S. Davis (p. 385) DUSKY-FRUITED MALACOTHRIX Ann ≤ 44 cm. **ST:** 1–many, spreading to ascending or erect, gen branched ± throughout, glabrous, ± glaucous. **LF:** basal ± fleshy, ± wide-lobed, lobes obtuse; proximal cauline oblong or oblanceolate to obovate, ± fleshy, with 3–8 pairs of ± equal lobes, ± white hair-tufted at lobe bases; distal reduced, ± 2–4-toothed at base. **INFL:** heads few to many in cyme- or panicle-like clusters; involucre 5–8 mm; outer phyllaries gen 1/2 inner, gen ovate, glabrous; receptacle glabrous. **FL:** corolla 5–8 mm, white, ligules of outermost fls gen exserted 1–3 mm. **FR:** 1.2–2 mm, ribs ± equal or 5 prominent; outer

pappus a scarious, rounded-toothed crown, bristles 0. $2n=14$. Diatomaceous shale of slides, burns, openings in chaparral, Bishop-pine/Douglas fir forest; 100–1300 m. SnFrB, SCoR, WTR. Apr–Jun ★

M. saxatilis (Nutt.) Torr. & A. Gray   Per, subshrub, ≤ 90(200) cm. **ST:** from rhizome or caudex, gen erect, gen branched distally, gen lfy proximally, glabrous or sparsely hairy to tomentose. **LF:** gen all cauline (basal withering early), linear to obovate, entire to irregularly lobed. **INFL:** heads few to gen many in cyme- or panicle-like clusters; involucre 10–12 mm; outer phyllaries gen 1/2 inner, gen lanceolate, glabrous to proximally tomentose; receptacle glabrous. **FL:** corolla 13–20 mm, white, ligules of outermost fls exserted 8–14 mm. **FR:** 1.3–2.5 mm, minutely spiny; outer pappus of irregular teeth, bristles 0. $2n=18$. Vars. intergrade.

var. ***arachnoidea*** (E.A. McGregor) E.W. Williams (p. 385)   CARMEL VALLEY MALACOTHRIX   Pl from rhizome or caudex, lfy, tomentose. **LF:** lanceolate to elliptic, gen entire or proximal toothed, tip acute. **INFL:** ± dense. Rocky, open banks, shale outcrops, cliff faces, coastal scrub, chaparral; 25–900 m. CCo, SCoRO, WTR. May–Aug(Oct) ★

var. ***commutata*** (Torr. & A. Gray) Ferris (p. 385)   Pl from rhizome or caudex, lfy, glabrous to ± hairy. **LF:** lanceolate to elliptic, gen entire, tip acute. **INFL:** dense. Crumbling shale, canyons, chaparral, foothill woodland; 200–1600 m. s SNF, Teh, SCoR, SCo, WTR. May–Aug(Oct)

var. ***implicata*** (Eastw.) H.M. Hall (p. 385)   Pl from rhizome or caudex, lfy, glabrous to lightly hairy. **LF:** 2-pinnately divided, lobes linear, tips obtuse. **INFL:** open to ± dense. ± Clay flats or rocky, canyon slopes; < 330 m. n ChI, s ChI (San Nicolas Island). Apr–Aug

var. ***saxatilis*** (p. 385)   CLIFF MALACOTHRIX   Pl from rhizome, lfy, glabrous to lightly hairy. **LF:** linear to ovate, entire or obscurely toothed (proximal sometimes lobed), tip obtuse or acute. **INFL:** ± dense. Coastal bluffs, on flats or in crevices; < 200 m. n SCo, sw WTR (Santa Barbara Co.). Apr–Aug ★

var. ***tenuifolia*** (Nutt.) A. Gray (p. 385)   Pl gen from caudex, sparsely lfy near heads, gen glabrous. **LF:** narrow, gen toothed, lobes sharp, tip acute. **INFL:** open. Canyons, coastal scrub, chaparral, foothill woodland; < 2000 m. s SNF, SCoRO, SCo, s ChI (Santa Catalina Island; introduced on San Clemente, San Nicolas islands), TR, PR. May–Aug

M. similis W.S. Davis & P.H. Raven   MEXICAN MALACOTHRIX   Ann ≤ 30 cm. **ST:** 1–11, erect or ascending, branched mostly from base, gen glabrous. **LF:** basal mostly lance-linear, ± pinnately lobed, not fleshy; cauline reduced, ± entire. **INFL:** heads gen many in cyme- or panicle-like clusters; involucre gen 6–10 mm; outer phyllaries ± 1/2 inner, glabrous; receptacle glabrous. **FL:** corolla 4–9 mm, gen yellow, ligules of outer fls exserted 2–4 mm; 70–100% of pollen 4-pored. **FR:** 1.4–1.7 mm; outer pappus of needle-like teeth, bristles 1. $2n=28$. Presumed extinct in CA; beaches, dunes; < 40 m. SCo (Hueneme Beach in 1925), n ChI (Santa Cruz Island in 1888); Mex. Apr–May ★

M. sonchoides (Nutt.) Torr. & A. Gray (p. 385)   Ann ≤ 50 cm. **ST:** 1–5, ascending to erect, branched throughout, glabrous, ± glaucous. **LF:** basal ± fleshy, often spreading in flat rosette, ± widely lobed, gen with minute, soft, spine-like teeth on lobes and midrib; cauline ± fleshy, narrowly oblong to elliptic, with 3–8 pairs of ± equal, oblong to triangular lobes; distal reduced, gen widest at base, ± clasping. **INFL:** heads few to many in open, cyme- or panicle-like clusters; involucre 7–13 mm; outer phyllaries ± 1/2 inner, ovate, margins ± with tack-shaped hairs; receptacle densely bristly. **FL:** corolla 10–14(16) mm, bright yellow, ligules of outermost fls exserted 6–10 mm. **FR:** 1.8–3 mm; outer pappus a scarious, rounded-toothed crown, bristles 0. $2n=14$. On dunes or in deep, fine sandy soils in arroyos, plains in Joshua-tree woodland, grassland; < 1400 m. SNE, DMoj; to WY, NM. Apr–Jun

M. squalida Greene (p. 385)   ISLAND MALACOTHRIX   Ann ≤ 30 cm. **ST:** stout, ascending to erect, branched at and above base, ± lfy, glabrous, ± glaucous. **LF:** basal not fleshy, widely lanceolate to ovate, sharply lobed, lobes long-acuminate, short-toothed; proximal ovate, 4–8 sharp-toothed-lobed; distal not notably reduced, narrowly ovate with 5–10 narrow, sharp teeth or lobes. **INFL:** heads gen few in open cyme- or panicle-like clusters; involucre 9–12 mm; outer phyllaries ± 2/3 inner, widely ovate, glabrous, mid-stripe ± red; receptacle gen not bristly. **FL:** corolla 12–19 mm, light yellow, ligules of outermost fls exserted 6–11 mm. **FR:** 1.3–2.1 mm; prominent ribs ± extending beyond tip; outer pappus minute, irregular teeth, bristles 0(1). $2n=28$. Shallow soils, open areas between shrubs, on ridges; < 200 m. n ChI (Anacapa, Santa Cruz islands). Mar–Jun ★

M. stebbinsii W.S. Davis & P.H. Raven (p. 385)   Ann ≤ 60 cm. **ST:** 1–5+, ± erect, branched distally, gen glabrous. **LF:** basal not fleshy, obovate, with 3–5 pairs of lobes; cauline reduced distally on st, proximally ± 2–4-lobed. **INFL:** heads gen few in open cyme- or panicle-like clusters; involucre 6–10 mm; outer phyllaries ± 1/2 inner, lanceolate to ovate, glabrous; receptacle glabrous. **FL:** corolla gen 6–8 mm, white or pale yellow, ligules of outermost fls exserted 1–2 mm. **FR:** 1.7–2.3 mm, ribs ± equal, 0 near tip; outer pappus of narrow teeth, bristles 1(2). $2n=28$. Gravelly soils on flats, gullies, streambeds, streambanks; 300–1300 m. SNE, D; NV, AZ. Mar–May

M. torreyi A. Gray (p. 385)   Ann < 25(40) cm. **ST:** 1–5, erect to ascending, branched at and above base, sparsely lfy, ± with tack-shaped hairs, ± glaucous. **LF:** basal ± fleshy, lobes 2–6 pairs, short-toothed, acute or obtuse; cauline ± fleshy, obovate to oblong, gen with 3–8 pairs of oblong or triangular to linear lobes, distal reduced, ± elliptic and lobed, to linear, bases strap-like; ± tack-shaped hairs on margins. **INFL:** heads gen few in open cyme-like clusters; involucre 8–14 mm; outer phyllaries ± 1/2 inner, gen lanceolate, ± tack-shaped hairs on mid-portions; receptacle glabrous or sparsely bristly. **FL:** corolla 14–20 mm, medium yellow, ligules of outermost fls exserted 7–10 mm. **FR:** ± 2.5–4 mm; prominent ribs gen wing-like, extending beyond tip, outer pappus of irregular teeth, bristles (0)1–6. $2n=14$. Coarse soils, dry sagebrush slopes, juniper grassland, desert scrub; 1500–1800 m. GB, n DMoj; to OR, ID, WY, CO, AZ. Apr–Jul

MALPERIA

David J. Keil & A. Michael Powell

1 sp. (Anagram of name of type collector, E. Palmer, 1831–1911) [Nesom 2006 FNANM 21:509–510]

M. tenuis S. Watson (p. 391)   BROWN TURBANS   Ann ≤ 40 cm, glabrous to puberulent. **ST:** erect, much-branched. **LF:** proximally opposite, distally alternate, gen sessile, ≤ 5 cm, linear, gen entire. **INFL:** heads discoid, many, 8–14 mm, in open, cyme-like clusters; involucre cylindric to narrowly bell-shaped, 5–6 mm wide; phyllaries 20–30+, graduated in 3–5 series, lanceolate; receptacle flat, epaleate. **FL:** 20–30; corolla 5–6 mm, cylindric, ± white, pink-tinged; anther base rounded, tip oblong; style branches long, club-shaped, rounded. **FR:** 3 mm, slender, 5-ribbed; pappus of 3 bristles, 5 mm, and 3 scales, ± 0.5 mm. $2n=20$. Sandy creosote-bush scrub; < 500 m. SCo (extirpated), s DSon; n Mex. Dec–Apr ★

MATRICARIA

Ann, sometimes aromatic. **ST:** branched, erect or decumbent. **LF:** alternate, irregularly 2–3-pinnately lobed; segments linear; petiole short or 0. **INFL:** heads radiate or discoid, 1–3; receptacle conic, epaleate; phyllaries in 2–3 unequal series, margins scarious. **RAY FL:** 0 or 10–25; ray white. **DISK FL:** many; corolla yellow, tubular, narrowed distally; anthers minute, tip

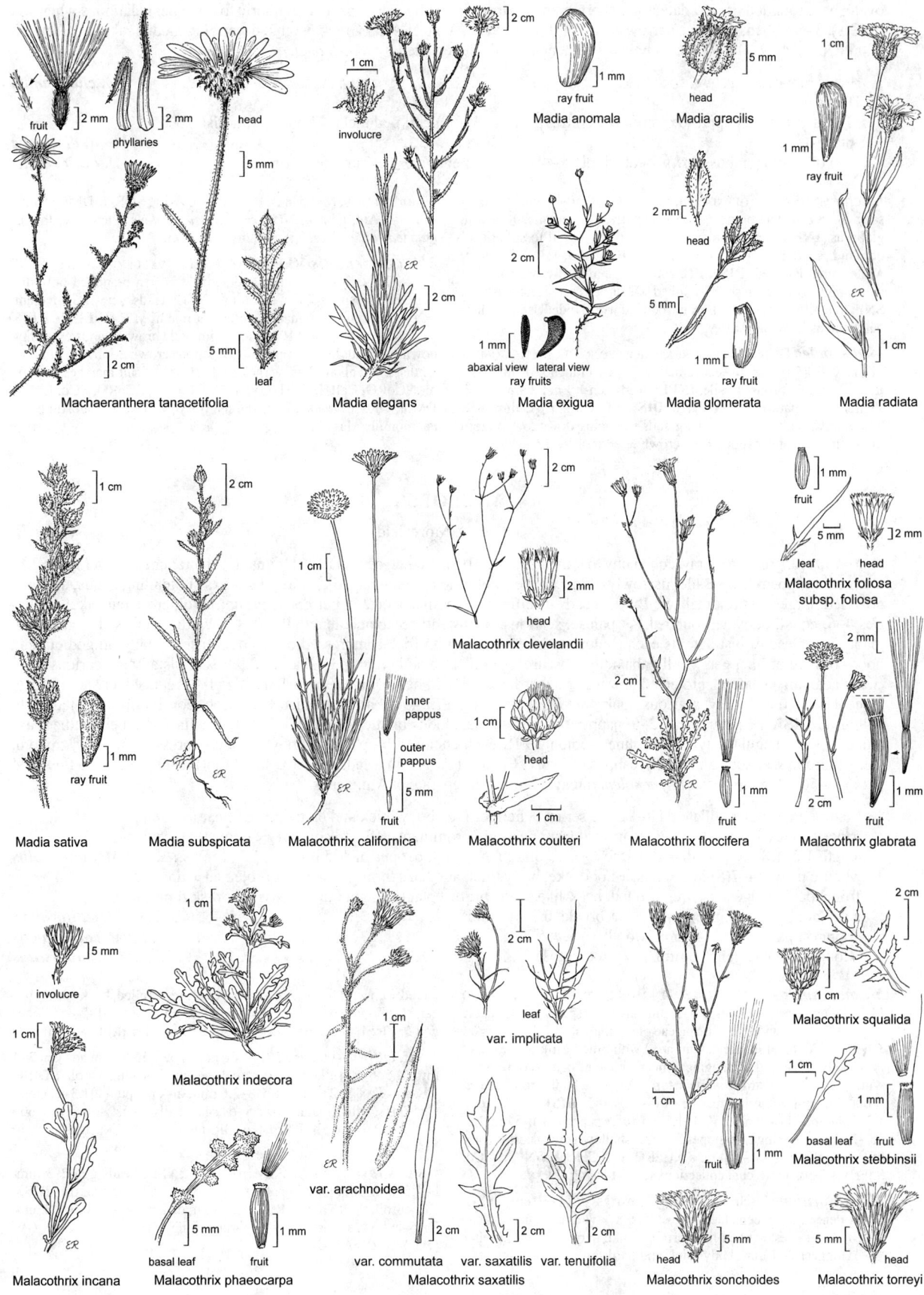

fruit

phyllaries

head

leaf

Machaeranthera tanacetifolia

involucre

ER

Madia elegans

ray fruit

Madia anomala

abaxial view   lateral view
ray fruits

Madia exigua

head

Madia gracilis

head

ray fruit

Madia glomerata

ray fruit

ER

Madia radiata

ray fruit

Madia sativa

ER

Madia subspicata

inner pappus

outer pappus

fruit

ER

Malacothrix californica

head

head

Malacothrix clevelandii

Malacothrix coulteri

fruit

ER

Malacothrix floccifera

fruit

leaf   head

**Malacothrix foliosa
subsp. foliosa**

fruit

fruit

Malacothrix glabrata

involucre

Malacothrix incana

ER

Malacothrix indecora

basal leaf   fruit

Malacothrix phaeocarpa

ER

var. arachnoidea

leaf

var. implicata

var. commutata   var. saxatilis   var. tenuifolia

Malacothrix saxatilis

fruit

head

Malacothrix sonchoides

Malacothrix squalida

basal leaf   fruit

Malacothrix stebbinsii

head

Malacothrix torreyi

ovate, base rounded or ± cordate; style short, branches truncate with bushy tips. **FR**: cylindric to obconic, gelatinous when wet (in CA), 3–5-ribbed; pappus a narrow crown or 0. 7 spp.: Eur, Asia, N.Am. (Greek: womb, describing medicinal use) [Brouillet 2006 FNANM 19:540–542] *Chamomilla* nomenclaturally superseded by *Matricaria*.

1. Heads radiate . . . . *M. chamomilla*
1′ Heads discoid
  2. Tip of fr entire, pappus 0 or crown ± entire; pl (1)10–30(50) cm; peduncle ≤ 1.5(2.5) cm; disturbed
     places. . . . . *M. discoidea*
  2′ Tip of fr and pappus crown 2-lobed; pl 15–45(70) cm; peduncle ≤ 6.5 cm; undisturbed wet places. . . . *M. occidentalis*

M. chamomilla L. GERMAN CHAMOMILE Pl 10–60 cm; strongly scented. **ST**: often branched only distally. **LF**: sessile, ≤ 8 cm, glabrous. **INFL**: heads radiate, gen 5–9 diam, ovoid to spheric or obovoid, shattering at maturity. **RAY FL**: 10–25, ray 7–8.5 mm, white, soon deflexed. **DISK FL**: corolla 1–2 mm. **FR**: 5-ribbed, tip and pappus 0 or crown entire or lobed. Disturbed roadsides; ± 100 m. SNF; to e N.Am; native to Eur; introduced worldwide. Used medicinally and as herb tea. Apr–May

M. discoidea DC.(p.391) PINEAPPLE WEED, RAYLESS CHAMOMILE Pl (1)10–30(50) cm; sweet-scented. **ST**: gen branched from base. **LF**: ≤ 5 cm, glabrous, sessile. **INFL**: heads gen ± 1 cm diam, conic, shattering at maturity. **RAY FL**: 0. **DISK FL**: corolla 1–2 mm. **FR**: 3–5-veined, with narrow brown glands extending down to ± bottom of fr; tip truncate; pappus 0 or crown ± entire. 2*n*=18. Abundant.

Disturbed sites, riverbanks; < 2250 m. CA-FP, SNE, w DMoj; native to nw N.Am, ne Asia. [*Chamomilla suaveolens* (Pursh) Rydb.; *M. matricarioides* (Less.) Porter, misappl.] Feb–Aug

M. occidentalis Greene (p. 391) VALLEY MAYWEED Pl 15–45(70) cm; not strongly scented. **ST**: often branched only distally. **LF**: sessile, ≤ 7 cm, glabrous. **INFL**: heads gen ≤ 1.5 cm diam, ± conic to spheric, remaining intact at maturity. **RAY FL**: 0. **DISK FL**: corolla 1–2 mm. **FR**: angled, with wide brown glands extending down to ± middle of fr; tip 2-lobed; pappus crown 2-lobed. Common. Undisturbed alkali flats, vernal pools, edges of salt marshes; < 2400 m. NCoRO, CaRH, SNH, GV, n CCo, SnFrB, SCoRO, SCo, ChI, w DMoj; OR. [*Chamomilla o.* (Greene) Rydb.] Used as substitute for chamomile. Mar–Aug

MICROPUS

James D. Morefield

Ann 1–50 cm, green to ± gray, cobwebby to tomentose. **ST**: 1(5), ± erect, central axis dominant, gen branched, ± evenly lfy. **LF**: alternate [opposite], ± sessile, narrowly oblanceolate to elliptic [spoon-shaped], entire; distal lvs subtending heads, scarcely crowded, largest > proximal lvs. **INFL**: heads disciform, ± sessile, 1 or 2–5 per dense group; involucre ± cup-like; phyllaries 4–6, equal, obovate, rounded, << paleae, scarious, persistent; receptacle length 0.5–1.8 × width, depressed or obovoid, glabrous; paleae of pistillate fls each enclosing fl, deciduous with fr, ± compressed-obovoid, obtuse, distally bulged out and hood- or helmet-like, gen woolly abaxially, obscurely parallel-veined (veins 5+), margin ± reflexed distally as scarious wing (wing adaxially ± lateral, projected upward or inward, ± visible in head); paleae of disk fls 0, or 1–3, reduced, erect, ± folded, lanceolate, acute, papery, scarious, glabrous. **PISTILLATE FL**: 4–12, all subtended by paleae; corolla obscure, narrowly cylindric. **DISK FL**: staminate, 2–5; pappus 0 or of 1–5 deciduous bristles (hidden in head); corolla 4–5 lobed; anther base tailed, tip ± triangular; style tips ± linear-oblong. **FR**: each enclosed by palea, ± obovoid, ± compressed laterally, smooth, shiny (corolla scar adaxial, lateral), pappus 0. 5 spp.: ± coastal sw N.Am, Medit. (Greek: small foot) [Morefield 2006 FNANM 19:454–456] Our spp. (sect. *Rhyncholepis*) may comprise a separate N.Am. genus.

1. Mature paleae of pistillate fls 8–12 in 2 series, ± helmet-like, body thick and cartilaginous near midvein,
    papery otherwise, wing projected inward from near top, prominent, ± flat to concave, ± obovate; receptacle
    length 1.2–1.8 × width; disk fl paleae 1–3, corolla lobes gen 4, pappus of 1–5 bristles, 1.7–2 mm . . . . *M. amphibolus*
1′ Mature paleae 4–7(8) in 1 series, ± hood-like, body thick and hard throughout, wing projected ± upward
    from adaxial edge, obscure, ± rolled, beak-like, ± linear; receptacle length 0.5–0.8 × width; disk fl paleae 0,
    corolla lobes gen 5, pappus 0 or of 1 bristle, 0.9–1.5 mm. . . . . *M. californicus*
  2. Longest paleae gen 3–4 mm, woolly . . . . var. *californicus*
  2′ Longest paleae gen 2–3 mm, silky-tomentose. . . . . var. *subvestitus*

M. amphibolus A. Gray (p. 391) MOUNT DIABLO COTTONSEED Pl 2–20 cm. **INFL**: heads 3.5–5 mm diam, ± spheric; receptacle length 1.2–1.8 × width; mature paleae of pistillate fls 8–12 in 2 series, ± helmet-like (profile like human head with billed cap), longest 2–3 mm, body thick and cartilaginous near midvein, papery otherwise, wing projected inward from near top, prominent, ± flat to concave, ± obovate, not withering; paleae of disk fls 1–3. **DISK FL**: corolla 1.2–1.9 mm, lobes gen 4. **FR**: 1–1.5 mm; disk pappus of 1–5 bristles, 1.7–2 mm. Openings on slopes, ridges, shallow soils, sedimentary or volcanic rocks; 40–900 m. s NCoRO, w NCoRI, n SNF, n CCo, SnFrB, s SCoRO. Recent collections few. Mar–Jun ★

M. californicus Fisch. & C.A. Mey. Q-TIPS Pl 1–50 cm. **INFL**: heads depressed; receptacle length 0.5–0.8 × width; mature paleae 4–7(8) in 1 series, ± hood-like (profile ± like harp or inverted letter "Q"), longest 2–4 mm, body thick and hard throughout, wing pro-

jected ± upward from adaxial edge, obscure, ± rolled, beak-like, ± linear, withering; paleae of disk fls 0. **DISK FL**: corolla 1–2 mm, lobes gen 5. **FR**: 1.4–2.6 mm; disk pappus 0 or of 1 bristle, 0.9–1.5 mm.

var. *californicus* (p. 391) COTTONTOP **INFL**: heads gen 3–4 mm, 4.5–6 mm diam; longest paleae gen 3–4 mm, woolly. **DISK FL**: corolla 1.3–2 mm. **FR**: 1.8–2.6 mm; disk pappus (0)1.3–1.5 mm. Clearings, often disturbed, dry or seasonally moist soils; 10–1600 m. CA-FP (exc e SNH, SnJt); w OR, nw Baja CA. Expected in SnJt. Mar–Jun

var. *subvestitus* A. Gray (p. 391) **INFL**: heads gen 2–3 mm, 3–4.5 mm diam; longest paleae gen 2–3 mm, silky-tomentose. **DISK FL**: corolla 1–1.5 mm. **FR**: 1.4–2.2 mm; disk pappus (0)0.9–1.3 mm. Dry, shallow, exposed soils (incl serpentine); 50–1100 m. c SNF, CW (exc SCoRI). Apr–Jun

MICROSERIS
Kenton L. Chambers

Ann, per, gen taprooted; mealy, hairs drying as minute white scales; sap milky. **LF**: basal and cauline, gen linear to lanceolate or oblanceolate, gen variably entire to pinnately lobed. **INFL**: heads 1 on long peduncle, liguliflorous, ± nodding in bud; involucre gen fusiform to spheric; phyllaries in 2–several series, outer overlapping, inner often ± black-hairy; receptacle flat to low-convex, epaleate, pitted. **FL**: 5–many; ligules ± equaling involucre in ann spp. to much exceeding involucre in per spp., white, yellow, or orange, abaxially often ± red or purple, readily withering. **FR**: cylindric to fusiform, gen square-topped, not beaked; ribs ± 10, ± scabrous (outer fr hairy in some); pappus of gen 5–many ± lanceolate, bristle-tipped scales. (Greek: small chicory) 14 spp.: w N.Am, S.Am, New Zealand, Australia. [Chambers 2006 FNANM 19:338–346] Hybridization common. Self-pollinating (ann) or self-sterile and ± complex (per).

1. Per; st simple or branched, scapose or ± lfy; ligules much exceeding involucre; pappus parts often > 5 (subg. *Scorzonella*)
  2. Pl from rhizome-like caudex; pappus bristles 24–48, ± brown, barbed, basal scales 0; lvs all basal; sphagnum bogs, NCoRO . . . . **M. borealis**
  2′ Pl taprooted; pappus bristles 5–30, each expanded at base into a small to large scale; lvs gen basal to cauline; various habitats
    3. Pappus scales 15–30, silvery; bristles plumose; widespread . . . . **M. nutans**
    3′ Pappus scales 5–24, often dull white or ± brown, bristles ± smooth or barbed to ± plumose
      4. Pappus scales 4–10 mm; bristles barbed to ± plumose
        5. Pappus scales 5–10, ± dull yellow-brown; st branched or simple, proximally lfy to mid-st; outer phyllary tips recurved; gen GV, adjacent foothills . . . . **M. sylvatica**
        5′ Pappus scales 9–19, silvery to dull white; st simple, lfy proximally; phyllary tips erect; n KR and adjacent OR, rare . . . . **M. laciniata** subsp. **detlingii**
      4′ Pappus scales 0.5–4 mm, bristles ± smooth to barbed
        6. Outer phyllaries often purple-spotted, lance-ovate to broadly ovate, 2.5–9 mm wide, gen glabrous . . . . **M. laciniata** subsp. **laciniata**
        6′ Outer phyllaries rarely purple-spotted, linear to ± deltate or lanceolate, 0.5–2.5 mm wide, often mealy, sometimes ± black-hairy
          7. Pappus scales 8–24, bristles white, barbed — KR and adjacent OR . . . . **M. laciniata** subsp. **siskiyouensis**
          7′ Pappus scales 5–10, bristles white, smooth or bristles ± brown, barbed
            8. Pappus scales 0.5–2.5 mm, bristles white, smooth; NW . . . . **M. laciniata** subsp. **leptosepala**
            8′ Pappus scales 2–4 mm, bristles ± brown, barbed; CCo, SnFrB . . . . **M. paludosa**
1′ Ann; st scapose; ligules ± equaling involucre; pappus parts < 6 (subg. *Microseris*)
  9. Fr gen < 3 mm
    10. Fr widest near middle, tapered to base and (slightly) to tip; pappus scales 1–4 mm; NCo, CCo, SnFrB . . . . [2]**M. bigelovii**
    10′ Fr widest at tip, tapered to base; pappus scales 0.2–2.5 mm; widespread . . . . **M. elegans**
  9′ Fr gen > 3 mm
    11. Pappus scales ≤ 1 mm . . . . **M. douglasii** subsp. **tenella**
    11′ Pappus scales averaging 1.5 mm or longer
      12. Pappus scales 3.5–11 mm, lance-linear, barely inrolled, midrib > 1/5 scale width . . . . **M. acuminata**
      12′ Pappus scales ≤ 7 mm (if lance-linear midrib < 1/5 scale width)
        13. Pappus scales curved only at base, midrib linear, thicker at base; NCo, CCo, SnFrB . . . . [2]**M. bigelovii**
        13′ Pappus scales becoming ± curved throughout, midrib evenly tapered from thick base
          14. Pappus scales 5, barely or not pigmented, becoming slightly inrolled; SNF, GV, e CW, SCo (rare) . . . . **M. campestris**
          14′ Pappus scales 5 or fewer, darkly to not pigmented, becoming strongly inrolled . . . . **M. douglasii**
            15. Pappus scales gen < fr (if > fr, then fr > 4.5 mm); widespread . . . . subsp. **douglasii**
            15′ Pappus scales ≥ fr; fr ≤ 4.5 mm; SCo, PR, s ChI . . . . subsp. **platycarpha**

M. acuminata Greene (p. 391) Ann 5–35 cm, scapose. **LF**: 3–20 cm, gen linear-lobed. **INFL**: involucre 10–22 mm; outer phyllaries deltate, << inner. **FL**: 5–50; ligule yellow. **FR**: body 4.5–7 mm, flared at tip, evenly brown, glabrous; pappus scales 5, 3.5–11 mm, barely inrolled, white to ± brown, hairy or not, midrib > 1/5 scale width, bristle-tips 4–7 mm, barbed. 2*n*=36. Grassy, open rocky or clay soil; < 760 m. NCoR, CaRF, n&c SNF, ScV, n SnJV, e SnFrB, SCoRO (rare); s OR. Apr–Jun

M. bigelovii (A. Gray) Sch. Bip. (p. 391) Ann 6–60 cm, scapose. **LF**: 2–25 cm, gen lobed; tip often wide, blunt. **INFL**: involucre 5–14 mm; outer phyllaries deltate, << inner. **FL**: 5–100; ligule yellow or orange. **FR**: body 2.5–5.5 mm, widest near middle, ± brown; outermost frs gen hairy; pappus silvery to ± black, scales 5, 1–4 mm, curved at base, glabrous, midrib linear, abruptly wider at base, bristles

3–8 mm, slightly barbed. 2*n*=18. Open sandy soil, or soil pockets on rocky coastal headlands; < 525 m. NCo, CCo, SnFrB; to BC. Apr–Jul

M. borealis (Bong.) Sch. Bip. (p. 391) NORTHERN MICROSERIS Per 15–70 cm, scapose, from rhizome-like caudex; roots several, fleshy. **LF**: basal, 6–30 cm, entire to few-toothed. **INFL**: involucre 10–18 mm; outer phyllaries narrow, << inner. **FL**: 18–50; ligule yellow. **FR**: body 4–8 mm, ± brown; ribs smooth; tip not flared; pappus scales 0; bristles 24–48, 5–10 mm, ± brown, barbed. 2*n*=18. Wet meadows, sphagnum bogs; 1000 m. NCoRO (Bald Mtn, Humboldt Co.); to AK. Jul–Sep ★

M. campestris Greene (p. 391) Ann 5–50 cm, scapose. **LF**: 3–20 cm, gen lobed. **INFL**: involucre 5–12 mm; outer phyllaries deltate, << inner. **FL**: 5–100+; ligule yellow or white. **FR**: 3–5+ mm, widest at tip, gray or pale brown, often dark-spotted; outermost gen gla-

brous; pappus scales 5, 1–4.5 mm, becoming curved throughout, barely inrolled, not or barely pigmented, glabrous, midrib tapered from base, bristles 3–5.5 mm, barbed. 2*n*=36. Open clay grassland, often near vernal pools; < 500 m. SNF, GV, e CW, SCo (rare). Derivative of *M. douglasii* × *M. elegans.* Apr–Jun

M. douglasii (DC.) Sch. Bip.   Ann 5–60 cm, scapose. **LF**: 3–25 cm. **INFL**: involucre 7–16 mm; outer phyllaries deltate, << inner. **FL**: 5–100+; ligule yellow or white. **FR**: gray to brown or ± black, often dark-spotted; outermost frs gen hairy; pappus scales ≤ 5, silvery to ± black, often hairy, bristles 1–8 mm, ± strongly barbed. 2*n*=18. Highly variable; subspp. intergrade.

   subsp. ***douglasii*** (p. 391)   **FR**: 4–10 mm, widest at tip; pappus scales 1–6.5 mm, becoming curved throughout and strongly inrolled, bristles 3–8 mm. Inland clay soils, grassland, often near vernal pools or serpentine outcrops (coastal in s CCo); < 1100 m. NCoR, CaRF, SNF, Teh, GV, CW, SCo (rare); s OR. Mar–Jun

   subsp. ***platycarpha*** (A. Gray) K.L. Chambers (p. 391)   SMALL-FLOWERED MICROSERIS   **FR**: 3–4.5 mm, widest at tip; pappus scales 2.5–7 mm (gen > fr), becoming curved and ± inrolled, bristles 1–4 mm. Clay soils, grassland, often near vernal pools or serpentine outcrops; < 1100 m. c&s SCo, s ChI, PR; Baja CA. Mar–May   ★

   subsp. ***tenella*** (A. Gray) K.L. Chambers (p. 391)   **FR**: 3–6.5 mm, gen widest near middle; pappus scales ≤ 1 mm, bristles 3–8.5 mm. Inland clay soils, grassland, often near vernal pools or serpentine outcrops (coastal in s CCo); < 500 m. NCoRH, NCoRI, ScV, w-c SnJV, CW, n SCo, n ChI. Hybridizes with *M. bigelovii*; often confused with *M. elegans.* Mar–Jun

M. elegans A. Gray (p. 391)   Ann 5–35 cm, scapose. **LF**: 2–20 cm, gen lobed. **INFL**: involucre 4–8(10) mm; outer phyllaries deltate, << inner. **FL**: 5–100; ligule yellow or orange. **FR**: 1.5–3 mm, widest at tip, evenly brown to ± black; outermost sometimes hairy; pappus scales 5, 0.2–2.5 mm, ± flat, white, glabrous, midrib dark, bristles 3–5 mm, barely barbed. 2*n*=18. Gen inland clay grassland, often near vernal pools (coastal in SW); < 700 m. NCoR, CaRF, SNF, GV, SnFrB, SCoR, SCo, ChI; Baja CA. Apr–Jun

M. laciniata (Hook.) Sch. Bip.   Per 15–120 cm, gen ± branched, proximally lfy. **LF**: 10–50 cm, entire to lobed. **INFL**: involucre 10–30 mm; outer phyllaries linear to ovate, < to << inner. **FL**: 13–100+; ligule yellow. **FR**: 3.5–8 mm, not or barely wider at tip, gray to brown, smooth or outermost scabrous on ribs; pappus scales 5–24, < 10 mm, silvery to white, bristles 4–12 mm, ± smooth to barbed. 2*n*=18.

   subsp. ***detlingii*** K.L. Chambers   DETLING'S SILVERPUFFS   Taproot slender, much elongated. **ST**: simple. **INFL**: outer phyllaries broadly lanceolate to elliptic, thin, glabrous, keel 0. **FL**: 18–85+. **FR**: pappus scales 9–19, 4–9 mm, bristles barbed. Clay slopes, grassland,

scrub, forest edges; 600–1000 m. ne KR (Iron Gate Reservoir, Siskiyou Co.); s OR. May–Jun   ★

   subsp. ***laciniata*** (p. 391)   Taproot thick, ± elongated. **ST**: branched. **INFL**: outer phyllaries lance-ovate to broadly ovate, thin, gen glabrous, keel 0. **FL**: 25–100+. **FR**: pappus scales 5–10, < 4 mm, bristles smooth proximally. Open grassland, meadows, rocky slopes, forest edge; < 2000 m. NW, CaRH, MP; to sw WA. Intergrades with *M. laciniata* subsp. *leptosepala* in NCoR. Apr–Aug

   subsp. ***leptosepala*** (Nutt.) K.L. Chambers   Taproot thick, ± elongated. **INFL**: outer phyllaries linear to lanceolate, mealy, ± black-hairy, keel fleshy. **FL**: 13–70. **FR**: pappus scales 5–10, < 2.5 mm; bristles smooth proximally. Open grassland, meadows, rocky slopes, forest edge; < 2000 m. NW; to sw WA. May–Aug

   subsp. ***siskiyouensis*** K.L. Chambers   Taproot thick, ± elongated. **INFL**: outer phyllaries deltate to lanceolate, mealy, keel fleshy. **FL**: 13–50. **FR**: pappus scales 8–24, < 2.5 mm, bristles barbed throughout. Open grassland, meadows, rocky slopes, forest edge; < 2000 m. KR; sw OR. [*M. howellii* A. Gray, misappl.] Intergrades with *M. laciniata* subsp. *leptosepala.* May–Jul

M. nutans (Hook.) Sch. Bip. (p. 391)   Per 10–70 cm, branched, proximally lfy. **LF**: 5–30 cm, entire to lobed. **INFL**: involucre 8–22 mm, gen ± mealy and black-hairy; outer phyllaries linear to deltate, << inner. **FL**: 13–75; ligule yellow. **FR**: 3.5–8 mm, not wider at tip, pale brown to ± red, smooth or outermost scabrous on ribs; pappus scales 15–30, 1–3 mm, silvery, bristles 4–7 mm, plumose. 2*n*=18. Common. Moist, rocky meadows, open conifer woodland, sagebrush scrub; 1000–3400 m. KR, NCoRH, CaRH, SNH, MP; to BC, SD, CO. Highly variable. Apr–Jul

M. paludosa (Greene) J.T. Howell (p. 391)   MARSH MICROSERIS   Per 15–70 cm, gen branched and lfy only near base. **LF**: 5–35 cm, entire to lobed. **INFL**: involucre 10–20 mm, gen mealy and ± black-hairy; outer phyllaries linear to ovate-deltate, tapered, not recurved, < inner. **FL**: 25–70; ligule yellow-orange. **FR**: 4–7 mm, not wider at tip, straw-colored or dull white, smooth or outermost scabrous on ribs; pappus scales 5–10, 2–4 mm, ± brown, bristles 6–9 mm, barbed. 2*n*=18. Moist grassland, open woodland; < 300 m. CCo, SnFrB. Like *M. laciniata* subsp. *leptosepala* exc pappus. Apr–Jun   ★

M. sylvatica (Benth.) Sch. Bip. (p. 391)   SYLVAN MICROSERIS   Per 15–75 cm, gen few-branched and lfy near base. **LF**: 10–35 cm, wavy-margined to lobed. **INFL**: involucre 12–25 mm, glabrous or mealy; outer phyllaries deltate to ovate, < inner, tips recurved. **FL**: 25–100; ligule yellow. **FR**: 5–12 mm, not wider at tip, straw-colored or dull white, smooth or outermost scabrous on ribs; pappus scales 5–10, 4–10 mm, yellow-brown, bristles 6–9 mm, ± plumose. 2*n*=18. Grassland, open woodland; < 1700 m. NCoRI, SNF, s SNH, Teh, ScV, e SnFrB, SCoR, WTR, w DMoj. Mar–Jun   ★

MONOLOPIA   HILLSIDE DAISY

Dale E. Johnson

Ann 5–80 cm, ± woolly. **ST**: erect to ascending, occ decumbent. **LF**: opposite proximally, alternate distally, linear to (ob)lanceolate or ovate, entire to wavy-dentate or shallowly lobed. **INFL**: heads 1 or in cyme-like clusters, radiate (± disciform); involucre hemispheric; phyllaries 4–11 in 1 series, free or fused into lobed cup, tips black-hairy; receptacle convex to conic, smooth or pitted, epaleate. **RAY FL**: 1 per phyllary; corolla gen yellow, with small lobe opposite ray. **DISK FL**: 20–100; corolla yellow; lobes 4–5, with glandless and/or glandular hairs; anther tip elliptic-oblong or ovate; style tips oblong, appendages ± deltate, papillate. **FR**: ± obconic, ray fr 3-angled, disk fr 2- or 4-angled; pappus 0 or 2–7 persistent scales. 5 spp.: CA. (Greek: single husk, from phyllaries) [Johnson 2006 FNANM 21:349–351]

1. Heads ± disciform; ray 0–0.5 mm; disk fr 2-angled; disk pappus of 2–7 scales . . . . *M. congdonii*
1′ Heads radiate; ray 2–20 mm; disk fr 4-angled; disk pappus 0
  2. Ray entire to slightly toothed, or middle lobe < outer
    3. Fr glabrous or sparsely hairy, ± 2 mm; branches ± spreading, gen near top of pl . . . . *M. gracilens*
    3′ Fr ± uniformly gray-strigose, 2.5–3 mm; branches ± erect . . . . *M. stricta*
  2′ Ray ± equally 3-lobed
    4. Fr ± uniformly gray-strigose; phyllaries free to ± 1/2 fused . . . . *M. lanceolata*
    4′ Fr ± glabrous or hairs concentrated at top; phyllaries fused into cup with triangular lobes . . . . *M. major*

M. congdonii (A. Gray) B.G. Baldwin (p. 391) SAN JOAQUIN WOOLLYTHREADS **ST**: decumbent to ascending, branched from base. **INFL**: heads ± disciform; peduncle 0.2–2 cm; involucre 4–5 mm; phyllaries 4–7, free, reflexed in fr. **RAY FL**: ray 0–0.5 mm, 3-lobed. **DISK FL**: corolla 1–1.5 mm, 4-lobed, bell-shaped; lobes glandular. **FR**: 2.5–3 mm; disk fr 2-angled, compressed front-to-back, fringed; disk pappus of 2–7 scales, 0.5–1 mm. $2n$=22. Grassland, sandy soils; 90–700 m. s SnJV. [*Lembertia c.* (A. Gray) Greene] Feb–May ★

M. gracilens A. Gray (p. 391) WOODLAND WOOLLYTHREADS **ST**: erect, branches mostly distal, ± spreading. **INFL**: heads radiate; peduncle 2–12 cm; involucre 5–7 mm; phyllaries 7–11, elliptic-oblanceolate, free. **RAY FL**: ray 5–10 mm, entire or slightly lobed. **DISK FL**: corolla 2.5–4 mm, 5-lobed, funnel-shaped; lobes with glandless hairs. **FR**: ± 2 mm, disk fr 4-angled, not compressed, ± glabrous. $2n$=26. Serpentine grassland, open chaparral, oak woodland; 100–1200 m. SnFrB, SCoR. Mar–Jul ★

M. lanceolata Nutt. (p. 391) COMMON HILLSIDE DAISY **ST**: erect, simple or branched ± throughout, branches ascending. **INFL**: heads radiate; peduncle 1–13 cm; involucre 6–10 mm; phyllaries ± 8, elliptic-oblanceolate, free or fused to ± 1/2. **RAY FL**: ray 10–20 mm, 3-lobed. **DISK FL**: corolla 3–4 mm, 5-lobed, funnel-shaped; lobes with glandular and glandless hairs. **FR**: 2–4 mm, disk fr 4-angled, compressed front-to-back, strigose. $2n$=20. Grassland, bare clay, open chaparral, woodland; 50–1600 m. s SNF, Teh, SnJV, SCoR, c SCo, WTR, SnGb, nw PR, w edge DMoj. Feb–Jun

M. major DC. (p. 391) **ST**: erect or ascending, simple or branched ± throughout. **INFL**: heads radiate; peduncle 1–13 cm; involucre 8–13 mm; phyllaries gen 8, fused into cup with triangular lobes. **RAY FL**: ray 8–20 mm, 3-lobed, sometimes cream. **DISK FL**: corolla 3.5–4 mm, 5-lobed, funnel-shaped; lobes with glandular and glandless hairs. **FR**: 2.5–4 mm, disk fr 4-angled, compressed front-to-back, ± glabrous or hairs concentrated at top. $2n$=24. Grassland, vertic clay, occ serpentine; 10–1100 m. NCoRI, sw ScV, n SnJV, SnFrB, n SCoR, n WTR. Feb–Jul

M. stricta Crum (p. 391) **ST**: gen erect, simple or branched ± throughout, branches ± erect. **INFL**: heads radiate; peduncle 3–5 cm; involucre 5–7 mm; phyllaries ± 8, free, oblanceolate. **RAY FL**: ray 2–7 mm (w SnJV) or 9–17 mm (s&e SnJV), entire or slightly lobed. **DISK FL**: corolla 3–4 mm, 5-lobed, funnel-shaped; lobes with glandular and glandless hairs. **FR**: 2.5–3 mm, disk fr 4-angled, not compressed, strigose. $2n$=26. Grassland, bare clay, open chaparral, woodland; 50–800 m. s SNF, Teh, w&s SnJV, SCoRI, n WTR. Feb–May

MONOPTILON   DESERT STAR

Ann, gen ± short-bristly. **ST**: ≤ ± 25 cm, gen several from base, tufted or gen prostrate to ascending. **LF**: alternate, often tufted below heads, linear or oblanceolate, entire. **INFL**: heads radiate, 1, ± sessile; involucre bell-shaped or hemispheric; phyllaries 10–14 in 1 series, equal, ± folded, gen acuminate, ± purple; receptacle flat or convex, pitted, epaleate. **RAY FL**: (7)12–21; ray white to purple, often dark-veined, esp abaxially. **DISK FL**: 28–40; corolla yellow; style tips short-triangular. **FR**: compressed, oblong to obovate, finely appressed-hairy, light brown; pappus of scales and gen 1 or more slender bristles. 2 spp.: sw N.Am. (Greek: 1 feather, from pappus of *Monoptilon bellidiforme*) [Nesom 2006 FNANM 20:210]

1. Pappus a minute crown of fused scales 0.1–0.2 mm and 1 plume-tipped bristle 3–4 mm . . . . ***M. bellidiforme***
1′ Pappus of (0)1–15 nonplumose bristles 1–2 mm and several narrow scales 0.5–1 mm ± dissected into bristles . . . . ***M. bellioides***

M. bellidiforme Torr. & A. Gray (p. 395) **ST**: ≤ ± 6 cm. **LF**: 4–10 mm, gen oblanceolate, obtuse. **INFL**: phyllaries 4–4.5 mm. **RAY FL**: corolla 5–7 mm; ray 3–5 mm. **DISK FL**: corolla 3–4.5 mm. **FR**: 1.5–2 mm. $2n$=16. Sandy flats, washes; 500–1450 m. DMoj; to sw UT, w AZ. Mar–Jun

M. bellioides (A. Gray) H.M. Hall (p. 395) **ST**: ≤ 25 cm. **LF**: 5–10 mm, linear to linear-oblanceolate, obtuse to ± acute. **INFL**: phyllaries 4–6 mm. **RAY FL**: corolla 6–11.5 mm; ray 5–8.5 mm. **FR**: 1.5–2 mm. $2n$=16. Sandy flats, washes; -60–1450 m. D; to sw UT, w AZ, nw Mex. Jan–May(Sep)

MUNZOTHAMNUS

L.D. Gottlieb

1 sp.: CA. (Greek: Munz's shrub, for P.A. Munz, CA botanist, 1892–1974) [Gottlieb 2006 FNANM 19:349–350]

M. blairii (Munz & I.M. Johnst.) P.H. Raven (p. 395) BLAIR'S MUNZOTHAMNUS Shrub 1–1.5 m; sap milky. **ST**: ± fleshy; branches to 1 cm diam, tomentose or becoming ± glabrous; with persistent petiole bases. **LF**: cauline, alternate, crowded or tufted at branch tips or not; petiole ± 1 cm, narrowly winged, base thickened; blade 4–6 cm, oblong-obovate, irregularly wavy-toothed or shallowly lobed, bright green, base tapered, tip rounded. **INFL**: heads liguliflorous, many in terminal panicle-like clusters; involucre 7–10 mm; phyllaries in 2 series, outer 7–10, unequal, < 1/2 × inner, glabrous to sparsely glandular-hairy, inner 8–10, lanceolate to linear, margins narrowly scarious; receptacle ± flat, pitted, epaleate. **FL**: 9–12; ligule white or rosy to ± purple. **FR**: 3–3.5 mm, cylindric, faces 5, each with 1–2 long, narrow, shallow grooves; pappus of 25–35 bristles, free, white, minutely barbed, deciduous. $2n$=16. Rocky canyon walls; ± 200 m. s ChI (San Clemente Island). [*Stephanomeria b.* Munz & I.M. Johnst.] Sep–Nov ★

NESTOTUS

Lowell E. Urbatsch

Per, subshrub, from woody caudex, forming mats < 5 m diam. **ST**: ± prostrate, stolon-like with short ± erect branches, often stalked-glandular. **LF**: alternate, gen crowded, persisting 2–3 yrs, linear to oblanceolate, gen rigid, tapered to indefinite petiole, ± clasping, entire. **INFL**: head radiate, 1; peduncle ≤ 5 cm, ± bractless; involucre hemispheric; phyllaries ± equal in 2–3 series, lf-like, linear to oblanceolate or ± ovate, acute to acuminate, faces often stalked-glandular; receptacle convex, pitted, epaleate. **RAY FL**: 5–8[11]; corolla yellow. **DISK FL**: 9–27; corolla funnel-shaped, yellow; anther tip triangular; style tips lance-triangular. **FR**: narrowly obconic, gen densely silky; pappus of soft, minutely barbed bristles. 2 spp.: w N.Am. (Anagram of *Stenotus*) [Urbatsch et al. 2006 FNANM 20:169–170]

N. stenophyllus (A. Gray) R.P. Roberts et al. (p. 395) **ST**: fl branches 2–12 cm. **LF**: ≤ 2 cm, 1-veined, ciliate, scabrous, faces densely stalked-glandular. **INFL**: involucre 5–9 mm. **RAY FL**: ray 7–12 mm. **DISK FL**: corolla 4.5–7.6 mm, tube hairy. **FR**: 3.5–6 mm; pappus bristles ± 30, ± white. Rocky soil, sagebrush scrub, juniper woodland; 900–2300 m. MP; to WA, ID, n NV. [*Stenotus s.* (A. Gray) Greene] Apr–May

NICOLLETIA

Bruce G. Baldwin, adapted from Strother (2006)

[Ann] per (5)10–29[50] cm, often glaucous, pungently scented. **ST**: erect to spreading, branched from base or throughout. **LF**: alternate exc 1–4 proximal pairs opposite, gen 1-pinnately lobed, lobes gen [3]5–11, bristle-tipped, each lobe with ± terminal, embedded oil gland. **INFL**: heads radiate, 1; bracts subtending heads 2–4[6], deltate to lanceolate, each with 1–5 oval to linear oil-glands; involucre 4–8(10) mm diam, bell-shaped to obconic, cylindric, or fusiform; phyllaries [6]8–12 in ± 2 series, free, linear to ovate, persistent, most with 1–5 oil-glands; receptacle convex to conic, ± pitted, epaleate. **RAY FL**: [7]8–12; corolla ± white to pink, ray sometimes with ± pink to ± purple stripes. **DISK FL**: [15]30–100+; corolla yellow, purple, or ± white, tube << cylindric throat, lobes deltate to lanceolate; anther tip ± lanceolate; style tips thread-like, minutely bristly. **FR**: club-shaped, sparsely puberulent, hairs gen ± red; pappus of 5(6) bundles of 7–15 minutely barbed bristles subtending and alternating with 5(6) lanceolate, 1-awned scales. 3 spp.: sw US, n Mex. (J.N. Nicollet, French astronomer, explorer, 1786–1843) [Strother 2006 FNANM 21:231–232]

N. occidentalis A. Gray (p. 395) HOLE-IN-THE-SAND PLANT Pl deeply taprooted. **LF**: 2–7 cm, 1° lf axis linear, lobes ± linear-cylindric to linear, 1° lf axis width mostly 2–3 × lobe width. **INFL**: peduncle gen 2–10 mm; bracts subtending head 4–8 mm; involucre 14–18 mm, obconic to cylindric. **RAY FL**: ray 4–9 mm, 2.5–4 mm wide, pale pink to bright pink-purple. **DISK FL**: (30)60–100+, corolla 8–9.5 mm, proximally yellow, distally purple to red. **FR**: 7–9 mm; pappus bristles 3–7 mm, scales 6–8 mm. 2*n*=20. Deep sandy soils, washes; 600–1525 m. e edge s SNH, DMoj; n Baja CA. Apr–Jun

NOTHOCALAIS FALSE-DANDELION

Kenton L. Chambers

Per from ± thick caudex; sap milky. **ST**: ± scapose. **LF**: basal, entire to pinnately lobed, ± glabrous or white-ciliate. **INFL**: heads liguliflorous, 1, erect; outer phyllaries gen ± = inner, glabrous or margins and midrib soft-white-hairy; receptacle flat, epaleate, pitted, glabrous. **FL**: 13–many; ligules much exceeding involucre, yellow, often abaxially ± red, readily withering. **FR**: cylindric, ± 10-ribbed, tapered distally, not beaked, brown or gray; pappus of many soft bristles or narrow, tapered, bristle-tipped scales. 4 spp.: N.Am. (Greek: false *Calais*, a synonym of *Microseris*) [Chambers 2006 FNANM 19: 335–337]

1. Outer phyllaries lanceolate to ovate, minutely and evenly purple-dotted; lvs linear to widely oblanceolate, entire to often lobed, gen glabrous; pappus of soft bristles . . . . ***N. alpestris***
1′ Outer phyllaries lance-linear, midrib gen purple-lined; lvs widely linear, entire to wavy-margined, often finely ciliate; pappus of narrow, bristle-tipped scales . . . . ***N. troximoides***

N. alpestris (A. Gray) K.L. Chambers (p. 395) ALPINE LAKE FALSE-DANDELION **ST**: 3–45 cm, glabrous below head. **LF**: 3–20 cm. **INFL**: involucre glabrous, 10–20 mm; outer phyllaries = inner but wider, faces glabrous. **FR**: 5–10 mm, ± narrowed in distal 1–2 mm; pappus 6–10 mm, of 30–50 soft, minutely barbed, white bristles. 2*n*=18. Uncommon. Rocky slopes, meadows, forest openings; 1300–3400 m. KR, n&c SNH; to WA. Often confused with *Agoseris monticola*. Jul–Sep

N. troximoides (A. Gray) Greene (p. 395) SAGEBRUSH FALSE-DANDELION **ST**: 5–40 cm, often tomentose below head. **LF**: 7–30 cm. **INFL**: involucre 14–30 mm; outer phyllaries ≤ inner, glabrous or soft-white-hairy, esp on margins and midrib, midrib gen purple-lined. **FR**: 7–13 mm, cylindric-fusiform; pappus 10–17 mm, of 10–30 narrow, tapered, bristle-tipped, silvery scales. 2*n*=18. Open, rocky places; 700–2000 m. KR, n SNH, MP; to s BC, MT. Apr–Jun

ONCOSIPHON

Ann, scented. **ST**: several from base, branched. **LF**: alternate, petioled or sessile, 2–3-pinnately dissected; segments linear, dotted with resin glands. **INFL**: heads discoid, 1–4 in cyme-like clusters; receptacle convex to conic, epaleate; phyllaries graduated in 3–4 series, margins scarious. **FL**: gen 100–250+; corolla yellow, tubular, 4-lobed, tube and proximal throat not externally differentiated, swollen; anthers minute, tip lanceolate, base rounded or ± cordate; style short, branches truncate, tips bushy. **FR**: cylindric or 4-angled and ribbed, faces resin-gland-dotted; pappus a narrow crown. 8 spp.: s Afr. (Greek: swollen tube, describing corolla) [Källersjö 1988 Bot J Linn Soc 96:299–322; Keil 2006 FNANM 19:539]

O. piluliferum (L. f.) Källersjö STINKNET Pls 15–45(70) cm, pungently scented. **ST**: branched distally or throughout. **LF**: 1–4+ cm, puberulent or minutely strigose. **INFL**: heads gen 5–8 mm diam, ± spheric. **FL**: corolla 1.5–2 mm. **FR**: 0.6–0.8 mm, tip and pappus crown 0.05–0.1 mm, entire or minutely toothed. Disturbed sites; coastal scrub, chaparral; < 500 m. SCo, PR, DSon; AZ; native to S.Afr; alien in Australia. [*Matricaria globifera* (Thunb.) Harv.] Mar–Jul

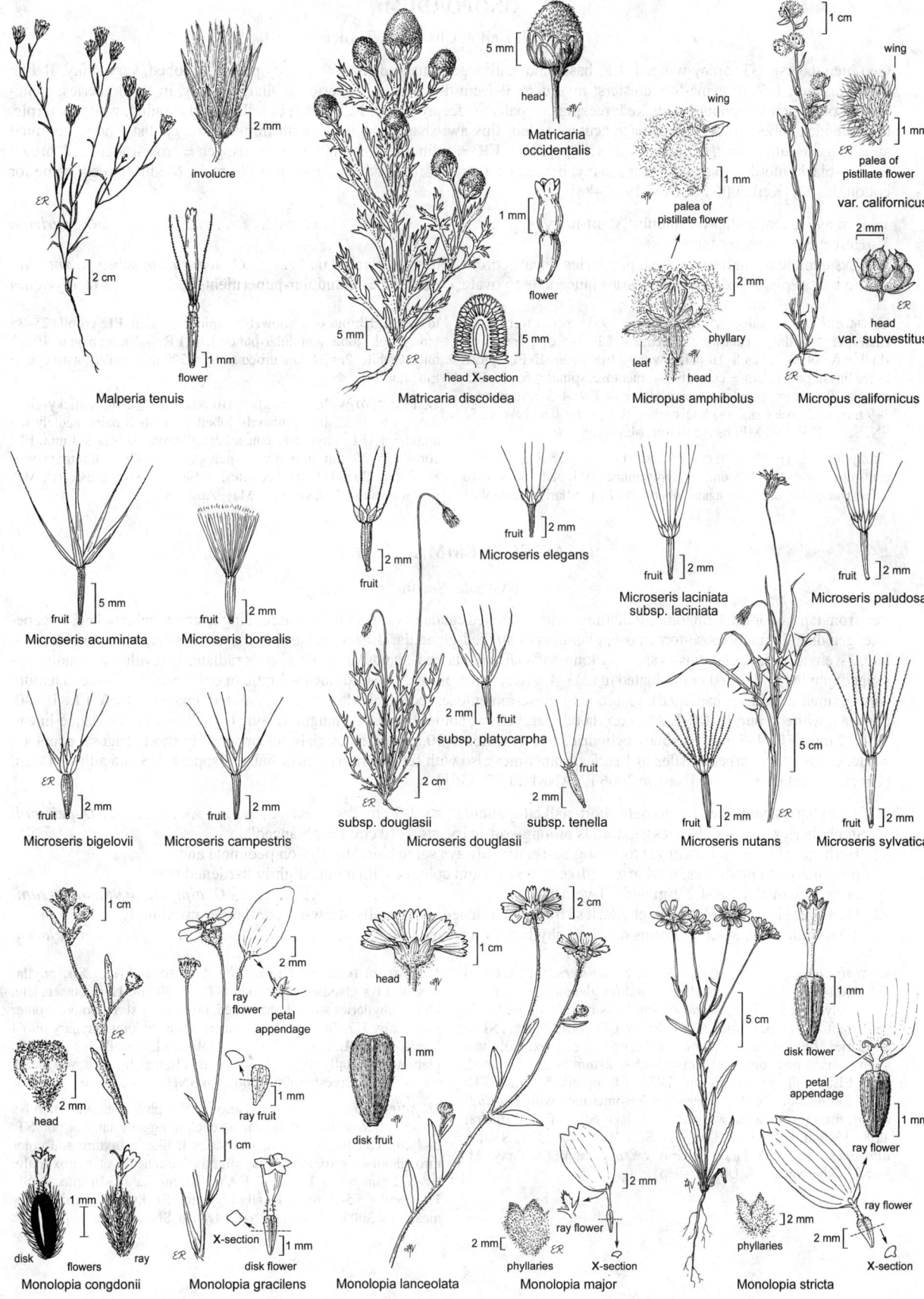

Malperia tenuis

involucre [2 mm

flower [1 mm

[2 cm

Matricaria occidentalis

head [5 mm

flower [1 mm

head X-section [5 mm

Matricaria discoidea

[5 mm

Micropus amphibolus

wing

palea of pistillate flower [1 mm

leaf phyllary head [2 mm

Micropus californicus

wing

[1 cm

palea of pistillate flower [1 mm

var. californicus

head [2 mm

var. subvestitus

Microseris acuminata

fruit [5 mm

Microseris borealis

fruit [2 mm

Microseris elegans

fruit [2 mm

Microseris bigelovii

fruit [2 mm

Microseris campestris

fruit [2 mm

subsp. platycarpha

fruit [2 mm

subsp. douglasii

[2 cm

Microseris douglasii

subsp. tenella

fruit [2 mm

Microseris laciniata subsp. laciniata

fruit [2 mm

Microseris paludosa

fruit [2 mm

Microseris nutans

fruit [2 mm

[5 cm

Microseris sylvatica

fruit [2 mm

Monolopia congdonii

[1 cm

head [2 mm

disk flowers ray [1 mm

Monolopia gracilens

ray flower petal appendage [2 mm

ray fruit [1 mm

[1 cm

disk flower X-section [1 mm

Monolopia lanceolata

head [1 cm

disk fruit [1 mm

Monolopia major

[2 cm

ray flower [2 mm

phyllaries [2 mm X-section

Monolopia stricta

[5 cm

disk flower [1 mm

petal appendage ray flower [1 mm

phyllaries [2 mm

ray flower X-section [2 mm

ONOPORDUM

David J. Keil & Charles E. Turner

Gen bien, coarse. **ST**: spiny-winged. **LF**: basal and cauline, alternate, dentate to deeply pinnately lobed, very spiny. **INFL**: heads discoid, 1–7, in cyme-like clusters; involucre 3–7 cm diam, (hemi)spheric; phyllaries many, in 8–10+ series, spine-tipped, outer gen spreading or reflexed; receptacle epaleate, deeply pitted. **FL**: many; corolla (nearly) radial, white to purple, tube slender, lobes linear; anther base acutely tailed, tips awl-shaped; style tip with minutely hairy distal node, terminal appendage ± entire, long, cylindric, minutely papillate. **FR**: ± cylindric, 4–5-angled, glabrous, gen ± cross-roughened, brown to gray-black, mottled; pappus bristles many, barbed or plumose, fused at base. ± 40 spp.: Eurasia, Medit. (Greek: name for cotton thistle) [Keil 2006 FNANM 19:87–88]

1. Herbage green, ± sticky-glandular, short-hairy . . . . ***O. tauricum***
1′ Herbage ± canescent-tomentose
  2. Lvs dentate to shallowly lobed; phyllaries linear; corolla lobes not glandular . . . . ***O. acanthium*** subsp. ***acanthium***
  2′ Lvs ± deeply 1–2 × lobed; phyllaries lanceolate to ovate; corolla lobes glandular-puberulent . . . . ***O. illyricum***

O. acanthium L. subsp. *acanthium* (p. 395) SCOTCH THISTLE Bien, per ≤ 30 dm, ± canescent-tomentose. **LF**: 1–5 dm, dentate to shallowly lobed, lobes 8–10 pairs, widely triangular. **INFL**: phyllaries linear, puberulent, ± cobwebby-tomentose, spines ≤ 5 mm. **FL**: corolla 20–25 mm, purple or white, glabrous. **FR**: 4–5 mm; pappus 7–9 mm, pink to ± red. 2*n*=34. Disturbed sites; < 1600 m. NW, CaR, SN, SnJV, CW, SW, MP; native to Eur. May–Aug ◆

O. illyricum L. (p. 395) ILLYRIAN THISTLE Pl ≤ 25 dm, ± canescent-tomentose. **LF**: 1–5 dm, ± deeply pinnately lobed, lobes 8–10 pairs, triangular, entire or again lobed. **INFL**: phyllaries lanceolate to ovate, glabrous or ± cobwebby, spines ≤ 5 mm. **FL**: corolla 25–35 mm, purple, lobes glandular-puberulent. **FR**: 4–5 mm; pappus 10–12 mm, ± white. 2*n*=34. Disturbed sites; < 500 m. s SnFrB; native to se Eur. Jun–Jul ◆

O. tauricum Willd. TAURIAN THISTLE Pl ≤ 2 m, ± sticky-glandular. **LF**: 1–2.5 dm, pinnately lobed, lobes 6–8 pairs, acutely triangular. **INFL**: phyllaries lanceolate, glabrous, spines ≤ 4 mm. **FL**: corolla 25–30 mm, purple-pink, gen glabrous. **FR**: 5–6 mm; pappus 8–10 mm. 2*n*=34. Disturbed sites; < 1400 m. SNF, n SCoRO, MP; CO; native to s Eur, sw Asia. May–Aug ◆

OREOSTEMMA ASTER

Guy L. Nesom

Per, from taproot or thick rhizome, sometimes with branched caudex. **ST**: decumbent-ascending to erect, unbranched, ± scape-like, gen distally finely loose-tomentose, glandless or stalked-glandular. **LF**: most basal in rosette, sessile, linear to oblanceolate, 3-veined, entire, glabrous to sparsely long-soft-hairy or stalked-glandular. **INFL**: heads radiate, 1; involucre broadly top-shaped; phyllaries ± equal or graduated in (2)3–4 series, gen ± purple, linear to linear-elliptic or oblanceolate, lf-like in texture (or proximal margins ± hardened), glabrous to loose-tomentose; receptacle flat, shallowly pitted, epaleate. **RAY FL**: 10–40; corolla ± white to purple. **DISK FL**: corolla tubular, yellow; anther tip ovate-triangular; style branch appendages lance-linear, 1.5–2.2 mm. **FR**: 4–5 mm, narrowly cylindric, ± brown, ribs 5–10, raised, faces glabrous or sparsely short-strigose; pappus ± white, of minutely barbed bristles in 1 series, sometimes also with few short bristles or hairs. 3 spp.: w US and adjacent Can. (Greek: mountain + crown) [Nesom 2006 FNANM 20:359–361]

1. St, phyllaries, and often lvs densely short-stalked-glandular; sts 2–7 cm; lvs linear . . . . ***O. peirsonii***
1′ St, phyllaries, and lvs glandless, glabrous or long-soft-hairy; sts 4–70 cm; lvs oblanceolate
  2. Herbage, phyllaries ± glabrous to ± long-soft-hairy, always some hairs distally on peduncle and proximally on phyllaries; phyllaries lf-like in texture and color or with a tan, slightly hardened base, outer proximally 0.8–1.2 mm wide, 1-veined . . . . ***O. alpigenum*** var. ***andersonii***
  2′ Herbage, phyllaries glabrous; phyllaries strongly hardened, proximally straw-colored, outer proximally 1.5–2 mm wide, 3-veined, veins dividing phyllary into 4 bands . . . . ***O. elatum***

O. alpigenum (Torr. & A. Gray) Greene var. *andersonii* (A. Gray) G.L. Nesom (p. 395) Sts, phyllaries, and lvs glandless, ± glabrous to densely long-soft-hairy, always some hairs distally on peduncle and proximally on phyllaries. **ST**: 4–40 cm. **LF**: oblanceolate. **INFL**: phyllaries lf-like in texture and color, slightly hardened, outer proximally 0.8–1.2 mm wide, 1-veined. **RAY FL**: corolla gen 10–16 mm. **DISK FL**: corolla 5–9 mm. **FR**: hairy throughout. 2*n*=18,36. Peatlands (sometimes with *Darlingtonia*), marshes, moist to wet meadows, lake edges, forest, tundra; 1200–3500 m. KR, NCoR, CaR, SN, SnJt, Wrn, W&I; OR, NV. [*A. alpigenus* (Torr. & A. Gray) A. Gray var. *andersonii* (A. Gray) M. Peck] 2 other vars. in w US. Jun–Sep

O. elatum (Greene) Greene TALL ALPINE-ASTER Sts, phyllaries, and lvs glandless, glabrous. **ST**: 35–70 cm. **LF**: oblanceolate. **INFL**: phyllaries strongly hardened, proximally straw-colored, outer proximally 1.5–2 mm wide, 3-veined, veins dividing phyllary into 4 bands. **RAY FL**: corolla 7–12 mm. **DISK FL**: corolla ± 7 mm. **FR**: glabrous or distally slightly hairy. Peatlands, marshy areas, wet meadows, montane forest; 1000–1500 m. n SNH. Jul–Aug ★

O. peirsonii (Sharsm.) G.L. Nesom Sts, phyllaries, and often lvs densely short-stalked-glandular, sparsely long-soft-hairy or not. **ST**: 2–7 cm. **LF**: linear. **INFL**: phyllaries lf-like in texture and color throughout or proximally tan, slightly hardened, outer proximally 0.8–1.2 mm wide, 1-veined. **RAY FL**: corolla 12–16 mm. **DISK FL**: corolla 5.5–8 mm. **FR**: slightly hairy. Rocky slopes, ridges, dry meadows; 3000–3800 m. s SNH. [*Aster p.* Sharsm.] Jul–Sep

OROCHAENACTIS

1 sp. (Greek: mountain *Chaenactis*) [Keil 2006 FNANM 21:414]

O. thysanocarpha (A. Gray) Coville (p. 395) Ann 1–25 cm, short-hairy to ± tomentose and glandular. **ST**: spreading to erect, simple or ± branched, slender. **LF**: proximal opposite, distal alternate, 1–4 cm, linear to narrowly oblong-oblanceolate, entire. **INFL**: heads discoid, 6–8 mm, sessile in small, lfy-bracted clusters at st tips; involucre cylindric; phyllaries 4–7 in 1 series, 3–6 mm, free, oblong, ± purple; receptacle flat, knobby, epaleate. **FL**: 4–9; corolla 3–5 mm, yellow, lobes short-triangular; anther tip short-triangular; style tips tapered. **FR**: 2.5–4 mm, club-shaped, ribbed, minutely puberulent; pappus of 11–17 scales, 1.5–2.5 mm, oblanceolate, obtuse, fringed, deciduous together in a ring. 2*n*=18. Open conifer forest, dry meadows, gravelly slopes; 1600–3800 m. s SNH. Jun–Sep

OSMADENIA

Robert L. Carr & Gerald D. Carr

1 sp. (Greek: odor gland) [Carr & Carr 2006 FNANM 21:269]

O. tenella Nutt. (p. 395) OSMADENIA Ann 0.5–4 dm, ± densely glandular, esp infl, ± scabrous and long-hairy, strongly scented; tacklike glands 0. **ST**: erect; branchlets many, thread-like, ± spreading. **LF**: mostly cauline, alternate, sessile; proximal (1)2–5 cm, linear, entire, scabrous, glandular, proximally ± ciliate or shaggy-hairy; distal reduced. **INFL**: heads many, radiate, in loose cyme-like clusters; peduncle spreading, bracts few, 5–12(20) mm, narrowly lance-linear; phyllaries 3–6 mm, partly enfolding ray fr, tips acute to attenuate; receptacle flat, glabrous; paleae between ray and disk fls, 3.5–7 mm, fused. **RAY FL**: 3–5; tube slender, > 1/2 ray; ray deeply 3-lobed, lobes ± = ray, 3–8 mm, white, aging ± red, occ red-splotched on adaxial face. **DISK FL**: 3–10; corolla gen colored like rays; anthers yellow, tips lanceolate; style tips linear. **FR**: ray fr 1.8–2.7 mm, back rounded to angled, rough, glabrous, adaxial face ± flat, beak short, adaxially off-center, pappus 0; disk fr ± angled, tapered at base, ± thinly hairy, pappus of ± 8 scales, long and short. 2*n*=18. Locally common. Barren, often rocky, sandy soils, coastal-sage scrub, grassland, oak savanna; 30–1200 m. SW (exc ChI); Baja CA. Self-sterile. May–Jul

PACKERA GROUNDSEL, RAGWORT, BUTTERWEED

Debra K. Trock

Bien or per 3–100+ cm from rhizome or taproot with thin branched fibrous roots, loosely hairy to glabrous. **ST**: 1–several. **LF**: pinnate or gen simple, basal and proximal cauline gen petioled, mid sessile, reduced, distal bract-like. **INFL**: heads discoid or radiate, in compact or open cyme-like clusters, rarely single; involucre gen bell-shaped (cylindric); main phyllaries gen 8, 13, or 21 in 1 series, reflexed in fr, green to ± red, linear, glabrous or hairy, subtended by a few reduced outer phyllaries, not black-tipped; receptacle epaleate. **RAY FL**: 0–13(21); corolla pale yellow to deep orange-red. **DISK FL**: 20–80+; corolla bell-shaped to tubular, lobes 5 erect to recurved, gen yellow (to deep orange-red); style tips truncate. **FR**: cylindric, gen prominently ribbed, glabrous or stiff-hairy; pappus of white minutely barbed bristles. 64 spp.: N.Am, 1–2 in Siberia. (J.G. Packer, Canadian botanist, b. 1929) [Trock 2006 FNANM 20:570–602] Formerly in *Senecio*. The common names groundsel, ragwort, and butterweed are applied to spp. of both *Packera* and *Senecio*.

1. Heads discoid
  2. Pls < 15 cm, ± scapose; cauline lvs much reduced; heads gen < 5 . . . . *2P. werneriifolia*
  2′ Pls gen > 15 cm; cauline lvs gradually reduced; heads gen > 5
    3. Caudex taprooted; lvs thin; heads gen > 10, phyllaries deep green or ± red-tipped; gen below subalpine
       elevations . . . . *2P. indecora*
    3′ Caudex weakly spreading and fibrous rooted; lvs thick, firm; heads gen 4–8, phyllaries deep red or
       green with ± red tips; often subalpine to alpine elevations. . . . . *2P. pauciflora*
1′ Heads radiate, some with rays barely exceeding phyllaries
  4. Basal lvs ± pinnately lobed
    5. Herbage glabrous or at most sparsely and irregularly hairy when young
      6. Caudex taprooted; peduncle conspicuously bracted; pls in SN and e . . . . *2P. multilobata*
      6′ Caudex fibrous rooted; peduncle not bracted or bracts inconspicuous; pls w of SNH
        7. Pls stout or slender, gen < 50 cm; basal lvs with 1–3 pairs of lateral lobes, terminal lobes ± round,
           ultimate margins evenly, regularly incised, sinuses rounded, bases ± cordate; heads (3)8–15+
           . . . . *P. bolanderi* var. *bolanderi*
        7′ Pls large, robust, gen > 50 cm; basal lvs with 2–6+ pairs of lateral lobes, terminal lobes ovate to
           oblong, ultimate margins irregularly incised, bases tapered; heads 15–50+ . . . . *P. breweri*
    5′ Herbage gen hairy or becoming glabrous near fl time
      8. Heads gen 3–6; pls gen ≤ 3 dm; basal lvs gen ≤ 3 cm, with 1–3 pairs of lateral lobes . . . . *2P. ionophylla*
      8′ Heads 5–30+; pls 2–5+ dm; basal lvs gen > 4 cm, with 2–6 pairs of lateral lobes
        9. Basal lvs persistent; heads small, rays 7–10 mm; herbage evenly but lightly hairy. . . . . *2P. multilobata*
        9′ Basal lvs withering before fl; heads large, rays 10–15 mm; herbage densely hairy to becoming
           glabrous with age . . . . *P. eurycephala*
          10. Lvs irregularly lobed or dissected, lateral lobes coarsely dentate, midribs winged; disturbed sites,
              woodland, roadsides . . . . var. *eurycephala*
          10′ Lf margins finely dissected, lateral lobes also dissected, midribs not winged; serpentine soils, rocky
              sites . . . . var. *lewisrosei*

4′ Basal lvs with margins various but not (or weakly) lobed
  11. Herbage hairy to becoming glabrous at fl time
    12. Rays red-orange; heads gen 1–3; pls gen on serpentine . . . . ***P. greenei***
    12′ Rays yellow; heads 2–many; pls of various soils
      13. Basal lvs round to widely ovate, obovate or spoon-shaped, petiole >> blade
        14. Fr 0.7–1 mm; phyllaries 6–8 mm; disk fls 35–50+ . . . . ***P. bernardina***
        14′ Fr 2–2.5 mm; phyllaries 7–10 mm; disk fls 60–75+ . . . . ²***P. ionophylla***
      13′ Basal lvs ovate, elliptic, lanceolate, oblanceolate or narrowly spoon-shaped, petiole < or > blade
        15. Pls gen 7–15 cm; heads gen < 5 . . . . ²***P. werneriifolia***
        15′ Pls 10–70+ cm; heads ≥ 5
          16. Herbage sparsely tomentose, becoming glabrous; ray 12–16 mm, pls only on serpentine . . . . ***P. layneae***
          16′ Herbage canescent, woolly, or densely tomentose, sometimes becoming glabrous; ray 8–10+ mm,
            pls of various habitats
            17. Lvs ovate, elliptic, or widely lanceolate; phyllaries tomentose, fr 2.5–3.5+ mm. . . . . ***P. cana***
            17′ Lvs lanceolate to oblanceolate; phyllaries glabrous, fr 1.5–2 mm. . . . . ***P. macounii***
  11′ Herbage glabrous or at most tomentose to woolly at fl on proximal st, in lf axils and/or at bases of heads
    18. Lvs thick, firm, glaucous . . . . ***P. clevelandii***
    18′ Lvs thick or thin, glabrous to sparsely tomentose
      19. Heads 6–20+
        20. Lvs widely lanceolate to ± hastate, margins sharply dentate, pls 5–7+ dm; heads 12–20+, phyllaries
          light green . . . . ***P. pseudaurea*** var. ***pseudaurea***
        20′ Lvs ovate, elliptic, oblanceolate, spoon-shaped, oblong or ± reniform, margins entire to dentate or
          weakly lobed, pls 1–8 dm; heads 6–20+, phyllaries dark green often red-tipped
          21. Caudex taprooted; lvs not noticeably firm; heads gen eradiate or with ray 3–5 mm . . . . ²***P. indecora***
          21′ Caudex stout and fibrous-rooted; lvs ± thick, firm; heads radiate, ray 5–10 mm
            . . . . ***P. streptanthifolia*** var. ***streptanthifolia***
      19′ Heads 1–6
        22. Lvs reniform to ± round; pls only in recently burned chaparral on gabbroic soils. . . . . ***P. ganderi***
        22′ Lvs elliptic, ovate, oblanceolate, obovate, spoon-shaped, or at most ± reniform; pls of various
          habitats, but not in chaparral
          23. Ray 5–7 mm, deep orange-yellow; heads gen > 4 . . . . ²***P. pauciflora***
          23′ Ray 6–12+ mm, yellow; heads gen 1–4
            24. Phyllaries densely tomentose; head (1)4; ray 6–10+ mm. . . . . ***P. hesperia***
            24′ Phyllaries glabrous; head 1(3); ray 7–12+ mm . . . . ***P. subnuda*** var. ***subnuda***

P. bernardina (Greene) W.A. Weber & Á. Löve (p. 399) SAN BERNARDINO RAGWORT Per 1.5–4 dm; caudex branched, ± erect, fibrous-rooted. **ST**: 1 per rosette, woolly-tomentose or becoming ± glabrous. **LF**: basal blade << petiole, 1–2.5 cm, 1–1.5 cm wide, widely spoon-shaped to obovate, tapered, ± entire or dentate near tip, woolly-tomentose or becoming ± glabrous; cauline sessile, spoon-shaped to linear. **INFL**: heads radiate, 2–8 in compact cluster; phyllaries 13 or 21, 6–8 mm, green, woolly-tomentose, distally becoming glabrous. **RAY FL**: 8(13); ray 8–10 mm. **DISK FL**: 35–50+. **FR**: 0.7–1 mm, glabrous or with sharp bristles, esp on ribs. 2*n*=46. Dry rocky slopes, duff of pine forest; 1700–2200 m. e SnBr. [*Senecio b.* Greene] May–Jul ★

P. bolanderi (A. Gray) W.A. Weber & Á. Löve var. ***bolanderi*** (p. 399) SEACOAST RAGWORT Per, 2–5(7) dm from fibrous-rooted creeping caudex. **ST**: 1 or 2–3 loosely clustered, glabrous. **LF**: basal petioled, 5–12 cm, 3–7 cm wide, round to oblong or fiddle-shaped, ± cordate at base, pinnately lobed, lateral lobes 1–3 pairs, shallowly to deeply, regularly incised, sinuses rounded, faces glabrous; cauline petioled or sessile and clasping, lanceolate, margins incised. **INFL**: heads radiate, (3)8–15+ in ± flat-topped cluster; phyllaries 13 or 21, 6–8 mm, proximally with ± brown multicelled hairs. **RAY FL**: 8 or 13; ray 10–15+ mm. **DISK FL**: 30–45+. **FR**: 2.5–3.5 mm, glabrous. 2*n*=46. Coastal forest, wet cliffs; < 300 m. NCo; to WA. [*Senecio b.* A. Gray] May–Jul ★

P. breweri (Burtt Davy) W.A. Weber & Á. Löve (p. 399) BREWER'S RAGWORT Per or bien 4–12+ dm, from fibrous-rooted, stout, erect caudex. **ST**: gen 1, glabrous. **LF**: basal and proximal cauline petioled, 10–30+ cm, 2–7 cm wide, spoon-shaped to obovate, abruptly contracted to tapered at base, ± pinnate, terminal lobe ovate to oblong, lateral lobes 2–6+ pairs, smaller, midrib narrowly winged, margins irregular, crenate or dentate to sharply cut, faces glabrous; mid and distal petioled or sessile, obovate, ± pinnate to irregularly incised, terminal lobe narrow. **INFL**: heads radiate, 15–50+ in ± flat-topped cluster; phyllaries 13 or 21, 7–10 mm, green, glabrous. **RAY FL**: 8–10+; ray 10–15+ mm. **DISK FL**: 45–60+. **FR**: 4–5 mm, glabrous. 2*n*=46. Common. Seasonally damp grassland, oak savanna, disturbed areas; 200–1500 m. s SNF, Teh, SnJV, CW, WTR. [*Senecio b.* Burtt Davy] Apr–Jun

P. cana (Hook.) W.A. Weber & Á. Löve WOOLLY GROUNDSEL Per 1–3+ dm from thick, ± erect, branched caudex. **ST**: 1 per rosette, densely woolly or canescent. **LF**: basal, proximal cauline 2.5–5+ cm, 1–3 cm wide, ovate or elliptic to widely lanceolate, tapered at base, entire or irregularly wavy to weakly dentate, abaxial face persistently woolly, adaxial woolly or becoming ± glabrous; cauline petioled or sessile and weakly clasping, elliptic to lanceolate, entire or weakly dentate. **INFL**: heads radiate, 8–15+ in ± flat-topped cluster; phyllaries 13 or 21, 5–8+ mm, green, densely tomentose. **RAY FL**: 8–10(13); ray 8–10+ mm. **DISK FL**: 35–50+. **FR**: 2.5–3.5+ mm, glabrous. 2*n*=46, 92. Locally abundant. Sagebrush scrub, rocky slopes, crevices; 1200–3500 m. KR, CaR, SN, GB; to BC, MB, MN, KS, NM. [*Senecio c.* Hook.] Jun–Aug

P. clevelandii (Greene) W.A. Weber & Á. Löve (p. 399) CLEVELAND'S RAGWORT Per 3–8+ dm, from stout taprooted caudex. **ST**: gen 1, glaucous. **LF**: basal and proximal cauline thick, firm, 3–10+ cm, 1–2 cm wide, lanceolate to narrowly oblanceolate, tapered at base, entire to crenate, dentate, or shallowly lobed, glaucous; cauline gradually reduced, distal ± sessile and weakly clasping, ± lanceolate. **INFL**: heads radiate, 12–20+ in ± flat-topped cluster; phyllaries (13)21, 7–10 mm, green with purple tips, glabrous. **RAY FL**: 8–10(13); ray 5–7 mm. **DISK FL**: 25–40+. **FR**: 1–2 mm, glabrous. 2*n*=46. Chaparral, woodland along streams in serpentine soils; 260–900 m. NCoR, c SNF. [*Senecio c.* Greene] May–Aug ★

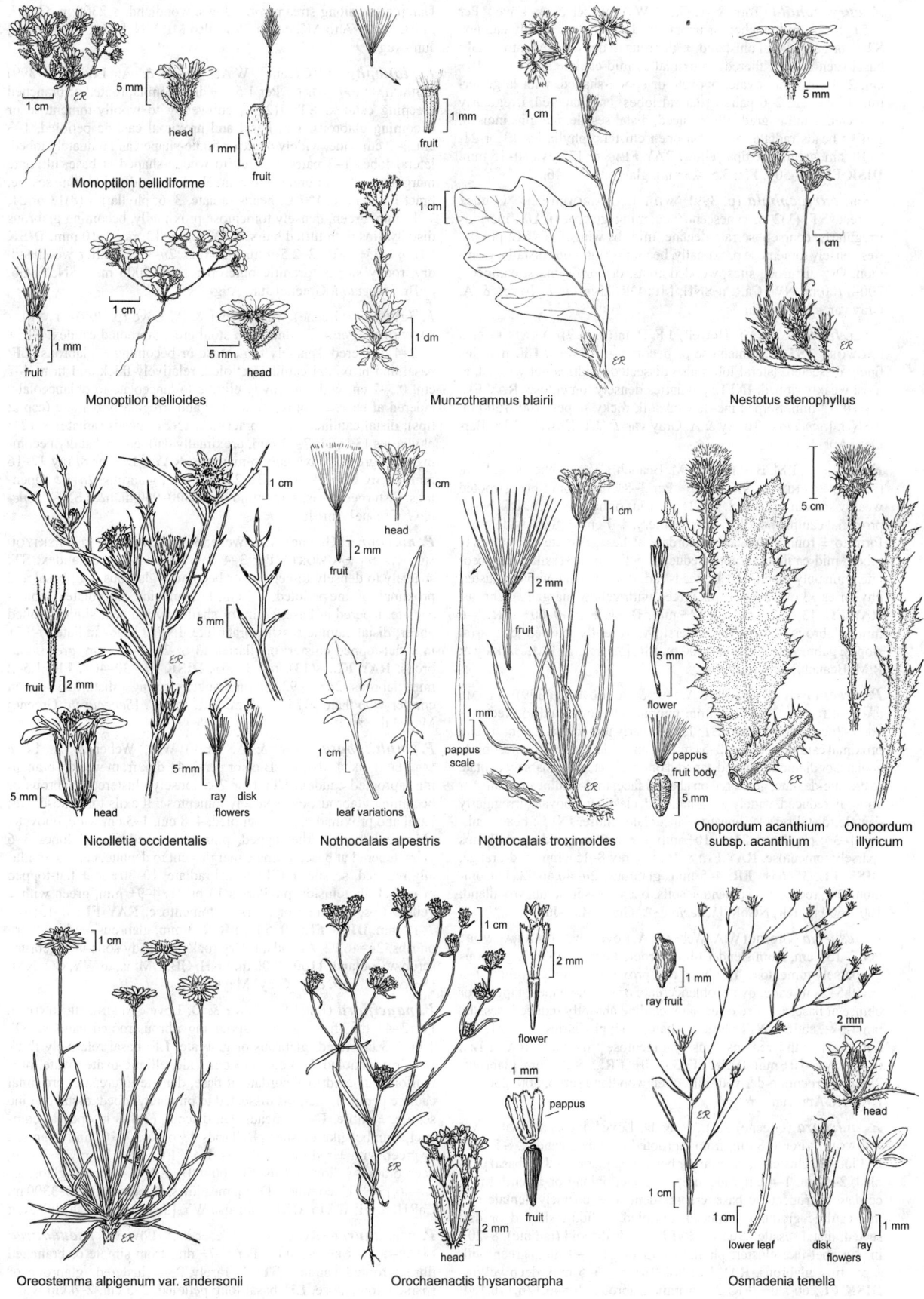

Monoptilon bellidiforme

head

fruit

1 cm

5 mm

1 mm

Monoptilon bellioides

fruit

head

1 cm

5 mm

1 mm

Munzothamnus blairii

fruit

1 cm

1 mm

1 dm

Nestotus stenophyllus

5 mm

1 cm

Nicolletia occidentalis

fruit

head

ray   disk
flowers

1 cm

5 mm

5 mm

5 mm

2 mm

Nothocalais alpestris

head

fruit

pappus
scale

leaf variations

1 cm

2 mm

1 cm

1 cm

Nothocalais troximoides

fruit

1 cm

2 mm

1 mm

Onopordum acanthium
subsp. acanthium

flower

pappus

fruit body

5 mm

5 mm

5 mm

Onopordum
illyricum

5 cm

Oreostemma alpigenum var. andersonii

1 cm

Orochaenactis thysanocarpha

flower

pappus

fruit

head

1 cm

2 mm

1 mm

2 mm

Osmadenia tenella

ray fruit

head

lower leaf

disk
flowers

ray

1 mm

2 mm

1 cm

1 cm

1 mm

P. eurycephala (Torr. & A. Gray) W.A. Weber & Á. Löve Per 2–5+ dm, from branched or unbranched, taprooted, woody caudex. **ST**: 1 or gen several clustered, ± glabrous to densely tomentose. **LF**: basal (gen soon withered), proximal to mid-cauline similar, 7–10+ cm, 2–3 cm wide, ovate, obovate or spoon-shaped, with large terminal lobe and 2–6 pairs of lateral lobes, base tapered, irregularly dissected; cauline gradually reduced, distal sessile, margins incised. **INFL**: heads radiate, 5–20+ in open clusters; phyllaries 13 or 21, 8–10 mm, green, occ tips yellow. **RAY FL**: 8 or 13; rays 10–15 mm. **DISK FL**: 35–50+. **FR**: 3.5–4.5 mm, glabrous. 2*n*=46.

var. ***eurycephala*** (p. 399) WIDEHEAD GROUNDSEL Caudex ± erect. **ST**: (1)2–6, canescent (becoming glabrous). **LF**: margins irregularly, coarsely serrate-dentate, midribs winged. **INFL**: phyllaries densely tomentose proximally, becoming glabrous distally. Common. Dry disturbed sites, wooded areas, dry or drying streambeds; 200–1700 m. NW, CaR, n SNH, MP; OR. [*Senecio e.* Torrey & A. Gray var. *e.*] May–Jun

var. ***lewisrosei*** (J.T. Howell) J.F. Bain (p. 399) LEWIS ROSE'S RAGWORT **ST**: 1, tomentose or becoming glabrous. **LF**: margins finely dissected, lateral lobes also dissected, midribs not winged, at most weakly ribbed. **INFL**: phyllaries densely tomentose. **RAY FL**: ray 10–13 mm. Serpentine-derived soil, rocky slopes; 100–1500 m. n SNH. [*Senecio e.* Torrey & A. Gray var. *l.* (J.T. Howell) T.M. Barkley] Apr–Jun ★

P. ganderi (T.M. Barkley & R.M. Beauch.) W.A. Weber & Á. Löve (p. 399) GANDER'S RAGWORT Per 3–8+ dm, from fibrous-rooted weakly spreading caudex. **LF**: basal, proximal cauline petioled, firm, leathery, 4–7 cm, 4–8 cm wide, reniform to ± round, truncate to cordate at base, dentate or shallowly lobed; mid-cauline abruptly reduced, petioled or sessile, oblanceolate, pinnately lobed. **INFL**: heads radiate, 3–6+ in compact cluster; phyllaries 13 or 21, 8–11 mm, green with yellow margins, glabrous. **RAY FL**: 13 or 21; ray 10–12+ mm. **DISK FL**: 40–60+. **FR**: 5–6 mm, glabrous. Chaparral understory, recently burned chaparral slopes, gabbroic soils; 700–1100 m. PR. [*Senecio g.* T.M. Barkley & R.M. Beauch.] Apr–May ★

P. greenei (A. Gray) W.A. Weber & Á. Löve (p. 399) FLAME RAGWORT Per 2–3+ dm, from unbranched horizontal to erect rhizomes (occ stoloniferous). **ST**: 1, irregularly tomentose. **LF**: basal and proximal cauline petioled, 2–5 cm, 2–4 cm wide, ± round, ovate, oblanceolate, or diamond-shaped, tapered or obtuse at base, coarsely dentate to crenate-dentate, gen ± red on abaxial face; mid-cauline gradually or abruptly reduced, widely wing-petioled, clasping, obovate, irregularly dentate; distal sessile, linear to lanceolate, entire. **INFL**: heads radiate, 1–3+; phyllaries 21, 8–10+ mm, green (tips occ deep red), tips sparsely tomentose. **RAY FL**: 8–10(13); ray 8–15+ mm, red-orange. **DISK FL**: 35–65+. **FR**: 4–5 mm, glabrous. 2*n*=40,46,92. Uncommon. Dry, rocky, gen serpentine soils, open areas in scrub, woodland; 100–1600 m. KR, NCoR. [*Senecio g.* A. Gray] May–Jul

P. hesperia (Greene) W.A. Weber & Á. Löve WESTERN RAGWORT Per 7–15+ cm, from slender fibrous-rooted caudex. **ST**: 1, glabrous or sparsely tomentose. **LF**: basal and proximal cauline petioled, 1–3 cm, 0.5–2 cm wide, ovate, oblanceolate or spoon-shaped, tapered or obtuse at base, ± entire or dentate; cauline abruptly reduced, sessile, bract-like, entire. **INFL**: heads radiate, (1)4; phyllaries 13 or 21, 6–9 mm, green with ± red tips, densely tomentose proximally. **RAY FL**: 8 or 13; ray 6–10+ mm. **DISK FL**: 35–50+. **FR**: 1.5–2.5 mm, glabrous. 2*n*=46. Serpentine-derived soils, open woodland scrub; 500–2500 m. KR; OR. Apr–Jun ★

P. indecora (Greene) Á. Löve & D. Löve RAYLESS MOUNTAIN RAGWORT Per 1–8 dm, from taprooted branched caudex. **ST**: 1 or 2–3 loosely clustered, glabrous or becoming glabrous. **LF**: basal petioled, 2–5 cm, 1–4 cm wide, elliptic-ovate, oblong or ± reniform, ± cordate or truncate at base, crenate-dentate to coarsely dentate-cut; mid-cauline gradually reduced, petioled, ± fiddle-shaped or dissected; distal sessile, ± entire. **INFL**: heads discoid (radiate), 8–20+ in ± umbel-like cluster; phyllaries 13 or 21, 7–9 mm, green with ± red tips, glabrous. **RAY FL**: 0(8–10); ray 3–5 mm, deep yellow. **DISK FL**: 60–70+. **FR**: 2–2.5 mm, glabrous. 2*n*=46,126,176,184.

Damp areas along streams, meadows, woodland; < 2300 m. CaRH, SNH, MP; WA to AK, e Can, WY; also MI, MN. [*Senecio i.* Greene] Jun–Aug ★

P. ionophylla (Greene) W.A. Weber & Á. Löve (p. 399) TEHACHAPI RAGWORT Per 1.5–3+ dm, from taprooted or branched creeping caudex. **ST**: 1(2–3), cobwebby to woolly-tomentose or becoming glabrous. **LF**: basal and proximal cauline petioled, 1–3 cm, 1–2 cm wide, widely ovate or fiddle-shaped and pinnately lobed, lateral lobes 1–3 pairs, tapered to wedge-shaped at base, ultimate margins ± entire or crenate to coarsely dentate; distal cauline sessile, bract-like, entire. **INFL**: heads radiate, 3–6; phyllaries (8)13 or 21, 7–10 mm, green, densely tomentose proximally, becoming glabrous distally, tips with tufted hairs. **RAY FL**: 8–13; ray 8–10 mm. **DISK FL**: 60–75+. **FR**: 2–2.5 mm, glabrous. 2*n*=46. Conifer woodland, dry, rocky slopes, granitic outcrops; 1400–3000 m. s SN, SnGb, SnBr. [*Senecio i.* Greene] Jun–Aug ★

P. layneae (Greene) W.A. Weber & Á. Löve (p. 399) LAYNE'S RAGWORT Per 4–7+ dm, from stout erect taprooted caudex. **ST**: 1 or 2–4 clustered, sparsely tomentose or becoming ± glabrous. **LF**: basal and proximal cauline petioled, relatively thick and firm, 4–7 cm, 0.5–2 cm wide, narrowly elliptic to lanceolate to oblanceolate, tapered at base, ± entire, or weakly and irregularly dentate (esp at tips); distal cauline sessile, bract-like. **INFL**: heads radiate, 5–12+; phyllaries 13 or 21, 7–11 mm, proximally dark green distally becoming light green, proximally tomentose. **RAY FL**: 5 or 8; ray 12–16 mm. **DISK FL**: 50–60+. **FR**: 2.5–3.5 mm, glabrous. 2*n*=92. Openings, disturbed areas, serpentine soils; 300–900 m. n&c SNF. [*Senecio l.* Greene] Apr–Jun ★

P. macounii (Greene) W.A. Weber & Á. Löve (p. 399) SISKIYOU MOUNTAINS RAGWORT Per 3–4 dm, from taprooted caudex. **ST**: sparsely to densely tomentose or becoming glabrous. **LF**: basal and proximal cauline petioled, 3–5 cm, 1–2 cm wide, lanceolate to oblanceolate, tapered at base, entire or shallowly toothed, slightly rolled under; distal cauline sessile, bract-like. **INFL**: heads radiate, 6–15+ in ± flat-topped cluster; phyllaries 13 or 21, 5–7+ mm, green, glabrous. **RAY FL**: 8(13); ray 8–10+ mm. **DISK FL**: 30–40+. **FR**: 1.5–2 mm glabrous. 2*n*=46,92. Streambanks, clearings, disturbed sites in conifer woodland; 375–1200 m. KR; to BC. [*Senecio m.* Greene] May–Jul ★

P. multilobata (Torr. & A. Gray) W.A. Weber & Á. Löve LOBELEAF GROUNDSEL Bien or per 2–4+ dm, from weakly branching taprooted caudex. **ST**: 1 or 2–5 loosely clustered, glabrous or becoming glabrous, occ sparsely tomentose, lf axils tomentose. **LF**: basal and proximal cauline petioled, 4–8 cm, 1–3 cm wide, obovate, oblanceolate or fiddle-shaped, pinnately lobed, lateral lobes 3–6 pairs, tapered at base, ultimate margins cut to dentate; cauline gradually reduced, sessile. **INFL**: heads radiate, 10–30+ in ± flat-topped or umbel-like cluster; phyllaries 13 or 21, 4–9+ mm, green with ± yellow tips, glabrous or sparsely tomentose. **RAY FL**: 8–13; ray 7–10 mm. **DISK FL**: 40–50+. **FR**: 2–3 mm, glabrous or short-hairy on ribs. 2*n*=46,92. Abundant. Dry rocky or sandy soils in sagebrush scrub, woodland; 1100–3000 m. SNH, GB, DMtns; to WY, CO, NM. [*Senecio m.* Torr. & A. Gray] May–Jul

P. pauciflora (Pursh) Á. Löve & D. Löve ALPINE GROUNDSEL Per 2–4+ dm, from weakly spreading fibrous-rooted caudex. **ST**: 1 or 2–3 clustered, glabrous or glabrate. **LF**: basal relatively thick and firm, petioled, 2–4 cm, 1–3 cm wide, elliptic-ovate, ovate, or ± reniform, tapered to ± cordate at base, dentate to crenate; proximal cauline petioled, margins dissected to pinnately lobed; distal cauline sessile, ± entire. **INFL**: heads gen discoid, (2)4–8 in open to compact ± umbel-like cluster; phyllaries 13 or 21, 7–10 mm, deep red or green with ± red tips, glabrous. **RAY FL**: 0(8–13); ray 5–7 mm, deep orange-yellow. **DISK FL**: 60–80+. **FR**: 1–1.5 mm, glabrous. 2*n*=46,130+. Uncommon. Damp meadows, woodland; 1800–3300 m. CaRH, SNH; WA to AK, e Can; also WY. [*Senecio p.* Pursh] Jul–Aug

P. pseudaurea (Rydb.) W.A. Weber & Á. Löve var. ***pseudaurea*** FALSE-GOLD GROUNDSEL Per 5–7+ dm, from simple or branched fibrous-rooted caudex. **ST**: 1, rarely 2–4 clustered, glabrous or sparsely tomentose. **LF**: basal long petioled, 2–5 cm, 2–4 cm wide,

widely lanceolate to ± hastate, truncate, ± cordate or obtuse at base, sharply dentate; proximal cauline gradually reduced, margins pinnately lobed to cut; distal becoming sessile. **INFL:** heads radiate, 12–20+ in open, ± flat-topped cluster; phyllaries (13)21(30+), 3–8 mm, light green, glabrous. **RAY FL:** 8 or 13, rarely 0; ray 6–10+ mm. **DISK FL:** 70–80+. **FR:** 1–1.5 mm, glabrous. $2n$=40,44,46,80. Damp streambanks, wet meadows, open, wet woodland; 1500–2000 m. CaRH, SN, Wrn; to BC, AB, SK, WY, NV. [*Senecio p.* Rydb. var. *p.*] Jun–Jul

P. streptanthifolia (Greene) W.A. Weber & Á. Löve var. ***streptanthifolia*** ROCKY MOUNTAIN GROUNDSEL Per 2–5+ dm, from stout fibrous-rooted caudex. **ST:** 1 or 2–5 clustered, glabrous or with tufted woolly hairs proximally and in axils. **LF:** basal and proximal cauline petioled, ± thick, firm, 2–4 cm, 1–3 cm wide, spoon-shaped to oblanceolate or ovate, tapered to abruptly contracted at base, entire, crenate, dentate, or weakly lobed; mid-cauline gradually reduced, becoming sessile, ± entire. **INFL:** heads radiate, 6–20+ in loose ± flat-topped cluster; phyllaries (8)13 or 21, 4–7+ mm, green or ± red-tipped, glabrous. **RAY FL:** 8 or 13; ray 5–10 mm. **DISK FL:** 35–60+. **FR:** 1–2.5 mm, glabrous. $2n$=46,92. Common. Forest, open meadows, dry or damp loamy soils; 1100–2900 m. CaR, SNH, MP; to YT, SK, NM. [*Senecio s.* Greene] May–Aug

P. subnuda (DC.) Trock & T.M. Barkley var. ***subnuda*** CLEFTLEAF GROUNDSEL Per 1–3+ dm, from fibrous rooted or creeping slender caudex. **ST:** 1(2–3 clustered), glabrous. **LF:** basal petioled, 2–4 cm, 1–3 cm wide, ovate, obovate or elliptic, tapered at base, ± entire to crenate-dentate. **INFL:** heads radiate, 1(3) subtended by 2 large bractlets; phyllaries (13)21, 5–8+ mm, green or ± red-tipped, glabrous. **RAY FL:** 13; ray 7–12+ mm. **DISK FL:** 40–55+. **FR:** 1.5–2.5 mm, glabrous. $2n$=46,90. Uncommon. Damp or marshy meadows, lakeshores, streambanks; 1500–3000 m. CaRH, n&c SNH, MP; to BC, MT, WY, NV. [*Senecio cymbalarioides* H. Buek] Jun–Sep

P. werneriifolia (A. Gray) W.A. Weber & Á. Löve HOARY GROUNDSEL Per, ± scapose, 7–15+ cm, from stout branching occ densely crowded rhizomes. **ST:** 1 or 3–5 clustered, woolly-tomentose or canescent, occ becoming glabrous. **LF:** basal lvs sessile or short-petioled, 1.5–4 cm, 0.5–2.5 cm wide, narrowly lanceolate to elliptic, tapered at base, entire or dentate toward the tips, margins often rolled under; cauline abruptly much-reduced and bract-like. **INFL:** heads radiate or discoid, 1–5(8); phyllaries 13 or 21, 4–10 mm, green or ± red-tipped, gen hairy. **RAY FL:** 0, 8 or 13; ray 5–10 mm. **DISK FL:** 30–50+. **FR:** 1.5–2 mm, glabrous. $2n$=44,46. Common. Rocky talus slopes, sandy soils in forest openings; 3000–3650 m. SNH, SNE; to MT, WY, CO, NM. [*Senecio w.* (A. Gray) A. Gray] Jul–Aug

PALAFOXIA

Ann, per, subshrub. **ST:** ascending or erect. **LF:** simple, proximal opposite, distal alternate, gen petioled; blade entire. **INFL:** heads discoid or radiant [radiate], in ± flat-topped clusters; involucre ± cylindric or ± obconic; phyllaries in 2–3 ± equal series; receptacle flat, epaleate. **RAY FL:** 0[13]; corolla white to ± purple. **DISK FL:** 9–40(90); corolla radial or ± bilateral, white to ± purple, lobes narrowly triangular; anther tip triangular; style branches linear. **FR:** 4-angled, narrowly obpyramidal; pappus of 3–10 scales, sometimes 0 in outer fr. 12 spp.: sw US, Mex. (J. Palafox, Spanish general, 1776–1847) [Strother 2006 FNANM 21:388–391]

P. arida B.L. Turner & M.I. Morris (p. 399) Ann (or short-lived per). **ST:** gen erect, much-branched. **LF:** 2–12 cm; petioles 5–15 mm; blade linear to lance-linear. **INFL:** heads discoid (± radiant); phyllaries linear, scabrous to densely glandular. **FL:** 9–40; corolla white to pink; anthers pink to purple. **FR:** 10–15 mm, ± strigose; pappus of outer fr 0 or of 3–8 scales of varying length; pappus of inner fr of 4 scales 8–12 mm, and 4 shorter scales. $2n$=24.

1. Pl gen 1–7 dm; main st diam ≤ 5 mm; heads
   (incl fls) gen 20–25 mm . . . . var. ***arida***
1′ Pl gen 9–20 dm; main st diam gen 5–10 mm;
   heads (incl fls) gen 28–35 mm . . . . var. ***gigantea***

var. ***arida*** **ST:** ± rough-hairy, glandular distally. **LF:** 2–10 cm, ± rough-canescent. **INFL:** heads gen 5–40; involucre 5–10 mm diam, cylindric; phyllaries 10–20 mm. **FL:** 9–20; corolla 9–11 mm, outer ± bilateral. **FR:** 10–15 mm. Washes, dunes, other sandy places in creosote-bush scrub; < 1000 m. D; NV, AZ, n Mex. Jan–Sep

var. ***gigantea*** (M.E. Jones) B.L. Turner & M.I. Morris GIANT SPANISH-NEEDLE **ST:** ± glabrous. **LF:** 6–12 cm, gen ± glabrous. **INFL:** heads gen 10–20; involucre 10–20 mm diam, obconic; phyllaries 16–25 mm. **FL:** 18–40; corolla 10–13 mm, radial. **FR:** 12–16 mm. Sand dunes; < 100 m. DSon (se Imperial Co.); sw AZ; SON. Mar–May ★

PECTIS

Ann [per] often scented. **ST:** prostrate to erect. **LF:** simple, opposite, sessile, narrow, bristly-ciliate, dotted with embedded oil glands. **INFL:** heads radiate, peduncled, [1] in lfy cyme-like clusters; involucre cylindric to bell-shaped; phyllaries in 1 series, free, gland-dotted; receptacle epaleate. **RAY FL:** as many as and subtended by phyllaries; corolla yellow. **DISK FL:** few to many; corolla yellow, [4]5-lobed, gen 2-lipped. **FR:** cylindric, gen puberulent; pappus of bristles, scales, or awns. ± 85 spp.: w US, Caribbean, S.Am. (Greek: comb, from ciliate lvs) [Keil 2006 FNANM 21:222–230]

P. papposa Harv. & A. Gray var. ***papposa*** (p. 399) CHINCH-WEED Ann 1–20 cm, mound-shaped; herbage spicy-scented. **ST:** 1–several from base, simple or much-branched. **LF:** narrowly linear, gland-dotted on margins. **INFL:** heads in ± dense cyme-like clusters, 6–10 mm diam; peduncle 3–10 mm; phyllaries 8, 3–5 mm, linear, each with ± terminal gland and several smaller ± marginal glands. **RAY FL:** 8; corolla 3–6 mm. **DISK FL:** 6–14; corolla 2–3.5 mm. **FR:** 2–4.5 mm; pappus of ray fr a low crown; pappus of disk fr ± 20 ± plumose bristles (rarely a crown). Arid plains, rocky slopes, in creosote-bush scrub; < 1500 m. W&I, D; to sw UT, sw NM, nw Mex. After summer rain, (Jun)Aug–Nov(Jan)

PENTACHAETA

David J. Keil & Meredith A. Lane

Ann from slender taproot, ± hairy (but appearing glabrous). **ST:** gen simple or branching proximally, erect, gen flexible, glabrous to hairy, green to ± red. **LF:** gen narrowly linear, ciliate, green. **INFL:** heads radiate, disciform, or discoid, nodding in bud, 1 or in open, ± flat-topped cluster; peduncle slender; involucre 3–7 mm, gen bell-shaped; phyllaries in 2–3(4) series, graduated or ± equal, lanceolate to (ob)ovate, green, margins widely scarious; receptacle flat to slightly convex, epaleate. **RAY FL:** present or 0, or corolla reduced to tube; corolla white, yellow, 2-colored yellow and white, or ± red. **DISK FL:** 4–90+; corolla yellow to ± red or maroon; style tips linear, acute. **FR:** 1.5–3 mm, oblong-fusiform, gen compressed, gen hairy; pappus

of 0–20 fragile, slender bristles. 6 spp.: CA, nw Baja CA. (Greek: five bristles, from pappus) [Keil 2007 Madroño 54:343–344; Nesom 2006 FNANM 20:46–48] Similar, closely related to *Rigiopappus, Tracyina*.

1. Well developed ray fls 0; ray of pistillate fls 0 or pistillate fls 0
2. Disk fls gen 4; corolla 3-lobed . . . . **P. alsinoides**
2′ Disk fls 6–34; corolla 5-lobed . . . . **P. exilis** subsp. **exilis**
1′ Well developed ray fls gen present, ray evident
  3. Ray fls white, ray sometimes ± red abaxially; pappus bristles present or 0
    4. Ray fls 7–16; peduncle glabrous to short-hairy. . . . . **P. bellidiflora**
    4′ Ray fls (0)1–3; peduncle soft-shaggy-hairy. . . . . **P. exilis** subsp. **aeolica**
  3′ Ray fls gen yellow or proximally yellow, distally white, sometimes ± brown-orange;
    pappus bristles present
    5. Ray fls 7–12, disk fls 10–23; largest lvs < 2.5 cm . . . . **P. fragilis**
    5′ Ray fls 14–52, disk fls 30–91; largest lvs 2.5–5.5 cm
      6. Phyllaries lanceolate, soft-shaggy-hairy; pappus bristles 8–12 . . . . **P. lyonii**
      6′ Phyllaries narrowly to widely-elliptic, glabrous or sparsely short-hairy; pappus bristles 5–8 . . . . **P. aurea**
        7. Ray proximally yellow, distally white . . . . subsp. **allenii**
        7′ Ray yellow to orange . . . . subsp. **aurea**

P. alsinoides Greene (p. 403)   **ST**: 3–14 cm, hairy below heads. **LF**: < 3.5 cm, 1 mm wide; adaxially hairy; abaxially gen glabrous. **INFL**: heads ≤ 200 per pl; peduncle hairy; involucre narrowly obconic; phyllaries elliptic, glabrous or sparsely hairy. **RAY FL**: gen 5; ray 0 or < 1 mm, corolla gen reduced to tube, light yellow, tips ± red. **DISK FL**: gen 4; corolla 3-lobed, colored as ray fls. **FR**: pappus bristles 3, slightly expanded at base. 2*n*=18. Grassy areas, open woodland, chaparral openings; < 550 m. NCoR, SNF, n CCo (San Francisco), SnFrB, SCoR, w WTR. Mar–Jun

P. aurea Nutt.   **ST**: 5–36 cm, short-hairy. **LF**: < 5.5 cm, 3 mm wide; adaxially glabrous to sparsely long-hairy; abaxially glabrous to densely hairy. **INFL**: heads ≤ 22 per pl; peduncle short-hairy; involucre widely obconic; phyllaries narrowly to widely elliptic, glabrous or sparsely short-hairy. **RAY FL**: ray 3–12 mm, yellow to ± brown-orange or proximally yellow and distally white. **DISK FL**: 30–90; corolla 5-lobed, yellow. **FR**: pappus bristles 5–8, expanded at base.

  subsp. **allenii** D.J. Keil ALLEN'S PENTACHAETA **RAY FL**: 30–45; ray yellow in proximal 1/3, white in distal 2/3. Grassy areas; < 500 m. s SCo, PR (Orange Co.). Mar–May ★

  subsp. **aurea** (p. 403) GOLDEN-RAYED PENTACHAETA **RAY FL**: 14–52; ray yellow to ± brown-orange. 2*n*=18. Grassy areas, scrub, woodland, conifer forest; < 2250 m. SCo, SnGb, SnBr, PR; Baja CA. Mar–Jul ★

P. bellidiflora Greene (p. 403) WHITE-RAYED PENTACHAETA **ST**: 6–17 cm, sparsely hairy. **LF**: < 4.5 cm, 1 mm wide, glabrous. **INFL**: heads ≤ 4 per pl; peduncle glabrous to short-hairy; involucre widely bell-shaped; phyllaries elliptic to obovate, glabrous. **RAY FL**: 7–16; ray 3–6 mm, white, sometimes ± red abaxially. **DISK FL**: 16–38; corolla 5-lobed, yellow. **FR**: pappus bristles 0 or 5, slightly expanded at base. 2*n*=18. Grassy or rocky areas; < 620 m. n CCo, SnFrB. Mar–May ★

P. exilis (A. Gray) A. Gray   **ST**: 2–15 cm, hairy at base and below heads. **LF**: < 3.2 cm, 1 mm wide; adaxially sparsely hairy; abaxially glabrous. **INFL**: heads ≤ 23 per pl; peduncle soft-shaggy-hairy; involucre ± bell-shaped; phyllaries elliptic to obovate, glabrous. **RAY FL**: 0–5; ray 0–2 mm, white or pink or corolla reduced to a ± yellow tube tinged with red. **DISK FL**: 6–34; corolla 5-lobed, yellow or ± red. **FR**: pappus bristles 0, 3, or 5, not expanded at base. 2*n*=18.

  subsp. **aeolica** Van Horn & Ornduff (p. 403) SAN BENITO PENTACHAETA **INFL**: heads gen radiate. **PISTILLATE FL**: (0)1–5; ray well developed, white. **DISK FL**: yellow. Grassland, woodland; < 750 m. SnFrB, SCoR. Mar–May ★

  subsp. **exilis** **INFL**: heads discoid or disciform. **PISTILLATE FL**: 0–3; ray 0 or vestigial, white to ± red, tube yellow or ± red. **DISK FL**: yellow or ± red. Grassland, foothill woodland; < 900 m. NCoR, SNF, GV, CW. Mar–May

P. fragilis Brandegee (p. 403) FRAGILE PENTACHAETA **ST**: 4–16 cm, glabrous to short-hairy (or hairy at base and proximal to heads). **LF**: < 2.5 cm, 3 mm wide; adaxially hairy; abaxially glabrous. **INFL**: heads ≤ 52 per pl; peduncles ± hairy; involucres widely bell-shaped; phyllaries elliptic to obovate, glabrous or hairy. **RAY FL**: 7–12; ray ≤ 5 mm, yellow. **DISK FL**: 10–23; corolla 5-lobed, yellow. **FR**: pappus bristles 8–20, not expanded at base. 2*n*=18. Grassy areas, chaparral, arid woodland, conifer forest; 50–2100 m. s SN, SnJV, s SCoR, WTR. Mar–Jun ★

P. lyonii A. Gray (p. 403) LYON'S PENTACHAETA · **ST**: 6–48 cm, hairy, esp proximal to heads and at nodes. **LF**: ≤ 5.5 cm, ≤ 6 mm wide; adaxially hairy; abaxially glabrous. **INFL**: heads ≤ 36 per pl; peduncles hairy; involucre widely bell-shaped; phyllaries lanceolate, soft-shaggy-hairy. **RAY FL**: 17–42; ray 3–8 mm, yellow. **DISK FL**: 21–91; corolla 5-lobed, yellow. **FR**: pappus bristles 8–12, slightly expanded at base. 2*n*=18. Coastal scrub, grassland, chaparral openings; < 400 m. c SCo (Ventura, Los Angeles cos.), s ChI (Santa Catalina Island), WTR. Mar–Aug ★

PERICOME

Per, subshrub. **ST**: many from base. **LF**: simple, gen opposite (or distal-most alternate), petioled, deltate-ovate, puberulent, gland-dotted; tip long-acuminate. **INFL**: heads discoid, small, 3–30+ in ± flat-topped cyme-like clusters, these often in lfy-bracted, compound cyme-like clusters; peduncle slender; involucre cylindric to bell-shaped; phyllaries in 1 series, linear, ± fused; receptacle rounded, epaleate. **FL**: 30–70; corolla 4-lobed, creamy yellow; anther tips triangular; style tips linear, tapered. **FR**: oblanceolate, flat, black, puberulent, margins ± thickened, densely ciliate; pappus a low crown of fringed scales, sometimes with 1–2 bristles. 2 spp.: sw US, n Mex. (Greek: hairs around, from ciliate fr margin) [Yarborough & Powell 2006 FNANM 21:334–335]

P. caudata A. Gray (p. 403)   **ST**: ≤ 2 m, much-branched, ± puberulent, resin-dotted. **LF**: many; petiole 1.5–5 cm; blade 3–12 cm, base rounded to cordate or hastate, margin entire or proximal 1/2 toothed or shallowly few-lobed. **INFL**: peduncle 5–30 mm; involucre 4.5–6 mm diam; phyllaries 4.5–7 mm, fused in proximal 1/2, margins trans-parent, tip soft-hairy. **FL**: corolla 3–5 mm. **FR**: 3.5–5 mm; pappus scales ± 1 mm, bristles 0–2, 1–4.5 mm, gen unequal. 2*n*=36. Canyons, rocky slopes, pinyon/juniper woodland, riparian; 1200–2550 m. s SN, SNE, DMtns; to CO, OK, TX, n Mex. Jul–Oct

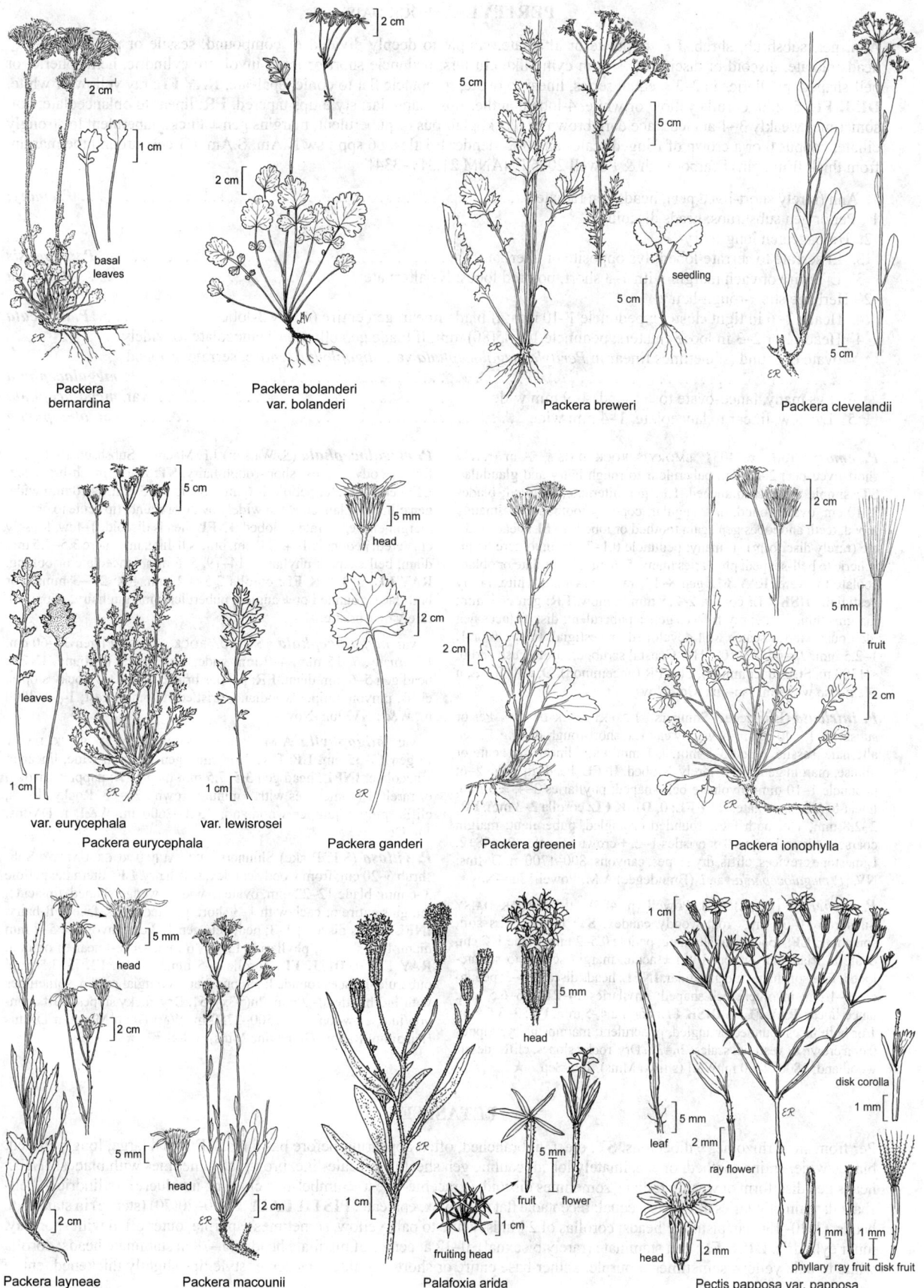

Packera
bernardina

Packera bolanderi
var. bolanderi

Packera breweri

Packera clevelandii

var. eurycephala

var. lewisrosei

Packera eurycephala

Packera ganderi

Packera greenei

Packera ionophylla

Packera layneae

Packera macounii

Palafoxia arida

Pectis papposa var. papposa

PERITYLE ROCK DAISY

Ann, per, subshrub, shrub. **LF:** opposite or alternate, simple to deeply divided or compound, sessile or petioled. **INFL:** heads radiate, discoid or disciform, 1 or in cyme-like clusters; peduncle short or long; involucre cylindric, hemispheric, or bell-shaped; phyllaries in 2–3 ± equal series, linear to ovate; receptacle flat to conic, epaleate. **RAY FL:** ray yellow or white. **DISK FL:** 5–200; corolla yellow or white, 4-lobed; anther tips triangular; style tips tapered. **FR:** linear to oblanceolate, flat, sometimes weakly 3–4-angled; face dark brown or black, glabrous or puberulent, margins gen ± thick, puberulent to strongly ciliate; pappus 0 or a crown of fringed scales and 0–2 slender bristles. 66 spp.: sw N.Am, S.Am. (Greek: around the margin, from thick fr margin) [Yarborough & Powell 2006 FNANM 21:317–334]

1. Ann (rarely short-lived per); heads gen radiate . . . . *P. emoryi*
1′ Per or gen subshrubs; heads discoid
  2. Herbage gen long-hairy
    3. Lf serrate to serrate-lobed; lvs opposite or alternate . . . . *P. inyoensis*
    3′ Lf entire or each margin with 1–3 short, pointed lobes; lvs alternate . . . . *P. villosa*
  2′ Herbage short-rough-hairy
    4. Heads 2–6 in tight clusters; peduncle 1–10 mm; lf blade linear, gen entire (rarely 3-lobed) . . . . *P. intricata*
    4′ Heads 1 or 2–3 in loose clusters; peduncle 10–45(80) mm; lf blade gen elliptic or lanceolate to widely ovate or round (sometimes linear in *Perityle megalocephala* var. *oligophylla*), entire, serrate or lobed
     . . . . *P. megalocephala*
      5. Lvs many, lance-ovate to ± round, 4–9 mm wide . . . . var. *megalocephala*
      5′ Lvs few, linear to lanceolate, 1–4 mm wide . . . . var. *oligophylla*

P. emoryi Torr. (p. 403) EMORY'S ROCK DAISY Ann (rarely short-lived per) 2–60 cm, puberulent to rough-hairy and glandular. **ST:** simple to much-branched. **LF:** gen alternate, petioled; blades 2–10 cm, ovate, round, or triangular, coarsely toothed to palmately lobed, teeth and lobes gen again toothed or lobed. **INFL:** heads radiate (rarely disciform), 1–many; peduncle 0.1–7 cm; involucre hemispheric to bell-shaped; phyllaries many, 5–6 mm, lanceolate or oblanceolate to ovate. **RAY FL:** gen 8–14; ray 1.5–4 mm, white, rarely vestigial. **DISK FL:** corolla 2–2.5 mm, yellow. **FR:** gen 2–3 mm; margins thin, ciliate; ray fr faces gen ± puberulent; disk fr faces gen glabrous; pappus scales well developed or vestigial, bristle 0 or 1, 1–2.5 mm. 2*n*=64–72, 100–116. Coastal scrub, creosote-bush scrub; < 1300 m. SCo (uncommon), ChI, PR (uncommon), D; to UT, AZ, n Mex; also w S.Am. Jan–Jun, Oct–Nov

P. intricata (Brandegee) Shinners DESERT ROCK DAISY Per or subshrub, 13–40 cm, from woody caudex, short-rough-hairy. **LF:** gen alternate, sessile, blade 3–8 mm, ± 1 mm wide, linear; tip acute or obtuse, margin gen entire, rarely 3-lobed. **INFL:** heads discoid, 2–6; peduncle 1–10 mm; involucre bell-shaped; phyllaries 3–4, ± 5 mm, linear to (ob)lanceolate. **RAY FL:** 0. **DISK FL:** corolla 2–3 mm. **FR:** 2–2.8 mm, 1 or both faces rounded or angled, puberulent; margin coarsely ciliate; pappus 0 or bristles 1–2, + crown of scales. 2*n*=38±2. Limestone crevices, cliffs, dry slopes, canyons; 800–1700 m. DMtns; NV. [*P. megalocephala* var. *i.* (Brandegee) A.M. Powell] Jun–Nov

P. inyoensis (Ferris) A.M. Powell (p. 403) INYO ROCK DAISY Subshrub 10–30 cm, from woody caudex. **ST:** many; hairs soft, spreading. **LF:** opposite or alternate; petiole 0.5–2 mm; blade 1–2 cm, ovate to triangular or round, tip ± acute, margin serrate to serrate-lobed, face soft-hairy and glandular. **INFL:** heads discoid, 1–3; peduncle 1–4 cm; involucre bell-shaped; phyllaries 14–21, 5.6–6.5 mm, lance-linear. **RAY FL:** 0. **DISK FL:** corolla 4–5 mm. **FR:** 3–3.5 mm, 1 or both faces rounded or angled, puberulent; margins hairy; pappus 0 or a crown of vestigial scales. 2*n*=36. Dry, rocky slopes, cliffs, desert woodland; 1800–2800 m. W&I (s Inyo Mtns). Jun–Sep ★

P. megalocephala (S. Watson) J.F. Macbr. Subshrub 15–60 cm, from woody caudex, short-rough-hairy. **ST:** many, much-branched. **LF:** gen alternate; petiole 1–6 mm; blade 7–15 mm, 1–10 mm wide, gen elliptic or lanceolate to widely ovate or round, tip acute to obtuse, margin entire, serrate or lobed. **INFL:** heads discoid, 1–few, loosely clustered; peduncle 1–4.5(8) cm, bracts lf-like; involucre 3.5–7.5 mm diam, bell-shaped; phyllaries 14–19, 5–6 mm, lanceolate or oblong. **RAY FL:** 0. **DISK FL:** corolla 3.5–4.2 mm. **FR:** 2.5–3 mm; 1 or both faces rounded or ± angled, puberulent; margin hairy; pappus 0, rarely of 1 bristle.

var. ***megalocephala*** NEVADA ROCK DAISY Pl gen 30–60 cm. **LF:** many, 7–15 mm, 4–9 mm wide, lance-ovate to round. **INFL:** head gen 5–6 mm diam. **FR:** pappus bristle 0. 2*n*=34. Rocky slopes, cliffs, pinyon/juniper woodland, bristlecone-pine forest; 1400–3000 m. W&I; NV. Jun–Nov

var. ***oligophylla*** A.M. Powell SMALL-LEAVED ROCK DAISY Pl gen 15–35 cm. **LF:** few, 7–17 mm, gen 1–4 mm wide, linear to lanceolate. **INFL:** head gen 3.5–7.5 mm diam. **FR:** pappus bristle 0 or rarely 1 (sometimes with a minute crown). 2*n*=68. Rocky slopes, cliffs, pinyon/juniper woodland; 1300–2600 m. W&I, n DMtns. Jul–Oct

P. villosa (S.F. Blake) Shinners HANAUPAH ROCK DAISY Subshrub 6–20 cm, from woody caudex, soft-hairy. **LF:** alternate; petiole 3–6 mm; blade 12–22 mm, ovate to widely wedge-shaped, tip acute, margins entire or each with 1–3 short, pointed lobes, face short-soft-hairy. **INFL:** heads discoid, 1–3; peduncle gen 1–2 cm; involucre 5–7 mm diam, bell-shaped; phyllaries 13–23, 6 mm, lance-linear to oblong. **RAY FL:** 0. **DISK FL:** corolla 4–5 mm, yellow. **FR:** 3–3.5 mm, puberulent; faces rounded; pappus 0 or a vestigial crown, sometimes with 1–2 bristles 1–2 mm. 2*n*(=3*x*)=51. Dry, rocky slopes, cliffs, pinyon/juniper woodland; 1500–2750 m. W&I (Inyo Mtns), n DMtns (Panamint Range, Grapevine Mtns). Jul–Sep ★

PETASITES

Per from stout rhizome; ± dioecious. **ST:** erect, unbranched, often appearing before basal lvs. **LF:** basal large, long-petioled, blades wide, entire, toothed, or ± palmately lobed; cauline gen sheathing, scale-like, proximal sometimes with blades. **INFL:** heads gen disciform or weakly radiate, sometimes discoid, in raceme-like to ± umbel-like clusters; involucre ± cylindric to bell-shaped; main phyllaries in 1 series, equal; receptacle flat to convex, epaleate. **PISTILLATE FL:** 0–20(70) (sterile) in staminate heads, (1)30–130+ in pistillate heads; corollas of 2 kinds, white to pale yellow, sometimes ± purple, outer often with short ray, inner cylindric. **DISK FL:** gen staminate (rarely bisexual), 0–12 at center of pistillate heads, 11–78 in staminate heads; corolla white to pale yellow, sometimes ± purple; anther base entire or short-sagittate, tips acute; style tips slightly thickened, entire

or slightly lobed; pappus reduced. **FR**: cylindric, 5–10-ribbed; pappus bristles 60–100+. 15–18 spp.: N.Am, Eurasia. (Greek: broad-brimmed hat, from large lvs) [Bayer et al. 2006 FNANM 20:635–640]

P. frigidus (L.) Fr. var. ***palmatus*** (Aiton) Cronquist (p. 403) WESTERN SWEET COLTSFOOT **ST**: 2–8 dm. **LF**: basal blade 10–40 cm wide, ± round to cordate or reniform, lobes palmate, coarsely toothed or again lobed, adaxially (±) glabrous, sometimes abaxially loosely tomentose. **INFL**: bracts 2–6 cm, entire to serrate, parallel-veined; heads gen 10–20; involucre 5–9 mm, bell-shaped, often ± purple; phyllaries linear. **PISTILLATE FL**: outer corollas with tube 4–6 mm, ray 2–7 mm, white to pale pink; inner corollas 3–5 mm, cylindric. **DISK FL**: corolla 3.5–5 mm, white to pale pink. **FR**: 3–4.5 mm; pappus 6–13 mm. 2*n*=60,61,62. Forest, streambanks, gen wet soil; < 1400 m. NW, n SNH, nw&n-c CW; to YT, e Can, ne US. Jan–Apr

PETRADORIA

Lowell E. Urbatsch & Meredith A. Lane

1 sp.: w N.Am. (Greek: rock goldenrod) [Urbatsch et al. 2006 FNANM 20:171–172]

P. pumila (Nutt.) Greene var. ***pumila*** (p. 403) ROCK-GOLDENROD Per (subshrub) < 3 dm, glabrous (hairy), taprooted from branched, woody base. **ST**: several, erect, striate, gummy, green, white to tan in age. **LF**: alternate, 2–12 cm, 1–12 mm wide, linear to (ob)lanceolate, leathery, resin-dotted, entire, 3–5-veined, margins scabrous. **INFL**: heads radiate, many, in dense, flat-topped clusters; involucre < 10 mm, < 3 mm diam, cylindric; phyllaries 10–21, graduated in 3–6 series, oblong to ovate, light yellow, tips green; receptacle convex, pitted, sometimes with 1–3 scales or awn-like appendages ≤ 1 mm, epaleate. **RAY FL**: (1)2–3; ray 2.5–4.5 mm, yellow. **DISK FL**: 2–4, staminate; corolla 4.5–6.2 mm, yellow, lobes 0.8–1.3 mm; anther tip ± lanceolate; style tips awl-shaped. **FR**: 4–5 mm, 6–9-veined, slightly compressed, glabrous; pappus bristles thread-like, tan to ± brown. 2*n*=18. Rocky soils, pine forest, juniper scrub, often on limestone; 1050–3400 m. DMtns; to ID, WY, NM. [*P. p.* subsp. *p.*] Other var. in AZ, NM, UT. Jul–Oct ★

PEUCEPHYLLUM PYGMY-CEDAR

1 sp.: sw N.Am. (Greek: fir-lf) [Strother 2006 FNANM 21:378]

P. schottii A. Gray (p. 403) Shrub or small tree ≤ 3 m, rounded. **ST**: densely lfy, green. **LF**: alternate, gen 1–2 cm, narrowly linear, thick, gen entire, glabrous, gland-dotted, resin-varnished. **INFL**: head 1, discoid; peduncle 8–25 mm, lfy-bracted; involucre cylindric or obconic to bell-shaped; phyllaries 9–18 in 1–2 series, 8–12 mm, linear to lanceolate, thick, acuminate, gland-dotted near tip, margins often scarious; receptacle flat, epaleate. **FL**: 12–21; corolla 6.5–8.5 mm, pale yellow, sometimes distally ± purple, tube << throat; anther base weakly tailed, tip lanceolate to ovate; style branches minutely papillate, rounded-truncate. **FR**: 3–4 mm, narrowly obconic, weakly angled, ± black, bristly; pappus of 30–60 fine bristles, 2–5 mm (sometimes also 15–20 slender scales 4–6 mm), straw-colored to red-brown. 2*n*=20. Rocky slopes, washes, creosote-bush scrub; < 1400 m. D; to UT, AZ, nw Mex. Dec–Jun

PHALACROSERIS

Kenton L. Chambers

1 sp. (Greek: bald chicory) [Chambers 2006 FNANM 19:374]

P. bolanderi A. Gray (p. 403) BOLANDER'S MOCK DANDELION Per from fleshy caudex and ± black taproot, glabrous; sap milky. **ST**: 10–45 cm, scapose. **LF**: basal, 6–20 cm, linear to oblanceolate, entire, tip obtuse to acuminate. **INFL**: heads liguliflorous; involucre 7–13 mm, green, glabrous; phyllaries lanceolate, ± equal or few outermost shorter, inner ± fused near base in fr, not reflexed; receptacle convex, epaleate, smooth. **FL**: 13–many; ligules 10–18 mm, much exceeding involucre, yellow, readily withering. **FR**: 3–4 mm, oblong, ± 3-angled, flat-topped, smooth, pale brown, dark-spotted; pappus 0 or a low crown. 2*n*=18. Wet meadows, sphagnum bogs; 1600–3000 m. SNH. Jun–Aug

PLECOSTACHYS

Shrub to subshrub, mildly scented. **ST**: slender, much-branched, thinly white-tomentose. **LF**: cauline, alternate, ± sessile, entire, adaxially thinly tomentose, abaxially densely gray-tomentose. **INFL**: heads disciform, gen in compact, rounded to ± flat-topped clusters; involucre ± cylindric; phyllaries graduated in 3–4 series, proximally tomentose, distally scarious, inner oblanceolate, tips white, spreading, ray-like; receptacle flat or convex, pitted, epaleate. **PISTILLATE FL**: < to > disk fls, corolla narrowly tubular, minutely 5-lobed. **DISK FL**: corolla ± yellow or lobes pink to purple, tube > narrowly cylindric throat; anther base short-tailed, tip triangular; style branches truncate. **FR**: ± cylindric, few-ribbed, puberulent, scar oblique; pappus of ± 20 fragile white bristles. 2 spp.: s Afr. (Greek: braided spike, from intricately branched habit) [Riefner & Nesom 2009 Phytologia 91:542–565]

P. serpyllifolia (P.J. Bergius) Hilliard & B.L. Burtt PETITE-LICORICE **ST**: 3–15 dm, spreading, lfy. **LF**: (4)5–9 mm, ± elliptic to round or ± obovate, adaxially ± glabrous in age. **INFL**: heads 20–50+; peduncle 0–3 mm; involucre ± 4 mm. **PISTILLATE FL**: 4–7(10); corolla ± 2.5 mm. **DISK FL**: 3–7; corolla ± 3 mm, tube ± 2 mm, lobes ± 0.3 mm. **FR**: 0.7–0.8 mm, ± brown; pappus 2–2.5 mm. Sea bluffs, grassy areas, coastal wetlands, strand, saline or alkaline soils, escaped from cult; < 350 m. CCo, SCo; Portugal; native to s Afr. Feb–Jun

PLEIACANTHUS

L.D. Gottlieb

1 sp.: W US. (Greek: unusually thorny) [Gottlieb 2006 FNANM 19:361]

P. spinosus (Nutt.) Rydb. (p. 403) THORNY SKELETONWEED
Subshrub 1–5 dm, from woody caudex; sap milky. **ST:** 1–4+, woolly-
hairy in axils of bud scales at and just below ground level, otherwise glabrous; branches rigidly spreading, thorn-tipped. **LF:** cauline;
proximal linear, entire, distal bract-like. **INFL:** heads liguliflorous, 1;
peduncle 1–4 mm; involucre 7–12 mm; phyllaries in 2 series, outer
4–6, unequal, < 1/2 × inner, inner 3–5, equal, lance-linear; receptacle
flat or convex, smooth, epaleate. **FL:** 3–5; ligules pink or lavender to
red-purple (white). **FR:** 6–8 mm; cylindric, faces 5, each with a long,
narrow, shallow, groove; pappus persistent, of 50+ free, minutely
barbed, light tan bristles of 2 lengths. 2*n*=16. Gravelly washes,
slopes, sagebrush scrub, pinyon/juniper woodland; 425–2900 m.
SNH, SnGb, SnBr, GB, DMtns; to OR, MT, UT, AZ. [*Stephanomeria
s.* (Nutt.) Tomb] Jun–Sep

PLEUROCORONIS ARROWLEAF

David J. Keil & A. Michael Powell

Per or subshrub. **LF:** proximally opposite, distally alternate, sometimes in axillary clusters. **INFL:** heads discoid, 1 or few
in cyme-like clusters; phyllaries graduated in 3–5 series, outer short, ovate, inner lanceolate; receptacle ± flat, epaleate. **FL:**
25–30; corolla narrowly cylindric, white or purple tinged; anther base ± rounded, tip ovate or oblong; style branches ± 1.5 mm,
club-shaped. **FR:** ellipsoid to obconic, 4–5-ribbed; sides densely hairy; pappus of bristles and scales. 3 spp.: sw US, nw Mex.
(Greek, Latin: side crown, from pappus) [Nesom 2006 FNANM 21:510–511]

P. pluriseta (A. Gray) R.M. King & H. Rob. (p. 409) Subshrub ≤
60 cm. **ST:** slender, much-branched, distally glandular. **LF:** thin; petiole thread-like, 20–60 mm, glandular, blade 3–10 mm, lanceolate to
± diamond-shaped, gen few-toothed, glabrous or ± glandular. **INFL:**
heads 6–11 mm; phyllaries ± 25, 2.5–6 mm, glandular, tips darker,
gen recurved. **FL:** corolla 4–5 mm. **FR:** 3–4 mm; pappus bristles
10–16, 2.5–5 mm, scales 10–12, 1–2 mm. 2*n*=18. Common. Rocky
canyons, wash banks, creosote-bush scrub; < 1700 m. D; to UT, AZ,
nw Mex. Oct–Jun

PLUCHEA

David J. Keil & G. Ledyard Stebbins

Ann to large shrub, stiff. **LF:** simple, alternate. **INFL:** heads disciform, many, in ± flat-topped or panicle-like clusters; involucre ± hemispheric; phyllaries graduated in 3–6 series, outer leathery distally, inner narrower, ± membranous; receptacle flat,
epaleate. **PISTILLATE FL:** many; corolla very slender, 4–5-lobed. **DISK FL:** few, bisexual or staminate; corolla 5-lobed,
pink or purple (in CA); anther bases short-tailed, tips ± ovate; style branches 0 to short, obtuse. **FR:** ± cylindric, grooved;
pappus 1 series of slender bristles. 40–60 spp.: trop, warm temp. (N.A. Pluche, 18th century French naturalist) [Nesom 2006
FNANM 19:478–484]

1. Ann or per, glandular; lvs 4–12 cm, ovate, toothed, not crowded . . . . *P. odorata* var. *odorata*
1′ Shrub or tree, nonglandular, finely silky-hairy; lvs 1–4 cm, linear to lanceolate or narrowly elliptic,
   entire, crowded . . . . *P. sericea*

P. odorata (L.) Cass. var. ***odorata*** (p. 409) SALTMARSH-
FLEABANE Pl 5–12+ dm, coarse, scented. **LF:** petioled. **INFL:**
involucre 4.5–5.5 mm; phyllaries lavender to bright purple. **PISTIL-
LATE FL:** corolla 3.5–4 mm, purple. **DISK FL:** corolla 4–5 mm,
purple. **FR:** ± 1 mm, minutely rough-hairy; pappus bristles slender to
tip. 2*n*=20. Moist, often saline valley bottoms; < 1450 m. GV, n CCo,
SW, D; to e US, Caribbean, n S.Am, Afr, Pacific islands. Jun–Nov

P. sericea (Nutt.) Coville (p. 409) ARROW-WEED Pl 1–5 m, not
scented. **LF:** sessile or subsessile. **INFL:** involucre 3.5–6 mm; phyllaries brown-purple. **PISTILLATE FL:** corolla 5–6.5 mm, pink to
deep rose. **DISK FL:** corolla 5–6 mm, pink to deep rose. **FR:** 0.5–1
mm, smooth. 2*n*=20. Forming thickets in stream bottoms, washes,
canyons, around springs, sometimes in saline areas; < 950 m. SnJV, s
CCo, SCoR, SW (exc n ChI), D; to UT, TX, nw Mex. Mar–Jul

POROPHYLLUM

Subshrub [ann, per]; strongly pungent-scented. **ST:** erect. **LF:** simple, alternate or opposite; blade dotted with embedded oil
glands. **INFL:** heads discoid, peduncled, 1 or in ± lfy cyme-like clusters; involucre ± cylindric; phyllaries in 1 series, equal,
free or fused at base, gland-dotted; receptacle flat, epaleate. **FL:** corolla white to green-yellow or purple; style branches long,
slender. **FR:** cylindric; pappus bristles 25–50+. ± 30 spp.: N.Am, S.Am. (Greek: pore-lf, from gland-dotted lvs) [Strother 2006
FNANM 21:233–235] *P. ruderale* (Jacq.) Cass. var. *macrocephalum* (DC.) Cronquist (ann, lvs narrowly elliptic to ovate or
obovate) collected as an agricultural weed in Seal Beach, Orange Co. (SCo).

P. gracile Benth. (p. 409) ODORA, SLENDER PORELEAF **ST:** 1–
many, 3–7 dm, glaucous, glabrous; branches slender, ascending. **LF:**
1–5 cm, entire, glaucous. **INFL:** heads 1–few; peduncle slender;
phyllaries 5, free, 10–16 mm, oblong, glaucous, dotted or streaked
with glands. **FL:** 20–30; corolla 7–9 mm, ± purple or ± white. **FR:**
8–10 mm; pappus 6–7 mm, dull white to ± brown. 2*n*=48. Rocky
slopes, washes, scrub, desert woodland; < 1600 m. PR, D; to NV,
TX, n Mex. Sep–Jun

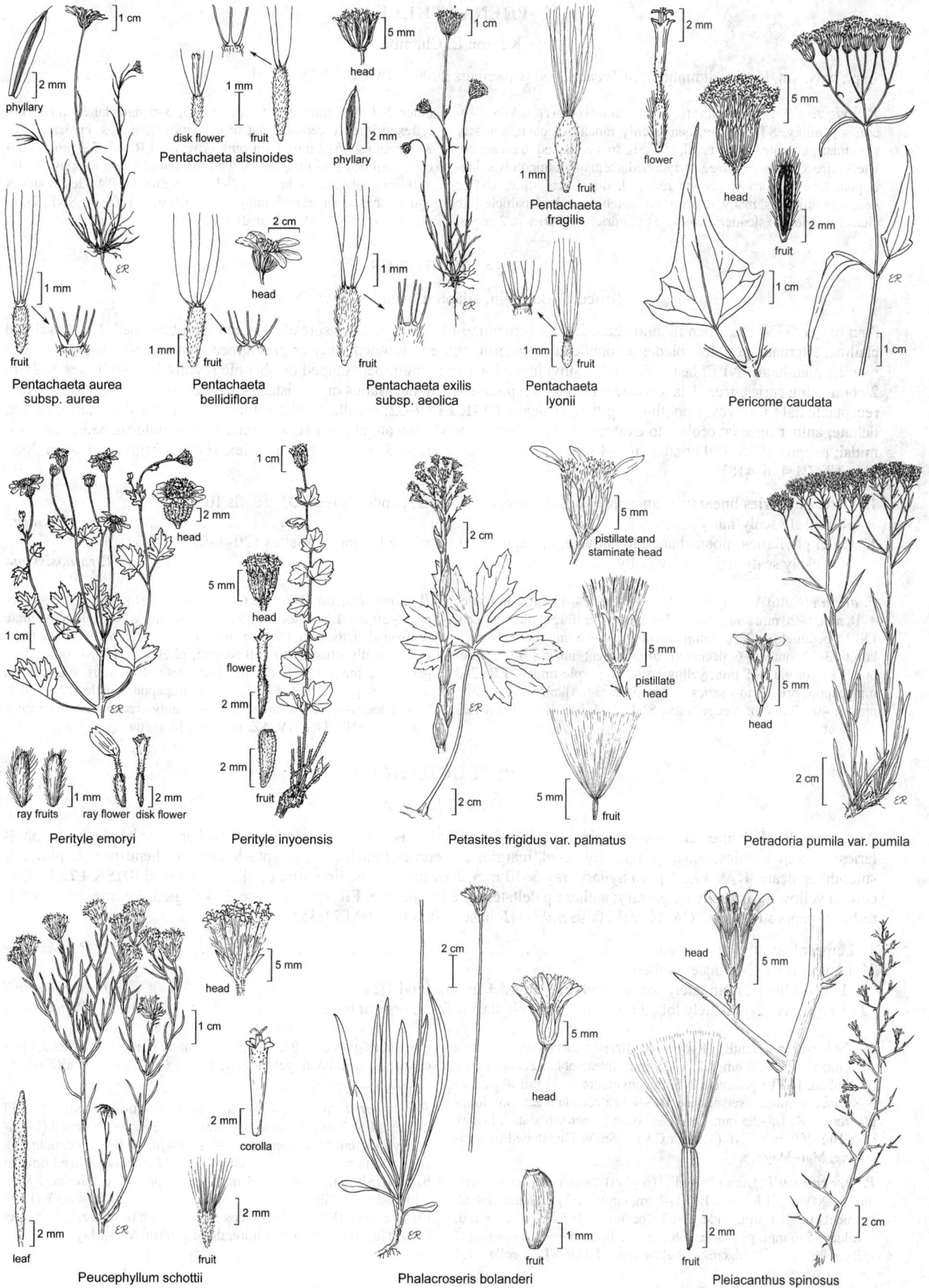

Pentachaeta alsinoides

phyllary

disk flower   fruit

Pentachaeta aurea
subsp. aurea

fruit

Pentachaeta
bellidiflora

head

fruit

Pentachaeta exilis
subsp. aeolica

fruit

phyllary

Pentachaeta
fragilis

fruit

Pentachaeta
lyonii

fruit

flower

head

fruit

Pericome caudata

Perityle emoryi

ray fruits   ray flower   disk flower

head

Perityle inyoensis

flower

fruit

Petasites frigidus var. palmatus

pistillate and
staminate head

pistillate
head

fruit

head

Petradoria pumila var. pumila

Peucephyllum schottii

head

corolla

leaf

fruit

Phalacroseris bolanderi

head

head

fruit

Pleiacanthus spinosus

head

fruit

PRENANTHELLA

Kenton L. Chambers

1 sp.: w N.Am. (Latin: diminutive of *Prenanthes*) [Chambers 2006 FNANM 19:359–360]

P. exigua (A. Gray) Rydb. (p. 409) BRIGHTWHITE Ann 5–30 cm; sap milky. **ST**: slender, gen openly much-branched, sparsely glandular-puberulent. **LF**: basal, 1–3 cm, spoon-shaped to oblanceolate, entire to dentate or irregularly lobed, teeth often minutely spine-tipped, faces glabrous; cauline reduced, distal scale-like. **INFL**: heads liguliflorous, many, in open, intricately-branched, panicle-like clusters; peduncle slender; involucre cylindric; phyllaries in 2 series, outer 2–3, < 1 mm, deltate, inner 3–5, 3–5 mm, lance-linear, not reflexed in fr; receptacle flat or ± convex, glabrous, epaleate. **FL**: 3–4; ligules 1.5–2 mm, light pink or white. **FR**: 2.5–3.5 mm, cylindric, 5-ribbed and grooved, ± white to pale brown; pappus of 80+ stiff, minutely barbed bristles, 2–3 mm, white. 2*n*=14. Desert slopes and washes to sagebrush/juniper woodland; < 1900 m. SNE, D; to OR, CO, w TX, n Mex. Mar–Jun

PSATHYROTES

Bruce G. Baldwin, adapted from Strother (2006)

Ann or per, 3–30 cm, often mound-shaped; odor turpentine-like. **ST**: erect or spreading, gen much-branched. **LF**: basal and cauline, alternate, petioled; blades ± rounded to reniform, entire to toothed, gray or gray-green, gen scaly- or soft-hairy to ± woolly, glandular. **INFL**: heads discoid, 1; involucre 3–6+ mm diam, bell-shaped or obconic [cylindric]; phyllaries 8–24 in 2 contrasting series, free, ± lanceolate, obovate, or spoon-shaped, deciduous or persistent, tips erect or spreading to reflexed; receptacle flat to convex, smooth or ± pitted, epaleate. **DISK FL**: 9–32; corolla ± yellow, tube << cylindric throat, lobes erect, deltate; anther tip ± lanceolate to ovate; style branches rounded-truncate at tip. **FR**: ± cylindric or spindle-shaped to obpyramidal; pappus of 35–150 bristles in 1–4 series, free or basally fused. 3 spp.: sw US, nw Mex. (Greek: brittle) [Strother 2006 FNANM 21:416–418]

1. Outer phyllaries linear to ± lanceolate, tips gen erect, like inner; pappus bristles 35–50; fls 10–20;
   pl sparsely scaly-hairy . . . . ***P. annua***
1′ Outer phyllaries spoon-shaped to obovate, tips spreading to reflexed; pappus bristles 120–140; fls 16–32;
   pl densely scaly-hairy to ± woolly . . . . ***P. ramosissima***

P. annua (Nutt.) A. Gray (p. 409) Ann (per), 4–15 cm. **LF**: blades 4–18 mm, 6–26 mm wide, rounded-deltate to reniform, often toothed. **INFL**: peduncle 1–22(28) mm; involucre 6–9 mm, obconic; phyllaries 13–19, outer 5–6, deciduous or persistent, inner 8–13, deciduous. **FL**: corolla 4–5 mm, yellow, often ± purple-tipped. **FR**: 2–3 mm; pappus bristles in 1 series, 1–4 mm. 2*n*=34. Alkali soils, washes, playas; 400–2000 m. n edge SnBr, SNE, DMoj; to ID, UT, nw AZ. Apr–Oct

P. ramosissima (Torr.) A. Gray (p. 409) TURTLEBACK Ann or per, 3–30 cm. **LF**: blades 8–25 mm, 8–30 mm wide, rounded-deltate to ± rounded, toothed. **INFL**: peduncle 3–35(50) mm; involucre 6–10 mm, broadly obconic to bell-shaped; phyllaries 15–24, outer 5–6, persistent, inner 10–18, distally lanceolate, deciduous. **FL**: corolla 4.5–5 mm, pale yellow. **FR**: 1.5–3.5 mm; pappus bristles 120–140 in 2–4 series, 3–4 mm, ± equal. 2*n*=34. Sandy creosote-bush scrub; < 1200 m. s SNE, D; s NV, AZ, nw Mex. Mar–Jun

PSEUDOBAHIA

Dale E. Johnson

Ann, ± woolly. **LF**: alternate, entire to pinnately lobed. **INFL**: heads 1, radiate; involucre bell-shaped or hemispheric; phyllaries 3–±8 in 1 series, equal, proximally fused, margins sometimes translucent; receptacle conic or hemispheric, pitted or smooth, epaleate. **RAY FL**: 1 per phyllary; ray 5–10 mm, ± ovate, yellow, tip entire or slightly toothed. **DISK FL**: 8–25+; corolla yellow, proximally long-hairy; anther tip deltate; style tips deltate. **FR**: oblanceolate, 3–4-angled, ± compressed front-to-back; pappus 0. 3 spp.: CA. (Greek: false *Bahia*) [Johnson 2006 FNANM 21:351–353]

1. Largest lvs entire or 3-lobed . . . . ***P. bahiifolia***
1′ Largest lvs 1–2-pinnately lobed
  2. Largest lvs 1(2)-pinnately lobed; phyllaries fused ± in proximal 1/2 . . . . ***P. heermannii***
  2′ Largest lvs 2-pinnately lobed (exc small pls); phyllaries fused only at base . . . . ***P. peirsonii***

P. bahiifolia (Benth.) Rydb. (p. 409) HARTWEG'S GOLDEN SUNBURST Pl 5–20 cm. **LF**: 8–25 mm, linear-oblanceolate, entire or 3-lobed. **INFL**: peduncle 2–5 cm; involucre 5–6 mm; phyllaries 3–8, lance-elliptic, fused at base. **DISK FL**: corolla ± 2.5 mm, lobes glabrous. **FR**: 1.5–2.5 mm. 2*n*=8. Grassland, open woodland, in clay soil; 100–200 m. c SNF (Madera Co.), e SnJV. Threatened by agriculture. Mar–May ★

P. heermannii (Durand) Rydb. (p. 409) FOOTHILL SUNBURST, BRITTLESTEM Pl 1–4 dm. **LF**: 1–4 cm; largest 1(2)-pinnately lobed, segments 0.5–1.5 mm wide. **INFL**: peduncle 2–7 cm, often ± red; involucre 5–6 mm; phyllaries ± 8, lance-elliptic, fused ± in proximal 1/2, enclosing frs, hard, crested below sinus. **DISK FL**: corolla ± 2.5 mm, lobes glabrous. **FR**: 2–2.5 mm. Sandy or rocky grassland, open chaparral, woodland, yellow-pine forest; 100–1600 m. CaRF (Butte Co.), SNF. Mar–Jun

P. peirsonii Munz (p. 409) SAN JOAQUIN ADOBE SUNBURST Pl 2–7 dm. **LF**: 2–6 cm, triangular-ovate, gen 2-pinnately lobed (1-pinnately in small pls); segments 1–5 mm wide. **INFL**: peduncle 2–8 cm; involucre 6–9 mm; phyllaries ± 8, oblanceolate, fused only at base. **DISK FL**: corolla ± 3 mm; lobes sparsely glandular. **FR**: ± 3 mm. 2*n*=16. Grassland, bare dark clay; 100–900 m. s SNF (Kern Co.), se SnJV (Fresno, Tulare cos.). Some ray fl tips reflect ultraviolet light. Threatened by agriculture, flood control. Mar–May ★

PSEUDOGNAPHALIUM CUDWEED, EVERLASTING

Guy L. Nesom

Ann to per, gen taprooted. **ST**: 1–several from base, gen woolly-tomentose, sometimes stalked- or sessile-glandular. **LF**: basal and cauline or mostly cauline, often ± clasping and/or decurrent, gen narrowly lanceolate to oblanceolate, entire, ± tomentose, sometimes adaxially glandular. **INFL**: heads disciform, gen in tight groups in flat-topped to cyme- or panicle-like cluster; involucre cylindric to gen urn-shaped, ± bell-shaped when pressed; phyllaries graduated in (2)3–7(10) series, persistent, ± spreading when dry, bases gen green, distally gen sessile-glandular, tips gen stiff-papery, opaque or clear, dull or shiny, ± white, rosy, tawny, or brown; receptacle flat to convex, epaleate, glabrous. **PISTILLATE FL**: many; corolla narrowly tubular, minutely lobed, ± yellow to ± red. **DISK FL**: few; corolla yellow or ± red; anther tip ± triangular; style branches truncate, hair-tufted. **FR**: oblong-compressed or cylindric, glabrous or ± papillate; pappus bristles in 1 series, deciduous, gen free, or weakly coherent. ± 100 spp.: worldwide, mostly Am, gen temp. (Greek: deceptively similar to *Gnaphalium*, a related genus) [Nesom 2006 FNANM 19:415–425]

1. Opposing lf faces strongly to weakly contrasting in color, adaxial green, not tomentose
  (exc *Pseudognaphalium biolettii*), sometimes glandular, abaxial gen gray to white, tomentose
  2. Lvs crowded, internodes 1–3(10) mm; lf blade linear to lance-linear, margin strongly rolled under;
    phyllaries bright white, opaque, dull; fr ridged, smooth . . . . ***P. leucocephalum***
  2′ Lvs not crowded, internodes gen > 5 mm; lf blade oblong or lanceolate to oblanceolate (distal sometimes
    linear), flat or margin slightly curled under; phyllaries straw-colored to shiny white (± pink), opaque or
    transparent, dull or shiny; fr papillate-roughened or ridged, smooth
    3. St not glandular; lf base not decurrent, clasping with ear-like basal lobes, opposing faces strongly
      contrasting in color . . . . ²***P. biolettii***
    3′ St stalked-glandular; lf base decurrent, weakly to not clasping, opposing faces strongly to weakly
      contrasting in color . . . . ***P. macounii***
1′ Opposing lf faces not or weakly contrasting in color, both gen gray to gray-green or ± green, tomentose,
  adaxially sometimes glandular beneath long hairs
  4. Lf base ± clasping, decurrent 1–2 mm or not (decurrent 2–15 mm in *Pseudognaphalium californicum*)
    5. Infl a dense terminal cluster or ± open, cyme-like; involucre 3–6 mm; phyllaries ± white, silver-gray to
      ± yellow or brown, transparent; pistillate fls 135–200; pappus bristles loosely coherent, shed in clusters
      or easily fragmented rings
      6. Involucre 3–4 mm; disk fls 4–10; pistillate corolla red-tipped; fr papillate . . . . ***P. luteoalbum***
      6′ Involucre 4–6 mm; disk fls gen 18–28; pistillate corolla evenly yellow; fr glabrous . . . . ***P. stramineum***
    5′ Infl ± flat-topped to panicle-like cluster; involucre 4–7 mm; phyllaries gen silvery white to white,
      sometimes pink, gen opaque; pistillate fls 29–140; pappus bristles distinct, released singly
      7. St stalked-glandular; abaxial lf face gen green, stalked-glandular; phyllaries in 7–10 series . . . . ²***P. californicum***
      7′ St not glandular; abaxial lf face gen white-tomentose or woolly-tomentose; phyllaries in 4–6 series
        8. Ann or short-lived per; involucre 4–4.5 mm . . . . ***P. roseum***
        8′ Per; involucre 5–6 mm
          9. Pl sharply scented; adaxial lf face densely stalked-glandular . . . . ²***P. biolettii***
          9′ Pl unscented; adaxial lf face not glandular . . . . ²***P. microcephalum***
  4′ Lf base not ± clasping, decurrent or not
    10. Lf base not decurrent
      11. Involucre 4–5 mm; phyllaries in 3–4 series; disk fls (1)2–5(6); st 2–3 mm diam near base; adaxial lf
        face occ sessile-glandular beneath long hairs . . . . ***P. canescens***
      11′ Involucre 5–6 mm; phyllaries in 4–6 series; disk fls 5–9; st 3–5 mm diam near base; adaxial lf face
        not glandular . . . . ²***P. microcephalum***
    10′ Lf base decurrent
      12. St not glandular, lvs sessile-glandular beneath long hairs . . . . ***P. thermale***
      12′ St and lvs stalked-glandular, glands sometimes obscured by nonglandular hairs
        13. Lf densely, loosely gray-tomentose . . . . ***P. beneolens***
        13′ Lf green or thinly gray-tomentose
          14. Lf gen narrowly lance-oblong, 5–10(20) mm wide; infl flat-topped cluster; involucre bell-shaped to
            ± spheric when pressed; phyllaries in 7–10 series, white to pink . . . . ²***P. californicum***
          14′ Lf linear to lanceolate, oblong, or narrowly spoon-shaped, 3–5(7) mm wide; infl panicle-like
            cluster, gen elongate to widely columnar; involucre short-cylindric to top-shaped when pressed;
            phyllaries in 4–5 series, gen ± pink, sometimes white to ± green . . . . ***P. ramosissimum***

P. beneolens (Davidson) Anderb.   Ann or short-lived per, scented. **ST**: 3–11 dm, persistently tomentose, obscurely glandular beneath long hairs. **LF**: 3–6 cm, 1.5–3.5 mm wide, ± smaller distally, gen linear, not clasping, decurrent 5–15 mm, flat, faces loosely gray-tomentose, obscurely glandular beneath long hairs. **INFL**: gen loose, panicle-like; involucre 5–6 mm, top- to bell-shaped when pressed; phyllaries in (4)5–6(7) series, ovate to ovate-oblong, dull to shiny, white, opaque, inner gen with thread-like keel and short point, glabrous. **PISTILLATE FL**: (39)44–69. **DISK FL**: 5–8(11). **FR**: ridged, smooth or weakly papillate-roughened. $2n=14$. Dry, open slopes and ridges, disturbed sites; < 850(1950) m. KR, NCoRO, NCoRI, SNF, s SNH, ScV, SnFrB, SCoRO, SCo, SnGb, PR; Baja CA. [*Gnaphalium canescens* DC. subsp. *b.* (Davidson) Stebbins & D.J. Keil] Cauline lvs becoming curving-coiling. Jun–Oct

P. biolettii Anderb. (p. 409)   Per, sharply scented. **ST**: 2–12 dm; sometimes ± woody near base, proximally becoming ± glabrous, distally persistently tomentose, at least near heads, not glandular; internodes gen > 5 mm, 1.5–5(8) cm. **LF**: not crowded, 4–10(15) mm wide, oblong to (ob)lanceolate, base clasping, not decurrent, flat or margin slightly curled under, often wavy, adaxial face ± bright green, tomentose or not, densely stalked-glandular, abaxially gen white-tomentose. **INFL**: ± flat-topped; involucre 5–5.5(6) mm, top-shaped to bell-shaped when pressed; phyllaries in 4–5 series, ovate to oblong-ovate or oblong, opaque, shiny, white or sometimes ± pink, often longitudinally wrinkled or grooved, glabrous. **PISTILLATE FL**: 41–73. **DISK FL**: 5–13. **FR**: ridged, smooth. Rocky slopes, roadsides, dunes, coastal scrub, chaparral, oak woodland; 5–600(1200) m. c&s SNF, CCo, SCoR, ChI, SnJt; Baja CA. [*Gnaphalium bicolor* Bioletti, illeg.] Apr–Jun

P. californicum (DC.) Anderb. (p. 409)   Ann to per, scented. **ST**: 2–13 dm, stalked-glandular, sometimes ± tomentose; internode gen 10–20 mm. **LF**: gen not crowded, 4–10 cm, 5–10(20) mm wide, reduced distally on st or not, oblanceolate to lanceolate, base ± clasping or not, decurrent 0–15 mm, flat, or margin slightly curled under, sometimes wavy, faces gen green, stalked-glandular, sticky, sometimes ± tomentose. **INFL**: ± flat-topped or rounded; involucre 5.5–7 mm, bell-shaped to round when pressed; phyllaries in 7–10 series, widely ovate to oblong-obovate, shiny or dull, white to pink, opaque, glabrous. **PISTILLATE FL**: 105–140. **DISK FL**: 7–12. **FR**: ridged, smooth. 2*n*=28. Sandy canyons, dry hills, coastal chaparral; 60–800 m. CA-FP (exc GV); OR, Baja CA. [*Gnaphalium c.* DC.] Lf bases (±) clasping and decurrent, or decurrent and non-clasping. Apr–Jul

P. canescens (DC.) Anderb.   Ann or per. **ST**: 2–10+ dm, persistently tomentose, not glandular. **LF**: gen 2–4(5) cm, 2–8(15) mm wide, ± oblanceolate, not clasping, not decurrent, flat, opposing faces weakly contrasting in color, tomentose, adaxially less densely so, occ also sessile-glandular. **INFL**: gen loose, rounded to ± flat-topped; involucre 4–5 mm, top- to bell-shaped when pressed; phyllaries in 3–4 series, narrowly lance-ovate, dull to shiny, white, opaque to transparent, glabrous. **PISTILLATE FL**: (16)24–44. **DISK FL**: (1)2–5(6), 5–6 more common in n part of range. **FR**: ridged, weakly papillate-roughened. 2*n*=28. Rocky sites, pine woodland; < 2500 m. c SNH, SCo, SnGb, SnBr, PR, DMtns; UT and AZ to OK, TX, Mex. [*Gnaphalium c.* DC.] Aug–Oct

P. leucocephalum (A. Gray) Anderb.   WHITE RABBIT-TOBACCO Bien or short-lived per, scented. **ST**: 3–7 dm, densely and persistently white-tomentose, gen with stalked-glandular hairs protruding through long hairs; internodes 1–3(10) mm. **LF**: crowded, 3–7 cm, 1–5(6) mm wide, linear to lance-linear, ± clasping, not decurrent, margin strongly curled under, adaxial face green, densely stalked-glandular, abaxially densely white-tomentose. **INFL**: rounded or ± flat-topped cluster; involucre 5–6 mm, widely bell-shaped when pressed; phyllaries in 5–7 series, oblong to oblong-ovate, bright white, opaque, dull, glabrous. **PISTILLATE FL**: 66–85. **DISK FL**: 29–44. **FR**: ridged, smooth. 2*n*=28. Sandy or gravelly benches, dry stream bottoms, canyon bottoms; < 500 m. SCo, SnBr, PR; AZ, NM, Mex. [*Gnaphalium l.* A. Gray] Jul–Oct   ★

P. luteoalbum (L.) Hilliard & B.L. Burtt (p. 409)   Ann, tap- or fibrous-rooted, unscented. **ST**: 1–6 dm, loosely tomentose, not glandular; internodes 1–10+ mm. **LF**: proximally ± crowded, distally ± widely spaced, 1–6 cm, 2–8(10) mm wide, linear to narrowly obovate or ± spoon-shaped, distal smaller, ± clasping, gen decurrent 1–2 mm, margin gen curled under, faces gen ± gray-tomentose or adaxial ± glabrous. **INFL**: gen ± long-peduncled, dense to ± openly cyme-like; involucre 3–4 mm, widely bell-shaped when pressed; phyllaries in 3–4 series, ovate to ovate-oblong, silvery-gray to ± yellow or brown, transparent, glabrous or proximally tomentose. **PISTILLATE FL**: 135–160. **DISK FL**: 4–10, corolla yellow or red-tipped. **FR**: papillate; pappus bristles loosely coherent, shed in clusters or easily fragmented rings. 2*n*=14,28. Disturbed sites, fields, streambeds, drying mud; < 850 m. NCo, SNF, GV, CW, SW; widely scattered worldwide; native to Eurasia. [*Gnaphalium l.* L.; *Gnaphalium luteo-album*, orth. var.] Apr–Aug

P. macounii (Greene) Kartesz   MACOUN'S EVERLASTING   Ann or bien, often scented. **ST**: 4–9 dm, stalked-glandular throughout, distally gen ± white-tomentose; internodes gen > 5 mm. **LF**: not crowded, 3–10 cm, 3–13 mm wide, lanceolate to oblanceolate, distal linear, clasping or not, decurrent 5–10 mm, flat or margin slightly curled under, adaxial face stalked-glandular, otherwise ± glabrous, abaxially tomentose. **INFL**: flat-topped cluster; involucre 4.5–5.5 mm, bell-shaped to ± round when pressed; phyllaries in 4–5 series, ovate to ovate-oblong, straw-colored to cream, transparent, shiny, glabrous. **PISTILLATE FL**: 47–101(156). **DISK FL**: 5–12. **FR**: not ridged, ± papillate-roughened. Open slopes, meadows, floodplains; 800–2300 m. s SNF; OR to s Can, e US, Mex. [*Gnaphalium m.* Greene] Jul–Oct

P. microcephalum (Nutt.) Anderb. (p. 409)   Per, unscented. **ST**: 3–10 dm, persistently tomentose, not glandular. **LF**: crowded or not, 2–5(8) cm, 5–10(18) mm wide, gradually smaller distally, narrowly oblanceolate, becoming lanceolate, clasping or not, not decurrent or decurrent 3–10 mm, flat, opposing faces weakly contrasting in color, tomentose, adaxial less densely so, not glandular. **INFL**: loose, flat-topped to panicle-like cluster; involucre 5–6 mm, top- to bell-shaped when pressed; phyllaries in 4–6 series, ovate to oblong-ovate, white, opaque, dull, tomentose at least at base, inner narrower, all gen with thread-like thickened keel and short point. **PISTILLATE FL**: 29–49. **DISK FL**: 5–9. **FR**: ridged, smooth to weakly papillate-roughened. 2*n*=28. Grassy hillsides, gravelly canyon bottoms, chaparral, coastal-sage scrub; < 2250 m. SnFrB, SCoRO, SCo, n ChI, WTR, PR; Baja CA. [*Gnaphalium canescens* subsp. *m.* (Nutt.) Stebbins & D.J. Keil] Jun–Aug

P. ramosissimum (Nutt.) Anderb. (p. 409)   Bien, scented. **ST**: 5–15 dm, ± tomentose, stalked-glandular. **LF**: crowded proximally or not, (1)3–7 cm, 3–5(7) mm wide, linear to lanceolate, oblong, or narrowly spoon-shaped, not clasping, decurrent 2–10 mm, margin curled under and closely wavy, faces ± green, loosely tomentose or not, stalked-glandular. **INFL**: panicle-like cluster; involucre 5–6 mm, short-cylindric to top-shaped when pressed; phyllaries in 4–5 series, ovate to ovate-oblong, acute to acuminate, gen ± pink, sometimes white or ± green, transparent, dull, loosely tomentose at base. **PISTILLATE FL**: 38–62. **DISK FL**: 2–7. **FR**: ridged, smooth. 2*n*=28. Dry, open slopes, woodland, sandy fields, dunes; < 600 m. w NW, w CW, SCo, n ChI, WTR, SnGb, PR. [*Gnaphalium r.* Nutt.] Jul–Sep

P. roseum (Kunth) Anderb. (p. 409)   Ann or short-lived per. **ST**: 5–20 dm, persistently woolly-tomentose, not glandular. **LF**: crowded proximally or not, 3–7 cm, (3)6–15(20) mm wide, lance-oblong to oblanceolate, mid-cauline base ± clasping, not decurrent, margin gen wavy, opposing faces not or weakly contrasting in color, gen white or woolly-tomentose, sometimes tardily ± glabrous adaxially, stalked- or sessile-glandular beneath long hairs. **INFL**: flat-topped cluster; involucre 4–4.5 mm, bell-shaped when pressed; phyllaries in 5–6 series, ovate to ovate-oblong, gen white, sometimes pink, opaque or transparent, dull to shiny, glabrous. **PISTILLATE FL**: 45–90(110). **DISK FL**: (5)6–12(18). **FR**: weakly ridged, smooth. 2*n*=18. Open, disturbed sites; 10–100 m. PR; Mex, C.Am. [*Gnaphalium r.* Kunth] Mar–Jun

P. stramineum (Kunth) Anderb. (p. 409)   Ann or bien, unscented. **ST**: 2–8+ dm, loosely to densely tomentose, not glandular. **LF**: ± crowded, 2–8(9.5) cm, 2–5(10) mm wide, reduced distally on st, oblong to narrowly oblanceolate or weakly spoon-shaped, ± clasping, gen not decurrent (decurrent 1–2 mm), flat or margin slightly curled under, faces loosely persistently gray-tomentose, not glandular. **INFL**: dense, 1–2 cm diam or ± cyme-like; involucre 4–6 mm ± spheric; phyllaries in 4–5 series, ovate to oblong-obovate, ± white, often ± yellow with age, transparent, shiny, glabrous. **PISTILLATE FL**: 160–200. **DISK FL**: gen 18–28. **FR**: weakly, if at all, ridged, otherwise smooth or papillate-roughened, glabrous; pappus bristles loosely coherent, shed in clusters or easily fragmented rings. 2*n*=28. Many habitats, dunes, chaparral slopes, roadsides; < 2500 m. NCo, CCo, SnFrB, SCoRO, SCo, ChI, SnBr; to BC, MT, NE, TX, Mex; also S.Am; naturalized in e US. [*Gnaphalium s.* Kunth] Mar–Aug

P. thermale (E.E. Nelson) G.L. Nesom (p. 409) Per, scented. **ST**: 2–7 dm, loosely tomentose, not glandular. **LF**: proximally or not crowded, 3–8 cm, 3–6 mm wide, reduced distally on st, narrowly oblanceolate, not clasping, decurrent 5–18 mm, flat, faces loosely tomentose, sessile-glandular beneath long hairs. **INFL**: loose to dense, flat-topped to panicle-like cluster; involucre (4)5–6 mm, top- to bell-shaped when pressed; phyllaries in 3–4(5) series, ovate to ovate-oblong, ± white, transparent or opaque, gen shiny, sometimes dull, outer widely acute, inner rounded to short-pointed, glabrous. **PISTILLATE FL**: 35–55. **DISK FL**: (2)4–7. **FR**: ridged, densely papillate-roughened. Openings in forest, riverbeds, banks, roadsides; 100–2500 m. KR, NCoRO, CaRH, c&s SNF, SNH, SnGb, SnBr, SnJt, PR (Cuyamaca Mtns), SNE; to BC, WY, UT. [*Gnaphalium canescens* subsp. *t.* (E.E. Nelson) Stebbins & D.J. Keil] Jun–Sep

PSILOCARPHUS WOOLLY-MARBLES, WOOLLYHEADS

James D. Morefield

Ann 1–15(20) cm, gray to green, cobwebby to woolly or tomentose. **ST**: 1, erect, or 2–10, ascending to prostrate, gen ± forked, ± lfless between forks. **LF**: gen opposite (distal ± alternate), ± sessile, linear to ovate or obovate, entire; distal lvs subtending heads, crowded, largest > proximal lvs. **INFL**: heads disciform, sessile, single or 2–4 per dense group, gen spheric; involucre 0, simulated by paleae and lvs; receptacle length 1–2 × width, ± obovoid, glabrous (gen not lobed); paleae of pistillate fls each enclosing fl, deciduous with fr, ± obovoid or cylindric, length gen < 3 × width, obtuse, bulged out and hood-like distally, tomentose to woolly abaxially, prominently net-veined, veins 5+, margin ± reflexed distally as scarious wing, wing adaxially lateral, gen at ≥ 2/3 palea, projected inward, hidden in head, ± beak-like; paleae of disk fls 0. **PISTILLATE FL**: (8)20–100+, all subtended by paleae; corolla obscure, narrowly cylindric. **DISK FL**: staminate, 2–10; pappus 0; corolla 4–5 lobed; anther base tailed, tip ± triangular; style tips ± linear-oblong. **FR**: each enclosed by palea, ± club-shaped (then ± compressed laterally) to ± cylindric, smooth, shiny, corolla scar ± terminal; pappus 0. 5 spp.: w N.Am, s S.Am. (Greek: slender chaff) [Morefield 2006 FNANM 19:456–460] Spp. sometimes appear to intergrade where ranges overlap; needs study.

1. Largest heads 6–14 mm diam; paleae gen ± hidden by hairs, longest 2.8–4 mm
    2. Largest heads 9–14 mm, ovoid, silky-tomentose, receptacle deeply lobed; paleae ± cylindric, length gen 3.5–6 × width, wing at ± 1/2 palea length . . . . ***P. brevissimus*** var. ***multiflorus***
    2′ Largest heads 6–9 mm, ± spheric, gen woolly, receptacle not or shallowly lobed; paleae obovoid, length 1.5–3 × width, wing at ≥ 2/3 palea length
        3. Distal lvs gen lanceolate to ovate, longest gen 8–15 mm, 1.5–4 × width; pls gen woolly; sts (1)2–10; fr ± club-shaped . . . . ***P. brevissimus*** var. ***brevissimus***
        3′ Distal lvs gen oblanceolate to linear, longest gen 17–35 mm, 4.5–9 × width; pls gen silky-tomentose; sts 1(3); fr ± cylindric . . . . ***P. elatior***
1′ Largest heads gen 3–6 mm diam; paleae ± visible through hairs, longest 1.5–2.7 mm
    4. Distal lvs linear to narrowly oblanceolate, gen 6–12 × width, (3)3.5–5 × heads; fr ± cylindric . . . . ***P. oregonus***
    4′ Distal lvs spoon-shaped to obovate or ovate, gen 1.2–5 × width, 1–2.5(3) × heads; fr ± club-shaped
        5. Distal lvs ± appressed to heads, ovate to widely elliptic, gen 1.2–1.8(2) × width; proximal st internodes (2)3–6 × lf lengths; disk fl corolla gen 4-lobed . . . . ***P. chilensis***
        5′ Distal lvs gen ± spreading, spoon-shaped to obovate, gen 2–5 × width; proximal st internodes gen 1–2(3) × lf lengths; disk fl corolla gen 5-lobed . . . . ***P. tenellus***

P. brevissimus Nutt. Pl ± green to ± gray, cobwebby to woolly. **ST**: proximal internodes gen 0.5–1.5(2) × lf lengths. **LF**: distal gen appressed to heads, lanceolate to ovate, longest 1–2.5(3) × heads. **INFL**: paleae hidden or visible through hairs, longest 2.8–4 mm. **DISK FL**: 0.8–1.6 mm; corolla gen 5-lobed. **FR**: 0.8–1.9 mm, ± club-shaped.

var. ***brevissimus*** (p. 413) DWARF WOOLLYHEADS Pl gen woolly. **ST**: (1)2–10. **LF**: longest gen 8–15 mm, 1.5–4 × width. **INFL**: largest heads 6–9 mm, ± spheric, gen woolly; receptacle not or shallowly lobed; paleae obovoid, length 1.5–3 × width, wing at ≥ 2/3 palea length. 2*n*=28. Drying edges of vernal pools, mud flats, drainages; 10–2500 m. CA-FP (exc c&s SNH, Teh, ChI, SnGb, SnJt), MP (exc Wrn); to BC, SK, WY, UT, nw Baja CA; also s S.Am. [*P. globiferus* Nutt.] May–Jun

var. ***multiflorus*** Cronquist DELTA WOOLLY-MARBLES Pl gen silky-cobwebby. **ST**: gen 1. **LF**: longest gen 14–25 mm, 3–6 × width. **INFL**: largest heads 9–14 mm, ovoid, silky-tomentose; receptacle deeply lobed; paleae ± cylindric, length gen 3.5–6 × width, wing at ± 1/2 palea length. **DISK FL**: corolla 5-lobed. Vernal pools, flats; 10–500 m. ScV, n SnJV, SnFrB. Like *P. elatior*. May–Jun ★

P. chilensis A. Gray (p. 413) ROUND WOOLLY-MARBLES Pl gen ± green, silky-cobwebby, coastal pls ± gray to white woolly. **ST**: (1)2–7; proximal internodes (2)3–6 × lf lengths. **LF**: distal ± appressed to heads, ovate to widely elliptic, longest 5–12 mm, gen 1.2–1.8(2) × width, 1–2(2.5) × heads. **INFL**: largest heads 3–5.5 mm diam; paleae ± visible through hairs, longest gen 1.5–2.7 mm. **DISK FL**: 0.8–1.3 mm; corolla gen 4-lobed. **FR**: 0.6–1.2 mm, ± club-shaped. Uncommon. Vernal pools, coastal interdunes; < 700 m. CaRF, SNF, GV, CCo, SCoRI, SCo; c Chile. [*P. tenellus* var. *globiferus* (DC.) Morefield; *P. tenellus* var. *tenuis* (Eastw.) Cronquist; *P. globiferus* Nutt., misappl.] A similar SnBr form seems best assigned to *P. brevissimus* var. *b.* Mar–Jul

P. elatior (A. Gray) A. Gray TALL WOOLLY-MARBLES Pl ± green to silvery, gen silky-tomentose. **ST**: 1(3); proximal internodes gen 0.5–1.5(2) × lf lengths. **LF**: distal appressed to heads, gen oblanceolate to linear, longest 17–35 mm, gen 4.5–9 × width, 2.5–5 × heads. **INFL**: largest heads 6–8 mm diam; paleae ± hidden by hairs, longest gen 2.8–3.8 mm. **DISK FL**: 1.3–1.9 mm; corolla 5-lobed. **FR**: 0.9–1.7 mm, ± cylindric. Drying, vernally wet or inundated places, often disturbed; 900–1700 m. CaRH, MP (exc Wrn); to sw BC, nw MT. Expected in NW. Jun–Jul ★

P. oregonus Nutt. (p. 413) OREGON WOOLLYHEADS Pl silvery to ± white, silky-tomentose. **ST**: (1)2–10; proximal internodes gen 0.5–1.5(2) × lvs. **LF**: distal appressed to heads, linear to narrowly oblanceolate, longest 12–20 mm, gen 6–12 × width, (3)3.5–5 × heads. **INFL**: largest heads 4–6 mm diam; paleae ± visible through hairs, longest 1.5–2.7 mm. **DISK FL**: 0.7–1.4 mm; corolla 4-lobed. **FR**: 0.6–1.2 mm, ± cylindric. Drying clay of vernal pools, drainages, moist rocky slopes; 10–1800 m. KR, NCoRO, NCoRI, CaR, n SN, c SNH, GV, CW, MP; to se WA, w ID, n NV. For s SN, SW pls, see under *P. tenellus*. Mar–Jul

P. tenellus Nutt. (p. 413) SLENDER WOOLLY-MARBLES Pl ± green to ± gray, cobwebby to silky-tomentose. **ST**: (1)2–10; proximal internodes gen 1–2(3) × lf lengths. **LF**: distal gen ± spreading, spoon-shaped to obovate, longest 6–15 mm, gen 2–5 × width, 1.5–2.5(3) × heads. **INFL**: largest heads 3–5.5 mm diam; paleae ± visible through hairs, longest gen 1.5–2.7 mm. **DISK FL**: 0.8–1.5 mm; corolla gen 5-lobed. **FR**: 0.7–1.2 mm, ± club-shaped. Common. Dry, seasonally moist slopes, flats, burns, trails, rarely vernal pools; < 2400 m. CA-FP (exc e SNH, Teh, SnBr, e PR, SnJt); to sw BC, n ID, nw Baja CA. Montane pls from s SN to SW and nw Baja CA narrowly lvd like *P. oregonus*, may be intermediate. Mar–Jul

PSILOSTROPHE PAPER-DAISY

Per, subshrub. **ST**: 1–many, erect, ± hairy. **LF**: simple, alternate, entire or pinnately lobed, ± soft-hairy; distal often ± glandular. **INFL**: heads radiate, 1 or in cyme-like clusters; peduncle short to long; involucre cylindric to bell-shaped; phyllaries 5–13 in 1–2 ± equal series; receptacle ± flat, epaleate. **RAY FL**: gen 2–6; corolla yellow, fading to cream, persistent on fr; ray ovate, 3-lobed, reflexed when dry. **DISK FL**: 5–25+; corolla yellow, densely hairy; anther tip triangular; style tips truncate. **FR**: cylindric (or ray fr slightly flat), ribbed, glabrous, glandular or soft-hairy; pappus of 4–6 unequal transparent scales (in CA). 7 spp.: sw US, n Mex. (Greek: naked turn, perhaps describing receptacle) [Strother 2006 FNANM 21:453–455]

P. cooperi (A. Gray) Greene (p. 413) Pl 2–9 dm. **ST**: densely white-tomentose, openly branched. **LF**: 1–8 cm, linear, entire, tomentose or becoming glabrous. **INFL**: heads 1–few; peduncle 3–8 cm; involucre 3–5 mm diam, cylindric; phyllaries in 2 series, outer 6–8 mm, lanceolate, ± soft hairy, inner shorter, membranous. **RAY FL**: 3–6; ray 8–20 mm. **DISK FL**: 10–25; corolla 4–5 mm. **FR**: 2–3 mm, glabrous or sparsely glandular; pappus scales 1–2.5 mm. 2*n*=32. Dry slopes, washes, desert scrub and woodland; 150–1950 m. e DMoj, n DSon; to UT, AZ, nw Mex. Mar–Jul, Oct–Jan

PULICARIA FALSE-FLEABANE
David J. Keil & G. Ledyard Stebbins

Ann to per. **LF**: basal or some cauline, alternate. **INFL**: heads radiate, 1–many; involucre ± hemispheric; phyllaries in 2–4+ series, narrowly membranous-margined; receptacle flat to convex, epaleate. **RAY FL**: (10)20–30+; ray linear, yellow. **DISK FL**: (9)40–100+; corolla yellow; anther base appendages bristle-like; style branches ± club-like, spreading. **FR**: ± cylindric, 5-ribbed, short-hairy; outer pappus a short crown of ± fused scales, inner of barbed bristles. ≥ 100 spp.: Eur, Asia, Afr, alien in N.Am; 1 cult for insecticidal properties. (Latin: flea, from use as flea repellent) [Preston 2006 FNANM 19:471–472]

P. paludosa Link Pl 6–12 dm, from short rhizome, stiff, ± soft-hairy. **LF**: 1–3(8) cm, linear to oblong, entire, often rolled under, ± clasping. **INFL**: heads many; peduncle 1–3 cm; involucre 4–5 mm, 6–10 mm diam; phyllaries 2.7–4.8 mm, ± equal in 2–3 series, narrowly linear. **RAY FL**: corolla 3.5–4 mm, tube 1.5–2 mm, ray 1.5–2 mm. **DISK FL**: corolla 2.2–3 mm. **FR**: ± 1 mm; outer pappus < 0.4 mm; inner pappus bristles 8–16+, 2–3 mm. 2*n*=18. Invasive weed in watercourses, moist soils, disturbed ground, esp in SW; < 950 m. w SnJV, SW, w DMtns, DSon; native to Eur. Jul–Oct

PYRROCOMA
David J. Keil & Gregory K. Brown

Per from woody taproot. **ST**: 1–many, decumbent to erect, gen red-tinged. **LF**: basal and cauline, alternate, entire, toothed, or sharply lobed, glabrous to tomentose or glandular; basal petioled; cauline gen clasping, ± reduced. **INFL**: heads gen radiate, 1–30+; involucre hemispheric to bell-shaped; phyllaries ± equal or ± graduated in 2–6 series, herbaceous; receptacle flat to weakly convex, epaleate. **RAY FL**: 10–80; corolla 2–35 mm, ray adaxially yellow, abaxially yellow to orange or red. **DISK FL**: 20–100+; corolla 5–15 mm, cylindric to funnel-shaped, yellow. **FR**: 3–4-angled, gen hairy; pappus bristles 15–60, gen rigid, unequal. ± 15 spp.: w N.Am. (Greek: ± red hair, describing pappus) [Bogler 2006 FNANM 20:413–424]

1. Ray fls obscure, often << disk fls (heads ± disciform); involucre subtended by lfy bracts
   .... ***P. carthamoides*** var. ***cusickii***
1′ Ray fls conspicuous, >> disk fls; involucre not subtended by lfy bracts
   2. Pl glandular
      3. Glands sessile; pl glabrous; heads crowded in spike-like clusters .... ***P. lucida***
      3′ Glands stalked; pl proximally gen hairy and red-tinged; heads in raceme- or panicle-like clusters
         4. Heads in panicle-like clusters; pl sometimes becoming proximally glabrous .... ***P. lanceolata*** var. ***subviscosa***
         4′ Heads in raceme-like clusters; pl proximally tomentose .... ***P. hirta***
            5. Pl densely glandular; phyllaries ± equal, barely overlapping, herbaceous nearly to base .... var. ***hirta***
            5′ Pl sparsely glandular; phyllaries unequal, much overlapping, herbaceous only at tip .... var. ***lanulosa***
   2′ Pl glandless
      6. Head gen 1
         7. Phyllaries ± oblong to oblanceolate, graduated in 3–4 series, glabrous; fr glabrous .... ***P. apargioides***
         7′ Phyllaries ± linear, in 2–3 series, barely or not overlapping, gen tomentose to woolly; fr silky .... ***P. uniflora***
            8. Herbage tufted-woolly; involucre 10–13 mm; phyllaries unequal .... var. ***gossypina***
            8′ Herbage tomentose, becoming glabrous; involucre 6–9 mm; phyllaries equal .... var. ***uniflora***
      6′ Heads (1)3–25
         9. Heads in ± flat-topped to panicle-like clusters
            10. Lvs minutely spiny-serrate .... ***P. lanceolata*** var. ***lanceolata***
            10′ Lvs entire or minutely dentate .... ²***P. racemosa*** var. ***paniculata***
         9′ Heads in ± narrow raceme- or ± spike-like clusters .... ***P. racemosa***

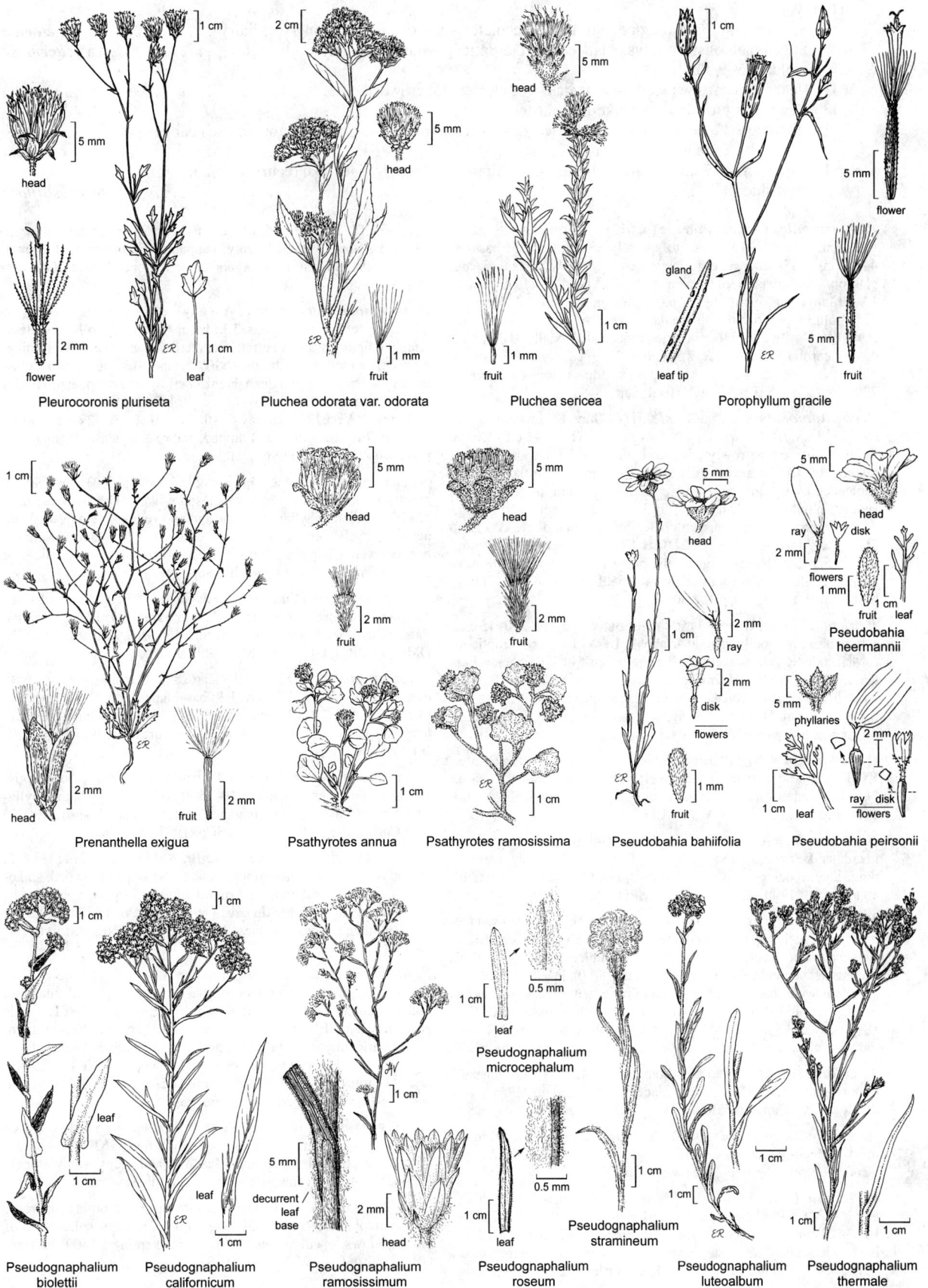

Pleurocoronis pluriseta

Pluchea odorata var. odorata

Pluchea sericea

Porophyllum gracile

Prenanthella exigua

Psathyrotes annua

Psathyrotes ramosissima

Pseudobahia bahiifolia

Pseudobahia heermannii

Pseudobahia peirsonii

Pseudognaphalium bioletti

Pseudognaphalium californicum

Pseudognaphalium ramosissimum

Pseudognaphalium microcephalum

Pseudognaphalium roseum

Pseudognaphalium stramineum

Pseudognaphalium luteoalbum

Pseudognaphalium thermale

11. Involucre 10–15 mm
 12. St cobwebby, hairs tangled; basal lvs lanceolate; phyllaries densely long-soft-hairy . . . . var. *pinetorum*
 12´ St gen glabrous; basal lvs oblanceolate to elliptic; phyllaries ciliate . . . . var. *racemosa*
11´ Involucre 5–9 mm
 13. Heads in a narrow raceme- to panicle-like cluster, not crowded . . . . ²var. *paniculata*
 13´ Heads gen crowded in a spike-like cluster
  14. Involucre 12–16 mm diam, phyllaries yellow-green throughout, sparsely tomentose near base or
      throughout, tip glandular-ciliate . . . . var. *congesta*
  14´ Involucre 5–7 mm diam, phyllaries green-tipped, leathery, ciliate (exc recurved tip), margin
      translucent . . . . var. *sessiliflora*

P. apargioides (A. Gray) Greene (p. 413) ALPINE FLAMES **ST:** 5–18 cm, 0–few-lvd, glabrous to sparsely tomentose. **LF:** basal ± 4–10 cm, ± (ob)lanceolate, leathery, gen ± coarsely dentate or cut, glabrous; cauline reduced, gen entire. **INFL:** head gen 1; involucre 13–20 mm diam, hemispheric; phyllaries graduated in 3–4 series, 6–10 mm, ± oblong to oblanceolate, green toward tip, glabrous. **RAY FL:** 11–40; ray 7–16 mm, often reddened abaxially. **DISK FL:** 45–90; corolla 5–7 mm. **FR:** 2.5–6 mm, 3-angled, glabrous; pappus 5–7.5 mm, tan. 2*n*=12. Rocky slopes, meadows, forest openings; 2200–3800 m. SNH, SNE; w NV. Jul–Sep

P. carthamoides Hook. var. ***cusickii*** (A. Gray) Kartesz & Gandhi (p. 413) LARGE-FLOWERED GOLDENWEED **ST:** 6–43 cm. **LF:** ± 5–21 cm; blade ± narrowly (ob)lanceolate, ± entire to spiny-serrate, puberulent. **INFL:** heads ± disciform, 1–4, in raceme-like cluster, subtended by lfy bracts; involucre 15–20 mm diam, narrowly bell-shaped; phyllaries ± lanceolate, in 3–5 series, ± not overlapping, entire to serrate, herbaceous with pale, thin, dry margin. **RAY FL:** < 25; ray 2–7 mm, gen ≤ pappus. **DISK FL:** many; corolla 9–13 mm. **FR:** 3–5.5 mm, 4-angled, glabrous; pappus ± 6–9 mm, tan to red-brown. 2*n*=12. Barren rocky areas, lava fields; 900–2800 m. CaRH, n SNH, MP; to WA, ID, n NV. Jun–Sep

P. hirta (A. Gray) Greene TACKY GOLDENWEED **ST:** 10–45 cm, ± tomentose to woolly, glandular above. **LF:** ± 1–19 cm, elliptic to (ob)lanceolate, gen toothed. **INFL:** heads gen 5–13, in raceme-like cluster; involucre 9–20 mm diam, hemispheric to bell-shaped; phyllaries in 2–3 series, ± equal or unequal, lance-linear. **RAY FL:** 10–30; ray ± 8–15 mm. **DISK FL:** many; corolla 5–8 mm. **FR:** 3-angled, ± 3–5 mm, silky; pappus 5–8 mm, tan.

var. ***hirta*** (p. 413) Pl densely glandular. **LF:** densely tomentose. **INFL:** phyllaries ± equal, barely overlapping, herbaceous ± to base. 2*n*=24. Meadows, open forest, sometimes in alkaline soils; 1200–1600 m. MP; to OR, ID, NV. Jul–Aug

var. ***lanulosa*** (Greene) G.K. Br. & D.J. Keil (p. 413) Pl sparsely glandular. **LF:** woolly. **INFL:** phyllaries unequal, obviously overlapping, herbaceous only at tip. Meadows, open forest, sometimes in alkaline soil; 1700–2200 m. n SNH, MP; to WA, ID. Jun–Aug

P. lanceolata (Hook.) Greene Pl proximally short tomentose or ± glabrous, distally ± glabrous or densely glandular. **ST:** 15–50 cm. **LF:** ± (ob)lanceolate, sometimes glandular; basal petioled, 6–30 cm, dentate; cauline sessile, reduced. **INFL:** heads (1)3–25(50), in ± flat-topped to panicle-like clusters; involucre 9–22 mm wide, hemispheric; phyllaries overlapping in 3–4 series, 6–11 mm, unequal, ± lanceolate. **RAY FL:** 18–40; ray 6–12 mm. **DISK FL:** 20–100; corolla 5–7 mm. **FR:** ± 4 mm, 4-angled, silky; pappus ± 5–7 mm, tan. Variable; needs study.

var. ***lanceolata*** Pl (±) glabrous. 2*n*=12,24,36. Alkaline meadows, marsh edges, hillsides, open places; 1300–2800 m. MP, n SNE; to OR, ID, s-c Can, c US. Jun–Sep

var. ***subviscosa*** (Greene) G.K. Br. & D.J. Keil (p. 413) Pl copiously stalked-glandular, esp infl. 2*n*=12. Dry slopes and alkaline meadows; 1500–2000 m. MP; nw NV. Jul–Aug

P. lucida (D.D. Keck) Kartesz & Gandhi (p. 413) STICKY PYRROCOMA Pl glabrous, ± shiny from gland-dots. **ST:** 25–75 cm. **LF:** 1–25 cm, (ob)lanceolate, entire or serrate. **INFL:** heads 12–30+, in crowded spike-like cluster; involucre 6–13 mm diam, bell-shaped; phyllaries 7–13 mm, overlapping, linear or lanceolate, herbaceous.

RAY FL: 12–20; ray 7–14 mm. **DISK FL:** 25–40; corolla 6–10 mm. **FR:** 2.5–4 mm, 4-angled, silky; pappus 5–7 mm, tan to ± brown. 2*n*=24. Alkaline clay flats, sagebrush scrub, open forest; 700–2050 m. n SNH. Jul–Sep ★

P. racemosa (Nutt.) Torr. & A. Gray CLUSTERED GOLDENWEED **ST:** 15–90 cm, gen glabrous. **LF:** basal 5–36 cm, (ob)lanceolate to widely elliptic, entire to serrate, petiole tomentose; cauline clasping, reduced, gen serrate, glabrous. **INFL:** heads 3–15+, in ± narrow clusters; involucre 5–18 mm diam, hemispheric or bell-shaped; phyllaries overlapping in 4–5 series, 6–13 mm, (ob)lanceolate to oblong, (±) glabrous. **RAY FL:** 7–28; ray 5–10 mm. **DISK FL:** 20–65; corolla 5–8 mm. **FR:** 2.5–5.5 mm, 4-angled, glabrous to densely tomentose; pappus 6–9 mm, tan to ± brown.

var. ***congesta*** (Greene) G.K. Br. & D.J. Keil (p. 413) DEL NORTE PYRROCOMA **INFL:** gen crowded, spike-like; involucre 5–8.5 mm, 12–16 mm diam; phyllaries herbaceous and yellow-green throughout, acute, sparsely tomentose (at least near base), tip glandular-ciliate, not recurved. Chaparral, conifer forest, boggy sites; serpentine soil; 200–1000 m. nw KR; sw OR. Jul–Sep ★

var. ***paniculata*** (Nutt.) Kartesz & Gandhi **ST:** glaucous. **INFL:** narrow; involucre 5–8.5 mm, 8–12 mm diam. 2*n*=12,24. Alkaline flats, saline meadows; 150–2500 m. c&s SNH, Teh, SCoRI, GB, DMoj; to OR, ID, UT. Highly variable. Jul–Oct

var. ***pinetorum*** (D.D. Keck) Kartesz & Gandhi (p. 413) PINE PYRROCOMA **ST:** cobwebby. **LF:** basal lanceolate. **INFL:** raceme-like; involucre 10–15 mm, 12–15 mm diam; phyllaries densely long-soft-hairy. Meadows, open conifer forest; 600–1700 m. KR. Jul–Sep ★

var. ***racemosa*** (p. 413) **LF:** basal oblanceolate to elliptic. **INFL:** raceme-like; involucre 10–15 mm, 11–18 mm diam; phyllaries ciliate. Coastal valleys, marshes, sometimes in saline soils; < 900 m. NCoR, CaRH, s ScV, n CCo, SCoRI; OR. Jun–Oct

var. ***sessiliflora*** (Greene) G.K. Br. & D.J. Keil (p. 413) **ST:** glaucous. **LF:** basal linear-oblanceolate. **INFL:** gen crowded, spike-like; involucre 5–8.5 mm, 5–7 mm diam; phyllary base pale, leathery, margin translucent, tip green, recurved. 2*n*=12. Dry alkaline flats, saline meadows, marshes; 300–2200 m. se SNE, DMoj; to UT. Jul–Oct

P. uniflora (Hook.) Greene PLANTAIN GOLDENWEED **ST:** 7–38 cm. **LF:** ± tomentose or woolly; basal 3–12 cm, (ob)lanceolate, sharply dentate to cut; cauline few, clasping, reduced. **INFL:** heads 1(4), in raceme-like cluster; involucre 6–13 mm, 11–20 mm diam, hemispheric; phyllaries barely or not overlapping in 2–3 series, 6–12 mm, ± linear, herbaceous, gen tomentose to woolly. **RAY FL:** 25–45; ray 7–11 mm. **DISK FL:** 35–60; corolla 5–8 mm. **FR:** ± 3–4 mm, 3–4-angled, silky; pappus ± 5–8 mm, tan.

var. ***gossypina*** (Greene) Kartesz & Gandhi (p. 413) BEAR VALLEY PYRROCOMA Herbage woolly-tufted. **INFL:** involucre 10–13 cm; phyllaries unequal. Meadows and seeps, stony slopes; 1600–2300 m. SnBr (near Baldwin Lake). Jul–Sep ★

var. ***uniflora*** (p. 413) Herbage tomentose, becoming glabrous. **INFL:** involucre 6–9 mm; phyllaries equal. 2*n*=12. Alkaline soils of mtn meadows, open conifer forest, near hot springs; 1400–2900 m. GB; to OR, n Can, WY, CO. May–Sep

RAFINESQUIA

L.D. Gottlieb

Ann; taprooted; sap milky. **ST**: 1–3, erect, simple or branched distally, hollow, glabrous. **LF**: basal and cauline, alternate; basal oblong to oblanceolate, pinnately lobed, ± petioled or sessile; cauline sessile, clasping; distal bract-like, entire or toothed. **INFL**: heads liguliflorous, 1 or in cyme- or panicle-like clusters; involucre cylindric to obconic; outer phyllaries 8–14, spreading to reflexed, unequal, < 1/2 × inner; main phyllaries ± 7–20, ± equal, lance-linear, scarious-margined, acuminate, reflexed in fr; receptacle flat to convex, minutely roughened, epaleate. **FL**: 15–30; ligule white or cream, often rose-tinged, esp abaxially, readily withering. **FR**: 9–18(20) mm, ± fusiform, tan to mottled gray-brown, body tapered to beak, ribs 5; outer frs with upward-pointing hairs, papillate, or scaly, inner frs mostly smooth to weakly wrinkled; pappus on small disk at beak tip, of 5–21 ± persistent, ± plumose bristles. 2 spp.: sw US, n Mex. (C.S. Rafinesque, naturalist, polymath, traveled widely in Am, 1783–1840) [Gottlieb 2006 FNANM 19:348–349]

1. Ligules surpassing phyllaries by 5–8 mm; fr beak slender, > body; pappus bristles wholly plumose; barbs straight, separate . . . . ***R. californica***
1′ Ligules surpassing phyllaries by 15–20 mm; fr beak stout, < body; pappus bristles plumose on proximal 65–80%; barbs crooked, entangled . . . . ***R. neomexicana***

R. californica Nutt. CALIFORNIA CHICORY **ST**: gen 1, 2–15+ dm, erect, branched chiefly in distal 1/2. **INFL**: heads few to many in cyme- or panicle-like clusters; involucre 12–20 mm. **FR**: 9–12 mm (body 4–5 mm, beak 5–7 mm); pappus bristles 6–10 mm. 2*n*=16. Open sites in scrub, woodland; often common after fire; 100–1500 m. NCo, NCoR, CaRF, SNF, s SNH, ScV (Sutter Buttes), CW, SW, SNE, D; OR, NV, AZ; Baja CA. Apr–Jul

R. neomexicana A. Gray (p. 413) DESERT CHICORY **ST**: 1–several, 1.5–6 dm, often supported by branches of shrubs. **INFL**: heads 1–few in cyme-like clusters; involucre 18–25 mm. **FR**: 11–16 mm (body 8–10 mm, beak 3–6 mm); pappus bristles 10–15 mm. 2*n*=16. Gravelly, sandy soils, scrub, woodland; 200–1500 m. s SNH, SnBr, W&I, D; to UT, NM, TX; n Mex. Feb–Jun

RAILLARDELLA

Bruce G. Baldwin

Per, from gen branched rhizome; rosettes often clumped. **LF**: most or all basal, cauline 0 or proximal only, gen opposite, sessile, lanceolate or oblanceolate to linear, entire or minutely dentate, glabrous, silky-hairy, or minutely coarse-hairy, and/or glandular. **INFL**: glandular, sometimes also hairy; heads radiate or discoid, gen 1; involucre 3–25+ mm diam, hemispheric or bell-shaped to cylindric; phyllaries as many as ray fls, each gen 1/2+ enveloping a ray ovary, falling with fr, coarse-hairy and ± glandular; receptacle flat or convex, glabrous or minutely bristly; paleae in 1 involucre-like series between ray and disk fls or peripheral in discoid heads, fused or free, ± equal, ciliate-hairy. **RAY FL**: 0–13; corolla yellow to red; ray often deeply lobed. **DISK FL**: 7–80+; corolla same color as ray fls, tube < throat, lobes deltate; anthers ± yellow, tips lance-ovate to ovate-oblong; style branches glabrous proximal to branches, tips awl-shaped, bristly. **FR**: ± cylindric, ± straight, ascending-hairy, black; ray and disk pappus of 8–30, awl-shaped or ± bristle-like, ciliate-plumose scales or ray pappus 0. 3 spp.: montane CA, OR, w NV. (Latin: small *Raillardia* or *Railliardia*) [Baldwin & Strother 2006 FNANM 21:256–257] Self-sterile.

1. Lvs glabrous or cauline (esp distal) sometimes glandular; corolla orange to ± red; ray fls 6–13; disk fls 45–80+ . . . . ***R. pringlei***
1′ Lvs silky-hairy or glandular and sometimes sparsely, minutely coarse-hairy; corolla yellow to yellow-orange; ray fls 0–5(7); disk fls 7–40+.
  2. Lvs silky-hairy, silvery, sometimes sparsely glandular, glands inconspicuous; ray fls 0; pl 1–15 cm . . . . . . ***R. argentea***
  2′ Lvs glandular, occ sparsely and minutely coarse-hairy; ray fls 0–5(7); pl 6–53 cm . . . . ***R. scaposa***

R. argentea (A. Gray) A. Gray (p. 413) SILKY RAILLARDELLA Pl 1–15 cm. **LF**: basal, 0.7–8 cm, entire or minutely dentate, gen oblanceolate, silky-hairy, minutely glandular or not. **INFL**: heads discoid, cylindric to bell-shaped; peduncle 0.1–12.5 cm, subtending bracts 0–1; phyllaries 0; paleae 5–15, 6–16 mm, ± fused. **RAY FL**: 0. **DISK FL**: 7–26; corolla 6–11 mm, yellow. **FR**: 5–9.5 mm, ± linear; pappus scales 16–30, 6–11 mm. 2*n*=34,36. Dry, open, gen gravelly sites in conifer forests, semi-barren subalpine and alpine slopes, flats; 1800–3900 m. KR, NCoRH (Yolla Bolly Mtns), CaRH, SNH, SnBr, Wrn, SNE; OR, w NV. Interfertile but not occurring with *R. pringlei*; hybridizes with *R. scaposa*. Jul–Sep

R. pringlei Greene (p. 413) SHOWY RAILLARDELLA Pl ± 25–50 cm. **LF**: basal or mostly opposite, 2.5–15 cm, linear to lanceolate or oblanceolate, entire or minutely dentate, glabrous or distal glandular. **INFL**: heads radiate, 1 or 2–3 in open, ± flat-topped cluster, bell-shaped to hemispheric; peduncle 10–28 cm, often subtended by 1–few gen alternate bracts; phyllaries 6–13; paleae 16–24, 10–14 mm, free or ± fused, sometimes overlapping. **RAY FL**: 6–13; corolla orange to ± red, ray 6–20+ mm. **DISK FL**: 45–80+; corolla 8–11.5 mm, same color as ray. **FR**: 5–8 mm, club-shaped; pappus scales ± 8–17, 7–11 mm. 2*n*=34. Wet meadows, streambanks, seeps, on serpentine-derived soils, in conifer forest; 1300–2200 m. KR (Trinity Alps, Scott Mtns). Jul–Oct ★

R. scaposa (A. Gray) A. Gray (p. 413) Pl 6–53 cm. **LF**: ± basal, 1–16 cm, linear to lanceolate or oblanceolate, entire, glandular, occ sparsely and minutely coarse-hairy. **INFL**: heads radiate or discoid, ± cylindric to hemispheric; peduncle 1–39 cm, gen subtended by 1–few bracts; phyllaries 0–5(7); paleae 8–20, 8–18 mm, free or fused, sometimes overlapping. **RAY FL**: 0–5(7); corolla yellow to orange-yellow, ray 5–25+ mm. **DISK FL**: 7–44; corolla 7.5–12 mm, same color as ray. **FR**: 4.5–10 mm, linear; pappus scales gen 8–30, 0 on some ray frs, 7.5–11.5 mm. 2*n*=68,70. Dry to wet, often sandy sites, of meadows, open conifer forest, semi-barren subalpine and alpine slopes, flats; 2000–3500 m. CaRH, SNH; OR, w NV. Hybridizes with *R. argentea*. Jun–Sep

RATIBIDA PRAIRIE CONE-FLOWER

Per from taproot, rhizome, or caudex. **ST**: 1–many from base, variously branched, gen erect, ribbed, strigose or rough-hairy. **LF**: basal and cauline, alternate, petioled, blade linear to ovate in outline, gen pinnately 1–2 × divided, the basal and distal-most sometimes entire; faces strigose to rough-hairy, gland-dotted. **INFL**: heads radiate, 1 or in open, few-headed clusters; peduncle slender, bracts 0–few; involucre rotate; phyllaries in 2 unequal series, linear or narrowly triangular, spreading in fl, reflexed in fr, outer gen longer; receptacle ± spheric to columnar, paleate; palea ovate, folded around fr, tip hair-tufted, margins each with oblong resin gland. **RAY FL**: 3–15+, sterile; corolla yellow, maroon, or 2-colored yellow and maroon; ray oblong to widely obovate. **DISK FL**: 50–400+; corolla yellow-green or distally ± purple; anthers gen brown-purple; style tips triangular, long papillate. **FR**: oblanceolate, ± 4-angled, compressed side-to-side, winged, ciliate on abaxial margin; pappus 0 or 1–2 tooth-like projections or a low crown. 7 spp.: N.Am. (Dacian name for a sp. of *Aster*) [Urbatsch & Cox 2006 FNANM 21:60–63]

R. columnifera (Nutt.) Wooton & Standl. **ST**: 3–10 dm, finely strigose. **LF**: cauline divided into linear to oblong or narrowly elliptic, lobes; faces strigose. **INFL**: heads 1 at st tips and from distal lf axils; peduncle 10–40 cm; receptacle cylindric. **RAY FL**: 4–12; ray 7–30 mm, ovate to obovate, 2–3-lobed. **DISK FL**: corolla 1–2.5 mm. **FR**: 2–3 mm; pappus of 0–2 minute teeth. 2*n*=28. Dry, rocky slopes, upland valleys, open pine forest; < 1600 m. PR, MP; native N.Am prairies, BC to ON, s to AZ, LA, n Mex. [*R. columnaris* (Pursh) D. Don var. *pulcherrima* (DC.) D.J.N. Hind] Incl in seed mixes; occ escapes or persists from cult. May–Aug

RHAGADIOLUS

David J. Keil & G. Ledyard Stebbins

Ann; sap milky. **ST**: 1, simple or branched. **LF**: basal and cauline, alternate, simple, linear or lanceolate, ± entire, toothed, or pinnately lobed, distal reduced. **INFL**: heads liguliflorous, few to many; phyllaries in 2 rows, outer short, inner fused at base, spreading in fr; receptacle ± flat, epaleate. **FL**: 5–10; ligules yellow, readily withering. **FR**: cylindric, of 2 kinds; outer 5–8 enclosed by rigidly spreading inner phyllaries, long-persistent; inner 0–2 deciduous; pappus 0. 2 spp.: Eur, w Asia. (Latin: crevice-like, from strongly folded inner phyllaries) [Strother 2006 FNANM 19:300–301]

R. stellatus (L.) Gaertn. Pls sparsely hairy throughout. **ST**: 0.7–4 dm. **LF**: 2.5–14 cm, oblong to obovate, sparsely toothed or lobed, terminal lobe largest. **INFL**: heads 5–8 mm in fl; outer phyllaries 5; inner phyllaries 5–8, hardened, hooked at tip in fr. **FL**: outer 5–8; inner 1–3; corolla 6–8 mm. **FR**: outer frs 10–15 mm, curved, spreading with attached phyllaries to form star-shaped structure. 2*n*=10. Uncommon. Disturbed places; < 350 m. s NCoRO (Napa Co.), n SnFrB; native to Eur. Apr–Jun ◆

RIGIOPAPPUS

Bruce G. Baldwin, adapted from Strother (2006)

1 sp.: w US. (Latin: rigid pappus) [Roberts & Urbatsch 2003 Taxon 52:209–228; Strother 2006 FNANM 20:48–49] Similar, closely related to *Pentachaeta*, *Tracyina*.

R. leptocladus A. Gray (p. 413) Ann 5–30+ cm, taprooted; gen self-pollinated. **ST**: gen erect, simple or with branches often exceeding main st, minutely soft- to coarse-hairy, glabrous in age. **LF**: basal and at fl mostly cauline, alternate, ± sessile, mostly 9–30+ mm, 1–2(5+) mm wide, narrowly oblanceolate to lance-linear or linear, entire, glabrous or sparsely and minutely coarse-hairy, 1(3)-veined. **INFL**: heads radiate, 1, at st tips; peduncle often wiry, ± minutely coarse- to soft-hairy near heads; involucre 4–7+ mm, 1–3(5+) mm wide, ± cylindric to obconic; phyllaries 11–20+ in 2+ series, free, ± equal, linear to lanceolate or awl-shaped, green, minutely coarse-hairy, inner gen each ± enclosing a fl, margins scarious; receptacle ± flat, faintly pitted, paleae between ray and disk fls (easily over-looked). **RAY FL**: 3–13+ (gen 3, 5, or 8); corolla ± yellow, often red- or purple-tinged, tube 1.5–2+ mm, ray inconspicuous, 0.3–1 mm, narrowly elliptic. **DISK FL**: 5–70+; corolla pale yellow, occ distally purple, tube 0.2–0.4 mm, throat 1–1.5 mm, lobes 2–4, 0.1–0.3 mm; anther tip thread-like; style tips long-tapering distally. **FR**: 4–5 mm, ± cylindric to spindle-shaped, barely or not compressed, minutely wrinkled, all minutely coarse-hairy or ray fr ± glabrous in age; pappus gen of 3–5 awl-shaped scales, 1–4+ mm or in rays occ much reduced or 0, persistent. 2*n*=18. Grassy sites, openings among shrubs or trees; 100–2200 m. NW (exc NCo), CaR, SN, GV, CW, WTR, SnGb, MP; to WA, ID, NV. Apr–Jul

RUDBECKIA CONE-FLOWER

Lowell E. Urbatsch

Ann to per. **ST**: erect, simple or sparingly branched. **LF**: simple, basal and cauline, alternate, petioled or distal sessile; blade entire, toothed or pinnately lobed. **INFL**: heads radiate or discoid, 1–few, terminal; peduncle long, stout; involucre disk-shaped; phyllaries ± equal or unequal in 1–3 series, spreading or reflexed; receptacle gen columnar to conic or ± spheric, paleate; paleae oblong, entire, folded around frs, hairy distally. **RAY FL**: 0 or 7–21, sterile; corolla yellow, ray 2–6 cm. **DISK FL**: many; corolla green-yellow to dark brown-purple, diam of tube and throat ± equal; anther tip triangular; style tips rounded or blunt-triangular to long-tapered. **FR**: oblong, ± 4-angled; pappus a crown of fused scales or 0. 23 spp.: N.Am. (O.J. Rudbeck, 1630–1702, and O.O. Rudbeck, 1660–1740, botany professors at Uppsala, Sweden) [Urbatsch & Cox 2006 FNANM 21:44–60]

1. Heads discoid. . . . . ***R. occidentalis***
1′ Heads radiate
  2. St hairy; pappus 0; disk corollas dark brown-purple; receptacle ± spheric or ovoid . . . . ***R. hirta*** var. ***pulcherrima***

413

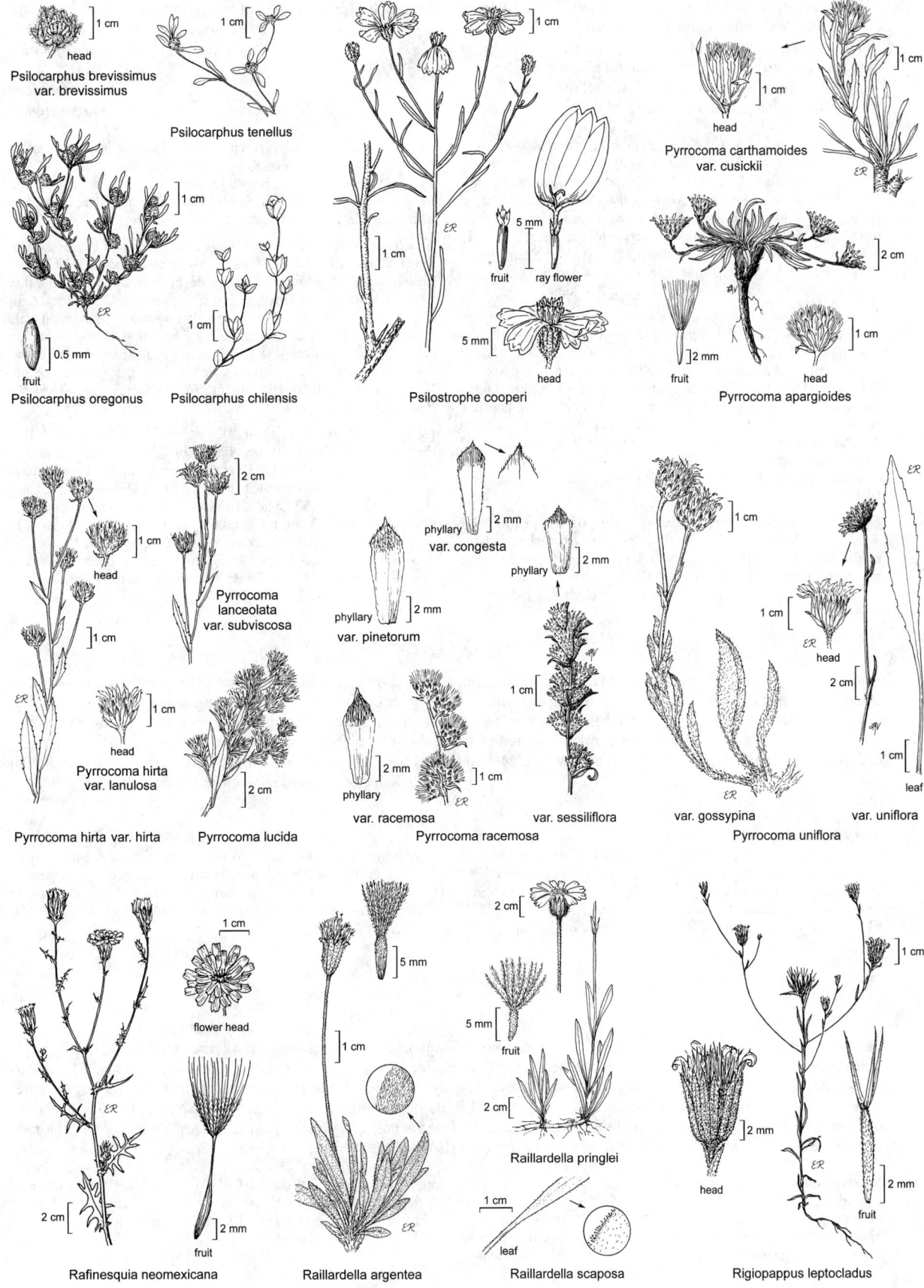

Psilocarphus brevissimus var. brevissimus

head

1 cm

Psilocarphus tenellus

1 cm

Psilocarphus oregonus

fruit

0.5 mm

1 cm

Psilocarphus chilensis

1 cm

Psilostrophe cooperi

1 cm

fruit

5 mm

ray flower

5 mm

head

Pyrrocoma carthamoides var. cusickii

head

1 cm

1 cm

Pyrrocoma apargioides

2 cm

fruit

2 mm

head

1 cm

Pyrrocoma lanceolata var. subviscosa

2 cm

Pyrrocoma hirta var. hirta

1 cm

head

1 cm

Pyrrocoma hirta var. lanulosa

head

1 cm

Pyrrocoma lucida

2 cm

Pyrrocoma racemosa

var. pinetorum

phyllary

2 mm

var. congesta

phyllary

2 mm

phyllary

2 mm

var. racemosa

phyllary

2 mm

1 cm

var. sessiliflora

1 cm

Pyrrocoma uniflora

var. gossypina

1 cm

var. uniflora

head

1 cm

2 cm

leaf

1 cm

Rafinesquia neomexicana

1 cm

flower head

fruit

2 mm

2 cm

Raillardella argentea

5 mm

1 cm

fruit

2 mm

Raillardella pringlei

2 cm

fruit

5 mm

2 cm

leaf

1 cm

Raillardella scaposa

Rigiopappus leptocladus

head

2 mm

fruit

2 mm

1 cm

2′ St glabrous or nearly so; pappus a crown of fused or ± separate scales; disk corollas green-yellow; receptacle ovoid or conic to columnar
 3. Lvs green; abaxial lf face hairy...*R. californica*
 3′ Lvs blue-green, glaucous, abaxial lf face glabrous
   4. Lf margins entire or with a few shallow teeth...*R. glaucescens*
   4′ Lf margins gen coarsely toothed or lobed..*R. klamathensis*

R. californica A. Gray CALIFORNIA CONE-FLOWER Per; rhizome stout. **ST:** 6–18 dm, gen unbranched, glabrous. **LF:** 10–60 cm, 5–15 cm wide, blade lanceolate to ovate or elliptic, entire, coarsely toothed or pinnately lobed; abaxially hairy. **INFL:** head radiate, 1; peduncle 2–6 dm; outer phyllaries 1–2 cm, linear-oblong, > inner; receptacle 1.5–6 cm, ovoid to columnar; paleae ± green. **RAY FL:** 8–21; ray 2–6 cm, often reflexed. **DISK FL:** corolla 4–5 mm, green-yellow. **FR:** 4–5 mm, glabrous; pappus scales 1–1.5 mm. 2*n*=36. Meadows, seeps, mires; 800–2600 m. SNH. Jul–Aug

R. glaucescens Eastw. (p. 419) WAXY CONE-FLOWER Per; rhizome stout. **ST:** 5–15 dm, gen simple to sparingly branched, glabrous, glaucous. **LF:** 20–50 cm, 3–10 cm wide, blade lanceolate to elliptic, entire or shallowly few-toothed; faces glabrous, glaucous, blue-green. **INFL:** head radiate, 1; outer phyllaries ≤ 1.7 cm, linear-oblong, > inner; receptacle 1.5–3.5 cm, conic to columnar; paleae ± brown. **RAY FL:** 7–15; ray 2–4 cm, often reflexed. **DISK FL:** corolla 3–4 mm, green-yellow. **FR:** 4–5.5 mm, glabrous; pappus scales ± 1.2 mm. 2*n*=36. Meadows, seeps, mires, streambanks, often on serpentine; 60–1300 m. n NCo, KR; sw OR. [*R. californica* var. *glauca* S.F. Blake] Jul–Sep

R. hirta L. var. **pulcherrima** Farw. (p. 419) BLACK-EYED SUSAN Ann to short-lived per. **ST:** 3–8 dm, simple or few-branched distally, stiffly hairy. **LF:** 8–15 cm, 1–2.5(5) cm wide, blade lanceolate to elliptic to oblanceolate, entire or shallowly crenate-serrate. **INFL:** heads radiate, 1–few; peduncle 5–20 cm; outer phyllaries 1–3 cm, lin-ear, ± = inner, stiffly hairy; receptacle ovoid to nearly spheric; paleae dark purple. **RAY FL:** gen 8–16; ray 2–4 cm, gen spreading. **DISK FL:** corolla 3–4 mm, dark brown-purple. **FR:** 1.5–2.7 mm; pappus 0. 2*n*=38. Meadows, fields, roadsides; 100–1200 m. c SNF, GV; US, s Can; native in c US, widely cult as orn. Jun–Aug

R. klamathensis P.B. Cox & Urbatsch KLAMATH CONE-FLOWER Per; rhizome stout. **ST:** 5–10 dm, few-branched distally, glabrous, ± glaucous. **LF:** 15–65 cm, 5–14 cm wide; blade ovate to elliptic, entire to coarsely toothed or lobed, glabrous, ± glaucous. **INFL:** head radi-ate, 1; outer phyllaries ≤ 2 cm, narrowly obovate, > inner; receptacle 1.5–3.5 cm, conic to columnar; paleae ± brown. **RAY FL:** 7–15; ray 2–3.5 cm, often reflexed. **DISK FL:** corolla 3–4 mm, green-yellow. **FR:** 4–6 mm, glabrous; pappus scales ± distinct, ± 1.5 mm. 2*n*=36. Meadows, seeps; 1000–1600 m. KR. [*R. californica* var. *intermedia* Perdue] Aug–Sep

R. occidentalis Nutt. (p. 419) WESTERN CONE-FLOWER Per; rhizome stout. **ST:** 6–20 dm, simple or few-branched distally, ± glabrous, glaucous. **LF:** 12–30 cm, 3–9 cm wide; blades elliptic to widely ovate, coarsely serrate to nearly entire, glabrous or short-hairy on 1 or both faces. **INFL:** heads discoid, 1–few; peduncle 15–45 cm; outer phyllaries 1–3 cm, linear-oblong, > or ± = inner; disk ovoid to columnar, 2–5 cm; paleae dark purple. **RAY FL:** 0. **DISK FL:** corolla 4–6 mm, dark brown-purple. **FR:** 3–5 mm; glabrous; pappus 0 or gen several scales to 1.3 mm. 2*n*=38. Meadows, seeps; 1200–2800 m. KR, NCoRH, n SNH; to WA, MT, WY, UT. Jun–Aug

SANVITALIA

Ann [per]. **ST:** simple to much-branched, prostrate to erect. **LF:** simple, opposite. **INFL:** heads radiate, [1] in few-headed cyme-like clusters; peduncle slender; involucre disk-like to hemispheric; phyllaries in 1–2 series; receptacle conic, paleate; paleae lanceolate, ± awn-tipped. **RAY FL:** 5–13; tube 0; ray 2–3-lobed, cream to orange, persistent on fr. **DISK FL:** many; corolla cream to yellow or brown; lobes minute; style tips triangular. **FR:** glabrous; ray fr thick, pappus of short, stout awns; disk fr short, pappus 0 or of 2 awns. 5 spp.: sw US, Mex, n C.Am, S.Am. (Either Sanvital, a Spanish botanist, or the Italian Sanvital family) [Strother 2006 FNANM 21:70–71] *S. procumbens* Lam. was a garden weed in Riverside (e SCo) in 1954 (apparently not persisting).

S. abertii A. Gray (p. 419) ABERT'S SANVITALIA Ann 2–29 cm. **ST:** spreading or erect, strigose. **LF:** sessile or short-petioled, 2–5 cm, linear to lanceolate or narrowly elliptic, acute, scabrous. **INFL:** heads gen few; peduncle 0–30 cm; phyllaries 5–11, prominently veined, acute, ± glabrous; awn-tips of paleae exceeding disk fls. **RAY FL:** 1 per phyllary; corolla yellow, drying cream; ray 2–3 mm, thick, ± leathery, gen 2-lobed. **DISK FL:** corolla 1–2 mm, cylindric, yel-low, drying cream. **FR:** ray fr 3–4 mm, straw-colored, pappus awns 3, ≤ 1 mm; disk fr 2.5–3.5 mm, brown, ± 4-angled, warty, pappus 0. 2*n*=22. Dry slopes, washes, scrub, woodland; 1450–1750 m. se DMtns (Clark, Mescal, New York mtns); to TX, n Mex. Aug–Oct ★

SAUSSUREA

David J. Keil & Charles E. Turner

Per. **LF:** basal and cauline, alternate. **INFL:** heads discoid, 1–many in cyme-like clusters; involucre ovoid to ± obconic; phyl-laries many, graduated in several series, obtuse to acute; receptacle flat or rounded, epaleate, sometimes bristly. **FL:** few to many; corolla white to blue or purple, tube slender, throat abruptly wider, lobes linear; anther base short-tailed, tip linear, acute; style branches oblong, short-papillate, tips minutely hairy. **FR:** oblong, ± angled, attached at base; pappus of 1–2 series, outer of short bristles or scales, inner of gen longer plumose bristles fused at base. ± 300 spp.: Eurasia, N.Am. (Theodore and Horace de Saussure, Swiss naturalists, 18th century) [Keil 2006 FNANM 19:165–168]

S. americana D.C. Eaton (p. 419) AMERICAN SAW-WORT **ST:** 3–12 dm, several to many from stout caudex, erect, lfy, ± loosely tomentose, ± glandular. **LF:** proximal blades 5–15 cm, lanceolate to triangular, sharply dentate, acute. **INFL:** peduncle 0–5 cm; involucre 10–15 mm; phyllaries pale green, ± loosely tomentose, margins ± membranous, tips gen dark. **FL:** 8–21; corolla ± 10 mm, white to dark purple. **FR:** 4–6 mm, glabrous or gland-dotted; pappus bristles ± brown, outer ≤ 7 mm, inner ± 10 mm. Meadows, slopes; 1700–1800 m. KR; to AK, MT. Jun–Aug ★

SCOLYMUS GOLDENTHISTLE

David J. Keil & G. Ledyard Stebbins

Ann to per, thistle-like, spiny; sap milky. **ST:** erect, ± branched distally. **LF:** basal and cauline, alternate, toothed or pinnately lobed, cauline decurrent on st as spiny wings; teeth and lobes spine-tipped. **INFL:** heads liguliflorous, several to many, in spike- to panicle-like clusters; phyllaries in 3+ series; receptacle paleate, paleae broad, scarious, slightly winged, tightly enclosing frs. **FL:** 30–60+; ligules yellow, readily withering. **FR:** flattened front-to-back, falling with paleae; pappus 0 or of a few rigid bristle-like scales. 3 spp.: Medit. (Greek: ancient name) [Strother 2006 FNANM 19:220]

S. hispanicus L. **ST:** 2–22 dm, stout. **LF:** basal to 45 cm, petioled, blade soft, ± deeply lobed, weakly spiny; cauline 4–20 cm, rigid, oblong to ovate, dentate or shallowly lobed, strongly veiny abaxially, spiny and ± hairy. **INFL:** heads sessile or short-peduncled, 1 in distal lf axils or > 1 in narrow panicle-like clusters; involucre 15–20 mm. **FR:** 3–5 mm, club-shaped; pappus present. 2*n*=20. Uncommon. Disturbed grassy areas; < 100 m. se ScV, n CCo, SnFrB; e US; native to Eur. May–Aug ◆

SCORZONERA

David J. Keil & G. Ledyard Stebbins

Per in CA; sap milky. **LF:** basal and cauline, entire to pinnately lobed. **INFL:** heads liguliflorous, large, 1–many; phyllaries graduated in 3–5+ series; receptacle ± flat, epaleate. **FL:** many; corolla yellow to white or purple; ligule readily withering. **FR:** cylindric, not beaked; pappus of several rows of plumose bristles. ± 150 spp.: Medit, Eurasia. (Old French: viper) [Strother 2006 FNANM 19:306–307]

S. hispanica L. SPANISH SALSIFY, VIPER'S GRASS **ST:** 2.5–10 dm from vertical rootstock, sparsely woolly at base, few-branched, lfy. **LF:** 15–40 cm, linear to ovate-elliptic, margins wavy with a few teeth. **INFL:** heads 2–3 cm, ≤ 4 cm in fr. **FL:** ligule yellow, some- times purple-tinged abaxially. **FR:** 10–15 mm; outer strongly ribbed, rough; inner weakly ribbed; pappus bristles ± = fr, dirty white. 2*n*=14. Open fields; < 500 m. KR, NCoRO; native to Eur. May–Jul

SENECIO RAGWORT, GROUNDSEL, BUTTERWEED

Debra K. Trock

Ann to shrub, from taproot, rhizome, or button-like caudex. **ST:** 1–many, simple or branched. **LF:** alternate; mostly basal to evenly distributed; proximal gen ± petioled; middle gen reduced, sessile, often clasping. **INFL:** heads radiate, disciform, or discoid, gen in cyme-like clusters; involucre cylindric to urn- or bell-shaped, main phyllaries gen 8, 13, or 21 in 1 series, subtended by few to many, gen much-reduced outer phyllaries, reflexed in fr, green, often black-tipped, linear to narrowly lanceolate, glabrous or hairy; receptacle epaleate. **RAY FL:** 0–21; ray gen yellow (white, pink-purple), occ much-reduced and scarcely exceeding phyllaries. **DISK FL:** 3–100+; corolla tubular to bell-shaped, lobes erect to recurved, pale to deep yellow; anther tip ± triangular-ovate; style branch tips obtuse or truncate. **FR:** cylindric, gen shallow-ribbed or -angled, glabrous or stiff-hairy; pappus of minutely barbed bristles, white to tan. 1000+ spp.: worldwide, esp abundant in warm temperate, subtrop and trop areas at mid to upper elevations. (Latin: old man, from white pappus) [Barkley 2006 FNANM 20:544–570] Many N.Am spp. formerly treated as *Senecio* now in *Packera*. The common names groundsel, ragwort, and butterweed apply to spp. of both genera. Neither *Pericallis hybrida* B. Nord. [*S. hybridus* Regel, illeg.] nor *S. squalidus* L. ◆ naturalized in CA; *S. hieraciifolius* L. var. *h.* [*Erechtites hieraciifolia* (L.) DC. var. *h.* (orth. var.)] not documented in CA.

1. Heads discoid, disciform, or minutely radiate, rays (if present) gen ≤ 2 mm, scarcely, if at all exceeding involucre
  2. Heads disciform or minutely radiate, pistillate fls present
   3. Pistillate fls fewer than disk fls
    4. Herbage densely curly-hairy; lvs 1–2-pinnately lobed or divided; disk fls 40–50 . . . . ***S. sylvaticus***
    4′ Herbage glabrous or sparsely hairy; lvs gen dentate or 1-pinnately lobed; disk fls 8–30
     5. Cauline lvs gen narrowed at base, sometimes weakly clasping; involucre narrowly urn-shaped, finely hairy; CW, SCo, ChI. . . . . ***S. aphanactis***
     5′ Mid and distal cauline lvs widest at base, truncate to cordate-clasping — 1–2+ cm wide — at base, often ± purple tinged; involucre cylindric or urn-shaped, glabrous; D . . . . ²***S. mohavensis***
   3′ Pistillate fls more numerous than disk fls
    6. Lvs pinnately lobed . . . . ***S. glomeratus***
    6′ Lvs dentate or minutely dentate
     7. Cauline lvs ± narrowly lanceolate, ± clasping, sharply and evenly fine-dentate, flat. . . . . ***S. minimus***
     7′ Cauline lvs linear or lance-linear, tapered to base, remotely minutely dentate or appearing entire, margins ± rolled under. . . . . ***S. quadridentatus***
 2′ Heads discoid, pistillate fls 0
  8. Ann
   9. Herbage glabrous, glaucous. . . . . ²***S. mohavensis***
   9′ Herbage glabrous or sparsely hairy when young, not glaucous . . . . ***S. vulgaris***
  8′ Bien or per
   10. Pl glabrous or minutely hairy among heads
    11. Caudex erect; herbage ± red-tinged, not glaucous; petioles winged, lvs firm, not fleshy, dentate or minutely dentate; wet hillsides and meadows . . . . ²***S. hydrophiloides***

11′ Caudex short, button-like; herbage blue-green, glaucous; petioles unwinged, lvs ± fleshy, entire to
    dentate; swamps and alkaline soils . . . . ²*S. hydrophilus*
10′ Pl ± hairy, occ becoming ± glabrous in age
  12. Fls 6–40 per head
    13. Phyllaries (4)5–12 mm; disk fls 35–45 . . . . ²*S. integerrimus* var. *exaltatus*
    13′ Phyllaries 3–5 mm; disk fls 10–20 . . . . ²*S. scorzonella*
  12′ Fls 40–90+ per head
    14. Phyllaries 8(13), 4–8 mm . . . . ²*S. aronicoides*
    14′ Phyllaries ± 21, 7–10 mm . . . . *S. astephanus*
1′ Heads radiate, rays 3–20 mm, gen well exceeding involucre
15. Shrub or subshrub with shrub-like branching; lvs or lf segments narrowly linear to thread-like
  16. Outer phyllaries prominent, often some > 1/2 inner . . . . *S. flaccidus*
    17. Herbage gen ± gray-tomentose . . . . var. *douglasii*
    17′ Herbage gen ± glabrous . . . . var. *monoensis*
  16′ Outer phyllaries inconspicuous or 0
    18. Lvs 1–2-pinnately divided into linear, entire lobes . . . . *S. lyonii*
    18′ Lvs gen entire or toothed (occ lobed in *Senecio spartioides*)
      19. Phyllaries 3–5 mm . . . . *S. linearifolius* var. *linearifolius*
      19′ Phyllaries 6–10 mm
        20. Lf 0.5–3 mm wide; phyllaries gen 13; involucre gen 7–9 mm wide when pressed; fr 3–5 mm;
          s CCo . . . . *S. blochmaniae*
        20′ Lf 1–6 mm wide; phyllaries 8(13); involucre gen 3–6 mm wide when pressed; fr 2–3.5 mm;
          SNH, SnBr, SNE, DMtns . . . . *S. spartioides*
15′ Ann or bien to per; lvs or lf segments of various shapes but gen wider than linear
21. Ann
  22. Rays yellow . . . . *S. californicus*
  22′ Rays red- or pink-purple . . . . *S. elegans*
21′ Bien or per
23. Lvs deeply lobed or divided
  24. Lvs sharply cut to pinnately lobed, lobes entire; phyllary tips green . . . . *S. clarkianus*
  24′ Lvs 1–3-pinnately lobed, lobes obovate to spoon-shaped, ultimate margin dentate; phyllary tips gen
    black . . . . *S. jacobaea*
23′ Lvs entire or toothed
25. Rosette of basal lvs 0, lvs well distributed along st
  26. Pl gen 40–200 cm
    27. Lvs ± linear to lanceolate, dentate to ± entire, tapered to ± sessile at base . . . . *S. serra* var. *serra*
    27′ Lvs narrowly to widely triangular, dentate (rarely ± entire), at least proximal truncate to cordate
      at base . . . . *S. triangularis*
  26′ Pl 2–30 cm
    28. Pls ≤ 10 cm from branching rhizomes; sts 1–3; lf entire, wavy or curled under; ray 5–10 mm
      . . . . *S. pattersonensis*
    28′ Pls 10–30 cm from ± rhizomatous spreading bases; sts 2–several; lf toothed to ± entire; ray
      8–12 mm . . . . *S. fremontii*
      29. Distal lvs well developed; phyllaries 7–10 mm . . . . var. *fremontii*
      29′ Distal lvs smaller, fewer; phyllaries 5–7 mm . . . . var. *occidentalis*
25′ Rosette of basal lvs present, lvs gen reduced distally on st
30. Pl with stout rhizome . . . . ²*S. scorzonella*
30′ Pl with short, erect, ± button-like caudex bearing many fleshy-fibrous roots
  31. Pl gen glabrous or nearly so
    32. Caudex erect; herbage ± red-tinged, not glaucous; petioles winged, lvs firm, not fleshy, dentate
      or minutely dentate; wet hillsides and meadows . . . . ²*S. hydrophiloides*
    32′ Caudex short, button-like; herbage blue-green, glaucous; petioles unwinged, lvs ± fleshy, entire
      to dentate; swamps, alkaline soils . . . . ²*S. hydrophilus*
  31′ Pl hairy when young, often ± glabrous in age
    33. Lf sharply toothed, minutely dentate, or ± entire; phyllaries 8 (rarely 13); ray fls 0 or 1–2 . . . . . ²*S. aronicoides*
    33′ Lf entire to ± dentate; phyllaries 13 or 21; ray fls 5, 8 or 13 . . . . *S. integerrimus*
      34. Ray fls white; basal lvs distinctly petioled, blades ovate to rounded-deltate, or rarely obovate
        to oblanceolate . . . . var. *ochroleucus*
      34′ Ray fls yellow; basal lvs distinctly petioled or not, blades elliptic to lanceolate or oblanceolate,
        occ rounded-deltate to ± round
        35. Pl densely or sparsely hairy; phyllaries lanceolate, tips black . . . . ²var. *exaltatus*
        35′ Pl distinctly and densely hairy; phyllaries linear to awl-like, tips gen green, rarely minutely
          black-tipped . . . . var. *major*

S. aphanactis Greene CHAPARRAL RAGWORT Ann 5–20+ cm, from short, thin taproot. **ST:** 1, often branched near base, thin, delicate, glabrous or sparsely hairy. **LF:** ± evenly distributed, sessile, 2–4 cm, 0.5–1 cm wide, oblanceolate to linear, narrowed or sometimes weakly clasping at base, pinnately lobed or dentate (± entire); distal bract-like. **INFL:** heads disciform or minutely radiate, 4–10+ in ± open clusters; involucre narrowly urn-shaped, phyllaries 8 or 13, 5–6 mm, tips green; outer phyllaries inconspicuous. **RAY FL:** ± 5; ray 0.5–1+ mm. **DISK FL:** 8–20. **FR:** 1.7–2.5 mm, densely hairy. 2*n*=40. Alkaline flats, dry open rocky areas; 10–550 m. CW, SCo, ChI; Baja CA. Feb–May ★

S. aronicoides DC. (p. 419) RAYLESS RAGWORT Bien or per 3–10 dm, from button-like caudex bearing fleshy, fibrous roots. **ST:** 1, sparsely cobwebby or becoming unevenly glabrous. **LF:** basal 7–20 cm, 2–3 cm wide, ovate or elliptic to oblanceolate or oblong, tapered to petiole, minutely dentate to sharply toothed or ± entire, mid and distal smaller, becoming sessile, weakly clasping. **INFL:** heads discoid or radiate, (6)15–30+ in ± flat-topped clusters; involucre bell-shaped, phyllaries 8(13), 4–8 mm, tips green or black; outer phyllaries 0–few, inconspicuous to > 1/2 inner. **RAY FL:** 0 or 1–2; ray 4–7 mm. **DISK FL:** 40–75. **FR:** 1.5–2.5 mm, glabrous. 2*n*=40. Dry or drying sites in open woodland, upper foothill, montane forest; 800–3020 m. NW, CaR, SN, ScV, CW, MP; s OR. Apr–Jul

S. astephanus Greene (p. 419) SAN GABRIEL RAGWORT Per 3–10 dm, from erect unbranched caudex. **ST:** 1–several, unevenly tomentose, becoming ± glabrous. **LF:** reduced distally on st, basal 10–25+ cm, 3–7+ cm wide, elliptic-ovate to lanceolate, tapered to petiole, dentate to minutely dentate, mid similar but sessile, distal bract-like. **INFL:** heads discoid, (5)10–20+ in loose or congested cyme-like cluster; involucre bell-shaped, phyllaries ± 21, 7–10 mm, green, tomentose or becoming ± glabrous; outer phyllaries ≤ 1/2 inner. **RAY FL:** 0. **DISK FL:** 70–90+. **FR:** 2–2.5 mm, glabrous. 2*n*=40. Steep rocky slopes in chaparral/coastal-sage scrub and oak woodland; 400–1500 m. SCoR, TR. Apr–Jun ★

S. blochmaniae Greene (p. 419) BLOCHMAN'S RAGWORT Subshrub 4–13 dm, from woody taproot. **ST:** multiple, erect or arching upward, glabrous or sparsely cobwebby, often proximally decumbent, rooting where buried in sand. **LF:** fleshy, evenly distributed, crowded, proximal often withered, 3–12 cm, 0.5–3 mm wide, linear to thread-like, entire. **INFL:** heads radiate, 5–20+ in ± flat-topped clusters; involucre bell-shaped, gen 7–9 mm wide when pressed, phyllaries gen 13, 7–10 mm, tips green; outer phyllaries inconspicuous, << inner. **RAY FL:** 5–8, ray 8–10 mm. **DISK FL:** 25–40+. **FR:** 3–5 mm, short-stiff hairy. 2*n*=40. Coastal sand dunes, sandy floodplains; < 150 m. CCo. May–Nov ★

S. californicus DC. (p. 419) CALIFORNIA RAGWORT Ann 1–5 dm, from taproot. **ST:** 1–6, branched from base or ± throughout, glabrous or sparsely hairy. **LF:** ± evenly distributed or distally reduced, proximal 2–8 cm, 0.5–3 cm wide, lance-linear to lanceolate, tapered to winged petiole or sessile, ± entire to dentate or pinnately lobed, ± fleshy, mid and distal sessile, clasping. **INFL:** heads radiate; (1)3–15+ in open clusters; involucre bell-shaped, phyllaries gen 21 (fewer), 5–8 mm, tips black; outer phyllaries inconspicuous or 0. **RAY FL:** 10–15; ray 8–12 mm. **DISK FL:** 45–55+. **FR:** 2–2.5 mm, short-stiff hairy. 2*n*=40. Sandy, dry or drying sites; < 1200 m. s SN, Teh, CW, SW, DSon; Mex. Mar–Jun

S. clarkianus A. Gray (p. 419) CLARK'S RAGWORT Per (5)6–10(12) dm, from fibrous-rooted caudex. **ST:** 1, sparsely hairy. **LF:** ± evenly distributed, proximal 9–16 cm, 1–8 cm wide, lanceolate to oblong, tapered to petiole, sharply cut to pinnately lobed, distal sessile, weakly clasping. **INFL:** heads radiate, 8–20 in ± flat-topped clusters; involucre bell-shaped, phyllaries 13 or 21, 5–8 mm, tips dark green; outer phyllaries conspicuous, ± = inner. **RAY FL:** 8 or 13; ray 8–12(20) mm. **DISK FL:** 35–45+. **FR:** 2.5–3 mm, glabrous. Wet meadows, conifer forest; 1200–2700 m. n SNH, c&s SN. Jul–Sep

S. elegans L. RED-PURPLE RAGWORT Ann 2–5+ dm, from taproot. **ST:** 1–several, branched distally or throughout, hairy, sticky-glandular. **LF:** ± evenly distributed, proximal 3–8 cm, 1.5–3.5 cm wide, obovate, abruptly narrowed or tapered to petiole, 1–2-pin-

nately lobed, ultimate margin crenate to dentate. **INFL:** heads radiate, 8–30+ in ± flat-topped clusters; involucre bell-shaped, phyllaries 13+, 6–10 mm, tips black; outer phyllaries conspicuous, many, graduated, ≤ ± 1/2 inner, black-tipped. **RAY FL:** 13–21; ray 10–15 mm, deep red- or pink-purple. **DISK FL:** 100+. **FR:** 2.5–3 mm, hairy on ribs. 2*n*=20. Disturbed coastal sites; < 100 m. CCo, SnFrB, SCo; s Afr. Mar–Jun ◆

S. flaccidus Less. THREADLEAF RAGWORT Shrub or subshrub (3)4–12+ dm, from woody taproot. **ST:** 1–many, erect or arching, branches few, glabrous or ± tomentose. **LF:** ± evenly distributed, sessile or short-petioled, 3–10(12) cm, narrowly linear to thread-like, or divided into gen entire linear to thread-like lobes. **INFL:** heads radiate, 5–10(20+) in ± flat-topped compound clusters; involucre bell-shaped, phyllaries, 7–10(12) mm, green, occ black tipped; outer phyllaries [0] ± prominent, often some > 1/2 inner. **RAY FL:** 8, 13, or 21; ray 10–15 (20) mm. **FR:** densely hairy.

var. ***douglasii*** (DC.) B.L. Turner & T.M. Barkley DOUGLAS' THREADLEAF RAGWORT Pl gen ± gray-hairy. **INFL:** main phyllaries 13 or 21. **DISK FL:** 40–50. **FR:** 4.5–5.5 mm. 2*n*=40. Gen disturbed dry, open, sandy or rocky sites; 30–1500 m. NCoR, CaR, SN (exc n SNH), GV, CW, SCo, ChI, TR, PR; Baja CA. Gen Jun–Oct

var. ***monoensis*** (Greene) B.L. Turner & T.M. Barkley SMOOTH THREADLEAF RAGWORT Pl gen ± glabrous. **INFL:** main phyllaries (13)21(30+). **DISK FL:** 45–55. **FR:** 4–5 mm. 2*n*=40. Desert basins, washes, rocky or sandy sites; 600–2000 m. SNE, D; to UT, TX, Baja CA. Apr–Jun, Sep–Nov

S. fremontii Torr. & A. Gray Per 1–3 dm, from ± rhizomatous spreading bases. **ST:** 2–several clustered, spreading at base, often purple-tinged, glabrous. **LF:** stiff, fleshy, ± evenly distributed, gen < 4 cm, 2 cm wide, ovate or obovate to oblanceolate, truncate to tapered at base, toothed to ± entire. **INFL:** heads radiate, (1)2–5(13) in ± flat-topped clusters; involucre bell-shaped, phyllaries 8 or 13, 5–10 mm; outer phyllaries few, inconspicuous, ≤ ± 1/2 inner. **RAY FL:** ± 8; ray 8–12 mm. **DISK FL:** 35–50. **FR:** 2.5–3.5 mm, glabrous.

var. ***fremontii*** (p. 419) DWARF MOUNTAIN RAGWORT Pl 1–3 dm. **LF:** distal well developed. **INFL:** heads (1)3–5(13); phyllaries 7–10 mm. **DISK FL:** 40–50. **FR:** 2.5–3 mm. 2*n*=40,40+,80. Talus slopes, rocky sites, subalpine conifer woodland; 2500–3500 m. CaRH, Wrn; to BC, AB, WY, UT, NV. Jun–Sep

var. ***occidentalis*** A. Gray (p. 419) WESTERN DWARF MOUNTAIN RAGWORT Pl 1–2 dm. **LF:** distal smaller, fewer. **INFL:** heads (1)2–4(5); phyllaries 5–7 mm. **DISK FL:** 35–45. **FR:** 3–3.5 mm. 2*n*=40,40+,80. Rocky subalpine, alpine sites; 2800–4000 m. SN (exc n SNH), SnBr, SNE; w NV. Jul–Sep

S. glomeratus Poir. (p. 419) CUTLEAF BURNWEED Ann or per (2)10–20 dm, from taproot with branched lateral roots. **ST:** 1–several, gray- to white-tomentose or unevenly hairy. **LF:** ± evenly distributed, proximal short-petioled, often withered before fl; mid sessile, ± clasping, 3–15 cm, 1–4 cm wide, linear to ± lanceolate, pinnately lobed, lobes dentate; distal bract-like. **INFL:** heads disciform, 20–120, grouped at branch tips of panicle-like or ± flat-topped cluster; involucre cylindric, phyllaries 13, 4.5–5.5 mm, light green, hairy at base; outer phyllaries inconspicuous, << inner. **PISTILLATE FL:** 25–35; corolla minutely lobed, ray 0. **DISK FL:** 8–10. **FR:** 0.75–1.5 mm, hairy in grooves between ribs. 2*n*=60. Disturbed sites, mostly coastal; < 300 m. NCo, NCoRO, w CW, n ChI; to WA; Australia, New Zealand. [*Erechtites g.* (Poir.) DC.; *Erechtites glomerata*, orth. var.] Apr–Sep ❖

S. hydrophiloides Rydb. (p. 419) SWEET MARSH RAGWORT Bien or per 3–10(14) dm, from erect button-like caudex, with fleshy-fibrous roots. **ST:** single or 2–4 clustered, ± red tinged, glabrous or ± hairy among heads when young. **LF:** gradually reduced distally on st, firm but not stiff, 5–15(20) cm, 2–7 cm wide, elliptic to broadly lanceolate, tapered to winged petiole, dentate to minutely dentate; distal sessile and bract-like. **INFL:** heads discoid or radiate, (6)15–30+ in congested ± flat-topped clusters; involucre widely cylindric or bell-shaped, phyllaries 8, 13, or 21, 4–9 mm, green, black-tipped; outer phyllaries ± inconspicuous, ≤ 1/2 inner, black-tipped. **RAY FL:** 0(5 or 8); ray 5–10 mm. **DISK FL:** 30–45. **FR:** 2–3 mm, glabrous.

$2n$=40. Damp hillsides, meadows, seeps; 1200–2200 m. CaRH, n SN, MP; to WA, MT, UT. Jun–Aug ★

S. hydrophilus Nutt. WATER RAGWORT, ALKALI-MARSH RAGWORT Bien or per 4–10(20) dm, from short button-like caudex, with fleshy-fibrous roots. **ST:** 1 or 2–4 loosely clustered, blue-green, glaucous. **LF:** ± fleshy, reduced distally on st, proximal 5–20+ cm, (1)2–10 cm wide, elliptic to oblanceolate, tapered to unwinged petiole, entire or minutely dentate, mid and distal sessile, occ clasping, becoming bract-like. **INFL:** heads discoid or radiate, 20–40(80+) in rounded or ± flat-topped clusters; involucre cylindric or narrowly urn-shaped, phyllaries 8 or 13, 5–8 mm, green, black tipped; outer phyllaries few, inconspicuous, << inner, black-tipped. **RAY FL:** 0 or 5; ray 3–5 mm. **DISK FL:** 20–35. **FR:** 2.5–3 mm, glabrous. $2n$=40. Marshes, swampy places, standing water, alkaline sites; 200–2500 m. NCoR, CaR, SN, GV, CW, GB; to BC, SD, CO, UT. May–Sep

S. integerrimus Nutt. Bien or per (1)2–7 dm, from button-like caudex; roots fleshy-fibrous. **ST:** 1, ± cobwebby, tomentose, or long-soft-wavy hairy, sometimes ± glabrous in age. **LF:** basal and proximal cauline ± petioled, 6–25 cm, 1–6 cm wide, elliptic to lanceolate or oblanceolate, occ rounded-deltate or ± round, tapered to truncate or cordate, entire to ± dentate, mid-cauline sessile, distal reduced, bract-like. **INFL:** heads gen radiate, (3)6–15(40+) in ± flat-topped clusters; involucre ± urn-shaped, phyllaries 13 or 21, (4)5–12 mm, linear or lanceolate, green- or black-tipped; outer phyllaries few, ≤ 2/3 inner. **RAY FL:** (0) gen 5, 8 or 13; ray 5–15(20) mm, yellow or white. **DISK FL:** 35–45. **FR:** 2.5–3 mm, gen glabrous (hairy on ribs). $2n$=40,80 (in CA).

var. **exaltatus** (Nutt.) Cronquist (p. 419) COLUMBIA RAGWORT **ST:** tomentose or long-soft-wavy hairy at fl. **LF:** basal and proximal cauline indistinctly petioled, elliptic to lanceolate or oblanceolate, occ rounded-deltate or ± round, base tapered to truncate or cordate. **INFL:** heads 6–15(30+); phyllaries (4)5–12 mm, lanceolate, tips black. **RAY FL:** gen 5(0); ray 6–15 mm, yellow. Open woodland, sagebrush scrub, grassland; 500–3200 m. CaR, SN, GB; to AB, WA, WY. May–Jul

var. **major** (A. Gray) Cronquist (p. 419) LAMBS-TONGUE RAGWORT **ST:** 1, densely cobwebby, tomentose or long-soft-wavy hairy at fl. **LF:** basal and proximal cauline petioled, lanceolate to oblanceolate, base tapered. **INFL:** heads 6–12(30+); phyllaries 8–12 mm, linear to awl-like, gen green-tipped, rarely minutely black-tipped. **RAY FL:** gen 5, ray 5–8 mm, yellow. Openings in conifer forest and sagebrush scrub; 100–3600 m. KR, NCoR, CaRH, SN, W&I; OR, NV. May–Aug

var. **ochroleucus** (A. Gray) Cronquist PALE-YELLOW RAGWORT **ST:** 1, unevenly becoming ± glabrous at fl. **LF:** basal and proximal cauline petioled, ovate to rounded-deltate, or obovate to oblanceolate, base tapered or truncate. **INFL:** heads (3)8–12(40+); phyllaries 7–10 mm, linear, green- or black tipped. **RAY FL:** gen 5, ray (5)10–15(20) mm, white. $2n$=40,80. Moist openings in conifer forest; 600–1700 m. CaRH; to BC, AB, MT. May–Jul

S. jacobaea L. (p. 425) TANSY RAGWORT Bien or per 2–9 dm, from taproot or taprooted branched caudex. **ST:** 1–10+, often purple-tinged, sparsely hairy to glabrous. **LF:** ± evenly distributed, basal (occ withered before fl) and proximal cauline 4–25 cm, 2–10 cm wide, ovate to broadly ovate, tapered to petiole, 1–3-pinnately lobed, lobes obovate to spoon-shaped, ultimate margin dentate, mid and distal sessile, smaller. **INFL:** heads radiate, 10–60+ in ± flat-topped clusters; involucre widely cylindric or urn-shaped, phyllaries 13, 3–5 mm, gen black-tipped; outer phyllaries gen inconspicuous, occ ≤ 2/3 inner. **RAY FL:** ± 13; ray 8–12 mm. **DISK FL:** 60–70. **FR:** 0.75–1.25 mm, sparsely hairy, or ray fr glabrous. $2n$=40. Pastures, disturbed areas; < 1500 m. NCo, KR, CaR, n SN, ScV, SnFrB; to BC, e N.Am; native to Eurasia. Toxic to livestock. Jun–Sep ◆

S. linearifolius A. Rich. var. **linearifolius** LINEAR-LEAVED AUSTRALIAN FIREWEED Subshrub 6–18 dm. **ST:** many from base, ± ascending, proximally woody, initially cobwebby, becoming glabrous. **LF:** ± evenly distributed, (proximal early-withered) ± sessile, 2–7 cm, 1–4 mm wide, narrowly oblanceolate or linear-oblong, tapered at base, often with triangular basal lobes, entire to finely ser-

rate, margin often rolled under. **INFL:** heads radiate, 10–50+ in ± flat-topped clusters; involucre cylindric to weakly bell-shaped, phyllaries 13, 3–5 mm, tips dark green; outer phyllaries inconspicuous. **RAY FL:** 4–8; ray 3–6 mm. **DISK FL:** 8–20. **FR:** 1.3–2.5 mm, glabrous or minutely strigose. Seasonal wetland, sandy soils, drainage areas, disturbed areas; 50–100 m. PR; Australia. May–Oct ◆

S. lyonii A. Gray (p. 425) ISLAND SENECIO Subshrub (2)4–15 dm, from woody taproot. **ST:** several from base, sparsely cobwebby or glabrous. **LF:** ± evenly distributed, (proximal early-withered), sessile or short-petioled, (4)6–10(13) cm, 1.5–10 cm wide, oblong or elliptic in outline, 1–2-pinnately divided, lobes 1–2 mm wide, linear, entire. **INFL:** heads radiate, 4–12 in cyme-like clusters; involucre bell-shaped, phyllaries 13 or 21, 5–8 mm, black-tipped; outer phyllaries inconspicuous, << inner. **RAY FL:** 13; ray 8–10 mm. **DISK FL:** 45–55. **FR:** 2.5–3 mm, densely hairy. $2n$=40. Ocean bluffs, open hillsides; < 500 m. s ChI; Baja CA. Feb–May

S. minimus Poir. (p. 425) COASTAL BURNWEED Ann or per (5)10–20 dm, from taproot with branched lateral roots. **ST:** 1–few from base, branched distally, sparsely and unevenly tomentose to glabrous. **LF:** ± evenly distributed (proximal withered before fl), 4–20 cm, 1–4 cm wide, ± narrowly lanceolate, tapered to short, ± clasping petiole or distal sessile, margin sharply and evenly fine-dentate. **INFL:** heads disciform, 40–200+, in rounded to ± flat-topped clusters; involucre cylindric, phyllaries 7–10, 4–6.5 mm, light yellow-green, gen minutely black-tipped; outer phyllaries inconspicuous, << inner. **PISTILLATE FL:** 9–20+ in 1–3 series; corolla minutely lobed, ray 0. **DISK FL:** 3–5. **FR:** 1.5–2.5 mm, hairy in grooves between the ribs, occ glabrous. $2n$=60. Disturbed coastal sites; < 200 m. NCo, NCoRO, w CW; to WA; Australia, New Zealand. [*Erechtites m.* (Poir.) DC.; *Erechtites minima*, orth. var.] Jun–Sep ❖

S. mohavensis A. Gray (p. 425) MOJAVE RAGWORT Ann (0.7)1–4 dm, from taproot. **ST:** 1, proximally or distally branched, often purple-tinged, glabrous, glaucous. **LF:** fleshy, glaucous, ± evenly distributed, proximal 2–6 cm, 0.5–4 cm wide, ovate to obovate, tapered to base, coarsely lobed or irregularly dentate, mid and distal similar, sessile, truncate to cordate-clasping. **INFL:** heads discoid, disciform, or minutely radiate, 3–15+ in loose cyme-like clusters; involucre cylindric or urn-shaped, phyllaries 8 or 13, 6–7 mm, green or red-tinged; outer phyllaries inconspicuous, << inner. **PISTILLATE FL:** 0–8; corolla 0 or narrowly tubular, ray 0–2 mm, barely surpassing phyllaries. **DISK FL:** 15–30. **FR:** 1.5–3 mm, hairy. $2n$=40. Sandy, rocky washes, desert flats; 100–700 m. D; NV, AZ, SON. Disjunct between w N.Am and sw Asia; populations from deserts of sw Asia have been treated as *S. mohavensis* subsp. *breviflorus* (Kadereit) M. Coleman. Mar–May

S. pattersonensis Hoover (p. 425) MOUNT PATTERSON SENECIO Per 2–10 cm, from branched rhizome. **ST:** 1–3, occ ± red tinged, glabrous. **LF:** ± evenly distributed, proximal withered early, petioled or sessile, 1–4 cm, 3–5 mm wide, oblanceolate to lance-linear, decurrent, weakly clasping, entire (or 1–2 lobed), wavy or margin rolled under. **INFL:** heads radiate, 1–4 at branch tips; involucre bell-shaped, phyllaries 12–18, 5–8 mm, light green; outer phyllaries inconspicuous, << inner. **RAY FL:** ± 8; ray 5–10 mm. **DISK FL:** 35–45. **FR:** 3.5–4.5 mm, glabrous, light green. Talus slopes; 3000–3700 m. c SNH, SNE; NV. Jul–Aug ★

S. quadridentatus Labill. COTTON BURNWEED Per 4–12 dm, from taproot. **ST:** 1–several from base, ± gray-tomentose, becoming ± glabrous. **LF:** ± evenly distributed, sessile, 2.5–20 cm, 1–5 mm wide, linear or lance-linear, base tapered, margin remotely minutely dentate, ± rolled under. **INFL:** heads disciform, (20)50–200 in ± flat-topped clusters; involucre cylindric, phyllaries 11–13, 6.5–10 mm, light yellow-green, occ minutely black-tipped; outer phyllaries inconspicuous, << inner. **PISTILLATE FL:** 12–30 in 1–3 series; corolla minutely lobed, ray 0. **DISK FL:** 5–10. **FR:** 2.5–3 mm, hairy in grooves between ribs. $2n$=40. Disturbed areas in chaparral, coastal scrub; < 570 m. WTR, SCo; Australia, New Zealand. [*Erechtites q.* (Labill.) DC.] Jan–Jul

S. scorzonella Greene (p. 425) SIERRA RAGWORT Per 1–4(5) dm, from stout rhizome. **ST:** 1–4, woolly to tufted-tomentose or

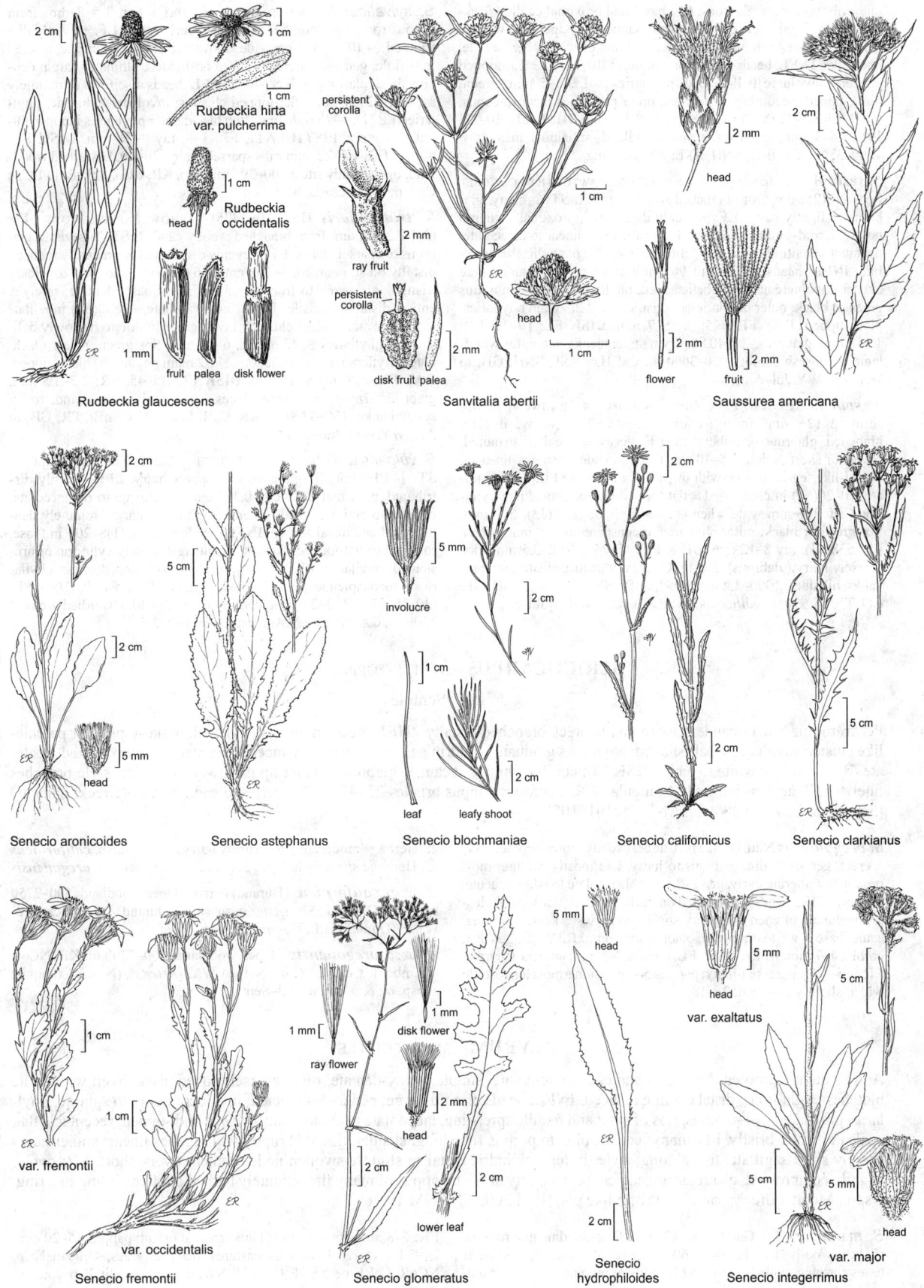

419

Rudbeckia glaucescens
2 cm · 1 cm · Rudbeckia hirta var. pulcherrima · 1 cm · 1 cm · head · Rudbeckia occidentalis · 1 mm · fruit · palea · disk flower

Sanvitalia abertii
persistent corolla · 1 cm · ray fruit · 2 mm · persistent corolla · 2 mm · disk fruit · palea · head · 1 cm · ER

Saussurea americana
2 mm · head · 2 cm · 2 mm · flower · 2 mm · fruit · ER

Senecio aronicoides
2 cm · 2 cm · ER · head · 5 mm

Senecio astephanus
5 cm · ER

Senecio blochmaniae
5 mm · involucre · 2 cm · 1 cm · leaf · leafy shoot · 2 cm

Senecio californicus
2 cm · 2 cm

Senecio clarkianus
2 cm · 5 cm · ER

Senecio fremontii
var. fremontii · 1 cm · 1 cm · ER · var. occidentalis · ER

Senecio glomeratus
1 mm · ray flower · 1 mm · disk flower · 2 mm · head · 2 cm · 2 cm · lower leaf · ER

Senecio hydrophiloides
5 mm · head · 2 cm · ER

Senecio integerrimus
var. exaltatus · 5 mm · head · 5 cm · var. major · 5 cm · 5 mm · head · ER

unevenly becoming glabrous. **LF**: basal and proximal cauline (4)6–12 cm, 1.5–3 cm wide, oblanceolate to lanceolate, tapered to winged petiole, dentate to minutely dentate; mid and distal smaller, sessile, bract-like. **INFL**: heads discoid or radiate, (5)10–24+ in ± flat-topped clusters; involucre bell-shaped, phyllaries ± 13, 3–5 mm, green, black-tipped, becoming ± glabrous; outer phyllaries inconspicuous, ≤ ± 1/2 inner. **RAY FL**: (0)5+; ray 5–8(10) mm. **DISK FL**: 10–20. **FR**: 2.5–3 mm, glabrous. Open woodland, subalpine meadows; 1600–3500 m. CaRH, SNH, SNE; NV. Jul–Aug

S. serra Hook. var. **serra** TALL RAGWORT, SAWTOOTH GROUNDSEL Per 4–10(25) dm, from branched woody caudex. **ST**: 1–many, glabrous or lightly hairy. **LF**: ± evenly distributed, proximal withered before fl, mid-cauline 5–20 cm, 1–4 cm wide, ± linear to lanceolate, dentate to ± entire, tapered to petiole or ± sessile base, distal ± bract-like. **INFL**: heads radiate, 40–90+ in ± flat-topped to panicle-like cluster; involucre narrowly bell-shaped, phyllaries ± 8, 4–6 mm, tips green or black; outer phyllaries ± conspicuous < 2/3 main phyllaries, black-tipped. **RAY FL**: ± 5; ray 5–7 mm. **DISK FL**: 10–20. **FR**: 1.5–2 mm, glabrous. 2*n*=40,80. Open streambanks in conifer woodland, sagebrush scrub; 1000–3000 m. CaRH, s SNF, SNH, GB; to WA, MT, WY. Jul–Aug

S. spartioides Torr. & A. Gray BROOM-LIKE RAGWORT Subshrub 2–12+ dm, from woody taproot. **ST**: 1–many, distally branched, glabrous or puberulent. **LF**: fleshy, ± evenly distributed, sessile or short-petioled, 5–10 cm, 1–6 mm wide, narrowly linear to thread-like, entire or occ with thread-like lobes. **INFL**: heads radiate, 10–20(60) in compound ± flat-topped clusters; involucre cylindric (gen 3–6 mm wide when pressed), phyllaries 8(13), 6–9 mm, tips green or black; outer phyllaries inconspicuous, << inner. **RAY FL**: 5–8(13); ray 8–12 mm. **DISK FL**: 15–25. **FR**: 2–3.5 mm, ribs sparsely hairy (glabrous). 2*n*=40. Open, dry disturbed sites, streambanks, hillsides; 1000–3500 m. SNH, SnBr, SNE, DMtns; to SD, NE, CO, TX. [*S. s.* var. *multicapitatus* (Rydb.) S.L. Welsh] Jul–Sep

S. sylvaticus L. WOODLAND RAGWORT Ann 1.5–8 dm, from fibrous-rooted taproot. **ST**: 1, densely curly-hairy. **LF**: ± evenly distributed, 3–10 cm, 1–4 cm wide, obovate to oblong in outline, tapered to petiole, gen 1–2-pinnately lobed or divided, ultimate margin dentate, distal clasping, ± bract-like. **INFL**: heads disciform or minutely radiate, 12–24+ in ± flat-topped clusters; involucre cylindric, phyllaries 13(21), 4–7+ mm, tips often black; outer phyllaries inconspicuous, << inner. **PISTILLATE FL**: 1–8; ray 0–2+ mm. **DISK FL**: 40–50. **FR**: 1.5–2.5 mm, ribs sparsely hairy. 2*n*=40. Disturbed woodland, open, sandy sites; 100–300 m. NCo, KR, CCo, SnFrB; to BC, e Can, ne US; Eurasia. Mar–Sep

S. triangularis Hook. (p. 425) ARROWLEAF RAGWORT Per (2)5–12(20) dm, from branched woody caudex. **ST**: 1–several, glabrous or sparsely hairy. **LF**: ± evenly distributed or gradually reduced distally on st, petioled, 3–10+ cm, 2–6 cm wide, narrowly to widely triangular, tapered to truncate or cordate at base, dentate, rarely ± entire, distal subsessile. **INFL**: heads radiate, 10–30(60) in ± flat-topped or raceme-like clusters; involucre cylindric to narrowly bell-shaped, phyllaries 8, 13, or 21, 6–10 mm, tips green, rarely black; outer phyllaries ± inconspicuous, ≤ 1/2 main phyllaries, black-tipped. **RAY FL**: ± 8; ray 9–15 mm. **DISK FL**: 35–45. **FR**: 2.5–3.5 mm, glabrous. 2*n*=40,80. Damp places, open conifer woodland, rocky streambanks; 100–3300 m. KR, CaR, SN, SnGb, SnBr, PR, GB; to AK, WY, NM. Jun–Sep

S. vulgaris L. COMMON GROUNDSEL Ann 1–6 dm, from taproot. **ST**: 1–10+, initially glabrous or sparsely hairy. **LF**: ± evenly distributed, proximal 2–10 cm, 0.3–4 cm wide, ovate to oblanceolate, tapered to petiole, lobed to dentate, ultimate margin minutely dentate, mid and distal sessile. **INFL**: heads discoid, (1)8–20+ in loose, rounded to flat-topped clusters; involucre ± widely cylindric or urn-shaped, phyllaries 13–21, 4–8 mm, gen black-tipped; outer phyllaries ± inconspicuous, black-tipped, << inner. **RAY FL**: 0. **DISK FL**: (30)55–65. **FR**: 2–2.5 mm, densely hairy. 2*n*=40. Disturbed areas; < 1300 m. CA-FP, GB; N.Am, Eur, Asia. Feb–Jul

SERICOCARPUS WHITE-TOPPED ASTER

John C. Semple

Per from rhizome or caudex, ascending to erect, branched distally. **INFL**: heads radiate, peduncled, in flat-topped or panicle-like cluster; involucre ± bell-shaped; phyllaries graduated, midrib gen ± swollen, translucent; receptacle convex, pitted, epaleate. **RAY FL**: ray white to pink. **DISK FL**: corolla white to cream, ± glabrous; anther tip narrowly triangular; style branches finely papillate, tips narrowly triangular. **FR**: fusiform; pappus bristles 25–45, in 2–4 series. 5 spp.: N.Am. (Greek: hairy fr) [Semple & Leonard 2006 FNANM 20:101–105]

S. oregonensis Nutt. (p. 425) Caudex woody, fibrous-rooted. **ST**: ± erect, gen 4–10 dm, glabrous to hairy. **LF**: mostly cauline; most proximal withering early, gen 3–8 cm, oblanceolate to elliptic, acute, ± hairy. **INFL**: heads ± congested on each branch of flat-topped cluster; peduncle of each head very short; phyllaries ± keeled, oblong, acute, base ± white, tip green, outer ± spreading. **RAY FL**: gen ± 5; corolla 4–7 mm, white. **DISK FL**: corolla 3–5 mm; anthers ± purple. **FR**: 2.5–5.5, appressed-hairy; pappus 5–7 mm, innermost longest, tip wider, slightly flattened. 2*n*=18.

1. Herbage moderately to densely hairy..... subsp. ***californicus***
1′ Herbage sparsely hairy............... subsp. ***oregonensis***

    subsp. ***californicus*** (Durand) Ferris Open woodland; 500–2750 m. NCoRO, CaRF, SN. [*Aster o.* subsp. *c.* (Durand) D.D. Keck; *S. o.* var. *c.* (Durand) G.L. Nesom] Jul–Sep

    subsp. ***oregonensis*** Open woodland; 100–2200 m. KR, NCoR (Humboldt Co.), CaR, n SNH; to WA. [*Aster o.* (Nutt.) Cronquist subsp. *o.*; *S. o.* var. *o.*] Jul–Sep

SILYBUM MILK THISTLE

Ann or bien, taprooted. **LF**: basal and cauline, alternate, simple, spiny-dentate, often coarsely lobed, dark green with white blotches, ± glabrous, distal cauline reduced. **INFL**: heads discoid, large; peduncles bracted; involucre ovoid to spheric; phyllaries graduated in 4–6 series, tips of outer and middle spreading, lanceolate to ovate, spiny-fringed and -tipped; receptacle flat, epaleate, white-bristly. **FL**: many; corolla pink to purple, tube long, slender, throat abruptly wider, lobes linear; anther bases sharply short-sagittate, tips oblong; style tip long-cylindric distal to slightly swollen node, branches very short. **FR**: ovoid, slightly compressed, glabrous, attachment scar slightly angled; pappus of many flat, minutely barbed bristles, falling in a ring. 2 spp.: Medit. (Greek: name for thistle-like pls) [Keil 2006 FNANM 19:164]

S. marianum (L.) Gaertn. (p. 425) **ST**: 2–30 dm, glabrous or slightly woolly. **LF**: basal 15–60+ cm; petiole winged; cauline lf bases clasping, coiled, spiny. **INFL**: body of involucre 2–6 cm diam. **FR**: 6–8 mm, tan to brown, black-spotted or not; pappus 15–20 mm. 2*n*=34. Invasive. Roadsides, pastures, disturbed areas; < 900 m. NCo, NCoR, CaRF, n&c SNF, GV, CW, SW; native to Medit. Feb–Jun ❖

SOLIDAGO GOLDENROD
John C. Semple

Per from woody caudex or rhizome, branched distally. **LF**: alternate, often sessile, resinous or not. **INFL**: heads radiate, few to many, in ± flat-topped to panicle-like, often ± 1-sided clusters; involucre cylindric to bell-shaped; phyllaries gen ± graduated in 3–5 series, midrib gen ± swollen, translucent, tip flat or slightly swollen; receptacle slightly convex, honeycombed, ridges uneven, epaleate. **RAY FL**: few to many; ray yellow. **DISK FL**: few to many; corolla yellow, gen glabrous; anther tip narrowly triangular; style branches finely papillate, tips triangular. **FR**: obconic, compressed; pappus of 25–45 long-barbed bristles in 1 series. ± 100 spp.: esp N.Am (S.Am, Eurasia). (Greek: make-well, for purported medicinal value) [Semple & Cook 2006 FNANM 20:107–166]

1. Mid-cauline lvs largest
  2. Peduncle bracts, and/or phyllaries sparsely to moderately minutely stalked-glandular; st gen glabrous
    proximally; infl pyramid-shaped, proximal branches not crowded; n SNH, MP . . . . ***S. lepida* var. *salebrosa***
  2′ Infl not glandular; st ± hairy to base, not waxy distally; infl pyramid- or club-shaped; widespread
    3. Infl ± widely pyramid-shaped; involucre 3–5 mm; lvs weakly serrate; disturbed sites . . . ***S. altissima* subsp. *altissima***
    3′ Infl ± club-shaped; involucre 2.5–3.5 mm; lvs gen serrate; ± undisturbed sites . . . . ***S. elongata***
1′ Proximal cauline or basal lvs largest
  4. Infl gen flat- to round-topped; phyllaries long-acuminate, not strongly graduated; lf gen long-ciliate . . . . ***S. multiradiata***
  4′ Infl gen panicle- or raceme-like, gen ± pyramid- or club-shaped, but not flat- or round-topped; phyllaries
    obtuse, rounded, acute or acuminate, ± strongly graduated; lf not long-ciliate
    5. Lvs and heads strongly resinous; NCo, NCoRO, NCoRI, CCo, SnFrB . . . . ***S. spathulata***
    5′ Lvs and heads not strongly resinous; widespread
      6. Herbage sparsely to densely short-hairy; distal nodes without axillary lf-clusters . . . . ***S. velutina***
        7. Herbage ± densely short-hairy; outer phyllaries strigose; disk corolla throat obscure . . . . subsp. *californica*
        7′ Herbage ± sparsely short-hairy; outer phyllaries ± glabrous; disk corolla throat ± obvious . . . . . . subsp. *sparsiflora*
      6′ Herbage ± glabrous; distal nodes often with axillary lf-clusters
        8. Phyllaries lanceolate to ± ovate, not inrolled, obtuse to acuminate; rays 6–21; KR, CaRH, n SNH
          (e slope), GB . . . . ***S. spectabilis***
        8′ Phyllaries very narrowly triangular, inrolled near tip, sharply acute; rays 3–13; Teh, CW, SW,
          W&I, n DMtns
          9. Lvs gen < 10 × longer than wide; infl gen club-shaped, panicle-like . . . . ***S. confinis***
          9′ Lf gen 10+ × longer than wide; infl gen narrow, sometimes spike-like — s SCoRI . . . . ***S. guiradonis***

S. altissima L. subsp. ***altissima*** LATE GOLDENROD **ST**: from rhizome, 2–15 dm, hairy to base. **LF**: 4–15 cm; mid-cauline largest, (ob)lanceolate, strongly 3-veined, ± densely scabrous to short-hairy. **INFL**: gen large, dense, ± widely pyramid-shaped; heads many; involucre 3–5 mm, phyllaries lanceolate, strongly graduated, outer 1/4–1/3 inner, margin ± strigose. **RAY FL**: 10–15; ray 1–2 mm. **DISK FL**: 3–7; corolla 2.5–4.5 mm. **FR**: 1–1.5 mm, ± strigose. $2n=36,54$. Uncommon. Disturbed sites; < 2800 m. NW, CaR, SN, CW, GB; native to c&e N.Am; widely cult, escaped elsewhere. [*S. a.* var. *a.*] Oct–Nov

S. confinis A. Gray (p. 425) SOUTHERN GOLDENROD Pl ± glabrous. **ST**: < 21 dm, often stout, from short, branched caudex. **LF**: proximal largest, 5–25 cm, gen < 10 × longer than wide, entire, often ± fleshy, base nearly sheathing, distal-most sometimes scale-like or with axillary lf-clusters. **INFL**: panicle-like, gen club-shaped; heads gen many; involucre 2.5–4 mm, phyllaries very narrowly triangular, inrolled near tip, sharply acute, ± strongly graduated, outer 1/3–2/3 inner; midrib gen enlarged, translucent. **RAY FL**: 3–13; ray 1–2.5 mm. **DISK FL**: 10–20; corolla ± 3–4 mm. **FR**: 1–1.5 mm, ± strigose. $2n=18$. Wet streambanks, springs, marshes; gen < 2500 m. Teh, CCo, SnFrB (extirpated), SCoR, SW, W&I, DMtns. Involucres of n DMtns pls like those of *S. spectabilis*. Apr–Oct

S. elongata Nutt. (p. 425) WEST COAST CANADA GOLDENROD Herbage ± sparsely strigose (more so distally). **ST**: 25–170 cm, rhizomed. **LF**: mid-cauline largest, 5–15 cm, ± lanceolate, gen 3-veined, toothed. **INFL**: panicle-like, ± club-shaped; heads many; involucre 2.5–3.5 mm, phyllaries lanceolate, strongly graduated, outer 1/4–1/3 length inner. **RAY FL**: 8–15; ray 1.5–2 mm. **DISK FL**: 5–12; corolla 2.5–3.5 mm. **FR**: 1–1.5 mm, ± strigose. $2n=18,36$. Meadows, thickets; < 2800 m. NW, CaR, SN, deltaic SnJV, CW, GB; to s BC. [*S. canadensis* L. subsp. *e.* (Nutt.) Keck.] Pls in CW have thicker, veinier lvs. *S. elongata* and *S. lepida* belong to the taxonomically difficult *S. canadensis* complex incl ± 20 taxa in N.Am. May–Oct

S. guiradonis A. Gray (p. 425) GUIRADO'S GOLDENROD Pl gen ± glabrous. **ST**: < 13 dm, from short caudex, often slender. **LF**: proximal largest, 5–20 cm, narrowly oblanceolate, entire, often ± fleshy, base nearly sheathing, distal-most reduced to linear scales, often with axillary lf-clusters. **INFL**: narrow; heads few to many; involucre 2.5–4 mm, phyllaries narrowly triangular, inrolled near tip, sharply acute, strongly graduated, outer 1/3–1/2 inner, midrib gen enlarged, translucent. **RAY FL**: 8–10; ray 1–2.5 mm. **DISK FL**: 10–20; corolla ± 3–4 mm. **FR**: 1–1.5 mm, ± strigose. $2n=18$. Perennial stream banks and seeps, serpentine; 600–900 m. s SCoRI (San Benito, Fresno cos.). Sep–Oct ★

S. lepida DC. var. ***salebrosa*** (Piper) Semple (p. 425) WESTERN GOLDENROD Herbage glabrous to ± strigose. **ST**: < 15 dm, rhizomed, sometimes ± waxy. **LF**: mid-cauline largest, 5–15 cm, narrowly elliptic to ± lanceolate, 3-veined, toothed. **INFL**: ± open, pyramid-shaped; bracts, branches sparsely to moderately minutely stalked-glandular; heads many; involucre 3–4 mm, phyllaries gen lanceolate, strongly graduated, outer 1/4–1/3 inner. **RAY FL**: 8–15; ray 2–2.5 mm. **DISK FL**: 7–12; corolla 3–4 mm. **FR**: 1–1.5 mm, sparsely strigose. $2n=18,36,54$. Moist streambanks, lakesides; 1000–2000 m. n SNH, MP; to AK, c&e Can, n Mex. [*S. canadensis* var. *s.* (Piper) M.E. Jones; *S. gigantea* Aiton, misappl.] *S. lepida* is the most widespread and variable sp. of the *S. canadensis* complex. *S. lepida* var. *l.* not documented in CA. May–Sep

S. multiradiata Aiton (p. 425) NORTHERN GOLDENROD Pl ± glabrous, hairier distally. **ST**: < 5 dm, from woody caudex. **LF**: proximal cauline < 12 cm, ± linear to spoon-shaped, distal reduced, ± linear to ovate, ± clasping, gen long-ciliate. **INFL**: flat- to round-topped or panicle-like with head-bearing branches from distal axils distinctly shorter than main st; heads few to many; involucre 4–7 mm, phyllaries ± lanceolate, gen long-acuminate, occ acute, not strongly graduated, outer gen 2/3 to = inner. **RAY FL**: 12–18; ray 2–4 mm. **DISK FL**: 12–35; corolla 3–5 mm. **FR**: 2–3 mm, ± sparsely strigose. $2n=18,36$. Slopes, meadows; 1250–3950 m. KR, CaRH, SN, n SNE, W&I; to AK, arctic Can, NM. Jun–Sep

S. spathulata DC. (p. 425)   COAST GOLDENROD   Pl ± glabrous, exc infl. **ST:** 1–5 dm, decumbent to erect, from short, woody caudex. **LF:** glabrous, gen resinous-sticky, proximal largest, 5–10 cm, spoon-shaped, crenate, obtuse, distal reduced, entire, acute. **INFL:** wand- to club-like, strigose; heads few to many; involucre 4–7 mm, phyllaries resinous-sticky, outer ovate, inner narrowly lanceolate, strongly graduated. **RAY FL:** 4–10; ray 2.3–4 mm. **DISK FL:** 10–18; corolla 4.3–6 mm. **FR:** 2–3 mm, strigose. 2*n*=18. Dunes, headlands; < 600 m. NCo, NCoRO, NCoRI, CCo, SnFrB. May–Nov

S. spectabilis (D.C. Eaton) A. Gray (p. 425)   SHOWY GOLDENROD   Pl ± glabrous. **ST:** < 18 dm, from short caudex. **LF:** basal largest, < 25 cm, oblanceolate, tapered, entire or serrate toward tip, ± fleshy, cauline reduced, entire, often with axillary lf-clusters. **INFL:** ± panicle-like, ascending to arching, tip sometimes 1-sided; heads gen many; involucre 3–4 mm, phyllaries lanceolate to ± ovate, obtuse to acuminate, strongly graduated. **RAY FL:** 6–21; ray 1.5–3.5 mm. **DISK FL:** 8–22; corolla 2.5–4.5 mm. **FR:** 1–2 mm, sparsely strigose. 2*n*=18. Bogs, alkaline meadows; 300–2300 m. KR, CaRH, n SNH (e slope), GB; to OR, NV. Jul–Sep

S. velutina DC.   VELVETY GOLDENROD   Herbage sparsely to densely short-soft-hairy. **ST:** 2–15 dm, short-rhizomed. **LF:** proximal < 14 cm, oblanceolate to obovate, serrate, sometimes 3-veined, base tapered, distal much reduced, entire. **INFL:** ± 1-sided, pyra-mid- or wand-shaped, gen many-headed (or short, raceme-like, few-headed); involucre 3–6 mm, phyllaries lanceolate or narrower, acute to obtuse, strigose, sometimes ± resinous, strongly graduated, outer 1/4–1/3 inner. **RAY FL:** 6–11; ray 3–6 mm. **DISK FL:** 5–17; corolla 3–5 mm. **FR:** 0.7–2.5 mm, ± moderately to densely strigose. 2*n*=18,36,54. Typical subsp. native to Mex.

subsp. ***californica*** (Nutt.) Semple (p. 425)   CALIFORNIA GOLDENROD   Herbage gen ± densely short-soft-hairy. **LF:** sometimes 3-veined. **INFL:** long, 1-sided, pyramid- or wand-shaped, many-headed (or short, raceme-like, few-headed); involucre 3–5 mm, phyllaries gen narrow, acute, outer strigose, not resinous. **RAY FL:** 6–11; ray 3–5 mm. **DISK FL:** 6–17; corolla 3–5 mm, throat obscure. **FR:** 0.7–1.5 mm, ± densely strigose. 2*n*=18,36. Woodland margins, grassland, disturbed soils; < 2500 m. CA-FP, MP, W&I; OR, Baja CA. [*S. c.* Nutt.] May–Nov

subsp. ***sparsiflora*** (A. Gray) Semple   Pl ± sparsely short-hairy. **LF:** gen 3-veined. **INFL:** pyramid-shaped, ± sparsely branched, 1-sided at tip; heads gen many; involucre 3–6 mm, phyllaries lanceolate, rounded or obtuse, outer ± glabrous, ± resinous. **RAY FL:** 6–9; ray 4–6 mm. **DISK FL:** 5–12; corolla 3.5–5 mm, throat ± obvious. **FR:** ± 1–2.5 mm, ± strigose. 2*n*=18,36,54. Woodland margins, grassland, disturbed soils; 500–2200 m. GB; to s ID, CO, w TX, n Mex. [*S. s.* A. Gray] Jul–Oct

SOLIVA

Linda E. Watson

Ann, often mat-forming, some with stolons, minutely strigose to long-soft-hairy or glabrous. **ST:** 1–many, prostrate to erect, simple or branched. **LF:** basal or basal and cauline, alternate, opposite, or whorled, sessile or petioled, bases ± sheathing, blades gen (1)2–3-pinnately divided into narrow lobes. **INFL:** heads disciform, 1, sessile near st base or at branch tips, ± enveloped by sheathing bases of opposite or whorled subtending lvs, overtopped by branches; involucre ± hemispheric; phyllaries persistent, 5–8+ in 1–2+ series, free, lanceolate to ovate, ± equal, margins, tips gen scarious; receptacle flat to hemispheric or conic, epaleate. **PISTILLATE FL:** 5–100+ in 1–8+ series; corolla 0; style base hard, spine-like. **DISK FL:** 2–8+, staminate, ovary reduced; corolla very short, ± yellow or ± white, tube wide-cylindric, << throat, lobes 3–4; anther tip narrowly triangular; style tip unbranched, tack-shaped. **FR:** ± oblanceolate to obovate, flattened front-to-back, winged or wingless, glabrous to densely short-hairy; pappus 0. 4–8 spp.: S.Am. (Salvador Soliva, 18th century Spanish physician) [Watson 2006 FNANM 19:545–546]

S. sessilis Ruiz & Pav. (p. 425)   Pl erect or spreading to mat-forming, stolons gen 0, soft-hairy or becoming ± glabrous. **ST:** 1–15 cm; branches arising below heads. **LF:** 1–2(7+) cm, cauline proximally alternate or gen ± whorled throughout. **INFL:** involucre 2–4(5) mm diam. **PISTILLATE FL:** 5–12(17+) in 1–2+ series. **DISK FL:** 4–8+; corolla 1.5–2.5 mm. **FR:** body (1.5)2.5–3+ mm, ± oblanceolate to obovate, wings (occ 0) entire or ± wavy to lobed, distal wing margins gen projecting as spine-like teeth; stylar spine erect, 1–2 mm; faces glabrous or minutely rough- or soft-hairy. 2*n*=110–120. Disturbed areas, esp hard-packed paths; < 1500 m. NW, SNF, CW, SW; widely naturalized; native to S.Am. Apr–Jul

SONCHUS   SOW THISTLE

David J. Keil & G. Ledyard Stebbins

Ann to per [shrub]; sap milky. **ST:** erect, smooth, distally branched. **LF:** basal and cauline, alternate, ± entire to toothed and coarsely pinnate-lobed; cauline gen sessile, clasping. **INFL:** heads liguliflorous, in cyme-like clusters; involucre swollen at base; phyllaries gen in 3 series, outer many, short-triangular, inner series linear, tapered; receptacle ± flat, epaleate. **FL:** many; ligule yellow, readily withering. **FR:** gen ± flat, beakless; pappus of many fine, white bristles. ± 55 spp.: Eurasia, Afr. (Ancient Greek name for a kind of thistle) [Hyatt 2006 FNANM 19:273–276]

1. Per from long rhizomes; fr ± compressed, 3–4-angled, 2-ribbed between angles . . . . ***S. arvensis***
  2. Peduncles and phyllaries stalked-glandular . . . . [subsp. ***arvensis***]
  2′ Peduncles and phyllaries gen sessile-glandular, rarely tomentose . . . . subsp. ***uliginosus***
1′ Ann from taproots; fr weakly to strongly flattened
  3. St slender; lvs deeply lobed, lobes toothed and often with smaller 2° lobes; fr only slightly flattened, transversely roughened; ligule > corolla tube; uncommon . . . . ***S. tenerrimus***
  3′ St stout; lvs toothed or with wide lobes; fr flat, with thin edges, smooth or cross-roughened; ligule ≤ corolla tube; widespread and common
    4. Proximal (clasping) lobes of lvs rounded, strongly curved to coiled; fr 3-ribbed per side, otherwise smooth; ligule < corolla tube . . . . ***S. asper*** subsp. ***asper***
    4′ Proximal (clasping) lobes of lvs acute; fr 2–4-ribbed and cross-wrinkled; ligule ± = corolla tube . . . . . . ***S. oleraceus***

S. arvensis L. PERENNIAL SOW THISTLE Per 4–15 dm, from long rhizomes producing fl and non-fl sts. **LF**: basal 5–15 cm, entire to deeply lobed, short-petioled; cauline coarsely lobed and toothed, basal clasping lobes curled. **INFL**: peduncle 2–8 cm, glandular-hairy; involucre 1.4–1.8 cm. **FL**: ligule ± = tube. **FR**: 2.5–3 mm, ± compressed, 2-ribbed on each face, rough on ribs, dark brown; pappus bristles ± 4 × fr. ◆

subsp. *uliginosus* (M. Bieb.) Nyman **INFL**: peduncle gen sessile-glandular; phyllaries gen sessile-glandular, rarely tomentose, longer 10–15 mm. $2n=36$. Uncommon. Damp soil; < 1800 m. nw MP, s CCo (Santa Barbara Co., extirpated); N.Am; native to Eur. Mar–Jun

S. asper (L.) Hill subsp. *asper* (p. 425) PRICKLY SOW THISTLE Ann 1–12 dm. **ST**: mostly unbranched proximal to infl. **LF**: basal tapered to gen sessile bases; cauline 6–30 cm, distal often widest near base, proximal clasping lobes rounded, strongly curved to coiled; blades dentate, sometimes ± lobed, teeth and lobes soft-spine-tipped. **INFL**: peduncle 0.5–5 cm, gen ± bristly-glandular; involucre 10–13 mm. **FL**: ligule ± 1/3 < tube. **FR**: 2–3 mm, flat, 3-ribbed per face, otherwise smooth, straw-colored to red-brown; pappus bristles ± 3 × fr. $2n=18$. Common. Slightly moist disturbed sites; along streams; < 1900 m. CA; N.Am; native to Eur. Much like *S. oleraceus*. All yr

S. oleraceus L. (p. 425) COMMON SOW THISTLE Ann 1–14 dm. **ST**: often proximally branched. **LF**: basal gen < cauline, gen tapered or abruptly wing-petioled; cauline 5–35 cm, distal often widest at base, proximal clasping lobes acute, not curled or coiled; blades nearly all lobed exc in dwarfed pls, lobes variable in width, terminal lobe often widely arrowhead-shaped. **INFL**: peduncle 0.5–7 cm, glabrous to bristly-glandular, sometimes cottony-tomentose just proximal to heads; involucre 10–13 mm. **FL**: ligule ± = tube. **FR**: 2.5–3.8 mm, flat, 2–4-ribbed, cross-wrinkled, dark brown; pappus bristles ± 2 × fr. $2n=32$. Abundant. Disturbed places; < 2500 m. CA; N.Am; native to Eur. Much like *S. asper*. All yr

S. tenerrimus L. Ann to per 1–8 dm. **ST**: slender, gen much-branched. **LF**: basal gen < cauline, tapered to clasping bases; cauline 3–20 cm, basal clasping lobes acute; blades deeply lobed, lobes linear to triangular, often secondarily lobed. **INFL**: peduncle 0.5–8 cm, gen stalked glandular, tomentose proximal to heads; involucre 10–12 mm. **FL**: ligule ± 1/3 > tube. **FR**: 2.5–3.5 mm, 1–3-ribbed per face, red-brown; ribs roughened; pappus bristles ± 2 × fr. $2n=14$. Uncommon. Disturbed sites; gen ≤ 500 m. GV, SCo, ChI, PR; native to s Eur. Mar–May

SPHAEROMERIA

Timothy K. Lowrey & Elizabeth McClintock

Per or subshrub, gen aromatic, silky; hairs basally forked, attached off-center; herbage dotted with resin glands. **LF**: alternate, often crowded at base; entire or 3-lobed to 1–3-pinnately dissected. **INFL**: heads disciform, 1–many; involucre hemispheric or bell-shaped; phyllaries graduated in 2–3 series, scarious-margined and -tipped; receptacle hemispheric, glabrous or hairy, epaleate. **PISTILLATE FL**: 4–15+, marginal; corolla narrowly tubular, lobes gen 3. **DISK FL**: 30–50; corolla tubular to bell-shaped; anther base rounded or ± cordate, tip awl-shaped to lanceolate or ovate; style tips truncate, brush-like. **FR**: cylindric, 2-3- or 5–10-ribbed; pappus gen 0 or a narrow crown of awl-shaped scales. 9 spp.: w N.Am. (Greek: sphere + division) [Lowrey & Shultz 2006 FNANM 19:499–502] Recent studies place *Sphaeromeria* in *Artemisia* (Garcia et al. 2011 Amer J Bot 98:638–653).

1. Subshrub; lvs entire or 3–4-lobed; disk corolla yellow-white; receptacle glabrous . . . . ***S. cana***
1′ Per from long, thick woody caudex; lvs 1–3-pinnately divided; disk corolla yellow; receptacle with white curly hairs 0.5–1 mm . . . . ***S. potentilloides***
  2. Sts nearly lfless; basal lvs gen 1-pinnately divided; heads 1–3(4) . . . . var. ***nitrophila***
  2′ Sts lfy; basal lvs gen 2–3-pinnately divided; heads (3)4–20 . . . . var. ***potentilloides***

S. cana (D.C. Eaton) A. Heller (p. 429) GRAY CHICKENSAGE **ST**: 15–60 cm, lfy, gen branched from base. **LF**: 10–40 mm, few basal, most cauline, sessile; basal and proximal cauline gen 3–4-lobed, distal entire. **INFL**: heads (1)3–12, in dense clusters. **DISK FL**: corolla lobes soft-hairy. **FR**: 1.8–1.9 mm, 10-ribbed; pappus 0. $2n=18$. Uncommon. Rocky places, ledges, ridges, talus; 1800–4000 m. c&s SNH, SNE, n DMtns; to OR, NV. Sage-like odor sometimes reported. Jul–Sep

S. potentilloides (A. Gray) A. Heller FIVEFINGER CHICKENSAGE **ST**: 3–30 cm, branched from base. **LF**: basal 20–70 mm, 1–3-pinnately divided; basal and proximal cauline petioled, distal sessile. **INFL**: heads (1)2–20, in open clusters. **DISK FL**: corolla lobes glabrous. **FR**: 0.8–1.5 mm, 5-ribbed, swelling when wet; pappus 0.

var. ***nitrophila*** (Cronquist) A.H. Holmgren et al. ALKALI TANSY-SAGE **ST**: 3–15 cm, nearly lfless. **LF**: basal lvs gen 1-pinnately divided; cauline gen 0. **INFL**: heads 1–3(4); involucre 3–4 mm. $2n=18$. Gen alkaline areas; 1600–2200 m. c SNH, SNE; to ID, NV. May–Jul ★

var. ***potentilloides*** (p. 429) **ST**: 10–30 cm, lfy. **LF**: basal gen 2–3-pinnately divided, cauline pinnately divided, rarely 3-lobed. **INFL**: heads (3)4–20; involucre 4–5.1 mm. $2n=18$. Uncommon. Mtn meadows, hot springs, gen not alkaline areas; 1300–2100 m. n&c SNH, GB; to OR, NV. May–Jul

STEBBINSOSERIS

Kenton L. Chambers

Ann, ± scapose; sap milky. **LF**: gen all basal, entire to pinnately lobed, glabrous or mealy, hairs drying as minute, white scales. **INFL**: heads, liguliflorous, 1 on naked peduncle, ± nodding in bud; involucre glabrous; phyllaries in 2–4 series, ± lanceolate, reflexed when dry, outer < 1/3 × inner; receptacle flat, epaleate, ± pitted. **FL**: 8–many; ligules equaling or slightly surpassing involucre, white or yellow, abaxially often ± red, readily withering. **FR**: 10-ribbed, gray to brown or ± purple; pappus scales 5, ± cut at tip, dull white (silvery) to ± brown, bristle-tip finely barbed. 2 spp.: w N.Am. (Greek: Stebbins' chicory, for G.L. Stebbins, Jr., Am geneticist, evolutionist, 1906–2000) [Chambers 2006 FNANM 19:346–347] Spp. derived independently by hybridization: *Uropappus lindleyi* × ann *Microseris*.

1. Fr spindle-shaped, ± dark brown to ± purple, 5–8 mm, tip not widened at base of pappus; pappus scales 3–5 mm; n&c CCo, rare . . . . ***S. decipiens***

1′ Fr spindle-shaped to columnar, sometimes narrowed and long-tapered above, gen gray to pale brown or violet (or ± dark purple in SCo) 4.5–12 mm, tip slightly widened at base of pappus; pappus scales 4–11 mm; widespread. . . . . ***S. heterocarpa***

S. decipiens (K.L. Chambers) K.L. Chambers (p. 429) SANTA CRUZ MICROSERIS Pl 10–60 cm. **LF:** 5–20 cm. **INFL:** involucre 6–19 mm. **FL:** ligules yellow. **FR:** pappus scales < fr, glabrous, midrib linear from near base, bristles 4–5 mm, hair-like, from irregularly cut scale tip. 2*n*=36. Open, sandy, shaly, or serpentine sites, coastal; 10–500 m. n&c CCo. Derived from *Microseris bigelovii* × *Uropappus lindleyi*. Apr–May ★

S. heterocarpa (Nutt.) K.L. Chambers (p. 429) GRASSLAND SILVERPUFFS Pl 8–60 cm. **LF:** 5–35 cm. **INFL:** involucre 6–30 mm. **FL:** ligules yellow or white. **FR:** pappus scales < to > fr, sometimes hairy, midrib thicker at base, tapered, bristle 3–8 mm, arising from notched scale tip. 2*n*=36. Open, sometimes disturbed sites, rocky to clay soils, gen inland exc SCo; < 1700 m. NCoR, SNF, GV (rare), CW, SCo, ChI, WTR, PR; n Mex, se AZ. Derived from *Microseris douglasii* × *Uropappus lindleyi*. Apr–Jun

STENOTUS

Lowell E. Urbatsch

Per to subshrub, taprooted, ± mat or mound-forming, from branched caudex; glabrous or scabrous to tomentose, sometimes ± stalked-glandular. **ST:** fl sts gen erect to spreading, gen unbranched, proximal 1/4–2/3 bearing often many, crowded lvs. **LF:** alternate, basal persisting 2–3 yrs, 2–10 cm, linear to oblanceolate, tapered to ± indefinite petiole, entire, 1–3-veined. **INFL:** heads radiate, gen 1 (2–4 in ± flat-topped cluster); peduncle < 15 cm; involucre hemispheric; phyllaries weakly graduated in 2–3 series, linear to ± ovate, ± lf-like, bases sometimes papery, tips acute to acuminate; receptacle convex, pitted, epaleate. **RAY FL:** 6–20; corolla yellow. **DISK FL:** 25–50; corolla 6–7.5 mm, funnel-shaped, yellow; anther tip narrowly triangular; style tips long tapered, appendage length 2–4 × width. **FR:** 2.4–6 mm, glabrous to densely silky; pappus of soft bristles ≤ 10 mm. ± 3 spp.: w N.Am. (Greek: narrowness, for lf width) [Morse 2006 FNANM 20:174–177] Other sp. of *Stenotus* in TJM (1993) [*S. stenophyllus* (A. Gray) Greene] now treated in *Nestotus*.

1. St and lvs glabrous or scabrous; glandless or stalked-glandular, sometimes sticky; taproot gen well developed, stout; lvs rigid . . . . ***S. acaulis***
1′ St and lvs ± tomentose with crinkled, shaggy hairs; glandless to sparsely stalked-glandular proximal to head; taproot weakly developed, branched; lvs ± flexible . . . . ***S. lanuginosus*** var. ***lanuginosus***

S. acaulis (Nutt.) Nutt. (p. 429) STEMLESS GOLDENWEED Cespitose or matted. **LF:** linear to widely oblanceolate. **INFL:** heads 1 or 2–4 per cluster; involucre 6–10 mm. **RAY FL:** 5–15; ray 6–12 mm. **DISK FL:** 17–40; corolla 5–9.5 mm, tube gen glabrous. **FR:** glabrous or densely silky; pappus white or tan. Dry, rocky open scrub, dry woodland, open forest, alpine; 1900–3660 m. CaR, SN, GB, DMtns; to OR, MT, WY, CO, AZ. May–Aug

S. lanuginosus (A. Gray) Greene var. ***lanuginosus*** WOOLLY STENOTUS Loosely cespitose. **LF:** on proximal 2/3 of st or occ mostly basal, linear to linear-oblanceolate. **INFL:** head 1; involucre 9–15 mm. **RAY FL:** 9–17; corollas 8–14 mm. **DISK FL:** 30–50, corolla 5.5–8.5 mm, tube thinly hairy. **FR:** thinly silky; pappus white. Shallow, rocky soils, sagebrush scrub, juniper woodland, dry meadows; ± 1500 m. s MP (Lassen Co.); to WA, MT, NV. Other var. in ID and MT. May–Jun ★

STEPHANOMERIA

L.D. Gottlieb

Ann to subshrub, glabrous or hairy; sap milky. **ST:** 1–5+, ascending or erect, simple or branched, 1–20 dm. **LF:** alternate, basal rosette and cauline, gen ± thin, linear to oblanceolate, entire to pinnately lobed, gen withered at fl in ann and some per; distal cauline gen reduced and ± bract-like. **INFL:** heads liguliflorous, 1 or in dense or panicle-like clusters from nodes along branches, ± sessile to peduncled; involucre cylindric; phyllaries gen in 2 series, outer < 1/4–1/2 × inner; appressed or reflexed; inner equal, glabrous or glandular-hairy; receptacle smooth or pitted, epaleate. **FL:** 4–21; ligules lavender, pink, or white, readily withering. **FR:** 1.9–6.5 mm; ± cylindric, 5-angled, each face gen with long, narrow, central groove; pappus of 5–40, white to tan bristles, wholly plumose or plumose at least on distal 1/2, free throughout or widened at bases and then basally fused, gen in groups of 2–4, deciduous entirely or only widened bristle bases persistent after distal portion breaks off. 16 spp.: w N.Am. (Greek: wreath division, alluding to plumose pappus bristles) [Gottlieb 2006 FNANM 19:350–359] Ann spp. complexly interrelated, distinguished by different combinations or expressions of same traits; mature frs and pappus needed for identification. *S. blairii* moved to *Munzothamnus*, *S. spinosa* to *Pleiacanthus*.

1. Per from woody caudex or from stout or slender rhizome
2. Fls 4–6(8); cauline lvs gen much reduced, bract-like, gen withered at fl
3. Pl from woody caudex; pappus bristles gen tan, plumose on distal 80%. . . . . ***S. pauciflora***
3′ Pl from rhizome; pappus bristles white, wholly plumose. . . . . ***S. tenuifolia***
2′ Fls 7–21; cauline lvs persistent, green at fl
4. Pl from woody caudex; lvs 10–20 cm; young herbage tomentose; phyllaries many, graduated in several series; receptacle pitted . . . . ***S. cichoriacea***
4′ Pl from rhizome; herbage glabrous or sparsely hairy; phyllaries in 2 series, inner long, outer < 1/2 × inner; receptacle smooth
5. Lf margins not thickened, spines 0; lvs 3–8 cm; pappus bristles wholly plumose; dry forest, n CA . . . . ***S. lactucina***
5′ Lf margins thickened, with minute spines; lvs 2–8 cm; pappus bristles plumose on distal 80%; SNE, DMoj. . . . . ***S. parryi***

425

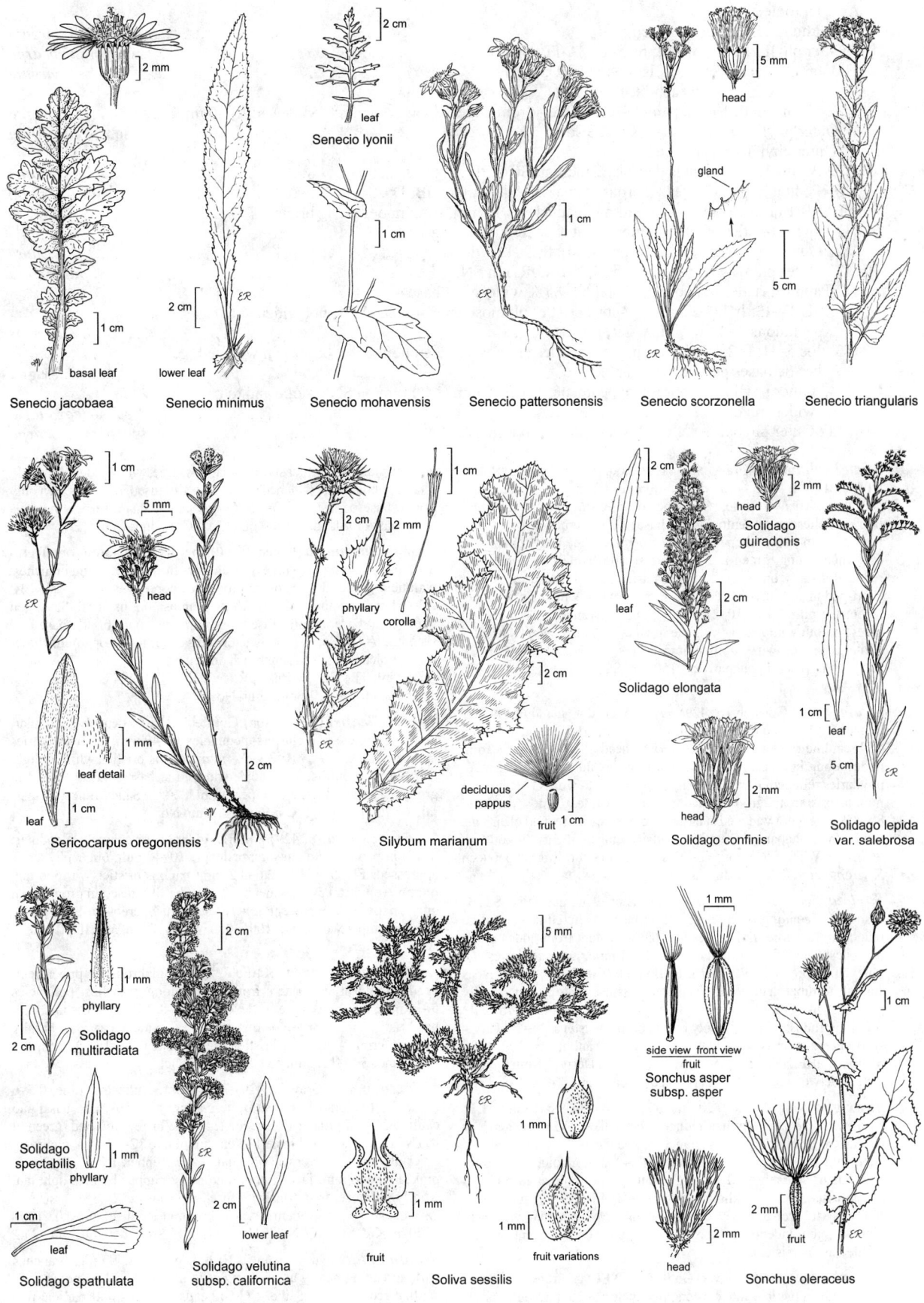

Senecio jacobaea
basal leaf

Senecio minimus
lower leaf
ER

Senecio lyonii
leaf

2 cm
1 cm

Senecio mohavensis

Senecio pattersonensis
ER
1 cm

Senecio scorzonella
head
gland
ER
5 mm
5 cm

Senecio triangularis

Sericocarpus oregonensis
ER
head
leaf
leaf detail
1 cm
5 mm
1 mm
2 cm

Silybum marianum
phyllary
corolla
deciduous pappus
fruit
1 cm
2 cm
2 mm
ER

Solidago elongata
leaf
2 cm

Solidago guiradonis
head
2 mm

Solidago confinis
head
2 mm

Solidago lepida var. salebrosa
leaf
1 cm
5 cm
ER

Solidago multiradiata
phyllary
1 mm
2 cm

Solidago spectabilis
phyllary
1 mm

Solidago spathulata
leaf
1 cm

Solidago velutina subsp. californica
lower leaf
ER
2 cm

Soliva sessilis
ER
fruit
fruit variations
5 mm
1 mm

Sonchus asper subsp. asper
side view front view
fruit
1 mm

Sonchus oleraceus
head
fruit
1 cm
2 mm
ER

1′ Ann, taprooted
  6. Fr without long groove on each face . . . . ***S. virgata***
    7. Outer phyllaries appressed; fls 5–6; CA-FP, GB. . . . . subsp. ***pleurocarpa***
    7′ Outer phyllaries reflexed; fls 8–9; SCoRO, SW . . . . subsp. ***virgata***
  6′ Fr with long groove on each face
    8. Heads in small, ± open panicle-like clusters from nodes along branches; peduncles 10–40 mm . . . . [2]***S. exigua***
      9. Outer phyllaries reflexed — CCo, SCoR . . . . subsp. ***carotifera***
      9′ Outer phyllaries appressed
        10. Peduncles, involucres densely glandular; SCo, PR . . . . subsp. ***deanei***
        10′ Peduncles, involucres glabrous or sparsely glandular; GB, D . . . . subsp. ***exigua***
    8′ Heads 1 or in clusters on spreading peduncles ≤ 10 mm, from nodes along branches
      11. Pappus bristles wholly plumose or at least to widened bases
        12. Outer phyllaries reflexed (appressed); fls 9–15; widespread . . . . ***S. elata***
        12′ Outer phyllaries appressed; fls 5; KR, CaRH, n SN, MP . . . . ***S. paniculata***
      11′ Pappus bristles plumose on distal 60–85%, widened at base or not
        13. Fls 11–13; fr 1.9–2.3 mm; pappus bristles plumose on distal 80–85%, not widened at bases, entirely
           deciduous — SCo, ChI, WTR, PR . . . . ***S. diegensis***
        13′ Fls 5–11; fr 2.3–6.5 mm; pappus bristles plumose on distal 60–85%, bases gen widened, gen only
           bristle bases persistent . . . . [2]***S. exigua***
          14. Outer phyllaries appressed; peduncles gen 2–5 mm (≤ 20 mm on SN e slope and to n and e);
             widespread . . . . subsp. ***coronaria***
          14′ Outer phyllaries reflexed; peduncles 5–10 mm; c&s SNF. . . . . subsp. ***macrocarpa***

S. cichoriacea A. Gray SILVER ROCK-LETTUCE, CHICORYLEAF WIRE-LETTUCE Per 4–10+ dm from large, ± branched, woody caudex with 1–many rosettes. **ST:** 1 per rosette, simple or with wand-like branches; ± tomentose. **LF:** basal 10–20 cm, narrowly (ob)lanceolate to spoon-shaped, entire or dentate, ± densely tomentose, esp when young; green at fl; cauline sessile, ± clasping, gradually reduced distally on st, becoming bract-like. **INFL:** heads sessile or short-peduncled; involucre 12–15 mm; phyllaries many, graduated; receptacle pitted. **FL:** 10–21; ligule pink or lavender. **FR:** 5–6 mm; faces smooth with grooves visible as fine lines or 0; pappus bristles white to pale brown, wholly plumose, free, persistent. $2n$=16. Rock faces, coastal scrub, chaparral; < 1500 m. SCoR, SCo, ChI, TR, PR. May–Nov

S. diegensis Gottlieb (p. 429) Ann 5–20 dm, glandular-puberulent. **ST:** 1, branches ascending or spreading. **LF:** basal withered at fl; cauline reduced, bract-like. **INFL:** heads 1 or in clusters from nodes along branches; peduncle 3–4 mm; involucre 7–9 mm; outer phyllaries reflexed. **FL:** 11–13; ligule pale pink to white. **FR:** 1.9–2.3 mm; faces smooth, grooved; pappus bristles white, plumose on distal 80–85%, free, not widened at bases, deciduous. $2n$=16. Old clearings, sand dunes, chaparral openings, roadside embankments; 20–600 m. SCo, ChI, WTR, PR; n Baja CA. Derived from hybridization between *S. virgata* and *S. exigua*. Jul–Jan

S. elata Nutt. Ann 5–15 dm, glabrous or glandular-hairy. **ST:** 1, branches ascending or spreading. **LF:** basal withered at fl; cauline reduced, bract-like. **INFL:** heads 1 or in clusters from nodes along branches; peduncle 3–7 mm; involucre 5–7 mm; outer phyllaries gen reflexed (appressed). **FL:** 9–15; ligule pink. **FR:** 2.8–4.5 mm; faces smooth to tubercled, grooved; pappus bristles white or tan, wholly plumose, in some populations free, deciduous, in others, basally fused in groups, wholly or only bristle base persistent. $2n$=32. Chaparral openings, grassy meadows, roadside embankments; 100–1400 m. NW, CaRF, SN, CW, w SCo, WTR; sw OR. Derived from hybridization between *S. exigua* and *S. virgata*. Jul–Nov

S. exigua Nutt. Ann 1–20 dm. **ST:** 1, branches spreading. **LF:** basal withered at fl; cauline reduced, bract-like. **INFL:** heads 1 or in clusters from nodes along branches, peduncles 2–10 mm, or in ± open panicle-like clusters, peduncles 10–40 mm; involucre 5–7 mm, glabrous or sparsely to densely glandular; outer phyllaries reflexed or appressed. **FL:** 5–11; ligule white to pink. **FR:** 2.1–6.5 mm; faces smooth to tubercled, grooved; pappus bristles white to tan, variously plumose, gen widened at base, entire pappus gen persistent or only bristle base persistent.

subsp. ***carotifera*** (Hoover) Gottlieb **INFL:** glabrous or puberulent; heads in panicle-like clusters; peduncle 10–25 mm; outer phyllaries reflexed. **FL:** 7–9. **FR:** 3.2–4.3 mm; pappus bristles white,

wholly plumose; in coastal populations, free, wholly deciduous; in inland populations, widened at base, basally fused in groups, gen only bristle base persistent. $2n$=16. Open, sandy, shale, serpentine soils near coast or inland; < 1000 m. CCo, SCoR. Jun–Oct

subsp. ***coronaria*** (Greene) Gottlieb **INFL:** glabrous or puberulent; in c&s CA heads single or in clusters from nodes along branches; peduncle gen 2–5 mm; outer phyllaries appressed. **FL:** 5–11. **FR:** 2.3–3.1 mm; pappus bristles white, plumose on distal 60–85%, gen widened at base, basally fused in groups, gen only bristle base persistent. $2n$=16. Coastal sandy sites, grassland, forest openings, limestone, volcanic soils in sagebrush desert; < 2800 m. SNH, Teh, SnJV, CCo, ChI, TR, GB, DMtns; OR, sw ID, NV. On SN e slope and to n and e, peduncles ≤ 20 mm. Jun–Nov

subsp. ***deanei*** (J.F. Macbr.) Gottlieb **INFL:** densely glandular; heads in panicle-like clusters; peduncles 10–40 mm; outer phyllaries appressed. **FL:** 7–9. **FR:** 2.1–2.4 mm; pappus bristles white to light tan, plumose on distal 55–60%, widened at bases, basally fused in groups, gen only bristle base persistent. $2n$=16. Sandy fields, chaparral; < 1700 m. SCo, PR; n Baja CA. Jun–Nov

subsp. ***exigua*** (p. 429) **INFL:** glabrous or sparsely glandular; heads in panicle-like clusters; peduncle 10–40 mm; outer phyllaries appressed. **FL:** 5–8. **FR:** 2.6–3.2 mm; pappus bristles white to tan, plumose on distal 50%, widened at base, basally fused in groups, gen only bristle base persistent. $2n$=16. Sagebrush, creosote-bush scrub, pinyon/juniper woodland; 100–2000 m. GB, D; to WA, ID, CO, TX; Baja CA. Apr–Jul

subsp. ***macrocarpa*** Gottlieb **INFL:** glabrous or puberulent; heads single or in clusters from nodes along branches, peduncles 5–10 mm; outer phyllaries reflexed. **FL:** 6–8. **FR:** 5.5–6.5 mm; pappus bristles tan, plumose on distal 60–70%, widened at base, basally fused in groups, gen only bristle base persistent. $2n$=16. Openings in woodland; 300–1200 m. c&s SNF. May–Oct

S. lactucina A. Gray (p. 429) Per from slender rhizome, 0.5–6 dm. **ST:** 1(4), branches erect or ascending; glabrous. **LF:** basal and cauline 3–8 cm, linear to lanceolate, entire or few-toothed, green at fl. **INFL:** heads 1; peduncle 1–5 cm; involucre 12–14 mm, glabrous; outer phyllaries appressed, unequal, < 1/2 × inner. **FL:** 7–10; ligule pink. **FR:** 5–6 mm; faces smooth, grooved; pappus bristles light tan, wholly plumose, free or basally fused in groups of 5–6+, persistent. $2n$=16. Sandy soils, openings in yellow-pine, red-fir forests; 1100–2300 m. KR, NCoRO, NCoRH, CaRH, n&c SNH, MP; OR. Jun–Aug

S. paniculata Nutt. Ann 2–10 dm, glabrous. **ST:** 1; branches stiff, spreading nearly at right angles to st. **LF:** basal withered at fl, cauline reduced, bract-like. **INFL:** heads 1 or in small panicle-like clusters; peduncle 2–10 mm; involucre 6–9 mm; outer phyllaries

appressed. **FL**: 5; ligule pale lavender. **FR**: 3.8–4.2 mm; faces tubercled, grooved; pappus bristles tan, wholly plumose, widened at base, basally fused in groups, entire pappus or only bristle base persistent. 2*n*=16. Sandy or volcanic soils, plains, foothills, roadsides; 200–1400 m. KR, CaRH, n SN, MP; OR, WA, ID. Jun–Oct

S. parryi A. Gray   Per 1–3 dm, from stout rhizome, glabrous. **ST**: 1–3; branches ascending. **LF**: 2–8 cm, linear to lanceolate, ± thick and firm, with weakly spine-tipped teeth or lobes; green at fl. **INFL**: heads 1 to few in cyme-like clusters; peduncle 2–30 mm; involucre 10–14 mm; outer phyllaries appressed, unequal, < 1/2 × inner. **FL**: 8–14; ligule white or pale lavender. **FR**: 4.5–6 mm; ribs well developed, faces bumpy, grooved; pappus bristles ± white to tan, plumose on distal 80%, widened at base, basally fused in groups, entirely persistent or only bristle base persistent. 2*n*=32. Gravelly slopes; 700–2000 m. SNE, DMoj; NV, UT, AZ. May–Jun

S. pauciflora (Torr.) A. Nelson (p. 429)   WIRE-LETTUCE   Per to subshrub, 2–10+ dm, from woody caudex, glabrous (tomentose). **ST**: 1–5+, intricately branched, branches ascending to widely spreading. **LF**: basal withered at fl; cauline reduced, deciduous in age, distal bract-like. **INFL**: heads 1; peduncle 0–10 mm; involucre 8–11 mm; outer phyllaries appressed. **FL**: 5–6(8); ligule white or lavender-pink. **FR**: 3.5–5 mm; faces tubercled, grooved; pappus bristles tan, plumose on distal 80%, widened at base, basally fused in groups, entirely persistent or only bristle base persistent. 2*n*=16. Sandy, gravelly washes, slopes, plains; 200–1500 m. s SNH, Teh, SnJV, SCoR,

SCo, TR, e PR, SNE, D; to WY, KS, TX, nw Mex. [*S. pauciflora* var. *parishii* (Jeps.) Munz] Mar–Nov

S. tenuifolia (Raf.) H.M. Hall   NARROW-LEAVED WIRE-LETTUCE   Per 2–7 dm, from stout rhizome, glabrous. **ST**: 1–5+; erect to ascending; branches few to many, internode lengths and branching angles variable. **LF**: proximal linear to thread-like, 5–8 cm, ± entire or few-toothed, withered at fl; cauline reduced, distal-most bract-like. **INFL**: heads 1, peduncle (0)1–5 cm; involucre 5–15 mm; outer phyllaries appressed. **FL**: 4–5; ligule light pink. **FR**: 3–6 mm; faces smooth, grooved; pappus bristles white, wholly plumose, free, persistent. 2*n*=16. Volcanic, granitic, sandstone outcrops, rocky ridges, cliff bases; 300–3000 m. CaRH, SNH, GB, DMoj; to WA, ND, CO, TX; nw Mex. Jul–Aug

S. virgata Benth.   Ann 5–20 dm, glabrous (± tomentose). **ST**: single, branches wand-like. **LF**: basal withered at fl; cauline reduced, bract-like. **INFL**: heads 1 or in clusters from nodes along branches; peduncle 3–10 mm; involucre 6–8 mm; outer phyllaries appressed or reflexed. **FL**: 5–9; ligule white to dark pink. **FR**: 2.2–3.6 mm; smooth to tubercled, not grooved; pappus bristles white, wholly plumose, free, deciduous.

subsp. **pleurocarpa** (Greene) Gottlieb (p. 429)   **INFL**: outer phyllaries appressed. **FL**: 5–6. 2*n*=16. Chaparral openings, grassland; < 1800 m. CA-FP, GB. Jun–Nov

subsp. **virgata**   **INFL**: outer phyllaries strongly reflexed. **FL**: 8–9. 2*n*=16. Chaparral openings, grassland; < 1800 m. SCoRO, SW; n Baja CA. Jun–Oct

STYLOCLINE   NESTSTRAW

James D. Morefield

Ann 1–10(20) cm, gen ± gray, cobwebby to tomentose. **ST**: 1, erect, or 2–10+, ascending to decumbent, ± forked, gen evenly lfy proximally, ± lfless between distal forks. **LF**: alternate, ± sessile, gen elliptic to oblanceolate, entire; distal lvs subtending heads, crowded, largest wider than proximal lvs. **INFL**: heads disciform, ± sessile, 2–10 per dense group; involucre 0 or vestigial, simulated by paleae, or ± cup-like (then phyllaries 1–4, ± unequal, like reduced paleae); receptacle length (2.8)4–8 × width, ± cylindric to club-shaped, glabrous; paleae of pistillate fls each gen enclosing fl, or outermost open, deciduous with fr, ± boat-shaped, acute to obtuse, ± woolly abaxially, obscurely parallel-veined (veins 5+), margin reflexed distally or fully as scarious wing, wing terminal, erect to curved inward, visible in head; paleae of disk fls gen 2–5, reduced, erect, ± folded, lanceolate, acute, papery, scarious, glabrous or cobwebby. **PISTILLATE FL**: 12–25+, all subtended by paleae; corolla obscure, narrowly cylindric. **DISK FL**: staminate (ovary partly developed in some spp.), 2–6; pappus of (0)1–10(13) free, deciduous bristles, hidden in head; corolla gen 5-lobed; anther base tailed, tip ± triangular; style tips ± linear-oblong. **FR**: each enclosed by palea, ± obovoid, variously compressed, smooth, shiny, corolla scar ± terminal, pappus 0. 7 spp.: sw N.Am. (Greek: column bed, for long receptacle) [Morefield 2006 FNANM 19:450–453] Close to *Logfia*. See also *Ancistrocarphus*, *Micropus*.

1. Longest paleae winged throughout, wing widest in proximal 2/3 of palea; phyllaries 1–3.5 mm, cordate or elliptic to obovate, ± persistent
   2. Wing of longest palea elliptic to ± obovate; disk fl ovary (0.2)0.3–0.6 mm; disk pappus bristles (5)6–12(13); heads woolly, dull, hairs evident . . . . ***S. citroleum***
   2′ Wing of longest palea ± cordate; disk fl ovary 0–0.2 mm; disk pappus bristles 1–5(6); heads ± cobwebby, often shiny, hairs hidden by palea wings . . . . ***S. gnaphaloides***
1′ Longest paleae winged distally, wing widest in distal 1/3 of palea; phyllaries 0 or ≤ 0.5 mm, ± awl-shaped, deciduous
   3. Receptacle 2.8–3.5 × width, club-shaped; disk fl ovary 0.3–0.6 mm; fr 0.6–0.8 mm; heads nearly spheric, largest 3–4 mm diam; longest paleae 1.9–3.1 mm; proximal lvs obtuse . . . . ***S. sonorensis***
   3′ Receptacle 4–8 × width, ± cylindric; disk fl ovary 0–0.3(0.4) mm; pl with > 1 of the following: fr 0.8–1.6 mm, heads ovoid to fusiform, largest heads 5–9 mm diam, longest paleae 3.4–4.5 mm, proximal lvs acute
      4. Heads ± spheric, woolly, largest 5–9 mm diam; longest paleae 3.4–4.5 mm; outermost paleae ± closed
         5. Bodies of longest paleae cartilaginous; fr compressed front-to-back; largest distal lvs 4–11 mm, all ± elliptic to oblanceolate, distal-most gen 0.8–1.2 × heads . . . . ***S. intertexta***
         5′ Bodies of longest paleae exc midvein papery; fr compressed laterally; largest distal lvs (7)11–17 mm, some or all awl-shaped to lanceolate, distal-most gen 1.5–2 × heads . . . . ***S. micropoides***
      4′ Heads ovoid to fusiform, tomentose, largest 1.5–4 mm diam; longest paleae 2–3.3 mm; outermost paleae open, concave
         6. Largest heads 1.5–2.5 mm diam; longest palea 2–2.7 mm; fr 0.7–1 mm; disk corolla 0.8–1.1 mm, gen 4-lobed; lvs obtuse, not mucronate . . . . ***S. masonii***
         6′ Largest heads 2.5–4 mm diam; longest palea 2.8–3.3 mm; fr 1.1–1.6 mm; disk corolla 1.1–1.7 mm, gen 5-lobed; lvs ± acute, mucronate . . . . ***S. psilocarphoides***

S. citroleum Morefield (p. 429) OIL NESTSTRAW Pl 2–9(13) cm. **LF:** ± acute, mucronate, longest 6–13 mm; largest distal lvs 4–12 mm, 2–3.5 mm wide, ± elliptic to widely oblanceolate. **INFL:** heads nearly spheric, woolly (hairs evident), dull, largest 4–5.5 mm, 3.5–5 mm diam; phyllaries 1.5–2.5 mm, elliptic to obovate, ± persistent; receptacle 1.5–2.5 mm, 4–6 × width, club-shaped; longest paleae 2.5–3.5 mm, winged throughout, wing elliptic to ± obovate, widest in median 1/3 of palea, palea body (exc midvein) papery; outermost paleae ± closed. **DISK FL:** corolla 1–1.6 mm; ovary (0.2)0.3–0.6 mm. **FR:** 0.8–1 mm, compressed laterally; disk pappus bristles (5)6–12(13), 1.4–1.8 mm. Open, stable, often crusted sand, clay, dry drainage edges, between *Atriplex* shrubs; 60–300 m. s SnJV, s SCo (extirpated). Some features suggest *S. gnaphaloides* × *Logfia filaginoides* origin; collections gen mixed with 1 or both. Mar–Apr ★

S. gnaphaloides Nutt. (p. 429) EVERLASTING NESTSTRAW Pl 1–15(20) cm. **LF:** gen obtuse, not or scarcely mucronate, longest 6–14 mm; largest distal lvs 4–13 mm, 1.5–3 mm wide, spoon-shaped to obovate or elliptic. **INFL:** heads ± spheric, ± cobwebby (hairs hidden by palea wings), often shiny, largest 3–6 mm diam; phyllaries 1–3.5 mm, ± ovate, ± persistent; receptacle 1.3–2.2 mm, 5–8 × width, cylindric; longest paleae 1.8–4.5 mm, winged throughout, wing ± cordate, widest in proximal 1/3 of palea, palea body (exc midvein) papery; outermost paleae ± closed. **DISK FL:** corolla 1–1.8 mm; ovary 0–0.2 mm. **FR:** 0.8–1 mm, compressed laterally; disk pappus bristles 1–5(6), 1.3–1.9 mm. 2*n*=28. Common. Open, gen sandy soil, often old disturbances; < 1200(1700) m. s NCoRI, SNF, SnJV, CW, SW, sw DMoj (exc DMtns), w edge DSon; s-c AZ, to nw Mex. [*Stylocline gnaphalioides*, orth. var.] Mar–May

S. intertexta Morefield (p. 429) MOJAVE NESTSTRAW Pl 2–8(11) cm. **LF:** acute, mucronate, longest 6–15 mm; largest distal lvs 4–11 mm, 1–2.5 mm wide, all ± elliptic to oblanceolate, distal-most gen 0.8–1.2 × heads. **INFL:** heads ± spheric, woolly, largest 5–6 mm diam; phyllaries 0 or ≤ 0.5 mm, ± awl-shaped, deciduous; receptacle 1.4–2.7 mm, 4–7 × width, cylindric; longest paleae 3.4–4.2 mm, winged distally, wing elliptic to ovate, widest in distal 1/3 of palea, palea body cartilaginous; outermost paleae ± closed. **DISK FL:** corolla 1.1–2.3 mm; ovary 0–0.3 mm. **FR:** 1–1.4 mm, compressed front-to-back; disk pappus bristles (0)1–4(8), 1.1–2 mm. Open, stable, rocky or sandy, often calcareous soils, rock bases, drip lines, dry drainages; 40–1400 m. n&e DMoj, DSon; to sw UT, w-most AZ. Feb–May

S. masonii Morefield MASON'S NESTSTRAW Pl 1–7(10) cm. **LF:** obtuse, not mucronate, longest 5–9 mm; largest distal lvs 2–5 mm, ± 1 mm wide, oblong to narrowly elliptic. **INFL:** heads ± fusiform, tomentose, largest 2–5 mm, 1.5–2.5 mm diam; phyllaries 0 or ≤ 0.5 mm, ± awl-shaped, deciduous; receptacle 2–3 mm, 5–8 × width, ±

cylindric; longest paleae 2–2.7 mm, winged distally, wing oblanceolate to teardrop-shaped, widest in distal 1/3 of palea, palea body cartilaginous; outermost paleae open, concave. **DISK FL:** ovary 0–0.1 mm; corolla 0.8–1.1 mm, gen 4-lobed. **FR:** 0.7–1 mm, compressed front-to-back; disk pappus bristle 0 or 1, 0.7–1 mm. Open loose sand of washes and flats; 100–1200 m. s SNF, s SnJV, SCoRO, WTR. Mar–Jun ★

S. micropoides A. Gray (p. 429) DESERT NESTSTRAW, WOOLLYHEAD FANBRACT Pl 2–14(20) cm. **LF:** acute, mucronate, longest 8–20 mm; largest distal lvs (7)11–17 mm, 1.5–2.5 mm wide, some or all awl-shaped to lanceolate, distal-most gen 1.5–2 × heads. **INFL:** heads ± spheric, woolly, largest 5–9 mm diam; phyllaries 0 or ≤ 0.5 mm, ± awl-shaped, deciduous; receptacle 2–3 mm, 5–8 × width, cylindric; longest paleae 3.4–4.5 mm, winged distally, wing elliptic to ovate, widest in distal 1/3 of palea, palea body exc midvein papery; outermost paleae ± closed. **DISK FL:** ovary 0–0.3 mm; corolla 1.2–1.9 mm. **FR:** 1–1.4 mm, compressed laterally; disk pappus bristles 2–5(10), 1.1–2 mm. 2*n*=28. Gen stable sand or gravel, often rock bases or drip lines; 70–1600 m. s SNE, n&e DMoj, DSon, DSon/PR; to s UT, w TX, n-most Mex. Feb–May

S. psilocarphoides M. Peck (p. 429) PECK NESTSTRAW Pl 1–8(18) cm. **ST:** ± lfless between proximal forks. **LF:** ± acute, mucronate, longest 8–18 mm; largest distal lvs 3–10 mm, 1–2 mm wide, elliptic to spoon-shaped. **INFL:** heads ovoid, tomentose, largest 3.5–5 mm, 2.5–4 mm diam; phyllaries 0 or ≤ 0.5 mm, ± awl-shaped, deciduous; receptacle 2–3 mm, 5–8 × width, ± cylindric; longest paleae 2.8–3.3 mm, winged distally, wing oblanceolate to teardrop-shaped, widest in distal 1/3 of palea, palea body cartilaginous; outermost paleae open, concave. **DISK FL:** ovary 0.1–0.3(0.4) mm; corolla 1.1–1.7 mm. **FR:** 1.1–1.6 mm, compressed front-to-back; disk pappus bristles (0)1–3, 1.1–1.5 mm. Open, gen stable, sandy or gravelly, rock bases, drip lines; 100–2000 m. n edge TR, SNE, DMoj, w edge DSon; to se OR, sw ID, sw UT. See also *S. masonii.* Feb–May

S. sonorensis Wiggins (p. 429) MESQUITE NESTSTRAW Pl 2–10(15) cm. **LF:** proximal obtuse, distal acute; mucronate, longest 6–13 mm; largest distal lvs 3–10 mm, 2–3 mm wide, ± elliptic to narrowly ovate. **INFL:** heads nearly spheric, woolly, largest 3.5–4.5 mm, 3–4 mm diam; phyllaries 0 or ≤ 0.5 mm, ± awl-shaped, deciduous; receptacle 1.2–2.2 mm, 2.8–3.5 × width, club-shaped; longest paleae 1.9–3.1 mm, winged distally, wing ± elliptic, widest in distal 1/3 of palea, palea body (exc midvein) papery; outermost paleae ± closed. **DISK FL:** ovary 0.3–0.6 mm; corolla 0.9–1.4 mm. **FR:** 0.6–0.8 mm, slightly compressed laterally; disk pappus bristles (1)3–8, 0.9–1.3 mm. Sandy drainages with *Prosopis*; ± 400 m. DSon (Hayfield, Riverside Co., Apr 1930, presumed extirpated); se AZ, ne Sonora. Like *Logfia depressa.* Mar–Apr ★

SYMPHYOTRICHUM AMERICAN-ASTER

Geraldine A. Allen

Ann or per from caudex or rhizome. **ST:** gen erect, 1–24 dm. **LF:** basal and/or cauline, alternate, gen entire; basal gen petioled. **INFL:** heads gen radiate, 1 or in cyme- or panicle-like cluster; involucre cylindric to hemispheric; phyllaries graduated or ± equal in 2–6 series, at least inner proximally with pale, papery margins; receptacle ± flat, pitted, epaleate. **RAY FL:** [8]14–100+; corolla violet to pink or white. **DISK FL:** 4–60+; corolla, anthers gen yellow, tube gen < throat; anther tip ± triangular; style branches flat on inner face, base ± warty, tip acute, hairy. **FR:** gen rounded, ± ribbed, ± brown; pappus of bristles, white to ± brown. ± 90 spp.: N.Am, S.Am, e Eurasia. (Greek: bristle hair) [Brouillet et al. 2006 FNANM 20:465–539] Reported records of *S. laeve* (L.) G.L. Nesom var. *geyeri* (A. Gray) G.L. Nesom from Del Norte Co. are *S. subspicatum* (Nees) G.L. Nesom.

1. Ann, taprooted; ray corolla < 8 mm, gen > disk fls by 2 mm or less
  2. Ray fls 90–110, rays ≤ 0.2 mm wide; disk fls > 35; phyllaries obtuse to rounded; lvs elliptic to obovate, ± obtuse . . . . ***S. frondosum***
  2′ Ray fls 21–54, rays ≥ 0.2 mm wide; disk fls 7–23; phyllaries acute to acuminate; lvs linear to oblanceolate, obtuse to acute . . . . ***S. subulatum***
    3. Ray corolla pink to lavender, gen 2.5–3.5 mm, drying in 2–3 coils; ray fls ≥ 30, disk fls 11–23 . . . . . . var. ***elongatum***
    3′ Ray corolla gen white to pink, 1.3–3 mm, drying in 1–2 coils or curling; ray fls 27–40, disk fls 7–15
      4. Phyllaries > 30; pappus bristles gen 3.5–4 mm, < ray corollas . . . . var. ***parviflorum***
      4′ Phyllaries < 30; pappus bristles gen 4.0–5.2 mm, > ray corollas . . . . var. ***squamatum***

429

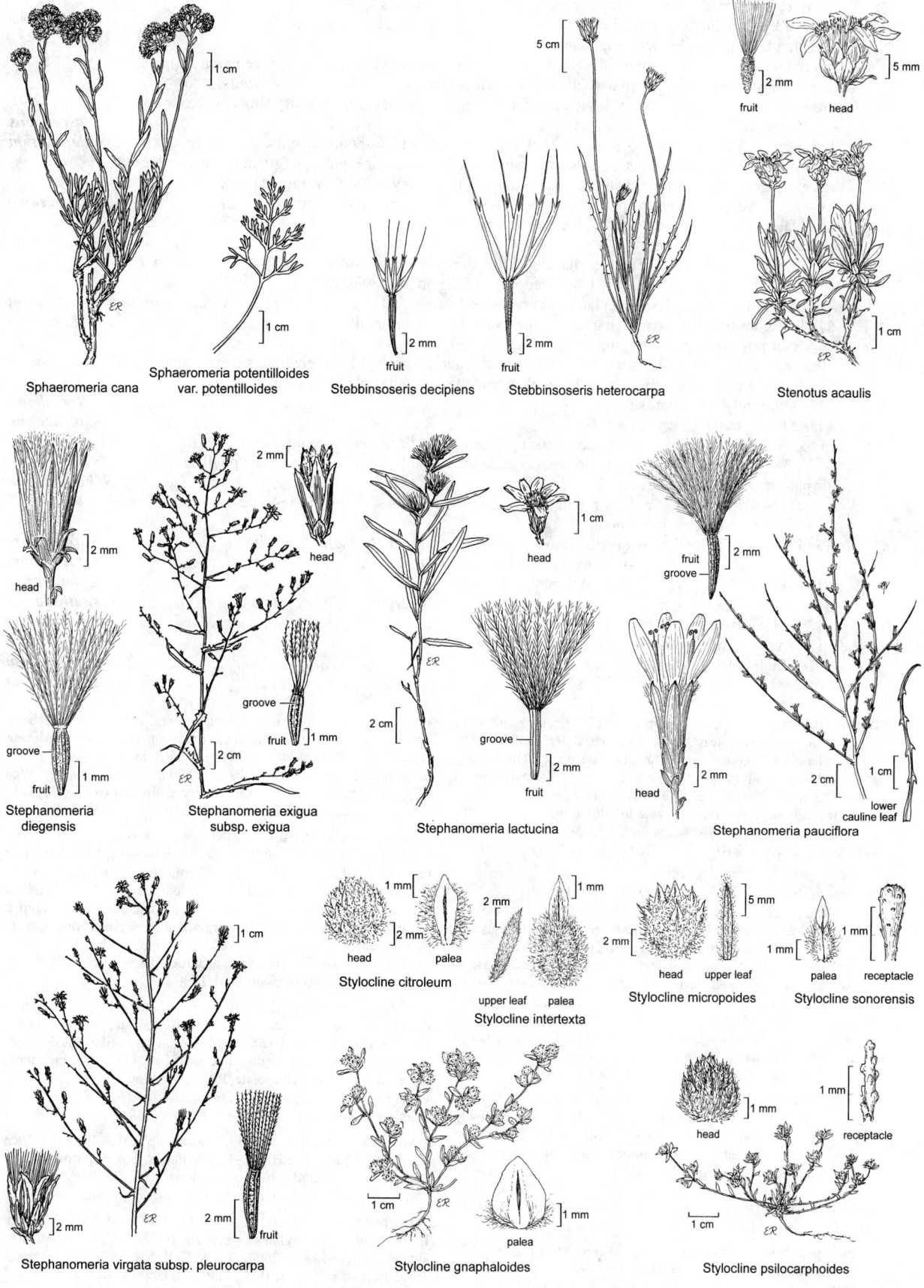

Sphaeromeria cana

Sphaeromeria potentilloides var. potentilloides

Stebbinsoseris decipiens

Stebbinsoseris heterocarpa

Stenotus acaulis

Stephanomeria diegensis

Stephanomeria exigua subsp. exigua

Stephanomeria lactucina

Stephanomeria pauciflora

Stephanomeria virgata subsp. pleurocarpa

Stylocline citroleum

Stylocline intertexta

Stylocline micropoides

Stylocline sonorensis

Stylocline gnaphaloides

Stylocline psilocarphoides

1′ Per, gen from rhizome or caudex; ray corolla 6–12 mm, conspicuous
   5. Involucre ± glandular — peduncle ± glandular . . . . ***S. campestre***
   5′ Involucre glabrous to ± hairy, not glandular
     6. Outer phyllaries obtuse to rounded, < inner ones, ± pale-margined at base, with green area at tip gen
        < 2.5 × longer than wide; st gen densely and uniformly strigose, especially below heads
       7. Infl open, branches gen > 5 cm; lvs glabrous to ± strigose, gen lacking short lfy shoots in axils;
          st 2–6 dm . . . . ***S. ascendens***
       7′ Infl narrow, branches gen < 5 cm; lvs strigose, gen with short lfy shoots in axils; st 4–10 dm . . . . . . ***S. defoliatum***
     6′ Outer phyllaries acute to obtuse, ≤ inner ones, green throughout, or ± pale-margined at base with green
        area at tip gen > 2.5 × longer than wide; st ± glabrous to ± hairy distally or with hairs gen ± in lines
       8. Lvs rough-hairy on faces, elliptic to obovate, cauline > basal; phyllaries acute . . . . ***S. greatae***
       8′ Lvs glabrous to ± soft-hairy, gen ± linear to narrowly elliptic (if wider, basal > cauline);
          phyllaries acute to obtuse
         9. Lvs ± equal or mid-cauline largest, basal gen 0 at fl; ray white to pink or violet; < 2000 m
           10. Outer phyllaries pale-margined > 1/2 length, length gen 3+ × width
             11. St ± hairy in lines; lvs lanceolate; infl bracts well developed . . . . ***S. lanceolatum*** var. ***hesperium***
             11′ St ± glabrous; lvs linear to narrowly lanceolate; infl bracts small . . . . ***S. lentum***
           10′ Outer phyllaries pale-margined < 1/2 length, length gen < 3 × width
             12. Ray gen pink to white; infl gen narrow, lateral branches gen < 10 cm; phyllaries gen ± equal . . . ***S. bracteolatum***
             12′ Ray violet; infl gen open, lateral branches gen 10+ cm; outer phyllaries gen < inner
               13. Outer phyllaries obtuse . . . . ***S. chilense***
               13′ Outer phyllaries ± acute . . . . ***S. subspicatum***
         9′ Basal or proximal cauline lvs largest, basal gen present at fl; ray violet to purple; > 1000 m
           14. St 7–15 dm; cauline lvs elliptic to oblanceolate base lobed, clasping; heads > 35; phyllaries ±
              pale-margined at base . . . . ***S. hendersonii***
           14′ St 0.5–8 dm; cauline lvs linear to narrowly elliptic or heads < 35; phyllaries pale-margined at base
              or outermost green throughout
             15. Lvs < 7 × longer than wide; outermost phyllaries gen green throughout, ± = inner . . . . ***S. foliaceum***
               16. Heads 1(2–4); st 0.5–2 dm; phyllaries ± purple-tinged . . . . var. ***apricum***
               16′ Heads 1–12 (gen > 1); st 1–4 dm; phyllaries green . . . . var. ***parryi***
             15′ Lvs 7+ × longer than wide; outermost phyllaries pale-margined at base, < inner . . . . ***S. spathulatum***
               17. Lvs ± linear . . . . var. ***yosemitanum***
               17′ Lvs lanceolate to narrowly elliptic
                 18. St 4–8 dm, heads gen 10–50 . . . . var. ***intermedium***
                 18′ St 2–6 dm, heads gen 3–10 . . . . var. ***spathulatum***

S. ascendens (Lindl.) G.L. Nesom (p. 435) Per; rhizome long. **ST:** 2–6 dm, strigose at least distally. **LF:** basal and cauline, 5–15 cm, smaller distally on st, oblong to oblanceolate, acute, glabrous to ± strigose. **INFL:** heads in open cyme-like cluster; phyllaries oblong to narrowly obovate, obtuse (outer) to ± acute (inner), green at tip, pale-margined at base, green zone obovate to elliptic. **RAY FL:** 15–40; corolla 8–15 mm, violet. **FR:** hairy. 2*n*=26,52. Meadows, disturbed places; 500–2500 m. SNH (e slope), Teh, SnGb, SnBr, GB; to w Can, CO, n AZ. [*Aster a.* Lindl.] Jul–Sep

S. bracteolatum (Nutt.) G.L. Nesom (p. 435) Per; rhizome short. **ST:** 4–10 dm, ± hairy distally. **LF:** basal and cauline (basal 0 at fl), sessile, 5–15 cm, narrowly lanceolate, acute, gen entire, glabrous to ± hairy. **INFL:** heads in narrow, many-headed cyme-like cluster; phyllaries ± equal, oblong to ovate, outer ± green throughout, obtuse to acute, inner pale-margined at base, acute, green zone oblanceolate to elliptic. **RAY FL:** 20–40, corolla 6–12 mm, white to pink or pink-purple. **FR:** hairy. 2*n*=16,32,48,64. Wet places; 500–2000 m. KR, CaR, SN, GB; to BC, CO. [*Aster eatonii* (A. Gray) Howell; *S. e.* (A. Gray) G.L. Nesom] The name *S. bracteolatum* has nomenclatural priority over *S. eatonii* (Brummitt 2011 Taxon 60:230). Jul–Aug

S. campestre (Nutt.) G.L. Nesom Per; rhizome long. **ST:** 1–4 dm, glandular and gen strigose. **LF:** cauline, 2–6 cm, linear to narrowly oblong, acute to obtuse, glabrous to ± short-hairy. **INFL:** heads in narrow cyme-like cluster; phyllaries linear to lanceolate, acute, green, ± pale-margined at base, ± glandular. **RAY FL:** 15–30; corolla 6–10 mm, violet. **FR:** hairy. 2*n*=10. Dry meadows; 1800–3050 m. SNH, GB; to e WA, UT. [*Aster c.* Nutt.; *S. c.* var. *bloomeri* (A. Gray) G.L. Nesom] Jul–Aug

S. chilense (Nees) G.L. Nesom (p. 435) Per; rhizome gen long. **ST:** 4–10 dm, hairy distally. **LF:** basal and cauline (basal 0 at fl), ± sessile, ± oblanceolate, acute, entire to finely serrate, ± hairy. **INFL:** heads in

cyme-like cluster; phyllaries oblong to ovate, outer green ± to base, obtuse, inner pale-margined, ± acute, green zone oblanceolate to linear. **RAY FL:** 15–40, corolla 8–12 mm, violet. **FR:** hairy. 2*n*=48,64. Grassland, salt marshes, disturbed places; < 500 m. w&c NW, w&c CW, n SCo, n ChI; to BC. [*Aster c.* Nees; *S. c.* var. *invenustum* (Greene) G.L. Nesom; *S. c.* var. *medium* (Jeps.) G.L. Nesom] Jun–Oct

S. defoliatum (Parish) G.L. Nesom SAN BERNARDINO ASTER Per; rhizome short to long. **ST:** 4–10 dm, strigose ± throughout. **LF:** basal and cauline (basal 0 at fl), 4–12 cm, narrowly oblong to oblanceolate, acute, strigose; smaller lvs tufted in axils. **INFL:** heads in narrow cyme-like cluster; phyllaries ± oblong, obtuse (outer) to ± acute (inner), green at tip, pale-margined at base, green zone obovate to elliptic. **RAY FL:** 15–40, corolla 8–12 mm, ± white to pale violet. **FR:** hairy. 2*n*=36. Grassland, disturbed places; < 2050 m. SnGb, SnBr, PR. [*Aster bernardinus* H.M. Hall] Jul–Nov ★

S. foliaceum (DC.) G.L. Nesom Per; rhizome long. **ST:** ± hairy distally. **LF:** basal and cauline, elliptic to obovate, acute to obtuse, ± entire, gen glabrous. **INFL:** heads 1 or in cyme-like cluster; phyllaries oblong to ovate, acute, outer ± lf-like, inner pale-margined, green zone elliptic to lanceolate. **RAY FL:** 15–60, corolla 12–25 mm, violet to purple. **FR:** gen hairy. 2*n*=16,32,48,64,80,96. [*Aster f.* DC.] Highly variable.

var. ***apricum*** (A. Gray) G.L. Nesom **ST:** 0.5–2 dm. **LF:** gen 3–10 cm. **INFL:** heads 1(2–4); phyllaries gen ± purple-tinged. 2*n*=16,32. Alpine and subalpine meadows; 2700–3200 m. KR, CaR, SN; to BC, MT, CO, NM. [*Aster f.* var. *a.* A. Gray] Aug

var. ***parryi*** (D.C. Eaton) G.L. Nesom **ST:** 1–4 dm. **LF:** 5–12 cm. **INFL:** heads 1–12; phyllaries green. 2*n*=16,32,64. Open woodland to alpine meadows; 1400–2800 m. NW, CaR, SN, SnBr, SnJt, Wrn; to MT, NM. [*Aster f.* var. *p.* (D.C. Eaton) A. Gray] Jul–Aug

S. frondosum (Nutt.) G.L. Nesom (p. 435)   Ann. **ST**: decumbent to erect, 2–6 dm, ± glabrous. **LF**: basal and cauline, sessile, gen 2–5 cm, elliptic to obovate, ± obtuse, glabrous to finely ciliate. **INFL**: heads in narrow cyme-like cluster; phyllaries oblanceolate to obovate, obtuse to rounded, green, inner ± pale-margined at base. **RAY FL**: 90–110, corolla < 8 mm, pink-purple, slightly > disk corolla. **FR**: ± hairy. 2*n*=14. Marshes, lake edges, often alkaline; 700–2450 m. KR, CaR, SN, n CW, TR, PR, GB; to BC, WY, AZ. [*Aster f.* (Nutt.) Torr. & A. Gray] May–Oct

S. greatae (Parish) G.L. Nesom   GREATA'S ASTER   Per; rhizome long. **ST**: ascending to erect, 5–12 dm, sparsely hairy. **LF**: basal and cauline (basal gen 0 at fl), 6–15 cm, elliptic to obovate, ± clasping, acute, entire to serrate, hairy. **INFL**: heads in open cyme-like cluster, gen lfy-bracted; phyllaries gen ± equal, lanceolate to oblong, acute, green, ± pale-margined at base, ± spreading, green zone lanceolate to narrowly elliptic. **RAY FL**: 15–40, corolla 8–15 mm, pale violet. **FR**: hairy. 2*n*=16. Damp places in canyons; 300–2000 m. SnGb (s slope). [*Aster g.* Parish] Formerly misspelled *Aster greatai.* Aug–Oct ★

S. hendersonii (Fernald) G.L. Nesom   HENDERSON'S ASTER   Per; rhizome long. **ST**: 7–15 dm, glabrous to ± hairy distally. **LF**: basal and cauline, 4–15 cm, ± smaller distally on st, oblanceolate to elliptic, acute, entire, ± glabrous. **INFL**: heads in open cyme-like cluster; phyllaries linear to oblong, acute, green at tip, ± pale-margined at base, green zone linear. **RAY FL**: 25–45, corolla 9–15 mm, violet. **FR**: hairy. 2*n*=16,32. Meadows, forest openings; 1000–1500 m. KR; to OR, ID. [*Aster foliaceus* var. *lyallii* (A. Gray) Cronquist] Variable. Jul–Aug

S. lanceolatum (Willd.) G.L. Nesom var. ***hesperium*** (A. Gray) G.L. Nesom (p. 435)   Per; rhizome long. **ST**: 6–16 dm; hairs ± in lines throughout. **LF**: basal and cauline (basal 0 at fl), 8–16 cm, lanceolate, acute, entire to ± serrate, ± hairy. **INFL**: heads in open, lfy-bracted cyme-like cluster; phyllaries linear to oblong, acute to acuminate, pale-margined at base to > 1/2 length, green zone ± lanceolate. **RAY FL**: 18–45, corolla 6–12 mm, ± white to violet. **FR**: hairy. 2*n*=64. Wet places; < 2000 m. SCo, PR, SNE; to AB, n-c US. [*Aster h.* A. Gray; *A. l.* subsp. *h.* (A. Gray) Semple & Chmiel.] Jul–Aug

S. lentum (Greene) G.L. Nesom (p. 435)   SUISUN MARSH ASTER   Per; rhizome long. **ST**: 4–15 dm, ± glabrous. **LF**: basal and cauline (basal gen 0 at fl), sessile, 5–15 cm, linear to narrowly lanceolate, acute, gen glabrous. **INFL**: heads at branch tips, in open cyme-like cluster; bracts small; phyllaries linear to oblong, acute to ± obtuse, ± pale-margined at base to > 1/2 length, green zone ± lanceolate. **RAY FL**: 20–45, corolla 8–14 mm, violet. **FR**: hairy. 2*n*=64. Marshes; < 300 m. s ScV, CCo, SnFrB. [*Aster l.* Greene] Grades into *S. chilense.* Threatened by habitat loss. May–Nov ★

S. spathulatum (Lindl.) G.L. Nesom   Per; rhizome long. **ST**: glabrous to ± hairy. **LF**: basal and cauline, 5–15 cm, ± smaller distally on st, acute, ± entire, ± glabrous. **INFL**: heads in open cyme-like cluster; phyllaries linear to oblong, acute or outer ± obtuse, green at tip, ± pale-margined at base, green zone elliptic to lanceolate. **RAY**

FL: 15–40, corolla 8–12 mm, violet. **FR**: hairy. [*Aster s.* Lindl.; *A. occidentalis* (Nutt.) Torr. & A. Gray] Variable.

var. ***intermedium*** (A. Gray) G.L. Nesom   **ST**: 4–8 dm. **LF**: lanceolate to narrowly elliptic. **INFL**: heads gen 10–50. 2*n*=32,48,64. Meadows; 1200–2000 m. NW, CaR, SN, GB; to BC, ID, NV. [*Aster occidentalis* var. *i.* A. Gray; *A. s.* var. *i.* (A. Gray) Cronquist] Jun–Aug

var. ***spathulatum*** (p. 435)   **ST**: 2–6 dm. **LF**: lanceolate to narrowly elliptic. **INFL**: heads gen 3–10. 2*n*=16,32,48,64. Meadows; 1200–2800 m. NW, CaR, SN, TR, SnJt, GB; to nw Can, MT, WY, CO, NM. [*Aster occidentalis* var. *o.*] Smaller forms intergrade with *S. foliaceum* var. *parryi.* Jun–Aug

var. ***yosemitanum*** (A. Gray) G.L. Nesom   WESTERN BOG ASTER   **ST**: 3–8 dm. **LF**: ± linear. **INFL**: heads gen 3–25. 2*n*=16,32. Uncommon. Meadows; 1200–2100 m. KR, CaR, SN; s OR. [*Aster occidentalis* var. *y.* (A. Gray) Cronquist] Jul–Sep

S. subspicatum (Nees) G.L. Nesom   DOUGLAS ASTER   Per; rhizome long. **ST**: 4–12 dm, ± hairy distally. **LF**: basal and cauline (basal 0 at fl), ± sessile, oblanceolate to elliptic, acute, entire to serrate, ± hairy. **INFL**: heads in ± open cyme-like cluster; phyllaries oblong or oblanceolate to linear, pale-margined or outer green ± to base, ± acute, green zone obovate to elliptic. **RAY FL**: 15–45, corolla 8–16 mm, violet. **FR**: hairy. 2*n*=48,64,80,96. Open disturbed places; < 1000 m. NW, CaR, CW, MP; to BC, AB, MT. [*Aster s.* Nees] Highly variable; grades into *S. chilense.* Jul–Sep

S. subulatum (Michx.) G.L. Nesom   ANNUAL SALTMARSH ASTER   Ann. **ST**: erect, 2–20 dm, ± glabrous. **LF**: basal and cauline (basal 0 at fl), ± sessile, 2–9 cm, linear to oblanceolate, obtuse to acute, ± glabrous. **INFL**: heads in open cyme-like cluster; phyllaries linear to lanceolate, acute to acuminate, green, ± pale-margined. **RAY FL**: 21–54, corolla gen < 7 mm, white to pink or lavender, barely > disk corolla. **FR**: ± hairy. 2*n*=10,20. [*Aster s.* Michx.] Var. *ligulatum* (Shinners) S.D. Sundb. [*Aster subulatus* var. *ligulatus* Shinners] not in CA.

var. ***elongatum*** (A.G. Jones & Lowry) S.D. Sundb. **ST**: 3–20 dm. **RAY FL**: 30–54, corolla gen 2.5–3.5 mm, pink to lavender, drying in 2–3 coils. 2*n*=20. Wet places; < 50 m. SCo; Baja CA; native to se US, West Indies. [*Aster s.* var. *e.* A.G. Jones & Lowry; *A. bahamensis* Britton] Aug–Oct

var. ***parviflorum*** (Nees) S.D. Sundb. (p. 435)   **ST**: 7–15 dm. **RAY FL**: 27–40, corolla gen 2–3 mm, white to pink, drying in 1–2 coils or curling. 2*n*=10. Marshes, disturbed places; < 1200 m. GV, CW, SW, DSon; to c&se US, Mex, n S.Am. [*Aster s.* var. *cubensis* (DC.) Shinners; *S. expansum* (Spreng.) G.L. Nesom] Jul–Oct

var. ***squamatum*** (Spreng.) S.D. Sundb. **ST**: 3–15 dm. **RAY FL**: 21–30, corolla gen 1.5–3 mm, ± white, ± curled when dry. 2*n*=20. Salt marshes, wet places, disturbed ground; ≤ 50 m. s ScV, CCo; se US, Mex, Eurasia, Australia; native to S.Am. [*Conyza squamata* Spreng.; *Aster squamatum* (Spreng.) G.L. Nesom] Jul–Oct

SYNTRICHOPAPPUS

Dale E. Johnson

Ann 1–11 cm, gen loosely woolly. **ST**: decumbent or erect, gen branched from base or ± throughout. **LF**: alternate (or proximal opposite), simple. **INFL**: heads 1, radiate; involucre ± cylindric; phyllaries 5–8 in 1 series, spreading in fr, oblanceolate, acute, alternate phyllaries scarious-margined; receptacle convex, ± knobby or smooth, epaleate. **RAY FL**: 1 per phyllary; ray 3-lobed, yellow or adaxially white and abaxially ± pink-purple with veins ± red. **DISK FL**: 10–20+, corolla narrowly funnel-shaped, yellow, glabrous; anther tip narrowly triangular; style tips narrowly triangular. **FR**: narrowly obconic; pappus 0, or of many minutely barbed bristles, slightly widened and fused at base. 2 spp.: sw US. (Greek: fused bristly pappus) [Johnson 1991 Novon 1:119–124; Johnson 2006 FNANM 21:379–380]

1. Ray yellow . . . . ***S. fremontii***
1′ Ray adaxially white, abaxially ± pink-purple with red veins . . . . ***S. lemmonii***

S. fremontii A. Gray (p. 435)   Pl ± decumbent to erect. **LF**: 5–20 mm, proximal wedge-shaped, distal spoon-shaped, tip gen 3-lobed or entire, margins rolled under. **INFL**: peduncle 3–25 mm, ± green; involucre 5–7 mm; phyllaries gen 5. **RAY FL**: ray 3–5 mm. **FR**:

2–3.5 mm, strigose; pappus bristles 30–40, ± 2 mm. 2*n*=12. Open, sandy to gravelly areas in scrub or woodland; 600–2500 m. DMoj, sw DSon; to sw UT, nw AZ, n Baja CA. Mar–Jun

S. lemmonii (A. Gray) A. Gray (p. 435) LEMMON'S SYNTRICHOPAPPUS Pl ± erect, sometimes ± glabrous in age. **LF**: 3–8 mm, linear to narrowly oblanceolate; tip entire, obtuse. **INFL**: peduncle 3–15 mm, ± red; involucre 4–5 mm; phyllaries 5–8. **RAY** **FL**: ray 2–3(4) mm. **FR**: 2–2.5 mm, ± glabrous; pappus 0 or bristles 20–30, ± 1 mm. 2*n*=14. Open, sandy to gravelly areas in chaparral or Joshua-tree or pinyon/juniper woodland; 900–1500 m. SCoRO (Monterey Co.), e WTR, SnGb, SnBr, SnJt. Apr–May ★

TAGETES MARIGOLD

Ann [per, shrub], gen scented. **ST**: erect. **LF**: pinnately divided or compound (in CA), opposite throughout or distal alternate, sessile or petioled, dotted with embedded oil glands. **INFL**: heads gen radiate, peduncled, 1 or in terminal cyme-like clusters; involucre cylindric to bell-shaped; phyllaries in 1 series, equal, fused, gland-dotted; receptacle flat or rounded, epaleate. **RAY** **FL**: (0)1–many; corolla white to yellow, orange, or brown; style tips long, tapered. **DISK FL**: 3–many; corolla yellow to orange or brown. **FR**: cylindric; pappus of scales, sometimes with 1 or more awns. ± 50 spp.: N.Am, S.Am. TOXIC to root-parasite nematodes. (Possibly for the Etruscan god Tages) [Strother 2006 FNANM 21:235–236] *T. erecta* L. (heads 1–few, gen large; involucre widely cylindric or bell-shaped, 13–19 mm; ray fls 8–many) a waif in disturbed places.

T. minuta L. WILD MARIGOLD Ann 2–10 dm, glabrous. **LF**: lflets serrate or dentate. **INFL**: peduncle 5–5.5 mm, slender; phyllaries 3–5, not splitting apart. **RAY FL**: corollas pale yellow; ray 1–2 mm, inconspicuous. **DISK FL**: 3–5; corolla 3–4 mm, yellow. **FR**: 4.5–7 mm; pappus of 1–2 acuminate scales 2–3 mm and 3–5 ovate to lanceolate scales 0.5–1 mm. 2*n*=24. Disturbed places; < 1000 m. s SNF, s SnJV, SnFrB (extirpated), s SCoRO, SCo, w WTR; e US, Afr, Australia; native to w S.Am. May cause contact dermatitis in susceptible individuals. Sometimes cult. Jun, Nov ◆

TANACETUM TANSY

Linda E. Watson

Ann, per; ≤ 150 cm, glabrous or hairy, often aromatic. **ST**: 1 or 2–5+, prostrate to erect, branched proximally and/or distally, glabrous or hairy. **LF**: basal and/or cauline, alternate, petioled or sessile, ovate or elliptic to obovate or spoon-shaped, gen 1–3-pinnately lobed, ultimate margins entire, crenate, or dentate. **INFL**: heads radiate or radiant to disciform [discoid], gen in ± flat-topped clusters, subsessile or peduncled; involucre gen hemispheric or wider; phyllaries 30–60+, ± equal or graduated in 3–5+ series, free, persistent, lanceolate to oblong or ± ovate, outer sometimes keeled, margins and tips scarious, pale to sometimes ± brown or black; receptacle flat to conic or hemispheric, epaleate, glabrous or hairy. **RAY FL**: 0 or 10–21, pistillate or sterile, ray oblong to fan-shaped, ± yellow or white [pink] (in disciform heads, peripheral pistillate fls 8–30+, corolla pale yellow, ± bilateral, 3–4-lobed). **DISK FL**: 60–300+; corolla yellow, tube < narrowly funnel-shaped throat, lobes (4)5, triangular; anther tip narrowly triangular; style tips truncate, brush-like. **FR**: obconic or ± cylindric, gen 5–10 ribbed, gen resin-gland-dotted; pappus a crown of short scales. 160 spp.: Eur, Asia, N.Am. (Possibly Greek through Latin: immortality) [Watson 2006 FNANM 19:489–491]

1. Lf blade gen 1–2-pinnately lobed, 1° lobes gen 3–5 pairs, oblong to ± ovate, at least abaxial face gen
   puberulent; ray fls gen 10–21+; rays white 2–8(12+) mm; pappus 0 or 0.1–0.2+ mm . . . . ***T. parthenium***
1′ Lf blade 2–3-pinnately lobed, 1° lobes 4–24+ pairs, linear to ± oblong or elliptic, gen cobwebby to
   long-soft-hairy, sometimes ± glabrous; ray fls 0 or 8–30+, rays 1–8+ mm, yellow; pappus 0.1–0.5+ mm
   2. Lf blade gen long-soft-hairy to tomentose, sometimes ± glabrous; heads disciform to ± radiant or ±
      radiate, (1)5–12(20+) in ± flat-topped cluster; involucre 8–22+ mm diam . . . . ***T. bipinnatum***
   2′ Lf blade glabrous or sparsely hairy; heads disciform, 20–200 in ± flat-topped cluster; involucre 5–10 mm
      diam . . . . ***T. vulgare***

T. bipinnatum (L.) Sch. Bip. (p. 435) DUNE TANSY Per from rhizome, 5–80 cm. **ST**: ± decumbent to ascending or erect. **LF**: basal and cauline, petioled or sessile, gen 7–25+ cm, 3–10+ cm wide, ovate to elliptic, obovate, or spoon-shaped, 2–3-pinnately lobed, 1° lobes gen 6–24+ pairs, linear to ± oblong to linear-elliptic, 2° lobes ± lanceolate to oblong or ovate, sometimes curled, ultimate margins entire or ± dentate, faces gen ± long-soft-hairy to tomentose, sometimes ± glabrous. **INFL**: heads disciform to ± radiant or ± radiate, (1)5–12(20+); involucre 8–22+ mm diam; receptacle flat to hemispheric. **RAY FL**: 8–21+ or rayless pistillate fls 15–30+; corolla yellow, ray 0–7+ mm, or rayless fls with ± bilateral, 3–5-lobed corolla. **DISK FL**: corolla, 2–4 mm. **FR**: 2–3(4) mm, weakly 5-ribbed or -angled. 2*n*=54. Uncommon. Coastal dunes; < 30 m. NCo, n CCo; to AK, n Can, n-c US; Eurasia. [*T. camphoratum* Less.] Jul–Oct

T. parthenium (L.) Sch. Bip. FEVERFEW Per from taproot, 20–80 cm. **ST**: erect or ascending, branched, gen glabrous proximally, puberulent distally. **LF**: mainly cauline, petioled, 4–10+ cm, 1.5–10 cm wide, oblong-ovate to deltate, gen 1–2(3)-pinnately lobed, primary lobes 3–5+ pairs, oblong to ± ovate, ultimate margins dentate to pinnately lobed, at least abaxial face gen puberulent. **INFL**: heads radiate, 5–20(30); involucre 5–7 mm diam, shallowly hemispheric.

RAY FL: 9–21+; ray 2–8(12+) mm, white. **DISK FL**: corolla 2–2.5 mm, yellow. **FR**: 1–2 mm, 5–10-ribbed. 2*n*=18. Occasional. Disturbed areas, fields, woodland; gen < 2100 m. KR, NCoR, n SN, GV, SnFrB, SCoRO, SCo, TR, SnJt; to BC, e US; native to Eur. Long cult in Eur and US as orn and for medicine. Jun–Aug

T. vulgare L. COMMON TANSY Per from rhizome, gen 40–150 cm. **ST**: erect, glabrous or sparsely hairy. **LF**: basal and cauline, petioled or sessile, 4–20 cm, 2–10 cm wide, widely oblong or oval to elliptic, pinnately divided, axis ± winged, 1° lobes 4–13 pairs, lance-linear to lanceolate or narrowly elliptic, often pinnately lobed or toothed, ultimate margins serrate, faces glabrous or sparsely hairy. **INFL**: heads 20–200 in compact flat-topped clusters; involucre 5–10 mm diam; receptacle convex to conic. **RAY FL**: 0; rayless pistillate fls ± 20; corolla tubular, yellow, 3–4-lobed. **DISK FL**: corolla 2–3 mm, yellow. **FR**: 1–2 mm, 4–5-angled or -ribbed. 2*n*=18. Uncommon. Escape from cult in disturbed areas; < 2000 m. NCo, NCoRO, CaRH, n SNH, SCoRO, SnBr, MP; to AK, n Can, e US; native to Eur. TOXIC: dried lvs and fls have been used medicinally, esp in home remedies; overdoses may be toxic; also causes contact dermatitis. Jun–Aug ❖

TARAXACUM DANDELION

Luc Brouillet

Per from taproot; sap milky. **ST**: 0. **LF**: simple, basal, blades oblong to obovate, oblanceolate or linear-oblanceolate, with large distal lobe or not, ± toothed or pinnately lobed. **INFL**: heads liguliflorous, 1, scapes unbranched, hollow; involucre cylindric to bell-shaped (urn-shaped in fr); outer phyllaries (6)8–18(20), graduated in 2–3 series, erect or reflexed, ovate to lanceolate, tips horned or not; main phyllaries 7–25 in 2(3) series, equal, erect, spreading to reflexed in fr, linear, tips horned or not, margin often scarious, irregularly toothed; receptacle ± flat, glabrous, epaleate. **FL**: (15)20–150; ligules yellow, readily withering. **FR**: oblanceoloid to obovoid, with slender beak >> body, 4–12-ribbed, ribs sharply roughened, glabrous; pappus of 50–150, white, slender, minutely barbed bristles. 60(2000) spp., many reproducing by asexually produced seeds: Eurasia, N.Am, S.Am, ± worldwide as weeds. (Arabic to Persian: bitter herb) [Brouillet 2006 FNANM 19:239–252]

1. Lvs toothed or shallowly lobed, sessile or petioles widely winged; outer phyllaries erect, lanceolate to widely ovate; main phyllaries (10)12–16
   2. Tips of outer and main phyllaries gen not horned; SnBr . . . . ***T. californicum***
   2′ Tips of outer and main phyllaries often with horn-like appendages; W&I . . . . ***T. ceratophorum***
1′ Lvs of fl pls gen sharply cut, lobes often ± toothed or with 2° lobes, petioles ± narrowly winged; outer phyllaries reflexed, lanceolate; main phyllaries 13–19
   3. Fr red to ± red-brown or purple; lvs gen sharply cut; petioles slightly winged distally . . . . ***T. erythrospermum***
   3′ Fr olive or olive-brown to straw-colored or ± gray; lvs ± widely lobed, or occ sharply cut; petioles ± winged . . . . ***T. officinale***

T. californicum Munz & I.M. Johnst. (p. 435) CALIFORNIA DANDELION **LF**: 10–20+, ± widely winged-petioled, oblanceolate, gen toothed, occ shallowly lobed proximally. **INFL**: involucre 11–16 mm; outer phyllaries erect, 5–7 mm, lance-ovate to widely ovate, tips hornless; main phyllaries 12–16, tips ± rounded, gen hornless. **FR**: pale brown; body sharply roughened in distal 2/3–3/4. 2n=16. Moist meadows; 1950–2400 m. SnBr. May–Aug ★

T. ceratophorum (Ledeb.) DC. HORNED DANDELION **LF**: ± 10, sessile or widely wing-petioled, oblanceolate, ± toothed or ± lobed. **INFL**: involucre (5)8–19(21) mm; outer phyllaries erect, 5–12 mm, lanceolate to ovate, tips often with horn-like appendages; main phyllaries (10)12–14(17), tips obtuse to rounded, horned. **FR**: straw-colored to olive or brown; body sharply rough in distal 1/3–1/2. 2n=16,32,40,48. Moist alpine meadows; 2900–3100 m. W&I; to AK, n Can, WY, NM; Eurasia. [*T. officinale* subsp. *c.* (Ledeb.) Thell.] Jun–Aug ★

T. erythrospermum Besser RED-SEEDED DANDELION **LF**: 20+, slightly wing-petioled distally, oblanceolate to obovate, fl pl lvs gen sharply cut, lobes recurved, often ± toothed or with 2° lobes. **INFL**: involucre 10–25 mm; outer phyllaries reflexed, 3.8–10 mm, narrowly lanceolate; main phyllaries 18–19, tips long-acuminate, gen some horned. **FR**: red to ± red-brown or purple; body sharply rough in distal 1/2. 2n=16,24,32. Abundant. Esp disturbed areas; 50–2250 m. CA-FP, GB; native to Eur. All yr

T. officinale F.H. Wigg. (p. 435) COMMON DANDELION **LF**: gen 20+, narrowly wing-petioled, oblanceolate to obovate, toothed or in fl pl sharply recurved-lobed, lobes ± wide, dentate or occ sharply cut. **INFL**: involucre 14–25 mm; outer phyllaries reflexed, 6–12 mm, lanceolate; main phyllaries 13–18, tips acuminate, hornless. **FR**: olive or olive-brown to straw-colored or ± gray; body sharply rough in distal 1/3. 2n=24,40. Abundant. Esp disturbed areas; < 3300 m. CA-FP, GB; native to Eur. All yr

TETRADYMIA COTTONTHORN, HORSEBRUSH

Shrub. **ST**: ± tomentose. **LF**: alternate and gen clustered in axils, linear to (ob)lanceolate, sometimes persisting as stiff spines, glabrous to tomentose. **INFL**: heads discoid, axillary or in ± rounded or flat-topped, terminal clusters; involucre cylindric to hemispheric; phyllaries 4–6 in 1–2 ± equal series, often keeled; receptacle flat, epaleate. **DISK FL**: 4–9; corolla cream to yellow, lobes long, spreading or recurved; anther base ± sagittate, tip obtuse or acute; style branches papillate to short-bristly, tips truncate to conic. **FR**: obconic or fusiform, often 5-angled; pappus 0 or of gen many bristles or slender scales. 10 spp.: w N.Am. (Greek: 4 together, from 4-fld heads of some spp.) Esp fl buds TOXIC to sheep (toxicity poorly understood). [Strother 2006 FNANM 20:629–632]

1. St unarmed (lvs not forming spines)
   2. Fls (5)6–9 per head; pappus 0 (but fr covered with long, pappus-like hairs); axillary lf clusters gen 0 . . . . . ²***T. comosa***
   2′ Fls 4(5) per head; pappus present; axillary lf clusters gen present
      3. Main lvs tomentose or silvery; clustered lvs ± (ob)lanceolate . . . . ***T. canescens***
      3′ Main lvs glabrous or sparsely tomentose; clustered lvs thread-like to linear-oblanceolate
         4. Main lvs narrowly awl-shaped, stiffly ascending or appressed, gen 5–10 mm; clustered lvs gen 3–10 mm, glabrous; fr gen 3–4 mm, short-stiff-hairy; pappus of ± 100 bristles . . . . ***T. glabrata***
         4′ Main lvs ± thread-like, soft, 10–40 mm; clustered lvs gen 10–20 mm, loosely tomentose; fr 5–6 mm, long-soft-hairy; pappus of ± 20 stiff bristles or slender scales (± hidden by fr hairs) . . . . ***T. tetrameres***
1′ St armed with spines derived from main lvs
   5. Fr glabrous, 2.5–3.5 mm . . . . ***T. argyraea***
   5′ Fr ± long-soft-hairy, 4–8 mm
      6. Main lvs narrowly (ob)lanceolate, finally deciduous, base not hardened; pappus 0 . . . . ²***T. comosa***
      6′ Main lvs tapered to tip, persistent as stout spines, base expanded, hardened; pappus present
         7. Clustered lvs tomentose or silvery-hairy; fr ± short-hairy; pappus of many fine bristles 9–12 mm . . . . . ***T. stenolepis***
         7′ Clustered lvs ± glabrous; fr densely long-hairy; pappus of ± 25 slender scales 6–9 mm
            8. Spines sharply recurved, 5–25 mm, tomentose; involucre gen 8–12 mm, ± bell-shaped; fr gen 6–8 mm . . . . ***T. spinosa***

8′ Spines straight, (1)2–5 cm, becoming glabrous; involucre gen 7–9 mm, obconic; fr gen 4–5 mm . . . . . . *T. axillaris*
9. Involucre and peduncle glabrous; fr hairs gen 6–8 mm . . . . var. *axillaris*
9′ Involucre and peduncle tomentose; fr hairs gen 9–11 mm . . . . var. *longispina*

T. argyraea Munz & J.C. Roos STRIPED HORSEBRUSH Pl ≤ 20 dm, spiny. **ST:** becoming ± glabrous in stripes below spines. **LF:** main lvs 1–2(3) cm, canescent or becoming glabrous, linear, forming straight or ± upturned spines; clustered lvs 3–8(20) mm, thread- to club-like, ± glabrous. **INFL:** heads gen 2–5; peduncles gen 1–4 mm, tomentose, bracts 0; involucre ± 7 mm, obconic; phyllaries 5, narrowly elliptic, tomentose. **FL:** 5; corolla ± 9 mm, pale yellow. **FR:** 2.5–3.5 mm, glabrous; pappus of many fine bristles, ± 8 mm. 2*n*=60. Blackbush or sagebrush scrub, desert woodland; 1200–2230 m. se DMtns. (May)Jul–Oct ★

T. axillaris A. Nelson Pl ≤ 15 dm, spiny. **ST:** evenly tomentose. **LF:** main lvs (1)2–5 cm, forming straight spines, tomentose, becoming glabrous; clustered lvs 2–12(20) mm, thread- to club-like, ± glabrous. **INFL:** heads 1–3 in axils of spines; peduncle gen 4–15 mm, gen bracted; involucre 7–9 mm, obconic; phyllaries 5, ovate. **FL:** 5–7; corolla 7.5–9 mm, pale yellow. **FR:** gen 4–5 mm, densely long-white-hairy; pappus of ± 25 slender scales, 6–7.5 mm.

var. ***axillaris*** **INFL:** peduncle and involucre glabrous. **FR:** hairs 6–8(10) mm. 2*n*=60. Desert scrub, woodland; 1150–2300 m. W&I, n DMtns; s NV. Apr–May

var. ***longispina*** (M.E. Jones) Strother (p. 435) **INFL:** peduncle and involucre tomentose. **FR:** hairs 9–11(14) mm. 2*n*=60,62. Desert scrub, woodland; 1200–2300 m. s SNF, e WTR, n SnGb, s SNE, DMoj; to sw UT. Apr–Jun

T. canescens DC. (p. 435) Pl 1–8 dm, unarmed. **ST:** unevenly tomentose, becoming ± glabrous in stripes below nodes. **LF:** main lvs ≤ 4 cm, ± (ob)lanceolate, tomentose to silvery; clustered lvs like (gen <) main lvs. **INFL:** heads gen 3–6(10) in flat-topped clusters; peduncle 5–15(25) mm, bracts 0; involucre 6–8(12) mm, cylindric to obconic; phyllaries 4, oblong to ovate. **FL:** 4; corolla 7–15 mm, creamy to bright yellow. **FR:** 2.5–5 mm, glabrous or short-stiff-hairy; pappus of many fine bristles, 6–11 mm. 2*n*=60,62,90,120. Sagebrush scrub, pinyon/juniper woodland, conifer forest; (700)1000–3400 m. CaRH, SNH, TR, s PR, GB, DMoj; to BC, MT, WY, NM. Jul–Oct

T. comosa A. Gray Pl 3–12 dm, ± spiny or not. **ST:** tomentose. **LF:** main lvs 2–6 cm, gen narrowly lanceolate, becoming rigid, spine-tipped but finally deciduous, tomentose; clustered lvs gen 0 (8–15 mm, ± oblanceolate). **INFL:** heads gen 3–6; peduncle gen 3–8 mm, tomentose, bract gen 1, narrow; involucre ± 8 mm, ± widely bell-shaped; phyllaries 5(6), elliptic to (ob)ovate, often unequal. **FL:** (5)6–9; corolla 8–9 mm, (± brown) yellow. **FR:** ± 4 mm; hairs long, white, pappus-like; pappus 0. 2*n*=60. Coastal scrub, chaparral; 100–1400 m. SCo, TR, PR; n Baja CA. Jun–Nov

T. glabrata Torr. & A. Gray (p. 435) Pl ≤ 12 dm, unarmed. **ST:** unevenly tomentose, becoming ± glabrous in stripes below nodes. **LF:** main lvs 5–10(20) mm, narrowly awl-shaped, ascending to appressed, sparsely woolly or becoming glabrous; clustered lvs 3–10(15) mm, thread- to ± club-like, glabrous. **INFL:** heads gen 3–7; peduncle gen 5–10 mm, bracts 0; involucre 7–10 mm, obconic; phyllaries 4, lanceolate to obovate. **FL:** 4; corolla 9–10 mm, cream to golden yellow. **FR:** gen 3–4 mm, short-stiff-hairy; ribs glandular; pappus of many fine bristles, 6–8 mm. 2*n*=60,62,120,180. Desert scrub, pinyon/juniper or Joshua-tree woodland; 700–2400 m. GB, w DMoj; to OR, ID, UT. Apr–Jul

T. spinosa Hook. & Arn. Pl ≤ 10 dm, spiny. **ST:** tomentose. **LF:** main lvs 5–25 mm, tomentose, forming rigid, recurved spines; clustered lvs 3–15(25) mm, thread-like to ± oblanceolate, ± glabrous. **INFL:** heads gen 1–2 in axils of spines; peduncle gen 5–30 mm, tomentose, bracted; involucre 8–12 mm, ± bell-shaped; phyllaries 4–6, oblong to ovate. **FL:** 5–8; corolla 6–10 mm, ± yellow. **FR:** gen 6–8 mm, densely long-white-hairy; pappus of ± 25 slender scales, 6–9 mm. 2*n*=60. Gen saltbush scrub; 800–2400 m. s MP, n SNE; to OR, MT, WY, n NM. Apr–Jun

T. stenolepis Greene (p. 435) Pl ≤ 12 dm, spiny. **ST:** unevenly tomentose, becoming ± glabrous in stripes below spines. **LF:** main lvs 2–3 cm, tomentose or becoming glabrous, forming ± straight spines; clustered lvs 10–30 mm, ± oblanceolate, tomentose or silvery-hairy. **INFL:** heads gen 4–7; peduncle gen 5–12 mm, tomentose, bracts 0; involucre 8–10 mm, narrowly obconic; phyllaries (4)5, narrowly ovate. **FL:** (4)5; corolla 10–12 mm, pale yellow. **FR:** 5–8 mm, ± long-soft-hairy; pappus of many fine bristles, 9–12 mm. 2*n*=60. Desert woodland, creosote-bush or saltbush scrub; 600–1700 m. s SNF, Teh, e WTR, n SnGb, SNE, DMoj; s NV. Jun–Nov

T. tetrameres (S.F. Blake) Strother DUNE HORSEBRUSH Pl ≤ 20 dm, unarmed. **ST:** tomentose. **LF:** sparsely tomentose; main lvs 1–4 cm, ± thread-like, soft; clustered lvs 10–20 mm, thread-like to linear-oblanceolate. **INFL:** heads gen 4–6 on short side branches; peduncle gen 1–3 mm, tomentose, bracts 0–2; involucre 8–9 mm, obconic; phyllaries 4(5), widely elliptic. **FL:** 4(5); corolla ± 8 mm, pale yellow. **FR:** 5–6 mm, densely long-soft-hairy; pappus of ± 20 stiff bristles or slender scales, 3–5 mm, ± hidden by fr hairs. 2*n*=60. Dunes, deep sand, sagebrush scrub; 1200–2100 m. n SNE; w NV. May–Sep ★

TETRANEURIS

Mark W. Bierner

Per [ann]; caudex woody, ± branched. **ST:** erect [± decumbent], unbranched [sparingly-branched distally], ± hairy. **LF:** all basal [basal and cauline], simple, entire [toothed or lobed], glabrous or ± hairy, ± gland-dotted [or not]. **INFL:** heads radiate [discoid], 1 [in panicle-like, ± flat-topped, or tightly congested clusters]; peduncle expanded distally, ± hairy; involucre hemispheric to bell-shaped; phyllaries in 3 series, free, outer and middle similar; receptacle hemispheric to conic, pitted, epaleate. **RAY FL:** corolla yellow, fading to cream, ± persistent on fr, ray fan-shaped to oblanceolate, gen 3-lobed. **DISK FL:** corolla yellow [distally ± purple]; anther tip ± triangular; style tips truncate. **FR:** obpyramidal to narrowly obpyramidal, gen 5-angled, ± hairy; pappus of membranous, gen awn-tipped scales. 9 spp.: w N.Am. (Greek: four nerves, from ray corolla) [Bierner & Turner 2006 FNANM 21:447–453]

T. acaulis (Pursh) Greene var. ***arizonica*** (Greene) K.F. Parker (p. 435) Pl (2)6–15(30+) cm; caudex branches notably thickened distally. **ST:** 1–20(35). **LF:** 2–6 cm, blades spoon-shaped or oblanceolate to linear-oblanceolate, densely gland-dotted. **INFL:** heads 1–20(35); peduncle (1)5–15(30) cm, occ woolly distally; involucre 7–12 mm, 12–16 mm diam; outer and middle phyllaries 3.9–6.5 mm, lanceolate or ovate to obovate. **RAY FL:** 9–15; corolla 11–17 mm. **DISK FL:** 50–150+; corolla 2.7–4.3 mm. **FR:** 3–3.7 mm; pappus scales 5–7, 2–3.5 mm, oblanceolate. 2*n*=28,30,56,60. Roadsides, hillsides, grassland, aspen meadows, forest edges; 1300–2900 m. SnJt, DMtns; to ID, CO, AZ. [*Hymenoxys a.* (Pursh) K.F. Parker var. *a.* (Greene) K.F. Parker] Apr–Sep

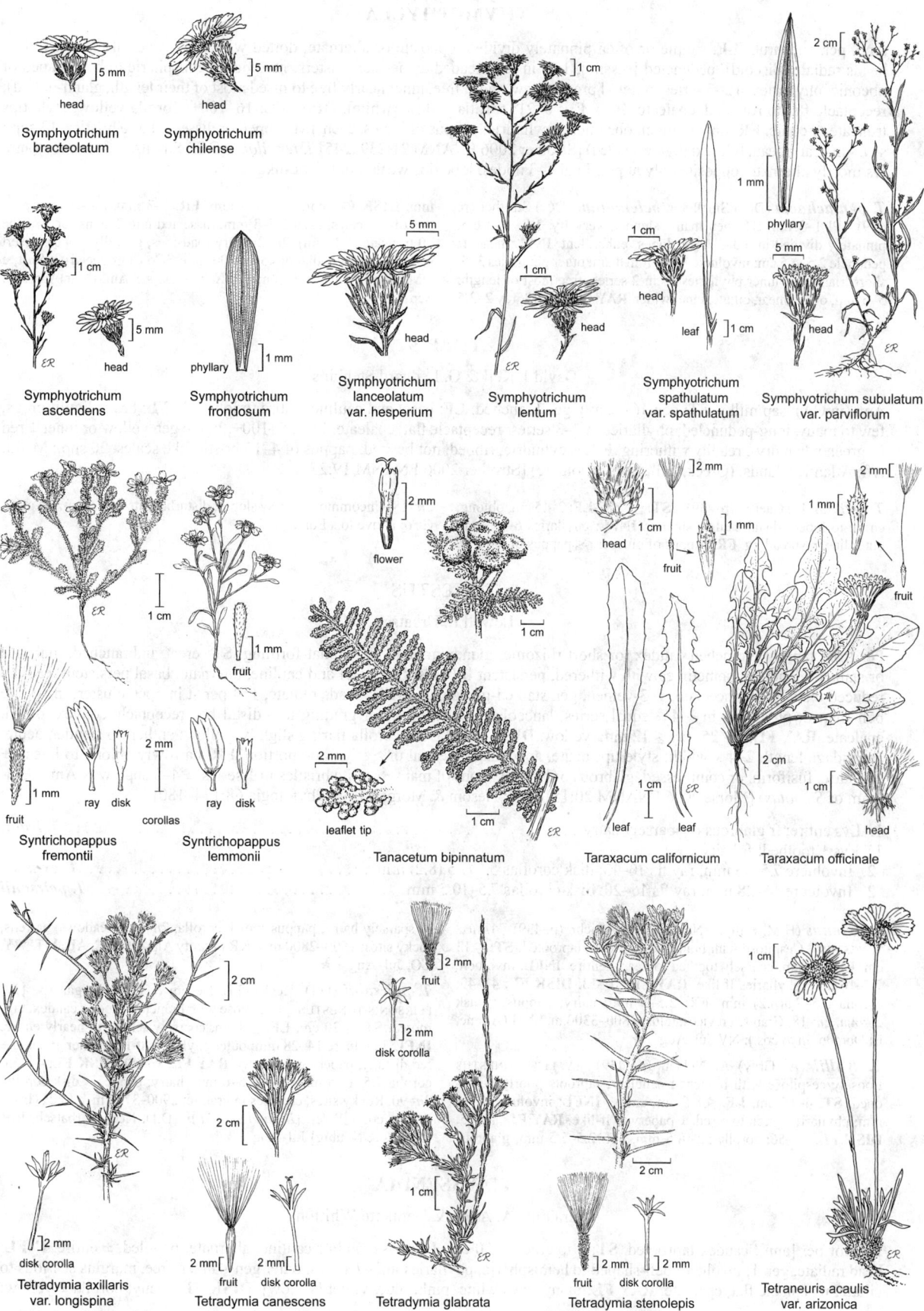

Symphyotrichum
bracteolatum
head
5 mm

Symphyotrichum
chilense
head
5 mm

Symphyotrichum
ascendens
head
1 cm
5 mm

Symphyotrichum
frondosum
phyllary
1 mm

Symphyotrichum
lanceolatum
var. hesperium
head
5 mm

Symphyotrichum
lentum
head
1 cm
head

Symphyotrichum
spathulatum
var. spathulatum
head
1 cm
leaf
1 cm
1 mm

Symphyotrichum subulatum
var. parviflorum
phyllary
2 cm
head
5 mm

Syntrichopappus
fremontii
fruit
1 mm
ray
disk
2 mm
1 cm

Syntrichopappus
lemmonii
fruit
1 mm
corollas
ray
disk

Tanacetum bipinnatum
flower
2 mm
leaflet tip
2 mm
1 cm

Taraxacum californicum
head
1 cm
fruit
2 mm
1 mm
leaf
leaf
1 cm

Taraxacum officinale
fruit
2 mm
1 mm
1 cm
head
2 cm

Tetradymia axillaris
var. longispina
disk corolla
2 mm
2 cm

Tetradymia canescens
fruit
disk corolla
2 mm
2 mm
2 cm

Tetradymia glabrata
fruit
2 mm
disk corolla
2 mm
1 cm

Tetradymia stenolepis
fruit
disk corolla
2 mm
2 mm
2 cm

Tetraneuris acaulis
var. arizonica
1 cm

THYMOPHYLLA

Ann, per, subshrub. **LF**: simple or often pinnately divided, opposite or alternate, dotted with embedded oil glands. **INFL**: heads radiate [discoid], peduncled [sessile], 1 or in few-headed cyme-like clusters; involucre hemispheric to bell-shaped or obconic; phyllaries in 2–3 series (outer, if present, few, short, free, inner nearly free to fused most of their length, gland-dotted); receptacle flat to rounded, epaleate. **RAY FL**: 0–21; corolla yellow [white]. **DISK FL**: 16–100+; corolla yellow; style tips truncate or conic. **FR**: obpyramidal, obconic, or cylindric; pappus of scales, each awn-tipped or dissected into bristles. 13 spp.: s N.Am, Caribbean. (Greek: thyme-leaved) [Strother 2006 FNANM 21:239–245] *T. tenuiloba* (DC.) Small var. *tenuiloba* (ann; lvs mostly alternate, opposite only at proximal 1–3 nodes, lobe tips weak, not bristle-like) a waif in SCo.

T. pentachaeta (DC.) Small var. *belenidium* (DC.) Strother (p. 439) Pl 1–3 dm. **ST**: gen many, slender, very lfy. **LF**: 1–2 cm, pinnately divided into 3–5 linear lobes, puberulent. **INFL**: head 1; peduncle 2.5–4.5 cm; involucre 3.5–6 mm diam; outer phyllaries 3–5, short-triangular; inner phyllaries 13 in 2 series, free most of length, 5–6 mm, outer linear, ciliate, inner wider. **RAY FL**: gen 13; ray 2–3.5 mm. **DISK FL**: corolla 2.5–3 mm. **FR**: 2–3 mm, puberulent; pappus of 10 scales, each 2.5–3 mm, dissected into 3 awns (or 5 of the 10 truncate, ± 1 mm). 2*n*=26. Dry roadsides, gravelly slopes, desert scrub, desert woodland; 450–1700 m. e SCo (introduced), e SnJt, e DMoj, se DMtns, w DSon; to TX, n Mex; s S.Am. (Mar)Apr–Jun, Sep–Oct

TOLPIS

David J. Keil & G. Ledyard Stebbins

Ann (in CA); sap milky. **ST**: gen 1(–many), gen branched. **LF**: basal and cauline, entire to lobed. **INFL**: heads liguliflorous, few to many, long-peduncled; phyllaries in 2–3 series; receptacle flat, epaleate. **FL**: 30–100+; ligule gen yellow or inner ± red (± green when dry), readily withering. **FR**: ± cylindric, ribbed, not beaked; pappus of 4–15 bristle-like scales. 20 spp.: Medit, Afr, Atlantic islands. (Greek: ball, from involucre) [Strother 2006 FNANM 19:277]

T. barbata (L.) Gaertn. (p. 439) **ST**: ≤ 7 dm. **LF**: 2–15 cm, oblong, entire to lobed; distal cauline smaller. **INFL**: phyllaries 6–15 mm, outer linear, spreading. **FR**: pappus of outer fr << pappus of inner fr. 2*n*=18. Uncommon. Grassy slopes, disturbed sites; < 600 m. NCoRO, CCo; native to s Eur. Apr–Jun

TONESTUS

Lowell E. Urbatsch

Per from taproot, branched caudex, or short rhizome, glandular, sometimes mat-forming. **ST**: erect, unbranched, glabrous or stalked-glandular, sometimes with withered, persistent lf bases. **LF**: basal and cauline, alternate, basal persistent, cauline reduced distally, oblanceolate, 1–3-veined, gen stalked-glandular. **INFL**: heads radiate, 1–4 per st in open cluster; involucre bell-shaped; phyllaries in 3–4 ± equal series, lanceolate, 1-veined, outer grading into distal lvs; receptacle convex, pitted, epaleate. **RAY FL**: 10–25; ray ≤ 12 mm, yellow. **DISK FL**: 25–65; corolla flaring slightly, yellow; anther tip slender, acute, appendage length 2–4 × width; style tips acute, appendage ± equal or > stigmatic portion. **FR**: narrowly oblong to linear, ± cylindric, fusiform, or compressed, glabrous or hairy; pappus of many ± white bristles in 1 series. ± 4–5 spp.: w N.Am. (Anagram of *Stenotus*) [Morse 2006 FNANM 20:181–184; Nesom & Morgan 1990 Phytologia 68:174–180]

1. Lvs entire; fr glabrous or sparsely hairy .... *T. lyallii*
1' Lvs ± toothed; fr hairy
  2. Involucre 7.5–16 mm; ray fls 10–13; disk corollas 5.5–7.5 (8.2) mm .... *T. eximius*
  2' Involucre 14–28 mm; ray fls 16–20; disk corollas 7.5–10.5 mm .... *T. peirsonii*

T. eximius (H.M. Hall) A. Nelson & J.F. Macbr. (p. 439) TAHOE TONESTUS Cespitose with branched caudices, taprooted. **ST**: ≤ 13 cm. **LF**: ≤ 5 cm, coarsely toothed to nearly entire. **INFL**: involucre 7.5–16 mm; phyllaries lf-like. **RAY FL**: 10–13. **DISK FL**: 41–49; corolla 5.5–7.5(8.2) mm. **FR**: 2.5–6 mm, hairy; pappus = disk corolla. 2*n*=18. Granite crevices, talus; 1800–3300 m. SNH (Alpine, El Dorado, Inyo cos.); NV. Jul–Aug ★

T. lyallii (A. Gray) A. Nelson (p. 439) LYALL'S TONESTUS Loosely cespitose with branched caudices, taproots poorly developed. **ST**: 4–15 cm. **LF**: ≤ 8.5 cm, entire. **INFL**: involucre 8–11 mm; phyllaries green to ± red, ± papery to lf-like. **RAY FL**: 11–25. **DISK FL**: 27–56; corolla 5.8–8.5 mm. **FR**: 2.5–5.5 mm, glabrous or sparsely hairy; pappus = disk corolla. 2*n*=18. Meadows, barrens, rocky sites; 1500–2800 m. c KR (Trinity Alps); to BC, AB, MT, WY, CO. Jul–Aug ★

T. peirsonii (D.D. Keck) G.L. Nesom & D.R. Morgan (p. 439) PEIRSON'S TONESTUS Cespitose with branched woody caudex, taprooted. **ST**: ≤ 20 cm. **LF**: ≤ 8 cm, coarsely serrate to nearly entire. **INFL**: involucre 14–28 mm; outer phyllaries lf-like; inner green to ± red distally, papery proximally. **RAY FL**: 16–20. **DISK FL**: 44–63; corolla 7.5–10.5 mm. **FR**: 2.5–6 mm, hairy; pappus < disk corolla. 2*n*=90. Rocky sites, crevices in granite; 2900–3700 m. c SNH (Inyo, Fresno cos.), W&I. [*Lorandersonia p.* (D.D. Keck) Urbatsch, R.P. Roberts & Neubig] Jul–Aug ★

TOWNSENDIA

Geraldine A. Allen & Jeannette Whitton

Bien or per [ann]; caudex taprooted. **ST**: 0 to erect, < 30 cm. **LF**: basal and/or cauline, alternate, petioled, ± entire. **INFL**: head radiate, gen 1; involucre bell-shaped to hemispheric; phyllaries in 2–7 series, outer gen < inner, free, margins scarious to ciliate; receptacle flat, epaleate. **RAY FL**: many; rays white, pink, blue, violet [yellow]. **DISK FL**: many; corolla ± yellow;

style branches flat, tip hairy. **FR**: ± compressed, brown, glabrous to ± hairy; pappus of minutely barbed to ± plumose, gen flat bristles (ray fr occ also with short outer series). 27 spp.: w N.Am. (D. Townsend, Am botanist, 1787–1858) [Strother 2006 FNANM 20:193–203] Some spp. reproduce by asexual seed. *T. parryi* D.C. Eaton not in CA.

1. Lvs long-soft-woolly; pappus readily deciduous. . . . . ***T. condensata***
1′ Lvs ± finely strigose; pappus ± persistent
  2. Pl ≤ 3 cm; peduncles 0–15 mm; fr glabrous. . . . . ***T. leptotes***
  2′ Pl 3–10 cm; peduncles > 30 mm; fr ± spreading-hairy . . . . ***T. scapigera***

T. condensata Parry (p. 439) CUSHION TOWNSENDIA Bien or per ≤ 3 cm. **LF**: basal, 0.5–1.5 cm, ± narrowly obovate, rounded at tip, long-soft-woolly. **INFL**: heads ± sessile or on lfy sts; phyllaries ± equal, lanceolate, ± acuminate, scarious-margined, ± long-hairy. **RAY FL**: rays ± white. **FR**: short-hairy; pappus readily deciduous. 2*n*=18. Gravelly slopes; 3200–3700 m. n SNE (Mono Co.); to sw Can, WY, CO. Jun–Aug ★

T. leptotes (A. Gray) Osterh. (p. 439) SLENDER TOWNSENDIA Per ≤ 3 cm. **LF**: basal, gen 1–2 cm, ± narrowly oblanceolate, gen acute, ± finely strigose. **INFL**: peduncle 0–15 mm; phyllaries lanceo-late, acuminate, margins scarious-ciliate, glabrous to ± hairy. **RAY FL**: rays white to pink or blue. **FR**: glabrous; pappus ± persistent. 2*n*=18. Rocky or sandy slopes; 3500–3800 m. W&I; to MT, NM. Jun–Jul ★

T. scapigera D.C. Eaton (p. 439) Pl 3–10 cm. **LF**: basal and cau-line, 1–5 cm (< petiole), oblanceolate to obovate, ± obtuse to rounded, strigose. **INFL**: peduncle > 30 mm; phyllaries narrowly ovate to lanceolate, ± acute, scarious-margined, ± hairy. **RAY FL**: rays white to pink or violet. **FR**: ± spreading-hairy; pappus ± persistent. Rocky slopes, openings in sagebrush; 1400–3600 m. GB; NV. May–Aug

TRACYINA

Bruce G. Baldwin

1 sp.: CA. (Joseph P. Tracy, nw CA botanist, 1879–1953) [Strother 2006 FNANM 20:50] Similar, closely related to *Pentachaeta*, *Rigiopappus*.

T. rostrata S.F. Blake (p. 439) BEAKED TRACYINA Ann 5–30+ cm, from slender taproot; gen self-pollinated. **ST**: erect, simple or branched distally, branches often exceeding main st, glabrous. **LF**: basal and at fl mostly cauline, alternate, ± sessile, 10–25(35+) mm, 1–2(5+) mm wide, narrowly lanceolate to linear, entire, ciliate, gla-brous or sparsely, minutely coarse-hairy. **INFL**: heads radiate, 1 at branch tips; peduncle ± minutely soft-hairy near heads; involucre 5–7+ mm, 1–3+ mm diam, ± cylindric to obconic; phyllaries (6)20–30+, graduated in (2)3–4+ series, appressed, linear to awl-shaped, green, flat, glabrous or sparsely and minutely coarse-hairy, persis-tent; receptacle convex, pitted or smooth, epaleate. **RAY FL**: (3)12–15(20+); corolla ± yellow, often purple-tinged, tube 2.5–3+ mm, ray inconspicuous, 1–1.5+ mm, narrowly elliptic. **DISK FL**: (4)15–25+; corolla pale yellow to ± purple, tube 1–1.5 mm, < throat, throat 2–2.5 mm, lobes erect, 0.1–0.3 mm, lance-deltate; anther tip reduced, ± triangular; style branch tips long-tapering distally. **FR**: 5–6 mm, ± spindle-shaped, long-tapered distally (almost beaked), glabrous or (ray fr) glabrous in age or (disk fr) ± minutely coarse-hairy; pappus of (12)30–40 long-tapered bristles in 1(2) series, 1–4+ mm, ± equal or outer shorter, minutely barbed, persistent, fragile. Grassy slopes; 100–400+ m. NCoRO, NCoRI. May–Jun ★

TRAGOPOGON   GOAT'S BEARD, SALSIFY

David J. Keil & G. Ledyard Stebbins

Ann, bien [per] from strong taproot; herbage ± glabrous in CA; sap milky. **ST**: branches few, stiffly ascending. **LF**: basal and cauline, alternate, sessile, sheathing, entire, grass-like. **INFL**: heads liguliflorous, 1, gen closed by mid-day; peduncle long, bractless; involucre cylindric, urn-shaped, or narrowly conic in bud, ± bell-shaped in fl; phyllaries in 1 series, linear to lanceolate, acute, reflexed in fr; receptacle flat to convex, pitted, epaleate. **FL**: corolla yellow to bronze or purple; ligules readily withering. **FR**: 2.5–3 cm, cylindric or ± fusiform, 5–10-ribbed, ribs gen roughened; beak stout, > body; pappus of stout plumose bristles, 2° bristles tangled, tips of a few 1° bristles exceeding others, unbranched; frs spreading, forming spheric ball 4–5 cm diam. ± 150 spp.: Eurasia, Medit, n Afr, widely naturalized (hybrid spp. have evolved from aliens in w N.Am). (Greek: goat's beard) [Soltis 2006 FNANM 19:303–306]

1. Corolla purple; phyllaries ≥ fls . . . . ***T. porrifolius***
1′ Corolla pale to bright yellow; phyllaries < to >> fls
  2. Peduncles in fl much wider distally; corolla pale lemon-yellow; ligules of outer fls much exceeded
    by phyllaries . . . . ***T. dubius***
  2′ Peduncles in fl not or slightly wider distally; corolla bright yellow; ligules of outer fls exceeding
    phyllaries or at same level . . . . ***T. pratensis***

T. dubius Scop. Ann, bien, 3–10 dm. **LF**: 2–5 dm, tips straight, faces ± minutely tomentose, soon glabrous. **INFL**: peduncle much wider distally in fl; involucre conic in bud, phyllaries 8–13, 2.5–4 cm in fl heads, much exceeding fls, 4–7 cm in fr heads. **FL**: ligule pale lemon-yellow. **FR**: 25–35 mm; pappus ± white. 2*n*=12. Uncommon. Disturbed places; < 2700 m. KR, CaRH, SN, GV, CW, SCo, TR, PR, GB; native to Eur. May–Sep

T. porrifolius L. (p. 439) SALSIFY, OYSTER PLANT Bien 4–10 dm. **LF**: 2–4 dm, tips straight, faces gen glabrous. **INFL**: peduncle much wider distally in fl; involucre conic in bud, phyllaries 5–11, 2.5–4 cm in fl heads, ≥ fls, 4–7 cm in fr heads. **FL**: ligule purple. **FR**: 2.5–4 cm; pappus ± brown. 2*n*=12. Common. Disturbed places; < 1300 m. NW, n SNF, GV, CW, SW, MP, DSon; native to Eur. Mar–Nov

T. pratensis L. MEADOW SALSIFY Bien 1.5–8 dm, ± hairy when young, becoming glabrous. **LF**: 2–4 dm, tip gen recurved or coiled, faces ± minutely tomentose, soon glabrous. **INFL**: peduncle not or slightly wider distally in fl; involucre urn-shaped in bud, phyllaries gen 8, 12–30 mm in fl heads, exceeded by outer fls or at same level, 18–45 mm in fr heads. **FL**: ligule bright yellow. **FR**: 15–25 mm; pap-pus ± white. 2*n*=12. Gen ± moist, disturbed places; 900–1800 m. KR, CaRH, SNH, MP; native to Eur. May–Aug

TRICHOCORONIS

David J. Keil & A. Michael Powell

Ann or per, ≤ 30 cm, ascending, hairy. **LF**: sessile; opposite or distal alternate, < 2.5 cm, ± oblong, serrate. **INFL**: open; heads discoid, 1–few, small; involucre hemispheric to bell-shaped; phyllaries ± 30, ± equal in 2–3 series; receptacle convex to conic, epaleate. **FL**: 75–125; corolla tube narrow; style branches slender. **FR**: 4–5-ribbed, minutely bristly on ribs; pappus 0.1–1.5 mm, of 2–6 scales or bristles. 2 spp.: sw US, Mex. (Greek: hair crown, from pappus) [Nesom 2006 FNANM 21:487–488]

T. wrightii (Torr. & A. Gray) A. Gray var. ***wrightii*** (p. 439) WRIGHT'S TRICHOCORONIS **INFL**: few-branched; heads ± 5 mm diam. **FL**: corolla ± 1 mm, throat white proximally, maroon distally, lobes white. **FR**: ± 1.5 mm. 2*n*=30. Moist places, drying riverbeds; < 500 m. GV, SCo (San Jacinto Valley); s TX, ne Mex. Considered naturalized in TJM (1993) but possibly native; study needed. May–Sep ★

TRICHOPTILIUM

1 sp. (Greek: feathery bristle, from pappus) [Strother 2006 FNANM 21:418]

T. incisum (A. Gray) A. Gray (p. 439) Ann, per, 5–25 cm, gen ± tomentose. **ST**: 1–several from base. **LF**: simple, basal and cauline, alternate or ± opposite, mostly clustered in proximal 1/2 of pl, 1–3(5) cm, sessile or tapered to a short, winged petiole; blade oblanceolate to spoon-shaped or ovate, acute, sharply dentate or shallowly lobed, densely tomentose, resin-dotted. **INFL**: head discoid or radiant, 1; peduncle 3–11 cm, slender, bractless, glandular-puberulent; involucre 6–12 mm diam, hemispheric or bell-shaped, often appearing ± rotate when pressed; phyllaries in 2 ± equal series, free, 5–7 mm, linear-elliptic, acute; receptacle rounded, epaleate. **FL**: 30–100+; corolla 4–7 mm, yellow (pink), outermost sometimes enlarged and ± bilateral; style-branches truncate. **FR**: 2–3 mm, obpyramidal, glabrous to densely strigose; pappus of 5 scales, each 3–5 mm, dissected into many bristles. 2*n*=26. Dry slopes, plains, creosote-bush scrub; < 1500 m. D; s NV, sw AZ, n Baja CA. Jan–May, Oct–Dec

TRIPLEUROSPERMUM

Ann, bien, per, gen not scented. **ST**: erect or ascending, gen branched. **LF**: basal and cauline, alternate, 1–3-pinnately lobed; segments linear or thread-like; petiole short or 0. **INFL**: heads radiate, 1 or in cyme-like clusters; phyllaries graduated in 2–5 series, margins scarious; receptacle convex to conic, epaleate. **RAY FL**: 10–35; ray white. **DISK FL**: 300–500; corolla gen yellow, ± tubular, 5-lobed, each lobe tip with an abaxial resin gland; anthers minute, tips triangular, bases rounded or ± cordate; style short, branches truncate with bushy tips. **FR**: 3-angled, ± compressed front-to-back, 3–5-ribbed, ± roughened between ribs, glabrous, abaxially with gen 2–3 subapical resin glands; pappus 0 or a minute lobed crown. 40 spp.: Eur, Asia, N.Am, n Afr. (Greek: three ribbed seeds, describing fr) [Brouillet 2006 FNANM 19:548–551]

T. inodorum (L.) Sch. Bip. SCENTLESS MAYWEED, FALSE CHAMOMILE Gen ann. **ST**: 3–6 dm, glabrous or sparsely hairy when young. **LF**: ≤ 8 cm, glabrous; ultimate segments thread-like, not fleshy. **INFL**: heads 3–4.5 cm diam; phyllaries with dark green or brown centers and narrow, colorless or light brown, scarious margins. **RAY FL**: 10–25, ray 8–12 mm, white, becoming reflexed. **DISK FL**: corolla 1–2.5 mm. **FR**: ± 2 mm, brown, minutely papillate; adaxial face with 3 wide white ribs; abaxial face with 2 ± round resin glands below tip. 2*n*=18, 36. Disturbed sites, roadsides; < 1500 m. NCo, CaR; to e N.Am; native to Eurasia. [*Matricaria i.* L.; *T. maritimum* (L.) W.D.J. Koch, misappl.] Similar to *Anthemis cotula*; differing by unscented herbage and epaleate receptacles. May–Aug

TRIXIS

Shrub in CA. **LF**: alternate, sessile or petioled; blade long, entire or toothed, gen ± glandular. **INFL**: heads 1 or in flat-topped or ± conic clusters, sessile or short-peduncled; involucre cylindric; phyllaries gen in 1 series, linear, keeled; receptacle flat, epaleate, short-hairy. **FL**: 4–25+; corolla yellow, 2-lipped, outer lip spreading, resembling ray of ray fl, 3-lobed, sometimes recurved, inner lip deeply 2-lobed, recurved or coiled; anthers exserted, base with tail-like appendage, tip oblong, acute; style branch tips ± truncate. **FR**: ± cylindric, 5-ribbed, black or brown, short-hairy and glandular; pappus of many finely barbed bristles. ± 65 spp.: N.Am, S.Am. (Greek: 3-fold, from outer corolla lip) [Keil 2006 FNANM 19:75–76]

T. californica Kellogg var. ***californica*** (p. 439) Pl glandular and short-hairy. **ST**: 2–20 dm, erect, stiff, much-branched. **LF**: many; petiole 1.5–5 mm, winged; blade 2–11 cm, lance-linear to lanceolate, acute at both ends, entire or serrate, margin often rolled under. **INFL**: peduncle gen ≤ 5 mm, bearing 5–7 bracts; phyllaries 8–10, 8–14 mm, green, tips acute. **FL**: 11–25; corolla glandular, tube 6–9 mm, outer lip 5–8 mm, spreading, inner lip 4–5.5 mm, gen coiled. **FR**: 6–10.5 mm; pappus 7.5–12 mm, dull white. 2*n*=54. Dry, rocky slopes, washes, desert flats; gen ≤ 1300 m. se DMoj, DSon; to TX, n Mex. Jan–May, Jul–Aug

UROPAPPUS SILVERPUFFS

Kenton L. Chambers

1 sp. (Latin: tailed pappus, from awn-tipped scales) [Chambers 2006 FNANM 19:322]

U. lindleyi (DC.) Nutt. (p. 445) Ann 5–70 cm, lfy sts erect or pls scapose; sap milky. **LF**: 5–30 cm, all basal or some cauline, alternate, ± linear, long-tapered, entire to narrowly lobed, ± soft-hairy, esp petiole base. **INFL**: heads liguliflorous, 1 on always-erect peduncle; involucre 10–40 mm, narrowly bell-shaped in fl, cylindric to ovoid in fr, glabrous; phyllaries in 3–4 series, narrowly lanceolate, reflexed in fr; receptacle flat or convex, epaleate, shallowly pitted. **FL**: 5–many; ligules 2–10 mm, equaling or slightly exceeding involucre, pale yellow, abaxially often ± red, readily withering. **FR**: 7–17 mm, slender-fusiform, tapered to beak-like tip in CA, 10-ribbed, gen ± black, scabrous; pappus scales 5, 5–15 mm, deciduous, smooth, silvery, bristle-tip 4–6 mm, slender, smooth, from notched scale tip.

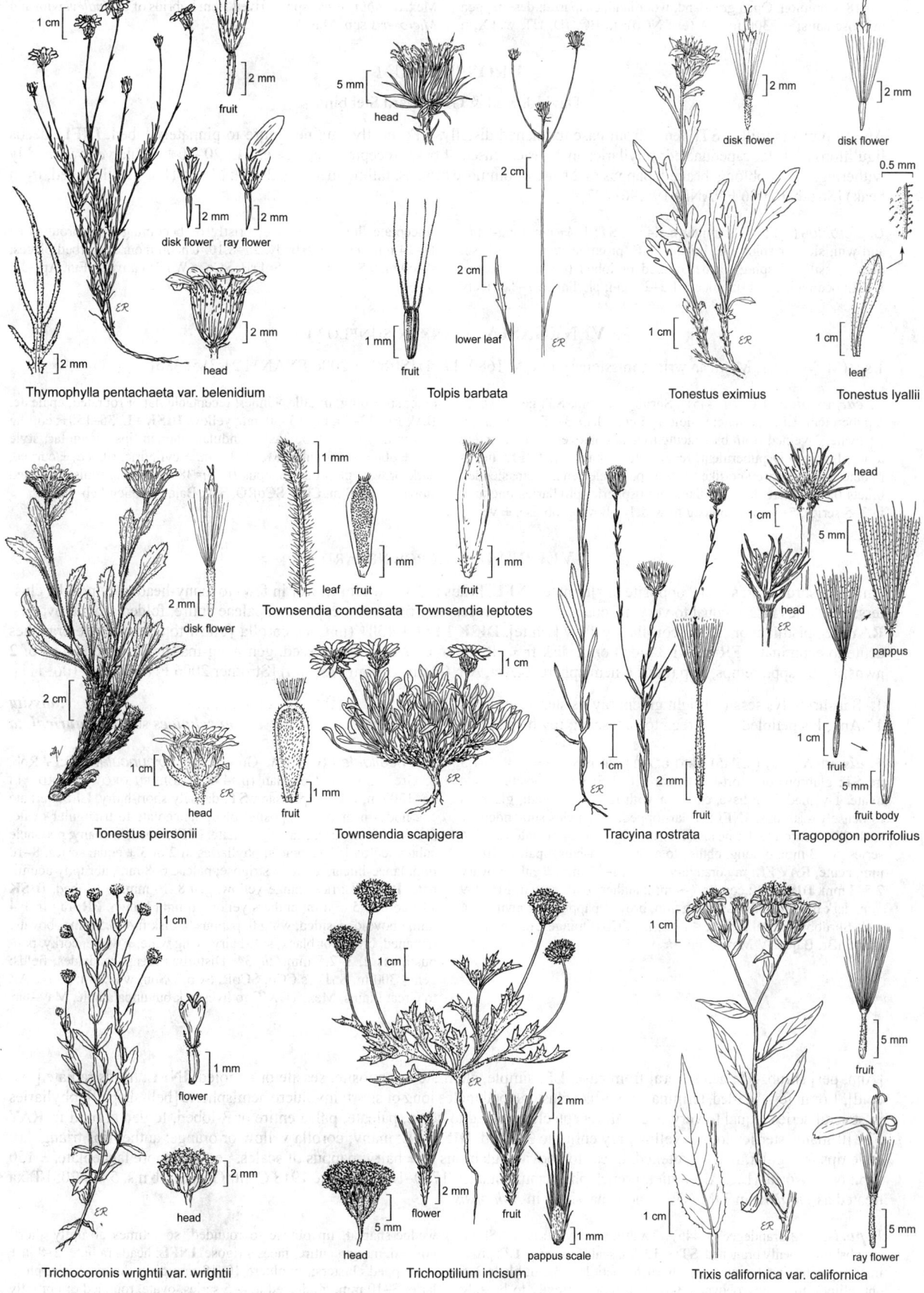

439

Thymophylla pentachaeta var. belenidium

Tolpis barbata

Tonestus eximius

Tonestus lyallii

Tonestus peirsonii

Townsendia condensata   Townsendia leptotes

Townsendia scapigera

Tracyina rostrata

Tragopogon porrifolius

Trichocoronis wrightii var. wrightii

Trichoptilium incisum

Trixis californica var. californica

2*n*=18. Common. Open grassland, woodland, chaparral, deserts, gen in loose soils; < 2300 m. CA (exc NCo); to BC, ID, UT, w TX, n Mex. *Stebbinsoseris* spp. derived from hybrids of *U. lindleyi* with ann *Microseris* spp. Mar–May

UROSPERMUM

David J. Keil & G. Ledyard Stebbins

Ann to per; sap milky. **ST**: gen 1 from base, branched distally. **LF**: mostly cauline, entire to pinnately lobed. **INFL**: heads liguliflorous, 1, long-peduncled; phyllaries in 1 series, fused at base; receptacle epaleate. **FL**: 20–50+; ligule yellow, readily withering. **FR**: ± oblong, beaked; pappus of 2 rows of plumose bristles, falling in a ring. 2 spp.: Medit. (Latin: tailed seed, from beak) [Strother 2006 FNANM 19:296–297]

U. picroides (L.) F.W. Schmidt (p. 445) **ST**: 1–4+ dm, long-hairy and with slender spine-like bristles. **LF**: proximal tapered to base; distal sessile, clasping; blade toothed or lobed (or distal entire). **INFL**: peduncle stout; involucre 1.3–2.2 cm; phyllaries 7–8, widely lanceolate, long-acuminate, bristly or becoming ± glabrous. **FR**: 11–14 mm; beak ± > body. 2*n*=8,10. Uncommon. Disturbed places; < 400 m. n SNF, n CCo, SnFrB, SCo, s WTR; native to Eur. Apr–Jul

VENEGASIA CANYON-SUNFLOWER

1 sp. (M. Venegas, Mexican writer, missionary to CA, 1680–1764) [Strother 2006 FNANM 21:385–386]

V. carpesioides DC. (p. 445) Shrub 5–15 dm. **ST**: gen several, ± puberulent. **LF**: alternate; petiole 1–5 cm; blade 3–15 cm, triangular-ovate, 3-veined from base, acute to acuminate, entire to coarsely toothed, minutely puberulent, resin-dotted abaxially. **INFL**: heads radiate, 1 or in ± lfy cyme-like clusters; peduncle gen 2–3 cm, slender, bracts 0; involucre 1–3 cm diam, hemispheric; phyllaries unequal, in 3–5 series, 5–14 mm, oblong to widely obovate, obtuse, ± veiny, outer spreading, middle > inner; receptacle flat or rounded, epaleate. **RAY FL**: 12–34; ray 15–30 mm, yellow. **DISK FL**: 35–150+; corolla 5–6 mm, yellow, tube base glandular; stamen tips triangular; style tips ± obtuse-triangular. **FR**: 2–3 mm, ± cylindric, ribbed, ± curved, dark brown, glabrous; pappus 0. 2*n*=38. Canyons, moist, wooded slopes; < 1000 m. CCo, SCoRO, SW; Baja CA. Gen Feb–Jul

VERBESINA CROWNBEARD

Ann, per, shrub. **LF**: simple, opposite or alternate. **INFL**: heads radiate [discoid], 1 or in few- to many-headed cyme-like clusters; phyllaries in 2–6 equal to very unequal series, free; receptacle flat to conic, paleate; paleae entire, folded around ovaries. **RAY FL**: pistillate or sterile; corolla ± yellow [white]. **DISK FL**: 60–150+ (in CA); corolla yellow to orange; style branches acute to acuminate. **FR**: ray fr 3-sided or 0; disk fr ± elliptic to obovate, compressed, gen wing-margined; pappus 0 or of 2 awns. ± 300 spp.: temps, trop mtns, n hemisphere, S.Am, Afr. (Derived from *Verbena*) [Strother 2006 FNANM 21:106–111]

1. Subshrub; lvs sessile, bright green; ray fls sterile, style 0 . . . . *V. dissita*
1′ Ann; lvs petioled, dull green to canescent; ray fls pistillate, style present . . . . *V. encelioides* subsp. *exauriculata*

V. dissita A. Gray (p. 445) BIG-LEAVED CROWNBEARD Pl 1–1.5 m. **ST**: glabrous or short-hairy near infl. **LF**: gen opposite; blade ovate, 1-veined from base, entire or with few, short teeth, glabrous to densely scabrous. **INFL**: ± flat-topped; peduncles subtended by scale-like entire bracts; heads (1)3–16; phyllaries graduated in 3–4 series, 5–13 mm, oblong, obtuse to acute, short-hairy; paleae 10–13 mm, acute. **RAY FL**: ray orange-yellow, 12–25 mm, ± entire; ovary 2.5–3 mm. **DISK FL**: corolla 7–9 mm; anthers dark brown. **FR**: 8–9 mm, dark brown, glabrous; wing thin, brown; pappus of 2 awns, 3–4 mm. Shrubby coastal slopes; < 200 m. s SCo (Orange Co.), naturalized SnBr; Baja CA. May–Aug ★

V. encelioides (Cav.) A. Gray subsp. *exauriculata* (B.L. Rob. & Greenm.) J.R. Coleman (p. 445) GOLDEN CROWNBEARD Pl 15–130 cm; odor unpleasant. **ST**: densely short-hairy. **LF**: alternate exc nodes near base opposite; blade lanceolate to triangular-ovate, 3-veined from base, coarsely dentate. **INFL**: heads 1–many; peduncle subtended by lf-like bracts; phyllaries in 2 or 3 ± equal series, 8–10 mm, lance-linear, acute, ± strigose; paleae 6–8 mm, abruptly acuminate. **RAY FL**: ray orange-yellow, gen 8–10 mm, ± 3-lobed. **DISK FL**: corolla 5–6 mm; anthers yellow to light brown. **FR**: ray fr 3–4 mm, obovoid, 3-sided, wing 0, pappus 0; disk fr 4–6.5 mm, obovate, flattened, brown or black, soft-hairy, wing wide, ± white, corky, pappus awns 2, 1–2.5 mm. 2*n*=34. Disturbed areas, roadsides, fields; gen ≤ 300 m. SnJV, s CCo, SCoR, SCo, DSon; w N.Am; native AZ to Great Plains, Mex. TOXIC to livestock but unpalatable. May–Jan

VIGUIERA

[Ann, per] shrub. **ST**: gen several from base. **LF**: simple, alternate or opposite, sessile or petioled. **INFL**: heads radiate [discoid], 1 or in few-headed, terminal cyme-like clusters; peduncles long or short; involucre hemispheric [bell-shaped]; phyllaries in several series, equal to very unequal; receptacle rounded to conic, paleate; palea entire or 3-lobed, folded around fr. **RAY FL**: [0]many, sterile; corolla yellow; ray entire to 3-lobed. **DISK FL**: many; corolla yellow or orange; anther tips triangular; style tips triangular. **FR**: ± flattened, linear to obovate, glabrous or ± hairy; pappus of scales, gen 1 or more lanceolate. ± 150 spp.: New World. (L.G.A. Viguier, French physician, botanist, 1790–1867) [Blake 1918 Contr Gray Herb n.s. 54:1–205] Taxa treated as *Viguiera* in TJM (1993) now classified in *Bahiopsis*.

V. purisimae Brandegee (p. 445) LA PURISIMA VIGUIERA Shrub or subshrub, openly branched. **ST**: < 12 dm, scabrous, gray. **LF**: proximal opposite, distal sometimes alternate; petiole ≤ 3 cm; blade 2–6 cm, elliptic to widely obovate, 3-veined, base narrowly to broadly wedge-shaped, tip obtuse to rounded, sometimes abruptly short-pointed, margin entire, faces strigose. **INFL**: heads radiate, 2–8 in ± flat-topped clusters; involucre 10–18 mm diam, hemispheric; phyllaries 3–10 mm, graduated in 4–5 series, ovate, rounded or abruptly

short-pointed, dark green to ± black, puberulent, ciliate; paleae 6–10 mm, linear-oblong, acuminate. **RAY FL**: 10–13; ray 10–12 mm, ovate, yellow. **DISK FL**: corolla ± 6 mm, yellow. **FR**: 2.5–3.3 mm, obovate, glabrous; pappus a crown of ± fused scales ± 1 mm and 2 awn-like scales, 2.5–4.5 mm. Coastal-sage scrub; < 750 m. PR (San Diego Co.); Baja CA. Apr–Oct ★

VOLUTARIA

Ann. **ST**: erect, openly branched. **LF**: basal and proximal cauline winged-petioled, mid and distal cauline sessile; entire to dentate or pinnately divided, long-soft hairy, minutely glandular. **INFL**: heads radiant, long-peduncled, 1 or in few-headed cyme-like cluster; involucre ovoid, 10–15 mm diam, phyllaries many, graduated in several series, appressed, ovate to lanceolate, entire, acute, tipped by ascending to reflexed, flattened spine; receptacle flat, epaleate, bristly. **FL**: corolla pink to purple [blue, yellow, or orange]; peripheral fls sterile, corolla enlarged, spreading, lobes (5–6), linear; inner fls bisexual, tube slender, throat narrowly cylindric, lobes linear-oblong; anther bases tailed, tips lanceolate; style tip with ± hairy node and short, linear terminal segment. **FR**: ± barrel-shaped, weakly compressed, ribbed, tip with prominent collar, faces pitted, attachment scar lateral within cavity with hard rim; pappus persistent, of several series of narrow scales, white to tan. 16 spp.: Medit, w Asia, Atlantic islands, ne Afr. (Latin: having twist, for spirally coiled corolla lobes of original sp.) [Keil 2006 FNANM 19:174–175] *V. canariensis* Wagenitz (disciform heads, white fls) established as weed in sw DSon.

V. muricata (L.) Maire MOROCCO KNAPWEED **ST**: ≤ 50 cm. **LF**: proximal oblanceolate, ± dentate to deeply lobed, distal smaller, linear, ± entire or lobed, long-shaggy-hairy (hairs jointed), minutely resin-gland-dotted. **INFL**: head 3–5 cm diam; peduncle 5–15 cm, distally lfless, finely cobwebby-tomentose, esp near tips; involucre 15–18 mm, phyllaries in 5–7 series, minutely spiny-serrate, finely soft hairy, outer and middle ovate, dark-margined, with spine tip 3–5 mm, inner lanceolate, with flattened, ± spineless tips. **FL**: sterile fl corolla lobes 10–15 mm, spreading; bisexual fl corolla tube 5–6 mm, throat 2–4 mm, lobes 5–6 mm. **FR**: pale gray-brown; pappus scales 1–2.5 mm, irregularly toothed, outer < inner. 2*n*=24. Disturbed sites; < 100 m. SCo; sw Eur, nw Afr. [*Centaurea m.* L.; *Cyanopsis m.* (L.) Dostál] Probable escape from cult; becoming problem weed in s SCo. Mar–May

WYETHIA MULE'S EARS

Per from stout taproot and caudex. **ST**: gen ± erect, branched basally or distally. **LF**: basal and cauline, alternate, basal largest, present on fl sts or 0, lance-linear to broadly ovate, cauline often sessile, distal reduced. **INFL**: heads gen radiate, 1–few, ± large; involucre hemispheric or bell-shaped; phyllaries ± many in 2–3 series, outer often ± lf-like; receptacle ± convex, paleate; paleae ± linear, entire, gen ± hairy. **RAY FL**: 0 or (1)5–25; ray gen yellow. **DISK FL**: 35–150+; corolla yellow or orange, lobes short; anthers brown or yellow, tips triangular; style tips linear, tapered. **FR**: pappus 0, of 1–several triangular or lanceolate scales, or a low crown; ray fr 3-angled; disk fr 4-angled, sometimes ± compressed. 10 spp.: w N.Am. (Nathaniel J. Wyeth, Am explorer, 1802–1856) [Weber 2006 FNANM 21:100–106] Some spp. sometimes treated in *Agnorhiza*.

1. Heads discoid (or ray fls rarely 1–3) . . . . *W. invenusta*
1′ Heads radiate, ray fls 5–21+
  2. Basal lvs 0 on fl sts (or gen < cauline lvs)
    3. Outer phyllaries many, gen not lf-like, gen < 6 mm wide; involucre hemispheric; pl gen > 5 dm
      4. Pl densely soft-hairy; fr 8–12 mm, glabrous or distally short-hairy; pappus (0.5)1–3 mm . . . . *W. elata*
      4′ Pl scabrous and sparsely bristly; fr 5–6 mm, glabrous; pappus 0 or minute . . . . *W. reticulata*
    3′ Outer phyllaries 4–6, ± lf-like, gen 6–17 mm wide; involucre bell-shaped; pl gen 1–3(5) dm
      5. Head 1; lvs glabrous; pappus 0 or a crown 0.1–0.3 mm . . . . *W. bolanderi*
      5′ Heads gen several; young lvs tomentose; pappus a crown of triangular scales 1–1.5 mm . . . . *W. ovata*
  2′ Basal lvs present on fl sts, gen > cauline lvs
    6. Outer phyllaries wide, lf-like, gen >> inner; head gen 1(2), large
      7. Pl shiny green, glabrous or sparsely short-hairy, glandular; phyllaries glabrous or glandular;
        fr 10–13 mm . . . . *W. glabra*
      7′ Pl densely tomentose, often becoming ± glabrous; phyllaries persistently tomentose;
        fr 12–15 mm . . . . *W. helenioides*
    6′ Outer phyllaries narrow, gen not or barely lf-like, gen barely or not > inner; heads 1–few
      8. Pl ± tomentose, often becoming glabrous; phyllaries gen few . . . . *W. mollis*
      8′ Pl glabrous to short-hairy; phyllaries gen many
        9. Phyllaries ± soft-hairy, ciliate; head gen 1 . . . . *W. angustifolia*
        9′ Phyllaries glabrous or puberulent, margins occ ± puberulent, not ciliate; heads 1–4 . . . . *W. longicaulis*

W. angustifolia (DC.) Nutt. Pl 3–9 dm, ± rough-hairy. **LF**: basal blades 10–50 cm, lance-linear to (ob)lanceolate or nearly deltate, short-strigose, scabrous, or short rough-hairy, occ appearing varnished. **INFL**: head gen 1; peduncle 20–30 cm, soft-hairy; involucre 2–4 cm diam, hemispheric; phyllaries 15–30 mm, linear to ± lanceolate, gen not lf-like, ± soft-hairy, ciliate; paleae 13–15 mm. **RAY FL**: 8–21; ray 15–45 mm. **DISK FL**: corolla 10–11 mm. **FR**: 7–8 mm, ± hairy; pappus of 1–4 awn-tipped scales, 7–9 mm (and occ a crown of smaller scales). 2*n*=38. Grassland; < 2050 m. NW, CaRF, SN, GV, CW, MP; to WA. Hybridizes with *W. helenioides* in SnFrB. Apr–Aug

W. bolanderi (A. Gray) W.A. Weber BOLANDER'S MULE'S EARS Pl 1–3 dm, ± sticky. **ST**: decumbent, seldom branched distally. **LF**: ± equal; blades 4–12 cm, widely ovate-cordate, obtuse, abaxially veiny, glabrous, gland-dotted, ± sticky, shining. **INFL**: peduncle 0–2 cm; involucre bell-shaped; outer phyllaries 4–6, 18–30 mm, obovate, veiny, lf-like, > inner; paleae ± 15 mm, tip tomentose. **RAY FL**: 8–13; ray 20–30 mm. **DISK FL**: corolla 10 mm. **FR**: 7–9 mm, glabrous; pappus 0 or a crown 0.1–0.3 mm. 2*n*=38. Grassland, chaparral; 300–1000 m. ne NCoRI, n&c SNF. [*Agnorhiza b.* (A. Gray) W.A. Weber] Mar–May

W. elata H.M. Hall HALL'S WYETHIA Pl 5–10 dm, densely soft-hairy. **ST**: erect, gen branched distally. **LF**: cauline; blades 8–20 cm, lance-ovate to triangular-ovate, base obtuse-cordate, short-soft-hairy or short-tomentose, gland-dotted. **INFL**: occ flat-topped, lfy-bracted;

peduncle 4–12 cm; involucre hemispheric; phyllaries 25–35 mm, lanceolate to ovate, acuminate, outer occ reflexed, gen not lf-like, inner erect, short-rough-hairy. **RAY FL**: 10–14(23); ray 50–60 mm. **DISK FL**: corolla 12 mm. **FR**: 8–12 mm, glabrous or distally short-hairy; pappus an uneven crown of scales, (0.5)1–3 mm. $2n=38$. Open woodland, forest; 500–1400 m. c SN. [*Agnorhiza e.* (H.M. Hall) W.A. Weber] May–Aug(Oct) ★

W. glabra A. Gray (p. 445) Pl 1–4(6) dm, glabrous or sparsely short-hairy, glandular. **LF**: basal blades 25–45 cm, oblong lanceolate to (ob)ovate, glabrous to finely stalked-glandular, occ short-soft-hairy, shiny. **INFL**: heads 1(2); peduncle 1–6 cm; outer phyllaries 40–70 mm, lanceolate, obtuse to acute, lf-like, spreading, gen > rays, inner shorter, ± erect; paleae 15–20 mm. **RAY FL**: (8)12–27; ray 25–50 mm. **DISK FL**: corolla 10–11 mm. **FR**: 10–13 mm, sparsely hairy and/or glandular; pappus scales 2–several, 1–5 mm, lanceolate to triangular, glabrous. Gen shady sites; < 950 m. KR, s NCoR, CW. Mar–Jun

W. helenioides (DC.) Nutt. Pl 2–7 dm, densely tomentose, often becoming ± glabrous. **LF**: basal blades 25–45 cm, elliptic to oblong-ovate, densely tomentose, becoming glabrous. **INFL**: heads gen 1; peduncle 1–6 cm; outer phyllaries 40–70+ mm, narrowly ovate, lf-like, spreading, often > rays, inner shorter, ± erect; paleae 15–20 mm. **RAY FL**: 13–21; ray 20–50 mm. **DISK FL**: corolla 10–11 mm. **FR**: 12–15 mm, distally minutely strigose; pappus scales 2–several, 1–5 mm, lanceolate to triangular, ± hairy. $2n=38$. Open grassland, woodland, scrub; < 2000 m. KR, NCoR, CaRF, n&c SN, ScV (Sutter Buttes), SnJV, CW. ± glabrous pls approach *W. glabra*. Hybridizes with *W. angustifolia* in SnFrB. Mar–May(Aug)

W. invenusta (Greene) W.A. Weber (p. 445) RAYLESS MULE'S EARS Pl (2)6–10 dm, short-rough-hairy, ± sticky. **ST**: erect, gen branched distally, distally densely glandular. **LF**: cauline, 7–20 cm, ovate to triangular; base obtuse to cordate, short- to long-stiff hairy, gen ± glandular. **INFL**: heads gen discoid; involucre 2–3 cm diam; phyllaries 20–35 mm, lanceolate to ovate, acuminate, striate, ciliate, outer not lf-like; paleae 12 mm. **RAY FL**: 0 (or 1–3); ray 6–10 mm. **DISK FL**: corolla 10 mm. **FR**: 7–8 mm, glabrous; pappus 0. Open forest, clearings; 600–2300 m. s SN. [*Agnorhiza i.* (Greene) W.A. Weber] Jun–Aug

W. longicaulis A. Gray HUMBOLDT COUNTY WYETHIA Pl 2–5(6) dm, glabrous or puberulent, ± gland-dotted. **LF**: basal blades 10–25(40) cm, oblong-oblanceolate, glabrous, gland-dotted, appear-ing varnished. **INFL**: heads 1–4; peduncle 3–8 cm; involucre 15–30 mm diam, bell-shaped; phyllaries ± equal, 15–30 mm, lanceolate to ovate, barely lf-like, glabrous or puberulent, margins occ ± puberulent; paleae 15–17 mm. **RAY FL**: 5–10; ray 20–30 mm. **DISK FL**: corolla 10–11 mm. **FR**: 6–10 mm, glabrous or distally puberulent; pappus scales < 1 mm, obtuse. Grassland, open forest; 750–1500 m. KR, n NCoR. May–Jul ★

W. mollis A. Gray (p. 445) Pl 3–5(10) dm, ± tomentose, often becoming glabrous. **LF**: basal blade 20–40 cm, oblanceolate to widely obovate, ± densely tomentose, gen glandular, becoming glabrous. **INFL**: heads 1–3; peduncle 1–10 cm, lfy-bracted; involucre ± 2 cm diam, ± bell-shaped; phyllaries gen few, ± equal, 15–35 mm, lance-oblong, acute to obtuse; paleae 15–16 mm. **RAY FL**: 6–15; ray 15–45 mm. **DISK FL**: corolla 10 mm. **FR**: 9–11 mm; distally puberulent to tomentose; pappus of scales ≤ 1 mm, occ also 1–few lanceolate scales ≤ 8 mm. $2n=38$. Open forest, meadows, dry rocky slopes, sagebrush scrub; 900–3400 m. KR, NCoRH, CaR, n&c SN, MP, n SNE; se OR, w NV. May–Jul(Sep)

W. ovata Torr. & A. Gray SOUTHERN MULE'S EARS Pl 1–3(5) dm, often branched, ± sticky-glandular. **ST**: decumbent, seldom distally branched. **LF**: ± equal; blades 7–20 cm, elliptic to broadly ovate or ± round, base acute to obtuse, truncate, or cordate, soft-silky-hairy or strigose. **INFL**: flat-topped, ± lfy-bracted; peduncle 1–20 cm; involucre 1.5–2 cm diam, bell-shaped; outer phyllaries 4–6, erect, 30–50 mm, ± linear to ovate, ± lf-like, inner shorter, narrower; paleae 13–15 mm. **RAY FL**: 5–8; ray 8–18 mm. **DISK FL**: corolla 8 mm. **FR**: 9–11 mm, glabrous or puberulent near tip; pappus scales 1–1.5 mm. $2n=38$. Grassland, open woodland and forest; 300–2750 m. s SN, SW; n Baja CA. [*Agnorhiza o.* (Torr. & A. Gray) W.A. Weber] May–Sep

W. reticulata Greene (p. 445) EL DORADO COUNTY MULE EARS Pl 4–7 dm, ± sticky-glandular. **ST**: erect, gen branched distally. **LF**: cauline; proximal > distal; blades 5–15 cm, lance-ovate to deltate, base broadly obtuse to ± cordate, abaxially veiny, sparsely bristly or scabrous, gland-dotted, appearing varnished. **INFL**: involucre 2–3 cm diam, hemispheric; phyllaries ± equal, 20–30 mm, oblong or lanceolate, gen not lf-like, outer often spreading or reflexed; paleae 12–13 mm. **RAY FL**: 10–21; ray 20–25 mm. **DISK FL**: corolla 10 mm. **FR**: 5–6 mm, glabrous; pappus 0 or crown 0.1–1 mm. $2n=38$. Wooded slopes, chaparral; 150–600 m. n SNF (El Dorado Co.). [*Agnorhiza r.* (Greene) W.A. Weber] Threatened by development. May–Aug ★

XANTHISMA

David R. Morgan

Ann, per, subshrub, from taproot or ± branched caudex. **LF**: simple, alternate, entire to deeply 1–2-pinnately lobed, teeth or lobes bristle-tipped, distal reduced. **INFL**: heads radiate [discoid], 1 or in cyme-like clusters; involucre hemispheric [obconic, bell-shaped]; phyllaries graduated in 2–8 series, proximally straw-colored, distally gen green, bristle-tipped; receptacle flat to convex, with short, triangular scales, glabrous, epaleate. **RAY FL**: [0]8–many; corolla yellow [white, pink, blue, or purple]. **DISK FL**: 10–many; corolla yellow; anther tip lanceolate; style tips triangular to linear. **FR**: elliptic, oblong, or obovoid, several- to many-ribbed, sparsely to densely hairy; pappus of many unequal bristles (ray pappus occ 0). 17 spp.: temp w N.Am. (Greek: yellow condition, describing fl heads of original sp.) [Hartman 2006 FNANM 20:383–393]

1. Ann. . . . . ***X. gracile***
1′ Per or subshrub
  2. Distal bracts of peduncle overlapping, grading into phyllaries; st nearly lfless above base. . . . . ***X. junceum***
  2′ Distal bracts of peduncle few, well separated, not grading into phyllaries; st gen lfy, at least in
     proximal 1/2 . . . . ***X. spinulosum*** var. ***gooddingii***

X. gracile (Nutt.) D.R. Morgan & R.L. Hartm. (p. 445) ANNUAL BRISTLEWEED Ann. **ST**: erect, 3–25 cm, lfy, branched at or above base, bristly throughout. **LF**: 1–3 cm, 3–7 mm wide; proximal oblanceolate, elliptic, or oblong in outline, pinnately lobed, ± appressed-hairy. **INFL**: involucre 6–7 mm, 7–12 mm wide; phyllaries in 4–6 series, lance-linear, hairy. **RAY FL**: 16–18; ray 7–12 mm, yellow. **DISK FL**: 44–65; corolla 4.5–5.5 mm. **FR**: 2.2–2.8 mm, canescent; pappus ≤ 5 mm, bristles proximally slightly widened, white to red-brown. $2n=4,8$. Sandy, clay, or rocky places, desert woodland or scrub; 900–1550 m. e DMoj; to CO, TX, n Mex. [*Machaeranthera g.* (Nutt.) Shinners] Apr–Sep ★

X. junceum (Greene) D.R. Morgan & R.L. Hartm. (p. 445) RUSH-LIKE BRISTLEWEED Per, subshrub. **ST**: spreading to erect, 4–10 dm, occ woody at base, sparsely strigose, slightly glandular near heads, nearly lfless above base. **LF**: 1–2 cm, ≤ 4 mm wide, proximal gen linear, pinnately lobed or serrate, glabrous or tomentose. **INFL**: peduncle long, bracts overlapping, grading into phyllaries; involucre 5–8 mm, 10–12 mm wide; phyllaries in 5–6 series, linear, glandular. **RAY FL**: 15–25; ray 5–6 mm, yellow. **DISK FL**: 25–40; corolla 5–6.5 mm. **FR**: 2.5–3 mm, hairy; pappus ± = disk corolla, tan. Dry hillsides; chaparral, coastal scrub; < 1000 m. PR; s AZ, n Mex. [*Machaeranthera j.* (Greene) Shinners] May–Oct ★

X. spinulosum (Pursh) D.R. Morgan & R.L. Hartm. var. ***gooddingii*** (A. Nelson) D.R. Morgan & R.L. Hartm. SPINY GOLDENWEED Per, subshrub. **ST:** erect or ascending, 2–6 dm, gen lfy, at least in proximal 1/2, ± glandular-puberulent to canescent. **LF:** 2–5 cm, proximal 1–2-pinnately lobed, glabrous or sparsely hairy, stalked-glandular. **INFL:** head 1, peduncle long, bracts few, well separated, not grading into phyllaries; involucre 6–9 mm, 10–18 mm wide, hemispheric; phyllaries lance-linear, prominently glandular-puber-ulent and scabrous. **RAY FL:** 30–45; ray 6–16 mm, yellow. **DISK FL:** many; corolla 4–6 mm. **FR:** 2–3 mm, appressed-hairy; pappus 3–5 mm, tan. Rocky places, washes; creosote-bush scrub; < 900 m. se DMoj, ne DSon; NV, AZ, nw Mex. [*Machaeranthera pinnatifida* (Hook.) Shinners var. *g.* (A. Nelson) B.L. Turner & R.L. Hartm.] Widespread and variable; 7 vars. in sw US and n Mex. Feb–May, Sep–Oct

XANTHIUM

Bruce G. Baldwin, adapted from Strother (2006)

Ann, coarse, 10–200+ cm. **ST:** erect, branched; nodal spines present or 0. **LF:** gen alternate (proximal 2–6 occ opposite), peti-oled or distal sessile, linear to nearly round, often ± palmately or pinnately lobed, ultimate margins entire or ± toothed, faces minutely coarse-hairy or ± strigose, gen gland-dotted. **INFL:** heads unisexual, 1, axillary or in raceme- to spike-like clusters, pistillate gen proximal, staminate gen distal. **PISTILLATE HEAD:** 2–5+ mm diam at fl, 6–20+ mm diam in fr, ± ellipsoid; bracts 30–75+ in 6–12+ series, all proximally fused or outer 5–8 free, tips free, mostly ± hooked exc distal 1–3 (gen longer, stouter), together becoming a hard, prickly bur. **STAMINATE HEAD:** 3–5 mm diam; involucre ± saucer-shaped, phyllaries 6–16+ in 1–2+ series, free, ± 1 mm; receptacle conic to columnar, paleae spoon-shaped to wedge-shaped or linear, membra-nous, distally ± soft-hairy or minutely coarse-hairy. **PISTILLATE FL:** 2: corolla 0, style tips linear, exserted. **STAMINATE FL:** 20–150+; corolla ± funnel-shaped, ± white, lobes 4–5, erect or reflexed, filaments fused, anthers free or weakly fused; style tip undivided. **FR:** fusiform, black, enclosed in obovoid to ellipsoid, 2-chambered bur; pappus 0. 2–3 spp.: ± worldwide; native to New World. (Greek: yellow, from fr-extract dye) [Strother 2006 FNANM 21:19–20]

1. Nodal spines present, gen 3-lobed, 15–30+ mm, golden; lf blade ± lanceolate to ovate or lance-linear, abaxially densely strigose, gray to white . . . . ***X. spinosum***
1′ Nodal spines 0; lf blade nearly round to ± pentagonal or deltate, abaxially minutely coarse-hairy, green . . . ***X. strumarium***

X. spinosum L. (p. 445) SPINY COCKLEBUR Pl 10–60(120+) cm. **LF:** petiole 1–15(25+) mm; blade 4–8(12+) cm, 1–3(5+) cm wide, often pinnately 3(7+)-lobed. **FR:** bur 10–12(15+) mm. 2*n*=36. Disturbed, seasonally wet, often alkaline sites, in grassland, marshes, watercourses; < 1350 m. CA-FP; worldwide, perhaps native to S.Am only. Jul–Oct

X. strumarium L. (p. 445) COCKLEBUR Pls 10–80(200) cm. **LF:** petiole 2–10(14+) cm; blade 4–12(18+) cm, 3–10(18+) cm wide, occ palmately 3–5-lobed. **FR:** bur 10–30+ mm. 2*n*=36. Dis-turbed, seasonally wet, often alkaline sites, in grassland, marshes, watercourses; gen < 1400 m. CA; worldwide, native to Am. Highly variable. Jul–Oct

XYLORHIZA DESERT-ASTER

Per to shrub. **ST:** gen ± white, glabrous or hairy. **LF:** all cauline, alternate, entire or toothed; midrib white. **INFL:** heads radiate, 1, sessile or peduncled; involucre bell-shaped or hemispheric; phyllaries graduated in 3–6 series; receptacle convex, epaleate. **RAY FL:** gen 12–60; corolla white to blue or purple. **DISK FL:** 30–140; corolla yellow; style tips linear, acute. **FR:** linear to club-shaped, weakly compressed, 4-ribbed, with long, ± appressed hairs; pappus of 30–45, unequal, barbed bristles. 10 spp.: w N.Am. (Greek: woody root) [Nesom 2006 FNANM 20:406–409]

1. Per or subshrub; st gen branched only in proximal 1/2; peduncle 8–25 cm . . . . ***X. tortifolia*** var. ***tortifolia***
1′ Shrub; st gen branched throughout; peduncle 0–11 cm
2. Younger sts and peduncles short-glandular; phyllaries loosely appressed, outermost glandular, innermost < immediately preceding series . . . . ***X. cognata***
2′ Younger sts and peduncles gen glabrous (rarely puberulent proximal to heads); phyllaries tightly appressed, ± glabrous; innermost ≥ immediately preceding series . . . . ***X. orcuttii***

X. cognata (H.M. Hall) T.J. Watson (p. 445) MECCA-ASTER Shrub ≤ 1.5 m, ill-scented. **ST:** gen branched throughout, gen short-glandular, glabrous in age. **LF:** 1–5 cm, oblanceolate to ovate, obtuse or acute at tip, ± spiny-dentate or entire, not reduced distally on st, glabrous to glandular. **INFL:** peduncle 0–11 cm; phyllaries 8–19 mm, 0.8–2.2 mm wide, ± glabrous to densely glandular, innermost < immediately preceding series. **RAY FL:** 20–30; tube 5–8 mm; ray 1.8–2.5 cm, lavender or light blue. **DISK FL:** 40–80; corolla 7–9 mm. **FR:** 3–4.5 mm; pappus bristles ≤ 9.5 mm. 2*n*=12. Arid canyons, washes; creosote-bush scrub; < 400 m. DSon (Imperial, Riverside cos.). Jan–Jul ★

X. orcuttii (Vasey & Rose) Greene (p. 445) ORCUTT'S WOODY-ASTER Shrub ≤ 1.5 m. **ST:** gen glabrous (rarely puberulent proximal to heads). **LF:** 2–6 cm, oblanceolate to oblong or lanceolate, obtuse or acute at tip, ± spiny-dentate or entire, not much reduced distally on st, glabrous or margins sparsely hairy. **INFL:** heads sessile (or peduncle ≤ 11 cm); phyllaries 5–20 mm, 1.5–3.5 mm wide, ± gla-brous, innermost ≥ immediately preceding series. **RAY FL:** 25–40; tube 4–6 mm; ray 1.2–3.2 cm, lavender or light blue. **DISK FL:** 55–140; corolla 8–10.5 mm. **FR:** 3–4 mm; pappus bristles ≤ 12 mm. 2*n*=12. Arid canyons, barren slopes; creosote-bush scrub; < 400 m. s DSon. Jan–May ★

X. tortifolia (Torr. & A. Gray) Greene var. ***tortifolia*** (p. 445) MOJAVE-ASTER Per, subshrub, 2–6(8) dm from much-branched caudex. **ST:** gen branched proximally, with long, nonglandular hairs and shorter, stalked glands. **LF:** 2.5–10 cm, reduced distally on st, linear to lanceolate, oblanceolate or elliptic, acute to spine-tipped, ± spiny-dentate, gen soft-hairy and glandular. **INFL:** peduncle 8–25 cm; phyllaries 5–25 mm, 0.7–2.5 mm wide, soft-hairy, glandular, innermost < to > immediately preceding series. **RAY FL:** 25–60; tube 4–6 mm; ray 1–3.3 cm, lavender, light blue, or white. **DISK FL:** 70–110; corolla 5.5–8.5 mm. **FR:** 3–6 mm; pappus bristles ≤ 9 mm. 2*n*=12,24. Desert slopes, canyons, woodland, creosote-bush and salt-bush scrub; 240–2000 m. s SNE, D; to sw UT, w AZ. Mar–Jun, Oct

BALSAMINACEAE TOUCH-ME-NOT or BALSAM FAMILY

Peter F. Zika

Ann, [per, subshrub, few epiphytes], glabrous [(hairy)]; roots fibrous [rhizomatous to tuberous]. **ST**: watery to fleshy. **LF**: gen alternate, simple, petioled, toothed or entire, veins pinnate, ± arched. **INFL**: cyme, umbel-like, terminal or axillary, 1–8-fld. **FL**: bisexual, bilateral, twisting 180°; sepals 3[5], free, upper becoming lower, gen with nectar spur [or spur 0]; petals 3 [5, free or] ± fused in 2s; stamens 5, filaments stout, forming column around pistil, ± united at tip, shed at base, anthers united into cap over stigma; ovary 1, superior, gen 5-chambered, stigmas 1–5. **FR**: capsule, explosive [or berry-like drupe]. 2 genera, ± 850 spp.: trop, warm temp; some orn. [Janssens et al. 2006 Syst Bot 31:171–180] Scientific Editor: Thomas J. Rosatti.

IMPATIENS TOUCH-ME-NOT, JEWELWEED

ST: lower gen with adventitious roots. **LF**: alternate, opposite, or whorled; stipules 0 or gland-like. **FL**: variously spotted or lined; sepals 3, lateral 2 greatly reduced, ± green, lower with nectar spur, not green; petals apparently 3, lateral 2 forming a lip, upper 1 a banner over mouth of tube. **FR**: green. **SEED**: 1–7, green to black, ridged or papillate. ± 850 spp.: trop, warm temp, Eurasia, Afr, boreal Am. (Latin: impatient, from explosive fr) [Zika 2006 Novon 16:443–448]

1. Fls yellow-orange (pink, cream), with dark red marks or not; lvs acute, entire to coarse-crenate; stipules 0 . . . [***I. capensis***]
1′ Fls white, pink, or purple, with yellow or dark purple marks or not; lvs acute to acuminate, serrate; stipules gland-like
  2. Taper to fl spur gradual, spur ± straight, 10–10 mm; lvs alternate; stipules dome-shaped, unlobed,
     < 1 mm . . . . ***I. balfourii***
  2′ Taper to fl spur abrupt, spur ± recurved, 5–6 mm; lvs opposite or whorled; stipules oblong, lobed or not,
     2–4 mm . . . . ***I. glandulifera***

I. balfourii Hook. f. (p. 451) KASHMIR BALSAM Pl 5–10 dm. **LF**: alternate; stipules gland-like, dome-shaped, unlobed, < 1 mm; blade 2–13 cm, lanceolate to wide-ovate, acuminate, serrate. **INFL**: 1–8(14)-fld, axillary, terminal. **FL**: white, pink, or purple, with yellow marks or not; spur ± straight, 10–20 mm, taper to spur gradual, lateral sepals 2–5 mm; self-pollinating fls 0. **FR**: club-shaped. 2*n*=14. Shores, streambanks; < 1000 m. NCo, n CCo, s SnFrB, ne SCo, SnBr; OR; native to Himalaya Mtns. Introduced in CA as orn. Apr–Sep

I. glandulifera Royle (p. 451) POLICEMAN'S HELMET Pl 6–25 dm. **LF**: opposite or whorled; stipules gland-like, oblong, lobed or not, 2–4 mm; blade 3–18 cm, gen narrow-lanceolate, ovate, or elliptic, acute to acuminate, serrate. **INFL**: gen 2–15-fld, axillary, terminal. **FL**: white or purple, gen with dark purple marks; spur ± recurved, 5–6 mm, taper to spur abrupt, lateral sepals 4–7 mm; self-pollinating fls 0. **FR**: club-shaped. 2*n*=18. Disturbed, moist, often shaded sites, streambanks; < 100 m. n NCo; to s BC, n ID, invasive along coast in OR, WA; native to Eurasia. Introduced in CA as orn. Apr–Sep

BATACEAE SALTWORT FAMILY

Steve Boyd & William J. Stone

Per, subshrub, salt-tolerant; dioecious [monoecious]. **LF**: simple, opposite, sessile, ± cylindric or ± 3-angled, narrowly elliptic to narrowly oblanceolate or linear, fleshy, base with triangular appendage, margins entire; stipules minute. **INFL**: dense, cone- or spike-like, axillary, ± sessile; fls unisexual. **STAMINATE FL**: 8–12, ± free, 4-ranked, subtended by ± widely ovate bract and involucre- or sac-like bractlets; calyx ± bilateral, cup- or bell-shaped, blunt-lobed; petals 4(5), alternate stamens; stamens 4(5), > petals, anthers opening by longitudinal slits. **PISTILLATE FL**: (2)4–12(24), free or fused, subtended by small bractlet, << fl; perianth 0; ovary superior, fleshy, chambers 4, ovule 1 per chamber, stigma 2-lobed, ± spheric, sessile. **FR**: drupe or drupe-like aggregate, floating. **SEED**: 1–4. *n*=11. 1 genus, 2 spp.: coastal, trop, subtrop Am, Pacific, Australia. [Bayer & Appel 2003 Fam Gen Vasc Pl 5:30–32] Staminate perianth variously interpreted as tepals, petals, or staminodes. Scientific Editor: Bruce G. Baldwin.

BATIS

2 spp. (Greek: name of a seashore plant) *Batis argillicola* P. Royen of New Guinea, ne Australia is monoecious.

B. maritima L. (p. 451) SALTWORT, BEACHWORT **ST**: prostrate to ascending, < 1.5 m; base woody. **LF**: (0.5)1–2 cm, linear to narrowly oblanceolate, ± cylindric, ± flattened above. **INFL**: staminate ± 6 mm, ovoid-cylindric, bracts or bractlets rounded; pistillate ± 6 mm, cylindric (obconic) in fr; fls fused in age. **STAMINATE FL**: petals white, blade ± triangular, clawed, stamens exserted. **FR**: ± 10 mm, fleshy, spongy or corky in age. Coastal salt marshes; < 10 m. SCo; to se US, Caribbean, n S.Am; HI. Apr–Sep

BERBERIDACEAE BARBERRY FAMILY

Michael P. Williams

Per, shrub, [tree], gen rhizomed, caudexed or not, glabrous, glaucous, or hairy. **ST**: spreading to erect, branched or not. **LF**: simple, 1–3-ternate, or pinnately compound, basal and cauline, gen alternate, deciduous or evergreen, petioled, stipuled. **INFL**: gen raceme, spike, or panicle, scapose, terminal, or axillary. **FL**: gen bisexual, radial; sepals 6–18 or 0, gen in whorls of 3; petals gen 6, in 2 whorls of 3, or 0; stamens 6–12(13), free or fused at base, in 2 whorls or not, anthers dehiscent by flap-like

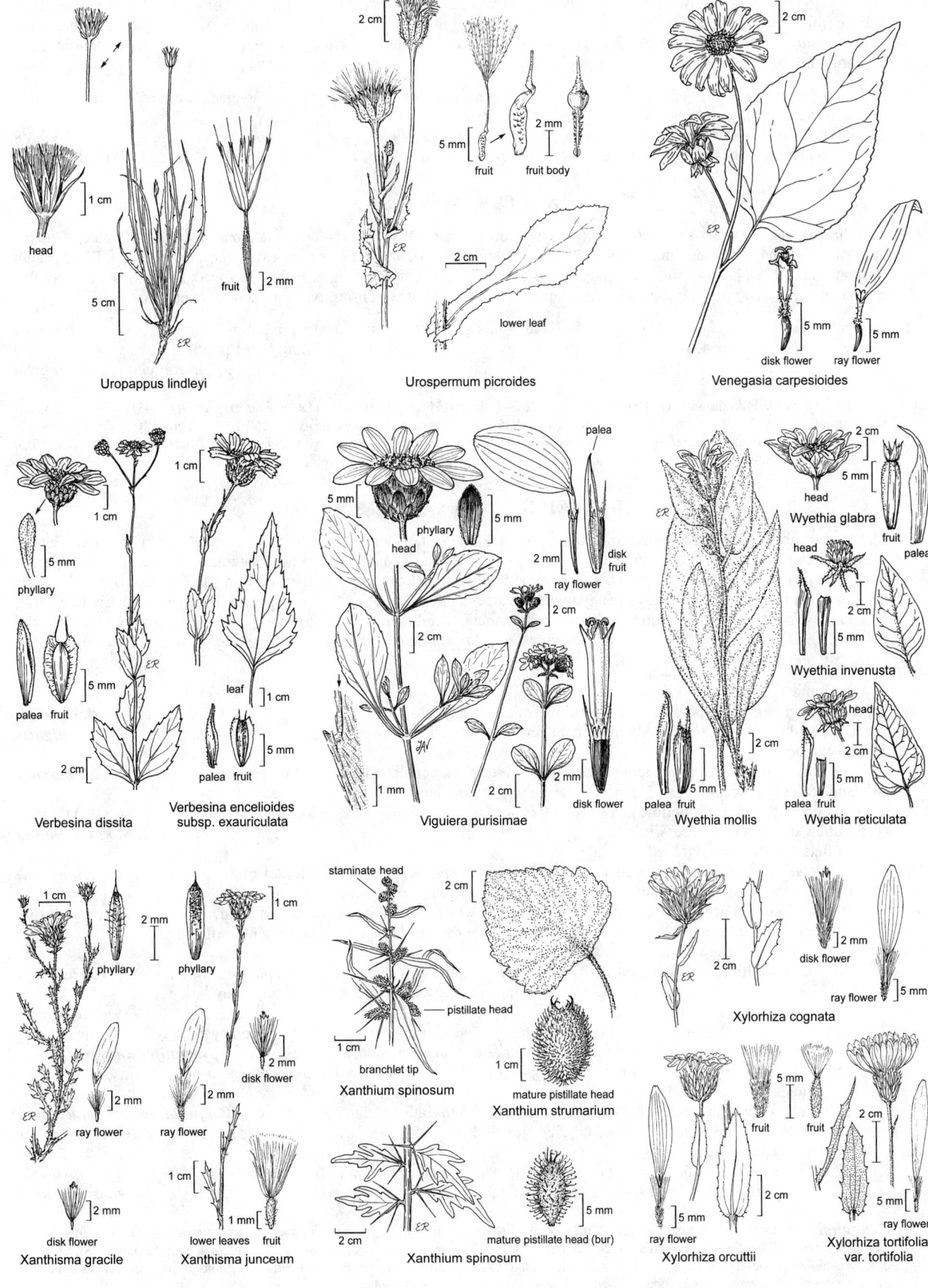

Uropappus lindleyi

Urospermum picroides

Venegasia carpesioides

Verbesina dissita

Verbesina encelioides subsp. exauriculata

Viguiera purisimae

Wyethia mollis

Wyethia glabra

Wyethia invenusta

Wyethia reticulata

Xanthisma gracile

Xanthisma junceum

Xanthium spinosum

Xanthium strumarium

Xanthium spinosum

Xylorhiza cognata

Xylorhiza orcuttii

Xylorhiza tortifolia var. tortifolia

valves or longitudinal slits; ovary superior, chamber 1, ovules gen 1–10, style 1 or 0, stigma flat or spheric. **FR**: berry, capsule, achene [follicle]. 16 genera, ± 670 spp.: temp, trop worldwide; some cult (*Berberis, Epimedium, Nandina* (heavenly bamboo), *Vancouveria*). [Wang 2007 Syst Bot 32:731–742] Lower sepals sometimes called "bracteoles", inner petals "staminodes". Scientific Editor: Thomas J. Rosatti.

1. Shrub; lvs simple or pinnately compound, gen leathery (exc *Berberis vulgaris*), gen spine-toothed; fr a berry ... **BERBERIS**
1′ Per; lvs 1–3-ternate, not leathery, not spine-toothed; fr a capsule or achene
  2. Infl a spike; lflets triangular to fan-shaped; perianth 0; fr an achene .... **ACHLYS**
  2′ Infl a raceme or panicle; lflets ovate to ± cordate; perianth parts 12+; fr a 2-valved capsule .... **VANCOUVERIA**

ACHLYS VANILLA LEAF

Per; rhizome scaly; caudex short, erect; wood, esp of roots, often yellow. **LF**: 1–few, basal, long-petioled, 1-ternate; lflets triangular to fan-shaped, bases tapered. **INFL**: ± scapose, spike, dense; lateral fls gen unisexual, terminal bisexual. **FL**: perianth 0; stamens (8)9(13), anther valves flap-like, curled inward; ovule 1, style 0, stigma ± flat, furrowed. **FR**: achene, curved, furrowed, brown to red-purple. 2 spp.: w N.Am, Japan. (Greek: thin mist or obscurity, from inconspicuous fls)

1. Central lflet (3)6–8(12)-lobed, 7–16 cm, 8–17 cm wide; stamens 4–6 mm; ovary 1.5–2 mm; fr brown .....*A. californica*
1′ Central lflet 3(5)-lobed, 4–11 cm, 4–8 cm wide; stamens 3–4 mm; ovary 1–1.5 mm; fr red-purple ....
....*A. triphylla* subsp. *triphylla*

A. californica I. Fukuda & H.G. Baker (p. 451) Pl 3–5 dm; rhizome internodes (7.5)9–10 cm. **INFL**: < 4 cm. **FR**: 3.5–5 mm. 2*n*=24. Moist, shaded sites, conifer forest; < 1500 m. NCo, w KR, w NCoRO; to BC. Apr–Jun

A. triphylla (Sm.) DC. subsp. ***triphylla*** (p. 451) Pl 2–4 dm; rhizome internodes 2.5–5 cm. **INFL**: 2.5–5 cm. **FR**: 3–4.5 mm. 2*n*=12. Moist shaded sites, conifer forest; < 1830 m. NCo, w KR, w NCoRO; to BC. Other subsp. in Japan. Apr–Jun

BERBERIS OREGON-GRAPE, BARBERRY

Shrub, gen rhizomed. **ST**: spreading to erect, branched, spiny or not, vine-like or not; inner bark, wood gen bright yellow; over-wintering bud scales deciduous or not. **LF**: simple or pinnately compound, cauline, alternate, gen leathery, gen persistent; lflets gen 3–11, ± round to lanceolate, gen spine-toothed. **INFL**: raceme, axillary or terminal. **FL**: sepals 9 in 3 whorls of 3; petals 6 in 2 whorls of 3, bases gen glandular; stamens 6; ovules 2–9, stigma ± spheric. **FR**: berry, spheric to elliptic, gen purple-black. ± 600 spp.: temp worldwide. (Latin: ancient Arabic name for barberry) [Kim 2004 J Pl Res 117:175–182] Roots often TOXIC: spines may inject fungal spores into skin. Contact with filament causes stamen to snap inward, possibly to deposit pollen on pollinator.

1. Lvs simple
  2. Lvs evergreen, with 6–8 coarse, gen terminal spines; petals orange .... **B. darwinii**
  2′ Lvs deciduous, with >> 8, small, marginal spines; petals yellow................................. **B. vulgaris**
1′ Lvs compound
  3. Bud scales persistent among upper lvs, 15–45 mm, thick, lanceolate; lflets ± palmately veined .... **B. nervosa**
  3′ Bud scales gen deciduous, < 5 mm, thin, ovate to deltate; lflets gen pinnately veined
    4. Infl open, fls gen < 10; petiole gen < 2 cm; terminal lflets gen lanceolate, oblong, or narrowly elliptic — SW, D
      5. Lflets gen 3, terminal sessile .... **B. harrisoniana**
      5′ Lflets 3–9, terminal stalked
        6. Lflets flat to ± wavy, not folded along midrib, serrate, spines gen > 8 per side, ± 1 mm............... **B. nevinii**
        6′ Lflets wavy, gen folded along midrib, ± lobed, spines 3–8 per side, 1–4 mm
          7. Terminal lflet gen < 2 × width; berries yellow-red when fresh, ± glaucous, not inflated.......... **B. higginsiae**
          7′ Terminal lflet gen > 2 × width; berries blue-black when fresh, heavily glaucous, not inflated or
             inflated only due to insect parasitism
            8. Terminal lflet gen lance-ovate, length gen < 3 × width; fls 8–12; fr yellow- to purple-red .... **B. fremontii**
            8′ Terminal lflet gen narrow-lanceolate, length gen > 3 × width; fls 3–5; fr red-brown to dark purple
               .... **B. haematocarpa**
    4′ Infl dense, fls > 10; petiole gen > 2 cm (0–3 cm in *Berberis pinnata*); terminal lflets gen ovate to wide-elliptic
      9. Lflet margin strongly wavy, thick, gen hard, spines 6–10(12) per side, 2–5 mm........ **B. aquifolium** var. **dictyota**
      9′ Lflet margin wavy to flat, ± thin, not hard, spines gen > 10 per side (to 0 in *Berberis pinnata*), 1–2 mm
        10. Petioles gen > 3 cm; lflets gen 5–9
          11. Sts ascending to erect, gen 1–2 m; lflets gen shiny adaxially .... **B. aquifolium** var. **aquifolium**
          11′ Sts spreading to erect, gen < 0.8 m; lflets gen dull adaxially .... **B. aquifolium** var. **repens**
        10′ Petioles gen 1–3 cm; lflets gen 7–11... **B. pinnata**
          12. Lflet margins gen flat, spines gen < 1 mm; upper sts reclining to weakly erect; pl 2–4 m ...... subsp. **insularis**
          12′ Lflet margins gen wavy, spines gen > 1 mm; upper sts gen erect; pl gen < 2 m............... subsp. **pinnata**

B. aquifolium Pursh **ST**: spreading to erect, 0.1–2 m; bud scales gen deciduous. **LF**: cauline, not crowded, 8–24 cm; petiole gen 3–6 cm; lflets 5–9, terminal 2–7.5 cm, 1.5–4.5 cm wide, ± round to elliptic, ± flat to strongly wavy, base ± lobed to wedge-shaped, tip acute to obtuse (exc tooth), margin serrate, spines 6–24(40) per side, 1–5 mm. **INFL**: 3–6 cm, dense; axis internodes 2–4 mm in fl, fr. **FR**: 4–7 mm diam, ovoid to obovoid, glaucous, blue to purple. **SEED**: 4–5 mm. Vars. intergrade, need study; abaxial papillae on lvs evidently of no taxonomic value.

var. *aquifolium* **ST**: ascending to erect, gen 1–2 m. **LF**: 10–24 cm; petiole 1–2.5(5) cm; lflets gen 5–9, 3–7.5 cm, 2–4.5 cm wide, ovate to elliptic, ± flat, gen shiny adaxially, base oblique to obtuse, spines 12–24 per side, 1–2 mm. **FR**: ovoid to obovoid, dark blue. 2*n*=28. Conifer forest; 400–2300 m. KR, NCoR, CaR, SN, SnGb, MP; to BC. [*B. piperiana* (Abrams) McMinn] Mar–Jun

var. *dictyota* (Jeps.) Jeps. (p. 451) **ST**: erect, < 1 m. **LF**: 9–15 cm; petiole 1–5 cm; lflets gen 7–9, 2–6 cm, 2–3.5 cm wide, ± round to ovate, base ± lobed to wedge-shaped, margin strongly wavy, thick, gen hard, spines 6–10(12) per side, 2–5 mm. **FR**: ± ovoid, blue-purple. 2*n*=28. Slopes, canyons, conifer forest, oak woodland, chaparral; 90–2200 m. NW, CaR, CW, SN, ScV (Sutter Buttes), SCoR, TR, PR. Possibly intergrades with *B. pinnata* subsp. *p.* in PR. Mar–May

var. *repens* (Lindl.) Scoggan (p. 451) **ST**: spreading to erect, gen < 0.8 m. **LF**: 8–18 cm; petiole 1–6 cm; lflets gen 5–7, 3–7 cm, 2.5–4.5 cm wide, round, wide-elliptic, or ovate, ± flat, gen dull adaxially, base oblique to ± rounded, spines 11–15 per side, 0.5–1 mm. **FR**: obovoid to elliptic, blue to dark blue. 2*n*=28. Slopes, canyons, conifer forest, oak woodland, chaparral; 150–2600 m. KR, NCoR, CaR, SN, PR, GB, DMoj; to BC, Great Plains. [*B. r.* Lindl.] Apr–Jun

B. darwinii Hook. **ST**: ascending to erect, 1.5–3 m; nodal spines 3-branched. **LF**: evergreen, 1.2–2.5 cm, simple, ovate to obovate, spines 1–6 per side, gen terminal; petiole 0.1–0.3 cm. **FR**: 4–7 mm, ovoid, red to dark purple-black. Coastal conifer forest, disturbed areas; 10–100 m. NCo, CCo; OR; native to s S.Am (Chile, Argentina). Mar–Jun

B. fremontii Torr. (p. 451) **ST**: erect, 0.1–4(5) m; bud scales < 5 mm, gen deciduous. **LF**: 3–6 cm, crowded on short lateral sts; petiole < 1 cm; lflets 3–7(9), terminal 1.5–2.5 cm, 1–1.5 cm wide, gen lance-ovate, wavy, gen folded along midrib, base truncate to wedge-shaped, tip gen acute, margin ± lobed, spines 3–8 per side, 2–3 mm. **INFL**: 4–5.5 cm, open; axis internodes 2–10 mm, 5–10 mm in fr; fls 8–12. **FR**: 6–15 mm diam, ± spheric, glaucous, yellow- to purple-red. **SEED**: 3–4 mm. Rocky slopes, pinyon/juniper woodland, chaparral; 900–1850 m. PR, e&s DMoj; to CO, NM, Mex. Intergrades with *B. haematocarpa*, esp in e DMoj; overlaps morphologically with *B. higginsiae* in PR, Baja CA. Larval exit holes suggest fr inflation may be due to parasitism by insects. Mar–May ★

B. haematocarpa Wooton **ST**: erect, 0.5–4 m; bud scales < 5 mm, gen deciduous. **LF**: 3–6 cm, crowded on short lateral sts; petiole < 1 cm; lflets 3–5, terminal 3–3.5 cm, 0.8–1.2 cm wide, gen narrow-lanceolate, wavy, gen folded along midrib, base truncate to wedge-shaped, tip gen acuminate, margin ± lobed, spines 3–8 per side, 2–3 mm. **INFL**: 2–3.5 cm, open; axis internodes 2–10 mm, 5–10 mm in fr; fls 3–5. **FR**: 8–10 mm diam, ± spheric, red-brown to dark purple. **SEED**: 3–4 mm. Rocky slopes, pinyon/juniper woodland, chaparral; 1000–1850 m. e&s DMoj; to TX, Mex. Perhaps best treated as synonym of *B. fremontii*. Mar–May

B. harrisoniana Kearney & Peebles KOFA MOUNTAIN BARBERRY **ST**: erect, 0.5–1.5 m; bud scales < 5 mm, gen deciduous. **LF**: 3–5 cm, crowded on short lateral sts; petiole 3–5 cm; lflets gen 3, palmate, sessile, terminal 3–5.5 cm, 2.5–4 cm wide, rhombic-oblong or triangular, leathery, base truncate to wedge-shaped (or asymmetric on lateral lflets), tip sharp-acuminate, margin divergently lobed, spine-tipped teeth 3–5, 3–5 mm. **INFL**: 1.5–2.5 cm, open; axis internodes 2–10 mm, 5–10 mm in fr; fls 6–11. **FR**: 5–6 mm diam, ± spheric to ovoid, blue-black. **SEED**: 3–4 mm. Rocky or talus slopes, thorn scrub; 750–850 m. e DSon (Whipple Mtns); to w AZ. Jan–Mar ★

B. higginsiae Munz **ST**: erect, 2–2.5 m; bud scales < 5 mm, gen deciduous. **LF**: 3–5.5 cm, crowded on short lateral sts; petiole < 1 cm; lflets 3–5(7), terminal 1.5–3.5 cm, 1.5–2.5 cm wide, ovate, oblong, or lanceolate, wavy, strongly folded along midrib, base truncate to wedge-shaped, tip gen acute, margin upturned, lobed, spines 2–5 per side, 2–4 mm. **INFL**: 3.5–6 cm, open; axis internodes 2–10 mm, 5–10 mm in fr; fls 5–9. **FR**: 6–8(10) mm diam, ± spheric, ± glaucous, yellow-red, drying ± black. **SEED**: 2.5–3 mm. Rocky slopes, pinyon/juniper woodland, chaparral; 700–1900 m. PR; n Baja CA. Intergradation with *B. fremontii*, *B. haematocarpa* needs study; lf, fr characters highly variable throughout range. Mar–Apr

B. nervosa Pursh **ST**: gen spreading to erect, 0.1–0.6(2) m; bud scales 15–45 mm, lanceolate, thick, persistent among upper lvs. **LF**: ± crowded distally, 12–45 cm; petiole gen 2–7 cm; lflets 7–23, terminal 2.5–8 cm, 1.5–3 cm wide, lanceolate to ovate, flat, base ± oblique to rounded, tip acute, margin serrate, spines 10–24 per side, 1–2 mm. **INFL**: 4–15 cm, ± open; axis internodes 2–8 mm, 4–10 mm in fr; fls > 20. **FR**: 8–12 mm diam, ovoid to obovoid, subglaucous, blue-purple. **SEED**: 4–6 mm. 2*n*=56. Conifer forest; < 2000 m. NW, n SNH (Sierra Co.), SnFrB, n SCoR; to BC, ID. Mar–Jun

B. nevinii A. Gray (p. 451) NEVIN'S BARBERRY **ST**: erect, 1–4 m; bud scales gen deciduous. **LF**: cauline or crowded on short lateral sts, 3.5–7(12) cm; petiole gen 0.5–2 cm; lflets gen 3–5, terminal 2.5–4 cm, 1.2–2 cm wide, narrow-elliptic to lanceolate, flat to ± wavy, base ± obtuse, tip acute to acuminate, margin serrate, spines 8–10 per side, ± 1 mm. **INFL**: 3.5–6.5 cm, open; axis internodes 5–10 mm in fl; fls 3–5. **FR**: 5–8 mm diam, spheric, ± red. **SEED**: 3.5–4 mm. Sandy to gravelly soils, washes, chaparral; < 650 m. SW. Often used in restoration with inadequate documentation, so that planted and natural populations confused; reports of elevations up to ± 1220 m based on plantings. Mar–May ★

B. pinnata Lag. **ST**: upper reclining to erect; bud scales gen deciduous. **LF**: cauline, not crowded, 9–20 cm; petiole gen 1–3 cm; lflets gen 7–11, terminal 3–7 cm, 2–4.5 cm wide, ovate to wide-elliptic, wavy, base ± lobed to truncate, tip acute to obtuse exc tooth, margin entire to gen dentate to serrate, spines 15–23 per side, 0.1–2 mm. **INFL**: 3–7.5 cm, dense; axis internodes 2–4 mm in fl. **FR**: 6–8 mm diam, ovoid to obovoid, glaucous, blue-purple. **SEED**: 3–4 mm. Relationship to *B. aquifolium* needs study.

subsp. *insularis* Munz (p. 451) ISLAND BARBERRY Pl 2–4 m. **ST**: upper reclining to weakly erect. **LF**: dentate, serrate, or subentire; lflet margins gen flat, spines gen < 1 mm. **INFL**: 3–7.5 cm. Closed-cone-pine forest, oak woodland, chaparral; < 350 m. n ChI. Feb–Apr ★

subsp. *pinnata* Pl gen < 2 m. **ST**: upper gen erect. **LF**: entire or gen serrate; lflet margins gen wavy, spines gen > 1 mm. **INFL**: 3.5–7.5 cm. Rocky slopes, conifer forest, oak woodland, chaparral; < 1900 m. NW, CW, WTR, SnGb, PR; to BC, Baja CA. Intergrades with *B. aquifolium* var. *dictyota*. Feb–May

B. vulgaris L. **ST**: ascending to erect, 1.5–3 m; single nodal spines present, 4–8 mm. **LF**: deciduous, 2–5 cm, simple, obovate to obovate-oblong, fine-dentate, spines 8–24 per side, < 1 mm; petiole 0.1–1 cm. **INFL**: 3–6 cm; fls 10–20. **FR**: 10–15 mm, elliptic, red-purple. Disturbed areas; 500–1000 m. NW; eastern US; native to Eur. Principal alternate host of wheat stem rust (*Puccinia graminis* Pers.). Mar–Jun

VANCOUVERIA INSIDE-OUT-FLOWER

Per, rhizomes extensive, scales brown. **ST**: 0. **LF**: basal, long-petioled, 2–3-ternate; lflet blades ovate to ± cordate, lobes 3, shallow, teeth 0 or shallow. **INFL**: raceme or panicle, ± scapose, open, long-peduncled; fls spreading to pendent. **FL**: sepals gen 12–15, 8–9 mm, outer 6–9 << inner 6, bract-like, deciduous, inner petal-like, persistent, in age reflexed; petals 6, < inner sepals, reflexed from base, distally glandular; stamens gen 6, held against ovary, style, anther valves flap-like, pointed tipward; ovules 2–10, style 1, < ovary, persistent, beak-like in fr, stigma cup-like. **FR**: capsule, 2-valved, gen elliptic. **SEED**: with oily body for ant dispersal. 3 spp.: temp w N.Am. (Captain George Vancouver, British explorer, 1757–1798) [Zhang 2007 Syst Bot 32:81–92] Pedicel appears to arise from inside fl, from tip instead of base, yielding an upside-down or "inside-out" fl.

1. Pedicel glabrous; lflet margin thin, membranous; petiole in age straw; lvs deciduous in fr . . . . *V. hexandra*
1′ Pedicel short glandular-hairy, at least at base; lflet margin thick, ± white; petiole in age red-brown; lvs persistent in fr
  2. Infl gen ± raceme, branched below or not; petals yellow, tips reflexed, hood-like; ovary, pedicel short
    glandular-hairy . . . . *V. chrysantha*
  2′ Infl panicle; petals white, tinged lavender or not, tips flat, notched; ovary glabrous, pedicel gen glabrous
    distally . . . . *V. planipetala*

V. chrysantha Greene (p. 451) SISKIYOU INSIDE-OUT-FLOWER **LF**: persistent in fr, 10–36 cm, adaxially glabrous to sparse-short-hairy, abaxially sparse- to dense-hairy; petiole sparse-hairy, in age red-brown. **INFL**: gen ± raceme, branched below or not; upper axis, pedicels short glandular-hairy. **FL**: outer sepals 2–4 mm, inner 6–10 mm; petals 4–6 mm, yellow, tip reflexed, hood-like; filaments glandular-hairy. **FR**: body 9–12 mm, short-glandular-hairy. $2n$=12. Dry sites, chaparral, conifer forest; < 1500 m. n NCo, w KR; sw OR. May–Jun ★

V. hexandra (Hook.) C. Morren & Decne. (p. 451) **LF**: deciduous in fr, 8–27 cm, adaxially glabrous, abaxially sparse-hairy; petiole gen glabrous, in age straw-colored. **INFL**: ± raceme, branched below or not; axis, pedicels glabrous. **FL**: outer sepals 2–4 mm, inner 5–7 mm;

petals 4–6 mm, white, tip strongly reflexed, ± hood-like; filaments red-glandular. **FR**: body 8–10(15) mm, short glandular-hairy. $2n$=12. Conifer forest; < 1900 m. NCo, w KR, n NCoRO; to w WA. May–Jul

V. planipetala Calloni (p. 451) REDWOOD IVY **LF**: persistent in fr, 10–30 cm, adaxially glabrous, abaxially glabrous to sparse-hairy; petiole sparse-hairy, in age glabrous, red-brown. **INFL**: panicle; upper axis short-glandular-hairy: pedicel short-glandular-hairy in lower 1/3. **FL**: outer sepals 2–4 mm, inner 4–5 mm; petals 3–4 mm, white or tinged lavender, tip flat, notched; filaments glabrous. **FR**: body 5–7 mm, glabrous. $2n$=24. Coastal conifer forest; < 1550 m. NW (exc NCoRI), SnFrB, SCoRO (Santa Lucia Range); sw OR. Late Apr–Jul

BETULACEAE BIRCH FAMILY

John O. Sawyer, Jr.

Shrub, tree; monoecious. **ST**: trunk < 35 m; bark smooth to scaly, peeling in thin layers or not, lenticels present or not. **LF**: simple, alternate, petioled, deciduous; stipules deciduous; blade ovate to elliptic, gen serrate, gen ± doubly so. **INFL**: catkin, gen appearing before lvs, often clustered; bracts each subtending 2–3 fls, 3–6 bractlets. **STAMINATE INFL**: pendent, ± elongate. **PISTILLATE INFL**: pendent or erect, developing variously in fr (see key to genera). **STAMINATE FL**: sepals 0–4, minute; petals 0; stamens 1–10; pistil vestigial or 0. **PISTILLATE FL**: sepals 0–4; petals 0; stamens 0; pistil 1, ovary inferior or superior, chambers 2, each 1-ovuled by abortion, stigmas 2. **FR**: achene, nut, winged or not, subtended or enclosed by 1–2 bracts. 6 genera, 155 spp.: gen n hemisphere; some cult. [Furlow 1997 FNANM 3:507–538] Scientific Editor: Thomas J. Rosatti.

1. Lf base oblique-cordate; frs 1–2 per catkin, not winged, each enclosed in a papery involucre of 2 fused
  bracts; pistillate calyx present, ovary inferior . . . . **CORYLUS**
1′ Lf base subcordate, rounded, ± truncate, or tapered; frs many per catkin, winged, each subtended but not
  enclosed by 1 bract; pistillate calyx 0, ovary superior
  2. Pistillate catkin cone-like, with woody bracts remaining attached after fr release . . . . **ALNUS**
  2′ Pistillate catkin not cone-like, with papery, lobed bracts released with but not attached to fr . . . . **BETULA**

ALNUS ALDER

ST: trunk < 35 m; bark smooth, gray to brown; twigs glabrous to fine-hairy, red-gray; lenticels small; winter buds stalked, 0–6-scaled. **LF**: glabrous to fine-hairy; blade 3–15 cm, cordate to elliptic or diamond-shaped. **STAMINATE INFL**: 5–20 cm; bracts each subtending 3 fls, 4 bractlets. **PISTILLATE INFL**: 5–20 mm; bracts each subtending 2 fls, 4 fused bractlets. **STAMINATE FL**: sepals 4; stamens 1–4. **PISTILLATE FL**: sepals 0. **FR**: many, in cone-like catkin, not enclosed by bract, winged, bracts 3 mm, woody, persistent. ± 25 spp.: n hemisphere, S.Am. (Latin: alder) Root nodules contain nitrogen-fixing bacteria; wood used for interior finishing, to smoke fish, meats.

1. Tree, > 9 m, not forming thickets
  2. Lf blade adaxially shiny, base cordate to truncate. . . . . *[A. cordata]*
  2′ Lf blade adaxially dull, base rounded to tapered
    3. Lf margin gen serrate, not tightly rolled under . . . . *A. rhombifolia*
    3′ Lf margin doubly serrate, tightly rolled under . . . . *A. rubra*
1′ Shrub, < 9 m, forming thickets
  4. Lateral buds stalked, scales 0; pedicels < cone-like catkins; lf thick, margin doubly serrate to crenate,
    teeth pointed, blunt, or rounded . . . . *A. incana* subsp. *tenuifolia*
  4′ Lateral buds sessile, scales present; pedicels > cone-like catkins; lf thin to firm, margin serrate, teeth
    pointed . . . . *A. viridis*
    5. Blade firm, adaxially dark green, margins gen densely serrate; coastal, 0–500 m . . . . subsp. *fruticosa*
    5′ Blade thin, adaxially yellow-green, margins coarsely doubly serrate; montane, 1000–2700 m . . . . subsp. *sinuata*

A. incana (L.) Moench subsp. *tenuifolia* (Nutt.) Breitung MOUNTAIN ALDER Shrub. **ST**: trunks < 9 m. **LF**: blade thick, base rounded to subcordate, tip rounded to acute, margin gen ± flat, adaxially dark green, dull, midrib, major veins gen indented, abaxially yellow-green. Wet places; 1200–2400 m. KR, NCoRH, CaRH, SNH; to AK, w Can, WY, NM. 2 other subspp. in AK, Can, Eur. Apr–Jun

A. rhombifolia Nutt. (p. 451) WHITE ALDER Tree. **ST**: trunk < 35 m. **LF**: blade thick, base rounded to tapered, tip rounded to acute, margin gen ± flat, adaxially green, midrib, major veins not indented, abaxially yellow-green. Along permanent streams; 100–2400 m. CA-FP, MP (uncommon), w DSon; to WA, ID. Jan–Apr

A. rubra Bong. (p. 451)   RED ALDER   Tree. **ST**: trunk < 25 m. **LF**: blade thick, base rounded to tapered, tip acute, margin tightly rolled under, adaxially gray-green, midrib, major veins indented, abaxially ± gray-green, rusty-hairy, or with rusty, sessile glands. Wet places, esp after logging; < 1000 m. NCo, w KR, NCoRO, CCo, SnFrB; to s AK, ID. Feb–Mar

A. viridis (Chaix) Lam. & DC.   GREEN ALDER   Shrub. **ST**: trunks < 8 m. **LF**: blade narrow- to broad-ovate, base tapered to cordate, tip acute or short-acuminate to tapered, margin ± flat, adaxially yellow-green to dark green, hairs sparse, esp on veins, moderately to heavily resinous. **INFL**: appearing with or before lvs. [*Betula v.* Chaix] 2 other subspp. in Can, Eur.

   subsp. ***fruticosa*** (Rupr.) Nyman   SIBERIAN ALDER   **ST**: trunks < 3 m. **LF**: blade firm, broad-ovate, base rounded to cordate, tip acute

to short-acuminate, gen densely serrate, adaxially dark green, shiny, abaxially green, hairs sparse, esp on veins, moderately to heavily resinous. Rocky to sandy coasts, streamsides, damp open areas; < 500 m. NCo; to AK, w Can, n Asia. [*A. f.* Rupr.] Spring

   subsp. ***sinuata*** (Regel) Á. Löve & D. Löve   THINLEAF OR SITKA ALDER   **ST**: trunks < 8 m. **LF**: blade thin, narrow- to broad-ovate, base tapered to subcordate, tip acute to tapered, coarsely doubly serrate, adaxially yellow-green, shiny, abaxially green, glabrous or hairs restricted to or denser in major vein axils, lightly to moderately resinous. Along creeks, seeps, meadow margins; 1000–2700 m. KR, NCoRO, NCoRH, w CaRH (Grizzly Peak, Shasta Co.); to AK, w Can. [*A. s.* (Regel) Rydb.] Spring

BETULA   BIRCH

ST: trunk < 30 m; bark smooth or scaly, aromatic, often peeling in thin layers; twigs puberulent, glandular, or both; lenticels prominent; winter buds sessile, 3-scaled. **LF**: glandular-hairy; blade 2–5 cm, wide-elliptic, base ± truncate to tapered. **STAMINATE INFL**: 2–7 cm; bracts each subtending 3 fls, 3 bractlets. **PISTILLATE INFL**: 2–3 cm; bracts each subtending 3 fls, 3 bractlets. **STAMINATE FL**: sepals 4; stamens 2. **PISTILLATE FL**: sepals 0. **FR**: many, in a non-cone-like catkin, not enclosed by bract, winged; bracts lobed, papery, released with but not attached to fr. 50 spp.: circumboreal. (Latin: birch) Important wildlife food; wood used for interior finishing; many spp. cult.

1. Pl < 3 m; lf leathery, tip round, margin crenate . . . .*B. glandulosa*
1′ Pl > 3 m; lf not leathery, tip acute, margin doubly serrate exc at base . . . .*B. occidentalis*

B. glandulosa Michx. (p. 451)   DWARF RESIN BIRCH   Shrub. **ST**: trunks < 2 m; bark brown to gray, not peeling; twigs waxy-gray, with resin glands. **LF**: petiole < 6 mm, hairy; blade 1–2 cm, elliptic to wide-ovate, glands on both surfaces, base tapered. **PISTILLATE INFL**: 1–3 cm; bracts resin-dotted. $2n=28$. Streams, meadow edges; 1300–2300 m. CaRH, Wrn; to AK, e N.Am. Furlow reports *B. pumila* L. from CA, but this author (Sawyer) assigns CA specimens, incl those identified as *B. pumila* var. *glandulifera* Regel, to *B. glandu-*

losa, reserving *B. pumila*, with *B. pumila* var. *glandulifera* as a synonym, for pls outside CA. May–Jun   ★

B. occidentalis Hook. (p. 451)   WATER BIRCH   **ST**: trunks to 10 m; bark red-brown to black, not peeling; twigs with large resin glands, hairy. **LF**: petiole < 15 mm, hairy; blade 2–5 cm, wide-ovate, glands esp adaxially, base ± truncate to tapered. **PISTILLATE INFL**: 3–5 cm; bract fringed with hairs. Streamsides, springs; 600–2900 m. KR, CaRH, SNH, GB, DMtns; scattered in w N.Am. Apr–May

CORYLUS   HAZELNUT

Shrub, small tree. **ST**: trunk < 6 m; bark smooth or scaly, dark brown; twigs glandular-hairy, becoming glabrous, brown; lenticels small; winter buds ciliate. **LF**: hairy; blade 4–10 cm, oblong to ovate, base oblique-cordate. **STAMINATE INFL**: 4–7 cm; bracts each subtending 3 fls, 3 bractlets. **PISTILLATE INFL**: < 1 cm, appearing as terminal bud; bracts each subtending 2 fls, 6 bractlets. **STAMINATE FL**: sepals 0; stamens 4. **PISTILLATE FL**: sepals 4; stigmas showy, red. **FR**: 1–2 per catkin, each enclosed in a papery involucre of 2 fused bracts, not winged. 15 spp.: n hemisphere. (Latin: hazelnut, filbert) [Furlow 1997 Syst Bot 26:283–298] Flexible sts used in basket-making; some cult as food crop.

C. cornuta Marshall subsp. ***californica*** (A. DC.) E. Murray (p. 461)   CALIFORNIA HAZEL   **ST**: trunks < 4 m. **LF**: petiole 5–10 mm; blade ± velvety-hairy, base cordate, tip acute to acuminate. **FR**: 2–3 cm, involucre vase-shaped. Common. Many habitats, esp moist, shady places; < 2100 m. NW, CaR, SN, CCo, SnFrB; to BC. [*C. cornuta* var. *californica* (A. DC.) Sharp] Jan–Mar

BIGNONIACEAE   TRUMPET-CREEPER FAMILY

Lúcia G. Lohmann

Shrub, tree [woody vine]; nectaries often on st, lf, calyx, ovary. **LF**: gen 1–3-pinnate or -palmate (or simple), gen opposite or whorled. **FL**: bisexual, showy; calyx gen 5-lobed, 2-lipped or not; corolla funnel- or bell-shaped, 5-lobed, gen 2-lipped; stamens epipetalous, gen 4, paired, a 5th vestigial (a staminode) [or 0]; ovary superior, chambers gen 2, placentas 4, axile (or chamber 1, placentas 2–4, parietal), style long, stigma 2-lobed. **FR**: gen capsule, long, cylindric to ± round, sometimes flat, 2-valved. **SEED**: many, flat, gen winged. 81 genera, 827 spp.: gen trop, esp S.Am; many orn (*Campsis*, trumpet creeper; *Catalpa*; *Jacaranda*). [Alcantara & Lohmann 2010 Amer J Bot 97:782–795] Scientific Editor: Thomas J. Rosatti.

1. Lvs 1- or 2-pinnate; calyx not 2-lipped; fr oblanceolate or ± round
  2. Woody vine; lvs 1-pinnate; corolla red-orange, 5–8 cm; staminode incl, glabrous; anther sacs 2;
     fr oblanceolate . . . . [*Campsis radicans*]
  2′ Tree; lvs 2-pinnate; corolla purple-blue, 3–4 cm; staminode exserted, glandular-hairy;
     anther sac 1; fr ± round. . . . . [*Jacaranda mimosifolia*]
1′ Lvs simple; calyx 2-lipped; fr linear
  3. Tree; lvs cordate; corolla white with yellow and purple markings or yellow stripes or brown dots . . . . **CATALPA**
  3′ Shrub; lvs linear; corolla white or light pink to lavender with yellow and purple markings . . . . **CHILOPSIS**

CATALPA SOUTHERN CATALPA

Tree, deciduous. **LF**: simple, opposite to whorled, cordate. **INFL**: panicle, terminal. **FL**: calyx inflated, 2-lipped, purple-green; corolla fragrant, lower petal notched; stamens incl, anther sacs 2; staminode incl, glabrous. **FR**: linear, not flat. **SEED**: 6–12 mm, oblong; glabrous. $2n=40$. 10 spp.: temp N.Am, e Asia. (Muscogee: winged head, for corolla)

1. Lf blade 10–26 cm, tips acute; corolla < 3.5 cm, white with yellow and purple markings in throat; calyx < 15 mm . . . . ***C. bignonioides***
1′ Lf blade 15–30 cm, tips acuminate; corolla 4–6 cm, white with yellow stripes or brown dots in throat; calyx 15–25 mm . . . . ***C. speciosa***

C. bignonioides Walter Pl ≤ 15 m. **FR**: 35 cm, < 1 cm diam. Streambanks, roadsides, railroad embankments; < 1500 m. n SNF, SnJV, SCo, expected elsewhere; OR to e N.Am. May–Jun

C. speciosa (Warder) Engelm. Pl 15–20 m. **FR**: 30 cm, > 1 cm diam. Disturbed places; < 1500 m. n SNF, ScV, PR; IN to n AR. Jun–Jul

CHILOPSIS DESERT-WILLOW

1 sp. (Greek: resembling lips, from fl shape)

C. linearis (Cav.) Sweet subsp. ***arcuata*** (Fosberg) Henrickson (p. 461) Shrub, 1.5–7 m, willow-like; deciduous. **LF**: simple, gen alternate (often some opposite to whorled on same pl); blade 10–26 cm, linear, curved. **INFL**: panicle or raceme, terminal. **FL**: calyx 8–14 mm, inflated, 2-lipped, gen soft-hairy, ± purple; corolla 2–5 cm, fragrant, white or light pink to lavender with yellow and purple markings in throat, lobes spreading, margins jagged, wavy; stamens incl, anther sacs 2; staminode incl, glabrous. **FR**: < 35 cm, linear, not flat. **SEED**: 6–12 mm, oblong; both ends long-hairy. $2n=40$. Common. Sandy washes; < 2100 m. SCo, e TR, e PR, D; to UT, NM, n Mex. May–Sep

BORAGINACEAE BORAGE or WATERLEAF FAMILY

Ronald B. Kelley, Robert Patterson, Richard R. Halse & Timothy C. Messick, family description, key to genera; treatment of genera by Ronald B. Kelley, except as noted

Ann to shrub, or non-green root parasite, gen bristly or sharp-hairy. **ST**: prostrate to erect. **LF**: cauline, often with basal rosette, gen simple, gen alternate. **INFL**: cymes, gen elongate, panicle-, raceme-, or spike-like, gen coiled in fl (often described as scorpioid), gen uncoiled in fr, or heads, spikes, or panicles, or fls 1–2 per axil. **FL**: bisexual, gen radial; sepals (4)5(10), fused at least at base, or free; corolla gen (4)5(10)-lobed, salverform, funnel-shaped, rotate, or bell-shaped, appendages 0 or 5 at top of tube, alternate stamens; stamens epipetalous; ovary superior, entire to 4-lobed, style 1(2), entire or 2-lobed or -branched. **FR**: nutlets 1–4, free (fused), smooth to roughened, prickly or bristly or not, or valvate or circumscissile capsule. ± 120 genera, ± 2300 spp.: trop, temp, esp w N.Am, Medit; some cult (*Borago, Heliotropium, Echium, Myosotis, Nemophila, Phacelia, Symphytum*). Many genera may be TOXIC from pyrrolizidine alkaloids or accumulated nitrates. [Olmsted et al. 2000 Molec Phylogen Evol 16:96–112] Recently treated to incl Hydrophyllaceae, Lennoaceae. Scientific Editors: Ronald B. Kelley, Robert Patterson, Bruce G. Baldwin, Thomas J. Rosatti.

1. Pls non-green, parasitic . . . . **PHOLISMA**
1′ Pls green, non-parasitic
2. Ovary deep-lobed, style base hidden within lobes
3. Nutlets spreading
4. Corolla blue to red-purple; nutlet ± spheric or disk-shaped, prickles barbed; bien, per . . . . **CYNOGLOSSUM**
4′ Corolla white; nutlet gen compressed, prickles straight or hooked at tip, not barbed; ann
5. Calyx lobes in fr unequal, upper 2 >> others, partly fused, arched over 1 nutlet, with 5–10 stout spines each with hooked bristles; nutlets 2 . . . . **HARPAGONELLA**
5′ Calyx lobes in fr ± equal or, if unequal, upper 2 > others, free, not arched over 1 nutlet, without spines but with hooked or straight prickles; nutlets gen 4 . . . . **PECTOCARYA**
3′ Nutlets ± erect
6. Corolla rotate; anthers 5–8 mm, closely appressed to style, separating in age . . . . **BORAGO**
6′ Corolla salverform, funnel-shaped, or ± tubular (exc *Symphytum*); anthers < 5 mm, gen not closely appressed to style
7. Fls 1–2 per lf axil; calyx lobes with 1–2 teeth in fr . . . . **ASPERUGO**
7′ Fls 1 per upper lf axil or in panicle-, raceme-, or spike-like cymes; calyx lobes without teeth in fr
8. Nutlet scar ± flat to strong-convex, with thick rim
9. Corolla appendages ovate to oblong, gen above anthers; corolla throat ± expanded above tube or tube and throat not differentiated . . . . **ANCHUSA**
9′ Corolla appendages lance-linear to lanceolate, at same level as anthers; corolla throat expanded above tube . . . . **SYMPHYTUM**
8′ Nutlet scar flat, curved, or grooved, without thick rim
10. Receptacle ± flat; nutlet scar gen basal (exc *Mertensia*)
11. Shrub, 1–3+ m . . . . **²ECHIUM**

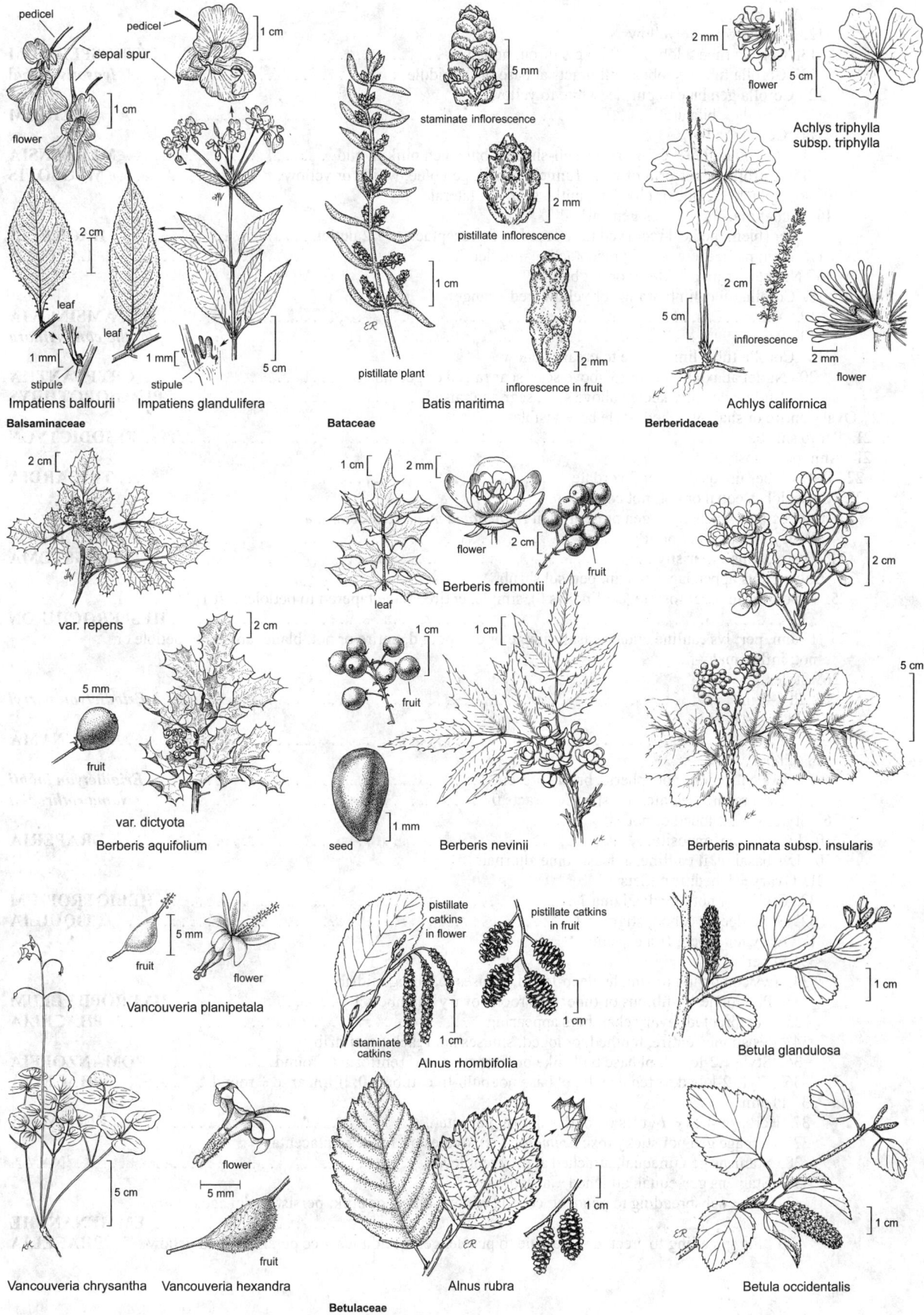

pedicel

pedicel

sepal spur

1 cm

1 cm

flower

2 cm

leaf

leaf

1 mm

stipule

1 mm

stipule

5 cm

Impatiens balfourii

Impatiens glandulifera

Balsaminaceae

5 mm

staminate inflorescence

2 mm

pistillate inflorescence

1 cm

2 mm

inflorescence in fruit

pistillate plant

Batis maritima

Bataceae

2 mm

flower

5 cm

Achlys triphylla subsp. triphylla

2 cm

5 cm

inflorescence

2 mm

flower

Achlys californica

Berberidaceae

2 cm

var. repens

2 cm

5 mm

fruit

var. dictyota

Berberis aquifolium

1 cm

2 mm

flower

2 cm

fruit

leaf

Berberis fremontii

1 cm

1 cm

fruit

1 mm

seed

Berberis nevinii

2 cm

5 cm

Berberis pinnata subsp. insularis

5 mm

fruit

flower

Vancouveria planipetala

pistillate catkins in flower

pistillate catkins in fruit

1 cm

staminate catkins

1 cm

Alnus rhombifolia

1 cm

Betula glandulosa

5 cm

Vancouveria chrysantha

5 mm

flower

5 mm

fruit

Vancouveria hexandra

1 cm

1 cm

Alnus rubra

1 cm

Betula occidentalis

Betulaceae

11′ Ann, bien, per
  12. Corolla white to yellow
    13. Corolla tube > lobes; infl bracts throughout . . . . **LITHOSPERMUM**
    13′ Corolla tube << lobes; infl bracts at base or to middle . . . . [*Myosotis verna*]
  12′ Corolla gen blue to purple (white to yellow)
    14. Corolla ± bilateral . . . . ^{2}**ECHIUM**
    14′ Corolla radial
      15. Corolla often ± cylindric or bell-shaped, blue, gen pink in bud . . . . **MERTENSIA**
      15′ Corolla salverform or wide-funnel-shaped, gen blue, white, or yellow . . . . **MYOSOTIS**
10′ Receptacle ± conic or elongate; nutlet scar gen lateral
  16. Nutlet margin prickles gen barbed
    17. Per (bien); pedicel recurved to reflexed in fr; receptacle ± 1/2 nutlet . . . . **HACKELIA**
    17′ Ann; pedicel erect in fr; receptacle ± = nutlet . . . . **LAPPULA**
  16′ Nutlet margin prickles 0 or not barbed
    18. Corolla tube, limb orange or yellow (red-orange)
      19. Ann . . . . **AMSINCKIA**
      19′ Per . . . . *Cryptantha confertiflora*
    18′ Corolla tube, limb white to cream-yellow
      20. Nutlet adaxially grooved above scar, scar raised or gen not . . . . **CRYPTANTHA**
      20′ Nutlet adaxially keeled above scar; scar gen raised . . . . **PLAGIOBOTHRYS**
2′ Ovary entire or shallow-lobed, style base visible
21. Per to shrub . . . . **ERIODICTYON**
21′ Ann, per, subshrub
  22. Calyx lobes unequal, outer 3 cordate . . . . **TRICARDIA**
  22′ Calyx lobes equal or not, not cordate
    23. Calyx sinus appendages gen present (0 in *Pholistoma membranaceum*); ann
      24. Ovary, fr hairy, not bristly . . . . **NEMOPHILA**
      24′ Ovary, fr hairy, bristly . . . . **PHOLISTOMA**
    23′ Calyx sinus appendages 0; ann, per, subshrub
    25. Pl per; lvs in basal rosette (cauline lvs 0), simple, entire, blade tapered to petiole; infl 1-fld
    . . . . **HESPEROCHIRON**
    25′ Pl ann, per; lvs cauline and/or basal, simple or compound, entire or not, blade tapered to petiole or
      not; infl 1–many-fld
    26. Styles 2
      27. Pl 1–3 m . . . . *Eriodictyon parryi*
      27′ Pl gen < 0.5 m
        28. Pl ann . . . . ^{2}**NAMA**
        28′ Pl per
          29. Lvs entire; infl not spheric, bracts present . . . . *Eriodictyon lobbii*
          29′ Lvs crenate-dentate; infl spheric, bracts 0 . . . . *Nama rothrockii*
    26′ Styles 0–1, 2-lobed or not
      30. Lvs cauline, opposite . . . . **DRAPERIA**
      30′ Lvs basal or, if cauline, at least some alternate
        31. Ovary ± lobed; fr nutlets
          32. Style 0 or not lobed; stigma 1 . . . . **HELIOTROPIUM**
          32′ Style deep-2-lobed; stigmas 2 . . . . **TIQUILIA**
        31′ Ovary not lobed; fr a capsule
          33. Pl per
            34. Lvs compound to simple, deep-lobed, sinuses reaching midrib
              35. Pl with fleshy-fibrous or tuber-like roots; ovary chambers 1 . . . . **HYDROPHYLLUM**
              35′ Pl taprooted; ovary chambers appearing 2 . . . . ^{3}**PHACELIA**
            34′ Lvs simple, entire, toothed, or lobed, sinuses not reaching midrib
              36. Style ± 2-lobed; pl base bulb-like or from tubers; lf reniform to round . . . . **ROMANZOFFIA**
              36′ Style 2-lobed, often deeply; pl base not bulb-like, tubers 0; lf linear to ± round . . . . ^{3}**PHACELIA**
          33′ Pl ann
            37. Herbage sticky; ovules on both sides of placenta . . . . **EUCRYPTA**
            37′ Herbage gen not sticky (exc *Emmenanthe*); ovules on 1 side of placenta
              38. Stamens gen unequal, attached at different levels . . . . ^{2}**NAMA**
              38′ Stamens gen equal, attached ± at same level
                39. Fls ± subspreading to pendent; corolla white, yellow, or pink, persistent, in age papery
                . . . . **EMMENANTHE**
                39′ Fls spreading to erect; corolla blue to purple, gen deciduous, occ persistent and yellow . . . ^{3}**PHACELIA**

AMSINCKIA FIDDLENECK

Ronald B. Kelley & Fred R. Ganders

Ann; hairs gen bristly, often bulbous-based. **ST:** gen erect, 2–12 dm, gen green. **LF:** basal and cauline, alternate, sessile or lower short-petioled, gen linear to narrow-lanceolate or -oblong, gen not succulent, ± entire. **INFL:** spike-like cymes, gen ± terminal, tip coiled. **FL:** gen radial; calyx lobes 5 or 2–4 (see key); corolla tube gen not constricted, gen orange or yellow (red-orange), appendages gen 0, throat gen open, glabrous, limb on large-fld taxa gen with 5 dark spots. **FR:** nutlets erect, ± triangular, adaxially gen with exposed elliptic attachment scar, gen with rounded or sharp tubercles. 14 spp.: w N.Am, sw S.Am, widely alien elsewhere. (W. Amsinck, patron of Hamburg Botanic Garden, 1752–1831) Self-compatible; often heterostylous; large-fld taxa gen cross-pollinated, small-fld self-pollinated.

1. Calyx lobes unequal in width, 2–4 from fusion below middle, notched at tip; corolla tube gen 20-veined near base; nutlet gen smooth, cobblestone-like, or round-tubercled (exc *Amsinckia spectabilis* var. *spectabilis*)
   2. St glabrous to sparse-coarse-hairy, glaucous, ± white to pink below; lf glaucous; nutlet smooth, adaxially with longitudinal slit-like groove, attachment scar hidden
      3. Corolla 12–22 mm, limb 8–14 mm diam, gen orange; gen heterostylous; nutlet 3–4 mm, ovate, groove forked at base . . . . *A. furcata*
      3′ Corolla 8–12 mm, limb 2–6(8) mm diam, yellow (to orange-yellow); homostylous; nutlet 4–6 mm, lanceolate, groove unforked at base . . . . *A. vernicosa*
   2′ St hairy, not glaucous, ± green below; lf not glaucous; nutlet smooth, cobblestone-like, or tubercled, adaxially without slit-like groove, attachment scar exposed
      4. Nutlet 1.5–2 mm, ± black; lf fine-toothed; st gen decumbent; coastal dunes or occ sandy bluffs . . . . ²*A. spectabilis* var. *spectabilis*
      4′ Nutlet 2.5–5 mm, gray to brown; lf entire; st erect to ascending; inland
         5. Corolla 14–22 mm, limb 10–16 mm diam; heterostylous
            6. Nutlet gray, dull, cobblestone-like; corolla yellow-orange; SCoR, w WTR . . . . *A. douglasiana*
            6′ Nutlet ± brown, smooth, shiny; corolla red-orange; nw SnJV . . . . *A. grandiflora*
         5′ Corolla 8–16 mm, limb 2–10 mm diam; homostylous . . . . *A. tessellata*
            7. Corolla 12–16 mm, limb 6–10 mm diam; anthers not appressed to, gen below stigma . . . . var. *gloriosa*
            7′ Corolla 8–12 mm, limb 2–6 mm diam; anthers appressed to stigma . . . . var. *tessellata*
1′ Calyx lobes ± equal in width, 5, not fused above base (exc *Amsinckia spectabilis* var. *spectabilis*); corolla tube 10-veined near base; nutlet gen sharp-tubercled, often ridged
   8. Nutlet 1–2 mm; lf ± succulent, fine-toothed . . . . *A. spectabilis*
      9. St erect; calyx lobes not fused above base; nutlet 1–1.5 mm; heterostylous . . . . var. *microcarpa*
      9′ St gen decumbent; calyx lobes gen 2–3, ± 1/2-fused; nutlet 1.5–2 mm; heterostylous or homostylous . . . . ²var. *spectabilis*
   8′ Nutlet 2–4 mm; lf not succulent, gen entire
      10. Corolla bilateral, tube bent, limb gen with 2 dark spots . . . . *A. lunaris*
      10′ Corolla radial, tube straight, limb with 0 or 5 dark spots
         11. Corolla throat ± closed by 5 hairy appendages, tube constricted in lower 1/2; stamens attached below middle of corolla tube, style incl . . . . *A. lycopsoides*
         11′ Corolla throat open, glabrous, appendages 0, tube not constricted; stamens gen attached above middle of corolla tube, if below then style exserted
            12. Corolla 10–20 mm, limb 8–14 mm diam; style gen ± exserted; anthers not appressed to, gen below stigma (if stigma at bottom 1/3 of tube, then anthers above stigma, at throat) . . . . *A. eastwoodiae*
            12′ Corolla 4–11 mm, limb 1–10 mm diam; style not exserted; anthers gen appressed to stigma
               13. St erect, ± gray, appressed to reflexed hairs below present; lf ± gray, hairs appressed to ascending; corolla ± yellow . . . . *A. retrorsa*
               13′ St decumbent to erect, green to brown, appressed to reflexed hairs below ± 0; lf green, hairs ± not appressed to ascending; corolla ± orange
                  14. St ascending to erect; corolla 7–11 mm, limb 4–10 mm diam, ± orange, tube exserted . . . . *A. intermedia*
                  14′ St decumbent to ± ascending; corolla 4–7 mm, limb 1–3 mm diam, yellow to orange-yellow, tube ± exserted . . . . *A. menziesii*

A. douglasiana A. DC. (p. 461) DOUGLAS' FIDDLENECK **ST:** slender, branched in ± upper 1/2. **LF:** occ ± narrow-spoon-shaped. **FL:** calyx lobes unequal in width, 2–4 from fusion below middle, notched at tip, margins coarse-dark-brown-hairy; corolla 14–22 mm, tube 20-veined near base, limb 10–16 mm diam, yellow-orange with 5 dark spots. **FR:** 3–5 mm, gray, dull, cobblestone-like. 2n=12. Unstable shaly sedimentary slopes; (100)150–1600 m. SCoR, w WTR. Heterostylous. Mar–Jun ★

A. eastwoodiae J.F. Macbr. (p. 461) EASTWOOD'S FIDDLENECK **FL:** calyx lobes 5, ± equal in width, not fused above base; corolla 10–20 mm, tube 10-veined near base, limb 8–14 mm diam, orange; stamens gen attached near tube top, anthers not appressed to, gen below stigma (if stigma at bottom 1/3 of tube, then anthers above

stigma, at throat); style gen ± exserted from throat. **FR:** 2.5–4 mm, ± sharp-tubercled, occ ridged. 2n=24. Open valleys, hills; 10–1500 m. NCo, NCoRI, SNF, Teh, GV, SCoR, SW. Gen homostylous. Mar–May

A. furcata Suksd. (p. 461) FORKED FIDDLENECK **ST:** glabrous to sparse-coarse-hairy, glaucous, ± white to pink below. **LF:** glaucous. **FL:** calyx lobes unequal in width, 2–4 from fusion below middle, notched at tip; corolla 12–22 mm, tube 20-veined near base, limb 8–14 mm diam, gen orange, with 5 dark spots. **FR:** 3–4 mm, ovate, ± gray, smooth, shiny, adaxially with longitudinal slit-like groove forked at base, attachment scar hidden. 2n=14. Semi-barren, loose, shaly slopes; 50–1000 m. w SnJV, SCoRI (Kings Co. n). [*A. vernicosa* var. *f.* (Suksd.) Jeps.] Gen heterostylous, homostylous at s end of range, Kings Co. Mar–May ★

A. grandiflora (A. Gray) Greene (p. 461) LARGE-FLOWERED FIDDLENECK **LF:** linear to narrow-ovate. **FL:** calyx lobes unequal in width, 2–4 from fusion below middle, notched at tip, hairs brown; corolla 14–20 mm, tube 20-veined near base, limb 10–15 mm diam, red-orange. **FR:** 3–4 mm, ± brown, smooth, shiny, abaxial longitudinal ridges 3. $2n=12$. Grassy slopes; < 300 m. nw SnJV (w San Joaquin Co.; presumed extinct in Contra Costa Co.). Heterostylous. Mar–May ★

A. intermedia Fisch. & C.A. Mey. COMMON FIDDLENECK **ST:** ascending to erect, green to brown, spreading-bristly, branches 0–many. **LF:** spreading, green, coarse-hairy to bristly. **FL:** calyx lobes 5, ± equal, not fused above base; corolla 7–11 mm, tube exserted, tube 10-veined near base, limb 4–10 mm diam, ± orange, gen with 5 dark spots; stamens attached above middle of tube, anthers gen appressed to stigma; style to throat. **FR:** 2–3.5 mm, ± sharp-tubercled, occ ridged. $2n=30,34,38$. Open, gen disturbed areas; < 1750(2350) m. CA (uncommon D); to BC, w MT, UT, w TX, Baja CA. [*A. menziesii* (Lehm.) A. Nelson & J.F. Macbr. var. *i.* (Fisch. & C.A. Mey.) Ganders, ined.] Homostylous. Perhaps better treated as part of *A. menziesii*. Hybridizes with *A. lycopsoides*, *A. retrorsa*; see note under *A. eastwoodiae*. Mar–Jun

A. lunaris J.F. Macbr. (p. 461) BENT-FLOWERED FIDDLENECK **ST:** erect, slender. **FL:** bilateral; calyx lobes 5, ± equal in width, not fused above base; corolla 7–10 mm, tube bent, 10-veined near base, limb 4–7 mm diam, orange, gen with 2 dark spots. **FR:** 2.5–4 mm, ± sharp-tubercled, occ ridged. $2n=8$. Gravelly slopes, grassland, openings in woodland, often serpentine; (5)50–800 m. NCoR, sw ScV, CCo (Marin, Santa Cruz cos.), SnFrB. Heterostylous, if homostylous then corolla smaller, anthers in upper, lower groups. Mar–Jun ★

A. lycopsoides Lehm. (p. 461) BUGLOSS-FLOWERED FIDDLENECK **FL:** calyx lobes 5, ± equal in width, not fused above base; corolla 7–11 mm, tube 10-veined near base, constricted in lower 1/2, throat ± closed by 5 well-developed hairy appendages, limb 4–10 mm diam, ± yellow-orange, often with 5 dark spots; stamens attached below constriction in lower 1/3 of tube; style short, incl in tube. **FR:** 2.5–3 mm, ± sharp-tubercled, gen not ridged. $2n=30$. Common. Open, grassland, foothill woodland, gen disturbed areas; 5–850(1770) m. NW, CaRF, SNF, GV, CW, SCo, ChI, PR; to BC, ID. Homostylous. Hybridizes with *A. menziesii* var. *intermedia*. Mar–Jun

A. menziesii (Lehm.) A. Nelson & J.F. Macbr. (p. 461) COMMON FIDDLENECK, SMALL-FLOWERED FIDDLENECK **ST:** decumbent to ± ascending, green to brown, spreading-bristly, branches ± many, from base. **LF:** spreading, green, coarse-hairy to bristly. **FL:** calyx lobes 5, ± equal, not fused above base; corolla 4–7 mm, tube ± exserted, tube 10-veined near base, limb 1–3 mm diam, yellow to orange-yellow, gen spotted; stamens attached above middle of tube, anthers gen appressed to stigma; style to throat. **FR:** 2–3.5 mm, ± sharp-tubercled, occ ridged. $2n=34$. Shade-tolerant, open, disturbed areas at forest/woodland edges; < 1700 m. CA-FP, MP; to AK, ID; naturalized in c&e US; also in S.Am. [*A. micrantha* Suksd.] Homostylous. May–Jul

A. retrorsa Suksd. RIGID FIDDLENECK Herbage ± gray, bristly, softer appressed- to reflexed-hairy below. **ST:** erect, branches 0 (few above). **FL:** calyx lobes 5, ± equal in width, not fused above base; corolla 5–8 mm, limb 1.5–3 mm diam, ± yellow, tube not exserted, 10-veined near base, limb gen unmarked; anthers gen appressed to stigma; style incl. **FR:** 2–3 mm, ± dense-sharp-tubercled, ridged or not. $2n=16,26$. Shade-intolerant, disturbed, dry sites, roadsides; < 1600(2250) m. CA (rare D); to se BC, ID, UT. [*A. helleri* Brand; *A. parviflora* A. Heller] Homostylous. Feb–May

A. spectabilis Fisch. & C.A. Mey. **LF:** ± succulent, fine-toothed. **FL:** calyx lobes 5, ± equal in width, not fused above base, or unequal in width, 2–3 ± 1/2 fused, with dark cilia; corolla 7–19 mm, limb 5–14 mm diam, yellow, tube 10-veined near base. **FR:** 1–2 mm, ± black, sharp-tubercled or obscure, occ ridged. Vars. intergrade in s CCo (near Morro Bay).

var. ***microcarpa*** (Greene) Jeps. & Hoover (p. 461) MESA FIDDLENECK **ST:** erect. **FL:** calyx lobes ± equal in width, not fused above base; corolla 12–19 mm, limb 8–14 mm diam. **FR:** 1–1.5 mm, not ridged; tubercles sharp. $2n=10,12$. Localized. Sandy mesas, stabilized dunes; < 180 m. s CCo. Heterostylous. Apr–Jun

var. ***spectabilis*** (p. 461) SEASIDE FIDDLENECK **ST:** gen decumbent. **FL:** calyx lobes gen 2–3, unequal in width, ± 1/2-fused; corolla 7–15 mm, limb 5–12 mm diam. **FR:** 1.5–2 mm, occ ridged; tubercles sharp or obscure. $2n=10$. Coastal dunes or sandy bluffs; < 170(300) m. NCo, CCo, SCo, ChI; to BC, Baja CA. [*A. s.* var. *sancti-nicolai* (Eastw.) I.M. Johnst.] Heterostylous or homostylous. Apr–Aug

A. tessellata A. Gray **FL:** calyx lobes unequal in width, reduced to 2–4 from fusion below middle, notched at tip; corolla 8–16 mm, yellow or orange, tube 20-veined near base, limb 2–10 mm diam. **FR:** 2.5–4 mm, gray, dull, cobblestone-like or round-tubercled, ridged or not. $2n=24$. Homostylous.

var. ***gloriosa*** (Suksd.) Hoover (p. 461) CARRIZO FIDDLENECK **FL:** calyx dense-brown-hairy; corolla 12–16 mm, limb 6–10 mm diam, orange; anthers not appressed to, gen below stigma. Sandy or shaly soils; 50–1950 m. NCoRI, Teh, sw SnJV, SnFrB, SCoR, WTR, w DMoj. Often forms expansive dense populations in the Carrizo Plain area, s SCoRI and sw SnJV. Mar–Jun

var. ***tessellata*** (p. 461) DESERT FIDDLENECK **FL:** calyx gen ± white hairy; corolla 8–12 mm, limb 2–6 mm diam, yellow; anthers appressed to stigma. Rocky or sandy soils; 50–2280 m. s SNF, Teh, SnJV, SnFrB, SCoR, TR (n slope), GB, D; to WA, ID, UT, AZ, Baja CA, S.Am. Feb–Jun

A. vernicosa Hook. & Arn. (p. 461) WAXY FIDDLENECK **ST:** (exc infl) ± glabrous, glaucous, ± white to pink below. **LF:** glaucous. **FL:** calyx lobes unequal in width and reduced to 2–4 from fusion below middle, notched at tip; corolla 8–12 mm, tube 20-veined near base, limb 2–6(8) mm diam, yellow (to orange-yellow). **FR:** 4–6 mm, lanceolate, gray, smooth, shiny, adaxially with a longitudinal slit-like groove, unforked at base, attachment scar hidden. $2n=14$. Uncommon. Gen shaly slopes; 50–1400 m. s SNF, w SnJV, SnFrB, SCoR, w WTR, DMoj. Homostylous. Mar–May

ANCHUSA ALKANET

[Ann] per (may fl 1st yr); hairs bristly, bases bulbous or not. **ST:** ± erect. **LF:** basal and cauline, petioled to sessile, clasping, lance-linear [oblong to oblanceolate], ± entire. **INFL:** axillary or terminal, gen spike-like cymes; tip coiled. **FL:** corolla funnel-shaped to salverform, pale blue to violet, appendages 5, ovate to oblong, ± puberulent; stamens incl to exserted. **FR:** nutlets 1–4, erect, ± ovoid, irregularly angled or wrinkled, exposed attachment scar basal or oblique, scar surrounded by thick rim. ± 35 spp.: Eurasia, Afr. Orn, cult for drugs, dyes. (Greek: ancient name for alkanet) *A. arvensis* (L.) M. Bieb. occ in orchards.

1. Calyx lobes >> tube, linear; nutlet straight, 5–10 mm .... [*A. azurea*]
1′ Calyx lobes 0.5 × tube, lanceolate; nutlet bent, 2–3 mm .... ***A. officinalis***

A. officinalis L. COMMON ALKANET **ST:** 4–7 dm. **LF:** upper 5–10 mm wide. **FL:** calyx 5–7 mm, 8–12 mm in fr, lobes lanceolate; corolla radial, salverform, tube 5–10 mm, straight, limb 6–12 mm diam, violet to blue-violet, (dark blue), appendages ± white; stamens on upper tube, ± exserted. **FR:** nutlet base 2–3.5 mm wide, tip incurved, scar basal. $2n=16$. Roadsides; 70–150 m. CaRF, ScV; widespread US; native to Eur. May–Jul

ASPERUGO MADWORT

1 sp. (Latin: rough, from hairs)

A. procumbens L. Ann, bristly, stout hairs curved to reflexed, base bulbous. **ST**: prostrate to weakly decumbent, 3–8 dm, angled, branches many. **LF**: ± cauline, opposite at base, alternate above, petioled to sessile above; blades 2–7 cm, oblanceolate to elliptic, scabrous. **INFL**: fls 1–2 per lf axil; pedicels reflexed in fr. **FL**: calyx 5-lobed, 2–3 mm, in fr expanded, compressed, 5–8 mm, 10–15 mm wide, papery, with 1–2 teeth per lobe, enclosing nutlets; corolla 2–3 mm, 2 mm diam, ± funnel-shaped, blue-violet; stamens incl, anthers free, sessile. **FR**: nutlets ± erect, 2–4, 2–3 mm, compressed, ovate, dark, shiny, minute-tubercled, scar lateral. 2*n*=48. Disturbed areas; 1000–1500 m. MP; widespread US; native to c&e Eur. Most collections historical. Apr–Jul

BORAGO BORAGE

Ann [per], bristly to rough-hairy. **ST**: ascending to erect. **LF**: cauline, ± petioled, ovate to oblanceolate, entire. **INFL**: cymes, terminal, 2–3-fld; pedicels ± spreading to pendent in fr. **FL**: calyx deep-5-lobed; corolla rotate to bell-shaped, lobes spreading, throat appendages erect, glabrous; stamens exserted, filament base dilated, anthers adherent around style, separating in age. **FR**: nutlets ± erect, 4, stout, obovoid, irregularly tubercled, exposed stipe-like basal attachment scar, rim of attachment scar thickened. 4 spp.: s Eur, n Afr. (Latin: ancient name)

B. officinalis L. (p. 461) Gen branching above; taprooted. **ST**: 2–7 dm. **LF**: lower blades 8–20 cm, 3–8 cm wide. **FL**: calyx lobes 8–10 mm, 10–15 mm in fr, linear; corolla rotate, bright blue, limb 18–20 mm diam; anthers 5–8 mm, dark brown. **FR**: nutlets 5–7 mm, ± oblong. 2*n*=16. Open, disturbed sites; < 550 m. CCo, SnFrB, SCo; widespread US; native to s Eur. Orn, cult for bees (nectar source), potherb. Jun–Aug

CRYPTANTHA

Ronald B. Kelley, Michael G. Simpson & Kristen E. Hasenstab-Lehman

Ann to per, gen erect. **ST**: branches 0 or gen ascending to erect, hairy. **LF**: gen sessile; basal whorled, cauline gen alternate, reduced above (opposite below); gen strigose, rough-hairy, or bristly, largest bristles gen bulbous-based. **INFL**: gen terminal, raceme- or gen spike-like cymes, in groups of 1–5 (> 5), gen coiled in bud, gen elongated in fr; bracts gen 0. **FL**: gen unscented, persistent or not; sepals fused at base; corolla tube gen 1–13 mm, limb 0.5–12 mm diam, gen white, appendages gen 5; anthers incl; ovary gen 4-lobed. **FR**: pedicel 0 or < 12 mm in fr; nutlets 1–4, gen gray to brown, smooth to granular, tubercled, or papillate, with abaxial, longitudinal ridge to not; margin rounded, sharp-edged, or a ± flat rim or wing; adaxially grooved above attachment scar, scar gen lateral, narrow, open to closed, raised or gen not, edges inrolled to sharp-angled, gen forked or flared open at base; central fr axis ("axis") not reaching to extending beyond fr. ± 200 spp.: w N.Am, w S.Am, ne Asia (1 sp.). (Greek: hidden fls, from cleistogamous fls of some S.Am spp.) Gen homostylous. The tissue between ovary lobes, interpreted as a receptacle and/or style (style sometimes 0, then stigma attached to top of receptacle), extends to various degrees in fr, forming what is often called the gynobase (here "fr axis"), to which the nutlets are laterally attached at maturity, leaving an attachment scar. Ann spp. without yellow corolla appendages gen self-pollinating; per spp. gen homostylous in CA. Some spp., e.g., *C. angustifolia*, *C. ambigua*, *C. barbigera*, *C. mariposae*, hybridize with co-occurring spp. Observation of nutlets, hairs best at 10+× gen critical for identification. Corolla limb diam gen < at end of fl period, esp for ann spp. *C. sobolifera* Payson does not occur in CA.

1. Bracts present; gen ann, gen wider than tall, often rounded to cushion-like; root gen red-purple, staining
   2. Per from a woody, branched caudex; root not red-purple . . . . *C. cinerea* var. *abortiva*
   2′ Ann; root gen red-purple, staining
      3. Calyx circumscissile below middle in fr, upper tube, lobes falling as 1 unit, lower tube scarious, persistent, cup-like in fr
         4. Corolla limb 3.5–6 mm diam, appendages larger than small to minute, yellow . . . . *C. similis*
         4′ Corolla limb gen 1–2 mm diam, appendages small to minute, white to pale yellow . . . . *C. circumscissa*
            5. St gen > 2 cm; calyx 3–4 mm in fr; fr axis ± = mature nutlets; nutlet beak 0; lf margin rolled under . . . . var. *circumscissa*
            5′ St 0.5–2 cm; calyx 2.5–3 mm in fr; fr axis < mature nutlets; nutlet strongly beaked; lf margin ± flat . . . . var. *rosulata*
      3′ Calyx not circumscissile in fr . . . . *C. micrantha*
         6. Corolla limb 1.5–4 mm diam, appendages larger than minute, yellow . . . . var. *lepida*
         6′ Corolla limb 0.5–1.2 mm diam, appendages minute, ± white . . . . var. *micrantha*
1′ Bracts gen 0; ann or per, gen taller than wide (rounded or cushion-like); root gen not red-purple
   7. Bien to per; lvs gen basal or tufted; nutlet wide-rounded to obtuse at tip; tip of attachment scar groove well below nutlet tip
      8. Corolla tube > calyx; fls heterostylous
         9. Corolla yellow . . . . *C. confertiflora*
         9′ Corolla white . . . . *C. flavoculata*
      8′ Corolla tube ≤ calyx; fls homostylous
         10. St 1–2(3) cm; basal lvs < 1.5 cm . . . . *C. roosiorum*
         10′ St gen > 2 cm; basal lvs gen > 1.5 cm

11. Nutlet abaxially ± smooth
    12. Nutlet 4.5–6 mm, margin winged . . . . *C. crymophila*
    12′ Nutlet gen ≤ 4.5 mm, margin a ± flat narrow rim or occ sharp-angled
        13. Caudex gen woody, gen branched; basal lvs sparse-hairy. . . . . [2]*C. nubigena*
        13′ Caudex not woody, unbranched; basal lvs dense-tomentose — e MP . . . . [2]*C. schoolcraftii*
11′ Nutlet abaxially granular, tubercled, or wrinkled
    14. Nutlet adaxially ± smooth, esp near attachment scar
        15. Nutlet attachment scar edges gapped . . . . *C. subretusa*
        15′ Nutlet attachment scar edges gen not gapped
            16. Caudex often woody, gen branched; basal lvs green, sparse-soft-bristly to dense-rough-hairy. . . . [2]*C. nubigena*
            16′ Caudex not woody, unbranched; basal lvs ± gray, dense-tomentose — e MP . . . . [2]*C. schoolcraftii*
    14′ Nutlet adaxially granular, tubercled, or wrinkled
        17. Pedicel in fr 0–2(3) mm
            18. Corolla limb 1–3 mm diam; 1 nutlet > 2–3. . . . . [2]*C. racemosa*
            18′ Corolla limb 3–10 mm diam; nutlets ± same size
                19. Corolla limb 7–10 mm diam; calyx in fr 7.5–10 mm; nutlet attachment scar edges flat; e Wrn . . . *C. celosioides*
                19′ Corolla limb 3–7 mm diam; calyx in fr 3.5–8 mm; nutlet attachment scar edges raised; W&I, n
                    DMtns . . . . *C. hoffmannii*
        17′ Pedicel in fr (2)3–12 mm
            20. Corolla late-deciduous to persistent; nutlet attachment scar edges flat
                21. Corolla cream, aging brown, late-deciduous; nutlet abaxial ridge low . . . . *C. humilis*
                21′ Corolla white, aging white, persistent; nutlet abaxial ridge 0 . . . . [2]*C. racemosa*
            20′ Corolla ± early-deciduous; nutlet attachment scar edges raised
                22. Pl long-lived per; caudex woody; nutlet abaxial ridge rounded, ± indistinct near base . . . . *C. tumulosa*
                22′ Pl bien or short-lived per; caudex not woody; nutlet abaxial ridge sharp-angled . . . . *C. virginensis*
7′ Gen ann; lvs gen cauline; nutlet narrow-acute to acuminate at tip; tip of attachment scar groove ± to nutlet tip
23. Nutlets ± smooth
    24. Nutlet asymmetric, attachment scar off-center
        25. Nutlets 4, attachment scar strongly off-center, appearing marginal; infl elongate spikes . . . . *C. affinis*
        25′ Nutlet 1(2), attachment scar off-center; infl spheric clusters . . . . *C. glomeriflora*
    24′ Nutlet symmetric, attachment scar central
        26. Calyx hairs curved or hooked and ± straight; nutlets 1
            27. Nutlet attachment scar edges widening into an open gap at base; fr axis > 1/2 nutlet
                28. Corolla limb 1–2 mm diam; calyx lobe midveins with many stout, hooked or curved bristles, occ
                    basally encrusted with exudate; nutlet 2.3–2.8 mm. . . . . *C. rostellata*
                28′ Corolla limb 3–6 mm diam; calyx lobe midveins with appressed to ascending curved hairs, with
                    hooked or curved bristles only at tip, not basally encrusted with exudate; nutlet 1.6–2 mm. . . . . . *C. spithamaea*
            27′ Nutlet attachment scar edges gen abutted to overlapped at base; fr axis < 1/2 nutlet
                29. Nutlet lance-ovate, swollen at base, margin rounded, tip acuminate, long-beaked; corolla limb
                    1–5(6) mm diam; calyx hairs gen basally encrusted with exudate . . . . *C. flaccida*
                29′ Nutlet ovate, not swollen at base, margin sharp-angled, tip ± acute, short-beaked; corolla limb
                    0.5–1(1.5) mm diam; calyx hairs not basally encrusted with exudate . . . . *C. sparsiflora*
        26′ Calyx hairs not hooked, ± straight (wavy); nutlets 1–4
            30. Calyx lobe tip bristles reflexed; corolla limb 0.5–1 mm diam — nutlets 1 or 2(4). . . . . *C. nemaclada*
            30′ Calyx lobe tip bristles gen spreading or ascending; corolla limb 0.5–8 mm diam
                31. Calyx lobes 6–10 mm in fr; longest calyx lobe bristles 3–4 mm in fr — nutlets 1 or 2 . . . . [2]*C. ganderi*
                31′ Calyx lobes 1–6 mm in fr; longest calyx lobe bristles 0 or < 3 mm in fr
                    32. Bracts present
                        33. St prostrate to decumbent, ± green in fr; nutlets (3)4, smooth. . . . . *C. leiocarpa*
                        33′ St erect to ascending, red-brown in fr; nutlet 1, smooth or 2, 1 smooth, 1 rough . . . . [2]*C. maritima*
                    32′ Bracts gen 0
                        34. Nutlet margin a ± flat narrow rim, esp near tip, to sharp-angled; nutlets 4
                            35. Corolla limb (4)5–7.5 mm diam; nutlets ± ovate. . . . . *C. mohavensis*
                            35′ Corolla limb 0.5–1.5 mm diam; nutlets ± lanceolate . . . . *C. watsonii*
                        34′ Nutlet margin gen rounded, occ sharp-angled; nutlets 1–4
                            36. Calyx (1)1.5–3 mm in fr
                                37. Nutlet 1.3–1.6 mm, abaxially ± round . . . . *C. microstachys*
                                37′ Nutlet 1.5–2.1 mm, abaxially ± flat
                                    38. Calyx obconic at base; corolla limb 1–2.5 mm diam; nutlets gen 1 . . . . *C. gracilis*
                                    38′ Calyx rounded at base; corolla limb 3–6 mm diam; nutlets gen 2 . . . . [2]*C. incana*
                            36′ Calyx (2.5)3–6.5(8) mm in fr
                                39. Nutlet abaxially ± flat. . . . . *C. torreyana*
                                    40. Calyx 1–1.5(2) mm, (2.5)3–4.5 mm in fr, lobe margin sparse-strigose; nutlet 1.3–1.5 mm . . . . var. *pumila*
                                    40′ Calyx 2–2.5 mm, 4–6.5(8) mm in fr, lobe margin dense-strigose; nutlet (1.6)1.7–2.2(2.5) mm
                                    . . . . var. *torreyana*

39′ Nutlet abaxially ± round
  41. Nutlet 1(2), adaxially ± rounded to convex, basally ± round in ×-section . . . . ***C. hispidula***
  41′ Nutlets 1–4, adaxially ± flat, basally not round in ×-section
    42. Corolla limb 0.5–1(1.5) mm diam; nutlets gen 4; st often stiffly branched above; W&I . . . . . ***C. fendleri***
    42′ Corolla limb 1–8 mm diam; nutlets 1–4; st 0–few-branched ± above middle or branched
      throughout; CA-FP
      43. St 0–few branched ± above middle . . . . ***C. dissita***
      43′ St branched throughout
        44. Calyx long-soft-hairy, midveins not thickened, not bristly; nutlets 1(2). . . . . ***C. milobakeri***
        44′ Calyx minute-hairy, midveins thickened, bristly; nutlets 1–4 . . . . ***C. clevelandii***
          45. Pedicel 0–0.5 mm; calyx ± appressed to st in fr, narrow at base, lobes linear; corolla
            limb 1–2.5 mm diam; nutlets 1–2 . . . . var. ***clevelandii***
          45′ Pedicel 0.5–1 mm; calyx not appressed to st in fr, ± wide at base, lobes ovate; corolla
            limb (1.5)2–5 mm diam; nutlets (1–2)3–4 . . . . var. ***florosa***
23′ Nutlets, or at least 1, rough
  46. At least 1 nutlet with all or part of margin a ± flat rim (occ seeming sharp-angled) or wing
    47. Pedicel 1–4(6) mm in fr . . . . ***C. holoptera***
    47′ Pedicel gen 0–1(1.5) mm in fr
      48. Nutlets 1 (2), margin a ± flat inward-bent narrow rim to basally sharp-angled, white-grainy
        papillate to minute-spiny; calyx narrow-urn-shaped in fr . . . . ***C. utahensis***
      48′ Nutlets gen 4(3), margin occ seeming sharp-angled, gen narrow- to wide-winged along entirety, not
        white-grainy papillate to minute-spiny; calyx wide-urn-shaped or ovoid in fr
        49. At least some nutlet margins with a strongly lobed, wide wing . . . . ***C. pterocarya***
          50. Nutlets similar in 1 fr, all winged . . . . var. ***cycloptera***
          50′ Nutlets dissimilar in 1 fr, 3 winged, 1 not
            51. Nutlet wing wide, gen > 0.5 mm . . . . var. ***pterocarya***
            51′ Nutlet wing narrow, gen < 0.5 mm . . . . [2]var. ***purpusii***
        49′ Nutlet margin occ seeming sharp-angled, gen a ± flat narrow, occ uneven or minute-toothed rim,
          not strongly lobed, not a wide wing
          52. Nutlets similar in 1 fr, not differing in persistence, size, or margin shape
            53. Corolla limb 1–2(2.5) mm diam; calyx 4–6 mm in fr; nutlets 1.2–1.8(2) mm, not compressed,
              abaxially strongly convex, dome-like; bracts present . . . . ***C. costata***
            53′ Corolla limb 4–9 mm diam; calyx 2.5–4 mm in fr; nutlets 1.8–2.5 mm, compressed, abaxially
              low-rounded; bracts 0 . . . . ***C. oxygona***
          52′ Nutlets dissimilar in 1 fr, 1 more persistent, also differing in size or margin shape
            54. Nutlet 1.3–1.7 mm, in 1 fr 1 > others, margins ± same; basal lf rosette gen 0 at fl; calyx lobe
              midvein dense-spreading-bristly . . . . ***C. inaequata***
            54′ Nutlet 2–2.5 mm, in 1 fr 1 not > others but with margin ± sharp-angled, (2)3 with margin a ±
              flat narrow rim or wing; basal lf rosette gen well developed at fl; calyx lobe midvein with 0–few
              bristles . . . . [2]***C. pterocarya*** var. ***purpusii***
46′ All nutlets with margin rounded or sharp-angled, not a ± flat rim or wing
  55. Nutlets 2–4, dissimilar in 1 fr, 1 more persistent, > other(s), of similar textures or not
    56. Nutlets similar in texture
      57. Pl gen erect, not sprawling; infl dense, fls ± 2-rowed, axis not ± flat; corolla limb 1–4 mm diam;
        nutlets 0.9–1.7 mm. . . . . ***C. angustifolia***
      57′ Pl gen decumbent, sprawling; infl not dense, fls not 2-rowed, axis ± flat; corolla limb 0.5–1 mm
        diam; nutlets 1.5–2.5 mm . . . . ***C. dumetorum***
    56′ Nutlets dissimilar in texture, largest ± smooth, with smaller papillae to tubercles
      58. Nutlets 2, 1.5–2 mm, lanceolate; bracts present; calyx 2–3 mm in fr, oblong, lobe midvein hairs
        straight. . . . . [2]***C. maritima***
      58′ Nutlets 4, 0.7–1 mm, triangular-ovate; bracts 0; calyx 1–1.5 mm in fr, ± spheric, lobe midvein
        hairs occ hooked . . . . ***C. micromeres***
55′ Nutlets 1–4, gen of similar persistence, size, texture
  59. Nutlets gen lanceolate to lance-ovate
    60. Nutlets gen 3–4
      61. Nutlets fine-papillate; spikes in 1s or 2s; ne SNE, W&I, ne DMoj . . . . ***C. scoparia***
      61′ Nutlets coarsely tubercled to papillate; spikes in 2s or 3s; widespread, exc ne SNE, incl W&I,
        ne DMoj. . . . . ***C. nevadensis***
        62. Pl ± sprawling; nutlets lanceolate; spikes not elongated in fr; calyx lobes recurved at tips;
          corolla limb 1–2 mm diam . . . . var. ***nevadensis***
        62′ Pl ± erect; nutlets lance-ovate; spikes elongated in fr; calyx lobes erect to ± recurved at tips;
          corolla limb 2–5 mm diam . . . . var. ***rigida***
    60′ Nutlets 1(2–3)
      63. Calyx in fr strongly recurved from st; nutlet incurved . . . . ***C. recurvata***

63′ Calyx in fr not recurved from st; nutlet(s) straight
  64. Nutlet attachment scar edges a large, wide, deep-concave gap; calyx ± not constricted above nutlet(s) . . . . *C. excavata*
  64′ Nutlet attachment scar edges a small, narrow, shallow groove; calyx ± tightly constricted above nutlet(s)
    65. Corolla limb 1–2.5 mm diam
      66. Corolla limb 1–1.5 mm diam, appendages 0; calyx lobe tips recurved to spreading . . . . *C. decipiens*
      66′ Corolla limb 1.5–2.5 mm diam, appendages minute, pale yellow; calyx lobe tips spreading to erect . . . . ²*C. ganderi*
    65′ Corolla limb 3–6 mm diam
      67. Calyx 1.5–2 mm, 3–3.5 mm in fr, not appressed to st in fr, lobes oblong, not spreading-fine-bristly; nutlets 2.2–2.5 mm . . . . *C. corollata*
      67′ Calyx 1–1.5 mm, 2.5–3(3.5) mm in fr, appressed to st in fr, lobes linear, outer 3 spreading-fine-bristly; nutlets 1.5–2 mm . . . . *C. rattanii*
59′ Nutlets lance-ovate, ovate, or triangular-ovate
  68. Nutlet abaxial ridge ± present
    69. Corolla limb 3–8 mm diam
      70. Calyx 1–1.5, 2–2.5 mm in fr, ± round; nutlet ± papillate . . . . ²*C. incana*
      70′ Calyx 1.5–2, 3–4 mm in fr, ovate; nutlet dense-sharp-papillate to tubercled . . . . . . *C. muricata* var. *muricata*
    69′ Corolla limb 1–3.5 mm diam
      71. St few-branched, without an evident central axis; nutlets 1.8–2.2 mm, ± tubercled . . . . *C. muricata* var. *denticulata*
      71′ St many-branched, with an evident central axis; nutlets 1.1–1.3(1.9) mm, sharp-papillate . . . . *C. muricata* var. *jonesii*
  68′ Nutlet abaxial ridge gen 0
    72. Corolla limb (2)3–9 mm diam
      73. Pedicel 2–3 mm in fr; calyx silky-hairy . . . . *C. crinita*
      73′ Pedicel 0–1.5(2) mm in fr; calyx rough-hairy to bristly
        74. Nutlets ovate, tip abruptly beaked; spikes gen in 1s to occ 3s, bracts gen sparse; lvs 1–2 cm, gen lowest 1–3 pairs opposite . . . . *C. mariposae*
        74′ Nutlets lance-ovate to ovate, tip gradually tapered or acute; spikes gen in 2–4(5)s, bracts gen 0; lvs 1–5(8) cm, alternate
          75. Calyx lobes narrow-oblong, midvein without coarse, spreading bristles; nutlets gen 1–2, ± ovate, sparse-low-tubercled (smooth) . . . . *C. intermedia* var. *hendersonii*
          75′ Calyx lobes lance-linear to lanceolate, midvein with few to many coarse, spreading bristles; nutlets gen 3–4, lance-ovate, dense-tubercled to warty
            76. Calyx 4.5–9 mm in fr; lvs oblong to lanceolate; infl gen in 2s; nutlet 1.9–2.5 mm . . . . *C. barbigera* var. *fergusoniae*
            76′ Calyx 4–5.5 mm in fr; lvs gen linear to lanceolate; infl gen in 3s; nutlet 1.4–2.1(2.2) mm . . . . *C. intermedia* var. *intermedia*
    72′ Corolla limb 0.5–2.5(3) mm diam
      77. Bracts present; nutlets 1.2–1.5 mm
        78. Calyx 4–6 mm in fr, lanceolate; fl persistent; lowest st lvs opposite; corolla limb gen < 1 mm diam; n&c SnJV . . . . *C. hooveri*
        78′ Calyx 3–4 mm in fr, oblanceolate; fl not persistent; lowest st lvs alternate; corolla limb 1.5–2 mm diam; s ChI . . . . *C. traskiae*
      77′ Bracts gen 0; nutlets 1.4–2.8 mm
        79. Nutlet attachment scar edges triangular-flare-gapped at base
          80. Nutlets 1.5–2 mm, tubercle tips not translucent; nutlets 1–4, lance-ovate to ovate . . . . *C. barbigera* var. *barbigera*
          80′ Nutlets ± 3 mm, tubercle tips translucent; nutlets 3–4, triangular-ovate . . . . *C. clokeyi*
        79′ Nutlet attachment scar edges wide-forked, but gen abutted, not gapped, at base
          81. Nutlet tubercles very numerous, narrow, ± longer than broad; pedicel < 0.5 mm in fr . . . . *C. echinella*
          81′ Nutlet tubercles fewer, wider than broad, rounded; pedicel 0.5–1 mm in fr
            82. Spikes gen in 1s; st branches 1–many; pl gen ± without main st . . . . *C. ambigua*
            82′ Spikes gen in 2s or 3s; st branches few; pl with main st . . . . *C. simulans*

C. affinis (A. Gray) Greene   SIDE-GROOVED CRYPTANTHA   Ann 5–30(40) cm. **ST:** branches 0 or gen few in upper 1/2; gen strigose, some hairs upcurved, some scattered, coarse, spreading. **LF:** few, 1–4(5) cm, oblanceolate to oblong or elliptic, lowest opposite, upper ± reduced; bristles sparse, minute-bulbous-based. **INFL:** in 1s or 2s, few-fld, slender; bracts gen 1–few at base, large, lf-like; pedicel 0.5–1 mm in fr, ascending. **FL:** calyx 1.5–2 mm, 2.5–3.5(4) mm in fr, ± constricted above, lobes erect to ± spreading, late-deciduous, ± lanceolate, bristles ascending to spreading, midvein ± thickened; corolla deciduous, limb 1–2 mm diam, appendages minute, white. **FR:** nutlets 4, 1.8–2.5 mm, similar, ovate, ± brown, gen mottled, ± smooth, shiny, margin rounded, base ± pointed; abaxially low-rounded, ridge 0; adaxially appearing deformed, 1 side much narrower, groove from attachment scars strongly off-center, ± curved, edges not raised, ± abutted entire length, occ short-forked-gapped at base; axis not to nutlet tips. Open areas, gen conifer forest, chaparral; 630–2600 m. KR, NCoR, CaR, SN, SCoRO, TR, PR, Wrn; to WA, MT, WY. May–Aug

C. ambigua (A. Gray) Greene   WILKES' CRYPTANTHA   Ann 10–30(45) cm, gen without a central st, occ erect. **ST:** 1–many, branches

from base, above, occ decumbent; ± short-strigose and spreading-fine-bristly. **LF:** 2–4(5) cm, gen oblong to linear, upper little reduced; ± sparse-strigose, bristles on margin, midvein, spreading to ascending ± bulbous-based. **INFL:** in 1s (2s); bracts gen 0; pedicel gen 0.5–1 mm in fr, ascending to spreading. **FL:** calyx 1.5–2.5 mm, 4–6(7) mm in fr, ± truncate at base, ± constricted above, late-deciduous, lobes erect to occ spreading, ± lance-linear, margin with ascending rough hairs, midvein ± thickened with ± tawny coarse bristles; corolla late-deciduous, limb 1–2 mm diam, appendages minute. **FR:** nutlets 3–4, (1.4)1.6–2 mm, wide-ovate, ± fine-grainy, sparse-low-tubercled, ± gray, ± shiny, margin rounded, base truncate, tip tapered; abaxially low-rounded, ridge gen 0; adaxially biconvex, attachment scar edges not raised, gen abutted entire length, occ narrow-gapped, wide-forked but gen not gapped at base; axis to nutlets or not. Dry, sandy or gravelly areas in sagebrush scrub to open conifer forest, occ pinyon/juniper woodland; 1100–2500 m. CaRH, n&c SNH, GB (exc W&I); to s BC, MT, WY, CO. May–Aug

C. angustifolia (Torr.) Greene NARROW-LEAVED CRYPTANTHA
Ann 5–20(30) cm; root rarely red-purple. **ST:** branches from base, decumbent to ascending; strigose and rough-hairy to bristly, coarse hairs spreading. **LF:** 1–4 cm, linear, scattered; rough-hairy, bristles bulbous-based. **INFL:** in gen 2s, dense, fls 2-rowed; bracts 0; pedicel not elongated in fr, < 0.5 mm, gen ascending. **FL:** calyx 1–2 mm, 2.5–4 mm in fr, persistent, lobes ± linear, dense-strigose, esp thickened midvein, gen 1 lobe ± longer, more densely strigose, tips ± erect; corolla limb 1–4 mm diam, appendages yellow. **FR:** nutlets (3)4, 0.9–1.2 mm (1 gen ± > others), lance-ovate, ± brown, pale-tubercled, margin gen rounded to occ ± sharp-angled, base ± rounded; abaxially rounded, gen dull, ridge 0; adaxially flat to ± rounded, ± shiny, attachment scar edges not raised, abutted near tip, narrow-gapped lower 1/2 to ± narrow-gapped entire length, narrow-flared-gapped at base; axis beyond nutlets. Common, sandy, occ silty to gravelly soils, creosote-bush scrub, desert woodland; < 1600 m. SNE, D; to w TX, n Mex. Jan–Jun

C. barbigera (A. Gray) Greene Ann 6–50 cm, erect, stout, white-hairy. **ST:** 1–few, branches 0 gen below middle; sparse-strigose and gen dense-bristly. **LF:** 1–5(8) cm, oblong to ± blunt-lanceolate; bristles dense, spreading, minute-bulbous-based. **INFL:** in (1s) 2s (3s), gen many-fld; bracts 0; pedicel 0–0.5 mm, ascending to spreading, gen spreading-soft-hairy. **FL:** calyx 3–6 mm, gen constricted above, 1 lobe gen longer, midvein thickened, spreading-bristly. **FR:** nutlets ovate to lance-ovate, dense-tubercled, ± brown, occ mottled, dull, margin ± rounded, base truncate, tip tapered to ± beaked; abaxially rounded, ridge 0; adaxially flat, attachment scar edges not raised, abutted entire length or variously gapped, triangular-flare-gapped at base; axis to nutlets or beyond.

var. **barbigera** BEARDED CRYPTANTHA **FL:** calyx 5–10 mm in fr, lobes gen spreading or recurved, gen lance-linear, margins gen dense-soft-hairy with spreading, fine bristles below middle; corolla deciduous, limb 1–2.5(3) mm diam, appendages 0 or minute, white. **FR:** nutlets 1–4, 1.5–2 mm. Open, sandy to rocky soils, gen in creosote-bush scrub, desert woodland; 15–1860(2250) m. s SN, Teh, s SnJV, s SNE, D; to UT, NM, Baja CA. Feb–Jun

var. **fergusoniae** J.F. Macbr. PALM SPRINGS CRYPTANTHA **LF:** 1–4(6) cm. **FL:** calyx 4.5–9 mm in fr, lobes gen erect, lanceolate, margins sparse-long-soft-hairy, rough-hairy to bristly below middle; corolla late-deciduous, limb (3)4–9 mm diam, appendages low, yellow. **FR:** (2)3–4, 1.9–2.5 mm. Gen granitic, often sandy substrates; 150–1500 m. s SNH (e slope), n SnBr, e PR, s SNE, D; AZ. Mar–May

C. celosioides (Eastw.) Payson COCKS-COMB CAT'S-EYE Bien to short-lived per, 15–50 cm; caudex not woody, gen unbranched. **ST:** 1–many, lateral 0 or gen < central st; strigose, spreading-bristly, esp lower 1/2. **LF:** 1–few basal rosettes; 2.5–9 cm; basal ± wide-spoon-shaped, thick, gray-green, dense-strigose to ± tomentose, bristles gen appressed, bulbous-based; cauline oblong, spreading-sharp-bristly. **INFL:** dense, gen cylindric; pedicel not elongated in fr, 0.5–2 mm. **FL:** calyx 3–5 mm, 7.5–10 mm in fr, dense-spreading-bristly; corolla deciduous, tube 3–5 mm, aging ± off-white, limb 7–10 mm diam, appendages yellow. **FR:** nutlets 2–4, 3.3–3.8 mm, ovate to elongate-ovate, ± dull, margin a ± flat narrow rim; abaxially ± flat, irregu-

larly tubercled, ridge narrow near tip, obscure near base; adaxially ± obscurely tubercled; attachment scar edges abutted to ± gapped entire length, edges not raised. Barren volcanic scree, talus slopes; 1550–1920 m. e Wrn; to e WA, MT, to Can, n Great Plains. Jun–Aug ★

C. cinerea (Greene) Cronquist var. abortiva (Greene) Cronquist (p. 461) BOW-NUT CRYPTANTHA Per 3–20 cm; caudex woody, branched. **ST:** prostrate to decumbent (ascending), ± many, lfy, branches from base; ± strigose. **LF:** basal rosettes gen few; 1.5–9 cm; cauline ± > basal, linear to linear-oblanceolate, abaxially with bulbous-based bristles, adaxially ± strigose. **INFL:** gen cylindric, ± not elongated in fr; bracts lf-like; pedicels short-elongated in fr, 1.5–4.5 mm. **FL:** calyx 2–4 mm, 4–7 mm in fr, appressed-soft-white-hairy, bristles few; corolla deciduous, aging off-white to tan, tube 2–4 mm, limb 5–8 mm diam, appendages yellow. **FR:** nutlets gen 4, 1.8–2.3 mm, ± ovate, ± gray, shiny, margins sharp-angled, strongly bowed in profile; abaxially smooth, ridge 0; adaxially smooth, attachment scar edges ± overlapped, not raised. Locally common. Gravelly soils; 1980–3560 m. SnBr, SNE, DMtns; s NV. May–Aug

C. circumscissa (Hook. & Arn.) I.M. Johnst. Ann 0.5–15 cm diam, rounded to cushion-like. **ST:** 0.5–8 cm, branches slender; strigose and bristly or rough-hairy, hairs gen ascending to occ spreading. **LF:** 0.3–1.5 cm; hairs rough, gen ascending to spreading. **INFL:** in 1s to 3s, 1–5-fld, dense in fr; bracts present; pedicel ± 0.5 mm, erect. **FL:** calyx circumscissile below middle, cup-like in fr, hairs ± like lvs; corolla limb gen 1–2 mm diam, appendages white to pale yellow. **FR:** nutlets ± compressed, base truncate; abaxial ridge 0; adaxial attachment scar edges not raised, abutted near tip.

var. **circumscissa** (p. 461) CUSHION CRYPTANTHA Gen 4–15 cm; root, occ pl base gen red-purple, esp when dry. **ST:** branches gen many, throughout, ± open at base. **LF:** linear to narrow-oblong, margin rolled under. **FL:** calyx 3–4 mm in fr. **FR:** nutlets 3–4, 1.2–1.7 mm (1 occ > others), triangular-ovate to lance-oblong, fine-grainy-dull or smooth-shiny, brown, or mottled brown, gray, margin sharp-angled, tip not beaked; abaxially rounded to ± flat; adaxially rounded, attachment scar edges variably gapped lower 1/2, forked at base; axis to nutlet tips. 2n=24,36. Sandy or silty flats, slopes, washes; 150–3650 m. SN, s SnJV, SCoRI, e SCo, TR, e PR, GB, D. [C. c. var. hispida (J.F. Macbr.) I.M. Johnst.] Mar–Aug

var. **rosulata** J.T. Howell ROSETTE CUSHION CRYPTANTHA Pl 0.5–3.5 cm, gen prostrate; root, pl base not red-purple. **ST:** 0.5–2 cm, branches few, gen compact; strigose, hairs gen ascending. **LF:** oblanceolate to narrow-oblong, margin ± flat. **FL:** calyx 2.5–3 mm in fr; corolla appendages minute, white. **FR:** nutlets (2)3–4, 1.5–1.8 mm, similar, lance-triangular, smooth, shiny, gray, margin a ± flat linear rim, tip strongly beaked; abaxially ± flat; adaxially biconvex, attachment scar edges occ gapped at base; axis not to nutlet tips. Barren granitic gravels; 2950–3650 m. SNH (Inyo, Tulare cos.). Jul–Aug ★

C. clevelandii Greene Ann 10–60 cm, gen erect, slender. **ST:** branches gen throughout, ascending to spreading. **LF:** 1–5 cm, scattered, reduced distally on st, ± linear to lance-linear. **FL:** calyx 1.5–2.5 mm, lobes linear to ovate, tips ± spreading, ± dense-strigose-bristly, midvein thickened. **FR:** nutlets 1.5–2 mm, lanceolate to lance-ovate, gen mottled, ± smooth, ± shiny, margin rounded, base ± rounded to narrow-truncate, tip ± beaked; abaxially low-rounded, ridge 0; adaxially ± flat, attachment scar edges abutted, forked at base, occ gapped.

var. **clevelandii** CLEVELAND'S CRYPTANTHA **ST:** sparse-strigose and fine-bristly. **LF:** hairs sparse, appressed, midvein, margin fine-bristly. **INFL:** in 1s or 2s; bracts gen 0; pedicel 0–0.5 mm, in age erect. **FL:** calyx (2.5)3–4 mm in fr, lobes linear, midvein with few conspicuous bristles at base; corolla limb 1–2.5 mm diam, appendages pale yellow. **FR:** nutlets gen 1–2; axis beyond nutlets. Loamy soils, slopes, grassland, coastal scrub; 50–150(230) m. SCo, s ChI; Baja CA. Mar–May

var. **florosa** I.M. Johnst. COASTAL CRYPTANTHA **ST:** strigose and ± spreading-soft-hairy. **LF:** appressed-hairy, occ fine-bristly. **INFL:** in 2s to 3s (4s); bracts 0 or occ 1–2 at base; pedicel 0.5–1 mm, ± spreading. **FL:** calyx 3–4(4.5) mm in fr, lobes ovate, midvein, margin with many spreading to reflexed bristles; corolla limb (1.5)2–5 mm diam, appendages yellow. **FR:** nutlets gen (1–2)3–4; axis not to

± beyond nutlets. Common; sandy or rocky soils, chaparral, coastal scrub, foothill woodland; 5–1500 m. Teh, CW, n SCo, ChI, WTR, s SnGb, e PR; Baja CA. Mar–Jun

C. clokeyi I.M. Johnst. CLOKEY'S CRYPTANTHA Ann 8–20(30) cm, stout. **ST:** branches gen throughout; gen strigose and sparse-rough-hairy. **LF:** 0.5–3 cm, oblong to lance-linear, reduced above; adaxially appressed- to ascending-rough-hairy, veins sunken; abaxially, marginally ascending- to spreading-fine-bristly, hairs throughout bulbous-based. **INFL:** in 1s (2s), gen few-fld; bracts gen 0 or 1+ at base; pedicel elongated in fr, 0.5–1.5(2) mm, ascending to spreading. **FL:** calyx 2–3 mm, 5–9(10) mm in fr, gen constricted above, late-deciduous, lobes lance-linear, tips ± erect, margin dense-strigose to occ ascending soft-hairy, midvein ± thickened, spreading-bristly, esp toward base; corolla late-deciduous, limb 1–2 mm diam, appendages minute. **FR:** nutlets 3–4, ± 3 mm, triangular-ovate, ± brown to mottled, shiny, tubercles ± dense, sharp, tips ± white-translucent, margin sharp-angled, base wide-truncate, tip acute; abaxially low-rounded, ridge 0; adaxially biconvex, attachment scar edges not raised, gen abutted entire length, rarely narrow-gapped, wide-forked, triangular-flare-gapped at base; axis beyond nutlets. Rocky to gravelly slopes, ridge crests, desert woodland; (850)1050–1650+ m. nw DMoj, n DMtns. Apr–May ★

C. confertiflora (Greene) Payson YELLOW-FLOWERED CRYPTANTHA Per 13–45 cm; caudex woody, branched. **ST:** erect, many, gray-green; tomentose below, strigose and sparse-bristly above. **LF:** gen many basal rosettes; 3–12 cm; basal lvs ± oblanceolate, gray-green, dense-strigose, abaxially with bulbous-based bristles. **INFL:** gen dense, ± head-like above, interrupted, with scattered smaller axillary clusters below; pedicel gen not elongated in fr, 0–1 mm. **FL:** calyx 6–10 mm, 9–14 in fr, strigose and spreading-bristly; corolla yellow, limb 8–12 mm diam, tube 9–13 mm. **FR:** nutlets gen 4, 3.5–4 mm, wide-ovate to triangular, smooth, gray, ± shiny, margins sharp-angled; abaxial ridge 0; adaxial attachment scar edges abutted to ± overlapped, edges not raised; axis 3–7 mm beyond nutlets. Common. Dry, rocky soils; 1050–3100(3350) m. s SNH, SnBr, SNE, DMtns; to ne AZ, sw UT. Heterostylous. May–Jul

C. corollata (I.M. Johnst.) I.M. Johnst. COAST RANGE CRYPTANTHA Ann 10–45 cm, erect, slender. **ST:** branched to not; sparse-bristly below, hairs appressed. **LF:** 1–4 cm, oblong to narrow-oblanceolate; margin appressed- to ± spreading-minute-soft-bulbous-based-bristly. **INFL:** in 2s or 3s; bracts 0; pedicel 0.5 mm, erect to ascending. **FL:** calyx 1.5–2 mm, 3–3.5 mm in fr, constricted above, late-deciduous, lobes erect to ± spreading, oblong, short-rough-hairy, hairs ascending to occ spreading, midvein not thickened; corolla limb 3–5 mm diam, appendages yellow. **FR:** nutlets gen 1(2), 2.2–2.5 mm, lance-ovate, grainy, dense-tubercled, ± dull, margin rounded, base truncate, tip short-beaked; abaxially ± rounded, spine wide, obscure toward base; adaxially ± flat, attachment scar edges raised toward tip, abutted at tip or entire length, flare-gapped at base; axis not to nutlet tips. Dry, rocky slopes, ridges, grassland, foothill woodland, occ on serpentine; 200–1450 m. s SNF, Teh, s SnJV, SnFrB, s SCoRO, SCoRI, WTR, SnGb, SnJt. Mar–Jun

C. costata Brandegee RIBBED CRYPTANTHA Ann 10–20 cm, coarse; root often red-purple. **ST:** branches few, stiff; dense-strigose and sparse-spreading-bristly. **LF:** 1–3(4) cm, linear to lanceolate, ± folded along midvein, drying long-acuminate or awl-shaped, dense-strigose and rough-hairy to sparse-spreading-bristly, hairs bulbous-based. **INFL:** in 1s or 2s, dense; bracts ± scattered, spreading-bristly, deciduous; pedicel 0–0.5 mm, stiff-spreading. **FL:** calyx 2.5–4 mm, 4–6 mm in fr, gen constricted above, persistent, lobes lance-linear, tips ± spreading, strigose on margins, midvein thickened, esp in age, dense-long-bristly; corolla limb 1–2(2.5) mm diam, appendages minute, pale yellow. **FR:** nutlets (3)4, 1.2–1.8(2) mm (1 ± > others), triangular or oblong-ovate, brown, shiny, margin occ sharp-angled or gen ± flat narrow rim, base wide-truncate; abaxially strongly convex, dome-like, minute-rippled, sparse-grainy to obscurely tubercled, ridge 0; adaxially smooth, flat to ± concave, attachment scar edges not raised, abutted near tip, gapped below, triangular-flare-gapped at base; axis beyond nutlets. Fine sand deposits (coarser soils), creosote-bush scrub; < 600(1000) m. e DMoj, DSon; s NV, AZ, Baja CA. Jan–May ★

C. crinita Greene (p. 461) SILKY CRYPTANTHA Ann 10–40 cm, erect. **ST:** branches many, gen throughout; appressed- to ascending-rough-hairy. **LF:** 1–4 cm, narrow-oblanceolate to oblong; rough-hairy to bristly, hairs appressed to ascending, some bulbous-based. **INFL:** raceme-like, in 2s or 3s; bracts 0; pedicel elongated in fr, 2–3 mm, long-soft-hairy. **FL:** calyx 2–4 mm, 4.5–6.5 mm in fr, deciduous, lobes linear-oblong, tips not constricted above, white, long silky-hairy, hairs appressed to ascending, midvein not thickened; corolla limb (3)4–9 mm diam, appendages yellow. **FR:** nutlet gen 1, 2.3–2.8 mm, lance-ovate, dense-grainy, tubercles few, ± dull, ± brown, margin rounded, base ± truncate, tip ± beaked; abaxially low-rounded, ridge 0; adaxially biconvex, attachment scar edges ± raised, abutted near tip, narrow-gapped below, triangular-flared-gapped at base; axis not to nutlet tips. Rocky volcanic soils, gravelly streambanks, gravel bars, gen foothill woodland; 90–1120 m. CaR (Shasta, Tehama cos.). Mar–Jun ★

C. crymophila I.M. Johnst. SUBALPINE CRYPTANTHA Per, tufted, 13–40 cm; caudex woody, branched. **ST:** sts many, lfy; sharp-bristly. **LF:** basal rosettes gen many; 4–10 cm; basal oblanceolate to acute-oblong; adaxially strigose; abaxially appressed-sharp-bulbous-based-bristly. **INFL:** ± head-like, dense above with axillary clusters below, not elongated in fr; pedicel gen ± not elongated in fr, 1–3.5 mm. **FL:** calyx 4–5 mm, 9–15 mm in fr, dense-sharp-± yellow-bristly; corolla deciduous, aging white to off-white, tube 3–5 mm, limb 4–7(8) mm diam, appendages yellow. **FR:** nutlets gen 4, 4.5–5.5(6) mm, ± elliptic, ± brown, ± dull, margin narrow-winged; abaxially ± flat, ± smooth, ridge 0; adaxially ± smooth, attachment scar edges gapped entire length, edges not raised; axis to nutlets or beyond. Rocky volcanic semi-barren soils, scree; 2600–3200 m. n&c SNH (Alpine, Tuolumne cos.). Jul–Aug ★

C. decipiens (M.E. Jones) A. Heller (p. 469) GRAVEL CRYPTANTHA Ann 10–40 cm, erect, slender. **ST:** branches 0 to throughout; strigose to rough-hairy, hairs appressed to occ spreading. **LF:** few, 0.5–3 cm, linear to narrow-lanceolate; hairs gen appressed to ± spreading, minute-bulbous-based. **INFL:** in (1s) 2s (3s); bracts 0; pedicel ± 0. **FL:** calyx 2–2.5 mm, 3–4 mm in fr, tightly constricted above, deciduous, lobes lance-linear, tips gen spreading to recurved, dense-rough-hairy to coarse-bristly, hairs ascending to spreading, midvein thickened; corolla limb 1–1.5 mm diam, appendages 0. **FR:** nutlets 1(2), 2.3–2.7 mm, lance-ovate, brown, fine-grainy, ± white-papillate, ± dull, margin rounded, base rounded, tip beaked; abaxially rounded, ridge 0; adaxially flat to low-biconvex, attachment scar edges ± raised, abutted to gapped entire length, triangular-flare-gapped at base; axis not to nutlet tips. Open, gravelly, occ sandy slopes, washes, creosote-bush scrub, Joshua-tree woodland; 60–1400 m. se SNH, Teh, n SnGb, ne SnBr, e PR, W&I, D; to sw UT. Mar–May

C. dissita I.M. Johnst. LAKE CRYPTANTHA Ann 8–25 cm, erect, tufted, stout. **ST:** branches 0–few, ± above middle; spreading- to ascending-long-soft-hairy. **LF:** crowded at base, scattered, ± not reduced distally on st; 0.5–2.5 cm, oblong to wide-linear; hairs long, soft, gen appressed to occ spreading, bulbous-based on margin, abaxial midvein. **INFL:** in (2s) 3s; bracts 0; pedicel not elongated in fr, ≤ 0.5 mm, ascending. **FL:** calyx 1.5–2.5 mm, 3.5–5.5(6) mm in fr, not constricted above, late-deciduous, lobes linear, dense-soft-hairy, midvein thickened, with few dark-tipped bristles; corolla deciduous, limb (4)5–8 mm diam, appendages pale yellow. **FR:** nutlet (1)2–4, 1.8–2.2 mm, lanceolate to lance-ovate, gen mottled, smooth, shiny, margin rounded, base truncate, tip ± beaked; abaxially rounded, ridge 0; adaxially ± flat, attachment scar edges not raised, ± abutted entire length, forked at base, occ gapped; axis to nutlets or ± beyond. Serpentine, rocky outcrops, gravelly slopes, chaparral, foothill woodland; 150–900 m. e KR, se NCoRO, s NCoRI. [*C. clevelandii* var. *d.* (I.M. Johnst.) Jeps. & Hoover] Mar–Jun ★

C. dumetorum (A. Gray) Greene (p. 469) SCRAMBLING CRYPTANTHA Ann 10–60 cm, prostrate to ascending, sprawling. **ST:** branches throughout, prostrate to ascending, ± brittle; strigose to appressed-stiff-hairy. **LF:** 0.5–3(4) cm, linear to lanceolate, ± succulent; sparse-strigose and rough-hairy, hairs ascending to ± appressed, bulbous-based. **INFL:** in 1s or 2s, ± sparse, fls not 2-rowed, axis ± flat; bracts gen 0 or occ 1–2 near base; pedicels ± 0. **FL:** calyx 1–1.5 mm, 2–3 mm in fr, persistent; lobes dissimilar, 2 ± partly fused,

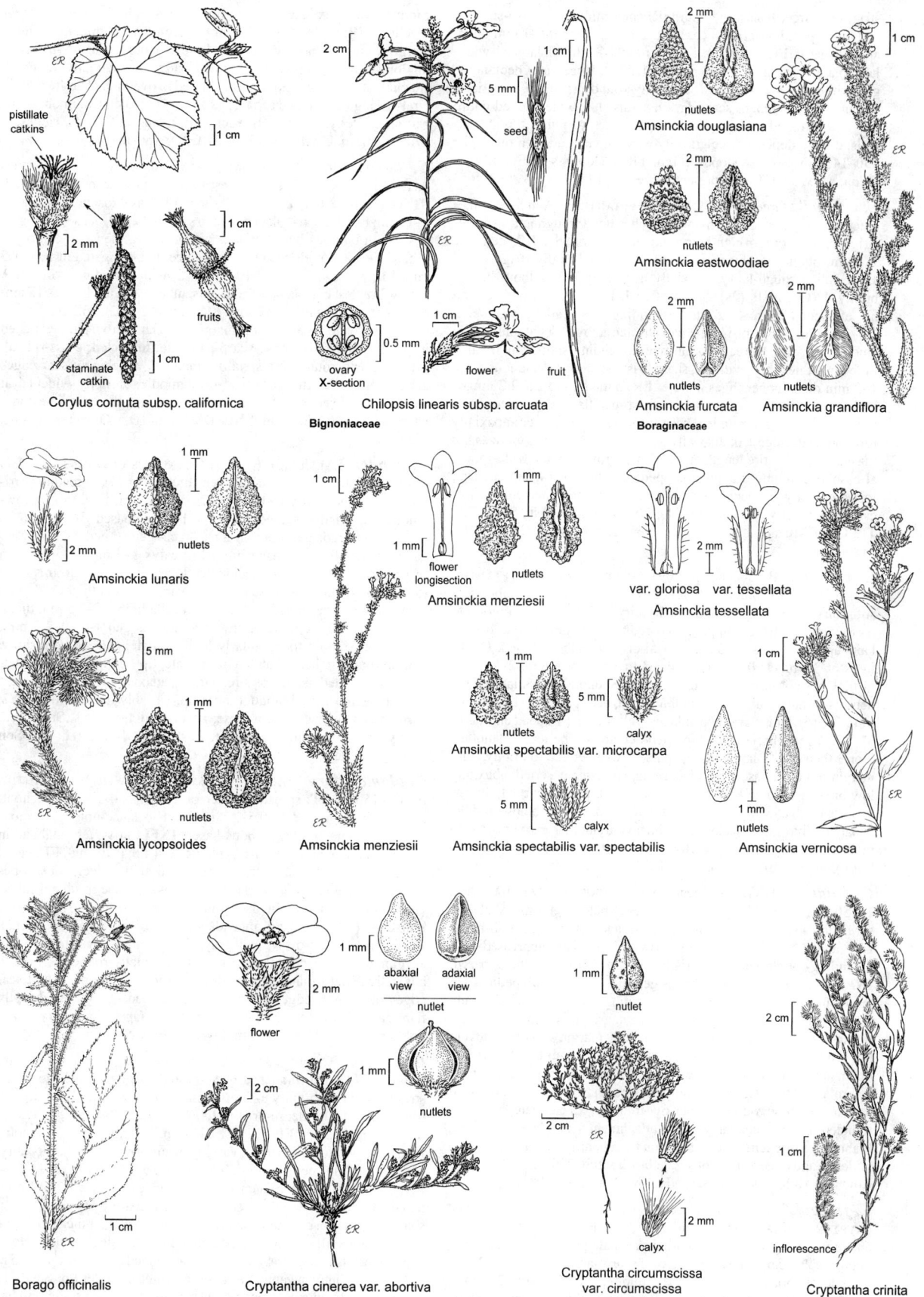

pistillate
catkins

staminate
catkin

fruits

1 cm

2 mm

1 cm

1 cm

Corylus cornuta subsp. californica

2 cm

5 mm

seed

1 cm

ovary
X-section

0.5 mm

1 cm

flower

fruit

Chilopsis linearis subsp. arcuata

Bignoniaceae

2 mm

nutlets

Amsinckia douglasiana

2 mm

nutlets

Amsinckia eastwoodiae

2 mm

nutlets

Amsinckia furcata

2 mm

nutlets

Amsinckia grandiflora

1 cm

Boraginaceae

1 mm

nutlets

2 mm

Amsinckia lunaris

1 cm

1 mm

flower
longisection

1 mm

nutlets

Amsinckia menziesii

1 mm

var. gloriosa   var. tessellata

2 mm

Amsinckia tessellata

5 mm

Amsinckia lycopsoides

1 mm

nutlets

1 mm

nutlets

5 mm

calyx

Amsinckia spectabilis var. microcarpa

5 mm

calyx

Amsinckia spectabilis var. spectabilis

1 cm

1 mm

nutlets

Amsinckia vernicosa

Amsinckia menziesii

Borago officinalis

1 cm

flower

2 mm

1 mm

abaxial
view

adaxial
view

nutlet

1 mm

nutlets

2 cm

Cryptantha cinerea var. abortiva

1 mm

nutlet

2 cm

2 mm

calyx

Cryptantha circumscissa
var. circumscissa

2 cm

1 cm

inflorescence

Cryptantha crinita

strigose, 3 free, lanceolate with thickened midvein, sparse-strigose and long-spreading-bristly; corolla deciduous, limb 0.5–1 mm diam, appendages 0. FR: nutlets 4, 1.5–2.5 mm (1 > others), larger 1 wide-lanceolate, ± rough, persistent, smaller 3 ± lanceolate, deciduous, gen white-sharp-grainy, dull; abaxially rounded, ridge 0, adaxially ± rounded, attachment scar edges not raised, ± wide-gapped entire length, not flared or forked at base; axis gen to nutlet tips. Sandy soils, dunes, slopes, or occ gravelly washes, gen under/on other pls; 250–1400 m. SNH (e slope), e Teh, TR (n slope), s SNE, DMoj, n DSon; s NV, sw UT. Gen present in wet yrs. Mar–May

C. echinella Greene HEDGEHOG CRYPTANTHA Ann 5–35 cm, ± erect, ± pale green. ST: gen few, branches 0 to throughout; ± sparse-strigose, spreading-rough-hairy and occ fine-bristly. LF: gen few 1–5 cm, oblong to oblanceolate, occ linear; adaxially strigose and ± ascending rough-hairy; abaxially ascending fine-bulbous-based-bristly. INFL: in 1s (2s); bracts often lf-like, at base; pedicel gen not elongated in fr, < 0.5 mm, ascending to spreading. FL: calyx 1.5–2.5 mm, 4–6 mm in fr, ± constricted above, deciduous, lobes lance-linear, tips ± erect, margin with ascending short, rough hairs, midvein ± thickened, with tan, sharp bristles; corolla deciduous, limb 1–2 mm diam, appendages minute. FR: nutlets 3–4, 1.5–2.2 mm, ± wide-ovate, fine-grainy, densely ± sharp-papillate, ± gray, occ mottled, ± dull, margin rounded, base truncate, tip ± beaked; abaxially low-rounded, ridge ± 0; adaxially gen flat, attachment scar edges not raised, abutted entire length or variably narrow-gapped, wide-forked at base, occ minute-triangular-gapped; axis to ± below nutlets. Dry, loose, gravelly, or rocky soils, disturbed areas, gen open conifer forest, occ juniper woodland or scrub; 700–3200 m. s SNF, SNH, Teh, TR, GB, DMtns (Panamint Range, New York Mtns); e OR, ID, NV. May–Aug

C. excavata Brandegee (p. 469) DEEP-SCARRED CRYPTANTHA Ann 5–30 cm. ST: branches 0–few, throughout; strigose to appressed-soft-hairy and occ spreading-rough-hairy. LF: 0.6–3 cm, oblong to narrow-spoon-shaped, upper much reduced; rough-hairy to bristly, hairs appressed to spreading, abaxial bristles bulbous-based. INFL: in 2s or 3s; bracts 0; pedicel ± 0. FL: calyx 1–1.5 mm, 2–2.5 mm in fr, ± not constricted above, ± erect, deciduous, lobes lanceolate, dense-soft-hairy, midvein ± not thickened, occ sparse-bristly; corolla limb 3–5(6) mm diam, appendages yellow. FR: nutlet gen 1(2 or 3), 2–2.4 mm, horizontal, ± perpendicular to lobe axis, (if nutlets 2 or 3 then erect), lance-ovate, sparse-low-tubercled to grainy, dull, margin rounded, base ± rounded, tip tapered, gen exserted; abaxially low biconvex, spine keel-like; adaxially flat, tip ± incurved, attachment scar edges ± inrolled, ± abutted near tip, gapped below middle, expanded into large, wide, deep-concave triangular gap; axis not to nutlet tips. Steep, sandy, gravelly slopes, soils, dry streambanks, foothill woodland; 100–600 m. s NCoRI. Mar–May ★

C. fendleri (A. Gray) Greene SAND DUNE CRYPTANTHA Ann 10–50 cm, erect. ST: gen 1, branches 0 below, gen many above; spreading to ascending, dense-strigose and spreading-rough-hairy. LF: many, 1–5 cm, linear to narrow-oblanceolate; appressed- and ascending-rough-hairy and bristly; abaxial bristles bulbous-based. INFL: in 1s or 2s, gen open; bracts gen 0 (1–2 near base); pedicel not elongated in fr, ± 0.5 mm, ± ascending. FL: calyx 1.5–2.5 mm, 3.5–6(7.5) mm in fr, deciduous, ± asymmetric, lobes linear to lance-linear, gen constricted above, tip spreading, margin strigose, midvein thickened, coarse-bristly; corolla ± deciduous, limb 0.5–1(1.5) mm diam, appendages 0 or minute, white. FR: nutlets gen 4, 1.4–2 mm, ± similar, lanceolate, gray or mottled, smooth, shiny, margin rounded to ± angled, base rounded to narrow-truncate, tip elongate; abaxially rounded, ridge 0; adaxially ± flat, attachment scar edges not raised, ± abutted entire length, widened to a triangular pit at base; axis ± to nutlets. Sand dune, sandy soils, sagebrush scrub; 2200 m. W&I (Little Cowhorn Valley); to WA, MT, NE, NM. Jun–Aug ★

C. flaccida (Lehm.) Greene WEAK-STEMMED OR PALE CRYPTANTHA Ann 15–50 cm, erect, slender, wiry. ST: branches above to throughout, appressed-hairy, ± pale green. LF: lowest occ opposite; 2–6 cm, ± linear, pale green; sparse-appressed-hairy. INFL: in 1s to 5s; bracts 0; pedicel < 0.5 mm, erect. FL: calyx 2–3 mm, 3–4.5(5) mm in fr, ± persistent, base asymmetric, lobes ± linear, margins strigose to ciliate, midvein thickened, with ± dense-tufted, spreading, conspicuous recurved or hooked bristles gen basally

encrusted with exudate; corolla limb 1–5(6) mm diam, appendages ± yellow. FR: nutlet 1, 1.8–2.3 mm, lance-ovate, ± smooth, ± shiny, margin rounded, basally swollen, base ± acute to narrow-truncate, tip long-beaked; abaxially rounded, ridge 0; adaxially rounded, attachment scar edges not raised, abutted to overlapped entire length, rarely triangular-gapped at base; axis to < 1/2 nutlet. Common; semi-barren, gravelly, loose soils, rocky sites, washes, slopes, ridges; < 1700(2250) m. CA-FP, w MP; to WA, ID. Apr–Jul

C. flavoculata (A. Nelson) Payson (p. 469) YELLOW-EYED CRYPTANTHA Per 5–35 cm, cespitose; caudex woody, branched. ST: gen many, ± equal; strigose, bristly. LF: basal rosettes gen many; 2–11 cm; basal linear-oblanceolate to spoon-shaped, dense-silky-strigose, with bulbous-based bristles or not. INFL: dense, cylindric or ± head-like, ± not elongated in fr; pedicel not elongated in fr, 1–1.5 mm. FL: calyx 4–7 mm, 6–10 mm in fr, with spreading, white or ± yellow bristles; corolla late-deciduous, tube 7–12 mm, limb 8–12 mm diam, appendages bright yellow. FR: nutlets 2–4, 2.5–3.2 mm, ovate to narrow-ovate, brown, dull, margin a ± flat narrow rim, ± arched in profile; abaxially dense-sharp-papillate, tubercled, cross-ridged, ridge present, rounded; adaxially tubercled, attachment scar edges gapped entire length, narrowed near middle, abruptly widened at base, edges raised; axis beyond nutlets. Common. Loose limestone-based soils; 1200–3200 m. SNE, DMtns; to ID, CO. Heterostylous. May–Jul

C. ganderi I.M. Johnst. (p. 469) GANDER'S CRYPTANTHA Ann 10–40 cm, stout. ST: branches gen throughout (0), spreading; strigose and rough-hairy to bristly. LF: 1–4(5) cm, linear to narrow-lanceolate, bristles spreading, some bulbous-based. INFL: in 1s, fls occ clustered, gen open in age; bracts gen 0; pedicel not elongated in fr, < 0.5 mm, ascending. FL: calyx 3–4 mm, 6–10 mm in fr, constricted above, late-deciduous, lobes linear-oblong, tips erect to spreading, dense-strigose to rough-hairy, midvein thickened, with many, 3–4 mm, ± spreading bristles; corolla limb 1.5–2.5 mm diam, appendages minute, pale yellow. FR: nutlets gen 1–2, 2–2.5 mm, lanceolate, smooth to irregularly low-papillate, gen mottled, shiny, margin rounded, base rounded to truncate, tip ± long-beaked; abaxially low-rounded, ridge obscure, narrow; adaxially ± flat, attachment scar edges not raised, abutted near tip, variably gapped lower 1/2, narrow-triangular-gapped at base; axis not to nutlet tips. Stabilized, ± silty, fine sand deposits, creosote-bush scrub; 140–250 m. w DSon (Borrego Valley area); s AZ, nw Mex. Jan–May ★

C. glomeriflora Greene CLUSTERED-FLOWER CRYPTANTHA Ann 3–15 cm. ST: spreading to erect, branches 0–few, throughout; gen dense-strigose. LF: 0.5–2 cm, linear to lance-oblong, strigose to bristly, some bristles bulbous-based. INFL: in 1s, terminal and in upper lf, branch axils, ± dense, spheric, 2–4-fld; pedicel 0. FL: calyx 1.5–2 mm, 2–2.5 mm in fr, ± constricted above, deciduous, lobes lance-linear, fused 1/3–1/2 length, strigose at base, midvein thickened, esp outermost lobe ascending coarse-bristly near tip; corolla limb 0.5–1(1.5) mm diam, appendages minute, white. FR: nutlet 1(2), 1.5–2 mm, ± wide-ovate, pale gray, smooth, shiny, margin rounded, asymmetric, base ± truncate, tip acute; abaxially rounded, ridge 0; adaxially ± flat to unevenly biconvex, groove from attachment scar edges off-center, edges abutted to overlapped near base, abruptly flare-gapped at base; axis not to nutlet tip(s). Open slopes, dry meadows, creekbeds; 1800–3750 m. SNH, SNE; w NV. Jun–Sep ★

C. gracilis Osterh. SLENDER CRYPTANTHA Ann 10–35 cm, slender. ST: erect, branches 0 or throughout, decumbent to erect; strigose and rough-hairy to short-bristly. LF: 1–3.5 cm, linear to narrow-oblanceolate, dense-bristly. INFL: in 1s or 2s (3s), gen dense; bracts 0; pedicels 0. FL: calyx 1.5–2 mm, 2–3 mm in fr, deciduous, lobes lanceolate to narrow-ovate, gen uniformly tufted soft-hairy, hairs ascending; corolla limb 1–2.5 mm diam, appendages yellow. FR: nutlet gen 1 (if 2–3, then 1–2 small, undeveloped), 1.7–2.1 mm, lanceolate to narrow-ovate, smooth, shiny, margin rounded to gen sharp-angled, base wide-truncate; abaxially ± flat, spine ± concave basally; adaxially biconvex, attachment scar edges abutted to ± gapped, esp basally, or even occ overlapped, edges narrow-forked at base, occ triangular-gapped at base; axis not to nutlet tip. Sandy to rocky soils, dry slopes, creosote-bush scrub, Joshua-tree and pinyon/juniper woodland; 730–2400 m. SNE, e DMoj; to e WA, s ID, w CO, n AZ. Mar–Jun

C. hispidula Brand NAPA CRYPTANTHA Ann 10–40(50) cm, slender. **ST:** gen erect, branches 0–many, spreading-ascending; sparse-strigose and short-spreading-rough-hairy. **LF:** 0.5–2 cm, linear to narrow-oblong, sparse-appressed- to -spreading-fine-bulbous-based-bristly. **INFL:** in 2s or 3s; bracts 0; pedicel ≤ 0.5 mm, ± ascending. **FL:** calyx 1.5–2 mm, (2.5)3–4.5 mm in fr, constricted above, lobes ± linear, tips spreading, dense-ascending-rough-hairy, midvein ± not thickened, bristly on 3 lobes; corolla late-deciduous, limb (2.5)3–6 mm diam, appendages yellow. **FR:** nutlet 1(2), 1.7–2.2 mm, lanceolate to ovate, gen mottled gray-brown, smooth, shiny, margin ± rounded, basally swollen, base ± truncate, tip ± abruptly long-beaked; abaxially rounded, ridge 0; adaxially low-rounded to biconvex, attachment scar edges ± abutted or occ ± gapped, short-forked-gapped at base; axis not to nutlet tip(s). Locally common. Semi-barren serpentine gravelly outcrops, slopes, chaparral, foothill woodland; 200–950 m. KR, NCoRI. Mar–May

C. hoffmannii I.M. Johnst. WHITE MOUNTAINS CRYPTANTHA Bien to short-lived per, 17–30(40) cm; caudex not woody, unbranched. **ST:** 1–many, central st gen > others, ± erect; conspicuously white-bristly. **LF:** basal rosette gen 1; gen 2–5 cm, spoon-shaped, thick, gray-green, dense-strigose to silky-tomentose, bulbous-based bristles dense abaxially, sparse adaxially. **INFL:** gen cylindric; pedicel not elongated in fr, 0.5–1.5(2) mm. **FL:** calyx 3–5 mm, 4.5–8 mm in fr, dense-white-bristly; corolla late-deciduous, in age ± off-white, limb 5–7 mm diam, appendages yellow. **FR:** nutlets 2–4, 2.7–3.5 mm, ovate, constricted above, ± brown, ± dull, margin a ± flat narrow rim; abaxially coarse-tubercled, wrinkled, ridge low, ± rounded, esp near base; adaxially ± tubercled, wrinkled, attachment scar edges raised, gapped ± entire length. Locally common. Rocky, loose soils, gen limestone or volcanic, gen pinyon/juniper woodland; 1740–2900(3100) m. W&I, n DMtns; w NV. Jun–Jul

C. holoptera (A. Gray) J.F. Macbr. (p. 469) WINGED CRYPTANTHA Ann (short-lived per), 10–60+ cm, erect, with central axis, coarse, occ woody at base. **ST:** gen 1, branches throughout; strigose and rough-bristly, bristles spreading to ascending, some bulbous-based. **LF:** 3–6 cm; basal, lower cauline gen few, elliptic to lanceolate, short-petioled, long-lived; upper sessile; short-rough-bulbous-based-bristly, bristles sparse abaxially. **INFL:** raceme-like, in (1s) 2s, fls ± few, not 2-rowed; bracts 0, occ 1–2 near base; pedicel ± elongated in fr, 1–1.5 mm, gen erect to ascending. **FL:** calyx 1.5–2 mm, 2.5–3.5 mm in fr, persistent, lobes dense-bristly to rough-hairy, hairs spreading, midvein thickened; corolla deciduous, limb 2–3 mm diam, flat, appendages yellow. **FR:** nutlets 4, (1.5)1.8–2.5 mm, ovate to triangular, dark with pale tubercles, ± dull, margin a ± flat narrow to wide non-papery wing; abaxially low-rounded, ridge 0; adaxially biconvex, attachment scar edges raised, abutted near tip, wide-triangular-flare-gapped at base; axis beyond nutlets. Gravelly to rocky soils, washes, slopes, ridges; 50–1220 m. e DMoj, DSon; w NV, w AZ, n Mex. Jan–May ★

C. hooveri I.M. Johnst. (p. 469) HOOVER'S CRYPTANTHA Ann 5–15 cm. **ST:** erect, branches 0 or at base; dense-strigose to rough-hairy. **LF:** ± many, scattered, lowest opposite; 1–2.5 cm, linear to narrow-spoon-shaped, gen folded to inrolled; strigose to ascending-hairy, hairs bulbous-based abaxially. **INFL:** head-like, only lowest fls in each developing fr; bracts scattered, thread-like; pedicel not elongated in fr, < 0.5 mm, ± spreading. **FL:** calyx 1.5–3 mm, 4–6 mm in fr, persistent, lobes ± linear, margins dense-strigose to ascending rough-hairy, midvein not thickened, long spreading-bristly; corolla limb gen ≤ 1 mm diam. **FR:** nutlets 3–4, 1.2–1.5 mm, triangular- to wide-ovate, gray or tan, occ mottled, fine-grainy, shiny, margin ± sharp-angled to obtuse or ± flat at base, base wide-truncate, tip acute; abaxially low-rounded, tubercled, ridge 0; adaxially biconvex, sparse-papillate, attachment scar edges not raised, abutted ± entire length, wide-triangular-flare-gapped at base; axis to nutlets or not. Dry, coarse sand, flats and hills; < 80 m. n&c SnJV. (Mar) Apr–May ★

C. humilis (A. Gray) Payson (p. 469) LOW CRYPTANTHA Per 5–30 cm, cespitose; caudex woody, branched. **ST:** many, erect, green; ± strigose and sparse-spreading-bristly. **LF:** rosettes gen many; 1.5–6 cm; basal spoon-shaped to oblanceolate, ± thin, green to gray-green, strigose to ± tomentose and appressed-bulbous-based-bristly. **INFL:** cylindric, dense, ± narrow, not or short-elongated in fr; pedicel 3–12

mm in fr. **FL:** calyx 3–4.5 mm, 7–11(13) mm in fr, ± green, conspicuously spreading-soft-bristly; corolla late-deciduous, cream aging brown, tube 3–5 mm, limb 7–10 mm diam, appendages yellow. **FR:** nutlets 2–4, 3–4 mm, ovate, ± uniformly grainy, tubercled, fine-wrinkled, dull, margin a ± flat narrow rim; abaxial ridge low, ± rounded; adaxial attachment scar edges abutted near tip, ± wide-gapped lower 1/2, edges not raised; axis beyond nutlets. Common. Gravelly soils, sagebrush to subalpine; 1700–3600 m. SNH (rare s), GB, n DMtns; to OR, ID, w CO. May–Aug

C. inaequata I.M. Johnst. PANAMINT CRYPTANTHA Ann 5–40 cm, erect; root occ red-purple. **ST:** branches 0 to throughout, esp from base, gen sparse, ascending; rough-hairy and strigose esp near base, hairs gen spreading. **LF:** many, scattered; 1.5–4.5(5) cm, narrow-oblanceolate to linear, strigose to ± rough-hairy with bulbous-based bristles, esp abaxially. **INFL:** in 1s or 2s, ± dense, fls ± 2-rowed; bracts 0 or occ few, esp near base; pedicels 0.5–1.5 mm, gen ascending. **FL:** calyx 1.5–2.5 mm, 2.5–4 mm in fr, persistent, symmetric, lobes wide-lanceolate, dense-spreading-bristly, esp on midvein, midvein ± thickened; corolla late-deciduous, limb 1.5–4 mm diam, ± flat, appendages yellow. **FR:** nutlets (3)4, 1.3–1.7 mm (1 > others), ovate to narrow-triangular, brown, fine-grainy, pale tubercled, shiny to ± dull, margin gen a ± flat narrow rim, occ sharp-angled; abaxially low-rounded, ridge 0; adaxially biconvex, ± sparse-tubercled, attachment scar edges not raised, abutted near tip, wide-flare-gapped at base; axis beyond nutlets. Gravelly to clay soils, rocky slopes, washes, creosote-bush scrub, desert woodland; < 1400 m. e D; to s NV, sw UT, nw AZ. Mar–May

C. incana Greene TULARE CRYPTANTHA Ann 15–50 cm, slender. **ST:** erect, branches gen throughout; thin-strigose to rough-hairy, some hairs ascending to spreading. **LF:** 1–3.5 cm, oblong to narrow-oblanceolate; abaxially bristly, some bristles bulbous-based; adaxially ± appressed-hairy. **INFL:** in (1s) 2s or 3s; bracts 0; pedicel < 0.5 mm, ± ascending. **FL:** scented; calyx 1–1.5 mm, 2–2.5 mm in fr, deciduous, lobes oblong, short rough-hairy, hairs ascending, midvein not thickened, bristles 2–4 per lobe, coarse, spreading; corolla deciduous, limb 3–6 mm diam, appendages prominent, deep yellow. **FR:** nutlets (1)2(3), 1.5–2 mm, lance-ovate, pale to dark gray, gen smooth or occ papillate, margin rounded to basally ± sharp-angled, base wide-truncate, tip ± beaked; abaxially ± flat, widened near base, sharp ridged at tip; adaxially low biconvex, attachment scar edges not raised, abutted toward tip, ± gapped ± lower 1/2 or abutted entire length, fork-gapped at base; axis to nutlets or shorter. Gravelly or rocky areas, open conifer forest, occ chaparral; 1770–3000 m. sw SNH. May–Aug ★

C. intermedia (A. Gray) Greene Ann 10–60+ cm, erect. **LF:** 1.5–5(7) cm, gen strigose and ± spreading-bristly on margins, midveins, occ ascending rough-hairy, bulbous-based hairs ± dense. **INFL:** bracts gen 0. **FL:** calyx 2–3 mm, lobes appressed, dense-rough-hairy, midvein ± thickened, bristles 0 or spreading; corolla deciduous, appendages ± yellow. **FR:** nutlets 1–4, fine-grainy to low-tubercled (smooth), ± rough, gen dull, abaxially ± rounded, ridge 0. $2n=24$.

var. *hendersonii* (A. Nelson) Jeps. & Hoover (p. 469) HENDERSON'S CRYPTANTHA **ST:** gen 1, branches 0–few, ± stiff; ascending- to spreading-rough-hairy. **LF:** linear to oblanceolate, ascending rough-hairy, some hairs bulbous-based. **INFL:** in 2s to 4s; pedicel ± 0.5 mm, ascending. **FL:** calyx 3.5–5 mm in fr, not appressed to st in fr, ovate, truncate at base, not constricted above, gen late-deciduous, lobes narrow-oblong, tips ± erect, hairs appressed or ascending to occ spreading, bristles 0; corolla limb 4–8 mm diam. **FR:** nutlet gen 1–2, 2.2–2.5 mm, ± ovate, low tubercled (smooth), gen dark ± gray, margin ± sharp-angled, occ rounded, base rounded, tip acute; adaxially biconvex to ± flat, attachment scar edges not raised, abutted or narrow-gapped entire length, wide-forked but not gapped at base; axis to nutlets or shorter. Loamy to rocky soils, often volcanic, gen open conifer forest, sagebrush scrub, occ grassland, chaparral; 150–1830 m. occ NW (exc NCo), CaR, n SNH (e slope), MP; to BC, w ID, nw NV. May–Jul

var. *intermedia* **ST:** 1–several, branches throughout, gen slender; ± long-spreading-fine-bristly. **LF:** 1.5–5 cm, linear to lanceolate, occ oblong. **INFL:** in (2s) 3s (5s); pedicel < 0.5 mm, erect to ascend-

ing. **FL:** calyx 4–5.5 mm in fr, ± ovate-oblong, ± narrow-rounded at base, ± constricted above, lobes lance-linear to occ lanceolate, tips erect to ± spreading, margin appressed-hairy, long-spreading-bristly, esp below middle; corolla limb 3–6 mm diam. **FR:** nutlet gen 3–4, 1.4–2.1(2.2) mm, ± dense-warty to tubercled (sharp-papillate or low-tubercled), gray-brown, often mottled, margin rounded to occ basally sharp-angled, base rounded to narrow-truncate, tip tapered, occ ± beaked; adaxially flat to occ low biconvex, attachment scar edges ± raised near tip, gen narrow-gapped near tip, abutted medially or ± abutted entire length, triangular-flare-gapped at base; axis gen to or beyond nutlets. Sandy to gravelly flats, slopes, gen granitic or serpentine-based, grassland, chaparral, foothill woodland, occ open conifer forest; < 2300 m. NW (exc NCo), SN (w slope), ScV (Sutter Buttes), CW (exc CCo), SW (common); to AZ, Baja CA. May–Jul

C. leiocarpa (Fisch. & C.A. Mey.) Greene BEACH CRYPTANTHA
Ann 5–20(40) cm, often ± mat-forming. **ST:** branches many, from base, becoming decumbent or prostrate; strigose to rough-hairy; hairs ascending to spreading. **LF:** 1–3.5 cm, linear to narrow-oblanceolate, strigose to rough-hairy or bristly, hairs ascending; some bristles bulbous-based. **INFL:** in 1s or 2s (3s), dense, 1–3 fl, not elongated in fr; bracts gen lf-like throughout; pedicel gen 0. **FL:** calyx 1.5–2 mm, 2.5–4 mm in fr, late-deciduous, lobes lance-linear, dense-strigose and rough-hairy, hairs ascending, midvein ± thickened, bristles spreading, ± tan; corolla late-deciduous, limb 1–2.5(3.5) mm diam, appendages pale yellow. **FR:** nutlets (3)4, 1.6–2 mm, narrow-ovate to ovate, gen mottled, ± smooth, shiny (dull), margin rounded, base gen rounded, tip acute; abaxially rounded, ridge 0; adaxially gen low-rounded, attachment scar edges not raised, abutted entire length or occ short forked-gapped at base; axis to or beyond nutlets. Sandy soils, coastal dunes, beaches; < 100(250) m. NCo, CCo, SCo; coastal s OR. Mar–Aug

C. mariposae I.M. Johnst. (p. 469) MARIPOSA CRYPTANTHA
Ann 8–25 cm. **ST:** branches 0 or gen few; sparse-strigose and gen ascending to occ spreading ± rough-hairy. **LF:** lowest 1–3 gen opposite; 1–2 cm, oblong to oblanceolate; sparse-short-bristly, bristles gen ascending, bulbous-based abaxially. **INFL:** in 1s (3s); bracts gen 1–few; pedicel ± elongated in fr, < 0.5 mm, ascending to erect. **FL:** calyx 1.5–3 mm, 4–6(7) mm in fr, late-deciduous, lobes ± lanceolate, margin ± strigose, midvein thickened, gen dense-spreading-bristly; corolla late-deciduous, limb (2)3–6 mm diam, appendages prominent, ± deep yellow. **FR:** nutlets (2)3–4, 1.9–2.2 mm, ovate, gen fine-grainy, ± tubercled, ± brown, dull to shiny, margin rounded, base wide-truncate, tip abruptly beaked; abaxially low-rounded, ridge 0; adaxially gen biconvex to occ ± flat, attachment scar edges not raised, abutted toward tip, ± gapped lower ± 1/2, ± flared- or wide-fork-gapped at base; axis beyond nutlets. Rocky, semi-barren serpentine ridges, slopes, gen chaparral; 200–650 m. n&c SNF. Apr–Jun ★

C. maritima (Greene) Greene (p. 469) GUADALUPE CRYPTANTHA
Ann 10–30(40) cm. **ST:** ascending to erect, red-brown in fr, branches 0 to throughout; gen strigose and spreading-rough-hairy, occ also sparse-long-white-spreading-soft-hairy. **LF:** 1–4 cm, linear to narrow-oblanceolate, ± rough, appressed-bristly, hairs ± bulbous-based. **INFL:** in 1s or 2s, gen not elongated in fr; bracts irregular, lf-like throughout; pedicel < 0.5 mm, ascending. **FL:** calyx 1.5–2 mm, 2–3 mm in fr, constricted above, late-deciduous, lobes lance-linear, dense-long-soft- to -rough-hairy, hairs ascending, occ spreading-bristly; corolla limb 0.5–1.5 mm diam, appendages minute, pale yellow. **FR:** nutlets 1–2, 1.5–2 mm, lanceolate, 1 smooth, shiny, brown, margin rounded, base ± rounded, a 2nd, if present, fine-tubercled, ± dull, ± gray, deciduous; abaxially rounded, ridge 0; adaxially rounded to biconvex, attachment scar edges abutted entire length to occ overlapped or ± gapped; axis to nutlet tips. Common, widespread. Sandy to gravelly soils; < 1250 m. s SCo (San Diego Co.), ChI, e SnBr, e PR, D; sw NV, AZ, Mex. Mar–May

C. micrantha (Torr.) I.M. Johnst. Ann, open, ± cushion-like; root, st base red-purple. **ST:** branches thread-like; strigose. **LF:** ± crowded at base, near infl; 0.3–0.7 cm, oblong to oblanceolate, hairs ± ascending, short-bristly. **INFL:** in 1s or 2s, axillary or in branch forks, dense in fr; bracts throughout; pedicel 0.5–0.8 mm, erect. **FL:** calyx ± 1–1.5 mm, 1.8–2.5 mm in fr, ± persistent, lobes linear-oblong, strigose. **FR:** nutlets (3)4, 1–1.3 mm, ± lanceolate, white-grainy to tubercled or

smooth, ± shiny, margin rounded, base rounded; abaxially rounded, ridge 0; adaxially ± flat, attachment scar edges narrow-gapped entire length, ± flared at base; axis beyond nutlets.

var. **lepida** (A. Gray) I.M. Johnst. MOUNTAIN RED-ROOT CRYPTANTHA Pl 8–15+ cm, gen taller than wide. **ST:** branches few. **FL:** corolla limb 1.5–4 mm diam, center yellow, appendages prominent, yellow. 2*n*=24. Mtn slopes, flats, valleys, granite-based gravelly soils, gen conifer forest, also chaparral, foothill woodland, Joshua-tree woodland; 300–2800 m. s SN, Teh, TR, PR, n SNE; Baja CA. Mar–Aug

var. **micrantha** (p. 469) RED-ROOT CRYPTANTHA Pl 2.5–10 cm, gen wider than tall. **ST:** branches many, spreading (erect). **FL:** corolla limb 0.5–1.2 mm diam, appendages minute, ± white. 2*n*=24. Desert flats, washes, sandy to fine-gravelly soils; < 1900 m. SW (exc ChI), SNE, D; to UT, TX, n Mex, Baja CA. Feb–Jun

C. micromeres (A. Gray) Greene (p. 469) MINUTE-FLOWERED CRYPTANTHA Ann 5–50 cm; slender. **ST:** branches gen throughout; short-rough-hairy, hairs spreading. **LF:** basal rosette gen well developed; 1–4.5 cm, linear to oblong, short-bristly, hairs abaxially sparse, bulbous-based. **INFL:** in 1s to 4s, slender; bracts 0; pedicel gen 0. **FL:** calyx < 1 mm, 1–1.5 mm in fr, ± dense-spreading-hairy, ± late-deciduous, lobes lance-ovate, tips constricted, together, margins ciliate, midvein ± thickened, dense-strigose and spreading-bristly, some bristles hooked at tip; corolla late-deciduous, limb ± 0.5 mm diam, appendages 0. **FR:** nutlets 4, 0.7–1 mm (1 ± > others), triangular-ovate, ± brown, gen shiny, margin rounded, base truncate, tip acute, ± larger 1 ± smooth, 3 smaller tubercled, white-papillate; abaxially low-rounded, ridge 0; adaxially biconvex, attachment scar edges ± raised, abutted to narrow-gapped entire length, triangular-forked-gapped at base; axis gen to nutlet tips. Open sites, disturbed, coarse soils, chaparral, woodland, burns; < 1350 m. n&c SNF, CW (exc SCoRI), SW; Baja CA. Mar–Jul

C. microstachys (A. Gray) Greene TEJON CRYPTANTHA Ann 10–50 cm. **ST:** ± red-brown, branches 0–many, above middle; strigose and spreading- to ascending-rough-hairy to bristly. **LF:** basal rosette gen well developed; 0.5–5(6) cm, linear to oblong, gen spreading-hairy (strigose), lower midvein, margin ± sparse-spreading-bristly. **INFL:** in 1s to 3s; bracts 0; pedicel 0. **FL:** calyx < 1 mm, (1)1.5–2 mm in fr, constricted above, late-deciduous, lobes ± linear, tips erect to ± spreading, rough-hairy, hairs spreading to ascending, midvein ± thickened, with coarse spreading bristles; corolla late-deciduous, limb 0.5–1.5(2.5) mm diam, appendages minute, ± white. **FR:** nutlets 1(2), 1.3–1.6 mm, lanceolate, smooth, shiny, margin rounded, ± plump, base ± rounded, tip beaked; abaxially rounded, ridge 0; adaxially low-rounded, attachment scar edges abutted entire length or short-forked-gapped at base; axis not to nutlet tips. Open sites, chaparral, woodland, burns; 50–1950 m. NCoRI, s SN, Teh, SnFrB, SCoR, SW (exc ChI); Baja CA. Apr–Jun

C. milobakeri I.M. Johnst. MILO BAKER'S CRYPTANTHA Ann 10–50 cm. **ST:** branches ± throughout; strigose and spreading-rough-hairy. **LF:** occ lowest opposite; 0.5–3 cm, linear-oblong to lance-linear, midveins sunken adaxially, dense-rough-hairy to bristly, hairs ascending. **INFL:** in 2s or 3s; bracts 0, occ 1 near base; pedicel < 0.5 mm, spreading. **FL:** calyx 1.5–2 mm, 3–4 mm in fr, deciduous, lobes linear-oblong, hairs ± tan, dense, ± ascending long-shaggy, esp on margin, midvein not thickened, sparse-hairy, not bristly; corolla limb 2–6 mm diam, appendages white or yellow. **FR:** nutlet 1(2), 1.5–2 mm, lance-ovate, dark brown, ± smooth, ± shiny, margin rounded, base wide-truncate, tip acute; abaxially low-rounded, ridge 0; adaxially ± flat, ridge sharp-angled, attachment scar edges abutted entire length, raised, forked at base; axis not to nutlet tip(s). Rocky or gravelly, gen non-serpentine soils, gen open conifer forest, chaparral; 120–1750 m. NW (exc NCo), CaR, n SNH (Butte Co.); sw OR. Apr–Jul

C. mohavensis (Greene) Greene MOJAVE CRYPTANTHA Ann 10–40 cm. **ST:** branches 0 to gen throughout; rough-hairy, hairs ascending to spreading. **LF:** 0.5–4 cm, linear to oblong, dense-spreading-bulbous-based-bristly. **INFL:** in 2s or 3s; bracts 0; pedicel 0.5 mm, ascending to spreading. **FL:** scented; calyx 1.5–2 mm, 3–5 mm in fr, late-deciduous, lobes lanceolate, dense-rough-hairy, hairs ascending, midvein ± thickened with few to many coarse, ± spread-

ing bristles; corolla deciduous, limb (4)5–7.5 mm diam, appendages prominent, deep yellow. **FR:** nutlets (3)4, 2–2.5 mm, ± ovate, ± smooth, margin sharp-angled or a ± flat linear rim esp near tip, base ± wide-truncate, tip ± acute; abaxially ± flat, widened near base, low-ridged near tip; adaxially low biconvex, attachment scar edges abutted toward tip, ± to wide-gapped lower 1/2, or abutted entire length, fork- to triangular-gapped at base; axis beyond nutlets. Open granitic-based gravelly or rocky slopes, gen pinyon/juniper woodland, occ creosote-bush scrub to open conifer forest; 500–2650 m. s SNH (e slope), Teh, n TR, w DMoj. Apr–Jul

C. muricata (Hook. & Arn.) A. Nelson & J.F. Macbr. Ann, gen stout, gen gray- to yellow-green; root occ red-purple. **LF:** 0.5–5+ cm, linear to narrow-oblanceolate. **INFL:** in (1s) 2s to 5s; bracts gen 0; pedicel not elongated in fr, < 0.5 mm, gen ascending, occ spreading. **FL:** calyx lobes ± lanceolate, tips ± constricted to erect, occ ± yellow, margins dense-ascending rough-hairy, midvein ± thickened. **FR:** nutlets (2)3–4, occ mottled, base truncate; abaxially gen low-biconvex (low-rounded), ridge rounded (obscure); adaxially ± biconvex (± flat), attachment scar edges not raised.

var. **denticulata** (Greene) I.M. Johnst. PRICKLY-NUT CRYPTANTHA Pl 10–40(50) cm. **ST:** gen 1, branches few, gen below middle; sparse-strigose, soft-hairy, and conspicuously spreading-rough-hairy. **LF:** gen few, scattered; ± sparse-appressed-soft-hairy and spreading-rough-hairy to bristly. **FL:** calyx 1.5–2.5 mm, 3–4.5 mm in fr, deciduous, lobe midvein dense-spreading-bristly; corolla late-deciduous, limb 1.5–3.5 mm diam, appendages pale yellow. **FR:** nutlets 1.8–2 mm, lance-ovate to ovate, gray, fine- to coarse-brown-tubercled, margin gen rounded at tip, sharp-angled at base, occ sharp-angled (rounded), tip ± beaked; attachment scar edges gen abutted near tip, ± gapped lower 1/2 or occ abutted entire length, triangular-flared at base; axis beyond not to nutlets. Locally common, open conifer forest, often sagebrush scrub; 1800–2770 m. e slope SNH, Teh, SnGb, SnBr, ne PR, n SNE, s W&I, D; w NV, AZ, Baja CA. Apr–Aug

var. **jonesii** (A. Gray) I.M. Johnst. JONES' CRYPTANTHA Pl 20–100 cm. **ST:** gen 1–several, central axis evident, branches many, stiff; dense-strigose, soft-hairy, and sparse-spreading-rough-hairy. **LF:** gen many, reduced above; densely ± appressed-soft-hairy and sparse-spreading-rough-hairy. **FL:** calyx 1–1.5 mm, 2–2.5(3.5) mm in fr, late-deciduous, lobe midvein sparse-spreading-rough-hairy; corolla deciduous, limb 1.5–2.5(3.5) mm diam, appendages pale yellow to white. **FR:** nutlets 1.1–1.3(1.9) mm, gen triangular-ovate, fine- to coarse-sharp-papillate, ± brown, shiny, margin sharp-angled, tip acute; attachment scar edges abutted entire length, triangular-gapped at base; axis to nutlets or beyond. Coastal scrub, chaparral, foothill woodland; < 2000 m. NCoR, SNF, Teh, e ScV, CW, SW, D; AZ, Baja CA. Mar–Jun

var. **muricata** (p. 469) SHOWY PRICKLY-NUT CRYPTANTHA Pl 10–60 cm; root occ red-purple. **ST:** gen 1, central axis evident, branches many, throughout; dense-strigose, soft-hairy, and conspicuously spreading-rough-hairy. **LF:** gen many, reduced above; densely ± appressed-soft-hairy and sparse- to dense-spreading-rough-hairy. **FL:** occ scented; calyx 1.5–2 mm, 3–4 mm in fr, deciduous, lobe tips gen erect, midvein ± sparse-spreading-rough-hairy, occ bristly; corolla late-deciduous, limb (3)4–8 mm diam, appendages yellow. **FR:** nutlets 1.6–2.1 mm, gen lance- to occ triangular-ovate, ± dense-sharp-papillate to tubercled, margin sharp-angled, tip ± acute; axis beyond nutlets. Coastal scrub, chaparral, foothill woodland, open conifer forest; 25–2130 m. s SnJV, CW (exc CCo), SW. Mar–Jul

C. nemaclada Greene (p. 469) COLUSA CRYPTANTHA Ann 10–35 cm, slender. **ST:** branches 0 or from base; strigose to ascending-rough-hairy. **LF:** 0.5–2.5(3) cm, linear to acutely-oblong, ascending-rough-hairy. **INFL:** in 1s or 2s; bracts 0; pedicel < 0.5 mm, spreading to ascending. **FL:** calyx 2–3 mm, 3–4.5(5) mm in fr, constricted above, deciduous, lobes linear-oblanceolate, tips gen spreading, midvein coarse-long-bristly, bristles spreading to reflexed, short bristles near tip strongly reflexed; corolla deciduous, limb 0.5–1 mm diam, appendages minute, white. **FR:** nutlets 1–2(4), 1.7–2 mm, lanceolate to lance-ovate, smooth, shiny, margin rounded; abaxially rounded, ridge 0; adaxially shallow-biconvex, attachment scar edges not raised, gen abutted, ± flare-gapped at base; axis not to nutlet

tips. Chaparral, foothill woodland, occ juniper woodland; < 200–1600(2000) m. s NCoRI, Teh, e CW (exc CCo), nw WTR. Apr–Jun

C. nevadensis A. Nelson & P.B. Kenn. Ann, gen ± gray. **ST:** 1–several, branches 0 to long, below middle. **LF:** 0.5–5 cm, ± dense-strigose and occ ± spreading-rough-hairy to fine-bristly esp on margin, hairs gen bulbous-based. **INFL:** in 2s or 3s; pedicel gen not elongated in fr, gen < 1 mm. **FL:** calyx late-deciduous, lobe midvein thickened; corolla late-deciduous. **FR:** nutlets 3–4 (1 occ late-deciduous), margin rounded to sharp-angled, base truncate to rounded, tip tapered; abaxially rounded, ridge 0; adaxially flat to low-rounded, attachment scar edges ± not raised, abutted to gapped entire length, triangular-flare-gapped at base; axis ± to nutlet tips.

var. **nevadensis** NEVADA CRYPTANTHA Pl 10–60 cm, becoming sprawling, often supported by other pls. **ST:** gen slender; occ sparse-spreading-rough-hairy. **LF:** linear to oblong. **INFL:** not elongate, clustered in fr; bracts occ. **FL:** calyx 2–3.5 mm, 6–11 mm in fr, gen ovate, constricted above, lobes linear, tips gen spreading, margin gen dense-strigose and ascending rough-hairy, spreading-coarse-bristly; corolla limb 1–2 mm diam, appendages minute, white, occ 0. **FR:** nutlets 1.8–2.3 mm, lanceolate, dull, unevenly warty, coarse-sharp-tubercled (1 occ ± smooth), base truncate. Sandy, gravelly soils, slopes, washes, gen creosote-bush scrub, mixed desert woodland; 60–2220(2750) m. e Teh, n TR, W&I, D (common). Feb–Jul

var. **rigida** I.M. Johnst. RIGID CRYPTANTHA Pl 5–35 cm, ± erect. **ST:** ± short and stiff and ± spreading-rough-hairy to fine-bristly, esp near base. **LF:** linear to oblanceolate. **INFL:** gen elongated in fr; bracts 0. **FL:** calyx 1.5–3 mm, 4.5–7.5 mm in fr, gen not constricted, lobes lanceolate, tips gen erect, margin sparse-strigose, midvein gen spreading-fine-bristly; corolla limb 2–5 mm diam, appendages pale yellow. **FR:** (1.6)1.7–2.1(2.5) mm, lance-ovate, dense-sharp-papillate, dull to shiny, base rounded. Sandy to gravelly soils, gen open slopes, open grassland, woodland; 80–2380 m. s SN, Teh, w SnJV, CW (exc CCo), TR, SnJt, w DMoj, DMtns (New York Mtns); s NV, w AZ. Mar–Jul

C. nubigena (Greene) Payson SIERRA CRYPTANTHA Bien to per 3–30 cm, tufted to cespitose; caudex gen woody, branched. **ST:** gen several, elongate, slender to stout; strigose and spreading-bristly, occ dense-tomentose to ± yellow rough-hairy. **LF:** basal rosettes gen several; 2–5 cm; basal narrow-oblanceolate to spoon-shaped, occ folded; strigose and sparse-soft-bristly to dense-rough-hairy, bristles often bulbous-based. **INFL:** interrupted with scattered smaller axillary clusters below, dense, ± head-like above, gen not elongated in fr; pedicel gen not elongated in fr, 0.5–1(3) mm. **FL:** occ scented; calyx (2)2.5–4 mm, (3.5)4.5–8 mm in fr, strigose, white (green) to dense-spreading-bristly, ± yellow; corolla (late-)deciduous, limb (3)3.5–6 mm diam, appendages yellow. **FR:** nutlets 2–4, 2.2–4.5 mm, lanceolate to wide-ovate, occ ± green, ± shiny (dull), margin a ± flat narrow rim; abaxially gen irregularly wrinkled to ± tubercled, occ smooth, ridge rounded; adaxially ± smooth, attachment scar occ off-center, edges gen abutted entire length (gapped at base), not raised; axis to or beyond nutlets. Slopes, ridges, gravel, scree, talus, occ dry meadows, open pine forest; (2400)2600–3900+ m. c&s SNH, ne SNE, W&I; w NV. Jun–Aug

C. oxygona (A. Gray) Greene SHARP-NUT CRYPTANTHA Ann 10–45 cm. **ST:** branches 0 or from base; strigose and ± sparse-ascending- to -spreading-rough-hairy, hairs ± bulbous-based. **LF:** 0.5–3(4) cm, narrow-oblanceolate to linear, hairs fine, ± appressed to ascending, occ spreading, ± bulbous-based. **INFL:** in 2s or 3s; bracts 0; pedicel ± 0.5 mm, ascending to spreading. **FL:** gen scented; calyx 1–1.5(2) mm, 2.5–4 mm in fr, occ ± constricted above, deciduous, lobes lanceolate, margins gen ascending soft-hairy, midvein ± thickened with few, coarse-bristles; corolla deciduous, limb 4–9 mm diam, appendages prominent, yellow. **FR:** nutlets 4, 1.8–2.5 mm, ± ovate, ± gray or mottled, dense-papillate to white-grainy, shiny, margin a ± flat narrow wing, occ ± toothed, wider near tip, base wide-truncate, tip ± acute; abaxially low-rounded, ridge 0; adaxially shallow-biconvex, attachment scar edges abutted near tip, lower 1/2 ± gapped or abutted entire length, triangular-gapped at base; axis gen beyond nutlets. Open, gravelly sites, slopes, washes, gen foothill woodland to pinyon/juniper woodland or sparse conifer forest, occ desert; 200–2450 m. c&s SNF, SNH (e slope), Teh, SnJV, SCoRI, SW (exc ChI), n SNE, w DMoj; w-c NV. Mar–Jul

C. pterocarya (Torr.) Greene Ann 10–40(50) cm. **ST**: branches 0–few, occ many, throughout; strigose and short-rough-hairy, hairs gen ascending, occ spreading. **LF**: 0.5–5 cm, linear to oblong below, ± lanceolate above, gen strigose to short rough-hairy, hairs occ spreading on margins, gen bulbous-based. **INFL**: in (1s) 2s or 3s; bracts gen 0; pedicel 0.5 mm, occ elongated in fr to 1(1.5) mm, spreading to occ ascending. **FL**: calyx ± constricted above, deciduous, lobes lanceolate, gen ± ovate in fr, margins sparse- to dense-strigose or ascending-rough-hairy, ± brown, midvein ± thickened with 0–few spreading bristles; corolla deciduous, appendages minute or 0. **FR**: nutlets (3)4, 2–2.5 mm, lanceolate to lance-oblong, gray, ± brown, or mottled gray and brown, gen papillate to white-tubercled, dull to shiny, base truncate, tip ± elongate; abaxially low-rounded, spine 0; adaxially biconvex, occ smooth, attachment scar edges not raised, abutted toward tip, ± gapped lower 1/2 or abutted entire length, ± triangular-gapped at base; axis gen to nutlet tips.

  var. *cycloptera* (Greene) J.F. Macbr. TUCSON CRYPTANTHA Pl gen stout, yellow-green, basal rosette gen not well developed. **FL**: calyx 2–3 mm, 4–5(6) mm in fr, late-deciduous, lobes sparse-rough-hairy, esp in fr; corolla limb 1–1.5 mm diam, appendages minute, pale yellow. **FR**: nutlets similar, margin a ± flat, lobed wing > 0.5 mm wide, ± brown or mottled gray and brown; abaxially low white-tubercled, shiny; adaxially smooth, dull. Gravelly to rocky soils, slopes, washes, often limestone-based, gen creosote-bush scrub, desert woodland, occ pinyon/juniper woodland; < 1400 m. s SNH, D; to s UT, AZ, w TX, n Mex. Mar–May

  var. *pterocarya* (p. 469) WINGED-NUT CRYPTANTHA Pl gen slender, green, basal rosette gen not well developed. **FL**: calyx 2–3 mm, 3–4(6) mm in fr, deciduous; lobes dense-strigose; corolla limb 1–2 mm diam, appendages 0 to minute, pale yellow to white. **FR**: nutlets dissimilar, 1 persistent, margin narrow, rounded, not a wing, (2)3 deciduous, margin a ± flat lobed wing > 0.5 mm wide, gray or mottled gray and brown, grainy to white-tubercled, shiny. Sandy to gravelly soils; (100)200–2630 m. s SN, Teh, w SnJV, n TR, e PR, GB, D; to WA, ID, CO, AZ. Mar–Jul

  var. *purpusii* Jeps. PURPUS' CRYPTANTHA Pl slender, green; basal rosette well developed. **ST**: branches gen many, gen from base. **LF**: basal, gen linear. **FL**: calyx 1.5–2.5 mm, 2.5–3(3.5) mm in fr, constricted above, deciduous, lobes not widening in fr, dense-strigose; corolla limb 1.5–2.5 mm diam, appendages minute, pale yellow or white. **FR**: nutlets gen dissimilar, 1 persistent, margin ± sharp-angled, rim 0, (2)3 deciduous, margin a ± flat linear rim or lobed wing < 0.5 mm wide, ± brown or mottled gray and brown, ± dense-papillate, shiny. Sandy to gravelly soils, desert slopes, gen juniper woodland; 900–2200 m. s SNH (e slope), n SnBr, e PR, W&I, n DMtns; n Baja CA. Mar–Jun

C. racemosa (A. Gray) Greene (p. 469) SHRUBBY CRYPTANTHA Per 20–100 cm, gen fl 1st yr, becoming rounded, shrubby, ± woody. **ST**: branches many, throughout, spreading to ascending, those distal ± wiry; canescent and sparse-spreading-bristly. **LF**: 1.5–4(6) cm, narrow-oblanceolate to linear, bulbous-based-bristly. **INFL**: raceme-like, gen in 1s, open, few-fld; bracts minute; pedicel elongated in fr, 1–4(6) mm, spreading. **FL**: calyx 1–2 mm, 2–4 mm in fr, ± persistent, lobes lance-linear, bristly in lower 1/2, bristles spreading to reflexed, midvein thickened; corolla persistent, limb 1–3 mm diam, flat, appendages yellow. **FR**: nutlets 3–4 (1 > others), 0.8–2 mm, persistent, long-ovate, dark gray with pale tubercles, ± shiny, margin a ± flat rim, wider near tip; abaxially low-rounded, sparse-fine-tubercled to white-grainy, ridge 0; adaxially biconvex, sparse-tubercled, attachment scar edges not raised, ± abutted near tip, wide-triangular-gapped; axis beyond nutlets. Rocky slopes, crevices, washes, canyons, desert scrub; < 1400(1670) m. n SnBr, e PR, n W&I, D; s NV, w AZ, Baja CA. Feb–May

C. rattanii Greene RATTAN'S CRYPTANTHA Ann 15–60 cm, slender. **ST**: branches 0 to throughout; bristles appressed to spreading. **LF**: few, 1.5–6 cm, linear to narrow-lanceolate, margin minute-appressed- to -spreading-bulbous-based-bristly. **INFL**: in 2s or 3s; bracts 0; pedicel ± 0. **FL**: calyx 1–1.5 mm, 2.5–3(3.5) mm in fr, deciduous, enlarged below, ± constricted above, lobes erect to ± spreading, linear, dense-rough-hairy to bristly, hairs ascending to spreading, midvein ± thickened; corolla limb 3–6 mm diam, appendages yellow. **FR**: nutlets 1–3, 1.5–2 mm, lanceolate, ± gray, fine-

grainy, ± dull, margin rounded, base rounded, tip ± beaked; abaxially rounded, ridge ± 0; adaxially ± flat, attachment scar edges ± raised, abutted entire length, narrow-gapped at base; axis not to nutlet tips. Rocky, gravelly slopes, grassland, coastal scrub, foothill woodland; 150–780 m. n SCoR. [*C. decipiens* (M.E. Jones) A. Heller, misappl.] Apr–Jul ★

C. recurvata Coville (p. 469) CURVED-NUT CRYPTANTHA Ann 5–40 cm, slender. **ST**: branches 0, from base, or throughout, coarse-strigose. **LF**: 1–3 cm, linear to narrow-oblanceolate, strigose with many appressed to spreading bulbous-based bristles. **INFL**: in 2s or gen 1s; bracts 0; pedicel gen 0. **FL**: calyx 1.5–2 mm, 2–3.5 mm in fr, recurved in fr, late-deciduous, ± constricted above, lobes narrow-linear, dense-ascending- to -spreading-tan-hairy, midvein thickened, spreading-bristly; corolla limb 0.5–1(1.5) mm diam, appendages 0. **FR**: nutlet 1(2, then 1 > other), 1.6–2 mm, lanceolate, incurved, brown, grainy-papillate, dull, margin rounded, base rounded, tip ± beaked; abaxially ± biconvex, ridge low; adaxially ± flat, incurved at tip, attachment scar edges, gen ± abutted near tip, flare-gapped at base; axis not to nutlet tip(s). Sandy to rocky soils, creosote-bush scrub, Joshua-tree woodland; 650–1700(1830) m. SNE, DMoj; to OR, ID, CO, AZ. Mar–Jun

C. roosiorum Munz (p. 469) BRISTLECONE CRYPTANTHA Per 1–2(3) cm; caudex woody, branched. **ST**: ± many, ± ≥ basal lvs, spreading-soft-bristly. **LF**: basal rosettes gen many; 0.5–1.2(1.5) cm, oblanceolate to narrow-spoon-shaped, ± thick, gray-green, ± dense-silky-hairy to tomentose, bristles mostly appressed, bulbous-based. **INFL**: dense, ± head-like, not elongated in fr; pedicel short-elongated in fr, 1.5–3 mm. **FL**: calyx 2.5–3 mm, 3.5–5(6) mm in fr, dense-strigose, ± spreading-bristly; corolla late-deciduous, aging ± brown, tube 2–3 mm, limb 4.5–5.5 mm diam, appendages yellow. **FR**: nutlets gen 1–3, 2–2.5 mm, lance-ovate to ovate, ± constricted above, ± dark gray, ± shiny to dull, margin sharp-angled or occ rounded; abaxially dense-tubercled with low-rounded wrinkles or papillate, ridge rounded; adaxially tubercled to papillate, attachment scar edges abutted near tip, ± wide-gapped lower 1/2, narrow-triangular, edges not raised; axis beyond nutlets. Rocky, dry meadows in open bristlecone pine-limber pine forest; 2570–3230 m. s W&I (n Inyo Mtns). Jun–Jul ★

C. rostellata (Greene) Greene RED-STEMMED CRYPTANTHA Ann 5–20+ cm, ± stout, stiff. **ST**: branches few, above to occ throughout; ± red, strigose and canescent. **LF**: gen several opposite pairs below infl; ascending, 1–1.5 cm, ± oblanceolate,± thick, persistent, hairs ascending to spreading. **INFL**: in 1s or 2s, ± stiff; bracts 0; pedicel < 0.5 mm, spreading to ascending. **FL**: calyx 2–3 mm, 3–3.5(4) mm in fr, ± persistent, lobes lanceolate, margins sparse-ciliate, midvein with many stout, hooked or curved bristles occ basally encrusted with exudate; corolla limb 1–2 mm diam, appendages minute, pale yellow. **FR**: nutlet 1, 2.3–2.8 mm, lance-ovate to lanceolate, shiny, margin rounded to sharp-angled, base wide-truncate, tip ± beaked; abaxially rounded, ridge 0; adaxially flat, attachment scar edges abutted near tip, ± gapped lower 1/3, wide-forked at base; axis not to nutlet tip. Open, rocky, dry sites, sparse grassland, chaparral, foothill woodland; 40–800 m. NCoRI, n SNF, ScV, nw MP; e WA, e OR. Apr–Jun

C. schoolcraftii Tiehm SCHOOLCRAFT'S CRYPTANTHA Bien (per) 6–25 cm, cespitose; caudex not woody, unbranched. **ST**: 1–many, central gen > lateral; ± gray, tomentose and spreading-bristly. **LF**: basal rosettes gen 1; 1.5–4 cm, ± spoon-shaped, thick, ± gray, dense-tomentose and appressed-bristly, abaxial bristles bulbous-based. **INFL**: dense, head-like to ± cylindric; pedicel not elongated in fr, 0–1 mm. **FL**: calyx 2–4 mm, 5–7 mm in fr, dense-bristly; corolla deciduous, limb 3–6 mm diam, appendages yellow. **FR**: nutlets gen 2–4, 2.2–3.2 mm, lanceolate to ovate, ± gray, ± shiny, margin a ± flat narrow rim; abaxially smooth to ± roughened, ridge rounded; adaxially smooth to ± tubercled, attachment scar edges abutted or occ ± gapped at base, edges not raised. Barren ash deposits, sagebrush scrub; 1700–1750 m. e MP (Coppersmith Hills, Lassen Co.); to se OR, nw NV. Jun–Jul ★

C. scoparia A. Nelson GRAY CRYPTANTHA Ann 5–30 cm, ± gray; central axis ± poorly defined. **ST**: slender, stiff, branches throughout; dense-strigose, occ spreading-rough-hairy. **LF**: 1–4(5) cm, linear to lance-linear, strigose, occ spreading-rough-hairy, hairs minute-bulbous-based. **INFL**: in 1s or 2s, occ crowded; bracts 0; pedicel ± 0. **FL**: calyx 1.5–2.5 mm, 4–6 mm in fr, not constricted

above, lobes ± erect, gen persistent, lance-linear, 1 gen longer, margin with ascending, short rough hairs, midveins thickened, gen coarse spreading-bristly, hairs occ bulbous-based; corolla late-deciduous, limb 1–1.5 mm diam, appendages 0. **FR:** nutlets gen 4, 1.6–2.1 mm, lanceolate, fine-sharp-papillate, mottled gray-brown, dull, margin ± sharp-angled, base truncate, tip tapered to ± beaked; abaxially low-rounded to ± flat, ridge 0; adaxially flat, attachment scar edges not raised, abutted entire length or narrow-gapped below middle, gradually forked, flare-gapped at base; axis gen to nutlet tips. Sandy, gravelly soils, open slopes, flats, gen sagebrush scrub, pinyon-pine woodland, occ Joshua-tree woodland; 1480–2740 m. ne SNE, W&I, ne DMoj; to WA, ID, UT. Apr–Jul ★

C. ***similis*** K. Mathew & P.H. Raven DOME CRYPTANTHA Ann, ascending to spreading, 3–12 cm diam, cushion-like; root, pl base gen red-purple. **ST:** 0.5–10 cm, branches few–many, throughout, slender; coarse-strigose, hairs gen ascending. **LF:** ± crowded at ends of branches; 0.3–1 cm, linear to oblanceolate, margin rolled-under, rough-hairy, hairs ± ascending to appressed. **INFL:** in 1s to 3s, in axils or branch forks, 1–5-fld, dense in fr; bracts throughout; pedicel < 0.5 mm, erect. **FL:** calyx 1.5–2 mm, 2.5–3 mm in fr, circumscissile below middle in fr, ± remaining green, lower tube persistent, scarious in age, cup-like in fr, hairs ± like lvs; corolla limb 3.5–6 mm diam, appendages prominent, yellow. **FR:** nutlets (2)3–4, 1.2–1.5 mm, triangular-ovate, brown, fine-grainy, shiny, margin sharp-angled, base truncate, tip not beaked; abaxially low-rounded, ridge 0; adaxially rounded, attachment scar edges ± abutted, forked at base; axis gen to nutlet tips. $2n=12$. Gravelly to coarse sandy soils, Joshua-tree woodland, occ pinyon/juniper woodland; 700–1870 m. n SnGb, nw SnBr, se PR, s DMoj. Mar–May

C. ***simulans*** Greene PINE CRYPTANTHA Ann 5–45 cm, slender, pale green. **ST:** 1, branches few; gen strigose, occ ascending rough-hairy. **LF:** scattered, lowest ± persistent during fl; 1–5(7) cm, linear to narrow-oblanceolate, abaxial, margin bristles ascending, fine, ± bulbous-based. **INFL:** gen in 2s or 3s, occ 1s, few-fld; bracts 0; pedicel elongated in fr, 0.5–1 mm, erect to ascending. **FL:** calyx 2–3 mm, 4–7(8) mm in fr, constricted above, deciduous, lobes lance-linear, tips spreading, margin ascending-fine-bristly, midvein ± thickened with many short, recurved, coarse bristles; corolla deciduous, limb 1.5–2 mm diam, appendages minute. **FR:** nutlets 3–4, 1.9–2.2(2.5) mm, wide-ovate, dense-grainy, sparse-tubercled, ± gray, dull, margin rounded, base truncate, tip ± beaked; abaxially low-rounded, ridge 0; adaxially flat, attachment scar edges not raised, abutted, wide-forked but not gapped at base; axis not to nutlet tips. Dry gravelly sites, disturbed areas, gen open conifer forest; (250)450–2600 m. KR, NCoRH, CaR, n SNF, SNH, Teh, TR, e PR, MP; to e WA, w ID. May–Aug

C. ***sparsiflora*** (Greene) Greene FEW-FLOWERED CRYPTANTHA Ann 10–30 cm, slender. **ST:** 1, branches few; sparse-strigose. **LF:** lower occ opposite; 1–3 cm, narrow-oblong to -oblanceolate, strigose. **INFL:** in 1s or 2s, few-fld, dense; bracts 0, occ 1–2 near base; pedicel not elongated in fr, < 0.5 mm, spreading to ascending. **FL:** calyx 1.5–2 mm, 2–3 mm in fr, lobes ovate to oblong, margins sparse-ciliate, midvein ± thickened, with short, ascending curved, ± hook-tipped bristles; corolla limb 0.5–1(1.5) mm diam, appendages minute, ± white. **FR:** nutlet 1, ± 2 mm, ovate, ± smooth, ± shiny, margin sharp-angled, base ± truncate, tip ± beaked; abaxially low-rounded, ridge 0; adaxially ± flat, attachment scar edges abutted, wide-forked at base; axis not to nutlet tip. Open, dry, rocky sites, sparse grassland, chaparral, foothill woodland; 300–1300 m. c&s SNF, Teh, e SnFrB, SCoR. Apr–May

C. ***spithamaea*** I.M. Johnst. Ann 8–20 cm. **ST:** branches ± ascending; dense-appressed-soft-hairy. **LF:** occ opposite below infl; 0.8–2.5 cm, linear to oblong, bristles ascending to spreading, occ bulbous-based. **INFL:** in 1s or 2s; bracts gen 0, occ at base; pedicel < 0.5 mm, ascending. **FL:** calyx 1.5–2.5 mm, 3.5–4.5 mm in fr, late-deciduous, lobes linear, tips ± spreading, soft-hairy, with hooked or curved bristles only at tip, midvein not thickened; corolla late-deciduous, limb 3–6 mm diam, appendages yellow. **FR:** nutlet 1(2), 1.6–2 mm, lanceolate, smooth, shiny, margin rounded, base truncate, tip ± beaked; abaxially rounded, ridge 0; adaxially ± flat, attachment scar edges abutted to occ overlapped near tip, ± gapped lower 1/3, triangular-gapped at base; axis not to nutlet tip(s). Open, semi-barren

sites, serpentine gravelly slopes, creekbeds, chaparral, foothill woodland; 270–760 m. n&c SNF. Apr–May

C. ***subretusa*** I.M. Johnst. CRATER LAKE CRYPTANTHA Per 5–20(25) cm, gen cespitose; caudex woody, gen branched. **ST:** ± many, stout, gray-green; ± tomentose with sparse, spreading bristles. **LF:** basal rosettes gen many; 1–4 cm, spoon-shaped, ± notched to obtuse at tip, thick, gray-green, strigose to ± tomentose and sparse-bristly, abaxial bristles appressed, bulbous-based. **INFL:** dense, ± cylindric; pedicel not elongated in fr, 0–1.5 mm. **FL:** calyx 2.5–4 mm, 5.5–7(8) mm in fr, dense-bristly; corolla deciduous, limb 4–6 mm diam, appendages yellow. **FR:** nutlets 2–4, 3–4.8 mm, narrow-ovate (lanceolate), ± shiny, margin a ± flat narrow rim; abaxially convex, tubercled, wrinkled (± roughened), ridge rounded, gen lighter in color; adaxially ± smooth, attachment scar edges gapped (abutted near tip), edges not raised. Pumice, ash, occ serpentine, gen conifer forest to subalpine; (1320)2050–2720 m. se KR (Scott Mtns to Mount Eddy), CaRH, MP; to e OR, n NV. Jul–Aug

C. ***torreyana*** (A. Gray) Greene Ann. **ST:** branched or not. **LF:** gen below middle of st; 2–5(7) cm, linear to oblanceolate, strigose to rough-hairy or bristly, hairs ± ascending, few ± bulbous-based. **INFL:** gen in 2s, gen congested at tips; bracts 0; pedicel ± 0.5 mm, gen not elongated in fr. **FL:** calyx gen constricted above, late-deciduous, lobe tips spreading, midvein ± thickened; corolla late-deciduous, limb 1–2 mm diam, appendages minute, white, or 0. **FR:** nutlets (2)3–4, wide-ovate, smooth (fine-grainy), gen mottled, shiny, margin rounded, base wide-rounded to truncate, tip ± acute; abaxially ± flat, esp toward base, ridge 0; adaxially biconvex, attachment scar edges abutted, occ raised near tip, forked but not gapped at base.

var. ***pumila*** (A. Heller) I.M. Johnst. DWARF CRYPTANTHA Gen 5–20(25) cm. **ST:** sparse-strigose and gen dense-spreading-rough-hairy to bristly. **FL:** calyx 1–1.5(2) mm, (2.5)3–4.5 mm in fr, lobes gen lanceolate, margins sparse-strigose, midvein dense-spreading-bristly. **FR:** nutlets 1.3–1.5 mm; axis gen not to nutlet tips. Gen chaparral, foothill woodland; 150–1280 m. n CW (exc CCo). Apr–Jun

var. ***torreyana*** (p. 469) TORREY'S CRYPTANTHA Gen (10)15–40 cm. **ST:** ± dense-strigose and sparse-spreading-rough-hairy. **FL:** calyx 2–2.5 mm, 4–6.5(8) mm in fr, lobes lance-linear, margin dense-strigose, midvein ± sparse-spreading-rough-hairy, occ bristly. **FR:** nutlets (1.6)1.7–2.2(2.5) mm; axis gen to nutlets or shorter. Slopes, disturbed soils, gen conifer forest, occ chaparral, woodland; 100–2650 m. NW (exc NCo), CaR, SN, GB; to s AK, MT, CO. Apr–Aug

C. ***traskiae*** I.M. Johnst. TRASK'S CRYPTANTHA Ann 5–20 cm, gen decumbent to ± erect, slender. **ST:** branches gen throughout; gen strigose, sparse-spreading-bristly at base. **LF:** 0.5–2 cm, linear to narrow-oblong, strigose to gen ascending-rough-hairy, abaxial bristles sparse (bulbous-based). **INFL:** in 1s or 2s, congested at tips; bracted below; pedicel gen 0. **FL:** calyx 1.5–2 mm, 3–4 mm in fr, deciduous, lobes lance-linear, tips gen spreading, margin dense-ascending rough-hairy, midvein ± thickened, ± dense-coarse-spreading-± yellow-bristly; corolla limb 1.5–2 mm diam. **FR:** nutlets (3)4, 1.2–1.4(1.5) mm, ± ovate, grainy, ± brown, gen mottled, dull, margin rounded, base truncate, tip ± beaked; abaxially low-rounded, ± tubercled above middle, ± ridged; adaxially ± flat, attachment scar edges unevenly raised, ± abutted, small-narrow-triangular-gapped at base; axis gen to nutlet tips. Rocky, gravelly, to sandy open sites, occ stabilized dunes; 5–200 m. s ChI (San Clemente, San Nicolas islands). Feb–Jun ★

C. ***tumulosa*** (Payson) Payson NEW YORK MOUNTAINS CRYPTANTHA Per 7–25 cm; caudex woody, branched. **ST:** gen many; dense-tomentose, hairs ± yellow, spreading. **LF:** basal rosettes gen many; 3–6 cm; basal oblanceolate to ± spoon-shaped, ± folded, thick, ± gray, dense-silky-tomentose, bristles appressed, bulbous-based. **INFL:** dense, cylindric, narrow, gen not elongated in fr; pedicel elongated in fr to 3–6 mm. **FL:** calyx 3.5–4.5 mm, 7–10 mm in fr, bristles dense, ± yellow, soft; corolla deciduous, tube 3.5–4.5 mm, limb 6–8 mm diam, appendages yellow. **FR:** nutlets gen 1–3, 3–4 mm, wide-ovate, occ constricted above, ± gray, dull, margin sharp-angled to a ± flat narrow rim; abaxially ± tubercled, ridge rounded, ± indistinct near base; adaxially ± papillate, tubercled, attachment scar edges wide-triangular-gapped, edges ± raised; axis beyond nutlet(s). Limestone, occ granitic gravel or clay soils, gen pinyon/juniper woodland; 1400–2100 m. e DMoj, n&e DMtns; sw NV (Spring, Charleston Mtns). Apr–Jun ★

C. utahensis (A. Gray) Greene SCENTED CRYPTANTHA Ann 10–30 cm. **ST:** branches 0 to gen throughout; ± dense-strigose. **LF:** 0.5–5 cm, linear to oblong; ± appressed-bristly, some bristles bulbous-based. **INFL:** in 2s or 3s, occ 1s; bracts 0; pedicel < 0.5 mm in fr, spreading to ± downcurved. **FL:** scented; calyx 2–2.5 mm, 2.5–3(4) mm in fr, ± tightly constricted above, deciduous, lobes elliptic to ovate, dense-strigose to rough-hairy near margin, hairs ascending, midvein thickened, ± brown; corolla limb (2.5)3–5 mm diam, appendages prominent, deep yellow. **FR:** nutlet 1(2), 1.7–2.5 mm, lanceolate to lance-ovate, white-grainy papillate to minute-spiny, ± dull, margin distally a ± flat inward bent narrow rim to basally sharp-angled, base rounded, tip ± acute; abaxially low-rounded, ridge 0; adaxially occ smooth, biconvex, attachment scar edges ± narrow-gapped, triangular-flare-gapped at base; axis gen to nutlet(s). Rocky, gravelly areas, slopes, washes, creosote-bush scrub, Joshua-tree and pinyon/juniper woodland; (80)300–2300(2600) m. se SN, ne SnGb, ne SnBr, e PR, s SNE, D; to sw UT, AZ. Mar–Jul

C. virginensis (M.E. Jones) Payson (p. 469) VIRGIN RIVER CRYPTANTHA Bien to short-lived per, 10–40+ cm; caudex not woody, unbranched. **ST:** 1–many, central > lateral; tomentose and with dense, spreading, ± yellow bristles. **LF:** basal rosette gen 1; 5–12 cm, oblanceolate to spoon-shaped, ± folded, thick, gray-green, dense-strigose to silky-tomentose, bristles bulbous-based, sparse, ± spreading abaxially, appressed adaxially. **INFL:** dense, gen cylindric, wider in fr; pedicel to 2–5 mm in fr. **FL:** scented; calyx 3–4 mm,

7–11 mm in fr, dense-spreading-yellow-bristly; corolla deciduous, tube 3–4 mm, limb 7–10 mm diam, appendages yellow. **FR:** nutlets gen 1–2, 3.3–4.5 mm, wide-ovate, occ constricted above, ± tubercled, irregularly wrinkled, gray to tan, ± shiny, margin a ± flat narrow rim to sharp-angled; abaxially rough, ridge sharp-angled; adaxial attachment scar edges wide-triangular-gapped, edges ± raised; axis beyond nutlet(s). Loose, limestone soils; 840–2450 m. SNE, DMoj; to nw AZ, sw UT. Apr–Jun

C. watsonii (A. Gray) Greene WATSON'S CRYPTANTHA Ann 10–35 cm, slender. **ST:** branches 0–few, throughout; thin-strigose and spreading-short-rough-hairy. **LF:** 0.5–4 cm, linear to oblanceolate, short-bulbous-based-bristly abaxially, densely, occ roughly strigose adaxially. **INFL:** in (1s) 2s, gen dense, open in fr; bracts gen 0, occ 1–few near base; pedicel not elongated in fr, < 0.5 mm, ± ascending. **FL:** calyx 1.5–2 mm, 2.5–3.5 mm in fr, deciduous, lobes lanceolate, dense-strigose, midvein ± not thickened with few, coarse, spreading bristles; corolla deciduous, limb 0.5–1.5 mm diam, appendages minute, pale yellow to white. **FR:** nutlets (3)4, 1.4–2 mm, ± lanceolate, ± compressed, gray or mottled, smooth (sparse-papillate), shiny, margin a ± flat narrow rim esp near tip to sharp-angled, base rounded, tip ± elongate; abaxially ± flat, ridge 0; adaxially biconvex, attachment scar edges not raised, ± abutted, short-forked at base; axis gen to nutlet tips. Rocky, gravelly soils, sagebrush scrub, pinyon/juniper woodland, occ conifer forest; 1250–3100(3300) m. SNH (e slope), GB; to WA, MT, CO. May–Aug

CYNOGLOSSUM HOUND'S TONGUE

Per, bien, ± hairy, taprooted. **ST:** erect. **LF:** entire, basal petioled, cauline petioled or not. **INFL:** panicle-like cymes, ± terminal, bracted or not. **FL:** calyx ± deep-5-lobed, enlarged in fr; corolla 5-lobed, funnel-shaped or salverform, blue to red-purple, appendages large; style entire. **FR:** nutlets gen 4, spreading, 5–8 mm diam, ± spheric or disk-shaped, short-barbed-prickly, adaxial attachment scar at tip. 80 spp.: worldwide. (Greek: dog tongue)

1. Infl ± among lvs; bracts ± lf-like; corolla lobes dull red-purple, appendages purple; nutlet disk-shaped, abaxially ± flat, margin raised . . . . ***C. officinale***
1′ Infl above lvs; bracts scale-like or 0; corolla lobes ± blue, appendages white; nutlet ± spheric, abaxially rounded, margin not raised
  2. St glabrous, ± glaucous; basal lf blades 3–10 cm wide, abruptly narrowed to petiole; cauline lvs few, petioled; corolla limb gen 10–15 mm diam . . . . ***C. grande***
  2′ St spreading-hairy, not glaucous; basal lf blade 1–4 cm wide, tapered to petiole; cauline lvs ± many, sessile; corolla limb 4–9 mm diam . . . . ***C. occidentale***

C. grande Lehm. (p. 469) GRAND HOUND'S TONGUE Per. **ST:** gen 1, 3–9 dm, glabrous, ± glaucous. **LF:** abaxially hairy, adaxially ± glabrous; basal petiole 8–15 cm, ± unwinged, blade 8–15 cm, 3–10 cm wide, ± ovate to elliptic, base truncate or cordate, abruptly narrowed to petiole; cauline few, petioled. **INFL:** above lvs; bracts scale-like or 0; pedicels 10–25 mm. **FL:** corolla 8–12 mm, ± salverform, tube gen violet, limb gen 10–15 mm diam, lobes ± blue, appendages white. **FR:** nutlet ± spheric, ascending-spreading, abaxially rounded, margin not raised. 2*n*=24. Chaparral, woodland; 10–1700 m. NW, CaR, SN, SnFrB, SCoR, PR; to BC. Feb–May

C. occidentale A. Gray (p. 469) WESTERN HOUND'S TONGUE Per. **ST:** 1–several, clustered, 1.5–5 dm, spreading-hairy, not glaucous. **LF:** rough-hairy; basal blade 5–15 cm, 1–4 cm wide, ± oblanceolate, base tapered, petiole 4–10 cm, winged; cauline ± many, sessile. **INFL:** above lvs; bracts scale-like or 0; pedicels 4–8 mm. **FL:**

corolla 6–9 mm, bell-shaped, limb 4–9 mm diam, tube, lobes ± blue, gen tinged rose to brown, appendages white. **FR:** nutlets ± spheric, ascending-spreading, abaxially rounded, margin not raised. Open, dry, conifer forest; 900–2600 m. KR, NCoRH, NCoRI, CaRH, SNH, MP; to c OR. May–Jul

C. officinale L. COMMON HOUND'S TONGUE Bien. **ST:** 1, 3–12 dm, ± soft-hairy. **LF:** soft-hairy; basal petiole 4–10 cm, unwinged, blade 8–20 cm, oblanceolate or narrow-elliptic, base tapered; cauline many, sessile. **INFL:** ± among lvs; bracts ± lf-like; pedicels 5–12 mm. **FL:** corolla 3–5 mm, ± salverform, limb 4–9 mm diam, tube, lobes dull red-purple, drying ± blue or not, appendages purple. **FR:** nutlets disk-shaped, descending-spreading, abaxially ± flat, margin raised. 2*n*=24. Disturbed areas, often associated with logging activities; 850–1850 m. KR, CaRH, n SNH; N.Am; native to Eurasia. May–Jul ❖

DRAPERIA

Robert Patterson & Richard R. Halse

1 sp. (J.W. Draper, Am historian, scientist, 1811–1882)

D. systyla (A. Gray) A. Gray (p. 473) Per; caudex woody; hairs soft, long. **ST:** decumbent to erect, occ rooting, 1–4 dm, slender. **LF:** cauline, opposite; lower petioled, upper sessile; blade 1–7 cm, ovate, entire. **INFL:** pedicels 1–2 mm. **FL:** calyx lobes 4–6 mm, 6–9 mm in fr, linear, hairy; corolla 7–14 mm, funnel-shaped, white to pink or

lavender, hairy outside; stamens incl, unequal, unequally attached; ovary chambers appearing 2, style 1, 3–4 mm, incl, lobes 2. **FR:** capsule, 2–3 mm wide, spheric, long-hairy. **SEED:** 1–4, ovoid, angled, dark brown, honeycombed. *n*=9. Woodland, talus, rock crevices; 200–3000 m. KR, CaRH, SN. May–Oct

469

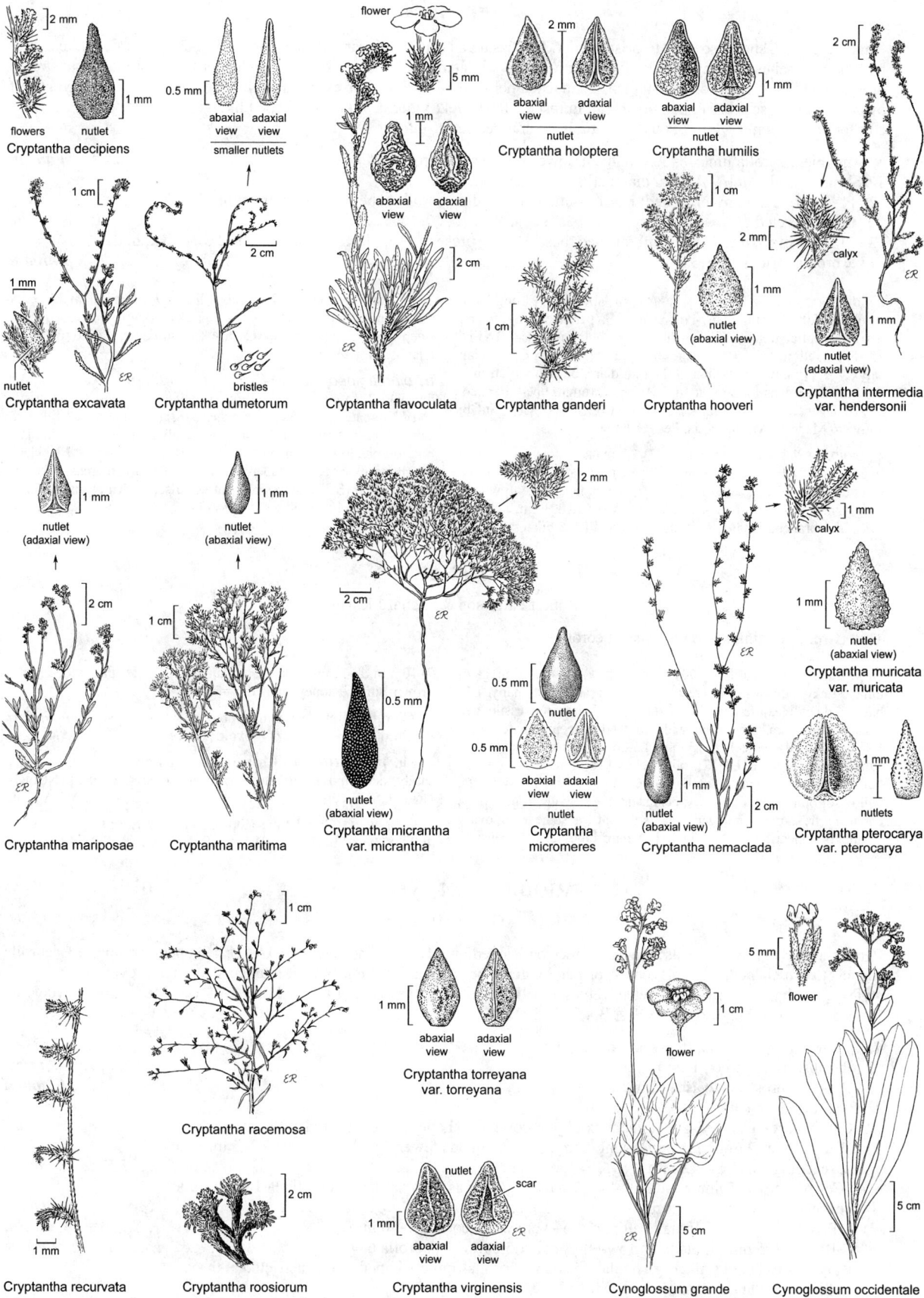

Cryptantha decipiens

flowers   nutlet

2 mm

1 mm

0.5 mm   abaxial   adaxial
          view   view

smaller nutlets

flower

5 mm

1 mm

abaxial   adaxial
view   view

2 mm

abaxial   adaxial
view   view

nutlet

Cryptantha holoptera

2 mm

abaxial   adaxial
view   view

nutlet

Cryptantha humilis

2 cm

1 cm

2 mm   calyx

1 mm

nutlet
(abaxial view)

Cryptantha hooveri

1 mm

nutlet
(adaxial view)

Cryptantha intermedia
var. hendersonii

1 cm

1 mm

nutlet

Cryptantha excavata

2 cm

bristles

Cryptantha dumetorum

2 cm

abaxial   adaxial
view   view

Cryptantha flavoculata

1 cm

Cryptantha ganderi

1 mm

nutlet
(adaxial view)

2 cm

Cryptantha mariposae

1 mm

nutlet
(abaxial view)

1 cm

2 cm

Cryptantha maritima

2 mm

2 cm

0.5 mm

nutlet
(abaxial view)

Cryptantha micrantha
var. micrantha

0.5 mm   nutlet

0.5 mm   abaxial   adaxial
          view   view

nutlet

Cryptantha
micromeres

1 mm   calyx

1 mm

nutlet
(abaxial view)

Cryptantha muricata
var. muricata

1 mm

1 mm   nutlet
        (abaxial view)

2 cm

nutlet
(abaxial view)

Cryptantha nemaclada

1 mm

nutlets

Cryptantha pterocarya
var. pterocarya

1 cm

Cryptantha racemosa

1 mm

Cryptantha recurvata

2 cm

Cryptantha roosiorum

1 mm

abaxial   adaxial
view   view

Cryptantha torreyana
var. torreyana

nutlet   scar

1 mm

abaxial   adaxial
view   view

Cryptantha virginensis

1 cm   flower

5 cm

Cynoglossum grande

5 mm   flower

5 cm

Cynoglossum occidentale

ECHIUM

Ann, bien, [per], shrub; strigose to bristly-hairy. **LF:** basal and cauline, linear to lanceolate, entire. **INFL:** panicle-like cymes, terminal; branches 3–many, ± spike-like. **FL:** radial to ± bilateral; calyx deep-lobed, often longer in fr; corolla throat straight or ± curved, lobes equal or not; stamens 5, attached below mid-tube, incl or exserted; style exserted. **FR:** nutlet erect, short, ovate, 3-angled, scar basal, flat. 40 spp.: s Eurasia, Afr. (Greek: viper, from nutlet shaped like viper's head) Several entities cult for orn, esp on CA coast, some potentially naturalized, some may be hybrids.

1. Ann, bien; corolla limb 18–30 mm diam; lower 2 stamens short-exserted . . . . *E. plantagineum*
1′ Shrub; corolla limb 9–13 mm diam; all stamens long-exserted
  2. Pl long-lived, many-branched; basal rosette 0; fl ± radial, corolla tube ± = calyx; nutlet black,
    fine-tubercled . . . . *E. candicans*
  2′ Pl short-lived, once-fl unbranched; basal rosette present; fl ± bilateral, corolla tube 2+ × calyx; nutlet
    ± brown, prickly . . . . *E. pininana*

E. candicans L. f. (p. 473) PRIDE OF MADEIRA Long-lived shrub 10–20(30) dm, many-branched. **LF:** persistent, 6–25 cm, narrow-elliptic, dense-strigose, few hairs bulbous-based. **INFL:** 15–40 cm, elliptic, dense; branches many, ± spreading. **FL:** ± radial; calyx 4–5(7) mm; corolla limb 9–11 mm diam, blue to violet, tube ± = calyx; stamens long-exserted. **FR:** black, rough, fine-tubercled. $2n$=16. Open, dry slopes, bluffs; < 500 m. CCo, SnFrB, SCo, s SnGb; native to Madeira, Macaronesia. Feb–Oct ❖

E. pininana Webb & Berthel. (p. 473) TOWER OF JEWELS Short-lived, once-fl shrub 20–30+ dm. **ST:** gen unbranched. **LF:** basal rosette lvs 25–50 cm; cauline many, 10–30 cm, lanceolate to widely lance-ovate, sparse-bulbous-based-bristly. **INFL:** 100+ cm, ± dense-cylindric; branches many, ascending, short. **FL:** ± bilateral; calyx 4–5(7) mm; corolla tube 2+ × calyx, limb 10–13 mm diam, blue; stamens long-exserted. **FR:** ± brown, prickly. $2n$=16. Disturbed areas, steep slopes; < 240 m. s NCo, c CCo, sporadic; native to Canary Islands. May–Aug

E. plantagineum L. (p. 473) SALVATION JANE Ann, bien, 4–8 dm. **LF:** 2–15 cm, narrow-elliptic to oblong, bulbous-based-bristly. **INFL:** spreading, 5–15 cm; branches 2–8, ascending. **FL:** ± bilateral; calyx 7–12(17) mm; corolla tube ± 2 × calyx, limb 18–30 mm diam, blue-purple; lower 2 stamens short-exserted, upper 3 incl. **FR:** black, rough, wrinkled to fine-tubercled. $2n$=16. Disturbed areas; < 200 m. c CCo, SCo (San Diego Co.); native to s Eur, widely naturalized. Potentially problematic weed. May–Jul

EMMENANTHE

Robert Patterson & Richard R. Halse

1 sp. (Greek: abiding fl, from persistent corolla)

E. penduliflora Benth. (p. 473) WHISPERING BELLS Ann, glandular, sticky, odorous. **ST:** erect, 5–85 cm, branches 0–many. **LF:** basal and cauline, alternate; lower short-petioled, upper sessile, gen clasping, 1–12 cm, gen < 3 cm wide, toothed to deep-pinnate-lobed. **INFL:** terminal; pedicels 5–15 mm, 12–25 mm in fr, thread-like, ± subspreading to recurved. **FL:** calyx lobes 4–11 mm, 1–4 mm wide, lanceolate to ovate, glandular; corolla 6–15 mm, bell-shaped, white, yellow, or pink, glandular-hairy, persistent in age, withering, papery, enclosing fr; stamens incl, gen equal, attached ± at same level; ovary chambers appearing 2, style incl, 1–4 mm, lobes 2. **FR:** capsule, 7–10 mm, 2–4 mm wide, glandular. **SEED:** 6–15, flat, wide-elliptic, brown; surface honeycombed. n=18.

1. Corolla yellow to cream; fr oblong . . . . var. ***penduliflora***
1′ Corolla white to pink; fr ovoid . . . . var. ***rosea***

  var. ***penduliflora*** Chaparral to creosote-bush scrub, rocky, sandy, decomposed granite, serpentine soils; < 2200 m. NCoRH, NCoRI, SNF, c&s SNH, SnJV, CW, SW, SNE, D; to UT, AZ. Apr–Jul

  var. ***rosea*** Brand Talus slopes, rocky, sandy, or serpentine soils; 400–1800 m. SnFrB, SCoRI, n WTR (Mount Pinos). Apr–Jun

ERIODICTYON YERBA SANTA

Gary L. Hannan

Per to shrub. **ST:** prostrate to ascending or erect; bark shredding. **LF:** cauline, alternate. **INFL:** gen open, terminal. **FL:** corolla funnel- to urn-shaped, white, lavender, or purple, gen hairy abaxially; stamens incl, filaments gen hairy; ovary chambers 2, styles 2, gen hairy. **FR:** 1–3 mm wide; valves 4. **SEED:** striate, dark brown or black. 11 spp.: sw US, Mex. (Greek: wool net, from abaxial lvs) [Ferguson 1998 Syst Bot 23:253–268]

1. Pl low per or subshrub, sts woody at base or not
  2. Pl subshrub, st erect, 1–3 m . . . . *E. parryi*
  2′ Pl per, st prostrate to ascending, 0.16–0.5 m . . . . *E. lobbii*
1′ Pl shrub, sts woody
  3. Calyx, corolla stalked-glandular abaxially; corolla urn-shaped, ± constricted at throat
    4. Calyx lobes 2.5–4 mm, dense-long-hairy, stalked glands fewer than hairs; corolla 3–5 mm, tube
      exceeding calyx by < 2 mm . . . . *E. tomentosum*
    4′ Calyx lobes 4–5 mm, moderate-hairy, stalked glands as many as hairs; corolla 6–10 mm, tube
      exceeding calyx by > 2 mm . . . . *E. traskiae*
      5. Petiole 10–20 mm; lf blade elliptic, 1.5–4 cm wide; corolla 7–10 mm . . . . subsp. ***smithii***
      5′ Petiole 2–5 mm; lf blade narrow-elliptic, 1–1.5 cm wide; corolla 6–7 mm . . . . subsp. ***traskiae***
  3′ Calyx, corolla not stalked-glandular abaxially; corolla funnel-shaped, not constricted at throat
    6. Lf linear to lance-linear, 0.2–1.1 cm wide
      7. Infl head-like; calyx lobes 6–8 mm, dense-long-hairy . . . . *E. capitatum*

7′ Infl open, branched; calyx lobes 2–5 mm, glabrous to sparse-hairy
  8. Lf abaxially with all but midvein obscured by hairs; pedicel, peduncle glabrous, sticky; calyx lobes
     3–5 mm, abaxially glabrous, sticky; corolla 11–16 mm, lavender . . . . *E. altissimum*
  8′ Lf abaxially with all but midvein not obscured by hairs; pedicel, peduncle sparse-stiff-hairy, not
     sticky; calyx lobes 2–3 mm, abaxially sparse-stiff-hairy, not sticky; corolla 3–7 mm, white. . . . . *E. angustifolium*
6′ Lf lanceolate to oblong or ovate, 1–5 cm wide
  9. Lf adaxially sparse-hairy to dense-tomentose. . . . . *E. crassifolium*
    10. Lf adaxially dense-gray- to -white-tomentose; lf ± ovate, not sticky; corolla 8–16 mm . . . . var. *crassifolium*
    10′ Lf adaxially dull ± green, sparse- to moderate-short-hairy; lf lanceolate, sticky between hairs;
       corolla 6–10 mm. . . . . var. *nigrescens*
  9′ Lf adaxially glabrous to sparse-long-wavy-hairy
    11. Calyx abaxially glabrous to sparse-hairy, adaxially short-stiff-hairy; corolla 8–12(17) mm, abaxially
       sparse-hairy; fl buds appearing dark, not gray-hairy. . . . . *E. californicum*
    11′ Calyx dense-long-hairy; corolla 6–8 mm, abaxially dense-long-hairy; fl buds appearing gray-hairy
      . . . . *E. trichocalyx*
      12. St sparse- to dense-long-wavy-hairy; lf adaxially sparse-long-wavy-hairy, abaxially dense-white-
         tomentose, all but midvein, secondary veins obscured . . . . var. *lanatum*
      12′ St glabrous to sparse-short-hairy; lf adaxially ± glabrous, sparse-short-hairy mostly on midvein,
         abaxially sparse-tomentose between yellow-green, glabrous veins in net-like pattern . . . . var. *trichocalyx*

E. altissimum P.V. Wells (p. 473) INDIAN KNOB MOUNTAINBALM **ST**: 2–4 m; twigs glabrous, sticky. **LF**: 5–9 cm, 0.2–0.4 cm wide, linear, sessile, entire, adaxially glabrous, sticky, abaxially white-tomentose, all but midvein obscured; margin strongly rolled under. **INFL**: pedicel, peduncle sticky, glabrous. **FL**: calyx lobes 3–5 mm, abaxially sticky, glabrous, adaxially glabrous; corolla 11–16 mm, funnel-shaped, lavender, abaxially sparse-stiff-hairy; style 5–7 mm. **SEED**: many. Sandstone ridges, chaparral; < 270 m. SCoRO (sw San Luis Obispo Co.). Mar–Jun ★

E. angustifolium Nutt. (p. 473) NARROW-LEAVED YERBA SANTA **ST**: < 2 m; twigs sticky, glabrous to sparse-hairy. **LF**: sessile to short-petioled; blade 5–10 cm, 0.2–1.1 cm wide, linear to lance-linear, entire to coarse-toothed, adaxially sticky, glabrous to sparse-hairy, abaxially white-hairy between veins in net-like pattern; margin rolled under. **INFL**: pedicel, peduncle sparse-stiff-hairy. **FL**: calyx lobes 2–3 mm, abaxially sparse-stiff-hairy, adaxially moderate-stiff-hairy; corolla 3–7 mm, funnel-shaped, white, abaxially moderate-stiff-hairy; style 2–3 mm. **SEED**: 1–8. *n*=14. Washes, slopes, pinyon/juniper woodland; 1460–1770 m. e DMtns (New York, Granite mtns); to UT, AZ, Baja CA. May–Aug ★

E. californicum (Hook. & Arn.) Torr. (p. 473) CALIFORNIA YERBA SANTA **ST**: 1–3 m; twigs glabrous, sticky (sparse-hairy). **LF**: short-petioled; blade ≤ 15 cm, 4–5 cm wide, lanceolate to oblong, entire to toothed, adaxially sticky, glabrous to sparse-hairy, abaxially hairy between veins in net-like pattern; margin rolled under. **FL**: buds appearing dark, not gray-hairy; calyx lobes 1–4 mm, abaxially glabrous to sparse-hairy, adaxially short-stiff-hairy; corolla 8–12(17) mm, funnel-shaped, white to purple, abaxially sparse-hairy; ovary dense-hairy, style 3–8 mm. **SEED**: 2–20. *n*=14. Slopes, fields, roadsides, woodland, chaparral; 20–1830 m. NW, CaR, SN, GV, CW; s OR. Apr–Jul

E. capitatum Eastw. (p. 473) LOMPOC YERBA SANTA **ST**: < 3 m; twigs sticky, glabrous. **LF**: 4–9 cm, 0.2–0.5(1) cm wide, linear, sessile, entire, adaxially sticky, glabrous to sparse-hairy, abaxially tomentose; margin strongly rolled under. **INFL**: head-like, to 2.5 cm wide. **FL**: calyx lobes 6–8 mm, linear, dense-long-hairy; corolla 6–15 mm, funnel-shaped, lavender, abaxially dense-hairy; styles 3–6 mm. **SEED**: 5. *n*=14. Ravines, mesas, chaparral, Bishop-pine woodland; 40–900 m. s CCo, s SCoRO, w WTR (Santa Barbara Co. endemic). Apr–Jul ★

E. crassifolium Benth. **ST**: 1–3 m; twigs dense-tomentose. **LF**: petiole 3–15 mm; blade 3–17 cm, 10–40 mm wide, lanceolate to ovate, ± entire to coarse-toothed, adaxially sparse-hairy to dense-tomentose, abaxially tomentose. **FL**: calyx lobes 2–4 mm, dense-long-hairy; corolla 6–16 mm, funnel-shaped, lavender, abaxially dense-hairy; styles 3–8 mm. **SEED**: 8–14. *n*=14. Vars. intergrade, ± distinguished by st, lf hairs.

var. ***crassifolium*** (p. 473) THICK-LEAVED YERBA SANTA **ST**: dense-gray- to -white tomentose, dark st obscured. **LF**: ± ovate, dense-gray- to -white-tomentose. **FL**: corolla 8–16 mm. Slopes, road-sides, washes, river bottoms, mesas, chaparral, woodland; 15–1520 m. SCo, WTR, SnGb, PR. Apr–Jun

var. ***nigrescens*** Brand BICOLORED YERBA SANTA **ST**: dull ± green, moderate-hairy, sticky, brown, hairs short. **LF**: lanceolate, sticky between hairs, adaxially sparse- to moderate-short-hairy, dull ± green, abaxially tomentose, ± green, net-veined. **FL**: corolla 6–10 mm. Slopes, roadsides, washes, river bottoms, mesas, chaparral, woodland; 100–2440 m. Teh (rare), SCoRO, SCo, WTR, SnGb, SnJt. Apr–Jul

E. lobbii (A. Gray) Greene (p. 473) MATTED YERBA SANTA Per, gen forming low mats > 1 m diam, rhizomed, short-glandular-hairy and dense-long-woolly-tomentose, some hairs stiff, spreading. **ST**: prostrate to ascending, 16–50 cm, many-branched, woody at base occ or not. **LF**: 0.5–6 cm, 2–15 mm wide, oblanceolate to obovate, adaxially gen sticky, strongly rolled under, smallest clustered in axils. **INFL**: pedicels 0–2 mm. **FL**: sepals 5–12 mm, linear to narrow-lanceolate; corolla 7–12 mm, wide-funnel- to narrow-bell-shaped, purple to pink, limb 7–9 mm wide, lobes 2–3 mm, 2–4 mm wide; stamens 4–8 mm, attached 2–4 mm above corolla base, filament gen with 1–4 short hairs; style 4–6 mm. **FR**: 2–4 mm, loculicidal and septicidal. **SEED**: 1–2 mm, gen elliptic-ovoid, gen angled below, black, papillate. 2*n*=28. Dry, sandy or rocky alluvial slopes, ridges, open pine forest; 900–2350 m. KR, CaRH, n SNH, nw MP, Wrn (rare); s OR, w NV. [*Nama l.* A. Gray] Move from *Nama* based on molecular data. Jun–Aug

E. parryi (A. Gray) Greene (p. 473) POODLE-DOG BUSH Sub-shrub, dense-glandular, sticky, strong-scented. **ST**: erect, 1–3 m, stout, branched from, gen woody at base. **LF**: dense, sessile; blade 4–30 cm, lanceolate, entire or toothed, margins of upper occ rolled under. **INFL**: terminal, branched; fls dense-clustered, short-pedicelled. **FL**: calyx lobes 3–6 mm, glandular, coarse-long-hairy; corolla 10–20 mm, shallow-lobed, funnel-shaped, blue, lavender, or purple, glandular, hairy abaxially; stamens incl, unequal; ovary chambers appearing 2, style 4–7 mm. **FR**: valves 4, 3–4 mm, ovoid, glandular-hairy. **SEED**: many, oblong-ovoid, angled, shiny black, fine-ridged, minute-net-sculptured. *n*=13. Gen disturbed areas, chaparral, dry granitic soils of slopes, ridges; often following fires; 120–2440 m. s SN, Teh, s SCoRO, TR, PR, DMtns (Panamint Range, Little San Bernardino Mtns), w edge DSon (rare); Baja CA. [*Turricula p.* (A. Gray) J.F. Macbr.] Move from *Turricula* based on molecular data; causes severe contact dermatitis in some people. May–Aug

E. tomentosum Benth. (p. 473) WOOLLY YERBA SANTA **ST**: 1–3 m; twigs dense-white-tomentose. **LF**: petiole 5–12 mm; blade 3–10 cm, 1–5 cm wide, oblanceolate to oblong, entire to coarse-toothed, dense-white-tomentose. **FL**: calyx lobes 2.5–4 mm, dense-long-hairy, stalked glands fewer than hairs; corolla 3–5 mm, urn-shaped, ± constricted at throat, white to lavender, abaxially sparse-hairy, glandular, tube exceeding calyx by < 2 mm; filaments glabrous; ovary glabrous, style 1–2 mm. **SEED**: 10–12. *n*=14. Slopes, ridges, ravines, disturbed areas, grassland, chaparral; 150–1400 m. SCoR. May–Jul

E. traskiae Eastw. **ST**: < 2 m; twigs tomentose. **LF**: petiole 2–20 mm; blade 3–14 cm, 1–4 cm wide, lanceolate to elliptic, entire to coarse-toothed, dense-tomentose, margin rolled under. **FL**: calyx lobes 4–5 mm, moderate-hairy, stalked glands as many as hairs; corolla 6–10 mm, urn-shaped, ± constricted at throat, white to lavender, abaxially sparse- to dense-hairy, glandular, tube exceeding calyx by > 2 mm; filaments glabrous; ovary gen glabrous, styles 1–2 mm. **SEED**: 2–4. *n*=14.

subsp. ***smithii*** Munz SMITH'S YERBA SANTA **LF**: petiole 10–20 mm; blade 8–14 cm, 1.5–4 cm wide, elliptic. **FL**: corolla 7–10 mm; style base glabrous to sparse-stiff-hairy. Slopes, disturbed areas, often following fires, chaparral; 150–1430 m. s SCoRO, w WTR. May–Jul

subsp. ***traskiae*** TRASK'S YERBA SANTA **LF**: petiole 2–5 mm; blade 3–6.5 cm, 1–1.5 cm wide, narrow-elliptic. **FL**: corolla 6–7 mm; style base glabrous. Slopes, chaparral; 150–460 m. s ChI (Santa Catalina Island). May–Jul

E. trichocalyx A. Heller **ST**: < 2 m; twigs glabrous to hairy. **LF**: petiole 4–13 mm; blade 3–14 cm, 1–4 cm wide, lance-linear to narrow-oblong, entire to toothed, abaxially sparse- to dense-tomentose, adaxially sticky, glabrous to sparse-hairy, margin rolled under. **FL**: buds appearing gray-hairy; calyx lobes 2–5 mm, dense-long-hairy; corolla 6–8 mm, funnel-shaped, white to lavender, abaxially dense-long-hairy; styles 1–6 mm. **SEED**: 4–8. *n*=14.

var. ***lanatum*** (Brand) Jeps. SAN DIEGO YERBA SANTA **ST**: sparse- to dense-long-wavy-hairy. **LF**: adaxially sparse-long-wavy-hairy, abaxially dense-white-tomentose, all but midvein, secondary veins obscured. Slopes, mesas, ravines, chaparral, woodland, open pine forest; 300–2200 m. PR, w edge DSon; Baja CA. Apr–Jun

var. ***trichocalyx*** (p. 473) HAIRY YERBA SANTA **ST**: sticky, glabrous to sparse-short-hairy. **LF**: adaxially ± glabrous, sparse-short-hairy mostly on midvein, abaxially sparse-tomentose between yellow-green, glabrous veins in net-like pattern. Slopes, mesas, ravines, chaparral, woodland, open pine forest; (30)120–2600 m. SCo, SnGb, SnBr, PR, w edge D (rare). Apr–Jul

EUCRYPTA

Robert Patterson & Richard R. Halse

Ann, glandular, sticky, odorous. **ST**: erect, much-branched. **LF**: 1–3-pinnate-lobed; lower cauline opposite, petioled, upper alternate, upward smaller, sessile, clasping; petioles gen narrow-winged, ciliate. **INFL**: terminal or axillary; pedicels thread-like, elongate in fr. **FL**: calyx < 1/2-fused, bell-shaped, glandular, lobes oblong to ovate to spoon-shaped, ciliate; corolla bell-shaped, gen ≥ calyx, with or without V-shaped transverse fold between each pair of filaments below throat; stamens incl, equal, equally attached; ovary chamber 1 (or appearing ± 5 from complex, enlarged placenta), ovules on both sides of placenta, style 1, stigmas 2. **FR**: capsule, ovoid to spheric, bristly. **SEED**: 5–15. 2 spp.: sw US. (Greek: well hidden, from seeds)

1. Calyx lobes erect, enclosing fr; seeds of 1 kind; lower lvs 1-pinnate-lobed . . . . ***E. micrantha***
1′ Calyx lobes spreading, not enclosing fr; seeds of 2 kinds; lower lvs 2–3-pinnate-lobed . . . . ***E. chrysanthemifolia***
  2. Corolla = calyx . . . . var. ***bipinnatifida***
  2′ Corolla > calyx . . . . var. ***chrysanthemifolia***

E. chrysanthemifolia (Benth.) Greene **ST**: erect to spreading, < 9 dm. **LF**: lower 2–10 cm, 1–5 cm wide, petioles < 1/2 blade, widened, clasping, blade oblong to wide-ovate, 2–3-pinnate-lobed, lobes obtuse, upper lvs smaller, narrower, less lobed, bases clasping. **INFL**: fls 4–15 per branch; pedicels gen recurved in fr. **FL**: calyx 2–4 mm, lobes spreading, not enclosing fr; corolla 2–6 mm, V-shaped fold 0, lobes hairy abaxially; style < 3 mm. **FR**: 2–4 mm wide. **SEED**: 6–8, dark brown, of 2 kinds, elliptic or round, disk-like, smooth and oblong-ovoid, wrinkled.

var. ***bipinnatifida*** (Torr.) Constance **ST**: openly spreading. **LF**: lower 2–7 cm, 1–4 cm wide, lobes 7–9. **INFL**: fls 4–8 per branch. **FL**: corolla 2–3 mm, 2–3 mm wide, = calyx, white or ± blue; style < 1 mm. *n*=10,20. Cliffs, rocky slopes, washes, crevices; 30–2300 m. s SNF, Teh, SnBr, e PR, SNE, D; NV, AZ, Baja CA. Mar–May

var. ***chrysanthemifolia*** (p. 477) **ST**: erect, stout. **LF**: lower 2–10 cm, 1–5 cm wide, lobes 9–13. **INFL**: fls 8–15 per branch. **FL**: corolla 3–6 mm, 4–8 mm wide, > calyx, yellow-white; style 1–3 mm. *n*=10. Roadsides, burns, coastal bluffs, ravines; < 1000 m. s SNF, Teh, s SnJV, CW, SW; Baja CA. Mar–Jun

E. micrantha (Torr.) A. Heller (p. 477) **ST**: weak, < 3 dm, gen stalked-glandular. **LF**: lower 1–5 cm, < 2 cm wide, petiole short, widened to clasping base, blade oblong or ovate, deep-1-pinnate-lobed, lobes 7–9, oblong or oblanceolate, straight or sickle-shaped, entire or few-toothed, upper lvs greatly reduced, lobed, toothed or entire. **INFL**: fls 4–12 per branch; pedicels gen erect in fr. **FL**: calyx 2–5 mm, lobes erect, enclosing fr, gen black-glandular; corolla 2–4 mm, white or blue-purple, tube yellow, V-shaped fold present; style 1–2 mm. **FR**: 2–3 mm wide. **SEED**: 7–15, oblong, in age incurved, worm-like, black or dark-brown, wrinkled. *n*=6,12. Canyons, hillsides, rocky crevices, washes, slopes; 60–2500 m. SnJt, SNE, D; to UT, TX, Mex. Mar–Jun

HACKELIA STICKSEED

Ronald B. Kelley & Robert L. Carr

Per (bien); hairs appressed to spreading; caudex gen branched in age, often ± woody, taprooted. **ST**: ascending or erect. **LF**: lowest petioles gen ± = blades, ± winged, others 0. **INFL**: coiled cymes, gen > 3, gen terminal and axillary, ± bracted; pedicel in fr elongated, recurved to reflexed. **FL**: calyx deep-5-lobed; corolla rotate-salverform, gen white with yellow patch adaxially, lobes appendaged near base. **FR**: nutlets erect, > style, attachment scar lateral-medial, gen with barb-tipped prickles abaxially and on margin. 40 spp.: gen w N.Am, se Asia. (J. Hackel, Czech botanist, 1783–1869) Values for corolla limb diam take into account shrinkage during fl period. Difficult, study needed, esp in n CA, se Asia; sometimes merged with *Lappula*.

1. Corolla pink or white
  2. Corolla pink, when dry ± blue, appendages pink . . . . ***H. mundula***
  2′ Corolla white, when dry ± brown, appendages white
    3. Nutlet 5–6 mm, abaxial prickles ± 10, << marginal prickles; calyx 3.5–5.5 mm; corolla limb 12–19 mm diam, appendages longer than wide; anthers at base of tube . . . . ***H. bella***
    3′ Nutlet 4–4.5 mm, abaxial prickles 10–25, ± = marginal prickles; calyx 1.5–3 mm; corolla limb (5)7–14 mm diam, appendages ± as long as wide; anthers ± at top of tube . . . . ***H. californica***

473

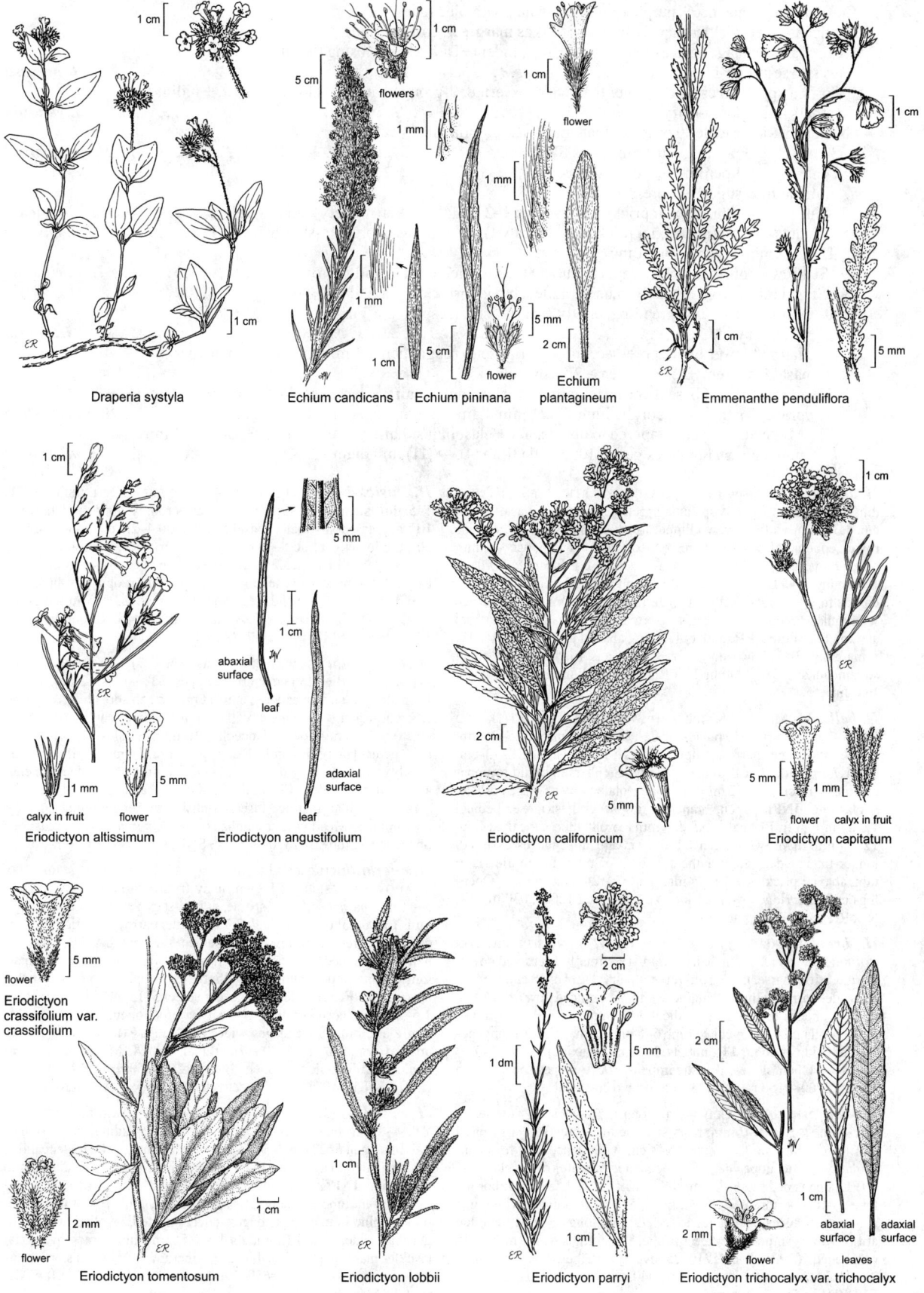

Draperia systyla

Echium candicans   Echium pininana   Echium plantagineum

flowers

flower

flower

Emmenanthe penduliflora

1 cm

5 cm

1 mm

1 mm

1 cm

1 cm

5 cm

5 mm

2 cm

1 cm

1 mm

5 cm

1 cm

1 cm

5 mm

calyx in fruit   flower

Eriodictyon altissimum

abaxial surface

leaf

adaxial surface

leaf

Eriodictyon angustifolium

Eriodictyon californicum

Eriodictyon capitatum

flower   calyx in fruit

1 cm

1 mm

5 mm

5 mm

1 cm

5 mm

2 cm

5 mm

1 cm

5 mm

1 mm

flower

Eriodictyon crassifolium var. crassifolium

Eriodictyon tomentosum

flower

Eriodictyon lobbii

Eriodictyon parryi

Eriodictyon trichocalyx var. trichocalyx

5 mm

1 cm

2 mm

2 cm

1 dm

5 mm

1 cm

1 cm

2 cm

1 cm

abaxial surface   adaxial surface

leaves

2 mm   flower

1′ Corolla gen ± blue, occ ± purple (pink), occ white with blue margins
  4. Nutlet abaxial prickles many, ± as numerous as marginal; corolla tube > to >> calyx
    5. Corolla appendages gen as long as wide, incl, indented at tip, not closing throat, not hiding anthers;
      lvs sparse-rough-hairy. . . . . *H. nervosa*
    5′ Corolla appendages much longer than wide, exserted, 2-pronged at tip, ± closing throat, ± hiding
      anthers; lvs velvety-hairy. . . . . *H. velutina*
  4′ Nutlet abaxial prickles 0 or fewer than marginal; corolla tube ≤ calyx
    6. Cymes (1)2–3 per infl, ± terminal — s SNH. . . . . *H. sharsmithii*
    6′ Cymes gen > 3 per infl, terminal and axillary
     7. Hairs at mid-st gen ± appressed
      8. Nutlets 2–3 mm, abaxial prickles 0–7; calyx 1–2 mm; lvs ± soft velvety-hairy; W&I . . . . *H. brevicula*
      8′ Nutlets 3–4.5 mm, abaxial prickles 8–15; calyx 4–5 mm; lvs appressed-stiff-hairy; MP. . . . . *H. cusickii*
     7′ Hairs at mid-st gen ± spreading
      9. St 1(few) from non-woody caudex; nutlet abaxial prickles 0(3); infl narrow . . . . *H. floribunda*
      9′ St gen many from ± woody caudex; nutlet abaxial prickles gen > 3; infl ± wide
       10. Hairs of lower lvs spreading, bristly, gen >> 1 mm, esp on margins; basal lvs gen < 22 cm, gen
         < 2.5 cm wide . . . . *H. setosa*
       10′ Hairs of lower lvs ± appressed, soft or strigose to rough, gen < 1 mm, occ > 1 mm on margins;
         basal lvs often > 22 cm, often > 2.5 cm wide
        11. Mid cauline lf base truncate to cordate, ± clasping; st hairs ± dense, gen < 1 mm; nutlet abaxial
          prickles gen 10–17; corolla limb 7–12+ mm diam . . . . *H. amethystina*
        11′ Mid cauline lf base tapered to obtuse, not ± clasping; st hairs gen 0 to ± sparse, often > 1 mm;
          nutlet abaxial prickles gen 4–10; corolla limb (4)5–8(11) mm diam . . . . *H. micrantha*

H. amethystina Eastw. (p. 477) AMETHYST STICKSEED **ST:** 4–8 dm; hairs ± dense, gen ± spreading, gen < 1 mm. **LF:** basal 10–30 cm, 1.3–4 cm wide, narrow-elliptic; lower cauline smaller, gen narrower, ephemeral, hairs ± appressed, soft, gen < 1 mm, occ > 1 mm on margins; mid-cauline gen lance-ovate, base truncate to cordate, ± clasping. **INFL:** open; pedicel 6–9 mm in fr. **FL:** calyx 2–4 mm; corolla tube gen = calyx, throat open, limb 7–12+ mm diam, blue with pink in center, appendages exserted, longer than wide; anthers at ± middle of tube. **FR:** nutlets 4–5 mm, abaxial prickles gen 10–17, < marginal. 2*n*=24. Meadows, forest clearings, roadsides, occ along streambanks; 1370–2200 m. NCoRH, n SNH (Plumas, Placer cos.). Jun–Jul ★

H. bella (J.F. Macbr.) I.M. Johnst. (p. 477) BEAUTIFUL STICKSEED **ST:** 5–7 dm; hairs gen spreading. **LF:** basal gen many, 15–26 cm, 2.2–4.5 cm wide, narrow-elliptic, obtuse; lower cauline 8–13 cm, 0.6–1.7 cm wide, linear to narrow-elliptic, ephemeral; mid to upper cauline 4–10 cm, 14–42 mm wide, lanceolate to ovate, base cordate, ± clasping. **INFL:** open, branches few–several, strigose; pedicel 12–23 mm in fr. **FL:** calyx 3.5–5.5 mm; corolla tube ± = calyx, limb 12–19 mm diam, white, gen ± brown in age, appendages exserted, longer than wide, white; anthers at base of tube. **FR:** nutlets 5–6 mm, abaxial prickles ± 10, << marginal. 2*n*=24. Uncommon. Openings in forest, ridgetops, roadsides, streambanks; 1200–2030 m. KR, NCoRH; sw OR. Jun–Jul

H. brevicula (Jeps.) J.L. Gentry (p. 477) POISON CANYON STICKSEED **ST:** 2–6 dm; hairs at mid-st ± strongly appressed downward, ± stiff, coarse. **LF:** ± soft-velvety-hairy; basal 6–18 cm, 0.5–1.8 cm wide, narrow-elliptic; cauline ± similar, reduced upward. **INFL:** open, branches few–several; pedicel 3–9+ mm in fr. **FL:** calyx 1–2 mm; corolla tube gen = calyx, limb 5–8 mm diam, pale blue, appendages as wide as long. **FR:** nutlets 2–3 mm, abaxial prickles 0–7, < marginal. Open slopes, dry streambeds, rocky slopes, open aspen stands; 2700–3150 m. W&I (Mono, n Inyo cos.). Jul ★

H. californica (A. Gray) I.M. Johnst. (p. 477) CALIFORNIA STICKSEED **ST:** 3–10 dm; hairs strigose to ± long, soft. **LF:** basal and lowest cauline 3–17 cm, 0.6–3 cm wide, lanceolate to oblanceolate; mid to upper cauline lanceolate to ovate, often clasping. **INFL:** open; pedicel 6–15 mm in fr. **FL:** calyx 1.5–3 mm; corolla tube ± = calyx, throat ± closed, limb (5)7–14 mm diam, white, gen ± brown in age, appendages exserted, ± as long as wide, swollen at tip, white; anthers at ± top of tube. **FR:** nutlets 4–4.5 mm, dull, roughened, abaxial prickles 10–25, evenly distributed, ± = marginal. 2*n*=24. Meadows, forest openings; 940–2490 m. NCoRH, CaRH, n SNH, MP; s OR. Jun–Aug

H. cusickii (Piper) Brand (p. 477) CUSICK'S STICKSEED **ST:** 1–5 dm, slender; hairs at mid-st gen strongly appressed upward. **LF:** appressed-stiff-hairy; basal 5–18 cm, 0.4–2.8 cm wide, narrow-elliptic; lower cauline 5–10 cm, 0.4–1.1 cm wide, narrow-elliptic to lanceolate. **INFL:** open, few-fld, strigose; pedicel 6–7 mm in fr. **FL:** calyx 4–5 mm; corolla tube gen = calyx, throat open, limb(5)7–13 mm diam, ± blue, appendages ± longer than wide. **FR:** nutlets 3–4.5 mm, abaxial prickles 8–15, < marginal. 2*n*=48. Under junipers; 1300–1950 m. MP; to e-c OR. May–Jul ★

H. floribunda (Lehm.) I.M. Johnst. (p. 477) MANY-FLOWERED STICKSEED Bien to occ short-lived per. **ST:** gen 4–12 dm, 1(few) from slender, simple caudex; hairs at mid-st gen spreading, ± coarse. **LF:** basal gen < cauline, ± withered at fl; lower cauline gen 5–24 cm, 0.5–3.5 cm wide, oblanceolate to narrow-elliptic. **INFL:** narrow; pedicel 4–10 mm in fr. **FL:** calyx 1.5–2.3 mm; corolla tube gen = calyx, throat ± closed, limb (3)4–7(8) mm diam, ± blue, appendages wider than long. **FR:** nutlets 2–4 mm, abaxial prickles gen 0(3), < marginal, marginal occ fused basally, ± forming a wing. 2*n*=±24. Uncommon. Meadows, streambanks, other vernally wet areas, occ open slopes, forests; 1370–3050 m. SNH, GB; w N.Am. Jun–Aug

H. micrantha (Eastw.) J.L. Gentry (p. 477) JESSICA'S STICKSEED, MEADOW STICKSEED **ST:** gen many from ± stout, woody caudex, 3–11 dm; hairs gen 0 to ± sparse, at mid-st gen ± spreading, gen > 1 mm. **LF:** basal 6–33 cm, 0.7–3.7 cm wide, narrow-elliptic to oblanceolate, ± green at fl; lower cauline gen 5–23 cm, 0.6–2.4 cm wide, hairs ± appressed, strigose to rough, gen < 1 mm, occ > 1 mm on margins; mid to upper cauline gen elliptic, base tapered to obtuse, not ± clasping. **INFL:** ± wide, ± strigose; pedicel 5–12 mm in fr. **FL:** calyx 1.5–2.7 mm; corolla tube gen = calyx, throat open, limb (4)5–8(11) mm diam, blue, appendages wider than long. **FR:** nutlets 3–5 mm, abaxial prickles gen 4–10, < marginal. 2*n*=24. Meadows, streambanks, shrubby slopes, open forest; 1200–3500 m. KR, NCoRH, CaRH, SNH, MP, n SNE (Sweetwater Mtns); w N.Am. Jun–Aug

H. mundula (Jeps.) Ferris (p. 477) PINK-FLOWERED STICKSEED **ST:** 4–8 dm; hairs ± dense, gen long-soft-spreading, occ strigose. **LF:** basal gen 6–22 cm, 0.5–2.8 cm wide, oblanceolate; lower cauline often smaller, in age ovate to lanceolate; mid to upper gen clasping to ± clasping. **INFL:** branched; pedicel 12–30 mm in fr. **FL:** calyx 2–3 mm; corolla tube ± = calyx, throat ± closed, limb 10–18 mm diam, pink, ± blue in age, appendages much-exserted, swollen at tip, longer than wide, pink. **FR:** nutlets 4.5–6.5 mm, shiny, smooth, abaxial prickles many, evenly distributed, ± = marginal. Dry open slopes, forest openings, roadsides; 1650–2900 m. e-c KR (Castle Crags), SNH; sw OR. May–Jul

H. nervosa (Kellogg) I.M. Johnst. (p. 477) SIERRAN STICKSEED **ST**: 4–7 dm; ± glabrous above, hairs gen sparse below, often ± spreading. **LF**: sparse-rough-hairy; basal 3–12 cm, 0.6–2.5 cm wide, ± oblong to oblanceolate; lower cauline gen smaller, gen sessile, ovate to lanceolate upward; uppermost ± clasping. **INFL**: open, branches few–several; pedicel 6–12 mm in fr. **FL**: calyx 1.5–3.5+ mm; corolla tube > calyx, ± white, throat open, limb (8)10–18 mm diam, ± blue, appendages gen as long as wide, indented at tip; anthers visible at throat. **FR**: nutlets 4–6 mm, dull, roughened, abaxial prickles many, evenly distributed, ± = marginal. Moist open slopes, openings in forest; 1800–2930 m. s CaRH, n&c SNH; w NV. May–Aug

H. setosa (Piper) I.M. Johnst. (p. 477) BRISTLY STICKSEED, SISKIYOU STICKSEED **ST**: gen many, 3–6 dm; mid-st ± spreading-bristly-hairy. **LF**: basal many, 9–22 cm, 1–2.5 cm wide, narrow-oblong-ovate; cauline 5–9 cm, 0.4–1 cm wide, linear to lanceolate or oblanceolate, base tapered, not clasping; lower cauline hairs spreading, bristly, gen >> 1 mm, esp on margins. **INFL**: few-branched, ± wide, strigose; pedicel 6–9 mm in fr. **FL**: calyx 2.5–3.5 mm; corolla tube < calyx, throat ± white, limb 8–13 mm diam, blue, appendages longer than wide. **FR**: nutlets 3.2–3.7 mm, abaxial prickles 9–13, < marginal. $2n=48$. Open, wooded ridges; (300)1200–1700 m. KR, NCoRO, n SNH (Butterfly Valley), s MP (Sierra Valley); sw OR. Jun–Jul

H. sharsmithii I.M. Johnst. (p. 477) SHARSMITH'S STICKSEED **ST**: 1–3 dm, ± strigose, hairs appressed downward below, upward above. **LF**: basal 3–14 cm, 0.6–2.3 cm wide, elliptic to lanceolate; cauline sessile, 1–4 cm, 0.5–1.6 cm wide, oblanceolate to ovate. **INFL**: (1)2–3, gen few-fld, ± terminal; pedicel 12–17+ mm in fr. **FL**: calyx 2–2.7 mm; corolla tube < calyx, limb 5.5–7(8) mm diam, ± white with pale blue margin, occ pale blue, appendages wider than long. **FR**: nutlets 2–3 mm, abaxial prickles 0–5, < marginal. Crevices in cliffs, talus slopes; 3150–3700 m. s SNH; c NV. Jul–Aug ★

H. velutina (Piper) I.M. Johnst. (p. 477) VELVETY STICKSEED **ST**: 4–8 dm; hairs gen spreading, ± dense below middle, ± dense to ± sparse above. **LF**: velvety-hairy; basal gen 5–17 cm, 0.5–2 cm wide, narrow-elliptic to oblanceolate; lower cauline reduced, ovate to narrow-lanceolate, ± clasping upward. **INFL**: ± dense, branches several, few-fld; pedicel 6–20 mm in fr. **FL**: calyx 2–3 mm; corolla tube gen >> calyx, gen ± purple, throat ± closed, limb 12–20 mm diam, gen blue to ± purple (pink), appendages long-exserted, much longer than wide, recurved, 2-pronged at tip, white; anthers hidden. **FR**: nutlets 4.5–6.5 mm, dull, roughened, abaxial prickles many, evenly distributed, ± = marginal. Dry, open slopes, forest clearings, roadsides; 1350–2750 m. SNH; w NV. Jun–Aug

HARPAGONELLA

Ronald B. Kelley & Timothy C. Messick

1 sp. (Latin: small grappling hook, from calyx spines)

H. palmeri A. Gray (p. 477) PALMER'S GRAPPLINGHOOK **ST**: ascending to erect, 3–30 cm. **INFL**: pedicels in fr 0.5–1 mm, twisted. **FL**: calyx > nutlets, upper 2 lobes >> others, partly fused, arched over 1 nutlet, ± bur-like, with 5–10 stout spines each with hooked bristles, lower 3 lobes distinct; corolla ± 1 mm diam, ± funnel-shaped, white. **FR**: nutlets 2, spreading, 1–4 mm, dissimilar, ± oblanceolate, margins entire. $2n=24$. Dry, semi-barren sites in chaparral, coastal scrub, grassland; < 1000 m. SCo, PR, sw DSon; sw AZ, nw Mex. May be indistinct from *Pectocarya*; needs study. Mar–Apr ★

HELIOTROPIUM HELIOTROPE

Ronald B. Kelley & Dieter H. Wilken

Ann, per [shrub], glabrous to bristly or strigose. **ST**: prostrate to erect, branched. **LF**: gen cauline, petioled to sessile, gen entire. **INFL**: fl 1 in axils or many in terminal coiled spike-like cymes. **FL**: corolla rotate to bell-shaped, white to purple; stamens attached on upper tube, incl, anthers ± sessile; style 0 or not lobed, stigma 1, linear to disk-like. **FR**: nutlets 2 or 4, erect, gen ovoid to spheric, smooth, roughened, or hairy, scar gen lateral. ± 250 spp.: temp, trop. Orn, cult for medicinal drugs. (Greek: sun turning, from some spp. fl at summer solstice)

1. Per
  2. Sts, lvs not fleshy, short-soft-hairy; corolla purple . . . . ***H. amplexicaule***
  2' Sts, lvs fleshy, glabrous; corolla white . . . . ***H. curassavicum*** var. ***oculatum***
1' Ann
  3. Corolla 8–15+ mm diam; fls 1 in axils; D . . . . ***H. convolvulaceum*** var. ***californicum***
  3' Corolla 3–5 mm diam; infl many in terminal coiled spikes; n&c SNF, GV, CCo, SnFrB, MP . . . . ***H. europaeum***

H. amplexicaule Vahl FRAGRANT HELIOTROPE Per, not fleshy. **ST**: decumbent to ascending, 2–6 dm, short-soft-hairy. **LF**: 4–9 cm, oblong to oblanceolate, short-petioled to subsessile, acute, short-soft-hairy. **INFL**: spike-like cymes 3–5. **FL**: gen fragrant; calyx lobes ± lance-linear, bristly; corolla 4–6 mm, 5–6 mm diam, bell-shaped, purple, throat yellow. **FR**: nutlets 2, irregularly roughened, faintly tubercled. $2n=26,28$. Open sites, fields; < 500 m. n&c SCo; native to Argentina. All yr

H. convolvulaceum (Nutt.) A. Gray var. *californicum* (Greene) I.M. Johnst. (p. 477) MORNING-GLORY HELIOTROPE Ann, tap-rooted. **ST**: ascending to erect, 7–18 cm, canescent. **LF**: 1–4 cm, elliptic to ovate, gen petioled, acute, dense-strigose. **INFL**: fls 1 in axils. **FL**: opening late afternoon, closing next morning, ± fragrant; calyx lobes lanceolate, long-tapered, dense-spreading-bristly; corolla 7–10 mm, 8–15+ mm diam, wide-bell-shaped to salverform, papery, white, tube long-exserted, constricted, throat swollen, green-yellow; anthers in throat. **FR**: nutlets 4, long-soft-hairy. $2n=42$. Sandy soils, dunes; < 700 m. D; w AZ, n Mex. Typical var. to Great Plains. Apr–Oct

H. curassavicum L. var. *oculatum* (A. Heller) Tidestr. (p. 477) SEASIDE HELIOTROPE, ALKALI HELIOTROPE Per, fleshy, occ from rhizome-like root. **ST**: prostrate to ± ascending, 1–6 dm, glabrous. **LF**: 1–6 cm, gen oblanceolate, short-petioled to subsessile, acute to obtuse, glabrous. **INFL**: spike-like cymes 2–4. **FL**: calyx lobes oblong to narrow-ovate, glabrous; corolla 3–5 mm, 3–5(7) mm diam, salverform to bell-shaped, white, throat gen blue-purple, upper tube ± yellow. **FR**: nutlets 4, smooth. $2n=26, 28$. Moist to dry, saline to alkaline soils, gen near water; < 2250 m. CA (exc KR, NCoRH, c SNH); s NV, sw UT, to w TX, n Mex. Feb–Oct

H. europaeum L. EUROPEAN HELIOTROPE Ann, taprooted. **ST**: ascending to erect, 5–40 cm, puberulent to short-soft-hairy. **LF**: 1.5–5 cm, elliptic to ovate, petioled, obtuse, appressed-short-hairy. **INFL**: spike-like cymes 2–4. **FL**: calyx lobes linear to lanceolate, bristly; corolla 2–4 mm, 3–5 mm diam, salverform, white. **FR**: nutlets 4, irregularly roughened, faintly tubercled. $2n=24,32$. Open, often disturbed sites; < 1400 m. n&c SNF, GV, CCo, SnFrB, MP; e US; native to s&e Eur, n Afr. May–Aug

HESPEROCHIRON

Robert Patterson & Richard R. Halse

Per, scapose; root caudex-like. **LF**: in basal rosette, spreading or ascending; blade tapered to petiole, gen entire, gen ciliate. **INFL**: fls 1; peduncle erect or spreading, 1–10 cm, slender. **FL**: calyx lobes 2–9 mm, gen not alike, glabrous to hairy, ciliate; corolla scales 0, tube gen dense-hairy inside, throat gen yellow, lobes glabrous to hairy, white or ± blue, gen tinged or marked with lavender or purple; stamens incl, gen unequal, filament base widened; ovary hairy, chamber 1, style 1, 2–5 mm, stigmas 2. **FR**: capsule, 5–11 mm, ovoid, hairy. **SEED**: many, ovoid, angular, red-brown, honeycombed or pitted. 2 spp.: w US, n Mex. (Greek: evening or western centaur)

1. Corolla bell- or funnel-shaped, lobes gen 0.8–1.5 × tube; lvs gen hairy both surfaces. . . . . ***H. californicus***
1′ Corolla rotate, lobes gen (1.5)2–4 × tube; lvs gen glabrous at least abaxially . . . . ***H. pumilus***

H. californicus (Benth.) S. Watson (p. 477)   Rhizomes 0. **LF**: gen > 6, < 8 cm, < 3 cm wide, oblanceolate to elliptic or ovate. **INFL**: fls gen > 5 per pl. **FL**: corolla 10–30 mm, limb 10–20 mm wide, lobes 3–10 mm, oblong. n=8. Wet meadows, flats, valleys; 770–2620 m. KR, CaRH, SNH, Teh, WTR, SnBr, GB, s DMoj (Rabbit Springs, San Bernardino Co.); to WA, MT, WY, UT, Baja CA. Apr–Jul

H. pumilus (Griseb.) Porter (p. 477)   Rhizomes gen slender. **LF**: gen 2–10, 1–7 cm, < 2 cm wide, linear-oblong to oblanceolate or oblong. **INFL**: fls gen 1–8 per pl. **FL**: corolla 5–15 mm, limb 7–30 mm wide, lobes 3–11 mm, rounded. n=8. Wet meadows, slopes, flats; 450–3000 m. KR, NCoRI, CaRH, SNH, Teh, WTR, GB (exc W&I), n DMoj (Death Valley); to WA, MT, UT, AZ. Apr–Jul

HYDROPHYLLUM   WATERLEAF

Robert Patterson & Richard R. Halse

Per; roots fleshy-fibrous or tuber-like, from rhizomes. **ST**: erect, fleshy. **LF**: simple, pinnate-lobed, or compound, basal or cauline, alternate; petiole widened, clasping; lflets toothed or lobed, hairy, gen paler abaxially. **INFL**: gen branched, gen head-like cymes; pedicels gen elongate, recurved in fr or not. **FL**: calyx bell-shaped, lobes linear to lanceolate, acute to obtuse, glabrous or hairy, gen ciliate; corolla lobed to middle, > calyx, bell-shaped, lobes hairy; stamens equal, exserted, filaments hairy; ovary chamber 1, style 1, exserted, stigmas 2, base persistent. **FR**: capsule, 3–5 mm, spheric; tip gen bristly, loosely enclosed by calyx. **SEED**: 1–4, oblong to spheric, brown, net-like. n=9. 8 spp.: N.Am. (Greek: water lf) [Constance 1942 Amer Midl Naturalist 27:710–731]

1. Infl near ground, << subtending lvs; anthers << 1 mm . . . . ***H. capitatum*** var. ***alpinum***
1′ Infl well above ground, gen ≥ subtending lvs; anthers 1–2 mm
  2. Lf wide-ovate to round; lflets gen 3–5 . . . . ***H. tenuipes***
  2′ Lf ± oblong; lflets 7–15
    3. Lflets acute to acuminate, teeth gen 4–8 per side . . . . ***H. fendleri*** var. ***albifrons***
    3′ Lflets obtuse to acute, teeth gen 2–4 per side . . . . ***H. occidentale***

H. capitatum Benth. var. ***alpinum*** S. Watson (p. 477)   WOOLEN-BREECHES   Rhizome short. **ST**: very short, spreading-hairy. **LF**: 4–12 cm; petiole 3–15 cm; blade ovate to oblong, deep-lobed, lflets 5–7, lanceolate to obovate, obtuse or acute, short-pointed, entire, lower pair gen distinct, terminal ± merged. **INFL**: near ground, << subtending lvs; peduncle 1–5 cm; pedicels 4–15 mm. **FL**: calyx lobes 3–4 mm, < 8 mm in fr; corolla 4–10 mm, lobes 2–6 mm, white to purple or white with lavender marks; anthers << 1 mm; style 7–10 mm. **SEED**: 1–3. Moist slopes, meadows, flats; 900–2500 m. CaRH, n SNH, MP; to OR, ID, UT. May–Jul

H. fendleri (A. Gray) A. Heller var. ***albifrons*** (A. Heller) J.F. Macbr. (p. 477)   Rhizome short. **ST**: 25–90 cm, with reflexed bristles. **LF**: 6–30 cm; petiole 3–18 cm; blade ± oblong, deep-lobed, lflets 7–11, lanceolate, acute to acuminate, teeth gen 4–8 per side, lower 2–3 lflet pairs gen distinct, upper deep-lobed. **INFL**: well above ground, gen > subtending lvs; peduncle 3–18 cm; pedicels 2–10 mm. **FL**: calyx lobes 3–6 mm, < 7 mm in fr; corolla 6–11 mm, lobes 4–5 mm, white, purple, or white with lavender marks; anthers 1–2 mm; style 9–14 mm. **SEED**: gen 2. Moist, shady, wooded slopes; 1100–2000 m. KR; to BC, ID. May–Jul

H. occidentale (S. Watson) A. Gray (p. 477)   Rhizome short. **ST**: 6–60 cm, short-hairy or with reflexed bristles. **LF**: 5–40 cm; petiole 2–15 cm; blade oblong to oblong-ovate, deep-lobed to compound, lflets 0 or 7–15, oblong, entire or deep-cut, obtuse or acute, teeth gen 2–4 per side, lower lflet pairs gen distinct, terminal widely merged. **INFL**: well above ground, gen > subtending lvs; peduncle 5–30 cm; pedicels 2–5 mm. **FL**: calyx lobes 3–4 mm, < 13 mm in fr; corolla 6–10 mm, lobes 4–6 mm, white to lavender or white with lavender marks; anthers 1–2 mm; style 7–19 mm. **SEED**: gen 2. Moist, shaded slopes, woodland, meadows, streambanks, chaparral; 600–3000 m. NW, CaRH, SN, ScV, SnFrB; to OR, UT, AZ. May–Jul

H. tenuipes A. Heller (p. 477)   Rhizome long. **ST**: 2–8 dm, with reflexed bristles. **LF**: 8–20 cm wide; petiole 5–30 cm; blade wide-ovate to round; lflets gen 3–5, lobed, coarse-serrate to cut, lowest pair(s) gen smaller, distinct, terminal merged or not (or appearing 3-lobed). **INFL**: well above ground, ≥ subtending lvs; peduncle 2–14 cm; pedicels 4–12 mm. **FL**: calyx lobes 4–7 mm, < 9 mm in fr; corolla 5–7 mm, lobes 3–4 mm, cream, ± green, purple, or blue; anthers 1–2 mm; style 9–14 mm. **SEED**: 1–3. Moist, shaded, wooded slopes, streambanks; < 1500 m. NCo, KR, NCoRO; to WA. May–Jul

LAPPULA

Ann, hairy, taprooted. **ST**: ± erect; branches 0–many. **LF**: basal and cauline, sessile, entire. **INFL**: raceme-like cymes, ± terminal; bracts lf-like; pedicels erect in fr. **FL**: calyx ± deep-5-lobed, enlarged in fr; corolla 5-lobed, funnel-shaped, appendages present; style entire. **FR**: nutlets gen 4, ovate, covered with ± long, barbed prickles, scar lateral. 12–14 spp.: n hemisphere, esp Asia. (Latin: little bur)

1. Nutlet margin prickles in 2(3) rows, bases not or ± widened, not fused . . . . [***L. squarrosa***]
1′ Nutlet margin prickles in 1 row, bases widened or ± so, gen fused . . . . ***L. redowskii***
  2. Nutlet margin prickles wider at base, fused into swollen crown . . . . var. ***cupulata***
  2′ Nutlet margin prickles ± wider at base, ± fused, not in crown . . . . var. ***redowskii***

477

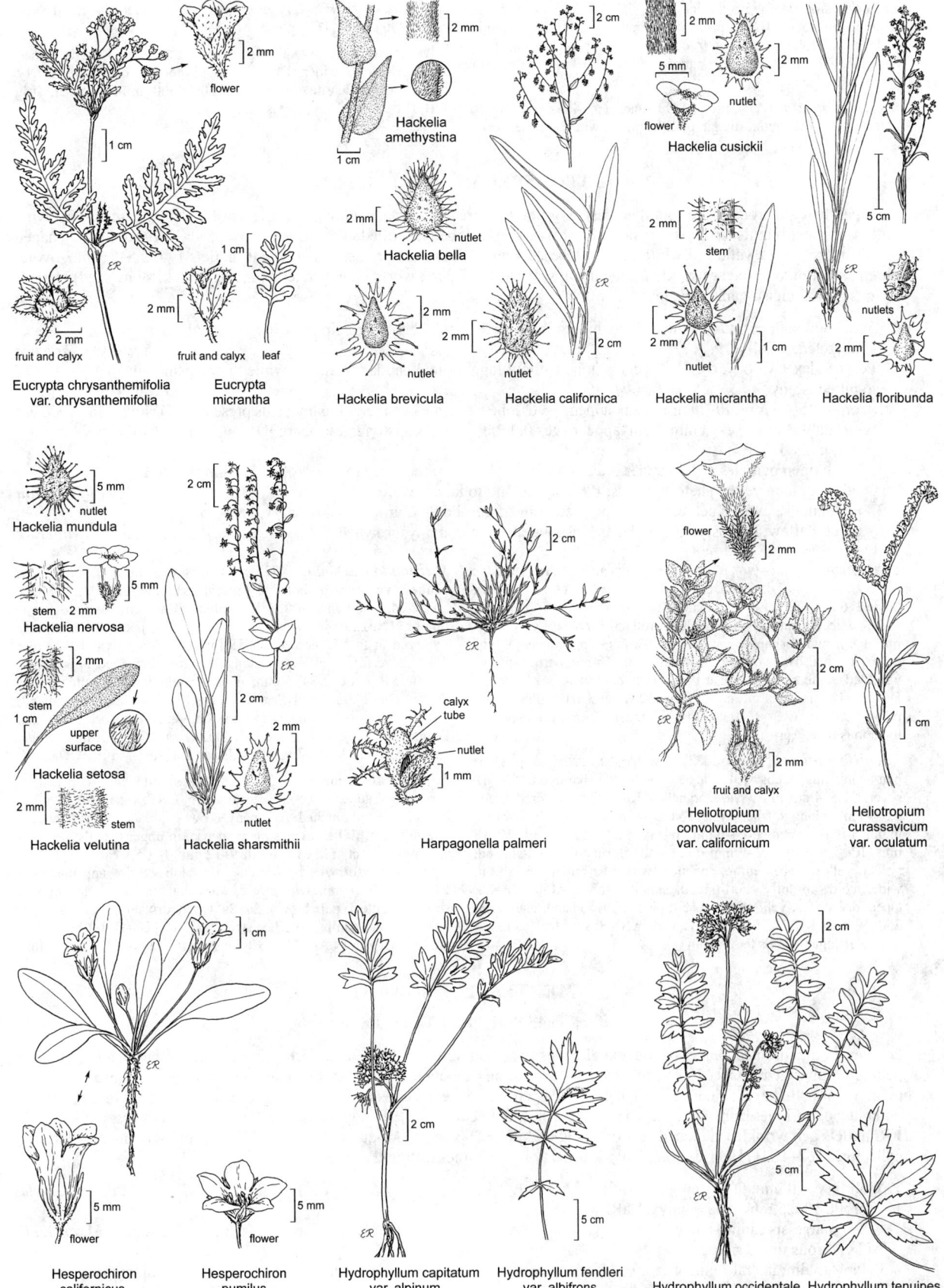

Eucrypta chrysanthemifolia var. chrysanthemifolia

flower

1 cm

ER

fruit and calyx

2 mm

Eucrypta micrantha

1 cm

2 mm

fruit and calyx

leaf

Hackelia amethystina

2 mm

1 cm

Hackelia bella

nutlet

2 mm

Hackelia brevicula

nutlet

2 mm

Hackelia californica

nutlet

2 mm

2 mm

ER

2 cm

2 cm

Hackelia cusickii

nutlet

flower

2 mm

5 mm

Hackelia micrantha

stem

2 mm

nutlet

2 mm

1 cm

Hackelia floribunda

ER

nutlets

2 mm

5 cm

Hackelia mundula

nutlet

5 mm

Hackelia nervosa

stem

2 mm

5 mm

Hackelia setosa

stem

2 mm

upper surface

1 cm

Hackelia velutina

stem

2 mm

Hackelia sharsmithii

2 cm

2 cm

nutlet

2 mm

ER

Harpagonella palmeri

2 cm

calyx tube

nutlet

1 mm

Heliotropium convolvulaceum var. californicum

flower

2 mm

2 cm

ER

fruit and calyx

2 mm

Heliotropium curassavicum var. oculatum

2 cm

1 cm

Hesperochiron californicus

1 cm

ER

flower

5 mm

Hesperochiron pumilus

flower

5 mm

Hydrophyllum capitatum var. alpinum

2 cm

ER

Hydrophyllum fendleri var. albifrons

5 cm

Hydrophyllum occidentale   Hydrophyllum tenuipes

2 cm

5 cm

ER

L. redowskii (Hornem.) Greene **ST**: 0.5–3.5 dm. **LF**: 1–4 cm, linear to lanceolate. **INFL**: pedicel 1–2 mm. **FL**: calyx 3–3.5 mm in fr, lobes lanceolate, ± erect in fr; corolla 1.5–2.5 mm diam, white to occ blue. **FR**: nutlet 2–3 mm, marginal prickles in 1 row, bases ± or much wider, gen fused.

var. ***cupulata*** (A. Gray) M.E. Jones (p. 483) CROWNED STICKSEED **FR**: nutlet margin prickles much wider at base, fused

into swollen crown. Very dry, open, rocky, often disturbed sites; 600–2200 m. GB, DMoj; w N.Am. Mar–Jun

var. ***redowskii*** (p. 483) WESTERN STICKSEED **FR**: nutlet marginal prickles ± wider at base, ± fused, not in crown. 2*n*=48. Dry, open, rocky, often disturbed sites; 1300–3300 m. SNH, SnBr, SnJt, GB, DMtns; w N.Am, Eurasia. Apr–Jul

LITHOSPERMUM STONESEED

Ann, per, hairy, taprooted, red root dye present or not. **ST**: erect. **LF**: gen cauline, ± sessile, entire. **INFL**: panicle-like cyme or fls 1 in upper lf axils; bracts throughout. **FL**: calyx deep-5-lobed, enlarged in fr, lobes equal; corolla 5-lobed, funnel-shaped or salverform, gen ± yellow (± white), tube > lobes, appendages present or 0; style entire. **FR**: nutlets 1–4, 2.5–6+ mm, ovoid, plump, smooth to pitted or wrinkled, attachment scar basal. 75 spp.: worldwide, gen temp, mtns. (Greek: stone seed) Heterostylous or not; cleistogamous fls present or 0.

1. Ann; caudex 0; corolla ± white (blue-white), tube with 5 vertical bands of hairs; fr ± brown, wrinkled, ± tubercled, ± dull . . . . ***L. arvense***
1′ Per; caudex ± woody; corolla ± yellow, tube without longitudinal bands of hairs; fr white to gray, smooth to pitted, shiny
  2. Corolla 15–35 mm, 10–20 mm diam, appendaged, lobe margin jagged; cleistogamous fls present; se DMtns . . . ***L. incisum***
  2′ Corolla 9–18 mm, 7–13 mm diam, appendages 0, lobes entire; cleistogamous fls 0; KR, NCoRH, NCoRI, CaR, n SNH, MP
    3. Fls in open panicles or 1 in upper lf axils; pedicels 4–7 mm, recurved in fr; corolla 12–18 mm, 1.5–2 × calyx, golden yellow; heterostylous; lf blades oblong to lance-ovate, few, scattered . . . . ***L. californicum***
    3′ Fls in dense cymes; pedicels 1–3 mm, ± erect in fr; corolla 9–12 mm, 1–1.5 × calyx, pale- to green-yellow; homostylous; lf blades lanceolate to linear, many, crowded . . . . ***L. ruderale***

L. arvense L. GROMWELL Ann, strigose; caudex 0; red root dye ± 0. **ST**: 1–few, 1–7 dm, ± not clustered, branched at base. **LF**: ± few; blade 2–6 cm, ± linear to lanceolate or oblong. **INFL**: fls 1 in upper lf axils; pedicels 1 mm, erect in fr. **FL**: corolla 5–8 mm, ± = calyx, 2–4 mm diam, funnel-shaped, ± white (blue-white), appendages 0, tube with 5 vertical bands of hairs. **FR**: nutlet 2.5–3 mm, narrow-ovoid, wrinkled, ± tubercled, ± dull, ± brown, late-dehiscent. 2*n*=14,28,42. Disturbed areas, open meadows; 100–1500 m. n SNH, SnFrB (extirpated), MP; to WA, e N.Am; native to Eurasia. Homostylous; cleistogamous fls 0. Apr–Jun

L. californicum A. Gray (p. 483) CALIFORNIA STONESEED Per; hairs spreading, ± coarse; caudex ± woody; red root dye 0. **ST**: 1–several, 1.5–4 dm, clustered, ± branched. **LF**: ± few, scattered; blade 2.5–5 cm, oblong to lance-ovate. **INFL**: open panicles ± few, or fls 1 in upper lf axils; pedicels 4–7 mm, recurved in fr. **FL**: corolla 12–18 mm, 1.5–2 × calyx, 7–9 mm diam, salverform to funnel-shaped, golden yellow, lobes entire, appendages 0. **FR**: nutlet 3.5–5 mm, wide-ovoid, abruptly short-tipped, smooth, shiny, white. 2*n*=28. Open, dry slopes, yellow-pine forest, pine/oak woodland, chaparral; 250–1900 m. KR, NCoRH, NCoRI, CaR, n SNH; sw OR. Heterostylous; cleistogamous fls 0. Apr–Jun

L. incisum Lehm. (p. 483) PLAINS STONESEED Per, strigose; caudex woody; red root dye present. **ST**: few–several, 1–3 dm, clustered, ± unbranched. **LF**: many; blade 1.5–6 cm, linear to linear-oblong. **INFL**: cymes many, in upper lf axils; pedicels 2–5 mm, ± recurved in fr. **FL**: corolla 15–35 mm, 2–3.5 × calyx, 10–20 mm diam, salverform, yellow, tube long, lobe margin jagged, appendages yellow. **FR**: nutlet 2.5–3.5 mm, ovoid, acutely tipped, ± pitted, shiny, ± gray. 2*n*=24,36. Sandy, rocky slopes, pinyon/juniper woodland; 1650–1700 m. se DMtns (Keystone Canyon, New York Mtns, San Bernardino Co.); s NV, to e BC, MT, Great Plains. Homostylous; small green fertile cleistogamous fls present. Apr–May ★

L. ruderale Lehm. (p. 483) WESTERN STONESEED Per; hairs ± spreading; caudex ± woody; red root dye 0. **ST**: 1–several, 2–5 dm, clustered, ± unbranched. **LF**: many, crowded; blade 3–8 cm, lanceolate to linear. **INFL**: cymes many, dense, in upper lf axils; pedicels 1–3 mm, ± erect in fr. **FL**: corolla 9–12 mm, 1–1.5 × calyx, 7–13 mm diam, ± salverform, pale- to green-yellow, lobes entire, appendages 0. **FR**: nutlet 5–6+ mm, wide-ovoid, ± attenuate into stout tip, smooth, shiny, ± white to pale brown. 2*n*=24. Open, dry slopes, plains, sagebrush steppe, conifer forest, chaparral; (750)1200–1800 m. CaRH, n SNH, MP; nw N.Am. Homostylous; cleistogamous fls 0. Apr–Jun

MERTENSIA BLUEBELL

Ronald B. Kelley & Elaine Joyal

Per, gen from taprooted, branched caudex; glabrous to spreading-hairy. **ST**: ± erect. **LF**: cauline and gen basal, alternate, gen petioled, upper gen sessile. **INFL**: gen panicle- or raceme-like cymes; bracts 0. **FL**: calyx gen deep-lobed; corolla often ± cylindric or bell-shaped, blue, gen pink in bud, tube gen well developed, exceeding calyx, abruptly expanded at throat, with or without ring of inner hairs, appendages present or not; filaments often ± flat, gen attached ± below appendages, anthers incl. **FR**: nutlets gen wrinkled, attached near or below middle. ± 50 spp.: N.Am, temp Eurasia. (F.C. Mertens, German botanist, pl collector, 1764–1831) Hybrids common; identification sometimes difficult, esp in MP.

1. Corolla 6–10 mm, tube < calyx; nw KR . . . . ***M. bella***
1′ Corolla 10–25 mm, tube > calyx; CaRH, SNH, GB
  2. Pl ± sparse-spreading-hairy . . . . ***M. cusickii***
  2′ Pl glabrous to ± strigose
    3. Pl 4–15 dm; lateral veins of cauline lvs conspicuous; wet places in mtns; fl Jun–Aug . . . ***M. ciliata*** var. ***stomatechoides***
    3′ Pl gen < 4(5) dm; lateral veins of cauline lvs obscure; gen spring-moist places of plains, foothills; fl Apr–Jun
      4. Basal lvs rare on fl pls; cauline lvs gen 1.5–4 × longer than wide; corolla limb 0.3–0.5(0.6) × tube; sts gen 1–2, from shallow, tuber-like root. . . . . ***M. longiflora***

4′ Basal lvs gen well developed on fl pls; cauline lvs gen 2.5–7 × longer than wide; corolla limb 0.5–0.8 × tube; sts many, from large fleshy taprooted caudex . . . . ***M. oblongifolia***
   5. Lf blade ± glabrous; st gen < 20 cm . . . . var. ***nevadensis***
   5′ Lf blade hairy 1 or both surfaces; st gen > 20 cm
      6. Lf blade hairy both surfaces . . . . var. ***amoena***
      6′ Lf blade hairy adaxially . . . . var. ***oblongifolia***

M. bella Piper (p. 483) OREGON LUNGWORT Pl 2–5 dm from spheric tuber-like root, glabrous or sparse-hairy. **ST:** 1, slender, branched or not. **LF:** lateral veins conspicuous; basal not persistent; cauline petioled, strigose adaxially. **INFL:** raceme-like, open. **FL:** calyx 2–3 mm, strigose; corolla 6–10 mm, wide-bell-shaped, tube rudimentary, 1–1.5 mm, < calyx, without ring of hairs inside, ± unappendaged; filaments slender, attached ± 1 mm above corolla base, ± > anthers; style incl. Wet meadows, springs, under taller pls; 1500–1800 m. nw KR (Siskiyou Co.); to sw OR, n ID, nw MT. May–Jul ★

M. ciliata (Torr.) G. Don var. ***stomatechoides*** (Kellogg) Jeps. (p. 483) STREAMSIDE BLUEBELL Pl 4–15 dm from thick branched thick-taprooted caudex, glabrous, occ glaucous. **ST:** clustered, lfy. **LF:** basal gen > cauline; cauline with conspicuous lateral veins, lower petioled; blades lanceolate to ovate, acute. **INFL:** panicle-like, open. **FL:** calyx 1.5–4 mm; corolla 10–17 mm, limb gen 0.8–1.2(1.5) × tube, tube > calyx, gen without with ring of hairs inside, appendaged; filaments wide, gen > anthers; style exserted 2–5 mm. 2*n*=24,48. Streamsides, wet meadows, damp thickets, wet cliffs; 1310–3380 m. s CaRH, SNH, MP, W&I; to w NV. Jun–Aug

M. cusickii Piper TOIYABE BLUEBELLS Pl 3–5 dm from taprooted branched caudex, ± sparse-spreading-hairy. **ST:** ± clustered. **LF:** basal gen few; cauline ± veiny, lower large, petioled, upper smaller, sessile. **INFL:** panicle-like, open or ± dense. **FL:** calyx 3–6 mm; corolla 10–16 mm, limb 0.8–1 × tube, tube > calyx, with ring of hairs inside near base, appendaged; filaments wide, flat, ± = anthers; style ± incl. Streamsides, dry drainage-bottoms, wooded slopes, drying meadows; ± 2650 m. Wrn; to se OR, sw ID, c NV. Intermediate in morphology, ecology between "short" and "tall" bluebells. May–Jun ★

M. longiflora Greene (p. 483) LONG BLUEBELLS Pl gen < 4 dm from tuber-like root, glabrous to ± strigose. **ST:** gen 1–2, easily detached. **LF:** basal rare on fl pls; cauline few, gen sessile, gen

1.5–4 × longer than wide, lateral veins obscure. **INFL:** ± panicle-like, dense. **FL:** calyx 3–6 mm; corolla 15–25 mm, limb 0.3–0.5(0.6) × tube, tube >> calyx, glabrous inside, appendaged; filaments wide, > anthers; style ± incl. Open, gen spring-moist, drying places of plains, foothills, esp with sagebrush or sparse ponderosa-pine forest; 1500–2200 m. MP; to BC, MT. Apr–Jun ★

M. oblongifolia (Nutt.) G. Don SAGEBRUSH BLUEBELL Pl gen < 4 dm from stout, deep, thick-taprooted, fleshy caudex, glabrous to strigose. **ST:** many, firmly attached. **LF:** basal gen well developed on fl pls; cauline gen 2.5–7 × longer than wide, lateral veins obscure, lower gen petioled. **INFL:** ± panicle-like, gen dense. **FL:** calyx 2.5–6 mm; corolla 10–20 mm, limb 0.5–0.8 × tube, tube > calyx, without or occ with sparse ring of hairs inside, appendaged; filaments wide, ± = anthers; style ± incl.

   var. ***amoena*** (A. Nelson) L.O. Williams BEAUTIFUL SAGEBRUSH BLUEBELLS **ST:** gen > 20 cm. **LF:** blade hairy both surfaces. **FL:** corolla tube hairy inside. 2*n*=48. Open slopes, drier meadows, gen spring-moist places, esp with sagebrush; 1700–2130 m. n SNH (rare), Wrn; to WA, WY, UT. Apr–Jun ★

   var. ***nevadensis*** (A. Nelson) L.O. Williams (p. 483) **ST:** gen < 20 cm. **LF:** blade ± glabrous, occ bumpy both surfaces. **FL:** corolla tube glabrous inside. 2*n*=24. Open slopes, drier meadows, gen spring-moist places, esp with sagebrush; 1760–2510 m. CaRH, n&c SNH, Wrn, n SNE (Sweetwater, Masonic mtns, Bodie Hills); to e OR, ID, CO. Most common var. in CA. Apr–Jun

   var. ***oblongifolia*** SAGEBRUSH BLUEBELLS **ST:** gen > 20 cm. **LF:** blade glabrous abaxially, hairy adaxially. **FL:** corolla tube hairy inside. 2*n*=48. Open slopes, drier meadows, gen spring-moist places, esp with sagebrush; 1580–2380 m. n SNH (rare), MP (esp Wrn); to e WA, WY, UT. Apr–Jun ★

MYOSOTIS FORGET-ME-NOT

Ronald B. Kelley & Elaine Joyal

Ann to per, glabrous to rough-hairy; roots gen fibrous. **ST:** decumbent to erect. **LF:** basal gen oblong or oblanceolate; cauline gen linear to elliptic. **INFL:** gen raceme-like cymes, coiled, in age ± open; bracts 0 (lf-like). **FL:** calyx lobes 5, tube hairs appressed to spreading, hooked at tip or not; corolla salverform or wide-funnel-shaped, gen blue, white, or yellow, appendages prominent or not; stamens incl; style gen incl. **FR:** nutlets gen 4, ± lens-shaped, smooth, shiny, each with raised outer margin, attachment scar adaxially, at base, small. 50 spp.: temp, boreal. (Greek: mouse ear, from lf) *M. arvensis* (L.) Hill reported from Orange Co., 1938, not persisting.

1. Calyx tube hairs appressed, not hooked at tip
  2. Corolla 2–5 mm diam; calyx tube ± ≤ lobes; style < nutlets; st often decumbent but base not creeping or stolon-like; ann to short-lived per . . . . *M. laxa*
  2′ Corolla 5–10 mm diam; calyx tube >> lobes; style ± ≥ nutlets; st base often creeping or stolon-like; per
    . . . . *M. scorpioides*
1′ Calyx tube hairs at least some spreading, some or all hooked
  3. Calyx lobes unequal, 2 longer; corolla 1–2 mm diam, white . . . . [*M. verna*]
  3′ Calyx lobes ± equal; corolla 1–10 mm diam, gen blue, occ initially yellow or pink
    4. Pedicel in fr ≥ calyx; corolla 5–10 mm diam; per . . . . *M. latifolia*
    4′ Pedicel in fr < to << calyx; corolla 1–3 mm diam; ann, occ bien
      5. Fls ± in upper 1/2 of pl; corolla yellow turning blue; style ≥ nutlets; abaxial lf hairs not hooked at tip . . . *M. discolor*
      5′ Fls ± throughout pl; corolla deep blue; style << nutlets; abaxial lf hairs hooked at tip . . . . *M. micrantha*

M. discolor Pers. CHANGING FORGET-ME-NOT Ann, occ bien, puberulent to rough-hairy; roots fibrous. **ST:** 1–5 dm, slender, branched or not. **LF:** sparse, gen 1–4 cm, 2–8 mm wide, abaxial hairs not hooked at tip; basal oblanceolate; cauline ± linear to oblong. **INFL:** fls ± in upper 1/2 of pl; bracts 0 (or 1–2 near base); pedicel in fr < calyx. **FL:** calyx 3–5 mm, tube hairs spreading, hooked at tips, also strigose or puberulent-strigose; corolla gen 1–3 mm diam, wide-

funnel-shaped, yellow turning blue, appendages prominent, yellow turning red. **FR:** nutlets gen ≤ style, dark brown or ± black. 2*n*=64. Roadsides, moist ground, wet meadows; < 1650 m. NW, CaR, n SN, c SNF, ScV, CCo, SnFrB, PR, MP; to e N.Am; native to Eur. Apr–Jul

M. latifolia Poir. (p. 483) BROADLEAVED FORGET-ME-NOT Per. **ST:** < 70 cm, base woody. **LF:** basal large, ovate; cauline oblong.

INFL: bracted at base; pedicel in fr ascending to spreading, 5+ mm, ≥ calyx. **FL**: calyx 3–6 mm in fr, tube hairs ± spreading, gen hooked at tips; corolla 5–10 mm diam, salverform, tube 2 × calyx, gen pink turning blue, appendages prominent, yellow. **FR**: nutlets 2–3 mm, > style, wide-ovate, dark brown. 2*n*=18. Moist, disturbed, shady places; < 460 m. NCo, CCo, SnFrB; S.Am; native to nw Afr. Feb–Jul ❖

M. laxa Lehm. (p. 483) BAY FORGET-ME-NOT Ann to short-lived per. **ST**: 1–4 dm, slender, weak, often decumbent but base not creeping or stolon-like; ± strigose. **LF**: 1.5–8 cm, 3–15 mm wide; basal oblanceolate; cauline oblong to lanceolate. **INFL**: bracted at base; pedicel spreading in fr, > calyx. **FL**: calyx 3–8 mm, tube hairs appressed-strigose, not hooked at tips, tube ± ≤ lobes, lobes equal or not, narrow-triangular; corolla 2–5 mm diam, ± salverform, blue, appendages prominent, yellow. **FR**: nutlets < 2 mm, > style, brown to black. 2*n*=±80. Moist soil to gen shallow water; < 2050(2400) m. NW (exc NCoRH), n SNH, s SNH, ScV; to s BC, w MT; ± circumboreal, S.Am. Highly localized in CA. May–Sep

M. micrantha Lehm. SMALL-FLOWERED FORGET-ME-NOT Ann or winter ann. **ST**: < 2 dm, branches 0 or from near base; ± puberulent or rough-hairy. **LF**: gen < 2 cm, 7 mm wide, abaxial hairs hooked,

esp on midrib; basal oblanceolate or wider; cauline oblong or elliptic. **INFL**: fls from near pl base, where 1 per axil; bracts in lower 1/2, lf-like; pedicel in fr ascending or ± spreading, << calyx. **FL**: calyx 3–5 mm, strigose at least above, tube hairs spreading, hooked at tips; corolla 1–2 mm diam, wide-funnel-shaped, deep blue, appendages minute, ± blue. **FR**: nutlets >> style, brown, occ paler. 2*n*=±36–40. Uncommon. Roadsides, streambanks, disturbed open areas; 500–1650 m. KR, NCoRO, MP; n US, s Can, native to Eurasia. Apr–Jun

M. scorpioides L. (p. 483) WATER FORGET-ME-NOT Per. **ST**: 2–6 dm, gen unbranched, base often creeping or stolon-like; ± strigose. **LF**: 2.5–8 cm, 7–20 mm wide; basal oblanceolate; cauline oblong or elliptic to lance-elliptic. **INFL**: bracts 0; pedicel in fr spreading, ± ≥ calyx. **FL**: calyx 3–5 mm, tube hairs sparse-appressed-strigose, not hooked at tip, lobes << tube, occ unequal, wide-triangular, in fr < 6 mm; corolla 5–10 mm diam, blue, salverform, blue, appendages prominent, yellow. **FR**: nutlets > 1.5 mm, ± ≤ style, ovate, ± black, marginal rim obscure. 2*n*=64. Shallow water, wet soil; (75)1050–2100+ m. SNH, Wrn; w OR, WA, to e N.Am; native to Eur. Other reported locations need documentation; most easily confused with *M. laxa*. Jun–Aug

NAMA PURPLE MAT

Sarah Taylor

Per or gen ann, hairy. **LF**: cauline, gen alternate, margin crenate-dentate or gen entire, wavy, flat, or rolled under. **INFL**: heads or gen clusters, terminal, not coiled, or fls 1–2 in lf axils; bracts 0 or gen lf-like. **FL**: calyx gen free from ovary; corolla salverform to bell-shaped, occ cylindric; stamens gen attached to corolla at different levels, gen unequal, part fused to corolla gen narrow-winged, scales at base 0; styles 1–2. **FR**: gen loculicidal, ovoid to elliptic. **SEED**: gen many, small, red-brown, brown, black, or yellow to orange. ± 55 spp.: sw US, trop Am, HI. (Greek: a stream) *N. lobbii* A. Gray now in *Eriodictyon*.

1. Per; fls in heads; lf crenate-dentate . . . . ***N. rothrockii***
1′ Ann; fls in clusters or 1–2 in lf axils; lf entire
  2. Calyx fused to ovary in lower 1/3–1/2; lf margins wavy, gen ± rolled under . . . . ***N. stenocarpa***
  2′ Calyx free from ovary; lf margins flat or rolled under, not wavy
    3. Corolla 1–3 mm, bowl- or bell-shaped, basal tube ± 0.5 mm; styles 2; seeds ≤ 4 . . . . ***N. californica***
    3′ Corolla > 2 mm, if bowl- or bell-shaped then basal tube 0 or indistinct; styles 1–2; seeds > 4
      4. Style 1, lobed < 1/2 length
        5. Corolla limb > 3.5 mm diam, lobes ≥ 2 mm wide . . . . ***N. aretioides***
          6. Style 1–2 mm . . . . var. *californica*
          6′ Style 3–7 mm . . . . var. *multiflora*
        5′ Corolla limb < 3.5 mm diam, lobes < 2 mm wide . . . . ***N. densa***
          7. Pl gray-canescent, hairs gen spreading, rough; style 0.3–1 mm . . . . var. *densa*
          7′ Pl green, hairs gen sparse-bristly-strigose; style (0.6)1–2.5 mm . . . . var. *parviflora*
      4′ Styles 2
        8. Corolla limb < 4 mm diam
          9. Calyx lobes glandular . . . . ***N. dichotoma*** var. *dichotoma*
          9′ Calyx lobes nonglandular
            10. Lf sessile; sepals gen pale green; corolla funnel-shaped or salverform; seeds oblong to ovoid,
              ± cross-ridged . . . . ***N. depressa***
           10′ Lf petioled; sepals white- or gray-canescent at least in basal 1/3–1/2; corolla cylindric to ±
              funnel-shaped; seeds irregular, ± net-like . . . . ***N. pusilla***
        8′ Corolla limb 5–14 mm diam
          11. St ascending to erect; seeds fusiform, yellow to orange . . . . ***N. hispida*** var. *spathulata*
          11′ St prostrate; seeds spheric to ovoid, brown to black . . . . ***N. demissa***
            12. Lf petioled, elliptic to diamond-shaped; pl gen gray-green . . . . var. *covillei*
            12′ Lf sessile, linear to spoon-shaped; pl gen green . . . . var. *demissa*

N. aretioides (Hook. & Arn.) Brand Hairs gen dense, coarse, appressed to spreading, gen swollen at base. **ST**: prostrate, 3–12 cm, repeatedly forked. **LF**: ± sessile. **INFL**: fls sessile. **FL**: calyx lobes narrowly linear to lanceolate; corolla 5–18 mm, salverform, limb > 3.5 mm diam, lobes ≥ 2 mm wide; fused parts of filaments unwinged; style 1, lobes 2, < 1/2 length. **FR**: 2–4 mm. **SEED**: gen compressed, irregularly elliptic-ovoid, brown to black, smooth to minute-cross-ridged, with depressions. 2*n*=14. Other vars. in ID, NV.

  var. ***californica*** (Brand) Jeps. **LF**: 4–20 mm, narrowly linear to spoon-shaped. **FL**: corolla 5–9 mm, white or pink; stamens 1–4

mm, attached 1–3 mm above corolla base; style 1–2 mm. **SEED**: ≤ 1 mm, black, smooth or minute-roughened. Sandy, loam flats, slopes; 1370–1500 m. s MP (exc Wrn); w NV. May–Jul

  var. ***multiflora*** (A. Heller) Jeps. (p. 483) **LF**: 7–30 mm, lanceolate to spoon-shaped. **FL**: corolla 9–18 mm, gen pink or purple; stamens 3–8 mm, attached 2–5 mm above corolla base; style 3–7 mm. **SEED**: 0.6–0.8 mm, brown to black, minute-cross-ridged. Dry, sandy areas; 1000–2150 m. n SNH, GB, n DMtns (Argus, Panamint ranges); NV. May–Jun

N. californica (A. Gray) J.D. Bacon (p. 483)   Gen puberulent to fine-strigose, hair bases gen swollen. **ST**: prostrate, 3–10 cm, forked. **LF**: gen sessile, 5–14 mm, 1–4 mm wide, oblanceolate, spoon-shaped, or elliptic. **INFL**: fls ± sessile. **FL**: calyx lobes 2–5 mm, lance-linear, silky-hairy; corolla 1–3 mm, bowl- or bell-shaped, white to pale pink, basal tube ± 0.5 mm, limb 1–2 mm diam, lobes 0.4–0.8 mm, 0.5–0.8 mm wide; stamens 0.8–1.3 mm, attached ≤ 0.5 mm above corolla base, free filament abruptly expanded above attachment; styles 2, 0.5–1 mm. **FR**: 2–2.5 mm. **SEED**: ≤ 4, 0.8–1 mm, ovoid, minute-cross-ridged, with depressions. 2*n*=14. Dry, sandy areas; 500–2500 m. NCoRI, s SN, Teh, SnJV, e SnFrB, SCoRI, TR, w edge DMoj. Apr–Jun

N. demissa A. Gray   Hairs gen dense, fine to coarse, gen mealy-glandular, bases swollen. **ST**: prostrate, 3–20 cm, forked. **INFL**: pedicels 0–5 mm. **FL**: corolla funnel-shaped to salverform; stamens attached 2–4 mm above corolla base; styles 2. **SEED**: ± 0.5 mm, spheric to ovoid, brown to black, ± cross-ridged, with depressions.

var. ***covillei*** Brand   Pl gen gray-green, hairs 0.5–1.2 mm, soft, shaggy. **LF**: petiole 1.5–5 mm, winged; blade 5–13 mm, elliptic to diamond-shaped. **FL**: calyx lobes 4–8 mm, linear to oblanceolate; corolla 8–12 mm, blue-violet to pink, limb 8–9 mm diam, lobes 2–4 mm, 3–4 mm wide; stamens 3–6 mm; styles 2–4 mm. **FR**: 2–5 mm. **SEED**: black. Dry, sandy flats, slopes; 30–900 m. n DMoj (Death Valley region, rare Ivanpah Valley). Feb–May

var. ***demissa*** (p. 483)   Pl gen green, hairs < 1 mm, fine-strigose. **LF**: sessile, 1–4 cm, linear to spoon-shaped. **FL**: calyx lobes 3–8 mm, lance-linear; corolla 7–14 mm, blue-purple to rose-pink, limb 5–9 mm diam, lobes 2–3 mm, 2–4 mm wide; stamens 4–7 mm; styles 3–6 mm. **FR**: 3–4 mm. **SEED**: elliptic-ovoid. 2*n*=14. Sandy or gravelly flats, slopes; 70–1780(2040) m. s SNH, Teh, SnGb, SnBr, e PR, SNE, D; NV, AZ, Mex. Mar–May

N. densa Lemmon   Hairs gen dense, stiff, gen appressed, bases gen swollen. **ST**: prostrate, forked. **LF**: sessile, lanceolate to oblanceolate. **INFL**: pedicels 0–1 mm. **FL**: calyx lobes lance-linear; corolla cylindric to funnel-shaped, white to pale purple, limb < 3.5 mm diam, lobes < 2 mm wide; fused part of filament unwinged; style 1, lobes 2, < 1/2 length. **FR**: 2–4 mm. **SEED**: 0.6–0.9 mm, gen elliptic to ovoid, angled on underside, brown to black, smooth to cross-ridged.

var. ***densa*** Pl gray-canescent, hairs gen spreading, rough. **FL**: corolla 2–5 mm, gen narrowed below limb; stamens attached ± 1 mm above corolla base; style 0.3–1 mm. **SEED**: gen shiny, gen smooth. 2*n*=14. Sandy or gravelly flats, slopes; 880–3560 m. e CaRH, n&c SNH, GB; OR, w NV. May–Jul

var. ***parviflora*** (Greenm.) C.L. Hitchc. (p. 491)   Pl green, hairs gen sparse-bristly-strigose. **FL**: corolla 3–8 mm, gen not narrowed below limb; stamens attached 1–2 mm above corolla base; style (0.6)1–2.5 mm. **SEED**: minute-cross-ridged, with depressions. 2*n*=14,28. Sandy or gravelly flats, slopes; 1340–1680 m. e CaRH, MP (exc Wrn); to WA, WY, w CO. Jun–Aug

N. depressa A. Gray   Pl puberulent, hairs appressed to ascending. **ST**: prostrate, 2–10 cm, forked, gen lfless in lower 1/2. **LF**: sessile, 2–16 mm, oblanceolate to spoon-shaped. **INFL**: fls pedicelled. **FL**: calyx lobes 3–5 mm, linear to ± spoon-shaped, gen pale green, nonglandular; corolla 3–6 mm, funnel-shaped or salverform, pink or white, limb 2–4 mm diam, lobes 0.6–1.2 mm, 0.8–1.2 mm wide; stamens 2–3 mm, attached 1–2 mm above corolla base; styles 2, 1–2 mm. **FR**: 2–4 mm. **SEED**: 0.4–0.7 mm, oblong to ovoid, brown, ± cross-ridged, gen with depressions. 2*n*=14. Dry, sandy or gravelly flats, slopes; 360–1700 m. s SNH, Teh, n SnBr, SNE, DMoj; sw NV. Apr–May

N. dichotoma (Ruiz & Pav.) Choisy var. ***dichotoma*** FORKED PURPLE MAT   Pl short-glandular-strigose and short-nonglandular-hairy, nonglandular hairs spreading or not. **ST**: gen erect, 5–20 cm, simple or forked. **LF**: sessile, 6–20 mm, linear to spoon-shaped. **INFL**: pedicels 1–2 mm, slender. **FL**: calyx lobes 4–10 mm, linear to spoon-shaped, glandular; corolla 3–8 mm, cylindric to bell-shaped, white to ± blue, limb 2–4 mm diam, lobes 1–2 mm, 1–1.5 mm wide, basal tube 0; stamens 2–4 mm, attached 0.5–1.2 mm above corolla base, filament wider just above attachment; styles 2, 1–3 mm. **FR**: 2–6 mm. **SEED**: 0.5–0.7 mm, irregularly oblong to ovoid, brown, cross-ridged, with depressions. 2*n*=28. Granite or limestone slopes, ridges; 1900–2200 m. e DMtns (New York Mtns); to CO, TX, Mex. Oct ★

N. hispida A. Gray var. ***spathulata*** (Torr.) C.L. Hitchc.   Pl gen mealy-glandular, hairs gen dense, fine to bristly-strigose, bases gen swollen. **ST**: ascending to erect, 7–30 cm, branches 0–many. **LF**: sessile, 1–5 cm, linear to spoon-shaped, ± rolled under. **INFL**: pedicels 0–4 mm. **FL**: calyx lobes 4–10 mm, ± linear to oblanceolate; corolla 10–15 mm, salverform to narrow-bell-shaped, blue to purple-lavender, basal tube 0, limb 5–14 mm diam, lobes 1–4 mm, 2–5 mm wide; stamens 2–6 mm, attached 1–4 mm from corolla base; styles 2, 1–4 mm. **FR**: 3–6 mm. **SEED**: 0.3–0.7 mm, fusiform, yellow to orange, net-like. 2*n*=14. Dry, sandy or gravelly flats, slopes; < 700(900) m. n SnBr, e DMoj, DSon; to w TX, n Mex. Mar–May

N. pusilla A. Gray   Pl gen dense-short-spreading-hairy, hairs slender, bases gen swollen. **ST**: prostrate, 2–6 cm, forked. **LF**: petiole 1–6 mm, winged; blade gen 2–11 mm, lanceolate to gen ovate, base abruptly narrowed. **INFL**: pedicels ≤ 4 mm. **FL**: calyx lobes 3–6 mm, linear to ± spoon-shaped, white- or gray-canescent at least in basal 1/3–1/2, nonglandular; corolla 3–5 mm, cylindric to ± funnel-shaped, white to pale pink, limb ≤ 2 mm diam, lobes ≤ 0.5 mm, 0.5–1 mm wide; stamens 1–3 mm, attached 1–2 mm above corolla base; styles 2, 1–2 mm. **FR**: 2–4 mm. **SEED**: ± 0.4 mm, irregular, brown-black, surface ± net-like, with depressions. 2*n*=14. Sandy to rocky flats; 200–1740 m. n SnBr, W&I, DMoj, n DSon; s NV, w AZ. Mar–May

N. rothrockii A. Gray (p. 491)   Per, rhizomed, in colonies, gen dense-short-glandular-hairy and long-bristly-nonglandular-hairy, longer hairs gen swollen at base. **ST**: erect, 20–30 cm, sticky, branches 0–few. **LF**: 2–6 cm, lanceolate to elliptic, crenate-dentate. **INFL**: heads; bracts 0; pedicels ± 1 mm. **FL**: calyx lobes 8–15 mm, lance-linear; corolla 13–18 mm, ± funnel-shaped, pink, purple, or pale blue, limb 8–10 mm diam, lobes 4–6 mm, 3–6 mm wide; stamens 6–11 mm, attached 4–7 mm above corolla base; styles 2, 7–11 mm. **FR**: 3–6 mm. **SEED**: 1–2 mm, ovoid, red-brown to brown, minute-pitted. 2*n*=34. Sandy alluvial flats, gravelly granitic slopes, meadows; 1600–3050 m. c&s SNH, SnBr, SNE; NV, nw AZ. Jul–Aug

N. stenocarpa A. Gray MUD NAMA   Pl short-soft-silky-hairy and short-glandular-hairy; some hairs stiff, swollen at base. **ST**: prostrate to ascending, 8–40 cm, branches many. **LF**: petiole 0(3) mm; 5–30 mm, oblanceolate, oblong, or spoon-shaped, base gen ± clasping st, margins wavy, gen ± rolled under. **INFL**: pedicels ≤ 5 mm. **FL**: calyx lobes 3–6 mm, to 9 mm in fr, oblanceolate to spoon-shaped, gen bristly, fused to ovary in lower 1/3–1/2; corolla 4–6 mm, ± funnel-shaped, white to cream, limb 3–5 mm diam, lobes 1–2 mm, 2–3 mm wide; stamens 2–4 mm, attached 0.5–1.5 mm above corolla base, filament gen toothed at attachment point; ovary partly inferior, style 1, 1–2 mm, lobes 2. **FR**: 3–9 mm. **SEED**: ± 0.5 mm, irregularly angled, tan to brown, honeycombed. 2*n*=14. Intermittently wet areas; < 810 m. SnJV, SCo, s ChI, w PR, se DSon; to TX, n Mex. Mar–Oct ★

NEMOPHILA

Robert Patterson & Richard R. Halse

Ann. **ST**: simple to branched, prostrate to erect, fleshy, brittle, angled or winged, glabrous to gen bristly (prickly). **LF**: cauline, lower gen opposite, upper opposite or alternate, gen reduced; petiole gen bristly-ciliate; blade pinnate-toothed or -lobed, gen bristly. **INFL**: fls 1 in lf axils or opposite lvs; pedicels longer in fr, recurved. **FL**: calyx bell-shaped to rotate, sinuses gen with spreading or reflexed appendages; corolla bell-shaped to rotate, white, blue, or purple, spotted or marked or not; stamens incl; ovary chamber 1, style 1, gen 1/3–1/2 forked. **FR**: gen 2–7 mm wide, spheric to ovoid, hairy, gen enclosed by calyx. **SEED**: ovoid, smooth, wrinkled or pitted, with a conic, colorless appendage at 1 end. 11 spp.: se US, w N.Am. (Greek: woodland-loving)

1. St with minute reflexed prickles; lower lvs alternate . . . . *N. breviflora*
1′ St without reflexed prickles; lower lvs opposite (exc *Nemophila pulchella* var. *gracilis*)
  2. Corolla ≥ 10 mm wide (6–40 mm wide in *Nemophila menziesii*)
   3. Seeds smooth or pitted; corolla with or without purple spot at lobe tips
    4. Corolla 10–50 mm wide; style 3–6 mm . . . . **N. maculata**
    4′ Corolla 2–10 mm wide; style < 2 mm . . . . [2]**N. spatulata**
   3′ Seeds wrinkled and tubercled; corolla without purple spot at lobe tips . . . . **N. menziesii**
    5. Upper, lower lvs dissimilar, lower lobed, upper entire to toothed
     6. Lower lvs 5–7-lobed; corolla 6–15 mm wide . . . . [2]var. **integrifolia**
     6′ Lower lvs 6–13-lobed; corolla 10–40 mm wide . . . . [2]var. **menziesii**
    5′ Upper, lower lvs similar, deep-5–13-lobed
     7. Corolla white, black-dotted from center ± to margin . . . . var. **atomaria**
     7′ Corolla blue with white center or blue-veined, gen black-dotted at center . . . . [2]var. **menziesii**
  2′ Corolla < 10(12) mm wide
   8. Corolla gen rotate; filaments > corolla tube
    9. Calyx appendages 1–4 mm in fr; seeds wrinkled and tubercled . . . . [2]**N. menziesii** var. **integrifolia**
    9′ Calyx appendages 0–0.5 mm in fr; seeds smooth or ± rough . . . . **N. pulchella**
     10. Corolla blue with white center, >> calyx; style 2–3 mm . . . . var. **pulchella**
     10′ Corolla white, ± ≥ calyx; style ≤ 2 mm
      11. Cauline lvs opposite; style < 1 mm; seeds 2–4 . . . . var. **fremontii**
      11′ Cauline lvs alternate; style 1–2 mm; seed gen 1 . . . . var. **gracilis**
   8′ Corolla bell- to bowl-shaped; filaments ≤ corolla tube
    12. Corolla gen dark-veined, dotted, lobes occ purple-spotted
     13. Lf oblong to ovate, deep-5–9-lobed . . . . **N. pedunculata**
     13′ Lf oblanceolate to spoon-shaped, shallow-3–5-lobed . . . . [2]**N. spatulata**
    12′ Corolla not dark-veined, dotted, lobes not purple-spotted
     14. Style 2–4 mm; lower lvs deep-lobed, lobes similar, gen well separated, stalked . . . . **N. heterophylla**
     14′ Style < 2 mm; lower lvs cut or shallow-lobed, lobes dissimilar, merging or not stalked . . . . **N. parviflora**
      15. Upper lvs alternate, lobes deep, gen acute . . . . var. **parviflora**
      15′ Upper lvs opposite, lobes shallow, rounded or obtuse
       16. Lf glabrous to bristly, blade base tapered . . . . var. **austiniae**
       16′ Lf long-hairy, blade base truncate or cordate . . . . var. **quercifolia**

N. breviflora A. Gray GREAT BASIN NEMOPHILA **ST:** prickles minute, reflexed. **LF:** lower alternate, blade 7–30 mm, 15–40 mm wide, lobes 3–6, acute, gen entire. **INFL:** pedicels < 5 mm, < 15 mm in fr. **FL:** calyx lobes 3–5 mm, stiff-ciliate, appendages 1–2 mm in fr; corolla 2 mm, 1–4 mm wide, bell-shaped, white or ± purple, tube > filaments; anthers << 1 mm; style < 1 mm, tip lobed. **SEED:** gen 1, ± red, smooth but regularly deep-pitted in rows. *n*=9. Streambanks, meadows, thickets; 1500–2200 m. MP; to BC, MT, CO. May–Jun ★

N. heterophylla Fisch. & C.A. Mey. (p. 491) **LF:** lower opposite, 1–4 cm, blade = petiole, oblong to ovate, lobes 5–7, deep, gen round, stalked, gen well separated, entire to 1–3-toothed; upper alternate, ± sessile, blade lanceolate to ovate, lobes 0 or 3–5, entire. **INFL:** pedicels < 10 mm, < 60 mm in fr. **FL:** calyx lobes 2–4 mm, appendages < 1 mm in fr; corolla 3–10 mm, 4–12 mm wide, bowl-shaped, white or ± blue, tube = filaments; anthers < 1 mm; style 2–4 mm. **SEED:** 2–5, yellow-brown, smooth or roughened. *n*=9. Forest, chaparral, roadsides, slopes, streambanks, talus; 30–1700 m. NW, CaR, SN, GV, CW; OR. Feb–Jun

N. maculata Lindl. (p. 491) FIVESPOT **LF:** opposite, petiole ≥ blade; lower blades 8–30 mm, 3–15 mm wide, oblong to ovate, lobes 5–9, entire or 1–3-toothed; upper oblanceolate to spoon-shaped, sessile, tip entire or 3-toothed. **INFL:** pedicels stout, 10–20 mm, < 70 mm in fr. **FL:** calyx lobes 4–9 mm, appendages 1–4 mm in fr; corolla 8–20 mm, 10–50 mm wide, bowl-shaped to rotate, white with dark veins, dots, purple-spotted at lobe tips, tube filaments; anthers 1–3 mm; style 3–6 mm. **SEED:** 2–12, green-brown, smooth or shallow-pitted. *n*=9. Meadows, roadbanks, woodland; 60–3100 m. SN, ScV. May–Jul

N. menziesii Hook. & Arn. BABY BLUE-EYES **LF:** opposite; lower 1–5 cm, blade = petiole, linear-oblong to ovate, lobes 5–13, obtuse, entire or 1–3-toothed; upper ± sessile, entire, toothed, or less lobed than lower. **INFL:** pedicels 20–60 mm, < 70 mm in fr. **FL:** calyx lobes 4–8 mm, appendages 1–4 mm in fr; corolla 5–20 mm, 6–40 mm wide, bowl-shaped to rotate, bright blue with white center to white, gen blue-veined, black-dotted, tube ≤ filaments; anthers 2–3 mm; style 2–7 mm. **FR:** 5–15 mm wide. **SEED:** 4–20, brown to black, wrinkled, tubercled. *n*=9. Highly variable; vars. intergrade.

var. ***atomaria*** (Fisch. & C.A. Mey.) Voss (p. 491) **LF:** all similar, deep-5–13-lobed. **FL:** corolla 6–12 mm, 10–30 mm wide, white, black-dotted from center ± to margin, ± blue-tinted or -veined or not, tube = filaments. **SEED:** 8–12. Coastal bluffs, grassy slopes; 15–1500 m. NW, CCo, SnFrB; OR. Feb–Jun

var. ***integrifolia*** Parish **LF:** lower 5–7-lobed, upper sessile, diamond-shaped, oblong, or oblanceolate, entire to shallowly few-toothed. **FL:** corolla 5–10 mm, 6–15 mm wide, gen rotate, blue, black-dotted at center or blue-veined, tube < filaments. **SEED:** 4–10. Grassland, canyons, woodland, burns, slopes; 100–1900 m. n SNH, SnJV, CCo, SCoR, SW, SNE, DMoj; Baja CA. Jan–Jul

var. ***menziesii*** (p. 491) **LF:** lower 6–13-lobed, upper fewer, short-petioled, toothed or lobes narrower. **FL:** corolla 5–20 mm, 10–40 mm wide, bright blue with white center or blue-veined, gen black-dotted at center, tube = filaments. **SEED:** 10–20. Meadows, grassland, chaparral, woodland, slopes, desert washes; 15–1600 m. CA-FP, DMoj. Feb–May

N. parviflora Benth. **ST:** glabrous or soft- to bristly-hairy. **LF:** lower cut or lobes shallow, dissimilar, merging or not stalked; upper opposite or alternate. **INFL:** pedicels 2–15 mm, < 30 mm in fr. **FL:** calyx lobes 1–3 mm, appendages < 1 mm in fr; corolla 2–4 mm, 1–5 mm wide, bell-shaped, white or blue, tube ≥ filaments; style < 2 mm. **SEED:** 2–4, yellow to brick-red, smooth but shallow-pitted. *n*=9.

var. ***austiniae*** (Eastw.) Brand (p. 491) **ST:** ± glabrous to bristly. **LF:** glabrous to bristly, blade base tapered; lower 1–2 cm, lobes 5–7, shallow, oblong or ovate, obtuse, entire; upper opposite, short-stalked. **FL:** corolla 1–3 mm wide. Meadows, streambanks, roadsides, forest, ridges; 1100–2300 m. KR, NCoRH, n SNH, MP; to WA, ID, UT. May–Jul

var. ***parviflora*** **ST:** ± glabrous to bristly. **LF:** upper alternate, gen sessile, 1–4 cm, asymmetric, blade base truncate or cordate, lobes

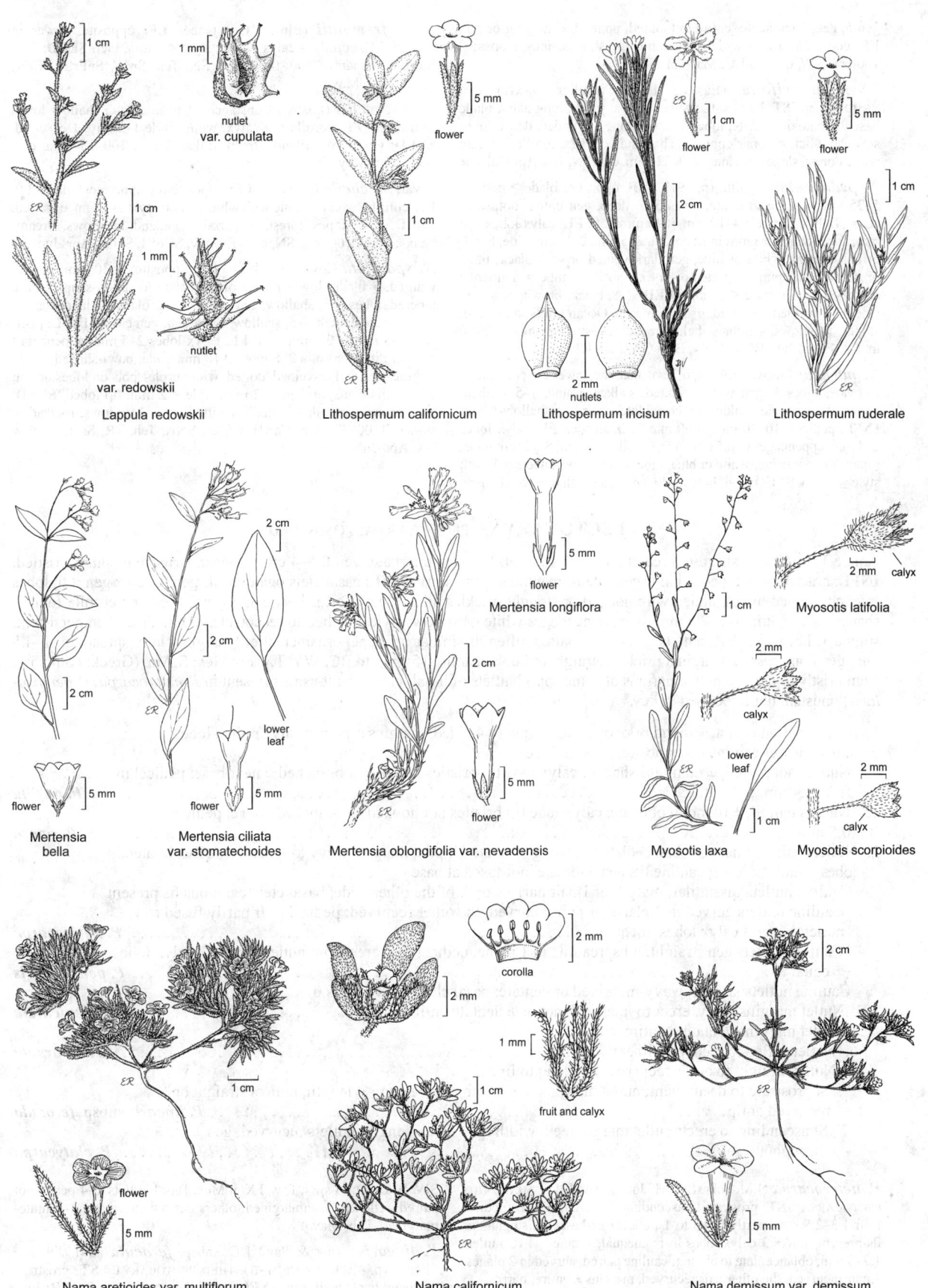

1 cm
1 mm
nutlet
var. cupulata
1 cm
ER
1 mm
nutlet
var. redowskii
Lappula redowskii

5 mm
flower
1 cm
ER
Lithospermum californicum

ER
flower
1 cm
2 cm
2 mm
nutlets
Lithospermum incisum

5 mm
flower
1 cm
ER
Lithospermum ruderale

2 cm
lower
leaf
2 cm
ER
2 cm
5 mm
flower
Mertensia
bella
5 mm
flower
Mertensia ciliata
var. stomatechoides

5 mm
flower
Mertensia longiflora
2 cm
5 mm
flower
Mertensia oblongifolia var. nevadensis

1 cm
calyx
lower
leaf
ER
1 cm
Myosotis laxa

2 mm
calyx
Myosotis latifolia
2 mm
calyx
2 mm
calyx
Myosotis scorpioides

ER
flower
5 mm
Nama aretioides var. multiflorum

2 mm
corolla
2 mm
1 mm
fruit and calyx
1 cm
ER
Nama californicum

2 cm
ER
5 mm
Nama demissum var. demissum

gen 5, deep, gen acute, entire to toothed, upper 3 ± merging or not. **FL**: corolla 1–5 mm wide. Woodland, forest, roadsides, slopes; < 1600 m. NW, CW; to BC. Mar–Jul

var. ***quercifolia*** (Eastw.) H.P. Chandler OAK-LEAVED NEMOPHILA **ST**: hairs soft, shaggy, or wavy. **LF**: long-hairy, blade base truncate or cordate, lobes 5–7, shallow, ± rounded, dense-long-soft-wavy-hairy; upper opposite, short-stalked. **FL**: corolla 3–5 mm wide. Forest, slopes, ravines; 700–2200 m. c&s SN, Teh. Apr–Jul ★

N. pedunculata Benth. (p. 491) **LF**: opposite; blade ≥ petiole, 5–35 mm, oblong to ovate, lobes 5–9, deep, gen entire, obtuse or acute. **INFL**: pedicels 4–12 mm, < 45 mm in fr. **FL**: calyx lobes 1–4 mm, appendages < 3 mm in fr; corolla 2–5 mm, 2–8 mm wide, bowl- or bell-shaped, white or blue, gen dark-veined or with black, blue, or purple dots, purple-spotted or not at lobe tips, tube = filaments; anthers < 1 mm; style < 2 mm. **SEED**: 2–8, black, brown, or green, smooth, wrinkled, or pitted. *n*=9. Common. Ocean bluffs, grassland, slopes, meadows, sandbars, fields, woodland, streambanks; < 2400 m. CA-FP, MP; to BC, ID, NV, Baja CA. Feb–Jul

N. pulchella Eastw. **LF**: opposite or alternate; lower 2–5 cm, oblong to ovate, lobes 5, gen well separated, stalked, round, 1–5-toothed; upper often sessile, oblong or lanceolate, entire to shallow-lobed. **INFL**: pedicels 10–30 mm, < 50 mm in fr, slender. **FL**: calyx lobes 2–4 mm, appendages 0–0.5 mm in fr; corolla 3–5 mm, 5–12 mm wide, rotate, >> to = calyx, white or blue, tube < filaments; anthers < 1 mm; style ≤ 3 mm. **SEED**: 1–4, brown or ± green, smooth or ± rough. *n*=9.

var. ***fremontii*** (Elmer) Constance **LF**: opposite, lower in rosette. **FL**: corolla = calyx, white; style < 1 mm, incl. **SEED**: 2–4. Slopes, chaparral; 400–1400 m. s SN, Teh, SnJV, SnFrB, SCoR, WTR. Mar–May

var. ***gracilis*** (Eastw.) Constance **LF**: alternate, shallow-lobed or toothed. **FL**: corolla ± > calyx, white; style 1–2 mm, ± exserted. **SEED**: gen 1. Woodland, streambanks, slopes; 100–1300 m. c&s SNF. Mar–May

var. ***pulchella*** (p. 491) **LF**: opposite (or uppermost alternate). **FL**: corolla >> calyx, blue with white center; style 2–3 mm, exserted. **SEED**: 2–4. Slopes, forest, chaparral, woodland, meadows, streambanks; 100–2100 m. s SN, c SNF, SnJV, SCoRI, SnGb. Apr–Jun

N. spatulata Coville (p. 491) **LF**: opposite, 5–30 mm, petiole winged, > blade; lower blades oblanceolate to spoon-shaped, base tapered, lobes 3–5, shallow, gen entire; upper blades oblanceolate to wide-tapered, teeth 3–5, shallow, triangular, gen entire. **INFL**: pedicels 3–8 mm, < 30 mm in fr. **FL**: calyx lobes 2–5 mm, appendages 1–2 mm in fr; corolla 2–8 mm, 2–10 mm wide, bowl-shaped, white or blue, gen darker-veined, dotted, with purple-spots on lobes or not, tube > filaments; anthers < 2 mm; style < 2 mm, tip lobed. **SEED**: 5–7, brown, smooth but shallow-pitted. *n*=9. Meadows, roadsides, slopes; 1100–3200 m. CaRH, s SNF, SNH, Teh, TR, SnJt, Wrn; w NV. Apr–Jul

PECTOCARYA PECTOCARYA, COMBSEED

Ann. **ST**: 2–40 cm, strigose, breaking at nodes or not. **LF**: gen alternate, gen 0.5–4 cm, ± linear, strigose to sharp-bristled. **INFL**: bracted or not; pedicel in fr gen free from nutlets, gen recurved. **FL**: basal cleistogamous fls gen 0; calyx gen < fr, lobes free, not arched over 1 nutlet, with hooked or straight prickles, in fr ± equal or, if unequal, upper 2 > others; corolla funnel-shaped, white, limb 0.5–2.5 mm diam, appendages white or yellow; style attached to receptacle, unbranched, gen persistent, stigma 1, head-like. **FR**: nutlets gen 4, gen paired, often dissimilar in shape, ornamentation, margin width, spreading, 1–4.5 mm, gen compressed, marginal prickles straight or hooked at tip. 15 spp.: to BC, WY, TX, nw Mex; S.Am. (Greek: comb nut, from bristly to dentate nutlet margins of some spp.) Nutlets of basal, cleistogamous fls (present in *P. heterocarpa*, *P. peninsularis*) unusual, not to be used in key.

1. Nutlets paired or not, ± round or obovate to ± equally 4-sided, margins ± entire; calyx radial, lobes > nutlets; lower cauline lvs opposite, fused at base
   2. Nutlets not paired, ± diamond-shaped; calyx lobe tip bristles hooked; st branched ± near base; pedicel in fr 1–2.3 mm. . . . . ***P. pusilla***
   2′ Nutlets paired, ± round to obovate; calyx lobe tip bristles not hooked; st branched above; pedicel in fr ± 0.5 mm . . . . ***P. setosa***
1′ Nutlets paired, linear or oblanceolate to oblong or ovate-elliptic, margins wavy to ± dentate; calyx bilateral, lobes < nutlets; lower cauline lvs gen alternate, not fused at base
   3. Cauline nutlets dissimilar, margin of 1 pair narrow or 0, of the other wide; basal cleistogamous fls present
      4. Cauline nutlets curved in 2 planes, 1 pair incurved, the other recurved; pedicel in fr partly fused to 1 nutlet; lower 3 calyx lobes unequal. . . . . ***P. heterocarpa***
      4′ Cauline nutlets gen straight, all spreading in 1 plane; pedicel in fr free from nutlets; lower 3 calyx lobes ± equal. . . . . ***P. peninsularis***
   3′ Cauline nutlets similar, wavy-margined or dentate; basal cleistogamous fls 0
      5. Nutlet margins wavy, erect to incurved, not or ± dentate entire length. . . . . ***P. penicillata***
      5′ Nutlet margins dentate ± entire length
         6. Nutlet coiled to recurved, linear . . . . ***P. recurvata***
         6′ Nutlet straight to occ ± recurved, ± oblong to linear
            7. St prostrate to decumbent; nutlet margin teeth width at base gen << length; nutlets straight or ± recurved at tip. . . . . ***P. linearis*** subsp. ***ferocula***
            7′ St ascending to erect; nutlet margin teeth width at base ± = length; nutlets recurved, gen ± so, ± throughout . . . . ***P. platycarpa***

P. heterocarpa (I.M. Johnst.) I.M. Johnst. (p. 491) MIXED-NUT PECTOCARYA **ST**: prostrate to ascending, 2–25 cm. **INFL**: pedicel in fr 1.3–2.8 mm, partly fused to 1 nutlet. **FL**: basal cleistogamous fls present; lower 3 calyx lobes in fr unequal, < others. **FR**: nutlets 1.2–3 mm, oblanceolate to oblong; cauline paired, curved in 2 planes, 1 pair incurved, the other pair recurved, margins ± entire, narrow- to wide-membranous, to dentate. 2*n*=24. Washes, roadsides, openings in creosote-bush scrub, Joshua-tree woodland; < 1600 m. s SNF, Teh,

SW, W&I, D; to sw UT, w TX, n Mex. Basal nutlets 2–4 per fl, not paired, reflexed, 1 unmargined, others narrow-entire- to ± -dentate-margined. Feb–May

P. linearis (Ruiz & Pav.) DC. subsp. ***ferocula*** (I.M. Johnst.) Thorne (p. 491) NARROW-TOOTHED PECTOCARYA **ST**: prostrate to decumbent, 6–26 cm. **INFL**: pedicel in fr 1.5–3 mm. **FR**: nutlets 2–3.8 mm, linear-oblong, straight or ± recurved near tip; margin

teeth free ± to base, width at base gen << length. 2*n*=48. Roadsides, grassy slopes, clearings; 5–2100 m. s SNF, Teh, GV, CW, SW, DMoj, w DSon; n Baja CA; also s S.Am. Apparently rare, scattered in D. Feb–May

P. penicillata (Hook. & Arn.) A. DC. (p. 491) NORTHERN PECTOCARYA **ST**: prostrate to decumbent, 2–25 cm. **INFL**: pedicel in fr ascending, 1.3–2.5 mm. **FR**: nutlets 1.6–2.4 mm, oblanceolate, straight, ± equal in width, bristled at tip to above ± middle; margins wavy, erect to incurved, not or ± dentate. 2*n*=24. Disturbed sites, roadsides; 90–2100 m. CA-FP (exc NCo, CCo), MP (exc Wrn), W&I, D; to BC, w WY, AZ, n Baja CA. D pls with margins of 2 nutlets narrower than other 2, bristles from nutlet base to tip, 2*n*=48, may be distinct tetraploid, as yet undescribed hybrid sp.; see Cronquist (1984) Intermountain Flora, Vol 4, for description, illustration. Feb–May

P. peninsularis I.M. Johnst. (p. 491) BAJA PECTOCARYA **ST**: ascending to erect, 2–24 cm. **INFL**: pedicel in fr 1–1.5 mm, free from nutlets. **FL**: basal cleistogamous fls present; lower 3 calyx lobes in fr ± equal. **FR**: nutlets 1.1–2 mm, ovate-elliptic; cauline paired, gen straight, all spreading in 1 plane, margins narrow- to wide-membranous, ± dentate. 2*n*=24. Washes, roadsides, clearings; 30–300 m. w DSon; Baja CA. Seldom collected. Basal nutlets 2–4, not paired, reflexed, 1–3 unmargined, others narrowly ± dentate-margined. Feb–Apr

P. platycarpa (Munz & I.M. Johnst.) Munz & I.M. Johnst. (p. 491) WIDE-TOOTHED PECTOCARYA **ST**: ascending to erect, 4–25 cm. **INFL**: pedicels in fr 2.5–4 mm. **FR**: nutlets 2.5–4.5 mm, linear to spoon-shaped-oblong, recurved or gen ± so, ± throughout; margin teeth fused at base, width at base ± = length. 2*n*=48. Washes, road-

sides in creosote-bush scrub, Joshua-tree woodland; < 1600 m. SW, W&I, D; to sw UT, w TX, Baja CA. Feb–May

P. pusilla (A. DC.) A. Gray (p. 491) LITTLE PECTOCARYA **ST**: ascending to erect, 3–20(38) cm, branched ± near base. **LF**: lower opposite, fused at base; upper alternate. **INFL**: pedicel in fr 1–2.3 mm. **FL**: calyx lobes in fr ± equal, > nutlets, appressed-short-hairy, tip with hooked-tipped bristles. **FR**: not paired, radially arranged; nutlets 1.5–3 mm, ± diamond-shaped; margins entire, not membranous-winged. 2*n*=24. Dry, semi-barren sites in grassland, chaparral, woodland, roadsides; 100–1800 m. NW, CaR, SNF, n&c SNH, CW; to WA; waif in Chile. Mar–Jun

P. recurvata I.M. Johnst. (p. 491) ARCHED-NUT PECTOCARYA **ST**: ascending to erect, 3.5–21 cm. **INFL**: pedicels in fr 2–3 mm. **FR**: nutlets 2.5–4 mm, linear, coiled to recurved; margin teeth free ± to base, linear (or width at base < length). 2*n*=24. Shelter of rocks, bases of shrubs, occ roadsides, creosote-bush scrub, Joshua-tree woodland; 10–1600 m. s SN, n TR, W&I, D; to s UT, sw NM, n Mex. Feb–May

P. setosa A. Gray (p. 491) ROUND-NUT PECTOCARYA **ST**: ascending to erect, 2–23 cm, branched above. **LF**: lower opposite, fused at base; upper alternate. **INFL**: pedicels in fr reflexed or ascending, ± 0.5 mm. **FL**: calyx lobes in fr ± equal, > nutlets, with appressed, short, and several spreading, long, stiff, straight-tipped bristles. **FR**: nutlets 1.5–4 mm, ± round to obovate; margins ± entire, membranous-winged, wide on 3 nutlets, narrow on 1, ± scattered hook-tipped bristles. 2*n*=24. Gravelly clearings in sagebrush scrub, creosote-bush scrub, pinyon/juniper woodland, grassland; 150–2300 m. s SN, CW, SW, se MP (exc Wrn), SNE, D; to e WA, s ID, w UT, AZ, n Baja CA. Mar–Jun

PHACELIA

Robert Patterson, Laura M. Garrison & Debra R. Hansen

Ann to per, gen glandular-hairy, taprooted or from ± thick caudex. **LF**: gen alternate, simple to 2-pinnately compound, gen ± reduced upward. **INFL**: cyme, gen dense, coiled, gen 1-sided; pedicel gen ≤ 5 mm. **FL**: calyx lobes gen 5, gen fused at base, gen persistent, enlarging in fr; corolla gen deciduous, at least some persistent and withering in fr in some spp., rotate to tubular or bell- or funnel-shaped, ± white, blue, purple, pink or yellow, tube and throat not always clearly differentiated, scales of tube base free or fused to filaments, or scales 0; stamens gen attached at same level, equal, gen exserted; ovary chamber 1, sometimes appearing as 2 due to intrusion of the 2 placentas, placentas parietal, enlarging and meeting in fr, style 2-lobed, gen hairy below lobes. **FR**: capsule, oblong to spheric. **SEED**: 1–many, oblong to spheric, gen brown; abaxial surface gen pitted or cross-furrowed. ± 175 spp.: Am; some cult for orn. (Greek: cluster, from dense infl) Bristly hairs may cause dermatitis. [Gilbert et al. 2005 Syst Bot 30:627–634] Some CA per spp. intergrade, hybridize, difficult to distinguish. *P. ixodes* Kellogg, incl in TJM (1993), not known from CA.

Key to Groups

1. Bien or per from ± woody taproot or branched caudex . . . . **Group 1**
1′ Ann from slender taproot (some key in both of the following groups)
  2. Most cauline lf blades deeply lobed to compound (some sinuses reaching midrib) . . . . **Group 2**
  2′ Most cauline lf blades entire to shallowly lobed (lowest sometimes deeply few-lobed at base, but sinuses gen not reaching midrib) . . . . **Group 3**

Group 1

1. Seeds 8–200
  2. Lf blade 5–25 mm, ± round; stamens incl; n W&I, n&e DMoj . . . . ***P. perityloides***
   3. Corolla 12–15 mm; pedicels in fr 10–30 mm, reflexed . . . . var. ***jaegeri***
   3′ Corolla 10–12 mm; pedicels in fr 5–15 mm, ascending to spreading . . . . var. ***perityloides***
  2′ Lf blade 15–120 mm, lanceolate or oblong to widely ovate; stamens gen exserted; NW, CaRH, SNH, MP
   4. Corolla persistent in fr; stamens 10–15 mm — n KR, Wrn . . . . ***P. sericea*** var. ***ciliosa***
   4′ Corolla deciduous; stamens 8–11 mm
    5. Infl head-like, dense; filaments glabrous; pl 10–30 cm . . . . ***P. hydrophylloides***
    5′ Infl ± elongate; filaments hairy; pls gen > 30 cm
     6. Corolla 10–12 mm, pale blue to purple . . . . ***P. bolanderi***
     6′ Corolla 3–7 mm, cream to ± green or brown-white . . . . ***P. procera***
1′ Seeds 1–4
  7. Lf compound; lflets coarsely toothed or lobed . . . . ***P. ramosissima***

7′ Lf gen simple (or basal lobed to compound); lflets entire (if any)
  8. Fls few; pedicel 10–20 mm; pl 5–15 cm — KR . . . . ***P. dalesiana***
  8′ Fls many; pedicel < 5 mm; pl (5)10–200 cm
    9. Basal lvs simple, entire or 3–7-lobed, gen silvery to ± white
      10. St densely glandular . . . . .²***P. corymbosa***
      10′ St gen not glandular
        11. Lf gen elliptic to obovate to ± round, veins deeply impressed; coastal dunes, n NCo, < 20 m . . . . . . ²***P. argentea***
        11′ Lf gen lanceolate to widely elliptic, veins prominent but not deeply impressed; shrubland, open
          woodland, widespread, 380–4000 m . . . . ²***P. hastata***
        12. St decumbent to ascending, 5–20(30) cm; hairs mostly spreading; calyx lobes gen glandular
          . . . . ²subsp. ***compacta***
        12′ St ascending to ± erect, 20–50 cm; hairs mostly ± appressed; calyx lobes gen not glandular . . . . ²subsp. ***hastata***
    9′ Basal lvs compound or dissected, lvs ± green (± gray) or silvery to ± white
      13. Lvs gen silvery to ± white . . . . ²***P. hastata***
      14. St decumbent to ascending, 5–20(30) cm; hairs mostly spreading; calyx lobes gen glandular
        . . . . ²subsp. ***compacta***
      14′ St ascending to ± erect, 20–50 cm; hairs mostly ± appressed; calyx lobes gen not glandular . . . . . ²subsp. ***hastata***
    13′ Lvs gen ± green (± gray)
      15. St densely glandular . . . . .²***P. corymbosa***
      15′ St gen not glandular
        16. Lf basally 2-lobed if not entire; coastal dunes — n NCo . . . . ²***P. argentea***
        16′ Lvs gen dissected or compound
          17. Basal lvs compound, lflets 3–7, lanceolate to ovate
            18. St finely stiff-hairy; pl 10–60 cm; inland mtns, 900–3500 m . . . . ***P. mutabilis***
            18′ St coarsely stiff-hairy; pls (15)50–200 cm; coastal foothills or mtns, < 1000 m
              19. Style 8–12 mm; corolla blue to lavender (white) . . . . ²***P. californica***
              19′ Style 6–9 mm; corolla green-white to ± yellow . . . . ***P. nemoralis***
                20. Mid-st gen < 7 mm diam; corolla 3.5–5 mm . . . . subsp. ***nemoralis***
                20′ Mid-st gen 7–10 mm diam; corolla 5–6 mm . . . . subsp. ***oregonensis***
          17′ Basal lvs dissected, segments 3–15, oblanceolate or lanceolate to triangular
            21. Central st erect, gen 1 . . . . ***P. heterophylla*** subsp. ***virgata***
            21′ Central sts gen decumbent to ascending, gen > 1
              22. Calyx lobes lanceolate to ovate, ± overlapping in fr . . . . ***P. imbricata***
                23. Lf segments 7–15; outer calyx lobes narrowly ovate, often glandular . . . . subsp. ***imbricata***
                23′ Lf segments 3–7; outer calyx lobes lanceolate, not glandular . . . . subsp. ***patula***
              22′ Calyx lobes linear to narrowly ovate or oblanceolate, not overlapping in fr
              24. Corolla blue to lavender (white); stamens 7–10 mm; calyx lobes 6–8 mm in fr . . . . ²***P. californica***
              24′ Corolla white to cream; stamens 9–14 mm; calyx lobes 8–12 mm in fr . . . . ***P. egena***

Group 2

1. At least some corollas persistent, withering in fr, ± enclosing fr
  2. Seeds 1–4; anthers exserted; corolla blue or purple to white
    3. Lvs gen simple, lobes 0–6
      4. Corolla purple to violet; largest lvs 1–2-lobed at base . . . . ***P. marcescens***
      4′ Corolla white to pale blue; largest lvs 2–6-lobed or -toothed at base . . . . ***P. stebbinsii***
    3′ Lvs gen compound, segments toothed or lobed
      5. Fr ± spheric, 2–3 mm; pedicels 1–3 mm in fr . . . . .²***P. distans***
      5′ Fr ovoid to elliptic, 2.5–4 mm; pedicels < 1 mm in fr
        6. St stiffly erect; lflet tips acute; infl dense, gen 2–4-branched; fr glabrous proximally . . . . ²***P. tanacetifolia***
        6′ St decumbent to weakly erect; lflet tips rounded; infl ± open proximally, dense distally, gen
          unbranched; fr ± puberulent proximally . . . . ²***P. umbrosa***
2′ Seeds 5–30; anthers incl; corolla gen yellow (lobes occ ± purple in *Phacelia adenophora*;
    ± white in *Phacelia tetramera*)
  7. Corolla tube puberulent inside; filaments puberulent
    8. Corolla 3.5–8 mm; style 1.5–3 mm; stamens 2.5–5 mm . . . . ***P. adenophora***
    8′ Corolla 2–4 mm; style 0.5–1.5 mm; stamens 1.5–3 mm . . . . ***P. monoensis***
  7′ Corolla tube glabrous inside; filaments glabrous
    9. Seeds cross-striate but not deeply furrowed; pls 10–40 cm . . . . ***P. inundata***
    9′ Seeds cross-furrowed, furrows 5–9; pls 2–15 cm
      10. Sepals, petals gen 5; corolla 2–3 mm . . . . ***P. inyoensis***
      10′ Sepals, petals gen 4; corolla 1.3–2 mm . . . . ***P. tetramera***
1′ Corolla gen deciduous in fr
  11. Seeds gen ≥ 7 per fr
    12. Corolla gen 2–5 mm, bell-shaped; stamens, styles ± 1–3 mm
      13. Distal st, infl axis, pedicels with dark, stalked glands . . . . ***P. glandulifera***

13′ Distal st, infl axis, pedicels short-hairy, glandular or not, glands not esp dark
  14. Calyx lobes gen linear to oblong, 3–6 mm in fr, minutely ciliate, not glandular . . . . ***P. ivesiana***
  14′ Calyx lobes oblanceolate, 4–10 mm in fr, short-stiff-hairy, ± glandular
    15. Pedicels straight; seeds cross-furrowed; s SCoRI, TR, PR, W&I, D . . . . ***P. affinis***
    15′ Pedicels curved; seeds pitted; SCo. . . . . ***P. stellaris***
12′ Corolla 5–20 mm, rotate to funnel- or bell-shaped; styles 2–8 mm; stamens 3–9 mm
  16. Seeds cross-furrowed
    17. Proximal lvs 1–2-compound; filaments puberulent . . . . ***P. bicolor*** var. ***bicolor***
    17′ Proximal lvs deeply lobed to 1-compound; filaments glabrous
      18. Corolla lobes white to pink; seed ± 0.5 mm. . . . . ***P. brachyloba***
      18′ Corolla lobes blue to pink or violet; seed 1–1.5 mm . . . . ***P. fremontii***
  16′ Seeds pitted
    19. St gen erect, 30–120 cm, 0–few-branched; lvs strongly glandular — s ChI. . . . . ***P. lyonii***
    19′ St decumbent, spreading to erect, 2–40 cm, gen branched at base; lvs short-spreading-hairy, not
      strongly glandular
      20. Fr 6–9 mm, obovoid, ± compressed below middle — NCo, CCo, n ChI . . . . ***P. insularis***
      21. St gen decumbent to ascending; style 2.5–4 mm; seed 1.5–2 mm; NCo, CCo . . . . var. ***continentis***
      21′ St gen erect; style 5–6 mm; seed 1–1.5 mm; n ChI . . . . var. ***insularis***
      20′ Fr 4–7 mm, ovoid, ± swollen below middle
        22. Corolla tube, throat white, lobes ± violet . . . . ***P. davidsonii***
        22′ Corolla light blue to purple throughout . . . . ***P. douglasii***
11′ Seeds 1–6 per fr
  23. Some lvs simple (esp distal cauline), entire or toothed
    24. Lf blade ± deltate to ± round; corolla cream-white. . . . . ***P. malvifolia***
    24′ Lf blade linear to ovate or oblanceolate; corolla ± blue to lavender
      25. Calyx lobes in fl 3–5 mm, becoming unequal in fr; stamens 2–4 mm; mtns s and e of GV . . . . ***P. austromontana***
      25′ Calyx lobes in fl 2–3 mm, subequal in fr; stamens 4–5 mm; mtns w of GV . . . . ***P. breweri***
  23′ Lvs lobed to compound, segments gen toothed to lobed
    26. Most calyx lobes pinnately lobed — s ChI . . . . ***P. floribunda***
    26′ Calyx lobes entire to crenate
      27. Calyx lobes puberulent, margin stiffly ciliate, midrib ± 2° veins raised
        28. Calyx lobes 4–6 mm, 6–10 mm in fr; corolla 8–10 mm; stamens 8–12 mm; style 6–8 mm;
        w CA-FP, c&s SN, Teh, GV . . . . ***P. ciliata***
        28′ Calyx lobes 3–4 mm, 5–7 mm in fr; corolla 3–4 mm; stamens 1–3 mm; style ± 2 mm; e CaR, MP
        . . . . ***P. thermalis***
      27′ Calyx lobes gen short- to long-hairy, margin ciliate but not stiffly so, only midrib raised
        29. Calyx lobes strongly unequal, 3 lobes gen ± lanceolate, entire, other 2 ovate to ± round, gen crenate;
        seed gen 1 . . . . ***P. platyloba***
        29′ Calyx lobes ± alike (if unequal not strongly), linear to oblong or oblanceolate, entire; seeds gen > 1
          30. Stamens, style branches incl; stamens 3–5 mm; style 3–5 mm. . . . . ***P. cryptantha***
          30′ Some stamens (and gen style branches) exserted; stamens 4–15 mm; style 5–15 mm
            31. Seed adaxial surface with central ridge separating 2 longitudinal grooves
              32. Sepals 1–1.5 mm > fr; pedicels ± thread-like, densely long-hairy, not glandular . . . . ***P. pedicellata***
              32′ Sepals ± = fr; pedicels stout, short-glandular-hairy
                33. Corolla lobes white; calyx lobes ± 1 mm wide; fr ovoid, sparsely stiff-hairy . . . . ***P. amabilis***
                33′ Corolla lobes blue-lavender or purple; calyx lobes ± 1.5 mm wide; fr ovoid to spheric,
                puberulent . . . . ***P. crenulata***
                  34. Corolla tube, throat white, lobes lavender to blue; calyx lobes 2–3 mm, 3–4.5 mm in fr. . . var. ***minutiflora***
                  34′ Corolla tube white at least at base, throat, lobes purple; calyx lobes 2.5–5.5 mm, 3.5–5.5 mm in fr
                    35. St glandular above; corolla 5–10 mm; stamens, style exserted ≥ 9 mm. . . . . var. ***ambigua***
                    35′ St gen glandular throughout; corolla 4.5–7 mm; stamens, style exserted 5.5–11 mm. . . . . . . var. ***crenulata***
            31′ Seed adaxial surface not clearly ridged and grooved
              36. Calyx lobes ± curved, not enclosing fr
                37. Fr ± puberulent, stiff hairs few, gen < 1 mm, not bulb-based . . . . ***P. vallis-mortae***
                37′ Fr with stiff, > 1 mm, bulb-based hairs . . . . ***P. cicutaria***
                  38. Corolla yellow-white; calyx lobes ± yellow . . . . var. ***cicutaria***
                  38′ Corolla lavender; calyx lobes ± gray. . . . . var. ***hispida***
              36′ Calyx lobes ± straight, ± enclosing fr
                39. Calyx lobes 7–12 mm in fr . . . . ***P. hubbyi***
                39′ Calyx lobes 4–8 mm in fr
                  40. Fr ± spheric, 2–3 mm; pedicels 1–3 mm in fr. . . . . [2]***P. distans***
                  40′ Fr ovoid to elliptic, 2.5–4 mm; pedicels < 1 mm in fr
                    41. St stiffly erect; lflet tips acute; infl dense, gen 2–4-branched; fr glabrous proximally . . . . [2]***P. tanacetifolia***
                    41′ St decumbent to weakly erect; lflet tips rounded; infl ± open proximally, dense distally,
                    gen unbranched; fr ± puberulent proximally. . . . . [2]***P. umbrosa***

Group 3

1. At least some corollas persistent, withering in fr, ± enclosing fr
  2. Seeds 1–4 per fr; stamens 3–7 mm, exserted; SN
    3. Fr ± spheric; calyx lobes 4–7 mm in fr, strongly unequal; seeds 2–4 . . . . *P. quickii*
    3′ Fr elliptic to ovoid, ± compressed; calyx lobes 3–4 mm in fr, ± equal; seeds 1–2
      4. Corolla violet to purple; stamens 5–6 mm; style 5–6 mm; seeds (1)2 . . . . *P. marcescens*
      4′ Corolla white to pale blue; stamens 6–7 mm; style 4–5 mm; seed gen 1 . . . . *P. stebbinsii*
  2′ Seeds gen 5–50 per fr; stamens 1–5 mm, incl; SNE, n DMoj
    5. Seeds 20–50, pitted . . . . *P. saxicola*
    5′ Seeds 5–25, cross-furrowed
      6. Corolla puberulent inside; filaments puberulent . . . . *P. monoensis*
      6′ Corolla tube glabrous inside; filaments glabrous
        7. Sepals, petals gen 5; corolla 2–3 mm . . . . *P. inyoensis*
        7′ Sepals, petals gen 4; corolla 1.3–2 mm . . . . *P. tetramera*
1′ Corolla deciduous in fr
  8. Seeds 1–4 per fr
    9. Cauline lf blade gen as wide as long
      10. Lvs, sts, and calyx lobes densely yellow-stiff-hairy; corolla 5–7 mm; stamens 5–10 mm . . . . ²*P. malvifolia*
      10′ Lvs, sts sparsely white- to clear-stiff-hairy; calyx lobes short-hairy; corolla 3–5 mm; stamens
        2–3 mm. . . . . *P. rattanii*
    9′ Cauline lf blade gen longer than wide
      11. Lf crenate to lobed (lobes gen ± toothed); adaxial seed surface with central ridge separating 2
        longitudinal grooves
        12. Corolla 3–7 mm, stamens 2–5 mm; style 2–5 mm
          13. Corolla 5–7 mm, widely bell-shaped; stamens 3–5 mm; style 3–5 mm; seeds 2.5–3.5 mm . . . . *P. anelsonii*
          13′ Corolla 3–5 mm, bell-shaped; stamens 2–3 mm; style 2–3 mm; seeds 2–3 mm . . . . *P. coerulea*
        12′ Corolla 6–10 mm; stamens 10–15 mm; style 12–15 mm . . . . *P. crenulata*
          14. Corolla ± 4 mm; stamens, style exserted < 2 mm . . . . var. *minutiflora*
          14′ Corolla ≥ 4.5 mm; stamens, style exserted ≥ 5.5 mm
            15. Stamens, style exserted ≥ 9 mm; st glandular distal to middle; corolla 5–10 mm . . . . var. *ambigua*
            15′ Stamens, style exserted 5.5–11 mm; st gen glandular throughout; corolla 4.5–7 mm . . . . var. *crenulata*
      11′ Lf blade entire (or lobes entire, obtuse to acute); adaxial seed surface not ridged and grooved
        16. Distal st, infl axis, pedicels not glandular or only sparsely so
          17. Proximal-most lvs opposite
            18. Filaments glabrous . . . . ²*P. racemosa*
            18′ Filaments short-hairy . . . . *P. humilis*
              19. Calyx lobes 8–12 mm in fr; stamens 6–8 mm, clearly exserted . . . . var. *dudleyi*
              19′ Calyx lobes 5–8 mm in fr; stamens 4–6 mm, barely exserted . . . . var. *humilis*
          17′ Proximal-most lvs alternate
            20. Calyx lobes 4–5 mm in fr, oblong; corolla 4–6 mm, light blue, lobes sparsely short-hairy outside;
              ne CW . . . . *P. breweri*
            20′ Calyx lobes 8–10 mm in fr, linear; corolla 3–4 mm, lavender, lobes glabrous outside; s SNH,
              w edge DMoj. . . . . ²*P. novenmillensis*
        16′ Distal st, infl axis, pedicels glandular
          21. Cauline lf blade linear to narrowly oblanceolate, gen tapered to base
            22. Proximal- to mid-cauline lvs opposite; KR, CaR, n&c SN . . . . ²*P. racemosa*
            22′ Proximal- to mid-cauline lvs alternate; KR, SnGb, SnBr
              23. Calyx lobes 2–3 mm, 3–5 mm in fr, subequal; corolla rotate, lobes violet to purple; serpentine
                soils, KR . . . . *P. greenei*
              23′ Calyx lobes 3–5 mm, 5–15 mm in fr, strongly unequal; corolla bell-shaped, lobes white to pale
                blue; non-serpentine soils, SnGb, SnBr . . . . ²*P. mohavensis*
          21′ Cauline lf blade elliptic to ovate, gen clearly petioled
            24. Calyx lobes in fr 2–3 mm wide; corolla 6–8 mm; stamens 7–8 mm; style 6–9 mm . . . . ²*P. purpusii*
            24′ Calyx lobes in fr gen < 2 mm wide; corolla 2–6 mm; stamens 2–5 mm; style 1–5 mm
              25. Calyx lobes in fr 8–10 mm, puberulent, stiffly ciliate, some hairs > 1 mm . . . . ²*P. novenmillensis*
              25′ Calyx lobes in fr 4–6 mm, short-stiff-hairy, ciliate, hairs < 1 mm
                26. Style gen divided 1/3–1/2 length. . . . . *P. austromontana*
                26′ Style divided ± to near base . . . . *P. eisenii*
  8′ Seeds gen ≥ 5 per fr
    27. Lf blades, esp proximal, gen as wide as long
      28. Corolla 7–40 mm; stamens (4)10–45 mm; style (3)8–45 mm
        29. Corolla lacking flap-like appendage fused to filament base
          30. Cymes dense; pls 30–200 cm; lvs not deeply lobed; corolla shallow bell-shaped to rotate, limb
            25–40 mm diam, lavender (white), often with lavender-spotted white throat . . . . *P. grandiflora*
          30′ Cymes lax; pls 10–100 cm; some basal lvs deeply lobed; corolla limb 10–25 mm diam, throat not
            spotted. . . . . *P. viscida*

31. Corolla white, rotate to bell-shaped, limb 10–15 mm diam; style incl, 3–10 mm — Santa Barbara, Ventura cos. . . . . var. *albiflora*
31′ Corolla rotate, throat white, limb 15–25 mm diam, pale lavender to bright blue; style incl or exserted, 5–16 mm . . . . var. *viscida*
29′ Corolla with flap-like appendage fused to filament base
  32. Filament flap hairy; corolla purple (white)
    33. Corolla purple (white) throughout, tube length > limb width, ± bell-shaped with slight constriction at throat . . . . ***P. minor***
    33′ Corolla purple with white spot at base of each corolla lobe, tube length < limb width, shallow-rotate to widely bell-shaped . . . . ***P. parryi***
  32′ Filament flap glabrous; corolla white or pink to bright blue
    34. Seeds 10–24 per fr; corolla white to pale magenta or pale blue, occ with diffuse area lacking pigment at base of sinuses, 7–12 mm . . . . ***P. longipes***
    34′ Seeds ≥ 40 per fr; corolla deep to sky blue, white spots at base of sinuses, 6–40 mm
      35. Style cleft 2/3 to 3/4; s SNH, Teh (e slope), w edge DMoj . . . . ***P. nashiana***
      35′ Style cleft 1/4 to 1/2; DMoj, n&w DSon. . . . . ***P. campanularia***
        36. Corolla rotate, gen < 25 mm, length ≤ width; w DSon . . . . subsp. *campanularia*
        36′ Corolla funnel-shaped, gen > 25 mm, length ≥ width; DMoj, n DSon . . . . subsp. *vasiformis*
28′ Corolla 3–12 mm; stamens 2–10 mm; style 1–12 mm
  37. Lvs, sts, calyx lobes densely yellow-stiff-hairy; fr 2–3 mm; corolla cream-white — NCo, CW. . . . . [2]***P. malvifolia***
  37′ Lvs, sts, calyx lobes ± short-hairy, sometimes glandular; fr 3–7 mm
    38. Seeds with 4–8 cross-furrows; calyx lobes 1.5–3 mm, 2–4 mm in fr; fr gen spheric
      39. Corolla 8–12 mm; stamens 5–6 mm; style 5–6 mm . . . . ***P. calthifolia***
      39′ Corolla 4–7 mm; stamens 2–4 mm; style 2–3 mm
        40. Corolla cream-white; infl axis among lvs, finely glandular. . . . . ***P. neglecta***
        40′ Corolla violet to purple; infl axis gen > lvs, ± dark-glandular. . . . . ***P. pachyphylla***
    38′ Seeds pitted, not cross-furrowed; calyx lobes 2–5 mm, 4–9 mm in fr; fr ovoid to ± oblong
      41. Corolla 6–12 mm; style 3–5 mm
        42. Corolla narrowly bell-shaped, limb gen 3–5 mm diam; upper infl ± short-glandular-hairy (some hairs < 0.5 mm) . . . . ***P. mustelina***
        42′ Corolla widely bell-shaped, limb gen > 5 mm diam; upper infl glandular-puberulent
. . . . ***P. pulchella*** var. ***gooddingii***
      41′ Corolla 3–6 mm; style 1–2 mm
        43. Lf blade widely elliptic to obovate, base tapered, margin entire to barely toothed. . . . . [2]***P. parishii***
        43′ Lf blade ± round, base truncate to slightly lobed, margin clearly toothed
          44. Calyx lobes 6–8 mm in fr, some 1–2 mm wide; lf blade crenate to irregularly toothed. . . . . . ***P. peirsoniana***
          44′ Calyx lobes 4–6 mm in fr, gen < 1 mm wide; lf blade dentate to weakly lobed, lobes obtuse . . . . ***P. rotundifolia***
27′ Lf blades gen longer than wide
  45. St prostrate to ± ascending; fl 1–2 mm; infl axillary, partly hidden by lvs; n CaRH. . . . . ***P. cookei***
  45′ St gen ascending to erect; fl ≥ 3 mm; infl gen terminal, not hidden by lvs
    46. Seeds 20–80+ per fr
      47. Lvs cauline, gradually reduced distally on st; blade slightly lobed to toothed; fr glandular-puberulent. . . ***P. lemmonii***
      47′ Lvs ± basal, ± abruptly reduced distally on st; blade entire to barely toothed; fr short-hairy . . . . [2]***P. parishii***
    46′ Seeds ≤ 20 per fr
      48. Proximal pedicels > distal (esp in fr), gen S-shaped or recurved in fr
        49. Fr obovoid, slightly flattened, 6–9 mm; NCo, CCo, n ChI . . . . ***P. insularis***
          50. St gen decumbent to ascending; style 2.5–4 mm; seed 1.5–2 mm; NCo, CCo. . . . . var. *continentis*
          50′ St gen erect; style 5–6 mm; seed 1–1.5 mm; n ChI . . . . var. *insularis*
        49′ Fr ovoid, swollen at base, 4–7 mm; SN, TR, PR, SNE, DMoj
          51. Corolla 4–8 mm; style 2–4 mm . . . . ***P. curvipes***
          51′ Corolla 7–15 mm; style 4–8 mm. . . . . ***P. davidsonii***
      48′ Pedicels gen equal, gen straight in fr (exc *Phacelia orogenes*)
        52. Lf linear, oblong, or narrowly (ob)lanceolate, tapered to indistinct petiole
          53. Proximal lvs opposite; calyx lobes 1–3 mm, 2–8 mm in fr; corolla 2–6 mm; stamens, style 2–5 mm
            54. Corolla 2–3 mm; stamens, style ± 2 mm; style glabrous — KR . . . . ***P. leonis***
            54′ Corolla 3–6 mm; stamens, style 3–5 mm; style short-hairy below branches
              55. Pedicels 2–7 mm, proximal S-shaped or recurved in fr, > distal pedicels; corolla tube white to ± yellow, lobes violet; s SNH . . . . ***P. orogenes***
              55′ Pedicels 1–3 mm, proximal straight in fr, gen = distal pedicels; corolla lavender; KR, CaRH . . . . ***P. pringlei***
          53′ Proximal lvs alternate; calyx lobes 3–5 mm, 5–15 mm in fr; corolla 5–10 mm; stamens, style 5–14 mm
            56. Pedicels < 2 mm in fr; calyx lobes subequal; fr 5–8 mm; ne CA . . . . ***P. linearis***
            56′ Pedicels gen 2–5 mm in fr; calyx lobes unequal (longest 1–2 mm > shortest); fr 3–5 mm; s CA
              57. Stamens incl or barely exserted, subequal, 5–9 mm, filaments gen white; corolla tube base white to lavender, lobes ± spreading, lavender . . . . ***P. exilis***
              57′ Stamens exserted, unequal, 5–14 mm, filaments violet; corolla tube base gen yellow, lobes ± erect, white, aging pale blue . . . . [2]***P. mohavensis***

52′ Lf blade elliptic or lanceolate to ovate, clearly petioled
  58. Seeds with 7–9 cross-furrows . . . . **P. gymnoclada**
  58′ Seeds pitted, without transverse furrows
    59. Corolla 6–15 mm
      60. Lf blade toothed to slightly lobed; calyx lobes short-hairy, not clearly ciliate; corolla tube
        yellow, lobes lavender to purple
        61. Corolla 9–14 mm; seeds 8–10; nw PR . . . . **P. keckii**
        61′ Corolla 7–11 mm; seeds 10–16; NCoRI, c SNF, CW . . . . **P. suaveolens**
      60′ Lf blade entire or few-lobed near base; calyx lobes puberulent to short-hairy, margin clearly
        stiff-ciliate; corolla lavender to violet
        62. Corolla 6–10 mm; stamens 3–5 mm; styles 3–5.5 mm . . . . **P. congdonii**
        62′ Corolla 10–15 mm; stamens 8–10 mm; styles 6–10 mm . . . . **P. divaricata**
    59′ Corolla 3–8 mm
      63. Stamens 7–8 mm; style 5–9 mm
        64. Sts stiff-hairy, many hairs > 1 mm; corolla gen white; SCoRO, WTR . . . . **P. grisea**
        64′ Sts puberulent to short-stiff-hairy, hairs < 1 mm; corolla lavender to violet; s CaRH, SN, MP . . . . [2]**P. purpusii**
      63′ Stamens 2–4 mm; style 1–4 mm
        65. Calyx lobes 4–6 mm in fr, short-glandular-hairy, stiff hairs on margin < 1 mm; infl open
          proximally; W&I, ne DMtns . . . . **P. barnebyana**
        65′ Calyx lobes 7–11 mm in fr, stiff-hairy, weakly glandular, stiff hairs on margin gen > 1 mm;
          infl dense; s CaR, n&c SN, e SnFrB, SCoRI
          66. Corolla white to light lavender, lobes violet-streaked, tube ± 1 mm wide; lf veins not
            impressed; fr obtuse; e SnFrB, SCoRI . . . . **P. phacelioides**
          66′ Corolla lobes gen purple, tube ± 2 mm wide, lavender; lf veins impressed; fr beaked; s CaR,
            n&c SN . . . . **P. vallicola**

P. adenophora J.T. Howell Ann 10–40 cm. **ST:** spreading to ascending, branched at base, short-hairy, glandular or not. **LF:** 10–30 mm; blade > petiole, oblong to ± ovate, gen deeply lobed, lobes entire or toothed. **FL:** calyx lobes 2–5 mm, 4–7 mm in fr, narrowly oblanceolate, short-hairy; corolla persistent in fr, 3.5–8 mm, narrowly bell-shaped, yellow or lobes occ ± purple, tube puberulent inside; stamens 2.5–5 mm, puberulent; style 1.5–3 mm, puberulent. **FR:** 2.5–6 mm, oblong, puberulent. **SEED:** < 15, 1–1.6 mm; cross-furrows 8–12. *n*=12. Open areas; 1200–2600 m. n SNH, MP; OR, w NV. May–Aug

P. affinis A. Gray (p. 491) Ann 6–30 cm. **ST:** erect, simple to branched at base, short-hairy; hairs spreading to reflexed, glandular-puberulent distally. **LF:** 8–70 mm; blade > petiole, narrowly oblong, deeply lobed to compound, lobes entire, lflets ± lobed. **INFL:** pedicel < 10 mm in fr, straight. **FL:** calyx lobes 4–5 mm, 6–10 mm in fr, oblanceolate, short-stiff-hairy, glandular; corolla deciduous, 3–5 mm, bell-shaped, tube yellow, lobes white to lavender; stamens 2–3 mm, glabrous; style 1–2 mm, gen glabrous. **FR:** 4–6 mm, oblong to elliptic, puberulent proximally, sparsely short-glandular-hairy distally. **SEED:** 15–30, ± 1 mm; cross-furrows 5–8. *n*=11,12. Open, sandy or gravelly areas; < 3400 m. s SCoRI (Caliente Mtn), TR, PR, W&I, D; to UT, NM, Baja CA. If recognized taxonomically, pls with spreading st hairs, glandular proximal lvs assignable to *P. affinis* var. *patens* J.T. Howell. Mar–Jun

P. amabilis Constance (p. 491) SALINE VALLEY PHACELIA Ann 10–60 cm. **ST:** decumbent to erect, few-branched, sparsely soft-hairy, glandular-puberulent. **LF:** 20–120 mm; blades > petioles, proximal gen oblong, deeply lobed, lobes crenate to toothed, distal ± ovate, coarsely toothed. **FL:** calyx lobes 3–4 mm, ± 1 mm wide, 4–5 mm in fr, oblong, short-hairy, glandular; corolla deciduous, 6–8 mm, bell-shaped, white; stamens 9–15 mm, glabrous; style 12–15 mm, puberulent proximally. **FR:** 3–4 mm, ovoid, sparsely stiff-hairy. **SEED:** 2–4, 3–4 mm; abaxial surface finely pitted; adaxial surface with central ridge separating 2 longitudinal grooves. Presumed extinct; gravelly soils, canyons; 500–700 m. DMoj. Apr–May ★

P. anelsonii J.F. Macbr. AVEN NELSON'S PHACELIA Ann 10–50 cm. **ST:** erect, gen simple, glandular-hairy; gland-tipped hairs dark. **LF:** 15–80 mm; blades > petioles, proximal oblong to oblanceolate, deeply lobed, lobes gen crenate, distal ± ovate, toothed. **FL:** calyx lobes 3–4 mm, 3.5–5.5 mm in fr, oblanceolate, sparsely short-hairy; corolla deciduous, 5–7 mm, widely bell-shaped, white,

pale blue, or lavender; stamens 3–5 mm, glabrous; style 3–5 mm, puberulent. **FR:** 2–3.5 mm, ovoid, glandular-puberulent. **SEED:** 2–4, 2.5–3.5 mm; abaxial surface pitted; adaxial surface with central ridge separating 2 longitudinal grooves. Sandy or gravelly soils, creosote-bush scrub, woodland; 1200–1500 m. e DMtns (New York Mtns); to sw UT. Apr–May ★

P. argentea A. Nelson & J.F. Macbr. (p. 491) SAND DUNE PHACELIA Per 10–45 cm, ± fleshy. **ST:** prostrate to ascending, ± stiff-hairy, not glandular. **LF:** thick; blade 20–120 mm, gen > petiole, elliptic to obovate to ± round, entire or 2-lobed at base, veins deeply impressed. **FL:** calyx lobes 3–4 mm, 6–7 mm in fr, oblong to lanceolate; corolla deciduous, 5–7 mm, bell-shaped, white to cream; stamens 6–9 mm, exserted, hairy; style 6–10 mm, exserted, hairy. **FR:** 3–4 mm, ovoid, stiff-hairy. **SEED:** 1–3, 1.5–3 mm, pitted in vertical rows. *n*=22. Sand dunes; < 20 m. n NCo (Del Norte Co.); OR. Mar–Sep ★

P. austromontana J.T. Howell (p. 491) Ann 5–27 cm. **ST:** decumbent to ascending, many-branched, puberulent, sparsely stiff-hairy, ± glandular. **LF:** 10–30 mm; blade ≥ petiole, narrowly elliptic to lanceolate, entire or few-lobed. **FL:** calyx lobes 3–5 mm, 4–6 mm and becoming unequal in fr, linear to narrowly oblanceolate, short-hairy, glandular; corolla deciduous, 3–6 mm, bell-shaped, pale blue to lavender; stamens 2–4 mm, glabrous to papillate; style 3–4 mm, gen divided 1/3–1/2 length, puberulent. **FR:** 3–5 mm, ovoid, beaked, puberulent. **SEED:** 2–4, 1–2 mm, pitted. *n*=9. Open, sandy to rocky areas; 1500–3000 m. c&s SN, TR, SnJt, W&I, DMtns; to sw UT. May–Jun

P. barnebyana J.T. Howell BARNEBY'S PHACELIA Ann 5–30 cm. **ST:** erect, simple to branched at base, glandular-puberulent. **LF:** 10–20 mm, distal little reduced; blade < petiole, gen ovate, entire to obscurely toothed. **INFL:** open proximally. **FL:** calyx lobes 2–4 mm, 4–6 mm in fr, narrowly oblanceolate, short-glandular-hairy; corolla deciduous, 3–5 mm, narrowly bell-shaped, tube white to yellow, lobes pale violet; stamens 2–3 mm, glabrous; style 1–2 mm, puberulent. **FR:** 3.5–5.5 mm, elliptic, puberulent. **SEED:** 15–20, ± 1 mm, pitted. Limestone scree; 1600–2700 m. W&I, ne DMtns (Clark Mtn Range); w NV. Probably closely related to *P. lemmonii*. May–Jul ★

P. bicolor S. Watson var. **bicolor** (p. 495) Ann 6–40 cm. **ST:** decumbent to erect, gen branched at base, short-hairy, glandular-puberulent. **LF:** 20–60 mm, distal little reduced, 1–2-compound;

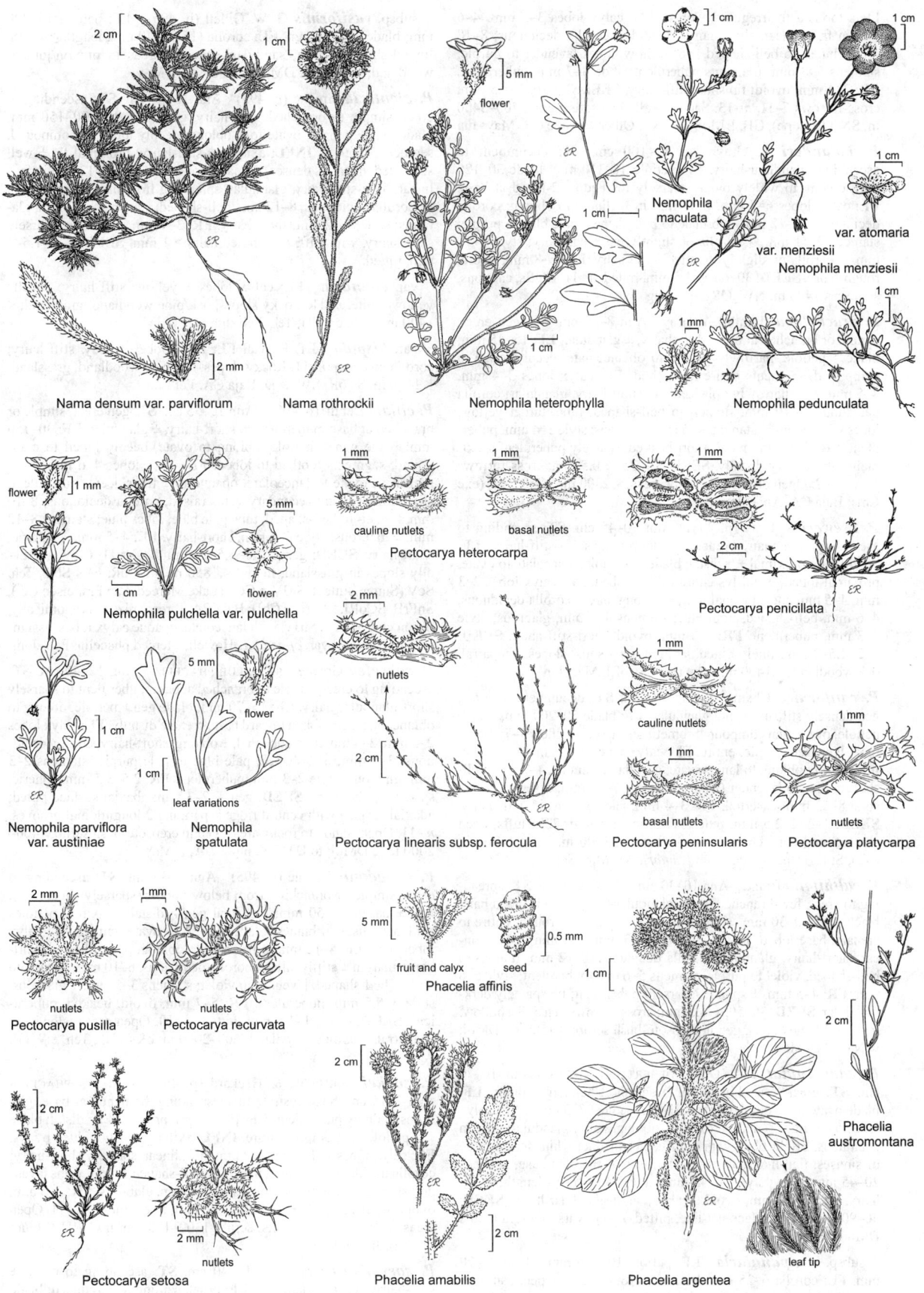

2 cm

1 cm

5 mm

flower

ER

1 cm

ER

1 cm

1 cm

1 cm

Nemophila maculata

var. menziesii

var. atomaria

Nemophila menziesii

1 mm

1 cm

ER

Nama densum var. parviflorum

Nama rothrockii

Nemophila heterophylla

1 cm

Nemophila pedunculata

flower

1 mm

5 mm

1 cm

flower

Nemophila pulchella var. pulchella

cauline nutlets

1 mm

basal nutlets

1 mm

Pectocarya heterocarpa

1 mm

2 cm

ER

Pectocarya penicillata

5 mm

flower

1 cm

1 cm

leaf variations

Nemophila parviflora var. austiniae

Nemophila spatulata

nutlets

2 mm

2 cm

ER

Pectocarya linearis subsp. ferocula

cauline nutlets

1 mm

basal nutlets

1 mm

Pectocarya peninsularis

nutlets

1 mm

Pectocarya platycarpa

2 mm

1 mm

nutlets

Pectocarya pusilla

nutlets

Pectocarya recurvata

5 mm

fruit and calyx

seed

0.5 mm

Phacelia affinis

1 cm

2 cm

Phacelia austromontana

2 cm

ER

Pectocarya setosa

2 mm

nutlets

2 cm

ER

2 cm

Phacelia amabilis

ER

leaf tip

Phacelia argentea

lflets toothed to irregularly lobed. **FL:** calyx lobes 3–5 mm, 4–6 mm in fr, ± linear, short-hairy, glandular; corolla deciduous, 8–18 mm, funnel- to bell-shaped, tube yellow, lobes lavender to purple; stamens 5–8 mm, filaments puberulent; style 3–7 mm, puberulent. **FR:** 3–4 mm, ovoid; tip short-stiff-hairy. **SEED:** 12–20, 1–1.5 mm; cross-furrows 7–11. *n*=13. Sandy or alkaline soils, scrub; 700–3400 m. SNH (e slope), GB, DMoj; OR, NV. Other var. in e OR. May–Jun

P. bolanderi A. Gray Per 12–100 cm. **ST:** decumbent to ascending, glandular-hairy. **LF:** petiole 10–110 mm; blade 30–120 mm, oblong to widely ovate, coarsely toothed or 2-lobed at base. **FL:** calyx lobes 6–7 mm, 8–11 mm in fr, linear to oblong; corolla deciduous, 10–12 mm, subrotate to bell-shaped, pale blue to purple; stamens 9–11 mm, = corolla or slightly exserted, hairy; style 9–11 mm, = corolla or slightly exserted, base hairy. **FR:** 6–8 mm, ovoid, rough-hairy. **SEED:** 30–60, 1–1.5 mm, pitted. *n*=11. Bluffs, canyons, slopes; < 1400 m. NW; OR. Apr–Aug

P. brachyloba (Benth.) A. Gray Ann 8–60 cm. **ST:** gen erect, simple or branched at base, short-hairy, ± glandular. **LF:** 15–45 mm; blade > petiole, narrowly elliptic to oblanceolate, deeply lobed to compound, segments entire to toothed. **FL:** calyx lobes 3–4 mm, 4–5 mm in fr, narrowly oblanceolate, short-hairy, glandular; corolla deciduous, 7–10 mm, funnel- to bell-shaped, tube, throat yellow, lobes white to pink; stamens 4–5 mm, glabrous; style 3–4 mm, puberulent. **FR:** 4–5 mm, ovoid, short-beaked, sparsely puberulent proximally, short-hairy distally. **SEED:** 10–25, ± 0.5 mm; cross-furrows 5–7. *n*=12. Open or burned, sandy areas; < 2300 m. SCoR, SW (exc ChI); Baja CA. Apr–May

P. breweri A. Gray (p. 495) Ann 10–45 cm. **ST:** spreading to ascending, gen many-branched, puberulent, short-stiff-hairy. **LF:** 10–40 mm; proximal ± = distal; blades > petiole, lanceolate to ovate, proximal lobed, distal lvs entire to few-lobed. **FL:** calyx lobes 2–3 mm, 4–5 mm and subequal in fr, ± oblong, hairy; corolla deciduous, 4–6 mm, bell-shaped, light blue; stamens 4–5 mm, glabrous; style 2–3 mm, puberulent. **FR:** 2–3 mm, ovoid, short-stiff-hairy. **SEED:** 1–2, 1.5–2 mm, finely pitted. *n*=11. Rocky soils, slopes, chaparral, oak woodland; < 1400 m. e SnFrB, n SCoRI. Mar–Jun

P. californica Cham. Per 15–90 cm. **ST:** decumbent to erect, gen densely stiff-hairy, not glandular. **LF:** blade 50–200 mm, < or > petiole, ovate, gen compound, sometimes dissected, lflets 3–7; distalmost lvs often simple, entire. **FL:** calyx lobes 3–5 mm, 6–8 mm in fr, narrowly oblong to lance-ovate; corolla deciduous, 4–7 mm, bell-shaped, blue to lavender (white); stamens 7–10 mm, exserted, hairy; style 8–12 mm, exserted. **FR:** 3–4 mm, narrowly ovoid, stiff-hairy. **SEED:** 1–2, 2–2.5 mm, pitted in vertical rows. *n*=22. Bluffs, open slopes, road cuts, chaparral, woodland; < 1000 m. NCo, NCoRO, CCo, SnFrB. Intergrades with *P. imbricata*. Mar–Sep

P. calthifolia Brand Ann 10–30 cm, fleshy, ± brittle. **ST:** spreading to erect, few-branched, short-glandular-hairy; gland-tipped hairs black. **LF:** 10–30 mm, ± basal; blade ± = petiole, ± round, entire to crenate, base lobed. **FL:** calyx lobes 2–3 mm, 3–4 mm in fr, ± linear, short-hairy, glandular; corolla deciduous, 8–12 mm, funnel- to bell-shaped, violet to purple; stamens 5–6 mm, puberulent; style 5–6 mm. **FR:** 4–5 mm, ± spheric, sparsely puberulent; tip sparsely dark-glandular. **SEED:** 30–50, 1–1.5 mm; cross-furrows gen 5–8, shallow. *n*=11,12. Sandy soils, gen in creosote-bush scrub; < 1500 m. DMoj; w NV. Mar–May

P. campanularia A. Gray DESERT BLUEBELLS Ann 10–70 cm. **ST:** erect, branching from base, short-hairy, glandular. **LF:** blade ovate to ± round, toothed. **INFL:** pedicel 7–30 mm. **FL:** calyx lobes 5–10 mm, ≤ mature capsule, oblong, hairy, glandular; corolla deciduous, bright blue, with a diffuse or distinct white spot at base of sinuses, flap-like glabrous scale fused to filament base; stamens 20–45 mm, hairy; style 20–45 mm, cleft 1/4 to 1/2 its length, short-hairy. **FR:** 7–15 mm, ovoid, beaked, short-glandular hairy. **SEED:** 40–90, 1.5 mm, surface net-like, pitted. *n*=11. Vars. intergrade in s DMtns.

subsp. ***campanularia*** **LF:** petiole 10–100 mm; blade 20–70 mm. **FL:** corolla 6–25 mm, length ≤ width, rotate. Open, sandy or gravelly areas to pinyon/juniper woodland; < 1600 m. w DSon. Much less common than *P. campanularia* subsp. *vasiformis*. Feb–May

subsp. ***vasiformis*** G. W. Gillett (p. 495) **LF:** petiole 10–200 mm; blade 20–100 mm. **FL:** corolla (15)25–40 mm, length ≥ width, funnel-shaped. Open, sandy or gravelly areas to pinyon/juniper woodland; < 1600 m. DMoj, n DSon. Feb–May

P. cicutaria Greene (p. 495) Ann 18–60 cm. **ST:** ascending to erect, simple to branched, stiff-hairy, glandular. **LF:** 20–150 mm; blade gen > petiole, ovate to ± oblong, deeply lobed to compound, segments toothed. **INFL:** axis long-stiff-hairy; proximal fls ± well separated; distal fls dense. **FL:** calyx lobes 6–8 mm, 9–12 mm in fr, ± linear, long-stiff-hairy, glandular, spreading from base of capsule in fr; corolla deciduous, 8–12 mm, bell-shaped; stamens 8–12 mm, glabrous; style 8–12 mm, short-hairy. **FR:** 3–4 mm, ± spheric, sparsely short-hairy, with stiff, bulb-based hairs > 1 mm. **SEED:** 2–4, 1.5–3 mm, pitted.

var. ***cicutaria*** **FL:** calyx lobes ± yellow, stiff-hairy; corolla yellow-white. *n*=11. Rocky slopes, oak/pine woodland, grassland; < 1400 m. SNF, c SNH, Teh. Feb–Jun

var. ***hispida*** J.T. Howell **FL:** calyx lobes ± gray, stiff-hairy; corolla lavender. *n*=11. Rocky slopes, oak/pine woodland, grassland; < 1400 m. SCoR, SW, w D; Baja CA. Feb–Jun

P. ciliata Benth. (p. 495) Ann 10–55 cm. **ST:** gen erect, simple or branched at base, puberulent to short-hairy, ± glandular. **LF:** 30–150 mm; blade gen > petiole, oblong to ovate, deeply lobed to compound, segments toothed to lobed. **FL:** calyx lobes 4–6 mm, 6–10 mm in fr, ovate to lanceolate, opaque to ± translucent, puberulent, ciliate, midrib and secondary veins raised; corolla deciduous, 8–10 mm, funnel- to bell-shaped, tube pale blue, lobes blue; stamens 8–12 mm, ± glabrous; style 6–8 mm, short-hairy. **FR:** 4–5 mm, ± spheric, short-hairy. **SEED:** gen 4, 2.5–3.5 mm, pitted. *n*=11. Clay or gravelly slopes in grassland, fields; < 1850 m. s NCoR, c&s SNF, Teh, ScV (Sutter Buttes), SnJV, n CCo (Lake Merced, San Francisco Co.), SnFrB, SCoRI, SW (exc ChI); Baja CA. If recognized taxonomically, pls on clay soils in SnJV, with lanceolate, opaque calyx lobes, assignable to *P. ciliata* var. *opaca* J.T. Howell, Merced phacelia. Feb–Jun

P. coerulea Greene SKY-BLUE PHACELIA Ann 12–40 cm. **ST:** ascending to erect, simple to branched at base, puberulent to sparsely short-glandular-hairy. **LF:** 15–70 mm; blade gen > petiole, oblong to oblanceolate, gen lobed toward base, crenate distally. **FL:** calyx lobes 2–3 mm, 3–4 mm in fr, subequal, ± oblong, short-hairy; corolla deciduous, 3–5 mm, bell-shaped, pale blue to pale purple; stamens 2–3 mm, glabrous; style 2–3 mm, puberulent. **FR:** 2.5–3.5 mm, spheric, sparsely puberulent. **SEED:** gen 4, 2–3 mm; abaxial surface pitted; adaxial surface with central ridge separating 2 longitudinal grooves. *n*=11. Open, sandy to rocky areas, gen in creosote-bush scrub; 1400–2000 m. e DMoj; to UT, TX, n Mex. Apr–May ★

P. congdonii Greene (p. 495) Ann 5–35 cm. **ST:** ascending to erect, simple or branching from below middle, sparsely short-hairy. **LF:** ± basal, 10–50 mm; proximal gen > distal; proximal petioles > blades, distal < blades; blade elliptic to ovate, entire. **FL:** calyx lobes 4–6 mm, 5–12 mm in fr, linear to oblong, puberulent to short-hairy, margin ± stiffly ciliate; corolla deciduous, 6–10 mm, funnel- to widely bell-shaped, lavender to violet; stamens 3–5 mm, ± glabrous; style 3–5.5 mm, puberulent. **FR:** 5–7 mm, ovoid, beaked, puberulent. **SEED:** 6–15, 1–1.5 mm, pitted. *n*=10. Open areas, slopes, in chaparral, oak/pine woodland; 600–2000 m. c&s SNF, Teh, e WTR. Apr–Jun

P. cookei Constance & Heckard (p. 495) COOKE'S PHACELIA Ann 2–15 cm. **ST:** prostrate to ± ascending, branched at base, glabrous to finely puberulent. **LF:** 10–20 mm; proximal ± = distal; blade ± = petiole, gen elliptic, entire. **INFL:** axillary, partly hidden by lvs. **FL:** calyx lobes 1–1.5 mm, ± 2 mm in fr, linear, sparsely short-hairy; corolla deciduous, 1–2 mm, narrowly bell-shaped, white, veins lavender; stamens < 1 mm, glabrous; style 0.5 mm, glabrous. **FR:** 2–3 mm, ovoid, short-hairy. **SEED:** 4–7, 1–1.5 mm, finely pitted. *n*=11. Open areas, volcanic, sandy soils, scrub; 1300–1700 m. n CaRH (Mount Shasta). Jun–Jul ★

P. corymbosa Jeps. Per 15–40 cm. **ST:** ascending to erect, ± stiff-hairy, densely glandular. **LF:** basal, cauline few, reduced; blade 20–150 mm, ≥ petiole, lanceolate to oblanceolate, entire to dissected,

segments 3–7. **FL:** calyx lobes 3.5–6 mm, 6–10 mm in fr, linear to narrowly ovate; corolla deciduous, 5–7 mm, ± cylindric, white; stamens 10–12 mm, exserted, gen hairy; style 10–12 mm, exserted. **FR:** 3–4 mm, narrowly ovoid, stiff-hairy. **SEED:** 1–2, 1.5–3 mm, pitted in vertical rows. *n*=11,22. Gen serpentine soils, slopes, flats, ridges; 120–2950 m. KR, NCoR, CaRH; OR. Intergrades with *P. egena.* Apr–Sep

P. crenulata S. Watson Ann 7–60 cm. **ST:** gen erect, 0–few-branched at base, stiff-hairy, glandular. **LF:** 20–80(120) mm, abruptly reduced distal to st base; blade > petiole, oblong to elliptic, crenate to deeply lobed. **FL:** calyx lobes ± 1.5 mm wide, oblong, puberulent to short-hairy, glandular; corolla deciduous, bell-shaped, tube white, throat white to purple; stamens 10–15 mm, glabrous; style 12–15 mm, glandular-puberulent. **FR:** 2.5–4 mm, ovoid to spheric, puberulent. **SEED:** gen 4, 2–3.5 mm; abaxial surface finely pitted; adaxial surface with central ridge separating 2 longitudinal grooves. Var. *ambigua,* var. *minutiflora* intergrade, perhaps better treated as vars. of *P. ambigua* M.E. Jones (Atwood 1975 Great Basin Naturalist 35:127–190); study needed.

var. ***ambigua*** (M.E. Jones) J.F. Macbr. **ST:** gen densely stiff-hairy throughout, glandular distally. **FL:** calyx lobes 3–5 mm, 4–5 mm in fr; corolla 5–10 mm, lobes purple; stamens and style exserted 9+ mm. **FR:** 3.3–3.5 mm, spheric. *n*=11. Sandy to gravelly washes, slopes; < 1600 m. D; to sw UT, AZ, nw Mex. Mar–May

var. ***crenulata*** **ST:** gen glandular- and stiff-hairy throughout. **FL:** calyx lobes 2.5–4 mm, 3.5–5.5 mm in fr; corolla 4.5–7 mm, lobes purple; stamens and style exserted 5.5–11 mm. **FR:** 2.5–4 mm, ovoid. *n*=11. Sandy to gravelly washes, slopes; < 2450 m. se MP, SNE, e DMoj; to UT. Mar–May

var. ***minutiflora*** (Munz) Jeps. **ST:** gen densely stiff-hairy throughout, glandular distally. **FL:** calyx lobes 2–3 mm, 3–4.5 mm in fr; corolla ± 4 mm, throat white, lobes lavender to blue; stamens and style exserted < 2 mm. **FR:** ± 3 mm, ± spheric. *n*=11. Sandy to gravelly washes, slopes; < 1500 m. DSon; AZ, Baja CA. Mar–Apr

P. cryptantha Greene Ann 16–50 cm. **ST:** gen erect, 0–few-branched, puberulent to sparsely stiff-hairy, glandular. **LF:** 20–150 mm; blade > petiole, elliptic to ovate, proximal lobed to compound, distal gen compound, segments toothed. **FL:** calyx lobes 4–7 mm, 8–10 mm in fr, ± linear, ciliate; corolla deciduous, 4–7 mm, narrowly bell-shaped, blue to lavender; stamens incl, 3–5 mm, glabrous; style incl, 3–5 mm, short-hairy. **FR:** 4–5 mm, ± spheric, short-stiff-hairy. **SEED:** gen 4, 1.5–3 mm, pitted. *n*=11. Gravelly or rocky slopes, canyons; < 1900 m. s SN, SCoRI, SnGb, SnBr, PR, SNE, DMoj; to UT. Mar–May

P. curvipes S. Watson Ann 4–15 cm. **ST:** spreading to ascending, many-branched, short-hairy, glandular distally. **LF:** 10–40 mm; proximal ± = distal; blade gen > petiole, elliptic to oblanceolate, entire. **INFL:** pedicels 1–6 mm, distal < 25 mm in fr. **FL:** calyx lobes 3–6 mm, 7–10 mm in fr, unequal, narrowly oblong, puberulent, stiffly ciliate; corolla deciduous, 4–8 mm, rotate to bell-shaped, tube, throat white, lobes blue to violet; stamens 2–6 mm, short-hairy; style 2–4 mm, short-hairy. **FR:** 4–5 mm, ovoid, appressed-puberulent. **SEED:** 6–16, 1–1.5 mm, pitted. *n*=11. Sandy to rocky slopes, chaparral, oak/pine woodland, conifer forest; 500–2700 m. s SN, Teh, TR, PR, SNE, DMoj; to sw UT, nw AZ. Apr–Jun

P. dalesiana J.T. Howell (p. 495) SCOTT MOUNTAIN PHACELIA Per 5–15 cm. **ST:** few, decumbent, densely glandular-hairy. **LF:** basal in rosette; cauline few, reduced; blade 10–50 mm, ± = petiole, oblong to elliptic, entire. **INFL:** fls few; pedicel 10–20 mm in fr. **FL:** calyx lobes 3–5 mm, 4–7 mm in fr, oblanceolate; corolla deciduous, 6–9 mm, rotate, white, throat purple-marked; stamens 6–8 mm, incl to slightly exserted, glabrous; style 6–7 mm, incl to = corolla. **FR:** 4 mm, ± spheric, hairy. **SEED:** 2–4, 2.5–4.5 mm; surface net-veined. *n*=8. Meadows, streambanks, conifer forest; 1500–2000 m. KR. May–Jul ★

P. davidsonii A. Gray Ann 5–20 cm. **ST:** decumbent to erect, 0–few-branched at base, puberulent to sparsely stiff-hairy. **LF:** 8–70 mm; blade > petiole, elliptic to oblanceolate; proximal deeply lobed to compound; distal entire to lobed, segments obtuse. **INFL:** proxi-

mal pedicels > distal (esp in fr). **FL:** calyx lobes 4–5 mm, 5–8 mm in fr, oblanceolate, short-hairy; corolla deciduous, 7–15 mm, rotate to bell-shaped, tube, throat white, lobes ± violet; stamens 3–6 mm, sparsely short-hairy; style 4–8 mm, short-hairy. **FR:** 4–7 mm, ovoid, puberulent and short-hairy. **SEED:** 7–15, 1–2 mm, pitted. *n*=10. Sandy to rocky soils, slopes, chaparral, conifer forest; 150–2500 m. c&s SN, Teh, SCoRO (Big Pine Mtn), TR, PR. Apr–Jun

P. distans Benth. (p. 495) Ann 5–80 cm. **ST:** decumbent to erect, simple to branched at base, puberulent, sparsely stiff-hairy, finely glandular distally. **LF:** 20–100(150) mm; blade > petiole, gen 1–2-compound, segments toothed. **FL:** calyx lobes 3–4 mm, 4–5 mm in fr, gen unequal, oblong to oblanceolate, densely hairy, glandular; corolla (at least some) ± persistent in fr, 6–9 mm, funnel- to bell-shaped, white to blue; stamens 8–12 mm, glabrous; style 7–12 mm, glabrous to ± hairy below. **FR:** 2–3.1 mm, ± spheric to oblong, puberulent. **SEED:** 2–4, 2–3.1 mm, pitted. *n*=11. Common. Clay to rocky soils, slopes; < 2700 m. s NCoR, n&s SNF, s SNH, ScV (Sutter Buttes), SnJV, CW, SW, SNE, D; s NV, n Mex. Mar–May

P. divaricata (Benth.) A. Gray Ann 9–40 cm. **ST:** decumbent to erect, simple to branched at base, puberulent to short-hairy. **LF:** 10–80 mm; blade gen > petiole, elliptic to narrowly ovate, entire to irregularly 1–2-lobed at base. **FL:** calyx lobes 5–7 mm, 8–11 mm in fr, unequal, lanceolate to obovate, puberulent, margin stiff-ciliate; corolla deciduous, 10–15 mm, funnel- to widely bell-shaped, lavender to violet; stamens 8–10 mm, glandular-puberulent; style 6–10 mm, short-hairy. **FR:** 5–10 mm, ovoid; tip short-hairy. **SEED:** 8–16, 1–1.5 mm, pitted. *n*=10. Open areas, chaparral, woodland, grassland; < 1500 m. s NW, CCo, SnFrB, n SCoRI. Apr–Jun

P. douglasii (Benth.) Torr. Ann 6–40 cm. **ST:** spreading to erect, branched at base, short-hairy, glandular. **LF:** 5–80 mm, ± basal; blade ± = petiole, oblanceolate to ovate, deeply lobed to compound, segments rounded to obtuse. **INFL:** pedicel 3–8 mm. **FL:** calyx lobes 2–5 mm, 4–7 mm in fr, oblanceolate, short-hairy; corolla deciduous, 6–12 mm, widely bell-shaped, light blue to purple; stamens 3–7 mm, glabrous to puberulent; style 2–7 mm, puberulent, glandular or not. **FR:** 5–7 mm, ovoid, short-hairy. **SEED:** 8–20, 0.5–1 mm, pitted. *n*=11. Open, gen sandy areas; < 1700 m. s SNF, Teh, s ScV, SnJV, CW, SCo, s ChI, WTR, SnGb, w DMoj. If recognized taxonomically, pls in w SnJV, SCoRI with glandular infl axis, glabrous style < 3.5 mm assignable to *P. douglasii* var. *petrophila* Jeps. Mar–May

P. egena (Brand) J.T. Howell (p. 495) Per 15–60 cm. **ST:** ascending, stiff-hairy, not glandular. **LF:** mostly basal; blade 100–250 mm, > petiole, lanceolate to oblanceolate; basal dissected, segments 7–11(15); cauline gen entire. **FL:** calyx lobes 4–6 mm, 8–12 mm in fr, linear to oblanceolate; corolla 5–9 mm, bell-shaped, white to cream; stamens 9–14 mm, exserted, hairy; style 9–15 mm, exserted. **FR:** ± 3 mm, ovoid, stiff-hairy. **SEED:** 1–2, 2.5–3 mm, pitted in vertical rows. *n*=22. Slopes, streambanks, flats, chaparral, woodland; 50–2500 m. NW (exc NCo), CaR, SN, GV, CW (exc CCo), WTR, SnGb. Intergrades with *P. imbricata* (esp in SnFrB), *P. corymbosa, P. hastata.* Some pls in NCoR mat-forming. Apr–Jul

P. eisenii Brandegee (p. 495) Ann 2–10(15) cm. **ST:** gen erect, gen simple, puberulent to short-stiff-hairy, glandular. **LF:** 10–25 mm; proximal-most opposite; blade ≥ petiole, elliptic to oblanceolate, gen entire. **FL:** calyx lobes 1–3 mm, 4–5 mm in fr, unequal, linear to narrowly oblanceolate, short-hairy; corolla deciduous, 2–4 mm, bell-shaped, white to lavender; stamens 3–5 mm, glabrous; style 1–4 mm, divided ± to near base, glabrous to glandular-puberulent. **FR:** 2–3 mm, ovoid to spheric, short-hairy. **SEED:** 2–4, 1–1.5 mm, pitted. *n*=9. Sandy or gravelly soils, conifer forest; 1300–3400 m. SN. If recognized taxonomically, pls with some proximal lvs lobed, corolla 3–4 mm, style 2–4 mm assignable to *P. eisenii* var. *brandegeeana* J.T. Howell. May–Jul

P. exilis (A. Gray) G.J. Lee TRANSVERSE RANGE PHACELIA Ann 5–25 cm. **ST:** decumbent to erect, simple to branched at base, short-hairy, ± glandular-puberulent. **LF:** 10–35 mm; blade ± narrowly (ob)lanceolate, tapered to petiole, entire. **FL:** calyx lobes 3–5 mm, 5–15 mm in fr, unequal, linear to oblanceolate, puberulent, short-hairy; corolla deciduous, 5–8 mm, bell-shaped, lobes lavender; stamens 5–9 mm, subequal, puberulent, filaments white; style 5–8 mm, puberu-

lent. **FR:** 3–5 mm, ovoid, short-hairy, gland-dotted. **SEED:** 4–8, 1–2 mm, pitted. *n*=9. Sandy or rocky slopes, flats, meadows; 1100–2700 m. s SN, WTR (Lockwood Valley), SnGb, SnBr. May–Aug ★

P. floribunda Greene (p. 495) MANY-FLOWERED PHACELIA Ann 25–60 cm. **ST:** erect, branched, soft-hairy, glandular-puberulent. **LF:** (30)50–180 mm; blade > petiole, ± ovate, ± compound, lflets scalloped to lobed. **FL:** calyx lobes 3.5–4.5 mm, 4–5.5 mm in fr, unequal, gen 3–5-lobed, 1–2 sometimes simple, oblong, short-hairy; corolla deciduous, 5–7 mm, bell-shaped, pale blue to lavender; stamens 5–6 mm, glabrous; style 2–4 mm, glabrous. **FR:** 2–3 mm, ovoid, short-hairy. **SEED:** 1–4, 1.5–2 mm, pitted. *n*=11. Ravines, gen coastal-sage scrub; < 500 m. s ChI (San Clemente Island); Baja CA (Guadalupe Island). Mar–May ★

P. fremontii Torr. (p. 495) Ann 7–30 cm. **ST:** decumbent to erect, branched at base, puberulent, gen glandular distally. **LF:** 15–50 mm, ± basal; blade > petiole, oblong to oblanceolate, deeply lobed to compound, segments gen rounded. **FL:** calyx lobes 3–5 mm, 4–6 mm in fr, subequal, linear to oblanceolate, short-hairy; corolla deciduous, 7–15(20) mm, funnel-shaped to bell-shaped, tube 0 throat yellow, sometimes distal throat dark violet, lobes blue to pink or violet; stamens 3–8 mm, glabrous; style 3–5 mm, short-hairy. **FR:** 4–6 mm, ovoid, puberulent below, short-stiff-hairy distally. **SEED:** 10–18, 1–1.5 mm; cross-furrows 6–9. *n*=13. Sandy or gravelly soils, scrub, grassland; < 2900 m. s SN, SnJV, e SnFrB, SCoRI, TR, SNE, DMoj; to sw UT, AZ. Mar–Jun

P. glandulifera Piper (p. 495) Ann 5–25 cm. **ST:** erect, 0–few-branched, short-hairy; gland-tipped hairs dark. **LF:** 10–50 mm; proximal ± = distal; blade gen > petiole, oblong to oblanceolate, deeply lobed to compound, segments obtuse or toothed. **FL:** calyx lobes 2–3 mm, 3–5 mm in fr, unequal, linear to oblanceolate, short-hairy; corolla deciduous, 2–5 mm, bell-shaped, tube yellow, lobes ± blue; stamens 1–2.5 mm, glabrous; style 1–3 mm. **FR:** 4–5.5 mm, elliptic, glabrous proximally, short-hairy distally. **SEED:** 7–14, 1–1.5 mm; cross-furrows 8–10. *n*=13. Gen sandy soils, scrub, juniper woodland; 800–2500 m. CaR, GB; to e WA, WY. May–Jun

P. grandiflora (Benth.) A. Gray Ann 30–200 cm. **ST:** erect, coarse, simple or branched distally, glandular-puberulent and sparsely stiff-hairy. **LF:** 30–200 mm; blade ≥ petiole, widely ovate to ± round, irregularly toothed. **INFL:** pedicel 2–10 mm. **FL:** calyx lobes 5–8 mm, ≥ mature capsule, oblanceolate, hairy, glandular; corolla deciduous, limb 25–40 mm diam, shallow-bell-shaped to rotate, lavender (white), often 2-colored with lavender-spotted white throat; stamens = corolla length, exserted from open throat, short-hairy, glandular, filaments dilated slightly at base; style 15–25 mm, short-hairy. **FR:** 8–14 mm, ovoid, beaked, puberulent; tip short-glandular-hairy. **SEED:** 80–120, 0.75–1.5 mm, pitted. *n*=11. Open, sandy, ± moist areas, chaparral and coastal-sage scrub, often in washes or burns; < 1150 m. SCo, ChI (rare), s slope WTR, s slope SnGb, w PR; Baja CA. Abundant after fire. Apr–Jun

P. greenei J.T. Howell (p. 495) SCOTT VALLEY PHACELIA Ann 2–15 cm, aromatic. **ST:** gen erect, simple to branched at base, ± short-glandular. **LF:** 8–30 mm; proximal-most opposite to subopposite; blade ± narrowly (ob)lanceolate, tapered to petiole, entire. **FL:** calyx lobes 2–3 mm, 3–5 mm in fr, narrowly oblanceolate, short-hairy; corolla deciduous, 5–6 mm, ± rotate, tube white to ± yellow, lobes violet to purple; stamens 4–6 mm; style 6–7 mm. **FR:** 3–4 mm, ± spheric, ± beaked, short-hairy. **SEED:** gen 2 per chamber, ± 3 mm, pitted. *n*=10. Serpentine soils, openings in conifer forest; 800–1800 m. KR. May–Jun ★

P. grisea A. Gray Ann (5)10–60 cm. **ST:** gen erect, branched, stiff-hairy, finely glandular. **LF:** 10–80 mm; blade > petiole, lanceolate to widely ovate, toothed to lobed. **FL:** calyx lobes 2.5–4 mm, 6–8 mm in fr, unequal, oblanceolate to obovate, densely short-hairy, glandular; corolla deciduous, 5–7 mm, widely bell-shaped, gen white to pale lavender; stamens 7–8 mm, papillate; style 5–9 mm, short-hairy. **FR:** 4–5 mm, ovoid, short-glandular-bristly. **SEED:** 5–10, 1–1.5 mm, pitted. *n*=9. Gravelly slopes, gen in chaparral; 300–1200 m. SCoRO, w WTR. If recognized taxonomically, pls < 15 cm in w WTR with ± crenate lf blades, subequal calyx lobes assignable to *P. hardhamiae* Munz. Apr–Jul

P. gymnoclada S. Watson NAKED-STEMMED PHACELIA Ann 5–20 cm. **ST:** spreading to ascending, branched at base, short-hairy, glandular-puberulent. **LF:** 10–40 mm, ± basal or reduced distally on st; blade < to > petiole, oblanceolate to ovate, wavy to obtusely lobed. **INFL:** pedicel 1–7 mm. **FL:** calyx lobes 3–5 mm, 5–8 mm in fr, ± linear, short-hairy; corolla deciduous, 5–11 mm, funnel- to bell-shaped, tube yellow, lobes lavender to purple; stamens 3–6 mm, short-hairy; style 2–5 mm, short-hairy. **FR:** 3–4 mm, ± oblong, sparsely short-hairy. **SEED:** 5–8, ovoid, 1–1.5 mm; cross-furrows 7–9. *n*=13. Clay to gravelly soils, gen in scrub; 180–2300 m. n SNE (Mono Co.); OR, NV. May–Jun ★

P. hastata Lehm. Per. **ST:** not glandular. **LF:** mostly basal; blade 15–120 mm, ≤ petiole, lanceolate to widely elliptic, gen entire (or 2–4-lobed or compound with 3–5 lflets). **FL:** calyx lobes 3–7 mm, 5–9 mm in fr, linear to lanceolate; corolla 4–7 mm, urn- to bell-shaped, white to lavender; stamens 6–10 mm, exserted, glabrous to hairy; style 7–10 mm, exserted. **FR:** 2–4 mm, ovoid, stiff-hairy. **SEED:** 1–3, 1.5–2.5 mm, pitted in vertical rows. Subspp. intergrade.

subsp. ***compacta*** (Brand) Heckard (p. 495) **ST:** decumbent to ascending, 5–20(30) cm; hairs mostly stiff, spreading. **FL:** calyx lobes gen glandular. *n*=22,33. Sandy to rocky slopes, flats, talus, conifer forest, alpine; 1500–4000 m. CaRH, SNH, Wrn, SNE; to WA, NV. Jul–Sep

subsp. ***hastata*** (p. 495) **ST:** ascending to ± erect, 20–50 cm; hairs mostly ± appressed, some stiff, spreading. **FL:** calyx lobes gen not glandular. *n*=11,22. Sandy to rocky slopes, scrub, conifer forest; 380–3100 m. KR, CaRH, SNH, SnBr, GB; to w Can, SD, CO. May–Jul

P. heterophylla Pursh subsp. ***virgata*** (Greene) Heckard (p. 495) Bien, weak per, 20–120 cm. **ST:** central erect, gen 1, > lateral, lateral ascending, stiff-hairy, not glandular. **LF:** mostly basal; blade 50–150 mm, gen > petiole, lanceolate to ovate, basal dissected, segments 5–9, distal simple to dissected. **FL:** calyx lobes 3–6 mm, 6–10 mm in fr, oblong to lanceolate; corolla 4–7 mm, bell-shaped, white to lavender; stamens 8–10 mm, exserted, hairy; style 8–10 mm, exserted. **FR:** 2.5–3 mm, ovoid, stiff-hairy. **SEED:** 1–3, 1.5–2.5 mm, pitted in vertical rows. *n*=11,22. Slopes, flats, roadsides; 100–2900 m. NW, CaR, n&c SN, GB; to OR, ID, WY. Sometimes intergrades with *P. mutabilis*. Other subsp. in Rocky Mtns. May–Jul

P. hubbyi (J.F. Macbr.) L.M. Garrison Ann 18–60 cm. **ST:** ascending to erect, simple to branched, stiff-hairy, glandular. **LF:** 20–150 mm; blade gen > petiole, ovate to ± oblong, deeply lobed to compound, segments toothed. **INFL:** axis hairs long, wavy; fls dense, proximal touching. **FL:** calyx lobes 6–8 mm, 9–12 mm in fr, ± linear, ± gray, shaggy-hairy, glandular, closely covering capsule in fr; corolla deciduous, 8–12 mm, bell-shaped, lavender; stamens 8–12 mm, glabrous; style 8–12 mm, short-hairy. **FR:** 3–4 mm, ± spheric, wavy-hairy. **SEED:** 2–4, 1.5–3 mm, pitted. Gravelly or rocky slopes, chaparral, grassland; < 1000 m. n SCo, n ChI (Santa Cruz Island), WTR. [*P. cicutaria* var. *h.* (J.F. Macbr.) J.T. Howell] Apr–Jul ★

P. humilis Torr. & A. Gray Ann 5–20 cm. **ST:** gen erect, simple to branched at base, short-stiff-hairy, sparsely glandular. **LF:** 10–40 mm; proximal-most opposite; blade gen >> petiole, elliptic to ovate, entire. **FL:** calyx lobes 2–3 mm, ± linear, densely white-hairy; corolla deciduous, 4–7 mm, bell-shaped, lavender to violet; stamens short-hairy; style 4–7 mm, glabrous to short-hairy. **FR:** 2–5 mm, ovoid, short-hairy. **SEED:** 1–4, finely pitted.

var. ***dudleyi*** J.T. Howell **FL:** calyx lobes 8–12 mm in fr; stamens 6–8 mm, clearly exserted. **SEED:** 2.5–3 mm. *n*=11. Flats, meadows; 800–2900 m. SNH (e slope, Mono Co.), Teh. Jun–Jul

var. ***humilis*** (p. 495) **FL:** calyx lobes 5–8 mm in fr; stamens 4–6 mm, barely exserted. **SEED:** 1.5–2.5 mm. *n*=11. Flats, meadows; 1500–2800 m. SNH (e slope), GB; c WA, se OR, nw NV. May–Jul

P. hydrophylloides A. Gray (p. 495) Per 10–30 cm. **ST:** decumbent to ascending, hairy, gen not glandular. **LF:** petiole 5–50 mm; blades 15–60 mm, oblong to ovate, gen coarsely toothed or lobed, 1–2 lobes often deep. **INFL:** head-like. **FL:** calyx lobes 3–5 mm, 7–10 mm in fr, narrowly oblong; corolla 5–8 mm, bell-shaped, white

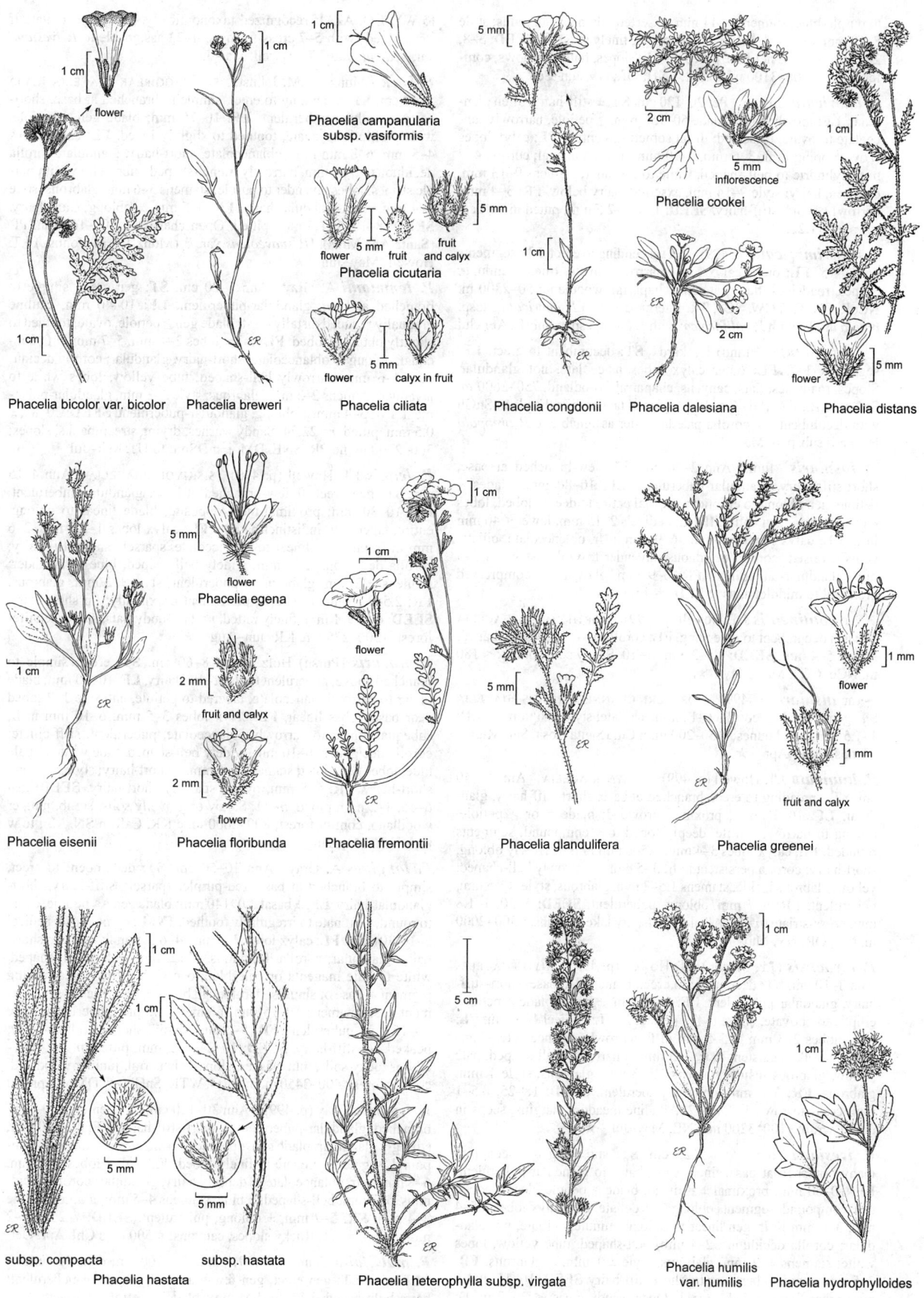

Phacelia campanularia subsp. vasiformis

Phacelia cicutaria
flower fruit fruit and calyx

Phacelia cookei
inflorescence

Phacelia bicolor var. bicolor   Phacelia breweri   Phacelia ciliata   Phacelia congdonii   Phacelia dalesiana   Phacelia distans

flower — Phacelia egena

Phacelia eisenii   Phacelia floribunda   Phacelia fremontii   Phacelia glandulifera   Phacelia greenei

fruit and calyx

subsp. compacta   subsp. hastata
Phacelia hastata   Phacelia heterophylla subsp. virgata   Phacelia humilis var. humilis   Phacelia hydrophylloides

to purple-blue; stamens 8–11 mm, exserted, filaments glabrous; style 8–10 mm, exserted. **FR:** 5–7 mm, ovoid, finely strigose. **SEED:** 3–8, 2–3 mm; surface net-like, pitted. *n*=11. Slopes, flats, meadows, conifer forest; 1500–3100 m. CaRH, SNH; OR, NV. Jun–Sep

P. imbricata Greene Per 20–120 cm. **ST:** ± stiff-hairy, often glandular. **LF:** mostly basal; blade 50–150 mm, ≥ petiole, narrowly lanceolate to ovate, dissected; distal sometimes entire. **FL:** calyx lobes ± overlapping in fr, 3–6 mm, 5–10 mm in fr, ± unequal; corolla 4–7 mm, cylindric to bell-shaped, white to lavender; stamens 9–13 mm, exserted, hairy; style 9–14 mm, exserted, hairy below. **FR:** 3–4 mm, narrowly ovoid, stiff-hairy. **SEED:** 1–3, 2–2.5 mm, pitted in vertical rows. *n*=11,22.

subsp. ***imbricata*** (p. 499) **ST:** ascending to erect. **LF:** segments gen 7–15. **FL:** outer calyx lobes narrowly ovate, often glandular. Slopes, roadsides, flats, canyons, chaparral, woodland; 50–2300 m. NCoR, SN, GV, CW, SCo, TR. Intergrades with *P. imbricata* subsp. *patula* in TR; with *P. californica*; with *P. egena* esp in SnFrB. Apr–Jul

subsp. ***patula*** (Brand) Heckard **ST:** decumbent to erect. **LF:** segments 3–7. **FL:** outer calyx lobes lanceolate, not glandular. Slopes, roadsides, flats, canyons, chaparral, woodland; 750–2600 m. SnGb, SnBr, PR; Baja CA. If recognized taxonomically, pls in SnGb with decumbent sts, corolla pale lavender assignable to *P. oreopola* Heckard subsp. *o.* May–Aug

P. insularis Munz Ann 2–20 cm. **ST:** few-branched at base, short-stiff-hairy, glandular-puberulent. **LF:** 10–80 mm; blades ≤ petiole, gen oblong to elliptic; proximal entire to deeply lobed, lobes unequal; distal gen entire. **INFL:** pedicels 2–15 mm, lower < 40 mm in fr. **FL:** calyx lobes 5–6 mm, 6–9 mm in fr, oblanceolate, ciliate, veins ± raised; corolla deciduous, lavender to violet; stamens 3–5 mm, glandular-puberulent. **FR:** 6–9 mm, obovoid, ± compressed proximal to middle, hairy. **SEED:** 8–15, pitted.

var. ***continentis*** J.T. Howell (p. 499) NORTH COAST PHACELIA **ST:** gen decumbent to ascending. **FL:** corolla 5–8 mm, ± bell-shaped; style 2.5–4 mm. **SEED:** 1.5–2 mm. *n*=10. Sandy soils, bluffs; < 180 m. NCo, CCo. Mar–May ★

var. ***insularis*** (p. 499) NORTHERN CHANNEL ISLANDS PHACELIA **ST:** gen erect. **FL:** corolla 6–12 mm, ± rotate; style 5–6 mm. **SEED:** 1–1.5 mm. Sand dunes; < 50–200 m. n ChI (Santa Rosa, San Miguel islands). Mar–Apr ★

P. inundata J.T. Howell (p. 499) PLAYA PHACELIA Ann 10–40 cm. **ST:** spreading to erect, branched at base, short-stiff-hairy, glandular. **LF:** 10–30 mm; proximal crowded; blade = or > petiole, oblong to narrowly ovate, deeply lobed to ± compound, segments rounded. **FL:** calyx lobes 3–4 mm, 5.5–8 mm in fr, narrowly oblong, short-hairy; corolla persistent in fr, 3–5 mm, ± narrowly bell-shaped, yellow, glabrous inside; stamens 1.5–3 mm, glabrous; style < 1.5 mm, puberulent. **FR:** 4–7 mm, oblong, puberulent. **SEED:** 5–30, 1–1.8 mm, cross-striate. *n*=12. Alkaline flats, dry lake margins; 1300–2000 m. MP; OR, NV. May–Aug

P. inyoensis (J.F. Macbr.) J.T. Howell (p. 499) INYO PHACELIA Ann 3–10 cm. **ST:** decumbent to erect, branched at base, short-stiff-hairy, glandular-puberulent. **LF:** 5–20 mm, ± basal; blade > petiole, elliptic to obovate, entire to few-lobed. **INFL:** pedicel 1–6 mm. **FL:** calyx lobes 2–3 mm, 3.5–5 mm in fr, narrowly oblanceolate, short-hairy; corolla persistent in fr, 2–3 mm, ± narrowly bell-shaped, pale yellow, glabrous inside; stamens 1.5–3 mm, glabrous; style 1 mm, glabrous. **FR:** 3–4 mm, oblong, puberulent. **SEED:** 18–25, 0.5–1 mm; cross-furrows 5–9. *n*=12. Alkaline meadow margins, seeps in desert scrub; 1100–3200 m. SNE. May–Jul ★

P. ivesiana Torr. Ann 5–25 cm. **ST:** spreading to ascending, many-branched at base, finely short-hairy to glandular-puberulent. **LF:** 10–60 mm; proximal ± = distal; blade > petiole, deeply lobed to ± compound, segments oblong to ± deltate. **FL:** calyx lobes 2.5–4 mm, 3–6 mm in fr, gen linear to oblong, minutely ciliate, not glandular; corolla deciduous, 2–4 mm, bell-shaped, tube yellow, lobes white; stamens 1–2 mm, glabrous; style ± 1 mm, ± glabrous. **FR:** 3–5 mm, ovoid, puberulent; tip short-stiff-hairy. **SEED:** 10–15, 1–1.5 mm; cross-furrows 5–12. *n*=11. Open, sandy areas; < 2000 m. D;

to WY, CO, AZ. If recognized taxonomically, pls with ± deltate lf lobes, seeds with 5–7 cross-furrows, *n*=23 assignable to *P. ivesiana* var. *pediculoides* J.T. Howell. Mar–Jun

P. keckii Munz & I.M. Johnst. SANTIAGO PEAK PHACELIA Ann 5–40 cm. **ST:** ascending to erect, simple to branched at base, short-hairy, glandular-puberulent. **LF:** 10–75 mm; blade gen > petiole, widely elliptic to ovate, toothed to slightly lobed. **FL:** calyx lobes 4–5 mm, 6–8 mm in fr, oblanceolate, short-hairy, glandular; corolla deciduous, 9–14 mm, narrowly funnel-shaped, tube yellow with purple stripes, lobes lavender to purple; stamens 3–6 mm, glabrous; style 3–4 mm, short-glandular-hairy. **FR:** 3–5 mm, ± oblong, short-hairy. **SEED:** 8–10, 1–1.5 mm, pitted. Open chaparral; 500–1600 m. PR (Santa Ana Mtns). [*P. suaveolens* var. *k.* (Munz & I.M. Johnst.) J.T. Howell] May–Jun ★

P. lemmonii A. Gray Ann 7–20 cm. **ST:** gen erect, simple to branched at base, glandular-puberulent. **LF:** 10–40 mm, cauline gradually reduced distally on st; blade gen ≥ petiole, ovate, toothed to slightly obtusely lobed. **FL:** calyx lobes 2–4 mm, 5–7 mm in fr, subequal, oblong to oblanceolate, short-hairy, glandular; corolla deciduous, 4–6 mm, narrowly bell-shaped, tube yellow, lobes white to lavender; stamens 2–3 mm, glabrous; style 2–3 mm, glandular-puberulent. **FR:** 3–4 mm, ± oblong, glandular-puberulent. **SEED:** 30–80, ± 0.5 mm, pitted. *n*=22,24. Sandy washes, drying streambanks, slopes; 300–2300 m. ne PR, SNE, DMoj, n DSon; to UT. Mar–Jul

P. leonis J.T. Howell (p. 499) SISKIYOU PHACELIA Ann 4–15 cm. **ST:** gen erect, 0–few-branched at base, glandular-puberulent. **LF:** 10–30 mm; proximal-most opposite; blade linear to oblong, entire, tapered to indistinct petiole. **FL:** calyx lobes 1–2.5 mm, 2–6 mm in fr, unequal, linear to oblanceolate, sparsely short-stiff-hairy; corolla deciduous, 2–3 mm, widely bell-shaped, blue to lavender; stamens ± 2 mm, glabrous to puberulent; style ± 2 mm, glabrous. **FR:** 2.5–3.5 mm, ± spheric, puberulent proximally; tip short-hairy. **SEED:** 6–9, ± 1 mm, finely pitted. *n*=11. Sandy flats, slopes, conifer forest; 1200–2750 m. KR. Jun–Aug ★

P. linearis (Pursh) Holz. Ann 8–60 cm. **ST:** erect, simple to branched above, puberulent to short-stiff-hairy. **LF:** 10–80 mm; blade linear to narrowly lanceolate, tapered to petiole, entire to 1–2-lobed near base, lobes linear. **FL:** calyx lobes 3–5 mm, 6–10 mm in fr, subequal, linear to narrowly oblanceolate, puberulent, stiff-ciliate; corolla deciduous, 6–10 mm, widely bell-shaped, tube white to pale blue, lobes gen violet; stamens 5–6 mm, ± short-hairy; style 5–8 mm, short-hairy. **FR:** 5–8 mm, ovoid, sparsely short-hairy. **SEED:** gen 6–15, 1–2 mm, pitted. *n*=11. Sandy or gravelly soils, scrub, juniper woodland, conifer forest; 800–2000 m. e KR, CaR, n SN, MP; to w Can, WY. May–Jul

P. longipes A. Gray Ann 10–50 cm. **ST:** decumbent to erect, simple to branched at base, red-purple, sparsely stiff-hairy, short-glandular-hairy. **LF:** ± basal, 20–140 mm; blade gen << petiole, ovate to round, ± crenate to irregularly toothed. **INFL:** cyme loose; pedicel 3–10(30) mm. **FL:** calyx lobes 3–4 mm, 4–6 mm in fr, ± linear, short-hairy, glandular; corolla deciduous, 7–12 mm, widely bell-shaped, white to pale magenta or pale blue, occ with diffuse area lacking pigment at base of sinuses, flap-like glabrous appendage fused to filament base; stamens 10–15 mm, filaments purple, puberulent; style 10–15 mm, puberulent. **FR:** 5–8 mm, ovoid, short-glandular-hairy; beaked, tip stiff-hairy. **SEED:** 10–24, 1–2 mm, pitted. *n*=11. Gravelly or rocky soils, mtn washes, slopes, chaparral, juniper woodland, conifer forest; 400–2450 m. SCoRO, WTR, SnGb, sw DMoj. Apr–Jul

P. lyonii A. Gray (p. 499) Ann 30–120 cm. **ST:** gen erect, 0–few-branched, glandular-puberulent to short-stiff-hairy. **LF:** 40–150 mm, strongly glandular; blade gen > petiole, ± ovate, deeply lobed to compound, segments crenate to finely lobed. **FL:** calyx lobes 4–5 mm, 6–8 mm in fr, oblanceolate, ± densely hairy, glandular; corolla deciduous, 5–7 mm, bell-shaped, ± blue; stamens 4–5 mm, glabrous; style 2–4 mm. **FR:** 5–7 mm, ± oblong, puberulent. **SEED:** 7–20, 1.5–2 mm, pitted. *n*=11. Rocky slopes, canyons; < 500 m. s ChI. Apr–Oct

P. malvifolia Cham. (p. 499) Ann 20–100 cm; densely yellow-stiff-hairy. **ST:** gen erect, gen few-branched, glandular-puberulent; hairs bulb-based. **LF:** 20–140 mm; blade ≥ petiole, ± deltate to ±

round, proximal lobed to compound (segments 3, again toothed to irregularly lobed), distal gen toothed or lobed. **FL:** calyx lobes 3–5 mm, 4–6 mm in fr, unequal, oblong to oblanceolate; corolla deciduous, 5–7 mm, widely bell-shaped, cream-white; stamens 5–10 mm, glabrous; style 8–12 mm, short-hairy. **FR:** 2–3 mm, spheric, puberulent. **SEED:** 2–6, 2–3 mm, pitted. *n*=11. Sandy or gravelly soils, scrub, conifer forest; < 1400 m. NCo, CW; OR. Apr–Jul

P. marcescens J.F. Macbr.   Ann 5–20 cm. **ST:** gen erect, simple to branched, short-glandular-hairy, sparsely stiff-hairy. **LF:** 10–50 mm, proximal-most opposite; blade ≥ petiole, elliptic to ovate, largest 1–2-lobed at base, smaller gen entire. **FL:** calyx lobes 2–3 mm, 3–4 mm in fr, subequal, linear to oblanceolate, short-hairy; corolla persistent in fr, 4–5 mm, widely bell-shaped, purple to violet; stamens 5–6 mm, glabrous; style 5–6 mm, short-hairy. **FR:** 2.5–3.5 mm, ovoid, ± compressed, short-hairy. **SEED:** (1)2, 1.5–2 mm, pitted. *n*=8. Sandy to gravelly soils, meadows, conifer forest; 1300–2600 m. n&c SN. Jun–Jul

P. minor (Harv.) F. Zimm. (p. 499)   WILD CANTERBURY BELLS   Pl 20–100 cm. **ST:** erect or reclining, 0–few-branched, short-glandular-hairy, sparsely stiff-hairy. **LF:** 20–110 mm; blade ≤ petiole, ovate to ± round, irregularly toothed. **INFL:** cyme lax, 15–30 cm; pedicel 10–30 mm. **FL:** calyx lobes 5–10 mm, 6–12 mm in fr, narrowly oblong, short-hairy, glandular; corolla deciduous, 10–40 mm, ± bell-shaped with slight constriction between tube and throat, purple (white), flap-like hairy appendage fused to filament base; stamens 15–35 mm, short-glandular-hairy, exserted; style 20–40 mm, short-hairy. **FR:** 7–13 mm, >> calyx, rounded at base, puberulent; beaked, tip short-stiff-hairy. **SEED:** 30–90, ± 1 mm, pitted. *n*=11. Exposed slopes, coastal-sage scrub, desert canyons; < 1850 m. SCo, e WTR, SnGb, SnBr, PR, w DSon; Baja CA. Abundant after fire. Mar–Jun

P. mohavensis A. Gray   MOJAVE PHACELIA   Ann 5–25 cm. **ST:** erect, 0–few-branched, glandular-puberulent, short-stiff-hairy. **LF:** 10–45 mm; blade linear to narrowly oblanceolate, entire, tapered to petiole. **FL:** calyx lobes 3–5 mm, 5–10(15) mm in fr, unequal, linear to oblanceolate, puberulent to short-hairy, glandular; corolla deciduous, 5–8 mm, bell-shaped, tube base gen ± yellow, lobes white, aging pale blue; stamens 5–14 mm, unequal, ± glabrous, filaments violet; style 5–8 mm, short-hairy. **FR:** 3–5 mm, ovoid, short-stiff-hairy, gland-dotted. **SEED:** 4–8, 1–2 mm, pitted. *n*=9. Sandy or gravelly soils, conifer forest; 900–2570 m. s SNH, SnGb, SnBr. Apr–Aug ★

P. monoensis Halse (p. 499)   MONO COUNTY PHACELIA   Ann 2–12 cm. **ST:** spreading to ascending, branched throughout, short-hairy, ± glandular. **LF:** 8–25 mm; blade gen = petiole, ± oblong to ovate, entire to lobed. **FL:** calyx lobes 2–4 mm, 4–6 mm in fr, narrowly oblanceolate, short-hairy; corolla persistent in fr, 2–4 mm, gen narrowly bell-shaped, yellow, puberulent inside; stamens 1.5–3 mm, puberulent; style 0.5–1.5 mm, puberulent. **FR:** 2.5–4 mm, ovoid, puberulent. **SEED:** < 10, 1–1.7 mm; cross-furrows 8–11. *n*=12. Fractured rhyolitic clay soils, sagebrush scrub; 1900–2900 m. n SNE (Mono Co.); w NV. May–Jul ★

P. mustelina Coville (p. 499)   DEATH VALLEY ROUND-LEAVED PHACELIA   Ann 6–30 cm. **ST:** decumbent to ascending, many-branched, ± short-glandular. **LF:** 10–40 mm; blade ≥ petiole, ± round, irregularly toothed. **FL:** calyx lobes 2–4 mm, 5–6 mm in fr, narrowly oblanceolate, short-glandular-hairy; corolla deciduous, 6–10 mm, limb gen 3–5 mm diam, narrowly bell-shaped, violet to purple; stamens 2–4 mm, short-hairy; style 3–5 mm, sparsely short-hairy. **FR:** 3–4 mm, ovoid, puberulent. **SEED:** 20–60, ± 0.5 mm, finely pitted. *n*=12. Gravelly or rocky slopes, creosote-bush scrub, pinyon/juniper woodland; < 2100 m. DMtns; w NV. Mar–Jun ★

P. mutabilis Greene (p. 499)   Bien or short-lived per, 10–60 cm. **ST:** decumbent to erect, finely stiff-hairy, not glandular. **LF:** blade 20–180 mm, gen = petiole, lanceolate to ovate, entire to compound (then segments 3, terminal largest). **FL:** calyx lobes 3–5 mm, 6–10 mm in fr, ± unequal, ± linear to oblanceolate; corolla deciduous, 4–6 mm, tubular to bell-shaped, white to ± yellow or lavender; stamens 6–8 mm, exserted, hairy; style 6–8 mm, exserted. **FR:** 2–3 mm, ovoid, stiff-hairy. **SEED:** 1–4, 1.5–2.5 mm, pitted. *n*=11,22. Roadsides, ridges, open forest; 900–3500 m. KR, NCoR, CaRH, SNH, SnJt, MP; OR, NV. Similar to *P. nemoralis*, intergrades with *P. hastata*. May–Oct

P. nashiana Jeps. (p. 499)   CHARLOTTE'S PHACELIA   Ann 3–18(40) cm. **ST:** ascending to erect, simple to branched at base, short-stiff-hairy; gland-tipped hairs black. **LF:** ± basal, 15–70 mm; blade gen < petiole, widely ovate to ± round, irregularly crenate to slightly lobed. **INFL:** pedicel 5–10 mm. **FL:** calyx lobes 4–10 mm, 5–12 mm in fr, oblong, short-hairy, glandular; corolla deciduous, 10–20 mm, rotate to widely bell-shaped, tube white, throat blue with ± white spot at base of each sinus, lobes bright blue, flap-like glabrous appendage fused to filament base; stamens 12–20 mm, short-hairy; style 10–18 mm, cleft 2/3 to 3/4 its length. **FR:** 7–14 mm, extending > 3 mm from calyx, ovoid, beaked, short-glandular-hairy. **SEED:** 40–90, ± 2 mm, pitted. *n*=11. Sandy to rocky, granitic e-facing slopes, gen Joshua-tree or pinyon/juniper woodland; < 2400 m. s SNH, Teh (e slope), w edge DMoj. Feb–Jun ★

P. neglecta M.E. Jones   Ann 3–20 cm. **ST:** ascending to erect, 0–few-branched, finely glandular. **LF:** ± basal, 10–50 mm; blade ≥ petiole, ± round, wavy to crenate. **INFL:** axis among lvs, pedicel 1–8 mm, < 12 mm in fr. **FL:** calyx lobes 1.5–2 mm, 2–3.5 mm in fr, narrowly ovate, densely short-hairy, glandular; corolla deciduous, 4–6 mm, cream-white, funnel- to bell-shaped; stamens 2–4 mm, puberulent; style 2–3 mm. **FR:** 3–4 mm, spheric, puberulent. **SEED:** > 60, ± 1 mm; cross-furrows 4–7. *n*=11. Clay or alkaline soils, flats, slopes; < 1000 m. D; s NV, AZ. Mar–May

P. nemoralis Greene   Per, short-lived, 50–200 cm. **ST:** ascending to erect, densely stiff-hairy, not glandular. **LF:** blade 40–150(250) mm, < to > petiole, widely lanceolate to ovate; proximal-most gen compound, lflets 3–7; distal-most often simple, entire. **FL:** calyx lobes 3.5–5 mm, 5–8 mm in fr, (ob)lanceolate; corolla bell-shaped; stamens 7–9 mm, exserted, hairy; style 6–9 mm. **FR:** 2–3 mm, ovoid to ± spheric, stiff-hairy. **SEED:** 1–4, 1.5–2 mm, pitted. *n*=11,22.

subsp. ***nemoralis***   **ST:** gen < 7 mm diam at midpoint. **LF:** basal gen with 3 lflets. **FL:** corolla 3.5–5 mm, green-white. *n*=11. Moist slopes, streambanks, mixed-evergreen or conifer forest; 50–1150 m. NCoRO, CW (exc SCoRI). Apr–Jul

subsp. ***oregonensis*** Heckard   **ST:** gen 7–10 mm diam at midpoint. **LF:** basal gen with 3–7 lflets. **FL:** corolla 5–6 mm, ± yellow. *n*=22. Moist slopes, streambanks, mixed-evergreen or conifer forest; < 800 m. NCo, NCoRO; to WA. Perhaps best treated as part of *P. californica*. May–Aug

P. novenmillensis Munz (p. 499)   NINE MILE CANYON PHACELIA   Ann 5–10 cm. **ST:** ascending to erect, 0–few-branched, short-soft-hairy, ± sparsely glandular-puberulent. **LF:** 20–80(110) mm; blade gen < petiole, (ob)lanceolate to narrowly elliptic, entire; proximal-most sometimes lobed to irregularly compound. **FL:** calyx lobes 2–4 mm, 8–10 mm in fr, subequal, linear, long-ciliate; corolla deciduous, 3–4 mm, bell-shaped, lavender; stamens 2–4.5 mm, glabrous; style 4–5 mm. **FR:** 2–3 mm, ovoid, short-hairy. **SEED:** 2–4, 1.5–2 mm, pitted. Open, sandy to gravelly soils, pinyon/juniper woodland, conifer forest; 1600–2200 m. s SNH (e slope), w edge DMoj. May–Jun ★

P. orogenes Brand (p. 499)   MOUNTAIN PHACELIA   Ann 2–10 cm. **ST:** erect, simple to branched, puberulent, sparsely glandular. **LF:** 5–30 mm; proximal opposite; blade linear to narrowly lanceolate, entire, tapered to petiole. **INFL:** pedicels 2–7 mm, proximal > distal. **FL:** calyx lobes 2–3 mm, 5–8 mm in fr, unequal, linear to oblanceolate, short-hairy; corolla deciduous, 4–6 mm, widely bell-shaped, tube white to ± yellow, lobes violet; stamens 3–5 mm, glabrous; style 3–4 mm, short-hairy. **FR:** 2.5–3.5 mm, ± spheric, short-hairy. **SEED:** 3–6, 1.5–2 mm, pitted. *n*=9. Gravelly slopes, meadow edges, conifer forest; 2060–3400 m. s SNH (Mineral King, Tulare Co.). Jul–Aug ★

P. pachyphylla A. Gray (p. 499)   Ann 4–17 cm. **ST:** erect, 0–few-branched, ± dark-glandular. **LF:** ± basal, 20–50 mm; blade gen < petiole, ± round, entire to slightly crenate. **INFL:** axis gen > lvs. **FL:** calyx lobes 2–3 mm, 3–4 mm in fr, oblong, densely short-hairy, glandular; corolla deciduous, 5–7 mm, funnel- to bell-shaped, violet to purple; stamens 2–4 mm, short-hairy; style 2–3 mm. **FR:** 5–7 mm, ± spheric, short-hairy, with some stalked glands. **SEED:** > 60, ± 1 mm; cross-furrows 6–8. *n*=11. Flats, ± alkaline soils, creosote-bush scrub; < 1300 m. D; Baja CA. Apr–May

P. parishii A. Gray (p. 499) PARISH'S PHACELIA Ann 5–15 cm. **ST**: ascending to erect, branched at base, glandular-puberulent. **LF**: ± basal, 8–30 mm; blade > petiole, ± widely elliptic to obovate, entire to barely toothed. **FL**: calyx lobes 3–5 mm, 6–8 mm in fr, unequal, ± linear to ovate, puberulent; corolla deciduous, 4–6 mm, narrowly bell-shaped, tube yellow, lobes lavender; stamens 2–4 mm, sparsely short-hairy; style 1–2 mm. **FR**: 3–5 mm, ± oblong, short-hairy. **SEED**: 20–40, 1–1.5 mm, finely pitted. Clay or alkaline soils, dry lake margins; 540–1200 m. w DMoj (nw San Bernardino Co.); NV. Apr–Jul ★

P. parryi Torr. Ann 10–100 cm. **ST**: 0–few-branched, reclining or erect, stiff-hairy, glandular. **LF**: 10–120 mm; blade ≤ petiole, oblong to ovate, irregularly toothed. **INFL**: pedicel 10–30 mm. **FL**: calyx lobes 5–10 mm, 6–12 mm in fr, ± linear, sparsely hairy, ± glandular; corolla deciduous, 10–20 mm, rotate to widely bell-shaped, tube length < limb width, tube ± proximal throat white to light violet, lobes violet to purple with basal crescent-shaped white spot, flap-like hairy appendage fused to filament base; stamens 10–20 mm, long-hairy; style 11–22 mm, far exserted, short-hairy. **FR**: 6–10 mm, >> calyx, rounded at base, beaked, short-stiff-hairy, gland-dotted. **SEED**: 40–90, ± 1 mm, pitted. *n*=11. Open gravelly areas, slopes, coastal-sage scrub, chaparral; < 2400 m. SCoRI, SW (exc ChI, w WTR), w edge D; Baja CA. Abundant after fire. Mar–May

P. pedicellata A. Gray (p. 499) Ann 12–50 cm. **ST**: erect, 0–few-branched above base, short-stiff-hairy, glandular. **LF**: 20–120 mm; blade gen = petiole, ovate to round, proximal compound (lflets 3–7, rounded or toothed), distal lobed to compound (segments 3, gen rounded). **INFL**: pedicel ± thread-like, densely long-hairy, not glandular. **FL**: calyx lobes 3–4.5 mm, 4–5.5 mm in fr, narrowly oblanceolate, sparsely hairy, ciliate, glandular; corolla deciduous, 5–7 mm, bell-shaped, pink to blue (white); stamens 6–8 mm, glabrous; style 6–8 mm. **FR**: 3–3.5 mm, ± spheric, puberulent. **SEED**: gen 4, ± 3 mm; abaxial surface pitted; adaxial surface with central ridge separating 2 longitudinal grooves. *n*=11. Sandy or gravelly washes, canyons; < 1400 m. D; s NV, AZ, Baja CA. Mar–May

P. peirsoniana J.T. Howell PEIRSON'S PHACELIA Ann 4–30 cm. **ST**: erect, simple to branched above base, glandular-puberulent, sparsely short-stiff-hairy. **LF**: 10–60 mm; blade gen < petiole, ± round, crenate to irregularly toothed. **FL**: calyx lobes 3–4 mm, 6–8 mm in fr, some 1–2 mm wide, oblong, short-hairy; corolla deciduous, 4–6 mm, narrowly bell-shaped, white to violet; stamens 2–3 mm, sparsely short-hairy; style 1–2 mm, short-hairy. **FR**: 4–6 mm, oblong, puberulent. **SEED**: 20–50, ± 1 mm, pitted. *n*=12. Rocky slopes, canyons, sagebrush scrub, pinyon/juniper woodland; 1500–2700 m. n SNE; w NV. May–Jun ★

P. perityloides Coville Per 5–40 cm. **ST**: pendant to spreading, glandular. **LF**: petiole 3–50 mm; blade 5–25 mm, ± round, irregularly toothed to lobed. **FL**: calyx lobes 3–6 mm, 4–6 mm in fr, oblong to oblanceolate; corolla narrowly funnel- to bell-shaped, tube ± yellow or aging purple, lobes white; stamens 3–6 mm, unequal, incl, glabrous; style 4–6 mm, incl, short-lobed. **FR**: 2.5–4 mm, narrowly ovoid, hairy. **SEED**: 50–200, ± 0.5 mm, angular, pitted. *n*=11.

var. ***jaegeri*** Munz JAEGER'S PHACELIA **ST**: base spreading-hairy, not clearly woolly. **INFL**: pedicel 10–30 mm in fr, reflexed. **FL**: calyx lobes < 1 mm wide; corolla 12–15 mm. Crevices on cliffs, rocky, often calcareous slopes; 1900–2300 m. e DMtns (Clark Mtn); NV. May–Jun ★

var. ***perityloides*** (p. 507) **ST**: base clearly woolly, some hairs spreading. **INFL**: pedicel 5–15 mm in fr, spreading to ascending. **FL**: calyx lobes 1–2 mm wide; corolla 10–12 mm. Crevices on cliffs, rocky, often calcareous slopes; 600–2200 m. n W&I, n DMoj. Mar–Jul

P. phacelioides (Benth.) Brand (p. 507) MOUNT DIABLO PHACELIA Ann 5–20 cm. **ST**: ascending to erect, 0–many-branched at base, puberulent, sparsely stiff-hairy. **LF**: 20–80(100) mm; blade ≤ petiole, elliptic to lanceolate, entire. **FL**: calyx lobes 4–6 mm, 7–11 mm in fr, ± oblong, densely stiff-hairy, weakly glandular, densely ciliate, stiff hairs on margin gen > 1 mm; corolla deciduous, 4–6 mm, narrowly bell-shaped, white to light lavender, lobes violet-

streaked; stamens 2–3 mm, short-hairy; style ± 2 mm, short-hairy. **FR**: 3.5–4 mm, ovoid, not beaked, short-hairy. **SEED**: 5–15, 1–1.5 mm, pitted. *n*=11. Open, rocky slopes; 500–1400 m. e SnFrB, SCoRI. Apr–Jun ★

P. platyloba A. Gray (p. 507) Ann 9–45 cm. **ST**: erect, simple to branched at base, puberulent, sparsely short-stiff-hairy. **LF**: 15–90 mm; blade > petiole, elliptic to ovate, 1–2-compound, lflets lobed to toothed. **FL**: calyx lobes 2–4 mm, ciliate, strongly unequal, 3 gen ± lanceolate, entire, 2 ovate to ± round, gen crenate, stalked; corolla deciduous, 4–5 mm, widely bell-shaped, ± blue to lavender; stamens 3–4 mm, glabrous; style 4–6 mm. **FR**: 2–3 mm, ± oblong, densely puberulent. **SEED**: gen 1, 2–3 mm, finely pitted. *n*=11. Gravelly or rocky soils, chaparral, woodland; 300–1200 m. c&s SNF. Apr–Jun

P. pringlei A. Gray (p. 507) Ann 2–18 cm. **ST**: erect, 0–few-branched, glabrous to glandular-puberulent proximally. **LF**: 7–30 mm; proximal gen opposite; blade linear to narrowly lanceolate, entire, tapered to petiole. **FL**: calyx lobes 1–2 mm, 2–6 mm in fr, gen unequal, linear to oblanceolate, short-hairy, glandular; corolla deciduous, 3–5 mm, ± rotate, lavender; stamens 3–4 mm, papillate; style 3–4 mm, short-hairy. **FR**: 2.5–3.5 mm, ± spheric, short-hairy. **SEED**: 3–8, 1.5–2 mm, pitted. *n*=11. Open, steep slopes, ridges, conifer forest; 900–2700 m. KR, CaRH. May–Aug

P. procera A. Gray Per 50–200 cm. **ST**: erect, hairy. **LF**: petiole < 40 mm; blade 50–120 mm, lanceolate to ovate, coarsely toothed to lobed. **INFL**: pedicel black-glandular. **FL**: calyx lobes 3–5 mm, 7–8 mm in fr, linear, black-glandular; corolla deciduous, 3–7 mm, ± bell-shaped, cream to ± green or brown-white; stamens 8–10 mm, exserted, hairy; style 6–10 mm, exserted. **FR**: 6–8 mm, ovoid, short-rough-hairy, glandular. **SEED**: gen 12–16, ± 2 mm, angled; surface net-like, pitted. *n*=11. Meadows, slopes, talus, conifer forest; 1200–2200 m. KR, NCoRH, CaRH, n SNH; to WA, ID. Jun–Aug

P. pulchella A. Gray var. ***gooddingii*** (Brand) J.T. Howell (p. 507) GOODDING'S PHACELIA Ann 5–20 cm. **ST**: ascending to erect, branched throughout, glandular-puberulent. **LF**: 5–40 mm; blade ≥ petiole, ovate to ± round, ± toothed. **FL**: calyx lobes 4–5 mm, 6–9 mm in fr, oblanceolate, glandular-puberulent; corolla deciduous, 6–12 mm, limb gen > 5 mm diam, bell-shaped, tube yellow, lobes lavender to violet; stamens 3–5 mm, puberulent; style 4–5 mm, short-hairy. **FR**: 3–5 mm, ± oblong, short-hairy. **SEED**: (25)30–50, ± 0.5 mm, pitted. Clay soils, flats; 800–1400 m. n DMoj; nw AZ. Other vars. in NV, s UT, n AZ. Apr–Jun ★

P. purpusii Brandegee Ann 10–40 cm. **ST**: gen erect, 0–few-branched, puberulent to short-stiff-hairy, glandular. **LF**: 12–50 mm; blade > petiole, elliptic to ovate, entire to toothed. **FL**: calyx lobes 3–4 mm, 4–7 mm in fr, unequal, oblanceolate to obovate, short-hairy, ciliate; corolla deciduous, 6–8 mm, bell-shaped, lavender to violet; stamens 7–8 mm, papillate; style 6–9 mm, short-hairy. **FR**: 3–5 mm, ovoid, short-glandular-hairy. **SEED**: 3–7, 1.5–2 mm, pitted. *n*=9. Sandy or gravelly soils, conifer forest; 700–2300 m. s CaRH, SN, MP. May–Jul

P. quickii J.T. Howell (p. 507) Ann 4–18 cm. **ST**: decumbent to erect, simple or branched, puberulent, glandular or not. **LF**: 8–50 mm; blade linear to oblanceolate, entire, tapered to petiole. **FL**: calyx lobes 2–4 mm, 4–7 mm in fr, strongly unequal, linear to oblanceolate, short-hairy; corolla late-deciduous, 3–5 mm, widely bell-shaped, white to blue; stamens 3–6 mm, glabrous, pollen yellow; style 3–6 mm, short-hairy. **FR**: 2–2.5 mm, ± spheric, short-hairy. **SEED**: 2–4, 1–1.5 mm, pitted. *n*=8. Open granitic areas; 1000–2400 m. n&c SN. May–Jul

P. racemosa (Kellogg) Brandegee Ann 4–18 cm. **ST**: erect, 0–few-branched, glabrous to glandular-puberulent. **LF**: 10–40 mm, proximal- to mid-cauline opposite; blade linear to narrowly oblanceolate, entire, gen tapered to base. **FL**: calyx lobes 1–2 mm, 2–3 mm in fr, ± linear, short-hairy; corolla deciduous, 2–4 mm, narrowly bell-shaped, pale blue; stamens 1–2 mm, glabrous; style 1–2 mm, short-hairy. **FR**: 2–3.5 mm, ± spheric, short-hairy. **SEED**: gen 4, 1.5–2.5 mm, finely pitted. *n*=7. Gravelly to rocky slopes, conifer forest; 1500–3000 m. KR, CaR, n&c SN. Jun–Sep

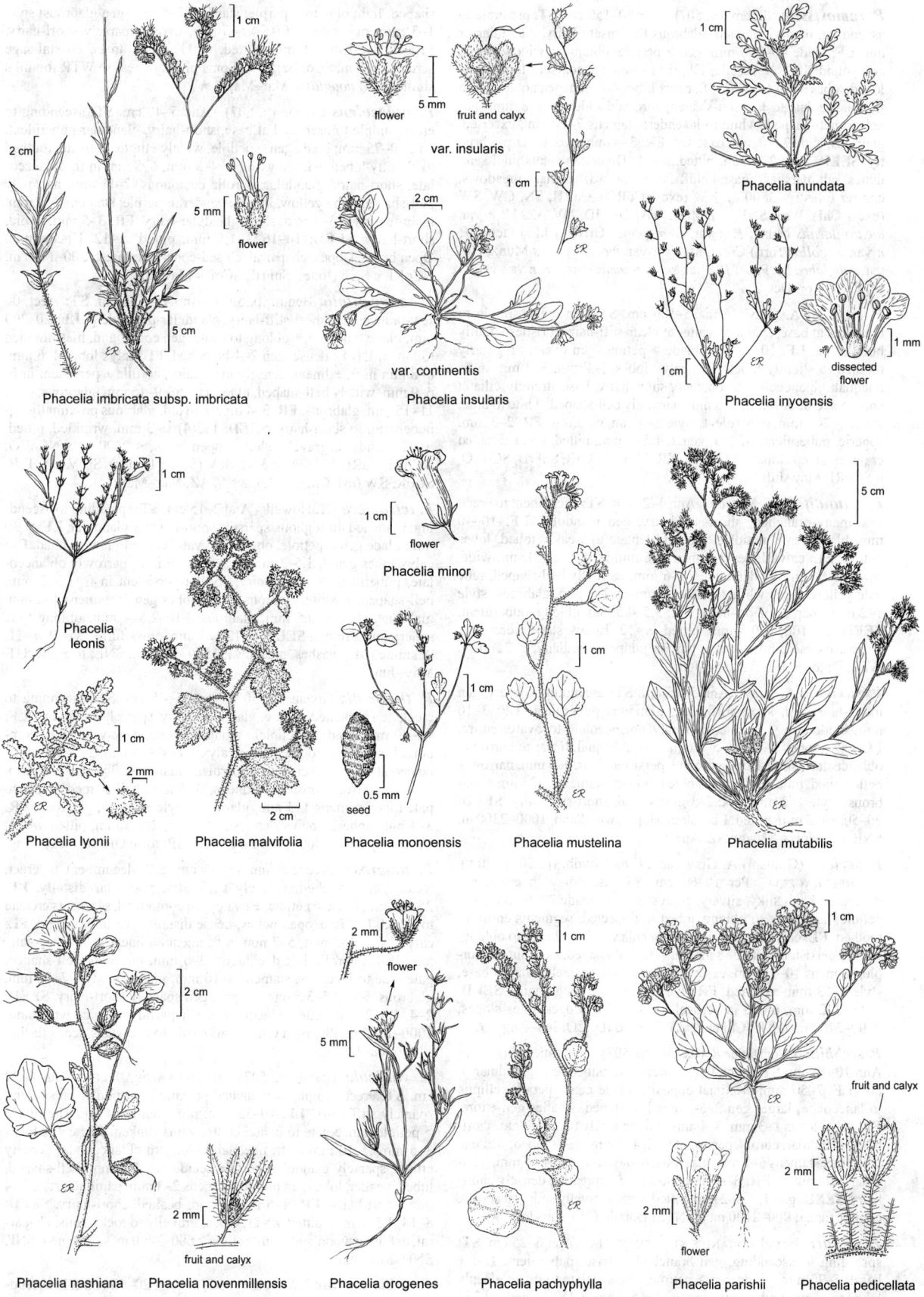

flower 5 mm fruit and calyx

var. insularis

var. continentis

Phacelia insularis

Phacelia imbricata subsp. imbricata

Phacelia inundata

Phacelia inyoensis

dissected flower

Phacelia leonis

Phacelia minor

Phacelia lyonii

Phacelia malvifolia

seed

Phacelia monoensis

Phacelia mustelina

Phacelia mutabilis

flower

Phacelia nashiana

Phacelia novenmillensis

flower

fruit and calyx

Phacelia orogenes

Phacelia pachyphylla

Phacelia parishii

fruit and calyx

flower

Phacelia pedicellata

P. ramosissima Lehm. (p. 507) Per 30–150 cm. **ST:** prostrate to ascending, many-branched, glabrous to densely hairy, glandular or not. **LF:** blade 40–200 mm, gen > petiole, oblong to widely ovate, compound; lflets ± sessile, elliptic to oblong, coarsely toothed or lobed, lobes often toothed. **FL:** calyx lobes 4–6 mm, gen not longer in fr, oblanceolate to ± spoon-shaped; corolla deciduous, 5–8 mm, funnel- to bell-shaped, white to lavender; stamens 7–10 mm, exserted, glabrous; style 7–10 mm, exserted. **FR:** 3–4 mm, ovoid, sharply bristly. **SEED:** 2–4, 2–3 mm, pitted. *n*=11. Diverse habitats, incl sand dunes, salt marshes, coastal bluffs, canyons, washes, flats, meadows, conifer forest; < 3800 m. NW (exc NCoRO), CaRH, SN, CW, SW (exc s ChI), Wrn, SNE, n DMtns; WA, OR, ID, NV, AZ. [*P. r.* var. *austrolitoralis* Munz; *P. r.* var. *eremophila* (Greene) J.F. Macbr.; *P. r.* var. *latifolia* (Torr.) Cronquist; *P. r.* var. *montereyensis* Munz; *P. r.* var. *subglabra* M. Peck; *P. r.* var. *suffrutescens* Parry; *P. r.* var. *valida* M. Peck] Apr–Oct

P. rattanii A. Gray Ann 15–100 cm. **ST:** gen erect, simple to branched at base, sparsely white- to clear-stiff-hairy; bristles slightly bulb-based. **LF:** 10–75 mm; blade > petiole, gen ovate, irregularly toothed to slightly lobed. **FL:** calyx lobes 2–4 mm, 3–7 mm in fr, unequal, oblanceolate to obovate, short-hairy, base strongly ciliate; corolla deciduous in fr, 3–5 mm, narrowly bell-shaped, white to blue; stamens 2–3 mm, puberulent; style 2–3 mm, glabrous. **FR:** 2–3 mm, spheric, puberulent. **SEED:** gen 2, 1.5–2 mm, pitted. *n*=11. Shaded crevices, steep slopes; < 1400 m. KR, NCoR, s CaR, SnFrB, SCoRO, n SCoRI. May–Jul

P. rotundifolia S. Watson Ann 4–28 cm. **ST:** decumbent to erect, few–many-branched, short-stiff-hairy, gen glandular. **LF:** 10–40 mm; blade gen < petiole, ± round, dentate to weakly lobed, lobes obtuse. **FL:** calyx lobes 2–4 mm, 4–6 mm in fr, gen < 1 mm wide, short-hairy; corolla deciduous, 3–6 mm, narrowly bell-shaped, tube pale yellow, lobes white to violet; stamens 2–3 mm, glabrous; style 1–2 mm, sparsely short-hairy. **FR:** 3.5–4.5 mm, oblong, puberulent. **SEED:** 50–100, ± 0.5 mm, pitted. *n*=12. Rocky slopes, crevices, ledges, creosote-bush scrub, pinyon/juniper woodland; < 2500 m. W&I, D; to sw UT, AZ. Apr–Jun

P. saxicola A. Gray Ann 5–15 cm. **ST:** ascending to erect, gen many-branched, short-stiff-hairy, glandular-puberulent. **LF:** 3–10 mm; blade gen = petiole, narrowly oblanceolate to ovate, entire. **FL:** calyx lobes 3–4 mm, 5–7 mm in fr, subequal, linear to narrowly oblanceolate, short-hairy; corolla ± persistent in fr, 3–4 mm, narrowly bell-shaped, tube white, lobes blue to violet; stamens 1–2 mm, ± glabrous; style 1–2 mm. **FR:** 2–3 mm, ovoid, short-stiff-hairy. **SEED:** 20–50, ± 0.5 mm, pitted. Limestone slopes, woodland; 1000–2300 m. SNE, n DMoj; s NV, nw AZ. Apr–Sep

P. sericea (Graham) A. Gray var. ***ciliosa*** Rydb. (p. 507) BLUE ALPINE PHACELIA Per 10–60 cm. **ST:** ascending to erect, not glandular; hairs silky, silvery, appressed. **LF:** blade 20–65 mm, ± = petiole, lanceolate to oblong, lobed to dissected; segments entire or toothed. **FL:** calyx lobes 2–7 mm, not enlarged in fr, linear to oblong; corolla persistent in fr, 5–8 mm, urn- to bell-shaped, dark blue to purple; stamens 10–15 mm, exserted, glabrous to sparsely hairy at base; style 6–13 mm, exserted. **FR:** 4–6 mm, ovoid, rough-hairy. **SEED:** 8–18, 1–2 mm, pitted in vertical rows. *n*=11. Ridges, talus slopes; 2100–2700 m. n KR (China Peak), Wrn; to ID, CO. Jun–Aug ★

P. stebbinsii Constance & Heckard (p. 507) STEBBINS' PHACELIA Ann 10–40 cm. **ST:** erect, 0–few-branched, puberulent, glandular or not. **LF:** 7–50 mm, proximal opposite; blade gen > petiole, elliptic to lanceolate, larger gen 2–6-toothed or -lobed, smaller gen entire. **FL:** calyx lobes 1–3 mm, 3–4 mm in fr, narrowly oblanceolate, short-hairy, glandular; corolla persistent in fr, 4–5 mm, bell-shaped, white to pale blue; stamens 6–7 mm, glabrous; ovules few, style 4–5 mm, glandular-puberulent. **FR:** 3–4 mm, ovoid, ± compressed, densely short-hairy. **SEED:** gen 1, 1.5–2 mm, pitted. *n*=8. Gravelly soils, meadows, conifer forest; 900–2100 m. n SN (El Dorado Co.). Jun–Jul ★

P. stellaris Brand BRAND'S STAR PHACELIA Ann 6–25 cm. **ST:** spreading to ascending, gen branched at base, puberulent. **LF:** ± basal, 5–70 mm; blade gen > petiole, oblanceolate to ovate, deeply lobed to compound, segments rounded to obtuse. **INFL:** pedicel 2–8 mm, curved. **FL:** calyx lobes 2–5 mm, 4–8 mm in fr, oblanceolate, short-stiff-hairy, glandular; corolla deciduous in fr, 3–5(7) mm, bell-

shaped, light blue to ± purple; stamens 2–4 mm, gen glabrous; style 1–3 mm, puberulent. **FR:** 4.5–6 mm, ovoid, sparsely short-hairy. **SEED:** 8–20, 0.5–1 mm, pitted. *n*=11. Open areas, coastal-sage scrub; < 400 m. SCo; Baja CA. Some pls from near se WTR foothills similar to *P. douglasii*. Mar–May ★

P. suaveolens Greene (p. 507) Ann 5–40 cm. **ST:** ascending to erect, simple to branched at base, short-hairy, glandular-puberulent. **LF:** 10–75 mm; blade gen > petiole, widely elliptic to ovate, toothed to slightly lobed. **FL:** calyx lobes 4–5 mm, 6–8 mm in fr, oblanceolate, short-hairy, glandular; corolla deciduous, 7–11 mm, narrowly bell-shaped, tube yellow, lobes lavender to purple; stamens 3–6 mm, glabrous; style 3–4 mm, short-glandular-hairy. **FR:** 3–5 mm, ovoid, short-hairy. **SEED:** 10–16, 1–1.5 mm, pitted. *n*=12. Uncommon. Open burns, slopes, chaparral, closed-cone-pine forest; 200–1700 m. NCoRI, c SNF (Ione), SnFrB, SCoRI. May–Jun

P. tanacetifolia Benth. (p. 507) Ann 15–100 cm. **ST:** erect, 0–few-branched, ± short-stiff-hairy, glandular-puberulent. **LF:** 20–200 mm; blade > petiole, ± oblong to ovate, gen compound, lflets toothed to lobed. **INFL:** dense, gen 2–4-branched. **FL:** calyx lobes 4–6 mm, 6–8 mm in fr, ± linear, densely long-hairy; corolla ± persistent in fr, 6–9 mm, widely bell-shaped, blue; stamens 9–15 mm, glabrous; style 11–15 mm, glabrous. **FR:** 3–4 mm, ± ovoid, glabrous proximally; tip puberulent to short-hairy. **SEED:** 1–2(4), 2–3 mm, wrinkled, pitted. *n*=11. Sandy to gravelly slopes, open areas; < 2500 m. s NCoRO, NCoRI, CaRF, SN (exc n SN), ScV (Sutter Buttes), SnJV, e SnFrB, SCoR, SW (exc ChI), DMoj; s NV, AZ. Mar–May

P. tetramera J.T. Howell Ann 2–15 cm. **ST:** spreading to ascending, branched throughout, sparsely short-hairy, ± glandular. **LF:** 5–30 mm; blade gen = petiole, oblong to ovate, entire to few-toothed. **FL:** calyx lobes gen 4, 1.5–3 mm, 3.5–4.5 mm in fr, narrowly oblanceolate, puberulent to short-hairy; corolla persistent in fr, 1.3–2 mm, bell-shaped, ± white, glabrous inside, lobes gen 4; stamens 1–2 mm, glabrous; style < 0.5 mm, glabrous. **FR:** 2.5–4 mm, oblong to ± spheric, puberulent. **SEED:** 5–12, ± 1 mm; cross-furrows 6–9. *n*=11. Alkaline flats, washes, meadows; 1500–2400 m. SNE; to e OR, UT. May–Jun

P. thermalis Greene (p. 507) Ann 5–45 cm. **ST:** ascending to erect, gen branched at base, glandular-hairy, sparsely stiff-hairy. **LF:** 12–80 mm; blade > petiole, oblong to ovate, deeply lobed to compound, segments toothed. **FL:** calyx lobes 3–4 mm, 5–7 mm in fr, lanceolate, entire to crenate, puberulent, margin stiffly ciliate, midrib, 2° veins raised; corolla deciduous, 3–4 mm, bell-shaped, white to pale blue; stamens 1–3 mm, glabrous; style ± 2 mm, glabrous. **FR:** 3–4 mm, spheric, puberulent. **SEED:** 2–4, 2–2.5 mm, pitted. *n*=11. Open clay flats; 1000–1700 m. e CaR, MP; to se OR, ID. May–Aug

P. umbrosa Greene Ann 15–45 cm. **ST:** decumbent to erect, weak, gen branched, sparsely stiff-hairy, glandular distally. **LF:** 25–90 mm; blade > petiole, ± ovate, gen compound, segments crenate to lobed. **INFL:** ± open below, dense distally, gen unbranched. **FL:** calyx lobes 3–5 mm, 5–7 mm in fr, unequal, linear to oblanceolate, long-hairy; corolla late-deciduous, 3–6 mm, narrowly bell-shaped, pale blue to lavender; stamens 4–10 mm, glabrous; style 7–12 mm, glabrous. **FR:** 2.5–3.5 mm, elliptic, gen sparsely short-hairy. **SEED:** 2–4, 1.5–2 mm, pitted. Uncommon. Chaparral, oak/pine woodland; 1000–1600 m. PR; Baja CA. Relationship to *P. distans* needs further study. Jun–Jul

P. vallicola Brand (p. 507) MARIPOSA PHACELIA Ann 6–27 cm. **ST:** erect, simple to branched proximal to middle, puberulent, sparsely stiff-hairy. **LF:** 10–35 mm; proximal-most opposite; blade > petiole, lanceolate to ovate, entire, veins sunken. **FL:** calyx lobes 3–5 mm, 7–10 mm in fr, unequal, oblong to oblanceolate, densely ciliate, sparsely glandular; corolla deciduous, 4–6 mm, bell-shaped, tube lavender, lobes gen purple; stamens 2–3 mm, papillate; style 2–4 mm, short-hairy. **FR:** 4–6 mm, ovoid, beaked, short-hairy. **SEED:** 8–14, 1.5–2 mm, pitted. *n*=11. Open, gravelly to rocky soils, chaparral, oak/pine woodland, conifer forest; 600–2400 m. s CaR, n&c SNF, SNH. Mar–Jun

P. vallis-mortae J.W. Voss (p. 507) Ann 20–60 cm. **ST:** ascending to erect, simple to branched, puberulent and sparsely stiff-reflexed-hairy. **LF:** 15–80 mm; blade > petiole, ± oblong, compound,

lflets toothed or slightly lobed. **FL:** calyx lobes 4–6 mm, 7–10 mm in fr, ± curved, not enclosing fr, linear to narrowly elliptic, long-hairy; corolla deciduous, 8–15 mm, funnel- to bell-shaped, lavender to violet; stamens 6–12 mm, glabrous; style 5–12 mm, short-glandular-hairy. **FR:** 3–4 mm, ovoid, ± puberulent, stiff hairs few, gen < 1 mm, not bulb-based. **SEED:** gen 4, 2.5–3 mm, pitted. *n*=11. Sandy to rocky soils, scrub; 90–2700 m. w SnJV, SNE, e DMoj, n DSon; to sw UT, nw AZ. If recognized taxonomically, pls in w SnJV with dark-veined corollas assignable to *P. vallis-mortae* var. *heliophila* (J.F. Macbr.) J.W. Voss; relationships to *P. cryptantha* or *P. cicutaria* need study. May–Jun

P. viscida (Lindl.) Torr.   Ann. **ST:** erect, 0–few branched, short-stiff-hairy, glandular. **LF:** 10–100 mm; blade > petiole, elliptic to broadly ovate, irregularly toothed, some basal lvs deeply lobed. **INFL:** cyme loose, pedicel 2–10 mm. **FL:** calyx lobes 5–8 mm, 5–12 mm in fr, gen oblong, short-hairy, glandular; corolla decidu-ous; stamens 8–16 mm, purple, short-hairy; style ± = corolla length. **FR:** 5–12 mm, ovoid, beaked, puberulent; tip short-glandular-hairy. **SEED:** 40–200, ≤ 1 mm, pitted. *n*=11. Glandular secretions stain clothing, may cause dermatitis.

var. ***albiflora*** (Nutt.) A. Gray   Pl 10–70 cm. **FL:** corolla limb 10–15 mm diam, rotate to bell-shaped, white; style incl, 3–10 mm. Exposed, ± moist areas, coastal-sage scrub, chaparral; < 1600 m. SCoRO, n SCo, WTR. Abundant in disturbed areas and after fires. Mar–Jun

var. ***viscida***   Pl 20–100 cm. **FL:** corolla limb 15–25 mm diam, rotate, tube, throat white or purple, lobes pale lavender to bright blue; style 5–16 mm. Exposed, ± moist areas, coastal-sage scrub, chaparral; < 1600 m. s CCo, SCoRO, SCo, ChI (Santa Catalina, Anacapa, Santa Cruz, Santa Rosa islands), WTR; Baja CA. Abundant in disturbed areas and after fires. Mar–Jun

PHOLISMA

George Yatskievych

Per, non-green, parasitic. **ST:** < 1.5 m. **LF:** 5–25 mm, linear to triangular, glandular. **FL:** 7–10 mm; ovary chambers 10–32. **FR:** capsule, ± circumscissile below middle. 3 spp.: s CA, w AZ, nw Mex. (Greek: scale, from scaly st)

1. Infl spike or panicle, dense; calyx lobe glandular hairs < 0.5 mm . . . . *P. arenarium*
1′ Infl head, concave; calyx lobe glandular hairs < 1.5 mm . . . . *P. sonorae*

P. arenarium Hook. (p. 507) **ST:** 3–8 dm, 1–2 cm diam. **FL:** calyx lobes linear to spoon-shaped; corolla lavender to blue-purple, abaxially puberulent, margin white; ovary chambers 10–20. 2*n*=36. Uncommon. Sandy soil, coastal dunes, chaparral, desert; < 1900 m. CCo, SCo, PR, D; w AZ, nw Mex. Parasitic on *Croton, Eriodictyon,* various shrubby Asteraceae. Apr–Jul, Oct

P. sonorae (A. Gray) Yatsk.   SAND FOOD   **ST:** 5–15 dm, 0.5–2 cm diam. **FL:** calyx lobes linear; corolla pink to purple, abaxially glabrous, margin white; ovary chambers 12–32. 2*n*=36. Dunes, sandy areas; < 200 m. DSon (se Imperial Co.); w AZ, nw Mex. Parasitic on *Eriogonum, Tiquilia, Ambrosia, Pluchea.* Threatened by off-road vehicles. Apr–May ★

PHOLISTOMA

Robert Patterson & Richard R. Halse

Ann, fleshy. **ST:** prostrate or reclined, many-branched, brittle; angles ± glabrous, bristly, or gen with hooked prickles. **LF:** cau-line, lower opposite, upper alternate; petioles gen winged, clasping; blade pinnate-lobed, uppermost reduced, short-petioled, gen deltate, 3-lobed, with small, sharp bristles both surfaces. **INFL:** terminal, axillary, opposite lvs, or fls 1; pedicels present. **FL:** calyx lobes hairy, bristly-ciliate; corolla rotate, lobes ± = tube, gen hairy; stamens incl, equal, equally attached; ovary bristly-hairy, chamber appearing 1, style 1, 2-lobed in distal 1/2. **FR:** capsule, spheric, stout-bristly-hairy. **SEED:** 1–8, spheric, brown, pitted or honeycombed. 3 spp.: CA, AZ, Baja CA. (Greek: scale mouth)

1. Calyx sinus appendages 0; calyx not enclosing mature fr . . . . *P. membranaceum*
1′ Calyx sinus appendages present; calyx enclosing mature fr
  2. Petiole narrow-winged; style 1–3 mm; seeds 1–2 mm . . . . *P. racemosum*
  2′ Petiole wide-winged; style 4–8 mm; seeds 2–3 mm . . . . *P. auritum*
    3. Corolla < 10 mm wide . . . . var. *arizonicum*
    3′ Corolla 10–30 mm wide . . . . var. *auritum*

P. auritum (Lindl.) Lilja   **ST:** 2–12 dm. **LF:** petiole wide-winged, clasping; lower lvs 4–16 cm, 1–8 cm wide, blade oblong to lance-ovate, base cordate, tip acuminate, lobes 5–13, oblong or lanceolate, obtuse or acute, entire or 1–5-toothed. **INFL:** fls 1 or 2–6 in cymes; pedicels 1–3 cm. **FL:** calyx lobes 3–9 mm, ± lanceolate, sinus appendages 1–4 mm; corolla 3–15 mm, 5–30 mm wide, blue to pur-ple with darker marks in throat; style 4–8 mm. **FR:** 5–10 mm wide, enclosed in calyx. **SEED:** 1–4.

var. ***arizonicum*** (M.E. Jones) Constance   ARIZONA PHOLISTOMA **LF:** lobes gen 5–11, obtuse. **FL:** corolla 3–7 mm, = calyx, < 10 mm wide. Desert scrub; 300–700 m. DSon; AZ. Mar–Apr ★

var. ***auritum*** (p. 507)   FIESTA FLOWER   **LF:** lobes 7–13, obtuse or acute. **FL:** corolla 7–15 mm, gen > calyx, 10–30 mm wide. *n*=9. Ocean bluffs, talus slopes, woodland, streambanks, canyons; < 1900 m. NCoRI, c SNF, s SN, Teh, SnJV, CW, SW. Mar–Jun

P. membranaceum (Benth.) Constance (p. 507) **ST:** 5–90 cm, gen glaucous. **LF:** petiole narrow-winged, not clasping; lower lvs 2–13 cm, 1–8 cm wide, blade oblong to ovate, base cordate or trun-cate, tip obtuse, lobes 5–11, oblong, obtuse, entire or 1-toothed. **INFL:** fls gen 2–10 in cymes; pedicel 5–20 mm. **FL:** calyx lobes 1–3 mm, oblong, sinus appendages 0; corolla 3–6 mm, < 10 mm wide, white, lobe gen with purple spot; style 1–3 mm. **FR:** 2–4 mm wide, not enclosed in calyx. **SEED:** 1–2. *n*=9. Beaches, bluffs, ravines, wooded slopes, desert washes; 40–1400 m. c&s SNF, Teh, SnJV, CW, SW, D; Baja CA. Feb–May

P. racemosum (A. Gray) Constance (p. 507)   **ST:** 2–6 dm. **LF:** petiole narrow-winged, clasping; lower lvs 3–10 cm, 2–8 cm wide, blade ovate to ± deltate, base cordate or truncate, tip obtuse, lobes 5–9, obtuse or acute, entire or 3–5-toothed. **INFL:** fls gen 2–6 gen in cymes; pedicel 5–30 mm. **FL:** calyx lobes 2–4 mm, ± lanceolate, sinus appendages < 2 mm; corolla 4–10 mm, 5–15 mm wide, white to blue; style 1–3 mm. **FR:** 5–8 mm wide, enclosed in calyx. **SEED:** 4–8. *n*=9. Moist shaded, areas, hillsides, ravines, ocean bluffs, coastal scrub; < 500 m. SCo, ChI, PR; Baja CA. Feb–May

PLAGIOBOTHRYS   POPCORNFLOWER

Ann (per), gen strigose to spreading-hairy; fibrous- to taprooted, staining red dye present or not. **ST:** branched at base or above, < 5 dm. **LF:** cauline or basal and cauline, 0.5–10 cm, gen smaller tipward, linear to oblanceolate. **INFL:** raceme- or spike-like cymes, coiled in bud, gen elongate in fr; bracts 0–many. **FL:** calyx lobes fused below middle, 2–10 mm in fr; corolla rotate to funnel-shaped or cylindric, white or white with yellow area, tube gen ± yellow inside, limb 1–12 mm diam, appendages prominent to minute, white to yellow. **FR:** nutlets gen 4, ± ovate (triangular to ± lanceolate), rarely on narrow stalk or short peg, variously roughened, abaxially gen with central ridge, lateral ridges, cross-ribs, gen tubercled, occ prickly or bristly; adaxially keeled above attachment scar, scar on side gen near middle to base, sometimes on bottom or oblique (on angle between side and bottom), gen raised. ± 65 spp.: temp w N.Am, w S.Am, ne Asia, Australia. (Greek: sideways pit, from position of nutlet attachment scar) Nutlet characters in key gen best for 3 nutlets farthest from st; yellow on corolla changes to white after pollination.

1. Nutlet scar on top of ± 0.5 mm angled narrow stalk . . . .*P. collinus*
  2. Pl cespitose; st 0.2–0.8 dm, erect; infl ± among lvs. . . . . var. *ursinus*
  2′ Pl not cespitose; st 1–4 dm, ± prostrate; infl exceeding lvs
    3. Corolla limb 4–8 mm diam; st hairs gen ± fine, soft . . . . var. *californicus*
    3′ Corolla limb 1–2 mm diam; st hairs gen coarse, rough
      4. Nutlets 1.5–2 mm; lf lanceolate, 3–5 mm wide. . . . . var. *fulvescens*
      4′ Nutlets 1–1.5 mm; lf linear, 2–2.5 mm wide . . . . var. *gracilis*
1′ Nutlet scar ± sessile, not on narrow stalk, or on short, ± basal peg
  5. Lower cauline lvs alternate, occ crowded, occ ± opposite, upper alternate; basal rosette present or 0; pl taprooted
    6. Nutlet scar long, narrow, elevated along ridge
      7. Corolla limb 1–3 mm diam, appendages white, minute; nutlet cobblestone-like, tubercles 0 . . . . *P. jonesii*
      7′ Corolla limb 4–10 mm diam, appendages yellow, prominent; nutlet irregularly ribbed, tubercled . . . . **P. kingii**
        8. Infl coiled in fr; sts 0.5–1.5 dm, ± equally tall . . . . var. *harknessii*
        8′ Infl ± elongate in fr; sts 1–4 dm, central ± taller. . . . . var. *kingii*
    6′ Nutlet scar short, ± round (elliptic), sunken below base of ridge
      9. Infl coiled in fr; nutlet scar gen above middle; basal rosette not persistent in fr . . . . *P. hispidus*
      9′ Infl gen elongate in fr; nutlet scar at or ± below middle; basal rosette gen present in fr
        10. Calyx gen circumscissile, lobes gen curved over fr; st, abaxial lf midveins, margins ± red, staining
          11. Bracts 0–few, near base of infl; corolla limb (3)4–9 mm diam; nutlet ± arched. . . . . *P. nothofulvus*
          11′ Bracts few–many, ± throughout infl; corolla limb 2–3 mm diam; nutlet gen arched or folded
            12. Nutlets gen 1–2, strongly attached; calyx ± 3 mm, circumscissile in fr, lobes curved over fr, dense-spreading-hairy . . . . *P. arizonicus*
            12′ Nutlets gen (2)3–4, weakly attached; calyx 3–5 mm, gen not circumscissile in fr, lobes erect to occ curved over fr, ± appressed-hairy . . . . ²*P. canescens* var. *catalinensis*
        10′ Calyx gen not circumscissile, lobes gen erect to ± spreading; st, abaxial lf midveins, margins green or ± red, staining or not
          13. Nutlet scar deep-concave, pit-like
            14. Bracts 0–few, near base of infl; basal rosette persistent; st, abaxial lf midveins, green, not staining; calyx hairs brown . . . . *P. fulvus* var. *campestris*
            14′ Bracts many, throughout infl; basal rosette short-lived, withered by fl; st, abaxial lf midveins, ± red, staining; calyx hairs tawny to ± white . . . . *P. infectivus*
          13′ Nutlet scar convex (to shallow-concave)
            15. Nutlet interspaces between cross-ribs gen wide, flat. . . . . *P. canescens*
              16. Lower st, abaxial lf midveins ± green, not staining. . . . . var. *canescens*
              16′ Lower st, abaxial lf midveins ± red, staining. . . . . ²var. *catalinensis*
            15′ Nutlet interspaces between cross-ribs gen narrow, groove-like, sometimes obscure
              17. Nutlet ± cross-shaped
                18. Infl ± not coiled toward tip, not elongate in fr; calyx 5–7 mm; nutlet 2–3 mm; fls gen < 10 per branch . . . . *P. shastensis*
                18′ Infl ± strongly coiled toward tip, elongate in fr; calyx 3–4(5) mm; nutlet 1–2 mm; fls gen > 10 per branch . . . . *P. tenellus*
              17′ Nutlet ± ovate
                19. Calyx hairs minutely hooked at tip; nutlet 1–1.3 mm. . . . . *P. uncinatus*
                19′ Calyx hairs not hooked at tip; nutlet 1.5–2.2 mm
                  20. Nutlet irregularly papillate, tubercled, cross-ribs 0; se SnFrB (Mount Hamilton Range). . . . . . *P. verrucosus*
                  20′ Nutlet not papillate or regularly papillate, cross-ribs prominent; c&s SN, SNH, Teh, s SCoRO, SnBr. . . . . *P. torreyi*
                    21. Nutlet abaxially with ± rows of round, translucent papillae, cross-ribs rounded; nutlet strongly arched, base pointed. . . . . var. *perplexans*
                    21′ Nutlet abaxially ± without papillae, cross-ribs flat; nutlet ± not arched, base rounded
                      22. St decumbent to ascending; cauline lvs, bracts many, crowded, ovate to elliptic . . . . var. *diffusus*
                      22′ St ± erect; cauline lvs, bracts ± few, scattered, ± oblong . . . . var. *torreyi*

5′ Lower cauline lvs opposite, upper gen alternate; basal rosette 0; pl fibrous-rooted, becoming taprooted
  23. Per; st stout; st, lf long-soft-spreading-hairy . . . . ***P. mollis***
    24. Nutlet gray, cross-ribs irregular; se CaRH, ne SNH, MP (exc Wrn) . . . . var. ***mollis***
    24′ Nutlet brown, cross-ribs netted; n SnFrB (near Petaluma, s Sonoma Co.) . . . . var. ***vestitus***
  23′ Ann; st slender; st, lf hairs not simultaneously long, soft, spreading
    25. Fls gen present near st base; calyx lobe midvein thickened; st prostrate
      26. Nutlet lance-ovate, dull, attached ± weakly; scar < 20% fr length, bordered by gen low ridge, funnel-
        like collar gen 0; lower lvs 3–8 cm . . . . ***P. humistratus***
      26′ Nutlet wide-ovate to deltate, shiny, attached strongly; scar 20–35% fr length, gen bordered by
        funnel-like collar; lower lvs 1–2 cm . . . . ***P. scriptus***
    25′ Fls gen 0 near st base; calyx lobe midvein occ thickened; st prostrate to erect
      27. Nutlet scar 25–60% fr length, ± deep-concave
        28. Nutlet abaxially with large prickles, barb-tipped or not, or rarely ± without
          29. Nutlet abaxial cross-ribs prominent, creating irregular net-like pattern . . . . ***P. acanthocarpus***
          29′ Nutlet abaxial cross-ribs 0 to obscure, net-like pattern 0
            30. Nutlet shiny; prickles (0)few, on ridges abaxially . . . . ***P. austiniae***
            30′ Nutlet ± dull; prickles many, throughout abaxially
              31. Nutlet 1.5–3 mm, prickles long, slender, not conic, ± barb-tipped, ± 0 adaxially; scar triangular . . . ***P. greenei***
              31′ Nutlet 1–1.5 mm, prickles short, stout, conic, blunt-tipped, present adaxially; scar narrow-
                triangular . . . . ***P. hystriculus***
        28′ Nutlet abaxially gen without large prickles, without barb-tipped projections (minute-bristled)
          32. Nutlet scar gen linear, wider near base or not; st, lvs ± glabrous . . . . ***P. strictus***
          32′ Nutlet scar ± pear-shaped or ± triangular; st, lvs strigose
            33. Nutlet scar < 25% fr length, concave, flanked by indistinct granular roughness . . . . [2]***P. trachycarpus***
            33′ Nutlet scar 40–60% fr length, pit-like, flanked by prominent, crowded wrinkles
              34. Calyx lobes ± not longer in fr; st prostrate to decumbent; lower lvs 0.5–2.5 cm; corolla limb
                1–1.5 mm diam, appendages white; nutlet ± 1 mm, cross-ribs wide, crowded, scar ± pear-shaped
                . . . . ***P. distantiflorus***
              34′ Calyx lobes longer in fr; st ascending to erect; lower lvs 4–8 cm; corolla limb > 2 mm diam,
                appendages yellow; nutlet 1.5–2 mm, cross-ribs narrow, not crowded, scar ± triangular . . . . ***P. glyptocarpus***
                35. Corolla limb 5–9 mm diam . . . . var. ***glyptocarpus***
                35′ Corolla limb 2–3 mm diam . . . . var. ***modestus***
      27′ Nutlet scar < 20% fr length, gen ± flat or ± concave (deep-concave in *Plagiobothrys diffusus*)
        36. Nutlet scar basal, gen on short peg; calyx lobe midvein ± stout
          37. St prostrate to decumbent; calyx lobes gen bent, ± same direction, gen upward; nutlet lanceolate
          . . . . ***P. leptocladus***
          37′ St ascending to erect; calyx lobes ± bent or straight, erect to ± spreading; nutlet lance-ovate
            38. Calyx base fleshy, lobes oblong to spoon-shaped; pedicel thick, hollow; nutlet adaxially fine- to
              obscure-granular to smooth, margins with conspicuous, entire border . . . . ***P. glaber***
            38′ Calyx base not fleshy, lobes linear to lanceolate; pedicel slender, gen solid; nutlet adaxially
              coarse-grainy, margins with obscure, not entire border . . . . ***P. stipitatus***
              39. Corolla limb 2–4 mm diam . . . . var. ***micranthus***
              39′ Corolla limb 5–12 mm diam . . . . var. ***stipitatus***
        36′ Nutlet scar near base or oblique, not on short peg; calyx lobe midvein gen slender
          40. Nutlet scar ± linear, slit- or ridge-like
            41. Nutlet ± not tubercled, shiny, 1.8–2.5 mm, abaxial cross-ribs 0; corolla limb 2–4 mm diam . . . . ***P. lithocaryus***
            41′ Nutlet ± tubercled, dull, 0.8–2 mm, abaxial cross-ribs few–many; corolla limb gen ≤ 2 mm or
             > 5 mm diam
              42. Corolla limb 1.5–2 mm diam; pedicel gen 0–1 mm; nutlet abaxially gen ± flat, interspaces
                narrower than or as wide as cross-ribs . . . . ***P. undulatus***
              42′ Corolla limb 5–10 mm diam; pedicel gen 2–15(30) mm; nutlet abaxially convex, interspaces
                wider than cross-ribs . . . . ***P. chorisianus***
                43. Lvs of lower pair gen fused at base, ± sheathing st; st erect to decumbent, branched from upper
                  axils; corolla limb 6–10 mm diam; pedicels gen > calyx . . . . var. ***chorisianus***
                43′ Lvs of lower pair gen ± not fused at base, not or ± sheathing st; st prostrate, branched from
                  lower axils; corolla limb 5–7 mm diam; pedicels gen < calyx . . . . var. ***hickmanii***
        40′ Nutlet scar oblong to triangular or elliptic to ovate, not slit- or ridge-like
          44. St hairs spreading
            45. Bracts few, near infl base; pedicels 1–3+ mm, slender; calyx ± deciduous; nutlet cross-ribs
              prominent, scar near base, ± oblong to occ narrow-triangular, gen sunken in groove . . . . ***P. parishii***
            45′ Bracts ± many, throughout infl; pedicels 0–1 mm, stout; calyx persistent; nutlet cross-ribs low,
              scar gen oblique, ± ovate, not sunken in groove . . . . ***P. salsus***
          44′ St hairs appressed
            46. Nutlet scar gen oblique at base
             47. Nutlet adaxial ridge, or at least lower part, sunken in groove . . . . ***P. reticulatus***

48. Nutlet ± ovate; adaxial ridge in wide, well-defined groove; trough around scar gen wide, distinct; abaxial cross-ribs gen irregularly net-like, interspace tubercles gen sparse or small . . . . var. ***reticulatus***

48′ Nutlet ± oblong-ovate; adaxial ridge in narrow, poorly defined groove; trough around scar gen narrow, ± obscure; abaxial cross-ribs ± few, prominent, scattered, irregular, interspace tubercles gen dense or large . . . . var. ***rossianorum***

47′ Nutlet adaxial ridge not sunken

49. Bracts 0–few, near infl base; calyx base ± 4- ridged in fr; corolla limb 3–9 mm diam
. . . . [2]***P. tener*** var. ***tener***

49′ Bracts many, ± throughout infl (exc bracts below infl middle in *Plagiobothrys bracteatus*); calyx base not ridged in fr; corolla limb 1–3 mm diam

50. Nutlet wide-ovate, ± symmetric . . . . ***P. bracteatus***

50′ Nutlet ± oblong-ovate, asymmetric. . . . . ***P. cognatus***

46′ Nutlet scar gen above base

51. Nutlet scar narrow-elliptic to obovate; corolla limb 1–2 mm diam

52. Nutlet plump, ± symmetric, sharp-roughened, often also minute-bristled or scabrous, dull; adaxial ridge at ± middle, gen not folded over to 1 side . . . . ***P. hispidulus***

52′ Nutlet ± flat, asymmetric, smooth to rounded-roughened, rarely bristled or scabrous, shiny; adaxial ridge beyond middle, ± folded over to 1 side below

53. Nutlet wrinkled, tubercled; s NCoRI, CaRH, n&c SNH, sw ScV, GB. . . . . ***P. cusickii***

53′ Nutlet not wrinkled, not tubercled; SNE (Deep Springs Lake, ne Inyo Co.). . . . . ***P. nitens***

51′ Nutlet scar ± ovate to triangular; corolla limb 1.5–9 mm diam

54. Bracts many, ± throughout infl; calyx base not ridged in fr; corolla limb 1.5–5 mm diam

55. Nutlet scar, adjacent ridge in deep trough. . . . . ***P. diffusus***

55′ Nutlet scar, adjacent ridge not in trough . . . . [2]***P. trachycarpus***

54′ Bracts 0–few, near base of infl; calyx base ± 4-ridged in fr; corolla limb 3–9 mm diam. . . . . ***P. tener***

56. Pl fleshy; st, lf ± glabrous, glaucous; calyx appressed to st, lobes erect in fr; gen on serpentine; NCoRI (Lake, n Napa cos.). . . . . var. ***subglaber***

56′ Pl not fleshy; st, lf sparse-hairy, not glaucous; calyx ± spreading from st, lobes ascending to ± spreading in fr; gen not serpentine; NCoR, CaR, w MP. . . . . [2]var. ***tener***

P. acanthocarpus (Piper) I.M. Johnst. (p. 507) ADOBE POPCORNFLOWER Ann, strigose; ± taprooted. **ST**: spreading to erect, 1–4 dm. **LF**: cauline; lower 2–6 cm. **INFL**: bracts throughout; pedicels 1–2 mm. **FL**: calyx 3–6 mm; corolla limb 1–2.5 mm diam, appendages pale yellow. **FR**: nutlet 1–1.8 mm, ovate, dull to shiny; abaxial cross-ribs prominent, irregular-net-like, prickles (0)few–many, slender, barb-tipped, ± in columns, 0 adaxially; scar near base, ± triangular, 25–60% fr length, deep-concave. Vernal pools, moist clay soil; < 700 m. CaRF, GV, e SnFrB, SCoRO, SCo; Mex. Intergrades with *P. greenei* in n. Mar–May

P. arizonicus (A. Gray) A. Gray (p. 507) ARIZONA POPCORNFLOWER Ann, ± red, staining. **ST**: ascending to erect, 1–4 dm; ± red, esp lower; hairs rough, sharp, ± spreading. **LF**: basal in rosette, 1.5–5 cm; cauline alternate; abaxial lf midveins, margins dark red. **INFL**: bracts ± throughout. **FL**: calyx ± 3 mm, circumscissile, lobes curved over fr, dense-spreading-hairy; corolla limb 2–2.5 mm diam, appendages pale yellow. **FR**: nutlets gen 2, ± 2 mm, wide-ovate, arched, strongly attached; abaxial ridge, lateral ridges, cross-ribs narrow, interspaces wide; scar near middle, round. Common. Dry, coarse soils in deserts, scrub, woodland; < 2100 m. e NCoRI, s SN, Teh, SnJV, e SnFrB, SCoRI, SW, W&I, D; to NM, n Mex. Intergrades with *P. canescens* var. *c.*, *P. nothofulvus*. Feb–May

P. austiniae (Greene) I.M. Johnst. (p. 507) AUSTIN'S SPINY-NUT POPCORNFLOWER Ann, strigose. **ST**: decumbent to erect, 1–4 dm. **LF**: cauline, lower 1–5 cm. **INFL**: bracts near base. **FL**: calyx 5–8 mm; corolla limb 1–2.5 mm diam, appendages yellow. **FR**: nutlet 2–3 mm, ovate, acuminate, ± granular, shiny; abaxial ridge, lateral ridges prominent, cross-ribs 0 to obscure, prickles (0 or) on ridges, few, stout, barb-tipped, ± 0 adaxially; scar near base, ± triangular, deep-concave. Common in n. Vernal pools, wet areas in thin rocky, clay soils; < 500 m. CaRF, n&c SNH, e ScV, ne SnJV; sw OR. [*Plagiobothrys austinae*, orth. var.] Mar–May

P. bracteatus (Howell) I.M. Johnst. (p. 507) BRACTED POPCORNFLOWER Ann, sparse- to dense-strigose. **ST**: gen ascending, occ decumbent, gen 1–4 dm. **LF**: basal lanceolate, hairs bulbous-based; lower cauline 3–10 cm. **INFL**: bracts below middle. **FL**: calyx 2–4 mm, strigose; corolla limb 1–3 mm diam, appendages pale yellow. **FR**: nutlet 1–1.8 mm, ± flat, lance-ovate, ± symmetric, ± dull;

abaxial ridge short, near tip, lateral ridges ± arched, cross-ribs high, ± crowded, esp toward tip, interspaces narrow, tubercled, gen glabrous (bristly); margins with narrow border; adaxial ridge near middle, not folded to 1 side, not in trough; scar oblique, ± ovate to wide-triangular; ± sunken by surrounding ridge. Common. Vernal pools, wet places in grassland, coastal-sage scrub, chaparral; < 2000 m. NW, CaR, SN (exc s SNH), ScV, n SnJV, CW, SCo, PR, MP, w DMoj; sw OR, nw Mex. Apr–Jun

P. canescens Benth. (p. 507) Ann, gen ± green, rarely red, staining on sts, lvs. **ST**: prostrate to erect, 1–6 dm; canescent, rarely rough or bristly. **LF**: basal in ± rosette, 1.5–5 cm; cauline alternate. **INFL**: 1+, bracts ± throughout. **FL**: calyx 3–6 mm, gen not circumscissile in fr, lobes erect to spreading (circumscissile, lobes curved over fr), ± appressed-hairy; corolla limb 2–3 mm diam, appendages pale yellow. **FR**: nutlets (2)3–4, 2–2.5 mm, round-ovate, gen arched, gen weakly attached; abaxial ridge, lateral ridges, cross-ribs narrow; interspaces wide, flat; scar near middle, round.

var. ***canescens*** VALLEY POPCORNFLOWER Pl gen green. **ST**: prostrate to erect, st ± green, not staining. **LF**: midveins, margins green. **INFL**: bracts ± throughout. **FL**: calyx 4–6 mm, lobes spreading to erect. Common. Grassland, woodland, coastal scrub, desert scrub, roadsides; gen < 1400 m. CaRF, n SNH (Plumas Co.), SNF, Teh, GV, CW, SW, w DMoj. Intergrades with *P. arizonicus*, *P. nothofulvus*. Mar–Jun

var. ***catalinensis*** (A. Gray) Jeps. SANTA CATALINA POPCORNFLOWER Pl ± red, staining. **ST**: ± prostrate to ascending; lower red. **LF**: abaxial midveins, margins red. **INFL**: bracts throughout. **FL**: calyx 3–5 mm, variable on 1 pl, lobes erect to occ over fr. Uncommon. Grassland, coastal scrub, desert scrub, roadsides; gen < 1400 m. SnJV, SCoRI, ChI, DMoj. Possible hybrid between *P. canescens* var. *c.* and *P. arizonicus* stabilized on Santa Catalina Island. Apr–Jun

P. chorisianus (Cham.) I.M. Johnst. ARTIST'S POPCORNFLOWER Ann, sparse-short-strigose. **ST**: prostrate to erect, 1–4 dm. **LF**: cauline, lower 3–7 cm. **INFL**: bracts near base; pedicels gen 2–15(30) mm. **FL**: calyx ± 4 mm; corolla limb 5–10 mm diam, appendages yellow. **FR**: nutlet 0.8–1.5 mm, ovate, ± dull, brown; abaxial ridge short,

near tip, lateral ridges 0, cross-ribs few, irregular to ± 0, interspaces wider than cross-ribs, granular, tubercled or netted; adaxial ridge 60–75% fr length; scar near base, linear, in groove. May intergrade with *P. undulatus* in SnFrB. Vars. intergrade.

var. *chorisianus* (p. 507) CHORIS' POPCORNFLOWER **ST:** decumbent to erect, branched from upper axils. **LF:** lower pair gen fused at base, ± sheathing st. **INFL:** pedicel gen >> calyx. **FL:** corolla limb 6–10 mm diam. Grassy, moist places, ephemeral drainages, coastal scrub, chaparral; < 650 m. KR, NCoRO, n CCo, w SnFrB. Mar–Jun ★

var. *hickmanii* (Greene) I.M. Johnst. HICKMAN'S POPCORNFLOWER **ST:** prostrate, branched from lower axils. **LF:** lower pair gen ± free at base, not or ± sheathing st. **INFL:** pedicel gen < calyx. **FL:** corolla limb 5–7 mm diam. Moist places, vernal pools, sandy deposits over clay pans; < 200 m. c CCo, sw SnFrB, n SCoRO. Apr–Jul ★

P. cognatus (Greene) I.M. Johnst. (p. 507) COGNATE POPCORNFLOWER Ann, strigose. **ST:** prostrate to occ ascending, 0.5–2+ dm. **LF:** cauline, lower 2–7 cm. **INFL:** bracts below middle to ± throughout. **FL:** calyx 2–4 mm, strigose to spreading-hairy; corolla limb 1–2 mm diam. **FR:** nutlet 1.2–1.8 mm, ± flat, oblong-ovate, asymmetric, dull or shiny, brown; abaxial ridge low, short, near tip, lateral ridges obscure, cross-ribs few, low, scattered, interspaces wide, tubercled or papillate-dentate and scabrous-bristled; adaxial ridge beyond middle, gen folded to 1 side below; scar oblique, triangular, solid, gen ± flat, not sunken. Moist places in meadows, sagebrush flats, forests; 1050–2520 m. KR, NCoRI, CaR, n&c SN, GB; to e WA, Rocky Mtns, AZ. May–Aug

P. collinus (Phil.) I.M. Johnst. (p. 507) Ann, hairs ± spreading, fine or coarse. **ST:** prostrate to erect, gen 0.2–4 dm. **LF:** cauline, lower gen opposite, 1–4 cm, upper alternate. **INFL:** gen elongate, bracts at least below; pedicel to 1.5 mm. **FL:** calyx ± 3 mm; corolla limb 1–8 mm diam, appendages gen inconspicuous, ± white. **FR:** nutlet 1–2 mm, ovate; abaxial ridge, lateral ridges, cross-ribs sharp; scar near middle, at tip of ± 0.5 mm, narrow, angled stalk.

var. *californicus* (A. Gray) Higgins CALIFORNIA POPCORNFLOWER Pl not cespitose. **ST:** ± prostrate, 1–4 dm, hairs gen ± fine, soft. **LF:** 1–3 cm, 2–5 mm wide, oblanceolate. **INFL:** elongate, exceeding lvs. **FL:** corolla limb 4–8 mm diam, appendages yellow. **FR:** nutlet 1.5–2 mm. Openings in coastal scrub, occ chaparral; gen < 700(2300) m. s SnJV, CW, SW, nw edge DSon (San Gorgonio Pass); Mex. Uncommon to rare outside of SW. Feb–May

var. *fulvescens* (I.M. Johnst.) Higgins ROUGH-STEMMED POPCORNFLOWER Pl not cespitose. **ST:** ± prostrate, 1–4 dm, hairs gen coarse, rough. **LF:** 1–3 cm, 3–5 mm wide, lanceolate. **INFL:** elongate, exceeding lvs. **FL:** corolla limb ± 2 mm diam. **FR:** nutlet 1.5–2 mm. Dry ± gravelly places, openings in chaparral, coastal scrub, conifer forest, occ coastal scrub; gen 300–1800+ m. SCoR, SW, w DSon; AZ, Mex, Chile. Mar–Jun

var. *gracilis* (I.M. Johnst.) Higgins SAN DIEGO POPCORNFLOWER Pl not cespitose. **ST:** ± prostrate, 1–3 dm, hairs gen coarse. **LF:** 1–4 cm, 2–2.5 mm wide, linear. **INFL:** elongate, exceeding lvs. **FL:** corolla limb 1.5–2 mm diam. **FR:** nutlet 1–1.5 mm. Dry places, ± clay openings in coastal scrub, occ chaparral; 5–640 m. SCo (Riverside, San Diego cos.), ChI, sw PR (w San Diego Co.). Jan–May

var. *ursinus* (A. Gray) Higgins BEAR VALLEY POPCORNFLOWER Pl cespitose. **ST:** ± erect, 0.2–0.8 dm, hairs gen coarse. **LF:** 1–2.5 cm, 3–5 mm wide, oblanceolate. **INFL:** short, among lvs, ± dense. **FL:** corolla limb 1–2 mm diam. **FR:** nutlet 1.5–2 mm. Sandy or gravelly granite-based soils, open conifer forest; gen 1100–2400 m. SnBr, SnJt, s PR (Tecate Mtn); Mex?. Apr–Jun

P. cusickii (Greene) I.M. Johnst. (p. 507) CUSICK'S POPCORNFLOWER Ann, ± strigose. **ST:** erect to occ prostrate, 0.5–3+ dm, ± branched above base. **LF:** cauline, lower 3–10 cm. **INFL:** bracts below middle to throughout. **FL:** calyx 1.5–4 mm, strigose; corolla limb 1–2 mm diam. **FR:** nutlet 1–1.6 mm, ± flat, ovate, asymmetric, shiny; abaxial ridge short, near tip, lateral ridges obscure, cross-ribs ± evenly spaced, rounded, interspaces ± tubercled

(bristled or scabrous); adaxial ridge beyond middle, ± folded to 1 side below; scar near base, occ oblique, gen hollow, ± oblong, with thin, ± incurved margins, sunken. Wet, muddy areas, flats in sagebrush scrub, conifer forest; (60)450–2100 m. s NCoRI, CaRH, n&c SNH, sw ScV, GB; to e WA, ID, NV. May–Aug

P. diffusus (Greene) I.M. Johnst. SAN FRANCISCO POPCORNFLOWER Ann, strigose. **ST:** prostrate to ± ascending, 0.5–2.5 dm. **LF:** cauline, lower 5–10 cm. **INFL:** bracts throughout, spreading. **FL:** calyx 2.5–5 mm, lobes ± appressed-hairy; corolla limb (2)3–5 mm diam, appendages yellow. **FR:** nutlet 1–1.5 mm, ovate, ± flat, tan to gray, ± dull; abaxial ridge obscure, rounded, lateral ridges 0 to obscure, cross-ribs low, ± net-like or ± tubercled; interspaces wide, ± smooth to ± tubercled; lower part of adaxial ridge in trough; scar near base, elliptic to ovate, deep-concave, sunken. Moist places, seeps; 30–150 m. c CCo, w-c SnFrB. Apr–Jun ★

P. distantiflorus (Piper) I.M. Johnst. (p. 507) MADERA POPCORNFLOWER Ann, slender, strigose. **ST:** prostrate to decumbent, 1–3 dm. **LF:** cauline, lower 0.5–2.5 cm. **INFL:** bracts ± throughout. **FL:** calyx 2.5–3 mm; corolla limb 1–1.5 mm diam, appendages minute, white. **FR:** nutlet 1 mm, elongate-ovate, narrow near tip, ± arched, ± dull; abaxial cross-ribs crowded, wide, wrinkled, irregularly toothed, occ bristled, interspaces narrow; scar near base, 50–60% fr length, ± pear-shaped, pit-like, flanked by crowded wrinkles. Uncommon. Moist places in grassland, open woodland, seeps, springs; 50–1040 m. n SNF (Placer Co.), c SNF. Mar–May

P. fulvus (Hook. & Arn.) I.M. Johnst. var. *campestris* (Greene) I.M. Johnst. (p. 511) FIELD POPCORNFLOWER Ann, tomentose, some hairs spreading, long, rough. **ST:** erect, 3–6 dm. **LF:** basal in persistent rosette, 2–10 cm, oblong to wide-oblanceolate; cauline alternate, few, elliptic. **INFL:** bracts 0–few, near base; pedicels 1–3 mm. **FL:** calyx 5–6 mm, lobes dense-brown-hairy; midvein green, obscured by hair; corolla limb 3–4 mm diam, appendages yellow. **FR:** nutlets 2–4, 2.5–3.5 mm, triangular-ovate, ± narrowed below tip, ± flat, ± brown; abaxial papillae 0–few, cross-ribs 0–few, narrow, straight, interspaces wide; scar near middle, ± round, deep-concave. Grassland, open woodland, loamy, sandy, or gravelly soil; < 500 m. NCoR, CaRF, SNF, SnFrB, SCoRO, SCo; s OR. May form dense populations. Mar–May

P. glaber (A. Gray) I.M. Johnst. (p. 511) HAIRLESS POPCORNFLOWER Ann, sparse-short-strigose. **ST:** ascending to erect, 1–2 dm, thick, hollow. **LF:** cauline, lower 2–11 mm. **INFL:** dense, bracts below; pedicels 0.5–2 mm, thick, hollow. **FL:** calyx 3 mm, 8–10 mm in fr, narrow, basal 2–3 mm fused into fleshy cylinder, lobes appressed, ± spreading at tip, oblong to spoon-shaped, midveins thick; corolla limb ± 3 mm diam. **FR:** nutlet 1.8–2 mm, lanceolate to lance-ovate, shiny, dark brown, attached firmly; abaxial ridge narrow, obscure below middle, cross-ribs above middle, few, rounded, wrinkled, ± flat below, ± tubercled; margins with entire border; adaxial ridge V-shaped, fine- to obscure-granular to smooth; scar basal, gen on peg. Presumed extinct. Wet, saline, ± alkaline soils in valleys, coastal marshes; < 100 m. CCo, s SnFrB (Hollister). Apr–May ★

P. glyptocarpus (Piper) I.M. Johnst. (p. 511) SCULPTURED-NUT POPCORNFLOWER Ann, strigose. **ST:** ascending to erect, 1–5 dm. **LF:** cauline, lower 4–8 cm. **INFL:** bracts near base. **FL:** calyx 3–5 mm, lobes longer in fr; corolla limb 2–9 mm diam, appendages yellow. **FR:** nutlet 1.5–2 mm, ± ovate, ± flat, ± dull; abaxial ridge, cross-ribs narrow, high, ± toothed, occ bristled, interspaces wide, granular; scar near base, ± 40% fr length, ± triangular, pit-like, flanked by crowded wrinkles.

var. *glyptocarpus* **FL:** corolla limb 5–9 mm diam. Common. Moist places, vernal pools, grassland, woodland, forest; 40–550 m. NCoRI, CaR, n&c SNF, n ScV; sw OR. Mar–May

var. *modestus* I.M. Johnst. CEDAR RIDGE POPCORNFLOWER **FL:** corolla limb 2–3 mm diam. Seeps, moist openings in grassland, ponderosa-pine forest; 50–870 m. NCoRI (Lake Co.), n SNF (Butte, Nevada cos.). May be minor variant or hybrid, possibly involving *P. distantiflorus*. Apr–May ★

P. greenei (A. Gray) I.M. Johnst. (p. 511) GREENE'S SPINY-NUT POPCORNFLOWER Ann, strigose. **ST:** decumbent to erect, 1–4 dm.

LF: cauline, lower 1–5 cm. **INFL:** bracts near base. **FL:** calyx 5–8 mm; corolla limb 1–2.5 mm diam, appendages pale yellow. **FR:** nutlet 1.5–3 mm, ovate, ± narrowed near tip, granular, ± dull; abaxial cross-ribs 0, prickles many, slender, ± barb-tipped, ± 0 adaxially; scar near base, triangular, deep-concave. Wet sites, grassland, woodland; < 900 m. NCoR, CaRF, SNF, n SNH, ScV, n SnJV; sw OR. Mar–May

P. hispidulus (Greene) I.M. Johnst. (p. 511) HARSH POPCORNFLOWER Ann, strigose. **ST:** prostrate to erect, 0.5–4 dm. **LF:** cauline, lower 1–5 cm. **INFL:** bracts gen below middle. **FL:** calyx 1.5–3 mm, lobes ± spreading; corolla limb 1–2 mm diam. **FR:** nutlet 1.2–1.8 mm, ovate, ± symmetric, dull, brown to black; abaxial ridge short, near tip, papillate-dentate, lateral ridges 0, cross-ribs narrow, scattered, papillate-dentate, interspace margins wide, granular to papillate-dentate; adaxial ridge gen not folded; scar near base, ± narrow-elliptic, ± sunken. Common. Moist to drying meadows, flats, conifer forest openings; 1200–3400 m. KR, NCoR, CaRH, SN, TR, PR, GB; to WA, WY, NV. May–Aug

P. hispidus A. Gray (p. 511) CASCADE POPCORNFLOWER, BRISTLY POPCORNFLOWER Ann, sharp-hairy and sparse-short-tomentose; taproot ± red. **ST:** erect, 0.5–2 dm. **LF:** gen cauline, alternate, 1.5–4 cm, hairs blister-based. **INFL:** short, coiled in fr, few-fld, bracts below; pedicel 0. **FL:** calyx 2 mm; corolla limb < 1.5 mm diam, appendages minute, white. **FR:** nutlet 1–4, 1–2.4 mm, ovate, not ± flat, dull, granular-roughened, occ tubercled; abaxial ridge ± prominent, lateral ridges low, wide, cross-ribs 0; scar gen above middle, ± round. Dry places, gen in sandy, gravelly soil; 1200–2800 m. KR, CaR, SNH, GB; to c OR and w NV, e WA. Jun–Aug

P. humistratus (Greene) I.M. Johnst. (p. 511) LOW POPCORNFLOWER Ann, sparse-strigose. **ST:** prostrate, 1–4 dm. **LF:** cauline, lower 3–8 cm. **INFL:** bracts ± throughout; fls gen present near st base; lower pedicels stout, < 2 mm, recurved. **FL:** calyx 6–10 mm, lobes longer in fr, midvein ± thick; corolla limb 1–2 mm diam, appendages pale yellow. **FR:** nutlet 2–2.5 mm, lance-ovate, often bristled, dull, attached ± loosely; abaxial ridge obscure, cross-ribs 0–low, irregular, interspaces wide; scar near base, < 20% fr length, triangular to ovate, gen concave, bordered by low ridge. Vernal pools, wet places, grassland; gen < 200 m. CaRF, SNF, GV, SCoRO, SCo. Mar–May

P. hystriculus (Piper) I.M. Johnst. (p. 511) BEARDED POPCORNFLOWER Ann, strigose. **ST:** ± prostrate to erect, 1–4 dm. **LF:** cauline, lower 2–6 cm. **INFL:** bracts throughout; pedicels 1–2 mm. **FL:** calyx 3–6 mm; corolla limb 1–2.5 mm diam, appendages pale yellow. **FR:** nutlet 1–1.5 mm, ovate, granular, dull; abaxial cross-ribs ± 0, prickles many, stout, conic, blunt-tipped, also adaxial; scar near base, narrow-triangular, deep-concave. Wet grassland, vernal pool margins; < 50 m. sw ScV. Mar–May ★

P. infectivus I.M. Johnst. DYE POPCORNFLOWER Ann, ± red, staining. **ST:** erect, 1–5 dm, ± canescent, ± spreading-hairy. **LF:** basal few, 2–8 cm, linear, in ± short-lived rosette; cauline alternate, many, oblong; abaxial lf midveins red, margins red. **INFL:** bracts throughout, occ sparse; pedicels 1–2 mm. **FL:** calyx 5–7 mm, lobes sparse-hairy, midveins dark red; corolla limb 3–4 mm diam, occ stained pink, appendages yellow. **FR:** nutlets 2–4, 2–3 mm, triangular-ovate, ± narrowed below tip, ± arched, pale gray-green, irregularly coarse-granular; abaxial ridge narrowly ± white-winged, cross-ribs 0–few, irregular; scar near middle, narrow-elliptic, deep-concave, crown 0. Open grassland, friable clay to sandy soils; 80–830 m. NCoRI, w SnJV, e SnFrB, SCoR. Mar–May

P. jonesii A. Gray (p. 511) MOJAVE POPCORNFLOWER Ann, sharp-spreading-bristly. **ST:** ascending to erect, < 5 dm. **LF:** cauline, alternate, lower 2–10 cm, hairs ± blister-based. **INFL:** bracts near base; pedicel 0–1 mm. **FL:** calyx 4–8 mm; corolla funnel-shaped, limb 1–3 mm diam, appendages minute, white. **FR:** nutlets 3–4, 2–3 mm, triangular-ovate, cobblestone-like, tubercles 0; abaxial ridge, lateral ridges weak, cross-ribs 0; scar near middle, long, narrow, on ridge. Sandy, gravelly, rocky slopes, creosote-bush scrub, pinyon/juniper woodland; < 1800 m. s SNH, SNE, D; to s UT, w AZ, Mex. Apr–May

P. kingii (S. Watson) A. Gray (p. 511) GREAT BASIN POPCORNFLOWER Ann, hairs coarse, stiff, spreading, unequal; taproot ± red. **ST:** ascending to erect, 0.5–4 dm. **LF:** cauline, alternate,

lower dense, < 6 cm, upper sparse. **INFL:** bracts near base; pedicel 0–1.5 mm. **FL:** calyx 3–4 mm; corolla rotate to bell-shaped, limb 4–10 mm diam, appendages yellow. **FR:** nutlet 2–3 mm, ovate, arched or bent; abaxial ridge, lateral ridges, cross-ribs irregular, coarse; scar near middle, long, narrow, on ridge. Vars. intergrade in SNE.

var. ***harknessii*** (Greene) Jeps. NORTHERN GREAT BASIN POPCORNFLOWER **ST:** gen several, 0.5–1.5 dm, ascending, ± equal. **LF:** < 1 cm wide. **INFL:** ± short, coiled in fr. Dry, open slopes, sagebrush scrub, saltbush scrub, juniper woodland; 1200–2300 m. c SNH, GB; to se OR, w UT. May–Jul

var. ***kingii*** SOUTHERN GREAT BASIN POPCORNFLOWER **ST:** erect at pl center to ascending at pl sides, 1–4 dm, central > lateral. **LF:** < 2 cm wide. **INFL:** ± elongate in fr. Dry, open slopes, sagebrush scrub, saltbush scrub, juniper woodland; 1200–2300 m. SNE, n DMoj; NV. May–Jun

P. leptocladus (Greene) I.M. Johnst. (p. 511) ALKALI PLAGIOBOTHRYS Ann, sparse- to dense-strigose; taprooted or not. **ST:** prostrate to decumbent, 1–3+ dm. **LF:** cauline, lower 3–10 cm. **INFL:** open, bracts sparse; pedicels 0–1+ mm, thick (elongating). **FL:** calyx 2 mm, 5–8 mm in fr, narrow, gen strongly bent upward, lobe midveins ± thick; corolla limb 1–2 mm diam, appendages pale yellow. **FR:** nutlet 1–1.8 mm, lanceolate, dull, gray to brown, attached ± loosely; abaxial ridge only at tip, cross-ribs 0–few, above middle, ± flat below, tubercled, often bristly, margins rounded, entire; adaxial ridge shallow-V-shaped, granular; scar basal, gen on peg. Gen alkaline or saline clay soils, vernal pools, wet places; < 1400 m. CaRF, GV, SnFrB, SCoR, SW, GB, w DMoj; to AK, c Can, NE, CO, Mex. See also *P. bracteatus.* Mar–Jul

P. lithocaryus (A. Gray) I.M. Johnst. (p. 511) MAYACAMAS POPCORNFLOWER Ann, short-strigose. **ST:** erect, 1–3 dm, ± not branched at base. **LF:** cauline, lower 3–6 cm. **INFL:** bracts throughout; pedicels 1–4 mm. **FL:** calyx 4 mm; corolla limb 2–4 mm diam. **FR:** nutlet 1.8–2.5 mm, ovate, shiny, ± tan; abaxial ridge rounded, cross-ribs 0, tubercles 0; scar near base, linear, at end of slit. Presumed extinct. Moist sites; 300–450 m. s NCoRI (valleys near Mayacamas Mtns). Apr–May ★

P. mollis (A. Gray) I.M. Johnst. CREEPING POPCORNFLOWER, SOFT POPCORNFLOWER Per, dense-long-soft-spreading-hairy. **ST:** prostrate to decumbent, 1–3 dm, rooting at nodes. **LF:** cauline, lower 4–8 cm, upper opposite or alternate. **INFL:** bracts 0. **FL:** calyx 4–5 mm; corolla limb 5–10 mm diam. **FR:** nutlet ± 1.5 mm, ovate; abaxial cross-ribs irregular or netted; scar near middle, ovate or triangular.

var. ***mollis*** (p. 511) **FR:** nutlet gray; abaxial cross-ribs irregular, ± = interspaces. Uncommon. Moist, alkaline places in sagebrush scrub; 1200–1700 m. se CaRH, ne SNH, MP (exc Wrn); s-c OR, nw NV. May–Aug

var. ***vestitus*** (Greene) I.M. Johnst. (p. 511) PETALUMA POPCORNFLOWER **FR:** nutlet brown; abaxial cross-ribs netted, gen < interspaces. Presumed extinct. Wet sites in grassland; < 50 m. n SnFrB (near Petaluma, s Sonoma Co.). May–Jul ★

P. nitens (Greene) I.M. Johnst. SHINY-NUTLET POPCORNFLOWER Ann, glabrous to sparse-short-strigose. **ST:** ascending to erect, 1–1.5 dm, gen branched above base. **LF:** cauline, lower 2–5 cm. **INFL:** bracts ± throughout. **FL:** calyx 1.5–3 mm; corolla limb 1–2 mm diam. **FR:** nutlet 1–1.5 mm, ovate, ± flat, asymmetric, shiny, gray to tan; abaxial ridge ± 0, lateral ridges 0, cross-ribs ± 0; adaxial ridge to beyond middle, ± folded to 1 side below; scar gen near base, elongate, with thin, incurved margins, ± sunken. Wet areas below springs; 1510 m. SNE (Deep Springs Lake, ne Inyo Co.); NV, n UT. [*P. scouleri* (Hook. & Arn.) I.M. Johnst. var. *penicillatus* (Greene) Cronquist, in part] May–Jul ★

P. nothofulvus (A. Gray) A. Gray (p. 511) RUSTY POPCORNFLOWER, FOOTHILL SNOWDROPS Ann, ± red, staining, hairs spreading, rough, sharp. **ST:** ± erect, 2–7 dm. **LF:** basal in rosette, 3–10 cm; cauline few, alternate. **INFL:** in 2s or 3s; bracts 0 or near base. **FL:** ± fragrant; calyx 2–3 mm, circumscissile, lobes gen curved over fr; corolla limb (3)4–9 mm diam, appendages yellow. **FR:** nutlets gen 3, ± 2 mm, round-ovate, abruptly narrowed below acute tip, ± arched; abaxial ridge, lateral ridges, cross-ribs narrow;

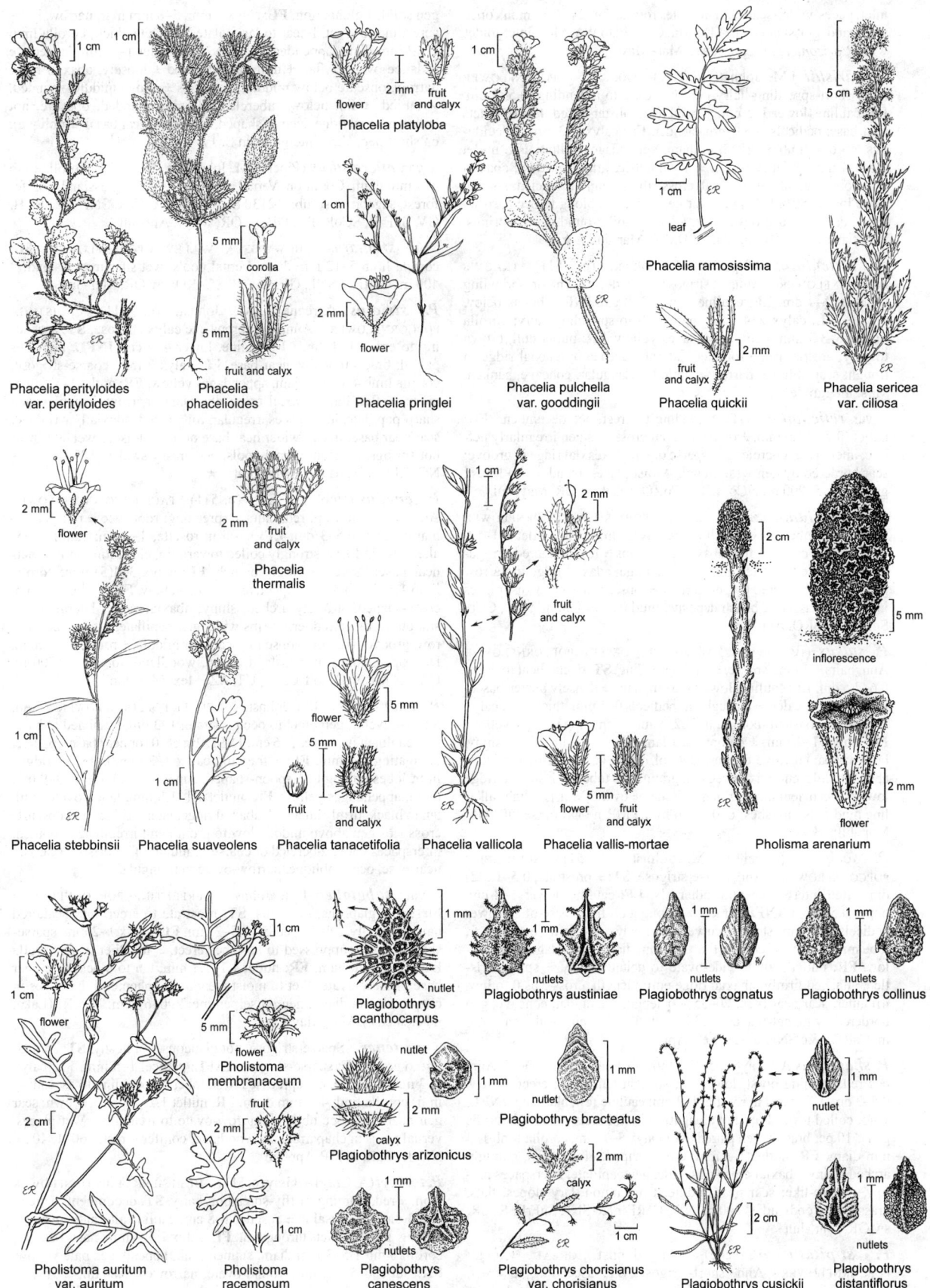

Phacelia perityloides var. perityloides

Phacelia phacelioides

Phacelia platyloba

flower

fruit and calyx

Phacelia pringlei

flower

Phacelia pulchella var. gooddingii

leaf

Phacelia ramosissima

fruit and calyx

Phacelia quickii

Phacelia sericea var. ciliosa

corolla

fruit and calyx

flower

Phacelia thermalis

fruit and calyx

Phacelia stebbinsii

Phacelia suaveolens

flower

fruit

fruit and calyx

Phacelia tanacetifolia

fruit and calyx

Phacelia vallicola

flower

fruit and calyx

Phacelia vallis-mortae

inflorescence

flower

Pholisma arenarium

flower

Pholistoma auritum var. auritum

flower

Pholistoma membranaceum

fruit

Pholistoma racemosum

nutlet

Plagiobothrys acanthocarpus

nutlets

Plagiobothrys austiniae

nutlets

Plagiobothrys cognatus

nutlets

Plagiobothrys collinus

nutlet

calyx

Plagiobothrys arizonicus

nutlet

Plagiobothrys bracteatus

nutlets

Plagiobothrys canescens

calyx

Plagiobothrys chorisianus var. chorisianus

Plagiobothrys cusickii

nutlet

nutlets

Plagiobothrys distantiflorus

interspaces wide; scar near middle, round. 2*n*=24. Common; open woodland, grassland; gen < 1550 m. CA-FP; to WA, Mex. Intergrades with *P. arizonicus*, *P. canescens*. Mar–May

P. parishii I.M. Johnst. (p. 511) PARISH'S POPCORNFLOWER Ann, short-spreading-hairy. **ST:** prostrate to ascending, 0.5–3 dm. **LF:** cauline, lower 1–5 cm, abaxial hairs blister-based. **INFL:** bracts near base; pedicels 1–3+ mm, slender. **FL:** calyx 2–3 mm, ± deciduous; corolla limb 3–7(8) mm diam, appendages yellow. **FR:** nutlet 0.8–1.4 mm, ± lance-ovate, ± asymmetric, gray to brown; abaxial ridge obscure, lateral ridges 0, cross-ribs strongly tubercled; adaxial ridge low, ± 80% fr length; scar near base, ± oblong to occ narrow-triangular, gen in groove. Wet, alkaline soil around desert springs, mud flats; 750–2210 m. SNE, c DMoj. Mar–Jun ★

P. reticulatus (Piper) I.M. Johnst. (p. 511) TRACY'S POPCORNFLOWER Ann, ± strigose. **ST:** decumbent or sprawling to erect, 1–4 dm. **LF:** cauline, lower 3–8 cm. **INFL:** bracts below middle. **FL:** calyx 2–4 mm, appressed- to spreading-hairy; corolla limb 1.5–3.5 mm diam, appendages yellow. **FR:** nutlet dull, brown to gray; abaxial ridge rounded, lateral ridges ± 0; adaxial ridge in trough; scar oblique, narrow-elliptic to triangular, concave, sunken. Vars. intergrade.

var. ***reticulatus*** **ST:** Ascending to erect, occ decumbent. **FR:** nutlet (0.7)1–1.6 mm, ± ovate; abaxial cross-ribs gen irregularly net-like, interspace tubercles gen sparse or small; adaxial ridge in groove; scar bordered by gen wide trough. Moist places, meadows in forest, grassland; 5–900 m. NCo, KR, NCoRO, CCo; sw OR. May–Jul

var. ***rossianorum*** I.M. Johnst. FORT ROSS POPCORNFLOWER **ST:** ± decumbent or sprawling, occ ascending. **FR:** nutlet 1.1–1.9 mm, oblong-ovate, ± flat; abaxial cross-ribs ± few, scattered, irregular, interspace tubercles gen dense or large; adaxial ridge in narrow groove; scar bordered by gen narrow, ± obscure trough. Moist places in coastal grassland, beach deposits, mud flats; < 800 m. NCo, CCo, SnFrB, SCoRO. Apr–Jun

P. salsus (Brandegee) I.M. Johnst. DESERT POPCORNFLOWER Ann, hairs ± 1 mm, sparse, ± stiff, spreading. **ST:** decumbent to erect, 0.6–2.5 dm. **LF:** cauline, lower 3–6 cm, hairs ± densely blister-based. **INFL:** bracts ± dense throughout; pedicels 0–1 mm, thick. **FL:** calyx 2–5 mm, persistent; corolla limb 2–5 mm diam, appendages yellow. **FR:** nutlet 1–2 mm, ± compressed-lanceolate, asymmetric, ± shiny, brown; abaxial ridge obscure, lateral ridges 0, cross-ribs few, low, above middle, curved or irregular, granular to tubercled; adaxial ridge low, entire fr length; scar gen oblique, ± ovate. Moist, saline, alkaline mud flats, marshes; 610–1410 m. ne MP, ne DMoj; se OR, NV. Mar–Jul ★

P. scriptus (Greene) I.M. Johnst. (p. 511) SCRIDGEE'S POPCORNFLOWER Ann, sparse-strigose. **ST:** ± prostrate, 0.5–1.5(2) dm, often interwoven with other pls. **LF:** cauline, lower 1–2 cm, upper 0.5–2 cm. **INFL:** bracts ± throughout; fls to base of st; lower pedicels < 2 mm, stout, recurved. **FL:** calyx 4–8 mm, longer in fr, lobe midvein thick; corolla limb ± 2 mm diam, appendages pale yellow. **FR:** nutlet 2 mm, wide-ovate to deltate, shiny, ± sparse-bristled, attached firmly; abaxial ridge entire length, cross-ribs 0 to low, irregular; scar near base, 20–35% fr length, triangular, concave, gen bordered by funnel-like collar. Moist thin, rocky, clay soil; gen < 150 m. CaRF, n&c SNF, n ScV. Feb–Apr

P. shastensis A. Gray (p. 511) SHASTA POPCORNFLOWER Ann, occ red, staining on st, lower lvs, straight-hairy. **ST:** erect, 1–few, 0.5–3 dm. **LF:** basal in rosette, 1–3 cm; cauline few, alternate. **INFL:** ± not coiled toward tip, not elongating in fr, bracts ± throughout; fls gen < 10 per branch, gen paired. **FL:** calyx 5–7 mm; corolla limb 1–3 mm diam. **FR:** nutlet 2–3 mm, cross-shaped, gray to tan, strongly arched, shiny; abaxial cross-ribs wide, gen papillate, interspaces narrow, groove-like; scar near middle. Uncommon. Dry slopes, flats, grassland, woodland; < 800 m. NW, CaRF, n&c SNF, SnFrB, SCoR; sw OR. Mar–Jun

P. stipitatus (Greene) I.M. Johnst. GREAT VALLEY POPCORNFLOWER Ann, short-strigose. **ST:** ascending to erect, 1–5 dm, often fleshy, ± hollow. **LF:** cauline, lower 2–11 cm, hairs ± blister-based. **INFL:** ± open, bracts below middle; pedicels 0–3 mm,

gen solid, ± elongating. **FL:** calyx 3 mm, 5–8 mm in fr, narrow, lobes spreading to erect, linear to lanceolate, midveins thick; corolla limb 2–12 mm diam, appendages yellow. **FR:** nutlet 1–2 mm, lanceolate to lance-ovate, ± flat, ± dull, tan, attached ± loosely; abaxial ridge narrow, obscure below middle, cross-ribs ± above middle, rounded, wrinkled, ± flat below, tubercled; margins rounded, obscure, not entire; adaxial ridge deep-V-shaped, tubercled; scar basal (± oblique), on short peg. Vars. intergrade in SnJV.

var. ***micranthus*** (Piper) I.M. Johnst. (p. 511) **FL:** corolla limb 2–4 mm diam. Common. Vernal pools, wet sites in grassland, conifer forest, sagebrush scrub; < 2130 m. KR, NCoR, CaR, SNF, n&c SNH, GV, SnFrB, SCoR, PR, GB; se OR, w NV. Apr–Jul

var. ***stipitatus*** SHOWY GREAT VALLEY POPCORNFLOWER **FL:** corolla limb 5–12 mm diam. Vernal pools, wet sites in grassland; < 400 m. CaRF, n SNF, GV, SnFrB, SCoRO; sw OR. Mar–Jun

P. strictus (Greene) I.M. Johnst. (p. 511) CALISTOGA POPCORNFLOWER Ann, ± glabrous exc calyx strigose. **ST:** ascending to erect, 1–4 dm. **LF:** cauline, lower 4–9 cm. **INFL:** branches paired; bracts 0 or few near base. **FL:** calyx 3 mm, coarse-strigose; corolla limb 4–6 mm diam, appendages yellow. **FR:** nutlet 1–1.5 mm, ovate, ± flat, dull; abaxial cross-ribs many, narrow, irregular, gen sharp-papillate, interspaces irregular, roughened; adaxially wrinkled, scar near base, linear, wider near base or not. Moist to wet sites near hot springs, adjacent vernal pools, occ grassy swales; 100–150 m. s NCoRI (nw Napa Co.). Mar–Jun ★

P. tenellus (Hook.) A. Gray (p. 511) PACIFIC POPCORNFLOWER Ann, rarely entire pl red, hairs ± spreading; roots occ ± red. **ST:** 1–many, erect, 0.5–3 dm. **LF:** basal in rosette, 1–5 cm; cauline few, alternate. **INFL:** ± strongly coiled toward tip, elongating in fr, bracts near base; fls gen > 10 per branch. **FL:** calyx 3–4(5) mm; corolla limb 1.5–3 mm diam, appendages ± pale yellow. **FR:** nutlet 1–2 mm, cross-shaped, strongly arched, shiny; abaxial ridge, lateral ridges smooth to tubercled, cross-ribs wide, gen papillate, interspaces narrow, groove-like, often obscure; scar near middle, ± round. Common. Dry, open slopes in grassland, scrub, woodland, forest; < 1700 m. CA-FP, MP, w D; to BC, ID, UT, AZ, Mex. Mar–Jun

P. tener (Greene) I.M. Johnst. SLENDER POPCORNFLOWER Ann. **ST:** 1–several, gen erect to occ prostrate, 1–3 dm, branched at base. **LF:** cauline few, lower 2–6 cm. **INFL:** bracts 0 (or near base); pedicel elongating < 5 mm. **FL:** ± fragrant; calyx 1–3 mm, base ± 4-ridged in fr, lobes oblong to ± spoon-shaped, cupped; corolla limb 3–9 mm diam, appendages yellow. **FR:** nutlet 1.1–1.8 mm, lance-ovate, dull, tan to black, firmly attached; abaxial ridge, lateral ridges gen obscure, cross-ribs gen above middle, low to prominent, irregular, ± toothed, interspaces wide, tubercled or coarse-granular, often bristly; scar gen near base, occ ± oblique, narrow-ovate to triangular.

var. ***subglaber*** I.M. Johnst. SERPENTINE POPCORNFLOWER Fleshy, ± glabrous, glaucous. **ST:** prostrate to erect. **LF:** scattered hairs abaxially. **INFL:** pedicel 0.5–2 mm. **FL:** calyx 1–2 mm, sparse-short-strigose, appressed to st, lobes erect, ± together in fr; corolla limb 3–7 mm diam. **FR:** nutlet 1.1–1.4 mm, tan to black; scar near base, narrow-ovate. Wet to moist meadows, ephemeral drainages in chaparral, woodland, gen on serpentine; 290–690 m. NCoRI (Lake, n Napa cos.). Mar–May

var. ***tener*** Sparse-strigose, not glaucous, not fleshy. **ST:** spreading to erect. **LF:** sparse-strigose. **INFL:** pedicel 1–5 mm. **FL:** calyx 1–3 mm, strigose, not appressed to st, lobes ± spreading to ascending in fr; corolla limb 4–9 mm diam. **FR:** nutlet 1.2–1.8 mm, brown; scar gen near base, occ oblique, narrow-ovate to triangular. Wet places, vernal pools in chaparral, oak woodland, conifer forest; 160–1340 m. NCoR, CaR, w MP. Apr–Jul

P. torreyi (A. Gray) A. Gray (p. 511) HIGH SIERRA POPCORNFLOWER Ann, ± red, staining, stiffly-spreading-hairy. **ST:** decumbent to erect, 0.5–3 dm. **LF:** basal in rosette, 1–8 cm; cauline alternate, ovate to oblong. **INFL:** bracts throughout. **FL:** calyx 1.5–4 mm, hairs straight; corolla limb 1.5–3 mm diam, stained pink, appendages pale yellow. **FR:** nutlet 1.5–2.2 mm, round-ovate, narrowed near tip, ± gray, ± dull; abaxial ridge, lateral ridges low, obscure, sparse-papillate, cross-ribs flat, ± smooth; scar near middle, ± round.

var. ***diffusus*** I.M. Johnst. **ST**: decumbent to ascending, few–many. **LF**: cauline dense, ovate to elliptic. **INFL**: bracts dense, ovate to elliptic. Common. Moist to dry meadows, flats, forest edges; 1200–3400 m. SNH, Teh, SnBr. Jun–Aug

var. ***perplexans*** I.M. Johnst. CHAPARRAL POPCORNFLOWER Ann, ± red, staining, hairs stiff, spreading. **ST**: ascending to erect, 0.5–3 dm. **LF**: basal in ± rosette, 1–8 cm; cauline narrow-ovate. **FL**: calyx 1.5–2.5 mm; corolla limb 2–3 mm diam. **FR**: nutlet 1.5–1.8 mm, pointed at base, gray to ± brown, arched, ± dull; abaxial ridge, lateral ridges tubercled, cross-ribs rounded, gen with rows of round, translucent papillae. Uncommon. Gravelly soils in chaparral, ± fire follower; 500–2100 m. c&s SN, s SCoRO. Apr–Jun

var. ***torreyi*** YOSEMITE POPCORNFLOWER **ST**: ± erect, few. **LF**: cauline ± sparse, ± oblong. **INFL**: bracts ± sparse, ± oblong. Forest edges, flats; 1200–2100 m. SNH (scattered). May–Jul ★

P. trachycarpus (A. Gray) I.M. Johnst. (p. 511) ROUGH-NUTLET POPCORNFLOWER Ann, ± dense-strigose. **ST**: prostrate to ascending, 0.5–4 dm. **LF**: cauline, lower 5–10 cm. **INFL**: bracts ± throughout, spreading. **FL**: calyx 4–6 mm; corolla limb 1.5–4 mm diam, appendages yellow. **FR**: nutlet 1–1.7 mm, wide-ovate, ± flat, dull; abaxial ridge, lateral ridges, cross-ribs gen narrow, tubercled or sharp-papillate, occ bristled, interspaces wide, irregular, granular-papillate; adaxial ridge > 75% fr length; scar near base, < 25% fr length, ovate to triangular, concave. Shallow vernal pools, wet places in grassland, scrub, chaparral, woodland; < 1390 m. s ScV, SnJV, SnFrB, SCoRO, SCo, WTR, PR. Mar–May

P. uncinatus J.T. Howell (p. 511) HOOKED POPCORNFLOWER Ann, ± red, staining, stiff spreading-hairy. **ST**: decumbent to erect, 1–3 dm. **LF**: basal in rosette, 1–4 cm; cauline alternate, ovate. **INFL**: bracts ± throughout. **FL**: calyx 1.5–2 mm, hairs dense, spreading, gen hooked at tip; corolla limb 1.5–2 mm diam, stained pink, appendages pale yellow. **FR**: nutlet 1–1.3 mm, ovate, base rounded, brown, ± arched; abaxial ridge, lateral ridges narrow, ± tubercled, with many irregular translucent papillae, occ wrinkled, cross-ribs 0, interspaces 0; scar near middle, ± round. Chaparral, canyon sides, rocky outcrops, ± fire follower; 300–600 m. SCoRO. Apr–May ★

P. undulatus (Piper) I.M. Johnst. (p. 511) WAVY-STEMMED POPCORNFLOWER Ann, sparse-short-strigose. **ST**: erect turning decumbent, 1–4+ dm. **LF**: cauline, lower 2–6 cm. **INFL**: bracts below middle, sparse; pedicels gen 0–1 mm. **FL**: calyx 2–3 mm; corolla limb 1.5–2 mm diam, appendages pale yellow. **FR**: nutlet 0.8–1.6 mm, ovate to lance-ovate, ± dull, brown; abaxial ridge, lateral ridges ± 0, cross-ribs ± many, low, ± curved, obscurely tubercled, interspaces narrower than or as wide as cross-ribs; adaxial ridge ± 75% fr length; scar near base, linear, in groove below ridge. Wet places, vernal pools; < 400 m. NCo, NCoRO, s ScV, CW (exc SCoRI), SCo, WTR, PR. Mar–Jun

P. verrucosus (Phil.) I.M. Johnst. FORGET-ME-NOT POPCORNFLOWER Ann, ± red, staining, hairs stiff, spreading. **ST**: erect, 0.5–2 dm. **LF**: basal in rosette, 1–3 cm; cauline alternate, ovate. **INFL**: bracts throughout. **FL**: calyx 1.5–2 mm, hairs dense, appressed to ascending, straight; corolla limb 1.5–3 mm diam, stained pink, appendages pale yellow. **FR**: nutlet ± 1.5 mm, ovate, brown, ± flat; abaxial ridge narrow, lateral ridges 0, cross-ribs 0, interspaces 0, surface with irregular papillae, tubercles, occ low wrinkles; scar near middle, ± round. 2*n*=24. Open chaparral, gravelly soil, ± fire follower; 700–850 m. se SnFrB (Mount Hamilton Range); s S.Am. [*P. myosotoides* (Lehm.) Brand, misappl.; *P. m.* subsp. *v.* (Phil.) N. Horn, ined.] Mar–May ★

ROMANZOFFIA

Robert Patterson & Richard R. Halse

Per [ann], ± scapose, base bulb-like or from tubers [taproot]; herbage ± glabrous to gen sparse-soft-hairy. **ST**: erect. **LF**: basal long-petioled, reniform to round, shallow-lobed or toothed; cauline few, reduced, alternate [gen opposite]. **INFL**: open; pedicels present. **FL**: corolla > calyx, bell- to funnel-shaped, white, gen yellow in throat; stamens incl, subequal; ovary chambers appearing 2, style 1, thread-like, incl, ± 2-lobed. **FR**: capsule, oblong to ovoid to obovoid. **SEED**: many, ovoid, angled, brown, pitted. 5 spp.: CA, AK, MT, w Can. (Count N.P. Romanzoff, 1754–1826, promoter of Russian expedition to CA in 1816)

1. Petioles overlapped; 1700–2100 m . . . . ***R. sitchensis***
1' Petioles not overlapped; < 800 m
  2. Infl exceeding lvs; pedicels in fr 1–3 cm . . . . ***R. californica***
  2' Infl not or ± exceeding lvs; pedicels in fr < 1 cm . . . . ***R. tracyi***

R. californica Greene (p. 511) Pl 10–40 cm; tubers clustered, ovoid, green to brown, tomentose. **LF**: petioles 2–12 cm, ± widened at base, not overlapped; blade 8–45 mm wide. **INFL**: exceeding lvs, glandular or not; pedicels slender, 1–3 cm in fr. **FL**: calyx lobes 2–5 mm, lance-linear, gen acute, occ glabrous, glandular or not; corolla 5–12 mm; ovary sparse-soft-hairy, glandular; style 4–7 mm. **FR**: 6–10 mm. *n*=11. Ocean bluffs, roadbanks, wet cliffs, moist rocky areas; < 800 m. NW, n CCo, SnFrB; to WA. Mar–May

R. sitchensis Bong. Pl 10–30 cm; tubers 0. **LF**: petioles 1–6 cm, widened at base, overlapped, forming bulbous base of st, gen ciliate; blades 1–4 cm wide. **INFL**: exceeding lvs, glandular or not; pedicels slender, 1–3 cm in fr. **FL**: calyx lobes 2–4 mm, oblong, gen obtuse, glabrous, glandular or not; corolla 5–11 mm; ovary sparse-soft-hairy, glandular, style 2–5 mm. **FR**: 4–7 mm. *n*=11. Uncommon. Moist clefts in rocks, mtns; 1700–2100 m. KR; to AK, w Can, MT. Jul–Aug

R. tracyi Jeps. TRACY'S ROMANZOFFIA Pl 2–12 cm, in rounded tufts; tubers clustered, ovoid, brown, tomentose. **LF**: petioles 1–8 cm, soft-hairy, widened at base, not overlapped; blades 10–35 mm wide, ± glabrous to soft-hairy. **INFL**: dense, not or ± exceeding lvs; pedicels stout, < 1 cm in fr. **FL**: calyx lobes 2–5 mm, lanceolate, acute, soft-hairy; corolla 6–8 mm; ovary soft-hairy, style 2–3 mm. **FR**: 5–8 mm. *n*=11. Rocky ocean bluffs; < 30 m. NCo; to BC. Mar–May ★

SYMPHYTUM COMFREY

Per; root thick, carrot-like. **ST**: ascending to erect, internodes winged or not, sharp-bristly. **LF**: gen cauline, sharp-bristly; lower petioled; upper short-petioled to sessile; blade lanceolate to ovate, base decurrent or not. **INFL**: terminal or axillary, gen peduncled, coiled. **FL**: calyx deep-lobed, bristly, expanded in fr; corolla bell- to ± urn-shaped, throat expanded above tube, appendages 5, alternate stamens, at same level at anthers, lance-linear to lanceolate [or not], papillate; stamens attached on upper tube; style exserted. **FR**: nutlets 1–4, ovoid; tip ± incurved; scar at base, ± flat with thick, ring-like, minute-toothed rim. 35 spp.: Eurasia. (Greek: growing together, from putative healing properties) Seeds, herbage TOXIC to humans, livestock from pyrrolizidine alkaloids. Orn, folk medicine, cult for forage.

1. St internodes winged; lf base decurrent; nutlets black, shiny, smooth . . . . [***S. officinale***]
1' St internodes not winged; lf base gen not decurrent; nutlets brown, dull, fine-grainy . . . . ***S. ×uplandicum***

S. xuplandicum Nyman RUSSIAN COMFREY **ST:** 6–10 dm, branched. **LF:** 5–15 cm. **FL:** calyx 2–4 mm, to 5.5 mm in fr, lobes gen linear-oblong, in fr triangular; corolla 13–16 mm, ± pink, turning purple to blue-purple. **FR:** nutlets 3–4 mm, 2–2.5 mm wide. $2n=36,40$. Wet, open sites; < 100 m. NCo, n NCoRO, ScV (waif in Yolo Co.); to BC, ne US; native to Eur. [*S. asperum* Lepechin, misappl.] May–Jul ◆

TIQUILIA

Ann to subshrub, variously hairy, glandular or not; ± taprooted, rhizome gen 0. **ST:** prostrate. **LF:** cauline, gen clustered, evergreen, petioled, margin rolled under, entire or ± crenate. **INFL:** ± axillary; fls 1 or clustered, sessile. **FL:** radial to ± bilateral; calyx lobes 5, not enlarged in fr; corolla gen ± funnel-shaped, tube yellow in youth, appendages 0; style 2-lobed, stigmas 2. **FR:** nutlets 1–4, not separate to base, 4-grooved to deep-4-lobed, ± tubercled or not. 27 spp.: w hemisphere deserts. (Native S.Am name for fl)

1. Branches alternate; lf veins obscure, blade ovate to narrow-elliptic; per, subshrub; style branched < 1/3 from tip; fr 4-grooved . . . . ***T. canescens***
  2. Corolla 4–7.5 mm, limb 2.5–4.5 mm diam . . . . var. ***canescens***
  2′ Corolla 8–12 mm, limb 5–8 mm diam. . . . . var. ***pulchella***
1′ Branches opposite; lf veins obvious, ± sunken, blade ovate, round, or obovate; ann, per; style branched 1/2–4/5 from tip; fr 4-lobed
  3. Rhizome present; st ± glandular; lf veins deeply sunken, lateral pairs 4–7 . . . . ***T. plicata***
  3′ Rhizome 0; st ± nonglandular; lf veins shallowly sunken, lateral pairs 2–3
    4. Ann, not woody; style < calyx; lf margin entire, lateral veins ± 30° from midvein; corolla pink to white; seed oblong-ovoid . . . . ***T. nuttallii***
    4′ Per, ± woody; style > calyx; lf margin ± crenate, lateral veins ± 45° from midvein; corolla blue, purple, or lavender; seed spheric . . . . ***T. palmeri***

T. canescens (DC.) A.T. Richardson SHRUBBY TIQUILIA Per, subshrub, woody. **ST:** branches alternate, many, hairs ± spreading. **LF:** occ clustered, white-tomentose; blade 5–13 mm, ovate to narrow-elliptic, veins obscure, margin entire, spiny-ciliate. **INFL:** fls 1 or clustered in lf axils; bracts 0. **FL:** calyx 3–5(6) mm, lobes 2/3–3/4 free; style short-exserted from calyx, branched < 1/3. **FR:** spheric, 4-grooved, not lobed; nutlets 2–2.5 mm, ovoid, minutely tubercled, hairy or not.

  var. ***canescens*** (p. 523) **FL:** corolla 4–7.5 mm, limb 2.5–4.5 mm diam, lavender, pink, or white. $n=9$. Slopes, ridges, occ washes; 300–1525 m. se DMoj, DSon; sw N.Am, TX, n Mex. Mar–May

  var. ***pulchella*** (I.M. Johnst.) A.T. Richardson **FL:** corolla 8–12 mm, limb 5–8 mm diam, blue or lavender. Slopes, ridges, occ washes; 250–600 m. DSon; sw AZ. Mar–May

T. nuttallii (Hook.) A.T. Richardson (p. 523) ANNUAL TIQUILIA Ann, rosette prostrate; rhizome 0. **ST:** branches opposite, ± nonglandular, hairs ± appressed. **LF:** clustered, hairs ± spreading; blade 3.5–9 mm, ovate to round, margin entire, lateral vein pairs 2–3, shallow-sunken, ± 30° from midvein. **INFL:** fls clustered in lf axils; bracted. **FL:** calyx 3–5 mm, lobes 2/3–3/4 free, hairs within short; corolla 3–4 mm, limb 2–2.5 mm diam, pink to white; style < calyx, branched 1/3–1/2. **FR:** 4-lobed; nutlets oblong-ovoid, smooth, shiny. $n=8$.

Sandy plains, pumice gravel, washes, slopes, saline flats; < 2750 m. s SN, Teh, e MP, SNE, DMoj; to WA, UT; also Argentina. May–Aug

T. palmeri (A. Gray) A.T. Richardson (p. 523) PALMER'S TIQUILIA Per, ± woody, bark white; rhizome 0. **ST:** branches opposite, ± nonglandular; hairs ± shaggy. **LF:** clustered, ± gray-strigose; blade 3.5–11 mm, ovate to round, margin ± crenate, lateral veins 2–3 pairs, shallow-sunken, ± 45° from midvein. **INFL:** fls clustered in lf axils; bracted. **FL:** calyx 2–3.5 mm, lobes ± 1/2 free, hairs within short or 0; corolla 5–9 mm, limb 4–5 mm diam, blue, purple, or lavender; style > calyx, branched 1/2. **FR:** deeply 4-lobed; nutlets spheric, smooth, shiny. $n=8,9$. Sandy gravel soils, on terraced flats; < 650 m. D (esp w edge DSon and near Colorado River); sw NV, w AZ, n Mex. Mar–Jun

T. plicata (Torr.) A.T. Richardson (p. 523) FAN-LEAVED TIQUILIA Per, not woody, matted; rhizomed. **ST:** branches opposite, ± glandular. **LF:** clustered, white-canescent; blade 3–12 mm, obovate to wide-ovate, margin entire, lateral veins 4–7 pairs, deeply sunken. **INFL:** fls clustered in lf axils; bracts 0. **FL:** calyx 2–3 mm, lobes ± free, hairs within long; corolla 4–6 mm, 2–3 mm diam, blue to lavender; style > calyx, branched 1/2–4/5. **FR:** deeply 4-lobed; nutlets ovoid, smooth, shiny. $n=8$. Dunes, sandy gravel flats; < 1100 m. D; w AZ, s NV, n Mex. Mar–Jul

TRICARDIA
Robert Patterson & Richard R. Halse

1 sp. (Greek: 3 hearts, from calyx)

T. watsonii S. Watson (p. 523) THREE HEARTS Per; herbage long-soft-hairy, in age ± glabrous; taproot woody, gen topped by a branched caudex covered by old petiole bases. **ST:** several from root crown, erect, 5–40 cm. **LF:** gen in basal rosette, petioled, 2–9 cm, 5–25 mm wide; cauline alternate, lower short-petioled, upper sessile, 6–20 mm, 3–15 mm wide, entire. **INFL:** open, terminal; fls pedicelled. **FL:** calyx lobes 5, unequal, not alike, enlarged in fr, outer 3 cordate, 5–9 mm, 9–25 mm in fr, scarious, veiny, green to ± purple, inner 2 ± 4 mm, linear; corolla 4–8 mm, bell-shaped to rotate, white to cream, gen marked lavender; stamens incl, unequal, equally attached; ovary chamber appearing 1, style 1, 3–4 mm, tip 2-lobed. **FR:** capsule, 7–9 mm, glabrous. **SEED:** 4–8, oblong, brown, rough. $n=8$. Sandy or gravelly desert slopes, flats, mtns, gen under shrubs; 100–2300 m. SnBr, SNE, D; to UT, AZ. Apr–Jun

511

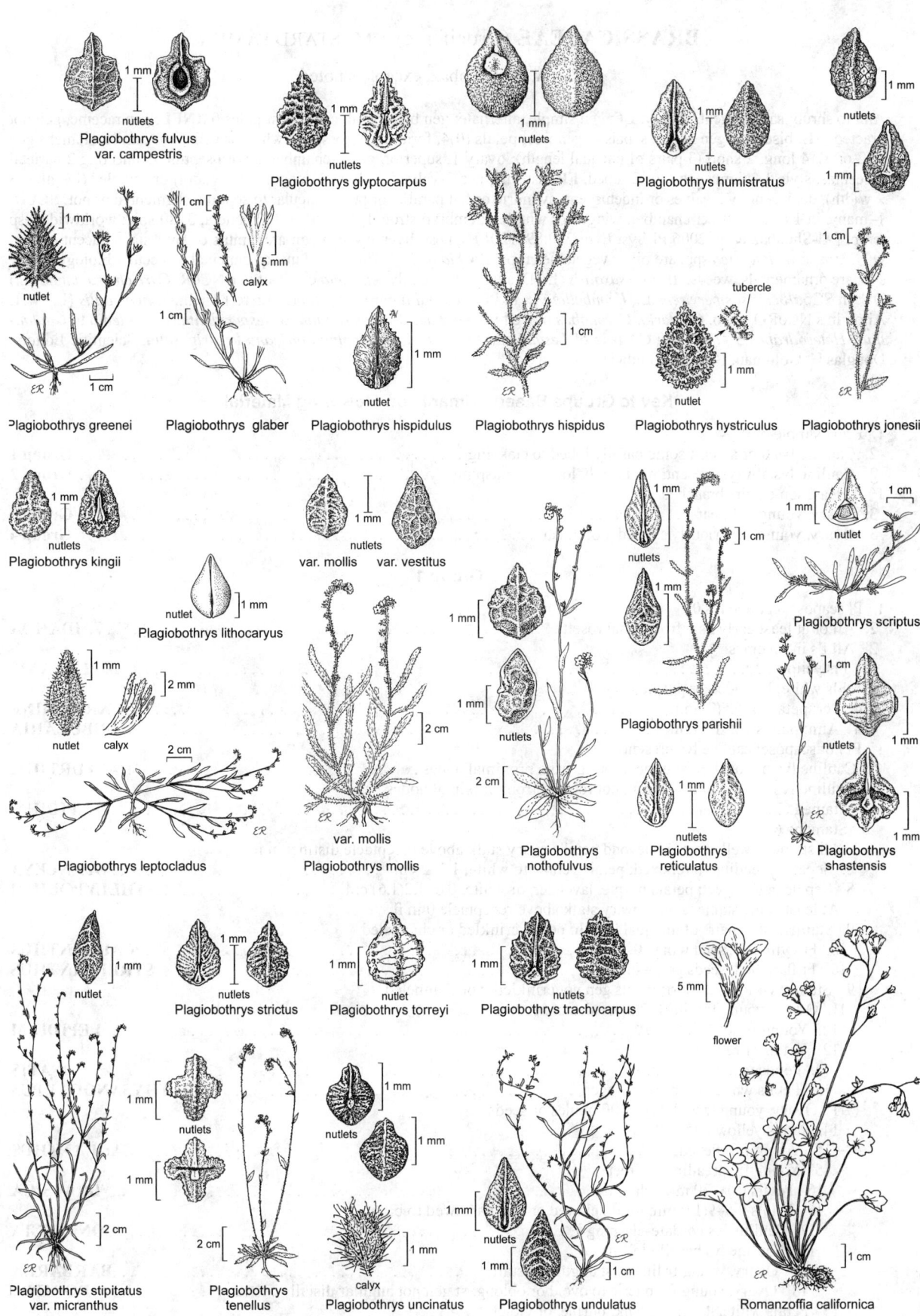

Plagiobothrys fulvus var. campestris

Plagiobothrys glyptocarpus

Plagiobothrys humistratus

Plagiobothrys greenei

Plagiobothrys glaber

Plagiobothrys hispidulus

Plagiobothrys hispidus

Plagiobothrys hystriculus

Plagiobothrys jonesii

Plagiobothrys kingii

Plagiobothrys lithocaryus

var. mollis   var. vestitus

Plagiobothrys parishii

Plagiobothrys scriptus

Plagiobothrys leptocladus

var. mollis

Plagiobothrys mollis

Plagiobothrys nothofulvus

Plagiobothrys reticulatus

Plagiobothrys shastensis

Plagiobothrys strictus

Plagiobothrys torreyi

Plagiobothrys trachycarpus

Plagiobothrys stipitatus var. micranthus

Plagiobothrys tenellus

Plagiobothrys uncinatus

Plagiobothrys undulatus

Romanzoffia californica

BRASSICACEAE (Cruciferae)   MUSTARD FAMILY

Ihsan A. Al-Shehbaz, except as noted

Ann to shrub; sap pungent, watery. **LF**: gen simple, alternate; gen both basal, cauline; stipules 0. **INFL**: gen raceme, gen not bracted. **FL**: bisexual, gen radial; sepals 4, gen free; petals (0)4, forming a cross, gen white or yellow to purple; stamens gen 6 (2 or 4), 4 long, 2 short (3 pairs of unequal length); ovary 1, superior, gen 2-chambered with septum connecting 2 parietal placentas; style 1, stigma entire or 2-lobed. **FR**: capsule, gen 2-valved, "silique" (length ≥ 3 × width) or "silicle" (length < 3 × width), dehiscent by 2 valves or indehiscent, cylindric or flat parallel or perpendicular to septum, segmented or not. **SEED**: 1–many, in 1 or 2 rows per chamber, winged or wingless; embryo strongly curved. ± 330 genera, 3780 spp.: worldwide, esp temp. [Al-Shehbaz et al. 2006 Pl Syst Evol 259:89–120] Highest diversity in Medit area, mtns of sw Asia, adjacent c Asia, w N.Am; some *Brassica* spp. are oil or vegetable crops; *Arabidopsis thaliana* used in experimental molecular biology; many spp. are ornamentals, weeds. *Aurinia saxatilis* (L.) Desvaux in cult only. *Aubrieta* occ waif in c NCoR, *Carrichtera annua* (L.) DC. in SCo, *Iberis sempervirens* L., *I. umbellata* L. in PR, *Teesdalia coronopifolia* (Bergeret) Thell., *T. nudicaulis* (L.) W.T. Aiton in s NCoRO, CCo. *Cardaria, Coronopus* moved to *Lepidium*; *Caulostramina* to *Hesperidanthus*; *Guillenia* to *Caulanthus*; *Heterodraba* to *Athysanus*; CA taxa of *Lesquerella* to *Physaria*; *Malcolmia africana* to *Strigosella*. Scientific Editors: Douglas H. Goldman, Bruce G. Baldwin.

Key to Groups Based Primarily on Flowering Material

1. Hairs simple or 0
  2. Cauline lvs 0 or at least some basally lobed to clasping . . . . **Group 1**
  2′ Cauline lvs always present, not basally lobed or clasping. . . . . **Group 2**
1′ At least some hairs branched
  3. Ovary, young fr linear . . . . **Group 3**
  3′ Ovary, young fr variously shaped, not linear . . . . **Group 4**

Group 1

1. Pl scapose; cauline lvs 0
  2. All or at least early fls 1 from basal rosette . . . . **IDAHOA**
  2′ All fls in racemes
    3. Fls yellow . . . . **DIPLOTAXIS**
    3′ Fls white, lavender, or purple
      4. Per; petals 4–5.5(7) mm . . . . **CARDAMINE**
      4′ Ann; petals 1.2–1.5 mm . . . . **SUBULARIA**
1′ Pl not scapose; cauline lvs present
  5. Cauline lvs petioled; pl aquatic, rooting from proximal nodes . . . . **NASTURTIUM**
  5′ Cauline lvs sessile; pl terrestrial, not rooting from proximal nodes
    6. Stamens 2 . . . . **²LEPIDIUM**
    6′ Stamens 6
      7. All stamens well exserted beyond petals; ovary stalk above receptacle distinct in fl
        8. Sepals spreading to reflexed; petals yellow to white, 1.3–2 cm . . . . **STANLEYA**
        8′ Sepals gen ± erect; petals purple, lavender, or white, 0.6–1.2(1.6) cm. . . . . **²THELYPODIUM**
      7′ At least some stamens incl; ovary stalk above receptacle 0 in fl
        9. Stamens in 3 pairs of unequal length; petals crinkled or channeled
          10. Fr cylindric; seed wings 0. . . . . **³CAULANTHUS**
          10′ Fr flattened; seeds gen winged . . . . **STREPTANTHUS**
        9′ Stamens 4 long, 2 short; petals gen not crinkled, not channeled
          11. Ovary, young fr with 1–2(4) ovules or seeds
            12. Young fr not reflexed . . . . **²LEPIDIUM**
            12′ Young fr reflexed
              13. Petals yellow . . . . **ISATIS**
              13′ Petals white or ± purple-tinged. . . . . **THYSANOCARPUS**
          11′ Ovary, young fr with (4)6–300 ovules or seeds
            14. Petals yellow
              15. Young fr reflexed . . . . **³CAULANTHUS**
              15′ Young fr spreading to erect
                16. Petals (6)8–30 mm, clawed . . . . **²BRASSICA**
                16′ Petals 0.5–9(12) mm (or 0), clawed or not, narrowed to base
                  17. Cauline lvs cordate-clasping . . . . **CONRINGIA**
                  17′ Cauline lvs basally lobed
                    18. Ovary, young fr linear; st angular distally . . . . **BARBAREA**
                    18′ Ovary, young fr spheric to ovoid or oblong; st gen not angular distally . . . . **RORIPPA**
            14′ Petals white, lavender, pink, purple, or violet

    19. Ovary, young fr ovate, round, or cordate
       20. Per; seeds smooth . . . . NOCCAEA
       20′ Ann; seeds striate . . . . THLASPI
    19′ Ovary, young fr linear
       21. Petals not clawed, gradually narrowed to base . . . . BOECHERA
       21′ Petals clawed
         22. Petals 15–30 mm . . . . [2]BRASSICA
         22′ Petal 4.5–12
           23. Young fr reflexed . . . . [3]CAULANTHUS
           23′ Young fr erect to spreading . . . . [2]THELYPODIUM

Group 2

1. Fls bilateral
  2. Ovary, young fr linear; ovules or seeds many . . . . [2]STREPTANTHUS
  2′ Ovary, young fr ovate to obcordate; ovules or seeds 2 . . . . [IBERIS]
1′ Fls radial
  3. Stamens 2 or 4
    4. Ovary linear . . . . [4]CARDAMINE
    4′ Ovary broader, not linear
      5. Ovary, young fr with 10–24 ovules or seeds . . . . HORNUNGIA
      5′ Ovary, young fr with 2–4 ovules or seeds. . . . . [2]LEPIDIUM
3′ Stamens 6
  6. Ovary, young fr ovate to ovoid, obovate, round, spheric, oblong, or elliptic, occ of 2 segments
    of such shapes
    7. Petals 15–30 mm, purple to lavender (white); style, stalk above receptacle well developed . . . . LUNARIA
    7′ Petals 0.5–12(15) mm, yellow to cream or white (lavender); style, stalk above receptacle gen 0
      8. Ovary, young fr with 24–80 ovules or seeds . . . . [2]RORIPPA
      8′ Ovary, young fr with 1–12(14) ovules or seeds
        9. Ovary, young fr 2-segmented
          10. Petals cream or pale yellow, veins dark brown or purple; ovules 4–6 . . . . [CARRICHTERA]
          10′ Petals white, lavender, or yellow, dark veins 0; ovules 1 or 2
            11. Pl stiff-hairy below; style distinct; petals yellow . . . . RAPISTRUM
            11′ Pl glabrous; style 0; petals white to lavender . . . . [2]CAKILE
        9′ Ovary, young fr unsegmented
          12. Ovules or seeds 1–2
            13. Young fr erect to spreading. . . . . [2]LEPIDIUM
            13′ Young fr reflexed . . . . THYSANOCARPUS
          12′ Ovules or seeds 6–14
            14. Per, roots fleshy; st 5–12(20) dm; basal lvs 1.5–12 dm. . . . . ARMORACIA
            14′ Bien or per, fleshy roots 0; st (0.1)0.5–3(4) dm; basal lvs 0.1–1 dm. . . . . COCHLEARIA
  6′ Ovary, young fr linear (narrowly lanceolate or awl-shaped)
    15. Cauline lvs compound
    16. Pl aquatic, rooting from proximal st nodes . . . . NASTURTIUM
    16′ Pl terrestrial, not rooting from proximal st nodes . . . . [4]CARDAMINE
    15′ Cauline lvs simple, occ shallowly divided
    17. Pl with multicellular stalked glands . . . . CHORISPORA
    17′ Glands 0
      18. Stamens ± equal in length, well-exserted beyond petals or all fl parts spreading
        19. Petals yellow to white; filament bases hairy or papillate . . . . STANLEYA
        19′ Petals white, lavender, or purple; filaments glabrous
          20. Fr ± sessile on receptacle . . . . [3]CAULANTHUS
          20′ Fr stalked above receptacle. . . . . THELYPODIUM
      18′ Stamens 4 long, 2 short, or in 3 pairs of unequal length
        21. Stamens in 3 pairs of unequal length
          22. Fr cylindric; seeds wingless . . . . [3]CAULANTHUS
          22′ Fr distinctly to ± flattened; seeds gen winged at least at tip
            23. Sepals 4–8 mm; sts (0.7)1–8 dm . . . . [2]STREPTANTHUS
            23′ Sepals 2.8–3.2 mm; st 0.5–2 dm
              24. Longest stamens with fused filaments and sterile anthers . . . . SIBAROPSIS
              24′ Longest stamens with free filaments and fertile anthers . . . . STREPTANTHELLA
        21′ Stamens 4 long, 2 short
         22. Petals dark-veined
          26. Per or subshrub, glabrous; lvs entire or dentate . . . . HESPERIDANTHUS

26′ Ann, gen hairy; lvs gen pinnately lobed
  27. Fr corky, indehiscent, lower segment rudimentary, seedless . . . . **RAPHANUS**
  27′ Fr not corky, dehiscent, lower segment well developed, seeded
    28. Fr valve 3-veined; terminal segment (0)1–6-seeded . . . . **COINCYA**
    28′ Fr valve 1-veined; terminal segment seedless . . . . **ERUCA**
25′ Petals not dark-veined
  29. Cauline lvs entire
    30. Bien or per with caudex; ovules 28–90 per ovary . . . . **BOECHERA**
    30′ Per with rhizomes or tubers; ovules 6–28 per ovary . . . . **⁴CARDAMINE**
  29′ At least proximal cauline lvs pinnately divided, lobed, wavy, or dentate
    31. Petals white, pink, lavender, or purple
      32. Ovules or seeds (1)2(4) . . . . **²CAKILE**
      32′ Ovules or seeds many
        33. Lf lateral lobes linear to thread-like, entire . . . . **SIBARA**
        33′ Lf lateral lobes not linear to thread-like, gen dentate
          34. At least most distal lvs entire or dentate . . . . **³CAULANTHUS**
          34′ All lvs pinnately lobed, divided or dissected
            35. Seeds wingless . . . . **⁴CARDAMINE**
            35′ Seeds winged . . . . **PLANODES**
    31′ Petals gen yellow
      36. Infl bracted at least below . . . . **[ERUCASTRUM]**
      36′ Infl gen not bracted
        37. Ovary, young fr 2-segmented
          38. Sepals gen erect to ascending; fr valves 1-veined . . . . **BRASSICA**
          38′ Sepals spreading to reflexed; fr valves 3–7-veined
            39. Stigma entire; seeds ovoid . . . . **HIRSCHFELDIA**
            39′ Stigma 2-lobed; seeds spheric . . . . **SINAPIS**
        37′ Ovary, young fr unsegmented
          40. Stigma entire . . . . **²RORIPPA**
          40′ Stigma 2-lobed
            41. Seeds in 2 rows; fr valves 1-veined . . . . **DIPLOTAXIS**
            41′ Seeds in 1 row; fr valves 3-veined . . . . **SISYMBRIUM**

Group 3

1. Infl bracted throughout . . . . **TROPIDOCARPUM**
1′ Infl not bracted or only most proximal few fls bracted
  2. Pl scapose; cauline lvs 0
    3. Sepals 4–5 mm; petioles of previous yr persistent . . . . **ANELSONIA**
    3′ Sepals 1.5–2.5 mm; petioles of previous yr not persistent . . . . **³DRABA**
  2′ Pl not scapose; at least 1 cauline lf present
    4. Cauline lvs sessile, basally lobed to sagittate or clasping
      5. All hairs tree-like . . . . **PHOENICAULIS**
      5′ Hairs various, not all tree-like
        6. Hairs stalked, stellate (few simple) . . . . **²ARABIS**
        6′ Hairs sessile, or if stalked then not stellate, or a mixture of > 1 type
          7. Ann; stigma 2-lobed; petals dark-veined . . . . **CAULANTHUS**
          7′ Bien or per; stigma entire; petals not dark-veined
            8. Petals ± white (pink or purple); young fr appressed to rachis; pls glaucous distally . . . . **TURRITIS**
            8′ Petals white, pink, lavender, or purple; young fr variously oriented (appressed);
               pls not glaucous distally
              9. Fr flattened . . . . **²BOECHERA**
              9′ Fr cylindric . . . . **TRANSBERINGIA**
    4′ Cauline lvs petioled, if sessile not basally lobed, sagittate, or clasping
      10. All hairs sessile . . . . **ERYSIMUM**
      10′ Hairs simple or stalked and branched
        11. Stigma 2-lobed
          12. Petals (6.5)8–10(12) mm . . . . **STRIGOSELLA**
          12′ Petals (13)15–30 mm
            13. Sepals 5–8 mm; hairs simple, stalked-forked . . . . **HESPERIS**
            13′ Sepals 10–15 mm; hairs many-branched . . . . **MATTHIOLA**
        11′ Stigma entire
          14. Cauline lvs 1–3-pinnately divided to dissected or comb-like
            15. Petals yellow; hairs many-branched . . . . **DESCURAINIA**

15′ Petals white to ± purple; hairs gen a mixture of simple, 2-forked, many-branched
  16. Per; lvs rigid . . . . **POLYCTENIUM**
  16′ Ann; lvs not rigid . . . . **SIBARA**
 14′ Cauline lvs entire, dentate, wavy (lobed)
  17. At least some lvs lobed or wavy . . . . **HALIMOLOBOS**
  17′ All lvs entire or dentate
   18. Petals yellow . . . . ³**DRABA**
   18′ Petals white, lavender, pink, or purple.
    19. Petals (4)5–20 mm; seeds winged
     20. Fr erect or ascending, gen < 2.5 mm wide; branched hairs on lf surface gen 2–4-rayed . . . . ²**ARABIS**
     20′ Fr gen reflexed, occ erect or ascending, ≥ 2 mm wide; branched hairs on lf surface 2–12-rayed
     . . . . ²**BOECHERA**
    19′ Petals 2–4(5) mm; seeds wingless
     21. Fr cylindric; seeds in 1 row . . . . **ARABIDOPSIS**
     21′ Fr flat; seeds in 2 rows . . . . ³**DRABA**

Group 4

1. Pl scapose; cauline lvs 0
 2. Ovules or seeds 1–4 . . . . ²**CUSICKIELLA**
 2′ Ovules or seeds ≥ 10
  3. Petals yellow to orange . . . . ³**DRABA**
  3′ Petals white, lavender, rose, or purple
   4. Sepals 4–5 mm; petioles of previous yr persistent . . . . **ANELSONIA**
   4′ Sepals 1.5–3 mm; petioles of previous yr not persistent . . . . ³**DRABA**
1′ Pl not scapose; at least 1 cauline lf present
 5. Cauline lvs sessile, basally lobed, sagittate, or clasping
  6. Per with well-developed woody caudex; hairs many-branched . . . . **PHOENICAULIS**
  6′ Ann or bien (per), woody caudex 0; hairs a mixture of different types
   7. Petals white (pink); at least some branched hairs sessile, stellate. . . . . **CAPSELLA**
   7′ Petals yellow (white); branched hairs stalked, not stellate . . . . **CAMELINA**
 5′ Cauline lvs petioled, if sessile then not basally lobed or clasping
  8. Infl bracted throughout . . . . **TROPIDOCARPUM**
  8′ Infl not bracted or occ most proximal fls bracted
   9. Petals strongly 2-lobed . . . . **BERTEROA**
   9′ Petals entire or notched
    10. Infl 1-sided; young fr reflexed; ann . . . . **ATHYSANUS**
    10′ Infl not 1-sided; young fr not reflexed; ann to subshrub
     11. Cauline lvs only with sessile, 2-rayed hairs . . . . **LOBULARIA**
     11′ Cauline lvs with various hair types but never only as above
      12. At least some filaments winged, dentate, or appendaged
       13. Ann; style 0.3–1.6 mm . . . . ²**ALYSSUM**
       13′ Per; style 4–12 mm . . . . **[AUBRIETA]**
      12′ All filaments wingless, not dentate, or unappendaged
       14. Ovules or seeds 1 or 2, near ovary or fr tip
        15. Petals (10)12–15 mm . . . . **DITHYREA**
        15′ Petals 0.9–3(4) mm
         16. Petals 0.9–1.3 mm; ovaries, young fr ovoid . . . . **EUCLIDIUM**
         16′ Petals 1.5–3(4) mm; ovaries, young fr round to obovate, flat. . . . . ²**ALYSSUM**
       14′ Ovules or seeds 4–100, parietal
        17. Sepals linear to oblong-linear, erect; petals narrowly oblanceolate to linear . . . . **LYROCARPA**
        17′ Sepals oblong to ovate, erect to reflexed; petals ovate, obovate, spoon-shaped, oblong, or oblanceolate
         18. Cauline lvs entire or dentate (wavy)
          19. Lf hairs stellate, ± sessile, webbed or not . . . . **PHYSARIA**
          19′ Lf hairs simple, forked, cross-shaped, tree-like, or a mixture of 2 or more of types, never only sessile-stellate
           20. Young fr with 1–4 seeds . . . . ²**CUSICKIELLA**
           20′ Young fr with > 4 seeds. . . . . ³**DRABA**
        18′ Cauline lvs 1–3-pinnately divided or comb-like
         21. Per with well-developed caudex
          22. Lvs rigid, lobes linear . . . . **POLYCTENIUM**
          22′ Lvs not rigid, lobes obovate or oblong . . . . **SMELOWSKIA**
         21′ Ann or bien
         23. Petals yellow; most or all hairs many-branched . . . . **DESCURAINIA**
         23′ Petals white; hairs simple or forked . . . . **HORNUNGIA**

Key to Groups Based on Fruiting Material

1. Fr silicle
  2. Hairs simple or 0 . . . . **Group 5**
  2′ At least some hairs branched . . . . **Group 6**
1′ Fr silique
  3. At least some hairs branched . . . . **Group 7**
  3′ Hairs simple or 0 . . . . **Group 8**

Group 5

1. Fr transversely 2-segmented
  2. Fr corky; glabrous; coastal, esp beaches . . . . **CAKILE**
  2′ Fr not corky; hairy; inland habitats
    3. Cauline lvs 1- or 2-pinnately divided; fr pedicels strongly recurved; terminal segment seedless, flattened
       . . . . **[CARRICHTERA]**
    3′ Cauline lvs dentate; fr pedicels erect to ascending, straight; terminal segment 1(3)-seeded, spheric or
       ovoid . . . . **RAPISTRUM**
1′ Fr unsegmented
  4. Fr flat parallel to septum, cylindric, or weakly flat perpendicular to septum
    5. Cauline lvs 0
      6. Fr pedicel 1 from basal rosette; seeds winged; pls terrestrial . . . . **IDAHOA**
      6′ Frs in racemes; seeds wingless; pls aquatic . . . . **SUBULARIA**
    5′ Cauline lvs 1–many
      7. Fr 3–5 cm, 2–3.5 cm wide; stalk above receptacle 0.7–1.8 cm . . . . **LUNARIA**
      7′ Fr ≤ 1.2 cm, ≤ 0.5 cm wide; stalk above receptacle 0(0.5) mm
        8. Fr dehiscent; ovules or seeds 5–90 per fr
          9. Fr valve midveins distinct; ovules, seeds 8–14 per fr; petals white . . . . **COCHLEARIA**
          9′ Fr valve midveins obscure; ovules, seeds 18–90 per fr; petals yellow . . . . **RORIPPA**
        8′ Fr indehiscent; ovules or seeds 1–2(4) per fr
          10. Fr ascending to spreading, cylindric, wings 0, radiating veins 0 . . . . ³**LEPIDIUM**
          10′ Fr reflexed, flat parallel to septum, winged, veins gen radiating . . . . **THYSANOCARPUS**
  4′ Fr clearly flat perpendicular to septum
    11. Cauline lvs sessile, basally lobed, sagittate, or clasping
      12. Fr indehiscent, reflexed, winged, with 1 seed at ± middle . . . . **ISATIS**
      12′ Fr gen dehiscent, erect to spreading to descending, not winged, 2–16-seeded
        13. Seeds 2 per fr . . . . ³**LEPIDIUM**
        13′ Seeds 2–16 per fr
          14. Seeds smooth (minutely net-like), yellow-brown to brown . . . . **NOCCAEA**
          14′ Seeds concentrically striate, ± black-brown . . . . **THLASPI**
    11′ Cauline lvs petioled (0), if sessile then not basally lobed, sagittate, or clasping
      15. Ovules or seeds 8–24 per fr
        16. Per 5–12(20) dm, roots fleshy . . . . **ARMORACIA**
        16′ Ann 0.2–3 dm, fleshy roots 0 . . . . **HORNUNGIA**
      15′ Ovules or seeds 2(4) per fr
        17. Raceme not elongated in fr; seeds winged . . . . **[IBERIS]**
        17′ Racemes elongated in fr; seeds wingless . . . . ³**LEPIDIUM**

Group 6

1. Pl scapose; cauline lvs 0
  2. Seed, ovules to 4 per fr . . . . **CUSICKIELLA**
  2′ Seed, ovules ≥ 10 per fr
    3. Seed with club-shaped hairs; fr valves with distinct net-like lateral veins . . . . **ANELSONIA**
    3′ Seed surface minutely net-like; fr valves obscurely lateral-veined . . . . ²**DRABA**
1′ Pl not scapose; at least 1 cauline lf present
  4. Cauline lvs sessile, basally lobed, sagittate, or clasping
    5. Fr obtriangular-obcordate, flat perpendicular to septum; style 0.2–0.7 mm . . . . **CAPSELLA**
    5′ Fr pear-shaped to obovoid, flat parallel to septum; style 1–3.5 mm . . . . **CAMELINA**
  4′ Cauline lvs petioled, if sessile then not basally lobed, sagittate, or clasping
    6. Fr flat perpendicular to septum, or 4-valved and septum 0
      7. Infl bracted throughout . . . . **TROPIDOCARPUM**
      7′ Infl not or only basally bracted
        8. Seeds 2 per fr; fr spectacle-shaped . . . . **DITHYREA**
        8′ Seeds 4–100 per fr; fr not spectacle-shaped
          9. Lvs gen comb-like, rigid . . . . **POLYCTENIUM**

   9′ Lvs entire or variously divided, not comb-like or rigid
     10. Fr obcordate; stigma prominently 2-lobed . . . . **LYROCARPA**
     10′ Fr not obcordate; stigma entire
       11. Style 2–9 mm; per . . . . ²**PHYSARIA**
       11′ Style ≤ 0.1 mm; ann . . . . **HORNUNGIA**
6′ Fr flat parallel to septum, cylindric, ovoid, or 4-angled
  12. Infl 1-sided; fr reflexed . . . . **ATHYSANUS**
  12′ Infl not 1-sided; fr not reflexed
   13. Cauline lvs 1–3-pinnately divided
    14. Ann or bien; petals yellow . . . . **DESCURAINIA**
    14′ Per; petals white, lavender, or purple . . . . **SMELOWSKIA**
   13′ Cauline lvs entire, crenate, dentate, or wavy
    15. Seeds 1–2(4) per fr
     16. Fr indehiscent, ovoid, occ 4-sided . . . . **EUCLIDIUM**
     16′ Fr dehiscent, flat parallel to septum
      17. Hairs stellate, occ also simple; petals yellow . . . . **ALYSSUM**
      17′ Hairs 2-rayed only; petals white to purple . . . . **LOBULARIA**
    15′ Seeds (4)6–many per fr
     18. Filaments appendaged; petals deeply 2-lobed . . . . **BERTEROA**
     18′ Filaments not appendaged; petals entire to notched
      19. Hairs ± sessile, stellate; fr septum tip gen with midvein . . . . ²**PHYSARIA**
      19′ Some hairs stalked, gen of > 1 type; fr septum tip without midvein . . . . ²**DRABA**

Group 7

1. Pl scapose; cauline lvs 0
  2. Seeds with club-shaped hairs; fr valves with distinct net-like lateral veins . . . . **ANELSONIA**
  2′ Seed surface minutely net-like; fr valves obscurely lateral-veined . . . . ²**DRABA**
1′ Pl not scapose; cauline lvs 1 to many
  3. Cauline lvs sessile, basally lobed to sagittate or clasping
   4. Fr flat parallel to septum
    5. Pl with many-branched, small, tree-like hairs; ovules or seeds (6)8–16(18) per fr . . . . **PHOENICAULIS**
    5′ Pl with stellate, forked, or few-rayed hairs, if hairs tree-like then ovules > 24
     6. Branched hairs stellate, 3–5-rayed; fr straight, erect, appressed to rachis or diverging, 0.6–2 mm wide . . . **ARABIS**
     6′ Branched hairs forked, few- to many-rayed, or tree-like; fr arc-shaped to straight, reflexed to pendent
       or spreading (erect), 1–6 mm wide . . . . ²**BOECHERA**
   4′ Fr cylindric or 4-angled
    7. Seeds in 1 row; stigma 2-lobed; branched hairs 2-rayed . . . . **CAULANTHUS**
    7′ Seeds in 2 rows; stigma entire; branched hairs tree-like or stellate
     8. Fr pedicels ascending; fr cylindric, not appressed to rachis . . . . **TRANSBERINGIA**
     8′ Fr pedicels erect; fr ± 4-angled, appressed to rachis . . . . **TURRITIS**
  3′ Cauline lvs petioled, if sessile then not basally lobed, sagittate, or clasping
  9. Infl bracted throughout . . . . **TROPIDOCARPUM**
  9′ Infl not or only basally bracted
   10. Pl hairs sessile, stellate and 2–5(8)-rayed, simple hairs 0 . . . . **ERYSIMUM**
   10′ Pl with other types of branched hairs, simple hairs gen present
    11. Stigma conic, prominently 2-lobed, lobes gen decurrent
     12. Seeds winged . . . . **MATTHIOLA**
     12′ Seed wings 0
      13. Fr glabrous, constricted between seeds; hairs simple, forked . . . . **HESPERIS**
      13′ Fr hairy, not constricted between seeds; pls with at least some many-branched hairs . . . . **STRIGOSELLA**
    11′ Stigma capitate, entire, if ± 2-lobed then lobes not decurrent
     14. Cauline lvs pinnately divided to dissected or comb-like
      15. Fr flat parallel to septum . . . . **SIBARA**
      15′ Fr cylindric or flat perpendicular to septum
       16. Lvs pinnately lobed or dissected, not rigid; fr cylindric . . . . **DESCURAINIA**
       16′ Lvs gen comb-like, rigid; fr ± flat perpendicular to septum . . . . **POLYCTENIUM**
     14′ Cauline lvs entire or dentate (wavy)
      17. Fr cylindric; seeds in 1 row
       18. Fr glabrous; cauline lvs entire . . . . **ARABIDOPSIS**
       18′ Fr hairy; at least some cauline lvs dentate to wavy . . . . **HALIMOLOBOS**
      17′ Fr gen flat parallel to septum; seeds in (1)2 rows
       19. Fr valves with elongate simple hairs mixed with stalked-forked or smaller stellate ones . . . . [**AUBRIETA**]
       19′ Fr valves glabrous or with hairs of 1 size or type
        20. Fr narrowly linear; seeds winged (wing 0) . . . . ²**BOECHERA**
        20′ Fr lanceolate to oblong (linear); seeds wingless — if winged then cauline lvs 1 . . . . ²**DRABA**

Group 8

1. Pl scapose; cauline lvs 0
  2. Per, cespitose; ovules or seeds 8–18 per fr; petals white . . . . **²CARDAMINE**
  2′ Ann (short-lived per); ovules or seeds 20–36 per fr; petals yellow . . . . **²DIPLOTAXIS**
1′ Pl not scapose; at least 1 cauline lf present
  3. Pl glandular . . . . **CHORISPORA**
  3′ Pl not glandular
   4. Fr indehiscent
    5. Fr winged, flat, lanceolate, not segmented, 1-seeded . . . . **ISATIS**
    5′ Fr wings 0, cylindric or angled, segmented, seeds gen ≥ 1
     6. Style ± absent; pl glabrous . . . . **CAKILE**
     6′ Style 1–5 mm; pl hairy . . . . **RAPHANUS**
   4′ Fr dehiscent
    7. Fr segmented
     8. Seeds in 2 rows
      9. Terminal fr segment not flattened; petal veins same color as blade . . . . **²DIPLOTAXIS**
      9′ Terminal fr segment flattened; petal veins darker than blade . . . . **ERUCA**
     8′ Seeds in 1 row
      10. Fr valves prominently 3–7-veined
       11. Fr (30)40–150-seeded; sepals erect . . . . **COINCYA**
       11′ Fr 4–16(24)-seeded; sepals spreading or reflexed . . . . **SINAPIS**
      10′ Fr valves obscurely veined or only midvein prominent
       12. Infl bracted . . . . **[ERUCASTRUM]**
       12′ Infl not bracted
        13. Fr spreading to erect, not appressed to rachis, terminal segment seedless or 1- or 2-seeded,
          if appressed then terminal segment seedless . . . . **BRASSICA**
        13′ Fr erect, appressed to rachis, terminal segment 1- or 2-seeded . . . . **HIRSCHFELDIA**
    7′ Fr unsegmented
     14. Fr flat parallel to septum
      15. Placenta margin flat in fr; fr valves coiled when dehiscent . . . . **²CARDAMINE**
      15′ Placenta margin not flat in fr; fr valves not coiled when dehiscent
       16. Seed wings 0
        17. Stalk above receptacle 6–28 mm . . . . **²STANLEYA**
        17′ Stalk above receptacle 1–4(5) mm
         18. All cauline lvs pinnately dissected, segments thread-like or linear . . . . **SIBARA**
         18′ At least some cauline lvs entire to dentate or pinnately lobed, segments not
           thread-like or linear
          19. Most distal lvs entire — stamens gen in 3 pairs of unequal length . . . . **²STREPTANTHUS**
          19′ Most distal lvs dentate or lobed
           20. Cauline lvs sessile, basally lobed or clasping; petals yellow . . . . **²BARBAREA**
           20′ Cauline lvs petioled, not basally lobed or clasping; petals white to purple . . . . **³THELYPODIUM**
       16′ Seeds winged at least at tip
        21. Stamens 4 long, 2 short; petals clawed
         22. At least some cauline lvs entire or dentate . . . . **BOECHERA**
         22′ All cauline lvs pinnately lobed . . . . **PLANODES**
        21′ Stamens in 3 pairs of unequal length; petals gen not clawed
         23. Fr strongly reflexed, tip indehiscent, forming a beak . . . . **STREPTANTHELLA**
         23′ Fr gen erect to spreading (reflexed), dehiscent throughout, beak 0.
          24. Sepals 2.8–3.2 mm; st 0.5–2 dm . . . . **SIBAROPSIS**
          24′ Sepals 4–8 mm; st (0.7)1–8 dm . . . . **²STREPTANTHUS**
     14′ Fr cylindric or 4-angled
      25. Cauline lvs pinnately compound; proximal nodes rooting . . . . **NASTURTIUM**
      25′ Cauline lvs simple; proximal nodes not rooting
       26. Stalk above receptacle 6–28 mm . . . . **²STANLEYA**
       26′ Stalk above receptacle 0–4(5) mm
        27. Cauline lvs basally lobed, sagittate, or clasping
         28. Cauline lvs dentate to pinnately lobed
          29. Seeds in 2 rows . . . . **²RORIPPA**
          29′ Seeds in 1 row
           30. Bien or per; petals not crinkled . . . . **²BARBAREA**
           30′ Ann; petals gen crinkled . . . . **³CAULANTHUS**
        28′ Cauline lvs entire or wavy
         31. Fr 4-angled; petals yellow . . . . **CONRINGIA**
         31′ Fr cylindric; petals white to purple . . . . **³THELYPODIUM**

27′ Cauline lvs not basally lobed, sagittate, or clasping
  32. Stigma prominently 2-lobed
    33. Fr valves 1-veined; petals purple to white. . . . . [3]**CAULANTHUS**
    33′ Fr valves distinctly 3-veined; petals yellow . . . . **SISYMBRIUM**
  32′ Stigma entire
    34. Per, occ subshrub, caudex woody . . . . **HESPERIDANTHUS**
    34′ Ann or bien
      35. Stalk above receptacle slender, 1–5 mm; infl spike-like. . . . . [3]**THELYPODIUM**
      35′ Stalk above receptacle stout, 0–1 mm; infl not spike-like
        36. Seeds in 1 row; petals (2.5)4–17 mm . . . . [3]**CAULANTHUS**
        36′ Seeds in 2 rows; petals 0.5–0.8 mm . . . . [2]**RORIPPA**

ALYSSUM

Ann [per]; hairs stellate, occ also simple. **ST:** 1–several. **LF:** entire. **FL:** sepals erect to spreading, bases not sac-like; petals yellow to cream, occ white in age; filaments slender or winged, toothed or not. **FR:** silicle, round (oblong), dehiscent, unsegmented, flat parallel to septum, glabrous or hairy; stigma entire. **SEED:** 1–2(4)[8]. ± 170 spp.: n N.Am, Eurasia, n Afr. (Greek: without rabies, supposed cure for hydrophobia and madness) *A. strigosum* Banks & Sol. reports in CA based on misidentified pls of *A. simplex*.

1. Sepals persistent in fr; filaments slender, not winged or toothed . . . . *A. alyssoides*
1′ Sepals early-deciduous; filaments winged and/or toothed
  2. Fr glabrous (sparsely hairy), 2.5–4(4.5) mm . . . . *A. desertorum*
  2′ Fr hairy, (3.5)4–6.5(7) mm. . . . . *A. simplex*

A. alyssoides (L.) L. (p. 523) Hairs gen stellate, appressed, some spreading and simple. **ST:** decumbent to erect, 0.5–3.5(5) dm. **FL:** sepals persistent; petals 2–3(4) mm, linear to narrowly oblanceolate, cream, in age ± white; filaments slender, not winged. **FR:** (2)3–4(5) mm, round, bulged over seed, hairy; style 0.3–0.6(1) mm, glabrous to hairy at base; pedicel spreading to ascending, 2–6 mm. **SEED:** 1–2 mm, oblong to ovate; wing to 0.1 mm. 2*n*=32. Disturbed areas, flats, ledges, bluffs; < 1800 m. e KR, n SNH; native to Eur, sw Asia, n Afr. May–Jul

A. desertorum Stapf Hairs stellate, appressed. **ST:** ascending to erect, (0.2)1–2(2.8) dm. **FL:** sepals early-deciduous; petals 2–2.5 mm, light yellow; filaments narrowly winged, not toothed. **FR:** 2.5–4(4.5) mm, round, shallowly notched, glabrous (sparsely hairy); style 0.3–0.7 mm, glabrous; pedicel ascending-spreading, 1–3.5(4.5) mm.

SEED: 1.2–1.5 mm, ovate, margined or not. 2*n*=32. Disturbed areas, rocky sagebrush flats; 1000–1500 m. SCoRI, MP; to WA, NE; native to Eur, c Asia. Apr–Jul

A. simplex Rudolphi Hairs stellate, coarse, appressed or not. **ST:** erect (decumbent), (0.3)0.7–3(4) dm. **FL:** sepals early-deciduous; petals 2–3.5 mm, light yellow; filaments winged, toothed. **FR:** (3.5)4–6.5(7) mm, round; hairs coarsely stellate, rays spreading; style 0.7–1.6 mm, hairy; pedicel spreading, 2–5(6) mm. **SEED:** 1.6–2 mm, ovate, margined or not. 2*n*=16. Roadsides, foothills, open rangeland; 1000–1500 m. NCoRI, SCoRI, MP; to MT, WY, n Mex; native to Eur, c&sw Asia, n Afr. [*A. minus* (L.) Rothm. var. *micranthum* (C.A. Mey.) T.R. Dudley; *A. minus* subsp. *micranthum* (C.A. Mey.) Breistr.] Apr–Jul

ANELSONIA

1 sp. (A. Nelson, Rocky Mtns botanist, 1859–1952)

A. eurycarpa (A. Gray) J.F. Macbr. & Payson (p. 523) Per, scapose, cespitose; caudex many-branched; hairs stalked, many-branched. **ST:** 1–4 cm, several from rosette. **LF:** basal, dense, overlapped, entire, canescent, linear to broadly oblanceolate, base tapered; cauline 0. **INFL:** scapose, umbel-like. **FL:** sepals erect, 4–5 mm, base not sac-like; petals 4.5–6 mm, white to ± purple. **FR:** 1.5–3 cm, 5–9 mm wide, lanceolate to oblong or ovate, ± purple, dehis-

cent, unsegmented, flat parallel to septum, glabrous; base ± obtuse or rounded, tip ± tapered, sharp-pointed; valves leathery; style 1–2 mm, stigma entire; pedicel erect to ascending, 4–15 mm. **SEED:** 10–24, 2–3 mm, ± flat, oblong to ovate, 2 rows per chamber; coat silvery, hairs dense, minute, club-shaped. 2*n*=14. Broken rock, talus, slopes, ridges; 1600–4100 m. SNH, SNE; to ID, NV. Jun–Jul

ARABIDOPSIS

Ann [per]; hairs simple, forked. **ST:** 1–few, branched or not, slender. **LF:** basal rosetted, petioled, entire to dentate [lobed]; cauline sessile, entire or toothed, base not lobed. **INFL:** terminal. **FL:** sepal bases not sac-like; petals white [pink or purple]. **FR:** silique, linear, glabrous, dehiscent, unsegmented, cylindric [flat parallel to septum]; pedicel ascending to spreading; stigma entire. **SEED:** many, 1 row per chamber, plump; wing 0. 10 spp.: Eur, Asia, N.Am. (Greek: resembling *Arabis*)

A. thaliana (L.) Heynh. (p. 523) MOUSE-EAR CRESS, THALE CRESS Branched. **ST:** erect, 0.5–3(5) dm; hairs below simple and forked, spreading, above 0. **LF:** basal, 0.8–3.5(4.5) cm, (1)2–10(15) mm wide, oblanceolate, entire or minutely dentate, hairs 2–4-rayed; cauline few, 0.4–1.8(2.5) cm, 1–6(10) mm wide, base tapered. **FL:** petals 2–3.5(4) mm, white. **FR:** 1–1.5 cm, 0.5–0.8 mm wide; pedi-

cel spreading, 3–10(15) mm, slender. **SEED:** 40–70, 0.3–0.5 mm, oblong. 2*n*=10. Disturbed ground, sandy areas, flats, fields; < 1600 m. NW, ScV, CaR, n SN, SnFrB, SCo, Wrn; to e US where abundant; native to Eurasia. Sp. widely used in experimental biology. Entire genome sequenced in 2000. Feb–May

ARABIS ROCKCRESS

Ann, bien, per; base not woody; hairs simple, forked, or stellate; caudex branched or not. **ST**: branched or not, lfy. **LF**: basal rosetted, petioled, entire or dentate; cauline sessile, entire or dentate, base lobed or not. **FL**: sepals erect or ascending, base of lateral pair sac-like or not; petals spoon-shaped to oblong, oblanceolate, or obovate, white to pink or deep purple. **FR**: silique, erect to spreading, linear, dehiscent, unsegmented, straight, flat parallel to septum, glabrous. **SEED**: 12–110, 1 row per chamber, flat, winged or margined. < 70 spp.: temp N.Am, Eurasia, n Afr. (Latin: of Arabia) Most spp. of *Arabis* in TJM (1993) moved to *Boechera*; *A. glabra* to *Turritis*. CA record of *A. aculeolata* Greene not confirmed.

1. Cauline lf bases lobed or sagittate
    2. Petals (5.5)6.5–9(10) mm; bases of lateral sepal pair sac-like; fr 1.2–2 mm wide; cauline lvs (2)4–12(18), gen well-spaced. . . . . ***A. eschscholtziana***
    2′ Petals 3.5–5(5.5) mm; sepal base not sac-like; fr 0.8–1(1.2) mm wide; cauline lvs (7)10–45(61), gen overlapping . . . . ***A. pycnocarpa*** var. ***pycnocarpa***
1′ Cauline lf bases tapered, not lobed or sagittate
    3. Hairs of basal lvs stellate, 3- or 4-rayed. . . . . ***A. modesta***
    3′ Hairs of basal lvs simple to 2(4)-rayed or 0
        4. Seeds ± round, 2–2.5 mm; fr 2–3 mm wide . . . . ***A. blepharophylla***
        4′ Seeds oblong, 1–1.3 mm wide; fr 1.5–2 mm wide
            5. Pl glabrous or with a few simple hairs at teeth tips of basal lvs; fr 2–4 cm . . . . ***A. mcdonaldiana***
            5′ Pl moderately to densely hairy, hairs simple, forked; fr (3)4.5–5 cm. . . . . ***A. oregana***

A. blepharophylla Hook. & Arn. (p. 523) COAST ROCKCRESS
Per; caudex branches 0–few; hairs simple to forked or (3)4-rayed. **ST**: 1–few, simple or few-branched distally, 0.6–2.5(3) dm, hairy (glabrous). **LF**: basal 2–8(12) cm, oblanceolate to obovate, entire or dentate, hairy or glabrous, margin hairy, tip obtuse; cauline 2–7, not basally lobed. **FL**: sepals 5–7 mm; petals (12)14–18 mm, 4–7 mm wide, widely spoon-shaped, rose-purple. **FR**: erect, 2–4 cm, 2–3 mm wide; style 0.2–1(1.5) mm; pedicel erect to ascending, (3)5–10(15) mm, slender, hairy. **SEED**: 20–28, 2–2.5 mm, ± round; wing 0.2–0.4 mm wide. 2*n*=16. Rocky outcrops, bluffs, grassy slopes; 50–300 m. CCo, SnFrB. Mar–Apr ★

A. eschscholtziana Ledeb. (p. 523) Bien or per; caudex branches 0–few; hairs simple to forked or 0. **ST**: 1–few, gen branched distally, 2–7(10) dm, hairy (glabrous). **LF**: basal 2–13 cm, narrowly oblanceolate to spoon-shaped, entire or dentate, hairy to glabrous; cauline (2)4–12(18), gen well spaced, basally lobed. **FL**: sepals 3.5–5 mm, base of lateral pair sac-like; petals (5.5)6.5–9(10) mm, 1.5–3 mm wide, narrowly spoon-shaped to linear-oblanceolate, white (pink). **FR**: erect to ascending, 3.5–6.5 cm, 1.2–2 mm wide; style (0.1)0.3–1 mm; pedicel erect to ascending, 3.5–10(15) mm, slender, glabrous or sparsely hairy. **SEED**: 54–80, 1–1.8 mm, oblong; wing to 0.2 mm wide, or 0. 2*n*=32,64. Gravelly soils, swales, open woodland, meadows, rock crevices, ledges; < 2800 m. KR, NCoRO, CaR, SN, SnBr, W&I; to AK, YT, WY, NV. [*A. hirsuta* (L.) Scop. var. *glabrata* Torr. & A. Gray] May–Jul

A. mcdonaldiana Eastw. MCDONALD'S ROCKCRESS Per; caudex branched. **ST**: few to many, simple, (0.6)1.5–3(4) dm, slender, glabrous. **LF**: basal 1–4(5.5) cm, entire, wavy-margined to few-dentate, glabrous or teeth bristle-tipped; cauline (2)3–5(6), well spaced, oblong, entire, base tapered, not basally lobed. **FL**: sepals 4–8 mm; petals 8–16 mm, 2–5 mm wide, spoon-shaped, purple. **FR**: erect to ascending, 2–4 cm, 1.5–2 mm wide, flat, glabrous; style 0.3–1.5 mm; pedicel ascending, 3–10(13) mm, glabrous. **SEED**: 20–34, oblong; wing 0.1–0.2 mm wide, at 1 end or 0. 2*n*=16. Deep ± red soils, steep slopes, dry ridges, serpentine areas; 200–1800 m. NCoRO. [*Arabis macdonaldiana* Eastw., orth. var.] May–Jun ★

A. modesta Rollins MODEST ROCKCRESS Per; caudex simple or few-branched; hairs 3- or 4-rayed, stalked. **ST**: 1 or few, simple or gen branched distally, 2.2–5.5(6.7) dm, hairy. **LF**: basal 2–8(11) cm, obovate to oblanceolate, margin entire to wavy or dentate; cauline (2)4–6(9), well spaced, base not lobed, oblong, shallowly dentate to entire. **FL**: sepals (4)5–6.5(8) mm; petals (10)12–16(20) mm, 4–6(7) mm wide, spoon-shaped, purple. **FR**: ascending to spreading, (2.8)3.5–6 cm, 1.5–2 mm wide, glabrous; style 0.5–1(1.5) mm; pedicel ascending to spreading, 0.7–1.8(2.5) cm. **SEED**: 20–34, 1.7–2.2, oblong; wing 0.2–0.5 mm wide, at 1 end . 2*n*=32. Deep soil on steep slopes, cliffs, shaded canyon ledges; 150–500 m. KR, se NCoRI; s OR. Isolated populations near Yolo-Napa Co. line (se NCoRI) need study. Mar–May ★

A. oregana Rollins OREGON ROCKCRESS Per; caudex branched or not; hairs simple and long-stalked-forked to 2 mm, some 3- to 4-rayed. **ST**: 1 or few, branched or simple, 2–5 dm; hairs simple and forked. **LF**: basal 1.5–8(12) cm, obovate to oblanceolate, entire to wavy-margined or dentate, hairy on surfaces or occ margins only; cauline 3–6(7), not basally lobed, oblong, entire or dentate. **FL**: sepals (5)6–8 mm; petals (10)12–15(16) mm, (3)4–5 mm wide, purple or pink. **FR**: erect to ascending, (3)4.5–5 cm, 1.5–2 mm wide, straight, glabrous; style 0.5–1 mm; pedicel ascending to erect, 5–10 mm, hairy to glabrous. **SEED**: 24–30, oblong; wing to 0.1 mm wide, at 1 end or 0. 2*n*=32. Rocky hillsides, steep banks, chaparral; 500–1400 m. KR, NCoRO; s OR. Apr–May ★

A. pycnocarpa M. Hopkins var. ***pycnocarpa*** Bien or weak per; caudex branched or not; hairs simple and short-stalked forked, coarse, spreading. **ST**: 1–several, simple or branched distally, erect, 1–8 dm. **LF**: basal short-petioled, 2–10 cm, oblong to obovate or spoon-shaped, entire to dentate; cauline (7)10–45(61), ovate to oblong or lanceolate, sagittate or lobed at base. **FL**: sepals 2–4 mm; petals 3.5–5(5.5) mm, 1–2(2.5) mm wide, white. **FR**: (3.5)4–6 cm, 0.8–1(1.2) mm wide; style (0.2)0.5–1 mm; pedicel 0.5–1.5 cm, glabrous to sparsely hairy. **SEED**: (54)60–86, oblong or ± round; wing to 0.2 mm, continuous. 2*n*=32. Gravelly soils, swales, disturbed sites, meadows, shady slopes, occ moist; < 2500 m. KR, NCoRO, CaR, SN, SnBr, W&I; Can, US exc AK, HI. [*A. hirsuta* (L.) Scop. var. *p.* (M. Hopkins) Rollins] Other var. in ON, c US. Mar–Jul

ARMORACIA

Per, glabrous; roots fleshy. **LF**: basal long petioled, simple, entire or crenate; cauline entire to pinnately lobed, gen not basally lobed. **INFL**: panicle. **FL**: sepal spreading to ascending, base not sac-like; petals short-clawed, white. **FR**: silicle, dehiscent, unsegmented, ovate to oblong or ± round, flat perpendicular to septum; stigma entire or ± 2-lobed. **SEED**: in 2 rows per chamber; wing 0. 3 spp.: Eurasia. (Greek: horse radish)

A. rusticana G. Gaertn. et al. (p. 523) HORSE RADISH Gen erect; roots deep. **ST**: branched distally, 5–12(20) dm. **LF**: basal 1.5–12 dm, wide-oblong to ovate, crenate; proximal cauline petioled, distal sessile, pinnately lobed or occ entire. **FL**: sepals 2–4 mm; petals obovate, 5–7(8) mm. **FR**: 4–6 mm, gen aborted; style ≤ 0.5 mm; pedicel ascending, 0.8–2 cm. **SEED**: 8–12, ovate; rare. 2*n*=32. Ditches, roadsides, moist places, fields; < 1100 m. GV, CCo, SnFrB; temp N.Am; native to Eurasia. Cult over 2000 yrs for roots. Mar–Jul

ATHYSANUS

Ann; hairs simple, or short-stalked with 2–4-rays. **ST**: simple or branched at base, lfy. **LF**: basal not rosetted, short petioled, entire or dentate; cauline short-petioled or sessile, base not lobed, entire or dentate. **INFL**: 1-sided; bracts 0. **FL**: sepals erect, early-deciduous, base not sac-like; petals < to > sepals or rudimentary, white. **FR**: silicle, reflexed, indehiscent or late-dehiscent, unsegmented, round or obovate to elliptic, flat parallel to septum, glabrous to hairy; pedicel recurved. **SEED**: 1 or 6–12, in 1 row, flat, not winged. 2 spp.: w N.Am, nw Mex. (Greek: without fr fringe)

1. Fr indehiscent, not twisted, hairs hooked if present; septum 0; seed 1 . . . . *A. pusillus*
1′ Fr late-dehiscent, gen twisted, hairs not hooked if present; septum present; seeds 6–12 . . . . *A. unilateralis*

A. pusillus (Hook.) Greene (p. 523)   Hairy at least proximally. **ST**: (2)5–30(50) cm. **LF**: basal 4–25 mm, 1–8 mm wide, oblanceolate to obovate or oblong, entire or dentate; cauline 1–6, like basal, sessile. **FL**: sepals 0.5–1 mm; petals 1–3 mm, spoon-shaped, occ 0. **FR**: 2–2.5 mm wide, ± round, smaller hairs branched; style to 0.2 mm; pedicel (1.5)2–4(6) mm. **SEED**: to 1.2 mm. 2*n*=26. Grassy, open slopes, rocky outcrops, chaparral, flats, floodplains, cliffs, ledges; < 2000 m. CA-FP, MP, DMtns (Granite Mtns); to BC, MT, NV, Mex. Feb–Jun

A. unilateralis (M.E. Jones) Jeps. (p. 523)   **ST**: hairy; branched near base, prostrate to ascending, (3)7–25(35) cm. **LF**: basal nearly sessile, 0.5–2.2 cm, oblanceolate to obovate or oblong, entire to each side 1-toothed; hairs 3–4-rayed; cauline 2–5, tapered at base. **FL**: sepals 0.6–1 mm; petals 1.3–1.7 mm. **FR**: 3–5 mm, obovate to elliptic or round; style 0.1–0.2 mm; pedicel 1–2(3) mm. **SEED**: 0.9–1.2 mm, plump, oblong. Uncommon. Grassy, open slopes, flats, clay soils, floodplains, gypsum-clay slopes; 100–900 m. CaRF, s SNF, Teh, GV, SCoR; OR, Baja CA. [*Heterodraba u.* (M.E. Jones) Greene] Feb–May

BARBAREA   WINTER CRESS, ROCKET

Bien, per, erect; hairs simple or 0. **ST**: angled. **LF**: basal petioled, rosetted, pinnately lobed, terminal lobe > lateral; cauline dentate or pinnately lobed; middle, distal sessile, base lobed. **INFL**: terminal; bracts 0. **FL**: sepals erect or spreading, base not sac-like; petals yellow, ± clawed. **FR**: silique, linear, dehiscent, unsegmented, cylindric to ± 4-sided or flat parallel to septum; valves strongly 1-veined, glabrous; stigma entire to ± 2-lobed. **SEED**: 10–40, in 1 row, wingless. 22 spp.: N.Am, Eurasia, Australia, n Afr. (Saint Barbara)

1. Fr 4.5–8 cm; seeds (34)38–48(52); pedicel width ± = fr; distal lvs pinnately lobed . . . . *B. verna*
1′ Fr (0.7)1.5–4(4.5) cm; seeds 16–36; pedicel width gen < fr; distal lvs undivided or lobed
  2. Style 0.2–1.2(2) mm, stout; tips of sepals or basal lobes of distal lvs gen with 1–many hairs . . . . *B. orthoceras*
  2′ Style (1)1.5–3.5 mm, slender; tips of sepals, basal lobes of distal lvs glabrous . . . . *B. vulgaris*

B. orthoceras Ledeb. (p. 523)   Glabrous exc sepals, basal lf lobe tips. **ST**: (1)2–6(10) dm, branched distally. **LF**: pinnately lobed; basal 1.5–12(25) cm, lateral lobe pairs (1)2–4(5), terminal lobe much larger, ovate, entire or irregularly toothed; cauline lobed (not), base strongly lobed. **FL**: sepals 2.5–3.5 mm; petals 5–7(8) mm, bright yellow. **FR**: erect to ascending, (2.5)3–4(4.5) cm; style 0.2–1.2(2) mm; pedicel (2)3–6(7) mm, narrower than fr. **SEED**: 24–36, ovate or oblong, 1.2–1.5 mm. 2*n*=16. Damp meadows, wet rocks, streambanks, moist woodland, grassland, scree, ledges; < 3400 m. CA-FP (exc GV), MP, n SNE, W&I; AK, Can, e N.Am, also c&e Asia. Mar–Jul

B. verna (Mill.) Asch. (p. 523)   EARLY WINTER CRESS   **ST**: (1)2.5–8 dm, branched, glabrous. **LF**: pinnately lobed; basal 2–18 cm, lateral lobe pairs (3)6–10; cauline lateral lobe pairs 1–4, base strongly lobed. **FL**: sepals 3–5 mm; petals (5)6–7(8.5) mm, yel-

low. **FR**: ascending, 4.5–8 cm, straight; style 0.2–1(2) mm; pedicel (2)3–6(7) mm, stout, width ± = fr. **SEED**: (34)38–48(52), 1.8–2.5 mm, oblong. 2*n*=16. Uncommon. Damp soils, fields, rocky outcrops, disturbed areas; < 1600 m. SCoRO (Monterey Co.); widespread to BC, e N.Am; native to Eurasia. Mar–Jul

B. vulgaris W.T. Aiton (p. 523)   YELLOW ROCKET   Glabrous to sparsely hairy. **ST**: (1.5)2–9(12) dm, branched distally. **LF**: basal lateral lobe pairs 1–3(5); distal cauline lvs coarsely dentate (entire). **FL**: sepals 3–5 mm; petals (5)6–9(10) mm, yellow. **FR**: erect to ascending, (1)1.5–3 cm, flat or 4-sided; style (1)1.5–3.5 mm; pedicel 3–7 mm, narrower than fr. **SEED**: 18–24(28), 1.2–1.5 mm, oblong. 2*n*=16. Uncommon. Disturbed areas, riverbanks, grassland, fields; 1000–3000 m. GB; to AK, Can, most of US; native to Eurasia. Apr–Jul

BERTEROA

Ann or bien [per]; hairs simple, stellate. **ST**: simple, branched distally. **LF**: basal petioled, entire to dentate, not rosetted; cauline sessile, not lobed at base. **INFL**: terminal; fls many. **FL**: sepals erect to spreading, base not sac-like; petals white, 2-lobed; lateral filaments basally appendaged. **FR**: silicle, dehiscent, oblong to elliptic [ovate, round], hairy or glabrous, unsegmented, flattened parallel to septum; stigma entire. **SEED**: 4–16, in 2 rows, plump or flattened; wing narrow or 0. 5 spp.: Eurasia. (C.G. Bertero, Italian-Chilean botanist, physician, 1789–1831)

B. incana (L.) DC.   Densely hairy throughout. **ST**: (2)3–8(11) dm, erect. **LF**: basal (2.5)3.5–8(10) cm, oblanceolate, entire or wavy. **FL**: sepals 2–2.5 mm; petals (4)5–6.5(8) mm, obcordate, lobes 1–3 mm. **FR**: (4)5–8.5(10) mm, 2–4 mm wide; style slender, 1–4 mm; pedicel

erect to ascending, (4)5–9(12) mm. **SEED**: 1–2.3 mm, lens-shaped or ovate-round, flattened, narrow-margined. 2*n*=16. Uncommon. Disturbed areas, fields; < 2800 m. MP, expected elsewhere; to BC, e N.Am; native to Eurasia. May–Sep   ◆

BOECHERA   ROCKCRESS

Michael D. Windham & Ihsan A. Al-Shehbaz

Per (bien); caudex simple or branched, persistent lf bases gen absent; rosetted or not; rosette at ground surface or elevated on woody base; hairs simple or 2–14-rayed, stalked or sessile. **ST**: simple or branched, lfy. **LF**: basal petioled, simple, gen entire

or dentate, gen hairy; cauline sessile, base gen lobed, entire or dentate. **INFL**: gen elongated. **FL**: sepals bases gen not sac-like; petals gen white, lavender, or purple, claw present or 0; pollen ellipsoid in sexual pls, spheric in pls with asexual seeds. **FR**: silique, dehiscent, gen linear, edges gen parallel, unsegmented, flat parallel to septum; stigma entire or 2-lobed. **SEED**: in 1 or 2 rows, winged or not. 110+ spp.: temp N.Am, Russian Far East. (T.W. Boecher, Danish cytogeneticist, 1909–1983) [Windham & Al-Shehbaz 2006–2007 Harvard Pap Bot 11:61–88, 11:257–274, 12:235–257] Some spp. with both fertile & sterile sts. Previously incl in *Arabis*, but the 2 genera in different tribes. *B. horizontalis* (Greene) Windham & Al-Shehbaz [*Arabis suffrutescens* S. Watson var. *horizontalis* (Greene) Rollins] not in CA.

1. Fr, gen ovary hairy
  2. Fr pedicel < 2.5 mm; petals ± yellow to brick-red . . . . ***B. yorkii***
  2′ Fr pedicel > 2.5 mm; petals purple, lavender, or white
   3. Fr pedicel base gen abruptly recurved; fr gen appressed to infl axis
    4. Sts elevated above ground surface on woody base; petals 9–16 mm; fr 2.5–4 mm wide; seeds 1.7–2.8 mm . . . . ***B. pulchra***
    4′ Sts from ± ground surface, from non-woody caudex; petals 4–8 mm; fr 0.9–1.8 mm wide; seeds 1–1.4 mm . . . . [3]***B. retrofracta***
   3′ Fr pedicel base not abruptly recurved; fr gen not appressed to infl axis
    5. Proximal sts with simple and 2-rayed hairs only
     6. Fr pedicel with branched hairs only . . . . [2]***B. arcuata***
     6′ Fr pedicel with gen simple hairs . . . . [2]***B. breweri*** subsp. ***breweri***
    5′ Proximal sts with at least some 3–12-rayed hairs
     7. Petals gen < 9 mm
      8. St proximally with 7–12-rayed hairs; seeds 1–1.3 mm, gen in 2 rows . . . . [2]***B. shockleyi***
      8′ St proximally with gen 3–7-rayed hairs; seeds 1.4–6 mm, in 1 row
       9. Basal lvs > 6 mm wide; fr erect to ascending; seeds 2.5–6 mm, wing 0.7–1.8 mm wide . . . . [2]***B. repanda***
       9′ Basal lvs < 6 mm wide; fr horizontal to pendent (appressed); seeds 1.4–2.5 mm, wing 0.05–0.3 mm wide
        10. Style < 0.2 mm; basal lvs gen dentate; fr 1.9–2.2 mm wide; seeds 1.4–1.8 mm . . . . ***B. puberula***
        10′ Style 1.5–2 mm; basal lvs entire; fr 2.5–3 mm wide; seeds 2–2.5 mm . . . . [2]***B. serpenticola***
     7′ Petals gen > 9 mm
      11. At least some basal lvs prominently dentate to ± pinnately lobed; style 0.5–1 mm; ovules < 60 . . . . ***B. subpinnatifida***
      11′ Basal lvs gen entire; style gen < 0.5 mm; ovules > 60
       12. Basal lvs 1–2 mm wide; infl 7–15-fld; fr pedicel gen ascending near base, recurved near tip . . . . ***B. lincolnensis***
       12′ Basal lvs 3–13 mm wide; infl > 15-fld; fr pedicel descending or horizontal near base
        13. Fr gen > 7 cm, sparsely hairy throughout; seeds 1.7–2 mm, gen in 1 row. . . . . [2]***B. californica***
        13′ Fr gen < 7 cm, hairy on distal 2/3, glabrous near base; seeds 2–2.5 mm, gen irregularly in 2 rows . . . ***B. xylopoda***
1′ Fr, ovary glabrous
 14. Cauline lf base not lobed
  15. Style > 2.2 mm
   16. Petals ± white, 6–8 mm; fr pendent, 4–7.5 cm . . . . ***B. constancei***
   16′ Petals lavender to purple, 8–13 mm; fr ascending, 1.5–2.5 cm . . . . ***B. parishii***
  15′ Style < 2.2 mm
   17. Basal lvs glabrous or hairs simple only
    18. Lf margins with minute simple hairs; petals 4–5 mm . . . . ***B. shevockii***
    18′ Lvs glabrous; petals 6–10 mm
     19. Caudex with crowded, persistent lf bases; fr ascending, not appressed to infl axis; seeds in 1 row . . . ***B. davidsonii***
     19′ Caudex without crowded, persistent lf bases; fr erect, appressed to infl axis; seeds gen in 2 rows. . . . . . [3]***B. lyallii***
   17′ Basal lvs with at least some branched hairs
    20. Pl gen glabrous throughout exc margins of basal lvs
     21. Fr gen > 3 mm wide; seeds 3–6 mm
      22. Margins of basal lvs with some 3–4-rayed hairs; ovules 18–30; pollen spheric. . . . . [3]***B. covillei***
      22′ Margins of basal lvs with simple and 2-rayed hairs; ovules 10–20; pollen ellipsoid . . . . [2]***B. howellii***
     21′ Fr gen < 3 mm wide; seeds 1.5–2.2 mm
      23. Petals 6–8.5 mm; fr pedicel erect; seeds gen in 2 rows. . . . . [3]***B. lyallii***
      23′ Petals 4–5 mm; fr pedicel ascending; seeds in 1 row . . . . [2]***B. tiehmii***
    20′ Pl at least hairy on surfaces of basal lvs, gen throughout pl
     24. Fr pendent or descending
      25. Fr 5–8 mm wide; seeds 5–6 mm . . . . ***B. glaucovalvula***
      25′ Fr < 3 mm wide; seeds < 2 mm
       26. Fr pedicel base gen abruptly recurved; fr reflexed, gen appressed to infl axis; cauline lvs 15–40; infl gen > 15-fld. . . . . [3]***B. retrofracta***
       26′ Fr pedicel base not abruptly recurved; fr wide-pendent, not appressed to infl axis; cauline lvs 2–17; infl < 15-fld
        27. Surfaces of basal lvs with simple and 2-rayed hairs. . . . . [2]***B. pendulina***
        27′ Surfaces of basal lvs with 4–8-rayed hairs. . . . . ***B. pendulocarpa***

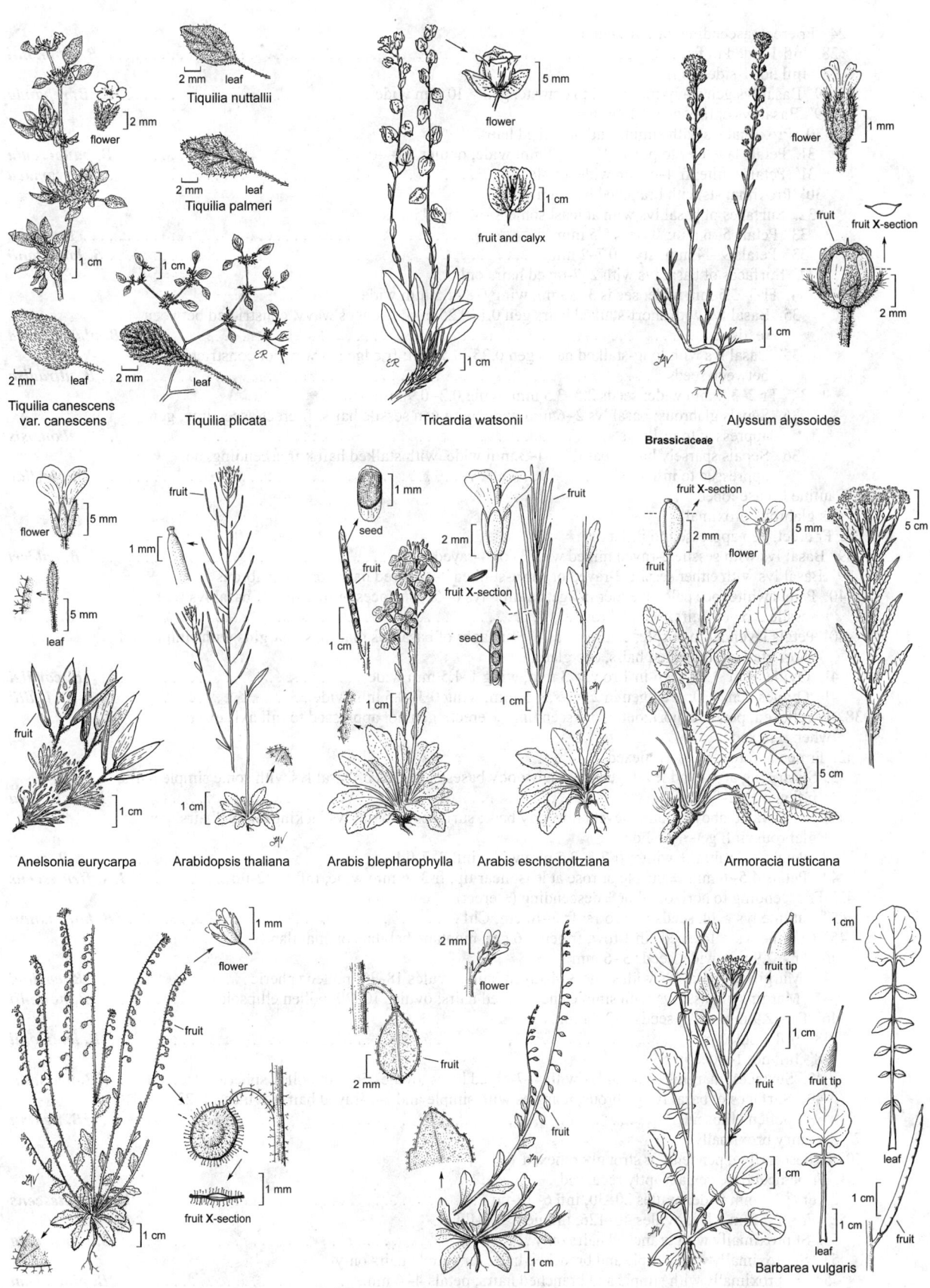

Tiquilia nuttallii

Tiquilia palmeri

Tiquilia canescens
var. canescens

Tiquilia plicata

Tricardia watsonii

Alyssum alyssoides

Brassicaceae

Anelsonia eurycarpa

Arabidopsis thaliana

Arabis blepharophylla

Arabis eschscholtziana

Armoracia rusticana

Athysanus pusillus

Athysanus unilateralis

Barbarea orthoceras

Barbarea vulgaris

Barbarea verna

24′ Fr erect, ascending, or horizontal
  28. Infl 1-sided in fr . . . . [3]***B. lemmonii***
  28′ Infl not 1-sided in fr
    29. Basal lvs gen wavy-margined to dentate, gen > 10 mm wide . . . . [2]***B. repanda***
    29′ Basal lvs entire, gen < 10 mm wide
      30. Proximal sts with simple and branched hairs
        31. Petals lavender to purple; fr 1.3–2 mm wide; ovules 24–40. . . . . [2]***B. paupercula***
        31′ Petals white; fr 4–5 mm wide; ovules 8–12 . . . . ***B. pygmaea***
      30′ Proximal sts with branched hairs only
        32. Surfaces of basal lvs with at least some 8–14-rayed hairs
          33. Petals 5–6 mm; style < 0.5 mm. . . . . ***B. dispar***
          33′ Petals 9–14 mm; style 0.7–2 mm. . . . . ***B. johnstonii***
        32′ Surfaces of basal lvs with 2–7-rayed hairs only
          34. Fr 3–5.5 mm wide; seeds 3–8 mm, wing 0.8–2.5 mm wide
            35. Basal lvs with short-stalked hairs gen 0.1–0.25 mm; fr edges wavy, constricted between
              seeds . . . . ***B. platysperma***
            35′ Basal lvs with long-stalked hairs gen 0.25–0.5 mm; fr edge not wavy or constricted
              between seeds . . . . ***B. ultraalsa***
          34′ Fr 2–3.2 mm wide; seeds 2.5–3.5 mm, wing 0.2–0.9 mm wide
            36. Sepals glabrous; basal lvs 2–6 mm wide, with gen sessile hairs; fr erect-ascending, gen
              appressed to infl axis . . . . [2]***B. elkoensis***
            36′ Sepals sparsely hairy; basal lvs 1–3 mm wide, with stalked hairs; fr ascending, not
              appressed to infl axis . . . . [2]***B. pinzliae***
14′ Cauline lf base lobed
 37. Sts glabrous proximally
  38. Fr erect, gen appressed to infl axis
    39. Basal lvs with sessile 2-rayed mixed with 3- or 4-rayed hairs . . . . ***B. calderi***
    39′ Basal lvs with either sessile 2-rayed or short-stalked 2–6-rayed hairs, or occ glabrous
      40. Petals white, occ pale lavender in age; cauline lvs 6–52; surfaces or margins of basal lvs with
        sessile 2 rayed hairs, occ glabrous . . . . [2]***B. stricta***
      40′ Petals lavender to purple; cauline lvs 1–7; surfaces of basal lvs glabrous, margins with simple and
        short-stalked, branched hairs, occ glabrous
        41. Ovules gen < 30; seed in 1 row, > 3 mm, wing 1–1.5 mm wide . . . . [3]***B. covillei***
        41′ Ovules gen > 30; seed gen in 2 rows, < 3 mm, wing 0.3–0.5 mm wide . . . . [3]***B. lyallii***
  38′ Fr reflexed, pendent, horizontal, or ascending (± erect), gen not appressed to infl axis exc gen
    when reflexed
    42. Fr pendent to strongly reflexed
      43. Sts arising at ground level, gen lacking woody bases; surfaces of basal lvs with some simple hairs;
        infl gen > 12-fld. . . . . [2]***B. rectissima***
      43′ Sts elevated above ground level on woody base; surfaces of basal lvs lacking simple hairs, gen
        glabrous; infl gen < 12-fld
        44. Petals 8–11 mm, ± white; fr 2.5–3.5 mm wide; infl 3–7-fld. . . . . ***B. rollei***
        44′ Petals 4.5–6 mm, ± purple or rose at least near tip; fr 3–6 mm wide; infl 6–12-fld . . . . [3]***B. suffrutescens***
    42′ Fr ascending to horizontal or ± descending (± erect)
      45. Cauline lvs > 14; seeds in 2 rows; fr 6–10 cm; ChI . . . . [2]***B. hoffmannii***
      45′ Cauline lvs < 14; seeds in 1 row; fr gen < 6 cm; montane habitats on mainland
        46. Fr > 2.9 mm wide; seeds 3–6 mm
          47. Margins of basal lvs with some 3–4-rayed hairs; ovules 18–30; pollen spheric . . . . [3]***B. covillei***
          47′ Margins of basal lvs with simple and 2-rayed hairs; ovules 10–20; pollen ellipsoid . . . . [2]***B. howellii***
        46′ Fr < 2.9 mm wide; seeds 1–2.2 mm
          48. Infl 1-sided in fr. . . . . [3]***B. lemmonii***
          48′ Infl not 1-sided in fr
            49. Surfaces, margins of basal lvs with 4–7-rayed hairs; ovules 56–80; pollen spheric. . . . . [2]***B. peirsonii***
            49′ Surfaces of basal lvs glabrous, margins with simple and 2–3-rayed hairs; ovules 14–22;
              pollen ellipsoid . . . . [2]***B. tiehmii***
 37′ Sts hairy proximally
  50. Fr descending, pendent, or strongly reflexed
    51. Fr pedicel base gen abruptly recurved
      52. Fr > 2.7 mm wide; ovules 20–30; infl 6–12-fld . . . . [3]***B. suffrutescens***
      52′ Fr < 2.7 mm wide; ovules 46–126; infl gen > 12-fld
        53. St proximally with branched hairs only . . . . [3]***B. retrofracta***
        53′ St proximally with simple and branched hairs, or simple hairs only
        54. St proximally with simple and branched hairs; petals 4–6 mm. . . . . [3]***B. pinetorum***
        54′ St proximally with simple hairs only; petals 3–4 mm. . . . . [2]***B. rectissima***
    51′ Fr pedicel base ± recurved or straight

55. Fr pedicel hairy
  56. St proximally with simple and branched hairs, or simple hairs only
    57. Petals gen > 8 mm, purple to magenta; sts elevated above ground level on woody caudex with
      crowded, persistent lf bases; pollen ellipsoid . . . . ³*B. koehleri*
    57′ Petals gen < 8 mm, lavender to white; sts from ± ground level, gen lacking woody caudex and
      crowded, persistent lf bases; pollen spheric
      58. Fr spreading-descending; fr pedicel with many 2–3-rayed hairs. . . . . ²*B. pauciflora*
      58′ Fr reflexed to ± appressed; fr pedicel with few simple and 2-rayed hairs only, occ glabrous. . . . ³*B. pinetorum*
  56′ St proximally with branched hairs only
    59. Petals gen > 9 mm; fr gen > 6 cm . . . . ²*B. californica*
    59′ Petals gen < 9 mm; fr gen < 6 cm
      60. Surfaces of basal lvs with hairs 0.1–0.2 mm; seed wing 0.25–0.5 mm wide; ovules 34–64;
        pollen ellipsoid . . . . ²*B. cobrensis*
      60′ Surfaces of basal lvs with hairs 0.2–0.7 mm; seed wing < 0.25 mm wide; ovules 74–134;
        pollen spheric . . . . ²*B. inyoensis*
55′ Fr pedicel glabrous (sparsely hairy)
  61. Petals gen > 8 mm, purple to magenta; caudex woody, lf bases crowded, persistent . . . . ³*B. koehleri*
  61′ Petals gen < 8 mm, white to lavender (rose-purple); caudex woody or not, crowded-persistent
    lf bases 0
    62. Fr > 2.7 mm wide; seed wing > 0.7 mm wide; sts elevated above ground surface on
      woody caudex . . . . ³*B. suffrutescens*
    62′ Fr < 2.7 mm wide; seed wing < 0.7 mm wide; sts gen not elevated above ground surface on
      woody caudex
      63. Fr reflexed to ± appressed; pollen spheric; KR, CaR, SN
        64. St proximally with stalked, 2–5-rayed hairs; seeds 1.5–1.8 mm . . . . ³*B. pinetorum*
        64′ St proximally with ± sessile, 2-rayed hairs; seeds 2–2.5 mm . . . . *B. tularensis*
      63′ Fr wide-pendent to spreading-descending, not appressed; pollen ellipsoid; SW, GB, D
        65. St proximally gen simple-hairy; seeds in 2 rows . . . . ²*B. pendulina*
        65′ St proximally gen branched-hairy; seeds in 1 row
          66. Basal lvs dentate; st proximally with 2–5-rayed hairs . . . . *B. perennans*
          66′ Basal lvs entire; st proximally with 4–10-rayed hairs . . . . ²*B. cobrensis*
50′ Fr erect, ascending, or horizontal (pendent)
  67. Lf hairs ± sessile, branches gen appressed; sepals gen glabrous
    68. Basal lvs with branched hairs 2-rayed only; petals white, occ pale lavender in age; pollen ellipsoid . . . ²*B. stricta*
    68′ Basal lvs with some 3–4-rayed hairs; petals purple to lavender (± white); pollen spheric
      69. Fr erect to ascending, gen appressed to infl axis
        70. Cauline lvs 3–6; seeds 2.5–3.5 mm . . . . ²*B. elkoensis*
        70′ Cauline lvs > 6; seeds 1.4–2 mm . . . . *B. pratincola*
      69′ Fr spreading-ascending to horizontal, not appressed to axis
        71. Cauline lvs gen > 10; fr 1.7–2.5 mm wide; seeds 1.4–2 mm . . . . *B. divaricarpa*
        71′ Cauline lvs gen < 10; fr 2.5–3.5 mm wide; seeds 2.5–3.2 mm . . . . *B. rigidissima*
  67′ Lf hairs stalked, branches not appressed; sepals gen hairy
    72. St proximally with some simple hairs
      73. Fr erect, appressed to infl axis; ovules 24–40 . . . . ²*B. paupercula*
      73′ Fr ascending to horizontal, not appressed to infl axis; ovules > 40
        74. Fr pedicel hairs 2–4-rayed only
          75. Fr 1.2–1.5 mm wide, ovules 46–52; lf surface hairs < 0.3 mm . . . . *B. evadens*
          75′ Fr 1.5–2.2 mm wide, ovules 80–250; lf surface hairs ≥ 0.3 mm
            76. Petals 9–14 mm, purple; fr pedicel gen > 10 mm; pollen ellipsoid . . . . ²*B. arcuata*
            76′ Petals 5–8 mm, lavender to ± white; fr pedicel gen < 10 mm; pollen spheric. . . . . ²*B. pauciflora*
        74′ Fr pedicel glabrous or with simple, occ some 2-rayed hairs
          77. Basal lvs gen ≤ 3 mm wide; caudex woody, with crowded, persistent lf bases . . . . ³*B. koehleri*
          77′ Basal lvs gen > 3 mm wide; caudex woody or not, crowded-persistent lf bases 0
            78. Distal cauline lvs glabrous, occ ciliate; cauline lf basal lobes gen > 3 mm; NW, CaR, SNE . . . *B. sparsiflora*
            78′ Distal cauline lvs hairy; cauline lf basal lobes gen < 3 mm; widespread
              79. St proximally with some 3 rayed, 0.1–0.5 mm hairs. . . . . *B. rubicundula*
              79′ St proximally with simple and 2-rayed hairs only (all simple), 0.4–1 mm . . . . *B. breweri*
                80. St 0.6–2 dm; fr pedicel gen < 9 mm, hairy . . . . ²subsp. *breweri*
                80′ St 1.8–4.5 dm; fr pedicel gen > 9 mm, glabrous or hairs few, scattered . . . . subsp. *shastaensis*
    72′ St proximally with branched hairs only
      81. Infl 1-sided in fr. . . . . ³*B. lemmonii*
      81′ Infl not 1-sided in fr
        82. St proximally with at least some 2–3-rayed hairs
          83. Petals 8–10 mm; infl 30–70-fld; pollen ellipsoid; ChI . . . . ²*B. hoffmannii*
          83′ Petals 4–7 mm; infl 5–25-fld; pollen spheric; mainland montane areas
            84. Petals white or lavender; fr < 1.9 mm wide; KR. . . . . *B. acutina*

84′ Petals purple; fr > 1.9 mm wide; SnBr, W&I
  85. Infl 12–25-fld; st proximally with hairs 0.4–0.6 mm only; seeds 1–1.5 mm; SnBr . . . . ²***B. peirsonii***
  85′ Infl 5–8-fld; st proximally with hairs 0.05–0.2 mm only; seeds 2.5–3.5 mm; W&I . . . . ²***B. pinzliae***
 82′ St proximally with 4–12-rayed hairs only
  86. Petals 3–5 mm
   87. Sts > 1.5 dm; fr 4–6.2 cm; ovules 48–68 . . . . ***B. bodiensis***
   87′ Sts gen < 1.5 dm; fr 1.7–4.5 cm; ovules 32–44 . . . . ***B. depauperata***
  86′ Petals 5–9 mm
   88. Style 1.5–2 mm; ovules 20–24 . . . . ²***B. serpenticola***
   88′ Style < 0.7 mm; ovules 74–190
    89. Petals 1.2–2 mm wide; surfaces of basal lvs with hairs 0.2–0.7 mm; seeds 1.7–2 mm;
     pollen spheric . . . . ²***B. inyoensis***
    89′ Petals 0.8–1.2 mm wide; surfaces of basal lvs with hairs 0.1–0.2 mm; seeds 1–1.3 mm;
     pollen ellipsoid . . . . ²***B. shockleyi***

B. acutina (Greene) Windham & Al-Shehbaz POINTED ROCKCRESS Short-lived per or bien; caudex not woody. **ST:** gen 1 per caudex branch, from center of basal rosette at ± ground surface, 1.5–6 dm; proximally with short-stalked, 2–4-rayed hairs 0.15–0.4 mm. **LF:** basal 1.5–6 mm wide, entire (minutely dentate), hairs 0.15–0.35 mm, short-stalked, gen 3–5-rayed; cauline gen 6–20, distal hairy or glabrous, basal lobes 0.3–2 mm. **INFL:** gen 5–20-fld, not 1-sided in fr; fr pedicels ascending to erect, 3–10 mm, gen straight, glabrous or hairs appressed, branched. **FL:** sepals hairy; petals 5–7 mm, 0.9–1.3 mm wide, white or lavender; pollen spheric. **FR:** ascending, not appressed, 2.5–7.5 cm, 1.2–1.8 mm wide, glabrous; style 0.3–1 mm; ovules 46–100. **SEED:** in 1 row, 1.8–2.2 mm; wing 0.1–0.4 mm wide. Gravelly slopes, in meadows, open forest; 1200–2250 m. KR; to OR, ID. Jun–Jul

B. arcuata (Nutt.) Windham & Al-Shehbaz (p. 533) ARCHING ROCKCRESS Caudex ± woody. **ST:** gen 1 per caudex branch, from center of basal rosette at ground surface or elevated on woody base; gen 3–8 dm, proximally with simple and short-stalked, 2-rayed hairs 0.4–1 mm. **LF:** basal 2–12 mm wide, gen entire, hairs gen short-stalked, 2–5-rayed, 0.4–0.8 mm; cauline 10–45, distal hairy, basal lobes 2–6 mm. **INFL:** 12–70-fld, 1-sided in fr; fr pedicel spreading-ascending (horizontal), ± recurved or straight, 8–22 mm, hairs ± appressed, 2–4-rayed. **FL:** sepals hairy; petals 9–14 mm, 2–4 mm wide, purple; pollen ellipsoid. **FR:** spreading-ascending (horizontal), not appressed, 6–13 cm, 1.5–2.2 mm wide, glabrous or with few hairs; style < 0.5 mm; ovules 90–250. **SEED:** gen in 1 row, 1.5–1.7 mm; wing 0.1–0.2 mm wide. Rocky hillsides, cliffs, in pine forest, chaparral; 300–2000 m. SNF, CW, SCo, TR. [*Arabis sparsiflora* Nutt. var. *a.* (Nutt.) Rollins] Mar–Jun

B. bodiensis (Rollins) Al-Shehbaz BODIE HILLS ROCKCRESS Caudex ± woody. **ST:** gen 1 per caudex branch, from center of basal rosette at ± ground surface; 1.5–3.5 dm, proximally with short-stalked, 5–10-rayed hairs to 0.7 mm. **LF:** basal 1–3 mm wide, entire, hairs short-stalked, 5–10-rayed, 0.08–0.25 mm; cauline 4–9, distal hairy, basal lobes 0.5–2 mm. **INFL:** 8–25-fld, not 1-sided in fr; fr pedicel ascending to spreading-ascending, straight, 3–7 mm, hairs appressed, branched. **FL:** sepals hairy; petals 4–5 mm, ± 1 mm wide, lavender to purple; pollen spheric. **FR:** spreading-ascending, not appressed, 4–6.2 cm, 1.2–1.8 mm wide, glabrous; style 0.1–0.2 mm; ovules 48–68. **SEED:** in 1 row, 1–1.5 mm; wing 0.1–0.15 mm wide. Rocky soil, igneous rock crevices; 2400–2900 m. SNE (n Mono Co.); w NV. [*Arabis b.* Rollins] Jul–Aug ★

B. breweri (S. Watson) Al-Shehbaz Caudex woody. **ST:** gen 1 per caudex branch, from center of basal rosette at ground surface or elevated on woody base; 0.6–4.5 dm, proximally with simple and long-stalked, 2-rayed hairs (simple only) 0.4–1 mm. **LF:** basal 3–11 mm wide (dentate), hairs long-stalked, 2–4-rayed, 0.4–0.8 mm; cauline 5–28, distal hairy, basal lobes gen < 3 mm. **INFL:** 7–30-fld, not 1-sided in fr; fr pedicel ascending to spreading-ascending, straight, 3–25 mm, glabrous or hairs spreading, simple and 2-rayed. **FL:** sepals hairy; petals 7–12 mm, 2–4 mm wide, purple (lavender); pollen ellipsoid. **FR:** ascending to spreading-ascending, not appressed, 3.5–10 cm, 1.5–2.2 mm wide, glabrous (sparsely hairy); style < 0.3 mm; ovules 48–96. **SEED:** in 1 row, 1.2–1.7 mm; wing 0.1–0.2 mm wide. [*Arabis b.* S. Watson]

subsp. ***breweri*** BREWER'S ROCKCRESS **ST:** 0.6–2 dm. **INFL:** fr pedicel gen < 9 mm, hairy. Rocky outcrops, ledges, talus; 300–2300 m. KR, NCoRH, NCoRI, CaR, n SNF, ScV (Sutter Buttes), SnFrB, SCoRI, w WTR; sw OR. [*Arabis b.* S. Watson var. *b.*] Mar–Jul

subsp. ***shastaensis*** Windham & Al-Shehbaz SHASTA ROCKCRESS **ST:** 1.8–4.5 dm. **INFL:** fr pedicel gen > 9 mm, glabrous or hairs few, scattered. 2*n*=14. Rocky areas in woodland; 300–1200 m. e KR, NCoRI, CaRF, n SNF, ScV (Sutter Buttes); sw OR. [*Arabis b.* S. Watson var. *austiniae* (Greene) Rollins] Mar–Jun

B. calderi (G.A. Mulligan) Windham & Al-Shehbaz CALDER'S ROCKCRESS Short-lived per; caudex not woody. **ST:** gen 1 per caudex branch, from center of basal rosette at ± ground surface; 1–4.5 dm, proximally glabrous. **LF:** basal 1.5–6 mm wide, entire, hairs sessile and 2–4-rayed, 0.15–0.4 mm; cauline 5–17, distal glabrous, basal lobes 1–3 mm. **INFL:** 10–25-fld, not 1-sided in fr; fr pedicel erect, straight, 5–10 mm, glabrous. **FL:** sepals glabrous; petals 6–9 mm, 1.5–2.5 mm wide, purple; pollen spheric. **FR:** erect, appressed, 3.5–6.5 cm, 1.8–2.5 mm wide, glabrous; style < 0.2 mm; ovules 64–134. **SEED:** gen in 2 rows, 1.8–2.2 mm; wing 0.2–0.5 mm wide. Exposed rocky ridges, meadows, open forest near timberline; 2050–3350 m. SNH, MP, W&I; to YT, WY. Jun–Aug

B. californica (Rollins) Windham & Al-Shehbaz (p. 533) CALIFORNIA ROCKCRESS Caudex woody. **ST:** gen 1 per caudex branch, from center of basal rosette on elevated woody bases; gen > 3.5 dm, proximally with short-stalked, 2–4 rayed hairs 0.3–0.9 mm. **LF:** basal 3–13 mm wide, entire (minutely dentate), hairs short-stalked, 4–8-rayed, 0.2–0.5 mm; cauline 12–55, distal hairy, basal lobes 1–6 mm. **INFL:** 30–120-fld, not 1-sided in fr; fr pedicel descending to horizontal, ± recurved (straight), 4–20 mm, hairs appressed, 3–7-rayed. **FL:** sepals hairy; petals 9–14 mm, 1.5–3 mm wide, purple (± pink); pollen spheric. **FR:** pendent, not appressed, 6–12 cm, 1.5–2.5 mm wide, glabrous or sparsely hairy throughout; style < 0.3 mm; ovules 140–180. **SEED:** gen in 1 row, 1.7–2 mm; wing 0.2–0.4 mm wide. Rocky slopes, gravelly soil, in chaparral, oak woodland; 350–2300 m. SW; Mex. [*Arabis sparsiflora* Nutt. var. *c.* Rollins] Mar–Jun

B. cobrensis (M.E. Jones) Dorn (p. 533) MASONIC ROCKCRESS Caudex ± woody. **ST:** gen 1 per caudex branch, from center of basal rosette at ± ground surface; gen > 2.5 dm, proximally with short-stalked, 4–10-rayed hairs 0.1–0.2 mm. **LF:** basal 1–4 mm wide, entire, hairs short-stalked, 4–10-rayed, 0.1–0.2 mm; cauline 5–10, distal ± hairy, basal lobes 1–1.5 mm. **INFL:** 10–25-fld, not 1-sided in fr; fr pedicel spreading-ascending to horizontal, gen straight near base, recurved or reflexed near tip, 4–17 mm, hairs appressed, branched (glabrous). **FL:** sepals hairy; petals 3.5–6 mm, 0.7–1 mm wide, white to lavender; pollen ellipsoid. **FR:** pendent, not appressed, 2.5–5.5 cm, 1.7–2.5 mm wide, glabrous; style < 0.2 mm; ovules 34–64. **SEED:** in 1 row, 1.4–1.8 mm; wing 0.25–0.5 mm wide. 2*n*=14. Sandy soil, under shrubs in semi-desert; 1350–3400 m. SNE, n DMtns; to OR, WY. [*Arabis c.* M.E. Jones] May–Jun ★

B. constancei (Rollins) Al-Shehbaz (p. 533) CONSTANCE'S ROCKCRESS Caudex woody. **ST:** gen 1 per caudex branch, from center of basal rosette elevated above ground surface on woody base; 1.2–3 dm, proximally glabrous. **LF:** basal 1.5–4 mm wide, entire,

margin hairs simple and short-stalked, 2-rayed, 0.3–0.8 mm, faces glabrous; cauline 6–12, distal glabrous, basal lobes 0. **INFL:** 5–15-fld, gen 1-sided in fr; fr pedicel arched, gen abruptly recurved near base, 4–12 mm, glabrous. **FL:** sepals glabrous; petals 6–8 mm, 1.5–2 mm wide, ± white; pollen ellipsoid. **FR:** pendent, not appressed, 4–7.5 cm, 3–3.5 mm wide, glabrous, edges ± wavy (non-parallel); style 2.5–5 mm; ovules 18–28. **SEED:** in 1 row, 3–4 mm; wing 0.5–1 mm wide. 2*n*=14. Serpentine slopes, ridges; 1100–1900 m. n SNH (Plumas Co.). [*Arabis c.* Rollins] May ★

B. covillei (Greene) Windham & Al-Shehbaz COVILLE'S ROCKCRESS Caudex ± woody. **ST:** gen 1 per caudex branch, from center of basal rosette at ± ground surface; 0.5–2.5 dm, glabrous. **LF:** basal 1.5–5 mm wide, entire, margin hairs simple and short-stalked, 2–4-rayed, 0.15–0.4 mm, faces glabrous; cauline 2–7, distal glabrous, basal lobes 0.2–1 mm (0). **INFL:** ascending to ± erect, straight, 2–9-fld, not 1-sided in fr; fr pedicel ascending to nearly erect 6–20 mm, glabrous. **FL:** sepals glabrous; petals 5–6 mm, 1–2 mm wide, lavender to ± purple; pollen spheric. **FR:** 3.5–5 cm, ascending to ± erect (appressed), 3–5 mm wide, glabrous, edges ± wavy (parallel); style 0.2–1 mm; ovules 18–30. **SEED:** in 1 row, 4–5 mm; wing 1–1.5 mm wide. Rocky slopes in alpine meadows, open conifer forest; 2200–3500 m. CaRH, SNH; OR, NV. Jul–Aug

B. davidsonii (Greene) N.H. Holmgren DAVIDSON'S ROCKCRESS Caudex woody; crowded, persistent lf bases abundant. **ST:** gen 1 per caudex branch, from center of basal rosette at ± ground surface; 0.6–2.3 dm, glabrous. **LF:** basal 3.5–14 mm wide, gen entire, glabrous; cauline 3–10, glabrous, basal lobes 0. **INFL:** 4–24-fld, not 1-sided in fr; fr pedicel ascending, straight, 3–18 mm, glabrous. **FL:** sepals glabrous; petals 6–10 mm, 2.5–4 mm wide, white to lavender; pollen ellipsoid. **FR:** ascending to spreading-ascending, not appressed, 2.5–7 cm, 1.5–2.5 mm wide, glabrous; style 0.1–0.8 mm; ovules 28–50. **SEED:** in 1 row, 1.8–2.2 mm; wing 0.1–0.5 mm wide. Ledges, rock outcrops; 1200–3400 m. SNH, Wrn; OR, NV. [*Arabis d.* Greene] Apr–Jul

B. depauperata (A. Nelson & P.B. Kenn.) Windham & Al-Shehbaz SOLDIER ROCKCRESS Caudex ± woody. **ST:** gen 1 per caudex branch, from center of basal rosette at ± ground surface; gen < 1.5 dm, proximally with short-stalked, 4–8-rayed hairs 0.07–0.4 mm. **LF:** basal 0.7–3(5) mm wide, entire, hairs short-stalked, 5–8-rayed, 0.05–0.25 mm; cauline 3–7, distal hairy (glabrous), basal lobes 0.5–1.5 mm. **INFL:** 9–23-fld, not 1-sided in fr; fr pedicel spreading-ascending, straight, 1.5–7 mm, hairs appressed, branched (glabrous). **FL:** sepals hairy; petals 3.5–5 mm, 0.8–1.2 mm wide, lavender to ± purple; pollen spheric. **FR:** spreading-ascending, not appressed, 1.7–4.5 cm, 1–1.5 mm wide, glabrous; style < 0.2 mm; ovules 32–44. **SEED:** in 1 row, 1.2–1.5 mm; wing 0.1–0.15 mm wide. Exposed ridges, talus slopes, in subalpine, alpine habitats; 2650–3900 m. s SNH, W&I; NV. [*Arabis lemmonii* S. Watson var. *d.* (A. Nelson & P.B. Kenn.) Rollins] Jun–Jul

B. dispar (M.E. Jones) Al-Shehbaz PINYON ROCKCRESS Caudex ± woody. **ST:** gen 1 per caudex branch, from center of basal rosette at ± ground surface; 0.9–3 dm, proximally with short-stalked, 5–12-rayed hairs 0.1–0.3 mm. **LF:** basal 2–5 mm wide, entire, hairs short-stalked, 5–16-rayed, 0.1–0.3 mm; cauline 1–5, distal hairy, basal lobes 0. **INFL:** 4–20-fld, not 1-sided in fr; fr pedicel ascending, straight, 4–15(25) mm, hairs appressed, branched. **FL:** sepals hairy; petals 5–6 mm, 1–1.5 mm wide, purple to lavender; pollen ellipsoid. **FR:** spreading-ascending, not appressed, 4–7.3 cm, 2.7–4 mm wide, glabrous; style < 0.5 mm; ovules 44–52. **SEED:** in 1 row, 1.9–2.3 mm; wing 0.3–0.5 mm wide. Rocky slopes, gravelly soil, in desert scrub, pinyon/juniper woodland; 1200–2500 m. SnBr, SNE, DMtns; sw NV. [*Arabis d.* M.E. Jones] Apr–May ★

B. divaricarpa (A. Nelson) Á. Löve & D. Löve (p. 533) SPREADINGPOD ROCKCRESS Short-lived per or bien; caudex not woody. **ST:** gen 1 per caudex branch, from center of basal rosette at ± ground surface; 3–9 dm, proximally with simple and ± sessile, 2–4-rayed hairs to 0.7 mm. **LF:** basal 2–10 mm wide, entire (minutely dentate), hairs sessile, 2–6-rayed 0.1–0.4 mm; cauline (10)15–56, distal glabrous, basal lobes 1–5 mm. **INFL:** 12–40(65)-fld, not 1-sided in fr; fr pedicel spreading-ascending to horizontal, straight,

5–12 mm, glabrous. **FL:** sepals hairy; petals 6–9 mm, 1.5–3 mm wide, purple (lavender); pollen spheric. **FR:** spreading-ascending to horizontal, not appressed, 4.5–11 cm, 1.7–2.5 mm wide, glabrous; style < 0.2 mm; ovules 114–142. **SEED:** gen in 1 row, 1.4–2 mm; wing 0.1–0.2 mm wide. 2*n*=21. Rock outcrops, talus slopes, gravelly hillsides; 900–2200 m. CaRH, MP; to WA, WY. [*Arabis ×d.* A. Nelson] May–Jul

B. elkoensis Windham & Al-Shehbaz ELKO ROCKCRESS Caudex ± woody. **ST:** 1–5 per caudex branch, from margin or center of basal rosette at ± ground surface; 1–3 dm, proximally with ± sessile, 2–5-rayed hairs 0.1–0.4 mm. **LF:** basal 2–6 mm wide, entire, hairs ± sessile, 2–5-rayed, 0.1–0.4 mm; cauline 3–6, distal glabrous or sparsely ciliate, basal lobes 0.2–0.5 mm (0). **INFL:** 4–11-fld, not 1-sided in fr; fr pedicel erect to ascending, straight, 4–8 mm, glabrous or hairs few, appressed, branched. **FL:** sepals glabrous; petals 4–7 mm, 1.2–2 mm wide, lavender (± white); pollen spheric. **FR:** erect to ascending, gen appressed, 3–7 cm, 2–3 mm wide, glabrous, edges gen ± wavy (non-parallel); style < 0.3 mm; ovules 26–42. **SEED:** in 1 row, 2.5–3.5 mm; wing 0.4–0.9 mm wide. Gravelly-rocky soil in open forest, subalpine meadows; 2050–3150 m. n&c SNH, n SNE; NV. [*Arabis platysperma* var. *p.*, in part] Jul

B. evadens Windham & Al-Shehbaz HIDDEN ROCKCRESS Caudex ± woody. **ST:** gen 1 per caudex branch, from center of basal rosette at ± ground surface; 1–2.5 dm, proximally with short-stalked, 2–5-rayed hairs 0.2–0.6 mm, and few simple hairs to 1 mm. **LF:** basal 1.5–3 mm wide, entire, hairs short-stalked, 4–8-rayed, 0.1–0.2 mm; cauline 3–5, distal hairy, basal lobes 0.5–1 mm. **INFL:** 10–22-fld, not 1-sided in fr; fr pedicel spreading-ascending, straight, 4–8 mm, hairs few, appressed, branched. **FL:** sepals hairy; petals 3–4 mm, 0.5–0.8 mm wide, white; pollen ellipsoid. **FR:** spreading-ascending, not appressed, 3–4 cm, 1.2–1.5 mm wide, glabrous; style ± 0.1 mm; ovules 46–52. **SEED:** in 1 row, 1–1.1 mm; wing ± 0.1 mm wide. Rock outcrops; ± 2600 m. s SNH. [*Arabis fernaldiana* Rollins var. *stylosa* (S. Watson) Rollins] Jun ★

B. glaucovalvula (M.E. Jones) Al-Shehbaz BLUEPOD ROCKCRESS Caudex gen not woody. **ST:** gen 1 per caudex branch, from center of basal rosette at ± ground surface; 1–4.5 dm, proximally with short-stalked, 4–8-rayed hairs 0.1–0.4 mm. **LF:** basal 2–6 mm wide, entire, hairs short-stalked, 4–8-rayed, 0.1–0.4 mm; cauline 6–10, distal hairy, basal lobes 0. **INFL:** gen 10–25-fld, gen 1-sided in fr; fr pedicel reflexed, strongly curved at base, 2–10 mm, hairs appressed, branched. **FL:** sepals hairy; petals 6–9 mm, 1.5–2.5 mm wide, light purple to lavender; pollen ellipsoid. **FR:** reflexed, occ appressed, 1.8–4.5 cm, 5–8 mm wide, glabrous; style 0.2–0.6 mm; ovules 24–62. **SEED:** in 2 rows, 5–6 mm; wing 1.8–2.5 mm wide. 2*n*=14. Rocky slopes, gravelly soil, gen under desert shrubs; 600–2000 m. SNE, DMoj; sw NV. [*Arabis g.* M.E. Jones] Mar–Apr

B. hoffmannii (Munz) Al-Shehbaz HOFFMANN'S ROCKCRESS Caudex woody; lf bases persistent, occ crowded. **ST:** gen 1 per caudex branch, from center of basal rosette gen elevated on woody base; 5–7 dm, proximally glabrous or with short-stalked, 2–3-rayed hairs 0.15–0.4 mm. **LF:** basal 3–8 mm wide, narrowly oblanceolate, coarsely dentate, hairs short-stalked, 2–7-rayed, 0.15–0.4 mm; cauline 15–65, most distal hairy, basal lobes 1–4 mm. **INFL:** 30–70-fld, not 1-sided in fr; fr pedicel spreading-ascending, straight, 10–45 mm, glabrous. **FL:** sepals hairy; petals 8–10 mm, 1.5–2 mm wide, white or pale lavender; pollen ellipsoid. **FR:** spreading-ascending to horizontal, not appressed, 6–10 cm, 2.5–3 mm wide, glabrous; style < 0.5 mm; ovules 170–220. **SEED:** in 2 rows, 1.2–1.6 mm; wing 0.1–0.2 mm wide. Thin soil over sandstone; < 500 m. n ChI (Santa Cruz Island). [*Arabis h.* (Munz) Rollins] Feb–Mar ★

B. howellii (S. Watson) Windham & Al-Shehbaz HOWELL'S ROCKCRESS Caudex ± woody. **ST:** gen 1 per caudex branch, from center of basal rosette at ± ground surface; 0.6–2(3) dm, glabrous. **LF:** basal 1–7 mm wide, entire, margin hairs simple and short-stalked, 2-rayed, faces glabrous; cauline 2–4, distal glabrous throughout or ciliate near base, basal lobes (0)0.2–1 mm. **INFL:** 2–5-fld, not 1-sided in fr; fr pedicel ascending, straight, 4–10 mm, glabrous. **FL:** sepals glabrous; petals 4–8 mm, 1–2.5 mm wide, white to ± purple; pollen ellipsoid. **FR:** ascending to ± erect, gen not appressed, 2.5–

6.5 cm, 3–7 mm wide, glabrous, edges ± wavy (non-parallel); style < 0.3 mm; ovules 10–20. **SEED:** in 1 row, 3–6 mm; wing 1.3–2.5 mm wide. Uncommon. Rock outcrops, talus slopes, gravelly soil, in alpine, subalpine habitats; 1800–3800 m. CaRH, SNH; NV, OR. [*Arabis platysperma* A. Gray var. *h.* (S. Watson) Jeps.] Jun–Aug

B. inyoensis (Rollins) Al-Shehbaz INYO ROCKCRESS Short-lived per; caudex not woody. **ST:** gen 1 per caudex branch, from center of basal rosette at ± ground surface; gen 2.5–6.5 dm, proximally with short-stalked, 7–12-rayed hairs 0.1–0.2 mm. **LF:** basal 1–4(8) mm wide, entire; hairs short-stalked, 3–10-rayed, 0.2–0.7 mm; cauline (7)12–35, distal hairy, basal lobes 0.5–2 mm. **INFL:** 10–65-fld, not 1-sided in fr; fr pedicel spreading-ascending to horizontal, ± straight, 5–15 mm, hairs few, appressed, branched. **FL:** sepals hairy; petals 5–8 mm, 1.2–2 mm wide, lavender to ± purple; pollen spheric. **FR:** spreading-ascending to widely pendent, not appressed, 3.7–6.5 cm, 1.5–2.2 mm wide, glabrous; style < 0.2 mm; ovules 74–134. **SEED:** in 2(1) rows, 1.7–2 mm; wing 0.1–0.2 mm wide. 2*n*=21. Limestone, volcanic outcrops, clay soils, in desert scrub, pinyon/juniper woodland; 1200–3500 m. e SNH, SNE, DMtns; to UT. [*Arabis i.* Rollins] Apr–Jun

B. johnstonii (Munz) Al-Shehbaz JOHNSTON'S ROCKCRESS Caudex woody. **ST:** gen 1 per caudex branch, from center of basal rosette at ± ground surface; 0.5–2 dm, proximally with short-stalked, 4–10-rayed hairs 0.07–0.15 mm. **LF:** basal 1.5–4 mm wide, entire; hairs short-stalked, 6–14-rayed 0.07–0.15 mm; cauline 4–10, distal hairy, basal lobes 0. **INFL:** 10–18-fld, not 1-sided in fr; fr pedicel spreading-ascending, straight, 5–14 mm, hairs appressed, branched. **FL:** sepals hairy; petals 9–14 mm, 2–4 mm wide, purple; pollen ellipsoid. **FR:** spreading-ascending, not appressed, 4–6 cm, 2.5–4 mm wide, glabrous; style 0.7–2 mm; ovules 26–34. **SEED:** in 1 row, 1.9–2.7 mm; wing 0.3–0.7 mm wide. Rocky areas, gravelly soil, in chaparral, grassland, open oak/pine woodland; 1300–1700 m. PR (Cuyamaca Mtns, SnJt). [*Arabis hirshbergiae* S. Boyd; *A. j.* Munz] Feb–Mar ★

B. koehleri (Howell) Al-Shehbaz KOEHLER'S ROCKCRESS Caudex woody; crowded, persistent lf bases present. **ST:** gen 1 per caudex branch, from center of basal rosette elevated on woody base; 0.8–4.5 dm, proximally with short-stalked, 2–4-rayed hairs to 0.5 mm, some simple hairs to 1 mm. **LF:** basal 1–3 mm wide, entire; hairs short-stalked, 2–4-rayed, 0.1–0.3 mm; cauline 3–30, distal glabrous, basal lobes 0.5–2.5 mm. **INFL:** 6–35-fld, not 1-sided in fr; fr pedicel spreading-ascending to horizontal, straight, 10–18 mm, glabrous or hairs few, simple and 2-rayed. **FL:** sepals hairy; petals 8–12 mm, 2.5–4 mm wide, purple to magenta; pollen ellipsoid. **FR:** spreading-ascending to pendent, not appressed, 5–7.5 cm, 1.8–2.5 mm wide, glabrous; style < 0.2 mm; ovules 58–94. **SEED:** in 1 row, 1.3–1.8 mm; wing 0.1–0.2 mm wide. Rock outcrops; 100–1000 m. KR; s OR. [*Arabis k.* Howell; *A. k.* var. *stipitata* Rollins] Apr–May ★

B. lemmonii (S. Watson) W.A. Weber LEMMON'S ROCKCRESS Caudex woody. **ST:** gen 1 per caudex branch, from center of basal rosette or laterally below sterile shoot, at ± ground surface; 0.5–2.5 dm, proximally glabrous or with short-stalked, 2–6-rayed hairs 0.1–0.2 mm. **LF:** basal 1.5–5 mm wide, entire (minutely dentate); hairs short-stalked, 3–9-rayed, 0.1–0.2 mm; cauline 2–12, distal glabrous or sparsely hairy, basal lobes 0.1–0.5 mm or 0. **INFL:** 3–17-fld, 1-sided in fr; fr pedicel ascending to horizontal, 2–6 mm, glabrous or hairs few, appressed, branched. **FL:** sepals glabrous or hairs few; petals 3.5–6 mm, 1–1.5 mm wide, purple to lavender; pollen ellipsoid. **FR:** spreading-ascending to horizontal, not appressed, 1.6–4.4 cm, 1.6–2.3 mm wide, glabrous; style 0.1–0.2 mm; ovules 28–44. **SEED:** in 1 row, 1.3–2 mm; wing 0.1–0.5 mm wide. 2*n*=14. Cliffs, talus slopes, gravelly soil, in alpine, subalpine habitats; 2000–4350 m. CaRH, SNH, Wrn, SNE; to AK, CO. [*Arabis l.* S. Watson var. *l.*] Jun–Aug

B. lincolnensis Windham & Al-Shehbaz LINCOLN ROCKCRESS Caudex woody. **ST:** gen 1 per caudex branch, from center of rosette elevated on woody base; 2–4.2 dm, proximally with short-stalked, 3–6-rayed hairs 0.1–0.4 mm. **LF:** basal 1–2 mm wide, entire; hairs short-stalked, 3–8-rayed, 0.1–0.4 mm; cauline 10–25, distal hairy, basal lobes 0 (to 1 mm). **INFL:** 7–15-fld, not 1-sided in fr; fr pedicel spreading-ascending and ± straight near base, recurved near tip,

10–25 mm, hairs appressed, branched. **FL:** sepals hairy; petals 10–12 mm, 2–3 mm wide, lavender to purple; pollen ellipsoid. **FR:** pendent, not appressed, 3.2–5.5 cm, 2–2.5 mm wide, hairy throughout; style 0.1–0.3 mm; ovules 86–120. **SEED:** in 2 rows, 1–1.5 mm; wing 0.07–0.12 mm wide. 2*n*=14. Rocky slopes, gravelly soil, sagebrush, shrubland; 1400–2000 m. W&I, DMtns; to UT. [*Arabis pulchra* S. Watson var. *munciensis* M.E. Jones] Apr–May ★

B. lyallii (S. Watson) Dorn LYALL'S ROCKCRESS Caudex woody. **ST:** gen 1 per caudex branch, from center of basal rosette at ± ground surface; 0.3–1.5(2) dm, glabrous. **LF:** basal 1–5(8) mm wide, entire, margins gen with simple or short-stalked, 2–3-rayed hairs 0.1–0.3 mm, occ glabrous, faces glabrous (rarely youngest lvs of sterile shoots with 4–6-rayed hairs 0.05–0.1 mm); cauline 1–5, glabrous, basal lobes (0)0.5–1.5 mm. **INFL:** 2–10(15)-fld, not 1-sided in fr; fr pedicel erect, straight, 3–8(15) mm, glabrous. **FL:** sepals glabrous; petals 6–8.5 mm, 1.5–3 mm wide, lavender to ± purple; pollen ellipsoid or spheric. **FR:** erect, appressed, 3–5.6 cm, 1.5–2.5 mm wide, glabrous; style 0.1–0.7 mm; ovules 34–64. **SEED:** gen in 2 rows, 1.5–2 mm; wing 0.3–0.5 mm wide. Uncommon. Cliffs, talus slopes, gravelly soil, in alpine, subalpine habitats; 2000–3900 m. CaRH, SNH, Wrn, W&I; to AK, UT. [*Arabis l.* S. Watson] Jun–Aug

B. parishii (S. Watson) Al-Shehbaz PARISH'S ROCKCRESS Caudex ± woody. **ST:** gen 1 per caudex branch, from center of basal rosette at ± ground surface; 0.3–1.4 dm, proximally with short-stalked, 2–8-rayed hairs 0.2–0.4 mm. **LF:** basal 0.5–2 mm wide, entire; hairs short-stalked, 6–12-rayed, 0.07–0.15 mm; cauline 2–8, distal hairy, basal lobes 0. **INFL:** 5–20-fld, not 1-sided in fr; fr pedicel ascending, straight, 3–7 mm, hairs appressed, branched. **FL:** sepals hairy; petals 8–13 mm, 2.5–4 mm wide, lavender to purple; pollen ellipsoid. **FR:** ascending, not appressed, 1.5–2.5 cm, 1.8–2.5 mm wide, glabrous; style 3–8 mm; ovules 12–20. **SEED:** in 1 row, 1.5–2 mm; wing 0.05–0.2 mm wide. 2*n*=14. Gravelly hillsides in sagebrush-juniper-pine associations; 1900–2800 m. SnBr. [*Arabis p.* S. Watson] Mar–May ★

B. pauciflora (Nutt.) Windham & Al-Shehbaz (p. 533) HAIRY STEM ROCKCRESS Caudex ± woody. **ST:** gen 1 per caudex branch, from center of basal rosette at ± ground surface; 3–8 dm, proximally with simple hairs 0.6–1.5 mm and stalked, 2–3-rayed hairs 0.2–0.4 mm. **LF:** basal 3–10 mm wide, dentate (entire); hairs stalked, 2–5-rayed, 0.3–0.6 mm; cauline (8)14–60, distal glabrous or ± hairy, basal lobes gen 3–10 mm. **INFL:** 17–60-fld, not 1-sided in fr; fr pedicel horizontal to spreading-descending, gen straight, 4–13 mm, hairs spreading, 2–3-rayed. **FL:** sepals hairy; petals 5–8 mm, 1–2 mm wide, lavender to ± white; pollen spheric. **FR:** horizontal to pendent, not appressed, 5.5–10.5 cm, 1.5–2.2 mm wide, glabrous; style < 0.5 mm; ovules 80–162. **SEED:** in 1 row, 1.4–1.8 mm; wing 0.1–0.25 mm wide. 2*n*=21. Rocky soil in sagebrush areas, scrub, conifer forest edges; 700–2500 m. CaRH, n&c SNH, GB; to BC, UT. [*Arabis sparsiflora* Nutt. var. *subvillosa* (S. Watson) Rollins] May–Jun

B. paupercula (Greene) Windham & Al-Shehbaz SMALL-FLOWERED ROCKCRESS Caudex ± woody. **ST:** gen 1 per caudex branch, from center of basal rosette at ± ground surface; 0.3–1.5 dm, proximally with simple and short-stalked, 2–4-rayed hairs 0.07–0.2 mm. **LF:** basal 1–5 mm wide, entire; hairs short-stalked, 2–6-rayed, 0.07–0.2 mm; cauline 2–6, distal glabrous, basal lobes 0 (0.2–1.5 mm). **INFL:** 3–8-fld, not 1-sided in fr; fr pedicel erect, straight, 3–9 mm, glabrous or hairy. **FL:** sepals glabrous or hairy; petals 4–7 mm, 1–2 mm wide, lavender to ± purple; pollen ellipsoid. **FR:** erect, appressed, 2.5–5.5 cm, 1.3–2 mm wide, glabrous; style 0.2–1 mm; ovules 24–40. **SEED:** gen in 1 row, 1.5–2.5 mm; wing 0.3–1 mm wide. Rock outcrops, talus, gravelly soil, in alpine, subalpine habitats; 2500–3700 m. CaRH, SNH, Wrn, W&I; to WA, WY. [*Arabis lyallii* S. Watson var. *nubigena* (J.F. Macbr. & Payson) Rollins; *A. microphylla* Nutt. var. *m.*, misappl.] Jun–Aug

B. peirsonii Windham & Al-Shehbaz PEIRSON'S ROCKCRESS Caudex woody. **ST:** gen 1 per caudex branch, from center of basal rosette at ± ground surface; 1–2.5 dm, proximally glabrous or with few short-stalked, gen 2–5-rayed hairs 0.4–0.6 mm. **LF:** basal 2.5–6 mm wide, entire; hairs short-stalked, 4–7-rayed, 0.3–0.5 mm; cauline 3–12, distal glabrous or ciliate, basal lobes 0.5–2 mm. **INFL:** 12–25-

fld, not 1-sided in fr; fr pedicel ascending to spreading-descending, straight, 2–6 mm, glabrous. **FL:** sepals hairy; petals 5–6 mm, 1.5–2 mm wide, purple; pollen spheric. **FR:** ascending to spreading-ascending, not appressed, 2–3.7 cm, 2–2.8 mm wide, glabrous; style 0.1–0.3 mm; ovules 56–80. **SEED:** in 1 row, 1–1.5 mm; wing 0.05–0.1 mm wide. Granitic ledges, talus slopes; 2700–3350 m. SnBr. [*Arabis breweri* S. Watson var. *pecuniaria* Rollins] Jun–Sep ★

B. pendulina (Greene) W.A. Weber RABBIT-EAR ROCKCRESS Caudex ± woody. **ST:** gen 2–6 per caudex branch, from margin of basal rosette, at ± ground surface; 0.6–3.7 dm, proximally with simple hairs 0.3–0.8 mm. **LF:** basal 1.5–6 mm wide, entire (dentate); hairs simple and short- to long-stalked, 2-rayed hairs 0.3–0.8 mm; cauline 2–13, distal ciliate to glabrous, basal lobes 0 (to 0.7 mm). **INFL:** 4–14-fld, not 1-sided in fr; fr pedicel ± horizontal, curved or angled downward near tip, 3–10 mm, glabrous or hairs few, simple. **FL:** sepals with few hairs (glabrous); petals 4–6 mm, 1–1.5 mm wide, ± white to pale lavender; pollen ellipsoid. **FR:** pendent, not appressed, 2.2–4 cm, 1.2–2.1 mm wide, glabrous; style 0.1–0.5 mm; ovules 40–90. **SEED:** in 2 rows, 0.9–1.2 mm, wing 0. 2*n*=14. Rock outcrops, open gravelly flats, hillsides; 2000–3000 m. W&I; to WY, NV, AZ. Apr–Jun ★

B. pendulocarpa (A. Nelson) Windham & Al-Shehbaz DROPSEED ROCKCRESS Caudex gen not woody. **ST:** gen 1 per caudex branch, from center of basal rosette at ± ground surface; 0.6–3 dm, proximally with simple and stalked, 2–4-rayed hairs 0.1–0.4 mm. **LF:** basal 1.5–5 mm wide, entire; hairs short-stalked, 4–8-rayed, 0.08–0.2 mm; cauline 6–17, distal glabrous or ± hairy, basal lobes 0. **INFL:** 4–11-fld, not 1-sided in fr; fr pedicel ± horizontal near base, ± recurved, 3–12 mm, glabrous or hairs few, appressed, branched. **FL:** sepals hairy; petals 4–6 mm, 1–1.5 mm wide, white or lavender; pollen ellipsoid. **FR:** pendent, not appressed, 2–3.8 cm, 1.5–2.2 mm wide, glabrous; style 0.3–0.5 mm; ovules 66–92. **SEED:** gen in 2 rows, 1–1.2 mm; wing 0.05–0.1 mm wide. 2*n*=14. Rock outcrops, gravelly slopes, in open meadows; 1800–3000 m. GB; to w Can, MT, UT. [*Arabis holboellii* Hornem. var. *p.* (A. Nelson) Rollins] Apr–Jul

B. perennans (S. Watson) W.A. Weber (p. 533) PERENNIAL ROCKCRESS Caudex woody. **ST:** gen 2–5 per caudex branch, arising laterally below sterile shoot or lf rosette, gen elevated on woody base; gen 2–7 dm, proximally with short-stalked, 2-rayed hairs 0.2–0.4 mm, gen mixed with 3–5-rayed (or simple) hairs. **LF:** basal gen 3–15 mm wide, dentate; hairs short-stalked, 3–6-rayed, 0.2–0.4 mm; cauline 4–17, distal glabrous, basal lobes 0.5–3.5 mm. **INFL:** 16–35-fld, not 1-sided in fr; fr pedicel ± horizontal near base, gen ± recurved near tip, 6–25 mm, glabrous or hairs few. **FL:** sepals hairy; petals 5–9 mm, 1–1.5 mm wide, white to ± purple; pollen ellipsoid. **FR:** pendent, not appressed, 3–7 cm, 1.7–2.1 mm wide, glabrous; style 0.05–0.4 mm; ovules 60–96. **SEED:** in 1 row, 1.1–1.5 mm; wing 0.1–0.2 mm wide. 2*n*=14. Rocky slopes, gravelly soil, desert, chaparral, low montane habitats; 500–2000 m. s SNH, SnGb, SnBr, e PR, SNE, D; to UT, NM, n Mex. [*Arabis p.* S. Watson] Feb–May

B. pinetorum (Tidestr.) Windham & Al-Shehbaz WOODLAND ROCKCRESS Short-lived per; caudex not woody. **ST:** gen 1 per caudex branch, from center of basal rosette at ± ground surface; 2–9.6 dm, proximally with simple and short-stalked, 2–5-rayed hairs to 1 mm. **LF:** basal 2–11 mm wide, entire (minutely dentate); hairs short-stalked, 2–5-rayed (few simple) hairs 0.2–0.5 mm; cauline 7–33, distal ± hairy or glabrous, basal lobes 1–3 mm. **INFL:** 15–63-fld, not 1-sided in fr; fr pedicel reflexed, gen abruptly recurved near base, 4–12 mm, hairs few spreading, simple and 2-rayed, or glabrous. **FL:** sepals hairy; petals 4–6 mm, 1–1.5 mm wide, white to lavender. **FR:** reflexed to ± appressed, 4.5–8.5 cm, 1.5–2 mm wide, glabrous; style 0.05–0.4 mm; ovules 70–110. **SEED:** in 1 row, 1.5–1.8 mm; wing 0.15–0.3 mm wide. Rock outcrops, gravelly soil, in meadows, open conifer forest; 1100–3200 m. KR, SNH; to OR, NV. [*Arabis holboellii* Hornem. var. *p.* (Tidestr.) Rollins] May–Jul

B. pinzliae (Rollins) Al-Shehbaz PINZL'S ROCKCRESS Caudex gen woody. **ST:** 1–3 per caudex branch, from margin of basal rosette at ± ground surface; 0.4–1.6 dm, proximally with short-stalked, 2–5-rayed hairs 0.05–0.2 mm. **LF:** basal 1–3 mm wide, entire; hairs short-stalked, 2–5-rayed, 0.05–0.2 mm; cauline 3–6, distal hairy,

basal lobes 0 (to 0.3 mm). **INFL:** 5–8-fld, not 1-sided in fr; fr pedicel ascending to spreading-ascending, straight, 2–6 mm, glabrous or hairs few, appressed, branched. **FL:** sepals sparsely hairy; petals 4–5 mm, 1–1.5 mm wide, purple; pollen spheric. **FR:** ascending to spreading-ascending, not appressed, 2.5–4.8 cm, 2.5–3.2 mm wide, glabrous; style 0.1–0.3 mm; ovules 26–34. **SEED:** in 1 row, 2.5–3.5 mm; wing 0.2–0.9 mm wide. Gravelly granitic soil in alpine, subalpine areas; 3000–3400 m. W&I; NV. [*Arabis p.* Rollins; *Arabis pinzlae*, orth. var.] Reported from Mammoth Mtn, c SNH (Constantine-Shull 2000 Madroño 47:209). Jul ★

B. platysperma (A. Gray) Al-Shehbaz (p. 533) PIONEER ROCKCRESS Caudex woody. **ST:** gen 1 per caudex branch, from center of basal rosette at ± ground surface or elevated on woody base; 0.6–3.5 dm, proximally with short-stalked, 2–5-rayed hairs gen 0.1–0.25 mm. **LF:** basal 3–10 mm wide, entire; hairs short-stalked, 2–5-rayed, 0.1–0.3 mm; cauline 3–12, distal glabrous to ± hairy, basal lobes 0. **INFL:** 2–7-fld, not 1-sided in fr; fr pedicel ascending, straight, 3–13 mm, glabrous or hairs few, appressed, branched. **FL:** sepals hairy; petals ± white to purple, 4–6 mm, 1–2 mm wide; pollen ellipsoid. **FR:** 2.5–8.5 cm, ascending, not appressed, 3–5.5 mm wide, glabrous, edges wavy (non-parallel); style 0.05–1 mm; ovules 16–44. **SEED:** in 1 row, 3–8 mm; wing 0.8–2.5 mm wide. 2*n*=14. Rock outcrops, gravelly soil, in dry pine forest, lodgepole-chaparral woodland; 1600–3000 m. KR, NCoRH, CaRH, SNH, SnGb, GB; OR. [*Arabis p.* A. Gray] Jun–Aug

B. pratincola (Greene) Windham & Al-Shehbaz MEADOW ROCKCRESS Short-lived per; caudex not woody. **ST:** gen 1 per caudex branch, from center of basal rosette at ± ground surface; 2–6 dm, proximally with sessile, 2–3-rayed hairs 0.15–0.6 mm. **LF:** basal 1.5–7 mm wide, entire (minutely dentate); hairs sessile, 2–6-rayed, 0.1–0.45 mm; cauline (7)13–42, distal occ hairy at tip, basal lobes 1–4 mm. **INFL:** (7)14–45-fld, not 1-sided in fr; fr pedicel erect, straight, 5–12 mm, glabrous. **FL:** sepals glabrous; petals 6–10 mm, 1.5–2 mm wide, lavender; pollen spheric. **FR:** erect, appressed, 4–6.5 cm, 1.5–2 mm wide, glabrous; style 0.1–0.3 mm; ovules 60–140. **SEED:** gen in 2 rows, 1.4–2 mm; wing 0.1–0.35 mm wide. Rocky slopes, in meadows, forest edges; 1900–3200 m. KR, CaRH, SNH; to OR, ID, NV. Jun–Aug

B. puberula (Nutt.) Dorn SILVER ROCKCRESS Short-lived per; caudex not woody. **ST:** gen 1 per caudex branch, from center of basal rosette at ± ground surface; gen 2–6.3 dm, proximally with short-stalked, 3–8-rayed hairs 0.1–0.5 mm. **LF:** basal 1.5–5 mm wide, dentate (entire); hairs short-stalked, 5–12-rayed, 0.05–0.2 mm; cauline 7–65, distal hairy, basal lobes 0.7–3 mm (occ 0). **INFL:** 10–64-fld, occ ± 1-sided in fr; fr pedicel arched, recurved distal to horizontal to ascending base, 4–10 mm, hairs appressed, branched. **FL:** sepals hairy; petals 5–9 mm, 0.8–1.8 mm wide, white to lavender; pollen ellipsoid. **FR:** closely pendent (appressed), 3–6.5 cm, 1.9–2.2 mm wide, hairy throughout; style 0.01–0.1 mm; ovules 38–64. **SEED:** in 1 row, 1.4–1.8 mm; wing 0.1–0.3 mm wide. 2*n*=14. Ledges, rocky slopes, gravelly hillsides; 1300–2900 m. e SNH, GB; to OR, UT. [*Arabis p.* Nutt.] Apr–Jul

B. pulchra (S. Watson) W.A. Weber (p. 533) BEAUTIFUL ROCKCRESS Caudex woody. **ST:** gen 1 per caudex branch, from center of rosette elevated on woody base; gen 3–7.5 dm, proximally with short-stalked, gen 4–7-rayed hairs 0.1–0.3 mm. **LF:** basal 1–3 mm wide, entire; hairs short-stalked, 4–9-rayed, 0.1–0.3 mm; cauline 10–30, distal hairy, basal lobes 0 (to < 0.5 mm). **INFL:** 8–25-fld, occ gen ± 1-sided in fr; fr pedicel reflexed, abruptly recurved at base, 8–16 mm, hairs appressed, branched. **FL:** sepals hairy; petals 9–16 mm, 2–5 mm wide, purple (white); pollen ellipsoid. **FR:** strongly reflexed, gen appressed, 3.3–8 cm, 2.5–4 mm wide, hairy throughout; style 0.1–0.3 mm; ovules 68–106. **SEED:** in 2 rows, 1.7–2.8 mm; wing 0.25–0.65 mm wide. Rocky, gravelly, sandy slopes in chaparral, sagebrush scrub, evergreen woodland; 600–2800 m. e&s SN, s GV, s SCoR, TR, PR, SNE, DMoj; NV, Baja CA. [*Arabis p.* S. Watson] Mar–Jun

B. pygmaea (Rollins) Al-Shehbaz (p. 533) TULARE COUNTY ROCKCRESS Caudex woody; gen with crowded, persistent lf bases. **ST:** gen 2–5 per caudex branch, from margin of basal rosettes, at ± ground surface; 0.2–0.8 dm, proximally with simple and short-

stalked, 2–3-rayed hairs < 0.4 mm. **LF:** basal 0.8–1.5 mm wide, entire; hairs short-stalked, 2–4-rayed, 0.05–0.4 mm; cauline 2–4, distal hairy, basal lobes 0. **INFL:** 2–5-fld, not 1-sided in fr; fr pedicel erect to ascending, straight, 2–7 mm, hairs ± appressed, branched, or glabrous. **FL:** sepals hairy; petals 3–5 mm, 0.7–1 mm wide, white; pollen ellipsoid. **FR:** erect to ascending, gen appressed, 1.3–3.3 cm, 4–5 mm wide, glabrous; style 0.05–0.4 mm; ovules 8–12. **SEED:** in 1 row, 3–5 mm; wing 0.8–2 mm wide. Barren gravel flats; 2100–3400 m. s SNH (Tulare Co.). [*Arabis p.* Rollins] May–Jul ★

B. rectissima (Greene) Al-Shehbaz BRISTLY LEAF ROCKCRESS Bien or short-lived per; caudex not woody. **ST:** gen 1 per caudex branch, from center of basal rosette at ± ground surface; 2–10 dm, proximally glabrous or with few simple hairs < 0.7 mm. **LF:** basal 3–13 mm wide, entire or denticulate; hairs simple, < 0.8 mm in, gen mixed with short-stalked, 2–3-rayed hairs 0.15–0.5 mm; cauline 6–45, distal ± hairy, basal lobes 1–1.5 mm. **INFL:** 10–50-fld, not 1-sided in fr; fr pedicel reflexed, abruptly recurved at base, 4–10 mm, glabrous. **FL:** sepals hairy; petals 3–4 mm, 0.7–1.2 mm wide, white; pollen ellipsoid. **FR:** strongly reflexed, gen appressed, 5–9 cm, 1.8–2.5 mm wide, glabrous; style 0.2–0.8 mm; ovules 46–80. **SEED:** in 1 row, 1.8–2.1 mm; wing 0.15–0.25 mm wide. Rocky slopes in open conifer forest; 1100–2800 m. CaRH, SNH, SnBr; OR. [*Arabis r.* Greene] May–Jul

B. repanda (S. Watson) Al-Shehbaz YOSEMITE ROCKCRESS Caudex gen not woody. **ST:** gen 1 per caudex branch, from center of basal rosette at ± ground surface; gen 2–9 dm, proximally with short- to long-stalked, 2–6-rayed hairs 0.2–0.5 mm (few simple hairs to 1.5 mm). **LF:** basal 7–50 mm wide, wavy-margined to coarsely dentate (entire); hairs short- to long-stalked, 3–6-rayed, 0.2–0.5 mm; cauline (3)8–30, distal ± hairy, basal lobes 0. **INFL:** 7–25-fld, not 1-sided in fr; fr pedicel erect to spreading-ascending, straight, 3–10 mm, glabrous or with spreading 2–5-rayed (simple) hairs. **FL:** sepals glabrous or ± hairy; petals 3.5–6 mm, 0.8–1 mm wide, white; pollen ellipsoid. **FR:** erect to spreading-ascending, not appressed, 3.5–13.5 cm, 2.5–4 mm wide, glabrous or hairy throughout; style 0.5–1.5 mm; ovules 34–50. **SEED:** in 1 row, 2.5–6 mm; wing 0.7–1.8 mm wide. 2*n*=14. Rock outcrops, talus, gravelly soil in meadows, open pine forest; 1400–3600 m. NCoRH, SNH, e WTR, SnGb, SnBr, SnJt; NV. [*Arabis r.* S. Watson; *A. r.* var. *greenei* Jeps.] Jun–Jul

B. retrofracta (Graham) Á. Löve & D. Löve (p. 533) REFLEXED ROCKCRESS Bien or short-lived per; caudex not woody. **ST:** gen 1 per caudex branch, from center of basal rosette at ± ground surface; 1.5–10 dm, proximally with short-stalked, 2–8-rayed hairs 0.1–0.2 mm. **LF:** basal 2–7 mm wide, entire or occ dentate; hairs occ short-stalked, 5–10-rayed, 0.05–0.2 mm; cauline 15–40, distal hairy, basal lobes 0.5–2.5 mm (0). **INFL:** 15–140-fld, occ ± 1-sided in fr; fr pedicel reflexed, abruptly recurved at base, 7–18 mm, hairs appressed. **FL:** sepals hairy; petals 4–8 mm, 0.8–2.2 mm wide, white to lavender; pollen ellipsoid. **FR:** strongly reflexed, gen appressed, 3.5–9 cm, 0.9–1.8 mm wide, glabrous or sparsely hairy throughout; style 0.05–0.3 mm; ovules 60–116. **SEED:** in 1 row, 1–1.4 mm; wing to 0.1 mm. 2*n*=14. Rock outcrops, sandy soil, in grassland, sagebrush steppe, open conifer forest; 900–2900 m. KR, NCoRH, CaRH, SNH, GB; AK, w Can, CO. [*Arabis holboellii* Hornem. var. *r.* (Graham) Rydb.] Apr–Aug

B. rigidissima (Rollins) Al-Shehbaz TRINITY MOUNTAIN ROCKCRESS Caudex woody. **ST:** gen 1 per caudex branch, from center of basal rosette at ± ground surface or on ± elevated woody base; 2–6 dm, proximally with ± sessile, 2–3-rayed (some simple) hairs 0.25–0.4 mm. **LF:** basal 2–6 mm wide, entire; hairs ± sessile, 2–5-rayed, 0.1–0.4 mm; cauline 5–10, distal glabrous, basal lobes 0.5–2 mm. **INFL:** 5–16-fld, not 1-sided in fr; fr pedicel spreading-ascending, straight, 4–10 mm, glabrous. **FL:** sepals glabrous; petals 6–8 mm, 1.5–2 mm wide, purple; pollen spheric. **FR:** spreading-ascending, not appressed, 4–7.6 cm, 2.5–3.5 mm wide, glabrous, edges ± wavy; style 0.2–0.8 mm; ovules 24–54. **SEED:** in 1 row, 2.5–3.2 mm; wing 0.3–1 mm wide. Rocky areas in open conifer forest; 1800–2100 m. KR; NV. [*Arabis r.* Rollins] Jul–Aug ★

B. rollei (Rollins) Al-Shehbaz (p. 533) ROLLE'S ROCKCRESS Caudex woody. **ST:** gen 1 per caudex branch, from center of basal rosette on elevated woody base; 1.5–2.5 dm, glabrous. **LF:** basal 3–8

mm wide, entire, margins with simple hairs 0.2–0.7 mm, surfaces glabrous or hairs short-stalked, 2–4-rayed, 0.2–0.4 mm; cauline 6–12, distal glabrous, lobes 0.5–2.5 mm. **INFL:** 3–7-fld, not 1-sided in fr; fr pedicel arched, strongly curved near base, 4–6 mm, glabrous. **FL:** sepals glabrous; petals 8–11 mm, 2–2.5 mm wide, ± white; pollen ellipsoid. **FR:** pendent, not appressed, 3.5–6 cm, 2.5–3.5 mm wide, glabrous, edges wavy; style 0.5–1 mm; ovules 14–22. **SEED:** in 1 row, 3–4 mm; wing 0.3–0.6 mm wide. Among rocks on sparsely forested slopes; 1600–1800 m. n KR (n Siskiyou Co.). [*Arabis r.* Rollins] Aug ★

B. rubicundula (Jeps.) Windham & Al-Shehbaz MOUNT DAY ROCKCRESS Caudex woody. **ST:** gen 1 per caudex branch, from center of basal rosette gen ± elevated on woody base; 1–5 dm, proximally with simple hairs 0.5–1 mm, mixed with long-stalked, 2–3-rayed hairs 0.1–0.5 mm. **LF:** basal 4–8 mm wide, minutely dentate; hairs long-stalked, 2–5-rayed, 0.1–0.5 mm; cauline 8–25, distal hairy, basal lobes 1–3 mm. **INFL:** 12–34-fld, not 1-sided in fr; fr pedicel spreading-ascending to horizontal, straight, 5–10 mm, hairs spreading, simple or 2-rayed. **FL:** sepals hairy; petals 6–8 mm, 2–2.5 mm wide, ± purple; pollen spheric. **FR:** ± horizontal, not appressed, 4–8 cm, 1.7–2.2 mm wide, glabrous; style 0.1–0.3 mm; ovules 70–102. **SEED:** in 1 row, 1.4–1.8 mm; wing 0.1–0.2 mm wide. Rocky slopes; ± 1200 m. SnFrB (Santa Clara Co.). Apr–May ★

B. serpenticola Windham & Al-Shehbaz SERPENTINE ROCKCRESS Caudex woody. **ST:** gen 1 per caudex branch, from center of basal rosette at ± ground surface; 0.5–1.8 dm, proximally with short-stalked, 5–8-rayed hairs 0.1–0.2. **LF:** basal 1–2 mm wide, entire; hairs short-stalked, 5–8-rayed, 0.1–0.2 mm; cauline 4–14, distal hairy, basal lobes 0.1–0.5 mm. **INFL:** 10–20-fld, not 1-sided in fr; fr pedicel horizontal, ± straight, 3–7 mm, hairs appressed, branched. **FL:** sepals hairy; petals 6–8 mm, 1.5–2.5 mm wide, purple; pollen ellipsoid. **FR:** horizontal (pendent), not appressed, 3–4.5 cm, 2.5–3 mm wide, sparsely hairy to ± glabrous; style 1.5–2 mm; ovules 20–24. **SEED:** in 1 row, 2–2.5 mm; wing to 0.2 mm wide. Serpentine ridges, talus; 1100, 2100 m. s KR. Mar–Jun ★

B. shevockii Windham & Al-Shehbaz SHEVOCK'S ROCKCRESS Caudex woody, crowded, persistent lf bases present. **ST:** gen 1 per caudex branch, from center of basal rosette at ± ground surface; 0.5–1 dm, proximally with simple hairs 0.02–0.10 mm. **LF:** basal 1–2 mm wide, entire, surfaces glabrous, margins with simple hairs near tip, 0.02–0.10 mm; cauline 3–7, distal glabrous, basal lobes 0. **INFL:** 4–7-fld, not 1-sided in fr; fr pedicel ascending, straight, 5–9 mm, glabrous. **FL:** sepals glabrous; petals 4–5 mm, 1–1.2 mm wide, lavender; pollen ellipsoid. **FR:** ascending, not appressed, 2.5–3 cm, 1–1.5 mm wide, glabrous; style ± 0.5 mm; ovules 30–34. **SEED:** in 1 row, ± 1.2 mm; wing ± 0.1 mm wide. Rock outcrop ledges; ± 2500 m. s SNH (Tulare Co.). Jun ★

B. shockleyi (Munz) Dorn SHOCKLEY'S ROCKCRESS Caudex ± not woody; crowded, persistent lf bases 0 (present). **ST:** gen 1 per caudex branch, from center of basal rosette at ± ground surface; gen 2–5 dm, proximally with short-stalked, 7–12-rayed hairs 0.1–0.2 mm. **LF:** basal 3–10 mm wide, entire; hairs short-stalked, 7–12-rayed, 0.1–0.2 mm; cauline 14–60, distal hairy, basal lobes 0.5–4 mm. **INFL:** 20–70-fld, not 1-sided in fr; fr pedicel spreading-ascending, straight, 7–28 mm, hairs appressed, branched. **FL:** sepals hairy; petals 6–9 mm, 0.8–1.2 mm wide, lavender; pollen ellipsoid. **FR:** spreading-ascending, not appressed, 4.5–11 cm, 1.5–2 mm wide, glabrous to ± hairy throughout; style 0.05–0.6 mm; ovules 140–190. **SEED:** gen in 2 rows, 1–1.3 mm; wing 0.05–0.1 mm wide (0). Rock outcrops, gravelly soil, gen dolomite; 1200–2500 m. SnBr, W&I, DMtns; to UT. [*Arabis s.* Munz] Apr–May ★

B. sparsiflora (Nutt.) Dorn (p. 533) SICKLEPOD ROCKCRESS Short-lived per to bien; caudex gen not woody. **ST:** gen 1 per caudex branch, from center of basal rosette at ± ground surface; 3–8 dm, proximally with simple and short-stalked, 2-rayed hairs 0.4–1.5 mm. **LF:** basal 3–12 mm wide, entire (dentate); hairs short-stalked, 2–5-rayed, 0.3–0.8 mm; cauline gen 15–35, distal glabrous (ciliate), basal lobes 3–10 mm. **INFL:** 12–50-fld, not 1-sided in fr; fr pedicel ascending, straight or ± recurved, 3–18 mm, ± hairs spreading, gen simple, or glabrous. **FL:** sepals hairy; petals 7–13 mm, 2–5 mm wide, lavender to purple; pollen ellipsoid. **FR:** spreading-ascending (horizontal), not appressed, 5–13 cm, 1.7–2 mm wide, glabrous; style 0.05–0.3

mm; ovules 90–170. **SEED**: in 1 row, 1.5–2 mm; wing 0.1–0.2 mm wide. 2*n*=14. Rocky slopes, sandy soil, in sagebrush steppe, open conifer communities; 200–2800 m. KR, NCoRH, CaR, n SNF, GB; to BC, ID, UT. [*Arabis s.* Nutt.] Apr–Jun

B. stricta (Graham) Al-Shehbaz (p. 533) DRUMMOND'S ROCKCRESS Bien or short-lived per; caudex gen not woody. **ST**: gen 1–4 per caudex branch, from center of basal rosette, at ± ground surface; 1.5–10 dm, proximally glabrous or with few sessile, 2-rayed hairs 0.3–0.7 mm. **LF**: basal 2–10 mm wide, entire; hairs with sessile, 2-rayed hairs 0.3–0.7 mm (glabrous); cauline 6–52, distal glabrous, basal lobes 0.5–3 mm. **INFL**: 8–80-fld, not 1-sided in fr; fr pedicel erect, straight, 5–25 mm, glabrous. **FL**: sepals glabrous; petals 5–11 mm, 1.5–2.7 mm wide, white, occ pale lavender in age; pollen ellipsoid. **FR**: erect, appressed, 4–10 cm, 1.5–3.5 mm wide, glabrous; style 0.05–0.3 mm; ovules 110–216. **SEED**: in 2 rows, 1.3–2.2 mm; wing 0.3–0.8 mm wide. 2*n*=14. Rocky slopes, gravelly soil, in open conifer forest, montane meadows; 1800–3400 m. SNH, GB; to AK, e N.Am, NM. [*Arabis drummondii* A. Gray] May–Aug

B. subpinnatifida (S. Watson) Al-Shehbaz KLAMATH ROCKCRESS Caudex woody, gen with crowded, persistent lf bases. **ST**: gen 1 per caudex branch, from center of basal rosette at ± ground surface; 1–5 dm, proximally with short-stalked, 2–6-rayed hairs 0.1–0.2 mm. **LF**: basal 1–5 mm wide, dentate to ± pinnately lobed on sterile shoots gen entire; hairs short-stalked, gen 4–9-rayed, 0.05–0.2 mm; cauline 10–60, distal hairy, basal lobes 0.5–3 mm. **INFL**: reflexed, strongly recurved, 8–30-fld, not 1-sided in fr; fr pedicel 5–15 mm, hairs appressed, branched. **FL**: sepals hairy; petals 9–14 mm, 1.5–3 mm wide, purple (lavender); pollen ellipsoid. **FR**: pendent, not appressed, 3.5–8 cm, 1.6–3 mm wide, hairy throughout; style 0.5–1 mm; ovules 24–42. **SEED**: in 1 row, 2.5–3.5 mm; wing 0.4–0.8 mm wide. 2*n*=14. Igneous, metamorphic rock outcrops, gravelly soil; 700–2400 m. KR, n NCoRO, NCoRH; to OR, UT. [*Arabis s.* S. Watson] Mar–May

B. suffrutescens (S. Watson) Dorn (p. 533) WOODY ROCKCRESS Caudex woody. **ST**: gen 1 per caudex branch, from center of basal rosette elevated on woody base; 1–5 dm, proximally glabrous or hairs short-stalked, 2-rayed, 0.1–0.3 mm. **LF**: basal 1.5–6 mm wide, entire; faces glabrous or hairs short-stalked, 2–6-rayed, 0.07–0.4 mm; cauline 4–12, distal glabrous, basal lobes 0.5–2 mm. **INFL**: 6–12-fld, gen 1-sided in fr; fr pedicel reflexed, gen abruptly recurved at base, 4–18 mm, glabrous. **FL**: sepals glabrous; petals 4.5–6 mm, 2–2.5 mm wide, at least near tip purple or rose; pollen ellipsoid or spheric. **FR**: reflexed to pendent, occ appressed, 3–8 cm, 3–6 mm wide, glabrous, edges wavy; style 0.4–1.2 mm; ovules 20–30. **SEED**: in 1 row, 2.5–5.5 mm; wing 0.8–1.5 mm wide. Rocky slopes, gravelly soil, gen with sagebrush; 1500–3000 m. KR, NCoRH, CaRH, SNH, MP; to WA, ID. [*Arabis s.* S. Watson] Jun–Aug

B. tiehmii (Rollins) Al-Shehbaz (p. 533) TIEHM'S ROCKCRESS Caudex woody, occ with crowded, persistent lf bases. **ST**: gen 2–5 per caudex branch, from margin of basal rosette at ± ground surface; 0.8–3 dm, proximally glabrous. **LF**: basal 2–7 mm wide, entire to minutely dentate, margins with simple and short-stalked, 2–3-rayed hairs 0.2–0.4 mm, faces glabrous; cauline 3–5, distal glabrous, basal lobes 0 (< 0.5 mm). **INFL**: 10–16-fld, not 1-sided in fr; fr pedicel ascending, straight, 3–7 mm, glabrous. **FL**: sepals glabrous; petals 4–5 mm, 1.2–1.7 mm wide, white to lavender; pollen ellipsoid. **FR**:

ascending, not appressed, 1.5–3.7 cm, 1.4–1.7 mm wide, glabrous; style 0.2–0.6 mm; ovules 14–22. **SEED**: in 1 row, 1.7–2.2 mm; wing < 0.5 mm wide or 0. Rock outcrops, gravelly soil; 3000–3600 m. c SNH; NV. [*Arabis t.* Rollins] Jun–Aug ★

B. tularensis Windham & Al-Shehbaz TULARE ROCKCRESS Bien or short-lived per; caudex not woody. **ST**: gen 1 per caudex branch, from center of basal rosette at ± ground surface; 2–7 dm, proximally with ± sessile, 2-rayed hairs 0.3–0.6 mm. **LF**: basal 3–7 mm wide, entire; hairs ± sessile, 2–5-rayed, 0.2–0.55 mm; cauline 7–17, distal sparsely hairy or glabrous, basal lobes 2–5 mm. **INFL**: 19–39-fld, not 1-sided in fr; fr pedicel reflexed, ± recurved near base, 5–13 mm, glabrous. **FL**: sepals hairy; petals 6–7 mm, 1.2–2 mm wide, white to pale lavender; pollen spheric. **FR**: reflexed (appressed), 4–8.5 cm, 2–2.3 mm wide, glabrous; style 0.3–0.7 mm; ovules 88–104. **SEED**: ± in 2 rows, 2–2.5 mm; wing 0.15–0.25 mm wide. Rocky slopes in montane, subalpine habitats; 2400–3200 m. s SNH. Jun–Jul ★

B. ultraalsa Windham & Al-Shehbaz SNOW MOUNTAIN ROCKCRESS Caudex woody. **ST**: 1 per caudex branch, from center of basal rosette at ± ground surface; ± 1 dm, proximally with long-stalked, 2–6-rayed hairs gen 0.25–0.5 mm. **LF**: basal 4–6 mm wide, entire; hairs long-stalked, 3–7 rayed, gen 0.25–0.5 mm; cauline 2–5, distal hairy, basal lobes 0. **INFL**: 3–4-fld, not 1-sided in fr; fr pedicel erect-ascending, straight, 4–5 mm, glabrous or hairs few, appressed, branched. **FR**: erect-ascending, occ appressed, 3–4 cm, ± 5 mm wide, glabrous; style 0.5–0.7 mm; ovules ± 16. **SEED**: in 1 row, 5.5–6.5 mm; wing 1–2 mm wide. Rocky soil; ± 1800 m. NCoRH (Snow Mtn, Lake Co.). [*Boechera ultra-alsa*, orth. var.] Fls not seen to date. Jun–Jul ★

B. xylopoda Windham & Al-Shehbaz BIGFOOT HYBRID ROCKCRESS Caudex woody. **ST**: gen 1 per caudex branch, from center of rosette elevated on woody base; 3–7.5 dm, proximally with short-stalked, 2–5-rayed hairs 0.25–0.9 mm. **LF**: basal 3–7 mm wide, entire to dentate; hairs short-stalked, 4–8-rayed, 0.15–0.5 mm; cauline 8–18, distal ± hairy, basal lobes 1–2 mm. **INFL**: 20–38-fld, not 1-sided in fr; fr pedicel descending-spreading, gen recurved, 9–20 mm, hairs ± appressed, branched (glabrous). **FL**: sepals hairy; petals 9–12 mm, 2–3.5 mm wide, purple; pollen spheric. **FR**: reflexed, not appressed, 5–7 cm, 2–2.5 mm wide, distal 2/3 hairy, glabrous proximally; style 0.1–0.5 mm; ovules 98–126. **SEED**: ± in 2 rows, 2–2.5 mm; wing 0.2–0.25 mm wide. Rock outcrops, gravelly slopes, gen under shrubs; 800–2500 m. e PR, GB, D; NV. [*Arabis pulchra* S. Watson var. *gracilis* M.E. Jones] Mar–May

B. yorkii S. Boyd LAST CHANCE ROCKCRESS Caudex woody. **ST**: gen 1 per caudex branch, from center of basal rosette at ± ground surface; 1–3 dm, proximally with short-stalked, 4–7-rayed hairs 0.1–0.5 mm, occ with simple and short- to long-stalked, 2–3-rayed hairs to 1.5 mm. **LF**: basal 1–3 mm wide, entire; hairs short-stalked, 4–7-rayed, 0.3–0.6 mm; cauline 9–17, distal hairy, basal lobes 0. **INFL**: 8–35-fld, not 1-sided in fr; fr pedicel reflexed, straight, 1–2 mm, hairs appressed, branched. **FL**: sepals hairy; petals 9–10 mm, 0.8–1 mm wide, ± yellow to brick-red; pollen ellipsoid. **FR**: immature fr reflexed, not appressed, ± 4 cm, hairy throughout; style ± 0.3 mm. Calcareous rock crevices, ledges; 2250–2400 m. DMtns (Last Chance Range, Inyo Co.). Mature fr, seed not seen to date. May ★

BRASSICA MUSTARD

Ann to per; hairs simple or 0. **ST**: erect, simple or branched, glabrous or hairy distally. **LF**: basal petioled, gen rosetted, dentate to pinnately lobed; cauline petioled or sessile, bases lobed or not. **INFL**: terminal. **FL**: sepals erect to ascending, base gen not sac-like; petals gen yellow, clawed. **FR**: silique, linear, dehiscent, segmented, round or 4-sided to flat parallel to septum; valves 1-veined, glabrous; terminal segment conic to cylindric, 0–3-seeded; stigma entire to ± 2-lobed. **SEED**: (4)10–50, in 1 row, spheric to ovoid. 35 spp.: Medit, Eurasia, some naturalized ± worldwide. (Latin: cabbage) *Brassica* incl most important vegetable, seed-oil, condiment crops in Brassicaceae.

1. Distal cauline lvs sessile, base lobed or clasping
  2. Lvs fleshy; petals (1.5)1.8–2.5(3) cm; terminal fr segment (3)4–11 mm . . . . ***B. oleracea***
  2′ Lvs not fleshy; petals 0.5–1.6 cm; terminal fr segment (5)9–22 mm
    3. Petals 10–16 mm; fls not overtopping buds . . . . ***B. napus***
    3′ Petals 6–11(13) mm; fls overtopping buds . . . . ***B. rapa***
1′ Distal cauline lvs short-petioled to sessile, base tapered

4. Pedicel, fr erect, ± appressed; pedicel (2)3–5(6) mm. . . . . *B. nigra*
4′ Pedicel, fr spreading to ascending, not appressed; pedicel (5)8–25 mm
  5. Basal lvs stiff-hairy, persistent, lateral lobe pairs 4–10; petals 4–7 mm, 1.5–2(2.5) mm wide . . . . *B. tournefortii*
  5′ Basal lvs ± glabrous, deciduous, lateral lobe pairs 1–3(4); petals 7–15 mm, 3–7.5 mm wide
    6. Fr 1.5–3 cm, 1.5–2 mm wide, stalk above receptacle 1–1.5 mm, terminal segment 3–6 mm . . . . *B. fruticulosa*
    6′ Fr (2)3–5(6) cm, 2–5 mm wide, stalk above receptacle 0, terminal segment (4)5–10(15) mm . . . . *B. juncea*

B. fruticulosa Cirillo Ann or per; glabrous. **ST:** 3–9 dm, branched. **LF:** basal petioled, to 10 cm, early-deciduous, pinnately lobed, lateral lobe pairs 1–3(4); cauline petioled, dentate to lobed. **FL:** sepals 3–8 mm; petals 7–15 mm, 3–4 mm wide. **FR:** spreading to ascending, 15–30 mm, 1.5–2 mm wide, cylindric, strongly narrowed between seeds; terminal segment 3–6 mm, conic, 0–1-seeded; pedicel spreading-ascending, 1–1.5(2.5) cm, slender. **SEED:** 10–26, 0.7–1.2 mm wide, spheric. 2*n*=16. Uncommon. Disturbed areas, fields; < 300 m. SCo; native to sw Eur, nw Afr. Dec–Mar

B. juncea (L.) Czern. INDIA MUSTARD Ann, glabrous, ± glaucous. **ST:** 2–10 dm, branched distally. **LF:** basal early-deciduous, 0.5–10 dm, lateral lobes in 1–3 pairs; cauline petioled or sessile, dentate to lobed, distal reduced, base tapered, not lobed. **FL:** sepals (3.5)4–6(7) mm; petals (7)9–13 mm, 5–7.5 mm wide. **FR:** ± spreading to ascending, (2)3–5(6) cm, 2–5 mm wide; terminal segment (4)5–10(15) mm, seedless; pedicel spreading to ascending, (0.5)1–1.5(2) cm, slender. **SEED:** 1.2–2 mm wide, spheric. 2*n*=36. Uncommon. Disturbed areas, fields; < 300 m. GV; widespread N.Am, native to Eurasia. Cult for seed oil used to make table mustard. May–Sep

B. napus L. SWEDE RAPE, RAPESEED Ann, bien, glabrous or hairy, glaucous. **ST:** 3–13 dm, branched distally. **LF:** basal pinnately lobed, 0.5–5.5 dm; middle, distal cauline sessile, base strongly lobed or clasping. **FL:** not overtopping buds; sepals (5)6–10 mm; petals 10–16 mm, (5)6–9(10) mm wide. **FR:** ascending, (3.5)5–10(11) cm, (2.5)3.5–5 mm wide; terminal segment (5)9–16 mm, seed 0(2); pedicel spreading to ascending, 1–3 cm, slender. **SEED:** 1.8–2.7(3) mm wide, spheric. 2*n*=38. Disturbed areas, fields; < 500 m. SnJV, SCo; widespread N.Am, native to Eur. Source of canola oil. May–Sep

B. nigra (L.) W.D.J. Koch (p. 537) BLACK MUSTARD Ann; hairs sparse to dense, stiff, esp below. **ST:** 3–20 dm, gen branched distally. **LF:** basal pinnately lobed, serrate-dentate; cauline similar to basal, distal lvs smaller, sessile, base tapered, not lobed. **FL:** sepals 4–6(7) mm; petals 7–11(13) mm, (2.5)3–4.5(5.5) mm wide. **FR:** erect, 1–2.5(2.7) cm, (1.5)2–3 mm wide, terminal segment (1)2–5(6)

mm, seedless; pedicel erect, ± appressed, (2)3–5(6) mm. **SEED:** 1.2–1.5(2) mm wide, spheric. 2*n*=16. Common. Disturbed areas, fields; < 1500 m. CA-FP; widespread N.Am, native to Eur. Apr–Sep ❖

B. oleracea L. CABBAGE Bien, per; glabrous, glaucous. **ST:** 5–10 dm, branched distally. **LF:** fleshy; basal pinnately lobed, 0.5–7.5 dm; middle, distal cauline sessile, base lobed or gen clasping. **FL:** sepals 8–15 mm; petals (1.5)1.8–2.5(3) cm, (6)8–12 mm wide, yellow or cream-white. **FR:** ascending, (2.5)5–8(10) cm, (2.5)3–4(5) mm wide; terminal segment (3)4–11 mm, seeds 0(2); pedicel spreading to ascending, slender, (0.8)1.4–2.5(4) cm. **SEED:** 1.7–2.5 mm wide, spheric. 2*n*=18. Se-facing sea cliffs; < 400 m. c&s NCo, n CCo; to WA, MT, TX, e N.Am, native to Eur. Most important vegetable crop in family. May–Aug

B. rapa L. (p. 537) TURNIP, FIELD MUSTARD Ann, bien, erect; hairs 0 or few. **ST:** 3–10 dm, simple or branched. **LF:** basal pinnately lobed, lateral lobes 2–4(6) pairs, terminal lobe obovate, wavy-dentate; middle, distal cauline sessile, base lobed, gen clasping. **FL:** overtopping buds; sepals (3)4–6.5(8) mm; petals 6–11(13) mm, (2.5)3–6(7) mm wide. **FR:** ascending to ± spreading, (2)3–8(11) cm, 2–4(5) mm wide; terminal segment 8–22 mm; pedicel ascending to spreading, (0.5)1–2.5(3) cm. **SEED:** 1.1–2 mm wide, spheric. 2*n*=20. Disturbed areas; < 1500 m. CA-FP, SNE; widespread N.Am, native to Eur. Jan–May ❖

B. tournefortii Gouan (p. 537) Ann, branched ± from base, st, lvs densely stiff-hairy. **ST:** (1)3–7(10) dm. **LF:** basal rosetted, persistent, petioled, pinnately lobed, serrate-dentate, lateral lobe pairs 4–10; cauline few, base tapered, not lobed. **FL:** sepals 2.5–4.5 mm; petals 4–7 mm, 1.5–2(2.5) mm wide. **FR:** spreading to ascending, 3–7 cm, 2–4(5) mm wide, cylindric, narrowed between seeds; terminal segment 10–20 mm, 1(3)-seeded; pedicel spreading, 8–15 mm. **SEED:** 1–1.2 mm wide, spheric. 2*n*=20. Roadsides, washes, open areas; < 800 m. SW, D; to s NV, TX; native to Medit, sw Asia. Locally abundant. Jan–Jun ❖

CAKILE SEA ROCKET

Ann (per), fleshy, many-branched, glabrous. **ST:** erect or decumbent. **LF:** fleshy, cauline, petioled or not, base not lobed. **INFL:** terminal, elongating. **FL:** sepal base not sac-like; petals white to lavender (0). **FR:** silique or silicle, fleshy, in age corky, indehiscent, segmented, segments 2, transversely jointed, each gen 1-seeded; stigma entire or ± 2-lobed. **SEED:** oblong. 7 spp.: shores, e N.Am, Eur, n Afr. (Arabic name) [Rodman 1974 Contr Gray Herb 205:3–146]

1. Lower fr segment lateral lobes or horns 0; petals 0 or 1.5–3 mm wide . . . . *C. edentula*
1′ Lower fr segment lateral lobes or horns 2; petals 3–6 mm wide . . . . *C. maritima*

C. edentula (Bigelow) Hook. (p. 537) **ST:** gen ascending to erect, ≤ 8 dm. **LF:** cauline ovate to spoon-shaped, petioled or not. **FL:** sepals 3.5–5 mm; petals 0 or 4.5–9.5 mm. **FR:** 1.2–2.9 cm, 5–9 mm wide; terminal segment 7–15 mm, tip ± flat. 2*n*=18. Beach dunes; < 50 m. NCo, CCo, SCo; native to e N.Am. Now less common in CA, being replaced by *C. maritima*. May–Nov

C. maritima Scop. (p. 537) **ST:** prostrate or mound-forming to erect, ≤ 8 dm. **LF:** cauline, broadly ovate to lanceolate, petioled. **FL:** sepals 4–5.5 mm; petals 8–14 mm. **FR:** 1.2–2.7 cm; terminal segment conic, gen 4-sided, tip acute to blunt; pedicel 1.5–7 mm. 2*n*=18. Beach dunes; < 100 m. NCo, CCo, SCo, ChI; to BC, e N.Am, Mex; native to Eur. May–Nov ❖

CAMELINA FALSE FLAX

Ann, bien; hairs simple, stalked-forked, ± stellate, or many-branched. **ST:** erect. **LF:** cauline, sessile, entire to dentate, base lobed or clasping. **INFL:** elongated; bracts 0. **FL:** sepals erect to ascending, base not sac-like; petals yellow (white); filaments in 3 pairs of unequal length. **FR:** silicle, obovoid to pear-shaped, dehiscent, unsegmented, ± flattened parallel to septum, ± stalked above receptacle, tip extending onto style; valves leathery; stigma entire. **SEED:** in 2 rows, oblong, plump, wing 0. 8 spp.: Eurasia. (Greek: low flax, for inhibiting growth of flax pls) *C. alyssum* (Mill.) Thell. an historical waif.

C. microcarpa DC. (p. 537) Ann, densely to moderately stiff-hairy. **ST:** (0.8)2–8(10) dm, simple or branched distally. **LF:** lanceolate to oblong, lobed at base. **FL:** sepals 2–3.5 mm; petals (2.5)3–4(6) mm, pale yellow. **FR:** 3.5–5(7) mm, 2–4(5) mm wide, pear-shaped;

style 1–3.5 mm, beak-like; pedicel ascending, 4–14(17) mm, slender. **SEED:** 0.8–1.5 mm. 2*n*=40. Fields, roadsides, slopes; < 1800 m. CA-FP, MP; N.Am; native to Eurasia. May–Jul

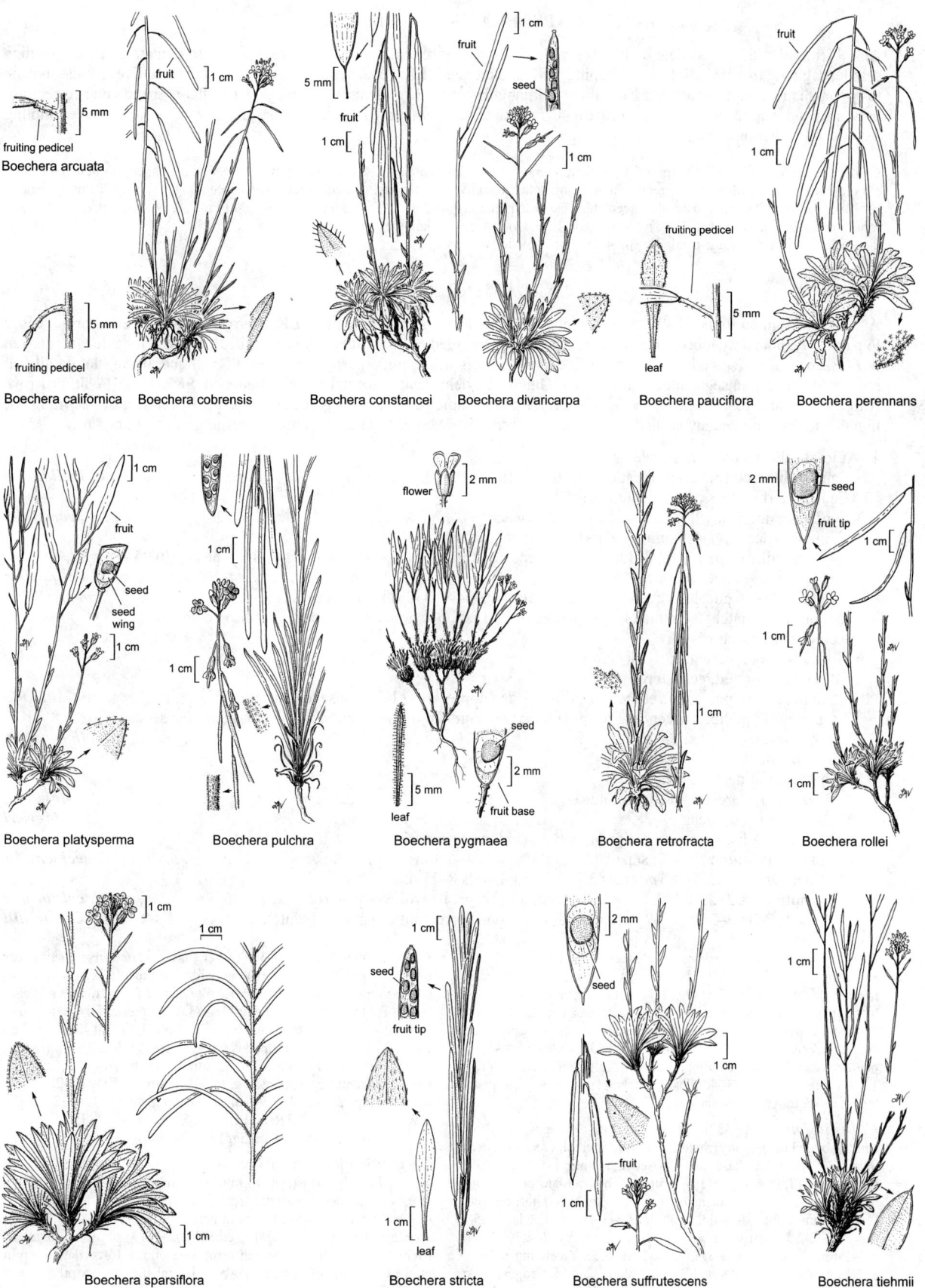

Boechera arcuata

fruiting pedicel

fruiting pedicel

Boechera californica

Boechera cobrensis

fruit

Boechera constancei

fruit

seed

fruit

Boechera divaricarpa

fruiting pedicel

leaf

Boechera pauciflora

fruit

Boechera perennans

fruit

seed

seed wing

Boechera platysperma

Boechera pulchra

flower

seed

fruit base

leaf

Boechera pygmaea

Boechera retrofracta

seed

fruit tip

Boechera rollei

Boechera sparsiflora

seed

fruit tip

leaf

Boechera stricta

seed

fruit

Boechera suffrutescens

Boechera tiehmii

CAPSELLA SHEPHERD'S PURSE

Ann, bien; hairs sessile, stellate, occ also simple. **LF:** basal rosetted, petioled, pinnately lobed to dentate or entire; cauline sessile, sagittate to basally lobed or clasping. **INFL:** elongated. **FL:** sepals erect to ascending, base sac-like; petals obovate to spoon-shaped, white or ± pink. **FR:** silicle, obtriangular-obcordate, dehiscent, unsegmented, flat perpendicular to septum; valves veined; stigma entire. **SEED:** (10)20–40, in 2 rows per chamber, wingless. 4 spp.: Eurasia. (Latin: little box resembling a medieval wallet or purse)

C. bursa-pastoris (L.) Medik. (p. 537)   Stellate hairs 3–5-rayed, simple hairs gen at pl base. **ST:** erect, simple or branched distally, (0.5)1–5(7) dm. **LF:** basal 3–6 cm, oblanceolate, ± entire to pinnately lobed or dissected; cauline 1–5.5(8) cm, oblong, linear to lanceolate. **FL:** sepals 1.5–2 mm; petals (1.5)2–4(5) mm, obovate, white (pink). **FR:** (3)4–9(10) mm, base wedge-shaped, tip notched or not, glabrous; pedicel spreading to ascending, (3)5–15(20) mm, slender. 2*n*=32. Disturbed areas; < 2800 m. CA; N.Am; native to Eurasia. Jan–Oct

CARDAMINE BITTER-CRESS

Ann to per, from taproots, fibrous roots, or tuber-like rhizomes; hairs 0 or simple. **LF:** alternate, opposite, or whorled; entire to palmately, pinnately lobed, or compound; cauline lvs petioled or 0, not lobed at base [lobed]. **INFL:** elongated, bracts gen 0. **FL:** sepals erect (spreading), bases sac-like or not; petals white, pink, purple, or violet. **FR:** silique, linear, flat parallel to septum, dehiscent, unsegmented; valves gen coiling when dehiscent; placental margins flattened. **SEED:** (4)10–80, in 1 row, wingless. ± 200 spp.: temp, worldwide. (Greek: for cress) Some N.Am spp. (e.g., *C. californica, C. nuttallii, C. pachystigma*) highly variable, more study needed; spp. treated conservatively here. *C. flexuosa* With. a waif in gardens, nurseries.

1. At least some cauline lvs simple or 0
  2. Caudex distinct; rhizomes 0; cauline lvs 0 or 1(2).. ***C. bellidifolia***
  2′ Caudex 0; rhizomes present; cauline lvs 2–23
    3. Rhizome not fleshy, 1–3 mm wide; rhizome lvs gen 0 .... ***C. cordifolia***
    3′ Rhizome fleshy, (3)4–18 mm wide; rhizome lvs present
      4. Petals white or pale rose, 8–13(15) mm; seeds 12–22; st (2)2.7–6(7) dm; lflets of rhizome lvs (0)3(5–7)
        .... ²***C. californica***
      4′ Petals pink to purple (white), 14–18 mm; seeds 10–14; st 1–3 dm; rhizome lvs simple............. ***C. pachystigma***
1′ Cauline lvs pinnately or palmately compound; lflets 3–25, occ pinnately divided, appearing compound
  5. Ann or bien; rhizomes 0
    6. Basal lvs not rosetted, early-deciduous... ***C. pensylvanica***
    6′ Basal lvs rosetted, persistent
      7. Stamens 4(6); pedicels erect to ascending; fr gen appressed, glabrous; seeds margined............... ***C. hirsuta***
      7′ Stamens 6; pedicels ascending to spreading; fr gen not appressed, sparsely hairy or glabrous; seeds
        not margined .... ***C. oligosperma***
  5′ Per; rhizomed
    8. Rhizome not fleshy, cylindric
      9. Rhizome lvs present; seeds 6–16(24).. ***C. angulata***
      9′ Rhizome lvs 0; seeds 14–28.. ***C. breweri***
    8′ Rhizome fleshy, segments fusiform, ovoid to oblong or spheric
      10. Lflets of cauline lvs 5–7; sepals 1.7–2 mm; petals 4–6 mm .... ***C. occidentalis***
      10′ Lflets of cauline lvs 3(5); sepals 3.5–5.5 mm; petals 8–15 mm
        11. Cauline lvs 2–5; st (2)2.7–6(7) dm; seeds 12–22; petals white or pale rose .... ²***C. californica***
        11′ Cauline lvs 1–3; st 0.5–2(3) dm; seeds 8–16; petals pale pink to purple (white).................. ***C. nuttallii***

C. angulata Hook.   Per; rhizome slender, ≤ 2 mm wide, cylindric. **ST:** simple, (1.5)2.5–8.5(10) dm; hairs at base sparse to dense. **LF:** lflets of rhizome lvs 3(5), petiole (2)4–12(14) cm; cauline lvs (3)4–8, lflets 3(5), ovate to lanceolate, margin dentate or entire. **FL:** sepals 2.5–4 mm; petals 8–15 mm, 4–8 mm wide, white to ± pink. **FR:** erect to ascending, 1.5–3.2 cm, 1.4–2 mm wide; style (0.5)1–4 mm; pedicel ascending to spreading, (0.9)1.2–2.5 cm. **SEED:** 6–16(24), 1.8–2.3 mm, oblong. 2*n*=40. Shady thickets, streambanks, forest; < 900 m. NW; to AK. Apr–Jun ★

C. bellidifolia L. (p. 537)   Per, ± cespitose, ± scapose, glabrous; caudex branches 0–many, rhizomes 0. **ST:** several, erect to ascending, 1–8(14) cm. **LF:** basal many, rosetted, simple, 1–5(7) cm; blade (0.5)0.8–1.7(2.5) cm, ovate to oblong (oblanceolate or obovate), margin entire (wavy or dentate); cauline 0 or 1(2), short-petioled. **INFL:** ± umbel-like, few-fld. **FL:** sepals 2–3(4) mm; petals 4–5.5(7) mm, 1.3–2(2.5) mm wide, white. **FR:** erect, (0.8)1.3–2.8(3.7) cm, 1.3–2 mm wide; style 0–3 mm, stout; pedicel ascending to erect, 3–6(8) mm. **SEED:** 8–18, 1.5–2 mm, oblong. 2*n*=16. Ledges, rocky, moist soil, alpine slopes, volcanic peaks, mossy places, streambanks; 1800–2850 m. CaRH; to AK, ne US, Greenland, n Eurasia. [*C. b.* var. *pachyphylla* Coville & Leiberg] Jun–Sep ★

C. breweri S. Watson (p. 537)   Per, glabrous, or sparsely hairy near base; rhizome slender, 1–3(4) mm wide, cylindric. **ST:** decumbent to erect, simple to branched, (0.6)1.5–6(7) dm. **LF:** rhizome lvs 0; cauline 3–8(11), 2.5–12(13.5) cm; lflets (1)3–5, terminal ovate to round (± cordate), margin crenate, dentate, or lobed, lateral lflets smaller. **FL:** sepals 2–3(3.8) mm; petals 3.5–6(7) mm, 1.5–2.5(3) mm wide, white. **FR:** erect, 1.5–3.5 cm, 1–1.5 mm wide; style 0.2–1.5(2.5) mm; pedicel ascending to spreading, (0.7)1–2 cm. **SEED:** 14–28, 1–1.5 mm, oblong. 2*n*=42–48. Wet places, conifer forest; 400–3200 m. n NW, CaRH, SNH, WTR, SnBr, Wrn, SNE; to BC, MT, CO. [*C. b.* var. *orbicularis* (Greene) Detling] Jun–Jul

C. californica (Nutt.) Greene (p. 537)   MILK MAIDS, TOOTH WORT   Per; rhizome (3)4–10 mm wide, tuber-like, spheric to ovoid or ± oblong, deeply buried; hairs 0 (minute, simple). **ST:** (2)2.7–6(7) dm, erect, simple. **LF:** lflets of rhizome lvs (0)3(5–7), 8–25(38) cm; terminal lflet (1.5)2.5–7.5(10) cm, (1.2)2–9(13) cm wide, ovate to round, widely cordate, or reniform, base obtuse to cordate, margin entire to dentate or shallowly wavy; lateral lflets when present ± = terminal lflet or smaller; cauline 2–5, petioled; lflets (0)3(5), widely ovate to ± round or lanceolate. **FL:** sepals 3.5–4.5(5.5) mm; petals 8–13(15) mm, 4–8 mm wide, white or pale rose. **FR:** 2.2–5(6) cm,

2–3 mm wide; style 3–6 mm; pedicel ascending, 1–3(4) cm. **SEED**: 12–22, 1.7–2.8 mm, oblong. 2*n*=32. Gen shaded sites, canyons, woodland; < 1400 m. CA-FP; Baja CA. [*Dentaria c.* Nutt.; *C. c.* var. *cardiophylla* (Greene) Rollins; *C. c.* var. *cuneata* (Greene) Rollins; *C. c.* var. *integrifolia* (Nutt.) Rollins; *C. c.* var. *sinuata* (Greene) O.E. Schulz; *C. pachystigma* var. *dissectifolia* (Detling) Rollins] Jan–May

C. cordifolia A. Gray   Per, glabrous or hairy at least at base; rhizome cylindric, 1–3 mm wide. **ST**: 2–7(10) dm, simple or branched above. **LF**: simple, 5–15 cm, cordate or reniform, petiole 2.5–12 cm; cauline (3)5–17(23), gen simple, fleshy, petioled, 2–10 cm, 1–5.5(8.5) cm wide, reniform to cordate or ovate, distal reduced. **FL**: sepals 2.5–4.5 mm; petals 7–12 mm, 4–6 mm wide, obovate, white. **FR**: ascending to erect, (2)2.5–3.7(4) cm, 1.2–2 mm wide; style 0.5–3(6) mm; pedicel ascending to spreading, (0.7)1–2 cm. **SEED**: 14–24, 1.6–2 mm, oblong. 2*n*=24. Streambanks, meadows, wet or moist areas, mixed-conifer forest; 600–3600 m. KR, CaR, n SNH; to BC, WY, NM. [*C. c.* var. *lyallii* (S. Watson) A. Nelson & J.F. Macbr.] May–Aug

C. hirsuta L.   Ann, stiff-hairy below. **ST**: erect, (0.3)1–3.5(4.5) dm. **LF**: basal rosetted, persistent; petioles ciliate; lflets (2)4–7(11) pairs, terminal > lateral, 0.6–3 cm wide, reniform to ± round, margin entire, wavy, dentate, or 3–5-lobed; cauline 1–4(6), similar and smaller than basal, petioled. **FL**: sepals 1.5–2.5; petals 2.5–4.5(5) mm, 0.5–1.1 mm wide, occ 0, white; stamens 4(6). **FR**: erect, (0.9)1.5–2.5(2.8) cm, (0.8)1–1.4 mm wide; style 0.1–0.6(1) mm; pedicel erect to ascending, (2)3–10(14) mm. **SEED**: 14–40, 0.9–1.3(1.5) mm, oblong, margined. 2*n*=16. Disturbed areas, roadsides, fields; < 800 m. KR, CCo; to BC, e N.Am; native to Eur. Feb–Jul

C. nuttallii Greene (p. 537)   Per, glabrous or sparsely hairy; rhizome slender, segments fleshy, 2–5 mm wide, ovoid to oblong or cylindric. **ST**: simple, 0.5–2(3) dm, glabrous, or hairy distally. **LF**: rhizome simple or compound, lflets 3(5), terminal lflet or simple lf (0.9)1.3–4(5.2) cm, reniform to ± round, ovate, or oblong, base cordate to obtuse, margin crenate to dentate or 5–7-lobed; petiole (2)3–18(21) cm; cauline 1–3, lflets 3(5), petioled, terminal 1–3.5(6) cm, widely ovate to oblong or linear. **FL**: sepals 3.5–5 mm; petals 1–1.5 cm, 4–7.5 mm wide, pale pink to purple (white). **FR**: 2.6–5.6 cm, 2–2.3 mm wide; style 4–8 mm; pedicel ascending to spreading, 1–3.5 cm. **SEED**: 8–16, 2–2.5 mm, oblong. Gen moist sites, canyons, forest; 150–2200 m. NW, n&c SNH; to BC. [*C. n.* var. *covilleana* (O.E. Schulz) Rollins; *C. n.* var. *dissecta* (O.E. Schulz) Rollins; *C. n.* var. *gemmata* (Greene) Rollins] Mar–May

C. occidentalis (S. Watson) Howell   Per, glabrous or stiff-hairy; rhizomes fleshy, 3–10 mm wide, ovoid to spheric. **ST**: erect or ascending, simple or branched above, 1–5 dm. **LF**: basal not rosetted, petioled, pinnately compound; lflets (3)5(7), terminal 0.5–2 cm, round or widely ovate or ± cordate, margin entire or wavy, > lateral; cauline 3–7, lflets 5–7, petioled, terminal obovate to oblanceolate, entire to dentate or wavy-margined. **FL**: sepals 1.7–2 mm; petals 4–6 mm, 1.5–2 mm, white. **FR**: 1.5–3.5 cm, 1.7–2.2 mm wide; pedicel spreading-ascending, 7–18 mm; style 0.5–1.5 mm. **SEED**: 18–40, 1–1.6 mm, ovate. 2*n*=64. Wet soils, lake margins, creeks; 150–1500 m. NW; to AK. Mar–May

C. oligosperma Nutt. (p. 537)   Ann, bien, hairy at least near base, taprooted. **ST**: erect to ascending, 1–several from base, (0.5)0.8–3(4) dm. **LF**: basal rosetted, pinnately compound, persistent; lflets 5–9(13), terminal 4–15(23) mm, round or ovate (oblong), entire or dentate (3–5-lobed), > lateral; cauline 3–8, similar to basal, petioled. **INFL**: 3–10 cm, elongated. **FL**: sepals 1.3–1.8(2) mm; petals 2.5–3.5 mm, 0.9–1.5 mm wide, white; stamens 6. **FR**: ascending to erect, (1.3)1.6–2.8 cm, 1–1.7 mm wide; hairs 0 or sparse; style 0.4–1(1.5) mm; pedicel ascending to spreading, 2–9(12) mm. **SEED**: 16–32(42), 1–1.6 mm, oblong, not margined. 2*n*=16. Wet meadows, shady banks, damp areas; 50–3300 m. CA-FP, W&I; to BC, MT, CO, Baja CA. Mar–Jul

C. pachystigma (S. Watson) Rollins   Per, glabrous; rhizome fleshy, (3)4–18 mm wide, ovoid to oblong. **ST**: erect, 1–3 dm, simple. **LF**: rhizome lvs simple, 1–2.5 cm, round to reniform, cordate, or wide-ovate, margin entire or wavy; petiole 6.7–19.5 cm; cauline near infl, 2–5, similar to rhizome lvs, petioled. **INFL**: short. **FL**: sepals 4–7 mm; petals 14–18 mm, 5–7 mm wide, pink to purple (white). **FR**: ± erect, 3.2–5.4 cm, 2–4 mm wide; style 4–7 mm; pedicel ascending to spreading, 0.7–2.4 cm. **SEED**: 10–14, 2–2.8 mm, ovate. Rocky or serpentine outcrops, slopes, cliffs, lava talus; 250–2900 m. KR, NCoR, n SNF, SNH, s SCoRO. Mar–May

C. pensylvanica Willd. (p. 537)   Ann, bien, rhizomes 0. **ST**: erect, simple or branched distally, (0.5)1–5.5(7) dm; hairs few or gen 0 near base. **LF**: basal not rosetted, early-deciduous; proximal cauline pinnately compound, lflets (5)7–13(19), terminal 1.3–3(4) cm, ± round to obovate, oblanceolate, or elliptic, margin entire, wavy, or 3–5-lobed, > lateral; distal cauline smaller, petioled. **FL**: sepals (1)1.3–2.3 mm; petals 2–3.5(4) mm, 0.8–1.5 mm, white. **FR**: ± erect, (1.4)1.7–2.7(3.2) cm, 0.8–1 mm wide; style 0.5–1 mm; pedicel ascending, (3)4–10(13) mm. **SEED**: 40–80, 0.7–1 mm, oblong. 2*n*=32, 64. Wet or muddy places; 200–2200 m. CaR, n SNF; to AK, e N.Am. Apr–Jul

CAULANTHUS   JEWELFLOWER

Ann to per, glabrous or hairs simple (forked). **LF**: basal rosetted or not, petioled, entire, dentate or pinnately lobed, gen deciduous; cauline petioled, or sessile with bases lobed or sagittate. **INFL**: elongated. **FL**: calyx urn-shaped or cylindric, sepals erect (spreading), base sac-like or not; petals yellow, purple, brown, or white, gen channeled, margin wavy or not; stamens in 3 pairs of equal length, or 4 long and 2 short, (all equal), free or filaments of longer pair(s) fused. **FR**: silique, linear, dehiscent, unsegmented, ± sessile, cylindric (flat perpendicular or parallel to septum); stigma entire or 2-lobed. **SEED**: 24–210, 1 row per chamber, oblong to ovate (± spheric), plump, wing 0. 17 spp.: sw US, nw Mex. (Greek: st fl, in reference to insertion of fls along st)

1. Cauline lvs petioled or occ ± sessile, bases not lobed or sagittate
    2. Per with woody caudex, glabrous or sparsely hairy
       3. St strongly inflated . . . . ***C. crassicaulis***
       3′ St not inflated (± inflated)
          4. Basal lvs not rosetted; pedicel 5–35 mm; stigma lobes opposite septum margins . . . . ***C. glaucus***
          4′ Basal lvs rosetted; pedicel 1–6 mm; stigma lobes opposite valves . . . . ***C. major***
    2′ Ann or bien (short-lived per), woody caudex 0; glabrous or moderately to densely stiff-hairy proximally.
       5. Sepals, petals, stamens widely spreading; filaments ± equal in length . . . . ***C. anceps***
       5′ Sepals, petals, stamens erect (sepals ascending to spreading); filaments in 2 or 3 pairs of unequal length
          6. Petals 2.5–5(6.5) mm, neither channeled nor wavy-margined, undifferentiated into blade and claw; calyx not urn-shaped . . . . ***C. lasiophyllus***
          6′ Petals (6)8–17.5 mm, channeled, gen wavy-margined, well differentiated into blade and claw; calyx gen urn-shaped
             7. Petals ± white or ± purple; seeds 152–198 . . . . ***C. pilosus***

7′ Petals creamy white (pink); seeds 44–96
  8. Stamens 4 long, 2 short; fr 3.8–7.7 cm; seeds 44–58 . . . . *C. flavescens*
  8′ Stamens in 3 pairs of unequal length; fr 6.5–12.5 cm; seeds 78–96 . . . . *C. hallii*
1′ Cauline lvs sessile, bases lobed, sagittate, or clasping
  9. St much inflated, to 4 cm wide . . . . *C. inflatus*
  9′ St not inflated
    10. Fr flat perpendicular to septum; seeds ± spheric . . . . *C. californicus*
    10′ Fr cylindric, flat parallel to septum, or 4-sided; seeds ovoid to oblong
      11. Infl with a terminal cluster of sterile fls; stamens in 3 pairs of unequal length, filaments of upper or
        lower pairs fused
        12. Fr reflexed (spreading); lf hairs 2-rayed forked and simple; cotyledons deeply 3-lobed; stigma lobes
          0.5–1.5 mm . . . . *C. coulteri*
        12′ Fr erect to ascending; lf hairs simple; cotyledons entire; stigma lobes 1–4 mm . . . . *C. lemmonii*
      11′ Infl without a terminal cluster of sterile fls; stamens gen 4 long, 2 short, filaments free
        13. Pl glabrous throughout; stamens in 3 pairs of unequal length; fr spreading to ascending . . . . *C. amplexicaulis*
        13′ Pl hairy at least basally; stamens 4 long, 2 short; fr reflexed (spreading)
          14. Hairs 0 or appressed, 2-rayed; distal st gen wavy, gen weak . . . . *C. cooperi*
          14′ Hairs spreading, simple; st straight, rigid
            15. Fr flat parallel to septum or 4-sided; cauline lvs lance-linear . . . . *C. heterophyllus*
            15′ Fr cylindric; cauline lvs ovate to oblong . . . . *C. simulans*

C. amplexicaulis S. Watson (p. 537) Ann, glabrous, glaucous. **ST:** erect, 0.4–11 dm, not inflated. **LF:** basal rosetted, 1.5–10 cm, obovate to oblanceolate, margin dentate; cauline sessile, bases lobed to clasping, dentate or entire. **INFL:** terminal sterile fl cluster 0. **FL:** sepals 4–9 mm, purple or ± yellow; petals 10–18 mm, tips gen reflexed, distal pair longer, ± purple, proximal gen straw-colored or paler purple, margins wavy; stamens in 3 pairs of unequal length, filaments free. **FR:** ascending to spreading, 4.5–14(16.7) cm, 1–1.5 mm wide, gen cylindric; style to 0.4 mm, stigma ± entire; pedicel spreading to ascending, 2.5–19 mm. **SEED:** 40–92, 1.4–2.2 mm, oblong; cotyledons entire. 2*n*=28. Chaparral, open, sandy or rocky areas, serpentine, granitic and shale scree; 800–2900 m. s SCoRO, TR. [*C. a.* var. *barbarae* (J.T. Howell) Munz] Apr–Aug

C. anceps Payson Ann, sparsely to densely hairy. **ST:** erect, 3.5–15 dm, simple or branched distally. **LF:** basal not rosetted, early-deciduous; cauline petioled, 2–13 cm, lanceolate to oblong, dentate, base not lobed. **INFL:** terminal sterile fl cluster 0. **FL:** sepals spreading, 3.5–5.5 mm; petals spreading, 4–8 mm, 2–4 mm wide, neither channeled nor wavy-margined, white to lavender; stamens ± equal, spreading. **FR:** ascending to reflexed, 3–6.7 cm, 1.2–2 mm, straight; style 1–4 mm, stigma ± entire; pedicel reflexed or ascending, 3–10 mm. **SEED:** 40–54, 1.4–1.8 mm, oblong to ovoid. 2*n*=28. Open slopes, plains, gen alkaline soil, roadsides, hillsides; 300–1700 m. s SNF, sw SnJV, se SCoRO, SCoRI, n WTR. [*Guillenia lemmonii*, ined.] Mar–May

C. californicus (S. Watson) Payson (p. 541) CALIFORNIA JEWELFLOWER Ann, ± glabrous, or sparsely bristly proximally. **ST:** decumbent to erect, 0.9–5.5 dm, gen branched distally. **LF:** basal rosetted, 1–11 cm, oblanceolate, coarsely dentate to shallowly lobed; cauline sessile, ovate to ± round, entire to coarsely dentate, base lobed to clasping. **INFL:** terminal sterile fl cluster present. **FL:** sepals erect to ascending, 4–9(11) mm, keeled; petals 5.5–12 mm, margins wavy, ± white, veins purple; filaments in 3 pairs of unequal length, free or longest pair fused. **FR:** ascending to reflexed, 1.7–5 cm, 3.5–6 mm wide, flat perpendicular to septum; style 0.2–2.7 mm, stigma strongly 2-lobed, lobes opposite valves; pedicel reflexed to ascending, 2–11 mm. **SEED:** 46–100, 1–1.6 mm wide, ± spheric; cotyledons deeply 3-lobed. 2*n*=28. Flats, slopes, gen in non-alkaline grassland; 70–1000 m. s SnJV, WTR. Formerly more widespread in s SnJV. Feb–Apr ★

C. cooperi (S. Watson) Payson Ann. **ST:** 1–8 dm, gen weak, not inflated, gen wavy and branched distally, glabrous or hairs appressed, 2-rayed, minute. **LF:** basal 1–8 cm, coarsely dentate to lobed (entire), oblanceolate to spoon-shaped; cauline sessile, lanceolate to oblong, entire to dentate, base lobed to clasping. **INFL:** terminal sterile fl cluster 0. **FL:** sepals 3–6.5 mm, ± purple or yellow-green; petals 4.5–9 mm, not wavy, ± purple or yellow-green; stamens 4 long, 2 short, filaments free. **FR:** spreading to reflexed, 2–6 cm, 1.5–2.5 mm wide, cylindric, gen curved; style 0.2–2.7 mm, stigma ± 2-lobed; pedicel reflexed, 1–4.5 mm. **SEED:** 24–48, 1–2 mm, oblong; cotyle-

dons entire. 2*n*=28. Common. Open, sandy, gravelly soil, gen among shrubs; 300–2500 m. n TR, e PR, s SNE, D; to sw UT, w AZ, Baja CA. Pls with st, fr ± green or straw-colored gen mixed with pls with st, fr dark purple. Mar–Apr

C. coulteri S. Watson (p. 541) Ann, bristly proximally, hairs simple and stalked 2-rayed. **ST:** 1–16 dm, not inflated. **LF:** basal 1–17 cm, narrowly oblong to oblanceolate, coarsely dentate to pinnately lobed, hairs simple and forked; cauline sessile, lanceolate, entire or dentate, base lobed to clasping. **INFL:** terminal sterile fl cluster present. **FL:** sepals erect to ascending, 5–15(19) mm, keeled, ± yellow-green, margins ± purple or brown; petals 8–25(30) mm, wavy-margined, white or ± purple, veins dark purple; filaments in 3 pairs of unequal length, free or longest pair fused. **FR:** spreading to reflexed, 3.5–15 cm, 2.2–3.5 mm wide, cylindric; style 1–4 mm, stigma strongly 2-lobed, lobes 0.5–1.5 mm, opposite valves; pedicel reflexed (spreading), 3–16 mm. **SEED:** 70–96, 1.5–3.5 mm, ovoid; cotyledons deeply 3-lobed. 2*n*=28. Dry, exposed slopes, chaparral, woodland, scrub, grassland; 80–2100 m. s SNF, Teh, sw SnJV, se SnFrB, e SCoRO, SCoRI, nw WTR, sw edge DMoj (Kern Co.). Mar–Jul

C. crassicaulis (Torr.) S. Watson (p. 541) THICK-STEM WILD CABBAGE Per from woody caudex; glabrous or sparsely hairy. **ST:** 2–10 dm, inflated to 3 cm wide, simple (branched). **LF:** basal rosetted 2–20 cm, obovate to oblanceolate, glabrous or hairy, entire to dentate or pinnately lobed, persistent; cauline petioled, linear to narrowly oblanceolate, entire, base not lobed. **INFL:** terminal sterile fl cluster present. **FL:** sepals 7.5–14 mm, creamy white to ± green or purple; petals 10–15 mm, margins not wavy, purple or brown; stamens 4 long, 2 short. **FR:** erect to ascending, 4.5–14 cm, 2–2.5 mm wide, cylindric; style 0.1–0.6 mm, stigma strongly 2-lobed, lobes to 1 mm, opposite valves; pedicel ascending, 1–5 mm. **SEED:** 98–126, 1.5–4 mm, oblong; cotyledons entire. 2*n*=28. Dry sagebrush scrub, pinyon/juniper woodland; 900–2900 m. W&I, n&c DMtns; to ID, WY, CO, AZ. [*C. c.* var. *glaber* M.E. Jones] Recognition of *C. crassicaulis* var. *glaber* based only on unreliable differences in hairs of lvs, sepals. Apr–Jul

C. flavescens (Hook.) Payson (p. 541) Ann, glabrous or sparsely to densely stiff-hairy. **ST:** erect, 7–12 dm, simple or branched distally. **LF:** basal lyre-shaped to pinnately lobed, early-deciduous; proximal cauline, petioled, 2–13.5 cm, lanceolate to oblong or oblanceolate; distal sessile. **INFL:** terminal sterile fl cluster 0. **FL:** sepals spreading to erect, 6–9 mm, cream to pale yellow or purple; petals 7–13 mm, channeled, wavy-margined, white or cream (± pink); stamens 4 long, 2 short. **FR:** ascending to reflexed, 3.8–7.7 cm, 1.4–2 mm; style 1–3.5 mm, stigma ± entire; pedicel ascending to strongly reflexed, 4.5–8 mm. **SEED:** 44–58, 1.3–1.8 mm, oblong. 2*n*=28. Dry, exposed slopes, open hillsides, vertic clay, often serpentine; 80–750 m. s NCoRI, ScV (Montezuma Hills), se SnFrB, SCoRI. [*Guillenia f.* (Hook.) Greene] Mar–May

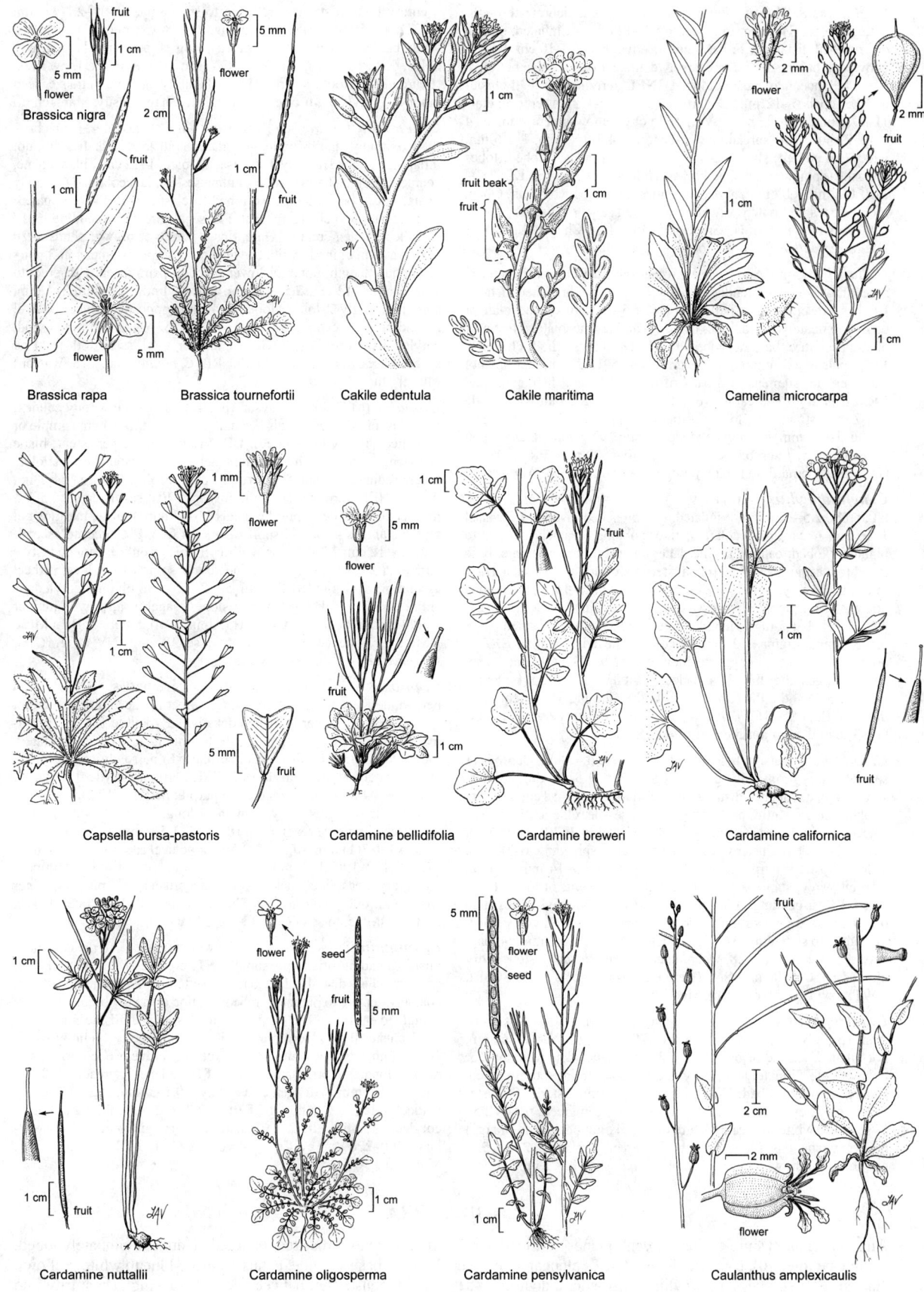

Brassica nigra

Brassica rapa

Brassica tournefortii

Cakile edentula

Cakile maritima

Camelina microcarpa

Capsella bursa-pastoris

Cardamine bellidifolia

Cardamine breweri

Cardamine californica

Cardamine nuttallii

Cardamine oligosperma

Cardamine pensylvanica

Caulanthus amplexicaulis

C. glaucus S. Watson   Per from woody caudex, glabrous throughout, glaucous. **ST:** erect to ascending, 3–12 dm, not inflated, simple or branched distally. **LF:** basal not rosetted, blade 2–10 cm, widely obovate to oblong, entire or coarsely dentate; distal cauline petioled, linear to lanceolate, base not lobed. **INFL:** terminal sterile fl cluster 0. **FL:** sepals 7–12 mm, ± purple to ± yellow-green, erect; petals 11–17.5 mm, margin not wavy, purple or ± yellow-green; stamens 4 long, 2 short. **FR:** spreading to ascending, 4.5–15.5 cm, 1–1.6 mm wide, gen curved; style 0.2–1 mm, stigma strongly 2-lobed, lobes opposite septum margins; pedicel ascending, 5–35 mm. **SEED:** 180–210, 1–2 mm, oblong; cotyledons entire. 2*n*=20. Uncommon. Open, rocky slopes and outcrops, gen in crevices, sagebrush scrub; 1400–2500 m. W&I, n DMtns (Grapevine Mtns, Last Chance Range); w NV. [*Streptanthus g.* (S. Watson) Jeps.] Apr–Jun

C. hallii Payson   Ann, sparsely to densely stiff-hairy. **ST:** erect to ascending, 2–12 dm, slender (± inflated), simple or branched distally. **LF:** basal rosetted, short-petioled, blade 1.5–11.5 cm, oblanceolate or oblong, pinnately lobed, lobes dentate; cauline lanceolate to linear or oblong, ± entire, base not lobed. **INFL:** terminal sterile fl cluster 0. **FL:** sepals 4–6.5 mm, erect, cream; petals 6–10.5 mm, margins not wavy, cream; stamens in 3 pairs of unequal length, filaments free. **FR:** ascending to spreading, 6.5–12.5 cm, 1.8–2.2 mm wide; style 0.5–2 mm, stigma ± 2-lobed; pedicel ascending, 9–25 mm. **SEED:** 78–96, 1–1.6 mm, oblong; cotyledons entire. Uncommon. Dry, open areas, chaparral, scrub, rocky places; 150–1800 m. e PR, s DMtns (Little San Bernardino Mtns), w edge DSon; n Baja CA. Apr–May

C. heterophyllus (Nutt.) Payson (p. 541)   Ann, bristly proximally. **ST:** erect, 2.5–12 dm, not inflated, gen branched distally. **LF:** basal 1–10 cm, linear-oblanceolate to linear-oblong, coarsely dentate to pinnately lobed; cauline sessile, lance-linear, entire or dentate, base lobed to clasping. **INFL:** terminal sterile fl cluster 0. **FL:** sepals 4–9 mm, purple or yellow to creamy white; petals 6–15 mm, margins not wavy, purple or ± yellow, veins darker purple. **FR:** reflexed, 4.5–10 cm, 1–1.5 mm wide, flat parallel to septum or 4-sided; style 0.5–3.5 mm, stigma ± 2-lobed; pedicel reflexed, 2–8 mm. **SEED:** 56–82, 1.2–2 mm, oblong; cotyledons entire. 2*n*=28. Dry, open scrub, chaparral, gen after fire, disturbance; < 1400 m. e SCoRO (La Panza Range), SCo and s PR (common in both), ChI and TR (uncommon in both); Baja CA. [*C. h.* var. *pseudosimulans*, ined.] 2 vars. in TJM (1993) based on minor fl color differences. Mar–May

C. inflatus S. Watson (p. 541)   DESERT CANDLE   Ann, glabrous or sparsely hairy proximally, glaucous. **ST:** erect, 1.5–9.7 dm, inflated, simple or occ branched distally. **LF:** basal rosetted, 2–18 cm, obovate to oblanceolate, entire or dentate; cauline sessile, oblong to ovate or lanceolate, entire, base strongly lobed to clasping. **INFL:** elongated; terminal sterile fl cluster 0. **FL:** sepals erect or spreading, 6–14 mm, at least tip dark purple, ± white basally; petals 9–14 mm, purple, margins wavy; stamens in 3 unequal pairs, filaments of longest pair fused. **FR:** ascending, 3.5–12.7 cm, 2–3 mm wide, cylindric; style 0.1–0.4 mm, stigma strongly 2-lobed, lobes opposite valves; pedicel ascending to spreading, 3–12 mm. **SEED:** 58–82, 1.5–3 mm, oblong, cotyledons entire. 2*n*=28. Open, sandy plains to rocky slopes, dry hillsides; 150–1500 m. s SNF (Kern Co.), w edge c&s SnJV, SCoRI, n WTR, sw DMoj (common). Mar–May

C. lasiophyllus (Hook. & Arn.) Payson   CALIFORNIA MUSTARD   Ann, sparsely to densely stiff-hairy. **ST:** erect, (0.8)2–10(16) dm, not inflated, simple or branched. **LF:** basal lanceolate to oblong or oblanceolate, pinnately lobed to dentate; distal cauline short petioled, smaller, base not lobed. **INFL:** elongated; terminal sterile fl cluster 0. **FL:** calyx cylindric, sepals erect, 2–4 mm, gen green; petals 2.5–5(6.5) mm, white to creamy white or ± pink, not channeled, margin not wavy; stamens 4 long, 2 short. **FR:** erect to reflexed, 2–4.8(5.7) cm, 0.7–1.2 mm wide, straight to outcurved; style 0.5–2 mm, stigma

± entire; pedicel strongly reflexed to spreading, (0.7)1–2.2(3) mm. **SEED:** 14–60, 0.9–1.5 mm, oblong. 2*n*=28. Common. Desert flats, sandy banks, gravelly or rocky areas, talus slopes, shrubland, grassy fields, disturbed sites; < 1400 m. CA (exc MP); to BC, UT, nw Mex. [*Guillenia l.* (Hook. & Arn.) Greene] Highly variable in fl size, seed number, lf division, fr orientation and size, hair density. Mar–Jun

C. lemmonii S. Watson   LEMMON'S JEWELFLOWER   Ann, bristly proximally, hairs simple, gen glabrous distally. **ST:** 1–8 dm, not inflated, branched distally. **LF:** basal petioled, 1–12 cm, oblanceolate, coarsely dentate, hairs simple; cauline sessile, lanceolate to narrowly ovate, entire or dentate, base lobed to clasping. **INFL:** elongated; terminal sterile fl cluster present. **FL:** sepals erect to ascending, 6–17 mm, keeled, gen creamy white, tips ± purple or brown; petals 8–20 mm, wavy-margined, white or veins dark purple; stamens in 3 pairs of unequal length, upper or lower filament pairs fused. **FR:** erect to ascending, 5–12 cm, 2.5–3.5 mm wide, cylindric; style 0.2–4 mm, stigma strongly 2-lobed, lobes 1–4 mm, opposite valves; pedicel ascending to spreading, 3–16 mm. **SEED:** 52–72, 2–3.5 mm, ovoid; cotyledons entire. 2*n*=28. Grassland, chaparral, scrub; 80–1100 m. sw SnJV, se SnFrB, e SCoRO, SCoRI. [*C. coulteri* var. *l.* (S. Watson) Munz] Mar–May   ★

C. major (M.E. Jones) Payson (p. 541)   Per from woody caudex, glabrous, or petioles, sepals occ hairy. **ST:** erect, 2–10 dm, simple or branched distally (± inflated). **LF:** basal rosetted, persistent; blade 1–14 cm, entire or dentate to pinnately lobed; distal short-petioled, much reduced, linear to narrowly oblanceolate, entire, base not lobed. **INFL:** terminal sterile fl cluster 0. **FL:** sepals erect, 6.5–9.5 mm, creamy white to purple; petals 11–17 mm, purple, margins not wavy; stamens 4 long, 2 short, filaments free. **FR:** erect to ascending, 4.5–12 cm, 2.2–2.8 mm wide, cylindric or only ± flattened; style 0.05–0.2(0.4) mm, stigma ± 2-lobed, lobes opposite valves; pedicel ascending, 1–6 mm. **SEED:** 46–58, 2–3.5 mm, oblong; cotyledons entire. 2*n*=28. ± dry, gen rocky slopes, sagebrush, pinyon/juniper woodland; 1500–3200 m. n SNH (Alpine Co.), SnGb, SnBr, s MP, e DMtns (Providence, New York mtns); se OR, n&w NV, e UT. [*C. m.* var. *nevadensis* Rollins] May–Jul

C. pilosus S. Watson (p. 541)   CHOCOLATE DROPS   Bien to weak per; moderately to densely hairy. **ST:** erect to ascending, 2–12 dm, not inflated, simple or branched distally. **LF:** petiole 1–8 cm; proximal blades 2–24 cm, oblanceolate to oblong in outline, pinnately lobed, lobes dentate; distal cauline linear to narrowly oblanceolate, entire or dentate, base not lobed. **INFL:** terminal sterile fl cluster 0. **FL:** sepals 4.5–9.5 mm, ± green to purple; petals 7–12 mm, ± white or ± purple, margins wavy; stamens 4 long, 2 short, filaments free. **FR:** ascending to spreading, 2–18 cm, 1–1.5 mm wide, gen curved; style 0.1–0.3(1) mm, stigma ± 2-lobed; pedicel ascending, 4–18 mm. **SEED:** 152–198, 1–2 mm, oblong; cotyledons entire. Uncommon. Open, dry areas, flats, rocky slopes, sagebrush scrub, pinyon/juniper woodland; 600–2800 m. c&s SNH (e slope), s MP (Honey Lake), SNE, n DMtns; to se OR, s ID, w UT. Mar–Jul

C. simulans Payson   PAYSON'S JEWELFLOWER   Ann, ± conspicuously spreading-bristly proximally. **ST:** erect, 1–7 dm, not inflated, gen branched distally. **LF:** basal rosetted, oblanceolate, 1–8 cm, coarsely dentate to pinnately lobed; cauline ovate to oblong, coarsely dentate to entire, sessile, base lobed to clasping. **INFL:** terminal sterile fl cluster 0. **FL:** sepals erect, 3–6.5 mm, keeled, yellow; petals 10–14 mm, creamy white to pale yellow, midvein occ purple; stamens 4 long, 2 short, filaments free. **FR:** reflexed (spreading), 3–7.5 cm, 1.2–1.5 mm wide, gen curved; style 0.1–3 mm, stigma 2-lobed; pedicel reflexed, 2–5 mm. **SEED:** 48–62, 1–2 mm, oblong to ovoid; cotyledons entire. 2*n*=28. Chaparral, scrub, pinyon/juniper woodland; 400–2200 m. e SCo (w Riverside Co.), e PR, w edge DSon. Mar–Jun   ★

CHORISPORA

Ann to per; hairs simple when present; glands multicellular, stalked. **LF:** basal rosetted, petioled, dentate to pinnately lobed; cauline petioled [0], bases not lobed. **INFL:** elongated. **FL:** sepals erect, base sac-like; petals ± purple-blue to white [yellow], clawed. **FR:** silique, linear, cylindric, sessile, indehiscent, segmented, constricted between seeds, breaking at maturity into 1-seeded corky segments, tapered at tip; style 2–15 mm, stigma strongly 2-lobed. **SEED:** 5–30, in 1 row, embedded in corky septum cover; wing 0. 11 spp.: Eurasia. (Latin: from breaking between seeds)

C. tenella (Pall.) DC. (p. 541) Ann, glandular, occ hairy. **ST:** erect to ascending, (0.5)1–4(5.6) dm, simple or branched. **LF:** basal, lower cauline (1.5)2.5–8(13) cm, oblong to oblanceolate, coarsely dentate; distal cauline similar, reduced. **FL:** sepals erect, (3)4–5(6) mm, free, forming tube, ± purple; petals 8–10(12) mm, oblanceolate, long-clawed, purple, lavender (white). **FR:** spreading to ascending, (1.4)1.8–2.5(3) cm, 1.5–2 mm wide; style (0.6)1–1.8(2.2) cm; pedicel spreading, (2)3–5 mm, stout. **SEED:** 10–24, 1–1.4 mm, oblong. 2*n*=14. Disturbed areas, pastures; < 2300 m. CaR, GV, SCo, GB; widespread in Can, US; native to Eurasia. Apr–Jul ◆

COCHLEARIA SCURVYGRASS

Ann to per, low, ± fleshy, glabrous. **LF:** basal simple, thick, petioled; cauline petioled or sessile, basally lobed, entire or dentate. **INFL:** elongate. **FL:** sepals ascending to spreading, base not sac-like; petals oblanceolate to oblong or elliptic, white. **FR:** silicle, round to oblong, ± inflated or flat perpendicular to septum, sessile, dehiscent, unsegmented; stigma entire. **SEED:** 5–32, 2 rows per chamber. 16 spp.: N.Am, Eur, Asia, nw Afr. (Latin: spoon, from basal lvs of some spp.)

C. groenlandica L. (p. 541) GREENLAND COCHLEARIA Bien, per. **ST:** spreading to erect, (0.1)0.5–3(4) dm. **LF:** petiole 1–7(10) cm; basal blades 0.7–2(2.5) cm, ovate to deltoid, margin entire to wavy or obscurely dentate, base truncate to tapered; upper cauline sessile, entire to obscurely dentate, not basally lobed. **FL:** sepals 1–2(3) mm; petals 2–4(5) mm, oblanceolate to spoon-shaped. **FR:** 3–5.5(7) mm, ovoid to obovoid; style 0.1–0.4 mm; pedicel ascending, (2)5–15(20) mm. 2*n*=14. Seabird nesting areas on offshore rocks, islands; < 100 m. n NCo (Del Norte Co.); to AK, e Can. [*C. officinalis* L. var. *arctica* (DC.) Gelert] Gen treated as var. of *C. officinalis*, but latter is a Eur tetraploid (2*n*=24), of a different lineage. Jun–Aug ★

COINCYA

Ann to per; hairs simple. **LF:** basal rosetted, petioled, dentate to pinnately lobed; cauline petioled, base not lobed. **INFL:** elongated. **FL:** sepals erect, base sac-like; petals long-clawed, yellow, veins dark brown to purple. **FR:** silique, sessile, linear, cylindric, dehiscent, segmented; valves 3(5)-veined; terminal segment indehiscent, (0)1–6-seeded; stigma 2-lobed. **SEED:** 4–100, in 1 row, spheric to oblong; wing 0. 6 spp.: Eur, nw Afr. (A.H.C. de Coincy, Spanish botanist, 1837–1903)

C. monensis (L.) Greuter & Burdet subsp. *recurvata* (All.) Leadlay (p. 541) STAR MUSTARD Ann, hairy. **ST:** erect to ascending, (0.8)1–10 dm, simple or branched. **LF:** basal long-petioled, blade (3)5–20 cm, gen pinnately lobed, lobe pairs 3–9(10); cauline petioled, similar to basal, lobes fewer. **FL:** sepals narrowly oblong, tips hairy; petals 1.3–2.2(2.6) cm, 2.5–7(9) mm wide. **FR:** spreading to ascending, (1)3–9 cm; lower segment (0.8)2.5–7.5 cm, 1.5–3 mm wide; terminal segment (0.5)0.7–2.3(3.4) mm. **SEED:** (15)20–75(90) in lower segment, (0)1–5 in terminal segment, 0.8–1.6 mm, ± spheric. 2*n*=24,48. Sandy, disturbed areas; 6 m. NCo (Manila, Humboldt Co.); e US, native to Eur, n Afr. May–Aug ◆

CONRINGIA HARE'S EAR

Ann [bien], glabrous, glaucous. **ST:** erect, simple or branched. **LF:** basal not rosetted, gen entire, ± fleshy; cauline cordate-clasping, entire. **INFL:** elongated. **FL:** sepals erect to ± ascending, oblong, lateral pair ± sac-like; petals narrowly obovate, clawed, yellow [white]. **FR:** silique, linear, dehiscent, unsegmented, cylindric or 4-angled; stigma entire. **SEED:** in 1 row, oblong, plump, wing 0. 6 spp.: Eurasia. (H. Conring, professor at Helmstedt, Germany, 1606–1681)

C. orientalis (L.) Dumort. (p. 541) Ann. **ST:** (1)3–7 dm, simple or branched distally. **LF:** basal 5–9 cm, oblanceolate to obovate, narrowed to base, ± entire; cauline (1)3–10(15) cm, sessile, lance-oblong. **FL:** sepals 6–8 mm; petals 7–12 mm, 2–3 mm wide, lemon to pale yellow, claws slender. **FR:** (5)8–14 cm, 2–2.5 mm wide, 4-angled to ± cylindric; style 0.5–4 mm; pedicel ascending, (0.8)1–1.5(2) cm. **SEED:** 2–3 mm, oblong. 2*n*=14. Disturbed areas, fields; < 3500 m. SCo, SNE, DSon; widespread in N.Am, native to Eurasia. Apr–Aug

CUSICKIELLA

Per, low-cespitose, cushion-like; caudex woody, many-branched; hairs simple or mixed with stalked forked or ± many-branched hairs. **ST:** scapose, erect, lvs 0–few. **LF:** basal rosetted at tip of caudex branches, sessile, entire, thickened at base; cauline few or 0, bases not lobed. **INFL:** ± elongated; bracts 0–few. **FL:** sepals erect or ascending, base not sac-like; petals white to pale yellow. **FR:** silicle, sessile, ovoid to ellipsoid, dehiscent, unsegmented; valves rounded or keeled, glabrous or hairy; stigma entire. **SEED:** 1–4, plump, ± ovoid to oblong; wing 0. 2 spp.: w US. (W.C. Cusick, OR pl collector, 1842–1922) [Rollins 1988 J Jap Bot 63:65–69]

1. Lf (3)5–12(14) mm; petals white; bracts 0; fr valves rounded . . . . *C. douglasii*
1′ Lf 1.5–4.5 mm; petals pale yellow; bracts subtending lower pedicels; fr valves keeled . . . . *C. quadricostata*

C. douglasii (A. Gray) Rollins (p. 541) Crisped-hairy. **ST:** 1–5 cm. **LF:** oblanceolate to oblong, thick, leathery, margin ciliate; cauline 0. **FL:** sepals 2.5–4 mm; petals 3.5–6 mm. **FR:** (2)3–5.5(7) mm, ovoid to ellipsoid; style (0.5)1–2(2.5) mm; pedicel spreading to ascending, 2–6(8) mm. **SEED:** 2–3 mm, ovoid to oblong. Rocky ridges, scree, loose volcanic slopes; 1500–2500 m. KR, NCoR, SN, SnBr, GB; to WA, ID, UT. May–Jun

C. quadricostata (Rollins) Rollins (p. 541) BODIE HILLS CUSICKIELLA Branched hairs curled, simple hairs stiff. **ST:** 2–6 cm, simple. **LF:** linear to oblong, not leathery, margin ciliate; cauline 1–5 as bracts. **FL:** sepals 2.5–4 mm; petals 3–4.5 mm. **FR:** 3–5 mm, ovoid, 4-sided; style 0.5–1 mm; pedicel spreading to ascending, 2–5 mm. **SEED:** 1.8–2.5 mm, ovoid to oblong. Rocky flats, sagebrush, slopes, pinyon/juniper woodland; 2300–2800 m. SNE (Mono Co.); w NV. May–Jun ★

DESCURAINIA   TANSY MUSTARD

Ann, bien (per); hairs minute, many-branched, tree-like, occ mixed with fewer simple hairs, club-shaped glandular papillae occ present. **ST**: gen branched distally. **LF**: petioled, finely 1–3-pinnately lobed or divided, basal gen early-deciduous; cauline similar to basal, less divided distally on st, base not lobed. **INFL**: elongating. **FL**: sepals erect to spreading, base not sac-like; petals obovate, yellow [± white]. **FR**: silique or silicle, dehiscent, linear, oblong, club-shaped, ellipsoid, or obovoid, not flattened, unsegmented; stigma entire. **SEED**: 5–100, in 1 or 2 rows, ellipsoid to oblong, plump; wing 0. 45–47 spp.: Eurasia, esp N.Am and S.Am, Canary Islands. (F. Descourain, French botanist, 1658–1740) May be TOXIC to livestock. [Detling 1939 Amer Midl Naturalist 22:481–520] Taxonomically difficult, most characters highly variable.

1. Fr sparsely hairy at least in youth . . . .*D. adenophora*
1′ Fr glabrous
  2. Fr fusiform, obovate, club-shaped, or ellipsoid (broadly linear, widest distally)
    3. Fr fusiform, obovate, or broadly ellipsoid; seeds 4–12, in 1(2) row per chamber; valve midvein obscure
      4. Fr fusiform, both ends long-acute; style (0.2)0.3–0.6(0.8) mm; st (1.3)2–10.5(13.5) dm, simple at base, branched distally . . . . *D. californica*
      4′ Fr obovate, tip obtuse, acute at base; style 0.05–0.3 mm; st (0.8)1.5–3.2(4.1) dm, branched throughout . . . . *D. paradisa*
    3′ Fr club-shaped (broadly linear, widest distally); seeds 16–40, in 2 rows per chamber; valve midvein prominent . . . . *D. pinnata*
      5. Infl axis glabrous, nonglandular . . . . subsp. *glabra*
      5′ Infl axis sparsely to densely hairy, glandular or not
        6. Pl sparsely to densely hairy (canescent), gen glandular; sepals yellow or lavender . . . . subsp. *brachycarpa*
        6′ Pl densely canescent, gen not glandular; sepals purple or rose . . . . subsp. *ochroleuca*
  2′ Fr linear
    7. Fr appressed; pedicel erect to erect-ascending . . . . *D. incana*
    7′ Fr not appressed; pedicel widely spreading to ascending
      8. Fr septum with 2- or 3-veined central band; lvs 2- or 3-pinnately lobed . . . . *D. sophia*
      8′ Fr septum not veined; lvs 1-pinnately lobed
        9. Fr 3–8(10) mm; petals 0.7–1.2 mm; seed 0.5–0.8 mm; pl not glandular . . . . *D. nelsonii*
        9′ Fr (8)12–35 mm; petals 1.7–3 mm; seed 0.7–1.5 mm; pl glandular or not
          10. Ultimate segment of cauline lvs oblong to lanceolate, dentate to incised (pinnately lobed); fr straight or strongly incurved. . . . . *D. incisa* subsp. *incisa*
          10′ Ultimate segment of cauline lvs linear, entire; fr straight or ± incurved . . . . *D. longipedicellata*

D. adenophora (Wooton & Standl.) O.E. Schulz   Bien, glandular at least distally, canescent. **ST**: branched distally, 4.5–13 dm. **LF**: petiole 1–3 cm; basal and most proximal cauline 1-pinnately lobed, 2–10 cm; lateral lobes 2–5 pairs, 4–12 mm, oblanceolate to lanceolate, entire or serrate to crenate; distal cauline similar, sessile or short-petioled. **FL**: sepals ascending, 2–2.9 mm, glandular; petals 1.8–2.6 mm, oblanceolate. **FR**: 8–16(20) mm, linear, straight, abruptly tapered at both end; valves sparsely hairy at least in youth; style 0.1–0.2 mm; pedicel ascending to spreading, 1.3–3.1 cm. **SEED**: 48–64, in 1 or 2 rows, 0.7–1 mm, ellipsoid. 2*n*=42. Gravelly flats, open woodland, lake margins; 900–2200 m. e PR, DMoj; to NM, sw TX, Mex. [*D. obtusa* (Greene) O.E. Schulz subsp. *a.* (Wooton & Standl.) Detling] May–Aug

D. californica (A. Gray) O.E. Schulz (p. 547)   Ann, bien, hairy, or glabrous distally; nonglandular. **ST**: (1.3)2–10.5(13.5) dm, branched distally, slender. **LF**: 1.5–6 cm, oblanceolate to obovate in outline, ± green, 1-pinnately lobed; lobes in 2–4(5) pairs, 0.5–2.2 cm, lanceolate, margins entire to crenate or dissected; distal cauline sessile or short-petioled, gen sparsely hairy. **FL**: sepals spreading, 0.9–1.5 mm, ± yellow; petals 1.1–1.8 mm, oblanceolate. **FR**: erect to spreading, (2)3–5(6) mm, fusiform, both ends long-acute; valves glabrous, midvein obscure; style (0.2)0.3–0.6(0.8) mm; pedicel spreading to ascending or erect, 3–9(11) mm, slender, straight. **SEED**: 4–12, in 1 row, 1–1.4 mm, ellipsoid. 2*n*=14. Open sites, sagebrush scrub, aspen groves, open woodland; 1700–3400 m. SNH, GB, DMtns; to WY, NM. May–Aug

D. incana (Fisch. & C.A. Mey.) Dorn (p. 547)   Bien, hairy, canescent to ± green (glandular). **ST**: erect, simple proximally, many-branched distally, (1.5)2.5–12 dm. **LF**: basal, proximal cauline 1.5–10(13) cm, widely lanceolate to oblanceolate or ovate, 1-pinnately lobed; ultimate segments linear to oblong or lanceolate, 3–10(15) mm, entire; distal cauline smaller, lobes narrower. **FL**: sepals 1–1.8 mm, ± yellow; petals 1.2–2 mm, oblanceolate. **FR**: erect, appressed, (4)5–10(15) mm, 0.7–1.2(1.5) mm wide, linear, straight; valves gla-

brous; septum midveined; style 0.1–0.4 mm; pedicel erect to erect-ascending, 2–8(11) mm, straight. **SEED**: 14–22, in 1 row, 0.8–1.2 mm, ellipsoid to narrowly oblong. 2*n*=14,28. Open sites, meadows, sagebrush scrub, open aspen groves, roadsides; 100–3500 m. KR, SN, SnBr, GB; to AK, WY, NM, QC, ne US. May–Sep

D. incisa (Engelm.) Britton subsp. ***incisa***   Ann; glabrous or hairy distally, glandular or not, ± green. **ST**: branched distally (proximally), (1.3)2–8.2(10.7) dm. **LF**: basal, proximal cauline 1-pinnate, 1.5–10.3 cm, obovate to oblanceolate in outline; lateral lobes (3)5–9 pairs; distal cauline lf lobes oblong to lanceolate. **INFL**: elongate; axis hairs minute. **FL**: sepals 1.6–2.4 mm, ± yellow; petals 1.7–2.8 mm. **FR**: erect to ascending, 0.8–2 cm, 0.9–1.3 mm wide, linear, gen straight; valves glabrous; septum not veined; style 0.1–0.3 mm; pedicel ascending to spreading, (3)5–10(12) mm, straight. **SEED**: 14–26, in 1 row, 0.9–1.3 mm, oblong. 2*n*=14. Dry creeks, streambanks, disturbed areas, meadows, sagebrush scrub, open woodland, talus, alpine areas; 900–2900 m. TR, PR, GB; to BC, YT, NM, n Mex. Jun–Aug

D. longipedicellata (E. Fourn.) O.E. Schulz   Ann, nonglandular (glandular), sparsely to moderately hairy, ± green, gen glabrous distally. **ST**: erect, simple or branched throughout, (1.5)3–6.2(8.5) dm. **LF**: basal, proximal cauline pinnate, 1.5–7 cm, ovate to oblanceolate in outline; ultimate divisions linear or oblong, entire to dentate or cut; distal cauline sessile or short-petioled, lobes entire, linear to thread-like. **FL**: sepals 1.5–2 mm, yellow; petals 1.7–2.6 mm. **FR**: ascending to gen erect, (0.9)1.2–1.7 mm, 0.8–1.1 mm wide, linear, straight or curved inward; valves glabrous; septum not veined; style 0.1–0.2 mm; pedicels spreading to ascending, (0.8)1–1.5(2) mm. **SEED**: 18–32, in 1 row, 1–1.3 mm, oblong. 2*n*=14. Shale soils below cliffs, dry washes, grassland, sagebrush scrub, pinyon/juniper woodland; 750–2500 m. CaR, GB; to BC, CO, AZ. [*D. incisa* subsp. *filipes* (A. Gray) Rollins] Apr–Jul

D. nelsonii (Rydb.) Al-Shehbaz & Goodson   Ann, ± green, sparsely to densely hairy (glabrous distally), nonglandular. **ST**: branched

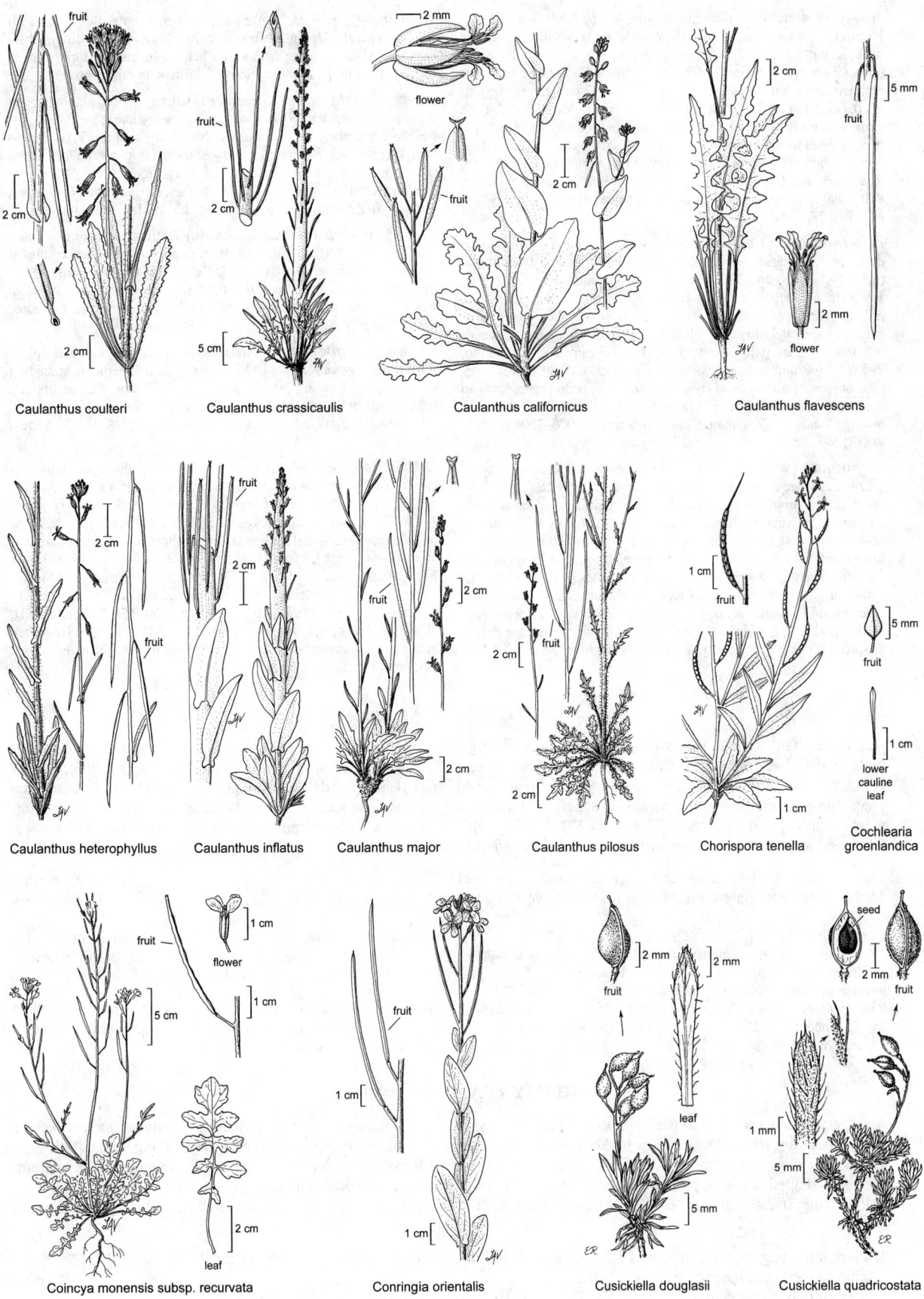

Caulanthus coulteri

Caulanthus crassicaulis

Caulanthus californicus

Caulanthus flavescens

Caulanthus heterophyllus

Caulanthus inflatus

Caulanthus major

Caulanthus pilosus

Chorispora tenella

Cochlearia groenlandica

Coincya monensis subsp. recurvata

Conringia orientalis

Cusickiella douglasii

Cusickiella quadricostata

distally, proximally, or throughout (simple), (0.7)0.9–3.2(4.5) dm. **LF**: basal, proximal cauline 1-pinnately lobed, 1.5–4 cm, oblong to ovate in outline; lateral lobes 2–5 pairs, dentate to pinnately lobed; ultimate segments entire or dentate; distal cauline sessile or short-petioled, lobes gen narrower. **FL**: sepals 0.7–1.2 mm, ± yellow; petals 0.7–1.2 mm. **FR**: 3–8(10) mm, 0.7–1 mm wide, linear; valves glabrous, midvein prominent; septum not veined; style 0.1–0.2 mm; pedicel ascending-spreading, at 20–45° angle, (1.5)2.5–7(10) mm, straight. **SEED**: 6–12, in 1 row, 0.5–0.8 mm, oblong. 2*n*=14. Roadsides, sagebrush scrub, dry washes, silty flats, gravelly ground; 800–3000 m. GB, DMoj (Calico Mtns), expected elsewhere; to WA, MT, WY. May–Jul

D. paradisa (A. Nelson & P.B. Kenn.) O.E. Schulz Ann, glandular or not, sparsely to densely hairy. **ST**: branched throughout, (0.8)1.5–3.2(4.1) dm. **LF**: basal, proximal cauline pinnately lobed, 1.5–3 cm, oblanceolate to obovate in outline; ultimate lobes 1–5 mm, oblong to linear or lanceolate, entire or dentate; distal cauline sessile, smaller, moderately to densely hairy. **FL**: sepals spreading to ascending, 0.8–1.2 mm, pale yellow; petals 0.9–1.3 mm. **FR**: 2.5–4 mm, 2–5 mm, 1–2 mm wide, obovate, obtuse at tip, acute at base; valves glabrous, midvein obscure; style 0.05–0.3 mm; pedicel spreading to ascending, 2.5–7(9) mm. **SEED**: 4–5, in 1 or 2 rows, 0.8–1.2 mm, oblong. Sandy washes, dunes, sagebrush scrub; 1000–2300 m. GB; to OR, NV. Apr–Jun

D. pinnata (Walter) Britton Ann, sparsely to densely hairy, ± green or canescent, glandular or not (glabrous distally). **ST**: (0.8)1.3–5.7(9.2) dm, simple or branched distally, proximally, or throughout. **LF**: basal, proximal cauline 1–15 cm, 1- or 2-pinnately lobed, oblanceolate or oblong to ovate in outline; lateral lobes 4–9 pairs, dentate to pinnately lobed; ultimate lobes oblanceolate to ovate, entire or dentate; distal cauline sessile or short-petioled, lobes gen narrower. **FL**: petals, bright yellow to cream. **FR**: 4–13(17) mm, 1.2–2.2 mm wide, club-shaped (broadly linear, widest distally); valves with prominent midvein, glabrous; style to 0.2 mm, gen ± absent; pedicel ascending to spreading or ± descending, 4–23 mm, straight or curved. **SEED**:

16–40, 0.6–0.9 mm, in 2 rows per chamber, oblong. Subsp. *halictorum* (Cockerell) Detling, subsp. *intermedia* (Rydb.) Detling, subsp. *menziesii* (DC.) Detling represent a heterogeneous mixture of intermediates within *D. pinnata* and its hybrids with other spp.

 subsp. ***brachycarpa*** (Richardson) Detling Pl gen glandular, gen ± green (canescent). **INFL**: sparsely to densely hairy, gen glandular; pedicels spreading to ascending at 20–60(80)° angle, (7)10–18(23) mm. **FL**: sepals 1.5–2.6 mm, yellow or lavender; petals (1.7)2–3 mm, 0.6–1 mm wide. 2*n*=14,28. Disturbed areas, sagebrush scrub, pinyon/juniper woodland, sandy fields, dry washes, streambanks, dry slopes, cliffs; 100–2200 m. CA; to BC, e Can, TX, n Mex. Mar–Jul

 subsp. ***glabra*** (Wooton & Standl.) Detling Pl nonglandular, not canescent. **INFL**: glabrous; pedicels spreading to descending at 70–90(100)° angle, 4–10(15) mm. **FL**: sepals 0.8–1.5 mm, at least tip ± pink; petals 1–1.8 mm, 0.3–0.7 mm wide. 2*n*=28. Sandy areas, washes, scrub, oak/pine woodland, floodplains, limestone outcrops; 200–2400 m. CA; to NV, NM, n Mex. Feb–Jun

 subsp. ***ochroleuca*** (Wooton) Detling Pl nonglandular (glandular), densely canescent. **INFL**: densely hairy; pedicels spreading to ascending, forming 30–60° angle, (4)6–12 mm. **FL**: sepals 1–2 mm, purple or ± pink; petals 1.5–2 mm, 0.3–0.5 mm wide. Gravelly hills, desert grassland, roadsides; 200–2400 m. se CA; to TX, n Mex. Mar–May

D. sophia (L.) Prantl (p. 547) Ann, hairy throughout or glabrous distally, nonglandular. **ST**: erect, simple or branched distally, (1)2–7(10) dm. **LF**: basal, proximal cauline 2- or 3-pinnately lobed, 1–10(15) cm, oblong to widely ovate; ultimate lobes 2–10 mm, linear or oblong, entire; distal cauline sessile or short petioled, ultimate segments narrower. **FL**: sepals 1.8–2.8 mm, ± yellow; petals 2–3 mm. **FR**: not appressed, (1.2)1.5–2.7(3) cm, 0.5–0.8(1) mm wide, linear; valves glabrous; septum with 2- or 3-veined central band; style 0.1–0.2 mm; pedicel spreading to ascending, (5)8–15(20) mm. **SEED**: 20–48, in 1 row, 0.7–1.3 mm, oblong. 2*n*=28. Common. Disturbed areas, fields, canyon bottoms, desert; < 3000 m. CA; throughout Can, US; native to Eurasia. May–Aug ❖

DIPLOTAXIS WALL ROCKET

Ann to per, hairs simple; base woody or not. **ST**: ascending to erect, lfy or not. **LF**: basal rosetted or not, pinnately lobed or toothed, petioled; cauline 0 or entire to lobed, petioled or sessile. **INFL**: elongated. **FL**: sepals ascending to spreading, base of lateral pair not sac-like; petals ± obovate, yellow [violet or white]. **FR**: silique, dehiscent, segmented, gen narrowed between seeds, linear, cylindric to ± flat parallel to septum, stalked above receptacle or sessile; valves 1-veined, glabrous; terminal segment seedless [seeded]; stigma 2-lobed. **SEED**: [12]20–46[276], in 2 rows. 25–30 spp.: Eur, n Afr, w Asia. (Greek: 2-rowed, from seed arrangement) *D. erucoides* (L.) DC. occ weed in botanic gardens.

1. Gen ann; st ± lfless, hairs reflexed; fr stalk above receptacle 0–0.5 mm . . . . ***D. muralis***
1′ Per; st lfy, gen glabrous adaxially; fr stalk above receptacle 0.5–3 mm . . . . ***D. tenuifolia***

D. muralis (L.) DC. (p. 547) Gen scapose, hairy. **ST**: (0.5)2–5(6) dm. **LF**: basal rosetted, pinnately lobed, lateral lobe pairs 2–4(6); cauline short-petioled to sessile if present. **FL**: sepals 3–5.5 mm, glabrous or hairs few; petals 5–8(10) mm. **FR**: (1.5)2–4 cm, 1.5–2.5 mm wide; pedicel (3)8–20(37) mm. **SEED**: 20–36, 0.9–1.3 mm, ovoid. 2*n*=42. Uncommon. Disturbed areas, dry streambeds; < 300 m. SCo; Can, US, native to Eur. May–Aug

D. tenuifolia (L.) DC. (p. 547) Gen glabrous above. **ST**: 2–7(10) dm, gen woody at base. **LF**: basal not rosetted, deeply pinnately lobed, lateral lobes 2–5; cauline short-petioled. **FL**: sepals 4–6 mm, glabrous or hairs few; petals 7–11(13) mm, 5–8 mm wide. **FR**: 2–5 cm, 1.5–2.5 mm wide; pedicel 8–35 mm. **SEED**: 20–32(46), 1–1.3 mm, ovoid. 2*n*=22. Uncommon. Disturbed areas, roadsides; < 300 m. ScV, SCo; Can, US, native to Eur. May–Aug

DITHYREA SPECTACLEPOD

Ann, per; hairs dense, stalked, stellate, occ simple. **LF**: basal rosetted, entire to dentate or pinnately lobed; cauline petioled or sessile, base not lobed. **INFL**: elongated or not. **FL**: sepals erect, forming a tube, linear to narrowly oblong, lateral pair base sac-like; petals tongue-shaped, long-clawed, white to lavender. **FR**: silicle, indehiscent, flattened perpendicular to septum, spectacle-shaped, splitting into 1-seeded halves, not segmented; valves keeled, enclosing seeds; stigma entire. **SEED**: 2, oblong, wing 0. 2 spp.: CA, Mex. (Greek: 2 shields, reference to spectacle-shaped fr) [Rollins 1979 Publ Bussey Inst Harvard 3–32]

1. Ann; infl elongated in fr; fr valves 3.5–5(6) mm, 4–6(7) mm wide, hairs 0 or occ on fr margin, branched, club-shaped papillae present . . . . ***D. californica***
1′ Per; infl not elongated in fr; fr valves 8–11 mm, 8–10 mm wide, hairs sessile, stellate, occ mixed with forked (simple) hairs, club-shaped papillae 0 . . . . ***D. maritima***

D. californica Harv. (p. 547) Pl densely hairy. **ST:** several from base, branched distally, (0.7)1–6(7) dm. **LF:** basal 2–11 cm, oblanceolate to widely ovate, dentate to shallowly lobed; cauline similar, smaller, entire to dentate. **FL:** sepals (6)7–9(10) mm; petals (1)1.2–1.5 cm, (1.5)2–3 mm wide, white to pale lavender. **FR:** style 0.1–0.5(1) mm; pedicel spreading, 1.5–2.5 mm. **SEED:** 2, 3–4 mm, oblong. $2n=20$. Abundant in sandy places, washes, scrub; 50–1400 m. D; NV, w AZ, nw Mex. Mar–May

D. maritima (Davidson) Davidson (p. 547) BEACH SPECTACLEPOD Pl rhizomed; densely hairy. **ST:** decumbent, several to many from base, simple or few-branched distally, 1–2.1 dm. **LF:** basal 2–11(14) cm, widely ovate to oblanceolate, wavy to dentate; distal cauline short-petioled to sessile, entire or wavy. **FL:** sepals (6)7–10 mm; petals 1.2–1.5 cm, 2.5–3.5 mm wide, white to ± purple. **FR:** pedicel spreading, 1.5–2.5 mm. **SEED:** 2, 3–4 mm, oblong. $2n=60$. Seashores, coastal sand dunes; < 50 m. s CCo, SCo, ChI; Baja CA. Mar–Aug ★

DRABA

Ann to per, gen cushion- or mat-forming, occ scapose, hairs simple, forked, or many-branched. **LF:** basal gen rosetted; cauline entire or shallowly toothed, base gen not lobed, occ 0. **INFL:** gen many-fld, elongated or not; bracts gen 0. **FL:** sepals bases equal; petals gen short-clawed, yellow or white (lavender or red). **FR:** silique or silicle, dehiscent, linear to lanceolate or ovate, occ ovoid or spheric, cylindric or flat parallel to septum, unsegmented; stigma entire. **SEED:** in 2 rows; wing gen 0. 370+ spp.: n hemisphere, S.Am mtns. (Greek: acrid, describing taste of crucifer lvs) [Al-Shehbaz & Windham 2007 Harvard Pap Bot 12:409–419]

1. Ann or bien
  2. Petals deeply 2-lobed; infl axis gen wavy in fr . . . . ***D. verna***
  2′ Petals not divided; infl axis straight in fr
    3. Style (0.8)1–3.5 mm . . . . [2]***D. corrugata***
    3′ Style ≤ 0.4 mm
      4. Infl umbel-like in fr . . . . **[*D. reptans*]**
      4′ Infl elongated to ± congested, not umbel-like in fr
        5. Infl axis glabrous
          6. Cauline lvs (0)1–5; pedicels 3–14 mm, ≤ fr . . . . [2]***D. albertina***
          6′ Cauline lvs 4–12(15); pedicels 7–28(35) mm, > fr . . . . **[*D. nemorosa*]**
        5′ Infl axis hairy
          7. Basal lvs not rosetted, late-season fls without petals; seeds 0.5–0.7 mm, ovoid . . . . ***D. cuneifolia***
          7′ Basal lvs rosetted, late-season fls with petals; seeds 0.8–1.1 mm, oblong . . . . [2]***D. praealta***
1′ Per
  8. Basal lvs abaxially glabrous (lf margins hairy)
    9. Style (1)1.6–3 mm
      10. Fr glabrous; seed 3–4.5 mm wide, round, winged . . . . ***D. carnosula***
      10′ Fr hairy; seed 1–1.6 mm, oblong, wingless . . . . ***D. howellii***
    9′ Style 0.1–0.8 mm
      11. Pl cushion-forming; lf midvein prominent. . . . . [2]***D. densifolia***
      11′ Pl. not cushion-forming; lf midvein obscure
        12. Pedicel base not decurrent; fr 1.5–3 mm wide, narrowly lanceolate . . . . ***D. cruciata***
        12′ Pedicel base decurrent; fr 2.2–4.5 mm wide, ovate to lance-elliptic . . . . [2]***D. incrassata***
  8′ Basal lvs abaxially hairy
    13. Pl cushion-forming; fls yellow; cauline lvs 0
      14. Style 1.5–3.8 mm; seeds winged . . . . ***D. pterosperma***
      14′ Style 0.1–0.8(1) mm; seeds wingless
        15. Lf hairs sessile, comb-like, 7–16-rayed, long axis parallel to midvein . . . . ***D. oligosperma***
        15′ Lf hairs stalked, not comb-like, 2–12-rayed, long axis not parallel to midvein
          16. Fr not symmetric, flat, gen twisted. . . . . ***D. sierrae***
          16′ Fr symmetric, flat or base inflated, not twisted
            17. Lvs abaxially sparsely hairy, hairs 2–4-rayed; fr flat. . . . . [2]***D. densifolia***
            17′ Lvs abaxially densely hairy, hairs 2–12-rayed; fr inflated proximally
              18. Infl gen elongated in fr; basal lvs ciliate, adaxial hairs simple, some 2-rayed . . . . ***D. novolympica***
              18′ Infl umbel-like in fr; basal lvs not ciliate, adaxial hairs 5–12-rayed. . . . . ***D. subumbellata***
    13′ Pl not cushion-forming; fls white (yellow); cauline lvs (0)1–many
      19. Infl axis glabrous
        20. Hairs on basal lvs 8–12-rayed on abaxial surface . . . . [2]***D. lonchocarpa***
        20′ Hairs on basal lvs 2–5-rayed on abaxial surface, occ also simple
          21. Fr base ± inflated; petals white; infl umbel-like in fr. . . . . [2]***D. monoensis***
          21′ Fr base flat; petals yellow; infl elongated in fr
            22. Seeds 8–12; pedicels decurrent at base . . . . [2]***D. incrassata***
            22′ Seeds (12)14–44; pedicels not decurrent at base
              23. Per, short-lived, not tufted; seeds 20–44; style ≤ 0.1 mm; petals 2–3 mm . . . . [2]***D. albertina***
              23′ Per, not short-lived, tufted; seeds 12–20; style (0.2)0.5–2 mm; petals 4–7 mm
                24. Seeds winged; basal lvs not ciliate; fr gen not twisted, straight, occ ± twisted . . . . ***D. asterophora***
                24′ Seeds wingless; basal lvs ciliate; fr twisted, curved . . . . ***D. sharsmithii***

19′ Infl axis hairy
  25. Basal lvs with simple hairs; petals persistent .... ***D. longisquamosa***
  25′ Basal lvs with some branched hairs; petals deciduous
    26. Style (0.8)1–3.5 mm; petals yellow
      27. Cauline lvs 0(1) .... ***D. saxosa***
      27′ Cauline lvs 4–33
        28. Fr oblong, not twisted; seeds 10–20; most proximal fls bracted .... ***D. aureola***
        28′ Fr elliptic, gen twisted; seeds 16–28; fls not bracted .... [2]***D. corrugata***
    26′ Style 0.1–0.6 mm; petals white (yellow)
      29. Lf hairs simple or 2-rayed
        30. Fls yellow; fr flat, ± twisted, 2.5–5 mm wide .... ***D. lemmonii***
        30′ Fls white; fr ± inflated proximally, not twisted, (1.2)1.5–2.5 mm wide .... [2]***D. monoensis***
      29′ Lf hairs gen 4–12-rayed
        31. Fr hairs 2–7-rayed only
          32. Infl not bracted; pl tufted; st (1)2–10(15) cm .... ***D. breweri***
          32′ Infl bracted proximally; pl not tufted; st (6)10–30(38) cm .... ***D. cana***
        31′ At least some fr hairs simple
          33. Infl 8–30(37)-fld; lvs dentate; seeds (24)30–52 .... [2]***D. praealta***
          33′ Infl 3–13-fld; lvs entire; seeds 16–32
            34. Lf hairs on both surfaces 4–8-rayed; infl axis not wavy in fr; sepals persistent .... ***D. californica***
            34′ Lf hairs on abaxial surface 8–12-rayed; infl axis ± wavy in fr; sepals deciduous .... [2]***D. lonchocarpa***

D. albertina Greene Ann to short-lived per. **ST**: (0.3)0.5–3(4.2) dm, branched distally, hairs simple proximally, fewer stalked 2-rayed, glabrous distally. **LF**: basal 3–28(35) mm, oblanceolate to obovate, entire to minutely dentate, ciliate, hairs 2–4-rayed abaxially, adaxially simple and forked or 0; cauline lvs (0)1–5. **INFL**: 5–30(50)-fld, elongated in fr; axis glabrous; pedicels 3–14 mm. **FL**: sepals 1.4–2.1 mm; petals 2–3 mm, 0.7–1.2 mm wide, yellow. **FR**: 4–12(15) mm, 1–2.1 mm wide, linear to narrowly elliptic or lanceolate, flat, not twisted, glabrous; style ≤ 0.1 mm. **SEED**: 20–44, 0.7–1 mm, oblong. $2n=24$. Moist meadows, streambanks, woodland, rocky knolls, disturbed areas; 900–3700 m. CaRH, SNH, SnGb, MP, n SNE, W&I; to AK, w Can, CO, NM. Jun–Aug

D. asterophora Payson (p. 547) Per, gen scapose, loosely tufted. **ST**: 3–11 cm, simple, glabrous or hairs sparse proximally, (2)4-rayed. **LF**: basal 4–14(17) mm, broadly obovate to spoon-shaped, entire, not ciliate, hairs stalked, (2)4(5)-rayed; cauline lvs 0(1). **INFL**: 5–20(27)-fld; axis glabrous; pedicels 3–9 mm. **FL**: sepals 3–4 mm; petals 5–7 mm, 1.5–2.5 mm wide, yellow. **FR**: 5–11(14) mm, 3.5–6 mm wide, lanceolate to broadly ovate or oblong, flat, occ ± twisted, glabrous or puberulent, hairs simple and fewer 2–3-rayed; style (0.2)0.5–1.6(2) mm. **SEED**: (12)14–18, 1.8–2.8 mm, ovate; wing 0.5–1 mm wide. $2n=40$. Rock crevices, alpine barrens, talus; 2600–3300 m. n&c SNH; w NV. [*D. a.* var. *macrocarpa* C.L. Hitchc.] Jun–Aug ★

D. aureola S. Watson (p. 547) GOLDEN ALPINE DRABA Short-lived per. **ST**: branched or not, 2.5–15 cm, hairs stiff, simple and long-stalked, 2–4-rayed. **LF**: basal (0.7)1–1.5(3) cm, oblanceolate to linear, ciliate, abaxial hairs stalked, 3–5-rayed, adaxial hairs simple and stalked, 2(5)-rayed; cauline lvs 5–33. **INFL**: 12–83-fld, dense, occ lower 1–9 fls bracted; axis as hairy as st; pedicels 3–12(19) mm. **FL**: sepals 2.5–4 mm; petals 4–6 mm, 0.5–1.2 mm wide, yellow. **FR**: 6–16 mm, 3–5 mm wide, oblong, not twisted, flat, hairs 2–4-rayed; style (0.8)1–2 mm. **SEED**: 10–20, 1.4–1.9 mm, oblong. $2n=20$. Scree, talus, gen volcanic substrates, alpine meadows, open conifer forest; 2250–3200 m. CaRH (Lassen Volcanic National Park), e KR (Mount Eddy); to WA. Jul–Aug ★

D. breweri S. Watson (p. 547) Per, tufted, gen canescent. **ST**: (1)2–10(15) cm, few to many from caudex, hairs stalked, 4–10-rayed. **LF**: basal (2.5)4–15(25) mm, oblanceolate to obovate, hairs on both surfaces dense, 4–10-rayed; cauline (0)1–3(6), entire; petiole ciliate. **INFL**: (5)7–18(24)-fld; axis as hairy as st; pedicels ± appressed, 1.5–3(4) mm, hairy. **FL**: sepals 1.2–2 mm; petals 2–3 mm, 0.7–1.1 mm wide, white. **FR**: 3.5–9(11) mm, 1.5–2.5 mm wide, lanceolate to oblong or linear, flat, gen twisted, hairs 2–5-rayed; style 0.1–0.4 mm. **SEED**: 28–40, 0.5–0.7 mm, ovoid. $2n=32$. Open, rocky areas, talus, exposed ridges, gen above timberline; 3100–4100 m. CaRH, SNH, n SNE, W&I. Jul–Aug

D. californica (Jeps.) Rollins & R.A. Price (p. 547) CALIFORNIA DRABA Short-lived per. **ST**: 1–several, (2)4–9(12) cm, hairs 4–8-rayed. **LF**: basal 6–20 mm, oblanceolate, entire, hairs on both surfaces 4–8-rayed; cauline lvs 0–3; petiole ciliate. **INFL**: 3–13-fld; axis hairy as st; pedicels (2)4–9(12) mm, hairy. **FL**: sepals 1.7–2.5 mm, persistent; petals 2–3 mm, 0.8–1 mm wide, white. **FR**: (5)6–9(11) mm, 1.8–2.5 mm wide, oblong to lanceolate, not twisted, ± flat, hairs 2–3 rayed and simple (0); style 0.1–0.4 mm. **SEED**: 22–32, 0.8–1 mm, oblong, flat. Open, rocky areas, grassy meadows; 3250–4000 m. n W&I (White Mtns, Mono Co.). Jun–Aug ★

D. cana Rydb. CANESCENT DRABA Per, not tufted. **ST**: 1–many from caudex, gen branched distally, (6)10–30(38) cm, hairs 4–10-rayed throughout, simple near base. **LF**: basal (5)8–20(35) mm, oblanceolate to linear, entire or dentate, ciliate, hairs on both surfaces short-stalked, 4–12-rayed; cauline lvs 3–10(17), bract-like proximally. **INFL**: (10)15–47(63)-fld, most proximal bracted; axis as hairy as st; pedicels erect or ascending, appressed, 2–5(10) mm. **FL**: sepals 1.5–2 mm; petals 3–5 mm, 0.7–1.7 mm wide, white. **FR**: 4–12 mm, 1.5–2.5 mm wide, lanceolate to oblong, flat, ± twisted or not, ± appressed, hairs 3–7-rayed; style 0.1–0.6 mm. **SEED**: 28–48, 0.5–0.9 mm, ovoid. $2n=32$. Subalpine to alpine meadows, tundra, rock crevices, outcrops; < 4100 m. s SNH (e slope, Inyo Co.); n&w US, Can. May–Aug ★

D. carnosula O.E. Schulz (p. 547) MOUNT EDDY DRABA Per, scapose, loosely tufted; caudex branches creeping. **ST**: 3–12 cm, glabrous throughout. **LF**: basal, 3–15 mm, oblanceolate to obovate, entire, surfaces glabrous, margin and petiole hairy, hairs short-stalked, 2–4-rayed or fewer simple; cauline lvs 0. **INFL**: 2–6-fld; axis glabrous; pedicels 3–8 mm, glabrous or hairs few, simple. **FL**: sepals 3–4 mm; petals 5–7 mm, 1.3–2 mm wide, yellow. **FR**: 1–2.3 cm, 4–6 mm wide, lanceolate, flat, not twisted, glabrous; style 2–3 mm. **SEED**: 8–12, 3–4.5 mm wide, round, flat; wing 1–1.5 mm wide. Rocky slopes; 2000–2750 m. KR (Mount Eddy area, Trinity, Siskiyou cos.). Jun–Jul ★

D. corrugata S. Watson Bien to short-lived per. **ST**: 5–13(17) cm, hairy throughout, hairs simple and long-stalked, 2-rayed hairs mixed with smaller 2–4-rayed hairs. **LF**: basal densely overlapping, (0.5)1–2.2(4.5) cm, oblong to narrowly oblanceolate, entire, abaxial hairs stalked, 2–4-rayed, simple hairs gen along midvein, adaxial hairs gen simple and long-stalked, 2-rayed; petiole and margin ciliate, hairs simple; cauline lvs (4)6–10(13). **INFL**: (10)18–55(67)-fld; axis as hairy as st; pedicels 2–6(8) mm, hairy. **FL**: sepals 2–2.7 mm; petals 2–3.5 mm, 0.2–0.5 mm wide, linear, yellow. **FR**: 5–20 mm, (1.5)2–3(4) mm wide, elliptic, flat, ± twisted, hairs stalked, (2)4-rayed; style (0.8)1–3.5 mm. **SEED**: 16–28, 1–1.2 mm, oblong. Rocky slopes, talus, pine woodland; 2000–3500 m. SnGb, SnBr; Baja CA. Jun–Jul

D. cruciata Payson (p. 547)   MINERAL KING DRABA   Per, tufted, scapose. **ST:** (3)5–14(18) cm, unbranched, glabrous or hairy proximally with stalked, (3)4(5)-rayed hairs, glabrous distally. **LF:** basal (4)6–11(16) mm, oblanceolate to ovate, entire to minutely dentate, not ciliate, midvein obscure, surfaces glabrous; cauline lvs 0. **INFL:** (3)5–18(22)-fld; axis glabrous; pedicels not decurrent at base, (4)5–10(13) mm, glabrous. **FL:** sepals 1.5–2 mm; petals 4–6 mm, 1.2–2 mm wide, yellow. **FR:** (4)6–12(16) mm, 1.5–3 mm wide, narrowly lanceolate, flat, not twisted, glabrous or puberulent, hairs simple and short-stalked, 2–4-rayed; style (0.1)0.3–0.8 mm. **SEED:** 6–10(12), 1–1.7 mm, ovoid. Gravelly slopes, subalpine areas; 2500–3050 m. s SNH (near Mineral King, Tulare Co.). Jul–Aug ★

D. cuneifolia Torr. & A. Gray (p. 547)   Ann, occ scapose. **ST:** (2)3–27(37) cm, simple, hairs 2–4(5)-rayed, base with simple hairs. **LF:** basal not rosetted, (0.4)1–3.5(5) cm, oblanceolate to broadly obovate, dentate, hairs stalked, 2–4-rayed, occ mixed with simple; cauline lvs 0–6. **INFL:** 10–50(70)-fld, elongated to ± congested, not umbel-like; axis hairs 2–4-rayed; pedicels (1)2–7(10) mm, hairy. **FL:** sepals 1.5–2.5 mm; petals (2)2.5–4.5(5) mm, 1–2 mm wide, white, or late-season fl petals 0. **FR:** (3)6–12(16) mm, 1.7–2.7(3) mm wide, oblong to linear or broadly ovate, flat, not twisted, hairs simple or 2–4-rayed (glabrous); style to 0.4 mm. **SEED:** (12)24–66(72), 0.5–0.7 mm, ovoid. 2*n*=16,32. Open or disturbed places; < 2100 m. s SN, SnJV, SW, W&I, D; w US, n Mex. [*D. c.* var. *c.*; *D. c.* var. *integrifolia* S. Watson; *D. c.* var. *sonorae* (Greene) Parish] Jan–May

D. densifolia Nutt. (p. 547)   Per, tufted, scapose, cushion-forming. **ST:** (0.5)1.5–10(17) cm, few to many, unbranched, glabrous (some simple and 2–4(5)-rayed hairs). **LF:** basal 2.5–9(14) mm, linear to ± oblanceolate, entire, ciliate; midvein prominent; glabrous or with 2–4-rayed hairs; cauline lvs 0. **INFL:** 2–10(22)-fld; axis glabrous or as hairy as st; pedicels (0.7)1.5–10(25) mm, glabrous (hairy). **FL:** sepals 2–3 mm; petals 2–5 mm, 1–1.7(2) mm wide, yellow. **FR:** (2.5)3–6(8) mm, 2–3 mm wide, ovate to ± lanceolate, flat, not twisted, hairs simple and short-stalked, 2–5-rayed; style 0.3–0.6(0.8) mm. **SEED:** 4–12, 1.2–2(2.6) mm. 2*n*=36. Alpine barrens, rocky slopes; 1900–3650 m. SNH, n SNE, W&I; nw US, w Can. Jun–Aug

D. howellii S. Watson   HOWELL'S DRABA   Per, loosely tufted. **ST:** unbranched, (2)4–11(15) cm, hairs 2–4-rayed and simple (0). **LF:** basal 4–16(25) mm, oblanceolate to obovate, entire, hairs stalked, (2–3)4-rayed, (glabrous with non-ciliate margin hairs); cauline lvs (0)1–3(4), entire. **INFL:** (5)7–18(25)-fld; axis as hairy as st or glabrous; pedicels (4)7–10 mm. **FL:** sepals 2.5–3.2 mm; petals 4–8 mm, 1–2 mm wide, yellow. **FR:** 6–11(15) mm, 3–5 mm wide, lanceolate to widely ovate, flat, not twisted, hairs simple and 2–4-rayed; style (1)1.6–3 mm. **SEED:** 8–22, 1–1.6 mm, oblong. Rock crevices; 1950–2650 m. KR; sw OR. Jul–Aug ★

D. incrassata (Rollins) Rollins & R.A. Price   SWEETWATER MOUNTAINS DRABA   Per, tufted, scapose. **ST:** few to many from caudex, unbranched, 2–8 cm, glabrous. **LF:** basal 3–10 mm, obovate to oblanceolate, entire, ciliate; midvein obscure; both surfaces glabrous or with simple and short-stalked 2-rayed hairs proximal to tip; cauline lvs 0. **INFL:** 8–22-fld; axis glabrous; pedicels 3–7(10) mm, base decurrent. **FL:** sepals 1.7–3 mm; petals 3–5 mm, 1.7–2.5 mm wide, yellow. **FR:** 3–7(10) mm, 2.2–4.5 mm wide, ovate to lance-elliptic, flat, not twisted, glabrous (puberulent, hairs simple); style 0.2–0.8 mm. **SEED:** 8–12, 1.5–1.8 mm, oblong. 2*n*=24. Alpine barrens, rocky slopes; 2500–3500 m. n SNE (Sweetwater Mtns, Mono Co.). Jun–Aug ★

D. lemmonii S. Watson (p. 547)   Per, tufted, scapose. **ST:** 3–10(15) cm, with simple hairs, fewer short-stalked, 2-rayed hairs. **LF:** basal 4–10(18) mm, oblanceolate to obovate, entire, ciliate, abaxial hairs gen stalked, 2-rayed, adaxial hairs ± only simple; cauline lvs 0. **INFL:** 4–15(21)-fld; axis hairy as st; pedicels 4–10(14) mm. **FL:** sepals 2–2.7 mm; petals 4–6 mm, 1.5–2.5 mm wide, yellow. **FR:** 4–9 mm, 2.5–5 mm wide, ovate, flat, ± twisted, hairs simple (or mixed with short-stalked, 2-rayed hairs); style 0.1–0.6 mm. **SEED:** 10–16, 1–1.4 mm, ovoid. 2*n*=50. Common. Talus, rock crevices, rocky meadows; 3050–4000 m. SNH. Jul–Aug

D. lonchocarpa Rydb. (p. 547)   SPEAR-FRUITED DRABA   Per, tufted. **ST:** few to many, (1)3–11 cm, glabrous or hairs minutely stalked, 8–12-rayed. **LF:** basal (1.5)3–15 mm, oblanceolate to obovate, entire, abaxial hairs short-stalked, 8–12-rayed, adaxially glabrous or hairs simple and long-stalked, branched, lf midvein obscure; cauline lvs 0 or 1(4), entire. **INFL:** 3–9-fld; axis glabrous or as hairy as st, ± wavy in fr; pedicels 2–9(15) mm, glabrous or hairy. **FL:** sepals 1.5–2 mm; petals 2–3.5 mm, 1–1.5 mm wide, white. **FR:** 6–15(18) mm, 1–2(3) mm wide, linear to narrowly lanceolate or oblong, flat, ± twisted or not, glabrous or puberulent, hairs simple and minutely stalked, 2-rayed; style 0.1–0.3 mm. **SEED:** 16–28, 0.7–1 mm, ovoid. 2*n*=16. Calcareous scree; 2800–4000 m. c SNH (Convict Creek Basin, Mono Co.), n W&I (White Mtns); to AK, CO; e Russia. Jun–Jul ★

D. longisquamosa O.E. Schulz   Per, tufted, scapose. **ST:** unbranched, 1.5–9 cm, stiff-simple-hairy throughout. **LF:** 5–20 mm, oblanceolate to obovate, entire, ciliate, both surfaces simple-hairy; cauline lvs 0. **INFL:** 4–16-fld; axis hairy; pedicels 3–7 mm. **FL:** sepals 1.5–2.2 mm, persistent; petals 3.5–5 mm, 1.5–2 mm wide, yellow, persistent. **FR:** 3.5–7 mm, 2.5–5 mm wide, ovate to ± round, flat, not twisted, hairs simple; style 0.4–1 mm. **SEED:** 10–16, 1–1.2 mm, ovoid. Gravelly areas; 3000–3900 m. s SNH (Fresno, Tulare cos.). [*D. lemmonii* S. Watson, misappl., in part] Jul ★

D. monoensis Rollins & R.A. Price (p. 551)   WHITE MOUNTAINS DRABA   Per, occ scapose. **ST:** unbranched, (0.5)1–4 cm, hairs gen wavy, simple and 2-rayed on both surfaces (0 adaxially). **LF:** (3)5–16(20) mm, narrowly oblanceolate, entire (minutely toothed near tip), both surfaces with simple and fewer stalked, 2-rayed hairs, margin ciliate or not; cauline lvs 0–3, entire. **INFL:** (3)6–13(17)-fld, umbel-like in fr; axis as hairy as st (glabrous); pedicels 1–2.5(4) mm. **FL:** sepals 1–1.5 mm; petals 1.5–2 mm, 0.5–0.6 mm wide, white. **FR:** (2)3–5 mm, (1.2)1.5–2.5 mm wide, ovoid to ellipsoid, ± inflated at base, not twisted, puberulent, hairs simple (0); style 0.1–0.2 mm. **SEED:** 12–20, 0.6–0.8 mm, ovoid. Moist gravel, rock crevices; 3600–4000 m. n W&I (White Mtns, Mono Co.). Jul–Aug ★

D. novolympica Payson & H. St. John (p. 551)   Per, tufted, cushion-forming, scapose. **ST:** unbranched, 0.5–3.5 cm, hairs dense, simple and stalked, 2–5-rayed. **LF:** basal 2–8 mm, oblong to linear-oblanceolate, ciliate, entire, abaxial hairs dense, stalked, 2–12-rayed, adaxial hairs simple, some 2-rayed; cauline lvs 0. **INFL:** 2–12-fld, gen elongated in fr; axis as hairy as st; pedicels 1–5 mm, hairy. **FL:** sepals 1.5–2.5 mm; petals 2–4 mm, 1.5–2 mm wide, yellow. **FR:** (2.5)3–4(5) mm, 1.5–3.5 mm wide, ovoid, ± inflated at base, not twisted, densely with 2–6-rayed hairs; style 0.2–0.6(0.8) mm. **SEED:** 4–8(12), 1.2–1.8 mm, oblong. 2*n*=42. Uncommon. Open, rocky slopes, talus, scree; 1500–3700 m. n SNH; w Can, nw US. [*D. paysonii* J.F. Macbr. var. *treleasei* (O.E. Schulz) C.L. Hitchc.] Jun–Aug

D. oligosperma Hook. (p. 551)   Per, tufted, scapose, cushion-forming. **ST:** unbranched, (1)2–6(10) cm, glabrous or hairs sessile, stellate. **LF:** (2)4–11(15) mm, linear to linear-oblanceolate, entire, midvein prominent, hairs sessile, comb-like, 7–16-rayed, long axis parallel to midvein and branches shorter toward hair tips, adaxial occ glabrous; cauline lvs 0. **INFL:** 4–12(17)-fld; axis glabrous or as hairy as st; pedicels (2)3–10(13) mm. **FL:** sepals 1.5–3 mm; petals 2.5–4 mm, 1.5–3 mm wide, yellow. **FR:** 3–6(7) mm, ovoid to lanceolate, inflated at least near base, not twisted, puberulent with simple and sessile, gen unequally 2-rayed hairs; style 0.1–0.8(1) mm. **SEED:** 6–12, 1.3–1.8 mm, ovoid. 2*n*=32,64. Common. Alpine barrens, dry slopes, tundra; 2000–3900 m. SNH, n SNE, W&I; w Can, w US. Gen produces seeds asexually. May–Aug

D. praealta Greene   TALL DRABA   Ann to short-lived per. **ST:** branched or not, (5)8–32(38) cm; simple, 2–5-rayed hairs mixed throughout. **LF:** basal rosetted, (0.7)1–3.5(4.5) cm, oblanceolate, dentate, ciliate, hairs stalked, 3–6-rayed, occ also simple and 2-rayed adaxially; cauline lvs (1)2–5(9), dentate. **INFL:** 8–30(37)-fld, elongated in fr; axis as hairy as st; pedicels (3)4–10(12) mm, hairy. **FL:** sepals 1.7–2.5 mm; petals 2.8–3.5(4) mm, 0.8–1.2 mm wide, white. **FR:** (5)7–12(15) mm, 1.5–2.5(3) mm wide, lanceolate to lance-linear, flat, not appressed, not twisted, puberulent with simple and short-stalked, 2–4-rayed hairs; style to 0.15 mm. **SEED:** (24)30–52, 0.8–1.1 mm, oblong. 2*n*=56. Montane or subalpine moist meadows, streambanks, forest, talus, shale cliffs; 2500–4100 m. c SNH (e slope, Mono, Inyo cos.); w US, Can; e Russia. Jun–Aug ★

D. pterosperma Payson WINGED-SEED DRABA   Per, scapose, cushion-forming. **ST:** unbranched, 3–11 cm, hairs gen wavy, 2–7–branched. **LF:** basal 2–7 mm, obovate to oblanceolate, entire, ciliate, surfaces with dense, wavy, stalked, 5–12-rayed hairs; cauline lvs 0. **INFL:** 4–12-fld; axis hairy as st; pedicels (3)4–9(12) mm, hairy. **FL:** sepals 3–4 mm; petals 6–7 mm, 1.5–2.5 mm wide, yellow. **FR:** 5–10(13) mm, (2.5)3.5–5(6) mm wide, widely ovate to lanceolate, flat, not twisted, hairs short-stalked, 2–7-rayed; style 1.5–3.8 mm. **SEED:** 8–12, 1.6–3 mm, ovate; wing 0.3–0.9 mm wide. Marble or limestone crevices, talus, scree; 1500–2450 m. c KR (Marble Mtns, Siskiyou Co.). Jun–Sep ★

D. saxosa Davidson SOUTHERN CALIFORNIA ROCK DRABA   Per, tufted, scapose. **ST:** unbranched, 5–15 cm; hairs simple, long-stalked and 2-rayed, mixed with smaller 2–4-rayed. **LF:** 1–3 cm, oblanceolate, entire, abaxial hairs dense, stalked, (2)4-rayed, adaxial hairs long-stalked, 2(4)-rayed; cauline lvs 0(1). **INFL:** 12–43-fld; axis as hairy as st; pedicels (6)10–17 mm, hairy. **FL:** sepals 2–2.5 mm; petals 3–5 mm, 0.5–1 mm wide, yellow. **FR:** 6–10 mm, 2.5–4 mm wide, oblong, flat, ± twisted or not, with short-stalked, (2)4-rayed hairs; style (0.8)1.2–3.5 mm. **SEED:** 8–24, oblong, 1.2–1.6 mm. Rocky slopes; 2400–3300 m. e PR (SnJt, Santa Rosa Mtns). [*D. corrugata* S. Watson var. *s.* (Davidson) Munz & I.M. Johnst.] Jun–Jul ★

D. sharsmithii Rollins & R.A. Price (p. 551) MOUNT WHITNEY DRABA   Per, densely tufted, scapose. **ST:** unbranched, (1.5)3–7(10) cm, glabrous or proximally with simple, 2-rayed hairs. **LF:** 4–7(10) mm, oblanceolate to oblong, entire, ciliate, abaxial hairs short-stalked, 2–5-rayed, adaxial hairs 0 or simple and short-stalked, 2-rayed; cauline lvs 0. **INFL:** 2–10(16)-fld; axis glabrous; pedicels (3)5–12(16) mm, glabrous. **FL:** sepals 2.8–3.5 mm; petals 4–6 mm, 2–3 mm wide, yellow. **FR:** (5)8–15(20) mm, 2–4(5) mm wide, oblong to lanceolate or ovate, flat, twisted, curved, gen glabrous; style (0.5)0.9–1.8(2) mm. **SEED:** 16–20, 1.2–1.7 mm, oblong. Rock crevices, slopes; 3300–3800 m. s SNH (Fresno, Inyo cos.). Jul–Aug ★

D. sierrae Sharsm. (p. 551) SIERRA DRABA   Per, tufted, cushion-forming, scapose. **ST:** (0.5)1–4 cm, with 2–4-rayed, branched hairs occ mixed with fewer simple ones. **LF:** 2–6(8) mm, oblong, entire, occ ciliate, hairs of both surfaces dense, stalked, 7–12-rayed; cauline lvs 0. **INFL:** 2–6(9)-fld; axis as hairy as st; pedicels 2–5(7) mm, hairy. **FL:** sepals 2.5–3.5 mm; petals 4–5 mm, 1.2–1.8 mm wide, yellow. **FR:** 4–8(10) mm, (2)3–4.5 mm wide, lanceolate to elliptic or widely ovate, flat, gen twisted, not symmetric, hairs short-stalked, 2–4-rayed, occ fewer simple ones; style (0.1)0.3–0.7(1) mm. **SEED:** 8–14(16), 1.2–1.5 mm, ovoid. Rock crevices, granite outcrops; 3500–3900 m. c SNH (Fresno, Inyo cos.). Jun–Aug ★

D. subumbellata Rollins & R.A. Price (p. 551) MOUND DRABA   Per, tufted, cushion-forming, scapose. **ST:** unbranched, (0.5)1–2.5 cm, hairs dense throughout, ± gray, stalked, tree-like, 5–12-rayed. **LF:** 1.5–4 mm, obovate to widely oblong, entire, not ciliate, hairs of both surfaces like those of st, adaxial hairs occ also long-stalked, spurred; cauline lvs 0. **INFL:** 2–5(10)-fld, umbel-like in fr; axis as hairy as st; pedicels 1.5–3(6) mm. **FL:** sepals 1.8–2.8 mm; petals 2.8–4 mm, 1–1.5 mm wide, yellow. **FR:** 2–5 mm, 2–3 mm wide, ovoid or lance-ovoid, inflated at base, flattened near tip, not twisted, hairs short-stalked, tree-like, 4–12-rayed; style 0.2–0.6 mm. **SEED:** 6–12, oblong, 1–1.2 mm. Talus, among rocks; 3300–4100 m. s SNH (e slope, nw Inyo Co.), n W&I (White Mtns, Mono Co.). Jul–Aug ★

D. verna L. (p. 551) Ann, scapose. **ST:** unbranched, (2)5–20(30) cm, hairs simple and 2(4)-rayed proximally, glabrous distally. **LF:** basal 2–18(30) mm, obovate to oblanceolate or linear, entire or 1–5-toothed, hairs of both surfaces simple, 2–4-rayed; cauline lvs 0. **INFL:** 4–20(30)-fld; axis glabrous, gen wavy in fr; pedicels (2)5–20(35) mm, glabrous. **FL:** sepals 1–1.5 mm, petals (1.5)2–4.5(6) mm, 1–2 mm wide, deeply 2-lobed, white. **FR:** (2.5)4–9(12) mm, 1.5–2.5(3.5) mm wide, obovate to oblanceolate, lanceolate, elliptic, oblong, or linear, flat, not twisted, glabrous; style ≤ 0.2 mm. **SEED:** (20)32–70(84), 0.3–0.6(0.8) mm, ovoid. $2n=14,16,20,24,28,30,32,$ $34,36,38,40,52,54,58,60,64.$ Open or disturbed areas; < 2500 m. CA-FP, MP; N.Am, native to Eurasia, n Afr. A highly variable, complex sp. occ split into many poorly defined taxa. Feb–May

ERUCA GARDEN-ROCKET

1 sp.: Eurasia. (Latin: perhaps burn, from spicy taste)

E. vesicaria (L.) Cav. subsp. *sativa* (Mill.) Thell. (p. 551)   Ann, hairs simple or 0. **ST:** (1)2–8(10) dm, hairs 0 or reflexed toward base. **LF:** basal petioled, (3)5–20(27) cm, widely oblanceolate, deeply pinnately lobed (dentate); lateral lobes 3–9 pairs, margin entire or dentate; cauline ± sessile, lobed or not. **INFL:** elongated. **FL:** sepals erect, (6)7–10(12) mm, linear to oblong, lateral pair base sac-like; petals (1.2)1.5–2(2.6) cm, (4)5–7(9) mm wide, widely obovate, yellow, veins dark brown to purple. **FR:** silique, (1)1.5–3.5(4) cm, (2.5)3–5 mm wide, linear to oblong or elliptic, dehiscent, segmented; terminal segment (4)5–10(11) mm, sword-like, flattened, seedless; pedicel erect to ascending, 2–8(10) mm. **SEED:** 10–50, in 2 rows, 1.6–2.5 mm, orange to brown. $2n=22.$ Disturbed areas; cult fields; < 1200 m. CaR, GV, SCoR, DSon; N.Am, native to Eurasia. [*E. s.* Mill.] May–Sep

ERYSIMUM WALLFLOWER

Ann to subshrub [shrub]; hairs sessile, appressed, 2–5(8)-rayed. **LF:** basal rosetted, petioled, entire, dentate, or pinnately lobed; cauline sessile or petioled, bases not lobed. **INFL:** elongated. **FL:** sepals oblong to linear, erect, base of lateral pair sac-like or not; petals clawed, yellow or orange (white, purple, or brown). **FR:** silique, dehiscent, linear, cylindric, 4-sided, or flat parallel or perpendicular to septum, unsegmented; stigma entire or 2-lobed. **SEED:** 15–100, in 1 or 2 rows, plump or flattened, oblong, winged or not. ± 150 spp.: N.Am, Eurasia, n Afr. (Greek: to help or save, from alleged medicinal properties of some spp.) [Rossbach 1958 Madroño 14:261–267] Fls, fr, basal lvs needed for identification. All native CA taxa related to *E. capitatum*; hybridization blurs limits of some spp.

1. Petals 3–9(12) mm, 1.5–3 mm wide
  2. Fr valves densely hairy inside; sepals 1.8–3.2 mm; petal claw 1.5–3.5 mm . . . . ***E. cheiranthoides***
  2′ Fr valves glabrous (sparsely hairy) inside; sepals 4–8 mm; petal claw 3–8 mm . . . . ***E. repandum***
1′ Petals (10)13–30(35) mm, 3–10(15) mm wide
  3. Most proximal sts woody; pl gen subshrub
    4. Fr hairs 2-rayed; stigma lobes 2, prominent, longer than wide; petals orange to yellow, brown, red, purple, violet, or white . . . . ***E. cheiri***
    4′ Fr hairs 2–4-rayed; stigma lobes obscure, as long as wide; petals yellow or cream
      5. Fr flat perpendicular to septum; lf hairs 2(3)-rayed . . . . ***E. insulare***
      5′ Fr flat parallel to septum or 4-angled; lf hairs 2–5-rayed
        6. Distal lvs petioled; fr flat parallel to septum; pedicel stout, 5–17(22) mm . . . . ***E. franciscanum***
        6′ Distal lvs sessile; fr 4-angled or ± flat parallel to septum; pedicels slender, (3)5–10 mm . . . . ***E. suffrutescens***

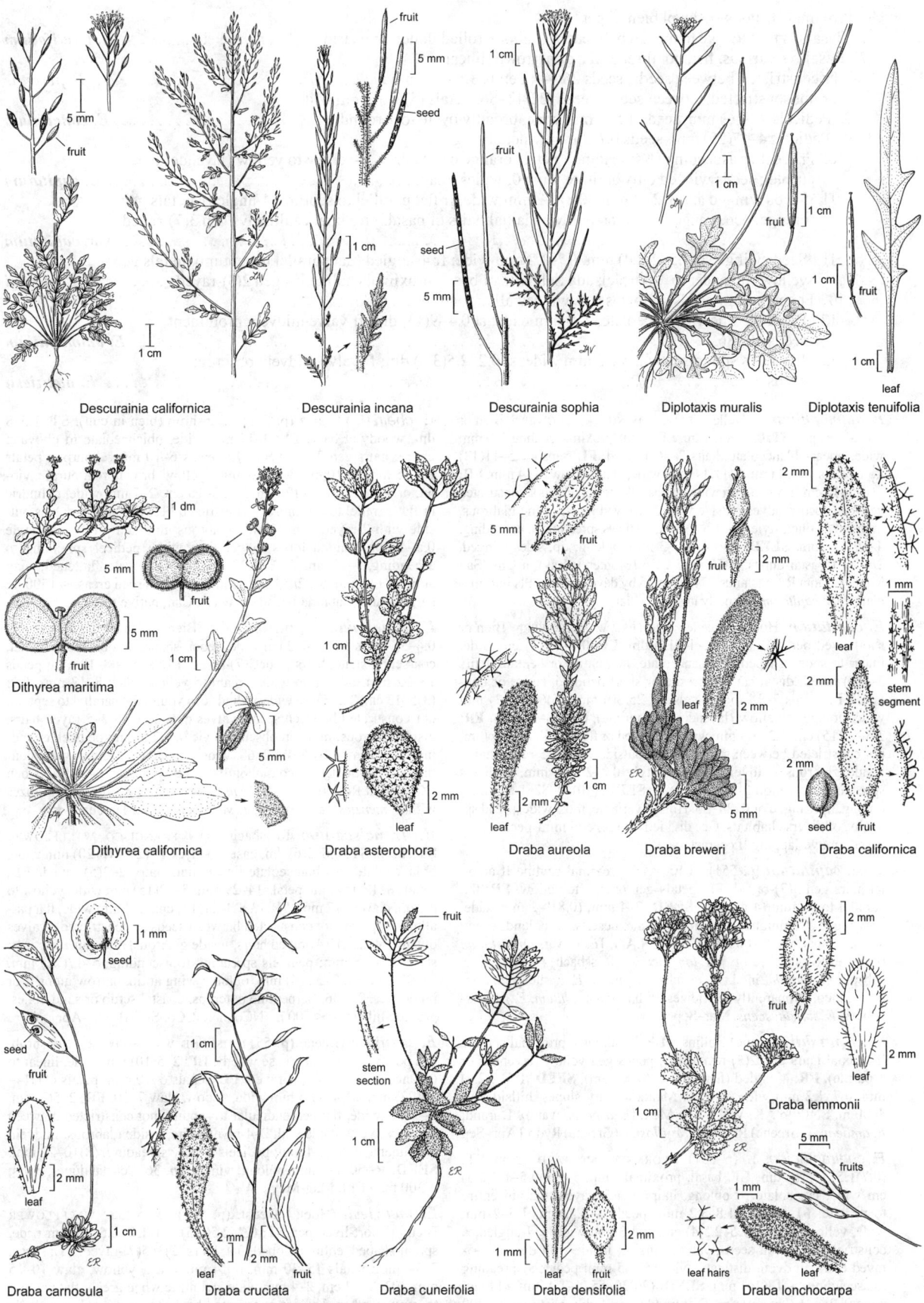

Descurainia californica

Descurainia incana

Descurainia sophia

Diplotaxis muralis

Diplotaxis tenuifolia

Dithyrea maritima

Dithyrea californica

Draba asterophora

Draba aureola

Draba breweri

Draba californica

Draba carnosula

Draba cruciata

Draba cuneifolia

Draba densifolia

Draba lemmonii

Draba lonchocarpa

3′ Proximal sts not woody; pl bien or per
  7. Basal lvs thread-like to narrowly linear, margins ± rolled under, appearing cylindric . . . . ***E. teretifolium***
  7′ Basal lvs various, flat, not thread-like or narrowly linear
    8. Fr constricted between seeds; seeds 26–44; petals 3.5–6 mm wide . . . . ***E. perenne***
    8′ Fr not constricted between seeds; seeds (32)42–86; petals (5)6–16 mm wide
      9. Pedicels 2–4(6) mm; seeds 1.5–3 mm wide, broadly ovate to ± round. . . . . ***E. concinnum***
      9′ Pedicels 4–17(25) mm; seeds 0.7–2 mm wide, oblong
        10. Fr erect or ascending (± spreading); petals orange or occ orange-yellow to yellow (lavender to ±
           purple); seeds winged only at tip or wing 0; widespread. . . . . ***E. capitatum***
          11. Seeds winged at tip, 2–4 mm, (0.8)1–2 mm wide; fr flat parallel to septum (4-angled); petals gen
             orange, occ orange-yellow or yellow; adaxial hairs of basal, proximal cauline lvs gen 3(7)-rayed
             . . . . var. ***capitatum***
          11′ Seeds wingless, 1.5–2(2.4) mm, 0.7–1.2 mm wide; fr 4-angled (flat parallel to septum); petals gen
             yellow (lavender to ± purple); adaxial hairs of basal, proximal cauline lvs gen 2(3)-rayed . . . . var. ***purshii***
        10′ Fr spreading; petals yellow; seeds winged all around; NCo, CCo
          12. Basal lvs linear-oblanceolate, 2–9 mm wide; st 0.4–9(13) dm; fr valve midvein prominent;
             seeds 50–86. . . . . ***E. ammophilum***
          12′ Basal lvs spoon-shaped, 5–15 mm wide; st 0.2–2.5(3.5) dm; fr valve midvein obscure;
             seeds 32–74. . . . . ***E. menziesii***

E. ammophilum A. Heller SAND-LOVING WALLFLOWER Bien or short-lived per. **ST:** 0.4–9(13) dm. **LF:** basal, proximal cauline 2–9 mm wide, linear-oblanceolate, hairs 2–4(7)-rayed. **FL:** sepals 7.5–11(13) mm; petals 14–24 mm, 6–11(14) mm wide, yellow, claw 8–14 mm. **FR:** (2)3.5–12 cm, 1.5–3.5 mm wide, flat parallel to septum, not constricted between seeds; valves outside with 2–4-rayed hairs, inside glabrous, midvein distinct; style 0.3–1.5(2) mm; pedicels spreading to ascending, 4–10(13) mm. **SEED:** 50–86, 1.5–3 mm, oblong; tip, sides winged. $2n=36$. Coastal dunes; < 50 m. c CCo (Monterey Bay), n ChI (San Miguel, Santa Rosa islands). Threatened by development. Pls intermediate to *E. capitatum* formerly in s SCo. Mar–Apr ★

E. capitatum (Hook.) Greene WESTERN WALLFLOWER Bien or short-lived per. **ST:** (0.5)1.2–10(12) dm. **LF:** 0.3–1.5(3) cm wide, linear to spoon-shaped or oblanceolate, flat, entire or dentate, hairs 2–4(7)-rayed; distal cauline sessile or short-petioled, entire or dentate. **FL:** sepals 7–14 mm; petals 12–25(30) mm, (5)6–10(13) mm wide, orange to yellow (lavender to ± purple), claw 8–16 mm. **FR:** 3.5–11(15) cm, 1.3–3.3 mm wide, 4-angled or flat parallel to septum, not constricted between seeds; valves outside with 2–5-rayed hairs, inside glabrous, midvein distinct; style 0.2–2.5(3) mm; pedicels spreading to ascending, 4–17(25) mm. **SEED:** (40)54–82, 1.5–4 mm, oblong, tip winged or 0. $2n=36$. Highly variable, most widespread sp. in CA, of diverse habitats. Occ divided into several infraspecific taxa based on overlapping lf characters.

  var. ***capitatum*** (p. 551) **LF:** basal, proximal cauline lf adaxial hairs gen 3(7)-rayed. **FL:** petals gen orange to yellow. **FR:** flat parallel to septum (4-angled). **SEED:** 2–4 mm, (0.8)1–2 mm wide, tip winged. Common. Open areas, alpine, deserts, woodland, sandy areas, chaparral; < 4000 m. CA; w N.Am. [*E. c.* var. *angustatum* (Greene) Rossbach; *E. c.* var. *lompocense* (Rossbach) Kartesz; *E. c.* subsp. *angustatum*, ined.; *E. c.* subsp. *c.*, ined.; *E. c.* subsp. *lompocense*, ined.] Apparently hybridizes with *E. ammophilum*, *E. franciscanum*, *E. suffrutescens*. Mar–Sep

  var. ***purshii*** (Durand) Rollins **LF:** basal, most proximal cauline lf adaxial hairs gen 2(3)-rayed. **FL:** petals gen yellow (lavender or ± purple). **FR:** 4-angled (flat parallel to septum). **SEED:** 1.5–2(2.4) mm, 0.7–1.2 mm wide, wing 0. Meadows, dry slopes, hillsides; < 4000 m. SNE; to AK, WY, TX, n Mex. [*E. asperum* var. *p.* Durand; *E. amoenum* (Greene) Rydb.; *E. argillosum* (Greene) Rydb.] Apr–Sep

E. cheiranthoides L. (p. 551) WORMSEED MUSTARD Ann. **ST:** (0.7)1.5–10(15) dm. **LF:** basal, proximal cauline (0.2)0.5–1.2(2.3) cm wide, lanceolate or oblong, hairs 3–4(5)-rayed, margin entire to dentate. **FL:** sepals 1.8–3.2 mm; petals 3–5.5 mm, 1.5–2 mm wide, yellow. **FR:** (1)1.5–2.5(4) cm, 1–1.3 mm wide, ± 4-angled, ± constricted between seeds; valves outside, inside with dense, 3–5-rayed hairs, midvein distinct; style 0.5–1.5 mm; pedicels spreading to ascending, 5–13(16) mm. **SEED:** (20)30–55, 1–1.4 mm, oblong; wing 0. $2n=16$. Uncommon. Disturbed areas, fields, pastures; < 3000 m. SN, GV; to e N.Am; circumboreal, native to Eurasia. Jun–Aug

E. cheiri (L.) Crantz (p. 551) Subshrub (bien in cult). **ST:** 1.5–8 dm, woody at base. **LF:** 3–15 mm wide, oblanceolate to obovate, entire; hairs gen 2–3-rayed. **FL:** sepals 6–10 mm, ± purple; petals 20–35 mm, 5–10 mm wide, orange, yellow, brown, red, purple, violet, or white, claw 7–12 mm. **FR:** 3–10 cm, 2–7 mm wide, cylindric or flat parallel to septum, not constricted between seeds; valves outside with 2-rayed hairs, inside glabrous, midvein prominent; style 0.5–4 mm, stigma lobes longer than wide; pedicel spreading to ascending, 7–13 mm. **SEED:** 32–44, 2–4 mm, ovate, flattened; wing at tip or continuous. $2n=12$. Uncommon. Disturbed areas; < 1500 m. s ChI (Santa Catalina Island), e WTR; Can; native to s Eur. Apr–Jul

E. concinnum Eastw. (p. 551) Bien or short-lived per. **ST:** 0.4–5(7) dm. **LF:** 0.4–2 cm wide, spoon-shaped to oblanceolate, flat, coarsely dentate, hairs 2- or 3(7)-rayed. **FL:** sepals 8–19 mm; petals 15–32 mm, 6–16 mm wide, cream to yellow, claw 8–12 mm. **FR:** (3)5–13 cm, 2.2–5 mm wide, cylindric in youth, flat parallel to septum, not constricted between seeds; valves outside with 2–5-rayed hairs, inside glabrous, midvein obscure; style 0.5–2.5 mm; pedicel ascending, 2–4(6) mm. **SEED:** 42–68, widely ovate to ± round; wing continuous. $2n=36$. Cliffs, coastal bluffs, dunes, prairies; < 400 m. NCo, n CCo (Point Reyes); sw OR. [*E. menziesii* subsp. *c.*, ined.] Hybridizes with *E. menziesii* at Fort Bragg, Mendocino Co. Mar–Jun ★

E. franciscanum Rossbach SAN FRANCISCO WALLFLOWER Subshrub. **ST:** 0.6–5(6) dm, base woody. **LF:** (2)3–16(20) mm wide, oblanceolate to oblanceolate-linear, flat; hairs 2–3(5)-rayed. **FL:** sepals 8–12(15) mm; petals 14–29 mm, 5–12(15) mm wide, yellow to cream, claw 9–17 mm. **FR:** (3.8)4–11(14) cm, 2–4 mm wide, flat parallel to septum, not constricted between seeds (± constricted); valves outside with (2)3(4)-rayed hairs, inside glabrous, midvein ± distinct; style 0.5–3.5 mm; pedicels spreading to ascending, 5–17(22) mm. **SEED:** 32–64, 2–3.5(4) mm, oblong; wing at tip, narrow along 1 or both sides. $2n=36$. Serpentine outcrops, coastal scrub or sand dunes, granitic hillsides; < 500 m. NCo, n&c CCo, SnFrB. Jan–Apr ★

E. insulare Greene (p. 551) ISLAND WALLFLOWER Subshrub, occ per. **ST:** 0.5–3 dm, base woody. **LF:** 2–5(10) mm wide, linear to oblanceolate; hairs 2(3)-rayed. **FL:** sepals 6–10 mm; petals (11)14–20(22) mm, (3)4–11.5 mm wide, yellow, claw 7–10. **FR:** 2–5(7) cm, 2–3 mm wide, flat perpendicular to septum, not constricted between seeds; valves outside with 2–4-rayed hairs, inside glabrous, midvein prominent; style 1–4 mm; pedicels widely spreading, (8)10–22 mm. **SEED:** 36–50, 1–2 mm, oblong, wing 0. $2n=36$. Coastal dunes, cliffs; < 300 m. n ChI. Mar–May ★

E. menziesii (Hook.) Wettst. (p. 551) MENZIES' WALLFLOWER Bien or short-lived per. **ST:** 0.2–2.5(3.5) dm. **LF:** 0.5–1.5 cm wide, spoon-shaped, entire to lobed, flat; hairs (2)3–5(7)-rayed. **FL:** sepals 7–14 mm; petals 15–30 mm, 6–14 mm wide, yellow, claw 10–15 mm. **FR:** 3–14 cm, 2–4 mm wide, cylindric when green, flat parallel to septum when dry, not constricted between seeds; valves outside with (2)3 or 4(6)-rayed hairs, inside glabrous, midvein obscure; style

0.3–2 mm; pedicel spreading, 4–15 mm. **SEED**: 32–74, 1.8–2.8(3.5) mm, oblong; wing widest at tip. 2*n*=36. Coastal dunes, headlands, cliffs; < 300 m. NCo, CCo; to s OR. [*E. m.* subsp. *eurekense*, ined.; *E. m.* subsp. *yadonii*, ined.] Threatened by development. Jan–Aug ★

E. perenne (Coville) Abrams (p. 551) Per (bien), caudex short. **ST**: 0.4–6.5 dm. **LF**: 3–10 mm wide, spoon-shaped to widely oblanceolate, ± entire to dentate, flat, hairs 2–5-rayed. **FL**: sepals 8–12 mm; petals 15–22 mm, 3.5–6 mm wide, yellow, claw 8–14 mm. **FR**: 3.8–14 cm, 1.2–3 mm wide, flat parallel to septum, constricted between seeds; valves outside with 2- or 3(4)-rayed hairs, inside glabrous, midvein distinct; style (1.5)2–5.5 mm; pedicel spreading-ascending, 4–12 mm. **SEED**: 26–44, 2–3.4 mm, ovoid, wing 0 (tip winged). 2*n*=36. Decomposing marble, gravelly ground, talus, granitic sand, alpine fell-fields; 2000–4000 m. KR, CaRH, SNH, n SNE (Sweetwater Mtns); OR, NV. [*E. capitatum* var. *p.* (Coville) R.J. Davis; *E. c.* subsp. *p.*, ined.] Jun–Sep

E. repandum L. (p. 551) Ann. **ST**: (0.4)1.5–4.5(7) dm. **LF**: (2)5–12(17) mm wide, linear to narrowly oblanceolate or oblong, wavy, dentate or shallowly wavy, flat, hairs 2(3)-rayed. **FL**: sepals 4–8 mm; petals 6–8 mm, 1.5–2 mm wide, light yellow, claw 3–8 mm. **FR**: (2)3–8(10) cm, 1.5–2 mm wide, 4-angled, ± constricted between seeds; valves outside with 2(3)-rayed hairs, inside glabrous or occ sparsely hairy, midvein distinct; style 1–4 mm; pedicel spreading, 2–4(6) mm, ± as wide as fr. **SEED**: (40)50–80(90), 1–1.5 mm, oblong; wing 0 (tip winged). 2*n*=16. Disturbed areas, fields; < 2100 m. n&c SnJV, MP; N.Am; native to Eur. Apr–Jun

E. suffrutescens (Abrams) Rossbach SUFFRUTESCENT WALLFLOWER Subshrub, occ per. **ST**: 1.5–8.1 dm, base woody. **LF**: 1.5–6(7) mm wide, linear to linear oblanceolate, flat, entire (minutely dentate), hairs 2–4-rayed. **FL**: sepal 6–11 mm; petals (11)14–20(22) mm, (3)4–11.5 mm wide, yellow, claw 8–13 mm. **FR**: (2)3–8.4(11) cm, 1.5–2.4(3.5) mm wide, 4-angled to ± flat parallel to septum, not constricted between seeds; valves outside with 2–4-rayed hairs, inside glabrous, midvein prominent; style 0.5–4 mm, stout; pedicel ascending, (3)5–10 mm. **SEED**: 48–82, 1.5–2.5(3), oblong; wing 0 or at tip. 2*n*=36. Stabilized coastal sand dunes, coastal scrub; < 150 m. s CCo, n SCo. [*E. insulare* subsp. *s.*, ined.] Hybridizes with *E. capitatum*. Dec–Aug ★

E. teretifolium Eastw. (p. 551) SANTA CRUZ WALLFLOWER Bien or short-lived per, caudex present. **ST**: (1.4)2.5–8(10) dm. **LF**: 0.3–3 mm wide, thread-like to narrowly linear, finely toothed, margin ± rolled under, appearing cylindric, hairs 2(3)-rayed. **FL**: sepals 7–11 mm; petals 15–20(25) mm, 5–10 mm wide, yellow-orange to yellow, claw 6–13 mm. **FR**: (4)7–12(15) cm, 1.2–2.2(2.5) mm wide, ± flat parallel to septum, ± constricted between seeds; valves outside with 2- and 3(4)-rayed hairs, inside glabrous, midvein prominent; style 0.5–2(2.5) mm, slender, cylindric, sparsely hairy; pedicel spreading to ascending, 5–14 mm. **SEED**: 40–72, 1.5–2.3(2.7) mm, ovoid; wing at seed tip. Sandy areas in coastal-sage scrub or chaparral; 100–400 m. sw SnFrB (Santa Cruz Sandhills, Santa Cruz Co.). Highly endangered by development, sand mining. Feb–May ★

EUCLIDIUM

1 sp.: Eurasia. (Greek: tightly shut, from indehiscent fr)

E. syriacum (L.) W.T. Aiton (p. 555) Ann, hairy throughout, hairs minutely stalked, 2(3)-forked, some simple. **ST**: (0.4)1–4(4.5) dm, rigid, branched proximally. **LF**: simple, not rosetted, petioled, base not lobed or clasping; basal, proximal cauline (1.5)2–9(12) cm, oblong to lance-oblong, entire, dentate, or wavy-margined; distal cauline ± sessile. **INFL**: spike-like, open. **FL**: sepals erect, 0.6–0.9 mm, base not sac-like; petals 0.9–1.3 mm, 0.1–0.2 mm wide, spoon-shaped, white. **FR**: silicle, sessile, appressed, 2–2.5 mm, 1.5–2 mm wide, ovoid, occ 4-sided; woody; style 1–1.8 mm, conical, stigma 2-lobed; pedicel erect, 0.5–1(1.2) mm, stout. **SEED**: 2(4), 1.3–1.7 mm, oblong. 2*n*=14. Disturbed pastures, roadsides; 1000–1500 m. MP; to WA, CO; native to Eurasia. May–Jun

HALIMOLOBOS

[Ann] per, hairs ± many-branched, tree-like. **LF**: basal simple, entire or dentate to deeply lobed; distal cauline sessile, base wedge-shaped to truncate (lobed). **INFL**: elongated or not. **FL**: sepals erect, base not sac-like; petals oblanceolate to spoon-shaped, not clawed, white. **FR**: silique [silicle], dehiscent, unsegmented, linear [ovate to oblong], flat parallel or perpendicular to septum, hairy, hairs small ± sessile, some long-stalked; stigma entire. **SEED**: 16–110, in 1[2] rows, oblong, wingless. 6 spp.: sw US, n Mex. (Greek: sea pod, from resemblance of fr hair cover to salt) [Bailey et al. 2007 Syst Bot 32:140–156] *H. virgata* moved to *Transberingia*.

H. jaegeri (Munz) Rollins (p. 555) Per; moderately to densely hairy. **ST**: many-branched distally, 1.5–7.5 dm, ± woody at base. **LF**: proximal, middle cauline petioled, (1.5)3–8(11.5) cm, oblanceolate to obovate, densely hairy, base tapered, margin coarsely dentate. **FL**: sepals 2–4 mm; petals (3.5)4.5–6 mm, 1–1.5 mm wide. **FR**: (1)1.5– 2.6 cm, 0.6–0.9 mm wide, cylindric, constricted between seeds; valves densely hairy; style 1–2.2 mm; pedicels spreading, (3)4–9(12) mm, densely hairy. **SEED**: 28–38, in 1 row, 0.9–1.2 mm, oblong. Limestone cliffs, steep rock outcrops, sagebrush-juniper areas; 1200–2600 m. W&I, DMoj; NV. May–Sep

HESPERIDANTHUS

Per or subshrub, glabrous, glaucous. **ST**: gen branched distally, woody at base. **LF**: all cauline, ± fleshy, petioled to ± sessile, base not lobed, [entire] dentate. **INFL**: elongated. **FL**: sepals erect to ascending, base not sac-like; petals obovate to spoon-shaped, white to yellow, lavender, or purple, gen not clawed. **FR**: silique, dehiscent, unsegmented, linear, cylindric or ± flat parallel to septum; stigma entire or 2-lobed. **SEED**: 8–110 in 1 row, wingless. 5 spp.: w US, n Mex. (Greek: resembling fls of *Hesperis*) [Al-Shehbaz 2005 Harvard Pap Bot 10:47–51]

H. jaegeri (Rollins) Al-Shehbaz (p. 555) Per, occ subshrub, woody, branched at base, 1–3 dm, deep-rooted. **ST**: wavy. **LF**: 3–7 per st, (2)3–6 cm, 1–3.5 cm wide, ovate, coarsely dentate; petiole 1–2 cm. **FL**: sepals 5–7 mm, oblong, ± purple; petals 9–14 mm, 2.5–4 mm wide, spoon-shaped, white to purple, purple-veined. **FR**: (2)3–5 cm, 1–1.2 mm wide; style 0.7–1.5 mm; pedicel spreading to ascending, 6–14 mm. **SEED**: 26–42, 1.2–1.5 mm, oblong. Rocky crevices, cliffs, limestone clefts; 1500–2800 m. s W&I (Inyo Mtns). [*Caulostramina j.* (Rollins) Rollins] Marble, Teufel canyons, Cerro Gordo Peak in Inyo Co. Apr–Jun ★

HESPERIS ROCKET

Bien, per; hairs simple, stalked-forked, gen mixed with some stalked glands. **LF:** gen rosetted, entire to dentate or lobed; cauline petioled or sessile [base occ lobed or clasping st]. **INFL:** elongated; bracts 0 [throughout]. **FL:** sepals erect, inner pair sac-like at base; petals white to purple, yellow, brown, or green, clawed. **FR:** silique, linear, cylindric to 4-sided or flat parallel to septum, late-dehiscent, unsegmented; valves 1–3-veined; stigma deeply 2-lobed. **SEED:** 4–40, in 1 row; wing 0. 25 spp.: Eurasia. (Greek: evening, when some fls are most fragrant)

H. matronalis L. (p. 555) DAME'S ROCKET Bien (per). **ST:** 4–8(11) dm; hairs simple and forked at least proximally; glands gen 0. **LF:** basal and proximal cauline (2)4–15(20) cm, 0.8–4(6) cm wide, oblong to lanceolate or obovate, entire or minutely dentate; distal cauline petioled or ± sessile, base not lobed. **FL:** sepals 5–8 mm; petals (1.3)1.5–2(2.2) cm, 3.5–9 mm wide, purple, rose, or white; claw 6–12 mm. **FR:** (4)6–10(14) cm, 2–2.5 mm wide, ± cylindric, constricted between seeds; pedicel spreading to ascending, (0.5)0.7–1.7(2.5) cm. **SEED:** (2.5)3–4 mm, oblong. 2*n*=24. Roadsides, slopes, woodland; 700–2200 m. KR (Trinity Co.), n SNH (Plumas Co.); Can, US; native to Eur. Apr–Jul

HIRSCHFELDIA PERENNIAL, SHORTPOD, or SUMMER MUSTARD

1 sp.: Eurasia. (C.C.L. Hirschfeldt, Austrian horticulturist, 1742–1792) Generic status uncertain, possibly best incl in *Erucastrum* or expanded *Brassica*.

H. incana (L.) Lagr.-Fossat (p. 555) Ann to per; hairs simple. **ST:** (2)4–15(20) dm, hairs reflexed. **LF:** basal rosetted, pinnately lobed, (3)5–25(40) cm, obovate to lanceolate; lateral lobes 1–6(9) pairs; cauline dentate to pinnately lobed, petioled or sessile, base not lobed. **INFL:** much elongated. **FL:** sepals spreading to reflexed, base not sac-like; petals obovate, yellow, clawed. **FR:** silique, dehiscent, erect, appressed, 0.7–1.5(1.7) cm, 1–1.7 mm wide, cylindric, segmented, glabrous (sparsely hairy); valves obscurely 3(7)-veined; terminal segment 3–6 mm, 1–2-seeded; pedicel erect, 2–4(5) mm, stout. **SEED:** 10–22, in 1 row, 0.9–1.5 mm, ovoid; wing 0. 2*n*=14. Disturbed areas; < 2000 m. NCo, KR, NCoRI, CaR, SNF, GV, CW, SCo, PR, DMoj; OR, NV; native to Medit. Apr–Oct ❖

HORNUNGIA

Ann [per]; hairs 0 or mixture of simple and branched. **LF:** basal rosetted or not, pinnately lobed, dentate, or entire, petioled or sessile, bases not lobed. **INFL:** open. **FL:** sepals spreading or reflexed, not sac-like at base; petals not clawed. **FR:** silicle, ovate to elliptic, flat perpendicular to septum, dehiscent, unsegmented; stigma entire. **SEED:** [4]10–24, gen in 2 rows; wings 0. 3 spp.: Eurasia. (E.G. Hornung, German pharmacist in Schwarzburg, 1795–1862) [Appel & Al-Shehbaz 1998 Novon 7:338–340] *Hutchinsia* used by Rollins (1993) and in TJM (1993), but name illegitimate.

H. procumbens (L.) Hayek (p. 555) Ann. **ST:** decumbent to erect, branched at base (simple), (2)5–22(30) cm. **LF:** basal, proximal cauline petioled, (0.4)0.7–1.5(2.5) cm, obovate to oblanceolate, entire to toothed or lobed; distal cauline sessile. **FL:** sepals 0.6–1.1 mm; petals 0.6–1.2 mm, spoon-shaped, white. **FR:** (2)3–4(4.5) mm, elliptic to obovate or oblong; valve with prominent midvein; style to 0.1 mm; pedicel spreading, 3–8(12) mm. **SEED:** 10–24, 0.5–0.6 mm, oblong. 2*n*=12, 24. Saline flats, shaded sites, woodland, desert, meadows, salt marshes, sagebrush scrub; < 2900 m. CA (exc KR, n&c SNH); to BC, e Can, CO; native to Eurasia. [*Hutchinsia p.* (L.) Desv.] Feb–Jul

IDAHOA

1 sp. (State of Idaho)

I. scapigera (Hook.) A. Nelson & J.F. Macbr. (p. 555) Ann, scapose, glabrous. **LF:** basal, rosetted, (3)7–20(25) mm, ovate, obovate to spoon-shaped, entire or lyre-shaped; petiole (0.2)0.5–2(3) cm. **INFL:** peduncles many, (2)3–13 cm, 1-fld, slender. **FL:** sepals spreading to ascending, 1.5–1.8 mm, not sac-like at base; petals 2–2.5 mm, white. **FR:** silicle, flat parallel to septum, dehiscent, unsegmented, (5)6–9(10) mm, round to wide-ovate; style 0.3–0.8 mm, stigma entire; pedicel (1.5)2.5–11(13) cm. **SEED:** 6–10(12), in 2 rows, 3.5–5 mm wide; wing 0.8–1.2 mm wide. 2*n*=16. Moist ledges, slopes, meadows, foothills; 200–1700 m. NCoRI (Mount Saint Helena), CaR, n SN, s SNF, SnFrB (Mount Hamilton), SCoRI, GB; to BC, MT. Mar–May

ISATIS

[Ann] bien [per]; hairs simple or 0. **LF:** rosetted or not, petioled, entire or lobed; cauline sessile, base lobed to clasping. **INFL:** elongated. **FL:** sepals erect to ascending, base not sac-like; petals yellow or cream, not clawed. **FR:** indehiscent, pendent, oblong to oblanceolate or round to ovate, flat perpendicular to septum, unsegmented, winged; style 0, stigma entire. **SEED:** 1(2), oblong, plump. ± 60 spp.: Eurasia. (Greek: pl with dark dye)

I. tinctoria L. (p. 555) WOAD Bien, glaucous, glabrous or hairy proximally. **ST:** (3)4–10(15) dm, branched distally. **LF:** basal (3)6–20(25) cm, oblong to oblanceolate, entire to wavy or dentate; cauline entire. **INFL:** panicle of racemes. **FL:** sepals 1.5–2.8 mm; petals 2.5–4 mm, 0.9–1.5 mm wide, oblanceolate, yellow. **FR:** (0.9)1.1–2(2.7) cm, 3–6(10) mm wide, oblong-oblanceolate to elliptic-obovate; pedicel reflexed, 5–10 mm, club-shaped near tip. **SEED:** 2.3–3.5(4.5) mm, oblong. 2*n*=14,28. Disturbed areas, pastures; 100–2200 m. KR, CaR, n SNH, MP; w&e N.Am, native to Eurasia. Once cult as source of the blue dye woad. Apr–Jul ◆

LEPIDIUM PEPPERGRASS, PEPPERCRESS

Ann to per (shrub); hairs 0 or simple. **LF:** basal rosetted or not, petioled, entire, dentate, to 1–3-pinnately lobed; cauline short-petioled to sessile, base occ lobed to clasping. **INFL:** elongated or congested. **FL:** sepals erect or spreading, oblong to ovate,

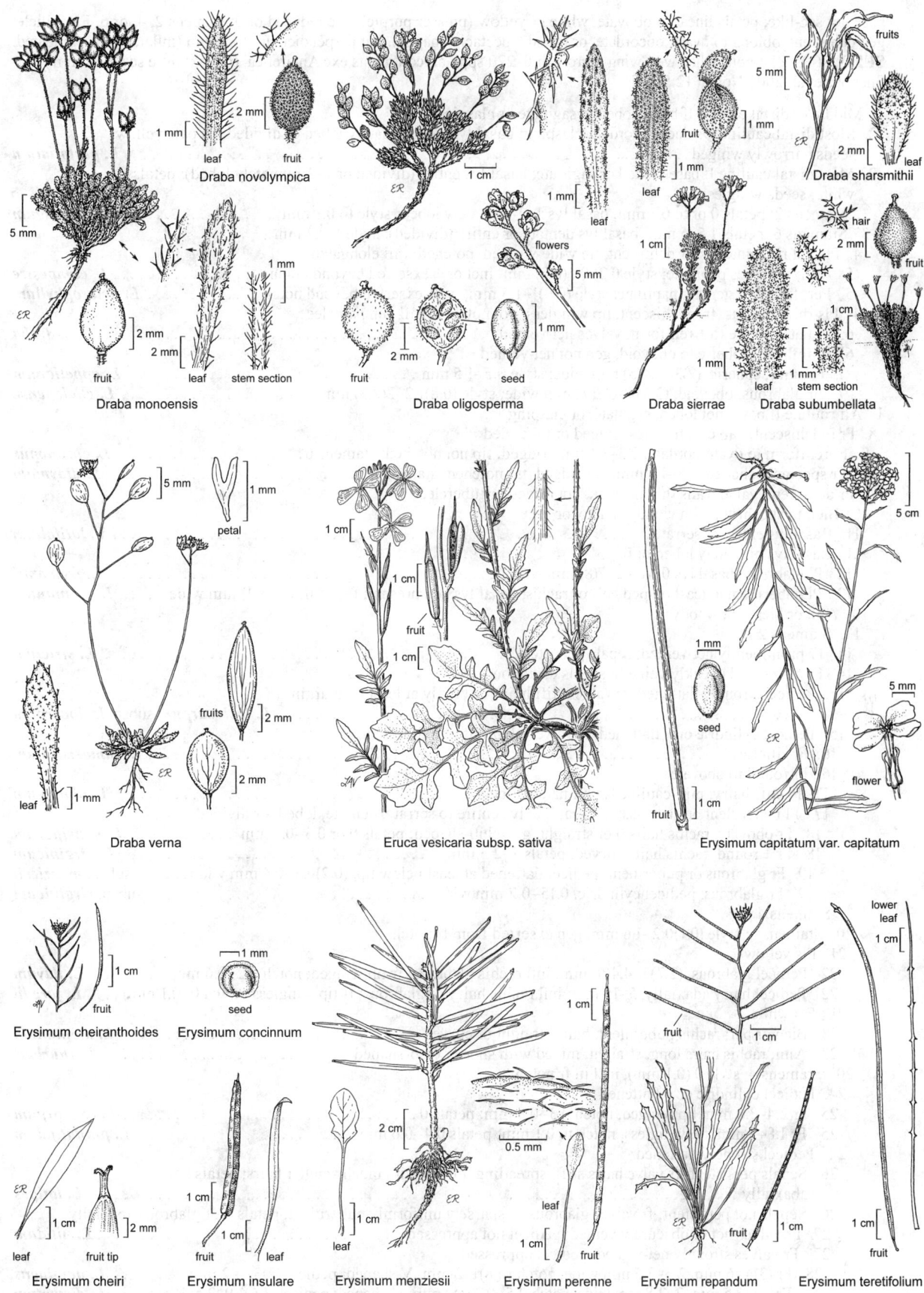

Draba novolympica

Draba monoensis

Draba oligosperma

Draba sierrae

Draba subumbellata

Draba sharsmithii

Draba verna

Eruca vesicaria subsp. sativa

Erysimum capitatum var. capitatum

Erysimum cheiranthoides

Erysimum concinnum

Erysimum cheiri

Erysimum insulare

Erysimum menziesii

Erysimum perenne

Erysimum repandum

Erysimum teretifolium

base not sac-like; petals linear to obovate, white or yellow (pink or purple), occ reduced or 0; stamens 2, 4, or 6. **FR**: silicle, gen dehiscent, oblong to ovate, obcordate, or round (spectacle-shaped), flat perpendicular to septum (inflated), unsegmented. **SEED**: 2(4), gelatinous when wet; wing narrow or 0. 220 spp.: all continents exc Antarctica. (Greek: little scale, from fr) [Al-Shehbaz et al. 2002 Novon 12:5–11]

1. Middle or distal cauline lf bases lobed to sagittate or clasping
  2. Most distal cauline lvs deeply cordate-clasping; basal lvs 2–3-pinnately lobed or divided; fls pale yellow;
    seeds narrowly winged . . . . **L. perfoliatum**
  2′ Most distal cauline lf bases lobed or sagittate; basal lvs entire (divided or 1–2-pinnately lobed); petals
    white; seeds wingless
    3. Stamens 2; petals 0 or to 0.7 mm; basal lvs 1–2-pinnately lobed; style to 0.1 mm . . . . ²**L. oblongum**
    3′ Stamens 6; petals 1.5–5 mm; basal lvs dentate or entire (divided); style ≥ 0.2 mm
      4. Pls not rhizomatous; fr dehiscent, tip wide-winged, notched; infl elongated
        5. Ann; fr valves papillate; style 0.2–0.5(0.7) mm, incl or ± exserted beyond notch . . . . **L. campestre**
        5′ Per; fr valves gen not papillate; style (0.6)1–1.5 mm, well-exserted beyond notch . . . . **L. heterophyllum**
      4′ Pls rhizomatous; fr indehiscent, tip wingless, not notched; infl gen panicles
        6. Fr flat, cordate to ± reniform; valves net-veined . . . . **L. draba**
        6′ Fr inflated, spheric to obovoid, gen not net-veined
          7. Fr hairy, (2)3–4.4(5) mm wide; style 0.5–1.5 mm . . . . **L. appelianum**
          7′ Fr glabrous, obovoid, (3.5)4–6.2(7) mm wide; style (0.8)1.2–2(2.3) mm . . . . **L. chalepense**
1′ All cauline lf bases not lobed, sagittate, or clasping
  8. Fr indehiscent, valve walls thick, ridged or tubercled
    9. Fr reniform to ovate-cordate, 2.3–3.4 mm, ridged, tip not notched; stamens 6 . . . . **L. coronopus**
    9′ Fr spectacle-shaped, 1.3–1.7 mm, not ridged, tip notched; stamens 2 . . . . **L. didymum**
  8′ Fr dehiscent, valve walls thin or thick, not ridged or tubercled
    10. Bien to subshrub, with woody caudex or base
      11. Basal lvs entire or serrate . . . . **L. latifolium**
      11′ Basal lvs pinnately lobed or 0
        12. Pl glabrous; basal lvs 0; fr 4.2–7(8) mm wide . . . . **L. fremontii**
        12′ Pl puberulent at least on pedicels or rachis; basal lvs gen present; fr (1.5)1.8–3.6(4) mm wide . . . . ²**L. montanum**
    10′ Ann or bien, not woody
      13. Stamens 2
        14. Fr prominently net-veined; sepals persistent . . . . **L. strictum**
        14′ Fr not veined (weakly veined); sepals deciduous
          15. Pedicel strongly flattened; fr valves stiff hairy or bristly at least on margin
          . . . . **L. lasiocarpum** subsp. **lasiocarpum**
          15′ Pedicel cylindric or ± flattened; fr valves glabrous or puberulent
            16. Fr elliptic . . . . **L. ramosissimum**
            16′ Fr round to obovate
              17. Pl stiff hairy; mid-cauline lvs pinnately lobed . . . . ²**L. oblongum**
              17′ Pl puberulent or glabrous; mid-cauline lvs entire to serrate, dentate, lobed, or divided
                18. Fr obovate; rachis hairs 0 or straight, gen club-shaped; petals 0 or 0.3–0.9 mm . . . . **L. densiflorum**
                18′ Fr ± round; rachis hairs curved; petals 1–2.5 mm . . . . **L. virginicum**
                  19. Fr glabrous or puberulent; pedicel flattened at least below tip, (0.2)0.3–0.4 mm wide . . . . . . subsp. **menziesii**
                  19′ Fr glabrous; pedicel cylindric, 0.15–0.2 mm wide . . . . subsp. **virginicum**
      13′ Stamens 4 or 6
        20. Stamens 6; style (0.1)0.2–1.6 mm, gen exserted from fr notch
          21. Fls yellow
             22. Pedicel glabrous, (2.7)3–4.4(5) mm; infl rachis glabrous; fr tip winged; notch 0.2–0.6 mm . . . . **L. flavum**
            22′ Pedicel hairy adaxially, 5–15 mm; infl rachis hairy or glabrous; fr tip wingless, notch 0 (0.1 mm) . . . . **L. jaredii**
          21′ Fls white
             23. Bien or per; rachis puberulent, hairs straight or curved . . . . ²**L. montanum**
            23′ Ann; rachis hairs long, straight, mixed with shorter club-shaped . . . . **L. thurberi**
        20′ Stamens 4; style 0 (0.1 mm), incl in fr notch
          24. Pedicel cylindric or ± flattened
             25. Fr 2.4–3.6 mm, tip winged, notch 0.3–0.8 mm; petals 0 . . . . **L. oxycarpum**
            25′ Fr 1.8–2 mm, tip wingless, notch to 0.1 mm; petals 0.4–0.6 mm . . . . **L. pinnatifidum**
          24′ Pedicel strongly flattened
             26. Sepals persistent; fr valve hairs stiff, spreading, mixed with much smaller hairs; petals hairy
              abaxially . . . . **L. latipes**
            26′ Sepals not persistent; fr valves glabrous or sparsely uniformly puberulent; petals 0 or glabrous abaxially
              27. Fr valves not or obscurely veined; pedicel not appressed . . . . **L. nitidum**
              27′ Fr valves strongly net-veined; pedicel appressed
                28. Fr (3)4–6 mm, 2.5–3.5 mm wide, notch (0.8)1–2 mm, V-shaped; pedicel (2)3–4.2 mm . . . . **L. acutidens**
                28′ Fr 2.5–3.5 mm, 2–2.8 mm wide, notch 0.5–0.7(0.8) mm, U-shaped; pedicel (1.6)1.9–2.5(3) mm . . . . **L. dictyotum**

L. acutidens (A. Gray) Howell Ann, puberulent or coarsely hairy. **ST**: erect to ascending, few to many from base, (0.5)0.8–3 dm, simple. **LF**: basal not rosetted, early-deciduous, 2–7 cm, linear and undivided, or pinnately divided; mid-cauline sessile, 1.2–5.8 cm, 0.5–2(3) mm wide, linear, entire, base tapered, not lobed. **INFL**: rachis hairs straight, cylindric. **FL**: sepals 0.7–1.1 mm; petals 0; stamens 4. **FR**: (3)4–6 mm, 2.5–3.5 mm wide, ovate to ovate-oblong, flat, tip winged; notch (0.8)1–2 mm, V-shaped; valves glabrous (hairy), strongly net-veined; style 0; pedicel erect to ascending, (2)3–4.2 mm, strongly flattened, puberulent. Alkaline flats, fields, stream beds; < 1500 m. NCoRI, CaR, GV, SW, MP; to OR, Baja CA. [*L. dictyotum* var. *a.* A. Gray] Feb–Apr

L. appelianum Al-Shehbaz (p. 555) WHITE-TOP Per, rhizomed, densely stiff-hairy. **ST**: erect or ascending, 1–several from base, (1)1.5–3.5(5) dm, branched distally. **LF**: basal not rosetted, early-deciduous, 1–7 cm, obovate to oblanceolate; mid-cauline 1–5(8) cm, 0.5–1.5(3) cm wide, oblong to lanceolate, base lobed to clasping, dentate to ± entire. **INFL**: panicle, gen not elongated; rachis hairs curved. **FL**: sepals 1.4–2 mm; petals (2.2)2.8–4 mm, 1–3 mm wide, widely obovate, white; stamens 6. **FR**: indehiscent, (2)3–4.4(5) mm wide, spheric, inflated, tip wingless, notch 0; valve walls thin, densely puberulent, not veined; style 0.5–1.5 mm; pedicel spreading to ascending, 3–9(12) mm, cylindric, hairy. **SEED**: 1.5–2 mm, ovoid. 2*n*=16. Saline soils, fields; 400–2400 m. CA; to e N.Am; native to c Asia. [*Cardaria pubescens* (C.A. Mey.) Jarm.] Apr–Sep ◆

L. campestre (L.) W.T. Aiton (p. 555) Ann; hairs dense, stiff. **ST**: (0.8)1.2–5(6.3) dm, erect, simple or branched distally. **LF**: basal rosetted, (1)2–6(8) cm, oblanceolate or oblong, entire to pinnately lobed; mid-cauline sessile, (0.7)1–4(6.5) cm, (2)5–10(15) mm wide, oblong to lanceolate, dentate or ± entire, base sagittate or lobed. **INFL**: much elongated, rachis stiff-straight-hairy. **FL**: 1–1.8 mm; petals (1.5)1.8–2.5(3) mm, (0.2)0.5–0.7 mm wide, spoon-shaped, white; stamens 6. **FR**: (4)5–6(6.5) mm, (3)4–5 mm wide, widely oblong to ovate, flat, broadly winged at tip, notch (0.2)0.4–0.6 mm; valve walls thin, papillate, not veined; style 0.2–0.5(0.7) mm, incl to ± exserted in notch; pedicel (3)4–8(10) mm, cylindric, spreading, hairy. **SEED**: 2–2.8 mm, ovoid. 2*n*=16. Common. Disturbed areas, fields, woodland, meadows; 100–2600 m. KR, CaR, n SN; widespread N.Am; native to Eur. Potentially problematic weed. May–Jul

L. chalepense L. (p. 555) LENS-PODDED HOARY CRESS Per, rhizomed, densely stiff-hairy, occ glabrous distally. **ST**: erect or decumbent proximally, (0.8)2.1–6.6(9.2) dm, many-branched distally. **LF**: basal not rosetted, early-deciduous, 2–9(14) cm, obovate to spoon-shaped or ovate; mid-cauline (1.5)2.6–9.3(13.2) cm, (0.7)1.2–3.1(4.5) cm wide, oblong to lanceolate, obovate, or oblanceolate, base lobed to clasping, entire to ± dentate. **INFL**: panicle, elongated; rachis glabrous or puberulent, hairs curved or straight. **FL**: sepals 1.7–3 mm; petals 3–5 mm, 1.2–2.4 mm wide, obovate, white; stamens 6. **FR**: 3.5–5.8(7) mm, (3.5)4–6.2(7) mm wide, obovoid, tip wingless, notch 0; valve walls thin, smooth, glabrous, gen not veined; style (0.8)1.2–2(2.3) mm; pedicel ascending to spreading, 5–16(19) mm, cylindric, glabrous or sparsely puberulent adaxially. **SEED**: 1.5–2.3 mm, ovate. 2*n*=48,80,128. Disturbed areas, pastures, fields, riverbanks; 300–4200 m. CA-FP, GB; Can, w&c US; native to w&c Asia. [*Cardaria c.* (L.) Hand.-Mazz.] May–Jun ◆

L. coronopus (L.) Al-Shehbaz (p. 555) SWINE CRESS Ann, glabrous or puberulent. **ST**: prostrate to decumbent, gen several from base, (0.3)0.6–2.5(3.5) dm, branched distally. **LF**: basal rosetted, (3)4–10(15) cm, 1–2-pinnately divided, lobes entire or dentate; mid-cauline petioled, pinnately divided, base cuneate, lobes entire or dentate. **INFL**: lf-opposed, not elongated, rachis glabrous. **FL**: sepals 1–1.5 mm, persistent; petals, white, obovate, 1–2 mm, 0.4–0.6 mm wide; stamens 6. **FR**: indehiscent, reniform to ovate-cordate, flattened, 2.3–3.4 mm, 3–4.4 mm wide, wingless, notch 0; valve walls thick, distinctly ridged, glabrous, prominently veined; style 0.2–0.7 mm; pedicel (0.7)1–2(2.4) mm, stout, cylindric, ascending, glabrous. **SEED**: 1.2–1.6 mm, oblong. 2*n*=32. Disturbed areas, fields, pastures; < 300 m. ScV, SnFrB, DSon; e Can, c&e US; native to s Eur, sw Asia, n Afr. [*Coronopus squamatus* (Forssk.) Asch.] Mar–Jun ◆

L. densiflorum Schrad. (p. 555) Ann, bien, puberulent or glabrous. **ST**: erect, gen 1, (1)2.5–5(6.5) dm, branched distally. **LF**: (1.5)2.5–8(11) cm, oblanceolate, spoon-shaped, or oblong, serrate or pinnately lobed, early-deciduous; mid-cauline petioled, (0.7)1.3–6.2(8) cm, (0.5)1.5–10(18) mm wide, narrowly oblanceolate or linear, tapered at base, not lobed, entire or dentate. **INFL**: much-elongated, rachis glabrous or puberulent, hairs club-shaped. **FL**: sepals 0.5–0.8(1) mm; petals 0 or thread-like, 0.3–0.9 mm, white; stamens 2. **FR**: (2)2.5–3(3.5) mm, 1.5–2.5(3) mm wide, obovate, flat, tip winged, notch 0.2–0.4 mm; valves not veined, glabrous or puberulent; pedicel (1.5)2–3.5(4) mm, cylindric, spreading to ± ascending, puberulent adaxially. **SEED**: 1–1.3 mm, ovate. Fields, pastures, meadows, disturbed sites, floodplains, chaparral; < 3500 m. KR, CaR, n&c SNH, GV, SCo, GB, DMoj; N.Am; naturalized in Eur, Asia, S.Am. [*L. d.* var. *elongatum* (Rydb.) Thell.; *L. d.* var. *macrocarpum* G.A. Mulligan; *L. d.* var. *pubicarpum* (A. Nelson) Thell.; *L. d.* var. *ramosum* (A. Nelson) Thell.] Highly variable. May–Jul

L. dictyotum A. Gray (p. 559) Ann; hairs stiff, spreading. **ST**: erect to ascending, 0.2–1.3(2) dm, few to many from base, outer decumbent, simple. **LF**: basal not rosetted, early-deciduous, (1.5)2.2–5.7(7) cm, pinnately lobed or divided, lobes linear to narrowly oblong, entire; mid-cauline sessile, blade 1–5 cm, 0.5–2 mm wide, linear, entire (linear lobed), base tapered, not lobed. **INFL**: elongate, rachis hairs stiff, cylindric, straight. **FL**: sepals 0.7–1.1 mm; petals 0; stamens 4. **FR**: 2.5–3.5 mm, 2–2.8 mm wide, ovate, flat, tip winged, notch 0.5–0.7(0.8) mm, U-shaped; valve walls thin, stiff-hairy (glabrous), strongly net-veined; style 0; pedicel (1.6)1.9–2.5(3) mm, strongly flattened, erect to ± ascending, stiff-hairy throughout (only adaxially). **SEED**: 1.2–1.8 mm, ovate. Saline soils, dry stream beds, roadsides, sandy flats, fields, meadows, dried pools; < 1600 m. NCoRI, SNF, GV, SCoRI, e SCo, n WTR, GB, w DMoj; to WA, UT, Baja CA. Mar–Jun

L. didymum L. (p. 559) LESSER SWINE CRESS Ann, fetid, glabrous or hairy. **ST**: erect to decumbent, few to many from base, 1–4.5(7) dm, branched distally. **LF**: basal not rosetted, early-deciduous, 1–6(8) cm, 1–2-pinnately divided, lobes entire to lobed; mid-cauline short-petioled to ± sessile, 1.5–3.5(4.5) cm, 0.5–1.2 cm wide, gradually reduced and less divided distally, lobes entire to serrate. **INFL**: elongated, rachis glabrous or hairs straight, cylindric. **FL**: sepals 0.5–0.7(0.9) mm; petals 0.4–0.5 mm, elliptic to linear, white; stamens 2. **FR**: 1.3–1.7 mm, 2–2.5 mm wide, spectacle-shaped, flat, wingless, notch 0.2–0.4 mm; valve walls thick, tubercled, glabrous, strongly veined; style 0; pedicel spreading, 1.4–2.5(4) mm, cylindric, glabrous or hairy adaxially. **SEED**: 1–1.2 mm, ovate. 2*n*=32. Common. Disturbed areas, fields, pastures; < 1000 m. CA; N.Am; native to S.Am, introduced worldwide. [*Coronopus d.* (L.) Sm.] Mar–Jul

L. draba L. (p. 559) HEART-PODDED HOARY CRESS Per, rhizomed, densely stiff-hairy or glabrous. **ST**: gen 1, (0.8)2–6.5(9) dm, many branched distally. **LF**: basal not rosetted, early-deciduous, (1.5)3–10(15) cm, obovate to spoon-shaped or ovate, margin dentate or entire; mid-cauline (1)3–9(15) cm, (0.5)1–2(5) cm wide, ovate to elliptic, lanceolate, or obovate, sessile, clasping or base lobed, dentate or entire. **INFL**: panicle, elongated or not; rachis glabrous or hairs curved to straight. **FL**: sepals 1.5–2.5 mm; petals (2.5)3–4(4.5) mm, (1)1.3–2(2.2) mm wide, obovate, white; stamens 6. **FR**: indehiscent, (2)2.5–3.7(4.3) mm, (3.2)3.7–5(5.6) mm wide, cordate to ± reniform, flat, tip obtuse to tapered, wingless, notch 0; valves glabrous, net-veined; style (0.6)1–1.8(2) mm; pedicel ascending to spreading, 5–10(15) mm, cylindric, glabrous or puberulent adaxially. **SEED**: 1.5–2.3 mm, ovate. 2*n*=32,64. Disturbed areas, saline soils, pastures, fields; < 3300 m. CA-FP, GB; Can, c&e US; native to Eurasia. [*Cardaria d.* (L.) Desv.] Apr–Aug ◆

L. flavum Torr. (p. 559) Ann, glabrous. **ST**: prostrate or decumbent, (0.2)0.4–3(4.6) dm, base branched. **LF**: basal rosetted, (0.7)1.3–5.2(6) cm, spoon-shaped to oblanceolate or linear, pinnately lobed, lobes ovate to oblong, entire; mid-cauline petioled, obovate to spoon-shaped or oblanceolate, (0.6)1–1.8(2.3) cm, base not lobed, dentate to entire. **INFL**: not elongated, rachis glabrous. **FL**: sepals 1–2 mm; petals 2–3 mm, 0.6–1 mm wide, spoon-shaped, yellow; stamens 6. **FR**: (2.2)2.5–3.8(4.2) mm, (1.6)2.2–3.2(3.5) mm wide, ovate (± round), flat, tip divergent-winged, notch 0.2–0.6 mm; valves glabrous (hairy), ± net-veined; style 0.7–1.6 mm, exserted beyond notch; pedicel spreading, (2.7)3–4.4(5) mm, cylindric, glabrous. **SEED**: 1–1.6 cm, ovate. Alkaline or sandy soils, sagebrush scrub,

mesas, floodplains, washes, roadsides; 600–1600 m. SNE, D; NV, Baja CA. [*L. f.* var. *felipense* C.L. Hitchc.] Mar–Jun

L. fremontii S. Watson (p. 559) Per, gen shrubby, glabrous, glaucous. **ST**: erect or ascending, several from woody base, 2–5.5(10) dm, many-branched distally. **LF**: mid-cauline (1.5)2.2–8.4(10.2) cm, pinnately 3–7(9)-lobed, or linear and undivided, base tapered, not lobed, lobes (0.7)1–2.8(4.2) mm wide, linear, entire; distal linear, entire. **INFL**: elongated panicle, rachis glabrous. **FL**: sepals 1.5–2.5(3) mm; petals 2.5–4.2 mm, 1.5–2.2 mm wide, spoon-shaped, white; stamens 6. **FR**: (4)4.5–7(8) mm, 4.2–7(8) mm wide, obovate to round, flat, tip winged, notch (0.1)0.2–0.5 mm; valves glabrous, not veined; style 0.2–0.8(1) mm, exserted beyond notch; pedicel spreading, (3.5)4.3–7.6(8.5) mm, cylindric, glabrous. **SEED**: 1.6–2.1 mm, ovate. Sandy or gravelly soils, washes, barren knolls, bluffs, pinyon/juniper woodland, rocky slopes; 450–2100 m. se SNH, SnGb, SnBr, s SNE, W&I, D; to s UT, w AZ. [*L. f.* var. *stipitatum* Rollins] Mar–Jun

L. heterophyllum Benth. Per, stiff-hairy, caudex branched. **ST**: erect to ascending, gen decumbent at base, 1–5 dm, simple or few-branched distally. **LF**: basal rosetted, oblanceolate or oblong-elliptic, 1–4.5 cm, entire to fine-dentate; mid-cauline oblong to lanceolate, 1–3.5 cm, 3–8 mm wide, dentate, base sagittate or lobed. **FL**: sepals 1.6–2.2 mm; petals 1.8–2.8 mm, 0.8–1.4 mm wide, spoon-shaped, white; stamens 6. **FR**: 4–5.5 mm, 3.5–4 mm wide, broadly oblong to ovate, flat, tip broadly winged, notch 0.2–0.3 mm; valves gen not papillate, veins 0; style (0.6)1–1.5 mm, well exserted beyond notch; pedicel spreading, 2.8–5 mm, cylindric, stiff hairy. **SEED**: 1.8–2.2 mm, ovoid. Uncommon. Fields, roadsides, open slopes; 1400–1500 m. KR, CaR; to BC, e N.Am; native to Eur. May–Jun

L. jaredii Brandegee (p. 559) JARED'S PEPPERGRASS Ann, sparsely soft-hairy. **ST**: erect to ascending, 1 to several from base, 1–6(7) dm, branched distally. **LF**: basal not rosetted, early-deciduous; mid-cauline 2–7.5(10) cm, 0.2–1 cm wide, lanceolate to linear, entire or few toothed near tip, base tapered, not lobed. **INFL**: much-elongated; rachis soft-curved-hairy, or glabrous. **FL**: sepals 1.8–2.5 mm; petals 2.8–4 mm, 1.2–1.8 mm wide, spoon-shaped, lemon yellow, ± white in age; stamens 6. **FR**: 3–3.8(4) mm, 2.8–3.2(3.5) mm wide, widely ovate, flat, tip wingless, notch 0 (0.1 mm); valves smooth or minutely papillate, glabrous, obscurely veined; style 0.3–0.8(1) mm; pedicel 5–15 mm, cylindric, spreading, hairy adaxially. **SEED**: 1.8–2.2 mm, oblong. 2*n*=16. Alkali bottoms, slopes, washes, dry hillsides, vertic clay, acidic, gypsiferous; 500–700 m. sw SnJV, se SCoRI. Mar–Apr ★

L. lasiocarpum Nutt. subsp. ***lasiocarpum*** (p. 559) Ann; hairs stiff, spreading. **ST**: erect to decumbent, 1 to several from base, (0.2)0.6–3(3.8) dm, branched distally. **LF**: basal not rosetted, late-deciduous, (0.7)1.5–4.5(7.5), spoon-shaped to oblanceolate, 1–2-pinnately lobed or dentate; mid-cauline short-petioled, (0.7)1.2–3.3(5) cm, (2)4–12 mm wide, lanceolate to oblanceolate, ± entire or dentate, not lobed. **INFL**: much-elongated; rachis spreading stiff-hairy or bristly. **FL**: sepals 1–1.3(1.5) mm, oblong; petals 0 or (0.3)0.6–1.5(2) mm, (0.1)0.2–0.5 mm wide, oblanceolate to linear, white; stamens 2. **FR**: 2.8–4(4.6) mm, 2.4–3.6(4) mm wide, ovate to ovate-round, flat, tip winged, notch (0.2)0.3–0.6(0.7) mm; valves stiff-hairy or bristly at least on margin, not veined; style 0–0.1 mm, incl in notch; pedicel spreading, (1.8)2–4(4.6) mm, strongly flattened, hairy at least adaxially. **SEED**: 1.4–2.2 mm, ovate. 2*n*=32. Dry flats, washes, streambeds, roadsides, sagebrush scrub, pinyon/juniper woodland; 50–2700 m. SW, W&I, D; to CO, TX, Mex. [*L. l.* var. *l.*] Mar–Jun

L. latifolium L. (p. 559) PERENNIAL PEPPERWEED Per, rhizomed, glabrous or hairy. **ST**: erect, (2)3.5–12(15) dm, branched distally. **LF**: basal not rosetted, (2)3.5–15(25) cm, elliptic-ovate to oblong, entire or serrate, leathery; mid-cauline sessile or short-petioled, (1)2–9(12) cm, 0.3–4.5 cm wide, oblong to lanceolate, entire or serrate, base not lobed. **INFL**: panicle, ± elongated or not; rachis glabrous or puberulent, hairs cylindric. **FL**: sepals 1–1.4 mm; petals 1.8–2.5 mm, 0.8–1.3 mm wide, obovate, white; stamens 6. **FR**: (1.6)1.8–2.4(2.7) mm, 1.3–1.8 mm wide, oblong-elliptic to broadly ovate or ± round, flat, wingless, notch 0 (0.1 mm); valves glabrous or soft-hairy, not veined; style 0.05–0.15 mm, exserted beyond notch; pedicel spreading to ascending, 2–5(6) mm, cylindric, glabrous or puberulent adaxially.

SEED: 0.8–1.2 mm, oblong. 2*n*=24. Pastures, disturbed areas, fields, grassland, saline meadows, streambanks, sagebrush scrub, pinyon/juniper woodland, edge of marshes; < 2500 m. CA (exc KR, D); s Can, US; native to Eurasia. Jun–Sep ◆

L. latipes Hook. (p. 559) Ann, puberulent or stiff-hairy. **ST**: erect decumbent, 1 or several from base, 0.2–1.5(3.8) dm, simple or branched. **LF**: basal not rosetted, early-deciduous, 2–10 cm, linear, entire, dentate, or pinnately divided into 2–10 lobe pairs; mid-cauline similar, smaller, entire, base tapered, not lobed. **INFL**: compact, cylindric to head-like, elongated or not; rachis puberulent, hairs cylindric. **FL**: sepals 1.1–1.4 mm, ovate; petals 1.9–3 mm, 0.8–1.3 mm wide, obovate-oblong, ± green, hairy abaxially, margin fringed; stamens 4. **FR**: 5–7 mm, 2.8–4 mm wide, oblong-ovate, tip winged, notch 1.4–2.8 mm; valve walls thick, with a mixture of long, much shorter hairs, strongly net-veined; style 0; pedicel 2.5–5 mm, strongly flattened, appressed, erect to ± ascending, puberulent throughout (only adaxially). **SEED**: 2–2.4 mm, oblong. Alkaline soils, vernal pool margins, salt marsh edges, pastures; < 700 m. NCo, NCoR, GV, CCo, SnFrB, SCoRI, SCo; Baja CA. [*L. l.* var. *heckardii* Rollins] Mar–Jun

L. montanum Nutt. Bien, per, occ cespitose or base woody, glabrous or hairy. **ST**: 1 to many from base, 0.4–5(7) dm, gen many-branched distally. **LF**: rosetted or not, (0.8)1.5–4(6) cm, (0)1–2-pinnately lobed or divided, lobes entire or dentate; mid-cauline short-petioled, similar to basal or undivided and linear, base tapered, not lobed. **INFL**: much-elongated; rachis puberulent, hairs straight or curved (glabrous). **FL**: sepals 1.2–1.8(2.1) mm; petals 2.2–3.7(4.3) mm, 1.3–1.8 mm wide, spoon-shaped to oblanceolate, white; stamens 6. **FR**: 2–4.3(5) mm, (1.5)1.8–3.6(4) mm wide, ovate to ± round (oblong), flat, tip winged, notch 0.1–0.3 mm; valves glabrous (puberulent), not veined; style 0.2–0.7(0.9) mm, exserted beyond notch; pedicel (2.7)3.3–8.5(10) mm, cylindric, spreading, puberulent adaxially. **SEED**: 1.2–1.8 mm, oblong. 2*n*=32. Sandy, gravelly, gen saline soils, pinyon/juniper woodland, sagebrush scrub, rocky hillsides, roadsides; 800–2700 m. CaRH, GB, DMoj; to OR, MT, AZ, n Mex. [*L. m.* var. *canescens* (Thell.) C.L. Hitchc.; *L. m.* var. *cinereum* (C.L. Hitchc.) Rollins] Highly variable, needs study. Many vars., subspp., based on trivial characters. Apr–Aug

L. nitidum Nutt. (p. 559) Ann, puberulent. **ST**: erect to decumbent, 1–many from base, (0.4)1–3.5(4.2) dm, occ branched distally. **LF**: basal not rosetted, early-deciduous, (0.8)1.5–7.3(8) cm, pinnately divided, lobes linear (oblong to lanceolate), entire (dentate); mid-cauline similar to basal but smaller, occ undivided, linear, petioled to sessile, base not lobed. **INFL**: much-elongated; rachis puberulent, hairs straight. **FL**: sepals (0.7)0.9–1.3(1.5) mm; petals 0 or (0.8)1.2–2(2.8) mm, 0.2–1(1.6) mm wide, oblanceolate, white; stamens 4. **FR**: (2.5)3–5.5(6.5) mm, (2)2.6–5(5.4) mm wide, round to broadly ovate, flat, tip winged, notch 0.3–0.7(1) mm; valves glabrous or margin sparsely puberulent, not or obscurely veined; style 0–0.1 mm, incl in notch; pedicel spreading (± erect), 2.5–5(6.5) mm, strongly flattened, puberulent or only adaxially. **SEED**: 1.6–2.6 mm, ovate. Alkaline soils, meadows, pastures, dry vernal pools, fields, beaches; < 1000 m. CA (exc e D); to WA, Baja CA. [*L. n.* var. *howellii* C.L. Hitchc.; *L. n.* var. *oreganum* (Greene) C.L. Hitchc.] Feb–Mar

L. oblongum Small Ann, stiff hairy. **ST**: erect to decumbent, gen several from base, (5)1–2.4(3.2) dm, branched distally. **LF**: basal not rosetted, 0.7–3.5 cm, 1–2-pinnately lobed, lobes entire or dentate; mid-cauline gen sessile, 0.8–2 cm, pinnately lobed, obovate to oblanceolate, base lobed or not. **INFL**: elongated; rachis hairs straight. **FL**: sepals 0.7–1 mm; petals 0 or 0.1–0.7 mm, 0.05–0.15 mm wide, linear-oblanceolate, white; stamens 2. **FR**: 2.2–3.5 mm, 2–3 mm wide, round to widely obovate or elliptic, flat, tip winged, notch 0.2–0.3 mm; valves glabrous or sparsely puberulent along margin, not veined; style to 0.1 mm, incl in notch; pedicel 2–3.5(5) mm, cylindric, spreading, puberulent adaxially (throughout). **SEED**: 1.2–1.6 mm, ovate. Bluffs, slopes, disturbed areas, roadsides, flats, pastures; 200–500 m. s CCo, SW; to c&s US, C.Am. [*L. o.* var. *insulare* C.L. Hitchc.] Mar–Aug

L. oxycarpum Torr. & A. Gray (p. 559) Ann, glabrous or puberulent. **ST**: erect to decumbent, several from base, 0.4–1.5(2) dm,

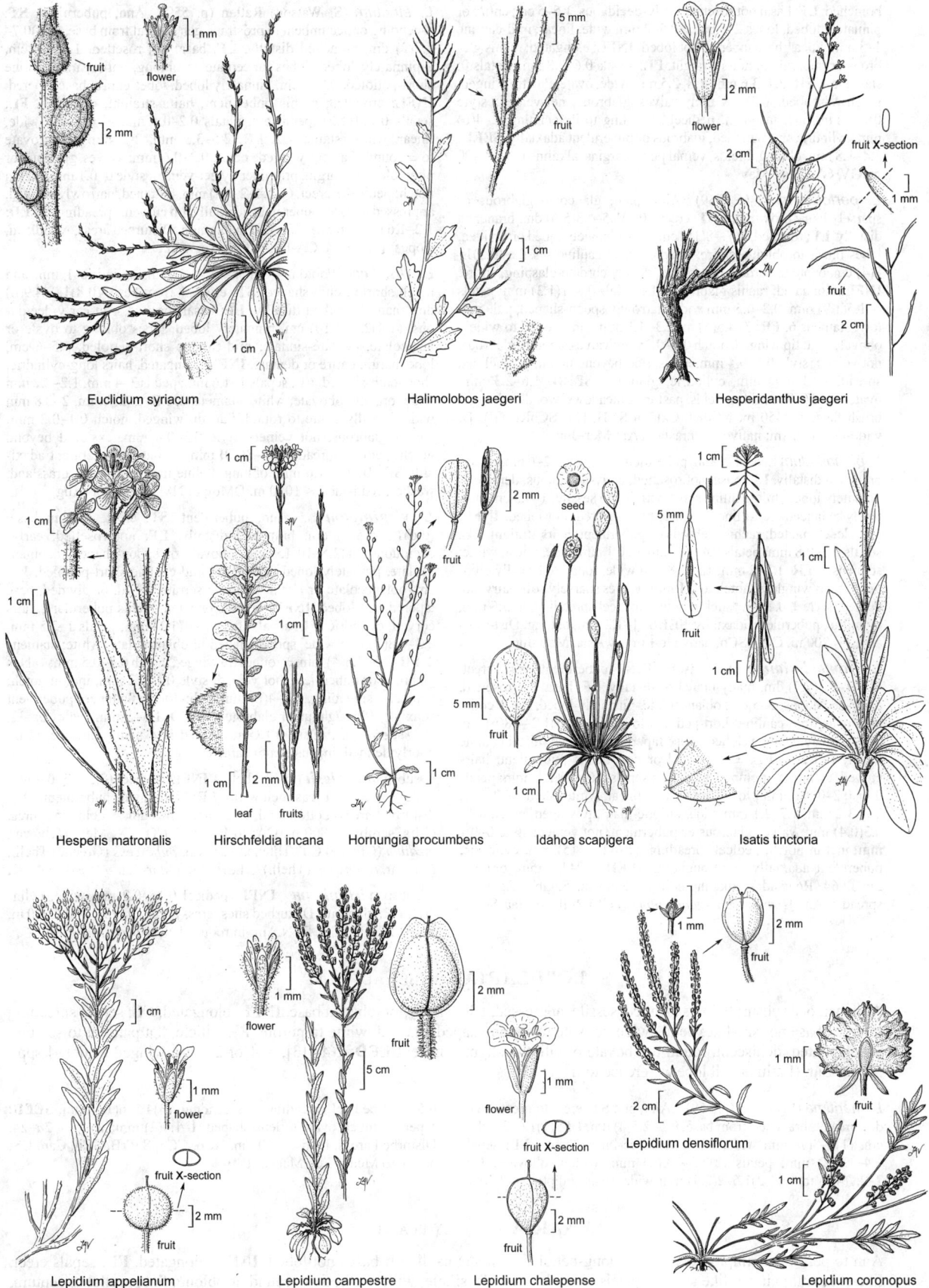

Euclidium syriacum

Halimolobos jaegeri

Hesperidanthus jaegeri

Hesperis matronalis

Hirschfeldia incana

Hornungia procumbens

Idahoa scapigera

Isatis tinctoria

Lepidium appelianum

Lepidium campestre

Lepidium chalepense

Lepidium densiflorum

Lepidium coronopus

branched. **LF**: basal not rosetted, early-deciduous, 1.5–5 cm, entire or pinnately lobed, lobes 2–5 pairs, 0.5–2 mm wide, linear; mid-cauline 1–3 mm, linear, base tapered, not lobed. **INFL**: elongated; rachis glabrous or puberulent, hairs straight. **FL**: sepals 0.6–0.8 mm; petals 0; stamens 4. **FR**: 2.4–3.6 mm, 1.8–2.5 mm wide, ovate, flat, tip winged, notch V-shaped, 0.3–0.8 mm; valves glabrous, net-veined; style 0–0.1 mm, incl in notch; pedicel spreading to descending, 2–4(6) mm, cylindric or ± flattened, glabrous or puberulent adaxially. **SEED**: 1.4–1.8 mm, oblong. Fields, vernal pool margins, alkaline flats; < 400 m. GV, CW. Mar–May

L. perfoliatum L. (p. 559) Ann, bien, glaucous, glabrous or sparsely hairy proximally. **ST**: erect, (0.7)1.5–4.3(5.6) dm, branched distally. **LF**: rosetted, (1)3–8(15) cm, 2–3-pinnately lobed or divided, lobes linear to oblong, entire; middle, distal cauline sessile, (0.5)1–3(4) cm, ovate to cordate or ± round, deeply cordate-clasping, entire. **INFL**: elongated; rachis glabrous. **FL**: sepals 0.8–1(1.3) mm; petals 1–1.5(1.9) mm, 0.2–0.5 mm wide, narrowly spoon-shaped, pale yellow; stamens 6. **FR**: 3–4.5(5) mm, 3–4.1 mm wide, round to widely obovate, flat, tip winged, notch 0.1–0.3 mm; valves smooth, glabrous, not veined; style 0.1–0.4 mm, ± exserted beyond notch; pedicel gen spreading, 3–6(7) mm, cylindric, glabrous. **SEED**: 1.6–2.3 mm, ovate. 2*n*=16. Roadsides, fields, pastures, meadows, woodland, sagebrush flats; < 2450 m. NCoRI, CaR, n SNH, GV, SCoRI, GB, D; widespread N.Am; native to Eurasia, n Afr. Mar–Jun

L. pinnatifidum Ledeb. Ann, puberulent. **ST**: erect, 2–6 dm, many-branched distally. **LF**: basal not rosetted, early-deciduous, dentate to pinnately lobed; mid-cauline short-petioled to sessile, 1–3.3 cm, narrowly oblanceolate to linear, entire, base tapered, not lobed. **INFL**: panicle, elongated; rachis glabrous or puberulent, hairs straight. **FL**: sepals 0.7–0.8 mm; petals 0.4–0.6 mm, to 0.1 mm wide, linear, white; stamens 4. **FR**: 1.8–2 mm, 1.7–1.8 mm wide, round to broadly elliptic, flat, tip wingless, notch to 0.1 mm; valves sparsely soft-hairy, not veined; style ± 0.1 mm, incl in notch; pedicel spreading, 2–3.5 mm, cylindric, puberulent adaxially. **SEED**: 1–1.2 mm, oblong. Disturbed areas; < 200 m. CCo, SCo; native to Eur, sw Asia. May–Jun

L. ramosissimum A. Nelson Bien, puberulent. **ST**: erect, (0.6)1–5.3(7.7) dm, many-branched distally. **LF**: basal not rosetted, early-deciduous, 2–5 cm, oblanceolate, pinnately lobed, lobes entire to dentate; mid-cauline short-petioled to sessile, (0.6)1.2–4.8(6) cm, oblanceolate, dentate (lobed), base tapered, not lobed; distal cauline linear, entire. **INFL**: ± elongated or not; rachis puberulent, hairs curved, cylindric or club-shaped. **FL**: sepals 0.6–0.9(1.1) mm; petals 0 or 0.2–0.8(1) mm, to 0.1 mm wide, linear, white; stamens 2. **FR**: 2.2–3.2 mm, 1.7–2.1 mm wide, elliptic, flat, tip winged, notch 0.1–0.3(0.4) mm; valves glabrous or puberulent, not veined; style 0–0.1 mm, incl in notch; pedicel spreading, (1.6)2–3.8(5) mm, cylindric, puberulent adaxially or throughout. **SEED**: 1.2–1.6 mm, oblong. 2*n*=32,64. Roadsides, alkaline soils; < 2900 m. SnGb, PR; widespread N.Am. [*L. r.* var. *bourgeauanum* (Thell.) Rollins] Mar–Sep

L. strictum (S. Watson) Rattan (p. 559) Ann, puberulent. **ST**: ascending or decumbent to prostrate, gen several from base, (0.4)0.7–1.7(2) dm, branched distally. **LF**: basal not rosetted, 1.5–5.6 cm, 2-pinnately lobed, lobes lanceolate to oblong, entire; mid-cauline short-petioled, 0.8–3 cm, pinnately lobed, lobes entire, base tapered. **INFL**: crowded; rachis puberulent, hairs straight, cylindric. **FL**: sepals 0.7–1(1.2), persistent; petals 0.2–0.5 mm, < 0.1 mm wide, linear, white; stamens 2. **FR**: 2.5–3.3 mm, 2–3 mm wide, ovate to ± round, flat, tip winged, notch 0.3–0.6 mm; valves glabrous or puberulent at margin, prominently net-veined; style 0–0.1 mm, incl in notch; pedicel ± erect, (1)1.4–2.5(3) mm, flattened, narrowly winged, appressed at base, puberulent adaxially, tip curved, spreading. **SEED**: 1.2–1.6 mm, oblong. 2*n*=32. Uncommon. Disturbed areas, woodland, slopes; < 1700 m. CA-FP; to OR. Apr–Jun

L. thurberi Wooton Ann, hairs ≤ 1 mm, cylindric 1 mm and much shorter, club-shaped. **ST**: erect, 1 from base, (0.8)1.2–4.9(6) dm, many-branched distally. **LF**: basal rosetted, gen early-deciduous, (1.4)2.2–7(10) cm, pinnately lobed, lobes oblong to ovate or lanceolate, dentate-sinuate; mid-cauline short-petioled, 1.5–6 cm, lobe margin entire or dentate. **INFL**: elongated, hairs long-cylindric, short-club-shaped. **FL**: sepals 1–1.6 mm; petals 3–4 mm, 1.2–2.2 mm wide, broadly obovate, white; stamens 6. **FR**: 2–2.9 mm, 2–2.8 mm wide, broadly ovate to round, flat, tip winged, notch 0.1–0.2 mm; valves glabrous, not veined; style 0.3–0.8 mm, exserted beyond notch; pedicel spreading, 4–8(10) mm, cylindric, puberulent adaxially. **SEED**: 1.3–1.6 mm, oblong. Saline flats, clay soils, grassland, washes, roadsides; < 1000 m. DMoj; to NM, Mex. Apr–Aug

L. virginicum L. Ann, puberulent. **ST**: erect, 1 from base, (0.6)1.5–5.5(7) dm, branched distally. **LF**: not rosetted, early-deciduous, (1)2.5–10(15) cm, obovate or spoon-shaped to oblanceolate, pinnately lobed to dentate; mid-cauline short-petioled, 1–6 cm, oblanceolate or linear, entire to serrate, lobed, or divided, base tapered, not lobed. **INFL**: much elongated; rachis puberulent, hairs curved, cylindric. **FL**: sepals (0.5)0.7–1(1.1) mm; petals 1–2.5 mm, 0.3–0.8(1) mm wide, spoon-shaped to oblanceolate, white; stamens 2. **FR**: 2.5–3.5(4) mm, ± round, tip winged, notch 0.2–0.5 mm; valves glabrous or puberulent, not veined; style 0.1–0.2 mm, incl in notch; pedicel ± spreading, 2.5–4(6) mm, cylindric or flattened, puberulent adaxially (throughout, or glabrous). **SEED**: 1.3–2 mm, ovate. 2*n*=32. [*L. ruderale* L., misappl.] Gen divided into many vars. based on poorly defined, inconsistent characters.

subsp. **menziesii** (DC.) Thell. **INFL**: pedicel (0.2)0.3–0.4 mm wide, flattened at least below tip. **FR**: glabrous or puberulent. Dry, disturbed areas, bottomland, riverbanks, meadows, fields, pastures, cliffs, scrub; < 2800 m. CA; to BC, MT, CO, TX, Mex. [*L. v.* var. *medium* (Greene) C.L. Hitchc.; *L. v.* var. *pubescens* (Greene) Thell.; *L. v.* var. *robinsonii* (Thell.) C.L. Hitchc.] Mar–Jun

subsp. **virginicum** **INFL**: pedicel 0.15–0.2 mm wide, cylindric. **FR**: glabrous. Disturbed sites, grassy areas, fields; 200–2900 m. SNH, TR, PR, GB, D; N.Am; introduced elsewhere. Mar–Sep

LOBULARIA SWEET ALYSSUM

Ann, per, occ subshrub; hairs 2-rayed, sessile, appressed. **LF**: entire, not lobed at base. **INFL**: elongated. **FL**: sepals spreading to erect, base not sac-like; petals wide-obovate to spoon-shaped, clawed, white to purple. **FR**: silicle, flat parallel to septum, unsegmented, dehiscent, round to obovate or elliptic; stigma entire. **SEED**: 2(4)[13], in 1 or 2 rows, winged or not. 4 spp.: Medit, Eur. (Latin: small lobe, reference to fr)

L. maritima (L.) Desv. (p. 559) Ann, per. **ST**: erect to prostrate or decumbent, branched from base, 0.5–2.5(4) dm. **LF**: (1)1.6–2.5(4.2) cm, (1)2–3(7) mm wide, linear to lance-linear, acute. **FL**: sepals 1.4–1.7(2) mm; petals (2)2.3–2.8(3) mm, widely obovate. **FR**: 2–2.7(4.2) mm, (1.2)1.5–2(2.9) mm wide, round-elliptic; style 0.4– 0.5 mm; pedicel spreading to ascending, (3)4.5–6(10) mm. **SEED**: 1 per chamber, ovate to lens-shaped, 1–1.4(2) mm; wing 0. 2*n*=24. Disturbed areas, fields; < 600 m. NCo, CCo, SnFrB, SCo; Can, US; native to Medit, Eur. Mar–Oct ❖

LUNARIA MONEY PLANT, HONESTY

Ann to per; hairs simple. **LF**: simple, long-petioled, dentate; cauline lf bases not lobed. **INFL**: elongated. **FL**: sepals erect, linear, lateral pair sac-like at base; petals long-clawed. **FR**: silicle, widely elliptic to round or oblong, flat parallel to septum, dehiscent, long-stalked above receptacle, unsegmented; style long, slender, stigma 2-lobed. **SEED**: 4–8, in 2 rows, flat, winged; stalks fused to septum. 3 spp.: Eur. (Latin: moon, from shiny fr septum)

L. annua L. (p. 559) (Ann) bien. **ST:** (3)4–10(12) dm, hairy at least proximally. **LF:** basal (1.5)3–12(18) cm, (1)2–8(12) cm wide, wide-cordate to cordate-ovate; petiole (1.5)3–10(17) cm; distal sessile. **FL:** sepals (5)6–9(10) mm; petals (1.5)1.7–2.5(3) cm, 5–10 mm wide, wide-obovate, purple to lavender (white); claw 5–10 mm. **FR:** 3–5 cm, 2–3(3.5) cm wide, oblong-round; stalk above receptacle 7–18 mm; style 4–10 mm; pedicel (0.7)1–1.5 cm. **SEED:** (6)7–10(12) mm, 5–9 mm wide, reniform. 2*n*=30. Disturbed areas; < 1000 m. CaR, CCo, SnFrB, SCo; Can, US; native to Eur. Apr–Jun

LYROCARPA

[Ann] per, occ subshrub; hairs dense throughout, many-branched. **LF:** petioled; basal not rosetted; cauline dentate to pinnately lobed, petioled, base not lobed. **INFL:** elongated. **FL:** sepals erect, linear to linear-oblong, base of lateral pair sac-like; petals linear to spoon-shaped, long-clawed, ± yellow to brown or purple. **FR:** dehiscent, obcordate [to lyre-shaped], flat perpendicular to septum, unsegmented; stigma prominently 2-lobed. **SEED:** 6–16[20], wingless. 3 spp.: sw US, Mex. (Greek: lyre fr, from fr shape)

L. coulteri Hook. & Harv. (p. 559) COULTER'S LYREPOD Per or subshrub. **ST:** (2)3–9(11) dm, straw-colored, woody proximally. **LF:** petiole 0.5–2.5(4) cm; blade (1)2–9(12) cm, (0.5)1–5(7) cm wide, lanceolate to ovate; distal less divided, dentate to wavy. **FL:** sepals (6)8–10(11) mm; petals (1.3)1.6–2.5 cm, 1–3 mm wide, claw 7–12 mm. **FR:** (0.8)1–2.5 cm, (0.8)1–1.6 cm wide, obcordate, base obtuse, tip notched to truncate; valves hairy; style 0–0.5 mm; stigma lobes spreading; pedicel ascending to spreading (reflexed), 2–7(11) mm. **SEED:** 6–16, 2–3 mm wide, round. 2*n*=36,42. Dry slopes, gravelly flats, washes, sandy plains, ravines, granite hillsides; < 1300 m. DSon; AZ, Baja CA. [*L. c.* var. *palmeri* (S. Watson) Rollins] Fr shape extremely variable. Sep–Apr ★

MATTHIOLA STOCK

[Ann] bien, per [occ subshrub], gen ± gray; hairs dense, many-branched (simple or forked); stalked multicellular glands present or 0. **LF:** entire or dentate to pinnately lobed; cauline petioled or sessile, base not lobed. **INFL:** elongated. **FL:** sepals erect, inner pair sac-like at base; petals long-clawed. **FR:** silique, linear, cylindric or flat parallel to septum, dehiscent, unsegmented; stigma conical, 2-lobed, with 2–3 horns. **SEED:** (5)15–60, in 1 row, winged [or not]. 50 spp.: Eurasia, Afr. (P.A. Matthioli, Italian botanist, 1500–1577) Other spp. cult, incl *M. longipetala* (Vent.) DC.

M. incana (L.) W.T. Aiton **ST:** (1)2.5–6(9) dm, occ woody at base. **LF:** linear to oblong or oblanceolate, entire to wavy-margined, proximal, middle (2.5)4–16(22) cm, (0.5)0.8–1.8(2.5) cm wide. **FL:** fragrant; sepals 1–1.5 cm; petals 2–3 cm, 0.7–1.5 cm wide, obovate to ovate, purple, violet, pink, or white, claw 1–1.7 cm. **FR:** (4)6–12(15) cm, (3)4–6 mm wide, flat parallel to septum, ± constricted between seeds; style 1–5 mm; pedicel ascending, (0.6)1–2(2.5) cm. **SEED:** 2.5–3.3 mm, ± round; wing 0.2–0.5 mm wide. 2*n*=14. Sandy areas, beaches, ocean bluffs; < 300 m. s NCo, CCo, SCo; Mex; native to Eur. Cult worldwide as orn. Mar–Jun

NASTURTIUM

Per, aquatic, rhizomatous; hairs simple or 0. **ST:** prostrate to erect, rooting at proximal nodes. **LF:** pinnately compound, petioled, base lobed or not, simple in deeply submersed pls. **INFL:** elongated; bracts 0 or subtending proximal fls. **FL:** sepals erect to ascending, base or lateral pair not sac-like; petals white (pink), not clawed. **FR:** silique, dehiscent, linear, cylindric, unsegmented; stigma entire. **SEED:** 24–60, in 1 or 2 rows; wing 0. 5 spp.: worldwide. (Latin: nose distortion, in reference to pl pungency) [Al-Shehbaz & Price 1998 Novon 8:124–126]

1. Seeds in 1 row, minutely net-veined; fr 1–1.2 mm wide . . . . *N. gambelii*
1′ Seeds in 2 rows, coarsely net-veined; fr (1.8)2–2.5(3) mm wide . . . . *N. officinale*

N. gambelii (S. Watson) O.E. Schulz (p. 563) GAMBEL'S WATER CRESS Pl gen hairy. **ST:** erect to decumbent, branched, 5–12 dm. **LF:** 3–10 cm; lflets 9–17, linear to narrowly oblong or lanceolate, 0.5–2.5 cm, 2–8(10) mm wide, dentate to ± wavy. **FL:** sepals 3–4 mm; petals 6–8 mm, 2–2.5 mm wide, oblanceolate, white. **FR:** 2–3 cm, straight to incurved, ± narrowed between seeds; style 1–2.5 mm; pedicel spreading to ± reflexed, 0.9–2.5 cm. **SEED:** 24–40, ovate, 1–1.2 mm. Marshes, streambanks, lake margins; < 350 m. s CCo, SCo; Mex. [*Rorippa g.* (S. Watson) Rollins & Al-Shehbaz] Intermediates known with *N. officinale*. May–Aug ★

N. officinale W.T. Aiton (p. 563) WATER CRESS Pl ± glabrous (sparsely hairy). **ST:** erect to decumbent, branched, 1–11(20) dm. **LF:** 2–15(20) cm; lflets 3–9(13), (0.4)1–4(5) cm, (0.3)0.7–2.5(4) cm wide, round to ovate, oblong, or lanceolate, dentate to entire. **FL:** sepals 2–3.5 mm; petals 2.8–4.5(6) mm, 1.5–2.5 mm wide, white. **FR:** (0.6)1–1.8(2.5) cm, narrowed between seeds; style 0.5–1(1.5) mm; pedicels spreading, 0.5–1.7(2.4) cm. **SEED:** (28)36–60, 0.8–1.1(1.3) mm. 2*n*=32. Streams, springs, marshes, lake margins, swamps; < 3000 m. CA-FP, MP, n SNE, W&I, DMtns; temp worldwide. [*Rorippa nasturtium-aquaticum* (L.) Hayek] Widely cult for edible greens. Mar–Nov

NOCCAEA

[Bien] per, gen glaucous; hairs simple or 0. **ST:** erect or decumbent. **LF:** basal rosetted, petioled, entire or dentate; cauline sessile, base lobed or clasping. **INFL:** elongated or not. **FL:** sepals erect, lateral pair not sac-like at base; petals white to ± purple, obscurely clawed. **FR:** silicle, flat perpendicular to septum, dehiscent, unsegmented; valves keeled, gen winged; stigma entire. **SEED:** 2–10[24], in 1 row, wingless. 80 spp.: temp, gen n hemisphere. (D. Nocca, botanist, director of botanical garden, Pavia, Italy 1758–1841) [Koch & Al-Shehbaz 2004 Syst Bot 29:375–384]

N. fendleri (A. Gray) Holub Pl ± glaucous; caudex branched or simple. **ST:** simple or branched distally. **LF:** petiole 0.4–7.3 cm; blade 0.4–3 cm; cauline ovate to ± oblong. **FL:** sepals 1.6–5.3 mm; petals spoon-shaped. **FR:** pedicel 2.5–15 mm. **SEED:** 2–10, 1.1–2.1 mm, ovate. 2*n*=28. 5 subspp.

1. Fr elliptic, tip acute, length 2.2–3.2 × width; pedicel ascending to ± spreading, forming 15–60(70)° angle with rachis . . . . subsp. *californica*
1′ Fr obovate to obcordate, tip obtuse to truncate or notched, length 1–2 × width; pedicel spreading to strongly descending, forming (60)70–130° angle with rachis . . . . subsp. *glauca*

    subsp. *californica* (S. Watson) Al-Shehbaz & M. Koch KNEELAND PRAIRIE PENNYCRESS **ST**: (5)9.5–11.5(20) cm. **LF**: basal 5–7(8) mm wide, spoon-shaped to obovate, base tapered; petiole 0.8–1.8 × > blade; cauline 2–5. **FL**: petals 6–8 mm, 1.6–2.5 mm wide, white. **FR**: 7–10.5 mm, 2.7–4 mm wide, wingless; style (1.3)1.5–2(2.4) mm. **SEED**: 2–6. Serpentine outcrops; 500–700 m. NCoRO (Kneeland Prairie). [*Thlaspi c.* S. Watson] May–Jun ★

    subsp. *glauca* (A. Nelson) Al-Shehbaz & M. Koch (p. 563) **ST**: (1.5)5–32(45) cm. **LF**: basal 4–9(15) mm wide, ovate to oblong, base tapered; petiole 0.8–1(2) × > blade; cauline (4)7–16(21). **FL**: petals (3.4)4–7(8.5) mm, (1)1.5–2.7(4.2) mm, white or ± pink-purple. **FR**: (2.5)5–8(12) mm, (1.5)2.6–4.5(6.6) wide, wingless; style (0.4)1–2.2(3) mm. **SEED**: 4–6. 2*n*=14, 28. Open alluvial flats or fans, scree, near snow banks, limestone cliffs, forest openings, (sub)alpine meadows, rocky or talus slopes, streambanks; 100–4400 m. KR, NCoRH, CaRH, n SNH, MP; to WA, WY, NM, Mex. [*Thlaspi montanum* L. var. *m.*] Apr–Aug

PHOENICAULIS

1 sp. (Greek: palm st, resemblance of st with petiole remnants to that of a palm)

P. cheiranthoides Nutt. (p. 563) Per, woody caudex well developed; hairs many-branched, tree-like. **ST**: simple (branched distally), (0.8)1.2–2.5(3) dm, glabrous. **LF**: rosetted, petioled, entire, persistent; basal (2)3–7(10) cm, (0.5)0.8–2(2.6) cm wide, linear-oblanceolate to obovate, petiole (1)2–5.5(8) cm; cauline 5–12, ovate to oblong or lance-linear, entire, sessile, base lobed. **INFL**: elongated. **FL**: sepals erect, 4–6 mm, base of lateral pair sac-like; petals 9–13(15) mm, 2.5–4 mm wide, obovate-oblanceolate, pink to purple, not clawed. **FR**: silique, dehiscent, 2–6(9) cm, (2)3–5(6) mm wide, lanceolate to linear, flat parallel to septum, unsegmented, glabrous; valves with prominent midvein; style 0.5–2 mm, stigma entire; pedicel spreading, (0.6)1–3(3.5) cm. **SEED**: (6)8–16(18), in 1 row, (2)2.5–3.2 mm, ovate to oblong. 2*n*=28. Volcanic, rocky areas, barren clay slopes, sandy banks, meadows, hillsides, sagebrush scrub; 1000–3200 m. KR, NCoRH, CaRH, SNH, GB; to WA, ID. Apr–Jun

PHYSARIA BLADDERPOD

Per with caudex (ann or bien); hairs stellate, ± sessile. **ST**: erect to decumbent or prostrate, simple or branched distally. **LF**: basal gen rosetted, simple, entire to wavy or dentate (pinnately lobed); cauline petioled or sessile, entire to wavy or dentate, base not lobed. **FL**: sepals oblong to ovate, erect or spreading, sac-like at base; petals yellow (white or ± purple), widely ovate to spoon-shaped, clawed or not. **FR**: silicle, dehiscent, spheric to ovoid, ellipsoid, oblong, or spectacle-shaped, inflated and bladdery or not, unsegmented, gen not flattened (flattened); stigma entire. **SEED**: 4–28(40), in 2 rows, wingless (narrowly winged). 105 spp.: w N.Am, S.Am. (Greek: bladder, from inflated fr) [Al-Shehbaz & O'Kane 2002 Novon 12:319–329]

1. Fr spectacle-shaped, tip deeply notched, strongly inflated, bladdery . . . . *P. chambersii*
1′ Fr spheric to obovoid, ellipsoid, or obcordate, tip obscurely notched or not, gen not inflated or bladdery
  2. Ann or bien, caudex 0 . . . . *P. tenella*
  2′ Per, with simple or branched caudex
    3. Fr strongly compressed
      4. Fr compressed perpendicular to septum, obcordate, ± inflated, tip notched to truncate . . . . *P. cordiformis*
      4′ Fr compressed parallel to septum, ellipsoid to obovoid, tip acute . . . . *P. occidentalis* subsp. *occidentalis*
    3′ Fr not or only ± compressed
      5. Basal, cauline lvs ± similar in shape, width; fr septum complete; valves hairy inside . . . . *P. ludoviciana*
      5′ Basal, cauline lvs different in shape, width; fr septum perforated; valves glabrous to sparsely hairy inside . . . . *P. kingii*
        6. Fr tip truncate or ± notched; valves gen sparsely hairy inside . . . . subsp. *kingii*
        6′ Fr tip round; valves glabrous inside
          7. Style 6–9 mm; seeds 4–8 . . . . subsp. *bernardina*
          7′ Style (2)2.5–6 mm; seeds 8–16 . . . . subsp. *latifolia*

P. chambersii Rollins (p. 563) CHAMBERS' PHYSARIA Cespitose with thick caudex, silvery-pubescent. **ST**: gen decumbent, many, 5–15 cm. **LF**: basal 3–6 cm, obovate to round, entire or dentate; cauline 1–2 cm, 3–6 mm wide, entire, spoon-shaped, gen acute. **FL**: sepals 5–8(9) mm; petals 9–12 mm, narrowly oblanceolate. **FR**: 1–1.5 cm, ± 2 cm wide, spectacle-shaped, greatly inflated into 2 bladdery, reniform halves, gen densely hairy, tip deeply notched; style (4)6–8 mm. **SEED**: gen 8. 2*n*=8,10,16,24. Clay hillsides, sagebrush and pinyon/juniper areas, limestone gravel, dolomite ridges, steep banks; 1500–2500 m. n DMtns (Clark, Grapevine mtns); to OR, UT, AZ. Apr–Jul ★

P. cordiformis Rollins Caudex simple or branched, densely hairy. **ST**: prostrate to decumbent, unbranched, 0.5–1.5 dm. **LF**: basal 2–4(6) cm, deltate to elliptic or ± round, entire or dentate; cauline 1–2(3) cm, 3–6 mm wide, oblanceolate to linear, entire. **FL**: sepals 3.5–6(8) mm; petals (5)7–8.5(10) mm, obovate to oblanceolate. **FR**: 3.3–5.6 mm, wider than long, obcordate, ± inflated, notched to truncate at tip, not bladdery, flattened perpendicular to septum; style 3–6.5 mm, gen hairy; pedicel 5–10 mm. **SEED**: 4–8. 2*n*=10. Steep hillsides, dry sandy, gravelly soils, juniper and sagebrush areas, rocky ridges, Bristlecone-pine forest; 3200–3660 m. W&I; NV. May–Aug

P. kingii (S. Watson) O'Kane & Al-Shehbaz Caudex simple or branched. **ST**: prostrate to decumbent or erect, few to numerous, 0.5–2(4) dm. **LF**: basal (1.2)2–6(8) cm, ± round to oblanceolate, wide-elliptic, or rhombic, entire to wavy or dentate; cauline 0.5–2 cm, obovate to elliptic or spoon-shaped, different in shape, width from basal. **FL**: sepals 4–6(7) mm; petals 6–13 mm, obovate to oblanceolate. **FR**: 3.2–9 mm, ± spheric to obovoid or ellipsoid, truncate to round-acute, ± compressed, densely hairy outside, glabrous to sparsely hairy inside; septum perforated; style 2–9 mm; pedicel 4.5–10(15) mm. **SEED**: 4–12(16), flat. [*Lesquerella k.* (S. Watson) S. Watson] 7 subspp. total. Highly variable, more study needed.

559

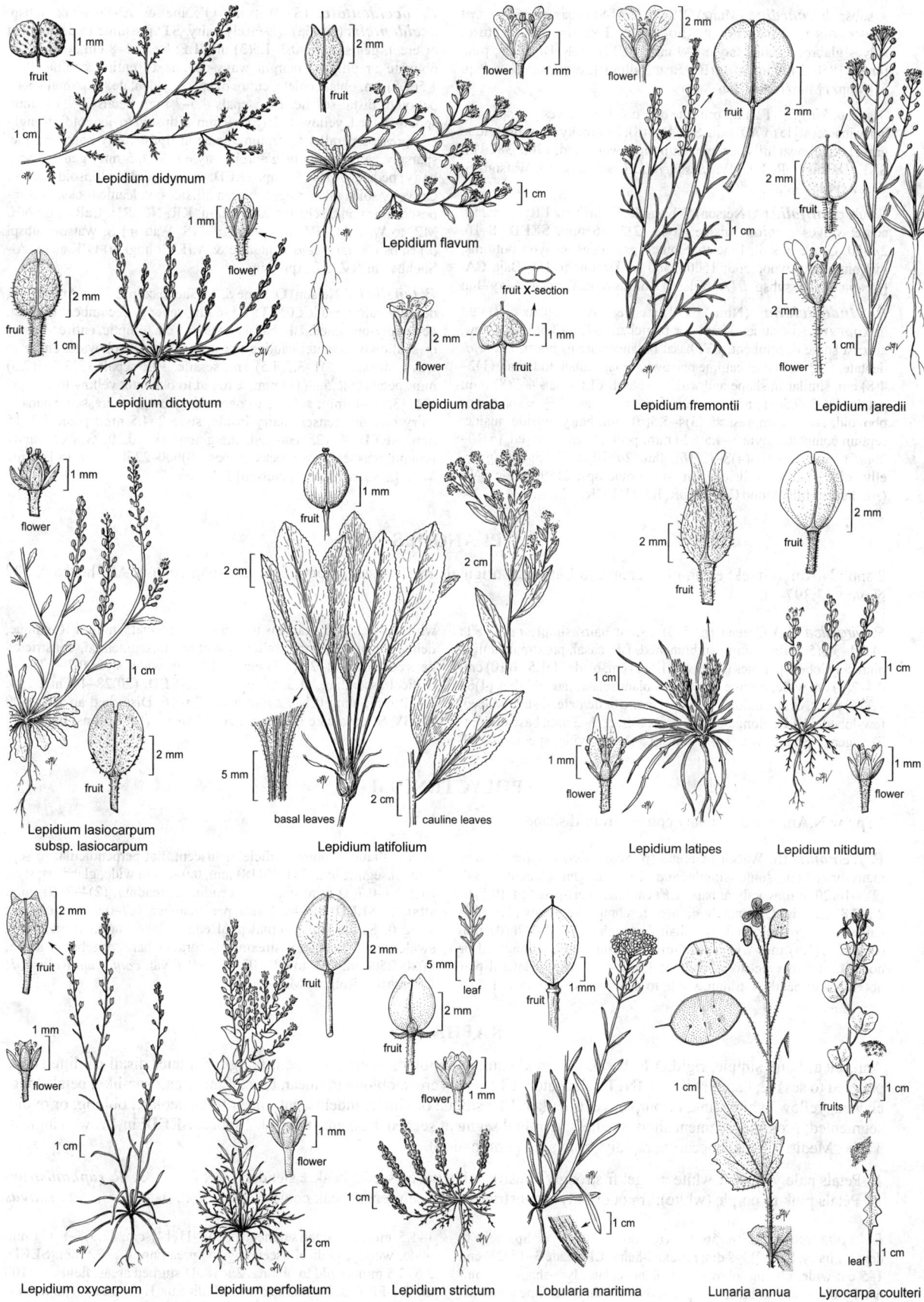

Lepidium didymum

Lepidium flavum

Lepidium fremontii

Lepidium jaredii

fruit X-section

Lepidium dictyotum

Lepidium draba

Lepidium lasiocarpum
subsp. lasiocarpum

Lepidium latifolium

basal leaves

cauline leaves

Lepidium latipes

Lepidium nitidum

Lepidium oxycarpum

Lepidium perfoliatum

Lepidium strictum

Lobularia maritima

Lunaria annua

Lyrocarpa coulteri

subsp. ***bernardina*** (Munz) O'Kane & Al-Shehbaz (p. 563) SAN BERNARDINO MOUNTAINS BLADDERPOD **FR**: tip round-obtuse; valves glabrous inside; style 6–9 mm. **SEED**: 4–8. Dry flats, pine forest; 1850–2400 m. SnBr (Big Bear Valley). [*Lesquerella k.* subsp. *b.* (Munz) Munz] May–Jun ★

subsp. ***kingii*** **FR**: tip truncate or ± notched; valves gen sparsely hairy inside; style to 7(9) mm. **SEED**: 4(8). Dry rocky soils, limestone gravel, sagebrush hillsides, pinyon/juniper woodland; 1700–3800 m. SNE, DMtns; OR, ID, NV. [*Lesquerella k.* subsp. *k.* (S. Watson) S. Watson] May–Jun

subsp. ***latifolia*** (A. Nelson) O'Kane & Al-Shehbaz **FR**: tip round-acute; valves glabrous inside; style (2)2.5–6 mm. **SEED**: 8–16. 2*n*=10. Gravelly soil, limestone outcrops, ridges, canyon bottoms, pinyon/juniper woodland; 1500–2500 m. DMtns; to UT, Baja CA. [*Lesquerella k.* subsp. *l.* (A. Nelson) Rollins & E.A. Shaw] May–Jun

P. ludoviciana (Nutt.) O'Kane & Al-Shehbaz SILVER BLADDERPOD Caudex simple or branched. **ST**: 1–3.5(5) dm, few, at least some decumbent. **LF**: basal oblanceolate to linear, entire or dentate, (1)2–6(9) cm; cauline narrowly oblanceolate to linear, (1)2–4(8) cm, similar in shape and width to basal. **FL**: sepals 4–7(8) mm; petals (5)6.5–9.5(11) mm, oblanceolate to obovate. **FR**: ± spheric or obovoid, occ ± compressed, (3)4–5.5(6) mm, hairy outside, inside; septum complete; style 3–4.5(6.5) mm; pedicel gen recurved, (5)10–20(25) mm. **SEED**: (4)8–12(16), flat. 2*n*=10,20,30. Sandy, gravelly soils, pastures, hillsides, limestone outcrops; 2150 m. e-c SNE (Anchorite Hills, Mono Co.); to OR, ID, MN, OK, NM. Apr–Aug ★

P. occidentalis (S. Watson) O'Kane & Al-Shehbaz subsp. ***occidentalis*** (p. 563) Densely hairy. **ST**: prostrate to decumbent or erect, gen simple, 0.3–1.5(3) dm. **LF**: basal 1–8 cm, ± round to obovate or elliptic, margin wavy-dentate to entire; cauline 0.5–1.5(2.5) cm, oblanceolate, entire or sparsely dentate, proximal short-petioled, distal sessile. **FL**: sepals 4.5–7 mm; petals 7–9(14) mm, spoon-shaped, yellow. **FR**: (5)6–9 mm, ellipsoid to obovoid, strongly compressed parallel to septum, acute at tip, densely hairy outside, sparsely hairy inside or glabrous; style (2)3–6.5 mm, gen sparsely hairy; pedicel 5–10(15) mm. **SEED**: 4–16, ovoid to ellipsoid. 2*n*=10. Gravelly soils, talus, ridges, barren hillsides, volcanic rocks, decomposed limestone, schist; 600–3350 m. KR, NCoRH, CaRH, n SNH, MP; to WA, ID, NV. [*Lesquerella o.* (S. Watson) S. Watson subsp. *o.*] Subsp. *cinerascens* (Maguire & A.H. Holmgren) O'Kane & Al-Shehbaz in NV, UT. Apr–Aug

P. tenella (A. Nelson) O'Kane & Al-Shehbaz (p. 563) Ann (bien), densely hairy; caudex 0. **ST**: 1.5–6 dm, several, decumbent to erect, gen many-branched. **LF**: basal (1.5)3–6.5 cm, elliptic, entire to wavy or shallowly dentate; cauline linear to elliptic or obovate, entire to wavy, distal (0.5)1–3.5(4.5) cm, sessile. **FL**: sepals (3)3.5–6(7.5) mm; petals (5)6.5–8(11) mm, ± round to obovate, yellow to orange. **FR**: (3.5)4–6 mm, spheric to obovoid, gen ± compressed, sparsely hairy outside, densely hairy inside; style 2–4.5 mm; pedicel 5–15 mm. **SEED**: 4–12, flattened, margined. 2*n*=10,20. Gravel, sandy ground, washes, loose rocky slopes; (0)600–2250 m. D; to UT, nw Mex. [*Lesquerella t.* A. Nelson] Feb–Apr

PLANODES

2 spp.: N.Am. (Greek: error, in reference to Linnaeus' original error in assignment of sp. to wrong genus) [Al-Shehbaz 2007 Novon 17:397–402]

P. virginica (L.) Greene (p. 563) Ann; hairs simple or 0. **ST**: (0.5)1–3.5(5.5) dm, simple or branched. **LF**: basal, proximal cauline pinnately lobed, petioles (0.3)0.8–21.5(2) cm, blade (1)1.5–7(10) cm, 0.4–2(3) cm wide, outline oblong to oblanceolate; lateral lobes (4)6–12(15) pairs, ovate, oblong, to linear, entire or dentate; distal reduced, few-lobed. **INFL**: elongated. **FL**: sepals erect, 1–2 mm, base not sac-like, gen ± purple; petals 2–3 mm, 0.5–1 mm wide, spoon-shaped, white, not clawed; stamens in 3 pairs of unequal length. **FR**: silique, dehiscent, linear, flat parallel to septum, unsegmented, constricted between seeds, 1–2.5(3.2) cm, 1–1.5 mm wide; style 0.2–0.7 mm; pedicel ascending, (1.5)2.5–6(8) mm. **SEED**: (20)28–44, in 1 row, 1–1.2 mm; wing 0.1–0.2 mm wide. 2*n*=16. Disturbed areas; < 500 m. GV, SCo; to c&e US, Baja CA. [*Sibara v.* (L.) Rollins] Feb–Apr

POLYCTENIUM

1 sp.: w N.Am. (Greek: many combs, from lf shape)

P. fremontii (S. Watson) Greene (p. 563) Per, cespitose; hairs many-branched, some simple and 2-rayed; gen glaucous. **ST**: (2)5–16(20) cm, woody at base. **LF**: cauline, short-petioled, (0.5)1–2.2(2.8) cm, rigid, pinnately divided to comb-like, lobes (2)3–4(5) pairs, linear, sparsely to densely hairy; terminal lobe (2)4–7(10) mm, (0.3)0.5–1(1.5) mm wide; distal-most sessile. **INFL**: elongated or not. **FL**: sepals erect to ascending, 1.5–2 mm, base of lateral pair not sac-like; petals 5–6 mm, white to pale purple, not clawed. **FR**: linear silique to oblong silicle, dehiscent, flat perpendicular to septum, unsegmented, (2)4–13(18) mm, 0.9–2 mm wide, glabrous; style (0.2)0.4–0.7(1) mm; pedicel ascending-spreading, (2)4–7(10) mm, straight. **SEED**: 30–46, 1 row per chamber, 0.7–0.9 mm, oblong; wing 0. Saline soils, vernal pool edges, lake margins, meadows, swales, mud flats, dry streambeds, gravel bars, sagebrush scrub; 1000–2500 m. GB; to OR, ID, NV. [*P. f.* var. *confertum* Rollins; *P. williamsiae* Rollins] May–Jul

RAPHANUS

Ann, bien; hairs simple, rigid. **LF**: basal, proximal cauline petioled, pinnately lobed, margin dentate; distal cauline short-petioled to sessile, base not lobed. **INFL**: elongated. **FL**: sepals erect, oblong to linear, base of inner pair sac-like; petals long-clawed, yellow, white, pink, or purple, veins darker. **FR**: silique or silicle, indehiscent, linear to lanceolate, oblong, or ovoid, segmented; proximal segment short, seedless; terminal segment seeded, beaked; stigma ± 2-lobed. **SEED**: in 1 row, wingless. 3 spp.: Medit. (Greek: appearing rapidly, from seed germination)

1. Petals pale yellow, ± white in age; fr strongly constricted between seeds; beak ± slender . . . . ***R. raphanistrum***
1′ Petals pink to purple (white); fr not or only ± constricted between seeds; beak conic . . . . ***R. sativus***

R. raphanistrum L. (p. 563) JOINTED CHARLOCK Sparsely to densely hairy. **ST**: (2)3–8 dm, reflexed-hairy. **LF**: blade 3–15(22) cm, 1–5 cm wide, oblong, obovate, or oblanceolate, lyre-shaped to pinnately lobed; lateral lobes 1–4 pairs, dentate; distal cauline ± sessile, dentate. **FL**: sepals 7–11 mm; petals 15–25 mm, 4–7 mm wide, claw 8–14 mm. **FR**: cylindric to narrowly lanceolate; proximal segment 1–1.5 mm; terminal segment (1.5)2–11(14) cm, (2.5)3–8(11) mm wide, woody; pedicel ascending to spreading, 0.7–2.5 cm. **SEED**: 2.5–3.5 mm, ovoid to oblong. 2*n*=18. Disturbed areas, fields; < 1100 m. CA-FP, GB; N.Am; native to Medit Eur. Hybridizes with *R. sativus* to produce swarms highly variable in fl color and fr constriction. Apr–Jul

R. sativus L. (p. 563) RADISH Sparsely hairy to glabrous. **ST:** branched distally, (1)4–13 dm. **LF:** basal blade 2–60 cm, 1–20 cm wide, oblong, obovate, oblanceolate, or spoon-shaped in outline, lyre-shaped to pinnately divided; lateral lobes 1–12, dentate. **FL:** sepals 5.5–10 mm; petals 15–25 mm, 3–8 mm wide. **FR:** fusiform or lanceolate, occ ovoid; proximal segment 1–3.5 mm; terminal segment (1)3–15(25) cm, (5)7–13(15) mm wide, corky; pedicel spreading to ascending, 0.5–4 cm. **SEED:** 2.5–4 mm, spheric to ovoid. 2*n*=18. Disturbed areas, fields; < 1500 m. CA-FP; N.Am, temp worldwide; native to Medit Eur. See note under *R. raphanistrum*. May–Jul ❖

RAPISTRUM WILD TURNIP

Ann [per]; hairs simple [or 0]. **LF:** proximal petioled, pinnately lobed; distal sessile or short-petioled, lobed or dentate, base not lobed. **INFL:** much elongated. **FL:** sepals ascending, base not sac-like; petals obovate, yellow, short-clawed. **FR:** silicle, indehiscent, segmented, strongly constricted between segments; proximal segment tardily dehiscent, 1(3)-seeded; terminal segment indehiscent, 1-seeded (seedless); stigma 2-lobed. **SEED:** 1–4, in 1 row, wingless. 2 spp.: Eur. (Latin: turnip-like)

R. rugosum (L.) All. (p. 563) Pl erect; hairs stiff. **ST:** (1)3–10(15) dm, branched distally. **LF:** basal 2–25 cm, oblanceolate to obovate, lateral lobe pairs 1–5. **FL:** sepals 2.5–5 mm; petals 6–11 mm, 2.5–4 mm, pale yellow. **FR:** hairy or glabrous; proximal segment 0.7–3 mm, ellipsoid; terminal segment 1.5–3.5 mm, 1–2.8 mm wide, spheric to ovoid, gen ribbed; style 1–3(5) mm; pedicel erect, appressed, 1.5–5 mm. **SEED:** 1.5–2.5 mm, ovoid. 2*n*=16. Disturbed areas, fields; < 200 m. CCo; to WA, ON, e US, Mex; native to s Eur. Doubtfully naturalized in CA. Apr–Jul

RORIPPA YELLOW CRESS

Ann to per, occ with caudex or rhizome; hairs simple or 0. **ST:** prostrate to erect, branched or not, lfy. **LF:** basal rosetted or not, simple, entire or dentate to 1–3-pinnately divided; cauline petioled or sessile, gen lobed to sagittate at base, entire to dentate or pinnately lobed. **INFL:** elongated or congested; bracts 0 [rarely throughout]. **FL:** sepals erect to spreading, base not sac-like, gen deciduous (persistent); petals present (vestigial or 0), yellow [white or pink], gen not clawed. **FR:** silique, linear or narrowly oblong, or silicle, spheric to ovoid or broadly oblong; dehiscent, unsegmented; stigma entire or ± 2-lobed. **SEED:** 10–300, 1(2) row(s) per chamber, gen wingless. 85 spp.: worldwide, on all continents exc Antarctica. (Latinized Old Saxon: for these or perhaps other crucifers) [Al-Shehbaz & Price 1998 Novon 8:124–126] Other taxa in TJM (1993) moved to *Nasturtium*.

1. Cauline lf bases not lobed or clasping . . . . **R. tenerrima**
1′ At least some cauline lf bases lobed or clasping
  2. Infl umbel-like in fr, congested; sepals persistent . . . . **R. subumbellata**
  2′ Infl elongated, not umbel-like in fr; sepals gen deciduous
    3. Pls papillate; hairs hemispheric, sac-like . . . . **R. sinuata**
    3′ Pls not papillate; glabrous or hairs cylindric
      4. Fr valves soft hairy; sepals persistent . . . . **R. columbiae**
      4′ Fr valves glabrous; sepals deciduous
        5. Fr ± spheric, 1.2–3 mm
          6. Per; pedicel 4–15 mm; petals 3–5 mm; style 1–1.5(2) mm . . . . **R. austriaca**
          6′ Ann or bien; pedicel 1.5–3.7(4.3); petals 0.6–1.2 mm; style 0.1–0.7(1) mm . . . . **R. sphaerocarpa**
        5′ Fr not spheric, (2)4–18 mm
          7. St erect, unbranched at base; petals (1.5)1.8–2.5(3) mm . . . . **R. palustris**
            8. St, abaxial lf surface densely bristly . . . . subsp. **hispida**
            8′ St, gen abaxial lf surface glabrous . . . . subsp. **palustris**
          7′ St erect, prostrate to decumbent, or ascending, branches gen few to several from base;
            petals 0.5–1.8(2) mm
          9. Fr ovoid to pear-shaped; petals erect . . . . **R. curvipes**
          9′ Fr linear to oblong; petals spreading . . . . **R. curvisiliqua**

R. austriaca (Crantz) Besser AUSTRIAN FIELD-CRESS Per, rhizomes short, thickened. **ST:** simple, many-branched distally, 4–11(18) dm, glabrous or hairy near base. **LF:** basal not rosetted, pinnately lobed; middle, distal cauline (2.5)4–12(15) cm, sessile, base lobed to clasping, lanceolate, entire or serrate. **INFL:** elongated. **FL:** sepals ascending, 2–3 mm, oblong; petals 3–5 mm, 1.7–2.5 mm wide, obovate. **FR:** silicle, rarely produced, 2.5–3 mm, 1.5–2.7 mm wide, ± spheric; valves glabrous; style 1–2 mm; pedicel spreading to ± ascending, 4–15 mm. **SEED:** 18–40, 0.7–0.9 mm, ovoid. 2*n*=16. Disturbed areas, mud flats, river banks, wet grassland; 1000–1900 m. MP; to Can, n-c&e US; native to Eur. May–Jul ◆

R. columbiae (S. Watson) Howell (p. 567) COLUMBIA YELLOW CRESS Per, soft-hairy, roots creeping. **ST:** ± erect or decumbent to prostrate, branched distally, 1–3.2(4) dm. **LF:** basal 0; mid-cauline 2.4–5.2 cm, short-petioled or sessile, lobed or not at base, oblanceolate to oblong, wavy to pinnately lobed, lateral lobes gen to midrib, oblong to ovate, entire or dentate; most distal cauline lobed at base. **INFL:** elongated. **FL:** sepals ascending, 2–3.5 mm, oblong, hairy, persistent; petals 2.7–4.2 mm, 0.7–1.7 mm wide, oblanceolate to spoon-shaped. **FR:** silicle, (1.5)2.5–5.5(7) mm, (1)1.7–2.8(3.5) mm wide, ± spheric to oblong-ellipsoid; valves densely hairy; style 0.7–3.2 mm; pedicels ascending, ± appressed, (3)4–10(12) mm, slender, densely hairy. **SEED:** 24–40, 0.7–0.9 mm, ovoid-spheric. Streambanks, lake or pond margins, meadows, wet fields; 1000–1800 m. MP; to WA. Jun–Aug ★

R. curvipes Greene (p. 567) Ann (± short-lived per), glabrous or hairy. **ST:** erect or ascending to decumbent or prostrate, gen few to several from base, 1–4.2(5) dm, proximally stiff-hairy. **LF:** not rosetted, early-deciduous; proximal, mid-cauline short-petioled to sessile, (2)3.5–10(12) cm, oblong or oblanceolate to obovate, base lobed to clasping, pinnately lobed to dentate or entire distally on st. **INFL:** elongated. **FL:** sepals erect, 0.8–1.8 mm, oblong; petals erect, 0.5–1.8 mm, 0.2–1 mm wide, oblanceolate to spoon-shaped. **FR:** silicle or silique, 2–8(8.8) mm, (0.5)1–2.5 mm wide, ovoid to pear-shaped, curved; valves glabrous; style 0.3–1 mm; pedicel spreading, (1.2)1.7–5(8) mm, slender. **SEED:** (20)30–80, 0.5–0.7 mm, ± reniform. 2*n*=16. Muddy shores, streambanks, meadows, seepage areas; 100–3500 m. CA-FP, SNE; N.Am. [*R. c.* var. *truncata* (Jeps.) Rollins] May–Sep

R. curvisiliqua (Hook.) Britton (p. 567) Ann, glabrous or sparsely hairy. **ST:** ascending or decumbent to prostrate, gen few to several from base, (0.5)1–4(6) dm. **LF:** basal rosetted or not, early-deciduous; proximal, mid-cauline (2)3–9(13) cm, petioled or sessile, base lobed, oblong to oblanceolate, spoon-shaped, or obovate, pinnately lobed or divided; lateral lobes linear to oblong or ovate, entire or dentate. **INFL:** elongated. **FL:** sepals ascending, 0.8–2(2.5) mm, oblong; petals spreading, 0.6–1.8(2) mm, 0.3–1.3 mm wide, oblong to oblanceolate. **FR:** silique, 4–13(18) mm, 1–2 mm wide, oblong to linear, curved upward; valves glabrous; style 0.1–0.8 mm; pedicel ± spreading, 1–4.5(9) mm, straight. **SEED:** (30)42–106, 0.5–0.7 mm. Uncommon. Streambanks, marshy ground, seepage areas, lake shores, mud flats, meadows; < 3500 m. CA-FP, MP; to AK, Rocky Mtns. May–Oct

R. palustris (L.) Besser Ann (short-lived per), glabrous or bristly. **ST:** erect, branched distally, (0.5)1–10(14) dm. **LF:** basal rosetted, early-deciduous, pinnately divided; cauline (1.5)2.5–10(18) cm, petioled or ± sessile, base lobed to clasping; lateral lobes oblong or ovate, ± entire to dentate. **INFL:** elongated. **FL:** sepals erect, 1.5–2.4(2.6) mm, oblong; petals (1.5)1.8–2.5(3) mm, 0.5–1.5(2) mm wide, spoon-shaped. **FR:** silicle (silique), (2.5)4–10 mm, (1.5)1.7–3(3.5) mm wide, oblong to ellipsoid, gen ± curved; valves glabrous; style 0.2–1(1.2) mm; pedicel spreading to reflexed, (2.5)3–10(14) mm, slender, straight or curved. **SEED:** 20–90, 0.5–0.9 mm, ovoid to ± spheric. 2*n*=32. Highly variable.

subsp. ***hispida*** (Desv.) Jonsell **ST:** densely bristly. **LF:** densely bristly abaxially. Gen wet places; 1200–1500 m. MP; N.Am. [*R. p.* var. *h.* (Desv.) Rydb.] Jun–Aug

subsp. ***palustris*** (p. 567) **ST:** glabrous (sparsely hairy proximally). **LF:** glabrous abaxially. Gen wet places; < 3200 m. CA; N.Am. [*R. p.* var. *occidentalis* (S. Watson) Rollins] Mar–Sep

R. sinuata (Nutt.) Hitchc. (p. 567) Per, sparsely to moderately hairy, hairs hemispheric; with rhizomes or creeping roots. **ST:** decumbent to prostrate, few to many from base, branched distally, 1–4.2(5) dm, proximal portions hairy. **LF:** basal not rosetted, early-deciduous; mid-cauline pinnately lobed to wavy, (1.5)2.5–6.5(9) cm, sessile, base gen lobed, oblong to oblanceolate or lanceolate, glabrous adaxially, hairs on abaxial veins sac-like; lateral lobes oblong or ovate, dentate to wavy or entire; distal cauline sessile, bases lobed. **INFL:** elongated. **FL:** sepals ascending, 2.2–3.7(4.5) mm, oblong; petals (2.7)3.2–5.3(6) mm, 1.5–2.5 mm wide, oblanceolate to spoon-shaped. **FR:** silique, (4)4.7–11.5(16) mm, (1)1.5–2.5 mm wide, oblong to lanceolate or linear, curved; valves glabrous or hairy; style (0.8)1–2.5(3.5) mm; pedicel spreading, 4–12(14.5) mm, slender, gen

recurved. **SEED:** (30)50–82(98), 0.7–1 mm, ± broadly reniform. 2*n*=16. Lake shores, streambanks, fields, moist ground; 900–2600 m. GB; to Can, c US, TX. Apr–Aug

R. sphaerocarpa (A. Gray) Britton Ann (bien). **ST:** decumbent or erect, 1–4(5.5) dm, simple or base few- to many-branched, branched distally; proximally stiff hairy. **LF:** basal rosetted, early-deciduous; proximal, mid-cauline pinnately lobed to divided, (3.5)4.5–9(12) cm, oblong to oblanceolate, short-petioled to sessile, base lobed; lateral lobes oblong to ovate, crenate to ± entire, smaller than terminal lobe. **INFL:** elongated. **FL:** sepals ascending, 0.7–1.3 mm, oblong to ovate; petals 0.6–1.2 mm, 0.2–0.5 mm wide, oblanceolate to spoon-shaped. **FR:** silicle, 1.2–2.5(3) mm wide, ± spheric, valves glabrous; style 0.1–0.7(1) mm; pedicels spreading to ± reflexed, 1.5–3.7(4.3) mm, slender, straight or recurved. **SEED:** 20–42, 0.5–0.7 mm, broadly reniform. Uncommon. Lake margins, muddy streambanks, moist ground; 1200–3300 m. SnBr, SnJt; to ID, CO, TX, n-c Mex. May–Aug

R. subumbellata Rollins (p. 567) TAHOE YELLOW CRESS Per, rhizomed, glabrous or wavy-hairy. **ST:** decumbent, many-branched distally, 0.5–2.5(3) dm. **LF:** basal not rosetted, early-deciduous; mid-cauline ± pinnately lobed to wavy, 1–3.2 cm, broadly oblanceolate to oblong, sessile to short-petioled, base not or minutely lobed; surfaces hairy or glabrous adaxially; lateral lobes oblong to ovate, gen entire. **INFL:** ± umbel-like, congested. **FL:** sepals erect, 2–3 mm, oblong or ovate, persistent; petals 2.5–3.5 mm, 1–1.7 mm wide, spoon-shaped to oblanceolate. **FR:** silicle, 3–5.5 mm, 2–3.5 mm wide, ± spheric to broadly oblong; valves glabrous; style 0.8–1.5 mm; pedicel erect to ascending, 3–7(9) mm, slender, straight, hairy. **SEED:** 30–44, ± angled, 0.8–1.1 mm. Sandy lake margins; 1800–2500 m. n SNH (Lake Tahoe Basin); w NV. Jun–Sep ★

R. tenerrima Greene Ann, glabrous. **ST:** prostrate to decumbent, base many-branched distally, 0.7–3.5(4) dm. **LF:** basal not rosetted, early-deciduous; proximal, mid-cauline pinnately lobed, 2–9(11) cm, oblong to oblanceolate or lanceolate, short-petioled, base not lobed; lateral lobes linear to oblong, ovate, or obovate, entire to dentate or wavy; distal sessile, base not lobed. **INFL:** elongated. **FL:** sepals ascending, 0.7–1.3 mm, oblong; petals 0.5–0.8 mm, 0.1–0.3 mm wide, oblong to oblanceolate or spoon-shaped. **FR:** silique or silicle, 3–7(9) mm, (0.8)1–1.7(2) mm wide, lanceolate to narrowly ovoid or lance-oblong, curved upward, middle gen ± constricted; valves papillate; style (0.2)0.5–1 mm; pedicel ascending to spreading, (1)1.5–3.2(4.2) mm, slender, straight. **SEED:** 20–80, 0.5–0.7 mm, broadly reniform. Marshes, wet meadows, streambanks; < 1600 m. PR; to BC, CO, TX, Mex. Jun–Oct

SIBARA

Ann, glabrous or with simple, forked, or many-branched hairs. **LF:** basal, proximal cauline finely pinnately divided; distal petioled, pinnately lobed, base not lobed. **INFL:** much elongated. **FL:** sepals erect, base not sac-like; petals spoon-shaped to oblanceolate, white to pink or purple, clawed. **FR:** silique, dehiscent, linear, flat parallel to septum or cylindric, unsegmented; stigma entire. **SEED:** 14–94, in 1 or 2 rows, wingless. 5 spp.: N.Am. (Anagram of *Arabis*) [Al-Shehbaz 2007 Novon 17:397–402] *S. virginica* now called *Planodes virginicum*.

1. Pl hairy; fr (0.8)1.2–2.5(3.2) cm, 1.2–1.5 mm wide, curved; petals 2–3.5 mm, 0.5–1 mm wide; seeds 16–24 ... *S. deserti*
1′ Pl glabrous; fr 2.5–4.1 cm, 0.7–0.9 mm wide, straight; petals 3.5–6 mm, 2–3 mm wide; seeds 32–40 .... *S. filifolia*

S. deserti (M.E. Jones) Rollins (p. 567) DESERT WINGED ROCKCRESS Hairs minute, forked or many-branched (simple). **ST:** 1–3.5(4.5) dm, distally branched or not. **LF:** basal early-deciduous; cauline 2–4 cm, deeply pinnately divided; lobes 2–20 mm, 0.3–3 mm wide, linear to lance-linear. **FL:** sepals 1.5–2 mm, oblong; petals white. **FR:** spreading to reflexed, flat; style 1–2.5(3) mm; pedicel spreading to reflexed, (2)3–5.5(10) mm, hairy or not. **SEED:** 0.9–1.4 mm, oblong. 2*n*=26,28. Washes, steep hillsides, dry flats, scree, calcareous rubble, rocky bluffs, exposed crevices; 50–1400 m. n&e DMoj; se NV. [*S. rosulata* Rollins] Fr length, orientation variable. Mar–Apr ★

S. filifolia (Greene) Greene (p. 567) SANTA CRUZ ISLAND ROCKCRESS **ST:** gen 1.5–3 dm, distally simple or few-branched. **LF:** basal early-deciduous; cauline 2–4 cm, pinnately divided; lobes 5–15 mm, 0.2–0.8 mm wide, linear to thread-like, entire. **FL:** sepals 2.2–3 mm, oblong; petals spoon-shaped, purple to lavender. **FR:** style 0.5–0.8 mm; pedicel ascending to ± spreading, (2)3–10(15) mm, straight, slender. **SEED:** 1–1.3 mm, oblong. Dry ridges; < 500 m. ChI (Santa Catalina, San Clemente, Santa Cruz islands). Apr ★

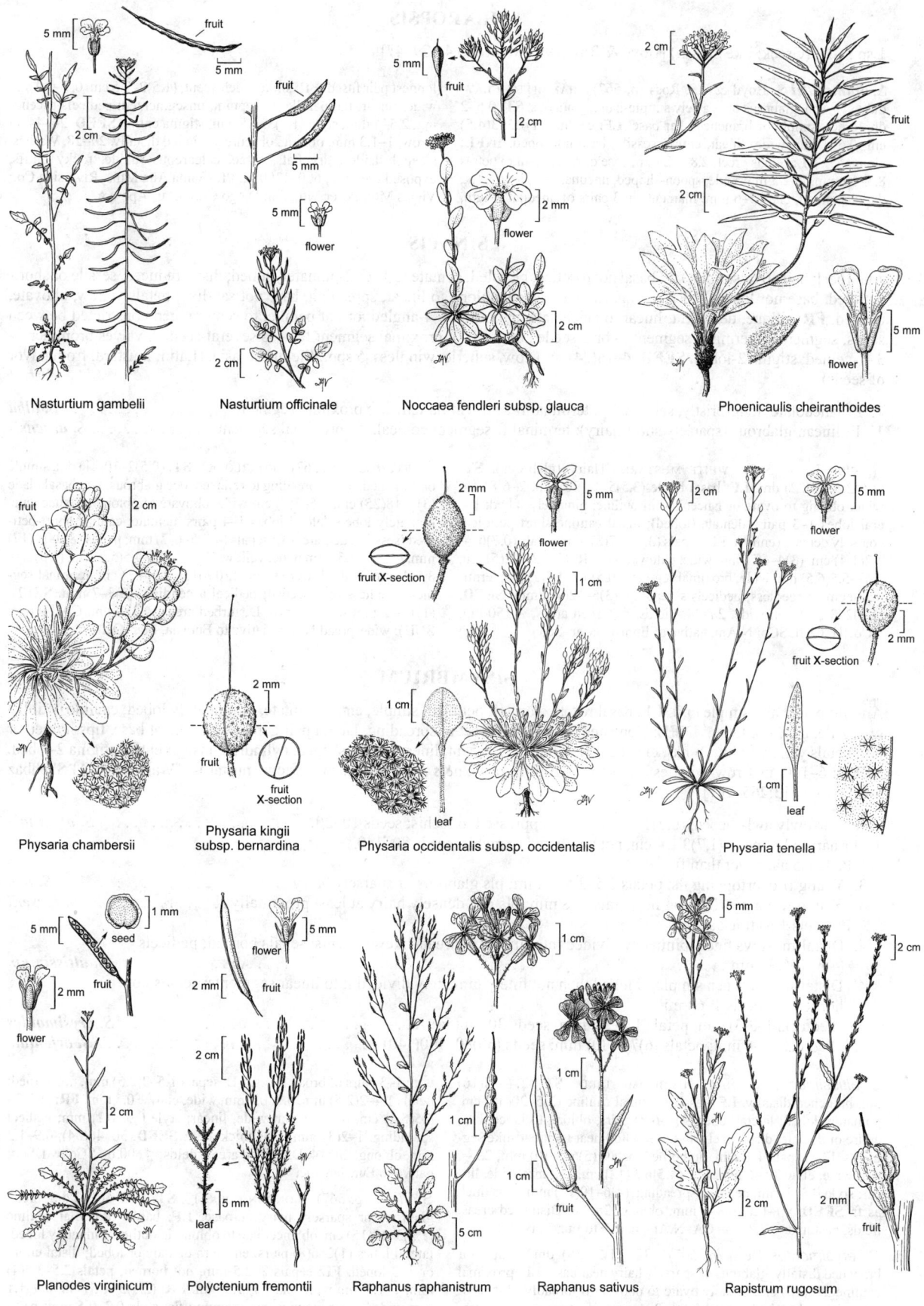

Nasturtium gambelii

Nasturtium officinale

Noccaea fendleri subsp. glauca

Phoenicaulis cheiranthoides

Physaria chambersii

Physaria kingii
subsp. bernardina

Physaria occidentalis subsp. occidentalis

Physaria tenella

Planodes virginicum

Polyctenium fremontii

Raphanus raphanistrum

Raphanus sativus

Rapistrum rugosum

SIBAROPSIS

1 sp.: CA. (Greek: like *Sibara*) [Boyd & Ross 1997 Madroño 44:29–47]

S. hammittii S. Boyd & T.S. Ross (p. 567) HAMMITT'S CLAY-CRESS Ann, glabrous or sparsely simple-hairy, glaucous. **ST**: 0.5–2 dm, erect, simple or branched near base. **LF**: cauline (1)1.5–3(4.5) cm, 0.5–1 mm wide, linear, entire, sessile, base not lobed. **INFL**: elongated. **FL**: sepals erect, 2.8–3.2 mm, base only ± sac-like; petals 8.5–10 mm, 2–2.5 mm wide, spoon-shaped, unequal, purple to pink, darker-veined, claw 5–6 mm; filaments in 3 pairs of unequal length, longest pair fused. **FR**: silique, dehiscent, (1.5)2–2.5 cm, 0.7–0.9 mm wide, linear, flat parallel to septum, unsegmented; pedicels ascending, 2.5–4 mm; style (1.5)3–4.5 mm, stigma entire. **SEED**: 24–44, in 1 row, 1–1.3 mm, oblong, obscurely winged distally. 2*n*=28. Washes, steep hillsides, dry flats, scree, calcareous rubble, rocky bluffs, exposed crevices; 600–1300 m. PR (Santa Ana Mtns, Riverside Co.; Viejas Mtn, Poser Mtn, San Diego Co.). Mar–Apr ★

SINAPIS

Ann [per]; hairs 0 or simple. **LF**: basal not rosetted, petioled, dentate to 1- or 2-pinnately lobed; distal reduced, sessile or short-petioled, base not lobed. **INFL**: elongated. **FL**: sepals oblong to linear, spreading, base not sac-like; petals yellow, obovate, clawed. **FR**: silique, dehiscent, linear to lanceolate, cylindric, ± 4-angled, or flat parallel to septum, gen constricted between seeds, segmented; terminal segment 1- or 2-seeded or seeds 0; proximal segment few- to several-seeded, valves prominently 3–7-veined; stigma 2-lobed. **SEED**: 4–16(24), in 1 row, spheric, wingless. 5 spp.: Medit, Eurasia. (Latin: mustard, from flavor of seeds)

1. Fr lanceolate, long-bristly, short-hairy; terminal fr segment flattened, ≥ proximal segment. . . . . ***S. alba***
1′ Fr linear, glabrous (sparsely short-hairy); terminal fr segment conical, << proximal segment . . . . ***S. arvensis***

S. alba L. (p. 567) WHITE MUSTARD Hairy (glabrous). **ST**: (1.5)2.5–10(22) dm. **LF**: basal blade (3.5)5–14(20) cm, 2–6(8) cm wide, oblong to ovate or lanceolate in outline, pinnately lobed; lateral lobes 1–3 pairs, dentate (lobed), cauline short-petioled, coarsely dentate (entire). **FL**: sepals (3.8)4–7(8) mm; petals (0.7)0.8–1.2(1.4) cm, (3)4–6(7) mm wide, pale yellow. **FR**: (1.5)2–4.2(5) cm, (2)3–5.5(6.5) mm wide; proximal segment (0.5)0.7–1.7(2) cm; terminal segment seedless; pedicels spreading, (3)6–12(17) mm. **SEED**: (1.7)2–3(3.5) mm wide. 2*n*=24. Fields, disturbed areas; < 1500 m. NCoRH, CCo, SCo; N.Am, native to Eurasia. Mar–Sep

S. arvensis L. (p. 567) CHARLOCK **ST**: (0.5)2–10(21) dm, simple or branched, hairs spreading to reflexed, occ glabrous. **LF**: basal blade (3)4–18(25) cm, 1.5–5(7) cm wide, obovate to oblong or lanceolate, pinnately lobed; lateral lobes 1–4 pairs, dentate; cauline short-petioled (sessile), dentate. **FL**: sepals (4.5)5–6(7) mm; petals (8)9–12(17) mm, (3)4–6(7.5) mm wide, yellow. **FR**: (1.5)2–4.5(5.7) cm, (1.5)2.5–3.5(4) mm wide; lower segment (0.6)1.2–3.5(4.3) cm; terminal segment seedless or 1-seeded; pedicel ascending, (2)3–7 mm. **SEED**: (1)1.5–2 mm wide. 2*n*=18. Disturbed areas; < 1800 m. CA-FP (exc SNH); widespread N.Am, native to Eurasia. Mar–Oct ❖

SISYMBRIUM

Ann [to per]; hairs simple or 0. **LF**: basal rosetted or not, petioled, simple, entire, dentate, or pinnately lobed; cauline petioled or sessile, base not lobed. **INFL**: elongated. **FL**: sepals erect to spreading, lateral pair gen not sac-like at base, tips horned or not; petals yellow [white, pink], clawed. **FR**: silique, dehiscent, linear or awl-shaped, cylindric, unsegmented; stigma 2-lobed. **SEED**: 6–160, in 1 row, wingless. 41 spp.: N.Am, Eurasia, n&s Afr. (Greek: for various mustards) [Warwick & Al-Shehbaz 2003 Novon 13:265–267]

1. Fr narrowly awl-shaped, (0.7)1–1.4(1.8) cm, appressed to rachis; seeds 10–20. . . . . ***S. officinale***
1′ Fr narrowly linear, (1.7)3–14 cm, not appressed to rachis; seeds 30–140
　2. Pedicels narrower than fr
　　3. Young fr overtopping fls; petals 2.5–3.5(4) mm; pls glabrous or sparsely hairy . . . . ***S. irio***
　　3′ Young fr not overtopping fls; petals 6–8 mm; pls gen densely hairy at least proximally . . . . ***S. loeselii***
　2′ Pedicels ± as thick as fr
　　4. Distal-most lvs finely pinnately divided into linear to thread-like segments; sepals horned; pedicels (4)6–10(13) mm . . . . ***S. altissimum***
　　4′ Distal-most lvs gen simple, if lobed then not finely pinnately divided into linear segments; sepals not horned; pedicels 1–6 mm
　　　5. Pedicels 1–2(3) mm; petals 1.4–2.5 mm; seeds 30–46(54) . . . . ***S. erysimoides***
　　　5′ Pedicels 3–6 mm; petals (6)7–9(10) mm; seeds (60)80–100(140) . . . . ***S. orientale***

S. altissimum L. (p. 567) TUMBLE MUSTARD **ST**: (2)4–12(16) dm, branched distally. **LF**: basal, proximal cauline (2)5–20(35) cm, pinnately lobed; lateral lobes (3)4–6(8) pairs, oblong to lanceolate, entire or dentate; distal finely dissected into linear to thread-like segments. **FL**: sepals 4–6 mm, tip horned; petals (5)6–8(10) mm, 2.5–4 mm wide, claw 3.5–6 mm. **FR**: (4.5)6–9(12) cm, 1–2 mm wide, linear; style 0.5–2 mm; pedicels spreading, (4)6–10(13) mm, ± as thick as fr. **SEED**: 90–120, 0.8–1 mm, oblong. 2*n*=14. Disturbed areas, fields, pastures; < 2700 m. CA; N.Am; native to Eur. May–Jul

S. erysimoides Desf. (p. 567) **ST**: (1)2–6(8) dm, simple or branched distally, glabrous or sparsely hairy near base. **LF**: proximal cauline 2–8(10) cm, broadly ovate to obovate or broadly oblanceolate, pinnately lobed; lateral lobes 2–4 pairs, dentate; distal dentate with 1–3 pairs of broad lobes. **FL**: sepals 1.5–2(2.5) mm, not horned; petals 1.4–2(2.5) mm, 0.2–0.5 mm wide, claw ± 0.5 mm. **FR**: (1.7)2–4.5(5.2) cm, 0.9–1.2 mm wide, linear; style 0.5–1(2) mm; pedicel spreading, 1–2(3) mm, ± as thick as fr. **SEED**: 30–46(54), 0.9–1.3 mm, oblong. 2*n*=14. Disturbed areas, fields; < 600 m. SCo, w DSon; native to Eur. Jan–Oct

S. irio L. (p. 567) LONDON ROCKET **ST**: (1)2–6(7.5) dm, erect, glabrous or sparsely hairy at base. **LF**: basal, proximal cauline (1.5)3–12(15) cm, oblanceolate to oblong in outline, pinnately lobed; lateral lobes (1)2–6(8) pairs, entire to dentate or lobed; distal entire or 1–3-lobed. **FL**: sepals 2–2.5 mm, not horned; petals 2.5–3.5(4) mm, 1–1.5 mm wide, claw 1–1.5 mm. **FR**: (2.5)3–4(5) cm, 0.9–1.1 mm wide, linear, younger overtopping fls; style 0.2–0.5 mm; pedi-

cel ascending to spreading, (5)7–12(20) mm, much narrower than fr. **SEED:** 40–90, 0.8–1 mm, oblong. $2n=14$. Disturbed areas, fields, pastures; < 1700 m. GV, CCo, SCoR, SW, W&I, D; to UT, TX, Baja CA, also e US; native to Eur. Jan–Apr ❖

S. loeselii L. (p. 567) **ST:** (2)3.5–12(17.5) dm, branched distally, gen densely reflexed-hairy proximally, glabrous distally. **LF:** basal, proximal cauline (1.5)2.5–8(12) cm; lateral lobes 2–4 pairs, entire or dentate; distal entire or dentate. **FL:** sepals 3–4 mm, not horned; petals 6–8 mm, 2–3 mm wide, claw 2.5–3.5 mm. **FR:** 2–3.5(5) cm, 0.9–1.1 mm wide, linear, younger not overtopping fls; style 0.3–0.7 mm; pedicel spreading to ascending, 5–12(15) mm, narrower than fr. **SEED:** 40–60, 0.7–1 mm, oblong. $2n=14$. Disturbed areas, fields, pastures; 1000–2400 m. GB; to n US, Can, native to Eur. May–Nov

S. officinale (L.) Scop. (p. 567) HEDGE MUSTARD **ST:** 2.5–7.5(11) dm, branched distally, reflexed-hairy. **LF:** basal, proximal cauline (2)3–10(15) cm, oblanceolate to oblong-obovate in outline, pinnately lobed; lateral lobes (2)3–4(5) pairs, entire to dentate or

lobed; distal lobed to dentate or entire. **FL:** sepals 2–2.5 mm; petals 2.5–4 mm, 1–2 mm wide. **FR:** ascending to erect, appressed, (0.7)1–1.4(1.8) cm, 1–1.5 mm wide, narrowly awl-shaped; style 0.8–1.5(2) mm; pedicel erect to ascending, 1.5–3(4) mm, stout. **SEED:** 10–20, 1–1.3 mm, oblong. $2n=14$. Disturbed areas, fields, pastures; < 2200 m. CA-FP; N.Am; native to Eurasia. Apr–Sep

S. orientale L. (p. 567) **ST:** (1)2–7(8.5) dm, soft-hairy proximally, gen glabrous distally. **LF:** 3–8(10) cm, broadly oblanceolate to oblong-oblanceolate in outline, pinnately lobed; lateral lobes 2–5 pairs, << terminal lobe; distal gen hastate, simple, narrowly lanceolate to linear. **FL:** sepals 3.5–5.5 mm, not horned; petals (6)7–9(10) mm, 2.5–4 mm wide, claw 3–5.5 mm. **FR:** (5)6–10(13) cm, 1–1.5 mm wide, linear; style 1–3(4) mm; pedicel ascending, 3–6 mm, ± as thick as fr. **SEED:** (60)80–100(140), 1–1.5 mm, oblong. $2n=14$. Disturbed areas, fields, roadsides; < 1300 m. GV, CW, SW, DMoj; to WA, TX, Baja CA, native to Eurasia. May–Jun

SMELOWSKIA

[Ann] per; gen cespitose, caudex branched; hairs many-branched, tree-like, occ mixed with simple and stalked-forked hairs. **LF:** basal rosetted, 1- or 2-pinnately lobed or divided (entire); cauline petioled or sessile, base not lobed. **INFL:** elongated or not; bracts 0 (present). **FL:** sepal ascending to spreading, base not sac-like; petals spoon-shaped to obovate, white to pink or purple (yellow). **FR:** dehiscent, cylindric to 4-angled or flat parallel or perpendicular to septum, unsegmented; stigma entire or ± 2-lobed. **SEED:** 4–30, in 1 row, wingless. 27 spp.: w N.Am, e&c Asia. (T. Smielowsky, Russian botanist, 1769–1815) [Al-Shehbaz & Warwick 2006 Harvard Pap Bot 11:91–99]

S. ovalis M.E. Jones (p. 567) Pl deep-rooted; caudex several-branched. **ST:** several to many from caudex, 0.3–1.8 dm, simple or branched distally, densely hairy, hairs simple, mixed with smaller, many-branched ones. **LF:** basal 0.5–2.5 cm, pinnately divided, obovate to ovate or oblong in outline; ultimate segments 2–10 mm, obovate or oblong; cauline short-petioled to sessile, reduced distally

on st. **FL:** sepals 2–2.5 mm, persistent; petals 3.5–4.5 mm, 1.5–2.5 mm wide, white or pink. **FR:** 2–6 mm, 2–3 mm wide, ovoid to ± oblong, cylindric, ± appressed, glabrous; style 0.2–1 mm; pedicel erect to ascending, 3–10 mm. **SEED:** 4–8, oblong, 1–1.5 mm. Loose talus, mica schist, moraines, rock crevices; 1500–3350 m. CaRH (Lassen Peak); to BC. [*S. o.* var. *congesta* Rollins] Jul–Aug ★

STANLEYA PRINCE'S PLUME

[Ann] per, subshrub; hairs 0 or simple, glaucous. **LF:** basal, proximal-most cauline petioled, simple to entire or 1(2)-pinnately lobed; middle, distal cauline petioled to sessile, base occ lobed or sagittate. **INFL:** dense, elongated. **FL:** sepals oblong to linear, spreading to reflexed, base not sac-like; petals yellow to white, long-clawed; filaments equal; anthers linear, coiled. **FR:** silique, dehiscent, linear, flat parallel to septum or cylindric, unsegmented; stalk above receptacle [0.4]0.6–2.8 cm; style 0 or short, stigma entire. **SEED:** 10–70, in 1 row, oblong, wingless. 7 spp.: w US. (E.S. Stanley, English ornithologist, 1775–1851) Concentrates selenium to TOXIC levels, rarely eaten.

1. Middle, distal cauline lvs sessile, base lobed to sagittate . . . .*S. viridiflora*
1′ Middle, distal cauline lvs petioled, base not lobed or sagittate
  2. Seeds 46–70; petal claw glabrous; filaments 5–13 mm; petal blade 0.3–1 mm wide, linear . . . . *S. elata*
  2′ Seeds 10–38; petal claw hairy; filaments 11–28 mm; petal blade 2–3 mm wide, oblanceolate to oblong
     . . . . *S. pinnata* var. *pinnata*

S. elata M.E. Jones Per, glaucous. **ST:** erect, 6–15(18) dm, simple or few-branched. **LF:** basal early-deciduous; proximal, mid-cauline (5.5)8–21(26) cm, 2–8(13) cm wide, broadly lanceolate or oblong to ovate, entire (proximally few-lobed), gradually smaller distally on st, base not lobed. **INFL:** 6–20 cm, dense. **FL:** sepals 7–11 mm; petals 8–13 mm, 0.3–1 mm wide, glabrous, yellow to ± white, claw 4–7 mm, glabrous; filament bases papillate. **FR:** 4–9(10.5) cm, 1.5–2 mm wide; stalk above receptacle 7–20 mm; style 0.2–1.5 mm; pedicel spreading to reflexed, (5)7–11(15) mm. **SEED:** 46–70, 1.5–2.6 mm, oblong. $2n=28$. Among boulders in canyons, scrub; 1000–2500 m. SNE, DMtns; NV, AZ. May–Jul

S. pinnata (Pursh) Britton var. *pinnata* (p. 571) Per to subshrub. **ST:** (1.2)3–12(15.3) dm, glaucous, hairs 0 or sparse; base branched, woody. **LF:** basal, proximal cauline 3–15 cm, 2–5 cm wide, oblanceolate to wide-lanceolate or ovate, pinnately lobed; distal cauline entire or few-lobed, base not lobed. **INFL:** 1–3 dm, dense; buds ± yellow. **FL:** sepals 8–16 mm; petals 8–20 mm, 2–3 mm wide, yellow, claw 4–10 mm, densely hairy inside; filament base hairy. **FR:** 3–9

cm, 1.5–3 mm; stalk above receptacle 7–28 mm; pedicel spreading, 3–11 mm, hairs 0 or few. **SEED:** (10)28–38, 2.5–4.5 mm, oblong. $2n=28,56$. Chaparral, open sites, slopes, canyons, desert scrub, woodland, dunes; < 2900 m. s SNF, SnJV, SCoR, SCo, WTR, SnGb, PR, GB, DMoj; to OR, MT, KS, NM. [*S. p.* var. *inyoensis* (Munz & J.C. Roos) Reveal] *S. pinnata* var. *inyoensis* based on subshrubby pls, but these sporadic throughout range. Apr–Sep

S. viridiflora Nutt. (p. 571) GREEN-FLOWERED PRINCE'S PLUME Per, glabrous throughout, glaucous. **ST:** (2.5)4–12(14) dm, simple or branched distally. **LF:** basal (2.2)5–18(22) cm, lanceolate to oblanceolate or ovate, entire to occ dentate (pinnately lobed); middle, distal cauline entire, sessile, base lobed to sagittate. **FL:** sepals 12–18 mm; petals 13–20 mm, 1–3 mm wide, lemon-yellow to white; claw 7–11 mm, glabrous; filaments glabrous. **FR:** 3–6(7) cm, 1.2–2 mm wide; stalk above receptacle (6)11–22(25) mm; style to 0.3 mm; pedicel spreading, 4–9(12) mm. **SEED:** 28–50, 2–3 mm, oblong. $2n=28$. Cliffs, shale, clay knolls, steep bluffs, white ash deposits; 1300–2700 m. s MP; to ID, MT, CO, UT. May–Jul ★

STREPTANTHELLA

1 sp. (Latin: small *Streptanthus*)

S. longirostris (S. Watson) Rydb. (p. 571)　Ann; gen glaucous, glabrous or hairy proximally. **ST**: (1.2)2–6(7.5) dm. **LF**: proximal cauline 2–5.5(6.5) cm, lanceolate to oblanceolate; distal cauline linear, sessile, entire. **FL**: sepals erect, 2–4(5) mm, lateral pair base ± sac-like; petals (3.5)4–6(7) mm, 0.7–1.1 mm wide, spoon-shaped, crinkled, clawed, white to ± yellow, purple-veined; stamens 4 long, 2 short. **FR**: silique, dehiscent exc near tip, (2.5)3.5–6(7) cm, 1.5– 2(2.2) mm wide, linear, flat parallel to septum, unsegmented, tip narrowed to beak (2)3.5–6(8) mm; stigma entire; pedicel recurved to reflexed, (1)2–5(7). **SEED**: (12)16–28(34), in 1 row, 2–3 mm, flat; wing 0.3–0.7 mm wide. $2n=28$. Common in sandy soils, desert scrub, woodland, slopes, chaparral; 50–2200 m. s SnJV, SCoRI, GB, D; to WA, CO, NM, Baja CA. Mar–Jun

STREPTANTHUS　JEWELFLOWER

Ann to per, gen ± glaucous; hairs simple or 0. **LF**: basal rosetted or not, petioled, entire or dentate to pinnately lobed or divided; cauline sessile, occ petioled, base gen lobed or clasping. **INFL**: elongated. **FL**: radial or bilateral; calyx urn- or occ bell-shaped, sepals erect, base ± sac-like, keeled or not; petal blade narrower to wider than proximal 1/2, gen channeled, margins ± crinkled or not; stamens in 3 pairs of unequal length, or 4 long and 2 short, longest filaments fused or free. **FR**: silique, dehiscent, linear, flat parallel to septum, unsegmented; stigma entire or 2-lobed. **SEED**: 10–120, in 1 row, gen winged. 35 spp.: sw US, n Mex. (Greek: twisted fl, from wavy-margined petals)

1. All cauline lvs petioled or sessile, base not lobed, sagittate, or clasping
  2. Ann; mid-cauline lvs ± linear; filaments of long stamen pairs fused; fr 1.2–2 mm wide . . . . ***S. barbiger***
  2′ Per; mid-cauline lvs round to broadly obovate; all filaments free; fr 2.5–3.2(3.5) mm wide. . . . . ***S. howellii***
1′ Middle or distal cauline lvs sessile, base lobed, sagittate, or clasping
  3. Raceme bracted at least below most proximal fls
    4. Basal, proximal cauline lvs 1–2-pinnately lobed or divided; fr not constricted between seeds (± constricted)
      5. Mid-cauline lf base not lobed, pinnately divided into thread-like segments; fr abruptly reflexed at stalk above receptacle. . . . . ***S. diversifolius***
      5′ Mid-cauline lf base lobed to dentate or entire (pinnately lobed, not thread-like-segmented); fr ascending to spreading (reflexed).
        6. Fr 6–12 cm, 1.7–2.5 mm wide, valve midvein distinct; seeds 60–100, 2–2.5 mm; petals white, veins violet . . . . ***S. farnsworthianus***
        6′ Fr 2–5 cm, 1.2–1.7 mm wide, valve midvein obscure; seeds 22–38, 1–1.5 mm; petals rose-purple . . . ***S. fenestratus***
    4′ Basal, proximal cauline lvs entire, wavy, dentate, or ± lobed; fr gen constricted between seeds
      7. Ann; sepals 4–5 mm; pedicel not expanded at receptacle. . . . . ***S. gracilis***
      7′ Bien or short-lived per; sepals 6–10(13) mm; pedicel expanded at receptacle
        8. Mid-cauline lvs lance-linear; longest filament pair fused, 13–16 mm; short filaments 7–9 mm. . . . ***S. oblanceolatus***
        8′ Mid-cauline lvs oblong to obovate or ± round; longest filament pair free, (5)7–11 mm; short filaments (1.5)3–5 mm . . . . ***S. tortuosus***
  3′ Raceme not bracted
    9. Raceme with terminal, sterile fl cluster; fr sparsely to densely stiff-hairy
      10. Fr 1.3–2.5 cm, 2.5–3.5 mm wide; seeds wingless; petals purple, veins darker, proximal 1/2 narrower than blade; Mount Hamilton Range, Santa Clara Co.. . . . . ***S. callistus***
      10′ Fr 3.5–11.4 cm, 1.5–2.5 mm wide; seeds winged; petals purple with white margins, ± purple-white, lemon-yellow, or ± yellow-white, proximal 1/2 at least as wide as blade; not Mount Hamilton Range
        11. Mid-cauline lf base not lobed; long filament pairs fused; Mount Diablo, Contra Costa Co.. . . . . ***S. hispidus***
        11′ Mid-cauline lf base lobed; only longest filament pair fused; Fresno, Merced, Monterey, San Benito cos. . . . . ***S. insignis***
          12. Sepals of fertile fls ± purple; terminal sterile fl cluster dark purple; petals ± purple-white; fr sparsely (densely) stiff hairy. . . . . subsp. ***insignis***
          12′ Sepals of fertile fls ± green-yellow or ± purple; terminal sterile fl cluster pale yellow, ± green-yellow, or ± purple; petals lemon yellow or ± yellow-white; fr moderately to densely stiff hairy . . . . subsp. ***lyonii***
    9′ Raceme terminal sterile fl cluster 0; fr gen glabrous
      13. Per, caudex gen woody; all stamens fertile, free; calyx ± bell-shaped; anthers 2.5–5 mm
        14. Distal lvs ± same size as mid-cauline; petals darker-veined, ± crinkled . . . . ***S. barbatus***
        14′ Distal lvs smaller than mid-cauline; petals not darker-veined, not crinkled
          15. Fr descending, recurved; petal proximal 1/2 yellow-green, blades purple or ± brown . . . . ***S. longisiliquus***
          15′ Fr spreading to ascending, straight or ± curved; petals gen all white, yellow, purple, or ± brown
            16. Stamens 4 long, 2 short; fr valve midveins obscure at least near tip; petal proximal halves white or pale yellow
              17. Sepals pale yellow to white; stigma ± unlobed; racemes open in bud; st 2.5–8.6 dm; woody caudex elevated . . . . ***S. bernardinus***
              17′ Sepals purple; stigma 2-lobed; racemes dense in bud; st (2.5)6–15(18) dm; woody caudex 0 . . . . ***S. campestris***

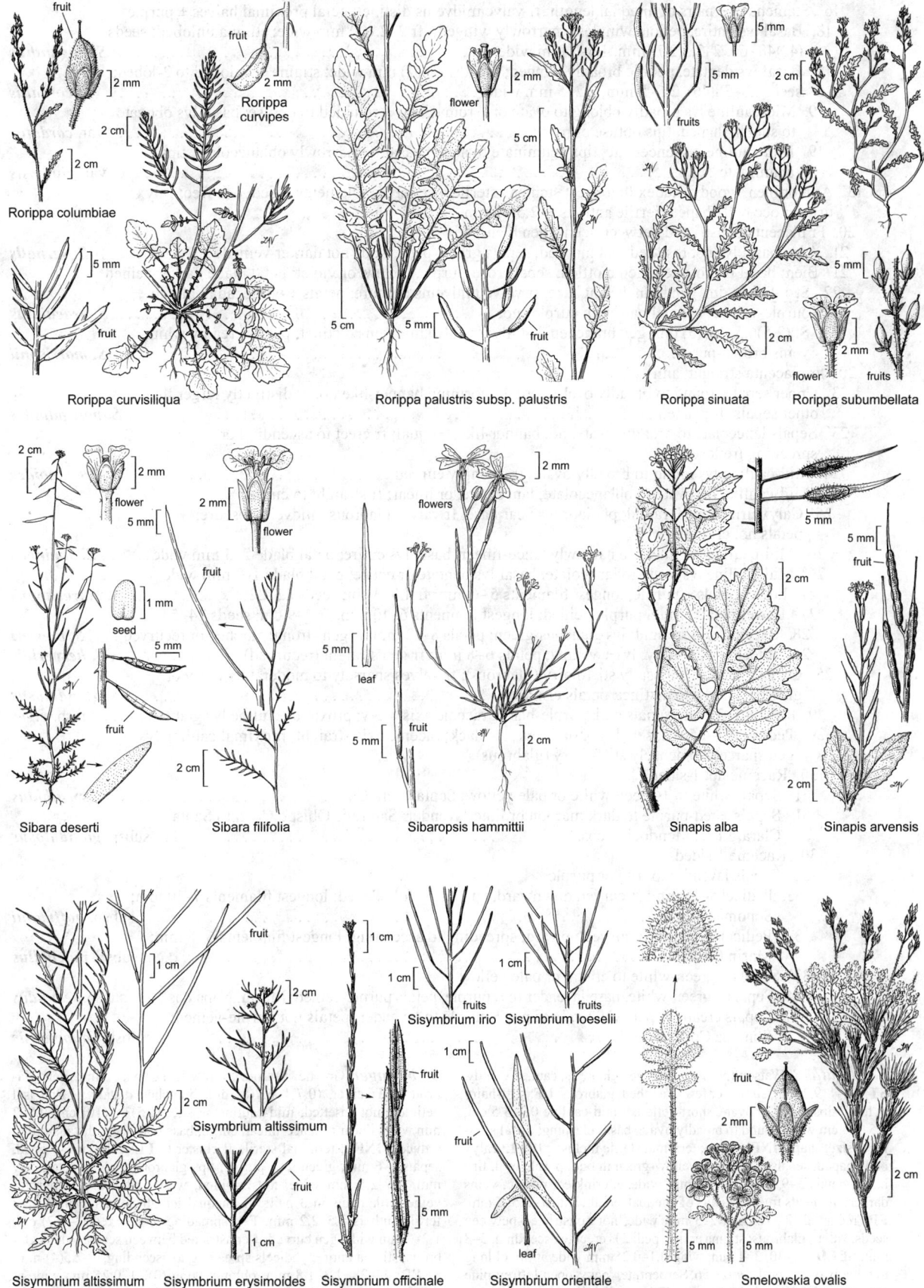

Rorippa columbiae

Rorippa curvipes

Rorippa curvisiliqua

Rorippa palustris subsp. palustris

Rorippa sinuata

Rorippa subumbellata

Sibara deserti

Sibara filifolia

Sibaropsis hammittii

Sinapis alba

Sinapis arvensis

Sisymbrium altissimum

Sisymbrium altissimum

Sisymbrium erysimoides

Sisymbrium irio

Sisymbrium loeselii

Sisymbrium officinale

Sisymbrium orientale

Smelowskia ovalis

16′ Stamens in 3 pairs of unequal length; fr valve midveins distinct; petal proximal halves ± purple
  18. Basal lvs entire, petiole wingless (narrowly winged); fr 2–2.7(3) mm wide; stigma unlobed; seeds
     (42)48–60, 2–2.5(2.7) mm, 1.5–2 mm wide . . . . ***S. oliganthus***
  18′ Basal lvs dentate, petiole broadly winged; fr (2.5)3–6(7) mm wide; stigma ± unlobed to 2-lobed;
     seeds 20–38(46), 2.5–5 mm, 2.2–5 mm wide. . . . . ***S. cordatus***
    19. Mid-cauline lvs broadly oblong to ovate or ± round, tips ± rounded to obtuse; basal lvs obovate
      to spoon-shaped, tips obtuse or round . . . . var. ***cordatus***
    19′ Mid-cauline lvs lanceolate, tips acuminate to acute; basal lvs narrowly oblanceolate, tips
      acuminate . . . . var. ***piutensis***
13′ Ann or bien, woody caudex 0; longest stamens sterile, occ fertile, filaments fused, occ free; calyx
  urn- or occ bell-shaped; fertile anthers 1–2.5(3) mm
  20. Fr placenta constricted between seeds; gen bien
    21. Ann; basal lvs not rosetted, not mottled; sepals green; petal blade not darker-veined. . . . . ***S. vernalis***
    21′ Bien; basal lvs rosetted, gen mottled; sepals rose-purple, yellow, or violet; petal blade darker-veined
      22. St 1.4–5(6) dm, gen branched at base; fr valve midveins obscure; petals ± white, adaxial pair ±
        purple-veined, abaxial pair with purple spot. . . . . ***S. brachiatus***
      22′ St (2.5)6.5–12(15) dm, gen branched distally; fr valve midveins distinct; petals creamy white,
        veins brown-purple. . . . . ***S. morrisonii***
  20′ Fr placenta straight; ann
    23. Upper sepal ± round to broadly ovate-cordate, forming banner-like hood, distinctly larger than
      other sepals; fr pendent. . . . . ***S. polygaloides***
    23′ Sepals lanceolate to broadly ovate, not banner-like, ± equal; fr erect to ascending or
      spreading (reflexed).
      24. Mid-cauline lvs round to broadly ovate; fr strongly curved . . . . ***S. drepanoides***
      24′ Mid-cauline lvs ovate to oblanceolate, lanceolate, or linear; fr straight to curved
        25. Calyx urn-shaped, radial; pl glabrous near base; fr valves glabrous, midvein obscure;
          petals not crinkled
          26. Mid-cauline lvs linear to narrowly lance-linear; basal lvs entire; petal blade 2–3 mm wide. . . . . . . ***S. vimineus***
          26′ Mid-cauline lvs ovate to lanceolate; basal lvs dentate or entire; petal blade 1–2 mm wide
            27. Lower petals ± purple; longest filaments 5–6 mm; fr 1.3–3 cm; seeds 12–22. . . . . ***S. batrachopus***
            27′ Lower petals white, purple-veined; longest filaments 6–10 mm; fr 3–9 cm; seeds 24–54
              28. Raceme axis straight; lvs glaucous-green; petals 8–12 mm; fr gen strongly arched or recurved . . . . ***S. breweri***
              28′ Raceme axis wavy; lvs ± yellow; petals 6–8 mm; fr gen straight (recurved) . . . . ***S. hesperidis***
        25′ Calyx bilateral; pl basally stiff-hairy (glabrous); fr valves sparsely to moderately hairy or
          glabrous, midvein distinct; petals crinkled . . . . ***S. glandulosus***
          29. Pedicel 1–3.2 cm; sepals dark purple-black; raceme axis wavy; proximal cauline lvs glabrous . . . . subsp. ***niger***
          29′ Pedicel 0.2–1.5 cm; sepals not dark purple-black; raceme axis straight; proximal cauline lvs
            gen sparsely to densely stiff-hairy (glabrous)
            30. Raceme not 1-sided
              31. Sepals white to ± green-white or pale yellow; Santa Clara Co. . . . . subsp. ***albidus***
              31′ Sepals ± red-purple to dark maroon or lilac-lavender; San Luis Obispo Co. n to Santa
                Clara, Lake, Mendocino cos. . . . . subsp. ***glandulosus***
            30′ Raceme 1-sided
              32. Sepals lavender to rose or purple
                33. Pedicel 5–15 mm; fr curved downward, spreading to reflexed; longest filaments 7–10 mm;
                  Sonoma Co. . . . . subsp. ***hoffmanii***
                33′ Pedicel 2–5 mm; fr curved upward, spreading to ascending; longest filaments 6–8 mm;
                  Marin Co. . . . . subsp. ***pulchellus***
              32′ Sepals ± green-white to cream or pale yellow
               34. Sepals ± green-white, base lavender to ± purple; petals purple-veined; Marin, Napa cos . . . . subsp. ***secundus***
              34′ Sepals cream or pale yellow, base not ± purple to lavender; petals not purple-veined;
                Sonoma Co. . . . . subsp. ***sonomensis***

S. barbatus S. Watson (p. 571)   Per, gen glabrous, caudex woody. **ST:** 1.5–8(9.2) dm, simple or few-branched, glabrous. **LF:** stiff-hairy at tooth tips; basal obovate, short-petioled; mid-cauline 0.7–3.5 cm, 0.5–2.4 cm wide, round to broadly ovate, bases clasping; distal similar in size, shape. **INFL:** open; terminal sterile fl cluster 0. **FL:** calyx bell-shaped, sepals 4–7 mm, ± yellow-green in bud, purple in fl, tips hairy; petals 5–9 mm, 0.3–0.7 mm wide, ± crinkled, purple, veins darker; filaments free, in 3 pairs of unequal length; anthers 2.5–5 mm. **FR:** recurved, 2–7(8) cm, 2–3 mm wide, not constricted between seeds, valves glabrous; stigma entire; pedicels erect to ascending, 3–8 mm. **SEED:** 16–30, 3–4 mm; wing 0.1–0.25 mm wide at tip, oblong to broadly ovate, or 0. 2*n*=28,56. Serpentine, rocky, open Jeffrey-pine forest; 800–2200 m. KR (Siskiyou, Tehama, Trinity cos.). Jun–Aug

S. barbiger Greene   BEARDED JEWELFLOWER   Ann, gen glabrous. **ST:** erect, (0.7)1–6.7(8) dm, branched distally. **LF:** basal petioled, not rosetted; mid-cauline sessile, (1.5)3–9(10) cm, 0.5–2 mm wide, linear to lance-linear, entire, base not lobed; distal similar, reduced. **INFL:** terminal sterile fl cluster 0. **FL:** calyx urn-shaped, sepals 4–6 mm, green to ± purple, tips glabrous (hairy); petals 5–9 mm, 1.5–2.5 mm wide, not crinkled, white, adaxial pair purple-veined; filaments in 3 pairs of unequal length, longer pairs fused; fertile anthers 1.5–2.2 mm. **FR:** spreading to reflexed, 2–6(7) cm, 1.2–2 mm wide, gen curved, ± constricted between seeds, valves glabrous; stigma entire; pedicels spreading to ascending, 1–2.5(4) mm. **SEED:** 22–38, 1.3–1.8 mm, oblong; wing (0)0.1–0.25 mm wide at tip. 2*n*=28. Serpentine barrens, chaparral; 200–1500 m. s NCoR, NCoRH. May–Aug   ★

S. batrachopus J.L. Morrison (p. 571) TAMALPAIS JEWELFLOWER
Ann, glabrous. **ST:** simple or branched near base, 0.3–1.7(2.8) dm.
LF: basal petioled, mottled, not rosetted; mid-cauline sessile, 0.5–2.5
cm, 1–7 mm wide, lanceolate, entire or dentate, base lobed; distal
similar, reduced, entire. **INFL:** terminal sterile fl cluster 0. **FL:** calyx
urn-shaped, sepals 3–5 mm, ovate, keeled, ± red-purple; petals 5–8
mm, 1–1.5 mm wide, not crinkled, adaxial pair ± white, veins purple,
lower ± purple; filaments in 3 pairs of unequal length, longest pair
fused, 5–6 mm; fertile anthers 1.4–1.7 mm. **FR:** spreading to reflexed,
1.3–3 cm, 1–1.5 mm wide, gen straight, ± constricted between seeds;
stigma entire; pedicels spreading, 1–2.5(4) mm. **SEED:** 12–22, 1.3–2
mm, oblong; wing (0)0.1–0.3 mm wide at tip. 2*n*=28. Serpentine barrens, chaparral; 100–600 m. s NCoRO, nw SnFrB (Mount Tamalpais). May–Jun ★

S. bernardinus (Greene) Parish LAGUNA MOUNTAINS
JEWELFLOWER Per, gen glabrous, caudex woody, elevated. **ST:**
erect, 2.5–8.6 dm, simple or few-branched. **LF:** basal oblanceolate
to spoon-shaped, not rosetted; petioles ciliate; mid-cauline 2–10
cm, 0.5–2.8 cm wide, ovate to lanceolate, sessile, entire, acute to
acuminate, base clasping; distal reduced. **INFL:** open in bud; terminal sterile fl cluster 0. **FL:** calyx bell-shaped, sepals 5–9 mm, not
keeled, pale yellow to white, tips hairy; petals 7–11 mm, 1–1.3 mm
wide, not crinkled, white; filaments free, 4 long, 2 short; anthers all
fertile, 3.5–4.5 mm. **FR:** ascending to spreading, 5–12.7 cm, 1.5–3
mm wide, straight or curved, not constricted between seeds; valve
midveins obscure; stigma ± unlobed; pedicels spreading, 4–7.5
mm. **SEED:** 36–56, 2–3 mm, oblong; wing 0.1–0.2 mm wide at tip.
2*n*=14. Montane conifer forest, chaparral; 1200–2500 m. e TR, PR;
Baja CA. Jun–Aug ★

S. brachiatus F.W. Hoffm. (p. 571) Bien, glabrous, glaucous. **ST:**
1.4–5(6) dm, branched at base (simple). **LF:** basal rosetted, broadly
obovate to ± round, dentate, fleshy, mottled; mid-cauline sessile, 0.7–
3.7 cm, 0.3–1.5 cm wide, ovate to cordate, serrate-dentate to entire,
base lobed to clasping; distal similar, reduced. **INFL:** open; terminal sterile fl cluster 0. **FL:** calyx urn-shaped, sepals 5–7 mm, ovate,
keeled, base rose-purple to ± yellow; petals 7–10 mm, 1–1.5 mm
wide, not crinkled, adaxial pair ± purple-veined, abaxial pair purple
spotted; filaments in 3 pairs of unequal length, longest pair fused,
8–10 mm; fertile anthers 2–2.5 mm. **FR:** spreading, 4–6 cm, 1–1.3
mm wide, gen straight, constricted between seeds and at placenta;
valves glabrous, midvein obscure; stigma entire; pedicels spreading-
ascending, 1–2 mm, straight. **SEED:** 22–30, 1.8–2.5 mm, oblong;
wing 0–0.1 wide at tip. 2*n*=28. Serpentine barrens, open chaparral or
woodland; 600–950 m. s NCoRI. [*S. b.* subsp. *hoffmanii* R.W. Dolan
& LaPré] Jun–Jul ★

S. breweri A. Gray (p. 571) Ann, glaucous, glabrous exc occ
sepals. **ST:** (0.5)1.5–6.5(8) dm, simple or branched at base. **LF:**
basal not rosetted, broadly ovate to obovate, dentate or entire; mid-
cauline 1.5–10 cm, 0.3–4.5 cm wide, ovate, sessile, base clasping;
distal much reduced, narrowly lanceolate, entire. **INFL:** open, axis
straight; terminal sterile fl cluster 0. **FL:** calyx urn-shaped, sepals 5–7
mm, ovate, keeled, purple or white; petals 8–12 mm, 1–2 mm wide,
not crinkled, white, purple-veined or adaxial pair white; filaments
in 3 pairs of unequal length, longest pair fused, 7–10 mm; fertile
anthers 2–3 mm. **FR:** erect to ascending, 3–9 cm, 1–1.5 mm wide,
gen strongly arched or recurved; valves glabrous, midvein obscure;
stigma entire; pedicel spreading to erect, 1.5–5(6) mm, straight.
SEED: 24–54, 1–1.5 mm, oblong; wing 0–0.1 wide at tip. Serpentine
barrens in chaparral or woodland; 250–2100 m. s-most KR, NCoRH,
c&s NCoRI, e SnFrB, SCoRI. May–Jul

S. callistus J.L. Morrison (p. 571) MOUNT HAMILTON
JEWELFLOWER Ann, bristly hairy. **ST:** 2–9 cm, simple or branched,
bristly esp proximally. **LF:** basal not rosetted, oblong to obovate, dentate; mid-cauline sessile, 0.8–1.7 cm, 0.4–1.3 cm wide, broadly ovate
to obovate, bristly, dentate, base clasping; distal similar, reduced.
INFL: open, axis straight; terminal sterile fl cluster present. **FL:**
calyx bell-shaped, sepals 3–5 mm, narrowly ovate, keeled, green to
± purple; petals 8–11 mm, 2.5–3.5 mm wide, not crinkled, purple,
veins darker; filaments in 3 pairs of unequal length, longest pair
fused, 5–6.5 mm; fertile anthers 1.4–1.8 mm. **FR:** spreading, 1.3–2.5
cm, 2.5–3.5 mm wide, curved upwards; valves stiff-hairy, midvein

prominent; stigma ± 2-lobed; pedicels spreading, 2–3 mm. **SEED:**
40–60, 1.2–1.5 mm, ovoid; wing 0. 2*n*=28. Open chaparral, gravelly sedimentary scree; 500–900 m. se SnFrB (Arroyo Bayo, Mount
Hamilton Range). Most endangered taxon in genus. Apr–May ★

S. campestris S. Watson (p. 571) SOUTHERN JEWELFLOWER
Per, woody caudex 0, glaucous, gen glabrous. **ST:** (2.5)6–15(18) dm,
glaucous, simple or few-branched. **LF:** glabrous exc margins ciliate, basal gen rosetted, oblanceolate to obovate; mid-cauline sessile,
3.5–11(15) cm, 0.6–1.4 cm wide, lanceolate to narrowly ovate, entire
to wavy (dentate), base lobed to clasping; distal reduced. **INFL:**
dense in bud, later open; terminal sterile fl cluster 0. **FL:** calyx bell-
shaped, sepals 7–10 mm, broadly ovate to oblong, not keeled, purple,
tips hairy or not; petals 9–12 mm, 0.5–1 mm wide, not crinkled,
light purple, proximal halves pale yellow; filaments free, 4 long, 2
short; anthers all fertile, 3–4 mm. **FR:** ascending to spreading, 6–14
cm, 2–3.5 mm wide, straight or ± curved, not constricted between
seeds; valves glabrous, midvein obscure; stigma 2-lobed; pedicels
spreading-ascending, 5–18 mm. **SEED:** 50–102, 2–3 mm, oblong;
wing 0.1–0.2 mm wide at tip. Open, rocky conifer forest, chaparral,
woodland; 900–2300 m. TR, PR; n Baja CA. May–Jun ★

S. cordatus Nutt. Per, caudex simple or branched, glaucous, gen
glabrous. **ST:** (1)3–9(11) dm, simple or few-branched. **LF:** basal
rosetted, spoon-shaped to narrowly oblanceolate; petiole ciliate,
broadly winged; mid-cauline sessile, 2–9 cm, 0.7–4.5(6) cm wide,
broadly oblong to ovate or ± round to lanceolate, entire or few-
toothed, base lobed to clasping; distal reduced. **INFL:** open; terminal
sterile fl cluster 0. **FL:** calyx bell-shaped, sepals 5–12 mm, broadly
oblong, not keeled, ± green-brown to purple, tips hairy or not; petals
9–15 mm, 0.7–1 mm wide, not crinkled, purple to ± brown; filaments
free, in 3 pairs of unequal length, longest pair 7.5–10 mm; anthers
all fertile, 2.5–5 mm. **FR:** ascending to ± spreading, 5–10.5(14.5)
cm, (2.5)3–6(7) mm wide, straight, not constricted between seeds;
valves glabrous, midvein distinct; stigma ± unlobed to 2-lobed; pedicels spreading-ascending, 3–11(14) mm. **SEED:** 20–38(46), 2.5–5
mm, broadly oblong to ± round; wing 0.1–0.9 mm wide, continuous.

var. ***cordatus*** (p. 571) **LF:** basal obovate to spoon-shaped, tip
obtuse to rounded; mid-cauline broadly oblong to ovate or ± round,
± round to obtuse at tip. **FR:** 2.5–6(7) mm wide; stigma ± entire to
prominently 2-lobed. 2*n*=28,56. Common. Rocky, sandy sagebrush
scrub, pinyon/juniper woodland, talus, calcareous outcrops; 1200–
3100 m. e CaRH, GB, e DMtns; to se OR, WY, n NM. [*S. c.* var.
duranii Jeps.] Apr–Jul

var. ***piutensis*** J.T. Howell PIUTE MOUNTAINS JEWELFLOWER
LF: basal narrowly oblanceolate, tip acuminate; mid-cauline lanceo-
late, tip acuminate or acute. **FR:** 3–4.5 mm wide; stigma ± entire.
Open chaparral, Piute-cypress stands; 1200–1700 m. s SNH (Piute
Mtns). Jun–Jul ★

S. diversifolius S. Watson (p. 571) VARIED-LEAVED JEWELFLOWER
Ann, glabrous throughout. **ST:** erect, (1)2–9(10.5) dm, branched distally. **LF:** basal not rosetted, pinnately divided; mid-cauline petioled,
2–12 cm, pinnately divided into thread-like segments 0.4–2.7 cm,
0.5–1.5 mm wide, lf base not lobed; distal cauline sessile, clasping,
lance-linear to ovate or cordate, entire. **INFL:** open; terminal sterile
fl cluster 0; bracted below or between most proximal 1–2 fls. **FL:**
calyx urn-shaped, sepals 5–7 mm, keeled, yellow or ± purple; petals 8–16 mm, 1.5–3.5 mm wide, recurved, crinkled, pale yellow to
± white, veins purple; filaments free, in 3 pairs of unequal length,
longest pair 4–8 mm; fertile anthers 3.5–4 mm. **FR:** abruptly reflexed
at stalk above receptacle, 3–9 cm, 1–1.5(2.2) mm wide, straight, not
constricted between seeds; valves glabrous, midvein obscure; stigma
± 2-lobed; pedicels spreading, 2–10 mm, straight. **SEED:** (22)38–80,
1–1.5(2.2) mm, oblong; wing 0–0.1(0.4) mm wide at seed tip. 2*n*=28.
Open woodland, rocky slopes; 200–1900 m. n SNF, c&s SN. Apr–Jul

S. drepanoides Kruckeb. & J.L. Morrison SICKLE-FRUIT
JEWELFLOWER Ann, gen glabrous. **ST:** erect, 0.4–3.5(4.5) dm,
base simple or branched. **LF:** basal not rosetted, round, teeth few,
blunt; mid-cauline sessile, 1.3–9 cm, 1–7.5 cm wide, round to broadly
ovate, succulent, entire to shallowly dentate, base lobed to cordate;
distal similar, reduced. **INFL:** dense; terminal sterile fl cluster 0.
FL: calyx urn-shaped, sepals 5–7 mm, ovate, ± green-yellow or ±

purple, tips glabrous or hairy; petals 7–10 mm, 1–1.5 mm wide, upper ± white or with purple-veins, lower ± purple; filaments in 3 pairs of unequal length; longest pair fused, 8–10 mm, lower pair fused ± proximally; fertile anthers 1.8–2.5 mm. **FR:** spreading, 3–9 cm, 1–1.2 mm wide, strongly curved, ± constricted between seeds or not; pedicels ± ascending, 1.4–5 mm, straight. **SEED:** 30–50, 1–1.5 mm, oblong; wing 0 or to 0.1 mm wide at seed tip. $2n=28$. Open chaparral or Jeffrey-pine woodland, on serpentine; 250–1800 m. s-most KR, NCoRH, n NCoRI, n SNF (Butte Co.). May–Jul ★

S. farnsworthianus J.T. Howell FARNSWORTH'S JEWELFLOWER Ann, glabrous, gen ± green-purple. **ST:** 2–10 dm, erect, simple or branched distally. **LF:** basal rosetted, early-deciduous, 2-pinnately lobed or divided; mid-cauline sessile, 3–16 cm, 1–4 cm wide, lanceolate, dentate distally or entire, base clasping; distal reduced, entire. **INFL:** open; terminal sterile fl cluster 0; bracted proximally or only between most proximal 1–2 fls. **FL:** calyx urn-shaped, sepals 6–10 mm, ± keeled or not, violet-purple; petals 8–11 mm, 1.5–2 mm wide, recurved, crinkled, white, veins violet; filaments free, in 3 pairs of unequal length, longest pair 7–9 mm; fertile anthers 3.5–5 mm. **FR:** ascending, 6–12 cm, 1.7–2.5 mm wide, straight to ± curved, not constricted between seeds; valves glabrous, midvein distinct; stigma ± 2-lobed; pedicels spreading, 4–15 mm, straight. **SEED:** 60–100, 2–2.5 mm, ovate or ± round; wing 0.1–0.3 mm wide at seed tip. $2n=28$. Foothill woodland; 400–1400 m. c SN (Madera, Fresno cos.), s SN. May–Jun ★

S. fenestratus (Greene) J.T. Howell TEHIPITE VALLEY JEWELFLOWER Ann, glabrous. **ST:** erect, (0.5)1–2(3.5) dm, simple or branched proximally. **LF:** rosetted, 1- or 2-pinnately divided; mid-cauline ovate to lanceolate, coarsely dentate or entire (pinnately lobed but not thread-like-segmented), lf base lobed; distal clasping, entire. **INFL:** open; terminal sterile fl cluster 0; bracted proximally or only between most proximal 1–2 fls. **FL:** calyx ± urn-shaped, sepals 5–7 mm, ± keeled or not, purple; petals 9–15 mm, 2.5–3.5 mm wide, not crinkled, rose-purple; filaments free, in 3 pairs of unequal length, longest pair 5–7 mm; fertile anthers 2–2.5 mm. **FR:** ascending to spreading (reflexed), 2–5 cm, 1.2–1.7 mm wide, not or ± constricted between seeds; valve midvein obscure; stigma entire; pedicels spreading, 2–7 mm, straight. **SEED:** 22–38, 1–1.5 mm, oblong; wing 0–0.1 mm wide at seed tip. $2n=28$. Granite ledges, sand, open mixed-conifer/oak woodland; 1050–1800 m. s SNH (Kings River Canyon, Fresno Co.). May–Jun ★

S. glandulosus Hook. Ann, stiff-hairy basally (glabrous throughout). **ST:** (0.8)1.5–9(12) dm, simple to branched throughout. **LF:** basal not rosetted, petioled, early-deciduous, coarsely dentate to ± lobed; mid-cauline sessile, 1–12 cm, lance-linear to oblanceolate, entire to coarsely dentate, lf base lobed to clasping; distal similar, reduced, gen entire. **INFL:** open, 1-sided or not; terminal sterile fl cluster 0. **FL:** calyx bilateral, sepals (3)5–10(13) mm, lanceolate to broadly ovate, white to yellow, rose, purple, or purple-black, glabrous or bristles sparse; petals 7–17 mm, 1–3 mm wide, ± equal or adaxial pair longer, crinkled; filaments in 3 pairs of unequal length; longest pair fused, 5–13 mm; fertile anthers 1–2.5 mm. **FR:** ascending to spreading or reflexed, 3–11 cm, 1.5–2.5 mm wide, straight or curved, not constricted between seeds; valve midveins distinct, glabrous or sparsely to moderately hairy; stigma ± entire; pedicels ascending to spreading, 0.2–3.2 cm. **SEED:** 22–70, 1.5–2.1 mm, ovate to oblong; wing continuous, 0.1–0.5 mm wide. 8 subspp. recognized, 7 in CA, most local.

subsp. ***albidus*** (Greene) Al-Shehbaz et al. (p. 571) METCALF CANYON JEWELFLOWER **ST:** 3–12 dm, sparsely hairy proximally, glabrous distally. **LF:** cauline glabrous; distal entire. **INFL:** not 1-sided, axis straight; pedicels 4–8 mm, glabrous. **FL:** sepals 7–11 mm, ± green-white to white or pale yellow, glabrous; petals 11–17 mm, white, veins light brown or ± purple; longest filaments 8–10 mm. **FR:** spreading to ascending, straight, glabrous. $2n=28$. Serpentine, grassy, barren slopes; 100–800 m. se SnFrB (Santa Clara Co.). [*S. a.* Greene subsp. *a.*] Apr–Jul ★

subsp. ***glandulosus*** (p. 571) **ST:** 1–10 dm, densely to moderately bristly proximally, less distally. **LF:** cauline densely to sparsely bristly; distal entire or minutely dentate. **INFL:** not 1-sided, axis

straight; pedicels 3–15 mm, sparsely hairy or glabrous. **FL:** sepals 5–13 mm, ± red-purple to dark maroon or lilac-lavender, sparsely hairy or glabrous; petals 8–17 mm, lavender to purple, veins darker or not; longest filaments 8–13 mm. **FR:** ascending to spreading or reflexed, straight to recurved, glabrous or sparsely hairy. Serpentine or metamorphic (Franciscan formation), rocky, gen barren slopes, chaparral openings, steep woodland; 150–1400 m. s NCoRO (uncommon), s NCoRH, c&s NCoRI, SnFrB, SCoRO (uncommon), n&c SCoRI. [*S. albidus* subsp. *peramoenus* (Greene) Kruckeb.; *S. g.* subsp. *arkii* M.S. Mayer; *S. g.* subsp. *raichei* M.S. Mayer] [Mayer & Beseda 2010 Ann Missouri Bot Gard 97:106–116] Apr–Jul

subsp. ***hoffmanii*** (Kruckeb.) M.S. Mayer & D.W. Taylor HOFFMAN'S BRISTLY JEWELFLOWER **ST:** 1.5–4.5(7) dm, densely bristly proximally, sparsely to moderately distally. **LF:** cauline moderately to sparsely bristly; distal flat, entire or minutely dentate. **INFL:** 1-sided, axis straight; pedicels 5–15 mm, sparsely hairy or glabrous. **FL:** sepals 6–7 mm, lavender or rose to purple, sparsely hairy or glabrous; petals 10–12 mm, lavender to ± purple; longest filaments 7–10 mm. **FR:** spreading to reflexed, recurved, sparsely hairy or glabrous. Serpentine outcrops; 125 m. sw NCoRO. May–Jul ★

subsp. ***niger*** (Greene) Al-Shehbaz et al. (p. 571) TIBURON JEWELFLOWER **ST:** 2–11 dm, glabrous. **LF:** cauline glabrous; distal entire. **INFL:** not 1-sided, axis wavy; pedicels 1–3.2 cm, glabrous. **FL:** sepals 7–8 mm, dark purple-black, glabrous; petals 10–12 mm, dark purple; longest filaments 8–10 mm. **FR:** 3.5–6.7 cm, spreading-ascending, straight, glabrous. $2n=28$. Serpentine outcrops in grassland; < 150 m. n CCo (Tiburon Peninsula, Marin Co.). [*S. n.* Greene] May–Jul ★

subsp. ***pulchellus*** (Greene) Kruckeb. MOUNT TAMALPAIS BRISTLY JEWELFLOWER **ST:** 1–3(3.7) dm, densely bristly proximally, glabrous distally. **LF:** cauline bristly; distal entire. **INFL:** 1-sided, axis straight; pedicels 2–5 mm, sparsely hairy or glabrous. **FL:** sepals 5–7 mm, rose to lavender or purple, glabrous or sparsely hairy; petals 8–11 mm, purple; longest filaments 6–8 mm. **FR:** curved upwards, spreading to ascending, glabrous or sparsely hairy. Dry, open grassland, chaparral, open conifer/oak woodland, occ on serpentine; 150–800 m. nw SnFrB (Marin Co.). [*S. g.* var. *p.* (Greene) Jeps.] May–Jun ★

subsp. ***secundus*** (Greene) Kruckeb. **ST:** 1.5–9.3 dm, densely stiff-hairy proximally, sparsely hairy distally. **LF:** cauline densely to moderately hairy; distal dentate. **INFL:** 1-sided, axis straight; pedicels 7–10 mm, sparsely hairy or glabrous. **FL:** sepals 6–7 mm, ± green-white, base lavender to ± purple, glabrous; petals 12–16 mm, veins purple; longest filaments 7–9 mm. **FR:** recurved, spreading to reflexed, glabrous. Rocky, open slopes, forest openings; 100–1200 m. s NCoRO, sw NCoRI, nw SnFrB. Apr–Jun

subsp. ***sonomensis*** (Kruckeb.) M.S. Mayer & D.W. Taylor UNCOMMON JEWELFLOWER **ST:** 2.3–7(9) dm, moderately to densely bristly proximally, glabrous distally. **LF:** cauline bristly; distal entire or finely dentate. **INFL:** 1-sided, axis straight; pedicels 3–10 mm, sparsely hairy or glabrous. **FL:** sepals 5–6 mm, cream or pale yellow, base not ± purple or lavender, glabrous or sparsely hairy; petals 10–17 mm, veins not purple; longest filaments 7–10 mm. **FR:** spreading to ± reflexed, arched, glabrous or hairs sparse. Serpentine outcrops; ± 300 m. NCoRO (c Sonoma Co.). May–Jul

S. gracilis Eastw. ALPINE JEWELFLOWER Ann, slender, glabrous throughout. **ST:** (0.6)1–3.5 dm, erect, gen branched at base. **LF:** basal rosetted, oblanceolate to spoon-shaped, wavy to dentate (lobed); mid-cauline short-petioled or sessile, 0.5–3 cm, oblong to ovate, entire or dentate, base lobed; distal reduced, clasping, entire. **INFL:** terminal sterile fl cluster 0; bracted proximally or between lowest 2 fls. **FL:** calyx urn-shaped, sepals 4–5 mm, not keeled, rose-purple; petals 7–10 mm, 1.5–2.5 mm wide, not crinkled, ± pink; filaments free, in 3 pairs of unequal length, longest pair 5–6 mm; anthers fertile, 1.5–2 mm. **FR:** ascending, 3–7 cm, 1–1.5 mm wide, constricted between seeds; valves glabrous, midvein obscure; stigma entire; pedicels spreading to ascending, 3–6 mm, straight or recurved, not expanded at receptacle. **SEED:** 24–52, 1–1.5 mm, oblong; wing 0–0.25 mm wide. Rocky slopes; 2600–3600 m. se SNH (Kings-Kern Divide region). Jun–Sep ★

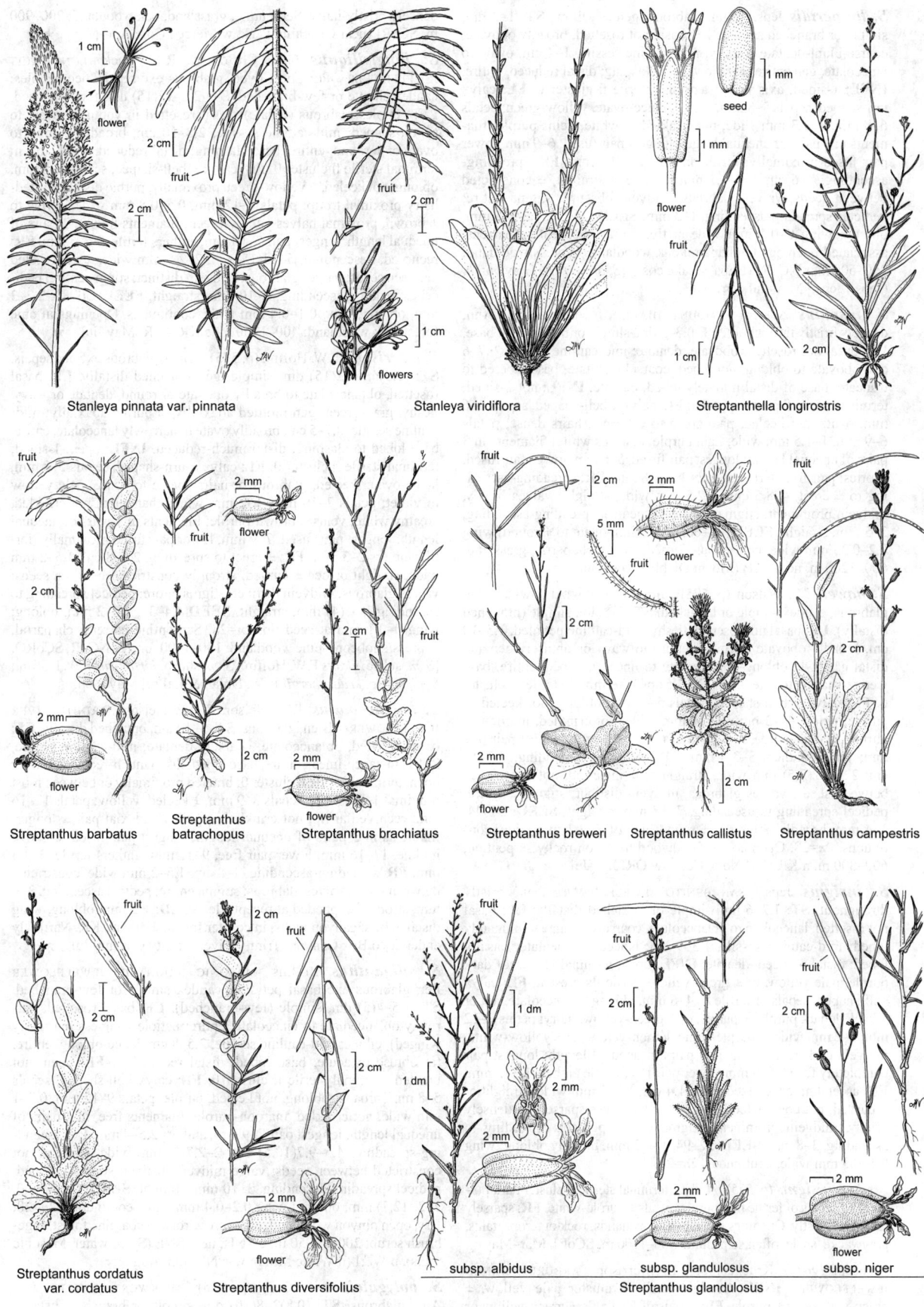

Stanleya pinnata var. pinnata

Stanleya viridiflora

Streptanthella longirostris

Streptanthus barbatus

Streptanthus batrachopus

Streptanthus brachiatus

Streptanthus breweri

Streptanthus callistus

Streptanthus campestris

Streptanthus cordatus var. cordatus

Streptanthus diversifolius

subsp. albidus

subsp. glandulosus

subsp. niger

Streptanthus glandulosus

S. hesperidis Jeps. Ann, glabrous, gen ± yellow. **ST:** 1–3 dm, simple or branched at base. **LF:** basal not rosetted, broadly obovate, coarse-blunt-dentate distally; mid-cauline sessile, 1–4 cm, ovate to lanceolate, gen entire, ± yellow, base clasping; distal reduced, entire. **INFL:** 1-sided, axis wavy; terminal sterile fl cluster 0. **FL:** calyx urn-shaped, sepals 5–8 mm, keeled, lance-ovate, yellow-green; petals 6–8 mm, 1–1.3 mm wide, not crinkled, ± white, veins purple; filaments in 3 pairs of unequal length; longest pair fused, 6–9 mm, lower pair fused proximally; fertile anthers 1.7–2.5 mm. **FR:** spreading-ascending, 3–6 cm, 0.9–1.1 mm wide, gen straight, ± constricted between seeds; valves glabrous, midvein obscure; stigma entire; pedicels spreading-ascending, 1–2 mm. **SEED:** 26–38, 1.2–1.5 mm, oblong; wing 0–0.1 mm wide at tip. 2*n*=28. Serpentine barrens, associated openings in chaparral/oak woodland, cypress woodland; 250–600 m. s NCoRI (Napa, s Lake cos.). [*S. breweri* A. Gray var. *h.* (Jeps.) Jeps.] May–Jul ★

S. hispidus A. Gray MOUNT DIABLO JEWELFLOWER Ann, densely bristly throughout. **ST:** 0.3–3 dm, simple or branched at base. **LF:** basal not rosetted, obovate, dentate; mid-cauline sessile, 0.7–6 cm, obovate to oblong, not lobed, coarsely dentate, base tapered to truncate; base of distal minutely lobed, dentate. **INFL:** not 1-sided; terminal sterile fl cluster present. **FL:** calyx ± bell-shaped, sepals 4–6 mm, ovate, not keeled, pale green to ± purple, hairs dense; petals 6–9 mm, 1–1.5 mm wide, light purple, margins white; filaments in 3 pairs of unequal length; longest pair fused, 5–6 mm, lower pair fused, shortest pair free; fertile anthers 1.5–1.8 mm. **FR:** spreading-ascending to ± erect, 4–8.5 cm, 2–2.5 mm wide, straight; valves bristly, midvein prominent; stigma ± 2-lobed; pedicel spreading-ascending, 2–5 mm, straight. **SEED:** 34–66, 1.6–2 mm, ovate to ± round; wing 0.2–0.35 mm wide, continuous. 2*n*=28. Rocky chaparral, grassland; 600–1200 m. ne SnFrB (Mount Diablo). Mar–Jun ★

S. howellii S. Watson (p. 575) HOWELL'S JEWELFLOWER Per, glabrous; caudex simple or branched. **ST:** 3–8 dm, simple (branched distally). **LF:** basal not rosetted, fleshy; mid-cauline sessile, 1.5–10 cm, broadly obovate to round, entire to wavy or obtusely dentate; distal narrowly oblong-oblanceolate to linear, unlobed, entire, base wedge-shaped to attenuate. **INFL:** open; terminal sterile fl cluster 0. **FL:** calyx ± bell-shaped, sepals 5–8 mm, oblong, not keeled, ± purple; petals 8–12 mm, 0.5–1 mm wide, not crinkled, maroon to purple; filaments free, in 3 pairs of unequal length, longest pair 6–7 mm; fertile anthers 3–3.5 mm. **FR:** spreading-ascending, 5.5–12 cm, 2.5–3.2(3.5) mm wide, straight to ± incurved, not constricted between seeds; valves glabrous, midvein distinct; stigma ± entire; pedicel spreading to ascending, 7–17 mm, straight. **SEED:** 24–44, 3–4 mm, broadly oblong to ± round; wing 0.5–1.1 mm wide, continuous. 2*n*=28. Open conifer/hardwood forest on rocky serpentine; 600–800 m. n KR (Del Norte Co.); sw OR. Jun–Jul ★

S. insignis Jeps. SAN BENITO JEWELFLOWER Ann, bristly throughout. **ST:** 1.2–6 dm, simple or branched distally. **LF:** basal not rosetted, lanceolate to oblanceolate, coarsely dentate to pinnately lobed; mid-cauline sessile, 1.3–9 cm, lanceolate, dentate, basally lobed; distal reduced, dentate. **INFL:** open; terminal cluster of dark purple, pale yellow, or ± green-yellow sterile fls present. **FL:** calyx bell-shaped, sepals of fertile fls 4–6 mm, oblong to lanceolate, not or ± keeled, dark purple, ± purple, or ± green-yellow, hairy; petals 7–12 mm, 1–2 mm wide, ± purple-white, lemon-yellow, or ± yellow-white, midvein darker; filaments in 3 pairs of unequal length; longest pair completely fused, 6–9 mm, lower pair free; fertile anthers 1.3–2 mm. **FR:** ascending or reflexed, 3.5–11.4 cm, 1.5–2 mm wide, straight or ± curved, not constricted between seeds; valves sparsely to densely bristly, midvein prominent; stigma entire; pedicels spreading to ascending, 3–8 mm. **SEED:** 32–94, 1.4–2 mm, broadly oblong; wing 0.1–0.3 mm wide, continuous. 2*n*=28.

subsp. ***insignis*** (p. 575) **INFL:** terminal sterile fl cluster dark purple. **FL:** sepals of fertile fls ± purple; petals ± purple-white. **FR:** sparsely (densely) bristly. Openings in chaparral, badlands, rock outcrops, talus, graywacke, shale, often serpentine; 300–1100 m. SCoRI. Mar–May

subsp. ***lyonii*** Kruckeb. & J.L. Morrison ARBURUA RANCH JEWELFLOWER **INFL:** terminal sterile fl cluster pale yellow, ± green-yellow, or ± purple. **FL:** sepals of fertile fls ± green-yellow or ± purple; petals lemon yellow or ± yellow-white. **FR:** moderately to

densely bristly-hairy. Serpentine, grassland, oak woodland; 200–900 m. SCoRI (near Ortigalita Peak, w Merced Co.). Apr–May ★

S. longisiliquus G.L. Clifton & R.E. Buck LONG-FRUIT JEWELFLOWER Per, short-lived, glabrous exc sepals, occ petioles; caudex simple or few-branched. **ST:** 2.2–12(15) dm, few-branched. **LF:** petioles glabrous (ciliate); basal rosetted in youth, obovate to spoon-shaped; mid-cauline sessile, 2.5–10 cm, broadly oblong to ovate or ± round, entire, base clasping; distal reduced. **INFL:** open; terminal sterile fl cluster 0. **FL:** calyx ± bell-shaped, sepals 6–8 mm, oblong, not keeled, ± yellow-green proximally, purple distally, tufted-hairy proximal to tip; petals 8–12 mm, 0.5–0.8 mm wide, purple to ± brown, proximal halves yellow-green; filaments free, in 3 pairs of unequal length, longest pair 7–10 mm; anthers fertile, 3.5–5 mm. **FR:** recurved, descending, 5–13(15) cm, 2–2.5 mm wide, ± constricted between seeds; valves glabrous, midvein distinct; stigma entire; pedicels spreading-ascending, 5–10 mm, straight. **SEED:** 50–82, 2.2–3 mm, oblong; wing 0.1–0.4 mm wide, continuous. Openings in pine forest, oak woodland; 400–1700 m. e KR, CaR. May–Jul ★

S. morrisonii F.W. Hoffm. (p. 575) Bien, glabrous exc occ sepals. **ST:** (2.5)6.5–12(15) dm, simple, gen branched distally. **LF:** basal rosetted, oblanceolate to broadly obovate or round, dentate or wavy, fleshy, gray-green, gen mottled adaxially, ± purple abaxially; mid-cauline sessile, 0.7–5 cm, broadly ovate to narrowly lanceolate, entire, base lobed to clasping; distal much-reduced. **INFL:** open, 1-sided; terminal sterile fl cluster 0. **FL:** calyx ± urn-shaped, sepals 5–8 mm, lance-ovate, keeled, glabrous or hairs sparse to dense, pale yellow to violet; petals 7–10 mm, 1–2 mm wide, channeled, not crinkled, creamy white, veins ± brown-purple; filaments in 3 pairs of unequal length; longest pair fused, 7–9 mm, lower pair fused proximally; fertile anthers 2–3 mm. **FR:** ± erect to spreading, 2.5–8 cm, 1.5–2 mm wide, straight or occ ± curved, strongly constricted between seeds; valves glabrous, midvein distinct; stigma entire; pedicel spreading to ascending, 1–3(4) mm, straight. **SEED:** 24–38, 1.5–2 mm, oblong; wing 0–0.1 mm at seed tip. 2*n*=28. Serpentine barrens, chaparral, cypress/knobcone-pine woodland; 150–1100 m. c&s NCoR, SCoRO. [*S. m.* subsp. *elatus* F.W. Hoffm.; *S. m.* subsp. *hirtiflorus* F.W. Hoffm.; *S. m.* subsp. *kruckebergii* R.W. Dolan & LaPré] May–Sep

S. oblanceolatus T.W. Nelson & J.P. Nelson TRINITY RIVER JEWELFLOWER Bien, glabrous. **ST:** 5–10 dm, branched distally. **LF:** basal rosetted, oblanceolate, coarsely dentate; mid-cauline sessile, 2–10 cm, lance-linear, entire, base lobed; distal much-reduced. **INFL:** open; terminal sterile fl cluster 0; bracted proximally or between most proximal 1–2 fls. **FL:** sepals 8–9 mm, ± keeled, yellow; petals 12–16 mm, recurved at tip, not crinkled, ± yellow, adaxial pair ± longer; filaments in 3 pairs of unequal length; longest pair fused to above middle, 13–16 mm, lower pair free, 9–11 mm; anthers fertile, 3–3.5 mm. **FR:** spreading-ascending, 4–8 cm, 1.5–2 mm wide, constricted between seeds, valves glabrous; stigma entire; pedicel ascending, 3–6 mm, strongly expanded at receptacle. **SEED:** ± 2 mm, oblong, wing distal. Cliffs, canyon walls, in conifer forest; ± 400 m. KR. Narrowly endemic to Box Canyon, Trinity River, Trinity Co. Jun–Jul ★

S. oliganthus Rollins MASONIC MOUNTAIN JEWELFLOWER Per, glabrous exc basal petioles; caudex simple or few-branched. **ST:** 1.5–4(5) dm, simple (few-branched). **LF:** basal rosetted, narrowly oblanceolate to lanceolate, entire; petiole wingless (narrowly winged), ciliate; mid-cauline sessile, 2.5–8 cm, lance-oblong, entire, tip obtuse to acute, base lobed; distal reduced. **INFL:** open, not bracted; terminal sterile fl cluster 0. **FL:** calyx bell-shaped, sepals 5–8 mm, broadly oblong, not keeled, purple; petals 9–12 mm, 0.7–1 mm wide, not crinkled, maroon-purple; filaments free, in 3 pairs of unequal length, longest pair 7–9 mm; anthers 2.5–4 mm. **FR:** ascending-spreading, 4.5–9.7(10.5) cm, 2–2.7(3) mm wide, straight, not constricted between seeds; valve midvein distinct; stigma unlobed; pedicel spreading-ascending, 3–10 mm, straight. **SEED:** (42)48–60, 2–2.5(2.7) mm, oblong; wing 0.2–0.4 mm wide, continuous. 2*n*=28. Dry, open pinyon woodland, pine forest, rocky subalpine forest, sagebrush scrub; 2000–3050 m. c SNH, n&c SNE (Sweetwater, Masonic mtns), n W&I (n White Mtns); w-c NV. Jun–Aug ★

S. polygaloides A. Gray (p. 575) MILKWORT JEWELFLOWER Ann, glabrous. **ST:** (0.8)2–8(10) dm, simple or branched distally. **LF:** basal rosetted, early-deciduous, wavy-dentate or 1- or 2-pin-

nately lobed, lobes broadly linear to thread-like lobes; mid-cauline sessile, 1–10 cm, linear, entire, base lobed; distal similar, reduced. **INFL**: open; terminal sterile fl cluster 0; fls clearly bilateral. **FL**: calyx urn-shaped, sepals 4–6 mm, not keeled, ± green-yellow or ± purple, upper 6–8 mm wide, ± round to broadly ovate-cordate, forming banner-like hood; lower 3–4 mm wide, broadly ovate, keeled; lateral pair 1.5–2 mm wide, lance-ovate; petals 5–8 mm, 0.7–1.2 mm wide, crinkled, channeled, white, veins ± brown; filaments in 3 pairs of unequal length; longest pair fused, 5–6 mm, lower pair free, 4–5 mm; fertile anthers 1.5–2 mm. **FR**: pendent, 2.4–5.6 cm, 1.2–1.7 mm wide, straight, not constricted between seeds; valves glabrous, midvein obscure or ± distinct; stigma entire; pedicel strongly recurved, 2–5 mm. **SEED**: (10)18–50, 1.7–2 mm, oblong; wing 0.2–0.3 mm wide at tip. $2n=28$. Serpentine barrens, chaparral openings, sparse pine/cypress woodland; 200–1100 m. SNF, n SNH. Accumulates high nickel concentrations. May–Jul

S. tortuosus Kellogg (p. 575)   MOUNTAIN JEWELFLOWER   Bien or short-lived per, glabrous throughout. **ST**: (0.5)1.5–12(15) dm, simple or many-branched at base, distally. **LF**: basal ± rosetted, early-deciduous, broadly ovate, obovate or oblong, entire to wavy or dentate distally; mid-cauline sessile, (0.7)1.5–6(9) cm, oblong to obovate or ± round, base lobed to clasping; distal round to oblong-ovate, clasping. **INFL**: open to dense, 1-sided or not; bracted proximally or between most proximal 1–2 fls; terminal sterile fl cluster 0. **FL**: calyx urn-shaped, sepals 6–10(13) mm, keeled or not, ± purple, gray-green, or ± yellow, tips recurved; petals (6)8–14 mm, 1–2.5 mm wide, not crinkled, ± purple or ± yellow-white, veins gen purple; filaments free, in 3 pairs of unequal length, longest pair (5)7–11 mm; fertile anthers (1.5)2.5–4.5(6) mm. **FR**: recurved-spreading to pendent, (3)4–13(16) cm, 1.5–2.5(3) mm wide, occ constricted between seeds; valves glabrous, midvein obscure or ± distinct; stigma entire; pedicel spreading to ascending, (2)3–12(17) mm, expanded at receptacle. **SEED**: 26–76(110), 1.5–2.5 mm, broadly oblong to ovate or round; wing 0.1–0.5 mm wide at tip, continuous. $2n=28$. Gen rocky to sandy soils, in open conifer forest, alpine areas, woodland; 200–4100

m. KR, NCoR, CaR, SN, n SnFrB, SCoRO; sw OR, NV. [*S. t.* var. *flavescens* Jeps.; *S. t.* var. *orbiculatus* (Greene) H.M. Hall; *S. t.* var. *suffrutescens* (Greene) Jeps.] Highly variable, needs study. Apr–Sep

S. vernalis R. O'Donnell & R.W. Dolan   EARLY JEWELFLOWER   Ann, glabrous. **ST**: 0.2–2 dm, simple or branched distally. **LF**: basal not rosetted, early-deciduous, broadly obovate to ± round, coarsely dentate, fleshy, not mottled; mid-cauline sessile, 5–18 mm, ovate to oblong, entire, base lobed to clasping; distal similar, much-reduced. **INFL**: open, occ 1-sided; bracts 0; terminal sterile fl cluster 0. **FL**: calyx urn-shaped, sepals 5–7 mm, green, lance-ovate, keeled; petals 6.5–8 mm, ± 1.5 mm, wide, not crinkled, white, veins not darker; filaments in 3 pairs of unequal length; longest pair completely fused, 6–8 mm, gen recurved, lower pair fused proximally; fertile anthers 1.5–2 mm. **FR**: 3–5 cm, 1.5–2 mm wide, spreading-ascending, straight, constricted between seeds; valves glabrous, midvein obscure; stigma entire; pedicels spreading-ascending, 1–2 mm, straight. **SEED**: 16–20, 1.6–2 mm, oblong; wing 0.1–0.2 mm wide at tip. Serpentine talus, gravel; ± 600–900 m. NCoRI (Three Peaks, Lake Co.). Mar–May   ★

S. vimineus (Greene) Al-Shehbaz & D.W. Taylor   Ann, glabrous exc occ sepals. **ST**: erect, 1–7.5 dm, branched above base. **LF**: basal not rosetted, early-deciduous, narrowly ovate to oblong, entire; mid-cauline sessile, 2–12 cm, linear to narrowly lance-linear, entire, base lobed; distal similar, reduced. **INFL**: open, gen 1-sided; terminal sterile fl cluster 0. **FL**: calyx urn-shaped, sepals 6–8 mm, lance-ovate, keeled, recurved at tip, ± green to ± purple (sparsely hairy); petals 8–12 mm, 2–3 mm wide, unequal, adaxial pair > lower, not crinkled, white, veins purple; filaments in 3 pairs of unequal length; longest pair completely fused, 7–11 mm, recurved, lower pair 4–7 mm, fused proximally; fertile anthers 2–2.7 mm. **FR**: spreading, 3.5–6.5 cm, 1–1.2 mm wide, recurved, ± constricted between seeds; valves glabrous, midvein obscure; stigma entire; pedicel spreading, 1–3(6) mm, straight. **SEED**: 28–40, 1–1.5 mm, narrowly oblong; wing 0–0.1 mm wide at tip. Serpentine grassland, ridges, barrens, openings in chaparral; 250–800 m. NCoRI (Colusa, Lake, Napa cos.). May–Jul

STRIGOSELLA

Ann; hairs simple and tree-like [stalked-forked]. **LF**: simple, dentate to pinnately lobed; cauline sessile or petioled, base not lobed. **INFL**: elongated. **FL**: sepals erect, lateral pair not sac-like at base; petals white to pink or purple, clawed. **FR**: silique, dehiscent, linear, cylindric or 4-angled, unsegmented, stigma conical, 2-lobed, lobes fused. **SEED**: 40–80, in 1 row, wingless. 20 spp.: Eurasia, n Afr. (Latin: with bristly hairs) [Warwick et al. 2007 Ann Missouri Bot Gard 94:56–78]

S. africana (L.) Botsch. (p. 575)   **ST**: (0.4)1.5–3(5) dm, simple or branched proximally. **LF**: proximal cauline (0.5)1.5–6(10) cm, (0.3)1–2.5(3.5) cm wide, oblanceolate to oblong, entire or dentate; distal smaller, sessile. **FL**: sepals (3.5)4–5 mm; petals (6.5)8–10(12) mm, 1–2 mm wide, violet to pink (white). **FR**: ascending, (2.5)3.5– 5.5(7) cm, 1–1.3 mm wide; style 0; pedicel 0.5–2(4) mm. **SEED**: 1–1.2 mm, oblong. $2n=14,28$. Disturbed areas, desert scrub, fields, flats; 600–2400 m. SNE, DMoj; to Rocky Mtns; native to Afr. [*Malcolmia a.* (L.) W.T. Aiton] May–Aug

SUBULARIA   AWLWORT

Ann, aquatic, scapose, hairs 0. **LF**: basal rosetted, sessile, erect, linear to narrowly awl-like, entire; cauline 0. **INFL**: ± elongated. **FL**: sepals erect to ascending, base not sac-like; petals white, not clawed. **FR**: silicle, dehiscent, obovoid to ellipsoid or oblong, ± flat perpendicular to septum; unsegmented; stigma entire. **SEED**: 4–18, in 2 rows, wingless. 2 spp.: n hemisphere. (Latin: awl, from lf shape) [Mulligan & Calder 1964 Rhodora 66:127–135]

S. aquatica L. subsp. ***americana*** G.A. Mulligan & Calder (p. 575)   WATER AWLWORT   **ST**: (0.5)1.5–15(23) cm. **LF**: (0.5)1–7(10) cm, narrowly awl-like. **FL**: sepals (0.5)0.7–1(1.3) mm; petals 1.2–1.5 mm, 0.2–0.5 mm wide. **FR**: (1.5)2–3.5(5.5) mm, 1.2–2(2.5) mm wide, obovoid to wide-ellipsoid; pedicels ascending to erect, 1–7(10) mm, straight. **SEED**: (4)6–12(18), 0.8–1 mm, oblong. Shallow lake margins, streambanks, wet sedge meadows, muddy flats, salt marshes; 1800–3200 m. SNH; to AK, Can, ne US. [*S. a.* var. *a.* (G.A. Mulligan & Calder) B. Boivin] Other subsp. native to n Eurasia. Jul–Oct   ★

THELYPODIUM

Ann to per; hairs 0 or simple. **LF**: basal rosetted, petioled, entire to pinnately lobed; mid-cauline petioled or sessile, base lobed to sagittate or wedge-shaped. **FL**: sepals erect to reflexed, bases sac-like or not; petals linear to oblanceolate, spoon-shaped, or obovate, clawed or not, white to lavender or purple; stamens free (± fused). **FR**: silique, dehiscent, linear, unsegmented, ± narrowed between seeds, cylindric or ± flat parallel to septum, stalked above receptacle; stigma entire. **SEED**: 1 row per chamber, ± flat; wing gen 0. 16 spp.: w N.Am. (Greek: female foot, from fr stalk above receptacle) [Al-Shehbaz 1973 Contr Gray Herb 204:1–148]

1. Mid-cauline lvs petioled, pinnately lobed to dentate
  2. Fr spreading; proximal sts solid; pedicel ± straight, spreading (ascending); petal blades linear; basal lf petioles not ciliate . . . . *T. laciniatum*
  2′ Fr ± erect, gen appressed; proximal sts hollow to infl; pedicel upcurved, tip erect; petal blades spoon-shaped to oblanceolate; basal lf petioles ciliate . . . . *T. milleflorum*
1′ Mid-cauline lvs sessile, entire
  3. Per, caudex thick; basal lvs persistent . . . . *T. flexuosum*
  3′ Bien, caudex 0; basal lvs withered by fr time
    4. Cauline lvs narrowed to base, not lobed or sagittate; basal lf petioles not ciliate . . . . *T. integrifolium*
    5. Petals white; fr pedicel 6–9(13) mm, ± white; stalk above receptacle 1–3 mm . . . . subsp. *affine*
    5′ Petals lavender to purple (± white); fr pedicel 2–5(6) mm, not ± white; stalk above receptacle 0.5–1 mm
      . . . . subsp. *complanatum*
    4′ Cauline lvs expanded to base, base lobed or sagittate; basal lf petioles ciliate
      6. Infl dense, spike-like, little expanded in fr; petals gen white (pale lavender), blades crinkled
        7. Fr pedicel 1–2 mm, spreading (± ascending), stout . . . . *T. brachycarpum*
        7′ Fr pedicel 2–5(10) mm, erect to erect-ascending, partly or fully appressed, slender. . . . . *T. crispum*
      6′ Infl open, panicle-like, much expanded in fr; petals lavender or purple (white), blades not crinkled or only at base
        8. Paired filaments partly to completely fused; petals narrowly oblanceolate, 0.5–1.2(3) mm wide; seeds 22–40 . . . . *T. howellii* subsp. *howellii*
        8′ Paired filaments free; petals linear, 0.3–0.5(0.8) mm wide; seeds 50–82 . . . . *T. stenopetalum*

T. brachycarpum Torr. (p. 575) SHORT-PODDED THELYPODIUM Bien, gen glaucous. **ST**: branched or not, (1.3)3–8(12) dm; hairs 0 or near base. **LF**: basal blade 3–14(20) cm, ± entire to pinnately lobed, withered by fr time, petiole ciliate; mid-cauline sessile, sagittate, entire. **INFL**: dense. **FL**: petals 8–12(16) mm, 0.3–0.5(1) mm wide, linear to narrowly oblanceolate, white, crinkled. **FR**: (0.8)1.2–2.7(3) cm, 1–1.5(2) mm wide, cylindric, narrowed between seeds; stalk above receptacle 1–2(5) mm, slender; style 0.5–1 mm; pedicel spreading (± ascending), 1–2 mm, stout. **SEED**: 12–26, (1.3)1.5–2 mm, plump. Alkaline soils, adobe flats, pond margins; 800–2320 m. KR, NCoR, CaR, GB; s OR. Apr–Aug ★

T. crispum Payson (p. 575) Bien, gen glaucous. **ST**: 1–7(12) dm; hairs 0 or toward base. **LF**: basal blade 2–15(25) cm, dentate to pinnately lobed (entire), withered by fr time, petiole ciliate; mid-cauline sessile, base lobed or sagittate, entire or dentate. **INFL**: dense. **FL**: petals 6–11(14.5) mm, 0.5–0.7(1) mm wide, linear to narrowly oblanceolate, white to pale lavender, crinkled. **FR**: 1–2.5(4) cm, 0.7–1(1.8) mm wide, cylindric, narrowed between seeds; stalk above receptacle 0.5–1.5(3.5) mm, slender; style 0.5–1.5(2.5) mm; pedicel erect to erect-ascending, partly or fully appressed, 2–5(10) mm, slender. **SEED**: 22–50, 1–1.5 mm, plump. 2*n*=26. Alkaline or sandy soils, lake margins, scrub; 1200–3000 m. SNH, GB; NV. Jun–Aug

T. flexuosum B.L. Rob. (p. 575) Per; caudex thick, woody; hairs 0. **ST**: ± decumbent to erect, 2–6(8) dm, slender, gen wavy. **LF**: basal blade 4–17(20) cm, entire, persistent, petiole glabrous; mid-cauline sessile, sagittate to clasping. **INFL**: open, much expanded in fr. **FL**: petals 6–10 mm, 1.5–3.5 mm wide, spoon-shaped, occ oblanceolate to obovate, lavender to white, not crinkled. **FR**: 1–2.5(4) cm, 0.8–1(1.5) mm wide, cylindric, narrowed between seeds; stalk above receptacle 0.5(1) mm, slender; style 1–2(3) mm; pedicel spreading, (2.5)4–9(16) mm, slender. **SEED**: 12–30, 1–1.5 mm, plump. 2*n*=26. Canyons, slopes, among shrubs; 900–1800 m. MP; OR, ID, NV. Apr–Jun

T. howellii S. Watson subsp. ***howellii*** HOWELL'S THELYPODIUM Bien, glaucous; hairs 0 exc on petiole. **ST**: 1–9 dm, simple or branched distally. **LF**: basal blade 2–10(13.5) cm, lyre-shaped (dentate or entire), withered by fr time, petiole ciliate; mid-cauline sessile, sagittate to clasping, entire. **INFL**: open, much expanded in fr. **FL**: petals 5–8(12) mm, 0.5–1.2(3) mm wide, narrowly oblanceolate, lavender to purple, not crinkled; paired filaments partly to completely fused. **FR**: 1.5–4.5(7) cm, cylindric, narrowed between seeds; stalk above receptacle 0.5–1(3.5) mm; style (0.5)1–2.7(4) mm; pedicel ascending, 3–8(14.5) mm, stout, straight or ± curved. **SEED**: 22–40, 1–1.7(2) mm, plump. 2*n*=26. Alkaline meadows, flats, sagebrush scrub; 1000–1600 m. MP; to WA. Other subsp. in e OR only. May–Jul ★

T. integrifolium (Nutt.) Walp. Bien, gen glaucous; hairs 0. **ST**: (2)4.5–17(28) dm, straight, glabrous, glaucous, branched distally. **LF**: basal 5–31(54) cm, entire, petiole glabrous; mid-cauline sessile,

tapered to base, base not lobed or sagittate. **INFL**: open, expanded in fr or not. **FL**: petals 6–9(13) mm, 0.8–1.5(2) mm wide, spoon-shaped to oblanceolate, purple to white, not crinkled. **FR**: cylindric, narrowed between seeds; style 0.5–1.5 mm; pedicel straight, spreading. **SEED**: 14–40, 1–2 mm wide, oblong. Only 2 of 5 subspp. occur in CA.

subsp. ***affine*** (Greene) Al-Shehbaz **FL**: petals white. **FR**: 2–4 cm, upcurved; stalk above receptacle 1–3 mm; pedicel 6–9(13) mm, stout, base ± flat. Among shrubs, low dunes, meadows; 700–1100 m. SNE, DMoj; to sw UT. Jun–Oct

subsp. ***complanatum*** Al-Shehbaz (p. 575) FOXTAIL THELYPODIUM **FL**: petals lavender to purple (± white). **FR**: 1.5–3 cm, straight or curved; stalk above receptacle 0.5–1 mm; pedicel 2–5(6) mm, ± stout, base flat. Alkaline or silty soils, woodland; 1100–2500 m. GB; to OR, nw UT. Jun–Aug ★

T. laciniatum (Hook.) Walp. (p. 579) Bien, glabrous throughout. **ST**: 2.5–10(14) dm, solid throughout. **LF**: basal blade (4)7–24(45) cm, pinnately lobed or dentate, petiole glabrous; middle petioled, dentate to pinnately lobed, base not lobed. **INFL**: dense, spike-like. **FL**: petals 7–18(20) mm, 0.3–0.8(1.5) mm wide, linear, white to purple, crinkled. **FR**: spreading, (2.5)3.5–10(12) cm, cylindric, ± narrowed between seeds; stalk above receptacle 1–5 mm; pedicel spreading (ascending), 3–7(15) mm, straight, stout; style 0.7–2.5(4) mm. **SEED**: 56–108, 1–1.5(1.8) mm, oblong. 2*n*=26. Rocky hillsides, basalt cliffs; 600–1900 m. CaRH, GB; to BC, ID. Apr–Aug

T. milleflorum A. Nelson (p. 579) MANY-FLOWERED THELYPODIUM Bien, glabrous exc petioles. **ST**: (2)4.5–13(21) dm, hollow to infl. **LF**: basal blade (4)6–23(28) cm, dentate (lobed), petiole ciliate; mid-cauline petioled, dentate, base not lobed. **INFL**: dense, spike-like. **FL**: petals (7)9–15(16) mm, 1–2 mm wide, oblanceolate to spoon-shaped, white. **FR**: ± erect, gen appressed, 3–8(10) cm, cylindric, narrowed between seeds; stalk above receptacle 1–4 mm, ± stout; style 0.5–1.5(3) mm, stout; pedicel upcurved, erect at tip, (1.5)2.5–5(7) mm, stout. **SEED**: 50–78, 1–1.5(2) mm, oblong. 2*n*=26. Sandy soils, scrub; 1300–2500 m. GB; to s BC, UT. Apr–Aug ★

T. stenopetalum S. Watson (p. 579) SLENDER-PETALED THELYPODIUM Bien, glaucous; hairs 0 exc on petiole. **ST**: 3–9 dm, simple or branched distally. **LF**: basal blade 1–4(6) cm, entire (wavy), withered by fr time, petiole ciliate; mid-cauline sessile, sagittate to clasping, entire. **INFL**: open, much expanded in fr. **FL**: petals 5–8(12) mm, 0.3–0.5(0.8) mm wide, linear, lavender (white), crinkled between blade and claw; paired filaments free. **FR**: 2.5–5(7.3) cm, cylindric, narrowed between seeds; stalk above receptacle 0.5–3.5(5) mm; style 1–2(2.5) mm; pedicel spreading, 4–8 mm, stout. **SEED**: 50–82, 1–15 mm, plump. Alkaline flats, lake shores; 1900–2200 m. SnBr (Big Bear Valley). May–Aug ★

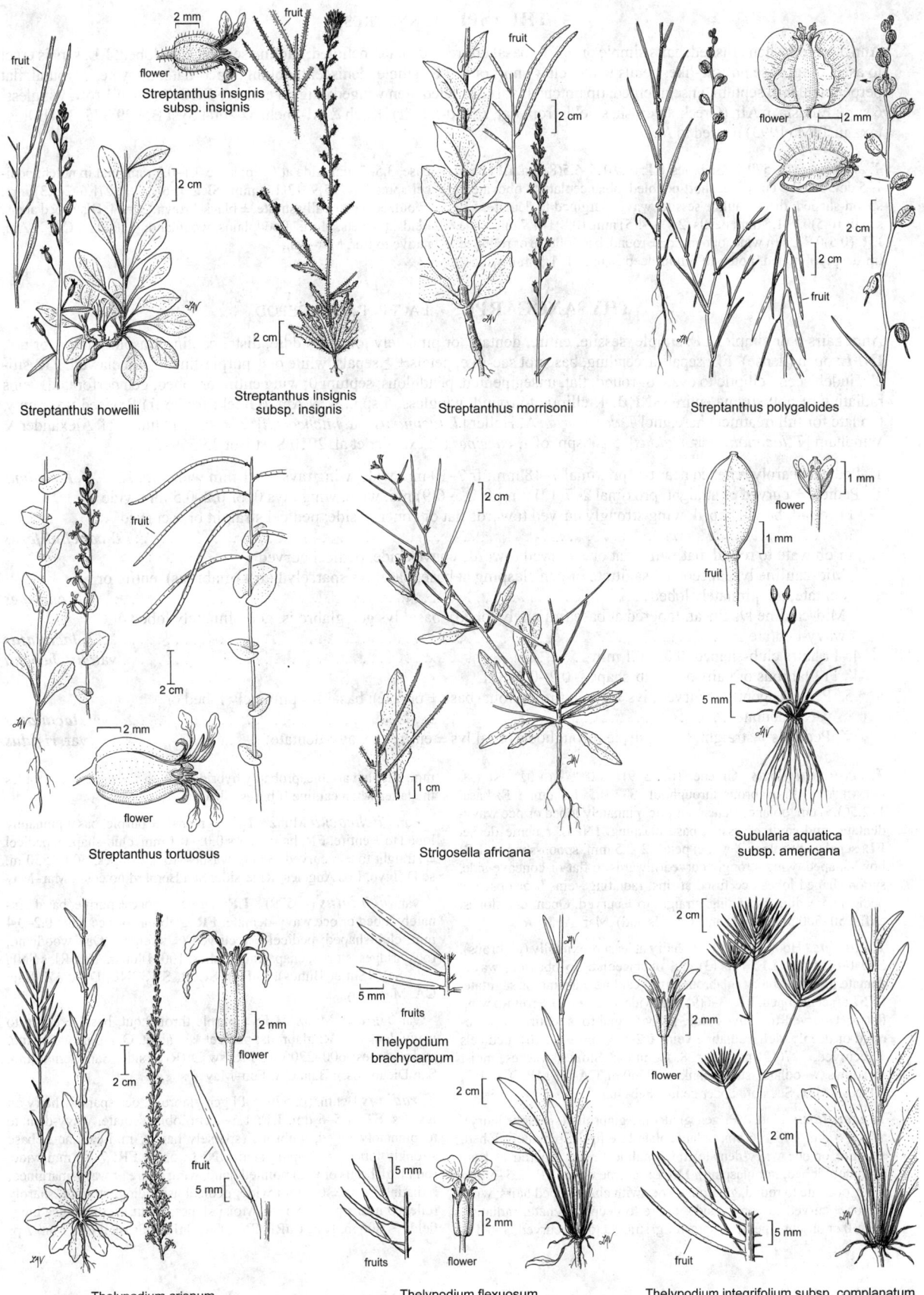

Streptanthus insignis
subsp. insignis

Streptanthus howellii

Streptanthus insignis
subsp. insignis

Streptanthus morrisonii

Streptanthus polygaloides

Streptanthus tortuosus

Strigosella africana

Subularia aquatica
subsp. americana

Thelypodium crispum

Thelypodium
brachycarpum

Thelypodium flexuosum

Thelypodium integrifolium subsp. complanatum

THLASPI PENNYCRESS

Ann, scented when crushed; hairs simple or 0. **LF:** basal entire to dentate, petioled; cauline sessile, base lobed. **FL:** sepals erect to ascending, base not sac-like; petals white, claw short or 0. **FR:** silicle, dehiscent, oblong, obcordate, obovate, or round, flat perpendicular to septum, unsegmented, tip notched; valves keeled, gen winged; stigma entire. **SEED:** 6–16 in 1 row, wingless. 6 spp.: Eurasia, n Afr. (Greek: to crush, shield, from flat, shield-like fr) [Koch & Al-Shehbaz 2004 Syst Bot 29:375–384] Other taxa in TJM (1993) moved to *Noccaea*.

T. arvense L. (p. 579)   Glabrous. **ST:** (0.9)1.5–5.5(8) dm. **LF:** basal 1–5 cm, 0.4–2.3 cm wide, short-petioled, oblanceolate to obovate or spoon-shaped; distal cauline sessile, wavy-margined to dentate. **FL:** sepals (0.5)0.9–1.5 mm; petals (2.4)3–4(5) mm, (0.8)1–1.7 mm wide. **FR:** (0.5)0.7–2 cm wide, obovate or ± round, base obtuse or rounded, tip deeply notched, notch to 5 mm deep; wings 1–1.5 mm wide at base, 3.5–5 mm wide at fr tip; style 0.1–0.3 mm, incl in notch; pedicel spreading, (5)9–13(15) mm. **SEED:** 6–16, (1.2)1.6–2(2.3) mm, ovoid, concentrically striate, ± black-brown. 2*n*=14. Disturbed areas, fields, thickets, bluffs, floodplains, woodland; < 2000 m. CA; N.Am; native to Eur. Mar–Aug

THYSANOCARPUS LACEPOD, FRINGEPOD

Ann; hairs 0 or simple. **LF:** simple, sessile, entire, dentate, or pinnately lobed; middle, distal cauline clasping, lobed or not. **INFL:** open (dense). **FL:** sepals ascending, base not sac-like; petals ± ≥ sepals, white or ± purple-tinged, not clawed. **FR:** silicle, indehiscent, elliptic to ovate or round, flat, unsegmented, pendulous; septum 0; wing entire or lobed, gen perforated, veins radiating or not; stigma entire. **SEED:** 1, elliptic to round, wingless. 5 spp.: w N.Am. (Greek: fringe fr) Revised taxonomy, too late for full treatment here, incl *T. desertorum* A. Heller [*T. laciniatus* var. *hitchcockii*], *T. rigidus* (Munz) P.J. Alexander & Windham [*T. laciniatus* var. *rigidus*], 5 subspp. of *T. curvipes* (Alexander et al. 2010 Syst Bot 35:559–577).

1. Pedicel sharply reflexed near tip, proximal 7–18 mm; fr 7–10 mm wide, wing rays ± 0.1 mm wide . . . .*T. radians*
1′ Pedicel ± curved or straight, proximal 2–7(12) mm; fr 2.5–6(9) mm wide, wing rays 0, or 0.2–0.5 mm wide
　2. Fr round, bowl-shaped, wing strongly curved towards flat or concave side; pedicel straight or ± curved
　. . . . . *T. conchuliferus*
　2′ Fr obovate to round, flat, wing flat or ± curved towards convex side; pedicel curved
　　3. Mid-cauline lvs lanceolate, sagittate or gen clasping at base; basal lvs sparsely hairy (glabrous), entire or
　　　dentate, not pinnately lobed. . . . . *T. curvipes*
　　3′ Mid-cauline lvs linear, tapered at base (minutely lobed); basal lvs gen glabrous, gen pinnately lobed or
　　　wavy-dentate . . . . *T. laciniatus*
　　　4. Fr hairs club-shaped, 0.05–0.1 mm . . . . var. *hitchcockii*
　　　4′ Fr glabrous or hairs occ club-shaped, 0.2–0.4 mm
　　　　5. Pedicels gently recurved; lvs ± green throughout (base ± purple); basal lvs pinnately lobed or
　　　　　wavy-dentate. . . . . var. *laciniatus*
　　　　5′ Pedicels ± straight; lvs ± purple throughout; basal lvs ± entire to wavy-dentate. . . . . var. *rigidus*

T. conchuliferus Greene (p. 579) SANTA CRUZ ISLAND FRINGEPOD   Pl glabrous throughout. **ST:** 0.5–1.5 dm. **LF:** basal 1–2.5(3.5) cm, oblanceolate to elliptic, pinnately lobed or occ wavy-dentate; mid-cauline sessile, base clasping. **INFL:** raceme dense. **FL:** sepals ± purple to ± white; petals 2–2.5 mm, spoon-shaped. **FR:** bowl-shaped, wing strongly curved inwards to flat or concave side, spoon-shaped lobes occ fused at tips, radiating veins 0 or obscure; pedicel spreading-ascending, straight to ± curved. Open, dry slopes, cliffs; 50–500 m. n ChI (Santa Cruz Island). Mar–Apr ★

T. curvipes Hook. (p. 579)   Pl hairy at least proximally (glabrous). **ST:** 1–6(8) dm. **LF:** 1–6(13) cm, oblanceolate to obovate, wavy-dentate to entire, hairy (glabrous); mid-cauline clasping or sagittate. **INFL:** raceme open. **FR:** 3–6(9) mm wide, obovate to ± round; wing flat or ± curved to convex side, hairy to glabrous, entire to perforated or deeply cleft, radiating veins 0.2–0.5 mm wide (0); pedicels ± recurved, 3–7(12) mm. 2*n*=28. Common. Slopes, washes, moist meadows, woodland, streambanks; < 2500 m. CA-FP, MP, D; to BC, ID, NM, Mex. See note under genus. Feb–Jun

T. laciniatus Nutt.   Pl gen glaucous, glabrous (sparsely hairy). **ST:** 1–6 dm. **LF:** 1–6 cm, oblanceolate to elliptic, pinnately lobed, ± entire or occ wavy-dentate; mid-cauline linear, tapered at base (minutely lobed, not clasping). **INFL:** raceme open. **FR:** 2.5–5 mm wide, obovate to round, glabrous or occ with club-shaped hairs; wing flat or ± curved to convex side, entire to deeply crenate, radiating veins 0 or obscure; pedicels spreading, straight to ± recurved, 3–6(10)

mm. Highly variable, probably hybridizes with *T. curvipes*, resembles the latter when cauline lf bases lobed.

var. *hitchcockii* Munz   **LF:** ± green or ± purple, basal pinnately lobed to ± entire. **FR:** hairy, hairs 0.05–0.1 mm, club-shaped; pedicel ± straight to ± recurved. Sandy washes, rocky slopes; 600–1850 m. se D (Inyo, Los Angeles, Riverside, San Bernardino cos.). Mar–May

var. *laciniatus* (p. 579)   **LF:** ± green or occ ± purple, basal pinnately lobed or occ wavy-dentate. **FR:** glabrous or occ hairs 0.2–0.4 mm, club-shaped; pedicel ± recurved (± straight). Oak woodland, rocky ridges, slopes, chaparral, washes; 100–1800 m. NCoRI, s SNF, Teh, ScV (Sutter Buttes), SnFrB, SCoR, SW, SNE, D; w AZ, Baja CA. Mar–May

var. *rigidus* Munz   **LF:** ± purple throughout, basal ± entire to wavy-dentate. **FR:** glabrous; pedicel ± straight. Oak/pine woodland, rocky slopes; 600–2200 m. PR, sw D (Riverside, San Bernardino, San Diego cos.); Baja CA. Feb–May ★

T. radians Benth. (p. 579)   Pl gen glabrous, occ sparsely hairy on lvs, frs. **ST:** 1.5–6 dm. **LF:** 1.5–4 cm, oblanceolate, wavy-dentate to pinnately lobed, glabrous (sparsely hairy); mid-cauline lf base strongly lobed or clasping stem. **INFL:** open. **FR:** 7–10 mm wide, round, glabrous or with pointed hairs; wing entire or wavy-margined, radiating veins ± 0.1 mm wide; pedicel ascending, straight, sharply reflexed near tip, 7–18 mm. Moist slopes, pastures, open meadows, fields; < 800 m. NW, CaRF, SNF, GV, SnFrB, SCoRI; OR. Mar–Apr

TRANSBERINGIA

1 sp.: N.Am, Russian Far East. (Latin: across the Bering Strait) [Al-Shehbaz & O'Kane 2003 Novon 13:396]

T. bursifolia (DC.) Al-Shehbaz & O'Kane subsp. ***virgata*** (Nutt.) Al-Shehbaz & O'Kane (p. 579) VIRGATE HALIMOLOBOS Per, with caudex; stiff-hairy proximally, hairs simple to forked or many-branched. **ST**: branched, (0.5)1–4.5(6.5) dm, distal st hairs appressed, occ simple; glabrous in age. **LF**: basal simple, petioled, rosetted, (0.5)1.2–6.5(10) cm, (2)3–8(17) mm wide, oblanceolate to obovate or lance-linear, entire or dentate; cauline sessile, (0.5)1–3(4.2) cm, (1)2.5–7(10) mm wide, lanceolate to lance-linear or oblong, entire, base strongly sagittate. **INFL**: elongated. **FL**: sepals erect, 1.2–2.5, base not sac-like; petals 2.5–4.5 mm, (0.8)1–1.5 mm wide, not clawed, white. **FR**: silique, dehiscent, (1.5)2–3.3(4) cm, (0.7)1–1.5(2) mm wide, linear, cylindric, unsegmented; style slender, 0.2–0.8(1) mm, entire; pedicels ascending, (3)5–14(18) mm. **SEED**: 70–150, in 2 rows, 0.8–1 mm, oblong, wingless. 2*n*=16. Meadows, near aspen groves, pinyon/juniper woodland; 2000–3700 m. SNE; to w Can, CO. [*Halimolobos v.* (Nutt.) O.E. Schulz] Other subsp. in Greenland, AK, n Can, Russia. May–Jul ★

TROPIDOCARPUM

Ann; hairs simple, or stalked-forked. **ST**: prostrate to erect, branched at least distally. **LF**: basal not rosetted; proximal cauline deeply pinnately lobed, lateral lobes entire to dentate or lobed; distal short-petioled to sessile, less divided, not basally lobed. **INFL**: bracted. **FL**: sepals oblong, erect to ascending, base not sac-like; petals obovate to oblanceolate or spoon-shaped, yellow or occ tinged purple, short-clawed. **FR**: silique, linear to oblong, or obconic silicle, dehiscent, unsegmented, flat perpendicular to septum (or septum 0); valves 2 or 4. **SEED**: 4–70, 1 or 4 rows per chamber; wing 0. 4 spp.: CA, Baja CA, Chile. (Greek: keeled fr) [Al-Shehbaz 2003 Novon 13:392–395]

1. Fr obconic, 0.4–0.5 cm, 4–8-seeded; valves thick-leathery, tubercled near tip . . . . ***T. californicum***
1′ Fr linear or oblong, 0.9–7 cm, 25–70-seeded; valves thin-leathery, smooth throughout
  2. Fr oblong, 0.9–2 cm, (3)4–5 mm wide, length (1.6)2.8–5 × width; valves (2)4 . . . . ***T. capparideum***
  2′ Fr narrowly linear, (2.5)3–6(7) cm, 1.5–2(3) mm wide, length 13–46 × width; valves 2 . . . . ***T. gracile***

T. californicum (Al-Shehbaz) Al-Shehbaz (p. 579) KINGS GOLD **ST**: 0.3–2.5 dm. **LF**: proximal cauline blade 2.5–4.5 cm; lateral lobes entire, pairs 2–4. **FL**: sepals 1.2–2 mm; petals 1.6–2.5 mm, 0.7–0.9 mm wide, oblanceolate to spoon-shaped, yellow. **FR**: silicle, 4–5 mm, width similar, obconic, hairy, hairs spreading or pointing distally; valves thick-leathery, tubercled near tip; style 0.5–3 mm; pedicels spreading to ascending, 3–10(28) mm. **SEED**: 4–8, 1.2–1.5 mm, oblong. ± Alkaline, sandy clay soil in *Atriplex* scrub; ± 65 m. SnJV (Kern, Kings cos.). [*Twisselmannia c.* Al-Shehbaz] Mar ★

T. capparideum Greene (p. 579) CAPER-FRUITED TROPIDOCARPUM **ST**: 1.5–7 dm. **LF**: proximal cauline blade 1.2–7 cm; lateral lobes entire, pairs 3–6. **FL**: sepals 2.5–3.5 mm; petals 3–5 mm, 1.5–2 mm wide, obovate to spoon-shaped, yellow, occ tinged ± purple. **FR**: silique, 0.9–2 cm, (3)4–5 mm wide, oblong, length (1.6)2.8–5 × > width; valves (2)4; style 1–2.6 mm; pedicels spreading to ascending, 5–17(25) mm. **SEED**: 25–40, 1.2–1.6 mm, oblong. Alkaline soils, low hills, valleys; < 400 m. nw SnJV, SCoRO. Mar–Apr ★

T. gracile Hook. (p. 579) **ST**: (0.4)1–4.5(6) dm. **LF**: blade (1.5)2.5–10(15) cm; lateral lobes entire to dentate or pinnately lobed, pairs 3–8(12). **FL**: sepals 2.5–4 mm; petals 3–6 mm, 1.5–4 mm wide, yellow, occ tinged purple. **FR**: silique, linear, (2.5)3–6(7) cm, 1.5–2(3) mm wide, length 13–46 × > width; valves 2; style (0.3)0.5–2.5(4) mm; pedicels erect to ascending, (4)6–17(35) mm. **SEED**: 30–70, 1.2–1.6 mm, oblong. 2*n*=16. Common. Grassy banks, open fields, roadsides, pastures; < 1450 m. s NCoRO, NCoRI, CaRF, SNF, Teh, GV, CW, SW, w DMoj; Baja CA. Mar–May

TURRITIS TOWER MUSTARD

Bien (per), glaucous distally; hairs on proximal parts simple to stalked-forked or many-branched. **LF**: basal rosetted, petioled, [entire or] dentate to pinnately lobed; cauline sessile, sagittate to clasping, dentate or entire. **INFL**: much elongated. **FL**: sepals erect, base not sac-like; petals ± white (pink or purple), not clawed. **FR**: silique, dehiscent, linear, cylindric or ± 4-angled, unsegmented; stigma entire. **SEED**: 130–200, 2 rows per chamber, wingless or only tip winged; cotyledons face-to-face. 2 spp.: N.Am, Eurasia, n Afr. (Latin: tower, from orientation of overlapping lvs, frs, giving pl a pyramidal shape) [Al-Shehbaz 2005 Novon 15:519–524]

T. glabra L. (p. 579) **ST**: simple, occ few-branched distally, (3)4–12(15) dm. **LF**: basal (4)5–12(15) cm, 1–3 cm wide, oblanceolate to spoon-shaped or oblong; cauline 2–9(12) cm, (0.5)1–2.5(4) cm wide, lanceolate to oblong-elliptic or ovate. **FL**: sepals 2.5–5 mm; petals 5–8.5 mm, 1.3–1.7 mm wide, linear-oblanceolate to narrowly spoon-shaped, cream (lilac or purple). **FR**: erect, appressed, (3)4–10(12.5) cm, 0.7–1.5 mm wide; pedicel 6–16(20) mm, glabrous; style 0.5–0.8(1) mm. **SEED**: 0.6–1.2 mm, oblong to ± round. Open fields, meadows, slopes; < 2800 m. CA-FP, MP; temp N.Am, Eurasia, n Afr. [*Arabis g.* (L.) Bernh.; *A. g.* var. *furcatipilis* M. Hopkins] Apr–Jul

BURSERACEAE TORCHWOOD FAMILY

Duncan M. Porter

Tree [shrub]; gen dioecious or monoecious. **ST**: gen erect, < 15 m. **LF**: simple or compound, cauline, gen alternate, deciduous, petioled; lf axis often winged; hairs dense to 0. **INFL**: gen panicle or fls 1. **FL**: radial; gen unisexual, disk ring- or cup-shaped; sepals 3–5, gen united below; petals [0–2]3–5; stamens gen 1–2 × number of petals; ovary superior, chambers 2–5, style 0 or 1. **FR**: drupe or capsule; stones 1–5, 1-seeded. 18 genera, ± 700 spp.: worldwide esp trop; some cult (*Boswellia*, frankincense; *Commiphora*, myrrh; *Bursera*, *Protium*, copal). [Weeks et al. 2005 Molec Phylogen Evol 35:85–101] Scientific Editor: Thomas J. Rosatti.

BURSERA ELEPHANT TREE, TOROTE, COPAL

Tree, aromatic; dioecious. **ST**: < 10 m; bark smooth, shedding. **INFL**: panicle. **FL**: petals inserted beneath disk; stamens 10; ovary chambers 2–3. **FR**: drupe-like when young, capsule-like in age, valves [2]3. ± 100 spp.: trop, subtrop Am. (Joachim Burser, German physician, botanist, 1583–1639) [Porter 1974 Madroño 22:273–276]

B. microphylla A. Gray (p. 579) LITTLELEAF ELEPHANT TREE **ST**: < 4 m; branches spreading, gen red; mature bark white. **LF**: odd-1-pinnate, 2–8 cm, glabrous; lf axis winged; lflets 7–33, 2–5[10] mm. **FL**: staminate fls 3–5-parted, pistillate fls 3-parted; sepals ± 5 mm; petals ± 4 mm, white to cream. **FR**: 5–8 mm; valves 3; stone yellow.

Rocky slopes; < 700 m. w edge DSon (e San Diego, w Imperial, Riverside cos.); to AZ, Mex. Populations local. *B. hindsiana* (Benth.) Engl. (some lvs simple; Baja CA Sur, w SON) reported but not confirmed for PR (s San Diego Co.). May–Jun ★

CACTACEAE CACTUS FAMILY

Bruce D. Parfitt, except as noted

Per, shrub, tree, gen fleshy. **ST**: cylindric to spheric, or flat; surface smooth, tubercled, or ribbed (grooved); nodal areoles bearing fls. **LF**: gen 0 or early-deciduous, flat to ± cylindric. **SPINES**: areoles gen with central, radial spines, occ with glochids. **FL**: gen 1 per areole, bisexual [unisexual], sessile, radial [bilateral]; perianth parts gen many [5], scale-like to petal-like; stamens many; ovary inferior [superior], style 1, stigma lobes gen several [many]. **FR**: dry to fleshy or juicy, indehiscent to variously dehiscent, spiny, scaly, or naked; tubercled or smooth. **SEED**: gen many, occ 0–few. ± 125 genera, ± 1800 spp.: Am (esp deserts), Afr; many cult, some edible. [Parfitt & Gibson 2004 FNANM 4:92–257] Spines smaller, fewer (0) in shade forms; yellow spines blacken in age. Introduced spp. increasingly escape cult. Hybridization common in some genera. Taxa of *Escobaria* in TJM (1993) moved to *Coryphantha*. Scientific Editors: Bruce D. Parfitt, Douglas H. Goldman, Bruce G. Baldwin.

1. St regularly segmented; areoles with glochids; seeds tightly encased within bone-like aril, ± white when dry
  2. St segments flat, tubercles 0 to ± developed; spine surface not separating . . . . **OPUNTIA**
  2′ St segments cylindric to club-like, gen tubercled; spine surface sheath-like, gen separating and
     late-deciduous
    3. Spines not flat; spine sheath 0 or fully separating; mature pl gen 0.3–4 m . . . . **CYLINDROPUNTIA**
    3′ Major spines flat; spine sheath separating only from tip; mature pl gen < 0.25 m . . . . **GRUSONIA**
1′ St not regularly segmented; glochids 0; dry seeds gen black to ± red-brown, lacking a ± white bone-like aril
  4. Largest spines with transverse ridges; spines gen 2–5 mm diam near base, rigid
    5. Ovary, gen st tip densely woolly, obscuring epidermis, ovary scales tapered to sharp tips . . . . . . **ECHINOCACTUS**
    5′ Ovary, st tip glabrous, only areoles woolly, ovary scales gen rounded . . . . **FEROCACTUS**
  4′ Largest spines smooth or longitudinally ridged; spines gen < 2 mm diam near base, rigid or not, thin,
    flexible if > 2 mm wide
    6. St ribs 0 or inconspicuous, tubercles prominent
      7. Spines all straight to ± curved; st tubercles with woolly, adaxial groove connecting spine cluster
        to axillary fl areole . . . . **CORYPHANTHA**
      7′ Some spines in each areole gen hook-shaped; st tubercles not adaxially grooved . . . . **MAMMILLARIA**
    6′ St ribs prominent, tubercles ± 0 or prominent on ribs
      8. St > 30 cm diam, branched but gen not in basal 1.5–2 m; perianth gen white . . . . **CARNEGIEA**
      8′ St < 15 cm diam, not branched, or branched in basal 1.5 m; perianth variously colored, but
        gen not white
        9. St length > 8 × diam; fr persistent when empty, densely spiny; coastal . . . . **BERGEROCACTUS**
        9′ St length < 6 × diam; fr not persistent when empty, spineless or spines deciduous; inland
          10. Fls lateral; ovary minutely scaly, spiny, spines deciduous in clusters; st soft-fleshy,
            branches 1–500 . . . . **ECHINOCEREUS**
          10′ Fls at st tip or nearly so; ovary scaly, spineless; st ± hard, branches 0–few (more if
            st damaged) . . . . **SCLEROCACTUS**

BERGEROCACTUS

1 sp. (A. Berger, German succulent specialist, 1871–1931)

B. emoryi (Engelm.) Britton & Rose (p. 583) GOLDEN-SPINED CEREUS Shrub, thicket-forming, ascending to sprawling or decumbent, many-branched. **ST**: < 2 m, 3–6 cm diam, cylindric, soft-fleshy, ± obscured by spines, gen not regularly segmented; ribs gen 12–18, prominent, tubercles ± 0. **SPINES**: 30–45 per areole, < 2 mm diam, needle-like, straight or curved, yellow, darker in age, smooth; central spines 1–3, decurved, longest < 60 mm; radial spines straight. **FL**: lateral to terminal, at distal margin of spine cluster, 3.5–5 cm, 2.5–4

cm diam; ovary glabrous, densely spiny, scales minute or none; outer perianth parts yellow, tips ± red, midveins green, inner perianth parts all yellow. **FR**: spheric, fleshy, dry in age, densely spiny, extruding seeds and pulp at tip. **SEED**: 3 mm, ± flat-obovoid, black, shiny. $2n$=44. Sandy open hills, coastal only; < 100 m. SCo (San Diego Co.), s ChI; Baja CA. Threatened by development, collecting, feral goats. May–Jun ★

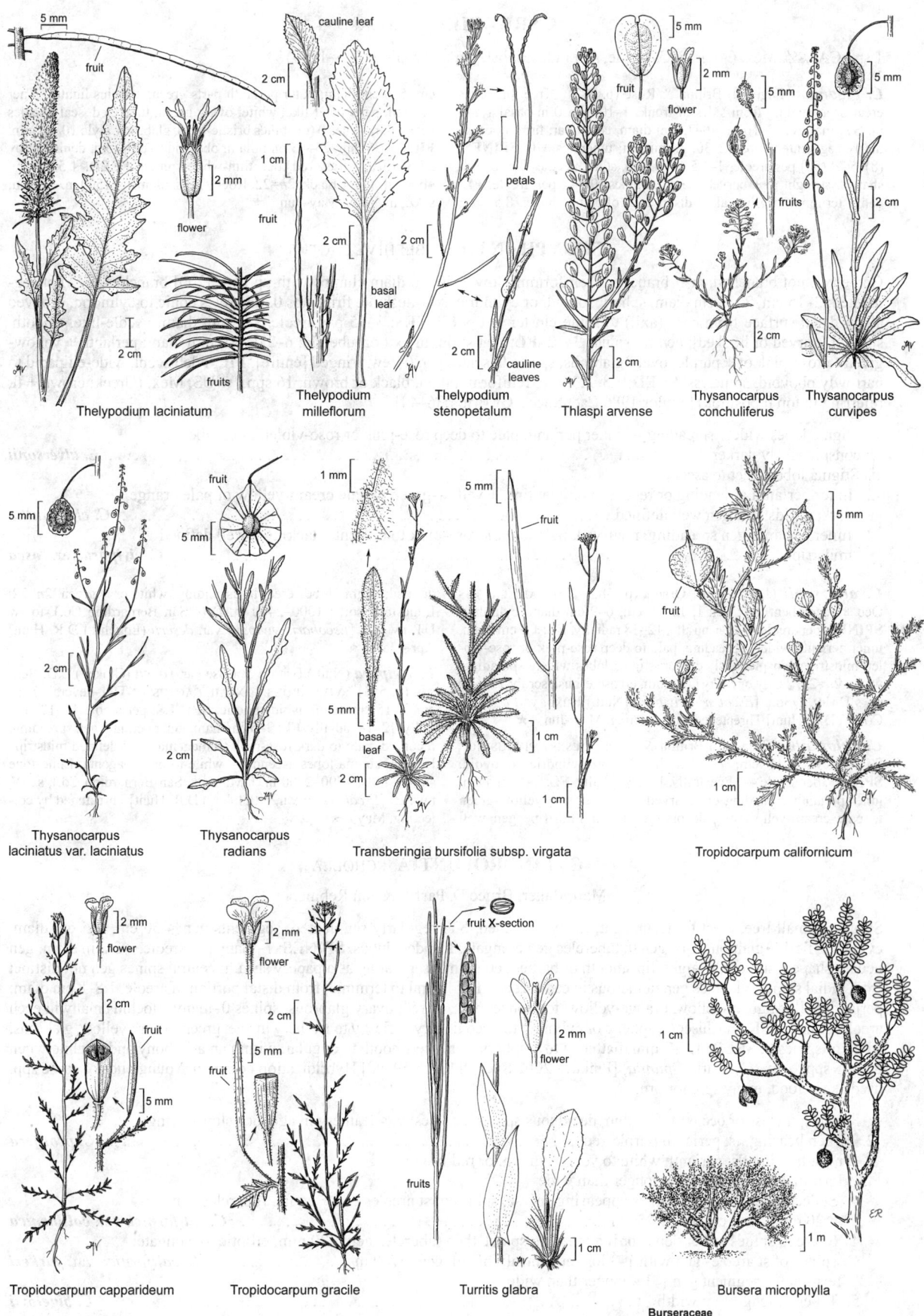

Thelypodium laciniatum

Thelypodium milleflorum

Thelypodium stenopetalum

Thlaspi arvense

Thysanocarpus conchuliferus

Thysanocarpus curvipes

Thysanocarpus laciniatus var. laciniatus

Thysanocarpus radians

Transberingia bursifolia subsp. virgata

Tropidocarpum californicum

Tropidocarpum capparideum

Tropidocarpum gracile

Turritis glabra

Bursera microphylla

Burseraceae

CARNEGIEA SAGUARO

1 sp.: CA, AZ, Mex. (Andrew Carnegie, Am industrialist, philanthropist, 1835–1919)

C. gigantea (Engelm.) Britton & Rose (p. 583)   Tree, branches erect or ascending from sides of trunk, 1–10, gen 0 in basal 1.5–2 m. **ST:** massive, 3–16 m, 30–75 cm diam, columnar, firm, not regularly segmented; ribs 12–30, prominent; tubercles ± 0. **SPINES:** (8)15–28(50) per areole, 1–1.5 mm diam, stout, needle-like to awl-shaped, straight to ± curved; central spines (0)4–7 per areole. **FL:** gen ± terminal, occ lateral, at distal edge of spine cluster, 8.5–12.5 cm, 5–6 cm diam; outer perianth parts green, margins lighter; inner perianth parts petal-like, white; ovary green, tubercled, scaly, scales triangular, spines 0 (or spines bristle-like), glabrous, style 10–15 mm. **FR:** 45–75 mm, 25–45 mm diam, obovoid to ellipsoid, dehiscent by 2+ vertical splits, red, occ thin-white-spined. **SEED:** 1.5–2 mm, obovoid, shiny, black. 2*n*=22. Rocky hills, plains; < 1500 m. e DSon; s AZ, nw Mex. May–Jun ★

CORYPHANTHA BEEHIVE CACTUS

Gen erect, not branched, or if branched then forming low ± 50 cm diam clumps with up to 12[200] branches. **ST:** not segmented; 2–15 cm, 2–15 cm diam, spheric, ovoid, or cylindric, soft-fleshy to firm, ribs 0, tubercles conic to cylindric, grooved along distal surface from base (axil) to spine cluster at tip. **SPINES:** 3–95 per areole, < 1 mm diam, needle-like, smooth, straight [curved or hooked]; central spines 3–12. **FL:** ± at st tip, in axil of tubercle; 6–37[100] mm diam; perianth ± yellow-green to rose-pink or ± purple; ovary glabrous, spines 0, scales gen 0–few, fringed [entire]. **FR:** indehiscent, wide-ellipsoid to narrowly obovoid, spineless. **SEED:** 1.3–2.3 mm, reniform, pitted, black or brown. 16 spp.: w US, Mex. (Greek: crown + fl, referring to top fl position) [Taylor 1986 Cact Succ J Gr Brit 4:36–44]

1. Stigma lobes widely spreading — inner perianth pale to deep rose-pink or rose-violet, midstripe
   conspicuously darker . . . .*C. alversonii*
1′ Stigma lobes erect to ascending
  2. Inner perianth ascending or recurved only at tips, ± yellow-green to pale cream-yellow or pale orange,
     darker midstripe gen well defined . . . . *C. chlorantha*
  2′ Inner perianth gen spreading, recurved, pale to dark rose-pink to magenta, lacking a well defined
     midstripe. . . . . *C. vivipara* var. *rosea*

C. alversonii (J.M. Coult.) Orcutt (p. 583)   FOXTAIL CACTUS   Occ ± decumbent. **ST:** gen 1, 10–27 cm, 6–9 cm diam, cylindric. **SPINES:** per areole 8–10 centrally, 12–18 radially. **FL:** ± 3 cm diam, inner perianth widely spreading, pale to deep rose-pink or rose-violet, midstripe conspicuously darker; stigma lobes widely spreading, white. 2*n*=22. Sandy or rocky alluvium, creosote-bush scrub; 75–600 m. s DMoj, DSon. [*Escobaria vivipara* (Nutt.) Buxb. var. *a.* (J.M. Coult.) D.R. Hunt] Threatened by collecting. May–Jun ★

C. chlorantha (Engelm.) Britton & Rose   DESERT PINCUSHION   **ST:** 1–few, 7–15 cm, 7–9 cm diam, short-cylindric to ovoid. **SPINES:** per areole 4–11 centrally, 12–33 radially. **FL:** 2–3 cm diam, inner perianth ascending or recurved only at tips, ± yellow-green to pale cream-yellow or pale orange, darker midstripe gen well defined; stigma lobes erect to ascending, white or ± green. 2*n*=22. Limestone soils; 1000–2400 m. D (e San Bernardino Co.); to sw UT, nw AZ. [*Escobaria vivipara* var. *deserti* (Engelm.) D.R. Hunt] Apr–May ★

C. vivipara (Nutt.) Britton & Rose var. *rosea* (Clokey) L.D. Benson (p. 583)   VIVIPAROUS FOXTAIL CACTUS   **ST:** 1–several, 7–18 cm, 7–15 cm diam, ovoid-spheric. **SPINES:** per areole 10–12 centrally, 12–18 radially. **FL:** 3–5 cm diam, inner perianth gen spreading, recurved, pale to dark rose-pink to magenta, well defined midstripe lacking; stigma lobes ascending, white to pale magenta. Limestone slopes, hills; 1500–2700 m. DMtns (ne San Bernardino Co.); s NV, nw AZ. [*Escobaria v.* var. *r.* (Clokey) D.R. Hunt] Threatened by collecting. May ★

CYLINDROPUNTIA CHOLLA

Marc Baker, Bruce D. Parfitt & Jon Rebman

Shrub or small tree, erect to decumbent, many-branched. **ST:** regularly segmented, segments gen < 50 cm, < 5 cm diam, cylindric, fleshy, glabrous; ribs gen 0; tubercles gen elongate. **LF:** deciduous. **SPINES:** 1–many per areole, < 2 mm diam, gen needle-shaped, smooth, straight, tip smooth or barbed, epidermis separating as a papery sheath; central spines gen not distinct from radial spines; glochids gen numerous in each areole. **FL:** lateral to terminal, from distal portion of areole, 1.8–8 cm diam; perianth yellow, green-yellow, orange-yellow, to bronze, pink, or red; ovary glabrous, spines 0–many, glochids many in each areole, scales 0. **FR:** indehiscent; spheric or cylindric to obconic, dry or fleshy to leathery in age, green to dark yellow, glabrous, spiny or spines 0. **SEED:** 1.9–7 mm, flattened to ± spheric, surface smooth to angular, within an aril, bony and ± white when dry. 35 spp.: Am. (Cylindric *Opuntia*) [Pinkava 2002 Succ Pl Res 6:59–98] Hybridization common. Young buds of some spp. used as for food, many spp. for orn.

1. Fr gen spineless or occ with few thin, deciduous spines, gen fleshy to leathery in age, occ slow-drying
  2. Fr gen bearing fls; perianth purple-red . . . . *C. prolifera*
  2′ Fr not bearing fls; perianth white to yellow, green, or red-brown
    3. Terminal st segment > 3 × longer than wide
      4. St decumbent to erect, spines appearing sparse, spines of st areoles gen ± equal, tubercles gen
         < 20 mm, oval; coastal, < 250 m . . . . [2]*C. californica* var. *californica*
      4′ St erect, spines rather dense or if appearing sparse then tubercles gen > 20 mm, elliptic to elongate,
         spines of st areoles gen with 1–3 longer central; inland, gen > 700 m . . . . [2]*C. californica* var. *parkeri*
    3′ Terminal st segment gen < 3 × longer than wide
      5. Tubercle length ± = width . . . . *C. bigelovii*

5′ Tubercle length ± 2 × width

  6. Spines 7–10 per tubercle; fr < 2.5 cm; seed 0 or irregularly shaped, < 3 mm . . . . ***C.* ×*fosbergii***

  6′ Spines 9–16 per tubercle; fr 2.5–3.5 cm; seed ± spheric, ± 3 mm . . . . ***C. munzii***

1′ Fr gen spiny, gen dry at maturity

  7. St < 1 cm diam, tubercle gen < 1 mm high . . . . ***C. ramosissima***

  7′ St > 1.5 cm diam, tubercle > 3 mm high

    8. Terminal st segment gen < 1 dm, tubercle length < 2 × width . . . . ***C. echinocarpa***

    8′ Terminal st segment gen > 1 dm, tubercle length > 3 × width

      9. St gen branched above only; trunk gen 1, inner perianth parts gen yellow . . . . . . ***C. acanthocarpa*** var. ***coloradensis***

      9′ St gen branched near base; trunks several to many; inner perianth parts green-yellow, yellow, bronze, to red

        10. Filaments red to magenta; perianth parts yellow, bronze, to red; e DSon (extreme se San Diego Co., sw Imperial Co.) . . . . ***C. wolfii***

        10′ Filaments green to yellow; perianth parts yellow to yellow-green; SCo, PR, DSon

          11. St decumbent to erect, spines appearing sparse, tubercles gen < 20 mm, oval; coastal, < 250 m . . . . [2]***C. californica*** var. ***californica***

          11′ St erect, spines rather dense or if appearing sparse then tubercles gen > 20 mm, elliptic to elongate; inland, gen > 250 m

            12. Spines gen not prominent, not obscuring young segments; st segments 14–43 cm, tubercles 16–35 mm . . . . [2]***C. californica*** var. ***parkeri***

            12′ Spines prominent, obscuring young segments; st segments 10–26 cm, tubercles 11–25 mm . . . . ***C. ganderi***

C. acanthocarpa (Engelm. & J.M. Bigelow) F.M. Knuth var. ***coloradensis*** (L.D. Benson) Pinkava (p. 583) BUCKHORN CHOLLA   Pl < 4 m. **ST:** trunk gen 1; main branches spreading to erect, few to several, gen long; terminal segments < 50 cm, 2–2.5 cm diam, firmly attached; tubercles 20–40 mm, < 7 mm high. **SPINES:** 12–21, < 5 cm, pale yellow- to red-brown, sheath pale yellow-brown. **FL:** inner perianth < 3.5 cm, yellow, tinted purple to brown-red; filaments purple-red. **FR:** dry, tubercled, proximal tubercles >> distal; base acute; spiny. **SEED:** < 6 mm, gen fertile. $2n=22$. Creosote-bush scrub, Joshua-tree woodland; < 1600 m. D; NV, AZ. [*Opuntia a.* Engelm. & J.M. Bigelow var. *c.* L.D. Benson] *Cylindropuntia* ×*deserta* (Griffiths) Pinkava, probably *C. acanthocarpa* × *C. echinocarpa*. May–Jun

C. bigelovii (Engelm.) F.M. Knuth (p. 583)   TEDDY-BEAR CHOLLA   Pl 1–2 m. **ST:** trunk gen 1; main branches few, short, spreading; terminal segments gen < 15 cm, 3–5 cm diam, easily detached; tubercles 4–11 mm, < 3 mm high. **SPINES:** 4–12, < 2.5 cm, pale yellow-brown, dark brown in age, sheath translucent to pale brown. **FL:** inner perianth < 4 cm, pale yellow or light green to ± white; filaments green. **FR:** leathery, yellow, tubercled, proximal tubercles ± = distal; base obtuse; spines 0 or few, thin, early-deciduous. **SEED:** < 4 mm, gen sterile. $2n=22,33$. Rocky fans, benches, creosote-bush scrub; < 1000 m. DMoj (Kelso Dunes), DSon; s NV, AZ, n Mex. [*Opuntia b.* Engelm.] Gen reproduces by rooting of detached st segments. Mar–May

C. californica (Torr. & A. Gray) F.M. Knuth   **ST:** trunks several to many; main branches decumbent to erect. **FL:** inner perianth < 3 cm, yellow to green-yellow, gen with purple tips; filaments green. **FR:** leathery to slow-drying, tubercled, proximal tubercles >> distal; base gen acute; spines 0–many. **SEED:** < 7 mm, gen fertile. $2n=22$.

var. ***californica***   SNAKE CHOLLA   Pl < 1.5 m. **ST:** terminal segments < 25 cm, 2–4 cm diam, firmly attached; tubercle 7–20 mm, < 5 mm high. **SPINES:** 6–15, gen < 2 cm, yellow to orange-brown, sheath translucent white to gold-brown. Coastal-sage scrub, coastal chaparral; < 250 m. s SCo; n Baja CA. [*Opuntia parryi* Engelm. var. *serpentina* (Engelm.) L.D. Benson] Apr–Jul   ★

var. ***parkeri*** (J.M. Coult.) Pinkava (p. 583)   CANE OR VALLEY CHOLLA   Pl < 3 m. **ST:** terminal segments gen 16–40 cm, 1.7–4 cm diam, firmly attached; tubercle 16–35 mm, < 7 mm high. **SPINES:** 0–20, gen < 3.5 cm, yellow to orange-brown, sheath translucent white to gold-brown. Chaparral, pinyon/juniper woodland; 700–1900 m. PR, w DSon; n Baja CA. [*Opuntia parryi* var. *p.*] Densely spined forms can be confused with *C. ganderi*. Apr–Jul

C. echinocarpa (Engelm. & J.M. Bigelow) F.M. Knuth (p. 583) SILVER OR GOLDEN CHOLLA   Pl < 3 m. **ST:** trunk gen 1–2, main branches spreading to curving upwards, few, gen short; terminal seg-ments gen < 10 cm, 2–3 cm diam, firmly attached; tubercle 6–15 mm, < 8 mm high. **SPINES:** 9–20, < 4 cm, pale gray to translucent yellow, sheath gen same color. **FL:** inner perianth < 2.5 cm, green-yellow (red-brown); filaments green to pale yellow. **FR:** dry, tubercled, prox-imal tubercles ± = distal; base obtuse; spines dense distally. **SEED:** < 6 mm, gen fertile. $2n=22$. Creosote-bush/white bur-sage, blackbush, saltbush, other desert and GB scrub; 300–1600 m. SNE, D; to UT, AZ, n Baja CA. [*Opuntia e.* Engelm. & J.M. Bigelow; *O. wigginsii* L.D. Benson] Mar–Jun   ★

C.* ×*fosbergii (C.B. Wolf) Rebman et al. (p. 583)   MASON VALLEY CHOLLA, PINK TEDDY-BEAR CHOLLA   Pl < 2.5 m. **ST:** trunk 1; branches gen several, long, gen curving upwards; terminal segments gen < 10 cm, 4–6 cm diam, easily detached; tubercle 10–20 mm, 3–5 mm high. **SPINES:** 7–10, < 2.5 cm, pale red-brown, sheath pale yel-low-brown. **FL:** inner perianth < 2.5 cm, pale red-brown; filaments green. **FR:** dry to leathery, tubercled, proximal tubercles ± = distal; base gen obtuse; spines 0 or few, thin, deciduous. **SEED:** < 3 mm, gen sterile. $2n=33$. Valley floors, alluvial fans; 300–450 m. DSon (e San Diego Co.). [*Opuntia f.* C.B. Wolf; *O.* ×*f.*] Mar–May   ★

C. ganderi (C.B. Wolf) Rebman & Pinkava   GANDER'S CHOLLA   Pl < 1.7 m. **ST:** trunks several to many; main branches ascending; terminal segments gen < 25 cm, 2.5–4.4 cm diam, firmly attached; tubercles 11–25 mm, 5–10 mm high. **SPINES:** 20–35, < 2.5 cm, pale yellow, light brown in age, sheath yellow-brown to golden. **FL:** inner perianth < 2.7 cm, yellow to yellow-green, gen red-tipped; filaments green. **FR:** dry, tubercled, proximal tubercles >> distal; base acute; densely-spined to bur-like. **SEED:** < 5 mm, gen fertile. $2n=22$. Desert, chaparral, pinyon/juniper woodland, sandy flats, rocky hill-sides, boulder fields; 200–1100 m. se PR, DSon (e San Diego Co., s Riverside Co.). Mar–May

C. munzii (C.B. Wolf) Backeb. (p. 583)   MUNZ'S CHOLLA   Pl < 2.4 m. **ST:** trunk 1; branches several, spreading to curving upwards; terminal segments gen < 10 cm, 3–5 cm diam, gen easily detached; tubercles 10–16 mm, 3–8 mm high. **SPINES:** 9–16, < 4 cm, light yellow to pale red-brown, sheath yellow-brown. **FL:** inner peri-anth < 2 cm, yellow-green to red-brown; filaments green. **FR:** dry, tubercled; proximal tubercles ± = to distal; base obtuse; gen with deciduous spines. **SEED:** gen fertile. $2n=22$. Gravelly or sandy soils of washes, canyon walls; 150–600 m. DSon (Chocolate, Chuckwalla mtns, Imperial, Riverside cos.); Baja CA. [*C.* × *munzii* (C.B. Wolf) Backeb.; *Opuntia m.* C.B. Wolf; *O.* ×m.]. Most localities in US ± inaccessible. Mar–May   ★

C. prolifera (Engelm.) F.M. Knuth   COAST CHOLLA   Pl < 2 m. **ST:** trunk gen 1, branches few to several, gen curving upwards; ter-minal segments < 13 cm, 3.5–5 cm diam, easily detached; tubercles 1.2–2.5 cm, 4–9 mm high. **SPINES:** 6–14, gen < 2 cm, pale red-

brown to dark brown, sheath pale yellow-brown. **FL:** produced from areoles of older fr; inner perianth < 2 cm, purple-red; filaments green, gen tinted purple. **FR:** fleshy, chained, green, smooth to shallowly tubercled; spines 0. **SEED:** < 4 mm, gen sterile. $2n=22,33$. Ocean bluffs, inland coastal scrub; < 450 m. SCo, ChI; Baja CA. [*Opuntia p.* Engelm.] Apr–Jul

C. ramosissima (Engelm.) F.M. Knuth (p. 583) DIAMOND CHOLLA, PENCIL CACTUS Pl < 1.5 m. **ST:** trunk 1–several, decumbent to erect, main branches spreading to ascending; terminal segments < 10 cm, 4–8 mm diam, firmly attached; tubercles 4.5–8.5 mm, ≤ 1 mm high. **SPINES:** gen 1, < 6 cm, pink-gray to dark brown, sheath ± white to pale yellow. **FL:** inner perianth < 6 mm, orange-pink to red-brown; filaments pale green. **FR:** dry, tubercled; proximal tubercles ± = distal; base obtuse to acute, gen continuous with st;

spines dense (0), bur-like. **SEED:** < 5 mm, gen fertile. $2n=22,44$. Creosote-bush/white bur-sage, saltbush, other desert scrub; < 1300 m. D; NV, AZ, n Mex (SON, Baja CA). [*Opuntia r.* Engelm.] See *C. echinocarpa.* Apr–Aug

C. wolfii (L.D. Benson) M.A. Baker WOLF'S CHOLLA Pl < 2 m. **ST:** trunk gen several to many, main branches gen long, erect; terminal segments < 40 cm, 2.5–4.2 cm diam, firmly attached; tubercles 15–25 mm, < 9 mm high. **SPINES:** 12–22, < 3 cm, pale to dark brown, sheath translucent to pale brown. **FL:** inner perianth < 3.5 cm, pale purple-brown; filaments red to magenta. **FR:** dry, tubercled, proximal tubercles ± = distal; base obtuse to acute; spines dense. **SEED:** < 6 mm, fertility unknown. $2n=66$. Dry places above valley floors; 300–1200 m. se PR, w DSon; n Baja CA. [*Opuntia w.* (L.D. Benson) M.A. Baker] Mar–May ★

ECHINOCACTUS CLUSTERED BARREL CACTUS

Erect, branched [or not], forming compact, ± 1 m diam mounds; branches 30–50(130). **ST:** not segmented, [4]15–40[250] cm, [8]15–30[80] cm diam, flat-topped spheric to short-cylindric, hard, tip densely woolly [to glabrous]; ribs [7]11–25[60+], prominent, tubercles indistinct. **SPINES:** (5)10–19 per areole, gen 2–5 mm diam near base, gen awl-shaped, gen flat, ringed with conspicuous ridges, straight to curved; central spines [1]4 per areole. **FL:** at st tip, near distal edge of spine cluster, 4–5 cm diam; perianth yellow tinged with pink; ovary densely long-woolly, spines 0, scaly, distal scales long-tapered, tips spine-like. **FR:** dehiscent via basal pore, ovoid, densely woolly, spineless, but distal scales spine-like at tip. **SEED:** 2.8–4.7 mm, spheric, or ± reniform to obovoid, shiny or dull, ± red-brown to black. 6 spp.: sw US, Mex. (Greek: spine + cactus) [Chamberland 1997 Syst Bot 22:303–313]

E. polycephalus Engelm. & J.M. Bigelow var. ***polycephalus*** (p. 583) **ST:** gen spheric. **SPINES:** red to gray (straw), spreading, initially canescent; radial spines 6–14, 3–4.5 cm, spreading, ± curved. **FL:** ± 5 cm, inner perianth bright yellow. **FR:** scales < dried perianth.

SEED: 2.8–4.7 mm, rounded or faceted, papillate-roughened, gen dull. $2n=22$. Rocky hills, silty valleys; < 1400 m. DMoj, n DSon; to AZ, Mex. Mar–Aug

ECHINOCEREUS HEDGEHOG CACTUS

Gen erect to ascending, [sprawling, pendent, or decumbent], branched or not, branches gen few–500, occ in dense mounds. **ST:** [2]5–60[200]cm, (1)4–15 cm diam, spheric to long-cylindric, soft, not regularly segmented; ribs prominent, 4–13[26], tubercles ± 0 along rib-crests. **SPINES:** [0]4–55 per areole, < 2 mm diam, needle- to dagger-like, glabrous to puberulent, straight, curved, or curly; central spines (0)1–6(9). **FL:** lateral, near distal margin of spine cluster; perianth purple to lavender, orange, or red [yellow or green]; ovary glabrous, spiny, scales minute. **FR:** spheric to obovoid, indehiscent or splitting laterally, densely spiny, spine clusters deciduous. **SEED:** 0.8–2 mm, obovoid to ± spheric, dull, wrinkled or tubercled, gen black. 49 spp.: sw US, Mex. (Greek: hedgehog + *Cereus*) [Taylor 1985 The Genus *Echinocereus*. Timber Press]

1. Perianth ± purple to lavender, inner perianth parts gen acuminate to bristle-tipped; anthers yellow; spines glabrous, at least 1 central spine gen angular to flat . . . . ***E. engelmannii***
1′ Perianth orange to red, inner perianth parts round to notched; anthers pink to purple; spines < 1 yr old minutely puberulent esp near tip, all ± angular . . . . ***E. mojavensis***

E. engelmannii (Engelm.) Lem. (p. 587) Clump-forming or loose, open mounds gen < 0.7 m diam. **ST:** < 60, 5–45(70) cm, 4–9 cm diam, cylindric; ribs 10–13; tubercles ± 0. **SPINES:** (8)15–20 per areole, color and shape variable; central spines 2–7, straight to twisted. **FL:** short-funnel- to bell-shaped. **FR:** 20–30 mm, spines glabrous. $2n=44$. Dry habitats; < 2400 m. SnBr, PR, W&I, D; to UT, AZ, Mex. Highly variable; occ unclearly divided into vars. Needs study. May–Jun

E. mojavensis (Engelm. & J.M. Bigelow) Rümpler (p. 587) Pl gen forming dense mounds to 1 m diam. **ST:** 1–500, 5–40 cm, 5–15 cm diam, ± spheric to cylindric; ribs 5–12; tubercles ± prominent on ribs. **SPINES:** 3–11 per areole, highly variable, gen gray; central spines (0)1–6, similar to radial spines, gen curved or curly and twisted. **FL:** narrowly funnel-shaped. **FR:** 20–25 mm, spines minutely puberulent. $2n=22$. Dry habitats; 150–3000 m. SnBr, W&I, D; to UT, AZ. [*E. triglochidiatus* Engelm.] Apr–Jun

FEROCACTUS BARREL CACTUS

Erect or leaning, branches 0, occ branched from tip-injury. **ST:** (0)10–200(300) cm, 10–35 cm diam, depressed-spheric to short-columnar, hard, glabrous, not segmented; ribs 13–31, prominent; tubercles not conspicuous on ribs. **SPINES:** [6]10–32 per areole, 2–4.5 mm wide, gen awl-shaped, gen flat, ringed with conspicuous ridges, straight to curved or ± hooked, some bristle-like; central spines gen 4 per areole. **FL:** ± terminal, near distal edge of spine cluster, 3–6 cm diam; perianth yellow to red [or purple, or white with ± purple midstripes]; ovary glabrous, spines 0, scales numerous, gen rounded, margins minutely fringed or toothed. **FR:** spheric, ovoid, or cylindric, glabrous, spineless, dehiscent by basal pore. **SEED:** [1]1.5–3 mm, spheric to subreniform, pitted, black. 25 spp.: sw US, Mex. (Latin: fierce cactus) [Taylor 1984 Bradleya 2:19–38]

1. Perianth gen maroon outside, gen yellow within; st gen > 0.4 m, 25–50 cm diam; ribs (18)21–31; inland
. . . . ***F. cylindraceus***
1′ Perianth ± all green-yellow; st gen < 0.2 m, 10–25 cm diam; ribs 13–21; coastal . . . . ***F. viridescens***

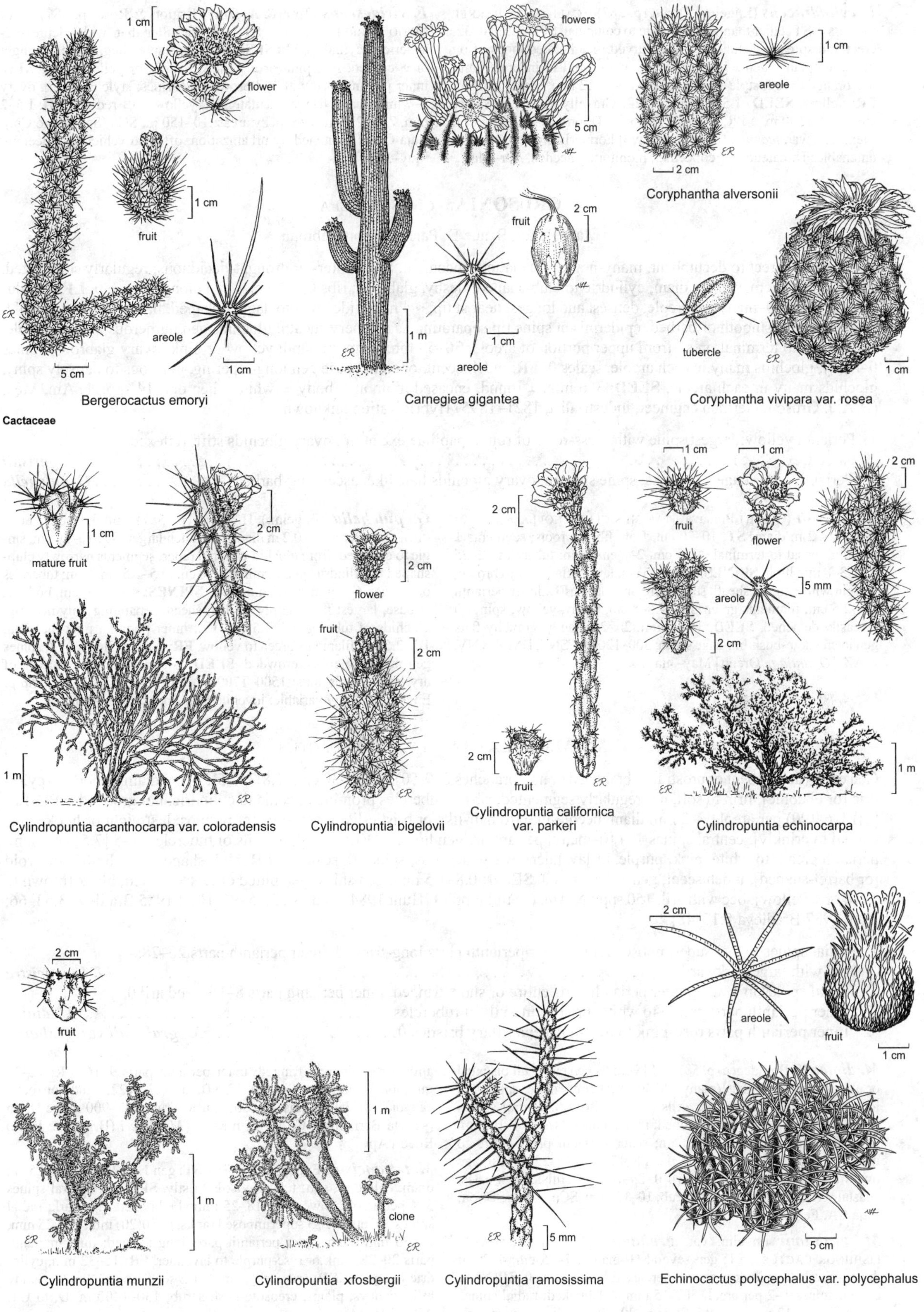

Bergerocactus emoryi

Cactaceae

Carnegiea gigantea

Coryphantha alversonii

Coryphantha vivipara var. rosea

Cylindropuntia acanthocarpa var. coloradensis

Cylindropuntia bigelovii

Cylindropuntia californica var. parkeri

Cylindropuntia echinocarpa

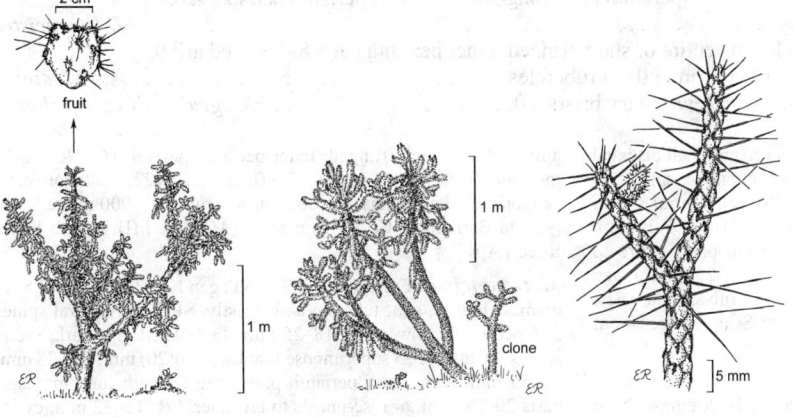

Cylindropuntia munzii

Cylindropuntia ×fosbergii

Cylindropuntia ramosissima

Echinocactus polycephalus var. polycephalus

F. cylindraceus (Engelm.) Orcutt (p. 587) CALIFORNIA BARREL
CACTUS   **ST:** taller than wide, spheric to columnar. **SPINES:** 10–32,
erect and spreading, longest gen recurved to ± hooked, hooked on
immature plants, gen ± red [yellow], gray in age. **FL:** inner perianth
occ orange to red; style 12–20 mm, ovary 9–12 mm, scales fringed.
FR: yellow. **SEED:** 1.5–3 mm. $2n=22$. Gravelly, rocky, or sandy
areas; 60–1500 m. e PR, D (esp e DMoj, w DSon); to sw UT, AZ, n
Mex. [*F. c.* var. *lecontei* (Engelm.) Bravo] Formerly recognized vars.
untenable. Threatened by collecting; monitoring needed. Apr–May

F. viridescens (Torr. & A. Gray) Britton & Rose (p. 587) SAN
DIEGO BARREL CACTUS   **ST:** gen not taller than wide, depressed-
spheric to cylindric. **SPINES:** 10–24, gen spreading, longest straight
or ± curved, red, pink or ± yellow, ± gray or yellow in age. **FL:**
inner perianth occ with red-brown midstripes; style 6–9 mm, ovary
5–7 mm, scales fringed-dentate. **FR:** yellow (to ± red). **SEED:** 1.5–2
mm. $2n=22$. Sandy to rocky areas; 10–150 m. SCo (San Diego Co.);
Baja CA. Threatened by urbanization, off-road vehicles, collecting.
May–Jun ★

GRUSONIA CLUB-CHOLLA

Marc Baker, Bruce D. Parfitt & Jon Rebman

Per or shrub, erect to decumbent, many-branched, matted, succulent. **ST:** winter- or drought-deciduous, regularly segmented,
segments < 30 cm, < 6 cm diam, cylindric to club-shaped, fleshy, glabrous; ribs 0, tubercles gen elongate, occ 0. **LF:** deciduous. **SPINES:** 0–many per areole, densest and longest near st tip, < 4 mm wide, awl- to dagger-shaped, flat to angular, straight,
roughened, tip smooth or barbed, epidermis at spine tip separating as a papery sheath; glochids gen numerous in each areole.
FL: lateral to terminal on st, from upper portion of areole, 30–50 mm diam; perianth yellow or pink; ovary glabrous, spines
0–many, glochids many in each areole, scales 0. **FR:** indehiscent, obconic, base gen long-tapering, glabrous to densely spiny,
glochids many in each areole. **SEED:** 3–6 mm, ± round, encased in an aril; bony, ± white when dry. 14 spp.: N.Am, Mex.
(H.A.J. Gruson, German engineer, industrialist, 1821–1895) Hybridization unknown.

1. Perianth yellow; largest spine with cross-rows of rough papillae exc at tip; ovary glochids stiff, reflexed-
   barbed. . . . . ***G. parishii***
1′ Perianth pink-magenta; largest spine smooth; ovary glochids hair-like, ascending-barbed . . . . ***G. pulchella***

G. parishii (Orcutt) Pinkava PARISH'S CLUB-CHOLLA Mats or
clumps < 2 m diam. **ST:** 10–20 cm, from fibrous roots; segments ±
obovoid, basal to terminal < 7.5 cm, 2–3 cm diam; tubercles 12–25
mm, 3–8 mm high. **SPINES:** < 21, < 5 cm, largest flat, gray to brown,
margin white, thick, sheath separating only at tip. **FL:** inner perianth
1.5–2.5 cm; filaments green. **FR:** 4.5–8 cm, fleshy, yellow; spines 0
or easily detached. **SEED:** 3–4.5 mm. $2n=22$. Sandy, gravelly flats
gen in creosote-bush/bur-sage scrub; 300–1200 m. SNE, DMoj, s NV,
w AZ. [*Opuntia p.* Orcutt] May–Jun ★

G. pulchella (Engelm.) H. Rob. (p. 587) BEAUTIFUL CLUB-
CHOLLA   Per gen < 0.2 m diam, occ much larger. **ST:** 10–20 cm, single to clumped, from glochid-covered tuber; segments narrowly club-
shaped to cylindric, gen terminal < 10 cm, 0.5–2.5 cm diam; tubercles
occ 0, gen 6–9 mm, < 1.5 mm high. **SPINES:** < 15, < 6 cm, bulbous
at base, largest flat, sharply angled; sheath separating only near tip;
glochids of tuber gen 1–1.5 cm. **FL:** inner perianth pink-magenta,
1.5–2.5 cm; filaments green to yellow. **FR:** 2–3 cm, fleshy, red; spines
gen thin, numerous, crowded. **SEED:** 3–6 mm. $2n=22$. Borders of
dry lakes, sandy flats; 1500–1700 m. SNE; NV, w UT. [*Opuntia p.*
Engelm.] Highly variable; juvenile forms occ fl. May–Jun ★

MAMMILLARIA FISHHOOK CACTUS

Gen erect (decumbent or prostrate), branched or not, branches 0–9(50). **ST:** 5–30 cm, [1.8]3–7.5[20] cm diam, spheric to cylindric [or obconic], firm to soft, not regularly segmented; ribs 0, tubercles prominent, conic to cylindric, not grooved. **SPINES:**
[2]14–64(90) per areole, < 2 mm diam, needle-like [to hair-like or bristle-like], glabrous [or plumose], straight or hooked [or
curved to crinkly]; central spines 1–4 [0–many] per areole, gen hooked. **FL:** lateral, in axils of tubercles, 1–5 [7.5] cm diam;
perianth cream to white, pink, purple, or lavender; ovary glabrous, spines 0, scales 0. **FR:** club-shaped or cylindric to ovoid
[or barrel-shaped], indehiscent, gen red, spines 0. **SEED:** 0.8–1.5 mm, gen shiny, gen pitted or raised-netted, black [brown to
± red or ± yellow], occ with aril. 150 spp.: N.Am. (Latin: nipple) [Hunt 1984 Bradleya 2:65–96; Hunt 1985 Bradleya 3:53–66;
Hunt 1987 Bradleya 5:17–48]

1. Radial spines in 2–3 superimposed ranks; outer perianth parts long-fringed, inner perianth parts 20–28;
   seed with large corky aril. . . . . ***M. tetrancistra***
1′ Radial spines in 1 rank; outer perianth parts entire or short-fringed, inner perianth parts 8–16; seed aril 0
  2. Inner perianth parts cream to white; bristles in axils of tubercles . . . . ***M. dioica***
  2′ Inner perianth parts rose-pink to rose-purple; axillary bristles 0. . . . . ***M. grahamii*** var. ***grahamii***

M. dioica K. Brandegee (p. 587) Pl with fls gen either all bisexual
or all pistillate. **ST:** gen 1(many), 5–30 cm, 3–7 cm diam, spheric to
long-cylindric, firm; tubercle axils bristly. **SPINES:** central spines
1–4 per areole, 8–15 mm, 1 hooked; radial spines 11–22, 4–10 mm.
FL: 10–22 mm, 20–40 mm diam; outer perianth parts entire to
minutely fringed; inner perianth parts 8–12. **FR:** 10–25 mm, in age
ovoid to club-shaped. **SEED:** aril 0. $2n=[44]66$. Hillsides, washes,
coastal scrub to creosote-bush scrub; 10–1500 m. SCo, w edge DSon;
Baja CA. Feb–Apr

M. grahamii Engelm. var. ***grahamii*** (p. 587) GRAHAM'S
FISHHOOK CACTUS   **ST:** gen several (1–many), 7–15 cm, 4–7 cm
diam, spheric to cylindric, firm; tubercle axillary bristles 0. **SPINES:**
central spines 1–2 per areole, 12–15 mm, ≥ 1 hooked; radial spines
17–28(35), 6–12 mm. **FL:** 15–25 mm, 20–30 mm diam; outer perianth parts minutely fringed; inner perianth parts 9–16. **FR:** 12–25
mm, long-club-shaped in age. **SEED:** aril 0. $2n=22$. Sandy or rocky
canyons, washes, plains, creosote-bush scrub; 300–900 m. ne DSon
(se San Bernardino Co.); AZ, n Mex. [*M. milleri* (Britton & Rose)
Boed.] Apr ★

M. tetrancistra Engelm. (p. 587)   **ST:** gen 1, 7–25 cm, 3.5–7.5 cm
diam, cylindric, soft; tubercle axils bristly. **SPINES:** central spines
3–4 per areole, lowermost 18–25 mm, 1+ hooked, tips dark; radial
spines 30–60, in 2–3 superimposed ranks, 6–10(20) mm. **FL:** 25 mm,
25–35 mm diam; outer perianth parts long-fringed; inner perianth
parts 20–28, pink or rose-purple to lavender. **FR:** 15–32 mm, cylindric in age. **SEED:** aril corky, tan, ≥ 1/2 seed length. $2n=22$. Sandy
hills, valleys, plains, creosote-bush scrub; 130–1400 m. D; to UT,
AZ, n Mex. Apr

OPUNTIA PRICKLY-PEAR

Shrub, tree; roots fibrous [tuberous]. **ST:** gen erect, < 6 [12] m; segments gen flat (± cylindric), gen firmly attached; tubercles 0 to ± developed; ribs 0. **LF:** small, conic, fleshy, deciduous, present on young sts, ovaries. **SPINES:** 0–many per areole, cylindric or flat, tip smooth or barbed, epidermis persistent; glochids gen many. **FR:** juicy, fleshy or dry; wall thick, bearing areoles; spiny or not. **SEED:** in a bony, ± white aril. ± 150 spp.: Am; *Opuntia ficus-indica* cult for food, others for orn. (Possibly from Papago Indian name ("opun") for this food pl; or for a spiny pl of Opus, Greece) Spines smaller, fewer in shade forms; yellow spines blacken in age. Spineless sts, ovaries, and fr gen with glochids, these occ long, conspicuous; hybridization common. Taxa with cylindric to club-shaped sts moved to *Cylindropuntia, Grusonia.*

1. St gen velvety to minutely papillate-hairy (puberulent), or if glabrous then perianth pink-magenta, stigma
   white, seeds ± spheric
   2. Tree < 6 m, trunks > 15 cm diam — sexually mature when young and ± shrubby . . . . *[O. leucotricha]*
   2′ Shrub gen < 1 m, all or most branches ± touching the ground
      3. Perianth yellow, stigma green; seed 2–3 mm . . . . *O. microdasys*
      3′ Perianth pink-magenta, stigma white; seed 6.5–9 mm . . . . *O. basilaris*
         4. Spines 2–8, rigid, < 2.6 cm — c Kern Co. . . . . var. *treleasei*
         4′ Spines 0, barbed bristles present
            5. St segment length < 2 × width, width > 5 × thickness . . . . var. *basilaris*
            5′ St segment length > 2 × width, width < 2 × thickness . . . . var. *brachyclada*
1′ St glabrous; if perianth pink-magenta then stigma green and seeds ± flat-sided
   6. Terminal st segment gen easily detached when fresh, 2–5.5 cm, width ± = thickness; w CaRH
      (Siskiyou Co.) . . . . *O. fragilis*
   6′ Terminal st segments firmly attached when fresh, > 4 cm, width > 4 × thickness; mostly SW, D
      7. Fr dry in age, tan, gen spiny; perianth yellow to pink-magenta; style slender, white, base ± thicker . . . *O. polyacantha*
         8. Spines gen pale, more numerous and hair-like, curling, reflexed near pl base; ovary areoles 20–33;
            fr gen a densely spiny bur . . . . var. *erinacea*
         8′ Spines gen brown to black, spreading to ascending, straight, stiff on all st segments; ovary areoles
            11–21; fr less spiny, not bur-like . . . . var. *hystricina*
      7′ Fr juicy, gen with some deep purple, spines 0; perianth yellow, orange, or red; style thick, white to red,
         base much thicker
         9. Trunk 1, erect; areoles gen > 38 per ovary
            10. Mature pl > 3 m; st bristles < 1 mm, inconspicuous; ovary tubercled when fresh . . . . *O. ficus-indica*
            10′ Mature pl < 2.5 m; st bristles 2–12 mm, conspicuous; ovary smooth
               11. Longest spines gen 25–50 mm; style white when fresh; seed 3 mm; e PR, DMoj (desert mtns and
                  lower montane slopes, canyons) . . . . *O. chlorotica*
               11′ Longest spines 19–25(50) mm; style red when fresh; seed 3.5–4 mm; SW . . . . *O. oricola*
         9′ Basal branches several, ± decumbent to ascending; areoles ≤ 36 per ovary
            12. Style and filaments white
               13. St segment < 14 cm wide; perianth base red; fr gen green inside . . . . *O. phaeacantha*
               13′ St segment > 15 cm wide; perianth base yellow; fr red inside
                  14. St segment lightly glaucous, gray-green, gen obovate, < 20 cm wide . . . . *O. engelmannii* var. *engelmannii*
                  14′ St segment strongly glaucous, ± silvery-blue, round, > 25 cm wide . . . . *[O. robusta]*
            12′ Style pink or filaments yellow to orange-yellow (or both)
               15. St segment > 15 cm wide . . . . *O. ×occidentalis*
               15′ St segment < 15 cm wide
                  16. St segment oblong-elliptic to narrowly obovate; major spines gen round, 4–11 per areole . . . . *O. littoralis*
                  16′ St segment gen obovate; major spines gen flat, gen 0–4 per areole . . . . *O. ×vaseyi*

O. basilaris Engelm. & J.M. Bigelow Shrub. **ST:** 7–40 cm, branches sprawling to ascending or erect; segments 5–21 cm, green to ± blue, gen papillate-puberulent. **SPINES:** gen 0(8) per areole, glochids many. **FL:** inner perianth ± 4 cm, pink-magenta; filaments deep magenta-red; style white or pink, stigma white. **FR:** 2–4 cm, dry in age, green and purple becoming tan, gen puberulent; areoles 24–76. **SEED:** 6.5–9 mm, ± spheric.

var. ***basilaris*** (p. 587) BEAVERTAIL **ST:** segments 8–21 cm, 5–13 cm wide, flat, ± obovate. **SPINES:** 0. 2*n*=22. Desert to pinyon/juniper woodland; 150–2200 m (higher n). s SN, Teh, SnGb, SnBr (and adjacent SCo), e PR, s SNE, D; to UT, AZ, Mex. Mar–Jun

var. ***brachyclada*** (Griffiths) Munz (p. 587) SHORT-JOINT BEAVERTAIL **ST:** segments 5–13 cm, 1.5–5 cm wide, ± flat, ± cylindric to ± club-shaped. **SPINES:** 0. 2*n*=22. Chaparral, oak/pine woodland; 1200–1800 m. SnGb, SnBr. Barely distinct from *O. basilaris* var. *b.* Threatened by collecting. Apr–Jun ★

var. ***treleasei*** (J.M. Coult.) Toumey (p. 587) BAKERSFIELD CACTUS **ST:** segments 9–20 cm, 5–7.5 cm wide, flat, gen obovate. **SPINES:** 2–8 per areole, 7–26 mm, spreading, gen straight, yellow. **FL:** spines 0–6, < 11 mm, pale yellow. 2*n*=22,33. Grassland; 120–150 m. Teh, se SnJV (Kern Co.). Threatened by habitat loss. Mar–Apr ★

O. chlorotica Engelm. & J.M. Bigelow PANCAKE PRICKLY-PEAR ± tree-like. **ST:** 1.5–2 m; branches ascending; segments 13–20 cm, gen round, glabrous. **SPINES:** 3–8 in at least distal areoles, longest 2.5–5 cm, gen straight, flat, ± reflexed, translucent yellow. **FL:** inner perianth 2–2.5 cm, yellow; filaments white; style white, stigma yellow-green. **FR:** ± 4 cm, juicy; exterior purple-red; middle layer white, seed-pulp pink; areoles 40–70. **SEED:** 3 mm. 2*n*=22. Pinyon/juniper woodland, desert scrub edge, chaparral; 600–1300 m. e PR, DMoj (desert mtns and lower montane slopes, canyons); to NV, NM, n Mex. If recognized taxonomically, stabilized hybrids with *O. phaeacantha* in DMtns (New York Mtns, e San Bernardino Co.), s NV, w AZ assignable to *O. ×curvispina* Griffiths (2*n*=44). May–Jun

O. engelmannii Engelm. var. ***engelmannii*** ENGELMANN PRICKLY-PEAR Shrub, mound-forming. **ST:** gen < 1 m; proximal branches gen decumbent, distal spreading to ascending; segments 15–25 cm, gen obovate; gray-green, glabrous. **SPINES:** 3–12 in all areoles, longest 4–5 cm, straight, spreading from areole, ± appressed

to st, ± flat, yellow, coated chalky-white, base often red-brown. **FL**: inner perianth 4–5 cm, yellow; filaments white; style white, stigma yellow-green to green. **FR**: 4–6.5 cm, juicy, red-purple throughout; areoles 20–32. **SEED**: 4–6 mm. 2*n*=66. Desert scrub, dry oak woodland; 900–1500 m. SnJt, e PR, DMtns; to NV, TX, Mex. Hybridizes with *O. phaeacantha*. Mar–May

O. ficus-indica (L.) Mill. (p. 587) MISSION PRICKLY-PEAR Treelike. **ST**: 4–5 m; branches gen ascending; segments 25–43 cm, gen elliptic-obovate, gray-green, glabrous. **SPINES**: gen 0(6) per areole, if present then longest ± 1–3 cm, flat, white-tan or brown-gray. **FL**: inner perianth yellow or orange; filaments pale green to pale pink; ovary tubercled, style white (pale pink), stigma green. **FR**: 6–9 cm, juicy, yellow-orange or purple; areoles 43–71. **SEED**: 4.5–5 mm. 2*n*=88. Dry coastal habitats; 6–450 m. s CCo, s SCoRO, SCo, s ChI, w WTR, w PR; cult in warm regions worldwide; native range unknown. Hybrids with *O. phaeacantha* uncommon. May–Jun

O. fragilis (Nutt.) Haw. (p. 587) BRITTLE PRICKLY-PEAR Shrub, low. **ST**: 6.5 cm; decumbent-sprawling; segments 2–5.5 cm, 2–3 cm wide, gen elliptic-obovate, thickness ± = width, ± tubercled, terminal gen easily detached, green, glabrous. **SPINES**: 3–7 in all areoles, round, longest 3.5 cm, ± rigid, straight, spreading, gray, brown at tip. **FL**: inner perianth yellow, base occ red; filaments white or red; style white, stigma green. **FR**: ± 1.3 cm, dry, tan; areoles ± 19–21, spines 1–5, 5–7 mm. 2*n*=66. Juniper woodland; 880 m. w CaRH (Shasta Valley, Siskiyou Co.); to n&e Can, e-c US, TX. Most northern range of any cactus, near Arctic Circle. Apr–Jul ★

O. littoralis (Engelm.) Cockerell Shrub. **ST**: ≤ 1.5 m, clumps < 9 m diam; branches spreading to sprawling; segments 15–22 cm, oblong-elliptic to narrowly obovate, gray-green, glabrous. **SPINES**: 4–11 in all areoles, longest 2–4 cm, gen round, gen straight, distal spreading, proximal ± reflexed, yellow, gen coated ± white, base yellow (brown). **FL**: inner perianth yellow to dull red; filaments orange-yellow; style pink or red, stigma yellow-green to green. **FR**: 3.5–5 cm, juicy, dark red-purple throughout; areoles 22–36. **SEED**: 3–4.5 mm. 2*n*=66. Coastal-sage scrub, chaparral; 8–400 m. SCo, ChI, s PR; Mex. Highly variable. Hybridizes with other spp. of same chromosome number. Apr–Jun

O. microdasys (Lehm.) Pfeiff. BUNNY EARS Shrub. **ST**: < 1(2) m, branches sprawling to erect; segments 7–15 cm, flat, ellipticobovate to round, light green, velvety pubescent. **SPINES**: 0, glochids abundant, yellow or red-brown. **FL**: inner perianth yellow; filaments white; style white, stigma green. **FR**: 2–3 cm, red, fleshy, pubescent; areoles 35–60, spines 0, glochids abundant, some to 13 mm. **SEED**: 2–3 mm, ± flattened. 2*n*=22. Desert, chaparral; 490 m. PR (San Diego Co.); AZ, Mex. Mar–Jun

O. ×occidentalis Engelm. & J.M. Bigelow Shrub. **ST**: < 1 m; branches sprawling to ± erect; segments 19–35 cm, gen obovate, gray-green, glabrous. **SPINES**: 3–6 in 90–100% of areoles, gen flat, gen straight, longest 2.5–5 cm, ± spreading, distal 1–2 yellow, coated ± white, base brown, proximal 2–4 shorter, ± reflexed. **FL**: inner perianth yellow to deep pink; filaments gen yellow; style pink to white, stigma green. **FR**: 4.5–5 cm, juicy, red-purple; areoles 24–30. **SEED**: 4–5.5 mm. 2*n*=66. Chaparral; 7–400 m. SCo, w edge PR. *O. littoralis* × (*O. engelmannii* × *O. phaeacantha*). Hybridizes with spp. of same chromosome number. Mar–May

O. oricola Philbrick (p. 587) Gen tree-like. **ST**: < 2.5 m; branches gen ascending to spreading; segments 16–25 cm, gen round, green, glabrous. **SPINES**: 5–13 in ± all areoles, flat, longest 2–2.5(5) cm, rigid, gen curved, reflexed, translucent yellow. **FL**: inner perianth yellow; filaments orange-yellow; style red, stigma green. **FR**: 3.7–6 cm, juicy, spheric; exterior purple-red; interior white-yellow; seed-pulp gen red; areoles 23–63. **SEED**: 3.5–4 mm. 2*n*=66. Coastal-sage scrub, chaparral; 3–450 m. SCo, ChI, WTR, w PR; Baja CA. Possible parent of *O. demissa* Griffiths, a putative hybrid. Apr–Jun

O. phaeacantha Engelm. (p. 587) BROWN-SPINED PRICKLY-PEAR Shrub. **ST**: 0.3–1 m; branches decumbent to ± spreading; segments 11–30 cm, gen obovate; gray-green, glabrous. **SPINES**: 1–4(6) per areole on distal 30–70% of segment, fewer proximally, largest 3–8 cm, gen flat, spreading, distal 1–2 red-brown near base, distally white or straw, smaller 1–3 ± reflexed, gen white or gray. **FL**: inner perianth 3.5–4 cm, yellow, base red; filaments white; style white, stigma yellow-green to green. **FR**: 2.5–6.5 cm, juicy, red-purple; interior gen green; areoles 15–32. **SEED**: 3–6 mm. 2*n*=66. Chaparral, Joshua-tree woodland, pinyon/juniper woodland; 45–2220 m. SCoRO, SnBr, e PR, DMtns, DSon; to SD, KS, OK, TX, Mex. Hybrids with *O. ficus-indica* uncommon or rare; stabilized hybrids with *O. chlorotica* in DMtns (New York Mtns, e San Bernardino Co.), s NV, w AZ called *O. ×curvispina* Griffiths (2*n*=44). May–Jul

O. polyacantha Haw. STARVATION PRICKLY-PEAR **ST**: gen < 0.5 m; branches decumbent or ascending to erect; segments 5.5–20 cm, elliptic to obovate, green, glabrous. **SPINES**: flat to round, gen ± white, base yellow-brown, surrounded by shorter, gen reflexed, whiter spines. **FL**: inner perianth 2–2.5 cm, yellow to pink-magenta; filaments gen white (magenta); style white, stigma green. **FR**: 2.5–4 cm, green, tinted red, in age dry, tan. **SEED**: 5–6.5 mm.

var. ***erinacea*** (Engelm. & J.M. Bigelow) B.D. Parfitt MOJAVE PRICKLY-PEAR, GRIZZLY BEAR PRICKLY-PEAR Shrub. **SPINES**: gen 4–24 per areole (0–3 in SnJt), gen pale, longest on oldest st segments, 1.7–18.5 cm, straight, ascending at tip, near pl base hair-like, curling, reflexed. **FR**: spiny exc sometimes in SnJt; areoles 20–33, each with 7–13 spines. 2*n*=44. Desert scrub, Joshua-tree woodland, pinyon woodland; 900–2200 m. se SNH, SnBr, SnJt, SNE, DMoj (esp DMtns); to UT, AZ. [*O. e.* Engelm. & J.M. Bigelow var. *e.*; *O. e.* Engelm. & J.M. Bigelow var. *utahensis* (Engelm.) L.D. Benson] May–Jun

var. ***hystricina*** (Engelm. & J.M. Bigelow) B.D. Parfitt PORCUPINE PRICKLY-PEAR Shrub, low. **SPINES**: (1)2–10 per areole, gen brown to black, (4)5–8 cm; straight, gen spreading, shorter spines gray. **FR**: spiny; areoles 11–21, each with 4–18 spines. 2*n*=44. Uncommon in CA. Desert grassland, pinyon/juniper woodland; 1500–2600 m. SNE; to CO, NM. May–Jun

O. ×vaseyi (J.M. Coult.) Britton & Rose Shrub. **ST**: < 1 m; branches sprawling to spreading; segments 9–22 cm, gen obovate, gray-green, glabrous. **SPINES**: gen 0–4 per areole, gen flat, longest gen 3–4.8 cm, gen straight, spreading to reflexed, yellow, gen ± white-coated, base brown or yellow, smaller spines 0–4, ± 5 mm. **FL**: inner perianth yellow, orange, or dull red; filaments orange-yellow; style gen pink, stigma green to yellow-green. **FR**: juicy, red-purple throughout; areoles 20–36. **SEED**: 4.5–6 mm. 2*n*=66. Chaparral, disturbed areas; 275–500 m. SCo, edges of adjacent zones. *O. littoralis* × *O. phaeacantha*. Apr–Jun

SCLEROCACTUS PINEAPPLE CACTUS

J. Mark Porter & Edward F. Anderson

Erect or ascending, branches gen 0. **ST**: 5–20 cm, 2–12 cm diam, ovoid to cylindric, not segmented, firm; ribs 8–21, prominent; tubercles distinct along ribs. **SPINES**: [2]10–24 per areole, 0.3–2.1 mm diam, needle-like or awl-shaped, straight to curved or hooked; central spines 1–11 per areole. **FL**: ± terminal, from upper edge of spine cluster, 25–75 mm diam; perianth ± green-yellow to magenta; ovary glabrous, spineless, scales sparse, rounded, ciliate at least near tip. **FR**: dehiscent by 2–4 short longitudinal slits, cylindric to ± spheric, spines 0. **SEED**: 2–3.7 mm, reniform, tubercled, glossy or shiny, black. ± 25 spp.: sw US, Mex. (Greek: hard or cruel cactus)

1. Central spines 4–8 per areole, all ± alike, robust, 1–1.5 mm wide at base, straight to curved but not hooked . . . ***S. johnsonii***
1′ Central spines 9–11 per areole, of 2 kinds, hooked spines 5–8 per areole, slender and flexible, 0.3–0.7 mm
     wide at base, distal-most spines 1–3 per areole, upright, straight, flat, ± white, 1–2.1 mm wide at base . . . ***S. polyancistrus***

587

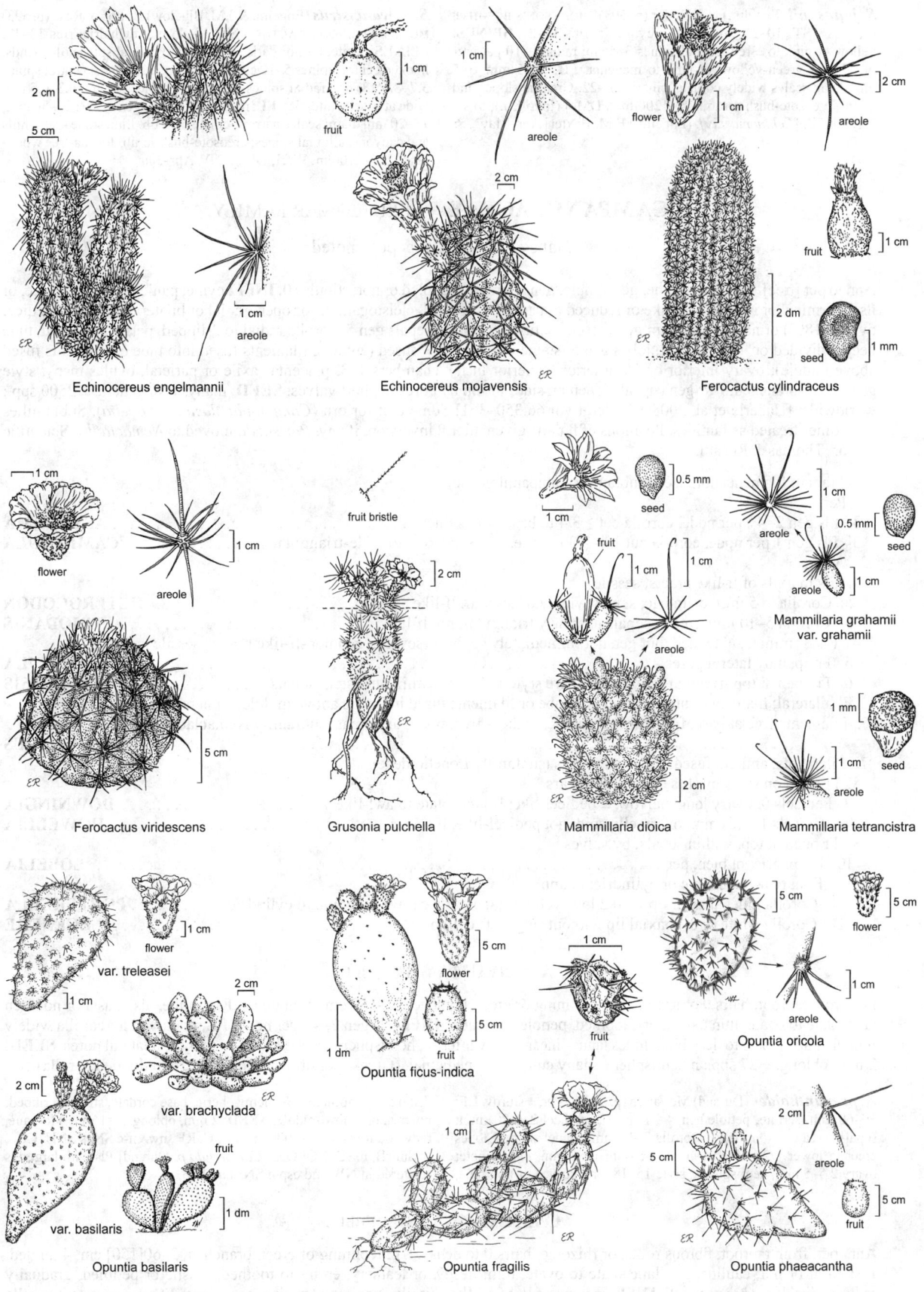

Echinocereus engelmannii

Echinocereus mojavensis

Ferocactus cylindraceus

fruit

areole

flower

areole

fruit

seed

flower

areole

fruit bristle

seed

areole

Mammillaria grahamii var. grahamii

seed

areole

Ferocactus viridescens

Grusonia pulchella

Mammillaria dioica

Mammillaria tetrancistra

var. treleasei

flower

var. brachyclada

flower

fruit

fruit

flower

areole

Opuntia oricola

var. basilaris

fruit

Opuntia basilaris

Opuntia ficus-indica

Opuntia fragilis

areole

fruit

Opuntia phaeacantha

S. johnsonii (Engelm.) N.P. Taylor (p. 593)   JOHNSON'S BEE-HIVE CACTUS   **ST**: 10–25 cm, ovoid to cylindric; ribs 17–21. **SPINES**: yellow or pink to ± red; central spines 3–4 cm; radial 9–10 per areole. **FL**: ± green-yellow, or pink to magenta. **FR**: 7–15 mm, 3–5 mm diam; scales widely cordate, ciliate. 2*n*=22. Granitic slopes and plains, creosote-bush scrub; 500–1200 m. n DMoj (Inyo Co.); to sw UT, nw AZ. [*Echinomastus j.* (Engelm.) E.M. Baxter] Apr–May   ★

S. polyancistrus (Engelm. & J.M. Bigelow) Britton & Rose (p. 593) MOJAVE FISHHOOK CACTUS   **ST**: 10–45 cm, cylindric; ribs 13–17. **SPINES**: white, red, or dark ± red-brown; central spines of 2 kinds, hooked central spines 5–10 cm, red-brown, distal-most central spines 3.7–8.6(13) cm; radial spines 10–15 per areole, 2–5 cm, 0.3–0.5 mm wide at base, white, flat. **FL**: rose-purple to magenta. **FR**: 20–30 mm, 15–20 mm diam; scales narrow, ciliate near tip. Limestone areas, hills and canyons, alluvial slopes; creosote-bush scrub, Joshua-tree woodland; 750–2100 m. W&I, DMoj; NV. Apr–Jun   ★

CAMPANULACEAE   BELLFLOWER FAMILY

Nancy R. Morin, except as noted

Ann to per [tree]. **LF**: gen cauline, gen simple, gen alternate, petioled or not; stipules 0. **INFL**: cyme, panicle, raceme, spike, or fls 1; terminal or in axils of lf-like or reduced bracts. **FL**: bisexual, cleistogamous or open, radial or bilateral, inverted (pedicel twisted 180°) or not; hypanthium gen present, ± fused to ovary; sepals gen 5; corolla radial to 2-lipped, petals gen fused, tube deeply divided on 1 side or not, lobes gen 5; stamens 5, free or ± fused (anthers, filaments fused into tube or filaments fused above middle); ovary inferior or 1/2 inferior (superior in fr), chambers 1–3, placentas axile or parietal, ovules many, style gen 1, 2–5-branched. **FR**: gen capsule, open on sides or top by pores or short valves. **SEED**: many. ± 90 genera, ± 2500 spp.: worldwide. [Haberle et al. 2008 J Molec Evol 66:350–361] Some cult for orn (*Campanula*, *Jasione*, *Lobelia*). Subfamilies sometimes treated as families. Positions of fl parts given after fl inversion, if any. *Parishella* moved to *Nemacladus*. Scientific Editor: Thomas J. Rosatti.

1. Fl radial; filaments not fused (subfamily Campanuloideae)
  2. Per
    3. Fls gen 2–4+ per node; corolla cut ≥ 3/4 to base, lobes linear . . . . **ASYNEUMA**
    3′ Fls gen 1 per node; corolla cut 1/4–2/3 to base, lobes narrow- to wide-triangular . . . . ²**CAMPANULA**
  2′ Ann
    4. Fls in axils of lf-like bracts, sessile
      5. Corolla 3–5 mm, cylindric; sepals widely triangular, lf-like . . . . **HETEROCODON**
      5′ Corolla 5–10 mm, rotate; sepals narrowly triangular, not lf-like . . . . **TRIODANIS**
    4′ Fls terminal, subtended, but gen not immediately so, by 1–several, gen non-lf-like bracts, sessile or not
      6. Fr open by lateral pores . . . . ²**CAMPANULA**
      6′ Fr open at top irregularly by tears where style falls off, within persistent sepals . . . . **GITHOPSIS**
1′ Fl bilateral; filaments, anthers fused into tube or filaments fused into tube above middle, anthers free
  7. Filaments free at base, fused above middle, anthers free; pls of dry areas (subfamily Nemacladoideae)
  . . . . **NEMACLADUS**
  7′ Filaments, anthers fused; pls of wet areas (subfamily Lobelioideae)
    8. Fr open on sides by slits or irregular tears
      9. Pedicels 0; ovary long, narrow, ± pedicel-like; lf lanceolate to awl-like . . . . **DOWNINGIA**
      9′ Pedicels 1–4(8) mm; ovary ellipsoid, not pedicel-like; lf linear. . . . . **HOWELLIA**
    8′ Fr open at top, within sepals, by valves
      10. Fr spheric; pl bien, per . . . . **LOBELIA**
      10′ Fr narrowly obconic or cylindric; pl ann
        11. Corolla blue, abaxial lip with 2 low, yellow ridges; fr narrowly obconic to cylindric . . . . **PORTERELLA**
        11′ Corolla white or 0, abaxial lip without ridges; fr cylindric. . . . . **LEGENERE**

ASYNEUMA   HAREBELL

Per, from taproot, hairs 0 or sparse. **ST**: reclining or erect, branched, 2–150 cm, 4-angled. **LF**: cauline, also basal or not, gen lanceolate to ovate, thin to leathery, toothed, petiole 0 or short. **INFL**: fls gen 2–4+ per node. **FL**: not inverted; corolla widely funnel-shaped, pale to deep blue, lobes lance-linear; ovary inferior, hemispheric, ribbed. **FR**: open by 2–3 lateral pores. **SEED**: 2 mm, oblong. ± 27 spp.: n hemisphere; many cult, some medicinal. (Greek: possibly not together, from divided corolla)

A. prenanthoides (Durand) McVaugh (p. 593)   Per, ± sturdy. **LF**: 10–60 mm, serrate; petiole gen < 5 mm. **FL**: pedicel 2–6(20) mm; sepals spreading to reflexed; corolla 7–14 mm, funnel-shaped, lobes erect in lower 1/2, reflexed in upper; stamens 6 mm, base ciliate; ovary 2.5–5 mm, hemispheric, style 15–18 mm, curved or not, blue, distal 55% papillate. **FR**: hemispheric; base cordate, strongly ribbed; pores at or below middle. **SEED**: 2 mm, oblong. *n*=16,17. Montane, redwood forests; 50–2000 m. NW, CaRH, nw&n-c SN, s SNH, CCo, w SnFrB, n SCoRO; OR. [*Campanula p.* Durand] Pls with large lvs scattered in NW and esp n SN. Jun–Sep

CAMPANULA   HAREBELL

Ann, per, from taproot, fibrous roots, or rhizome, hairs 0 to dense. **ST**: reclining or erect, branched, 2–60[150] cm, 4-angled. **LF**: basal or not, cauline, gen lanceolate to ovate, thin, fleshy, or leathery, entire to toothed, sessile or petioled, gradually reduced distally to bracts in infl. **INFL**: raceme and/or fls 1 [head, spike, panicle]; terminal or axillary. **FL**: not inverted; corolla cylindric to funnel- or bell-shaped, white to deep blue, cut 1/4–2/3 [to all the way] to base, lobes narrow- to wide-triangular;

ovary inferior, hemispheric, spheric, or oblong to obconic. **FR**: open by 2–3 lateral pores. **SEED**: 0.6–3.5 mm, oblong or fusiform. ± 400 spp.: n hemisphere; many cult, some medicinal. (Latin: little bell, from corolla shape) [Roquet et al. 2008 Syst Bot 33:203–217] *C. prenanthoides* moved to *Asyneuma*.

1. Ann, in dry places
  2. Lf oblong-ovate; ovary obovoid to spheric . . . . *C. angustiflora*
  2′ Lf ± linear or lanceolate; ovary oblong
    3. Corolla < 4 mm . . . . *C. griffinii*
    3′ Corolla > 7 mm
      4. Upper lvs, bracts ± linear; ovary (hypanthium surface) hairy but not papillate . . . . *C. exigua*
      4′ Upper lvs, bracts widely lanceolate; ovary papillate . . . . *C. sharsmithiae*
1′ Per, in dry places or not
  5. Style >> corolla, well exserted . . . . *C. scouleri*
  5′ Style ≤ corolla, not exserted
    6. Pl hairy throughout
      7. Lf ± 30 mm, entire . . . . *C. scabrella*
      7′ Lf 6–7 mm, with few large teeth . . . . *C. shetleri*
    6′ Pl hairs 0 or only on lf margins, st angles, sepals
      8. Lf margin entire; cauline lf 30–60 mm . . . . *C. rotundifolia*
      8′ Lf margin ± sharp-toothed or crenate; cauline lf < 20 mm
        9. Lf ovate, margin crenate, with recurved, stiff hairs; corolla bell-shaped; coastal . . . . *C. californica*
        9′ Lf narrowly oblong, margin with few sharp teeth, glabrous; corolla funnel-shaped; inland . . . . *C. wilkinsiana*

C. angustiflora Eastw. (p. 593) Ann, stiffly hairy. **ST**: erect, 5–20 cm. **LF**: sessile, 4.5–9 mm, oblong-ovate, leathery, few-toothed. **FL**: pedicel 3–20 mm; sepals erect, converging in fr; corolla 2.5–6 mm, cylindric, pale blue to white, lobes suberect; stamens 2–2.5 mm, base sparsely ciliate; ovary 2–3 mm, obovoid to spheric, style 2–3 mm, white, distal 50% papillate. **FR**: spheric, strongly ribbed; pores near middle. $n=15$. Chaparral, burns, serpentine soil; 30–1000 m. NCo, NCoRI, SnFrB, SCoRO. May–Jun

C. californica (Kellogg) A. Heller (p. 593) SWAMP HAREBELL Per, hairs stiff, recurved, on st angles, lf margins, ovary ribs. **ST**: clambering, 10–30 cm. **LF**: 10–20 mm, ovate, thin, crenate, petiole 0 or short. **FL**: pedicel 1–20 mm; sepals spreading; corolla 8–15 mm, bell-shaped, pale blue, lobes reflexed; stamens 5 mm, base sparsely ciliate; ovary 2–3 mm, hemispheric, style ± 8 mm, white, distal 95% papillate. **FR**: spheric, weakly ribbed; pores basal. Marshy areas; ± 5–400 m. s NCo, n CCo. Jun–Sep ★

C. exigua Rattan (p. 593) CHAPARRAL HAREBELL Ann, hairs 0 or stiff. **ST**: erect, 5–20 cm. **LF**: 5–11 mm, ± linear, leathery, few-toothed, sessile. **FL**: pedicel 3–20 mm; sepals erect; corolla 7–18 mm, funnel-shaped, pale blue to white, lobes ascending to suberect, tips spreading; stamens 4–6 mm, base ciliate; ovary 2–3 mm, oblong, base tapered, style 6–8 mm, blue, distal 66% papillate. **FR**: oblong, strongly ribbed; pores near middle. $n=17$. Talus slopes, gen serpentine soil; 300–1250 m. e SnFrB, n SCoRI. May–Jun ★

C. griffinii Morin (p. 593) Ann, hairs 0 or stiff. **ST**: erect, 2–20 cm. **LF**: 2–9 mm, ± linear, leathery, serrate, sessile. **FL**: pedicel 1–6 mm; sepals erect; corolla 2.2–3.7 mm, cylindric, white, lobes spreading; stamens 2 mm, base glabrous; ovary 1.7–3.6 mm, oblong, base tapered, style 2.5 mm, white, distal 33% papillate. **FR**: oblong, strongly ribbed; pores near middle. $n=17$. Chaparral, serpentine soil; 30–1350 m. NCoRI, SnFrB, n SCoR. May–Jun

C. rotundifolia L. (p. 593) Per, glabrous. **ST**: erect, 10–60 cm. **LF**: 30–60 mm, linear to lanceolate, thin, entire; basal petioled, cauline sessile. **FL**: pedicel 10–20 mm; sepals spreading to ascending; corolla 12–20 mm, bell-shaped, pale to deep blue [white], lobes reflexed; stamens 6 mm, base ciliate; ovary 4–6 mm, oblong, base rounded, style 11–12 mm, white to blue, distal 60% papillate. **FR**: hemispheric, weakly ribbed; pores near base. $n=34$. Moist slopes; 1200–2500 m. KR, CaRH; circumboreal. Jul–Sep

C. scabrella Engelm. (p. 593) ROUGH HAREBELL Per, cespitose, densely short-appressed-hairy. **ST**: < 6 cm. **LF**: sessile, blade ± 30 mm, ± linear to ovate, stiff, entire, narrowed to base. **FL**: pedicel 10 mm; sepals ascending; corolla 7.5–10 mm, funnel-shaped, powder

blue, lobes ascending to suberect; stamens ± 8 mm, base ciliate; ovary 2.5–4 mm, obconic, style 7.5–9.5 mm, blue, distal 50% papillate. **FR**: elongate, weakly ribbed; pores between middle, top. Bare talus slopes; 2100–2800 m. KR, CaRH, Wrn; to WA, MT. Jul–Aug ★

C. scouleri A. DC. (p. 593) SCOULER'S HAREBELL Per, hairs 0 or short. **ST**: reclining to erect, 20–30 cm. **LF**: petiole narrow, winged, 1–2 cm; blade 10–60 mm, widely lanceolate to ± round, thin to leathery, serrate. **FL**: pedicel 5–20 mm; sepals spreading to ascending; corolla 8–15 mm, widely bell-shaped, pale blue, lobes reflexed; stamens 4–6 mm, base ciliate; ovary 3 mm, hemispheric to obconic, style 12–25 mm, straight, blue, distal 20–40% papillate. **FR**: obconic, weakly ribbed; pores near middle. Shaded woodland, streamsides; 100–1900 m. KR, n NCoRO, CaRH (Mount Shasta, Lassen Peak), n SNH; to AK. Historical collections also from n SNF (n Butte, w Sierra cos.). Jun–Aug

C. sharsmithiae Morin (p. 593) SHARSMITH'S HAREBELL Ann, stiffly hairy. **ST**: erect, 5–25 cm. **LF**: 5–11 mm, widely lanceolate, fleshy, serrate, sessile. **FL**: pedicel 1–3 mm; sepals ascending to suberect; corolla 7–16 mm, funnel- to bell-shaped, deep purple, lobes recurved; stamens 4–6 mm, base ciliate; ovary 2–4.5 mm, oblong, with conic papillae, style ± 4.5–5.5 mm, blue, distal 75% papillate. **FR**: oblong, papillate, strongly ribbed; pores near middle. $n=17$. Talus slopes; 400–1000 m. s SnFrB (Mount Hamilton Range). Apr–Jun ★

C. shetleri Heckard (p. 593) CASTLE CRAGS HAREBELL Per, cespitose, densely short-appressed-hairy. **ST**: < 5 cm. **LF**: 6–7 mm, ± 6-sided, leathery; teeth few, large, petiole 1–2 mm. **FL**: pedicel 1–4 mm; sepals ascending to erect (erect in fr); corolla 7–10.5 mm, pale blue to white, lobes recurved to spreading; stamens 4–6.5 mm, base ciliate; ovary 2.5–4 mm, oblong, style 7–8 mm, blue, distal 33% papillate. **FR**: cup-shaped; base 3-lobed, weakly ribbed; pores between middle, top. $n=17$. Rock crevices; 1300–1500 m. e KR (Castle Crags). Jun–Sep ★

C. wilkinsiana Greene (p. 593) WILKINS' HAREBELL Per, forming dense colonies, glabrous. **ST**: reclining to erect, 5–30 cm. **LF**: 12–20 mm, narrowly oblong, thin, teeth few, sharp; sessile. **FL**: pedicel 10–30 mm; sepals ascending to suberect (erect in fr); corolla 12–15 mm, funnel-shaped, deep blue, lobes recurved to ascending; stamens 6 mm, base ciliate; ovary 4–5 mm, narrowly obconic, style ± 12 mm, blue, distal 50% papillate. **FR**: narrowly obconic, weakly ribbed; pores near top. Wet meadows, streamsides; 1800–2600 m. KR, CaRH (Mount Shasta). Jul–Sep ★

DOWNINGIA

Lisa M. Schultheis

Ann, glabrous. **ST**: decumbent to erect, (10)20–40 cm. **LF**: cauline, often deciduous before fl, 0.5–2(4) mm wide, lanceolate to awl-like (uppermost wider or not), sessile, gen entire. **INFL**: spike; terminal fls often aborted, overtopped by fertile; pedicels 0. **FL**: bilateral, gen inverted at full bloom by twisted ovary; corolla gen >> calyx, blue to pink or white, gen with a symmetric white or yellow spot on lower lip, tube entire, limb strongly 2-lipped, upper lip lobes 2, lower lip gen with 2 low ridges, these often with 2 knob-like projections near throat, lobes 3, > upper, gen obovate, obtuse-mucronate; stamens fused (filaments, anthers in tubes), gen 2 smallest anthers each with terminal tuft of bristles, 1 bristle triangular or horn-like, gen 0.2–0.5 mm, others linear, shorter; ovary inferior, long, narrow, ± pedicel-like, chambers 1–2, placentas parietal or axile. **FR**: dehiscent on sides by 3–5 sometimes translucent slits, tardily so or not. 15 spp.: w N.Am, Chile. (A.J. Downing, Am horticulturist, 1815–1852) [Schultheis 2001 Syst Bot 26:603–621] Fl part positions ("upper" is adaxial; "lower" is abaxial) given at full bloom. Corolla measurements are from base of tube to tip of longest lobe, color incl albino for ± all spp. Based on geog, interfertility, cytology, and sequences of nuclear as well as chloroplast DNA, *D. yina* Applegate (as recognized in TJM (1993)) split here into 3 morphologically indistinguishable spp. whose limits need further study, only 2 of which are documented for CA: *D. pulcherrima, D. willamettensis.*

1. Anthers abruptly bent, > 70° to filaments; lower corolla lip lobes ± parallel
   2. Corolla with 1 purple band or 3 distinct spots near throat; ovary chambers 2 . . . . *D. insignis*
   2′ Corolla without purple near throat, or, if present, diffuse; ovary chambers 1
      3. Corolla 3-colored (blue, white, yellow); lower corolla lobes obtuse, mucronate . . . . *D. bacigalupii*
      3′ Corolla 2-colored (blue, white); lower corolla lobes acute . . . . *D. elegans*
1′ Anthers not or ± bent, < 45° to filaments; lower corolla lip lobes divergent, not parallel
   4. Corolla ≤ 7 mm; fl not inverting, upper corolla lip 3-lobed
      5. Corolla 4–7 mm, with blue or purple spots or band near throat; seeds with longitudinal lines . . . . *D. laeta*
      5′ Corolla 2.5–4 mm, without blue or purple spots near throat; seeds with spiral lines, appearing twisted . . . *D. pusilla*
   4′ Corolla gen ≥ 7 mm; fl inverting, upper corolla lip 2-lobed
      6. Sinus between 2 upper corolla lobes with a backward-projecting horn . . . . *D. ornatissima*
         7. Upper corolla lobes densely hairy adaxially near tips. . . . . var. *eximia*
         7′ Upper corolla lobes glabrous or sparsely hairy adaxially near tips . . . . var. *ornatissima*
      6′ Sinus between 2 upper corolla lobes without a horn
         8. 2 largest bristles at tips of lower anthers twisted together (or not or 0) . . . . *D. bicornuta*
            9. 2 largest bristles at tips of lower anthers gen ≥ anthers; corolla tube light blue to pale yellow, with
               prominent blue veins adaxially . . . . var. *picta*
            9′ 2 largest bristles at tips of lower anthers < anthers; corolla tube blue, without prominent veins
               adaxially. . . . . var. *bicornuta*
         8′ 2 largest bristles at tips of lower anthers not twisted
            10. Corolla without purple near throat; seeds with spiral lines, appearing twisted . . . . *D. cuspidata*
            10′ Corolla with 1 purple, ± square spot or 2–3 purple, non-square spots or 1 purple band near throat;
               seeds with longitudinal lines
               11. Corolla with 1 purple, ± square spot (2-lobed or not) near throat; upper corolla lobes
                  ciliate-scabrous . . . . *D. concolor*
                  12. Mature fr gen 12–30 mm, readily dehiscent, lines translucent. . . . . var. *brevior*
                  12′ Mature fr (25)30–50 mm, tardily dehiscent, lines not translucent . . . . var. *concolor*
               11′ Corolla with 2–3 purple, non-square spots or 1 purple band near throat; upper corolla lobes glabrous
                  13. Anthers exserted, ± tapered, pointed at tip . . . . *D. pulchella*
                  13′ Anthers incl to ± exserted, blunt at tip
                     14. Ovary chambers 2; placentas axile; GV, SCo, n WTR. . . . . *D. bella*
                     14′ Ovary chambers 1; placentas parietal; CaR, SN
                        15. Upper 3 anthers densely hairy at tips; upper 3 sepals >> lower 2 . . . . *D. montana*
                        15′ Upper 3 anthers glabrous or ± hairy at tips; upper 3 sepals ± > lower 2
                           16. Pls at 500–1550 m in KR, CaR . . . . *D. pulcherrima*
                           16′ Pls at gen < 200 m in NW (650 m, Lake Co.) . . . . *D. willamettensis*

D. bacigalupii Weiler (p. 593) **FL**: corolla 5–18 mm, lateral sinuses >> upper, lower lobes blue with central white area incl 2 oblong, orange-yellow spots and occ diffuse purple spots near throat, ridges obscure or 0; anthers ± 90° to filaments; ovary glabrous to minutely papillate-scabrous, 1-chambered, placentas parietal. **FR**: (15)25–45(55) mm; lateral walls papery, easily ruptured when dry, lines translucent. **SEED**: with longitudinal lines. *n*=12. Vernal pools, grassy meadows; < 2000 m. KR, CaR, n SNH, MP; to s OR, sw ID, ne NV. Occ hybridizes with *D. pulcherrima.* Apr–Aug

D. bella Hoover (p. 593) **FL**: corolla 10–12 mm, glabrous, lateral sinuses ± = upper, lower lip blue with central white area incl 2 yellow spots gen alternate 2–3 purple spots near throat; anthers incl to ± exserted, < 45° to filaments; ovary glabrous to minutely spiny-scabrous, 2-chambered, placentas axile. **FR**: 18–60 mm, tardily dehiscent, lateral walls tough, lines not translucent. **SEED**: with longitudinal lines. *n*=11. Uncommon. Vernal pools; < 200 m (GV), 610–655 m (PR), 1400–1600 m (WTR). GV, PR (Santa Rosa Plateau, Riverside Co.), n WTR. Pls in PR, WTR (fr < 30 mm, nectar guides ± unusual) possibly distinct. Mar–May

D. bicornuta A. Gray **FL**: corolla 7–19 mm, densely white-hairy in tube, lateral sinuses > upper, appearing hinged, lower lip blue-purple, with central white area incl yellow-green spots, base purple with 2 prominent knob-like projections near throat; anthers < 45° to filaments, 2 larger bristles at tips of 2 lower, smaller anthers 0.6–2.7

mm, gen tightly twisted together (or 0 or not twisted); ovary glabrous to minutely spiny-scabrous, 2-chambered, placentas axile. **FR**: 35–65(90) mm, tardily dehiscent, lines not translucent. **SEED**: with longitudinal lines. *n*=11. Occ intermediates between *D. bicornuta* var. *b.* and *D. bicornuta* var. *picta* in San Joaquin, Stanislaus cos.

var. ***bicornuta*** (p. 593) **FL**: corolla 9–19 mm, tube (2.5)3–4(4.5) mm, blue, without prominent veins adaxially, 2 upper lobes erect or recurved, not crossed at tips, lower lip gen ± flat; 2 largest bristles at tips of 2 lower, smallest anthers 0.6–1.5 mm, < anthers. Vernal pools, roadside ditches, lake margins; < 1700 m. NCoRI, s CaR, n SN, c SNF, ScV, n SnJV, MP; to s OR, sw ID, w NV. Apr–Jul

var. ***picta*** Hoover (p. 593) **FL**: corolla 7–10 mm, tube (1.5)2–3 mm, light blue or pale yellow, with prominent blue veins adaxially, upper lobes reflexed, ± appressed to tube, crossed at tips, lower lip strongly concave, lateral sinuses >> upper; 2 largest bristles at tips of 2 lower, smallest anthers 1.6–2.7 mm, gen ≥ anthers. Vernal pools, roadside ditches, lake margins; < 100 m. n SNF, ScV, n SnJV. Apr–Jul

D. concolor Greene **FL**: corolla 7–13 mm, glabrous exc upper lobes ciliate-scabrous, lateral sinuses ± ≥ upper, lower lip blue with central white area incl 1 purple, ± square spot (2-lobed or not) near throat; anthers incl to ± exserted, < 45° to filaments; ovary glabrous to spiny-scabrous, 2-chambered, placentas axile. **SEED**: with longitudinal lines.

var. ***brevior*** McVaugh (p. 593) CUYAMACA LAKE DOWNINGIA **FR**: gen 12–30 mm, readily dehiscent, lines translucent. *n*=9. Lakeshores; 1400 m. c PR (Cuyamaca Lake). Threatened by development, grazing, recreation. May–Jul ★

var. ***concolor*** **FR**: (25)30–50 mm, tardily dehiscent, lines not translucent. *n*=8,9. Vernal pools, mud flats, pond and lake margins, roadside ditches; < 550 m. s NCoR, s ScV (Solano Co.), SnFrB, n SCoRI (n San Benito Co.). Apr–Jul

D. cuspidata (Greene) Rattan (p. 593) **FL**: corolla 7–15 mm, glabrous, lateral sinuses ± = upper, lower lip pale to bright blue or lavender with central white area incl 2 yellow spots near throat, ridges obscure; anthers incl, < 45° to filaments; ovary glabrous to minutely spiny-scabrous, 2-chambered, placentas axile. **FR**: (16)20–72 mm, lateral walls tough or papery, lines translucent or not. **SEED**: with spiral lines, appearing twisted. *n*=11. Vernal pools, lake margins, meadows; < 1700 m. NCoR, CaR, SNF, GV, SnFrB, SCoRO, PR, MP; Mex. Variable, needs study. Mar–Jun

D. elegans (Lindl.) Torr. (p. 593) **FL**: corolla 5–18 mm, lateral sinuses >> upper, lower lobes acute, blue with central white area, ridges obscure or 0; anthers ± 90° to filaments; ovary glabrous to minutely papillate-scabrous, 1-chambered, placentas parietal. **FR**: (15)25–45(55) mm; lateral walls papery, easily ruptured when dry, lines gen not translucent. **SEED**: with longitudinal lines. *n*=10. Vernal pools, grassy meadows; < 1600 m. NW; to BC, ID. [*D. e.* var. *corymbosa* (A. DC.) A. Gray] Occ hybridizes with *D. willamettensis*. Jun–Sep

D. insignis Greene (p. 593) **FL**: corolla 9–15 mm, glabrous, lateral sinuses > upper, lower lip sky-blue with prominent dark blue veins, central white area incl 2 yellow-green, ovate spots and 1 purple band or 3 distinct spots near throat; anthers ± 90° to filaments; ovary glabrous to minutely spiny-scabrous, 2-chambered, placentas axile. **FR**: (25)45–80 mm, tardily dehiscent, lateral walls tough, lines not translucent. **SEED**: with longitudinal lines. *n*=11. Vernal pools, roadside ditches, lake margins; < 1650 m. ScV, n SnJV, MP; se OR, w NV. Mar–May

D. laeta (Greene) Greene (p. 593) GREAT BASIN DOWNINGIA **FL**: not inverting; corolla 4–7 mm, ≤ calyx, glabrous, upper lip 3-lobed, pale blue or lavender with central white area incl 2 yellow spots alternate 3 purple spots or with 1 purple band near throat, lower lip 2-lobed, lateral sinuses ± = lower, lobes narrowly triangular, acute; anthers < 45° to filaments, tapered (becoming blunt as stigma elongates); ovary glabrous, minutely spiny-scabrous or not, 2-chambered, placentas axile. **FR**: 21–43 mm, tardily dehiscent, lateral walls tough, lines not translucent. **SEED**: with longitudinal lines. *n*=11. Ditches, ponds, streams, vernal pools; 1200–2200 m. MP; to s-c Can, MT, WY, UT. May–Jul ★

D. montana Greene (p. 593) **FL**: upper 3 sepals >> lower 2, occ minute-ciliate; corolla 9–12 mm, > calyx, glabrous, lateral sinuses > upper, lower lip pale lavender-blue to blue with central white area incl 2 yellow spots and purple markings near throat, the purple gen on small knob-like projections; anthers incl to ± exserted, < 45° to filaments, upper, larger 3 densely hairy at tips; ovary glabrous to minutely spiny-scabrous, 1-chambered, placentas parietal. **FR**: 15–35(45) mm; lateral walls firm, lines 3, translucent. **SEED**: with longitudinal lines. *n*=11. Grassy meadows, roadside ditches in pine forest; 300–1700 m. n CaRF (near Redding), CaRH, n&c SN. May–Aug

D. ornatissima Greene **FL**: corolla (7)8–13 mm, lateral sinuses ± = upper, sinus between 2 upper lobes with a backward-projecting horn, lower lip lavender-blue with central white area incl 2 yellow and 2–4 purple spots near throat; anthers < 45° to filaments; ovary minutely spiny- to papillate-scabrous, 2-chambered, placentas axile. **FR**: 25–65 mm, not easily ruptured even when dry, lateral walls tough, lines not translucent. **SEED**: with longitudinal lines. *n*=12. The 2 vars. occ in same population.

var. ***eximia*** (Hoover) McVaugh (p. 593) **FL**: upper corolla lobes densely hairy adaxially near tips, tips reflexed or not but not curved into a ring. Vernal pools, roadside ditches; < ± 160(230) m. s ScV (Sacramento Co.), SnJV. Apr–May

var. ***ornatissima*** (p. 593) **FL**: upper corolla lobes glabrous or sparsely hairy adaxially near tips, tips curled outward, backward into a ring or strongly recurved. Vernal pools, roadside ditches; < ± 150 m. n SNF, ScV, n SnJV. Apr–May

D. pulchella (Lindl.) Torr. (p. 593) **FL**: corolla 7–19 mm, glabrous, lateral sinuses ± ≥ upper, upper lobes widely divergent, lower lip deep blue with central white area incl 2 ovate yellow spots alternate 3 purple ones or with 1 purple band near throat, lower lobes widely obovate, obtuse to ± truncate, mucronate; anthers exserted, < 45° to filaments, ± tapered, pointed at tip; ovary glabrous to minutely spiny-scabrous, 2-chambered, placentas axile. **FR**: 35–75 mm, tardily dehiscent, lateral walls tough, lines not translucent. **SEED**: with longitudinal lines. *n*=11. Vernal pools, roadside ditches; < 400 m (1300 m, s Teh). NCoRI, s Teh, c&s ScV, SnJV, CCo, s SnFrB, n SCoRI. Apr–Jun

D. pulcherrima M. Peck **FL**: upper 3 sepals ± > lower 2; corolla 7–10 mm, glabrous, lateral sinuses gen > upper, lower lip blue with a central white area incl 2 yellow spots alternate 3 purple spots near throat; anthers < 45° to filaments, 3 upper, larger glabrous to ± hairy at tips; ovary glabrous to minutely papillate-scabrous, 1-chambered, placentas parietal. **FR**: 20–30 mm; lateral walls thin, easily fractured, lines translucent or not. **SEED**: with longitudinal lines. *n*=12. Edges of lakes, ponds, vernal pools, roadside ditches; 500–1560 m. KR, CaR; to WA, gen e of Cascade Range. [*D. yina* Applegate var. *major* McVaugh, in part] See note at genus. Apr–Jul

D. pusilla (A. DC.) Torr. (p. 593) DWARF DOWNINGIA **FL**: not inverting; corolla 2.5–4 mm, ≤ calyx, lateral sinuses ± = lower, glabrous, upper lip 3-lobed, with 2 yellow spots near throat, ridges obscure or 0, lobes white or blue, narrowly triangular, acute; anthers < 45° to filaments; ovary glabrous to minutely spiny-scabrous, 2-chambered, placentas axile. **FR**: 15–27 mm, lateral walls tough, lines translucent. **SEED**: with spiral lines, appearing twisted. *n*=11. Vernal pools, roadside ditches; < ± 150 m (488 m, NCoRI). s NCoRO, NCoRI, ScV, n&c SnJV, n SnFrB; Chile. Mar–May ★

D. willamettensis M. Peck (p. 593) **FL**: upper 3 sepals ± > lower 2; corolla 7–10 mm, glabrous, lateral sinuses gen > upper, lower lip blue with central white area incl 2 yellow spots alternate 3 purple spots near throat; anthers < 45° to filaments, 3 upper, larger glabrous to ± hairy at tips; ovary glabrous to minutely papillate-scabrous, 1-chambered, placentas parietal. **FR**: 20–30 mm; lateral walls thin, easily fractured, lines gen not translucent. **SEED**: with longitudinal lines. *n*=6,8,10. Edges of lakes, ponds, vernal pools; gen < 200 m (650 m, Lake Co.). NW; to WA, w of Cascade Range. [*D. yina* var. *major* McVaugh, in part] See note at genus. Jun–Jul

GITHOPSIS BLUECUP

Ann, glabrous to hairy; roots fibrous. **ST:** erect, 2–40 cm, 4-angled, branched or not. **LF:** cauline, widely linear to ovate, serrate, sessile. **INFL:** fls terminal, bracts 2–4, linear, lanceolate, oblanceolate, to ovate. **FL:** not inverted, pedicelled or not; sepals 0.5–3 × ovary, linear to narrowly triangular; corolla cylindric, funnel-, or bell-shaped, throat white, lobes linear to widely ovate, white to deep purple; ovary inferior, obconic or cylindric, narrowed near middle or not. **FR:** open at top irregularly by tears where style falls off, within persistent sepals. **SEED:** ± 1 mm, angular-fusiform. 4 spp.: w N.Am. (Greek: *Githago* -like) [Morin 1983 Syst Bot 8:436–468] Width in length-to-width ratios of corollas (or ovaries) measured at tube (or ovary) tops.

1. Corolla cylindric, 2.5–3 mm; fls gen cleistogamous; bracts ovate . . . . ***G. tenella***
1′ Corolla funnel- to bell-shaped, gen > 3 mm; fls open; bracts narrower than ovate
   2. Corolla showy, 1–1.5 × longer than wide; filament base wider than tip, ciliate . . . . ***G. pulchella***
      3. Ovary 2.5–5.5 mm; corolla gen < 12 mm . . . . subsp. ***serpentinicola***
      3′ Ovary ≥ 5.5 mm; corolla gen > 12 mm
         4. Ovary obconic, hairs dense, recurved, stiff; upper branches 15–30 mm . . . . subsp. ***campestris***
         4′ Ovary cylindric to narrowly obconic, hairs 0, minute, fine, appressed, or coarse; upper branches
            > 32 mm . . . . subsp. ***pulchella***
            5. St, ovary hairs 0 or fine, appressed; corolla 10–15 mm. . . . . var. ***glabra***
            5′ St, ovary puberulent or hairs rough, coarse; corolla 13–24 mm. . . . . var. ***pulchella***
   2′ Corolla not showy, 2–4.5 × longer than wide; filament base no wider than tip, glabrous or sparsely ciliate
      6. Ovary obconic, ± narrowed at top, base long-tapered; pedicel rarely 0; corolla 4.5–14 mm . . . . ***G. specularioides***
      6′ Ovary cylindric to obconic, narrowed near middle, base swollen; pedicel 0; corolla 1.5–7.5 mm . . . . ***G. diffusa***
         7. Corolla lobes white, pale blue, or pale pink, darker when dry
            8. Upper st 0.4–1 mm wide; corolla 5–7.5 mm . . . . subsp. ***candida***
            8′ Upper st 0.2–0.4 mm wide; corolla 1.5–5 mm . . . . subsp. ***filicaulis***
         7′ Corolla lobes light violet, violet-blue, or light to deep blue
            9. Ovary 5–6.5 × longer than wide . . . . subsp. ***diffusa***
            9′ Ovary 2–5 × longer than wide. . . . . subsp. ***robusta***

G. diffusa A. Gray   Pl glabrous to hairy. **ST:** clambering to erect, 2–30 cm. **LF:** 3–15 mm. **INFL:** bracts 2.5–10 mm, linear or oblanceolate, < 5 mm apart. **FL:** pedicel 0; sepals 1.5–2 × ovary; corolla 1.5–7.5 mm, 2.5–4.5 × longer than wide, narrowly funnel-shaped, lobes ≤ ovary, white to deep blue; filament base narrow, glabrous; ovary 4–9 mm, cylindric to obconic, 2–6.5 × longer than wide, narrowed near middle, base swollen, ribs 10, those at sinuses narrower, style 2–4.5 mm, distal 35–75% papillate. *n*=10,20.

subsp. ***candida*** (Ewan) Morin   **ST:** erect, 4–20 cm; upper 0.4–1 mm wide. **LF:** 6–13 mm. **INFL:** bracts 4–10 mm. **FL:** corolla 5–7.5 mm, widely funnel-shaped, lobes 2–5 mm, white or pale pink; ovary 5 × longer than wide. *n*=10. Chaparral, burned areas, gen on serpentine or similar soils; 480–1700 m. PR; Baja CA. May–Jun

subsp. ***diffusa*** (p. 599)   **ST:** often decumbent, 3–30 cm; upper 0.4–0.8 mm wide. **LF:** 4–10 mm. **INFL:** bracts 4–7 mm. **FL:** corolla 3–5 mm, base narrow, throat funnel-shaped, lobes ± 1.5 mm, light to deep blue; ovary 5–6.5 × longer than wide. *n*=10. Moist, disturbed areas; 200–1500 m. s SNF, SCoR (Monterey Co.), n ChI, TR, PR; Baja CA. Apr–Jun

subsp. ***filicaulis*** (Ewan) Morin   MISSION CANYON BLUECUP **ST:** decumbent, 8–25 cm; upper 0.2–0.4 mm wide. **LF:** 3–10 mm. **INFL:** bract 1–5 mm. **FL:** corolla 1.5–5 mm, funnel-shaped, lobes 0.7–2.5 mm, white or pale blue; ovary 3–4 × longer than wide. *n*=10. Moist or disturbed areas; 220–800 m. PR (San Diego, Riverside cos.). May ★

subsp. ***robusta*** Morin   **ST:** erect, 2–25 cm; upper 0.4–1 mm wide. **LF:** 5.5–15 mm. **INFL:** bracts 4–10 mm. **FL:** corolla 3–7 mm, funnel-shaped, lobes 1–3 mm, light violet or violet-blue; ovary 2–5 × longer than wide. *n*=20. Shaded or disturbed areas, burns; 50–2000 m. KR, NCoR, CaRF, n&c SNF, SnFrB, SCoR. Intergrades with *G. diffusa* subsp. *d.* in Monterey Co., *G. specularioides* in Butte Co. Apr–Jun

G. pulchella Vatke   Pl glabrous to hairy. **ST:** erect, 4–40 cm. **LF:** 3–20 mm. **INFL:** bracts 2–15 mm, lanceolate, 2–30 mm apart. **FL:** pedicel 0–10 mm; sepals 2 × ovary; corolla 7–24 mm, 1–1.5 × longer than wide, narrowly to widely bell-shaped, lobes < to > tube, light violet to deep blue; filament base wide, ciliate; ovary 2.5–10.5 mm, cylindric to narrowly obconic, 2–5 × longer than wide, base long-tapered, top round, ribs 10, equal, style 4–12 mm, distal 66–90% papillate. *n*=10.

subsp. ***campestris*** Morin   **ST:** 8–13 cm; upper branches 1.5–3 cm. **LF:** 8–13 mm. **INFL:** bracts < 5 mm apart. **FL:** pedicel 0; corolla 11–15 mm, 7–10 mm wide at tube top, widely bell-shaped, lobes 7–10 mm, bright blue; ovary 5.5–7 mm, obconic, deeply ribbed in fr, hairs dense, recurved, stiff, style 4–9 mm, 0.3–0.6 mm wide, distal 66% papillate. Uncommon. Open places, volcanic soil; 60–1000 m. CaR, n&c SN. May–Jun

subsp. ***pulchella*** (p. 599)   **ST:** 10–40 cm; upper branches > 3.2 cm. **LF:** 7–18 mm. **INFL:** bracts 5–13 mm apart. **FL:** pedicel 0–7 mm; corolla 10–24 mm, 7–14 mm wide at tube top, widely bell-shaped, lobes 5–19 mm, deep blue; ovary 5.5–10.5 mm, cylindric to narrowly obconic, deeply ribbed in fr, style 5–12 mm, 0.5–1 mm wide, distal 75–90% papillate. The subsp. incl the following 2 vars.

var. ***glabra*** (Jeps.) Morin   **ST:** hairs 0 or finely appressed. **FL:** pedicels 0–7 mm; sepals 5–10 mm; corolla 10–15 mm; ovary 5.5–7 mm, ± cylindric to narrowly obconic, hairs 0 or finely appressed. Grassy openings on serpentine outcrops; 180–730 m. n&c SNF. May–Jun

var. ***pulchella***   **ST:** puberulent or hairs rough, coarse (0). **FL:** pedicels gen 0; sepals 8–16 mm; corolla 13–24 mm; ovary 7.5–10.5 mm, narrowly cylindric, hairs minute or coarse. Grassy openings on serpentine, non-serpentine soils; 30–1220 m. n&c SN. May–Jun

subsp. ***serpentinicola*** Morin   SERPENTINE BLUECUP **ST:** 4–25 cm; upper branches 2–6 cm; hairs fine or 0. **LF:** 3.5–9 mm. **INFL:** bracts 5–15 mm apart. **FL:** pedicel 0–10 mm; corolla 7–13 mm, 4–8 mm wide at sinus, narrowly bell-shaped, lobes 3–6 mm, light to deep blue-violet; ovary 2.5–5.5 mm, obconic, shallowly ribbed in fr, style 5–9 mm, 0.2–0.6 mm wide, distal 70–80% papillate. Serpentine, similar outcrops, and Ione Formation; 300–640 m. n&c SNF. May–Jun ★

G. specularioides Nutt. (p. 599)   Pl glabrous to hairy. **ST:** erect, 2–40 cm. **LF:** 4–20 mm. **INFL:** bracts 3–12 mm, lanceolate to oblong, > 5 mm apart. **FL:** pedicel rarely 0; sepals 1–3 × ovary; corolla 4.5–14 mm, 2 × longer than wide, funnel-shaped, lobes < to ± = tube, deep blue; filament base narrow, sparsely ciliate; ovary 4–14 mm, obconic, ± 3 × longer than wide, top ± narrowed, base long-tapered, ribs 10, equal, style 2.5–6.5 mm, distal 50% papillate. *n*=19,20. Chaparral, oak woodland; 60–1500 m. KR, NCoR, CaR, SN, ScV, SnFrB, s SCoRI (San Luis Obispo, Kern cos.), e SCo (San Bernardino Co.); to BC. Apr–May

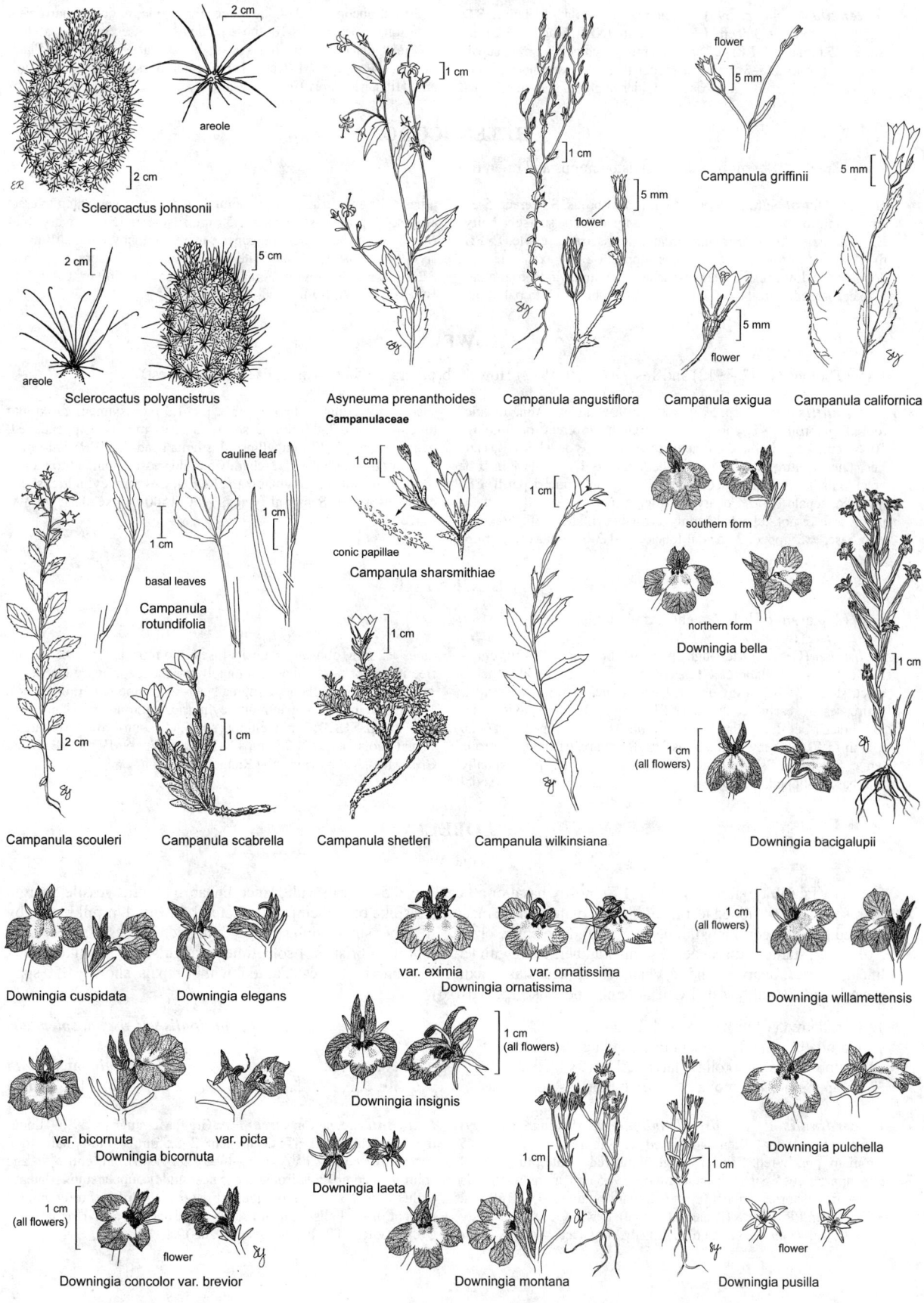

Sclerocactus johnsonii

areole

Sclerocactus polyancistrus

areole

Asyneuma prenanthoides
Campanulaceae

Campanula angustiflora

flower

Campanula griffinii

Campanula exigua

flower

Campanula californica

cauline leaf

Campanula
rotundifolia

basal leaves

conic papillae

Campanula sharsmithiae

southern form

northern form
Downingia bella

Campanula scouleri

Campanula scabrella

Campanula shetleri

Campanula wilkinsiana

Downingia bacigalupii

1 cm
(all flowers)

Downingia cuspidata

Downingia elegans

var. eximia   var. ornatissima
Downingia ornatissima

1 cm
(all flowers)

Downingia willamettensis

var. bicornuta   var. picta
Downingia bicornuta

1 cm
(all flowers)

Downingia insignis

Downingia laeta

Downingia pulchella

1 cm
(all flowers)

flower

Downingia concolor var. brevior

Downingia montana

flower

Downingia pusilla

G. tenella Morin (p. 599) DELICATE BLUECUP Pl hairy. **ST:** clambering to erect, 3–9 cm. **LF:** 4–6.5 mm. **INFL:** bracts 2–5.6 mm, ovate, < 5 mm apart. **FL:** gen cleistogamous; pedicel 0–2 mm; sepals = ovary; corolla 2.5–3 mm, 2 × longer than wide, cylindric, lobes < tube, deep blue; filament base no wider than tip, glabrous; ovary 3–5 mm, obconic, ± 2–4.5 × longer than wide, middle ± narrowed, base long-tapered, ribs 10, those ending at sinuses ± raised, style ± 2–3 mm, distal 25% papillate. n=9. Moist places in oak woodland; 1100–1900 m. s SNH (Kern, Tulare cos.), possibly s SCoRI (Cholame Hills, Monterey Co.). May–Jun ★

HETEROCODON

1 sp. (Greek: different bell, from cleistogamous and open fls)

H. rariflorum Nutt. (p. 599) Ann; roots fibrous. **ST:** erect, 5–30 cm, simple or branched from base, 4-angled, thin, sparsely hairy. **LF:** cauline, 2–10 mm, round-cordate, thin, serrate, sessile. **INFL:** fls 1, sessile, terminal and at nodes, opposite or in axil of lf. **FL:** not inverted, lower cleistogamous; sepals 2–4 mm, widely triangular, lf-like, toothed; corolla 3–5 mm, cylindric, tube white to pale blue, lobes ± 1.5 mm, triangular, ascending to suberect, spreading to erect at tips, deep blue; stamens ± 1.5 mm, filaments linear; ovary inferior, 2–3 mm, short-oblong, papery, gen with many long, stiff hairs, style 3 mm, distal 25% papillate. **FR:** open by lateral pores near base. **SEED:** 0.5 mm, elliptic. Vernally wet places; < 2500 m. CA-FP; to BC, MT, NV. Apr–Jul

HOWELLIA

1 sp. (Thomas (1842–1912) and Joseph (1830–1912) Howell, brothers and botanists in Pacific Northwest)

H. aquatilis A. Gray (p. 599) WATER HOWELLIA Ann, aquatic, rooted, glabrous. **ST:** submersed, ascending to erect, or floating; 10–60 cm. **LF:** alternate (sometimes ± opposite or whorled), cauline, linear, entire or minutely toothed, sessile. **INFL:** fls 1 in lf or bract axils, scattered; pedicels stout, 1–4(8) mm; bract 1 per fl. **FL:** inverted; sepals 1/6–2/5 ovary, narrowly triangular; corolla white, 0 in lower fls or not, tube 2–2.7 mm, narrowly cylindric, split adaxially ± to base, ± 2-lipped, 2 adaxial lobes ± 1–1.3 mm, narrowly strap-shaped, spreading, 3 abaxial lobes ± 1–1.3 mm; stamens fused into tube, anthers 0.2–0.9 mm, 2 shorter anthers minutely appendaged; ovary inferior, 5–13 mm, ellipsoid, stigma head-like. **FR:** 5–13 mm, 1–2 mm diam, ellipsoid, 1-chambered, depressed-conic at top, open on sides by slits or irregular tears. **SEED:** 2–4 mm, cylindric, shiny, chestnut-brown. Seasonal ponds; 1100–1500 m. NCoRH; to WA, MT. Jun–Aug ★

LEGENERE

1 sp. (Anagram of E.L. Greene, Am botanist, 1843–1915)

L. limosa (Greene) McVaugh (p. 599) LEGENERE Ann, emergent or terrestrial, glabrous. **ST:** reclining, 10–30 cm; lateral branches erect, slender, stiff, fleshy or not. **LF:** cauline, narrowly triangular, entire, sessile, early-deciduous. **INFL:** raceme, terminal; axis ± zigzag; bract 1 per fl, 6–12 mm, lf-like, spreading; pedicels 6–20 mm in fr. **FL:** not inverted; sepals 1/6–1/4 ovary, triangular; corolla white, 0 in lower fls or not, tube ± 1.5 mm, cylindric, split abaxially ± to base, 2-lipped, 2 abaxial lobes ± 2 mm, narrow, erect, 3 adaxial lobes ± 2 mm, obovate; stamens fused into tube, anthers 0.5–1 mm, free in age or not, 2 shorter minutely appendaged; ovary inferior, ± 3.5 mm, narrowly obconic, stigma head-like, smooth. **FR:** 6–10 mm, 1–2 mm diam, ± > hypanthium, cylindric, top rounded, chamber 1; open at top. **SEED:** 1 mm, elliptic, shiny, chestnut-brown. Wet areas, vernal pools, ponds; < 950 m. s NCoR, s ScV, n SnJV, SnFrB (Santa Cruz Mtns, Mount Hamilton Range). May–Jun ★

LOBELIA

Tina Ayers

Bien, per [shrub], glabrous or hairy. **LF:** mostly basal or all cauline, 0.5–1.5 cm wide, lance-linear to elliptic, sessile, margin with small, gland-tipped teeth; cauline alternate. **INFL:** raceme [or spike or panicle]. **FL:** bilateral, inverted in full bloom by twisted pedicel; corolla red, blue (or white), tube entire or with an upper sinus, limb strongly 2-lipped, 2 lobes of upper lip < 3 of lower; stamens fused, gen 2 smaller anthers each with terminal tuft of bristles, 1 sometimes triangular or horn-like, others linear, shorter; ovary ± spheric, chambers 2, placentas 2, axile. **FR:** spheric, valves 2, at top, within sepals, short. ± 350 spp.: ± worldwide. (Matthias de l'Obel, Flemish botanist, 1538–1616)

1. Corolla red (white); st erect. 4–20 dm . . . . ***L. cardinalis*** var. ***pseudosplendens***
1′ Corolla blue; st decumbent to sprawling, < 8.5 dm
  2. Upper sinus of corolla = lateral; fl bracts lf-like . . . . ***L. dunnii*** var. ***serrata***
  2′ Upper sinus of corolla >> lateral; fl bracts not lf-like . . . . [***L. erinus***]

L. cardinalis L. var. ***pseudosplendens*** McVaugh (p. 599) CARDINAL FLOWER Bien, short-lived per. **ST:** erect, 4–20 dm, < 1.5 cm diam, purple-red. **FL:** corolla glabrous, red (white), tube 15–20 mm, upper sinus >> lateral; anther tube 3.5–4.5 mm, triangular bristle at tips of 2 shorter anthers 0. n=7. Stream bottoms; 450–1600 m. SnGb, SnBr, PR, DMtns (Panamint Range); to w TX, Mex. Seriously TOXIC, esp when used as a home remedy. Aug–Oct

L. dunnii Greene var. ***serrata*** (A. Gray) McVaugh (p. 599) Long-lived per; rhizomed. **ST:** decumbent, 2–8.5 dm, < 4 mm diam, light green. **FL:** corolla hairy, blue, tube 12(14)–19 mm, entire, in age splitting incompletely from base to near middle, upper sinus = lateral; anther tube 2.3–3 mm, triangular bristle at tips of 2 shorter anthers present. n=7. Falls, seeps of cliffs; 30–1850 m. c SNH (Mariposa Co.), SCoRO, TR, PR; n Baja CA. Jul–Oct

NEMACLADUS

Ann, from taproot. **ST**: prostrate, decumbent, or erect; base gen ± brown or ± purple; branches 0 or below middle. **LF**: basal; petiole short or 0. **INFL**: ± raceme-like; bract 1 per fl, small; pedicel gen thread-like. **FL**: inverted or not; sepals linear to triangular; corolla ± radial or 2-lipped, lobes 5; filaments free at base, fused into tube around style distally, appendages attached to a stalk or directly on 2 adjacent filaments, each with 2–12 cells, anthers free, all alike; ovary superior to 1/2 inferior, hemispheric to obconic, nectary glands 3, mounded or donut-like, on free part of ovary, stigma 2-lobed, papillate. **FR**: gen > hypanthium, hemispheric to fusiform, top pointed or rounded, chambers 2; open at top gen by 2 valves (or circumscissile). **SEED**: elliptic to oblong. 18 spp.: sw US, nw Mex. (Greek: thread-like branch) [Morin 2008 J Bot Res Inst Texas 2:397–400] In descriptions, "filaments" incl both free and fused parts thereof.

1. Fr circumscissile; pls prostrate . . . .*N. californicus*
1′ Fr valvate; pls erect or ascending
  2. Ovary superior in fr; fr 2–3 × > sepals . . . . *N. longiflorus*
    3. Corolla 3–3.5 mm, tube 1.5–2.5 mm; filaments ± 2–3 mm . . . . var. *breviflorus*
    3′ Corolla 5–10 mm, tube 2.5–5.5 mm; filaments 3.5–7.5 mm . . . . var. *longiflorus*
  2′ Ovary partly to completely inferior in fr; fr < or ± > sepals
    4. St < 1 cm; infl gen head-like . . . . *N. twisselmannii*
    4′ St gen > 1 cm, spreading, decumbent, or erect; infl an open raceme
      5. St below branches silver-gray, rarely dark; corolla white or cream with yellow and maroon marks or
        yellow, pink, or ± brown tips
        6. Fl not inverted, adaxial corolla lobes 3, abaxial 2; filaments declined, straight or tips ± curved;
          lvs entire . . . .*N. rubescens*
        6′ Fl inverted, adaxial corolla lobes 2, abaxial 3; filaments arched over, strongly curved; lvs toothed or
          pinnately lobed . . . . *N. tenuis*
          7. Corolla lobes dissimilar, adaxial 2 linear, arched, maroon, abaxial lobes yellow, central oblong-
            elliptic, flanking abaxial ones ± reniform . . . . var. *aliformis*
          7′ Corolla lobes similar, oblong-elliptic, white, tips pink or ± yellow . . . . var. *tenuis*
      5′ St below branches ± red, brown, or dark purple, not silver-gray; corolla white, cream, pink, lavender,
        or ± purple, with dark maroon marks or not
        8. Corolla white or cream with darker marks, divided ± to base, tube 0
          9. Fr base rounded, top acute; sepals spreading or reflexed
            10. Corolla lobes 1.3–2.5 mm . . . . *N. orientalis*
            10′ Corolla lobes 0.2–0.5 mm . . . . *N. rigidus*
          9′ Fr base acute, top acute, ± pointed, or rounded; sepals erect
            11. Pedicel straight, base to tip; sepals ± 0.5 mm; corolla lobes 0.7 mm . . . . *N. capillaris*
            11′ Pedicel straight near base, curved at tip; sepals ± 2 mm; corolla lobes 2–4 mm
              12. Filaments 2 mm; fr 2–2.5 mm; seed 0.5 mm, with clearly pitted rows; lf blades irregularly serrate . . . *N. interior*
              12′ Filaments 3 mm; fr 2.5–3 mm; seed 1 mm, with deeply impressed, vertical lines; lf blades entire
                or obscure-toothed . . . . *N. montanus*
        8′ Corolla white or pale lavender, midvein on lobes sometimes pink, lavender, or maroon, otherwise
        without darker marks, divided 1/3–1/2 to base, tube bell-shaped or cylindric
          13. Corolla tube narrow- or wide-cylindric . . . . *N. secundiflorus*
            14. Corolla tube 0.5–0.8 mm; lf 2–3 mm, elliptic . . . . var. *robbinsii*
            14′ Corolla tube 2.5–3.5 mm; lf 3–6 mm, oblanceolate . . . . var. *secundiflorus*
          13′ Corolla tube bell-shaped
            15. Corolla lobes all held on adaxial side of fl, bases of 2 lowest fused into spur. . . . . *N. calcaratus*
            15′ Corolla lobes radiating or ± held on abaxial side of fl, bases of 2 lowest ± fused, not forming spur
              16. Infl axis strongly zigzag
                17. Pedicel straight exc curved up below tip; lvs toothed or pinnately lobed; bracts linear to
                  lanceolate . . . . *N. glanduliferus*
                17′ Pedicel S-curved; lvs entire or irregularly dentate; bracts ovate . . . . *N. sigmoideus*
              16′ Infl axis straight or weakly zigzag
                18. Anthers 0.5 mm, filaments ± 2 mm; bracts 2–4 mm, recurved; corolla white to lavender,
                  midvein on lobes lavender; fr base acute, tip rounded . . . . *N. gracilis*
                18′ Anthers 0.1–0.3 mm, filaments 0.7–2 mm; bracts 2–9 mm, erect or appressed; corolla white or ±
                  pink; fr bulging near base or not, tip acute
                  19. Bracts ascending, appressed to pedicel base; pedicel straight, tip hooked upward abruptly; fr
                    base, tip acute. . . . . *N. pinnatifidus*
                19′ Bracts erect, not appressed to pedicel base; pedicel ± S-curved or straight, tip erect; fr bulging
                  near base . . . . *N. ramosissimus*

N. calcaratus Morin   CHIMNEY CREEK NEMACLADUS   Erect, 2–4 cm, branches from base or 0.5–1 cm above. **ST**: ascending; base dull, red-brown. **LF**: 2–5 mm, narrowly lanceolate to ± spoon-shaped or elliptic, remotely toothed, hairy, narrowed to wide petiole. **INFL**: axis ± zigzag; bracts 1–1.2(2) mm, lance-ovate, clasping pedicel at base, reflexed above; pedicels 10–12 mm, ± 0.1 mm diam, spreading, straight or arched. **FL**: not inverted; hypanthium ± 0.5 mm; sepals 0.8–1.5 mm, lanceolate, lowest reflexed, others ascending to erect;

corolla divided 2/3 to base, white with pink or red midvein, tube widely cup-shaped, lobes all held on adaxial side of fl, bases of 2 lowest fused into spur; filaments slightly upcurved, 1.8–2 mm, tip ± curved, long-hairy, appendages stalk-like, attached at base of ovary, cells narrow, pointed, anthers 0.3–0.4 mm; ovary ± 1/2-inferior. **FR:** ± 3 mm, base, tip rounded. **SEED:** 1–1.2 mm, oblong, impressed vertical lines crossed by fine transverse lines. Decomposed granite flats; 1900–2100 m. s SNH (Chimney Creek). Jun ★

N. californicus (A. Gray) Morin (p. 599) Prostrate, rosettes 1 or linked by naked, creeping st, 1–10 cm diam. **ST:** reclined. **LF:** 5–20 mm, oblanceolate, entire, hairy, narrowed to slender petiole. **INFL:** head-like, axis compressed; bracts 3–9 mm, oblanceolate or spoon-shaped, spreading; pedicels 2–5 mm, 0.1–0.2 mm diam, spreading, straight, tip not curved. **FL:** not inverted; hypanthium ± 1 mm; sepals 2–6 mm, oblanceolate, erect; corolla not conspicuously 2-lipped, divided 1/2 to base, white, tube bell-shaped, lobes spreading, elliptic or oblong, acute, 1.5–2 mm; filaments declined, curled up, 1–3 mm, few hairs at tip, appendage attached at base of filament, paddle-shaped, cells tube-like, blunt, 0.1 mm, anthers 0.5–0.6 mm; ovary 1/2–2/3-inferior. **FR:** 4–5 mm, base acute, tip dome-like; circumscissile. **SEED:** 0.5–0.7 mm, widely elliptic, pitted. Sandy or gravelly soils; 650–1900 m. se SCoRO, s SCoRI (Caliente Range), ne WTR (n Ventura Co.), DMoj (San Bernardino Co.). [*Parishella c.* A. Gray] Apr–May

N. capillaris Greene (p. 599) Erect, 7–18 cm, 1st branches gen 1.5 cm from base. **ST:** stiffly ascending. **LF:** 3–15 mm, ovate, entire, glabrous or hairy, narrowed abruptly to short petiole. **INFL:** axis zigzag, esp in fr; bracts 1–3 mm, lanceolate to elliptic or ovate, spreading to erect, appressed to pedicel; pedicel 8–12 mm, 0.1–0.15 mm diam, spreading, straight or ± curved, tip not curved. **FL:** not inverted; hypanthium ± 1 mm; sepals ± 0.5 mm, elliptic to lanceolate, erect; corolla conspicuously 2-lipped, divided ± to base, white, lobes oblanceolate, glabrous, 3 adaxial erect, with ± pink area at tip and sometimes other darker marks as well, ± 0.7 mm, 2 abaxial spreading, 1 mm; filaments declined, ± 1 mm, tip ± curved, glabrous, appendage cells 2, directly attached to base of filaments, 0.1 mm, blunt, anthers 0.1–0.2 mm; ovary 1/2-inferior. **FR:** 1.5–2.5 mm, base acute, tip rounded. **SEED:** 0.5–0.7 mm, widely elliptic, narrowly zigzag-ridged. n=9. Dry slopes, burned areas, volcanic outcrops; 400–2100 m. KR, NCoR, CaR, n SNF, SNH, SCoRO, WTR, MP; OR. May–Jul

N. glanduliferus Jeps. (p. 599) Erect, 5–25 cm, branches from base. **ST:** stiffly ascending. **LF:** 3–16 mm, oblanceolate to elliptic, toothed or pinnately lobed, hairy, tapered to petiole. **INFL:** axis strongly zigzag; bracts 2–5(10) mm, spreading or recurved, linear to lanceolate; pedicel 6–13(20) mm, 0.2 mm diam, spreading, straight exc curved below tip. **FL:** not inverted; hypanthium ± 1 mm; sepals 1.3–3 mm, lance-linear to ± deltate, spreading; corolla not conspicuously 2-lipped, white, tube bell-shaped, lobes elliptic, spreading, 0.5–2 mm, lower ciliate or all densely hairy; filaments 1.6–2.3 mm, entire length curved, tip glabrous, appendages on long narrow stalk attached to base of filaments, anthers 0.2–0.4 mm; ovary 1/2-inferior. **FR:** ± 2–4 mm, base, tip rounded. **SEED:** ± 0.5 mm, cylindric, with impressed, vertical lines crossed by fine transverse lines. Sandy or gravelly soils, canyons; 150–1900 m. PR, s SNE, D; NV, AZ, Baja CA. Mar–May

N. gracilis Eastw. (p. 599) SLENDER NEMACLADUS Erect, 2.5–10 cm, branches from base or 0.5–1.2 cm above. **ST:** spreading to ascending; base dull red-brown. **LF:** 2.5–8 mm, oblanceolate to oblong or spoon-shaped, irregularly dentate to ± pinnately lobed, hairy, narrowed to wide petiole. **INFL:** axis straight or weakly zigzag; bracts 2–4 mm, linear-oblong, tips recurved; pedicels (5)10–18 mm, 0.1 mm diam, reflexed, ± S-curved, tip erect. **FL:** inverted; hypanthium 0.5 mm; sepals ± 1 mm, narrowly deltate, spreading; corolla not conspicuously 2-lipped, divided 1/3–1/2 to base, white to lavender, lavender-veined, sparsely hairy, tube bell-shaped, lobes spreading; filaments ± 2 mm, tip curved, finely hairy, appendages attached at base, cells slender, pointed, anthers ± 0.5 mm; ovary 1/2-inferior. **FR:** ± 1.5 mm, base acute, tip rounded. **SEED:** ± 0.5 mm, widely elliptic, with vertical zigzag ridges alternate clearly pitted rows. Rocky slopes, sandy washes; < 1900 m. s SNF (Kern Co.), SnJV (Kern Co.), SCoRI, SCoRO (San Rafael Mtns). Mar–Apr ★

N. interior (Munz) G.T. Robbins Erect, 7–25 cm, 1st branches gen 3–6 cm above base. **ST:** stiffly ascending. **LF:** 10–20 mm, oblanceolate to elliptic, irregularly serrate, glabrous, sessile or abruptly narrowed to petiole. **INFL:** axis strongly zigzag; bracts 1–3 mm, linear to lanceolate, appressed, enfolding pedicel base or not; pedicels 7–13 mm, 0.1–0.2 mm diam, ascending, straight, tip curved. **FL:** not inverted; hypanthium 0.5 mm; sepals ± 2 mm, narrowly triangular, erect; corolla conspicuously 2-lipped, divided ± to base, white, 3 adaxial lobes erect, 2.5 mm, with yellow spot at base, 2 maroon arches above, glabrous, 2 abaxial spreading, 3 mm; filaments declined, ± 2 mm, tip ± curved, glabrous, appendages attached to ± 0.4 mm stalk, wide, blunt, anthers 0.4–0.8 mm; ovary 1/2-inferior. **FR:** 2–2.5 mm, base, tip acute. **SEED:** ± 0.5 mm, elliptic, with clearly pitted rows. Dry, gravelly slopes, yellow-pine forest; 150–2700 m. SNF, SNH, Teh, SnJV, SnBr. May–Jul

N. longiflorus A. Gray Erect, 2.5–21 cm, branches gen from base. **LF:** 3–12 mm, oblanceolate to ovate or oblong, entire to finely crenate, hairy, narrowed to a winged petiole. **INFL:** axis ± zigzag; bracts 2–4 mm, elliptic to ovate, appressed to pedicel; pedicels 6–23 mm, 0.1 mm diam, ascending, S-curved, tip abruptly curved. **FL:** inverted; hypanthium ± 0; sepals 1–2 mm, elliptic, erect; corolla salverform, conspicuously 2-lipped, divided < 1/2 to base, white, deep pink abaxially or on veins, tube cylindric, lobes 1.2–3.5 mm, elliptic, 2 adaxial erect, puberulent, 3 abaxial spreading, yellow spot on throat, glabrous; filaments arched, 2–7.5 mm, tip abruptly curved, finely hairy, appendage stalk attached 2/3 from base, cells slender, acute, ± 0.15 mm, anthers 0.2–0.6 mm; ovary superior in fr. **FR:** ± 2.5–5 mm, base, tip acute. **SEED:** 0.2–0.5 mm, widely elliptic to ± round, with obscure, lengthwise, wavy ridges.

var. ***breviflorus*** McVaugh **FL:** corolla 3–3.5 mm, tube 1.5–2.5 mm; filaments ± 2–3 mm. **SEED:** < 0.5 mm, ± round. Sandy or gravelly slopes, washes; 800–1300 m. s SNH, SnGb, SnBr, PR, sw DMtns (Little San Bernardino Mtns); UT, AZ, Baja CA. Apr–Jun

var. ***longiflorus*** (p. 599) **FL:** corolla 5–10 mm, tube 2.5–5.5 mm; filaments 3.5–7.5 mm. **SEED:** ± 0.5 mm, widely elliptic. Sandy or gravelly slopes, washes; 300–2400 m. SCo, SnGb, SnBr, PR, W&I; Baja CA. Apr–Jun

N. montanus Greene Erect, 8–18 cm, 1st branches gen 2–2.5 cm from base. **ST:** ascending. **LF:** 6–18 mm, oblanceolate to elliptic, entire or obscurely toothed, hairy, narrowed to short, wide petiole. **INFL:** axis strongly zigzag; bracts 1–3 mm, linear to lanceolate, spreading; pedicels 10–15 mm, 0.1 mm diam, ascending, gen straight, tip upturned. **FL:** not inverted; hypanthium ± 1 mm; sepals ± 2 mm, triangular, erect; corolla conspicuously 2-lipped, divided ± to base, white, glabrous or ciliate, 3 adaxial lobes erect, 2 mm, yellow spot at base and 2 maroon arches above, 2 abaxial spreading, 3–4 mm, 2 maroon lines near base; filaments declined, ± 3 mm, tip ± curved, glabrous, appendage pad at base of filaments, cells obtuse, anthers ± 0.5 mm; ovary 1/2-inferior. **FR:** 2.5–3 mm, base acute, tip ± pointed. **SEED:** ± 1 mm, elliptic, with deeply impressed, vertical lines. n=9. Serpentine soils; 300–1200 m. NCoRI, NCoRH, n ScV, se SnFrB (Mount Hamilton). May–Jul

N. orientalis (McVaugh) Morin Erect, 5–25 cm, branched from base. **ST:** stiffly ascending. **LF:** 3–16 mm, oblanceolate to elliptic, toothed or pinnately lobed, hairy, tapered to petiole. **INFL:** axis ± straight; bracts 1–6 mm, spreading, lanceolate to ovate; pedicels 6–16 mm, 0.1 mm diam, stiffly ascending, straight to tip. **FL:** not inverted; hypanthium ± 1 mm; sepals 0.8–2.3 mm, linear-elliptic to ± deltate, erect; corolla 2-lipped, divided ± to base, white, lobes oblanceolate, 3 adaxial erect, 1.3–2.5 mm, maroon at tips, ciliate, 2 abaxial 1.3–2.5 mm; filaments 1–2 mm, declined, tip ± curved, glabrous, appendages stout stalks, cells blunt, anthers 0.2–0.4 mm; ovaries 1/3–1/2-inferior. **FR:** ± 2–4 mm, base rounded, tip acute. **SEED:** ± 0.5 mm, cylindric, with impressed, vertical lines crossed by fine transverse lines. Dry slopes, sandy soils, washes; < 2400 m. SCoRI, SnGb, n SnBr, SNE, D; to UT, NM, Baja CA. [*N. glanduliferus* Jeps. var. *o.* McVaugh] Mar–May

N. pinnatifidus Greene (p. 599) Erect, 6–20 cm, branches many from base to below middle. **ST:** stiffly erect. **LF:** 5–20 mm, oblanceolate, deeply pinnately lobed, toothed or entire, glabrous, narrowed to a long petiole. **INFL:** axis ± straight; bracts 2–5 mm, linear to

elliptic, appressed to pedicel base, proximal pinnately lobed; pedicels 5–15 mm, < 0.1 mm diam, ascending to spreading, straight, tip hooked upward abruptly. **FL:** not inverted; hypanthium 0.5–0.7 mm; sepals 1–1.2 mm, linear to narrowly triangular, separated by a sinus, erect; corolla not conspicuously 2-lipped, divided ± 1/3 to base, glabrous, white or faintly pink-tinged, tube bell-shaped, 1–1.4 mm, lobes elliptic, 0.5–0.8 mm, adaxial 2 reflexed, 3 abaxial spreading, or all reflexed; filaments arched, ± 0.7 mm, tip strongly curved, glabrous, appendage continuous with filament base, cells slender, acute, anthers 0.1–0.2 mm; ovary 1/2-inferior in fr. **FR:** 3–4 mm, base, tip acute. **SEED:** 0.5 mm, elliptic, with widely spaced zigzag ridges. Dry washes, burned areas, chaparral; 300–1520 m. SCo, SnGb, PR, DSon; Baja CA. May–Jun

N. ramosissimus Nutt. Erect, 5–32 cm, branched from base only. **ST:** ascending. **LF:** 3–18 mm, oblanceolate, irregularly toothed or ± pinnately lobed, hairy on margins, base, narrowed to slender or wide petiole. **INFL:** axis straight; bracts 2–9 mm, linear, erect; pedicels 6–22 mm, 0.7 mm diam, ± spreading, ± S-curved or straight, tip erect. **FL:** gen not inverted; hypanthium 0.5 mm; sepals ± 0.5 mm, deltate, erect; corolla not conspicuously 2-lipped, divided > 1/2 to base, white, glabrous, tube bell-shaped, lobes elliptic, 1 mm, spreading to ascending, midvein dark; filaments 1–2 mm, curled upward, appendage cells extended from base of filaments, narrow, blunt, anthers 0.2–0.3 mm; ovary 1/2-inferior. **FR:** ± 1.5–2.5 mm, bulging near base (base asymmetrically narrowed, tip rounded). **SEED:** ± 0.5 mm, ± spheric, with clearly pitted rows. Dry, sandy or gravelly soils, burned areas, chaparral; 30–1600 m. SCoRO, SCo, WTR, SnGb, PR. Apr–May

N. rigidus Curran (p. 599) Spreading to decumbent, 2–4(9) cm, branched from base. **ST:** spreading, stout; base shiny, purple. **LF:** 5–10 mm, elliptic to oblanceolate, fleshy, entire to scalloped, hairy, narrowed to wide petiole. **INFL:** axis strongly zigzag; bracts 2–3 mm, widely elliptic, spreading or reflexed; pedicels 5–12 mm, 0.2 mm diam, spreading, straight, in age curved. **FL:** not inverted; hypanthium 1 mm; 2 adaxial, middle abaxial sepals linear-elliptic, 0.7 mm, 2 flanking abaxial sepals widely triangular, 1.4 mm; corolla divided to base, 2-lipped, white, 3 adaxial lobes erect, 0.2–0.4 mm, ovate, glabrous, veins and tips maroon, 2 abaxial 0.4–0.5 mm, oblong, acute; filaments declined, 1.2–1.6 mm, tip ± curved, glabrous, appendage pad near base of filaments, cells wide, blunt, anthers 0.2–0.3 mm; ovary ± inferior in fr. **FR:** 3–4 mm, base oblique, tip pointed. **SEED:** 0.6–0.7 mm, elliptic, with wide zigzag ridges alternate pitted rows. Bare soil, sand; 200–2500 m. GB; e OR, ID, NV. May–Jun

N. rubescens Greene Erect, branches many, gen 2–2.5 cm from base, 5–20 cm. **ST:** ascending, base silver-gray. **LF:** 5–20 mm, elliptic, blunt, entire, glabrous or coarsely hairy, narrowed abruptly to winged petiole; basal ± yellow or yellow-green in age, not ± purple. **INFL:** axis weakly zigzag; bracts 1–2(3) mm, lanceolate to ovate, appressed to pedicel base; pedicels 8–15 mm, 0.1–0.2 mm diam, spreading to ascending, straight, tip not curved. **FL:** not inverted; hypanthium 0.2–0.3 mm; sepals 0.7–1.5 mm, elliptic to deltate, erect; corolla conspicuously 2-lipped, divided ± to base, cream, 3 adaxial lobes 1.5–2.5 mm, widely oblong-acute, with ± brown chevron at tip, middle adaxial erect, flanking spreading, ± ciliate, 2 abaxial declined, 2–3 mm, narrowly oblong, ± brown-tipped; filaments 1.6–3 mm, declined, straight or tip ± curved, appendage stalk continuous with base of filaments, cells widely tubular, blunt, anthers 0.5–0.7 mm; ovary 1/4–1/2-inferior. **FR:** 2–2.5 mm, base narrowed, tip rounded. **SEED:** 0.4 mm, widely elliptic, with wavy ridges alternate weakly pitted rows. Dry, sandy or gravelly soils; < 1600 m. PR (e slope), SNE, D; NV, AZ, Baja CA. Apr–May

N. secundiflorus G.T. Robbins Erect, 2.5–12.5(16) cm, branched from base or gen 2 cm above. **ST:** spreading to ascending; base dull, red-brown. **LF:** 2–6 mm, narrowly oblanceolate to ± spoon-shaped or elliptic, irregularly dentate, hairy, narrowed to wide petiole. **INFL:** ± 1-sided or not; axis straight or zigzag; bracts 1–3 mm, widely lanceolate to ovate or oblong, erect or reflexed; pedicels 8–15 mm, ± 0.1 mm diam, spreading, ± S-curved, tip erect. **FL:** inverted; hypanthium 0.5 mm; sepals ± 0.8–2.5 mm, linear or deltate, erect; corolla salverform, divided 1/3 to base, lobes all held on lower side of fl; filaments arched, 0.8–2.5 mm, tip ± curved, long-hairy, appendages wide pads attached on filaments near tip of ovary, cells slender, pointed, anthers

0.1–0.7 mm; ovary ± 1/4-inferior. **FR:** ± 2–2.5 mm, base, tip rounded. **SEED:** 0.3–0.5 mm, ± spheric, with zigzag ridges alternate clearly pitted rows.

var. ***robbinsii*** Morin ROBBINS' NEMACLADUS **LF:** 2–3 mm, elliptic. **INFL:** not 1-sided; axis zigzag; bracts 1–1.5 mm, ovate to lanceolate, reflexed. **FL:** sepals 0.8–1 mm, deltate; corolla pale lavender or white, tube 0.5–0.8 mm, narrowly cylindric, 2 adaxial lobes spreading, 0.3–0.5 mm, glabrous, 3 abaxial reflexed, 0.3–0.5 mm; filaments ± 0.8 mm, anthers 0.1 mm. Dry, gravelly slopes; 350–1700 m. s SNH, SCoRI, WTR. Apr–May ★

var. ***secundiflorus*** LARGE-FLOWERED NEMACLADUS **LF:** 3–6 mm, narrowly oblanceolate to ± spoon-shaped **INFL:** ± 1-sided or not; axis straight; bracts 2–3 mm, widely lanceolate to oblong, erect or reflexed; pedicels 9–14 mm, ± 0.1 mm diam, horizontal, ± S-curved, tip erect. **FL:** sepals ± 0.8–2.5 mm, linear; corolla white, with yellow blotch at lobe base and lavender veins, tube 2.5–3.5 mm, widely cylindric, 2 adaxial lobes spreading, 2–2.5 mm, glabrous to hairy, 3 abaxial reflexed, 2.5–3 mm; filaments ± 2–2.5 mm, anthers 0.5–0.7 mm. Dry, gravelly slopes; 200–2000 m. s SNH, SCoR. Apr–May ★

N. sigmoideus G.T. Robbins (p. 603) Erect, 4–12 cm, branches from base or 1.5–2 cm above. **ST:** spreading, base purple-brown. **LF:** 1.5–10 mm, ovate to elliptic, entire or irregularly dentate, short-hairy, sessile. **INFL:** axis strongly zigzag; bracts 0.8–1.5 mm, ovate; pedicels 10–18 mm, < 0.1 mm diam, spreading, S-curved, tip erect. **FL:** inverted; hypanthium 0.5 mm; sepals ± 1.5 mm, lance-deltate, erect, spreading in fr; corolla 2-lipped or all lobes held on abaxial side, divided 1/3 to base, white, gen yellow at lobe tips, hairy, tube bell-shaped, 2 adaxial lobes divergent, elliptic, 1–1.5 mm, 3 abaxial spreading, 1.5 mm; filaments ± 1.5 mm, erect, exserted, tip curved, appendages continuous with base of filaments, cells narrow, attenuate, anthers ± 0.3 mm; ovary ± 1/2 inferior. **FR:** ± 2 mm, base oblique, tip acute. **SEED:** ± 0.5 mm, widely elliptic, with zigzag ridges alternate clearly pitted rows. Sandy or gravelly soils, Joshua-tree woodland; 50–2300 m. s SN, Teh, TR, e PR, SNE, DMoj, nw DSon; NV, AZ, Baja CA. Pl in SNE with corolla to 3 mm. Apr–Jun

N. tenuis (McVaugh) Morin Erect, 9–20 cm, branches many, from base or 1.5–3.5 cm above. **ST:** spreading, base silver-gray. **LF:** 5–15 mm, lanceolate, toothed or pinnately lobed, glabrous or coarsely hairy, narrowed to short petiole; basal in age ± yellow or yellow-green, not ± purple. **INFL:** axis weakly zigzag; bracts 0.5–3 mm, lance-linear to ovate, spreading to reflexed or appressed to pedicel; pedicels 8–15(18) mm, ± 0.1 mm diam, spreading to ascending, double-curved, upturned at tip. **FL:** inverted; sepals acute, erect; corolla conspicuously 2-lipped or lobes ± alike, tube bell-shaped; filaments arched, appendage a narrow stalk 0.2–0.4 mm attached 1/2 to base on proximal free part of filament, cells tubular, blunt, anthers 0.4–0.5 mm; ovary 1/4–1/2-inferior. **FR:** 1–2.5 mm, base narrowed, tip rounded. **SEED:** 0.4 mm, widely elliptic, with wavy ridges alternate weakly pitted rows. [*N. rubescens* var. *t.* McVaugh] Vars. differ in fl morphology to an extent suggesting sp., were it not for intermediates.

var. ***aliformis*** Morin **INFL:** bracts 1–3 mm, lance-linear spreading to reflexed; pedicels 8–15(18) mm, 0.1 mm diam. **FL:** hypanthium 0.5–1 mm; sepals 1–2.5 mm, narrowly deltate; corolla divided ± to base, 2 adaxial lobes 0.9–2.5 mm, 0.3–0.5 mm wide, maroon or ± brown, linear, divergent, arched, 3 abaxial 0.9–2.2 mm, white with yellow and ± brown marks, ciliate; filaments 2–3.5 mm, anthers 0.5–0.6 mm. **FR:** 2–2.5 mm. **SEED:** 0.4 mm, widely elliptic, with wavy ridges alternate weakly pitted rows. Dry, sandy or gravelly soils; < 1600 m. D. Apr–May

var. ***tenuis*** **INFL:** bracts 0.5–1 mm, lance-linear to ovate, appressed to pedicel; pedicels 8–12 mm, < 0.1 mm diam. **FL:** sepals 0.7–1 mm, linear; corolla divided 1/2 to base, lobes ± alike, 0.9–2 mm, white with pale or deep pink or ± yellow tips; filaments 1–2 mm, arched, anthers 0.4–0.5 mm. **FR:** 1–2 mm. Dry, sandy or gravelly soils; < 1600 m. D; Baja CA. Apr–May

N. twisselmannii J.T. Howell (p. 603) TWISSELMANN'S NEMACLADUS Pl 0.5–1 cm, cushion-like, branches several from base. **ST:** base dull, red-brown. **LF:** 2–3 mm, ± spoon-shaped, entire,

hairy, narrowed to wide petiole. **INFL**: ± head-like; axis obscure; bracts 1–1.5 mm, oblong, flat; pedicels 2–4 mm, 0.2 mm diam, erect, straight. **FL**: not inverted; hypanthium 0.5 mm; sepals ± 1 mm, linear, erect; corolla divided 1/2 to base, white, hairy, tube bell-shaped, 2 adaxial lobes erect, 0.5–1.5 mm, 3 abaxial spreading, 0.5–1.5 mm; filaments ± 1.5 mm, tip ± curved, glabrous, cells attached at base of filaments, long-pointed, anthers 0.3 mm; ovary 2/3-inferior in fr. **FR**: ± 3.5 mm, base rounded, tip ± pointed. **SEED**: 0.7–0.8 mm, elliptic to elliptic-oblong, with deeply impressed, vertical lines crossed by fine transverse lines. Granitic sands, rocks, yellow-pine forest; 2240 m. s SNH (n Kern Co.). Jul ★

PORTERELLA

1 sp. (Thomas C. Porter, US botanist, 1822–1901)

P. carnosula (Hook. & Arn.) Torr. (p. 603)   Ann, 2–30 cm, emergent or terrestrial, glabrous. **ST**: erect, branches from base or 0. **LF**: cauline, sessile; blade (5)12–15 mm, (1)3–5 mm wide, narrowly ovate (aerial) to narrowly triangular (submersed), entire or few-toothed. **INFL**: raceme; bract 1 per fl, lf-like, ascending; pedicels in fr 1–3 cm. **FL**: inverted, fragrant; hypanthium 1–2.5 mm; sepals ≥ fr, narrowly triangular; corolla >> calyx, blue, with a symmetric yellow spot on lower lip, tube entire, 4–5 mm, cylindric, strongly 2-lipped, adaxial lobes 2, 1–2 mm, narrowly triangular, erect, abaxial lip with 2 low ridges, 3–8 mm, ± round, spreading; stamens fused, anthers ± 2 mm, tufted, 2 short anthers also with horn-like appendage ± 0.5 mm; ovary 2-chambered, stigma cup- or plate-like, papillate. **FR**: 5–10 mm, narrowly obconic to cylindric, valvate; top acute, ± pointed, or rounded. **SEED**: 1 mm, smooth exc finely striate. *n*=12. Moist, grassy roadsides, lake and pond edges; 1300–3500 m. CaRH, SNH, GB; to WY, AZ. Jun–Aug

TRIODANIS   VENUS LOOKING-GLASS

Ann; roots fibrous; hairs sparse, stiff, reflexed. **ST**: erect, gen 5–40 cm, 4-angled, branches from base or 0. **LF**: cauline, thin, serrate, sessile exc lowest ± short-petioled. **INFL**: fls 1–several in axils of bracts, sessile. **FL**: lower cleistogamous; sepals spreading, narrowly triangular, not lf-like; corolla rotate, lobes deep blue to blue-violet; stamen base wide, ciliate; ovary inferior, elliptic to obovoid, distal 50% papillate. **FR**: opening by lateral pores. **SEED**: ± 0.5 mm, widely elliptic. 7–8 spp.: N.Am, 1 Medit. (Greek: 3 teeth)

1. Ovary elliptic to ovoid, narrowed at top; lf widely lanceolate to ovate, base not clasping st . . . .*T. biflora*
1′ Ovary oblong to obovoid, not narrowed at top; lf round-cordate, base gen clasping st . . . . *T. perfoliata*

T. biflora (Ruiz & Pav.) Greene (p. 603)   **LF**: 5–15 mm, abaxial veins inconspicuous, tip acute. **FL**: open in upper 1–3 bract axils; corolla 5–9 mm, lobes 4–7 mm; stamens ± 2.5 mm; ovary 4.5–7 mm, style 3–3.5 mm. **FR**: pores near top. Disturbed areas; < 2000 m. CA-FP (exc CaR); to c&s US, S.Am. Apr–Jun

T. perfoliata (L.) Nieuwl. (p. 603)   **LF**: gen 8–11 mm, abaxial veins conspicuous, tip blunt. **FL**: open in upper >> 3 bract axils; corolla 8–10 mm, lobes 6–8 mm; stamens 3.5 mm; ovary 5–10 mm, style 4.5 mm. **FR**: pores near or below middle. Disturbed areas; < ± 1500 m. NW, n SNH (Plumas Co.), ScV (Sutter Buttes), SW (uncommon); to e US, S.Am. Jun–Jul

CANNABACEAE   HEMP FAMILY

Alan T. Whittemore, except as noted

Tree, shrub, or erect or twining per; dioecious or fls staminate and bisexual, wind-pollinated; epidermis with stiff hairs, glandular or not; sap watery. **LF**: petioled; simple, unlobed or palmately lobed or compound, all alternate or lower opposite. **INFL**: terminal or axillary, unisexual or with both male and bisexual fls. **FL**: perianth parts 4–6, free or fused; stamens 0 or 4–6; ovary 0 or 1, superior, chamber 1, ovule 1, style 1, short, stigmas 2, slender, plumose. **FR**: drupe or achene, occ ± enclosed in persistent perianth. 11 genera, ± 100 spp.: temp, trop areas worldwide. [Sytsma et al. 2002 Amer J Bot 89:1531–1546] Scientific Editors: Douglas H. Goldman, Bruce G. Baldwin.

1. Tree or shrub; lvs alternate, simple, unlobed; fls staminate and bisexual; fr a drupe; perianth not persistent . . . . . **CELTIS**
1′ Ann, per; lower lvs opposite, palmately lobed or compound (upper unlobed); fls staminate or pistillate;
     fr an achene, gen ± enclosed in persistent perianth.
   2. St erect; lf gen palmately compound; pistillate infl ± spike-like, erect to spreading . . . . **CANNABIS**
   2′ St twining; lf simple, palmately lobed; pistillate infl head- or cone-like, gen pendent (erect). . . . . **HUMULUS**

CANNABIS   HEMP

Alan T. Whittemore & Elizabeth McClintock

1 sp. (Greek, Latin: hemp) [Small & Cronquist 1976 Taxon 25:405–435]

C. sativa L. (p. 603)   HEMP, MARIJUANA   Ann, erect; hairs unbranched. **ST**: branched, < 4 m; inner bark fibrous. **LF**: lower opposite, palmately compound, upper alternate, palmately compound to occ simple; lflets gen 3–7, < 15 cm, narrowly lanceolate, coarsely serrate. **STAMINATE INFL**: panicle- or spike-like, > 15 cm, ± open. **PISTILLATE INFL**: erect to spreading, > 2 cm, dense. **PISTIL-** **LATE FL**: perianth parts fused into a short, unlobed tube or ring. **FR**: achene, 1–2 mm, gen ± enclosed in persistent perianth. Disturbed areas; gen < 600 m. CA-FP; possibly native to c Asia, but cult since pre-history. Highly variable. Psychoactive resin (THC) concentrated in pistillate infls; used in medicine; st fibers for rope, fabric, paper, etc. Jun–Oct

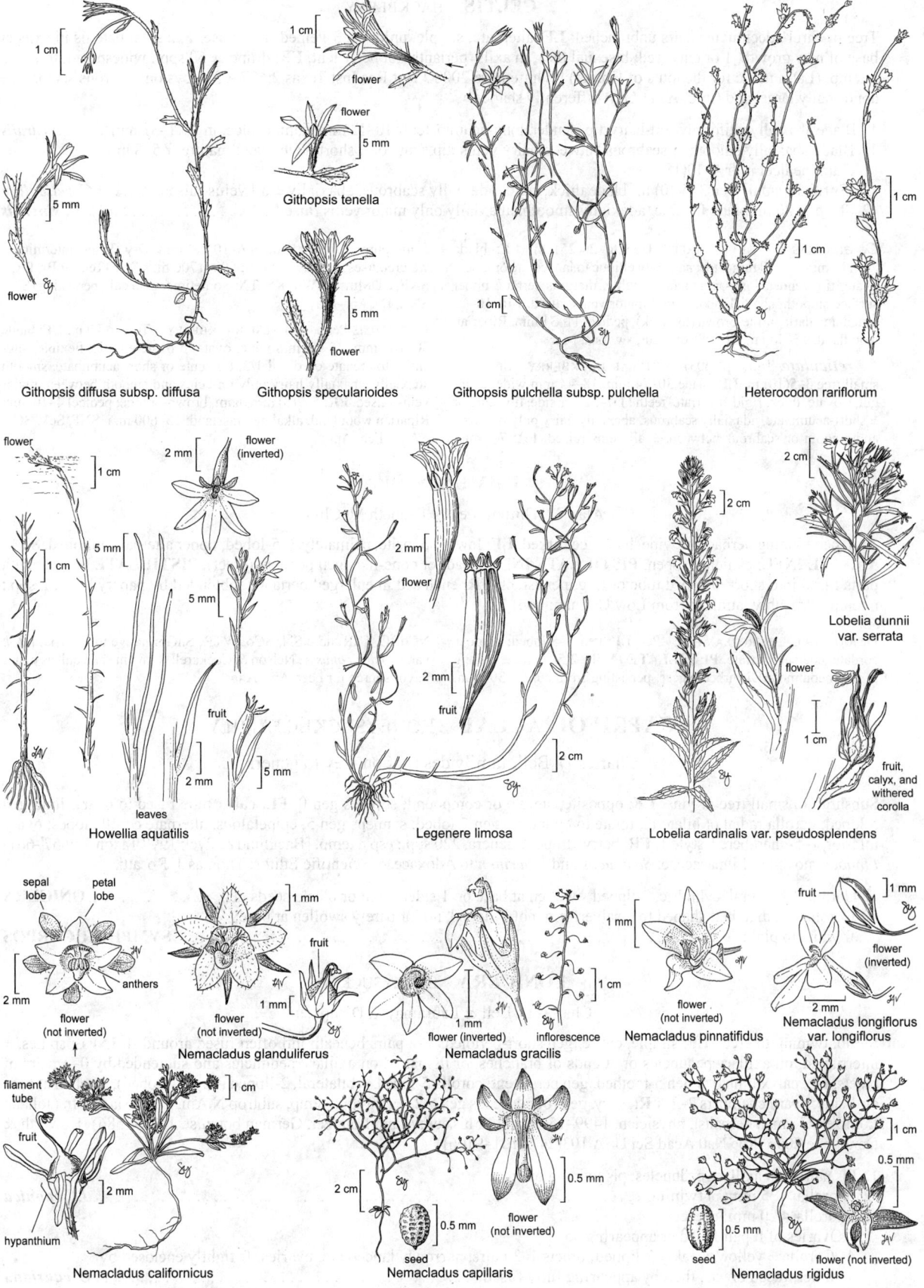

Githopsis tenella

Githopsis diffusa subsp. diffusa

Githopsis specularioides

Githopsis pulchella subsp. pulchella

Heterocodon rariflorum

Howellia aquatilis

Legenere limosa

Lobelia dunnii var. serrata

Lobelia cardinalis var. pseudosplendens

Nemacladus glanduliferus

Nemacladus gracilis

Nemacladus pinnatifidus

Nemacladus longiflorus var. longiflorus

Nemacladus californicus

Nemacladus capillaris

Nemacladus rigidus

CELTIS HACKBERRY

Tree or shrub, deciduous; hairs unbranched. **LF**: alternate, simple, unlobed, 3-veined from base. **FL**: staminate fls in axils at base of new growth, 1 or clustered; bisexual fls 1, in axils, perianth not persistent. **FR**: drupe. ± 60 spp.: widespread, trop and n temp. (Latin name for the lotus of Homer) [Whittemore 2008 J Bot Res Inst Texas 2:627–632] Lvs on vigorous sts may be abnormally large, and (in *C. reticulata*) differently shaped.

1. Blade abaxially uniformly soft-hairy, tip slender, long-acuminate; fr 10–12 mm diam; pedicel in fr 11–31 mm . . . *C. australis*
1′ Blade abaxially smooth or scabrous between hairy veins, tip acute or ± short-acuminate; mature fr 5–8 mm
   diam; pedicel in fr 4–10(18) mm
   2. Shrub or small tree 3–5(10) m; blade thick, rigid, adaxially scabrous, abaxially ± all veins raised . . . . . . *C. reticulata*
   2′ Tree to 20 m; blade flexible, adaxially smooth, abaxially only major veins raised . . . . *C. sinensis*

C. australis L. NETTLE-TREE Tree to 20(25) m. **LF**: blade 70–110 mm, 30–50 mm wide, narrowly elliptic to lanceolate or lance-ovate, tip slender, long-acuminate, flexible, margins serrate, upper surface smooth, abaxially soft-hairy, major veins raised. **FR**: 10–12 mm diam, dark purple-brown (to black); pedicel 11–31 mm. Riparian woodland; 15–500 m. ScV, SCo; s Eur, sw Asia. Mar

C. reticulata Torr. (p. 603) NETLEAF HACKBERRY Shrub or small tree 3–5(10) m. **LF**: blade 30–74 mm, 18–44 mm wide, triangular-ovate, thick, rigid, ± serrate, teeth (1)7–32 per side, tip acute or ± short-acuminate, adaxially scabrous, abaxially hairy only on veins and smooth or scabrous between, ± all veins raised. **FR**: 7–8 mm

diam, purple-brown; pedicel 6–10(18) mm. Dry slopes, intermittent watercourses, in gravelly soil; 500–1700 m. s SNF, Teh, SnBr, PR, s SNE, e DMtns; to WA, KS, TX. Sometimes also cult, persisting (ScV, SNE, D). Apr–May

C. sinensis Pers. CHINESE HACKBERRY Tree to 20 m. **LF**: blade 30–100 mm, 35–60 mm wide, ovate to ovate-elliptic, flexible, subentire to crenate on distal 1/2, tip acute or short-acuminate, smooth adaxially, abaxially hairy only on veins and smooth between, major veins raised. **FR**: 5–7(8) mm diam, brown-orange; pedicel 4–10 mm. Riparian woodland, alkaline grassland; 15–600 m. n SNF, ScV, SCo; e Asia. Feb–Apr

HUMULUS HOP

Alan T. Whittemore & Elizabeth McClintock

Ann, per, twining herbaceous vine, hairs occ forked. **LF**: lower opposite, palmately 3–5-lobed, upper alternate, unlobed. **STAMINATE INFL**: panicle, ± open. **PISTILLATE INFL**: head- or cone-like, gen pendent (erect). **PISTILLATE FL**: perianth parts fused into short, unlobed tube or ring, persistent. **FR**: enclosed in enlarged perianth subtended by papery bract. 2 spp.: n temp. (Probably latinized from Low German: hop)

H. lupulus L. EUROPEAN HOP Per. **LF**: gen 3–5-lobed; blade ± cordate, coarsely serrate. **PISTILLATE INFL**: 2.5–5 cm, ± oblong in fr. Uncommon. Disturbed places, persisting from cult; < 3000 m.

NCoRO, CaR, n&c SN, SCo, WTR, SnGb; native to Eurasia. [*H. l.* var. *neomexicanus* A. Nelson & Cockerell] Orn vine and cult as major flavor source for beer. Apr–Aug

CAPRIFOLIACEAE HONEYSUCKLE FAMILY

Charles D. Bell, family description, key to genera

Subshrub to small tree or vine. **LF**: opposite, simple or compound; stipules gen 0. **FL**: calyx tube fused to ovary, limb gen 5-lobed; corolla radial or bilateral, rotate to cylindric, gen 5-lobed; stamens gen 5, epipetalous, alternate corolla lobes; ovary inferior, 1–5-chambered, style 1. **FR**: berry, drupe. 5 genera, 220 spp.: esp n temp. [Backlund & Pyck 1998 Taxon 47:657–661] *Linnaea* moved to Linnaeaceae; *Sambucus* and *Viburnum* to Adoxaceae. Scientific Editor: Thomas J. Rosatti.

1. Corolla ± bilateral, cylindric, 2-lipped, swollen at base on 1 side; fr red or black; seeds gen > 2 . . . . **LONICERA**
1′ Corolla ± radial, bell-shaped to ± salverform, not 2-lipped, not or barely swollen at base;
   fr white to pink; seeds 2 . . . . **SYMPHORICARPOS**

LONICERA HONEYSUCKLE

Charles D. Bell & Lauramay T. Dempster

Shrub, twining to erect. **LF**: simple, entire, gen short-petioled; 1–2 pairs beneath infl often fused around st. **INFL**: spikes, ± interrupted, on axillary peduncles or at ends of branches, or fls paired on axillary peduncles and subtended by 0–3 pairs of bracts. **FL**: calyx limb 0 or gen 5-toothed, gen persistent; corolla 5-lobed, ± bilateral, 2-lipped (upper 4-lobed), swollen at base on 1 side; ovary chambers 2–3. **FR**: berry, gen round; seeds gen > 2. ± 200 spp.: temp, subtrop N.Am, Eur, Asia, n Afr. (Johann Lonitzer, German herbalist, physician, 1499–1569, and/or his son, Adam Lonitzer, German botanist, 1528–1586) [Howarth & Donoghue 2006 Proc Natl Acad Sci USA 103(24):9101–9106]

1. Fls paired on axillary peduncles; pls twining or erect
   2. Corolla > 25 mm; pl twining . . . . *L. japonica*
   2′ Corolla < 20 mm; pl erect
      3. Ovaries of fl pair fused or appearing so
         4. Corolla ± yellow, weakly 2-lipped; bracts 1–3 pairs, narrowly lanceolate; ovaries, fr tightly enclosed by
            inner, fused bracts, thereby appearing fully fused . . . . *L. cauriana*
      4′ Corolla dark red, strongly 2-lipped; bracts 0 or minute; ovaries, frs not enclosed by bracts, fused gen
         > 1/2 . . . . *L. conjugialis*

3′ Ovaries of fl pair free
    5. Bracts not lf-like, not forming involucre, not enveloping ovaries . . . . ***L. tatarica***
    5′ Bracts lf-like, forming a conspicuous involucre, ± enveloping ovaries . . . . ***L. involucrata***
      6. Pl 6–9 dm; corolla yellow, tube ± wider upward; stigma exserted; KR, NCoRH, CaRH, SN, MP . . . var. ***involucrata***
      6′ Pl 15–36 dm; corolla yellow, strongly tinged orange or red, tube cylindric; stigma incl or ± exserted;
        NCo, NCoRH, NCoRI, CCo . . . . var. ***ledebourii***
1′ Fls in ± interrupted spikes on axillary peduncles or at ends of branches; pls twining or trailing
   (exc *Lonicera interrupta*)
    7. Lf 5–10 cm; corolla 15–50 mm; infl a dense, short spike; upper lf pair fused around st; NW, CaR
      8. Corolla orange, weakly 2-lipped; stamens, style ± exserted . . . . ***L. ciliosa***
      8′ Corolla yellow-white tinged ± purple, strongly 2-lipped; stamens, style well exserted . . . . [***L. etrusca***]
    7′ Lf 1–8 cm; corolla gen < 15 mm; infl a ± long, interrupted spike; upper lf pair fused around st or not;
     widespread
      9. Upper lf pairs fused around st; corolla glabrous or not
        10. Stipules ± conspicuous, at least near infl; corolla glandular-hairy or hairy . . . . ***L. hispidula***
        10′ Stipules 0; corolla glabrous . . . . ***L. interrupta***
      9′ Upper lf pairs not fused around st; corolla often hairy . . . . ***L. subspicata***
        11. Lf < 2 × longer than wide; n SNH (Butte Co.), Teh, CW, SW. . . . . var. ***denudata***
        11′ Lf 3–4 × longer than wide; s ChI (Santa Catalina Island), WTR . . . . var. ***subspicata***

L. cauriana Fernald (p. 603)   Erect, 3–9 dm; herbage puberulent. **LF:** 2–5 cm; blade oblong-ovate, ciliate, base tapered, tip round or obtuse. **INFL:** fls paired; peduncle ± 2 mm; bracts 1–3 pairs, narrowly lanceolate, inner fused, tightly enclosing ovaries. **FL:** calyx limb exserted from bracts; corolla 6–9 mm, ± yellow, bell-shaped, weakly 2-lipped, divided 1/2; anthers exserted; ovaries (and fr) tightly enclosed by inner, fused bracts, thereby appearing fully fused; style ± = corolla, glabrous. **FR:** pair appearing fully fused, ± 8 mm, red. Bogs, wet meadows; 2200–3200 m. CaRH, SNH; also OR to AK, ID. May–Jul

L. ciliosa (Pursh) Poir. (p. 603)   ORANGE HONEYSUCKLE   Trailing to twining; herbage glabrous or soft-puberulent. **ST:** 3–30 dm. **LF:** deciduous, 6–10 cm; blade oval or ovate, ciliate, base tapered, tip round to sharp; upper 1–2 pairs fused around st. **INFL:** spike, short, dense; fls ± 20 in 2–4 whorls. **FL:** corolla 16–40 mm, ± cylindric, weakly 2-lipped, divided 1/6–1/4, orange; stamens, stigma ± exserted. **FR:** ± 8 mm, red, ± glaucous. Forest, thickets; 700–1700 m. NW, CaR; to BC, MT. May–Jun

L. conjugialis Kellogg (p. 603)   Erect, slender, 6–18 dm; herbage puberulent. **LF:** 2–8 cm; blade elliptic to round, base ± tapered, tip round to acute. **INFL:** fls paired; peduncle 14–30 mm, slender; bracts 0 or minute. **FL:** calyx limb 0 or inconspicuous; corolla 4–7 mm, strongly 2-lipped, dark red, hairs 0 or sparse, upper lip erect, shallowly 4-lobed, lower lip downturned; anthers 3 in upper lip, 2 exserted from deepest corolla sinuses; ovaries (and fr) fused gen > 1/2, style hairy well below stigma; stamens, stigma exserted. **FR:** pair fused gen > 1/2, 6–8 mm, bright red, translucent. Streambanks, moist places in conifer forest, open rocky slopes, talus; 140–3300 m. NW, CaR, SN, MP; to WA, w NV. May–Jul

L. hispidula (Lindl.) Torr. & A. Gray (p. 603)   Sprawling to twining, 18–60 dm; herbage puberulent. **LF:** 4–8 cm; blade oblong to ovate, base truncate or subcordate, tip gen obtuse; upper pairs fused around st, others gen with green or ± scale-like stipules. **INFL:** spike, long, interrupted, densely glandular, esp in fr. **FL:** corolla 12–16 mm, strongly 2-lipped, gen pink, glandular-hairy or hairy, upper lip shallowly 4-lobed; stamens, stigma exserted. **FR:** ± 8 mm, red. Canyons, streamsides, woodland; < 1100 m. NW, SN, CW, SW; s OR. [*L. h.* var. *californica* (Torr. & A. Gray) Rehder, nom. superfl.; *L. h.* var. *vacillans* A. Gray] May–Jun

L. interrupta Benth.   Erect; herbage glabrous or puberulent. **ST:** trunk rigid, woody, ± 3 dm; branches climbing or sprawling. **LF:** 2–2.5 cm; stipules 0; blade elliptic to round, base gen tapered to round, tip gen ± round; upper 1–3 pairs fused around st. **INFL:** spike, long, interrupted. **FL:** corolla 8–10 mm, strongly 2-lipped, deeply divided, cream-yellow, glabrous; stamens, stigma well exserted. **FR:** ± 10 mm, red. Dry slopes, ridges, floodplains, oak woodland, chaparral; 240–1400 m. NW, w CaR, SN, n ScV, CW, SW; AZ. Apr–May

L. involucrata (Richardson) Spreng.   TWINBERRY   Erect, 6–30 dm; herbage with stalked glands, sparsely hairy. **LF:** 3–12 cm; petiole 2–10 mm; blade elliptic to ovate, base rounded or tapered, tip acute to acuminate. **INFL:** fls paired; peduncle 12–32 mm; bracts in 2 pairs, lf-like, forming a conspicuous involucre, ± enveloping ovaries, outer opposite fls, 7–12 mm, wide-ovate, gen acute, inner 4–8 mm, alternate fls, deeply 2-lobed to entire, truncate, densely glandular. **FL:** calyx limb 0; corolla 12–18 mm, ± cylindric to narrowly bell-shaped, ± yellow, hairy, lobes ± 2 mm, subequal, ± spreading; stamens gen incl; ovaries of fl pair free. **FR:** 6–10 mm, black, in enlarged, purple or red, spreading or reflexed involucre.

var. ***involucrata*** (p. 603)   Pl 6–9 dm. **LF:** not leathery. **FL:** corolla yellow, tube ± wider upward; stigma exserted. Moist places; 600–2900 m. KR, NCoRH, CaRH, SN, MP; to AK, e N.Am, Mex. May–Jun

var. ***ledebourii*** (Eschsch.) Jeps. (p. 603)   Pl 15–36 dm. **LF:** leathery. **FL:** corolla yellow, strongly tinged orange or red, tube cylindric; stigma incl or ± exserted. Moist places; < 1500 m. NCo, NCoRH, NCoRI, CCo; s OR. May–Jul

L. japonica Murray   JAPANESE HONEYSUCKLE   Twining; herbage glabrous or soft-hairy. **LF:** gen 3–8 cm; blade oblong to ovate, base rounded, tip ± acute. **INFL:** fls paired, each pair subtended by 2 lf-like bracts, 4 ± round bractlets, bractlets ± 1/2 ovary; peduncle 5–10 mm. **FL:** corolla 25–40 mm, strongly 2-lipped, white turning yellow, often tinged ± purple, tube hairy; stamens, stigma exserted. **FR:** black. 2*n*=18. Disturbed places; gen < 1000 m. Deltaic GV, n CCo, SCo, SnGb, expected elsewhere; abundant in se US; native to Asia. Sporadic escape from cult. May–Jul ❖

L. subspicata Hook. & Arn.   Gen twining or reclining on shrubs, 9–24 dm; herbage glabrous to puberulent. **LF:** gen 1–4 cm; blade base round or ± tapered, tip round or obtuse; upper pairs not fused around st. **INFL:** spike, long, interrupted, often ± glandular-hairy. **FL:** corolla 8–12 mm, strongly 2-lipped, pale yellow, often hairy; stamens, stigma exserted. **FR:** ± 8 mm, red or yellow.

var. ***denudata*** Rehder (p. 603)   **LF:** wide-elliptic to ± round, < 2 × longer than wide; Chaparral slopes; < 1800 m. n SNH (Butte Co.), Teh, CW, SW. Jun–Jul

var. ***subspicata*** (p. 603)   SANTA BARBARA HONEYSUCKLE   **LF:** narrowly elliptic, 3–4 × longer than wide. Chaparral; < 1000 m. s ChI (Santa Catalina Island), WTR. Apr–May ★

L. tatarica L.   Erect, < 3 m; herbage glabrous or soft-hairy. **LF:** gen 3–6 cm; blade ovate, base round or ± cordate, tip obtuse; upper pairs not fused around st. **INFL:** fls paired, pedicels 0; peduncle ± 15 mm, slender; bracts not lf-like, not forming involucre, lower lanceolate, spreading, upper round-ovate, erect. **FL:** calyx limb ± deep-lobed; corolla ± 15 mm, white or pink fading ± yellow, glabrous, weakly 2-lipped, lobes ± = tube, subequal, obovate; stamens well exserted; ovaries of a fl pair free, stigma exserted. **FR:** ± 1 cm, red, orange, yellow. Disturbed places; 700–1100 m. w CaR; native to Siberia. Berries possibly TOXIC, attractive to children. Sporadic escape from cult. Jun–Jul

SYMPHORICARPOS WAXBERRY, SNOWBERRY

Charles D. Bell & Lauramay T. Dempster

Shrub. **ST:** decumbent to erect, slender. **LF:** simple, deciduous, short-petioled; blade gen elliptic to round, some often ± lobed. **INFL:** gen raceme, gen ± terminal, gen few-fld; fl subtended by 2 fused bractlets. **FL:** ± radial; hypanthium ± spheric; calyx with 5-toothed, persistent limb; corolla bell-shaped to ± salverform, gen 5-lobed, white or pink, often ± hairy inside; nectary glands [1]5, ± basal; stamens gen incl; ovary chambers 4, styles gen incl, stigma head-shaped. **FR:** drupe, gen berry-like, white to pink. **SEED:** 2 (1 per lateral ovary chamber), ± oblong, planoconvex. ± 10 spp.: N.Am, 1 in China. (Greek: to bear fr together, berries borne in clusters)

1. Corolla lobes hairy inside, ± = throat (subg. *Symphoricarpos*)
   2. Pl 6–18 dm, erect; corolla swollen on 1 side, with 5 nectary glands within swelling; infl gen 8–16-fld
      . . . . **S. albus** var. **laevigatus**
   2′ Pl 1.5–6 dm, sprawling; corolla scarcely or not swollen, nectary glands 5, 1 below each corolla lobe; infl
      gen 2–8-fld . . . . **S. mollis**
1′ Corolla lobes glabrous inside, << throat or tube (subg. *Anisanthus*)
   3. Corolla ± salverform, 8–15 mm, tube slender, glabrous inside, 3–4 × spreading lobes; nectary 1; lf ±
      similar adaxially, abaxially, blade gen 4–12 mm. . . . . **S. longiflorus**
   3′ Corolla narrowly bell-shaped, 6–10 mm, tube wide, ± hairy inside, 2–3 × erect lobes; nectaries 5;
      lf more obviously veined abaxially, blade gen 8–20 mm . . . . **S. rotundifolius**
      4. Pl trailing; corolla tube inside sparsely hairy throughout; s SNH, SW, SNE, DMtns. . . . . var. **parishii**
      4′ Pl ± erect; corolla tube inside hairy in middle 1/3, otherwise glabrous; CaR, SN, GB. . . . . var. **rotundifolius**

S. albus (L.) S.F. Blake var. ***laevigatus*** (Fernald) S.F. Blake (p. 609) SNOWBERRY Pl erect, 6–18 dm, glabrous or puberulent. **ST:** branches stiff, spreading; new shoots erect, unbranched, often with infl and larger, more variable lvs. **LF:** blade gen 1–3 cm, to 6 cm on new shoots. **INFL:** fls 8–16. **FL:** calyx limb ± spreading, divided 1/2; corolla 4–6 mm, bell-shaped, pink, swollen on 1 side, with 5 nectary glands within swelling, lobes ± 1/2 corolla, ± erect, lobes and upper throat ± densely hairy inside. **FR:** 8–12 mm, round. **SEED:** 4–5 mm. Shady woodland, streambanks, n slopes; < 1200 m. NW, w edge CaR, n SNF, CW, SW; to AK, MT; naturalized in e US. Fr may be TOXIC to humans. May–Jul(Sep)

S. longiflorus A. Gray (p. 609) FRAGRANT SNOWBERRY Pl 9–12 dm, stiff, glabrous or puberulent, often dotted with minute glands. **ST:** branches often spiny; young bark red or brown, old ± white, shredding. **LF:** blade 0.5–2 cm, entire, ± thick, lanceolate or not, ± blue, veins abaxially not prominent. **INFL:** fls 1 in axils or not. **FL:** fragrant; calyx limb ± erect, unevenly, often shallowly lobed, sinuses often round; corolla 8–15 mm, ± salverform, pink or cream, tube slender, often red or purple outside, glabrous inside, lobes 1/5–1/4 corolla, spreading; nectary 1, long, slender; style gen hairy above middle. **FR:** ± 7 mm, narrowly elliptic, dry. **SEED:** ± 5 mm. Among rocks; 1350–1600 m. GB, DMtns; to CO, TX. May–Jun

S. mollis Nutt. (p. 609) CREEPING SNOWBERRY, TRIP VINE Pl sprawling, 1.5–6 dm, ± glabrous to soft-hairy. **ST:** branches often rooting; root-crowns, old nodes often becoming swollen. **LF:** blade 0.5–3 cm. **INFL:** fls 2–8. **FL:** calyx limb spreading, divided 1/2; corolla ± 4 mm, bell-shaped, pink, often red outside, lobes ± erect, 1/2 corolla, hairy inside; nectary glands 5, 1 below each corolla lobe. **FR:** ± 8 mm, round. **SEED:** 2–4 mm. Ridges, slopes, open places in woodland; 9–3000 m. NW, CaR, SN, CW, SW, MP; to BC, ID, NM. Apr–May

S. rotundifolius A. Gray Pl 6–12 dm, stiff, puberulent. **ST:** old bark shredding. **LF:** blade 8–20 mm, abaxially paler, more prominently veined. **INFL:** fls 1–2 in axils. **FL:** calyx limb flaring, lobes deep, irregular, margin gen transparent; corolla 6–10 mm, narrowly bell-shaped, pink or white, ± hairy inside, lobes ± erect to ± spreading, ± 1/5–1/3 corolla; nectary glands 5, 1 below each corolla lobe; style glabrous. **FR:** 8–12 mm, ovoid. **SEED:** 4–6 mm.

var. ***parishii*** (Rydb.) Dempster Pl trailing, 3–6 dm. **ST:** branches often arched, tips rooting; twigs gen ± straight-hairy. **FL:** corolla 6–9 mm, tube inside sparsely hairy throughout. Slopes, ridges; 1100–3300 m. s SNH, SW, SNE, DMtns; NM. Difficult to distinguish from *S. rotundifolius* var. *r.* in Mono Co. Jun–Aug

var. ***rotundifolius*** (p. 609) Pl ± erect, 6–12 dm. **ST:** branches not arched; twigs fine-puberulent. **FL:** corolla 7–10 mm, tube inside hairy in middle 1/3, otherwise glabrous. Rocky or sandy slopes, open places in conifer forest; 1200–3200 m. CaR, SN, GB; to WA, WY, CO, w TX. Jun–Aug

CARYOPHYLLACEAE PINK FAMILY

Ronald L. Hartman & Richard K. Rabeler, except as noted

Ann to per; rarely dioecious (*Silene*), taprooted or rhizome gen slender. **LF:** simple, gen opposite (subwhorled), entire, pairs at nodes often ± connected at bases; stipules gen 0; petiole gen 0. **INFL:** gen cyme, gen open; fls 1–many; involucre gen 0 (present in *Dianthus*, *Petrorhagia*). **FL:** gen bisexual, radial; hypanthium often present but obscure; sepals (4)5, ± free or fused into a tube, margins gen scarious, more so on inner 2 or not, tube gen not scarious, awns gen 0; petals (4)5 or 0, gen tapered to base (or with claw long, limb expanded), entire to 2–several-lobed, limb gen without scale-like appendages adaxially, gen without ear-like lobes at base; stamens gen 10, gen fertile, gen free, gen from ovary base; nectaries 0 or 5; ovary superior, gen 1-chambered, placentas basal or free-central, styles 2–5 with 0 branches or 1 with 2–3 branches. **FR:** capsule or utricle (rarely ± dehiscent), gen sessile. **SEED:** appendage gen 0 (present in *Moehringia*). 83 or 89 genera, 3000 spp.: widespread, esp arctic, alpine, temp n hemisphere; some cult (*Agrostemma*, *Arenaria*, *Cerastium*, *Dianthus*, *Gypsophila*, *Lychnis*, *Sagina*, *Saponaria*, *Silene*, *Vaccaria*). [Rabeler & Hartman 2005 FNANM 5:3–215] Apetalous Caryophyllaceae can also be keyed in Rabeler & Hartman 2005 FNANM 5:5–8. Scientific Editor: Thomas J. Rosatti.

1. Fr utricle (± dehiscent in *Achyronychia*, *Scopulophila*); petals 0 or < 0.6 mm, scale-like; stamens on hypanthium rim; stipules present (exc *Scleranthus*), scarious
   2. Stipules 0; hypanthium in fr hard. . . . . **SCLERANTHUS**
   2′ Stipules present; hypanthium in fr ± hard or not (subfamily Paronychioideae)

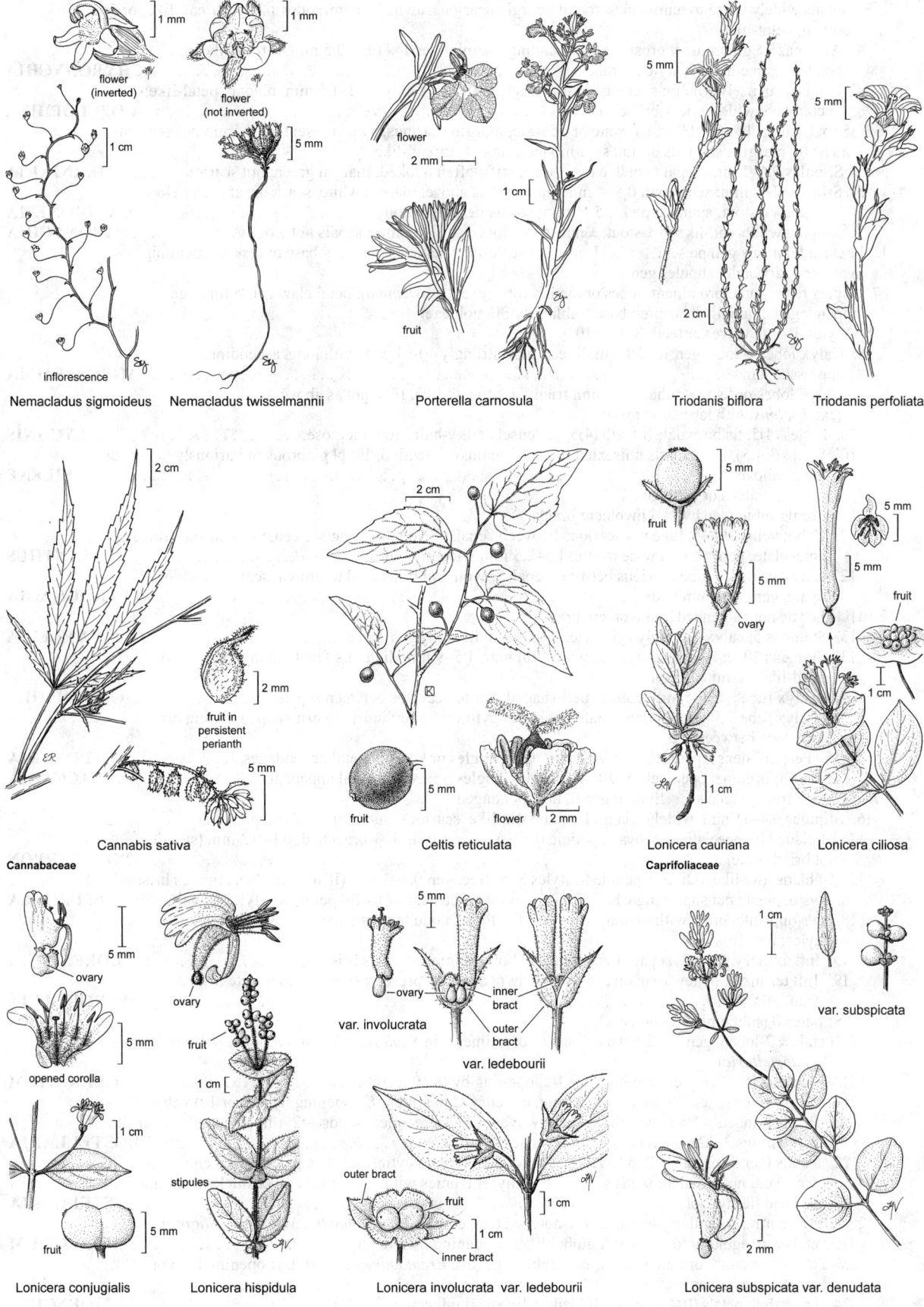

flower
(inverted)

1 mm

flower
(not inverted)

1 mm

1 cm

5 mm

flower

2 mm

1 cm

5 mm

5 mm

fruit

inflorescence

2 cm

Nemacladus sigmoideus   Nemacladus twisselmannii   Porterella carnosula   Triodanis biflora   Triodanis perfoliata

2 cm

fruit

5 mm

2 cm

5 mm

ovary

5 mm

5 mm

fruit

fruit in
persistent
perianth

2 mm

5 mm

5 mm

fruit

flower

2 mm

1 cm

1 cm

Cannabis sativa   Celtis reticulata   Lonicera cauriana   Lonicera ciliosa

Cannabaceae

Caprifoliaceae

ovary

5 mm

ovary

fruit

5 mm

5 mm

inner
bract

1 cm

var. subspicata

opened corolla

var. involucrata

outer
bract

var. ledebourii

1 cm

1 cm

outer bract

fruit

stipules

inner bract

5 mm

1 cm

fruit

2 mm

Lonicera conjugialis   Lonicera hispidula   Lonicera involucrata var. ledebourii   Lonicera subspicata var. denudata

3. Sepals widely ovate to reniform or round, margin scarious, awns 0; staminodes 14–19, thread-like, or 5, oblong, petal-like
  4. Ann, base ± glabrous; st prostrate to ascending; staminodes 14–19, ± 0.5 mm, thread-like, not petal-like; nectaries 0; style 2-branched . . . . **ACHYRONYCHIA**
  4′ Per, base densely woolly; st erect; staminodes (pistillate fls only) 5, 1–1.5 mm, oblong, petal-like; nectaries 5, wide; style 3-branched . . . . **SCOPULOPHILA**
3′ Sepals lanceolate to oblong (if ovate or obovate, margin scarious, awns present, or margin not scarious, awns 0), margins scarious or not; staminodes 0 or 4–5, thread-like
  5. Stipules 0.4–1 mm; sepal awn 0, hairs ± long, stiff, often hooked, margin green, not scarious . . . . . . **HERNIARIA**
  5′ Stipules 1–8 mm; sepal awn 0.5–4 mm, hairs not as above, margin white, scarious, at least below
    6. Sepal awn stout, spine-tipped, 1.5–4 mm; sepals densely woolly . . . . **CARDIONEMA**
    6′ Sepal awn thread-like to ± stout, ± not spine-tipped, 0.5–1.5 mm; sepals not woolly . . . . **PARONYCHIA**
1′ Fr capsule; petals gen present, > (0.6)1 mm, not scale-like; stamens on ovary base or disk surrounding ovary or ovary stalk; stipules gen 0
7. Sepals fused, tube prominent, lobes or teeth < tube (exc *Agrostemma*); petal claw gen > limb, gen appendaged adaxially near limb base (subfamily Silenoideae)
  8. Styles 3–5; fr valves or teeth 3–6 or 10
    9. Calyx lobes > tube, gen 12–50 mm, linear, tube strongly 10-ribbed, with long, ascending, appressed hairs . . . . **AGROSTEMMA**
    9′ Calyx lobes or teeth < tube, < 13 mm, triangular to awl-like, tube not as above (exc *Lychnis* with lobes 4–7 mm)
      10. Styles (4)5, fls bisexual; fr teeth (4)5; pl densely silky-hairy to tomentose. **LYCHNIS**
      10′ Styles 3(4,5), if 5 then fls unisexual (taxa dioecious); fr teeth 6, 10; pl glabrous or variously hairy but not as above . . . . **SILENE**
  8′ Styles 2; fr valves or teeth gen 4
    11. Fl or fls subtended by 2–6 involucre bracts
      12. Calyx veins 20–45, tube not scarious between sepals; involucral bracts green, not scarious, linear to lanceolate, gen < 1 mm wide (ovate, 1.5–2.5 mm wide in *Dianthus deltoides*) . . . . **DIANTHUS**
      12′ Calyx veins 15, tube scarious between sepals; involucral bracts red to brown, scarious, widely ovate, gen 5–12 mm wide. . . . . **PETRORHAGIA**
    11′ Fl or fls not subtended by involucre bracts
      13. Stamens 5; calyx narrowly cylindric, 0.8–1 mm diam; fls gen 1 in axils . . . . **VELEZIA**
      13′ Stamens 10; calyx cup-shaped to widely tubular, 1.5–9 mm diam; fls few to many, in open to head-like, terminal cyme
        14. Calyx tube ± 1.3–5 mm, cup- to bell-shaped, white-scarious between sepals . . . . **GYPSOPHILA**
        14′ Calyx tube 7.5–20 mm, lanceolate to oblong-cylindric or cylindric to urn-shaped, green, not scarious between sepals
          15. Per; infl dense; pedicels gen 0–3 mm; calyx angles or keels 0; petal appendages 2 . . . . **SAPONARIA**
          15′ Ann; infl open; pedicels 5–40+ mm; calyx angles or keels 5; petal appendages 0 . . . . **VACCARIA**
7′ Sepals ± free; petal claw < limb or gen 0, unappendaged
16. Stipules 0.4–11 mm, widely triangular to bristle-like, scarious (subfamily Polycarpoideae)
  17. Lf blade oblanceolate to obovate, petiole gen ± present; style 1, 3-branched, 0.1–0.3 mm (stipules not bristle-like) . . . . **POLYCARPON**
  17′ Lf blade awl-like to linear, petiole 0; styles 3–5, free, gen 0.3–3 mm (if shorter, then stipules bristle-like)
    18. Lvs opposite but appearing whorled due to axillary clusters of 16–30 per node; styles, fr valves 5. . . . **SPERGULA**
    18′ Lvs opposite, often with axillary clusters of < 16 per node but not appearing whorled; styles, fr valves 3
      19. Infl axillary, fls 1–2; sepals awned; petals 0 or vestigial; stipules bristle-like . . . . **LOEFLINGIA**
      19′ Infl terminal, fls few to many; sepals not awned; petals present; stipules lanceolate to widely triangular . . . . **SPERGULARIA**
16′ Stipules 0 (subfamily Alsinoideae)
20. Petals ± 2-lobed, gen > 1/2 to base (petals sometimes 0 in *Stellaria*, *Cerastium*; lobes < 1/5 to base in *Pseudostellaria*)
  21. Fr cylindric, often ± curved in upper 1/2, opening by (8)10 teeth. . . . . **CERASTIUM**
  21′ Fr ± ovoid or spheric to cylindric-oblong, not curved in upper 1/2, opening by 6 (8 or 10) valves
    22. Petals 2-lobed < 1/5 to base; fr spheric, valves ± 2–3 × recoiled; seeds 1–2; rhizomes with tuber-like thickenings 3–12 mm diam or with ± vertical fleshy roots . . . . **PSEUDOSTELLARIA**
    22′ Petals 0 or 2-lobed > 1/2 to base; fr ± ovoid or spheric to cylindric-oblong, valves ascending to recurved, not recoiled; seeds several to many; rhizomes with neither tuber-like thickenings nor vertical fleshy root . . . . **STELLARIA**
20′ Petals entire, irregularly toothed, or ± notched (sometimes 0 in *Minuartia*, *Moenchia*, *Sagina*)
23. Petals ± irregularly toothed; infl umbel-like; fr opening by 6 teeth . . . . **HOLOSTEUM**
23′ Petals 0 or entire or notched; infl not umbel-like (exc *Eremogone congesta*); fr opening by 6 or 8 teeth or 3–6 valves
  24. Fr teeth 8; petals (0)4; pl erect; lf blade ± linear, rigid, acute . . . . **MOENCHIA**

24′ Fr valves or teeth 3–6; petals 0 or 5 (if 4 then pl not erect, lf blade not linear, not rigid, not acute)
  25. Styles 4–5; fr valves 4–5. . . . . **SAGINA**
  25′ Styles 3; fr valves 3 or 6 or teeth 6
    26. Ovary sutures 3; fr valves 3 . . . . **MINUARTIA**
    26′ Ovary sutures 6; fr teeth or valves 6
      27. Lf blades needle-like to narrowly linear, often congested at or near base of fl sts, tips gen
        sharp-pointed. . . . . **EREMOGONE**
      27′ Lf blades ovate to lanceolate (narrowly so or not), not congested at or near base of fl sts, tips
        gen acute to acuminate
        28. Seed appendage 0; rhizome 0 . . . . **ARENARIA**
        28′ Seed appendage ± 0.7 mm, ± elliptic, spongy; rhizome slender, branched. . . . . **MOEHRINGIA**

ACHYRONYCHIA

1 sp. (Greek: chaff fingernail, from silvery, chaffy sepals) [Hartman 2005 FNANM 5:46–47] Fl, seed suggest possible unification with *Scopulophila*.

A. cooperi Torr. & A. Gray (p. 609) ONYX FLOWER, FROST-MAT Ann, prostrate to ascending, glabrous to ± hairy, taprooted. **ST:** many, 3–17 cm. **LF:** stipules 0.1–0.2 mm, ± ovate, scarious, ± fringed, white; blade 3–20 mm, oblanceolate, ± fleshy; vein 1. **INFL:** axillary; fls 20–60+; pedicels 0.5–2.5 mm. **FL:** 2.5–3 mm; hypanthium ± 10-ribbed, in fr ± cylindric; calyx abruptly expanded above; sepals 5, free, ± 1.2–1.5 mm, ovate to reniform, green, fleshy, margin wide, scarious, ± jagged, white, deciduous; petals 0; fertile stamens 1–4, staminodes 14–19, ± 0.5 mm, thread-like, on hypanthium rim; style 2-branched in upper 1/2, 0.3–0.4 mm. **FR:** utricle, ± dehiscent, ovoid; teeth 8–10, minute. **SEED:** 1, ± 1 mm, ovoid, ± compressed, tan, red dot near narrow end. Sandy slopes, flats, washes; 50–700 m. D; AZ, Mex. Jan–May

AGROSTEMMA CORN-COCKLE

Ann, erect, taprooted. **LF:** blade linear to narrowly lanceolate; vein 1 or lateral pair faint. **INFL:** terminal; fls 1–few; peduncles, pedicels 4–20+ cm. **FL:** sepals 5, fused, hairs long, ascending, appressed, tube prominent, 12–17 mm, 7–12 mm diam, ovoid to widely cylindric, round in ×-section, strongly 10-ribbed, lobes gen 12–50 mm, > tube, linear; petals 5, 24–40 mm, claw long, limb entire or notched; styles 5, 10–12 mm. **FR:** capsule, ovoid; teeth 5, ascending. **SEED:** many, black. 2 spp.: Medit Eur. (Greek: field garland) [Thieret 2005 FNANM 5:214–215]

A. githago L. var. ***githago*** Pl 30–90+ cm; hairs dense, long, silky, ± appressed. **ST:** simple or sparingly branched above. **LF:** 5–15 cm. **INFL:** lfy. **FL:** sepals green; petals exserted 10–20 mm, obovate, rounded to truncate, purple-red; stamens exserted 8–10 mm. **SEED:** 3–3.5 mm, widely ovate; tubercles thin, triangular. 2*n*=48. Disturbed areas; < 1000 m. NCoRI, n SN, ScV, SnFrB, SCo; to WA, MT, e N.Am; native to s Eur. Evidently eradicated from grain fields. Spring–summer

ARENARIA SANDWORT

Ann, per, erect to mat-forming, taprooted. **LF:** not congested at base of fl sts; blades narrowly lanceolate to ovate; veins 1–5. **INFL:** terminal or axillary; fls 1–many; peduncles, pedicels 1–50 mm. **FL:** sepals 5, ± free, 1.5–4 mm, ± lanceolate to widely ovate, glabrous to glandular-hairy; petals 0 or 5, 1.5–6 mm, entire; stamens 10; styles 3, 0.5–2 mm. **FR:** capsule, ovoid to urn-shaped; teeth 6, ascending to recurved. **SEED:** 8–20, gray- or dark brown. 210 spp.: n temp, esp mtns, S.Am, Eurasia. (Latin: sand, a common habitat) [Hartman, Rabeler, & Utech 2005 FNANM 5:51–56] Based in part on molecular evidence, most taxa moved to *Eremogone*.

1. Ann; lf blade ± ovate, veins gen 3–5 . . . . *A. serpyllifolia* var. *serpyllifolia*
1′ Per; lf blade lanceolate to oblanceolate to linear or needle-shaped, vein ± 1
  2. Fls few to many in cyme, terminal or axillary; petals 1.5–3.5 mm; st rounded, dull, hairs minute, down-curved. . . . . *A. lanuginosa* var. *saxosa*
  2′ Fl 1, axillary; petals 5–6 mm; st angled or grooved, shiny, glabrous exc at nodes . . . . *A. paludicola*

A. lanuginosa (Michx.) Rohrb. var. ***saxosa*** (A. Gray) Zarucchi et al. (p. 609) ROCK SANDWORT, SPREADING SANDWORT Per, tufted or sts trailing, green. **ST:** 10–40 cm, rounded, dull; hairs minute, down-curved, ± in lines. **LF:** 8–22 mm, 2–6 mm wide, gen narrowly lanceolate to oblanceolate, obtuse to acute; vein 1. **INFL:** terminal or axillary; fls few to many; pedicels 3–25 mm. **FL:** sepals 1.5–2.8 mm, in fr < 3.5 mm, acute to acuminate; petals 1.5–3.5 mm. **SEED:** 8–12, 0.7–0.8 mm, ± round, compressed, smooth, dark brown. 2*n*=44. Moist, sandy soil along streams; 1800–2600 m. SnBr, PR; to CO, TX, Mex. [*A. l.* subsp. *s.* (A. Gray) Maguire] Spring–summer ★

A. paludicola B.L. Rob. (p. 609) MARSH SANDWORT Per, erect or not, often supported by nearby pls, green. **ST:** 25–90 cm, angled or grooved, shiny, glabrous exc at nodes. **LF:** 20–55 mm, some 2–7 mm wide, ± lanceolate, narrowly acute; vein 1. **INFL:** fl 1, axillary; pedicels 15–50 mm. **FL:** sepals 2.8–3.5 mm, in fr < 4 mm, obtuse to rounded; petals 5–6 mm. **SEED:** 15–20, 0.8–0.9 mm, widely reniform, ± compressed, smooth, dark brown. Wet meadows, marshes; < 300 m. s CCo (Nipomo Mesa, San Luis Obispo Co.), SnFrB (extirpated), SCo (Santa Ana River); Mex. Threatened by development. Late spring–summer ★

A. serpyllifolia L. var. ***serpyllifolia*** (p. 609) THYMELEAF SANDWORT Ann, tufted or sts trailing, green. **ST:** 3–25 cm, dull; hairs minute, down-curved. **LF:** 2–7 mm, 1–4 mm wide, ± ovate, acute to acuminate; veins gen 3–5. **INFL:** terminal; fls few to many; pedicels 1–12 mm. **FL:** sepals 2.5–3 mm, in fr < 4 mm, narrowly acute to acuminate; petals 0.6–2.7 mm. **SEED:** 10–15, 0.5–0.6 mm, widely reniform, plump, ± gray; tubercles low, elongate. 2*n*=40. Disturbed areas, sand, gravel bars, dry woodland; 150–1800 m. NW, n&c SNF, SCo, MP; ± N.Am; native to Eur. [*A. s.* subsp. *s.*] Spring–summer

CARDIONEMA　SANDMAT

Per, ± prostrate, taprooted. **LF**: stipules 4–8 mm, lanceolate to ovate, scarious, ± entire, white; blade needle-like; vein 1. **INFL**: axillary; fls 1–5, ± sessile. **FL**: hypanthium cup-shaped, not abruptly expanded above; sepals 5, free, 1.2–2.8 mm (exc awn), oblong to obovate, densely woolly, margin scarious below, awn 1.5–4 mm, stout, spine-tipped; petals 5, 0.3–0.5 mm, scale-like; stamens 3–5, on hypanthium rim; styles 2, 0.2 mm. **FR**: utricle, elliptic. **SEED**: 1, tan. 6 spp.: w N.Am, w S.Am. (Greek: heart thread, from stamen shape) [Hartman 2005 FNANM 5:45–46]

C. ramosissimum (Weinm.) A. Nelson & J.F. Macbr. (p. 609)　**ST**: 5–30+ cm, often ± covered by stipules. **LF**: 5–13 mm, finely spine-tipped, glabrous. **FL**: sepal margin incurved near tip. **SEED**: 1.4–1.6 mm, narrowly ovate, not compressed. Sandy beaches, hills, dunes, bluffs; < 400 m. NCo, CCo, SCo, n ChI, PR; to WA, Mex; S.Am. Spring–early summer

CERASTIUM　MOUSE-EAR CHICKWEED

Ann, per, erect to mat-forming; taproot or rhizomes present. **LF**: blade linear to ovate; vein 1; axillary lf clusters gen 0. **INFL**: terminal or axillary; fls few to many, open to dense; pedicels 1–36+ mm. **FL**: sepals (4)5, 3–12 mm, free, lanceolate to ovate, hairy to glandular-hairy, hairs gen not exceeding tip; petals 0 or (4)5, 2.5–15 mm, ± 2-lobed, white [purple tinged]; stamens (4,5)10; styles (4)5, 0.5–3.3 mm. **FR**: capsule, cylindric, often ± curved in upper 1/2; teeth (8)10, spreading to recurved. **SEED**: several to many, pale to red-brown. ± 180 spp.: worldwide, esp n temp. (Greek: horn, from fr shape) [Morton 2005 FNANM 5:74–93]

1. Ann (or appearing so); sts gen all fl, ascending to erect
2. Calyx 8.5–12 mm; fr 14–22 mm; seeds 0.9–1.1 mm . . . . *C. dichotomum*
2′ Calyx 3–7 mm; fr 3.5–11 mm; seeds ± 0.4–1 mm
  3. Upper bracts scarious-margined; scarious margin of most outer sepals 0.2–0.6 mm wide
  . . . . ²*C. fontanum* subsp. *vulgare*
  3′ Upper bracts not scarious (rarely with minute scarious tip); scarious margin of most outer sepals < 0.1 mm wide
    4. Fl parts 4(5); sepals with 0 hairs exceeding tip; bracts not scarious or rarely with minute scarious tip; pedicels in fr gen > sepals . . . . **[*C. diffusum*]**
    4′ Fl parts 5; sepals with some hairs exceeding tip by 0.2–0.8 mm; bracts not scarious; pedicels in fr (exc lowest) < sepals. . . . . *C. glomeratum*
1′ Per; sts both non-fl (mat-forming) and fl (± erect)
5. Petals 1.5 mm < to 0.5 mm > sepals. . . . . ²*C. fontanum* subsp. *vulgare*
5′ Petals ± 1.2–6 mm > sepals
  6. Lvs not clustered in fl st axils; petals 1.2–2.8 mm > sepals; fr 5.5–8.5 mm; seeds 0.7–0.8 mm; 2900–4300 m. . . . . *C. beeringianum*
  6′ Lvs clustered in fl st axils, esp below; petals 3–6 mm > sepals; fr (7.5) 9–16 mm; seeds ± 1–1.5 mm; < 2500 m
    7. Sts 5–20(30) cm; petals 7.5–9 mm; sepals 4.2–6(7) mm . . . . *C. arvense* subsp. *strictum*
    7′ Sts 15–45 cm; petals 10–15 mm; sepals 6–9 mm. . . . . *C. viride*

C. arvense L. subsp. *strictum* Gaudin (p. 609)　FIELD MOUSE-EAR CHICKWEED　Per, gen not fl 1st yr, 5–20(30) cm, glandular-hairy above, hairs ± longer below. **ST**: non-fl (mat-forming) and fl (± erect). **LF**: on fl st gen 8–25 mm, linear or lanceolate, ± glabrous or not; axillary lf clusters present, esp below. **INFL**: bract margins gen scarious in distal 1/4; pedicels in fr 1–4 × sepals. **FL**: parts 5; calyx 4.2–6(7) mm, glandular-hairy, rarely with hairs exceeding tip, scarious margin of outer sepals < 0.2 mm wide; petals 7.5–9 mm, 3–6 mm > sepals. **FR**: 7.5–11 mm. **SEED**: ± 1–1.2 mm. 2*n*=36. Moist seeps, shaded areas, grassy, gen rocky or sandy slopes; 600–1850 m. KR, c SNF, n&c SNH, SnFrB; N.Am (exc se); S.Am, Eur. Can be difficult to distinguish from *C. viride* (see Morton). Spring

C. beeringianum Cham. & Schltdl. (p. 609)　BERING MOUSE-EAR CHICKWEED　Per, gen not fl 1st yr, 1.5–10 cm, glandular-hairy. **ST**: non-fl (mat-forming) and fl (± erect). **LF**: on fl st 5–15 mm, lanceolate to elliptic. **INFL**: bracts ± completely not scarious; pedicels in fr 0.3–2.2 × sepals. **FL**: parts 5; calyx 4.5–6 mm, ± glandular-hairy, scarious margin of outer sepals < 0.2 mm wide; petals 6–12 mm, 1.2–2.8 mm > sepals. **FR**: 5.5–8.5 mm. **SEED**: 0.7–0.8 mm. 2*n*=72. Moist, rocky areas, grassy meadows, open slopes; 2900–4300 m. n&c SNH, W&I; to AK, MT, CO, NM. [*C. b.* var. *capillare* Fernald & Wiegand] Jul–Aug

C. dichotomum L.　FORKED MOUSE-EAR CHICKWEED　Ann 7–18 cm, glandular-hairy. **ST**: ascending to erect. **LF**: 10–35 mm, lanceolate. **INFL**: bracts not scarious; pedicels in fr 0.3–0.9 × sepals. **FL**: parts 5; calyx 8.5–12 mm, glandular-hairy, tip ± glabrous, scarious margin of outer sepals < 0.2 mm wide; petals 8–10 mm, 1.5–3 mm < sepals. **FR**: 14–22 mm. **SEED**: 0.9–1.1 mm. 2*n*=38. Fields, roadsides, disturbed areas; 750–850 m. CaRH (Siskiyou Co.); to WA; native to sw Eur. Spring–early summer

C. fontanum Baumg. subsp. *vulgare* (Hartm.) Greuter & Burdet　COMMON MOUSE-EAR CHICKWEED　Per, often fl 1st yr, appearing ann, 6–35 cm; hairs nonglandular. **ST**: gen non-fl (mat-forming) and fl (± erect). **LF**: on fl st 8–25 mm, lanceolate to oblong, ± widely so. **INFL**: bract margins gen scarious; pedicels in fr 1–4 × sepals. **FL**: fl parts 5; calyx 4.5–7 mm, hairy, ascending, scarious margin of most outer sepals 0.2–0.6 mm wide; petals 4–5 mm, 1.5 mm < to 0.5 mm > sepals. **FR**: 6.5–11 mm. **SEED**: 0.6–0.7 mm. 2*n*=126,136, gen 144. Disturbed areas, grassy slopes, damp woodland, marshy ground; < 2200 m. NW, CaRH, c SNF, n&c SNH, SCoRO?, SCo, SnBr, PR; N.Am; native to Eur. Spring

C. glomeratum Thuill. (p. 609)　STICKY MOUSE-EAR CHICKWEED　Ann, 3–40 cm, hairy (± glandular). **ST**: ascending to erect. **LF**: on fl st 5–35 mm, lanceolate or oblanceolate to ovate. **INFL**: bracts not scarious; pedicels gen 0.5–0.9 × sepals (exc lowest). **FL**: fl parts 5; calyx 3.5–5 mm, glandular-hairy, with some hairs exceeding tip by 0.2–0.8 mm, scarious margin of most outer sepals < 0.1 mm wide; petals 3–5 mm, 1.5 mm < to 1 mm > sepals, often 0 on lateral branches. **FR**: 3.5–8 mm. **SEED**: 0.4–0.6 mm. 2*n*=72. Dry hillsides, grassland, chaparral, disturbed areas; < 1600 m. CA-FP (exc s SNH), MP, DSon; to AK, e N.Am (exc ± n-c); native to Eur. Spring

C. viride A. Heller　WESTERN FIELD MOUSE-EAR CHICKWEED　Per, gen not fl 1st yr, 15–45 cm, glandular-hairy above, hairs ± longer below. **ST**: non-fl (mat-forming) and fl (± erect). **LF**: on fl st gen

10–45 mm, lance-ovate to narrowly oblong, ± glabrous below or not; axillary lf clusters present, esp below. **INFL:** bract margins gen scarious in distal 1/4 of pl; pedicels in fr 2–4+ × sepals. **FL:** parts 5; calyx 6–9 mm, glandular-hairy, rarely with hairs exceeding tip, scarious margin of outer sepals < 0.2 mm wide; petals 10–15 mm, 3–6 mm >

sepals. **FR:** 9–16 mm. **SEED:** 1–1.5 mm. $2n=72$. Coastal grassland, dunes, rocky slopes; < 500 m. NCo, CCo, SnFrB; to OR. [*C. arvense* var. *maximum* Hollick & Britton] Can be difficult to distinguish from *C. arvense* subsp. *strictum* (see Morton). Spring–early summer

DIANTHUS   CARNATION, PINK

Ann to per, erect, taprooted or rhizomed. **LF:** linear to oblanceolate; vein 1 or lateral 2 less prominent. **INFL:** terminal; few to many-fld, dense, or 1–few-fld, open; involucral bracts 2–6, linear to ovate; pedicels 0–25 mm. **FL:** sepals 5, fused, glabrous to hairy, tube prominent, 1.3–2.2 cm, 1.8–3.3 mm diam, ± cylindric, veins 20–45[60], lobes 3–8 mm, < tube, triangular to lanceolate; petals 5, 13–24 mm, claw long, limb irregularly toothed or divided to narrow segments, unappendaged; stamen bases fused with petal bases to ovary stalk; styles 2, 5–12 mm. **FR:** capsule, ± tubular; stalk 1–4 mm; teeth 4, ascending. **SEED:** many, black. 320 spp.: Eurasia, s Afr. (Greek: divine fl, from beauty or fragrance of fl) [Rabeler & Hartman 2005 FNANM 5:159–162]

1. Fls few to many; pedicels 0–3 mm; bracts linear to lanceolate, mostly ≥ calyx tube, long-tapered
  2. Calyx moderately hairy; cauline lf ± linear; fr stalk ± 1 mm . . . . ***D. armeria*** subsp. ***armeria***
  2′ Calyx glabrous; cauline lf ± lanceolate; fr stalk 3–4 mm . . . . ***D. barbatus*** subsp. ***barbatus***
1′ Fls 1–few; pedicels mostly 5–25 mm; bracts ovate or obovate, 1/4–1/2 × calyx tube, acute to short-tapered
  3. Petal limb 4–9 mm, tip irregularly toothed . . . . ***D. deltoides*** subsp. ***deltoides***
  3′ Petal limb 8–15 mm, tip divided to narrow segments . . . . [***D. plumarius*** subsp. ***plumarius***]

D. armeria L. subsp. ***armeria***   GRASS PINK, DEPTFORD PINK   Ann, bien, 15–70 cm; taproot slender. **LF:** basal lanceolate to oblanceolate; cauline ± linear. **INFL:** ± open, fls few to several; bracts mostly ≥ calyx tube, linear to lanceolate, long-tapered; pedicels 0–3 mm. **FL:** calyx 1.5–2 cm, moderately hairy, hairs long, ± appressed, ribs 20–25, lobes long-tapered; petal limb 4–5 mm, pink or rose with white dots. **FR:** stalk ± 1 mm. $2n=30$. Disturbed areas; 400–2000 m. KR, NCoRO, NCoRI, CaR, SNH, PR, SNE; to BC, AB, e N.Am, TX; native to s Eur. Spring–summer

D. barbatus L. subsp. ***barbatus***   SWEET-WILLIAM   Per 30–60 cm; rhizome ± stout. **LF:** basal lanceolate to oblanceolate; cauline ± lanceolate. **INFL:** dense, fls many; bracts mostly ≥ calyx tube, ± linear, long-tapered; pedicels 0–3 mm. **FL:** calyx 1.5–1.8 cm, glabrous, ribs

40, lobes acute to short-tapered; petal limb 6–10 mm, white to pink, purple, violet, or 2-colored. **FR:** stalk 3–4 mm. $2n=30$. Disturbed areas; < 1500 m. NW, CaRF, CCo; to BC, AB, e N.Am, TX; native to s Eur. Spring–early summer

D. deltoides L. subsp. ***deltoides***   MAIDEN PINK, MEADOW PINK   Per 18–40 cm; rhizomes slender. **LF:** basal oblanceolate; cauline linear to lance-linear. **INFL:** fls 1–few; bracts 1/3–1/2 × calyx tube, ovate, short-tapered; pedicels mostly 5–10+ mm. **FL:** calyx 1.3–1.7 cm, hairs 0 to minute adaxially, ribs 25–30, lobes linear to ± triangular; petal limb 4–9 mm, deep pink with darker zigzag band near base. **FR:** stalk ± 3 mm. $2n=30$. Wet meadows, disturbed areas; < 2500 m. CaRH, n&s SNH, SnJV, MP; WA, BC, AB to CO, e N.Am; native to Eur. Early summer

EREMOGONE   SANDWORT

Per, prostrate (non-fl sts) or ascending to erect to mat-forming, taprooted. **LF:** needle-like to narrowly linear; vein 1. **INFL:** terminal, open to head- or umbel-like; fls 1–many; peduncles, pedicels 0–55 mm. **FL:** hypanthium present; sepals 5, ± free, 3–7.2 mm, lance-linear to ovate, glabrous to glandular-hairy; petals 5, 2–18 mm, entire or ± notched; stamens on hypanthium; ovary ± superior, styles 3, 2.5–3 mm. **FR:** capsule, ovoid to urn-shaped; teeth 6, ascending to recurved. **SEED:** 1–9, ± gray, dark brown, red-brown, yellow-tan, black-purple, or ± black. 90 spp.: n temp, esp w N.Am, Eurasia. (Greek: solitary or deserted + seed, allusion uncertain) [Hartman & Rabeler 2004 Sida 21:237–241] Based in part on molecular evidence (Harbaugh et al. 2010 Intl J Pl Sci 171:185–198), 2 subgenera of *Arenaria* treated here as *Eremogone*.

1. Infl gen head- or umbel-like, dense to ± open; pedicels gen < 7 mm . . . . ***E. congesta***
  2. Pedicels 0–0.2(2) mm; infl dense
    3. Lf gen 3–8 cm, ± thread-like, not fleshy; NCoRH, CaRH, n SNH, Wrn . . . . var. ***congesta***
    3′ Lf 2–3 cm, ± thread- or needle-like, ± fleshy; KR, NCoRH, MP . . . . var. ***crassula***
  2′ Pedicels 1–12 mm; infl ± open
    4. Infl umbel-like, bracts gen at base; sepals 3–4 mm, obtuse . . . . var. ***suffrutescens***
    4′ Infl not umbel-like, bracts scattered among fls; sepals 3.5–6.5 mm, acute
      5. Sepals 3.5–4.5 mm . . . . var. ***subcongesta***
      5′ Sepals 4.5–6.5 mm
        6. Lf 1–2 cm; sepals 4.5–5.5 mm; DMtns . . . . var. ***charlestonensis***
        6′ Lf 2–3.5 cm; sepals 5.5–6.5 mm; n SNH, MP . . . . var. ***simulans***
1′ Infl gen ± open cyme or fls 1–2; pedicels some or all > 7 mm (± 0.5–1.5 mm in *Eremogone ursina*)
  7. Sepals gen obtuse to rounded or broadly acute, abruptly pointed or not
    8. Lf blade 0.5–1 cm; sepals in fl 1.8–3 mm; SnBr . . . . ***E. ursina***
    8′ Lf blade 1–6 cm; sepals in fl 3–5.5(6.5) mm; CaR, n SN, MP
      9. Petals 12–18 mm; st glabrous; lvs not glaucous; fr 7.5–9 mm . . . . ***E. cliftonii***
      9′ Petals 4.5–10 mm; st glandular-hairy; lvs gen glaucous; fr 5–7 mm . . . . ***E. aculeata***
  7′ Sepals acute to acuminate, ± spine-tipped or not
    10. St 1–20 (30) cm; lvs 0.3–2 cm . . . . ***E. kingii*** var. ***glabrescens***

10′ St gen 20–40 cm; lvs 2–6 cm
  11. Sepals 3–4.3 mm, in fr < 5.5 mm; infl branches spreading; nectaries rounded, 0.3–0.4 mm . . . . *E. ferrisiae*
  11′ Sepals 4.5–7.2 mm, in fr < 8 mm; infl branches erect to ascending; nectaries rectangular,
       2-lobed or truncate, 0.7–1.5 mm . . . . *E. macradenia*
    12. Cauline lvs downcurved, 0.8–2 mm wide . . . . var. *arcuifolia*
    12′ Cauline lvs ± ascending, 0.8–1.2 mm wide . . . . var. *macradenia*

E. aculeata (S. Watson) Ikonn. (p. 609) PRICKLY SANDWORT Per, mat-forming, gen glaucous. **ST:** 7–15(20) cm, ± dull, glandular-hairy. **LF:** 10–35 mm, 0.5–1.5 mm wide, sharp-pointed, vein 1. **INFL:** terminal, open; fls few to many; pedicels 3–25 mm. **FL:** sepals 3–4(4.5) mm, in fr < 6 mm, gen obtuse to rounded, rarely abruptly acute; petals 4.5–10 mm; nectaries < 0.8 mm, rounded, grooved above. **FR:** 5–7 mm. **SEED:** 6–8, 2–2.5 mm, elliptic-oblong, compressed, yellow-tan to gray; tubercles low, rounded, elongate. 2*n*=22. Rocky slopes, alluvium, volcanic areas; 2100–2600 m. CaR, MP; to WA, MT, UT. [*Arenaria a.* S. Watson] Jun–Jul

E. cliftonii Rabeler & R.L. Hartm. CLIFTON'S EREMOGONE, CLIFTON'S SANDWORT   Per, tufted, green, not glaucous. **ST:** 15–35 cm, rounded, ± dull, glabrous. **LF:** 10–60 mm, 0.7–1.1 mm wide, sharp-pointed, vein 1. **INFL:** terminal, open; fls several to many; pedicels (5)10–25 mm, glandular-hairy. **FL:** sepals 3–5.5(6.5) mm, in fr < 6 mm, gen obtuse to rounded (esp inner sepals) or broadly acute, often abruptly soft-pointed; petals 12–18 mm; nectaries 0.5–0.6 mm, rounded. **FR:** 7.5–9 mm. **SEED:** 5–9, 2–3 mm, ovate, compressed, black; tubercles low, rounded, often elongate. Decomposing granite in meadows, oak/conifer woodland; 455–1770 m. n SN (Butte, Plumas cos.). Apr–Jul   ★

E. congesta (Nutt.) Ikonn. BALLHEAD SANDWORT Per, tufted, ± green. **ST:** 8–40 cm, ± dull, often glandular-hairy. **LF:** 10–80 mm, 0.5–2 mm wide, ± fleshy or not, sharply acute to spine-tipped, vein 1. **INFL:** terminal, gen head- or umbel-like; fls few to many, dense to ± open; pedicels gen < 7 mm. **FL:** sepals 3–6 mm, in fr < 6.5 mm, rounded to acute; petals 5–8 mm; nectaries < 0.2 mm, rounded. **FR:** 3.5–6 mm. **SEED:** 4–8, 1.4–3 mm, widely elliptic to ovate, compressed, red-brown; tubercles low, rounded, often elongate. [*Arenaria c.* Nutt.] Vars. often intergrade.

  var. *charlestonensis* (Maguire) R.L. Hartm. & Rabeler CHARLESTON SANDWORT   **LF:** 10–20 mm, < 1 mm wide. **INFL:** ± open; bracts closely enveloping sepals; pedicels (1)2–4 mm. **FL:** sepals 4.5–5.5 mm, acute. Sandy ridges; 2200–2500 m. DMtns (New York Mtns, Panamint Range); sw NV. [*Arenaria congesta* var. *charlestonensis* Maguire] Jun   ★

  var. *congesta* (p. 609) **LF:** gen 30–80 mm, 0.5–1 mm wide. **INFL:** dense; bracts closely enveloping sepals; pedicels 0–0.2(2) mm. **FL:** sepals 3.5–4.2 mm, obtuse. 2*n*=22. Gravelly or sandy soil, in open; 1300–2000 m. NCoRH, CaRH, n SNH, Wrn; to WA, MT, CO. [*Arenaria congesta* var. *congesta*] Jun–Aug

  var. *crassula* (Maguire) R.L. Hartm. & Rabeler **LF:** 20–30 mm, 1–2 mm wide, ± fleshy. **INFL:** dense; bracts closely enveloping sepals; pedicels 0–0.1 mm. **FL:** sepals 3.5–4 mm, obtuse. Dry ridges, rock crevices; 2200–2500 m. KR, NCoRH, MP; sw OR. [*Arenaria congesta* var. *crassula* Maguire] Jun–Jul

  var. *simulans* (Maguire) R.L. Hartm. & Rabeler **LF:** 20–35 mm, ± 0.5 mm wide. **INFL:** ± open; bracts scattered among fls; pedicels 1–5 mm. **FL:** sepals 5.5–6.5 mm, acute. Open, rocky slopes; 1300–1700 m. n SNH, MP; nw NV. [*Arenaria congesta* var. *simulans* Maguire] Apr–Jul

  var. *subcongesta* (S. Watson) R.L. Hartm. & Rabeler (p. 609) **LF:** 10–30 mm, 0.5–1 mm wide. **INFL:** ± open; bracts scattered among fls; pedicels 1–5 mm. **FL:** sepals 3.5–4.5 mm, acute. Open rocky slopes, flats, often on volcanics; 1350–2750 m. NCoRH, CaRH, n SNH, GB, DMtns; to UT. [*Arenaria congesta* var. *subcongesta* (S. Watson) S. Watson] Jun–Jul

  var. *suffrutescens* (A. Gray) R.L. Hartm. & Rabeler SUFFRUTESCENT SANDWORT   **LF:** 15–80 mm, 0.5–1.5 mm wide. **INFL:** umbel-like, ± open; bracts gen at base; pedicels 1–12 mm. **FL:** sepals 3–4 mm, obtuse. Rocky slopes, outcrops; 1200–3300 m. KR, CaR, c SNF, SNH, DMtns. [*Arenaria congesta* var. *suffrutescens* (A. Gray) B.L. Rob.] Jun–Aug

E. ferrisiae (Abrams) R.L. Hartm. & Rabeler (p. 613) FERRIS' SANDWORT   Per, tufted, green. **ST:** (10)20–40(100) cm, rounded, ± dull, glabrous to glandular-hairy. **LF:** 20–60 mm, 0.5–1 mm wide, subsucculent or not succulent, spine tipped, vein 1. **INFL:** cymes, terminal, open; fls many; pedicels 15–55 mm. **FL:** sepals 3–4.3 mm, in fr < 5.5 mm, acute to acuminate; petals 6–9 mm; nectaries rounded, 0.3–0.4 mm. **FR:** 6–7 mm. **SEED:** 4–5, 1.3–3.2 mm, ± round to ovate, compressed, red-brown to ± black; tubercles low, rounded to conic. Pine, oak woodland, granitic alluvium on foothills, mtn slopes; 1450–2900 m. c&s SNH, SnJt, SNE; to UT. [*Arenaria macradenia* subsp. *f.* Abrams; *E. m.* var. *f.* (Abrams) R.L. Hartm. & Rabeler] Elevation to sp. rank based primarily on nectary morphology (see key); for discussion, see Hartman et al. (2005 FNANM 5: 65). Apr–Aug

E. kingii (S. Watson) Ikonn. var. *glabrescens* (S. Watson) Dorn (p. 613) KING'S SANDWORT   Per, tufted to funnel-shaped, green. **ST:** 1–20(30) cm, ± dull, glandular-hairy or glabrous. **LF:** 3–20 mm, 0.5–1.2 mm wide, gen sharp-pointed, vein 1. **INFL:** terminal, open; fls few to many (1–2 in alpine); pedicels 2–15 mm. **FL:** sepals 2.5–4(5.5) mm, in fr enlarging little, acute to acuminate; petals 4–7(10) mm; nectaries < 0.5 mm, rounded. **FR:** 4.5–7 mm. **SEED:** 2–5, 1.2–1.8(2.5) mm, elliptic-oblong to ovate, compressed, red-brown to dark purple or black; tubercles low, rounded, often elongate. Rocky slopes, summits, canyon floors; 2100–4050 m. CaRH, SNH, SNE, DMtns (San Bernardino Co.); to OR, ID, UT. [*Arenaria k.* (S. Watson) M.E. Jones subsp. *compacta* (Coville) Maguire; *A. k.* var. *g.* (S. Watson) Maguire] Pls from Diamond Mtns (Plumas, Lassen cos.), atypical in habit (funnel-like), height (< 30 cm), sepal length (< 5.5 mm), petal length (< 10 cm), seed length (< 2.5 mm), need study. Jun–Sep

E. macradenia (S. Watson) Ikonn. MOJAVE OR DESERT SANDWORT   Per, tufted, green. **ST:** 20–40 cm, rounded, ± dull, glandular-hairy or not. **LF:** 20–60 mm, 0.5–2 mm wide, blunt to sharp-pointed, vein 1. **INFL:** terminal, gen open; fls several to many; pedicels 3–55 mm. **FL:** sepals 4.5–7.2 mm, in fr < 8 mm, acute to acuminate; petals 6–11 mm; nectaries rectangular, 2-lobed or truncate, 0.7–1.5 mm. **FR:** 6–7 mm. **SEED:** 4–9, 1.8–2.7 mm, ± spheric to ovate, compressed, red-brown to ± black; tubercles low, rounded to conic. [*Arenaria m.* S. Watson]

  var. *arcuifolia* (Maguire) R.L. Hartm. & Rabeler (p. 613) **LF:** downcurved, 0.8–2 mm wide. **INFL:** branches (and sepals) ± glabrous to densely glandular-hairy. **FL:** sepals 4.5–7 mm, in fr < 8 mm. Dry, often gravelly (decomposing granite) canyon slopes, dry yellow-pine and oak forests, ridges, summits; 650–2400 m. s SN, WTR, SnGb. [*Arenaria m.* var. *a.* Maguire; *A. m.* var. *kuschei* (Eastw.) Maguire] Based on recent collections, *E. macradenia* var. *kuschei* (Eastw.) R.L. Hartm. & Rabeler, recognized by densely glandular-hairy infl and sepals, intergrades completely with *E. macradenia* var. *arcuifolia*, previously considered largely glabrous (Ross & Boyd 1996 Crossosoma 22: 65–71). Jun–Jul

  var. *macradenia* (p. 613) **LF:** ± ascending, 0.8–1.2 mm wide. **INFL:** branches (and sepals) ± glabrous. **FL:** sepals 4.5–7.2 mm, in fr < 8 mm. Open woodland, sagebrush flats, dry rocky slopes, alluvial deposits, often on carbonates; 1100–2200 m. s SN, SnGb, SnBr, SNE, DMoj; to UT, AZ. [*Arenaria m.* var. *m.*] Apr–Jun

E. ursina (B.L. Rob.) Ikonn. (p. 613) BEAR VALLEY SANDWORT Per, tufted, green. **ST:** 10–18 cm, dull to ± shiny, often glandular-hairy. **LF:** 5–10 mm, 0.5–1 mm wide, sharp-pointed, vein 1. **INFL:** terminal, ± open; fls few to many; pedicels ± 0.5–1.5 mm. **FL:** sepals 1.8–3 mm, in fr < 4.2 mm, obtuse to rounded; petals 2–4.5 mm; nectaries < 0.5 mm, rounded. **FR:** 4.5–6 mm. **SEED:** 1–2, 2.2–2.5 mm, ± spheric to widely elliptic, compressed, dark purple; tubercles low, rounded, often elongate. Rocky soil, pinyon/juniper woodland; 1950–2100 m. e SnBr (Bear Valley). [*Arenaria u.* B.L. Rob.] Threatened by development, grazing, vehicles. May–Jun   ★

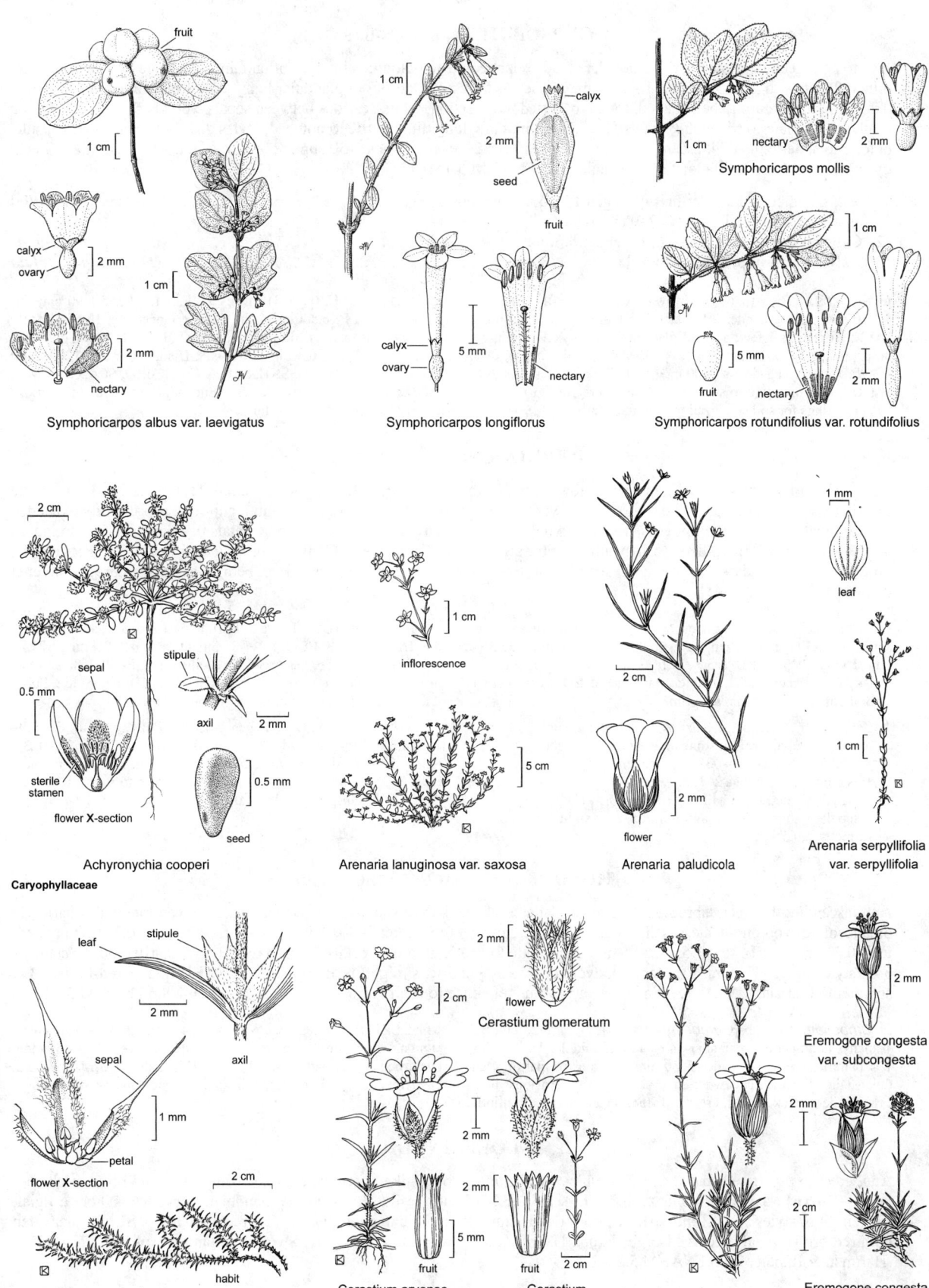

Symphoricarpos albus var. laevigatus

Symphoricarpos longiflorus

Symphoricarpos mollis

Symphoricarpos rotundifolius var. rotundifolius

Achyronychia cooperi

Arenaria lanuginosa var. saxosa

Arenaria paludicola

Arenaria serpyllifolia var. serpyllifolia

Caryophyllaceae

Cardionema ramosissimum

Cerastium glomeratum

Cerastium arvense subsp. strictum

Cerastium beeringianum

Eremogone aculeata

Eremogone congesta var. subcongesta

Eremogone congesta var. congesta

GYPSOPHILA   BABY'S-BREATH

Ann to per, erect, taprooted or rhizomed. **LF:** blade ± lanceolate to oblong; veins 1–3, often faint. **INFL:** gen panicle-like, terminal; fls ± few to many; pedicels 1–35+ mm. **FL:** sepals 5, fused, glabrous or glandular-hairy, tube ± prominent, ± 1.3–5 mm, 0.8–2 mm diam, cup- to bell-shaped, round to angled in ×-section, white-scarious between sepals, veins ± 5, teeth 0.2–1 mm, < tube, lanceolate to triangular; petals 5, ± 2.2–9 mm, claw long, limb entire to notched; styles 2, 1.2–2.5 mm. **FR:** capsule, oblong to spheric; teeth 4, ascending to recurved. **SEED:** 2–several, black. 150 spp.: temp Eurasia, n Afr, Australia. (Greek: gypsum lover, from habitat of 1 sp.) [Pringle 2005 FNANM 5:153–156]

1. Sepals, pedicels glandular-hairy; lvs gen 10–35 mm wide . . . . *[G. scorzonerifolia]*
1′ Sepals, pedicels glabrous; lvs 2–9 mm wide
  2. Calyx 3–5 mm; ann, bien, from slender taproot . . . . **G. elegans**
  2′ Calyx 1–3 mm; per, from stout rhizome . . . . **G. paniculata**

G. elegans M. Bieb. (p. 613)   SHOWY BABY'S-BREATH   Ann, bien 15–50 cm; taproot slender. **LF:** blade 2–5 mm wide, ± lance-linear. **INFL:** ± open; fls ± few; pedicels glabrous. **FL:** calyx 3–5 mm, glabrous; petals 2–5 × calyx, white to pink, veins purple. 2*n*=34. Open pine/fir forest, roadsides; < 2500 m. c SNH (Mammoth Lakes), GV, WTR; ± N.Am; native to sw Eur, se Asia. Source in CA may have been plantings for soil stabilization. Summer–early fall

G. paniculata L. (p. 613)   BABY'S-BREATH   Per 50–90 cm; rhizome stout. **LF:** blade 2–9 mm wide, ± lanceolate. **INFL:** openly branched; fls many; pedicels glabrous. **FL:** calyx 1–3 mm, glabrous; petals 1.5–2 × calyx, white. 2*n*=34,68. Disturbed areas; 1200–2100 m. NCoRI, CaRH, n&c SNH, SnJV, CCo, SCoRO, SCo, GB, DMoj; to BC, ne N.Am; native to c&e Eur, adjacent Asia. Infestations widely scattered. Summer–fall ❖

HERNIARIA   RUPTUREWORT

Ann, ± prostrate, taprooted. **LF:** opposite below, alternate above, oblanceolate to obovate; vein 0–1; stipules 0.4–1 mm, ovate to deltate, scarious, ciliate, white. **INFL:** axillary; fls 3–10, dense, ± sessile. **FL:** hypanthium cup-like, not abruptly expanded above; sepals 5, 0.6–1.2 mm, free, lanceolate to oblong, hairy, margin entire, not scarious; petals 0; stamens 2–5, staminodes 4–5, ± 0.5 mm, ± thread-like, on hypanthium rim; styles 2, or 2-branched, 0.1–0.4 mm, united in lower 1/3. **FR:** utricle, obovoid. **SEED:** 1, dark red-brown. 45 spp.: S.Am, Eur, s Asia, Afr. (Latin: rupture, 1 sp. being a supposed cure) [Thieret et al. 2005 FNANM 5:43–45]

H. hirsuta L.   Hairs dense, short, ± straight (exc sometimes at tip), spreading. **ST:** gen 4–20 cm. **LF:** stipules 0.5–1.3 mm; blade 3–12 mm. **INFL:** fls 3–8. **FL:** sepals in fr ± equal or not; stamens 2–5; styles 2 or 2-branched. **FR:** minutely papillate. **SEED:** ± compressed, smooth; margin with prominent rim.

1. Sepals in fr ± unequal; fl hairs of 2 kinds, long hairs 1/2–2/3 × sepals, tips of some or all hooked or tightly coiled, short hairs 1/4–1/3 × sepals, gen on hypanthium, tips recurved; stamen 2–3 . . . . var. *cinerea*
1′ Sepals in fr ± equal; fl hairs of ± 1 kind, 1/5–1/3 × sepals, ± shorter on hypanthium, tips ± straight; stamens gen 5 . . . . var. *hirsuta*

var. *cinerea* (DC.) Loret & Barrandon (p. 613)   **ST:** gen 5–20 cm. **INFL:** fls 3–8. **FL:** ± 1.2–1.8 mm; styles 2, 0.2–0.4 mm. **SEED:** 0.5–0.6 mm. 2*n*=36. Disturbed areas, alkaline hills, clay flats; < 800 m. s SNF, Teh, SnJV, SnFrB, SCo; OR, AZ, MD; native to s Eur, n Afr, sw Asia. [*H. h.* subsp. *c.* (DC.) Cout.] Spring–fall

var. *hirsuta* (p. 613)   **ST:** gen 4–15 cm. **INFL:** fls 3–6. **FL:** 0.9–1.1 mm; style 2-branched, < 0.1 mm. **SEED:** 0.6–0.7 mm. 2*n*=18,36. Sandy flats, roadsides, woodland; 200–1750 m. KR, NCoRI, n&c SNF, n SNH, GV, SnGb, PR; MA, MD; native to s Eur, n Afr, sw Asia. [*H. h.* subsp. *h.*] Spring–fall

HOLOSTEUM   JAGGED CHICKWEED

Ann, ascending to erect, taprooted. **LF:** basal blades oblanceolate to spoon-shaped, short-petioled; cauline blades narrowly oblanceolate to oblong-ovate, sessile; vein 1. **INFL:** terminal, umbel-like; fls 3–12; involucral bracts 5–15, ± 0.3–1.8 mm; pedicels 0.5–2.5 cm. **FL:** sepals 5, 2.5–4.5 mm, lanceolate to ovate, glabrous; petals 5, 3.3–5 mm, ± irregularly toothed; stamens 3–5; styles 3, 0.5–1.5 mm. **FR:** capsule, ovoid or ovoid-cylindric, straight; teeth 6, recurved. **SEED:** many, red-brown. 3–4 spp.: temp Eurasia, n Afr. (Greek: all bone, humorous reference to frailty of pl) [Rabeler & Hartman 2005 FNANM 5:94–95]

H. umbellatum L. subsp. *umbellatum* (p. 613)   Pl 5–25 cm. **ST:** 1–many, unbranched, glandular-hairy ± near middle. **LF:** cauline 1–4 pairs in lower 1/2; blades 2–30 mm, glabrous or margin glandular-hairy. **INFL:** bracts whorled, 1–2 mm, partly scarious; pedicels slender, reflexed in fl, erect in fr. **FL:** petals narrowly elliptic, white. **SEED:** 0.7–0.8 mm, compressed, with marginal ridge opposite elongate depression; tubercles low, rounded. 2*n*=20,40. Uncommon. Disturbed areas, roadsides; < 1270 m. KR, CaR, CCo?, MP; to BC, e N.Am; native to Eur; c&sw Asia; n Afr. Late winter–early spring

LOEFLINGIA

Ann, erect to prostrate, taprooted. **LF:** stipules 0.4–1.5 mm, bristle-like, scarious, entire, ± white; blade awl-like to oblong; vein 0–1. **INFL:** axillary, sessile; fls 1–2. **FL:** sepals 5, 2.7–6 mm, ± free, ± lanceolate, glandular-hairy; petals 0 or vestigial; stamens 3–5; styles 3, < 0.1 mm. **FR:** capsule, 1.5–3.7 mm, lanceoloid to ovoid; valves 3, ± recurved at tip. **SEED:** many, tan with red-brown band on curved edge. 7 spp.: N.Am, Medit, sw Asia. (P. Loefling, Swedish botanist, explorer, 1729–1756) [Hartman & Rabeler 2005 FNANM 5:26–27]

L. squarrosa Nutt. (p. 613)   SPREADING PYGMYLEAF   Pl 1–12 cm, much-branched at base, glandular-hairy, ± fleshy. **ST:** stiff. **LF:** blade 2–7 mm, erect to ± recurved, bristly; bases gen fused into a short scarious sheath; tip blunt to spine-tipped. **FL:** cleistogamous; sepals awned, becoming hardened, margin often scarious, lateral spurs often present, bristly. **FR:** 0.5–0.8 × sepals, 3-angled. **SEED:** 0.4–0.7 mm, minutely papillate on flat edge. Sand, gravel of hills, mesas, dunes, disturbed areas; < 1200 m. Teh, SnJV, SnFrB, SCo, PR, GB, DMoj; to OR, NE, AR, TX, n Baja CA. [*L. s.* var. *artemisiarum* (Barneby & Twisselm.) Dorn] Spring–summer

LYCHNIS CAMPION

Bien, per, erect, taprooted or roots fibrous. **LF**: petioled or not; oblanceolate to narrowly elliptic; vein 1, prominent. **INFL**: terminal; fls few; pedicels 10–55+ mm. **FL**: sepals 5, fused, densely silky-hairy to tomentose, tube prominent, 12–14 mm, 7–10 mm diam, elliptic to ovoid, rounded, strongly 10-ribbed (obscured by hairs), lobes 4–7 mm, < tube, linear to ± lanceolate; petals 5, 22–30 mm, claw > limb, limb widely notched, appendages 2; styles (4)5, 16–18 mm. **FR**: capsule, ovoid; stalk 0–0.5 mm; teeth (4)5, ascending. **SEED**: many, gray to black-purple. 30 spp.: n temp Eurasia, n Afr. (Greek: lamp, from flame-colored fl of some spp.) See note at *Silene*.

L. coronaria (L.) Desr. ROSE CAMPION Pl densely silky-hairy to tomentose. **ST**: sparingly branched above. **LF**: basal many, 9–15 cm, oblanceolate; cauline fewer, 5–15 cm, gen narrowly elliptic. **INFL**: lfy; branches, pedicels spreading, often arching upward. **FL**: calyx lobes twisted; petals obovate, red-purple, appendages 2–4 mm, awl-like, thickened. **SEED**: 1–1.5 mm, ± spheric; tubercles rounded, elongate. 2n=24. Disturbed areas, open slopes, redwood/Douglas-fir forests; 100–1220 m. KR, NCoR, CaRF (Butte Co.), CaRH (Plumas Co.), n SNF, n SNH (Butte Co.), c SNH, SnFrB, SCo, WTR; to BC, ID, UT, also e N.Am; native to se Eur. Summer

MINUARTIA SANDWORT

Ann, per, erect to mat-forming, taprooted or rhizomed. **LF**: blade thread-like to awl-shaped or narrowly oblong; veins or ribs 1–3. **INFL**: terminal or axillary, open to ± dense; fls 1–many; peduncles, pedicels 0.5–35+ mm. **FL**: hypanthium short, obscure; sepals 5, ± free, 1.9–7 mm, ± lanceolate to ovate, glabrous to glandular-hairy; petals 5 or 0, 0.7–10 mm, entire or notched; stamens on an obscure to prominent disk; styles 3, 0.3–2 mm. **FR**: capsule, narrowly ovoid to widely elliptic; valves 3, ascending to recurved. **SEED**: 1–many, red-tan to red-, purple-, or black-brown. 175 spp.: arctic to Mex, n Afr, s Asia. (J. Minuart, Spanish botanist, pharmacist, 1693–1768) [Rabeler et al. 2005 FNANM 5:116–136]

1. Ann, not cespitose, not mat-forming
 2. Infl ± glandular-hairy; lf (5)10–30 mm; seed 1.3–2 mm
  3. Lvs ± evenly spaced, thread-like, gen 0.3 mm wide, flexible, becoming curled; seeds red-brown, margin thin, wing-like . . . . *M. douglasii*
  3′ Lvs mostly near base, lance-linear, 1–1.5(2.5) mm wide, rigid, recurved; seeds brown to black-brown, margin thick, not wing-like . . . . *M. howellii*
 2′ Infl glabrous; lf 1.5–7(10) mm; seed 0.4–1 mm
  4. Petals < sepals or 0; pl 1–5 cm; lf, sepal with gen 1, often faint vein . . . . *M. pusilla*
  4′ Petals 1–1.8 × sepals; pl (1)2–20(25) cm; lf abaxially with 1 faint vein or 3 veins with 1 prominent, sepal with ± 3 faint veins or 3 or 5 prominent ribs
   5. Petals 1.5–1.8 × sepals; lf vein 1, faint abaxially; sepal with 3 faint veins, lateral pair rarely extending to tip . . . . *M. californica*
   5′ Petals 1–1.3 × sepals; lf veins 3, 1 prominent abaxially; sepal with 3 or 5 prominent ribs, lateral pair converging with midrib at tip . . . . *M. cismontana*
1′ Per, cespitose to mat-forming
 6. Sepal tip narrowly rounded, margin incurved . . . . *M. obtusiloba*
 6′ Sepal tip acute to acuminate, margin not incurved
  7. Pl ± densely cespitose, st decumbent to erect; taproot < 1.5 mm diam
   8. Petals ± 0.8–1 × sepals; pl glandular-hairy . . . . *M. rubella*
   8′ Petals 0; pl glabrous . . . . *M. stricta*
  7′ Pl mat-forming, rhizomes elongate and st or stolons trailing; taproot > 3 mm diam
   9. Pl densely glandular-hairy . . . . *M. nuttallii*
    10. Lf recurved; sepal ribs (1)3, ± equally prominent . . . . var. *fragilis*
    10′ Lf straight or ± recurved; sepal ribs 1 or 3, lateral pair 0 or less prominent
     11. Petals 0.7–0.9 × sepals; KR, CaRH, SNH, SnGb, SNE . . . . var. *gracilis*
     11′ Petals 1.1–1.6 × sepals; NW, CaRH, c SNH, Wrn . . . . var. *gregaria*
   9′ Pl gen glabrous or ± sparsely glandular-hairy esp in infl; st, lvs ± glabrous
    12. Sepals 5–6 mm, in fl gen 3-ribbed; petals 0.7–0.9 × sepals . . . . *M. decumbens*
    12′ Sepals 2.5–4.8 mm, in fl faintly 1–3-veined; petals 1.4–2.2 × sepals
     13. Pl glaucous; fl st from branching rhizomes; lvs > internodes; axillary lvs well developed . . . . *M. rosei*
     13′ Pl gray-green; fl st from stolons; lvs often < internodes; axillary lvs weakly developed . . . . *M. stolonifera*

M. californica (A. Gray) Mattf. (p. 613) CALIFORNIA SANDWORT Ann, simple or much-branched, (1)2–12 cm, glabrous, green; taproot thread-like. **ST**: widely spreading to erect. **LF**: 2–5 mm, 0.2–1.5 mm wide, << internode, linear to awl-shaped or narrowly oblong, ± straight, flexible, ± evenly spaced; axillary lvs 0. **FL**: sepals 2.5–2.8 mm, broadly lanceolate to elliptic, margin not incurved, veins 3, faint, lateral pair rarely extending to tip; petals 1.5–1.8 × sepals. **SEED**: 0.4–0.5 mm; margin thick, red-brown. 2n=26. Gravelly, sandy slopes, grassy ridges, chaparral, serpentine or not; < 1500 m. NW, CaR, SNF, c SNH, ScV, CW; s OR. Spring–summer

M. cismontana Meinke & Zika CISMONTANE MINUARTIA Ann, much-branched, (5)8–20(25) cm, glabrous, green or red-purple; taproot thread-like. **LF**: 2–7(9) mm, 0.5–1.2(1.8) mm wide, >> internode, lanceolate and long-pointed to linear, ± straight to outwardly curved, flexible, ± evenly spaced; axillary lvs occ present. **FL**: sepals 3.2–5.5 mm, lance-linear to lanceolate, margins not incurved, ribs 3 or 5, prominent, lateral pair converging with midrib at tip; petals 1–1.3 × sepals. **SEED**: 0.7–1 mm, margin thick, brown to ± red. Dry woodland, chaparral, often serpentine; (100)400–1700 m. KR, NCoRO, NCoRI, CaRH, n SNF, n&c SNH, SnJV, SCoRO; OR. Spring–summer

M. decumbens T.W. Nelson & J.P. Nelson (p. 613) THE LASSICS SANDWORT Per, mat-forming, 4–15 cm, gen ± glabrous, green; taproot > 3 mm diam. **ST**: in fl ascending to erect; trailing st < 30+ cm. **LF**: 3–6(9) mm, 0.7–2 mm wide, < to > internodes, needle-like to awl-shaped, ascending, ± rigid, ± evenly spaced; axillary lvs well developed. **INFL**: ± sparsely glandular-hairy. **FL**: sepals 5–6 mm, acute to acuminate, margin not incurved, ribs gen 3; petals 0.7–0.9 × sepals. **SEED**: 1.8–2.2 mm; margin thick, purple-brown. Serpentine soils in Jeffrey-pine forest; 1500–1600 m. n NCoRH (Mule Ridge, Trinity Co.). Spring–summer ★

M. douglasii (Torr. & A. Gray) Mattf. (p. 613) DOUGLAS' STITCHWORT Ann, simple or often branched from base, 4–30 cm, finely glandular-hairy at least above, green or purple; taproot thread-like. **ST:** erect to spreading. **LF:** 5–30 mm, gen 0.3 mm wide, thread-like, becoming curled, flexible, ± evenly spaced; axillary lvs 0. **FL:** sepals 2.5–3.7 mm, obtuse to acute, margin not incurved, ribs gen 3; petals 1.7–2.1 × sepals. **SEED:** 1.3–2 mm; margin thin, wing-like, red-brown. Rocky, sandy slopes, flats in chaparral, oak and pine woodland, often serpentine; 100–1800 m. NW, CaR, SNF, c SNH, GV, CW, SW, MP; s OR, AZ. Spring–early summer

M. howellii (S. Watson) Mattf. (p. 613) HOWELL'S SANDWORT Ann, simple or often branched from base, 12–30 cm, finely glandular-hairy, green or in fr purple; taproot < 2 mm. **ST:** erect to spreading. **LF:** (5)10–15 mm, 1–1.5(2.5) mm wide, lance-linear, recurved, rigid, mostly near base of st; axillary lvs 0. **FL:** sepals 1.9–3 mm, ± acute, margin not incurved, veins or ribs near base ± 3; petals 1.8–2.3 × sepals. **SEED:** 1.4–1.7 mm; margin thick, brown to black-brown. Chaparral, Jeffrey-pine/oak woodland, serpentine; 550–1000 m. KR; s OR. Spring–summer ★

M. nuttallii (Pax) Briq. NUTTALL'S SANDWORT Per, mat-forming, 2–20 cm, densely glandular-hairy, ± green; taproot > 3 mm diam; rhizomes, trailing sts < 60+ cm. **ST:** in fl ascending to erect. **LF:** 5–12(15) mm, ± 0.3–1.1 mm wide, > internodes, needle-like to awl-shaped, straight to recurved, ± rigid, ± evenly spaced; axillary lvs well developed. **FL:** sepals 3.5–7 mm, acute to acuminate, margin not incurved, ribs 1 or 3; petals 0.7–1.6 × sepals. **SEED:** 1.5–2.2 mm; margin thick, red-brown to dark brown. 2*n*=36. 1 other var., extending sp. range to BC, AB.

var. *fragilis* (Maguire & A.H. Holmgren) Rabeler & R.L. Hartm. (p. 613) **LF:** recurved. **FL:** sepal ribs (1)3, ± equally prominent; petals 0.8–1.1 × sepals. Uncommon. Basins, limestone talus; 1650–2400 m. GB; OR, NV. [*M. n.* subsp. *f.* (Maguire & A.H. Holmgren) McNeill] May–Jul

var. *gracilis* (B.L. Rob.) Rabeler & R.L. Hartm. (p. 613) **LF:** straight or ± recurved. **FL:** sepal ribs 1 or 3, lateral pair 0 or less prominent; petals 0.7–0.9 × sepals. Loose talus, sandy flats, gravelly areas, barren rock; 2600–3800 m. KR, CaRH, SNH, SnGb, SNE; w NV. [*M. n.* subsp. *g.* (B.L. Rob.) McNeill] Jul–Aug

var. *gregaria* (A. Heller) Rabeler & R.L. Hartm. (p. 613) **LF:** straight or slightly recurved. **FL:** sepal ribs 1 or 3, lateral pair 0 or less prominent; petals 1.1–1.6 × sepals. Sandy, rocky slopes, ridges, scree, barren rock, serpentine, chaparral, open Jeffrey-pine woodland; 650–3200 m. NW, CaRH, c SNH, Wrn; s OR. [*M. n.* subsp. *g.* (A. Heller) McNeill] Spring–summer

M. obtusiloba (Rydb.) House (p. 613) ALPINE SANDWORT Per, cespitose to mat-forming, 1–12 cm, glandular-hairy, green; taproot gen > 3 mm diam. **ST:** in fl ± erect; trailing st 2–20+ cm. **LF:** 1.5–8 mm, 0.4–1 mm wide, > internodes, needle- to awl-shaped, ± straight, flexible, ± evenly spaced; axillary lvs well developed. **FL:** sepals 2.9–4 mm, tip narrowly rounded; margin incurved, hood-like, ribs gen 3; petals 1.2–1.5 × sepals. **SEED:** 0.6–0.7 mm; margin thick,

red-tan. 2*n*=26,±52,78. Among dwarf willows, unglaciated granitics, metamorphics, alpine; 3150–3700 m. c&s SNH; to AK, MT, CO, NM. Summer ★

M. pusilla (S. Watson) Mattf. (p. 617) ANNUAL SANDWORT Ann, simple or branched, 1–5 cm, slender, glabrous, green; taproot thread-like. **ST:** spreading to erect. **LF:** 1.5–5 mm, 0.2–1.5 mm wide, << internode, awl-shaped to lanceolate, ± straight, flexible, ± evenly spaced, vein gen 1, often faint; axillary lvs 0. **FL:** sepals 1.5–3.5 mm, acute to acuminate, margin not incurved, vein 1, often faint; petals < sepals or 0. **SEED:** 0.4–0.7 mm; margin thick, purple-brown. Plains, open pine forest, chaparral slopes; < 2400 m. NCo, NCoRI, SNH, SnJV, CW, WTR, PR; to WA, CO. Spring–summer

M. rosei (Maguire & Barneby) McNeill (p. 617) PEANUT SANDWORT Per, mat-forming, 5–20 cm, gen glabrous, glaucous; taproot > 3 mm diam; rhizomes, trailing sts 5–20+ cm. **ST:** in fl ascending to erect. **LF:** 4–15 mm, ± 0.5–1.2 mm wide, > internodes, needle-like, straight or curved, ± flexible, ± evenly spaced; axillary lvs well developed. **INFL:** often glandular-hairy. **FL:** sepals 2.5–4 mm, acute to acuminate, margin not incurved, vein 1, faint; petals 1.4–2.2 × sepals. **SEED:** 2.3–2.8 mm; margin thick, red-brown to brown. Open serpentine slopes with scattered oak, Jeffrey pine; 750–1350 m. KR, NCoRH. Spring–summer ★

M. rubella (Wahlenb.) Hiern (p. 617) REDDISH SANDWORT Per, cespitose, 2–8(10) cm, densely glandular-hairy, green; taproot < 1.5 mm diam; rhizomes, trailing sts 0. **ST:** ascending to erect. **LF:** 1.5–10 mm, 0.3–0.8 mm wide, < to > internodes, needle-like, ± straight, flexible, mostly near base; axillary lvs well developed. **FL:** sepals 2.5–3.2 mm, acute to acuminate, margin not incurved, ribs 3; petals ± 0.8–1 × sepals. **SEED:** 0.4–0.5 mm; margin thick, red-brown. 2*n*=24. Rocky ridges, slopes, unglaciated metamorphics, granitics; 2400–3800 m. KR, SNH, n CCo, W&I; to AK, MT, CO, NM; circumboreal. Summer

M. stolonifera T.W. Nelson & J.P. Nelson (p. 617) SCOTT MOUNTAIN SANDWORT Per, mat-forming, 10–20 cm, ± glabrous to sparsely glandular-hairy, gray-green; taproot > 3 mm diam; trailing sts or stolons 6–20+ cm. **ST:** in fl ± erect. **LF:** 5–9 mm, 0.5–0.9 mm wide, often < internodes, needle-like, ± straight, rigid; axillary lvs weakly developed. **FL:** sepals 3.5–4.8 mm, narrowly acute to acuminate, margin not incurved, veins 1–3, faint in fl; petals 1.6–1.8 × sepals. **SEED:** 2–2.4 mm; margin thick, red-brown to brown. Serpentine soils, Jeffrey-pine forest; 1250–1400 m. s KR (Scott Mtn, Siskiyou Co.). Spring–summer ★

M. stricta (Sw.) Hiern (p. 617) BOG SANDWORT Per, cespitose, 0.8–2.5 cm, glabrous, green; taproot < 1.5 mm diam; rhizomes, stolons 0. **ST:** decumbent to erect. **LF:** 2–9 mm, 0.3–0.6 mm wide, > internodes, needle-like, ± straight, flexible, mostly near base; axillary lvs well developed. **FL:** sepals 2–3.2 mm, acute to acuminate, margin not incurved, vein 1 or 3, often prominent; petals 0. **SEED:** 0.5–0.6 mm; margin thick, red-brown. 2*n*=22,26,30. Granitic gravels, sandy wet spots, sedge meadows, alpine; 3500–3900 m. c&s SNH, W&I; ± n Can, CO; circumboreal. Aug ★

MOEHRINGIA

Per, ascending to erect, rhizomed. **LF:** not congested at base of fl sts; petiole short or 0; blade ± lanceolate or elliptic, acute; vein 1. **INFL:** terminal or axillary; fls 1–5; pedicels 2–25+ mm. **FL:** hypanthium short, obscure; sepals 5, ± free, 2.8–6 mm, ± ovate, minutely ciliate; petals 5, 2–8 mm, entire; stamens on disk; styles 3, 1.5–1.8 mm. **FR:** capsule, widely ovoid; valves 6, ± recurved. **SEED:** few, red-brown to ± black; appendage ± 0.7 mm, ± elliptic, spongy. 25 spp.: n temp. (P.H.G. Moehring, Danzig naturalist, 1710–1791) [Rabeler & Hartman 2005 FNANM 5:70–72]

M. macrophylla (Hook.) Fenzl (p. 617) LARGE-LEAVED SANDWORT **ST:** simple to branched, 2–18 mm, ± angled or grooved; hairs minute, peg-like; rhizome slender, branched. **LF:** 1.5–5 cm, ± evenly spaced, ± thin; margin smooth to minutely granular, gen ciliate in basal 1/2. **INFL:** bracts 1–4 mm, margin widely scarious, minutely ciliate; pedicels ascending to erect, ± spreading in fr or not.

FL: sepals acute to acuminate, margin scarious, midrib ± keeled; petals ± round. **FR:** black. **SEED:** 1.5–2.2 mm, widely elliptic; tubercles minute, low, rounded. 2*n*=48. Moist, shaded slopes, rocky ridges, summits, pine, oak forests, serpentine; 300–1800 m. NW, n&s SNH, SnFrB, SCoRO, PR; to BC, MT, CO, NM, also ± n Can, ne N.Am; Asia. Late spring–summer

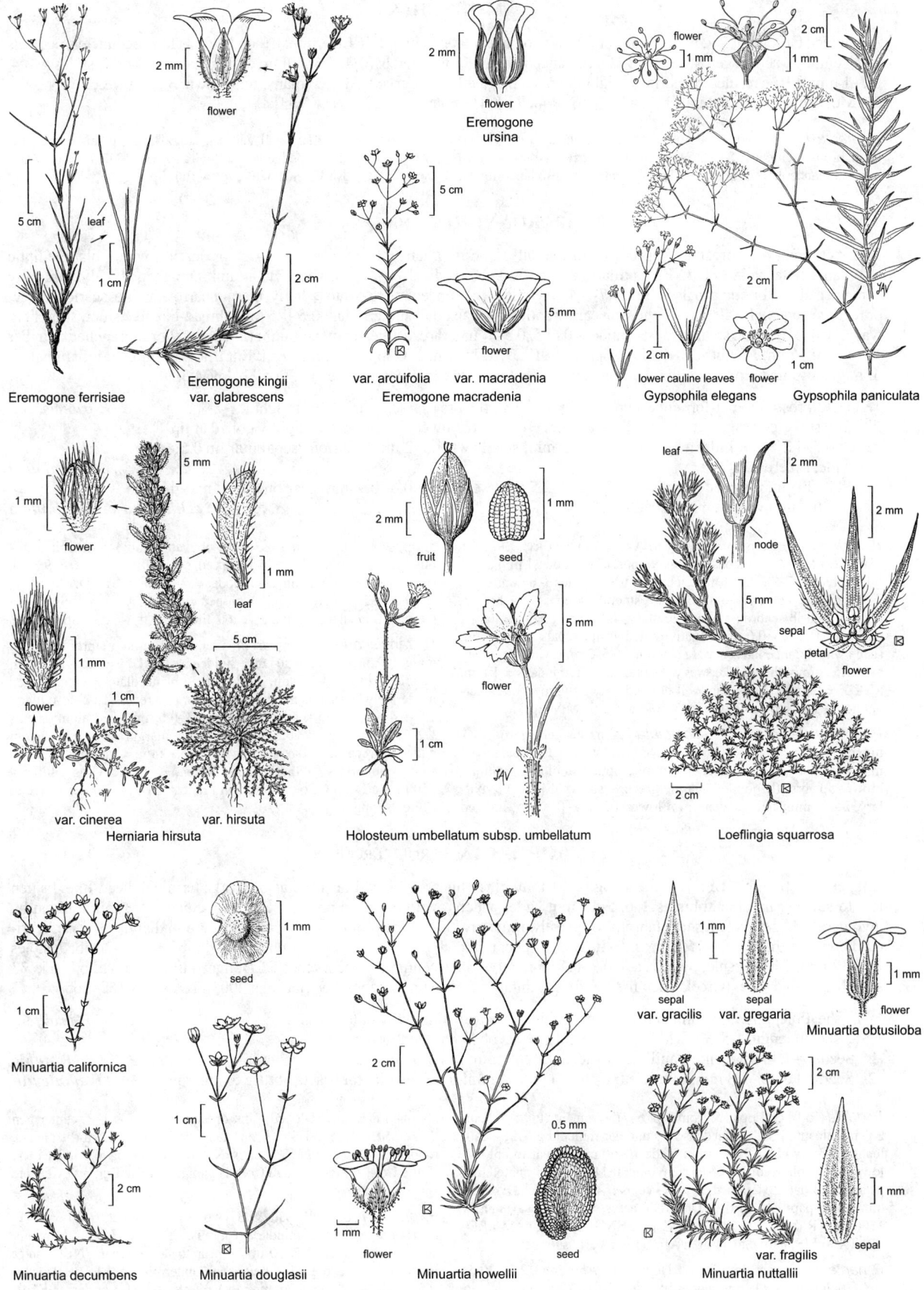

Eremogone ferrisiae

Eremogone kingii var. glabrescens

flower 2 mm

flower 2 mm
Eremogone ursina

var. arcuifolia   var. macradenia
Eremogone macradenia

flower

Gypsophila elegans   Gypsophila paniculata

lower cauline leaves   flower

var. cinerea   var. hirsuta
Herniaria hirsuta

flower
leaf
flower

fruit   seed
flower
Holosteum umbellatum subsp. umbellatum

leaf
node
sepal
petal   flower
Loeflingia squarrosa

Minuartia californica

seed

Minuartia decumbens

Minuartia douglasii

Minuartia howellii

flower   seed

sepal var. gracilis   sepal var. gregaria

flower
Minuartia obtusiloba

var. fragilis
Minuartia nuttallii
sepal

MOENCHIA

Ann, erect, taprooted. **LF**: petioled or not, linear to oblanceolate; vein 1. **INFL**: terminal, ± open; fls 1–few; peduncles, pedicels 1–7+ cm. **FL**: sepals gen 4, free, 3.8–7 mm, lanceolate, glabrous; petals (0)4, 2.5–6 mm, entire; stamens 4 or 8; styles gen 4, 0.7–1 mm. **FR**: capsule, cylindric, straight in upper 1/2; teeth 8, recurved. **SEED**: many, red-brown. 3 spp.: w&c Eur, Medit. (C. Moench, German naturalist, 1744–1805) [Rabeler & Hartman 2005 FNANM 5:93]

M. erecta (L.) G. Gaertn. et al. subsp. *erecta* UPRIGHT CHICKWEED Pl glabrous, glaucous. **ST**: 2.5–15 cm. **LF**: basal ± petioled, 5–18 mm, ± oblanceolate; cauline sessile, 3–12 mm, linear to lance-linear, ascending, rigid. **FL**: sepal veins 3, lateral faint; petals lanceolate, white. 2*n*=38. Disturbed areas; 200–600 m. n&c SNF, n CCo (Marin Co.); to BC, also IL, SC; native to sw Eur. Spring

PARONYCHIA NAILWORT

Ann, per, erect or ± prostrate, taprooted. **LF**: stipules 1–6 mm, lanceolate to ovate, scarious, ± entire, white; blade elliptic to oblanceolate; vein ± 1. **INFL**: axillary, dense; fls 1–12; pedicels 0–2 mm. **FL**: hypanthium cup-shaped; calyx abruptly expanded above or not; sepals 5, free, 0.7–4.4 mm (exc awn), lanceolate to ovate, ± hairy, margin narrow, white, scarious, erect or recurved adaxially at tip (awn then appearing at tip); awn abaxial, subterminal, 0.5–1.5 mm, thread-like to ± stout, straight to wavy; petals 0; fertile stamens 5, staminodes 0 or 5, 0.5–1 mm, thread-like, on hypanthium rim; styles 2 or 2-branched in upper 1/2, 0.2–0.5 mm. **FR**: utricle, ovoid to spheric. **SEED**: 1, brown. 110 spp.: N.Am, S.Am, Eurasia, Afr. (Greek: inflammation of finger, esp under nail [whitlow], ailment pl was believed to cure) [Hartman et al. 2005 FNANM 5:30–43]

1. Per; st prostrate, mat-forming; taproot > 3 mm diam; fl hairs sparse, near tip, ± straight. . . . . *P. franciscana*
1′ Ann; st ± erect; taproot < 1 mm diam; fl hairs ± dense, on lower 1/2, hooked or tightly coiled at tip
  2. St 0.5–1.2 cm, ± hidden by lvs; fl 1, 4.2–5 mm; sepal awn 1.5–2 mm; scarious sepal margin 0.5–2 mm wide, tip erect . . . . *P. ahartii*
  2′ St 2–20 cm, not hidden by lvs; fls 3–12, 2–2.5 mm; sepal awn 0.6–0.9 mm; scarious sepal margin 0.1–0.2 mm wide, tip recurved adaxially. . . . . *P. echinulata* var. *echinulata*

P. ahartii Ertter (p. 617) AHART'S NAILWORT, AHART'S PARONYCHIA Ann, inconspicuous, ± spheric; taproot < 1 mm diam. **ST**: ± erect, 0.5–1.2 cm, ± hidden by lvs. **LF**: stipules 3–6 mm; blade 2–7.5 mm, ± narrowly oblanceolate, smooth, green; tip a bristle; margin, midrib scabrous. **FL**: 1, axillary, 4.2–5 mm; hairs ± dense, on lower 1/2, 0.3–0.6 mm, tightly coiled at tip; sepals 3.8–4.4 mm (exc awn), ± lanceolate, margin scarious, 0.5–2 mm wide, tip erect, awn 1.5–2 mm, thread-like, wavy, ± spreading; staminodes ± 1 mm. **SEED**: ± 1.3 mm, lenticular. Well-drained, rocky outcrops, often vernal pool edges, volcanic upland; < 500 m. CaRF, ScV. Spring ★

P. echinulata Chater var. *echinulata* Ann, slender; taproot < 1 mm diam. **ST**: ± erect, 2–20 cm, not hidden by lvs. **LF**: stipules 1–3 mm; petiole ± present; blade 3–7 mm, elliptic to oblong, granular, brown; tip abruptly pointed; margin green, scabrous. **FL**: 3–12, axillary, 2–2.5 mm; hairs ± dense, on lower 1/2, 0.1–0.2 mm, hooked at tip, sepals 0.7–1 mm (exc awn), obovate to oblong, margin scarious, 0.1–0.2 mm wide, tip recurved adaxially, awn 0.6–0.9 mm, straight, stout, spreading; staminodes 0.5–0.6 mm. **SEED**: ± 0.7 mm, ± spheric. 2*n*=10,14,24,28. Disturbed clay soil; ± 25 m. n SnJV (La Grange, Stanislaus Co.); native to s Eur, nw Afr, w Asia. Spring

P. franciscana Eastw. (p. 617) Per, mat-forming; taproot > 3 mm diam. **ST**: prostrate, 5–50 cm, ± hidden by lvs. **LF**: stipules 3–6 mm; petiole 0; blade 5–10 mm, ± elliptic to oblanceolate, ± smooth, gen moderately hairy, green; tip a bristle; margin green. **FL**: 2–6, axillary, 1.9–2.4 mm; hairs sparse, near tip; 0.3–0.6 mm, ± straight; sepals 1.2–1.3 mm (exc awns), oblong to ovate, margin scarious, ± 1 mm wide, tip recurved adaxially, awn 0.5–0.7 mm, ± straight, slender, erect; staminodes 0. **SEED**: 1.2–1.3 mm, ± spheric. Grassy hills; < 250 m. s NCo, n CCo (around San Francisco Bay), SnFrB; native to Chile. Spring

PETRORHAGIA PROLIFEROUS PINK

Ann, erect, taprooted. **LF**: base sheathing, 1–9+ mm; blade linear to lance-linear; veins 3. **INFL**: terminal, head-like; fls gen few to several; involucral bracts 2–6, 5–12 mm wide, widely ovate, ± red to brown, scarious; pedicels 0–3 mm, hidden by involucre. **FL**: sepals 5, fused, glabrous to sparsely, minutely hairy, tube prominent, 8–14 mm, 1–3 mm diam, cylindric, scarious between sepals, veins 15, lobes 0.5–1.8 mm, < tube, rounded; petals 5, 10–14 mm, claw long, limb entire or 2-lobed; styles 2, 9–12 mm. **FR**: capsule, ovoid; stalk 0.2–0.7 mm; valves 4, ascending to recurved. **SEED**: many, black-brown to black. 33 spp.: Medit to c Asia. (Greek: rock fissure, from habitat of some spp.) [Rabeler & Hartman 2005 FNANM 5:162–165]

1. Lf sheath length ± = width, gen 1–2 mm; petals truncate or shallowly notched . . . . *P. prolifera*
1′ Lf sheath length 1.5–3 × width, gen 3–9 mm; petals obcordate to 2-lobed
  2. Seeds 1–1.4 mm, conic-papillate; lf sheath (3)4–9 mm; inner involucral bracts mucronate . . . . *P. dubia*
  2′ Seeds (1.3)1.5–1.8 mm, tubercled; lf sheath (2)3–4 mm; inner involucral bracts obtuse to mucronate. . . . . *P. nanteuilii*

P. dubia (Raf.) G. López & Romo (p. 617) Ann, erect, 9.5–60 cm. **ST**: middle internodes often densely glandular-tomentose. **LF**: sheath length 2–3 × width, (3)4–9 mm; blade 10–60 mm, linear to oblong, lowermost oblanceolate. **INFL**: involucral bracts mucronate. **FL**: petals obcordate to 2-lobed, ± pink, veins 5–6, dark. **SEED**: 1–1.4 mm, conic-papillate. 2*n*=30. Disturbed areas, woodland savanna; < 1800 m. KR, NCoR, CaRF, n SNF, n&c SNH, ScV, CCo, SCoRO, PR; TX to MS; native to s Eur, n Afr. Spring–early summer

P. nanteuilii (Burnat) P.W. Ball & Heywood Ann, erect, 21–52(65) cm. **ST**: lower, middle internodes minutely hairy. **LF**: sheath length 1.5–2 × width, (2)3–4 mm; blade 10–25 mm, gen linear. **INFL**: outer involucral bracts mucronate, inner obtuse to mucronate. **FL**: petals obcordate to ± 2-lobed, pink or ± purple, veins 1–3, center gen darker. **SEED**: (1.3)1.5–1.8 mm, tubercled. 2*n*=60. Disturbed areas, roadsides; 200–550(1417) m. s NCoRO (Sonoma Co.), NCoRH (Lake Co.), PR (San Diego Co.); BC; native to sw Eur, nw Afr. Late spring–summer

P. prolifera (L.) P.W. Ball & Heywood Ann, erect, 6–60 cm. **ST**: internodes glabrous or middle ± scabrous. **LF**: sheath length ± = width, gen 1–2 mm; blade 12–30 mm, linear to lance-linear. **INFL**: outer involucral bracts obtuse to mucronate, inner obtuse. **FL**: petals truncate or shallowly notched, pink to ± purple, vein 1, not dark. **SEED**: 1.1–1.8 mm, surface net-like. 2*n*=30. Disturbed areas; ± 400 m. c SNF (Mariposa Co.); BC, OR, ID, e N.Am; native to c&s Eurasia. Summer

POLYCARPON POLYCARP

Ann, matted or tufted, taprooted, glabrous. **LF**: opposite, appearing whorled or not; stipules 0.4–2.8 mm, lanceolate to triangular, scarious, entire to irregularly toothed or cut, white; petiole gen ± present; blade oblanceolate to obovate; vein 1. **INFL**: axillary, open to dense; fls few to many; pedicels 0.2–2 mm. **FL**: sepals 5, ± free, 1–2.2 mm, lanceolate to ovate, glabrous, margin scarious, white, awn conic to widely triangular; petals 5, 0.5–1.1 mm, entire or notched; stamens 3–5, ± fused at base; style 1, 3-branched, 0.1–0.3 mm. **FR**: capsule, ovoid to spheric; valves 3, margin rolled inward. **SEED**: several, brown. 15 spp.: worldwide. (Greek: many fr, from capsule number) [Thieret & Rabeler 2005 FNANM 5:25–26]

1. Stipules 0.4–1.2 mm; lvs opposite; sepals 1–1.5 mm, obscurely keeled, awn ± 0.1 mm . . . . ***P. depressum***
1′ Stipules 1.8–2.8 mm; lvs opposite, often appearing to be in whorls of 4; sepals 1.5–2.5 mm,
   prominently keeled, awn 0.3–0.7 mm . . . . ***P. tetraphyllum*** var. ***tetraphyllum***

P. depressum Nutt. (p. 617) **ST**: prostrate, much-branched, 1–6 cm. **LF**: stipules ovate to widely triangular; petiole slender; blade 3–7(9) mm, oblanceolate to obovate. **FL**: sepals elliptic to ovate, awn conic; petals linear to oblong. **SEED**: 0.4–0.5 mm, obliquely triangular, minutely granular. Bluffs, gravelly or sandy soil, chaparral, fields, disturbed areas; < 500 m. SCoRO, SCo, s ChI (Santa Catalina Island); Baja CA. Spring–summer

P. tetraphyllum (L.) L. var. ***tetraphyllum*** FOUR-LEAVED ALLSEED **ST**: prostrate to erect, often much-branched, esp above, 3–17 cm. **LF**: stipules lanceolate to widely triangular; petiole 0 or tapered to blade; blade 4–12 mm, obovate. **FL**: sepals lanceolate to ovate, awn widely triangular; petals linear to elliptic. **SEED**: 0.4–0.5 mm, obliquely triangular, granular. 2*n*=32,48,64. Disturbed shaded areas, roadsides; < 450 m, 1180 m. NCoRO, n SNF, c SN, ScV, CCo, SCo, SnGb; BC, TX, se N.Am; native to s Eur. Spring–fall

PSEUDOSTELLARIA

Per, sprawling to erect, rhizomed. **LF**: blade linear to ± lanceolate or elliptic; vein 1. **INFL**: terminal or axillary; fls 1–many. **FL**: hypanthium 0; sepals 5, free, 3–7 mm, lanceolate to ± ovate, glandular-hairy or glabrous, margin scarious; petals 5, 5–9.5 mm, 2-lobed < 1/5 to base; stamens on ovary base; styles 3, 2–4.5 mm. **FR**: capsule, spheric; valves 3, ± 2–3 × recoiled. **SEED**: 1–2, red-brown to brown. 21 spp.: w US, Eur, c&e Asia. (Latin: false *Stellaria*) [Hartman & Rabeler 2005 FNANM 5:114–116]

1. Pedicels, sepals glandular-hairy, often densely so; fls in lfy cymes; stamens 10, anthers purple . . . . ***P. jamesiana***
1′ Pedicels, sepals glabrous; fls 1, gen terminal; stamens 5, anthers yellow. . . . . ***P. sierrae***

P. jamesiana (Torr.) W.A. Weber & R.L. Hartm. (p. 617) TUBER STARWORT Pl 12–45 cm, glandular-hairy, at least above; rhizomes with spheric to elongate, tuber-like thickenings 3–12 mm diam. **ST**: simple to much-branched, ascending to erect, 4-angled. **LF**: 15–150 mm, ± smaller above, thick; margin ± smooth to roughened. **INFL**: pedicels in fr recurved to reflexed from base. **SEED**: 2–3.4 mm. 2*n*=96. Meadows, sagebrush-grassland, dry understory of conifer forest; 1400–2700 m. KR, NCoRH, CaRH, SN (exc n SNF), WTR, MP; to WA, MT, CO, TX. Summer

P. sierrae Rabeler & R.L. Hartm. SIERRA STARWORT Pl 9–27 cm, glabrous; rhizomes with ± vertical, fleshy roots enlarged toward tips. **ST**: simple or branched, ascending to erect, round (angled or grooved when pressed). **LF**: 7–35 mm, ± equal throughout, thick; margin ± smooth. **INFL**: pedicels in fr abruptly bent downward near upper end. **SEED**: 3–3.4 mm. Meadows, dry understory of mixed oak or conifer forest; 1400–2000 m. n&c SNH. First collected 1878; named, described 2002. Summer ★

SAGINA PEARLWORT

Ann, per, tufted to matted, taprooted. **LF**: linear to awl-shaped, gen not fleshy; vein 0–1. **INFL**: terminal or axillary; fl 1; pedicels 2–30 mm. **FL**: sepals 4–5, free, 1.3–3.5 mm, lanceolate to ovate, glabrous to glandular-hairy; petals 0 or 4–5, 1–3 mm, entire or notched; stamens 4, 5, 8, 10; styles 4–5, 0.1–0.6 mm. **FR**: capsule, ovoid; valves 4–5, spreading to recurved. **SEED**: many, gen obliquely triangular, ± compressed, brown or red-brown. ± 25 spp.: n temp, trop mtns. (Latin: fatten, from early use as forage) [Crow 2005 FNANM 5:140–147]

1. Ann, often glandular-hairy on pedicels, sepals, or both; sterile basal rosettes 0; st thread-like;
   pedicels ± straight
   2. Upper lvs minutely ciliate near base; sepals 4(5); petals 0(4), minute, < sepals. . . . . ***S. apetala***
   2′ Upper lvs glabrous; sepals gen 5; petals gen 5, ± = sepals . . . . ***S. decumbens*** subsp. ***occidentalis***
1′ Per, gen glabrous; sterile basal rosettes often present; st not thread-like; pedicels often recurved in fl,
   straight in fr
   3. Sepals 4(5), spreading to ascending in fr; petals 0 or 1/4–1/2 × sepals. . . . . ***S. procumbens***
   3′ Sepals 5, ± appressed in fr; petals 3/4–1 × sepals
      4. Lvs fleshy, widely linear; sepals gen 2.5–3.5 mm. . . . . ***S. maxima*** subsp. ***crassicaulis***
      4′ Lvs not fleshy, narrowly linear; sepals 1.5–2.5 mm
         5. Lf margins glabrous, base pairs not forming cup . . . . ***S. saginoides***
         5′ Lf margins minutely glandular, base pairs forming cup. . . . . [***S. subulata***]

S. apetala Ard. (p. 617) DWARF PEARLWORT Ann (1.5)3–8 cm, ± glandular-hairy distally; sterile basal rosettes 0. **ST**: thread-like, erect to decumbent. **LF**: upper minutely ciliate near base; blade 3–9(12) mm, narrowly linear. **INFL**: pedicels 2–8(12) mm, thread-like, gen straight, glandular-hairy. **FL**: sepals 4(5), ± appressed in fr, 1.5–2 mm, ± glandular-hairy; petals 0(4), minute, < sepals; stamens gen 4. **FR**: 1–1.2 × sepals. **SEED**: 0.2–0.3 mm, roughened or papillate,

brown; back grooved. 2*n*=12. Sandy disturbed areas, river bars, streamsides; < 700 m. NW, CaR, n&c SN, GV, CCo, SCoRO, SCo, n ChI, PR; BC, KS, formerly elsewhere in e N.Am; native to Eur. Spring–early summer

S. decumbens (Elliott) Torr. & A. Gray subsp. ***occidentalis*** (S. Watson) G.E. Crow WESTERN PEARLWORT Ann (2)4–16 cm,

glabrous or ± glandular-hairy distally; sterile basal rosettes 0. **ST**: thread-like, gen erect or ascending. **LF**: glabrous; blade 4–20 mm, narrowly linear. **INFL**: pedicels 2–14(20) mm, thread-like, gen straight, glabrous or ± glandular-hairy. **FL**: sepals gen 5, ± appressed in fr, 1.7–2.1(2.5) mm, ± glandular-hairy; petals gen 5, ± = sepals; stamens 5 or 10. **FR**: 1.2–1.7 × sepals. **SEED**: ± 0.4 mm, smooth to slightly roughened, brown; back grooved. Dry streams, chaparral, grassy areas, rock outcrops, vernal pools; < 2000 m. NW, n&c SN, GV, CCo, SnFrB, SCo, ChI, PR, MP; to BC, Baja CA. Spring–early summer

S. maxima A. Gray subsp. **crassicaulis** (S. Watson) G.E. Crow (p. 617) THICK-STEMMED PEARLWORT Per 3–18 cm, glabrous; sterile basal rosettes present. **ST**: not thread-like, spreading to decumbent. **LF**: fleshy; blade 7–22 mm, widely linear. **INFL**: pedicels 5–25 mm, slender to stout, gen straight. **FL**: sepals 5, ± appressed in fr, gen 2.5–3.5 mm; petals 5, 3/4–1 × sepals; stamens 10. **FR**: 1.3–1.6 × sepals. **SEED**: 0.4–0.5 mm, ± reniform, plump, smooth or slightly roughened, red-brown; back not grooved. 2n=46,66. Sandy bluffs, rock crevices; < 30 m. NCo, CCo; to AK. Spring–early fall

S. procumbens L. MATTED PEARLWORT Per 2–18 cm, glabrous; sterile basal rosettes often present. **ST**: slender, gen decumbent to ascending. **LF**: blade 3–10(20) mm, linear. **INFL**: pedicels 5–20(30) mm, thread-like, gen recurved in fl, straight in fr. **FL**: sepals 4(5), spreading to ascending in fr, 1.3–2 mm; petals 0,4(5), 1/4–1/2 × sepals; stamens gen 4. **FR**: 1.2–1.4 × sepals. **SEED**: 0.3–0.4 mm, smooth or slightly roughened, brown; back grooved. 2n=22. Wet, gravelly or sandy soil, roadsides, disturbed areas; < 15 m. NCo, CCo; to AK, MT, CO, e N.Am, Mex; native to Eur. Late spring–early fall

S. saginoides (L.) H. Karst. (p. 617) ARCTIC PEARLWORT Per (1)2–12 cm, glabrous; sterile basal rosettes often present. **ST**: slender, ascending or decumbent. **LF**: blade (3)5–15 mm, narrowly linear. **INFL**: pedicels 10–30 mm, thread-like, recurved in fl, straight in fr. **FL**: sepals 5, ± appressed in fr, 1.5–2.5 mm; petals 5, 3/4–1 × sepals; stamens gen 10. **FR**: 1.5–2 × sepals. **SEED**: 0.2–0.4 mm, smooth or slightly roughened, brown; back grooved. 2n=22. Moist banks, streamsides, dry creeks; (100)1000–3800 m. KR, NCoRO, NCoRH, CaRH, SN, TR, PR, GB; to AK, MT, WY, NM, Mex; circumboreal. May–Sep

SAPONARIA SOAPWORT

Per, erect, rhizomed. **LF**: petioled or not; blade oblanceolate to ovate; veins 3. **INFL**: terminal, dense; fls 20–40+; pedicels gen 0–3 mm. **FL**: sepals 5, fused, tube prominent, 15–20 mm, 4–8 mm diam, lanceolate to oblong-cylindric, rounded, veins 20, obscure, lobes 5, 1.5–5 mm, triangular-acuminate; petals 5, 25–40 mm, claw long, limb entire to ± obcordate, appendages 2; stamens fused with petals to ovary stalk; styles 2, 12–15 mm. **FR**: capsule, ± ovoid; stalk 2–3 mm; valves 4, ascending to recurved. **SEED**: many, ± purple to black. 40 spp.: Eurasia, n Afr. (Latin: soap, from sap lathering with water) [Thieret & Rabeler 2005 FNANM 5:157–158] *S. ocymoides* L. not in CA.

S. officinalis L. (p. 617) Pl 3–9+ dm, ± glabrous. **ST**: simple or branched above. **LF**: 3–10 cm. **INFL**: peduncles ascending or erect, bracted. **FL**: calyx tube obdepressed around pedicel, ± narrowed near base, lobes unequal, thinly scarious; petals pink, limb 8–12 mm wide, obovate, appendages 1–2 mm, lance-linear. **SEED**: 1.5–2 mm, ± spheric, notched, ± compressed, tubercled. 2n=28. Roadsides, oak woodland, streambeds, disturbed areas; < 1850 m. NW, CaRH, n SNF, SnFrB, SCoRO, SCo, PR, GB; N.Am; native to s Eur. "Double-fld" cultivars also naturalized. Jun–Sep ❖

SCLERANTHUS KNAWEL

Ann, bien, prostrate to erect, taprooted. **LF**: blade needle-like; vein 1. **INFL**: axillary, dense; fls 1–5, ± sessile. **FL**: hypanthium widely obovate to urn-shaped, abruptly expanded distally, in fr hard; sepals 5, free, 1.5–4 mm, narrowly triangular to awl-shaped, glabrous, margin thinly scarious; petals 0; stamens 2–10, on hypanthium rim; styles 2. **FR**: utricle, ovoid. **SEED**: 1. 10 spp.: Eurasia, Afr, Australia. (Greek: hard fl, from hypanthium in fr) [Thieret & Rabeler 2005 FNANM 5:149–151]

S. annuus L. subsp. **annuus** (p. 623) **ST**: prostrate to erect, much-branched, gen 4–20 cm, rigid; hairs ± in lines, fine, recurved. **LF**: 4–20+ mm; sheath scarious, ciliate; tip sharp-pointed. **INFL**: 3–15 mm diam. **FL**: 3–4.2 mm; hypanthium 10-ribbed; sepals narrowly triangular to awl-shaped, erect to spreading; styles ± 0.8–1 mm. **SEED**: 1.4–1.6 mm, widely ovoid, tan exc red crescent near acute tip. 2n=44. Meadows, stream margins, serpentine areas, disturbed areas; 300–1200 m. KR, NCoRO, NCoRH, CaR, n SN, c SNF, PR; to BC, SK, e N.Am; native to Eur. All yr

SCOPULOPHILA

Per, erect, taprooted; dioecious. **LF**: stipules 0.8–3.5 mm, triangular, scarious, jagged to ciliate, white; blade linear to lanceolate or oblanceolate; vein 1, obscure. **INFL**: axillary; fls 1–4, sessile. **FL**: unisexual (appearing bisexual); hypanthium in fr conic to urn-shaped, abruptly expanded above; sepals 5, free, 1.1–2.1 mm, elliptic to round, ± glabrous, margin wide, scarious, white; petals 0; stamens 5, sterile in pistillate fl, 1–1.5 mm, oblong, petal-like, on hypanthium rim; nectaries wide; ovary sterile in staminate fl, style 3-branched in upper 1/3, ± 1.5 mm. **FR**: utricle, ± dehiscent, ovoid; valves 3, minute. **SEED**: 1, tan. 2 spp.: sw US, Mex. (Greek: fond of high places) [Hartman 2005 FNANM 5:47–48] Fl, seed suggest possible unification with *Achyronychia*.

S. rixfordii (Brandegee) Munz & I.M. Johnst. (p. 623) RIXFORD ROCKWORT Pl glabrous exc base densely woolly. **ST**: many, branched above, 10–30 cm. **LF**: blade 8–25 mm, ± fleshy. **FL**: 2.2–4.2 mm; hypanthium green, in age brown, thickened, ± hard, ± angled; sepals erect to spreading, ± concave, often unequal, central portion linear to oblong, fleshy, green, margin much wider, entire to irregular, white, deciduous or not. **SEED**: 0.9–1.1 mm, ovoid, ± compressed, red dot near narrow end. Uncommon. Limestone outcrops; 1200–1550 m. SNE, n DMoj, DSon; w NV, w AZ. Apr–Jul

SILENE CATCHFLY, CAMPION

Ronald L. Hartman, Richard K. Rabeler & Dieter H. Wilken

Ann to per, ± erect, from caudex, taproot, or rhizome; rarely dioecious. **LF**: petioled or not; linear to oblanceolate, vein 1. **INFL**: gen terminal, open to dense; fls few to many, pedicels gen 5–40+ mm. **FL**: gen erect, gen bisexual; sepals 5, fused, tube prominent, 4–38 mm, 2–13 mm diam, cylindric to bell-shaped, rounded, hairs various or 0 (walls between hair cells gen

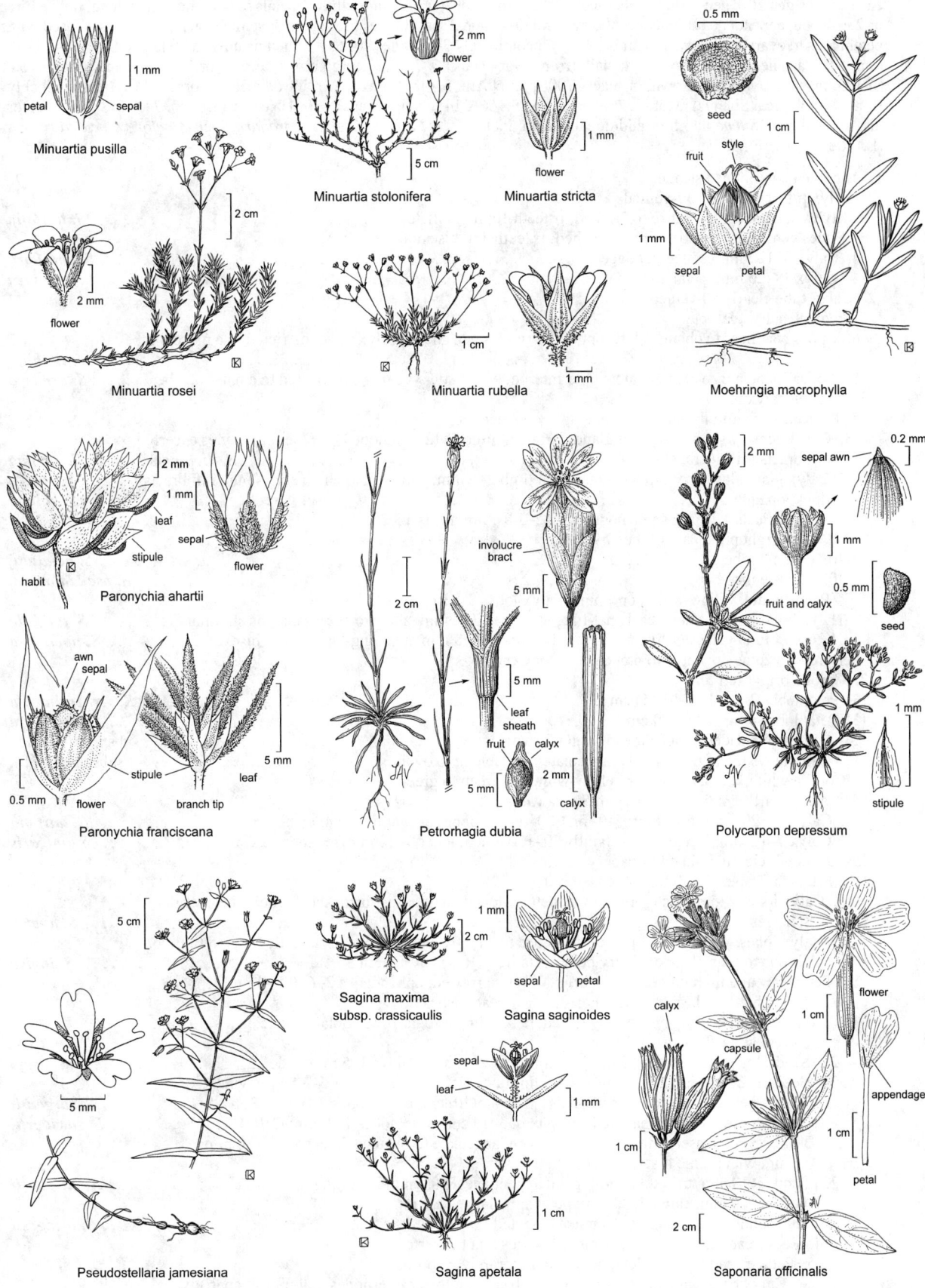

Minuartia pusilla

petal · sepal
1 mm

Minuartia stolonifera

2 mm
flower

5 cm

Minuartia stricta

flower
1 mm

Minuartia rosei

2 cm

flower
2 mm

Minuartia rubella

1 cm

1 mm

Moehringia macrophylla

seed
0.5 mm

fruit
style
sepal · petal
1 mm

1 cm

Paronychia ahartii

leaf
2 mm
1 mm
stipule
sepal
flower
habit

Paronychia franciscana

awn
sepal
stipule
flower
0.5 mm
leaf
branch tip
5 mm

Petrorhagia dubia

involucre
bract
5 mm
2 cm
5 mm
leaf
sheath
fruit · calyx
5 mm
2 mm
calyx

Polycarpon depressum

2 mm
sepal awn
0.2 mm
1 mm
fruit and calyx
0.5 mm
seed
1 mm
stipule

Pseudostellaria jamesiana

5 cm
5 mm

Sagina maxima
subsp. crassicaulis

2 cm
1 mm

Sagina saginoides

sepal · petal
1 mm

sepal
leaf
1 mm

1 cm

Sagina apetala

Saponaria officinalis

calyx
capsule
flower
1 cm
appendage
1 cm
petal
1 cm
2 cm

clear), veins gen 10+, gen dark, lobes or teeth 1–13 mm, < tube, triangular to linear; petals 5, 6–62 mm, claw long, limb entire or 2–6-lobed, appendages at junction of claw, limb 0–6, gen 2, basal lobes present or 0; stamens gen fertile, bases fused with petal bases to ovary stalk; ovary chamber 1 or ± incompletely 3–5, styles 3(4,5; if 5 then fls unisexual, taxon dioecious), 1–35 mm. **FR**: capsule, cylindric to ovoid; stalk (from ovary stalk) 0–7 mm, gen glabrous; teeth 6 or 10, ascending to recurved. **SEED**: many, gray to red, brown, or black. 700 spp.: N.Am, S.Am, Eurasia, Afr, introduced ± worldwide. (Greek: probably from mythological Silenus) [Morton 2005 FNANM 5:166–214] Oxelman et al. (2001 Nordic J Bot 20: 743–748) incl data for disarticulation of *Silene* into four additional genera, incl for CA *Lychnis* (*Lychnis coronaria*) and *Atocion* (*Atocion armeria* (L.) Raf., as *S. armeria* here).

1. Ann, bien, taproot gen slender
  2. Calyx tube clearly 16–30-veined
    3. Styles 5; petal limb ± entire to 2-lobed; fr teeth 10; fl pistillate . . . . ³*S. latifolia*
    3′ Styles gen 3; petal limb entire to notched; fr teeth 3; fl bisexual
      4. Calyx 8–12 mm; petal appendages 0 . . . . *S. coniflora*
      4′ Calyx 18–26 mm; petal appendages 2–5 mm . . . . *S. conoidea*
  2′ Calyx tube clearly 10-veined
    5. Pedicel, calyx glabrous
      6. Upper lvs linear to oblanceolate; upper internodes gen sticky; calyx not constricted to a narrow tube
      . . . . *S. antirrhina*
      6′ Upper lvs oblanceolate to ovate; upper internodes not sticky; calyx constricted to a narrow tube. . . . . *S. armeria*
    5′ Pedicel, calyx hairy
      7. Pedicels 0–5 mm; fls gen 1 per node; infl ± 1-sided
        8. Calyx densely short-hairy, not glandular; petal limb 5–12 mm, lobed ≥ 1/2 length; styles exserted; infl bracts thin, ± translucent . . . . *S. dichotoma* subsp. *dichotoma*
        8′ Calyx glandular-hairy esp on veins; petal limb 2–5 mm, entire to notched; styles incl; infl bracts thick, opaque . . . . *S. gallica*
      7′ Some pedicels > 5 mm; some nodes with 2–3 fls; infl not 1-sided
        9. Petals bright pink; calyx lobes 2–3 mm
          10. Petals 2-lobed . . . . [*S. pendula*]
          10′ Petals entire. . . . . [*S. pseudatocion*]
        9′ Petals white, yellow-white, or ± pink; calyx lobes 3–13 mm
          11. Calyx tube veins not clearly net-like, lobes gen 3–6 mm, acute to acuminate; fls staminate. . . . . ³*S. latifolia*
          11′ Calyx tube veins net-like above middle, lobes 6–13 mm, gen long-tapered; fls bisexual . . . . *S. noctiflora*
1′ Per, from rhizome or simple to branched, woody caudex
  12. Corolla bright red
    13. Petal lobes 2–4; pls 4–10(15) cm . . . . *S. serpentinicola*
    13′ Petal lobes 4–6; pls 20–70 cm . . . . *S. laciniata*
      14. Cauline lvs widely lanceolate to ovate; fr ovoid . . . . subsp. *californica*
      14′ Cauline lvs linear to narrowly lanceolate; fr ± oblong to ovoid . . . . subsp. *laciniata*
  12′ Corolla white, ± red, purple, or yellow-white or -green to green
    15. Calyx clearly ± inflated (papery in *Silene vulgaris*)
      16. Calyx 12–20 mm, short-hairy, 10- or ± 20-veined, lobes 3–6 mm; fr ovoid . . . . ³*S. latifolia*
      16′ Calyx 7–10 mm, gen glabrous, faintly 10–15-veined, lobes 1–3 mm; fr ± spheric . . . . *S. vulgaris*
    15′ Calyx not clearly inflated or papery in fr
      17. Petal claw tapered to limb; appendages 0
        18. Calyx lobes 4–5 mm, slightly < tube; petal 4-lobed, claw ± fusiform, short-woolly below middle; s SNH . . . . *S. aperta*
        18′ Calyx lobes 1–2 mm, << tube; petals entire to slightly notched, claw gradually tapered to limb, glabrous to sparsely short-hairy below middle; KR, s CaRH, c&n SNH. . . . . *S. invisa*
      17′ Petal claw and limb junction gen abrupt, often narrowed; appendages 2–6, 0.5–4 mm
        19. Petal limb 2-lobed, sometimes short-toothed near base
          20. Basal, lower cauline lvs densely tufted, fleshy, linear to narrowly oblanceolate, gen < 40 mm, < 5 mm wide; fl sts gen < 30 cm
            21. Seeds 2–3 mm; calyx 8–10 mm; petal limb 3–5 mm; basal lvs 1–5 mm wide . . . . ²*S. grayi*
            21′ Seeds 1–2 mm; calyx 9–15 mm; petal limb 2.5–3.5 mm; basal lvs 1.5–3.5 mm wide
              22. Walls between calyx hair cells clear; pl 10–15(20) cm; calyx lobes 2–3 mm; SNH, SNE . . . . *S. sargentii*
              22′ Walls between calyx hair cells purple; pl 3–10 cm; calyx lobes 1–2 mm; CaRH. . . . . *S. suksdorfii*
          20′ Basal lvs not densely tufted, not fleshy, gen lanceolate to oblanceolate, some > 30 mm, > 5 mm wide; some fl sts > 30 cm
            23. Petals < 10 mm, calyx 5–7 mm; petal appendages < 0.5 mm . . . . ²*S. menziesii*
            23′ Petals > 10 mm, calyx 7–20 mm; petal appendages > 1 mm
              24. Fls nodding; stamens, styles much exserted . . . . *S. bridgesii*
              24′ Fls spreading to gen erect; stamens, styles ± not exserted
                25. Pedicel, calyx puberulent, not glandular, rarely glabrous
                  26. Calyx base truncate to rounded in fr; basal lvs, lower internodes glabrous to sparsely puberulent . . . . *S. douglasii* var. *douglasii*

26′ Calyx base narrowed in fr; basal lvs, lower internodes moderately to densely puberulent,
± finely scabrous or not . . . . ²*S. verecunda*
25′ Pedicel, calyx glandular-puberulent to -hairy
  27. Basal lvs mostly > 10 mm wide, oblanceolate, lanceolate, or elliptic
    28. Cauline lvs << basal lvs; cauline lf pairs < 3; fr stalk 1–2 mm, glabrous or sparsely puberulent;
      calyx lobe veins wider than tube veins; > 1200 m . . . . *S. nuda*
    28′ Cauline lvs gradually reduced upward; cauline lf pairs > 4; fr stalk 3–6 mm, puberulent
      to woolly; calyx lobe veins ± as wide as tube veins; < 300 m . . . . *S. scouleri* subsp. *scouleri*
  27′ Basal, lower cauline lvs 0 or withering (if present, most < 10 mm wide, linear to
    narrowly oblanceolate)
    29. Middle cauline lvs (3)5–10 mm wide; calyx lobe margins membranous . . . . *S. marmorensis*
    29′ Middle cauline lvs < 6 mm wide; calyx lobe margin gen translucent . . . . ²*S. verecunda*
19′ Petal limb gen 4–8-lobed
  30. Calyx 15–40 mm; corolla 20–30 mm
    31. Petals pink to rose-red, limbs 14–20 mm; s CaRH . . . . ²*S. occidentalis*
    31′ Petals yellow-white, limbs 7–10 mm; SnGb, SnBr, e PR . . . . *S. parishii*
  30′ Calyx 5–20(25) mm; corolla gen < 20 mm (< 50 mm in *Silene occidentalis*)
    32. Calyx tube gen ≤ 10 mm
      33. Calyx tube cylindric, length > width; stamens, styles > petal claws; petal claw short-woolly . . . . . *S. lemmonii*
      33′ Calyx tube bell-shaped or elliptic, length ± = width; stamens, styles = to ± > petal claws; petal
        claw glabrous or ciliate at base
        34. Petals 5–11 mm, limb 2–4-lobed; fr stalk glabrous; fls gen ascending to erect; calyx lobes gen < 2 mm
          35. Lvs linear to oblanceolate; internodes 2–3 × > lvs; calyx clearly veined; petal lobes 4, outer gen
            > inner . . . . ²*S. grayi*
          35′ Lvs lanceolate to elliptic; internodes < lvs; calyx faintly veined; petal lobes gen 2 . . . . ²*S. menziesii*
        34′ Petals 12–20 mm, limb 4–6(8)-lobed; fr stalk puberulent; fls nodding to ± spreading; calyx
         lobes 2–5 mm . . . . *S. campanulata*
          36. Lvs 2–10 mm wide, linear to lanceolate . . . . subsp. *campanulata*
          36′ Lvs 10–30 mm wide, lanceolate to ± round . . . . subsp. *glandulosa*
    32′ Calyx tube gen > 10 mm
      37. Upper cauline lvs gen overlapping or exceeding pedicels; calyx canescent to soft-white-hairy;
        petal limb ± = claw, lobes > 10 mm
        38. Sts glandular-puberulent above; petal lobes equal or outer < inner, ± linear or lanceolate to
          broadly oblong, 5–10 mm, appendages 2, linear; stamens incl . . . . *S. hookeri*
        38′ Sts soft-shaggy-hairy or canescent above; petal lobes equal, linear, 15–25 mm,
          stamens long-exserted.
          39. Petal limb white with ± green center or ± pink, appendages 0 . . . . *S. bolanderi*
          39′ Petal limb salmon-orange, appendages present . . . . *S. salmonacea*
      37′ Fls located well above cauline lvs; calyx short-glandular-hairy; petal limb 1/3–1/2 × claw,
        lobes < 8 mm
        40. Fr stalk 4–7 mm; petal limb 7–17 mm; calyx > 15 mm . . . . ²*S. occidentalis*
        40′ Fr stalk 2–5 mm; petal limb 3–7 mm; calyx gen < 15(18) mm
          41. Petal appendages 2, deeply cut or fringed; petal claw, stamen base ciliate . . . . *S. bernardina*
          41′ Petal appendages 4–6, entire; petal claw, stamen base glabrous . . . . *S. oregana*

S. antirrhina L. (p. 623) SLEEPY CATCHFLY Ann 12–80 cm. **ST:** erect, glabrous or puberulent; upper internodes gen sticky. **LF:** gradually reduced upward, 1–3(6) cm, 3–5 mm wide; lower ± oblanceolate; upper linear to oblanceolate. **FL:** calyx 4–9 mm, glabrous, 10-veined, lobes 1–2 mm; petal claw glabrous, appendages 0 or to 0.4 mm, limb ± 2.5 mm, white to red, lobes 2; stamens incl; styles 3, ± incl. **FR:** ± ovoid; stalk ± 1 mm. **SEED:** < 1 mm, gray-black. $2n=24$. Open areas, burns; < 1800 m. CA-FP, MP, D (uncommon); ± N.Am. Apr–Aug

S. aperta Greene (p. 623) NAKED CATCHFLY, TULARE CAMPION Per 15–60 cm; caudex branches many. **ST:** erect, puberulent. **LF:** ± abruptly reduced gen above middle; basal, middle 5–12 cm, 1–4 mm wide, linear to oblanceolate; upper few, < 8 cm, 1–2 mm wide, linear. **FL:** calyx 6–10 mm, puberulent, 10-veined, lobes 4–5 mm; petal 12–20 mm, claw short-woolly below middle, appendages 0, limb 4-lobed, white to yellow-green; stamens ± = petal claws; styles 3, ± = petal claws. **FR:** ovoid; stalk 1–2 mm, puberulent. **SEED:** < 1 mm, brown. $2n=48$. Open areas, conifer forest; 1800–2800 m. s SNH. Jun–Sep ★

S. armeria L. SWEET-WILLIAM CATCHFLY Ann, bien (10)20–40 cm. **ST:** erect, glabrous. **LF:** gradually reduced upward, 1–6 cm, 5–25 mm wide; lower oblanceolate to elliptic or ovate; upper oblanceolate to ovate. **FL:** calyx 13–17 mm, glabrous, gen pink or purple,

10-veined, constricted to a narrow tube, lobes 1 mm; petal claw glabrous, appendages 2, limb ± 5 mm, pink or purple, lobes 0 or 2, shallow; stamens ± exserted; styles 3(4), ± exserted. **FR:** oblong; stalk 6–7 mm. **SEED:** 0.5–0.7 mm, brown. $2n=24$. Disturbed areas; 1500 m. SnGb?, PR (San Diego Co.), SNE (Mono Co.); to BC, UT, e N.Am; native to Eur. In CA, possibly only a waif; possibly from sown wildflower seed mixtures. Summer

S. bernardina S. Watson (p. 623) PALMER'S CATCHFLY Per 15–55 cm; caudex branches gen few. **ST:** erect, puberulent to short-hairy, glandular above or throughout. **LF:** gradually reduced upward; lower 2–8 cm, 2–6 mm wide, linear to oblanceolate; upper 1–6 cm, 1–4 mm wide, ± linear. **INFL:** axillary and terminal. **FL:** calyx 12–15 mm, glandular-puberulent, 10-veined, lobes 2–3.5 mm; petal claw ciliate at base, appendages 2, limb 4–6 mm, white, pink, or purple, lobes gen 4; stamens ± exserted; styles 3–4, ≥ stamens. **FR:** ± elliptic; stalk 2–5 mm, puberulent. **SEED:** 1.5–2 mm, brown. $2n=48$. Rocky slopes, scrub, conifer forest, alpine; 1350–3600 m. KR, NCoR, CaR, SN, GB, n DMtns; to WA, ID, NV; Baja CA. Jun–Aug

S. bolanderi A. Gray BOLANDER'S CATCHFLY Per 5–20 cm; caudex branches many. **ST:** decumbent to erect, soft-wavy-hairy. **LF:** slightly reduced upward, 2–8(9) cm, 8–25 mm wide, oblanceolate. **FL:** ascending to erect; calyx 12–20 mm, canescent to soft-white-

hairy, faintly 10-veined, lobes 4–7 mm; petal claw ± ciliate at base, appendages 0, limb white with ± green center or ± pink, lobes 15–25 mm, equal, linear; stamens exserted; styles 3, ± exserted. **FR:** oblong to ovoid; stalk 2–5 mm, glabrous to puberulent. **SEED:** ± 2 mm, black. Serpentine and non-serpentine soils, oak, conifer woodland; < 1000 m. NCoRO; OR. [*S. hookeri* Nutt. subsp. *b.* (A. Gray) Abrams] Segregation from *S. hookeri* discussed by Nelson et al. (2006 Madroño 53:72–76). Spring

S. bridgesii Rohrb. BRIDGES' CATCHFLY Per 16–50 cm; caudex branches 1 or few. **ST:** decumbent to erect, puberulent, sticky or glandular above. **LF:** gradually reduced above middle; lower gen withering; middle 2–6(8) cm, 6–15 mm wide, oblanceolate to elliptic; upper 1–5 cm, 3–12 mm wide, lanceolate, elliptic, or oblong. **INFL:** axillary and terminal. **FL:** nodding; calyx 7–14 mm, glandular-puberulent, 10-veined, lobes 2–3 mm; petal 14–28 mm, claw gen glabrous, appendages gen 2, limb 2-lobed, white; stamens exserted; styles 3, much exserted. **FR:** ovoid; stalk 2–3 mm, puberulent. **SEED:** 1–1.5 mm, brown. 2*n*=48. Open areas, conifer forest; 600–2400 m. NW, s CaR, SN; OR. Summer

S. campanulata S. Watson Per 5–40 cm; caudex branches gen many. **ST:** erect, glabrous to puberulent, glandular or not. **LF:** slightly reduced upward, 1–5 cm, 2–30 mm wide; lower lanceolate to ± round; upper linear to ovate. **INFL:** axillary and terminal; pedicel ± 0 to short. **FL:** nodding to ± spreading; calyx 6–10 mm, puberulent, glandular or not, faintly 10-veined, lobes 2–5 mm; petal to 2 × calyx, claw glabrous to ciliate, appendages 2, limb 4–6(8)-lobed, white to ± green or ± pink, lobes linear; stamens exserted; styles 3, exserted. **FR:** ovoid; stalk 1–2.5 mm, puberulent. **SEED:** 2–2.5 mm, brown.

subsp. *campanulata* (p. 623) RED MOUNTAIN CATCHFLY Pl 5–20 cm. **ST:** puberulent. **LF:** 1.5–4 cm, 2–10 mm wide, linear to lanceolate. Serpentine, chaparral, conifer forest; 500–1000 m. s KR, c NCoR (Red Mtn, Mendocino Co.; Cook's Springs, Colusa Co.). Pls intermediate to *S. campanulata* subsp. *glandulosa* need study. Summer ★

subsp. *glandulosa* C.L. Hitchc. & Maguire (p. 623) BELL CATCHFLY Pl 15–40 cm. **ST:** glabrous to puberulent, glandular or not. **LF:** 1–3(5) cm, 10–30 mm wide, lanceolate to ± round. 2*n*=48. Open or shaded areas, conifer forest; 300–1900 m. KR, NCoR; OR. Summer

S. coniflora Otth (p. 623) MULTINERVED CATCHFLY Ann 20–65 cm. **ST:** erect, gen glandular-short-hairy. **LF:** gradually reduced upward; lower 3–8(10) cm, 5–13 mm wide; upper 0.8–3 cm, 2–8 mm wide, lanceolate. **INFL:** axillary and terminal. **FL:** calyx 8–12 mm, glandular-short-hairy, 20–25-veined, lobes 1–3 mm; petal claw glabrous, appendages 0, limb 1–3 mm, white to pink, ± notched; stamens incl; styles 3, incl. **FR:** ovoid; stalk 1–1.5 mm. **SEED:** 0.5–1 mm, black. 2*n*=20. Open areas, burns; < 2000 m. s NCoRO, CW (exc SCoRI), SW; Baja CA; native to Asia. [*S. multinervia* S. Watson] Spring–early summer

S. conoidea L. LARGE SAND CATCHFLY Ann 50–80 cm. **ST:** erect, puberulent, glandular above. **LF:** gradually reduced upward; lower 5–12 cm, 5–10 mm wide, lanceolate to oblanceolate; upper 1–8 cm, 3–8 mm wide, lanceolate. **FL:** calyx 18–26 mm, puberulent, glandular or not, ± 30-veined, lobes 5–10 mm; petal claw glabrous, appendages 2, limb 8–12 mm, white, pink, or ± purple, entire, fine-toothed, or notched; stamens exserted; styles 3, ± exserted. **FR:** ovoid to conic; stalk 1–2.5 mm. **SEED:** 1–1.5 mm, gray-brown. 2*n*=20,24. Uncommon. Disturbed, open areas; < 500 m. n SN, SCo, DSon; to BC, SK, MO, TX; native to Eurasia. Early summer

S. dichotoma Ehrh. subsp. *dichotoma* FORKED CATCHFLY Ann 20–80 cm. **ST:** erect, gen short-rough-hairy. **LF:** gradually reduced upward; lower 6–8 cm, 15–30 mm wide, lanceolate to oblanceolate; upper 2–5 cm, 3–20 mm wide, gen lanceolate. **INFL:** pedicel ± 0. **FL:** gen spreading; calyx 9–14 mm, densely short-hairy, 10-veined, lobes 2–4 mm; petal claw glabrous, appendages 2, minute, limb 5–9 mm, white to red, lobes 2, ≥ 1/2 limb; stamens exserted; styles 3, exserted. **FR:** ovoid; stalk 1–2 mm. **SEED:** 1–1.5 mm, dark brown to black. 2*n*=24. Fields, roadsides; < 1000 m. CaR, SnFrB, SCo; to BC, SK e N.Am; native to Eur. Summer

S. douglasii Hook. var. *douglasii* (p. 623) DOUGLAS' CATCHFLY Per 10–40(70) cm; caudex branches few to many. **ST:** decumbent

to erect, puberulent. **LF:** ± gradually reduced upward; lower 2–6 cm, 2–8 mm wide, lanceolate to oblanceolate; upper 1–4 cm, 1–5 mm wide, ± linear. **FL:** calyx 10–14 mm, puberulent (glabrous), 10-veined, lobes 1.5–2.5 mm; petal claw ciliate at base, appendages 2, limb 4–11 mm, white to pink or ± purple, lobes 2, ≤ 1/2 limb; stamens ± = petal claws; styles 3–5, ± not exserted. **FR:** oblong to ovoid; stalk 3–4 mm, puberulent. **SEED:** 1–1.5 mm, brown. 2*n*=48. Open areas, scrub, oak woodland, conifer forest; 1400–2900 m. KR, n NCoR, CaR, SN, MP; to BC, MT, WY. Difficult to separate from *S. verecunda*; variation, esp in SN, needs study. Summer

S. gallica L. SMALL-FLOWER CATCHFLY, WINDMILL PINK Ann 10–40 cm. **ST:** decumbent to erect, short-rough-hairy to minutely bristly. **LF:** gradually reduced upward; lower 1–3.5 cm, 3–5 mm wide, oblanceolate; upper 0.8–2.5 cm, 2–4 mm wide, oblanceolate to oblong. **INFL:** axillary and terminal; pedicel 0 to short. **FL:** ascending to erect; calyx 6–10 mm, glandular-hairy esp on veins, 10-veined, lobes 2–2.5 mm; petal claw glabrous, appendages 2, limb < 6 mm, white, pink, or lavender, entire to notched; stamens incl; styles 3, incl. **FR:** ovoid; stalk < 1 mm, puberulent. **SEED:** ± 1 mm, red-brown. 2*n*=24. Fields, disturbed areas; < 1000 m. CA-FP; to BC, ID, AZ, e N.Am; native to Eur. Spring–early summer

S. grayi S. Watson GRAY'S CATCHFLY Per 10–20(30) cm; caudex branches many. **ST:** decumbent to erect, puberulent, glandular above. **LF:** abruptly reduced upward, fleshy; basal tufted, 1.5–4 cm, 1–5 mm wide, narrowly oblanceolate; cauline 0.4–1 cm, 1–3 mm wide, linear to oblanceolate. **FL:** nodding or not; calyx 8–10 mm, short-glandular-hairy, 10-veined, lobes 1–2 mm; petal claw ciliate at base, appendages gen 2, limb 3–5 mm, pink to ± purple, lobes 4, outer gen < inner; stamens ± = petals; styles 3, ± = petals. **FR:** ± ovoid; stalk 1.5–3 mm, puberulent. **SEED:** 2–3 mm, brown. 2*n*=48. Chaparral, conifer forest, alpine; 1200–2900 m. KR, n CaR; OR, NV. Summer

S. hookeri Nutt. (p. 623) HOOKER'S CATCHFLY Per 5–20 cm; caudex branches many. **ST:** decumbent to erect, appressed hairy below, often glandular-puberulent above. **LF:** ± reduced upward, 2–8(9) cm, 8–25 mm wide, oblanceolate. **FL:** ascending to erect; calyx 12–20 mm, canescent to soft-white-hairy, faintly 10-veined, lobes 4–7 mm; petals claw ± ciliate at base, appendages 2, limb lobes ± linear or lanceolate to broadly oblong, pink or white or purple, lobes 4, equal or outer < inner; stamens ± = petal claws; styles 3, ± exserted. **FR:** oblong to ovoid; stalk 2–5 mm, glabrous to puberulent. **SEED:** ± 2 mm, black. 2*n*=72. Serpentine soils, dry rocky ground, talus, or pine or oak forest; < 1400 m. KR, NCoR; OR. [*S. h.* subsp. *pulverulenta* (M. Peck) C.L. Hitchc. & Maguire] See note under *S. bolanderi*. Spring–early summer

S. invisa C.L. Hitchc. & Maguire (p. 623) SHORT-PETALED CAMPION Per 10–40 cm; caudex branches 0–many. **ST:** erect, puberulent, ± glandular above. **LF:** ± reduced upward, 1.5–5 cm, 2–6 mm wide, linear to narrowly oblanceolate. **FL:** calyx 7–9 mm, glandular-puberulent, 10-veined, lobes 1–2 mm; petal claw glabrous to sparsely short-hairy below middle, appendages 0, limb 1–2 mm, cream to pink, entire to ± notched; stamens incl; styles 3, = petal claws. **FR:** ± ovoid to cylindric; stalk < 1 mm, sparsely puberulent. **SEED:** ± 1 mm, brown. 2*n*=48. Open areas, conifer forest; 900–2800 m. KR, s CaRH, n&c SNH. Summer

S. laciniata Cav. Per 20–70 cm; caudex branches 0–few. **ST:** ± prostrate, reclining, or decumbent, ± puberulent to glandular-puberulent. **LF:** ± reduced upward, lower sometimes withering, 1–10 cm, 2–25 mm wide; lower oblanceolate to ovate, withering or not; upper ± linear or widely lanceolate to ovate. **FL:** ascending to erect; calyx 12–26 mm, glandular-puberulent, faintly 10-veined, lobes 3–6 mm; petal claw glabrous to ciliate, appendages 2, limb 6–15 mm, bright red, lobes 4–6; stamens ± exserted; styles 3, gen exserted. **FR:** oblong to ovoid; stalk 2–4 mm, glabrous to puberulent. **SEED:** 1–2.5 mm, red-brown. See Morton 2005 FNANM 5:189 for discussion of inclusion of *S. californica.*

subsp. *californica* (Durand) J.K. Morton (p. 623) CALIFORNIA PINK **LF:** widely lanceolate to ovate. **FR:** ovoid. **SEED:** 2–2.5 mm, red-brown. 2*n*=48,72. Chaparral, oak woodland, conifer forest, serpentine or not; < 2200 m. NW, CaR, SN, CCo, SnFrB, n SCoRO, SCo, WTR (n slope), w SnGb, PR; Mex. [*S. c.* Durand] Pls from Teh, WTR, w SnGb intermediate to (and sometimes difficult to separate from) *S. laciniata* subsp. *l.* Spring–summer

subsp. ***laciniata*** (p. 623) MEXICAN PINK **LF:** linear to narrowly lanceolate. **FR:** ± oblong to ovoid. **SEED:** 1–1.5 mm, red-brown. 2*n*=96. Chaparral, oak woodland; < 1200 m. s SCoRO, SW; Baja CA. [*S. l.* subsp. *brandegeei* C.L. Hitchc. & Maguire; *S. l.* subsp. *major* C.L. Hitchc. & Maguire; *S. l.* var. *angustifolia* C.L. Hitchc. & Maguire; *S. l.* var. *latifolia* C.L. Hitchc. & Maguire] Spring–summer

S. latifolia Poir. WHITE CAMPION Bien, per 30–100 cm; dioecious (fls appear bisexual); caudex branches few. **ST:** erect, gen rough-hairy, ± glandular above. **LF:** gradually reduced upward; lower 5–10 cm, 6–25 mm wide, oblanceolate; upper 1.5–8 cm, 3–15 mm wide, lanceolate. **INFL:** pedicel ± 0 to short. **FL:** ascending to erect; calyx 12–20 mm, short-hairy, lobes 3–6 mm; staminate calyx 10-veined; pistillate calyx ± 20-veined, much inflated in fr; petal ± 24–40 mm, claw glabrous, appendages 2, limb white, ± entire to 2-lobed; stamens ≥ petal claws; styles 5, exserted. **FR:** ovoid; stalk 1–2 mm. **SEED:** 1–2 mm, gray-brown. 2*n*=24. Fields, roadsides; < 1000 m. NCo, NCoRO, s SNF, n SNH?, GV, SnFrB, SCo?; N.Am; native to Eur. [*S. l.* subsp. *alba* (Mill.) Greuter & Burdet] Often mistaken for *S. noctiflora* L. Summer–fall

S. lemmonii S. Watson (p. 623) LEMMON'S CATCHFLY Per 15–45 cm; caudex branches gen few. **ST:** decumbent to erect, short-hairy below, glandular above; some short branches at base spreading to decumbent, only vegetative. **LF:** ± abruptly reduced upward, elliptic to oblanceolate; basal 2–3.5 cm, 8–10 mm wide; cauline few, 0.5–2.5 cm, 3–5 mm wide. **FL:** nodding; calyx 6–10 mm, ± short-glandular-hairy, 10-veined, lobes 1–2 mm; petal claw short-woolly, appendages 2, limb yellow-white to pink, lobes 4, 4–8 mm, linear; stamens exserted; styles 3, much exserted. **FR:** oblong to ovoid; stalk 2–3 mm, puberulent. **SEED:** 1–1.5 mm, rusty brown. 2*n*=48. Oak woodland, conifer forest; 850–2800 m. NW, CaR, SN, w SnFrB (Santa Cruz Mtns), SCoR, TR, PR, MP; OR. Spring–summer

S. marmorensis Kruckeb. (p. 623) MARBLE MOUNTAIN CAMPION Per 25–40 cm; caudex branches 0–few. **ST:** erect, puberulent, glandular above. **LF:** gradually reduced above, lanceolate; lower withering; middle 2–4.5 cm, (3)5–10 mm wide; upper 1.5–4 cm, 2–8(10) mm wide. **INFL:** axillary and terminal. **FL:** spreading to erect; calyx 12–14 mm, glandular-puberulent, faintly 10-veined, lobes 3–5 mm; petal claw glabrous, appendages 2, limb 4–6 mm, pale pink, lobes 2, deep; stamens ± = petals; styles 3, ± = petals. **FR:** ovoid; stalk 3–4 mm, puberulent. **SEED:** 2–3 mm, black. 2*n*=48. Oak woodland, conifer forest; 850–1000 m. KR (Humboldt, Siskiyou cos.). [Kruckeberg 1960 Madroño 15:172–177] Closely related to *S. bridgesii*. Summer ★

S. menziesii Hook. (p. 623) MENZIES' CATCHFLY Per 5–20(50) cm, ± mat-like, gen dioecious (fls appear bisexual); caudex branches many. **ST:** decumbent to erect, puberulent to short-hairy, gen glandular above. **LF:** slightly reduced upward, 2–6 cm, 3–20 mm wide, lanceolate to elliptic. **FL:** ascending to erect; calyx 5–7 mm, puberulent, ± glandular, faintly 10-veined, lobes 1–2 mm; petal claw glabrous, appendages 2, limb 1.5–3 mm, white, lobes gen 2; stamens << (pistillate fl) or ± = (staminate fl) petal claws; styles 3(4), << (staminate fl) or > (pistillate fl) petal claws. **FR:** ovoid; stalk 1–2 mm. **SEED:** 0.5–1 mm, black. 2*n*=24,48. Conifer forest, pinyon/juniper woodland; 900–2900 m. KR, NCoR, CaR, SN, SnBr, GB; to AK, MB, KS, NM. Jun–Jul

S. noctiflora L. NIGHT-FLOWERING CATCHFLY Ann 20–60(80) cm. **ST:** gen erect, rough-hairy, glandular above. **LF:** gradually reduced upward; lower 6–12(14) cm, 20–45 mm wide, elliptic to oblanceolate; upper 1–7 cm, 3–12 mm wide, lanceolate. **INFL:** fls pedicelled. **FL:** ascending to erect; calyx 14–22 mm, glandular-hairy, 10-veined, veins net-like above middle, lobes 6–13 mm; petal claw glabrous, appendages 2, limb white to ± pink, lobes 2; stamens gen incl; styles 3, incl. **FR:** ovoid; stalk 1–3 mm. **SEED:** ± 1 mm, red-brown. 2*n*=24. Open, disturbed areas, fields; < 1900 m. CaR, SNF, SCo, expected elsewhere; to AK, SK, NM, e N.Am; native to Eur. Often mistaken for *S. latifolia*. Summer

S. nuda (S. Watson) C.L. Hitchc. & Maguire STICKY CATCHFLY Per 15–50 cm; caudex branches 0–few. **ST:** erect, glandular-hairy. **LF:** abruptly reduced above base; basal 6–15 cm, 10–30 mm wide, oblanceolate to elliptic; cauline pairs < 3, 0.8–4 cm, 3–8 mm wide, linear to lanceolate. **INFL:** fls pedicelled or subsessile. **FL:** calyx 12–16 mm, glandular-puberulent, 10-veined, lobe veins wider than

tube veins, lobes 2–5 mm; petal claw glabrous to ciliate at base, appendages 2, limb 5–10 mm, pink, lobes 2; stamens ± incl; styles 3, incl. **FR:** conic to elliptic; stalk 1–2 mm. **SEED:** 1–1.5 mm, brown. 2*n*=48. Shrubland, juniper woodland, conifer forest; 1200–1900 m. n&c SNH, MP; s OR, w NV. Summer

S. occidentalis S. Watson (p. 623) WESTERN CAMPION Per 30–60 cm; caudex branches 0–few. **ST:** erect, short-soft-hairy, glandular above. **LF:** gradually reduced upward; lower 5–12 cm, 7–20 mm wide, gen oblanceolate; upper 2–8.5 cm, 3–18 mm wide, oblanceolate. **INFL:** axillary and terminal. **FL:** ascending to erect; calyx 15–38 mm, short-glandular-hairy, 10-veined, lobes 2–4 mm; petal claw 12–40 mm, ciliate at base, appendages 2, limb 7–20 mm, pink to rose-red, lobes 4; stamens exserted; styles 3, ± = petal claws. **FR:** oblong to ovate; stalk 4–18 mm, short-hairy. **SEED:** 1–1.5 mm, gray-brown. 2*n*=48. Chaparral, conifer forest; 700–2300 m. s CaRH, n SNH, MP. [*S. o.* subsp. *longistipitata* C.L. Hitchc. & Maguire] Summer ★

S. oregana S. Watson OREGON CAMPION Per 30–50(70) cm; caudex branches 0–few. **ST:** erect, puberulent, glandular above. **LF:** gradually reduced upward; lower 5–8 cm, 7–12 mm wide, oblanceolate; upper 1–6(8) cm, 2–6 mm wide, lanceolate. **INFL:** axillary and terminal. **FL:** calyx 9–15 mm, glandular-puberulent, 10-veined, lobes 2–3 mm; petal claw glabrous, appendages 4–6, limb white to pink, lobes 4–6; stamens incl to ± exserted; styles 3(4), gen incl. **FR:** elliptic to ovoid; stalk 3–4 mm, puberulent. **SEED:** 1–2 mm, brown. 2*n*=48. Sagebrush scrub, subalpine conifer forest; 1500–2500 m. CaRH, n&c SNH, MP, W&I; to WA, MT, WY. Summer ★

S. parishii S. Watson (p. 623) PARISH'S CATCHFLY Per 10–40 cm; caudex branches gen many. **ST:** ascending to erect, strigose to short-hairy, glandular below or throughout. **LF:** slightly reduced upward, 1.5–6 cm, 5–15 mm wide, lanceolate to ± ovate. **INFL:** pedicel ± 0 to short. **FL:** calyx 24–29 mm, short-glandular-hairy, 10-veined, lobes 4–8 mm; petal claw ciliate to puberulent, appendages 2, limb 7–10 mm, yellow-white, lobes ± 6; stamens ± = petals; styles 3, ± exserted. **FR:** ovoid to elliptic; stalk 2–3 mm. **SEED:** 1.5–2 mm, brown. 2*n*=48. Open, rocky to gravelly slopes, conifer forest, alpine; 1800–3350 m. SnGb, SnBr, e PR (SnJt, Santa Rosa Mtns, Hot Springs Mtn). Spring–summer

S. salmonacea T.W. Nelson et al. KLAMATH MOUNTAIN CATCHFLY, SALMON-FLOWERED CATCHFLY Per 5–14 mm; rhizome branched. **ST:** erect, gray-green, canescent. **LF:** lowest much-reduced, 2.5–3.5 cm, 0.4–0.8 cm wide, spoon-shaped to oblanceolate. **FL:** ascending to erect; calyx 18–23 mm, canescent, gray-green, distinctly 10-nerved, lobes 4–7 mm; petal claw glabrous, green becoming white at base, appendages 2, limb salmon-orange, lobes 4, equal; stamens long-exserted; styles 3, incl. **FR:** only immature seen; ± spheric; stalk ± 5 mm, glabrous. **SEED:** ± 2.2 mm, ± red. Serpentine and iron-rich soils in openings or mixed-evergreen forest; 760–1050 m. KR. Jun ★

S. sargentii S. Watson SARGENT'S CATCHFLY Per 10–15(20) cm; caudex branches many. **ST:** decumbent to erect, puberulent and gen glandular above. **LF:** ± abruptly reduced upward; basal tufted, fleshy, 1.5–3 cm, 1–3 mm wide, oblanceolate; cauline 1–2.5 cm, 1–2 mm wide, ± linear. **FL:** calyx 9–15 mm, glandular-puberulent, 10-veined, lobes 2–3 mm; petal claw ciliate at base, appendages 2, limb white to red-purple, lobes 2–4(6), outer gen < inner; stamens ± = petal claws; styles 3(4), incl. **FR:** ± ovoid; stalk 1.5–3 mm, woolly-puberulent. **SEED:** 1–2 mm, brown. 2*n*=48. Subalpine forest, alpine; 2400–3800 m. SNH, SNE (Sweetwater, White mtns); WA, ID, NV. Jul–Aug

S. scouleri Hook. subsp. ***scouleri*** SCOULER'S CATCHFLY Per 15–70 cm; caudex branches gen 0. **ST:** gen erect, densely puberulent, glandular above. **LF:** gradually reduced upward; lower 5–10 cm, 12–20 mm wide, oblanceolate to elliptic; upper 2–6 cm, 3–15 mm wide, lanceolate to ovate. **INFL:** axillary and terminal; fls subsessile to pedicelled. **FL:** ascending to erect; calyx 12–16 mm, short-glandular-hairy, 10-veined, lobes 2–4 mm; petal claw ± ciliate at base, appendages 2, limb white to rose, lobes 2–4(6), outer gen < inner; stamens < to ± = petal claws; styles 3(4), ± = petal claws. **FR:** elliptic to ovoid; stalk 3–6 mm, puberulent to woolly. **SEED:** 1–1.5 mm, gray-brown. 2*n*=48. Rocky slopes, coastal bluffs; < 300 m. NCo, CCo, SnFrB; to BC. [*S. s.* subsp. *grandis* (Eastw.) C.L. Hitchc. & Maguire] Summer

S. serpentinicola T.W. Nelson & J.P. Nelson SERPENTINE CATCHFLY Per 4–10(15) cm; thin branching rhizome from deep taproot. **ST**: erect, canescent. **LF**: gradually reduced upward, 2.5–4.5 cm, 5–15 mm, oblanceolate to obovate or spoon-shaped. **INFL**: terminal, fls 1–3(4). **FL**: ascending; calyx 13–17 mm, densely glandular-hairy, 10-veined, lobes 5–6 mm; petal claw glabrous, appendages 2, limb bright red, lobes 2–4, outer < inner; stamens exserted; styles 3, exserted. **FR**: ovoid to oblong; stalk 0.5–1 mm. **SEED**: 1.8–2 mm, dark brown. 2*n*=72. Serpentine soils, chaparral, conifer forest; 100–800 m. KR. Early summer ★

S. suksdorfii B.L. Rob. (p. 623) CASCADE ALPINE CAMPION, SUKSDORF'S CATCHFLY Per 3–10 cm; caudex branches gen many. **ST**: decumbent to erect, puberulent, glandular above. **LF**: ± abruptly reduced upward; basal tufted, fleshy, 0.5–4.5 cm, 1.5–3.5 mm wide; cauline 0.5–1.5 cm, 1–2 mm wide, ± linear. **FL**: calyx 9–12 mm, short-glandular-hairy (walls between hair cells purple), 10-veined, lobes 1–2 mm; petal claw ciliate at base, appendages 2, limbs white to ± purple, lobes 2; stamens ± = petal claws; styles 3(4), ± = petal claws. **FR**: ± ovoid; stalk 2–3.5 mm, puberulent. **SEED**: 1–2 mm, brown. 2*n*=48. Rocky slopes, alpine; 2400–3100 m. CaRH (Mount Shasta, Lassen Peak); to WA. Summer ★

S. verecunda S. Watson (p. 627) SAN FRANCISCO CAMPION Per 10–55 cm; caudex branches few to many. **ST**: erect, ± scabrous to puberulent, glandular above or not. **LF**: ± gradually reduced upward, stiff to flexible; lower 3–9 cm, 2–9 mm wide, gen lanceolate; middle spreading to erect; upper 1–4.5 cm, 2–6 mm wide, linear to lanceolate. **FL**: calyx 10–15 mm, ± densely puberulent to glandular-puberulent, 10-veined, lobes 2–5 mm; petal claw ciliate throughout or at base, appendages 2, limb white to rose, lobes 2; stamens ± = petal claws; styles 3(4), exserted. **FR**: oblong to ovoid; stalk 2–5 mm, puberulent. **SEED**: 1–1.5 mm, dark brown to black. 2*n*=48. Open areas, chaparral, sagebrush, oak woodland, pinyon/juniper woodland, conifer forest; < 3400 m. c&s NCoR, SN (exc n SNF, Teh), ScV (Sutter Buttes), CW (exc s CCo), SW, W&I, DMtns; to OR, UT, AZ, Baja CA. [*S. v.* subsp. *andersonii* (Clokey) C.L. Hitchc. & Maguire; *S. v.* subsp. *platyota* (S. Watson) C.L. Hitchc. & Maguire] Summer

S. vulgaris (Moench) Garcke BLADDER CAMPION Per 20–80 cm; caudex branches few; rhizome short. **ST**: decumbent to erect, glabrous to glaucous. **LF**: gradually reduced upward; lower 4–8 cm, 5–20 mm wide, lanceolate to oblanceolate; upper 3–4.5 cm, 5–15 mm wide, lanceolate to ovate. **INFL**: gen flat-topped. **FL**: calyx 7–10 mm, gen glabrous, inflated and papery in fr, faintly 10–15-veined; petal claw glabrous, appendages 0 or 2, minute, limb white, lobes 2, gen > 1/2 to base; stamens exserted; styles 3, exserted. **FR**: ± spheric; stalk 2–3 mm. **SEED**: 1–1.5 mm, ± black. 2*n*=24. Open areas, fields; < 1200 m. CaR, GV, SnFrB, SCo, expected elsewhere; to YT, SK, ID, AZ, e N.Am; native to Eur. Summer–fall

SPERGULA SPURREY

Ann, ascending to erect, taprooted. **LF**: opposite but appearing whorled due to axillary clusters of 16–30 per node; stipules 1–2 mm, ovate to triangular, entire, scarious, white; vein 1. **INFL**: terminal; fls several to many; pedicels 40+ mm. **FL**: sepals 5, ± free, 2.5–5 mm, elliptic to ± ovate, glandular-hairy; petals 5, 2.5–4 mm, entire, white; stamens 5 or 10; styles 5, 0.3–0.6 mm. **FR**: capsule, ovoid; valves 5, spreading to ± recurved. **SEED**: several, ± black. 5 spp.: temp Eurasia. (Latin: to scatter, from sowing seeds for early forage in Eur) [Hartman & Rabeler 2005 FNANM 5:14–16]

S. arvensis L. (p. 627) STICKWORT, STARWORT Pl glabrous or gen glandular-hairy. **ST**: 10–40+ cm; base ± branched. **LF**: 1–5 cm, ± linear; tip blunt to abruptly pointed; margin often strongly rolled under. **INFL**: bracts like stipules, often ± purple; pedicels erect to ascending, in fr spreading to reflexed. **FL**: sepals ± acute to rounded, margin widely scarious, ribs often 3, weak; petals ovate, persistent in fr. **SEED**: 1–1.5 mm diam, ± spheric, with ± white, club-shaped papillae or minutely roughened. 2*n*=18,36. Open slopes, pine woodland, sand dunes, fields, disturbed areas; < 200 m. NCo, NCoRO, n SNH, GV, CW (exc SCoRI), SCo; scattered in N.Am, native to Eur. Spring–early summer

SPERGULARIA SAND-SPURREY

Ann, per, erect to sprawling; taprooted. **LF**: thread-like to linear, vein 1; stipules 1–11 mm, lanceolate to widely triangular, scarious, ± entire or splitting ± at tip, white to tan. **INFL**: terminal, open to dense; fls few to many; pedicels 0.5–28+ mm. **FL**: sepals 5, united in basal 1/5, 1.5–10 mm, lanceolate to ovate, glabrous to glandular-hairy; petals 5, 0.6–9 mm, entire; stamens 2–10; styles 3, 0.3–1.9 mm. **FR**: capsule, ovoid; valves 3, spreading, tip recurved. **SEED**: few to many, dark brown, red-brown, or black, often winged. 60 spp.: w N.Am, w S.Am, Medit. (Latin: derivative of *Spergula*) [Hartman & Rabeler 2005 FNANM 5:16–23]

1. Pl strongly per; stamens 7–10
  2. Seeds 0.4–0.5 mm; calyx lobes 2.5–4 mm, in fr < 5 mm; styles 0.4–0.6 mm . . . . ***S. villosa***
  2′ Seeds 0.6–0.9 mm; calyx lobes 4.5–7 mm, in fr < 8 mm; styles 0.6–3 mm . . . . ***S. macrotheca***
    3. Petals pink to rosy or blue; styles 0.5–1.2 mm . . . . var. ***macrotheca***
    3′ Petals white; styles 1.2–3 mm
      4. Fr 1.2–1.4 × calyx; styles 1.2–1.9 mm . . . . var. ***leucantha***
      4′ Fr 0.8–1 × calyx; styles 2–3 mm . . . . var. ***longistyla***
1′ Ann (*Spergularia rubra*, *Spergularia media* short-lived per); stamens 2–10
  5. Seeds black, not papillate . . . . ***S. atrosperma***
  5′ Seeds light to dark brown or red-brown, if nearly black, seeds covered with gland-tipped papillae
    6. Stamens 6–10
      7. Seeds 0.8–1.1 mm, smooth, not papillate, gen winged; fr (4.5)5.5–8 mm; pl stout, lower main st gen 1–3 mm diam; lvs clearly fleshy . . . . ***S. media*** var. ***media***
      7′ Seeds 0.4–0.6 mm, roughened or variously sculptured, papillate, not winged; fr 2.7–5.4 mm; pl ± delicate, lower main st gen 0.3–1 mm diam; lvs barely to ± fleshy
        8. Lvs gen 0–2 per axillary cluster, ± fleshy; seeds light brown; stipules gen 1.5–4.5 mm, mostly deltate, dull, white to tan, gen inconspicuous, tip acute to short-acuminate . . . . ***S. bocconi***
        8′ Lvs 2–4+ per axillary cluster, barely fleshy; seeds red-brown to dark brown; stipules gen 3.5–5 mm, lanceolate, shiny, white, conspicuous, tip ± long-acuminate . . . . ***S. rubra***

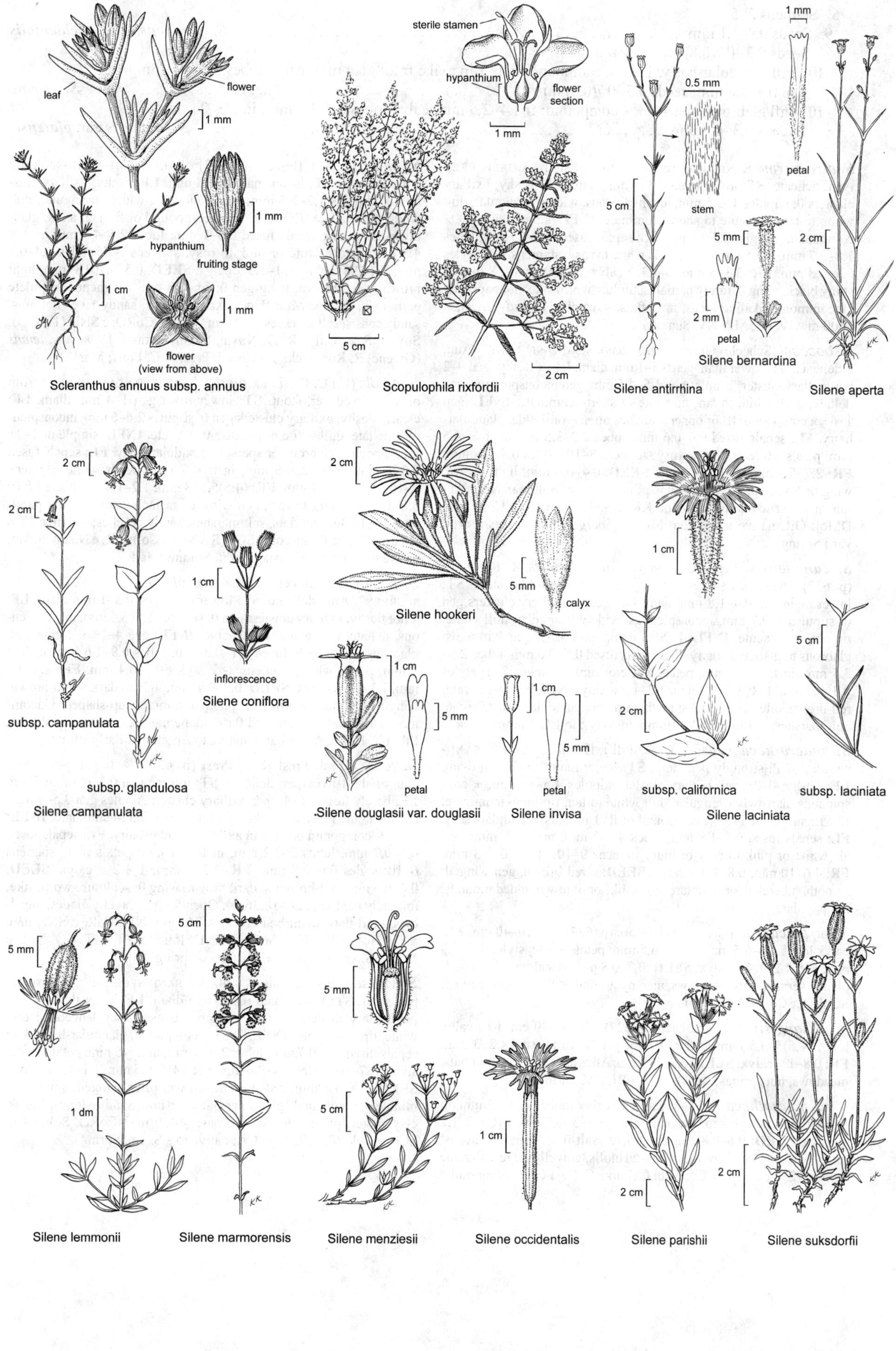

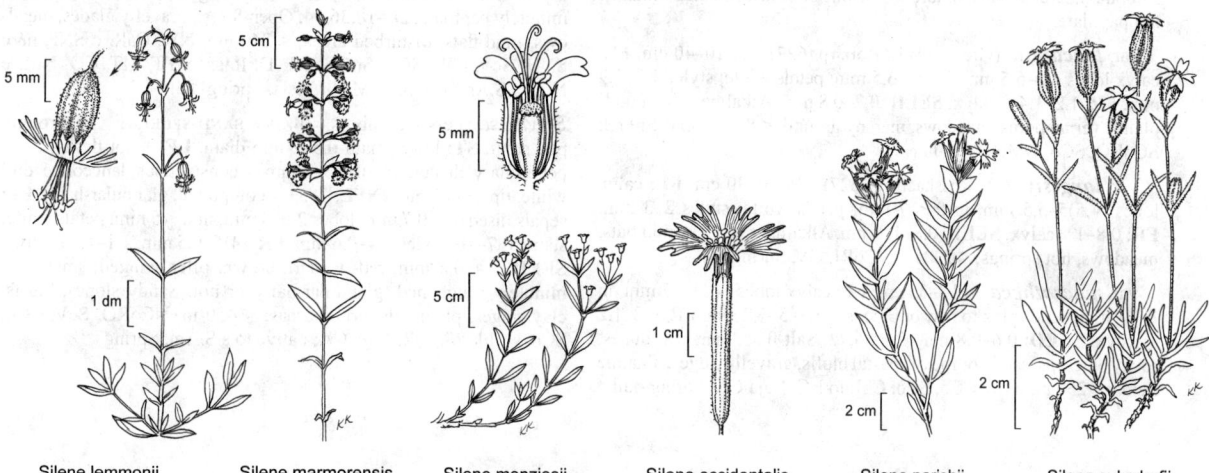

leaf
flower
1 mm
hypanthium
1 mm
fruiting stage
1 cm
1 mm
flower
(view from above)
Scleranthus annuus subsp. annuus

sterile stamen
hypanthium
flower
section
1 mm
5 cm
Scopulophila rixfordii
5 cm
2 cm

0.5 mm
stem
5 cm
5 mm
petal
2 mm
petal
Silene antirrhina

1 mm
petal
2 cm
Silene aperta

2 cm
2 cm
1 cm
inflorescence
Silene coniflora
subsp. campanulata
subsp. glandulosa
Silene campanulata

2 cm
5 mm
calyx
Silene hookeri
1 cm
5 mm
petal
.**Silene douglasii var. douglasii**
1 cm
5 mm
petal
Silene invisa

1 cm
2 cm
subsp. californica
Silene laciniata
5 cm
subsp. laciniata

5 mm
1 dm
Silene lemmonii
5 cm
Silene marmorensis
5 mm
5 cm
Silene menziesii
1 cm
Silene occidentalis
2 cm
Silene parishii
2 cm
Silene suksdorfii

6′ Stamens 2–5
  9. Seeds 0.9–1.1 mm. . . . . ***S. canadensis*** var. ***occidentalis***
  9′ Seeds 0.3–0.7(0.8) mm
    10. Infl glandular-hairy, 1–3+ × compound or fls 1 in axils; fr 2.8–6.4 mm; calyx lobes 2.5–4.5 mm,
      in fr < 4.8 mm; seeds 0.5–0.7(0.8) mm . . . . ***S. marina***
    10′ Infl glabrous, gen 4–7+ × compound; fr 1.4–2.6 mm; calyx lobes 0.9–1.6 mm, in fr < 2 mm;
      seeds 0.3–0.4 mm. . . . . ***S. platensis*** var. ***platensis***

S. atrosperma R. Rossbach (p. 627) BLACK SEED SAND-SPURREY Ann, delicate. **ST:** lower main 0.3–1 mm diam. **LF:** fleshy; axillary clusters 0; stipules 1–2.5 mm, inconspicuous, widely triangular, dull white to tan, tip acute to short-acuminate. **INFL:** simple or 1–3+ × compound, gen glandular-hairy. **FL:** sepals fused 0.2–0.5 mm, lobes 1.8–2.7 mm, < 3.5 mm in fr; petals white to rosy; stamens 4–8; styles 0.4–0.8 mm. **FR:** 3.3–5 mm, 1–1.3 × calyx. **SEED:** 0.6–0.8 mm, ± shiny black, wing 0, rarely partial; sculpture worm-like, not papillate. Uncommon. Alkaline areas, mud flats, streambeds, sandy areas; < 1500 m. GV, PR, MP; NV. Summer

S. bocconi (Scheele) Graebn. BOCCONE'S SAND-SPURREY Ann, ± delicate. **ST:** lower main gen 0.5–1 mm diam. **LF:** ± fleshy; gen 0–2 per axillary cluster; stipules gen 1.5–4.5 mm, gen inconspicuous, gen deltate, dull, white to tan, tip acute to short-acuminate. **INFL:** gen 1–6+ × compound; fls of upper branches often ± on 1 side, glandular-hairy. **FL:** sepals fused 0.4–0.6 mm, lobes 2.2–3.5 mm, in fr < 4.5 mm; petals white or pink to rosy; stamens 8–10; styles 0.4–0.6 mm. **FR:** 2.7–5.3 mm, 1–1.2 × calyx. **SEED:** 0.4–0.6 mm, light brown, wing 0; ± sculptured, minutely papillate. 2*n*=36. Salt marshes, alkaline areas, sandy soils; < 400 m. KR, c SNF, GV, CCo, SCo, s ChI, DMoj; OR; native to sw Eur, Medit. [*Spergularia bocconei*, orth. var.] Spring

S. canadensis (Pers.) G. Don var. ***occidentalis*** R. Rossbach (p. 627) WESTERN SAND-SPURREY Ann, delicate to ± stout. **ST:** lower main gen 0.4–1.8 mm diam. **LF:** fleshy; axillary clusters gen 0; stipules 1–2.7 mm, inconspicuous, widely triangular, dull white, tip obtuse to acute. **INFL:** 1–2+ × compound or fls gen 1 in axils; glabrous to glandular-hairy. **FL:** sepals fused 0.5–0.6 mm, lobes 2.5–3.5 mm, in fr < 4.5 mm; petals white or pink; stamens 2–4; styles 0.3–0.7 mm. **FR:** 3.5–5.3 mm, 1.2–1.3 × calyx. **SEED:** 0.9–1.1 mm, red-brown, often winged; ± smooth, minutely glandular-hairy. 2*n*=36. Salt marshes; < 3 m. n NCo (Humboldt Bay); to BC. Summer ★

S. macrotheca (Cham. & Schltdl.) Heynh. STICKY SAND-SPURREY Pl strongly per, stout. **ST:** lower main 0.8–3 mm diam. **LF:** fleshy; 0–2+ per axillary cluster; stipules 4.5–11 mm, ± conspicuous, narrowly triangular, dull white to tan, tip long-acuminate. **INFL:** simple or 1–3+ × compound or fls 1 in axils; glandular-hairy. **FL:** sepals fused 0.5–1.8 mm, lobes 4.5–7 mm, in fr < 8 mm; petals white or pink to rosy or blue; stamens 9–10; styles 0.5–3 mm. **FR:** 4.6–10 mm, 0.8–1.4 × calyx. **SEED:** ± red-brown, gen winged; smooth, tubercled, or sculpture worm-like or of low rounded mounds, not papillate.

var. ***leucantha*** (Greene) B.L. Rob. (p. 627) Pl 10–40 cm. **FL:** calyx lobes 4.5–5.5 mm, in fr < 6.5 mm; petals white; styles 1.2–1.9 mm. **FR:** 1.2–1.4 × calyx. **SEED:** 0.7–0.8 mm. Alkaline soils, floodplains, vernal pools, meadows, marshy ground; < 800 m. GV, SnFrB, SCoRO, SCo, DMoj. Apr–Jun

var. ***longistyla*** R. Rossbach (p. 627) Pl 10–30 cm. **FL:** calyx lobes (4.5)5–5.5 mm, in fr < 7 mm; petals white; styles 2–3 mm. **FR:** 0.8–1 × calyx. **SEED:** 0.8–1.2 mm. Alkaline marshes, mud flats, meadows, hot springs; < 200 m. NCoRI, GV. Spring

var. ***macrotheca*** Pl 5–35 cm. **FL:** calyx lobes (4.5)5–7 mm, in fr < 8 mm; petals pink to rosy or blue; styles 0.5–1.2 mm. **FR:** 0.8–1.2 × calyx. **SEED:** 0.6–0.8 mm. 2*n*=36,72. Salt flats, marshes, dunes, rocky outcrops, sandy or rocky coastal bluffs, gravelly ridges, alkaline fields; < 250 m. NCo, CCo, SCo, ChI; to BC, Baja CA. Spring–fall

S. marina (L.) Besser (p. 627) SALTMARSH SAND-SPURREY Ann, delicate. **ST:** lower main 0.6–2 mm. **LF:** fleshy; axillary clusters 0; stipules 1.2–3.5 mm, inconspicuous, widely triangular, dull white, tip ± acute. **INFL:** 1–3+ × compound or fls 1 in axils; glandular-hairy. **FL:** sepals fused 0.5–1 mm, lobes 2.5–4.5 mm in fr < 4.8 mm; petals white or pink to rosy; stamens 2–5; styles 0.4–0.7 mm. **FR:** 2.8–6.4 mm, 1–1.5 × calyx. **SEED:** 0.5–0.7(0.8) mm, light brown to red-brown, wing gen 0; smooth or ± roughened, papillate or not. 2*n*=18,36. Mud flats, alkaline fields, sandy river bottoms, sandy coasts, salt marshes; < 700 m. NCo, NCoRO, c SNF, GV, CCo, SnFrB, SCo, ChI, PR, D; N.Am, S.Am; Eurasia. [*S. m.* var. *tenuis* (Greene) R. Rossbach; *S. salina* J. Presl & C. Presl] Mar–Sep

S. media (L.) C. Presl var. ***media*** GREATER SEA-SPURREY Ann or short-lived per, stout. **ST:** lower main gen 1–4 mm diam. **LF:** clearly fleshy; axillary clusters gen 0; stipules 2.6–6 mm, inconspicuous, deltate, dull white, tip obtuse to ± acute. **INFL:** simple or 1–2+ × compound, glabrous or sparsely glandular-hairy. **FL:** sepals fused 0.3–1 mm, lobes 2.5–5 mm, in fr < 7 mm; petals white; stamens 9–10; styles 0.5–1 mm. **FR:** (4.5)5.5–8 mm, 1.2–1.4 × calyx. **SEED:** 0.8–1.1 mm, dark brown, gen winged; surface ± smooth, not papillate. 2*n*=18,36. Salt flats, salt marshes, sandy beaches; < 2 m. n CCo (Marin, Contra Costa cos.), SCo (Riverside Co.); OR, e N.Am; S.Am; native to coastal Eur, Asia, Medit. Summer–fall

S. platensis (Cambess.) Fenzl var. ***platensis*** LA PLATA SAND-SPURREY Ann, delicate. **ST:** lower main 0.3–1 mm diam. **LF:** ± not fleshy; axillary clusters gen 0; stipules 1.5–3.5 mm, inconspicuous, deltate, dull white, tip ± acute. **INFL:** gen 4–7+ × compound, glabrous. **FL:** sepals fused 0.1–0.3 mm, lobes 0.9–1.6 mm, in fr < 2 mm; petals white; stamens 3–5; styles 0.3–0.4 mm. **FR:** 1.4–2.6 mm, 1.3–1.5 × calyx. **SEED:** 0.3–0.4 mm, light to dark or red-brown, wing 0; sculpture worm-like, papillae minute, cup-shaped. Uncommon. Dried or brackish mud flats, adobe mesas; < 400 m. ScV, SCo, PR; TX, Baja CA; probably native to Argentina. Early spring

S. rubra (L.) J. Presl & C. Presl (p. 627) RED SAND-SPURREY Ann or short-lived per, delicate. **ST:** lower main 0.3–0.5 mm diam. **LF:** barely fleshy, 2–4+ per axillary cluster; stipules gen 3.5–5 mm, conspicuous, lanceolate, shiny, white, tip ± long-acuminate. **INFL:** 1–3+ × compound or fls 1 in axils; glandular-hairy. **FL:** sepals fused 0.5–0.7 mm, lobes 2–3.2 mm, in fr < 4 mm; petals pink; stamens 6–10; styles 0.6–0.8 mm. **FR:** 3.5–5 mm, 1–1.2 × calyx. **SEED:** 0.4–0.6 mm, red-brown to dark brown, wing 0; sculpture worm-like, minutely papillate. 2*n*=18,36,54. Open forest, gravelly glades, meadows, mud flats, disturbed areas; < 2400 m. NW, CaR, c SNF, n&c SNH, ScV, CW, SCo, SnGb, PR, DMtns; to YT, MT, CO, NM, e N.Am, S.Am; native to Medit, Asia. Spring–fall

S. villosa (Pers.) Cambess. HAIRY SAND-SPURREY Pl strongly per, stout. **ST:** lower main 0.5–2 mm diam. **LF:** ± not fleshy, 2–4+ per axillary cluster; stipules 3–8 mm, ± conspicuous, lanceolate, dull white, tip acuminate. **INFL:** 1–3+ × compound, glandular-hairy. **FL:** sepals fused 0.5–0.7 mm, lobes 2.5–4 mm, in fr < 5 mm; petals white; stamens 7–10; styles 0.4–0.6 mm. **FR:** (4)5–6.5 mm, 1.1–1.3 × calyx. **SEED:** 0.4–0.5 mm, red- to dark brown, often winged; smooth or minutely roughened, glandular-hairy or not. Sandy slopes, bluffs, clay ridges, plains, disturbed areas; < 450 m. NCoRO, ScV, CCo, SCo, s ChI, PR; OR, Baja CA; native to s S.Am. Spring

STELLARIA CHICKWEED, STARWORT

Ann, per, erect to prostrate; taprooted, rhizomed. **ST**: 4-angled or round. **LF**: petioled or not; linear to ovate, vein 1. **INFL**: terminal or axillary, umbel-like or not, open to dense; fls 1–many (if 1, axillary); peduncles, pedicels 0.8–50+ mm. **FL**: sepals (4)5, free, 1.5–5.5 mm, lanceolate to ovate, glabrous to glandular-hairy; petals 0 or (1)5, 0.8–7 mm, 2-lobed > 1/2 to base; stamens 10 or fewer; styles 3(4–5 in *Stellaria calycantha*), 0.2–2.8 mm. **FR**: capsule, ± ovoid or spheric to cylindric-oblong; valves 6(8,10), ascending to recurved. **SEED**: several to many, brown to ± yellow, ± red, or purple-brown. ± 190 spp.: worldwide. (Latin: star, from fl shape) [Morton 2005 FNANM 5:96–114] Presence of papillae on lf margins determined at 20×.

1. Ann, from slender to thread-like taproot
  2. Lvs crowded near base, upper lance-linear (to ovate); internode hairs 0 or scattered; bracts scarious . . . . *S. nitens*
  2′ Lvs ± evenly spaced; ± ovate; internode hairs in line; bracts lf-like
    3. Sepals 2–3 mm, < 4 mm in fr; petals 0; seeds 0.7–0.8 mm, yellow- to light red-brown . . . . *S. pallida*
    3′ Sepals 3–6 mm, > 6 mm in fr; petals gen present; seeds 0.9–1.7 mm, dark red- or purple-brown to brown
      4. Sepals 3–4.5 mm; seeds 0.9–1.3 mm, tubercles rounded . . . . *S. media*
      4′ Sepals 5–6 mm; seeds 1.1–1.7 mm, tubercles conical, esp on margin . . . . *S. neglecta*
1′ Per, from slender rhizomes
  5. Pl covered (exc ± lvs) with long wavy hairs — coastal ± marshes, bluffs, NCo, CCo . . . . *S. littoralis*
  5′ Pl glabrous or hairs restricted to st or lf margin
    6. Petals ≥ sepals
      7. Infl ± narrow; pedicels ascending to erect; fls gen 1–7 . . . . *S. longipes* subsp. *longipes*
      7′ Infl open, branches widely spreading; pedicels spreading or reflexed; fls gen many
        8. Sepals 4–5.5 mm, margin densely ciliate or not; infl terminal; seeds with prominent, elongate tubercles; lf margin ± not papillate, shiny . . . . *S. graminea*
        8′ Sepals 3–4 mm, margin ± glabrous; infl lateral; seeds minutely roughened; lf margin papillate, dull . . . . *S. longifolia*
    6′ Petals < 0.8 × sepals or 0
      9. Fls few (alpine) to many, in umbel-like cyme; bracts scarious; pedicels in fr spreading to reflexed; petals 0 . . . . *S. umbellata*
      9′ Fls 1 in axils or few to many in terminal cyme; bracts lf-like; pedicels in fr spreading to recurved or reflexed; petals present or 0
        10. Fls 1 in axils; petals gen 0; lf ± ovate
          11. Sepals gen 5, acute or acuminate, ribs in fr 3; lf margin ± wavy, glabrous; seeds 0.8–1 mm . . . . *S. crispa*
          11′ Sepals gen 4, ± obtuse, ribs in fr 1, obscure; lf margin ± flat, gen ciliate near base; seeds 0.6–0.7 mm . . . . *S. obtusa*
        10′ Fls gen few to many in terminal or axillary cymes; petals gen present; lf lanceolate to ovate
          12. Lf ± lanceolate to widely so, primary blades mostly > 2 cm, margin papillate, dull; internodes gen finely papillate; sepals 3–3.5 mm, in fr < 4.5 mm, with gen 3 ribs; styles gen 0.9–1.6 mm, slender, straight (often contorted on dried specimens) . . . . *S. borealis* subsp. *sitchana*
          12′ Lf elliptic to ovate, primary blades mostly < 2 cm, margin ± not papillate, shiny; internodes not papillate; sepals 1.5–2.5 mm, in fr < 3 mm, with gen 1–3 obscure veins; styles gen 0.5–0.9 mm, thick, outcurved . . . . *S. calycantha*

S. borealis Bigelow subsp. ***sitchana*** (Steud.) Piper & Beattie (p. 627) SITKA STARWORT Per, sprawling to erect, 15–50 cm, ± glabrous; rhizome white. **ST**: internodes gen finely papillate. **LF**: ± evenly spaced; blade 15–45 mm, ± lanceolate to widely so; margin papillate, flat, dull, sometimes ciliate near base. **INFL**: terminal or axillary, several to many-fld; bracts lf-like; pedicels erect to ascending, in fr curved to reflexed. **FL**: sepals 5, 3–3.5 mm, < 4.5 mm in fr, lanceolate, acute, glabrous, margin ± widely scarious, ribs in fr gen 3, prominent; petals (0)1–5, 0.5–0.8 × sepals; styles gen 0.9–1.6 mm, slender, straight (often contorted on dried specimens). **SEED**: 1–1.2 mm, dark red-brown, tubercles low, elongate. 2n=52. Sedge stands, meadows, streambanks, swamps, moist woodland; < 2100 m. NCo, NCoRO, CaRH, c&s SNF, SNH, CCo; to AK, MT, SD, UT. Pls from > 1000 m in c&s SNF, SNH often have ovate-elliptic lvs and resemble *S. crispa. S. borealis* subsp. *b.* erroneously reported for CA in FNANM (Morton 2005). May–Sep

S. calycantha (Ledeb.) Bong. (p. 627) NORTHERN STARWORT Per, prostrate to erect, 5–25 cm, often glabrous; rhizome white. **ST**: internodes glabrous or with wavy scattered hairs, not papillate. **LF**: ± evenly spaced; blade 3–25 mm, elliptic to ovate; margin ± not papillate, flat to wavy, shiny, ciliate or not. **INFL**: terminal or axillary, 1–few-fld; bracts lf-like; pedicels ascending to erect, in fr often recurved. **FL**: sepals 5, 1.5–2.5 mm, < 3 mm in fr, ovate to elliptic, ± acute, glabrous, margin scarious, veins gen 1–3, obscure; petals (0)1–5, 0.3–0.5 × sepals; styles gen 0.5–0.9 mm, thick, outcurved.

SEED: 0.7–0.9 mm, red-brown, ± smooth or minutely roughened. 2n=26. Mossy banks, bogs, dry creeks, wet meadows, shaded areas; 1700–3800 m. KR, NCoRO, NCoRI, CaRF, SN, SnGb, SnBr, SnJt, Wrn, W&I; to AK, MT, NM; ne Asia. Jun–Aug

S. crispa Cham. & Schltdl. (p. 627) CRISP STARWORT Per, prostrate to trailing, 10–40 cm, gen glabrous; rhizome white. **ST**: internodes rarely with scattered, wavy hairs. **LF**: ± evenly spaced; blade 8–20 mm, ± ovate; margin ± not papillate, ± wavy, shiny, glabrous. **INFL**: fls 1 in axils; pedicels ascending, in fr spreading to recurved. **FL**: sepals gen 5, 2.5–4 mm, lanceolate, acute to acuminate, glabrous, margin scarious, ribs in fr 3; petals gen 0. **SEED**: 0.8–1 mm, redbrown, tubercles low, elongate. 2n=26,52. Shaded, damp areas; < 1700 m. NCo, KR, CaRH, n SNH, CCo; to AK, MT. Summer

S. graminea L. COMMON STARWORT Per, sprawling to erect, 10–60 cm, gen glabrous; rhizome white. **ST**: internodes glabrous. **LF**: ± evenly spaced; blade 10–35 mm, linear to lanceolate; margin ± not papillate, flat, shiny, ± ciliate near base. **INFL**: terminal; fls many; bracts scarious; pedicels spreading to erect, in fr gen widely branched. **FL**: sepals 5, 4–5.5 mm, ± lanceolate, acute, margin widely scarious, densely ciliate or not, ribs in fr 3, prominent; petals 5, 1–1.4 × sepals. **SEED**: 0.9–1.1 mm, dark brown, tubercles prominent, elongate. 2n=26,39,52. Disturbed areas; < 400, and 1220 m. c SNH, SnJV, SnFrB, SCo; to BC, MT, e N.Am; native to Eur. Late spring–early summer

S. littoralis Torr. (p. 627) BEACH STARWORT Per, sprawling, 10–60 cm, covered (exc ± lvs) with long wavy hairs; rhizome white. **ST:** ± uniformly hairy. **LF:** ± evenly spaced; blade 10–45 mm, ± ovate; margin flat to wavy, ± shiny, densely ciliate. **INFL:** terminal; fls few to many; bracts lf-like; pedicels ascending to erect, in fr spreading to reflexed. **FL:** sepals 5, 2.8–5 mm, lanceolate, sharply acute, hairy, margin widely scarious, ciliate near base, ribs in fr 1 or 3; petals 5, 1–1.2 × sepals. **SEED:** ± 1 mm, red-brown, minutely roughened. Coastal ± marshes, bluffs; < 40 m. NCo, CCo. *S. littoralis* closely resembles *S. dichotoma* L. of e Asia; further study needed to determine if the 2 are conspecific and our pls thus introduced to CA. Spring ★

S. longifolia Willd. (p. 627) LONG-LEAVED STARWORT Per, sprawling to ascending, 15–40 cm, gen glabrous; rhizome white. **ST:** internodes ± scabrous. **LF:** ± evenly spaced; blade 15–35 mm, linear to lance-linear; margin papillate, flat, dull, ciliate near base or not. **INFL:** lateral, several- to gen many-fld; bracts scarious; pedicels spreading to erect, in fr widely branched. **FL:** sepals 5, 3–4 mm, lanceolate to narrowly elliptic, ± acute, ± glabrous, margin widely scarious, ribs in fr ± 3 near base; petals 5, 1–1.2 × sepals. **SEED:** 0.8–0.9 mm, red-brown, minutely roughened. 2*n*=26. Moist areas; ± 900 m. CaRH, n SNH; to AK, AZ, n-c US, e N.Am; circumboreal. Late spring–summer ★

S. longipes Goldie subsp. ***longipes*** (p. 627) GOLDIE'S STARWORT Per, ascending to erect, 5–35 cm, gen glabrous; rhizomes white. **ST:** internodes glabrous or hairs scattered, wavy. **LF:** ± evenly spaced; blade 10–40 mm, linear to lance-linear; margin not papillate, flat, shiny, sometimes ciliate near base. **INFL:** terminal or axillary, gen 1–7-fld, ± narrow; bracts lf-like or ± scarious; pedicels ascending to erect, in fr ± straight. **FL:** sepals 5, 3–5.5 mm, lanceolate to ± ovate, ± acute, glabrous, margin widely scarious, in fr ± 3-ribbed; petals 5, 1–1.2 × sepals. **SEED:** 0.7–1 mm, red-brown, minutely roughened. 2*n*=gen 52,78,104. Streambanks, moist to boggy meadows, seeps; (0)1250–3500 m. KR, NCoRI, CaRH, SN, CCo (San Luis Obispo Co.), SnBr, Wrn, n SNE, W&I; to AK, MT, NM; circumboreal. [*S. l.* var. *l.*] May–Aug

S. media (L.) Vill. COMMON CHICKWEED Ann but often over-wintering, prostrate to erect, 7–50 cm; taproot slender. **ST:** internode hairs in line. **LF:** ± evenly spaced; blade 8–45 mm, ± ovate; margin ± not papillate, ± flat, shiny, often ciliate near base. **INFL:** terminal or axillary, few-fld, ± dense; bracts lf-like; pedicels spreading to erect, in fr curved to reflexed. **FL:** sepals 5, 3–4.5 mm, > 6 mm in fr, lanceolate to ovate, acute to obtuse, glabrous or ± glandular-hairy, margin ± widely scarious, ribs often 1 or 3 near base; petals 5, 0.7–0.9 × sepals. **SEED:** 0.9–1.3 mm, dark red-brown to brown, tubercles rounded. 2*n*=40,42,44. Oak woodland, meadows, disturbed areas; < 1500 m. NW, CaRH, n&c SNF, n SNH, GV, CCo, SnFrB, SCo, ChI, DSon; N.Am; native to sw Eur. Often a pernicious urban weed; sometimes difficult to distinguish from *S. neglecta, S. pallida.* Feb–Sep

S. neglecta Weihe GREATER CHICKWEED Ann but often over-wintering, ascending to erect, 30–60(80) cm; taproot slender. **ST:** internode hairs in line. **LF:** ± evenly spaced; blade 8–50 mm, broadly ovate; margin ± not papillate, ± flat, shiny, often ciliate near base.

INFL: terminal or axillary, few-fld, ± dense; bracts lf-like; pedicels spreading to erect, in fr curved to reflexed. **FL:** sepals 5, 5–6 mm, > 6 mm in fr, lanceolate to lance-oblong, acute, ± hairy and glandular, margin ± widely scarious, ribs often 1 or 3 near base; petals 5, 0.5–0.7 × sepals. **SEED:** 1.1–1.7 mm, brown or purple-brown, tubercles conical, esp on margin. 2*n*=20, 22. Disturbed areas, fields, cult areas; < 400(600) m. NCoRO, Teh, GV, SnFrB, SCo, SnGb, PR; se N.Am; native to Eur. [*S. media* (L.) Vill. subsp. *n.* (Weihe) Gremli] Mar–May

S. nitens Nutt. (p. 627) SHINING CHICKWEED Ann, ascending to erect, 3–25 cm, glabrous or sparsely hairy; taproot thread-like. **ST:** internode hairs 0 or scattered. **LF:** crowded near base; blade 5–15 mm, margin not papillate, shiny, often ciliate; lower oblanceolate to obovate; upper lance-linear (to ovate). **INFL:** terminal, few- to several-fld; bracts scarious; pedicels ascending to erect, in fr ± straight. **FL:** sepals 5, 2.8–4.2 mm, ± lanceolate to needle-like, long-acuminate, glabrous, margin widely scarious, ribs in fr ± 3; petals 5, 0.2–0.5 × sepals or 0. **SEED:** 0.5–0.7 mm, brown, tubercles minute. 2*n*=20, 40. Sand dunes, streambanks, open woodland, beneath boulders, disturbed areas; < 1500 m. NW, n&c SNF, SNH, ScV (Sutter Buttes), SnJV, CW, SW; to BC, MT, AZ; Mex. Spring

S. obtusa Engelm. (p. 627) OBTUSE STARWORT, ROCKY MOUNTAIN STARWORT Per, gen prostrate, 4–20 cm, gen glabrous; rhizome white. **ST:** internodes glabrous. **LF:** ± evenly spaced; blade 5–12 mm, ± ovate; margin ± not papillate, ± flat, shiny, gen ciliate near base. **INFL:** fls 1 in axils; pedicels ascending, in fr spreading to reflexed. **FL:** sepals gen 4, 1.5–3.5 mm, ± ovate, ± obtuse, glabrous, margin ± thinly scarious, rib 1, obscure; petals 0. **SEED:** 0.6–0.7 mm, dark brown, tubercles low, ± elongate. 2*n*=26,52,±65,±78. Moist areas in woodland, shaded edges of creeks; 1600–2000 m. NCoR, CaR, c SNF, n&c SNH, MP; to BC, MT, CO. Late spring–summer ★

S. pallida (Dumort.) Crép. LESSER CHICKWEED Ann but often over-wintering, prostrate to erect, 7–50 cm; taproot slender. **ST:** internode hairs in line. **LF:** ± evenly spaced; blade 8–45 mm, ± ovate, margin ± not papillate, ± flat, shiny, often ciliate near base. **INFL:** terminal or axillary, few-fld, ± dense; bracts lf-like; pedicels spreading to erect, in fr curved to reflexed. **FL:** sepals 5, 2–3 mm, < 4 mm in fr, lanceolate to ovate, obtuse to acute, glabrous or ± hairy and glandular, margin ± thinly scarious, ribs often 1 or 3 near base; petals 0. **SEED:** 0.7–0.8 mm, yellow- to light red-brown, tubercles conical. 2*n*=22. Oak woodland, streambanks, grassy hills, flats, disturbed areas; < 450(1500) m. s NCo, n&c SNF, Teh/WTR, ScV, CCo, SCoRO, s ChI, PR; WA, CO, AZ, Baja CA, to e N.Am; native to sw Eur. Spring

S. umbellata Kar. & Kir. (p. 627) UMBELLATE STARWORT Per, prostrate to erect, 2–20 cm, glabrous; rhizome white. **ST:** internodes glabrous. **LF:** ± evenly spaced; blade 5–20 mm, elliptic to ovate, margin not papillate, gen flat, shiny, glabrous. **INFL:** terminal, gen umbel-like, few-(alpine) to many-fld; bracts scarious; pedicels erect to spreading, in fr spreading to reflexed. **FL:** sepals 5, 1.5–2.5 mm, lanceolate to ovate, ± acute, margin widely scarious, ribs in fr 3, weak; petals minute or 0. **SEED:** 0.6–0.7 mm, brown, ± smooth. 2*n*=26. Moist meadows, rocky summits, streamsides; 1800–2300 m. NCoRH, NCoRI, CaRH, SNH, Wrn; to AK, MT, NM; ne Asia. Summer

VACCARIA COW-HERB

1 sp. (Latin: cow, from use as fodder or prevalence in pastures) [Thieret & Rabeler 2005 FNANM 5:156]

V. hispanica (Mill.) Rauschert (p. 627) Ann (8)20–100 cm, glabrous, glaucous, taprooted. **LF:** 2–12 cm, petioled or not; blade lanceolate to ovate, base rounded to cordate-clasping; vein 1. **INFL:** terminal, ± flat-topped, open; fls 10–70+; bracts paired, lf-like; pedicels 5–40+ mm. **FL:** sepals 5, fused, glabrous, tube prominent, 7.5–17 mm, 1.5–9 mm diam, cylindric to urn-shaped, angles or keels 5, each with ± green wing < 1 mm wide; veins 5, teeth 1.5–3 mm, ovate to triangular; petals 5, 15–25 mm, claw long, limb oblanceolate to obovate or obcordate, pink to ± red, unappendaged; styles 2, 10–12 mm. **FR:** capsule, ovoid; stalk 0.5–1 mm; teeth 4, ascending to recurved. **SEED:** many, 1.6–1.8 mm, ± spheric, red-brown to black; tubercles fine, low. 2*n*=30. Disturbed areas; < 2800 m. KR, NCoR, CaRH, c SNF, n SNH, Teh, ScV, CW, SCo, PR, GB; ± N.Am; native to Eurasia, Medit. Likely eradicated from grain fields. May–Aug

VELEZIA

Ann, erect, taprooted. **LF:** blade linear to awl-shaped; veins 3. **INFL:** axillary; fl gen 1; pedicels 1.5–3.5 mm. **FL:** sepals 5, fused, tube prominent, 10–14 mm, 0.8–1 mm diam, narrowly cylindric, rounded, ribs 15, teeth 1–1.2 mm, lanceolate-acuminate; petals 5, 11–16 mm, claw ± = calyx, limb entire or notched, appendages 6–8, linear to lanceolate; stamens 5; styles

627

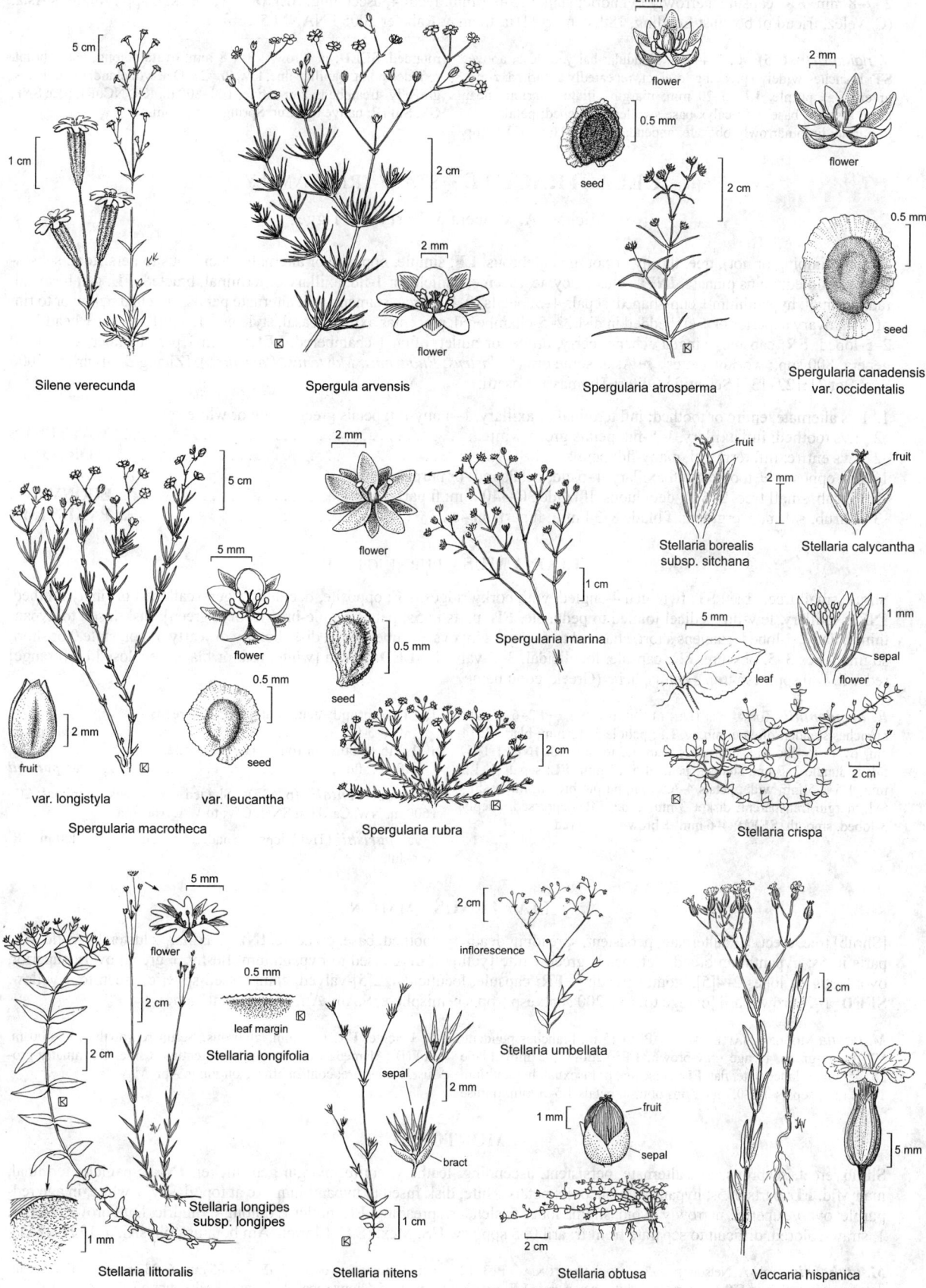

Silene verecunda

Spergula arvensis

Spergularia atrosperma

Spergularia canadensis var. occidentalis

var. longistyla

var. leucantha

Spergularia macrotheca

Spergularia rubra

Spergularia marina

Stellaria borealis subsp. sitchana

Stellaria calycantha

Stellaria crispa

Stellaria longifolia

Stellaria longipes subsp. longipes

Stellaria littoralis

Stellaria nitens

Stellaria umbellata

Stellaria obtusa

Vaccaria hispanica

2, 7–8 mm. **FR**: capsule, narrowly cylindric; stalk 0.2–0.7 mm; teeth 4, ascending. **SEED**: 6–8, black. 6 spp.: Medit, s Asia. (C. Velez, friend of botanist Loefling, 18th century) [Hartman & Rabeler 2005 FNANM 5:166]

V. rigida L. (p. 635) Pl 7–40 cm, glandular-hairy, at least above. **ST**: branches widely spreading to erect, repeatedly 2-forked, rigid, green to ± purple. **LF**: 5–20 mm; margin ciliate, scarious near base, fused at base. **FL**: calyx base swollen, hardened; petals pink to purple, limb narrowly obovate, appendages 0.4–0.6 mm. **FR**: tip rounded. **SEED**: in 1 row, 1.3–1.8 mm, ovate-oblong, with abrupt, rounded point; papillae fine, low. $2n=28$. Oak woodland, open ridges, gravelly streambeds, serpentine; 100–800 m. KR, NCoRI, n&c SNF, GV, SnFrB; native to s Eur. Spring–early summer

CELASTRACEAE STAFF-TREE FAMILY

Michael A. Vincent & Barry A. Prigge

Shrub (climbing or not), tree, thorny or not, gen glabrous. **LF**: simple, opposite or alternate, deciduous to persistent, subsessile or petioled; veins pinnate. **INFL**: cluster, cyme, raceme, panicle, or 1-fld, axillary or terminal, bracted. **FL**: gen bisexual, radial, small; hypanthium ± cup-shaped; sepals 4–5; petals (0)4–5, free; stamens 4–5, alternate petals, attached below or to rim of disk; ovary superior or ± embedded in disk, 2–5-chambered, placentas axile or basal, style gen 1, short, stigma ± head-like, 2–5-lobed. **FR**: capsule, winged achene, berry, drupe, or nutlet, often 1-chambered. **SEED**: gen 1 per chamber, arilled. 50 genera, 800 spp.: worldwide, esp se Asia; some orn (*Celastrus, Euonymus, Maytenus, Paxistima*). [Zhang & Simmons 2006 Syst Bot 31:122–137] Scientific Editor: Thomas J. Rosatti.

1. Lvs alternate, entire or toothed; infl terminal or axillary, 1–many-fld; petals green-white or white
  2. Lvs toothed; infl axillary, 1–5-fld; petals green-white. . . . . **MAYTENUS**
  2′ Lvs entire; infl terminal, many-fld; petals white . . . . **MORTONIA**
1′ Lvs opposite, ± toothed; infl axillary, 1–5-fld; petals red- or purple-brown
  3. Shrub, small tree, 2–6 m, deciduous; lf blade 30–140 mm; fl parts in 5s. . . . . **EUONYMUS**
  3′ Shrub, < 1 m, evergreen; lf blade 8–34 mm; fl parts in 4s . . . . **PAXISTIMA**

EUONYMUS BURNING BUSH

Shrub, small tree, erect. **ST**: twig gen 4-angled, with corky ridges. **LF**: opposite, deciduous, gen scalloped or finely toothed. **INFL**: axillary, few-fld; pedicel jointed to peduncle. **FL**: parts in 5s; petals purple-brown [or ± green]; disk fused to hypanthium, flat, ± 5-lobed; stamens short, attached to disk margin; ovary embedded in disk, bumpy or warty or not, style 0 or short, stigma lobes 3–5, obscure. **FR**: capsule, loculicidal, 3–5-valved. **SEED**: brown [white, red, or black], enclosed by [orange] red aril. 180 spp.: esp trop s hemisphere. (Greek: good name)

E. occidentalis Torr. WESTERN BURNING BUSH Pl 2–6 m. **ST**: branches slender, often climbing. **LF**: petiole 3–15 mm; blade 3–14 cm, ovate to obovate, thin, base truncate to tapered. **INFL**: 1–5-fld; peduncle 2–7 cm, slender; pedicel 5–15 mm. **FL**: sepals 1–1.5 mm, 1.5–2.5 mm wide; petals 4–6.5 mm, purple-brown, finely dotted, margin transparent; disk ± 3 mm wide. **FR**: depressed, deeply 3-lobed, smooth. **SEED**: 4–6 mm, ± brown; aril ± red.

1. Lf tip abruptly acuminate; twigs ± green;
  infl 1–5-fld. . . . . var. ***occidentalis***
1′ Lf tip ± obtuse or round; twigs ± white;
  infl 3–5-fld. . . . . .var. ***parishii***

  var. ***occidentalis*** (p. 635) Shaded streambanks, canyons; 20–1600 m. NW, CaRH, n SNH, CW; to WA. Apr–Jun

  var. ***parishii*** (Trel.) Jeps. Shaded canyons; 1300–2000 m. PR. Apr–Jun

MAYTENUS MAITEN

[Shrub] tree, erect. **LF**: alternate, persistent, spreading, leathery, toothed; base, tip acute. **INFL**: axillary clusters, 1–5-fld. **FL**: parts in 5s; hypanthium broad, flat; petals green-white [yellow]; disk fused to hypanthium, fleshy, ± green; ovary superior, ovoid, stigma lobes 2–4[5], stout, spreading. **FR**: capsule, loculicidal, 2[5]-valved, compressed-spheric, green to ± yellow. **SEED**: 1–2, brown; aril [orange to] red. 200 spp.: esp trop s hemisphere, S.Am, Afr. (Aboriginal name)

M. boaria Molina MAITEN TREE Pl to 15 m, branches pendent. **ST**: twigs green, in age gray-brown. **LF**: petiole 2–5 mm; blade 15–45 mm, ± lanceolate, flat. **FL**: unisexual or bisexual, hypanthium 1–1.2 mm; sepals 0.5–0.7 mm, tips obtuse; petals 1.5–3 mm, obtuse or ± acute. **FR**: 6–9 mm, glabrous. Scrub regrowth; 130–400 m. SnFrB (Alameda Co.); native to Argentina, Chile. Germinates profusely after fire; control efforts ongoing. Apr–May

MORTONIA

Shrub, erect, scabrous. **LF**: alternate, persistent, ascending, leathery, entire; margin gen thicker. **INFL**: panicle, terminal, many-fld. **FL**: parts in 5s; hypanthium obconic; petals white; disk fused to hypanthium exc at top, fleshy, ± white, in age red-purple; ovary superior, narrowly ovoid, stigma lobes 5, slender, spreading. **FR**: nutlet 1, oblong-cylindric, light brown. **SEED**: 1, straw-colored, difficult to separate from fr; aril 0. 5 spp.: sw US, Mex. (S.G. Morton, Am botanist, physician, 1799–1851)

M. utahensis (Trel.) A. Nelson (p. 635) UTAH MORTONIA Pl 3–12 dm, coarsely scabrous. **ST**: twigs cream-white, in age gray. **LF**: petiole ± 0–1 mm; blade 6–16 mm, ovate to round, abaxially concave transversely, convex longitudinally, base rounded to tapered, tip rounded to acute, mucronate or not. **INFL**: 8–65 mm, 6–23 mm wide. **FL**: hypanthium 1.5–2 mm; sepals 1–2.3 mm, keeled, tips often acute, keeled; petals 2.2–3 mm, ovate. **FR**: 5–7 mm, glabrous. Limestone slopes, canyon bottoms; 350–2100 m. n DMoj; to sw UT. Mar–May ★

PAXISTIMA

Shrub, prostrate to ascending. **ST**: twig 4-angled, with corky ridges. **LF**: opposite, evergreen, leathery, finely toothed. **INFL**: axillary, 1–3-fld. **FL**: parts in 4s; petals ± white to red-brown; disk fused to hypanthium, ± square, fleshy; stamens short, attached to disk margin; ovary embedded in disk, stigma lobes 2, obscure. **FR**: capsule, loculicidal, 2-valved. **SEED**: shiny, dark brown to black; aril surrounding base and 1 side of seed, white, thin, fringed. 2 spp.: N.Am. (Greek: thick stigma, possibly)

P. myrsinites (Pursh) Raf. (p. 635)   OREGON BOXWOOD   Pl 3–10 dm. **ST**: prostrate to spreading, ± stiff, densely branched. **LF**: petiole ± 1 mm; blade 8–34 mm, ovate to oblanceolate, base tapered, tip rounded to acute. **INFL**: peduncle 2–3 mm. **FL**: petals ± 1 mm, ovate. **FR**: 4–7 mm, obovoid. Shaded places; 180–2120 m. NW, CaR, n&c SN, n SnFrB (Marin Co.), MP; to BC, Rocky Mtns, n Mex. May–Jul

CHENOPODIACEAE   GOOSEFOOT FAMILY

Mihai Costea, family description, key to genera

Ann to shrub; hairs simple, stellate or glandular; pls gen scaly, mealy, or powdery from collapsed glands; gen monoecious. **ST**: occ fleshy. **LF**: blade simple, gen alternate, occ fleshy or reduced to scales, veins pinnate; stipules 0. **INFL**: raceme, spike, catkin-like, spheric heads, or fls 1; bracts 0–5, herbaceous, gen persistent or strongly modified in fr, wings, tubercles or spines present or 0. **FL**: bisexual or unisexual, small, green; calyx parts (1)3–5, or 0 in pistillate fls, free or fused basally, lf-like in texture, membranous or fleshy, deciduous or not, gen strongly modified in fr; corolla 0; stamens 1–5, opposite to calyx parts, filaments free, equal; anthers 4-chambered; ovary superior (1/2-inferior), chamber 1; ovule 1; styles, stigmas 1–4. **FR**: achene or utricle, gen with persistent calyx or bracts. **SEED**: 1, small, lenticular to spheric; seed coat smooth to finely dotted, warty, net-like, or prickly, margin occ winged. 100 genera, 1500 spp.: worldwide, esp deserts, saline or alkaline soils; some cult for food (*Beta vulgaris* subsp. *vulgaris*, beet, Swiss chard; *Spinacea oleraceae* L., spinach; *Chenopodium quinoa* Willd., quinoa); and some worldwide, naturalized ruderal or noxious agricultural weeds. *Nitrophila* treated in Amaranthaceae, *Sarcobatus* treated in Sarcobataceae. Scientific Editors: Douglas H. Goldman, Bruce G. Baldwin.

1. Sts jointed, ± fleshy or succulent; lvs scale-like
 2. Lvs alternate; shrub . . . . **ALLENROLFEA**
 2′ Lvs opposite; ann or subshrub
  3. Fls sunken in fleshy bracts of distal internode, adherent to each other and to bracts, forming a 3-parted cavity at fl-fall . . . . **SALICORNIA**
  3′ Fls sunken in axis, free, not forming a 3-parted cavity at fl-fall . . . . **ARTHROCNEMUM**
1′ Sts not jointed, gen not fleshy; lvs well developed
 4. Lvs or infl bracts spine- or bristle-tipped
  5. Lvs, infl bracts cylindric, abruptly bristle-tipped; calyx tip winged in fr . . . . **HALOGETON**
  5′ Lvs thread-like to ± cylindric, acute; bracts awl-like to ovate, gen spine-tipped; calyx in fr abaxially gen tubercled to winged . . . . **SALSOLA**
 4′ Lvs or infl bracts not spine- or bristle-tipped
  6. Pl with at least some stellate hairs, hairs gen golden-brown in age . . . . **KRASCHENINNIKOVIA**
  6′ Pl glabrous to puberulent, hairy, minutely scaly, or powdery, green or ± gray
   7. Lf blades gen fleshy, ± semicylindric
    8. Calyx in fr with horizontal, membranous wings or tubercles; lvs herbaceous . . . . **³KOCHIA**
    8′ Calyx in fr without wings or tubercles or occ horned or wing-margined; lvs fleshy . . . . **SUAEDA**
   7′ Lf blades weakly fleshy, flattened
    9. Pistillate fl calyx gen 0, gen enclosed in fr by 2 enlarged, free or fused bracts
     10. Pl gen with glands that collapse and become mealy or scaly; fr bracts variously shaped, gen thickened, tubercled and appendaged or not . . . . **ATRIPLEX**
     10′ Pl glabrous or with soft, branched hairs; fr bracts ± round, flattened and winged, appendages 0 . . . . . . **GRAYIA**
    9′ Pistillate, bisexual fls gen with calyx (sometimes 0 if pistillate in *Monolepis*), not enclosed by 2 bracts
     11. Calyx horizontally winged in fr
      12. Lf wavy-dentate; pl long-shaggy-hairy to tomentose, gen glabrous in age . . . . **CYCLOLOMA**
      12′ Lf entire; pl gen glabrous, occ ± hairy . . . . **³KOCHIA**
     11′ Calyx not horizontally winged in fr
      13. Calyx parts tubercled or with curved to hooked appendages
       14. Calyx parts densely hairy, with curved to hooked appendages . . . . **BASSIA**
       14′ Calyx parts glabrous to ciliate, tips tubercled . . . . **³KOCHIA**
      13′ Calyx parts not tubercled, appendages 0
       15. Ovary 1/2-inferior, embedded in receptacle; calyx thick, hard in fr; frs clustered . . . . **BETA**
       15′ Ovary superior; calyx thin in fr; frs not adherent
        16. Distal lvs cordate-clasping; style branches 3 . . . . **APHANISMA**
        16′ Distal lvs not clasping; style branches gen 2
         17. Calyx parts, stamens gen 5
          18. Pl glandular . . . . **DYSPHANIA**
          18′ Pl glabrous or powdery . . . . **CHENOPODIUM**

17′ Calyx parts (0)1–3, stamens 1–3
  19. Lvs gen linear to lance-linear or thread-like; calyx parts 1; stamens gen 3 . . . . **CORISPERMUM**
  19′ Lvs elliptic, ± oblong, triangular-lanceolate, oblanceolate or spoon-shaped; calyx parts 0
      or 1–3; stamens 1–2
    20. St many-branched throughout, ultimate branches thread-like; fls 1–3 per cluster . . . **MICROMONOLEPIS**
    20′ St simple or branched from base, not branched distally, ultimate branches not thread-like;
      fls 4–15+ per cluster. . . . . **MONOLEPIS**

ALLENROLFEA   IODINE BUSH

Margriet Wetherwax, Leila M. Shultz & Dieter H. Wilken

1 sp.: N.Am, S.Am. (R.A. Rolfe, English botanist, 1855–1921) [Shultz 2004 FNANM 4:321]

A. occidentalis (S. Watson) Kuntze (p. 635)   Shrub, 3–15 dm, glabrous; deciduous. **ST:** erect or decumbent, many-branched, woody proximally, distally fleshy in age; jointed; joints (2)3–5(10) mm, 1–4.5 mm wide. **LF:** clasping, 2–4 mm, 2–3 mm wide, scale-like, triangular, margins entire, tip acute. **INFL:** spike, 6–25 mm, cylindric; fls spirally arranged in 3s or 5s; bracts deciduous, fleshy, peltate. **FL:** bisexual, sessile; calyx 1–1.5 mm, 4–5-lobed, persistent, enclosing fr; stamens 1–2, exserted; stigmas 2(3). **FR:** utricle, ± 1 mm, ovoid. **SEED:** erect, red-brown. n=9. Flats, hummocks, in alkaline soils; < 1450 m. GV, e SnFrB, s SNE, D; to OR, ID, TX, Mex. Jun–Aug

APHANISMA

Margriet Wetherwax, Leila M. Shultz & Dieter H. Wilken

1 sp. (Greek: inconspicuous) [Shultz 2004 FNANM 4:261]

A. blitoides Moq. (p. 635)   APHANISMA   Ann, glabrous. **ST:** 1–6 dm, decumbent to erect, branched from base, fleshy. **LF:** alternate, gradually reduced distally on st; proximal sessile; distal cordate-clasping; blade 2–5 cm, elliptic to ovate, entire. **INFL:** axillary, sessile; fls 1–3(5). **FL:** bisexual; calyx parts 3(5), ± 1 mm, ± equal, concave; stamen 1; style branches 3. **FR:** 1.2–2 mm, depressed-spheric; equatorial margin ± thick, ring-like. **SEED:** 1–1.5 mm, round, horizontal, black. Coastal scrub, bluffs, saline sand; < 200 m. s CCo, SCo, ChI; Baja CA. Jun–Sep ★

ARTHROCNEMUM   PICKLEWEED

Peter W. Ball

Subshrub, glabrous. **ST:** gen many-branched, appearing jointed, green and fleshy when young, woody and not jointed in age. **LF:** opposite, sessile, base fused, decurrent, forming fleshy internode, tip obtuse to ± acute, soft. **INFL:** spike, terminal, cylindric, dense; bracts lf-like, fleshy; fls 3(5) per axil, sessile, sunken in axis, free, not forming a 3-parted cavity at fl-fall. **FL:** calyx fleshy, 3–4-lobed, deciduous in fr; stamens 1–2; stigmas 2–3. **FR:** wall free from seed, membranous. **SEED:** vertical, seed coat hard, dark brown to black, tubercled; storage tissue (perisperm) abundant. 2 spp.: Medit, CA, nw Mex. (Greek: jointed leg)

A. subterminale (Parish) Standl. (p. 635)   Pl 10–30 cm, clumps to 1 m diam. **ST:** spreading to erect. **INFL:** larger spikes 5–40 mm, 2–3 mm wide, fertile nodes < 20, non-fl nodes distally gen < 14. **FL:** anther 0.5–1 mm, dehiscing after exsertion. **SEED:** dark brown, 1–1.4 mm. Salt marshes, alkaline flats; < 800 m. GV, CCo, SnFrB, SCo, ChI, w DMoj, DSon; n Mex. [*Salicornia s.* Parish] Apr–Sep

ATRIPLEX   SALTBUSH, ORACH

Elizabeth H. Zacharias

Gen monoecious ann, to gen dioecious shrub, gen scaly. **LF:** gen alternate, distal ± reduced; blade entire to variously dentate or lobed. **INFL:** axillary or terminal. **STAMINATE INFL:** spheric cluster to spike-like or panicle; bracts 0. **PISTILLATE INFL:** cluster to spike- or panicle-like, occ 1; bracts 2 per fr, enlarged in age, free to variously fused, gen compressed, gen sessile, falling with fr (or not). **STAMINATE FL:** calyx lobes 3–5; stamens 3–5. **PISTILLATE FL:** calyx gen ± 0; stigmas 2. **SEED:** gen erect. ± 250 spp.: temp to subtrop worldwide. (Latin: name derived from Greek) [Welsh 2003 FNANM 4:322–381] Gen in alkaline or saline soils; some weedy; some accumulate selenium. Bract descriptions refer to 2 bracts surrounding fl, enlarging in fr. Australian *A. crassipes* J.M. Black possibly in SCo. In revised taxonomy, too late for full treatment here, *A. californica*, *A. joaquinana* moved to *Extriplex*, *A. covillei* to *Stutzia*, both new genera [Zacharias & Baldwin 2010 Syst Bot 35(4):839–857].

1. Shrub or subshrub; main st clearly woody
  2. Lf length gen > 5 × width, margins entire; fr bracts together 4-winged or tips long, tubercles in
     vertical rows; branch tips not spiny
    3. Fr bract tips long, tubercles in vertical rows; shrub or subshrub 1–6 dm . . . . ***A. gardneri*** var. ***falcata***
    3′ Fr bract together 4-winged from base to tip, tubercles 0; shrub 3–25 dm
      4. Twigs gen slender; shrub 10–25 dm; lvs 10–35 mm, 1–3 mm wide; fr bracts 4–8 mm, 4–6 mm wide
        . . . . ***A. canescens*** var. ***linearis***
      4′ Twigs gen not slender; shrub 3–20 dm; lvs 10–50 mm, 2–8 mm wide; fr bracts 4–25 mm,
        4–12 mm wide
        5. Fr bracts entire to dentate; widespread . . . . ***A. canescens*** var. ***canescens***

5′ Fr bracts irregularly to deeply sharp-dentate; s CA
  6. Pl gen 10–20 dm; fr bract 6–12 mm, deeply sharp-dentate. . . . . ***A. canescens*** var. ***laciniata***
  6′ Pl gen 3–15 dm; fr bract 4–8 mm, irregularly dentate . . . . ***A. canescens*** var. ***macilenta***
2′ Lf length < 5 × width, margins entire to dentate; fr bracts not 4-winged, tubercled or not;
  branch tips spiny or not
  7. Branch tips not spiny
    8. Lf margin sharply dentate; shrub 3–10 dm; fr bracts free — SNE, D . . . . ***A. hymenelytra***
    8′ Lf margin entire to wavy or irregular-sparsely dentate; shrub or subshrub ≤ 40 dm; fr bracts free to
      fused above middle
      9. Shrub or subshrub < 5 dm; branches ascending to erect; lf sessile — SCo, WTR; st prostrate . . . . ***A. glauca***
      9′ Shrub ≤ 40 dm; branches spreading to erect; lf sessile to petioled
        10. Pl < 15 dm; lf blade narrowly elliptic to ovate, oblong to hastate . . . . ***A. amnicola***
        10′ Pl 8–40 dm; lf blade ovate to deltate (± hastate)
          11. Fr bracts 4.5–9(12) mm, round to ± reniform, gen tubercled. . . . . ***A. nummularia***
          11′ Fr bracts 2–6 mm, ovate to round or ± spheric, smooth
            12. Twigs gen smooth, not striate . . . . [2]***A. lentiformis***
            12′ Twigs gen sharp-angled, striate . . . . [2]***A. torreyi*** var. ***torreyi***
  7′ Branch tips spiny — gen stiffly spreading
    13. Fr bracts 5–20(24) mm, margins wide, lf-like
      14. Fr bracts ± free, wide-elliptic to ± round, few-dentate proximally, entire distally. . . . . ***A. confertifolia***
      14′ Fr bracts fused ± to middle around fr, entire and spheric proximally, compressed and irregularly
        dentate distally . . . . ***A. spinifera***
    13′ Fr bracts 2–6 mm, margins not wide and lf-like
      15. Distal lvs gen sessile to clasping . . . . ***A. parryi***
      15′ Distal lvs gen ± sessile to short-petioled
        16. Lf blade oblong to narrowly oblanceolate, 2–4 mm wide . . . . ***A. polycarpa***
        16′ Lf blade ovate to deltate or ± hastate, 10–50 mm wide
          17. Twigs gen smooth, not striate . . . . [2]***A. lentiformis***
          17′ Twigs gen sharp-angled, striate . . . . [2]***A. torreyi*** var. ***torreyi***
1′ Per or ann; main st herbaceous to ± woody at base
  18. Per
    19. St gen erect; lvs entire
      20. Fr bracts 2–3 mm, obovate; coastal . . . . [2]***A. coulteri***
      20′ Fr bracts 3–5 mm, obovate to fan-shaped to ± spheric, inland. . . . . [3]***A. fruticulosa***
    19′ St gen prostrate to decumbent; lvs entire to dentate, wavy-dentate, or crenate
      21. Lf bases long-tapered to short-petioled, entire to wavy-dentate, dentate, or crenate
        22. Fr bracts thin; lvs gen serrate. . . . . [2]***A. suberecta***
        22′ Fr bracts spongy or fleshy; lvs entire to wavy-dentate
          23. Fr bract spongy . . . . [2]***A. lindleyi***
          23′ Fr bract fleshy, ± red. . . . . ***A. semibaccata***
      21′ Lvs gen sessile, entire
        24. Fr bracts ± free, entire . . . . ***A. californica***
        24′ Fr bracts fused proximally to distally, entire to dentate or crenate
          25. Fr bracts dentate distally; lvs lanceolate to elliptic
            26. Fr bracts 2–3 mm, obovate; coastal . . . . [2]***A. coulteri***
            26′ Fr bracts 3–5 mm, obovate to fan-shaped to ± spheric; inland . . . . [3]***A. fruticulosa***
          25′ Fr bracts entire to shallowly dentate throughout; lvs elliptic to ovate
            27. Fr bracts ± spongy, ± spheric; not mat-like . . . . ***A. leucophylla***
            27′ Fr bracts thin, not spongy, ovate to diamond-shaped; gen mat-like . . . . ***A. watsonii***
  18′ Ann
    28. Lvs gen green, glabrous to sparsely powdery or fine-scaly
      29. Pistillate fls of 2 kinds; fr bracts wide-elliptic to ± round, smooth, net-veined; seeds vertical between
        bracts, or horizontal in calyx . . . . ***A. hortensis***
      29′ Pistillate fls of 1 kind; fr bracts gen not wide-elliptic to ± round, smooth to tubercled or ribbed,
        not net-veined; seeds vertical between bracts
        30. Fr bracts fused proximally, dentate distally; staminate fls in clusters of elongated terminal
          spikes or panicles . . . . [2]***A. serenana*** var. ***serenana***
        30′ Fr bracts free to fused proximally, entire to dentate proximally; staminate, pistillate fls mixed
          or staminate fls in short spikes
          31. Fr bracts entire, ± round . . . . ***A. micrantha***
          31′ Fr bracts entire to variously dentate, not ± round
            32. Fr bracts entire, lobes basal . . . . ***A. covillei***
            32′ Fr bracts entire to few-dentate or dentate
               33. Fr bracts few-dentate to dentate
                 34. Lvs triangular-hastate; fr bract surfaces gen tubercled . . . . [2]***A. prostrata***

34′ Lvs linear to triangular-ovate; fr bracts smooth or 2-tubercled

   35. Brown seeds elliptic, wider than long; fr bracts smooth or 2-tubercled — lvs gen thickened, ± scaly . . . ***A. dioica***

   35′ Brown seeds round; fr bracts smooth . . . . ***A. patula***

33′ Fr bracts entire

   36. Seeds of 1 kind; fr bract ribbed — fr bracts 2–4 mm; lvs narrowly elliptic to deltate . . . . [2]***A. joaquinana***

   36′ Seeds of 2 kinds; fr bract not ribbed

      37. Fr bracts 4–12(25) mm, margins entire; lvs linear to lanceolate . . . . ***A. gmelinii*** var. ***gmelinii***

      37′ Fr bracts 3–7 mm, margins entire to finely dentate; lvs triangular-hastate . . . . [2]***A. prostrata***

28′ Lvs ± gray- or ± white-green, densely fine-scaly

  38. Fr bracts free to fused proximally — st gen striate; fr bracts ribbed; infl spike- or panicle-like,
     terminal, dense . . . . [2]***A. joaquinana***

  38′ Fr bracts fused ± proximally to near tip

   39. Fr bract widest proximally, gen ovate to deltate or diamond-shaped

    40. Lvs coarsely wavy-dentate, distal entire — fr bracts hard; lvs densely fine-scaly abaxially, ± green
      adaxially, red in age . . . . ***A. rosea***

    40′ Lvs gen all entire

     41. St 0–few-branched, ascending to erect, 1–8 dm; fr bracts 2.5–5 mm

      42. Base of proximal lf blades rounded; fr bracts 2.5–3.5 mm; s SnJV (Kern Lake bed) . . . . ***A. tularensis***

      42′ Base of proximal lf blades cordate; fr bracts 3–5 mm; GV . . . . ***A. cordulata***

       43. Fr bracts deltate-ovate, central tooth gen > others; GV . . . . var. ***cordulata***

       43′ Fr bracts deltate to fan-shaped, central tooth gen ≤ others; SnJV . . . . var. ***erecticaulis***

     41′ St many-branched, prostrate to erect, < 4 dm; fr bracts 1–4 mm

      44. Fr bracts 1–2 mm, entire, surfaces smooth; GB — sts < 1.5 mm thick . . . . ***A. pusilla***

      44′ Fr bracts 2–4 mm, entire to few-dentate, surfaces smooth or tubercled; GV, SCo, PR

       45. Staminate fl clusters terminal, pistillate fl clusters axillary . . . . [2]***A. persistens***

       45′ Staminate fl clusters axillary, or staminate and pistillate fls mixed, gen in axillary clusters

        46. St woolly near tips; lvs gen opposite; SCo, PR . . . . ***A. parishii***

        46′ St glabrous to densely scaly near tips; lvs alternate or opposite; GV, SnJV

         47. Lvs gen alternate; fr bracts smooth; SnJV . . . . ***A. minuscula***

         47′ Lvs gen opposite; fr bracts gen tubercled; GV

          48. Fr bracts gen tubercled on side facing st . . . . ***A. depressa***

          48′ Fr bracts gen tubercled on both surfaces — branches widely spreading . . . . ***A. subtilis***

   39′ Fr bracts gen widest at middle or distally, fan- to wedge-shaped or ± round

    49. Staminate fls in terminal clusters, pistillate fls in axillary clusters

     50. Pl prostrate to decumbent; fr bracts 1–1.5 mm . . . . ***A. pacifica***

     50′ Pl erect to decumbent (spreading); fr bracts ≥ 2 mm

      51. Pl < 3(5) dm; fr bracts 3–5 mm; lvs entire

       52. Lvs ovate to elliptic, 2–5 mm . . . . [2]***A. persistens***

       52′ Lvs narrowly lanceolate to elliptic, 5–12(24) mm . . . . [3]***A. fruticulosa***

      51′ Pl 3–10 dm; fr bracts 2–3.5 mm; lvs irregularly dentate (entire) . . . . ***A. serenana***

       53. Lf blades 5–20 mm; staminate infl spheric, short; s SCo . . . . var. ***davidsonii***

       53′ Lf blades 10–50 mm; staminate infl spike or panicle, elongate; widespread . . . . [2]var. ***serenana***

    49′ Staminate fls in axillary clusters, or staminate and pistillate fls mixed, gen in axillary clusters

     54. Fr bracts spongy, 5–12 mm, obconic to ± spheric — s SCo . . . . [2]***A. lindleyi***

     54′ Fr bracts not spongy, 2–6(8) mm, wedge-shaped to round, gen compressed, or if spheric, then
       densely tubercled

      55. Fr bracts round, crenate, ± dentate to cut . . . . ***A. elegans***

       56. Fr bracts deeply dentate to cut, teeth 0.5–1 mm wide . . . . var. ***elegans***

       56′ Fr bracts minutely crenate to finely dentate, teeth 0.3–0.5 mm wide . . . . var. ***fasciculata***

      55′ Fr bracts gen wedge-shaped to obovate or ± spheric, entire proximally, dentate distally

       57. Fr bract tip ± truncate to notched, 3–5-dentate — gen SNE, DMoj, incl edges . . . . ***A. truncata***

       57′ Fr bract tip acute to rounded in outline

        58. Lf margin coarsely serrate . . . . [2]***A. suberecta***

        58′ Lf margin entire, wavy or flat

         59. Lf blade base ± obtuse to tapered; fr bracts 2.5–6 mm, fused proximally . . . . ***A. coronata***

         60. Fr bracts 2.5–4 mm, ± 2.5–4 mm wide . . . . var. ***vallicola***

         60′ Fr bracts 4.5–6 mm, 3–5 mm wide

          61. St decumbent to ascending; fr bracts together ± compressed, tubercles gen few, gen <
            marginal teeth . . . . var. ***coronata***

          61′ St ± erect; fr bracts together ± spheric, tubercles dense, gen ± = marginal teeth . . . . var. ***notatior***

        59′ Lf blade base gen ± hastate to tapering; fr bracts 4–8 mm, fused to near tip . . . . ***A. argentea***

        62. St decumbent to erect, 15–30 cm; fr bracts sessile

         63. Lvs sessile or proximal petioled; sts, lvs scaly, inflated hairs not elongate . . . . var. ***hillmanii***

         63′ Lvs gen sessile; sts, lvs, fr bracts densely scaly, with elongate inflated hairs . . . . var. ***longitrichoma***

        62′ St erect, 30–80 cm; fr bracts short-stalked

         64. Distal lvs short-petioled; most proximal lvs alternate; GB, DMoj . . . . var. ***argentea***

         64′ Distal lvs sessile; most proximal lvs opposite; GV, CCo, e SnFrB, SW, D . . . . var. ***expansa***

A. amnicola Paul G. Wilson SWAMP SALTBUSH Shrub < 15 dm; branches many, spreading; monoecious. **LF:** petiole short; blade 10–30 mm, narrowly elliptic to ovate, entire to irregularly dentate, base tapered to lobed. **STAMINATE INFL:** clusters in terminal spikes. **PISTILLATE INFL:** axillary clusters, short terminal spikes; bracts in fr 4–6 mm, ovate to fan-shaped, fused proximally to distally, both sides convex, entire to irregularly dentate, smooth. Urban drainages, beaches; < 15 m. SCo; native to Australia. Feb–Oct ◆

A. argentea Nutt. SILVERSCALE Ann 1.5–8 dm, decumbent to erect, densely branched, herbage gray-scaly. **ST:** stout, angled, peeling. **LF:** blade 7–40 mm, elliptic to deltate, wavy-margined to entire, base gen ± hastate to tapering. **PISTILLATE INFL:** bracts in fr 4–8 mm, fused to near tip, deltate to ± spheric, base tapering, gen tubercled, margins green, dentate. **SEED:** 1.5–2 mm, brown.

var. *argentea* (p. 635) **ST:** 3–8 dm, erect; branches ascending. **LF:** short-petioled; blade 7–20 mm, elliptic to deltate. 2n=18. Saline soils; 900–1500 m. GB, DMoj; to c US, n Mex. Jun–Sep

var. *expansa* (S. Watson) S.L. Welsh & Reveal (p. 635) **ST:** 3–8 dm, erect; branches ascending. **LF:** gen curled toward st, blade 7–40 mm, lanceolate to ovate or deltate, entire to wavy-margined; proximal-most opposite, petioled; distal sessile. 2n=36. Saline soils; < 1500 m. GV, CCo, e SnFrB, SW, D; to TX, n Mex. [*A. a.* Nutt. var. *mohavensis* (M.E. Jones) S.L. Welsh] Jul–Nov

var. *hillmanii* M.E. Jones HILLMAN'S SILVERSCALE **ST:** 1.5–2(3) dm, decumbent to erect; branches spreading to ascending. **LF:** sessile or proximal petioled; blade 8–20(30) mm, ovate to elliptic or obovate, base round to tapered or ± hastate. **PISTILLATE INFL:** bracts in fr smooth, or flattened outgrowths few. 2n=18,36,54. Saline or clay valley bottoms; 875–1700 m. n SNH, e&s MP, n DMoj; se OR, w NV. Jun–Sep ★

var. *longitrichoma* (Stutz et al.) S.L. Welsh PAHRUMP ORACH **ST:** 1–3 dm, erect; branches decumbent to ascending, scaly. **LF:** gen sessile; blade 10–35 mm, elliptic to ovate, base tapered. 2n=36. Saline soils; 700–850 m. DMoj (exc DMtns); sw NV. Apr–Jun ★

A. californica Moq. CALIFORNIA ORACH Per, spreading to decumbent, < 3 dm, < 8 dm wide; taproot thick, ± fleshy; monoecious. **ST:** ± many from base; branches decumbent to ascending. **LF:** proximal opposite; blade 5–24 mm, lanceolate to elliptic, gray-scaly. **PISTILLATE INFL:** bracts in fr 3–4.5 mm, ± free, ovate to ± round, ± thin, entire, smooth. **SEED:** ± 2 mm, black. 2n=18. Sandy soils, coastal dunes, sea bluffs, scrub, salt marshes; < 200 m. s NCo, CCo, SCo, ChI; Baja CA. Apr–Oct

A. canescens (Pursh) Nutt. FOUR-WING SALTBUSH Shrub 3–25 dm, erect; branches many. **ST:** branches spreading to ascending. **LF:** blade linear to oblanceolate, densely white-scaly. **PISTILLATE INFL:** terminal; bracts in fr 4–25 mm, gen fused to near tip, ovoid to spheric, hard, wings 4, 3–6 mm wide, entire to wavy or deeply sharp-dentate. **SEED:** 1.5–2.5 mm. Vars. intergrade.

var. *canescens* (p. 635) Pl 8–20 dm. **LF:** blade 15–50 mm, linear to oblanceolate; **PISTILLATE INFL:** bracts in fr 6–25 mm, stalked, wings gen 3–6 mm wide, entire to dentate. 2n=36+. Clay to gravelly flats, slopes, scrub; < 2100 m. SNH (e slope), Teh, SCoRI, SCo, n TR, PR, GB, D. [*A. c.* subsp. *c.*] Jun–Aug

var. *laciniata* Parish CALEB SALTBUSH Gen 10–20 dm. **LF:** blade 10–25 mm, oblanceolate, base tapered, tip obtuse. **PISTILLATE INFL:** bracts in fr 6–12 mm, wings deeply sharp-dentate. Saline desert flats, alluvial fans; < 1500 m. WTR, D (exc DMtns); NV, Mex. Polyploid, probably of hybrid origin. Mar–Oct

var. *linearis* (S. Watson) Munz SLENDERLEAF SALTBUSH Pl 10–25 dm. **LF:** blade 10–35 mm, 1–3 mm wide, linear to linear-elliptic, tip acute. **PISTILLATE INFL:** bracts in fr ± sessile, 4–8 mm, tips free, exceeding wings, wings gen < 3.5 mm wide, margins dentate. 2n=18. Sandy soils, dunes, flats; < 800 m. PR/D, D; AZ; n Mex. [*A. c.* (Pursh) Nutt. subsp. *l.* (S. Watson) H.M. Hall & Clem.; *A. l.* S. Watson] May–Jul

var. *macilenta* Jeps. SALTON SALTBUSH Gen 3–15 dm. **LF:** blade 10–25 mm, oblanceolate to linear, base tapered, tip obtuse. **PISTILLATE INFL:** bracts in fr 4–8 mm, wings irregularly dentate.

Saline desert flats, alluvial fans; < 160 m. DSon. Polyploid, probably of hybrid origin. Mar–Oct

A. confertifolia (Torr. & Frém.) S. Watson (p. 635) SHADSCALE Shrub < 10 dm, gen rounded, ± erect, many-branched. **ST:** twigs gen spreading, stiff, spine-like in age. **LF:** ± sessile to short-petioled; blade 8–24 mm, elliptic to obovate or wide-ovate, firm, densely gray-scaly. **PISTILLATE INFL:** occ terminal; bracts in fr 5–12(24) mm, ± free, wide-elliptic to ± round, entire to few-dentate, smooth. **SEED:** 1.5–2 mm. 2n=18,36,54,72,90,108. Alkaline flats, gravelly slopes in scrub, pinyon/juniper woodland; < 2400 m. SNH (e slope), GB, SCoRI, DMoj; to OR, n-c US, n Mex. May hybridize with *A. canescens*. Apr–Jul

A. cordulata Jeps. HEARTSCALE Ann, erect, 1–5 dm. **ST:** 1–many from base, rigid; branches ascending to erect, gray-scaly, tips tomentose. **LF:** blade 6–15(20) mm, ovate, gray-scaly, proximal blade bases cordate, distal blade base rounded. **PISTILLATE INFL:** bracts in fr 3–5 mm, fused proximally, ovate to ± round, smooth to ± tubercled, margin thin, dentate distally. **SEED:** ± 1.5–2 mm, red-brown.

var. *cordulata* (p. 635) **INFL:** bracts in fr deltate-ovate, central tooth gen > others. 2n=36. Saline or alkaline soils; < 70 m. GV. Jun–Jul ★

var. *erecticaulis* (Stutz et al.) S.L. Welsh EARLIMART ORACH **INFL:** bracts in fr deltate to fan-shaped, central tooth gen ≤ others. 2n=54. Saline or alkaline soils; < 100 m. SnJV. Aug–Sep ★

A. coronata S. Watson Ann 1–3 dm. **ST:** 1–few from base; branches decumbent to erect, stiff, gen gray-scaly, in age glabrous, straw-colored. **LF:** blade 8–20 mm, elliptic to ovate, base tapered to ± obtuse; distal lvs gray-scaly. **PISTILLATE INFL:** bracts fused proximally, wide-deltate to ± round. **SEED:** 1–1.5 mm, dark brown.

var. *coronata* (p. 637) CROWNSCALE **ST:** branches decumbent to ascending. **PISTILLATE INFL:** bracts in fr 4.5–6 mm, 3–3.5 mm wide, ± compressed, tubercles gen few, gen < marginal teeth. 2n=36. Fine, alkaline soils; < 200 m. s ScV, SnJV, e SCoRI. Mar–Oct ★

var. *notatior* Jeps. (p. 637) SAN JACINTO VALLEY CROWNSCALE **ST:** branches ± erect. **PISTILLATE INFL:** bracts in fr 4.5–5 mm, 3–5 mm wide, ± spheric, tubercles dense, gen ± = marginal teeth. Alkaline flats; 400–500 m. e SCo (San Jacinto Valley, Riverside Co.). Apr–Aug ★

var. *vallicola* (Hoover) S.L. Welsh (p. 637) LOST HILLS CROWNSCALE **ST:** branches ascending to erect. **PISTILLATE INFL:** bracts in fr 2.5–4 mm, 2.5–4 mm wide, ± compressed to spheric, dentate, smooth to tubercled. **SEED:** 1–1.5 mm, dark brown. 2n=18. Dried ponds, alkaline soils; < 430 m. SnJV. [*A. v.* Hoover] Apr–Sep ★

A. coulteri (Moq.) D. Dietr. COULTER'S SALTBUSH Per < 5 dm; monoecious. **ST:** 1–several from base, decumbent to erect, branches gen many, spreading to ascending. **LF:** blade 7–20 mm, narrowly elliptic to lanceolate, gray-scaly esp abaxially. **STAMINATE INFL:** gen spike, terminal. **PISTILLATE INFL:** bracts in fr 2–3 mm, fused proximally, obovate, sharply dentate distally, smooth or few-tubercled. **SEED:** 1–1.5 mm, brown. 2n=18. Alkaline or clay soils, open sites, scrub, coastal bluff scrub; < 500 m. SCo, ChI; Baja CA. Pls from e SCo (Hemet) may be an undescribed taxon. Mar–Oct ★

A. covillei (Standl.) J.F. Macbr. (p. 637) COVILLE'S ORACH Ann 1–4.5 dm. **ST:** many-branched from base, ± striate, green, glabrous to sparsely fine-scaly. **LF:** blades 10–40 mm, lanceolate to deltate, fleshy, brittle in age, base tapered to hastate. **PISTILLATE INFL:** bracts in fr 5–20 mm, fused proximally, lanceolate to deltate, smooth or tubercled, base 2-lobed. **PISTILLATE FL:** calyx lobes(1)3(5), translucent. **SEED:** 1–1.5 mm, brown. 2n=18. Saline soils, flats; < 2100 m. SnJV, SNE, DMoj; NV. [*A. phyllostegia* (Torr. ex S. Watson) S. Watson sensu TJM (1993), in part] Apr–Aug

A. depressa Jeps. BRITTLESCALE Ann < 3 dm. **ST:** prostrate-decumbent to ascending, glabrous to scaly, gen brittle, ± red, peeling. **LF:** gen opposite, blade 2.5–8(10) mm, ovate to cordate, gen densely white-scaly, tip acute. **PISTILLATE INFL:** bracts in fr 2–3.5 mm, fused to near tip, diamond-shaped, entire to few-dentate, gen white-

scaly, gen tubercled on side facing st. **SEED**: ± 1–1.2 mm, ± red-brown. 2*n*=18. Alkaline or clay soils; < 320 m. GV. [*A. parishii* var. *d.* (Jeps.) S.L. Welsh] Pls from sw SnJV (Carrizo Plain) an undescribed taxon, pl green-yellow-scaly; lvs petioled, blades ovate; fr bracts ± 3–4 mm, smooth. Jun–Oct ★

A. dioica Raf. THICKLEAF ORACH Ann 5–15 dm. **ST**: branched at base, gen erect; branches ascending, ± red to straw-colored at base, green-striate distally. **LF**: petioles 10–30 mm; blade 20–80(190) mm, lanceolate to triangular-ovate, entire to dentate, green to ± red, sparsely fine-scaly, glabrous in age, base gen hastate. **PISTILLATE INFL**: bracts in fr 3–7 mm, fused at base, ovate to wide-triangular, spongy, few-dentate, smooth or tubercles 2, base truncate to obtuse. **SEED**: of 2 kinds; 1.5–2 mm wide, black, or 2–3 mm wide, brown. 2*n*=36,54. Moist, saline or alkaline soils; < 200 m. s ScV, SnJV, CCo, SnFrB, SCo; to e N.Am. [*A. subspicata* (Nutt.) Rydb.] Jul–Nov

A. elegans (Moq.) D. Dietr. WHEELSCALE Ann ± 1–5 dm. **ST**: decumbent to ascending, ± branched, fine-scaly, glabrous in age. **LF**: blade 3–25(30) mm, elliptic to oblanceolate or oblong, tapered to base, entire to irregularly dentate, densely white-scaly abaxially. **PISTILLATE INFL**: bracts in fr 2–4 mm, fused to near tip, round, compressed, dentate, smooth or 1–2-tubercled. **SEED**: 1–1.5 mm, brown. 2*n*=18. Vars. intergrade.

var. ***elegans*** (p. 637) **LF**: blade entire to irregularly dentate. **PISTILLATE INFL**: bract in fr deeply dentate, teeth 0.5–1 mm wide. ± saline creosote-bush scrub; < 800 m. e D; to TX, n Mex. Apr–May

var. ***fasciculata*** (S. Watson) M.E. Jones (p. 637) MECCA ORACH **LF**: blade gen entire. **PISTILLATE INFL**: bract in fr minutely crenate to finely dentate, teeth 0.3–0.5 mm wide. Saline or alkaline soils, dry lakes; < 880 m. s SNE, D; NV, AZ, n Mex. Mar–Jul

A. fruticulosa Jeps. (p. 637) BALLSCALE, LITTLE OAK ORACH Per, occ ± ann, ± decumbent to erect, < 3(5) dm; monoecious. **ST**: simple proximally; branches many distally, scaly, glabrous in age. **LF**: blade 5–12(24) mm, narrowly lanceolate to elliptic, densely gray-scaly. **STAMINATE INFL**: spike, terminal. **PISTILLATE INFL**: bracts in fr 3–5 mm, fused proximally to distally, wide-obovate to fan-shaped or ± spheric, ± hard, irregularly dentate distally, smooth to few-tubercled proximally. **SEED**: 1.5–1.75 mm, red- to dark-brown. 2*n*=18. Clay or alkaline soils, open sites, scrub; < 1000 m. NCoRI, Teh, GV, SnFrB, SCoRI, w DMoj. Apr–Nov

A. gardneri (Moq.) D. Dietr. var. *falcata* (M.E. Jones) S.L. Welsh (p. 637) FALCATE SALTBUSH Subshrub or shrub, decumbent to erect, 1–6 dm. **ST**: several to many from base, gen unbranched, gray-scaly. **LF**: blade 15–45 mm, oblong to narrowly oblanceolate, gray-green, densely scaly. **PISTILLATE INFL**: bracts ± sessile, in fr 4–8 mm, fused to near tip, fusiform to ovoid, smooth to few-tubercled, tip long-tapered, margin entire to dentate or tubercled. **SEED**: 1–2 mm. 2*n*=18,36. Open, gen alkaline soils, sagebrush scrub, chenopod scrub; 1200–1700 m. MP; to WA, MT, WY. May–Aug ★

A. glauca L. Subshrub or shrub < 5 dm. **ST**: prostrate; branches erect to ascending, scaly. **LF**: blade 5–20 mm, lance-oblong to ovate-round, entire to dentate, silvery gray-green. **INFL**: bracts in fr 4–5 mm, fused at base, ovate-deltate to diamond-shaped, entire to dentate, smooth to tubercled, base gen 2-lobed. **SEED**: 2 mm, light brown. 2*n*=18. Beaches, coastal bluffs, disturbed places, ann grassland; < 1000 m. SCo, WTR; native to Eur. Mar–Jul

A. gmelinii Bong. var. ***gmelinii*** (p. 637) GMELIN'S SALTBUSH Ann 0.5–7.5 dm. **ST**: 1+ from base, sprawling to erect, striate; branches ascending to erect. **LF**: ascending, petioled; blade 10–70 mm, linear to lanceolate, green, glabrous, base tapered to rounded or lobes ≥ 2. **PISTILLATE INFL**: bracts in fr 4–12(25) mm, free, diamond-shaped or ovate to lance-oblong, entire, smooth (tubercled). **SEED**: of 2 kinds; 1–2 mm, black, or 2–3 mm, brown. 2*n*=36,54. Salt marshes, beaches; < 50 m. NCo, CCo, SnFrB; to AK. [*A. patula* var. *obtusa* (Cham.) M. Peck] Jun–Aug

A. hortensis L. GARDEN ORACH Ann, erect, 5–18(25) dm. **ST**: gen 1, branches ascending to erect, striate, sparsely scaly. **LF**: petioled; gen alternate exc proximal opposite; blade 12–150 mm, ovate

to deltate, entire to wavy or dentate, green, glabrous to sparsely scaly; distal lvs abruptly reduced. **PISTILLATE INFL**: terminal; bracts in fr 0 or 2, free, 8–18 mm, wide-elliptic to ± round, smooth, clearly net-veined, entire. **PISTILLATE FL**: calyx lobes 5 if bracts 0, or 0 if bracts 2. **SEED**: of 2 kinds; gen horizontal, 1–2 mm, black if bracts 0; or vertical, 2–4.5 mm, brown if bracts 2. 2*n*=18. Open, disturbed places; < 1300 m. NCo, SnFrB, SCo, MP, expected elsewhere; to e N.Am; native to Eurasia. Jul–Nov

A. hymenelytra (Torr.) S. Watson (p. 637) DESERT-HOLLY Shrub 3–10 dm, rounded, silver-scaly. **ST**: simple proximally, erect; branches many, spreading to ascending. **LF**: petioled, blade 12–45 mm, wide-ovate to round, thick, irregular-sharply dentate. **PISTILLATE INFL**: occ terminal; bracts in fr sessile to short-stalked, free, 6–15(20) mm, round to ± reniform, entire to ± crenate. **SEED**: ± 2 mm. 2*n*=18. Slopes, washes, scrub; < 1500 m. SNE, D; to sw UT, AZ, n Mex. Jan–Apr

A. joaquinana A. Nelson (p. 637) SAN JOAQUIN SPEARSCALE Ann, erect, 1–10 dm. **ST**: gen striate; branches gen ascending, sparsely scaly, glabrous in age. **LF**: petioled, blade 10–70 mm, ovate to deltate, fine-scaly in youth, glabrous in age, gen irregularly wavy-dentate, base truncate to tapered; abruptly reduced distally on sts. **INFL**: spike- or panicle-like, terminal, dense; staminate calyx lobes 4; bracts in fr 2–4 mm, free to fused proximally, narrowly elliptic to deltate, entire, ribbed. **SEED**: ± 1–1.5 mm, dark brown to black. 2*n*=18. Alkaline soils; < 350(840) m. NCoRI, GV, CCo, SnFrB, SCoRI (e slope). [*Atriplex joaquiniana* A. Nelson, orth. var.] Apr–Sep ★

A. lentiformis (Torr.) S. Watson (p. 637) BIG SALTBUSH Shrub, erect, 8–40 dm, gen wider than tall; monoecious or dioecious. **ST**: branches many, spreading to ascending; twigs occ spine-like in age, densely fine-scaly, glabrous in age. **LF**: petioled, blade 12–60 mm, ovate to oblong-elliptic or deltate, entire to wavy, base truncate to ± hastate. **PISTILLATE INFL**: ± panicle-like, terminal; bracts in fr 2–6 mm, fused to middle, wide-ovate to round, entire to minutely crenate. **SEED**: ± 1.5 mm, dark brown. 2*n*=18. Alkaline or saline washes, dry lakes, scrub; < 1200 m. s SN, deltaic GV, SCoRI, SCo, s ChI, n WTR, PR, SNE, D; to sw UT, n Mex. If recognized taxonomically, coastal, ChI pls with large lvs, frs assignable to *A. lentiformis* subsp. *breweri* (S. Watson) H.M. Hall & Clem. Jul–Oct

A. leucophylla (Moq.) D. Dietr. (p. 637) BEACH SALTBUSH Per, prostrate to decumbent, < 3 dm, densely white-scaly; gen monoecious. **ST**: 3–10 dm, branches decumbent to erect. **LF**: blade 9–40 mm, elliptic to wide-ovate. **PISTILLATE INFL**: bracts in fr 3–6 mm, fused proximally, ± spheric, ± spongy, entire to shallowly dentate, low-tubercled. **SEED**: 2–3 mm, red-brown. 2*n*=36. Sandy soils, dunes; < 60 m. NCo, CCo, SnFrB, SCo, ChI; Baja CA. Apr–Oct

A. lindleyi Moq. Ann or short-lived per 1–4 dm. **ST**: branches decumbent to erect, brittle, sparsely white-scaly, glabrous in age. **LF**: sessile, distal short-petioled; blade 8–30 mm, oblanceolate to diamond-shaped, entire to coarsely wavy-dentate, green to finely white-scaly, base tapered. **PISTILLATE INFL**: bracts in fr 5–12 mm, fused to near tip, obconic to ± spheric, spongy. **SEED**: of 2 kinds; ± 1 mm, black; or 1.5 mm, red-brown. 2*n*=18. Open, disturbed places, fields; < 40 m. s SCo; native to Australia. Mar–May

A. micrantha Ledeb. RUSSIAN ATRIPLEX Ann, erect, 5–15 dm. **ST**: stiff, branches gen ascending, striate, green, sparsely scaly. **LF**: proximal opposite, distal alternate; blade 10–65 mm, triangular to lanceolate, green, entire to irregularly wavy-dentate, glabrous to sparsely fine-scaly, base hastate or tapering; distal lvs abruptly reduced. **INFL**: panicle-like; branches spike-like, terminal or axillary. **PISTILLATE INFL**: bracts in fr of 2 sizes, 2–2.5 mm or 5–7 mm; free, ± round, entire, smooth, not net-veined. **SEED**: of 2 kinds; 1–1.5 mm, spheric, black; or 2–3 mm, flattened, yellow-brown. 2*n*=36. Open, gen disturbed places; < 2000 m. KR, CaRF, GV, WTR, GB (exc W&I); to e US; native to Eurasia. [*A. heterosperma* Bunge] May–Oct

A. minuscula Standl. LESSER SALTSCALE Ann < 4 dm. **ST**: many from base, ascending to erect; branches spreading, brittle, ± red, peeling. **LF**: gen alternate; blade 4–10 mm, ovate to cordate, white-

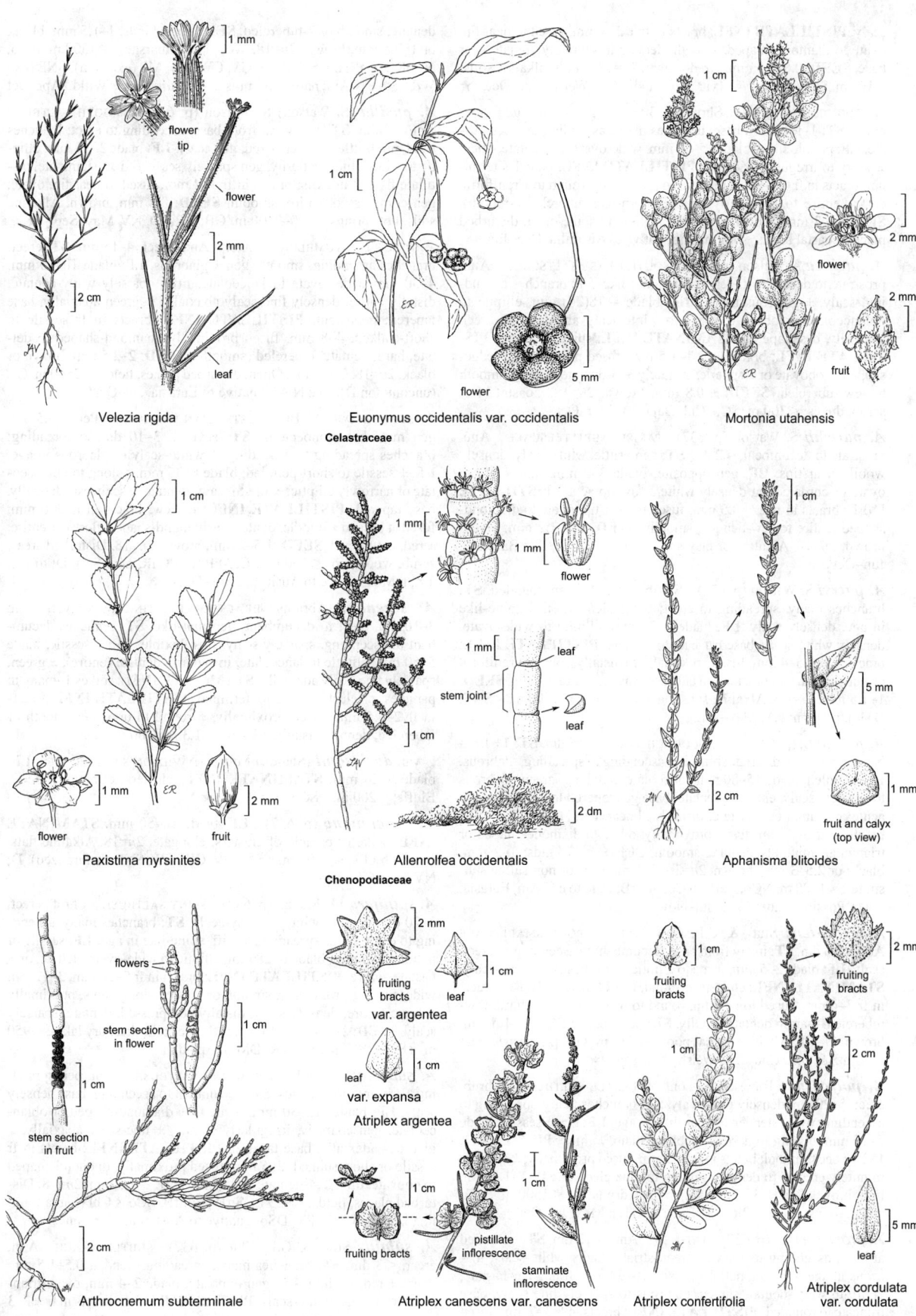

Velezia rigida

Euonymus occidentalis var. occidentalis
Celastraceae

flower tip

flower

leaf

1 mm

2 mm

2 cm

1 cm

flower

5 mm

Mortonia utahensis

1 cm

flower

2 mm

fruit

2 mm

Paxistima myrsinites

1 cm

flower

1 mm

fruit

2 mm

Allenrolfea occidentalis
Chenopodiaceae

flower

1 mm

1 mm

stem joint

leaf

leaf

1 mm

1 cm

2 dm

Aphanisma blitoides

1 cm

leaf

5 mm

fruit and calyx
(top view)

1 mm

2 cm

Arthrocnemum subterminale

flowers

stem section
in flower

1 cm

stem section
in fruit

1 cm

2 cm

fruiting
bracts

2 mm

leaf

1 cm

var. argentea

leaf

1 cm

var. expansa

Atriplex argentea

fruiting bracts

1 cm

pistillate
inflorescence

staminate
inflorescence

1 cm

Atriplex canescens var. canescens

fruiting
bracts

1 cm

1 cm

Atriplex confertifolia

fruiting
bracts

2 mm

2 cm

leaf

5 mm

Atriplex cordulata
var. cordulata

scaly. **PISTILLATE INFL**: bracts in fr ± 2–3 mm, fused to near tip, ovate to diamond-shaped, smooth, dentate to minutely crenate near base. **SEED**: 0.8–0.9 mm, ± red-brown. 2*n*=18. Sandy, alkaline soils; < 100 m. SnJV. [*A. parishii* var. *m.* (Standl.) S.L. Welsh] Apr–Oct ★

A. nummularia Lindl. Shrub 15–30+ dm, gray-scaly; gen dioecious. **ST**: 1–few from base; branches many, ascending to erect, striate. **LF**: petiole short, blade 15–65 mm, wide-ovate to ± deltate, thick, ± wavy to irregularly dentate. **PISTILLATE INFL**: panicles terminal; bracts in fr 4.5–9(12) mm, fused proximally, round to ± reniform, corky, entire to irregularly dentate, horn-like tubercled (smooth). **SEED**: ± 2 mm, brown. 2*n*=36,54,72. Sandy soils, open, disturbed places, coastal bluffs; < 40 m. SCo; native to Australia. Dec–Jun

A. pacifica A. Nelson (p. 637) SOUTH COAST SALTSCALE Ann, prostrate to decumbent, gen 1–4 dm, mat-like. **ST**: branches ascending, scaly, ± glabrous in age. **LF**: blade 4–18(25) mm, elliptic to oblanceolate or obovate, gray- to white-scaly abaxially, ± green adaxially, base tapered. **STAMINATE INFL**: spike, terminal. **PISTILLATE INFL**: bracts in fr 1–1.5 mm, fused proximally, wedge-shaped to obovate or ± spheric, minutely 3–5-dentate distally, smooth to few-tubercled. **SEED**: ± 0.8 mm, brown. 2*n*=18. Coastal bluff scrub, dunes; < 300 m. SCo, ChI; Baja CA. Mar–Oct ★

A. parishii S. Watson (p. 637) PARISH'S BRITTLESCALE Ann, prostrate to decumbent, < 2 dm. **ST**: gen brittle, white-scaly, densely woolly near tips. **LF**: gen opposite, blade 3–9 mm, lanceolate to ovate or cordate, gen densely white-scaly, tip acute. **PISTILLATE INFL**: bracts in fr 2.5–3.5 mm, fused to near tip, ovate to diamond-shaped, entire to few-dentate, smooth. **SEED**: ± 1–1.2 mm, ± red-brown. 2*n*=18. Alkaline or clay soils; < 470 m. SCo, PR; Baja CA. Jun–Oct ★

A. parryi S. Watson (p. 637) Shrub, erect, 2–5 dm, rounded. **ST**: branches many, spreading to erect; twigs slender, stiff, spine-like in age, densely scaly. **LF**: blade 5–20 mm, elliptic to wide-ovate, densely white-scaly, base truncate to cordate. **PISTILLATE INFL**: bracts in fr 2.5–4 mm, fused proximally to distally, round to reniform or fan-shaped, smooth, margin entire to wavy-dentate distally. **SEED**: 1–1.5 mm. 2*n*=18. Alkaline soils, flats, dry lakes; < 1500 m. s SNE, DMoj, n DSon; NV. May–Aug

A. patula L. (p. 637) SPEAR ORACH Ann 3–15 dm. **ST**: 1+ from base, erect, angled, striate; branches ascending to spreading, glabrous. **LF**: petioled, blade 15–80 mm, lanceolate to oblong, green, glabrous to sparsely scaly, entire to dentate, base gen tapered to ± rounded, tip acute; proximal occ hastate; distal lvs linear. **PISTILLATE INFL**: bracts in fr 3–8 mm, fused proximally, ovate to diamond-shaped or triangular, minutely dentate, smooth. **SEED**: of 2 kinds; 1–2 mm, black; or 2.5–3 mm, brown. 2*n*=36. Salt marshes or non-saline substrates; < 1400 m. NCo, CaRF, CCo, SnFrB; Can, to e N.Am, Eurasia, n Afr. Possibly naturalized. Jul–Nov

A. persistens Stutz & G.L. Chu VERNAL POOL SMALLSCALE Ann < 2.5 dm. **ST**: many from base, decumbent to ascending or erect, scaly. **LF**: blade 2–5 mm, ovate to elliptic, densely silvery gray-scaly. **STAMINATE INFL**: clusters terminal. **PISTILLATE INFL**: bracts in fr 3–4 mm, fused to near tip, round to ovate-oblong, ± hard, few-tubercled, margin dentate distally. **FR**: persistent. **SEED**: 1–1.5 mm, brown. 2*n*=18. Alkaline vernal pools; < 115 m. GV. [*A. parishii* var. *p.* (Stutz & G.L. Chu) S.L. Welsh] Jun–Sep ★

A. polycarpa (Torr.) S. Watson ALLSCALE SALTBUSH Shrub, erect, 5–20 dm, densely gray-scaly. **ST**: branches many, spreading to ascending; twigs slender, ± spine-like in age. **LF**: gen ± sessile, blade 3–25 mm, oblong to narrowly oblanceolate, ± thick. **PISTILLATE INFL**: occ terminal; bracts in fr 2–3 mm, fused proximally, ± spheric, minutely crenate to dentate, smooth to tubercled. **SEED**: 1–1.5 mm, light brown. 2*n*=18,36,72. Alkaline flats, dry lakes; < 1500 m. SnJV and margins, n TR, e PR, s SNE, D; to UT, n Mex. Jul–Oct

A. prostrata DC. (p. 637) FAT-HEN Ann 1–12 dm. **ST**: branched at base, ascending to ± erect, green-striate, finely white-scaly, glabrous in age. **LF**: gen petioled, blade 10–90 mm, triangular-hastate, entire to wavy-dentate, green, glabrous to sparsely fine-scaly, base truncate to rounded. **PISTILLATE INFL**: bracts in fr 3–7 mm, fused at base, ovate to deltate, base truncate to obtuse, entire to finely-

dentate, smooth or 2-tubercled. **SEED**: of 2 kinds, 1–1.5 mm, black, or 1–2.5 mm, brown. 2*n*=18. Wet places, marshes; < 1300 m. NCo, NCoRO, CaRH, n SNF, Teh, GV, CW, SW, MP (exc Wrn), SNE (exc W&I); to e N.Am; native to Eurasia. [*A. triangularis* Willd.] Apr–Oct

A. pusilla (S. Watson) S. Watson (p. 637) SMOOTH SALTBUSH Ann < 3 dm. **ST**: 1–several from base, ascending to erect; branches rigid, not brittle, slender, ± red, gen scaly. **LF**: blade 3–15 mm, elliptic to ovate, thick, ± fleshy, gen sparsely scaly. **INFL**: staminate, pistillate fls 1–2 in axils; bracts in fr 1–2 mm, fused to near tip, ovate, gen compressed, entire, smooth. **SEED**: 0.8 mm, brown. Alkaline soils, hot springs; 1300–2100 m. GB; OR, ID, NV. May–Sep ★

A. rosea L. TUMBLING ORACH Ann, erect, 4–15 dm. **ST**: erect; branches ascending, smooth, gen˙± glabrous. **LF**: blade 10–60 mm, 4–30 mm wide, ovate to lanceolate, firm, coarsely wavy-dentate, distally entire, densely fine-scaly abaxially, ± green adaxially, base tapered; persistent. **PISTILLATE INFL**: bracts in fr sessile to short-stalked, 4–8 mm, fused proximally, diamond-shaped to deltate, hard, dentate, tubercled (smooth). **SEED**: 2–2.5 mm, brown or black. 2*n*=18. Common. Open, disturbed places, fields; < 2500 m. CA (uncommon D); to e N.Am; native to Eurasia. Jul–Oct

A. semibaccata R. Br. AUSTRALIAN SALTBUSH Per, < 3.5 dm, gen mat-like; monoecious. **ST**: several, 3–10 dm, ± spreading; branches spreading to ascending, ± white-scaly or glabrous in age. **LF**: ± sessile to short-petioled; blade 8–35 mm, oblong to oblanceolate or narrowly elliptic, entire to wavy-dentate, ± scaly esp abaxially, base tapered. **PISTILLATE INFL**: fls few; bracts in fr 3–6 mm, fused at base to ± middle, ovate to ± diamond-shaped, fleshy, ± entire, ± red, net-veined. **SEED**: 1.5–2 mm, brown. 2*n*=18. Disturbed areas, scrub, woodland; < 1000 m. CA-FP (exc CaR, n&c SN), D; to UT, TX, n Mex; native to Australia. Apr–Dec ❖

A. serenana Abrams BRACTSCALE, STINKING ORACH Ann 3–10 dm, erect to decumbent, gen mat-like. **ST**: branches decumbent to ascending, sparsely scaly; tips flexible. **LF**: ± sessile, blade 5–50 mm, elliptic to lanceolate, irregularly dentate (entire), ± green, sparsely fine-scaly adaxially. **STAMINATE INFL**: spikes 1–many in panicles, or clusters spheric, terminal. **PISTILLATE INFL**: bracts in fr 2–3.5 mm, fused proximally, ± round to obovate, smooth or tubercled, dentate distally. **SEED**: 1–1.3 mm, brown. 2*n*=18.

var. *davidsonii* (Standl.) Munz DAVIDSON'S SALTSCALE **LF**: blade 5–20 mm. **STAMINATE INFL**: clusters terminal, spheric. Bluffs; < 200 m. s SCo. Apr–Oct ★

var. *serenana* (p. 637) **LF**: blade 10–50 mm. **STAMINATE INFL**: spike or panicle of clusters, elongate. 2*n*=18. Alkaline flats, coastal bluffs; < 2100 m. s SN, GV, CW, SW, SNE (naturalized), D; NV, Baja CA. Apr–Oct

A. spinifera J.F. Macbr. (p. 637) SPINY SALTBUSH Shrub, erect, 3–20 dm, densely white- to gray-scaly. **ST**: branches many, ascending to erect; twigs spreading, ± stiff, spine-like in age. **LF**: sessile or petioles 1–3 mm, blade 6–28 mm, elliptic to wide-ovate, deltate, or ± hastate, entire. **PISTILLATE INFL**: bracts in fr 5–20 mm, 3–11 mm wide, fused ± proximally, smooth or crested along veins, proximally spheric, entire, densely scaly, distally compressed, dentate, sparsely scaly. **SEED**: 2–3 mm, red-brown. Saline soils, flats, dry lakes; < 950 m. s SnJV, SCoRI, n WTR, DMoj. Apr–Jun

A. suberecta I. Verd. SPRAWLING SALTBUSH Ann, per, sprawling, 2–6 dm; monoecious. **ST**: decumbent to ascending, base densely scaly. **LF**: blades 12–30 mm, ovate to ± diamond-shaped or oblanceolate, gen coarsely, irregularly serrate, ± fine-scaly abaxially, ± glabrous adaxially, base tapered. **PISTILLATE INFL**: bracts in fr sessile or short-stalked, 2–5 mm, fused proximally, diamond-shaped to obovate, thin, 2–4-dentate distally. **SEED**: 1–1.5 mm. 2*n*=18. Disturbed places, fields; < 925 m. SnJV, SnFrB, SCo, s ChI (Santa Catalina Island), WTR, PR, DSon; native to Australia. Mar–Jun

A. subtilis Stutz & G.L. Chu (p. 637) SUBTLE ORACH Ann, erect, < 3 dm. **ST**: branches many, spreading, slender, 0.5–1.5 mm diam, ± red, peeling. **LF**: gen opposite, blade 2–4 mm, ovate-triangular to cordate, white-scaly. **PISTILLATE INFL**: bracts in fr ± 2–3 mm, fused to near tip, deltate, irregularly dentate, both surfaces gen

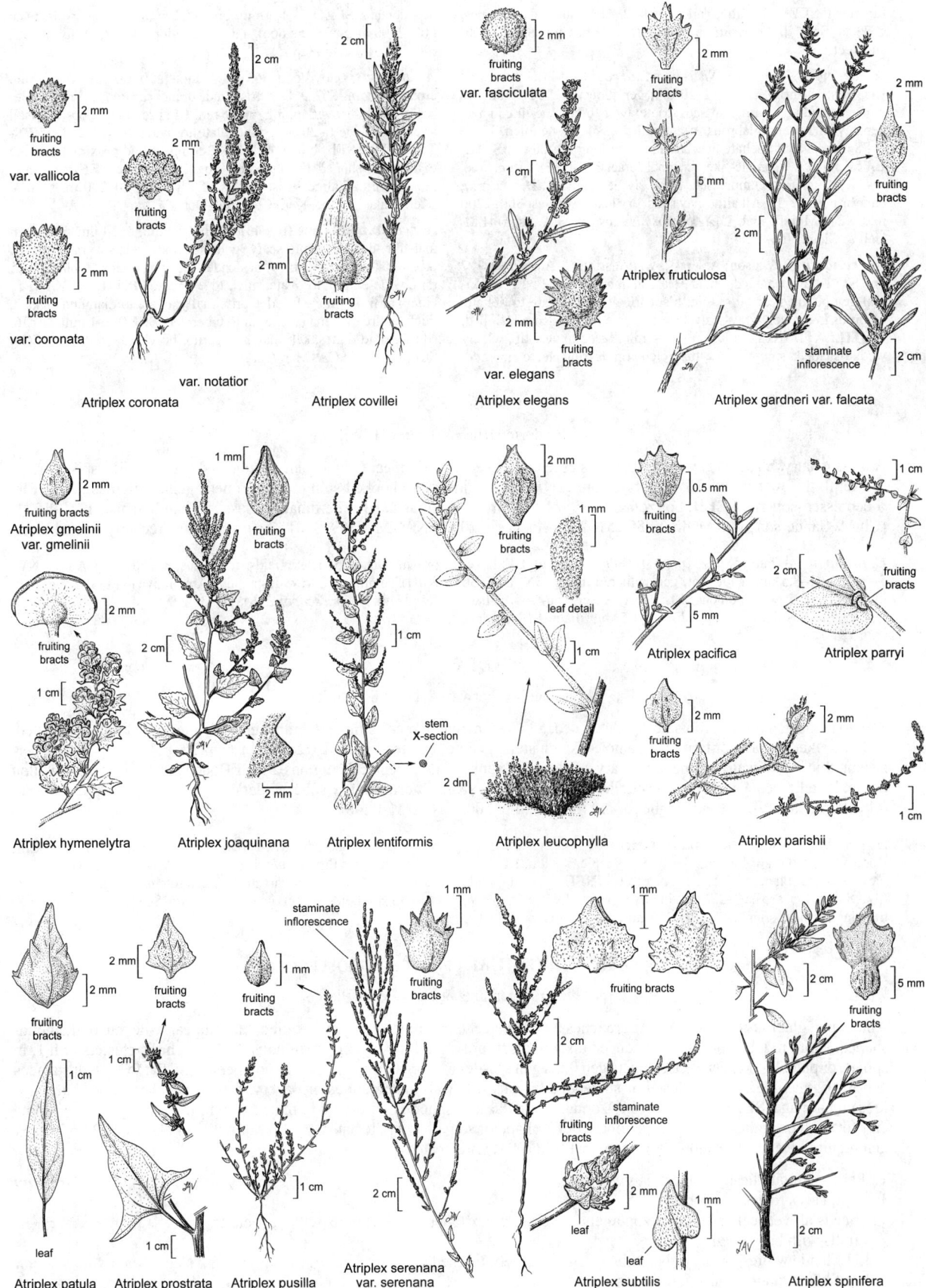

fruiting bracts

2 mm

var. vallicola

fruiting bracts

2 mm

var. coronata

fruiting bracts

2 mm

var. notatior

Atriplex coronata

2 cm

2 mm

fruiting bracts

2 mm

fruiting bracts

2 mm

Atriplex covillei

fruiting bracts

2 mm

var. fasciculata

1 cm

2 mm

fruiting bracts

var. elegans

Atriplex elegans

fruiting bracts

2 mm

5 mm

Atriplex fruticulosa

2 mm

fruiting bracts

2 cm

staminate inflorescence

2 cm

Atriplex gardneri var. falcata

fruiting bracts

2 mm

Atriplex gmelinii var. gmelinii

fruiting bracts

2 mm

1 cm

Atriplex hymenelytra

1 mm

fruiting bracts

2 cm

2 mm

Atriplex joaquinana

1 cm

stem X-section

2 dm

Atriplex lentiformis

2 mm

fruiting bracts

1 mm

leaf detail

1 mm

1 cm

Atriplex leucophylla

fruiting bracts

2 mm

0.5 mm

5 mm

Atriplex pacifica

fruiting bracts

2 mm

fruiting bracts

2 mm

1 cm

Atriplex parishii

1 cm

2 cm

fruiting bracts

Atriplex parryi

fruiting bracts

2 mm

2 mm

1 cm

1 cm

leaf

Atriplex patula

Atriplex prostrata

1 cm

Atriplex pusilla

staminate inflorescence

fruiting bracts

1 mm

1 mm

fruiting bracts

1 cm

2 cm

Atriplex serenana var. serenana

1 mm

fruiting bracts

2 cm

fruiting bracts

staminate inflorescence

leaf

2 mm

leaf

1 mm

Atriplex subtilis

2 cm

5 mm

fruiting bracts

2 cm

Atriplex spinifera

tubercled. **SEED:** 1.2 mm, dark-brown. $2n=18$. Saline depressions; < 70 m. SnJV. [*A. parishii* var. *s.* (Stutz & G.L. Chu) S.L. Welsh] Jun–Oct ★

A. torreyi (S. Watson) S. Watson var. ***torreyi*** (p. 643) TORREY'S SALTBUSH Shrub, erect, 12–20 dm; gen dioecious. **ST:** branches many, spreading, sharply angled, striate; twigs occ spine-like in age, densely fine-scaly, glabrous in age. **LF:** sessile to petioled, blade 7–38 mm, ovate to deltate or ± hastate, densely gray-scaly. **PISTILLATE INFL:** ± panicle-like, terminal; bracts in fr 2–4.5 mm, free, sessile, ± spheric, compressed, minutely dentate. **SEED:** 1.2 mm, dark brown. $2n=18$. Alkaline clay soils, dry lakes, washes; 300–2200 m. SNE, DMoj; to sw UT. [*A. lentiformis* subsp. *t.* (S. Watson) H.M. Hall & Clem.] Jun–Oct

A. truncata (S. Watson) A. Gray WEDGESCALE Ann, erect, < 7 dm. **ST:** 1–few, ± angled; branches ascending, gray-scaly. **LF:** proximal gen petioled, distal sessile; blade 10–40 mm, ovate to deltate, glabrous to sparsely gray-scaly, base truncate to rounded or ± lobed. **PISTILLATE INFL:** bracts in fr 2–4 mm, fused to near tip, widely wedge-shaped, smooth to ± tubercled, tip ± truncate to notched,

3–5-dentate. **SEED:** 1–1.5 mm, brown. $2n=18$. Alkaline soils, flats; 600–2500 m. SNH (e slope), TR (desert slopes), GB, DMoj; to BC, SK, CO, NM. Jun–Sep

A. tularensis Coville (p. 643) BAKERSFIELD SMALLSCALE Ann, erect, 1–8 dm. **ST:** 0–few-branched, branches rigid, brittle, white-scaly, distal sts overlapping, tips ± red. **LF:** proximal opposite, distal alternate; blade 6–20 mm, lanceolate to ovate, base rounded. **PISTILLATE INFL:** bracts in fr 2.5–3.5 mm, fused proximally, ovate to diamond-shaped, tip acute, few- to many-dentate. **SEED:** ± 1 mm, red-brown. Alkaline soils, shores of dry lake; 90–200 m. s SnJV (Kern Lake Bed, sw Kern Co.). Jun–Oct ★

A. watsonii Abrams (p. 643) Per < 10 dm, 5–30 dm diam, gen mat-like, densely white-scaly; gen dioecious. **ST:** several to many, prostrate to decumbent; twigs slender. **LF:** gen opposite, blade 8–25 mm, wide-elliptic to ovate, thick to ± fleshy. **PISTILLATE INFL:** bracts in fr 4–8 mm, fused ± proximally, ovate to diamond-shaped, thick, entire to crenate, smooth to tubercled. **SEED:** ± 1 mm. $2n=36$. Bluffs, sand dunes, salt marshes, scrub, beaches; < 170 m. s CCo, SCo, ChI; Baja CA. Mar–Oct

BASSIA

G. Frederic Hrusa & Dieter H. Wilken

Ann, gen hairy. **ST:** axis gen erect; branches ascending to erect. **LF:** linear to lanceolate, reduced distally. **INFL:** spike; bracts lf-like; fls 1–few per axil. **FL:** gen bisexual; calyx lobes 5, incurved, hooked-spiny in fr; stamens gen 5; stigmas gen 2. **FR:** ± depressed-spheric. **SEED:** horizontal. ± 10 spp.: warm temp Eurasia, Afr. (Ferdinando Bassi, Italian botanist, 1710–1774) [Chu & Sanderson 2008 Madroño 55:251–256; Mosyakin 2003 FNANM 4:309–310] *Kochia scoparia* recently treated here.

B. hyssopifolia (Pall.) Kuntze (p. 643) Pl gen < 1.5 m. **LF:** lower 5–60 mm, 1–3.5 mm wide, flat, gen withered in age. **INFL:** 5–50 mm; bracts 2–5 mm, ± oblong. **FL:** calyx densely tan-woolly, base leathery in fr, spines ± 1 mm. **FR:** 1–1.5 mm diam. **SEED:** dark brown. Disturbed sites, fields, roadsides; < 1200 m. CA (exc NW, SNH), common s CA; widespread N.Am; native to Eurasia. Occ confused with *Kochia scoparia*. May–Nov ❖

BETA BEET

Margriet Wetherwax & Leila M. Shultz

Ann to per, gen glabrous; roots fleshy, thickened. **ST:** decumbent to erect, simple to branched. **LF:** alternate, ± entire, petioled. **INFL:** spike, axillary, or terminal, panicle-like clusters, gen not bracted in distal 1/2. **FL:** bisexual; sepals 3–5, < 3 mm, persistent, thickened in age; stamens 5; ovary 1/2-inferior, sunken into receptacle, stigmas 2(3). **FR:** achene, enclosed by swollen perianth and receptacle, ± circumscissile, hard, clustered. **SEED:** horizontal, ± spheric, dark brown. x=9. ± 5 spp.: Eurasia. (Greek: probably from Celtic name for red root) [Shultz 2003 FNANM 4:266]

B. vulgaris L. subsp. ***maritima*** (L.) Arcang. (p. 643) SEA BEET Roots fibrous, occ swollen, not fleshy. **ST:** simple, < 8 dm. **LF:** petiole ± = blade; blade < 10 cm, oblanceolate. **INFL:** 1–3-fld, bracts (0)2–8 mm. **FL:** sepals 2–2.5 mm, incurved, abaxially keeled in age, margin scarious; stigmas 2. **FR:** 3–5 mm diam, 5–11 per cluster. Moist sandy places, disturbed areas; < 300 m. CCo, SnFrB, SCo, ChI, WTR, PR; NJ; s Eur. *B. vulgaris* subsp. *v.* [*B. macrocarpa* Guss.], cult beet, Swiss chard, occ waif in drainage ditches, irrigation channels close to cult fields, DSon; gen in e US. Feb–Sep

CHENOPODIUM PIGWEED, GOOSEFOOT

Steven E. Clemants & Nuri Benet-Pierce

Ann or per, glabrous or powdery. **ST:** branches 0 to gen erect (spreading). **LF:** gen petioled; blade linear to deltate or diamond-shaped, entire to lobed or toothed, reduced distally on st; proximal lvs gen early-deciduous. **INFL:** spheric clusters or fl 1, in spikes, or panicle-like, gen dense; bracts gen 0; fls gen sessile. **FL:** sepals gen 5, fused or not, persistent, flat to keeled; stamens gen 5; stigmas 2(5). **FR:** enclosed or subtended by calyx; fr wall membranous or papery, free or attached to seed and gen loosening in age. **SEED:** vertical or horizontal, lenticular to ± spheric, red-brown to black; wall thin. ± 100 spp.: temp; some cult for food or grain. (Greek: goose foot, from lf shape of some spp.) [Clemants & Mosyakin 2003 FNANM 4:275–299] Fr gen required for identification. Other spp. in TJM (1993) now treated in *Dysphania*.

1. Per from stout, fleshy caudex; calyx tube gen > lobes . . . . ***C. californicum***
1′ Ann; calyx tube gen < lobes
  2. Seeds all vertical or both horizontal and vertical; calyx lobes gen 3(4) in fls with vertical fr, 4–5 in fls with horizontal fr
    3. Lf blade white-powdery abaxially, margins acutely-toothed . . . . ***C. glaucum*** var. ***salinum***
    3′ Lf blade glabrous abaxially at least in age, margins entire to variously toothed
      4. Lf blade thick to fleshy, margin gen serrate, abaxially ± densely powdery and becoming glabrous
. . . . ***C. macrospermum***

4′ Lf blade thin, margin entire or toothed, gen glabrous

  5. Fl clusters 3–10 mm diam, in unbranched terminal and/or axillary spikes; sepals gen red, fleshy in fr; all seeds vertical

    6. Fl clusters subtended by lfy bracts throughout infl; stamens gen 1; fr maturing from base to tip of pl . . . ***C. foliosum***

    6′ Fl clusters subtended by lfy bracts only in proximal 1/2 of infl; stamens 3; fr maturing from tip to base of pl. . . . . ***C. capitatum***

      7. Fl clusters 6–10 mm diam; fr red, fleshy; lf blade base truncate to cordate (hastate), margin gen sharply toothed . . . . var. ***capitatum***

      7′ Fl clusters 3–5 mm diam; fr green to red, not fleshy; lf blade base tapered to truncate, margin entire or ± toothed . . . . var. ***parvicapitatum***

  5′ Fl clusters 2–4 mm diam, in axillary branched spikes; sepals green, membranous; seeds gen vertical, occ horizontal

    8. Calyx tube > lobes; stamen 1 . . . . ***C. chenopodioides***

    8′ Calyx tube < lobes; stamens 2–3 . . . . ***C. rubrum***

      9. St prostrate to spreading; lf margin entire or shallowly toothed. . . . . var. ***humile***

      9′ St erect to ascending; lf margin deeply toothed . . . . var. ***rubrum***

2′ Most or all seeds horizontal; calyx lobes gen 5

10. Fls 1 or in cymes, arranged in axillary and terminal panicles or branched spikes; lf blade glabrous to ± powdery abaxially . . . . ***C. simplex***

10′ Fl clusters in terminal and/or axillary spikes and/or panicles; lf blade gen powdery

  11. Primary lvs not fleshy, gen 2-lobed, gen toothed on margin

    12. Fr wall pitted at 20×

      13. Pl scented; proximal lf blades entire to 2-lobed or toothed near base, densely powdery; fr wall, seed coat ± white. . . . . ***C. watsonii***

      13′ Pl gen not scented; all lf blades entire, or 2-lobed near base and serrate or irregularly toothed, powdery; fr wall brown, seed coat black. . . . . ***C. berlandieri***

        14. Style base not yellow in fr; seed 1–1.3 mm diam. . . . . var. ***sinuatum***

        14′ Style base yellow in fr; seed 1.2–1.5 mm diam . . . . var. ***zschackei***

    12′ Fr wall not pitted at 20×

      15. Lf powdery in youth, esp abaxially, or gen glabrous; fr wall scale-like at 20×; seed margin sharply angled . . . . ***C. murale***

      15′ Lf powdery; fr wall smooth or papillate at 20×; seed margin rounded or ridged

        16. Lf blade sides ± parallel from near base, gen not lobed, gen finely serrate; fr wall smooth, gen attached to seed in age . . . . ***C. strictum*** var. ***glaucophyllum***

        16′ Lf blade sides not parallel, gen lobed or toothed; fr wall papillate or smooth, gen free from seed in age

          17. Pl to 1 m; infl branches straight; fl clusters gen in axillary and terminal branched spikes. . . . . ***C. album***

          17′ Pl to 2 m; infl branches ± curved or pendent; fl clusters in axillary and terminal spikes or panicles . . . . ***C. missouriense***

  11′ Primary lvs ± fleshy, entire or 1–2 pairs of lobes or teeth near base

    18. Lf blade 1-veined from base, entire

      19. Lf blade 4–6 mm wide, gen linear to narrowly lanceolate (elliptic), powdery abaxially, tip gen acuminate; infl dense . . . . [2]***C. desiccatum***

      19′ Lf blade gen < 3.5 mm wide, linear, densely powdery abaxially, tip ± round; infl open . . . . ***C. leptophyllum***

    18′ Lf blade 3-veined from base, entire or 1–2 pairs of lobes or teeth near base

      20. Lf blade length 3–many × width

        21. Fr wall attached to seed. . . . . ***C. hians***

        21′ Fr wall free from seed

          22. Pl erect to spreading; lf blade length 3–many × width, entire; calyx lobes gen enclosing fr in age; seed coat warty . . . . [2]***C. desiccatum***

          22′ Pl gen erect; lf blade length 3–5 × width, entire or gen 1–2 lobes near base; calyx lobes ± spreading in age; seed coat wrinkled. . . . . ***C. pratericola***

      20′ Lf blade length to 3 × width

        23. Lf blade length 2–3 × width, elliptic, lanceolate, or oblong

          24. Pl erect, branched; lf base gen rounded to tapered, tip gen obtuse to round, petiole gen strongly upcurved; fr wall smooth . . . . ***C. atrovirens***

          24′ Pl prostrate to ascending, branched from base; lf tip acute to round, petiole not upcurved; fr wall papillate . . . . ***C. littoreum***

        23′ Lf blade length gen < 2 × width, ± diamond-shaped to ovate, deltate, or elliptic

          25. Fr wall attached to seed

            26. Pl not ill-smelling, spheric in fr, many-branched distally, branches forked in pairs; lvs entire or gen 2-lobed, occ toothed; seed < 0.8 mm. . . . . ***C. nevadense***

            26′ Pl ill-smelling, not spheric in fr, few-branched from base, branches not forked in pairs; lvs entire; seed > 0.9 mm . . . . ***C. vulvaria***

25′ Fr wall free from seed

27. Infl clusters in spikes; lvs ± powdery abaxially; lobes near lf base spreading; fr visible between
calyx lobes in age; seed margin thick, square in ×-section .... *C. fremontii*

27′ Infl clusters in panicles; lvs moderately to densely powdery abaxially; lobes near lf base
occ spreading; fr hidden by calyx lobes in age; seed margin obtuse in ×-section
.... *C. incanum* var. *occidentale*

C. album L. (p. 643)   LAMB'S QUARTERS   Ann 18–100 cm, many-branched. **LF:** blade 15–70 mm, ovate to ± diamond-shaped, irregularly toothed (exc entire distally), dull-green adaxially, powdery abaxially. **INFL:** clusters or fl 1, in axillary and terminal branched spikes, branches straight. **FL:** sepals ovate, white-margined, gen keeled, powdery, gen enclosing fr, tip obtuse; stamens 5. **FR:** .1–1.5 mm diam; wall gen free, or free in age, gen papillate (smooth). **SEED:** horizontal, ± spheric or oval-lenticular, margin rounded; seed coat gen grooved, netted, occ smooth. 2*n*=54. Common. Disturbed areas, fields; < 1800 m. CA; ± temp worldwide; probably native to Eur. Widespread, polymorphic sp., probably with several distinct vars. Resembles *C. berlandieri, C. missouriense, C. strictum*. Jun–Oct

C. atrovirens Rydb. (p. 643)   Ann, erect, 7–60 cm, many-branched gen from base. **LF:** petiole gen strongly upcurved; blade 9–25 mm, length 2–3 × width, narrowly oblong or elliptic, gen prominently 3-veined, moderately to densely powdery, entire or occ 1–2 pairs of lobes near base, base gen rounded to tapered, tip gen obtuse to round, occ small-pointed at tip. **INFL:** clusters in axillary and terminal spikes and panicles. **FL:** sepals obovate, ± keeled, sparsely powdery, not enclosing fr in age, tip rounded. **FR:** 1–1.5 mm diam; fr wall smooth, attached to seed, loose in age. **SEED:** horizontal; seed coat wrinkled abaxially; margin thinly ridged. 2*n*=18. Open places, scrub, woodland, conifer forest; 300–3500 m. NW, CaR, SN, SnBr, PR, GB, n DMoj; to WY, c US, NM. ± like *C. pratericola*. Jul–Sep

C. berlandieri Moq. (p. 643)   PITSEED GOOSEFOOT   Ann 10–60 cm, gen not scented. **LF:** blade 15–30(40) mm, lanceolate to ± diamond-shaped, ovate, or ± deltate, gen with 2 lobes near base and serrate to irregularly toothed or entire, powdery. **INFL:** clusters, in branched spikes 5–17 cm. **FL:** sepals ovate to deltate, powdery, strongly keeled abaxially, enclosing fr in age. **FR:** 1–1.5 mm diam; wall attached to seed, brown, honeycomb-pitted at 20×. **SEED:** horizontal; seed coat black, honeycomb-pitted. 2*n*=36. Gen confused with *C. album*.

var. *sinuatum* (Murr) Wahl   **FR:** style base not yellow. **SEED:** 1–1.3 mm diam. Disturbed areas, fields, riverbanks; < 2000 m. CA-FP, D; OR, sw US. Jul–Sep

var. *zschackei* (Murr) Graebn.   **FR:** style base yellow. **SEED:** 1.2–1.5 mm diam. Disturbed areas, ocean bluffs, sandy washes; < 2200 m. CA-FP; to Can, w&c US. Jul–Sep

C. californicum (S. Watson) S. Watson (p. 643)   CALIFORNIA GOOSEFOOT   Per 20–90 cm; caudex stout, fleshy. **ST:** several from base, decumbent to ascending. **LF:** blade 40–100 mm, broadly deltate, coarsely dentate to wavy-toothed, base truncate to hastate or cordate, tip acute. **INFL:** clusters < 10 mm diam, in terminal, interrupted spikes 8–20 cm. **FL:** calyx tube gen > lobes, enclosing fr, lobes 4(5), oblong to elliptic, ± erect, flat, ± glabrous. **FR:** 1.5–2 mm diam; wall attached to seed. **SEED:** vertical. Gen open sites, sandy to clay soils; < 2000 m. s NCo, NCoRO, NCoRI, c&s SNF, Teh, GV, CW, SW, s SNE, w DMoj; Baja CA. Mar–Sep

C. capitatum (L.) Ambrosi   STRAWBERRY-BLITE   Ann 15–100 cm. **LF:** blade 25–100 mm, gen deltate, entire to sharply toothed, glabrous, base tapered to hastate, tip acute to acuminate. **INFL:** spheric clusters 3–10 mm diam, in unbranched terminal spikes 5–20 cm; lfy-bracted only in proximal 1/2 of infl. **FL:** sepals 3, subtending fr, smooth, glabrous, ± fleshy or membranous in fr; stamens 3. **FR:** ± 1 mm diam; wall strongly attached to seed, smooth, red to dark red or green; maturing from top to base of pl. **SEED:** vertical. 2*n*=18. Similar to *C. foliosum*.

var. *capitatum*   **LF:** blade base truncate to cordate (hastate), margin gen sharply toothed. **INFL:** spheric clusters 6–10 mm diam. **FR:** fleshy, red. Moist to dry, sandy or grassy meadows, thickets, open woodland; < 3000 m. c SNH, PR, W&I; to n Can, e US; native to Eur. Jun–Aug

var. *parvicapitatum* S.L. Welsh   **LF:** blade base tapered to truncate, margin entire or ± toothed. **INFL:** spheric clusters 3–5 mm diam. **FR:** membranous, not fleshy, green to red. Aspen groves, conifer forest, pinyon/juniper woodland, meadows, riverbanks; 1400–3000 m. c SNH, MP, W&I; to MT, CO, NM. Jun–Aug

C. chenopodioides (L.) Aellen (p. 643)   Ann 4–35 cm, prostrate to erect, densely branched, glabrous. **LF:** blade 8–35 mm, ± diamond-shaped to deltate, entire to irregularly few-toothed. **INFL:** clusters 3–4 mm diam, in axillary branched spikes. **FL:** calyx 3-lobed in vertically seeded fls, 4–5-lobed in horizontally seeded fls; calyx tube enclosing fr in vertical-seeded fls, lobes deltate, ± glabrous; stamen 1. **FR:** 0.5–1 mm diam; wall free from seed. **SEED:** gen vertical, ovoid. Saline soils, drying ponds, mudflats; < 2500 m. NCoRO, SN, SnBr, PR, GB; to WA; native to S.Am. Jul–Oct

C. desiccatum A. Nelson (p. 643)   Ann, erect to spreading, 12–40 cm. **ST:** gen several from base. **LF:** blade 15–25 mm, 4–6 mm wide, length 3–many × width, gen linear to narrowly lanceolate (elliptic), ± fleshy, (1)3-veined, entire, tip gen acuminate, densely powdery abaxially, less so adaxially. **INFL:** clusters densely packed in axillary and terminal panicles. **FL:** sepals obovate, keeled, densely powdery, gen enclosing fr in age, tip obtuse; stamens 5. **FR:** ± 1 mm diam; wall free from seed. **SEED:** horizontal, flat adaxially, black, warty, shiny. 2*n*=18. Uncommon. Open places, scrub, conifer forest; < 2900 m. SN, SnJV, TR, GB, n DMoj; to WY, MO, TX. Jul–Sep

C. foliosum (Moench) Asch. (p. 643)   Ann 5–60 cm. **LF:** blade 7–75 mm, lanceolate to narrowly triangular or deltate, base 2-lobed to hastate; lvs oblong to lanceolate, glabrous, gen deeply and irregularly toothed, base tapered. **INFL:** spheric clusters 3–8 mm diam, in unbranched terminal and axillary spikes 5–30 cm, lfy-bracted throughout. **FL:** sepals 3(4), lobes obovate, ± red in age, smooth, gen glabrous, enclosing fr only at base; stamen gen 1. **FR:** 1–1.5 mm diam; wall attached to seed; maturing from base to top of pl. **SEED:** vertical, round, dark red or brown, margin ridged. Open, gravelly or sandy soils, disturbed ground; < 1800 m. CaRH, c SNH, MP, w DMoj; to w Can, ne US; native to Eur. Jun–Aug

C. fremontii S. Watson (p. 643)   Ann 10–70 cm, gen branched from base. **LF:** blade thin; petiole to 2.5 cm; blade 0.5–2.5 cm, broadly triangular to ovate or elliptic, gen 1–2 lobed near base, lobes spreading ± powdery abaxially, glabrous adaxially, base ± truncate to round, tip obtuse to round. **INFL:** clusters in axillary and terminal spikes. **FL:** sepals ovate, keeled, ± powdery; fr visible in age. **FR:** 1–1.4 mm diam; wall free from seed. **SEED:** horizontal, margin ridged, thick, square in ×-section; seed coat black, ± warty, streaked. 2*n*=18. Gen shaded places, scrub, conifer forest; 700–3100 m. SN, TR, PR, GB, DMtns; to e N.Am, n Mex. ± like *C. incanum*. Jun–Oct

C. glaucum L. var. *salinum* (Standl.) B. Boivin   Ann 8–20 cm. **ST:** proximal gen prostrate; distal ascending. **LF:** blade 5–35 mm, lanceolate to oblong, glabrous to sparsely powdery adaxially, white-powdery abaxially, margins wavy-dentate, teeth acute; tip acute. **INFL:** axillary, ± 2 mm diam, ± spheric, lfy-bracted throughout. **FL:** sepals gen 3, abaxially rounded, glabrous, ± enclosing fr; stamen 1. **FR:** wall free from seed. **SEED:** horizontal, vertical, 0.9–1 mm. Open places, gen saline soils, drying ponds, streambanks; < 2200 m. CaR, SNH (e slope), SnBr, GB; to Can, c US. E US native *C. glaucum* var. *g.*, infl not lfy-bracted, lf tip obtuse, expected in CA. Jul–Oct

C. hians Standl. (p. 643)   Ann 30–80 cm, gen sparsely branched, scented. **LF:** blade 8–40 mm, 3–8(15) mm wide, length 3–many × width, narrowly lanceolate or elliptic-oblong, prominently 3-veined, entire or 2-lobed, ± densely powdery abaxially, glabrous to sparsely powdery adaxially; base tapered, tip acute or rounded. **INFL:** sparse; clusters widely spaced in axillary and terminal panicles, bracts lf-like. **FL:** sepals elliptic-oblong to narrowly ovate, ± keeled, powdery, not enclosing fr in age. **FR:** 1–1.5 mm diam; fr wall glandular-rough-

ened, attached to seed. **SEED**: horizontal, widely conic, flat adaxially, margin round. 2*n*=18. Open places, scrub, woodland; 300–2700 m. NW, CaR, SN, SCoRI, TR, PR, GB; to WY, NM. [*C. incognitum* Wahl] ± like *C. atrovirens, C. desiccatum, C. leptophyllum.* Jul–Aug

C. incanum (S. Watson) A. Heller var. ***occidentale*** D.J. Crawford (p. 643) Ann 8–60 cm, erect or spreading. **LF**: blade 6–18 mm, ± triangular-diamond-shaped to ovate, gen 2-lobed at base, moderately powdery to glabrous adaxially, powdery abaxially, base tapered, tip acute or rounded. **INFL**: widely spaced clusters in axillary and terminal panicles. **FL**: sepals ovate, strongly keeled, powdery, gen enclosing fr. **FR**: 1–1.2 mm diam; wall free from seed. **SEED**: horizontal, black, margin obtuse in ×-section. Open places, sandy or gravelly soils; 700–2300 m. SNH (e slope), GB, DMoj; to ID, sw UT. Other vars. in Rocky Mtns, c US; ± like *C. fremontii.* Apr–Aug

C. leptophyllum (Moq.) S. Watson Ann 10–40 cm, branched from base. **LF**: petiole 3–4 mm, gen < 1/4 blade; blade 8–25 mm, gen < 3.5 mm wide, linear, entire, densely powdery abaxially, less so adaxially, tip ± round. **INFL**: clusters widely spaced, in panicles. **FL**: sepals (4)5, lanceolate to elliptic, partially reflexed in age, densely powdery. **FR**: 0.5–0.7 mm diam; wall loosely attached to seed, white or ± yellow, papillate or smooth. **SEED**: horizontal, lenticular, margin rounded; seed coat black, finely netted adaxially. 2*n*=18. Open, gravelly soils, scrub; 300–3200 m. s NCoRH, c&s SN, W&I, n DMoj; to e US, n Mex. ± like *C. hians.* Jul–Sep

C. littoreum Benet-Pierce & M.G. Simpson Ann 5–40 cm, prostrate to ascending, mat-forming. **ST**: several from base. **LF**: blade 12–20 mm, length 2–3 × width, elliptic (lanceolate), fleshy, entire, ± glabrous adaxially, gen sparsely powdery abaxially, base tapered, tip acute to round, gen with small tooth. **INFL**: clusters in axillary and terminal spikes 5–15 cm. **FL**: sepals elliptic, ± flat, powdery, ± enclosing fr. **FR**: ± 1 mm diam; wall gen free from seed, papillate. **SEED**: horizontal, dark brown, finely wrinkled, margin round. Gen sandy soils, dunes; < 40 m. s CCo. [*Chenopodium carnosulum* Moq. var. *patagonicum* (Phil.) Wahl, misappl.] Jun–Oct ★

C. macrospermum Hook. f. Ann 8–60 cm. **ST**: proximal branches decumbent, distal ascending. **LF**: blade 8–60 mm, ± deltate, ± diamond-shaped, thick to fleshy, margin gen serrate, adaxially glabrous or sparsely powdery, powdery abaxially, glabrous in age, base rounded to tapered. **INFL**: ± spheric clusters 3–5 mm diam, in dense terminal and axillary spikes. **FL**: calyx 3-lobed in vertically seeded fls, 4–5-lobed in horizontally seeded fls; calyx enclosing fr, ± glabrous, tube > lobes, lobes deltate, lobe tips rounded to acute; stamen 1. **FR**: ± 1 mm diam; wall attached to seed. **SEED**: vertical, horizontal, ± red; seed coat netted. Wet places, marshes; < 100 m. NCo, deltaic GV, CCo, SnFrB, SCo; to WA; probably native to S.Am. [*C. m.* var. *halophilum* (Phil.) Standl.] Jul–Oct

C. missouriense Aellen Ann 40–200 cm. **LF**: petiole to 2.5 cm; blade 1.5–8(10) cm, ovate to ± triangular- or lanceolate-diamond-shaped, powdery, irregularly wavy-dentate to deeply toothed, occ lobed near base, base tapered. **INFL**: infl branches ± curved or pendent; clusters gen dense (or sparse if of late germination), in axillary and terminal spikes or panicles. **FL**: calyx lobes elliptic to ovate, not enclosing fr in age. **FR**: 1–1.5 mm diam; wall free or occ ± loosely attached to seed and free in age, thick, brown-black, papillate, shining. **SEED**: horizontal, ± spheric or oval, or a few vertical or oblique; seed coat black, occ ± red, netted, radially grooved or smooth, margin rounded. 2*n*=54. Open, gen disturbed areas; < 1000 m. n SNF, s SN, SnJV, SCo, PR; native to c&e US. Similar to *C. album.* Sep–Oct

C. murale L. Ann 15–50(80) cm. **LF**: 0.8–4 cm, deltate or ± diamond-shaped to ovate, toothed, dark green, powdery in youth, esp abaxially, or gen glabrous, shiny, base wedge-shaped. **INFL**: fls 1 or in clusters, in axillary and terminal panicles. **FL**: sepals ovate, keeled, red, powdery, tip acute to obtuse, ± enclosing fr. **FR**: 1–1.5 mm diam; wall attached to seed, scale-like. **SEED**: horizontal, 1–1.5 mm, not pitted, margin sharply angled. 2*n*=18. Common. Disturbed areas, fields; < 2900 m. CA-FP, D (uncommon); to Can, e US, n Mex; native to Eur. All yr

C. nevadense Standl. (p. 643) Ann 15–40 cm, many-branched distally, branches forked in pairs; spheric in fr. **LF**: blade 6–18 mm, ovate to ± diamond-shaped or elliptic, entire or gen 2-lobed, occ some teeth near base, lobes spreading; powdery abaxially, occ ± glabrous.

INFL: axillary and terminal, panicle-like. **FL**: calyx lobes ovate, abaxially ± keeled, tip obtuse ± enclosing fr. **FR**: wall minutely papillate, attached to seed. **SEED**: horizontal, < 0.8 mm, black. Washes, scrub; 1400–2000 m. c SNH (e slope), SNE; se OR, NV. Jun–Aug

C. pratericola Rydb. Ann, gen erect, 16–65 cm, simple or branched. **LF**: blade 1.5–4.2(6) cm, length 3–5 × width, elliptic to narrowly lanceolate, oblong-elliptic, or lanceolate, gen 1–2-lobed near base, 3-veined, thin, glabrous to sparsely powdery adaxially, densely powdery abaxially, tip acute or occ acuminate. **INFL**: clusters in terminal and axillary panicles. **FL**: calyx lobes (4)5, oblong-ovate, abaxially keeled, moderately to densely powdery, tip obtuse, ± spreading in fr. **FR**: 1–1.5 mm diam; wall free from seed. **SEED**: horizontal, lenticular, black, shiny, finely wrinkled. 2*n*=18. Open, dry places; < 2500 m. CaR, SN, GV, e SnFrB, SCoR, WTR, SNE, D; to e US. ± like *C. atrovirens.* Jun–Sep

C. rubrum L. RED PIGWEED Ann 10–50(70) cm. **LF**: blade 15–90 mm, deltate to ± diamond-shaped, entire to deeply toothed, glabrous, base gen tapered. **INFL**: clusters ± spheric, 3 mm wide, in axillary branched spikes. **FL**: sepals gen 3(4), fused only at base, lobes lanceolate to elliptic, flat (keeled) abaxially, gen glabrous to sparsely powdery; ± enclosing fr; stamens 2–3. **FR**: 0.5–1 mm diam; wall free from seed. **SEED**: vertical and occ horizontal, ± red. 2*n*=18.

var. ***humile*** (Hook.) S. Watson Ann, prostrate to spreading. **LF**: margin entire or shallowly toothed. **SEED**: 0.6–0.8 mm diam. Open, saline places, drying mudflats; <1100 m. NCoRO, NCoRI, CaR, Teh, GV, CCo, SnFrB, SCo, GB, DMoj; e US. Aug–Oct

var. *rubrum* Ann, erect to ascending. **LF**: margin deeply toothed. **SEED**: 0.8–1(1.2) mm diam. Open, saline places, drying mud flats; < 1100 m. NCoRO, NCoRI, CaR, Teh, GV, CCo, SnFrB, SCo, GB, DMoj; to Can, e US, Mex, Eurasia. Aug–Oct

C. simplex (Torr.) Raf. (p. 647) LARGE-SEEDED GOOSEFOOT Ann 35–150 cm, gen few-branched distally. **LF**: blade 25–150 mm, 10–80 mm wide, broadly ovate to deltate, wavy-lobed, dark green, lobes 3–5(7), glabrous to sparsely powdery, base truncate to ± cordate, tip long-tapered to acute or occ acuminate. **INFL**: fls 1 or in cymes, in terminal and axillary panicles. **FL**: sepals not enclosing fr, calyx lobes ovate to lanceolate, ± keeled abaxially, glabrous to sparsely powdery, tip acute to obtuse. **FR**: 1.5–3 mm diam, wall weakly attached to seed, smooth, netted. **SEED**: horizontal, 1–1.5 mm diam, lenticular, margin broadly ridged; seed coat netted, radially grooved, shiny. 2*n*=18. Disturbed or open places, scrub, conifer forest; 1400–2400 m. n SN, MP, W&I; to e US. Jul–Oct ★

C. strictum Roth var. *glaucophyllum* (Aellen) Wahl Ann 45–100(120) cm; proximal branches ascending. **LF**: 15–25(30) mm, lanceolate to lance-oblong, sides ± parallel from near base; gen finely serrate, base tapered to ± truncate, tip obtuse. **INFL**: clusters in simple and occ branched spikes. **FL**: sepals ovate, weakly keeled, white-margined, sparsely powdery, not enclosing fr at maturity. **FR**: wall gen attached to seed, thin, smooth. **SEED**: horizontal, ± 1.2 mm diam, lenticular, oval, margin ridged; seed coat black, smooth, shiny. 2*n*=36. Open, disturbed places; < 1700 m. CaRF, SN, GV, SnFrB, SCoRO, SCo, SnBr, DMtns (uncommon); to Can; native to e US. Aug–Oct

C. vulvaria L. Ann 5–40 cm, few-branched from base, odor unpleasant. **ST**: erect to ascending, powdery. **LF**: blade 3–20 mm, ovate to ± diamond-shaped, entire, glabrous to sparsely powdery adaxially, densely powdery abaxially, base obtuse, broadly tapered or rounded, tip gen acuminate. **INFL**: small clusters in axillary spikes or terminal panicles. **FL**: calyx tube ± enclosing fr, lobes abaxially ± flat, powdery. **FR**: ± 1 mm diam, wall attached to seed, minutely papillate. **SEED**: horizontal, > 0.9 mm, lenticular-oval, black, wrinkled. 2*n*=18. Uncommon. Open, disturbed areas; < 1400 m. CaRF, s SN, GV, CW, WTR; to e US; native to Eurasia. Jun–Oct

C. watsonii A. Nelson Ann 10–45 cm, scented. **ST**: erect or ascending, base branched. **LF**: petiole 1/2 to = blade; blade 10–26 mm, broadly rounded-ovate, rounded-diamond-shaped, or ovate, entire to 2-lobed near base, both surfaces powdery, base rounded or tapered to ± truncate, tip rounded to obtuse, acute, or small-toothed. **INFL**: lfy, clusters in panicles of spikes. **FL**: calyx tube short, sepals ovate, keeled abaxially, powdery, enclosing fr. **FR**: 1–1.3 mm diam, wall attached to seed, ± white, deeply pitted. **SEED**: horizontal. 2*n*=18. Woodland, scrub; < 3200 m. SnGb; to s Can, c US. Jul–Oct

CORISPERMUM BUGSEED

G. Frederic Hrusa

Ann, gen erect, branched, glabrous to sparsely long-hairy, gen glabrous in age. **ST**: branches 0–few, spreading to ascending. **LF**: gen linear. **INFL**: spike, terminal; bracts lf-like, reduced distally. **FL**: bisexual; perianth parts 0–5, scarious; stamens 1–3(5); ovary chamber 1, stigmas 2. **FR**: 1.6–5.2 mm, elliptic to obovate. **SEED**: vertical, wings 0 to narrow. ± 65 spp.: n temp. (Latin, Greek: leathery seed) [Mosyakin 2003 FNANM 4:313–321]

C. americanum (Nutt.) Nutt. var. *americanum* (p. 647) AMERICAN BUGSEED Pl well branched, 3–15 cm; glabrous. **LF**: 9–25 mm, 1–2.5 mm wide. **INFL**: 1–4 cm, narrow; bracts in fr 3–20 mm, gen covering fr, margin scarious. **FL**: perianth parts 1, stamens gen 3. **FR**: body 2–3.5 mm, ± obovate, ± yellow-green to brown, gen red-spotted or warty, convex abaxially, flat to ± convex adaxially; wing ± opaque, 0–0.2 mm, entire or ± cut, if ± 0 gen with remnant style-base < 0.1 mm beyond summit. Sandy soils, dunes; 900–1200 m. n DMoj (Eureka Valley); N.Am exc se, n Mex. [*C. hyssopifolium* L., misappl.] In CA apparently rare, seldom collected. Other var. in SW N.Am, Mex; CA material ± atypical, study needed. May ★

CYCLOLOMA WINGED PIGWEED

G. Frederic Hrusa & Dieter H. Wilken

1 sp. (Greek: circular wing, from calyx in fr) [Mosyakin 2003 FNANM 4:264–265]

C. atriplicifolium (Spreng.) J.M. Coult. (p. 647) Ann 12–75 cm, many-branched, rounded. **ST**: spreading, slender, striate, finely long-shaggy-hairy to tomentose, gen glabrous in age. **LF**: gradually reduced distally on st; petiole 0–12 mm; blade 5–65 mm, lanceolate to ovate, wavy-dentate; densely to sparsely long-shaggy-hairy, gen glabrous in age. **INFL**: panicle-like, terminal, open in fr; bracts 0; fls sessile. **FL**: bisexual or pistillate; calyx enclosing fr, 2–3 mm diam in fr, round-winged, lobes 5, ± keeled; stamens 5; ovary densely, finely tomentose, style deeply 2–3-lobed. **FR**: ± 2 mm diam. **SEED**: 1.5–2 mm diam, lens-shaped, long-shaggy-hairy, horizontal, dull black. 2*n*=36. Fields, disturbed areas, gen sandy; < 1250 m. GV, s SCo, w PR, DMoj; Can to n Mex; native to c N.Am. May–Sep

DYSPHANIA MEXICAN TEA, WORMWOOD

Steven E. Clemants & Nuri Benet-Pierce

Ann to per, glandular, ± strongly scented. **ST**: gen ± branched. **LF**: alternate, gen petioled; blade linear to ovate, entire to lobed, dentate or serrate, base gen tapered. **INFL**: spikes, panicles, or dense axillary spheric clusters; bracts lf-like, reduced, or 0. **FL**: gen sessile; calyx lobes 1–5, fused or not, flat to keeled, persistent; stamens 1–5; stigmas 1–3. **FR**: achene, ± 1 mm; fr wall free or attached to seed, thin, smooth to papillate, occ densely glandular. **SEED**: vertical or horizontal, red-brown to black. ± 32 spp.: temp; some cult for food, medicine. (Greek: obscure, apparently for inconspicuous fls) [Clemants & Mosyakin 2003 FNANM 4:267–275] Fr gen required for identification.

1. Pl ± sprawling; calyx lobes fused ± to tip, surface netted, tube enclosing fr. . . . . ***D. multifida***
1′ Pl gen erect; calyx deeply 5-lobed, ± spheric, surface not netted, lobes gen folded over fr
  2. Lf blade lobed, 3–65 mm; sepals glandular; fr wall not glandular, adherent to seed; seed margin gen ridged
    3. Lf blade ovate to elliptic, margin irregular-wavy to deeply pinnately lobed; infl spike or panicle-like, occ terminal; sepals densely stalked-glandular . . . . ***D. botrys***
    3′ Lf blade elliptic to lanceolate, margin coarsely wavy-dentate, lobes obtuse; infl ± spheric axillary clusters; sepals sessile-glandular. . . . . ***D. pumilio***
  2′ Lf blade coarsely dentate to jagged-cut or occ entire, 15–100 mm; sepals sparsely glandular; fr wall glandular, gen free from seed; seed margin rounded
    4. Infl bracts 0 or reduced, lf-like, < 2.5 mm . . . . ***D. anthelmintica***
    4′ Infl bracts lf-like, 3–25 mm
      5. Lvs, petioles gen densely glandular, st glabrous or gland-dotted . . . . ***D. ambrosioides***
      5′ Lvs, petioles, st glandular, soft-long-wavy-hairy . . . . ***D. chilensis***

D. ambrosioides (L.) Mosyakin & Clemants MEXICAN TEA Pl 25–130 cm. **ST**: glabrous or gland-dotted. **LF**: blade 15–100 mm, ovate to lance-oblong, entire to coarsely dentate, serrate, or jagged-cut, gen densely glandular. **INFL**: clusters, ± spheric, in terminal and axillary spikes; bracts 3–25 mm. **FL**: calyx deeply 5-lobed, lobes smooth, enclosing fr, sparsely glandular. **FR**: ± 0.5 mm diam; wall free from seed, glandular. **SEED**: vertical and horizontal, ovoid, ± red-brown, smooth. 2*n*=32. Disturbed places; < 1400 m. CA-FP; to Can, e US; native to trop Am. [*Chenopodium a.* L.] Jul–Sep

D. anthelmintica (L.) Mosyakin & Clemants WORMSEED Pl 37–75 cm. **LF**: blade 50–70 mm, lanceolate, dentate, with wide, straight teeth, gen densely glandular. **INFL**: spikes or panicles, axillary, terminal; bracts 0 or < 2.5 mm. **FL**: calyx deeply 5-lobed, lobes enclosing fr, glabrous. **FR**: ± 0.6 mm diam; wall free from seed, smooth, glandular. **SEED**: vertical and horizontal, black, bumpy. Disturbed areas, riverbanks; < 1100 m. ScV, SnFrB, SCoRO, SCo; to Can, e US; native to e US. Jun–Oct

D. botrys (L.) Mosyakin & Clemants (p. 647) JERUSALEM OAK Ann 14–65 cm. **ST**: glandular. **LF**: blade 3–65 mm, ovate to elliptic, wavy to pinnately lobed, lobes acute to obtuse, densely short-stalked-glandular abaxially. **INFL**: arching; clusters in axillary or terminal panicles. **FL**: short-pedicelled; calyx lobes distinct, weakly enclosing fr, ± flat, densely short-stalked-glandular. **FR**: ± 0.5 mm diam, margin gen acute; wall adherent to seed, white-blotchy, minutely papillate. **SEED**: horizontal, bumpy. 2*n*=18. Disturbed areas; < 2100 m. CA; to Can, e US; native to Eur, Asia. [*Chenopodium b.* L.] Jun–Oct

D. chilensis (Schrad.) Mosyakin & Clemants Pl 40–60 cm. **ST**: glandular, long-soft-wavy-hairy. **LF**: blade 20–90 mm, lanceolate, dentate with straight, widely spaced teeth, gen densely glandular, tip acuminate. **INFL**: spikes, axillary and terminal; bracts 10–18 mm. **FL**: calyx lobes ± = tube, tips long-soft-wavy-hairy and sparsely glandular in fl, in age enclosing fr and dry. **FR**: ± 0.8 mm diam; wall free from seed, glandular. **SEED**: gen vertical, ovoid. Disturbed areas, streambeds; < 500 m. NCoRO, n SNF, ScV, SnFrB; OR; native to s S.Am. Aug–Sep

Atriplex torreyi var. torreyi

1 cm

stem X-section

Atriplex watsonii

staminate inflorescence

1 cm

2 mm

fruiting bracts

Atriplex tularensis

1 dm

1 cm

5 mm

leaf

2 mm

fruiting bracts

Bassia hyssopifolia

2 cm

1 mm

fruit and calyx (side view)

Beta vulgaris subsp. maritima

5 cm

fruit and calyx

2 mm

5 mm

flower

2 mm

5 cm

2 mm

Chenopodium album

1 mm

top view   side view

fruit

1 mm

flower   fruit and calyx

2 cm

Chenopodium atrovirens

2 cm

2 cm

Chenopodium berlandieri

1 mm

fruit (top view)

1 mm

fruit and calyx

1 cm

Chenopodium californicum

2 cm

flower

1 mm

Chenopodium chenopodioides

0.5 mm

fruit and calyx

2 cm

Chenopodium desiccatum

2 cm

0.5 mm

fruit and calyx (top view)

Chenopodium foliosum

1 mm

fruit and calyx (side view)

2 cm

Chenopodium fremontii

1 cm

0.5 mm

fruit and calyx (top view)

Chenopodium hians

1 cm

1 cm

1 mm

fruit and calyx (top view)

Chenopodium incanum var. occidentale

1 cm

seed (top view)

0.5 mm

1 mm

0.5 mm

seed   seed

top view   side view

fruit

0.5 mm

fruit and calyx

Chenopodium nevadense

5 cm

2 cm

0.5 mm

fruit and calyx

D. multifida (L.) Mosyakin & Clemants (p. 647)   Pl 8–50 cm. **ST**: branches spreading to decumbent, glandular. **LF**: sessile; blade 3–45 mm, oblong to elliptic, dentate or deeply irregularly divided with narrow linear lobes, minutely glandular abaxially. **INFL**: clusters, < 5 mm diam, in axillary and terminal spikes, 10–25 mm. **FL**: calyx urn-shaped, lobes fused ± to tip, tube enclosing fr, papery, surface netted. **FR**: ovoid, ± 1 mm diam; wall loosely attached to seed. **SEED**: vertical, dark brown-black, warty, shiny. $2n=32$. Disturbed areas; < 1400 m. NW, c SNF, n ScV, CW, SCo, ChI (San Nicolas, Santa Rosa islands), s WTR; to e N.Am; native to S.Am. [*Chenopodium m.* L.] Jul–Oct

D. pumilio (R. Br.) Mosyakin & Clemants (p. 647)   Ann, prostrate to ascending, 12–45 cm. **ST**: glandular. **LF**: blade 4–25 mm, elliptic to lanceolate, coarsely wavy-dentate, lobes obtuse; base gen tapered, glandular abaxially. **INFL**: axillary, dense ± spheric clusters, 3–6 mm diam. **FL**: calyx lobes distinct ± to base, gen keeled, sessile-glandular, papery in age, weakly enclosing fr; stamens 0–1. **FR**: ± 0.5 mm diam; wall smooth, adherent to seed, margin acute or occ round. **SEED**: gen vertical. $2n=16,18$. Disturbed areas; < 3000 m. KR, NCoR, CaR, SN, GV, SnFrB, SCo, TR, w PR, MP; to e US, n Mex; native to Australia. [*Chenopodium p.* R. Br.] Jul–Aug

GRAYIA   HOP-SAGE

Margriet Wetherwax & Dieter H. Wilken

1 sp. (Asa Gray, eminent Am botanist, Harvard University, 1810–1888) [Holmgren 2003 FNANM 4:306–307]

G. spinosa (Hook.) Moq. (p. 647)   Shrub rounded; scaly-puberulent, hairs branched, glabrous in age; gen dioecious. **ST**: gen 3–10(15) dm, branches many, stiff; bark red-brown, ± white-ribbed, peeling in strips, older bark gray; twigs spine-like in age. **LF**: alternate, 5–25(40) mm, gen spoon-shaped to oblanceolate, flat, entire, tapered to short-petioled, blade green, tip gen ± white. **STAMINATE INFL**: spike-like, terminal, 7–18 mm; bract ± lf-like; fls 2–5 per cluster. **PISTILLATE INFL**: ± spike-like, axillary or terminal, 6–18 cm in fr; fls 1–few per cluster; bracts 3–10 mm, ± lf-like; fr bracts 2, 7–15 mm, fused, together sac-like, ± round, flat, winged, white to red-tinged, margins entire. **STAMINATE FL**: calyx lobes 4, 1.5–2 mm, enclosing stamens; stamens 4–5. **PISTILLATE FL**: stigmas 2, exserted. **FR**: gen 1.5–2 mm, brown. $2n=36$. Sandy to gravelly soils in scrub, pinyon/juniper woodland; 300–2900 m. SNH (e slope), Teh, se SnJV, WTR (n slope), GB, DMoj, nw DSon; to WA, MT, NM. Mar–Jun

HALOGETON

Margriet Wetherwax & Dieter H. Wilken

Ann, ± glabrous, glaucous, ± fleshy. **ST**: prostrate to erect, gen branched at base. **LF**: ± cylindric, fleshy, abruptly pointed, bristle- or spine-tipped, base clasping. **INFL**: axillary; fls densely clustered; bracts 0–2, lf-like. **FL**: bisexual or pistillate; perianth persistent, deeply 5-parted, gen enclosing fr, tip winged in fr; stamens 2–5; stigmas 2. **FR**: round, wall ± adherent to seed. **SEED**: vertical, brown to black. 5 spp.: Eurasia. (Greek: salty neighbor, from habitat) [Holmgren 2003 FNANM 4:403–404]

H. glomeratus (M. Bieb.) C.A. Mey. (p. 647)   SALTLOVER   **ST**: 6–40 cm, erect, lateral sts decumbent to spreading from base, lfy. **LF**: 4–17 mm, 1–1.5 mm wide, withered or deciduous in fr; bristle 1–2 mm, stiff. **INFL**: bracts 1.5–2 mm; fl clusters many. **FL**: perianth parts clawed, claw 2–3 mm, blade 2–4 mm, fan-like, membranous, veiny; stamens 3–5, filaments fused into 2 clusters of 2–3. **FR**: 0.5–2 mm. $2n=18$. Alkaline soils, open flats, scrub; 600–1800 m. CaR, GB, DMoj; to MT, NM, WA; native to Eurasia. TOXIC to livestock from concentrated oxalates. Jul–Aug   ◆

KOCHIA

G. Frederic Hrusa & Dieter H. Wilken

Ann to subshrub, gen erect, glabrous to tomentose. **LF**: alternate or lower ± opposite, ± thread-like to lanceolate, flat to cylindric, fleshy or not. **INFL**: spikes, simple or branched; bracts lf-like; fls 1–7 per axil. **FL**: bisexual or pistillate, sessile; calyx lobes 5, incurved, keeled; tubercled, winged or not in fr; stamens 5; stigmas 2–3. **FR**: ± compressed-spheric. **SEED**: horizontal. ± 15 spp.: w N.Am, Eurasia. (Wilhelm D. Koch, German physician, botanist, 1771–1849) [Chu & Sanderson 2008 Madroño 55:251–256] Native spp. recently treated in *Neokochia*, *K. scoparia* in *Bassia*.

1. Ann; lower cauline lvs narrowed to base or short-petioled, gen 3–5-veined below middle . . . . ***K. scoparia*** subsp. ***scoparia***
1′ Per or subshrub; lvs sessile, vein 1 or obscure
  2. Sts many from base, gen simple, glabrous to finely white-tomentose; lvs gen overlapping . . . . ***K. americana***
  2′ Sts 1–few from base, gen branched throughout, gray- to brown-puberulent; lvs not overlapping . . . . . . ***K. californica***

K. americana S. Watson (p. 647)   Subshrub 8–40 cm, root-sprouting. **ST**: many from base, gen simple, ascending to erect, gen finely white-tomentose, occ glabrous in age. **LF**: 5–20 mm, 1–2 mm wide, gen overlapping, ± cylindric to flat, ± fleshy, glabrous to spreading-long-hairy; vein obscure or 1. **INFL**: fls 1–3 per axil; bracts 4–15 mm, 0.5–1 mm wide, gen ascending, spreading in age. **FL**: calyx lobes gen white-tomentose, wings fan-shaped, in fr < 2 mm. **FR**: < 4 mm wide. Alkaline soils, flats, dry lake margins; 600–2200 m. GB, DMoj; to WA, MT, TX. May–Aug

K. californica S. Watson (p. 647)   Per or subshrub, 20–60 cm, root-sprouting. **ST**: 1–few from base, erect, gen branched throughout, densely gray- or brown-puberulent to long-spreading-hairy. **LF**: 3–12 mm, 1–3 mm wide, gen well spaced, gen flat, ± fleshy, ± appressed silky-hairy; vein obscure or 1. **INFL**: fls 1–2(5) per axil, bracts gen spreading. **FL**: calyx lobes densely short-hairy, wings in fr ± 1–2 mm. **FR**: < 3 mm wide. Alkaline soils, flats; < 1000 m. s SnJV, DMoj; s NV. May–Sep

K. scoparia (L.) Schrad. subsp. *scoparia*   Ann 20–120 cm. **ST**: simple to much-branched, glabrous to spreading-hairy. **LF**: 8–50 mm, 1–6 mm wide, flat, glabrous to appressed-hairy, gen 3–5-veined below middle. **INFL**: branched spike, short- to densely long-hairy, hairs < to > fls; fls 1–7 per axil; hairs gen hiding fls in immature infl. **FL**: calyx glabrous to thinly appressed-hairy, lobe margins gen bristly (glabrous); bisexual fls with tubercles or wings < 2 mm in fr. Disturbed places, fields, roadsides; < 2300 m. CaR, SN, GV, n SnFrB, SCo, SnBr, GB, D; to e US; native to Eurasia. Immature pls much like *Bassia hyssopifolia* (Pall.) Kuntze. If recognized taxonomically, a cult form with linear to thread-like lvs, short internodes, infl hairs < fls, rarely escaping in SnJV, SCo (expected elsewhere), assignable to *K. scoparia* subsp. *culta* (Voss) O. Bolòs & Vigo. Other subspp. in Eurasia. Aug–Nov   ❖

KRASCHENINNIKOVIA

Margriet Wetherwax & Dieter H. Wilken

Subshrub, gen erect, densely tomentose, hairs stellate; monoecious or dioecious. **LF**: petioled, linear to lanceolate, flat, entire. **INFL**: spike-like, terminal; staminate fls distal to pistillate; pistillate fls few, clustered, subtended by 2 ± lf-like, densely long-hairy bracts ± fused at base. **STAMINATE FL**: perianth 4-lobed; stamens 4. **PISTILLATE FL**: perianth lobes 0; stigmas 2. **FR**: ovate, flat, fr wall free. **SEED**: vertical, brown, white-hairy. ± 3 spp.: w N.Am, n Medit, temp Asia. (S. P. Krascheninnikov, Russian botanist, 1711–1755) [Holmgren 2003 FNANM 4:307–308]

K. lanata (Pursh) A. Meeuse & A. Smit (p. 647) WINTER FAT Gen 5–10 dm; hairs white, ± rust-colored in age. **LF**: 1–4 cm, 1.5–5 mm wide, margins gen inrolled; petiole 1.5–3.5 mm. **INFL**: 3–19 cm; staminate fls many; pistillate fls 1–4 in proximal axils. **STAMINATE FL**: bracts 0; perianth lobes 1–2 mm, densely hairy; stamens exserted. **PISTILLATE FL**: bracts 4–7.5 mm in fr, densely hairy; stigmas exserted. **FR**: 2.5–3.5 mm, white-hairy. $2n=18,36$. Rocky to clay soils, flats, gentle slopes; 100–2700 m. Teh, s SnJV, WTR (n slope), GB, DMoj; to WA, n-c US, NM, n Mex. May–Jul

MICROMONOLEPIS

Bruce G. Baldwin, adapted from Holmgren (2003)

1 sp.: w US. (Greek: small *Monolepis*) [Holmgren 2003 FNANM 4:301–302]

M. pusilla (S. Watson) Ulbr. (p. 647) DWARF MONOLEPIS Ann 4–14(20) cm, rounded, ± red, ± powdery, glabrous in age. **ST**: 1–5 from base, ± intricately branched throughout, ultimate branches thread-like. **LF**: 3–6 mm, elliptic to ± oblong, entire or hastate, tip obtuse to rounded. **INFL**: axillary or terminal; clusters 2–3-fld, or fl 1. **FL**: sepals 1–3, oblanceolate. **STAMINATE FL**: stamens 1–2. **PISTILLATE FL**: style 1, stigmas 2. **FR**: 0.5–1 mm, ± flattened; wall minutely tubercled, ± free from seed. **SEED**: vertical, lenticular, smooth, dark brown. Alkali flats; 1000–2100 m. n SNH (e slope), MP, n SNE; to WA, CO. [*Monolepis p.* S. Watson] May–Aug ★

MONOLEPIS POVERTY WEED

Bruce G. Baldwin, adapted from Holmgren (2003)

Ann, gen glabrous, gen powdery when young. **ST**: prostrate to ascending, simple or branched from base, not branched distally, ultimate branches not thread-like. **LF**: gen reduced distally on st, ± lanceolate to oblanceolate or spoon-shaped, entire or not, tip obtuse to rounded. **INFL**: clusters gen axillary, 4–15-fld; bracts lf-like. **FL**: bisexual or pistillate; sepals (0)1–3, ± green; stamens 0–1(2); stigmas 2, fused at base. **FR**: ± flattened; wall pitted to tubercled, free from seed or not. **SEED**: gen vertical, lenticular, smooth, brown to black. 5 spp.: temp w N.Am, S.Am, c&ne Asia. (Greek: 1 scale, from sepal number in most spp.) [Holmgren 2003 FNANM 4:300–301] *M. pusilla* moved to *Micromonolepis*.

1. Proximal lvs with ≥ 2 teeth near base, hastate or not, distal ± entire; fr 1.1–1.5 mm, wall minutely pitted, adherent to seed . . . . ***M. nuttalliana***
1′ All lvs entire; fr 0.5–0.7 mm, wall minutely papillate, free from seed . . . .*M. spathulata*

M. nuttalliana (Schult.) Greene (p. 647) NUTTALL'S POVERTY WEED Pl 5–50 cm, powdery, glabrous in age. **ST**: prostrate to ascending. **LF**: 10–30(40) mm, ± lanceolate, fleshy, proximal with ≥ 2 teeth near base, hastate or not, distal ± entire. **INFL**: fls 5–15+ per cluster. **FL**: sepal 1, 1 mm, oblanceolate to obovate, membranous; stamen 1. **FR**: 1.1–1.5 mm; wall minutely pitted, adherent to seed. $n=9$. Gen moist, ± alkaline clay soils in disturbed areas; < 3700 m. CA (exc NW); to AK, c N.Am, n Mex. Apr–Sep

M. spathulata A. Gray (p. 647) Pl 2–20 cm, glabrous to sparsely powdery. **ST**: decumbent to erect. **LF**: 3–25 mm, narrowly oblanceolate to spoon-shaped, fleshy, entire. **INFL**: fls 4–15+ per cluster. **FL**: outer pistillate, sepals 0; central bisexual, sepals 2–3, 0.5 mm, spoon-shaped to obovate; stamens 1–2. **FR**: 0.5–0.7 mm; wall minutely papillate, free from seed. $n=9$. Moist, ± alkaline streambanks and meadows; 2000–2700 m. SNH, SnBr, SNE; to OR, NV, Baja CA. Jun–Sep

SALICORNIA PICKLEWEED

Peter W. Ball

Ann or subshrub, glabrous. **ST**: gen many-branched, appearing jointed when young; internodes green to glaucous, fleshy when young. **LF**: opposite, sessile, decurrent; lf pairs fused at base, enclosing st. **INFL**: spike, terminal, cylindric, dense; bracts lf-like; fls gen 3 per axil, sessile, sunken in fleshy bracts of distal internode, adherent to each other and to bracts, forming a 3-parted cavity at fl-fall. **FL**: calyx fleshy, 3–4-lobed at tip, ± deciduous in fr; stamens 1–2; stigmas 2–3. **FR**: wall membranous, free from seed. **SEED**: vertical; seed coat membranous, pale brown, hairy [papillate]. ± 50 spp.: ± worldwide. (Greek: salt horn) [Kadereit at al. 2007 Taxon 56:1143–1170] Needs study. *S. subterminalis* moved to *Arthrocnemum*.

1. Subshrub, some sts vegetative only; central fls of a node separating lateral fls, fls in a horizontal row
  2. Woody sts spreading to erect; longest spikes with 12–40 fertile nodes . . . . ***S. pacifica***
  2′ Woody sts creeping; longest spikes with 7–14 fertile nodes . . . .*S. perennis*
1′ Ann, all sts terminated by infl; lateral fls of a node gen meeting below central fl, fls in a triangle.
  3. Lf, bract tips gen acute, sharp-mucronate; infl internode length < width . . . . *S. bigelovii*
  3′ Lf, bract tips obtuse to ± acute, not sharp; infl internode length ≥ width
    4. Infl internodes ± cylindric; central fl ± > lateral fls; anthers 0.3–0.5+ mm, exserted before dehiscence . . . *S. depressa*
    4′ Infl internodes widest distally; central fl >> lateral fls; anthers 0.2–0.4 mm, gen dehiscing inside fl, gen not exserted . . . .*S. rubra*

S. bigelovii Torr. (p. 647) Ann 9–60 cm, slender. **ST**: erect, simple or branching above middle; lf, bract tips acute, gen sharp-mucronate. **INFL**: 15–90 mm, 4–6 mm wide; fertile nodes 5–25, internodes 4–5 mm, 4.5–6.2 mm wide, width > length; lateral fls meeting below central fl. **FL**: anthers ± 0.6–0.7 mm, dehiscing after exsertion. **SEED**: 1–1.5 mm, hairs ± 0.1 mm, curved, tip hooked. Salt marshes; < 20 m. CCo (Morro Bay), SCo; to e US, Caribbean, Mex. Jul–Nov

S. depressa Standl. (p. 647) Ann 10–70 cm. **ST**: erect to prostrate, gen branched; lf, bract tips obtuse to ± acute, not sharp. **INFL**: 20–80 mm, 2.5–5 mm wide, fertile nodes 5–25; fertile internodes 3.5–7.5 mm, 2.5–5 mm wide, length ≥ width, ± cylindric; central fl ± > lateral fls, gen meeting below central fl. **FL**: anthers 0.3–0.5+ mm, dehiscing after exsertion. **SEED**: ± 1–2 mm, hairs ± 0.1 mm, hooked, curved. 2*n*=36. Salt marshes, alkaline flats; < 610 m. CCo, SCo, MP, w edge DMoj; to AK, e US, Baja CA. [*S. europaea* L., misappl.] Jul–Sep

S. pacifica Standl. (p. 647) Subshrub 10–70 cm. **ST**: spreading to erect, occ rooting at base; few- to many-branched. **INFL**: spikes 20–85 mm, 2.5–5 mm wide, longest with 12–40 fertile nodes. **FL**: central fls 1–2.5 mm wide, separating lateral fls; anthers 0.7–1 mm, dehiscing after exsertion. **SEED**: 1.2–1.5 mm, hairs 0.1–0.2 mm,

hooked, curved. Salt marshes, alkaline flats; < 800 m. NCo, SnJV, CCo, SnFrB, SCo, s ChI, PR, DMoj; Baja CA. [*Sarcocornia p.* (Standl.) A.J. Scott; *S. virginica* L., misappl.] Occ misidentified as *S. utahensis* Tidestr., not present in CA. Jul–Nov

S. perennis Mill. Subshrub. **ST**: woody st ± prostrate, matted to 1 m diam; fl sts erect, few-branched, 10–20 cm. **INFL**: spikes 10–25 mm, 2.9–4.4 mm wide, longest with 7–14 fertile nodes. **FL**: central fls 1.3–2.7 mm wide, separating lateral fls; anthers 0.8–1 mm wide, dehiscing after exsertion. **SEED**: 1.1–1.3 mm, hairy, longest hairs > 0.1 mm, hooked, curved. 2*n*=18. Salt marshes; < 20 m. NCo; to AK. [*Sarcocornia p.* (Mill.) A.J. Scott] Possibly a northern variant of *S. pacifica.* Occ misidentified as *S. virginica.* Aug–Oct

S. rubra A. Nelson Ann 5–25 cm. **ST**: ± erect, gen branched; lf, bract tips obtuse to ± acute, not sharp. **INFL**: 5–30(50) mm, 1.8–3.2 mm wide, fertile nodes 4–10(19); internodes widest distally, length ≥ width; lateral fls meeting below central fl, << central fl. **FL**: anthers 0.2–0.4 mm, gen dehiscing inside fl, gen not exserted. **SEED**: 1–2 mm, hairs ± 0.1 mm, hooked, curved. 2*n*=18. Salt marshes, alkaline soils; < 100 m. CCo; to AK, MI, NM, e Can. Occ misidentified as *S. europaea.* Aug–Oct

SALSOLA

G. Frederic Hrusa

Ann to shrub. **ST**: simple to many-branched. **LF**: gen reduced distally along st, thread-like to ± cylindric, spine-tipped, in age gen thick, rigid. **INFL**: axillary; bracts 1–2; fls gen 1 per axil. **FL**: bisexual; sepals 4–5, thickened in fr, persistent, gen tubercled to winged; stamens gen 5, exserted, style branches 2, exserted. **FR**: spheric to obovoid; tip ± depressed. **SEED**: horizontal. ± 100 spp.: ± worldwide. (Latin: salty, from habitats) [Mosyakin 2003 FNANM 4:398–403; Hrusa & Gaskin 2008 Madroño 55:113–131] An alternative treatment as separate genera *Kali* (*S. australis, S. gobicola, S. paulsenii, S. ryanii, S. tragus*), *Caroxylon* (*S. damascena*), and *Salsola* (*S. soda*) has been proposed (Akhani et al. 2007 Int J Plant Sci 168:931–956).

1. Shrub; lvs 3–9 mm, not reduced upward, oblong to ovate, gen puberulent . . . . ***S. damascena***
1′ Ann; lvs 5–55 mm, gen reduced upward, thread-like to lanceolate, glabrous to hairy or papillate.
  2. Pl strongly fleshy in fr; sepal wings in fr ± tubercled, thickened, rudimentary; lf and bract narrowed abruptly to short-pointed tip; st glabrous . . . . ***S. soda***
  2′ Pl not or ± fleshy in fr; sepal wings in fr thin, membranous, well developed on at least some frs; lf and bract tips narrowed gradually to sharp spine; st glabrous to bristly, papillate-hairy, or with mixed hair types
    3. St papillate-hairy or with mixed hair types
      4. Sepals 5, gen all winged, sepal tips in fr soft to spiny; st hairs mixed bristly- and papillate-hairy, or papillate-hairy only; bract wing 0.3–0.6 mm . . . . [2]***S. gobicola***
      4′ Sepals 5, 3 segments gen winged, 2 rudimentary to ± narrowly winged, sepal tips in fr hardened into sharp 3–5-parted spine; st hairs papillate; bract wing gen < 0.3 mm or 0 . . . . ***S. paulsenii***
    3′ St glabrous to bristly, bristles occ few or short
      5. Smallest 2 sepal wings fan-shaped to broadly obovate, length 1–2 × width; anthers ≤ 0.7 mm . . . . ***S. australis***
      5′ Smallest 2 sepal wings oblong, spoon-shaped, lanceolate, linear, or rudimentary, gen not ± fan-shaped, length > 3 × width; anthers ≥ 0.6 mm
        6. Bract wing narrow, 0.3–0.6 mm, opaque-white to ± translucent, base not surrounding fr; sepal wings ± translucent, smallest 2 narrowly oblanceolate to linear, or 1 (both) rudimentary . . . . [2]***S. gobicola***
        6′ Bract wing broad, gen > 0.4 mm, ± translucent, base surrounding fr; sepal wings opaque to translucent, smallest 2 narrowly winged
          7. Smallest 2 sepal wings narrower at base, broadened gradually or sharply to gen rounded tips; st hairs sparse, short; anthers 0.6–0.9 mm . . . . ***S. ryanii***
          7′ Smallest 2 sepal wings with ± parallel sides, tip unevenly lobed, or wing linear and tip ± truncate; st glabrous to bristly; anthers 0.6–1.3 mm . . . . ***S. tragus***

S. australis R. Br. (p. 653) Ann < 2 m, branched, not readily breaking at base, glabrous to ± minutely scabrous (short-bristly). **ST**: brittle, blue-glaucous to green, occ red-striped. **LF**: opposite or alternate below, alternate above, 8–52 mm, blade ± deciduous, base broader in age, margin broad, translucent, tip sharp-pointed to spiny. **INFL**: open to dense, prickly, gen not rigid; bract not surrounding fr, subcylindric, weakly spiny, narrowly wing-margined in age, lower margin ± translucent. **FL**: sepals 2.5–3 mm, lobes soft in fr; anthers 0.5–0.7 mm. **FR**: deciduous in age; 4.8–7.9 mm diam incl wings; developed wings 5, opaque, veins few, gen dark, margin gen smooth, smallest wings fan-shaped to broadly obovate. 2*n*=18. Disturbed places, road banks, open slopes, railroad tracks, shorelines; < 700 m. GV, CW,

SCo, D (rare DMoj); sw N.Am, Mex, Afr; possibly native to Australia. [*S. kali* L. subsp. *pontica* (Pall.) Mosyakin, misappl.; *S. tragus,* misappl., in part] Mar–Jan ◆

S. damascena Botsch. Shrub 60–100+ cm, gen puberulent. **ST**: branched ± throughout, branches ascending to erect, not ribbed longitudinally, not red-striped. **LF**: alternate, 3–9 mm, lower oblong, upper ovate, gen puberulent, tip obtuse to acute. **INFL**: bracts ovate, lf-like, narrowly wing-margined. **FL**: sepals 2–3 mm, wing ± 2 mm. **FR**: 7–11 mm diam incl wings; developed wings 5, opaque, veins numerous. Clay soils, flats; ± 1000 m. se SCoRI (Temblor Range); native to Medit, sw Asia. [*S. vermiculata* L., in part] Jun–Oct ◆

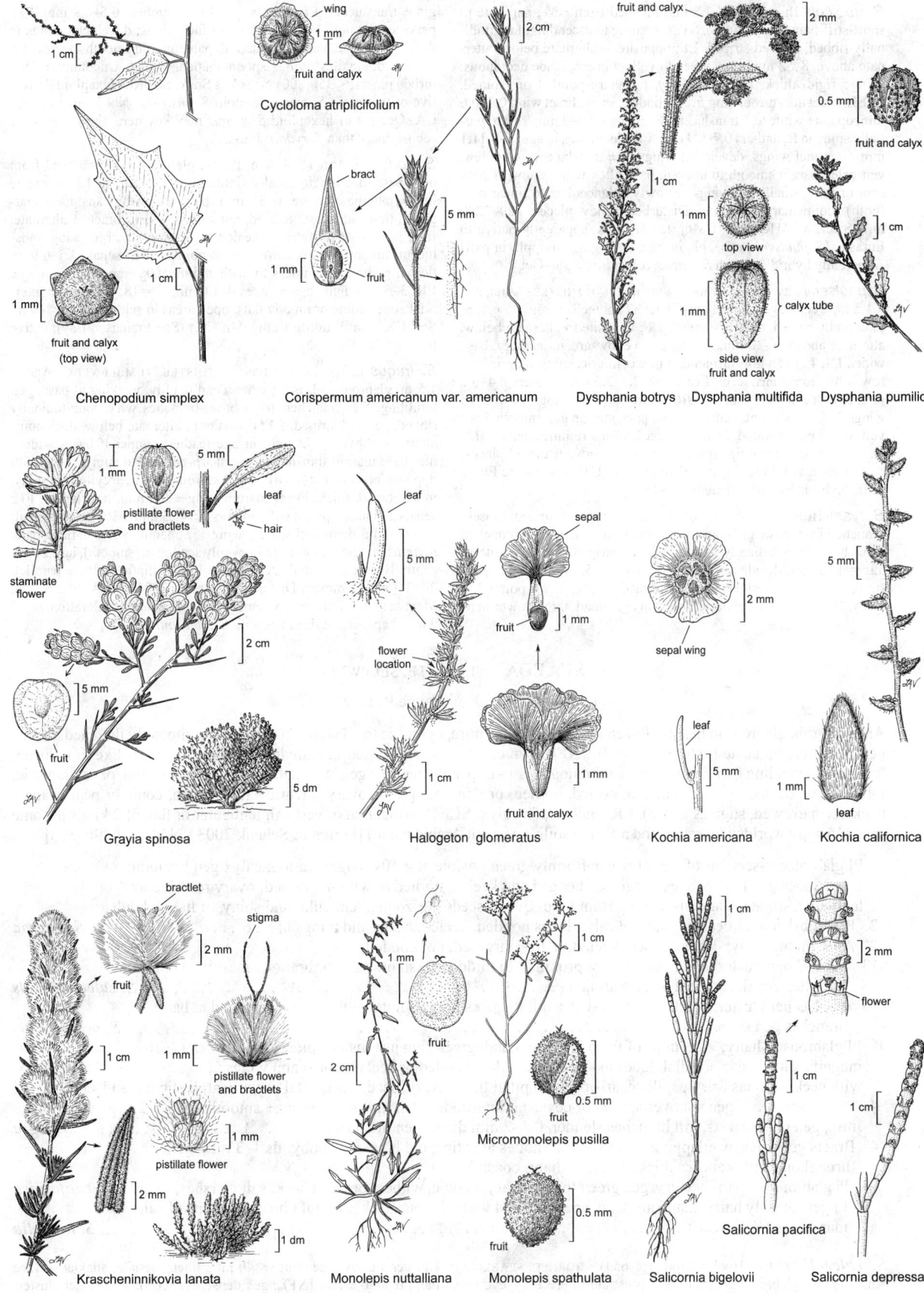

Cycloloma atriplicifolium

wing
1 mm
fruit and calyx

Chenopodium simplex

1 mm
fruit and calyx
(top view)

Corispermum americanum var. americanum

bract
fruit
1 mm
5 mm
2 cm

Dysphania botrys

fruit and calyx
2 mm

0.5 mm
fruit and calyx

1 cm

top view
1 mm

side view
fruit and calyx
calyx tube
1 mm

Dysphania multifida

1 cm

Dysphania pumilio

staminate
flower
pistillate flower
and bractlets
1 mm
leaf
hair
5 mm
fruit
2 cm
5 mm
5 dm

Grayia spinosa

leaf
5 mm
flower
location
fruit
1 mm
fruit and calyx
1 mm
1 cm

Halogeton glomeratus

sepal
fruit
1 mm
sepal wing
2 mm

Kochia americana

leaf
5 mm

5 mm
leaf
1 mm

Kochia californica

bractlet
fruit
2 mm
stigma
pistillate flower
and bractlets
1 mm
pistillate flower
1 mm
1 cm
2 mm
1 dm

Krascheninnikovia lanata

fruit
1 mm
2 cm

Monolepis nuttalliana

1 cm
fruit
0.5 mm

Micromonolepis pusilla

fruit
0.5 mm

Monolepis spathulata

1 cm

1 cm

Salicornia bigelovii

2 mm
flower
1 cm

Salicornia pacifica

1 cm

Salicornia depressa

S. gobicola Iljin   Ann < 1.5 m, branched from base, papillate to short-stiff-hairy (± glabrous). **ST:** spreading to ascending, longitudinally ribbed, occ red-striped. **LF:** opposite to alternate below, alternate above, 8–52 mm, gray-green to yellow-green, blade deciduous; in age rigid, thick, broad, leathery, tip sharp-pointed or -spined. **INFL:** bract not surrounding fr, ± cylindric, spiny, bract wing 0.3–0.6 mm, opaque-white to ± translucent. **FL:** sepals 2.5–3 mm, soft to partially spiny in fr; anthers 0.7–0.9 mm. **FR:** deciduous in age; 5.4–11.1 mm diam incl wings, developed wings 5, ± translucent, veins few, gen pale, margin smooth to uneven, base often with yellow or pink spot or both, smallest 2 wings narrowly oblanceolate to linear or 1 (both) rudimentary. Common. Disturbed, sandy, places; 120–2200 m. s SnJV, n WTR, SNE, DMoj; to UT, Mex, apparently native to Eurasia. [*S. paulsenii*, misappl., in part; *S. tragus*, misappl., in part] Apparently hybridizes with *S. tragus, S. paulsenii*. Jul–Oct ◆

S. paulsenii Litv. (p. 653)   BARBWIRE RUSSIAN THISTLE   Ann, gen < 1.5 m, ± conic, gen densely papillate. **ST:** branched from base, longitudinally ribbed, occ red-striped. **LF:** opposite to alternate below, alternate above, 5–32 mm; in age gen yellow-green, leathery, base wider. **INFL:** bract not surrounding fr, ± cylindric, spiny, margin narrow, white to ± translucent. **FL:** sepals 2.5–3.5 mm; anthers 0.4–0.7 mm. **FR:** deciduous in age; 6.6–10.7 mm diam incl wings; developed wings 3–5, translucent, veins few, gen pale, margin gen smooth, base yellow- to pink-spotted, in fr smallest 2 wings rudimentary or 1(2) narrowly linear; sepal tips spiny. Common. Sandy, disturbed places; < 1000 m. n WTR (se Cuyama Valley), D; to UT; native to se Eur, c Asia. Hybridizes with *S. tragus*. Jul–Oct ◆

S. ryanii Hrusa & Gaskin (p. 653)   Ann < 2 m, ± rounded, loosely branched from base, glabrous to sparsely short-bristly. **ST:** branched from base, longitudinally ribbed, occ red-striped. **LF:** opposite to alternate below, deciduous, alternate above, 5–55 mm, in age upper rigid, base wider, leathery, margin translucent, tip sharp-pointed to spiny. **INFL:** bract ± cylindric, spiny, in age broad, thick, lower mar-gin ± translucent. **FL:** sepals 2.5–3 mm; anthers 0.6–0.9 mm. **FR:** persistent or not; 4.6–8.1 mm diam incl wings; developed wings in fr 5, gen opaque, veins few, dark to pale, margin smooth to unevenly scalloped, smallest wings spoon-shaped. 2*n*=54. Uncommon. Disturbed places; < 350(800) m. ScV, s SnJV, SCoRI. Hexaploid derivative of tetraploid *S. tragus*, diploid *S. australis*. Not known outside CA. Occ pls of hexaploid *S. tragus* may key here, these gen more robust, hairy than *S. ryanii*. Jun–Oct

S. soda L.   Ann 15–45 cm, fleshy, glabrous. **ST:** branched from near base, not longitudinally ribbed, not red-striped. **LF:** opposite below, alternate above, 6–55 mm, base widened, translucent-margined, tip rigid, short, pointed, not spined. **INFL:** bracts ± alternate, free, lanceolate to ovate, ± keeled, base broad, margin wing translucent, tip abruptly narrowed, short-pointed. **FL:** sepals 3.5–6 mm, fleshy, glabrous, wingless or with rudimentary appendages in age. **FR:** 3–6 mm diam, appendages 0–1.6 mm. 2*n*=18. Uppermost intertidal zone, saline or muddy flats, open areas in salt marshes; < 40 m. ScV (Delevan Wildlife Refuge), n CCo (San Francisco Bay); native to s Eur. Jul–Oct ❖

S. tragus L. (p. 653)   RUSSIAN THISTLE, TUMBLEWEED   Ann < 1.5 m, glabrous to bristly; when dead readily breaking at base, gen tumbling. **ST:** branched from base, branches wiry, longitudinally ribbed, gen red-striped. **LF:** opposite to alternate below, deciduous, alternate above, 8–52 mm; in age leathery, upper lf bases widening, base margin translucent, tip sharp-pointed to spiny, fused with opposite bract or not. **INFL:** bract surrounding fr, ± cylindric, spiny, in age broad, thick, lower margin wing ± 0.5 mm, translucent. **FL:** sepals 2–5 mm, tips not stiff; anthers 0.6–1.3 mm. **FR:** gen persistent; 2.9–8.4 mm diam incl wings; wings 5, opaque, veins dark to pale, margins minutely toothed to unevenly scalloped (smooth), largest gen centrally notched, smallest linear to blunt-elliptic, sides ± parallel. 2*n*=36(54). Common. Disturbed places; < 2800 m. CA; to e N.Am, Mex; native to Eurasia. Extremely variable in habit, coloration, sepal wing shape, etc. Hybridizes with *S. paulsenii*. Jul–Oct ◆

SUAEDA   SEABLITE, SEEPWEED

H. Jochen Schenk & Wayne R. Ferren, Jr.

Ann to shrub, glabrous to hairy. **LF:** gen alternate; blade entire, cylindric to adaxially flattened or completely flattened, fleshy, gen glaucous, tip acute [obtuse to round]. **INFL:** cyme; clusters sessile, gen in panicles of spikes; bracts lf-like to reduced; bractlets subtending fls 1–3, minute, membranous; fls 1–12 per cluster. **FL:** gen bisexual; calyx radial, bilateral, or asymmetric, lobes 5, gen fleshy, rounded, hooded, keeled, horned, or wing-margined; ovary ± lenticular, rounded, conic or pear-shaped, neck occ narrowed, stigmas 2–4(5). **FR:** enclosed in calyx. **SEED:** horizontal or vertical, lenticular or flat, of 2 kinds in some spp. 115 spp.: worldwide, saline and alkaline soils. (Ancient Arabic name) [Ferren & Schenk 2003 FNANM 4:390–398]

1. Pl glabrous; ×-sections of fresh lvs ± uniformly green, visible at ≥ 10× magnification; calyx gen bilateral, 1 or 3 lobes gen larger, lobes in age gen hooded and keeled, horned or wing-margined; ovary rounded to lenticular, stigmas glabrous, arising from tip of ovary; seeds horizontal, lenticular and shiny, or flat and dull
  2. Per, subshrub, or occ ann; mature calyx lobes hooded, keeled, horns and wings 0; SCo . . . . *S. esteroa*
  2′ Ann; mature calyx lobes horned, keeled, wing-margined, not hooded
    3. Fls 3–7 per cluster; bracts gen widest proximal to middle, fresh bracts membranous-margined at base; branches gen decumbent to ascending . . . . *S. calceoliformis*
    3′ Fls 1–3 per cluster; bracts gen widest at ± middle, fresh bracts not membranous-margined at base; branches gen spreading . . . . *S. occidentalis*
1′ Pl glabrous to hairy; ×-sections of fresh lvs with a dark green ring just inside epidermis, visible at ≥ 10× magnification; calyx ± radial, lobes in age gen rounded, hooded or keeled, horns and wings 0; ovary occ with neck, stigmas hairy-papillate, arising from pit at tip of ovary; seeds horizontal or vertical, lenticular and shiny
  4. Bracts gen < lvs, gen not overlapping or covering internodes at st tips; old lf scars ± smooth; fls 0.7–2 mm, gen on distal sts; infl branches slender, 0.4–2 mm diam; gen inland . . . . *S. nigra*
  4′ Bracts gen = lvs, overlapping, covering internodes at st tips; old lf scars knobby; fls 1–3 mm, gen throughout; infl branches thick, 2–4 mm diam; coastal
    5. Pl glabrous to sparsely hairy, gen green to red; ovary ± conic, without obvious neck; salt marshes, CCo . . . *S. californica*
    5′ Pl gen densely hairy, glaucous; ovary pear-shaped with obvious neck; coastal bluffs, margins of salt marshes, s CCo, SCo, ChI . . . . *S. taxifolia*

S. calceoliformis (Hook.) Moq. (p. 653)   HORNED SEABLITE   Ann < 8 dm, glabrous, glaucous. **ST:** prostrate to erect, 1–several; branches gen decumbent to ascending, green to dark red, gen striped.

LF: gen tightly ascending, < 40 mm, linear, sessile; adaxial surface flat, green to ± red. **INFL:** gen dense, branched; fls 3–7 per cluster; bracts subtending branches = lvs; bracts subtending fls < lvs, gen

widest proximal to middle, fresh bracts membranous-margined at base. **FL:** bilateral, 1–4 mm incl horns; mature calyx lobes horned, ± keeled, wing-margined; ovary rounded to lenticular, stigmas gen 2, glabrous. **SEED:** horizontal; lenticular seeds 0.8–1.7 mm, shiny, gen black; flat seeds 1–1.5 mm, dull, brown. $2n=36,54$. Saline or alkaline, wetland soils, gen dried; < 2200 m. CA; to AK, e Can, TX, Mex. Jul–Oct

S. californica S. Watson (p. 653) CALIFORNIA SEABLITE Shrub 3–8 dm, mound-like, glabrous to sparsely hairy. **ST:** decumbent, several from base, dull gray-brown, old lf scars knobby; branches spreading, many, herbaceous sts pale green to ± red. **LF:** overlapping, ± sessile; petioles ± 1 mm; blades 5–35 mm, ± lanceolate, ± cylindric to flat, green. **INFL:** clusters scattered throughout pl; branches thick, 2–4 mm diam; fls 1–5 per cluster; bracts gen = lvs, densely overlapping at branch tips. **FL:** bisexual or lateral pistillate, radial, 2–3 mm; calyx lobes rounded to hooded, glabrous; ovary ± conic, without obvious neck, stigmas 3, hairy-papillate. **SEED:** horizontal or vertical, 1.5–2 mm, lenticular, shiny, black. Margins of coastal salt marshes; < 5 m. CCo. Jul–Oct ★

S. esteroa Ferren & S.A. Whitmore (p. 653) ESTUARY SEABLITE Per, subshrub, or occ ann, 1–6 dm, glabrous, gen glaucous. **ST:** decumbent to erect; branches gen ascending, straw-colored, gen exfoliating. **LF:** ascending, sessile; distal < 60 mm, lance-linear, adaxial surface flat, overlapping, green to ± red, gen glaucous; proximal = mid-st, straw-colored, gen withered, at least vascular strands persistent. **INFL:** clusters on distal sts; fls gen 3–5 per cluster; bracts < lvs. **FL:** bilateral, 1.5–3.5 mm; calyx lobes hooded, keeled; ovary rounded to lenticular, stigmas 2–3, linear, glabrous. **SEED:** horizontal; lenticular form 1–1.3 mm, shiny, black to ± red; flat form 1.2–2 mm, dull, brown. Coastal salt marshes; < 5 m. SCo; n Mex. May–Oct ★

S. nigra (Raf.) J.F. Macbr. (p. 653) BUSH SEEPWEED Subshrub, shrub, or occ ann, 2–15 dm, glabrous to hairy, glaucous. **ST:** spreading to erect, several from base, base gen woody; branches spreading, herbaceous sts shiny, green to yellow-brown or red. **LF:** ascending

to wide-spreading, gen not overlapping; petiole 0–1 mm; blade 5–30 mm, ± cylindric to flat, linear to narrowly lanceolate, base narrow, yellow-green to red. **INFL:** gen open, branches thin, 0.4–2 mm diam; bracts gen < lvs; fls 1–12 per cluster, gen on distal sts. **FL:** gen bisexual, radial, 0.7–2 mm; calyx lobes rounded; ovary ± pear-shaped, stigmas 2–3, hairy-papillate. **SEED:** horizontal or vertical, 0.5–2 mm, lenticular, shiny, black. $2n=18$. Alkaline, saline habitats in interior and desert, occ coastal; < 1600 m. GV, CCo, SnFrB, SW, GB, D; to w Can, TX, Mex. [*S. moquinii* (Torr.) Greene] May–Sep

S. occidentalis (S. Watson) S. Watson WESTERN SEABLITE Ann < 3.5 dm, glabrous, glaucous. **ST:** erect, gen branched near base, branches gen spreading, green to dark red. **LF:** ascending to spreading, < 30 mm, linear, sessile; adaxial surface flat; green to ± red. **INFL:** branched, gen spreading; fls 1–3 per cluster; bracts subtending branches = lvs, bracts gen widest at ± middle, fresh bracts not membranous-margined at base. **FL:** bilateral, 1–3 mm incl horns; mature calyx lobes horned, keeled, wing-margined; ovary spheric to lenticular, stigmas gen 2, glabrous. **SEED:** horizontal; lenticular form 1–1.5 mm, shiny, gen black; flat form 1–1.5 mm, dull, brown. Dry, saline or alkaline wetlands; < 2200 m. GB, DMoj; to WA, WY. Jul–Sep ★

S. taxifolia (Standl.) Standl. (p. 653) WOOLLY SEABLITE Subshrub or shrub, < 15 dm, glabrous to gen densely hairy, glaucous. **ST:** spreading to erect, several from base, dull, gray-brown, old lf scars knobby; branches spreading, herbaceous branches pale green to red. **LF:** ascending to wide-spreading, ± sessile; blades < 30 mm, lanceolate to short-elliptic, ± cylindric or adaxial surface flat, blue-green, yellow-green, or red. **INFL:** clusters gen throughout; branches thick, 2–4 mm diam; fls gen 1–3 per cluster; bracts gen = lvs, overlapping. **FL:** bisexual or lateral pistillate, radial, 1–3 mm; calyx lobes rounded to hooded, gen hairy; ovary pear-shaped, stigmas 3–4, hairy-papillate. **SEED:** horizontal or vertical, ~2 mm, lenticular, shiny, black or brown. Coastal bluffs, margins of salt marshes; < 15 m. s CCo, SCo, ChI; Baja CA. [*S. californica* var. *pubescens* Jeps.; *S. c.* var. *t.* (Standl.) Munz] All yr ★

CISTACEAE ROCK-ROSE FAMILY

John W. Thieret & Elizabeth McClintock, final revision by Thomas J. Rosatti & Bruce G. Baldwin

Ann to shrub, aromatic, of sunny areas, often sandy or chalky substrates; hairs nonglandular, in stellate clumps or not, peltate or not, and/or glandular. **LF:** simple, alternate or opposite [whorled], often ± reduced, entire or not, petioled or not, stipuled or not. **INFL:** raceme- or panicle-like cymes or fls 1. **FL:** gen bisexual, ± radial; sepals 3 or 5 (outer 2 often narrower), free, often persistent in fr, 3 twisted in direction opposite that of petals; petals [0(3)]4–5, gen ephemeral; stamens (3–10) many, free, often sensitive to touch, ± persistent in fr or not; ovary superior, chambers 1 (or ± 3–12 from intruded parietal placentas), style 0–1, stigma 1(3), lobes 0 or 3–12. **FR:** loculicidal capsule, valves 3–12. **SEED:** 3–many. 8 genera, ± 175 spp.: temp, esp se US, Medit; some cult (*Cistus; Helianthemum; Tuberaria*). [Arrington 2004 Ph.D. Dissertation Duke Univ] Fls open in sunshine for < 1 day. Scientific Editors: Thomas J. Rosatti, Bruce G. Baldwin.

1. Petals white (often drying ± yellow) or rose to purple, with red or yellow near base or not; lvs gen linear or elliptic to ovate, gen opposite; fr valves 5–12 . . . . **CISTUS**
1′ Petals yellow; lvs gen linear to lanceolate or oblanceolate, alternate or opposite; fr valves gen 3
  2. Per or subshrub (sometimes fl in 1st yr); basal lvs 0, cauline gen alternate; style < 2 mm . . . . **HELIANTHEMUM**
  2′ Ann; basal lvs in rosette, often withering early, proximal cauline gen opposite, distal 0–few, alternate; style 0 . . . . **TUBERARIA**

CISTUS ROCK-ROSE

Shrub, evergreen. **ST:** < 2.5 m. **LF:** gen opposite, petioled or not; stipules 0. **INFL:** panicle-like or fls 1. **FL:** sepals 3 or 5; petals 4–5, white (often drying ± yellow) or rose to purple, with red or yellow near base or not, often ± wrinkled; ovary 1-chambered, placentas 5, style 0 or 1, stigma large, hemispheric, 5–12-lobed. **FR:** valves 5–12. ± 20 spp.: Medit; cult as orn. (Ancient Greek name) [Guzman & Vargas 2005 Molec Phylogen Evol 37:644–660] Pls incl hybrids sometimes escape cult; scented resin from some spp. may be myrrh of biblical, other references.

1. Sepals 3; fls 1 at ends of lateral branchlets . . . . *C. ladanifer*
1′ Sepals 5; fls gen in panicle-like cymes
  2. Style thread-like, ± = stamens; petals rose to purple . . . . *C. incanus*
  2′ Style ± 0; petals white (often drying ± yellow), with red or yellow near base or not
    3. Lf petioled, blade ovate to elliptic, main veins from base gen 1; pedicel 10–100 mm . . . . *C. salviifolius*
    3′ Lf sessile, linear to lance-linear or linear- to ovate-oblong, main veins from base gen 3; pedicel 5–15 mm . . . . *C. monspeliensis*

C. incanus L. (p. 653)   **ST:** < 1.3 m; hairs gen glandular and non-glandular. **LF:** petiole 0–15 mm; blade elliptic, tapered to base, ± wrinkled adaxially, main veins from base gen 1, margin ± wavy. **INFL:** (1)10-fld; pedicel gen < 2 cm. **FL:** sepals 5, lance-ovate, acuminate; petals 2–3 cm, rose to purple; style thread-like, ± = stamens. **FR:** valves 5. 2*n*=18. Uncommon. Disturbed places; < 1000 m. n&s CCo, n SCo, SnGb, WTR (Liebre Mtns), PR; native to s Eur. Possibly naturalizing in KR on I-5 median north of Shasta Lake. [*C. creticus* L.; *C. i.* subsp. *corsicus* (Loisel.) Heywood; *C. villosus* L., incl vars.] Jan–Aug

C. ladanifer L. (p. 653)   GUM CISTUS   **ST:** < 2.5 m, shiny, sticky; hairs ± 0. **LF:** sessile [or not], 3–8 cm, lance-linear to narrow-oblong, main veins from base gen 3. **INFL:** fls 1 at ends of lateral branchlets. **FL:** sepals 3; petals 3–5 cm, white, gen with red or yellow near base; style ± 0. **FR:** valves 6–12. 2*n*=18. Uncommon. Disturbed places; < 800 m. s CCo, s SCo, WTR, SnGb; native to sw Eur. Mar–Aug

C. monspeliensis L. (p. 653)   **ST:** < 1 m; hairs gen glandular and nonglandular. **LF:** sessile, 15–60 mm, linear to lance-linear, wrinkled adaxially, main veins from base gen 3; margin gen rolled under, ± not ciliate. **INFL:** 1–11-fld; pedicel 5–15 mm. **FL:** sepals 5, outer rounded to broadly wedge-shaped at base; petals 1–2 cm, white; style ± 0. **FR:** valves 5. 2*n*=18. Uncommon. Disturbed places; < 300 m. NCo, CCo (Carmel Highlands); native to s Eur, n Afr. Apr–Oct

C. salviifolius L. (p. 653)   **ST:** < 1 m; hairs gen nonglandular. **LF:** petiole 2–10 mm; blade 1–4 cm, ovate to elliptic, wrinkled adaxially, main veins from base gen 1. **INFL:** 1(4)-fld; pedicel 10–100 mm. **FL:** sepals 5; petals 1.5–2.5 cm, white, with yellow near base or not; style ± 0. **FR:** valves 5. 2*n*=18. Uncommon. Disturbed places; < 1000 m. s CCo, s SCo, SnGb; native to s Eur. [*Cistus salvifolius*, orth. var.] Apr–May

HELIANTHEMUM   SUN-ROSE, RUSH-ROSE

Per or subshrub, evergreen; hairs gen in stellate clumps, rarely glandular (exc infl), sparse to dense. **ST:** gen erect, ± broom-like. **LF:** basal 0; cauline gen alternate [opposite], gen linear to lanceolate or oblanceolate, ± sessile. **INFL:** raceme- or panicle-like. **FL:** sepals 5, outer 2 gen narrower; petals yellow; stamens 5–many; style < 2 mm, stigma ± hemispheric. **FR:** valves gen 3. **SEED:** 3–many. ± 120 spp.: ± range of family. (Greek: sun fl) CA spp. esp abundant after fire, here retained in *Helianthemum* (polyphyletic) in lieu of combination in *Crocanthemum* (or *Hudsonia*) for *H. greenei*.

1. Infl hairs dense, gen 0.5–0.7 mm, most glandular, dark red, fewer nonglandular, white; outer 2 sepals
   lanceolate . . . . ***H. greenei***
1' Infl hairs sparse to dense, gen < 0.5 mm, most nonglandular, white, fewer glandular, dark red;
   outer 2 sepals linear . . . .***H. scoparium***

H. greenei B.L. Rob.   ISLAND RUSH-ROSE   **ST:** 15–30 cm. **LF:** 7–30 mm, 1–4 mm wide. **FL:** outer sepals 2.5–4 mm, 0.5–1 mm wide, lanceolate; inner sepals 4.5–7 mm, 3–4 mm wide, ovate, acuminate; petals 5–8 mm, obovate; stamens 20–25. **FR:** 4–6 mm, ovoid. **SEED:** ± 15. Dry, rocky slopes, ridges, chaparral; < 500 m. ChI. Threatened by grazing, feral animals. Mar–Oct ★

H. scoparium Nutt. (p. 653)   PEAK RUSH-ROSE   **ST:** 12–65 cm. **LF:** 5–40 mm, 0.5–6 mm wide. **FL:** outer sepals 0.5–4.5 mm, < 0.5

mm wide, linear; inner sepals 2.5–7 mm, 2–3.5 mm wide, ovate, acuminate; petals 3–11 mm, obovate; stamens 5–45. **FR:** 2.5–4 mm, ovoid. **SEED:** 4–10. Dry sandy or rocky soil of hills, slopes, ridges; < 1500 m. NCoR, n&c SNF, n SNH, SnJV, CW, SW. [*H. suffrutescens* B. Schreib., Bisbee Peak rush-rose] Threatened by mining. Pls with evidently 10 stamens (± appearing as 5 united pairs) reported from Mount Diablo. Feb–Sep

TUBERARIA

Ann [per]; hairs nonglandular, spreading, long, white, red below or not, or some hairs short, red. **ST:** erect. **LF:** basal in rosette, often withering early; proximal cauline gen opposite, distal 0–few, alternate; petioled or not; stipuled or not. **INFL:** raceme-like. **FL:** sepals 5, persistent, outer 2 gen narrower; petals 5, yellow; stamens 10–many; style 0, stigma large, hemispheric. **FR:** valves 3. 8–12 spp.: Eur, n Afr. (Latin: from tuber-like swellings on roots) [Sales & Hedge 1995 Taxon 44:437–438]

T. guttata (L.) Fourr. (p. 653)   **ST:** 5–40 cm. **LF:** 1–6 cm, ± linear, wider below; margin ± rolled under or not. **FL:** pedicel 7–15 mm, slender; petals 7–10 mm, gen with red to purple near base. 2*n*=36. Uncommon. Disturbed places; 80–150 m. n&c SNF, e edge ScV; native to Eur, n Afr. Cleistogamous fls, possibly without petals, sometimes form, set fr. Apr–May

CLEOMACEAE   SPIDERFLOWER FAMILY

Robert E. Preston & Staria S. Vanderpool

Ann, per, shrub, often ill-smelling. **LF:** gen 1-palmate, gen alternate, gen petioled; stipules gen minute, often bristle-like or hairy; lflets 0 or 3–7. **INFL:** raceme, head, or fls 1, expanded in fr; bracts gen 3-parted below, simple above, or 0. **FL:** gen bisexual, radial to ± bilateral; sepals gen 4, free or fused, gen persistent; petals gen 4, free, ± clawed; stamens gen 6, free, exserted, anthers gen coiling at dehiscence; ovary superior, gen on stalk-like receptacle, chamber gen 1, placentas gen 2, parietal, style 1, persistent, stigma gen minute, ± head-like. **FR:** 2 nutlets or gen capsule, septicidal; valves gen 2, deciduous, leaving septum (frame-like placentas) behind; pedicel gen ± reflexed to spreading. 17 genera, ± 150 spp.: widespread trop to arid temp. [Iltis & Cochrane 2007 Novon 17:447–451] Treated as Capparaceae in TJM (1993). Scientific Editor: Thomas J. Rosatti.

1. Shrub; fr a capsule, inflated; petals 4–5 mm wide . . . . ***Peritoma arborea***
1' Ann, short-lived per; fr a capsule, not or ± inflated, or 2 nutlets; petals < 4 mm wide
  2. Fls in axillary heads; style in fr stout, spine-like; fr 2 nutlets . . . . **OXYSTYLIS**
  2' Fls in ± terminal racemes or 1 in lf axils; style in fr stout or not but not spine-like; fr 2 nutlets or a capsule
    3. Stamens 8–32; ovary ± sessile; fr a capsule, valves persistent . . . . **POLANISIA**

3′ Stamens 6; ovary on stalk-like receptacle; fr 2 nutlets or a capsule, valves deciduous

  4. Fr 2 nutlets, each 1(3)-seeded; septum 1 mm wide . . . . **WISLIZENIA**

  4′ Fr a capsule, 2–many-seeded; septum 1–25 mm wide

    5. Fr 2–6 mm, often wider than long; septum elliptic to round . . . . **CLEOMELLA**

    5′ Fr gen 12–55 mm, longer than wide; septum linear to oblong

      6. Infl open, few-fld; receptacle in fr 2–5 mm; anthers 3–6 mm. . . . . **CARSONIA**

      6′ Infl dense, many-fld; receptacle in fr 5–18 mm; anthers 1.8–2.6 mm. . . . . **PERITOMA**

CARSONIA

1 sp. (Named for Carson Desert, NV)

C. sparsifolia (S. Watson) Greene (p. 653) Per 1–9 dm, densely branched, glabrous, glaucous. **LF**: ± sparse; lflets 3 below, 1 (lvs simple) above, 0.4–1.5 cm, obovate. **INFL**: raceme, open, few-fld, 2–10 cm, not much expanded in fr. **FL**: sepals free, deciduous, 1.4–3 mm, ovate, acuminate, minutely serrate, brown-green; petals 9–13 mm, strap-shaped, recurved, yellow with brown central streak; stamens 9–15 mm, yellow, anthers 3–6 mm, brown; style 0.1–0.4 mm. **FR**: 15–45 mm, 1–3 mm wide, smooth; receptacle 2–5 mm. 2*n*=32. Sand dunes, sandy areas on alkali lake margins; 850–2030 m. SNE, DMoj (Eureka Valley); w NV. [*Cleome s.* S. Watson] May–Aug

CLEOMELLA

Ann, gen glabrous. **ST**: gen ascending to erect, sometimes prostrate when older, gen branched from base, often red-tinged. **LF**: gen many; petiole gen 7–20 mm; lflets gen 3. **INFL**: raceme, ± terminal, fls 1 in lf axils, or both; pedicel gen 4–25 mm. **FL**: parts gen yellow; sepals fused in basal 1/3, gen entire; petals ± sessile, upper 2 often recurved. **FR**: 2–6 mm, often wider than long; septum elliptic to round; receptacle stalk-like. **SEED**: < 10. ± 10 spp.: arid w N.Am. (Diminutive of *Cleome*) [Holmgren 2004 Brittonia 56:103–106]

1. Receptacle in fr 0.3–3 mm; petals < 3 mm

  2. Fls 1 in lf axils; receptacle reflexed in fr . . . . ***C. brevipes***

  2′ Fls in racemes, sometimes also 1 in lf axils; receptacle spreading to ascending in fr . . . . ***C. parviflora***

1′ Receptacle in fr 6–15 mm; petals 3.5–9 mm

  3. Pl hairy; older sts prostrate . . . . ***C. obtusifolia***

  3′ Pl glabrous; older sts ascending to erect

    4. Branches 0 or gen from lower nodes; lf elliptic to ovate . . . . ***C. hillmanii*** var. ***hillmanii***

    4′ Branches gen from upper nodes; lf linear-elliptic . . . . ***C. plocasperma***

C. brevipes S. Watson SHORT-PEDICELLED CLEOMELLA Pl glaucous. **ST**: 5–45 cm, rough. **LF**: petiole 0.5–3 mm; lflets 5–15 mm, linear to obovate, fleshy. **INFL**: fls 1 in lf axils, incl those near base; pedicels 1.5–3 mm. **FL**: sepals 0.8–1.2 mm, ovate, acuminate; petals 1.5–2 mm, pale yellow; stamens 1.5–2.2 mm, anthers 0.3–0.5 mm; style 0.1–0.3 mm. **FR**: 2–3 mm, 2–3.2 mm wide, round; valves ± conic; receptacle 0.5–3 mm, reflexed. Alkaline marshes, wet, salt-encrusted soil around thermal springs; 400–2100 m. SNE (exc W&I), DMoj (exc DMtns); w NV. May–Oct ★

C. hillmanii A. Nelson var. ***hillmanii*** HILLMAN'S CLEOMELLA **ST**: 5–55 cm, erect, smooth, branches 0 or gen from lower nodes. **LF**: petiole 2–6 cm; lflets 0.8–2.4 cm, elliptic to ovate. **INFL**: raceme 3–10 cm. **FL**: sepals 1–2 cm, lance-ovate; petals 4–8 mm, yellow; stamens 8–12 mm, anthers 2.5–3 mm; style 1–2 mm. **FR**: 3.5–6 mm, 4–10.5 mm wide; valves conic; receptacle 7–15 mm, gen spreading. Alkaline meadows and flats; 1220–1462 m. MP (Lassen Co.); to OR, ID, NV. May ★

C. obtusifolia Torr. & Frém. (p. 657) MOJAVE STINKWEED Pl hairy. **ST**: 1–9 dm, rough, younger ascending to erect, older prostrate, in ± round mat. **LF**: lflets 5–15 mm, obovate. **INFL**: on older sts raceme, 1–10 cm, dense, on younger sts fls 1 in lf axils. **FL**: sepals 1–1.5 mm, ovate, green, hairy, margin long-hairy; petals 4–6(9) mm, dark yellow, abaxially hairy; stamens 8–14 mm, anthers 1.5–2.3 mm; style 1.5–5 mm. **FR**: 3–4 mm, hairy, striate; valves conic to horn-shaped; receptacle 6–8 mm, reflexed. Desert scrub, sandy, rocky alkaline flats; < 1800 m. SNE, D; w NV. Variable; needs study. Apr–Oct

C. parviflora A. Gray **ST**: branched gen from base, 3–45 cm, smooth. **LF**: few; lflets 5–35 mm, linear-elliptic, ± fleshy. **INFL**: raceme, 0.5–30 cm, fls sometimes also 1 in lf axils. **FL**: sepals 0.5–1 mm, lanceolate; petals 1.8–2.2 mm, pale yellow; stamens 1.9–2.5 mm, anthers 0.4–0.5 mm; style < 0.2 mm, stigma 2-lobed, 0.3 mm, purple. **FR**: 3–4 mm; valves ± conic; receptacle 0.3–0.8 mm, spreading to ascending. Wet, alkaline meadows near thermal springs in sagebrush desert; 350–2140 m. GB, DMoj; w NV. Often occurs with *C. brevipes*, *C. plocasperma*. May–Sep

C. plocasperma S. Watson (p. 657) **ST**: branched gen from upper nodes, 10–55(80) cm, smooth. **LF**: lflets 15–45 mm, linear-elliptic. **INFL**: raceme, 1–20 cm. **FL**: sepals 0.9–2.2 mm, lanceolate; petals 3.5–7 mm; stamens 8–12 mm, anthers 1.5–1.9 mm; style 0.8–1.2 mm. **FR**: 4–5 mm; valves ± hemispheric to horn-shaped; receptacle 6–10 mm, spreading to ascending. Wet, alkaline meadows, greasewood flats, near thermal springs; 875–1710 m. ne SnBr, GB (exc Wrn, W&I), DMoj (exc DMtns); to OR, ID, UT. May–Oct

OXYSTYLIS

1 sp. (Greek: sharp style)

O. lutea Torr. & Frém. (p. 657) Ann, ± glabrous. **ST**: branched from base, 5–15 dm. **LF**: petiole 2.5–7 cm; lflets 3, 2–6 cm, ± elliptic, thick, firm. **INFL**: axillary head, 0.5–3 cm, ± spheric, ± sessile, not elongating in fr; pedicels 1–3 mm, thicker, reflexed in fr. **FL**: radial; sepals free, 1–2 mm, lanceolate, green; petals 2–3 mm, elliptic, straw-yellow, ± sessile; stamens 3–5 mm, yellow, anthers 1–1.2 mm; ovary 1–1.5 mm, ± sessile, lobes 2, ± separate, each 1-ovuled, style 2–3 mm, wide, fleshy at base. **FR**: 2 nutlets, each 2.5 mm, ± spheric, smooth, white to deep purple; receptacle < 2 mm, ± stalk-like; style 4–11 mm, stout, spine-like. **SEED**: 1 per nutlet. 2*n*=20,40. Desert wash, alkaline sink scrub, alkaline flats, in sandy or alkaline soil; < 945 m. n&e DMoj; w NV (Amargosa Desert). Mar–Oct

PERITOMA BEE PLANT

Ann, shrub, glabrous to densely glandular-hairy. **ST**: gen branched from upper nodes. **LF**: petiole 5–45 mm; lflets 3–5. **INFL**: raceme, terminal, gen 1–4 cm, gen 5–40 cm in fr; pedicels 4–20 mm. **FL**: often ± unisexual (stamens or pistils vestigial), ± bilateral, parts gen yellow; sepals free or fused, deciduous above base; petal claw 0 to short. **FR**: longer than wide; septum linear to oblong; receptacle stalk-like, reflexed to ascending. **SEED**: 10–40. 6 spp. (Greek: cutting all-around, from calyx) [Iltis & Cochrane 2007 Novon 17:447–451]

1. Shrub; fr inflated; petals 4–5 mm wide . . . .*P. arborea*
   2. Fr fusiform . . . . var. *angustata*
   2′ Fr obovoid or spheric
      3. Fr obovoid. . . . . var. *arborea*
      3′ Fr spheric . . . . var. *globosa*
1′ Ann; fr not inflated; petals < 4 mm wide
   4. Pl densely glandular-hairy; fr oblong, ± flat; sepals awl-shaped, gen 4–6 mm . . . . *P. platycarpa*
   4′ Pl glabrous or ± so; fr ± linear, not ± flat; sepals ovate to lanceolate, gen 1.6–3 mm
      5. Lflets gen 3; petals purple (white) . . . . *P. serrulata*
      5′ Lflets gen 5; petals golden or pale yellow
         6. Fr 40–60 mm; receptacle in fr 15–25 mm; petals golden yellow, 10–13 mm; stamens 20–30 mm; pl 5–10(20) dm . . . . *P. jonesii*
         6′ Fr 15–40 mm; receptacle in fr 5–17 mm; petals pale yellow, 5–8 mm; stamens 10–15 mm; pl (1.5)2.5–3 dm. . . *P. lutea*

P. arborea (Nutt.) H.H. Iltis BLADDERPOD Shrub, much branched, gen 5–20 dm, minutely hairy. **LF**: petiole 1–3 cm; lflets gen 3, 15–45 mm, oblong-elliptic. **INFL**: 1–30 cm; pedicels 8–15 mm, thicker in fr. **FL**: sepals fused in basal 1/2, 4–7 mm, ± entire, green; petals 8–14 mm, 4–5 mm wide, yellow, claw 0; stamens 15–25 mm, yellow, anthers 2–2.5 mm; style 0.9–1.2 mm or pistil aborting in bud. **FR**: capsule, 3–6 cm, 1–2.5 cm wide, inflated, smooth, leathery, light brown, tardily dehiscent; valves 2–3; receptacle stalk-like, 1–2 cm, stout, reflexed or not. 2*n*=34,40. [*Isomeris a.* Nutt.] Vars. ± distinct.

var. *angustata* (Parish) H.H. Iltis (p. 657) **FR**: fusiform, tapered to base, tip. Slopes, washes, desert scrub; 60–1220 m. s DMoj, DSon; AZ, Mex. [*Isomeris a.* var. *angustata* Parish] Nov–Jun

var. *arborea* **FR**: obovoid, tapered to receptacle, tapered to tip or abruptly narrowed to short beak. Slopes, grassland, coastal scrub; 5–1200 m. SCo, s ChI, PR; Baja CA. [*Isomeris a.* var. *a.*] All yr

var. *globosa* (Coville) H.H. Iltis (p. 657) **FR**: spheric, base cordate or rounded, tapered to receptacle, or abruptly narrowed to stalk-like base similar to receptacle, tip abruptly narrowed to short beak. Slopes, roadcuts, grassland, coastal scrub; 15–1830 m. s SNF, Teh, s SnJV, SCoR, SCo, ChI, TR, w DMoj. [*Isomeris a.* var. *g.* Coville; *I. a.* var. *insularis* Jeps.] Oct–Jul

P. jonesii (J.F. Macbr.) H.H. Iltis JONES' BEE PLANT Ann, sparsely branched from upper nodes, 5–10(20) dm, glabrous. **LF**: lflets 5, 1.5–6 cm, linear to elliptic. **FL**: sepals fused in basal 1/2, 1.6–2.6 mm, lanceolate, minutely dentate, yellow, persistent; petals 10–13 mm, oblong to ovate, golden yellow; stamens 20–30 mm, yellow, anthers 1.9–2.6 mm. **FR**: 40–60 mm, 2–5 mm wide, ± linear, not ± flat, striate; receptacle 15–25 mm. Dry, sandy flats, desert scrub,

weedy roadsides; 300–1200 m. w DSon; AZ, Baja CA. CA specimens not seen; treatment ± based on FNANM 7:205, 207. Summer

P. lutea (Hook.) Raf. (p. 657) YELLOW BEE PLANT Ann, sparsely branched from upper nodes, (1.5)2.5–3 dm, ± glabrous. **LF**: lflets gen 5, 1.5–6 cm, linear to elliptic. **FL**: sepals fused in basal 1/2, 1.6–2.6 mm, lanceolate, minutely dentate, yellow, persistent; petals 5–8 mm, oblong to ovate, pale yellow; stamens 10–15 mm, yellow, anthers 1.9–2.6 mm. **FR**: 15–40 mm, 2–5 mm wide, ± linear, ± round in ×-section, striate; receptacle 5–17 mm. 2*n*=34. Dry, sandy flats, desert scrub, weedy roadsides; 1100–2400 m. SNE, DMtns; to n WA, NE, NM, Baja CA. [*Cleome l.* Hook.] May–Aug

P. platycarpa (Torr.) H.H. Iltis (p. 657) GOLDEN BEE PLANT Ann, densely branched from base, 1–6 dm, green with purple, densely glandular-hairy. **LF**: lflets 3, 1–3.5 cm, ovate to obovate. **FL**: sepals free, gen 4–6 mm, awl-shaped, entire, yellow, deciduous; petals 6–12 mm, oblong, golden yellow; stamens 10–17 mm, yellow, anthers 1.8–2 mm; style 1–3 mm. **FR**: 12–25 mm, 8–12 mm wide, oblong, ± flat, hairy; receptacle 10–18 mm. 2*n*=40. Alkaline, clay soils, volcanic tuff, dry foothills, often sagebrush scrub; 1200–1500 m. CaRH (Shasta Valley), MP; OR, w NV. [*Cleome p.* Torr.] May–Jul

P. serrulata (Pursh) DC. ROCKY MOUNTAIN BEE PLANT Ann, open, 3–8 dm, ± glabrous. **LF**: lflets gen 3, 2–6 cm, elliptic. **FL**: sepals fused in basal 1/2, gen 1.7–3 mm, ovate, acuminate, minutely dentate, purple to green, persistent; petals 7–12 mm, oblong to ovate, purple (white); stamens 18–24 mm, purple, anthers 2–2.3 mm, green; style 0.1–0.5 mm. **FR**: 30–55 mm, 3–6 mm wide, ± linear, ± round in ×-section, smooth; receptacle 10–15 mm. 2*n*=34,60. Roadsides, disturbed areas; 305–1700 m. n SNH, SnJV, CCo, SCo, SnGb, PR, GB (exc W&I); s BC, Great Plains. [*Cleome s.* Pursh] Evidently wide-ranging but in CA rarely collected, perhaps only a waif. May–Jul

POLANISIA

Ann, sticky, glandular-hairy. **LF**: lflets 3–5, oblanceolate to ovate. **INFL**: raceme, terminal, dense. **FL**: bilateral; sepals ± free, deciduous; petals gen unequal, obovate, clawed, notched at tip; stamens 8–32, unequal, anthers not coiling; ovary ± sessile, subtended on 1 side by a prominent nectary. **FR**: dehiscent in upper 1/2; valves persistent; receptacle not stalk-like. **SEED**: many. 5 spp.: N.Am. (Greek: many, unequal, from stamens) [Inda et al. 2008 Pl Syst Evol 274:111–126] Taxonomic position problematic.

P. dodecandra (L.) DC. subsp. *trachysperma* (Torr. & A. Gray) H.H. Iltis (p. 657) WESTERN CLAMMYWEED **LF**: many; petiole 1.5–4.5 cm; lflets 1.6–4.5 cm. **INFL**: appearing after lvs, 5–30 cm; pedicels 1–2.5 cm, green to purple. **FL**: sepals 4–7 mm, purple; petals 6–13 mm, white, claw 4–5 mm; stamens 10–20, 5–40 mm, filaments purple, anthers 1–1.3 mm; nectary bright orange, spheric, concave at top, 2–2.3 mm; style 5–17 mm, deciduous, stigma purple. **FR**: 4–7 cm, 5–9 mm wide, green. 2*n*=20. Gravel bars, roadsides, disturbed areas; 30–1680 m. NCoRI, CaR, n&c SNF, ScV, MP; to e Can, se US. Jun–Sep

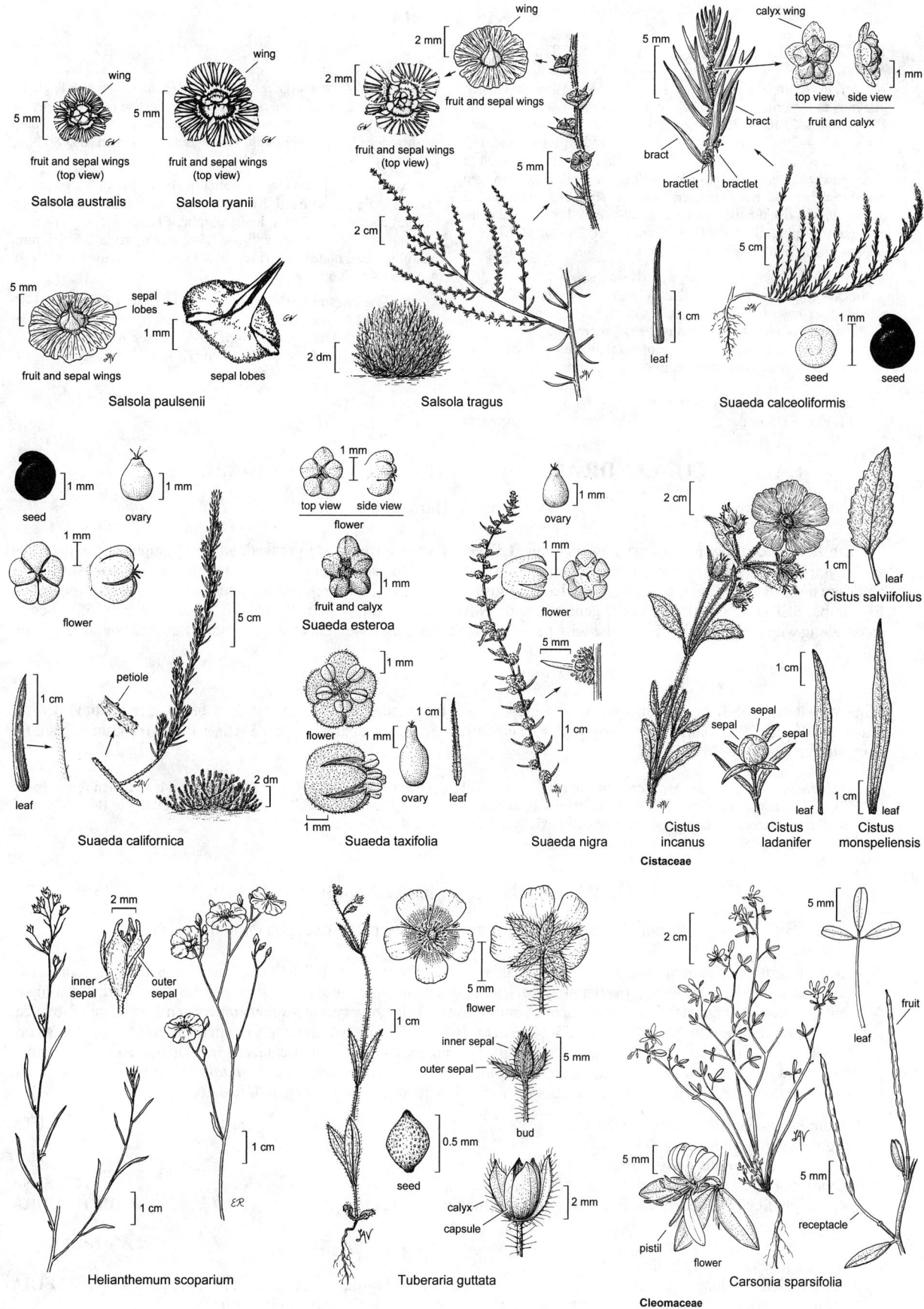

wing
5 mm
fruit and sepal wings (top view)
Salsola australis

wing
5 mm
fruit and sepal wings (top view)
Salsola ryanii

wing
2 mm
2 mm
fruit and sepal wings
fruit and sepal wings (top view)
5 mm
2 cm
2 dm
Salsola tragus

5 mm
sepal lobes
1 mm
fruit and sepal wings
sepal lobes
Salsola paulsenii

calyx wing
5 mm
bract
bract
bractlet
bractlet
top view
side view
fruit and calyx
5 cm
1 cm
leaf
1 mm
seed
seed
Suaeda calceoliformis

1 mm
seed
1 mm
ovary
1 mm
flower
1 cm
petiole
5 cm
leaf
2 dm
Suaeda californica

1 mm
top view
side view
flower
1 mm
fruit and calyx
Suaeda esteroa
1 mm
flower
1 mm
ovary
1 cm
leaf
1 mm
Suaeda taxifolia

1 mm
ovary
1 mm
flower
5 mm
1 cm
Suaeda nigra

2 cm
1 cm
leaf
Cistus salviifolius

1 cm
sepal
sepal
sepal
sepal
leaf
leaf
Cistus incanus
Cistus ladanifer
1 cm
leaf
Cistus monspeliensis
Cistaceae

2 mm
inner sepal
outer sepal
1 cm
1 cm
Helianthemum scoparium

5 mm
flower
1 cm
inner sepal
outer sepal
5 mm
bud
0.5 mm
seed
calyx
capsule
2 mm
Tuberaria guttata

5 mm
2 cm
leaf
fruit
5 mm
5 mm
pistil
flower
receptacle
Carsonia sparsifolia
Cleomaceae

WISLIZENIA

1 sp. (F.A. Wislizenus, pl collector in sw US, 1810–1889)

W. refracta Engelm. JACKASS-CLOVER Ann, short-lived per, glabrous to puberulent. **ST:** much-branched from base, 0.5–24 dm. **LF:** petiole 3–25 mm; lflets gen 3. **INFL:** raceme, terminal, 1–3 cm, dense, in fr 4–20 cm; pedicels 5–10 mm. **FL:** radial; sepals free, 1–2 mm, ± entire, green; petals 2.5–6.3 mm, elliptic, yellow, tapered to base, claw ± 0; stamens 8–14 mm, yellow; ovary 0.3–0.6 mm, gen exserted, lobes 2, ± free, each gen 1-ovuled, style 2–5.5 mm. **FR:** 2 nutlets; valves deciduous; receptacle stalk-like, reflexed; style elongate but not spine-like. **SEED:** 1(3) per nutlet. 2*n*=40. TOXIC but seldom eaten. Valuable honey pl.

1. Pl short-lived per; lflets 3 below, 1 (lf simple) above, linear-elliptic, 3–8 × longer than wide; sepals ovate, < 1.75 × longer than wide . . . . .subsp. *palmeri*
1′ Pl ann; lflets 3, ovate to obovate, 2–3 × longer than wide; sepals lanceolate, > 1.75 × longer than wide
　2. Lflet 2 × longer than wide; anthers 1.6–2.1 mm; SnJV . . . . .subsp. *californica*
　2′ Lflet 2–3 × longer than wide; anthers 0.9–1.2 mm; s DMoj, n DSon . . . . subsp. *refracta*

subsp. **californica** (Greene) S. Keller (p. 657) Ann. **ST:** green, tan. **LF:** lflets 3, 4–20 mm, obovate. **FL:** receptacle 6–10 mm; sepals 1–2 mm, lanceolate; anthers 1.6–2.1 mm; style 3–5 mm. Sandy washes, weedy roadsides, cult, fallow fields; < 125 m. SnJV. [*W. c.* Greene] Jul–Nov

subsp. **palmeri** (A. Gray) S. Keller (p. 657) PALMER'S JACKASS-CLOVER Per, short-lived. **ST:** brown-gray. **LF:** lflets 3 below, 1 (lf simple) above, 17–35 mm, linear-elliptic. **FL:** receptacle 5–14 mm; sepals 1–1.7 mm, ovate; anthers 1.5–2.3 mm; style 2.5–5.5 mm. Sandy washes, dunes, desert scrub; < 130 m. DSon; nw Mex. [*W. p.* A. Gray] Apr–Nov ★

subsp. **refracta** (p. 657) Ann. **ST:** green, tan. **LF:** lflets 3, 7–30 mm, ovate. **FL:** receptacle 3–6 mm; sepals 1–1.5 mm, lanceolate; anthers 0.9–1.2 mm; style 2–5 mm. Sandy washes, roadsides, alkaline flats; 90–1160 m. s DMoj, n DSon; to UT, w TX, Mex. Apr–Oct ★

COMANDRACEAE BASTARD TOADFLAX FAMILY

Danica T. Harbaugh

Per, subshrub, green root-parasite; sts, lvs glabrous. **LF:** entire, simple [reduced to scales], alternate; stipules 0. **INFL:** gen small, simple or compound cymes, axillary or terminal. **FL:** bisexual [unisexual]; calyx 0; petals 4–6; stamens opposite petals, on fleshy disk; ovary inferior, chamber 1, ovules 2–4, suspended from top of free-central placenta, style 1, stigma lobes 2–4. **FR:** drupe. **SEED:** 1, spheric [ovoid]. 2 genera, 2 spp.: N.Am, Eur, Medit. [Nickrent et al. 2010 Taxon 59:538–558] Segregated, along with other families, from otherwise paraphyletic Santalaceae (Nickrent et al.). Scientific Editor: Thomas J. Rosatti.

COMANDRA BASTARD TOADFLAX

Rhizome extensive. **ST:** green, blue-green, or ± gray, striate. **LF:** ± sessile. **FL:** subtended by bractlet; calyx tube bell- or urn-shaped, lobes 5(6); stamen base hair-tufted. **FR:** crowned by persistent calyx. 4 spp.: 3 Am, 1 Eur. (Greek: hair, man, for hairy stamen bases)

C. umbellata (L.) Nutt. subsp. **californica** (Rydb.) Piehl (p. 657) **ST:** many from rhizome, 10–40 cm, branched, glaucous. **LF:** 15–55 mm, lanceolate, acute to sharp-pointed, abaxially paler or not. **FL:** 2–3.5 mm; calyx lobes lanceolate to narrow-ovate, ± white; anthers ± 0.5 mm. **FR:** 5–7.5 mm, oblong-ovoid. Gen dry, ± rocky areas; 300–3000 m. KR, CaR, SN, SNE, n DMoj; to BC, sw NV, nw AZ. Apr–Aug

CONVOLVULACEAE MORNING-GLORY FAMILY

Robert E. Preston & Lauramay T. Dempster, except as noted

Ann, per, subshrub, gen twining or trailing. **LF:** 0 or alternate. **INFL:** cyme or fls 1 in axils; bracts subtending fls 0 or 2. **FL:** bisexual, radial; sepals (4)5, ± free, overlapping, persistent, often unequal; corolla gen showy, gen bell-shaped, ± shallowly 5-lobed, gen pleated and twisted in bud; stamens 5, epipetalous; pistil 1, ovary superior, chambers gen 2, each gen 2-ovuled, styles 1–2. **FR:** gen capsule. **SEED:** 1–4(6). 55–60 genera, 1600–1700 spp.: warm temp to trop; some cult for food or as orn (*Ipomoea*). [Stefanović et al. 2003 Syst Bot 28:791–806] Monophyletic only if Cuscutaceae incl, as treated here. *Ipomoea cairica* (L.) Sweet, *I. hederacea* Jacq. [*I. nil* L., misappl.], *I. indica* (Burm.) Merr. (incl *I. mutabilis* Ker Gawl.), *I. purpurea* (L.) Roth, *I. triloba* L., all incl in TJM (1993), not naturalized. Scientific Editor: Thomas J. Rosatti.

1. Pls non-green, parasitic, without lvs . . . . .CUSCUTA
1′ Pls green, photosynthetic, with lvs
　2. Styles 2, free or united only at base
　　3. Ovary not lobed; lf elliptic; pl tufted. . . . . CRESSA
　　3′ Ovary shallowly to deeply 2-lobed; lf reniform; pl matted . . . . DICHONDRA
　2′ Style 1
　　4. Stigma unlobed, head-like. . . . . [IPOMOEA]
　　4′ Stigma lobes 2, oblong to linear
　　　5. Stigma lobes oblong, tips obtuse; calyx 7–25 mm; corolla 20–75 mm. . . . . CALYSTEGIA
　　　5′ Stigma lobes linear to narrowly spoon-shaped, tips acute; calyx 3–10 mm; corolla ≤ 40 mm
　　　　. . . . . CONVOLVULUS

CALYSTEGIA MORNING-GLORY
R.K. Brummitt

Per, subshrub from caudex or rhizome, glabrous to tomentose. **ST**: short to high-climbing, gen twisting, twining. **LF**: gen > 1 cm, linear to reniform or sagittate to hastate (deeply divided). **INFL**: peduncle gen 1-fld; bracts gen ± opposite, lobed or not, > 1 mm below calyx, not hiding it, small, to < 1 mm below calyx, hiding it or ± so, large. **FL**: gen showy; corolla glabrous, white or yellow to pink or purple; ovary chamber 1, style 1, stigma lobes 2, oblong, tips obtuse. **FR**: ± spheric, ± inflated. **SEED**: gen ± 4. ± 25 spp.: temp, worldwide. (Greek: hiding calyx, by bracts of some) [Brummitt 2002 Madroño 49:130–131] Intermediates common, often difficult to identify. Molecular evidence indicates close relationship with *Convolvulus* (Carine et al. 2004 Amer J Bot 91:1070–1085). Bracts qualify as bractlets by some definitions. Lf blade length measured along midrib.

1. Bracts > 1 mm below calyx, not hiding it, ≤ 5 mm wide or variously toothed or lobed like lvs
  2. Pl glabrous or nearly so
    3. Bracts elliptic to widely elliptic-oblong, entire...........................²*C. peirsonii*
    3′ Bracts gen linear, entire or lobes basal
      4. St ± stiffly erect, intertwined; lf linear to narrowly triangular, lobes 0 or ± linear, sinus rounded....... **C. longipes**
      4′ St trailing to strongly climbing, ± not intertwined; lf triangular to reniform, lobes wider than linear, 2(3)-tipped, sinus V-shaped or ± closed ....**C. purpurata**
        5. St strongly climbing, > 1 m; lf triangular, tip acute, sinus V-shaped, bracts rarely lobed....... subsp. **purpurata**
        5′ St trailing to weakly climbing, gen < 1 m; lf ovate-triangular to reniform, tip gen rounded to notched, sinus gen ± closed; bracts gen with small lobes................................ subsp. **saxicola**
  2′ Pl puberulent to tomentose at least around lf sinus or at peduncle tip
    6. Bracts entire, not lobed like lvs, ± sessile.............................. **C. occidentalis** subsp. **occidentalis**
    6′ Bracts toothed or lobed like lvs, stalked
      7. Lf lobes 7–9, linear, finger-like .... **C. stebbinsii**
      7′ Lf ± entire or lobes 2, not linear, not finger-like
        8. Lf blade widely triangular to ± reniform, tip acute to acuminate or notched, lobes gen 2-tipped; hairs ± densely spreading....................................... **C. malacophylla** subsp. **malacophylla**
        8′ Lf blade gen narrowly triangular, tip ± acute, lobes gen 1-tipped; hairs minute or fine .... **C. occidentalis** subsp. **fulcrata**
1′ Bracts < 1 mm below calyx, ± hiding it, often > 4 mm wide, not lobed like lvs
  9. Bracts < 4 mm wide
    10. Pl hairs dense, not appressed................................... **C. collina** subsp. **tridactylosa**
    10′ Pl hairs 0 or minutely appressed
      11. Lf blade narrowly triangular, lobes clearly defined, directed ± basally, base with deep sinus ....²*C. peirsonii*
      11′ Lf blade gen triangular-hastate to reniform, lobes not clearly defined, directed ± laterally, base truncate to wedge-shaped or widely rounded, occ with shallow sinus
        12. Bracts ± obtuse; pl glabrous. ²*C. atriplicifolia* subsp. **buttensis**
        12′ Bracts acute; pl hairs gen minutely appressed......................... ²*C. subacaulis* subsp. **episcopalis**
  9′ Bracts > 4 mm wide
    13. St trailing or weakly to strongly climbing, ≥ 1 m
      14. Pl from rhizomes, st not woody; peduncle 1-fld; marshes, riverbanks, or disturbed areas
        15. Bracts strongly sac-like at base, margins overlapped; corolla gen 55–75 mm...... **C. silvatica** subsp. **disjuncta**
        15′ Bracts not strongly sac-like at base, margins not overlapped; corolla 30–58 mm................. **C. sepium**
          16. Bracts < sepals; corolla gen ± 40 mm .... subsp. **binghamiae**
          16′ Bracts gen > sepals; corolla 30–58 mm .... subsp. **limnophila**
      14′ Pl from woody caudex, lower st often woody; peduncle often > 1-fld; gen dry places .... **C. macrostegia**
        17. Bracts 16–30 mm wide, strongly keeled to ± sac-like at base
          18. Bracts 19–37 mm; calyx 16–25 mm; corolla 47–68 mm; stamens 23–35 mm; s ChI......... subsp. **amplissima**
          18′ Bracts 13–26 mm; calyx 10–22 mm; corolla 36–60 mm; stamens 17–26 mm; n ChI .... subsp. **macrostegia**
        17′ Bracts 4–16 mm wide, flat or ± keeled
          19. Bracts obtuse.. subsp. **cyclostegia**
          19′ Bracts acute
            20. Lvs hairy... subsp. **arida**
            20′ Lvs glabrous to puberulent
              21. Lf blade triangular, > 7 mm wide exc basal lobes .... subsp. **intermedia**
              21′ Lf blade linear to narrowly triangular, < 7 mm wide exc basal lobes .... subsp. **tenuifolia**
    13′ St not or only weakly climbing, < 1 m
      22. Pl hairs 0 or sparsely appressed-minute and lf triangular-hastate to reniform
        23. Corolla deep pink or ± purple; lf reniform, glabrous, ± fleshy; ocean beaches, dunes .... **C. soldanella**
        23′ Corolla white or ± yellow (pink- or purple-tinged); lf ± reniform or not, glabrous or not, not fleshy; inland
          24. Bracts obtuse; pl gen glabrous .... ²*C. atriplicifolia* subsp. **buttensis**
          24′ Bracts acute; pl gen minutely appressed-hairy .... ²*C. subacaulis* subsp. **episcopalis**
      22′ Pl tomentose or hairs densely spreading, or lf not triangular-hastate to reniform

25. St well developed, < 100 cm, without basal rosette of lvs
   26. Pl tomentose; lf margin wavy . . . . ²*C. collina* subsp. *venusta*
   26′ Pl tomentose or hairs densely spreading; lf margin not wavy . . . . **C. malacophylla**
     27. Lf widely triangular to ± reniform, tip gen obtuse to notched, lobes gen 2-tipped; hairs
        brown or golden-brown (± gray); CaR, SN . . . . subsp. **malacophylla**
     27′ Lf ± narrowly triangular, tip gen acute, lobes gen 1-tipped (2nd tip 0 or poorly developed);
        hairs ± gray (± brown); SnFrB, SCoR, WTR . . . . subsp. **pedicellata**
25′ St poorly developed, < 30 cm, ≤ basal rosette of lvs
   28. Pl hairs sparse to ± dense, short, spreading to reflexed; lf margin not or ± wavy
     . . . . **C. subacaulis** subsp. **subacaulis**
   28′ Pl ± tomentose; lf margin wavy . . . . **C. collina**
     29. Sepals tomentose . . . . ²subsp. **venusta**
     29′ Sepals glabrous or strigose at center
        30. Lf tip obtuse, lobes gen indistinct, blade triangular to reniform . . . . subsp. **collina**
        30′ Lf tip acute, lobes distinct, blade triangular . . . . subsp. **oxyphylla**

C. atriplicifolia Hallier f. subsp. ***buttensis*** Brummitt (p. 657)
BUTTE COUNTY MORNING-GLORY   Per from rhizome, gen glabrous.
ST: decumbent to ± erect, 10–50 cm. **LF:** blade gen 2–4 cm, gen
triangular-hastate to reniform; lobes not clearly defined, directed ±
laterally, truncate to wedge-shaped or widely rounded, occ with shallow sinus. **INFL:** peduncle < 5 cm, gen < subtending lf; bracts not
fully hiding calyx, 8–12 mm, 3–9 mm wide, ± cordate-elliptic, entire,
± obtuse. **FL:** sepals 10–15 mm; corolla 30–45 mm, white (pink-
tinged). Dry, rocky places in open forest, chaparral; 600–1200 m.
KR, CaRH, SnFrB. Scattered, uncommon, variable; intergrades with
C. occidentalis subsp. *o.* Pls intermediate to *C. atriplicifolia* subsp.
a. of OR, WA, in KR; pls ± intermediate to *C. subacaulis* in SnFrB.
May–Jul ★

C. collina (Greene) Brummitt   Per from rhizome, tomentose. **ST:**
decumbent, not or weakly climbing. **LF:** gen < 3 cm, reniform (basal
lobes indistinct) to distinctly lobed; margin gen wavy. **INFL:** peduncle < 6 cm, > subtending lf or not; bracts partly to entirely hiding
calyx, entire. **FL:** sepals 8–13 mm; corolla 25–55 mm, white.

   subsp. ***collina*** (p. 657)   **ST:** 8–15 cm. **LF:** reniform to triangular;
lobes gen indistinct. **INFL:** bracts ± hiding calyx, 8–15 mm, 6–14
mm wide, widely ovate. **FL:** sepals glabrous or strigose at center;
corolla 30–54 mm. Open grassy or rocky places or in open oak/
pine woodland, often serpentine; < 600 m. NCoR, SnFrB, SCoRO.
Apr–Jun

   subsp. ***oxyphylla*** Brummitt (p. 657)   MOUNT SAINT HELENA
MORNING-GLORY   **ST:** 8–15(20) cm. **LF:** lobes distinct. **INFL:**
bracts ± hiding calyx, 8–17 mm, 5–11 mm wide, lanceolate to widely
ovate. **FL:** sepals glabrous or strigose at center; corolla 27–53 mm.
Open grassy or rocky places or in open oak/pine woodland, often
serpentine; < 600 m. NCoR (Lake, Sonoma, Napa cos.). Intermediate between *C. collina* subsp. *c.* and *C. collina* subsp. *tridactylosa.*
Apr–Jun ★

   subsp. ***tridactylosa*** (Eastw.) Brummitt (p. 657)   THREE-
FINGERED MORNING-GLORY   **ST:** gen 15–30 cm. **LF:** lobes distinct;
margin ± wavy. **INFL:** bracts not fully hiding calyx, 7–10 mm, 2–3.5
mm wide, linear-elliptic to lanceolate. **FL:** outer sepals tomentose;
corolla 27–33 mm. Open grassy or rocky places or in open oak/pine
woodland, often serpentine; < 600 m. NCoRO, NCoRI (Mendocino,
Lake cos.). May intergrade with *C. occidentalis* subsp. *o.*, *C. malacophylla* subsp. *m.* Apr–Jun ★

   subsp. ***venusta*** Brummitt (p. 657)   SOUTH COAST RANGE
MORNING-GLORY   Pl densely spreading-hairy or tomentose. **ST:**
8–30 cm. **LF:** ± triangular to ± reniform; lobes distinct or not. **INFL:**
bracts spreading, not hiding calyx, 8–16 mm, 4–10 mm wide, lanceolate to widely ovate, ± flat. **FL:** sepals finely tomentose; corolla
25–44 mm. Open grassy or rocky places or in open oak/pine woodland, often serpentine; < 600 m. SCoR. Intergrades with *C. subacaulis*
subsp. *episcopalis*, *C. malacophylla* subsp. *pedicellata.* Apr–Jun ★

C. longipes (S. Watson) Brummitt (p. 657)   Subshrub from woody
caudex, ± hemispheric, glabrous. **ST:** ± stiffly erect or intertwining,
3–10 dm. **LF:** < 6 cm, linear to narrowly triangular; lobes 0 or ± linear; sinus rounded; upper lvs less lobed. **INFL:** peduncle gen 1-fld, <
20 cm, >> subtending lf; bracts gen alternate, not hiding calyx, 3–17

mm, 0.2–3 mm wide, linear, often with basal lobes. **FL:** sepals 8–11
mm; corolla 28–36 mm, white or cream to pale pink or lavender. Dry,
rocky places, desert scrub; 600–1300 m. s CA-FP, D; NV, AZ. Intergrades with *C. macrostegia* subsp. *tenuifolia* in San Diego Co., *C.
malacophylla* subsp. *pedicellata* in San Luis Obispo Co., *C. peirsonii*
in Los Angeles Co. May–Jul

C. macrostegia (Greene) Brummitt   Per, subshrub from woody
caudex, glabrous to densely short-hairy. **ST:** slender, weakly climbing to woody, or strongly climbing, 1–9 m. **LF:** < 13 cm, gen widely
triangular, lobed. **INFL:** peduncle 1- to several-fld, gen > subtending
lf; bracts ± hiding calyx, 6–37 mm, 4–30 mm wide, lanceolate to ±
round, entire, flat to sac-like. **FL:** sepals 7–25 mm; corolla 22–68
mm, white or fading pink.

   subsp. ***amplissima*** Brummitt (p. 657)   ISLAND MORNING-
GLORY   Pl ± glabrous in age. **LF:** lobes 2–3-tipped; sinus widely
rounded. **INFL:** bracts 19–37 mm, 17–30 mm wide, widely ovate,
strongly keeled to sac-like, tip round to acuminate. **FL:** sepals 16–25
mm; corolla 47–68 mm. Rocky slopes, canyon walls; gen < 100 m. s
ChI. Most like, but larger than, *C. macrostegia* subsp. *m.* Feb–Jul ★

   subsp. ***arida*** (Greene) Brummitt (p. 657)   Pl hairy. **ST:** slender,
trailing or weakly climbing. **LF:** lobes 1-tipped; sinus rounded to
square. **INFL:** bracts 10–21 mm, 6–10 mm wide, lanceolate, not or
± keeled, tip acute. **FL:** sepals 9–12 mm; corolla 24–34 mm. Coastal
scrub, chaparral; < 1000 m. SnGb, SnBr, n PR (incl SnJt). Intergrades
with *C. macrostegia* subsp. *intermedia*, *C. macrostegia* subsp. *tenuifolia*, *C. occidentalis* subsp. *fulcrata*, probably *C. peirsonii.* May–Jun

   subsp. ***cyclostegia*** (House) Brummitt (p. 657)   Pl glabrous to
puberulent. **ST:** trailing or climbing, 1–2.5 m. **LF:** triangular; lobes
gen ± 2-tipped; sinus acute, rounded, or ± square. **INFL:** bracts 6–20
mm, 7–16 mm wide, widely ovate to ± round, flat to ± sac-like, tip
notched to acuminate. **FL:** sepals 9–15 mm; corolla 28–52 mm.
Coastal scrub; < 350 m. CCo, n SCo. Intergrades with *C. macrostegia* subsp. *intermedia*, *C. macrostegia* subsp. *m.*, *C. purpurata* subsp.
p. Mar–Aug

   subsp. ***intermedia*** (Abrams) Brummitt (p. 657)   Pl glabrous to
puberulent. **ST:** trailing or climbing. **LF:** lobes rounded to 2-angled;
sinus ± acute to rounded or square. **INFL:** bracts 10–20 mm, 6–12
mm wide, lanceolate, not or ± keeled. **FL:** sepals 9–16 mm; corolla
32–40 mm. Coastal or inland hills; gen < 200 m. SCo, s ChI, s WTR,
SnGb, n PR (incl SnJt). Intergrades with other subspp., probably with
C. peirsonii. Mar–Aug

   subsp. ***macrostegia***   Pl ± glabrous in age. **ST:** climbing, often
> 4 m. **LF:** lobes gen 2-tipped; sinus widely rounded. **INFL:** bracts
13–26 mm, 16–27 mm wide, widely ovate to ± round, strongly keeled
to sac-like, tip notched to acuminate. **FL:** sepals 10–22 mm; corolla
36–60 mm. Coastal scrub; gen < 500 m. n ChI; Guadalupe, San Martin islands, Baja CA. Intergrades with *C. macrostegia* subsp. *amplissima*, *C. macrostegia* subsp. *cyclostegia*, *C. macrostegia* subsp. *intermedia.* Apr–Jul

   subsp. ***tenuifolia*** (Abrams) Brummitt (p. 657)   Pl glabrous to
puberulent. **ST:** trailing or weakly climbing. **LF:** blade linear to narrowly triangular; lobes linear; sinus rounded or square. **INFL:** bracts

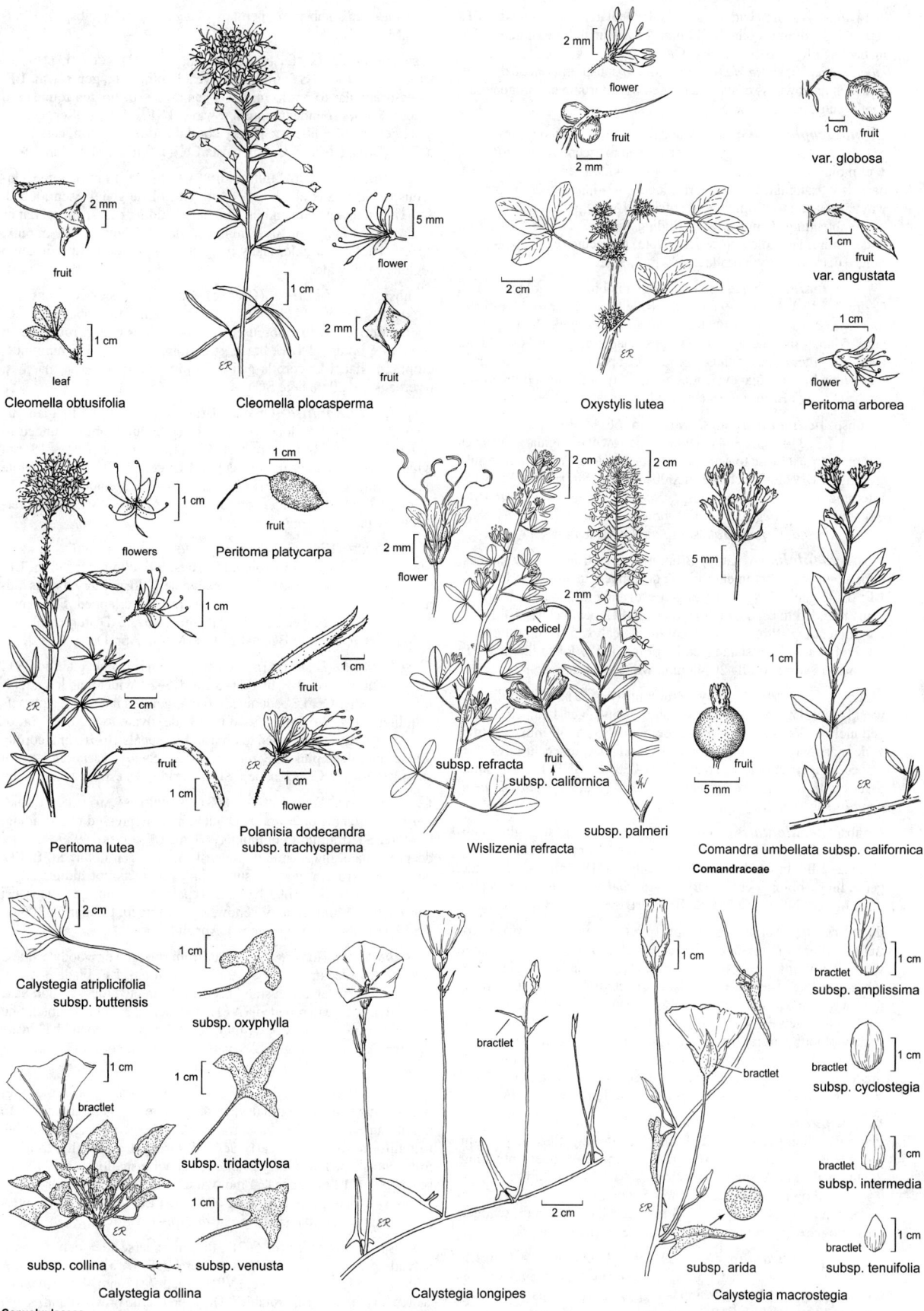

Cleomella obtusifolia

Cleomella plocasperma

Oxystylis lutea

var. globosa

var. angustata

Peritoma arborea

flowers

Peritoma platycarpa

Peritoma lutea

Polanisia dodecandra
subsp. trachysperma

Wislizenia refracta

subsp. refracta

subsp. californica

subsp. palmeri

Comandra umbellata subsp. californica

Comandraceae

Calystegia atriplicifolia
subsp. buttensis

subsp. oxyphylla

subsp. tridactylosa

subsp. venusta

subsp. collina

Calystegia collina

Calystegia longipes

subsp. arida

Calystegia macrostegia

subsp. amplissima

subsp. cyclostegia

subsp. intermedia

subsp. tenuifolia

Convolvulaceae

8–14 mm, 4–8 mm wide, narrowly lanceolate, flat, tip acute. **FL:** sepals 7–10 mm; corolla 22–40 mm. Dry, ± inland areas; gen < 500 m. c&s SCoR, SCo, w PR; Baja CA. Intergrades with *C. longipes*, *C. macrostegia* subsp. *arida*, *C. macrostegia* subsp. *intermedia*; extent to which narrowness of lvs determined by environment (vs genetics) needs study, esp in Mex. May

C. malacophylla (Greene) Munz   Per from rhizome, tomentose or hairs densely spreading. **ST:** decumbent to ascending, 10–100 cm, ± climbing or not. **LF:** < 6 cm, < 9 cm wide (gen much smaller), narrowly triangular to ± reniform; lobes 1–2-tipped; margin gen not wavy; tip acute to acuminate or notched. **INFL:** peduncle 6–9 cm, < or > subtending lf; bracts gen ± hiding calyx, 7–20 mm, 5–15 mm wide, entire, lanceolate to widely ovate, gen acute. **FL:** sepals 9–15 mm, ± densely hairy; corolla 20–45 mm, white.

subsp. *malacophylla* (p. 663)   Pl hairs brown or golden-brown (± gray). **LF:** 3–6 cm, 4–9 cm wide, widely triangular to ± reniform; lobes gen 2-tipped; tip acute to acuminate or notched. **INFL:** peduncle 3–5 cm, < subtending lf. Dry slopes, chaparral; gen 1000–2400 m. CaR, SN. Intergrades with *C. occidentalis*, perhaps *C. collina* subsp. *tridactylosa*. Molecular evaluation under way of *C. malacophylla* var. *berryi* (Eastw.) Brummitt (bracts not hiding calyx; s SN). Jun–Aug

subsp. *pedicellata* (Jeps.) Munz (p. 663)   Pl hairs ± gray (± brown). **LF:** 3–4.5 cm, 3–4 cm wide, ± narrowly triangular; lobes gen 1-tipped; tip ± obtuse to acute. **INFL:** peduncle 6–9 cm, often > subtending lf. Dry slopes, chaparral; 300–1900 m. SnFrB, SCoR, WTR. Intergrades with *C. collina* subsp. *venusta*, *C. longipes*, *C. macrostegia* subsp. *cyclostegia*, *C. malacophylla* subsp. *m.*, *C. occidentalis* subsp. *fulcrata*, *C. purpurata* subsp. *p.*, *C. subacaulis*. Apr–Jul

C. occidentalis (A. Gray) Brummitt   Per from woody caudex; puberulent to finely tomentose. **ST:** decumbent to strongly climbing. **LF:** blade gen 1.5–4 cm; lobes gen ± distinct, rounded to 2-tipped; sinus rounded to ± square or tapered. **INFL:** peduncle 1–4-fld, < to ± > subtending lf; bracts gen ± not hiding calyx, 5–12 mm, linear to ± round, ± sessile or stalked, entire or toothed or lobed like lvs. **FL:** sepals 9–15 mm; corolla 20–48 mm, white to creamy yellow.

subsp. *fulcrata* (A. Gray) Brummitt (p. 663)   **ST:** trailing or weakly climbing, gen < 1 m. **LF:** lobes distinct, gen 1-tipped. **INFL:** peduncle 1-fld; bracts 2–15 mm below calyx, 5–30 mm, 2–7 mm wide, narrowly lanceolate to widely triangular, lobed like lvs. Dry slopes, chaparral, pine woodland; 300–2700 m. SNF, TR, PR. Intergrades with *C. macrostegia* subsp. *arida*, *C. malacophylla*, *C. occidentalis* subsp. *o.* May–Aug

subsp. *occidentalis* (p. 663)   **ST:** decumbent to strongly climbing, gen > 1 m. **LF:** lobes gen indistinct, gen 2-tipped. **INFL:** peduncle (1)2–4-fld; bracts 1–7 mm below calyx, 4–18 mm, 1–5 mm wide, gen ± linear-oblong, entire. Dry slopes, chaparral, pine woodland; < 1200 m. KR, NCoR, CaRH, SNH, SnFrB, MP; OR. May–Jul

C. peirsonii (Abrams) Brummitt (p. 663)   PEIRSON'S MORNING-GLORY   Per from rhizome, glabrous, ± glaucous. **ST:** decumbent to weakly climbing, < 0.4 m. **LF:** blade < 2 cm, narrowly triangular; lobes clearly defined, 1/3–1/2 blade, directed ± basally; sinus deep, narrowly rounded to ± square. **INFL:** peduncle 2–8 cm, gen < subtending lf; bracts ± hiding calyx, 3–7 mm, 2.5–4 mm wide, elliptic to widely oblong, entire, flat. **FL:** calyx 9–13 mm; corolla 25–40 mm, white. Rocky slopes; 1000–1500 m. n SnGb, adjacent DMoj. Intergrades with *C. longipes* (coastal mainland), *C. macrostegia*, *C. occidentalis* subsp. *o.*, perhaps *C. sepium*. May–Jun   ★

C. purpurata (Greene) Brummitt   Per from woody caudex, glabrous, often glaucous. **ST:** trailing to strongly climbing, < 7 m. **LF:** blade gen 1.5–5 cm, triangular to reniform; lobes spreading, 2(3)-tipped; sinus V-shaped or ± closed. **INFL:** peduncle 1–5-fld, gen >> subtending lf; bracts gen not hiding calyx, 2–16 mm, 0.4–1.5 mm wide, gen linear. **FL:** sepals 7–14 mm; corolla 23–52 mm, white or cream to purple (often ± purple-striped). Subspp. intergrade.

subsp. *purpurata* (p. 663)   **ST:** strongly climbing, > 1 m. **LF:** triangular; sinus V-shaped; tip acute; lobes gen strongly angled; margin not wavy. **INFL:** bracts opposite, gen entire. Chaparral, coastal scrub; < 300 m. NCo, NCoRO, ScV (Sutter Buttes), CW (exc SCoRI), n SCo, w WTR. Intergrades with *C. macrostegia* subsp. *cyclostegia*,

C. occidentalis subsp. *o.*, perhaps *C. malacophylla* subsp. *pedicellata*. May–Jun

subsp. *saxicola* (Eastw.) Brummitt (p. 663)   COASTAL BLUFF MORNING-GLORY   **ST:** trailing to weakly climbing, gen < 1 m. **LF:** ovate-triangular to reniform; sinus gen ± closed; tip gen rounded to notched; lobes rounded; margin ± wavy. **INFL:** bracts often ± alternate, gen lobed ± like lvs. Rocky coastal scrub; < 100 m. c&s NCo, n CCo (Brooks Island, Contra Costa Co.), n SnFrB. May–Jun   ★

C. sepium (L.) R. Br.   HEDGE BINDWEED   Per from rhizome, glabrous to hairy. **ST:** climbing, < 4 m. **LF:** blade gen 4–8 cm, lobed. **INFL:** peduncle < subtending lf; bracts ± hiding calyx, entire, flat or keeled. **FL:** sepals gen 10–18 mm; corolla 30–58 mm, white or pink. $2n=22$. Highly variable, many geog subspp.; other subspp. in temp regions worldwide.

subsp. *binghamiae* (Greene) Brummitt   SANTA BARBARA MORNING-GLORY   Pl glabrous to sparsely hairy. **LF:** lobes ± spreading or not, ± entire to obscurely 2-tipped; sinus widely rounded; tip acute to ± obtuse. **INFL:** bracts 8–12 mm, < sepals, 4–8 mm wide, elliptic, ± flat. **FL:** corolla gen ± 40 mm, white. Coastal marshes, riverbanks; < 20 m. n&c SCo. Apr–Jun   ★

subsp. *limnophila* (Greene) Brummitt (p. 663)   Pl glabrous (densely hairy). **LF:** lobes ± abruptly spreading; sinus rounded to square; tip acute. **INFL:** bracts 13–28 mm, gen > sepals, 8–18 mm wide, narrowly ovate, flat or keeled. **FL:** corolla 30–58 mm, white or pink-tinged. Marshes, riverbanks; < 500 m. Deltaic GV, SnFrB, SCoR, TR, e DMoj (Amargosa River, 500 m); to e US, n Mex. Intergrades with *C. peirsonii*. May–Jul

C. silvatica (Kit.) Griseb. subsp. *disjuncta* Brummitt   LARGE BINDWEED   Per from rhizome, glabrous. **ST:** climbing, ≤ 6 m. **LF:** blade gen 5–12 cm, lobed. **INFL:** peduncle 1-fld, axillary; bracts hiding calyx, strongly sac-like at base, margins overlapped. **FL:** sepals 15–23 mm; corolla gen 55–75 mm, white. $2n=22$. Disturbed places; < 50 m. NCo, CCo; to BC; native to w Medit. Apr–Oct

C. soldanella (L.) R. Br. (p. 663)   Per from rhizome, glabrous. **ST:** decumbent, < 0.6 m. **LF:** blade 1–3 cm, 1.5–2 × wider than long, reniform, ± fleshy. **INFL:** peduncle 3–6 cm, gen > subtending lf; bracts ± hiding calyx, 7–16 mm, 5–10 mm wide, ovate to ± round, flat or enfolding calyx, tip notched to obtuse. **FL:** sepals 10–16 mm; corolla 32–52 mm, deep pink or ± purple. $2n=22$. Sandy seashores, coastal strand; < 50 m. NCo, CCo, n&c SCo; worldwide, esp temp. Apr–Aug

C. stebbinsii Brummitt (p. 663)   STEBBINS' MORNING-GLORY   Per from rhizome or ± woody caudex; hairs appressed to spreading, ± white. **ST:** trailing to climbing, < 1 m. **LF:** lobes 7–9, 7–55 mm, deep, palmate, linear, middle longest; margin gen upturned. **INFL:** peduncle 3–13 cm, gen >> subtending lf; bracts not hiding calyx, < 1.8 cm, 3–9-lobed like lvs. **FL:** sepals 7–11 mm, gen glabrous; corolla 30–35 mm, creamy yellow, ± pink-tinged. Chaparral; 300 m. n SNF (El Dorado, Nevada cos.). Apr–Jul   ★

C. subacaulis Hook. & Arn.   Per from rhizome or woody caudex; ± hairy. **ST:** decumbent or ascending, 2–20 cm. **LF:** blade 3–4 cm, ± triangular-hastate, with small backward-directed lobes or not; base tapered; tip acute to rounded. **INFL:** peduncle < 3.3 cm, < subtending lf; bracts ± hiding calyx, 7–17 mm, gen ± 4–9 mm wide, widely ovate to lanceolate or oblong. **FL:** sepals 10–13 mm; corolla 33–62 mm, white or cream (pink- or purple-tinged). Subspp. intergrade.

subsp. *episcopalis* Brummitt   CAMBRIA MORNING-GLORY   Pl gen minutely appressed-hairy; basal rosette 0. **ST:** decumbent to ascending, ± 20 cm. **LF:** basal rosette 0; blade gen triangular-hastate to reniform; lobes not clearly defined, directed ± laterally, truncate to wedge-shaped or widely rounded, occ with shallow sinus. **INFL:** bracts gen ± 12 mm, gen ± 4 mm wide, lanceolate, acute. Dry, open scrub, woodland; < 500 m. c SCoRO (San Luis Obispo Co.). Intergrades with *C. collina* subsp. *venusta*. Apr–Jun   ★

subsp. *subacaulis* (p. 663)   Pl hairs sparse to ± dense, short, spreading to reflexed. **ST:** gen ± 2 cm. **LF:** basal rosette > fls; ± widely rounded to acuminate. **INFL:** bracts 7–17 mm, 4–9 mm wide, widely ovate to oblong, rounded. Dry, open scrub or woodland; < 500 m. s NCoR, SnFrB. Intergrades with *C. malacophylla* subsp. *pedicellata*, *C. occidentalis* subsp. *o.* Apr–Jun

CONVOLVULUS MORNING-GLORY, BINDWEED

Ann, per from caudex or rhizome, gen ± glabrous. **ST**: gen trailing to high-climbing, gen twisting, twining. **LF**: gen > 1 cm, gen petioled, gen cordate or hastate. **INFL**: bracts gen 2, > 1 mm below calyx, not hiding it. **FL**: gen showy; corolla gen funnel-shaped, pleated, 5-angled or -lobed; stamens incl; ovary chambers 2, septa complete, stigma lobes 2, linear to narrowly spoon-shaped, tips acute. **FR**: spheric, ± inflated. **SEED**: gen 4. 250 spp.: gen temp. (Latin: entwine) Not easily distinguished from *Calystegia*.

1. Upper lvs deeply 3–many-lobed, lobes ± linear . . . . *[C. althaeoides]*
1′ Upper lvs entire to hastate, lobes wider than linear
  2. Lf hastate, petioled; corolla white, ± pink abaxially, esp at folds . . . . *C. arvensis*
  2′ Lf obovate to oblanceolate, ± sessile, narrowed to base; corolla pale-pink or -blue or white with yellow
    base, blue tips
    3. Corolla ± 0.6 cm, pale-pink or -blue . . . . *C. simulans*
    3′ Corolla 1.4–4 cm, center yellow, perimeter blue-purple, white between. . . . . *[C. tricolor]*

C. arvensis L. (p. 663) BINDWEED, ORCHARD MORNING-GLORY Per from deep, persistent root. **ST**: prostrate, tufted, or twining. **LF**: gen 2–3 cm, hastate, tip gen rounded. **INFL**: peduncles gen 2.5–6 cm, 1–several-fld, in fr gen reflexed; bracts ± linear. **FL**: calyx ± 5 mm, lobes oblong; corolla 2–2.5 cm, funnel-shaped, white, ± pink abaxially, esp at folds, margin entire. **FR**: ± 8 mm. *n*=24. Roadsides, open areas in many pl communities; < 2610 m. CA (rare GB, D); native to Eur. Mar–Oct ◆

C. simulans L.M. Perry (p. 663) SMALL-FLOWERED MORNING-GLORY Ann, strigose. **ST**: tufted, diffusely branched, 10–40 cm. **LF**: entire, gen < 6 cm, oblanceolate, ± sessile, narrowed to base. **INFL**: peduncles 1-fld, short, ± sharply nodding in fr; bracts 3–7 mm, linear or narrowly oblanceolate, 2–5 mm below calyx. **FL**: calyx lobes 3–4 mm, oblong-obovate; corolla ± 0.6 cm, bell-shaped, pale-pink or -blue, lobes ± 1/2 tube, ± ascending. **FR**: ± 7 mm. Clay substrates, occ serpentine, ann grassland, coastal-sage scrub, chaparral; 30–875 m. s SNF, SnJV/SCoRI, SnFrB, s SCoRO, SCo, ChI, WTR, PR; AZ, Baja CA. Apr–Jun ★

CRESSA

Per, subshrub, canescent. **ST**: prostrate to erect, not twining. **LF**: entire. **INFL**: fls 1 in upper lf axils, on ± 1 side of st. **FL**: calyx ± erect, hiding corolla tube; corolla lobes ± = tube; styles 2, stigma head-like. 4 spp.: trop & subtrop. (Greek: a Cretan woman) [Austin 2000 Bot J Linn Soc 133:27–39]

C. truxillensis Kunth (p. 663) ALKALI WEED Pl tufted. **ST**: 7–25 cm, much-branched from base, densely silky-canescent. **LF**: < 1 cm, ± sessile, elliptic. **INFL**: peduncle short, bracted. **FL**: 5–8 mm; sepals elliptic; corolla white, persistent, lobes ovate, acute; stamens, styles exserted. **SEED**: often 1, by abortion. 2*n*=28. Saline and alkaline substrates; < 1500 m. GV, CW (exc SCoRO), SCo, ChI, GB (exc Wrn, W&I), D (exc DMtns); to OR, TX, Mex, S.Am. May–Oct

CUSCUTA DODDER

Mihai Costea & Saša Stefanović

Vine, ann (per if on perennial host), not in contact with ground, attached to, holoparasitic on host by many small, specialized roots (haustoria) along st, gen glabrous. **ST**: thread-like, ± green, yellow, orange, or ± red. **LF**: 0 or scale-like, alternate, ± 2 mm. **INFL**: gen cyme, head- to panicle-like (fls 1), subtended by 0–3 bracts. **FL**: bisexual, radial, parts gen in 4s or 5s; calyx gen divided 2/5–3/5, persistent gen ± cream-white; corolla gen ± white, persistent (withered in fr) or not, tube cup-shaped to cylindric, bulged or horizontally ridged below lobes or gen not, gen with scales subtending stamens, lobes alternate stamens, erect to reflexed; ovary superior, chambers 2, each 2-ovuled, styles 2, gen free, persistent, stigmas 2, gen spheric, persistent. **FR**: capsule, gen indehiscent to irregularly dehiscent (or circumscissile near base), spheric to ovoid, depressed or not, thickened and/or raised around gen inconspicuous opening between styles or not. **SEED**: 1–4; coat papillate when hydrated, honeycombed when dry, (rarely neither, with cells ± rectangular, in ± jigsaw-puzzle-like arrangement); embryo gen slender, 1–3-coiled. ± 200 spp.: cosmopolitan, esp warmer regions of W. Hemisphere and Polynesia. (Aramaic, Hebrew: to cover, from habit) [Costea & Stefanović 2009 Syst Bot 34:570–579] By persistent, withered corolla, fr may be "capped" (corolla on top of fr), "surrounded" (fr at least in part visible, corolla ± loosely around fr), or "enclosed" (fr not visible, corolla ± tightly around fr). *C. pentagona* Engelm. excluded.

1. Styles 2, fused (subg. *Monogynella*)
  2. Style ≥ stigmas . . . . *[C. japonica* var. *formosana]*
  2′ Style < stigmas. (Extirpated) . . . . *[C. reflexa]*
1′ Styles 2, free
  3. Stigmas elongate; calyx lobe tip with a short, obtuse, fleshy appendage (subg. *Cuscuta*) . . . . *C. approximata*
  3′ Stigmas spheric to spheric-depressed; calyx lobe tip without appendage (subg. *Grammica*)
    4. Corolla scales 0
      5. Corolla lobes 1/3–1/2 tube, ovate-triangular, tips incurved . . . . *C. jepsonii*
      5′ Corolla lobes 3/4 to > tube, narrowly lanceolate to lanceolate, tips not incurved
        6. Styles 0.5–1(1.5) mm; anthers 0.25–0.5 mm, wide-elliptic; fr translucent, surrounded by corolla. . . *C. occidentalis*
        6′ Styles 1.2–3 mm; anthers 0.7–1.1 mm, oblong to linear; fr not translucent, ± enclosed by corolla
          (top not or barely visible)

7. Calyx 1/2–1/4 corolla tube; corolla lobes ≤ tube; filaments 0.2–0.6 mm . . . . *C. brachycalyx*
7′ Calyx 3/4 to = corolla tube; corolla lobes ≥ tube; filaments 0.6–1.1 mm . . . . *C. californica*
    8. Ovary, fr ovoid-conic, tip pointed; seed 1 . . . . var. *apiculata*
    8′ Ovary, fr spheric to spheric-depressed; seeds (1)2–4
        9. Perianth, pedicels dense-papillate . . . . var. *papillosa*
        9′ Perianth, pedicels not papillate. . . . . var. *californica*
4′ Corolla scales present
  10. Corolla scales gen reduced, shallowly finely dentate (rarely 2-lobed, ± fringed into 1–3 short units on
      each side of filaments) . . . . *C. suksdorfii*
  10′ Corolla scales well developed, fringed
    11. Calyx, corolla lobe tips dissimilar, calyx lobe tips obtuse to rounded, corolla lobe tips acute
      12. Calyx 1/2–3/4 corolla tube. (Extirpated) . . . . *C. suaveolens*
      12′ Calyx gen ± = corolla tube. . . . . *C. campestris*
    11′ Calyx, corolla lobes tips similar, both either acute to long-acuminate or obtuse to rounded
      13. Calyx, corolla lobes obtuse to rounded
        14. Calyx veined, shiny; calyx, corolla lobes finely dentate; seeds 1, spheric to spheric-ovoid. . . . . . . *C. denticulata*
        14′ Calyx not veined, not shiny; calyx, corolla lobes entire or calyx lobes finely serrate; seeds (1)2–4
          15. Fl parts in (3s)4s(5s); fr capped by corolla . . . . *C. cephalanthi*
          15′ Fl parts in (4s)5s; fr surrounded by corolla. (Extirpated) . . . . *C. obtusiflora* var. *glandulosa*
      13′ Calyx, corolla lobes acute to long-acuminate
        16. Corolla lobe tips incurved. . . . . *C. indecora* var. *indecora*
        16′ Corolla lobe tips straight or recurved but not incurved
          17. Embryo enlarged on 1 end
            18. Fls 1–2(5), (2.8)3–4(5) mm; calyx lobes lanceolate, ± entire; anthers 0.4–0.7 mm, filaments
              (0)0.1–0.3 mm; desert, on shrubs in *Atriplex, Ambrosia, Psorothamnus, Xylorhiza*. . . . . *C. nevadensis*
            18′ Fls (1)2–5, 1.9–3.2 mm; calyx lobes oblong to obovate-rhombic, ± irregularly finely dentate;
              anthers 0.3–0.4 mm, filaments 0.2–0.4 mm; desert, on *Bursera, Schinus*. (Extirpated) . . . . *C. veatchii*
          17′ Embryo not enlarged on 1 end
            19. Fl 5–7(9) mm; calyx gen 1/2 corolla tube, lobe bases overlapped; corolla lobes 1/4–1/3 tube;
              anthers 0.8–2 mm . . . . *C. subinclusa*
            19′ Fl 2.5–5(6) mm; calyx gen ± = corolla tube, lobe bases gen not overlapped; corolla lobes gen ± =
              tube; anthers 0.3–0.7 mm
              20. Fls in host infl, parts in 4s(5s); calyx, corolla lobes acuminate to long-acuminate; fr spheric to
                spheric-depressed, not thickened, not raised around opening between styles; seeds 1–4 . . . . . *C. howelliana*
              20′ Fls not in host infl, parts in 5s; calyx, corolla lobes acute to acuminate; fr elliptic-ovoid, ±
                thickened, ± raised around opening between styles; seeds 1–2
                21. Pedicels (0.5)1–5 mm; fl 2.5–4.5 mm; corolla tube ± cylindric, lobes reflexed to spreading,
                  ovate- to lance-oblong; inland salt flats. . . . . *C. salina*
                21′ Pedicels 0.5–2 mm; fl 3.5–6 mm; corolla tube bell-shaped, lobes spreading to erect, widely
                  ovate; coastal salt marshes, interdune depressions, tidal flats . . . . *C. pacifica*
                  22. Pedicels, calyx papillate; interdune depressions. . . . . var. *papillata*
                  22′ Pedicels, calyx not papillate; coastal salt marshes, tidal flats. . . . . var. *pacifica*

C. approximata Bab. (p. 663) ALFALFA DODDER **INFL:** head-like, fls 8–40; pedicels gen 0. **FL:** 3–4.2 mm, fleshy, not papillate, parts in 5s; calyx ± = corolla tube, bell-shaped, divided 1/3–1/2, veined, shiny, lobes widely triangular, bases overlapped, margins entire, tip with a short, obtuse, fleshy appendage; corolla tube 1.5–2 mm, bell-shaped (spheric or urn-shaped in fr), scales ± reaching stamen bases, oblong to obovate, rounded, shallowly fringed in distal 1/2, lobes spreading, gen = tube, ovate-round, margins entire, tip obtuse, straight; filaments 0.3–0.5 mm, anthers exserted, 0.3–0.5 mm, ovate; styles 0.8–1.6 mm, stigmas elongate. **FR:** circumscissile, 1.5–2.3 mm, 1.6–2 mm wide, depressed-spheric, not thickened or raised around opening between styles, translucent, capped by corolla. **SEED:** 4, 0.9–1.1 mm, 0.6–0.75 mm wide, ovate to widely elliptic. 2*n*=14,28. Uncommon. Esp on Fabaceae (e.g., alfalfa, clover); gen < 1500 m. NCoR, GV, MP; to w Can, UT; Asia, Eur, n Afr. May–Sep

C. brachycalyx (Yunck.) Yunck. **INFL:** panicle-like, fls 1–9; pedicels 1–6 mm. **FL:** 4.5–6 mm, not papillate, parts in 5s; calyx 1/4–1/2 corolla tube, bell- or cup-shaped, gen divided 2/3, not veined, not shiny when membranous or fleshy, lobes triangular-ovate, bases ± overlapped, margins entire, tip acute; corolla tube 2–3.5 mm, narrowly bell- to urn-shaped, bulged below lobes, scales 0, lobes erect, reflexed in age, ≤ tube, triangular-lanceolate, margins entire, tips acute, straight; anthers exserted, 0.7–1.1 mm, oblong to linear; styles 1.2–3 mm, ≥ ovary. **FR:** 1.6–2.3 mm, 1.8–2.5 mm wide, spheric,

spheric-depressed or obovoid, tip not pointed, not thickened or raised around opening between styles, not translucent, enclosed by corolla (top not visible). **SEED:** 2–4, 0.9–1.4 mm, 0.9–1.2 mm wide, widely elliptic to obovate. On herbs, chaparral, grassland, yellow-pine, red-fir forests; gen < 2500 m. CA-FP; to OR. Apr–Sep

C. californica Hook. & Arn. CHAPARRAL DODDER **INFL:** head- to panicle-like, fls 3–20; pedicels (0.5)1–2.5(3) mm. **FL:** 3–5(5.5) mm, papillate or not, parts in 5s; calyx 3/4 to = corolla tube, bell-shaped, gen divided 1/2–2/3, not veined, shiny when membranous, not shiny when fleshy, lobes triangular to lanceolate, bases overlapped, margins entire, tip acute to acuminate; corolla tube 1.6–2.4 mm, bell-shaped, bulged below lobes or not, scales 0, lobes erect, reflexed in age, ≥ tube, narrowly lanceolate, margins entire, tips acute to acuminate, straight; anthers exserted, 0.6–1 mm, oblong to linear; styles 1.2–2.2 mm, ≥ ovary. **FR:** 1.5–2.2 mm, 1.2–2.5 mm wide, spheric or ovoid-conic, tip pointed or not, not thickened or raised around opening between styles, not translucent, ± enclosed by corolla (top barely visible). **SEED:** 1–4, 0.9–1.3 mm, 0.8–1.1 mm wide, widely elliptic to obovate.

var. *apiculata* Engelm. **FL:** perianth (and pedicel) not papillate; filaments 0.5–1.1 mm; ovary (and fr) ovoid-conic, tip pointed. **SEED:** 1. On herbs; gen < 500 m. e DSon (near Colorado River); NV, Baja CA. Mar–Sep ★

var. ***californica*** (p. 663) **FL**: perianth (and pedicel) not papillate; filaments 0.5–1.1 mm; ovary (and fr) spheric to spheric-depressed. **SEED**: (1)2–4. On herbs, shrubs, roadsides, chaparral, grassland, yellow-pine forest; gen < 2500 m. CA-FP, SNE; to WA, CO, Mex. May–Sep

var. ***papillosa*** Yunck. **FL**: perianth (and pedicel) densely papillate; filaments 0.4–1 mm; ovary (and fr) spheric to spheric-depressed. **SEED**: (1)2–4. *n*=14. On herbs, shrubs in chaparral; < 1500 m. NCoRI, SNF, GV, SCoRO, SCo. Mar–Sep

C. campestris Yunck. FIELD DODDER **INFL**: head- to raceme-like, fls 6–25(30); pedicels 0.3–2.5(3.5) mm. **FL**: (1.9)2.1–3.6 mm, membranous, not papillate, parts gen in 5s; calyx gen ± = corolla tube, cup-shaped, divided 2/5–3/5, lobes ovate-triangular, veined, shiny, bases overlapped, margins entire, tips obtuse to rounded; corolla tube (1.1)1.5–1.9 mm, bell-shaped, scales ≥ corolla tube, oblong-ovate to spoon-shaped, rounded, uniformly densely fringed, lobes spreading, triangular-lanceolate, ± = tube, margins entire, tips acute to acuminate, incurved; filaments 0.4–0.7 mm, anthers exserted, (0.3)0.4–0.5 mm, widely elliptic; styles 0.8–1.6 mm, gen = ovary. **FR**: 1.3–2.8 mm, 1.9–3.8 mm wide, spheric-depressed to depressed, not thickened or raised around opening between styles, translucent or not, corolla enveloping 1/3 or less of fr. **SEED**: 4, 1.1–1.5 mm, 0.9–1.1 mm wide, ± spheric to widely elliptic. 2*n*=56. Gen on herbs, many crops, roadsides, fields; < 500 m. NCo, NCoRI, SNF, GV, CCo, SCo; worldwide. Treated as synonym of *C. pentagona* Engelm. in TJM (1993), but more recent work has shown that *C. campestris* and *C. pentagona* are distinct spp., and that the latter does not occur in CA. May–Nov

C. cephalanthi Engelm. (p. 663) BUTTONBUSH DODDER **INFL**: spike- or panicle-like, fls 3–18; pedicels 0–1 mm. **FL**: 2–3 mm, membranous, not papillate, parts in (3s)4s(5s); calyx gen 1/2 corolla tube, shallowly cup-shaped, divided 2/3, not veined, not shiny, lobes oblong-ovate, bases ± overlapped, margins entire to finely serrate, tip obtuse; corolla tube 1.8–2.2 mm, narrowly bell-shaped to cylindric, scales nearly reaching stamen bases, oblong, rounded, sparsely fringed laterally, more densely at tip, lobes erect to reflexed, 1/4–1/3 tube, ovate, margins entire, tip obtuse, straight; filaments 0.3–0.6 mm, anthers incl to ± exserted, 0.3–0.5 mm, ovate; styles (0.6)1–2 mm, ≥ ovary. **FR**: 2.5–3.2(4) mm, 2–4 mm wide, depressed-spheric to spheric, not thickened or raised around opening between styles, not translucent, capped by corolla. **SEED**: (1)2–3, 1.5–2 mm, 1.3–1.45 mm wide, widely ovate. 2*n*=60. On herbs, woody pls, near streams, rivers, lakes; < 1500 m. KR, CaR, Wrn; to Can, also e&s US. Jun–Oct

C. denticulata Engelm. SMALL-TOOTH DODDER **INFL**: head-like, 2–7(12); pedicels (0)0.5–2.2 mm. **FL**: 1.8–3.1 mm, membranous, not papillate, parts gen in 5s; calyx ± = corolla tube, bell- to urn-shaped, gen divided 2/3, veined, shiny, lobes obovate-round, bases overlapped, margins finely dentate, tip rounded; corolla tube 0.6–1.5 mm, bell-shaped, scales ± = corolla tube, oblong-ovate, rounded, uniformly finely dentate or fringed, lobes reflexed, ± = tube, ovate to widely elliptic, margins finely dentate, tip rounded, straight; filaments 0.2–0.4 mm, anthers incl to ± exserted, 0.25–0.4 mm, ± round to elliptic; styles 0.3–0.5 mm, < ovary. **FR**: 1.3–2.1 mm, 1–2 mm wide, spheric-ovoid, not thickened or raised around opening between styles, translucent, capped by corolla. **SEED**: 1, 0.85–1.1 mm, 0.8–1.1 mm wide, spheric to spheric-ovoid, embryo enlarged on 1 end. 2*n*=30. On herbs, esp shrubs in creosote-bush scrub, Joshua-tree woodland; gen < 1300 m. W&I, D; to UT, AZ, Baja CA. May–Oct

C. howelliana P. Rubtzov BOGGS LAKE DODDER **INFL**: in infl of host, head-like, fls 3–30; pedicels 0–0.6 mm. **FL**: 3–4.5 mm, membranous, papillate, parts in 4s(5s); calyx ≥ corolla tube, bell-shaped, divided 1/2–2/3, finely veined, shiny, lobes triangular-ovate, bases not overlapped, margins entire, tip acuminate to long-acuminate, recurved; corolla tube 1.5–2.2 mm, narrowly bell- to urn-shaped, scales ± 1/2 corolla tube, oblong-ovate, rounded, uniformly densely fringed, lobes ± erect to spreading, gen = tube, triangular-ovate, margins entire to long-acuminate, recurved; filaments 0.1–0.4 mm, anthers incl, 0.4–0.7 mm, elliptic; styles 0.4–1.1 mm, 1/4 to gen = ovary. **FR**: 1.2–1.5 mm, 0.8–1.2 mm wide, spheric to spheric-depressed, not thickened or raised around opening between styles, not translucent, enclosed by corolla, capped by it in age. **SEED**: 1–4,

0.9–1.4 mm, 0.8–1.3 mm wide, ± round to widely elliptic. Margins of vernal pools gen on *Eryngium, Navarretia, Polygonum polygaloides* subsp. *kelloggii, Epilobium campestre*; 140–1650 m. KR/CaRH, NCoRI, CaRF, n SNF, GV, MP. Aug–Sep

C. indecora Choisy var. ***indecora*** (p. 663) LARGE-SEEDED DODDER **INFL**: raceme- or panicle-like, fls 5–40; pedicels 0.5–6 mm. **FL**: 2–5.3 mm, fleshy, papillate, parts in 5s; calyx 1/2–3/4 corolla tube, cup-shaped, divided 1/3–2/3, not veined or shiny, lobes triangular-ovate to lanceolate, bases overlapped or not, margins entire, tip acute; corolla tube 1–3.2 mm, bell- or ± urn-shaped to ± spheric, scales reaching stamen bases, ± spoon-shaped, rounded, rarely truncate or 2–3(4) lobed, uniformly densely fringed, lobes ± erect, 1/3 to = tube, triangular-ovate, margins entire, tip acute, incurved; filaments 0.3–0.7 mm, anthers incl to ± exserted, 0.3–0.5 mm, ovate-elliptic to oblong; styles 1–2.5 mm, gen = ovary. **FR**: 2–3.5 mm, 1.9–4(5) mm wide, spheric to subspheric, thickened, raised around opening between styles, translucent, surrounded or capped by corolla. **SEED**: 2–4, 1.4–1.9 mm, 1.25–1.6 mm wide, widely elliptic to oblique, shape varied on 1 pl . 2*n*=30. Common. On herbs, woody pls, often in moist fields, roadsides; gen < 1500 m. NCo, NCoR, SN, GV, D; to Can, c&se US, Caribbean, Mex, S.Am. [*C. i.* var. *neuropetala* (Engelm.) Hitchc.] Jun–Nov

C. jepsonii Yunck. **INFL**: head- or raceme-like, fls 3–7; pedicels 0.5–1.5 mm. **FL**: 2–2.7(3) mm, fleshy, papillate, parts in 5s; calyx gen 1/2 corolla tube, shallowly cup-shaped, divided 1/2, not veined or shiny, lobes triangular, bases not overlapped, margins entire, tips acute; corolla tube 1.3–2 mm, bell or ± urn-shaped to spheric to, scales 0, lobes erect, 1/3–1/2 tube, ovate-triangular, margins entire, tip acute, incurved; filaments 0.2–0.3 mm, anthers incl, widely elliptic, 0.2–0.3 mm; styles 0.4–0.8 mm, < ovary. **FR**: ± spheric to ± depressed-spheric, 2–3 mm, 2–3.5 mm wide, thickened, raised around opening between styles, translucent, surrounded by corolla. **SEED**: 2–4, 0.9–1.1 mm, 0.8–1 mm wide. On *Ceanothus diversifolius, Ceanothus prostratus*; 1200–2300 m. KR, NCoRH, CaR, SNH. Possibly extinct; differs from *C. indecora*, with which it is sometimes united, in absence of corolla scales, smaller anthers. Jul–Sep ★

C. nevadensis I.M. Johnst. **INFL**: umbel-like, fls 1–2(5); pedicels (0.5)2.5–4 mm. **FL**: (2.8)3–4(5) mm, membranous exc receptacle, base of perianth fleshy or not, parts gen in 5s; calyx ± = corolla tube, narrowly bell-shaped, divided 2/3, gen veined, shiny, lobes lanceolate, bases overlapped, margins ± entire, tip acute to acuminate; corolla tube 1.3–2.5 mm, bell-shaped, scales reaching stamen bases, widely ovate to ± spheric, ± truncate, margins uniformly short-fringed, lobes ± erect to reflexed, = to ± > tube, ovate to oblong, margins entire to irregularly finely dentate, tip acute, straight; filaments (0)0.1–0.3 mm, anthers incl or ± exserted, 0.4–0.7 mm, elliptic to oblong; styles 0.5–1 mm, 1/2 to = ovary. **FR**: 1.4–2.1 mm, 1.2–2 mm wide, spheric-ovoid to ovoid, not thickened or raised around opening between styles, translucent, capped by corolla. **SEED**: 1, 0.9–1.2 mm, 0.85–1.1 mm wide, spheric-ovoid; embryo enlarged on 1 end. Desert, on shrubs in *Atriplex, Ambrosia, Psorothamnus, Xylorhiza*; 400–1500 m. DMtns; NV. Gen misidentified as *C. denticulata, C. salina*. May–Oct

C. obtusiflora Kunth var. ***glandulosa*** Engelm. **INFL**: spike- or panicle-like, fls 5–18; pedicels 0–1 mm. **FL**: 1.8–2.5 mm, membranous, not papillate, parts in (4s)5s; calyx gen = corolla tube, shallowly cup-shaped, divided 1/2, not veined, not shiny, lobes ovate, bases ± overlapped, margins entire, tip obtuse; corolla tube 1–1.5 mm, bell-shaped, scales reaching stamen bases, oblong, rounded, fringed laterally, densely in apical 2/3, lobes erect to spreading, gen = tube, ovate to ovate-oblong, margins entire, tip obtuse, straight; filaments 0.4–0.6 mm, anthers exserted, 0.3–0.4 mm, ovate to widely elliptic; styles 0.4–1.1 mm, ≤ ovary. **FR**: 1.5–3 mm, 2.5–4 mm wide, depressed-spheric, not thickened or raised around opening between styles, not translucent, surrounded by corolla. **SEED**: gen 4, 1.4–1.6 mm, 1.2–1.35 mm wide, widely ovate to widely elliptic. On herbs incl *Alternanthera, Dalea, Lythrum, Polygonum, Xanthium*; ± < 500 m. Extirpated, formerly sporadically collected in San Bernardino Co., 1890–1898, Sonoma, Butte, Merced cos., 1940s; e&s US, West Indies, Mex. Jul–Oct ★

C. occidentalis Millsp. **INFL**: head-like, fls 3–30; pedicels 0–0.5(1.5) mm. **FL**: 2.7–3.4 mm, membranous, gen not papillate, parts in 5s; calyx ± = corolla tube, bell-shaped, divided 2/5–1/2, not veined, gen shiny, lobes lanceolate to narrowly ovate, bases not overlapped, margins entire, tip acuminate; corolla tube 1.4–2.1 mm, cylindric to bell-shaped, bulged below lobes, scales 0, lobes erect, spreading (reflexed) in age, 3/4 to = tube, lanceolate, margins entire, tip acuminate, straight; filaments 0.3–0.6 mm, anthers ± exserted, 0.25–0.5 mm, widely elliptic; styles 0.5–1(1.5) mm, < ovary. **FR**: 1.8–2.2 mm, 2–2.6 mm wide, spheric to spheric-depressed, ± thickened, not raised around opening between styles, translucent, surrounded by corolla. **SEED**: 2–4, 0.85–1.3 mm, 0.8–1.1 mm wide, ± spheric to widely elliptic. Gen on herbs; < 2000 m. NW, c SNH, GV, CW, SNE; to WA, CO, Mex. [*Cuscuta californica* var. *breviflora* Engelm.] Mar–Sep

C. pacifica Costea & M. Wright GOLDENTHREAD **INFL**: umbel- to ± head-like, fls 2–17; pedicels 0.5–2 mm. **FL**: 3.5–6 mm, membranous, papillate or not, parts in 5s; calyx gen = corolla tube, bell- to cup-shaped, gen divided 1/3, not veined, not shiny, lobes ovate-triangular, bases ± overlapped, margins entire, tip acute to acuminate; corolla tube 1.5–2.6 mm, bell-shaped, scales 1/2–2/3 corolla tube, oblong, rounded, short-fringed, lobes spreading to erect, gen = tube, widely ovate, margins entire, tip acute to acuminate, straight; filaments 0.3–0.6 mm, anthers incl, 0.35–0.5 mm, widely elliptic to ± round; styles 0.4–1 mm, gen < ovary. **FR**: 2–3.6 mm, 1.4–2.1 mm wide, elliptic-ovoid, ± thickened, ± raised around opening between styles, not translucent, surrounded or capped by corolla. **SEED**: 1–2, 1.45–1.95 mm, 1.25–1.43 mm wide, ± spheric to widely elliptic.

var. ***pacifica*** (p. 663) **FL**: pedicels, calyx not papillate. 2*n*=30. Gen on *Salicornia, Jaumea* in coastal salt marshes, tidal flats; gen 0 m. NCo, CCo, SCo; to BC. [*C. salina* Engelm. var. *major* Yunck.] Jul–Oct

var. ***papillata*** (Yunck.) Costea & M. Wright **FL**: pedicels, calyx papillate. On herbs in coastal interdune depressions; 3–7 m. NCo (Mendocino Co.). [*C. salina* var. *papillata* Yunck.] Jul–Oct ★

C. salina Engelm. (p. 663) SALT DODDER **INFL**: umbel-like, fls 2–16; pedicels (0.5)1–5 mm. **FL**: 2.5–4.5 mm, membranous, papillate or not, parts in 5s; calyx gen = corolla tube, ± cylindric, gen divided 1/2, not veined, ± shiny, lobes lance-ovate to lanceolate, bases not overlapped, margins entire, tip acute to acuminate; corolla tube 1.2–2 mm, ± cylindric, scales 4/5–9/10 corolla tube, oblong to ± obovate, rounded, uniformly densely fringed, lobes reflexed to spreading, gen = tube, ovate- to lance-oblong, margins entire, tip acute to acuminate, straight; filaments 0.3–0.7 mm, anthers exserted, 0.3–0.7 mm, elliptic; styles 0.4–0.9 mm, gen < ovary. **FR**: 1.6–2.5 mm, 1.7–2.2 mm wide, elliptic-ovoid, ± thickened, ± raised around opening between styles, not translucent, surrounded or capped by corolla. **SEED**: 1, 1.35–1.6 mm, 1.25–1.43 mm wide, widely elliptic to ± spheric. 2*n*=30. Gen on herbs on inland salt flats; 50–100 m. KR, GV, SnFrB, W&I; to WA, UT, NM, Baja CA. May–Nov

C. suaveolens Ser. FRINGED DODDER **INFL**: ± umbel-like, fls 5–18; pedicels 2–5(7) mm. **FL**: 3–5 mm, gen membranous, parts in 5s; calyx 1/2–3/4 corolla tube, cup-shaped, gen divided 1/2, not veined, not shiny, lobes triangular ovate, bases ± overlapped, margins entire, rolled under, tip obtuse; corolla tube 2.5–3 mm, bell-shaped, scales gen = corolla tube, oblong-ovate, rounded, densely fringed,

lobes erect to spreading, ± 1/2 tube, ovate-triangular, not papillate, margins entire, tip acute, incurved; filaments 0.25–0.4 mm, anthers exserted, 0.6–0.8 mm, oblong-ovate; styles 1.6–5.5 mm. **FR**: 2.5–3.5 mm, 2.5–3.5 mm wide, spheric, thickened, not thickened or raised around opening between styles, not translucent, 3/4 surrounded by corolla. **SEED**: 2–4, 1.3–2 mm, 1–1.7 mm wide, ± spheric to widely elliptic. Gen on herbs, crops, roadsides, fields; 25–300 m. SnJV (extirpated); AL, NM, OH, SD, TX, Eur, native to S.Am. Jul–Oct

C. subinclusa Durand & Hilg. **INFL**: ± head-like, fls 5–20(45); pedicels 0–1 mm. **FL**: 5–7(9) mm, gen membranous, parts in 5s; calyx gen 1/2 corolla tube, bell-shaped, divided 3/5–2/3, finely veined, shiny, lobes widely ovate to lanceolate, bases overlapped, margins entire, tip acute (abruptly pointed); corolla tube 2.5–3.5(4.5) mm, cylindric, gen horizontally ridged below lobes, scales ± 1/2 corolla tube, oblong to spoon-shaped, rounded, irregularly short-fringed, lobes spreading to reflexed, 1/4–1/3 tube, ovate-triangular, papillate, margins entire, tip acute to ± acuminate, straight; filaments 0–0.1 mm, anthers incl to ± exserted, 0.8–2 mm, linear; styles 1–1.5 mm. **FR**: 1.5–3 mm, 1.2–2.5 mm wide, ± ovoid, thickened, raised around opening between styles, not translucent, capped by corolla. **SEED**: 1, 1.3–1.7 mm, 1.2–1.5 mm wide, ± spheric to widely ovoid. Gen on herbs, shrubs, in forests near streams, river canyon bottoms, salt marshes; < 2000 m. NCoR, SN, GV, SnFrB, SCoR, SCo, SnJt, GB; OR, Baja CA. Mar–Oct(Dec?)

C. suksdorfii Yunck. (p. 663) MOUNTAIN DODDER **INFL**: umbel-like, fls 2–7; pedicels 0–2 mm. **FL**: 2.8–3.3 mm, membranous, not papillate, parts in 4s or 5s; calyx > corolla tube, widely bell-shaped, divided 1/2–3/5, not veined, not shiny, lobes ovate, bases not overlapped, margins entire, tip long-acuminate; corolla tube 1.2–1.5 mm, bell-shaped, scales gen reduced, 1/2–3/4 tube, oblong, shallowly finely dentate (2-lobed, ± fringed into 1–3 short units on each side of filaments), lobes ± erect, > tube, triangular-ovate, margins entire, tip long-acuminate, straight; filaments 0.2–0.5 mm, anthers incl or ± exserted, 0.2–0.4 mm, widely elliptic; styles 0.3–0.7 mm, gen 1/4 ovary. **FR**: 2–3.2 mm, 2–3.6 mm wide, elliptic-ovoid, ovoid-conic, spheric to spheric-depressed, not thickened or raised around opening between styles, translucent, lower 1/2 surrounded by corolla. **SEED**: 2–4, 0.8–1.1 mm, 0.8–1.02 mm wide, ± spheric. Gen on herbs in Asteraceae, *Calyptridium, Trifolium*, in mtn meadows; 1500–2600 m. KR, n&c SNH, SnBr; to WA. Jul–Sep

C. veatchii Brandegee **INFL**: umbel-like, fls 1(2–5); pedicels 0.5–2.5(3.5) mm. **FL**: 1.9–3.2 mm, membranous exc receptacle, base of perianth fleshy or not, parts in (4s)5s; calyx ± ≥ corolla tube, ± bell-shaped, divided ± 2/3, gen veined, shiny, lobes oblong to obovate-rhombic, bases overlapped, margins ± irregularly finely dentate, tip acute to acuminate; corolla tube 0.8–1.6 mm, bell-shaped, scales reaching stamen bases, oblong, truncate, margins uniformly densely fringed, lobes suberect to reflexed, = to ± > tube, ovate to oblong, margins entire to irregularly finely dentate, tip acute, straight; filaments 0.2–0.4 mm, anthers incl or ± exserted, 0.3–0.4 mm, ± elliptic; styles 0.3–0.8 mm, 1/3 to 1/2 ovary. **FR**: 1.4–2.2 mm, 1.2–1.8 mm wide, spheric-ovoid to ovoid, not thickened or raised around opening between styles, translucent, capped by corolla. **SEED**: 1, 0.8–1.15 mm, 0.8–1.1 mm wide, ± spheric-ovoid; embryo enlarged on 1 end. Desert, on *Bursera, Schinus*; ± 400–1500 m. DSon (San Diego Co.; extirpated); Baja CA. Mar–Oct(Dec)

DICHONDRA

Per, matted from creeping stolons; hairs equally forked. **LF**: reniform; petiole > blade. **INFL**: fls 1 in axils; bracts 0; upper peduncle recurved in fr. **FL**: inconspicuous; calyx lobes 5, deep, ± equal, ovate to obovate; corolla ± > calyx, lobes > tube; ovary shallowly to deeply 2-lobed, styles 2, free or united only at base, stigmas head-shaped. **FR**: capsule or separating into nutlets, spheric to ± 2-lobed. 15+ spp.: temp & trop. (Greek: double grain, from deeply lobed fr of some) [Austin 1998 Econ Bot 52:88–106]

1. Fr spheric to ± notched capsule, each valve often 2-seeded — s CCo, SCo, ChI, PR . . . . ***D. occidentalis***
1′ Fr gen deeply 2-lobed, gen separating into 2, gen 1-seeded nutlets
  2. St 1–2 mm thick; calyx in fr ≥ 2.5 mm; NCo, w KR, n SNF, CCo, SnFrB . . . . ***D. donelliana***
  2′ St < 1 mm thick; calyx in fr < 2.5 mm; n SNF, SCo, expected elsewhere . . . . ***D. micrantha***

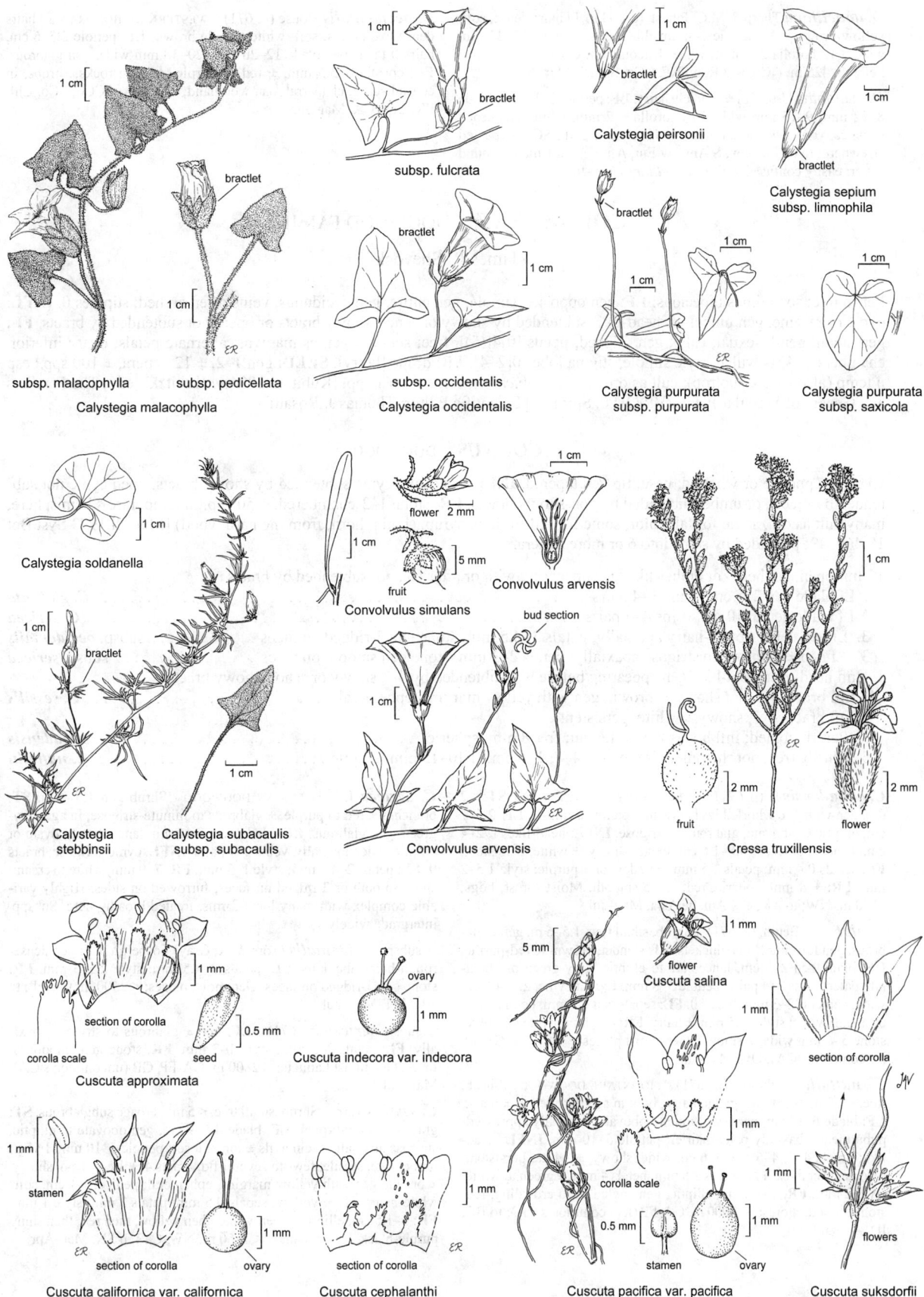

subsp. malacophylla

subsp. pedicellata

Calystegia malacophylla

subsp. fulcrata

bractlet

subsp. occidentalis

Calystegia occidentalis

bractlet

Calystegia peirsonii

bractlet

Calystegia sepium
subsp. limnophila

bractlet

Calystegia purpurata
subsp. purpurata

Calystegia purpurata
subsp. saxicola

Calystegia soldanella

Calystegia
stebbinsii

bractlet

Calystegia subacaulis
subsp. subacaulis

flower

fruit

Convolvulus simulans

Convolvulus arvensis

bud section

Convolvulus arvensis

fruit

flower

Cressa truxillensis

section of corolla

corolla scale

seed

Cuscuta approximata

ovary

Cuscuta indecora var. indecora

flower

Cuscuta salina

corolla scale

stamen

ovary

section of corolla

Cuscuta pacifica var. pacifica

stamen

section of corolla

ovary

Cuscuta californica var. californica

section of corolla

Cuscuta cephalanthi

flowers

Cuscuta suksdorfii

D. donelliana Tharp & M.C. Johnst. (p. 671) Pl hairs dense, soft, yellow-brown. **LF**: petiole 2–4 cm; blade 10–15 mm, 15–25 mm wide. **FL**: corolla 2–3 mm, white. Uncommon. Open slopes, moist fields; < 425 m. NCo, w KR, n SNF, CCo, SnFrB. Mar–Jun

D. micrantha Urb. Pl ± sparsely hairy. **LF**: petiole 2–3.5 cm; blade 8–12 mm, 9–15 mm wide. **FL**: corolla ± 2 mm, white. Disturbed sites, canyons near urban areas; < 1000 m. n SNF, SCo, expected elsewhere; to e US, Mex, S.Am; sw Eur, Afr, s Asia. Cult as ground cover; easily confused with *D. donelliana*. Apr–Jun

D. occidentalis House (p. 671) WESTERN DICHONDRA Pl hairs dense exc on lvs, soft, white to pale brown. **LF**: petiole 2.5–6 cm, hairs 0 to sparse; blade 12–20 mm, 20–30 mm wide, gen glabrous. **FL**: corolla 3–3.5 mm, ± red to purple. Among rocks, shrubs, in coastal scrub, chaparral, oak woodland; < 520 m. s CCo, SCo, ChI, PR; Baja CA. Mar–Jun ★

CORNACEAE DOGWOOD FAMILY

James R. Shevock

Per to tree; sometimes dioecious. **LF**: gen opposite, simple, gen entire, gen deciduous, veins often arched; stipules 0. **INFL**: cyme or raceme, gen umbel- or head-like, subtended by showy or ± non-showy bracts or open, not subtended by bracts. **FL**: gen small, gen bisexual; calyx gen 4-lobed; petals [0]4[(5)], free; stamens gen as many as, alternate petals; ovary inferior, chambers 1–4, 1-ovuled, style simple, stigma lobes 0[2–4]. **FR**: drupe [berry]. **SEED**: gen 1–2. ± 12 genera, ± 100 spp.: esp n temp (also s trop, subtrop); cult as orn (*Cornus*, *Aucuba*); some timber spp. [Kubitzki 2004 *in* Kubitzki (ed.), The Families and Genera of Vascular Plants. VI:82–90. Springer] Scientific Editor: Thomas J. Rosatti.

CORNUS DOGWOOD

LF: gen opposite or whorled; base, tip gen tapered. **INFL**: head-like cyme subtended by showy bracts, open cyme not subtended by bracts, or umbel subtended by non-showy bracts. **FR**: stone 1–2-chambered. ± 50 spp.: n temp, rare s hemisphere; many cult as orn, some for fall color; some fr used for jam, syrup. (Latin: horn, from the hard wood) [Murrell 1993 Syst Bot 18:469–495] Divided by some into 6 or more genera.

1. Infl open, not head- or umbel-like, gen appearing with or after lvs, not subtended by bracts
  2. Lf blade gen 2–5 cm, veins 3–4 pairs . . . . ***C. glabrata***
  2′ Lf blade gen 5–10 cm, veins 4–7 pairs. . . . . ***C. sericea***
    3. Lf gen dense-rough-hairy abaxially; petals 3–4.5 mm; stone gen 3-ridged on faces . . . . subsp. ***occidentalis***
    3′ Lf gen ± glabrous to strigose abaxially; petals 2–3 mm; stone gen smooth on faces . . . . subsp. ***sericea***
1′ Infl head- or umbel-like, gen appearing before lvs, subtended by 4–7 showy or ± non-showy bracts
  4. Infl bracts 4, ± not showy, ± brown, gen with yellow margins, ephemeral. . . . . ***C. sessilis***
  4′ Infl bracts 4–7, showy, ± white, persistent
    5. Per, rhizomed; infl bracts 4, 0.8–1.6 cm; fr ± 8 mm, spheric . . . . ***C. canadensis***
    5′ Shrub, tree, not rhizomed; infl bracts 4–7, 4–6 cm; fr 10–15 mm, elliptic . . . . ***C. nuttallii***

C. canadensis L. (p. 671) BUNCHBERRY Per, rhizomed. **ST**: < 2 dm, gen with 4–6 whorled lvs below infl, pair near mid-st. **LF**: 2.5–7 cm, elliptic to obovate, glabrous to strigose. **INFL**: head-like, 1, 2–4 cm, slender; bracts 4, 0.8–1.6 cm, ovate, showy, ± white, persistent. **FL**: sepals 0.4 mm; petals 1.5 mm, ± yellow or ± purple; style 1.5–2 mm. **FR**: ± 8 mm, spheric, red; stone smooth. Moist forest, bogs; 1100 m. NW; to AK, e N.Am; ne Asia. May–Jul

C. glabrata Benth. (p. 671) Shrub, small tree, 1.5–6 m, gen ± glabrous, gen forming clonal thickets. **ST**: slender, brown- to red-purple. **LF**: blade gen 2–5 cm, lanceolate to elliptic, gray-green or abaxially paler, veins 3–4 pairs; petiole 3–8 mm. **INFL**: cyme, 2.5–4.5 cm wide; pedicels 2–3 mm; bracts 0. **FL**: sepals gen < 1 mm; petals 4.5–5 mm, dull white; styles 3.5 mm, ± hairy. **FR**: 8–9 mm, white to ± blue; stone 5–6 mm wide, ± smooth. Gen moist places; < 1550 m. CA-FP (uncommon s CA); OR. May–Jun

C. nuttallii Audubon (p. 671) MOUNTAIN DOGWOOD Shrub, tree, < 25 m. **ST**: twigs green, hairy; bark in age dark red to ± black. **LF**: blade 6–12 cm, narrow-elliptic to obovate, adaxially appressed-puberulent, abaxially paler, hairier; petiole 5–10 mm. **INFL**: head-like; bracts 4–7, 4–6 cm, 3–6 cm wide, showy, ± white, persistent; receptacle convex. **FL**: sepals 2.5 mm; petals 4 mm, ± green to white; style 2 mm. **FR**: 1–1.5 cm, elliptic, gen angled from crowding, red; stone smooth. Forest; < 2000 m. CA-FP (less common s CA); to BC, ID. Apr–Jul

C. sericea L. AMERICAN DOGWOOD Shrub gen 1.5–4 m. **ST**: branches ± red to purple, ± glabrous to minute-strigose, in age gray-green, gen glabrous. **LF**: blade gen 5–10 cm, lanceolate to ovate or elliptic, paler abaxially, veins 4–7 pairs. **INFL**: cyme, strigose; bracts 0. **FL**: petals 2–4.5 mm; style 1–3 mm. **FR**: 7–9 mm, white to cream; stone smooth or 3-ridged on faces, furrowed on sides. Highly variable complex with many local forms, treated broadly here. Subspp. intergrade widely.

  subsp. ***occidentalis*** (Torr. & A. Gray) Fosberg **LF**: gen dense-rough-hairy abaxially. **FL**: petals 3–4.5 mm; style 2.5–3 mm. **FR**: stone gen 3-ridged on faces. Gen moist places; < 2500 m. CA-FP; to AK, MT. May–Jul

  subsp. ***sericea*** (p. 671) **LF**: gen ± glabrous to strigose abaxially. **FL**: petals 2–3 mm; style 1–2 mm. **FR**: stone gen smooth on faces. Gen moist habitats; < 2800 m. CA-FP, GB (uncommon s CA). May–Jul

C. sessilis Torr. Shrub, small tree, < 5 m; herbage subglabrous. **ST**: gray or yellow-brown. **LF**: blade 4.5–9 cm, gen obovate to elliptic, strigose abaxially (vein axils ± tomentose); petiole 5–10 mm. **INFL**: umbel-like, sessile, few- to several-fld; bracts 4, ± 1 cm, ± not showy, ± brown, gen with yellow margins, ephemeral; pedicel ± 1 cm, soft-white-hairy. **FL**: ± yellow; sepals 0.5 mm; petals 3 mm; style 1 mm. **FR**: 1–1.5 cm, elliptic, green-white, then yellow, then red, then shiny purple-black. Streambanks; < 1550 m. NW, CaR, n SN. Mar–Apr

CRASSULACEAE STONECROP FAMILY

Steve Boyd, except as noted

Ann to shrub [(± tree-like or climbing)], fleshy. **LF**: gen simple, alternate or opposite, in dense to open, basal (or terminal) rosettes or basal and cauline, not in rosettes, reduced distally or not, margin often ± red. **INFL**: gen cyme, gen bracted. **FL**: gen bisexual; sepals gen 3–5, gen ± free; petals gen 3–5, ± free or fused; stamens >> to = sepals, epipetalous or not; pistils gen 3–5, simple, fused at base or not, ovary 1-chambered, placenta 1, parietal, ovules 1–many, style 1. **FR**: follicles, gen 3–5. **SEED**: 1–many, small. ± 33 genera, ± 1400 spp.: ± worldwide, esp dry temp; many cult for orn. [Eggli (ed.) 2003 Illus Handbook Succulent Pls 6 (Crassulaceae). Springer] Water-stressed pls often ± brown or ± red. Consistent terminology regarding lvs, bracts difficult; in taxa with rosettes (e.g., *Aeonium*, *Dudleya*, some *Sedum*), structures in rosettes are lvs, those on peduncles are bracts, and those subtending fls are fl bracts; in taxa where infl is terminal, rosette lvs may "become" bracts as st rapidly elongates to form infl. Seed numbers given per follicle. Scientific Editor: Thomas J. Rosatti.

1. Ann; lf < 1 cm; fl 1–5 mm
    2. Basal, cauline lvs opposite, bases fused, ± sheathing; fls 1 in axils of lvs; petals < 2 mm .... ³**CRASSULA**
    2′ Basal lvs opposite, cauline alternate, bases free, not sheathing; fls 1 or ≥ 2 in cyme, terminal; petals
        1.3–5 mm .... **SEDELLA**
1′ Gen per to shrub; lf > 1 cm; fl gen > 10 mm (if ann, fl > 4 mm)
    3. Subshrub to shrub
        4. Lvs alternate
            5. Lvs ciliate, rosettes terminal; sepals 7–11 .... **AEONIUM**
            5′ Lvs not ciliate, rosettes 0; sepals 5 .... ²**SEDUM**
        4′ Lvs opposite
            6. Lvs gen 2–4 pairs; petals fused, tube >> sepals .... **COTYLEDON**
            6′ Lvs gen > 4 pairs; petals free or ± fused at base, tube 0 .... ³**CRASSULA**
    3′ Per (ann or bien in *Sedum radiatum*)
        7. Infls axillary .... **DUDLEYA**
        7′ Infls terminal
            8. Sepals, petals gen 5 .... ²**SEDUM**
            8′ Sepals, petals gen 4
                9. Fls bisexual; stamens = sepals in number .... ³**CRASSULA**
                9′ Fls ± unisexual; stamens 2 × sepals in number .... **RHODIOLA**

AEONIUM

[Per] to shrub. **LF**: rosettes near st tips, lvs alternate, gen obovate to oblanceolate, ciliate [or not]. **INFL**: terminal; branches many, fls on 1 side. **FL**: erect, calyx, corolla not circumscissile in fr; sepals [6]7–11[16], ± fused basally; petals = sepals in number, ± free, > sepals, cream or bright yellow; stamens 2 × sepals in number; pistils erect, ± free. **SEED**: many, 0.4–0.6 mm, ellipsoid, striate, brown. 31 spp.: Afr, Yemen, Madeira, Cape Verde, esp Canary Islands. (Latin: name given to *Aeonium arboreum* by Dioscorides, Greek botanist)

1. Lvs 50–75, green or brown-purple, ± shiny; sts few-branched, fleshy, bark ± smooth; adventitious prop roots
    0; fl rotate, petals 9–11, yellow; pedicel, calyx puberulent .... *A. arboreum* var. *arboreum*
1′ Lvs 15–25, gray-green, ± dull; sts much-branched, woody, bark rough; adventitious prop roots many; fl
    bell-shaped, petals 7–9, cream; pedicel, calyx glabrous .... *A. haworthii*

A. arboreum (L.) Webb & Berthel. var. *arboreum* Subshrub 1(2) m, ± open. **ST**: 10–40 mm diam, lf scars distinct. **LF**: rosettes gen dense, 10–20 cm diam, lvs 5–9(15) cm, 1.5–3 mm thick, oblong-oblanceolate. **INFL**: dense, conic to ovoid. **FL**: ± 2 cm wide; sepals 2–3.5 mm, 1.1–1.4 mm wide; petals narrow-oblong to lanceolate, 5–7 mm, 1.5–2 mm wide. 2*n*=[36],72. Bluffs, dunes, or persisting from cult; < 100 m. SCo; native to Canary Islands. Oct–Apr

A. haworthii Webb & Berthel. [Sub]shrub ± 7 dm, densely hemispheric. **ST**: 3–10 mm diam. **LF**: rosettes gen open, < 10 cm diam, lvs 3–6 cm, 2–5 mm thick, obovate. **INFL**: open, flat-topped to hemispheric. **FL**: ± 1 cm wide; sepals 3–4 mm, 1.5–2 mm wide; petals lanceolate, 7–9 mm, 1.2–1.8 mm wide. 2*n*=72. Sea cliffs, dunes, or persisting from cult; < 100 m. c&s SCo; native to Canary Islands. Apr–Jul

COTYLEDON

Shrub. **LF**: opposite, sessile; margin entire. **INFL**: terminal. **FL**: pendent, calyx, corolla circumscissile at base in fr; sepals 5, ± free; corolla lobes 5, tube > sepals, often > lobes; stamens 10, epipetalous on tube; pistils erect, ± free. **SEED**: many, elliptic, ridged. 9 spp.: esp s Afr; some cult for orn. (Greek: cavity)

C. orbiculata L. var. *oblonga* (Haw.) DC. Pl < 1 m, erect, branched below, glabrous [or not]. **LF**: gen 2–4 pairs, ± clustered near st tips or not, 5–18 cm, obovate to wedge-shaped, glaucous; tip rounded to obtuse, abruptly pointed. **INFL**: 10–30-fld, peduncle 20–40(50) cm, thick, ± bractless; pedicels pendent in fl, erect in fr. **FL**: sepals 2–5 mm; corolla orange to red or pink, tube (15)20–25 mm, cylindric, with hairy ring in proximal 1/4, lobes 10–15 mm, recurved. 2*n*=18. Coastal bluffs, dunes, or persisting from cult; < 100 m. c&s CCo, SCo; native to s Afr. Highly TOXIC to sheep and goats but rarely eaten. Jun–Jul

CRASSULA

Ann, per, shrub, glabrous (hairy). **ST:** erect to decumbent, branched or not. **LF:** opposite, 0.1–7 cm, linear to deltate or obovate, bases fused, ± sheathing; margins gen entire. **INFL:** terminal panicle or fls 1 in axils of lvs, either 2 per node, axillary, or 1 per node, terminal but appearing axillary by overtopping of main axis. **FL:** erect, sepals 3–5, ± fused at base; petals 3–5, spreading or recurved, free or ± fused at base; stamens = sepals in number; pistils 3–5. **FR:** spreading to erect. **SEED:** 0.2–0.6 mm, elliptic to elliptic-oblong (spheric, reniform), gen with longitudinal lines, sometimes ± smooth or papillate, red-brown. x=(7)8. ± 250 spp.: esp Afr, ann ± worldwide. (Latin: diminutive of thick) *C. argentea* Thunb., a synonym of *C. ovata* (Mill.) Druce, a waif.

1. Per to shrub; lvs gen > 10 mm; fls in terminal infl, many
   2. Lvs linear to narrow-deltate, ± cylindric . . . . ***C. tetragona***
   2′ Lvs obovate to ovate or wide-elliptic, ± flat
      3. Per; sts erect or decumbent; lf ovate to wide-elliptic, petiole 5–20 mm; pedicel 3–8 mm;
         petals gen 4, 3–4 mm . . . . [***C. multicava*** subsp. ***multicava***]
      3′ Shrub; sts erect; lf obovate, petiole ± 0; pedicel 8–12 mm; petals gen 5, 7–10 mm . . . . [***C. ovata***]
1′ Ann; lvs gen < 6 mm; fls in lf axils, 1–2 per node, or terminal, 1
   4. Fls 1 per node (or terminal); sepals ± 1/2 petals; seeds ≥ 3; pls gen aquatic or on wet substrates
      (e.g., seeps, vernal pools, ditches); sts ± decumbent to ascending, gen rooting at nodes
      5. Seeds with ± regular, continuous, longitudinal lines at 20×, ± dull or shiny but not glistening as if wet;
         follicles subtruncate at tip, suture abruptly outcurved in distal 1/4; sepals 0.5–1.5 mm, ovate to oblong
         . . . . ***C. aquatica***
      5′ Seeds with irregular, interrupted, longitudinal lines at 20×, shiny, glistening as if wet; follicles
         oblique-acute at tip, suture gradually outcurved in distal 1/2; sepals 0.4–1 mm, deltate-ovate to lanceolate . . . . ***C. solieri***
   4′ Fls gen 2 per node; sepals ≥ petals; seeds (1)2; pls gen terrestrial but not on wet substrates; sts ± erect,
      not rooting at nodes
      6. Fl parts gen in 3s; pedicels gen < 0.5 mm . . . . ***C. tillaea***
      6′ Fl parts gen in 4s or 5s; pedicels gen > 1.5 mm
         7. Fl parts gen in 5s; lvs gen 4–5 mm, tip acute, with short awn or point; petals ± 1.2 mm, lanceolate
            . . . . ***C. colligata*** subsp. ***lamprosperma***
         7′ Fl parts gen in 4s; lvs gen < 4 mm, tip acute to rounded, without awn or point; petals gen < 1.2 mm,
            narrow-deltate . . . . ***C. connata***

C. aquatica (L.) Schönl. (p. 671)   Ann, gen aquatic or on wet substrates. **ST:** decumbent, ± erect if stranded, gen branched at base, rooting at nodes. **LF:** 2–6 mm, oblanceolate to linear, tip acute. **INFL:** fl 1 per node, terminal; pedicel 0.5–20 mm. **FL:** parts in 4s; sepals ± 0.5–1.5 mm, ovate to oblong, rounded to obtuse; petals > sepals, 1–2 mm, ovate to oblong. **FR:** erect, oblong, tip subtruncate; suture straight, abruptly outcurved in distal 1/4. **SEED:** 6–17, elliptic-oblong, ± dull or shiny but not glistening as if wet, with ± regular, continuous, longitudinal lines at 20×. 2*n*=42. Salt marshes, vernal pools, margins of lakes, ponds; < 3000 m. NCo, CaRF, SNF, GV, CW, SW (exc n ChI); N.Am, Mex, n Eurasia. Mar–Jun(Aug)

C. colligata Toelken subsp. ***lamprosperma*** Toelken   Ann, terrestrial. **ST:** erect, to 16 cm, branched or not, not rooting at nodes, red-brown in age. **LF:** 4–5 mm, ovate to oblong; tip acute, with short awn or point. **INFL:** fls 2 per node; pedicel ± 1.5 mm. **FL:** parts in 5s; sepals ± 1.5 mm, lanceolate, mucronate; petals < sepals, ± 1.2 mm, lanceolate. **FR:** erect or ± recurved, lance-oblong. **SEED:** (1)2, elliptic, shiny, ± smooth. Open, gravelly alluvial bench; 150–200 m. SCo (San Gabriel River near Irwindale); native to s Australia. Jan–May

C. connata (Ruiz & Pav.) A. Berger (p. 671)   PYGMY-WEED   Ann, terrestrial. **ST:** erect, 2–6(10) cm, branched or not, not rooting at nodes, red in age. **LF:** 1–3(6) mm, ovate to oblong; tip acute to rounded, without awn or point. **INFL:** fls (1)2 per node; pedicel 0.2–6 mm. **FL:** parts gen in 4s; sepals 0.5–2 mm, lanceolate, acute to acuminate; petals gen < sepals, 0.6–1.2(1.5) mm, narrow-deltate. **FR:** ascending, ovoid. **SEED:** 1–2, elliptic, shiny, with ± wavy longitudinal lines at 20×. 2*n*=16. Open areas; < 1500 m. NW, CaRF, SNF, GV, CW, SW, w DMoj, DSon; to WA, s BC, TX, n C.Am; also w S.Am. Locally abundant. Feb–May

C. solieri (Gay) F. Meigen (p. 671)   Ann, gen aquatic or on wet substrate. **ST:** decumbent, or erect when stranded, 2–7 cm, ± branched. **LF:** 1–5 mm, oblong to linear; tip ± obtuse. **INFL:** fl 1 per node, terminal; pedicel 0.5–6(10) mm. **FL:** parts in 4s; sepals ± 0.4–1 mm, deltate-ovate to lanceolate, rounded to obtuse; petals > sepals, ± 1–1.5 mm, lanceolate. **FR:** ascending, tip lanceolate, oblique-acute; suture gradually outcurved in distal 1/2. **SEED:** 6–14, elliptic-oblong, shiny, glistening as if wet, with irregular, ± interrupted, longitudinal lines at 20×. Vernal pools, margins of lakes, ponds; < 2100 m. NCoRI, CaRF, n SNF, GV, SCo, PR; to OR, MT, WY, TX, Baja CA; also Chile. Similar to, often occurring with *C. aquatica*. Mar–Jun

C. tetragona L.   MINIATURE OR CHINESE PINE-TREE   Subshrub. **ST:** gen erect, 30–50(100) cm, 1–5(20) mm diam. **LF:** cauline, gen > 4 pairs, subsessile, 10–30 mm, linear to narrow-deltate, ± cylindric, pale green, mucronate. **INFL:** terminal, ± flat-topped, open; pedicels 3–5 mm. **FL:** parts in 5s; sepals erect, ± 1 mm, deltate, acute; petals 2–3 mm, narrow-deltate, elliptic, to oblanceolate, cream to white. **FR:** erect, ovoid. **SEED:** 6–10, elliptic, with indistinct rows of pointed papillae. Persisting from cult and at dump sites, wildland-urban interface; < 400 m. CCo, SCo, SnGb; native to s Afr. Jun–Sep

C. tillaea Lest.-Garl.   Ann, gen terrestrial. **ST:** gen erect, 1–6 cm, branched or not, not rooting at nodes, red in age. **LF:** 1–3 mm, oblong; tip ± acute. **INFL:** fls gen 2 per node; pedicel gen < 0.5 mm. **FL:** parts in 3s; sepals 1–1.5 mm, lanceolate, acuminate; petals < sepals, 0.5–1 mm, narrow-lanceolate. **FR:** ascending, ovoid. **SEED:** (1)2, elliptic, shiny, with ± wavy longitudinal lines at 20×. 2*n*=64. Open, gravelly sites; < 700 m. NCoRI, s CaRF, n&c SNF, GV, CW, SW; to WA, BC; native to Medit. Feb–May

DUDLEYA   DUDLEYA, LIVEFOREVER

Stephen Ward McCabe

Per, fleshy, glabrous, bisexual. **ST:** gen caudex- or corm-like, branched or not, ± covered with dried lvs. **LF:** in rosettes, evergreen or ± deciduous in summer (withering, falling or not), waxy or not, base wounding purple-red (yellow) or gen not. **INFL:** cyme; fl bracts ± subtending pedicels, < bracts; bracts alternate. **FL:** sepals 5, fused below; petals 5, fused at base, erect to

spreading above; stamens 10, epipetalous; carpels 5, ± fused below. **FR**: follicles 5, erect to spreading, many-seeded. **SEED**: < 1 mm, narrowly ovoid, brown, striate. ± 46 spp.: sw N.Am; some used as groundcover or cult for orn. (W.R. Dudley, 1st head of Botany Department, Stanford University, 1849–1911) [Thiede 2003 *in* Eggli (ed.) Illus Handbook Succulent Pls 6 (Crassulaceae):85–103. Springer] Fr just before opening gen most reliable for orientation; insect damage may cause branching in taxa characterized as non-branching.

1. St below surface, not elongate, gen simple; lvs vernal, ± petioled (basally gen < 1 cm wide, gen narrowed to < 4 mm wide) — petals, fr gen spreading (subg. *Hasseanthus*)
  2. Fl odor 0; petals yellow; fr spreading
    3. Lvs 4–15 cm, linear, ± narrowed above base, tip sharply acute; petals fused 1–2 mm . . . . ***D. multicaulis***
    3′ Lvs 1–7 cm, oblanceolate to spoon-shaped, strongly narrowed above base (to gen 0.5–3 mm wide) tip acute to obtuse; petals fused 0.5–1 mm . . . . ***D. variegata***
  2′ Fl odor musky-sweet; petals white to pale yellow; fr ascending to spreading
    4. Petals ascending, 7–14 mm, 3.5–5.5 mm wide, fused 1–2 mm; fr ascending; lf base 3–12 mm wide . . . . ***D. nesiotica***
    4′ Petals spreading, 5–10 mm, 2–4 mm wide, fused gen < 1 mm; fr spreading; lf base 1–4 mm wide
      5. Lower bracts < 1.5 × longer than wide; lvs 7–15 mm, ± spheric to spoon-shaped; petiole narrow . . . . . . ***D. brevifolia***
      5′ Lower bracts > 2 × longer than wide; lvs 10–60 mm, ± oblanceolate to club-shaped; petiole ± narrow
        . . . . ***D. blochmaniae***
        6. Lvs gen < 12, not to ± glaucous; s CCo, SCo, Santa Cruz Island . . . . subsp. ***blochmaniae***
        6′ Lvs gen > 15, glaucous or ± so; n ChI (Santa Rosa Island) . . . . subsp. ***insularis***
1′ St above surface or ± buried by moving substrate, often elongate, often branched; lvs gen evergreen, ± not petioled (basally gen 1–5 cm wide, gen not narrowed to < 4 mm wide)
  7. Petals erect below, ± spreading from near middle; fr ascending to spreading
    8. St gen < 3 cm between branches; pl gen < 2 dm across; mainland
      9. Lvs sticky, appearing oily, odor resinous. . . . . ***D. viscida***
      9′ Lvs not sticky, not appearing oily, odor 0 or not resinous
        10. Lvs covered with white powdery wax, appearing white; styles 2–3 mm; pedicels 2–5 mm; SnGb (Los Angeles Co.). . . . . ***D. densiflora***
        10′ Lvs lightly covered with thin, translucent wax, appearing olive-green; styles 1–2 mm; pedicels 1–2 mm; SCo, PR . . . . ***D. edulis***
    8′ St gen > 3 cm between branches; pl 0.5–10 dm across; mainland or ChI
      11. Lvs in ×-section round nearly to base, 2–4 mm thick . . . . ***D. attenuata*** subsp. ***attenuata***
      11′ Lvs in ×-section ± elliptic or in upper 1/2 round, 4–6 mm thick . . . . ***D. virens***
        12. Lvs green, rarely glaucous; outer lvs lax to erect; dry peduncle easily broken; s ChI (San Clemente Island). . . . . subsp. ***virens***
        12′ Lvs white, gray, or green, occ with red or pink, gen glaucous; outer lvs ascending to erect; dry peduncle not easily broken; SCo (near San Pedro, Los Angeles Co.), s ChI (San Nicolas, Santa Catalina, San Clemente islands)
          13. Lvs ± round in ×-section in upper 1/2, 15–30, 3–10(15) cm, 5–10(15) mm wide; peduncle 10–30 cm, ascending or erect. . . . . subsp. ***hassei***
          13′ Lvs ± wide-elliptic in ×-section in upper 1/2, 20–50, 6–25 cm, 10–32 mm wide; peduncle 6–70 cm, erect. . . . . subsp. ***insularis***
  7′ Petals erect below, erect to ascending from near middle, ± spreading at tips; fr gen erect to ± spreading
    14. Petals fused ± 1/2 length; pedicels in fr often sharply bent; rosette white, 7–60 cm across
      15. St 1–4 cm wide; lvs 15–25, gen 5–15 cm; pedicels 5–15(20) mm; fls gen erect. . . . . ***D. arizonica***
      15′ St 4–9 cm wide; lvs 40–60, 8–25(27) cm; pedicels 5–30(35) mm; fls pendent . . . . ***D. pulverulenta***
    14′ Petals fused < 1/3 length; pedicels in fr not sharply bent; rosette green or white, 0.5–55 cm across
      16. Petals adaxially cream, white, or pale yellow, margins ± jagged or not; petal keel ± not raised, ± marked with red-pink or purple
        17. St often > 8 cm, 10–80 mm wide; peduncle 4.5–54 cm; lf 2.5–22 cm
          18. St swollen at base; lf ± thin; rosettes gen 1 . . . . ***D. candelabrum***
          18′ St not swollen at base; lf ± thick; rosettes 1–100+
            19. Lf tip gen obtuse, margin with ≥ 2 angles between ad-, abaxial surfaces; upper margins of adjacent petals not touching; NCo, n&c CCo . . . . ***D. farinosa***
            19′ Lf tip gen acute, margin with 0–1(2) angles between ad-, abaxial surfaces; upper margins of adjacent petals gen touching; n ChI . . . . ***D. greenei***
        17′ St < 8 cm, 1.5–30 mm wide; peduncle 2–25(40) cm; lf 0.5–11 cm
          20. Lf wounding red or purple-red at base; infl 1° branches gen simple
            21. Lf ± deciduous in summer, ± round in ×-section; petals often purple-tinged near tip, with few purple flecks; s CCo, s SCoRO (San Luis Obispo Co.) . . . . ***D. abramsii*** subsp. ***bettinae***
            21′ Lf gen dry but ± not deciduous in summer; ± elliptic in ×-section; petals not purple-tinged near tip, with purple-flecks throughout or not
              22. Lower bracts 10–30 mm; s SCoRO (San Luis Obispo Co.); on serpentine . . . . ***D. abramsii*** subsp. ***murina***
              22′ Lower bracts 4–15 mm; s SNH, PR; not on serpentine, gen on granite . . . . ***D. abramsii*** subsp. ***abramsii***
          20′ Lf gen not wounding red or purple-red at base; infl 1° branches simple to branched

23. Lf triangular to triangular-ovate; petals pale yellow to medium yellow; insular (Santa Rosa Island) ... [2]***D. gnoma***
23′ Lf not triangular to triangular-ovate; petals ± pale yellow; not insular
  24. Rosettes 1–few; not on granite or limestone, often on serpentine or volcanics
    25. Infl 1° branches gen simple; pedicels 2–7 mm; peduncle 5–20 cm; se SnFrB (s Santa Clara Valley); on serpentine .... ***D. abramsii*** subsp. ***setchellii***
    25′ Infl 1° branches often branched; pedicels 3–12 mm; peduncle 5–25(40) cm; SnJV, SnFrB, SCoRI (Contra Costa Co. to w Fresno Co.); on serpentine or not .... ***D. cymosa*** subsp. ***paniculata***
  24′ Rosettes 1–50; on granite or limestone
    26. Lvs lance-oblong to lanceolate to subcylindric; infl 1° branches gen branched; s SNH
      .... ***D. abramsii*** subsp. ***calcicola***
    26′ Lvs elliptic to oblanceolate; infl 1° branches gen simple; SnBr............... ***D. abramsii*** subsp. ***affinis***
16′ Petals adaxially red, orange, yellow, yellow-green, or bright yellow-green, margins entire; petal keel often raised, rarely marked with red-pink or purple
  27. Pl stoloned or st branched; fr ascending
    28. Pl stoloned; c SCo (San Joaquin Hills, Orange Co.) .... ***D. stolonifera***
    28′ Pl st branched; ChI (Santa Barbara Co.) .... ***D. traskiae***
  27′ Pl simple or branched by sts; fr ± erect
    29. Pl gen < 10 cm wide (exc in deep shade or yrs with high rainfall); lvs 0.8–3 × longer than wide
      30. Peduncle erect, infl ± symmetric radially, 1° branches gen 3
        31. Internodes between bracts < 5 mm................................[2]***D. cymosa*** subsp. ***crebrifolia***
        31′ Internodes between bracts gen > 5 mm .... ***D. cymosa*** subsp. ***pumila***
      30′ Peduncle ascending, infl ± asymmetric radially, sometimes by pedicels turning to sun or away from cliff, 1° branches (0)2–3 (or, if peduncle erect and infl symmetric radially, 1° branches gen 2)
        32. Lvs in summer ± deciduous, papery when dry; bracts plump............... ***D. cymosa*** subsp. ***marcescens***
        32′ Lvs in summer ± not deciduous, ± tough when dry; bracts not or ± plump
          33. Lvs adaxially dark green, abaxially ± red; bracts ± spreading .... ***D. cymosa*** subsp. ***ovatifolia***
          33′ Lvs adaxially, abaxially ± equally colored; bracts ascending
            34. Lvs oblong to lanceolate, tips acute; petals bright yellow; infl branches 2(3); s WTR (Santa Monica Mtns)..[2]***D. cymosa*** subsp. ***agourensis***
            34′ Lvs ovate or deltate to oblanceolate or spoon-shaped, tips acuminate to mucronate; petals bright yellow, orange, or red; infl branches gen ≥ 3; NCoR, CaR, SN, CW, s WTR (Santa Monica Mtns).. ***D. cymosa*** subsp. ***cymosa***
29′ Pl 0.5–100 cm+ wide; lvs (2.5)3–10 × longer than wide
  35. St gen 0.15–2 cm wide; 1° infl branches gen 2, branched gen 0–1 ×
    36. Lvs deciduous in summer; n-facing volcanic cliffs, adjacent grassland; WTR (n of Santa Monica Mtns) .... ***D. parva***
    36′ Lvs dry in summer but ± not deciduous; level or sloping habitats of various orientations; n ChI, WTR (Santa Monica Mtns)
      37. Lvs white, 5–13 mm; n ChI (Santa Rosa Island)................................[2]***D. gnoma***
      37′ Lvs gray-white or white with some purple-red to gray-green showing through wax, 20–100 mm; mainland
        38. Lvs gen twisted when dry, base wounding purple-red; fl bud tips angled ± 35°; petals green- to lemon-yellow .... ***D. verityi***
        38′ Lvs occ twisted when dry, base not or ± wounding purple-red; fl bud tips angled ≥ 50°; petals bright yellow ....[2]***D. cymosa*** subsp. ***agourensis***
  35′ St 1–10 cm wide; 1° infl branches gen ≥ 3, branched 0–3 ×
    39. Rosettes 5–40, dense; lvs 2.5–8 mm wide, linear to linear-oblanceolate .... ***D. cymosa*** subsp. ***costatifolia***
    39′ Rosettes 1–100, gen not dense; lvs gen > 8 mm wide, lanceolate to oblanceolate
      40. St of older pls often exposed between dried lvs; lvs adaxially gen convex (transversely), tip margins gen not angled between ad-, abaxial surfaces; petals bright yellow (orange-yellow)... ***D. caespitosa***
      40′ St of older pls covered by dried lvs; lvs adaxially flat or convex (transversely), tip margins often angled between ad-, abaxial surfaces; petals bright yellow, red, tinged with red, red over yellow but appearing orange-pink, ± green, or green-yellow
        41. St (2)4–10 cm wide; peduncle 15–85 cm; petals red over yellow but appearing orange-pink...... ***D. palmeri***
        41′ St 1–3 cm wide; peduncle 5–95 cm; petals bright yellow, red over yellow but appearing orange-pink, mustard-, or green-yellow
          42. Internodes between bracts < 5 mm................................[2]***D. cymosa*** subsp. ***crebrifolia***
          42′ Internodes between bracts > 5 mm
            43. Lvs 5–30 cm, 1–4 cm wide; peduncle 1.5–9.5 dm; petals fused 1–2 mm................. ***D. lanceolata***
            43′ Lvs 3–15 cm, 0.3–2.5 cm wide; peduncle 0.5–5.1 dm; petals fused 1–4 mm .... ***D. saxosa***
              44. St 1–3 cm wide; lvs 4–15 cm; peduncle 1–5.1 dm; petals ± green, green-yellow, or bright yellow, rarely red-tinged .... subsp. ***aloides***
              44′ St 1–1.5 cm wide; lvs 3–9 cm; peduncle 0.5–2 dm; petals bright yellow, gen red-tinged ... subsp. ***saxosa***

D. abramsii Rose Cespitose or not; rosettes 1–many, 0.5–15 cm wide. **LF:** gen dry but ± not deciduous in summer, 2–30 mm, 3–20 mm wide, lance-oblong or lanceolate (elliptic to oblanceolate), gen glaucous, adaxially gen ± flat. **INFL:** peduncle 2–25 cm, 1–6 mm wide; lower bracts 4–40 mm; pedicels 0.5–7(11) mm. **FL:** sepals 2–5 mm, deltate; petals 8–13 mm, 1.5–3.5 mm wide, fused 1–4.5 mm, elliptic, acute, pale- or cream-yellow, margin often jagged, keel gen with fine, purple to red lines. 2n=34.

subsp. **abramsii** Rosettes few to many, 4–15 cm wide. **ST:** 10–15 mm wide. **LF:** 1–11 cm, 3–20 mm wide, oblong to lance-oblong or lanceolate, base wounding red. **INFL:** peduncle 2–18 cm, 1–5 mm wide; lower bracts 4–15 mm; 1° branches 2–3, ascending, gen simple, 2–15-fld; pedicels 0.5–7 mm. **FL:** petals fused 2–4.5 mm, keel gen red-lined. Uncommon. Outcrops, gen granite; 750–1750 m. s SNH, PR; n Baja CA. May–Jul

subsp. **affinis** K.M. Nakai (p. 671) SAN BERNARDINO MOUNTAINS DUDLEYA Rosettes gen 1–few, 3–6 cm wide. **ST:** 10–15 mm wide. **LF:** 2–4 cm, 7–15 mm wide, elliptic to oblanceolate, base gen not wounding purple-red. **INFL:** peduncle 3–8 cm, 1–3 mm wide; lower bracts 5–6 mm; 1° branches 0 or gen 2–3, ascending, gen simple, 3–8-fld. **FL:** petals fused 1.5–2.5 mm, keel gen red-lined. Outcrops, granitic or quartzite, rarely limestone; 1800–2600 m. SnBr. May–Jul ★

subsp. **bettinae** (Hoover) Bartel BETTY'S DUDLEYA Rosettes 5–40, 0.5–8 cm wide. **ST:** 3–20 mm wide. **LF:** ± deciduous in summer, 2–7 cm, 2–7 mm wide, ± round in ×-section, ± glaucous, base wounding purple-red, tip acute. **INFL:** peduncle 3–15 cm, 2–5 mm wide; lower bracts 10–20 mm; 1° branches 0 or gen 2(3), ascending, gen simple, 2–15-fld. **FL:** petals fused 1–2 mm, often purple-tinged near tip, with few purple flecks. Rocky outcrops in serpentine grassland; 50–180 m. s CCo, s SCoRO (San Luis Obispo Co.). Hybrids with *D. abramsii* subsp. *murina* likely. May–Jun ★

subsp. **calcicola** (Bartel & Shevock) K.M. Nakai (p. 671) LIMESTONE DUDLEYA Rosettes 1–50, 1–9 cm wide. **ST:** 0.3–2 cm wide. **LF:** 1–6(10) cm, 2–9 mm wide, lance-oblong to lanceolate to subcylindric, tip acute to subacuminate. **INFL:** peduncle 3–18 cm, 1.5–4.5 mm wide; 1° branches 2–4, gen spreading, branched (0)3 ×, 2–8-fld; pedicels 2–14 mm. **FL:** sepals 2.5–6 mm, triangular-ovate to lanceolate; petals 9–15 mm, 3.5–5 mm wide, fused 1–3 mm, lanceolate, narrowly acute, keel darker yellow (to red-tinged). Open, rocky, granite or gen limestone outcrops; 500–2600 m. s SNH. [*D. c.* Bartel & Shevock] May–Jun(Jul) ★

subsp. **murina** (Eastw.) Moran MOUSE-GRAY DUDLEYA Rosettes 1–3(10), 3–12 cm wide. **ST:** 10–30 mm wide. **LF:** gen dry but ± not deciduous in summer, 3–11 cm, 7–20 mm wide, lance-oblong, ± elliptic in ×-section, ± glaucous, base wounding purple-red, adaxially ± flat, tip acute to ± obtuse. **INFL:** peduncle 5–25 cm, 2–6 mm wide; lower bracts 10–30 mm; 1° branches 0 or gen 2–3; gen simple, ascending; pedicels ± 0.5–7 mm. **FL:** petals fused 1.5–3 mm, purple-flecked throughout, keel purple. Serpentine outcrops; 120–300 m. s SCoRO (San Luis Obispo Co.). Hybrids with *D. abramsii* subsp. *bettinae* likely. May–Jun ★

subsp. **setchellii** (Jeps.) Morin (p. 671) SANTA CLARA VALLEY DUDLEYA Rosettes 1–few, 2–9 cm wide. **ST:** 10–20 mm wide. **LF:** 3–8 cm, 7–15 mm wide, oblong-triangular to lance-elliptic to lanceolate, glaucous. **INFL:** peduncle 5–20 cm, 2–4 mm wide; lower bracts 10–25 mm; 1° branches 2–3, gen simple, ascending; pedicels 2–7 mm. **FL:** sepals 2–5 mm, deltate; petals 8–13 mm, 2.5–3.5 mm wide, fused 1–2.5 mm, elliptic, pale yellow, margin jagged or not, tip acute. Rocky outcrops in serpentine grassland; 120–300 m. se SnFrB (s Santa Clara Valley). [*D. s.* (Jeps.) Britton & Rose] Similar to *D. abramsii* subsp. *murina, D. cymosa* subsp. *paniculata.* May–Jun ★

D. arizonica Rose Rosettes 1(few), 10–25(30) cm wide, white. **ST:** 1–4 cm wide. **LF:** evergreen, 15–25, gen 5–15 cm, 1–5 cm wide, 2–4 mm thick, oblong to oblong-obovate, glaucous, base 1–3.5 cm wide, tip gen long-acuminate. **INFL:** peduncle 15–60 cm, 2–6 mm wide 1° branches 3–6, ascending, 4–27 cm, 3–6-fld; pedicels 5–15(20) mm; fls gen erect. **FL:** petals 9–15 mm, fused 4–8 mm, red or yellow, lobes 1.5–2 mm wide. 2n=34. Rocky slopes; 600–1500 m.

PR, DMtns, DSon; to UT, AZ, nw Mex. [*D. pulverulenta* subsp. *a.* (Rose) Moran] May–Jul

D. attenuata (S. Watson) Moran subsp. **attenuata** (p. 671) Rosettes 3–50, 2–15 cm wide. **ST:** 3–10(15) mm wide. **LF:** evergreen, 2–10 cm, 2–5 mm wide, 2–4 mm thick, in ×-section round ± to base, linear to linear-oblanceolate, glaucous, base wounding purple-red, 5–15 mm wide. **INFL:** peduncle 5–25 cm, 1–3 mm wide; 1° branches (0)2–3, simple, ascending, 2–11 cm, 3–15-fld; pedicels 0.5–3 mm. **FL:** sepals 1.5–4 mm, deltate-ovate; petals 6–10 mm, fused 0.5–3 mm, elliptic, acute, spreading from middle, white, often rose-flushed, red-lined. **FR:** 30–45° < erect, slightly swollen near base. 2n=34[68]. Coastal bluffs; < 50 m. s SCo (s San Diego Co.); n Baja CA. [*D. a.* subsp. *orcuttii* (Rose) Moran] May–Jun ★

D. blochmaniae (Eastw.) Moran Rosettes 1(5), 0.5–7 cm wide. **LF:** deciduous in summer, 1–6 cm, ± oblanceolate to club-shaped, base wounding purple-red, 1–4 mm wide, tip acute to rounded; petiole ± narrow. **INFL:** lower bracts > 2 × longer than wide; 1° branches 2–3, branched 0–1 ×, 1–6 cm, 3–10-fld; pedicels < 1 mm. **FL:** sepals 1.5–4 mm, deltate-ovate; petals spreading from base, 5–10 mm, elliptic, acute, white, keel often red-lined. **FR:** spreading.

subsp. **blochmaniae** BLOCHMAN'S DUDLEYA Rosettes 1–7 cm wide. **ST:** 7–25 mm, 4–15 mm wide, ± spheric to fusiform. **LF:** gen < 12, 1–6 cm, 3–8 mm wide, 2–4 mm thick, not to ± glaucous. **INFL:** peduncle (0.9)3–12(22) cm, 0.5–2 mm wide. 2n=34,[68,102]. Open, rocky slopes, often serpentine or clay-dominated; < 450 m. s CCo, SCo, n ChI (Santa Cruz Island); n Baja CA. Hybrids with *D. edulis* suspected. Apr–Jun ★

subsp. **insularis** (Moran) Moran (p. 671) SANTA ROSA ISLAND DUDLEYA Rosettes 0.5–4 cm wide. **ST:** 1–2 cm, 5–20 mm wide, ± spheric to oblong. **LF:** gen > 15, 1–3.5 cm, 2–7 mm wide, 1–3 mm thick, ± glaucous. **INFL:** peduncle 3–7 cm, 0.5–2 mm wide. 2n=34. Coastal bluffs; 3–90 m. n ChI (Santa Rosa Island). Apr–Jun ★

D. brevifolia (Moran) Moran (p. 671) SHORTLEAVED DUDLEYA Rosettes 1(3), 0.5–4 cm wide. **ST:** 1.5–3.5 cm, 1–6 mm wide, oblong. **LF:** deciduous in summer, 5–15, 0.7–1.5 cm, 2–7 mm wide, 2–4 mm thick, ± spheric to spoon-shaped distally, base wounding purple-red; petiole narrow. **INFL:** peduncle 2–11 cm, 0.5–1 mm wide; lower bracts < 1.5 × longer than wide; branches 0–1, 1–4 cm, 3–10-fld. **FL:** petals white (pale yellow). 2n=34. Bare sandstone terraces; < 250 m. s SCo (sw San Diego Co.). [*D. blochmaniae* subsp. *brevifolia* Moran] Hybrids with *D. edulis* suspected. Apr–Jun ★

D. caespitosa (Haw.) Britton & Rose (p. 671) Rosettes 1–100, gen not dense, (3)8–32 cm wide. **ST:** 1.5–4 cm wide, not swollen at base, those of older pls often exposed between dried lvs. **LF:** evergreen, 15–30, 5–20 cm, 1–2 cm wide, 3–8 mm thick, lance-oblong (extremely variable), adaxially gen convex (transversely), glaucous or not, base wounding purple-red or not, tip gen acute, margins gen not angled between ad-, abaxial surfaces. **INFL:** peduncle 10–60 cm, 3.5–10 mm wide; 1° branches 3–5, branched 0–2 ×; 3–15 cm, 4–15-fld; pedicels 1–6 mm. **FL:** sepals 2–5 mm, deltate-ovate, acute; petals 8–16 mm, 3–5 mm wide, fused 1.5–2.5 mm, elliptic, bright yellow (orange-yellow), tips erect, acute. 2n=102,136. Common. Coastal, rock, sand; gen < 100 m. s NCo, CCo, SCo, n ChI. [*Dudleya cespitosa,* orth. var.] Difficult complex of several entities that may not be separable; evidently intergrades with *D. cymosa, D. farinosa, D. palmeri, D. greenei, D. lanceolata.* Apr–Aug

D. candelabrum Rose (p. 671) CANDLEHOLDER DUDLEYA Rosette gen 1, (10)15–50 cm wide. **ST:** (2)3–8 cm wide, swollen at base. **LF:** evergreen, 20–45, 6–22 cm, 3–7 cm wide, obovate to oblong-oblanceolate, ± thin, ± glaucous or not, base wounding purple-red, 2–4.5 cm wide, tip acuminate. **INFL:** peduncle 15–54 cm, 6–11 mm wide; lower bracts 1–4+ cm, reflexed; 1° branches gen 3–7, branched 1–2 ×; terminal branches 2.5–13 cm, spreading, 5–25-fld; pedicels 2–6 mm. **FL:** sepals 5–8 mm, deltate-ovate, acute; petals 8–12 mm, 2.5–3.5 mm wide, fused 1.5–3.5 mm, oblong, acute, pale yellow. 2n=34. Open rocky places and n-facing slopes; < 380 m. n ChI. Hybrids with *D. greenei* suspected. Apr–Jul ★

D. cymosa (Lem.) Britton & Rose Rosettes 1–many, 0.5–30 cm wide. **ST:** 0.2–3.5 cm wide. **LF:** gen evergreen, 1.5–17 cm, 2.5–50

mm wide, gen oblanceolate to spoon-shaped (ovate or linear-oblanceolate), glaucous or not, tip acute or often acuminate to mucronate. **INFL**: peduncle 3–30(50) cm, 1–10 mm wide; 1° branches gen 2–4, gen spreading to ascending, branched 0–3 ×; branches 1–5(17) cm, 2–10(20)-fld; pedicels gen 5–15 mm. **FL**: sepals 1.5–5 mm, deltate-ovate, acute; petals 7–14 mm, 1.5–3.5 mm wide, fused 1–3 mm, elliptic to lanceolate, narrowly acute, yellow to red. 2*n*=34.

subsp. ***agourensis*** K.M. Nakai AGOURA HILLS DUDLEYA Cespitose; rosettes 1–6+, (2)5–10 cm wide. **ST**: gen 1–2 cm wide. **LF**: 3–10 cm, 1–1.5 cm wide, oblong to lanceolate, gen ± glaucous, gray-purple, base not or ± wounding purple-red, occ twisted when dry, tip acute. **INFL**: ± asymmetric radially when infl branches 2; peduncle 6–17 cm, 2–3 mm wide, ± erect; lower bracts ascending; 1° branches 2(3), ascending, simple; pedicels 2.5–9 mm. **FL**: bud tips angled ≥ 50°; petals bright yellow. Open, rocky volcanic slopes; < 460 m. s WTR (Santa Monica Mtns). May–Jun ★

subsp. ***costatifolia*** Bartel & Shevock PIERPOINT SPRINGS DUDLEYA Cespitose; rosettes 5–40, dense, 1–5 cm wide. **ST**: 1.5–2 cm wide. **LF**: (1)2–8 cm, 2.5–8 mm wide, linear to linear-oblanceolate, glaucous. **INFL**: peduncle 5–15(20) cm, 1.5–3.5 mm wide; 1° branches 2–4, ascending, branched 0–3 ×, terminal branches 1–4 cm, 2–7-fld; pedicels 2.5–9 mm. **FL**: petals bright yellow. Limestone outcrops; 1450–1600 m. s SNH (Tule River, Tulare Co.). [*D. abramsii* subsp. *c.* (Bartel & Shevock) Morin; *Dudleya cymosa* subsp. *costafolia*, orth. var.] May–Jun ★

subsp. ***crebrifolia*** K.M. Nakai & Verity SAN GABRIEL RIVER DUDLEYA Not cespitose, few lvs green in droughts; rosettes 1(few), (3)5–12 cm wide. **ST**: 1–2 cm wide. **LF**: 4–10 cm, 20–50 mm wide, elliptic to spoon-shaped, gen glaucous, tip ± acuminate. **INFL**: ± symmetric radially; peduncle gen 10–30(50) cm, 4–5 mm wide, erect; lower bracts 20–50, internodes < 5 mm; 1° branches (2)3, branched 0–2 ×, terminal branches 3–15 cm, 2–10(20)-fld; pedicels 3–8 mm. **FL**: petals fused 1–1.5 mm, mustard-yellow. Granitic slopes; 400 m. SnGb (Fish Canyon, Los Angeles Co.). Jun–Jul ★

subsp. ***cymosa*** (p. 671) Gen not cespitose; rosettes 1–several, 3–10(30) cm wide. **ST**: 1–3.5 cm wide. **LF**: 3–17 cm, 10–60 mm wide, ovate or deltate to oblanceolate or spoon-shaped, with broad base, glaucous or not, ± tough when dry, margin ± up-folded at widest point, tip ± recurved, acuminate to mucronate. **INFL**: ± asymmetric radially by pedicels turning to sun or away from cliff; peduncle 5–30(45) cm, 2–8 mm wide, ascending; lower bracts not or ± plump, ascending; 1° branches gen ≥ 3, ascending, terminal branches 1–17 cm, 4–20-fld. **FL**: petals bright yellow, orange, or red. Rocky outcrops, talus slopes, less often shaded canyon slopes; 100–2700 m. NCoR, CaR, SN, CW, s WTR (Santa Monica Mtns). [*D. c.* subsp. *gigantea* (Rose) Moran] Variable; needs study. Hybrids with *D. farinosa, D. lanceolata, D. palmeri* suspected. May–Jul

subsp. ***marcescens*** Moran (p. 671) MARCESCENT DUDLEYA Gen not cespitose; rosettes 1(15), 0.5–4 cm wide. **ST**: 2–10 mm wide. **LF**: ± deciduous in summer, 1.5–4 cm, 5–12 mm wide, elliptic to elliptic-ovate, ± glaucous or not, base wounding purple-red or gen not, papery when dry, tip acute. **INFL**: ± asymmetric radially; peduncle 3–10 cm, 1–3 mm wide; lower bracts plump; 1° branches 0–2, simple, 1–3 cm, 3–5-fld. **FL**: petals 2.5–3.5 mm wide, bright yellow (orange or red-marked). Shaded, rocky volcanic outcrops and slopes; 150–500 m. s WTR (Santa Monica Mtns). May–Jun ★

subsp. ***ovatifolia*** (Britton) Moran SANTA MONICA DUDLEYA Rosettes 1–few, 1–6 cm wide. **ST**: 1–1.5 cm wide. **LF**: gen 6–10, 2–5 cm, 15–25 mm wide, oblong to ovate or ovate-deltate with wide base, dark green adaxially, ± red abaxially, not glaucous, not papery when dry, tip acuminate to mucronate. **INFL**: ± asymmetric radially; peduncle 3–15 cm, 3–3.5 mm wide, ± ascending; lower bracts ± spreading, not or ± plump; 1° branches 3–4, gen simple, 1–3 cm, 3–5-fld; pedicels 3 mm, 1 mm wide. **FL**: petals bright yellow (pls with orange or red-marked petals rare, may be hybrids). Shaded, rocky outcrops and slopes; 150–500 m. s WTR (Santa Monica Mtns), PR (Santa Ana Mtns). May–Jun ★

subsp. ***paniculata*** (Jeps.) K.M. Nakai Not cespitose; rosettes 1–few, 4–11 cm wide. **ST**: 1–2 cm wide. **LF**: 3–10 cm, 5–20 mm

wide, lance-oblong, green to ± white, glaucous to not, tip acute. **INFL**: peduncle 5–25(40) cm, 1° branches 2–3, often branched, terminal branches 1–5 cm, 4–10-fld; pedicels 3–12 mm. **FL**: petals 1.5–2.5 mm wide, pale yellow-white (to pale yellow-pink in SnJV). Uncommon. Rocky outcrops, canyons; 30–1200 m. SnJV, SnFrB, SCoRI. Contra Costa to w Fresno cos. in SCoRI. May–Jun

subsp. ***pumila*** (Rose) K.M. Nakai Not cespitose; rosettes 1–few, 4–10(30) cm wide. **ST**: 1–2(3.5) cm wide. **LF**: few green in droughts, 1.5–5 (14) cm, 10–30 mm wide, diamond-shaped-oblanceolate to spoon-shaped, glaucous or not, margins up-folded at widest point, tip gen short-acuminate to mucronate, ± recurved. **INFL**: ± symmetric radially; peduncle 5–30 cm, 3–10 mm wide, erect; lower bracts ± not plump, internodes gen > 5 mm; 1° branches gen 3, terminal branches 1–3 cm, 3–6-fld. **FL**: petals bright yellow to red. Rocky cliffs, slopes; 50–2600 m. SCoRO, TR, PR. Hybrids with *D. lanceolata, D. caespitosa* suspected. May–Jul

D. densiflora (Rose) Moran (p. 671) SAN GABRIEL MOUNTAINS DUDLEYA Cespitose, gen < 3 cm between branches; rosettes 1–8, 7–25 cm wide. **ST**: 1–2.5 cm wide. **LF**: evergreen, 6–15 cm, 6–12 mm wide, ± cylindric exc at base, linear, covered with white powdery wax, appearing white, base ± wounding yellow, 1–2 mm wide, tip acute. **INFL**: peduncle 10–30 cm, 2–6 mm wide; 1° branches 3–several, branched 1–2 ×; terminal branches 2–4 cm, 2–8-fld; pedicels 2–5 mm. **FL**: sepals 1.5–2.5 mm, deltate-ovate; petals 5–10 mm, 2–3 mm wide, fused 0.5–2 mm, narrowly ovate, acute, white or pink; styles 2–3 mm. **FR**: 45–80° < erect, strongly swollen near base. 2*n*=34. Steep canyon walls; 300–520 m. SnGb (Los Angeles Co.). Jun ★

D. edulis (Nutt.) Moran (p. 671) Cespitose, gen < 3 cm between branches; rosettes 1–16, 5–10(20) cm wide. **ST**: 1.5–4.5 cm wide. **LF**: evergreen, 8–20 cm, 4–10 mm wide, ± cylindric exc at base, lightly covered with thin, translucent wax, appearing olive-green, base ± wounding yellow, 1.5–3 cm wide, tip acute. **INFL**: open, peduncle 10–50 cm, 3–10 mm wide; 1° branches several, branched 1–2 ×; terminal branches 4–10 cm, 3–11-fld; pedicels 1–2 mm. **FL**: sepals 2.5–4.5 mm, oblong-ovate; petals 7–10 mm, 2–3 mm wide, fused 1–2 mm, elliptic-oblong, acute, spreading or ± reflexed from middle, white–cream; styles 1–2 mm. **FR**: 45–60° < erect, strongly swollen at base. 2*n*=34. On soil, rocky slopes, ledges; < 1300 m. SCo, PR; n Baja CA. Hybrids with *D. stolonifera, D. blochmaniae* suspected. May–Jul

D. farinosa (Lindl.) Britton & Rose (p. 677) BLUFF LETTUCE Rosettes 4–60, 4–25 cm wide. **ST**: 1–3 cm wide, often elongate, not swollen at base, older parts gen not visible between dried lvs. **LF**: evergreen, 2.5–6 cm, 1–2.5 cm wide, 5–9 mm thick, oblong-ovate, glaucous or not, base 1–2.5 cm wide, tip gen obtuse, margin with ≥ 2 angles between ad-, abaxial surfaces. **INFL**: peduncle 10–35 cm, 3–8 mm wide; 1° branches 3–5, close-set, branched 0–2 ×; terminal branches 1–3.5 cm, ascending in age, 3–11-fld; pedicels 1–3 mm. **FL**: sepals 3–7 mm, deltate-ovate; petals 10–14 mm, 3–5 mm wide, fused 1–2 mm, oblanceolate, acute to obtuse, pale yellow, upper margins of adjacent petals not touching. 2*n*=34. Common: Coastal soils and cliffs; gen < 100 m (< 600 m in San Mateo Co.). NCo, n&c CCo; sw OR. Hybrids with *D. caespitosa, D. cymosa* subsp. *c.* suspected. Jun–Aug

D. gnoma S.W. McCabe MUNCHKIN DUDLEYA To 10 cm wide; rosettes 1–24, 0.8–5.1 cm wide. **ST**: (0.15) 1.2–2 cm wide. **LF**: gen dry but ± not deciduous in summer, 0.5–1.3 cm, 0.6–2.5 cm wide, triangular to triangular-ovate, white, glaucous. **INFL**: peduncle 2.5–13 cm, 2–5 mm wide; 1° branches 2(3), branched 0–1 ×, 1–10-fld; pedicels 1–3 mm. **FL**: sepals 3.5–4 mm, 2–3 mm wide; petals 8–9(11) mm, 3 mm wide, fused 1–2 mm, pale yellow to medium yellow, keel ± glaucous. 2*n*=68. Rocky slopes with shallow volcanic soils; < 80 m. n ChI (Santa Rosa Island). May–Jun ★

D. greenei Rose GREENE'S DUDLEYA Pl ± cespitose, to 1 m wide; rosettes 1–100+; (3.5)5–46 cm wide. **ST**: 2–5 cm wide, often elongate, not swollen at base. **LF**: evergreen, 3–22 cm, 1–3.5 cm wide, 4–8 mm thick, variable in shape, glaucous or not, base wounding red or yellow or not, 1–3 cm wide, tip gen acute, margin with 0–1(2) angles between ad-, abaxial surfaces. **INFL**: peduncle 4.5–50 cm, 3–5(14) wide; 1° branches 3–6, branched 0–2 ×; terminal

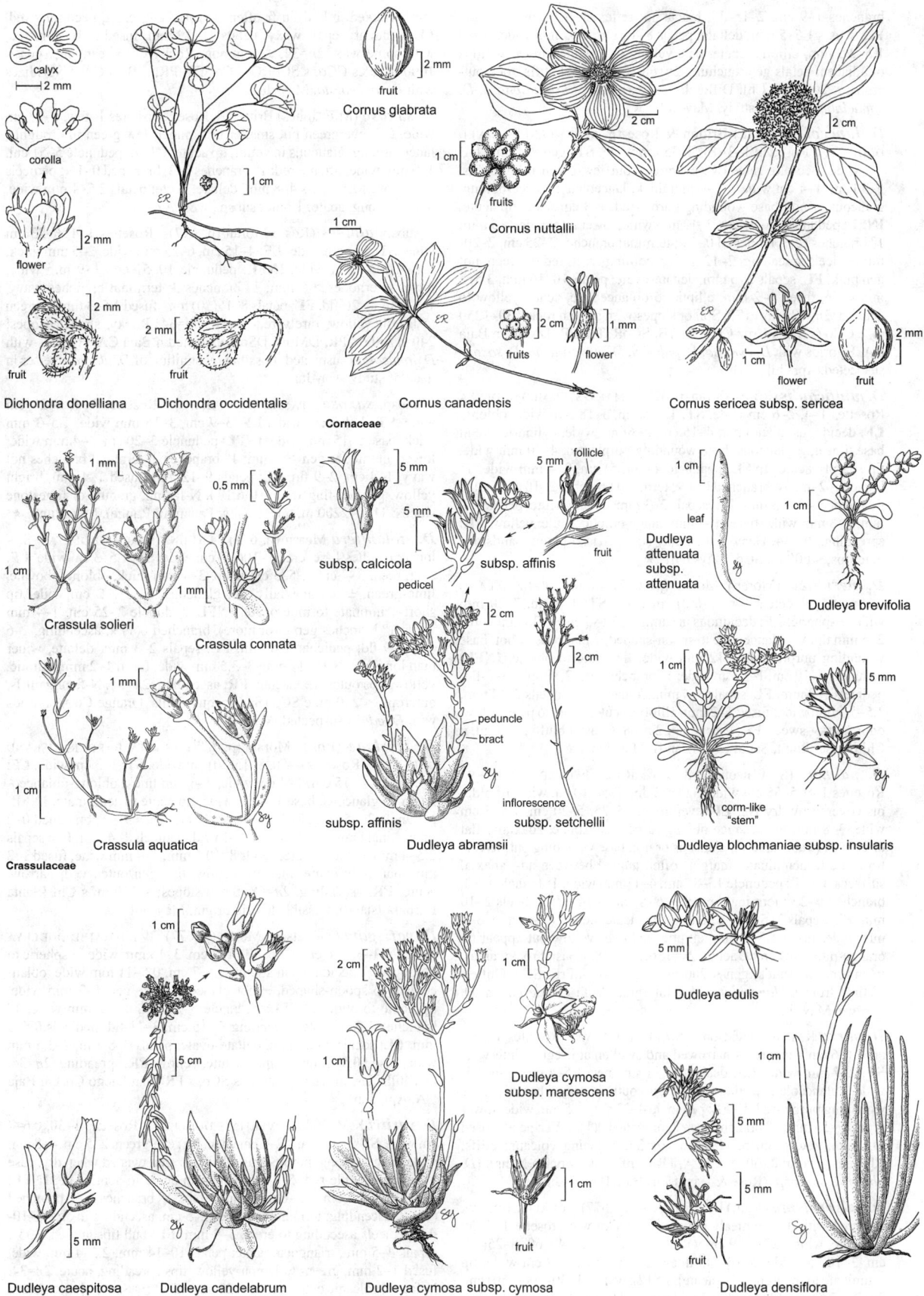

calyx

2 mm

corolla

2 mm

flower

2 mm

fruit

fruit

Dichondra donelliana **Dichondra occidentalis**

1 cm

ER

Cornaceae

Cornus glabrata

2 mm

fruit

1 cm

fruits

Cornus nuttallii

1 cm

ER

fruits

2 cm

flower

1 mm

Cornus canadensis

2 cm

ER

1 cm

flower

fruit

2 mm

Cornus sericea subsp. sericea

1 mm

0.5 mm

1 cm

Crassula solieri

1 mm

Crassula connata

5 mm

subsp. calcicola

5 mm

pedicel

5 mm

follicle

fruit

subsp. affinis

leaf

1 cm

Dudleya attenuata subsp. attenuata

1 cm

Dudleya brevifolia

1 mm

1 cm

Crassula aquatica

Crassulaceae

2 cm

peduncle

bract

subsp. affinis

2 cm

inflorescence

subsp. setchellii

Dudleya abramsii

1 cm

5 mm

corm-like "stem"

Dudleya blochmaniae subsp. insularis

1 cm

5 cm

5 mm

Dudleya caespitosa

Dudleya candelabrum

2 cm

1 cm

1 cm

fruit

Dudleya cymosa subsp. cymosa

1 cm

Dudleya cymosa subsp. marcescens

1 cm

fruit

5 mm

Dudleya edulis

5 mm

5 mm

fruit

1 cm

Dudleya densiflora

branches 1–9 cm, 2–15-fld; bracts not reflexed; pedicels 1–5 mm. **FL:** sepals 1.5–5 mm, deltate; petals 8–12 mm, 3–5 mm wide, fused 1.5–2.5 mm, elliptic, acute, pale yellow or ± white, upper margins of adjacent petals gen touching. 2*n*=68,102. Coastal cliffs, rock outcrops; < 200 m. n ChI. Difficult to separate from *D. caespitosa, D. candelabrum;* needs study. May–Jul ★

D. lanceolata (Nutt.) Britton & Rose (p. 677) LANCE-LEAVED DUDLEYA Rosettes 1–3 (8), 3–35 cm wide. **ST:** gen < 4 cm, 1–3 cm wide, occ elongate. **LF:** evergreen but few green in droughts, 5–30 cm, 1–4 cm wide, 1.5–6 mm thick, lanceolate to lance-oblong, glaucous or not, base wounding purple-red, 1–3 cm wide, tip acute. **INFL:** peduncle 15–95 cm, 3–12 mm wide; lower internodes > 5 mm; 1° branches (2)3, branched 0–1 ×; terminal branches 2–25 cm, 2–20-fld; pedicels spreading, 2–12 mm, becoming erect, red or green, not gen pink. **FL:** sepals 3–6 mm, deltate-ovate; petals 10–16 mm, 3.5–5 mm wide, fused 1–2 mm, elliptic to oblanceolate, acute, yellow to gen red. 2*n*=68,136,±170. Soil or slopes with broken rocks; 30–1250 m. c CCo (s Santa Cruz Co.), SnFrB, SCoR, TR, PR, DMtns; n Baja CA. Hybrids with *D. cymosa, D. palmeri, D. pulverulenta, D. saxosa* suspected. Apr–Jul

D. multicaulis (Rose) Moran (p. 677) MANY-STEMMED DUDLEYA Rosettes 1–4, 2–6 cm wide. **ST:** 1.5–5 cm, 3–18 mm wide, oblong. **LF:** deciduous in summer, 4–15 cm, 2–6 mm wide, cylindric exc at base, linear, ± glaucous, base wounding purple-red, 4–10 mm wide, tip sharply acute. **INFL:** peduncle 4–20(35) cm, 2–4 mm wide; 1° branches 2–many, branched 0–1 ×; terminal branches 2–10 cm, 3–15-fld; pedicels 0.5–3 mm. **FL:** sepals 2–3 mm, deltate-acute; petals 5–9 mm, 2–3 mm wide, fused 1–2 mm, lance-elliptic, acute, yellow. **FR:** spreading. 2*n*=34. Heavy, often clay soils, coastal plains, sandstone outcrops; < 600 m. SCo. May–Jun ★

D. nesiotica (Moran) Moran (p. 677) SANTA CRUZ ISLAND DUDLEYA Rosettes 1–6, 1–6(8) cm wide. **ST:** 1–3 cm, 7–20 mm wide, ± spheric. **LF:** deciduous in summer, 2.5–5 cm, 5–25 mm wide, 2–5 mm thick, oblanceolate to spoon-shaped, ± glaucous or not, base wounding purple-red, 3–12 mm wide, tip acute to obtuse. **INFL:** peduncle 3–10 cm, 1–3 mm wide; 1° branches gen 2, simple, 3–8-fld; pedicels 1–2 mm. **FL:** sepals 3–4 mm, deltate-ovate; petals 7–14 mm, 3.5–5.5 mm wide, fused 1–2 mm, elliptic, acute, white to pale yellow, odor musky-sweet. **FR:** ascending. 2*n*=68. Coastal bluffs; < 50 m. n ChI (Fraser Point, Santa Cruz Island). Mar–Jun ★

D. palmeri (S. Watson) Britton & Rose PALMER'S DUDLEYA Rosettes 1–8, 5–55 cm wide. **ST:** < 2 dm, (2)4–10 cm wide, of older pls covered by dry lvs. **LF:** evergreen, 15–25, 5–20 cm, 1.5–5 cm wide, 3–8 mm thick, lance-oblong to gen lanceolate, adaxially flat or convex (transversely), glaucous or not, base wounding purple-red, tip acute to acuminate, margins often angled between ad-, abaxial surfaces. **INFL:** peduncle 15–85 cm, 4–11 mm wide; 1° branches ± 3, branched 0–2 ×; terminal branches 5–8 cm, 5–14-fld; pedicels 2–10 mm. **FL:** sepals 3–5 mm, deltate-ovate, acute; petals 11–16 mm, 3–5 mm wide, fused 1.5–2 mm, elliptic, red over yellow but appearing orange-pink, tips erect, acute. 2*n*=136,170,238. Coastal rocky areas, inland marine sands; gen < 200 m. c&s CCo, SnFrB, SCo. Only ± distinct from *D. lanceolata.* May intergrade with *D. caespitosa, D. cymosa.* May–Jul

D. parva Rose & Davidson CONEJO DUDLEYA Rosettes 1–several, 1–6 cm wide; roots narrowed and swollen at irregular intervals. **ST:** 2–7 mm wide. **LF:** deciduous in summer, 1.5–4 cm, 3–6 mm wide, oblanceolate, ± glaucous esp in youth, papery when dry, base wounding purple-red. **INFL:** peduncle 4–23 cm, 1–5 mm wide; lower bracts 5–15 mm; 1° branches gen 2, branched 0(1) ×. **FL:** petals fused 1–2 mm, keel often red-flecked. 2*n*=34. N-facing volcanic cliffs, adjacent grassland; 60–450 m. WTR (n of Santa Monica Mtns). [*D. abramsii* subsp. *p.* (Rose & Davidson) Bartel] May–Jul ★

D. pulverulenta (Nutt.) Britton & Rose (p. 677) CHALK DUDLEYA Covered with dense, mealy powder or chalky wax; rosette 1, 7–60 cm wide, white. **ST:** 4–9 cm wide. **LF:** evergreen, 40–60, 8–25(27) cm, 3–10 cm wide, 3–10 mm thick, oblong, base 3–8 cm wide, tip acuminate to mucronate [to acute]. **INFL:** peduncle 30–100(150) cm, 5–20 mm wide; 1° branches 2–6, branched 0–1 ×; terminal branches twisted at base, nodding in youth, spreading in age; pedicels 5–30(35)

mm, reflexed in bud, in fr often sharply bent, erect, becoming red. **FL:** pendent; sepals waxy; petals 11–19 mm, fused 6–10 mm, red, with some wax. 2*n*=34. ± common. Rocky cliffs, canyons; gen < 1000 m. c&s CCo, s SCoRO, SCo, TR, PR; n Baja CA. Hybridizes with *D. lanceolata.* May–Jul

D. saxosa (M.E. Jones) Britton & Rose Rosettes 1–few, 6–23 cm wide. **LF:** evergreen but smaller in summer, few green in droughts, lance-oblong, glaucous in youth, tip acute. **INFL:** peduncle 5–51 cm, 1–9 mm wide, often ± red; 1° branches 2–3, branched 0–1 ×; pedicels 5–20 mm. **FL:** sepals 4–8 mm, deltate, acute; petals 2.5–4 mm wide, lance-oblong, acute. 1 other subsp., w&c AZ.

subsp. **aloides** (Rose) Moran (p. 677) Rosettes 1–4, 6–23 cm wide. **ST:** 1–3 cm wide. **LF:** 4–15 cm, 6–25 mm wide, 2–5 mm thick, base 10–25 mm wide. **INFL:** peduncle 10–51 cm, 1–9 mm wide; lower internodes > 5 mm; 1° branches 3; terminal branches wavy, 1–12 cm, 2–20-fld. **FL:** petals 8–15(20) mm, fused 1.5–3 mm, ± green to bright yellow, rarely red-tinged. 2*n*=34. Rocky, shaded slopes; 240–1700 m. PR, DMtns, DSon; expected n Baja CA. Hybrids with *D. lanceolata* suspected. Possible recognition of *D. alainae* Reiser in need of study. Apr–Jun

subsp. **saxosa** PANAMINT DUDLEYA Rosettes 1–4, 6–20 cm wide. **ST:** 1–1.5 cm wide. **LF:** 3–9 cm, 3–15 mm wide, 1.5–3 mm thick, base 5–15 mm wide. **INFL:** peduncle 5–20 cm, 2–4 mm wide; lower internodes gen > 5 mm; 1° branches 3, terminal branches not wavy, 1–4 cm, 2–9-fld. **FL:** petals 9–12 mm, fused 1–4 mm, bright yellow, gen red-tinged. 2*n*=136,170. N-facing, granitic or limestone slopes; 1100–2200 m. n DMtns (w Panamint Range). May–Jun ★

D. stolonifera Moran (p. 677) LAGUNA BEACH DUDLEYA Stoloned; rosettes 1–several, 1–10(15) cm wide. **ST:** 1.5–3 cm wide. **LF:** evergreen, 3–7 cm, 1.5–3 cm wide, 3–4 mm thick, oblong-obovate, lime-green, ± red abaxially, not glaucous, base 1–2 cm wide, tip short-acuminate to mucronate. **INFL:** peduncle 2–25 cm, 1–4 mm wide; 1° branches gen 2 (or more), branched 0(1) ×, ascending, 1–6 cm, 3–9-fld; pedicels 5–8 mm. **FL:** sepals 2–3 mm, deltate, wider than long; petals 10–11 mm, 3–3.5 mm wide, fused 1–2 mm, elliptic, yellow, tips outcurved, acute. **FR:** ascending. 2*n*=34. N-facing cliffs, outcrops; < 250 m. c SCo (San Joaquin Hills, Orange Co.). Hybrids with *D. edulis* suspected. May–Jul ★

D. traskiae (Rose) Moran (p. 677) SANTA BARBARA ISLAND DUDLEYA Rosettes 4–50, 8–15(20) cm wide. **ST:** 1–3 cm wide. **LF:** evergreen, 4–15 cm, 1–4 cm wide, 4–6 mm thick, oblong-oblanceolate, gen glaucous, base 1–4 cm wide, tip acute to acuminate. **INFL:** peduncle 20–50 cm, 3–8 mm wide; 1° branches 3–4+, branched 0–2 ×; terminal branches 4–10 cm, 7–15-fld; pedicels 1–4 mm. **FL:** sepals 2.5–4 mm, deltate, acute; petals 8–10.5 mm, 3–4 mm wide, fused 1–2 mm, narrowly ovate, mustard-yellow, tips gen outcurved, narrow, acute. **FR:** ascending. 2*n*=68. Steep slopes; < 110 m. s ChI (Santa Barbara Island). Possible hybrid origin. May–Jul ★

D. variegata (S. Watson) Moran (p. 677) VARIEGATED DUDLEYA Rosettes 1–3, 2–6 cm wide. **ST:** 1–3 cm, 3–15 mm wide, ± spheric to oblong. **LF:** deciduous in summer, 1–7 cm, 0.5–11 mm wide, oblanceolate to spoon-shaped, ± not glaucous, base gen 1–3 mm wide, tip acute to obtuse. **INFL:** peduncle 5–20 cm, 0.5–2 mm wide; 1° branches 2–3, simple, ascending, 2–15 cm, 3–11-fld; pedicels 0.5–3 mm. **FL:** sepals 2–3.5 mm, deltate-ovate; petals 5–8 mm, 2–3.5 mm wide, fused 0.5–1 mm, elliptic, acute, yellow. **FR:** spreading. 2*n*=34. Dry hillsides, mesas; < 300 m. s SCo, s PR (San Diego Co.); n Baja CA. Apr–Jun ★

D. verityi K.M. Nakai VERITY'S DUDLEYA Rosettes 1–30, 0.6–7 cm wide. **ST:** 2–10 cm, 0.2–1 cm wide. **LF:** evergreen, 2–5 cm, 4–8 mm wide, lance-oblong, purple-gray, glaucous, gen twisted when dry, base wounding purple-red, 5–8 mm wide, tip acute to acuminate. **INFL:** peduncle 3–15 cm, 1.5–6 mm wide, erect; 1° branches 2–3, branched (0)1 ×, ascending; terminal branches 2–5 cm, ascending in age, 2–10-fld; pedicels ascending to erect, 3–5 mm. **FL:** bud tips angled ± 35°; sepals 4–5 mm, triangular, acute; petals 10–14 mm, 2.5–4 mm wide, fused 1–2 mm, green- to lemon-yellow, tips spreading, acute. 2*n*=34. N-facing volcanic outcrops; 60–120 m. s WTR (w Santa Monica Mtns, Ventura Co.). Hybrids with *D. lanceolata* suspected. May–Jun ★

D. virens (Rose) Moran   Rosettes many, 4–25 cm wide. **ST**: 1–8 cm wide. **LF**: evergreen, 3–25 cm, 15–30 mm wide, oblong, 4–6 mm thick, in ×-section ± elliptic or in upper 1/2 round, glaucous or not, ± fragrant or not. **INFL**: peduncle 6–70 cm, 1.5–15 mm wide; 1° branches 2–many, branched 0–3 ×; terminal branches 1–10 cm, 3–20-fld; pedicels 1–5 mm. **FL**: sepals 2–4 mm, acute; petals 7–10 mm, 2–3 mm wide, fused 1–2 mm, elliptic, acute, white to pale yellow, some with red, keels rose or orange-brown. 1 other subsp., Guadalupe Island, Mex.

subsp. ***hassei*** (Rose) Moran   Rosettes 4–8 cm wide. **ST**: 1–3(4) cm wide. **LF**: 15–30, outer ascending to erect, 3–10(15) cm, 5–10(15) mm wide, 2–4 mm thick, ± round in ×-section in upper 1/2, linear-oblong to -lanceolate, base 8–15 mm wide, gen glaucous. **INFL**: peduncle 10–30 cm, 1.5–5 mm wide, ascending or erect, not easily broken when dry; 1° branches 2–4, branched 0–1 ×; terminal branches 2–10 cm, 3–15(20)-fld; pedicels gen 1–2(5) mm. **FL**: petals (7)8–10 mm, fused 1.5–2 mm, spreading from middle, white, keels marked rosy or not. 2*n*=68. Rocks, cliffs; < 200 m. s ChI (Santa Catalina Island). [*D. h.* (Rose) Moran] Hybrids with *D. virens* subsp. *insularis* suspected. Apr–Jun   ★

subsp. ***insularis*** (Rose) Moran   ISLAND GREEN DUDLEYA   Rosettes 10–25 cm wide. **ST**: 2–6(8) cm wide. **LF**: 20–50, outer ascending to erect, 6–25 cm, 10–32 mm wide, 2–4 cm wide at base, 2–4 mm thick, triangular-lanceolate, ± wide-elliptic in ×-section in upper 1/2, gen glaucous, tip ± obtuse. **INFL**: peduncle 6–70 cm, 5–15 mm wide, erect, not easily broken when dry; 1° branches several, branched 0–1 ×, terminal branches 1–5 cm, 3–8-fld; pedicels 2–3 mm. **FL**: sepals 2–3 mm; petals 8–10 mm, fused 1.5–2 mm, spreading from middle, white, keels with red to orange-brown. 2*n*=34. Rocks, cliffs; < 200 m. SCo (near San Pedro, Los Angeles Co.), s ChI (San Nicolas, Santa Catalina islands). Hybrids with *D. virens* subsp. *hassei* suspected. Apr–Jun   ★

subsp. ***virens***   BRIGHT GREEN DUDLEYA   Rosettes 5–10(18) cm wide. **ST**: 1–3.2 cm wide. **LF**: 20–50, outer lax to erect, 3–10 cm, 7–16 mm wide, 2–4 mm thick, elliptic to ± round in ×-section in upper 1/2, triangular-lanceolate to lance-linear, green, rarely glaucous, base 8–15 mm wide, tip acute. **INFL**: peduncle 7–46 cm, 4–7 mm wide, easily broken when dry; 1° branches often several, branched 0–3 ×; terminal branches 3–5 cm long, 5–12-fld; pedicels 2–4 mm. **FL**: petals 8–10 mm, spreading from middle, white, keels rarely with red to orange-brown. 2*n*=34. Rocks, cliffs, coastal flats; < 500 m. s ChI (San Clemente Island). May comprise 2 entities; study needed. Apr–Jun   ★

D. viscida (S. Watson) Moran (p. 677)   STICKY DUDLEYA   Cespitose, gen < 3 cm between branches; rosettes 1–16(32), 7–20 cm wide. **ST**: 1–4 cm wide. **LF**: evergreen, 6–15 cm, 5–15 mm wide, 3–5 mm thick, ± elliptic to wide-elliptic in ×-section in upper 1/2, linear-deltate, sticky, appearing oily, odor resinous, base 1–2 cm wide, tip acute. **INFL**: peduncle 15–70 cm, 2–10 mm wide; 1° branches 3–many, branched 1–2 ×; terminal branches 2–6 cm, 3–10-fld; pedicels 1–4 mm. **FL**: sepals 1.5–4 mm, ovate, acute; petals 6–9 mm, 2.5–3.5 mm wide, elliptic, acute, spreading from middle, pink (to white-pink). **FR**: ascending. 2*n*=34. Bluffs, rocky cliffs; < 450 m. s SCo (Orange, San Diego cos.). Hybrids with *D. edulis* suspected. May–Jun   ★

RHODIOLA

Per from short, scaly caudex, glabrous [hairy]; dioecious or not. **LF**: cauline, sessile, alternate, entire to toothed. **INFL**: gen hemispheric, gen dense. **FL**: ± unisexual; sepals, petals 4–5[6], not circumscissile in fr; sepals fused at base; petals ± free; stamens 2 × sepals in number, epipetalous; pistils 4–5(6), free or fused below. **FR**: erect. **SEED**: many, < 3 mm, ± fusiform, brown, striate. ± 40 spp.: gen montane to arctic; N temp, esp Asia. (Greek: rose, for scent of roots)

R. integrifolia Raf. subsp. ***integrifolia*** (p. 677)   WESTERN ROSEROOT   Pl (2)3–15(50) cm; caudex short, thick, fleshy, branched. **LF**: 5–30(50) mm, 2–15(20) mm wide, oblanceolate to obovate or elliptic, entire or toothed, tip acute to obtuse, gen green, ± glaucous or not. **INFL**: ± 1–3 cm, 7–50-fld, dense. **FL**: most or all unisexual; sepals, petals 4(5); sepals 1.5–3 mm, lanceolate to ovate; petals free, 2–4 mm, dark red to deep red-purple (narrowly spoon-shaped, > sepals, ± spreading in staminate, to linear or awl-like, ≤ sepals, erect in pistillate); anthers light brown to red-purple. **FR**: 3–6(10) mm. **SEED**: 1.4–2 mm. *n*=18. Cliffs, talus, alpine ridges, margins of meadows, streams; 1800–4000 m. KR, SNH, Wrn, W&I; w N.Am. [*Sedum rosea* (L.) Scop. subsp. *i.* (Raf.) Hultén; *Sedum roseum* subsp. *integrifolium*, orth. var.] May–Aug

SEDELLA

Ann, erect, glabrous, branches 0 or near base. **LF**: early-deciduous, sessile, 0.4–0.7 cm, oblong-elliptic to ovoid (obovoid), basal opposite, free, not fused around st, cauline alternate, entire, tip rounded to obtuse. **INFL**: terminal, fls 1–2+ in 0–3-branched cyme, subsessile. **FL**: sepals 5; petals 5, ± fused at base, linear to narrow-ovate, pale to bright or green-yellow, midrib often ± red; stamens 5 or 10, anthers 0.2–0.4 mm, yellow or red-brown; pistils 5, oblong, bases rounded, styles 0.2–1.2 mm, erect or recurved, stigmas ± 0.1 mm diam. **FR**: ± indehiscent, utricle-like, erect to outcurved, glabrous or glandular. **SEED**: 1, 0.7–2 mm, club-like, brown. 3 spp.: CA. (Latin: diminutive of *Sedum*) [Moran 1997 Haseltonia 5:53–60] *Parvisedum* is a superfluous name for *Sedella*.

1. Stamens 10 . . . . ***S. pumila***
1′ Stamens 5
  2. Calyx base tapered to pedicel; petals 2.5–3.8 mm; fr 1.5–2.5 mm, ± papillate; seed 1.2–1.5 mm . . . . ***S. leiocarpa***
  2′ Calyx base abruptly narrowed to pedicel; petals 1.3–2 mm; fr 1–1.5 mm, densely stipitate-glandular;
     seed < 1 mm . . . . ***S. pentandra***

S. leiocarpa H. Sharsm. (p. 677)   LAKE COUNTY STONECROP   Pl 1–3.5(5) cm, branches 0–3, ascending. **LF**: 2–4.5(7) mm, ± 1–2(4) mm wide. **FL**: calyx base tapered to pedicel, sepals 0.6–1 mm, ± 0.5 mm wide; petals ascending to ± erect in fl, appressed in fr, 2.5–3.8 mm, 0.5–0.9 mm wide, narrow-lanceolate, subacuminate, ± yellow becoming pale, finely red-lined exc margins; stamens 5, anthers ± 0.2 mm, yellow, pistils 1.5–2 mm, subglabrous, styles erect, 0.2–0.4 mm. **FR**: 1.5–2.5 mm, ± erect, ± papillate. **SEED**: 1.2–1.5 mm. Dry vernal pools, rocky depressions; 500–600 m. s NCoRI (Lake Co.). [*Parvisedum l.* (H. Sharsm.) R.T. Clausen] Apr–May   ★

S. pentandra H. Sharsm.   Pl 2–13 cm, simple, erect, or branches 1–4, ascending, exceeded by main axis. **LF**: 4–7 mm, 2–3 mm wide. **FL**: calyx base abruptly narrowed to pedicel, sepals 0.5–0.8 mm, 0.3–0.5 mm wide; petals ascending in fl, erect with tips ± touching in fr, 1.3–2 mm, 0.5–0.8 mm wide, lance-ovate, pale green-yellow, with red streaks abaxially or not; stamens 5, anthers ± 0.2 mm, yellow; pistils 1 mm, stipitate-glandular, styles erect, 0.2–0.4 mm. **FR**: 1–1.5 mm, erect, densely stipitate-glandular. **SEED**: 0.7–0.9 mm. *n*=9. Compacted ground, slate, shale, sandstone or serpentine outcrops; 300–700 m. NCoRI, e ScV, ne SnJV, s SCoRO, SCoRI. [*Parvisedum p.* (H. Sharsm.) R.T. Clausen] Mar–Jun

S. pumila (Benth.) Britton & Rose (p. 677) Pl (1)2–17 cm, branches gen several (0), erect or spreading, at same level as main axis. **LF**: (2)4–7 mm, 1–3 mm wide. **FL**: calyx base tapered to pedicel, sepals 0.5–0.8 mm, 0.3–0.4 mm wide; petals spreading in fl, erect in fr, (2)2.5–5 mm, 0.3–1.2 mm wide, elliptic to lanceolate, acute, pale to bright yellow; stamens 10, anthers 0.2–0.4 mm, yellow or red-brown; pistils 1–2 mm, stipitate-glandular near suture and on angles, often with fringed row of papillae on suture, styles erect, or, when short, often recurved, 0.3–1.2 mm. **FR**: 1.2–2.5 mm, erect to ascending (spreading), glabrous to glandular adaxially. **SEED**: 0.8–1.5 mm. $n=9$. Open, often wet sites, rock outcrops, clay soils, vernal pools; 30–1500 m. se NCoRO, NCoRI, CaRF, SNF, c SNH, GV, SCoRI; OR. [*Parvisedum p.* (Benth.) R.T. Clausen; *P. congdonii* (Eastw.) R.T. Clausen; *S. c.* (Eastw.) Britton & Rose] Mar–May

SEDUM STONECROP

Steve Boyd & Melinda F. Denton

Per (ann, bien, subshrub), gen from rhizomes or stout, scaly caudex, gen glabrous; rosettes 0 or open to dense. **LF**: sessile, gen alternate, gen obovate to spoon-shaped. **INFL**: terminal, gen raceme- to panicle-like. **FL**: sepals, petals gen 5, free to fused at base, sepals < petals, obtuse to long-tapered; petals erect to spreading; stamens 8 or 10, in 2 whorls, epipetalous or not; pistils 4–5, free or fused below. **FR**: free or fused at base, erect or spreading. **SEED**: many, elliptic, often winged at both ends. ± 450 spp.: temps, trop mtns, N.Am, Mex, C.Am, Eur, Asia, n&e Afr, Atlantic islands, Indian Ocean islands; cult as orn, green roofs. (Latin: to assuage, from healing properties of houseleek, to which *Sedum* was sometimes applied.) *S. roseum* moved to *Rhodiola*.

1. Subshrub . . . . ***S. praealtum***
1′ Ann to per
  2. Sts densely glandular-hairy toward base . . . . ***S. album***
  2′ Sts glabrous, glaucous, or papillate, not glandular-hairy toward base
    3. Lvs gen widest ± below middle, tapered to tip (exc *Sedum niveum*); lvs ± = bracts (exc *Sedum lanceolatum*)
      4. Petals white, pink-streaked, fused > 1 mm; fls 1–9; fr erect. . . . . ***S. niveum***
      4′ Petals yellow or white, ± free; fls 1–44; fr erect to spreading
        5. Lvs red-scarious in age
          6. Ann, bien; lf tip obtuse to acute; fr 3–5 mm, spreading to strong-reflexed; sts without plantlets in place of distal lvs, fls . . . . ***S. radiatum***
          6′ Per; lf tip long-tapered; fr 5–8 mm, ± erect; sts often with plantlets in place of distal lvs, fls . . . . ***S. stenopetalum***
        5′ Lvs staying green, fleshy
          7. Bracts ± elliptic to round, surfaces convex; fr spreading . . . . ***S. divergens***
          7′ Bracts lanceolate, surfaces ± flat; fr ± erect, tips outcurved. . . . . ***S. lanceolatum***
    3′ Lvs gen widest above middle, tapered to base; lvs > bracts
      8. Infl gen ± 3-branched; petals long-acuminate and frs erect, or petals acute and frs gen outcurved
        9. Petals long-acuminate, 8–13 mm, fused < 3 mm; fr erect; outer rosette lvs = inner . . . . ***S. oreganum***
        9′ Petals acute, (5)7–8(10) mm, ± free; fr gen outcurved; outer rosette lvs ≫ inner. . . . . ***S. spathulifolium***
      8′ Infl gen > 3-branched; petals obtuse to acute and frs erect
        10. Rosette open, internodes visible, some 3–5 mm
          11. Petals rounded to obtuse; anthers yellow (red-brown); infl 2–8 cm, 13–40(72)-fld . . . ***S. obtusatum*** subsp. ***retusum***
          11′ Petals acute; anthers yellow or red-brown; infl 1–14 cm, (12)20–70(120)-fld. . . . . ***S. oregonense***
        10′ Rosette dense, internodes not visible, < 3 mm
          12. Infl strongly white-glaucous; rosette lvs 11–60 mm, gen > 3.5 × longer than wide; anthers gen yellow (turning red-brown)
            13. Bracts 14–30 mm; rosettes 3–9 cm diam, lvs 14–60 mm, 7–16 mm wide, widest 6–9 mm below tip; petals ± pale yellow; fr free . . . . ***S. albomarginatum***
            13′ Bracts 7–13 mm; rosettes 1–6 cm diam, lvs 11–35 mm, 3–9 mm wide, widest 2–3 mm below tip; petals ± cream; fr fused at base . . . . ***S. oblanceolatum***
          12′ Infl not or ± glaucous; rosette lvs 6–50 mm, gen < 3 × longer than wide (or anthers red-purple); anthers yellow to purple-black
            14. Petals dark pink to dark red; bract surfaces convex, base cordate, not clasping. . . . ***S. laxum*** subsp. ***eastwoodiae***
            14′ Petals white to yellow or pink; bract surfaces flat, base obtuse to truncate, not clasping, or cordate, clasping
              15. Petals white to pink, or pale yellow . . . . ***S. laxum***
                16. Bract length ≫ width; pl 7–40 cm; rosette lvs 10–50 mm . . . . subsp. ***laxum***
                16′ Bract length < to > width; pl 9–24 cm; rosette lvs 9–32 mm
                  17. Petals pale yellow; bract base often truncate, not clasping; sepals < 1/3 petals. . . . . subsp. ***flavidum***
                  17′ Petals white to pink; bract base cordate, clasping; sepals 1/3–2/3 petals . . . . subsp. ***heckneri***
              15′ Petals cream or deep yellow, pink-tinged or not, red-veined or not (pale yellow or pink-white in *Sedum obtusatum* subsp. *boreale*) . . . . ***S. obtusatum***
                18. Petals cream, not ± pink- or red-tinged or -veined; sepals 3.5–6 mm, acute to long-tapered . . . subsp. ***paradisum***
                18′ Petals yellow, pale yellow, or white, ± pink- or red-tinged or -veined at least in age; sepals 1.8–5 mm, obtuse to acute
                  19. Bracts 8–15 mm; petals pale yellow or white, gen pink-tinged or -veined; infl 15–38(60)-fld . . . subsp. ***boreale***
                  19′ Bracts gen 4–9(13) mm; petals yellow, ± red-tinged or -veined in age; infl 8–26(46)-fld . . . subsp. ***obtusatum***

S. albomarginatum R.T. Clausen (p. 677) FEATHER RIVER
STONECROP Pl 14–25 cm, glaucous; rosettes dense, 3–9 cm diam,
internodes not visible, < 3 mm. **LF:** 14–60 mm, widest 6–9 mm
below tip, ± 1–4 mm thick, tip rounded or ± notched. **INFL:** 4–10
cm, 20–55-fld; bracts 14–30 mm, base truncate or obtuse. **FL:** petals
± 7–10 mm, obovate, pale yellow, obtuse or minutely mucronate;
anthers yellow, some turning red-brown. **FR:** free, 7–9 mm, erect.
SEED: 1–1.2 mm. *n*=15. Steep serpentine slopes; 300–900 m. n SNH
(Plumas, Butte cos.). Jun ★

S. album L. WHITE STONECROP St creeping to short-ascending,
erect in fl, densely glandular-hairy toward base; rosettes 0. **LF:**
4–20(25) mm, linear to ovate, green or ± red, hairs 0 or sparse;
tip obtuse to rounded. **INFL:** 5–18(30) cm, 15–50-fld. **FL:** sepals
(0.5)1–1.5(2) mm, hairs 0 or sparse; petals spreading, 2–4.5 mm,
lanceolate, ± acute, white (± pink). **FR:** free, 3–4 mm, erect. *n*=17.
Disturbed sites; gen < 1350 m. SN, CCo, SCo, SnBr, PR; n&w N.Am;
native to Eur. Jul–Sep

S. divergens S. Watson (p. 677) CASCADE STONECROP Pl 5–12
cm, matted; ascending sterile shoots many; rosettes 0. **LF:** opposite,
3–8 mm, obovate to ± spheric, glabrous or fine-papillate, tip gen
rounded. **INFL:** gen 1–5 cm, 3–17-fld. **FL:** petals 5–7 mm, lance-
oblong, acute or obtuse, mucronate, yellow; anthers yellow. **FR:**
fused at base, 5–7 mm, spreading. **SEED:** ± 0.5 mm. *n*=8. Gravelly
flats, slopes; 1600–2000 m. KR; to BC. Jul–Aug ★

S. lanceolatum Torr. (p. 677) SPEARLEAF STONECROP Pl 3–20
cm; rosettes gen dense, 0.3–1 cm diam, internodes not visible, ±
1 mm. **LF:** 5–30 mm, 1.5–2 mm thick, linear to ovate, tip acute.
INFL: 1–4 cm, 3–24-fld; bracts 3–10 mm, most fallen by fl. **FL:**
petals 5–8 mm, lanceolate, acute to acuminate, yellow, midrib often
± red; anthers yellow. **FR:** fused at base, 4–9 mm, ± erect, tips outcurved. **SEED:** ± 1 mm. *n*=8,16,24. Granite outcrops, rocky soils;
1800–2800 m. KR, CaR, SN; to AK, SD, NM. May–Aug

S. laxum (Britton) A. Berger Pl 7–40 cm, glaucous or not;
rosettes dense, 1–6 cm diam, internodes not visible, < 3 mm. **LF:**
9–50 mm, 1–5 mm thick, gen widest 3–8 mm below tip, tip rounded
or ± notched. **INFL:** 2–11 cm, 12–80-fld, flat-topped or not; bracts
5–20 mm. **FL:** sepals gen 1/3–2/3 petals, gen acute; petals 6–13 mm,
obovate, obtuse or acute. **FR:** free, 6–12 mm, erect. **SEED:** 1–2 mm.

  subsp. ***eastwoodiae*** (Britton) R.T. Clausen (p. 679) RED
MOUNTAIN STONECROP Pl 7–19 cm. **LF:** 10–29 mm. **INFL:** bracts
4–17 mm, base cordate, not clasping. **FL:** petals acute, dark pink to
dark red; anthers light red or ± red-purple. *n*=30. Serpentine soils
among rocks; 600–1200 m. c NCoRO (Red Mtn, Mendocino Co.). [*S.
e.* (Britton) A. Berger; *S. l.* var. *e.* (Britton) H. Ohba] Jul ★

  subsp. ***flavidum*** Denton (p. 679) PALE-YELLOW STONECROP
Pl 9–24 cm. **LF:** 9–25 mm. **INFL:** bract length < to > width, base
often truncate, not clasping. **FL:** sepals < 1/3 petals, obtuse; petals
obtuse, pale yellow; anthers yellow to ± red-brown. *n*=30. Serpentine
or basalt outcrops; 800–2000 m. KR, n NCoRH. [*S. l.* var. *f.* (Denton)
H. Ohba] May–Jul ★

  subsp. ***heckneri*** (M. Peck) R.T. Clausen (p. 679) HECKNER'S
STONECROP Pl 9–24 cm. **LF:** 11–32 mm. **INFL:** bract length < to
> width, base cordate, clasping. **FL:** petals obtuse, white to pink;
anthers ± red-brown to purple-black. *n*=15. Gen steep serpentine or
gabbro outcrops; 100–1800 m. KR, n NCoRO; sw OR. [*S. l.* var. *h.*
(M. Peck) H. Ohba] May–Sep ★

  subsp. ***laxum*** (p. 679) ROSEFLOWER STONECROP Pl 7–40 cm.
LF: 10–50 mm. **INFL:** bract length >> width, base obtuse, not clasping. **FL:** petals obtuse, gen white to light pink (light yellow); anthers
± red-brown to purple-black. *n*=15. Serpentine, occ basalt; 50–2000
m. NCo, KR; sw OR. [*S. l.* var. *l.*; *S. laxum* var. *latifolium* (R.T. Clausen) H. Ohba] May–Jul

S. niveum Davidson (p. 679) DAVIDSON'S STONECROP Pl
3–9 cm, matted; rosettes dense, 0.6–1.7 cm diam, internodes gen
not visible, gen < 1 mm. **LF:** 5–9 mm, 1–3 mm thick, oblong to
spoon-shaped, tip rounded to wide-acute. **INFL:** 1–2 cm, 1–9-fld.
FL: petals spreading to reflexed, 5–8 mm, lanceolate, acute, white,
pink-streaked; anthers red to black. **FR:** fused at base, 5–7 mm, erect.

SEED: ± 0.5 mm. *n*=16. Rocky ledges, crevices; 2100–3000 m.
SnBr, e PR (Santa Rosa Mtns), DMtns (New York Mtns); Baja CA.
S. pinetorum Brandegee, *Congdonia pinetorum* (Brandegee) Jeps. are
possible synonyms of *S. niveum.* Jun–Aug ★

S. oblanceolatum R.T. Clausen (p. 679) APPLEGATE STONECROP
Pl 7–14 cm, strongly glaucous; rosettes gen dense, 1–6 cm diam,
internodes gen not visible, < 3 mm. **LF:** 11–35 mm, widest 2–3 mm
from tip, 2–3 mm thick, tip obtuse to notched. **INFL:** 1–5 cm, 7–50-
fld, gen flat-topped; bracts 7–13 mm. **FL:** petals 7–11 mm, obovate,
± mucronate, pale yellow or cream; anthers yellow. **FR:** fused at base,
6–10 mm, erect. **SEED:** 0.9–1.5 mm. *n*=15. Rocky slopes; 400–2000
m. n KR (nw Siskiyou Co.); sw OR. Jul ★

S. obtusatum A. Gray SIERRA STONECROP Pl 3–22 cm, glaucous; rosettes gen dense, 1–6 cm diam, internodes gen not visible,
gen < 3 mm. **LF:** 6–33 mm, 1–4 mm thick, obovate to oblanceolate
or spoon-shaped, widest 2–8 mm below tip, tip rounded or obtuse
to ± notched. **INFL:** 2–12 cm, often flat-topped; bracts 4–19 mm,
obtuse or truncate. **FL:** sepals acute to long-tapered; petals 3.5–11
mm, obovate, rounded to obtuse, gen mucronate; anthers ± yellow to
dark red-brown. **FR:** free, 5–10 mm, erect.

  subsp. ***boreale*** R.T. Clausen (p. 679) **INFL:** 2–12 cm,
15–30(60)-fld; bracts 8–15 mm. **FL:** sepals 3.5–4.5 mm; petals ±
6–9 mm, pale yellow or white, gen pink-tinged or -veined. *n*=15.
Uncommon. Outcrops; 1400–2000 m. CaRH, SN. [*S. o.* var. *b.* (R.T.
Clausen) H. Ohba] Jun–Aug

  subsp. ***obtusatum*** (p. 679) **INFL:** 2–7(12) cm, 8–26(46)-fld;
bracts gen 4–9(13) mm. **FL:** sepals 1.8–5 mm; petals 4–10 mm, yellow, ± red-tinged or -veined in age. *n*=15. Outcrops; 1200–3700 m.
KR, SN. [*S. o.* var. *o.*] Jun–Aug

  subsp. ***paradisum*** Denton (p. 679) CANYON CREEK STONECROP
Pl 10–20 cm. **LF:** 10–33 mm, tip obtuse to notched. **INFL:** gen 3–9
cm, 10–58-fld; bracts 9–19 mm, base obtuse to truncate. **FL:** sepals
3.5–6 mm, acute to long-tapered, petals 7–10 mm, cream, not pink- or
red-tinged or -veined. **FR:** 9–10 mm. **SEED:** 1–1.6 mm. *n*=15. Granite outcrops, meta-volcanic outcrops, siltstone; 300–1400 m. se KR
(Trinity Co.). [*S. o.* var. *p.* (Denton) H. Ohba; *S. p.* (Denton) Denton,
ined.] Reported also from Shasta Co., and from 1850 m. Jun ★

  subsp. ***retusum*** (Rose) R.T. Clausen (p. 679) Rosette open,
internodes visible, some 3–5 mm. **INFL:** 2–8 cm, 13–40(72)-fld;
bracts (5)8–17 mm. **FL:** sepals 3 mm, obtuse; petals 6–9 mm, ± pale
yellow or cream. *n*=30. Outcrops; 400–2300 m. NCoRH, KR; sw
OR. [*S. laxum* subsp. *r.* (Rose) R.T. Clausen; *S. o.* var. *r.* (Rose) H.
Ohba] Jun–Aug

S. oreganum Nutt. (p. 679) OREGON STONECROP Pl 6–15 cm,
rhizomed; rosettes gen dense, 1.2–1.6 cm diam, internodes gen not
visible, gen 1–2 mm. **LF:** 5–25 mm, ± 2 mm thick, obovate to spoon-
shaped, widest 1–3 mm below tip, tip rounded to ± notched; outer
rosette lvs = inner. **INFL:** dense, gen ± 3-branched, 4–12 cm, 3–16-
fld; bracts 4–12 mm, base truncate. **FL:** petals 8–13 mm, narrow-
lanceolate, long-acuminate, yellow (midvein red); anthers light yellow-brown to red-brown. **FR:** fused at base, 5–8 mm, erect. **SEED:**
± 1 mm. *n*=12. Rocky ledges, gravelly ridges; 1100–1800 m. n KR;
to AK. Jul

S. oregonense (S. Watson) M. Peck (p. 679) CREAM STONECROP
Pl 4–21 cm; rosettes open, 1–8 cm diam, internodes gen visible,
gen 3–5 mm. **LF:** 12–41 mm, widest 3–9 mm below tip, 1–4 mm
thick, tip rounded or ± notched. **INFL:** gen panicle-like, 1–14 cm,
(12)20–70(120)-fld; bracts 5–16 mm, base obtuse or truncate. **FL:**
petals ± 5–11 mm, obovate, obtuse and gen ± mucronate, pale yellow
adaxially, gen red-marked or ± white abaxially; anthers yellow or
red-brown. **FR:** free, 6–9 mm, erect. **SEED:** 1–1.8 mm. *n*=45. Rock
outcrops; 900–2200 m. KR; OR. Jun–Aug

S. praealtum A. DC. GREEN COCKSCOMB Subshrub, sts gen
erect; rosettes 0. **LF:** 4–8 cm, elliptic-oblanceolate or -oblong, green,
tinged or spotted with red or not, shiny, tip rounded. **INFL:** 10–50
cm. **FL:** sepals unequal, 1.5–9.6 mm; petals spreading, ± 7.5 mm,
lanceolate, acute or obtuse, yellow. **FR:** ± free, 5–6 mm, spreading.
n=34. Disturbed sites; < 100 m. CCo (mouth of San Lorenzo River,

Santa Cruz Co.); Italy, Australia; native to c Mex, Guatemala. [*S. dendroideum* DC. subsp. *p.* (A. DC.) R.T. Clausen] Mar–Jun

S. radiatum S. Watson   COAST RANGE STONECROP   Ann, bien 4–20 cm, glabrous; rosettes dense, 0.5–0.7 cm, internodes not visible, gen < 1 mm. **LF:** 5–11 mm, oblong to narrow-ovate, ± flat adaxially, veins in age red-scarious, tip obtuse to acute. **INFL:** 2–4 cm, 1–44-fld. **FL:** petals 5–10 mm, lanceolate, acuminate, yellow or white, reflexed; anthers yellow to yellow-brown. **FR:** fused at base, 3–5 mm, spreading to strong-reflexed. **SEED:** ± 1 mm. *n*=8. Rocky ledges, gravelly serpentine slopes; 60–2100 m. KR, NCoR, CaRF, SN, SnFrB, SCoRO; sw OR. [*S. r.* var. *depauperatum* (R.T. Clausen) H. Ohba; *S. stenopetalum* subsp. *r.* (S. Watson) R.T. Clausen] Apr–Jul

S. spathulifolium Hook. (p. 679)   BROADLEAF STONECROP   Pl 5–22 cm, glabrous, often glaucous; rosettes dense, 1–6 cm diam, internodes not visible, gen 1–2 mm. **LF:** 11–22 mm, widest 2–5 mm below tip, 1–2 mm thick, tip rounded to obtuse; outer rosette lvs

>> inner. **INFL:** 3–8 cm, 5–48-fld; bracts 6–11 mm, elliptic, base truncate. **FL:** petals (5)7–8(10) mm, ± erect to spreading, lanceolate, acute, yellow; anthers yellow or red-brown. **FR:** fused at base, 4–8 mm, ± erect until mature, then spreading. **SEED:** ± 1 mm. *n*=15. Outcrops, often in shade; 50–2500 m. NW, CaR, SN, CW, TR, PR; to BC. [*S. s.* var. *pruinosum* (Britton) B. Boivin] Variable, intergrading complex. Apr–Aug

S. stenopetalum Pursh (p. 679)   Pl 5–27 cm, glabrous, often with plantlets in place of distal lvs, fls; rosettes gen dense, 0.7–2 cm diam, internodes gen not visible, ± 1 mm. **LF:** 6–16 mm, gen ± lanceolate, red-scarious in age, base often ± 3-lobed, tip long-tapered. **INFL:** gen raceme-like, gen ± 1.5 cm, 1–44-fld. **FL:** petals 4–10 mm, lanceolate, long-tapered, yellow (midrib ± red); anthers yellow or light brown. **FR:** ± fused at base, 5–8 mm, ± erect. **SEED:** ± 1 mm. *n*=8,16,24. Well-drained, rocky or gravelly soil; 1400–2600 m. NW, CaR, SN, SnFrB, MP; to BC, MT. [*S. s.* var. *monanthum* (Suksd.) H. Ohba] May–Aug

CROSSOSOMATACEAE   CROSSOSOMA FAMILY

Robert E. Preston & James R. Shevock

Shrub, small tree. **ST:** gen glabrous; branchlets or twigs gen thorny. **LF:** gen deciduous, simple, gen small, gen alternate, entire; stipules minute or 0. **INFL:** fls 1. **FL:** gen bisexual, radial; hypanthium short; sepals, petals gen 5(3–6), free; petals gen white, ephemeral; stamens 4–50, on or around disk; pistils 1–9, simple, styles short, stigmas head-like, ovules gen 2–many. **FR:** follicles, 1–9. **SEED:** brown to black, arilled. 4 genera, 9 spp.: w US, Mex. [Sosa & Chase 2003 Syst Bot 28:96–105] Scientific Editor: Thomas J. Rosatti.

1. Fls terminal; stamens 15–50; fr 1–9, 8–20 mm, stalked . . . . **CROSSOSOMA**
1′ Fls axillary; stamens 4–10, fr 1–3, 1–5 mm, sessile . . . . **GLOSSOPETALON**

CROSSOSOMA

LF: deciduous (or dry when dormant); petiole 0–short. **INFL:** peduncle bracted or not. **FL:** petals 9–15 mm, ± round or oblong, gen white. **FR:** cylindric, surface smooth or net-like, beak 1–2 mm. **SEED:** gen > 2 per fr, black, shiny, round to flat; aril conspicuous, fringed, ± yellow. 2 spp.: s CA to AZ, Mex. (Greek: fringe body, from aril)

1. Lvs 5–15 mm, clustered, glaucous; sepals 4–5 mm . . . . *C. bigelovii*
1′ Lvs 25–90 mm, gen not clustered, not glaucous; sepals ± 10 mm . . . . *C. californicum*

C. bigelovii S. Watson (p. 679)   Pl 1–2 m. **ST:** branchlets thorny. **LF:** elliptic to obovate or spoon-shaped. **FL:** sepals ± round; petals 9–12 mm, oblong, white to ± purple, clawed. **FR:** 1–3, 8–10 mm, ± straight. **SEED:** gen 2–5, ± 2 mm diam. Dry, rocky slopes, canyons; 150–1280 m. e SnBr, e PR, D; NV, AZ, Baja CA. Feb–Apr

C. californicum Nutt. (p. 679)   CATALINA CROSSOSOMA   Pl 1–5 m. **ST:** branchlets ± thorny. **LF:** oblong (to obovate or spoon-shaped). **FL:** sepals round-ovate; petals 12–15 mm, ± round, white, ± clawed. **FR:** 2–9, 15–20 mm, ± recurved at tip. **SEED:** gen > 20, ± 2.5 mm diam. Dry, rocky slopes, canyons; < 610 m. c SCo (Palos Verdes Peninsula), s ChI (San Clemente, Santa Catalina islands); Baja CA (Guadalupe Island). Feb–May   ★

GLOSSOPETALON

Small shrub, gen densely branched. **ST:** angled, ± green, ± thorny at tips or not, hairs 0 to sparse. **LF:** small, gen deciduous, ± entire, ± sessile. **FL:** petals narrow-oblanceolate, white. **FR:** ovoid, gen striate, gen beaked. **SEED:** gen 1–2, gen brown; aril ± inconspicuous, gen ± white. ± 5 spp.: w US, n Mex, esp limestone in desert mtns. (Greek: tongue petal, from petal shape) [Yatskievych 2007 Novon 17:529–530]

1. Lf spine-tipped; stipules 0; fls terminal; fr 1, < 1 mm . . . . *G. pungens*
1′ Lf tip rounded or short-pointed; stipules as 2 minute bristles near lf base; fls axillary; fr 1–2, 3–5.5 mm
. . . . *G. spinescens* var. **aridum**

G. pungens Brandegee   PUNGENT GLOSSOPETALON   Pl 5–20 cm, matted. **ST:** not thorny. **LF:** 6–10 mm, narrow-elliptic to oblanceolate; veins prominent abaxially. **FL:** sepals acuminate, 2–3 spine-tipped; petals 6–8 mm; stamens 10, longer opposite sepals. **SEED:** 1, < 1 mm, light brown. Limestone cliffs; 1700–2075 m. e DMtns (Clark Mtn Range); s NV. May–Jun   ★

G. spinescens A. Gray var. **aridum** M.E. Jones (p. 679)   NEVADA GREASEWOOD   Pl < 2 m, ± erect. **ST:** ± thorny at tips. **LF:** 5–17 mm, oblong to obovate; veins ± inconspicuous abaxially. **FL:** sepals ± rounded, 0 spine-tipped; petals 3–7 mm; stamens 6–10, ± equal. **SEED:** 1–2, 2–3 mm, gen shiny brown. Limestone; 850–2720 m. s KR, s SNH (Piute Mtns, Kern Co.), SnBr (n base), W&I, DMtns; to WA, WY, TX, n Mex. Apr–May

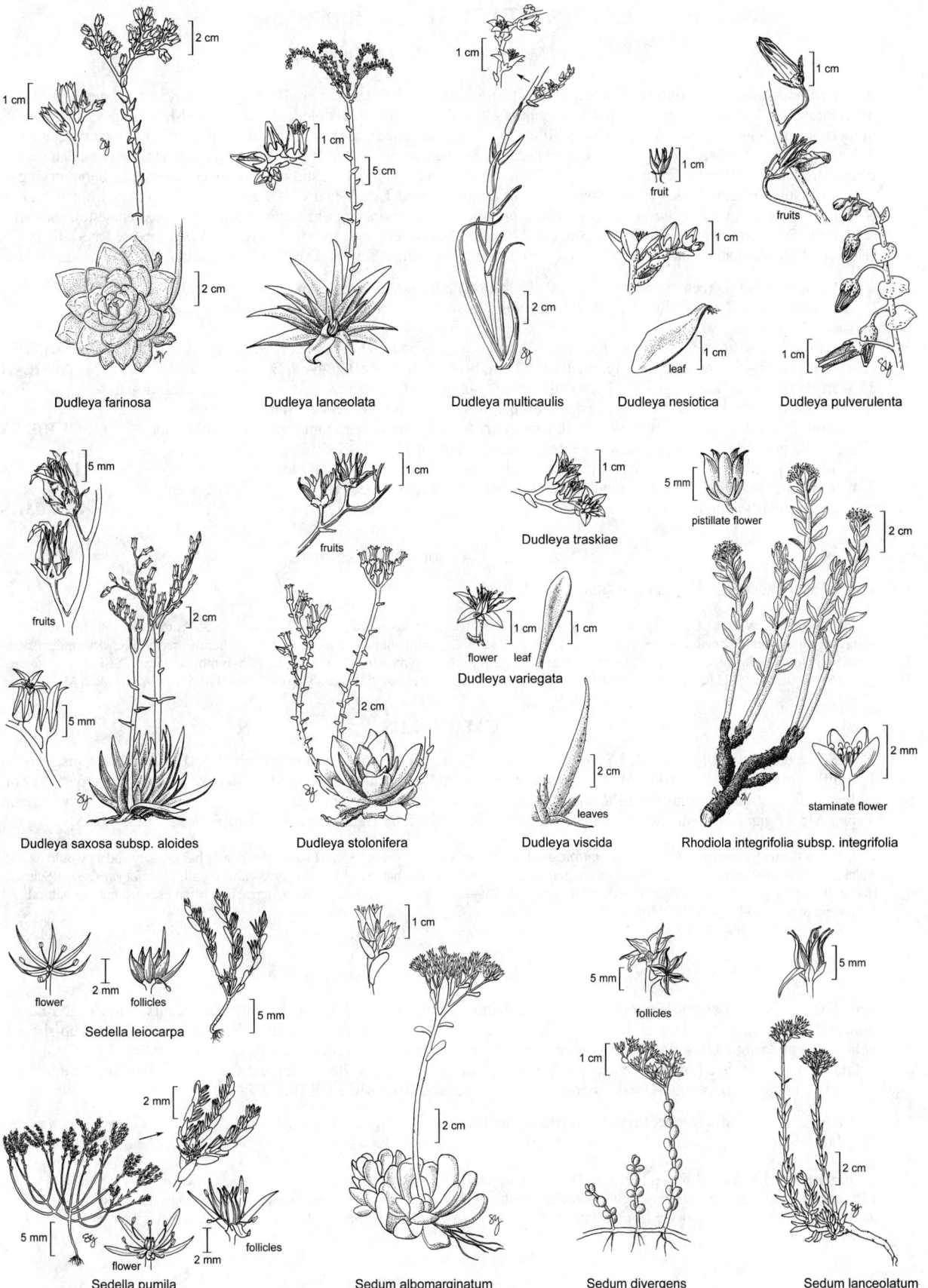

Dudleya farinosa

Dudleya lanceolata

Dudleya multicaulis

fruit

leaf

Dudleya nesiotica

fruits

Dudleya pulverulenta

fruits

Dudleya saxosa subsp. aloides

fruits

Dudleya stolonifera

Dudleya traskiae

flower   leaf

Dudleya variegata

leaves

Dudleya viscida

pistillate flower

staminate flower

Rhodiola integrifolia subsp. integrifolia

flower   follicles

Sedella leiocarpa

flower   follicles

Sedella pumila

Sedum albomarginatum

follicles

Sedum divergens

Sedum lanceolatum

CUCURBITACEAE GOURD FAMILY

John M. Miller & Robert L. Schlising

Ann, per; hairs often hard from calcium deposits; gen monoecious. **ST**: trailing or climbing, 1–many; tendril gen 1 per node, often branched. **LF**: gen simple, alternate, gen palmate-lobed and -veined, petioled; stipules 0. **INFL**: at nodes; staminate fls in racemes, panicles, small clusters, (or 1); pistillate fls gen 1. **FL**: unisexual [bisexual], radial; hypanthium > ovary; calyx gen 5-lobed (or ± 0); corolla rotate to cup- or bell-shaped, gen 5-lobed; stamens 3–5 (or ± 1–3 from fusion), anthers twisted together, often > filaments; ovary ± inferior, chambers 3–5, placentas parietal, styles 1–3, stigmas gen lobed, large. **FR**: berry, drying or not, or capsule, irregularly dehiscent, often gourd- or melon-like. **SEED**: 1–many. 100 genera, 700 spp.: esp trop; some cult (*Citrullus*, watermelon; *Cucumis*, cucumber; *Cucurbita*, gourd, pumpkin, squash; *Luffa*, loofah; *Sechium*, chayote). [Schaefer et al. 2009 Proc Roy Soc London Ser B, Biol Sci 276:843–851] Several cult spp. incl *Bryonia dioica* reported as waifs in CA (Howell 1958 Wasmann J Biol 16:1–157), but none recently. Scientific Editors: Douglas H. Goldman, Bruce G. Baldwin.

1. Corolla white to ± green or cream, < 2 cm wide; staminate fls gen in racemes, panicles, or small clusters, pistillate fls 1 or in small clusters, gen at same nodes as staminate; fr gen prickly, irregular-dehiscent or not; seeds gen not flat
  2. Staminate fl < 3 mm wide; fr asymmetric, < 1 cm, beak ± = body; D . . . . **BRANDEGEA**
  2′ Staminate fl ≥ 3 mm wide; fr ± symmetric, > 2 cm, beak << body or 0; CA-FP, D . . . . **MARAH**
1′ Corolla yellow to orange, gen 2–12 cm wide; staminate, pistillate fls 1–few, gen at different nodes; fr gen unarmed (prickly in some *Cucumis*), gourd- or melon-like, indehiscent; seeds ± flat
  3. Corolla > 3 cm wide, deeply cup-shaped, fused part > 3 cm; ann or per from large, tuber-like root . . . . **CUCURBITA**
  3′ Corolla gen < 3 cm wide, shallowly rotate to cup-shaped, fused part < 1 cm; ann
    4. Tendril branched; fr > 6 cm wide; lf ± palmately lobed, 1° lobes ± pinnately lobed. . . . . **CITRULLUS**
    4′ Tendril unbranched; fr 2–6 cm wide; lf angular or ± palmately lobed, 1° lobes, if present, ± entire or irregularly lobed . . . . **CUCUMIS**

BRANDEGEA DESERT STAR-VINE

1 sp. (T.S. Brandegee, CA botanist, engineer, 1843–1925)

B. bigelovii (S. Watson) Cogn. (p. 683) Per, taprooted. **ST**: ± glabrous; tendril unbranched. **LF**: gen round to cordate or ± square, gen deeply lobed, adaxially with white glands, terminal lobe gen longest. **INFL**: staminate fls few in axils, pistillate 1 per node, at same nodes as staminate. **FL**: corolla 1.5–3 mm wide, rotate or shallowly cup-shaped, cream; stigma 1, hemispheric. **FR**: dry, indehiscent, asymmetric, ± prickly; body 5–6 mm, ± = beak. **SEED**: 1. Canyons, washes, flats, sandy slopes; < 900 m. D; sw AZ, n Mex. Mar–May

CITRULLUS

Ann. **ST**: ± hairy; tendril branched. **LF**: deeply, ± palmately lobed; 1° lobes ± pinnately lobed. **INFL**: staminate, pistillate fls 1, at different nodes. **FL**: corolla gen < 3 cm wide, rotate or shallowly cup-shaped, yellow, deeply 5-lobed, fused part < 1 cm; anthers 3, free; stigmas 3, reniform. **FR**: melon-like, indehiscent, spheric to oblong; rind hard, smooth. **SEED**: many, ± 1 cm. 4 spp.: Afr, s Asia, Eur. (Latin: diminutive of citrus) [Dane & Liu 2007 Genet Resources Crop Evol 54:1255–1265]

C. lanatus (Thunb.) Matsum. & Nakai var. *citroides* (L.H. Bailey) Mansf. WILD WATERMELON **FR**: large, green, gen striped or mottled with darker green; pulp white. **SEED**: white to black. 2*n*=22. Disturbed areas; < 300 m. SW, DMoj; w Mex; native to s Afr. [*C. colocynthis* var. *l.* (Thunb.) Matsum. & Nakai, ined.] *C. lanatus* var. *caffrorum* (Alef.) Fosberg evidently has priority and so would be correct name; widely cult; hybridizes with *C. colocynthis* (L.) Schrad.; *C. lanatus* var. *l.*, cult watermelon, often escaped but not naturalized in CA. May–Aug

CUCUMIS CUCUMBER, MELON

Ann. **ST**: gen scabrous; tendril unbranched. **LF**: angular to ± palmately lobed; 1° lobes ± entire to irregularly lobed. **INFL**: staminate fls 1–several per node; pistillate fls gen 1, at different nodes. **FL**: corolla 2–3 cm wide, rotate or shallowly cup-shaped, yellow, deeply 5-lobed, fused part < 1 cm; anthers 3, free; stigmas 3–5, ± reniform. **FR**: gourd- or melon-like, indehiscent, cylindric to spheric; rind firm, net-veined, prickles 0 or weak; pulp green-white to orange. **SEED**: many, < 1 cm, ± flat; margin smooth. ± 40 spp.: Afr, s Asia. (Greek: cucumber) [Renner et al. 2007 BMC Evol Biol 7:58–69]

1. Lf angular or shallowly lobed; fr 3–6 cm wide, prickles 0 . . . . *C. melo* var. *dudaim*
1′ Lf deeply lobed; fr 2–3 cm wide, prickles weak . . . . *C. myriocarpus*

C. melo L. var. *dudaim* (L.) Naudin (p. 683) DUDAIM MELON **FR**: cylindric to ± spheric, orange, irregular-blotched or striped or not. Fields, roadsides; < 200 m. se DSon (Imperial Co.); native to Afr, s Asia. Mar–Sep ◆

C. myriocarpus Naudin (p. 683) PADDY MELON **FR**: ± spheric, yellow-green, green-striped. 2*n*=24. Fields, disturbed areas; < 300 m. s SCoRO (Santa Barbara Co.); native to s Afr. Mar–Sep ◆

CUCURBITA GOURD, FIELD PUMPKIN, SQUASH

Ann, per, from large, fleshy, tuber-like root. **ST**: glabrous to scabrous; tendril branched or not. **LF**: lanceolate to round, entire to deeply lobed. **INFL**: fls 1 per node, staminate, pistillate at different nodes. **FL**: corolla > 2 cm wide (staminate gen < pistillate),

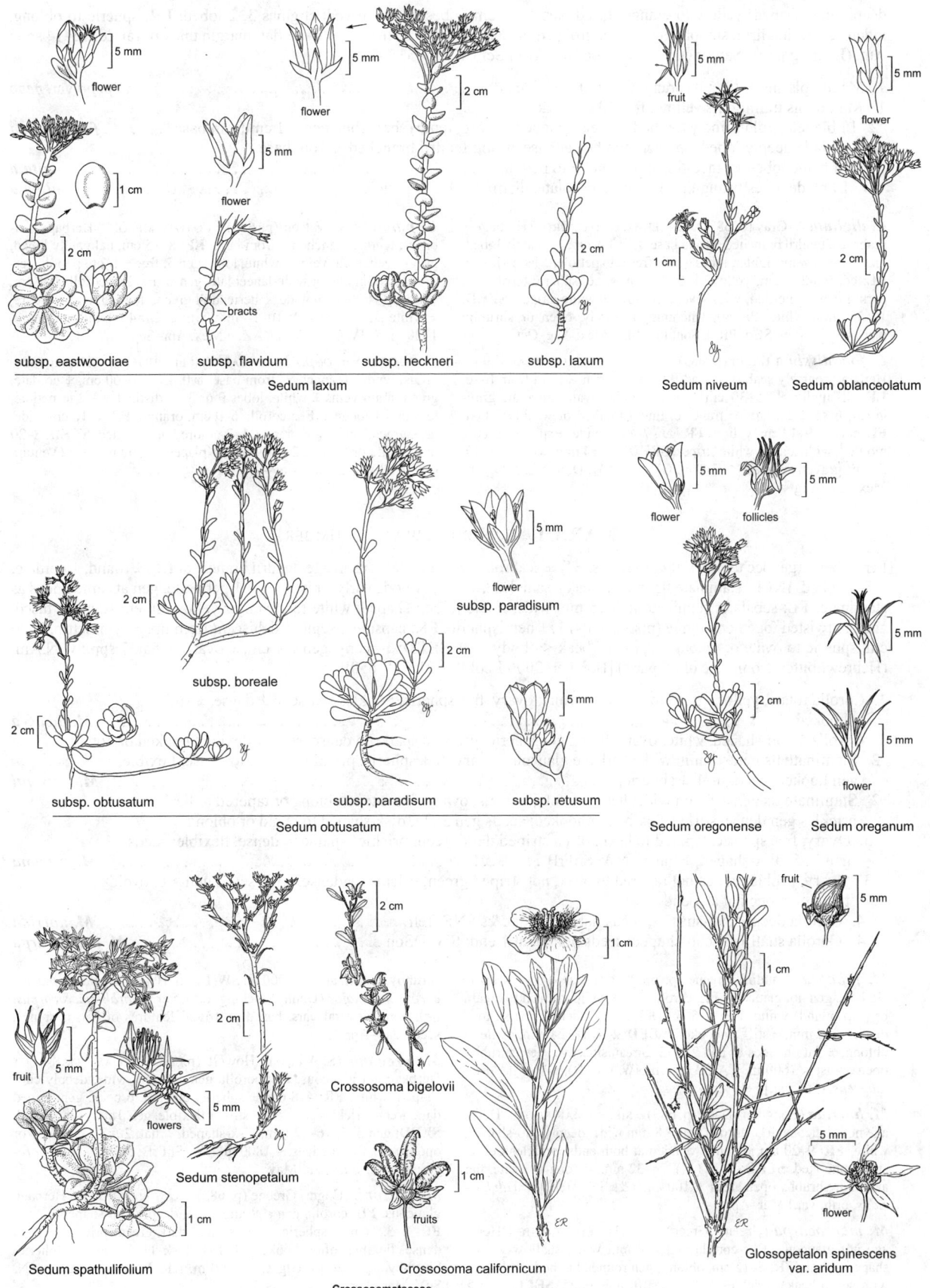

subsp. eastwoodiae subsp. flavidum subsp. heckneri subsp. laxum

Sedum laxum

Sedum niveum Sedum oblanceolatum

subsp. boreale subsp. paradisum subsp. retusum

subsp. obtusatum

Sedum obtusatum

Sedum oregonense Sedum oreganum

Sedum stenopetalum

Sedum spathulifolium

Crossosoma bigelovii

Crossosoma californicum

Crossosomataceae

Glossopetalon spinescens
var. aridum

deeply cup-shaped, yellow to orange, fused part 4–12 cm, lobes gen recurved; stigmas 3, 2-lobed. **FR:** spheric to oblong, indehiscent; rind firm, smooth to rough or grooved. **SEED:** many, < 20 mm, ± ovate, ± flat; margin thick or raised. 12–14 spp.: Am. (Latin: gourd) [Sanjur et al. 2002 Proc Natl Acad Sci USA 99:535–540]

1. Mature pls shrub-like; fls orange; fr > 10 cm wide . . . . ***C. pepo*** var. ***pepo***
1′ Mature pls trailing; fls yellow; fr < 10 cm wide
  2. Lf blade angular, finely toothed or weakly lobed at base; tendril branched gen > 1 cm from base . . . . . ***C. foetidissima***
  2′ Lf blade deeply lobed, coarsely toothed at base or not; tendril branched ± from base
    3. Lf blade lobes ± lance-linear, distinct ± to petiole . . . . ***C. digitata***
    3′ Lf blade lobes triangular or wide-lanceolate, distinct ± 1/2 to petiole . . . . ***C. palmata***

C. digitata A. Gray (p. 683) FINGER-LEAVED GOURD Herbage ± scabrous; tendril branched ± from base. **LF:** 3–9 cm, palmately lobed, green, main veins lighter, lobes gen 5, free ± to petiole, ± lance-linear. **FL:** corolla 3–5 cm, yellow. **FR:** 7–8 cm wide, spheric to oblong, dark green, ± mottled, with several narrow, ± white stripes. **SEED:** 10–11 mm, white. 2*n*=40. Uncommon. Sandy, open or shrubby places; < 1200 m. SCo, PR, DSon; to NM, n Mex. Aug–Oct

C. foetidissima Kunth (p. 683) BUFFALO GOURD, CALABAZILLA Herbage coarsely scabrous; tendril branched gen > 1 cm from base. **LF:** ill-smelling, 15–30 cm, gen triangular-ovate, angular, gray-green, base ± cordate or truncate, finely toothed or weakly lobed. **FL:** corolla 9–12 cm, yellow. **FR:** gen 7–8 cm wide, ± spheric, green, mottled, with coarse, white stripes. **SEED:** 12–14 mm, white. 2*n*=40. Sandy, gravelly places; < 1300 m. GV, CW, SW, D; to NE, MO, TX, Mex. Jun–Aug

C. palmata S. Watson (p. 683) COYOTE MELON Herbage scabrous; tendril branched ± from base. **LF:** 8–15 cm, palmately lobed, gray-green, main veins ± white, lobes gen 5, free ± 1/2 to petiole, terminal 3 triangular or wide-lanceolate, gen entire. **FL:** corolla 6–8 cm, yellow. **FR:** 8–9 cm wide, spheric, dull green, mottled, with obscure, ± white stripes. **SEED:** 10–14 mm, white. 2*n*=40. Sandy places; < 1300 m. SnJV, CW, SW, D; AZ, n Mex. Apr–Sep

C. pepo L. var. ***pepo*** CALABAZA, FIELD PUMPKIN Herbage scabrous; tendril branches ± from base or 0. **LF:** 20–60 cm, ± cordate, green, main veins ± white, lobes 0 or 3–5, distinct ± 1/2 to petiole, terminal 1 longest. **FL:** corolla 6–9 cm, orange. **FR:** > 10 cm wide, dark green, yellow, or orange, 1–2-colored or speckled. **SEED:** 8–20 mm, white or brown. 2*n*=40. Sandy places; < 100 m. SCo (Ventura Co.); cult worldwide. May–Sep

MARAH MAN-ROOT, WILD CUCUMBER

Per; tuber large; occ temporarily dioecious. **ST:** ± scabrous or hairy, glabrous in age; tendril branched. **LF:** ± round, ± cordate, ± 5–7-lobed. **INFL:** staminate fls in racemes or panicles (or 1 fl per node early); pistillate fl 1 per node, gen at same nodes as staminate. **FL:** sepals 0; staminate fl 3–15 mm wide, rotate to cup-shaped, white to cream or yellow-green; stamens fused, anthers twisted together; stigma (pistillate fls) 1, ± hemispheric. **FR:** capsule, irregular-dehiscent from tip, ± symmetric, 3–20 cm, spheric to ovate or oblong, ± prickly, beak << body or 0. **SEED:** 1–many, gen > 1 cm, ± ovate, ± flat. 7 spp.: w N.Am. (Hebrew: bitter, from taste of all parts) [Borchert 2006 Ecol Res 21:641–650]

1. Corolla rotate, yellow-green to cream or white; ovary, fr ± spheric, prickles sparse to ± dense, ± stiff, unhooked . . . . ***M. fabacea***
1′ Corolla ± cup-shaped, white; ovary, fr gen not spheric, prickles sparse to dense, stiff or flexible, hooked or not
  2. Staminate fls gen < 8 mm wide; herbage glaucous; ovary, fr ± spheric, prickles sparse to dense, flexible, gen hooked; seeds 1–4, ± spheric . . . . ***M. watsonii***
  2′ Staminate fls gen > 8 mm wide; herbage not glaucous; ovary, fr ovate to oblong or tapered to beak, prickles gen dense, stiff or flexible, unhooked; seeds gen 3–24, disk-shaped to ovoid or oblong
    3. Ovary, fr ± spheric, tapered to beak, often striped dark green, prickles sparse to dense, flexible; seeds gen 3–6, disk-shaped, ± flat — NW, SnFrB . . . . ***M. oregana***
    3′ Ovary, fr oblong, gen not tapered to beak, not striped green, prickles ± dense, stiff; seeds gen > 6, ovoid to oblong, occ flat at 1 end
      4. Corolla deeply cup-shaped; seeds flat at 1 end; c&s SNF, Teh . . . . ***M. horrida***
      4′ Corolla shallowly cup-shaped; seeds not flat at 1 end; SW, DSon . . . . ***M. macrocarpa***

M. fabacea (Naudin) Greene (p. 683) CALIFORNIA MAN-ROOT Herbage gen not glaucous. **FL:** corolla rotate, yellow-green to cream or (esp inland) white. **FR:** 4–5 cm, ± spheric; prickles sparse to ± dense, < 12 mm, ± stiff, unhooked. **SEED:** 2–4, 18–24 mm, ovate to oblong, ± flat on sides or not. 2*n*=32. Streamsides, washes, shrubby open areas; < 1600 m. CA-FP (exc n NW, n CaR), DMoj. [*Marah fabaceus*, orth. var.] Feb–Apr

M. horrida (Congdon) Dunn (p. 683) SIERRA MAN-ROOT Herbage not glaucous. **FL:** corolla gen > 8 mm wide, deeply cup-shaped, white. **FR:** 9–20 cm, oblong, rounded at both ends; prickles dense, stiff, unhooked. **SEED:** 6–16 (24), 26–32 mm, oblong or ovate, flat at 1 end. Shrubby open areas; < 1000 m. c&s SNF, Teh. [*Marah horridus*, orth. var.] Mar–Apr

M. macrocarpa (Greene) Greene (p. 683) CHILICOTHE Herbage not glaucous. **FL:** corolla gen > 8 mm wide, shallowly cup-shaped, white. **FR:** 5–12 cm, oblong, gen rounded at both ends (occ with sharp beak); prickles ± dense, stiff, unhooked. **SEED:** gen > 6, 13–33 mm, ovoid to oblong, not flat at 1 end. 2*n*=32,64. Washes,

shrubby or open areas; < 900 m. SW, DSon; Baja CA. [*Marah macrocarpus* var. *major* (Dunn) Stocking, orth. var.; *Marah macrocarpus*, orth. var.] Several vars. based on trivial features of lvs, fr spines, seeds. Jan–Apr

M. oregana (S. Watson) Howell (p. 683) COAST MAN-ROOT Herbage not glaucous. **FL:** corolla gen > 8 mm wide, deeply cup-shaped, white. **FR:** 4–8 cm, ± spheric, tapered to beak, gen striped dark green; prickles gen dense, stiff or flexible, gen 0 at tip, unhooked. **SEED:** gen 3–6, 16–22 mm, disk-shaped, ± flat. 2*n*=32. Shrubby or open areas, forest edges; < 1800 m. NW, SnFrB; to BC. [*Marah oreganus*, orth. var.] Mar–May

M. watsonii (Cogn.) Greene (p. 683) TAW MAN-ROOT Herbage glaucous. **FL:** corolla gen < 8 mm wide, deeply cup-shaped, white. **FR:** 2–3.5 cm, ± spheric, often striped dark green; prickles ± 0 to dense, flexible, often hooked. **SEED:** 1–4, 11–14 mm, ± spheric. Shrubby areas, forest edges; < 1200 m. KR, NCoRI, s CaR, n SN, ScV. Mar–Apr

DATISCACEAE DATISCA FAMILY

Steve Boyd & William J. Stone

Per, glabrous (sparsely short-hairy); pls with either male or bisexual fls [dioecious]. **LF:** alternate, pinnately lobed or compound. **INFL:** axillary spikes or small clusters. **BISEXUAL FL:** calyx lobes 3; stamens [0]2–4; ovary inferior, ovoid, 3-angled, chamber 1, placentas 3, parietal, styles 3, thread-like, 2-forked. **STAMINATE FL:** calyx lobes 4–9, unequal; corolla 0; stamens 8–12, filaments short. **FR:** capsule, opening at top between styles. **SEED:** many, minute. 1 genus, 2 spp.: w N.Am, Asia. [Zhang et al. 2006 Molec Phylogen Evol 39:305–322] With nitrogen-fixing root nodules. Scientific Editor: Bruce G. Baldwin.

DATISCA

(Name used by Dioscorides)

D. glomerata (C. Presl) Baill. (p. 683) DURANGO ROOT **ST:** gen clustered, erect, branched, 1–2 m. **LF:** alternate above, ± opposite or ± whorled below, ± 15 cm, ovate to lanceolate, acuminate, coarsely and unequally pinnately lobed; petioles 2–3(4) cm. **FL:** staminate fl calyx 2 mm; anthers 4 mm, ± sessile, yellow; bisexual fl calyx 5–8 mm; styles ± 6 mm. **FR:** ± 8 mm. **SEED:** ± 1 mm, light brown, pitted in rows. Dry streambeds or washes; < 2000 m. CA-FP, s SNE, DMoj (Victorville, Cottonwood Mtns); w NV, Baja CA. All parts TOXIC. May–Jul

DIPSACACEAE TEASEL FAMILY

Charles D. Bell & Elizabeth McClintock

Ann to per [(shrub)], armed or not. **ST:** gen branched. **LF:** simple, gen in basal rosettes and cauline, gen opposite, ± fused around st, entire, toothed, or pinnately lobed or dissected, petioled or not; stipules 0. **INFL:** head, terminal, on long peduncle, many-fld, dense, ± spheric or cylindric, subtended by involucre; each fl gen ± enclosed by involucel of 1–2 gen fused bractlets, this gen expanded above or in fr, gen subtended by 1 receptacle bract. **FL:** bisexual, ± bilateral, esp outermost; calyx limb cup-shaped or divided into 4–5(10) linear or bristle-like segments; corolla ± funnel-shaped, lobes 4–5, < tube, gen unequal; stamens gen 4, attached to corolla tube, alternate lobes; ovary inferior, 1-chambered, style ± exserted from corolla, stigma gen 2-lobed. **FR:** achene, enclosed by sometimes enlarged involucel, gen topped by persistent calyx. 10–11 genera, 270–350 spp.: Eur to e Asia, c&s Afr; several cult for orn. [Caputo et al. 2004 Pl Syst Evol 246:163–175] Scientific Editor: Thomas J. Rosatti.

1. Pl armed with prickles or spines; involucre bracts > fls, stiff; calyx, corolla 4-parted . . . . **DIPSACUS**
1′ Pl unarmed; involucre bracts gen < fls, flexible; calyx, corolla 5-parted . . . . **SCABIOSA**

DIPSACUS TEASEL

Bien, armed with prickles or spines ± throughout. **ST:** erect, gen < 2 m, stout, few-branched, rough-hairy. **LF:** gen < 5 dm, entire or toothed. **INFL:** gen ± cylindric; involucre bracts unequal, linear, receptacle bract ending in a spine. **FL:** calyx limb cup-shaped, persistent; corolla gen lavender (white), tube long, lobes unequal; stamens 4. **FR:** 4-angled, hairy. ± 15 spp.: Eur, w Asia, Afr. (Greek: thirst, from lf bases that in some spp. hold water)

1. Receptacle bract ending in a straight, ± flexible spine; involucre bract gen erect or upcurved. . . . . ***D. fullonum***
1′ Receptacle bract ending in a recurved, stiff spine; involucre bract gen ± spreading or reflexed . . . . ***D. sativus***

D. fullonum L. (p. 683) WILD TEASEL **LF:** pairs gen narrowly fused around st. **INFL:** 5–10 cm, ovoid-cylindric. **FR:** 6–8 mm. 2*n*=16,18. Roadsides, pastures, fields, sometimes moist sites; < 1700 m. NCo, KR, NCoRO, c&s SNF, CCo, SnFrB; native to Eur. Apr–Aug ❖

D. sativus (L.) Honck. (p. 683) FULLER'S TEASEL **LF:** pairs fused to widely fused around st. **INFL:** 5–10 cm, ovoid to ovoid-cylindric. **FR:** ± 6–8 mm. 2*n*=16,18. Disturbed areas, fields, vacant lots, pastures; < 800 m. NCo, NCoRO, CW, PR; native to Eur. Fr head long used to raise nap on woolen cloth. May–Jul ❖

SCABIOSA PINCUSHION FLOWER

Ann to per, unarmed. **ST:** erect, < 6 dm, branched, glabrous or hairy. **LF:** gen < 1 dm, pinnately lobed or dissected. **INFL:** ± spheric; outermost fls gen larger; peduncle often long; involucre bracts ± equal, lanceolate, flexible; involucel tubular, 8-ribbed, with expanded, fan-like, many-veined limb or not. **FL:** calyx limb divided into 5 bristles; corolla gen blue, purple, pink, or white, lobes 5, upper 2 smaller; stamens 4(2). **FR:** ± angled, gen hairy. ± 80 spp.: temp Eurasia, Afr. (Latin: itch, from medicinal use)

1. Involucel tube without openings; st ± glabrous; corolla purple, white, or pink; pl gen bien . . . . ***S. atropurpurea***
1′ Involucel tube with 8 valve-like openings; st shaggy-hairy; corolla blue; pl gen ann . . . . [***S. stellata***]

S. atropurpurea L. **ST:** < 6 dm. **LF:** cauline pinnately dissected. **INFL:** < 30 mm wide, elongating at maturity; limb of involucel < 1 mm, ± urn-shaped. **FR:** calyx bristles >> limb of involucel. Disturbed areas; < 600 m. NCoRO, NCoRI, CCo, SnFrB, SCoRO, SnBr, PR; native to s Eur. Mar–Nov

DROSERACEAE SUNDEW FAMILY

Elizabeth L. Painter & William J. Stone

Ann, per, [subshrub], carnivorous; roots weak. **LF**: gen basal rosette, often coiled in bud; blade with insect-catching hairs adaxially, hairs gland-tipped and sticky, [sensitive bristles]. **INFL**: cyme, raceme-like, [fls solitary]; fls [1] few, on long peduncle. **FL**: bisexual, radial; calyx lobes gen 5; petals gen 5, free or ± fused; stamens (4)5 [(10)20]; pistil 1, ovary superior, chamber 1, placentas gen 3(5), parietal, style gen 3(5), each gen 2-lobed. **FR**: capsule, loculicidal; valves gen 3(5). **SEED**: gen many, spindle-shaped. 3 genera (2 with 1 sp. each), 170+ spp.: temp, trop, esp Australia, S.Am, s Afr; esp in bogs, swamps; some cult as novelties (*Dionaea*, Venus' fly-trap, of se US). Scientific Editor: Thomas J. Rosatti.

DROSERA SUNDEW

Ann, per, often ± brown or ± red. **LF**: petiole long; adaxial blade hairs gland-tipped. **INFL**: raceme-like cyme. **FL**: sepals, petals, stamens gen 5; petals white, pink, or purple; styles, placentas, valves 3. 170+ spp.: range of family. (Greek: dewy) Insects and other organisms trapped by sticky fluid secreted by lf glands are secured by lf folding around them and digested by bacteria as well as additional lf secretions (enzymes, ribonucleases), providing nutrition; many cult and/or non-native taxa, not all documented by specimens, persisting in NCo (Mendocino Co., incl *D. tracyi* Macfarl.), n SNH (Plumas Co.) after reported, ill-advised plantings.

1. Lf ascending to erect, blade ± oblanceolate to linear-spoon-shaped; seed black . . . . *D. anglica*
1′ Lf gen spreading, blade ± round; seed light brown . . . . *D. rotundifolia*

D. anglica Huds. (p. 683) ENGLISH SUNDEW Per. **LF**: blade 15–50 mm, 2–7 mm wide. **INFL**: peduncle gen 1, 6–25 cm. **FL**: calyx 4–6 mm, ± 1/3 fused; petals 8–12 mm, white; style lobes 2, ≤ 2/3 to base. **SEED**: 1–1.5 mm, longitudinally striate-netted. 2*n*=40. Swamps, peatlands, often with *Sphagnum*; 1300–2000 m. KR, CaR, n SNH (n of Lake Tahoe), s Wrn; circumboreal. Lvs often esp long in CA (Barry Rice, pers. comm.). Sterile hybrids with *D. rotundifolia*, gen more clumped than *D. anglica* but distinguished only by fr, may be called *D.* ×*obovata* Mert. & W.D.J. Koch. Jun–Aug ★

D. rotundifolia L. (p. 683) ROUND-LEAVED SUNDEW Per. **LF**: blade 3–12 mm, 4–20 mm wide, base ± cordate or not. **INFL**: peduncles 1–several, 5–35 cm. **FL**: calyx 4–6 mm, fused at base; petals 4–6 mm, white to pink; style lobes 2, ± to base. **SEED**: 1–1.5 mm, finely, regularly longitudinally striate. 2*n*=20. Uncommon. Swamps, wet meadows, forests, peatlands, often with *Sphagnum*; < 2700 m. NW (esp near coast), CaR, SNH; to e US, circumboreal. Jun–Sep

ELAEAGNACEAE OLEASTER FAMILY

Elizabeth McClintock, except as noted

Shrub, tree, gen thorny, gen densely silvery-hairy throughout; hairs gen scale-like; occ ± dioecious. **LF**: simple, alternate or opposite, gen deciduous, entire; petiole gen short; stipules 0. **INFL**: gen umbel-like, axillary; fls 1–few. **FL**: radial; hypanthium rotate to salverform, fleshy in age, lower part gen constricted, persistent, with a nectary disk; sepals gen 4, ± petal-like; petals 0; anthers 4 or 8, ± sessile; ovary superior but appearing inferior, chamber 1, style 1. **FR**: achene enclosed in fleshy hypanthium, appearing drupe- or berry-like. 3 genera, ± 45 spp.: N.Am, Eur, Asia, e Australia; esp temp, subtrop. [Graham 1964 J Arnold Arbor 45:274–278] Scientific Editors: Douglas H. Goldman, Bruce G. Baldwin.

1. Lvs alternate; fls bisexual; stamens 4 . . . . **ELAEAGNUS**
1′ Lvs opposite; fls unisexual; stamens 8 . . . . **SHEPHERDIA**

ELAEAGNUS

FL: hypanthium bell-shaped to salverform, lobes 4; stamens ± exserted; disk flask-shaped, enclosing base of style; stigma ± elongate, on 1 side of style. ± 40 spp.: N.Am, s Eur, Asia. (Greek: olive, chaste-tree)

E. angustifolia L. (p. 683) RUSSIAN OLIVE Pl < 7 m, occ thorny. **LF**: 4–8 cm, lanceolate to oblong, more silvery on abaxial surface. **FL**: 5–10 mm, ± as wide at top, fragrant; hypanthium ± dark yellow inside, tube ± = lobes. **FR**: 10–20 mm, elliptic in outline, yellow. Uncommon. Disturbed, sometimes moist places; gen < 2050 m. ScV, SnFrB, SnGb, SnBr, PR, GB, DMoj; native to temp Asia. [*Elaeagnus angustifolius* L., orth. var.] Cult as orn. May–Jun ❖

SHEPHERDIA

Alan T. Whittemore

Shrub; dioecious. **FL**: small, clustered in lf axils. **STAMINATE FL**: hypanthium rotate; disk lobes alternate with stamens. **PISTILLATE FL**: hypanthium urn-shaped, ± closed distally; disk lobes 8, ± meeting above ovary; stigma cap-like. 3 spp.: N.Am. (John Shepherd, curator of Liverpool Botanic Garden, 1764–1836)

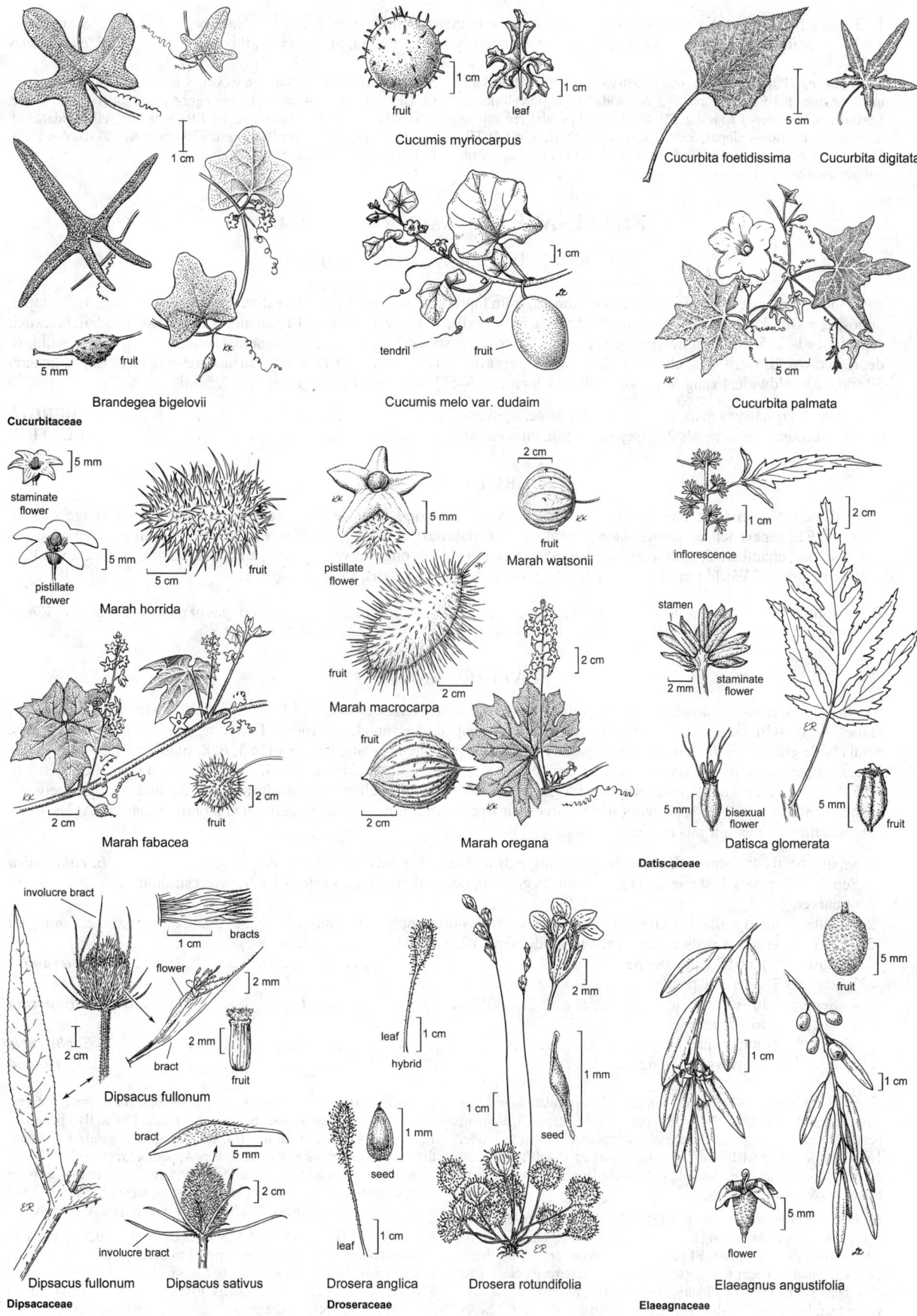

fruit

1 cm

leaf

1 cm

Cucumis myriocarpus

5 cm

Cucurbita foetidissima

Cucurbita digitata

1 cm

Brandegea bigelovii

5 mm

fruit

tendril

fruit

1 cm

Cucumis melo var. dudaim

5 cm

Cucurbita palmata

Cucurbitaceae

staminate
flower

5 mm

pistillate
flower

5 mm

fruit

5 cm

Marah horrida

Marah fabacea

2 cm

fruit

2 cm

pistillate
flower

5 mm

fruit

2 cm

Marah watsonii

fruit

Marah macrocarpa

2 cm

fruit

2 cm

Marah oregana

inflorescence

1 cm

2 cm

stamen

staminate
flower

2 mm

bisexual
flower

5 mm

fruit

5 mm

Datisca glomerata

Datiscaceae

involucre bract

bracts

1 cm

flower

2 mm

bract

2 mm

fruit

2 mm

2 cm

Dipsacus fullonum

bract

5 mm

involucre bract

2 cm

Dipsacus fullonum

Dipsacus sativus

Dipsacaceae

leaf

hybrid

1 cm

seed

1 mm

leaf

1 cm

2 mm

seed

1 mm

1 cm

Drosera anglica

Drosera rotundifolia

Droseraceae

fruit

5 mm

1 cm

1 cm

flower

5 mm

Elaeagnus angustifolia

Elaeagnaceae

1. Twigs silvery white; lvs abaxially silvery white; some branches thorn-tipped . . . . *S. argentea*
1′ Twigs orange-brown; lvs abaxially ± white, conspicuously brown-dotted; branches not thorny . . . . *S. canadensis*

S. argentea (Pursh) Nutt. (p. 693) BUFFALO-BERRY Pl 2–7 m, much-branched. **LF**: 1.3–6 cm, 1–2 cm wide, oblong or elliptical, adaxially silver-green. **FL**: white. **FR**: 5–7 mm, ellipsoidal, red. Along streams, river bottoms, slopes; 650–2100 m. e KR, SNH, SCoR, WTR, SnBr, MP, n SNE; to Can, c US. Fr sour, made into sauce eaten with buffalo meat along Overland Trail; occ cult as orn. Apr–May

S. canadensis (L.) Nutt. CANADIAN BUFFALO-BERRY Pl 0.7–4 m. **LF**: 4–6 cm, 2–4 cm wide, oblong or ovate to triangular-ovate, adaxially green. **FL**: brown-scaly. **FR**: 8–9 mm, ellipsoidal, red. Streambanks, slopes, conifer forest; 1700 m. e KR; to Alaska, e US. Fr bitter, inedible. Apr–May ★

ELATINACEAE WATERWORT FAMILY

Gordon C. Tucker & Evanielis U. Grissom

Ann, short-lived per, in or near water; roots fibrous, from taproot or not, gen from lower lf axils also. **ST**: gen soft. **LF**: simple, opposite, ± 4-ranked; stipules scarious. **INFL**: fls in upper axils, 1 or few in clusters. **FL**: small, inconspicuous, radial, bisexual; sepals, petals 2–5, equal in number, gen free; ovary spheric, styles 3–5, short. **FR**: capsule, septicidal, ± spheric, ovoid, or depressed-ovoid, walls thin; chambers 2–5, each several to many-seeded. **SEED**: small; surface net-like or glossy. 2 genera, 50 spp.: ± worldwide. [Yang & Tucker 2007 Fl China 13:55–56] Scientific Editor: Thomas J. Rosatti.

1. Pl glandular-hairy; sepals, petals 5; sepals acute, midvein rib-like . . . . **BERGIA**
1′ Pl glabrous; sepals, petals 2–4; sepals obtuse, midvein not apparent . . . . **ELATINE**

BERGIA WATERFIRE

ST: prostrate to ascending; branches 0 to many; base ± woody or not. **LF**: blade > petiole, elliptic, serrate, base wedge-shaped, tip acute. **FL**: sepals deltate; petals oblong, membranous, glabrous; stamens 5–10, filaments ± = petals, anthers elliptic; styles 5. **FR**: ovoid; chambers 5, each many-seeded. **SEED**: not visible through fr wall, oblong, ± curved, brown; surface ± net-like. ± 25 spp.: gen Old World trop, 1 N.Am. (P.J. Bergius, student of Linnaeus, 1730–1790)

B. texana (Hook.) Walp. (p. 693) **FL**: Sepals acute, midvein green, margins scarious; petals glabrous, ± white. Moist, disturbed soils, sandbars along rivers, margins of pools; < 200 m. GV, SCo; to e WA, c&s US, nw Mex. Jul–Aug

ELATINE WATERWORT

ST: Erect underwater, ± prostrate on wet ground, branched or not; base not woody. **LF**: petiole < 1/3 blade, flat, ± blade-like; blades narrow-elliptic to ± round, ± entire, bases wedge-shaped to ± round, tips round. **INFL**: fls 1(2) per node. **FL**: sepals, petals wide-elliptic, membranous; sepals pale green; petals pale green-white; stamens (1) 3, 6, 8, filaments 1/2 × to ± equal petals, anthers wide-ovoid; styles 3–4. **FR**: ± spheric or depressed-ovoid; chambers 3–4, each 3–15-seeded; pedicel gen ± 0. **SEED**: ± visible through fr wall, elliptic, straight or curved, brown to yellow-brown; surface net-like due to ridges between ± linear rows of pits. ± 25 spp.: worldwide. (Greek: fir tree, from a Eur sp. that suggests such a pl in miniature) At least 20× magnification needed for pits on seeds.

1. Sepals, petals 4; stamens 8; pedicels elongating in fr to 2–3 × fr; seeds curved 90–180° . . . . *E. californica*
1′ Sepals 2–3, petals 3; stamens (1) 3, 6; pedicels gen ± 0, occ ± elongating in fr to ≤ 1/2 fr; seeds straight
    or curved ≤ 30°
    2. Fl subsessile to stalked, ± erect; fr pedicel turned to 1 side; sepals 3, ± equal . . . . *E. ambigua*
    2′ Fl sessile, erect; fr pedicel not turned to 1 side; sepals 2, ± equal, or 3, 1 < others
        3. Stamens (1) 6 (if 3, opposite petals) . . . . *E. heterandra*
        3′ Stamens 3, alternate petals
            4. Seed pits 10–15 per row . . . . *E. brachysperma*
            4′ Seed pits 16–35 per row
                5. Seed pits wider than long . . . . *E. chilensis*
                5′ Seed pits ± as wide as long . . . . *E. rubella*

E. ambigua Wight **ST**: ± erect, 1–8 cm. **LF**: lanceolate to oblong-elliptic; petiole 1/4 blade. **INFL**: fls 1 per node. **FL**: sepals 3, ± equal; petals 3, equal, ± > sepals. **FR**: chambers 3; pedicel turned to 1 side, 1/4–1/2 fr length. **SEED**: narrow-oblong, curved 15–30°; pits ± 20 per row, wider than long. Vernal pools, wet places; 10–100 m. GV; native to e&s Asia. Jun–Aug

E. brachysperma A. Gray (p. 693) **ST**: decumbent to erect, 1–5(12) cm. **LF**: narrow-oblong to ovate; petiole 1/4–1/3 blade or indistinct. **INFL**: fls 1 per node. **FL**: sepals 2, ± equal, or 3, 1 < others; petals 3, equal, ± wider than sepals; stamens 3, alternate petals. **FR**: chambers 3. **SEED**: wide-oblong, curved ± 15°; pits 10–15 per row, wider than long. Muddy shores, shallow pools; 50–500 m. CA; c&s US, Mex. Apr–Sep

E. californica A. Gray (p. 693) **ST**: decumbent to erect, 1–5 cm. **LF**: oblong, tapered to base; petiole ± 1/2 blade. **INFL**: fls 1 per node. **FL**: sepals 4, fused in basal ± 1/3, separating in fr; petals 4, >, wider than sepals; stamens 8. **FR**: chambers 4; pedicel recurved, 2–3 × fr. **SEED**: oblong to elliptic, curved 90–180°; pits ± 15 per row, as wide as or wider than long. Pools, ponds, rice fields, streambanks; 50–1900 m. s NCoRH (Snow Mtn), GV, SnFrB; to WA, Baja CA. Mar–Aug

E. chilensis Gay (p. 693) **ST**: decumbent to erect, 0.5–10 cm. **LF**: narrow-oblong to wide-elliptic, tapered to base; petiole < 1/4 blade. **INFL**: fls 1(2) per node. **FL**: sepals 2, ± equal, or 3, 1 < others; petals 3, ovate; stamens 3, alternate petals. **FR**: chambers 3. **SEED**: 20–40 per chamber, elliptic, ± straight; pits 25–35 per row, wider than long. Muddy shores of ponds; ± 600–1700 m. n SNH, PR, MP (Madeline

Plains, Lassen Co.; Sierra Valley, Plumas Co.); temp S.Am; s Asia (introduced, India). Apr–Sep

E. heterandra H. Mason **ST**: decumbent, 2–5 cm. **LF**: obovate to wide-oblong-elliptic; petiole ± 1/4 blade. **INFL**: fls 1 per node. **FL**: sepals 2, ± equal; petals 3, ovate; stamens (1) 6 (if 3, opposite petals). **FR**: chambers 3. **SEED**: wide-elliptic, curved 15–30°; pits 12–16 per row, wider than long. Pond edges; 1000–1500 m. NCoR, SN. Apr–Jul

E. rubella Rydb. (p. 693) **ST**: prostrate to erect, 3–6(15) cm, often tinted ± red. **LF**: lance-oblong; petiole < 1/4 blade; tip blunt to notched. **INFL**: fls 1–2 per node. **FL**: sepals 2, ± equal, or 3, 1 < others; petals 3, equal, wide-elliptic; stamens 3, alternate sepals. **FR**: chambers 3. **SEED**: narrow-oblong, straight or curved < 15°; pits 16–35 per row, ± as wide as long. Muddy shores, shallow vernal pools, rice fields; < 500 m. CA; w US. Apr–Jul

ERICACEAE HEATH FAMILY

Gary D. Wallace, except as noted

Per, shrub, tree. **ST**: bark often peeling distinctively. **LF**: simple or 0, gen cauline, alternate, opposite (whorled), evergreen or deciduous, often leathery, petioled or not; stipules 0. **INFL**: raceme, panicle, cyme, or fls 1, terminal or axillary, gen bracted; pedicel often with 2 bractlets. **FL**: gen bisexual, gen radial, bell-shaped, cylindric, or urn-shaped; sepals gen (0)4–5, gen free; petals gen (0)4–5, free or fused; stamens (2–5)8–10, free, filaments rarely appendaged, anthers dehiscing by pores or slits, awns 0 or 2(4), seemingly abaxial, reduced or elongate, gen curved; nectary gen present at ovary base, gen disk-like; ovary superior or inferior, chambers gen 1–5, placentas axile or parietal, ovules 1–many per chamber, style 1, stigma head- to funnel-like or lobed. **FR**: capsule, drupe, berry. **SEED**: gen many, winged or not. ± 100 genera, 3000 spp.: gen worldwide exc deserts; some cult, esp *Arbutus*, *Arctostaphylos*, *Rhododendron*, *Vaccinium*. [Kron et al. 2002 Bot Rev 68:335–423] Monophyletic only if Empetraceae incl, as treated here. *Ledum* incl in *Rhododendron*. Non-green pls obtain nutrition from green pls through fungal intermediates. Scientific Editors: Gary D. Wallace, Thomas J. Rosatti.

1. Pl non-green; lvs 0 or scale-like
  2. Ovary chamber 1 or appearing > 1 by intrusion of parietal placentas; fr a berry; fr axis fleshy, not long-persisting
    3. Petals fused; anthers elongate, dehiscing by separate slits . . . . **HEMITOMES**
    3′ Petals free; anthers elongate or not, dehiscing by separate or fused slits
      4. Corolla densely hairy inside, ± glabrous outside; anthers not elongate; stigma < 5 mm wide, ± funnel-shaped, subtended by ring of hairs . . . . **PITYOPUS**
      4′ Corolla glabrous; anthers elongate; stigma 1.5–2.5 mm wide, crown-like, not subtended by hairs
      . . . . **PLEURICOSPORA**
  2′ Ovary chambers 4–5; fr a capsule; fr axis fibrous, gen persisting long after seed dispersal
    5. Petals free; fr dehiscing tip to base; seed fusiform
      6. Infl emerging from ground erect, with red or maroon stripes; pedicel not recurved when anthers open; sepals 0(2–4), petals 5; stamens exserted . . . . **ALLOTROPA**
      6′ Infl emerging from ground nodding, without red or maroon stripes; pedicel recurved to spreading when anthers open; sepals, petals gen 5; stamens incl . . . . **MONOTROPA**
    5′ Petals fused; fr dehiscing base to tip or indehiscent; seed ovate, winged or not
      7. Pl ± pink, in age brown; infl 1.5–17 dm, axis < 1.5 cm wide just below lowest fl; corolla 6–9 mm, cream to ± yellow; anthers awned; fr dehiscing; seed winged . . . . **PTEROSPORA**
      7′ Pl ± red; infl gen 1.5–3 dm, axis gen > 1 cm wide just below lowest fl; corolla 12–18 mm, red; anthers unawned; fr indehiscent; seed unwinged . . . . **SARCODES**
1′ Pl green; lvs lf-like (exc some *Pyrola*)
  8. Pl not or ± woody; lvs basal or cauline but on short sts, alternate to whorled; petals free
    9. Fls 1 or to 10 in ± head- or umbel-like raceme; fr dehiscing tip to base
      10. Fls in raceme; style in depression, stout; stigma not crown-like; nectary present; lvs cauline, leathery, elliptic to oblanceolate . . . . **CHIMAPHILA**
      10′ Fls 1; style not or in slight depression, slender; stigma crown-like; nectary 0; lvs ± basal, not leathery, ovate to obovate . . . . **MONESES**
    9′ Fls few–many in elongate raceme; fr dehiscing base to tip
      11. Infl 1-sided, arched, in fr ± erect; petal tubercles 2, basal, adaxial; nectary present . . . . **ORTHILIA**
      11′ Infl not 1-sided, ± erect; petal tubercles 0; nectary 0 . . . . **PYROLA**
  8′ Pl woody; lvs cauline, alternate (opposite, whorled); petals 0, free and reduced, or gen fused, or calyx, corolla ± indistinguishable, perianth parts 3–6, ± free
    12. Fls reduced, wind-pollinated, unisexual . . . . **EMPETRUM**
    12′ Fls not reduced, not wind-pollinated, bisexual
      13. Ovary inferior; fr a berry . . . . **VACCINIUM**
      13′ Ovary 1/2-inferior or gen superior; fr a berry, drupe, or capsule
        14. Fr gen ± enclosed by fleshy calyx; lvs evergreen . . . . **GAULTHERIA**
        14′ Fr not enclosed by calyx; lvs evergreen or deciduous
          15. Corolla gen persistent in fr; lf needle-like, gen < 1 mm wide . . . . **ERICA**
          15′ Corolla not persistent in fr; lf linear or wider, gen not needle-like, gen > 1 mm wide
            16. Fr a drupe or berry; pls gen of low elevations, dry habitats (exc *Arbutus*, some *Arctostaphylos*)
              17. Lvs opposite or whorled, ± linear; fr a drupe, stones fused, each gen 2-seeded . . . . **ORNITHOSTAPHYLOS**
              17′ Lvs alternate (exc some *Xylococcus*), wider than linear; fr a berry or drupe, stones free or fused, each 1–few-seeded

18. Ovary, fr smooth, not papillate; fr not juicy; infl gen dense
   19. Lf margin gen flat; filament swollen at base; fr flesh gen mealy, occ 0, stones free to all fused
     . . . . **ARCTOSTAPHYLOS**
   19′ Lf margin rolled under; filament not swollen at base; fr flesh dry, stones fused . . . . **XYLOCOCCUS**
18′ Ovary, fr papillate; fr ± juicy; infl gen open
   20. Fr a berry; seeds few per chamber . . . . **ARBUTUS**
   20′ Fr a drupe; seeds 1 per stone . . . . **COMAROSTAPHYLIS**
16′ Fr a capsule; pls gen of high elevations or moist habitats (some *Rhododendron, Menziesia* from
   low elevations, moist habitats)
  21. Anthers awned or not; fr loculicidal
    22. Lvs boat-shaped, opposite, 2–5 mm, appressed; fls 1 in upper lf axils; anthers awns elongate . . . . **CASSIOPE**
    22′ Lvs not boat-shaped, alternate, 15–80 mm, spreading; fls in racemes; anthers awns 0 or vestigial
     . . . . **LEUCOTHOE**
  21′ Anthers unawned; fr septicidal
    23. Fr dehiscing base to tip; petals fused at base or gen free . . . . ²**RHODODENDRON**
    23′ Fr dehiscing tip to base; petals ± fused
     24. Corolla gen rotate to cup-shaped, with pockets holding anthers until dehiscence . . . . **KALMIA**
     24′ Corolla gen not cup-shaped to rotate, without pockets
      25. Lf < 5 mm wide, margin strongly rolled under, abaxial surface < 1/3 visible; bud scales 0
       . . . . **PHYLLODOCE**
      25′ Lf > 5 mm wide, margin not strongly rolled under, abaxial surface > 1/3 visible (exc
       sometimes in bud); buds scales deciduous or ± persistent
       26. Corolla urn-shaped to ± spheric, 6–7(13) mm; lvs deciduous. . . . . **MENZIESIA**
       26′ Corolla widely bell- to funnel-shaped, 10–50 mm; lvs deciduous or evergreen . . . . ²**RHODODENDRON**

ALLOTROPA   SUGAR STICK

1 sp. (Greek: different turned, from erect infl)

A. virgata Torr. & A. Gray (p. 693)   Per, non-green, glabrous, rhizomed, roots brittle. **ST**: 0. **LF**: 0. **INFL**: raceme-like, < 5 dm, white with red or maroon stripes, emerging from ground erect, persistent after seed dispersal; bracts < 3 cm; bractlets 0; pedicel not recurved when anthers open. **FL**: sepals 0(2–4); corolla cup-shaped, white, petals 5, free, concave; stamens 10, exserted, anthers maroon, dehiscing by short separate slits; nectary lobes 10, short; ovary superior, chambers 5, placentas axile, style < 2 mm, stigma disk-like. **FR**: capsule, loculicidal, dehiscing tip to base. **SEED**: many per chamber, fusiform. 2*n*=26. Oak, mixed, or conifer forest; 75–3000 m. NW, CaRH, SNH; to BC, MT, NV. Petals incorrectly considered sepals by some. Jun–Aug

ARBUTUS   MADRONE

Shrub, tree, glabrous to hairy, burled or not. **ST**: erect; bark smooth at first, then shredding or fissured. **LF**: alternate, evergreen, leathery. **INFL**: panicle, bracted; bractlets 2. **FL**: sepals 5, fused at base; corolla urn-shaped, petals 5, fused; stamens 10, anthers dehiscing by short separate gaping slits, awns elongate; ovary superior, papillate, chambers 5. **FR**: berry. **SEED**: few per chamber. 20 spp.: N.Am, C.Am, w Eur, Medit, w Asia. (Latin: name for *Arbutus unedo* L., strawberry tree)

A. menziesii Pursh (p. 693)   PACIFIC MADRONE   **ST**: < 40 m, bark ± red, twigs stout. **LF**: blade < 12 cm, ovate to oblong, glabrous, rounded to pointed at tip, entire to minutely serrate, abaxially ± white, adaxially bright green. **FL**: < 8 mm; corolla yellow-white or ± pink. **FR**: < 12 mm, spheric, orange-red, papillate. 2*n*=26. Conifer, oak forests; 100–1500 m. NW, CaR, n SNF, n&c SNH, ScV, CW, n ChI (Santa Cruz Island), WTR, SnGb, PR; to BC, Baja CA. Mar–May

ARCTOSTAPHYLOS   MANZANITA

V. Thomas Parker, Michael C. Vasey & Jon E. Keeley

Shrub to small tree, prostrate to erect. **ST**: old sts gen ± red, smooth, bark gen thin, peeling, or gen ± gray or red-gray, shredding and rough, burls at base, woody, sprouting after fire, or gen 0; twig hairs 0 or gen ± like those on infl axes, bracts. **LF**: alternate, evergreen; blade flat to convex, base lobed to wedge-shaped, clasping st or not, margins gen flat, surfaces with stomata gen both abaxially, adaxially, alike in color, hairiness, less often only or fewer abaxially, gen differing in color, hairiness. **INFL**: ± raceme (gen 0–1-branched) or panicle (gen 2–10-branched), terminal, nascent infl present following st growth, gen late spring through winter, remaining dormant 4–6 months prior to fl (exc in *Arctostaphylos pringlei* subsp. *drupacea*); branches 0 or raceme-like; fl bracts lf-like, gen flat, or scale-like, often folded, keeled, tips rounded to acute to awl-shaped. **FL**: radial; sepals 5(4), free, persistent; corolla conic to urn-shaped, lobes in number = sepals, short, rounded, curved back, white to pink; stamens 2 × number of sepals, incl, filaments swollen, gen hairy at base, anthers dark red, awns elongate; ovary superior, on disk, 4–10-chambered, ovule 1 per chamber. **FR**: drupe, gen ± depressed-spheric to spheric; flesh gen thick, ± mealy, occ 0; stones 2–10, free, fused, or some fused. ± 62 spp.: N.Am (esp CA) to C.Am, Eurasia. (Greek: bear berries) [Keeley 1997 Madroño 44:109–111; Parker et al. 2007 Madroño 54:148–155]

1. Burls at base of main st, also above or not; pl sprouting after fire
   2. Lvs with stomata gen only abaxially, surfaces gen differing in color and/or hairiness
      3. Old st bark persistent, gray, shredding . . . . **A. tomentosa**
         4. Twig with glands . . . . subsp. **bracteosa**
         4′ Twig without glands
            5. Twig short-tomentose and long-nonglandular-hairy . . . . subsp. **daciticola**
            5′ Twig short-tomentose or short-nonglandular-hairy
               6. Lvs glabrous abaxially . . . . subsp. **hebeclada**
               6′ Lvs tomentose abaxially . . . . subsp. **tomentosa**
      3′ Old st bark gen smooth or peeling, ± red
         7. Pl prostrate; burls at base of main st or occ above; fr spheric; lf blades oblanceolate to obovate or narrowly elliptic; widespread . . . . [2]**A. uva-ursi**
         7′ Pl erect; burls at base of main st; fr depressed-spheric; lf blades oblong-ovate to lance-oblong; NCoRI, CW, n ChI . . . . **A. crustacea**
            8. Twig, nascent infl axes densely glandular-hairy . . . . subsp. **subcordata**
            8′ Twig, nascent infl axes glabrous or nonglandular-hairy, occ sparsely glandular-hairy
               9. Twig with short and long hairs
                  10. Lf abaxially densely nonglandular-hairy . . . . subsp. **crinita**
                  10′ Lf abaxially ± nonglandular-hairy, in age glabrous . . . . subsp. **crustacea**
               9′ Twig with short hairs
                  11. Lf abaxially densely short-nonglandular-hairy . . . . subsp. **insulicola**
                  11′ Lf abaxially glabrous
                     12. Pedicel, ovary glabrous . . . . subsp. **eastwoodiana**
                     12′ Pedicel, ovary gen short-nonglandular-hairy . . . . subsp. **rosei**
   2′ Lvs with stomata on both surfaces, occ fewer adaxially, surfaces gen the same in color and/or hairiness (exc *Arctostaphylos pacifica*)
      13. Old st bark persistent, shredding, gray . . . . [2]**A. rudis**
      13′ Old st bark smooth or peeling, ± red
         14. Pl prostrate to mounded; infl raceme, 0–1-branched
            15. Lf entire; fr dark brown; KR, NCoRO . . . . **A. nevadensis** subsp. **knightii**
            15′ Lf serrate; fr ± red; CCo (San Bruno Mtn) . . . . **A. pacifica**
         14′ Pl erect; infl panicle
            16. Twig and/or infl axes and/or bracts glandular
               17. Glands golden, on short hairs; burl 0 or often flat, obscure; NCoRH, CaRH, SNH, SnGb, SnBr, SnJt . . . . [3]**A. patula**
                  18. Lvs strongly white-glaucous . . . . subsp. **leucophylla**
                     19. Twig gen glandular; KR, NCoR, SnFrB, SCoR, PR . . . . subsp. **glandulosa**
                     19′ Twig not glandular; SCoRO (c&e Santa Lucia Range) . . . . subsp. **howellii**
                  18′ Lvs green to gray-green, ± glaucous
               17′ Glands black or clear (pink), on short and long hairs; burl hemispheric; KR, NCoR, SnFrB, SCoR, w SnBr, PR . . . . [2]**A. glandulosa** (in part)
            16′ Twig, infl axes, bracts not glandular (exc twig minutely glandular, infl axes sparsely short glandular in *Arctostaphylos rainbowensis*)
               20. Fr spheric (depressed-spheric in *Arctostaphylos mewukka*)
                  21. Stones gen free; fr dark mahogany-brown; SNF, n SNH . . . . **A. mewukka** subsp. **mewukka**
                  21′ Stones fused; fr dark red to dark brown with ± purple tinge; SnBr, PR
                     22. Twig minutely glandular; PR (San Diego, Riverside cos.) . . . . **A. rainbowensis**
                     22′ Twig nonglandular; SnBr, PR (San Jacinto, Santa Rosa, San Ysidro mtns) . . . . **A. parryana**
                        23. Lvs gray-glaucous; PR (San Jacinto, Santa Rosa, San Ysidro mtns) . . . . subsp. **desertica**
                        23′ Lvs green; SnBr . . . . subsp. **tumescens**
               20′ Fr depressed-spheric, (to ± spheric in *Arctostaphylos patula*)
                  24. Nascent infl bracts scale-like or awl-shaped
                     25. Twig hairs 0 or nonglandular; lvs dull to bright green; burl prominent; NCoRI, CaRF (Butte Co.)
                        . . . . **A. manzanita** subsp. **roofii**
                     25′ Twig hairs glandular or nonglandular; lvs bright green; burl 0 or often flat, obscure; NCoRH, CaRH, SNH, SnGb, SnBr, SnJt . . . . [3]**A. patula**
                  24′ Nascent infl bracts at least occ lf-like . . . . [2]**A. glandulosa** (in part)
                     26. Lvs strongly white-glaucous; SW (exc ChI) . . . . subsp. **adamsii**
                     26′ Lvs green, yellow-, or gray-green; KR, NCoR, SnFrB, SCoR, SCo (San Diego Co.), TR, PR
                        27. Twig hairs soft-wavy . . . . subsp. **mollis**
                        27′ Twig hairs not soft-wavy
                           28. Lvs shiny bright green; fr depressed-spheric, stones fused or in 2(3) fused units . . . . subsp. **gabrielensis**
                           28′ Lvs not shiny or bright green; fr depressed-spheric, stones gen free
                              29. Lf blades dark green, margins often ± red; fr markedly depressed-spheric; SCo (San Diego Co.) . . . . subsp. **crassifolia**
                              29′ Lf blades green or gray-green; fr not markedly depressed-spheric; KR, NCoR, SnFrB, SCoR, PR . . . . subsp. **cushingiana**

1′ Burls 0; pl not sprouting after fire
  30. Lvs with stomata abaxially, surfaces gen different in color and/or hairiness
    31. Fl parts in 4s; fr 3–4 mm wide
      32. Old st bark smooth, ± red; w SnFrB (Bolinas Ridge, Mount Tamalpais, Santa Cruz Mtns) . . . . ***A. sensitiva***
      32′ Old st bark gen rough or shredding, gray or red-gray; c NCo (Mendocino Co.), c&s NCoRO . . . . . . ***A. nummularia***
        33. Pl prostrate to mounded; lf blade 0.5–1.2 cm, 0.3–0.7 cm wide, oblong-elliptic; pygmy forest; c NCo
          (Mendocino Co.) . . . . subsp. ***mendocinoensis***
        33′ Pl prostrate to erect; lf blade 1–2.2 cm, 0.8–1.8 cm wide, round to round-ovate; coastal prairie,
          maritime chaparral, closed-cone forest; c NCo/NCoRO (Mendocino Co.) . . . . subsp. ***nummularia***
    31′ Fl parts in 5s; fr 5–15 mm wide
      34. Lf abaxially gen tomentose, sparsely so in age
        35. Lf blade oblong-ovate to -elliptic, 1.5–3 cm, base ± truncate to ± cordate; pl erect; s CCo (s Morro
          Bay, San Luis Obispo Co.) . . . . ***A. morroensis***
        35′ Lf blade narrowly obovate to oblanceolate, 1–2 cm, base wedge-shaped; pl prostrate to mounded; c
          CCo (around Monterey Bay, Monterey Co.) . . . . ***A. pumila***
      34′ Lf abaxially glabrous or sparsely nonglandular-hairy
        36. Pl gen prostrate to low-mounded
          37. Lvs oblanceolate to obovate, occ narrowly elliptic, base wedge-shaped; NCo, c SNH (above Convict
            Lake, Mono Co), CCo . . . . [2]***A. uva-ursi***
          37′ Lvs elliptic to round-ovate, base truncate to ± lobed; c CCo (nw Monterey Co.). . . . . ***A. edmundsii***
        36′ Pl erect
          38. Petioles 4–8 mm; lf oblong-elliptic, base rounded, tip obtuse; n ChI (Santa Cruz Island) . . . . ***A. insularis***
          38′ Petioles 1–2(4) mm; lf oblong to triangular-ovate, base lobed, tip acute; n-c CCo, s&w SnFrB
            39. Old st bark smooth, ± red; lf light green; w SnFrB (Santa Cruz Mtns). . . . . ***A. andersonii***
            39′ Old st bark persistent, shredding, gray or ± red; lf blue-green; n-c CCo, s SnFrB (Pajaro Hills)
              . . . . ***A. pajaroensis***
  30′ Lvs with stomata on both surfaces, surfaces similar in color and/or hairiness
    40. Pl prostrate to prostrate-mounded, 0.1–0.5(3) m
      41. Petioles to 2 mm; lf base lobed, clasping; nascent infl bracts lf-like
        42. Twig, nascent infls, pedicel, ovary nonglandular; < 150 m, c CCo (s Monterey, nw San Luis Obispo
          cos.). . . . . ***A. cruzensis***
        42′ Twig, nascent infl, pedicel, ovary glandular; 200–400 m, CCo (San Bruno Mtn). . . . . ***A. imbricata***
      41′ Petioles 1–12 mm; lf base rounded or obtuse to wedge-shaped, not clasping; nascent infl bracts
        scale-like (basal-most lf-like in *Arctostaphylos franciscana*)
        43. Twig, infl axes, bracts glandular; e KR (Scott Mtn Divide, Slate Mtn) . . . . ***A. klamathensis***
        43′ Twig, infl axes, bracts gen nonglandular; Sonoma Co. to San Luis Obispo Co.
          44. Infl panicle, 3–5-branched; lf blade gen 3–5 cm — s NCoRO (near Healdsburg and Santa Rosa,
            Sonoma Co.) . . . . ***A. stanfordiana*** subsp. ***decumbens***
          44′ Infl ± raceme or panicle, 0–3(5)-branched; lf blade gen 0.8–3 cm
            45. Fr 3–5 mm wide
              46. Lf blade ± elliptic, 0.8–1.2 cm, 0.4–0.7 cm wide; fr 3–5 mm wide; s CCo (n San Luis Obispo Co.)
                . . . . ***A. hookeri*** subsp. ***hearstiorum***
              46′ Lf blade round to round-elliptic, 1–2 cm, 1–1.5 cm wide; fr 4–5 mm wide; n CCo
                (San Francisco Presidio) . . . . ***A. montana*** subsp. ***ravenii***
            45′ Fr 6–8 mm wide
              47. Twig gray-tomentose; bracts 3–4 mm; c CCo (San Francisco) . . . . ***A. franciscana***
              47′ Twig sparsely short-nonglandular-hairy; bracts near infl base occ 5–10 mm; KR, NCoRH, CaRH,
                SNH, MP, n SNE (Sweetwater Mtns). . . . . ***A. nevadensis*** subsp. ***nevadensis***
    40′ Pl gen erect, occ mounded to erect, (0.1)0.3–8 m
      48. Nascent infl bracts fleshy, scoop-shaped or gen scale-like, deltate to awl-shaped, keeled or not
        49. Infl gen ± raceme, occ 1–2-branched
          50. Old st bark persistent, shredding, gray — s CCo, s SCoRO (Nipomo, Burton mesas, Point Sal; sw
            San Luis Obispo, nw Santa Barbara cos.) . . . . [2]***A. rudis***
          50′ Old st bark smooth, dark red with ± gray or ± glaucous patches or ± red
            51. Twig glandular-hairy; old sts dark red with ± gray or ± glaucous patches; n SNF (Amador,
             Calaveras cos.). . . . . ***A. myrtifolia***
            51′ Twig sparsely short-nonglandular-hairy; old sts ± red, without ± gray or ± glaucous patches; n
             CCo, s SnFrB (s Santa Cruz Mtns), SCoR, SnBr, PR, e DMtns
             52. Nascent infl spheric, cup-like, axis 0.3–1 cm, crook-necked, bracts ± spreading; fr 3–8 mm wide;
              n CCo, s SnFrB (s Santa Cruz Mtns) . . . . ***A. hookeri*** subsp. ***hookeri***
             52′ Nascent infl club-like, axis 0.5–1.5 cm, not crook-necked, bracts recurved; fr 8–14 mm wide;
              SCoR, SnBr, PR, e DMtns . . . . ***A. pungens***
      49′ Infl panicle, (1)3–8-branched
        53. Lf blade ± white, white-glaucous, gray-green, gray-glaucous
          54. Fr (10)12–16 mm wide, spheric or ± spheric
            55. Stones free, occ some fused; nascent infl bracts appressed; fr glabrous; s CaRF, CaRH
             (near Paradise), n SNF . . . . ***A. mewukka*** subsp. ***truei***

55′ Stones fused; nascent infl bracts spreading; fr glabrous or sticky; SnFrB, SCoR, TR, PR, sw
    DMtns (Little San Bernardino Mtns)
   56. Petiole 1–4 mm; lf glabrous or gen short-nonglandular-hairy, base lobed; ovary, fr glabrous, not
      sticky; n SCoRI (n&c Gabilan Range, San Benito, Monterey cos.) . . . . *A. gabilanensis*
   56′ Petiole 7–15 mm; lf glabrous, base rounded, truncate, or ± lobed; ovary, fr glandular-sticky;
      SnFrB, SCoR, TR, PR, sw DMtns (Little San Bernardino Mtns) . . . . *A. glauca*
54′ Fr 6–12 mm wide, depressed-spheric
  57. Nascent infl bracts nonglandular; KR, NCoR, CaRF, SN . . . . *A. viscida*
    58. Twig densely glandular-hairy; lf sparsely short-glandular-hairy, ciliate, papillate, scabrous; SN
      . . . . subsp. *mariposa*
    58′ Twig glabrous to sparsely short-nonglandular-hairy; lf glabrous
     59. Ovary, fr rough, sticky-glandular; KR, NCoR . . . . subsp. *pulchella*
     59′ Ovary, fr smooth, glabrous, not sticky; KR, NCoRI, CaRF, SN . . . . subsp. *viscida*
  57′ Nascent infl bracts glabrous or nonglandular- or glandular-hairy; KR, NCoRI (Sonoma, Colusa,
     Shasta, Trinity cos.), c&s NCoRO
    60. Lf blade glaucous, canescent or densely white-tomentose, in age glabrous; ovary densely
      nonglandular-hairy; KR, NCoRI (Sonoma, Colusa, Shasta, Trinity cos.). . . . . *A. malloryi*
    60′ Lf ± glaucous, glabrous or scabrous; ovary glabrous (minutely glandular-hairy); c&s NCoRO
      . . . . *A. manzanita* subsp. *glaucescens*
53′ Lf blade green, shiny or dull both surfaces, not ± white or ± gray due to ± glaucous or hairy surfaces
 61. Twig glandular
   62. Nascent infl spreading to erect, axes < 1 mm wide; bracts appressed, ± = buds; buds appearing as
      scattered beads
    63. Lf blade 1.5–3 cm, dull green; corolla white; twig finely glandular-hairy; KR, NCoRO . . . . *A. hispidula*
    63′ Lf blade 3–5 cm, bright green to ± glaucous; corolla white to pink; twig short-glandular-hairy;
      s NCoRI (s Mendocino, w Lake cos.). . . . . *A. stanfordiana* subsp. *raichei*
   62′ Nascent infl pendent, axes gen > 1 mm wide; bracts gen ± appressed, > buds, buds not appearing as beads
    64. Lf blade gen 2–6 cm, at least some > 3 cm, round to widely ovate or oblong
     65. Glands of twig, nascent infls dark or clear; lf dull green, scabrous, glandular; NCoRH, CaRF
      (Shasta, Tehama cos.) . . . . *A. manzanita* subsp. *wieslanderi*
     65′ Glands of twig, nascent infls gold or 0; lf bright green, glabrous; NCoRH, CaRH, SNH, SnGb,
      SnBr, SnJt . . . . [3]*A. patula*
    64′ Lf blade 1.5–3 cm, elliptic to ovate
     66. Twig with occ long glandular hairs; siliceous shale, sw SnFrB (n Ben Lomond Mtn, nw Santa
      Cruz Co.) . . . . *A. ohloneana*
     66′ Twig with glandular hairs or sessile glands; serpentine, s NCoRO (Sonoma Co.) . . . . *A. bakeri*
      67. Twig with glandular hairs; s NCoRO (central Sonoma Co. s of Guerneville) . . . . subsp. *bakeri*
      67′ Twig with sessile glands; s NCoRO (Sonoma Co. from The Cedars to Healdsburg) . . . . . . subsp. *sublaevis*
 61′ Twig nonglandular
  68. Nascent infl axis gen < 1 mm wide
    69. Lf blade 1–2.5 cm; bracts > buds; buds not appearing as beads; pl mounded to erect; s NCoRO
      (Vine Hill near Forestville, Sonoma Co.). . . . . *A. densiflora*
    69′ Lf blades 3–5 cm; bracts ± = buds; buds appearing as scattered beads; pl erect; NCoRO, c&s
      NCoRI . . . . *A. stanfordiana* subsp. *stanfordiana*
  68′ Nascent infl axis > 1 mm wide
    70. Lf blade 1–2.5 cm . . . . *A. montana* subsp. *montana*
    70′ Lf blade 1.5–5 cm
     71. Ovary, fr minutely glandular-hairy; pl erect; KR, NCoR. . . . . *A. manzanita* subsp. *elegans*
     71′ Ovary, fr glabrous; pl mounded or erect; KR, NCoRO, NCoRI, s CaRF, n&c SNF, n SNH, ScV,
      n&e SnFrB, TR
      72. Fr spheric, stones fused; TR . . . . *A. parryana* subsp. *parryana*
      72′ Fr depressed-spheric, stones free; KR, NCoRO, NCoRI, s CaRF, n&c SNF, n SNH, ScV, n&e
      SnFrB. . . . . *A. manzanita* (in part)
       73. Lf blade bright green; infl 2–4-branched, axis > 1 mm wide; nascent infl branches to 15 mm;
         s NCoRI (Vaca Mtns), e SnFrB (Mount Diablo) . . . . subsp. *laevigata*
       73′ Lf blade dull to bright green; infl 4–7-branched, axis ± 1 mm wide; nascent infl branches gen
         15–25 mm; KR, NCoRO, NCoRI, s CaRF, n&c SNF, n SNH, ScV, n SnFrB . . . . subsp. *manzanita*
48′ Nascent infl bracts gen lf-like, flat
 74. Old st bark persistent, gray, shredding (exc occ smooth on *Arctostaphylos osoensis*)
   75. Lf gray, dull, sparsely appressed-puberulent, in age glabrous, base rounded to wedge-shaped, not
      clasping; n SN (Placer, El Dorado cos.), c SNF (Tuolumne Co.). . . . . *A. nissenana*
   75′ Lf dark green, ± shiny, sparsely short-nonglandular-hairy or glabrous, base lobed, clasping; s CCo
      (w Los Osos Valley, San Luis Obispo Co.) . . . . *A. osoensis*
 74′ Old st bark smooth, ± red

76. Lf gray-canescent in youth, glabrous in age
77. Lf base lobed, clasping; petiole to 4 mm
78. Ovary, fr glabrous; SCoRO (se of Cuesta Pass, San Luis Obispo Co.) . . . . *A. luciana*
78′ Ovary, fr nonglandular-hairy or glandular- and nonglandular-hairy; e&sw SnFrB
79. Ovary, fr nonglandular-hairy; e SnFrB (Mount Diablo and vicinity) . . . . *A. auriculata*
79′ Ovary, fr glandular- and nonglandular-hairy; sw SnFrB (n Ben Lomond Mtn,
    nw Santa Cruz Co.). . . . . *A. glutinosa*
77′ Lf base wedge-shaped, rounded, truncate, or ± lobed, not clasping; petiole 3–10 mm
80. Ovary glabrous; corolla white
81. Infl 2–4-branched panicle; pedicel 8–10 mm; SCoRO (c&s Santa Lucia Range) . . . . *A. obispoensis*
81′ Infl 0–1-branched, ± raceme; pedicel 5–7 mm; sw SnFrB (s Santa Cruz Mtns) . . . . *A. silvicola*
80′ Ovary glandular- or nonglandular-hairy; corolla pink or white . . . . *A. canescens*
82. Ovary nonglandular-hairy; KR, NCoR, n CCo (Tassajara, Santa Lucia Range), w SnFrB (Mount
    Tamalpais; s Santa Cruz Mtns). . . . . subsp. *canescens*
82′ Ovary glandular-hairy; w KR, NCoRO, w SnFrB (Shasta-Santa Cruz cos.) . . . . subsp. *sonomensis*
76′ Lf green to glaucous, glabrous to variously hairy but not gray-canescent
83. Lf base lobed (truncate to lobed in *Arctostaphylos viridissima*), often clasping (exc *Arctostaphylos*
    hooveri with longer petiole)
84. Nascent infl nonglandular-hairy
85. Pedicel nonglandular-hairy; n ChI (e Santa Cruz Island). . . . . *A. viridissima*
85′ Pedicel glabrous or sparsely nonglandular-hairy; s CCo, s SCoRO/WTR
86. Lf blade 2–5 cm; fr 8–12 mm wide; stones gen fused; s CCo (Morro Bay to Avila Beach,
    San Luis Obispo Co.). . . . . *A. pechoensis*
86′ Lf blade 1–2.5 cm; fr 5–8 mm wide; stones free; s CCo, s SCoRO/WTR (w Santa Barbara Co.)
    . . . . *A. purissima*
84′ Nascent infl glandular-hairy
87. Fr 10–15 mm wide, spheric; stones fused; s SCoRO/w WTR (Santa Ynez Mtns). . . . . *A. refugioensis*
87′ Fr 6–10(15) mm wide, depressed-spheric; stones free; CCo, SnFrB, SCoRO
88. Lf blade gen ovate, 2.5–4.5 cm, not boat-shaped, sparsely glandular or not but not sticky
89. Lf blade bright green, not glaucous; twig long-glandular-hairy; CCo (San Bruno Mtn),
    w SnFrB (Montara Mtn) . . . . *A. montaraensis*
89′ Lf blade green, glaucous; twig short- and long-nonglandular-hairy; e SnFrB (Sobrante,
    Huckleberry ridges; Alameda, Contra Costa cos.). . . . . *A. pallida*
88′ Lf blade oblong-ovate, 3–6 cm, boat-shaped, glandular-sticky in youth, glabrous but possibly
    still sticky in age
90. Petiole 3–6 mm; lvs not clasping; SCoRO (Santa Lucia Range) . . . . *A. hooveri*
90′ Petiole < 3 mm; lvs clasping; w SnFrB (n Santa Cruz Mtns). *A. regismontana*
83′ Lf base rounded, truncate, wedge-shaped, or ± lobed, not clasping (exc occ in suspected hybrids of
    Arctostaphylos montereyensis and *Arctostaphylos pajaroensis*)
91. Lf-like bracts deep pink, glandular, deciduous after fl; SnBr, PR . . . . *A. pringlei* subsp. *drupacea*
91′ Lf-like bracts green, glandular or not, persistent after fl; NCo, nw&w KR, NCoRO, CCo,
    nw SnFrB, SCoRO, ChI, sw PR
92. Fr spheric; stones free or fused
93. Fr 8–15 mm wide; lf papillate, scabrous, base truncate to ± lobed; s ChI (Santa Catalina Island)
    . . . . *A. catalinae*
93′ Fr 6–8 mm wide; lf not papillate, not scabrous, base wedge-shaped, rounded, or truncate;
    sw PR. . . . . *A. otayensis*
92′ Fr depressed-spheric; stones free
94. Infl ± raceme, 0–1-branched
95. Twig, infl axes, bracts glandular; n CCo, nw SnFrB (Marin Co.) . . . . *A. virgata*
95′ Twig, infl axes, bracts nonglandular (exc occ minutely glandular in *Arctostaphylos nortensis*);
    nw KR, s CCo, s SCoRO
96. Pedicel, ovary, fr nonglandular-hairy; 500–940 m, nw KR (n Del Norte Co.) . . . . *A. nortensis*
96′ Pedicel, ovary, fr glabrous; 30–1250 m, s CCo (Pismo Beach vicinity), s SCoRO (s Santa
    Lucia, La Panza ranges and s). . . . . *A. pilosula*
94′ Infl panicle, 3–10-branched
97. Lvs overlapped; pl shrub, prostrate to mounded or erect, 0.1–2 m; n ChI (Santa Rosa Island)
    . . . . *A. confertiflora*
97′ Lvs not overlapped; pl shrub or tree, erect, 1–5 m; NCo, w KR, NCoRO, c CCo, n SCoRO
98. Lf lanceolate to ovate, dark green; nascent infl gen short- and long-nonglandular-hairy, occ
    glandular-hairy; NCo, w KR, NCoRO . . . . *A. columbiana*
98′ Lf round to oblong-ovate, bright green or ± glaucous; nascent infl glandular-hairy; c CCo
    (Fort Ord), n SCoRO (Mount Toro, nw Monterey Co.). . . . . *A. montereyensis*

A. andersonii A. Gray (p. 693) ANDERSON'S MANZANITA Tree-like, 2–5 m. **ST:** twig (and nascent infl axis) densely tomentose or short-nonglandular-hairy and long-glandular-hairy. **LF:** overlapped; petiole < 4 mm; blade 4–7 cm, 1.5–2.5 cm wide, oblong, boat-shaped, both surfaces light green, dull, appearing glabrous, base lobed, clasping, tip acute, margin entire; stomata abaxial. **INFL:** panicle, 4–6-branched; nascent infl pendent, axis 2–3 cm, > 1 mm wide; bracts 8–15 mm, ± lf-like (occ reduced), lanceolate; pedicel 6–8 mm. **FL:** ovary (and fr) glandular-hairy. **FR:** 6–8 mm wide, ± depressed-spheric, sticky; stones free. 2*n*=26. Open sites or forest edge, redwood or mixed-evergreen forest, occ in chaparral near coast; < 800 m. w SnFrB (Santa Cruz Mtns). Jan–Mar ★

A. auriculata Eastw. (p. 693) MOUNT DIABLO MANZANITA Shrub, erect, 1–4.5 m. **ST:** twig white-tomentose. **LF:** overlapped; petiole < 2 mm; blade 1.5–4.5 cm, 1.5–3 cm wide, oblong- to round-ovate, white-gray, canescent, base lobed, clasping, tip acute, margin entire, flat. **INFL:** panicle, 3–5-branched; nascent infl axis pendent, 1–1.5 cm, > 1 mm wide, tomentose; bracts 5–15 mm, lf-like, ovate to lance-ovate, canescent; pedicel 4–10 mm. **FL:** ovary (and fr) densely short-white-nonglandular-hairy. **FR:** depressed-spheric, 5–10 mm wide; stones free. 2*n*=26. Sandstone, upland chaparral near coast; 150–650 m. e SnFrB (Mount Diablo and vicinity). Feb–Mar ★

A. bakeri Eastw. Shrub, erect, 1–3 m. **LF:** erect; blade 2.5–3 cm, 1.5–2 cm wide, elliptic to ovate, dark green, base gen wedge-shaped, occ truncate to rounded, tip acute, margin entire, flat. **INFL:** panicle, 2–5-branched; nascent infl pendent, axis and bracts glandular-hairy, puberulent, or tomentose; bracts 2–4 mm, scale-like, deltate to awl-shaped; pedicel 3–5 mm, glabrous. **FL:** ovary (and fr) glabrous. **FR:** 8–10 mm wide, depressed-spheric; stones free. 2*n*=52.

subsp. ***bakeri*** (p. 693) BAKER'S MANZANITA **ST:** twig glandular-hairy. **LF:** petiole 3–6 mm; blade glandular-hairy, papillate, scabrous. **INFL:** nascent infl axes 10–15 mm, > 1 mm wide. Serpentine chaparral near coast; 75–300 m. s NCoRO (s of Guerneville, central Sonoma Co.). Feb–Apr ★

subsp. ***sublaevis*** P.V. Wells (p. 693) THE CEDARS MANZANITA **ST:** twig sparsely to densely short-nonglandular-hairy with sessile glands. **LF:** petiole 4–8 mm; blade appressed-puberulent, ± sparsely papillate. **INFL:** nascent infl axes 15–20 mm, > 1 mm wide. Serpentine chaparral near coast; 300–600 m. s NCoRO (from The Cedars to Healdsburg, Sonoma Co.). Feb–Apr ★

A. canescens Eastw. Shrub, erect, 0.3–3 m. **ST:** twig canescent to white-tomentose. **LF:** erect; petiole 3–10 mm; blade 2–5 cm, 1–3 cm wide, round-ovate, ovate, or elliptic, white-gray, canescent, base rounded to wedge-shaped, tip acute to abruptly soft-pointed, margin entire, flat. **INFL:** panicle, 1–3-branched; nascent infl pendent, bell-shaped, axis 1–2 cm, > 1 mm wide, ± obscured, ± canescent to white-tomentose; bracts 6–20 mm, lf-like, widely lanceolate, canescent; pedicel 5–9 mm, densely hairy, glandular or not. **FL:** ovary densely (fr sparsely) glandular- or nonglandular-hairy. **FR:** 5–10 mm wide, depressed-spheric; stones free. 2*n*=26.

subsp. ***canescens*** (p. 693) HOARY MANZANITA **ST:** twig short-canescent, nonglandular. **INFL:** nascent infl axis canescent; pedicel hairy. **FL:** ovary and young fr densely nonglandular-hairy. **FR:** glabrous in age. Chaparral, open forest; 200–2000 m. KR, NCoR, w SnFrB (Mount Tamalpais; s Santa Cruz Mtns); s OR. Jan–May ★

subsp. ***sonomensis*** (Eastw.) P.V. Wells SONOMA CANESCENT MANZANITA **ST:** twig canescent, glandular and not. **INFL:** bracts, pedicel glandular-hairy in youth. **FL:** ovary and young fr glandular-hairy. **FR:** gen ± glabrous in age. Chaparral, open forests; 60–1700 m. w KR, NCoRO, w SnFrB (Mount Tamalpais; s Santa Cruz Mtns). Mar–May ★

A. catalinae P.V. Wells (p. 693) SANTA CATALINA ISLAND MANZANITA Shrub, small tree, erect, 2–5 m. **ST:** twig (and infl bract) densely glandular- and long-white-nonglandular-hairy. **LF:** overlapped; petiole 2–6 mm; blade 2–5 cm, 1.5–3 cm wide, lance-ovate to elliptic, green-glaucous, glandular-bristly, papillate, scabrous, base truncate to ± lobed, tip acute, margin entire or ± serrate. **INFL:** panicle, 4–10-branched; nascent infl pendent, axis 2–3 cm, > 1 mm wide; bracts gen overlapped in distal 1/2 of infl, 6–10(15) mm, lf-like, lanceolate to narrowly ovate, acute, green, persistent; pedicel

2–5 mm, tomentose. **FL:** ovary densely (fr sparsely) white-nonglandular hairy, sparsely glandular. **FR:** 8–15 mm wide, spheric; stones free. 2*n*=26. Volcanic outcrops, ridges, maritime chaparral; 100–600 m. s ChI (Santa Catalina Island). Dec–Feb ★

A. columbiana Piper (p. 693) COLUMBIA MANZANITA Shrub, small tree, erect, (1)2–5 m. **ST:** twig gen white-tomentose, glandular or not. **LF:** spreading; petiole 4–10 mm; blade 4–6 cm, 2–3 cm wide, lanceolate to ovate, flat, dark green, dull, sparsely glandular-hairy, ± papillate, scabrous, base wedge-shaped to ± rounded, tip acute, margin entire. **INFL:** panicle, 3–8-branched; nascent infl pendent, axis 1.5–2.5 cm, > 1 mm wide; bracts lf-like, green, persistent, stiff, densely short- and long-white-nonglandular-hairy, occ glandular-hairy; pedicel 2–4 mm, glandular-hairy. **FL:** ovary densely white-nonglandular-hairy, sparsely glandular. **FR:** 8–11 mm wide, depressed-spheric, sparsely nonglandular-hairy; stones free. 2*n*=26. Rocky coastal uplands, maritime chaparral, conifer forest; < 800 m. NCo, w KR, NCoRO; to BC. [*A. c.* f. *tracyi* (Eastw.) P.V. Wells] *Arctostaphylos* ×*media* Greene (prostrate to mounded shrub < 1 m; infl ± raceme, 0–1-branched) presumed hybrid with *A. uva-ursi*. Mar–May

A. confertiflora Eastw. (p. 693) SANTA ROSA ISLAND MANZANITA Shrub, prostrate to mounded or erect, 0.1–2 m. **ST:** twig densely short-nonglandular- and -long-white-glandular-hairy. **LF:** overlapped; petiole 4–10 mm; blade 4–6 cm, 2–3 cm wide, ovate to elliptic, light green, dull, glandular-puberulent, ± papillate, scabrous, ± glandular-hairy near base, base wedge-shaped to ± rounded, tip rounded or abruptly soft-pointed, margin cupped. **INFL:** panicle, 3–5-branched; nascent infl pendent, axis 1.5–2 cm, > 1 mm wide; bracts ± densely overlapped near infl tip, 8–18 mm, lf-like, ovate to oblanceolate, green, densely short- and long-white-glandular-hairy, persistent; pedicel 3–5 mm, glandular-hairy. **FL:** ovary densely (and fr sparsely) white-nonglandular-hairy, sparsely glandular. **FR:** 8–11 mm wide, depressed-spheric; stones free. 2*n*=26. Sandstone outcrops, maritime chaparral; < 500 m. n ChI (Santa Rosa Island). Feb–Mar ★

A. crustacea Eastw. Erect, 1–3 m; burl prominent. **ST:** twig glabrous or tomentose, glandular or not. **LF:** spreading; petiole 2–5 mm; blade 2–5 cm, 1.5–2.5 cm wide, oblong-ovate to lance-oblong, abaxially glabrous to densely short-nonglandular-hairy, light green, adaxially ± glabrous (nonglandular-hairy), dark green, base truncate to ± lobed, tip acute, margin entire, occ toothed, cupped or ± rolled; stomata abaxial. **INFL:** panicle, 2–8-branched; nascent infl pendent, axis 1–2.5 cm, > 1 mm wide, glabrous, nonglandular-hairy, or glandular-hairy; lower bracts lf-like, upper often scale-like, tomentose, glandular or not; bracts 3–8 mm, scale-like and awl-shaped to deltate or lf-like and lanceolate, glabrous to hairy; pedicel 2–5 mm, sparsely nonglandular-hairy to glandular-hairy. **FL:** ovary glabrous to short-hairy, gen nonglandular. **FR:** 6–10 mm wide, depressed-spheric, glabrous to sparsely nonglandular-hairy; stones free. 2*n*=52.

subsp. ***crinita*** (J.E. Adams) V.T. Parker et al. (p. 693) CRINITE MANZANITA **ST:** twig (and nascent infl) short- and long-stiff-nonglandular-hairy. **LF:** densely nonglandular-hairy abaxially, occ adaxially. **FL:** ovary short- nonglandular-hairy. Chaparral, conifer forest; < 700 m. CCo (Fort Ord, Mount Toro), sw SnFrB (Santa Cruz Mtns). [*A. tomentosa* subsp. *crinita* (J.E. Adams) Gankin] Feb–Apr

subsp. ***crustacea*** **ST:** twig (and nascent infl) short- and long-stiff-nonglandular-hairy, occ sparsely glandular. **LF:** abaxially ± nonglandular-hairy, in age glabrous. **FL:** ovary short-nonglandular-hairy. Chaparral, conifer forest; < 1100 m. NCoRI, CW, n ChI. [*A. tomentosa* subsp. *c.* (Eastw.) P.V. Wells] Feb–Apr

subsp. ***eastwoodiana*** (P.V. Wells) V.T. Parker et al. EASTWOOD'S BRITTLE-LEAF MANZANITA **ST:** twig gen densely short-nonglandular-hairy. **LF:** blade abaxially glabrous. **INFL:** pedicel glabrous. **FL:** ovary glabrous. Chaparral, closed-cone conifer forest; < 650 m. s CCo/SCoRO (nw Santa Barbara Co.). [*A. tomentosa* subsp. *e.* P.V. Wells] Feb–Apr ★

subsp. ***insulicola*** (P.V. Wells) V.T. Parker et al. ISLAND MANZANITA **ST:** twig (and lf abaxially, nascent infl, ovary) gen short-nonglandular-hairy. Chaparral, conifer forest; 5–700 m. n ChI (Santa Cruz, Santa Rosa islands). [*A. tomentosa* subsp. *i.* P.V. Wells] Feb–Apr ★

subsp. *rosei* (Eastw.) V.T. Parker et al. ROSE'S MANZANITA ST: twig (and nascent infl, pedicel, ovary) gen short-nonglandular-hairy. LF: blade abaxially glabrous. Chaparral, conifer forest, coastal bluffs; < 500 m. CCo (Lake Merced, San Francisco Co.; Santa Lucia Range, Monterey Co.). [*A. tomentosa* subsp. *r.* (Eastw.) P.V. Wells] Feb–Apr

subsp. *subcordata* (Eastw.) V.T. Parker et al. (p. 693) SANTA CRUZ ISLAND MANZANITA ST: twig (and nascent infl) densely glandular-hairy. LF: abaxially ± tomentose, sparsely glandular-hairy, ± papillate. FL: ovary white-nonglandular-hairy, sparsely glandular. Chaparral, conifer forest; 100–730 m. n ChI (Santa Cruz, Santa Rosa islands). [*A. tomentosa* subsp. *s.* (Eastw.) P.V. Wells] Feb–Apr ★

A. cruzensis Roof (p. 697) ARROYO DE LA CRUZ MANZANITA Shrub, prostrate, 0.1–1 m. ST: twig (and nascent infl axis, bract, pedicel) sparsely nonglandular-hairy. LF: overlapped; petiole < 2 mm; blade 1.5–3 cm, 1–25 cm wide, oblong-ovate, bright green, tomentose, glabrous in age, smooth, base lobed, clasping, tip acute, margin entire or toothed, hairy-ciliate near base. INFL: panicle, 1–3-branched; nascent infl pendent, bell-shaped, crowded, hidden by large lower bracts, axis 0.5–1.5 cm, > 1 mm wide; bracts 5–15 mm, lf-like, lanceolate to lance-ovate; pedicel 4–5 mm. FL: ovary densely white-nonglandular-hairy. FR: 8–10 mm wide, depressed-spheric, glabrous; stones free. 2*n*=26. Sandy bluffs, maritime chaparral, coastal prairie; < 150 m. c CCo (s Monterey, nw San Luis Obispo cos.). Jan–Mar ★

A. densiflora M.S. Baker (p. 697) VINE HILL MANZANITA Shrub, mounded to erect, 1 m. ST: twig (and nascent infl axis) sparsely short-nonglandular-hairy. LF: erect; petiole 4–5 mm; blade 1–2.5 cm, 0.5–1.5 cm wide, elliptic to narrowly lanceolate, bright green, shiny, sparsely puberulent, smooth, base wedge-shaped to obtuse, tip acute, margin entire, flat. INFL: panicle, 3–5-branched; nascent infl pendent, axis 1–1.5 cm, < 1 mm wide; bracts 1.5–3 mm, > buds, scale-like, awl-shaped to deltate, glabrous; pedicel 4–5 mm, glabrous. FL: ovary glabrous. FR: 5–6 mm wide, ± spheric, glabrous; stones free. 2*n*=26. Marine acid sandy-clay soils, maritime chaparral; ± 100 m. s NCoRO (Vine Hill near Forestville, Sonoma Co.). Mar–Apr ★

A. edmundsii J.T. Howell (p. 697) LITTLE SUR MANZANITA Shrub, prostrate to mounded, 0.2–1.5 m. ST: twig (and nascent infl axis) sparsely short-nonglandular-hairy. LF: spreading; petiole 2–3 mm; blade 1–2.5 cm, 1–1.5 cm wide, elliptic to round-ovate, dark green, shiny, abaxially glabrous or sparsely nonglandular-hairy, adaxially sparsely puberulent, base truncate to ± lobed, tip abruptly soft-pointed, margin entire, flat; stomata abaxial. INFL: panicle, 2–5-branched; nascent infl pendent, axis 0.5–1 cm, > 1 mm wide; bracts 3–5 mm, scale-like, deltate to awl-shaped, (and pedicel) glabrous or sparsely nonglandular-hairy; pedicel 3–4 mm. FL: ovary sparsely nonglandular-hairy or glabrous. FR: 6–8 mm wide, ± spheric, glabrous; stones free. 2*n*=26. Sandy terraces, bluffs, maritime chaparral; < 100 m. c CCo (nw Monterey Co.). Nov–Dec ★

A. franciscana Eastw. (p. 697) FRANCISCAN MANZANITA Prostrate to prostrate-mounded, 0.2–1.5 m. ST: twig gray-tomentose, nonglandular. LF: spreading; petiole 3–5 mm; blade 1.5–2 cm, 0.5–1 cm wide, oblanceolate, ± puberulent, glabrous in age, smooth, bright green, base wedge-shaped, tip acute, margin entire, flat. INFL: panicle, 1–3-branched; nascent infl pendent, axis 0.5–1 cm, > 1 mm wide, gray-nonglandular-hairy, bracts near base lf-like, above scale-like, 3–4 mm, awl-shaped, concave when fresh, fleshy, nonglandular; pedicel 2–6 mm, glabrous. FL: ovary canescent. FR: 6–8 mm wide, depressed-spheric, glabrous; stones free. 2*n*=26. Serpentine outcrops, chaparral; < 300 m. n CCo (San Francisco). [*A. hookeri* subsp. *f.* (Eastw.) Munz] 1 pl in wild, others in cult. Jan–Apr ★

A. gabilanensis V.T. Parker & M.C. Vasey (p. 697) GABILAN MOUNTAINS MANZANITA Erect, 1–5 m. ST: twig (and lower infl bracts) sparsely short-nonglandular-hairy, with sessile glands. LF: overlapped; petiole 1–4 mm; blade 1.5–3.5 cm, 1.1–2.4 cm wide, ovate, gray-green, glabrous or gen short-nonglandular-hairy, base lobed, clasping, tip obtuse, margin entire, flat. INFL: panicle, 1–4-branched; nascent infl pendent, axis 1–3 cm, > 1 mm wide; bracts 3–5 mm, fleshy, deltate, nonglandular-hairy, lower lf-like; pedicel 8–10 mm, glabrous or nonglandular-hairy. FL: ovary glabrous. FR: 10–15 mm wide, spheric, glabrous; stones fused. 2*n*=26.

Open granitic outcrops, chaparral, Coulter-pine/chaparral woodland; < 710 m. n SCoRI (n&c Gabilan Range, San Benito, Monterey cos.). Feb–Apr ★

A. glandulosa Eastw. (p. 697) Shrub, erect, 1–4 m; burl hemispheric. ST: twig (and nascent infl axis, pedicel) glandular- or non-glandular-hairy, glands black, clear, or 0. LF: ascending or erect; petiole 5–10 mm; blade 2–4.5 cm, 1–2.5 cm wide, elliptic to ovate, bright green to gray-green or strongly glaucous, shiny or dull, glandular-puberulent to -hairy, papillate, scabrous or not, puberulent to tomentose, or in age glabrous, base wedge-shaped to rounded, occ ± lobed or truncate, tip acute, margin entire or toothed; stomata both surfaces or adaxial fewer than abaxial. INFL: panicle, 3–6-branched; nascent infl pendent, ± crowded, hairy, glandular-hairy or not; axis 1–3 cm, > 1 mm wide; pedicel 3–10 mm. FR: 6–10 mm wide, depressed-spheric, glabrous, sticky, or glandular-hairy; stones gen free. 2*n*=52. 2 more subspp. in Baja CA.

subsp. *adamsii* (Munz) Munz ST: twig sparsely short-nonglandular-hairy. LF: blade glabrous, strongly white-glaucous, base rounded, occ ± lobed to truncate. INFL: nascent infl axis, bract, pedicel, ovary densely short-white-nonglandular-hairy, bracts occ lf-like. Chaparral, conifer forest; 1500–2200 m. SW (exc ChI); n Baja CA. Jan–Apr ★

subsp. *crassifolia* (Jeps.) P.V. Wells (p. 697) DEL MAR MANZANITA ST: twig sparsely short-nonglandular-hairy. LF: blade smooth, dark green, margins often ± red. INFL: nascent infl axis, bract, pedicel, ovary densely white-nonglandular-hairy, bracts occ lf-like. Chaparral; < 100 m. SCo (San Diego Co.); Baja CA. Dec–Feb ★

subsp. *cushingiana* (Eastw.) J.E. Keeley et al. CUSHING MANZANITA ST: twig sparsely short-nonglandular-hairy. LF: blade smooth, green or gray-green. INFL: nascent infl axis, bract, pedicel, ovary densely white-nonglandular-hairy, some bracts lf-like. Chaparral, conifer forest; 50–1900 m. KR, NCoR, SnFrB, SCoR, PR; n Baja CA. [*A. g.* Eastw. f. *c.* (Eastw.) P.V. Wells] Jan–Apr

subsp. *gabrielensis* (P.V. Wells) J.E. Keeley et al. SAN GABRIEL MANZANITA ST: twig sparsely short-nonglandular-hairy. LF: shiny bright green, sparsely puberulent, in age glabrous, base wedge-shaped to rounded. INFL: nascent infl axis, bract, pedicel, ovary densely white-nonglandular-hairy, some bracts lf-like. FR: stones fused or in 2(3) fused units. Chaparral on granitic soils; 950–2000 m. s SCoRO (Sierra Madre Mtns), SnGb. [*A. gabrielensis* P.V. Wells] Jan–Apr ★

subsp. *glandulosa* (p. 697) ST: twig gen glandular- and short-nonglandular-hairy, glands black or clear (0 or pink). LF: blade glandular-puberulent, papillate, scabrous, or sparsely puberulent, smooth, green to gray-green, ± glaucous. INFL: nascent infl axis, bract, pedicel, ovary glandular-hairy. Chaparral, conifer forest; 30–1900 m. KR, NCoR, SnFrB, SCoR, PR; n Baja CA. [*A. g.* subsp. *zacaensis* (Eastw.) P.V. Wells] Jan–Apr

subsp. *howellii* (Eastw.) P.V. Wells ST: twig sparsely to densely short-nonglandular-hairy. LF: blade glandular-hairy, papillate, scabrous or sparsely nonglandular-hairy, smooth, green to gray-green, ± glaucous. INFL: infl axis, bract sparsely to densely glandular-hairy, pedicel sparsely nonglandular-hairy. FL: ovary densely white-nonglandular-hairy (glandular). Chaparral, conifer forest; 400–1800 m. SCoRO (c&e Santa Lucia Range). Jan–Apr

subsp. *leucophylla* J.E. Keeley et al. ST: twig long-glandular-hairy, glands black or clear (pink). LF: blade glandular-hairy, scabrous or sparsely puberulent, strongly white-glaucous, base rounded, occ ± lobed to truncate. INFL: nascent infl axis, bract, pedicel, ovary densely white-nonglandular-hairy. Chaparral, conifer forest; 200–1930 m. w SnBr, PR; n Baja CA. Jan–Apr

subsp. *mollis* (J.E. Adams) P.V. Wells ST: twig short- and long-white-soft-wavy-nonglandular-hairy. LF: blade shiny, smooth, bright green. INFL: nascent infl axis, bract, pedicel, ovary densely white-nonglandular-hairy, bracts occ lf-like. Chaparral, conifer forest; 60–2000 m. TR, PR; Baja CA. [*A. g.* subsp. *glaucomollis* P.V. Wells] Jan–Apr

A. glauca Lindl. (p. 697) Shrub, small tree, erect, 1–8 m. ST: twig glabrous. LF: erect; petiole 7–15 mm; blade 2.5–5 cm, 2–4 cm wide, oblong-ovate to ± round, white-glaucous, dull, glabrous,

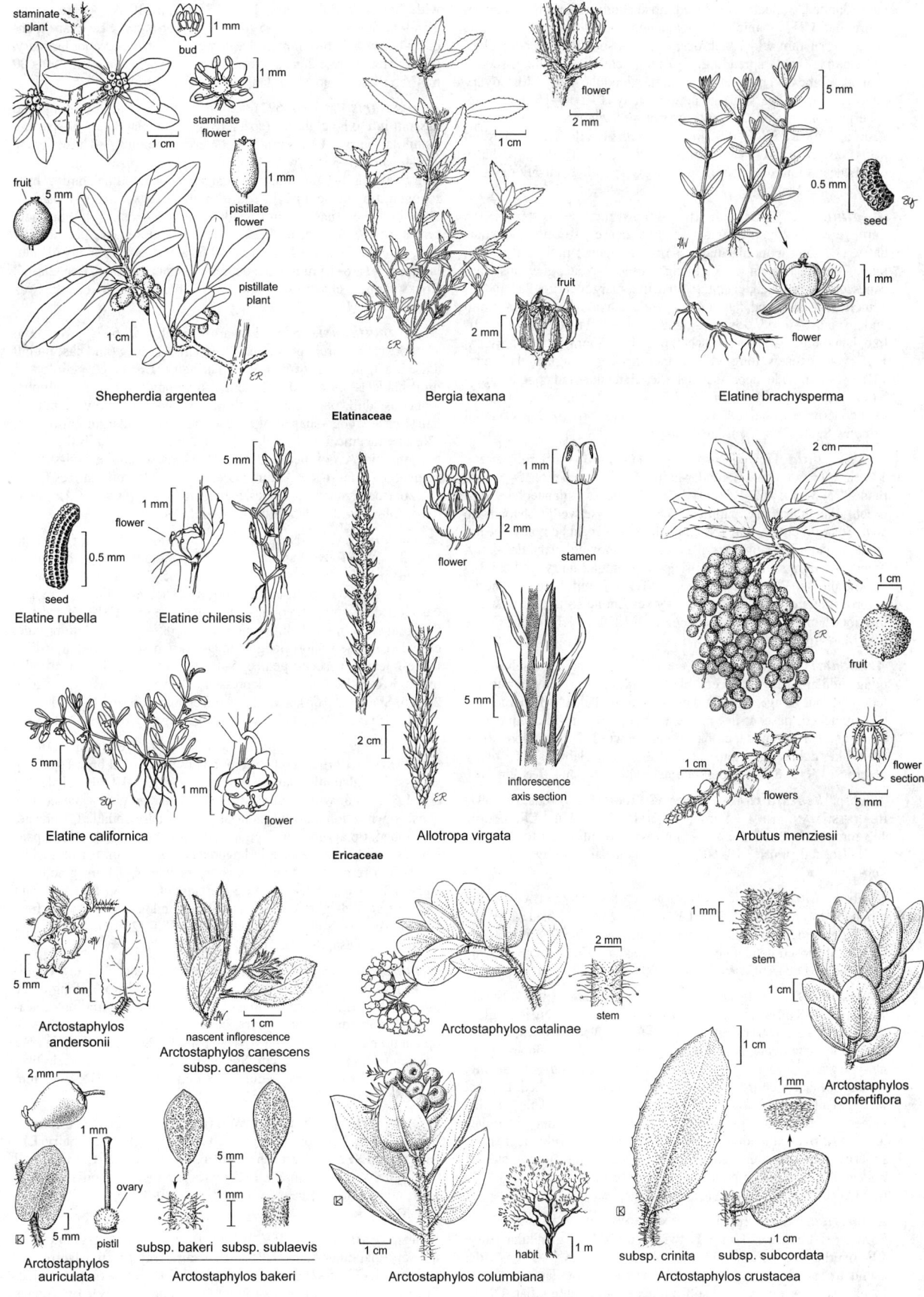

693

staminate plant

bud

1 mm

1 mm

staminate flower

1 mm

1 mm

pistillate flower

fruit

5 mm

pistillate plant

1 cm

1 cm

Shepherdia argentea

flower

2 mm

1 cm

fruit

2 mm

Bergia texana

Elatinaceae

5 mm

0.5 mm

seed

1 mm

flower

Elatine brachysperma

5 mm

1 mm

flower

0.5 mm

seed

Elatine rubella

Elatine chilensis

5 mm

1 mm

flower

Elatine californica

1 mm

2 mm

flower

stamen

2 cm

5 mm

inflorescence axis section

Allotropa virgata

Ericaceae

2 cm

1 cm

fruit

1 cm

flowers

flower section

5 mm

Arbutus menziesii

5 mm

1 cm

Arctostaphylos andersonii

nascent inflorescence
Arctostaphylos canescens subsp. canescens

2 mm

1 mm

ovary

5 mm

pistil

Arctostaphylos auriculata

5 mm

1 mm

subsp. bakeri subsp. sublaevis

Arctostaphylos bakeri

2 mm

stem

Arctostaphylos catalinae

1 cm

habit

1 m

Arctostaphylos columbiana

1 mm

stem

1 cm

Arctostaphylos confertiflora

1 cm

1 mm

1 cm

subsp. crinita subsp. subcordata

Arctostaphylos crustacea

base rounded, truncate, or ± lobed, tip abruptly soft-pointed, margin entire, flat. **INFL:** panicle, 4–8-branched; nascent infl pendent, axis 2–3 cm, > 1 mm wide, glabrous, occ short-nonglandular- or glandular-hairy; bracts spreading, 3–6 mm, scale-like, deltate, scoop-shaped, glabrous; pedicel 8–10 mm, glandular-hairy. **FL:** ovary densely (and fr sparsely) glandular-sticky. **FR:** 10–15 mm wide, spheric; stones fused. 2*n*=26. Rocky slopes, chaparral, woodland; < 2200 m. SnFrB, SCoR, TR, PR, sw DMtns (Little San Bernardino Mtns); nw Baja CA. Twigs short-nonglandular- or -glandular-hairy on pls near some other spp., presumably from hybridization; study needed. Dec–Mar

A. glutinosa B. Schreib. (p. 697) SCHREIBER'S MANZANITA Shrub, erect, 1–2 m. **ST:** twig sparsely to densely soft-nonglandular-hairy, occ long-glandular-hairy. **LF:** overlapped; petiole < 4 mm; blade 2–5 cm, 1–3 cm wide, oblong to oblong-ovate, glaucous, dull, canescent, base lobed, clasping, tip acute, margin entire, flat. **INFL:** panicle, 2–4-branched; nascent infl pendent, axis 1.5–2.5 cm, > 1 mm wide, sparsely to densely soft-glandular-hairy; bracts 5–15 mm, lf-like, lance-oblong, gray-canescent; pedicel 5–8 mm, glandular-hairy. **FL:** ovary densely long-white-glandular- and nonglandular-hairy. **FR:** 7–14 mm wide, depressed-spheric, glandular- and nonglandular-hairy, sticky; stones free. 2*n*=26. Siliceous shale outcrops, chaparral, knobcone-pine woodland; 180–650 m. sw SnFrB (n Ben Lomond Mtn, nw Santa Cruz Co.). Jan–Mar　★

A. hispidula Howell HOWELL'S MANZANITA Shrub, erect, 1–3 m. **ST:** twig fine-glandular-hairy. **LF:** erect; petiole 3–6 mm; blade 1.5–3 cm, 0.5–1.5 cm wide, elliptic to oblanceolate, dull green, glandular-hairy, papillate, scabrous, base wedge-shaped, tip acute, margin entire, flat. **INFL:** panicle, 3–6-branched; nascent infl spreading to ascending, axis 1–2 cm, < 1 mm wide, glandular-hairy; bracts 2–4 mm, scale-like, awl-shaped, glandular-hairy; pedicel 3–5 mm, glabrous. **FL:** ovary glabrous. **FR:** 5–7 mm wide, ± spheric, glabrous; stones free. 2*n*=26. Rocky serpentine soils or sandstone, interior chaparral, open woodland; 100–1250 m. KR, NCoRO; sw OR. Mar–Apr　★

A. hookeri G. Don **ST:** twig (and nascent infl axis) sparsely short-nonglandular-hairy. **LF:** erect; blade bright green, shiny, ± puberulent, glabrous in age, tip acute, margin entire, flat. **INFL:** ± raceme, 0–1-branched; nascent infl pendent, spheric or cup-like, dark, axis 0.3–1 cm, > 1 mm wide, crook-necked; bracts 2–5 mm, ± spreading, scale-like, deltate, glabrous; pedicel 2–6 mm, glabrous. **FL:** ovary glabrous. **FR:** 3–8 mm wide, spheric, glabrous; stones free. 2*n*=26.

　subsp. ***hearstiorum*** (Hoover & Roof) P.V. Wells (p. 697) HEARSTS' MANZANITA Shrub, prostrate, 0.1–0.2 m. **LF:** petiole 1–3 mm; blade 0.8–1.2 cm, 4–7 mm wide, ± elliptic. **FR:** 3–4 mm wide. Coastal prairie; 10–50 m. s CCo (n San Luis Obispo Co.). Feb–Apr　★

　subsp. ***hookeri*** (p. 697) HOOKER'S MANZANITA Shrub, mounded to erect, 0.5–1.5 m. **LF:** petiole 4–8 mm; blade 2–3 cm, 1–1.5 cm wide, lanceolate to widely elliptic. **FR:** 3–8 mm wide. Chaparral, closed-cone pine forest; < 200 m. n CCo, s SnFrB (s Santa Cruz Mtns). Feb–Apr　★

A. hooveri P.V. Wells (p. 697) HOOVER'S MANZANITA Shrub, small tree, 2–8 m. **ST:** twig (and nascent infl axis) densely long-glandular-hairy. **LF:** overlapped; petiole 3–6 mm; blade 4–6 cm, 2–3 cm wide, oblong to ovate, boat-shaped, glaucous, dull, glandular-sticky-hairy, ± glabrous in age, papillate, scabrous, base lobed, not clasping, tip acute, margin entire, flat. **INFL:** panicle, 4–6-branched; nascent infl pendent, axis 1.5–2.5 cm, > 1 mm wide; bracts 8–20 mm, lf-like, lanceolate, acuminate, glandular-hairy; pedicel 8–15 mm, glandular-hairy. **FL:** ovary glandular-hairy. **FR:** 10–15 mm wide, depressed-spheric, glandular-hairy, sticky; stones free. 2*n*=26. Rocky slopes, upland chaparral, open ponderosa-pine forest near coast; 450–1100 m. SCoRO (Santa Lucia Range). Feb–Apr　★

A. imbricata Eastw. (p. 697) SAN BRUNO MOUNTAIN MANZANITA Shrub, prostrate, 0.1–1 m. **ST:** twig densely long-glandular-hairy. **LF:** overlapped; petiole < 2 mm; blade 2.5–4 cm, 2–3 cm wide, round to round-ovate, light green, sparsely glandular-hairy, base lobed, clasping, tip abruptly soft-pointed, margin entire, flat. **INFL:** panicle, 3–5-branched; nascent infl pendent, axis 0.5–1 cm, > 1 mm

wide, long-glandular-hairy; bracts 5–10 mm, lf-like, ovate, acute, glandular-hairy; pedicel 3–5 mm, glandular-hairy. **FL:** ovary glandular-hairy. **FR:** 6–7 mm wide, depressed-spheric, glandular-hairy, sticky; stones free. 2*n*=26. Sandstone outcrops, chaparral; 200–400 m. CCo (San Bruno Mtn). Jan–Mar　★

A. insularis Parry (p. 697)　Shrub, erect, 2–5 m. **ST:** twig (and nascent infl axis) glabrous (sparsely short-nonglandular- or sparsely glandular-hairy). **LF:** spreading or erect; petiole 4–8 mm; blade 2.5–4.5 cm, 1–3 cm wide, oblong-elliptic, cupped, bright green, shiny, abaxially glabrous or sparsely nonglandular-hairy, base rounded, tip obtuse, margin entire; stomata abaxial. **INFL:** panicle, 5–10-branched; nascent infl pendent, axis 2–3 cm, > 1 mm wide; bracts 1.5–3 mm, scale-like, deltate, glandular-hairy; pedicel 6–9 mm, glandular-hairy. **FL:** ovary (and fr) glabrous or sparsely glandular-hairy. **FR:** 8–15 mm wide, depressed-spheric; stones free. 2*n*=26. Rocky slopes, chaparral, woodland; < 400 m. n ChI (Santa Cruz Island). Jan–Mar

A. klamathensis S.W. Edwards et al. (p. 697) KLAMATH MANZANITA Shrub, prostrate, 0.1–0.5 m. **ST:** twig (and nascent infl axis, bract, pedicel, ovary) glandular-hairy. **LF:** erect; petiole 4–7 mm; blade 1–3.5 cm, 0.5–2.5 cm wide, obovate to widely elliptic, glaucous, dull, papillate, ± scabrous, midvein sparsely glandular-hairy, base wedge-shaped or obtuse, tip obtuse, margin entire, flat. **INFL:** ± raceme, 0–1-branched; nascent infl pendent, axis 0.5–1 cm, > 1 mm wide, crook-necked; bracts 2–5 mm, scale-like, awl-shaped; pedicel 3–6 mm. **FR:** 6–7 mm wide, spheric, glabrous; stones fused. 2*n*=26. Rocky outcrops, slopes, subalpine forest; 1600–2000 m. e KR (Scott Mtn Divide, Slate Mtn). May–Jul　★

A. luciana P.V. Wells SANTA LUCIA MANZANITA Shrub, small tree, 2–3 m. **ST:** twig sparsely short-nonglandular-hairy. **LF:** overlapped; petiole < 2 mm; blade 2–4 cm, 1.5–2.5 cm wide, ovate to ± round, glaucous, dull, appressed-gray-canescent, glabrous in age, base lobed, clasping, tip obtuse, margin entire, flat. **INFL:** ± raceme, 0–1-branched; nascent infl pendent, axis 0.5–1 cm, > 1 mm wide, crowded, sparsely short-nonglandular-hairy; bracts 5–10 mm, lf-like, lance-linear, canescent; pedicel 5–10 mm, glabrous. **FL:** ovary glabrous. **FR:** 6–12 mm wide, depressed-spheric, glabrous; stones free. 2*n*=26. Shale outcrops, slopes, upland chaparral near coast; 100–800 m. SCoRO (se of Cuesta Pass, San Luis Obispo Co.). Jan–Mar　★

A. malloryi (W. Knight & Gankin) P.V. Wells MALLORY'S MANZANITA Shrub, erect, 1–3 m. **ST:** twig (and infl bract) sparsely short-sticky-glandular-hairy. **LF:** erect; petiole 5–10 mm; blade 2–3 cm, 1.5–2.5 cm wide, round to ovate, glaucous, dull, canescent or densely white-tomentose, glabrous in age, base rounded, truncate, or ± lobed, tip abruptly soft-pointed, margin entire, flat. **INFL:** panicle, 2–5-branched; nascent infl pendent, axis 1–2 cm, > 1 mm wide; bracts 3–5 mm, scale-like, appressed; pedicel 6–9 mm, glandular-hairy. **FL:** ovary densely white-nonglandular-hairy. **FR:** 7–9 mm wide, depressed-spheric, glabrous to nonglandular-hairy; stones free. 2*n*=26. Volcanic soils, interior chaparral; 650–1200 m. KR, NCoRI (Sonoma, Colusa, Shasta, Trinity cos.). Feb–Apr　★

A. manzanita Parry　Erect, 2–8 m; burled or not. **LF:** erect; petiole 6–12 mm; blade 2.5–5 cm, 1–3.5 cm wide, wide- or oblong-ovate to obovate, bright green, shiny, or glaucous, dull, glabrous or scabrous, veins nonglandular-hairy, base rounded to ± wedge-shaped, tip acute, margin entire, flat. **INFL:** panicle, 2–7-branched; nascent infl pendent, axis 1.5–4.5 cm; bracts appressed, 2–4 mm, scale-like, deltate to sharp-pointed; pedicel 3–8 mm, glabrous. **FR:** 8–12 mm wide, depressed-spheric. 2*n*=52.

　subsp. ***elegans*** (Jeps.) P.V. Wells (p. 697) KONOCTI MANZANITA Burl 0. **ST:** twig glabrous to sparsely short-nonglandular-hairy. **LF:** dull green. **INFL:** 4–7-branched; infl axis 1 mm wide. **FL:** ovary (and fr) minutely glandular-hairy. **FR:** stones gen fused. Woodland, chaparral, conifer forest, gen volcanic soils; 220–1850 m. KR, NCoR. Feb–May　★

　subsp. ***glaucescens*** P.V. Wells　Burl 0. **ST:** twig glabrous to sparsely glandular-hairy. **LF:** ± glaucous, gray-green, glabrous or scabrous. **INFL:** 5–7-branched, gen glandular-hairy; axis 1 mm wide. **FL:** ovary (and fr) glabrous (minutely glandular-hairy). **FR:** stones free. Chaparral, conifer forest; 150–600 m. c&s NCoRO. Feb–May

subsp. *laevigata* (Eastw.) Munz CONTRA COSTA MANZANITA Burl 0. **ST**: twig sparsely short-nonglandular-hairy. **LF**: bright green, shiny. **INFL**: 2–4-branched; axis > 1 mm wide. **FL**: ovary (and fr) glabrous. **FR**: stones free. Chaparral, rocky outcrops; 240–1100 m. s NCoRI (Vaca Mtns), e SnFrB (Mount Diablo). Feb–May ★

subsp. *manzanita* (p. 697) Burl 0. **ST**: twig glabrous to sparsely short-nonglandular-hairy. **LF**: dull to bright green. **INFL**: 4–7-branched; axis ± 1 mm wide. **FL**: ovary (and fr) glabrous. **FR**: stones free. Woodland, chaparral, conifer forest; 30–1200 m. KR, NCoRO, NCoRI, s CaRF, n&c SNF, n SNH, ScV, n SnFrB. Feb–May

subsp. *roofii* (Gankin) P.V. Wells Burl prominent. **ST**: twig glabrous to appressed-nonglandular-hairy. **LF**: dull to bright green. **INFL**: 4–7-branched; axis > 1 mm wide; bracts awl-shaped. **FL**: ovary (and fr) gen minutely glandular-hairy. **FR**: stones free or partly fused. Chaparral, conifer forest; 121–1800 m. NCoRI, CaRF (Butte Co.). Feb–May

subsp. *wieslanderi* P.V. Wells (p. 697) Burl 0. **ST**: twig (and infl axis) short-glandular-hairy, glands dark or clear. **LF**: dull green, scabrous, glandular-hairy. **INFL**: axis 1 mm wide. **FL**: ovary (and fr) glabrous to hairy. **FR**: stones free. Chaparral, conifer forest; 300–1500 m. NCoRH, CaRF (Shasta, Tehama cos.). Feb–May

A. mewukka Merriam Shrub, erect, 1–3 m; burled or not. **ST**: twig glabrous to ± nonglandular-hairy. **LF**: blade 3–7 cm, dull, glabrous, base rounded, margin entire, flat. **INFL**: panicle 2–7-branched; nascent infl pendent, axis 2–3 cm, > 1 mm wide, glabrous to ± nonglandular-hairy, occ glandular-hairy; bracts 3–6 mm, scale-like, deltate to lance-linear, glabrous; pedicel 4–8 mm, glabrous. **FL**: ovary glabrous. **FR**: 10–16 mm wide, depressed-spheric, glabrous, dark mahogany-brown. 2*n*=52.

subsp. *mewukka* (p. 701) Burled. **LF**: petiole 6–10 mm; blade 2–4 cm wide, oblong- to lance-ovate, gray-glaucous, tip acute. **INFL**: nascent axes 3–5; bracts spreading, exposing buds, ± flat, tip acute. **FR**: stones gen free. Chaparral, forest openings; 333–1800 m. SNF, n SNH. Mar–Apr

subsp. *truei* (W. Knight) P.V. Wells (p. 701) TRUE'S MANZANITA Burl 0. **LF**: petiole 10–15 mm; blade 4–7 cm wide, round to round-ovate, white-glaucous, tip obtuse. **INFL**: nascent axes 4–7, bracts appressed, hiding buds, keeled, tip acuminate. **FR**: stones free, occ some fused. *n*=26. Chaparral, forest openings; 290–1350 m. s CaRF, CaRH (near Paradise), n SNF. Mar–Apr ★

A. montana Eastw. Burl 0. **LF**: erect; petiole 2–6 mm; blade puberulent, glabrous in age, dark green, shiny, base rounded to wedge-shaped, tip acute, margin entire, flat. **INFL**: panicle, 1–3-branched; nascent infl pendent, axis 0.5–1.5 cm, > 1 mm wide, densely short-white-nonglandular-hairy; bracts appressed, scale-like, ovate, acuminate, 2–5 mm, glabrous; pedicel 3–6 mm, glabrous. **FL**: ovary glabrous. **FR**: spheric, glabrous; stones free. 2*n*=52.

subsp. *montana* (p. 701) MOUNT TAMALPAIS MANZANITA Mounded to erect, 0.5–2 m. **ST**: twig densely short-white-nonglandular-hairy. **LF**: blade 1–2.5 cm, 1–1.5 cm wide, round-elliptic to elliptic. **FR**: 6–8 mm wide. Serpentine chaparral; 250–800 m. nw SnFrB (Mount Tamalpais, Marin Co.). [*A. hookeri* subsp. *m.* (Eastw.) P.V. Wells] Feb–Apr ★

subsp. *ravenii* (P.V. Wells) V.T. Parker et al. (p. 701) PRESIDIO MANZANITA Prostrate, 0.1–0.3 m. **ST**: twig short-gray-nonglandular-hairy. **LF**: blade 1–2 cm, 1–1.5 cm wide, round to round-elliptic. **FR**: 4–5 mm wide. Serpentine chaparral; 60–95 m. n CCo (San Francisco Presidio). [*A. hookeri* subsp. *r.* P.V. Wells] Feb–Apr ★

A. montaraensis Roof (p. 701) MONTARA MANZANITA Mounded to erect, 0.5–5 m (< 0.5 m on exposed granite outcrops). **ST**: twig (and nascent infl axis) densely long-glandular-hairy. **LF**: overlapped; petiole < 2 mm; blade 2.5–4.5 cm, 1.5–2.5 cm wide, ovate, bright green, sparsely glandular-puberulent, base lobed, clasping, tip acute, margin entire, flat. **INFL**: panicle, 4–6-branched; nascent infl pendent, axis 1–1.5 cm, > 1 mm wide; bracts 6–9 mm, lf-like, lanceolate, acuminate, glandular-hairy; pedicel 5–6 mm, glandular-hairy. **FL**: ovary glandular-hairy. **FR**: 6–7 mm wide, depressed-spheric, glandular-hairy, sticky; stones free. 2*n*=26. Granite, sandstone out-

crops, chaparral, coastal scrub; 200–500 m. CCo (San Bruno Mtn), w SnFrB (Montara Mtn). Jan–Mar ★

A. montereyensis Hoover TORO MANZANITA Erect, 1–3 m. **ST**: twig (and lvs, nascent infl axis, bract, pedicel, ovary, fr) glandular-hairy. **LF**: erect, not overlapped; petiole 4–6 mm; blade 2–3 cm, 1.5–3 cm wide, round to oblong-ovate, bright green or ± glaucous, dull, glandular-hairy, papillate, scabrous, base gen rounded, truncate, or ± lobed, tip abruptly soft-pointed, margin entire, flat. **INFL**: panicle, 4–10-branched; nascent infl pendent, axis 1–2 cm, > 1 mm wide; bracts 3–12 mm, lf-like, green, persistent, lanceolate, acuminate; pedicel 5–6 mm. **FR**: 8–12 mm wide, depressed-spheric, sticky; stones free. 2*n*=26. Sandstone soils (stabilized dunes), chaparral; < 350 m. c CCo (Fort Ord), n SCoRO (Mount Toro, nw Monterey Co.). Jan–Mar ★

A. morroensis Wiesl. & B. Schreib. (p. 701) MORRO MANZANITA Erect, 1–4 m. **ST**: old sts gray, bark shredding; twig (and nascent infl axis) short- and long-white-nonglandular-hairy. **LF**: overlapped; petiole 2–5 mm; blade 1.5–3 cm, 1–2 cm wide, oblong-ovate to -elliptic, dark green, shiny, abaxially gray-tomentose, adaxially glabrous, base ± truncate to ± cordate, tip abruptly soft-pointed, margin entire, cupped; stomata abaxial. **INFL**: panicle, 2–5-branched; nascent infl pendent, bell-shaped, axis 0.5–0.8 cm, > 1 mm wide; bracts 5–8 mm, lf-like, lance-linear, acuminate, puberulent; pedicel 4–6 mm, glabrous or nonglandular-hairy. **FL**: ovary densely (fr sparsely) white-nonglandular-hairy. **FR**: 7–10 mm wide, depressed-spheric; stones free. 2*n*=26. Stabilized sand dunes, sandstones, chaparral; < 200 m. s CCo (s Morro Bay, San Luis Obispo Co.). Jan–Mar ★

A. myrtifolia Parry (p. 701) IONE MANZANITA Mounded to erect, 0.5–1.5 m. **ST**: old sts smooth, red with ± gray or ± glaucous patches; twig glandular-hairy. **LF**: spreading; petiole 1–3 mm; blade 0.6–1.5 cm, 0.3–0.8 cm wide, narrowly elliptic, bright green, shiny, sparsely glandular-hairy, papillate, base wedge-shaped, tip acute, margin entire, flat. **INFL**: ± raceme, 0–1-branched; nascent infl pendent, axis 0.5–1 cm, > 1 mm wide, glandular-hairy; bracts 1–2 mm, scale-like, deltate, glabrous; pedicel 1–3 mm, glabrous. **FL**: ovary white-nonglandular-hairy. **FR**: 3–4 mm wide, ± spheric, glabrous; stones free. 2*n*=26. Acidic sandy or clay soils, chaparral, woodland; 70–770 m. n SNF (Amador, Calaveras cos.). Jan–Feb ★

A. nevadensis A. Gray Prostrate to mounded, 0.1–0.5 m; burled or not. **ST**: twig (and nascent infl axis) sparsely short-nonglandular-hairy. **LF**: erect; petiole 3–7 mm; blade 1–3 cm, 1–1.5 cm wide, obovate or oblanceolate, bright green, shiny, puberulent, glabrous in age, base wedge-shaped, tip obtuse or acute, margin entire, flat. **INFL**: ± raceme, 0–1-branched; nascent infl pendent, spheric, axis 0.5–1 cm, > 1 mm wide; bracts 2–3(5–10) mm, scale-like, linear or lance-linear, acuminate, glabrous to sparsely nonglandular-hairy (glandular-hairy); pedicel 3–5 mm, glabrous. **FL**: ovary glabrous. **FR**: 6–8 mm wide, spheric, glabrous, dark brown; stones free. 2*n*=52.

subsp. *knightii* (Gankin & W.R. Hildreth) P.V. Wells KNIGHT'S MANZANITA Burl present, occ obscure. Rocky outcrops, conifer forests; 200–2600 m. KR, NCoRO. [*A.* ×*k.* Gankin & W.R. Hildreth] May–Jul

subsp. *nevadensis* (p. 701) PINE-MAT MANZANITA Burl 0. **INFL**: bracts near base occ 5–10 mm. Rocky outcrops, conifer forest; 130–3200 m. KR, NCoRH, CaRH, SNH, MP, n SNE (Sweetwater Mtns); to WA. May–Jul

A. nissenana Merriam (p. 701) NISSENAN MANZANITA Erect, 0.2–1.5 m. **ST**: old st bark persistent, gray, shredding; twig canescent. **LF**: erect; petiole 1–3 mm; blade 1–2 cm, 0.8–1.5 cm wide, elliptic to oblong-elliptic, glaucous, gray, dull, sparsely appressed-puberulent, glabrous in age, base wedge-shaped to rounded, tip acute, margin entire, flat. **INFL**: ± raceme, 0–1-branched; nascent infl pendent, axis 0.2–0.5 cm, > 1 mm wide, canescent; bracts 3–5 mm, lf-like, lanceolate, glabrous to sparsely nonglandular-hairy; pedicel 5–7 mm, glabrous to sparsely nonglandular-hairy. **FL**: ovary white-nonglandular-hairy. **FR**: 3–4 mm wide, ± cylindric, glabrous, mature splitting; stones free. 2*n*=26. Open, rocky shale ridges, chaparral, woodland; 450–1650 m. n SN (Placer, El Dorado cos.), c SNF (Tuolumne Co.). Feb–Mar ★

A. nortensis (P.V. Wells) P.V. Wells DEL NORTE MANZANITA Erect, 2–5 m. **ST:** twig (and infl bract) long-nonglandular-bristly, occ minutely glandular-hairy. **LF:** erect; petiole 4–5 mm; blade 2–4 cm, 1–2 cm wide, narrowly elliptic-ovate, glaucous, gray, dull, white-canescent, in age dark green, dull, glabrous, base truncate to rounded, tip acute, margin entire, flat. **INFL:** ± raceme, 0–1-branched; nascent infl pendent, axis 0.5–1.2 cm, > 1 mm wide, soft-nonglandular-hairy and long-nonglandular-bristly, occ minutely glandular-hairy; bracts 5–10 mm, lf-like, ovate, acute, green, persistent; pedicel 2–4 mm, sparsely white-nonglandular-hairy. **FL:** ovary (and fr sparsely) white-nonglandular-hairy. **FR:** 6–8 mm wide, depressed-spheric; stones free. 2*n*=26. Rocky slopes, occ serpentine, chaparral, conifer forest; 500–940 m. nw KR (n Del Norte Co.). Mar–May ★

A. nummularia A. Gray **ST:** old st bark persistent, gray or red-gray; twig (and nascent infl axis) densely short-nonglandular- and long-glandular-hairy. **LF:** spreading; blade abaxially light green, shiny, glabrous, midvein hairy, adaxially dark green, tip obtuse, margin entire, cupped; stomata abaxial. **INFL:** panicle, 4–12-branched; nascent infl pendent, axis 0.5–1 cm, < 1 mm wide; bracts appressed, 0.5–2 mm, scale-like, deltate, sharp-pointed, glabrous; pedicel 3–5 mm, glabrous. **FL:** fl parts in 4s; ovary white-nonglandular-hairy. **FR:** 3–4 mm wide, ± cylindric, glabrous, mature splitting; stones free. 2*n*=26.

  subsp. ***mendocinoensis*** (P.V. Wells) V.T. Parker et al. PYGMY MANZANITA Prostrate to mounded, 0.1–0.5 m. **ST:** old st bark red-gray, gen rough or shredding. **LF:** petiole < 1 mm; blade 0.5–1.2 cm, 0.3–0.7 cm wide, oblong-elliptic, base rounded to wedge-shaped. Pygmy pine forest, chaparral; 50–200 m. c NCo (Mendocino Co.). [*A. m.* P.V. Wells] Mar–May ★

  subsp. ***nummularia*** (p. 701) Prostrate to erect, 0.5–5 m. **ST:** old st bark gray, shredding. **LF:** petiole 1–3 mm; blade 1–2.2 cm, 0.8–1.8 cm wide, round to round-ovate, base truncate to ± lobed. Coastal prairie, maritime chaparral, closed-cone forest; 15–400 m. c NCo/NCoRO (Mendocino Co.). Mar–May

A. obispoensis Eastw. BISHOP MANZANITA Erect, 1–4 m. **ST:** twig (and nascent infl axis) sparsely short-nonglandular-hairy. **LF:** erect; petiole 5–7 mm; blade 2–4.5 cm, 1–2.5 cm wide, oblong- to lance-ovate, glaucous-gray, dull, appressed-canescent, in age glabrous, base rounded to truncate or ± lobed, tip acute, margin entire, flat. **INFL:** panicle, 2–4-branched; nascent infl pendent, bell-shaped, axis 1–2.5 cm, > 1 mm wide; bracts 7–14 mm, lf-like, lance-linear, acuminate, appressed-canescent to glabrous; pedicel 8–10 mm, glabrous. **FL:** ovary glabrous. **FR:** 9–14 mm wide, depressed-spheric, glabrous; stones free. 2*n*=26. Rocky, gen serpentine soils, chaparral, open closed-cone forest near coast; 60–950 m. SCoRO (c&s Santa Lucia Range). Feb–Mar ★

A. ohloneana M.C. Vasey & V.T. Parker OHLONE MANZANITA Erect, 1–2 m. **ST:** twig (and nascent infl axis) sparsely short-nonglandular- and occ -long-glandular-hairy. **LF:** erect; petiole 3–5 mm; blade 1.5–3 cm, 1–1.5 cm wide, elliptic to ovate-elliptic, puberulent and short-glandular-hairy, in age glabrous, light green, dull, base wedge-shaped, tip acute, margin entire, flat. **INFL:** panicle, 3–5-branched; nascent infl pendent, axis 1.5–2.5 cm, > 1 mm wide; bracts 1–3 mm, scale-like, ovate to deltate, sharp-pointed, nonglandular-hairy; pedicel 3–8 mm, glabrous. **FL:** ovary glabrous. **FR:** 5–8 mm wide, depressed-spheric, glabrous; stones free. 2*n*=26. Siliceous shale outcrops, chaparral, knobcone-pine woodland; 400–500 m. sw SnFrB (n Ben Lomond Mtn, nw Santa Cruz Co.). Feb–Mar ★

A. osoensis P.V. Wells OSO MANZANITA Erect, 1–4 m. **ST:** old st bark persistent, gray, shredding (smooth); twig (and nascent infl axis) sparsely short-nonglandular-hairy. **LF:** overlapped; petiole < 2 mm; blade 1.5–3 cm, 1.5–2.5 cm wide, ovate to round-ovate, dark green, ± shiny, sparsely short-nonglandular-hairy or glabrous, base deeply lobed, clasping, tip obtuse, margin entire, flat. **INFL:** ± raceme, 0–1-branched; nascent infl pendent, axis 0.5–1 cm, > 1 mm wide; bracts 4–8 mm, lf-like, lanceolate to ovate, glabrous; pedicel 8–9 mm, glabrous. **FL:** ovary glabrous. **FR:** 5–8 mm wide, depressed-spheric, glabrous; stones free. 2*n*=26. Dacite (volcanic) outcrops, chaparral; 50–375 m. s CCo (w Los Osos Valley, San Luis Obispo Co.). Dec–Feb ★

A. otayensis Wiesl. & B. Schreib. (p. 701) OTAY MANZANITA Erect, 1–3 m. **ST:** twig (and nascent infl axis) sparsely short-non-glandular- and long-glandular-hairy. **LF:** erect; petiole 5–8 mm; blade 2–3 cm, 1–2 cm wide, narrowly or oblong-elliptic to ovate, gray-glaucous, dull, appressed-tomentose, in age ± glabrous, base wedge-shaped, rounded, or truncate, tip abruptly soft-pointed, margin entire, flat. **INFL:** panicle, 4–7-branched; nascent infl pendent, axes 1.5–2.5 cm, > 1 mm wide; bracts spreading, 6–12 mm, lf-like, rigid, lance-linear, acuminate, green, long-glandular-hairy, persistent; pedicel 3–8 mm, glandular-hairy. **FL:** ovary densely nonglandular-hairy, sparsely glandular. **FR:** 6–8 mm wide, spheric, sparsely nonglandular-hairy to glabrous; stones free or fused. 2*n*=26. Volcanic rock outcrops, chaparral, woodland; 280–1700 m. sw PR. Jan–Mar ★

A. pacifica Roof PACIFIC MANZANITA Prostrate, 0.1–0.6 m; burls on sts. **ST:** twig (and nascent infl axis) short-nonglandular-hairy. **LF:** spreading; petiole 2–4 mm; blade 1–2 cm, 0.5–1 cm wide, elliptic, abaxially light green, adaxially dark green, sparsely non-glandular-hairy, base wedge-shaped, tip acute, margin serrate. **INFL:** ± raceme, 0–1-branched; nascent infl pendent, axis 0.5–1 cm, > 1 mm wide; bracts 0.5–1 mm, scale-like, awl-shaped, glabrous; pedicel 3–5 mm, glabrous. **FL:** ovary glabrous. **FR:** 6–8 mm wide, spheric, ± red, glabrous; stones free. 2*n*=52. Sandstone outcrops, chaparral; 300 m. CCo (San Bruno Mtn). [*A.* ×p. Roof] Jan–Mar ★

A. pajaroensis (J.E. Adams) J.E. Adams (p. 701) PAJARO MANZANITA Erect, 1–4 m. **ST:** old sts gray, bark shredding; twig (and nascent infl axis) short-nonglandular-hairy, hairs occ long. **LF:** overlapped; petiole < 2 mm; blade 2–4 cm, 1–2 cm wide, ovate to triangular-ovate, abaxially light green, glabrous, midvein nongland-ular-hairy, adaxially dark blue-green, glaucous, base lobed, clasping, tip acute, margin entire, cupped, tinged with red; stomata abaxial. **INFL:** panicle, 2–5-branched; nascent infl pendent, axis 1–1.5 cm. > 1 mm wide; bracts 5–10 mm, lf-like, lance-linear, acuminate, gla-brous; pedicel 5–8 mm, nonglandular-hairy or not. **FL:** ovary densely white-nonglandular-hairy. **FR:** 6–8 mm wide, depressed-spheric, sparsely nonglandular-hairy to glabrous; stones free. 2*n*=26. Sand-stone outcrops, chaparral; < 755 m. n-c CCo, s SnFrB (Pajaro Hills). Dec–Feb ★

A. pallida Eastw. (p. 701) PALLID MANZANITA Erect, 2–4 m. **ST:** twig densely short- and long-nonglandular-hairy. **LF:** overlapped; petiole < 2 mm, nonglandular-hairy; blade 2.5–4.5 cm, 2–3 cm wide, gen ovate, glaucous-green, dull, glabrous, base lobed, clasping, tip acute, margin entire, flat. **INFL:** panicle, 3–5-branched; nascent infl pendent, axis 0.5–1 cm, > 1 mm wide, densely short-nonglandular-hairy, occ with long glandular hairs; bracts 5–9 mm, lf-like, widely lanceolate, acute, glandular-hairy; pedicel 8–12 mm, glandular-hairy. **FL:** ovary glandular-hairy. **FR:** 8–10 mm wide, depressed-spheric, sticky; stones free. 2*n*=26. Siliceous shales, slopes, ridges, chapar-ral; 200–460 m. e SnFrB (Sobrante, Huckleberry ridges, Alameda, Contra Costa cos.). Jan–Mar ★

A. parryana Lemmon **ST:** twig (and nascent infl axis) sparsely to moderately short-nonglandular-hairy. **LF:** erect; petiole 5–10 mm; blade 1.5–5 cm, 1.5–2.5 cm wide, oblong-ovate to widely elliptic, gla-brous or sparsely nonglandular-hairy, base obtuse to truncate, tip acute, margin entire, flat. **INFL:** panicle, 3–5-branched; nascent infl pendent, axis 0.5–1.5 cm, > 1 mm wide; bracts ± appressed, 2–4 mm, scale-like, deltate to wide ovate, glabrous; pedicel 4–9 mm, glabrous. **FL:** ovary glabrous. **FR:** 7–10 mm wide, spheric, glabrous; stones fused.

  subsp. ***desertica*** J.E. Keeley et al. DESERT MANZANITA Erect, 1–2 m; burled. **LF:** gray-glaucous. 2*n*=26,52. Chaparral; 1200–2300 m. PR (San Jacinto, Santa Rosa, San Ysidro mtns). Jan–Mar

  subsp. ***parryana*** PARRY MANZANITA Mounded, gen 0.5–1 m. **LF:** green. 2*n*=52. Openings in chaparral, forest; 1500–2400 m. TR. Mar–May

  subsp. ***tumescens*** J.E. Keeley et al. INTERIOR MANZANITA Erect, 1–2 m; burl prominent. **LF:** green. 2*n*=52. Montane chaparral; 2100–2300 m. SnGb, SnBr. Mar–Apr ★

A. patula Greene (p. 701) GREENLEAF MANZANITA Erect, 1–3 m; burl 0 or often flat, obscure. **ST:** twig (and nascent infl axis) short-gold-glandular- or densely short-nonglandular-hairy. **LF:** erect; petiole

697

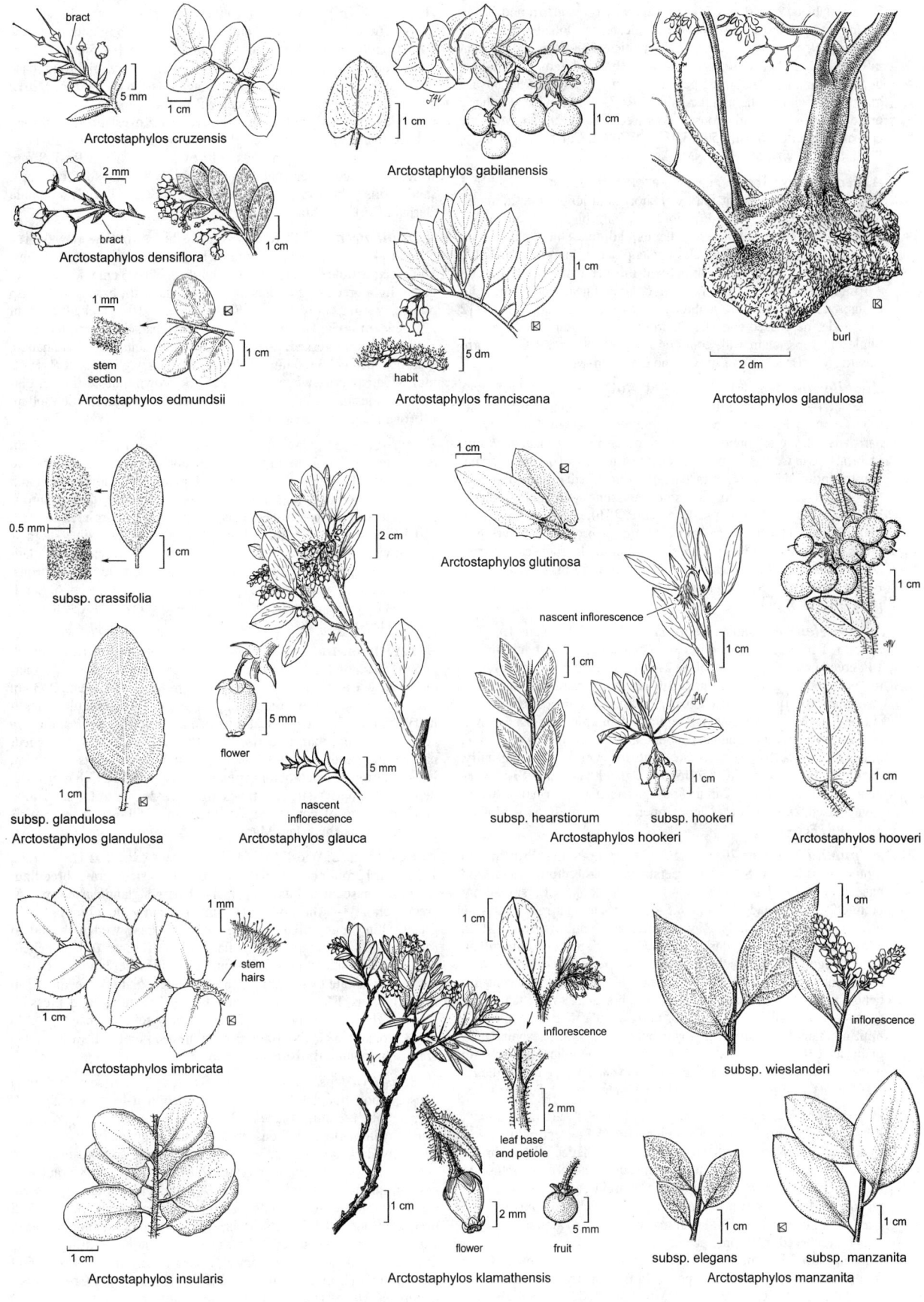

bract

5 mm

1 cm

Arctostaphylos cruzensis

2 mm

bract

1 cm

Arctostaphylos densiflora

1 mm

stem
section

1 cm

Arctostaphylos edmundsii

1 cm

1 cm

Arctostaphylos gabilanensis

1 cm

5 dm

habit

Arctostaphylos franciscana

burl

2 dm

Arctostaphylos glandulosa

0.5 mm

1 cm

subsp. crassifolia

1 cm

subsp. glandulosa
Arctostaphylos glandulosa

2 cm

5 mm

flower

5 mm

nascent
inflorescence

Arctostaphylos glauca

1 cm

Arctostaphylos glutinosa

nascent inflorescence

1 cm

1 cm

subsp. hearstiorum

1 cm

subsp. hookeri
Arctostaphylos hookeri

1 cm

1 cm

1 cm

Arctostaphylos hooveri

1 mm

stem
hairs

1 cm

Arctostaphylos imbricata

1 cm

Arctostaphylos insularis

1 cm

inflorescence

2 mm

leaf base
and petiole

1 cm

2 mm

flower

5 mm

fruit

Arctostaphylos klamathensis

1 cm

inflorescence

subsp. wieslanderi

1 cm

subsp. elegans

1 cm

subsp. manzanita
Arctostaphylos manzanita

7–15 mm; blade 2.5–6 cm, 1.5–4 cm wide, widely ovate to round, bright green, shiny, glabrous, base rounded, truncate, or ± lobed, tip abruptly soft pointed, margin entire, flat. **INFL**: panicle, 2–8-branched; nascent infl pendent, axis 1.5–3 cm, > 1 mm wide; bracts 4–6 mm, scale-like, deltate, acuminate, glabrous; pedicel 2–7 mm, glabrous. **FL**: ovary glabrous or white-nonglandular-hairy. **FR**: 7–10 mm wide, ± spheric or depressed-spheric, glabrous; stones free (fused). $2n=26$. Montane chaparral, conifer forest; 750–3350 m. NCoRH, CaRH, SNH, SnGb, SnBr, SnJt; to WA, MT, CO, NM, Baja CA. Apr–Jun

A. pechoensis (Abrams) Eastw. PECHO MANZANITA   Erect, 1–5 m. **ST**: twig (and nascent infl axis) short- and long-white-nonglandular-hairy. **LF**: overlapped; petiole 1–4 mm; blade 2–5 cm, 1–2.5 cm wide, oblong-ovate, green-glaucous, dull to ± shiny, puberulent, in age glabrous, base lobed, clasping, tip acute, margin entire, flat. **INFL**: panicle, 1–4-branched; nascent infl pendent, axis 1–1.5 cm, > 1 mm wide; bracts 8–15 mm, lf-like, lance-linear, acuminate, glabrous; pedicel 5–10 mm, glabrous. **FL**: ovary glabrous. **FR**: 8–12 mm wide, depressed-spheric, glabrous; stones gen fused. $2n=26$. Shale outcrops, chaparral, conifer forest; < 500 m. s CCo (Morro Bay to Avila Beach, San Luis Obispo Co.). Jan–Mar ★

A. pilosula Jeps. & Wiesl. (p. 701) SANTA MARGARITA MANZANITA   Erect, 1–5 m. **ST**: twig densely long-nonglandular-bristly-hairy. **LF**: erect; petiole 4–8 mm; blade 1–3 cm, 1–2 cm wide, narrowly elliptic to round-ovate, dark green to gray-glaucous, dull, glabrous, base wedge-shaped, truncate to ± lobed, tip obtuse or acute, margin entire, flat. **INFL**: ± raceme, 0–1-branched; nascent infl pendent, axis 1–2 cm, > 1 mm wide, short- and long-nonglandular-hairy; bracts ascending, 8–15 mm, lanceolate, lf-like, green, persistent, long-nonglandular-hairy; pedicel 2–5 mm, glabrous. **FL**: ovary glabrous. **FR**: 8–10 mm wide, depressed-spheric, glabrous; stones free. $2=26$. Shale outcrops, slopes, chaparral; 30–1250 m. s CCo (Pismo Beach vicinity), s SCoRO (s Santa Lucia, La Panza ranges and s). [*A. wellsii* W. Knight] Dec–Mar ★

A. pringlei Parry subsp. ***drupacea*** (Parry) P.V. Wells (p. 701) Erect, 1–5 m. **ST**: twig (and infl axis) densely short-glandular-hairy. **LF**: erect; petiole 5–10 mm; blade 2–5 cm, 1–4 cm wide, ovate, elliptic, or ± round, gray-glaucous, glandular-hairy, papillate, scabrous, base rounded, truncate, or ± lobed, tip abruptly soft-pointed, margin entire, flat. **INFL**: ± raceme, 0–1-branched; infl forming just before fl (unlike most or rest of genus), pendent, axis 1–1.5 cm, > 1 mm wide; bracts 6–10 mm, lf-like, lanceolate, acute, deep pink, glandular-hairy, deciduous after fl; pedicel 5–15 mm, glandular-hairy. **FL**: ovary glandular-hairy. **FR**: 6–12 mm wide, spheric, glandular-hairy, sticky; stones fused. $2n=26$. Rocky slopes, open conifer forest; 530–2400 m. SnBr, PR. Feb–Apr

A. pumila Nutt. (p. 701) SANDMAT MANZANITA   Prostrate to mounded, 0.1–1 m. **ST**: ± red, persistent, ± shredding; twig (and nascent infl axis) short-gray-nonglandular-hairy. **LF**: spreading; petiole 2–3 mm; blade 1–2 cm, 0.5–1.5 cm wide, narrowly obovate to oblanceolate, abaxially gray-tomentose in youth, sparsely so in age, adaxially convex, dark green, ± shiny, sparsely puberulent, base wedge-shaped, tip obtuse, margin entire, cupped, tinged with red; stomata abaxial. **INFL**: ± raceme, 0–1-branched; nascent infl pendent, axis 0.3–0.5 cm, > 1 mm wide; bracts 2–3 mm, scale-like, acute to sharp-pointed, glabrous or nonglandular-hairy; pedicel 3–4 mm, nonglandular-hairy. **FL**: ovary white-nonglandular-hairy or ± glabrous. **FR**: 5–6 mm wide, spheric, sparsely nonglandular-hairy to glabrous; stones free. $2n=26$. Sandy soils in chaparral, oak woodland; < 200 m. c CCo (around Monterey Bay, Monterey Co.). Feb–Apr ★

A. pungens Kunth (p. 701)   Erect, 1–3 m. **ST**: twig (and nascent infl axis) sparsely short-nonglandular-hairy. **LF**: erect; petiole 4–8 mm; blade 1.5–4 cm, 1–1.8 cm wide, elliptic to lance-elliptic, bright or dark green, shiny, tomentose, glabrous in age, base obtuse to wedge-shaped, occ rounded, tip acute, margin entire, flat. **INFL**: ± raceme, 0–1-branched; nascent infl pendent, club-like, axis 0.5–1.5 cm, > 1 mm wide; bracts 2–4 mm, scale-like, ovate-deltate, acuminate, recurved, glabrous; pedicel 5–10 mm, glabrous. **FL**: ovary glabrous. **FR**: 5–8 mm wide, depressed-spheric, glabrous; stones free (fused). $2n=26$. Rocky slopes, ridges, chaparral, conifer forest; 180–2300 m. SCoR, SnBr, PR, e DMtns; to UT, TX, Mex. Feb–Mar

A. purissima P.V. Wells LA PURISIMA MANZANITA   Erect, 1–4 m. **ST**: twig densely short- and sparsely long-nonglandular-hairy. **LF**: strongly overlapped; petiole < 2 mm; blade 1–2.5 cm, 1–2 cm wide, round-ovate to ± round, bright green, shiny, glabrous, base lobed, clasping, tip abruptly soft-pointed, margin entire, flat. **INFL**: ± raceme, 0–2-branched; nascent infl pendent, axis 0.5–1 cm, > 1 mm wide, densely nonglandular-hairy; bracts overlapped, 5–8 mm, lf-like, ovate to lance-ovate, glabrous; pedicel 3–5 mm, sparsely nonglandular-hairy or glabrous. **FL**: ovary glabrous. **FR**: 5–8 mm wide, depressed-spheric, glabrous; stones free. Sandstone outcrops, sandy soils, chaparral; < 300 m. s CCo, s SCoRO/WTR (w Santa Barbara Co.). Jan–Mar ★

A. rainbowensis J.E. Keeley & Massihi RAINBOW MANZANITA   Erect, 1–4 m; burled. **ST**: twig minutely glandular, appearing glabrous. **LF**: erect; petiole 6–12 mm; blade 3.5–5 cm, 2–3.5 cm wide, elliptic-ovate, light green to gray-glaucous, dull, glabrous, base rounded, tip obtuse, margin entire, flat. **INFL**: panicle, 4–10-branched; nascent infl pendent, axis 2–3 cm, > 1 mm wide, sparsely short-glandular-hairy; bracts appressed, 2–4 mm, scale-like, deltoid to awl-shaped, glabrous; pedicel 5–10 mm, glabrous. **FL**: ovary glabrous. **FR**: 8–12 mm wide, spheric, white-glaucous, dark brown, purple-tinged, glabrous; stones fused. $2n=26$. Granitic outcrops, chaparral; 150–800 m. PR (San Diego, Riverside cos.). Jan–Feb ★

A. refugioensis Gankin REFUGIO MANZANITA   Erect, 2–4 m. **ST**: twig (and nascent infl axis) densely short-nonglandular- and -long-glandular-hairy. **LF**: overlapped; petiole < 2 mm; blade 3–4.5 cm, 2–3 cm wide, wide- to oblong-ovate, glaucous, dull, glabrous or midrib ± hairy, base lobed, clasping, tip acute, margin entire, flat. **INFL**: panicle, 5–10-branched; nascent infl pendent, axis 2–3 cm, > 1 mm wide; bracts 4–10 mm, lf-like, ovate to widely elliptic, glandular-hairy; pedicel 7–9 mm, glandular-hairy. **FL**: ovary glabrous. **FR**: 10–15 mm wide, spheric, glabrous; stones fused. $2n=26$. Sandstone outcrops, chaparral; 300–820 m. s SCoRO/w WTR (Santa Ynez Mtns). Dec–Feb ★

A. regismontana Eastw. KINGS MOUNTAIN MANZANITA   Erect, 2–5 m. **ST**: twig (and nascent infl axis, bract, pedicel, ovary, fr) glandular-hairy. **LF**: overlapped; petiole < 3 mm; blade 3–6 cm, 2–3 cm wide, oblong-ovate, boat-shaped, pale green, dull, glandular-sticky-hairy, glabrous in age, ± scabrous or not, base lobed, clasping, tip acute, upcurved, margin entire, flat. **INFL**: panicle, 4–6-branched; nascent infl pendent, axis 1.5–2 cm, > 1 mm wide; bracts 5–12 mm, lf-like, narrowly lance-oblong; pedicel 6–10 mm. **FR**: 6–8 mm wide, depressed-spheric, sticky; stones free. $2n=26$. Granite, sandstone outcrops, edge of conifer forest, chaparral; 150–780 m. w SnFrB (n Santa Cruz Mtns). Jan–Mar ★

A. rudis Jeps. & Wiesl. (p. 701) SAND MESA MANZANITA   Erect, 1–3 m; burl prominent (0). **ST**: old st bark persistent, gray, shredding; twig (and nascent infl axis) sparsely short-nonglandular-hairy. **LF**: erect; petiole 3–8 mm; blade 1–3 cm, 1–2 cm wide, elliptic, bright green, shiny, puberulent, in age glabrous, base wedge-shaped to rounded, tip acute, margin entire, flat. **INFL**: ± raceme, 0–1-branched; nascent infl pendent, axis 0.5–1 cm, > 1 mm wide; bracts 2–6 mm, scale-like, deltate to acuminate, sharp-pointed, glabrous; pedicel 3–6 mm, glabrous. **FL**: ovary glabrous. **FR**: 8–14 mm wide, depressed-spheric, glabrous; stones free. $2n=26$. Sandy soils, chaparral; < 380 m. s CCo, s SCoRO (Nipomo, Burton mesas, Point Sal, sw San Luis Obispo, nw Santa Barbara cos.). Nov–Feb ★

A. sensitiva Jeps.   Erect, 1–2 m. **ST**: twig (and nascent infl axis) densely short-nonglandular- and -long-glandular-hairy. **LF**: spreading; petiole 1–3 mm; blade 1–2.2 cm, 0.8–1.8 cm wide, round to round-ovate, shiny, glabrous, midvein hairy, abaxially light green, adaxially dark green, base truncate to ± lobed, tip obtuse, margin entire, cupped; stomata abaxial. **INFL**: panicle, 4–8-branched; nascent infl pendent, axis 0.5–1 cm, > 1 mm wide; bracts appressed, 0.5–2 mm, scale-like, deltate, sharp-pointed, glabrous; pedicel 3–5 mm, glabrous. **FL**: fl parts in 4s; ovary white-nonglandular-hairy. **FR**: 3–4 mm wide, subcylindric, glabrous, mature splitting; stones free. $2n=26$. Rocky sites, chaparral, closed-cone conifer forest; < 600 m. w SnFrB (Bolinas Ridge, Mount Tamalpais, Santa Cruz Mtns). [*A. nummularia* var. *s.* (Jeps.) McMinn] Jan–Apr

A. silvicola Jeps. & Wiesl. (p. 701) BONNY DOON MANZANITA
Erect, 1–3(5) m. **ST:** twig (and nascent infl axis) sparsely to densely
soft-nonglandular-hairy. **LF:** erect; petiole 3–8 mm; blade 1.5–3.5
cm, 1–1.5 cm wide, narrowly obovate or oblong-elliptic, gray, dull,
appressed-canescent, in age glabrous, glaucous, base wedge-shaped,
tip obtuse, margin entire, flat. **INFL:** ± raceme, 0–1-branched;
nascent infl pendent, axis 0.5–1 cm, > 1 mm wide; bracts 5–12 mm,
lf-like, lanceolate, canescent to glabrous; pedicel 5–7 mm, sparsely
nonglandular-hairy or glabrous. **FL:** ovary glabrous. **FR:** 6–12 mm
wide, spheric, glabrous; stones free. 2*n*=26. Weathered sandstone
soils in chaparral, conifer forest; < 600 m. sw SnFrB (Santa Cruz
Sandhills, Santa Cruz Co.). Feb–Mar ★

A. stanfordiana Parry **LF:** erect; blade gen 3–5 cm, 1.5–2.5
cm wide, elliptic, oblong, or oblanceolate, glabrous, bright green to
± glaucous, ± shiny, glabrous (glandular-puberulent), base wedge-
shaped, tip abruptly soft-pointed, margin entire, flat. **INFL:** panicle,
3–5-branched; nascent infl erect to pendent, axis 2–2.5 cm, < 1
mm wide; bracts 1–1.5 mm, scale-like, ovate to deltate, acuminate, gen
glabrous; pedicel 5–10 mm, glabrous. **FL:** ovary glabrous. **FR:** 6–8
mm wide, depressed-spheric, glabrous; stones free. 2*n*=26.

subsp. ***decumbens*** (P.V. Wells) P.V. Wells RINCON RIDGE
MANZANITA Prostrate to mounded, 0.1–1 m. **ST:** twig sparsely
short-nonglandular-hairy. **LF:** glabrous, petiole 4–8 mm. **INFL:**
axes glabrous or sparsely short-nonglandular-hairy. Chaparral;
100 m. s NCoRO (near Healdsburg and Santa Rosa, Sonoma Co.).
Feb–Apr ★

subsp. ***raichei*** W. Knight (p. 701) RAICHE'S MANZANITA
Erect, 1–3 m;. **ST:** twig short-glandular-hairy. **LF:** papillate, ± rough,
petiole 4–8 mm. Chaparral; 400–945 m. s NCoRI (s Mendocino,
w Lake cos.). Feb–Apr ★

subsp. ***stanfordiana*** (p. 701) Erect, 1–3 m. **ST:** twig glabrous
(sparsely short-nonglandular-hairy). **LF:** glabrous, petiole 8–12 mm.
INFL: axes glabrous or sparsely short-nonglandular-hairy; bracts
appressed, ± = buds, short-glandular-hairy. Chaparral; 150–1000 m.
NCoRO, c&s NCoRI. Feb–Apr

A. tomentosa (Pursh) Lindl. Erect, 1–3 m; burl prominent. **ST:**
old sts gray, bark shredding. **LF:** spreading; petiole 2–5 mm; blade
2–5 cm, 1.5–2.5 cm wide, oblong-ovate to lance-oblong, abaxially
dull, adaxially dark to bright green, ± shiny, base truncate to ± lobed,
tip acute, margin entire, occ toothed, cupped or ± rolled; stomata
abaxial. **INFL:** panicle, 2–8-branched; nascent infl pendent, axis
1–2.5 cm, > 1 mm wide; bracts 8–15 mm, lf-like, lanceolate; pedicel
2–5 mm. **FL:** ovary densely white-nonglandular or -glandular-hairy.
FR: 6–10 mm wide, depressed-spheric, nonglandular or glandular-
hairy; stones free. 2*n*=52.

subsp. ***bracteosa*** (DC.) J.E. Adams (p. 701) **ST:** twig densely
glandular-bristly. **LF:** abaxially sparsely glandular-hairy, papillate,
scabrous or ± glabrous. **INFL:** densely glandular-bristly or tomentose. Chaparral, closed-cone conifer forest; 10–300 m. c CCo (Jacks
Peak and Fort Ord, Monterey Co.). Dec–Mar

subsp. ***daciticola*** P.V. Wells DACITE MANZANITA **ST:** twig
sparsely short-tomentose and long-white-nonglandular-hairy. **LF:**
abaxially tomentose. **INFL:** sparse- short- and long-white-nonglandular-hairy. Chaparral; 200–300 m. s CCo (w Los Osos Valley, San
Luis Obispo Co.). Dec–Mar ★

subsp. ***hebeclada*** (DC.) V.T. Parker et al. **ST:** twig sparsely
short-nonglandular-hairy. **LF:** abaxially glabrous. **INFL:** sparsely
short-nonglandular-hairy. Chaparral, closed-cone conifer forest;
100–300 m. c CCo (Jacks Peak, Monterey Peninsula). [*A. t.* var. *h.*
(DC.) McMinn] Dec–Mar

subsp. ***tomentosa*** (p. 701) **ST:** twig sparsely short-tomentose.
LF: abaxially tomentose. **INFL:** sparsely short-nonglandular-hairy.
Chaparral, closed-cone conifer forest; 10–500 m. c&s CCo (Mon-
terey, San Luis Obispo cos.). Dec–Mar

A. uva-ursi (L.) Spreng. (p. 705) BEAR-BERRY, KINNIKINNICK
Prostrate to mounded, 0.1–0.5 m; burled or not. **ST:** twig (and
nascent infl axis) sparsely short-nonglandular-hairy, occ long-
nonglandular- and/or short-glandular-hairy. **LF:** spreading; petiole
2–4 mm; blade 1–2.5 cm, 0.5–1.5 cm wide, oblanceolate to obovate,
occ narrowly elliptic, abaxially light green, shiny, sparsely puberulent, in age glabrous, adaxially dark green, base wedge-shaped, tip
obtuse, occ acute, margin entire, often cupped; stomata abaxial.
INFL: ± raceme, 0–1-branched; nascent infl pendent, axis 0.3–1
cm, > 1 mm wide; bracts 2–6 mm, scale-like, narrowly deltate, acuminate, glabrous; pedicel 2–4 mm, glabrous. **FL:** ovary glabrous.
FR: 6–12 mm wide, spheric, glabrous; stones free. 2*n*=26,52.
Rocky outcrops, slopes, stabilized dunes, closed-cone conifer forest, grassy coastal headlands, chaparral, subalpine forest; gen < 100
m (2400–3300 m in c SNH). NCo, c SNH (above Convict Lake,
Mono Co), CCo; to AK, Greenland, VA, CO, NM; also Guatemala,
circumboreal. Jan–Jun

A. virgata Eastw. MARIN MANZANITA Erect, 1–5 m. **ST:** twig
(and nascent infl axis) short- and long-glandular-hairy. **LF:** ascending, overlapped; petiole 2–4 mm; blade 3–5 cm, 1–2.5 cm wide,
narrowly oblong-ovate to lance-oblong, sparsely glandular-hairy,
papillate, bright green, ± shiny, base truncate to ± lobed, tip acute,
margin entire, flat. **INFL:** ± raceme, 0–1-branched; nascent infl
pendent, axis 1.5–2 cm, > 1 mm wide; bracts persistent, 8–20 mm,
lf-like, lance-linear, acuminate, long-glandular-hairy, green; pedicel
3–8 mm, glandular-hairy. **FL:** ovary glandular-hairy. **FR:** 6–8 mm
wide, depressed-spheric, glandular-hairy, sticky; stones free. 2*n*=26.
Sandstone, granite outcrops in chaparral, conifer forest; < 500 m. n
CCo, nw SnFrB (Marin Co.). Nascent infl possibly 0; needs study.
Dec–Feb ★

A. viridissima (Eastw.) McMinn (p. 705) WHITE-HAIRED
MANZANITA Erect, 1–4 m. **ST:** twig (and nascent infl axis) short-
and long-white-nonglandular-hairy. **LF:** overlapped; petiole 1–4 mm;
blade 2–3.5 cm, 1.5–2.5 cm wide, narrowly to oblong-ovate, puberulent, glabrous in age, bright green, shiny, base truncate to lobed, ±
clasping or not, tip abruptly soft-pointed, margin entire, flat. **INFL:**
± raceme, 0–1-branched; nascent infl pendent, axis 1–1.5 cm, > 1
mm wide; bracts 6–10 mm, lf-like, lanceolate, acute, ± nonglandular-hairy; pedicel 2–3 mm, nonglandular-hairy. **FL:** ovary densely
white-nonglandular-hairy. **FR:** 10–15 mm wide, ± spheric, sparsely
nonglandular-hairy to glabrous; stones free. 2*n*=26. Shale outcrops,
chaparral, closed-cone conifer forest; 100–550 m. n ChI (e Santa
Cruz Island). Jan–Mar ★

A. viscida Parry Erect, 1–3 m. **LF:** erect; petiole 5–12 mm; blade
2–5 cm, 2–4 cm wide, ovate to ± round, glabrous or densely glandular-hairy, papillate, scabrous, white-glaucous, dull, base rounded,
truncate or ± lobed, tip abruptly soft-pointed, margin entire, flat.
INFL: panicle, 4–7-branched; nascent infl erect to pendent, axis 1–3
cm, > 1 mm wide, densely glandular-sticky-hairy; bracts appressed,
3–4 mm, scale-like, deltate, acute to acuminate, glaucous, sparsely
short-nonglandular-hairy; pedicel 6–10 mm, glandular-hairy. **FR:**
6–8 mm wide, depressed-spheric; stones free. 2*n*=26.

subsp. ***mariposa*** (Dudley) P.V. Wells (p. 705) MARIPOSA
MANZANITA **ST:** twig densely glandular-hairy. **LF:** sparsely short-
glandular-hairy, ciliate, papillate, scabrous. **FL:** ovary (and fr)
glandular-bristly. Openings in chaparral, forest; 400–2200 m. SN.
Feb–Apr

subsp. ***pulchella*** (Howell) P.V. Wells **ST:** twig glabrous to
sparsely short-nonglandular-hairy. **LF:** glabrous. **FL:** ovary (and fr)
sticky-glandular, rough. Openings in chaparral, forest; 150–1400 m.
KR, NCoR; sw OR. Feb–Apr

subsp. ***viscida*** **ST:** twig glabrous to sparsely short-nonglandular-hairy. **LF:** glabrous, entire, smooth. **FL:** ovary (and fr) smooth,
glabrous. *n*=13. Openings in chaparral, forest; 30–2000 m. KR,
NCoRI, CaRF, SN; sw OR. Feb–Apr

CASSIOPE MOSS HEATHER

Shrub, small, glabrous to hairy. **ST**: decumbent or prostrate, often rooting. **LF**: opposite, appressed, evergreen, leathery or thin. **INFL**: fls 1 in upper lf axils; bracts 0; pedicel jointed to fl, bractlets 4–6. **FL**: sepals 4–5, free; petals 4–5, ± 2/3 fused, gen white; stamens 10, anthers dehiscing by gaping pores, awns elongate; ovary superior, chambers 5, placentas near top. **FR**: capsule, loculicidal. **SEED**: several per chamber. ± 14 spp.: s&e Asia, N.Am. (Greek: mother of Andromeda)

C. mertensiana (Bong.) G. Don (p. 705) WHITE HEATHER Pl low, densely branched. **ST**: < 3 dm, glabrous or finely hairy. **LF**: sessile, overlapping, 2–5 mm, boat-shaped, elliptic, concave, leathery, glabrous, abaxially not grooved; margin entire, ciliate, or minutely glandular, not rolled under. **INFL**: pedicel glabrous or hairy. **FL**: corolla widely bell-shaped, white, lobes 5; filaments glabrous. Moist, subalpine slopes, around rocks, areas of late snow; 1800–3505 m. KR, CaRH, SNH; to AK, w Can, MT. If recognized taxonomically, pls with lf 3–5 mm, margin minutely glandular-ciliate, in s CaRH (Lassen Peak), SNH, assignable to *C. mertensiana* subsp. *californica* Piper, pls with lf 2–3 mm, margin with white, ephemeral hairs, in KR, assignable to *C. mertensiana* subsp. *ciliolata* Piper. Jul–Aug

CHIMAPHILA PRINCE'S PINE, PIPSISSEWA

Gary D. Wallace & Erich Haber

Per, ± woody, evergreen, rhizomed. **LF**: cauline [basal], ± whorled, lanceolate to oblanceolate, leathery, gen prominently toothed, petioled. **INFL**: terminal, ± head- or umbel-like raceme; fls 1–10; peduncle gen papillate to glandular-hairy; bracts narrowly lanceolate to widely ovate. **FL**: radial, nodding, parts in 5s, free; petals spreading; stamens 10, filaments widened at base, ± hairy, anther pores on tubes; nectary present; ovary superior, style in depression, stout, stigma wide, peltate, lobes 5, ± flat, spreading. **FR**: capsule, erect; valves opening tip to base, margins not fibrous. 4–5 spp.: circumboreal, N.Am, C.Am, Eurasia. (Greek: winter loving, from evergreen habit)

1. Lf lanceolate to elliptic; infl 1–3-fld; sepals ± 5 mm, ovate; bracts widely ovate to obovate, persistent . . . . . . *C. menziesii*
1′ Lf oblanceolate; infl 3–10-fld; sepals < 2 mm, widely ovate; bracts narrowly lanceolate, deciduous . . . . *C. umbellata*

C. menziesii (D. Don) Spreng. (p. 705) LITTLE PRINCE'S PINE **ST**: < 15 cm, slender. **LF**: 1–several per node, gen 1–3(5) cm, toothed or entire; main veins ± white-bordered. **INFL**: ± glabrous to minutely papillate. **FL**: petals white, turning pink; filament base hairy. Uncommon. Montane conifer forest; 1000–2500 m. KR, NCoR, CaRH, n SNF, SNH, SCoRO, SnGb, SnBr, PR, MP; to BC, MT. Jun–Aug

C. umbellata (L.) W.P.C. Barton (p. 705) PRINCE'S PINE **ST**: < 30 cm, stout. **LF**: > several per node, gen 3–7 cm, toothed, esp toward tip; veins not bordered. **INFL**: ± densely glandular-hairy. **FL**: petals pink to red; filament base margins hairy. 2*n*=26. Common. Dry conifer forest; 300–2900 m. KR, NCoR, CaRH, SN, SnBr, SnJt, MP; to e N.Am, C.Am; Eurasia. Jun–Aug

COMAROSTAPHYLIS

Shrub, small tree, gen hairy to glandular, densely, rigidly branched, burled. **ST**: bark often shredding. **LF**: alternate, evergreen, leathery, entire or serrate. **INFL**: raceme or panicle, bracted; bractlets 2. **FL**: calyx lobes (4)5, > tube; corolla urn-shaped, petals (4)5, fused; stamens (8)10, anthers dehiscing by short separate slits, awned; ovary superior, chambers 4–6, placentas pendent, axile. **FR**: drupe, juicy, papillate, red or black; stones 4–6, fused. **SEED**: 1 per stone. 10 spp.: subtrop, trop Am. (Greek: arbutus grape cluster, from strawberry-tree-like frs)

C. diversifolia (Parry) Greene **ST**: erect, < 5 m; bark shredding; twigs gray-tomentose. **LF**: obovate, entire or serrate. **INFL**: raceme, gen gray-tomentose; bracts < 10 mm, lance-linear to oblong-ovate. **FL**: calyx lobes lanceolate or narrowly triangular. **FR**: red.

1. Lf margin rolled under; bracts < 3 mm, lance-linear
. . . . subsp. *diversifolia*

1′ Lf margin not or inconspicuously rolled under;
bracts gen 3–10 mm, oblong-ovate . . . . subsp. *planifolia*

subsp. *diversifolia* (p. 705) SUMMER HOLLY **INFL**: 3.5–8 cm. Chaparral; 100–550 m. SCo, PR; to n Baja CA. May–Jun ★

subsp. *planifolia* (Jeps.) G.D. Wallace (p. 705) **INFL**: 6–14 cm. Uncommon. Chaparral; 100–600 m. ChI (Santa Rosa, Santa Cruz, Santa Catalina islands), WTR. Mar–May

EMPETRUM

Kathleen A. Kron

Subshrub to low shrub, evergreen; gen dioecious. **ST**: spreading to decumbent, densely branching. **LF**: alternate or whorled, linear to ± oblong, stiff, strongly rolled under, apparent groove abaxially. **INFL**: axillary or terminal, fls 1 or in few-fld clusters, ± sessile. **FL**: unisexual [bisexual], reduced, perianth parts 3–6, ± free, dark purple-red, subtended by scaly bracts. **FR**: drupe, gen juicy. **SEED**: stones 2–9. 2–4 spp.: circumboreal, s Chile, Tierra del Fuego. (Greek: on rocks, from habitat) [Kron 1996 Ann Bot 77:293–304] Wind-pollinated.

E. nigrum L. (p. 705) BLACK CROWBERRY **ST**: gen decumbent, branches 15–40 cm. **LF**: crowded, 3–6 mm, glabrous exc along groove. **STAMINATE FL**: stamens (2)3(4). **PISTILLATE FL**: ovary superior, chambers 2–9, style deeply 2–9-lobed. **FR**: 4–6 mm wide, black or purple-black, occ red. Coastal cliffs on rocks; < 200 m. n NCo (Del Norte, Humboldt cos.); to AK; circumboreal. If recognized taxonomically, pls with bisexual fls assignable to *E. nigrum* var. *hermaphroditum* (Hagerup) T.J. Sørensen. Apr–May ★

ERICA HEATH

Shrub, glabrous to hairy. **ST**: gen erect. **LF**: alternate, opposite, or gen whorled, evergreen, leathery; margin entire or minutely dentate, gen rolled under, hiding abaxial surface. **INFL**: gen raceme, panicle, or umbel, gen bracted; pedicel not jointed to

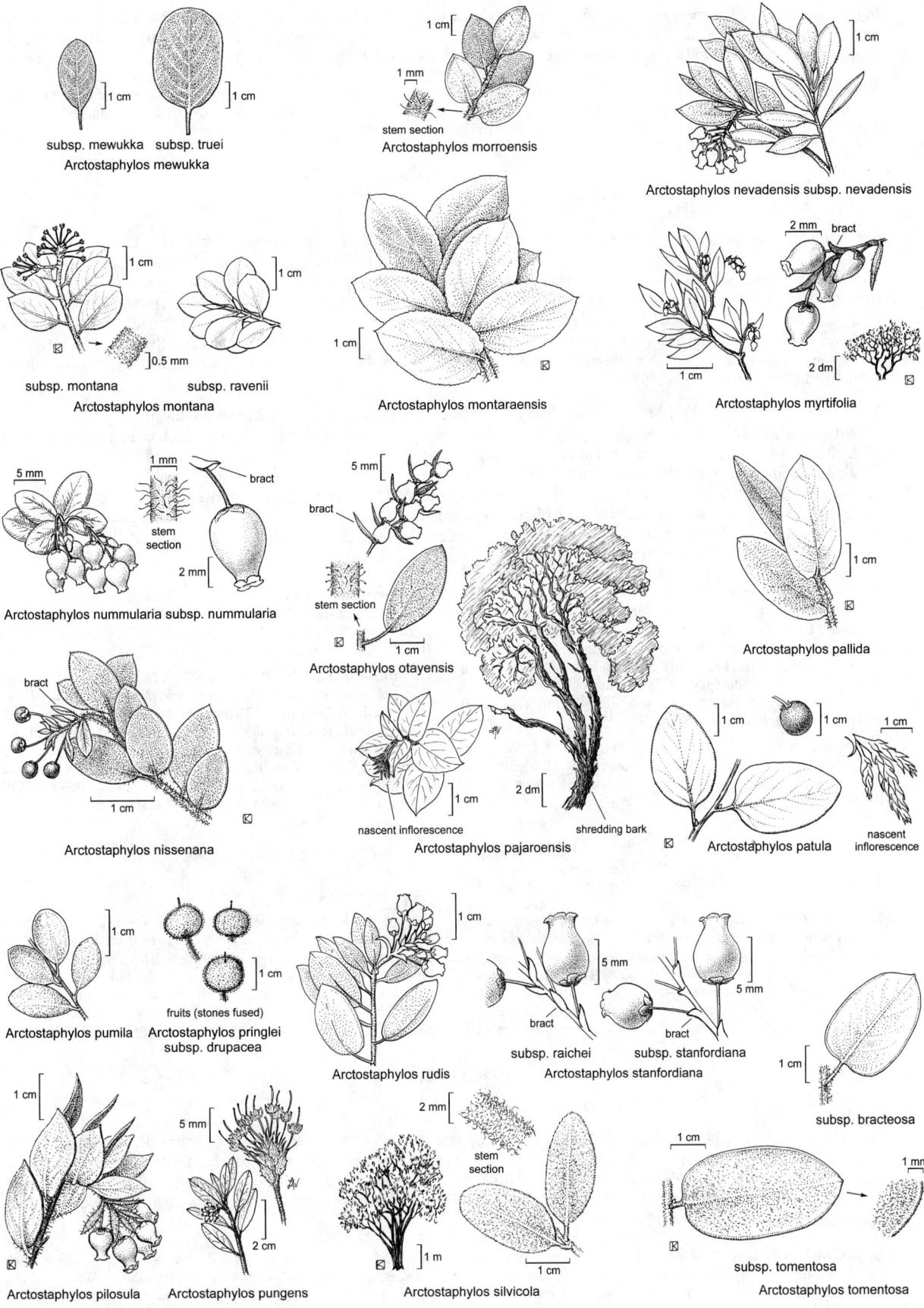

subsp. mewukka   subsp. truei
Arctostaphylos mewukka

Arctostaphylos morroensis

stem section

Arctostaphylos nevadensis subsp. nevadensis

subsp. montana   subsp. ravenii
Arctostaphylos montana

Arctostaphylos montaraensis

bract

Arctostaphylos myrtifolia

bract

stem
section

Arctostaphylos nummularia subsp. nummularia

bract

stem section

Arctostaphylos otayensis

Arctostaphylos pallida

bract

Arctostaphylos nissenana

nascent inflorescence

shredding bark

Arctostaphylos pajaroensis

Arctostaphylos patula

nascent
inflorescence

Arctostaphylos pumila   fruits (stones fused)
Arctostaphylos pringlei
subsp. drupacea

Arctostaphylos rudis

bract

subsp. raichei   subsp. stanfordiana
Arctostaphylos stanfordiana

bract

subsp. bracteosa

Arctostaphylos pilosula   Arctostaphylos pungens

stem
section

Arctostaphylos silvicola

subsp. tomentosa
Arctostaphylos tomentosa

fl, bractlets gen 3. **FL**: sepals gen 4, gen free; corolla spheric to tubular, petals gen 4, fused, gen persistent in fr; stamens gen 8, anthers gen 2-lobed, dehiscing gen by pores or separate slits, gen awned; nectary gen disk-like; ovary superior, chambers 4(8). **FR**: capsule, loculicidal, dehiscing tip to base. **SEED**: gen several per chamber. ± 630 spp.: s&e Afr, Eur. (Latin: heath)

E. lusitanica Rudolphi **ST**: < 2 m, branchlets short, hairs dense. **LF**: < 1 cm, gen < 1 mm wide, needle-like, bright green. **INFL**: fls 1 in lf axils. **FL**: sepals fused in basal 1/3, petal-like; petals white to ± pink. **FR**: > 4 mm. Disturbed, open, sandy areas; < 50 m. NCo (Humboldt Co.), expected CCo; native to sw Eur. Mar–Jul

GAULTHERIA

Walter S. Judd

Shrub, glabrous or short- or long-hairy, glandular or not, often smelling of wintergreen, gen rhizomed. **ST**: prostrate to erect, rooting at nodes or not. **LF**: alternate, evergreen, leathery, entire to serrate. **INFL**: raceme, each fl with 1 bract, 2 bractlets, or fls 1 in lf axils, each with 4–10 bractlets; pedicel jointed to fl. **FL**: sepals gen 5, fused; petals 5, fused, cylindric, urn-, or bell-shaped, white to red; stamens (5,8)10, anthers dehiscing by 2 short, slit-like to rounded pores, awns (0,2)4, sometimes reduced; ovary superior or 1/2-inferior, chambers (4)5, placentas axile, at top. **FR**: capsule, loculicidal or irregularly dehiscing, gen ± enclosed by fleshy, colorful calyx (or a berry and/or with non-fleshy calyx). **SEED**: few to many per chamber, appendages 0. ± 130 spp.: circum-Pacific, e N.Am, e Brazil, Himalayas. (J.F. Gaulthier, botanist, physician, Quebec, 1708–1756) [Middleton 1991 Bot J Linn Soc 106:229–258]

1. Lf (3)5–13 cm; st gen 20–200 cm, coarse; fls in axillary racemes, each with 1 bract, 2 bractlets; corolla urn-shaped, glandular-hairy; filaments puberulent, anther awns 4; calyx dark purple to ± blue-black in fr . . . . . ***G. shallon***
1′ Lf 0.5–3.5(4.5) cm; st < 35 cm, slender; fls 1 in lf axils, each with 4–10 bractlets; corolla bell-shaped, glabrous; filaments papillate, anther awns 0; calyx red in fr
  2. Sepals glabrous; st glabrous to puberulent, long glandular hairs 0 to moderate; lf minutely serrate or ± entire (or appearing so), esp near base, tip obtuse to rounded. . . . . ***G. humifusa***
  2′ Sepals glandular-hairy; st puberulent and moderate- to densely long-hairy, glandular or not; lf minutely serrate, tip short-acuminate or acute to obtuse . . . . ***G. ovatifolia***

G. humifusa (Graham) Rydb. (p. 705) ALPINE WINTERGREEN **ST**: low, < 2 dm, glabrous to puberulent, long, glandular hairs 0 to moderate. **LF**: 0.5–3 cm, minutely serrate or ± entire (or appearing so), esp near base; base rounded to truncate; tip obtuse to rounded. **INFL**: fls 1 in lf axils. **FL**: sepals glabrous, in fr red; corolla bell-shaped, glabrous, white (pink-tinged); filaments papillate, anther awns 0. **FR**: red. Wet subalpine forests; 1350–4000 m. KR, CaRH, SNH; to BC, CO. Jun–Aug

G. ovatifolia A. Gray SLENDER WINTERGREEN **ST**: low, < 3.5 dm; puberulent and moderate- to densely long-hairy, glandular or not. **LF**: 1–3.5(4.5) cm, minutely serrate; base rounded to truncate or ± cordate; tip short-acuminate or acute to obtuse. **INFL**: fls 1 in lf axils.

FL: sepals glandular-hairy, in fr red; corolla bell-shaped, glabrous, white (pink-tinged); filaments papillate, anther awns 0. **FR**: red. Wet fir forests; 400–1900 m. KR, NCoRO, n NCoRH, CaRH, n SNH; to BC, ID. Jun–Jul

G. shallon Pursh (p. 705) SALAL **ST**: erect, gen 2–20 dm, clumped, nonglandular-hairy or not, sparsely to moderately long-glandular-hairy. **LF**: (3)5–13 cm, minutely serrate; base obtuse to cordate; tip short-acuminate to acuminate. **INFL**: raceme, axillary, glandular-hairy. **FL**: sepals glandular-hairy; corolla urn-shaped, glandular-hairy, white to pink; filaments puberulent, anther awns 4. **FR**: dark purple to ± blue-black. Moist forest margins; < 1060 m. NCo, KR, NCoRO, CCo, SnFrB, s SCoRO; to AK. Apr–Jul

HEMITOMES GNOME PLANT

1 sp. (Greek: 1/2 eunuch, from 1 anther sac thought sterile)

H. congestum A. Gray (p. 705) Per, non-green, fleshy, gen glabrous, rhizomed; roots brittle. **ST**: 0. **LF**: 0. **INFL**: gen dense cyme, raceme, or fls 1; 2–10 cm, gen pink, cream, emerging from ground erect, not persisting after seed dispersal; bracts < 2 cm, margins ciliate; bractlets gen 0(1–2). **FL**: sepals (2)4, free, lateral 2 often folded, clasping corolla, other 2 flat if present; corolla cylindric to flask-shaped, inside densely hairy, petals 4(5), ± 2/3 fused, cream or gen pink; stamens gen 8, filaments densely hairy, anthers 1–2 mm, elongate, dehiscing by separate longitudinal slits, unawned; nectary 8–10-lobed; ovary superior, chamber 1 or appearing > 1 by intrusion of parietal placentas, style < 5 mm, hairy, stigma 1.5–2.5 mm wide, disk-like, yellow, subtended by dense hairs. **FR**: berry, < 1 cm. **SEED**: many, ovate. Uncommon. Mixed or conifer forests; 30–2700 m. NCo, KR, NCoRO, CaRH, n&s SNH, CCo, SnFrB, SCoRO; to BC. May–Jul

KALMIA

Kathleen A. Kron

Shrub [small tree]; evergreen. **ST**: prostrate to erect. **LF**: opposite [alternate, whorled], margins flat or rolled under. **INFL**: fls 1 in lf axils [raceme], pedicel jointed to fl. **FL**: sepals 5, fused near base; petals 5, fused, corolla gen rotate to cup-shaped with pockets holding anthers until dehiscence; stamens 10, filaments recurved toward corolla; ovary superior, chambers 5, placenta axile. **FR**: septicidal capsule, dehiscing tip to base, valves 5. **SEED**: small. 9–10 spp.: N.Am, Cuba. (P. Kalm, student of Linnaeus, collector of e N.Am pls, 1716–1779)

K. polifolia Wangenh. (p. 705) SWAMP LAUREL, BOX LAUREL Pl 1–7 dm, mat-forming. **ST**: ascending, glabrous or sparsely hairy, young sts with 2 edges. **LF**: 4–60 mm, 3–25 mm wide, linear to ovate or oblong, abaxially pale green to white, finely canescent. **INFL**: bracts deciduous. **FL**: corolla 7–11 mm, pink to rose-purple. **FR**: 4–5(7) mm wide. $2n=24,48$. Peat bogs, moist meadows, rock crevices; 1000–3500 m. KR, NCoRO, CaRH, SNH, Wrn; n N.Am. [*K. microphylla* (Hook.) A. Heller; *K. p.* subsp. *m.* (Hook.) Calder & Roy L. Taylor] 2 subspp. previously recognized evidently based on variation caused directly by altitude. Jun–Aug

LEUCOTHOE

Walter S. Judd

Shrub, small tree, coarsely branched, non- and/or glandular-hairy. **ST**: erect. **LF**: alternate, evergreen, serrate. **INFL**: raceme, in fall, bracted; pedicel jointed to fl, bractlets 2, ± basal. **FL**: sepals 5, fused at base; petals 5, fused, [cylindric to] urn-shaped; stamens 10, anthers dehiscing by pores, awns 4, [elongate to] vestigial, filaments ± straight, with hairs, papillae [or only papillae]; ovary superior, chambers 5, placentas axile, stigma truncate [or head-like/peltate]. **FR**: capsule, loculicidal, erect, persistent after seed dispersal, dehiscing tip to base. **SEED**: many per chamber, small, gen winged. 5 spp.: US, Asia. (Greek: name for a princess of Babylon) [Waselkov & Judd 2008 Brittonia 60:382–397]

L. davisiae A. Gray (p. 705)   SIERRA LAUREL   **ST**: < 1.5 m. **LF**: 1.5–8 cm, oblong to elliptic or ovate, leathery. **INFL**: in upper lf axils, < 15 cm, many-fld; pedicel recurved. **FL**: corolla 5.5–8.5 mm, >> calyx, white. **FR**: < 6 mm wide, thin-walled. Uncommon. Bogs, wet areas; 400–3000 m. KR, CaRH, SNH, Wrn; OR. Jun–Aug

MENZIESIA

Kathleen A. Kron

Shrub, branches straggling. **ST**: bark finely shredding. **LF**: alternate, often crowded toward branch tips, deciduous, papery, elliptic, flat-brown-hairy on midrib abaxially, margin rolled under in bud. **INFL**: terminal, umbel-like; pedicel spreading-downcurved, bud scales light brown, early-deciduous. **FL**: ± bilateral, sepals 4[5], fused 3/4; petals 4[5], fused; stamens 5[8,10], anthers dehiscing by short slits, unawned; ovary superior, chambers 4[5], placentas axile. **FR**: capsule, septicidal, dehiscing tip to base. **SEED**: many, small, fusiform. 4–5 spp.: temp Asia, Am. (A. Menzies, naturalist on Vancouver expedition, 1754–1842)

M. ferruginea Sm. (p. 707)   MOCK AZALEA   Herbage glabrous to flat-brown-hairy and/or glandular-hairy. **ST**: sprawling to erect, 1–4 m. **LF**: 1.5–4 cm, tip mucronate, margins glandular-ciliate. **INFL**: terminal on last yrs shoots, occ appearing axillary; pedicel straightening in fr. **FL**: corolla 6–7(13) mm, urn-shaped to ± spheric, ± yellow to orange-bronze, lobes short. **FR**: < 8 mm wide, valves 4. **SEED**: many, elongate, often with short tails, coat loose. Moist woodland, gen acidic or peaty soils; < 300 m. NCo, KR, NCoRO; to AK, MT, WY; AL. Jun–Jul

MONESES

Gary D. Wallace & Erich Haber

1 sp. (Greek: single delight, from single fl)

M. uniflora (L.) A. Gray (p. 707)   WOODNYMPH   Per < 10 cm, evergreen, rhizomes 0, roots rhizome-like. **LF**: ± basal, < 3.5 cm, ovate to obovate, not leathery, finely crenate or sharp-toothed, petioled. **INFL**: fl 1; peduncle minutely papillate above. **FL**: radial, nodding, parts in 5s, free; sepals fringed; petals < 1 cm, spreading, entire or minutely fringed, waxy-white to ± pink; stamens 10, filaments ± widened at base, glabrous, anther pores on tubes; nectary 0; ovary superior, style not or in slight depression, straight, slender, stigma crown-like, lobes 5, prominent. **FR**: capsule, erect; valves opening tip to base, margins not fibrous. 2*n*=26. Moist, mossy conifer forests; 100–1000 m. n NCo, w KR, n NCoRO, CaRH, c SNH (Fresno Co.); N.Am, Eurasia, circumboreal. May–Jul   ★

MONOTROPA

Per, non-green, glabrous to glandular-hairy; roots brittle, main often elongate. **ST**: 0. **LF**: 0. **INFL**: raceme-like or fls 1; emerging from ground nodding, erect in fr, persistent after seed dispersal, bracted; pedicel gen recurved to spreading when anthers open, erect in fr, jointed to fl, bractlets 1–2. **FL**: sepals gen 5, free; petals gen 5, free, oblong-cup-shaped, ± bulged at base; stamens gen 10, incl, anthers dehiscing by 1 or 2 slits, unawned; nectary lobes (8)10, ± clasping stamen bases; ovary superior, lines of dehiscence evident, chambers (4)5, placentas axile. **FR**: capsule, loculicidal, erect, dehiscing tip to base. **SEED**: many per chamber, fusiform. 2 spp.: n hemisphere. (Greek: 1 direction, from 1-sided infl)

1. Infl raceme-like, fls rarely 1; pl ± yellow to pink or red; stigma shallowly depressed, often subtended by ring
   of bristly hairs; nectary lobes short, stout; fr segments thin-walled, often irregularly deciduous . . . . ***M. hypopitys***
1′ Fls 1; pl white (pink or red-orange); stigma widely funnel-like, not subtended by hairs; nectary lobes
   elongate, slender; fr segments thick-walled, persistent . . . . ***M. uniflora***

M. hypopitys L. (p. 707)   PINESAP   **FL**: sepals ± unlike petals; style 1–2 mm wide. Uncommon. Mixed or conifer forest; 120–2200 m. NW (exc NCoRI), CaRH; to AK, e US and adjacent Can; also Mex, Eurasia. Jul–Aug

M. uniflora L.   GHOST-PIPE   **FL**: sepals ± like petals; style 2–5 mm wide. Low mixed or conifer forest; < 200 m. NCo, KR, NCoRO; to BC, e N.Am; also C.Am, n S.Am, e Asia. Jun–Jul   ★

ORNITHOSTAPHYLOS

1 sp. (Greek: bird grape cluster, for obscure reasons)

O. oppositifolia (Parry) Small (p. 707)   BAJA CALIFORNIA BIRDBUSH   Shrub, rigidly branched, glabrous to hairy, burled. **ST**: erect, < 2 m; bark thin. **LF**: opposite or whorled, 2.5–8 cm, 3–6 mm wide, ± linear, evergreen, leathery; margins rolled under. **INFL**: panicle, bracted; pedicel not jointed to fl, bractlets 2. **FL**: sepals (4)5, ± 1/2 fused; corolla ± spheric to urn-shaped, petals (4)5, ± 2/3

fused, white; stamens gen 10, anthers dehiscing by separate gaping slits, awned; ovary superior, chambers 5, placentas axile. **FR:** drupe; stones 5, fused. **SEED:** gen 2 per stone. Chaparral; 100–800 m. SCo (San Diego Co.); n Baja CA. Part of known population transplanted away from construction of US/Mex border fence. Jan–Apr ★

ORTHILIA

Gary D. Wallace & Erich Haber

1 sp. (Greek: straight spiral, from 1-sided raceme) Once placed in *Pyrola*.

O. secunda (L.) House (p. 707) ONE-SIDED WINTERGREEN · Per, ± shrubby or not, < 20 cm, evergreen, rhizomed. **LF:** ± cauline, gen near base, 1.5–6 cm, ovate-elliptic, leathery or not, entire to finely crenate, petioled. **INFL:** raceme, elongate, 1-sided, arched, ± erect in fr; peduncle densely papillate; bracts several, gen lanceolate. **FL:** radial, ± closed, parts in 5s, free; petals with 2 basal tubercles adaxially, ± green to cream-white; stamens 10, filaments ± narrow throughout, glabrous, anther pores not on tubes; nectary present; ovary superior, style straight, exserted, stigma peltate, lobes 5, shallow, domed. **FR:** capsule, pendent; valves opening base to tip, margins fibrous. $2n=38$. Dry, shady, conifer forests; 1000–3200 m. KR, NCoRH, NCoRO, s CaRH, SNH, SnBr, SnJt, MP, SNE (exc W&I); circumboreal, subarctic, N.Am, C.Am, Eurasia. Jul–Sep

PHYLLODOCE MOUNTAIN HEATHER

Shrub, gen matted, glabrous to glandular, gen rhizomed. **ST:** decumbent, rooting, rough from persistent, decurrent lf bases. **LF:** alternate, crowded, gen < 5 mm wide, needle-like, evergreen, leathery, abaxially channeled; margin strongly rolled under, abaxial surface < 1/3 visible. **INFL:** raceme, bracts lf-like; pedicel jointed to fl, bractlets 2. **FL:** sepals 5, fused at base; corolla [urn- or] cup-shaped, petals 5, ± 4/5 fused; stamens gen 10, anthers elongate, dehiscing by short separate slits, unawned; ovary superior, chambers 5, placentas axile. **FR:** capsule, septicidal, dehiscing tip to base. **SEED:** many per chamber, narrowly winged. 4–7 spp.: circumboreal. (Greek: sea nymph; possibly lf similar, from resemblance to *Erica*)

1. Stamens exserted; sepals, corolla lobes ± = corolla tube . . . . ***P. breweri***
1′ Stamens incl; sepals, corolla lobes < corolla tube . . . . ***P. empetriformis***

P. breweri (A. Gray) Maxim. (p. 707) **FL:** sepals < 4.5 mm; corolla pink to rose-purple, tube < 4.5 mm, lobes < 3.5 mm; filaments > 2 × anthers. Moist rocky slopes, meadows, subalpine; 1200–3500 m. s CaRH (Lassen Peak and nearby), SNH, SnBr; NV. Jul–Aug

P. empetriformis (Sm.) D. Don (p. 707) **FL:** sepals < 2.5 mm; corolla pink to rose-purple, tube < 5 mm, lobes < 2 mm; filaments < 2 × anthers. $2n=24$. Moist slopes, meadows, subalpine; 1450–2650 m. KR, n CaRH (Mount Shasta); to AK, MT, WY. Jul–Aug

PITYOPUS

1 sp. (Greek: pine foot, from habitat)

P. californicus (Eastw.) H.F. Copel. (p. 707) CALIFORNIA PINEFOOT · Per, non-green, fleshy; roots brittle. **ST:** 0. **LF:** 0. **INFL:** raceme or fls 1, 1–10 cm, cream to ± yellow, emerging from ground erect, not persistent after seed dispersal, bracted; bractlets 0. **FL:** sepals 4(5), free, lateral 2 often folded, clasping corolla, others flat against corolla; corolla cylindric, outside ± glabrous, inside densely hairy, petals 4(5), free, cream to ± yellow; stamens gen 8, anthers erect, horseshoe-shaped, dehiscing by 1 unified slit, unawned; nectary lobes 8–10, among stamen bases; ovary superior, chamber 1 or appearing > 1 by intrusion of parietal placentas, style < 5 mm, stigma < 5 mm wide, ± funnel-shaped, ± yellow, subtended by ring of hairs. **FR:** berry, < 1 cm. **SEED:** many, ovate. Mixed or conifer forest; < 1800 m. NCo, KR, NCoRO, s SNH, CCo, SnFrB; to WA. May–Jul ★

PLEURICOSPORA

1 sp. (Greek: seeds at side, from parietal placentas)

P. fimbriolata A. Gray (p. 707) Per, non-green, fleshy, glabrous; roots brittle. **ST:** 0. **LF:** 0. **INFL:** raceme, gen 6–10 cm, yellow-cream, emerging from ground erect, not persistent after seed dispersal, bracted, drying brown or black at tips; bractlets 0. **FL:** sepals 4, free, 5–8 mm; corolla cylindric, glabrous, petals 4(5), free, 8–10 mm, yellow-cream, margins jagged; stamens gen 8, anthers 3–4 mm, elongate, dehiscing by long, separate lateral slits, unawned; ovary superior, chamber 1, placentas parietal, style < 3 mm, stigma 1.5–2.5 mm wide, not subtended by hairs. **FR:** berry, < 1 cm wide, cream to ± white. **SEED:** many, ovate. $2n=52$. Mixed or conifer forest; 150–2800 m. NCo, KR, NCoRO, CaRH, SNH, SnFrB; to BC. Jun–Aug

PTEROSPORA PINEDROPS

1 sp. (Greek: winged seed)

P. andromedea Nutt. (p. 707) Per, non-green, ± pink, brown in age, densely sticky-glandular; roots brittle. **ST:** 0. **LF:** 0. **INFL:** raceme, 1.5–17 dm, pink to ± red, emerging from ground erect, persistent after seed dispersal, bracted, axis < 1.5 cm wide just below lowest fl; bractlets 0. **FL:** pendent; sepals 5, free; corolla urn-shaped, petals 5, ± 4/5 fused, 6–9 mm, cream to ± yellow, lobes recurved; stamens 10, anthers dehiscing by separate slits, awned; ovary superior, chambers 5, placentas axile, style < 3 mm, jointed to ovary, stigma gen 1.5 mm wide, disk-like. **FR:** capsule, pendent, < 1.3 cm wide, loculicidal, dehiscing base to tip. **SEED:** many per chamber, < 0.2 mm wide, ovate; wing terminal, < 1 mm wide, membranous. Mixed or conifer forest; 60–3700 m. KR, NCoRH, CaR, SNH, Teh, TR, PR, MP, SNE (exc W&I); to BC, Mex; also e N.Am. Jun–Aug

705

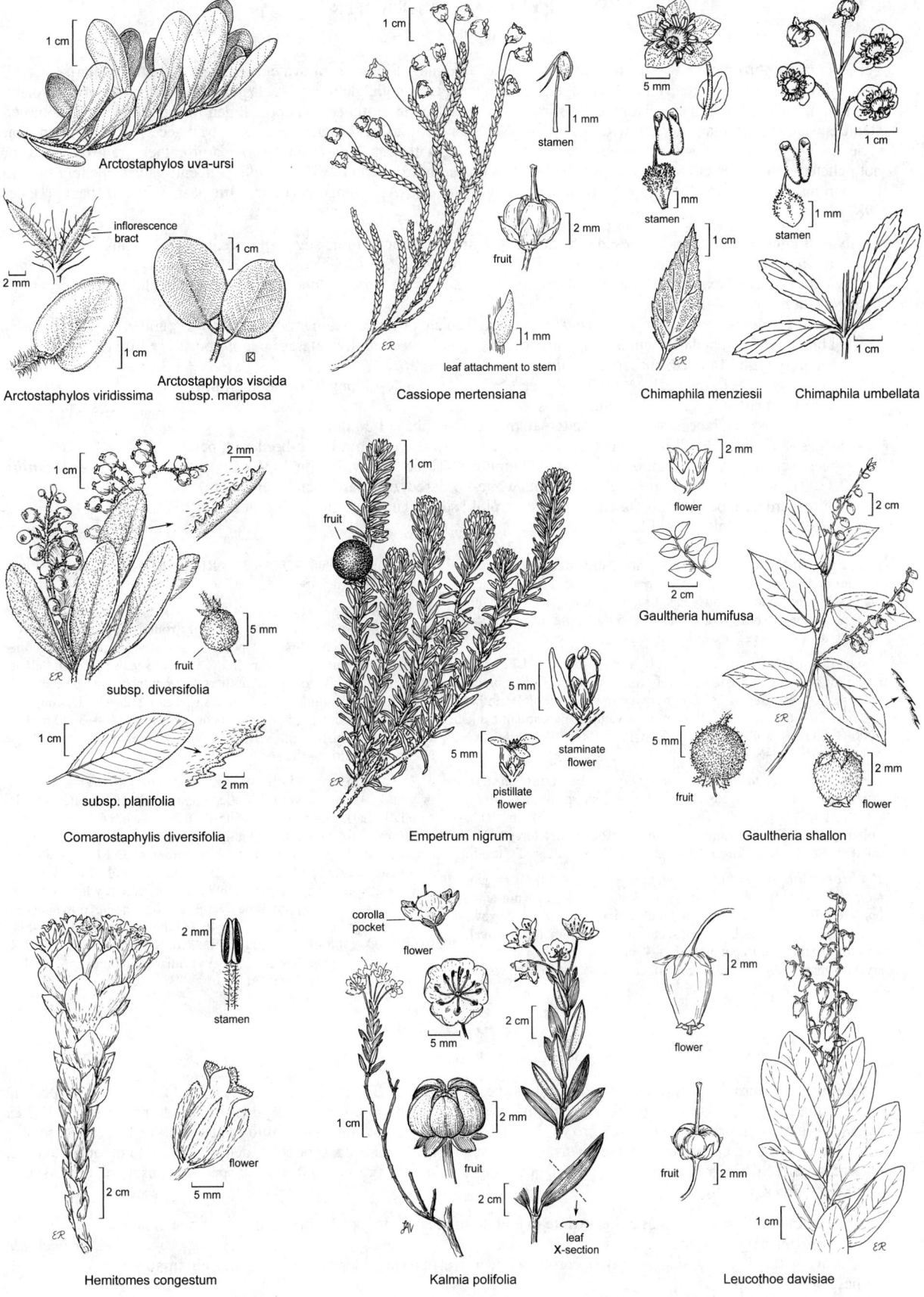

Arctostaphylos uva-ursi

1 cm

inflorescence bract

2 mm

Arctostaphylos viridissima

1 cm

Arctostaphylos viscida subsp. mariposa

1 cm

1 cm

Cassiope mertensiana

1 cm

stamen

1 mm

fruit

2 mm

leaf attachment to stem

1 mm

5 mm

mm

stamen

1 cm

Chimaphila menziesii

1 cm

stamen

1 mm

stamen

1 cm

Chimaphila umbellata

1 cm

1 cm

2 mm

subsp. diversifolia

5 mm

fruit

subsp. planifolia

1 cm

2 mm

Comarostaphylis diversifolia

fruit

1 cm

5 mm

staminate flower

5 mm

pistillate flower

Empetrum nigrum

flower

2 mm

2 cm

2 cm

Gaultheria humifusa

5 mm

fruit

flower

2 mm

Gaultheria shallon

2 mm

stamen

flower

5 mm

2 cm

Hemitomes congestum

corolla pocket

flower

5 mm

1 cm

fruit

2 cm

2 cm

2 cm

leaf X-section

Kalmia polifolia

flower

2 mm

fruit

2 mm

1 cm

Leucothoe davisiae

PYROLA WINTERGREEN

Gary D. Wallace & Erich Haber

Per, evergreen, rhizomed. **LF**: ± basal, reniform, ovate, ± round, elliptic, or obovate, ± entire to round- or sharp-toothed, petioled. **INFL**: raceme, ± erect, not 1-sided, elongate; peduncle smooth, glabrous; bracts gen 1–several, ovate or lanceolate. **FL**: radial, ± closed or bilateral, ± open, parts in 5s, free; petals without tubercles, upper 2 gen forming hood over upturned stamens; stamens 10, filaments gen widened at base, smooth, glabrous, anther pores gen on tubes; nectary 0; ovary superior, style straight, ± incl, or downcurved, exserted, stigma peltate, with 5 spreading lobes above a prominent, reflexed collar or not peltate, with 5 ± erect lobes projecting beyond a delicate, reflexed collar. **FR**: capsule, pendent; valves opening base to tip, margins fibrous. ± 15–20 spp.: gen circumboreal, high mtns of C.Am, Sumatra. (Latin: little pear, ± from lf shape) [Haber 1987 Syst Bot 12:324–335]

1. Style straight, ± incl; fl ± radial, ± closed; anthers < 1 mm, pore not on tube; stigma peltate, lobes spreading
(subg. *Amelia*) . . . . *P. minor*
1′ Style downcurved, exserted; fl bilateral, ± open; anthers 2–5.5 mm, pore on tube; stigma not peltate, lobes ± erect (subg. *Pyrola*)
  2. Fl bract gen >> pedicel, ovate; sepals ovate to gen lanceolate or lance-oblong; pore tube < ± 1/5 anther. . . *P. asarifolia*
    3. Fl bract < ± 1.5 × pedicel; sepals 2–3.5 mm, ovate to gen lanceolate; lf ovate, round, or obovate, entire to obscurely round-toothed; anthers 2–3 mm . . . . subsp. *asarifolia*
    3′ Fl bract gen >> 2 × pedicel; sepals 3–5.8 (gen >> 3.5) mm, lance-oblong; lf ovate to elliptic, gen sharp-toothed; anthers 2.5–3.5 mm. . . . . subsp. *bracteata*
  2′ Fl bract < pedicel, lanceolate; sepals deltate to ovate; pore tube ± 1/3 anther
    4. Lf gen < 4 cm, ovate-elliptic, obovate, or reduced, bract-like, veins not white-bordered; petals pale green; sepals gen < 1.8 mm, deltate-ovate, ± blunt; n SNH (near Downieville, Sierra Co.) . . . . *P. chlorantha*
    4′ Lf ± 0 or gen 4–10 cm, ovate to elliptic, veins white-bordered, or oblanceolate, veins not white-bordered; petals ± green, cream-white, or pink; sepals gen >> 2 mm, ovate, acute; NW, CaR, SN, CCo, SnFrB, SCoRO, TR, PR, MP . . . . *P. picta*

P. asarifolia Michx. **LF**: < 10 cm, abaxially often purple. **INFL**: < 6 dm incl peduncle; bract gen >> pedicel, ovate. **FL**: bilateral, ± open; sepals ovate to gen lanceolate or lance-oblong, acute to acuminate; petals pink to deep red; anthers 2–3.5 mm, pore tube < ± 1/5 as long; style downcurved, exserted, stigma lobes ± erect.

  subsp. ***asarifolia*** (p. 711) BOG WINTERGREEN **LF**: ovate, round, or obovate, entire to obscurely round-toothed. **INFL**: bract < 1.5 × pedicel. **FL**: sepals 2–3.5 mm, ovate to gen lanceolate; anthers 2–3 mm. 2*n*=46. Common. Moist forest, swamps, bogs, streambanks; 1000–3000 m. KR, CaRH, SNH, TR, MP, SNE (exc W&I); to AK, e N.Am; e Asia. Jul–Sep

  subsp. ***bracteata*** (Hook.) Haber (p. 711) LONG-BRACTED WINTERGREEN **LF**: ovate to elliptic, gen sharp-toothed. **INFL**: bract gen >> 2 × pedicel. **FL**: sepals 3–5.8 (gen >> 3.5) mm, lance-oblong; anthers 2.5–3.5 mm. Uncommon. Moist to dry forests; 100–2000 m. NCo, KR, NCoRO, NCoRH, n SNH; to s AK, MT. Jun–Jul

P. chlorantha Sw. GREEN-FLOWERED WINTERGREEN **LF**: gen < 4 cm, ovate-elliptic, obovate, or reduced, bract-like; veins not white-bordered. **INFL**: < 3 dm (incl peduncle); bracts < pedicel, narrowly lanceolate. **FL**: bilateral, ± open; sepals gen < 1.8 mm, deltate-ovate, ± blunt; petals pale green; anthers 2–4 mm, pore tube ± 1/3 as long; style downcurved, exserted, stigma lobes ± erect. 2*n*=46. Gen conifer,

mixed forest; ± 900–2200 m. NCoRH, n SNH; circumboreal, N.Am, Eurasia. Jul–Aug ★

P. minor L. (p. 711) LESSER WINTERGREEN **LF**: < 5 cm, ovate to oblong-obovate, entire to obscurely round-toothed. **INFL**: < 2 dm (incl peduncle); bract >> pedicel, lance-ovate, gen larger, wider at base of peduncle. **FL**: ± radial, ± closed; sepals 1.5 mm, deltate, acute; petals white to ± pink; anthers < 1 mm, pore not on tube; style straight, ± incl, stigma peltate, lobes spreading. 2*n*=46. Uncommon. Moist, mossy sites, high montane conifer forest; 2400–3000 m. KR, SNH, SnBr, SnJt, Wrn, SNE (exc W&I); circumboreal, to AK, e N.Am, also Eurasia. Jul–Aug

P. picta Sm. (p. 711) WHITE-VEINED WINTERGREEN **LF**: ± 0 or gen 4–10 cm, ovate to elliptic, ± entire, abaxially often purple, adaxially dark green, veins white-bordered, or oblanceolate, entire to prominently toothed, abaxially often ± blue-waxy when young, adaxially dull green, veins not white-bordered. **INFL**: < 4 dm incl peduncle; bract < pedicel, lanceolate. **FL**: bilateral, ± open; sepals gen >> 2 mm, ovate, acute; petals ± green, cream-white, or pink; anthers 2–5.5 mm, pore tube ± 1/3 as long; style downcurved, exserted, stigma lobes ± erect. 2*n*=46. Common. Dry ponderosa-pine forest; 400–2400 m. NW, CaR, SN, CCo, SnFrB, SCoRO, TR, PR, MP; to sw Can, NM. Variable, may hybridize with other spp. (pollen, seeds sometimes abortive). Jun–Aug

RHODODENDRON

Kathleen A. Kron & Walter S. Judd

Shrub to tree, glabrous, hairy, and/or with glandular scale-like hairs. **ST**: prostrate to erect, bark thin, sometimes peeling or shredding. **LF**: alternate, evergreen or deciduous, ovate to obovate to elliptic, margin entire, flat to rolled under. **INFL**: raceme, 1–many-fld, bracts green to red-brown. **FL**: sepals, petals gen 5, corolla radial to bilateral, 1–5 cm, petals free to ± fused, spots or blotch present or 0; anther awns 0; ovary superior. **FR**: capsule, septicidal, dehiscing base to tip or tip to base, placentas axile. **SEED**: many, fusiform, ± flat to not, wings and/or tails present or 0. ± 1000 spp.: n hemisphere, trop Asia, to Australia. (Greek: rose tree)

1. Lvs deciduous, thin; stamens 5; corolla white to pink to salmon, upper petal with yellow to orange blotch at base; fr dehiscing tip to base . . . . *R. occidentale*
1′ Lvs evergreen, leathery; stamens (8)10; corolla white to cream or to pink or purple; fr dehiscing tip to base or base to tip
  2. Pl with ± flat glandular scales at least abaxially on lf; petals white to cream; fr dehiscing base to tip . . . . *R. columbianum*
  2′ Pl without ± flat glandular scales; petals white to pink or purple; fr dehiscing tip to base . . . . *R. macrophyllum*

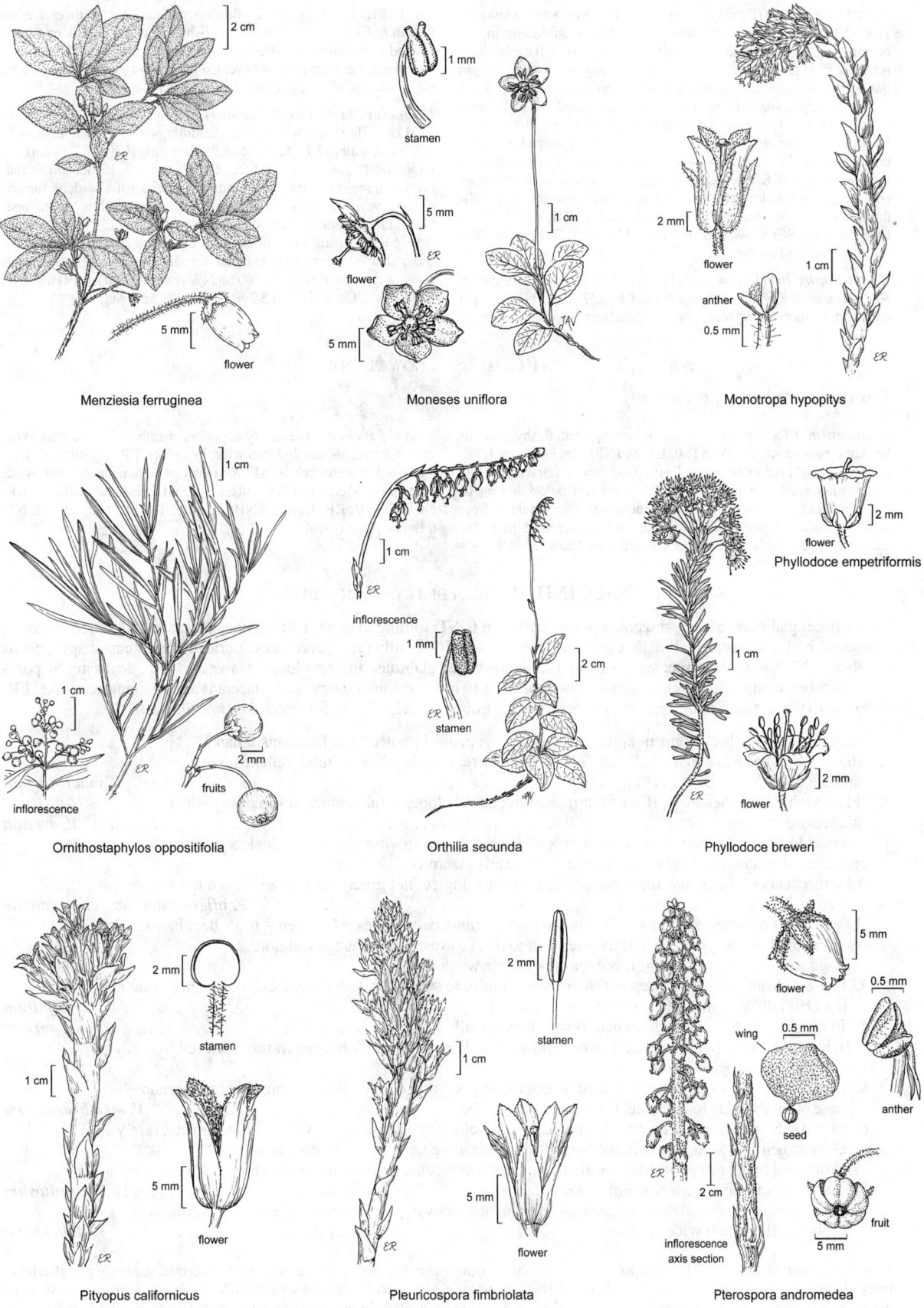

Menziesia ferruginea

Moneses uniflora

Monotropa hypopitys

Phyllodoce empetriformis

Ornithostaphylos oppositifolia

Orthilia secunda

Phyllodoce breweri

Pityopus californicus

Pleuricospora fimbriolata

Pterospora andromedea

R. columbianum (Piper) Harmaja (p. 711) WESTERN LABRADOR TEA **ST:** erect, < 2 m, bark smooth, peeling or shredding in age or not, twigs with unicellular hairs, papillae, and/or ± flat glandular scales. **LF:** (1)2–8 cm, 1.5–3 cm wide, leathery, evergreen, margin flat or ± rolled under, abaxially with sparse to dense papillae, ± flat glandular scales, adaxially with scattered ± flat glandular scales and/or papillae. **INFL:** ± rounded, bracts, bractlets with ± flat glandular scales, margins ciliate. **FL:** corolla widely bell-shaped, white to cream; stamens (8)10, ± equal. **FR:** 8–10 mm wide, ± longer, dehiscing base to tip. **SEED:** ± fusiform, coat ± elongated beyond narrow end. Coast, higher elevations inland, bogs, stream margins, occ well-drained sites; < 3630 m. NCo, KR, s NCoRO, CaRH, s SNF, SNH, CCo, SnFrB, n SCoRI, Wrn, SNE; to BC, MT, WY, UT. [*Ledum glandulosum* Nutt.] May–Aug

R. macrophyllum G. Don (p. 711) CALIFORNIA RHODODENDRON **ST:** < 4 m, coarse-branched, twigs stout. **LF:** (6)7–12(17) cm, 3–5(7) cm wide, leathery, evergreen, glabrous, midvein impressed, margin

flat. **INFL:** 1.5–3 cm, 10–20-fld, bracts deciduous; pedicel ± elongate in fr. **FL:** corolla widely funnel-shaped, white to pink or purple, adaxially brown-yellow-flecked; stamens 10, unequal. **FR:** longer than wide, dehiscing tip to base. Conifer forest margins; < 1515 m. NCo, KR, NCoRO, CCo, SnFrB; to BC. Apr–Jul

R. occidentale (Torr. & A. Gray) A. Gray (p. 711) CALIFORNIA AZALEA Hairs sparse to dense, glandular or not. **ST:** ≤ 8 m, densely branched, hairy. **LF:** (2.5)3.5–8.2(10.8) cm, (0.8)1.2–2.9(3.6) cm wide, deciduous, margin ciliate. **INFL:** short, 3–15-fld, bracts red-brown, margins ciliate or glandular. **FL:** corolla widely funnel-shaped, white to pink to salmon, upper petal yellow- to orange-blotched; stamens 5, ± equal. **FR:** longer than wide, dehiscing tip to base. **SEED:** ovate to fusiform, coat expanded around seed. 2*n*=26. Streambanks, ocean bluffs, moist wooded slopes and canyon bottoms, serpentine ridges; < 2700 m. NW (exc NCoRH), CaRH, n SNF, SNH, GV, CCo, SnFrB, n SCoRI, PR; OR. Apr–Aug

SARCODES SNOW PLANT

1 sp. (Greek: flesh-like, from red infl)

S. sanguinea Torr. (p. 711) Per, non-green, ± red, fleshy, glandular-hairy; roots thick, brittle. **ST:** 0. **LF:** 0. **INFL:** raceme, gen 1.5–3 dm, stout, bright red to orange-red, emerging from ground erect, persistent after seed dispersal; axis gen > 1 cm just below lowest fl; bracts < 8 cm, margins long-ciliate; bractlets 0. **FL:** sepals 5, free; corolla urn-shaped, petals 5, ± 3/4 fused, 12–18 mm, red; stamens 10, anthers < 4 mm, dehiscing by short separate slits, unawned; nectaries

± visible at ovary base; ovary superior, chambers 5, placentas axile, style < 8 mm, stigma 2–3 mm wide, head-like. **FR:** capsule, < 2.5 cm wide, indehiscent, brittle. **SEED:** many per chamber, < 1 mm wide, ovate, unwinged. 2*n*=64. Conifer or mixed forest; 700–3100 m. KR, NCoRO, NCoRH, CaRH, SNH, SCoRO, TR, PR, MP; sw OR, NV, n Baja CA. May–Jul

VACCINIUM BLUEBERRY, HUCKLEBERRY

Shrub, [tree] glabrous to hairy, rhizomed or not, burls gen 0. **ST:** trailing to erect. **LF:** cauline, alternate. **INFL:** raceme or fls 1, bracted; bud scales present; bractlets gen 2. **FL:** sepals 4–5, 2/3 to fully fused; corolla cylindric to urn- or cup-shaped, petals gen 4–5, ± 2/3 fused, gen white; stamens 8 or 10, filaments gen glabrous, anthers elongate, awned or not, dehiscing by pores on small tubes; ovary inferior, chambers 4–5, or appearing 10 by intrusion of ovary wall, placentas axile, stigma head-like. **FR:** berry. **SEED:** gen many. 400+ spp.: temp n hemisphere, trop mtns, Afr. (Latin: for *Vaccinium myrtillus* L.)

1. Lvs evergreen, leathery, veins not prominent abaxially; pedicel jointed to fl; filaments ± hairy
    2. Pl < 1.5 dm, branches slender; lf gen 7–17 mm, ± entire; corolla lobes >> tube, reflexed when anthers open . . . . ***V. macrocarpon***
    2′ Pl 5–30 dm, branches stout; lf 20–50 mm, serrate; corolla lobes < tube, erect to spreading when anthers open . . . . ***V. ovatum***
1′ Lvs deciduous (rarely evergreen in young pls of *Vaccinium parvifolium*), thin or ± thick, veins gen prominent abaxially; pedicel not jointed to fl; filaments glabrous
    3. Lf entire; calyx lobes gen > tube, persistent; twigs not angled, not green; fls 1–4 on lfless older shoots
        . . . . ***V. uliginosum*** subsp. ***occidentale***
    3′ Lf serrate or minutely so (but see *Vaccinium parvifolium*); calyx lobes ± 0 or gen < tube, deciduous; twigs angled or not, green or not; fls often 1 in axils of lowest lvs of youngest shoots
        4. Twigs strongly angled, green; fr red (or dark purple when dry)
            5. Pl erect shrub, 10–40 dm, gen not rhizomed; lf entire to serrate or with only deciduous, sharp point at tip, abaxially (esp midvein) puberulent . . . . ***V. parvifolium***
            5′ Pl brushy shrub, < 5 dm, rhizomed; lf serrate, abaxially glabrous. . . . . ***V. scoparium***
        4′ Twigs not or weakly angled, ± green or yellow-green but not green; fr not red (rarely dark red in *Vaccinium membranaceum*)
            6. Pl gen 5–15 dm; twigs weakly angled; lf gen ovate to elliptic or obovate, 2–5 cm, thin, membranous, base often rounded to truncate, tip acute. . . . . ***V. membranaceum***
            6′ Pl gen < 5 dm; twigs not or weakly angled; lf gen oblong or elliptic to obovate or oblanceolate, rarely elliptic, gen 1–3.5 cm, gen thin but not membranous, base tapered, tip seldom acute
                7. Youngest twigs gen puberulent or glandular; lf not glaucous, gen oblong or obovate to elliptic; corolla narrowly urn-shaped; fr < 9 mm wide . . . . ***V. cespitosum***
                7′ Youngest twigs gen glabrous, glaucous; lf glaucous, obovate to oblanceolate (elliptic); corolla ± spheric; fr > 9 mm wide. . . . . ***V. deliciosum***

V. cespitosum Michx. (p. 711) DWARF BILBERRY Shrub, gen hairy, rhizomed. **ST:** prostrate to erect, < 5 dm, gen rooting; twigs not or weakly angled, ± green, youngest gen puberulent or glandular. **LF:** deciduous, gen 1–3 cm, gen oblong or obovate to elliptic,

gen thin, not membranous, minutely serrate, abaxially gen glandular, base tapered, tip seldom acute. **INFL:** fls 1 in axils of lowest lvs of youngest shoots; pedicel not jointed to fl, bractlets 0. **FL:** calyx lobes ± 0; corolla < 6 mm, narrowly urn-shaped, ± white to ± pink; anthers

awned. **FR**: < 9 mm wide, gen blue-glaucous. 2*n*=24. Margins of wet meadows, mtn slopes; < 3400 m. NW (exc NCoRI), CaRH, SNH, CCo, SnFrB, Wrn; to AK, MT; also ne US, adjacent Can. [*Vaccinium caespitosum*, orth. var.] May–Jul

V. deliciosum Piper (p. 711)   CASCADE BILBERRY   Shrub, matted, glabrous, rhizomed. **ST**: < 4 dm, gen rooting; twigs weakly or gen not angled, youngest gen glabrous, glaucous. **LF**: deciduous, gen 1.5–3.5 cm, obovate to oblanceolate (elliptic), glaucous, gen thin, not membranous, gen serrate in upper 2/3, abaxially not glandular, base tapered, tip seldom acute. **INFL**: fls 1 in lf axils; pedicel not jointed to fl. **FL**: calyx lobes ± 0; corolla < 6 mm, ± spheric, ± pink; anthers awned. **FR**: > 9 mm wide, gen blue-glaucous. 2*n*=48. Alpine meadows, subalpine conifer forest, near coast; 600–2000 m. KR, n SNH, Wrn; to BC. Jun–Jul

V. macrocarpon Aiton   CRANBERRY   Shrub, glabrous to hairy, rhizomed. **ST**: prostrate and erect, < 1.5 dm, slender, rooting; twigs not angled, ± brown. **LF**: evergreen, gen 7–17 mm, narrowly elliptic to oblong, leathery, ± entire, abaxially glaucous. **INFL**: fls 1 in axils of reduced lowest lvs of youngest shoots; pedicel recurved, jointed to fl, bractlets 2, lf-like. **FL**: sepals 4, fused; petals 4, fused at base, lobes >> tube, reflexed when anthers open; filaments ± hairy, anthers awns 0. **FR**: 9–14 mm diam, red. 2*n*=24. Boggy soil at abandoned placer mine; ± 900 m. n SNH (near North Columbia, n Nevada Co.); native to e N.Am. Jun–Jul

V. membranaceum Torr. (p. 711)   THINLEAF HUCKLEBERRY   Shrub, ± glabrous, rhizome gen 0. **ST**: erect, gen 5–15 dm, not rooting; twigs weakly angled, yellow-green. **LF**: deciduous, 2–5 cm, gen ovate to elliptic or obovate, thin, membranous, serrate, teeth gen with a gland-tipped hair, abaxially with prominent veins, base often rounded to truncate, tip acute. **INFL**: fls 1 in axils of lowest lvs of youngest shoots; pedicel not jointed to fl, bractlets 2, scale-like. **FL**: calyx lobes ± 0; corolla < 6 mm, cylindric to urn-shaped, ± pink; anthers awned. **FR**: 9–11 mm diam, black (dark red). Wet meadows, mtn slopes; 1100–2200 m. KR, NCoRH, n SNH, Wrn; to AK, SD. Study needed of *V. coccineum* Piper (Siskiyou Mountains huckleberry), a sporadic red-fruited form not treated by Vander Kloet. Jun–Jul

V. ovatum Pursh (p. 711)   CALIFORNIA HUCKLEBERRY   Shrub, hairy, rhizome 0. **ST**: erect, 5–30 dm, stout, gen not rooting; twigs not angled, ± gray. **LF**: evergreen, 2–5 cm, elliptic to lanceolate, leathery,

serrate, abaxially with sparse, dark glandular hairs, veins not prominent. **INFL**: umbel-like, axillary, dense; pedicel jointed to fl, bractlets 2. **FL**: sepals 5, fused at base, lobes deltate; corolla < 8 mm, urn-shaped, lobes < tube, erect to spreading when anthers open; filaments ± hairy, anthers often short-awned. **FR**: 6–9 mm, black, glaucous or not. 2*n*=24. Edges, clearings in conifer forest; 3–800 m. NCo, KR, NCoRO, NCoRI, CW, n ChI (Santa Cruz, Santa Rosa islands), WTR, PR (uncommon); to BC. Mar–May

V. parvifolium Sm. (p. 711)   RED HUCKLEBERRY   Shrub, glabrous to puberulent, rhizome gen 0. **ST**: erect, 1–4 m, gen not rooting; twigs strongly angled, green. **LF**: deciduous, rarely evergreen in young pls, 10–25 mm, elliptic to ovate, thin, entire to serrate or with only deciduous, sharp point at tip, abaxially (esp midvein) puberulent, veins prominent. **INFL**: fls 1 in axils of lowest lvs of youngest shoots; pedicel not jointed to fl, bractlets 0. **FL**: calyx lobes ± 0 or < tube, rounded, deciduous; corolla < 5 mm, ± green or ± pink; anthers awned. **FR**: 6–10 mm diam, bright red. 2*n*=24. Moist, shaded woodland; 3–1400 m. NCo, KR, NCoRO, n&c SNH, SnFrB; to se AK. May–Jun

V. scoparium Coville   LITTLE-LEAVED HUCKLEBERRY   Shrub, brushy, glabrous, rhizomed. **ST**: gen erect, < 5 dm, rooting; twigs strongly angled, green. **LF**: deciduous, 8–15 mm, ovate to elliptic, serrate, abaxially glabrous. **INFL**: fls 1 in axils of lowest lvs of youngest shoots; pedicel not jointed to fl, bractlets 0. **FL**: calyx lobes ± 0 or < tube, rounded, deciduous; corolla < 4 mm, urn-shaped, pink; anthers awned. **FR**: 3–6 mm diam, red. Rocky subalpine woodland; 1800–2200 m. KR, NCoRO, CaRH; to w Can, SD, UT. Jun–Jul ★

V. uliginosum L. subsp. ***occidentale*** (A. Gray) Hultén (p. 711)   WESTERN BLUEBERRY   Shrub, glabrous, rhizomed or not. **ST**: erect, < 6 dm, or prostrate, gen rooting; twigs not angled, not green. **LF**: deciduous, 1–2 cm, elliptic to ovate, ± thick, entire, glaucous or not, abaxially visibly but not prominently veined. **INFL**: gen ± raceme, fls 1–4 on lfless older shoots; pedicel not jointed to fl, bractlets 0. **FL**: calyx lobes 4–5, gen > tube, triangular, persistent; anthers awned. **FR**: < 6 mm diam, blue-black, glaucous. Bogs, wet meadows; < 3400 m. NCo, KR, NCoRH, CaR, SNH; to BC, MT, NV. Part of circumboreal complex needing study. Some NCo (Big Lagoon, Humboldt Co.) pls with prominently veined lvs, fr > 6 mm diam may be *V. uliginosum* subsp. *u.*, native to lowland n Eur. Jun–Jul

XYLOCOCCUS

1 sp. (Greek: wood berry, from stone of fr)

X. bicolor Nutt. (p. 711)   Shrub, burled. **ST**: erect, < 2.5 m; bark shredding; twigs canescent. **LF**: gen alternate, evergreen, 2.5–4.5 cm, elliptic to oblong, leathery; margin entire, rolled under, abaxially densely white- to gray-hairy, adaxially glabrous, dark green. **INFL**: panicle, ± reflexed, dense, bracted; pedicel not jointed to fl, bractlets 2. **FL**: sepals gen 5; corolla urn-shaped, hairy, petals 5, fused, < 9 mm, white or ± pink; stamens gen 10, filaments woolly in lower 1/2, not swollen at base, anthers dehiscing by separate slits, awned; ovary superior, hairy, chambers gen 5, placentas axile. **FR**: drupe, < 9 mm wide, smooth, flesh dry; stones (3)5, fused into a smooth unit. 2*n*=26. Chaparral; < 650 m. SCo, s ChI (Santa Catalina Island), PR (exc SnJt); n Baja CA. Dec–Feb

EUPHORBIACEAE   SPURGE FAMILY

Mark H. Mayfield and Grady L. Webster, except as noted

Ann to shrub, tree [vine]; monoecious or dioecious. **ST**: gen branched [fleshy or spiny]. **LF**: gen simple, alternate or opposite, gen stipuled, petioled; blade entire, toothed, or palmately lobed. **INFL**: terminal or axillary panicle, raceme or spike, or (*Chamaesyce*, *Euphorbia*) a compact unit enclosed by an involucre appearing fl-like, terminal or axillary, 1 or in whorled, umbel-like, or cyme-like arrays. **FL**: unisexual, ± radial; sepals gen 3–5, free or fused; petals gen 0; stamens 1–many, free or filaments fused; ovary superior, chambers 1–4, styles free or fused, simple or lobed. **FR**: gen capsule. **SEED**: 1 per chamber; seed scar appendage sometimes present, pad- to dome-like. 218 genera, 6000+ spp.: ± worldwide esp trop; some cult (*Aleurites*, tung oil; *Euphorbia* spp.; *Hevea*, rubber; *Ricinus*). Many spp. ± highly TOXIC. [Wurdack & Davis 2009 Amer J Bot 96:1551–1570] *Eremocarpus* moved to *Croton*. *Tetracoccus* moved to Picrodendraceae. Scientific Editor: Bruce G. Baldwin.

1. Lvs opposite or 3-whorled
  2. Sts prostrate or erect, sap milky; lf base oblique; infl enclosed by involucre, fl-like, bisexual . . . . . . **CHAMAESYCE**
  2′ Sts erect, sap clear; lf base not oblique; infl not enclosed by involucre, not fl-like, unisexual . . . . . . **MERCURIALIS**

1′ Lvs alternate (sometimes opposite in *Euphorbia*)
  3. Infl ± fl-like, enclosed by involucre, bisexual . . . . **EUPHORBIA**
  3′ Infl not fl-like, not enclosed by involucre, unisexual or staminate and pistillate fls gen separated
    4. Lf blades palmately lobed, 1–5 dm; infl panicle, terminal . . . . **RICINUS**
    4′ Lf blades pinnately veined, entire or toothed, << 1 dm; infl various
      5. St and lf hairs 2-branched or stellate, or, if 0, petals 5
        6. Hairs 0 or 2-branched; petals present; filaments fused; seeds ± striate to pitted . . . . **DITAXIS**
        6′ Hairs stellate; petals 0; filaments free; seeds gen smooth, occ ± ridged
          7. Lf margin bluntly toothed; stipules persistent; staminate infl axillary . . . . **BERNARDIA**
          7′ Lf margin entire; stipules inconspicuous; staminate infl terminal . . . . **CROTON**
      5′ St and lf hairs simple or, if 0, petals 0
        8. Pistillate bract toothed, > fr . . . . **ACALYPHA**
        8′ Pistillate bract entire, << fr, or 0
          9. Lvs with stinging, nettle-like hairs; stamens 3–6; base of pistillate bracts not glandular; st sap clear . . . **TRAGIA**
          9′ Lvs glabrous or without nettle-like hairs; stamens 2–3; base of pistillate bracts glandular; st sap milky or clear
            10. Ann or per; stamens 2; seeds 2–3.5 mm . . . . **STILLINGIA**
            10′ Tree; stamens 2–3; seeds 6–9 mm . . . . **TRIADICA**

ACALYPHA

[Ann, per] shrub; sap clear; gen monoecious. **ST**: central erect, gen much-branched; lateral spreading to ascending. **LF**: cauline, alternate; hairs simple, sometimes glandular. **INFL**: spike, terminal or axillary; staminate bracts minute; pistillate bracts lf-like, toothed. **STAMINATE FL**: sepals 4; stamens 4–8, filaments free or fused at base; nectary disk 0. **PISTILLATE FL**: sepals 3(5); nectary 0; ovary 3-chambered, styles 3, deeply cut. **FR**: ± spheric, smooth or ± lobed. **SEED**: smooth to pitted; scar appendage minute. ± 400 spp.: trop, warm temp worldwide. (Greek: ancient name for a kind of nettle)

A. californica Benth. (p. 717) Pl < 1.5 m, hairy, ± glandular. **LF**: stipules 2–5 mm, linear; petiole < 1.5 cm; blade 1–2 cm, ovate to ± deltate, base truncate to ± lobed, margin crenate. **STAMINATE INFL**: 1.5–4 cm, slender. **PISTILLATE INFL**: < 2 cm; bracts together cup-like, hairy, margin glandular. **STAMINATE FL**: sepals ± 0.5 mm, puberulent; stamens >> sepals. **PISTILLATE FL**: sepals ± 1 mm, puberulent; ovary ± 1 mm diam, puberulent, styles ± red. **FR**: 1–3 mm diam, puberulent. Rocky slopes, chaparral, oak woodland; < 1300 m. s SCo (near San Diego), PR, w DSon; Baja CA. Jan–Jun

BERNARDIA

Shrub; sap clear; [monoecious] dioecious. **ST**: erect, gen much-branched. **LF**: cauline, alternate; hairs simple or stellate. **STAMINATE INFL**: spike or raceme, axillary. **PISTILLATE INFL**: terminal; fl occ 1. **STAMINATE FL**: sessile or short-pedicelled; calyx splitting into 3–4 parts; stamens 3–25, filaments free; nectar disk minute or 0. **PISTILLATE FL**: sessile; sepals 4–6; nectar disk 0; ovary 3-chambered, styles 3, free, 2-lobed or -toothed. **FR**: 3-lobed. **SEED**: scar appendaged. 30–40 spp.: trop, subtrop Am. (Bernard de Jussieu, French taxonomist, 1699–1776) *B. myricifolia* (Scheele) S. Watson not in CA.

B. incana C.V. Morton (p. 717) Pl < 2.5 m, hairy. **LF**: stipules ± 1 mm, deciduous; petiole 1–5 mm; blade 0.5–3 cm, elliptic, tip obtuse or rounded, margin crenate. **STAMINATE INFL**: raceme; pedicel 3–4 mm. **PISTILLATE INFL**: fl 1, sessile. **STAMINATE FL**: stamens 12–15; nectar disk of small glands. **PISTILLATE FL**: sepals 5, ± 2 mm, unequal; ovary tomentose, styles jagged. **FR**: 8–10 mm diam, tomentose. **SEED**: 5 mm, smooth; back ribbed. Washes, rocky canyons; < 1200 m. s DMoj, DSon; to TX, Mex. Apr–May, Oct–Nov

CHAMAESYCE   PROSTRATE SPURGE

Daryl Koutnik

Ann, per, glabrous to hairy; sap milky; gen monoecious. **ST**: prostrate to erect, < 5 dm; branches alternate. **LF**: cauline, opposite, short-petioled; blade base gen asymmetric, veins dark green. **INFL**: fl-like, gen 1 per node; involucre ± bell-shaped, bracts 5, fused; glands gen 4, distal appendages gen colorful, petal-like; fls central. **STAMINATE FL**: 3–many, gen in 5 clusters around pistillate fl, each fl a stamen. **PISTILLATE FL**: 1, central, stalked; ovary chambers 3, styles 3, separate or fused at base, divided to entire. **FR**: capsule, round to 3-angled or -lobed in ×-section. **SEED**: gen 4-angled, smooth or sculptured. ± 250 spp.: dry temp, subtrop worldwide, esp Am. (Greek: ancient name for kind of prostrate pl) [Wheeler 1941 Rhodora 43:97–154, 168–286]

1. Infls gen in dense axillary cyme-like clusters, occ 1 per node; lvs 8–35 mm; st gen erect (decumbent) . . . . *C. nutans*
1′ Infls 1 per node, occ crowded on lateral branches; lvs 1–28 mm; st prostrate to ascending, sometimes erect
  2. Involucre, fr hairy; st, lvs gen hairy or becoming glabrous
    3. Per
      4. Involucre urn-shaped; staminate fls ≤ 10 — DSon . . . . *C. arizonica*
      4′ Involucre ± bell-shaped; staminate fls ≥ 15
        5. Seed 3-angled, transversely 4–5-ridged, ridges rounded . . . . *C. pediculifera*

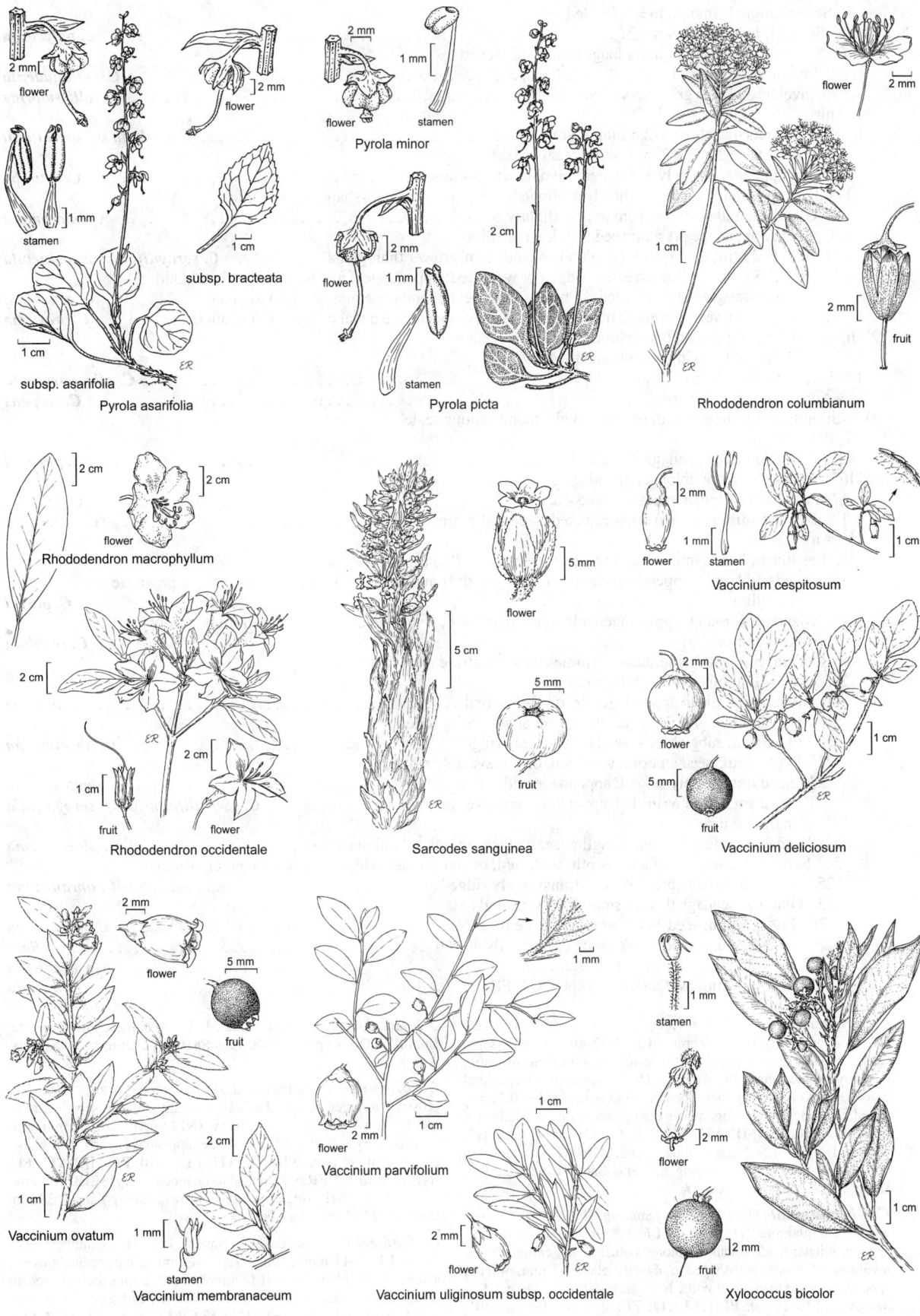

flower

stamen

2 mm

1 mm

subsp. bracteata

1 cm

subsp. asarifolia

Pyrola asarifolia

Pyrola minor

flower   stamen

flower   stamen

Pyrola picta

flower

fruit

Rhododendron columbianum

Rhododendron macrophyllum

flower

flower   fruit

Rhododendron occidentale

flower

fruit

Sarcodes sanguinea

flower   stamen

Vaccinium cespitosum

flower

fruit

Vaccinium deliciosum

flower

fruit

Vaccinium ovatum

stamen

Vaccinium membranaceum

flower

Vaccinium parvifolium

flower

Vaccinium uliginosum subsp. occidentale

stamen

flower

fruit

Xylococcus bicolor

    5′ Seed 4-angled, smooth to ± wrinkled
      6. St and lf hairs short, straight . . . . ²*C. polycarpa*
      6′ St and lf tomentose or hairs long, appressed, dense
        7. Involucre glands red . . . . *C. melanadenia*
        7′ Involucre glands green to yellow . . . . *C. vallis-mortae*
  3′ Ann
    8. Staminate fls 40–60; fr 2–2.5 mm; n ScV . . . . *C. ocellata* subsp. *rattanii*
    8′ Staminate fls < 20; fr ≤ 2 mm; widespread
      9. Gland appendage deeply 3–5-lobed; involucre urn-shaped . . . . *C. setiloba*
      9′ Gland appendage entire, shallowly scalloped, or 0; involucre bell-shaped or obconic
        10. Gland appendage 0; seed smooth to slightly wrinkled . . . . ²*C. micromera*
        10′ Gland appendage present; seed wrinkled or ridged
          11. Fr 1.5–2 mm; seed wrinkled; gland appendage narrower than gland . . . . *C. serpyllifolia* subsp. *hirtula*
          11′ Fr ≤ 1.5 mm; seed transversely ridged or wrinkled; appendage equal to or wider than gland
            12. Seeds transversely wrinkled; fr evenly strigose; appendage equal in width to gland . . . . *C. maculata*
            12′ Seeds transversely ridged; fr hairy on lobes only; appendage equal or greater in width than gland . . . *C. prostrata*
  2′ Involucre, fr glabrous; st, lvs glabrous, sometimes hairy
    13. Stipules fused into wide, membranous scale
      14. Per; staminate fls ≥ 15 . . . . *C. albomarginata*
      14′ Ann; staminate fls 5–10 . . . . *C. serpens*
    13′ Stipules separate or fused, but not a wide, membranous scale
      15. Per
        16. Glands round, appendage 0; lvs 2–4 mm . . . . *C. parishii*
        16′ Glands elliptic or oblong, appendage present; lvs 1–11 mm
          17. Fr ≥ 2 mm; stipules separate; seed ≥ 2 mm . . . . *C. fendleri*
          17′ Fr 1–1.5 mm; proximal stipules fused; seed 1–1.5 mm . . . . ²*C. polycarpa*
      15′ Ann
        18. Lvs linear, base symmetric; st prostrate to erect — PR, D
          19. Glands 1–4, oval, appendage narrower than gland; fr and seed > 1.5 mm; seed smooth; st prostrate
            to ascending . . . . *C. parryi*
          19′ Glands ± 4, round, appendage wider than gland or 0; fr and seed ≤ 1.5 mm; seed transversely
            ridged; st erect . . . . *C. revoluta*
        18′ Lvs lanceolate to round, base asymmetric; st prostrate to decumbent
          20. Lf margin toothed, at least toward tip
            21. Style entire; gland appendages deeply 3–5-lobed . . . . *C. hooveri*
            21′ Style divided; gland appendages entire or shallowly lobed
              22. Pl hairy to subglabrous; seed transversely ridged . . . . ²*C. abramsiana*
              22′ Pl glabrous; seed smooth, wrinkled, or transversely ridged
                23. Seed transversely ridged; appendage wider than gland . . . . *C. glyptosperma*
                23′ Seed smooth to wrinkled; appendage narrower than gland . . . . *C. serpyllifolia* subsp. *serpyllifolia*
          20′ Lf margin entire
            24. Seed ± 2 angled, flattened lengthwise, surface smooth; gland ovate . . . . *C. platysperma*
            24′ Seed 3–4-angled, surface smooth, wrinkled, or transversely ridged; glands round or elliptic
              25. Gland appendage present; seed transversely ridged . . . . ²*C. abramsiana*
              25′ Gland appendage 0; seed smooth or wrinkled
                26. Fr < 1.5 mm; seed 1–1.5 mm; staminate fls 2–5 . . . . ²*C. micromera*
                26′ Fr 2–2.5 mm; seed 1.5–2 mm; staminate fls 40–60 . . . . *C. ocellata*
                  27. Lf ≤ 15 mm, lanceolate to ovate; D . . . . subsp. *arenicola*
                  27′ Lf < 10 mm, ovate, sickle-shaped; CA-FP . . . . subsp. *ocellata*

C. abramsiana (L.C. Wheeler) Koutnik ABRAMS' SPURGE Ann. **ST:** prostrate, hairy to subglabrous. **LF:** 2–12 mm; stipules separate, 2–5-parted; blade ovate to elliptic-oblong, obtuse at tip, entire to finely toothed, hairy to glabrous. **INFL:** dense on short lateral branches; involucre < 1 mm, obconic, glabrous; gland < 0.5 mm, round or elliptic; appendage wider than gland, entire or shallowly 2-lobed, white. **STAMINATE FL:** 3–5. **PISTILLATE FL:** style divided 1/2. **FR:** 1.5–2 mm, oblong, round, glabrous. **SEED:** 1–1.5 mm, ovoid, transversely 4–6-ridged, white. Sandy flats; < 200 m. DSon; to AZ, Mex. Sep–Nov ★

C. albomarginata (Torr. & A. Gray) Small (p. 717) RATTLESNAKE WEED Per, glabrous. **ST:** prostrate. **LF:** 3–8 mm; stipules fused, triangular, ciliate; blade round to oblong, obtuse at tip, entire. **INFL:** involucre < 2.5 mm, bell-shaped to obconic; gland < 1 mm, oblong; appendage wider than gland, entire to slightly scalloped, white. **STAMINATE FL:** 15–30. **PISTILLATE FL:** style divided 1/2. **FR:**

2–2.5 mm, ovoid, 3-angled. **SEED:** 1–2 mm, oblong, smooth, white. Common. Dry slopes; < 2300 m. s SnJV, SW, D; to UT, TX, Mex. Apr–Nov

C. arizonica (Engelm.) Arthur ARIZONA SPURGE Per. **ST:** prostrate to erect, hairy. **LF:** 2–10 mm; stipules gen separate, minute; blade ovate, acute, entire, hairy. **INFL:** involucre < 2 mm, urn-shaped, hairy; gland < 0.5 mm, ovate; appendage wider than gland, entire, white to pink. **STAMINATE FL:** 5–10. **PISTILLATE FL:** style divided 1/2. **FR:** < 2 mm, spheric, lobed, hairy. **SEED:** ± 1 mm, ovoid, transversely ridged, white to brown. Sandy flats; < 300 m. DSon; to TX, Mex. Mar–Apr ★

C. fendleri (Torr. & A. Gray) Small Per. **ST:** decumbent, glabrous. **LF:** 3–11 mm; stipules separate, linear, entire; blade ovate, acute, entire, glabrous. **INFL:** involucre < 2 mm, bell-shaped to obconic, glabrous; gland < 1 mm, elliptic; appendage narrower than gland, scalloped, white. **STAMINATE FL:** 25–35. **PISTILLATE**

FL: style divided 1/2. **FR:** 2–2.5 mm, ovoid, lobed, glabrous. **SEED:** 2–2.5 mm, ovoid, smooth to slightly wrinkled, white. Uncommon. Dry slopes, woodland; 1500–2300 m. W&I, DMtns; to NE, TX, Mex. May–Oct

C. glyptosperma (Engelm.) Small   Ann. **ST:** prostrate, glabrous. **LF:** 3–15 mm; stipules separate, thread-like; blade ovate to ovate-oblong, rounded at tip, finely toothed, glabrous. **INFL:** involucre < 1 mm, obconic, glabrous; gland < 0.5 mm, elliptic; appendage wider than gland, scalloped, white. **STAMINATE FL:** 1–5. **PISTILLATE FL:** style divided < 1/2. **FR:** 1.5–2 mm, ovoid, lobed, glabrous. **SEED:** 1–1.5 mm, ovoid, transversely 3–4-ridged, white to light brown. Uncommon. Dry ground; 200–300 m. n CaRF (near Redding); to BC, e N.Am. Jul–Sep

C. hooveri (L.C. Wheeler) Koutnik (p. 717)   HOOVER'S SPURGE Ann, glabrous. **ST:** prostrate to decumbent. **LF:** 2–5 mm; stipules fused, fringed; blade round, rounded at tip, coarsely toothed, papillate. **INFL:** involucre ± 2 mm, bell-shaped; gland < 1 mm, round; appendage wider than gland, deeply 3–5-lobed, white. **STAMINATE FL:** 30–35. **PISTILLATE FL:** style entire. **FR:** 1.5–2 mm, spheric, lobed. **SEED:** ± 1.5 mm, ovoid, widely 4-angled, shallowly ridged, white. Vernal pools; < 250 m. GV (Butte, Tehama, Tulare cos.). Threatened by habitat loss. Jul–Sep ★

C. maculata (L.) Small (p. 717)   SPOTTED SPURGE   Ann. **ST:** prostrate, hairy. **LF:** 4–17 mm; stipules separate, fringed; blade ovate to oblong, acute to obtuse, finely toothed, hairy or becoming glabrous. **INFL:** dense on short, lateral branches; involucre < 1 mm, obconic, hairy; gland < 0.5 mm, elliptic; appendage width = gland width, scalloped, white to pink. **STAMINATE FL:** 4–5. **PISTILLATE FL:** style divided < 1/2. **FR:** < 1.5 mm, ovoid, lobed, strigose. **SEED:** < 1.5 mm, ovoid, transversely wrinkled, light brown. Disturbed places; < 2000 m. CA-FP, DSon; native to e US. Apr–Oct

C. melanadenia (Torr.) Millsp.   Per. **ST:** decumbent to ascending, tomentose or becoming glabrous. **LF:** 2–9 mm; stipules separate, linear; blade ovate, acute, entire, tomentose. **INFL:** involucre 1–1.5 mm, bell-shaped, tomentose; gland < 1 mm, oblong; appendage width = gland width, scalloped, white. **STAMINATE FL:** 15–20. **PISTILLATE FL:** style divided > 1/2. **FR:** 1.5–2 mm, ovoid, lobed, tomentose. **SEED:** 1–1.5 mm, ovoid, slightly wrinkled, white. Dry, stony slopes or flats; < 1300 m. SW, DSon; AZ, Baja CA. Dec–May

C. micromera (Boiss.) Wooton & Standl.   Ann, glabrous to hairy. **ST:** prostrate. **LF:** 2–7 mm; stipules fused below, separate above, triangular, ciliate; blade ovate to oblong, acute to obtuse, entire. **INFL:** involucre < 1 mm, bell-shaped, glabrous to hairy; gland << 0.5 mm, round, red or pink; appendage 0. **STAMINATE FL:** 2–5. **PISTILLATE FL:** style divided 1/2. **FR:** < 1.5 mm, spheric, angled. **SEED:** 1–1.5 mm, ovoid, smooth to slightly wrinkled, white to brown. Sandy places; < 1000 m. D; to UT, TX, n Mex. Apr–Jun, Sep–Dec

C. nutans (Lag.) Small   Ann. **ST:** decumbent to gen erect, sparsely hairy or becoming glabrous. **LF:** 8–35 mm; stipules fused, triangular; blade oblong, obtuse at tip, finely toothed, glabrous to hairy. **INFL:** gen in dense, axillary cyme-like clusters, occ 1 per node; involucre 1–2 mm, obconic, glabrous; gland < 0.5 mm, oblong; appendage wider than gland, entire, white to red. **STAMINATE FL:** 5–11. **PISTILLATE FL:** style divided 1/2. **FR:** 2–2.5 mm, ovoid, lobed, glabrous. **SEED:** 1–1.5 mm, ovoid, shallowly wrinkled, black to brown. Disturbed areas; < 300 m. GV, CW, SW, DSon; native to se US, S.Am. Apr–Oct

C. ocellata (Durand & Hilg.) Millsp.   Ann. **ST:** prostrate, glabrous to hairy. **LF:** stipules separate, thread-like; blade lanceolate to ovate, acute to obtuse, margin entire, rolled under, faces glabrous to hairy. **INFL:** involucre 1.5–2 mm, obconic to bell-shaped, glabrous to hairy; gland < 1 mm, round; appendage wider than gland or 0. **STAMINATE FL:** 40–60. **PISTILLATE FL:** style divided 1/2. **FR:** 2–2.5 mm, spheric, lobed, glabrous to hairy. **SEED:** 1.5–2 mm, ovoid, widely 3-angled, smooth to shallowly wrinkled, white.

subsp. ***arenicola*** (Parish) Thorne   **ST:** glabrous. **LF:** ≤ 15 mm; blade lanceolate to ovate, acute, glabrous. **INFL:** involucre glabrous; gland appendage 0. **FR:** glabrous. **SEED:** smooth. Sandy places; < 800 m. D; to UT, AZ. May–Sep

subsp. ***ocellata***   **ST:** glabrous. **LF:** < 10 mm; blade ovate, sickle-shaped, glabrous. **INFL:** involucre glabrous; gland appendage 0. **FR:** glabrous. Dry, sandy places; < 500 m. CA-FP (exc PR). May–Sep

subsp. ***rattanii*** (S. Watson) Koutnik   STONY CREEK SPURGE **ST:** hairy. **LF:** < 10 mm; blade ovate, acute, hairy. **INFL:** involucre hairy; gland appendage wider than gland, white. **FR:** hairy. Sandy or stony ground; < 100 m. n ScV (Glenn, Tehama cos.). May–Sep   ★

C. parishii (Greene) Millsp.   Per. **ST:** prostrate. **LF:** 2–4 mm; stipules separate, linear, ciliate; blade ovate, abruptly pointed, entire, glabrous. **INFL:** involucre ± 1 mm, bell-shaped, glabrous; gland < 0.5 mm, round, yellow to red; appendage 0. **STAMINATE FL:** 40–50. **PISTILLATE FL:** style divided 1/2. **FR:** < 2 mm, spheric, lobed, glabrous. **SEED:** < 1.5 mm, ovoid, 4-angled, slightly wrinkled, white. Uncommon. Sandy washes; < 1000 m. D; NV. Apr–Oct

C. parryi (Engelm.) Rydb.   PARRY'S SPURGE   Ann. **ST:** prostrate to ascending, glabrous. **LF:** 5–28 mm; stipules separate, linear; blade linear, acute to obtuse, entire, glabrous. **INFL:** involucre 1.5–2 mm, bell-shaped, glabrous; glands 1–4, < 0.5 mm, oval; appendage narrower than gland, entire, white. **STAMINATE FL:** 40–55. **PISTILLATE FL:** style divided 1/2. **FR:** 2 mm, spheric, lobed, glabrous. **SEED:** ± 2 mm, ovoid, 3-angled, smooth, brown to white. Sand dunes; < 700 m. DMoj; to Rocky Mtns, TX, Mex. May–Jun   ★

C. pediculifera (Engelm.) Rose & Standl.   Per. **ST:** prostrate to erect, hairy or becoming glabrous. **LF:** 2–20 mm; stipules separate, thread-like; blade ovate to spoon-shaped, acute, entire, hairy to glabrous. **INFL:** involucre 1.5–2 mm, bell-shaped, hairy to glabrous; gland 0.5 mm, oblong; appendages unequal, entire to almost 0 on some glands, largest wider than gland. **STAMINATE FL:** 22–25. **PISTILLATE FL:** style divided to base. **FR:** 2 mm, ovoid, lobed, hairy. **SEED:** 1–1.5 mm, ovoid, white, 3-angled, transversely 4–5-ridged, ridges rounded. Uncommon. Dry slopes; < 500 m. DSon; AZ, Mex. Jan–Apr

C. platysperma (Engelm.) Shinners   FLAT-SEEDED SPURGE Ann. **ST:** prostrate, glabrous. **LF:** 5–10 mm; stipules separate, 2–3-lobed; blade oblong to obovate, obtuse to rounded at tip, entire, glabrous. **INFL:** involucre 1.5–2 mm, bell-shaped, glabrous; gland 1 mm, ovate, glabrous; appendage 0. **STAMINATE FL:** ± 50. **PISTILLATE FL:** style divided to base. **FR:** < 4.5 mm, widely ovoid, slightly lobed, glabrous. **SEED:** 2.5–3 mm, ovoid, flattened lengthwise, smooth, white. Sandy soil; < 100 m. DSon (Coachella Valley); to sw AZ, n SON. May   ★

C. polycarpa (Benth.) Millsp. (p. 717)   Per. **ST:** prostrate to ascending, glabrous to hairy. **LF:** 1–10 mm; stipules separate or proximal fused, triangular; blade round to ovate, acute to obtuse, entire, glabrous to hairy. **INFL:** involucre 1–1.5 mm, bell-shaped, glabrous to hairy; gland < 1 mm, oblong; appendage wider to narrower than gland, entire to scalloped, white to red. **STAMINATE FL:** 15–32. **PISTILLATE FL:** style divided > 1/2. **FR:** 1–1.5 mm, spheric, lobed, glabrous to hairy. **SEED:** 1–1.5 mm, ovoid, smooth, white to light brown. Common. Dry, sandy slopes and flats; < 1000 m. SW, D; NV, Mex. All yr

C. prostrata (Aiton) Small (p. 717)   Ann. **ST:** prostrate, hairy or becoming glabrous. **LF:** 3–11 mm; stipules separate, linear; blade ovate to elliptic, obtuse at tip, finely toothed, glabrous to hairy. **INFL:** gen on short, lateral branches; involucre < 1.5 mm, obconic, hairy to glabrous; gland < 0.5 mm, oval; appendage = to or wider than gland, entire to scalloped, white. **STAMINATE FL:** 4. **PISTILLATE FL:** style divided 1/2. **FR:** 1–1.5 mm, spheric, lobed, lobes hairy. **SEED:** ± 1 mm, ovoid, transversely ridged, white to gray. Disturbed areas; < 250 m. s CA-FP; to e US; native to West Indies, S.Am. Aug–Sep

C. revoluta (Engelm.) Small   REVOLUTE SPURGE   Ann. **ST:** erect, glabrous. **LF:** 3–26 mm; stipules separate, linear; blade linear, acute to obtuse, margin entire, rolled under, faces glabrous. **INFL:** involucre < 1.5 mm, obconic, glabrous; gland < 0.5 mm, round; appendage wider than gland or 0, entire, white. **STAMINATE FL:** 5–10. **PISTILLATE FL:** style divided 1/2. **FR:** slightly < 1.5 mm, spheric, lobed, glabrous. **SEED:** 1–1.5 mm, ovoid, 3-angled, transversely 2–3-ridged, white to gray. Rocky slopes; < 3100 m. PR, D; to Rocky Mtns, Mex. Aug–Sep   ★

C. serpens (Kunth) Small Ann. **ST**: prostrate, rooting at nodes; glabrous. **LF**: 2–7 mm; stipules fused, triangular; blade ovate to oblong, obtuse at tip, entire, glabrous. **INFL**: involucre < 1.5 mm, obconic, glabrous; gland < 0.5 mm, oblong; appendage wider than gland, scalloped, white. **STAMINATE FL**: 5–10. **PISTILLATE FL**: style divided 1/2. **FR**: < 1.5 mm, spheric, lobed, glabrous. **SEED**: < 1.5 mm, ovoid, smooth, white to brown. Disturbed areas; < 200 m. SnJV, s SCoR, SW, expected elsewhere; to e US; native to S.Am. May–Sep

C. serpyllifolia (Pers.) Small THYME-LEAFED SPURGE Ann. **ST**: prostrate to ascending, glabrous to hairy. **LF**: 3–14 mm; stipules separate, linear; blade oblong to ovate, rounded at tip, finely toothed, glabrous to hairy. **INFL**: involucre ± 1 mm, bell-shaped, glabrous to hairy; gland < 0.5 mm, oblong; appendage narrower than gland, entire to scalloped, white. **STAMINATE FL**: 5–18. **PISTILLATE FL**: style divided 1/2. **FR**: 1.5–2 mm, ovoid, lobed, glabrous to hairy. **SEED**: 1–1.5 mm, ovoid, smooth to wrinkled, white to brown.

subsp. *hirtula* (S. Watson) Koutnik **ST**: hairy. **LF**: 3–7 mm, hairy. **INFL**: involucre hairy. **FR**: hairy. **SEED**: wrinkled. Dry places, woodland; < 2500 m. c SN, SW; Baja CA. Aug–Oct

subsp. *serpyllifolia* **ST**: glabrous. **LF**: glabrous. **INFL**: involucre glabrous. **FR**: glabrous. **SEED**: smooth to wrinkled. Common. Dry habitats; < 2500 m. CA; to BC, e N.Am, Mex. Aug–Oct

C. setiloba (Engelm.) Parish (p. 717) Ann. **ST**: prostrate, hairy. **LF**: 2–7 mm; stipules separate, thread-like; blade oblong to ovate, acute, entire, hairy. **INFL**: involucre < 1.5 mm, urn-shaped, hairy; gland < 0.5 mm, oblong; appendage wider than gland, 3–5-lobed, white. **STAMINATE FL**: 3–7. **PISTILLATE FL**: style divided to base. **FR**: < 1.5 mm, spheric, lobed, hairy. **SEED**: ± 1 mm, ovoid, wrinkled, white to brown. Uncommon. Sandy places; < 1500 m. D; to TX, Mex. All yr

C. vallis-mortae Millsp. DEATH VALLEY SANDMAT Per. **ST**: prostrate to decumbent, tomentose or becoming glabrous. **LF**: 4–6 mm; proximal stipules fused; distal stipules separate, thread-like, tomentose; blade oblong to ovate, obtuse at tip, entire, tomentose. **INFL**: gen clustered at branch tips; involucre < 2.5 mm, bell-shaped, tomentose; gland < 1 mm, oblong; appendage ≥ gland in width, entire to scalloped, white. **STAMINATE FL**: 17–22. **PISTILLATE FL**: style divided 1/2. **FR**: 2 mm, ovoid, 3-lobed, tomentose. **SEED**: < 2 mm, ovoid, smooth, white. Dry, sandy places; < 1300 m. SNE, DMoj. May–Oct ★

CROTON

Ann to shrub [tree]; sap clear or colored; monoecious or dioecious. **ST**: spreading to erect. **LF**: cauline, alternate, entire in CA; hairs gen stellate. **INFL**: cyme, spike, or raceme, gen terminal. **STAMINATE FL**: gen pedicelled; sepals gen 5; petals 5 or 0; stamens 8–50(300), filaments free, bent inward in bud; nectar disk gen divided. **PISTILLATE FL**: pedicel short or 0, becoming longer in fr; sepals gen 5, entire to lobed; petals gen 0; nectar disk entire; ovary 1–3-chambered, styles 2-lobed, ± dissected, or simple. **FR**: spheric or 3-lobed, smooth or tubercled. **SEED**: smooth to ribbed or pitted; scar appendaged. 1200–1300 spp.: trop, warm temp, worldwide. (Greek: tick, for resemblance of seed) [Berry et al. 2005 Amer J Bot 92:1520–1534]

1. Ann; monoecious; ovary 1-chambered, style 1, simple . . . . *C. setiger*
1′ Per to shrub; dioecious; ovary 3-chambered, styles 3, 2-lobed, lobes 2-forked
  2. Seed 3.5–5.5 mm; pedicels in fr 1–1.5(3) mm; staminate sepals ± 2–2.5 mm; CCo, SCo, s ChI, D . . . . *C. californicus*
  2′ Seed 6.5–7 mm; pedicels in fr 4–7 mm; staminate sepals 2.5–3 mm; se DSon . . . . *C. wigginsii*

C. californicus Müll. Arg. (p. 717) Per to subshrub; < 1 m; dioecious; hairs stellate, scale-like. **LF**: petiole 1–4 cm; blade 2–5.5 cm, elliptic to narrowly oblong, tip rounded to obtuse. **INFL**: raceme. **STAMINATE FL**: pedicel 1–5.5(7) mm; petals 0; stamens 10–15. **PISTILLATE FL**: pedicel ≤ 1 mm, 1–1.5(3) mm in fr; sepals ± 2 mm; ovary 3-chambered, styles 3, ± dissected. **SEED**: 3.5–5.5 mm, smooth. Sandy soils, dunes, washes; < 900 m. CCo, SCo, s ChI (Santa Catalina Island), D; AZ, Baja CA. Apr–Jul

C. setiger Hook. (p. 717) TURKEY-MULLEIN Ann < 2 dm, < 8 dm wide, mound-like; monoecious. **ST**: spreading to ascending. **LF**: petiole 1–5 cm; blade 1–6 cm, ovate, tip rounded. **STAMINATE INFL**: cyme, terminal. **PISTILLATE INFL**: cyme, axillary, proximal to staminate infl; fls 1–3. **STAMINATE FL**: pedicel 2–3 mm; receptacle finely bristly; sepals 5–6; petals 0; stamens 6–10. **PIS-**

TILLATE FL: pedicel ± 0; sepals 0; glands below ovary 4–5; ovary 1-chambered, style 1, simple. **FR**: ± 4 mm diam. **SEED**: 1, 3–4 mm, smooth or ± ridged; scar not appendaged. 2*n*=20. Dry, open, often disturbed areas; < 1000 m. CA-FP, w D; to WA. Herbage TOXIC to livestock, esp in hay. [*Eremocarpus s.* (Hook.) Benth.] May–Oct

C. wigginsii L.C. Wheeler (p. 717) WIGGINS' CROTON Subshrub to shrub, < 1 m; dioecious; hairs stellate, scale-like. **LF**: petiole 1–4 cm; blade 2–8.5 cm, narrowly elliptic to linear-oblong, tip rounded to obtuse. **INFL**: raceme. **STAMINATE FL**: pedicel 1–5.5(7) mm; petals 0; stamens 10–15. **PISTILLATE FL**: pedicel < 2 mm, 4–7 mm in fr; sepals ± 2 mm; ovary 3-chambered, styles 3, 2-lobed, lobes 2-forked. **SEED**: 6.5–7 mm, smooth. Sand dunes; < 100 m. se DSon (se Imperial Co.); AZ, nw Mex. Mar–May ★

DITAXIS

Ann to subshrub; sap clear; gen monoecious; hairs 0 or gen 2-branched, gen appressed. **ST**: spreading to erect, 1–10 dm. **LF**: alternate, stipuled. **INFL**: raceme, axillary; staminate fls gen distal to pistillate fls; axis appressed- to spreading-hairy; bracts entire. **STAMINATE FL**: sepals 5, edges abutting in bud; petals 5; stamens 5–15, gen in 2 sets, some > others, filaments fused into a column, staminodes 0–3 at column tip. **PISTILLATE FL**: sepals 5, overlapping in bud; petals 5; nectar disk ± dissected; ovary 3-chambered, styles 3, 2-lobed. **FR**: smooth. **SEED**: surface ± striate to pitted [net-like]; scar not appendaged. ± 50 spp.: trop, warm temp Am. (Greek: 2-ranked, from 2 sets of anthers)

1. Pl glabrous . . . . *D. serrata* var. *californica*
1′ Pl hairy, hairs 2-forked and appressed, occ simple or glandular
  2. Margin of stipules, bracts, and pistillate sepals stalked-glandular; infl axis minutely spreading-hairy . . . . *D. claryana*
  2′ Margin of stipules, bracts, and pistillate sepals glabrous or faintly glandular, glands not stalked; infl axis appressed-hairy
    3. Subshrub, sts brittle; pistillate sepals ± = petals; fr appressed-hairy; style lobes expanded . . . . *D. lanceolata*
    3′ Ann or per (occ woody at base), sts not brittle; pistillate sepals clearly > petals; fr ± spreading-hairy; style lobes not expanded
      4. Seeds angled in ×-section, clearly pitted; lvs lanceolate, not densely hairy, entire to faintly toothed . . . *D. neomexicana*
      4′ Seeds round in ×-section, ± striate, not clearly pitted; lvs widely elliptic to ovate, densely hairy, clearly toothed distally . . . . *D. serrata* var. *serrata*

D. claryana (Jeps.) Webster (p. 717) GLANDULAR DITAXIS Ann or per. **ST**: 1–5 dm; some hairs simple and spreading, others 2-forked and appressed. **LF**: 1–4 cm; stipules 1.5–3 mm, gland-toothed; blade lanceolate, abaxially hairy, margin finely gland-toothed. **STAMINATE FL**: sepals 3.5–5 mm, hairy; petals ± = sepals, glabrous or hairy; stamen column 1.5–2 mm. **PISTILLATE FL**: sepals 3.5–5.5 mm, unequal, faces glabrous or hairy, margins stalked-glandular; petals ± = sepals; ovary sparsely hairy, styles fused proximally, lobe tips expanded. **FR**: ± 4.5 mm. **SEED**: ± 2 mm, angled, faintly pitted. Sandy soils, creosote-bush scrub; < 100 m. DSon (Coachella Valley). [*Ditaxis clariana* (Jeps.) Webster, orth. var.] Dec–Mar ★

D. lanceolata (Benth.) Pax & K. Hoffm. (p. 717) Subshrub. **ST**: gen erect, 1–5 dm, brittle, appressed-hairy. **LF**: 2–6 cm; stipules ± 1 mm, entire; blade lanceolate, densely hairy, entire. **STAMINATE FL**: sepals 2.5–3 mm, hairy; petals 3–3.5 mm, abaxially hairy; stamen column ± 1.5 mm. **PISTILLATE FL**: sepals 3–4 mm, entire; petals ± = sepals, lanceolate to ovate, abaxially hairy; ovary densely appressed-hairy, styles gen free, lobe tips expanded. **FR**: 3–5 mm. **SEED**: 2–2.5 mm, angled, pitted. Rocky soils, slopes, canyons; < 600 m. DMoj (Eagle Mtn), DSon; AZ, Mex. Mar–May

D. neomexicana (Müll. Arg.) A. Heller (p. 717) Ann or per. **ST**: 1–3.5 dm, densely appressed-hairy. **LF**: 1–3.5 cm; stipules 1–1.5 mm, entire; blade lanceolate, ± hairy, entire to faintly toothed. **STAMINATE FL**: sepals 2–2.5 mm; petals ± 2 mm, glabrous; stamen column ± 1 mm. **PISTILLATE FL**: sepals 3–4 mm, abaxially hairy, margin ± entire; petals ± 2.5 mm, lanceolate, glabrous or appressed-hairy, hairs not exceeding petal tip; ovary stiff-hairy, styles free, lobe tips not expanded. **FR**: 3–4 mm. **SEED**: ± 2 mm, angled, pitted. Slopes, creosote-bush scrub; < 1000 m. s edge DMoj, DSon; to TX, Mex. Mar–Dec

D. serrata (Torr.) A. Heller Ann or per. **ST**: 1–5 dm. **LF**: 1–5 cm; stipules ≤ 1.5 mm; blade elliptic or lanceolate to ovate, ± toothed, at least distally. **STAMINATE FL**: sepals, petals ≤ 2.5 mm; stamen column ± 1 mm. **PISTILLATE FL**: sepals > petals; styles gen free, lobe tips not expanded. **SEED**: ± 2 mm.

var. ***californica*** (Brandegee) V.W. Steinm. & Felger (p. 717) CALIFORNIA DITAXIS **ST**: 1.5–5 dm, glabrous. **LF**: 1–5 cm; stipules ± 1 mm, faintly gland-toothed; blade lanceolate to elliptic, glabrous, finely toothed. **STAMINATE FL**: sepals 1.5–2.5 mm; petals 2–2.5 mm. **PISTILLATE FL**: sepals 2.5–4.5 mm, faintly gland-toothed; petals ± 1.5 mm, glabrous; ovary glabrous. **FR**: ± 3.5 mm. **SEED**: ± angled, ± pitted. Washes, canyons; 50–1000 m. DMoj (Eagle Mtn), nw DSon (Coachella Valley). [*D. c.* (Brandegee) A. Heller] Apr–Nov ★

var. ***serrata*** (p. 717) **ST**: 1–3.5 dm, densely appressed-hairy. **LF**: 1–3 cm; stipules 1–1.5 mm; blade widely elliptic to ovate, densely hairy, clearly toothed distally. **STAMINATE FL**: sepals 2–2.5 mm; petals ± 2 mm, glabrous. **PISTILLATE FL**: sepals 3–4 mm, abaxially hairy; petals ± 2.5 mm, obovate, abaxially clearly appressed-hairy, hairs exceeding petal tip; ovary stiff-hairy. **FR**: ± 2 mm. **SEED**: rounded, ± striate, not clearly pitted. Sandy or rocky soils, creosote-bush scrub; < 200 m. D; AZ, Mex. Apr–Nov

EUPHORBIA SPURGE

Mark H. Mayfield

Ann, per, glabrous or hairy; gen monoecious. **ST**: ascending to erect, gen < 1 m; branches equally forked. **LF**: cauline, gen alternate, occ opposite; stipules 0 or gland-like; gen ± sessile; lf base symmetrical. **INFL**: ± fl-like, gen clustered; clusters whorled in umbel-like, or cyme-like, bracted arrays; involucre gen < 5 mm, ± bell-shaped, bracts 5, fused, glands 4–5, alternating with bracts, appendages gen 0; fls central. **STAMINATE FL**: 5–many, in 5 clusters around pistillate fl; sepals 0; stamens 1. **PISTILLATE FL**: 1, central, stalked; sepals 0; ovary chambers 3, styles 3, gen fused at base, divided or simple. **FR**: capsule, round to 3-angled or -lobed in ×-section. **SEED**: round or angled in ×-section; surface smooth or sculptured, gen with a knob-like structure at attachment scar. ± 1500 spp.: warm temp to trop, worldwide. (Euphorbus, physician to the King of Mauritania, 1st century) [Wheeler 1936 Bull S Calif Acad Sci 35:127–147] Forms monophyletic group with *Chamaesyce*, often incl in *Euphorbia*. *E. myrsinites* L., *E. rigida* L. Bieb. occ in gardens but not considered naturalized; *E. serrata* L. ◆ considered extirpated from CA; *E. marginata* Pursh occ persisting from gardens, but recent records lacking. *E. graminea* Jacq. ◆ an urban weed.

1. Involucre with white petal-like appendages on glands
  2. Lvs opposite, margin serrate; seed with a minute knob at tip . . . . *E. exstipulata* var. *exstipulata*
  2′ Lvs alternate, margin entire; seed without a knob . . . . *E. misera*
1′ Involucre ± without obvious white petal-like appendages on glands
  3. Infl clustered at branch tips; involucre glands 1–3(5), cupped
    4. Involucre and fr hairy; seed flattened, oblong . . . . *E. eriantha*
    4′ Involucre and fr gen glabrous; seed ellipsoid to ovoid
      5. Bracts red at base, tip green; st glabrous . . . . [*E. cyathophora*]
      5′ Bracts uniformly green or ± white; st hairy
        6. Hairs on distal sts and abaxial lf faces, recurved; seed width ≥ 2.2 mm, surface coarsely tubercled . . . . [*E. davidii*]
        6′ Hairs on distal sts and abaxial lf faces ± straight, spreading; seed width ≤ 2.1 mm, surface finely tubercled . . . . [*E. dentata*]
  3′ Infl open, whorled at branch tips, alternate proximal to whorl; involucre glands 4(5), flat
    7. Lvs opposite throughout, 4-ranked; fr ≥ 8 mm . . . . *E. lathyris*
    7′ Lvs alternate proximal to infl; fr ≤ 6 mm
      8. Lf margin finely to coarsely toothed
        9. Lf tip acute . . . . *E. terracina*
        9′ Lf tip rounded, obtuse or notched
          10. Per; st densely hairy . . . . *E. oblongata*
          10′ Ann; st glabrous or sparsely hairy
            11. Whorled cyme branches 5; staminate fls 8–12; style divided < 1/2 . . . . *E. helioscopia*
            11′ Whorled cyme branches 3; staminate fls 5–10; style ± free, divided nearly to base . . . . *E. spathulata*
      8′ Lf margin gen entire (finely crenate)
        12. Shrub . . . . *E. dendroides*
        12′ Ann to per
          13. Lf linear, lanceolate to oblanceolate; seed surface smooth . . . . *E. virgata*

13′ Lf elliptic to (ob)ovate or spoon-shaped; seed surface gen pitted or low ridged, occ ± smooth

    14. Seed 1–1.5 mm, surface with broad pits . . . . **_E. peplus_**

    14′ Seed ≥ 2 mm, surface with low network of ridges to ± smooth; fr lobes smooth

        15. Ann (bien); gland horns > gland width . . . . **_E. crenulata_**

        15′ Per; gland horns < gland width

            16. Lf sessile; st ascending to erect; gland margin 2-horned, entire to scalloped between horns . . . . **_E. lurida_**

            16′ Lf narrowed to short petiole; st ascending to decumbent; gland margin truncate, scalloped . . . . **_E. schizoloba_**

E. crenulata Engelm. (p. 727) CHINESE CAPS Ann (bien). **ST:** ascending to erect, 1.2–6 dm, glabrous. **LF:** 1.5–3.5 cm, gen sessile, proximal-most petioled; blade elliptic to oblanceolate or ± spoon-shaped, obtuse to abruptly pointed at tip, entire to finely crenate, glabrous. **INFL:** involucre 2 mm, bell-shaped, glabrous; gland 1–2 mm, crescent-shaped, 2-horned. **STAMINATE FL:** 10–20. **PISTILLATE FL:** style divided < 1/2. **FR:** 3–3.5 mm, oblong, lobed, glabrous. **SEED:** 2–2.5 mm, oblong-ellipsoid, round; surface sculptured, net-like. Common. Dry places; < 1600 m. CA-FP; OR, s CO, n NM. Mar–Aug

E. dendroides L. Shrub. **ST:** spreading, 10–20 dm, glabrous, marked with persistent lf scars. **LF:** crowded, 4–8 cm, sessile; blade narrowly elliptic to narrowly oblanceolate, entire, glabrous. **INFL:** involucre 4–5 mm, broadly goblet-shaped, glabrous; gland ± 3 mm, oval, margin incised. **STAMINATE FL:** 10–15. **PISTILLATE FL:** style divided < 1/4. **FR:** 5–6 mm, spheric, deeply lobed, glabrous. **SEED:** ± 3.5 mm, oblong, radially flattened, elliptic in ×-section, surface smooth. Disturbed areas; < 3055 m. SCo, WTR; native to Medit, Canary Islands. All yr

E. eriantha Benth. (p. 727) BEETLE SPURGE Ann. **ST:** erect, branches ascending, 1.5–5 dm, glabrous or hairy and becoming glabrous. **LF:** 2–7 cm, short-petioled; blade linear, acute to obtuse and abruptly pointed, entire, sparsely hairy. **INFL:** 1 or few clustered at branch tips; involucre 1.5–2 mm, obconic, densely hairy; glands 1–3, ± 1.5 mm, round, cupped, lobes 5–7, curved over gland. **STAMINATE FL:** 23–36. **PISTILLATE FL:** style simple. **FR:** 4–5 mm, oblong, lobed, hairy. **SEED:** 3.5–4 mm, oblong, flattened, 4-angled; surface tubercled. Canyons, rocky slopes; < 1000 m. DSon; to TX, Mex. Mar–Apr

E. exstipulata Engelm. var. ***exstipulata*** CLARK MOUNTAIN SPURGE Ann. **ST:** erect, branches spreading from base, < 2.5 dm, minutely hairy. **LF:** opposite, 2–4 cm, petioled; stipules gland-like; blade linear to lanceolate, acute, serrate, sparsely hairy. **INFL:** 1 per node; involucre 1–2 mm, bell-shaped, minutely hairy; gland < 0.5 mm, oblong; appendage white, petal-like, ≥ gland in width, 2–4-lobed. **STAMINATE FL:** 8–14. **PISTILLATE FL:** styles ± free, divided nearly to base. **FR:** 2.5–3.4 mm, spheric, lobed, puberulent on angles. **SEED:** 2–3 mm, pyramidal, 3-angled, transversely 2–3-ridged, tubercled. Rocky slopes; 1800–2000 m. e DMtns (Clark Mtn Range, New York Mtns); to TX, Mex. Other vars. in sw US, Mex. Sep–Oct ★

E. helioscopia L. WARTWEED Ann. **ST:** erect, 1–5 dm, glabrous or sparsely hairy. **LF:** 1–3 cm, sessile; blade obovate to spoon-shaped, rounded to notched at tip, finely toothed, glabrous. **INFL:** whorled cyme branches 5; involucre ± 2 mm, obconic, glabrous; gland < 1 mm, oblong, entire. **STAMINATE FL:** 8–12. **PISTILLATE FL:** style divided < 1/2. **FR:** 2.5–3 mm, spheric, lobed, glabrous. **SEED:** 2–2.5 mm, ovoid, flattened, biconvex; surface ridges thin, net-like. Disturbed areas; < 200 m. n&c NCo, n&c NCoRO, SnFrB, SCo; native to Eur. Apr–Jul

E. lathyris L. (p. 727) CAPER SPURGE Bien. **ST:** 5–15 dm, glabrous. **LF:** opposite, 5–15 cm, sessile; blade linear to lanceolate, ± clasping at base, acute, entire, glabrous,. **INFL:** involucre 2.5–4 mm, bell-shaped, glabrous; gland ± 2 mm, crescent-shaped, 2-horned. **STAMINATE FL:** 15–40. **PISTILLATE FL:** style divided ± 1/2. **FR:** 8–15 mm, spheric, glabrous. **SEED:** 4–5 mm, oblong, round; surface shallowly net-like, brown; knob 0. Disturbed areas; < 200 m. NCo, GV, CCo, SCo, expected elsewhere; native to Eur. All yr

E. lurida Engelm. Per. **ST:** ascending to erect, several from base, 1–3.5 dm, glabrous. **LF:** 0.5–2 cm, sessile; blade obovate to spoon-shaped, obtuse to rounded at tip, entire, glabrous. **INFL:** involucre < 3 mm, obconic, glabrous; gland < 1.5 mm, oblong, margin entire to scalloped, shortly 2-horned. **STAMINATE FL:** < 20. **PISTILLATE FL:** style divided < 1/2. **FR:** ± 4 mm, ovoid, round, glabrous. **SEED:** ± 2.5 mm, oblong, round; surface shallowly net-like. Common. Dry slopes, flats; 1300–2800 m. SW; to UT, NM. [*E. palmeri* S. Watson] Apr–Jul

E. misera Benth. CLIFF SPURGE Shrub. **ST:** 5–10 dm, hairy, glabrous in age. **LF:** 0.4–1.5 cm, petioled; stipule thread-like; blade ovate to round, rounded at tip, entire, hairy. **INFL:** 1 at branch tip; involucre 2–3 mm, bell-shaped, hairy; glands 5, 1.5–2 mm, oblong; appendage petal-like, ≥ gland in width, scalloped, white. **STAMINATE FL:** 30–40. **PISTILLATE FL:** styles ± free, divided 1/2. **FR:** 4–5 mm, spheric, lobed, becoming glabrous. **SEED:** 2.5–3 mm, ovoid, round; surface wrinkled, white to gray. Rocky slopes, coastal bluffs; < 500 m. SCo, s ChI, w DSon; Baja CA. Jan–Aug ★

E. oblongata Griseb. Per. **ST:** several from base, 5–8 dm, densely hairy. **LF:** 4–6.5 cm, sessile; blade oblong to lanceolate, obtuse at tip, finely toothed, glabrous. **INFL:** involucre 1.5–2.5 mm, bell-shaped, glabrous; gland ± 1 mm, oval. **STAMINATE FL:** 20–40. **PISTILLATE FL:** style divided < 1/2. **FR:** 3–4.5 mm, spheric, tubercled. **SEED:** 2.5 mm, ovoid, smooth, brown. Disturbed areas; < 200 m. GV, SnFrB, expected elsewhere; native to Eur. Jun–Aug ◆

E. peplus L. (p. 727) PETTY SPURGE Ann. **ST:** 1–4.5 dm, glabrous. **LF:** 1–3.5 cm, petioled; blade obovate to oblanceolate (elliptic), obtuse at tip, entire, glabrous. **INFL:** involucre 1–1.5 mm, bell-shaped, glabrous; gland < 0.5 mm, crescent-shaped, 2-horned. **STAMINATE FL:** 10–15. **PISTILLATE FL:** styles ± free, divided nearly to base. **FR:** ± 2 mm, spheric, glabrous, lobed, lobes 2-keeled. **SEED:** 1–1.5 mm, oblong, 4-angled; surface pitted, white to gray. Common. Disturbed areas; < 300 m. CA-FP, DSon; to Can, e US; native to Eur. Feb–Aug

E. schizoloba Engelm. MOJAVE SPURGE Per. **ST:** ascending to decumbent, numerous, 1–4 dm, intricately woven, glabrous. **LF:** 0.6–2 cm, narrowed to short petiole; blade elliptic to obovate, acute to abruptly pointed, entire, glabrous. **INFL:** involucre 2–3 mm, bell-shaped, glabrous; gland 1–2 mm, crescent-shaped, margin scalloped. **STAMINATE FL:** < 20. **PISTILLATE FL:** style divided < 1/2. **FR:** 4–5 mm, oblong, lobed, glabrous. **SEED:** 2–3 mm, oblong-ellipsoid, round, white to gray; surface net-like to ± smooth. Rocky or sandy slopes; 1000–2300 m. W&I, DMtns; NV, AZ. [*E. incisa* Engelm.] Mar–May

E. spathulata Lam. (p. 727) Ann. **ST:** 1.5–4.5 dm, glabrous. **LF:** 1–3 cm, petioled to sessile; blade obovate to spoon-shaped, obtuse to notched at tip, finely toothed, glabrous. **INFL:** whorled cyme branches 3; involucre < 1.5 mm, bell-shaped, glabrous; gland < 1 mm, oblong. **STAMINATE FL:** 5–10. **PISTILLATE FL:** styles ± free, divided nearly to base. **FR:** 2–3 mm, spheric, lobed, tubercled. **SEED:** 1.5–2 mm, ovoid, tangentially flattened, biconvex; surface net-like. Open, gen disturbed places; < 1300 m. CA-FP; to WA, e N.Am, S.Am. Mar–Jun

E. terracina L. GERALDTON CARNATION WEED Ann or per. **ST:** ascending to erect, several from base, 3–8 dm, glabrous. **LF:** 2.5–6 cm, sessile; blade linear to narrowly lanceolate, acute, finely serrate, often with larger primary marginal teeth, glabrous. **INFL:** involucre < 2 mm, glabrous; glands < 1.5 mm, crescent-shaped, 2-horned, horns 1–1.5 mm, thread-like. **STAMINATE FL:** 10–15. **PISTILLATE FL:** style simple. **FR:** 4–5 mm, depressed spheric, deeply lobed, glabrous. **SEED:** 2.1–2.3, oblong, round; surface smooth, gray to ± white. Disturbed areas near habitations, sandy soil, seeps, ocean bluffs; < 160 m. SCo; native to Eur. Mar–Jul ◆

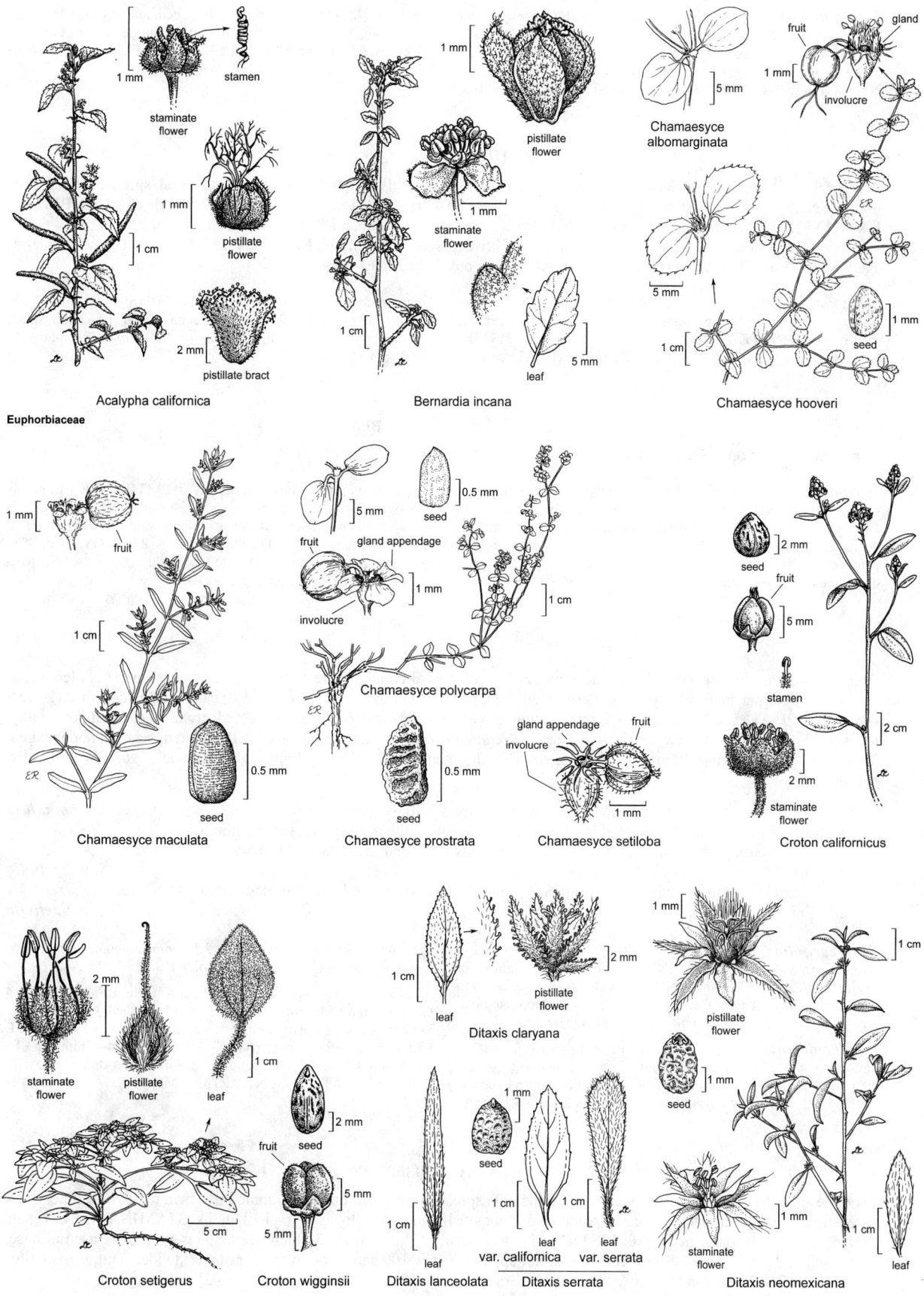

Acalypha californica

Euphorbiaceae

stamen
staminate flower
pistillate flower
pistillate bract

Bernardia incana
pistillate flower
staminate flower
leaf

Chamaesyce albomarginata
fruit
gland
involucre
Chamaesyce hooveri
seed

Chamaesyce maculata
fruit
seed

Chamaesyce polycarpa
fruit
gland appendage
involucre
seed

Chamaesyce prostrata
seed

Chamaesyce setiloba
gland appendage
involucre
fruit

Croton californicus
seed
fruit
stamen
staminate flower

Croton setigerus
staminate flower
pistillate flower
leaf

Croton wigginsii
fruit
seed

Ditaxis claryana
leaf
pistillate flower

Ditaxis lanceolata
leaf

Ditaxis serrata
seed
leaf
var. californica
leaf
var. serrata

Ditaxis neomexicana
pistillate flower
seed
staminate flower
leaf

E. virgata Waldst. & Kit.   LEAFY SPURGE   Per. **ST**: erect, several from base, 3–8 dm, glabrous to hairy. **LF**: 2–6 cm, sessile; blade linear to (ob)lanceolate, acute, entire, glabrous. **INFL**: involucre 1.5–2.5 mm, bell-shaped, glabrous; gland 1.5–2 mm, crescent-shaped, 2-horned. **STAMINATE FL**: 11–21. **PISTILLATE FL**: style divided ± 1/2. **FR**: 3–4 mm, spheric, lobed, smooth. **SEED**: 2–2.5 mm, oblong, transversely rounded; surface smooth. Fields, pastures; < 1400 m. s NCo (Sonoma Co.), e KR, CaR, MP; to e US; native to Eur. [*E. esula* L., misappl.] May–Aug   ◆

MERCURIALIS   MERCURY

Ann [per], gen glabrous; sap clear; (monoecious) dioecious. **ST**: central erect, branched; lateral < central, spreading to ascending. **LF**: cauline, opposite; stipules persistent; hairs simple. **STAMINATE INFL**: spike-like, axillary; fls clustered. **PISTILLATE INFL**: fls clustered, axillary. **STAMINATE FL**: calyx 3-lobed; stamens 8–15(20), free; nectary 0. **PISTILLATE FL**: sepals 3; staminodes 2, elongated; ovary 2-chambered, bristly, styles free, simple. **FR**: bristly. **SEED**: gen pitted; scar appendaged. 8 spp.: Eurasia, n Afr. (Geronimo Mercuriali, Italian physician, 1530–1606)

M. annua L. (p. 727)   Ann, 1–3 dm. **LF**: stipules 1–1.5 mm, lance-deltate; petiole 0.5–2(2.5) cm; blade 2–5 cm, ± ovate, margin serrate, ciliate. **STAMINATE INFL**: < 2 cm, short-peduncled. **PISTILLATE INFL**: fls 2–3; pedicel < 5 mm. **STAMINATE FL**: calyx ± 1 mm; stamens exserted. **PISTILLATE FL**: calyx ± 1 mm; ovary strigose, styles ± 1 mm. **FR**: 2–3 mm diam. **SEED**: 1.5–2 mm, shiny, pitted. 2*n*=16,32. Open, disturbed areas; < 200 m. SnFrB; native to Eur. Feb–Oct

RICINUS   CASTOR BEAN

1 sp. (Latin: tick, from seed shape)

R. communis L. (p. 727)   Shrub, occ tree-like, 1–3 m, ± glabrous; sap clear; monoecious. **ST**: trunk ascending to erect, branched above. **LF**: cauline, alternate, peltate; stipules fused, 1–1.5 cm, sheath-like, deciduous; petiole 1–3 dm, glandular distally; blade 1–5 dm, ± round, palmately 7–11-lobed, sharply toothed. **INFL**: panicle, terminal, 1–3 dm; staminate fls proximal to pistillate fls. **STAMINATE FL**: sepals 3–5; stamens many, clustered; nectary 0. **PISTILLATE FL**: sepals 3–5; ovary 3-chambered, bristly, styles 2-lobed, plumose, ± red. **FR**: 1.2–2 cm diam, ± spiny. **SEED**: 9–22 mm, smooth, shiny, mottled; scar appendaged. 2*n*=20. Disturbed areas; < 300 m. GV, CCo, SCo, expected elsewhere; e US; native to Eur. Highly TOXIC: seeds attractive to children, fatal when ingested. All yr   ❖

STILLINGIA

Ann, per, < 2 m; sap clear or milky; monoecious. **ST**: erect. **LF**: alternate, entire or toothed; stipules minute, petioled; blade base gen with 2 glands. **INFL**: spike, axillary or terminal; bracts glandular. **STAMINATE FL**: calyx 2-lobed; stamens 2; nectary disk 0. **PISTILLATE FL**: sepals 3, overlapping in bud, reduced, or 0; ovary 3-chambered, styles free, fused proximally, lobes 0. **FR**: gen 3-lobed, separating into 3 1-seeded segments; central axis persistent. **SEED**: pointed; scar not appendaged. 30 spp.: trop, warm temp. (Benjamin Stillingfleet, British botanist, 1702–1771) [Johnston & Warnock 1963 SW Naturalist 8:100–106]

1. Lf blades elliptic to ovate, sharply toothed, 3-veined; spikes axillary; seeds striate. . . . .*S. spinulosa*
1′ Lvs linear, entire or sparsely toothed in proximal 1/2, 1-veined; spikes terminal; seeds smooth
  2. Spikes exceeding lvs, open in proximal 1/3, pistillate fls well separated; base of lf blade entire; seeds ± 2 mm . . . .*S. linearifolia*
  2′ Spikes not exceeding lvs, dense in proximal 1/3, pistillate fls crowded; base of lf blade few-toothed; seeds 2.5–3 mm . . . .*S. paucidentata*

S. linearifolia S. Watson (p. 727)   Per < 7 dm. **LF**: blade 1–4 cm, < 2 mm wide, linear, entire. **INFL**: 2–7 cm, terminal; glands of pistillate bracts stalked, ± 1 mm. **PISTILLATE FL**: 3–6 per infl, well separated; styles ± 1 mm. **FR**: ± 3.5 mm. **SEED**: ± 2 mm, smooth. Dry slopes, washes; < 1500 m. SW, D; AZ, Mex. Mar–May

S. paucidentata S. Watson (p. 727)   Per < 5 dm. **LF**: blade 3–8 cm, 1.5–5 mm wide, linear, few-toothed in proximal 1/2, teeth ± spiny. **INFL**: 2–7 cm, terminal; glands of pistillate bracts subsessile, ± 1–1.5 mm. **PISTILLATE FL**: 2–3 per infl, crowded; styles ± 2.5 mm. **FR**: 4–4.5 mm. **SEED**: 2.5–3 mm, smooth. Slopes, flats, creosote-bush scrub; < 1500 m. DMoj, n DSon; AZ. Apr–Jun

S. spinulosa Torr. (p. 727)   Ann or per < 10 dm. **LF**: blade gen 2–4 cm, 5–12 mm wide, elliptic to ovate, sharply toothed. **INFL**: 1–2 cm, axillary; glands of pistillate bracts stalked, ± 2 mm. **PISTILLATE FL**: 1–2 per infl, ± open; styles 3–3.5 mm. **FR**: 4–5 mm. **SEED**: 3–3.5 mm, striate, minutely roughened. Sandy soils, dunes, creosote-bush scrub; < 900 m. D; s NV, AZ. Mar–May

TRAGIA   NOSEBURN

Per < 0.5 m; hairs stinging, nettle-like; monoecious. **ST**: spreading to erect, branched, sometimes twining. **LF**: cauline, alternate; stipules persistent. **INFL**: raceme, terminal or opposite lf; staminate fls distal to pistillate fls. **STAMINATE FL**: sepals [3]4–5; stamens 3–6[50]; nectary 0. **PISTILLATE FL**: sepals 5[4–8]; ovary 3-chambered, styles simple, ± fused at base. **FR**: ± spheric. **SEED**: smooth or ± rough; scar not appendaged. ± 100 spp.: trop, warm temp worldwide. (Tragus, name for Hieronymus Bock, German herbalist, 1498–1554) [Miller & Webster 1967 Rhodora 69:241–305]

T. ramosa Torr. (p. 727) DESERT TRAGIA Pl rough-hairy. **ST:** 1–3 dm. **LF:** stipules 1–4.5 mm, lanceolate to ovate; petiole 2–20 mm; blade 1–2 cm, lanceolate to ovate, base truncate to ± lobed, margin coarsely, sharply toothed. **INFL:** 0.5–1 cm, ± spreading; pedicels 1–2 mm; staminate fls 2–4; pistillate fl 1. **STAMINATE FL:** sepals 4–5, ± 1 mm, recurved; stamens 3–6, filaments ± flattened. **PISTIL-** **LATE FL:** sepals 5, 1.5–2 mm; ovary < 2 mm diam, puberulent to finely bristly, styles fused in proximal 1/3. **FR:** 3–4 mm, 6–8 mm wide, depressed-spheric, sparsely and finely bristly. **SEED:** 2.5–3.5 mm, ± spheric. Dry, rocky slopes, scrub, pinyon/juniper woodland; 900–1900 m. DMtns; to c US, TX, Mex. [*T. stylaris* Müll. Arg.] Apr–May ★

TRIADICA TALLOWTREE

Bruce G. Baldwin

Tree [shrub]; sap milky; monoecious. **ST:** branches slender, glabrous. **LF:** cauline, alternate; petiole tip with adaxial pair of large, ± spheroid glands; blade entire, adaxially glabrous, abaxially papillate-glaucous, with 0–few glands near margins. **INFL:** spike-like panicles or racemes, terminal or axillary, elongate, yellow-green; bracts 1–2 mm, with pair of large glands; pedicels 2–3 mm in ours; pistillate fls basal, 1/bract; staminate fls distal or throughout, gen 5–8/bract. **STAMINATE FL:** sepals 3, fused; stamens 2 or 3, filaments gen < 1 mm; nectary disk 0. **PISTILLATE FL:** sepals 3, fused; nectary disk 0; ovary 3-chambered, styles 3, fused proximally, simple. **FR:** ± spheric or ± 3-lobed, smooth. **SEED:** ± white, often staying attached to persistent fr axis. 3 spp.: native to e & se Asia. (Greek: 3, for lobes of calyx, ovary, fr) [Bower et al. 2009 Inv Pl Sci Managem 2:386–395; Esser 2002 Harvard Pap Bot 7:17–21]

T. sebifera (L.) Small CHINESE TALLOWTREE Pl to 13 m; deciduous. **LF:** petiole 2–7 cm; blade 3–9 cm, widely elliptic to ± (ob)ovate or triangular-(ob)ovate, acuminate. **INFL:** 4–16 cm. **PISTIL-** **LATE FL:** 0–17. **FR:** ± 13 mm; pedicel 4–13 mm. **SEED:** 6–9 mm, ± spheric. Stream edges; 10–300 m. n SNF, GV; native to Asia. [*Sapium s.* (L.) Roxb.] Spring–summer ❖

FABACEAE (Leguminosae) LEGUME FAMILY

Martin F. Wojciechowski, except as noted

Ann to tree. **LF:** gen alternate, gen compound, gen stipuled, gen entire, pinnately veined **INFL:** gen raceme, spike, umbel or head; or fls 1–few in axils. **FL:** gen bisexual, gen bilateral; hypanthium 0 or flat to tubular; sepals gen 5, gen fused; petals gen 5, free, fused, or lower 2 ± united into keel (see 3, Key to Groups, for banner, wings); stamens 10 or many (or [1], 5, 6, 7, 9), free or fused or 10 with 9 filaments at least partly fused, 1 (uppermost) free; pistil 1, ovary superior, gen 1-chambered, ovules 1–many, style, stigma 1. **FR:** legume, incl a stalk-like base (above receptacle) or not. **SEED:** 1–many, often ± reniform, gen hard, smooth. ± 730 genera, 19400 spp.: worldwide; with grasses, requisite in agriculture, most natural ecosystems. Many cult, most importantly *Arachis*, peanut; *Glycine*, soybean; *Phaseolus*, beans; *Medicago*, alfalfa; *Trifolium*, clovers; many orns. [Lewis et al. (eds) 2005 Legumes of the World. RBG, Kew] Unless stated otherwise, fr length incl stalk-like base, number of 2° lflets is per 1° lflet. Upper suture of fr adaxial, lower abaxial. *Anthyllis vulneraria* L. evidently a waif, a contaminant of legume seed from Eur. *Laburnum anagyroides* Medik., collected on Mount St. Helena in 1987, may be naturalized. *Ceratonia siliqua* L., carob tree (Group 2), differs from *Gleditsia triacanthos* L. in having evergreen (vs deciduous) lvs that are 1-pinnate (vs 1-pinnate on spurs on old sts, 2-pinnate on new sts) with 2–5(8) (vs 7–17) 1° lflets, commonly cult, now naturalized in s CA. *Aeschynomene rudis* Benth. ❖, *Halimodendron halodendron* (Pall.) Voss ❖ (possibly extirpated), *Lens culinaris* Medik. are agricultural weeds. *Caragana arborescens* Lam. only cult. *Ononis alopecuroides* L. ❖, *Sphaerophysa salsula* (Pall.) DC. ❖ all evidently extirpated. *Cercidium* moved to *Parkinsonia*; *Chamaecytisus* to *Cytisus*; *Psoralidium lanceolatum* to *Ladeania*. Scientific Editors: Martin F. Wojciechowski, Thomas J. Rosatti.

Key to Groups

1. Fl radial; calyx, corolla gen inconspicuous; petals free or fused, lobes not overlapped in bud; stamens 10 or gen many, often long-exserted; lf 2-pinnate (simple in alien *Acacia*). . . . . **Group 1: Mimosoideae**
1′ Fl gen bilateral, less often ± radial; calyx, corolla gen conspicuous; petals overlapped in bud, free or 2 lowermost ± fused; stamens 10 (or 5, 6, 7, 9), gen ± incl; lf 1- or 2-pinnate (less often simple or palmately compound)
  2. Fl ± radial to ± bilateral; sepals fused only at very base; lf 1- or 2-pinnate . . . **Group 2: Caesalpinioideae** (exc *Cercis*)
  2′ Fl bilateral; sepals free to ± entirely fused; lf simple, 1-pinnate, or palmate
    3. Upper petal (banner) inside lateral ones (wings) in bud; stamens free; lf simple, reniform or bi-lobed
    . . . . **Group 2: Caesalpinioideae** (*Cercis* only)
    3′ Upper petal (banner) outside lateral ones (wings) in bud (only banner present in *Amorpha*, petal position not evident in some *Dalea*); stamens gen with all or 9 filaments fused (free in *Thermopsis*, *Pickeringia*, *Calia*); lf 1-pinnate (lflets often 3), some palmately compound (esp *Lupinus*) or simple . . . **Group 3: Papilionoideae**

Group 1

Mimosoideae

1. Stamens 10; 1° lflets gen 2–4. . . . . **PROSOPIS**
1′ Stamens > 10; 1° lflets gen > 4, or lf simple

2. Filaments fused below, free above; lf 2-pinnate
  3. Infl axillary spike-like racemes or ± terminal panicle of head-like racemes; petiole with a gland; fr not
     or slowly dehiscent, valves not recurving; ScV, CCo, SnFrB, SCoRO, SW . . . . **ALBIZIA**
  3′ Infl axillary heads; petiole without a gland; fr dehiscent, valves recurving; DSon . . . . **CALLIANDRA**
2′ Filaments free; lf 2-pinnate or simple
  4. Pl unarmed . . . . **²ACACIA**
  4′ Pl armed with prickles or stipular spines
    5. Pl armed with prickles . . . . **SENEGALIA**
    5′ Pl armed with stipular spines
      6. Lf simple . . . . **²ACACIA**
      6′ Lf 2-pinnate . . . . **VACHELLIA**

Group 2

Caesalpinioideae

1. Lf simple; corolla pink-purple . . . . **CERCIS**
1′ Lf compound; corolla gen ± yellow
  2. Lf 1-pinnate, pl unarmed or main lf axis gen > 2 cm, a weak spine at tip; corolla gen yellow . . . . **SENNA**
  2′ Lf gen 2-pinnate (both 1- and 2-pinnate in some), pl armed, main lf axis gen < 2 cm (exc *Gleditsia*), a
     strong spine at tip, or (if lf interpreted as 1-pinnate; see *Parkinsonia*), lf subtended by a strong spine gen
     < 2 cm and main lf axis not a spine at tip; corolla ± yellow to orange, less often ± red or with orange to
     ± red marks
    3. Fl ± 3–5 mm; sepals, petals ± alike in color, texture; pl tree, gen armed . . . . **GLEDITSIA**
    3′ Fl > 5 mm; sepals, petals not alike in color, texture; pl per, shrub, small tree, armed or not
      4. Pl unarmed
        5. Shrub, small tree, < 4 m; 2° lflets 14–22 . . . . **²CAESALPINIA**
        5′ Per, shrub, < 30 cm or < 2.5 m; 2° lflets 3–13 . . . . **HOFFMANNSEGGIA**
      4′ Pl armed
        6. Pl with scattered prickles; sepals not alike; fr indehiscent . . . . **²CAESALPINIA**
        6′ Pl with stipular spines at nodes or thorns in lf axils (see lf scars) or main lf axis a strong spine; sepals
           alike; fr indehiscent to partly dehiscent late . . . . **PARKINSONIA**

Group 3

Papilionoideae

1. Shrub to tree
  2. Pl gen armed with prickles, spines, or thorns
    3. Lvs simple, sometimes small or falling early
      4. Pl gland-dotted; corolla indigo blue to pink-purple; seed 1 . . . . **⁴PSOROTHAMNUS**
      4′ Pl not gland-dotted; corolla yellow to ± red; seeds (1)2–several
        5. Corolla red-purple; fr narrowed between seeds, glabrous . . . . **ALHAGI**
        5′ Corolla yellow; fr not narrowed between seeds, densely hairy . . . . **ULEX**
    3′ Lvs all or mostly compound, gen persistent
      6. Lf simple or palmately compound; all filaments free . . . . **PICKERINGIA**
      6′ Lf pinnately compound; all or 9 filaments fused
        7. Pl gland-dotted; fr indehiscent; seed 1 . . . . **⁴PSOROTHAMNUS**
        7′ Pl not gland-dotted; fr dehiscent (sometimes slowly so); seeds gen several
          8. Lf even-pinnate; corolla wings purple-pink suffused with white, banner, keel yellow-white to pink or
             purple; fr not flat, gen narrowed between seeds . . . . **OLNEYA**
          8′ Lf odd-pinnate; corolla white or pink; fr flat, not narrowed between seeds . . . . **²ROBINIA**
  2′ Pl unarmed
    9. Petal 1 (the banner) — infl spike-like; fr indehiscent; seed 1 . . . . **AMORPHA**
    9′ Petals 5
      10. Filaments all fused; lvs simple or ternately or palmately compound
        11. Lvs simple, < 2.5 cm; sts gen ± lfless . . . . **SPARTIUM**
        11′ Lvs simple or gen compound, gen > 2.5 cm; sts lfy
          12. Lf of 3–17 lflets, palmately compound; corolla blue, purple, white, or yellow, banner glabrous to
             densely hairy, keel gen beaked . . . . **²LUPINUS**
          12′ Lf of 1–3 lflets, appearing simple or not; corolla yellow or white, banner gen glabrous, keel obtuse
            13. Style gen abruptly curved at ± middle or gently curved ± throughout; upper lip of calyx ± 2-lobed . . . . **CYTISUS**
            13′ Style ± abruptly bent at tip; upper lip of calyx 2-lobed . . . . **GENISTA**
      10′ Filaments all free or 9 fused, 1 (uppermost) free or 0; lvs gen odd-pinnately compound (or lflet
         number, arrangement gen irregular, as sometimes in *Acmispon*)
        14. Infl gen an umbel or fls 1–2; corolla gen yellow . . . . **³ACMISPON**

14′ Infl a raceme; corolla yellow or not
  15. Corolla yellow with dark marks; fr papery, inflated . . . . **COLUTEA**
  15′ Corolla white or pink to purple; fr not papery, not inflated
    16. Fl 6–10 mm; pl gland-dotted; fr indehiscent; seed 1 (see also *Marina*) . . . . ⁴**PSOROTHAMNUS**
    16′ Fl 14–25 mm; pl not gland-dotted; fr dehiscent or not; seeds several
      17. Fr dehiscent, flat, not narrowed between seeds; filaments 9 fused, 1 free; corolla ± white to pink. . . ²**ROBINIA**
      17′ Fr indehiscent, not flat, ± narrowed between seeds; filaments all free; corolla ± blue-purple
      . . . . [*Calia secundiflora*]
1′ Ann, bien, per, vine, subshrub
18. Lflets 0, but stipules lflet-like — lf axis ending as a tendril . . . . [*Lathyrus aphaca*]
18′ Lflets 2–many
  19. Lf palmately compound, lflets gen 3–9
    20. All filaments free; corolla yellow, 15–25 mm . . . . **THERMOPSIS**
    20′ All or 9 filaments fused; corolla not yellow or, if so, gen < 15 mm
      21. Lflets gland-dotted; fr indehiscent or transversely dehiscent; seed 1
        22. Lvs ± basal or clustered near st tips; lflets elliptic to oblanceolate or widely obovate; fr incl in calyx
          exc for beak. . . . . **PEDIOMELUM**
        22′ Lvs cauline; lflets obovate to linear; fr exserted from calyx. . . . . **LADEANIA**
      21′ Lflets not gland-dotted; fr indehiscent or dehiscent through longitudinal sutures; seeds gen several
        23. Filaments of all stamens fused; lflets gen 5–9, entire . . . . ²**LUPINUS**
        23′ Filaments of 9 stamens fused, the 10th (uppermost) free; lflets 3–5, entire, toothed, or wavy
          24. Lflets 3–9, lower 2 in stipular position or not, others ± palmately arranged, stipules gland-like,
            reduced to bumps, or inconspicuous, infl an umbel or 1–3-fld, corolla gen yellow (see also *Acmispon*). . . **LOTUS**
          24′ Lflets gen 3, lower 2 not in stipular position, stipules gen papery or membranous, not reduced to
            bumps, not inconspicuous, rarely lflet-like, if so then infl not an umbel and corolla not yellow
            25. Lflet entire; fr not enclosed in corolla. . . . . ³**ASTRAGALUS**
            25′ Lflet ± toothed or wavy; fr gen enclosed in corolla . . . . ²**TRIFOLIUM**
  19′ Lf pinnately to subpalmately compound (axis apparent beyond lowermost lflets), lflets 2–many
    26. Lflets 2; main lf axis ending as tendril or bristle. . . . . ²**LATHYRUS**
    26′ Lflets ≥ 3; main lf axis ending as lflet, tendril, or bristle
      27. Lflets 3; fr gen indehiscent (gen dehiscent in *Phaseolus*)
        28. Keel petals spirally coiled; lflet gen lobed — trailing or twining vine. **PHASEOLUS**
        28′ Keel petals not spirally coiled; lflet entire, toothed, or wavy
          29. Lflet toothed or wavy — fr ovate or reniform, gen 1-seeded, or ± coiled, several-seeded
            30. Corolla persistent, enclosing fr, yellow, 3.5–5 mm . . . . ²**TRIFOLIUM**
            30′ Corolla deciduous, yellow or not, 3.5–5 mm or not
              31. Fr spirally coiled (or sickle-shaped or straight), gen prickly; seeds 1–several. . . . . ²**MEDICAGO**
              31′ Fr not spirally coiled, not prickly; seeds 1–2
                32. Fr reniform, ridges gen net-like; corolla 2–3 mm. . . . . ²**MEDICAGO**
                32′ Fr ovate, ridges transverse to finely net-like; corolla 2.5–7 mm . . . . **MELILOTUS**
          29′ Lflet not toothed, not wavy
            33. Pl not gland-dotted; seeds gen 2–several; fr exserted from calyx or incl exc for beak
              34. Infl an umbel; fr exserted from calyx or incl exc for beak. . . . . ³**ACMISPON**
              34′ Infl a raceme; fr exserted from calyx . . . . ³**ASTRAGALUS**
            33′ Pl ± gland-dotted; seed 1; fr gen incl in calyx exc for beak
              35. Corolla cream to yellow; calyx conspicuously swollen in fr . . . . **RUPERTIA**
              35′ Corolla at least partly blue to purple; calyx swollen or not in fr
                36. Calyx swollen in fr; bracts at each node, esp lower, of infl united into fan-shaped, 3–5 toothed
                  blade. . . . . **BITUMINARIA**
                36′ Calyx not (or only ±) swollen in fr; bracts at nodes of infl not united . . . . **HOITA**
    27′ Lflets > 3 on all or most lvs; fr dehiscent or not
      37. Lf even-pinnate, main axis ending as a bristle or tendril or not
        38. Lflets ± 20–60; fr 4–8 cm or 15–20 cm; pl ann, shrub, small tree . . . . **SESBANIA**
        38′ Lflets < 30; fr gen < 8 cm; pl ann, per
          39. Stipules lflet-like, often > lflets; style longitudinally folded; lflets 4–6, glabrous . . . . **PISUM**
          39′ Stipules gen not lflet-like but sometimes ± = lflets; style not longitudinally folded; lflets 4–many,
            hairy or glabrous
            40. Style ± flat, puberulent near ± middle for ± 1/3–1/2 length adaxially; lflets ± rolled in bud . . . . ²**LATHYRUS**
            40′ Style gen not ± flat, puberulent at tip, all around or esp abaxially; lflets folded in bud . . . . **VICIA**
      37′ Lf odd-pinnate, main axis ending as a lflet
        41. Corolla wings << keel; lflets adaxially finely red-dotted; fr 1-seeded, leathery, strongly
          net-ridged . . . . **ONOBRYCHIS**
        41′ Corolla wings ± = keel; lflets adaxially dark gland-dotted or not, not red-dotted; fr not
          simultaneously 1-seeded, leathery, and strongly net-ridged

  42. Pl gland-dotted on sts, lflets, or both; fr indehiscent
    43. Fr several-seeded, long-exserted from calyx, glabrous or with bristles or prickles; lflets 6–10
      mm wide . . . . **GLYCYRRHIZA**
    43′ Fr 1-seeded, incl in calyx, glandular; lflets gen < 6 mm wide
      44. Petals from receptacle; stamens 10; infl head-like or not. ⁴**PSOROTHAMNUS**
      44′ Petals, exc banner, from side or top of column of fused filaments; stamens 5 or 9–10; infl not head-like
        45. Pl prostrate to decumbent
          46. St gland-dotted. . . . . ²**DALEA**
          46′ St not gland-dotted. . . . . ³**MARINA**
        45′ Pl ascending to erect
          47. Infl a dense spike; stamens 5 . . . . ²**DALEA**
          47′ Infl an open raceme; stamens 9–10 . . . . ³**MARINA**
  42′ Pl not obviously gland-dotted; fr dehiscent or not
    48. Fls 1–2 or several to many in umbel
      49. Stipules gland-like, often not apparent; fr exserted or not, ovate to oblong . . . . ³**ACMISPON**
      49′ Stipules conspicuous, scarious or lflet-like; fr exserted, gen linear to lanceolate
        50. Infl gen 10–20-fld umbel; fr indehiscent, segments 1–12. . . . . **CORONILLA**
        50′ Infl several-fld umbel or 1–2-fld; fr dehiscent, segments 0 . . . . **HOSACKIA**
    48′ Fls (1–2 or) many in spike to raceme that is head-like or not
      51. Seed 1; lflet tip with gland; infl a spike, dense, ± 1 cm. . . . . ³**MARINA**
      51′ Seeds several; lflet tip without gland; infl a spike or raceme, dense or not, gen > 1 cm
        52. Fr indehiscent, breaking into 1-seeded segments; corolla 6–9 mm; lflet gen < 6 mm . . . . . **ORNITHOPUS**
        52′ Fr gen dehiscent (sometimes slowly so), not breaking into 1-seeded segments; corolla often >
          9 mm; lflet often > 6 mm
          53. Style tip or stigma finely hairy; stipules spiny, free . . . . **PETERIA**
          53′ Style tip and stigma glabrous; stipules not spiny, free or fused
            54. Keel tip rounded to acute (or short-beaked); fr 1-chambered, if ± 2-chambered then septum
              from lower suture, rarely fusing with narrow flange from upper suture . . . . ³**ASTRAGALUS**
            54′ Keel tip beaked; fr ± 2-chambered, septum from upper suture, partial to complete. . . . . . . **OXYTROPIS**

ACACIA

David Seigler & John E. Ebinger

Shrub, tree, armed or not; gen evergreen. **LF**: even-2-pinnate or, if simple, true blades 0, petioles, main axes blade-like, with 1 prominent midvein or ≥ 2 gen prominent longitudinal veins; gen alternate, gen with a swollen, joint-like thickening at base that governs orientation, main axis with raised glands or not. **INFL**: head, gen axillary, 1 or in raceme or panicle, or fls in spike; staminate fls often present. **FL**: radial; sepals, petals 4–5, inconspicuous; stamens many, conspicuous, exserted, free; ovary simple. **FR**: gen dehiscent, occ tardily so, flat or ± cylindric. **SEED**: aril gen enlarged, forming cap or completely encircling seed. ± 960 spp.: trop, subtrop, esp Australia. (Greek: sharp point) [Orchard & Wilson 2001a, 2001b, (eds) Fl Australia. Vol 11. Mimosaceae, *Acacia*, part A and B. ABRS] Recognition of *Acacia*, *Senegalia* (incl *A. greggii*), *Vachellia* (incl *A. farnesiana*) current consensus; many Australian spp. cult, incl *A. cultriformis* G. Don, *A. elata* Benth., some naturalized, spreading in CA.

1. Lvs 2-pinnate
  2. 1° lflets 2–5 pairs . . . . *A. baileyana*
  2′ 1° lflets gen > 6 pairs (3–13)
    3. Twigs winged; 2° lflets ≥ 10 mm. . . . . [*A. decurrens*]
    3′ Twigs angled; 2° lflets < 7 mm
      4. Lf main axis with raised glands at each 1° lflet pair; fr glabrous, silver-blue . . . . *A. dealbata*
      4′ Lf main axis with raised glands at most 1° lflet pairs and gen between; fr ± hairy, dark brown. . . . . . . [*A. mearnsii*]
1′ Lvs simple
  5. Lf < 40 mm
    6. Stipular spines present; heads 1. . . . . [*A. paradoxa*]
    6′ Stipular spines 0; heads in racemes
      7. Lf > 4 mm wide, tip not spine-like. . . . . [*A. podalyriifolia*]
      7′ Lf ≤ 1.5 mm wide, tip spine-like . . . . [*A. verticillata*]
  5′ Lf gen > 40 mm
    8. Lf with 1 prominent midvein
      9. Petiole base 4–8 mm; raceme, occ panicle of 10–30 heads. . . . . [*A. pycnantha*]
      9′ Petiole base ≤ 4 mm; raceme of 2–9 heads
        10. Lf narrowly lanceolate; head 5–7 mm wide; gland above petiole base not obvious, < 1 mm wide . . . [*A. retinodes*]
        10′ Lf linear to narrowly elliptic; head 8–12 mm wide; gland above petiole base obvious, 1–2 mm wide . . . *A. saligna*
    8′ Lf with ≥ 2 ± prominent longitudinal veins
      11. Fls in spikes . . . . *A. longifolia*

11′ Fls in heads, these 1 or in racemes or panicles
  12. Lvs, sts resinous, vanilla-scented when crushed . . . . ***A. redolens***
  12′ Lvs, sts not resinous, not vanilla-scented when crushed
    13. Shrub, branches dense, spreading from base; lf narrowly oblong to obovate, gen 4–15 mm wide . . . . . ***A. cyclops***
    13′ Tree, gen with 1 trunk; lf (exc juvenile) lanceolate to oblanceolate, gen 6–30 mm wide . . . . ***A. melanoxylon***

A. baileyana F. Muell. COOTAMUNDRA WATTLE Shrub, small tree < 10 m, unarmed. **ST:** twig angled, silver-blue, ± hairy. **LF:** 2-pinnate, < 4 cm, silver-blue; petiole < 2 mm, glabrous; axis with raised glands at distal pairs of 1° lflets; 1° lflets 2–5 pairs, 10–28 mm; 2° lflets 8–24 pairs, 3.5–7.5 mm, 0.8–1.3 mm wide, linear. **INFL:** raceme, occ panicle of 8–35 heads, > lf. **FL:** bright yellow. **FR:** 3–15 cm, 7–13 mm wide, straight, flat, leathery, dark brown, silver-blue when young, glabrous. **SEED:** aril yellow, club-shaped, forming cap on seed. Uncommon. Roadsides, disturbed areas; < 300 m. CCo, SnFrB, SCo, WTR, SnGb; native to se Australia. Feb–Apr

A. cyclops G. Don WESTERN COASTAL WATTLE Shrub < 6 m, unarmed. **ST:** twig ± angled, glabrous. **LF:** simple, 4–10 cm, gen 4–15 mm wide, narrowly oblong to obovate, ± straight; petiole base 1–2 mm; tip ± oblique, with blunt point; longitudinal veins 3–5, ± prominent. **INFL:** ± raceme of 2 heads (1 often aborted), < 1/4 lf. **FL:** golden yellow. **FR:** 5–13 cm, 9–17 mm wide, sickle-shaped, flat, leathery to ± woody, dark brown, glabrous, often persistent after seed release. **SEED:** aril orange to scarlet, encircling seed in double fold. Uncommon. Disturbed areas, coastal dunes; < 100 m. SCo, PR; native to sw Australia. Often cult along highways. Dec–Mar

A. dealbata Link SILVER WATTLE Tree < 30 m, unarmed. **ST:** twig angled, silver-blue, short-hairy. **LF:** 2-pinnate, < 17 cm, silver-blue; petiole 8–22 mm, short-hairy; main axis with raised glands at each pair of 1° lflets; 1° lflet 6–30 pairs, 15–55 mm; 2° lflets 15–70 pairs, 2–5 mm, 0.4–0.8 mm wide, linear. **INFL:** raceme, occ panicle of 11–30 heads, gen = lf. **FL:** pale yellow to cream. **FR:** 2–11 cm, 6–14 mm wide, ± straight, flat, ± leathery, silver-blue, glabrous. **SEED:** aril light yellow, club-shaped, forming cap. Locally common. Disturbed areas, often roadsides; < 500 m. NCoRO, ScV, CCo, SnFrB, SCo, WTR, SnGb; native to se Australia. [*A. decurrens* Willd. var. *d.* (Link) Maiden] Often reported as *A. decurrens* Willd. Feb–Apr ❖

A. longifolia (Andrews) Willd. SYDNEY GOLDEN WATTLE Shrub, small tree < 10 m, unarmed. **ST:** twig angled, glabrous. **LF:** simple, 5–15 cm, 10–25 mm wide, narrowly elliptic to lance-linear, straight; petiole base 2–5 mm; longitudinal veins 2–4. **INFL:** spike, 2–5 cm, ± sessile, < lf. **FL:** bright yellow. **FR:** 5–15 cm, 5–9 mm wide, ± straight, ± cylindric, leathery, light to dark brown, glabrous. **SEED:** aril light yellow, cup-shaped, covering tip. Uncommon. Disturbed places, esp sandy coastal areas; < 150 m. CCo, SnFrB, SCo, SnGb, PR; native to se Australia, New Guinea, Kei Islands. Jan–Apr

A. melanoxylon R. Br. BLACKWOOD ACACIA Tree < 30 m, unarmed; root suckers common. **ST:** twig ± angled, glabrous. **LF:** juvenile (on seedlings, young branches) 2-pinnate; adult simple (occ 2-pinnate at tips), 4–14 cm, gen 6–30 mm wide, lanceolate to oblanceolate, ± straight; petiole base 2–5 mm; longitudinal veins 3–5, ± prominent. **INFL:** raceme of 2–8 heads < lvs. **FL:** pale yellow. **FR:** 5–15 cm, 4–8 mm wide, curved, twisted, flat, leathery, dark brown, glabrous. **SEED:** aril yellow to red, encircling seed in an irregular, double fold. Uncommon. Disturbed areas; < 200 m. NCo, CCo, SnFrB, SCo, n ChI, PR; native to se and e Australia. Feb–Mar ❖

A. redolens Maslin VANILLA-SCENTED WATTLE Shrub, small tree < 5 m, unarmed; lvs, sts resinous, vanilla-scented when crushed. **ST:** twig ± angled, hairy. **LF:** simple, (2)4–7 cm, 5–13 mm wide, oblanceolate, ± straight; petiole base 1–3 mm; longitudinal veins 5–10, ± prominent. **INFL:** head 1 or raceme of 2–6 heads, ± = lf. **FL:** light yellow. **FR:** 3–6 cm, 2–4 mm wide, sickle-shaped, flat, ± leathery, dark brown, glabrous. **SEED:** aril cream-white, enlarged, forming cap. Uncommon. Disturbed areas; occ cult along highways and for land reclamation; < 200 m. CCo, SCo, PR; native to sw Australia. Feb–May

A. saligna (Labill.) H.L. Wendl. GOLDEN WREATH WATTLE Shrub, small tree < 5 m, unarmed. **ST:** twig ± angled, glabrous, silver-blue. **LF:** simple, 7–21 cm, 4–25 mm wide, linear to narrowly elliptic, ± straight; petiole base 1–4 mm, gland above obvious, 1–2 mm wide; midvein prominent. **INFL:** raceme of 2–8 heads, < lf; head 8–12 mm wide. **FL:** golden yellow. **FR:** 8–12 cm, 4–7 mm wide, straight, flat, papery, dark brown, glabrous. **SEED:** aril light yellow, club-shaped, forming cap. Uncommon. Disturbed areas, coastal dunes; < 100 m. SCo, WTR; native to sw Australia. Mar–May

ACMISPON DEERVETCH, DEERWEED

Luc Brouillet

Ann, per, shrub, unarmed. **LF:** gen odd-1-pinnate (or ± palmately compound, rarely some or most simple); stipules often gland-like, bump-like, or conic, often not apparent; lflets 3–9, gen irregularly arranged, lowest not stipular in position. **INFL:** umbel or 1–2-fld, axillary, gen peduncled, often bracted. **FL:** corolla gen yellow (white, pink), fading darker; 9 filaments fused, 1 free. **FR:** dehiscent or not, exserted from calyx or not, ovoid to oblong, ± beaked. **SEED:** 1–several. ± 23 spp.: sw Can, w US, Mex, 1 sp. in Chile. (Greek: tip, probably for hooked-tipped fr) [Brouillet 2008 J Bot Res Inst Texas 2:387–394] Intermediates may be hybrids.

1. Fr dehiscent, gen flat, gen ± entirely exserted from calyx body, straight or ± curved, gen with straight or
  curved, 0.5–1.5 mm beak; ann, per
  2. Per (exc *Acmispon strigosus*); fr straight or ± curved; wings gen > keel; stigma ± glabrous or finely hairy
    3. Ann, gen prostrate, mat-forming; infl gen 1–2-fld; corolla 5–10 mm; lf axis flat, ± blade-like; fr gen
      curved near tip . . . . [2]***A. strigosus***
    3′ Per or shrub-like, prostrate (mat-forming or not) to erect; infl 1–9-fld; corolla 7–25 mm; lf axis not flat
      or blade-like; fr gen straight (or curved throughout, rarely only at or near tip)
      4. Lf axis incl petiole 10–35 mm; lflets 7–9, 1–2.2 cm, length 1.5–3 × width . . . . ***A. grandiflorus***
        5. Pl puberulent or soft-hairy; lflets gen 1–1.5 cm . . . . var. ***grandiflorus***
        5′ Pl ± glabrous, finely strigose, or puberulent; lflets gen 1.6–2.2 cm . . . . var. ***macranthus***
      4′ Lf axis incl petiole 1–8 mm; lflets 3–5, < 1 cm and/or length > 3 × width
        6. Shrub-like, ascending, (2)5–15 dm, finely strigose; corolla 12–24 mm . . . . ***A. rigidus***
        6′ Per, prostrate or low-ascending, 1–3 dm, silvery-silky or gray-puberulent; corolla 6–12 mm . . . . ***A. argyraeus***
          7. St gen prostrate, mat-forming; calyx lobes ± 2 mm; SnBr, SnJt, DMtns . . . . var. ***argyraeus***

7′ St decumbent to low-ascending; calyx lobes 2–3 mm; DMtns
   8. Lflet oblanceolate to obovate, length ± 3–4 × width; New York Mtns . . . . var. ***multicaulis***
   8′ Lflet obovate, length ± 2 × width; Providence Mtns . . . . var. ***notitius***
2′ Ann; fr gen straight; wings gen ± ≤ keel; stigma glabrous or puberulent
  9. Fls gen 1 per lf axil; peduncle << 1 cm, bracts 0
    10. Corolla white, pale yellow, or pink, often darkening in age; pl gen ascending to erect and calyx lobes
      < 2 × tube, or rarely prostrate and calyx lobes gen > 2 × tube
      11. Pl gen 1–4 dm; calyx lobes ± 1.5 × tube . . . . ***A. denticulatus***
      11′ Pl < 1 dm; calyx lobes ± 2 × tube . . . . ***A. rubriflorus***
    10′ Corolla yellow, reddening in age; pl prostrate to low-ascending; calyx lobes 0.8–2 × tube
      12. Calyx lobes 1–2 × tube; fr gen 3–4 mm wide; pl gen with soft, spreading hairs . . . . ***A. brachycarpus***
      12′ Calyx lobes ± 0.8–1.2 × tube; fr gen 2.3–3 mm wide; pl ± strigose or hairs soft, spreading . . . . . . ***A. wrangelianus***
  9′ Fls 1–several per lf axil; peduncle gen > 1 cm, gen bracted
    13. Infl gen 2–4-fld (1st-formed often 1–2-fld); lflets 3–7, obovate to ± round, terminal gen largest . . . . . ***A. maritimus***
    14. Corolla 3.5–5 mm, keel > other petals; fr 1–1.5 cm, becoming narrowed between seeds . . . . var. ***brevivexillus***
    14′ Corolla 6–10 mm, keel ± = other petals; fr 1.5–3 cm, not narrowed between seeds . . . . var. ***maritimus***
    13′ Infl gen 1-fld; lflets 1–9, lanceolate to obovate, ± equal
      15. Calyx lobes >> tube; lflets gen 3 (upper lvs often simple) . . . . ***A. americanus*** var. ***americanus***
      15′ Calyx lobes ± ≤ tube; lflets 3–9
        16. St ascending to erect; lf axis sometimes ± flat; corolla pink or salmon, quickly fading; fr straight,
          margin often wavy . . . . ***A. parviflorus***
        16′ St prostrate, often mat-forming; lf axis flat; corolla yellow; fr gen curved near tip, margin not wavy
        . . . . [2]***A. strigosus***
1′ Fr indehiscent, gen not flat, gen incl or ± exserted (sometimes exserted) from calyx body, curved most of
  length, gen with curved, 2–3 mm beak; gen per
17. Ann, prostrate, decumbent, or ascending; young growth ± not hairy; corolla 3–7 mm; fr exserted; SCoR,
  SW, gen s of Los Angeles
  18. Infl ± sessile; s SCoR, SW . . . . ***A. micranthus***
  18′ Infl peduncled; s SCo (San Diego Co.) . . . . ***A. prostratus***
17′ Per, shrub, prostrate to erect; young growth hairy if pl blooming 1st yr or fls < 7 mm; fr incl or exserted;
  widespread, incl SW
19. Per, shrub; ChI
  20. Calyx shaggy-hairy; fr incl (exc slowly elongating beak); lf gen silvery-silky with fine, straight,
    ultimately wavy or tangled hairs . . . . [2]***A. argophyllus***
    21. Infl peduncled, not crowded; calyx lobes 1.5–2.5 mm; st prostrate to ascending . . . . var. ***argenteus***
    21′ Infl ± sessile, crowded at st tip; calyx lobes 2.5–5 mm; st ascending to erect
      22. St gen erect; lvs overlapping; calyx lobes ± 2.5–4 mm; s ChI (San Clemente Island) . . . . var. ***adsurgens***
      22′ St gen ascending; lvs not overlapping; calyx lobes 2.5–5 mm; n ChI (Santa Cruz Island) . . . . var. ***niveus***
  20′ Calyx ± glabrous or strigose; fr soon-exserted; lf green or gray, sparsely or densely strigose or silky
    . . . . ***A. dendroideus***
    23. Lflets 3, densely strigose or ± silky, gen gray; n ChI (San Miguel Island) . . . . var. ***veatchii***
    23′ Lflets 3–5, finely or sparsely strigose, green; ChI (exc San Miguel Island)
      24. Fr 1–1.5 cm; peduncle bract 0; ChI (exc San Miguel, San Clemente Islands) . . . . var. ***dendroideus***
      24′ Fr 2.5–5 cm; peduncle bract gen pres; s ChI (San Clemente Island) . . . . var. ***traskiae***
19′ Per (exc *Acmispon glaber*); mainland
25. Lflets 3; infl 1–3-fld, ± sessile
  26. Lflet 2–5 mm; fr 6–9 mm; uncommon, se PR, sw DSon . . . . ***A. haydonii***
  26′ Lflet 4–15 mm; fr 10–15 mm; widespread, incl PR
    27. Calyx glabrous (see couplet 31 for vars.) . . . . [2]***A. glaber***
    27′ Calyx strigose . . . . ***A. procumbens***
      28. Corolla 9–12 mm; calyx 4–6 mm, lobes ± = tube; s SN . . . . var. ***jepsonii***
      28′ Corolla 6–8 mm; calyx 2–3 mm, lobes << tube; not s SN . . . . var. ***procumbens***
25′ Lflets ≥ 3; infl 2–15-fld, sessile or peduncled
  29. Lflet glabrous or finely strigose, gen green; hairs straight or ± wavy
    30. Pls gen ascending to erect (some on immediate coast prostrate), bushy-branched, 5–20 dm;
      st sparsely lfy; corolla yellow; calyx lobes gen 1–2 mm, not curved outward; coastal and
      inland . . . . [2]***A. glaber***
      31. Keel prominent, > wings; corolla gen 8–9 mm; gen inland, SW, DSon . . . . var. ***brevialatus***
      31′ Keel not prominent, ± = wings; corolla 7–12 mm; coastal and inland, NCo, NCoR, n SNF, CCo,
        SnFrB, SCo, WTR, PR . . . . var. ***glaber***
    30′ Pls prostrate or low-ascending, < 10 dm; st lfy; corolla pink-white or yellow; calyx lobes gen < 1
      mm or curved outward; esp coastal
      32. Calyx lobes narrow, some or all curved outward or hooked; corolla gen white to ± pink (to
        brick-red when dry) . . . . ***A. cytisoides***
      32′ Calyx lobes wide, straight, not hooked; corolla yellow . . . . ***A. junceus***

33. Peduncles 8–25 mm; st often wiry, gen decumbent; fr well exserted . . . . var. ***biolettii***
33′ Peduncles 1–5 mm; st gen stout, prostrate to ascending; fr ± exserted . . . . var. ***junceus***
29′ Lflet hairy, green, gray, or silvery; hairs straight or wavy
34. St, esp near tip, gen with spreading or obliquely directed, often straight, stiff hairs < 0.5 mm;
corolla 4–6 mm; coastal and inland . . . . ***A. heermannii***
35. Corolla gen 4–5 mm; ovary gen finely strigose; SCo, SnBr, PR, DSon . . . . var. ***heermannii***
35′ Corolla gen 5–6 mm; ovary gen soft spreading-hairy; NCo, NCoRO, CCo, SCoRO. . . . . var. ***orbicularis***
34′ St gen strigose or with ± spreading, often wavy hairs 0.2–0.4 mm; corolla 5–12 mm; esp inland
36. Fr exserted; lflet hairs gen wavy, not obscuring surface; lvs gray or green . . . . ***A. nevadensis***
37. Fls gen 3–5 per infl; calyx 6–7 mm, appearing blocky; banner abruptly upcurved 90° TR, ne PR
. . . . var. ***davidsonii***
37′ Fls gen 5–12 per infl; calyx 5–10 mm, not appearing blocky; banner upcurved 30–90° KR,
NCoR, CaR, SN, CCo, SCo, w PR, DMtns . . . . var. ***nevadensis***
36′ Fr incl or ± exserted; lflet hairs gen straight, tangled in age, gen obscuring surface; lvs
silky-silvery . . . . ²***A. argophyllus***
38. Infl gen < 1 cm wide, 4–8-fld; corolla 6–10 mm; esp c SN and s . . . . var. ***argophyllus***
38′ Infl often ± 1.5 cm wide, 10–15-fld; corolla 8–12 mm; n SN . . . . var. ***fremontii***

A. americanus (Nutt.) Rydb. var. ***americanus*** (p. 727)   Ann, gen hairy. **ST:** prostrate to erect, simple or openly branched, 0.5–6 dm. **LF:** pinnate or ± simple; stipules gland-like; lflets gen 3, gen 10–20 mm, lanceolate to elliptic; axis not flat. **INFL:** 1-fld, peduncle (3–10)15 mm, bracts simple. **FL:** calyx 2.5–6.5 mm, lobes >> tube, hairs soft-shaggy; corolla 5–9 mm, ± white or yellow to pink, wings ± = keel, stigma glabrous. **FR:** dehiscent, spreading or pendent, ± entirely exserted, 1.5–3 cm, oblong, ± straight, gen flat, beak curved, 0.9–1.7 mm. **SEED:** 3–8. 2*n*=14. Coast, chaparral, mtn forest, water courses, roadsides, other disturbed areas; < 2400 m. CA (exc DSon); to Can, c US, Mex. [*Lotus purshianus* Clem. & E.G. Clem. var. *p.*] Many races, ecological forms. May–Oct

A. argophyllus (A. Gray) Brouillet   Per, ± woody or not, gray or silvery-silky with fine, straight, ultimately wavy or tangled hairs 0.2–0.4 mm. **ST:** prostrate to erect, 1–6(10+) dm. **LF:** irregularly pinnate to ± palmate; stipules gland-like; lflets 3–7, 6–12 mm, obovate to lance-elliptic, hairs gen obscuring surface. **INFL:** head-like, 4–15(20)-fld, peduncle ± 0 or 1–3(10) mm, bracted or not. **FL:** calyx 4–8 mm, lobes ± ≤ tube, densely shaggy-hairy; corolla 6–12 mm, wings ± = keel, stigma glabrous. **FR:** indehiscent, horned, incl or ± exserted (exc beak), half-ovate, ± curved, gen not flat, beak curved, 2–3 mm. **SEED:** gen 1(2). 2*n*=14. [*Lotus argophyllus* (A. Gray) Greene] Major vars. geog distinct; mainland vars. intergrade with related spp.

var. ***adsurgens*** (Dunkle) Brouillet   SAN CLEMENTE ISLAND BIRD'S-FOOT TREFOIL   Pl ± woody, silvery. **ST:** gen erect; branches slender. **LF:** overlapping **INFL:** < 1 cm wide, 4–15-fld, subsessile, densely crowded at st tips. **FL:** calyx lobes ± 2.5–4 mm; corolla 6–10 mm. Dry, rocky slopes, bluffs; < 300 m. s ChI (San Clemente Island). [*Lotus argophyllus* var. *adsurgens* Dunkle] Related to *A. argophyllus* var. *niveus*. Mar–Jun   ★

var. ***argenteus*** (Dunkle) Brouillet   Pl ± woody, light gray. **ST:** prostrate to ascending. **LF:** not dense. **INFL:** 1.5–2 cm wide, 12–20-fld, peduncled, not crowded. **FL:** calyx lobes 1.5–2.5 mm; corolla 10–12 mm. Chaparral, bluffs; < 400 m. ChI (exc Santa Cruz Island). [*Lotus argophyllus* var. *argenteus* Dunkle] Like *A. argophyllus* var. *a.* Mar–Jun

var. ***argophyllus*** (p. 727)   Pl not woody, silky-canescent to silvery strigose. **ST:** prostrate or decumbent-ascending. **LF:** not dense. **INFL:** < 1 cm wide, 4–8-fld, sessile or not, ± crowded. **FL:** calyx lobes 1.5–3.5 mm; corolla 6–10 mm. Dry slopes in chaparral, canyons; < 1600 m. c&s SN, SCoR, SCo, SnGb, SnBr, PR. [*Lotus argophyllus* var. *argophyllus*] Variable, intergrading with other spp. in SnGb, SnBr. Apr–Jul

var. ***fremontii*** (A. Gray) Brouillet (p. 727)   Pl not woody, pubescent-appressed. **ST:** prostrate or decumbent-ascending. **LF:** not dense. **INFL:** often ± 1.5 cm wide (spheric), 10–15-fld, gen sessile, crowded at st tips. **FL:** calyx lobes 3–5 mm; corolla 8–12 mm. Openings along rivers, trails, on canyon slopes; 600–1200 m. n SN. [*Lotus argophyllus* var. *fremontii* (A. Gray) Ottley] Apr–Jul

var. ***niveus*** (Greene) Brouillet   SANTA CRUZ ISLAND BIRD'S-FOOT TREFOIL   Pl ± woody, silky. **ST:** gen ascending. **LF:** not overlapping. **INFL:** < 1 cm wide, 4–15-fld, ± sessile, densely crowded at st tips. **FL:** calyx lobes 2.5–5 mm; corolla 6–10 mm. Rocky slopes, dry riverbeds; < 300 m. n ChI (Santa Cruz Island). [*Lotus argophyllus* var. *niveus* (Greene) Ottley] Mar–Jun   ★

A. argyraeus (Greene) Brouillet   Per, hairy, silvery-silky or gray-puberulent. **ST:** 1–3 dm. **LF:** irregularly pinnate or ± palmate; stipules gland-like; lflets 3–5, 4–12 mm; axis (incl petiole) 1–8 mm, not flat. **INFL:** 1–3-fld, peduncle 10–20 mm, bract 0 or small. **FL:** calyx 4.3–7 mm, lobes ± ≤ tube, puberulent; corolla yellow, orange-red in age, wings unequal, > keel; stigma finely puberulent. **FR:** dehiscent, divergent, exserted, oblong, beak 0.5–1.5 mm. **SEED:** 2–several. [*Lotus a.* (Greene) Greene] Vars. similar, geog distinct.

var. ***argyraeus*** **ST:** gen prostrate, mat-forming. **LF:** canescent; lflets obovate. **FL:** calyx lobes ± 2 mm; corolla 7–10 mm. **FR:** 1.5–2 cm, gen straight. Open granitic slopes or with pine; 1500–2400 m. SnBr, SnJt, DMtns; to Mex. [*Lotus a.* var. *a.*] May–Aug

var. ***multicaulis*** (Ottley) Brouillet   SCRUB LOTUS   **ST:** decumbent to low-ascending. **LF:** green or canescent; lflets oblanceolate to obovate, length ± 3–4 × width. **FL:** calyx lobes 2.5–3 mm; corolla 8–12 mm. **FR:** 2–2.5 cm, curved at tip. Pinyon/juniper woodland; 1200–1500 m. DMtns (New York Mtns). [*Lotus a.* var. *m.* (Ottley) Isely] May–Aug   ★

var. ***notitius*** (Isely) Brouillet   PROVIDENCE MOUNTAINS LOTUS   **ST:** low-spreading or -ascending. **LF:** green or canescent; lflets obovate, length ± 2 × width. **FL:** calyx lobes 2–3 mm; corolla 8–12 mm. **FR:** 1–2.5 cm, straight. Pinyon/juniper woodland; 1200–2000 m. DMtns (Providence Mtns). [*Lotus a.* var. *n.* Isely] May–Aug   ★

A. brachycarpus (Benth.) D.D. Sokoloff (p. 731)   Ann, often ± fleshy, ashy green or ± green, soft-spreading-hairy. **ST:** mat-forming to ascending, 0.5–4 dm. **LF:** subpinnate or palmate; stipules gland-like; lflets gen 4, 4–12 mm, elliptic to obovate; axis flat, ± blade-like. **INFL:** axillary, 1-fld, ± sessile, bract 0. **FL:** calyx 3–6 mm, lobes 1–2 × tube, hairy; corolla 5–9 mm, yellow, reddening in age, wings ± = keel, stigma glabrous. **FR:** dehiscent, ascending, exserted, 6–12 mm, gen 3–4 mm wide, oblong, gen straight, gen flat, beak bent, 0.5–1.5 mm. **SEED:** 2–5. 2*n*=12. Abundant. Grassland, oak and pine woodland, desert flats and mtns, roadsides; < 1700 m. CA-FP, D; to sw UT, w NM, Mex. [*Lotus humistratus* Greene] Mar–Jun

A. cytisoides (Benth.) Brouillet   Per, glabrous or finely strigose, hairs straight or ± wavy. **ST:** mat-forming or low-ascending, not woody, lfy, 1–8 dm. **LF:** irregularly pinnate or palmate, gen green; stipules gland-like; lflets 3–5, 5–12 mm, elliptic to obovate. **INFL:** 3–10-fld, peduncle 2–12(25) mm, bracted or not. **FL:** calyx 3.5–6 mm, lobes < tube, narrow, gen spreading or curved outward or hooked, thinly minutely strigose to ± glabrous; corolla 8–10 mm, gen white to ± pink (to brick-red when dry, wings cream), often dark-striate, stigma minutely strigose. **FR:** indehiscent, spreading to upcurved, incl (exc beak), 7–10 mm, curved, gen not flat, beak gen

curved, 2–3 mm. **SEED**: 1–2. 2*n*=14. Coastal dunes, slopes, bluffs; < 200 m. CCo, SnFrB, SCoRO. [*Lotus benthamii* Greene] Evidently hybridizes with other spp. Mar–Jul

A. dendroideus (Greene) Brouillet Per, shrub, sparsely or densely strigose or ± silky, green or gray. **ST**: decumbent to ascending, 5–20 dm. **LF**: irregularly pinnate; stipules gland-like; lflets 7–15 mm, elliptic to obovate. **INFL**: 3–10-fld, peduncle 0 or 2–10 mm. **FL**: calyx 4–6 mm, lobes < tube, thinly strigose to ± glabrous; corolla 8–12 mm, yellow, wings ± = keel, stigma glabrous. **FR**: indehiscent, divergent to descending, soon-exserted, oblong-tapering, straight or curved, not flat; beak narrow, abruptly curved, 2–3 mm. 2*n*=14. [*Lotus d.* (Greene) Greene] Island-to-island variants recognized variously; the following distinct in morphology, geog.

var. ***dendroideus*** (p. 731) ISLAND BROOM **LF**: lflets 3–5, finely or sparsely strigose, ± green. **INFL**: peduncle bract 0. **FR**: 1–1.5 cm. Bluffs, inland canyons, open sites near ocean; < 350 m. ChI (exc San Miguel, San Clemente islands). [*Lotus d.* var. *d.*] Jan–Aug ★

var. ***traskiae*** (Abrams) Brouillet SAN CLEMENTE ISLAND LOTUS **LF**: lflets 3–5, finely or sparsely strigose, ± green. **INFL**: peduncle gen bracted. **FR**: 2.5–5 cm. Bluffs, inland canyons, open sites near ocean; < 350 m. s ChI (San Clemente Island). [*Lotus d.* var. *t.* (Abrams) Isely] Feb–Aug ★

var. ***veatchii*** (Greene) Brouillet SAN MIGUEL ISLAND DEERWEED **LF**: lflets 3, densely strigose or ± silky, gen gray. **INFL**: peduncle bracted. **FR**: 2.5–5 cm. Bluffs, inland canyons, open sites near ocean; < 350 m. n ChI (San Miguel Island). [*Lotus d.* var. *v.* (Greene) Isely] Mar–Jun ★

A. denticulatus (Drew) D.D. Sokoloff (p. 731) Ann, ± glabrous or puberulent. **ST**: decumbent to erect, often 1, 1–4 dm. **LF**: subpinnate or palmate; stipules gland-like or 0; lflets 2–4, 8–12 mm, elliptic to obovate; axis flat, ± blade-like. **INFL**: axillary, 1–2-fld, peduncle ± 0, bract 0. **FL**: calyx 3–5 mm, lobes ± 1.5 × tube, hairy; corolla 6–8 mm, cream-white or pale yellow, wings ± = keel; stigma glabrous. **FR**: dehiscent, erect or spreading, exserted, 0.5–1.5 cm, widely oblong, straight, gen flat, gen strigose, beak ± curved, 0.5–1.5 mm. **SEED**: 2–3. 2*n*=12. Open meadows, slopes, streambanks; < 1400 m. NCo, KR, NCoRO, CaRF, n SNF, ScV, e SnFrB, MP; to BC, ID, UT. [*Lotus d.* (Drew) Greene] May–Jul

A. glaber (Vogel) Brouillet DEERWEED, CALIFORNIA BROOM Subshrub, glabrous or finely strigose, hairs straight or ± wavy. **ST**: gen ascending to erect (prostrate and mat-forming or not), clustered, ± woody (yet green), bushy-branched, 5–20 dm. **LF**: ± pinnate, well spaced, often deciduous; stipules gland-like or 0, black; lflets 3–6 (gen 3 on upper st), 6–15 mm, elliptic, gen green. **INFL**: 2–7-fld, gen sessile, bract 0. **FL**: calyx 2.5–5 mm, lobes gen 1–2 mm, < tube, erect, glabrous; corolla yellow; stigma glabrous. **FR**: indehiscent, spreading or pendent, much exserted, 1–1.5 cm, oblong, curved, not flat; beak curved, 2–3 mm. **SEED**: gen 2. 2*n*=14. [*Lotus scoparius* (Torr. & A. Gray) Ottley] May hybridize with *A. cytisoides, A. junceus.*

var. ***brevialatus*** (Ottley) Brouillet **FL**: corolla gen 8–9 mm, keel > wings. Desert slopes, flats, washes; < 1400 m. SCo, TR, PR, DSon; AZ, Mex. [*Lotus scoparius* var. *b.* Ottley] Intergrades with *A. glaber* var. *g.* in Los Angeles Co. Mar–Aug

var. ***glaber*** (p. 731) **FL**: corolla 7–12 mm, keel ± = wings. Chaparral, roadsides, coastal sands; common; < 1500 m. NCo, NCoR, n SNF, CCo, SnFrB, SCo, WTR, PR; Baja CA. [*Lotus scoparius* var. *s.*] Gen erect, may be trailing in shade or mat-forming on beaches. Island forms here referred to *Lotus dendroideus.* Mar–Aug

A. grandiflorus (Benth.) Brouillet Per. **ST**: decumbent to erect, 1–4(15) dm. **LF**: irregularly pinnate; stipules gland-like, conic, black; lflets 7–9, elliptic to obovate, length 1.5–3 × width; axis (incl petiole) 10–35 mm, not flat. **INFL**: 3–9-fld, peduncle gen 10–80 mm, bract near tip. **FL**: calyx 6–9 mm, lobes ± ≤ tube, long-shaggy-hairy; corolla 15–25 mm, green-white or yellow, wings ± ≥ keel; stigma puberulent. **FR**: dehiscent, exserted, 2.5–6 cm, oblong, gen straight, gen flat, occ with small horn-like processes; beak ± straight or basally

curved, 0.5–1.5 mm. **SEED**: several. 2*n*=14. [*Lotus g.* (Benth.) Greene] Vars. are geog separated.

var. ***grandiflorus*** (p. 731) Pl puberulent or soft-hairy. **LF**: lflets gen 1–1.5 cm. Gen dry, open, disturbed sites, chaparral to yellow-pine forest; 300–1500 m. NW, s SN, CW, SW. [*Lotus g.* var. *g.*] Hairiness variable. Apr–Jul

var. ***macranthus*** (Greene) Brouillet Pl ± glabrous, finely strigose, or puberulent. **LF**: lflets gen 1.6–2.2 cm. Yellow-pine forest, moist river bottoms; 900–1800 m. KR, n&c SN. [*Lotus g.* var. *m.* (Greene) Isely] Apr–Jul

A. haydonii (Orcutt) Brouillet PYGMY LOTUS Per, finely strigose. **ST**: ascending or sprawling, bushy, rush-like, 1–20 dm. **LF**: sparse, subpalmate, early-deciduous; stipules gland-like or 0; lflets 3, 2–5 mm, elliptic. **INFL**: 1–2-fld, peduncle 0 or 1–3 mm, bract 0. **FL**: calyx 2.5–3 mm, lobes < tube, ± appressed-hairy; corolla 4–5 mm, yellow, keel > other petals. **FR**: indehiscent, ascending or reflexed, exserted, 6–9 mm, oblong, curved, not flat, beak curved, 2–3 mm. **SEED**: 1–2. Creosote-bush scrub, pinyon/juniper woodland; 600–1200 m. se PR, sw DSon (esp San Diego Co.); Baja CA. [*Lotus h.* (Orcutt) Greene] Local, little known. Mar–Jun ★

A. heermannii (Durand & Hilg.) Brouillet Per (or fl 1st yr, appearing ann), hairs spreading or obliquely directed, often straight, stiff, < 0.5 mm, esp near st tip. **ST**: prostrate, often mat-forming, 3–10 dm. **LF**: irregularly subpalmate; stipules gland-like; lflets 4–6, 4–16 mm, ovate to obovate, green, gray, or silver; axis sometimes flat, ± blade-like. **INFL**: 3–8-fld, peduncle gen 1–5 mm, bracted. **FL**: calyx 2–4 mm, lobes < tube, gen long-shaggy-hairy; corolla yellow to ± red, dark-tipped, wings ≥ other petals. **FR**: indehiscent, ascending to divergent, exserted, ± 5 mm, narrowly oblong, gen curved, gen flat, tapered to long, beak 2–3 mm, curved. **SEED**: 1–2. [*Lotus h.* (Durand & Hilg.) Greene] Vars. based esp on habitat, geog.

var. ***heermannii*** **FL**: corolla gen 4–5 mm; ovary gen finely strigose. Washes, riverbanks, chaparral; < 2000 m. SCo, SnBr, PR, DSon. [*Lotus h.* var. *h.*] Mar–Oct

var. ***orbicularis*** (A. Gray) Brouillet (p. 731) **FL**: corolla gen 5–6 mm; ovary gen soft-spreading-hairy. Coastal scrub, chaparral; < 250 m. NCo, NCoRO, CCo, SCoRO. [*Lotus h.* var. *o.* (A. Gray) Isely] Mar–Oct

A. junceus (Benth.) Brouillet Subshrub (or fl 1st yr, appearing ann), finely strigose, hairs straight. **ST**: not woody, 0.8–4 dm, lfy. **LF**: irregularly subpalmate; stipules gland-like, black; lflets 3–5, 5–10 mm, oblanceolate to obovate, gen green; axis sometimes flat, blade-like. **INFL**: 2–8-fld, bract present or 0. **FL**: calyx 3–5 mm, lobes gen < 1 mm, < tube, straight, finely strigose; corolla 6–8 mm, yellow, keel < other petals; stigma glabrous. **FR**: indehiscent, ascending to divergent, 6–8 mm, oblong-tapered, gen arched to ± straight, gen not flat; beak curved, 2–3 mm. **SEED**: 1–2. [*Lotus j.* (Benth.) Greene]

var. ***biolettii*** (Greene) Brouillet **ST**: gen decumbent, often wiry. **INFL**: peduncle 8–25 mm. **FR**: well exserted. Coastal sand, chaparral, disturbed areas; < 500 m. NCo, NCoRO, CCo, SCoRO. [*Lotus j.* var. *b.* (Greene) Ottley] Apr–Jul

var. ***junceus*** **ST**: prostrate to ascending, gen stout. **INFL**: peduncle 1–5 mm. **FR**: ± exserted. Chaparral, on serpentine or not; < 500 m. CCo, SCoRO. [*Lotus j.* var. *j.*] Apr–Jul

A. maritimus (Nutt.) D.D. Sokoloff Ann, often fleshy, glabrous or strigose. **ST**: clustered, prostrate or ascending, 0.5–5 dm. **LF**: irregularly pinnate; stipules gen not apparent; lflets 3–7, 5–15 mm, obovate to ± round; axis flat, ± blade-like. **INFL**: gen 2–4-fld (1st formed often 1–2-fld, fr often only 1), peduncle 5–15 mm, bracted or not. **FL**: calyx lobes < tube, strigose; corolla bright yellow; stigma glabrous. **FR**: dehiscent, ascending, exserted, gen 1.5–3 cm, narrowly oblong, often curved, flat, occ with small horn-like processes; beak hooked, 0.5–1.5 mm. **SEED**: 5–9. 2*n*=14. [*Lotus salsuginosus* Greene]

var. ***brevivexillus*** (Ottley) Brouillet **FL**: calyx ± 2.5 mm; corolla 3.5–5 mm, keel > other petals. **FR**: 1–1.5 cm, narrowed between seeds in age. Deserts, incl mtns; < 1850 m. D; NV, AZ, Mex. [*Lotus salsuginosus* var. *b.* Ottley] Mar–Jun

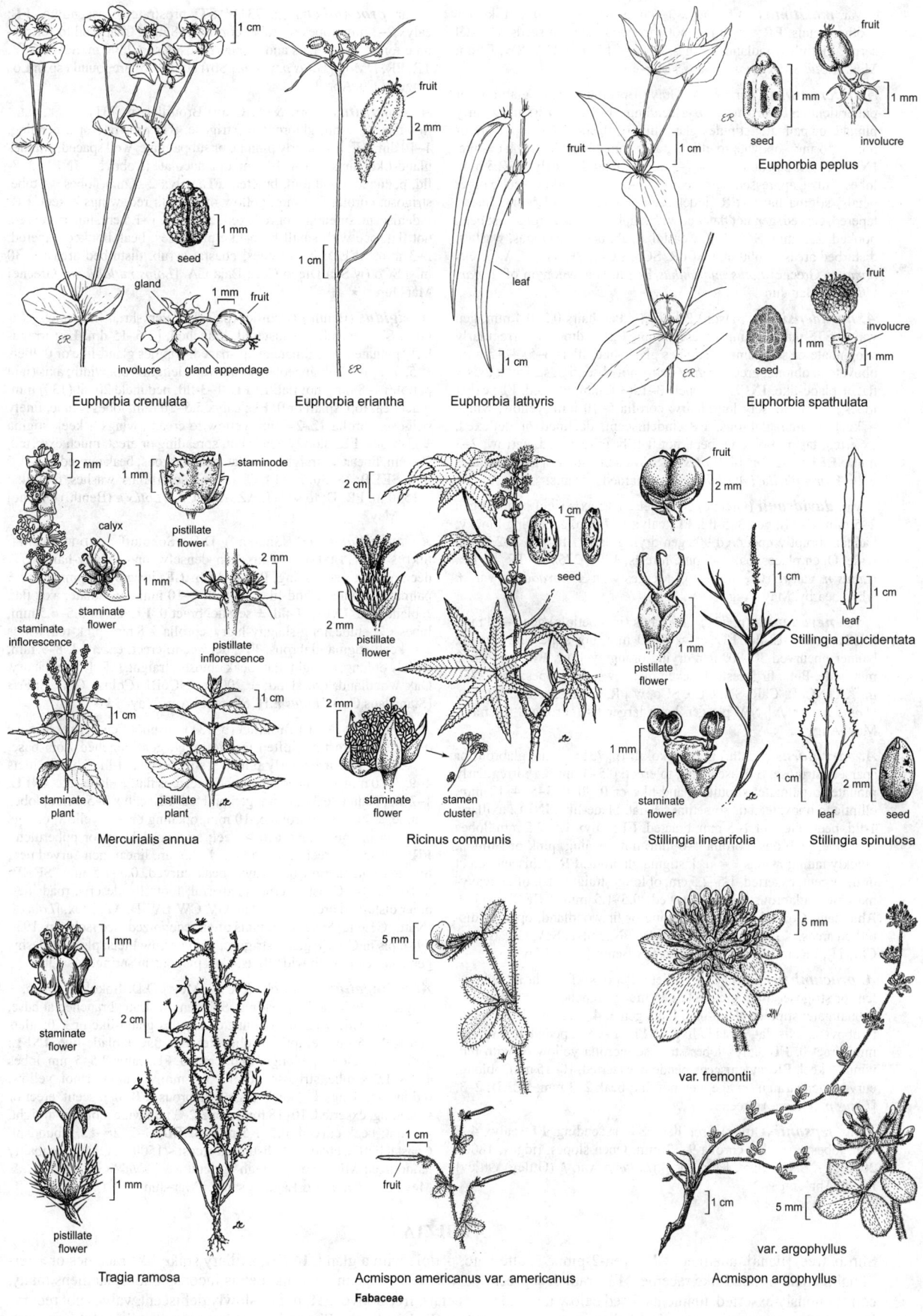

Euphorbia crenulata

Euphorbia eriantha

Euphorbia lathyris

Euphorbia peplus

Euphorbia spathulata

Mercurialis annua

Ricinus communis

Stillingia linearifolia

Stillingia paucidentata

Stillingia spinulosa

Tragia ramosa

Acmispon americanus var. americanus

Fabaceae

var. fremontii

var. argophyllus

Acmispon argophyllus

var. ***maritimus*** **FL**: calyx 3–4 mm; corolla 6–10 mm, keel ± = other petals. **FR**: 1.5–3 cm, not narrowed between seeds. Coastal scrub, foothill woodland, washes, talus; < 1200 m. CW, SW, DSon; Mex. [*Lotus salsuginosus* var. *s.*] Mar–Jun

A. micranthus (Torr. & A. Gray) Brouillet Ann, glabrous or puberulent. **ST**: prostrate to ascending, 1–8 dm. **LF**: irregularly pinnate or palmate; stipules gland-like or 0; lflets 4–7, opposite or not, 7–15 mm, obovate to elliptic; axis sometimes flat, ± blade-like. **INFL**: 2–5-fld, peduncle 0 or 1–5 mm, bract 0. **FL**: calyx 2–2.5 mm, lobes < tube, appressed-hairy; corolla 3–5 mm, ± yellow, keel > other petals, stigma hairy. **FR**: indehiscent, exserted, 1–1.5 cm, linear, tapered, curved, gen not flat, occ with small horn-like processes; beak hooked, 2–3 mm. **SEED**: 2. Coastal scrub, desert canyons, washes, disturbed areas; < 600 m. s SCoR, SCo, s ChI, PR; Baja CA. [*Lotus hamatus* Greene] *Lotus micranthus* Benth. is a synonym of *A. parviflorus*. Mar–Jun

A. nevadensis (S. Watson) Brouillet Per, hairs 0.2–0.4 mm, gen wavy. **ST**: mat-forming or ascending, 0.5–1 dm. **LF**: irregularly subpinnate or subpalmate; stipules gland-like; lflets 3–5, 4–12 mm, obovate to oblong, green or gray (hairs not obscuring surface); axis ± flat, ± blade-like. **INFL**: peduncle 3–15(30) mm, bracted. **FL**: calyx lobes < tube, loosely long-hairy; corolla 5–10 mm, yellow, wings > keel; stigma glabrous. **FR**: indehiscent, declined or deflexed, exserted, tapered-oblong, bent, not flat; beak recurved, narrow, 2–3 mm. **SEED**: 1–2. 2*n*=14. [*Lotus n.* (S. Watson) Greene] Intermediates with *A. argophyllus*, *A. glaber*. Many named variants.

var. ***davidsonii*** (Greene) Brouillet **LF**: axis (incl petiole) gen 2–5 mm. **INFL**: gen 3–5-fld. **FL**: calyx 6–7 mm, appearing blocky; banner abruptly upcurved 90° gen drying dark. **FR**: gen 2–2.2 mm wide. Open places, oak and pine forests; 1200–2750 m. TR, ne PR. [*Lotus n.* var. *d.* (Greene) Isely] Merges with *A. nevadensis* var. *n.* when nearby. May–Aug

var. ***nevadensis*** (p. 731) **LF**: axis (incl petiole) gen 5–10 mm. **INFL**: gen 5–12-fld. **FL**: calyx 5–10 mm, not appearing blocky; banner upcurved 30–90° gen drying orange-yellow. **FR**: gen 1.8–2 mm wide. Pine, fir forests, bracken meadows, dry slopes; 850–2750 m. KR, NCoR, CaR, SN, CCo, SCo, w PR, DMtns; to BC, ID, NV. [*Lotus n.* var. *n.*] NW pls gen have larger fls (corolla 8–10 mm). May–Aug

A. parviflorus (Benth.) D.D. Sokoloff (p. 731) Ann, glabrous or sparsely strigose. **ST**: ascending to erect, 0.5–4 dm. **LF**: irregularly pinnate to palmate; stipules gland-like or 0; lflets 3–5, 4–12 mm, elliptic to obovate; axis sometimes ± flat, blade-like. **INFL**: axillary, 1-fld, peduncle 0 or 1–5 mm, bracted. **FL**: calyx 1.5–2.5 mm, lobes ± = tube, glabrous; corolla 4–6 mm, not opening, pink or salmon, quickly fading, wings ± = keel; stigma glabrous. **FR**: dehiscent, erect or divergent, exserted, 1.5–2.5 cm, oblong, straight, flat, often wavy-margined, glabrous; beak ± curved, 0.5–1.5 mm. **SEED**: (3)5–9. Abundant. Coastal bluffs to oak/pine or fir woodland, open or disturbed areas; < 1300 m. NW, CaR, n&c SN, s SNF, ScV, CW, SCo, n ChI, TR, PR; to BC. [*Lotus micranthus* Benth.] Mar–May

A. procumbens (Greene) Brouillet Per or stiff subshrub, puberulent or strigose, often gray. **ST**: gen much-branched, 1–10 dm. **LF**: subpalmate; stipules gland-like; lflets gen 3, 4–12 mm, oblanceolate to obovate; axis flat, ± blade-like. **INFL**: 1–3-fld, peduncle 0 or 1–3 mm, bract 0. **FL**: calyx lobes strigose; corolla yellow or with red, wings > keel. **FR**: indehiscent, pendent, exserted, 10–15 mm, oblong, curved, gen straight in age, gen not flat; beak 2–3 mm. **SEED**: 2–3. [*Lotus p.* (Greene) Greene]

var. ***jepsonii*** (Ottley) Brouillet **ST**: ascending. **FL**: calyx 4–6 mm, lobes ± = tube; corolla 9–12 mm. Open slopes, ridges; 1800–2000 m. s SN (Tulare, Kern cos.). [*Lotus p.* var. *j.* (Ottley) Ottley] Local. Apr–Jun

var. ***procumbens*** (p. 731) **ST**: prostrate to ascending. **FL**: calyx 2–3 mm, lobes << tube; corolla 6–8 mm. Chaparral to Jeffrey-pine forest, sandy flats and slopes, roadsides; < 2300 m. ScV, SCoR, TR, PR, DMoj. [*Lotus p.* var. *p.*] Stiff subshrubs are found esp in Los Angeles Co. Apr–Jun

A. prostratus (Torr. & A. Gray) Brouillet (p. 731) NUTTALL'S ACMISPON Ann, glabrous or strigose. **ST**: prostrate or ascending, 1–10 dm. **LF**: irregularly pinnate or subpalmate, well spaced; stipules gland-like; lflets 3–6, 4–10 mm, oblanceolate to obovate. **INFL**: 3–8-fld, peduncle 8–30 mm, bracted. **FL**: calyx 2–3 mm, lobes << tube, strigose; corolla 5–7 mm, yellow often with red, wings ≥ keel. **FR**: indehiscent, spreading or reflexed, exserted, 1–1.5 cm, linear, curved, not flat, occ with small horn-like processes; beak hooked, tapered, 2–3 mm. **SEED**: 2. Beaches, coastal scrub, disturbed areas; < 30 m. s SCo (w San Diego Co.); Baja CA. [*Lotus nuttallianus* Greene] Mar–Jun ★

A. rigidus (Benth.) Brouillet (p. 731) Per, shrub-like, finely strigose. **ST**: ascending, clustered, branched, (2)5–15 dm. **LF**: irregularly pinnate to ± palmate, well spaced; stipules gland-like or 0; lflets 3–5, 5–17 mm, oblanceolate to obovate, length > 3 × width; axis (incl petiole) 1–8 mm, not flat. **INFL**: 1–3-fld, peduncle 20–60(130) mm, bract near top, small or 0. **FL**: calyx 5.5–10 mm, lobes < tube, finely strigose; corolla 12–24 mm, yellow to cream, wings > keel; stigma ± glabrous. **FR**: slowly dehiscent, spreading or erect, much exserted, 2–4 cm, linear, ± straight, gen flat, gen glabrous; beak curved, 0.5–1.5 mm. **SEED**: 18–30. 2*n*=14. Chaparral, desert flats, washes, foothills; < 1550 m. PR, D; to se UT, AZ, Baja CA. [*Lotus r.* (Benth.) Greene] Mar–May

A. rubriflorus (H. Sharsm.) D.D. Sokoloff RED-FLOWERED BIRD'S-FOOT TREFOIL Ann, gen densely long-shaggy-hairy. **ST**: decumbent to ascending, 0.4–0.9 dm. **LF**: irregularly pinnate or ± palmate; stipules gland-like; lflets 4, 3–10 mm, lanceolate; axis flat, ± blade-like. **INFL**: 1-fld, ± sessile, bract 0. **FL**: calyx 5.5–6.5 mm, lobes ± 2 × tube, long-shaggy-hairy; corolla 5–8 mm, pink-red, wings ± = keel; stigma glabrous. **FR**: dehiscent, erect, exserted, 8–9 mm, widely oblong, straight, flat; beak obtuse, straight, 0.5–1.5 mm, hairy. Oak woodland, grassland; ± 200 m. NCoRI (Colusa Co.), SnFrB (Stanislaus Co.). [*Lotus r.* H. Sharsm.] Apr–May ★

A. strigosus (Nutt.) Brouillet (p. 731) Ann, often fleshy, strigose or not. **ST**: prostrate, often mat-forming, gen branched from base, 0.3–5 dm. **LF**: irregularly pinnate; stipules gland-like, black; lflets 4–9, 3–10 mm, oblanceolate to obovate; axis flat, ± blade-like. **INFL**: 1–2-fld, peduncle 3–25 mm, gen bracted. **FL**: calyx 3–5.5 mm, lobes < tube, ± strigose; corolla 5–10 mm, opening or not, yellow, orange or ± red in age, wings gen > keel; stigma glabrous or puberulent. **FR**: dehiscent, erect or spreading, 1–3.5 cm, linear, gen curved near tip, gen flat, margin not wavy; beak curved, 0.5–1.5 mm. **SEED**: 5–10. 2*n*=14. Coastal scrub, chaparral, foothills, deserts, roadsides, other disturbed areas; < 2300 m. GV, CW, SW, D; AZ, Mex. [*Lotus s.* (Nutt.) Greene] Several variants often recognized (see Isely, pp. 193–198); pls in CA-FP gen ± strigose, with narrow lflets; pls in D fleshy, gen canescent, with wide lflets. Conspicuous in spring. Mar–Jun

A. wrangelianus (Fisch. & C.A. Mey.) D.D. Sokoloff Ann, ± strigose or hairs soft, spreading. **ST**: gen prostrate, branched at base, 0.5–3 dm. **LF**: irregularly pinnate; stipules gland-like or ± 0; lflets gen 4, 4–15 mm, elliptic to obovate; axis flat, ± blade-like. **INFL**: 1-fld, peduncle ± 0, ± longer in fr, bract 0. **FL**: calyx 2.5–5 mm, lobes ± 0.8–1.2 × tube, strigose; corolla 5–9 mm, opening or not, yellow, red in age, wings ± ≤ keel; stigma glabrous. **FR**: dehiscent, erect or spreading, exserted, 10–18 mm, gen 2.2–3 mm wide, oblong, straight, gen flat; beak curved, 0.5–1.5 mm. **SEED**: 3–7. 2*n*=12. Abundant. Coastal bluffs, chaparral, disturbed areas; < 1500 m. CA-FP, probably naturalized MP, DSon through agriculture. [*Lotus w.* Fisch. & C.A. Mey.; *L. subpinnatus* Lag., misappl.] Mar–Jun

ALBIZIA

Shrub, tree, [liana], unarmed. **LF**: even-2-pinnate, alternate; petiole with a gland. **INFL**: axillary spike-like racemes or ± terminal panicle of head-like racemes. **FL**: radial, ± green, yellow, white, or pink; sepals, petals inconspicuous; stamens many, conspicuously exserted, filaments fused below into tube ± = petals, free above. **FR**: not or slowly dehiscent, valves not recurving. ± 140 spp.: temp to trop Afr, Madagascar, s Asia, esp Am. (F. del Albizzi, Florentine nobleman, naturalist, 18th century)

1. Infl ± terminal panicle of head-like racemes; stamens 20–38 mm, filaments pink . . . . ***A. julibrissin***
1′ Infl axillary spike-like racemes; stamens 13–16 mm, filaments green to yellow . . . . ***A. lophantha***

A. julibrissin Durazz. SILK TREE, MIMOSA Tree, broad-crowned. **LF**: deciduous; 1° lflets 8–24; 2° lflets 26–60, < 1.5 cm. **FL**: sessile; corolla ± green, 7–11 mm; stamens 20–34. **FR**: 10–20 cm, oblong, glabrous, gray-brown. **SEED**: 8–15. Gen disturbed, coastal, riparian areas; < 500 m. ScV, SW, expected elsewhere; native to warm temp, trop Asia, widely cult in trop, warm temp. May–Jul

A. lophantha (Willd.) Benth. PLUME ACACIA Shrub, small tree. **LF**: evergreen; 1° lflets 14–30; 2° lflets ± 50, < 1 cm. **INFL**: axis 3–6 cm; fls 40–70. **FL**: corolla ± green, 5–7.2 mm; stamens 95–150. **FR**: 5–10 cm, oblong, glabrous, brown. **SEED**: 8–11. Gen disturbed, coastal areas; < 300 m. CCo, SnFrB, SCoRO, SCo, ChI, WTR, PR; native to se Asia, sw Australia. [*Paraserianthes l.* (Willd.) I.C. Nielsen] Widely cult in trop, warm temp. May–Jul

ALHAGI
Martin F. Wojciechowski & Duane Isely

Shrub, ± glabrous; thorns axillary; rhizome spreading. **LF**: simple, small. **INFL**: axillary raceme, open fls 2–7; main axis a thorn. **FL**: 9 filaments fused, 1 free. **FR**: indehiscent, oblong, round in ×-section, narrowed between seeds. ± 3 spp.: Medit, w Asia. (Arabic: pilgrim)

A. maurorum Medik. (p. 731) CAMEL THORN **ST**: 3–10 dm, much-branched, ± green. **LF**: 7–20 mm, elliptic or obovate. **FL**: corolla 8–9 mm, red-purple. **FR**: 1–3 cm, narrowed between seeds, glabrous; stalk-like base 2–3 mm. 2*n*=16. Uncommon. Arid agricul-tural areas; esp < 700 m. GV, s SNE, D; sporadic to w TX; native to w Asia. [*A. pseudalhagi* (M. Bieb.) Desv.] Desert forage, source of manna. Most infestations have been eradicated. Jun–Jul ◆

AMORPHA FALSE INDIGO
Steve Boyd & Duane Isely

Shrub, unarmed, gland-dotted. **LF**: odd-1-pinnate; stipules bristle-like, ephemeral; lflets with minute stipule-like appendages. **INFL**: raceme, spike-like, terminal. **FL**: petal 1 (banner), ± purple; stamens 10, exserted, filaments fused near base; style puberulent. **FR**: indehiscent. **SEED**: 1. ± 15 spp.: N.Am. (Greek: deformed, from 1 petal) [McMahon & Hufford 2004 Amer J Bot 91:1219–1230]

1. Main lf axis without prickle-like glands; lflet midrib gen ending as awn-like gland 0.5–1.5 mm . . . . ***A. fruticosa***
1′ Main lf axis with prickle-like glands; lflet midrib gen ending as sessile, head-like gland ≤ 0.5 mm . . . . ***A. californica***
  2. Pl hairy; longest calyx lobe (exc gland at tip) 1–2 mm . . . . var. ***californica***
  2′ Pl ± glabrous; longest calyx lobe (exc gland at tip) 0.5–1 mm — s NCoR, n SnFrB (Marin Co.) . . . . var. ***napensis***

A. californica Nutt. **LF**: main axis with prickle-like glands; lflet midrib gen ending as sessile, head-like gland ≤ 0.5 mm. **INFL**: gen 1 per st tip. 2*n*=20.

  var. ***californica*** (p. 731) Pl hairy. **FL**: longest calyx lobe (exc gland at tip) 1–2 mm. Wooded, shrubby, or open slopes; < 2300 m. KR, NCoRI, CaRF, n SNF, ScV, SCoR, SCo, TR, n&e PR. May–Jul

  var. ***napensis*** Jeps. NAPA FALSE INDIGO Pl ± glabrous. **FL**: longest calyx lobe (exc gland at tip) 0.5–1 mm. Chaparral; < 800 m. s NCoR (Napa, Sonoma cos.), n SnFrB (Marin Co.). May–Jul ★

A. fruticosa L. (p. 731) **LF**: main axis without prickle-like glands; lflet midrib gen ending as awn-like gland 0.5–1.5 mm. **INFL**: gen > 1 per st tip. 2*n*=20,40. Streambanks, canyons; < 1200 m. ScV, c&s SCo, SnBr, PR, ne DSon; US, e Can, n Mex. Highly variable, esp outside CA. May–Jul

ASTRAGALUS LOCOWEED, MILKVETCH
Martin F. Wojciechowski & Richard Spellenberg

Ann, per from crown, gen unarmed; hairs gen present, simple or branches 2, from base, parallel to lf surface, unequal or not. **ST**: 0 or prostrate to erect. **LF**: odd-1-pinnate (or palmately compound); lflets gen jointed to midrib, entire; stipules membranous, lower fused around st into sheaths (stipule sheaths) or not. **INFL**: raceme, head- or umbel-like or not, axillary; fls 2–many. **FL**: bilateral; keel petals with small protrusion at base locking into pit on adjacent wing; 9 filaments fused, 1 free; ovary (and fr) gen sessile, style slender, stigma minute. **FR**: gen 1- or ± 2-chambered, often mottled, gen ± dry in age, sometimes deciduous (falling from pl with or without pedicel, calyx, receptacle) before dehiscence. **SEED**: 2–many, smooth, compressed, ± notched at attachment scar. > 2500 spp.: ± worldwide (380 in N.Am, 97 in CA, incl many rare taxa). (Greek: ankle-bone or dice, perhaps from rattling of seeds within fr) Difficult; fl and fr needed for identification; fr said to be "deciduous" dehisce only after fr has separated from pl; many good spp. appear similar; some spp. complexes need study. Taxa near province boundaries may appear in > 1 key. Vars. keyed under spp. for simplicity; spp. with vars. so identified in key. Fr length incl beak and any stalk-like base unless fr body specified; fr depth is suture-to-suture axis. *A. tephrodes* A. Gray var. *brachylobus* (A. Gray) Barneby in sw UT, AZ, near CA.

1. Corolla bright red, banner 35–41 mm; calyx 18–24 mm; fr 25–40 mm, densely woolly . . . . ***A. coccineus***
1′ Corolla not red, banner < 35 mm; calyx < 18 mm; fr often < 25 mm, often not woolly
  2. Lflet ± 1 mm wide, gen spine-tipped; raceme 1–3-fld, ± incl in axil; peduncle inconspicuous or 0; fl, fr gen 3–10 mm. . . . . ***A. kentrophyta*** vars.
  2′ Lflet gen > 1 mm wide, not spine-tipped; raceme often > 3-fld, well exserted from axil; peduncle conspicuous; fl 3–30 mm, fr 2–60 mm
    3. Pl gen of CA-FP (also Wrn, adjacent GB, D) . . . . **Group 1**
    3′ Pl gen of GB, D, or both, occ adjacent CA-FP

4. Pl gen of D . . . . **Group 2**
4′ Pl gen of GB . . . . **Group 3**

Group 1

1. Ann
 2. Banner gen 2.5–3.3 mm; fls early dense, ascending, then spaced, reflexed; fr reflexed, 2.8–4.2 mm, width > depth — frs overlapped, ovate or ± round in top view. . . . . *A. gambelianus*
 2′ Banner gen > 3 mm; fls remaining dense, erect if banner ± 3 mm, otherwise erect, spreading, or reflexed; fr erect to pendent, gen > 4 mm, if not then width ± = to or << depth
  3. Fl, fr gen strongly ascending or spreading, in dense, oblong or ± spheric, head-like racemes; fr 2–4 or >> 4 mm, 1–2-chambered, not or 2-lobed in ×-section
   4. Corolla white-lavender to purple; fr 2–4 mm, 2-chambered, 2-lobed in ×-section . . . . *A. didymocarpus*
   4′ Corolla gen white; fr >> 4 m, 1-chambered, gen unlobed in ×-section . . . . [2]*A. hornii* var. *hornii*
  3′ Fl, fr often ascending early, then spreading or reflexed, often in loose racemes; fr > 4 mm, gen unlobed in ×-section, 1–2-chambered
   5. Infl 10–40-fld; fr inflated, papery, 1-chambered, 4–17 mm wide, sides gen strongly convex . . . . [2]*A. palmeri*
   5′ Infl 2–14-fld; fr not inflated, ± 2-chambered, ≤ 4 mm wide, sides parallel or ± convex
    6. Pl silvery-canescent; local, Cushenbury Canyon (ne SnBr, adjacent DMoj) . . . . [2]*A. albens*
    6′ Pl ± green, ± hairy; widespread, not Cushenbury Canyon
     7. Fr body 5–10 mm, tapered to spine-like or hooked beak ± = body; ovules 2–6; fr in head- or umbel-like racemes. . . . . *A. breweri*
     7′ Fr body > 10 mm or beak < 1/4 body; ovules 5–20; fr in head- or umbel-like racemes or not
      8. Keel > wings; fr base stalk-like, receptacle with peg-like extension (evident after fr drop) . . . . *A. claranus*
      8′ Keel << wings; fr base ± stalk-like or not, receptacle without peg-like extension
       9. Infl 2–7-fld, open, axis in fr 7–20 mm; fr ± purple or ± purple-mottled . . . . *A. pauperculus*
       9′ Infl 2–12-fld, ± dense, axis in fr 1–8 mm; fr pale yellow or black
        10. Fr on erect peduncle; seeds pitted; NCoR. . . . . *A. rattanii* vars.
        10′ Fr on spreading or reflexed peduncle; seeds smooth; GV, CCo, SnFrB, SCo . . . . *A. tener* vars.
1′ Per
 11. Hairs branched; corolla cream or green-white (tinged dull purple); fr erect; n GB, adjacent CA-FP . . . . *A. canadensis* var. *brevidens*
 11′ Hairs simple; corolla often pink-purple, white, or yellow; fr erect to reflexed; gen not n GB, adjacent CA-FP
  12. Calyx 10–12 mm, base with pouch adaxially; fls reflexed — petals dull yellow; fr hanging from stalk-like base; n SNH . . . . *A. gibbsii*
  12′ Calyx gen < 10 mm, base gen without pouch; fls ascending or spreading if calyx > 10 mm
   13. Terminal lflet not or ± jointed to midrib, joint unlike that of lateral lflets; lateral lflets often reduced or 0, spaces between gen >> 1.5 × lflet width; lvs few, sparse
    14. Calyx 8–10 mm; corolla ± white; fr ± 4-sided. . . . . [2]*A. bicristatus*
    14′ Calyx 4–5 mm; corolla red-pink and yellow; fr 2-sided . . . . *A. inversus*
   13′ Terminal lflet jointed to midrib, joint like that of lateral lflets; lateral lflets present, spaces between rarely > 1.5 × lflet width; lvs gen many, not sparse
    15. Pl tufted or matted; st < 15 cm (gen much shorter), from crown at surface; fls strongly ascending, pl hairs dense
     16. Banner, wings hairy on outside; fr 5–7 mm, ± incl in calyx . . . . *A. austiniae*
     16′ Banner, wings glabrous on outside; fr ≥ 7 mm, well exserted from calyx
      17. Calyx 4–6 mm; fr 3-sided in ×-section, gen < 4 mm wide; hairs of st, lf sparse, ± straight, gen < 0.5 mm. . . . . [2]*A. obscurus*
      17′ Calyx 6–14 mm; fr ± round or 2-lobed in ×-section, gen > 4 mm wide; hairs of st, lf dense, curly, often > 0.5 mm
       18. Fr hairs gen > 1.5 mm, hiding surface . . . . *A. purshii* vars.
       18′ Fr hairs gen <<, rarely > 1 mm, hiding surface or not
        19. Petals ± white with some lilac, banner 11–13.3 mm; fr 8–15 mm; lflets 7–13. . . . . *A. subvestitus*
        19′ Petals pink-purple, banner 16–18.5 mm; fr 13–25 mm; lflets 7–19. . . . . *A. leucolobus*
    15′ Pl gen open; st gen > 15 cm, gen from crown at or below surface; fls in age spreading or reflexed, pl hairs sparse, or both, if pl tufted or matted
     20. Lower stipules fused around st into sheath or low and ± inconspicuous
      21. Lflets of at least some lvs > 22; petals ± white or cream
       22. Banner 7–10 mm; fr 6–11 mm, beak long, hooked . . . . *A. pycnostachyus* vars.
       22′ Banner 10–19 mm; fr often > 11 mm, beak not long, not hooked
        23. Pl gray-white-woolly; stipules finely tomentose; ChI . . . . *A. miguelensis*
        23′ Pl ± green even if hairy; stipules gen glabrous or sparsely hairy or densely strigose; ChI or mainland
         24. Fr base not stalk-like; sts not more densely hairy than lvs; gen coastal. . . . . *A. nuttallii* vars.
         24′ Fr base stalk-like; sts more densely hairy than lvs or not; often non-coastal
          25. Fls early spreading, then reflexed; stalk-like fr base 5–40 mm, ± strigose
           26. Stalk-like fr base 14–40 mm . . . . *A. asymmetricus*
           26′ Stalk-like fr base 5–17 mm . . . . [2]*A. trichopodus* vars.

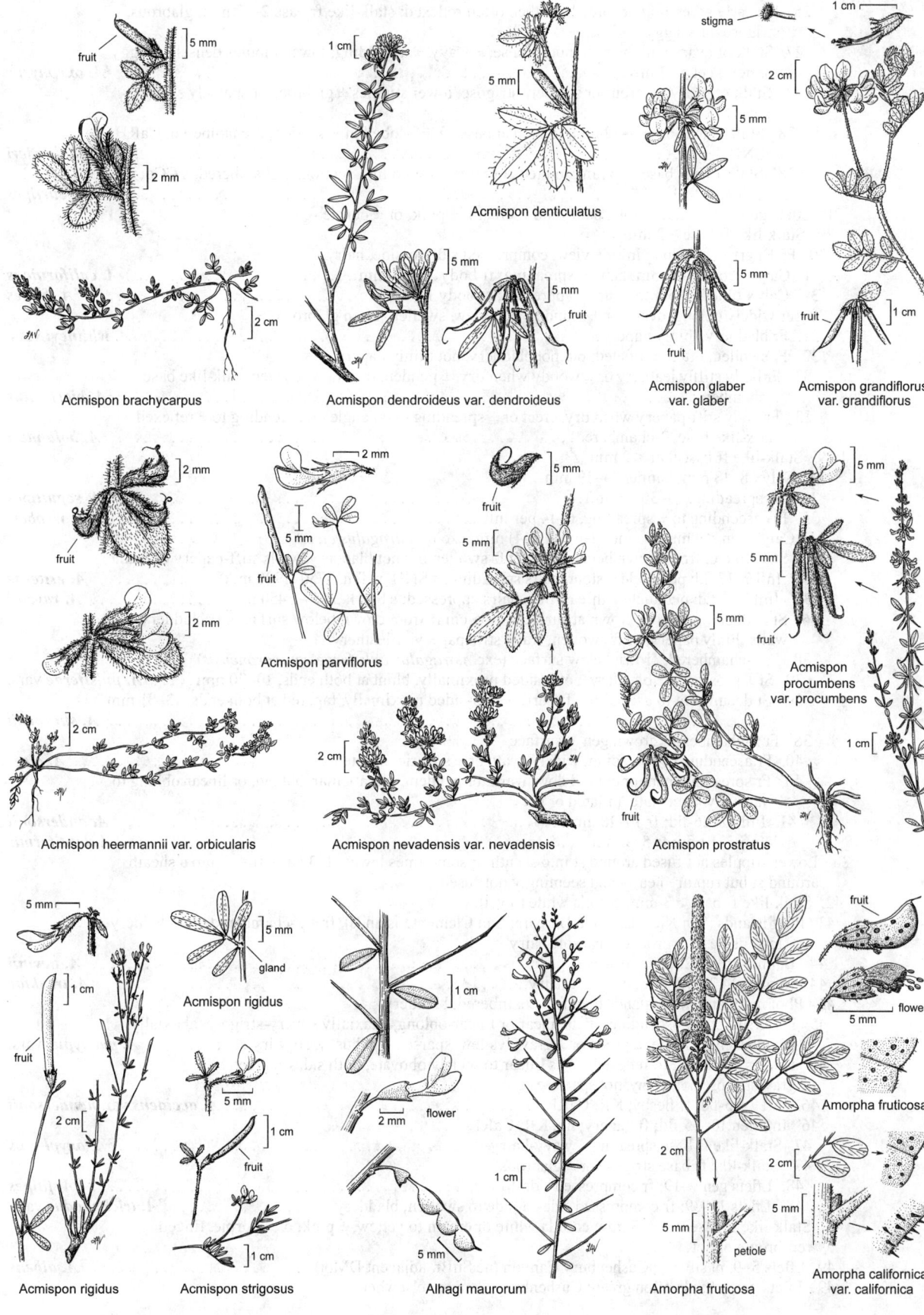

fruit

5 mm

2 mm

2 cm

Acmispon brachycarpus

1 cm

5 mm

fruit

Acmispon dendroideus var. dendroideus

Acmispon denticulatus

stigma

1 cm

2 cm

5 mm

5 mm

fruit

Acmispon glaber
var. glaber

fruit

1 cm

Acmispon grandiflorus
var. grandiflorus

2 mm

fruit

2 mm

2 mm

2 mm

fruit

Acmispon parviflorus

5 mm

fruit

5 mm

5 mm

5 mm

5 mm

fruit

Acmispon
procumbens
var. procumbens

1 cm

2 cm

Acmispon heermannii var. orbicularis

2 cm

Acmispon nevadensis var. nevadensis

fruit

1 cm

Acmispon prostratus

5 mm

5 mm

gland

Acmispon rigidus

fruit

1 cm

2 cm

Acmispon rigidus

5 mm

fruit

1 cm

Acmispon strigosus

1 cm

flower

2 mm

5 mm

1 cm

Alhagi maurorum

fruit

flower

5 mm

Amorpha fruticosa

2 cm

5 mm

petiole

Amorpha fruticosa

2 cm

5 mm

Amorpha californica
var. californica

25′ Fls early ascending or spreading, then often reflexed; stalk-like fr base 2–12 mm, glabrous, strigose, or shaggy

    27. Stalk of ovary with hairs minute, dense, ± wavy, ± ascending; lower stipules densely shaggy or not; keel ≥ 13 mm. . . . . ²*A. oxyphysus*

    27′ Stalk of ovary glabrous or sparsely strigose; lower stipules ± glabrous or sparsely shaggy; keel < 13 mm

        28. Stalk-like fr base 5–12 mm, jointed at base; fr ± 2-lobed in ×-section, 2-chambered; CaRH, SNH . . . . ²*A. bolanderi*

        28′ Stalk-like fr base 2–6 mm, jointed at top; fr ± round in ×-section, 1-chambered; s CCo, w SCoR, n ChI . . . . *A. curtipes*

21′ Lflets gen < 22; petals white, pale yellow, cream, ± pink, or ± purple

  29. Stalk-like fr base ≥ 2 mm

    30. Fr ± narrowly oblong in side view, compressed side-to-side, hairy

      31. Calyx tube 5.2–7 mm, hairs ± spreading; fr body 27–43 mm . . . . *A. californicus*

      31′ Calyx tube 3–3.5 mm, hairs ± appressed; fr body 17–30 mm . . . . ²*A. filipes*

    30′ Fr widely oblong, ovate, or ± round in side view, swollen, often glabrous

      32. Fr bladdery, thinly papery. . . . . *A. whitneyi* vars.

      32′ Fr swollen, often ± inflated, but not bladdery, not thinly papery

        33. Fr body stiffly leathery or ± woody when dry, ± pendent on a downcurved stalk-like base, 1-chambered. . . . . ²*A. bicristatus*

        33′ Fr body stiff-papery when dry, erect or ± spreading — at angle on ascending to ± reflexed stalk-like base, 2-chambered. . . . . ²*A. bolanderi*

  29′ Stalk-like fr base 0 or < 2 mm

    34. Calyx 8–13 mm; banner 14–19 mm

      35. Fls spreading, 10–30 per infl. . . . . ²*A. sepultipes*

      35′ Fls ascending to ± spreading, 6–14 per infl . . . . *A. webberi*

    34′ Calyx gen < 9 mm; banner gen 5–12(14) mm, exc in *Astragalus cicer*

      36. St 1–10 cm, from crown below surface; fr swollen but not bladdery, walls stiff-papery or leathery

        37. Infl 7–17-fld; peduncle ± stout, hairs spreading; s SNH, n Teh, 1750–1900 m. . . . . ²*A. ertterae*

        37′ Infl 2–8-fld; peduncle ± thread-like, hairs appressed; s SNH, 3400–3450 m . . . . *A. ravenii*

      36′ St > 10 cm, gen from crown at surface (1–35 cm if from crown below surface); fr bladdery, walls thinly papery, or fr swollen, walls stiff-papery or leathery

        38. Fr 1-chambered; crown below surface (exc *Astragalus pulsiferae var. coronensis*)

          39. St ± prostrate; fr bladdery, not 3-sided proximally, blunt at both ends, 10–20 mm . . . . . *A. pulsiferae* vars.

          39′ St decumbent to ascending; fr narrow, ± 3-sided proximally, tapered at both ends, 13–31 mm . . . . *A. shevockii*

        38′ Fr 2-chambered; crown gen at surface

          40. Fr ascending or spreading, widely ovate or ± spheric, inflated . . . . *A. cicer*

          40′ Fr spreading, bent or curved downward, or pendent, ± lenticular, oblong, or linear-oblong to narrowly lanceolate, inflated or not

            41. Infl 12–26-fld; fr 10–18 mm. . . . . *A. andersonii*

            41′ Infl 5–10-fld; fr 5–9 mm. . . . . *A. lentiformis*

20′ Lower stipules not fused around st into sheath or sometimes lowest 1–3 pairs fused into a sheath around st but ruptured early and seemingly not fused

  42. Stalk-like fr base > 3 mm; corolla white to yellow

    43. Pl of islands (San Nicolas, Santa Barbara, San Clemente islands), fr ± 2-chambered, not bladdery — lower side of fr grooved; lvs soft-hairy

      44. Banner 10.6–12.7 mm; ovary, fr glabrous . . . . *A. nevinii*

      44′ Banner 14.2–17.5 mm; ovary, fr hairy . . . . *A. traskiae*

    43′ Pl of mainland, if of islands then fr 1-chambered, bladdery

      45. Fls, frs strongly ascending; lflets linear or linear-oblong, adaxially silvery-strigose, abaxially less densely strigose, ± green or ± gray; lvs few, sparse — pl bushy, sts wiry . . . . *A. pachypus* vars.

      45′ Fls, frs spreading to reflexed; lflets linear to widely obovate, both sides ± green or ± gray-strigose; lvs many, not sparse

        46. Infl 7–15-fld; fr fleshy; KR, CaRF . . . . *A. accidens* var. *hendersonii*

        46′ Infl gen 10–65-fld; fr papery; not KR, CaRF

          47. Stalk-like fr base spreading-, wavy-hairy. . . . . ²*A. oxyphysus*

          47′ Stalk-like fr base strigose or ± glabrous

            48. Lflets gen 9–19; fr compressed side-to-side . . . . ²*A. filipes*

            48′ Lflets 15–39; fr compressed side-to-side to swollen, bladdery. . . . . ²*A. trichopodus* vars.

  42′ Stalk-like fr base 0 or < 3 mm; corolla white or cream to yellow, ± pink, or ± purple, tinged red-brown or not

    49. Lflets 5–9; pl silvery; Cushenbury Canyon (ne SnBr, adjacent DMoj) . . . . ²*A. albens*

    49′ Lflets gen > 9; pls often green; Cushenbury Canyon, elsewhere

50. Fr 4–7 mm, gen 2-chambered; calyx 3–4 mm; banner 4–7 mm
  51. St 3–10 dm; fls reflexed, 20–100 per infl. . . . . ***A. clevelandii***
  51′ St 1–4(5) dm; fls ascending, 2–13 per infl . . . . ***A. lemmonii***
50′ Fr 6.5–60 mm, 2- or often 1-chambered; calyx 4–13 mm; banner gen > 7 mm
  52. Fr bladdery, walls gen thinly papery
    53. Middle of fr 2-chambered, beak compressed side-to-side, triangular in side view,
        1-chambered; banner recurved 30–50° . . . . [2]***A. lentiginosus*** vars.
    53′ Middle of fr 1-chambered, beak often not or ± compressed side-to-side, not triangular in side
        view, 2-chambered or not; banner recurved 40–90°, often > 60°
      54. Fls, frs crowded in head-like clusters 2.5–3.5 cm wide; fr 12–18 mm, ascending, hairs ±
          spreading . . . . [2]***A. hornii*** var. ***hornii***
      54′ Fls, frs, if crowded, in clusters > 3.5 cm wide; fr gen > 18 mm, often spreading, hairs
          ascending or ± appressed or 0 in age
        55. Banner gen > 11 mm, recurved ± 40–45°; lflets often > 25 (see also *Astragalus macrodon*, at 57.)
          56. Lflets 19–29, not clearly smaller from lf base to tip; lflet midrib prominently raised
              abaxially; ovules 29–40; 250–300 m; PR (sw San Diego Co.) . . . . ***A. deanei***
          56′ Lflets 25–41, clearly, gradually smaller from lf base to tip; lflet midrib not prominently
              raised abaxially; ovules 34–55; 50–700 m; s SCoRO, SCo, PR, nw Baja CA . . . . ***A. pomonensis***
        55′ Banner gen < 11 mm, recurved 60–90°; lflets gen < 25 (see also *Astragalus deanei*, at 56.)
          57. Lf, fr minutely strigose or in age glabrous; lflet margin, midrib green; calyx lobes 0.7–2.6
              mm; ovules 51–71; s SN, Teh, GV, CW, SW . . . . ***A. douglasii*** vars.
          57′ Lf, fr hairs ± sparse to dense, ± ascending or spreading to ± incurved; lflet margin, midrib
              gen purple; calyx lobes 2.5–4.3 mm; ovules 29–52; c&s SCoR. . . . . ***A. macrodon***
  52′ Fr narrow or swollen, ± bladdery or inflated, walls stiff-papery, leathery, or woody
    58. St 3.5–10 cm, spreading-hairy, from crown below surface; s SNH, n Teh . . . . [2]***A. ertterae***
    58′ St gen > 10 cm, ascending- or appressed-strigose or ± glabrous, gen from crown at surface; s
        SNH, n Teh, elsewhere
      59. St erect to widely ascending, 60–150 cm; sw SW
        60. Fl reflexed; corolla dull lilac; fr reflexed, deciduous, < 10 mm, densely wavy-hairy . . . . . . ***A. brauntonii***
        60′ Fl widely ascending; corolla cream; fr erect, persistent, 15–25 mm, ± glabrous. . . . . ***A. oocarpus***
      59′ St decumbent to erect, 30–90 cm; widespread, incl sw SW
        61. Fr 1-chambered, thickly papery, ± strongly inflated; fls spreading; corolla often pink-purple
            (or cream); w edge DSon, adjacent foothills of SnBr, PR . . . . [2]***A. palmeri***
        61′ Fr ± 2-chambered, at least in middle, thinly papery to stiffly leathery or woody, linear
            to swollen or ± bladdery-inflated; fls ascending to reflexed; corolla white, cream or
            pink-purple; widespread (incl places at 61.)
          62. Beak of fr triangular in side view, compressed strongly side-to-side, 1-chambered; fr > 5
              mm wide, swollen, both sutures ± sunken; corolla ± purple or ± white . . . . [2]***A. lentiginosus*** vars.
          62′ Beak of fr acuminate to ± triangular in side view, gen not compressed strongly, gen
              2-chambered; fr gen < 5 mm wide, ± 3-sided, lower suture gen sunken, upper raised;
              corolla pale to dark lilac, pink- or green-white, or ± white
            63. Fr, ovary glabrous
              64. Pl ± glabrous; fr body 14–24 mm, spreading or reflexed; KR, NCoRO . . . . ***A. umbraticus***
              64′ Pl minutely strigose, esp above; fr body 20–42 mm, ascending; SnBr, DMtns
                65. Lflets 7–19; banner 7–10.2 mm; calyx tube 2.7–4.1 mm . . . . ***A. bernardinus***
                65′ Lflets 17–27; banner 12.6–15.7 mm; calyx tube 4.1–5 mm . . . . ***A. tricarinatus***
            63′ Fr, ovary hairy
              66. Fls gen 3–14 per infl, ascending; fr erect; st 5–15 cm; se KR, CaR (also MP) . . . . [2]***A. obscurus***
              66′ Fls gen (8)10–40 per infl, spreading or reflexed; fr ascending or reflexed; st > 15 cm;
                  SNE, SNF, NCoRO
                67. Pl ± glabrous, esp at base, ± green; calyx 6.5–9 mm; corolla white; NCoRO. . . . . ***A. agnicidus***
                67′ Pl hairy, gray to silver; calyx ± 5–13 mm; corolla pink- to lilac-white, wing tips ±
                    white, or ± cream; c&s SNF, e edge SNH, SNE
                  68. Herbage silvery-silky, hairs appressed; calyx 8.5–13 mm; corolla pink- to lilac-white,
                      wing tips ± white; e edge SNH, SNE . . . . [2]***A. sepultipes***
                  68′ Herbage gray-hairy, hairs ± spreading; calyx 5–8 mm; corolla ± cream; c&s SNF. . . . ***A. congdonii***

Group 2

1. Terminal lflet gen not jointed to midrib, if jointed, ± so and unlike that of lateral lflet; lateral lflets often
   reduced or 0, spaces between gen >> 1.5 × lflet width; lvs few, sparse
  2. Banner 10–14.2 mm; lflets 3–13; fr half-ovate to ± spheric, bladdery . . . . ***A. magdalenae*** var. ***peirsonii***
  2′ Banner 17–26 mm; lflets 5–11; fr plump-oblong, not bladdery. . . . . [2]***A. serenoi*** var. ***shockleyi***
1′ Terminal lflet jointed to midrib, gen like that of lateral lflet; lateral lflets present, spaces between rarely >
   1.5 × lflet width; lvs gen many, not sparse
  3. Hairs dense, branched; lflets 1–7, crowded near lf tip . . . . ***A. calycosus*** var. ***calycosus***

3′ Hairs 0 to dense, simple; lflets often > 7, gen not crowded near lf tip
　4. Ann, or fl 1st yr and appearing so
　　5. Fl, fr strongly ascending to erect, in dense, head-like racemes; fr ± spheric, 2–4 mm . . . . *A. didymocarpus* vars.
　　5′ Fl, fr ± spreading to ascending or reflexed, in often open racemes; fr linear to ± ovate-inflated, > 4 mm
　　　6. Keel 10–21 mm
　　　　7. Pl densely shaggy-hairy; st gen << 7 cm, ± prostrate, often ± 0; fr 1-chambered . . . . [2]*A. tidestromii*
　　　　7′ Pl ± glabrous to strigose; st > 9 cm, prostrate to erect; fr 1- or 2-chambered
　　　　　8. Keel 17–21 mm; fr 1-chambered . . . . [2]*A. crotalariae*
　　　　　8′ Keel 10–13 mm; fr 2-chambered at middle . . . . [4]*A. lentiginosus* vars.
　　　6′ Keel 2–10.5 mm
　　　　9. Fr bladdery, walls thinly papery, translucent
　　　　　10. Infl gen 10–35-fld
　　　　　　11. Lflets gen 15–33; fr 1-chambered . . . . *A. hornii* var. *hornii*
　　　　　　11′ Lflets gen 11–27; fr 2-chambered below beak . . . . [4]*A. lentiginosus* vars.
　　　　　10′ Infl gen 2–10-fld
　　　　　　12. Corolla ± white; lflets 11–19; upper suture of fr convex; ovules 13–21 . . . . [2]*A. allochrous* var. *playanus*
　　　　　　12′ Corolla ± purple; lflets 7–19; upper suture of fr straight or convex; ovules 7–24
　　　　　　　13. Keel 4.8–6 mm; lflets gen 11–19; upper suture of fr much less convex than lower; ovules 7–14
　　　　　　　　. . . . . *A. insularis* var. *harwoodii*
　　　　　　　13′ Keel 5.9–6.6 mm; lflets gen 7–13; upper suture of fr ± equally convex as lower; ovules 19–24 . . . . [2]*A. nutans*
　　　　9′ Fr linear to incurved or swollen, ± inflated, walls papery, ± translucent, or leathery
　　　　　14. Infl 20–40-fld . . . . [2]*A. palmeri*
　　　　　14′ Infl < 20-fld
　　　　　　15. Fr 1-chambered, swollen, half-ovate or -elliptic in side view, ± ovate or round in x-section
　　　　　　　16. Pl, fr hairs appressed, ± straight, ± obscuring surface; ovules 3–7 . . . . *A. aridus*
　　　　　　　16′ Pl, fr hairs ascending or spreading, ± wavy, not obscuring surface; ovules 10–19 . . . . *A. sabulonum*
　　　　　　15′ Fr 2-, rarely 1-chambered, then often ± linear, curved, ± 2- or 3-sided in x-section
　　　　　　　17. Lflets adaxially sparsely hairy; infl 1–6-fld; keel < 6 mm; fr diam often < 3 mm
　　　　　　　　18. Lflets of upper lvs gen blunt, notched at tip; fr early-deciduous, maturing pale brown . . . . . . *A. acutirostris*
　　　　　　　　18′ Lflets of upper lvs gen acute at tip; fr persistent, maturing dark brown or ± black . . . . *A. nuttallianus*
　　　　　　　17′ Lflets adaxially silvery-canescent; infl 3–16-fld; keel > 6 mm; fr diam often > 3 mm
　　　　　　　　19. Fr 10–18 mm, 2.8–3.5 mm wide, incurved, ± 3-sided; ovules 8–11; Cushenbury Canyon
　　　　　　　　　(ne SnBr, adjacent DMoj) . . . . [2]*A. albens*
　　　　　　　　19′ Fr 13–32 mm, 3.5–8.5 mm wide, straight, ± 2-sided, or incurved, ± 3-sided; ovules 20–30;
　　　　　　　　　widespread in DMoj . . . . [2]*A. mohavensis* vars.
　4′ Per
　　20. Fr, gen ovary, glabrous (*Astragalus douglasii*, *Astragalus nutans*, under 20′, glabrous or not)
　　　21. Lflet hairs 0 or few, restricted to margins, midrib on 1 or both surfaces; fr base stalk-like
　　　　22. Fls early ascending, then reflexed or not; keel 9.5–10.6 mm; fr ± ascending to spreading, ±
　　　　　2-chambered . . . . *A. cimae* vars.
　　　　22′ Fls strongly ascending; keel 11–19 mm; fr erect, 1-chambered . . . . [2]*A. preussii* vars.
　　　21′ Lflet hairs ± throughout 1 or both surfaces; fr base stalk-like or not
　　　　23. Keel 6–9.4 mm; fr 2-chambered, at least at middle
　　　　　24. Pl often coarse, lfy, gen in open areas; st not wiry; stalk-like fr base 0 — fr swollen, beak
　　　　　　compressed side-to-side, triangular in side view, 1-chambered . . . . [4]*A. lentiginosus* vars.
　　　　　24′ Pl slender, sparsely lfy, often in shelter of shrubs; st ± wiry; stalk-like fr base 1–5 mm
　　　　　　25. Fls 10–25 per infl, ascending; fr ascending, 3-sided . . . . *A. bernardinus*
　　　　　　25′ Fls 5–15 per infl, early ascending, then reflexed; fr pendent, ± 2-sided . . . . *A. jaegerianus*
　　　　23′ Keel 9.7–19 mm; fr 1- or 2-chambered
　　　　　26. Calyx glabrous; banner recurved ± 85°; fr bladdery, 1-chambered, base stalk-like . . . . *A. oophorus* var. *oophorus*
　　　　　26′ Calyx strigose; banner recurved < 50°; fr ovate or narrow, not bladdery, ± 2-chambered, base
　　　　　　stalk-like or not
　　　　　　27. Corolla pink-purple; lflets narrow, adaxially often paler than abaxially (silvery vs ± green); fr
　　　　　　　erect, stalk-like base 0; n Inyo Co. . . . . [2]*A. serenoi* var. *shockleyi*
　　　　　　27′ Corolla white or cream to lemon yellow; lflets narrow or wide, adaxially not paler than adaxially;
　　　　　　　fr ascending to pendent, stalk-like base 1–10 mm; w or sw edge of D
　　　　　　　28. Banner 15–22 mm; fr 15–27 mm, swollen, ± compressed side-to-side; lflets > 17 or not . . . *A. pachypus* vars.
　　　　　　　28′ Banner 12.6–15.7 mm; fr 24–42 mm, ± linear, 3-sided; lflets 17–27 . . . . *A. tricarinatus*
　20′ Fr, ovary with at least some hairs (*Astragalus douglasii*, *Astragalus nutans* glabrous or not)
　　29. Pl woolly- or wavy-hairy; sts < 9 cm, gen densely tufted
　　　30. Keel 18–28 mm; fr 1-chambered, densely white-hairy
　　　　31. Calyx with more black than white hairs; fr 25–50 mm; lf bases not notably persistent on crown;
　　　　　ne DMtns (e of Death Valley) . . . . *A. funereus*
　　　　31′ Calyx gen with more white than black hairs; fr 13–28 mm; lf bases persistent on crown;
　　　　　widespread, incl DMtns . . . . *A. newberryi* var. *newberryi*

30′ Keel gen < 20 mm; fr 2-chambered if densely white-hairy
   32. Infl 10–45-fld; corolla 2-colored, ± white with keel tip, wing tips, and sometimes banner tip ± purple; fr hairs spreading — pls from deep rhizomes (± stless pls of *Astragalus minthorniae* will key out here; see 36.) . . . . ²*A. layneae*
   32′ Infl 3–16-fld; corolla pink-purple or ± 2-colored (± as at 33.); fr hairs spreading or not
      33. Keel 11.5–20.8 mm; petals pink-purple; fr often ± 2-chambered, hairs spreading or not
. . . . *A. purshii* var. *tinctus*
      33′ Keel 10–12 mm; petals ± white, tinged dull purple; fr 1-chambered, hairs gen appressed . . . . . ²*A. tidestromii*
29′ Pl hairy to ± glabrous; sts gen > 9 cm (if < 9 cm, pl gen not woolly- or wavy-hairy exc sometimes in *Astragalus layneae*), tufted or not
   34. St ≤ 16 cm, from rhizome well below surface; corolla 2-colored, ± white with keel tip, wing tips, and sometimes banner tip ± purple . . . . ²*A. layneae*
   34′ St often > 15 cm, from crown at or ± below surface; corolla 1- or 2-colored (as at 35.).
      35. Lflet hairs 0 or few on margins, midrib on 1 or both sides; stalk-like fr base 0 or 2–7 mm . . . . ²*A. preussii* vars.
      35′ Lflet hairs ≥ few, ± throughout 1 or both sides; stalk-like fr base 0 or < 3 mm
         36. St, lf hairs ± spreading, curly or not, 0.8–1.5 mm; fr shaggy-hairy, ± erect . . . . . . *A. minthorniae* var. *villosus*
         36′ St, lf hairs appressed, not curly, if spreading then < 0.8 mm; fr often strigose or silky-canescent, erect to reflexed or pendent
            37. St slender, wiry, incurved, gen < 15 cm; petioles persistent, wiry; infl 1–4-fld — pls of limestone, resembling an unkempt nest . . . . *A. panamintensis*
            37′ St slender or coarse, if wiry then not incurved, often > 15 cm; petioles deciduous, not wiry; infl often > 4-fld
               38. Petals dingy white at least at base, dull lilac on margins or tips; wings ± twisted, tips shortly fringed or notched; fr pendent, ± compressed side-to-side, 2-chambered. . . . . *A. atratus* var. *mensanus*
               38′ Petals pale yellow or white to ± purple; wings plane or curved, tips not fringed, not notched; fr erect to reflexed, ± 3-sided or round in ×-section, sometimes plump and widely grooved on upper side, 1- or 2-chambered
                  39. Fr 2-chambered in middle, beak often 1-chambered; banner recurved 30–50°
                     40. Lflets gen > 11; fr in ×-section ± round, upper suture gen ± sunken in a wide channel; pl gen robust — fr beak widely triangular, 1-chambered . . . . ⁴*A. lentiginosus* vars.
                     40′ Lflets 3–11; fr ± 2- or 3-sided, upper suture raised; pl robust or delicate
                        41. Fr 10–18 mm, 2.8–3.5 mm wide, incurved, ± 3-sided; ovules 8–11; Cushenbury Canyon (ne SnBr, adjacent DMoj) . . . . ²*A. albens*
                        41′ Fr 13–32 mm, 3.5–8.5 mm wide, straight, ± 2-sided, or incurved, ± 3-sided; ovules 20–30; widespread in DMoj . . . . ²*A. mohavensis* vars.
                  39′ Fr 1-chambered; banner recurved 45–90°
                     42. Of the following, all true: most lvs with > 19 lflets; fr bladdery; infl gen > 12-fld . . . . . . *A. douglasii* vars.
                   42′ Of the following, 1 true: most lvs with ≪ 19 lflets; fr swollen but not bladdery; infl < 12-fld
                      43. Banner > 11 mm; calyx gen > 7 mm; fr stiff-papery or leathery
                        44. Keel < 14 mm; fr pendent to reflexed, 5–10 mm wide; SNE, DMtns (Inyo Co.) . . . . *A. casei*
                        44′ Keel > 16 mm; fr ascending or spreading, 10–14 mm wide; DSon . . . . ²*A. crotalariae*
                     43′ Banner < 11 mm; calyx 4–7 mm; fr ± papery, ± bladdery, or inflated
                      45. Infl gen 10–40-fld; petals gen bright pink-purple; w edge DSon, adjacent foothills of SnBr, PR . . . . ²*A. palmeri*
                      45′ Infl often < 10-fld; petals ± white or pink-purple; c&e DSon, e DMoj
                        46. Petals ± white; lflets 11–19; fr beak short, obscure . . . . ²*A. allochrous* var. *playanus*
                        46′ Petals pink-purple; lflets 7–13; fr beak deltate, compressed side-to-side
                          47. Banner 6.1–8 mm; ovules 8–12; n DMtns (Panamint Range), > 2000 m . . . . *A. gilmanii*
                          47′ Banner 7.8–10.4 mm; ovules 19–24; se DMtns, DSon, ≤ 2000 m. ²*A. nutans*

Group 3

1. Terminal lflet not or ± jointed to midrib, joint unlike that of lateral lflet; lateral lflets often reduced or 0, spaces between gen ≫ 1.5 × lflet width; lvs few, sparse
  2. Fr compressed side-to-side, base stalk-like; banner 9.4–12.2 mm . . . . *A. inversus*
  2′ Fr cylindric or ± compressed side-to-side, base not stalk-like; banner 17–26 mm. . . . . *A. serenoi* var. *shockleyi*
1′ Terminal lflet jointed to midrib, joint like that of lateral lflet; lateral lflets not reduced or 0, spaces between rarely > 1.5 × lflet width; lvs gen many, not sparse
  3. Ann, slender, rarely persisting to 2nd season
    4. Banner 3–5 mm; fr ≤ 4 mm. . . . . *A. didymocarpus* var. *dispermus*
    4′ Banner 4.7–7.6 mm; fr ≥ 12 mm
      5. Fr not inflated, 2–3 mm wide; lflet adaxially ± appressed-hairy . . . . *A. acutirostris*
      5′ Fr inflated, 6–10 mm wide; lflet adaxially ± glabrous . . . . *A. geyeri* var. *geyeri*
  3′ Per, often coarse — fl, fr > 5 mm (fr < 5 in *Astragalus anxius*)

6. Lf hairs branched
  7. Lflets 1–7 . . . . ***A. calycosus*** var. ***calycosus***
  7′ Lflets 7–25 . . . . ***A. canadensis*** var. ***brevidens***
6′ Lf hairs simple
  8. Fr densely white-hairy, resembling a cotton ball; st 0–14 cm; pl hairs dense, of 2 kinds or not
    9. Sts forming thickened crown covered by persistent lf bases; longer hairs of lf gen straight, some
      spreading; fr hairs of 2 kinds, some curly, short, some ± straight, long (see also *Astragalus*
      platytropis) . . . . ***A. newberryi*** var. ***newberryi***
    9′ Sts tufted or matted, lf bases not persistent; longer hairs of lf wavy, tangled; fr hairs of 1 kind, ± wavy
      or straight . . . . ***A. purshii*** vars.
  8′ Fr silvery-hairy or not, not resembling a cotton ball; st often > 10 cm; pl hairs sparse or dense, gen not of 2 kinds
    10. Calyx base strongly asymmetric, pedicel attached at lower side, upper ± pouched; petals white to ±
      yellow; fr curved 1/4 to ± full circle, base stalk-like
      11. St hairs subappressed or incurved; fr 2.5–4.5 mm wide; lowest stipules not fused
          . . . . ***A. curvicarpus*** var. ***curvicarpus***
      11′ St hairs gen spreading, straight; fr 4–8 mm wide; lowest stipules fused around st into sheath . . . . ***A. gibbsii***
    10′ Calyx base ± symmetric, pedicel attached ± at middle, upper side ± not pouched; petals white, ±
      yellow, or purple-pink; fr straight to curved ± full circle, base stalk-like or not
      12. Fr spreading or pendent, early ± flat, then bulged around seeds, base stalk-like; corolla white, pale
        yellow, or cream
        13. Calyx, lf hairs wavy, ± ascending or spreading; calyx > 6 mm . . . . ***A. californicus***
        13′ Calyx, lf hairs ± straight, appressed; calyx gen < 6 mm . . . . ***A. filipes***
      12′ Fr erect to pendent or reflexed, early ± swollen, then ± round, triangular, or 2-lobed in ×-section,
        base stalk-like or not; corolla ± white, ± yellow, or ± purple
        14. Stipule sheaths (on herbarium specimens or not) ruptured by st expansion or not
          15. Stalk-like fr base ≥ 0.5 mm
            16. Banner < 6 mm; fr body narrow, 7–11 mm, 3-sided. . . . . ***A. johannis-howellii***
            16′ Banner > 8 mm; fr body bladdery, > 15 mm, ± round in ×-section . . . . ***A. whitneyi*** vars.
          15′ Stalk-like fr base 0 (see also *Astragalus johannis-howellii*, at 16.)
            17. Banner gen < 9 mm (< 10 mm in *Astragalus anxius*); fr 1-chambered; st delicate, 1–20 cm
              18. Fr gen compressed side-to-side, ± 3-sided, 3.5–4.5 mm, hairs subappressed, ± straight;
                banner 6.5–10 mm, recurved ± 60–80°. . . . . ***A. anxius***
              18′ Fr bladdery, ± round in ×-section, 10–20 mm, hairs ± spreading; banner 5.2–8.5 mm,
                recurved 90–100° . . . . [2]***A. pulsiferae*** vars.
            17′ Banner often > 9 mm; fr ± 2-chambered; st often coarse, 0–35 cm
              19. Herbage sparsely hairy or ± glabrous; fls, frs strongly ascending, crowded . . . . [2]***A. agrestis***
              19′ Herbage densely hairy; fls ± ascending to spreading, frs ± ascending, spreading, reflexed, or
                pendent, crowded or not
                20. Pls cespitose; st < 2 cm; fr bladdery . . . . ***A. platytropis***
                20′ Pls tufted or loose-matted; st > 5 cm; fr not bladdery
                  21. Crown at surface; lowest few internodes above surface, densely white-tomentose; fr ×-section
                    blunt-triangular . . . . ***A. andersonii***
                  21′ Crown below surface; lowest few internodes below surface, ± glabrous; fr ×-section oblong
                    or blunt-triangular
                    22. Keel 6.7–8 mm; fr oblong in ×-section, not fully 2-chambered, ± narrowed at sutures . . . . ***A. monoensis***
                    22′ Keel 10–12.2 mm; fr bluntly 3-sided in ×-section, fully 2-chambered, not narrowed at
                      sutures . . . . ***A. sepultipes***
        14′ Stipule sheaths ± 0 (lower stipules not or appearing not to be fused around st into sheaths)
          23. Fls 1–4 per infl, strongly ascending; corolla pink-purple; sts low, tufted or matted, strigose to
            spreading-hairy . . . . ***A. argophyllus*** var. ***argophyllus***
          23′ Fls gen > 4 per infl, strongly ascending to reflexed; corolla ± red-purple to white; sts low or not,
            tufted or not, rarely matted, ± glabrous to strigose to spreading-hairy
            24. Fr ± 1-chambered at mid-section, stalk-like base gen > 3 mm or 0
              25. Stalk-like fr base 2–12 mm
                26. Stalk-like fr base 3.5–12 mm; ovary, fr glabrous . . . . ***A. oophorus*** vars.
                26′ Stalk-like base of fr 2–5 mm; ovary, fr ± hairy . . . . ***A. inyoensis***
              25′ Stalk-like base of fr 0
                27. Keel < 6 mm; fr 10–20 mm, bladdery . . . . [2]***A. pulsiferae*** vars.
                27′ Keel ≥ 8 mm; fr 12–55 mm, swollen but not bladdery
                  28. St ascending or erect, not mat-forming; lflets > 4 × longer than wide, spaces between >>
                  width; keel > 10 mm . . . . ***A. casei***
                28′ St ± prostrate to decumbent, ± mat-forming; lflets < 2.5 × longer than wide, spaces << or ± =
                  width; keel < 10 mm
                  29. Herbage, fr hairs 0 or stiff, straight, appressed, < 0.7 mm; among sagebrush, woodland
                    to barren areas on gravels, sands, volcanic ash . . . . ***A. iodanthus*** vars.
                29′ Herbage, fr hairs soft, wavy, spreading, 0.7–1.2 mm; sandy flats, dunes . . . . [2]***A. pseudiodanthus***

24′ Fr ± 2-chambered at mid-section, stalk-like base gen ≤ 3 mm or 0

30. Banner 4.8–6.1 mm; peduncles often in 2s or 3s in upper axils . . . . ***A. lemmonii***

30′ Banner > 7 mm; peduncles gen in 1s in upper axils

31. Stipules 7–17 mm, ± white or pale yellow; fr, herbage hairs ± 1.5–2.5 mm, spreading. . . . . ***A. malacus***

31′ Stipules < 6 mm, often ± green; fr, herbage hairs 0 or gen < 1.5 mm, ± appressed

32. Fls, frs in head-like racemes, strongly ascending; herbage sparsely hairy or ± glabrous . . . . . . ²***A. agrestis***

32′ Fls, frs not in head-like racemes, gen spreading or reflexed; herbage ± glabrous to densely hairy

33. Fr erect, < 4 mm wide, straight, beak slender, ± tooth-like. . . . . ***A. obscurus***

33′ Fr spreading or reflexed, gen > 4 mm wide, straight to curved, beak ± triangular

34. Crown at surface; fr not or ± curved, gen ± as deep as wide . . . . ***A. lentiginosus*** vars.

34′ Crown below surface; fr curved at least 1/2 circle, wider than deep . . . . ²***A. pseudiodanthus***

A. accidens S. Watson var. ***hendersonii*** (S. Watson) M.E. Jones (p. 741) Per, ± bushy; hairs sparse, minute, curved. **ST:** ± ascending, 3–6 dm. **LF:** 3–12 cm; lflets 15–29, 6–22 mm, ± oblong, tips gen notched. **INFL:** fls 7–15, ± spreading. **FL:** petals ± white, banner 13.8–19.3 mm, recurved ± 45°, keel 9.5–12.3 mm. **FR:** pendent; body 16–25 mm, 8–12 mm wide, plump, fleshy, glabrous, drying leathery or woody; stalk-like base 6–12 mm; upper, lower sutures thick, raised; beak sharp; chambers 2. 2*n*=26. Shrubland, open woodland, roadbanks; < 1250 m. KR, CaRF; sw OR. Apr–Jul

A. acutirostris S. Watson (p. 741) Ann; hairs ± appressed, ± curved. **ST:** prostrate or ascending, 2–30 cm. **LF:** 1–4 cm; lflets 7–15, 2–8 mm, ± oblong, tips gen notched. **INFL:** fls 1–6, ± spreading. **FL:** petals ± white, banner 4.7–7 mm, recurved ± 45°, keel 4.3–5.8 mm. **FR:** 12–30 mm, 2–3 mm wide, gently curved, ± 3-sided, pale brown, thinly walled but not bladdery, early-deciduous; chambers 2. Sandy or gravelly areas; 200–2100 m. SNE, DMoj, w DSon; NV. Often confused, growing with *A. nuttallianus* (fr persistent, in age dark brown or ± black, curved most strongly near base; at least upper lflets acute). Apr–May

A. agnicidus Barneby (p. 741) HUMBOLDT COUNTY MILKVETCH Per, coarse, lfy; hairs ± 0 at base, sparse, fine, spreading, wavy above. **ST:** erect, 3–9 dm. **LF:** 4–16 cm; lflets 13–27, 3–22 mm, ± oblong to ovate. **INFL:** fls 10–40, dense, soon reflexed. **FL:** calyx 6.5–9 mm; petals white, banner 9.1–11 mm, recurved 45°, keel 7–7.4 mm. **FR:** reflexed, 11–15 mm, ± 3 mm wide, ± 3-sided (lower narrow, grooved), papery in age, hairs sparse, ascending, curved; chambers 2. Open soil in woodland; 300–750 m. NCoRO (s Humboldt, Mendocino cos.). Like *A. umbraticus* (more n, fr glabrous, lvs ± glabrous). May–Aug ★

A. agrestis G. Don (p. 741) FIELD MILKVETCH Per; ± glabrous to sparsely strigose. **ST:** decumbent or ascending, 4–30 cm, underground for 0–9 cm, forming small patches, often supported by grasses. **LF:** 2–10 cm; stipule sheaths (on herbarium specimens or not) ruptured by st expansion or not; lflets 9–23, 5–20 mm, ± lanceolate to ovate, tips acute to notched. **INFL:** head-like, 0.5–2.5 cm, ovate; fls 5–15, ascending. **FL:** petals pink-purple to ± white, banner 15–22 mm, recurved ± 25°, keel 11.4–14 mm. **FR:** erect, 7–10 mm, 3–4 mm wide, ± ovate, pointed at tip, ± 3-sided (lower deeply grooved), stiff-papery, turning black, hairs white; chambers 2. 2*n*=16. Vernally moist soil, with sagebrush; 1600–2200 m. MP (Lassen Co.), SNE (Mono Co.); to MN, NM, esp Rocky Mtns; also Asia. May–Aug ★

A. albens Greene (p. 741) CUSHENBURY MILKVETCH Ann, ± per, delicate; hairs dense, appressed, branched, flat, silvery. **ST:** ± prostrate, 2–30 cm, loose-matted. **LF:** 1–5.5 cm; lflets 5–9, 2–10 mm, ovate to obovate, tips blunt, ± notched. **INFL:** fls 5–14, spreading or reflexed. **FL:** petals pink-purple, banner 7.3–9.5 mm, recurved ± 40°, keel ± = banner, > wings. **FR:** 10–18 mm, 2.8–3.5 mm wide, crescent-shaped, ± 3-sided (lower grooved), densely strigose, stiff-papery; chambers 2. Rocky areas; 1200–1900 m. Cushenbury Canyon (ne SnBr, adjacent DMoj). Threatened by limestone mining. Mar–May ★

A. allochrous A. Gray var. ***playanus*** (M.E. Jones) Isely (p. 741) PLAYA MILKVETCH Ann, per, lfy; sparsely minutely strigose. **ST:** ± prostrate to erect, 1–5 dm. **LF:** 2–12 cm; lflets 11–19, 5–20 mm, ± oblanceolate or oblong, tips shallowly notched or obtuse. **INFL:** fls 4–10, early ascending, then spreading or reflexed. **FL:** petals ± white, banner 5–7 mm, recurved ± 45°, keel 4–6 mm. **FR:** spreading, 15–30 mm, 12–20 mm wide, bladdery, minutely strigose, papery; beak short, obscure; chamber 1. 2*n*=22. Sandy flats; 600–1950 m. e DMoj; to TX, Mex. Mar–Jul ★

A. andersonii A. Gray (p. 741) Per, loose-tufted; hairs spreading or ascending, wavy, ± gray, on lower internodes dense, interwoven, ± white. **ST:** decumbent to ascending, 7–20 cm. **LF:** 2–10 cm; stipule sheaths scarious; lflets 9–21, 3–14 mm, elliptic to ± oblanceolate. **INFL:** fls 12–26, early ascending, then reflexed. **FL:** petals white, tinged dull purple or not, banner 9.5–14.5 mm, recurved ± 45°, keel 6.6–9 mm. **FR:** spreading or pendent, 10–18 mm, 3–5 mm wide, widely oblong, ± 3-sided (lower narrow, shallowly grooved), long-wavy-hairy, stiff-papery; upper suture thick, raised; chambers 2. 2*n*=24. Gen disturbed flats, slopes; 1300–2450 m. MP, n w NV. Apr–Jul

A. anxius Meinke & Kaye (p. 741) ASH VALLEY MILKVETCH Per, delicate, ± matted; hairs sparse, finely wavy. **ST:** prostrate, 3–20 cm, slender. **LF:** 1–3.5 cm; stipule sheaths ± present; lflets 9–15, 2.5–12 mm, obovate, tips blunt or notched. **INFL:** fls 7–15, spreading to reflexed. **FL:** petals purple and white, with pale lilac veins, banner 6.5–10 mm, recurved 60–80°, keel 3.5–3.7 mm. **FR:** spreading or reflexed, 3.5–4.5 mm, 3.2–4.2 mm wide, ovate, ± compressed side-to-side, ± 3-sided (lower blunt-angled), sparsely strigose, thinly papery; chamber 1. Gravelly volcanic soil among pines, sagebrush; 1550 m. MP (Lassen Co.). [*A. tegetarioides* var. *a.* (Meinke & Kaye) S.L. Welsh] Jun–Jul ★

A. argophyllus Nutt. var. ***argophyllus*** (p. 741) SILVER-LEAVED MILKVETCH Per, cespitose, from crown; hairs on lvs dense, appressed or ascending, soft-gray to silvery. **ST:** ± prostrate, tufted or matted, < 15 cm. **LF:** 2–15 cm; lflets 9–21, 4–15 mm, ± elliptic or ovate, tips acute or obtuse. **INFL:** fls 1–4, ascending. **FL:** petals bright pink-purple, banner 22–24 mm, keel 17–20 mm. **FR:** 15–25 mm, 7–12 mm wide, ± widely lanceolate, straight or curved, densely loose-strigose, early fleshy, then stiff-leathery; chamber 1. Heavy alkaline or saline soil; 1280–1350 m. MP (e Lassen Co.), SNE (nw Inyo Co.); to MT, WY, UT. Apr–Aug ★

A. aridus A. Gray (p. 741) Ann, silvery-silky-canescent. **ST:** decumbent or ascending, 3–30 cm. **LF:** 2–9 cm; lflets 7–17, 4–16 mm, ± oblong or oblanceolate, tips acute to notched. **INFL:** fls 3–9, ascending. **FL:** petals ± white, pink-tinged, or ± tan-pink, banner 3.3–6.5 mm, recurved ± 40°, keel 3.5–5 mm. **FR:** ascending, 10–17 mm, 4–7 mm wide, ± half-elliptic in side view, round in ×-section, ± gray, appressed-hairy, thick-papery; beak triangular, compressed side-to-side; chamber 1. Sandy places; < 350 m. DSon; AZ, nw Mex. Highly variable in stature: low, slender (drier spring seasons) or robust, coarse (moist seasons). Feb–Apr

A. asymmetricus E. Sheld. (p. 741) Per, clumped; finely silky-strigose, some hairs near top longer, more spreading. **ST:** erect, 5–12 dm, stout, often hollow. **LF:** 5–20 cm; midrib stiff, often curved; stipule sheaths present; lflets 17–35, 6–25 mm, linear to ± elliptic, tips obtuse or shallowly notched. **INFL:** fls 15–45, early spreading, quickly reflexed. **FL:** calyx tube 5–7 mm; petals ± cream, banner 12.6–17.6 mm, ± recurved, margins abruptly folded back, keel 11.5–14.7 mm. **FR:** pendent; body 25–40 mm, 13–18 mm wide, bladdery; stalk-like base 1.4–4 cm, arched, sparsely strigose; upper suture ± concave to ± convex, lower strongly convex; beak triangular; chamber 1; ovules 16–30. 2*n*=22. Grassy areas, open woodland, disturbed sites; 50–900 m. NCoRI, GV, SnFrB, SCoR. Like *A. curtipes, A. oxyphysus, A. trichopodus*, but differs in calyx length, stalk-like base of fr (length, hairs, presence or position of joint), ovule number. Weedy, poisonous to stock. Apr–Jul

A. atratus S. Watson var. ***mensanus*** M.E. Jones (p. 741) DARWIN MESA MILKVETCH Per, wiry, loose-matted or scrambling through scrub; minutely strigose. **ST:** 3–25 cm. **LF:** 1.5–15 cm; lflets 7–15, spaced, 3–16 mm, linear to ± ovate, tips acute to shallowly notched. **INFL:** fls 4–18, reflexed. **FL:** petals dingy white at least at base, dull lilac on margins or tips, banner 9.8–13.4 mm, recurved ± 90°, blade pinched to appear fiddle-shaped, wings irregularly toothed, ± twisted, keel 8.2–10 mm. **FR:** pendent, 16–22 mm, ± 4 mm wide, ± linear-oblong, ± compressed side-to-side, stiff-papery or leathery, minutely strigose; stalk-like base gen 0; upper suture more convex than lower; tip ± reflexed; chambers 2. Open foothills with pinyon, sagebrush; 1700–2350 m. DMtns (n and w of Panamint Valley, Inyo Co.). Other vars. in OR, ID, NV. May–Jul ★

A. austiniae A. Gray (p. 741) Per, dwarf, cespitose; hairs dense, wavy, silvery. **ST:** < 11 cm. **LF:** 1–5 cm; stipule sheaths overlapped; lflets 5–13, 1–7 mm, ± elliptic to oblanceolate, keeled abaxially. **INFL:** ± head-like; fls 4–14, erect to ascending. **FL:** petals ± white, dull lilac-tinged, banner 8.4–11.3 mm, recurved ± 35°, banner, wings finely hairy on outside, keel 6.2–8.1 mm. **FR:** ascending or spreading, ± incl in calyx, 5–7 mm, 3–4 mm wide, oblong-ovate, finely tomentose; chambers ± 2 in lower 2/3. Exposed ridges, meadows, above timberline; 2400–2750 m. n SNH (near Lake Tahoe). Jul–Sep

A. bernardinus M.E. Jones (p. 741) Per, often twining among sagebrush, wiry, sparsely lfy; minutely strigose, esp above. **ST:** 1–5 dm, slender. **LF:** 3–14 cm; lflets 7–19, ± spaced, 4–20 mm, ± lanceolate, tips acute to notched. **INFL:** ± among lvs; fls 10–25, spaced, ascending. **FL:** calyx tube 2.7–4.1 mm; petals pale to dark lilac, banner 7–10.2 mm, recurved 45–90°, keel 6.8–9.4 mm, < 0.5 mm < banner. **FR:** ascending, 20–30 mm, 4–5 mm wide, narrowly oblanceolate in side view, straight or ± curved, 3-sided (lower shallowly channeled or ± flat), pale, glabrous, papery; chambers 2. Stony areas among desert shrubs, junipers; 900–2300 m. SnBr, DMtns (New York, Ivanpah mtns). [*Astragalus bernardianus*, orth. var.] Apr–Jun ★

A. bicristatus A. Gray (p. 741) CRESTED MILKVETCH Per, sparsely lfy; often ± gray minutely strigose. **ST:** ascending or sprawling, < 5 dm. **LF:** 3–14 cm; stipule sheaths present; lflets 11–23, often ± spaced, 5–20 mm, linear to narrowly oblong, terminal not or ± jointed to midrib, tips gen obtuse or notched. **INFL:** fls 5–20, ascending. **FL:** petals ± white, banner 15–19 mm, recurved ± 50°, keel 12–13 mm. **FR:** ± pendent; body 20–43 mm, 6–9 mm wide, incurved, glabrous, early ± round in ×-section, fleshy, then ± 4-sided (prominently flanged on upper, lower), stiffly leathery or ± woody; stalk-like base downcurved, 6–10 mm, stout; chamber 1. Open, rocky areas in pine forest; 1700–2750 m. SCo (w Riverside, San Bernardino cos.), SnGb, SnBr, PR. May–Aug ★

A. bolanderi A. Gray (p. 741) Per, stiff, sparsely lfy; hairs fine, wavy or curly. **ST:** ± decumbent to erect, 1.5–4 dm, basally lfless. **LF:** 3–16 cm; stipule sheaths ± loose; lflets 13–27, 5–20 mm, ± oblanceolate to ± oblong, midrib prominent abaxially, extending to tip as small, hard point. **INFL:** fls 7–18, ± crowded, spreading. **FL:** petals ± white, ± faintly lilac-tinged, banner 13–17.6 mm, recurved ± 40°, keel 9.8–12.4 mm. **FR:** body erect or ± spreading, at angle to stalk, 10–30 mm, 7–12 mm wide, swollen, incurved, glabrous, stiff-papery; stalk-like base ascending to ± reflexed, 5–12 mm, jointed at base; chambers 2 below beak, a wing-like outgrowth protruding into each from central septum. 2*n*=22. Dry sandy, rocky areas on margins of meadows; 1400–3300 m. CaRH, SNH; w-c NV. Jun–Sep

A. brauntonii Parish (p. 741) BRAUNTON'S MILKVETCH Per, coarse; hairs ± white, dense, tangled, some longer, spreading. **ST:** ± erect, 7–15 dm, stout, hollow. **LF:** 3–16 cm; lflets 25–33, 3–20 mm, ± obovate, tips acute to obtuse. **INFL:** spike-like; fls 35–60, overlapped, reflexed. **FL:** petals dull lilac, persisting withered, banner 9.1–11.7 mm, recurved ± 40°, keel 6.4–8.5 mm. **FR:** reflexed, ± 1/2 incl in calyx, 6.5–9 mm, 3–4 mm wide, oblong in side view, bluntly 3-angled, not bladdery, hairs dense, wavy; chambers 2 in lower 1/2. Disturbed areas in chaparral; < 650 m. w WTR, SnGb, SnGb/SCo?, n PR. Threatened by development. Mar–Jul ★

A. breweri A. Gray (p. 741) BREWER'S MILKVETCH Ann, sparsely lfy; thinly minutely strigose. **ST:** few, 4–30 cm, slender.

LF: 1.5–7.5 cm; lflets 7–13, spaced, 3–12 mm, narrowly to widely ± obovate, tips notched. **INFL:** head- or umbel-like; fls 4–10, ascending and spreading. **FL:** petals pale yellow to white, ± streaked with lavender or not, banner 7.8–11.4 mm, recurved 35–45°, keel 5–7.8 mm. **FR:** ascending and spreading; body 5–10 mm, 2.5–4 mm wide, ovate; beak spine-like or hooked, ± = body; chambers 2; ovules 2–6. Open slopes, grassy areas, on serpentine or not; < 950 m. c&s NCoR, n SnFrB. In fl, like *A. rattanii* (ovules 8–20). Mar–Jun ★

A. californicus (A. Gray) Greene (p. 741) KLAMATH MILKVETCH Per, robust, open, widely branched; hairs wavy, ± gray. **ST:** clumped, ± ascending, 2–5 dm. **LF:** 3–8.5 cm; stipule sheaths fragile, papery; lflets 13–23, 5–20 mm, ± narrowly oblanceolate, tips obtuse or blunt, shallowly notched. **INFL:** fls 10–30, not crowded, in age reflexed. **FL:** petals cream, banner 11.5–17.4 mm, recurved ± 45°, keel 9.5–12.4 mm. **FR:** pendent; body 27–43 mm, 3.5–5 mm wide, linear oblong, compressed side-to-side, stiff-papery, sparsely to densely minutely strigose; stalk-like base 7–15 mm; chamber 1. Dry, open areas in scrub, woodland; gen < 1550 m. e KR, n CaRH, MP; s OR. Apr–Jul

A. calycosus S. Watson var. ***calycosus*** (p. 741) TORREY'S MILKVETCH Per, ± stless, tufted; silvery-strigose, hairs branched. **LF:** 1–7 cm; lflets 1–7, gen 3, crowded near lf tip, 5–19 mm, elliptic to obovate, tips gen obtuse or acute. **INFL:** fls 1–8, ascending or spreading. **FL:** petals ± white to bright purple, wing tips white, notched, banner 10–13 mm, keel 7.4–9.4 mm. **FR:** ascending, 10–25 mm, 3–4 mm wide, oblong, ± 3- sided (lower narrow, grooved), strigose; chambers ± 2. 2*n*=22. Rocky areas, sagebrush scrub, pine forest; 1500–3550 m. SNE, n DMtns; to ID, WY. Apr–Jul

A. canadensis L. var. ***brevidens*** (Gand.) Barneby (p. 741) Per from rhizome, lfy; ± strigose, hairs branched. **ST:** 1.5–5.5 dm. **LF:** 5–23 cm, stipules fused around st into sheath; lflets 7–25, 5–40 mm, widely lanceolate or elliptic, ± glabrous, tips obtuse or shallowly notched, small-pointed. **INFL:** spike-like; fls 20–many, reflexed, overlapped. **FL:** petals cream or green-white (tinged dull purple), banner 11.7–17.5 mm, recurved 40–90°, keel 8.9–13.6 mm. **FR:** erect; body 10–15 mm, 3–4 mm wide, ± cylindric, gen minutely strigose, stiff-papery; beak recurved, ± 3 mm, stiff; chambers 2. Heavy soil, moist at least in spring; 1400–2550 m. n SNH, MP, n SNE; to WA, MT, WY, CO. Jun–Sep

A. casei A. Gray (p. 741) Per, slender, open, widely branched, wiry, sparsely lfy; minutely strigose. **ST:** ascending or erect, 1–4 dm, often zigzag. **LF:** 3–10 cm, midrib rigid, tapered; lflets 5–15, spaced, 3–25 mm, linear to oblanceolate, tips gen obtuse or notched. **INFL:** fls 8–25, spaced, reflexed in age. **FL:** petals pink-purple, wing, keel tips white (petals all white), banner 12–18 mm, recurved 45°, keel 10.6–13.3 mm. **FR:** sharply pendent (reflexed); 20–55 mm, 5–10 mm wide, ± incurved, ± half-lanceolate, wider than deep, minutely strigose, early pulpy, then tough, wrinkled; beak sharp, rigid, compressed side-to-side; sutures keel-like; chamber 1. Dry, gravelly soils or dunes, with sagebrush or pinyon; 1200–2400 m. SNE, DMtns (Inyo Co.); w NV. Apr–Jun

A. cicer L. CHICKPEA MILKVETCH Per, diffuse, lfy. **ST:** prostrate or ± ascending, 3–7 dm, hairs ± 0 or short, stiff, appressed; stipules sheathing at base. **LF:** 4–21 cm, petioles gen 0; lflets elliptic to oblong, 5–35 mm, tips acute. **INFL:** dense, fls 6–30, ascending, in oblong or spheric heads, pedicels ascending. **FL:** petals white, banner 12–16 mm, ± straight, keel obtuse, 9–10.5 mm, wings 1–2 mm shorter; calyx, ovary with white, black hairs. **FR:** ascending or spreading, firmly attached to receptacle; body widely ovate or ± spheric, inflated, 6–14 mm, 5–10 mm wide, thinly fleshy, firm, green, leathery or stiff-papery in age, brown to black; stalk-like base < 1 mm; chambers 2. 2*n*=64. Moist grassy areas to open woodland, disturbed areas; < 2000 m. c SNF (Tuolumne Co.), n SNH (Nevada Co.); NV, UT; native to Eur, sw Asia. Introduced as forage, cover crop. May–Aug

A. cimae M.E. Jones Per, ± coarse; ± fleshy; hairs ± 0. **ST:** ± spreading, 2–25 cm. **LF:** 4.5–11 cm; lflets 11–23, 5–20 mm, obovate to ± round, tips obtuse or notched. **INFL:** fls 10–25, early ascending, then reflexed or not. **FL:** petals red-purple, white- or pale-tipped, banner 12–15 mm, keel 9.5–10.6 mm. **FR:** ± ascending to spreading;

body ± oblong in side view, bladdery or not, glabrous; stalk-like base 5–12 mm, tapered to narrow base; chambers ± 2.

1. Fr 8–12 mm wide, sides drying stiff-leathery or woody
. . . . var. *cimae*
1′ Fr 13–21 mm wide, sides drying thinly papery . . var. *sufflatus*

var. *cimae* (p. 741)   CIMA MILKVETCH   **FR:** body 15–25 mm, gen incurved > 90° (beak perpendicular to infl axis), fleshy when green; stalk-like base 6–8 mm. Gen among sagebrush; 1250–1850 m. e DMtns (e San Bernardino Co.); NV. Apr–Jun   ★

var. *sufflatus* Barneby   INFLATED CIMA MILKVETCH   **FR:** body 20–37 mm, straight to incurved 90° (beak ± parallel to infl axis), bladdery, in age papery; stalk-like base 5–12 mm. Calcareous substrates, with pinyon pine, gen sagebrush; 1500–2100 m. W&I (e slope Inyo Mtns). May   ★

A. claranus Jeps. (p. 743)   CLARA HUNT'S MILKVETCH   Ann, slender, sparsely lfy; sparsely minutely strigose. **ST:** ascending, 3–12 cm. **LF:** 1.5–5 cm; lflets 5–9, 2–10 mm, obovate, tips deeply notched. **INFL:** head-like, black-hairy; fls 2–7, early spreading, then reflexed. **FL:** petals ± white, banner, keel bright purple-tipped, banner 8.9–12 mm, recurved ± 20°, keel 7.4–9.1 mm, > wings. **FR:** spreading or reflexed, ephemeral; body 17–25 mm, 1.6–3.1 mm wide, crescent-shaped, tapered to ends, finely strigose, stiff-papery; stalk-like base attached to peg-like, 1.5–2.5 mm extension of receptacle (evident in calyx after fr drop); beak prominent, ± curved; chambers 2. Open grassy areas, thin clay soil; 100–200 m. s NCoR (Sonoma, Napa cos.), w ScV (Solano Co.). [*Astragalus clarianus* Jeps., orth. var.] Threatened by development. Apr–May   ★

A. clevelandii Greene (p. 743)   CLEVELAND'S MILKVETCH   Per, ± robust, bushy, lfy. **ST:** erect, 3–10 dm, glabrous. **LF:** 2–14 cm; lflets 13–27, 3–23 mm, elliptic to oblanceolate, ± glabrous adaxially, sparsely finely strigose abaxially, tips ± obtuse. **INFL:** spike-like; fls 20–100, reflexed. **FL:** petals white or cream, banner 4.8–6 mm, recurved ± 45°, keel 3.8–4.6 mm. **FR:** reflexed, 4.5–7 mm, ± 2 mm wide, incurved, half-ovate in side view, 3-sided, glabrous, stiff-papery; upper suture raised, lower in groove; chambers 2. 2*n*=26. Moist serpentine areas; 100–1500 m. NCoRI, NCoRH. Jun–Sep   ★

A. coccineus Brandegee (p. 743)   SCARLET MILKVETCH   Per, tufted, ± stless; hairs dense, ± white. **LF:** 3–10 cm; lflets 7–15, 3–14 mm, ± oblanceolate, tips ± acute. **INFL:** fls 3–10, ascending. **FL:** calyx 18–24 mm; petals scarlet, banner 35–41 mm, recurved 20–30°, keel 35–40 mm. **FR:** 25–40 mm, 10–12 mm wide, plump, ± narrowly ovate in side view, often curved, ± compressed top-to-bottom, hairs dense, shaggy, white or tawny, early fleshy, then leathery; chamber 1. 2*n*=22. Gravelly places, gen sagebrush scrub, pinyon woodland; 650–2450 m. SnBr, PR, SNE, DMoj; s NV, w AZ, n Baja CA. Differs in fr from *A. newberryi* ± by its longer fr, remnants of fl. Mar–Jun

A. congdonii S. Watson (p. 743)   Per, often coarse; hairs abundant, ± spreading. **ST:** ± ascending, 2–7 dm. **LF:** 3–14 cm; lflets 11–37, 3–15 mm, ± elliptic to round, tips blunt or notched. **INFL:** fls 8–35, spaced, spreading to reflexed. **FL:** petals ± cream, banner 10.4–16.6 mm, recurved ± 45°, keel 7.4–12.7 mm. **FR:** reflexed, 15–35 mm, ± 3 mm wide, linear, straight or curved, ± 3-sided (lower grooved), drying stiff-papery, hairs sparse to dense, appressed and spreading; chambers 2. 2*n*=26. Open, disturbed sites in scrub, woodland; 150–850 m. c&s SNF. Mar–Jun

A. crotalariae (Benth.) Torr. (p. 743)   SALTON MILKVETCH   Per, often fl 1st season, seemingly ann, bushy-clumped, coarse, ± ill-scented; ± strigose. **ST:** ± erect, 1.5–6 dm, often hollow. **LF:** 5–16.5 cm; lflets 9–19, 5–35 mm, ± obovate to round, flat, thick, tips notched. **INFL:** fls 10–25, ascending or spreading. **FL:** calyx 7.6–12.3 mm; petals bright red-purple or white, banner 21–28 mm, recurved ± 40°, keel 17–21 mm. **FR:** body 20–30 mm, 10–14 mm wide, inflated, ovate in side view, sparsely to densely strigose, with fine, net-like pattern, drying thick-papery; stalk-like base 1–1.5 mm, stout; sutures ± raised; beak erect or incurved, tipped by persistent style-base; chamber 1. 2*n*=24. Valleys, washes, in desert foothills or open, sandy, gravelly areas; -60–300 m. DSon; sw AZ, n Baja CA. Jan–Apr   ★

A. curtipes A. Gray (p. 743)   MORRO MILKVETCH   Per; minutely strigose and ± tomentose, often ± gray. **ST:** ascending, clumped,

2–4 dm. **LF:** 4–16 cm; stipule sheaths sparsely hairy; lflets 25–39, 2–25 mm, linear-oblong to narrowly obovate, tips blunt or shallowly notched. **INFL:** fls 15–35, ± dense, early spreading, then reflexed. **FL:** calyx tube 4–5 mm; petals cream or keel faintly lilac-tipped, banner 13–16 mm, recurved ± 50°, keel 10.7–12.7 mm. **FR:** ascending or ± spreading; body 23–36 mm, 12–20 mm wide, bladdery, ± half-ovate in side view, minutely sparsely strigose, thinly papery; stalk-like base 2–6 mm, stiff, sparsely strigose, jointed at top; chamber 1; ovules 26–37. Shrubland, grassy or disturbed areas near coast; < 450 m. s CCo, SCoR, SCo, n ChI. See note under *A. asymmetricus*. Phase with fr body 40–70 mm uncommon in San Benito Co. Feb–Jun

A. curvicarpus (A. Heller) Macbr. var. *curvicarpus* (p. 743)   SICKLE MILKVETCH   Per, lfy; hairs subappressed, incurved, curly or not, ± spreading, often ± gray. **ST:** decumbent to ascending, 1.5–4 dm, often stout. **LF:** 2.5–9 cm; lflets 7–21, 3–23 mm, obovate, tips obtuse or notched. **INFL:** fls 5–35, not crowded, reflexed. **FL:** petals early white, then pale yellow, banner 15–21 mm, recurved ± 45°, keel 11–15.2 mm. **FR:** pendent; body 20–35 mm, 2.5–4.5 mm wide, narrowly oblong, curved 1/4 to ± full circle, ± compressed side-to-side, sparsely wavy-hairy (glabrous), drying stiff-papery; stalk-like base 9–20 mm, jointed to receptacle; chamber 1. Loose soil, often with sagebrush; 1000–2900 m. CaRH, GB; to OR, ID, NV. Apr–Jul

A. deanei (Rydb.) Barneby (p. 743)   DEAN'S MILKVETCH   Per, coarse; hairs ± 0. **ST:** ± erect, 3–6 dm. **LF:** 8–18 cm; lflets 19–29, 4–21 mm, lanceolate to oblong, midrib raised abaxially, esp in basal 1/2, tips ± obtuse. **INFL:** fls 15–25, ± spreading, not crowded; pedicels slender, in fr thick. **FL:** petals ± white, banner 9.5–15.2 mm, recurved ± 45°, keel 7.8–10.5 mm. **FR:** ± ascending, 15–30 mm, 10–20 mm wide, bladdery, thinly minutely strigose, drying papery; upper suture ± straight, lower strongly convex; chamber 1; ovules 29–40. Open shrubby slopes in chaparral; 250–800 m. PR (sw San Diego Co.). Threatened by development. Mar–May   ★

A. didymocarpus Hook. & Arn.   TWO-SEEDED MILKVETCH   Ann, gen slender; ± gray ± minutely strigose. **ST:** prostrate to ± erect, 2.5–45 cm. **LF:** 0.8–7.5 cm; lflets 9–17, 2–14 mm, linear to oblanceolate, tips notched. **INFL:** head-like; fls 5–30, erect or ascending. **FL:** < 9 mm; calyx hairs ± mixed black, white; petals ± white, purple-tinged. **FR:** ascending, ± incl in calyx, 2–4 mm, ± 2 mm wide, ± spheric, 2-lobed in ×-section, ± minutely strigose (glabrous), coarsely wrinkled, drying stiff-papery; chambers 2. Vars. intergrade.

1. Keel 2.4–4.5 mm; s ± 1/2 CA
  2. Calyx hairs mostly black, lobes gen 0.8–1.5 mm, <(=) tube;
     lflets ± glabrous adaxially; st ± erect . . . . . var. *didymocarpus*
  2′ Calyx hairs mostly white, lobes gen 1.5–2.4 mm, > tube;
     lflets ± canescent adaxially; st ± prostrate . . . . var. *dispermus*
1′ Keel 4.7–7.2 mm; s CCo, s SCoRO, SW (exc WTR)
  3. Calyx hairs mostly black; keel abruptly curved,
     tip short, blunt . . . . var. *milesianus*
  3′ Calyx hairs mostly white; keel crescent-shaped,
     tip narrowly triangular, acute . . . . var. *obispoensis*

var. *didymocarpus* (p. 743)   **ST:** ± erect, 25–45 cm; herbage ± green, sparsely ± strigose. **INFL:** fls 5–25. **FL:** calyx hairs mostly black, lobes gen 0.8–1.5 mm; banner 2.8–6.1 mm, keel 2.4–4.5 mm, abruptly curved, tip bluntly pointed. 2*n*=24. Grassy areas; < 1350 m. c&s SNF, Teh, GV, CW, SW, DMoj; s NV. Feb–May

var. *dispermus* (A. Gray) Jeps.   **ST:** ± prostrate, 15–27 cm; herbage ± gray-hairy. **INFL:** fls 7–20. **FL:** calyx hairs mostly white, lobes gen 1.5–2.4 mm; banner 3.4–5.4 mm, keel 3.4–4.5 mm, abruptly incurved, tip bluntly or sharply triangular. 2*n*=26. Sandy or gravelly areas; -50–1500 m. D. Feb–May

var. *milesianus* (Rydb.) Jeps.   MILES' MILKVETCH   **ST:** ± erect, 25–45 cm; herbage ± green, sparsely ± strigose. **INFL:** fls 5–25. **FL:** calyx hairs mostly black, lobes 1–2 mm; banner 7.5–10 mm, keel 4.7–6.9 mm, abruptly incurved, tip short, blunt. Grassy areas near coast; < 400 m. s CCo. Mar–May   ★

var. *obispoensis* (Rydb.) Jeps.   **ST:** ascending, 2.5–25 cm; herbage ± gray-hairy. **INFL:** fls 5–30. **FL:** calyx hairs mostly white, lobes 1–2.5 mm; banner 6.2–8.6 mm, keel 5.2–7.2 mm, crescent-shaped, tip narrowly triangular, acute. Grassy hills, openings in chaparral; 400–1150 m. s SCoRO, SW (exc WTR); Baja CA. Mar–May

A. douglasii (Torr. & A. Gray) A. Gray   DOUGLAS MILKVETCH
Per, lfy; ± minutely strigose. **ST:** gen many, ± decumbent or erect,
2–10 dm. **LF:** 5–18 cm; lflets 7–25, 5–25 mm, elliptic, ovate, or
obovate, midrib often raised abaxially, margins ± green, tips obtuse
or shallowly notched. **INFL:** fls 10–30, spaced, spreading or ascending. **FL:** calyx green, lobes 0.7–2.6 mm; petals ± white to pale yellow,
banner 8–13 mm, recurved 60–90°, keel 6.2–10.8 mm, keel, wing
claws 2.2–4.9 mm, keel blade 4.3–6.4 mm, > keel claw. **FR:** spreading or ± ascending, 25–60 mm, 12–32 mm wide, bladdery, sparsely
hairy to glabrous, drying thinly papery; beak erect, pointed, compressed side-to-side; chamber 1; deciduous. Much like *A. macrodon*
(hairs ± spreading, ± wavy; lflet margins purple); like *A. pomonensis*
(lflets 25–41; keel blade ≤ keel claw); vars. ± intergrade.

1. Calyx tube ± evenly hairy throughout, lobes longer
   than wide, gen 1.4–2.5 mm (intergrades in SnBr
   with var. *parishii*) . . . . var. ***douglasii***
1′ Calyx tube most densely hairy between lobes,
   lobes ± as wide as long, gen 0.7–2.2 mm
   2. St prostrate to ± ascending; peduncle incurved-
      ascending, in fr often prostrate . . . . var. ***parishii***
   2′ St ± stiffly erect; peduncle erect, in fr erect. . . var. ***perstrictus***

   var. ***douglasii*** (p. 743)   **ST:** prostrate to ± ascending, 2–7 dm.
LF: lflets 11–25. **FL:** calyx tube ± evenly hairy throughout, lobes
gen 1.4–2.5 mm, longer than wide, pointed. **FR:** 25–60 mm. 2*n*=22.
Open areas; < 2500 m. c&s SN, Teh, s ScV, SnJV, CCo, SCoR, TR,
DMoj. Apr–Jul

   var. ***parishii*** (A. Gray) M.E. Jones   **ST:** prostrate to ± ascending,
2–6 dm. **LF:** lflets 15–25. **FL:** calyx tube most densely hairy between
lobes, lobes gen 0.7–2.2 mm, triangular, ± as wide as long. **FR:** 25–50
mm. 2*n*=22. Open areas; 100–2350 m. SCo, SnBr, PR. May–Aug

   var. ***perstrictus*** (Rydb.) Munz & McBurney   JACUMBA
MILKVETCH   **ST:** ± stiffly erect, 4–10 dm. **LF:** lflets 13–19. **FL:**
calyx tube most densely hairy between lobes, lobes gen 0.7–2.2 mm,
triangular, ± as wide as long. **FR:** 35–60 mm. 2*n*=22. Rocky areas
in open oak woodland; 850–1200 m. e PR; n Baja CA. Apr–Jun   ★

A. ertterae Barneby & Shevock (p. 743)   WALKER PASS MILKVETCH
Per; hairs spreading, long, soft, shaggy. **ST:** 3.5–10 cm, below surface
for ± 1/2 length, ± prostrate above. **LF:** 4–5, 3–6.5 cm, crowded distally; stipules below surface, bladeless, fused around st into sheath;
lflets 9–13, 6–13 mm, ± oblanceolate, tips blunt or shallowly notched.
INFL: fls 7–17, crowded, ascending. **FL:** petals cream, banner 10–12
mm, recurved ± 45°, keel 8.5–9 mm. **FR:** spreading, 16–22 mm,
7–9 mm wide, oblong, gently incurved, swollen, abruptly narrowed
at ends, bluntly 3-sided in ×-section (lower depressed), glabrous, in
age stiffly leathery; upper suture prominently raised, lower a low,
wavy ridge; chamber 1. Open areas with sandy, granitic soil, pine/oak
woodland; 1750–1900 m. s SNH (Kern Co.). Apr–May   ★

A. filipes A. Gray (p. 743)   BASALT MILKVETCH   Per, sparsely lfy;
± glabrous to densely strigose. **ST:** clumped, 3–9 dm. **LF:** 2.5–12
cm; stipule sheaths low, papery; lflets (5)9–19(23), ± spaced, 3–25
mm, linear or narrowly oblong, tips ± obtuse or notched. **INFL:** fls
(4)10–30, not crowded, gen reflexed. **FL:** petals ± dull white or pale
yellow, banner 10–15.5 mm, recurved 50–85°, keel 6.7–12 mm. **FR:**
spreading or pendent; body 17–30 mm, 3–6.5 mm wide, gen flat,
± narrowly oblong in profile, tapered at ends, glabrous or minutely
strigose, in age papery; often deciduous, dehiscing from base or not;
stalk-like base 6–16 mm, hairy; chamber 1. 2*n*=24. Dry, open areas in
sagebrush or pine (MP) or chaparral (SW); 750–2150 m. WTR, SnBr,
SnJt, MP; to BC, ID, NV, n Baja CA. May–Jul

A. funereus M.E. Jones (p. 743)   BLACK MILKVETCH   Per; hairs
± gray, dense, stiff, some short, ± wavy. **ST:** prostrate, loose-tufted,
2–8 cm. **LF:** 2.5–7 cm; lflets 7–17, ± crowded, 3–12 mm, obovate,
tips blunt or notched. **INFL:** fls 3–10, ascending. **FL:** calyx hairs
black or mixed black, white; petals pink-purple, banner 22–29 mm,
recurved ± 40°, keel 21–28 mm. **FR:** ascending, 25–50 mm, 10–15
mm wide, ± lanceolate in side view, straight, ± compressed top-to-
bottom at base, gently incurved near tip, leathery, hairs dense, 1.5–
2.5 mm, wavy, white; chamber 1. Gravelly, clayey, or rocky areas;
1050–1600 m. ne DMtns (Grapevine Mtns, e of Death Valley); NV.
[*A. purshii* var. *f.* (M.E. Jones) Jeps.] Apr–May   ★

A. gambelianus E. Sheld. (p. 743)   GAMBEL MILKVETCH, LITTLE
BLUE LOCO   Ann, slender; hairs minute, ± incurved. **ST:** erect to ±
decumbent, 2–30 cm, slender. **LF:** 1–4 cm; lflets 7–15, 1–9 mm, ±
oblanceolate, tips blunt or notched. **INFL:** black-hairy; fls 4–15, early
dense, ascending, then spaced, reflexed. **FL:** petals ± white, purple-
tinged, banner 2.5–3.3(5.5–6.5) mm, recurved ± 25°, keel 2–2.5(4.2)
mm. **FR:** reflexed, 2.8–4.2 mm, 2.4–3.6 mm wide, ovate or ± round in
top view, ± wrinkled on edges, minutely hairy; chambers ± 2. 2*n*=22.
Open, grassy areas, scrub; < 1450 m. CA-FP; sw OR, n Baja CA. Fr
hairs near coast straight, elsewhere incurved. Mar–Jul

A. geyeri A. Gray var. ***geyeri*** (p. 743)   GEYER'S MILKVETCH
Ann, or into 2nd season, often slender; minutely strigose. **ST:** prostrate to ascending, 1–20 cm. **LF:** 1.5–10 cm; lflets 3–13, ± spaced,
5–15 mm, ± linear to oblong, terminal often >> others. **INFL:** among
lvs; fls 3–8, spaced, early ascending, then reflexed. **FL:** petals ± white,
lilac-blushed or not, keel tip purple, banner 5.2–7.6 mm, recurved ±
45(80)°, keel 3.8–4.8 mm. **FR:** 15–25 mm, 6–10 mm wide, inflated,
half-ovate in side view, minutely strigose, thinly papery; beak triangular; upper suture straight or ± concave; chamber 1. Sandy areas;
± 1200 m. MP, SNE (Owens Valley); to WA, WY, UT. Apr–Jul   ★

A. gibbsii Kellogg (p. 743)   GIBBS' MILKVETCH   Per, lfy; hairs
0.5–0.8 mm, gen ± spreading, straight to wavy, ± gray. **ST:** many, ±
prostrate, 1.5–3.5 dm. **LF:** 1.5–9.5 cm; stipule sheaths small; lflets
7–19, 4–20 mm, obovate or oblong. **INFL:** fls 10–30, reflexed. **FL:**
calyx base with pouch on upper side; petals ± yellow, wings, banner
± equal, banner 14–18 mm, shallowly S-shaped, keel 12–15 mm. **FR:**
pendent; body 22–30 mm, 4–8 mm wide, oblong, incurved 1/4–1/2
circle, swollen but ± compressed side-to-side, densely wavy-hairy,
early fleshy, then leathery; stalk-like base 7–22 mm; sutures prominent, raised, keel-like; chamber 1. Among sagebrush or pines; 1250–
2400 m. n SNH, GB; w NV. May–Jul

A. gilmanii Tidestr. (p. 743)   GILMAN'S MILKVETCH   Ann, per;
hairs ± spreading or curved, minute. **ST:** ± ascending, 5–25 cm. **LF:**
1.5–7.5 cm; lflets 7–13, 2–12 mm, ± oblanceolate, margins often
purple, tips notched or not. **INFL:** ± among lvs; fls 4–9, spaced, early
ascending, then reflexed. **FL:** petals pink-purple, banner 6.1–8 mm,
recurved ± 45°, keel 4.7–6.1 mm. **FR:** 14–25 mm, 8–16 mm wide,
± ovate, bladdery, papery; hairs minute, curved; chamber 1; ovules
8–12. Gravelly areas; 2000–3050 m. n DMtns (Panamint Range); NV.
May–Jun   ★

A. hornii A. Gray var. ***hornii*** (p. 743)   HORN'S MILKVETCH
Ann, long-lived, appearing per or not, open, widely branched; hairs
± appressed or ascending. **ST:** 3–12 dm, slender-solid or stout-hol-
low. **LF:** 1.5–13 cm, often reflexed; lflets (11)15–33, 5–20 mm, ±
elliptic. **INFL:** head-like; fls 10–35, spreading and ascending. **FL:**
petals white to pale lilac, banner 7.8–10.2 mm, recurved ± 40°, keel
5.9–8.4 mm. **FR:** crowded, spreading, in ± cylindric or ± spheric
heads 2.5–3.5 cm wide, 12–18 mm, 7–9 mm wide, ± ovate, inflated,
bladdery, papery, hairs ± spreading, coarse; beak prominent, pointed;
chamber 1. Salty flats, lake shores; 60–300 m. s SnJV, SCo, WTR,
w edge DMoj; w-c NV. May–Sep   ★

A. insularis Kellogg var. ***harwoodii*** Munz & McBurney (p. 743)
HARWOOD'S MILKVETCH   Ann; ± gray-strigose. **ST:** decumbent to
ascending, 5–40 cm, slender. **LF:** 2–12 cm; lflets (9)11–19(21), ±
spaced, 4–20 mm, ± narrowly elliptic or oblong, tips gen notched.
INFL: among lvs; fls 4–9, spaced, early spreading, then reflexed.
FL: petals pink-violet, banner 5.5–7.4 mm, recurved ± 50–60°, keel
4.8–6 mm. **FR:** spreading or ± reflexed, 15–24 mm, 5–15 mm wide,
half-ovate, bladdery, sparsely strigose, papery; beak conspicuous;
upper suture straight or ± convex, lower strongly convex; chamber 1;
ovules 7–14. Sandy or gravelly areas; < 500 m. DSon; AZ, nw Mex.
Jan–May   ★

A. inversus M.E. Jones (p. 743)   SUSANVILLE MILKVETCH   Per,
wiry, sparsely lfy; thinly minutely strigose. **ST:** prostrate to spreading,
2–5 dm, slender. **LF:** 3–12 cm; lflets 5–11, spaced, 4–20 mm, narrow,
arched, terminal not or ± jointed to midrib. **INFL:** fls 5–12, spaced,
reflexed. **FL:** petals red-pink, banner white-tipped, wings, keel with
buff-yellow, banner 9.4–12.2 mm, recurved 35–45°, keel 8.2–10 mm.
FR: pendent; body 20–35 mm, 4–5 mm wide, straight or downcurved,
compressed side-to-side, minutely strigose or glabrous, papery; stalk-

741

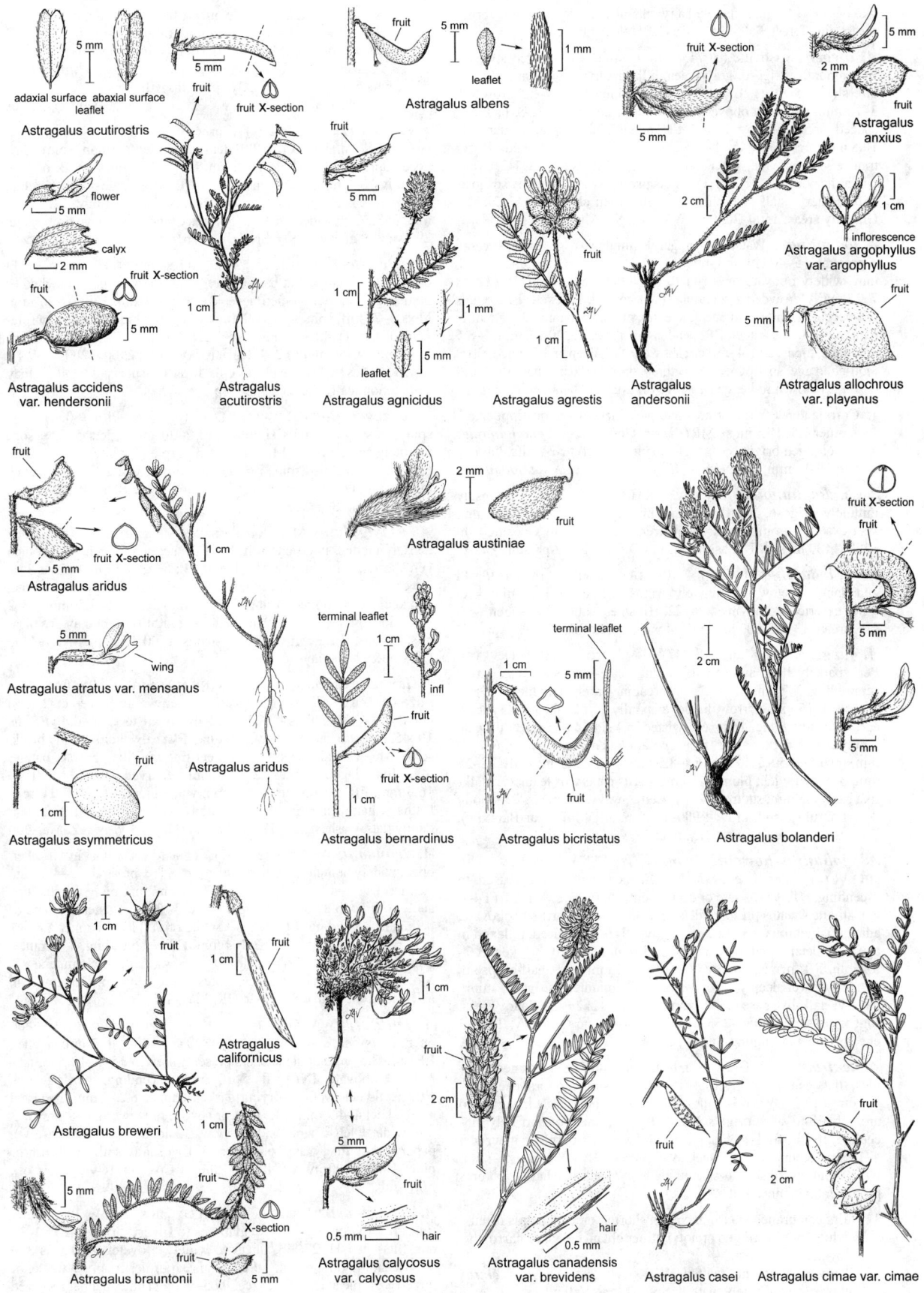

adaxial surface abaxial surface
leaflet
Astragalus acutirostris

5 mm
fruit
fruit **X**-section

flower
5 mm
calyx
2 mm
fruit
fruit **X**-section
**Astragalus accidens
var. hendersonii**

fruit
5 mm
fruit **X**-section
1 cm
**Astragalus
acutirostris**

fruit
5 mm
1 cm
leaflet
5 mm
1 mm
Astragalus agnicidus

fruit
1 cm
Astragalus agrestis

fruit
Astragalus albens
5 mm
leaflet
1 mm

fruit **X**-section
5 mm
2 cm
**Astragalus
andersonii**

fruit
5 mm
2 mm
fruit
**Astragalus
anxius**

1 cm
inflorescence
**Astragalus argophyllus
var. argophyllus**

fruit
5 mm
**Astragalus allochrous
var. playanus**

fruit
fruit **X**-section
5 mm
1 cm
Astragalus aridus

5 mm
wing
Astragalus atratus var. mensanus

fruit
1 cm
Astragalus asymmetricus

Astragalus aridus

2 mm
fruit
Astragalus austiniae

terminal leaflet
1 cm
infl
fruit
fruit **X**-section
1 cm
Astragalus bernardinus

terminal leaflet
1 cm
5 mm
fruit
Astragalus bicristatus

2 cm
fruit **X**-section
fruit
5 mm
5 mm
Astragalus bolanderi

1 cm
fruit
fruit
1 cm
**Astragalus
californicus**

Astragalus breweri
1 cm
fruit
5 mm
fruit
X-section
fruit
5 mm
Astragalus brauntonii

1 cm
fruit
5 mm
fruit
0.5 mm
hair
**Astragalus calycosus
var. calycosus**

fruit
2 cm
hair
0.5 mm
**Astragalus canadensis
var. brevidens**

fruit
Astragalus casei

fruit
2 cm
Astragalus cimae var. cimae

like base 6–14 mm, minutely hairy; chamber 1. Dry soils, sagebrush scrub, pine forest; 950–1950 m. e CaRH, MP. Jun–Aug ★

A. inyoensis Coville (p. 743)   INYO MILKVETCH   Per, sparsely lfy; minutely strigose, gray-green. **ST:** prostrate, loose-matted, 1–6 dm, slender, zigzag. **LF:** 1.5–4.5, spreading; lflets 9–21, crowded, 3–10 mm, narrowly obovate, tips blunt or notched. **INFL:** fls 6–15, spaced, spreading to reflexed. **FL:** petals pink-purple, banner 8.6–10.8 mm, recurved ± 45°, keel 8.2–9.6 mm. **FR:** pendent; body 12–15 mm, 6–8 mm wide, lanceolate, strongly incurved 1/8–1/2 circle, wider than deep, ± 3-sided (lower deeply grooved), minutely strigose, stiff-leathery; stalk-like base 2–5 mm, stout; chamber ± 1. 2*n*=22. Gravelly areas; 1500–2650 m. W&I; s-c NV. May–Jun ★

A. iodanthus S. Watson   Per; gen ± minutely strigose. **ST:** several, ± prostrate, 5–40 cm, few-branched. **LF:** 2–7 cm; lflets 9–21, 3–18 mm, widely obovate or ± round, tips blunt or notched. **INFL:** fls 7–25, early crowded, ascending, then spaced, reflexed. **FL:** petals red-purple, ± white with purple keel tip, or cream, banner 9–15.5 mm, recurved ± 45°, ± ≥ keel. **FR:** pendent or spreading, 20–40 mm, 5–8.5 mm wide, incurved 1/4–full circle, wider than deep or ± 3-sided, firm, mottled in age, stiff-papery or leathery, dark; stalk-like base 0; chambers ± 2 below middle, gen 1 above (or 2 up to 1-chambered beak).

1. Corolla gen ± white or cream to pale lilac-tinged or -tipped, banner 7.5–11 mm, se MP (Lassen Co.).... var. *diaphanoides*
1′ Corolla gen bright red-purple (drying violet) to ± white, banner gen 11–15 mm, SNE (Mono Co.) . . . . var. *iodanthus*

var. ***diaphanoides*** Barneby   SNAKE MILKVETCH   Pl densely minutely strigose, esp lflet margin, midrib. **FL:** calyx 3.7–5.3 mm; keel exserted from wings. Dry barren areas, sand to volcanic ash; 1200–1400 m. se MP (Lassen Co.); to OR, ID, NV. Apr–Jun ★

var. ***iodanthus*** (p. 743)   HUMBOLDT RIVER MILKVETCH   Pl gen glabrous below, infl minutely strigose. **FL:** calyx 3.4–5 mm; keel ± not exserted from wings. 2*n*=22. Hillsides, valley floors, gen with sagebrush; 1300–2600 m. SNE (Mono Co.); to OR, ID, UT. Apr–Jun

A. jaegerianus Munz (p. 747)   LANE MOUNTAIN MILKVETCH   Per, sparsely lfy; hairs minute, ± flat, scale-like. **ST:** weak, often scrambling, 3–7 dm. **LF:** 2–5 cm; spreading or reflexed; lflets 7–15, ± spaced, 3–15 mm, narrow, hairier adaxially. **INFL:** fls 5–15, spaced, early ascending, then reflexed, twisted to 1 side. **FL:** petals dull pale ± purple with darker veins, fading cream, often dingy, banner 6.5–10 mm, recurved 50–75°, keel 6.4–8.5 mm. **FR:** pendent; body 18–25 mm, 3–5 mm wide, plump, glabrous, stiff-papery or leathery; stalk-like base 3–5 mm; sutures thick, raised, ± wavy; chambers 2. Among desert shrubs, sand, gravel; 900–1200 m. c DMoj (near Barstow). Threatened by military activity, grazing, vehicles. Apr–Jun ★

A. johannis-howellii Barneby (p. 747)   LONG VALLEY MILKVETCH   Per, open, widely branched; hairs ± appressed to ascending. **ST:** ± prostrate or decumbent, 3–20 cm, slender. **LF:** 4–6 cm; stipule sheaths present; lflets 13–23, 2–6 mm, narrowly obovate, adaxially glabrous. **INFL:** ± among lvs; fls 6–12, spaced, reflexed in age. **FL:** petals ± white, banner 5–5.5 mm, recurved 90°, keel 3.3–3.9 mm. **FR:** pendent; body 7–11 mm, ± 3 mm wide, half-ellipsoid, 3-sided (lower deeply, openly grooved), minutely strigose, thinly papery; stalk-like base 0.5–2.5 mm; chambers 2. 2*n*=22. Sandy areas, sagebrush scrub; 2050–2550 m. SNE (sw Mono Co.); NV. Threatened by grazing, mining. Jun–Aug ★

A. kentrophyta A. Gray   Per, tufted or matted, gen ± armed (stipules, lflets gen spine-tipped); strigose, hairs simple or branched. **ST:** < 3 dm. **LF:** 2–26 mm; stipules like lflets or not, at least lower fused around st into sheath; lflets 3–9, 1–17 mm, linear or narrowly lanceolate. **INFL:** fls 1–3. **FL:** petals white to pink-purple, banner gen 4–10 mm, recurved ± 45°, keel 2.9–6.3 mm. **FR:** 4–9 mm, 1.5–4 mm wide, compressed side-to-side, finely hairy; stalk-like base 0; sutures prominent; chamber 1.

1. Hairs gen branched (1 branch gen shorter, or ± 0); petals often ± white, purple-tinged or not; banner oblanceolate or narrowly obovate; ovules gen 2–4
2. Pl not prostrate, not matted; W&I . . . . var. *elatus*
2′ Pl prostrate, densely matted; SNE (Mono Valley)
   . . . . var. *ungulatus*

1′ Hairs simple; petals gen pale purple to purple; banner widely obovate, obcordate, or round; ovules gen 5–8
3. Lflets 3 (5 on lower lvs or not), rigid, spine-tipped; pl a dense rounded cushion . . . . var. *danaus*
3′ Lflets 5–9, ± soft, minutely spine-tipped; pl a ± dense, flat mat . . . . var. *tegetarius*

var. ***danaus*** (Barneby) Barneby (p. 747)   SWEETWATER MOUNTAINS MILKVETCH   Pl a dense, rounded cushion, spiny; strigose, hairs simple. **ST:** gen < 5 cm. **LF:** 4–20 mm; lflets 3, or 5 on lower lvs, 3–7 mm, stiff, spine-tipped. **FL:** petals pale purple (± white with purple keel-tip), banner 4–5.6 mm, keel 3.3–4.1 mm. **FR:** 3.5–5 mm, 2–2.5 mm wide, ovules 5–8. Rocky places at, above timberline; 2900–4000 m. c&s SNH, SNE. Jul–Sep ★

var. ***elatus*** S. Watson (p. 747)   SPINY-LEAVED MILKVETCH   Pl erect, often branched at base, spiny; strigose, hairs gen branched (1 branch gen shorter, sometimes ± 0). **ST:** ± 1 dm. **LF:** 10–26 mm; lflets 3–7, stiff, spine-tipped. **FL:** petals gen white or faintly purple-veined, banner 4.8–6.2 mm, keel 3.7–4.1 mm. **FR:** gen 4–7 mm, 1.5–2 mm wide, ovules 2–4. Open rocky areas; 2900–3200 m. W&I; to WY, CO, NM. From e pls, CA pls differ in aspects of habit, fr, may be taxonomically distinct. Jun–Sep ★

var. ***tegetarius*** (S. Watson) Dorn (p. 747)   Pl a ± flat mat, ± spiny; ± strigose, hairs simple. **LF:** 2–20 mm; lflets 5–9, ± soft, minutely spine-tipped. **FL:** petals gen purple (± white), banner 3.9–6.2 mm, keel 2.9–4 mm. **FR:** gen 4–8 mm, gen 2–2.5 mm wide, ovules 5–8. Open, rocky areas; 2700–3600 m. c SNH, W&I; to OR, MT, NM. Jun–Sep

var. ***ungulatus*** M.E. Jones   SPINY MILKVETCH   Pl prostrate, densely matted, internode < lf; hairs dense, gen branched (1 branch gen shorter, sometimes ± 0), silvery. **LF:** 5–13 mm; lflets 5, or 3 on lower lvs, 3–9 mm, linear elliptic, spine-tipped. **FL:** petals ± white, gen keel purple-tinged, banner 5–6.5 mm. **FR:** gen 5–7.5 mm, 1.6–2 mm wide, ovules 2–3(4). Valley knolls, foothills, light clay or volcanic ash soils, in sagebrush/juniper stands; 1500–2200 m. SNE (Mono Co.); ne&c NV. May–Jul ★

A. layneae Greene (p. 747)   LAYNE MILKVETCH   Per from deep rhizome (gen overlooked); hairs coarse, gen ± gray. **ST:** erect, 2.5–16 cm. **LF:** 4–16 cm; lflets 11–23, 5–23 mm, ovate to ± round. **INFL:** fls 10–45, early ascending, then spreading. **FL:** calyx hairs mostly black, some white; petals ± white with keel tip, wing tips purple, banner often lilac-tinged, banner 12.5–18 mm, recurved ± 50°, keel 10.4–16.5 mm. **FR:** 20–65 mm, 3.5–8 mm wide, incurved 1/4–full circle, leathery, hairs spreading, wavy; chambers ± 2 in lower 1/2. 2*n*=44. Sandy flats, washes; 25–1750 m. W&I, DMoj; s NV, nw AZ. Mar–Jun

A. lemmonii A. Gray (p. 747)   LEMMON'S MILKVETCH   Per, open, widely branched; sparsely strigose. **ST:** ± prostrate, 1–4(5) dm, slender. **LF:** 1–4.5 cm; lflets 7–15, 2–11 mm, narrowly elliptic, tips acute. **INFL:** in 2s or 3s in upper axils; fls 2–13, clustered at end of peduncle, ascending. **FL:** petals ± white or dull lilac, banner 4.8–6.1 mm, recurved 45–85°, keel 3.4–4 mm. **FR:** ± spreading, 4–7 mm, ± 2 mm wide, elliptic, 3-sided (lower narrow), papery; chambers ± 2. Moist, alkaline meadows, lake shores; 1300–2900 m. GB (lower in MP), adjacent edge c SNH; s-c OR. May–Jul ★

A. lentiformis A. Gray (p. 747)   LENS-POD MILKVETCH   Per, open; hairs fine, spreading, ± gray. **ST:** spreading, 1–2 dm, slender. **LF:** 1.2–3.5 cm; stipule sheaths present; lflets 7–15, 2–10 mm, narrowly ± obovate. **INFL:** fls 5–10, early ascending, then reflexed. **FL:** petals pale yellow, drying darker, banner 6.2–7 mm, recurved ± 50°, keel 4–4.9 mm. **FR:** bent or curved downward, 5–9 mm, ± 3 mm wide, ± lens-shaped in side view, 3-sided (lower side narrow), papery, hairs fine, wavy; chambers 2. Dry sandy soil, sagebrush or pine; 1450–1800 m. n SNH (se Plumas Co., possibly Sierra Co.). May–Jul ★

A. lentiginosus Douglas   FRECKLED MILKVETCH   Per, sometimes fl 1st yr or ann, ± lfy; ± glabrous to silvery-strigose. **LF:** 1–15 cm; lflets gen 11–27(29), linear to widely ± ovate. **INFL:** fls 3–± 50, ascending or spreading. **FL:** petals ± purple, cream, ± white, or mixed ± purple and ± white, keel 0.65–0.8 × banner, banner recurved 30–50°. **FR:** variable but gen ovate or spheric, widely grooved above and below, gen ± bladdery, ± papery, deciduous; stalk-like base 0;

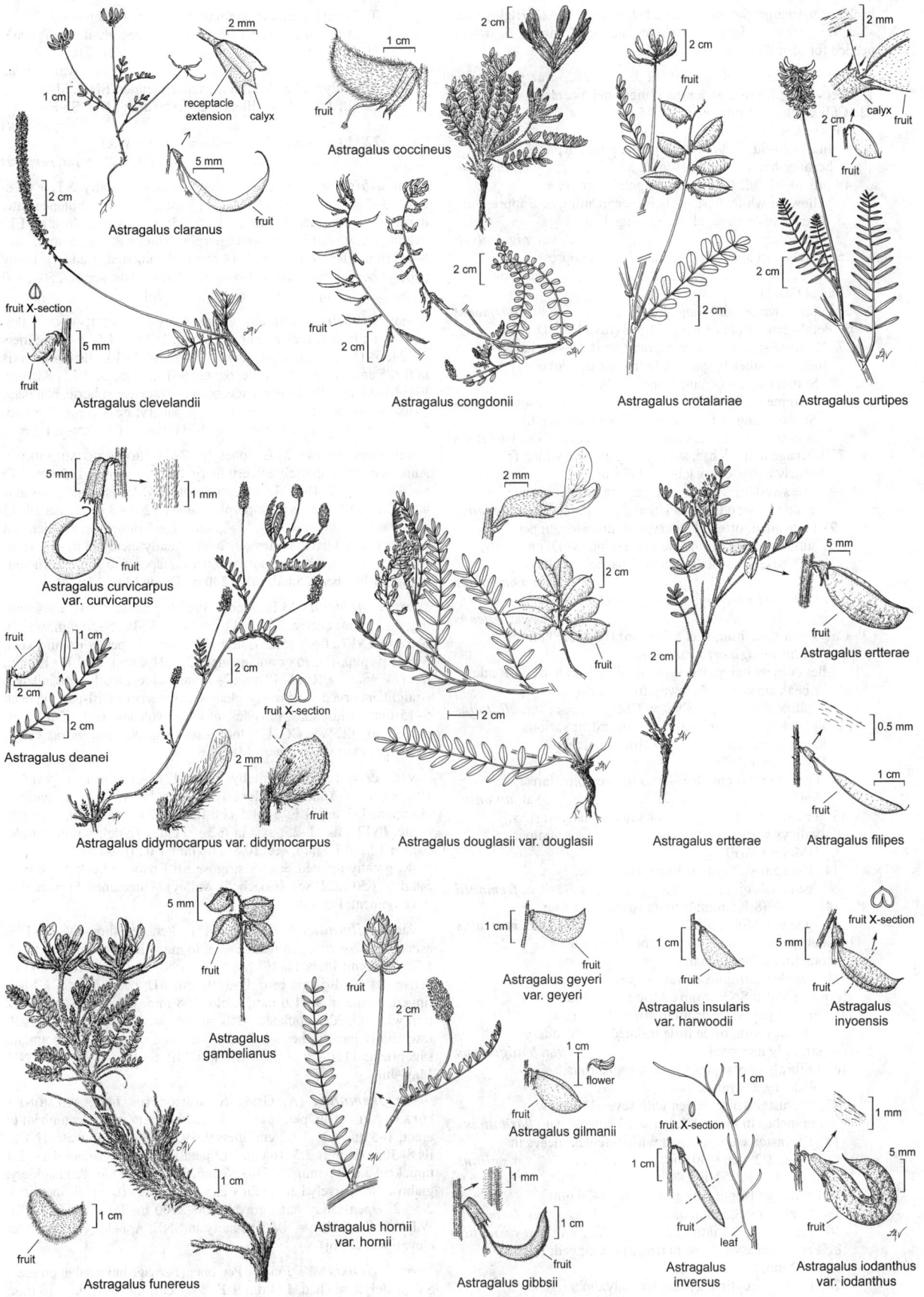

Astragalus claranus

receptacle extension   calyx

fruit

Astragalus coccineus

fruit X-section

fruit

Astragalus clevelandii

Astragalus congdonii

Astragalus crotalariae

calyx   fruit

fruit

Astragalus curtipes

fruit

Astragalus curvicarpus var. curvicarpus

fruit

Astragalus deanei

fruit X-section

fruit

Astragalus didymocarpus var. didymocarpus

fruit

Astragalus douglasii var. douglasii

fruit

Astragalus ertterae

Astragalus ertterae

fruit

Astragalus filipes

fruit

Astragalus gambelianus

fruit

Astragalus funereus

fruit

Astragalus hornii var. hornii

fruit

Astragalus geyeri var. geyeri

flower

fruit

Astragalus gilmanii

fruit

Astragalus gibbsii

Astragalus insularis var. harwoodii

fruit

fruit X-section

fruit

Astragalus inyoensis

fruit X-section

fruit   leaf

Astragalus inversus

fruit

Astragalus iodanthus var. iodanthus

beak gen triangular, flat; chambers 2 below beak. 2*n*=22. Highly variable; vars. often distinct, yet intermediates are common; fl, fr both needed for identification.

1. Lflets 3–5, linear-oblanceolate . . . . var. *piscinensis*
1′ Lflets gen ≥ 7, often wider than linear-oblanceolate
  2. Calyx gen > 6.5 mm; keel 8.4–15 mm;
    3. Pls of CA-FP
      4. Infl 7–20-fld, 1–4 cm in fr; petals pink-purple; herbage hairs subappressed. . . . . var. *idriensis*
      4′ Infl 10–32-fld, 4–15 cm in fr; petals purple, ± yellow, or white; herbage hairs ± spreading or ± appressed
        5. Petals cream to ± yellow; herbage hairs ± spreading . . . . var. *nigricalycis*
        5′ Petals purple (white); herbage hairs ± appressed . . . . [4]var. *variabilis*
    3′ Pls of GB, D
      6. Petals ± white, pink-tinged; MP . . . . var. *chartaceus*
      6′ Petals gen purple or ± pink; SNE (Inyo Co.), D
        7. Herbage sparsely hairy, ± green, or if densely hairy then gen either fr sparsely hairy or calyx lobes < 1.4 mm
          8. St prostrate, > 60 cm; moist alkaline flats, extreme n DMoj. . . . . var. *sesquimetralis*
          8′ St ascending or spreading, < 50 cm; gen sand, widespread. . . . . [4]var. *variabilis*
        7′ Herbage hairs dense, silvery, ± gray, or ± white, fr densely hairy, calyx lobes > 1.4 mm
          9. Fr ± swollen, < 7 mm wide, upper suture concave in side view; e DMoj, s DSon . . . . var. *borreganus*
          9′ Fr swollen, often bladdery, > 8 mm wide, upper suture convex in side view; n DMoj, nw DSon
            10. Pl base not woody; DSon (Coachella Valley) . . . . var. *coachellae*
            10′ Pl base ± woody; n DMoj (Eureka Valley) . . . . var. *micans*
2′ Calyx gen < 6.5 mm; keel 5.5–10 mm (8.4–12.3 mm in *Astragalus lentiginosus* var. *variabilis*)
  11. Lflet densely hairy, ashy-gray or silvery, at least on 1 side
    12. Fr beak curved down, away from st; clayey, alkaline flats, seeps; SNE, w DMoj . . . . var. *albifolius*
    12′ Fr beak straight or curved up, toward st; various habitats, gen exc clayey, alkaline flats, seeps; DMoj, CA-FP
      13. Fr cluster < 4 cm; dry slopes in open pine forest; SnGb. . . . . var. *antonius*
      13′ Fr cluster > 4 cm; gravel or sand on flats, washes, valleys with creosote, sagebrush, pinyon/juniper; DMoj, s CA-FP
        14. Keel 5.6–8.5 mm; fr hairs 0 to sparse; gen e DMoj. . . . . [2]var. *fremontii*
        14′ Keel > (8)8.5 mm; fr hairs sparse to dense; gen w DMoj . . . . [4]var. *variabilis*
  11′ Lflet sparsely hairy or ± glabrous, green (sometimes densely hairy, ± gray in *Astragalus lentiginosus* var. *ineptus*)
    15. Pls of MP, n SNE (from c Mono Co.), ± ne CA-FP
      16. Fr stiff-papery, ± opaque, minutely hairy at least in youth, often little inflated, not bladdery, strongly incurved. . . . . var. *lentiginosus*
      16′ Fr thinly papery, translucent, ± glabrous, bladdery, ± straight
        17. Fr cluster dense; st gen with several to many branches in lower 1/2, gen > 25 cm. . . . var. *floribundus*
        17′ Fr cluster ± open; st gen with 0–few branches in lower 1/2, gen < 30 cm . . . . var. *salinus*
    15′ Pls of s CA-FP, SNE, D
      18. Petals gen bright pink-purple; fr 12–36 mm
        19. Keel gen < 8.5 mm. . . . . [2]var. *fremontii*
        19′ Keel gen > 8.5 mm . . . . [4]var. *variabilis*
      18′ Petals gen ± white, sometimes pink-tinged; fr 6–22 mm
        20. Fr beak ± cylindric, slender; calyx lobes gen < 1.2 mm, acute, slender — s SNH . . . . . . var. *kernensis*

        20′ Fr beak flat, ± triangular; calyx lobes often > 1.2 mm, acute, slender (if < 1.2 mm, then blunt, triangular)
          21. Calyx lobes < 1.2 mm, blunt, triangular; TR . . . . var. *sierrae*
          21′ Calyx lobes > 1.2 mm, acute, slender; SNH, SNE
          22. Lf 1.5–5.5 cm, lflets 9–21, crowded; SNH, SNE . . . . var. *ineptus*
          22′ Lf 4–9 cm, lflets 13–27, ± spaced; W&I . . . . var. *semotus*

var. *albifolius* M.E. Jones Per, gen densely hairy. **ST:** ± prostrate, 3–7 dm. **LF:** lflets 9–21, 3–18 mm, narrowly oblanceolate, densely strigose. **INFL:** dense; fls 9–35; axis in fr 5–40 mm. **FL:** petals gen ± white with some purple, banner 8.2–11.5 mm, keel 6–8.5 mm. **FR:** 10–17 mm, 8–14 mm wide, inflated, bladdery, thinly papery; beak curved down, 3–5 mm. Alkaline flats, seeps; 250–2300 m. SNE, w DMoj (Los Angeles Co.). Apr–Jul

var. *antonius* Barneby SAN ANTONIO MILKVETCH Per, densely hairy. **ST:** prostrate or spreading, 1–3 dm. **LF:** 3–8 cm; lflets 11–21, 3–11 mm, ± obovate, densely strigose. **INFL:** fls 10–15; axis in fr < 5 cm. **FL:** petals purple, banner 9–10.5 mm, keel 7.2–8.2 mm. **FR:** 14–30 mm, 10–18 mm wide, plump-ovate to ± spheric, bladdery, sparsely strigose, pale yellow, ± shiny, papery; beak erect, 3–6 mm. Dry slopes in open pine forest; 1400–2600 m. SnGb. Apr–Jul ★

var. *borreganus* M.E. Jones (p. 747) BORREGO MILKVETCH Ann, occ per, ± densely silvery-hairy. **ST:** ascending, 1–3 dm. **LF:** 6–16 cm; lflets 7–19, 4–21 mm, ± obovate. **INFL:** fls 13–50; axis in fr 4.5–26 cm. **FL:** petals pink-purple, banner 12–14.8 mm, keel 10–13 mm. **FR:** erect to ascending, 15–23 mm, 4.5–6 mm wide, swollen, not bladdery, lanceolate or narrowly ovate, gently incurved in side view, papery, silky-hairy; upper suture concave, tapered to short, ± triangular, tooth-like beak. Sand; -67–1200 m. D; nw Mex. Mar–May ★

var. *chartaceus* M.E. Jones Per, ± glabrous. **ST:** ± ascending, 1–3.5 dm, coarse. **LF:** 4–11 cm; lflets 7–19, 5–20 mm, widely obovate. **INFL:** fls 5–15; axis in fr 1–3.5 cm. **FL:** petals ± white, keel, wing tips pink-tinged or not, banner 12.6–21.4 mm, keel 11–15 mm, rarely > wings. **FR:** 13–48 mm, 7–14 mm wide, greatly or ± inflated, straight or curved 3/4 circle, ± glabrous, leathery or stiff-papery; beak 5–15 mm. Arid plains, hillsides, often on volcanic soil; 1200–1750 m. MP; to OR, WY, CO, UT. Intergrades with *A. lentiginosus* var. *l.*, *A. lentiginosus* var. *salinus*. May–Jul

var. *coachellae* Barneby (p. 747) COACHELLA VALLEY MILKVETCH Ann, per, densely silvery-hairy. **ST:** ascending, clumped, 1–3 dm. **LF:** 5–11.5 cm; lflets 7–21, 5–17 mm, ± widely ovate. **INFL:** fls 11–25; axis in fr 3–10 cm. **FL:** petals pink-purple, banner 12.7–14.5 mm, keel 10.8–11.6 mm. **FR:** 16–21 mm, 9–14 mm wide, greatly inflated, ± gray strigose, stiff-papery; beak 3.5–6 mm. Sand; < 650 m. DSon (Coachella Valley). Threatened by vehicles, development. Feb–May ★

var. *floribundus* A. Gray (p. 747) Per, gen short-lived. **ST:** ± ascending, 2–5 dm, gen with several to many branches in lower 1/2. **LF:** 3–11 cm; lflets 11–19, 5–15 mm, obovate, ± glabrous. **INFL:** dense; fls 10–40; axis in fr 1–4(10) cm. **FL:** petals white, ± lilac-tinged, banner 8.8–11.6 mm, keel 6.6–8 mm. **FR:** 8–21 mm, 6–12 mm wide, bladdery-inflated, glabrous or sparsely strigose, pale yellow, thinly papery; beak 3–7 mm, incurved or erect. Often among sagebrush; 1150–2000 m. n SNH, s MP, SNE; s-c OR, w-c NV. May–Jul

var. *fremontii* (A. Gray) S. Watson (p. 747) FREMONT'S MILKVETCH Ann, per, sparsely to densely hairy. **ST:** decumbent to erect, 1–5 dm. **LF:** 3–12 cm; lflets 9–19, 5–19 mm, obovate. **INFL:** fls 8–30; axis in fr 2.5–16 cm. **FL:** petals ± purple, banner 9.1–12.4 mm, keel 5.6–8.5 mm. **FR:** 14–36 mm, 5–18 mm wide, gen bladdery, glabrous to sparsely hairy, thinly papery; beak 2–10 mm, ± incurved. 2*n*=22. Open sandy flats, gravel; 400–2900 m. Teh, SnJV, SCoRI, WTR, SNE, DMoj; s NV. Esp hairy in SNE. Apr–Jul(Oct, at lower elevations, DMoj)

var. *idriensis* M.E. Jones Per, open, herbage hairs subappressed. **ST:** widely branched, 1–4 dm. **LF:** 2–11 cm; lflets 7–29, 3–18 mm, widely obovate. **INFL:** fls 7–20; axis in fr 1–4 cm. **FL:** petals pink-

purple, banner 12–20 mm, keel 9–14 mm. **FR**: 12–30 mm, 5–16 mm wide, ± half-ovate, often incurved, greatly or ± inflated, minutely strigose, ± fleshy, drying thick-papery or ± leathery; beak 3–10 mm. Dry, grassy hillsides, often with oak; 700–2150 m. SCoRI, WTR. Highly variable. Apr–Jun

var. *ineptus* (A. Gray) M.E. Jones (p. 747) Per; hairs ± ascending, ± gray. **ST**: decumbent, 1–3 dm. **LF**: 1.5–5.5 cm; lflets 9–21, crowded, 2–10 mm, obovate. **INFL**: fls 10–21; axis in fr 10–25 mm. **FL**: calyx lobes > 1.2 mm; petals cream, banner 8.8–12.2 mm, keel 7.2–9.3 mm. **FR**: 10–18 mm, 6–12 mm wide, bladdery, minutely strigose (glabrous), thinly papery; beak erect or incurved. Open, gravelly places, talus slopes; 1250–3700 m. SNH (± n SNH), SNE. Intergrades with *A. lentiginosus* var. *semotus*. Jun–Aug

var. *kernensis* (Jeps.) Barneby KERN PLATEAU MILKVETCH Per; minutely strigose. **ST**: prostrate or decumbent, 2–12 cm, slender. **LF**: 1–5 cm; lflets 7–19, 2–7 mm, ± ovate, adaxially ± glabrous. **INFL**: fls 2–9; axis in fr 3–15 mm. **FL**: calyx lobes gen < 1.2 mm; petals ± white, banner 9.3–11.3 mm, keel 7.3–8.7 mm. **FR**: 6–10 mm, 6–9 mm wide, ± spheric, bladdery, thinly papery; beak ± cylindric, slender. Sandy areas; 2350–2750 m. s SNH; s NV. Jun–Jul ★

var. *lentiginosus* Per; sparsely strigose. **ST**: prostrate or decumbent, 1–5 dm. **LF**: 3–10 cm; lflets 5–19, 5–18 mm, widely obovate. **INFL**: fls 8–22; axis in fr 5–35 mm. **FL**: petals cream, keel tip, wing bases ± lilac-tinged, banner 7.4–11 mm, keel 6.3–8.4 mm, rarely > wings. **FR**: 10–23 mm, 4–10 mm wide, lanceolate or ovate in side view, strongly incurved, often little inflated, ± glabrous, stiff-papery, ± opaque; beak 4–9 mm. 2*n*=22. Dry, open areas, sagebrush or pines; 900–1800 m. MP; to WA, ID, NV. May–Jul

var. *micans* Barneby SHINING MILKVETCH Per; hairs dense, silvery or white, silky. **ST**: clumped, ascending or erect, 2–4 dm. **LF**: 4.5–9.5 cm; lflets 11–17, 5–14 mm, widely ± ovate. **INFL**: fls 12–35; axis in fr 4–10 cm. **FL**: petals white, tips tinged pink-lavender, banner 12.2–14.3 mm, keel 9.6–10 mm. **FR**: 15–20 mm, 8–10 mm wide, inflated, densely silky-hairy, stiff-papery; beak 2.5–4 mm. Dunes; 850–1200 m. n DMoj (Eureka Valley). Threatened by vehicles. More strongly per, hairs longer (1.1–2 mm) than similar pls of *A. lentiginosus* var. *variabilis*. Apr–Jun ★

var. *nigricalycis* M.E. Jones Per; hairs ± spreading, curly or wavy, ± black among fls, frs. **ST**: ± ascending, 2.5–5 dm, stout. **LF**: 4–16 cm; lflets 19–25, 6–25 mm, oblong to obovate. **INFL**: fls 10–32; axis in fr 4–11 cm. **FL**: petals cream to ± yellow, yellow when dry, banner 12–18.5 mm, keel 10.8–13 mm. **FR**: 20–35 mm, 10–20 mm wide, greatly inflated, bladdery, papery, hairs ± ascending; beak 4–8 mm. Dry, grassy areas, roadcuts; 100–1250 m. s SNF, SnJV, SCoRI, WTR. Mar–May(fall)

var. *piscinensis* Barneby (p. 747) FISH SLOUGH MILKVETCH Per, sparsely lfy, ± canescent. **ST**: prostrate, < 1 m. **LF**: 3–4 cm, petiole ± 0; lflets 3–5, linear-oblanceolate, lateral 7–20 mm, terminal 14–32 mm. **INFL**: fls 5–12; axis in fr 1.5–4 cm. **FL**: petals lavender, banner 13 mm, keel 9 mm. **FR**: 20–24 mm, 8–12 mm wide, strongly inflated, stiff-papery, densely strigose; beak 4.5–7 mm. Moist alkaline soil banks; 1300 m. c SNE (near Bishop, Mono, Inyo cos.). Jun–Jul ★

var. *salinus* (Howell) Barneby (p. 747) Per, gen short-lived; glabrous to sparsely strigose. **ST**: ascending, gen with 0–few branches in lower 1/2, 1–3 dm. **LF**: 4–10 cm; lflets 9–19, 5–20 mm, widely obovate. **INFL**: fls 10–25; axis in fr 1.5–9 cm. **FL**: petals ± white, keel, wing tips lilac-tinged or not, banner 9.5–13.3 mm, keel 6–9.6 mm. **FR**: 14–30 mm, 6–14 mm wide, bladdery, gen glabrous, thinly papery; beak 3–9 mm. 2*n*=22. Gen sagebrush; 1050–1600 m. MP; to OR, WY, UT. Intergrades with *A. lentiginosus* var. *floribundus*, *A. lentiginosus* var. *l.* Apr–Jun

var. *semotus* Jeps. Per; ± strigose, hairs ± appressed. **ST**: loose-tufted, < 15 cm. **LF**: 4–9 cm; lflets 13–27, 2–9 mm, ± spaced, ± narrowly elliptic, adaxially ± glabrous. **INFL**: fls 6–10; axis in fr 1–3 cm. **FL**: calyx lobes > 1.2 mm; petals ± white, banner 10.4–12 mm, keel 7.7–8.5 mm. **FR**: 10–20 mm, 6–12 mm wide, bladdery, mottled, sparsely papery; beak 4–7 mm, incurved. 2*n*=22. Dry sandy or gravelly flats, hillsides, sagebrush, pine forest; 2200–3500 m. W&I. Intergrades with *A. lentiginosus* var. *ineptus*. Jun–Aug

var. *sesquimetralis* (Rydb.) Barneby SODAVILLE MILKVETCH Per; sparsely strigose. **ST**: prostrate, 6–8 dm. **LF**: 2–5 cm; lflets 7–17, 6–18 mm, oblanceolate, adaxially ± glabrous. **INFL**: fls 5–12; axis in fr 1–2.5 cm. **FL**: petals purple, banner 12–14.5 mm, keel 9.3–9.5 mm. **FR**: 12–26 mm, 5–12 mm wide, ± inflated, sparsely strigose, stiff-papery; beak 4–8 mm. Moist, alkaline flats; 950 m. n DMoj (n Death Valley, e slope Last Chance Range); NV. May–Jun ★

var. *sierrae* M.E. Jones BIG BEAR VALLEY MILKVETCH Per, ± loose-matted; sparsely strigose. **ST**: ± prostrate, widely branched, 1–3.5 dm. **LF**: 2–5 cm; lflets 15–21, gen crowded, 3–8 mm, obovate, midribs gen ± arched. **INFL**: fls 5–15; axis in fr 1–3 cm. **FL**: calyx lobes < 1.2 mm; petals ± white, tips ± pink-tinged, banner 10.4–14.5 mm, keel 8.1–9.8 mm. **FR**: 15–22 mm, 8–15 mm wide, plump-ovate, bladdery, sparsely strigose, papery; beak 3–6 mm. Rocky meadows, pine woodland; 1700–3200 m. TR. May–Jul ★

var. *variabilis* Barneby Ann, per, gen robust, coarse; hairs sparse, appressed, straight to ± dense, spreading, wavy. **ST**: ascending, 1–4 dm, fl-st sometimes 1, weak. **LF**: 2.5–13 cm; lflets 11–25, 4–17 mm, obovate. **INFL**: fls 10–30; axis in fr 4–17 cm. **FL**: petals purple (white), banner 11.1–15 mm, keel 8.4–12.3 mm. **FR**: 12–30 mm, 8–15 mm wide, bladdery, sparsely to densely strigose or ± wavy-hairy, ± firm-papery; beak 3–9 mm, gently incurved. 2*n*=22. Sandy areas, esp with *Larrea*; 140–1850 m. s SNF, Teh, s SnJV, s-most SNE, w&s DMoj. Intergrades with *A. lentiginosus* var. *fremontii*, *A. lentiginosus* var. *micans* to n; *A. lentiginosus* var. *coachellae*, *A. lentiginosus* var. *nigricalycis* to w. Mar–Jun

A. leucolobus M.E. Jones (p. 747) BIG BEAR VALLEY WOOLLYPOD Per, ± cespitose; herbage hairs gen 0.6–1.3 mm, straight and wavy, tangled, ± gray. **ST**: < 7 cm. **LF**: 1.5–9 cm; lflets 7–19, 3–13 mm, ± widely obovate. **INFL**: fls 5–13, ascending. **FL**: petals pink-purple, banner 16–18.5 mm, ± > wings, recurved ± 40°, keel 14.3–16.8 mm. **FR**: ascending, 13–25 mm, 5–8 mm wide, ± incurved, ± compressed top-to-bottom, abruptly bent at tip, leathery, hairs dense, gen 0.7–1.5 mm, white; chambers ± 2. Dry, rocky areas, sagebrush or pines; 1450–2900 m. Teh, TR, SnJt. Much like vars. of *A. purshii* (banner >> wings, fr hairs > 1.5 mm, or banner 10–15 mm). May–Jul ★

A. macrodon (Hook. & Arn.) A. Gray (p. 747) SALINAS MILKVETCH Per; hairs ± ascending or spreading, wavy or curved. **ST**: clumped, ± ascending, 5–10 dm. **LF**: 5.5–15 cm; lflets 11–29, 7–20 mm, ± elliptic, margins, midribs gen purple. **INFL**: fls 8–35, early spreading, then reflexed. **FL**: calyx lobes 2.5–4.3 mm, purple; petals ± cream, tips red-brown-tinged, banner 8.3–11.4 mm, red-brown-streaked, recurved ± 90°, keel 7.5–9.1 mm. **FR**: ± spreading or reflexed, 20–40 mm, 14–20 mm wide, widely ovate, bladdery, thinly papery, hairs ± spreading or ascending; chamber 1. Eroded pale shales or sandstone, serpentine alluvium; 200–1550 m. c&s SCoR. See note under *A. douglasii*. Apr–Jun ★

A. magdalenae Greene var. *peirsonii* (Munz & McBurney) Barneby (p. 747) PEIRSON'S MILKVETCH Ann, per, silvery-canescent. **ST**: ascending or erect, 2–9 dm. **LF**: 1–15 cm; midrib ± flat; lflets (3)9–13, ± spaced, 2–8 mm, narrow, oblong, terminal not jointed to midrib. **INFL**: fls 5–20, ascending or spreading. **FL**: petals pink-purple, tips often white, banner 10–14.2 mm, recurved ± 40°, keel 8.5–10 mm. **FR**: spreading, 20–35 mm, 10–20 mm wide, half-ovate to ± spheric, bladdery, finely strigose, papery; chamber 1. Sand dunes; 50–250 m. DSon (Algodones Dunes); nw Mex. Seeds to 5 mm, largest of Am spp. of *Astragalus*. Dec–Apr ★

A. malacus A. Gray (p. 747) Per; hairs 1.5–2.5 mm, spreading. **ST**: ascending or erect, 1–4 dm. **LF**: 4–15 cm; lflets 7–21, 5–20 mm, elliptic to obovate, tips ± notched. **INFL**: fls 9–35, reflexed in age. **FL**: petals red-violet, banner 15–21 mm, recurved ± 45°, keel 12.3–16 mm. **FR**: pendent; body 18–38 mm, 5–6 mm wide, incurved, ± 3-sided, stiff-papery, hairs long; stalk-like base 1–3 mm, stout; chambers 2. Dry rocky or stiff soils, sagebrush, pinyon/juniper woodland; 900–2350 m. e MP, SNE, e DMoj; to OR, ID, NV. Apr–Jun

A. miguelensis Greene (p. 747) SAN MIGUEL ISLAND MILKVETCH Per, open; hairs dense, short, woolly, ± gray. **ST**: ± ascending, 1–4 dm. **LF**: 2.5–12 cm; stipule sheath finely tomentose; lflets 17–27, 6–22 mm, narrowly oblong or ± obovate, tips notched or blunt. **INFL**: fls

10–30, dense, ± spreading. **FL:** petals ± white to ± yellow, banner 12.5–16 mm, recurved ± 45°, keel 9–12 mm. **FR:** spreading, 16–26 mm, 13–23 mm wide, bladdery, papery, hairs fine; chamber 1. 2*n*=22. Slopes, bluffs, coastal beaches; < 500 m. s CCo, ChI. Mar–Jul ★

A. minthorniae (Rydb.) Jeps. var. ***villosus*** Barneby (p. 747) Per, robust or ± cespitose; hairs gen dense, ± spreading, curly or not, coarse, gray. **ST:** ascending or erect, 3–30 cm. **LF:** 4–17 cm; stipule sheath ± 0, lower stipules long-woolly; lflets 7–17, 8–25 mm, ± obovate. **INFL:** fls 7–35, ascending to reflexed. **FL:** petals cream (keel tip purple) or ± purple (wing tips pale), banner 12–18 mm, recurved ± 45°, keel 9.5–13 mm. **FR:** ± erect, 15–30 mm, 4–6 mm wide, ± compressed side-to-side but sides convex, leathery, hairs ± spreading; chambers 2. Rocky, calcareous hillsides, washes, gen pinyon/juniper woodland; 1350–2300 m. SnBr, W&I, DMtns (Inyo, San Bernardino cos.); s NV. Apr–Jun

A. mohavensis S. Watson Ann, ± per, tufted or open, widely branched; gray- or silvery-strigose or -canescent. **ST:** ± ascending, 5–35 cm. **LF:** 2–12.5 cm; lflets 3–11, 3–18 mm, ovate to ± round, tips gen blunt, sometimes lower, rarely upper shallowly notched. **INFL:** fls 3–15, early ascending, then reflexed. **FL:** petals pink-purple, banner 7–12.5 mm, recurved ± 45°, keel 6.4–10.5 mm. **FR:** reflexed, 13–32 mm, 3.5–8.5 mm wide, densely minutely strigose, stiff-leathery; chambers ± 2 below, 1 near tip; ovules 20–30.

1. Fr incurved 1/4–1/2 circle, ± 3-sided, upper suture raised, lower in wide, shallow groove; DMtns .... var. ***hemigyrus***
1′ Fr straight or incurved gen < 1/4 circle, ± compressed side-to-side or 3-sided, upper, lower sutures gen both raised; DMoj .... var. ***mohavensis***

var. ***hemigyrus*** (Clokey) Barneby (p. 747) CURVED-POD MILKVETCH Limestone hillsides, *Larrea* zone; 350–2750 m. D; s NV (Clark Co.). Previously thought extinct, re-collected near type locality. Apr–Jun ★

var. ***mohavensis*** Gen limestone; 750–2300 m. DMoj. Apr–Jun

A. monoensis Barneby (p. 747) MONO MILKVETCH Per from rhizome, crown below surface; hairs dense, silky, wavy. **ST:** 7–20 cm, 2–6 cm buried, ± glabrous, above ground decumbent, hairy. **LF:** 7–30 mm; stipule sheaths present, esp underground; lflets 9–15, crowded, 2–8 mm, ovate. **INFL:** head-like; fls 6–12, spreading. **FL:** petals ± pale pink-tinged but drying ± yellow, banner 10–13 mm, recurved ± 50°, keel 6.7–8 mm. **FR:** spreading or ascending, 15–20 mm, 6–9 mm wide, widely incurved-lanceolate, papery; hairs short, wavy; not fully 2 chambered; ovules 18–20. 2*n*=22. Open areas, pumice sand, gravel; 2100–3400 m. SNE (Mono, n Inyo cos.). Threatened by vehicles, road maintenance, grazing. Jun–Aug ★

A. nevinii A. Gray (p. 747) SAN CLEMENTE ISLAND MILKVETCH Per; hairs fine, soft, kinky, woolly, tangled. **ST:** ascending, bushy-branched, 1–3 dm. **LF:** 2–8 cm; lflets 11–25, 3–12 mm, ± oblong or obovate, tips blunt or notched. **INFL:** fls 15–30, dense, reflexed in age. **FL:** petals cream, banner 10.6–12.7 mm, recurved 30–40°, keel 9–10 mm. **FR:** pendent; body 14–20 mm, 3–5 mm wide, incurved, 3-sided (lower openly grooved), glabrous, stiff-papery; stalk-like base 5–10 mm, slender; beak stout; chambers ± 2. Sandy flats, dunes; probably < 200 m. ChI (San Clemente Island). Feb–Jul ★

A. newberryi A. Gray var. ***newberryi*** (p. 747) NEWBERRY'S MILKVETCH Per, ± cespitose, old lf bases persistent; hairs silky, longer ± straight, partly spreading. **LF:** 1.5–15 cm; lflets 3–15, 5–20 mm, obovate, tips acute or blunt, notched or not. **INFL:** fls 3–8, ascending. **FL:** calyx with more white than black hairs; petals pink-purple or ± white, tips pink-purple, banner 21.5–30 mm, recurved ± 40°, keel 18.5–26 mm. **FR:** 13–28 mm, 7–13 mm wide, ovate, gen incurved, hairs dense, longer 2–4.5 mm, white, woolly, some curly, some ± straight; chamber 1. 2*n*=22. Rocky areas, esp pinyon/juniper woodland; 1300–2350 m. SNE, ne DMoj; to OR, UT, NM. Like *A. purshii* (hairs extremely fine, tangled). Apr–Jun

A. nutans M.E. Jones (p. 747) PROVIDENCE MOUNTAINS MILKVETCH Ann, ± per; minutely strigose. **ST:** prostrate to erect, 6–15 cm. **LF:** 2–8 cm; lflets 7–13, 5–15 mm, ± narrowly elliptic or obovate, tips acute or shallowly notched. **INFL:** fls 6–10, ascending to reflexed. **FL:** petals pink-purple, wing tips often paler, banner

7.8–10.4, recurved ± 90°, keel 5.9–6.6 mm. **FR:** spreading, 15–25 mm, 11–15 mm wide, ovate, bladdery, sparsely strigose, rarely ± glabrous, thinly papery; beak compressed side-to-side, widely triangular; chamber 1; ovules 19–24, suspended from flange 1–2.5 mm wide. 2*n*=22. Sandy flats, washes of desert foothills, with *Larrea*, *Yucca*; 450–2000 m. se DMtns, DSon. Mar–Jun(Oct) ★

A. nuttallianus DC. SMALL-FLOWERED MILKVETCH Ann, slender; minutely strigose. **ST:** prostrate or ± ascending, 4–45 cm. **LF:** 1.5–6.5 cm; lflets 5–13, 2–10 mm, at least upper elliptic, acute at tips, lower blunt or not, notched at tip. **INFL:** fls 1–4. **FL:** 4–7 mm; corolla ± white, faintly lilac-tinged (± purple). **FR:** 10–20 mm, 2–3 mm wide, linear in side view, gently curved near base, ± straight toward tip, ± 3-sided (laterals ± convex, lower grooved), maturing dark brown or ± black, glabrous or minutely strigose; chambers ± 2. See note under *A. acutirostris*; vars. intergrade; *A. nuttallianus* var. *austrinus* (Small) Barneby (calyx lobes gen 1.8–2.8 mm, hairs 0.7–1.2 mm, spreading, stiff) near CA, in w AZ.

1. Lflet tips notched on lower lvs, acute on upper; calyx tube 1.4–1.7 mm .... var. ***cedrosensis***
1′ Lflet tips acute; calyx tube 1.9–2.8 mm ...... var. ***imperfectus***

var. ***cedrosensis*** M.E. Jones **LF:** lflet tips notched on lower lvs, acute on upper. **FL:** calyx tube 1.4–1.7 mm, ± as wide, lobes (0.7)1–1.6 mm, hairs ascending, short, not stiff. Rocky flats, washes; < 1400 m. DSon; AZ, nw Mex. Dec–Apr

var. ***imperfectus*** (Rydb.) Barneby (p. 747) **LF:** lflet tips acute. **FL:** calyx tube 1.9–2.8 mm, ± 1/2 as wide, lobes 1–1.7(2) mm, hairs ascending, short, not stiff. 2*n*=22. Sandy or gravelly flats, washes; 250–2150 m. DMoj, w DSon; to UT, AZ. Mar–May

A. nuttallii (Torr. & A. Gray) J.T. Howell NUTTALL'S MILKVETCH Per, gen robust, lfy; hairs fine, incurved-ascending or curly, or ± 0. **ST:** prostrate to erect, often in dense tangles, 2–10 dm. **LF:** 2.5–17 cm, gen spreading, arched; stipules fused around st into sheath, sometimes early ruptured, seemingly free; lflets 21–43, gen ± crowded, 3–25 mm, ± obovate or oblong, tips notched. **INFL:** fls 20–125, dense, early reflexed. **FL:** petals ± cream, lavender-tinged or not, banner 10–15 mm, recurved ± 40°, wings 1 mm < to 1.5 mm > banner, keel 9.7–14 mm. **FR:** 20–60 mm, 15–27 mm wide, bladdery, glabrous to sparsely hairy, papery; chamber 1; ovules 14–38.

1. Lflets gen gray-hairy adaxially, esp in s; c&s CCo .... var. ***nuttallii***
1′ Lflets gen glabrous adaxially; c&s NCo, n CCo .... var. ***virgatus***

var. ***nuttallii*** (p. 747) OCEAN BLUFF MILKVETCH **ST:** gen prostrate or decumbent (ascending if sheltered). **FR:** ± hairy, at least in youth; ovules 22–38. 2*n*=22. Rock, sandy areas, bluffs; < 250 m. CCo. All yr ★

var. ***virgatus*** (A. Gray) Barneby **ST:** ascending or erect, at coast prostrate or not. **FR:** glabrous or sparsely hairy; ovules 14–21. Sandy soil, bluffs; < 150 m. c&s NCo, n CCo. Apr–Jul(all yr)

A. obscurus S. Watson (p. 751) Per; minutely strigose. **ST:** ± prostrate, tufted or open, widely branched, < 15 cm. **LF:** 2.5–10 cm; lflets 5–15, ± spaced, 2–15 mm, narrowly to widely elliptic or oblong, ± thick. **INFL:** fls (3)6–14, ± dense, ascending. **FL:** petals ± graduated, cream or dirty white, dirty lilac-tinged or not, banner 7–10.5 mm, recurved ± 45°, keel 6.3–10 mm. **FR:** erect, 10–25 mm, 2.4–3.3 mm wide, linear-oblong, ± bluntly 3-sided (lower grooved), finely strigose, leathery; upper suture a thick ridge; chambers ± 2. Rocky, basalt, granite; 800–2150 m. se KR, CaR, MP; to OR, ID, NV. May–Jul

A. oocarpus A. Gray (p. 751) SAN DIEGO MILKVETCH Per, stout, lfy; hairs ± 0. **ST:** widely ascending to erect, 6–13 dm, hollow. **LF:** 4.5–17 mm; midrib grooved; lflets 17–35, 6–33 mm, widely ± lanceolate, glabrous (exc sparsely strigose margins, midveins), prominently veined, midvein raised abaxially, in groove adaxially. **INFL:** fls 20–75, widely ascending. **FL:** calyx 4.5–6 mm; petals cream, banner 10.5–12.5 mm, recurved 70–90°, keel 9–10.8 mm. **FR:** erect, 15–25 mm, 10–16 mm wide, greatly inflated, glabrous or minutely strigose, stiff-papery, persistent; chamber 1. 2*n*=22. Openings in chaparral, oak woodland; 400–1700 m. PR (c San Diego Co.). Like *A. douglasii* (all tissues thinner; fr less persistent). May–Aug ★

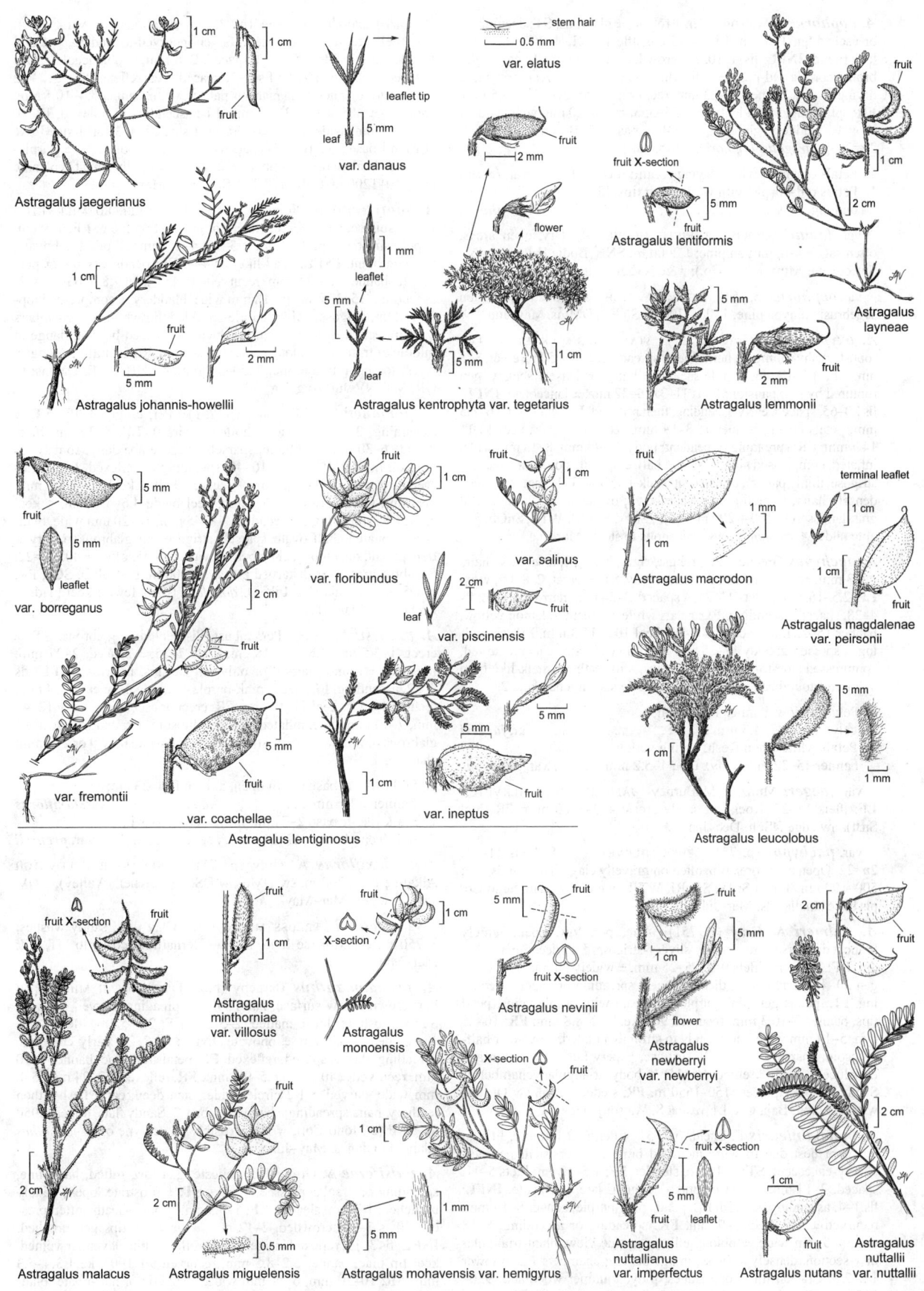

Astragalus jaegerianus

var. danaus

var. elatus

leaflet tip

leaf

fruit

fruit X-section

Astragalus lentiformis

Astragalus layneae

Astragalus johannis-howellii

Astragalus kentrophyta var. tegetarius

Astragalus lemmonii

var. borreganus

leaflet

var. floribundus

var. salinus

var. piscinensis

leaf

Astragalus macrodon

terminal leaflet

Astragalus magdalenae var. peirsonii

var. fremontii

var. coachellae

var. ineptus

Astragalus lentiginosus

Astragalus leucolobus

fruit X-section

Astragalus minthorniae var. villosus

Astragalus monoensis

X-section

Astragalus nevinii

fruit X-section

Astragalus newberryi var. newberryi

flower

Astragalus malacus

Astragalus miguelensis

Astragalus mohavensis var. hemigyrus

Astragalus nuttallianus var. imperfectus

Astragalus nutans

Astragalus nuttallii var. nuttallii

A. oophorus S. Watson   Per, ± robust; ± glabrous. **ST**: decumbent or ascending, 1–3 dm. **LF**: 5–15 cm; lflets 7–21, 4–20 mm, ovate to ± round. **INFL**: fls 4–10, not crowded, spreading. **FL**: calyx glabrous; corolla red-purple with white wing tips, or cream, or white drying cream, banner 16–23 mm, recurved ± 85°, keel 10–16.5 mm. **FR**: spreading or pendent; body 25–55 mm, 10–20 mm wide, widely ± ovate, bladdery, glabrous; stalk-like base 3.5–10 mm; chamber 1. 2*n*=24. [*Astragalus oophorous*, orth. var.]

1. Petals cream or white drying cream; lflets 7–11 . . . var. *lavinii*
1′ Petals red-purple with white wing tips; lflets > 11
  on longer lvs . . . . var. *oophorus*

  var. ***lavinii*** Barneby   LAVIN'S MILKVETCH   Dry, open areas, often sagebrush, pinyon pine; ± 2000 m. SNE (Bodie Hills, possibly Sweetwater Mtns, Mono Co.); w&c NV. Jun–Jul   ★

  var. ***oophorus*** (p. 751)   EGG MILKVETCH   Dry, open areas, often sagebrush, pinyon pine; 1500–3300 m. SNE, n DMtns. May–Jun   ★

A. oxyphysus A. Gray (p. 751)   STANISLAUS MILKVETCH   Per, robust, bushy-clumped; hairs fine, ± spreading, wavy. **ST**: ± erect, 3–8 dm. **LF**: 4.5–17 cm; stipule sheaths ± hairy or densely shaggy, gen ruptured by st expansion; lflets 11–31, 5–32 mm, ± lanceolate. **INFL**: fls 20–65, spaced, early ascending, then reflexed. **FL**: calyx tube 6–8.5 mm; petals cream, banner 15.3–19 mm, recurved ± 50°, keel 13.1–14.7 mm. **FR**: spreading or pendent; body 25–45 mm, 8–15 mm wide, inflated, compressed side-to-side, ± half-elliptic in side view, sparsely strigose, thinly papery, translucent; stalk-like base 3–11 mm, minutely, densely hairy, jointed at top; chamber 1; ovules 8–18. 2*n*=22. Arid grassland, scrub; 100–1200 m. s SNF, SnJV, SCoRI, WTR, SnGb. See note under *A. asymmetricus*. Poisonous to stock. Mar–Jun

A. pachypus Greene   Per, robust, rigid, bushy, sparsely lfy; hairs < 0.3 mm, appressed, ± scale-like, ± gray. **ST**: ± erect, 2–8 dm, wiry. **LF**: 2.5–16.5 cm; lflets 11–27, ± spaced, 3–34 mm, narrow. **INFL**: fls 4–28, spaced, ascending. **FL**: petals white or cream to lemon yellow, banner 15–22 mm, recurved ± 45°, keel 10.7–15.3 mm. **FR**: ascending or spreading; body 12–28 mm, 4–8 mm wide, straight or ± curved, compressed side-to-side, gen glabrous, stiff-leathery; stalk-like base 4–8 mm, stout; beak short, sharp, rigid, persistent; chambers 2.

1. Petals yellow, banner 15–17 mm;
  calyx tube 3.7–4.3 mm . . . . var. *jaegeri*
1′ Petals white when fresh, drying yellow,
  banner 15–22 mm; calyx tube 4–5.2 mm . . . . . . var. *pachypus*

  var. ***jaegeri*** Munz & McBurney   JAEGER'S BUSH MILKVETCH   **LF**: lflets 15–25. Rocky or sandy areas; 450–1200 m. n PR (incl SnJt), nw edge DSon. Dec–Jun   ★

  var. ***pachypus*** (p. 751)   BUSH MILKVETCH   **LF**: lflets 11–21. 2*n*=22. Open areas or scrub, often on gravelly clay, shale, sandstone; 500–1900 m. Teh, s SnJV, SCoRI, WTR, w edge D. San Diego Co. pls have smaller fls. Mar–Jul

A. palmeri A. Gray (p. 751)   Ann, per, low, open, widely branched; sparsely to densely silvery-strigose. **ST**: decumbent, 2–5 dm. **LF**: 2–16 cm; lflets 9–21, 5–25 mm, ± widely elliptic. **INFL**: fls gen 20–40, early dense, then ± spaced, spreading or widely ascending. **FL**: petals gen pink-purple, or cream with purple veins, petal tips, banner 7–10.3 mm, recurved 90°, keel 6.2–8.8 mm. **FR**: 10–25 mm, 5–14 mm wide, moderately to strongly inflated, ± ovate or half-ellipsoid, sparsely to densely strigose, papery but not esp thin and translucent; beak ± erect, 1/5–1/3 × body, triangular; chamber 1. Sandy or rocky places; 150–1650 m. PR, s edge DMoj, sw DMtns, w edge DSon; Baja CA. [*A. vaseyi* S. Watson] Dec–Jun

A. panamintensis E. Sheld. (p. 751)   Per, mat-forming, like an unkempt nest due to wiry, incurved branches, persistent petioles, silvery-canescent. **ST**: 1–15 cm, slender. **LF**: 1.5–12 cm; lflets 5–11, spaced, 2–14 mm, linear-elliptic, deciduous late, tips acute. **INFL**: fls 1–4, ascending, spaced. **FL**: petals pink-purple, banner 8–14 mm, recurved ± 45°, keel 7–9 mm. **FR**: spreading or ascending, 8–18 mm, 3–5 mm wide, ± oblong-elliptic in side view, blunt-triangular in ×-section, densely strigose, in age papery; chambers ± 2 in lower 1/2–2/3. Cracks, limestone ledges, pinyon/juniper woodland; 950–2150 m. n DMtns (Inyo Co.). Apr–Jun

A. pauperculus Greene (p. 751)   DEPAUPERATE MILKVETCH   Ann, delicate; finely strigose. **ST**: gen incurved-ascending, < 1 dm. **LF**: 1.5–5 cm; lflets 5–11, ± spaced, 2–8 mm, ± oblanceolate, tips notched or blunt. **INFL**: fls 2–7, spaced, gen reflexed in age. **FL**: petals purple, inner wing margins pale or white, banner 5.4–10.5 mm, recurved ± 40°, keel 4.3–5.8 mm. **FR**: ascending to reflexed, 12–20 mm, ± 3 mm wide, narrowly crescent-shaped, ± 3- or 4-sided exc at round base, glabrous to sparsely strigose, ± purple or ± purple-mottled; beak short; chambers ± 2. Open, vernally moist, volcanic clay; 40–1200 m. CaR, n SNF, n ScV. Mar–May   ★

A. platytropis A. Gray (p. 751)   BROAD-KEELED MILKVETCH   Per, cespitose; hairs dense, silvery or gray. **ST**: < 2 cm. **LF**: 1–9 cm; stipule sheaths ± present; lflets 5–15, 4–11 mm, elliptic to obovate, tips gen blunt. **INFL**: head-like; fls 4–9. **FL**: petals ± white or pale purple, banner 7.2–9.5 mm, recurved 30–45°, keel 7.8–8.6 mm. **FR**: ascending, 15–33 mm, 10–18 mm wide, bladdery, ± compressed top-to-bottom, strigose, valves purple-speckled, brown in age; chambers 2 by inflection of lower suture, which meets seed-bearing flange in middle of fr, so seeds along center of partition. Rocky hilltops, ridges, forest to above timberline; 2350–3500 m. c SNH, SNE, DMtns; to OR, MT, NV. Jul–Aug   ★

A. pomonensis M.E. Jones (p. 751)   Per, clumped, lfy. **ST**: ± ascending, 2.5–8 dm, stout, hollow; hairs 0. **LF**: 5–20 cm; lflets 25–41, 6–30 mm, ± elliptic, sparsely strigose abaxially, sometimes only on midrib. **INFL**: fls 10–45, spreading or reflexed. **FL**: petals cream, banner 11.1–15.3 mm, recurved ± 40°, keel 9.4–13.2 mm, keel and wing claws 5.1–6.9 mm, keel blade 4.6–6.9 mm, ≤ keel claw. **FR**: spreading or ascending, 18–45 mm, 10–20 mm wide, bladdery, ± ovate to half-ovate, sparsely strigose or ± glabrous, papery, ± transparent, deciduous; chamber 1; ovules 34–55, often ± 40. 2*n*=22. Shrubby, grassy, or disturbed areas; < 1200 m. s SCoRO, SCo, PR, w DSon; nw Baja CA. Like *A. douglasii* (lflets fewer; keel blade > keel claw). Mar–May

A. preussii A. Gray   Per, robust, ill-scented; ± glabrous. **ST**: ± erect, 1–3.5 dm. **LF**: 3.5–18 cm; lflets 7–25, ± spaced, 2–27 mm, linear to ± round, hairs 0 or only on margins, midribs. **INFL**: fls 4–22, ascending. **FL**: petals pink-purple or ± pale, banner 14–24 mm, recurved ± 40°, keel 11–19 mm. **FR**: erect or ascending; body 12–40 mm, 7–13 mm wide, inflated, oblong-ellipsoid, ± round in ×-section, glabrous or minutely hairy, stiff-papery; stalk-like base 0 or 2–7 mm; chamber 1.

1. Stalk-like fr base 0; infl open, axis in fr 4–23 cm;
  banner ± 14 mm . . . . var. *laxiflorus*
1′ Stalk-like fr base 2–7 mm; infl ± dense, axis in
  fr 1–9 cm; banner 17–24 mm . . . . var. *preussii*

  var. ***laxiflorus*** A. Gray (p. 751)   LANCASTER MILKVETCH   Alkaline flats; 700 m. sw DMoj, w DSon (Coachella Valley); s NV, sw UT, w AZ. Mar–May   ★

  var. ***preussii*** PREUSS' MILKVETCH   Clay flats, sandy washes; ± 750 m. e DMoj (se Inyo, ne San Bernardino cos.); to UT, AZ. Mar–May   ★

A. pseudiodanthus Barneby (p. 751)   TONOPAH MILKVETCH   Per, crown below surface; hairs soft, ± spreading, curved or curly. **ST**: ± prostrate, loose-matted, 2–3 dm. **LF**: 2.5–5 cm; lflets 7–19, ± crowded, 3–10 mm, ± obovate. **INFL**: fls 7–25, early crowded, spreading, then ± spaced, reflexed. **FL**: petals red-lilac, banner 9–10 mm, recurved ± 40°, keel 8.5–9.3 mm. **FR**: reflexed, 12–24 mm, 5–8 mm wide, curved > 1/2 circle, wider than deep, early fleshy, then leathery, hairs spreading; chambers 1 or ± 2. Sandy flats, dunes; 2050 m. SNE (e Mono Co.); w NV. Possibly an ecotype of *A. iodanthus* adapted to dunes. May–Jun   ★

A. pulsiferae A. Gray   Per, delicate, gen low, tufted; hairs fine, spreading or ± subappressed, ± gray. **ST**: ± prostrate, open, widely branched, 1–3 dm, slender. **LF**: 1–5.5 cm; stipule sheaths often present; lflets 3–13, crowded, 2–12 mm, ± obovate, tips gen notched. **INFL**: fls 3–13, reflexed in age. **FL**: corolla ± white, lavender-veined, keel tip lilac, banner 5.2–8.5 mm, recurved 90–100°, keel 3.4–5.3 mm. **FR**: 10–20 mm, 6–11 mm wide, ± spheric or half-ovate, bladdery, thinly papery, translucent, hairs ± sparse, long, wavy; chamber 1.

1. Crown at surface; st hairy; stipules distinct;
   fr hairy, hairs 1–1.7 mm . . . . var. ***coronensis***
1′ Crown below surface; st strigose to hairy;
   stipules distinct or lowermost fused around st;
   fr strigose to hairy, hairs 0.4–0.9 mm
   2. St, main lf axis, peduncle hairy, hairs ±
      spreading to ascending . . . . var. ***pulsiferae***
   2′ St, main lf axis, peduncle ± strigose, hairs
      ascending and subappressed . . . . var. ***suksdorfii***

var. ***coronensis*** S.L. Welsh et al. MODOC PLATEAU MILKVETCH
Sandy or gravelly soils, often with juniper, pines, sagebrush;
1300–1900 m. n SNH (Plumas Co.), e MP; w NV (Washoe Co.).
May–Jul ★

var. ***pulsiferae*** (p. 751) PULSIFER'S MILKVETCH 2*n*=24.
Sandy or rocky soil, often with pines, sagebrush; 1300–1900 m. n
SNH, e MP; NV. May–Aug ★

var. ***suksdorfii*** (Howell) Barneby SUKSDORF'S MILKVETCH
Loose, often rocky soil, often with pines, sagebrush; 1300–1600 m. s
CaRH; s WA. May–Jul ★

A. purshii Douglas PURSH'S MILKVETCH Per, sparsely to
densely cespitose; herbage hairs gen 1–2.3 mm, fine, cottony, dense,
tangled, silvery or gray. ST: 0–14 cm. LF: 1–15 cm; lflets 3–17, 2–20
mm, narrowly elliptic to ± round, tips blunt to notched. INFL: ±
among lvs; fls 1–11, ascending. FL: petals white, cream, pink-purple,
or purple, banner 9–26 mm, recurved ± 40°, keel 8–21.2 mm. FR:
ascending, 7–27 mm, 4–13 mm wide, ovate or widely lanceolate in
side view, hairs gen dense, gen 1.5–5 mm, all ± wavy or all straight,
white (fr resembling a cotton ball); chambers 1 or 2. Locally and
regionally variable. Like *A. newberryi* var. *n.* (longer, ± straight,
partly spreading hairs).

1. Calyx < 10 mm; fr 7–17 mm
   2. Fr incurved at least 1/2 circle; ne MP . . . . var. ***lagopinus***
   2′ Fr incurved near beak; SNH, SnBr,
      perhaps w edge SNE . . . . var. ***lectulus***
1′ Calyx > 10 mm; fr 13–27 mm
   3. Petals white or cream, tip pale lilac or not;
      infl 1–6-fld . . . . var. ***purshii***
   3′ Petals pink-purple or purple; infl 3–11 fld . . . . . . var. ***tinctus***

var. ***lagopinus*** (Rydb.) Barneby (p. 751) ST: 0–8 cm. LF: 1–7
cm; lflets 3–11, 5–15 mm. INFL: fls 2–7. FL: calyx 5.5–9 mm; petals
pink-purple, tips pale purple or not, banner 9–13.2 mm, keel 8–11.3
mm. FR: 7–17 mm, 3.8–7 mm wide, incurved at least 1/2 circle;
chamber 1; ovules 14–20. Dry plains, slopes, often on basalt or pumice, often with sagebrush; 1200–1850 m. ne MP; OR, NV. May–Jun

var. ***lectulus*** (S. Watson) M.E. Jones ST: 0–10 cm. LF: 1–5 cm;
lflets 3–11, 2–10 mm. INFL: fls 1–5. FL: calyx 5.6–8.8 mm; petals
pink or pale purple, banner 10.3–15 mm, keel 9.4–11.7 mm. FR:
7.5–15 mm, 4–8 mm wide, incurved near beak; chamber 1; ovules
24–32. Dry, open flats, slopes, often with juniper, pines, rocky slopes
above timberline; 1500–3650 m. SNH, SnGb, SnBr, w edge SNE.
May–Aug

var. ***purshii*** ST: 0–1 cm. LF: 1.5–15 cm; lflets 5–17, 2–10 mm.
INFL: fls 1–6. FL: calyx 12–19 mm; petals white or cream, tips pale
lilac or not, banner 19–26 mm, keel 15–21.2 mm. FR: 13–27 mm,
5–13 mm wide, straight or curved near tip; chamber 1; ovules 20–34.
2*n*=22. Dry flats, often in sagebrush; 1050–2100 m. MP; to sw Can,
ND, CO. Apr–Jun

var. ***tinctus*** M.E. Jones (p. 751) ST: 0–10 cm. LF: 2–11 cm;
lflets 3–17, 2–14 mm. INFL: fls 3–11. FL: calyx 12–19 mm; petals
pink-purple or purple, banner 14.6–25 mm, keel 11.5–20.8 mm. FR:
13–27 mm, 5–13 mm wide; chambers often ± 2 (esp in DMoj); ovules
18–46. Gravelly, sandy flats, slopes, often with pines or sagebrush;
450–3000 m. KR, NCoRH, CaRH, SNH, Teh, SCoRI, WTR, SnBr,
GB, DMoj; OR, NV. Apr–Jun

A. pycnostachyus A. Gray Per, stout, clumped, lfy; gen canescent or white-woolly. ST: ± erect, 4–9 dm, hollow, sparsely hairy
esp near base, ± red. LF: 3–15 cm; stipule sheaths present; lflets
23–41, crowded, 5–30 mm, narrow. INFL: fls many, dense, overlapped, reflexed. FL: petals green-white or cream, banner 7–10 mm,
recurved ± 35°, keel 7.1–9.1 mm. FR: reflexed, 6–11 mm, 3.5–6 mm
wide, ovate, ± inflated, papery, persistent, hairs 0 or sparse; stalk-like
base 0; beak 5–8 mm, stiff, slender, hooked; chamber 1.

1. Peduncle 2–4 cm; calyx tube 3–4 mm; s SCo
   . . . . var. ***lanosissimus***
1′ Peduncle 4–10 cm; calyx tube 3.5–5 mm;
   NCo, n CCo . . . . var. ***pycnostachyus***

var. ***lanosissimus*** (Rydb.) Munz & McBurney (p. 751)
VENTURA MARSH MILKVETCH Pl white-woolly. FR: gen sparsely
strigose; ovules 8–12. Disturbed areas, open, sand to gravel; < 100
m. c SCo. Sent ± to extinction by urbanization, rediscovered in 1997.
Jul–Oct ★

var. ***pycnostachyus*** COASTAL MARSH MILKVETCH Pl canescent. FR: glabrous; ovules 2–5. Coastal marshes, seeps, adjacent
sand; < 150 m. NCo, n CCo. Jun–Sep ★

A. rattanii A. Gray Ann; sparsely minutely strigose, hairs often
± black above. ST: decumbent to erect, 4–30 cm, gen slender. LF:
1.5–5 cm; lflets 5–13, 2–12 mm, narrowly to widely ± obovate, tips
gen blunt or notched. INFL: head-like; fls 2–10, spreading. FL: petals pink-purple, wing tips paler, banner 7.2–12 mm, recurved ± 40°,
keel 3.8–8.1 mm. FR: ascending, 15–57 mm, 2–3 mm wide, narrowly
linear, ± round in ×-section, tapered to a narrow, sharp beak, finely
strigose, papery; chambers ± 2; ovules 8–20. SEED: pitted.

1. Calyx 2.5–3.4 mm, tube 2–2.5 mm;
   fr 1.5–3 cm . . . . var. ***jepsonianus***
1′ Calyx 3.7–5 mm, tube 2.6–3.5 mm;
   fr 2.1–5.7 cm . . . . var. ***rattanii***

var. ***jepsonianus*** Barneby (p. 751) JEPSON'S MILKVETCH LF:
lflets gen 7–9. INFL: fls 4–9. FL: petals gen white exc ± purple banner tip, keel. Grasslands, grassy openings in woodland and chaparral,
vertic clay, often serpentine; 150–700 m. s NCoRI. Apr–Jun ★

var. ***rattanii*** (p. 751) RATTAN'S MILKVETCH LF: lflets 5–13.
INFL: fls 2–10. FL: petals pink-purple, claws, wing tips paler (white).
Often riverbanks, sandbars; 50–1500 m. n&c NCoR. Apr–Jul ★

A. ravenii Barneby (p. 751) RAVEN'S MILKVETCH Per, delicate,
open, widely branched; minutely strigose. ST: prostrate, 1–10 cm,
weak, slender, buried 1–6 cm. LF: 0.5–3 cm; stipule sheaths present;
lflets 7–13, 1–4 mm, ovate to ± round, tips notched. INFL: fls 2–8,
spreading. FL: petals ± white, veins faint lilac, keel tip lilac, banner
5.5–8.4 mm, recurved ± 85°, keel 4.5–5.5 mm. FR: ascending, 8–17
mm, 5–8 mm wide, ± incurved, swollen, sparsely strigose, papery;
chamber 1. 2*n*=22. Gravel; 3400–3450 m. s SNH (Fresno, Inyo cos.).
Closely related to *A. monoensis*. Jun–Sep ★

A. sabulonum A. Gray (p. 751) Ann, low, small or coarse, lfy;
hairs ± dense, ascending or spreading, ± wavy. ST: erect or decumbent,
2–26 cm. LF: 1.5–6.5 cm; lflets 5–15, 2–13 mm, oblanceolate, tips
blunt, ± notched. INFL: fls 2–7, spaced, spreading or ascending. FL:
petals dingy cream, lilac-tinged, banner 5.2–7.2 mm, recurved 50–70°,
keel 5–6.5 mm. FR: 9–20 mm, 5–11 mm wide, ± ovate, incurved,
often abruptly so near tip, leathery, hairs ± dense, stiff, spreading,
wavy; chamber 1. 2*n*=24. Sand, gravel; -50–900 m. D; to UT, NM, n
Mex. ≤ 2000 m outside CA. Feb–Jul (Nov–Apr in se CA) ★

A. sepultipes (Barneby) Barneby (p. 751) BISHOP MILKVETCH
Per, open, widely branched, loose-clumped; hairs gen ± dense, ±
appressed, silvery- or gray-silky. ST: ascending, 1.5–3.5 dm, ± glabrous, buried 1–5 cm. LF: 2–8 cm; stipule sheaths present; lflets
7–17, 3–14 mm, obovate, tips notched or ± blunt. INFL: fls 10–30,
spreading. FL: petals pale pink- or lilac-white, wing tips ± white,
banner 12.7–17.5 mm, recurved ± 40°, keel 10–12.2 mm. FR: spreading to reflexed, 15–20 mm, 3–6 mm wide, ± incurved, ± 3-sided, stiff-
papery, hairs fine, shaggy; chambers 2; ovules 14–20. Dry, granitic
sand, in sagebrush in pinyon forest; 1450–2000 m. e edge c&s SNH,
SNE (Inyo Co.). May–Jul

A. serenoi (Kuntze) E. Sheld. var. **shockleyi** (M.E. Jones) Barneby
(p. 751) SHOCKLEY'S MILKVETCH Per, bushy-clumped; sparsely
minutely strigose, often ± gray. ST: 1.5–4.5 dm. LF: 5–15 cm; lflets
5–11, spaced, 5–30 mm, narrow, adaxially more densely strigose,

paler than abaxially, terminal lflet not or ± jointed to midrib. **INFL:** fls 3–25, spaced, ascending. **FL:** petals strongly graduated, ± purple, wing tips pale, banner 17–26 mm, recurved 35–40°, keel 12.4–18.5 mm. **FR:** erect, 17–31 mm, 7–12 mm wide, plump-oblong, leathery, glabrous; beak stout, sharp; chambers ± 2. 2*n*=22(24?). Open, dry, alkaline gravelly clay, gen sagebrush, pinyon; 1150–2300 m. SNE, n DMtns (Inyo Co.); NV. May–Jun ★

A. shevockii Barneby (p. 751) SHEVOCK'S MILKVETCH Per, wiry, open; hairs sparse. **ST:** decumbent to ascending, 1–3.5 dm, slender; hairs at base spreading, above appressed. **LF:** 2.5–6.5 cm; stipule sheaths present; lflets 9–17, spaced, 2–10 mm, ± elliptic. **INFL:** fls 2–13, spaced, ascending. **FL:** petals ± cream, banner 9–9.8 mm, recurved ± 75°, keel 9–9.8 mm. **FR:** ascending, 13–31 mm, 3–4 mm wide, tapered at ends, incurved, ± 3-sided proximally, hairy, papery; upper suture a low keel, lower a keel near tip, depressed in a wide groove in lower 1/2; chamber 1. Granitic sand, Jeffrey-pine forest; ± 1900 m. s SNH (se Tulare Co.). Jun–Aug ★

A. subvestitus (Jeps.) Barneby (p. 751) KERN COUNTY MILKVETCH Per, cespitose or matted, lfy; herbage hairs dense, gen 0.7–1.2 mm, woolly, ± curly, ± gray. **ST:** prostrate, 1–8 cm. **LF:** 1.5–6.5 cm; lflets 7–13, 2–9 mm, ± elliptic or obovate. **INFL:** among lvs; fls 3–8. **FL:** petals ± white, with some lilac, keel tip pink, banner 11–13.3 mm, recurved ± 40°, keel 9.5–10.7 mm. **FR:** ascending, 8–15 mm, 4–7 mm wide, ± ovate, compressed top-to-bottom in lower 1/2, stiff-papery; hairs dense, gen < 1 mm, curly; beak erect or incurved, triangular, compressed side-to-side; chamber 1; ovules 11–14. Gravel, sand, in sagebrush; 1500–2650 m. s SNH (Tulare, Kern cos.). Jun–Jul ★

A. tener A. Gray Ann, delicate; ± sparsely strigose to ± glabrous. **ST:** erect or ascending, 2–30 cm. **LF:** 2–9 cm; lflets 7–17, ± spaced, 3–16 mm, lanceolate to obovate, glabrous adaxially, tip notched or pointed. **INFL:** dense; fls 2–12, spreading. **FL:** petals pink-purple, banner 5.2–11.8 mm, recurved 35–40°, keel 3.4–6.4 mm. **FR:** reflexed, 6–50 mm, 1.7–3.5 mm wide, ± narrowly lanceolate, straight or curved, openly grooved on lower side, glabrous, stiff-papery; base ± stalk-like or not; chambers ± 2. **SEED:** smooth.

1. Banner 5.2–6 mm; ovules 5–11; gen coastal . . . . var. *titi*
1′ Banner 7.8–11.8 mm; ovules 8–16; gen inland
  2. Fr 2.7–5 cm, incurved, base ± stalk-like, 3–5 mm
    . . . . var. *ferrisiae*
  2′ Fr 1–2.5 cm, ± incurved, base not stalk-like . . . . . . var. *tener*

var. *ferrisiae* Liston (p. 755) FERRIS' MILKVETCH **ST:** 6–26 cm. **LF:** 2–6 cm; lflets 7–15. **INFL:** fls 3–12. **FL:** banner 7.8–9.6 mm, keel 4.2–5.1 mm. **FR:** 2.7–5 cm, incurved; base ± stalk-like, 3–5 mm; ovules 10–16. Alkaline flats, vernally moist meadows; < 60 m. n ScV. Mar–Jun ★

var. *tener* (p. 755) ALKALI MILKVETCH **ST:** 4–30 cm. **LF:** 2–9 cm; lflets 7–17. **INFL:** fls 3–12. **FL:** banner 8.2–11.8 mm, keel 4.7–6.4 mm. **FR:** 10–25 mm, ± incurved; base not stalk-like; ovules 8–14. Alkaline flats, vernally moist meadows; < 60 m. s ScV, n SnJV, e SnFrB (where mostly extirpated), SCoRI. Mar–Jun ★

var. *titi* (Eastw.) Barneby COASTAL DUNES MILKVETCH **ST:** 2–12 cm. **LF:** 2–7 cm; lflets 7–13. **INFL:** fls 2–7. **FL:** banner 5.2–6 mm, keel 3.4–3.9 mm. **FR:** 6–14 mm, straight or outcurved; base not stalk-like; ovules 5–11. Moist sandy depressions (vernal pool) near coast, coastal bluffs, dunes; < 20 m. c CCo (Monterey Co.), SCo (extirpated?). Threatened by urbanization, recreation, alien pls. Mar–Jun ★

A. tidestromii (Rydb.) Clokey (p. 755) TIDESTROM'S MILKVETCH Per, or fl 1st yr, appearing ann, tufted; hairs dense, ± stiff, often curly, tangled, shaggy. **ST:** 0 or ± prostrate, < 7 cm. **LF:** 3–15 cm; lflets 7–19, 4–17 mm, widely obovate. **INFL:** fls 5–16, spaced, spreading-ascending. **FL:** petals ± white, dull purple-tinged, wings, keel tips dark purple, banner 12–17.7 mm, recurved ± 40°, keel 10–12 mm. **FR:** ascending, 15–55 mm, 6–16 mm wide, ± lanceolate, curved 1/4–5/4 circle, ± 4-sided, compressed side-to-side at ends, top-to-bottom in middle, minutely strigose, stiff-leathery; beak long, narrowly triangular; chamber 1. Open, calcareous gravel, foothills; 600–1600 m. DMoj (Cushenbury Canyon), e-c DMtns; s NV. Mar–May ★

A. traskiae Eastw. (p. 755) TRASK'S MILKVETCH Per, open, widely branched, lfy; hairs dense, ± gray. **ST:** several ± ascending, 1–4 dm. **LF:** 4–10 cm; lflets 21–29, 5–15 mm, ± obovate or ovate. **INFL:** fls 12–30, spaced, ascending or spreading. **FL:** petals cream, banner 14.2–17.5 mm, recurved 30–40°, keel 10.2–12.9 mm. **FR:** pendent or ascending; body 8–16 mm, 3–5.5 mm wide, ± half-elliptic in side view, ± 3-sided (lower grooved), thinly fleshy, then leathery; hairs minute, wavy; stalk-like base 4–9 mm; chambers ± 2. Sandy coastal bluffs, dunes; gen < 200 m. s ChI (San Nicolas, Santa Barbara islands). Mar–Jul ★

A. tricarinatus A. Gray (p. 755) TRIPLE-RIBBED MILKVETCH Per, loose-tufted; finely strigose. **ST:** ± erect, 5–25 cm, stiff. **LF:** 7–20 cm; lflets 17–27, spaced, 3–12 mm, ± narrowly obovate, silvery-strigose adaxially, sparsely hairy, ± green abaxially. **INFL:** fls 5–15, spaced, spreading-ascending. **FL:** calyx 6.1–7.6 mm, mixed black, white hairs, tube 4.1–5 mm; petals cream, banner 12.6–15.7 mm, recurved ± 45°, keel 9.7–11 mm. **FR:** ascending; body 24–42 mm, 3.5–5.5 mm wide, ± linear, 3-sided, glabrous, thinly papery; stalk-like base 1–2.5 mm, stout, jointed at top; upper suture a narrow ridge; chambers 2. **SEED:** 20–24. Exposed rocky slopes, canyon walls along desert washes; 450–1250 m. e SnBr (Whitewater, Morongo Valley), adjacent edges D. Feb–May ★

A. trichopodus (Nutt.) A. Gray Per, robust, bushy-branched; ± minutely strigose, gen also spreading-hairy. **ST:** ± erect, 2–10 dm, base hollow. **LF:** 2.5–20 cm, gen subsessile; stipule sheaths rarely present; lflets 15–39, 2–25 mm, ± lanceolate. **INFL:** fls 10–50, early spreading, then reflexed. **FL:** calyx 5–8.7 mm, tube 3.6–5.4 mm; petals ± cream, faintly lilac-blushed or -lined or not, banner 11.3–19 mm, recurved 40–45°, keel 8.6–13.7 mm. **FR:** dehiscent from tip, along top suture; pendent, persistent; body 13–45 mm, 4.8–21 mm wide, linear-elliptic, compressed side-to-side to ovate, bladdery, thinly papery, hairs sparse, minute; stalk-like base 5–17 mm, slender, strigose or ± glabrous, not jointed; chamber 1; ovules 10–30. See note under *A. asymmetricus*.

1. Fr compressed side-to-side, not bladdery,
  sides ± flat or ± convex . . . . var. *phoxus*
1′ Fr bladdery, ± not compressed side-to-side, sides convex
  2. Fr 8–21 mm wide, hairy n of San Diego, sometimes
    glabrous s, upper suture much less convex than lower
    . . . . var. *lonchus*
  2′ Fr 5–13 mm wide, glabrous, upper suture ± as
    convex as lower . . . . var. *trichopodus*

var. *lonchus* (M.E. Jones) Barneby (p. 755) **INFL:** fls 12–36. **FL:** banner 11.3–19 mm, keel 8.6–13 mm. **FR:** body 17–45 mm, 8–21 mm wide, ovate to half-ellipsoid, bladdery, with convex sides, hairy n of San Diego, sometimes glabrous s; upper suture much less convex than lower. 2*n*=22. Coastal bluffs, fields; < 1100 m. CCo, SCo, n ChI, w PR, sw DMoj; Baja CA. Feb–Jun

var. *phoxus* (M.E. Jones) Barneby ANTISELL MILKVETCH **INFL:** fls 10–50. **FL:** banner 11.4–16.7 mm, keel 9.3–12.7 mm. **FR:** body 15–36 mm, 5–9 mm wide, compressed side-to-side, with ± flat or ± convex sides, stalk glabrous (minutely strigose). Gen inland, grassy or shrubby hillsides, washes, openings in chaparral; < 3050 m. s SNH, Teh, s SnJV, s SCoR, n SCo (Santa Barbara, Ventura cos.), WTR, PR, w edge DMoj. Feb–Jun

var. *trichopodus* SANTA BARBARA MILKVETCH **INFL:** fls 10–50. **FL:** banner 11.5–15.4 mm, keel 9.5–11.5 mm. **FR:** body 13–35 mm, 5–13 mm wide, bladdery, with convex sides, gen glabrous; upper suture ± as convex as lower. Coastal bluffs, low grassy hills; < 350 m. SCo, n PR. Mar–Jun

A. umbraticus E. Sheld. (p. 755) BALD MOUNTAIN MILKVETCH Per, open, widely branched; ± glabrous. **ST:** several, diffuse or not, ascending or not, 2–5 dm. **LF:** 4–12 cm, ± petioled; lflets 11–23, 4–20 mm, widely oblong to ± round, tips notched. **INFL:** fls 10–25, ± spreading. **FL:** petals green-white, banner 10–14 mm, recurved 40–60°, keel 7.7–10 mm. **FR:** spreading or reflexed; body 14–24 mm, 2.5–3.5 mm wide, lance-linear, incurved 1/4–1/2 circle, glabrous, firm-papery, turning black; stalk-like base 0.8–1.9 mm; chambers 2; ovules 10–15. Dry, open woodland; 200–1250 m. w KR, n NCoRO (Humboldt Co.); w OR. May–Jul ★

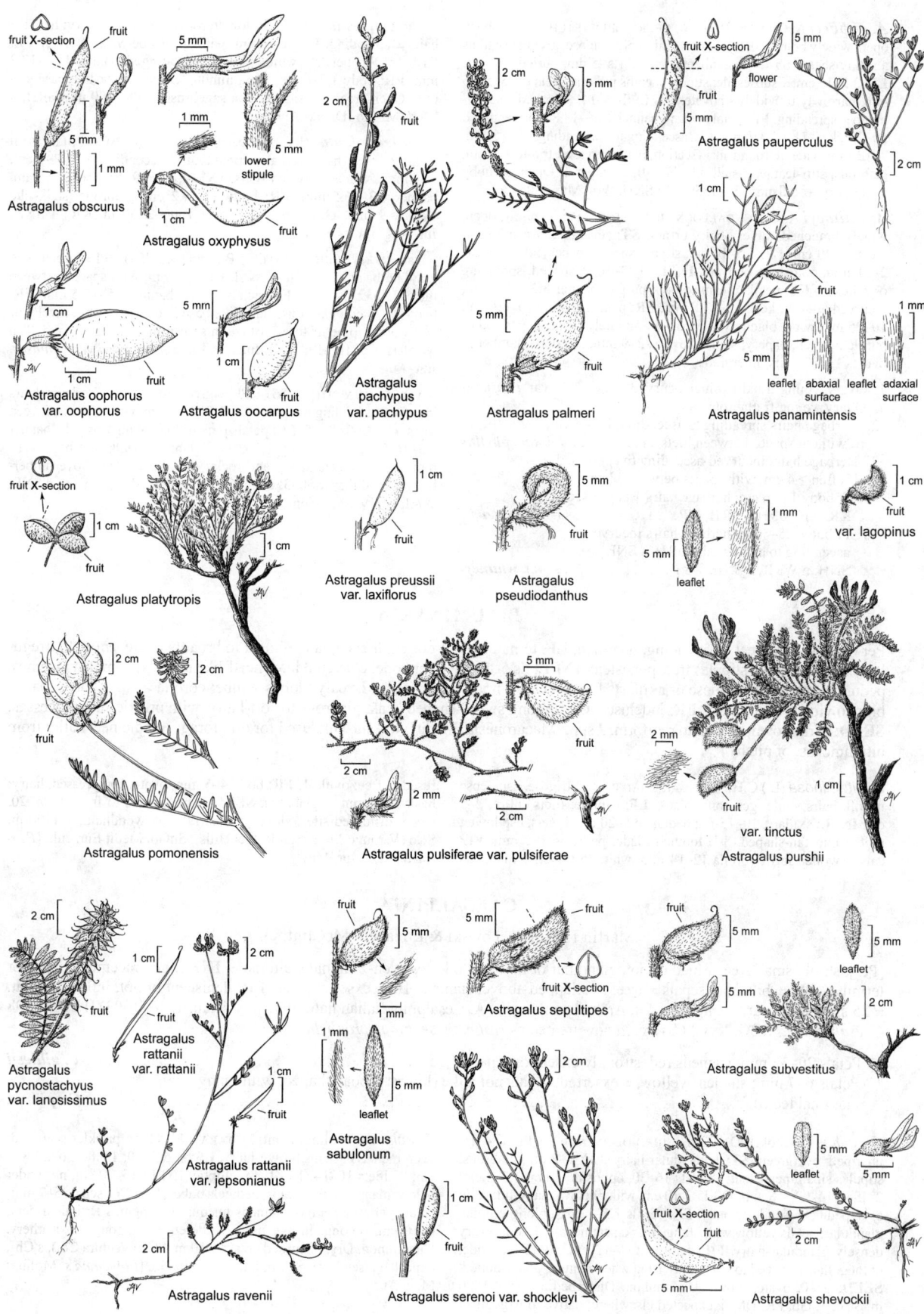

fruit X-section
fruit
5 mm
1 mm
Astragalus obscurus

5 mm
1 mm
lower stipule
5 mm
1 cm
fruit
Astragalus oxyphysus

1 cm
1 cm
fruit
Astragalus oophorus var. oophorus

5 mm
1 cm
fruit
Astragalus oocarpus

2 cm
fruit
Astragalus pachypus var. pachypus

2 cm
fruit
5 mm
5 mm
fruit
Astragalus palmeri

fruit X-section
flower
fruit
5 mm
5 mm
5 mm
2 cm
Astragalus pauperculus

1 cm
fruit
5 mm
leaflet abaxial surface
leaflet adaxial surface
1 mm
Astragalus panamintensis

fruit X-section
fruit
1 cm
Astragalus platytropis

1 cm
fruit
Astragalus preussii var. laxiflorus

5 mm
fruit
Astragalus pseudiodanthus

1 cm
fruit
var. lagopinus

5 mm
leaflet
1 mm

2 cm
2 cm
fruit
Astragalus pomonensis

5 mm
fruit
2 cm
2 mm
2 cm
Astragalus pulsiferae var. pulsiferae

2 mm
fruit
1 cm
var. tinctus
Astragalus purshii

2 cm
1 cm
fruit
Astragalus pycnostachyus var. lanosissimus

fruit
Astragalus rattanii var. rattanii

1 cm
fruit
Astragalus rattanii var. jepsonianus

2 cm
Astragalus ravenii

fruit
5 mm
1 mm
Astragalus sabulonum
1 mm
leaflet
5 mm

fruit
5 mm
fruit X-section
Astragalus sepultipes

2 cm
1 cm
fruit
Astragalus serenoi var. shockleyi

fruit
5 mm
leaflet
5 mm
5 mm
2 cm
Astragalus subvestitus

fruit
5 mm
leaflet
5 mm
5 mm
fruit X-section
fruit
5 mm
2 cm
Astragalus shevockii

A. webberi A. Gray (p. 755) WEBBER'S MILKVETCH Per, robust, open, widely branched, lfy; densely minutely strigose, giving dry lflets a satiny sheen. **ST**: ± decumbent, slender 1.5–5 dm, buried 0–6 cm. **LF**: 2.5–15 cm, ± subsessile; stipule sheaths present; lflets 9–25, 5–35 mm, narrowly to widely obovate. **INFL**: fls 6–14, ± spaced, ascending to ± spreading. **FL**: petals cream, banner 15.4–18.8 mm, recurved 50–75°, keel 11.4–13.7 mm. **FR**: ascending to ± spreading, 20–35 mm, 7–12 mm wide, ± round in ×-section, ± compressed top-to-bottom, glabrous, stiff-leathery; stalk-like base 0; chamber 1. Open, shrubby slopes, dry woodland; 850–1550 m. n SNH, Wrn. May–Jul ★

A. whitneyi A. Gray BALLOON MILKVETCH Per, gen low, open, widely branched, silvery-hairy or not. **ST**: several, ± ascending or erect, 4–40 cm. **LF**: 1.5–10 cm; stipule sheaths papery; lflets 5–21, 2–21 mm, oblong to obovate. **INFL**: fls 3–16, ± spaced, spreading or reflexed. **FL**: petals cream to pink-purple, banner 8.3–17.2 mm, recurved 50–80°, keel 7.3–13.8 mm. **FR**: pendent; body 15–60 mm, 10–25 mm wide, bladdery, glabrous to minutely strigose, gen red or purple mottled, papery; stalk-like base 2–9 mm, slender; chamber 1; ovules 13–37. Vars. intergrade.

1. Ovary, young fr hairs minute, curved . . . . var. ***confusus***
1′ Ovary, young fr glabrous
  2. Herbage hairs spreading to ascending; lf 2–4 cm,
    ± without spaces between lflets . . . . var. ***lenophyllus***
  2′ Herbage hairs incurved-ascending to appressed;
    lf often > 4 cm, with spaces between lflets
    3. Fr body 1.5–6 cm; herbage hairs ± appressed;
      KR, n NCoRH, CaRH, MP . . . . var. ***siskiyouensis***
    3′ Fr body 1.5–3 cm; herbage hairs incurved-
      ascending to appressed; CaRH, s SNF,
      SNH, n WTR, n SNE, W&I . . . . var. ***whitneyi***

var. ***confusus*** Barneby Pl low to robust, gray or silvery; herbage hairs ± spreading. **LF**: 2–10 cm, with spaces between lflets. **INFL**: fls 5–16. **FL**: petals ± white, pink- or lilac-tinged, banner 12.8–17.2 mm. **FR**: body 1.7–6 cm, hairs minute, curved; stalk-like base 4–9 mm. Open areas, hillsides, often sagebrush; 800–2300 m. CaRH, n SNH, MP; to ID, NV. Apr–Jul

var. ***lenophyllus*** (Rydb.) Barneby WOOLLY-LEAVED MILKVETCH Pl gray-green; herbage hairs spreading to ascending. **LF**: 2–4 cm, ± without spaces between lflets. **INFL**: fls 5–9. **FL**: petals cream, banner 8.3–16.5 mm. **FR**: body 1.5–4.2 cm, glabrous; stalk-like base 3–9 mm. Open, rocky places; 1500–2750 m. KR, n SNH. Jul–Aug ★

var. ***siskiyouensis*** (Rydb.) Barneby (p. 755) Pl ± green or ± gray; herbage hairs ± appressed. **LF**: 3–9 cm, with spaces between lflets. **INFL**: fls 4–16. **FL**: petals cream, banner 9.5–13.5 mm. **FR**: body 1.5–6 cm, glabrous; stalk-like base 3–6 mm. Open, gravelly or rocky areas, pine/fir forest, often on serpentine; 750–2750 m. KR, n NCoRH, CaRH, MP; s OR. Intergrades to *A. whitneyi* var. *confusus*. Jun–Aug

var. ***whitneyi*** (p. 755) Pl ± green to ± gray; herbage hairs incurved-ascending to appressed. **LF**: 3–11 cm, with spaces between lflets. **INFL**: fls 3–15. **FL**: petals pink or lilac, wing tips pale, banner 8.3–16.5 mm. **FR**: body 1.5–3 cm, glabrous; stalk-like base 3–15 mm. $2n=22$. Open, sandy and rocky slopes esp at or above timberline, to foothills; 1550–3500 m. CaRH, s SNF, SNH, n WTR, n SNE, W&I; w NV. May–Sep

BITUMINARIA

Per to subshrub, erect to spreading, unarmed. **LF**: ternate, gland-dotted; lflets 3, lance-linear to broadly ovate, entire to irregularly finely dentate; stipules free, persistent. **INFL**: fls 5–15, 2–3 per node, clustered into [head-like umbels] compact racemes, peduncles elongate, scapose or axillary; bracts at each node gen united. **FL**: calyx lobes > tube; corolla > calyx lobes, partly blue to violet; ovary hairy. **FR**: indehiscent, with long sword-shaped beak, glabrous to soft-hairy, with prickle-like processes. **SEED**: dark brown, oblong to reniform. 2 spp.: Macaronesia, Medit, sw Asia, cult, used for orn, forage, medicine. (Latin: from bituminosus, of pitch)

B. bituminosa (L.) C.H. Stirt. ARABIAN PEA Strigose, esp dense in infl, hairs white, gen some black. **LF**: cauline, lflets entire, 2–5 cm. **INFL**: axillary, 10–15-fld; peduncle 5–20 cm; bracts, esp lower, united into fan-shaped, 3–5 toothed blade; pedicels 1–2 mm. **FL**: calyx swollen in fr; corolla 12–14 mm, white to violet; 9 filaments fused, 1 free; ovule 1. **FR**: body 4–5 mm, ovate, compressed, hairy; beak 7–10 mm, ± glabrous. **SEED**: 4 mm, fused to fr wall. $2n=20$. Open, disturbed sites, slopes, chaparral, oak woodland; < 1100 m. SCo (Verdugo Mtns, San Rafael Hills), SnGb; Medit Eur, cult. [*Psoralea b.* L.] Apr–May

CAESALPINIA

Martin F. Wojciechowski & Elizabeth McClintock

[Per], shrub, small tree, armed or not, glandular or not. **LF**: odd- or even-2-pinnate, alternate. **INFL**: gen raceme, axillary or terminal. **FL**: ± bilateral; sepals ± free, overlapped above; stamens 10, ± exserted, free. **FR**: dehiscent or not, inflated or flat. ± 25 spp.: trop, warm temp Am, Afr, Arabia; some cult. (A. Cesalpino, Italian naturalist, physician, 1519–1603) [Lewis 1998 *Caesalpinia* .... RBG, Kew] *C. virgata* now treated as *Hoffmannseggia microphylla*.

1. Petals 20–35 mm; stamens red, strongly exserted; sepals alike . . . . ***C. gilliesii***
1′ Petals 6–7 mm; stamens yellow, ± exserted; sepals not alike (lowest 1 boat-shaped, with many long,
  marginal teeth). . . . . ***C. spinosa***

C. gilliesii (Hook.) D. Dietr. BIRD-OF-PARADISE Pl < 4 m, unarmed; evergreen. **ST**: ± glandular-hairy. **LF**: 1–2 dm, glabrous; stipules small, persistent; 1° lflets 16–30, opposite or not, 1.5–3 cm; 2° lflets 14–22, < 8 mm. **INFL**: < 10 cm, wider below, many-fld; main axis, pedicels glandular-hairy. **FL**: sepals 1.5–2 cm, oblong-elliptic, glandular; petals yellow with orange marks; stamens 8–10 cm; ovary densely glandular-hairy. **FR**: dehiscent, 6–12 cm, 1.9–2 cm wide, oblong, flat, ± curved to straight, twisted when mature, gland-dotted. **SEED**: 6–10, ovate, brown. Uncommon. Disturbed areas; < 1000 m. SCo, SnGb, PR, DMoj, expected elsewhere; native to Argentina, Uruguay, widely cult in arid temp, trop. Fr, seeds TOXIC. May–Aug

C. spinosa (Molina) Kuntze TARA Pl < 2 m; prickles scattered; evergreen. **ST**: twigs brown-hairy. **LF**: stipules 0; 1° lflets 6–20, < 8 cm; 2° lflets 10–14, 1.5–4 cm, elliptic. **INFL**: 15–20 cm, not wider below, many-fld; main axis, pedicels puberulent. **FL**: sepals 5–7 mm; petals yellow to ± red; stamens unequal in length. **FR**: indehiscent, 6–10 cm, oblong, in age more withered, dry, spongy to leathery. Uncommon. Dry, disturbed areas; < 50 m. SCo (Ventura Co.), s ChI, expected elsewhere; S.Am, introduced in US. [*Poinciana s.* Molina] May–Aug

CALLIANDRA FAIRY-DUSTER

Martin F. Wojciechowski & Elizabeth McClintock

Per, shrub, [tree], unarmed. **LF**: even-2-pinnate, alternate. **INFL**: head, axillary, few-fld. **FL**: radial, pink [purple-red to white]; sepals, petals inconspicuous; stamens many, strongly exserted, filaments fused basally. **FR**: dehiscent, valves recurving. ± 135 spp.: sw US to Uruguay. (Greek: beautiful stamens)

C. eriophylla Benth. (p. 755)   PINK FAIRY-DUSTER   Shrub < 60 cm; branches many, spreading from base. **LF**: deciduous; 1° lflets gen 4–8; 2° lflets gen 14–18, ± 2–3.5 mm. **FL**: stamens 18–22 mm. **FR**: erect, 4–8 cm, flat. **SEED**: 3–8. Sandy washes, slopes, mesas; ± 1500 m. DSon; AZ to TX, n Mex. Feb–Apr, after Sep–Oct rain   ★

CERCIS REDBUD

Martin F. Wojciechowski & Elizabeth McClintock

Shrub, tree, unarmed; deciduous. **LF**: simple, alternate, cordate to reniform, ± leathery, glabrous. **INFL**: umbel-like, axillary on short spur or ± sessile on woody branches. **FL**: bilateral, gen appearing before lvs; sepals fused at base; petals pink-purple, banner inside wings in bud, keel petals free; stamens 10, gen incl, free. **FR**: dehiscent, oblong, flat. **SEED**: 2–8. 10 spp.: n hemisphere. (Greek: applied perhaps to a poplar, but also to *Cercis siliquastrum* L., Judas tree)

C. occidentalis A. Gray (p. 755)   WESTERN REDBUD   Shrub, tree < 7 m, glabrous. **LF**: < 10 cm; petiole 15–20 mm. **INFL**: 2–5-fld. **FL**: keel 12–13 mm, > wings, banner. **FR**: 5–8 cm. Dry, shrubby slopes, canyons, streambanks, chaparral, foothill woodland, yellow-pine forest; 100–1500 m. NW, CaR, SN, GV, SnFrB, SCoRO, SnGb, SnBr, PR, MP; to sw OR, UT, TX. [*C. orbiculata* Greene] Mar–May

COLUTEA BLADDER SENNA

Martin F. Wojciechowski, Duane Isely & Elizabeth McClintock

Shrub, unarmed [armed]. **LF**: odd-1-pinnate, alternate; lflets 5–13. **INFL**: raceme, axillary, few-fld. **FL**: bilateral; sepals 5; petals 5, gen yellow [to ± red]; 9 filaments fused, 1 free. **FR**: indehiscent, ellipsoid, inflated, bladder-like, base stalk-like, walls papery. **SEED**: many. ± 28 spp.: c&s Eur, Medit, n Afr to w China. (Greek: ancient name) Lvs with cathartic properties.

C. arborescens L. (p. 755)   Pl < 3 m, rounded; branches many; deciduous;. **LF**: lflets 9–13, 1.5–3 cm, elliptic, sparsely hairy. **FL**: corolla 1.5–2 cm, yellow with dark marks. **FR**: body 2–3 cm, glabrous, stalk-like base 5–9 mm. 2*n*=16. Uncommon. Roadsides, woodland, disturbed areas; < 1600 m. n SNF, SnGb; introduced in US; native to s Eur. May–Aug

CORONILLA

[Ann] per, small shrub, unarmed. **LF**: alternate, odd-1-pinnate (simple); stipules gen conspicuous, free or fused; lflets gen 3–several. **INFL**: axillary umbel, peduncled, fls gen 10–20. **FL**: corolla white or pink to purple to yellow; 9 filaments fused, 1 free; style glabrous. **FR**: indehiscent, linear to lanceolate, segments 1–12. **SEED**: several. ± 25 spp.: Medit, n Afr, c&s Eur to w Asia, naturalized elsewhere. (Latin: small crown, for shape of infl) [Sokoloff 2003 Bot Zhurn 88:108–113] *C. valentina* L. subsp. *glauca* (L.) Batt. (per, small shrub; corolla yellow) is a waif in PR.

C. varia L.   CROWN VETCH   Erect to spreading or vine-like, to 1 m. **ST**: angled, hairs 0 to sparse. **LF**: stipule tips dark; petiole 0 to short; lflets 11–19, 1–2 cm, obovate to oblong, glabrous. **INFL**: head-like. **FL**: calyx 2-lipped, upper lip 1-lobed, lower 3-lobed, lobes < tube; corolla 1 cm, glabrous, wings white, enclosing keel, keel acute, tip purple. **FR**: with 3–12 segments, gen 2–6 cm, linear, longitudinally ridged, 4-angled. 2*n*=24. Disturbed sites, fields, planted as cover along roadsides; < 2200 m. n SN, ScV, expected elsewhere; widely cult, established, esp e US; native to Eurasia. [*Securigera v.* (L.) Lassen] May–Jul

CYTISUS

Shrub, [small tree], unarmed. **ST**: often ribbed, green, to 5 m. **LF**: 1-compound (gen ternate), gen alternate, petioled; stipules free or 0. **INFL**: gen terminal racemes, or axillary, peduncled clusters of 1–4(7) fls. **FL**: calyx bell-shaped to cylindric, 2-lipped, upper lip ± 2-lobed, lower gen 3-lobed; petals 5, gen yellow or white, gen not hairy, keel oblong-sickle-shaped to ± 1/2 circular (curve abaxial), claw ± 1/4 keel; stamens 10, filaments fused; style gen abruptly curved at ± middle or gently curved ± throughout. **FR**: dehiscent, gen oblong, papery to ± leathery; pedicel short. **SEED**: few to many, gen arilled. 65 spp.: Eur, w Asia, n Afr, Canary Islands; some cult. (Greek: name for several woody Fabaceae) *Chamaecytisus* (30 spp.) often segregated but recent work (Cristofolini 1991 Webbia 45:187–219; Cubas et al. 2002 Pl Syst Evol 233:223–244) supports treatment as 1 monophyletic group.

1. Corolla gen white, with dark or purple lines or spots or not; banner < 10 or 10–15 mm
  2. Banner < 10 mm; lflets < 10 mm . . . . *C. multiflorus*
  2′ Banner 10–15 mm; lflets 10–30 mm . . . . *C. proliferus*
1′ Corolla pale or golden yellow; banner gen > 10–25 mm
  3. Calyx glabrous; fr flat, glabrous exc margin; branches gen 5-angled . . . . *C. scoparius*
  3′ Calyx appressed-hairy; fr ± inflated, densely white-hairy; branches gen 8–10-angled . . . . *C. striatus*

C. multiflorus (L'Hér.) Sweet   SPANISH OR PORTUGUESE BROOM   Pl 3–4+ m; branches many, 5-angled, flexible, broom-like, gen lfless in fl, silvery-silky-hairy in youth, then ± glabrous. **LF**: on lower branches petioled, lflets 3, on upper 0 or sessile, lflets 1; lflets < 10 mm, lance-linear or oblong, silvery-silky-hairy. **INFL**: axillary clusters of 1–3 fls, pedicels < 10 mm. **FL**: calyx ± 5 mm, silky-hairy; corolla white, banner < 10 mm, dark-lined in lower center, gen ± reflexed. **FR**: 2.5–3 cm, linear-oblong, appressed-hairy. **SEED**: gen 4–6. Uncommon. Disturbed roadsides; ± 600 m. SCo (Los Angeles Co.); native to Spain, Portugal. Mar–Apr

C. proliferus L. f. TREE LUCERNE Pl < 5 m. **ST**: twigs round in ×-section, hairy. **LF**: petiole gen 5–12 mm, hairy; lflets 10–30 mm, lanceolate to ovate, adaxially gen glabrous or sparsely hairy, abaxially silky-hairy. **INFL**: axillary clusters of 3–7 fls, pedicels hairy. **FL**: calyx cylindric, 8–9 mm, silky-hairy, tube 5–6 mm, upper lip gen widely 2-lobed, gen ± 1/2 tube, lower shallowly 3-lobed; corolla white, purple-lined or -spotted inside, banner 10–15 mm, gen obovate, reflexed or not, keel oblong-sickle-shaped (curve abaxial), claw gen < 1/2 keel, ± hairy outside. **FR**: 3.5–5 cm, silky-hairy, black. Uncommon. Disturbed places; < 100 m. CCo, SnFrB; native to Canary Islands. [*Chamaecytisus p.* (L. f.) Link] Transfer from *Chamaecytisus* (Cristofolini 1991 Webbia 45: 187–219) supported by molecular data (Cubas et al. 2002 Pl Syst Evol 233: 223–242). Mar–May

C. scoparius (L.) Link (p. 755) SCOTCH BROOM Pl 2–2.5 m. **ST**: branches gen 5-angled, green, hairy in youth, then gen glabrous. **LF**: on younger sts sessile, simple; on older sts petioled, lflets 3, 5–20

mm, obovate to oblong; hairs appressed or 0. **INFL**: axillary clusters of 1–2 fls, pedicels < 10 mm, glabrous. **FL**: calyx < 5 mm, glabrous; corolla golden yellow, banner gen 15–20 mm, reflexed or not. **FR**: 2.5–4 cm, flat, brown or black, glabrous exc margin. **SEED**: 5–12. Common. Disturbed places; < 1000 m. NW, CaRF, n&c SNF, n SNH, GV, SnFrB, SCo, SnBr; native to s Eur, n Afr. Apr–Jul ◆

C. striatus (Hill) Rothm. Pl 2–3 m. **ST**: branches many, slender, gen 8–10-angled, silky-hairy in youth, then ± glabrous. **LF**: lflets 1–3, 5–15 mm, obovate, adaxially glabrous, glaucous, abaxially silky-hairy; on upper branches sessile, lflets 1 or 3, on lower branches petioled, lflets 3. **INFL**: cluster, axillary, gen 1–2-fld; pedicels 5–10 mm. **FL**: calyx ± 5 mm, appressed-hairy; corolla pale yellow, banner 10–25 mm, not reflexed. **FR**: 1.5–4 cm, ± inflated, densely white-hairy. **SEED**: several. Uncommon but locally abundant. Disturbed places; < 300 m. SnFrB, SCoRO, SCo, SnGb, PR, expected elsewhere; native to Spain, Portugal. Confused with *C. scoparius*. May–Aug ❖

DALEA

Michelle M. McMahon & Duane Isely

Ann, per, unarmed, gland-dotted. **LF**: gen odd-1-pinnate; stipules inconspicuous, thread-like or glandular. **INFL**: spike [or not], dense; bracts gen ± conspicuous. **FL**: calyx tube 10-ribbed; banner from receptacle, other petals from side or top of filament column; stamens 5 or 9–10, filaments fused; ovules 2. **FR**: indehiscent, incl in or ± exserted from calyx. **SEED**: 1. ± 165 spp.: Am. (T. Dale, English botanist, 18th century) Incl spp. sometimes put in *Parosela*, *Petalostemon*; exc other spp. here put in *Marina*, *Psorothamnus*.

1. Pl hairy, gen prostrate, < 3.5 dm; stamens 9–10
  2. Calyx 3–7 mm, lobes ± = tube; infl 8–15 mm wide; corolla ± ≥ calyx; lflet margin gen entire . . . . ***D. mollis***
  2′ Calyx ± 7–8 mm, lobes gen > tube; infl gen 14–16 mm wide; corolla < calyx; lflet margin shallowly
     lobed or wavy . . . . ***D. mollissima***
1′ Pl (exc infl) glabrous, ascending, gen 2.5–5 dm; stamens 5
  3. Infl (exc corollas) ± 12–16 mm wide, axis not visible between frs; calyx tube thinly papery, hairs long,
     silky, ascending. . . . . ***D. ornata***
  3′ Infl (exc corollas) 8–12 mm wide, axis visible between frs; calyx tube firm, ± puberulent. . . . . ***D. searlsiae***

D. mollis Benth. (p. 755) Ann, gen < 1(3.5) dm, mat-forming or not, hairy. **LF**: lflets 8–12, 3–7 mm, ± round to obovate-oblong, flat or folded, gen entire. **INFL**: 8–15 mm wide, ovoid or short-cylindric; bracts < ± 1 mm wide. **FL**: calyx 3–7 mm, lobes ± = tube, needle-like, shaggy-hairy; corolla ± ≥ calyx, ± white or tinged lavender; stamens 9–10. 2*n*=16. Common. Creosote-bush flats, washes, roadsides; < 800 m. D; to AZ, Mex. Mar–May

D. mollissima (Rydb.) Munz (p. 755) Ann, per, 0.5–3.5 dm, mat-forming or not, hairy. **LF**: lflets 8–14, 3–10 mm, ± round to obovate-oblong, gen folded, shallowly lobed or wavy. **INFL**: gen 14–16 mm wide, ovoid or short-cylindric; bracts gen 1–1.5 mm wide. **FL**: calyx ± 7–8 mm, lobes gen > tube, needle-like, shaggy-hairy; corolla < calyx, ± white or tinged lavender; stamens 9–10. 2*n*=16. Common. Desert flats, washes; < 900 m. D; s NV, w AZ, Mex. Mar–May

D. ornata (Hook.) Eaton & J. Wright ORNATE DALEA Per, glabrous. **ST**: clustered, ascending, 2.5–5 dm. **LF**: lflets 5–7, 10–20 mm, widely ovate to elliptic. **INFL**: (exc corollas) ± 12–16 mm wide, ovoid to cylindric, compact, axis not visible between frs. **FL**: calyx 4–6 mm, tube not slit, thinly papery, hairs long, ascending, silky; corolla pale lavender to rose-purple, banner 7–9 mm, other petals 4–6 mm; stamens 5. Open, rocky hillsides; 1400 m. MP (Shaffer Mtn, Lassen Co.); to WA, ID, NV. Jun ★

D. searlsiae (A. Gray) Barneby (p. 759) Per, glabrous. **ST**: clustered, ascending, 3–5 dm. **LF**: lflets 5–7, 7–16 mm, obovate to oblong. **INFL**: (exc corollas) 8–12 mm wide, oblong or narrowly oblong, dense, axis visible between frs. **FL**: calyx 3.5–4.5 mm, tube recessed or slit on upper side, firm, ± puberulent; corolla lilac-pink to rose-purple, banner 5–7 mm, other petals 3–5 mm; stamens 5. Juniper/sagebrush scrub, slopes, bluffs; 1200–2000 m. W&I, DMtns; to UT, AZ. May–Jun

GENISTA BROOM

Shrub, spiny or unarmed; gen deciduous. **ST**: gen ribbed or angled, green. **LF**: gen alternate, ternately 1-compound or simple, petioled; stipules fused to lf bases (0). **INFL**: axillary or terminal, racemes, heads, or fls in clusters on short-shoots. **FL**: bilateral; calyx gen < corolla, 2-lipped, upper 2-lobed, lobes ± 1/2 tube, lower gen 3-toothed, < upper lobes; petals 5, gen yellow, banner gen ovate or rounded, outside gen glabrous or variously hairy, keel narrowly oblong to obtuse, ± straight abaxially, often silky-hairy; stamens 10, filaments fused; style ± abruptly bent at tip. **FR**: gen dehiscent, narrowly oblong, compressed, or curved, ± inflated; pedicel < 7 mm. **SEED**: 1–several-seeded, gen arilled. 90 spp.: Eur, w Asia, n Afr, Canary Islands. (Latin: from *planta genista*, from which English Plantagenet monarchs took their name) Generic circumscription difficult, but Pardo et al. (2004 Pl Syst Evol 244:93–119) suggest recognizing *Genista* in broad sense (i.e., incl *Retama*, *Teline*, *Ulex*). Many naturalized CA pls are hybrids involving *G. canariensis*, *G. monspessulana*, and *G. stenopetala* Webb & Berthel. (native of Canary Islands; not in CA in pure form), although determining parentage in gen often difficult.

1. Lvs simple; corolla white, fr indehiscent, gen inflated . . . . ***G. monosperma***
1′ Lvs ternately 1-compound, simple on young sts; corolla yellow, fr dehiscent, not inflated
  2. Lflet linear or narrow-oblong to -oblanceolate, length gen ≥ 5 × width, margins rolled under; banner
     silky-hairy. . . . . ***G. linifolia***

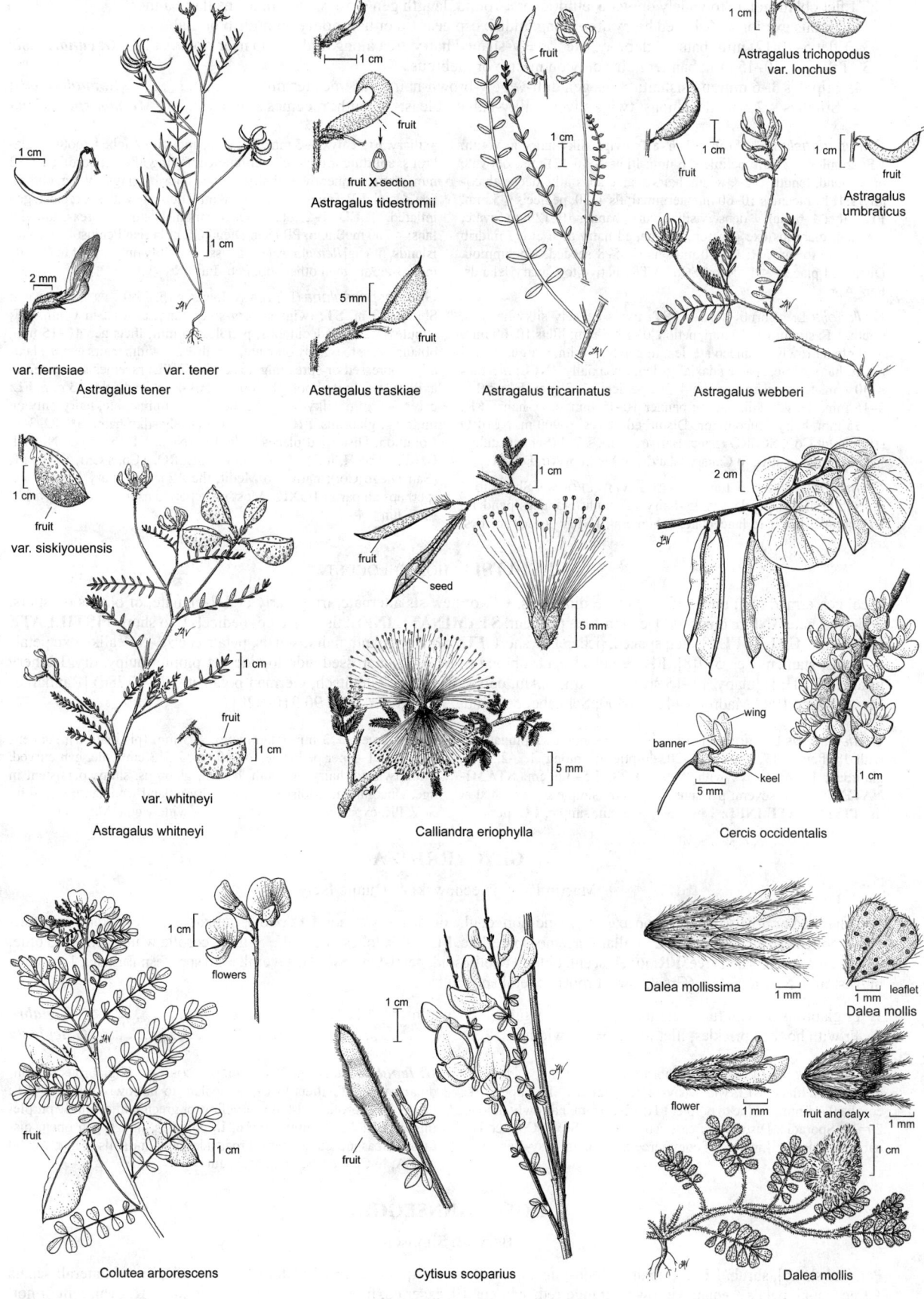

Astragalus tidestromii

fruit X-section

var. ferrisiae   var. tener

Astragalus tener

Astragalus traskiae

Astragalus tricarinatus

Astragalus trichopodus
var. lonchus

Astragalus umbraticus

Astragalus webberi

var. siskiyouensis

var. whitneyi

Astragalus whitneyi

Calliandra eriophylla

Cercis occidentalis

Colutea arborescens

Cytisus scoparius

Dalea mollissima

Dalea mollis

Dalea mollis

2′ Lflet oblanceolate to widely obovate, elliptic, or ± round, length gen < 5 × width, margins flat; banner glabrous exc for ± V-shaped hairy area along midrib esp near tip or hairy only on midrib or glabrous

    3. Lflet 5–10(12) mm; banner glabrous exc for ± V-shaped hairy area along midrib esp near tip. . . . . **G. canariensis**

    3′ Lflet gen 10–15 mm; banner hairy only on midrib or glabrous

       4. Stipules 3–6 mm, persistent; twigs gen dull-yellow-brown-hairy; racemes terminal . . . . [**G. maderensis**]

       4′ Stipules < 2 mm, deciduous; twigs silvery-silky-hairy, at least in youth; racemes axillary . . . . . . **G. monspessulana**

G. canariensis L. Shrub < 3 m. **ST**: twigs silky-hairy in youth. **LF**: stipules < 2 mm; petiole < 6 mm; lflets 5–10(12) mm, obovate or ± round, length 1–2 × width, hairs sparse adaxially, dense abaxially. **INFL**: racemes 10–60 mm, terminal; fls 4–20; pedicels < 5 mm. **FL**: calyx 4–6 mm, ± densely silky-hairy; banner 10–12 mm, ovate, notched, glabrous exc for ± V-shaped hairy area along midrib from base to tip. **FR**: 15–30 mm, hairy, 5–8 seeded. Uncommon. Disturbed places; < 1000 m. SCo, WTR; native to Canary Islands. Feb–Apr

G. linifolia L. Shrub < 2.5 m. **ST**: twigs densely silky-hairy in youth. **LF**: stipules 1–10 mm; petiole 0 or < 5 mm; lflets 10–60 mm, linear to narrowly oblanceolate, length gen 5 × width, margins rolled under, hairs white, sparse adaxially, dense abaxially. **INFL**: racemes < 30 mm, terminal, dense; fls 4–20; pedicels 2–7 mm. **FL**: calyx 5–15 mm, densely silky-hairy; banner 10–18 mm, silky-hairy. **FR**: 15–35 mm, hairy. Uncommon. Disturbed places; < 900 m. NCoRO (Humboldt Co.), SCoRO (Santa Barbara Co.), s ChI (Santa Catalina Island), SnGb; native to Canary Islands, w Medit, n Afr. Feb–Aug

G. monosperma (L.) Lam. BRIDAL VEIL BROOM Shrub to 4 m. **ST**: branches pendent, silky-hairy in youth, branches pendent. **LF**: simple, linear to lanceolate, ephemeral. **INFL**: loose racemes, axillary. **FL**: calyx 3–5 mm, 2-lipped, upper lip 2-lobed, toothed, glabrous, splitting around circumference, falling after fl; corolla 10–12 mm, white, banner ovate, hairy only on midrib, wings oblong, obtuse, gen < keel. **FR**: indehiscent, 14–18 mm, obovate with short beak, gen inflated. **SEED**: 1–2, black. Uncommon. Disturbed areas, alluvial fans; < 200 m. SnGb, PR (San Diego Co.); Iberian Peninsula, Canary Islands, n Afr. [*Retama m.* (L.) Boiss.] Possibly only a waif in CA; fls more fragrant than other spp. Feb–Jun ❖

G. monspessulana (L.) L.A.S. Johnson (p. 759) FRENCH BROOM Shrub < 3 m. **ST**: twigs silvery-silky-hairy, at least in youth. **LF**: stipules < 2 mm, deciduous; petiole < 5 mm; lflets gen 10–15 mm, oblanceolate to widely obovate, length ± 2 × width, hairs gen 0 adaxially, appressed or spreading abaxially. **INFL**: racemes short, dense, axillary, on short-shoots, 15–60 mm; fls 4–10; pedicels 1–3 mm. **FL**: calyx 5–7 mm, silky-hairy; banner 10–15 mm, ovate, hairy only on midrib or glabrous. **FR**: 15–25 mm, densely silky-hairy. **SEED**: 3–6. Common. Disturbed places; < 900 m. NCo, KR, NCoRO, NCoRI, CaRF, s CaRH, n SN, CCo, SnFrB, SCoRO, SCo, s ChI, WTR, PR (San Diego Co.); native to Medit, the Azores, Canary Islands. Fls (perhaps all parts) TOXIC. Most pls reported as this may be hybrids. Mar–Jun ◆

GLEDITSIA HONEY LOCUST

Tree, gen armed (gen unarmed in cult); ± dioecious. **LF**: of new sts alternate, irregularly odd-2-pinnate, of old sts on spurs, odd-1-pinnate. **INFL**: on spurs, ± catkin-like, pendent. **STAMINATE INFL**: fls clustered, pedicels 0 to short. **PISTILLATE OR BISEXUAL INFL**: fls gen spaced, pedicels > short. **FL**: radial, perianth hairy, sepals, petals each 3–5, ± alike exc petals > sepals; stamens gen 5–7[8]. **FR**: ± indehiscent, oblong to ovate, compressed side-to-side yet plump, pulpy, dry, leathery in age. **SEED**: [1]many. 13–16 spp.: e N.Am, S.Am, e&se Asia. (J.G. Gleditsch, German botanist, 1714–1786) [Randall & Meyers-Rice 1997 Madroño 44:399–400; Schnabel & Wendel 2003 Amer J Bot 90:310–320]

G. triacanthos L. HONEY LOCUST **LF**: deciduous; 2-pinnate lvs with 1° lflets 7–17, 2° lflets 4–20, elliptic to oblong, 1.3–2.5 cm, glabrous; 1-pinnate lvs with lflets 20–28, 1.5–3.5 cm. **STAMINATE INFL**: 1–several per spur, 3.5–8 cm, simple or branched at tip. **PISTILLATE INFL**: 1 per spur, 3–5 cm, simple. **FL**: perianth yellow-green, ± 3 mm (staminate), 4–5 mm (pistillate fl); stamens 5–7. **FR**: 1–3 per peduncle, 20–40 cm, 2.5–3 cm wide, gen curved, often twisted, hairy in youth, brown, glabrous, shiny, persistent in age. Uncommon. Moist riparian to dry upland woodland; < 800 m. ScV, PR, expected elsewhere; c&e US, widely cult. May–Jun

GLYCYRRHIZA

Martin F. Wojciechowski & Duane Isely

Per, unarmed or bristly or prickly on axes, fr, gland-dotted, glandular-hairy or not. **LF**: odd-1-pinnate; stipules deciduous; lflets gland-dotted esp abaxially. **INFL**: axillary raceme, spike-like. **FL**: calyx lobes unequal, ± ≤ tube; corolla white-yellow to blue; 9 filaments fused, 0 or 1 free. **FR**: indehiscent, oblong to ellipsoid, persistent. **SEED**: several. ± 20 spp.: esp Eurasia, 1 sp. each in Australia, N.Am, S.Am. (Greek: sweet root) Several spp. cult.

1. Fr glabrous or with fine, straight bristles; lflet length ± 2 × width . . . . **G. glabra**

1′ Fr with hooked prickles; lflet length ± 3 × width . . . . **G. lepidota**

G. glabra L. LICORICE Pl ± glabrous or finely hairy, ± glandular or not. **LF**: lflets 9–13, widely ovate or elliptic. **INFL**: open. **FL**: corolla 9–11 mm, ± blue to purple. **FR**: 12–30 mm, narrowly oblong. $2n=16$. Sporadic. Disturbed areas; esp < 500 m. ScV, SCo; occ US; native to Eurasia. Licorice of commerce from roots. May–Sep

G. lepidota Pursh (p. 759) WILD LICORICE Pl glabrous to glandular-hairy. **LF**: lflets 9–19, lanceolate to narrowly ovate. **INFL**: dense. **FL**: corolla 9–14 mm, ± yellow or green-white, faintly purple-tinged. **FR**: 12–20 mm. $2n=16$. In colonies, moist, gen open, disturbed areas incl streambanks, roadsides, alkaline soils or not; < 2000 m. CA; to Can, c US, Mex. May–Jul

HOFFMANNSEGGIA

Beryl B. Simpson

Per, [subshrub], shrub. **LF**: 2-pinnate, lf odd-pinnate, 1° lflets even-pinnate. **INFL**: lateral or terminal. **FL**: ± bilateral; sepals ± free, equal; petals ± equal, yellow to orange-red; stamens 10, exserted, free, filaments often glandular. **FR**: dehiscent or not. **SEED**: several. 21 spp.: s US, Mex, also Peru to Argentina. (J. Centurius, Count of Hoffmannsegg, Germany, 1766–1849) [Simpson 1999 Lundellia 2:14–54]

1. Per; 1° lflets 3–13, terminal ≤ lateral; fr indehiscent . . . . *H. glauca*
1′ Shrub; 1° lflets 3, terminal > lateral; fr dehiscent . . . . *H. microphylla*

H. glauca (Ortega) Eifert (p. 759) PIG-NUT, HOG POTATO Pl erect, < 30 cm; roots deep, tuberous. **LF:** 5–12 cm; 2° lflets 4–13, 2–6 mm. **INFL:** 5–23 cm, often glandular. **FL:** petals spreading, 10–18 mm wide, gen yellow, with some orange-red. **FR:** 1.5–4 cm, ± curved. Uncommon. Dry, alkaline flats in deserts, disturbed areas; < 900 m. SnJV, SCoRO, SCo, WTR, D; to TX, Mex, S.Am. Apr–Jun

H. microphylla Torr. (p. 759) Pl erect, < 2.5 m; taproot woody. **LF:** 2–5 cm; 2° lflets 3–7, 1.5–3.5 mm. **INFL:** 9–16 cm, strigose. **FL:** petals spreading, 6–12 mm wide, yellow, banner with red markings. **FR:** 18–23 mm, sickle-shaped. Common in sandy soils; < 1000 m. s SNE, DSon; AZ, Mex. [*Caesalpinia virgata* Fisher] Apr

HOITA LEATHER ROOT

Martin F. Wojciechowski & James W. Grimes

Per, unarmed; gland-dotted, hairs glandular, nonglandular, or both, at least above; caudex woody. **ST:** prostrate (incl stolon) to erect; base green to gray-brown. **LF:** odd-1-pinnate, cauline; stipules free; lflets 3. **INFL:** spike-like raceme, axillary, with 1 deciduous bract and 2–3 fls per node. **FL:** calyx not (or only ±) swollen in fr, lobes > tube, lowest ± keeled, > others; corolla at least partly blue to purple; 9 filaments fused, 1 less so or free; ovary hairy, ovule 1, style tip curved or bent, stigma feathery. **FR:** indehiscent, unevenly elliptic, brown to black, hairy; veins obvious. **SEED:** 1, ± reniform. 3 spp.: CA, Baja CA. (Native American name)

1. St prostrate (incl stolon) to decumbent, erect parts < 6.5 cm high; lflet obovate to round . . . . *H. orbicularis*
1′ St erect, < 2 m high; lflet lanceolate to round
  2. Fl 9–10 mm . . . . *H. macrostachya*
  2′ Fl 13–19 mm . . . . *H. strobilina*

H. macrostachya (DC.) Rydb. **ST:** erect, < 2 m, much-branched, gen striate; base hollow. **LF:** stipule 1.5–5 mm; petiole 2–10 cm; lflet 2–10 cm, lanceolate, both surfaces glandular, hairy or not. **INFL:** 5–15 cm; bract 3–11 mm; peduncle 4–10 cm. **FL:** 9–10 mm; calyx 4.5–9 mm; banner 5.5–10 mm. **FR:** 6–8 mm, brown; veins prominent. **SEED:** 5–7 mm. Streamsides, marshes, spring-moist places; < 2500 m. CA-FP; Baja CA. May–Aug

H. orbicularis (Lindl.) Rydb. (p. 759) **ST:** prostrate (incl stolon) to decumbent, faintly striate or not, 10–60 cm. **LF:** stipule 4.5–10 mm; petiole 10–50 cm; lflet 3–11 cm, obovate to round, both surfaces brown-glandular, hairy or not. **INFL:** 6–35 cm; bract 7–27 mm; peduncle 20–70 cm. **FL:** 12–23 mm; calyx 15–19 mm; banner 10–16 mm. **FR:** 6–9 mm, brown; veins obvious. **SEED:** 4–5 mm. Meadows, streamsides, moist hillsides; < 2250 m. CA-FP (exc GV); Baja CA. Apr–Aug

H. strobilina (Hook. & Arn.) Rydb. LOMA PRIETA HOITA **ST:** erect, rarely to 1 m, smooth to striate. **LF:** stipule 7–16 mm, becoming reflexed; petiole 3–7 cm; lflet 4.5–8 cm, lanceolate to round. **INFL:** 3–8 cm; bract 15–21 mm; peduncle 4–6 cm. **FL:** 13–19 mm; calyx 13–17 mm; banner 12–13 mm. **FR:** ± 10 mm, dark brown to black; veins obvious. **SEED:** 6–7 mm. Chaparral, oak woodland; < 600 m. SnFrB. Jun–Aug ★

HOSACKIA LOTUS

Luc Brouillet

Per, unarmed. **LF:** odd-1-pinnate; stipules scarious or lflet-like, fragile or not, early-deciduous; lflets 3–15, often ± opposite. **INFL:** several-fld umbel or 1–2-fld, axillary, gen peduncled, often bracted. **FL:** corolla yellow, white, or pink, fading darker; 9 filaments fused, 1 free. **FR:** dehiscent, exserted from calyx, linear to oblong, ± beaked. **SEED:** few–several. ± 11 spp.: sw Can, w US, Mex to Guatemala, esp CA. (D. Hosack, NY physician, botanist, mineralogist, 1769–1835) [Brouillet 2008 J Bot Res Inst Texas 2:387–394] Pollen apertures 3. Intermediates may be hybrids.

1. Lflets 3–5; corolla 8–10 mm; pl decumbent; NCoR . . . . *H. yollabolliensis*
1′ Lflets (3)5–15; corolla gen 8–16 mm; pl gen ascending or erect; if lflets < 5, not of NCoR
  2. Lf silky or canescent, gray or silvery; fr glabrous . . . . *H. incana*
  2′ Lf glabrous or hairy, rarely gray or silvery; fr gen hairy if lf hairy
   3. Peduncle bract 0 or just below umbel, gen simple or 2–3-parted; fr 1.5–3 mm wide
    4. Petal claw incl to ± exserted from calyx tube; pl often hairy . . . . *H. oblongifolia*
     5. Lflets 3–7, glabrous; corolla 8–9 mm; n SNF (Calaveras Co.), s SNH (Tulare Co.), SNE . . . . var. *cuprea*
     5′ Lflets 7–11, hairy or glabrous; corolla 9–13 mm; CA-FP, MP, DMoj . . . . var. *oblongifolia*
    4′ Petal claw exserted from calyx tube; pl ± glabrous
     6. Corolla wings pink-purple, fading white; bracts gen 3-parted . . . . *H. gracilis*
     6′ Corolla wings white; bracts 0 or gen simple . . . . *H. pinnata*
   3′ Peduncle bract gen well below umbel, gen 3–5-parted, sometimes fragile, deciduous; fr 3–5 mm wide
    7. Hairs spreading, glandular or not; stipules often wide, lflet-like, persistent, in age scarious or not; corolla pink to red-purple . . . . *H. stipularis*
     8. St wiry; stipules wide, clasping st . . . . var. *ottleyi*
     8′ St often fleshy; stipules narrow or wide, not or ± clasping st . . . . var. *stipularis*
    7′ Hairs gen ± 0 or appressed, rarely glandular; stipules scarious, fragile, not lflet-like, gen some deciduous; corolla initially white, pink, or yellow-green
     9. Fls gen 6–10; corolla 10–12 mm, white or ± pink, in age striate, darker; pedicel 2–3 mm, ± longer in fr; lflet length 2–3 × width — NCo, NCoRO, n SNH . . . . *H. rosea*
     9′ Fls gen 12–20; corolla 12–17 mm, yellow-green, dark-blotched in age; pedicel gen 3–6 mm, > 3 mm at least in fr; lflet length gen 1–2.5 × width . . . . *H. crassifolia*
      10. Pl glabrous or strigose; widespread exc s PR . . . . var. *crassifolia*
      10′ Pl soft-hairy; s PR (Otay Mtn, San Diego Co.) . . . . var. *otayensis*

H. crassifolia Benth.   Often robust. **ST:** sprawling or erect, 7–15 dm, base hollow. **LF:** stipules scarious, inconspicuous; lflets 9–15, 2–3 cm, elliptic or obovate, length gen 1–2.5 × width, abaxially pale. **INFL:** gen 12–20-fld; peduncle bract 0 or gen well below umbel, (1)3–5-parted; pedicels gen 3–6 mm, > 3 mm at least in fr. **FL:** calyx 5–8 mm, lobes < tube; corolla 12–17 mm, yellow-green, dark-blotched in age, claw exserted from calyx tube. **FR:** 3.5–7 cm, 3–5 mm wide, oblong, glabrous. **SEED:** several. 2*n*=14.

var. ***crassifolia*** (p. 759)   Glabrous or strigose. **FL:** ovules ± 14–35 (> 24 in s). Common. Chaparral, pine or mixed woodland, roadsides, disturbed places; 300–2100 m. KR, NCoR, CaRH, SN, SnFrB, SCoR, SCo, TR, PR, MP; to WA, NV, Baja CA. [*Lotus c.* (Benth.) Greene var. *c.*] Conspicuous. May–Aug

var. ***otayensis*** (Isely) Brouillet   OTAY MOUNTAIN HOSACKIA   Hairs long, soft, wavy. **FL:** ovules 14–20. Disturbed areas; 900–1000 m. s PR (Otay Mtn, San Diego Co.); Baja CA. [*Lotus c.* var. *o.* Isely] May–Aug ★

H. gracilis Benth. (p. 759)   HARLEQUIN LOTUS   Pl ± glabrous, stoloned or rhizomed. **ST:** sprawling to ascending, base often spongy, 1–5 dm. **LF:** stipules large, triangular, ± scarious, fragile; lflets 3–7, ± opposite, 6–20 mm, elliptic or obovate. **INFL:** 3–9-fld; peduncle bract just below umbel, gen 3-parted. **FL:** calyx 5–6 mm, lobes ± ≤ tube; corolla 10–16 mm, banner yellow, wings pink-purple, fading white, claw exserted from calyx tube. **FR:** 2–3 cm, 2–3 mm wide, oblong, glabrous. **SEED:** few. In water, springy areas, shores, meadows, roadside ditches; < 700 m. NCo, NCoRO, n CCo, SnFrB, n SCoRO; to BC. [*Lotus formosissimus* Greene] Mar–Jul ★

H. incana Torr.   Silky or canescent, gray or silvery. **ST:** erect, gen clustered, 1–2.5(3) dm. **LF:** stipules scarious; lflets 7–13, 0.8–1.6 cm, elliptic to obovate. **INFL:** 3–8-fld; peduncle bract separated from umbel or not. **FL:** calyx 5.5–7.5 mm, lobes < tube; corolla 12–15 mm, 2-colored (red and white), claw exserted from calyx tube. **FR:** 1.5–3.5 cm, linear, glabrous. **SEED:** few–several. Dry slopes, open pine forest; 800–1700 m. n&c SN. [*Lotus i.* (Torr.) Greene] May–Jun

H. oblongifolia Benth.   Often hairy. **ST:** sprawling or ascending, 1–6 dm. **LF:** stipules scarious, fragile; lflets 3–11, ± opposite, 1–2.5 cm, gen elliptic to oblong. **INFL:** 2–6-fld; peduncle bract 0 or just below umbel, simple or divided. **FL:** calyx 4–6 mm, lobes < tube; corolla 8–13 mm, gen white-yellow, claw incl to ± exserted from calyx tube. **FR:** 2.5–5 cm, 1.5–2 mm wide, oblong, glabrous. **SEED:** few.

var. ***cuprea*** (Greene) Brouillet   COPPER-FLOWERED BIRD'S-FOOT TREFOIL   **LF:** lflets 3–7, glabrous. **FL:** corolla 8–9 mm. Meadows, open pine woodland; 2400–2800 m. n SNF (Calaveras Co.), s SNH (Tulare Co.), SNE. [*Lotus o.* var. *c.* (Greene) Ottley] Jul–Aug ★

var. ***oblongifolia***   **LF:** lflets 7–11, hairy or glabrous. **FL:** corolla 9–13 mm. Locally common. Open, moist forest, river bottoms, marshy meadows; 200–2400 m. CA-FP, MP, DMoj; s OR, w&s NV, AZ, Mex. [*Lotus o.* (Benth.) Greene var. *o.*] May–Sep

H. pinnata (Hook.) Abrams (p. 759)   Pl ± glabrous. **ST:** 1, decumbent or ascending, base often spongy-thickened, 1.5–5 dm. **LF:** stipules scarious; lflets 5–7, ± opposite, 1–2.5 cm, elliptic or obovate. **INFL:** 4–10-fld; peduncle bract 0 or just below umbel, gen simple. **FL:** calyx 5–7 mm, lobes < tube; corolla 10–13 mm, banner bent or recurved 90–180°, yellow, other petals white, claw ± exserted from calyx tube. **FR:** 3–5 cm, 1.5–2.5 mm wide, linear, glabrous. **SEED:** few. Wet meadows, bogs; 600–1700 m. KR, NCoR, CaR, n SN, c SNH, CCo; to WA, ID. [*Lotus p.* Hook.] May–Jul

H. rosea Eastw.   Glabrous or puberulent. **ST:** spreading or erect, 1–7 dm. **LF:** stipules scarious, inconspicuous, fragile; lflets 9–15, 1.5–3 cm, elliptic, length 2–3 × width. **INFL:** 6–10-fld; peduncle bract 0 or well below umbel, 1–5-parted, pedicels 2–3 mm, ± longer in fr. **FL:** calyx 4–6 mm, lobes < tube; corolla 10–12 mm, white or ± pink, striate, darker in age, claw often exserted from calyx tube. **FR:** 3–5 cm, 3–4 mm wide, oblong, glabrous. **SEED:** several. Banks, streamsides, burns, logged areas; < 800 m. NCo, NCoRO, n SNH; to WA. [*Lotus aboriginus* Jeps.] May–Jul

H. stipularis Benth.   Gen spreading soft-hairy, glandular or not. **ST:** gen erect, 1.5–5(10) dm. **LF:** stipules often wide, lflet-like, persistent, in age reduced, scarious or not; lflets 9–15, 5–20 mm, oblong to ovate. **INFL:** 4–9-fld; peduncle bract well below umbel, (1)3(7)-parted, small. **FL:** calyx 5–6.5 mm, lobes < tube; corolla 10–12 mm, pink to red-purple, claw exserted from calyx tube. **FR:** 2–2.5 cm, 3–4 mm wide, widely oblong, ± hairy to ± glabrous. **SEED:** many. [*Lotus s.* (Benth.) Greene] Vars. geog distinct.

var. ***ottleyi*** (Isely) Brouillet   **ST:** wiry. **LF:** stipules wide, clasping st. Open pine forest, streambeds; 600–1200 m. KR, CaR, n&c SN, s SNH. [*Lotus s.* var. *o.* Isely] Apr–Jun

var. ***stipularis***   **ST:** often fleshy. **LF:** stipules narrow or wide, not or ± clasping st. Thickets, chaparral, logged areas; 200–1000 m. NCo, NCoR, CCo, SCoR. [*Lotus s.* var. *s.*] Variable; some pls intermediate to *H. rosea*. Apr–Jun

H. yollabolliensis (Munz) D.D. Sokoloff (p. 759)   YOLLA BOLLY MOUNTAINS BIRD'S-FOOT TREFOIL   Glabrous. **ST:** decumbent, clustered, 0.5–1.5 dm. **LF:** gen palmate; stipules inconspicuous, scarious; lflets 3–5, 3–10 mm, oblanceolate to obovate. **INFL:** 1–3-fld; peduncle bracted. **FL:** calyx 2.5–4 mm, lobes < tube; corolla 8–10 mm, white or ± yellow, claw ± = calyx tube, banner margin fringed. **FR:** 1.5–2.5 cm, narrowly oblong, glabrous. **SEED:** few. Open, dry slopes, fir forest; 1700–2100 m. n NCoRH (Yolla Bolly Mtns, South Fork Mtn, Trinity, Humboldt cos.). [*Lotus y.* Munz] Jun–Aug ★

LADEANIA

Per, unarmed, gland-dotted, glabrous to ± sparsely hairy; rhizomes or roots (or both) woody. **ST:** erect, < 7.5 dm, green or yellow toward base. **LF:** palmately compound, cauline; stipules free; lflets 3[5]. **INFL:** raceme, axillary, each node with 1 deciduous bract, 1–3 fls; peduncle gen >> subtending lf. **FL:** calyx flaring back, tearing along 1 lateral sinus in fr; corolla white, yellow, or purple; 9 filaments fused, 1 less so or free; ovary glabrous to hairy, ovule 1, style tip bent, stigma head-like. **FR:** indehiscent, ± spheric. **SEED:** 1, elliptic to round. 2 spp.: w N.Am, s Can to Mex. (LaDean Egan, b. 1949) [Egan & Reveal 2009 Novon 19:310–314]

L. lanceolata (Pursh) A.N. Egan & Reveal (p. 763)   **LF:** petiole 9–21 mm; lflets 17–33 mm, linear to oblanceolate. **FL:** 4–7 mm; calyx 2–2.5 mm; petals white to purple-blue. **FR:** 4–6 mm, papil-late-glandular to glandular, ± hairy. **SEED:** 4–5 mm, smooth, shiny. Alluvial plains, sand; < 2500 m. GB; to c Can, c US. [*Psoralidium l.* (Pursh) Rydb.] May–Jul ★

LATHYRUS   WILD PEA

Kelly Steele & Duane Isely

Ann, per, unarmed, glabrous or hairy (glandular), gen rhizomed. **ST:** sprawling, climbing, or erect; angled, flanged, or winged. **LF:** gen even-1-pinnate; stipules persistent, upper lobe > lower; main axis ending as tendril or short bristle; lflets ± rolled in bud, 0–16 (if 0, stipules lflet-like), ± opposite or alternate, linear to widely ovate. **INFL:** raceme, gen axillary, 1–many-fld. **FL:** upper calyx lobes gen <, wider than lower; corolla 8–30 mm, pink-purple or pale, occ white or yellow; 9 filaments fused,

759

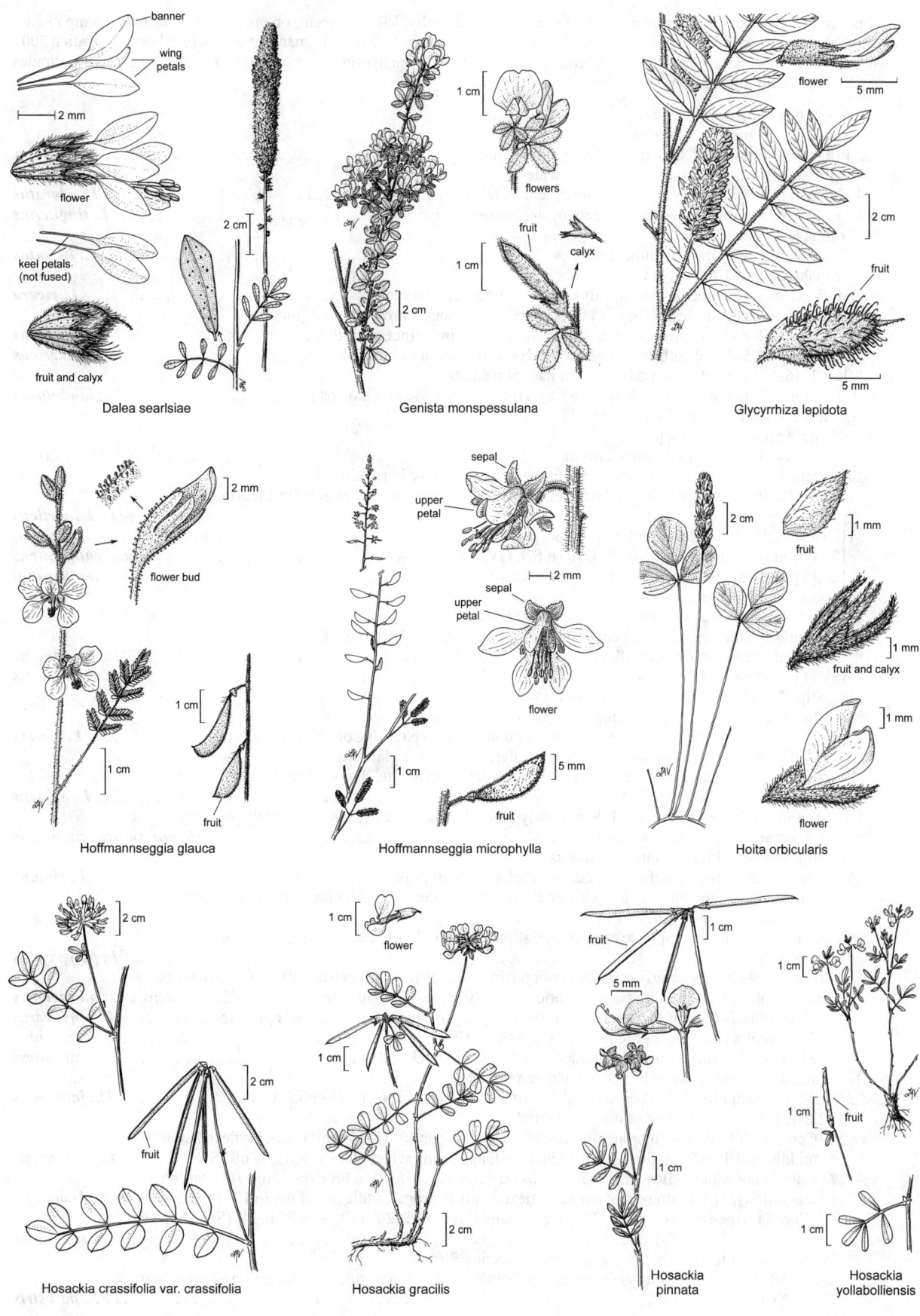

Dalea searlsiae

Genista monspessulana

Glycyrrhiza lepidota

Hoffmannseggia glauca

Hoffmannseggia microphylla

Hoita orbicularis

Hosackia crassifolia var. crassifolia

Hosackia gracilis

Hosackia pinnata

Hosackia yollabolliensis

1 free; style ± flat, puberulent near ± middle for ± 1/3–1/2 adaxially. **FR**: dehiscent, oblong, ± flat. ± 150 spp.: temp N.Am, S.Am, Medit, Eurasia. (Ancient Greek name) Seeds of most alien spp. TOXIC to humans, livestock (esp horses). [Broich 2007 Madroño 54:63–71] Some spp. variable, intergrading with others; some hybridization probable. *L. aphaca* L. (lflets 0, stipules lflet-like) not naturalized in CA.

1. Lflets 2; gen of disturbed habitats
 2. Corolla 20–30 mm
  3. Infl 4–15-fld; wings of st, petiole often ≥ 2 mm wide . . . . ***L. latifolius***
  3′ Infl 2–4-fld; wings of st, petiole < 2 mm wide
   4. Fr hairy; lflets ovate to narrowly elliptic, length 1.2–3 × width . . . . ***L. odoratus***
   4′ Fr glabrous; lflets lanceolate to widely ovate, length 1.2–6.5 × width . . . . ***L. tingitanus***
 2′ Corolla 8–15 mm
  5. Fr hairy; infl 1–2-fld; calyx tube ± = lobes . . . . ***L. hirsutus***
  5′ Fr glabrous; infl 1-fld; calyx tube ± ≤ lobes
   6. Infl axis not extended beyond fl; calyx tube << lobes; corolla red-purple . . . . ***L. cicera***
   6′ Infl axis extended as bristle beyond fl; calyx tube ± ≤ lobes; corolla not red-purple
    7. Peduncle 2–8 cm; corolla lavender or purple; fr veins indistinct, netted . . . . ***L. angulatus***
    7′ Peduncle 0.5–2 cm; corolla orange-red; fr veins distinct, longitudinal . . . . ***L. sphaericus***
1′ Lflets 2–16, > 2 at least on some lvs; gen of natural habitats
 8. Lflet, fr gland-dotted; lflets ± 14–16 — NCoR (Humboldt, n Mendocino cos.) . . . . ***L. glandulosus***
 8′ Lflet, fr not gland-dotted; lflets often < 14
  9. St with wings ± ≥ 1 mm wide
   10. Lflets (4)6(8), opposite; immediate coast, n NCo . . . . [2]***L. palustris***
   10′ Lflets gen 8–16, irregularly arranged; immediate coast only in n CCo
    11. Corolla 10–15 mm, white, gen lavender-striate, fading yellow-tan; nw KR (Del Norte,
      w Siskiyou cos.) . . . . [2]***L. delnorticus***
    11′ Corolla 15–20 mm, gen pink to pink-purple . . . . ***L. jepsonii***
     12. Pl gen puberulent; KR, NCoR, CaR, n SN, GV, CW, SCoRO . . . . var. ***californicus***
     12′ Pl glabrous; GV (esp Solano Co., nearby CCo) . . . . var. ***jepsonii***
  9′ St with angles or flanges < 1 mm wide
   13. Tendrils not coiled or branched, often a short bristle
    14. Stipule ± ≥ lflet; gen dunes, beaches of immediate coast — NCo, n&c CCo
     15. Pl glabrous or puberulent; lflet 2.5–4.5 cm . . . . [2]***L. japonicus***
     15′ Pl gray-hairy; lflet 1–2 cm . . . . ***L. littoralis***
    14′ Stipule < lflet; not of immediate coast
     16. Infl 1–2-fld; lflet gen 0.5–2 cm; fr glabrous or puberulent
      17. Lflets 10–15, elliptic to ovate, 1–2 cm; fr puberulent — NCo, NCoR, SnFrB. . . . . ***L. torreyi***
      17′ Lflets 4–8, gen oblong or narrowly lanceolate to linear; fr glabrous
       18. Lflets 4, lance-oblong, gen 0.5–1(2) cm; corolla green-white, dark-striate; NCoRH
         (Red Mtn, Humboldt Co.) . . . . ***L. biflorus***
       18′ Lflets of at least some lvs 4–8, narrowly lanceolate to linear, 1.5–8 cm; corolla pale
         lavender to purple; KR, SNH, MP . . . . [2]***L. lanszwertii*** var. ***aridus***
     16′ Infl 2–16-fld; lflet 1–8 cm; fr glabrous
      19. Sts erect, many, clustered; infl dense; corolla white to pink, 17–23 mm; MP (Modoc Co.) . . . . ***L. rigidus***
      19′ Sts erect or sprawling, 1–few, clustered; infl open or dense; corolla lavender to pink-purple or,
        if white, < 17 mm; n CA-FP
       20. Lflets 10–16, subopposite to alternate; stipules in width often ± = lflets; pl glabrous;
         KR, n&c NCoR . . . . [3]***L. polyphyllus***
       20′ Lflets 4–10, irregularly arranged or opposite; stipules in width gen ± ≤ lflets; pl often hairy
        21. Corolla 13–25 mm; lflets gen elliptic or widely ovate, gen opposite . . . . [2]***L. nevadensis*** var. ***nevadensis***
        21′ Corolla 7–12 mm; lflets gen linear to lanceolate, irregularly arranged or opposite . . . . ***L. lanszwertii***
         22. Corolla pale lavender to purple; KR, SNH, MP . . . . [2]var. ***aridus***
         22′ Corolla white to cream, banner often dark-striate; KR, NCoR . . . . [2]var. ***tracyi***
   13′ Tendrils, at least on some lvs, coiled and gen branched
    23. Lflets ± = stipules, 6–8; coastal beaches, dunes — NCo (Del Norte, Humboldt cos.) . . . . [2]***L. japonicus***
    23′ Lflets gen > stipules, 6–8 or not; gen inland
     24. Corolla white-yellow to bronze-orange (drying tan or dark), 11–14 mm; banner reflexed above
       middle; infl 10–15-fld, dense, gen 1-sided — fl appearing wide, blocky; KR, NCoR, SN. . . . . ***L. sulphureus***
     24′ Corolla not white-yellow (exc *Lathyrus lanszwertii* var. *tracyi*) to bronze-orange (occ fading
       tan-yellow), 11–14 mm or not; banner reflexed or bent from middle or below; infl 1–15-fld, gen ± open (1-sided)
      25. Corolla wine-red to crimson, 25–35 mm; banner reflexed 120–180° — s SCo, PR (San Diego Co.)
        . . . . ***L. splendens***
      25′ Corolla white to purple, 8–25 mm; banner bent ± 90°
       26. Lflets (4)6(8), opposite, glabrous or ± puberulent, fls 3–6 per infl, 15–20 mm; immediate coast,
         n NCo. . . . . [2]***L. palustris***

26′ Lflets, fls not entirely as above; immediate coast only in n CCo
27. Stipules > 1 cm wide, lflets 10–16, pls of KR, NCoR. . . . . ³*L. polyphyllus*
27′ Stipules < 1 cm wide and lflets ≤ 12 or pls not of KR, NCoR
   28. Infl gen 2-fld; ne DMoj (Grapevine Mtns, Inyo Co.) . . . . *L. hitchcockianus*
   28′ Infl occ 2–16-fld; not DMoj
     29. Corolla white or cream, fading yellow-tan (occ to lavender in *Lathyrus vestitus* var. *ochropetalus*) — NW
       30. St flanged or narrowly winged; stipules widely triangular, 0.5–0.8 cm wide;
         nw KR (Del Norte, w Siskiyou cos.) . . . . ²*L. delnorticus*
       30′ St angled or ± flanged; stipules gen lanceolate, < 0.5 mm wide; NW
         31. Lflets gen 4–6, glabrous or puberulent; corolla 9–12 mm. . . . . ²*L. lanszwertii* var. *tracyi*
         31′ Lflets gen 8–12, glabrous; corolla 14–16 mm . . . . *L. vestitus* var. *ochropetalus*
     29′ Corolla purple, lavender, or ± red (occ to white in *Lathyrus nevadensis* var. *nevadensis*)
       32. Stipules > 1 cm wide; lflets 10–16 — KR, NCoR. . . . . ³*L. polyphyllus*
       32′ Stipules < 1 cm wide; lflets 4–12
         33. Fls often crowded, gen 8–14, corolla 14–20 mm; lflets 8–12, gen subopposite to alternate;
          w CA-FP . . . . *L. vestitus*
           34. Corolla wine red to dark purple, 16–20 mm; banner gen reflexed > 90°; SCo, s ChI, PR
            . . . . var. *alefeldii*
           34′ Corolla pale lavender to purple, 14–18 mm; banner bent ± 90°; NCo, NCoR, CW, SW . . . . . var. *vestitus*
         33′ Fls gen spaced, either 2–10 and corollas 8–16 mm or 2–4 and corollas 13–25 mm; lflets
          4–10, opposite or subopposite; esp KR, CaR, SN
           35. Lflets gen opposite, gen elliptic or widely ovate; infl 2–4-fld; corolla 13–25 mm
            . . . . ²*L. nevadensis* var. *nevadensis*
          35′ Lflets gen subopposite or irregularly arranged, ovate to linear; infl 2–10-fld; corolla 8–16 mm
           36. Stipules widely ovate-triangular, often toothed, conspicuous; calyx glabrous exc lobes
            often finely ciliate; corolla gen 10–14 mm, banner upcurved above other petals. . . . . *L. brownii*
           36′ Stipules lanceolate, not toothed, often inconspicuous; calyx gen hairy;
            corolla 10–16 mm, banner gen more reflexed than other petals . . . . *L. lanszwertii* var. *lanszwertii*

L. angulatus L. Ann, glabrous. **ST**: often flanged, ± winged. **LF**: stipules small, narrow; lflets 2, 2–6 cm, thread-like to narrowly lanceolate; tendril coiled. **INFL**: 1-fld; peduncle 2–8 cm, axis extended beyond fl as bristle. **FL**: calyx lobes subequal, ± ≥ tube; corolla 9–12 mm, lavender or purple. **FR**: glabrous, faint-net-veined. 2*n*=14. Uncommon. Disturbed places; esp < 300 m. CA-FP; OR; native to Eur. Apr–Jun

L. biflorus T.W. Nelson & J.P. Nelson (p. 763) TWO-FLOWERED PEA Per, small, puberulent. **ST**: wiry, angled or flanged, not winged. **LF**: stipules small; lflets 4, gen 0.5–1(2) cm, lance-oblong; tendril not branched or coiled, bristle-like. **INFL**: 2-fld. **FL**: calyx tube > lobes; corolla 8–10 mm, green-white, dark-striate. **FR**: glabrous. Serpentine; ± 1300 m. NCoRH (Red Mtn, Humboldt Co.). Jun–Aug ★

L. brownii Eastw. (p. 763) Per, glabrous. **ST**: angled or flanged, not winged. **LF**: stipules widely ovate-triangular, often toothed; lflets 6–10, 1–5 cm, gen ovate or elliptic, occ lance-oblong; tendril gen branched, coiled, at least on some lvs. **INFL**: 2–6-fld. **FL**: calyx lobes < tube, often finely ciliate; corolla gen 10–14 mm, blue-purple to ± pink, petals upcurved, banner to ± 90°, above other petals, keel ± = wings. **FR**: glabrous. Dry, open woodland, streambanks; 1000–1700 m. NCo, KR, CaR, n&c SN; s OR. Apr–Jul

L. cicera L. (p. 763) Ann, glabrous. **ST**: narrowly winged. **LF**: stipules lanceolate; lflets 2, 3–6 cm, linear to lance-linear; tendril branched, coiled. **INFL**: 1-fld; axis not extended beyond fl. **FL**: calyx tube << lobes; corolla 10–15 mm, red-purple. **FR**: glabrous; upper suture furrowed. 2*n*=14. Uncommon. Disturbed areas; esp < 500 m. CA-FP; native to Eurasia. Apr–Jun

L. delnorticus C.L. Hitchc. (p. 763) DEL NORTE PEA Per, glabrous. **ST**: flanged or narrowly winged. **LF**: stipules small, 0.5–0.8 cm wide, widely triangular, toothed; lflets 10–16, gen subopposite, 3–5 cm, elliptic or lanceolate; tendril branched, coiled. **INFL**: 8–10-fld, ± dense. **FL**: calyx tube gen < lower lobes; corolla 10–15 mm, white, gen lavender-striate, fading yellow-tan. **FR**: glabrous. 2*n*=14. Streambanks, on serpentine; 250–700 m. nw KR (Del Norte, w Siskiyou cos.); OR. May–Jul ★

L. glandulosus Broich (p. 763) STICKY PEA Per, glandular-puberulent. **ST**: often strongly angled or flanged, gen not winged. **LF**: stipules small, narrow; lflets ± 14–16, gen subopposite to alternate, 3–5 cm, ovate to lanceolate, conspicuously gland-dotted abaxially; tendril branched, coiled. **INFL**: 5–12-fld. **FL**: calyx tube > upper lobes, ± ≤ lower; corolla 10–14 mm, lavender to purple. **FR**: gland-dotted. 2*n*=14. Roadsides, oak woodland; < 800 m. NCoR (Humboldt, n Mendocino cos.). Apr–Jun ★

L. hirsutus L. (p. 763) CALEY PEA Ann, ± glabrous, exc fr. **ST**: winged. **LF**: stipules narrow; lflets 2, 3–6 cm, narrowly elliptic to lance-oblong; tendril branched, coiled. **INFL**: 1–2-fld; axis not extended beyond fls. **FL**: calyx tube ± = lobes; corolla 9–14 mm, pink to blue-purple or 2-colored. **FR**: hairs with bulbous base. 2*n*=14. Uncommon. Disturbed places, wet meadows, creekbeds; < 1000 m. CA-FP (sporadic); to se US; native to Eurasia. Cult. May–Aug

L. hitchcockianus Barneby & Reveal BULLFROG MOUNTAIN PEA Per, glabrous or puberulent. **ST**: angled or flanged, not winged. **LF**: stipules small; lflets ± 4–6, 1–1.4 cm, linear to lanceolate; tendril coiled, occ branched. **INFL**: gen 2-fld, open. **FL**: calyx tube >> lobes; corolla 8–12 mm, lilac to purple. **FR**: glabrous. Washes, desert scrub; 1500 m. ne DMoj (Grapevine Mtns, Inyo Co.); w NV. Jun ★

L. japonicus Willd. SEASIDE PEA Per, glabrous or puberulent. **ST**: sprawling, angled or flanged, not winged. **LF**: stipules ± = lflets; lflets 6–8, 2,5–4.5 cm, ovate or elliptic, fleshy; tendril coiled, branched, or neither. **INFL**: 3–8-fld. **FL**: calyx tube < lower lobes; corolla 1.8–2.2 mm, blue-purple or 2-colored. **FR**: puberulent. 2*n*=14. Coastal beaches, dunes; < 30 m. NCo (Del Norte, Humboldt cos.); to AK, circumboreal, also Chile, Argentina. Jun–Aug ★

L. jepsonii Greene Per. **ST**: climbing, winged. **LF**: stipules small, gen narrow; lflets 10–16, gen subopposite to alternate, 3.5–5.5 cm, lanceolate or lance-oblong; tendril branched, coiled. **INFL**: 6–15-fld. **FL**: calyx tube > upper lobes, ± = lower; corolla 15–20 mm, gen pink to pink-purple. **FR**: glabrous. 2*n*=14.

var. *californicus* (S. Watson) Hoover Pl gen puberulent, occ glabrous, ± not robust, with st << 2.5 m. Forest, open areas; < 1500 m. KR, NCoR, CaR, n SN, GV, CW, SCoRO. Intermediates to *L. vestitus* may be hybrids. Apr–Aug

var. *jepsonii* (p. 763) DELTA TULE PEA Pl glabrous, often robust, with st < 2.5 m. Coastal, estuarine marshes; < 30 m. GV (esp Solano Co., nearby CCo). Apr–Aug ★

L. lanszwertii Kellogg   Per. **ST:** angled or ± flanged, not winged. **LF:** stipules small, gen narrow; lflets 4–10, opposite or subopposite, 1.5–8 cm; tendril branched, coiled, a bristle, or 0. **INFL:** 2–10-fld. **FL:** calyx tube gen > upper lobes; corolla 7–16 mm. **FR:** glabrous.

var. ***aridus*** (Piper) Jeps. (p. 763)   Pl gen puberulent. **ST:** gen ascending, not climbing. **LF:** lflets narrowly lanceolate to linear; tendril often a bristle or 0. **FL:** corolla 7–11 mm, pale lavender to purple. 2*n*=14. Open, dry woodland, meadows; 1200–2000 m. KR, SNH, MP; to WA. Intergrades with *L. lanszwertii* var. *l.* in CA. May–Jul

var. ***lanszwertii*** (p. 763)   Pl gen puberulent. **ST:** climbing or ascending. **LF:** lflets lance-elliptic to lanceolate; tendril gen coiled. **FL:** corolla 10–16 mm, purple to lavender. 2*n*=28. Open slopes, pine forest; 1200–2000 m. KR, GB; to WA, ID, UT. May–Jul

var. ***tracyi*** (Bradshaw) Isely   Pl glabrous or puberulent. **ST:** climbing or erect. **LF:** lflets linear to elliptic; tendril well developed or a bristle. **FL:** corolla 9–12 mm, white to cream, banner often dark-striate. 2*n*=14. Conifer forest; 200–1100 m. KR, NCoR. May–Jul

L. latifolius L. (p. 763)   PERENNIAL SWEET PEA   Per, glabrous, often robust. **ST:** wings often ≥ 2 mm wide. **LF:** stipules small or large; lflets 2, 5–14 cm, lanceolate to ovate; tendril branched, coiled. **INFL:** 4–15-fld. **FL:** calyx tube > lobes; corolla 20–25 mm, pink, pink-purple, or red. **FR:** glabrous. 2*n*=14. Disturbed areas, esp roadsides; gen < 2000 m. CA-FP; sporadic to e US; native to Eur. Cult as orn. Locally conspicuous. Apr–May

L. littoralis (Nutt.) Walp. (p. 763)   Per, densely gray-hairy. **ST:** prostrate or ascending, angled or flanged, not winged. **LF:** stipules ≥ lflets; lflets 4–8, 1–2 cm, obovate to short-oblong; tendril bristle-like. **INFL:** 4–8-fld, dense. **FL:** calyx tube < to > lobes; corolla 15–18 mm, 2-colored (pink-purple, white). **FR:** elliptic to oblong, hairy. Open coastal dunes; < ± 5 m. NCo, n&c CCo; to BC. Apr–Jul

L. nevadensis S. Watson var. ***nevadensis*** (p. 763)   Per, glabrous or puberulent. **ST:** angled or flanged, not winged. **LF:** stipules small, narrow or wide; lflets 4–8, gen opposite, 1.5–± 4 cm, gen elliptic or widely ovate; tendril branched and coiled or bristle-like. **INFL:** 2–4-fld. **FL:** calyx tube > lobes; corolla 13–25 mm, white to blue-purple. **FR:** glabrous. 2*n*=28. Conifer, mixed forests; 450–2300 m. KR, NCoR, CaR, n&c SN; to WA. Infraspecific taxa based on tendrils, lflet number, fl size possibly not warranted in CA. Apr–Jul

L. odoratus L.   SWEET PEA   Ann, hairy. **ST:** sprawling or climbing, gen winged. **LF:** stipules small; lflets 2, 3–6 cm, ovate to narrowly elliptic, length 1.2–3 × width; tendril branched, coiled. **INFL:** 2–4-fld, open. **FL:** calyx tube ± = lobes; corolla 20–30 mm, white, pink, or purple. **FR:** < 6 cm, hairy. 2*n*=14. Uncommon. Disturbed areas; esp < 1000 m. CA-FP; to e US; native to Eur. Widely cult. Apr–Jun

L. palustris L. (p. 763)   MARSH PEA   Per, gen glabrous or ± puberulent. **ST:** angled, flanged, or narrowly winged. **LF:** stipules lanceolate, lower lobe conspicuous; lflets (4)6(8), opposite, 2.5–5.5 cm, elliptic to lance-oblong; tendril branched, coiled. **INFL:** 3–6-fld. **FL:** calyx tube ± = lower lobes; corolla 15–20 mm, banner bent ± 90°, pink-purple or purple (white). **FR:** glabrous or initially glandular. 2*n*=14. Moist or wet coastal areas; gen < 100 m. n NCo; to AK, ne US, circumboreal. May–Aug ★

L. polyphyllus Nutt. (p. 763)   Per, glabrous, often robust. **ST:** angled or flanged, not winged. **LF:** stipules gen conspicuous, gen widely ovate, > 1 cm wide; lflets 10–16, subopposite to alternate, 3–6 cm, elliptic to ovate; tendril branched, coiled, or a bristle. **INFL:** 6–12-fld, ± 1-sided. **FL:** lower calyx lobe narrow, ± = tube, lateral triangular, not widened above base; corolla 16–20 mm, purple. **FR:** glabrous. 2*n*=14. Forest; < 1500 m. KR, n&c NCoR; to WA. Apr–Aug

L. rigidus T.G. White (p. 763)   RIGID PEA   Per, glabrous. **ST:** erect, many, clustered, angled or flanged, not winged. **LF:** stipules gen narrow, ± 1/2 lflet; lflets 6–10, 1–3 cm, lanceolate to elliptic; tendril bristle-like. **INFL:** 2–5-fld, dense. **FL:** calyx tube gen ± = lower lobes, > upper; corolla 17–23 mm, white to pink. **FR:** glabrous. 2*n*=14. Sagebrush scrub, disturbed areas; 800–1200 m. MP (Modoc Co.); to OR, ID, NV. Apr–Jul ★

L. sphaericus Retz.   Ann, glabrous. **LF:** stipules small, narrow; lflets 2, 2–6 cm, linear to narrowly lanceolate; tendril gen coiled. **INFL:** 1-fld; peduncle 0.5–2 cm, axis extended beyond fl as short bristle. **FL:** calyx lobes ± ≥ tube; corolla 8–14 mm, orange-red. **FR:** glabrous, longitudinally veined. 2*n*=14. Uncommon. Disturbed places; < 1000 m. CA-FP; OR; native to Eurasia, Afr. Apr–May

L. splendens Kellogg   PRIDE-OF-CALIFORNIA   Per, glabrous or puberulent. **ST:** gen climbing, angled or flanged, not winged. **LF:** stipules gen small, occ wide, wavy-margined; lflets gen 6–8, 2–4 cm, linear to widely ovate; tendril coiled. **INFL:** 4–6-fld, open. **FL:** calyx tube ± = lower lobe, > others; corolla 25–35 mm, wine-red to crimson, banner reflexed 120–180°; ovary glandular-puberulent. **FR:** glabrous. 2*n*=14. Chaparral; < 1050 m. s SCo, PR (San Diego Co.); Baja CA. Apr–Jun ★

L. sulphureus A. Gray   Per, glabrous. **ST:** angled or flanged, not winged. **LF:** stipules 1–2.5 cm, gen wide; lflets 6–12, often alternate, 1.5–4 cm, elliptic to ovate; tendril branched, coiled. **INFL:** 10–15-fld, dense, gen 1-sided. **FL:** calyx tube > lobes; corolla 11–14 mm, white-yellow to bronze-orange, drying tan or dark, appearing wide, blocky, banner reflexed above middle. **FR:** glabrous. 2*n*=14. Foothill woodland to fir forest; 600–2500 m. KR, NCoR, SN; sw OR. [*L. s.* var. *argillaceus* Jeps.] Apr–Jul

L. tingitanus L.   TANGIER PEA   Ann, glabrous. **ST:** winged. **LF:** stipules small or conspicuous, wide; lflets 2, 2–6 cm, lanceolate to widely ovate, length 1.2–6.5 × width; tendril branched, coiled. **INFL:** 2–3-fld, open. **FL:** calyx tube > lobes; corolla 25–30 mm, maroon to crimson. **FR:** < 10 cm, glabrous. 2*n*=14. Gen disturbed areas, esp coastal; gen < 500 m. CA-FP; OR, sporadic in w US; native to Eur, n Afr. Cult as orn. Apr–Jul

L. torreyi A. Gray (p. 763)   Per, puberulent. **ST:** angled or flanged, not winged. **LF:** stipules small, narrow or wide; lflets 10–15, 1–2 cm, elliptic to ovate; tendril a bristle or 0. **INFL:** 1–2-fld. **FL:** calyx tube gen < lower lobes; corolla 8–10 mm, lilac to pale purple-blue. **FR:** puberulent. 2*n*=14. Open woodland; < 800 m. NCo, NCoR, SnFrB; to WA. Apr–Jul

L. vestitus Nutt.   Per. **ST:** often sharply angled or flanged. **LF:** stipules small, entire to wide, wavy-margined; lflets gen 8–12, 2–4.5 cm, linear to elliptic; tendril branched, coiled. **INFL:** 8–15-fld, often dense. **FL:** calyx tube < or > lower lobes (lower lobes in some phases ± wider above base); banner bent or reflexed ± 90° or more. **FR:** glabrous or initially puberulent. 2*n*=14. Intergrading complex of taxa, local variants (see Broich, 1987).

var. ***alefeldii*** (T.G. White) Isely   Pl glabrous. **FL:** corolla 16–20 mm, wine red to dark purple, banner gen reflexed > 90°. Chaparral; < 1200 m. SCo, s ChI, PR. Intergrades with *L. vestitus* var. *v.* Feb–Jul

var. ***ochropetalus*** (Piper) Isely   Pl glabrous. **FL:** corolla 14–16 mm, lavender or white. Conifer or mixed forest; gen < 150(400) m. NCo, KR, NCoRO. Feb–Jul

var. ***vestitus*** (p. 763)   Pl glabrous or puberulent. **FL:** corolla 14–18 mm, pale lavender to purple, banner bent ± 90°. Conifer forest in n to chaparral and oak woodland in s; < 1500 m. NCo, NCoR, CW, SW. Variable; may hybridize with *L. jepsonii* var. *californicus.* Feb–Jul

LOTUS

Luc Brouillet

Ann, per, unarmed. **LF:** gen odd-1-pinnate; stipules gland-like, reduced to bumps, or inconspicuous; lflets 3–9, lower 2 in stipular position or not, others ± palmately arranged. **INFL:** umbel or 1–3-fld, axillary, peduncled, bract 1–3-parted. **FL:** corolla gen yellow, in age darkening; 9 filaments fused, 1 free. **FR:** dehiscent, exserted from calyx, linear to narrowly oblong, ± beaked. **SEED:** few to several. ± 125 spp.: Eur, Afr, to e Asia, Australia, New Caledonia; cult as forage, ground cover, orn. (Greek: derivation unclear) [Brouillet 2008 J Bot Res Inst Texas 2:387–394] Other taxa in TJM (1993) moved to *Acmispon* and *Hosackia.*

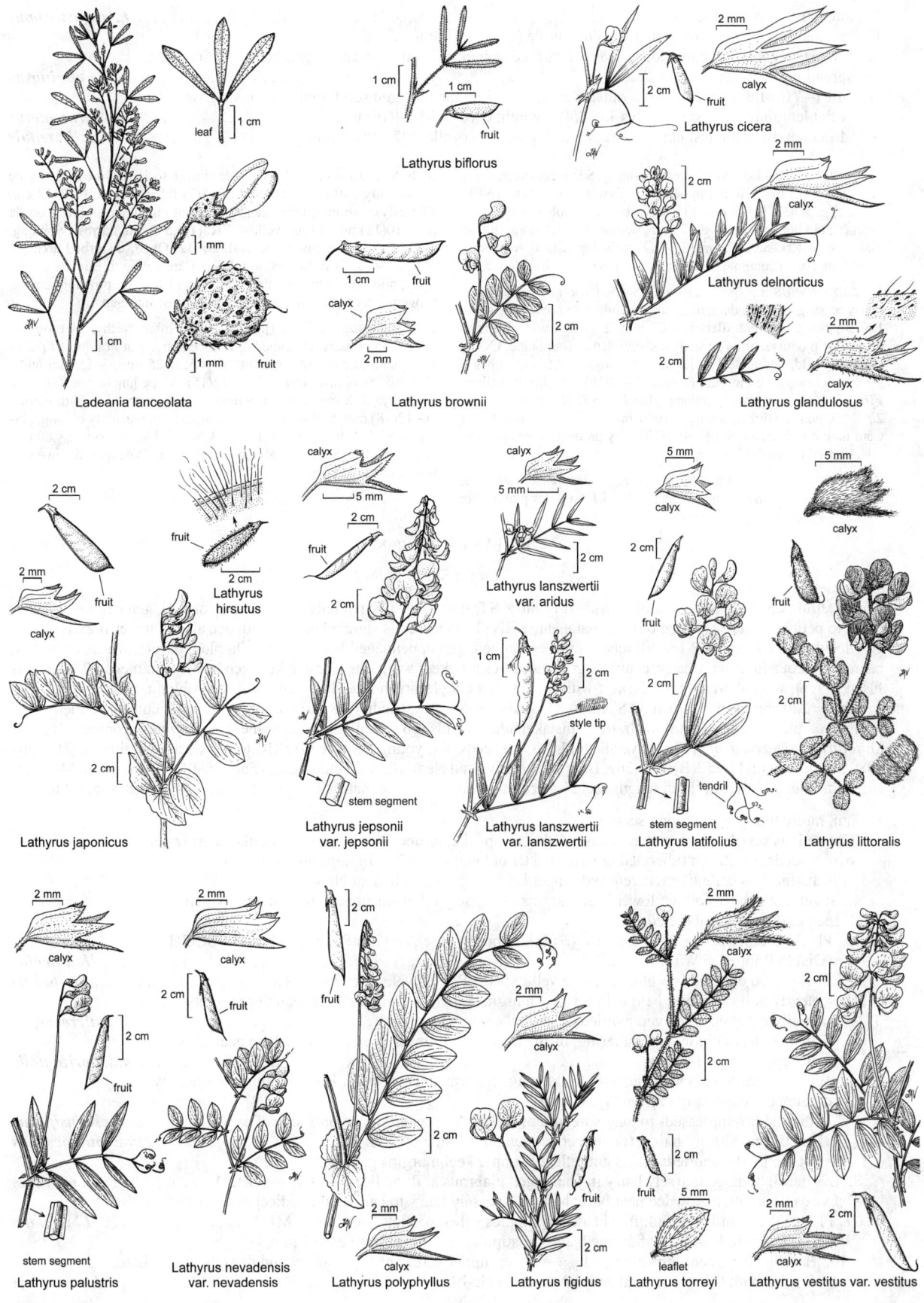

Ladeania lanceolata

leaf — 1 cm
fruit — 1 mm
1 cm

Lathyrus biflorus

1 cm
1 cm
fruit

Lathyrus cicera

2 mm
2 cm
fruit
calyx

Lathyrus brownii

1 cm fruit
calyx
2 mm
2 cm

Lathyrus delnorticus

2 cm
2 mm
calyx

Lathyrus glandulosus

2 cm
2 mm
calyx

Lathyrus japonicus

2 cm
2 mm
fruit
calyx

Lathyrus hirsutus

fruit
2 cm

Lathyrus jepsonii var. jepsonii

calyx
5 mm
2 cm
fruit
2 cm
stem segment

Lathyrus lanszwertii var. aridus

calyx
5 mm
2 cm

Lathyrus lanszwertii var. lanszwertii

1 cm
2 cm
style tip

Lathyrus latifolius

5 mm
calyx
2 cm
fruit
2 cm
tendril
stem segment

Lathyrus littoralis

5 mm
calyx
2 cm
fruit
2 cm

Lathyrus palustris

2 mm
calyx
2 cm
fruit
stem segment

Lathyrus nevadensis var. nevadensis

2 mm
calyx
2 cm
fruit

Lathyrus polyphyllus

2 cm
fruit
2 mm
calyx
2 cm
2 mm
calyx

Lathyrus rigidus

fruit
2 cm

Lathyrus torreyi

2 cm
fruit
5 mm
leaflet

Lathyrus vestitus var. vestitus

2 mm
2 cm
calyx
2 mm
calyx
2 cm

1. Ann, hairy; infl 1–3-fld; corolla 5–8(12) . . . . ***L. angustissimus***
1′ Per, glabrous or hairy; infl gen 3–12-fld; corolla (6)7–14(18) mm
  2. Infl (5)8–12(15)-fld; calyx lobes often outcurved in bud; rhizome present; st gen hollow, hairs often
    spreading. . . . . ***L. uliginosus***
  2′ Infl 1–7(10)-fld; calyx lobes not outcurved in bud; rhizome 0; st gen solid, glabrous or strigose
    3. Lflet length of longest upper lvs 1.6–3(4) × width; corolla 10–14(16) mm. . . . . ***L. corniculatus***
    3′ Lflet length of longest upper lvs (2.5)3.2–5 × width; corolla (6)7–10(12) mm. . . . . ***L. tenuis***

L. angustissimus L.   Ann, spreading-hairy. **ST**: ± prostrate, solid. **LF**: lflets 8–12 mm, elliptic to narrowly obovate, tip obtuse. **INFL**: 1–3-fld; peduncle 0.3–2 cm. **FL**: calyx 6–7 mm, lobes > tube, not outcurved in bud; corolla 5–8(12) mm, yellow. **FR**: 1.5–3 cm, linear, glabrous. **SEED**: several. 2*n*=12,24. Disturbed grassland, roadsides; < 1000 m. NCo (Sonoma, Mendocino cos.); native to Eur. Jun–Aug

L. corniculatus L. (p. 771) BIRD'S-FOOT TREFOIL Per, glabrous or strigose. **ST**: decumbent or ascending to erect, gen solid. **LF**: stipules gland-like; lflets 5, 4–22 mm, 2–11 mm wide, ovate to obovate, tip obtuse, often abruptly soft-pointed or acuminate. **INFL**: (1)3–7(10)-fld; peduncle 1.5–12 cm. **FL**: calyx 5–7.5 mm, lobes ± = tube, not outcurved in bud; corolla 10–14(16) mm, bright yellow. **FR**: 1.5–3(4) cm, narrowly oblong, glabrous. **SEED**: few to several. 2*n*=24. Open, disturbed areas; < 1000 m. CA-FP, GB; to n US, s Can; native to Eurasia. Some pls TOXIC by production of cyanide-releasing compounds. Jun–Sep

L. tenuis Willd.   Per, ± glabrous or young nodes sparsely strigose. **ST**: decumbent or ascending to erect, solid. **LF**: stipules gland-like; lflets 5, 5–15 mm, 1–4 mm wide, linear to lance-linear, tip acute or abruptly soft-pointed. **INFL**: 1–4(7)-fld; peduncle (1)2–12 cm. **FL**: calyx 4–6 mm, lobes gen < tube, not outcurved in bud; corolla (6)7–10(12) mm, bright yellow. **FR**: (1)1.5–3 cm, narrowly oblong, glabrous. **SEED**: few to several. 2*n*=12. Open, disturbed areas; < 600 m. NCo, CCo, SnFrB; to n US, s Can; native to Eur. [*L. glaber* Mill., nom. rej.] Diploids hard to distinguish from, probably among progenitors of allotetraploid *L. corniculatus*. Jun–Sep

L. uliginosus Schkuhr (p. 771) Per, often fleshy, glabrous or long-shaggy-hairy, rhizomed. **ST**: ascending, gen hollow. **LF**: stipules gland-like or not apparent; lflets 5, 8–25 mm, 3–15 mm wide, obovate, tip obtuse. **INFL**: (5)8–12(15)-fld; peduncle gen 8–14 cm. **FL**: calyx 7–8 mm, lobes ± = tube, often outcurved in bud; corolla 10–12(18) mm, yellow. **FR**: (1.5)2–3(3.5) cm, narrowly oblong, glabrous. **SEED**: few to several. 2*n*=24. Wet fields, roadsides; < 800 m. NCo, NCoRO, SnFrB, expected elsewhere; to WA, sporadic in e US; native to Eur. Jun–Sep

LUPINUS LUPINE

Teresa Sholars

Ann to shrub; cotyledons gen petioled, withering early. **ST**: gen erect. **LF**: palmately compound [or not], gen cauline; stipules fused to petiole; lflets 3–17, gen oblanceolate, entire. **INFL**: raceme, fls spiraled or whorled, occ also in lower lf axils; bracts gen deciduous. **FL**: calyx 2-lipped, lobes entire or toothed, gen appendaged between; corolla blue, purple, white, or yellow, banner glabrous to densely hairy, centrally grooved, sides reflexed, wing tips ± fused, keel gen beaked; stamens 10, filaments fused, 5 long with short anthers, 5 short with long anthers; style brush-like. **FR**: dehiscent, gen oblong. **SEED**: 2–12, gen smooth. ± 220 spp.: esp w N.Am, w S.Am to e US, also trop S.Am, Medit to w Asia, e trop Afr.; some cult for fodder, green manure, edible seed, orn. (Latin: wolf, from mistaken idea that pls rob soil of nutrients) Some (e.g., *Lupinus arboreus*, *Lupinus latifolius*, *Lupinus leucophyllus*) have alkaloids (esp in seeds, frs, young herbage) TOXIC to livestock (esp sheep). [Barneby 1989 Intermountain Flora 3(B):237–267; Isely 1998 Native and Naturalized Leguminosae (Fabaceae) US. M.L. Bean Museum, Brigham Young University] Infl length excludes peduncle; some CA spp. naturalized in e N.Am, S.Am, Australia, s Afr;.

1. Ann, rarely living > 1 growing season
  2. Cotyledons sessile, persistent at pl base, disk- or cup-like, or deciduous, leaving circular scar; fr gen
    ovoid; seeds gen 2, gen tubercled or wrinkled (fr oblong; seeds 2–6 in *Lupinus odoratus*)
    3. Fls distinctly whorled; bracts reflexed; upper keel margins ciliate near claw
      4. St gen not clearly hollow; lower keel margins as densely ciliate as upper; lflets adaxially gen hairy;
        seed dark brown, tubercled
        5. Pl 30–75 cm; petals gen pale yellow (± pink or bright blue); NW, CaRF, e SnFrB and n SCoRI
          (Diablo Range), w WTR. . . . . ***L. luteolus***
        5′ Pl 100–200 cm; petals ± blue, in age ± yellow; NCoRO (Covelo, Mendocino Co.). . . . . ***L. milo-bakeri***
      4′ St clearly hollow at least below; lower keel margins glabrous or less ciliate than upper; lflets
        adaxially glabrous; seed gen mottled, wrinkled or smooth . . . . ***L. microcarpus***
        6. Wings widely elliptic, not withering, in age translucent; lower keel margins ciliate near claw;
          calyx appendages 1–2 mm. . . . . var. ***horizontalis***
        6′ Wings linear to oblanceolate, withering, in age not translucent; lower keel margins occ sparsely
          ciliate near claw; calyx appendages gen 0
          7. Bracts short-appressed- to long-spreading-hairy; fr ± spreading, gen on 1 side of axis. . . . . var. ***densiflorus***
          7′ Bracts long-shaggy-hairy; fr gen ± erect, gen on > 1 side of axis . . . . var. ***microcarpus***
  3′ Fls not distinctly whorled; bracts not reflexed; upper keel margins glabrous
    8. Lvs basal; herbage sparsely hairy in youth, gen glabrous at fl; pedicel 3–5 mm — GB, D. . . . . ***L. odoratus***
    8′ Lvs cauline, often crowded near base; herbage sparsely hairy to canescent; pedicel gen < 3 mm
      9. Pl 1–2(3) cm; infls (1)2-fld; free blades of stipules < 1 mm; lflets 2–7 mm — MP . . . . ***L. uncialis***
      9′ Pl 3–20 cm; infls 3–many fld; free blades of stipules > 1 mm; lflets gen > 7 mm
        10. Herbage canescent; upper calyx lip ± = lower; upper suture of fr gen wavy and densely stiff-ciliate;
          fr sides with hairs short, inflated, when dry scale-like — D. . . . . ***L. shockleyi***

10′ Herbage sparsely to densely ± long-hairy; upper calyx lip < lower; upper suture of fr not obviously
wavy and densely stiff-ciliate; fr sides with hairs long, not inflated.
  11. Peduncle 0–1 cm; infl < lvs; fr oblong, narrowed between seeds — GB (Modoc, Inyo cos.)
        . . . . ***L. pusillus*** var. ***intermontanus***
  11′ Peduncle 2–5(10) cm; infl > lvs; fr ovate, not narrowed between seeds
    12. Infl 1–2.5 cm; pedicels 0.5–1.5 mm; calyx appendages present; seed smooth;
      s MP, W&I, e DMoj. . . . . ***L. brevicaulis***
    12′ Infl 3–10 cm; pedicels 2–3 mm; calyx appendages 0; seed wrinkled; W&I, e DMoj. . . . . ***L. flavoculatus***
2′ Cotyledons petioled, gen withering, gen deciduous, not leaving circular scar; fr oblong; seeds gen > 2, smooth
  13. Lower and often upper margins of keel ciliate near claw, both glabrous near tip
    14. Fls whorled — petals gen blue-purple (white, pink, lavender); c&s NW, GV, CW, SW . . . . [2]***L. succulentus***
    14′ Fls spiralled (if appearing ± whorled, pedicel bases clearly spiralled, not whorled)
      15. Infl < peduncle — banner yellow, wings gen pink (± white), keel white; SN, n SCoRO
        (Monterey Co.), SnGb, SnBr . . . . ***L. stiversii***
      15′ Infl gen > peduncle
        16. Petals gold-yellow or white; pedicel recurving . . . . ***L. citrinus***
          17. Petals gold-yellow, drying translucent, ± purple; c SNF (Fresno, Madera cos.) . . . . var. ***citrinus***
          17′ Petals white, pink- or lavender-tinged or not, banner occ drying translucent, yellow;
            c SNF (Mariposa Co.). . . . . var. ***deflexus***
        16′ Petals blue or dark pink to magenta; pedicel not recurving
          18. Herbage short-appressed- and stiff-spreading-stinging-hairy; lflets 10–20 mm wide —
            c&s CW (SnFrB, SCoRO, possibly others), SW. . . . . ***L. hirsutissimus***
          18′ Herbage not stiff-spreading-stinging-hairy; lflets gen < 10 mm wide
            19. Petiole flat, lflet-like; keel tip stout, blunt, upper margins ciliate middle to claw —
              c&s CW, SW . . . . ***L. truncatus***
            19′ Petiole not flat, not lflet-like; keel tip slender, pointed, upper margins gen glabrous
              20. Pedicel 5–9 mm; bract 10–15 mm, >> fl bud — SNF, Teh, deltaic GV, SCoRO. . . . . ***L. benthamii***
              20′ Pedicel < 5 mm; bract 3–8 mm, < to ± > fl bud
                21. Lflets 5–10 mm wide, adaxially glabrous; petals dark pink to magenta; e DMoj, DSon. . . . . . . ***L. arizonicus***
                21′ Lflets 2–4 mm wide, adaxially hairy at least near margins; petals gen blue (± pink);
                  s SCoR, SW, DMoj . . . . ***L. sparsiflorus***
13′ Lower, upper keel margins glabrous or upper ciliate only near tip
  22. Fls spiralled; bracts gen persistent; upper keel margins gen glabrous
    23. Pl with central peduncle erect, lateral often decumbent; open or disturbed areas, burns,
      s SN, c&s CW, SW, D. . . . . ***L. concinnus***
    23′ Pl with peduncles decumbent; stable dunes, s CCo (Nipomo Dunes, sw San Luis Obispo Co.). . . . . ***L. nipomensis***
  22′ Fls whorled at some or all nodes; bracts deciduous; upper keel margins gen ciliate near tip
    24. Banner longer than wide; pedicel gen < 3 mm
      25. Fr gen 3–6 mm wide; keel gen ciliate on upper margins near tip; abundant, < 1600 m, CA-FP,
        DMoj. . . . . ***L. bicolor***
      25′ Fr 6–9 mm wide; keel glabrous; uncommon, < 600 m, NCoRO, CaRF, SNF, SnFrB, SCoRO . . . . . ***L. pachylobus***
    24′ Banner as wide as or wider than long; pedicel gen > 3 mm
      26. Upper keel margins with (occ inconspicuous) tooth near middle — NCo, NCoR, CCo, SnFrB . . . . ***L. affinis***
      26′ Upper keel margins without tooth
        27. Fr 4–7 mm wide — < 1300 m. . . . . ***L. nanus***
        27′ Fr 8–10 mm wide
          28. Pedicel 4–5 mm; herbage sparsely hairy; ChI (San Clemente Island) . . . . ***L. guadalupensis***
          28′ Pedicel 6–8 mm; herbage densely hairy; c SNF (Mariposa, Tuolumne cos.). . . . . ***L. spectabilis***
1′ Per to shrub
  29. Upper keel margin ciliate from claw to middle, glabrous from middle to tip
    30. Pl fleshy; infl 9–15 cm; open or disturbed areas; fr 35–50 mm . . . . [2]***L. succulentus***
    30′ Pl not fleshy; infl 16–60 cm; open to shady, moist areas; fr 20–45 mm . . . . ***L. latifolius***
      31. Fl 13–18 mm
        32. St densely hairy; SnFrB . . . . var. ***dudleyi***
        32′ St subglabrous to sparsely strigose; c SNH, s SN, SW . . . . var. ***parishii***
      31′ Fl 8–14 mm
        33. Fl 8–10 mm
          34. St glabrous to strigose; MP . . . . var. ***barbatus***
          34′ St strigose; KR, CaR, MP . . . . var. ***viridifolius***
        33′ Fl 10–14 mm
          35. Wings covering most of keel; pl 4–24 dm; SN, SNE. . . . . var. ***columbianus***
          35′ Wings not covering most of keel; pl 3–12 dm; NW, CW, SW. . . . . var. ***latifolius***
  29′ Upper keel margin glabrous or variously ciliate, but not as at 29.
    36. Calyx with spur 1–3 mm
      37. Wings with dense patch of hair outside near tip; lf green, strigose; CaRH, SNH, SnGb, GB. . . . . ***L. arbustus***
      37′ Wings glabrous; lf silver-silky; c SNH, GB . . . . ***L. argenteus*** var. ***heteranthus***

36′ Calyx with bulge or spur 0–1 mm
  38. Banner back gen hairy (best seen in buds)
    39. Upper keel margin ± glabrous
      40. St prostrate to decumbent, gen < 4 dm . . . . [3]***L. albifrons*** var. ***collinus***
      40′ St erect, gen > 4 dm
        41. Shrub — coastal strand, dunes . . . . ***L. chamissonis***
        41′ Per
          42. Lf green
            43. Petals pale yellow to orange-yellow; CaRH, n&c SNH, GB . . . . ***L. angustiflorus***
            43′ Petals white to purple; n SNH . . . . ***L. apertus***
          42′ Lf silver- to white-hairy
            44. Petals gen yellow; n SNH (Plumas Co.) . . . . ***L. dalesiae***
            44′ Petals cream-yellow to -white, pale yellow, lavender, or blue; s SNH, TR, SNE;
              45. Petals lavender to blue; 1500–3000 m; s SNH, TR . . . . [2]***L. elatus***
              45′ Petals cream to pale yellow; gen > 2500 m; gen c SNH (n Inyo Co.) . . . . ***L. padre-crowleyi***
  39′ Upper keel margin ciliate
    46. Subshrub or shrub
      47. Keel gen lobed near base; fl 9–18 mm; s SNH, Teh, SW, SNE, D . . . . ***L. excubitus***
        48. Shrub, gen > 7 dm
          49. Fl 9–13 mm; lf silver-hairy; DMoj, adjacent s SNH, SNE . . . . var. ***excubitus***
          49′ Fl 14–18 mm; lf green-hairy; SnGb, SnBr, PR . . . . var. ***hallii***
        48′ Subshrub, gen < 7 dm
          50. Infl 14–40 cm; Teh, SnGb, SnBr . . . . [2]var. ***austromontanus***
          50′ Infl 3–14 cm; SnGb, SnBr, DSon
            51. Petiole < 12 cm; SnGb, SnBr . . . . var. ***johnstonii***
            51′ Petiole gen > 12 cm; sw DSon . . . . var. ***medius***
      47′ Keel gen unlobed near base; fl 9–16 mm; CA-FP . . . . ***L. albifrons***
        52. Subshrub, gen < 5 dm, base, some lower branches woody
          53. Lf woolly to shaggy-hairy; SCoRO (Santa Lucia Range) . . . . [2]var. ***abramsii***
          53′ Lf appressed-hairy, not woolly or shaggy-hairy; NW, SNF, CW . . . . [3]var. ***collinus***
        52′ Shrub, gen > 5 dm, woody ± throughout
          54. Lf woolly to shaggy-hairy; SCoRO (Santa Lucia Range) . . . . [2]var. ***abramsii***
          54′ Lf hairy but not woolly or shaggy-hairy; NW, SN, CW, s SCoRO, SCo, ChI, WTR, SnGb, PR
            55. Bract 4–8 mm; NW, SN, CW, SCo, ChI, WTR, SnGb, PR . . . . var. ***albifrons***
            55′ Bract 10–24 mm; CCo, nw SnFrB, s SCoRO, ChI, w WTR . . . . var. ***douglasii***
  46′ Per
    56. Lflet 10–30 mm at widest point
      57. Petals yellow — SnGb . . . . [2]***L. peirsonii***
      57′ Petals pink to ± purple or blue (or pale yellow)
        58. Peduncle 13–20 cm; lf long-spreading-hairy; petals light blue, pink, or straw (often drying pale
           yellow); SCoRO (Santa Lucia Range) . . . . [2]***L. cervinus***
        58′ Peduncle 8–15 cm; lf short-appressed-hairy; petals purple to violet; s NCoRI (Napa, Lake,
           Sonoma cos.) . . . . [2]***L. sericatus***
    56′ Lflet gen < 10 mm at widest point
      59. Pl < 1 dm
        60. Fls in many crowded whorls; 2000–3500 m; SNH, SnBr, SNE . . . . ***L. breweri*** var. ***grandiflorus***
        60′ Fls in few spaced whorls; 1500–3000 m; e KR (Mount Eddy, Siskiyou Co.),
          NCoRO (King Range, sw Humboldt Co.) . . . . [2]***L. lapidicola***
      59′ Pl > 1 dm
        61. Calyx gen bulged or spurred < 1 mm; CaR, n SNH, GB, DMtns . . . . ***L. argenteus*** (in part)
          62. Lvs basal and cauline; petioles (5)7–15 cm — c SNH, SNE . . . . var. ***montigenus***
          62′ Lvs cauline; petioles 1–10 cm
            63. St hairs appressed; lf gen appearing green; fl (8)10–12 mm; CaR, n SNH, GB, DMtns . . . . . . var. ***argenteus***
            63′ St hairs spreading; lf gray to silver; fl 8–10 mm; CaRH, SNH, Wrn, SNE . . . . var. ***palmeri***
        61′ Calyx not bulged or spurred
          64. Lf hairs appressed, occ sparse
            65. Bracts deciduous; lflets 20–50 mm, silver-silky; NW, CaR, n SNH . . . . ***L. obtusilobus***
            65′ Bracts persistent; lflets 30–80(130) mm, green, hairy;
              SNE (Big Pine Creek, Inyo Co.) . . . . ***L. pratensis*** var. ***eriostachyus***
          64′ Lf hairs ± spreading, dense
            66. Fl gen 12–15 mm; pl 2–3.5 dm; st prostrate to matted — SN . . . . [2]***L. grayi***
            66′ Fl gen 10–13 mm; pl 3–9 dm; st erect
              67. Petals lavender to purple, often turning brown; banner back densely hairy;
                KR, CaRH, MP . . . . ***L. leucophyllus***
              67′ Petals ± blue to purple; banner back glabrous to ± hairy; s SCoRO (San Luis
                Obispo Co.) . . . . [2]***L. ludovicianus***

38′ Banner back ± glabrous
  68. Subshrub or shrub
    69. Shrub, gen 5–20 dm; woody sts erect
      70. Infl 10–30 cm; petals gen yellow (lilac to purple, esp n of c NCo); NCo, CCo, SnFrB, SCo, ChI .... **L. arboreus**
      70′ Infl 20–45 cm; petals violet to blue; SW .... **L. longifolius**
    69′ Subshrub, < 5 dm; woody sts prostrate to decumbent
      71. Fl 14–18 mm — Teh, SnGb, SnBr .... ²**L. excubitus** var. **austromontanus**
      71′ Fl 10–16 mm
        72. Keel ± unlobed near base — fl 10–15 mm; NW, SNF, CW.................... ³**L. albifrons** var. **collinus**
        72′ Keel ± lobed near base
          73. Fl 10–13 mm; petals purple exc banner spot ± white; roots gen yellow; petioles gen < 2 × lflets,
            2–6 cm; n&c NCo, CCo (Point Reyes) .... ²**L. littoralis**
          73′ Fl 11–16 mm; petals white, yellow, rose, or purple, often on 1 pl; roots not yellow; petioles
            gen 2 × lflets, 4–10 cm; NCo, n&c CCo .... ²**L. variicolor**
  68′ Per (rarely woody at base)
    74. Lf adaxially glabrous to sparsely hairy, green (exc *Lupinus elmeri*)
      75. Upper keel margin ciliate (occ sparsely so in *Lupinus onustus*)
        76. Pl 3.5–10 dm; lvs cauline — NCo.. **L. rivularis**
        76′ Pl < 3.5 dm; lvs gen basal or cauline, clustered near base
          77. Fl 8–11 mm, bract 3–4 mm; lf abaxially silky-hairy; KR, CaRH, n SNH .... **L. onustus**
          77′ Fl 12–18 mm, bract 4–7 mm; lf abaxially long-stiff-hairy; MP .... **L. saxosus**
      75′ Upper keel margin gen glabrous (occ a few hairs)
        78. Petals ± yellow — NW, CaR
          79. Bract 2–7 mm; pl 4–6 dm; petals bright yellow to orange-yellow; KR, CaR ..... ²**L. croceus**
          79′ Bract 7–14 mm; pl 6–9 dm; petals pale yellow; nw NCoRH (South Fork Mtn, Humboldt,
            Trinity cos.) .... **L. elmeri**
        78′ Petals not yellow
          80. St slender; lvs cauline; lflets 10–40 mm; dry areas — KR .... **L. tracyi**
          80′ St stout; lvs basal and cauline; lflets 40–150 mm; moist to wet areas.................... **L. polyphyllus**
            81. Lflet adaxially puberulent to minutely rough-hairy, appearing ± glabrous; KR, NCoRH, CaR ... ²var. **pallidipes**
            81′ Lflet adaxially glabrous; CA-FP, GB
              82. Lflets 5–11; KR, CaR, SNH, SnGb, SnBr, SnJt, WTR, GB....................... var. **burkei**
              82′ Lflets 9–17; NW, CW (exc SCoRI) .... var. **polyphyllus**
    74′ Lf adaxially hairy, green-gray to silver
      83. Infl gen dense, bracts gen persistent (exc *Lupinus constancei*)
        84. Infl gen < lvs; pl matted — fls in 2–3 whorls; W&I (White Mtns) .... **L. lepidus** var. **utahensis**
        84′ Infl gen > lvs; pl matted or not
         85. Lvs cauline, appearing ± basal
          86. Bracts deciduous; petals pink; serpentine barrens — ne NCoRO (Lassics Range,
            se Humboldt, sw Trinity cos.) .... ³**L. constancei**
          86′ Bracts persistent; petals pink, violet, or blue; montane to alpine sites .... ²**L. lepidus** (in part)
            87. Infl ± head-like, 2–8 cm, gen < some lvs; pl < 1 dm; 2000–3500 m .... var. **lobbii**
            87′ Infl elongate, 4.5–11 cm, > lvs; pl 1.2–3.5 dm; 1000–2500 m .... var. **sellulus**
        85′ Lvs cauline, ± along st
          88. Pl ± prostrate, matted; petals pink — ne NCoRO (Lassics Range, se Humboldt,
            sw Trinity cos.) .... ³**L. constancei**
          88′ Pl gen erect, gen not matted; petals blue or violet (occ pink)
            89. Largest lflets > 40 mm
              90. Lvs cauline, strigose to shaggy-hairy, hairs > 1 mm; infl open; petals light blue; c&s SNH ... ²**L. covillei**
              90′ Lvs·basal and cauline, strigose, hairs < 1 mm; infl dense; petals violet to dark blue;
                c&s SNH, SNE .... **L. pratensis** var. **pratensis**
            89′ Largest lflets ≤ 40 mm.. ²**L. lepidus** (in part)
              91. Infl dense, 5–30 cm, whorls > 7, ± crowded; lvs cauline — meadows, vernally moist areas;
                CaRH, SNH, WTR, SnBr, GB, DMtns .... var. **confertus**
              91′ Infl gen open, 2–12 cm, whorls 3–7, ± well spaced; lvs, or at least some, basal
                92. Lflets gen 10–30 mm; fl 9–11 mm; 2500–3000 m, s SNH (Kaweah River, Tulare,
                  Fresno cos.)... var. **culbertsonii**
                92′ Lflets gen 5–15(23) mm; fl gen 7–9(12) mm; 3000–4000 m, c&s SNH, SNE............. var. **ramosus**
    83′ Infl ± open, bracts gen ± deciduous (exc *Lupinus covillei*, *Lupinus gracilentus*)
      93. Upper keel margin gen glabrous
        94. Pl < 2 dm; fl 4–11 mm; keel ± straight
          95. Petioles (2)3–6(8) cm; pl tufted; stipule 6–11 mm; pedicel (2)4–5 mm; pumice gravel flats —
            SNE (Mono Co.) .... **L. duranii**
          95′ Petioles 1–3(4) cm; pl matted or tufted; stipule 2–5 mm; pedicel 1–3(4) mm; montane.... **L. breweri** (in part)
            96. Fl 6–9 mm; lflet 6–20 mm; 1000–4000 m .... var. **breweri**
            96′ Fl 4–7 mm; lflet 3–5 mm; 2500–4000 m .... var. **bryoides**

94′ Pl > 2 dm; fl gen > 9 mm; keel gen upcurved
  97. Stipules green, some lf-like; fls 10–14 mm — SNH... ***L. fulcratus***
  97′ Stipule green to silvery, not lf-like
    98. Petals pale yellow or bright yellow to orange-yellow
      99. Fl 9–12 mm; petals pale yellow; NCoR, SnFrB, SN.. [2]***L. adsurgens***
      99′ Fl 12–15 mm; petals bright yellow to orange-yellow; KR, CaR .... [2]***L. croceus***
    98′ Petals ± yellow-white to white, ± blue, lavender, violet, purple, or pink
      100. Lvs basal and cauline; moist to wet places — KR, NCoRH, CaR .... [2]***L. polyphyllus*** var. ***pallidipes***
      100′ Lvs cauline; dry places
        101. Banner narrow, wings narrow, not covering keel tip; KR, NCoRH, CaRH, SNH, WTR .... ***L. albicaulis***
        101′ Banner wide, wings wide, covering keel tip
          102. Lf green, sparsely to densely hairy
            103. Fl 9–12 mm; NW, SNH, WTR, SnBr, SNE... ***L. andersonii***
            103′ Fl 13–16 mm; SnGb, SnBr, PR.. ***L. hyacinthinus***
          102′ Lf dull green- or gray-hairy to silver-silky
            104. Pl resprouting below ground, rhizomed; gen valleys .... ***L. formosus***
              105. Fl 10–14 mm; st 3–4 mm diam.. var. ***formosus***
              105′ Fl 16–18 mm; st 5–7 mm diam.. var. ***robustus***
            104′ Pl resprouting above ground, not rhizomed; mtns
              106. Petals white (banner spot turning tawny); seeds 7–11 mm — NCoRH (Anthony Peak,
                Mendocino Co.)... ***L. antoninus***
              106′ Petals not white; seeds 4–6 mm
                107. Lflets widest above middle, appressed-hairy to ± silky; petals pale yellow to lavender
                  or violet; NCoR, SN, SnFrB .... [2]***L. adsurgens***
                107′ Lflets widest below middle, densely silver-silky; petals lavender to blue; s SNH, TR.... [2]***L. elatus***
93′ Upper keel margin ciliate
 108. Pl ± prostrate to decumbent, gen < 3 dm
  109. Petals pink; ne NCoRO (Lassics Range, se Humboldt, sw Trinity cos.) .... [3]***L. constancei***
  109′ Petals purple, ± violet, lavender, rose, light blue, yellow, or white; KR, NCo, NCoRO, SN, CCo, SNE
    110. Lvs gen basal (if cauline, clustered near base); KR, NCoRO, SN, SNE
      111. Infl 10–16 cm; fl 10–16 mm; SN... [2]***L. grayi***
      111′ Infl 2–7 cm; fl 9–12 mm; e KR (Mount Eddy, Siskiyou Co.), NCoRO (King Range,
        sw Humboldt Co.)... [2]***L. lapidicola***
    110′ Lvs cauline; NCo, CCo
      112. Lflets 3–5; st weak — s NCo (Sonoma Co.), n&c CCo (Marin, Monterey cos.) .... ***L. tidestromii***
      112′ Lflets 5–9; st not weak
        113. Fl 10–13 mm; petals purple exc banner spot ± white; roots yellow; petioles gen < 2 × lflets,
          2–6 cm; n&c NCo, CCo (Point Reyes).. [2]***L. littoralis***
        113′ Fl 11–16 mm; petals white, yellow, rose, or purple, often on 1 pl; roots not yellow;
          petioles gen 2 × lflets, 4–10 cm; NCo, n&c CCo.................................... [2]***L. variicolor***
108′ Pl gen erect, gen > 2 dm
 114. Lf densely woolly
  115. St hairs < 1 mm, not sharp, not stiff; petiole 5–12 cm; fl 10–15 mm; s SCoRO (San Luis
    Obispo Co.)... [2]***L. ludovicianus***
  115′ St hairs sometimes 1–3 mm, sharp, stiff; petiole 6–30 cm; fl 10–18 mm;
    se SNH, SNE, DMoj... ***L. magnificus***
   116. Fl 10–13 mm... var. ***glarecola***
   116′ Fl (13)16–18 mm... var. ***magnificus***
 114′ Lf occ densely hairy but not woolly
  117. Lvs clustered near base
   118. Petals yellow; SnGb... [2]***L. peirsonii***
   118′ Petals not yellow, occ pale yellow or drying so; NCoRI, SCoRO
    119. Peduncle 13–20 cm; petals light blue, pink, or pale yellow (often drying pale yellow);
      SCoRO (Santa Lucia Range).. [2]***L. cervinus***
    119′ Peduncle 8–15 cm; petals purple to violet; s NCoRI (Napa, Lake, Sonoma cos.) .... [2]***L. sericatus***
  117′ Lvs clustered near base or not, some along st
   120. Fl 5–7(10) mm; n&c SNH, SNE... ***L. argenteus*** var. ***meionanthus***
   120′ Fl 8–18 mm; c&s SNH, GB, DMtns
    121. Lflets 30–110 mm
     122. Pl strigose to shaggy-hairy; lflet yellow-green, 5–11 mm wide; lower petioles 5–10 cm;
       bracts 7–15 mm, persistent, conspicuous; c&s SNH................................. [2]***L. covillei***
     122′ Pl puberulent to hairy; lflet green, 2–5 mm wide; lower petioles (3)5–14 cm;
       bracts 4–10 mm, ± deciduous, inconspicuous; c SNH (Rock Creek, Inyo, Mono cos.,
       to Yosemite National Park) .... ***L. gracilentus***
    121′ Lflets 15–50 mm
     123. Pl 4–7 dm; fl 13–15 mm; ne DMtns (Last Chance Range, Grapevine Mtns, Inyo Co.) .... ***L. holmgrenianus***
     123′ Pl 1–4 dm; fl 10–12 mm; GB, DMtns .... ***L. nevadensis***

L. adsurgens Drew (p. 771) Per 2–6 dm, hairy, silver to dull green. **ST:** erect. **LF:** cauline; stipules 5–17 mm; petiole 2–6 cm; lflets 6–9, 20–50 mm. **INFL:** 2–23 cm, ± open, fls spiralled to ± whorled; peduncle 2–8 cm; pedicels 2–6 mm; bract 2–8 mm. **FL:** 9–12 mm; calyx upper lip 4–6.5 mm, 2-toothed, lower 3–7 mm, entire to minutely 3-toothed; petals pale yellow to lavender or violet, banner back glabrous, spot yellow to white, keel upcurved, glabrous. **FR:** 2–4 cm, silky. **SEED:** 3–6, 4–6 mm, mottled brown. 2*n*=48. Dry slopes, montane forest; 500–3500 m. NCoR, SN, SnFrB (Santa Clara Co.); s OR. May–Jul

L. affinis J. Agardh (p. 771) Ann 2–5(6) dm, hairy. **LF:** petiole 3–10 cm; lflets 5–8, 20–50 mm, 4–11 mm wide. **INFL:** 4–20 cm, fls whorled; peduncle 5–18 cm; pedicels 3–6 mm; bract 5–7.5 mm. **FL:** 8–12 mm; calyx 5–7 mm, lips ± equal; petals blue, banner spot white, upper keel margins with (occ inconspicuous) tooth near middle, ciliate from tooth to near tip. **FR:** 3–5 cm, 5–9 mm wide, coarsely hairy. **SEED:** 5–8. Uncommon. Open areas; < 800 m. NCo, NCoR, CCo, SnFrB. Intergrades with *L. nanus.* Mar–May

L. albicaulis Douglas (p. 771) Per 3–12 dm; hairs puberulent to silky-appressed. **ST:** erect. **LF:** cauline; stipules 5–18 mm; petiole 2–7 cm; lflets 5–10, 20–70 mm. **INFL:** 10–44 cm, open, fls gen whorled; peduncle 2–12 cm; pedicels 2–7 mm; bract 6–16 mm. **FL:** (8)12–16 mm; calyx upper lip 6–12 mm, 2-toothed, lower 7–13 mm, entire to 3-toothed; petals gen purple (yellow-white), banner back glabrous, spot indistinct, keel strongly upcurved, glabrous, tip not covered. **FR:** 2–5 cm, silky. **SEED:** 3–7, 4–7 mm, gray to tan, mottled tan. *n*=24. Dry slopes, openings, ± montane; 500–3000 m. KR, NCoRH, CaRH, SNH, WTR; to WA. Doubtfully distinct from *L. andersonii.* May–Jul

L. albifrons Benth. Subshrub, shrub, < 50 dm, hairy, gen silver (± green). **ST:** decumbent to erect. **LF:** cauline, clustered near base or not, hairy; stipules 6–20 mm; petiole 1–8 cm; lflets 6–10, 10–45 mm. **INFL:** 4–30 cm, fls gen not to loosely whorled; peduncle 5–13 cm; pedicels 3–10 mm; bract 4–15 mm. **FL:** 9–16 mm; calyx upper lip 6–8 mm, deeply divided, lower 6–10 mm, entire to 3-toothed; petals violet to lavender, banner back gen hairy, spot gen yellow (to white) turning purple, keel gen unlobed near base, upper margins gen ciliate middle to tip, lower glabrous. **FR:** 3–5 cm, hairy. **SEED:** 4–9, 4–6 mm, mottled tan. Variable, ± indistinct from *L. excubitus.*

var. ***abramsii*** (C.P. Sm.) Hoover (p. 771) ABRAMS' LUPINE Subshrub, shrub, 2–10 dm, woolly to shaggy-hairy. **ST:** decumbent to erect. **LF:** stipules 9–10 mm; petiole 3–5 cm; lflets 7–10, 10–30 mm. **INFL:** 15–25 cm; peduncle 6–10 cm; pedicels 4–7 mm; bract 10–15 mm. **FL:** 11–16 mm; banner back hairy to glabrous. Open woodland; 600–2000 m. SCoRO (Santa Lucia Range). May–Jun ★

var. ***albifrons*** (p. 771) Shrub 5–50 dm, gen with distinct trunk, green to silvery. **ST:** erect. **LF:** stipules 7–10 mm; petiole 2–5 cm; lflets 6–10, 10–30 mm. **INFL:** 8–30 cm; peduncle 5–13 cm; pedicels 4–9 mm; bract 4–8 mm. **FL:** 9–14 mm. Common. Chaparral, foothill woodland; < 1500 m. NW, SN, CW, SCo, ChI, WTR, SnGb, PR; s OR. Mar–Jun

var. ***collinus*** Greene (p. 771) Subshrub 2–4 dm, appressed-silvery. **ST:** prostrate to decumbent. **LF:** cauline, clustered near base; stipules 6–8 mm; petiole 3–8 cm; lflets 6–9, 10–20 mm. **INFL:** 4–14 cm; peduncle 6–10 cm; pedicels 3–5 mm; bract 7–9 mm. **FL:** 10–15 mm; banner back gen hairy, occ ± glabrous, keel upper margin glabrous or ciliate. Cliffs, forest openings; < 2000 m. NW, SNF, CW; s OR. Mar–Jun

var. ***douglasii*** (J. Agardh) C.P. Sm. (p. 771) Shrub 10–20 dm, gen silver-silky. **ST:** erect. **LF:** stipules 10–20 mm; petiole 1–4 cm; lflets 7–9, 25–45 mm. **INFL:** 10–15 cm; peduncle 9–10 cm; pedicels 6–10 mm; bract 10–24 mm. **FL:** banner back sparsely hairy. Common. Coastal scrub, chaparral, open woodland; < 500 m. CCo, nw SnFrB, s SCoRO, ChI, w WTR. Mar–Jun

L. andersonii S. Watson (p. 771) Per 2–9 dm, green, hairy. **ST:** erect. **LF:** cauline; stipules 3–15 mm; petiole 2–6 cm; lflets 6–9, 20–60 mm. **INFL:** 2–23 cm, open, fls ± whorled; peduncle 1–8.5 cm; pedicels 1.5–5 mm; bract 2–10 mm. **FL:** 9–12 mm; calyx upper lip 5–7 mm, 2-toothed, lower 3–8 mm, 2–3-toothed; petals gen lavender to purple (± yellow), banner back glabrous, spot white turning purple, keel glabrous. **FR:** 2–4.5 cm, silky. **SEED:** 4–6, 4–6 mm, mottled tan, brown. Dry slopes; 1500–3000 m. NW, SNH, WTR, SnBr, SNE; s OR, w NV. ± indistinct morphologically from *L. albicaulis.* Jun–Sep

L. angustiflorus Eastw. (p. 771) Per 5–12 dm, green, glabrous to hairy. **ST:** erect. **LF:** cauline; stipules 5–13 mm; petiole ± 1–5 cm; lflets 6–9, 20–60 mm. **INFL:** 6–34 cm, open; peduncle ± 1–8 cm; pedicels 2–4 mm; bracts 3–7 mm, ± persistent. **FL:** 8–10(12) mm; calyx upper lip 4–8 mm, 2-toothed, lower 4–9 mm, entire to 3-toothed; petals pale yellow to orange-yellow, banner back gen hairy, spot orange to yellow, keel glabrous, tip lavender. **FR:** 2.5–4 cm, hairy. **SEED:** 4.5–5.5 mm, speckled tan, brown. Gen volcanic soils; 1000–3500 m. CaRH, n&c SNH, GB. Jun–Sep

L. antoninus Eastw. ANTHONY PEAK LUPINE Per 2–5 dm, gray-to silver-hairy. **ST:** erect. **LF:** cauline; stipules 10–12 mm; petiole 1–2 cm; lflets 6–7, 15–25 mm. **INFL:** 4–20 cm, open, fls spiralled; peduncle ± 1–4 cm; pedicels 3–4 mm; bracts 7–8 mm, ± deciduous. **FL:** 12–14 mm; calyx upper lip 6–8 mm, 2-toothed, lower 6–8 mm, 3-toothed; petals white, banner back glabrous, spot turning tawny, keel glabrous. **FR:** 2.5–3.5 cm, 14–17 mm wide, silky. **SEED:** 4–5, 7–11 mm, mottled brown. Open fir forest; ± 2000 m. NCoRH (Anthony Peak, Mendocino Co.). Jun–Jul ★

L. apertus A. Heller Per 2–6 dm, green, puberulent to sparsely appressed-hairy. **ST:** erect. **LF:** cauline; stipules 5–10 mm; petiole 2–5 cm; lflets 7–9, 25–55 mm. **INFL:** 8–11 cm, fls ± whorled to not; peduncle ± 1–8 cm; pedicels 3–6 mm; bract 3.5–5 mm. **FL:** 10–12 mm; calyx upper lip 3.5–6 mm, 2-toothed, lower 4.5–7 mm, entire to 3-toothed; petals white to purple, banner back hairy, spot gen white, keel glabrous. **FR:** 2–3 cm, hairy. **SEED:** 3–4, 5–6 mm. Dry, rocky soil; 1500–3000 m. n SNH. Reportedly TOXIC. Jun–Jul

L. arboreus Sims (p. 771) YELLOW BUSH LUPINE Shrub < 20 dm, green-glabrous to silver-hairy. **ST:** erect. **LF:** cauline; stipules 8–12 mm; petiole 2–3(6) cm; lflets 5–12, 20–60 mm. **INFL:** 10–30 cm, fls whorled or not; peduncle 4–10 cm; pedicels 4–10 mm; bract 8–10 mm. **FL:** 14–18 mm; calyx upper lip 5–9 mm, 2-toothed, lower 5–7 mm, entire; petals gen yellow (lilac to purple, esp n of c NCo), banner back glabrous, spot darker or not to white, keel upper margins ciliate claw to tip, lower glabrous. **FR:** 4–7 cm, brown to black, hairy. **SEED:** 8–12, 4–5 mm, black to tan, often striped lighter. Coastal bluffs, dunes, or more inland; < 100 m. NCo, CCo, SnFrB, SCo, ChI (probably naturalized n of Sonoma Co.). Grades ± into *L. rivularis* in NCo. Pls with yellow petals, sweet-smelling fls widely cult as sand binder. If recognized taxonomically, hairier pls from w SnFrB (yellow banner, blue wings) assignable to *L. arboreus* var. *eximius* (Burtt Davy) C.P. Sm.; pls with glabrous lflets, purple petals assignable to *L. propinquus* Greene (study needed). Apr–Jul

L. arbustus Douglas (p. 771) SPUR LUPINE Per 2–7 dm, green or gray-silky. **ST:** erect. **LF:** cauline and occ basal; stipules 4–9 mm; petiole 2–16 cm; lflets 7–13, 20–70 mm. **INFL:** 3–18 cm, open, fls whorled; peduncle 2–5 cm; pedicels 1–7 mm; bract 3–6 mm. **FL:** 8–14 mm; calyx spur distinct, 1–3 mm, upper lip 2–4 mm, 2-toothed, lower 2.5–5 mm, 3-toothed; petals blue, purple, pink, white, or ± yellow, banner back hairy, spot white, ± yellow, or 0, wings with dense patch of hair outside near tip, keel upper margins ciliate, lower glabrous. **FR:** 2–3 cm, silky. **SEED:** 3–6, 5–6 mm, tan. Open sagebrush scrub, mixed-conifer forest; 1500–3000 m. CaRH, SNH, SnGb, GB; to OR, ID, UT. Like *L. argenteus* exc wing hairs. May–Jul

L. argenteus Pursh Per, subshrub, green-glabrous to silver-hairy. **ST:** erect. **LF:** basal to cauline; stipules 2–12 mm; petiole gen 1–15 cm; lflets 5–9, 10–60 mm, < 10 mm wide, abaxially hairy, adaxially glabrous to hairy. **INFL:** 5–16(25) cm, fls whorled to not; peduncle 1–10 cm; pedicels 1–6 mm; bracts gen deciduous. **FL:** 5–14 mm; calyx upper lip 4–8 mm, entire to 2-toothed, lower 4–8 mm, entire to 3-toothed, bulge or spur 0–3 mm (may be variable on 1 pl); petals blue, violet, or white, banner back gen hairy, spot ± yellow to ± white to 0, keel upper margins ciliate, lower glabrous. **FR:** 1–3 cm, hairy or silky. **SEED:** 2–6, tan, brown, or red. Highly variable. vars. intergrade. *L. argenteus* var. *rubricaulis* (Greene) S.L. Welsh probably not in CA.

var. *argenteus* (p. 771) Pl 2–15 dm; st, lf hairs appressed. **LF:** cauline, green. **FL:** (8)10–12 mm; calyx bulge < 1 mm; petals blue or purple to white. Esp dry sagebrush scrub; 1000–2000 m. CaR, n SNH, GB, DMtns; to s Can, SD, NM. Jun–Oct

var. *heteranthus* (S. Watson) Barneby (p. 771) Pl 2–8 dm. **LF:** basal or some cauline, densely silver-silky. **FL:** 8–14 mm; calyx spur 1–2 mm; petals violet or blue to white, banner back silky, wings glabrous. Dry, open slopes, sagebrush scrub, pinyon/juniper woodland; 1000–3000 m. c SNH, GB; to OR, ID, UT. May–Sep

var. *meionanthus* (A. Gray) Barneby (p. 771) Pl 2–9 dm. **LF:** cauline, appressed-silvery to gray-green. **FL:** 5–7(10) mm; petals dull blue to lilac, banner back glabrous, spot yellow. Dry banks; 1500–3500 m. n&c SNH, SNE; NV. Jul–Aug

var. *montigenus* (A. Heller) Barneby (p. 771) Pl < 4 dm, densely silver-hairy. **LF:** basal and cauline; petioles (5)7–15 cm. **FL:** 9–12(14) mm; calyx bulge < 1 mm; petals blue to violet, banner spot yellow to cream. Dry, open montane forest, sagebrush scrub; 2500–3500 m. c SNH, SNE; NV. Jul–Aug

var. *palmeri* (S. Watson) Barneby (p. 771) Pl 3–6 dm, st hairs spreading. **LF:** cauline, densely gray-spreading-hairy and silver-silky. **FL:** 8–10 mm; calyx bulge or spur < 1 mm; petals blue, banner back hairy. Dry, open montane forest; 2000–2500 m. CaRH, SNH, Wrn, SNE; to WA, UT, NM. Like *L. argenteus* var. *a.* exc hairs. May–Jun

L. arizonicus (S. Watson) S. Watson (p. 771) ARIZONA LUPINE Ann 1–5 dm, short-appressed- and long-spreading-hairy. **LF:** petiole 2.5–7 cm; lflets 6–10, 10–40 mm, 5–10 mm wide, adaxially glabrous. **INFL:** 4–24 cm, fls spiraled, occ appearing ± whorled; peduncle 2–6 cm; pedicels 2–4 mm; bracts 4–8 mm, gen persistent. **FL:** 7–10 mm; calyx 3–6 mm, lips ± equal, upper deeply lobed; banner, wings dark pink to magenta, drying blue-purple or ± white, banner spot ± yellow, in age darker magenta, keel upper margins glabrous, lower ciliate near claw. **FR:** often on 1 side of infl, 1–2 cm, ± 5 mm wide, coarsely hairy. **SEED:** 4–6. Locally common. Sandy washes, open areas; < 1100 m. e DMoj, DSon; NV, AZ, Mex. Mar–May

L. benthamii A. Heller (p. 771) SPIDER LUPINE Ann 2–7 dm, short-appressed- and long-spreading-hairy. **LF:** petiole 3–12 cm; lflets 7–10, 20–50 mm, 1.5–3.5 mm wide, linear, adaxially glabrous. **INFL:** 6–30 cm, fls ± whorled or not; peduncle 4–7 cm; pedicels 5–9 mm; bract 10–15 mm, >> bud. **FL:** 10–18 mm; calyx 5–6.5 mm, lips ± equal, upper deeply lobed; petals bright blue, banner spot ± white, magenta in age, keel upper margins glabrous, lower ciliate near claw. **FR:** ± 3 cm, 5 mm wide, coarsely hairy. **SEED:** 5–8. Locally common. Rocky slopes, open areas; < 1500 m. SNF, Teh, deltaic GV, SCoRO. Mar–May

L. bicolor Lindl. (p. 771) MINIATURE LUPINE Ann (may live 2 seasons in NCo), 1–4 dm, hairy. **LF:** petiole 1–7 cm; lflets 5–7, 10–40 mm, 1–5 mm wide, occ linear, adaxially gen ± glabrous. **INFL:** 1–8 cm, fls in (0)5 whorls; peduncle 3–10 cm; pedicels 1–3.5 mm; bract 4–6 mm. **FL:** 4–10 mm; calyx upper lip 2–4 mm, deeply lobed, lower 4–6 mm; petals gen blue (light blue, pink, or white), banner spot white, in age magenta, keel ± white, gen pointed, upper margins gen ciliate near tip. **FR:** 1–3 cm, gen 3–6 mm wide, hairy. **SEED:** 5–8. 2*n*=48. Abundant. Open or disturbed areas; < 1600 m. CA-FP, DMoj; to BC, Baja CA; naturalized in AZ. Highly variable, needs study; named subspp. and vars. ± indistinct in geog, morphology. Vigorous pls with larger fls may be confused with *L. nanus*. Mar–Jun

L. brevicaulis S. Watson (p. 775) SAND LUPINE Ann 0.3–1 dm, hairy; cotyledons disk-like, persistent. **LF:** cauline, crowded near base; petioles 2–5 cm; lflets 6–8, 10–15 mm, 3–6 mm wide, linear to oblanceolate, adaxially glabrous. **INFL:** 1–2.5 cm, exceeding lvs, dense, fls spiraled; peduncle 2–5 cm; pedicels 0.5–1.5 mm; bracts 2–3 mm, straight, persistent. **FL:** 6–8 mm; calyx upper lip ± 3 mm, 2-toothed, lower ± 6 mm, appendages present; petals bright blue, banner spot white or yellow, keel glabrous. **FR:** ± 1 cm, ± 5 mm wide, ovate, hairy. **SEED:** 1–2, smooth. Sandy washes, open areas; < 2300 m. s MP, W&I, e DMoj; to OR, CO, Mex. Like small-fld *L. flavoculatus.* May–Jun

L. breweri A. Gray Per, subshrub, < 2 dm, matted or tufted, silver-silky. **ST:** prostrate, base ± woody. **LF:** cauline, clustered near base; stipules 2–5 mm; petiole 1–5(6) cm; lflets 5–10, 3–20 mm. **INFL:** 1–10 cm, ± dense; peduncle 1–8 cm; pedicels 1–3(4) mm; bract 3–5 mm. **FL:** 4–11 mm; calyx upper lip 4–7 mm, 2-toothed, lower 4–6 mm, entire to 3-toothed; petals blue to violet, banner back glabrous to densely hairy, spot white or yellow, keel straight, upper margins glabrous or ciliate, lower glabrous. **FR:** 1–2 cm, silky. **SEED:** 3–4, 3–4 mm, mottled tan, brown.

var. *breweri* Per. **LF:** lflets 6–20 mm. **INFL:** 1–2 cm; peduncle 1–3 cm. **FL:** 6–9 mm; keel glabrous, banner back ± glabrous. Common. Gen open montane forest; subalpine to alpine; 1000–4000 m. KR, NCoRH, CaRH, SNH, TR, GB; s OR, w NV. Jun–Aug

var. *bryoides* C.P. Sm. (p. 775) Per. **LF:** lflets 3–5 mm. **INFL:** < 2.5 cm; peduncle < 2 cm. **FL:** 4–7 mm; keel, banner glabrous. Gen open montane forest; 2500–4000 m. s SNH, WTR, SNE; NV. Jul–Aug

var. *grandiflorus* C.P. Sm. (p. 775) Per, matted. **LF:** lflets 6–11 mm. **INFL:** 2–10 cm; peduncle 2–8 cm. **FL:** 6–11 mm; keel ciliate, banner back ± silky. Volcanic sand; 2000–3500 m. SNH, SnBr, SNE. Jun–Aug

L. cervinus Kellogg (p. 775) SANTA LUCIA LUPINE Per 1.5–3 dm, gray-green, spreading-hairy. **ST:** erect. **LF:** cauline, clustered near base; stipules 5–6 mm; petiole 13–15 cm; lflets 4–8, 40–80 mm, 10–30 mm wide. **INFL:** < 20 cm, open, fls ± whorled to not; peduncle 13–20 cm; pedicels 3–6 mm; bract 3–4 mm. **FL:** 14–16 mm; calyx lips entire to 2-toothed, upper 6–7 mm, lower 8–10 mm; petals light blue, pink, or pale yellow (often drying pale yellow), banner back glabrous to ± hairy, spot yellow, keel upper margins ciliate, lower ciliate near claw. **FR:** 3–6 cm, silky. **SEED:** 4–8, 2–4 mm, light brown with brown line or mottled tan. Dry sites in forest; (305)500–1000(1350) m. SCoRO (Santa Lucia Range). May–Jun ★

L. chamissonis Eschsch. (p. 775) Shrub 5–20 dm, silvery, densely appressed-hairy. **ST:** erect. **LF:** cauline; stipules 8–10 mm; petiole 1–3.5 cm; lflets 5–9, 10–25 mm. **INFL:** 5–20 cm, fls ± whorled; peduncle 2–6 cm; pedicels 4–8 mm; bract 7–10 mm. **FL:** 8–16 mm; calyx upper lip 5–7 mm, deeply lobed, lower 7–9 mm, entire; petals light violet to blue, banner back densely hairy, spot yellow, keel upper margins glabrous, lower ± ciliate. **FR:** 2.5–3.5 cm, hairy. **SEED:** 4–8, 4–5 mm, mottled brown. Coastal strand, dunes; < ± 10 m. s NCo, CCo, SCo. Mar–Jul

L. citrinus Kellogg Ann 1–6 dm; hairs < 2 mm, soft, white, matted or not. **LF:** petiole 2–7 cm; lflets 6–9, 15–35 mm, 3–10 mm wide. **INFL:** 4–15 cm, fls spiralled or appearing ± whorled; peduncle 1–9 cm; pedicels 2.5–5 mm, recurving; bract 2.5–5 mm. **FL:** 8.5–12 mm; calyx 3–5 mm, lips ± equal; petals golden-yellow or white, keel upper margins glabrous, lower short-ciliate near claw. **FR:** 1–2 cm, 3–5 mm wide, glabrous or becoming so. **SEED:** 3–8, like bits of granite.

var. *citrinus* (p. 775) ORANGE LUPINE **FL:** petals gold-yellow, drying translucent, ± purple. Granitic soils, open yellow-pine forest; 600–1700 m. c SNF (Fresno, Madera cos.). Apr–Jul ★

var. *deflexus* (Congdon) Jeps. MARIPOSA LUPINE **FL:** petals white, pink- or lavender-tinged or not, banner occ drying translucent, yellow. Granitic soils; 400–600 m. c SNF (Mariposa Co.). Apr–May ★

L. concinnus J. Agardh (p. 775) BAJADA LUPINE Ann 1–3 dm, hairy. **ST:** decumbent or erect. **LF:** petiole 2–7 cm; lflets 5–9, 10–30 mm, 1.5–8 mm wide, occ linear. **INFL:** 1.5–9 cm, often dense, fls spiraled, gen also in lower lf axils; peduncle ≤ 8 cm, central erect, lateral often decumbent; pedicels 0.7–2 mm; bracts 2.5–4 mm, straight, persistent. **FL:** 5–12 mm; calyx 3–5 mm, lips ± equal, upper deeply lobed; petals pink to purple (white), banner spot white or ± yellow, keel gen glabrous. **FR:** 1–1.5 cm, 3–5 mm wide, hairy. **SEED:** 3–5. 2*n*=48. Common. Open or disturbed areas, burns; < 1700 m. s SN, c&s CW, SW, D; to UT, TX, Mex. Highly variable, gen self-pollinated, needs study; named vars. ± indistinct; pls in D with linear, coarsely hairy lflets, ± ciliate lower keel margins may be confused with *L. sparsiflorus.* Mar–May

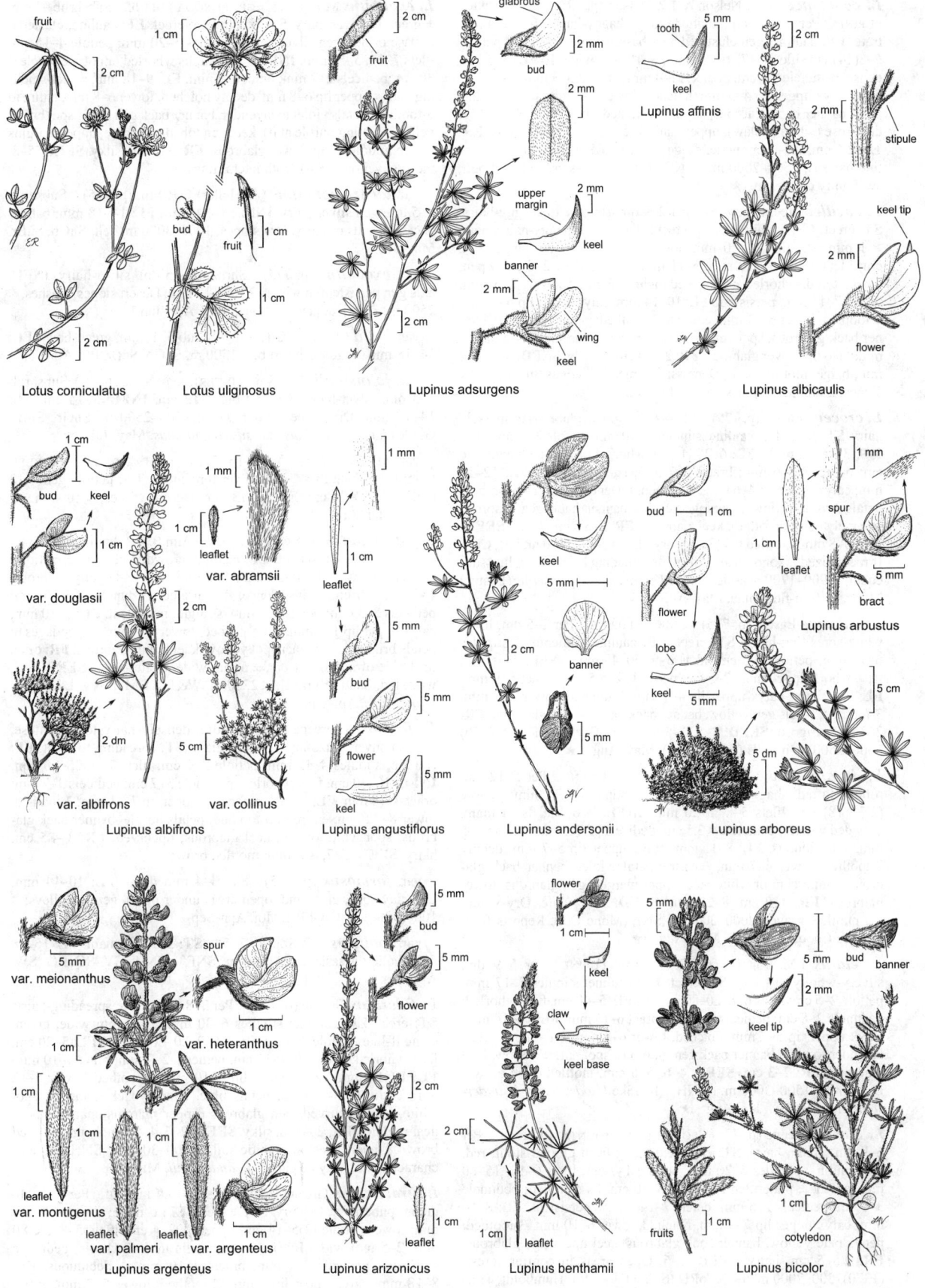

Lotus corniculatus

Lotus uliginosus

Lupinus adsurgens

Lupinus albicaulis

Lupinus affinis

var. douglasii

var. abramsii

var. albifrons

var. collinus

Lupinus albifrons

Lupinus angustiflorus

Lupinus andersonii

Lupinus arbustus

Lupinus arboreus

var. meionanthus

var. heteranthus

var. montigenus

var. palmeri

var. argenteus

Lupinus argenteus

Lupinus arizonicus

Lupinus benthamii

Lupinus bicolor

L. constancei T.W. Nelson & J.P. Nelson (p. 775)   THE LASSICS LUPINE   Per < 1.5 dm, matted, long-shaggy-hairy. **ST:** ± prostrate. **LF:** cauline, gen clustered near base; stipules < 6 mm; petiole 6–8(14) cm; lflets 6–7, 10–20 mm, 8–10 mm wide. **INFL:** 3–5 cm, dense; peduncle < 4 cm; pedicels 1–4 mm; bract 2.5–3 mm. **FL:** 8–12 mm; calyx upper lip 4–5 mm, notched, lower 4–5 mm, entire; petals pink, banner back glabrous, strongly reflexed, spot light yellow, keel dark rose (white at claw), upper margins ciliate, lower glabrous. **FR:** 1.5–2.5 cm, 0.5–1 cm wide, shaggy-hairy. **SEED:** 3–5, tan. Serpentine barrens; 1500–2000 m. ne NCoRO (The Lassics, se Humboldt, sw Trinity cos.). Jul ★

L. covillei Greene (p. 775)   Per 2–9 dm, strigose to shaggy-hairy. **ST:** erect. **LF:** cauline, yellow-green, strigose to shaggy-hairy, hairs > 1 mm; stipules 12–30 mm; lower petiole 5–10 cm, upper 0–2 cm; lflets 4–9, 30–110 mm, 5–11 mm wide. **INFL:** 2–6 cm, open, fls scattered, whorled or not; peduncle 2–6 cm; pedicels 2–5 mm; bracts 7–15 mm, persistent. **FL:** 10–14 mm; calyx upper lip 6–8 mm, 2-toothed, lower 6–11 mm, entire to 3-toothed; petals light blue, banner back glabrous, spot yellow, keel upper margins sparsely ciliate ± middle to tip, lower glabrous. **FR:** 2.5–4 cm, woolly. **SEED:** 4–6, 3–4 mm, beige, mottled dark. Depressions, meadow edges, moist, rocky slopes; 2500–3500 m. c&s SNH. Jul–Sep

L. croceus Eastw. (p. 775)   Per 4–6 dm, green, glabrous to sparsely hairy. **ST:** erect. **LF:** cauline; stipules 4–10 mm; petiole 2–8 cm; lflets 5–9, 30–60 mm. **INFL:** 6–28 cm, fls whorled or not; peduncle 2–6 cm; pedicels 3–6 mm; bracts 2–7 mm, late-deciduous. **FL:** 12–15 mm; calyx upper lip 4–6 mm, 2-toothed, lower 6–7 mm, 2–3-toothed; petals bright yellow to orange-yellow, banner back gen glabrous (sparsely hairy on ridge), keel glabrous. **FR:** 2–3.5 cm, hairy. **SEED:** 3–5, 6–8 mm, mottled tan. Rocky dry places; 900–2700 m. KR, CaR. If recognized taxonomically, pls with spreading hairs, calyx lips subequal, at 900–1700 m assignable to *L. croceus* var. *pilosellus* (Eastw.) Munz, saffron-fld lupine. May–Aug

L. dalesiae Eastw. (p. 775)   QUINCY LUPINE   Per 2–5 dm, long-white-spreading-hairy. **ST:** erect. **LF:** cauline, tomentose; stipules 6–16 mm; petiole 1–3 cm; lflets 6–9, 20–45 mm. **INFL:** 5–16 cm, fls ± whorled; peduncle 2–5 cm; pedicels 2–5.5 mm; bract 5–9 mm. **FL:** 9–12 mm; calyx upper lip 4–7 mm, 2-toothed, lower 3–7 mm, 3-toothed; petals gen yellow, banner back hairy, keel ± glabrous. **FR:** 2–3 cm, strigose. **SEED:** 3–5, 3–5 mm, tan. Dry pine forest; (850)1000–2500 m. n SNH (Plumas Co.). May–Aug ★

L. duranii Eastw. (p. 775)   MONO LAKE LUPINE   Per 5–12 cm, robust, tufted, shaggy-hairy. **LF:** basal; stipules 6–11 mm; petiole (2)3–6(8) cm; lflets 5–8, 5–20 mm. **INFL:** 2–6 cm, fls in many crowded whorls; peduncle < 3.5 cm; pedicels (2)4–5 mm; bracts 4–5 mm, ± deciduous. **FL:** 8–11 mm; calyx upper lip 5–7 mm, deeply 2-toothed, lower 6–7 mm, ± entire; petals violet, banner back glabrous, spot cream or white, keel upper margins gen glabrous, lower glabrous. **FR:** 1–2 cm, 8–20 mm. **SEED:** 3–5, white. Dry volcanic pumice, gravel; 2000–3000 m. SNE (Mono Co.). Reports from Madera Co. questionable. May–Aug ★

L. elatus I.M. Johnst. (p. 775)   SILKY LUPINE   Per 5–9 dm, silvery-woolly to -silky. **ST:** erect. **LF:** cauline; stipules 5–17 mm; petiole 2–5 cm; lflets 6–8, 20–80 mm. **INFL:** 5–40 cm, fls ± whorled; peduncle 2–8 cm; pedicels 2–4 mm; bract 6–11 mm. **FL:** 10–14 mm; calyx upper lip 5–7 mm, notched, lower 6–8 mm, 3-toothed; petals lavender to blue, banner back gen glabrous, spot pale ± yellow, keel glabrous. **FR:** 2–3 cm. **SEED:** 4–6, 5–6 mm, mottled olive-brown. Dry forest; 1500–3000 m. s SNH, TR. Like *L. adsurgens*, *L. andersonii*; needs study. Jun–Aug ★

L. elmeri Greene (p. 775)   SOUTH FORK MOUNTAIN LUPINE   Per 6–9 dm, green, hairy. **ST:** erect, emerging from ground stout, red. **LF:** cauline; stipules 6–20 mm; petiole 1–7 cm; lflets 6–10, 15–60 mm, green, ± puberulent. **INFL:** 15–20 cm, fls spiralled; peduncle 3–9 cm; pedicels 2–6 mm; bracts 7–14 mm, ± persistent. **FL:** 8–14 mm; calyx upper lip 7–9 mm, notched, lower 6–10 mm, 3-toothed; petals pale yellow, banner back glabrous, keel upcurved, glabrous. **FR:** 2.5–5 cm, hairy. **SEED:** 3–6. Open areas in conifer forest; (1370)1500–2000 m. nw NCoRH (South Fork Mtn, Humboldt, Trinity cos.). Jun–Jul ★

L. excubitus M.E. Jones   GRAPE SODA LUPINE   Subshrub, shrub, ± green to silver-hairy. **ST:** prostrate to erect. **LF:** cauline, clustered at base or not, gen silver-hairy; stipules 5–20 mm; petiole 4–15 cm; lflets 7–10, 5–50 mm. **INFL:** < 70 cm, fls whorled or not; peduncle < 30 cm; pedicels 2–7 mm; bract 8–9 mm. **FL:** 9–18 mm, sweet-smelling; calyx upper lip 6–8 mm, deeply notched, lower 6–8 mm, entire to 3-toothed; petals violet to lavender, banner back gen hairy, spot bright yellow (turning purple at fl), keel gen lobed near base, upper margins ciliate middle to tip, lower glabrous. **FR:** 3–5 cm, silky. **SEED:** 5–8, mottled yellow-brown with lateral lines.

var. ***austromontanus*** (A. Heller) C.P. Sm. (p. 775)   Subshrub 2–5 dm, gen silver-hairy. **INFL:** 14–40 cm. **FL:** 14–18 mm; banner back glabrous or hairy. Dry slopes; 1000–3000 m. Teh, SnGb, SnBr. May–Jul

var. ***excubitus*** (p. 775)   Shrub 10–15 dm, silver-hairy. **INFL:** axis gen persistent in winter. **FL:** 9–13 mm. Desert slopes, washes; < 2500 m. DMoj, adjacent s SNH, SNE. Apr–Jun

var. ***hallii*** (Abrams) C.P. Sm.   Shrub 5–15 dm, green-hairy. **FL:** 14–18 mm. Sagebrush scrub; < 1500 m. SnGb, SnBr, PR. Apr–Jun

var. ***johnstonii*** C.P. Sm.   INTERIOR BUSH LUPINE   Subshrub 1–3 dm, silver-hairy. **LF:** petiole < 12 cm. **INFL:** 3–14 cm. **FL:** 14–18 mm. Dry slopes under pines; 1500–2500 m. SnGb, SnBr. Grades into *L. excubitus* var. *austromontanus*. May–Jul ★

var. ***medius*** (Jeps.) Munz (p. 775)   MOUNTAIN SPRINGS BUSH LUPINE   Subshrub < 7 dm, silver-tomentose. **LF:** petiole gen > 12 cm. **INFL:** 3–14 cm. **FL:** 10–13 mm. Desert washes; < 1000 m. sw DSon. Mar–Apr ★

L. flavoculatus A. Heller (p. 775)   Ann 0.5–2 dm, hairy; cotyledons disk-like, persistent. **LF:** cauline, crowded near base; petioles 3–6 cm; lflets 7–9, 10–20 mm, 5–8 mm wide, adaxially glabrous. **INFL:** 3–10 cm, > lvs, dense, fls spiraled; peduncle 2–5(10) cm; pedicels 2–3 mm; bracts 2–3 mm, straight, persistent. **FL:** 7–10 mm; calyx upper lip 1–3 mm, deeply lobed, lower 4–5 mm, appendages 0; petals bright blue, banner spot yellow, keel blunt, glabrous. **FR:** often on 1 side of infl, 0.5–1 cm, ± 5 mm wide, ovate, hairy. **SEED:** 1–2, wrinkled. Sand or gravel; < 2200 m. W&I, e DMoj; NV. Like hairy *L. odoratus*. Apr–Jun

L. formosus Greene   Per 2–8 dm, densely hairy to tomentose, gray to silver; rhizomes 3–7 mm diam. **ST:** spreading to erect. **LF:** cauline; stipules 4–15 mm; petiole 2–7 cm; lflets 7–9, 25–70 mm. **INFL:** 10–30 cm, fls ± whorled; peduncle 3–7 cm; pedicels 3–7 mm; bract 4–14 mm. **FL:** 10–18 mm; calyx upper lip 7–11 mm, 2-toothed, lower 8–12 mm, entire to 3-toothed; petals purple, banner back glabrous, spot white or not, keel glabrous, upcurved. **FR:** 3–4.5 cm, hairy. **SEED:** 5–7, 4–7 mm, mottled brown.

var. ***formosus*** (p. 775)   **ST:** 3–4 mm diam. **FL:** 10–14 mm. Dry clay soils, grassland, open areas under pines, gen in valleys; < 1000(3000) m. CA-FP, DMoj. Apr–Sep

var. ***robustus*** C.P. Sm. (p. 775)   **ST:** 5–7 mm diam. **FL:** 16–18 mm. Valley grassland; < 1500 m. SNF, c SNH, GV, SCoRO, SW. Apr–Jun

L. fulcratus Greene (p. 775)   Per 3–8 dm, green, spreading-hairy. **ST:** erect. **LF:** cauline; stipules 6–30 mm, 2–10 mm wide, green, some lf-like; petiole 3–6 cm; lflets 6–9, 20–60 mm. **INFL:** 3–20 cm, fls ± whorled; peduncle 1–11 cm; pedicels 2–7 mm; bract 4–10 mm. **FL:** 10–14 mm; calyx upper lip 5–10 mm, 2-toothed, lower 5–12 mm, entire to 3-toothed; petals blue, banner back glabrous, spot white, keel upcurved, gen glabrous (upper margins sparsely hairy near middle). **FR:** 2–4 cm, silky. **SEED:** 2–6, 4–5 mm, beige, mottled brown. Under pines, in granitic soils; 1500–3000 m. SNH. Stipules characteristic. May be part of *L. andersonii*. May–Sep

L. gracilentus Greene (p. 775)   SLENDER LUPINE   Per 2–8 dm, green, puberulent to hairy. **ST:** ± erect. **LF:** cauline; stipules 10–15 mm; lower petiole (3)5–14 cm, upper (1)2–4 cm; lflets 5–8, 35–80 mm, 2–5 mm wide. **INFL:** 6–20 cm, fls in 4–8 whorls; peduncle 6–12 cm; pedicels 2–4 mm; bracts 4–10 mm, ± deciduous. **FL:** 8–18 mm; calyx upper lip 7 mm, 2-toothed, lower 5–7 mm, entire

to 2–3-toothed; petals blue, banner back glabrous, spot white to ± yellow, keel upper margins sparsely ciliate, lower glabrous. **FR:** 2–3 cm, densely hairy. **SEED:** 6–8. Open moist sites, Subalpine forest; 2500–3500 m. c SNH (Rock Creek, Inyo, Mono cos., to Yosemite National Park). In need of study. Jul–Sep ★

L. grayi S. Watson (p. 775) Per 2–3.5 dm, spreading-tomentose to -woolly. **ST:** prostrate to matted. **LF:** gen basal; stipules 4–10 mm; petiole 5–12 cm; lflets 5–11, 10–35 mm. **INFL:** 10–16 cm, fls ± whorled; peduncle 3–15 cm; pedicels 2–4 mm; bract 4–5(10) mm. **FL:** 10–16 mm; calyx upper lip 5–10 mm, deeply 2-toothed, lower 7–12 mm, entire to 3-toothed; petals deep purple to light blue, banner back glabrous to hairy, spot yellow turning ± red, keel upper margins densely hairy, lower gen ciliate near base. **FR:** 2–3.5 cm, hairy. **SEED:** 4–6, 3–4 mm, mottled gray-brown with dark lateral line. Common. Open forest slopes; 500–2500 m. SN. Occ fragrant. May–Jul

L. guadalupensis Greene (p. 777) GUADALUPE ISLAND LUPINE Ann 2–6 dm, sparsely hairy. **LF:** petiole 3–7 cm; lflets 7–9, 20–50 mm, 3–5 mm wide, occ linear. **INFL:** 5–15 cm, fls gen whorled; peduncle 5–8 cm; pedicels 4–5 mm; bract 8–10 mm. **FL:** 10–12 mm; calyx 6–10 mm, lips ± equal, upper deeply lobed; petals blue, banner spot white, keel upper margins ± ciliate near tip. **FR:** 3–6 cm, ± 10 mm wide, densely hairy. **SEED:** 6–8. Sand or gravel; < 300 m. s ChI (San Clemente Island); Baja CA (Guadalupe Island). Intergrades with *L. nanus*. Feb–Apr ★

L. hirsutissimus Benth. (p. 777) STINGING LUPINE Ann 2–10 dm (often > 10 dm after fire), short-appressed- and stiff-spreading-blister-based-stinging-hairy. **LF:** petiole 4–9 cm; lflets 5–8, 20–50 mm, 10–20 mm wide. **INFL:** 10–30 cm, fls spiralled; peduncle 5–8 cm; pedicels 2–5 mm; bracts 4–5 mm, gen persistent. **FL:** 12–18 mm; calyx 6–10 mm, lips ± equal, upper deeply lobed; petals dark pink to magenta, drying ± purple, banner spot ± yellow, in age magenta, keel upper margins glabrous, lower densely ciliate middle to near claw. **FR:** 2–4 cm, ± 8 mm wide, coarsely hairy. **SEED:** 3–6. Locally common. Dry, rocky areas, burns; < 1400 m. c&s CW (SnFrB, SCoRO, possibly others), SW; Baja CA. Mar–May

L. holmgrenianus C.P. Sm. (p. 777) HOLMGRENS' LUPINE Per 4–7 dm, long-hairy. **ST:** erect. **LF:** basal and cauline; stipules 5–20 mm; lower petioles 2–17 cm, > upper; lflets 4–7, 15–50 mm. **INFL:** 10–26 cm, open to ± dense, fls spiralled; peduncle 3–10 cm; pedicels 6–10 mm; bracts 8–10 mm, gen deciduous. **FL:** 13–15 mm; calyx bulge or spur < 1 mm, upper lip 6–7 mm, 2-toothed, lower 7–9 mm, entire; petals violet, banner back glabrous, spot yellow, keel upper margins ± ciliate, lower glabrous. **FR:** 4–5 cm, hairy. **SEED:** 5–7. Dry desert slopes; 1500–2500 m. ne DMtns (Last Chance Range, Grapevine Mtns, Inyo Co.); w NV. [*Lupinus holmgrenanus*, orth. var.] May–Jun ★

L. hyacinthinus Greene Per 4–10 dm, green, ± hairy. **ST:** erect. **LF:** cauline; stipules 5–16 mm; petiole 3–6 cm; lflets 7–12, 30–80 mm, 4–8 mm wide. **INFL:** 4–22 cm, fls ± whorled; peduncle 3–12 cm; pedicels 2–6 mm; bract 5–9 mm. **FL:** 13–16 mm; calyx upper lip 6–10 mm, 2-toothed, lower 7–11 mm, entire to 3-toothed; petals light blue to purple, banner back glabrous, spot ± yellow to white, keel glabrous. **FR:** 3–4 cm, silky. **SEED:** 3–7, 4–6 mm, beige, speckled brown. Dry slopes, under pines; 2000–3500 m. SnGb, SnBr, PR. Jun–Aug

L. lapidicola A. Heller HELLER'S MOUNT EDDY LUPINE Per < 1 dm, silver-silky. **ST:** ± prostrate, short. **LF:** cauline, clustered near base; stipules 4–5 mm; petiole 2–4.5 cm; lflets 6–8, 10–20 mm. **INFL:** 2–7 cm, fls in few spaced whorls; peduncle 5–10 cm; pedicels 2–4 mm; bracts 4–5 mm, gen deciduous. **FL:** 9–12 mm; calyx upper lip 4–5 mm, notched, lower 5–6 mm, ± 3-toothed; petals ± violet, banner back gen hairy, spot yellow, keel upper margins ciliate, lower glabrous. **FR:** not seen. Dry granite gravel; 1500–3000 m. e KR (Mount Eddy, Siskiyou Co.), NCoRO (King Range, sw Humboldt Co.). Jul ★

L. latifolius J. Agardh Per 3–24 dm, green, glabrous to hairy. **ST:** erect. **LF:** cauline; stipules 5–10 mm; petiole 4–20 mm; lflets 5–11, 40–100 mm, abaxially ± hairy, adaxially glabrous to hairy. **INFL:**

16–60 cm, open, fls whorled or not; peduncle 8–20 cm; pedicels 2–12 mm; bract 8–12 mm. **FL:** 8–18 mm; calyx upper lip 5–10 mm, entire to 2-toothed, lower 4–8 mm, entire or notched; petals blue or purple to white, banner back glabrous, spot gen white to ± yellow turning purple, keel upper margins ciliate claw to middle, lower gen ciliate. **FR:** 2–4.5 cm, ± densely hairy. **SEED:** 6–10, 3–4 mm, mottled dark brown. TOXIC: causes birth defects in livestock. [*L. rivularis* Lindl., in part] Other vars. to BC, UT, AZ, Baja CA.

var. ***barbatus*** (L.F. Hend.) Munz BEARDED LUPINE **ST:** glabrous to strigose. **FL:** 8–10 mm. Wet places; 1500–2500 m. MP. Jun–Jul ★

var. ***columbianus*** (A. Heller) C.P. Sm. (p. 777) Pl 4–24 dm. **ST:** subglabrous to strigose. **FL:** 10–14 mm; wings covering most of keel. Moist slopes, stream sides; 1000–3500 m. SN, SNE; to WA. May–Sep

var. ***dudleyi*** C.P. Sm. (p. 777) **ST:** densely hairy. **FL:** 13–16 mm. Chaparral; < 1000 m. SnFrB. Apr–May

var. ***latifolius*** (p. 777) Pl 3–12 dm. **ST:** glabrous to puberulent. **FL:** 10–14 mm; wings not covering most of keel. Moist areas, open woodland; < 2500 m. NW, CW, SW; to WA. Apr–Jul

var. ***parishii*** C.P. Sm. **ST:** ± glabrous to sparsely strigose. **FL:** 14–18 mm. Moist areas; < 3500 m. c SNH, s SN, SW. May–Aug

var. ***viridifolius*** (A. Heller) C.P. Sm. (p. 777) **ST:** strigose. **FL:** 8–10 mm. Moist areas; 1000–2000 m. KR, CaR, MP. Jun–Aug

L. lepidus Lindl. DWARF LUPINE Per < 6 dm, matted or not, hairy. **ST:** 0 or prostrate to ± erect. **LF:** gen basal; stipules 3–25 mm; petiole 2–10 cm; lflets 5–8, 5–40 mm. **INFL:** < 30 cm, gen dense; peduncle < 14 cm; pedicels 1–3 mm; bracts 4–15 mm, persistent. **FL:** 6–11 mm; calyx upper lip 3–7 mm, entire to 2-toothed, lower 4–7 mm, entire to 3-toothed; petals pink, violet, or blue, banner back glabrous, keel upper margins ciliate, lower glabrous. **FR:** 1–2 cm, hairy. **SEED:** 2–4, 2–4 mm, ± mottled tan or green to brown. Variable; vars. defined by habit, infl, bracts, habitats. Other vars. in w N.Am.

var. ***confertus*** (Kellogg) C.P. Sm. (p. 777) Pl 25–60 cm, hairy. **ST:** decumbent to erect. **LF:** cauline. **INFL:** 5–30 cm, > lvs, fls in > 7 ± crowded whorls; bract 5–14 mm. **FL:** 7–9 mm; banner spot ± yellow fading brown to red. 2*n*=48. Common. Meadows, vernally moist areas; 1500–3000 m. CaRH, SNH, WTR, SnBr, GB, DMtns; NV. Jun–Aug

var. ***culbertsonii*** (Greene) C.P. Sm. (p. 777) HOCKETT MEADOWS LUPINE Pl 15–40 cm, ± green. **ST:** 0 to short. **LF:** gen basal; lflets gen 10–30 mm. **INFL:** 4–12 cm, > lvs, gen open, fls in 3–7 spaced whorls; bract 4–5 mm. **FL:** 9–11 mm; banner spot white to light yellow. Rocky slopes; 2500–3000 m. s SNH (Kaweah River, Tulare, Fresno cos.). Jul–Aug ★

var. ***lobbii*** (S. Watson) C.L. Hitchc. (p. 777) Pl < 10 cm, hairy to shaggy-hairy. **ST:** prostrate. **LF:** gen basal. **INFL:** ± head-like, 2–8 cm, gen < some lvs; bract 5–6 mm. **FL:** 6–10 mm; banner spot white. Dry rocks, meadows; 2000–3500 m. KR, NCoR, CaRH, SNH, Wrn, SNE; to WA, NV. Jun–Aug

var. ***ramosus*** Jeps. Pl 13–30 cm, shaggy-hairy. **ST:** decumbent to erect. **LF:** basal to cauline; lflets gen 5–15(23) mm. **INFL:** 2–10 cm, open, fls in 3–7 ± spaced whorls; bract 4–9 mm. **FL:** gen 7–9(12) mm, fragrant; banner spot white to yellow. 2*n*=48. Subalpine; 3000–4000 m. c&s SNH, SNE. Jul–Aug

var. ***sellulus*** (Kellogg) Barneby (p. 777) Pl 12–35 cm. **ST:** short, prostrate to ± erect. **LF:** ± basal; lflets 10–30 mm. **INFL:** 4.5–11 cm, > lvs; bract 4–8 mm. **FL:** 8–9 mm; banner spot yellow to white turning red. 2*n*=48. Dry rocks, open woodland; 1000–2500 m. NW, CaRH, SNH, GB; to OR, ID, NV. Jun–Aug

var. ***utahensis*** (S. Watson) C.L. Hitchc. (p. 777) STEMLESS LUPINE Pl 10–25 cm, short-lived, matted, densely hairy. **ST:** short. **LF:** appearing basal. **INFL:** 3–6 cm, gen < lvs, fls in 2–3 whorls; bract 8–15 mm. **FL:** 7–10 mm; banner spot white, keel upper margins ciliate near tip. 2*n*=48. Sand or rocks, with sagebrush, lodgepole pine; 1500–3500 m. W&I (White Mtns); to OR, MT, CO. Jun–Jul ★

L. leucophyllus Lindl. (p. 777) Per 4–9 dm, white-woolly and long-stiff-hairy. **ST:** erect, branched from just above ground. **LF:** cauline, some clustered at base; stipule lobes 6–15 mm; petiole 6–20 cm; lflets 7–11, 30–70 mm. **INFL:** 8–30 cm, dense; peduncle 2–8 cm; pedicels 1–2 mm, stout; bracts 3–12 mm, gen persistent. **FL:** 10–13 mm; calyx upper lip 3–6 mm, 2-toothed, lower 3–8 mm, entire; petals lavender to purple, often turning brown, banner back densely hairy, spot yellow to brown, keel upper margins ciliate, lower glabrous. **FR:** 2–3 cm, hairy. **SEED:** 3–6, mottled gray-tan. *n*=24,48. Grassy hillsides, sagebrush flats; 500–2000 m. KR, CaRH, MP; to WA, MT, UT. May–Aug

L. littoralis Douglas (p. 777) Per, subshrub, < 3 dm, ± green to silvery, strigose to shaggy-hairy, long-hairy esp at nodes; roots yellow. **ST:** prostrate to decumbent. **LF:** cauline; stipules 8–16 mm; petiole 2–6 cm; lflets 5–9, 10–35 mm. **INFL:** 12–15 cm, ± open, fls whorled; peduncle 3–9 cm; pedicels 3–5 mm; bract 5–9 mm. **FL:** 10–13 mm; calyx upper lip 5–6 mm, 2-toothed, lower 6–7 mm, entire to 3-toothed; petals purple, banner back glabrous, spot ± white, keel upper margins ciliate, lower glabrous. **FR:** 3–4 cm, hairy. **SEED:** 8–12. Coastal sand; < 100 m. n&c NCo, CCo (Point Reyes); to BC. Doubtfully distinct from *L. variicolor*. May–Aug

L. longifolius (S. Watson) Abrams (p. 777) Shrub 5–15 dm, silvery to green, soft-short-hairy. **ST:** erect. **LF:** cauline; stipules 5–14 mm; petiole 4–10 cm; lflets 5–10, 30–60 mm. **INFL:** 20–45 cm, fls ± whorled or not; peduncle 5–12 cm; pedicels 5–10 mm; bract 4–11 mm. **FL:** 12–18 mm; calyx upper lip 8–10 mm, 2-toothed, lower 10–15 mm, entire; petals violet to blue, banner back glabrous, spot ± yellow to white or 0, keel upper margins ciliate middle to tip, lower glabrous. **FR:** 4–6 cm, dark, hairy. **SEED:** 6–8, 5–6 mm, ± brown to gray. Scrub, canyons; < 500 m. SW. Apr–Jun

L. ludovicianus Greene (p. 777) SAN LUIS OBISPO COUNTY LUPINE Per 3–6 dm, woolly-tomentose. **ST:** erect, branched just above ground, hairs < 1 mm, not sharp, not stiff. **LF:** cauline, clustered at base; stipules 7–12 mm; petiole 5–12 cm; lflets 5–9, 15–40 mm, oblanceolate. **INFL:** 10–40 cm, fls ± whorled or not; peduncle < 16 cm; pedicels 2–5 mm, stout; bract 7–8 mm. **FL:** 10–15 mm; calyx upper lip 6–7 mm, deeply notched, lower 6–8 mm, 3-toothed; petals ± blue to purple, banner back glabrous to ± hairy, spot yellow turning purple to white, keel upper margins ciliate middle to tip, lower glabrous. **FR:** 2–3 cm, hairy. **SEED:** 3–4, 4–7 mm, mottled ± gray. Open, grassy areas, on limestone, in oak woodland; 50–500 m. s SCoRO (San Luis Obispo Co.). Apr–Jul ★

L. luteolus Kellogg (p. 777) BUTTER LUPINE Ann 3–7.5 dm, sparsely hairy or in age glabrous, appearing glaucous; cotyledons disk-like, persistent, or leaving circular scar. **ST:** hard, rigid. **LF:** petiole 2–5 cm; lflets 7–9, 10–30 mm, 4–9 mm wide, adaxially gen hairy (exc e KR). **INFL:** 5–22 cm, fls in gen crowded whorls; peduncle 4–15 cm; pedicels 1–3 mm; bracts 5–11 mm, reflexed, hairy, persistent. **FL:** 10–16 mm; calyx upper lip 3–5 mm, lower 6–10 mm, appendages gen 0; petals gen pale yellow (± pink or bright blue), wings gen ciliate on upper (lower) margins near claw, keel upper, lower margins equally ± densely ciliate. **FR:** spreading, 1–1.5 cm, ± 10 mm wide, ovate, hairy. **SEED:** 2, dark brown, tubercled. Clearings, open or disturbed areas; < 1900 m. NW, CaRF, e SnFrB and n SCoRI (Diablo Range), w WTR; OR. May–Aug

L. magnificus M.E. Jones (p. 777) Per 6–12 dm, white-woolly. **ST:** erect, branched from base, hairs sharp, stiff, sometimes 1–3 mm. **LF:** gen basal; stipules 10–24 mm; petiole 6–30 cm; lflets 5–9, 20–55 mm, 6–15 mm wide. **INFL:** 20–50 cm, fls whorled or not; peduncle 10–50 cm; pedicels 2–8 mm; bract 4–5 mm. **FL:** 10–18 mm, fragrant; calyx upper lip 5–9 mm, 2-toothed, lower 5–11 mm, entire; petals lavender to rose, banner back glabrous, spot yellow turning purple, keel upper margins ciliate middle to tip, lower glabrous. **FR:** 3–7 cm, densely hairy. **SEED:** 5–8, 3–4 mm, tan. If recognized taxonomically, straight-keeled pls from SNE assignable to *L. magnificus* var. *hesperius* (A. Heller) C.P. Sm., McGee Meadows lupine.

var. ***glarecola*** M.E. Jones COSO MOUNTAINS LUPINE **FL:** 10–13 mm. Desert slopes, washes; 1500–2500 m. se SNH, DMoj. Apr–Jun ★

var. ***magnificus*** PANAMINT MOUNTAINS LUPINE **FL:** (13)16–18 mm. Desert slopes, washes; 1500–2500 m. SNE, n DMoj. May–Jun ★

L. microcarpus Sims CHICK LUPINE Ann 1–8 dm, sparsely to densely hairy; cotyledons disk-like, persistent, or leaving circular scar. **ST:** clearly hollow, at least below. **LF:** petiole 3–15 cm; lflets 5–11, gen 9, 10–50 mm, 2–12 mm wide, occ linear, adaxially glabrous. **INFL:** 2–30 cm; peduncle 2–30 cm; pedicels 0.5–5 mm; bracts 3.5–12 mm, reflexed, persistent. **FL:** 8–18 mm; calyx upper lip 2–6 mm, lower 5–10 mm, appendages gen 0; petals white to dark yellow, pink to dark rose, or lavender to purple, wings gen ciliate on upper (less often lower) margins near claw, keel upper margins ciliate, lower less so or glabrous near claw. **FR:** erect to spreading, often on 1 side of infl, 1–1.5 cm, ± 10 mm wide, ovate, hairy. **SEED:** 2, tan to brown, gen mottled, wrinkled or smooth. 2*n*=48. Highly variable; vars. intergrade.

var. ***densiflorus*** (Benth.) Jeps. (p. 777) **INFL:** bract short-appressed- to long-spreading-hairy. **FL:** calyx ± sparsely appressed- to -spreading-hairy, appendages gen 0; petals gen white to yellow (rose or purple), pink- or lavender-tinged or not, wings oblanceolate, withering, upper margins (rarely lower near claw) gen ciliate, keel lower margins occ sparsely ciliate near claw. **FR:** ± spreading, gen on 1 side of axis. Abundant. Open or disturbed areas, occ seeded on roadbanks; < 1600 m. NW (exc Siskiyou Co.), SNF, Teh, GV, CW, e SCo, TR, PR, MP, DMtns, DSon. Apr–Jun

var. ***horizontalis*** (A. Heller) Jeps. (p. 777) **INFL:** bract short- to long-spreading-hairy. **FL:** calyx appressed- to spreading-hairy, appendages 1–2 mm; petals lavender to purple, in age translucent, wings widely elliptic, persistent, ciliate on upper (and gen lower) margins near claw, keel lower margins ciliate near claw. Washes, sand, gravel; < 1500 m. s SNF, s SnJV, WTR, e DMoj. Apr–May

var. ***microcarpus*** (p. 777) **INFL:** bract long-shaggy-hairy. **FL:** calyx long-shaggy-hairy, appendages gen 0; petals gen pink to purple (± yellow or white), wings linear to lanceolate, withering, upper margins (rarely lower) gen ciliate near claw, keel lower margins gen glabrous near claw. **FR:** gen ± erect, gen on > 1 side of axis. Abundant. Open or disturbed areas, occ seeded on roadbanks; < 1600 m. CA-FP, MP, w DMoj; to BC, Baja CA, S.Am. Mar–Jun

L. milo-bakeri C.P. Sm. MILO BAKER'S LUPINE Ann 10–20 dm, sparsely hairy or glabrous in age, appearing glaucous; cotyledons disk-like, persistent, or leaving circular scar. **ST:** hard, rigid, with pith until old. **LF:** petiole 2–5 cm; lflets 7–9, 10–30 mm, 4–9 mm wide, adaxially gen hairy **INFL:** 5–22 cm, fls in gen crowded whorls; peduncle 4–15 cm; pedicels 1–3 mm; bracts 5–11 mm, reflexed, hairy, persistent. **FL:** 10–16 mm; calyx upper lip 3–5 mm, lower 6–10 mm, appendages gen 0; petals ± blue, in age ± yellow, wings gen ciliate on upper margins (rarely lower) near claw, keel upper, lower margins equally ± densely ciliate. **FR:** spreading, 1–1.5 cm, ± 10 mm wide, ovate, hairy. **SEED:** 2, dark brown, tubercled. Clearings, open or disturbed areas; < 1900 m. NCoRO (Covelo, Mendocino Co.). Jun–Sep ★

L. nanus Benth. (p. 777) Ann 1–6 dm, hairy. **LF:** petiole 2–8.5 cm; lflets 5–9, gen 7, 10–40 mm, 1–12 mm wide, occ linear. **INFL:** 2–20 cm, fls gen whorled; peduncle 2–15 cm; pedicels 2.5–7 mm; bract 4–12 mm. **FL:** 6–15 mm; calyx 4–8 mm, lips ± equal, upper deeply lobed; petals blue (light blue), lavender, pink, or white, banner spot white, keel upper margins ciliate near claw, lower glabrous. **FR:** 2–4 cm, 4–7 mm wide, hairy. **SEED:** 4–12. 2*n*=48. Abundant. Open or disturbed areas; < 1300 m. CA-FP (exc s SW); to BC. Mar–Jun

L. nevadensis A. Heller NEVADA LUPINE Per 1–4 dm, long-hairy. **ST:** erect. **LF:** basal and cauline; stipules 8–10 mm; basal petioles < 14 cm, cauline < 4 cm; lflets 6–10, 20–50 mm. **INFL:** 5–17 cm, fls spiraled; peduncle 3–6 cm; pedicels 4–8 mm; bract 4–5 mm. **FL:** 10–12 mm; calyx upper lip 3–4 mm, 2-toothed, lower 4–5 mm, 3-toothed; petals blue, banner back glabrous, spot white to ± yellow, keel strongly upcurved, upper margins ciliate, lower glabrous. **FR:** 2.5–4 cm, densely hairy. **SEED:** 3–4, light. Hillsides, valleys, with sagebrush; 1000–3000 m. GB, DMtns; OR, NV. Apr–Jun ★

Lupinus breweri
var. grandiflorus

bud — banner

keel

Lupinus chamissonis

2 cm

5 mm

keel

fruit

1 mm

Lupinus citrinus var. citrinus

fruits

calyx appendage

5 mm

5 mm

1 cm

5 mm

banner patch

5 mm

2 cm

cotyledons

Lupinus
brevicaulis

bud

keel

2 mm

Lupinus breweri
var. bryoides

1 mm

2 cm

1 mm

5 dm

1 cm

axillary fruit

Lupinus cervinus

Lupinus chamissonis

Lupinus concinnus

Lupinus croceus

2 cm

5 mm

Lupinus covillei

flower

5 mm

fruit

1 cm

5 cm

2 mm

bract

bud

1 cm

5 mm

flower

Lupinus elmeri

1 dm

5 mm

2 cm

5 mm

leaflet

Lupinus constancei

leaf

Lupinus dalesiae

2 cm

5 mm

flower

5 mm

keel

Lupinus duranii

5 mm

1 cm

2 mm

stem

leaflet

Lupinus elatus

bud

1 cm

flower

1 cm

keel

5 mm

var. excubitus

5 mm

flower

5 cm

5 cm

var. austromontanus

5 cm

1 dm

var. medius

Lupinus excubitus

flower

5 mm

2 mm

seed

2 cm

Lupinus
flavoculatus

flower

5 mm

keel

var. robustus

5 cm

2 mm

leaflet

1 cm

var. formosus

Lupinus formosus

flower

5 mm

bract

flower

Lupinus gracilentus

2 cm

stipule

Lupinus fulcratus

flower

2 cm

5 mm

keel

1 mm

leaflet

1 cm

Lupinus grayi

L. nipomensis Eastw. (p. 777) NIPOMO MESA LUPINE Ann 1–2 dm, hairy. **ST:** decumbent. **LF:** petiole 2–3 cm; lflets 5–7, 10–15 mm, 5–6 mm wide. **INFL:** 1–5 cm, dense, fls spiralled; peduncle 2–3.5 cm, decumbent; pedicels 1–1.5 mm; bracts 3–3.5 mm, gen persistent. **FL:** 6–7 mm; calyx 4–5.5 mm, lips ± equal, upper deeply lobed; petals pink, banner spot white or ± yellow, keel glabrous. **FR:** 1.5–2 cm, 5 mm wide, hairy or glabrous in age. **SEED:** 3–4. Stable dunes; < 25 m. s CCo (Nipomo Dunes, sw San Luis Obispo Co.). Intergrades with *L. concinnus*. Threatened by coastal development. Mar–May ★

L. obtusilobus A. Heller (p. 777) Per 1.5–3 dm, appressed-silvery-silky. **ST:** decumbent to erect. **LF:** cauline; stipules 7–14 mm; petiole 2–5 mm; lflets 6–7, 20–50 mm. **INFL:** 3–7 cm, fls ± whorled; peduncle 2–4 cm; pedicels 2–5 mm; bract 3–4 mm. **FL:** 11–13 mm; calyx upper lip 6–7 mm, 2-toothed, lower 6–7 mm, 3-toothed; petals blue to lilac, banner back hairy, spot yellow, keel upper margins ciliate, lower glabrous. **FR:** 2.5–4 cm, silky. **SEED:** 4–5, 3–4 mm, mottled brown. Gravelly summits; 2500–3500 m. NW, CaR, n SNH. Jun–Sep

L. odoratus A. Heller (p. 777) MOJAVE LUPINE Ann 1–3 dm, sparsely short-hairy in youth, glabrous at fl; cotyledons disk-like, persistent. **LF:** basal; petiole 2–12 cm; lflets 7–9, 10–20 mm, 3–8 mm wide, occ obovate, bright green. **INFL:** 4–13 cm, fls spiraled; peduncle 6–15 cm; pedicels 3–5 mm; bracts 2–4 mm, straight, persistent, tips sparsely ciliate. **FL:** with violet odor, 7–10 mm; calyx tips gen glabrous (ciliate), upper lip 3–3.5 mm, rounded or 2-toothed, lower 4–5 mm; petals deep blue-purple, banner spot white or yellow, in age magenta, keel glabrous. **FR:** 1.5–2.5 cm, ± 8 mm wide, oblong, upper suture wavy, densely long-ciliate, sides glabrous or hairs few, short, when dry scale-like. **SEED:** 2–6, wrinkled. Sandy flats, open areas; < 1600 m. GB, D; NV, AZ. May be confused with *L. flavoculatus*. Apr–May

L. onustus S. Watson (p. 777) Per 2–3 dm, green, silky. **ST:** short, decumbent. **LF:** cauline, clustered near base, abaxially silky, adaxially gen glabrous; stipules 8–10 mm; petiole 5–13 mm; lflets 5–9, 15–50 mm. **INFL:** 5–15 cm, fls spiralled; peduncle 4–8 cm; pedicels 3–5 mm; bract 3–4 mm. **FL:** 8–11 mm; calyx upper lip 2–5 mm, 2-toothed, lower 3.5–6 mm, entire; petals violet, banner back glabrous, keel upper margins ciliate, occ sparsely so, lower glabrous. **FR:** 3–4.5 cm, hairy. **SEED:** 5–6, 6–7 mm, brown. Dry banks, open forest, on serpentine; 1000–2000 m. KR, CaRH, n SNH; sw OR. Apr–Sep

L. pachylobus Greene (p. 777) BIG POD LUPINE Ann 1.5–4 dm, hairy. **LF:** petiole 4–8 cm; lflets gen 7, 20–25 mm, 2–5 mm wide. **INFL:** 1–4 cm, fls gen whorled; peduncle 3–12 cm; pedicels 1–2.5 mm; bract ± 6 mm. **FL:** 7–9 mm; calyx 4.5–6 mm, lips ± equal; petals blue, banner spot white, in age dark magenta, keel blunt, glabrous. **FR:** 3 cm, 6–9 mm wide, densely hairy, ± fleshy. **SEED:** gen 5. Uncommon. Open or disturbed areas; < 600 m. NCoRO, CaRF, SNF, SnFrB, SCoRO. Occurs, intergrades with *L. bicolor*. Mar–May

L. padre-crowleyi C.P. Sm. (p. 777) FATHER CROWLEY'S LUPINE Per 5–7.5 dm, silver- to white-woolly. **ST:** erect, from a woolly mat. **LF:** basal and cauline; stipules 5–11 mm; petiole 2–3 cm; lflets 6–9, 25–75 mm. **INFL:** 7–21 cm, fls ± whorled; peduncle 2–5.5 cm; pedicels 2–3.5 mm; bracts 4–9 mm, deciduous or not. **FL:** 10–14 mm; calyx upper lip 5–7 mm, 2-toothed, lower 5.5–8 mm, 3-toothed; petals cream to pale yellow, banner back gen hairy, keel glabrous. **FR:** 2–3 cm, silky. **SEED:** 2–3, 4–5 mm, white, mottled black. Decomposed granite; 2500–4000 m. c SNH (s Mono Co., Inyo Co.), s SNH (c Tulare Co.), SNE (s Mono Co., n Inyo Co.). Jun–Sep ★

L. peirsonii H. Mason (p. 777) PEIRSON'S LUPINE Per 3–6 dm, silver-silky. **ST:** erect, branched just above ground. **LF:** cauline, clustered at base, ± fleshy; stipules 15–20 mm; petiole 2–15 mm; lflets 5–8, 25–70 mm, widely oblanceolate. **INFL:** < 10 cm, fls ± whorled; peduncle < 24 cm; pedicels 1–2 mm; bract 5–7 mm. **FL:** 10–12 mm; calyx upper lip 4–6 mm, ± 2-toothed, lower 5–7 mm, entire; petals yellow, banner back gen hairy, keel upper margins ciliate middle to tip, lower glabrous. **FR:** 3–4 cm, silky. **SEED:** 3–5. Loose gravel; 1000–2000 m. SnGb. Apr–May ★

L. polyphyllus Lindl. Per 2–15 dm, green, glabrous or sparsely hairy. **ST:** erect, stout, gen hollow. **LF:** basal and cauline; stipules 5–30 mm; petioles 3–45 cm, upper shorter; lflets 5–17, 40–150 mm.

INFL: 6–40 cm, open, fls ± whorled; peduncle 3–13 cm; pedicels 3–15 mm; bract 7–11 mm. **FL:** 9–15 mm; calyx lips 4–7 mm, entire; petals violet to lavender to pink to white, banner back glabrous, spot yellow to white occ turning red-purple, keel upcurved, gen glabrous. **FR:** 2.5–4 cm, hairy. **SEED:** 3–9. 2*n*=48.

var. ***burkei*** (S. Watson) C.L. Hitchc. (p. 781) **LF:** lflets 5–11, abaxially occ sparsely hairy, adaxially glabrous. Wet places; 1500–3000 m. KR, CaR, SNH, TR, SnJt, GB; OR, ID, NV. May–Aug

var. ***pallidipes*** (A. Heller) C.P. Sm. **LF:** lflets 9–17, adaxially puberulent to minutely rough-hairy, appearing ± glabrous. Moist to wet places; < 2500 m. KR, NCoRH, CaR; to BC. May–Jun

var. ***polyphyllus*** (p. 781) **LF:** lflets 9–17, abaxially ± strigose, adaxially glabrous. Moist areas to bogs; < 3000 m. NW, CW (exc SCoRI); to BC. May–Aug

L. pratensis A. Heller Per 3–7 dm, green, hairy. **ST:** erect, hollow. **LF:** basal and cauline, green, strigose, hairs < 1 mm; stipules 5–20 mm; basal petioles 10–25 cm, cauline 1–4 cm; lflets 5–10, 30–80(130) mm. **INFL:** 5–28 cm, dense; peduncle 4–17 cm; pedicels 1–3 mm; bracts 5–10 mm, persistent. **FL:** 10–12 mm; calyx upper lip 4–7 mm, 2-toothed, lower 5–6 mm, entire; petals violet to dark blue, banner back glabrous or hairy, spot orange to red, keel upper margins densely ciliate, lower glabrous. **FR:** 1.5–2 cm, hairy to woolly. **SEED:** 4–6, 3–4 mm, mottled tan, brown. 2*n*=48.

var. ***eriostachyus*** C.P. Sm. **FL:** banner back hairy. Moist places; 1500–3500 m. SNE (Big Pine Creek, Inyo Co.). May–Sep

var. ***pratensis*** **FL:** banner back glabrous. Meadows, streambanks; 1000–3500 m. c&s SNH, SNE. May–Sep

L. pusillus Pursh var. ***intermontanus*** (A. Heller) C.P. Sm. (p. 781) INTERMONTANE LUPINE Ann 0.5–1.2 dm, hairy; cotyledons disk-like, persistent. **LF:** cauline, crowded near base; petioles 3–6 cm; lflets 5–6, 10–20 mm, ± 5 mm wide, adaxially glabrous. **INFL:** ± 3 cm, < lvs, dense, fls spiraled; peduncle 0–1 cm; pedicels < 1 mm; bracts ± 3 mm, straight, persistent. **FL:** 6 mm; calyx upper lip ± 2.5 mm, lower ± 5 mm; petals pale blue, fading ± pink or ± white, banner spot white or yellow, keel glabrous. **FR:** 1.5 cm, 6 mm wide, oblong, narrowed between seeds, hairy. **SEED:** 2, wrinkled, margin ridged. 2*n*=48. Open, sandy areas; < 1600 m. GB (Modoc, Inyo cos.); to WA, c US, AZ. May–Jun ★

L. rivularis Lindl. (p. 781) Per (subshrub), 3.5–10 dm, green, ± glabrous. **ST:** erect, dark brown to red, gen hollow. **LF:** cauline; stipules 7–15 mm; petiole 3–5 cm; lflets 5–9, 20–40 mm, adaxially glabrous. **INFL:** 15–50 cm, open, fls ± whorled or not; peduncle 3–15 cm; pedicels 5–10 mm; bract 8–10 mm. **FL:** 12–16 mm; calyx upper lip 7–8 mm, 2-toothed, lower 7–9 mm, entire to ± 3-toothed; petals violet, banner back glabrous, keel upper margins ciliate claw to tip, lower glabrous. **FR:** 3–7 cm, dark, sparsely hairy. **SEED:** 7–8, 3–4 mm, mottled brown with black line. Sand or gravel; < 500 m. NCo; to BC. Grades ± into blue-fld *L. arboreus*, but fls earlier (late winter, spring), not sweet-smelling. Mar–Jun

L. saxosus Howell (p. 781) Per 2–3 dm, green. **ST:** erect. **LF:** gen basal; stipules 10–20 mm; petiole 4–12(14) cm; lflets 7–12, 10–40 mm, abaxially long-stiff-hairy, adaxially glabrous. **INFL:** 5–20 cm, dense, fls ± whorled or not; peduncle 2–15 cm; pedicels 4–8 mm; bracts 4–7 mm, ± deciduous. **FL:** 12–18 mm; calyx upper lip 5–7 mm, 2-toothed, lower 5–6 mm, 3-toothed; petals blue, banner back glabrous, spot yellow turning violet, keel upper margins ± sparsely ciliate, lower glabrous. **FR:** 2–4 cm, shaggy-hairy. **SEED:** 4–6, brown. Sagebrush scrub, open areas; 1000–2500 m. MP; to WA, NV. May–Jun

L. sericatus Kellogg (p. 781) COBB MOUNTAIN LUPINE Per 1.5–5 dm, silver to gray-green, short-appressed-hairy. **ST:** gen erect. **LF:** cauline, clustered near base; stipules 2–7 mm; petiole 5–15 cm; lflets 4–7, 30–50 mm, widely spoon-shaped. **INFL:** 10–30 cm, open to dense, fls ± whorled; peduncle 8–15 cm; pedicels 4–6 mm; bract 3–4 mm. **FL:** 12–16 mm; calyx upper lip 6–10 mm, 2-toothed, lower 7–10 mm, 3-toothed; petals purple to violet, banner back glabrous to ± hairy, keel upper margins ciliate claw to tip, lower gen ± glabrous. **FR:** 2–3 cm, hairy. **SEED:** 3–7, 3–4 mm, light brown. Open wooded slopes; 500–1500 m. s NCoRI (Napa, Lake, Sonoma cos.). Mar–Jun ★

777

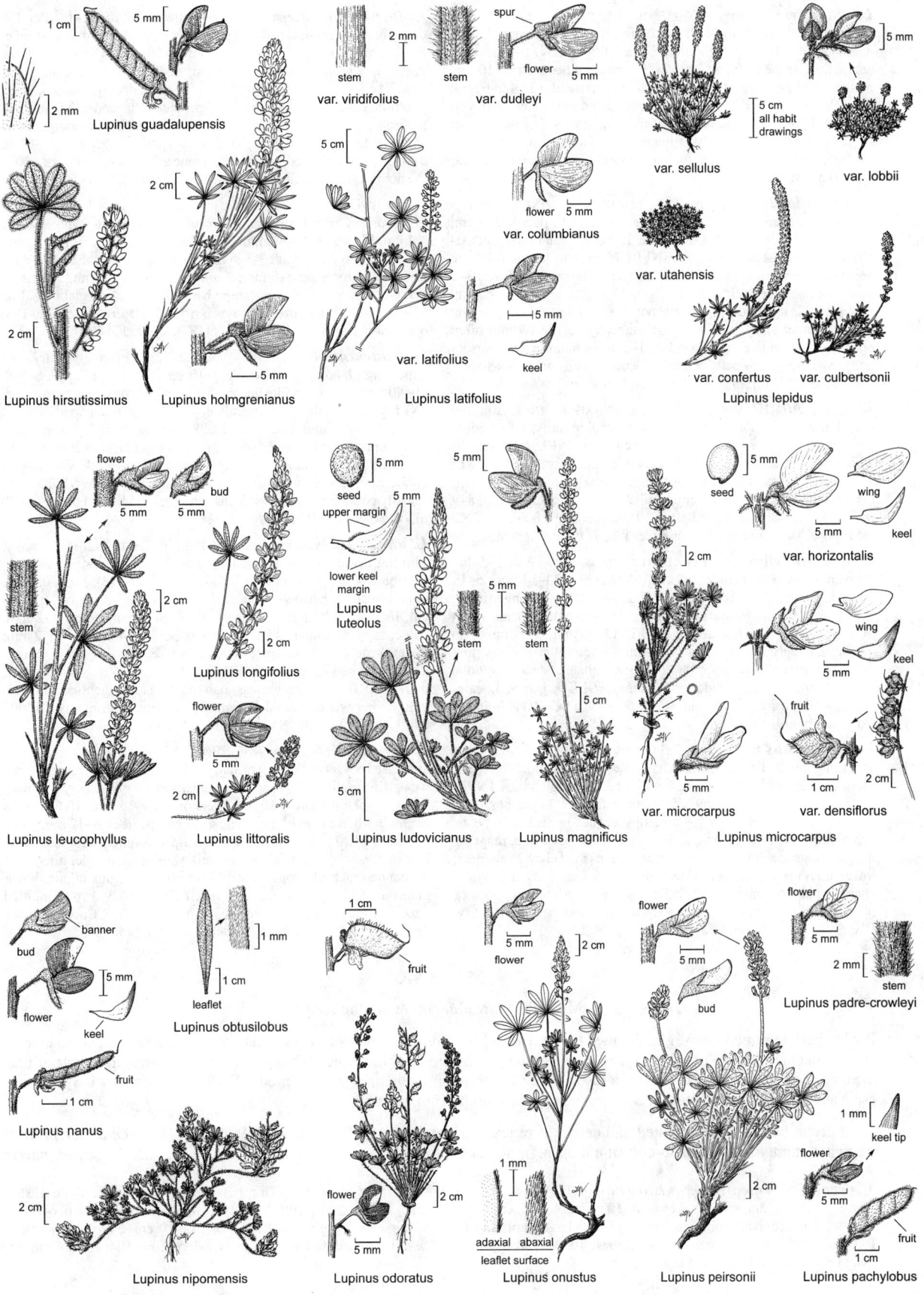

Lupinus guadalupensis

var. viridifolius   var. dudleyi

spur
flower

var. columbianus

var. latifolius

Lupinus latifolius

var. sellulus   var. lobbii

var. utahensis

var. confertus   var. culbertsonii

Lupinus lepidus

Lupinus hirsutissimus   Lupinus holmgrenianus

flower   bud

stem

Lupinus longifolius

seed
upper margin

lower keel margin

Lupinus luteolus

stem   stem

flower

Lupinus leucophyllus   Lupinus littoralis

Lupinus ludovicianus   Lupinus magnificus

seed   wing   keel

var. horizontalis

wing   keel

fruit

var. microcarpus   var. densiflorus

Lupinus microcarpus

banner
bud
flower
keel

fruit

Lupinus nanus

leaflet

Lupinus obtusilobus

fruit

flower

adaxial abaxial
leaflet surface

Lupinus nipomensis   Lupinus odoratus   Lupinus onustus

flower
bud

flower

Lupinus padre-crowleyi

stem

keel tip

flower

fruit

Lupinus peirsonii   Lupinus pachylobus

L. shockleyi S. Watson (p. 781) DESERT LUPINE Ann 0.5–3 dm, canescent; cotyledons disk-like, persistent. **LF:** cauline, crowded near base; petioles 4–12 cm; lflets 8–10, 10–30 mm, 4–10 mm wide, adaxially glabrous. **INFL:** 2–6 cm, fls spiraled; peduncle 1–10 cm; pedicels 1–3 mm; bracts 2–4 mm, straight, persistent. **FL:** 4.5–6 mm; calyx 3–6 mm, lips ± equal; petals dark blue-purple, banner spot yellow, keel blunt, glabrous. **FR:** 1.5–2 cm, 8–12 mm wide, upper suture gen wavy, densely stiff-long-ciliate, sides with hairs short, inflated, scale-like when dry. **SEED:** 2, wrinkled. Open, sandy areas; < 1200 m. D; NV, AZ. Apr–May

L. sparsiflorus Benth. (p. 781) COULTER'S LUPINE Ann 2–4 dm, short-appressed- and long-spreading-hairy. **LF:** petiole 3–4 cm; lflets 7–11, 15–30 mm, 2–4 mm wide, linear to oblanceolate, adaxially hairy at least near margins. **INFL:** 15–20 cm, fls spiraled, occ appearing ± whorled; peduncle 2–4 cm; pedicels 2–4 mm; bracts 3–5 mm, < buds, gen deciduous. **FL:** gen 10–12 mm; calyx 3–6 mm, lips ± equal, upper deeply lobed; petals gen blue (± pink), drying darker, banner spot ± white, in age magenta, keel lower and often upper margins ciliate near claw. **FR:** 1–2 cm, ± 5 mm wide, coarsely hairy. **SEED:** 4–5. Locally common. Washes, sandy areas; < 1300 m. s SCoR, SW, DMoj; to UT, AZ, n Mex. Mar–May

L. spectabilis Hoover (p. 781) SHAGGYHAIR LUPINE Ann 2–6 dm, dense-short-appressed and -long-spreading-hairy. **LF:** petiole 4–9 cm; lflets gen 9, 10–40 mm, 4–9 mm wide. **INFL:** 5–30 cm, fls whorled; peduncle 5–12 cm; pedicels 6–8 mm; bract 8–9 mm. **FL:** 11–14 mm; calyx 4–7 mm, lips ± equal; petals blue (white), banner spot white, keel upper margins ciliate near tip. **FR:** 3–5 cm, 8–9 mm wide, densely hairy. **SEED:** 5–10. Serpentine; < 800 m. c SNF (Mariposa, Tuolumne cos.). Intergrades with *L. nanus*. Apr–May ★

L. stiversii Kellogg (p. 781) HARLEQUIN LUPINE Ann 1–5 dm, sparsely hairy. **LF:** petiole 2–8 cm; lflets gen 7, 20–50 mm, 5–15 mm wide, adaxially bright green. **INFL:** 3–8 cm (often longer in fr), < peduncle, dense, fls spiralled; peduncle 5–10 cm; pedicels 1.5–4 mm; bracts 3–5 mm, late-deciduous. **FL:** 13–18 mm; calyx upper lip 4–6 mm, deeply lobed, lower 5–7 mm; banner yellow, wings pink (± white), keel white, upper, lower margins ciliate claw to ± middle. **FR:** ± 2 cm, 5–7 mm wide, gen glabrous. **SEED:** gen 5. Locally common. Clearings, open areas; < 2100 m. SN, n SCoRO (Monterey Co.), SnGb, SnBr. Apr–Jun

L. succulentus K. Koch (p. 781) ARROYO LUPINE Ann, often appearing per, 2–10 dm, sparsely hairy, fleshy. **LF:** petiole 6–15 cm; lflets 7–9, 20–60 mm, 7–20 mm wide, adaxially glabrous. **INFL:** 9–15 cm, fls whorled; peduncle 5–9 cm; pedicels 3–7 mm; bract 3–5 mm. **FL:** 12–18 mm; calyx 4–7 mm, lips ± equal, upper lobed; petals gen blue-purple (white, pink, lavender), banner spot white, magenta in age, wings sparsely ciliate on upper margins near claw, keel upper, lower margins ciliate near claw. **FR:** 3.5–5 cm, 8–10 mm wide, coarsely hairy to tomentose. **SEED:** 6–9. $2n=48$. Abundant. Open or disturbed areas, often seeded on roadbanks; < 800 m. c&s NW, GV, CW, SW; Baja CA. Feb–May

L. tidestromii Greene (p. 781) TIDESTROM'S LUPINE Per 1–3 dm, white-shaggy-hairy. **ST:** ± prostrate, weak. **LF:** cauline; stipules 8–12 mm; petiole 1–3 cm; lflets 3–5, 5–20 mm. **INFL:** 2–10 cm, open, fls whorled; peduncle 4–8 cm; pedicels 3–5 mm; bract 4–5 mm. **FL:** 11–13 mm; calyx upper lip 5–6 mm, deeply notched, lower 5–6 mm, entire to notched; petals light blue to lavender, banner back glabrous, spot white to yellow turning violet, keel upper margins ciliate claw to tip, lower glabrous. **FR:** 2–3 cm, shaggy-hairy. **SEED:** 5–8, 3–4 mm, mottled brown, tan. Dunes, beaches; < 100 m. s NCo (Sonoma Co.), n&c CCo (Marin, Monterey cos.). May–Jun ★

L. tracyi Eastw. (p. 781) TRACY'S LUPINE Per 4–7 dm, glabrous, glaucous. **ST:** erect, slender. **LF:** cauline; stipules 7–9 mm; petiole < 1 cm; lflets 6–7, 10–40 mm. **INFL:** 4–16 cm, fls ± whorled or not; peduncle 2–6 cm; pedicels 5–6 mm; bract 8–10 mm. **FL:** 8–10(12) mm; calyx upper lip 3–8 mm, 2-toothed, lower 3–5 mm, 3-toothed; petals ± white to dull blue, banner back glabrous, keel glabrous. **FR:** 1.5–2.5 cm, white-hairy, dark when dry. **SEED:** 3–4, 4–5 mm. Dry, open montane forest; 1500–2000 m. KR; s OR. Jul ★

L. truncatus Nutt. (p. 781) Ann 2–3(5) dm, finely hairy, gen appearing glabrous. **LF:** petiole 3–10 cm, flat, lflet-like; lflets 5–8, 20–40 mm, 2–5 mm wide, linear, adaxially glabrous, tip gen truncate. **INFL:** 3–25 cm, fls sparse, spiralled; peduncle 3–10 cm; pedicels 2–4 mm; bracts 2–5 mm, persistent. **FL:** 8–13 mm; calyx 3–4 mm, lips ± equal, upper deeply lobed; banner, wings magenta, banner spot ± yellow, dark magenta in age, keel tip stout, blunt, upper, lower margins ciliate middle to claw. **FR:** ± 3 cm, 5 mm wide, hairy. **SEED:** 6–8. Locally common. Openings in chaparral or woodland, burns; < 1000 m. c&s CW, SW; Baja CA. Mar–May

L. uncialis S. Watson LILLIPUT LUPINE Ann, 1–2(3) cm, hairy, forming dense lfy tuft; cotyledons disk-like, persistent. **LF:** stipule free blades < 1 mm; petiole 0.4–1.5 cm; lflets, (3)5, 0.2–0.7 cm, oblanceolate to narrowly obovate, hairy. **INFL:** among lvs, 0.5–1.5 cm, fls (1)2; peduncle < 1 cm; pedicels 4–6 mm; bracts 1–2 mm, ovate, early-deciduous. **FL:** 5–6 mm; upper lip 2-toothed, 1.5–2 mm; lower 3-toothed, 3–3.5 mm; petals 2-colored, banner white, wings, keel ± purple, keel glabrous. **FR:** 0.6–1 cm, 3.5–5 mm wide, ovate, hairy. **SEED:** 1–2. Open areas, barrens, talus in sagebrush and pinyon/juniper woodland, on limestone, rhyolite, volcanic ash; 1400–2400 m. MP; OR, ID, NV. May–Jul ★

L. variicolor Steud. (p. 781) Per, subshrub, 2–5 dm, gen dense-appressed or -spreading-silver-hairy. **ST:** prostrate to decumbent, not weak. **LF:** cauline, often appearing clustered near base 1st yr; stipules 7–8 mm; petiole gen 4–10 cm; lflets 6–9, 20–35 mm. **INFL:** 6–15 cm, fls ± whorled or not; peduncle 4–12 cm; pedicels 4–12 mm; bract 4–7 mm. **FL:** 11–16 mm; calyx upper lip 7–8 mm, 2-toothed, lower 8–9 mm, gen entire; petals white, yellow, rose, or purple, often on 1 pl, banner back glabrous, spot 0, keel upper margins ciliate, lower glabrous. **FR:** 3–4 cm, dark, hairy. **SEED:** 7–9, 3–4 mm, mottled dark. Coastal terraces, beaches; < 500 m. NCo, n&c CCo. Doubtfully distinct from *L. littoralis*. Apr–Jul

MARINA

Michelle M. McMahon & Duane Isely

Per [or not], unarmed, hairy, gland-dotted or not. **LF:** odd-1-pinnate; stipules obscure; lflets opposite. **INFL:** open raceme or dense spike, terminal. **FL:** calyx lobes < tube; corolla gen 2-colored (blue-violet, white), petals (exc banner) from side of filament column; stamens 9–10, filaments fused; ovule 1. **FR:** indehiscent, incl or ± exserted. **SEED:** 1. 38 spp.: sw US to C.Am, esp Mex. (Marina, interpreter for Mexican conqueror Cortez, 16th century)

1. St decumbent, not gland-dotted; infl dense, fls reflexed or spreading; e PR (Santa Rosa Mtns). . . . . ***M. orcuttii*** var. ***orcuttii***
1′ St ascending or erect, gland-dotted; infl open, fls spreading to ascending; D . . . . ***M. parryi***

M. orcuttii (S. Watson) Barneby var. ***orcuttii*** (p. 781) CALIFORNIA MARINA Pl < 2 dm, strigose, gen gray. **LF:** lflets 11–15, crowded, 2–5 mm, oblanceolate to obovate, folded, tip with large gland. **INFL:** 1–2 cm. **FL:** keel 5–5.5 mm. Rocky slopes; 1050–1150 m. e PR (Santa Rosa Mtns); Baja CA. May–Oct ★

M. parryi (Torr. & A. Gray) Barneby (p. 781) Pl 2–8 dm, strigose. **LF:** well spaced; lflets 11–23, 0.5–6 mm, oblong obovate to ± round, flat, ± gland-dotted. **INFL:** 2–10 cm. **FL:** keel 5–7 mm. $2n=20$. Open desert washes, rocky slopes, roadsides; < 800 m. D; AZ, Mex. Feb–Jun

MEDICAGO ALFALFA, MEDICK

Martin F. Wojciechowski & Duane Isely

Ann, per, unarmed, gen hairy. **ST:** prostrate to erect. **LF:** subpalmately compound or gen odd-1-pinnate; stipules ± fused to petiole, entire or deeply cut; lflets 3, gen dentate near tip. **INFL:** axillary or terminal, raceme, gen umbel- or ± head-like, 1–many-fld. **FL:** calyx lobes ± equal or not; corolla yellow or purple; 9 filaments fused, 1 free. **FR:** indehiscent, reniform or gen spirally coiled 1.5–8 turns (or sickle-shaped or straight), gen prickly. **SEED:** 1–several. 83 spp.: Medit to w&c Asia; several cult, naturalized in warm temp. (Greek: Medice, now Media, Asia Minor, source of alfalfa) *M. muricata* possibly naturalized in Carrizo Plain.

1. Fr 1-seeded, reniform, not prickly, black; fl 2–3 mm . . . . ***M. lupulina***
1′ Fr 1- to several-seeded, spirally coiled (or straight to sickle-shaped), prickly or not, tan or gray to black; fl 2–11 mm
  2. Per, gen erect; corolla 6–11 mm, violet or violet- to yellow-green (yellow); fr not prickly . . . . ***M. sativa***
  2′ Ann, gen prostrate, decumbent, or low; corolla gen 2–7 mm, yellow to orange-yellow; fr prickly or not
    3. Fr not prickly, ± 1–1.5 cm wide . . . . [***M. scutellata***]
    3′ Fr gen prickly, < 1 cm wide
      4. Stipules gen entire to ± dentate; pl hairs gen dense
        5. Lvs densely hairy; fr sparsely hairy . . . . ***M. minima***
        5′ Lvs, fr hairy, often glandular-hairy (or fr glabrous) . . . . ***M. rigidula***
      4′ Stipules deeply toothed or cut; pl hairs gen ± 0 to sparse (strigose in *M. praecox*)
        6. Lflet gen with dark central spot, obovate to obcordate, length ± = or 1–2 × width; lf pinnate or subpalmate; stipules gen cut 1/4–1/2 width
          7. Lflet gen obcordate, length ± = width; lf pinnate or ± palmate; fr a loose spiral, glabrous . . . . ***M. arabica***
          7′ Lflet obovate (obcordate), length 1–2 × width; lf pinnate; fr gen a tight spiral, sparsely (densely) hairy . . . . ***M. truncatula***
        6′ Lflet without dark central spot, gen wedge-shaped or obovate to obcordate, length 1–2 × width; lf pinnate; stipules gen cut ± ≥ 1/2 width
          8. Lflet gen 10–20 mm, gen glabrous; corolla 3.5–6 mm; CA-FP; common . . . . ***M. polymorpha***
          8′ Lflet 3–10 mm, gen strigose; corolla 2.5–4 mm; NCoRI, CaRF, SN, SnFrB, SCo; uncommon . . . . ***M. praecox***

M. arabica (L.) Huds. BURCLOVER, SPOTTED BURCLOVER Ann, hairs 0 to sparse. **ST:** gen sprawling, 1–4 dm. **LF:** pinnate or subpalmate; stipules gen cut 1/4–1/2 width; lflets gen 1–2.5 cm, gen obcordate, length ± = width, with dark central spot. **INFL:** 2–4-fld. **FL:** calyx 2–2.5 mm; corolla 4–5 mm, yellow. **FR:** a loose spiral, coiled 4–6 turns, spheric or short-cylindric; tan to brown, prickles curved, hooked, hairs 0. **SEED:** several. 2*n*=16. Disturbed and agricultural areas, fields, woodland; < 2000 m. NCo, NCoR, CaRF, n SN, ScV, CW; abundant esp in se US, elsewhere; native to Medit. Mar–Jun

M. lupulina L. BLACK MEDICK Ann, hairy or glandular-hairy. **ST:** prostrate to ascending, gen 1–4 dm, angled. **LF:** stipules entire or toothed; lflets gen 1–1.5 cm, obovate to oblong. **INFL:** 10–20-fld. **FL:** calyx 1–1.5 mm; corolla 2–3 mm, yellow. **FR:** 2–2.5 mm, reniform, veins prominent, concentric; black, prickles 0, hairs 0. **SEED:** 1. 2*n*=16,32. Disturbed and agricultural areas, forest, mtns; < 2500 m. CA-FP, GB, DMtns (Panamint Range); most of US, widespread elsewhere; native to Eur. Apr–Jul

M. minima (L.) Bartal. BURCLOVER Ann, hairs gen dense. **ST:** decumbent or ascending, 1–4 dm. **LF:** stipules lanceolate, gen entire to ± dentate; lflets 4–12 mm, elliptic or obovate, gen dentate near tip. **INFL:** 3–6-fld. **FL:** calyx 2–2.5 mm; corolla 3–5 mm, yellow. **FR:** gen clustered, coiled 3–4 turns, spheric to short-cylindric; tan to brown, prickles hooked, hairs sparse. **SEED:** several. 2*n*=16. Disturbed areas; < 1000 m. CA-FP; to OR, s-c US; sporadic in e US; native to Medit. Apr–Jul

M. polymorpha L. (p. 781) CALIFORNIA BURCLOVER Ann, gen glabrous. **ST:** prostrate, mat-forming, or ascending, 1–5 dm. **LF:** stipules gen cut ± ≥ 1/2 width; lflets gen 10–20 mm, wedge-shaped to obcordate, length 1–2 × width. **INFL:** 2–6-fld. **FL:** calyx ± 3 mm; corolla 3.5–6 mm, yellow. **FR:** a loose spiral, coiled 2–6 turns, ovate to short-cylindric; gray to black, prickly, hairs 0. **SEED:** several. 2*n*=14. Common. Chaparral, oak woodland, streambanks, roadsides, disturbed areas; < 1500 m. CA-FP; widespread in s US, warm-temp; native to Medit, Eur. Provides spring pasture in CA; bur-like frs lodge in fur. Mar–Jul ❖

M. praecox DC. Ann, gen strigose. **ST:** decumbent, 1–3 dm. **LF:** stipules lanceolate, gen cut ± 1/2 width; lflets 3–10 mm, wedge-shaped or obovate to obcordate, length 1–2 × width. **INFL:** gen axillary, 1–3-fld. **FL:** calyx 1.5–3 mm; corolla 2.5–4 mm, yellow, ephemeral. **FR:** a loose spiral, coiled 3–5 turns, spheric or disk-like; tan to dark, prickles slender, hooked, hairs sparse. **SEED:** few. 2*n*=14. Uncommon. Creek beds, oak woodland, disturbed areas; < 1500 m. NCoRI, CaRF, SN, SnFrB, SCo; native to Medit. Apr–May

M. rigidula (L.) All. TIFTON MEDICK, TIFTON BURCLOVER Ann, hairy, often glandular-hairy. **ST:** prostrate to ascending, gen 1–7 dm. **LF:** stipules entire or deeply toothed; lflets gen 1–1.5 cm, wide-obovate to -obcordate, entire to dentate near tip. **INFL:** 1–6-fld, peduncle > (≤) petiole. **FL:** calyx 3–5 mm; corolla 6–7 mm, yellow. **FR:** 4.5–9 mm, disk-like, cylindric to spheric, loosely coiled 3–7 turns, veins strongly curved; light yellow to dark gray, prickly, hairy, often glandular-hairy (or glabrous). **SEED:** 1–3. 2*n*=14. Uncommon. Disturbed and agricultural areas; < 400 m. PR; native to Eur, n Afr. Mar–Apr

M. sativa L. (p. 781) ALFALFA Per, ± glabrous or puberulent. **ST:** gen erect, 2–8 dm. **LF:** stipules lanceolate, entire to sharply toothed; lflets 1–2.9 cm, narrowly lanceolate to obovate. **INFL:** spike-like, 8–30-fld, longer in fr. **FL:** calyx 4–4.5 mm; corolla 6–11 mm, violet or violet- to yellow-green (yellow). **FR:** gen coiled 2–3 turns (or straight or sickle-shaped); light to dark yellow-brown, leathery, prickles 0, hairs 0. 2*n*=16,32. Disturbed, agricultural areas; < 2450 m. CA-FP, GB, DMtns (Panamint Range); US exc se; native to Eurasia. Cult; highly variable, polyploid complex in US, incl genetic components from several spp.; often divided into several spp. or subspp., none tenable. Apr–Oct

M. truncatula Gaertn. BARREL MEDICK, BARREL CLOVER Ann, sparsely hairy. **ST:** prostrate to ascending, gen 1–8 dm. **LF:** stipules deeply toothed; lflets gen 1–2.5 cm, obovate (obcordate), dentate near tip, length 1–2 × width, gen with dark central spot. **INFL:** 1–3(5)-fld, peduncle < (≥) petiole. **FL:** calyx 2.5–4 mm; corolla 5–7 mm, yellow. **FR:** 5–8 mm, cylindric, gen tight spiral, coiled 3–6 turns, veins not prominent; light yellow to dark gray, sparsely (densely) hairy, prickles 1–4 mm, straight to curved. **SEED:** 1–2. 2*n*=16. Uncommon. Disturbed and agricultural areas; < 800 m. SCoRO, PR; native to Medit. [*M. muricata* All., misappl.] Mar–Apr

MELILOTUS SWEETCLOVER

Kelly Steele & Duane Isely

Ann, bien, unarmed. **ST**: gen erect. **LF**: odd-1-pinnate; stipules gen narrow or bristle-like, bases fused to petiole; lflets 3, margin toothed or wavy. **INFL**: raceme, axillary or terminal, slender or short-cylindric, many-fld. **FL**: calyx lobes ± equal; corolla yellow or white; 9 filaments fused, 1 free. **FR**: indehiscent, 2–4 mm, ovate, compressed but thick, leathery, bumpy or not, ridges transverse to finely net-like. **SEED**: 1–2. 20 spp.: temp Eur, esp Medit, subtrop Asia, n Afr, widely introduced, naturalized; several spp. widely cult for forage, green manure, soil improvement. (Greek: honey-lotus) TOXIC: inclusion in hay enhances production of mold toxins that may cause cattle death.

1. Corolla 2.5–3 mm; fr bumpy or with faint lines . . . . ***M. indicus***
1′ Corolla 3–7 mm; fr irregularly cross-ridged or with network of lines
  2. Corolla white, 3.5–5 mm; fr with network of lines . . . . ***M. albus***
  2′ Corolla yellow, 4.5–7 mm; fr irregularly cross-ridged . . . . ***M. officinalis***

M. albus Medik. (p. 785) WHITE SWEETCLOVER Pl ± glabrous or strigose. **ST**: 0.5–2 m. **LF**: lflets 1–2.5 cm, elliptic-oblong to obovate, ± toothed. **INFL**: slender; axis gen 3–8 cm when fls open. **FL**: calyx ± 2 mm; corolla 3.5–5 mm, white. **FR**: 3–5 mm, with network of lines. **SEED**: 1. 2*n*=16,24. Locally abundant. Pastures, open disturbed sites; < 1900 m. CA; most of n US, adjacent Can; native to Eurasia. [*Melilotus alba*, orth. var.] Indistinguishable from *M. officinalis* prior to fl. May–Sep

M. indicus (L.) All. (p. 785) SOURCLOVER Ann, ± glabrous. **ST**: spreading or erect, 1–6 dm. **LF**: lflets 1–2.5 cm, oblanceolate to wedge-shaped-obovate, gen sharply dentate. **INFL**: slender, compact; axis gen 1–2 cm when fls open. **FL**: calyx 1–1.5 mm; corolla 2.5–3 mm, yellow. **FR**: 2–3 mm, bumpy or with faint lines. **SEED**: 1. 2*n*=16. Open, disturbed areas; < 1700 m. CA, more common s; to s OR, se US; native to Medit. [*Melilotus indica*, orth. var.] Apr–Oct

M. officinalis Lam. (p. 785) YELLOW SWEETCLOVER Bien, ± glabrous to strigose. **ST**: 0.5–2 m. **LF**: lflets 1–2.5 cm, elliptic-oblong to obovate, ± toothed. **INFL**: slender; axis gen 3–8 cm when fls open, longer in fr. **FL**: calyx 2–2.5 mm; corolla 4.5–7 mm, yellow. **FR**: 3–5 mm, plump, irregularly cross-ridged. **SEED**: 1. 2*n*=16. Open fields, disturbed sites, cult; < 2300 m. CA; widely distributed in temp areas, esp n&c US; native to Eurasia. May–Aug

OLNEYA IRONWOOD

Matt Lavin & Duane Isely

1 sp. (S.T. Olney, Am botanist, 1812–1878) [Lavin & Sousa 1995 Syst Bot Monogr 45:117–124]

O. tesota A. Gray (p. 785) IRONWOOD Shrub, tree, gen armed, canescent. **LF**: even-1-pinnate, alternate or clustered; stipular spines breaking off, leaving scar, occ 0; lflets 8–21, ± opposite, obovate or elliptic, thick; axis extending beyond lflets, pointed. **INFL**: raceme, axillary, 2–many-fld. **FL**: corolla 10–14 mm, wings purple-pink suf-fused with white, banner, keel yellow-white to pink or purple; 9 filaments fused, 1 free. **FR**: slowly dehiscent, oblong or elliptic, plump, gen narrowed between seeds, persistent. **SEED**: 1–3. 2*n*=18. Often abundant in washes; < 1300 m. PR, DSon; AZ, Mex. Apr–May

ONOBRYCHIS

Martin F. Wojciechowski & Duane Isely

Per [to small shrub], unarmed. **LF**: odd-1-pinnate; stipules papery, lower clasping st. **INFL**: raceme, spike-like. **FL**: calyx lobes > tube, narrowly lanceolate; corolla wings << keel; 9 filaments fused, 1 free. **FR**: indehiscent, 1-seeded. ± 130 spp.: Eurasia, n Afr. (Greek: ancient name)

O. viciifolia Scop. SANFOIN Pl ascending, ± hairy. **LF**: lflets 15–21, 1–2.5 cm, narrowly elliptic to obovate, adaxially finely red-dotted. **INFL**: dense early, peduncle 1–2 dm. **FL**: corolla ± 1 cm, pink. **FR**: ascending, ± 6–7 mm, obovate, short-prickly near margin, strongly net-ridged, leathery, stalk-like base 0. Uncommon. Disturbed places; 1000–1550 m. n SNH, MP; to BC, sporadic in US; native to c&se Eur. Cult for forage, soil improvement. Jun–Aug

ORNITHOPUS

Martin F. Wojciechowski & Duane Isely

Ann, unarmed, ± glabrous or hairy. **ST**: sprawling or ascending, 2–7 dm. **LF**: odd-1-pinnate; stipules small or 0; lflets many, mucronate. **INFL**: axillary, head-like, 1–5-fld, subtended by cluster of bracts or not. **FL**: calyx lobes ± equal, < tube; corolla 6–9 mm; 9 filaments fused, 1 free. **FR**: indehiscent, linear or oblong, flat or not, separating into 1-seeded segments, surface strongly net-like. ± 5 spp.: w Eur, Medit, to sw Asia, 1 in S.Am. (Greek: birdfoot)

1. Fr not flat, narrowed ± between seeds or not; infl not subtended by bracts; lflets 7–15 . . . . ***O. pinnatus***
1′ Fr flat, narrowed between seeds; infl subtended by cluster of lf-like bracts; lflets ± 19–37 . . . . ***O. sativus***

O. pinnatus (Mill.) Druce Pl glabrous or puberulent. **LF**: lflets linear to oblanceolate. **FL**: corolla yellow. **FR**: 2–3.5 cm, linear, gen curved. 2*n*=14. Coastal prairies, disturbed areas; 150–200 m. CCo (Santa Cruz Co.); native to Eur. Jun–Jul

O. sativus Brot. COMMON BIRDSFOOT Pl hairy. **LF**: lflets ovate to lanceolate. **FL**: corolla white or pink. **FR**: 1–4 cm, oblong, straight. Grassland, disturbed areas; 150–200 m. CCo (Santa Cruz Co.); native to sw Eur. Cult as fodder, forage pl. May–Jun

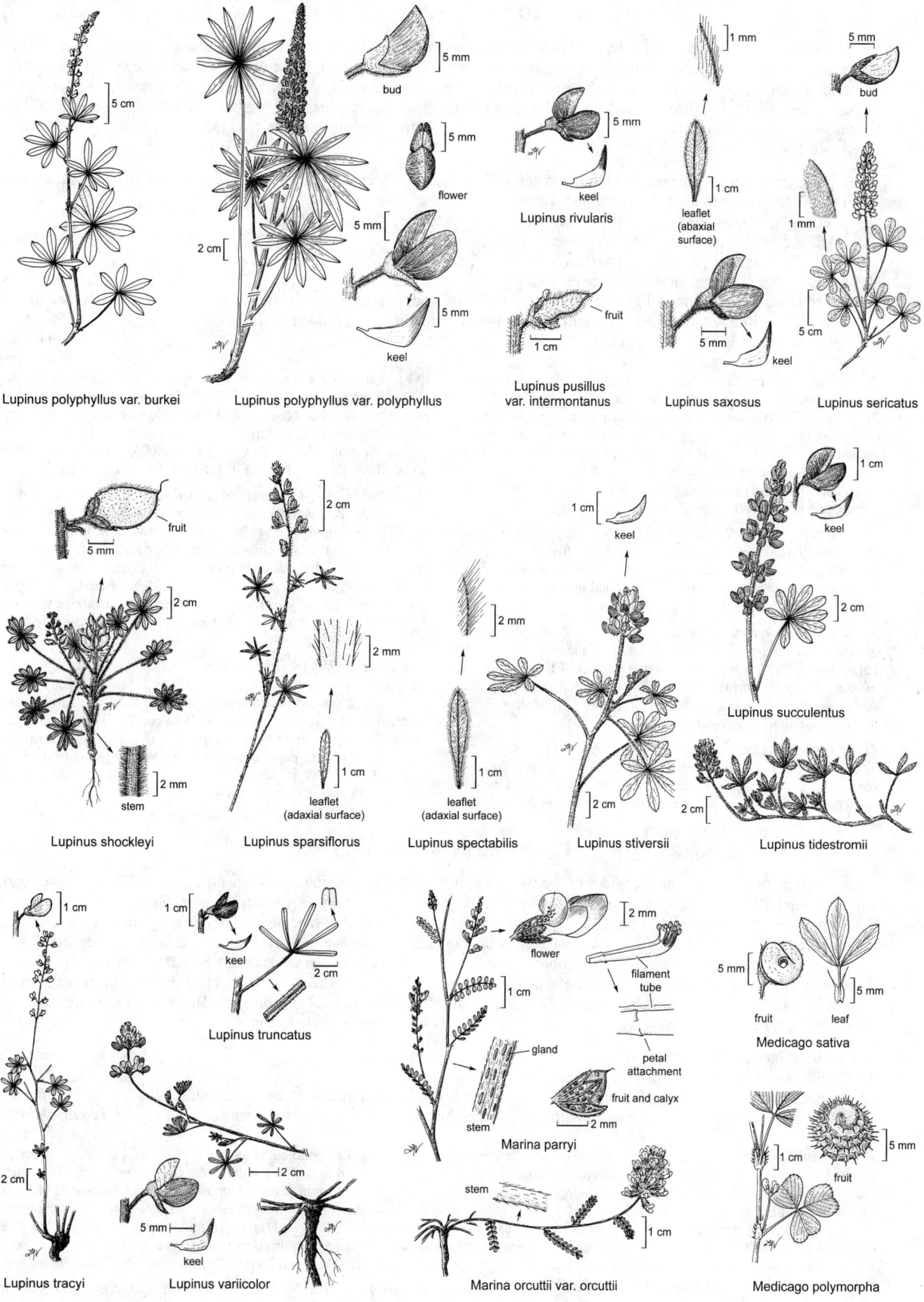

Lupinus polyphyllus var. burkei

Lupinus polyphyllus var. polyphyllus
bud
flower
keel

Lupinus rivularis
keel

leaflet (abaxial surface)

Lupinus pusillus var. intermontanus
fruit

Lupinus saxosus
keel

Lupinus sericatus
bud

Lupinus shockleyi
fruit
stem

Lupinus sparsiflorus
leaflet (adaxial surface)

Lupinus spectabilis
leaflet (adaxial surface)

Lupinus stiversii
keel

Lupinus succulentus
keel

Lupinus tidestromii

Lupinus truncatus
keel

Lupinus tracyi

Lupinus variicolor
keel

Marina parryi
flower
filament tube
petal attachment
fruit and calyx
gland
stem

Marina orcuttii var. orcuttii
stem

Medicago sativa
fruit
leaf

Medicago polymorpha
fruit

OXYTROPIS LOCOWEED, OXYTROPE

Per, unarmed, hairy. **LF**: odd-1-pinnate, gen basal; stipules gen partly fused to petiole, initially forming a sheath, or free. **INFL**: raceme, gen scapose, spike- or head-like or not, or 1–2-fld; bracts gen persistent. **FL**: calyx lobes < tube; corolla pink-purple, white, or ± yellow, keel tip beaked; 9 filaments fused, 1 free; style, stigma glabrous. **FR**: erect to reflexed, gen persistent, oblong to lanceolate, ± inflated, ± 2-chambered, septum from upper suture, partial to complete. ± 350 spp.: Eurasia, N.Am. (Greek: sharp keel) Seriously TOXIC: causes "staggers" in livestock, mostly outside CA. [Welsh 2001 Revision N Amer *Oxytropis*. E.P.S.]

1. Fl quickly reflexed, fr reflexed; lvs basal and gen 1–3 cauline; infl open, spike-like . . . . ***O. deflexa* var. *sericea***
1' Fl and fr ascending to erect; lvs basal; infl dense, spike- or head-like
  2. Lflets 7–17; pl silvery or gray, silky or tomentose, not glandular; corolla 7–12 mm
    3. Fls 2–12 per infl; fr ovate-inflated, 7–9 mm wide, thinly papery. . . . . ***O. oreophila* var. *oreophila***
    3' Fls 1–3 per infl; fr lanceolate to narrowly ovate, 5–6 mm wide, leathery . . . . ***O. parryi***
  2' Lflets 21–39; pl green, puberulent, gen ± sticky-glandular; corolla 10–20 mm . . . . ***O. borealis***
    4. Corolla white to cream; infl gen head-like, incl to ± exserted; pl sticky-glandular. . . . . var. ***australis***
    4' Corolla ± white with purple keel-tip, lilac, or red-purple, drying blue; infl spike-like, exserted;
      pl ± sticky-glandular . . . . var. ***viscida***

O. borealis DC. (p. 785) BOREAL LOCOWEED Pl gen cespitose, green, puberulent, gen ± sticky-glandular. **LF**: basal; lflets 21–39, 2–15 mm, reduced distally on st, ovate to lance-oblong. **INFL**: head- or spike-like raceme, gen exserted, longer in fr or not; fls 4–many, ascending to erect. **FL**: corolla 10–20 mm, white or ± yellow to red-purple (drying blue). **FR**: ascending to erect, 8–14[20] mm, 5–6 mm wide, lanceolate to ovate, ± inflated, papery, ± incompletely 2-chambered; stalk-like base 0.

  var. ***australis*** S.L. Welsh Pl sticky glandular. **LF**: basal; lflets 23–39, 3–10 mm, lance-oblong to ovate, flat or folded. **INFL**: gen head-like, incl to ± exserted. **FL**: corolla 12–18 mm, white to cream. Pine and aspen meadows to alpine ridges and outcrops; 3300–3900 m. s SNH, W&I; to UT; circumboreal. Jul–Aug

  var. ***viscida*** (Nutt.) S.L. Welsh STICKY OXYTROPE Pl ± sticky-glandular. **LF**: basal; lflets 23–39, 3–10 mm, lance-oblong to ovate, flat or folded. **INFL**: spike-like, exserted, fls 5–20. **FL**: corolla 12–18 mm, ± white with purple keel-tip, lilac, or red-purple, drying blue. Aspen meadows to alpine; 3300–3900 m. s SNH, W&I; ne OR, MT to CO; circumboreal. Jul–Aug

O. deflexa (Pall.) DC. var. ***sericea*** Torr. & A. Gray (p. 785) BLUE PENDENT-POD OXYTROPE Pl green or gray. **ST**: few to several. **LF**: basal and gen 1–3 cauline; stipules free to fused around petiole, 7–20 mm; lflets 15–31, 3–20 mm, lanceolate to ovate, flat or folded.

INFL: spike-like, open; fls 10–25, quickly reflexed; peduncle gen > 15 cm, often curved. **FL**: corolla 5–10 mm, dull white to pale lilac or blue, ± ≥ calyx. **FR**: reflexed, 3–4.5 cm, elliptic or oblong, membranous or papery, incompletely 2-chambered; stalk-like base gen short. $2n=16$. Moist meadows, forest openings; 2800–3200 m. W&I (White Mtns, Mono Co.); to AK, MT, UT, n NM. Jun–Aug ★

O. oreophila A. Gray var. ***oreophila*** (p. 785) ROCK-LOVING OXYTROPE Pl cespitose, silvery or gray, silky. **LF**: basal; lflets 7–17, 2–10 mm, elliptic to oblong, folded. **INFL**: head-like, incl or exserted; fls 2–12, ascending or erect. **FL**: corolla 7–10 mm, pink-purple or white. **FR**: ascending or erect, 9–15 mm, 7–9 mm wide, ovate-inflated, thinly papery, incompletely 2-chambered; gen sessile. $2n=16$. Open gravelly or rocky ground, talus, at or above treeline; 2700–3800 m. SnBr; to UT, AZ. Variable; alpine pls only a few cm tall. Jul ★

O. parryi A. Gray (p. 785) PARRY'S OXYTROPE Pl densely cespitose, gray, silky or tomentose, hairs white, black. **LF**: basal; lflets 11–15, 2–12 mm, ovate to oblong, folded. **INFL**: head-like, gen exserted; fls 1–3, ascending to erect. **FL**: corolla 7–10 mm, purple. **FR**: ascending to erect, 15–20 mm, 5–6 mm wide, lanceolate to narrowly ovate, leathery, ± 2-chambered, sessile. Near timberline, above; 3100–3800 m. c&s SNH, SNE (White, Inyo, Sweetwater mtns); to ID, WY, CO, NM. Jun–Jul ★

PARKINSONIA PALO VERDE

Tree, shrub, with stipular spines at nodes or thorns in lf axis (see lf scars) or main lf axis a strong spine. **ST**: ± zigzag; bark smooth, green. **LF**: alternate or in clusters of 1–6 in spine axils, gen even-2-pinnate, sometimes appearing 1-pinnate, main axis flat, lflets alternate, falling early or not; 1° lflets 2–6; 2° lflets 4–many. **INFL**: raceme, axillary, gen < 7-fld. **FL**: ± bilateral; sepals ± free, alike, gen reflexed; petals ± equal, gen yellow or cream-white; stamens 10, yellow to orange, free, exserted. **FR**: indehiscent to partly late-dehiscent, gen flat, oblong, ± inflated, narrowed between seeds or not. **SEED**: 1–several. 11–12 spp., 2 named hybrids: Am, s US to Argentina, Afr; cult. (J. Parkinson, London, apothecary, author, 1567–1650) [Haston et al. 2005 Amer J Bot 92:1359–1371; Hawkins 1996 Ph.D. Dissertation, Univ Oxford, UK] Haston et al., Hawkins support treatment of *Cercidium* and *Parkinsonia*, both recognized in TJM (1993), as single, monophyletic genus.

1. Pl with spines at nodes; lvs 10–30 cm; 2° lflets 30–60 . . . . ***P. aculeata***
1' Pl with thorns in lf axils or thorn-tipped branches; lvs 1–5 cm or 0; 2° lflets 2–6 or 8–16
  2. Lf petioled; 2° lflets 2–6, 4–8 mm; pl with thorns in lf axils; fr ± not narrowed between seeds. . . . . ***P. florida***
  2' Lf sessile; 2° lflets 8–16, 1–5 mm; pl with thorn-tipped branches; fr narrowed between seeds . . . . ***P. microphylla***

P. aculeata L. MEXICAN PALO VERDE Tree < 12 m, with spines at nodes. **LF**: ± sessile; 1° lflets gen 2–4, main axis < 30 cm, conspicuous, persistent ± as ribbon; 2° lflets 30–60, scattered, 3–5 mm, 1–1.5 mm wide, elliptic, ephemeral. **FL**: ± 2 cm wide; sepals reflexed; petals ± round, banner red-spotted at base, all red in age. **FR**: 4–12 cm, ± thickened, leathery. Uncommon. Disturbed, dry places; < 800 m. SnJV, SCo, WTR, PR, DSon; native to deserts of AZ, Baja CA, S.Am, widely cult in trop. Gen Apr–Oct

P. florida (A. Gray) S. Watson (p. 785) BLUE PALO VERDE Tree gen < 8 m, with thorns in lf axils; branches spreading, ± zigzagged,

± glabrous. **LF**: petioled, blue-green; 1° lflets 1 pair, axis < 1 cm; 2° lflets 2–6, 4–8 mm. **FL**: corolla 2–2.5 cm, banner 9–15 mm, widely ovate, yellow, orange-dotted or not. **FR**: indehiscent, 3–10 cm, flat, tan, ± not narrowed between seeds; tip beak-like. Uncommon. Washes, floodplains; ± 1100 m. D; to AZ, nw Mex. [*Cercidium f.* A. Gray, incl subsp. *f.*] Subspp. not recognized by Hawkins (2006, pers. comm.). Apr–May (gen 2 weeks before *Parkinsonia microphylla*)

P. microphylla Torr. (p. 785) LITTLE-LEAVED PALO VERDE Shrub, small tree 3–4(9) m, with thorn-tipped branches; branches gen ascending or spreading, broom-like, hairy. **LF**: sessile, yellow-green;

1° lflets 1 pair, axis 2–4 cm, gen early-deciduous; 2° lflets 8–16, 1–5 mm. **FL:** corolla 12–14 mm, banner < 10 mm, widely ovate, gen cream-white. **FR:** late dehiscent, < 11 mm, brown, narrowed between seeds; tip beak-like, gen ending in spine. Rock slopes; ± 600 m. D; to AZ, nw Mex. [*Cercidium m.* (Torr.) Rose & I.M. Johnst.] Branches used as livestock feed; seeds edible. Hybrids with *P. florida* reported [Jones et al. 1998 Madroño 45:110–118]. Apr–May (gen 2 weeks after *Parkinsonia florida*) ★

PEDIOMELUM BREADROOT

Martin F. Wojciechowski & James Grimes

Per, unarmed, gland-dotted, hairs glandular, nonglandular, or both; roots deep, woody, enlarged near ground surface. **ST:** main axis erect, nearly 0 to short; branches short, decumbent to ascending, underground or not. **LF:** ± palmately compound, ± basal or clustered near st tips; stipules at base of pl fused, those above free; lflets 5–7, elliptic to oblanceolate or widely obovate. **INFL:** raceme, basal, axillary, or terminal on branches, with 1 sometimes late-deciduous bract and 2–3 fls per node. **FL:** calyx base swollen on top, tube enlarging in fr; corolla at least partly blue to purple; 9 filaments fused, 1 less so or free; ovary ± hairy, ovule 1, style tip curved to bent, stigma head-like. **FR:** incl in calyx exc for beak, transversely dehiscent, hairy, rarely glandular. **SEED:** 1, elliptic, smooth or ridged. 21 spp.: s Can to c Mex. (Greek: plain apple) *P. mephiticum* (S. Watson) Rydb. incorrectly reported for s CA.

1. Lvs on pl all palmately compound, lflets 15–28 mm; pedicel 4–6 mm; seed smooth . . . . *P. californicum*
1′ Lvs on pl both palmately and subpalmately compound (1 lflet extended beyond others), lflets 25–42 mm; pedicel 0–1.5 mm; seed ridged . . . . *P. castoreum*

P. californicum (S. Watson) Rydb. (p. 785) **LF:** stipule 7–10 mm; petiole 8–11 cm; lflets 5–7. **INFL:** bract 6.5–8 mm. **FL:** 8–12 mm; calyx 9–10.5 mm; banner 10–11 mm. **FR:** ovate to round in outline; body 4–9 mm; beak 1–4 mm, straight, linear. **SEED:** 5–5.5 mm, reniform, red-brown. 2n=22. Open chaparral, woodland; 1000–2500 m. NCoRI, s SNF (Kern Co.), CW, TR, PR; Baja CA. Apr–Jul

P. castoreum (S. Watson) Rydb. (p. 785) BEAVER DAM BREADROOT **LF:** stipule 5–13.5 mm; petiole 6.8–15 cm; lflets 5–6. **INFL:** bract 3.5–8 mm. **FL:** 9–13 mm; calyx 10–12 mm; banner 9–13 mm. **FR:** ovate to elliptic in outline; body 6–8 mm; beak 8–11 mm, straight to curved, triangular. **SEED:** 6 mm, reniform, gray. Open areas, roadcuts; < 1750 m. DMoj (San Bernardino Co.); NV, AZ. Apr–May ★

PETERIA

Matt Lavin & Duane Isely

Per, armed or not. **ST:** decumbent to erect. **LF:** odd-1-pinnate; stipules spiny, free. **INFL:** raceme, spike-like, terminal. **FL:** calyx tube bulged on upper side near base or not, lobes < or > tube, upper pair fused ≥ 1/2; corolla pink or white; 9 filaments fused, 1 free; style tip or stigma finely hairy. **FR:** dehiscent, oblong, ± flat but plump, leathery; pedicel 3–4 mm. 4 spp.: s US, Mex. (R. Peter, 19th cent KY botanist) [Lavin & Sousa 1995 Syst Bot Monogr 45:124–135]

P. thompsoniae S. Watson (p. 785) SPINE-NODED MILKVETCH Roots spreading, tuberous. **ST:** 2–10 dm. **LF:** lflets 13–31, elliptic or obovate. **INFL:** often glandular-hairy. **FL:** calyx 7–11 mm, tube cylindric, gen dark-glandular-puberulent; corolla gen 15–20(25) mm; style or stigma hairs gen hidden by pollen. **FR:** 4–6 cm, glabrous. Sandy alluvial fans; 800–1800 m. DMtns (California Valley, se Inyo Co.); to ID, UT, sw CO, AZ. May–Jun ★

PHASEOLUS WILD BEAN

Alfonso Delgado-Salinas

[Ann], per, vine, unarmed; hairs gen incl minute, hooked ones. **LF:** odd-1-pinnate; axis extended beyond basal lflets, persistent; lflets 3, entire or lobed. **INFL:** raceme-like (fls > 1 per node) [to panicle-like], nodes 1–60, not swollen; bracts, bractlets gen persistent. **FL:** calyx lobes < (>) tube; corolla incurved, sickle-shaped in bud, banner oblong to round, in fl recurved 90°, wings twisted into a platform position, keel incurved, tightly, spirally coiled 1.5[2] turns; 9 filaments fused, 1 free; style thickened, ± bristly distally, stigma lateral, turned inward. **FR:** gen dehiscent, linear to oblong rarely rhombic, mostly flat in ×-section. **SEED:** 1–20, oblong to reniform or rarely discoid. ± 70 spp.: neotrop, warm regions. (Classical name, presumably for a bean) [Delgado-Salinas et al. 2006 Syst Bot 31:779–791] *P. coccineus* L., *P. lunatus* L., *P. vulgaris* L. in CA possibly as waifs from cult.

P. filiformis Benth. (p. 785) SLENDER-STEM BEAN Short-lived per, trailing or twining vine; taproot slender. **ST:** cylindric or angled, branches many at base, lowermost often opposite. **LF:** lflets gen 1–5 cm, 1–4.5 cm, wide ovate-triangular, gen lobed. **INFL:** raceme-like, peduncle 1.2–16 cm, nodes 2–6, 2-fld. **FL:** corolla ± 1 cm, pink-purple, keel coil diam ± 2.5 mm. **FR:** 2.5–3.5 cm, 4–5 mm wide, oblong, gen curved, pendant, ± glabrous. **SEED:** 4–6(7), 2–4 mm, oblong to reniform, wrinkled, net-like. Washes; ± 125 m. DSon (Coachella Valley, Riverside Co.); to w TX, n Mex. Gen Oct–Dec ★

PICKERINGIA CHAPARRAL PEA

Martin F. Wojciechowski & Duane Isely

1 sp. (C. Pickering, Am naturalist, 1805–1875)

P. montana Nutt. CHAPARRAL PEA Shrub, rhizomed; fls gen appearing before lvs on new growth; thorns terminal or axillary. **ST:** intricately branched, 1–3 m. **LF:** simple or palmately compound, evergreen; stipules 0; lflets 0 or 2–3, 1–2 cm, elliptic or ovate to obovate; petiole ± 0. **INFL:** raceme, terminal or axillary; bracts gen lf-like. **FL:** calyx 5–7 mm, barely lobed; corolla 1.5–1.8 cm, gen purple, keel petals free; stamens 10, free. **FR:** dehiscent, 3–6 cm, 4–5 mm wide, oblong, narrowed between seeds, margin often wavy. **SEED:** 1–8. 2n=28. Vars. intergrade but are geog separate.

1. Pl glabrous to inconspicuously strigose . . . . var. *montana*
1′ Pl strigose to tomentose . . . . var. *tomentosa*

var. *montana* (p. 785) **FR**: rarely abundant. Chaparral, open woodland; gen < 660 m (1700 m in s). NCoR, n SNF, SnFrB, SCoR, n ChI, TR. May–Aug

var. *tomentosa* (Abrams) I.M. Johnst. (p. 785) WOOLLY CHAPARRAL PEA **FR**: abundant. Chaparral, washes; < 1700 m. SnBr, PR; Baja CA. May–Aug ★

PISUM PEA

Martin F. Wojciechowski & Duane Isely

Ann, unarmed, glabrous. **LF**: even-1-pinnate; stipules like, often > lflets; lflets 4–6, opposite; main axis ending in tendril. **INFL**: axillary raceme, 1–3-fld. **FL**: corolla white or 2-colored; 9 filaments fused, 1 free; style longitudinally folded, puberulent on concave side. **FR**: dehiscent, oblong, flat. **SEED**: few to several. 2 spp.: Medit, w Asia, cult worldwide. (Ancient name)

P. sativum L. COMMON PEA Often climbing. **LF**: stipule margins toothed or wavy; lflets 2–4 cm, ovate or elliptic. **FL**: 10–25 mm, white, pink to purple, or both. 2*n*=14. Uncommon. Disturbed areas; < 1000 m. NCo, NCoRO, GV, SCoRO, SCo, PR, expected elsewhere; to e US; native to Medit, w Asia. Cult in temp worldwide. Feb–May

PROSOPIS MESQUITE

Martin F. Wojciechowski & Elizabeth McClintock

Shrub, tree. stipule spines gen 2 per node; roots long, spreading. **LF**: even-2-pinnate, alternate, deciduous; 1° lflets gen 2–4, opposite; 2° lflets gen many, opposite. **INFL**: axillary, head or spike-like raceme, many-fld. **FL**: radial, small, green-white or yellow; calyx shallowly bell-shaped, lobes short; petals gen inconspicuous; stamens 10, exserted, free; style exserted, gen appearing before stamens. **FR**: indehiscent, ± flat, ± narrowed between seeds or tightly coiled, pulpy when young, then woody. **SEED**: several. ± 44 spp.: esp Am (also sw Asia, n Afr). (Greek: burdock, for obscure reasons) Cult, naturalized worldwide.

1. Fr tightly coiled; petals fused
  2. Infl spike-like raceme, 4–8 cm; lf hairy . . . . **P. pubescens**
  2′ Infl head to spike-like raceme, < 4 cm; lf glabrous or short-hairy . . . . **[P. strombulifera]**
1′ Fr not coiled; petals free
  3. Pl ± glabrous; 2° lflets (10)15–25 mm, length 7–9 × width . . . . **P. glandulosa** var. **torreyana**
  3′ Pl with short, dense hairs; 2° lflets 4–15 mm, length 3–4 × width . . . . **P. velutina**

P. glandulosa Torr. var. *torreyana* (L.D. Benson) M.C. Johnst. (p. 785) HONEY MESQUITE Shrub, tree < 7 m; crown often wider than tall. **ST**: branches arched, crooked; spines 5–40 mm. **LF**: glabrous; 1° lflets gen 1 pair, 6–17 cm; 2° lflets 14–34, (10)15–25 mm, oblong, length 7–9 × width. **INFL**: spike-like raceme, 6–10 cm. **FL**: petals 2.5–3.5 mm. **FR**: 5–20 cm, linear, ± narrowed between seeds, glabrous. **SEED**: gen 5–18, 6–7 mm, oblong. Common. Grassland, alkali flats, washes, bottoms, sandy alluvial flats, mesas; < 1700 m. SnJV, SCo, SnGb, SnBr, PR, D; to TX, Baja CA, n Mex. Apr–Aug

P. pubescens Benth. SCREW BEAN, TORNILLO Shrub, tree < 10 m; crown gen ± narrow. **ST**: branches ascending; spines 4–12 mm. **LF**: hairy; 1° lflets 2 or 4, 3–5 cm; 2° lflets 10–16, 2–10 mm, oblong, length 2–3 × width. **INFL**: 4–8 cm. **FL**: petals 2–3 mm. **FR**: 3–5 cm, tightly coiled. **SEED**: gen 3 mm, ovate. 2*n*=28. Creek, river bottoms, sandy or gravelly washes or ravines; < 1300 m. SnJV, SnBr, D; sw US, n Mex. Fr used for food and as coffee substitute. Apr–Sep

P. velutina Wooton Shrub, tree < 15 m; crown ± spreading, rounded. **ST**: often crooked; spines 10–20 mm. **LF**: 1° lflets 2–4, 2–9 cm; 2° lflets gen 30–60, 4–15 mm, oblong, length 3–4 × width. **INFL**: spike-like raceme, 5–15 cm. **FL**: petals 2–3 mm. **FR**: 8–15 cm, linear, ± flat, ± narrowed between seeds. **SEED**: 5–7 mm, ovate. 2*n*=28. Uncommon. Sandy or rocky soils in canyons, washes; < 1000 m. SnJV, CCo, SCo, PR; native to sw US (exc CA), n Mex. Apr–Jun

PSOROTHAMNUS INDIGO-BUSH

Martin F. Wojciechowski & Duane Isely

Per to small tree, gen thorny, gland-dotted, esp sts, gen hairy. **ST**: gen intricately branched. **LF**: simple or gen odd-1-pinnate, lflets 1–3, gen more. **INFL**: axillary or terminal, raceme, spike- or head-like or not; pedicel bractlets (0)2. **FL**: calyx lobes gen unequal, upper 2 often largest; petals from receptacle, indigo blue to pink-purple; stamens 10, filaments partly fused; ovules gen 2. **FR**: indehiscent, incl in or exserted from calyx, gen glandular. **SEED**: 1. 9 spp.: deserts of sw US, Mex, basins of CO Plateau. (Greek: scabshrub)

1. Lf simple, linear or oblanceolate (or lflets 2–3)
  2. Lvs persistent; infl axis not ending as thorn . . . . **P. schottii**
  2′ Lvs deciduous by early summer; infl axis ending as thorn . . . . **P. spinosus**
1′ Lvs odd-1-pinnate or some simple
  3. Fr 2–4.5 mm, incl; infl dense; pedicel bractlets 0
    4. Terminal lflets > lateral or not; glands of twigs << 0.5 mm wide . . . . **P. emoryi**
    4′ Terminal lflets ± = lateral; glands of twigs ± 0.5 mm wide . . . . **P. polydenius**
  3′ Fr 7–10 mm, exserted; infl ± open; pedicel bractlets 2
    5. Fr glands many, small, in longitudinal lines esp distally; lvs gen silvery-strigose; DMoj, e DSon . . . . . . **P. fremontii**
      6. Lflet linear, < ± 1 mm wide; e DSon . . . . var. **attenuatus**
      6′ Lflet narrowly elliptic to ovate or obovate, ± 1.5–3 mm wide; DMoj . . . . var. **fremontii**
    5′ Fr glands few to several, large, scattered; lvs variously hairy, gen not strigose; SNE to SnBr, D . . . . . **P. arborescens**
      7. Lflets gen continuous with axis, silky-hairy . . . . var. **simplicifolius**
      7′ Lflets gen (or at least some) jointed to axis, glabrous to hairy, incl silky-hairy or not
        8. Calyx 7–9 mm, sparsely to densely hairy . . . . var. **arborescens**
        8′ Calyx 5–7 mm, hairs 0 or sparse . . . . var. **minutifolius**

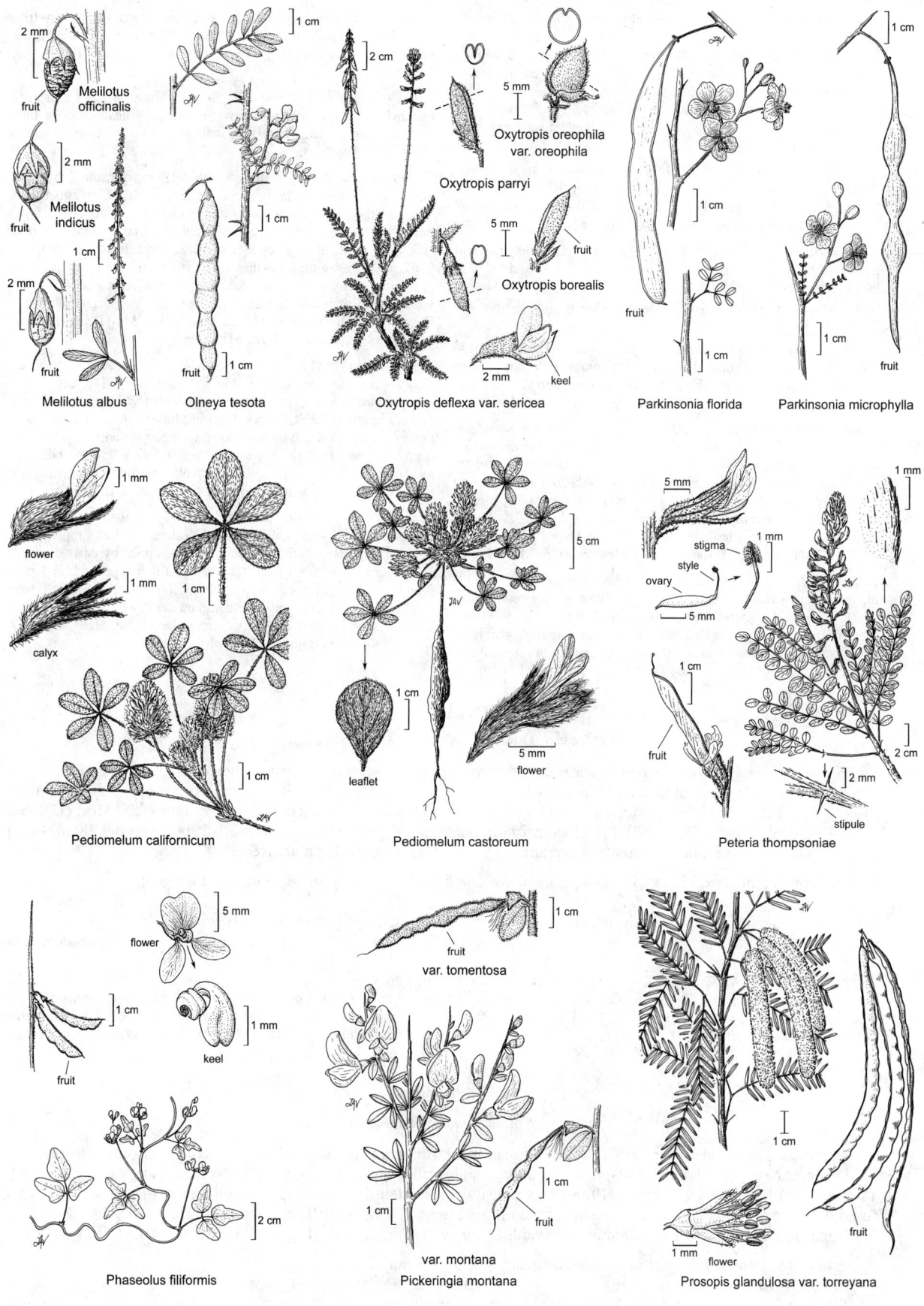

Melilotus officinalis

Melilotus indicus

Melilotus albus

Olneya tesota

Oxytropis oreophila var. oreophila

Oxytropis parryi

Oxytropis borealis

Oxytropis deflexa var. sericea

Parkinsonia florida

Parkinsonia microphylla

Pediomelum californicum

Pediomelum castoreum

Peteria thompsoniae

Phaseolus filiformis

var. tomentosa

var. montana

Pickeringia montana

Prosopis glandulosa var. torreyana

P. arborescens (A. Gray) Barneby MOJAVE INDIGO-BUSH Shrub < 1 m, armed or not, glabrous to ± puberulent. **LF**: lflets gen 5–7, 3–14 mm, terminal (occ all) often continuous with axis. **INFL**: raceme, open; pedicel bractlets 2. **FL**: calyx lobes ± equal, gen < tube; corolla 6–10 mm, indigo-blue to violet-purple, glands 0. **FR**: exserted, 7–10 mm, ovoid-ellipsoid, hairs 0 to fine, glands few to several, scattered, large, ± yellow. Possibly best united with *P. fremontii*; vars. similar morphologically, distinct geog.

var. ***arborescens*** (p. 793) **LF**: lflets lanceolate to ovate, larger 2.5 mm or more wide, terminal often continuous with axis. **FL**: calyx 7–9 mm, sparsely to densely hairy. Desert; 400–800 m. sw DMoj (Kern, San Bernardino cos.); Mex. Apr–May ★

var. ***minutifolius*** (Parish) Barneby (p. 793) **LF**: lflets lanceolate to ovate, larger 2.5 mm or more wide, gen jointed to axis, glabrous to loosely hairy. **FL**: calyx 5–7 mm, hairs 0 or sparse. Desert mtn slopes, canyons, talus; 150–1900 m. c&s SNE, n&c DMoj; sw NV. Gen May–Oct

var. ***simplicifolius*** (Parish) Barneby (p. 793) **LF**: lflets linear-oblanceolate, gen 1–2 mm wide, gen continuous with axis, silky-hairy. **FL**: calyx 5–7 mm, puberulent. Lower desert mtn slopes, flats, washes; 100–1100 m. SnBr, adjacent D. Apr–May

P. emoryi (A. Gray) Rydb. (p. 793) Subshrub < 1 m, < 2 m wide, gen densely canescent, occ glabrous in age, twig glands << 0.5 mm wide, red-orange. **LF**: lflets of middle lvs 5–9, 2–10 mm, narrowly oblong to obovate, terminal > lateral or not. **INFL**: spike-like, ovoid or spheric, dense; pedicel bractlets 0. **FL**: calyx 4–7 mm, lobes ± ≥ tube; corolla 4–8 mm, 2-colored (purple, white), puberulent. **FR**: incl, ± 2.5–3 mm, glabrous at base, glandular-hairy distally. 2*n*=20. Desert flats, washes, dunes; < 700 m. s DMoj, DSon; sw AZ, n Mex, Baja CA. Mar–May

P. fremontii (A. Gray) Barneby FREMONT'S DALEA Shrub < 1 m, gen silvery-strigose. **LF**: lflets exc uppermost gen jointed to axis. **INFL**: ± open, 10–25-fld; pedicel bractlets 2. **FL**: calyx 5–9 mm, lobes unequal, < tube; corolla 7–9.5 mm, magenta-purple, glands 0. **FR**: exserted, 7–10 mm, hairs 0, glands many, small, in longitudinal lines esp distally, orange. 2*n*=20. Like *P. arborescens* exc in fr, geog.

var. ***attenuatus*** Barneby NARROW-LEAVED PSOROTHAMNUS **LF**: lflets gen 5–25 mm, < ± 1 mm wide, linear. Granite, volcanic slopes, flats, canyons; 450–900 m. e DSon; to s NV, nw AZ. Apr–May ★

var. ***fremontii*** (p. 793) **LF**: lflets 3–15 mm, ± 1.5–3 mm wide, narrowly elliptic to ovate or obovate. Granite, volcanic slopes, flats, canyons; gen on sedimentary formations; 150–1350 m. DMoj; to s UT, ne AZ. Apr–May

P. polydenius (Torr.) Rydb. Shrub < 1.5 m, spreading, thorny in age, young branches, twigs finely reflexed-strigose or puberulent, with glands dense, ± 0.5 mm wide, orange. **LF**: lflets 7–13, 1–4.5 mm, obovate to ± round, tip gen notched, terminal ± = lateral. **INFL**: spike-like, ovate or short-cylindric, dense; pedicels < 1 mm, bractlets 0. **FL**: calyx 5–6 mm, lobes unequal, ± = tube, with longitudinal rows of glands; corolla 4.5–6 mm, pink-purple, persistent after fl, banner, wings gland-tipped. **FR**: incl, ± 2.5 mm. 2*n*=20. Locally abundant. Desert flats, hills, washes; 900–2250 m. SNE, DMoj, DSon (Imperial Co.); to e UT. [*Dalea p.* Torr.] May–Sep

P. schottii (Torr.) Barneby INDIGO-BUSH Shrub < 2 m, green, gray-strigose in age. **LF**: simple (or lflets 2–3), 1–3 cm, linear, densely puberulent, gland-dotted, persistent. **INFL**: open, 5–20-fld; pedicel bractlets 2. **FL**: calyx 4–6 mm, lobes < tube, glabrous to ± puberulent, glands gen many, brown to ± purple; corolla 7–10 mm, deep blue, banner with ± yellow spot, glands 0. **FR**: exserted, 7–10 mm; glands spaced, large. 2*n*=20. Gravelly to sandy slopes, benches, washes; < 600 m. DSon; AZ, nw Mex, Baja CA. Mar–May

P. spinosus (A. Gray) Barneby (p. 793) SMOKE TREE Shrub to small tree 1.5–8 m, gen lfless, gray-canescent. **LF**: simple, 0.5–2 cm, oblanceolate, thick, gland-dotted, deciduous by early summer. **INFL**: gen open, 5–15-fld; axis ending as thorn; pedicels ± 1 mm, bractlets 2. **FL**: calyx 4.5–5 mm, lobes < tube, glands forming lateral ring; corolla 6–8 mm, indigo-blue, glands ± 0. **FR**: ≥ calyx, plump-obovate, gland-dotted. 2*n*=10. Common. Desert washes; < 400 m. D; to c AZ, nw Mex, Baja CA. Jun–Jul(Oct–Nov)

ROBINIA LOCUST

Matt Lavin, Duane Isely & Elizabeth McClintock

Shrub, tree, armed, gen spreading from underground parts. **LF**: odd-1-pinnate, alternate, deciduous; stipular spines gen not gland-dotted. **INFL**: raceme, axillary. **FL**: calyx bell-shaped, lobes 5; petals 5, white or pink, banner reflexed; 9 filaments fused, 1 free. **FR**: flat [or plump], not narrowed between seeds, dehiscent. 4 spp.: temp US to subtrop sw US, Mex. (J. Robin (1550–1629), V. Robin (1579–1662), French botanists who introduced genus to Eur) [Lavin et al. 2003 Syst Bot 28:387–409] *R. hispida* L. spreading from planted pls in Sacramento River Delta, potentially naturalized.

1. Corolla pink; infl ascending to drooping, glandular-hairy; fr glandular-hairy, upper margin not winged; e DMtns . . . . ***R. neomexicana***
1′ Corolla white; infl pendent, finely, nonglandular-hairy; fr glabrous, upper margin narrowly winged; CA-FP, GB . . . . ***R. pseudoacacia***

R. neomexicana A. Gray (p. 793) NEW MEXICO LOCUST Shrub, small tree. **FL**: corolla 2–2.5 cm. Canyons in pinyon/juniper woodland, roadsides; < 1800 m. e DMtns; to s UT, w TX, n Mex. Apr–Aug ★

R. pseudoacacia L. (p. 793) BLACK LOCUST Tree. **FL**: corolla 1.5–2 cm. 2*n*=20. Locally common near abandoned houses, roadsides, canyon slopes, streambanks; 50–1900 m. CA-FP, GB; native to e US, widely cult. TOXIC: ingested seeds, lvs, bark may be fatal to humans, livestock. May–Jun ❖

RUPERTIA RUPERT'S SCURF-PEA

Martin F. Wojciechowski & James W. Grimes

Per, unarmed, gland-dotted (esp lvs), hairy or not; caudexed, rhizomed, or stoloned, roots deep, woody, extensive. **ST**: ± decumbent at base or erect. **LF**: odd-1-pinnate, cauline; stipules reflexed, deciduous; lflets 3. **INFL**: raceme, gen axillary, with 1 deciduous bract and 2–3 fls per node. **FL**: pedicelled; calyx conspicuously swollen in fr; corolla cream to yellow; 9 filaments fused, 1 less so or free; ovary ± hairy, ovule 1, style tip bent, stigma head-like. **FR**: indehiscent, elliptic to depressed-obovate in outline, beaked or not. **SEED**: 1, reniform, smooth. 3 spp.: w N.Am, esp CA. (Rupert C. Barneby, botanist, 1911–2000)

1. Stipules 13–15 mm, widely elliptic to obtriangular; bracts 9–13 mm . . . . ***R. hallii***
1′ Stipules 4–10 mm, ± elliptic, lance-linear to -oblanceolate, or triangular; bracts 3–7 mm
  2. Calyx in fl 6–8 mm; banner 10–14 mm; fr 4–7 mm, mucronate . . . . ***R. physodes***
  2′ Calyx in fl 9–12 mm; banner 14–15 mm; fr 9–13 mm, with widely attached, < 3 mm beak . . . . ***R. rigida***

R. hallii (Rydb.) J.W. Grimes (p. 793) HALL'S RUPERTIA Pl hairy or not. **ST:** erect, < 1 m. **LF:** stipule 13–15 mm, widely elliptic to obtriangular; petiole 10–30 mm; main axis 2–2.8 cm; lflets 4–9 cm, lanceolate to widely ovate, with glands but no hairs on both surfaces. **INFL:** bract 9–13 mm, late-deciduous; pedicels 2 mm. **FL:** calyx 14–15 mm; banner 11–12 mm. **FR:** 7–10 mm, elliptic, with hairs and sparse, minute, golden glands that fade with age; beak 1–3 mm, widely attached. **SEED:** 6–7 mm, red-brown. Woodland openings; < 2250 m. CaRH (Butte, Tehama cos.). Fr rare. Jun–Aug ★

R. physodes (Douglas) J.W. Grimes (p. 793) Pl sparsely hairy, stoloned or not. **ST:** erect or decumbent, ± 0.5 m. **LF:** stipule 4–10 mm, ± elliptic or lance-linear to -oblanceolate; petiole 11–65 mm; main axis 9–21 mm; lflet 3.5–7 cm, triangular to lanceolate, with glands, sparse hairs adaxially, becoming ± glabrous abaxially. **INFL:** bract 3–7 mm, deciduous; pedicels 1.5–2.5 mm. **FL:** calyx 6–8 mm; banner 10–14 mm. **FR:** 4–7 mm, depressed-obovate, golden-red, faintly net-sculptured, with red-brown hairs, mucronate. **SEED:** 5–6.5 mm, dark red-brown. 2*n*=22. Woodland; < 2500 m. KR, NCoR, CW, SCo, SnBr, PR; to BC, ID. [*Psoralea p.* Douglas] May–Sep

R. rigida (Parish) J.W. Grimes PARISH'S RUPERTIA Pl hairy; caudex woody. **ST:** erect, < 0.75 m; base purple. **LF:** stipule 4–10 mm, lance-linear or triangular; petiole 40–60 mm; lflet 35–65 mm, lanceolate, with glands and hairs on both surfaces, much denser on upper. **INFL:** bract 3–7 mm, deciduous; 1.5–2 mm. **FL:** calyx 9–12 mm; banner 14–15 mm. **FR:** 9–13 mm, elliptic, golden-brown, smooth to faintly net-sculptured, with glands and red-brown hairs; beak < 3 mm, widely attached. **SEED:** 6.5–7 mm, red-brown. Woodland, chaparral, lower montane conifer forest; < 2500 m. SCo, SnBr, PR; Baja CA. May–Jun ★

SENEGALIA
David Seigler & John E. Ebinger

Shrub, tree, armed with prickles, stipular spines 0. **LF:** even-2-pinnate, gen alternate, gen deciduous; petiole, main axis gen with raised glands. **INFL:** [head] spike, gen 1 (rarely in raceme, panicle). **FL:** radial; sepals, petals 4–5, inconspicuous; stamens many, conspicuous, exserted, free; ovary simple. **FR:** gen dehiscent, gen flat. **SEED:** aril gen 0. ± 200 spp.: trop, subtrop Am, Afr, Asia, Australia. [Seigler et al. 2006 Phytologia 88:38–93]

S. greggii (A. Gray) Britton & Rose (p. 793) CATCLAW, DEVIL'S CLAW Shrub, < 7 m, ± hairy. **ST:** twigs ± ridged; prickles recurved. **LF:** clustered on short-shoots or not, < 2.5 cm; petiole 2–10 mm, ± hairy; 1° lflets 1–3 pairs, 7–16 mm; 2° lflets 3–6 pairs, 2.8–6.3 mm, 0.6–2.5 mm wide, oblong to obovate, ± hairy. **INFL:** ≥ 1 clustered with lvs on short-shoots, stalked, 1–4 cm, gen > lf. **FL:** light yellow to cream. **FR:** 5–15 cm, 10–25 mm wide, curved or twisted, papery, light brown, glabrous. Uncommon; dry slopes in chaparral, flats, washes, disturbed areas; 100–1400 m. e PR, D; to TX, n Mex incl Baja CA. [*Acacia g.* A. Gray] Apr–Jun

SENNA
Martin F. Wojciechowski & Elizabeth McClintock

Per to tree, unarmed or main lf axis a weak spine at tip or branches a weak thorn at tip. **LF:** even-1-pinnate, alternate; stipules small or not, ephemeral or not; lflets 4–20(36). **INFL:** axillary or terminal, raceme or panicle. **FL:** gen ± bilateral, gen showy; sepals ± free; petals free, gen yellow; stamens free, 7 fertile, 3 sterile, or 10 fertile, anthers gen > filaments, opening by terminal pores. **FR:** dehiscent or not. **SEED:** few to many. ± 300 spp.: trop, esp Am, Afr, Australia, also warm temp, deserts. (Arabic: Sana) [Randell & Barlow 1998 Fl Australia 12:89–138] Some cult as orns; dried lvs of some cathartic, laxative.

1. Per to subshrub; lflets gen 4–18
  2. Lflets 4–8; gland(s) on main lf axis between lflet pairs; pl densely hairy . . . . ***S. covesii***
  2′ Lflets 12–18; gland(s) near base of petiole; pl glabrous . . . . [***S. marilandica***]
1′ Shrub to small tree; lflets gen 4–20
  3. Shrubs of dry places; lflets linear or gen ovate, gen flat, persistent or < 1 cm, ephemeral
    4. St, lf ± glabrous; lflets 4–8, 4–6 mm . . . . ***S. armata***
    4′ St, lf sparsely to densely hairy; lflets 6–16, 10–25 mm . . . . ***S. artemisioides***
  3′ Shrubs to small trees gen not of dry places; lflets oblong to elliptic, not ephemeral
    5. Stipules 6–17 mm, ovate-cordate, gen persistent; uppermost, unopened fls covered by bracts;
      fr flat, dehiscent . . . . ***S. didymobotrya***
    5′ Stipules 3.5–7 mm, lanceolate, ephemeral; uppermost, unopened fls not covered by bracts;
      fr not flat, indehiscent . . . . ***S. multiglandulosa***

S. armata (S. Watson) H.S. Irwin & Barneby (p. 793) SPINY SENNA Shrub, armed, ± lfless most of yr, ± glabrous. **ST:** 0.5–1 m; branches from base, grooved, a weak thorn at tip or not, green. **LF:** stipules 0 or minute; lflets 4–8, not overlapped, ± opposite, ± sessile, 4–6 mm, asymmetric, gen ovate, ephemeral; main axis elongating after lflets fall, a weak spine at tip. **INFL:** terminal raceme. **FL:** petals 8–12 mm, obovate, yellow to salmon-red. **FR:** dehiscent, 2.5–4 cm, lanceolate, straight. **SEED:** few. Uncommon. Sandy or gravelly washes; 100–1400 m. D; NV, AZ, Baja CA. Mar–Jul

S. artemisioides (DC.) Randell SILVER SENNA Shrub < 2 m, unarmed, sparsely to densely hairy. **LF:** 3–6 cm incl 5–15 mm petiole with glands 1–3 below lowest lflets; stipules small, ephemeral; lflets 6–16, not overlapped, sessile, 10–25 mm, 1 mm wide, linear. **INFL:** clustered towards upper axils, raceme, 4–12-fld; peduncles 5–40 mm, pedicels 5–25 mm. **FL:** petals 7–10 mm, glabrous, yellow; stamens 10, unequal. **SEED:** several. Dry, sandy washes to rocky slopes, disturbed areas; 100–1500 m. SCo, SnGb, SnBr, DSon; Native to Australia, cult, naturalized in sw US. [*Cassia a.* DC.] Jan–May

S. covesii (A. Gray) H.S. Irwin & Barneby COUES' CASSIA Subshrub, unarmed, lfy, dense-white to -gray-hairy. **ST:** 3–6 dm. **LF:** stipules bristle-like, some persistent; lflets 4–8, overlapped, opposite, short-stalked, 10–25 mm, elliptic. **INFL:** axillary raceme, 5–15 mm, few-fld. **FL:** petals ± 12 mm, oblong-obovate, prominently veined, golden-yellow. **FR:** erect, dehiscent, 2–5 cm, wide, ± straight, persistent. **SEED:** several. Dry, sandy desert washes, slopes; 330–760 m. DSon; to s NV, sw NM, Baja CA. Mar–Apr(fall) ★

S. didymobotrya (Fresen.) H.S. Irwin & Barneby Shrub, unarmed, lfy, hairy, gen strongly scented. **LF:** stipules gen persistent or late-deciduous, 6–17 mm, ovate-cordate; lflets 14–20(36), ± overlapped, opposite, 20–65 mm, sessile, oblong to elliptic, tips abrupt, slender. **INFL:** clustered in upper axils, raceme, many-fld; bracts in cone-shaped cluster covering uppermost, unopened fls, 1–2 cm, widely ovate, brown- to black-green, falling after fls open. **FL:** petals concave, longest 1.5–2.5 cm, obovate. **FR:** dehiscent, 8–12 cm, oblong, flat; transverse partitions between seeds. **SEED:** many. Uncommon. Disturbed areas; < 100 m. s CCo, expected elsewhere; native to trop Afr. [*Cassia d.* Fresen.] Mar–Oct

S. multiglandulosa (Jacq.) H.S. Irwin & Barneby   Shrub, small tree, unarmed, lfy, densely hairy. **LF:** stipules 3.5–7 mm, lanceolate, ephemeral; lflets 12–16, gen not overlapped, opposite, short-stalked, 25–45 mm, oblong to narrowly elliptic, abaxially more densely hairy, paler. **INFL:** axillary raceme, 5–15-fld; bracts falling before fls open.

FL: petals ± flat, longest 12–19 mm, obovate. **FR:** indehiscent, 8–12 cm, oblong, ± inflated. **SEED:** many. Disturbed areas; < 500 m. CCo, SnFrB; native to Mex, Guatemala, S.Am. Often planted along highways; potentially problematic weed. May–Jun

SESBANIA

Thomas J. Rosatti

Ann, shrub, small tree, unarmed. **LF:** even-1-pinnate; stipules gen deciduous; lflets gen many; main axis ending as bristle or not. **INFL:** raceme, axillary; bractlets occ appressed to calyx base. **FL:** calyx lobes subequal, < tube; corolla gen yellow, gen with dark spots on banner; 9 filaments fused, 1 free. **FR:** slowly dehiscent, linear, inflated, 4-angled or -winged. **SEED:** 2–many. ± 75 spp.: trop, warm temp. (Ancient Arabic name) [Lavin & Sousa 1995 Syst Bot Monogr 45:39–45]

1. Ann (in age ± woody at st base), (0.5)1–3(4) m; fr 3–5 mm wide, 4-angled; corolla 1–1.5(2) cm, ± yellow,
   banner ± red- or ± brown-spotted, esp on outside, less often ± red ± throughout; seeds (15)30–40 . . . . *S. herbacea*
1′ Shrub, small tree, 1–2 m; fr 10–15 mm wide, 4-winged; corolla 1.5–2.5 cm, scarlet or orange-red;
   seeds 4–8 . . . . *S. punicea*

S. herbacea (Mill.) McVaugh (p. 793)   BIGPOD SESBANIA, COFFEE WEED, COLORADO RIVER HEMP   **LF:** lflets 30–60, 1–2.5 cm, oblong. **INFL:** 2–6-fld. **FR:** 15–20 cm, linear. 2*n*=12. Along streams, other moist sites, often in cult or old fields; < 500 m. DSon, probably elsewhere; s US, Mex. [*S. exaltata* (Raf.) Cory] Apr–Oct

S. punicea (Cav.) Benth.   SCARLET SESBAN   **LF:** lflets ± 20–34, 0.8–2.5 cm, elliptic to elliptic-oblong. **INFL:** 5–15-fld. **FR:** 4–8 cm, oblong. 2*n*=12. Along streams, lake shores, other moist sites, roadsides, often cult as orn; < 200 m. NCoRO, GV, SnFrB, probably elsewhere; se US; native to S.Am. [*S. tripetii* F.T. Hubb.] Jun–Sep   ◆

SPARTIUM

Martin F. Wojciechowski & Elizabeth McClintock

1 sp.: Medit, Canary, Madeira, Azores islands. (Greek: ancient name)

S. junceum L. (p. 793)   SPANISH BROOM   Shrub, unarmed. **ST:** erect; branches few, < 3 m, rush-like, not angled, striate, green, gen ± lfless. **LF:** simple, alternate to subopposite, ephemeral, < 2.5 cm, linear to lanceolate; adaxially ± glabrous; abaxially appressed-hairy. **INFL:** raceme, terminal, open; fls several. **FL:** 2–3 cm, ± fragrant; calyx split adaxially ± to base (rarely ± 2-lipped, 5-lobed); petals yellow; stamens 10, filaments fused. **FR:** dehiscent, 5–10 cm, ± 5 mm wide. **SEED:** 4–18. Common. Disturbed areas; < 900 m. NCoRO, CaRF, SNF, ScV, SnFrB, SCoRO, SCo, s ChI, WTR; native to Medit. Cult as orn, readily invasive, becomes weedy; sts used for fiber, fls for yellow dye. Apr–Jun   ◆

THERMOPSIS   GOLDEN PEA, FALSE-LUPINE

Dieter H. Wilken

Per, unarmed, gen hairy, rhizomed. **ST:** erect; lower nodes lfless. **LF:** palmately compound; lower stipules clasping st, scarious, others not clasping st, green; lflets 3, widely ovate to oblanceolate. **INFL:** raceme, terminal; bracts like stipules, ± persistent. **FL:** calyx lobes 5, upper 2 ± fused; corolla yellow; stamens 10, free. **FR:** erect or spreading, slowly dehiscent, oblong, flat; base stalk-like base present; margin often wavy. **SEED:** 3–10. ± 20 spp.: temp N.Am, c&e Asia. (Greek: like a lupine) [Chen et al. 1994 Ann Missouri Bot Gard 81:714–742]

1. Lflets, pedicels, frs glabrous to sparsely long-soft-hairy; fl 1–2 per node . . . . *T. gracilis*
1′ Lflets, pedicels, frs long-soft-hairy; fls gen 2–5 per node
  2. Pl 8–25 dm; st stout, erect
    3. Fr ascending, straight; lower pedicels 2–4 mm; lflet lateral veins 12–16; WTR (Santa Ynez Mtns) . . . *T. macrophylla*
    3′ Fr spreading, curved; lower pedicels 4–8 mm; lflet lateral veins 18–24; KR, NCoRO . . . . *T. robusta*
  2′ Pl 3–8 dm; st slender, ascending to erect . . . . *T. californica*
    4. Stipules ovate to widely ovate; infl bracts < 15 mm wide; lflet hairs not appressed; NCo, NCoR, CW
      . . . . var. *californica*
    4′ Stipules lanceolate to ovate; infl bracts < 6 mm wide; lflet hairs appressed; CaR, WTR, s PR (San Diego Co.), MP
      5. Lflet silvery, hairs straight; CaR, WTR, MP . . . . var. *argentata*
      5′ Lflet velvety, hairs ± wavy; s PR (San Diego Co.) . . . . var. *semota*

T. californica S. Watson   Pl 3–8 dm. **ST:** ascending to erect, slender, branches irregular. **LF:** lflets gen 3–7 cm, hairy, lateral veins 12–18. **INFL:** 7–30 cm; pedicels 2–6 mm; fls 2–5 per node. **FR:** straight, gen ascending, gen densely hairy.

   var. *argentata* (Greene) C.J. Chen & B.L. Turner (p. 793)   SILVERY FALSE-LUPINE   Pl silvery-hairy. **ST:** gen 3–5 dm. **LF:** stipules lanceolate to ovate; hairs long, soft, straight, appressed. **INFL:** bracts < 5 mm wide. 2*n*=18. Chaparral, conifer forest; 1200–2200 m. CaR, WTR, MP. [*T. macrophylla* var. *a.* (Greene) Jeps.] May–Jun   ★

   var. *californica* (p. 793)   Pl gray-hairy. **ST:** gen 3–8 dm. **LF:** stipules ovate to widely ovate; hairs soft, not appressed. **INFL:** bracts < 15 mm wide. Meadows, chaparral, pine/oak woodland; < 1000 m. NCo, NCoR, CW. May–Jun

   var. *semota* (Jeps.) C.J. Chen & B.L. Turner (p. 793)   VELVETY FALSE-LUPINE   Pl velvety-hairy. **ST:** gen 3–5 dm. **LF:** stipules lanceolate to ovate; hairs long, soft, ± wavy, appressed. **INFL:** bracts < 6 mm wide. Meadows, pine/oak woodland; 1000–1500 m. s PR (San Diego Co.). [*T. macrophylla* var. *s.* Jeps.] Apr–Jun   ★

T. gracilis Howell   Pl 3–8 dm, green. **ST**: branches 0 to irregular. **LF**: lflets 3–7 cm, glabrous to sparsely long-soft-hairy, lateral veins 12–18. **INFL**: 12–20 cm; bracts < 5 mm wide; pedicels 6–16 mm; fls 1–2 per node. **FR**: straight, ± spreading, glabrous to sparsely long-soft-hairy. 2*n*=36. Open sites, gen mixed-evergreen forest; 100–1200 m. KR, NCoR; w OR. [*T. macrophylla* var. *venosa* (Eastw.) Isely] Apr–Jun ★

T. macrophylla Hook. & Arn.   SANTA YNEZ FALSE-LUPINE   Pl 12–25 dm, gray-hairy. **ST**: erect, stout, branches at base. **LF**: lflets 4–10 cm, densely long-soft-wavy-hairy, lateral veins 12–16. **INFL**: 25–60 cm; bracts < 5 mm wide; pedicels 2–4 mm; fls 3–5 per node. **FR**: straight, ascending, densely hairy. 2*n*=18. Sandstone, chaparral; 1000–1400 m. WTR (Santa Ynez Mtns). [*T. m.* var. *agnina* J.T. Howell] May–Jun ★

T. robusta Howell   ROBUST FALSE-LUPINE   Pl 8–18 dm, green to gray-hairy. **ST**: erect, stout, branches 0–few at base. **LF**: lflets 6–11 cm, densely long-soft-wavy-hairy, lateral veins 18–24. **INFL**: 20–45 cm; bracts < 8 mm wide; pedicels 4–8 mm; fls gen 3–5 per node. **FR**: curved, spreading, densely hairy. Shale, serpentine, open sites, forest; 150–1500 m. KR, NCoRO. May–Jun ★

TRIFOLIUM   CLOVER

Michael A. Vincent & Duane Isely

Ann, per, unarmed. **LF**: gen palmately compound; stipules conspicuous, partly fused to petiole, gen papery or membranous; lflets gen 3, occ 5–9, ± toothed or wavy. **INFL**: raceme (gen umbel-like), head, or spike, axillary or terminal, gen many-fld, gen involucred, gen peduncled; infl bracts 0 or forming vestigial ring or involucre; fl bracts present or not. **FL**: gen spreading to erect, gen becoming reflexed; corolla gen purple to pale lavender, occ yellow, persistent after fl; 9 filaments fused, 1 free. **FR**: gen indehiscent but gen breaking, short, plump, gen enclosed in corolla; base gen stalk-like. **SEED**: 1–6. (Latin: 3 lvs) ± 300 spp.: temps, trop mtns, n hemisphere, S.Am, Afr.; foodplant for lepidopterans, cult as green manure in crop rotation, fodder. [Ellison et al. 2006 Molec Phylogen Evol 39:688–705; Vincent 2009 Madroño 56:208]

1. Involucre bracts gen fused, gen > 1 mm, gen forming cup or bowl about base of infl; infl not sessile above 1–2 lvs . . . . **Group 1**
1′ Involucre bracts 0 or occ < 1 mm or forming a vestigial ring; infl occ ± sessile above 1–2 ± reduced lvs (stipules of which occ involucre-like) . . . . **Group 2**

Group 1

1. Calyx or banner soon inflated in fr; involucre bracts gen ± free
  2. Calyx hairy, inflated in fr; infl in fr ± 2 cm wide, spheric, fuzzy; banner not inflating in fr; per, stoloned . . . . [2]***T. fragiferum***
  2′ Calyx glabrous, not inflated in fr; infl not as above; banner inflating in fr; ann, not stoloned
    3. Corolla 10–20 mm . . . . ***T. fucatum***
    3′ Corolla 4–9 mm
      4. Involucre bracts < 1 mm, fused at base; infl gen 1–1.5 cm wide; corolla 6.5–9 mm; ScV, nw SnJV, CW . . . . ***T. hydrophilum***
      4′ Involucre bracts 2–2.5 mm, ± free to ± 1/2 fused; infl 0.5–1 cm wide; corolla 4.5–7.5 mm; NCoR, CaRF, SNF, Teh, GV, CW, SCo, ChI, PR . . . . ***T. depauperatum***
        5. Fr 3–4 mm, oblong, stalk-like base 0 or short; involucre bracts ± free or ± 1/2 fused, margins widely scarious, toothed near tip; corolla 5–6 mm; GV, SnFrB, SCoR . . . . var. ***amplectens***
        5′ Fr 2–3 mm, ovate or obovate, stalk-like base 0.5–1 mm; involucre bracts ± free, margins ± membranous; corolla 4.5–7.5 mm; NCoR, CaRF, SNF, Teh, GV, CW, SCo, ChI, PR . . . . var. ***truncatum***
1′ Calyx, corolla not inflated in fr (exc lower part of banner in *Trifolium barbigerum*); involucre bracts gen fused into a cup-, bowl-, or wheel-shaped, toothed or cut ring (exc gen ± free in *Trifolium monanthum*)
  6. Involucre cup- or bowl-shaped, when pressed gen partly hiding fls (exc in *Trifolium grayi*)
    7. Calyx lobes < 1/2 tube, not bristle-tipped . . . . ***T. microdon***
    7′ Calyx lobes > 1/2 tube, bristle-tipped
      8. Calyx lobes, involucre lobes ± entire; pl hairs gen finely wavy . . . . ***T. microcephalum***
      8′ Calyx lobes, involucre lobes, or both toothed, cut, or branched; pl hairs 0 or finely wavy or not
        9. Bristle-tips of lower 3 calyx lobes gen forked; involucre margin wavy, finely toothed . . . . ***T. cyathiferum***
        9′ Bristle-tips of lower 3 calyx lobes not divided; involucre margin wavy or not, gen coarsely toothed or cut 1/2 to involucre base
          10. Calyx lobes finely toothed, bristle at tip 1–1.5 mm, glabrous; involucre gen glabrous; banner not inflated . . . . ***T. buckwestiorum***
          10′ Calyx lobes entire, bristle at tip 3–4 mm, plumose (or lower lobes divided to near base into 2–3 bristles, each 3–4 mm, gen plumose); involucre gen puberulent; banner inflated below, persistent as twisted beak above
            11. Corolla gold- to sulphur-yellow; lvs, sts glabrous; seeds > 3 mm; ScV (North Table Mtn Plateau, Butte Co.). . . . . ***T. jokerstii***
            11′ Corolla purple, pink-purple, white, or 2-colored; lvs, sts gen hairy; seeds < 3 mm; NW, SN, GV, CW
              12. Corolla 5–10 mm, ≤ (rarely ± >) calyx; stalk-like fr base < 1 mm; infl gen bristly from calyx lobes . . . . ***T. barbigerum***
              12′ Corolla 8–16 mm, > calyx; stalk-like fr base 1–3 mm; infl not bristly . . . . ***T. grayi***
  6′ Involucre wheel-shaped or bracts ± free, when pressed hiding only bases of fls
    13. Involucre vestigial or bracts 2–5, inconspicuous, gen ± free, 1–3 mm; infl 1–6-fld . . . . ***T. monanthum***
      14. Rhizomes thin, white; pl mat-forming, not from central crown or thickened taproot; pl glabrous to ± hairy

15. Fls 1–5, peduncle bent upward at tip; lflet tips gen acute; SnGb, SnBr, SnJt . . . . subsp. ***grantianum***
15′ Fls 1–4, peduncle straight; lflet tips gen rounded or truncate; CaR, SN, SNE. . . . . subsp. ***monanthum***
14′ Rhizomes 0; pl not mat-forming, from central crown or thickened taproot; pl densely hairy
  16. Lflet tips gen rounded, truncate, or notched . . . . subsp. ***parvum***
  16′ Lflet tips acute to acuminate . . . . subsp. ***tenerum***
13′ Involucre gen conspicuous, bracts indefinite in number, gen > 5 mm, fused, if inconspicuous (e.g., 2–3
  mm in *Trifolium oliganthum*) fused at least at base; infl gen 8–many-fld
17. Per; involucre gen 2–3 cm wide; seeds 2–4
  18. Thickened fusiform root present; rhizomes 0; KR . . . . ***T. siskiyouense***
  18′ Thickened fusiform root 0; rhizomes present; NW, CaR, SN, SnJV, CW, SCo, PR, SNE . . . . ***T. wormskioldii***
17′ Ann; involucre < 2 cm wide; seeds gen 1–2
  19. Calyx lobes ± ≤ tube, often toothed or shouldered below tapered tip; calyx tube splitting between upper lobes
    20. Fl 5–8 mm; involucre 2–3 mm, gen cut > 1/2 to base; infl 6–10 mm wide, 5–15-fld . . . . ***T. oliganthum***
    20′ Fl 9–18 mm; involucre > 3 mm, cut < 1/2 to base; infl gen 12–30 mm wide, often > 15-fld
      21. Calyx glandular-hairy or bumpy (occ in age glabrous), lobes gen entire; lflets sharply serrate,
        longer teeth ± 1 mm . . . . ***T. obtusiflorum***
      21′ Calyx glabrous, not bumpy, lobes entire or 3-toothed; lflets serrate, teeth < 1 mm . . . . ***T. willdenovii***
  19′ Calyx lobes gen > tube, gen entire (occ ± forked in *Trifolium trichocalyx*), not shouldered below
    tapered tip; calyx tube not splitting between upper lobes
    22. Calyx hairy; involucre small, cut past middle — CCo (Monterey Peninsula) . . . . ***T. trichocalyx***
    22′ Calyx gen ± glabrous; involucre gen well developed, cut to middle
      23. Calyx tubular, lobes ± 3-parted — CCo (Monterey Peninsula) . . . . ***T. polyodon***
      23′ Calyx bell-shaped, lobes entire or toothed. . . . . ***T. variegatum***
        24. Infl 1–5-fld; corolla 3.5–9 mm . . . . var. ***geminiflorum***
        24′ Infl 5–many-fld; corolla 6–16 mm
          25. Infl 1.5–3 cm, 10–many-fld, corolla 9–17 mm . . . . var. ***major***
          25′ Infl 1–1.5 cm, 5–10-fld, corolla 6–10 mm. . . . . var. ***variegatum***

Group 2

1. Lflets 3–9 (at least some lvs with > 3 lflets)
  2. Lflets gen 7–9; infl 3–5 cm wide . . . . ***T. macrocephalum***
  2′ Lflets 3–7; infl 1–2.5 cm wide
    3. Pl silvery or gray, soft-hairy or tomentose; calyx lobe hairs ± 1 mm. . . . . ***T. andersonii***
      4. Longer petiole hairs 1.5–2 mm; peduncles gen << petioles, infl not or ± exceeding lvs . . . . subsp. ***andersonii***
      4′ Longer petiole hairs 0.4–1.2 mm; peduncles gen > petioles; infl much exceeding lvs . . . . subsp. ***beatleyae***
    3′ Pl green, strigose, puberulent, or hairy; calyx lobe hairs < 1 mm
      5. Lflet gen elliptic, serrate, teeth 0.1–0.3 mm; infl gen incl, 1–1.5 cm wide. . . . . . ***T. gymnocarpon*** subsp. ***plummerae***
      5′ Lflet obovate, coarsely serrate, larger teeth 0.4–0.5 mm; infl gen exserted, 1.5–2.5 cm wide . . . . ***T. lemmonii***
1′ Lflets 3
  6. Fls of infl, exc for outer 2–8, sterile, with calyx stalk-like, without petals, forming bur, delivered to
    ground by growing peduncle; pl prostrate or creeping . . . . ***T. subterraneum***
  6′ Fls of infl fertile, with calyx not stalk-like, with petals, not forming bur, not delivered to ground by
    peduncle; pl prostrate or creeping or not
    7. Calyx inflated in fr; infl in fr spheric or star-shaped, red, brown, green, or white, ± woolly, 0.8–2.5 cm
      8. Per, creeping and rooting or cespitose; bracts of lower fls forming involucre, ± 2 mm, green; corolla
        not becoming reflexed . . . . [2]***T. fragiferum***
      8′ Ann, decumbent to ± erect, not rooting; bracts of lower fls gen not forming involucre, < 1 mm, white;
        corolla becoming reflexed
        9. Infl star-shaped, 1–2.5 cm; fl bracts reduced. . . . . [*T. resupinatum*]
        9′ Infl spheric, 0.6–1.2 cm; fl bracts cup-shaped. . . . . ***T. tomentosum***
    7′ Calyx rarely inflated in fr (if inflated then << 2 cm wide); infl in fr not as at 7.
      10. Lf pinnately compound; corolla bright yellow
        11. Corolla striate; infl 0.8–1.3 cm wide, gen > 20-fld; petioles of mid-lvs > lflets . . . . ***T. campestre***
        11′ Corolla weak-striate; infl 0.4–0.8 cm wide, 5–10(20)-fld; petioles of mid-lvs gen << lflets . . . . ***T. dubium***
      10′ Lf palmately compound; corolla not bright yellow (exc *Trifolium aureum*)
        12. Lflets 4–10 cm, stipules 1–2.5 cm, green; calyx glabrous; corolla green-white to pink . . . . ***T. howellii***
        12′ Lflets or stipules (or both) << above, stipules gen not green; calyx hairy or glabrous; corolla
          green-white to pink or not
          13. Fls gen 1–5 per infl, bracts gen inconspicuous; corolla white to lavender-striate; pl gen small
            (see Group 1, couplet 13 for vars.) . . . . [2]***T. monanthum***
          13′ Fls > 5 per infl unless pl stressed; corolla white to lavender-striate or not; pl small or not
            14. Pedicels 1–6 mm, fls gen becoming reflexed
              15. Corolla bright yellow, 5–7 mm; calyx 1.5–2 mm; stalk-like fr base 2–3 mm . . . . ***T. aureum***
              15′ Corolla not bright yellow, gen > 6 mm; calyx gen > 2 mm; stalk-like fr base 0 or < 2 mm
                16. Lflet lanceolate; lvs mostly basal, occ 2–3 cauline; pls of s SNH, SNE . . . . ***T. kingii*** subsp. ***dedeckerae***

16′ Lflet wider than lanceolate, if lanceolate then lvs not mostly basal or pls not of s SNH, SNE

  17. St creeping, rooting, turf-forming; petioles from ground level, ± equal; corolla white . . . . **_T. repens_**

  17′ Pl not creeping, not rooting, not turf-forming; petioles not or some from ground level, lower >
upper; corolla white or not

    18. Calyx tube puberulent or hairy, occ ± so

      19. Corolla 5–8 mm; pedicel 2–5 mm . . . . **_T. breweri_**

      19′ Corolla 10–18 mm; pedicel 1–2 mm . . . . ²**_T. longipes_**

        20. Longer petals tapered to beak; 1400–2700 m — MP . . . . subsp. **_multipedunculatum_**

        20′ Longer petals acute to acuminate but not beaked; 1200–2700 m

          21. Sts gen well developed, not cespitose; petals gen white, short-tapered, acute, or obtuse;
1200–1800 m; KR, NCoR . . . . subsp. **_oreganum_**

          21′ Sts often poorly developed, gen cespitose; petals lavender to purple, longer ones acute
to acuminate; 1400–2700 m; KR, CaR . . . . subsp. **_shastense_**

    18′ Calyx tube glabrous

      22. Calyx lobes ciliate with short, flat bristles (or, through breakage, bristle bases); bracts
inconspicuous, often in a ring at base of infl. . . . . **_T. ciliolatum_**

      22′ Calyx lobes not ciliate; bracts subtending individual fls or 0

        23. Lvs basal exc 1 pair cauline; corolla 10–17 mm . . . . ²**_T. beckwithii_**

        23′ Lvs cauline; corolla 5–11 mm

          24. Infl 1.5–3 cm wide; bracts conspicuous but soon falling; calyx lobes sharply tapered but
not bristle-tipped . . . . **_T. hybridum_**

          24′ Infl 0.5–1.5 cm wide; bracts 0; calyx lobes bristle-tipped

            25. Calyx lobes ± 0.5 mm wide at base, glabrous, tube 1.5–2.5 mm

              26. Lflet obovate to obcordate, length 1.5–2.5 × width; CA-FP (exc ChI), DMoj . . . . **_T. gracilentum_**

              26′ Lflets narrowly elliptic or lanceolate, length 3.5–5 × width; ChI . . . . **_T. palmeri_**

            25′ Calyx lobes ± 0.2 mm wide at base, with gen few hairs, tube 0.8–1.4 mm . . . . **_T. bifidum_**

              27. Lflet 2.5–5.5 × longer than wide, notched at tip 0.1–0.4 × length, otherwise gen entire . . . var. **_bifidum_**

              27′ Lflet 1.5–2.5 × longer than wide, truncate or notched at tip < 0.2 × length, otherwise
dentate or entire . . . . var. **_decipiens_**

14′ Pedicels < 2 mm, fls becoming reflexed or not

  28. Infl sessile at end of st or in lf axils, or in lf axils on short peduncles 6–10 mm, << lvs;
corolla 4–6 mm

    29. Infl axillary or at end of st, peduncles 0 . . . . **_T. glomeratum_**

    29′ Infl axillary, peduncles short or short below, 0 above

      30. Corolla > calyx lobes; banner tip notched; peduncles short . . . . **_T. cernuum_**

      30′ Corolla < calyx lobes; banner tip occ ± toothed but not notched; peduncles short below,
0 above . . . . **_T. retusum_**

  28′ Infl peduncled (or immediately above 1–2 reduced lvs), peduncle gen > 10 mm, gen > lvs;
corolla gen > 5 mm

    31. Banner inflated in fr; involucre vestigial, ring-like; corolla gen 6–9 mm
. . . . **_T. depauperatum_** var. **_depauperatum_**

    31′ Banner not inflated in fr; involucre 0; corolla 6–9 mm or not

      32. Fls gen reflexed soon after opening; infl often turned to side or downward

        33. Calyx hairy, lobes twisted or bent, plumose . . . . **_T. eriocephalum_** subsp. **_eriocephalum_**

        33′ Calyx glabrous, lobes not twisted or bent, not plumose

          34. Infl 1–2 cm wide; calyx 3.5–4.5 mm, gen dark or purple-black; corolla 10–12 mm . . . . **_T. bolanderi_**

          34′ Infl 1.5–3 cm wide; calyx 3–7 mm, not dark or purple-black; corolla 10–17 mm

            35. Calyx 4.5–7 mm; infl erect, axis not exserted . . . . ²**_T. beckwithii_**

            35′ Calyx 3–4 mm; infl gen turned to side or downward, axis exserted, gen forked . . . . **_T. productum_**

      32′ Fls gen not reflexed after opening; infl gen ± erect

        36. Infl ± sessile above a pair of reduced lvs, stipules, or both (in *Trifolium macraei*, occ 1 of a
pair of infls short-peduncled)

          37. Infl gen paired, ± sessile or occ 1 short-peduncled above (1)2 lvs . . . . **_T. macraei_**

          37′ Infl single above subtending lvs

             38. Corolla 2-colored (purple with white tips); locally uncommon, s NCoR, n CCo, SnFrB . . . ²**_T. amoenum_**

            38′ Corolla not 2-colored (red-purple or pink); sporadic or locally common, CA-FP, GB

              39. Calyx lobes densely to sparsely plumose, lower one ± 1–1.5 × others; corolla pink

                40. Infl spheric, ± 1.5–2 cm; calyx lobes >> tube, densely plumose, 4–6 mm . . . . **_T. hirtum_**

                40′ Infl ± short-cylindric, 0.5–1.5 cm; calyx lobes ≤ tube exc lower one, sparsely plumose,
1–2.5 mm . . . . **_T. striatum_**

             39′ Calyx lobes sparsely plumose, straight, lower one ± 2 × others, or calyx lobes glabrous,
curved, ± equal; corolla red-purple or pink

              41. Lflets elliptic to obovate, ± entire; calyx hairy, veins 10, lobes straight; corolla gen
red-purple . . . . **_T. pratense_**

              41′ Lflets gen lanceolate to elliptic, teeth bristle-like; calyx glabrous, veins 24, lobes
straight to curved; corolla pink to purple . . . . **_T. vesiculosum_**

36′ Infl peduncled
  42. Corolla ≤ calyx; calyx lobes > tube, plumose
    43. Lflets oblanceolate to obovate, length 1.2–3 × width; corolla 2-colored (purple with white tips)
      44. Calyx 4–8 mm, lobes 3–6.5 mm . . . . ***T. albopurpureum***
      44′ Calyx 8–14 mm, lobes 6.5–12 mm . . . . ***T. olivaceum***
    43′ Lflets linear to narrow-lanceolate, -oblong, or -oblanceolate, length 3–8 × width; corolla
        1-colored (pale pink to white)
      45. Corolla 10–12 mm, ± = calyx . . . . ***T. angustifolium***
      45′ Corolla 3–4 mm, < or ≪ calyx . . . . ***T. arvense***
  42′ Corolla > calyx; calyx lobes > tube or not, plumose or not
    46. Per; calyx lobes glabrous or hairs few, curved or wavy
      47. Infl < 1 cm wide, fls gen 1–5, erect (see Group 1, couplet 13, for vars.) . . . . ²***T. monanthum***
      47′ Infl 1.5–3 cm wide, fls > 5, ascending-spreading . . . . ²***T. longipes***
        48. Calyx lobes 3–5 × tube; lflet length gen 6–10 × width; KR, NCoR, CaR . . . . subsp. ***elmeri***
        48′ Calyx lobes 1–3 × tube; lflet length 2–6 × width in CA-FP, to 10 × in GB; KR, NCoR,
          CaR, SN, SnBr, SnJt, GB
          49. Roots thickened, fusiform; rhizomes ± 0; calyx lobes densely long-hairy . . . . subsp. ***atrorubens***
          49′ Roots gen not thickened, not fusiform; rhizomes present; calyx ± glabrous . . . . subsp. ***hansenii***
    46′ Ann; calyx lobes ± plumose, hairs gen straight
      50. Corolla 2-colored (purple with white tips)
        51. Corolla 12–16 mm; calyx 10–12 mm . . . . ²***T. amoenum***
        51′ Corolla 7–12 mm; calyx 4–8 mm . . . . ***T. dichotomum***
      50′ Corolla 1-colored (purple, crimson, pink, white, or cream)
        52. Corolla cream; infl ovate to spheric, 1–2.5 cm; lflets elliptic to oblanceolate;
          stipules long-tapered . . . . [***T. alexandrinum***]
        52′ Corolla crimson or white; infl ± cylindric or ± conic, 2–6 cm; lflets obovate to
          obcordate; stipules widely rounded . . . . ***T. incarnatum***

T. albopurpureum Torr. & A. Gray (p. 793) Ann, hairy. **ST:** decumbent to erect. **LF:** cauline; stipules small; lflets 1–3 cm, oblanceolate to obovate. **INFL:** spike, 5–20 mm wide, ovate to short-cylindric. **FL:** calyx 4–8 mm, lobes 3–6.5 mm, > tube, tapered or bristle-like, plumose; corolla 5–8 mm, ≤ calyx, 2-colored (purple and white). **SEED:** 1–2. 2*n*=16. Abundant. Coastal dunes, grassland, wet meadows, open slopes, oak chaparral, pine woodland, roadsides, disturbed areas; < 2100 m. CA-FP; to BC, Baja CA. Mar–Jun

T. amoenum Greene (p. 793) TWO-FORK CLOVER Ann, gen robust, hairy. **ST:** erect. **LF:** cauline; stipules gen conspicuous; lflets widely obovate. **INFL:** head, ovate to spheric. **FL:** calyx 10–12 mm, lobes > tube, slender, plumose; corolla 12–16 mm, purple, tips white. **SEED:** 1–2. Moist, heavy soils, disturbed areas; < 100 m. s NCoR, n CCo, SnFrB. Apr–Jun ★

T. andersonii A. Gray Per, short-tufted or cushion-forming, soft-hairy or tomentose, silvery or gray. **ST:** 0 (peduncle ascending or erect). **LF:** basal; stipules entire, persistent; lflets 3–7, 5–20 mm, oblanceolate to obovate, entire. **INFL:** head-like, 1.5–2.5 cm wide; pedicels 0.5–1 mm. **FL:** calyx 8–10 mm, lobes slender, > tube, plumose, hairs ± 1 mm; corolla 10–15 mm, pink-purple or 2-colored. **SEED:** 1–2. 2*n*=16.

  subsp. ***andersonii*** (p. 797) ANDERSON'S CLOVER **LF:** lflets gen 1–2 cm; longer petiole hairs 1.5–2 mm. **INFL:** not or ± exceeding lvs; peduncle 1–4 cm, gen ≪ petiole. Rocky slopes, meadows, esp yellow-pine forest; 900–2000 m. n SNH, MP; NV. May–Aug ★

  subsp. ***beatleyae*** J.M. Gillett BEATLEY'S FIVE-LEAVED CLOVER **LF:** lflets gen 0.5–1.4 cm; longer petiole hairs 0.4–1.2 mm. **INFL:** much exceeding lvs; peduncle 2–10 cm, gen > petiole. Washes, talus, pine forest to alpine slopes; 1300–4000 m. SNE (Mono Co.); NV. [*T. a.* var. *b.* (J.M. Gillett) Isely; *T. a.* subsp. *monoense* (Greene) J.M. Gillett] May–Aug

T. angustifolium L. (p. 797) NARROW-LEAVED CLOVER Ann, hairy. **ST:** gen erect. **LF:** cauline; stipules bristle-tipped; lflets 2–4.5 cm, linear to narrowly lanceolate. **INFL:** spike, 1–5 cm, cylindric. **FL:** calyx 10–12 mm, 10- ribbed, lobes spreading, needle-like, plumose, hardening in fr, unequal, all or lower > tube; corolla 10–12 mm, ± = calyx, pale pink. **SEED:** 1. 2*n*=16. Disturbed areas; esp < 200 m. CCo, SCoR; OR; native to Medit. Late spring

T. arvense L. (p. 797) RABBITFOOT CLOVER Ann, hairy. **ST:** gen erect. **LF:** cauline; stipules lanceolate, occ bristle-tipped; lflets 0.5–2 cm, narrow-oblong or -oblanceolate. **INFL:** spike, 1–3 cm, ovate to short-cylindric, dull green to gray. **FL:** calyx 4.5–7.5 mm, lobes > tube, needle-like, plumose; corolla 3–4 mm, < or ≪ calyx, pale pink to white. **SEED:** 1. 2*n*=14,28. Disturbed areas; esp < 300 m. CA-FP; to BC, se US; native to Eur. Jul

T. aureum Pollich LARGE HOP CLOVER Ann, bien, glabrous or puberulent. **ST:** ascending or erect. **LF:** cauline; stipules lanceolate to ovate; lflets 1–2.5 cm, oblanceolate or obovate. **INFL:** head, 1–1.5 cm wide, ovate; fls quickly reflexed. **FL:** calyx 1.5–2 mm, 5-veined, glabrous, lobes unequal, ± = tube; corolla 5–7 mm, striate, initially bright yellow. **FR:** stalk-like base 2–3 mm; style 1–2 mm, persistent. **SEED:** 1. 2*n*=14. Disturbed areas; esp < 200 m. SnJV; to BC, ne US; native to Eur. [*T. agrarium* L., misappl.] Jul–Aug

T. barbigerum Torr. BEARDED CLOVER Ann, glabrous or puberulent. **ST:** decumbent to erect. **LF:** ± cauline; stipules conspicuous, lanceolate to ovate, lower gen overlapped, upper gen sharply cut; lflets 1.5–2.5 cm, oblanceolate to obovate, tips occ notched. **INFL:** head, 5–25 mm wide, 5–many-fld; involucre bowl-shaped, sharply toothed or deeply cut, gen puberulent. **FL:** calyx tube 5-veined, gen 2–3 mm, < lobes, lobes simple, entire, terminal bristle 3–4 mm, plumose, or lower lobes divided near base into 2–3 bristles, each 3–4 mm, gen plumose; corolla 5–10 mm, ≤ (rarely ± >) calyx, pink-purple (white) or 2-colored, lower part of banner inflated in fr. **FR:** stalk-like base < 1 mm. **SEED:** 1–2. 2*n*=16. Wet meadows, open, disturbed areas; gen < 700 m. NW, n&c SN, GV, CW; OR. [*T. minutissimum* D. Heller & Zohary] Apr–Jun

T. beckwithii S. Watson (p. 797) BECKWITH'S CLOVER Per, glabrous. **ST:** ascending. **LF:** basal exc 1 pair; cauline stipules conspicuous, green; lflets 1–4 cm, oblong to ovate. **INFL:** ± scapose, head-like, 2–3 cm wide; axis not exserted; pedicels 1–2 mm; fls soon reflexed. **FL:** calyx 4.5–7 mm, lobes unequal, ± = tube; corolla 10–17 mm, pink to purple or 2-colored. **SEED:** 2–4. 2*n*=48. Meadows, evergreen forest; 1200–2100 m. n SNH, MP; to SD. May–Aug

T. bifidum A. Gray PINOLE CLOVER Ann, glabrous or sparsely finely hairy. **ST:** sprawling to erect. **LF:** cauline; stipules short-awned; lflets 1–2 cm, oblanceolate to obovate, tip gen notched. **INFL:** head- or umbel-like, 7–15 mm wide, gen 5–many-fld; axis tip gen ± exserted; pedicels 1–2 mm; fls becoming reflexed. **FL:** calyx 3.5–5 mm, tube 0.8–1.4 mm, glabrous, lobes ≫ tube, ± 0.2 mm wide at base, awn-like, bristle-tipped, hairs gen few, fine; corolla 6–9 mm, dull yellow to pink-purple. **SEED:** 1–2. 2*n*=16.

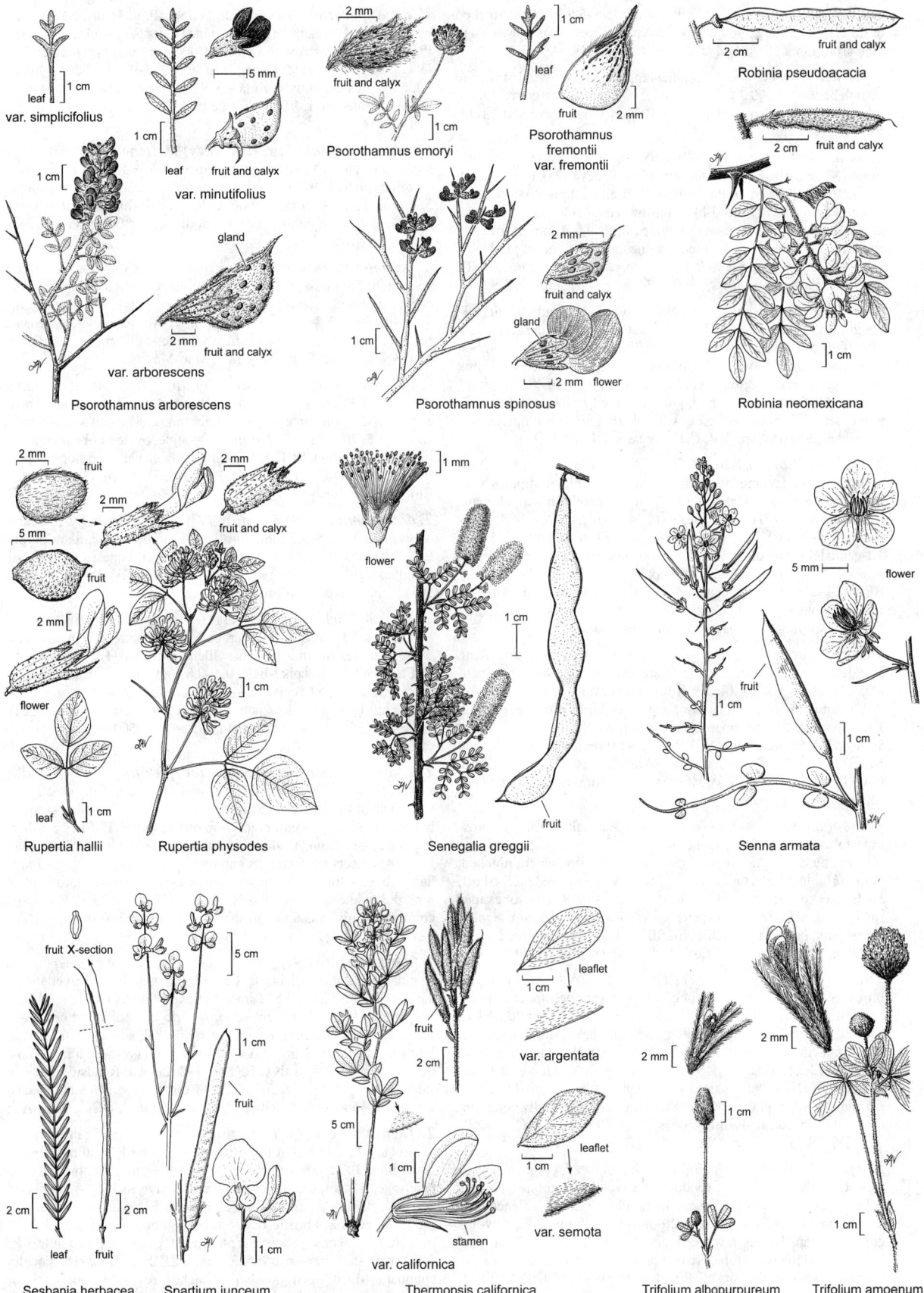

leaf
1 cm
var. simplicifolius

5 mm

leaf

1 cm

1 cm

fruit and calyx
var. minutifolii

1 cm

gland

var. arborescens

2 mm
fruit and calyx

Psorothamnus arborescens

2 mm
fruit and calyx
Psorothamnus emoryi

1 cm

1 cm
leaf

fruit
Psorothamnus
fremontii
var. fremontii
2 mm

2 mm
fruit and calyx

gland

2 mm   flower

1 cm

Psorothamnus spinosus

2 cm
fruit and calyx
Robinia pseudoacacia

2 cm   fruit and calyx

1 cm

Robinia neomexicana

2 mm
fruit

2 mm

5 mm

fruit

2 mm

2 mm
fruit and calyx

flower

leaf
1 cm

Rupertia hallii

1 cm

Rupertia physodes

1 mm

flower

1 cm

fruit

Senegalia greggii

1 mm

1 cm

fruit

1 cm

Senna armata

5 mm
flower

5 mm

fruit

flower

fruit X-section

5 cm

1 cm

fruit

2 cm

2 cm

leaf   fruit

Sesbania herbacea   Spartium junceum

fruit

2 cm

5 cm

1 cm

stamen
var. californica

1 cm
leaflet

var. argentata

1 cm
leaflet

var. semota

Thermopsis californica

2 mm

2 mm

1 cm

1 cm

Trifolium albopurpureum   Trifolium amoenum

var. ***bifidum*** (p. 797) **LF**: lflet length 2.5–5.5 × width, notched at tip 0.1–0.4 × length, otherwise gen entire. Open, grassy areas, oak chaparral, forest; < 760 m. NCoR, n SNF, GV, CW. Apr–Jun

var. ***decipiens*** Greene **LF**: lflet length 1.5–2.5 × width, truncate or notched at tip < 0.2 × length, otherwise dentate & entire. Open, grassy areas, oak chaparral, forest; < 1000 m. CA-FP (exc SNH, Teh); to s WA. Apr–Jun

T. bolanderi A. Gray (p. 797) BOLANDER'S CLOVER Per, glabrous. **ST**: clumped, short, ascending; internodes 1–2. **LF**: gen basal; stipules ovate, papery or thick; lflets 7–15 mm, elliptic to obcordate, finely serrate. **INFL**: head-like, gen turned to side, 1–2 cm wide; axis not exserted; peduncle 6–15 cm; pedicels 0.5–1 mm; fls quickly reflexed. **FL**: calyx 3.5–4.5 mm, 5-veined, gen dark or purple-black, lobes unequal, ± ≤ tube; corolla 10–12 mm, pale purple to rose. **SEED**: 2. Moist montane meadows; 2000–2300 m. c SNH. Jun–Jul ★

T. breweri S. Watson BREWER'S CLOVER Per, puberulent. **ST**: decumbent to ascending. **LF**: cauline; stipules sheathing to lanceolate; lflets 5–20 mm, gen obovate, ± entire to serrate. **INFL**: umbel-like, gen turned to side, 1–2 cm wide, 5–15-fld; pedicels 2–5 mm; fls or some becoming reflexed. **FL**: calyx 3.5–5 mm, ± 10-veined, lobes > tube; corolla 5–8 mm, yellow-white to pink-lavender. **FR**: gen exserted from corolla. **SEED**: 1. 2*n*=16. Dry forest, open areas, roadsides; 200–1800 m. KR, CaR, n&c SN; OR. May–Aug

T. buckwestiorum Isely (p. 797) SANTA CRUZ CLOVER Ann, glabrous. **ST**: decumbent to ascending. **LF**: cauline; stipules with many bristle-tipped teeth; lflets 0.5–1.5 cm, elliptic to obovate, finely serrate. **INFL**: head-like, 8–12 mm wide, 5–many-fld; involucre bowl-shaped, irregularly toothed, cut. **FL**: calyx tube 4–5 mm, 10-veined, lobes < tube, with 3–5 inconspicuous, lateral teeth, bristle at tip 1–1.5 mm; corolla 6–7 mm, pale pink or white, keel darker. **SEED**: 1–2. Grassy or disturbed areas; < 710 m. sw SnFrB (Mendocino, Monterey, Santa Cruz cos.). First heads enclosed in stipules, reduced, seeming cleistogamous. May–Jun ★

T. campestre Schreb. (p. 797) HOP CLOVER Ann, puberulent. **ST**: decumbent to erect. **LF**: 1-pinnate, cauline; stipules ovate; petioles of mid-lvs > lflets; lflets 0.6–1.5 cm, obovate. **INFL**: head-like, 8–13 mm wide, ovate to spheric, gen > 20-fld; fls quickly reflexed. **FL**: calyx 1.5–2 mm, upper lobes < lower; corolla 4–5.5 mm, bright yellow, striate, brown in age. **FR**: fragile; style persistent, < 1 mm. **SEED**: 1. 2*n*=14. Disturbed areas, roadsides; esp < 300 m. Gen n CA-FP; most US exc Rocky Mtns, sw; native to Eur. [*T. procumbens* L., nom. rej.] Apr–May

T. cernuum Brot. NODDING CLOVER Ann, glabrous. **ST**: prostrate to ascending. **LF**: cauline; stipules lance-triangular, tip long-pointed; lflets 0.4–1.5 cm, obovate to obcordate, tip blunt or notched, veins prominent abaxially. **INFL**: axillary, 1 cm; peduncle short; bracts lanceolate; fls quickly reflexed. **FL**: calyx 4 mm, lobes narrowly triangular, ± = tube, upper 2 > lower 3; corolla 4–5 mm, > calyx lobes, pink, banner tip notched. **SEED**: 1–4. 2*n*=16. Disturbed areas; < 150 m. NCoRO, n ScV, CCo, SnFrB; US; native to Medit. May

T. ciliolatum Benth. (p. 797) FOOTHILL CLOVER Ann, ± glabrous. **ST**: erect; internodes long. **LF**: gen cauline; upper stipules bristle-tipped; lflets 1–3 cm, oblanceolate to obovate, serrate. **INFL**: head-like, 7–20 mm wide, ovate or spheric, bracts inconspicuous, often in a ring at base; axis occ exserted; pedicels in age 3–6 mm; fls soon reflexed. **FL**: calyx 5–6 mm, tube glabrous, lobes unequal, ciliate with short, flat bristles (or, through breakage, bristle bases); corolla ± ≥ calyx, pink to purple. **SEED**: 1. 2*n*=16. Locally common. Grassland, chaparral, disturbed areas; 150–1700 m. KR, NCoR, CaR, SN, GV, CW, SW, MP exc Wrn; to WA, Baja CA. Mar–Jun

T. cyathiferum Lindl. (p. 797) BOWL CLOVER Ann, small to robust, ± glabrous. **ST**: ascending to erect. **LF**: cauline; stipules entire or deeply toothed, tip acuminate; lflets 1–3 cm, oblanceolate to obovate. **INFL**: head-like, 6–20 mm wide, 3–many-fld; involucre bowl-shaped, margin wavy, finely toothed. **FL**: calyx 7–11 mm, lower, lateral lobes ± = tube, bristle tips pen forked; corolla ± ≤ calyx, white or ± yellow, pink-tipped. **FR**: stalk-like base short. **SEED**: 1–2. 2*n*=16. Gen spring-moist valleys, chaparral, roadcuts, forest; < 2500 m. NW, CaR, n&c SN, GV, n SNE; to BC, ID, NV. May–Aug

T. depauperatum Desv. Ann, gen small, glabrous. **ST**: decumbent to erect. **LF**: cauline; lower stipules oblong, upper bristle-tipped; lflets 0.5–2 cm, narrowly oblong to obovate, entire or toothed, occ lobed, tip gen truncate. **INFL**: head-like, 0.5–1 cm wide, 3–many-fld. **FL**: calyx glabrous; corolla 4.5–9 mm, pink-purple, white-tipped, banner inflated in fr. **FR**: stalk-like base 0 to short; style gen persistent. 2*n*=16.

var. ***amplectens*** (Torr. & A. Gray) McDermott (p. 797) PALE SACK CLOVER **INFL**: involucre bracts 4–5, ± free or ± 1/2 fused, margins widely scarious, toothed near tip. **FL**: calyx 3–4 mm; corolla 5–6 mm. **FR**: 3–4 mm, > style, oblong; stalk-like base 0 or short. **SEED**: 2–5. Grassland, coastal woodland; < 800 m. GV, SnFrB, SCoR. Apr–Jun

var. ***depauperatum*** (p. 797) DWARF SACK CLOVER **LF**: lflets occ notched at tip or lobed. **INFL**: involucre vestigial, ring-like. **FL**: calyx 2.5–3.5 mm; corolla gen 6–9 mm. **FR**: > style, ovate or oblong; stalk-like base 0.5–1.5 mm. **SEED**: 2–4. Wet meadows, grassland, roadsides, open spring-moist, heavy soils; < 900 m. NCoR, CaRF, n&c SN, GV, SnFrB; to BC; Chile. Mar–May

var. ***truncatum*** (Greene) Isely (p. 797) TRUNCATE SACK CLOVER **LF**: lflet tip gen truncate, toothed. **INFL**: involucre bracts ± free, 2–2.5 mm, margins ± membranous. **FL**: calyx 2.5–3 mm; corolla 4.5–7.5 mm. **FR**: 2–3 mm, ± = style, ovate or obovate; stalk-like base 0.5–1 mm. **SEED**: 1–2. Grassy flats, disturbed slopes, openings in woodland; < 800 m. NCoR, CaRF, SNF, Teh, GV, CW, SCo, ChI, PR. Apr–Jun

T. dichotomum Hook. & Arn. (p. 797) Ann, erect; hairy. **FL**: calyx 4–8 mm, lobes 2.5–5.5 mm; corolla 7–12 mm, > calyx, dark purple, tips white. **SEED**: 1. 2*n*=32. Coastal dunes, open slopes, meadows, oak woodland, disturbed areas; < 1300 m. NW, c SN, GV, SnFrB, SCoR; to WA. [*T. albopurpureum* var. *d.* (Hook. & Arn.) Isely] Apr–Jun

T. dubium Sibth. (p. 797) LITTLE HOP CLOVER Ann, sparsely puberulent. **ST**: decumbent to erect. **LF**: 1-pinnate, cauline; stipules ovate; petioles of mid-lvs gen < lflets; lflets 0.6–1.2 cm, obovate. **INFL**: head- to ± spike-like, 0.4–0.8 cm wide, 5–10(20)-fld; fls quickly reflexed. **FL**: calyx 1.5–2 mm; corolla 3.5–4 mm, bright yellow, in age brown, weak-striate. **FR**: style persistent, < 1 mm. **SEED**: 1. 2*n*=14,16,28. Agricultural, disturbed areas; < 500 m. CA-FP; much of US; native to Eur. Apr–Jul

T. eriocephalum Nutt. subsp. ***eriocephalum*** (p. 797) HAIRY HEAD CLOVER Per, hairy. **ST**: gen ascending or erect, gen unbranched. **LF**: basal and cauline; lower stipules sheathing st, upper lanceolate; lflets 1–4 cm, elliptic-oblong to ovate. **INFL**: head-like, 1.5–3 cm, gen turned to side or downward; peduncle 5–15 cm; involucre 0; pedicels ± 0.5 mm; fls gen soon-reflexed. **FL**: calyx 5–9 mm, hairy, lobes > tube, ± linear, twisted or bent, plumose; corolla 8–14 mm, dull white or ± yellow. **FR**: hairy. **SEED**: 1–3. 2*n*=16. Locally common. Moist meadows to dry, open slopes; 300–1500 m. NW, CaR, n SN; OR. [*T. e.* var. *e.*] May–Aug

T. fragiferum L. (p. 797) STRAWBERRY CLOVER Per, creeping and rooting or cespitose, glabrous. **LF**: ± basal; stipules gen overlapped; petiole >> blade; lflets 0.5–2 cm, elliptic to obovate. **INFL**: head-like, 8–12 mm wide, in fr ± 2 cm wide, spheric, ± woolly, red or brown; involucre bracts ± 2 mm, ± free, green. **FL**: calyx hairy, in fr inflated, lobes linear, lower 2 bristle-like, > upper 3; corolla 5–6 mm, pink, quickly ± hidden by calyx. **SEED**: 1–2. 2*n*=16. Roadsides, gen in saline soil; < 1500 m. NCoR, ScV, SCoR, SW, MP, W&I, expected elsewhere; sporadic in much of US; native to Eur, Afr. Cult. May–Aug

T. fucatum Lindl. (p. 797) BULL CLOVER Ann, gen robust, ± glabrous. **ST**: decumbent to erect, gen hollow. **LF**: cauline; stipules wide; lflets 1–2.5 cm, elliptic to ± round, ± entire to toothed. **INFL**: head-like, 5–many-fld; involucre bracts ± free to 1/2 fused, margin scarious; fls quickly spreading or reflexed. **FL**: calyx 4–7 mm, glabrous, longer lobes bristle-like, > tube, gen branched; corolla 10–20 mm, dull white or ± yellow, purple-tipped, banner inflated in fr. **FR**: 5–10 mm; stalk-like base 0.5–2 mm. **SEED**: 2–4. 2*n*=16. Locally abundant. Moist, open grassland, marshes, roadsides, occ saline or serpentine soils; < 1100 m. NCo, NCoR, CaR, GV, SnFrB, SCo, ChI; OR. Vars. in need of study. Apr–Jun

T. glomeratum L. (p. 797) CLUSTERED CLOVER Ann, ± glabrous. **ST:** decumbent to ascending. **LF:** cauline; stipules striate, short-bristle-tipped; lflets 5–12 mm, obovate. **INFL:** sessile at end of sts or in succession of lf axils, head-like, 7–10 mm wide, many-fld; peduncle 0; pedicels < 1 mm. **FL:** calyx 3–5 mm, 10-ribbed, lobes triangular, recurved in fr; corolla ± > calyx, pink. **SEED:** 1–2. 2*n*=14,16. Uncommon. Disturbed areas; < 300 m. CA-FP; US; native to Eur, Afr. Mar–May

T. gracilentum Torr. & A. Gray (p. 797) PINPOINT CLOVER Ann, glabrous or ± hairy. **ST:** prostrate to erect. **LF:** cauline; stipules ovate-tapered, lower gen inconspicuous, < 1 cm; lflets 5–15 mm, obovate to obcordate, length 1.5–2.5 × width, teeth acute, tip shallowly notched. **INFL:** 1–1.5 cm wide, 3–many-fld, gen turned to side; axis gen ± exserted; bracts 0; pedicels 1–2 mm; fls in age recurved. **FL:** calyx 4–6 mm, glabrous, tube 10-veined, lobes > tube, ± 0.5 mm wide at base, short-bristle-tipped; corolla 5–7 mm, pink to pink-purple. **FR:** 4–6 mm, > petals. **SEED:** 1–2. 2*n*=16. Open, disturbed, moist or dry places, occ serpentine; < 1800 m. CA-FP (exc ChI), DMoj; to WA, AZ. [*T. g.* var. *inconspicuum* Fernald] Mar–Jun

T. grayi Lojac. (p. 797) GRAY'S CLOVER **INFL:** not bristly. **FL:** calyx tube 3–5 mm, gen ± = lobes; corolla 8–16 mm, > calyx, gen purple, tips white. **FR:** stalk-like base 1–3 mm. Wet meadows, foothill slopes, pine woodland; < 600 m. NCoRO, n&c SN, ScV, SnFrB, SCoRO. [*T. barbigerum* var. *andrewsii* A. Gray] Like *T. barbigerum*. Apr–Jun

T. gymnocarpon Nutt. subsp. *plummerae* (S. Watson) J.M. Gillett (p. 797) PLUMMER'S CLOVER Per, mat- or cushion-forming, hairy. **LF:** ± basal; lflets 3–5, 0.8–20 mm, gen elliptic, thick, serrate, teeth 0.1–0.3 mm. **INFL:** gen incl, 10–15 mm wide, 4–many-fld; peduncle 1–4 cm; fls spreading-reflexed. **FL:** calyx 4–6.5 mm, tube 10-veined, lobes < to > tube, narrow; corolla 7–11 mm, ± pink to ± tan. **FR:** 4–5 mm, exserted from corolla; stalk-like base ± 1 mm. **SEED:** 1. With sagebrush, juniper; 1500–1800 m. MP; to OR, MT, NM. [*T. g.* var. *p.* (S. Watson) J.S. Martin] Merges clinally into *T. g.* subsp. *g.* of Rocky Mtns. May–Jun ★

T. hirtum All. (p. 797) ROSE CLOVER Ann, hairy. **ST:** ascending or erect. **LF:** cauline; stipules bristle-tipped; lflets 1–2.5 cm, obovate. **INFL:** head, ± 1.5–2 cm wide, many-fld; sessile above 1–2 ± reduced lvs with involucre-like stipules; bristly in fr. **FL:** calyx 7–9 mm, lobes 4–6 mm, >> tube, subequal, bristle-like, densely plumose; corolla 11–15 mm, gen pink. **SEED:** 1. 2*n*=10. Disturbed areas, roadsides; < 2060 m. CA-FP; sporadic to e US; native to Eurasia, n Afr. Apr–May ❖

T. howellii S. Watson (p. 797) HOWELL'S CLOVER Per, gen robust, glabrous. **ST:** erect. **LF:** gen cauline; stipules of mid-lvs 1–2.5 cm, green; lflets 4–10 cm, elliptic to ovate. **INFL:** head- or short-raceme-like, 1.5–3 cm; axis gen exserted 2–3 mm; fls becoming reflexed. **FL:** calyx 3.5–4.5 mm, tube 10-veined, lobes slender, ± ≥ tube; corolla 9–14 mm, green-white to pink. **FR:** stalk-like base ± 1 mm. **SEED:** 1–3. Wet or shady places, meadows with sedges, alder swamps; 800–1800 m. n KR, n NCoR, n CaR; OR. Jul–Aug ★

T. hybridum L. (p. 797) ALSIKE CLOVER Ann, per, ± glabrous. **ST:** sprawling to erect. **LF:** cauline; stipules lance-ovate, prominently veined; lflets 1–4 cm, elliptic to obovate. **INFL:** 1.5–3 cm wide; fl bract < 1 mm, soon falling; pedicels 2–6 mm; fls becoming reflexed. **FL:** calyx 3–5 mm, tube 10-veined, lobes ± ≥ tube, lanceolate, sharply tapered, not bristle-tipped; corolla 6–11 mm, pink. **FR:** stalk-like base ± 0. **SEED:** 2–4. 2*n*=16. Disturbed areas; < 1500 m. NW, sporadic elsewhere; US; native to Eur. Cult. May–Oct

T. hydrophilum Greene (p. 801) SALINE CLOVER Ann, gen fleshy. **INFL:** gen 1–1.5 cm wide; involucre bracts basally fused, < 1 mm. **FL:** calyx 2.5–5 mm; corolla 6.5–9 mm, striate. **FR:** ± = style, ovate to oblong; stalk-like base 0.5–1 mm. **SEED:** 1–2. Salt marshes, open areas in alkaline soils; < 300 m. ScV, nw SnJV, CW. [*T. depauperatum* var. *h.* (Greene) Isely] Apr–Jun ★

T. incarnatum L. (p. 801) CRIMSON CLOVER Ann, hairy. **ST:** erect, gen unbranched. **LF:** cauline; stipules ovate to oblong, prominently veined; lflets 1–2.5 cm, obovate to obcordate. **INFL:** spike, 2–6 cm, ± cylindric or ± conic. **FL:** calyx 7–10 mm, tube prominently 10-veined, lobes ± ≥ tube, bristle-like; corolla 10–14 mm, crimson or white. **SEED:** 1. 2*n*=14. Uncommon. Disturbed areas; esp < 300 m. CA-FP; naturalized in se US; native to s Eur. May–Aug

T. jokerstii Vincent & Rand. Morgan BUTTE COUNTY GOLDEN CLOVER Ann, glabrous exc calyx. **ST:** decumbent to erect. **LF:** ± cauline; stipules conspicuous, lanceolate to ovate, lower gen overlapped, ± serrate; lflets 0.5–3.2 cm, elliptic to obovate, gen with prominent white to purple chevron. **INFL:** head, 12–30 mm wide, 5–many-fld; involucre bowl-shaped, glabrous, lobes rounded, toothed. **FL:** calyx tube 5-veined, lobes simple, entire, bristle at tip 3–4.5 mm, plumose; corolla gold- to sulphur-yellow, lower part of banner inflating. **SEED:** 1–2, > 3 mm. Vernal pool; < 400 m. ScV (North Table Mtn Plateau, Butte Co.). Mar–May ★

T. kingii S. Watson subsp. *dedeckerae* (J.M. Gillett) D. Heller (p. 801) DEDECKER'S CLOVER Per, cespitose, glabrous. **ST:** ± ascending. **LF:** mostly basal, occ 2–3 cauline; basal stipules clasping st; petiole gen >> blade; lflets 0.5–4 cm, lanceolate, thick, ± serrate. **INFL:** head-like, 1.5–3 cm wide; pedicels 1–2 mm; fls soon reflexed. **FL:** calyx 5–8 mm, glabrous, lobes gen > tube, unequal, narrow or bristle-like; corolla 12–15 mm, pink or pale violet. **FR:** stalk-like base < 2 mm. **SEED:** probably 2. Pinyon woodland to alpine crests, rock crevices; 2100–3500 m. s SNH, SNE. [*T. macilentum* var. *d.* (J.M. Gillett) Barneby] Basal lvs occ ± deteriorated on herbarium sheets. May–Jul ★

T. lemmonii S. Watson (p. 801) LEMMON'S CLOVER Per, cespitose, strigose. **LF:** basal and cauline; basal stipules papery; lflets 3–7, 1–2 cm, obovate, thick, coarsely serrate, larger teeth 0.4–0.5 mm. **INFL:** umbel, 1.5–2.5 cm wide, ± spheric, gen many-fld; peduncle gen bent or curved; fls soon reflexed. **FL:** calyx 4–5 mm, lobes > tube, slender, ± bristle-like; corolla 10–13 mm, pink-white. **FR:** stalk-like base 1–1.5 mm. **SEED:** 1–2. Pine forest, sagebrush flats; 1500–1800 m. n SNH; NV. May–Jul ★

T. longipes Nutt. LONG-STALKED CLOVER Per, rhizomed or not, cespitose or not, gen puberulent. **ST:** decumbent to erect, occ ± 0. **LF:** basal and cauline; stipules < 2 cm, lance-oblong to ovate; lflets 2–5 cm, linear to obovate. **INFL:** head-like, incl or exserted, 1.5–3 cm wide; peduncle gen bent or curved at tip; pedicels 1–2 mm. **FL:** calyx 5–10 mm, ± puberulent, lobes gen > tube, lanceolate to bristle-like; corolla 10–18 mm, dull white, purple, or 2-colored. **FR:** stalk-like base 0–1 mm. **SEED:** gen 1. 2*n*=16,24,32,48.

subsp. *atrorubens* (Greene) J.M. Gillett Roots thickened, fusiform; rhizomes ± 0. **LF:** lflet short-elliptic, obtuse, length 2.5–6 × width. **INFL:** pedicels 0.3–1 mm; fls spreading to ascending. **FL:** calyx lobes 1–3 × tube, densely long-hairy; corolla white to purple. Dry or boggy meadows, open slopes, woodland, subalpine; 1100–3000 m. SN, SnBr, SnJt. [*T. l.* var. *a.* (Greene) Jeps.] Jun–Sep

subsp. *elmeri* (Greene) J.M. Gillett (p. 801) ELMER'S CLOVER **LF:** lflet length gen 6–10 × width. **INFL:** pedicels 0.3–1 mm; fls spreading to ascending. **FL:** calyx lobes 3–5 × tube; corolla ± white. Moist, shaded areas, streambanks, meadows; 600–1700 m. KR, NCoR, CaR; OR. [*T. l.* var. *e.* (Greene) McDermott] Jun–Sep

subsp. *hansenii* (Greene) J.M. Gillett (p. 801) Rhizomed. **LF:** lflet length 2.5–6 × width in CA-FP, to 10 × in GB. **INFL:** pedicels 0.3–1 mm; fls spreading to ascending. **FL:** calyx ± glabrous, lobes 1–3 × tube; corolla white to purple. Dry or boggy meadows, open slopes, woodland, subalpine; 1100–3000 m. KR, NCoR, CaR, SN, GB; OR, NV. [*T. l.* var. *nevadense* Jeps.] Intergrades in s SN with *T. longipes* subsp. *atrorubens*, which is isolated geog in SnBr, SnJt (hairier calyx, darker corolla). Pls smaller at high elevations. Jun–Sep

subsp. *multipedunculatum* (P.B. Kenn.) J.M. Gillett Gen cespitose. **LF:** hairy abaxially, glabrous adaxially; lflets lanceolate, elliptic, or oblanceolate, length 1.5–4 × width. **INFL:** pedicels 1–2 mm; fls some or most becoming reflexed. **FL:** calyx lobes 1–3 × tube; corolla lavender to purple, longer petals tapered to beak. Conifer forest to alpine slopes; 1400–2700 m. MP; to WA. [*T. l.* var. *m.* (P.B. Kenn.) Isely] Jun–Sep

subsp. *oreganum* (Howell) J.M. Gillett Not cespitose. **LF:** lflet length 2–5 × width. **INFL:** pedicels 1–2 mm; fls soon reflexed. **FL:** calyx lobes 1–2 × tube; petals short-tapered, acute, or obtuse, gen white. Forested slopes, gravelly meadows, serpentine; 1200–1800 m. KR, NCoR; OR. [*T. l.* var. *o.* (Howell) Isely] Jun–Aug

subsp. ***shastense*** (House) J.M. Gillett   Gen cespitose. **LF:** lflet length 1.5–4 × width. **INFL:** pedicels 1–2 mm; fls some or most becoming reflexed. **FL:** calyx lobes 1–3 × tube; corolla 1–1.2 cm, lavender to purple, longer petals acute to attenuate. Conifer forest to alpine slopes; 1400–2700 m. KR, CaR. [*T. l.* var. *s.* (House) Jeps.] Jun–Sep

T. macraei Hook. & Arn. (p. 801)   MACRAE'S CLOVER   Ann. **ST:** decumbent to erect. **LF:** cauline; stipules ovate-acuminate or bristle-tipped; lflets 1–2 cm, narrowly elliptic to ovate. **INFL:** heads gen paired, ± sessile or occ 1 short-peduncled above (1)2 lvs, 5–15 mm wide, few–many-fld, ovate or spheric; peduncle 0 or 1; pedicels 0. **FL:** calyx 4–7 mm, hairy, lobes >> tube, tapered or bristle-like, plumose; corolla 5–9 mm, purple or 2-colored. **SEED:** 1. 2*n*=16. Disturbed areas, dunes; < 600 m. NCo, NCoRO, CCo, SnFrB, ChI; S.Am. Mar–May

T. macrocephalum (Pursh) Poir. (p. 801)   LARGE-HEAD CLOVER Per, rhizomed, hairy. **ST:** ascending. **LF:** basal and cauline; basal stipules brown-papery, others green, 1–3.5 cm; lflets gen 7–9, 1–2.5 cm, obovate, thick. **INFL:** head- or raceme-like, 3–6 cm, 3–5 cm wide. **FL:** calyx 1.5–2 cm, tube 10-veined, lobes >> tube, bristle-like, plumose; corolla 2–2.7 cm, light pink-purple or 2-colored. **SEED:** 1–3. 2*n*=32,48. Rocky flats or slopes, with sagebrush, juniper, to mtn ridges; 650–2500 m. KR, CaR, n SNH, MP; to WA, ID, NV. Apr–May

T. microcephalum Pursh (p. 801)   SMALL-HEAD CLOVER   Ann; hairs gen finely wavy. **ST:** decumbent to erect. **LF:** cauline; stipules gen ± bristle-tipped; lflets 0.8–2 cm, gen obovate, tip notched. **INFL:** head, 7–many-fld, gen bur-like in fr; involucre bowl-shaped, lobes ± entire. **FL:** calyx 4–6 mm, tube 10-veined, lobes > 1/2 tube, entire, bristle-tipped; corolla 4–7 mm, pink to lavender. **FR:** rupturing corolla. **SEED:** 1–2. 2*n*=16. Streambanks, moist, disturbed areas, roadsides, serpentine, conifer forest; 0–2700 m. NW, SN, GV, SnFrB, SCoRO, SCo, ChI, WTR, SnBr, PR, DMoj; to BC, ID, AZ, Mex. Apr–Aug

T. microdon Hook. & Arn. (p. 801)   THIMBLE CLOVER   Ann, puberulent. **ST:** decumbent to erect. **LF:** cauline; stipules wide, upper deeply cut or not; lflets 0.5–1.5 cm, oblanceolate to obovate, tip truncate or notched. **INFL:** head-like, 0.7–1.4 cm wide, 5–many-fld; involucre cup-shaped, divisions bristle-tipped. **FL:** calyx 2–3 mm, tube 10-veined, < 1/2 tube, triangular, not bristle-tipped; corolla 4–6 mm, white to pink. **SEED:** 1–2. 2*n*=16. Common locally. Open, moist or dry, gen disturbed areas; 10–1100 m. NCo, NCoR, n&c SNF, GV, CW, ChI, PR; to BC; S.Am. Mar–Jun

T. monanthum A. Gray   Per, small, cespitose, glabrous or puberulent. **ST:** slender or reduced. **LF:** gen basal; stipules lanceolate to ovate; lflets 2–12 mm, elliptic-oblanceolate to widely obovate. **INFL:** reduced head, incl or exserted from lvs, 1–6-fld; involucre vestigial or bracts 2–5, inconspicuous, gen ± free, 1–3 mm. **FL:** calyx 4–5 mm, lobes ± = tube, bristle-tipped; corolla 7–12 mm, white or lavender-striate. **FR:** occ rupturing corolla. **SEED:** 1–2. 2*n*=16.

subsp. ***grantianum*** (A. Heller) J.M. Gillett   GRANT'S CARPET CLOVER   Mat-forming, glabrous to ± hairy; roots thin, rhizomes thin, white. **LF:** lflet length 4–6 × width, tip gen acute. **INFL:** 1–5-fld; peduncle bent upward at tip. Mtn forest near streams; 1700–3000 m. SnGb, SnBr, SnJt; Mex. [*T. m.* var. *g.* (A. Heller) Parish] Jun–Aug

subsp. ***monanthum*** (p. 801)   CARPET CLOVER   Mat-forming, glabrous to ± hairy; roots thin, rhizomes thin, white. **LF:** lflet length 2–4 × width, tip gen rounded or truncate. **INFL:** 1–4-fld; peduncle straight. Pine belt upwards, wet meadows with aspen or willows, rocky slopes, alpine; 1700–3900 m. CaR, SN, SNE; NV. Jun–Aug

subsp. ***parvum*** (Kellogg) J.M. Gillett   SMALL CARPET CLOVER Not mat-forming, densely hairy; from thickened taproot, rhizomes 0. **LF:** lflet length 2–4 × width, tip gen rounded, truncate, or notched. **INFL:** 1–9-fld, peduncle strongly bent at tip. Pine forest, wet meadows and along streams; 1500–2900 m. CaR, SNH, SNE. [*T. m.* var. *p.* (Kellogg) McDermott] Jun–Aug

subsp. ***tenerum*** (Eastw.) J.M. Gillett   DELICATE CARPET CLOVER Not mat-forming, densely hairy; from central crown or thickened taproot, rhizomes 0. **LF:** lflet length 2–5 × width, tip acute to acuminate. **INFL:** 1–7-fld, peduncle strongly bent at tip. Pine forest, wet meadows, moist rocky slopes; 1600–3300 m. SNH. [*T. m.* var. *eastwoodianum* J.S. Martin] Jun–Aug

T. obtusiflorum Hook. (p. 801)   CLAMMY CLOVER   Ann, gen robust, glandular-hairy, gen sticky. **ST:** erect. **LF:** cauline; stipules deeply cut; lflets 2–4 cm, ± narrowly elliptic, sharply serrate, longer teeth ± 1 mm. **INFL:** head-like, 1.5–3 cm wide; involucre cut < 1/2 to base. **FL:** calyx 9–11 mm, glandular-hairy or bumpy (occ in age glabrous), tube slitting between upper lobes, lobes ± ≤ tube, gen entire, bristle-tipped; corolla 14–18 mm, pale lavender to dull purple, tips white. **SEED:** 1–2. 2*n*=16. Moist disturbed areas, gravel bars, marshes; 30–1800 m. NW, n&c SN, CW, SCo, SnGb, PR; sw OR. Apr–Jul

T. oliganthum Steud. (p. 801)   FEW-FLOWERED CLOVER   Ann, gen glabrous. **ST:** ascending to erect. **LF:** cauline; stipules toothed or deeply cut; lflets 1–2 cm, linear to obovate. **INFL:** head-like, 6–10 mm wide, 5–15-fld; involucre 2–3 mm, wheel-shaped, inconspicuous, gen cut > 1/2 to base into narrow lobes. **FL:** calyx 5–7 mm, tube slitting between upper lobes, lobes < tube, tapered to occ ± forked bristle; corolla 5–8 mm, ± ≥ calyx. **SEED:** 1–2. 2*n*=16. Woody or shrubby slopes, roadsides; < 1000 m. KR, NCoR, SN, GV, SnFrB, SCoRO; to BC. Mar–Jun

T. olivaceum Greene (p. 801)   OLIVE CLOVER   Ann, ascending to erect, hairy. **LF:** cauline; stipules lance-ovate, entire; lflets 1–3 cm, obovate, minutely serrate. **INFL:** spike, 5–20 mm wide, ovate to short-cylindric, peduncled; involucre 0. **FL:** calyx 8–14 mm, lobes 6.5–12 mm, > tube, plumose; corolla 4–7 mm, << calyx, 2-colored (purple and white). **SEED:** 1. Grassy slopes, valley meadows, disturbed areas; < 800 m. NCoRI, CaRF, n SNF, GV, SnFrB. [*T. albopurpureum* Torr. & A. Gray var. *o.* (Greene) Isely] Once abundant, esp GV. Apr–May

T. palmeri S. Watson   SOUTHERN ISLAND CLOVER   Ann, erect, glabrous. **LF:** cauline; lower stipules gen conspicuous, 15–20 mm, fibrous; lflets 10–30 mm, narrowly elliptic or lanceolate, length 3.5–5 × width, teeth bristle-like, tip acute. **INFL:** 1–1.5 cm wide, 3–many-fld, gen turned to side; axis gen ± exserted; bracts 0; pedicels 1–2 mm; fls in age recurved. **FL:** calyx 4–6 mm, tube 10-veined, lobes ± 0.5 mm wide at base, > tube, narrowly triangular-linear; corolla 5–9 mm, > calyx, purple or pink. **FR:** 4–6 mm. **SEED:** 1–2. Grassy places near ocean; 0–300 m. ChI; Baja CA. [*T. gracilentum* var. *p.* (S. Watson) McDermott] Mar–May   ★

T. polyodon Greene   PACIFIC GROVE CLOVER   Ann, decumbent-ascending, glabrous. **LF:** cauline; stipules 1 cm, widely ovate, upper deeply cut; lflets 5–20 mm, widely elliptic to obovate, tips rounded. **INFL:** head-like, incl or exserted from lvs, 0.5–2.5 cm wide, 1–many-fld; involucre wheel-shaped, gen well developed, cut to middle. **FL:** calyx 4–7 mm, tube 10–many-veined, lobes ± ≤ tube, ± 3-parted; corolla 8–10 mm, pink to white with purple tip. **FR:** stalk-like base short or 0. **SEED:** 2. Closed-cone-pine forest; moist meadows, streamsides; < 300 m. CCo (Monterey Peninsula). [*T. variegatum* phase 4 of TJM (1993)] Considered by some of hybrid origin. Seriously threatened by urbanization. Apr–Jun   ★

T. pratense L. (p. 801)   RED CLOVER   Per, gen hairy. **ST:** ascending. **LF:** cauline; stipules bristle-tipped; lflets 1.5–3.5 cm, elliptic to obovate, ± entire. **INFL:** terminal, above pair of reduced lvs, head-like, 2–3 cm wide; peduncle 0. **FL:** calyx 4.5–7.5 mm, tube 10-veined, lobes ± ≥ tube, bristle-like, straight, sparsely plumose; corolla 11–15 mm, gen red-purple. **FR:** circumscissile. **SEED:** 1–2. 2*n*=14, 28. Disturbed areas; 0–2900 m. CA-FP, n GB; US, esp n, Can; native to Eur. Important forage crop. Apr–Oct

T. productum Greene (p. 801)   PRODUCTIVE CLOVER   Per, glabrous. **ST:** ascending; internodes 1–3 below peduncle. **LF:** basal and cauline; lower stipules papery; lflets 1–5.4 cm, lanceolate or elliptic, coarsely serrate. **INFL:** head-like or short spike, gen turned to side or downward, 1.5–3 cm wide; axis exserted, gen forked; pedicels < 1 mm; fls quickly reflexed. **FL:** calyx 3–4 mm, glabrous, tube gen 10-veined, lobes ± = tube, slender; corolla 10–14 mm, red-purple or pink. **FR:** stalk-like base 2–4 mm. 2*n*=16. Conifer forest, meadows, to open ridges; 1100–2400 m. KR, CaRH, n&c SNH, MP; OR. [*T. kingii* var. *p.* (Greene) Jeps.] Jun–Aug

T. repens L. (p. 801)   WHITE CLOVER   Per, gen ± glabrous. **ST:** creeping, rooting, turf-forming. **LF:** cauline, alternate or clustered; stipules white-membranous; petioles >> blades, ± equal; lflets 0.5–

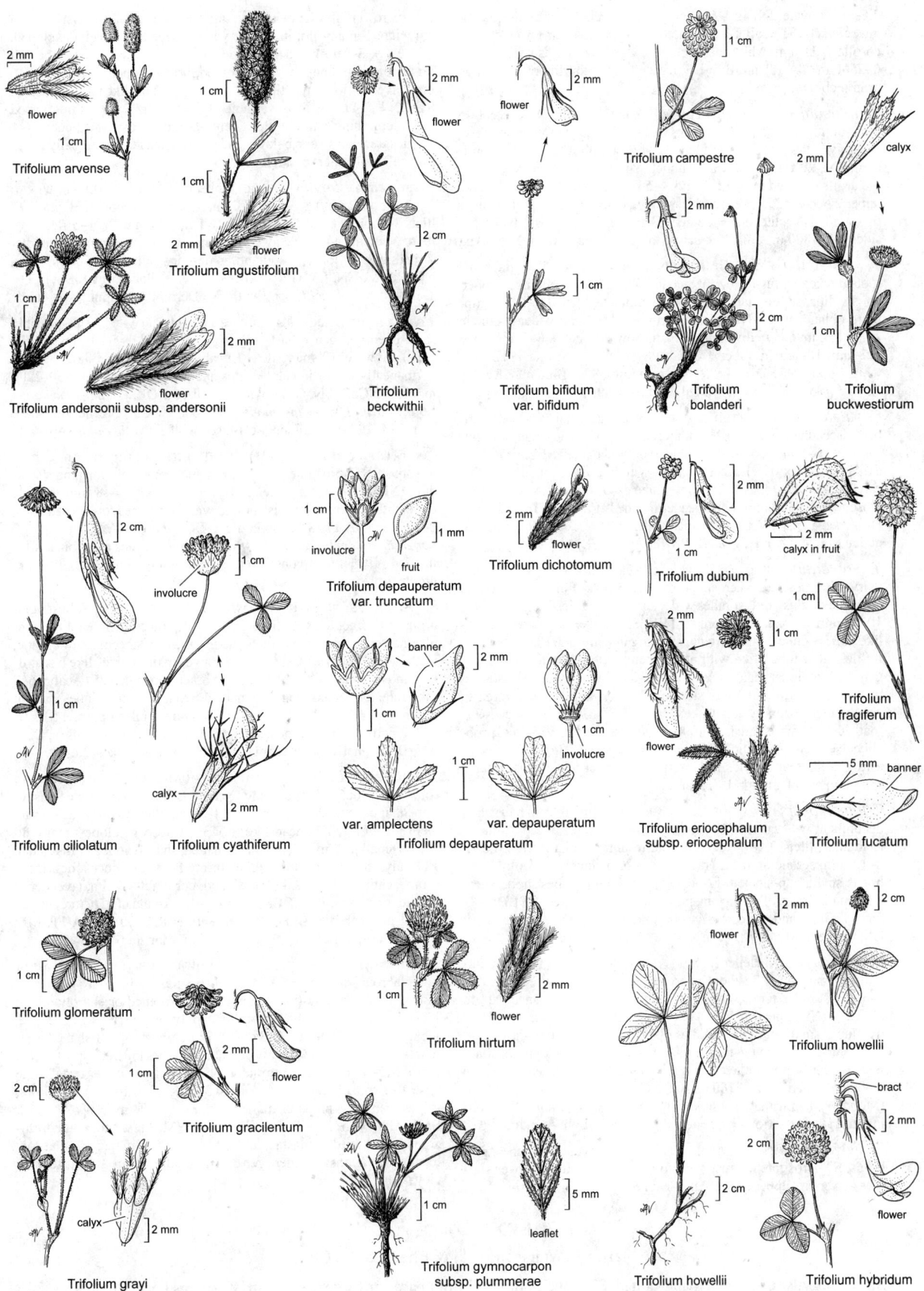

Trifolium arvense

Trifolium angustifolium

Trifolium andersonii subsp. andersonii

Trifolium beckwithii

Trifolium bifidum var. bifidum

Trifolium campestre

Trifolium bolanderi

Trifolium buckwestiorum

Trifolium ciliolatum

Trifolium cyathiferum

Trifolium depauperatum var. truncatum

Trifolium dichotomum

Trifolium dubium

var. amplectens

var. depauperatum

Trifolium depauperatum

Trifolium eriocephalum subsp. eriocephalum

Trifolium fragiferum

Trifolium fucatum

Trifolium glomeratum

Trifolium gracilentum

Trifolium hirtum

Trifolium howellii

Trifolium grayi

Trifolium gymnocarpon subsp. plummerae

Trifolium howellii

Trifolium hybridum

2.5 cm, obovate. **INFL:** 1–2.5 cm wide; pedicels 1–3 mm; fls becoming reflexed. **FL:** calyx 3–6 mm, lobes tapered, all or lower > tube; corolla 7–11 mm, white. **SEED:** 3–4. $2n=16,28,48$. Disturbed areas; 0–2500 m. CA-FP, n GB; n US; native to Eurasia. Important forage crop. Feb–Dec

T. retusum L. TEASEL CLOVER Ann, glabrous. **ST:** erect to ascending. **LF:** cauline; stipules 0.5–1.5 cm, triangular at base, abruptly awl-shaped at tip; lflets 0.7–1.9 cm, obovate, veins prominent. **INFL:** axillary, 1 cm; peduncle short below, 0 above; bracts linear; fls reflexed in fr. **FL:** calyx 4–5 mm, 10-veined, lobes > tube, upper 2 > lower 3, curved; corolla < calyx lobes, white or pink, banner tip rounded to acute, occ ± toothed. **SEED:** 2. $2n=16$. Disturbed roadsides; 150–800 m. CaR; US; native to s Eur, Medit, n Afr. May–Aug

T. siskiyouense J.M. Gillett SISKIYOU CLOVER Per, glabrous; roots thickened, fusiform, rhizome 0. **ST:** erect to ascending, slender. **LF:** cauline; stipules lanceolate, tips acuminate; lflets 0.8–3 cm, elliptic to oblanceolate. **INFL:** head, terminal, 1.5–2 cm wide; peduncle 1–3.5 cm; involucre divisions deep, narrow. **FL:** calyx 6–7 mm, tube ± 2 mm, 10-veined, lobes ± equal, lower longer; corolla 7–12 mm, white-cream. **SEED:** 2–4. Wet mtn meadows; 800–1400 m. KR; OR. [*T. wormskioldii* var. *s.* (J.M. Gillett) Isely] Jun–Jul ★

T. striatum L. KNOTTED CLOVER Ann, bien, hairy. **ST:** decumbent, ascending, or erect. **LF:** cauline; stipules ovate, tips triangular-linear; lflets 6–16 mm, obovate-oblong to oblanceolate. **INFL:** head-like, axillary, sessile, 0.5–1.5 cm, ± short-cylindric; fls many, dense. **FL:** calyx 10-nerved, lobes 1–2.5 mm, ≤ tube exc lower one, sparsely plumose; corolla 5–7 mm, pink, banner notched at tip. **SEED:** 1. $2n=14$. Disturbed areas; 0–685 m. NCoRI, CaRF, n SNF, CCo, SCoRO; US; native to Eur, n Afr. [*T. gemellum* Willd., misappl.] May–Jun

T. subterraneum L. (p. 801) SUBTERRANEAN CLOVER Ann, hairy. **ST:** prostrate or creeping, rooting. **LF:** cauline from ground, alternate or clustered; stipules wide, tapered; petioles > blades; lflets 10–15 mm, obovate or obcordate. **INFL:** head-like, ± 1 cm wide, burlike in fr; peduncle in fr curving, growing into ground. **FL:** outer 2–8, fertile, calyx tube often with purple band at top, corolla 8–14 mm, gen ± white; inner many, sterile, calyx stalk-like, with 4–5 bristles at tip, elongating, recurving to surround frs, forming bur, delivered to ground by growing peduncle, corolla 0. **SEED:** 1. $2n=16$. Meadows, roadsides, disturbed areas; < 1000 m. NCoR, n SN, GV, SCoR, probably elsewhere; sporadic US; native to s Eur, n Afr. TOXIC in excess: estrogen-like compounds may sterilize livestock. Fr of this, related spp. unique in family in US. Mar–Apr

T. tomentosum L. WOOLLY CLOVER Ann, glabrous to hairy. **ST:** decumbent to ± erect. **LF:** cauline; stipules ovate, free part lance-triangular; lflets 0.4–1.2 cm, obovate to obtriangular. **INFL:** heads 6–12 mm, sessile above 2 bract-like lvs or peduncled, woolly in fr; bracts small, cup-shaped. **FL:** calyx hairy on top, lobes short, linear inflating in fr; corolla 3–6 mm, pink, becoming reflexed. **SEED:** 1–2. $2n=16$. Grassland, roadsides; < 400 m. NCoR, n SNF, ScV, SnFrB; US; native to Medit. Apr–Jun

T. trichocalyx A. Heller (p. 801) MONTEREY CLOVER Ann, initially sparsely hairy, in age ± glabrous below infl. **ST:** spreading. **LF:** cauline; stipules toothed or lobed; lflets 5–10 mm, oblanceolate to obovate. **INFL:** head-like, incl or exserted, 4–14 mm wide, 1–10-fld; involucre small, cut past middle, glabrous or hairy. **FL:** calyx 6–7 mm, hairy, lobes gen > tube, bristle-tipped, occ ± forked; corolla 6 mm, ± ≤ calyx, pink or lavender. **FR:** not seen. Open closed-cone pine woodland, roadsides; < 100 m. w-c NCoRO (Mendocino Co.), CCo (Monterey Peninsula). Possible hybrid of *T. variegatum* and a non-involucrate sp.; seriously threatened by urbanization. Apr–Jun ★

T. variegatum Nutt. Ann, possibly short-lived per, gen ± glabrous. **ST:** prostrate to erect, wiry to fleshy. **LF:** cauline; lower stipules gen entire, upper deeply cut; lflets gen obovate or wedge-shaped,

occ narrower. **INFL:** head-like, incl or exserted from lvs, 0.5–2.5 cm wide, 1–many-fld; involucre wheel-shaped, gen well developed. **FL:** calyx 3–10 mm, bell-shaped, tube 10–many-veined, lobes all or some gen > tube, entire or toothed, bristle-tipped; corolla 3.5–16 mm, lavender to purple, tips gen white. **FR:** stalk-like base short or 0. **SEED:** 1–2. $2n=16$. Most variable of CA clovers; ± 30 entities named but intergrading without clear delimitations; most conspicuous CA entities treated as vars. below; keel shape seems taxonomically insignificant; molecular research needed.

var. *geminiflorum* (Greene) Vincent (p. 801) SMALL-FLOWERED VARIEGATED CLOVER Ann. **ST:** mat-forming or not, 1–30 cm. **LF:** lflets 3–8 mm, gen wedge-shaped. **INFL:** < 1 cm wide, 1–5-fld; involucre occ reduced. **FL:** calyx 3–5 mm; corolla 3.5–9 mm. Open, disturbed, grassy slopes, woodland, gen in dry soil; 200–2500 m. NCo, NCoR, n&c SN, CW, PR; to WA, ID. [*T. v.* phase 5 in TJM (1993); *T. v.* Nutt. var. *pauciflorum* (Nutt.) McDermott, misappl.] Apr–Jul

var. *major* Lojac. LARGE VARIEGATED CLOVER Ann, possibly short-lived per, gen robust. **ST:** gen ascending, 1–5(10) dm. **INFL:** 1.5–3 cm wide, 10–many-fld. **FL:** calyx 6–10 mm; corolla 9–17 mm. Permanently wet or inundated sites, incl meadows, marshes; 50–2200 m. NCo, NCoR, SN, GV, CW, SW, SNE; sw OR. [*T. appendiculatum* Lojac. var. *a.*; *T. v.* var. *melananthum* (Hook. & Arn.) Greene; *T. v.* phase 2 in TJM (1993)] Commonly confused with *T. wormskioldii*. Apr–Jul

var. *variegatum* (p. 801) VARIEGATED CLOVER Ann, small to robust. **ST:** prostrate to erect, occ mat- or tangle-forming, 0.3–5 dm. **INFL:** 1–1.5 cm wide, 5–10-fld. **FL:** calyx 4–8 mm; corolla 6–10 mm. Open gen moist fields, wet forest meadows, roadsides; 0–2500 m. CA-FP, SNE; sporadic to BC, MT, CO, AZ, Baja CA. [*T. appendiculatum* Lojac. var. *rostratum* (Greene) Jeps.; *T. v.* phases 1,3 in TJM (1993)] Gen considered typical; intergrading with *T. v.* var. *major*. Apr–Jun

T. vesiculosum Savi ARROWLEAF CLOVER Ann, robust, glabrous. **ST:** erect to ascending, grooved, 15–60 cm. **LF:** cauline; stipules narrow, awl-shaped; lflets 1.5–3 cm, gen lanceolate to elliptic, teeth bristle-like. **INFL:** head, gen above a pair of bract-like lvs, axillary and terminal, 4–6 cm, many-fld; bracts lanceolate. **FL:** calyx 6–7 mm, tube 24-veined, veins connected by lateral veins, lobes straight to curved, inflated in fr; corolla 12–15 mm, pink to purple. **SEED:** 2–3. $2n=16$. Disturbed roadsides; < 300 m. NCo, NCoRI, s ScV, CCo, SnFrB; US; native to Medit. Used in hydro-seed mixes. Jul–Nov

T. willdenovii Spreng. (p. 801) TOMCAT CLOVER Ann, small to robust, glabrous. **ST:** sprawling to erect. **LF:** cauline; stipules bristle-tipped, upper gen sharply cut; lflets 1–5 cm, linear to obovate, serrate, teeth < 1 mm. **INFL:** head-like, 1.5–3 cm wide, gen lobed, many-fld exc on small pls; involucre wheel-shaped, sharply lobed or dissected. **FL:** calyx 6–10 mm, shiny, tube slitting between upper lobes, lobes < tube, entire 3-toothed, bristle-tipped; corolla 8–15 mm, lavender to purple, tip gen white. **SEED:** 1–2. $2n=16$. Abundant. Disturbed, gen spring-moist, heavy soils, occ serpentine; 0–2500 m. CA-FP, MP; sporadic to BC, ID, NM, Baja CA, S.Am. Mar–Jun

T. wormskioldii Lehm. (p. 801) COW CLOVER Per, cespitose or not, glabrous, rhizomed. **ST:** decumbent or ascending. **LF:** gen basal; lower stipules bristle-tipped, upper wide, toothed or sharply lobed; lflets 1–3 cm, narrowly elliptic to widely ovate. **INFL:** head-like, incl or exserted from lvs, 2–3 cm wide; involucre wheel-shaped, segments or lobes many. **FL:** calyx 7–11 mm, lobes tapered, tips bristled; corolla 12–16 mm, pink-purple or magenta, tip white. **FR:** stalk-like base 0–1 mm. **SEED:** 2–4. $2n=16,32$. Beaches to mtn meadows, ridges, gen open moist or marshy places; < 3200 m. NW, CaR, SN, SnJV, CW, SCo, PR, SNE; to BC, WY, NM, Mex. Incl matted, rhizomed form in dry coastal sands; lush, long-stemmed form at low to mid elevations; slender, gen much smaller form at mid to high elevations. May–Oct

ULEX GORSE, FURZE

Martin F. Wojciechowski & Elizabeth McClintock

Shrub, heavily armed, not gland-dotted. **ST:** much-branched from base, stiffly spreading, striate; twigs becoming thorns. **LF:** simple, alternate; juvenile (on seedlings, young shoots near ground) linear; adult awl-like, stiff, becoming spines. **INFL:** gen cluster, axillary near twig tips, few-fld. **FL:** calyx 2-lipped, membranous, yellow, persistent; petals ± equal, yellow, persistent.

FR: ± exserted from calyx, ovate or oblong, explosively dehiscent. **SEED**: 1–3, with small basal outgrowth. ± 20 spp.: w Eur, n Afr. (Latin: ancient name)

U. europaeus L. (p. 801) **ST**: < 3 m; twigs hairy when young, stiff, intricately branched in age. **INFL**: fls 1 per axil. **FL**: densely hairy; calyx 15 mm; petals < 20 mm. **FR**: 1–2 cm, densely hairy. Common. Disturbed places, esp old fields, pastures; < 400 m. NCo, NCoRO, CCo, SCo; native to w Eur. [*Ulex europaea*, orth. var.] Old pls flammable; twigs, lvs ending in hard tips, also similar in color, texture. Apr–Jul ◆

VACHELLIA

David Seigler & John E. Ebinger

Shrub, tree, armed with stipular spines, prickles 0. **LF**: even-2-pinnate, gen alternate, gen deciduous; petiole, main axis gen with raised glands. **INFL**: head [spike], gen 1 (or in raceme, panicle). **FL**: radial; sepals, petals in 4–5, inconspicuous; stamens many, conspicuous, exserted, free; ovary simple. **FR**: dehiscent or not, flat or ± cylindric. **SEED**: rarely arilled. ± 160 spp.: trop, subtrop. (Rev. G.H. Vachell) [Seigler & Ebinger 2005 Phytologia 87:139–178]

V. farnesiana (L.) Wight & Arn. SWEET ACACIA Shrub, small tree < 8 m; stipular spines 7–30(50) mm, straight, white in age. **ST**: twig ± ridged, ± hairy. **LF**: clustered on short-shoot or not, deciduous, < 5 cm; petiole 3–16 mm, ± hairy, gland present; 1° lflets 2–6 pairs, 6–33 mm; 2° lflets 8–19 pairs, 1.6–6.3 mm, 0.5–1.7 mm wide, oblong. **INFL**: head, gen 1–3 per axil, also clustered with lvs on short-shoots, stalked, < lf. **FL**: bright yellow to dull orange. **FR**: indehiscent, 9–18 mm wide, ± straight, cylindric, leathery, dark brown, glabrous. **SEED**: ± embedded in a sweet pulp. Introduced in trop, subtrop worldwide, often a troublesome, invasive weed. Cult in s Eur for fl oils used in perfumes.

1. Fr 3–9 cm; seeds in 2+ rows . . . . var. *farnesiana*
1′ Fr 10–17 cm; seeds in 1 row . . . . var. *minuta*

var. *farnesiana* **LF**: gen > 30 mm; petiole ± puberulent. Uncommon. Disturbed areas, chaparral, dry scrub, forest; < 300 m. SCo, D; native to trop Am from FL to s TX, s AZ, Mex. [*Acacia f.* (L.) Willd. var. *f.*] Nov–Apr

var. *minuta* (M.E. Jones) Seigler & Ebinger **LF**: 15–30 mm; petiole gen densely puberulent. Disturbed areas, chaparral, dry scrub, forest, washes; < 300 m. SCo (San Diego Co.), expected PR; Baja CA. [*Acacia m.* (M.E. Jones) R.M. Beauch.; *A. f.* subsp. *m.* (M.E. Jones) Ebinger et al.] Nov–Apr

VICIA VETCH

Robert E. Preston & Duane Isely

Ann, per, unarmed. **ST**: gen sprawling or climbing, ridged or angled. **LF**: even-1-pinnate; stipules with an upper and smaller lower lobe, entire to dentate; lflets 4–many, alternate to opposite (often on 1 pl), linear to ovate; main axis gen ending as tendril. **INFL**: raceme or cluster, axillary; peduncle or pedicels present; bracts small or 0. **FL**: corolla gen lavender to purple, occ white or yellow; 9 filaments fused, 1 free; style gen not ± flat, puberulent at tip, all around or esp abaxially. **FR**: dehiscent, gen ± oblong, gen flat; base stalked or not. **SEED**: ≥ 2. ± 160 spp.: N.Am, Eurasia, S.Am, Afr. (Latin: vetch) [Steele & Wojciechowski 2003 Adv Legume Syst 10:355–370]

1. Lflet 5–12 cm; pl erect; tendrils 0 . . . . [*V. faba*]
1′ Lflet gen 1–4 cm; pl sprawling or climbing; tendrils gen present
  2. Infl a cluster of 1–4 fls, peduncle ± 0, < pedicels
    3. Corolla 5–6 mm . . . . [*V. lathyroides*]
    3′ Corolla 10–30 mm
      4. Corolla pink-purple to ± white; banner glabrous; fr hairy initially, glabrous at maturity . . . . **V. sativa**
        5. Calyx tube 4.5–5.5 mm, lobes 3–4.5 mm; corolla 10–18 mm, pink-purple to ± white . . . . subsp. **nigra**
        5′ Calyx tube 6–7 mm, lobes 5–11 mm; corolla 18–30 mm, gen pink-purple . . . . subsp. **sativa**
      4′ Corolla yellow or yellow-white, often purple-blotched or -tinged; banner puberulent abaxially or glabrous; fr hairy
        6. Banner glabrous; fr hairs from conspicuous tubercles . . . . **V. lutea**
        6′ Banner puberulent abaxially; fr hairs not from conspicuous tubercles . . . . [*V. pannonica*]
  2′ Infl a raceme, peduncled
    7. Corolla 3–9 mm
      8. Fr 6–10 mm, with spreading hairs, stalk-like base 0; lflets 10–16, fls gen 2–8 . . . . **V. hirsuta**
      8′ Fr 10–30 mm, glabrous or ± strigose, stalk-like base ± 1–2 mm; lflets, fls various, incl as at 8.
        9. Fr 10–13 mm, tip rounded, style reflexed; lflets 4–10, linear to elliptic, peduncle thread-like, 1–3 cm, often curved, 1–3-fld . . . . **V. tetrasperma**
        9′ Fr 10–30 mm, tip tapered to style, straight to curved; lflets or peduncle not as at 9.
          10. Lflets ± 12–20; fls 2–6; seeds 2 . . . . [*V. disperma*]
          10′ Lflets 4–10; fls 1–3; seeds gen > 2
            11. Style hairs at tip, gen abaxial; lflet tip gen truncate, notched, or 1–5-toothed; fls separated or 1 . . . . . . **V. hassei**
            11′ Style hairs at tip, below, all around; lflet tip acute, 1-toothed, or occ truncate; fls gen crowded or 1
            . . . . **V. ludoviciana** subsp. **ludoviciana**
    7′ Corolla > 9 mm
      12. Lf gen 1–1.5 dm; lflets 16–24, larger 2.5–4 cm — fr often upcurved, ± 12–13 mm wide . . . . **V. gigantea**
      12′ Lf rarely > 1 dm; lflets either fewer or smaller than at 12.
        13. Fls 1–3, crowded at tip of infl axis; fr hairy, margins ciliate; stipules unlobed, dentate . . . . [*V. bithynica*]

13′ Fls 3–25, ± evenly spaced along infl axis; fr glabrous to strigose, margins not ciliate; stipule lobes
   entire or lower lobe dentate
   14. Fls gen spaced, on > 1 side of axis (exc occ when pressed), 15–25 mm, length when pressed 2.5–3.5
      × width; stalk-like fr base 2–5 mm. . . . . ***V. americana*** subsp. ***americana***
   14′ Fls not spaced, gen on 1 side of axis, gen < 18 mm, length when pressed gen 4–6 × width; stalk-like
      fr base 1.5–3 mm
      15. Infl gen ± = subtending lf, 3–12-fld; corolla dark or red-purple; fr densely strigose . . . . ***V. benghalensis***
      15′ Infl gen > subtending lf, gen 10–25-fld; corolla blue-purple or lavender to ± white; fr glabrous or
         with short, sparse appressed hairs
         16. Calyx gen obliquely attached but only ± lopsided-swollen at base; fl length when pressed ± 3–3.5
            × width; fr 6–7 mm wide. . . . . **[*V. cracca*]**
         16′ Calyx both obliquely attached and lopsided-swollen at base; fl length when pressed 4–6 × width;
            fr 6–10 mm wide . . . . ***V. villosa***
            17. Hairs on sts, lvs 0 or sparse, appressed, ± 1 mm; lower calyx lobe narrowly lanceolate, gen
               1–2.5 mm . . . . subsp. ***varia***
            17′ Hairs on upper sts, lvs not sparse, spreading, 1–2 mm; lower calyx lobe linear, 2–4 mm . . . . . subsp. ***villosa***

V. americana Willd. subsp. ***americana*** (p. 805) AMERICAN
VETCH Per, hairy or glabrous. **ST:** sprawling or short, erect, to 1 m.
LF: stipules gen sharply lobed; lflets 8–16, 1–3.5 cm, widely elliptic,
wedge-shaped, to narrowly oblong, tip acute, truncate, notched, or
1–5-toothed. **INFL:** ± = subtending lf; fls 3–9, gen spaced, on > 1 side
of axis (exc occ when pressed). **FL:** calyx attachment oblique, lobes
unequal, lower 1.5–4 mm, > upper; corolla 15–25 mm, gen blue-
purple to lavender, length when pressed 2.5–3.5 × width. **FR:** 2.5–3
cm, 5–7 mm wide, glabrous or hairy; stalk-like base 2–5 mm. 2*n*=14.
Gen open, moist forest, along streams, disturbed areas; < 2400 m.
CA-FP (exc NCo, SCoRI, s ChI), GB (exc Wrn, W&I); N.Am (exc
se US). Often mistaken for *Lathyrus*. Mar–Jun

V. benghalensis L. PURPLE VETCH Ann, hairy. **ST:** sprawling or
climbing, 1–2 m. **LF:** stipules widely lanceolate, lower lobe entire or
few-toothed; lflets 10–16, 1.5–3 cm, elliptic to oblong. **INFL:** gen ±
= subtending lf; fls 3–12, gen on 1 side of axis. **FL:** calyx lopsided-
swollen at base or not, attachment oblique, tube 4–5 mm, lobes linear,
plumose, lower 3–8 mm, > upper; corolla (12)15–18 mm, dark or red-
purple, length when pressed 4–6 × width. **FR:** 2.5–3.5 cm, 8–12 mm
wide, densely strigose; stalk-like base ± 1.5 mm. 2*n*=14. Grassland,
roadsides, disturbed areas; < 1372 m. NCo, NCoRO, NCoRI, ScV,
CW (exc SCoRI), SW (exc SnBr, SnJt); sporadic to e US; native to
Eur. Cult. Mar–Jun

V. gigantea Hook. (p. 805) GIANT VETCH Per, robust, glabrous
or puberulent. **ST:** sprawling or climbing, 1–2 m. **LF:** stipules toothed
or not; lflets 16–24, 1.5–4 cm, elliptic or oblong, tip rounded to acute.
INFL: gen < subtending lf; fls 6–15, crowded, gen on 1 side of axis.
FL: calyx attachment ± oblique, tube 4 mm, lower lobe lanceolate,
± 4 mm, upper lobes deltate, < lower; corolla 12–14 mm, red-purple
or variegated to pale yellow. **FR:** 2–4.5 cm, ± 1.2–1.3 cm wide, often
upcurved, glabrous, black; stalk-like base 1–2 mm. Coastal shrub,
coastal forest, chaparral; < 305 m. NCo, w NCoRO, CCo, SnFrB, w
SCoRO, SCo (Santa Barbara Co.); to AK. [*V. nigricans* Hook. & Arn.
var. *g.* (Hook.) Broich] Mar–Aug

V. hassei S. Watson (p. 805) SLENDER VETCH Ann, glabrous to
sparsely hairy. **ST:** decumbent to erect, branched at base, 0.2–0.7
m. **LF:** stipule gen entire; lflets 4–10, 5–25 mm, narrowly oblong to
elliptic, tip gen truncate, notched, or 1–5-toothed. **INFL:** < subtending lf; fls 1–2, at tip. **FL:** calyx attachment basal, tube ± 2 mm, lobes
lanceolate, lower ± 1.5 mm, > upper, margins with loosely appressed
hairs; corolla 6–8 mm, lavender to white; style hairs at tip, gen abaxial. **FR:** 2–3 cm, 5–6 mm wide, oblong to saber-shaped, initially with
a few hairs; stalk-like base 1–2 mm. 2*n*=14. Coastal scrub, chaparral, oak woodland; < 1200 m. NCoRO, CW, SW (exc SnGb, SnBr,
SnJt); to s OR, Baja CA. Easily confused with *V. ludoviciana* in CA.
Mar–May

V. hirsuta (L.) Gray (p. 805) Ann, ± glabrous or finely hairy. **ST:**
decumbent or climbing, slender, 0.2–0.7 m. **LF:** stipules linear, occ
toothed; lflets 10–16, 0.5–1.5 cm, linear to narrowly elliptic, tip gen
truncate, notched, or 1–5-toothed. **INFL:** gen < subtending lf; fls gen
2–8, crowded near tip. **FL:** calyx attachment basal, tube 1–1.5 mm,
lobes 1–1.5 mm, linear, subequal; corolla 3–4.5 mm, dull white or

pale blue. **FR:** 6–10 mm, 3–5 mm wide, widely oblong or elliptic,
hairs spreading; stalk-like base 0. 2*n*=14. Open, disturbed sites in
scrub, woodland, forest; < 1372 m. NCo, NCoRO, n&c SNF, CCo,
SnFrB, SW; to BC, se US; native to Eur. Apr–Jul

V. ludoviciana Torr. & A. Gray subsp. ***ludoviciana*** (p. 805)
DEERPEA VETCH Ann, glabrous or hairy. **ST:** sprawling or low-
climbing, 3–10 dm. **LF:** stipules small; lflets 4–10, 1–2.5 cm, narrowly oblong to elliptic, tip acute, 1-toothed, or occ truncate. **INFL:**
< subtending lf; fls 1–3, near tip, gen crowded. **FL:** calyx attach-
ment basal, tube 1–2 mm, lobes awl-like to lanceolate, lower = tube,
> upper; corolla 4.5–7 mm, pale blue; style hairs at tip, below all
around. **FR:** 1.5–2.5 cm, 4–7 mm wide, oblong or saber-shaped, glabrous; stalk-like base ± 1–1.5 mm. 2*n*=14. Slopes, canyons, streambanks, in grassland, coastal-sage scrub, chaparral, oak woodland,
riparian woodland; < 1220 m. s CCo, SCoRO, SW (exc SnBr); to se
US, Mex. [*V. l.* var. *l.*] CA pls of this variable sp. ± identical to race in
s TX; closely related to *V. hassei*. Mar–Jun

V. lutea L. YELLOW VETCH Ann, glabrous or sparsely hairy. **ST:**
ascending or climbing, 2–6 dm. **LF:** stipules lobed or not; lflets 8–16,
1–2 cm, oblong or those of uppermost lvs linear, tip rounded or truncate, 1-toothed. **INFL:** << subtending lf; fls 1–3, crowded. **FL:** calyx
attachment oblique, tube 4–6.5 mm, lobes linear, lower 8–11.5 mm,
>> upper; corolla 20–30 mm, yellow, often purple-tinged, banner gla-
brous. **FR:** 2.5–3.5 cm, 7–14 mm wide, elliptic-oblong; hairs from
conspicuous tubercles; stalk-like base 1–2 mm. 2*n*=14. Roadsides,
disturbed areas; < 732 m. NCo, NCoRO, NCoRI, SnFrB; sporadic to
s US; native to Eur. May–Jul

V. sativa L. Ann, glabrous or hairy. **ST:** decumbent to ascending,
1–6 dm. **LF:** stipules gen toothed; lflets 8–14, 1.5–3.5 cm, tip acute,
truncate, or notched, often with 1 slender tooth. **INFL:** << subtending
lf; fls in sessile or ± peduncled clusters of 1–3, pedicels short. **FL:**
calyx attachment basal, lobes linear, ± equal; banner glabrous. **FR:**
2.5–6 cm, 2.5–8 mm wide, hairy initially, glabrous at maturity; stalk-
like base 0. 2*n*=12,14. Subspp. occ treated as spp.

   subsp. ***nigra*** (L.) Erhart (p. 805) NARROW-LEAVED VETCH **LF:**
   lflets 5–7 mm wide, linear to gen lance-oblong, length 4–10 × width.
   FL: calyx tube 4.5–5.5 mm, lobes 3–4.5 mm; corolla 10–18 mm,
   pink-purple to ± white. **FR:** black. **SEED:** 2.5–4 mm wide, spheric.
   Roadsides, disturbed areas, grassland, open areas in oak woodland,
   riparian woodland; < 1608 m. NCo, NCoRO, NCoRI, n&c SNF, GV,
   CW (exc SCoRI), SCo, ChI, PR; to se US; native to Eur. Mar–Jun

   subsp. ***sativa*** SPRING VETCH **LF:** lflets 4–10 mm wide, wedge-
   shaped to oblong, length 2–6 × width. **FL:** calyx tube 6–7 mm, lobes
   5–11 mm; corolla 18–30 mm, gen pink-purple. **FR:** brown to black.
   SEED: 6–8 mm wide, gen lens-shaped. Roadsides, disturbed areas,
   grassland, open areas in oak woodland, riparian woodland; < 1266 m.
   NCo, NCoRO, NCoRI, CaRF, n SNF, GV, CW (exc SCoRI), SW (exc
   n ChI, SnGb, SnJt); to e US; native to Eur. Mar–Jun

V. tetrasperma (L.) Schreb. (p. 805) SPARROW VETCH Ann, ±
glabrous. **ST:** slender, decumbent or climbing, 1–5 dm. **LF:** stipules
entire; lflets 4–10, 0.6–2 cm, linear to elliptic. **INFL:** > subtending

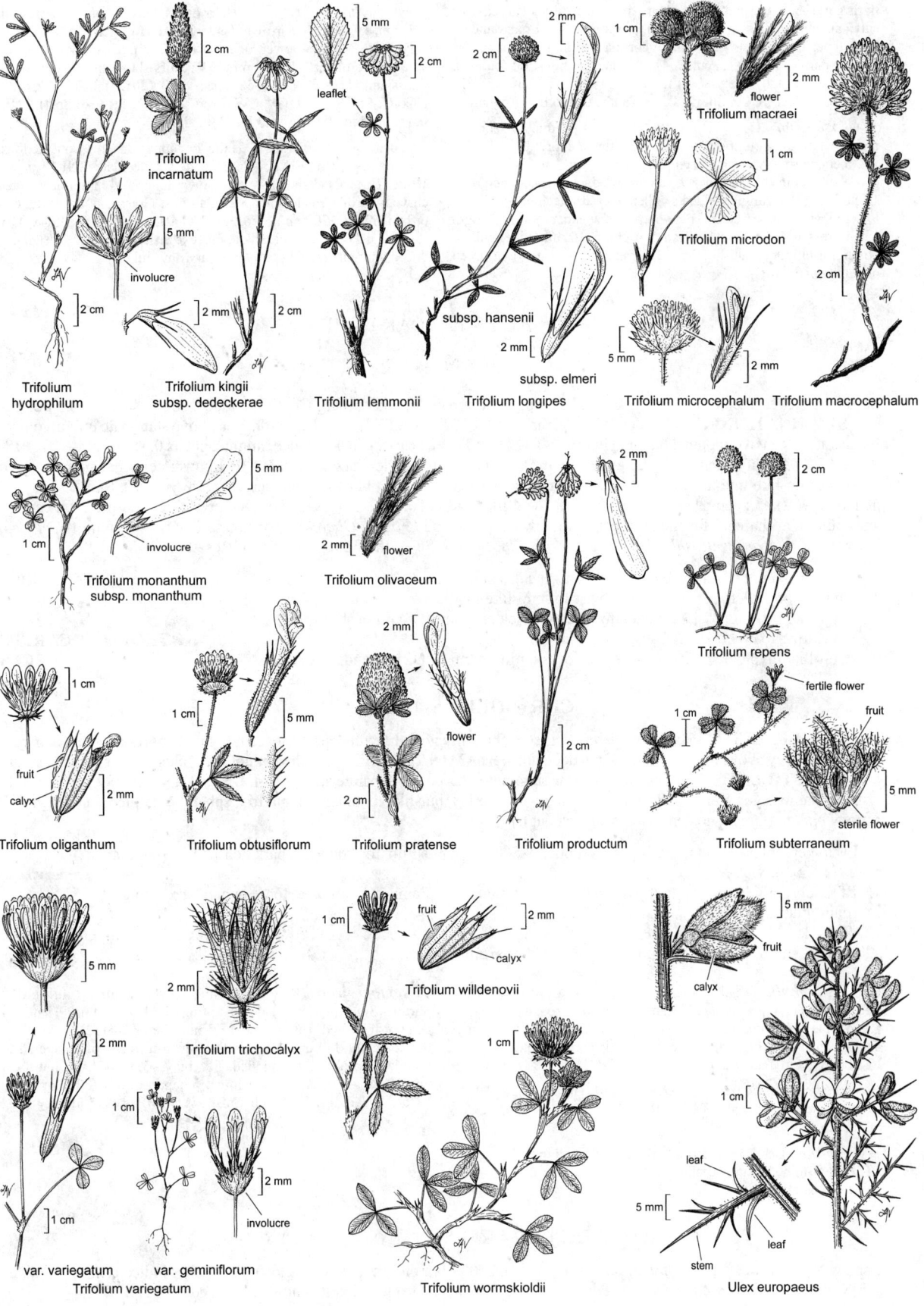

Trifolium incarnatum

leaflet

2 cm

5 mm

2 cm

2 mm

involucre

Trifolium hydrophilum

Trifolium kingii subsp. dedeckerae

2 cm

2 mm

2 cm

Trifolium lemmonii

subsp. hansenii

subsp. elmeri

2 mm

Trifolium longipes

flower

Trifolium macraei

1 cm

2 mm

Trifolium microdon

1 cm

5 mm

2 mm

Trifolium microcephalum

Trifolium macrocephalum

2 cm

Trifolium monanthum subsp. monanthum

5 mm

1 cm

involucre

Trifolium olivaceum

2 mm flower

2 mm

2 mm

Trifolium repens

2 cm

Trifolium oliganthum

fruit

calyx

1 cm

2 mm

Trifolium obtusiflorum

1 cm

5 mm

Trifolium pratense

2 mm

flower

2 cm

Trifolium productum

2 mm

2 cm

fertile flower

fruit

sterile flower

Trifolium subterraneum

1 cm

5 mm

var. variegatum

var. geminiflorum

Trifolium variegatum

5 mm

2 mm

1 cm

Trifolium trichocalyx

2 mm

1 cm

2 mm

involucre

fruit

calyx

Trifolium willdenovii

1 cm

2 mm

1 cm

Trifolium wormskioldii

fruit

calyx

leaf

leaf

stem

5 mm

1 cm

5 mm

Ulex europaeus

lf; fls gen 1–3, at tip, peduncle 1–3 cm, thread-like, often curved. **FL:** calyx attachment basal, tube 1–2 mm, lower lobe 0.5–1 mm, lanceolate, upper lobes << lower, deltate; corolla 4–6 mm, pale lavender to light purple. **FR:** 1–1.3 cm, 3–4 mm wide, glabrous; tip rounded, style reflexed; stalk-like base ± 1 mm. 2*n*=14. Disturbed areas; < 610 m. NCo, NCoRO, CCo, SnFrB; US, s Can; native to Eur. Apr–Jul

V. villosa Roth   HAIRY VETCH, WINTER VETCH   Ann. **ST:** sprawling, ascending, or climbing, to 1 m. **LF:** stipule entire; lflets 12–18, 1–2.5 cm, narrowly oblong to elliptic, tip rounded, 1-toothed. **INFL:** gen > subtending lf; fls gen > 9, gen crowded, occ spaced, gen on 1 side of axis. **FL:** calyx lopsided-swollen at base, attachment oblique, lower lobes >> upper; corolla blue-purple or lavender to white, length when pressed 4–6 × width. **FR:** 1.5–4 cm, 6–10 mm wide, widely oblong or elliptic; stalk-like base 2–3 mm. 2*n*=14. Important green manure, cover crop.

subsp. *varia* (Host) Corb.   Hairs on sts, lvs 0 or sparse, appressed, ± 1 mm. **LF:** stipule narrowly lanceolate. **INFL:** gen 10–20-fld. **FL:** calyx tube 3–4 mm, lower lobe gen 1–2.5 mm, narrowly lanceolate, straight; corolla 10–14(16) mm. **FR:** hairs short, sparse appressed. Grassland, roadside, disturbed areas; < 1200(2072) m. NCo, NCoRO, NCoRI, CaRF, n SN, GV, CW (exc CCo), TR, PR (exc SnJt), SNE; introduced in US; native to Eur. Mar–Jun

subsp. *villosa* (p. 805)   Hairs on upper sts, lvs conspicuous, spreading, 1–2 mm. **LF:** stipule widely lanceolate. **INFL:** gen > 19-fld. **FL:** calyx tube 2.3–4 mm, lower lobes 2–4 mm, linear, often curved; corolla 14–18 mm. **FR:** glabrous. Grassland, roadside, disturbed areas; < 2090 m. NCo, KR, NCoRI, c SNF, n SNH, GV, CW (exc SCoRI), SCo, n ChI, SnGb, PR (exc SnJt), GB (exc W&I); introduced, cult in w&s US; native to Eur. May–Jul

FAGACEAE   OAK FAMILY

John M. Tucker

Shrub, tree, evergreen or not; monoecious. **LF:** simple, alternate, petioled; margin entire to lobed; stipules small, gen deciduous. **STAMINATE INFL:** catkin or stiff spike, many-fld. **PISTILLATE INFL:** 1–few-fld, gen above staminate infl; involucre bracts many, gen overlapping, flat or cylindric. **STAMINATE FL:** calyx gen 4–6-lobed, minute; petals 0; stamens 4–12+. **PISTILLATE FL:** calyx gen 6-lobed, minute; petals 0; ovary inferior, style branches gen 3. **FR:** 1 nut subtended, partly enclosed by scaly, cup-like involucre or 1–3 nuts subtended, enclosed by spiny, bur-like involucre; mature yrs 1–2. **SEED:** gen 1. 7 genera, ± 900 spp.: gen n hemisphere. [Li et al. 2004 Int J Pl Sci 165:311–324] Wood of *Quercus* critical for pre-20th century ship-building, charcoal for metallurgy; some now supply wood (*Fagus*, *Quercus*), cork (*Quercus suber*), food (*Castanea*, chestnut). *Lithocarpus densiflorus* moved to *Notholithocarpus*. Scientific Editor: Thomas J. Rosatti.

1. Nuts 1–3, ± angled, enclosed by spiny, bur-like involucre . . . . **CHRYSOLEPIS**
1′ Nuts 1, not angled, partly enclosed by scaly, cup-like involucre
  2. Cup-like involucre with reflexed to spreading scales; staminate infl a stiff spike,
    spreading to erect . . . . **NOTHOLITHOCARPUS**
  2′ Cup-like involucre with appressed scales; staminate infl a catkin, pendent. . . . . **QUERCUS**

CHRYSOLEPIS   CHINQUAPIN

Evergreen. **LF:** leathery, adaxially ± glabrous, green, abaxially golden, with densely spaced, minute, appressed scales, margin entire or ± wavy above middle; stipules gen deciduous. **STAMINATE INFL:** branched or not, stiff, clustered or not, ascending to erect. **PISTILLATE INFL:** clustered below staminate on same or separate stalk, 1–3-fld. **STAMINATE FL:** sepals gen 6, minute; stamens gen 8–10+. **FR:** nuts 1–3, enclosed by spiny, bur-like involucre, ovoid to ± spheric, ± angled; mature yr 2. 2 spp.: w N.Am. (Greek: golden scale, from abaxial lf)

1. Lf tip gen obtuse to rounded, blade ± elliptic, 2–8(12) cm; pl shrub, top rounded; bark gen ± thin, ± smooth
  . . . . *C. sempervirens*
1′ Lf tip abruptly long-tapered, blade lanceolate to oblong, 5–15 cm; pl shrub, tree, top ± conic; bark ± thick,
  rough, furrowed. . . . . *C. chrysophylla*
  2. Tree, 15–30(45) m; lf blade ± flat. . . . . var. *chrysophylla*
  2′ Shrub, small tree, < 5(10) m; lf blade ± folded, margins upturned . . . . var. *minor*

C. chrysophylla (Hook.) Hjelmq.   GIANT CHINQUAPIN   Shrub, tree < 30(45) m; top ± conic. **ST:** trunk bark ± thick, rough, furrowed. **LF:** petiole 5–12 mm; blade 5–15 cm, lanceolate to oblong, adaxially dark green, abaxially golden, base tapered, tip abruptly long-tapered. **FR:** bur 3–5 cm diam; nut 6–15 mm.

var. *chrysophylla* (p. 805)   Tree 15–30(45) m. **LF:** blade ± flat. Conifer forest; < 2000 m. NW, n CaR, n SNH (El Dorado Co.), CCo, SnFrB; to WA. Jun–Sep

var. *minor* (Benth.) Munz (p. 805)   Shrub, small tree < 5(10) m. **LF:** blade ± folded, margins upturned. Conifer forest, closed-cone-pine forest, chaparral; < 1800 m. NW, CW (exc SCoRI); sw OR. Jun–Sep

C. sempervirens (Kellogg) Hjelmq. (p. 805)   BUSH CHINQUAPIN   Shrub < 3(10) m; top rounded. **ST:** trunk bark gen ± thin, ± smooth, gen not furrowed. **LF:** petiole 4–15 mm; blade 2–8(12) cm, ± elliptic, adaxially dull green, abaxially golden to rusty, base tapered to rounded, tip obtuse to rounded. **FR:** bur 2–3.5 cm diam; nut 8–13 mm. Rocky slopes, conifer forest, chaparral; 700–3300 m. KR, NCoRH, CaRH, SNH, SnGb, SnBr, PR, w MP; s OR. Jul–Aug

NOTHOLITHOCARPUS   TAN OAK, TANBARK OAK

1 sp.: w N.Am. (Greek: false *Lithocarpus*) [Manos et al. 2008 Madroño 55:181–190] Molecular studies indicate closer relationship with *Castanea*, *Castanopsis*, *Quercus* than with *Lithocarpus*, from which it is now segregated.

N. densiflorus (Hook. & Arn.) Manos et al.   Shrub, tree < 30(45) m, evergreen; trunk bark gray-brown. **LF:** petioles 10–25 mm; blade 3–14 cm, oblong to ± ovate, adaxially sparsely stellate-hairy, ± glabrous in age, abaxially fine-woolly, in age ± glabrous, base ± rounded, tip obtuse, margin entire to serrate; stipules early-deciduous. **STAMINATE INFL:** spike, stiff, spreading to erect, many-fld, dense. **PISTILLATE INFL:** below staminate infl on same or separate stalk, 1-fld. **STAMINATE FL:** sepals 5–6, minute; stamens 10–12. **FR:** mature yr 2; nut 1, 20–35 mm, ovoid to ± spheric, partly enclosed by cup-like involucre (cup), remnants of perianth and style persistent as small point at tip; cup (1.5)2–3 cm diam, saucer-shaped, scales slender, ± tapered, reflexed to spreading. [*Lithocarpus d.* (Hook. & Arn.) Rehder]

1. Tree < 30(45) m; lf gen serrate, prominently
    veined abaxially . . . . var. ***densiflorus***
1′ Shrub < 3 m; lf entire to few-toothed,
    obscurely veined abaxially . . . . var. ***echinoides***

    var. ***densiflorus*** (p. 805)   **LF:** 4–14 cm, 12–40 mm wide. Redwood, mixed-evergreen forest; < 1500 m. NW, CaR, SN, CW, WTR; s OR. [*Lithocarpus d.* (Hook. & Arn.) Rehder var. *d.*] Jun–Oct

    var. ***echinoides*** (R. Br. ter) Manos et al. (p. 805)   **LF:** 3–5(8) cm, 10–30 mm wide. Mixed-conifer and red-fir forest; 600–2000 m. KR, CaR, SN; s OR. [*Lithocarpus d.* (Hook. & Arn.) Rehder var. *e.* (R. Br. ter) Abrams] Jun–Aug

QUERCUS   OAK

Evergreen or not. **LF:** stipules small, gen early-deciduous. **STAMINATE INFL:** catkins, 1–several, pendent, slender, proximal on twig. **PISTILLATE INFL:** in upper lf axils, short-stalked; fl gen 1. **STAMINATE FL:** stamens 4–10. **PISTILLATE FL:** calyx minute, gen 6-lobed; ovary enclosed by involucre. **FR:** nut 1, partly enclosed by cup-like involucre (cup) with appressed scales (nut and cup = acorn); scales tubercled to not; mature yrs 1 (on younger sts) or 2 (on older sts). $2n=24$. ± 600 spp.: n hemisphere, to n S.Am, India. (Latin: ancient name for oak) [Manos et al. 1999 Molec Phylogen Evol 12:333–349] Many named hybrids; those (3) treated here form widespread populations; most others occur as single individuals, and some but not all of these are mentioned here, under the first parent treated (alphabetically). Reproduction of many spp. declining due to habitat degradation or loss as well as disease.

1. Acorn cup scales thin, not tubercled, shell hairy to woolly inside, at least at tip; bark dark gray or gray-brown to black
  2. Lf moderately to deeply lobed, lobes 1–4-toothed, teeth gen bristle-tipped; pl deciduous . . . . ***Q. kelloggii***
  2′ Lf entire to toothed, teeth abruptly pointed to spine-tipped, not bristle-tipped; pl evergreen
    3. Lf adaxially convex, margin inrolled or not, obscuring marginal teeth, blade gen widely elliptic to
        round; fr mature yr 1 . . . . ***Q. agrifolia***
      4. Lf abaxially glabrous to sparsely hairy exc vein axils hair-tufted . . . . var. ***agrifolia***
      4′ Lf abaxially densely tomentose . . . . var. ***oxyadenia***
    3′ Lf adaxially ± flat, margin not inrolled, blade gen lanceolate to oblong; fr mature yrs 1–2
      5. Lf abaxially ± yellow or ± white woolly; fr mature yr 1 . . . . ***Q. ilex***
      5′ Lf abaxially glabrous; fr mature yr 2
        6. Lf 2–5 cm, abaxially gen ± shiny, yellow-green; nut gradually tapered from below middle to tip . . . . . . ***Q. wislizeni***
          7. Shrub 2–4(6) m . . . . var. ***frutescens***
          7′ Tree gen 10–22 m . . . . var. ***wislizeni***
        6′ Lf 3–9(14) cm, abaxially gen ± dull . . . . ***Q. parvula***
          8. Tree < 17 m — s NCo, NCoRI, CW (exc SCoRI), WTR . . . . var. ***shrevei***
          8′ Shrub 1–6 m
            9. Lf margin gen entire; SCoRO (Santa Barbara Co.), n ChI (Santa Cruz Island), e WTR . . . . var. ***parvula***
            9′ Lf margin long-tapered dentate; SnFrB (Mount Tamalpais) . . . . var. ***tamalpaisensis***
1′ Acorn cup scales gen thick, gen tubercled to ± not, shell glabrous to woolly inside; bark light gray to ± white
  10. Acorn nut shell ± woolly inside; fr mature yr 2; pls evergreen
    11. Lf gen 3–8 cm; acorn cup > 1.5 cm wide; gen tree, > 7 m
      12. Lf adaxially with lateral veins ± not impressed, abaxially golden-puberulent, glabrous in age . . . . . . ***Q. chrysolepis***
      12′ Lf adaxially with lateral veins impressed, abaxially densely tomentose, sparsely tomentose in age —
          ChI . . . . ***Q. tomentella***
    11′ Lf gen 1–3.5 cm; acorn cup < 1.5 cm wide (if > 1.5 cm wide, lvs ≤ 3 cm); gen shrub, < 7 m
      13. Twigs rigid; lf margin wavy, strongly spine-toothed; blade elliptic to round-ovate; pls 2–6 m, erect . . . . ***Q. palmeri***
      13′ Twigs flexible; lf margin entire to irregularly serrate, teeth mucronate or spine-tipped; blade oblong
          or lanceolate to ovate; pls to 5 m, prostrate to erect
        14. Lf margin entire or with few irregular spine-tipped teeth; dry chaparral, 100–1800 m, sw PR . . . . ***Q. cedrosensis***
        14′ Lf margin entire to mucro-toothed; montane conifer zone, 900–2800 m, KR, NCoRH, NCoRI,
            CaRH, SNH, MP . . . . ***Q. vacciniifolia***
  10′ Acorn nut shell glabrous inside; fr mature yr 1; pls evergreen or not
    15. Lf lobed, margins gen without spines; pls deciduous
      16. Lf shallowly lobed (sinuses gen < 1/2 distance lobe tip to midrib), adaxially gen dull, green to blue-green
        17. Lf adaxially green; acorn cup scales tubercled; ChI . . . . ***Q. ×macdonaldii***
        17′ Lf adaxially blue- or gray-green; acorn cup scales ± tubercled; mainland
          18. Shrub, tree < 3 m, evergreen to deciduous; lvs 1.5–5 cm, gen irregularly coarsely toothed, adaxially
              blue- or gray-green . . . . [2]***Q. ×alvordiana***
          18′ Tree, deciduous; lvs 3–6(8) cm, ± entire, wavy, or ± lobed, adaxially blue-green . . . . ***Q. douglasii***
      16′ Lf moderately to deeply lobed (sinuses gen > 1/2 distance lobe tip to midrib), adaxially ± shiny, dark green
        19. Tree; acorn cup 10–30 mm deep, all scales tubercled; nut 30–50 mm, gen long-conic, tip pointed . . . . . . ***Q. lobata***
        19′ Shrub, tree; acorn cup 4–9 mm deep, lower scales ± tubercled, upper ± not; nut 20–30 mm, ovoid to
            ± spheric, tip rounded . . . . ***Q. garryana***

20. Tree, 8–20 m, gen with 1 trunk; terminal buds 5–12 mm, fusiform, hairs dense, ± yellow or ± white
. . . . var. *garryana*
20′ Shrub, small tree, often < 5 m, gen with > 1 trunk; terminal buds 2–5 mm, hairs sparse, brown or ± red
21. Rays of stellate hairs on lf abaxially gen 4–6, 0.25–0.5 mm; KR, NCoRH, CCo . . . . var. *breweri*
21′ Rays of stellate hairs on lf abaxially gen 6–8, < 0.3 mm; KR, NCoRI, CaR, SN, SnFrB, SCoRO, ne
WTR, MP . . . . var. *semota*
15′ Lf gen entire to toothed, not lobed, teeth with spines or not; pls gen evergreen
22. Lvs 7–11(18) cm, serrate, lateral veins of larger lvs gen 20–28, gen straight and prominent; stipules
persistent, > 10 mm, silky-hairy . . . . *Q. sadleriana*
22′ Lvs 1–6 cm, entire to dentate, lateral veins of larger lvs gen < 20, gen not straight and prominent;
stipules early-deciduous, < 10 mm, not silky-hairy
23. Shrub, tree gen > 5 m; lvs 2–6 cm, oblong to obovate, entire or ± toothed, adaxially green to blue-green
24. Shrub, small tree 5–6(10) m; lf blade (10)20–35(60) mm, adaxially shiny green to dull blue-green,
margin entire, dentate, or wavy-dentate . . . . [2]*Q. ×acutidens*
24′ Tree 5–18 m; lf blade (20)30–60(80) mm, adaxially blue-green, dull, margin wavy-dentate to
gen entire . . . . *Q. engelmannii*
23′ Shrub gen < 5 m; lvs 1–5 cm, elliptic to ovate or ± round, variously toothed, adaxially gen green
25. Lf 2-colored, adaxially yellow- to gray-green, abaxially ± white, densely fine-tomentose,
hairs obscuring lateral veins. . . . . *Q. cornelius-mulleri*
25′ Lf 1–2-colored, adaxially green to gray- or blue-green, abaxially gen green, not densely tomentose,
hairs not obscuring lateral veins
26. Lf gen oblong to obovate, margin wavy to obtuse-toothed; PR . . . . [2]*Q. ×acutidens*
26′ Lf elliptic to ovate or ± round, margin toothed, teeth abruptly pointed to spine-tipped
27. Acorn cup thin, scales ± tubercled to not; lf adaxially gray- or blue-green, dull
28. Acorn stalk 10–15 mm, nut cylindric-ovoid to elliptic, abruptly tapered to tip, 12–23 mm, gen
yellow-brown; lvs oblong to elliptic, gen regularly spine-toothed. . . . . *Q. turbinella*
28′ Acorn stalk 0, nut conic-ovoid, gradually tapered to tip, 20–40 mm, gen dark brown; lvs oblong
to elliptic or obovate, irregularly toothed, teeth blunt to weakly spiny
29. Shrub, tree < 3 m; lvs 1.5–5 cm, toothed but not spine-toothed, adaxially blue- to gray-green
. . . . [2]*Q. ×alvordiana*
29′ Shrub; lvs 1.3–2.8 cm, spine-toothed, adaxially gray-green . . . . *Q. john-tuckeri*
27′ Acorn cup thick, scales tubercled; lf adaxially gen green, ± shiny (dull or not in *Quercus durata*)
30. Lf adaxially dull, convex, margin often inrolled, obscuring teeth . . . . *Q. durata*
31. Lf adaxially strongly convex, abaxially short-hairy when young; gen on serpentine, NCoR,
n SN, n CCo, SnFrB, SCoR . . . . var. *durata*
31′ Lf adaxially ± convex, abaxially densely short-hairy when young; not on serpentine,
se WTR, SnGb (s slope). . . . . var. *gabrielensis*
30′ Lf adaxially ± shiny, flat to ± wavy, not convex, margin gen not inrolled
32. Lf abaxially with spreading, 2–6-rayed hairs — SCo, PR, chaparral or coastal-sage scrub,
< 200 m . . . . *Q. dumosa*
32′ Lf abaxially with appressed, 4–10-rayed hairs
33. Lf base wedge-shaped or gen truncate or rounded; margin mucro- or spine-toothed;
widespread on mainland . . . . *Q. berberidifolia*
33′ Lf base gradually tapered, wedge-shaped, or rounded; margin entire, wavy, or ± toothed,
teeth gen mucronate; ChI. . . . . *Q. pacifica*

Q. ×acutidens Torr. (p. 805)   Shrub, small tree 5–6(10) m, evergreen. **LF:** 2–6 cm; petiole 3–7 mm; blade gen oblong to obovate, ± leathery, adaxially shiny green to dull blue-green, abaxially ± densely puberulent, glabrous in age, dull, pale green, tip obtuse to short-toothed, margin entire, dentate, or wavy-dentate. **FR:** cup 10–18 mm wide, 6–9 mm deep, bowl-shaped, scales ± tubercled; nut 20–25 mm, oblong to ovoid, tip ± obtuse, shell glabrous inside; mature yr 1. Slopes, chaparral, woodland; < 1720 m. PR. Hybrids involving *Q. cornelius-mulleri*, *Q. engelmannii*, considered sp. by Torrey. Feb–May

Q. agrifolia Née   COAST LIVE OAK, ENCINA   Tree (6)10–25 m, evergreen; top wide; trunk bark furrowed in age, widely ridged, checkered, dark gray. **LF:** 2.5–6(9) cm; petiole 4–15 mm; blade gen widely elliptic to round, adaxially convex, ± dull green, abaxially glabrous to densely tomentose, dull, pale green, tip rounded to spine-toothed, margin inrolled or not, weakly spine-toothed. **FR:** cup 10–16 mm wide, 8–15 mm deep, obconic, scales thin, ± not tubercled, ± glabrous, ± brown; nut 25–35 mm, slender, ovoid, tip pointed, shell woolly inside; mature yr 1.

var. ***agrifolia*** (p. 805)   **LF:** abaxially glabrous to sparsely hairy exc vein axils hair-tufted. Valleys, slopes, mixed-evergreen forest,

woodland; < 1440 m. NCoRO, NCoRI, ScV, CW, SW; Baja CA. [*Q. a.* var. *frutescens* Engelm.] Hybridizes with *Q. kelloggii* (*Q. ×chasei* McMinn et al.), *Q. parvula* var. *shrevei*, *Q. wislizeni*. Mar–Apr

var. ***oxyadenia*** (Torr.) J.T. Howell (p. 805)   **LF:** abaxially densely tomentose. Gen granitics; 300–1500 m. SnGb, SnBr, PR; Baja CA. Hybridizes with *Q. kelloggii* (*Q. ×ganderi* C.B. Wolf). Mar–Apr

Q. ×alvordiana Eastw. (p. 805)   Shrub, tree < 3 m, evergreen to deciduous. **LF:** 1.5–5 cm; petiole 2–5 mm; blade gen oblong to widely elliptic, adaxially dull, blue- to gray-green, abaxially fine-hairy, dull, pale green, tip obtuse to abruptly pointed, margin gen irregularly coarsely toothed. **FR:** cup 10–16 mm wide, 8–10 mm deep, cup- to bowl-shaped, scales ± tubercled, light brown; nut 20–40 mm, gen narrowly ovoid, tapered to pointed tip, shell glabrous inside; mature yr 1. Dry slopes, hills; 180–1300 m. Teh, SCoR, WTR. Hybrids involving *Q. douglasii*, *Q. john-tuckeri*, considered sp. by Eastwood. Jan–Mar

Q. berberidifolia Liebm. (p. 805)   SCRUB OAK   Shrub 1–3 m or ± tree > 3 m, evergreen. **LF:** 1.5–3 cm; petiole 2–4 mm; blade oblong, elliptic, or ± round, adaxially ± flat to wavy, ± shiny, green, abaxially with minute appressed stellate hairs, dull, pale green, tip gen rounded,

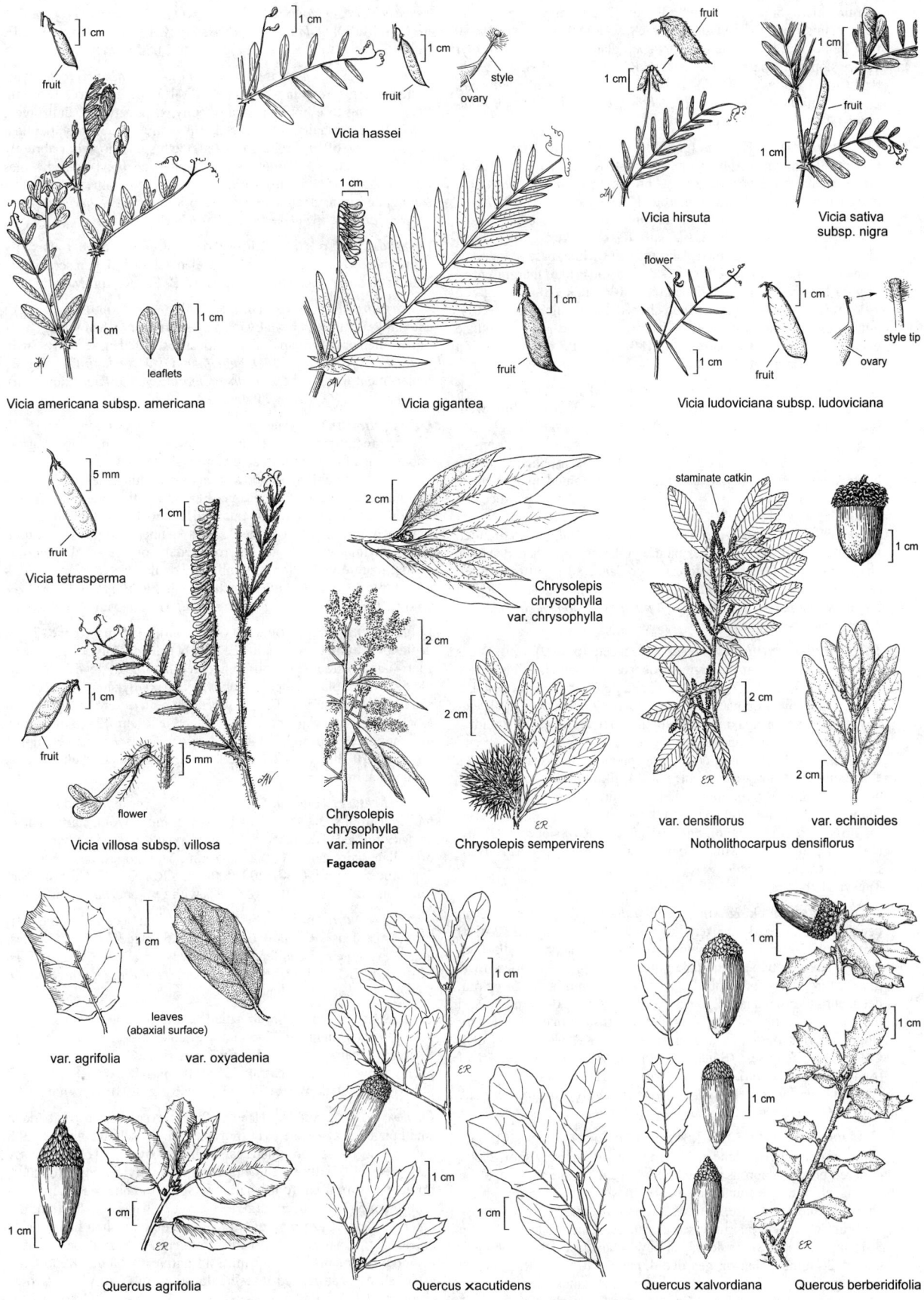

Vicia hassei

Vicia hirsuta

Vicia sativa
subsp. nigra

Vicia americana subsp. americana

leaflets

Vicia gigantea

fruit

flower

fruit

style tip

ovary

Vicia ludoviciana subsp. ludoviciana

Vicia tetrasperma

fruit

Vicia villosa subsp. villosa

flower

Chrysolepis
chrysophylla
var. chrysophylla

Chrysolepis
chrysophylla
var. minor

Fagaceae

Chrysolepis sempervirens

staminate catkin

var. densiflorus

var. echinoides

Notholithocarpus densiflorus

var. agrifolia

var. oxyadenia

leaves
(abaxial surface)

Quercus agrifolia

Quercus ✕acutidens

Quercus ✕alvordiana

Quercus berberidifolia

margin mucro- or spine-toothed. **FR:** cup 12–20 mm wide, 5–10 mm deep, hemispheric to bowl-shaped, thick, scales tubercled; nut 10–30 mm, gen ovoid, tip obtuse to acute, shell glabrous inside; mature yr 1. Dry slopes, chaparral; 100–1800 m. KR, NCoR, CaRH, SNF, Teh, ScV (Sutter Buttes), CW, SW; Baja CA. Hybridizes with *Q. durata*, *Q. engelmannii*, *Q. garryana* (*Q.* ×*howellii* J.M. Tucker), *Q. john-tuckeri*, *Q. lobata*. Feb–Apr

Q. cedrosensis C.H. Mull. (p. 809) CEDROS ISLAND OAK Tree to 5 m, decumbent shrub 2–3 m, or prostrate shrub to 2 dm, evergreen; trunk bark flaky, gray; twigs brown, hairy, dark gray in age; buds 1 mm, widely ovoid or subround, light brown, sparsely hairy. **LF:** 0.6–2(3.5) cm; petiole 1.5–2.5 mm; blade lanceolate, ovate, oblong, elliptic, or subround, adaxially flat or convex, glossy green, glabrous, abaxially glaucous, glabrous, veins white, base rounded or cordate, tip acute to widely rounded, gen spine-like, margin entire or with few irregular spine-tipped teeth. **FR:** stalk ± 0–10 mm; cup 7–12 mm wide, 5–6 mm deep, cup-shaped, scales thickened basally; nut 15–22 mm, narrowly ovoid to fusiform, tip acute, shell tomentose inside; mature yr 2. Chaparral; 100–1800 m. sw PR (Otay Mtn, San Diego Co.); Baja CA. Apr–May ★

Q. chrysolepis Liebm. (p. 809) MAUL OAK, CANYON LIVE OAK Shrub, tree < 20 m, evergreen; trunk bark in age narrowly furrowed, scaly, pale gray; twigs golden-tomentose, ± glabrous in age. **LF:** (1.5)3–6 cm, leathery; petiole 3–10 mm; blade oblong to oblong- or round-ovate, adaxially dark green, abaxially golden-puberulent, glabrous in age, dull, ± gray, tip acute to abruptly pointed, margin entire or spine-toothed. **FR:** cup 17–30 mm wide, 5–10 mm deep, saucer- to bowl-shaped, scales thick, ± tubercled to not, golden-tomentose; nut 25–30 mm, 14–20 mm wide, ± ovoid, oblong, or elliptic, tip rounded to pointed, shell woolly inside; mature yr 2. Canyons, shaded slopes, chaparral, mixed-evergreen forest, woodland; 30–2750 m. CA-FP (exc GV), e DMtns; OR, AZ, Baja CA. [*Q. c.* var. *nana* (Jeps.) Jeps.] Highly variable. Hybridizes with *Q. palmeri*, *Q. tomentella*, *Q. vacciniifolia*. Apr–May

Q. cornelius-mulleri Nixon & K.P. Steele (p. 809) MULLER'S OAK Shrub 1–2.5 m, evergreen, densely branched; twigs finely tomentose. **LF:** 2.5–3.5 cm, leathery; petiole 2–5 mm; blade oblong, ovate, or narrowly obovate, adaxially sparsely puberulent, dull, yellow- to gray-green, abaxially densely fine-tomentose, ± white, midrib yellow, tip acute to rounded, margin entire or 4–6-toothed. **FR:** cup 12–20 mm wide, 5–8 mm deep, hemispheric to cup-shaped, scales ± not tubercled, gray-canescent; nut 20–30 mm, elliptic to widely conic, tip obtuse, puberulent, shell glabrous inside; mature yr 1. Slopes, gen granitic soils, chaparral, pinyon woodland; 300–2140 m. s SNH, SnGb (n slope), SnBr (n slope), PR (e slope), s DMtns (Little San Bernardino Mtns), DSon (Eagle Mtns); Baja CA. Hybridizes with *Q. engelmannii* (*Q.* ×*acutidens*), *Q. lobata* (*Q.* ×*munzii* J.M. Tucker). Feb–Apr

Q. douglasii Hook. & Arn. (p. 809) BLUE OAK Tree 6–20 m, deciduous; trunk bark checkered into thin scales, ± gray. **LF:** 3–6(8) cm; petiole 3–9 mm; blade oblong to obovate, adaxially dull, blue-green, abaxially puberulent, pale blue-green, tip gen rounded, margin ± entire, wavy, or ± lobed. **FR:** cup 12–20 mm wide, 6–10 mm deep, cup- to bowl-shaped, scales ± tubercled; nut 20–35 mm, ovoid to subcylindric, tip pointed, shell glabrous inside; mature yr 1. Dry slopes, interior foothills, woodland; < 1590 m. KR, NCoRO, NCoRI, CaRF, SNF, Teh, ScV (Sutter Buttes), SnJV, SnFrB, SCoR, s ChI (Santa Catalina Island), WTR (n slope), MP. Hybridizes with *Q. garryana* (*Q.* ×*eplingii* C.H. Mull.), *Q. john-tuckeri* (*Q.* ×*alvordiana*), *Q. lobata* (*Q.* ×*jolonensis* Sarg.). Apr–May

Q. dumosa Nutt. (p. 809) NUTTALL'S SCRUB OAK Shrub 1–3 m, gen evergreen; twigs slender, 1–1.5 mm diam, sparsely short-hairy, dark red-brown, glabrous in age. **LF:** 1–2.5 cm; petiole < 5 mm; blade oblong, elliptic, or ± round, adaxially ± convex or not, ± shiny, green, abaxially fine-tomentose, in age glabrous, dull, pale green, tip obtuse to abruptly pointed, margin ± wavy or not, ± spine-toothed. **FR:** cup 8–15 mm wide, 5–8 mm deep, gen bowl-shaped, scales ± tubercled; nut 10–20 mm, ± slender, gen ovoid, tapered to tip, shell glabrous inside; mature yr 1. Gen sandy soils near coast, sandstone, chaparral, coastal-sage scrub; < 200 m. SCo, PR; Baja CA. Hybridizes with *Q.*

berberidifolia, *Q. douglasii*, *Q. engelmannii*, *Q. garryana* (*Q.* ×*eplingii* C.H. Mull.), *Q. lobata* (*Q.* ×*kinseliae* (C.H. Mull.) Nixon & C.H. Mull., *Q. dumosa* var. *kinseliae* C.H. Mull.). Mar–May ★

Q. durata Jeps. LEATHER OAK Shrub 1–3 m, evergreen; twigs tomentose, glabrous in age or not. **LF:** 1.5–3 cm; petiole < 5 mm; blade oblong to elliptic, adaxially convex, puberulent, dull green, abaxially short-hairy, pale green, tip spiny or abruptly pointed, margin wavy, often inrolled, toothed, teeth spine-tipped or abruptly pointed. **FR:** cup 12–18 mm wide, 4–6 mm deep, bowl-shaped, scales tubercled; nut 15–25 mm, ovoid to cylindric, tip obtuse or rounded, shell glabrous inside; mature yr 1. Hybridizes with *Q. berberidifolia*, *Q. garryana* (*Q.* ×*subconvexa* J.M. Tucker).

var. **durata** (p. 809) **LF:** adaxially strongly convex; abaxially short-hairy when young; margin inrolled. Chaparral, gen serpentine; 150–1500 m. NCoR, n SN, n CCo, SnFrB, SCoR. Apr–May

var. **gabrielensis** Nixon & C.H. Mull. SAN GABRIEL OAK **LF:** adaxially ± convex; abaxially densely short-hairy when young; margin inrolled or not. Chaparral, granitics; 450–1000 m. se WTR, SnGb (s slope). Hybridizes with *Q. engelmannii* (*Q.* ×*ewanii* I.M. Turner, replacement name for *Q.* ×*grandidentata* Ewan, a later homonym), *Q. lobata* Née (*Q.* ×*townei* Palmer). Apr–May ★

Q. engelmannii Greene (p. 809) ENGELMANN OAK Tree 5–18 m, evergreen; trunk bark in age narrowly furrowed, scaly, ± gray; young twigs finely tomentose, in age glabrous. **LF:** 2–6 cm; petiole 3–7 mm; blade oblong to obovate, adaxially dull blue-green, abaxially soft-hairy, glabrous in age, pale blue-green, tip obtuse to rounded, margin gen entire to wavy-dentate. **FR:** cup 10–15 mm wide, 6–8 mm deep, cup- to bowl-shaped, scales ± tubercled; nut 15–25 mm, oblong-cylindric to ovoid, tip rounded to obtuse, shell glabrous inside; mature yr 1. Slopes, foothills, woodland; < 1300 m. SCo, s ChI (1 tree on Santa Catalina Island), SnGb, PR; Baja CA. Hybridizes with *Q. berberidifolia*, *Q. cornelius-mulleri*. Apr–May ★

Q. garryana Hook. OREGON OAK Shrub 1–5 m to tree 8–20 m, deciduous; trunk bark thin, in age widely ridged, scaly, ± gray; twigs short-hairy, ± green, glabrous in age, red-brown. **LF:** 5–15 cm; petiole 5–25 mm; blade elliptic to obovate, adaxially shiny, dark green, abaxially short-hairy, dull, light green, tip obtuse to rounded, margin lobes 5–7 per lf, deep, entire or 2-toothed. **FR:** cup 12–25 mm wide, 4–9 mm deep, cup- to bowl-shaped, lower scales ± tubercled, upper ± not; nut 20–30 mm, ovoid to ± spheric, tip rounded, shell glabrous inside; mature yr 1.

var. **breweri** (Engelm.) Jeps. (p. 809) Shrub, spreading, with multiple trunks. **ST:** terminal buds 3–5 mm, ovoid, hairs sparse, brown or ± red. **LF:** blade 5–9 cm, abaxially with erect hairs, rays of stellate hairs gen 4–6, 0.25–0.5 mm. Mtn slopes, conifer forest, maritime chaparral; (150)1400–2000 m. KR, NCoRH, CCo (nw San Luis Obispo Co.); to OR. Hybridizes with *Q. sadleriana*. Apr–Jun

var. **garryana** (p. 809) Tree. **ST:** terminal buds 5–12 mm, fusiform, hairs dense, ± yellow or ± white. **LF:** blade 7–14 cm. Slopes, mixed-evergreen or conifer forest; 90–2140 m. NW, CaRF, SnFrB, SCoRO; to BC. Hybridizes with *Q. berberidifolia*, *Q. douglasii*, *Q. durata*, *Q. lobata*. Apr–Jun

var. **semota** Jeps. Shrub, small tree to 5 m, gen with multiple trunks. **ST:** terminal buds 2–5 mm. **LF:** blade 5–9 cm, abaxially with ± spreading hairs, rays of stellate hairs gen 6–8, < 0.3 mm. Dry slopes, open conifer forest, chaparral; 725–1800 m. KR, NCoRI, CaR, SN, SnFrB, SCoRO, ne WTR (Sierra Pelona Ridge), MP. Apr–Jun

Q. ilex L. HOLLY OAK Tree to 20 m, evergreen; trunk bark in small plates, gray-black; twigs gray-brown, densely woolly. **LF:** 3–8 cm; petiole 10–25 mm; blade lanceolate to ovate, adaxially glossy dark green, abaxially ± yellow or ± white woolly, base gen rounded, tip acute, margin entire to shallowly spine-toothed. **FR:** cup 7–15 mm wide, 5–7 mm deep, thin, top- or cup-shaped, scales thin, not tubercled; nut 15–25 mm, ovoid to conic-oblong, shell hairy inside at least at tip; mature yr 1. Escaped from cult, canyon slopes, washes near developments; 50–400 m. SW; native to Medit. Reportedly established as naturalized at several locations in s CA. Widely grown as an orn in CA. May–Aug

Q. john-tuckeri Nixon & C.H. Mull. (p. 809) TUCKER'S OAK
Shrub 2–5 m or ± tree < 7 m, evergreen; young twigs finely tomentose. **LF:** 1.3–2.8 cm; petiole 1–4 mm; blade oblong, elliptic, or
obovate, adaxially dull, gray-green, abaxially fine-hairy, pale gray-
green, base rounded to widely wedge-shaped, tip obtuse to rounded,
margin irregularly spine-toothed. **FR:** cup 10–15 mm wide, 5–7
mm deep, thin, obconic to hemispheric, scales ± tubercled to not;
nut 20–30 mm, ovoid to conic, tapered to tip, shell glabrous inside;
mature yr 1. Slopes on desert borders, chaparral, pinyon/juniper
woodland; 900–2090 m. s SN, Teh (e slope), SCoRI, WTR (n slope),
SnGb (n slope), sw edge DMoj. Hybridizes with *Q. berberidifolia*,
Q. douglasii. Feb–Apr

Q. kelloggii Newb. (p. 809) CALIFORNIA BLACK OAK Tree < 25
m, deciduous; trunk bark deeply furrowed in age, checkered, dark
gray-brown to black. **LF:** (6)9–20 cm; petiole (3)10–40 mm; blade
widely elliptic, obovate, or ± round, adaxially glabrous, bright green,
abaxially finely tomentose, ± glabrous in age, pale green, tip gen
acute, bristled, margin lobes gen 6 per lf, with 1–4 coarse, gen bristle-
tipped teeth. **FR:** cup 16–25 mm wide, 15–25 mm deep, gen cup-
shaped, scales thin, not tubercled, glabrous to puberulent; nut 20–35
mm, oblong-ovoid, puberulent, tip gen obtuse, shell woolly inside;
mature yr 2. Slopes, valleys, woodland, conifer forest; 30–2660 m.
CA-FP (exc GV, SCo, ChI), MP; OR, Baja CA. Hybridizes with
Q. agrifolia, *Q. parvula* var. *p.*, *Q. parvula* var. *shrevei*, *Q. wislizeni*
(*Q. ×morehus* Kellogg). Apr–May

Q. lobata Née (p. 809) VALLEY OAK, ROBLE Tree < 35 m, deciduous; trunk bark in age deeply checkered into ± square sections, ±
light gray. **LF:** 5–12 cm; petiole 5–12 mm; blade obovate, adaxially often ± shiny, dark green, abaxially fine-tomentose, dull to pale
green, tip obtuse to rounded, margin lobes 6–10 per lf, deep, obtuse,
gen coarsely 2–3-toothed at tip. **FR:** cup 14–30 mm wide, 10–30
mm deep, hemispheric, scales tubercled; nut 30–50 mm, 12–20 mm
wide, gen long-conic, tip pointed, shell glabrous inside; mature yr 1.
Slopes, valleys, savanna; < 1830 m. KR, NCoR, CaRF, SNF, s SNH,
Teh, GV, SnFrB, SCoR, nw SCo, ChI (Santa Cruz, Santa Catalina
islands), WTR, w SnGb. [*Q. l.* var. *turbinata* Jeps.; *Q. l.* var. *walteri*
Jeps.] Hybridizes with *Q. berberidifolia*, *Q. cornelius-mulleri*, *Q.
douglasii*, *Q. engelmannii*, *Q. garryana*, *Q. john-tuckeri*, *Q. pacifica*
(*Q. ×macdonaldii*). Mar–Apr

Q. ×macdonaldii Greene (p. 809) MACDONALD OAK Tree 5–15
m, deciduous; trunk bark scaly, ± gray; twigs tomentose. **LF:** 4–7
cm; petiole 3–10 mm; blade oblong to obovate, adaxially glabrous to
sparsely hairy, green, ± shiny, abaxially densely appressed-stellate-
hairy, ± pale green, tip obtuse to rounded, margin lobes 2–6(8) per lf,
shallow, gen pointed. **FR:** cup 10–20 mm wide, 6–10 mm deep, hemispheric, scales tubercled, canescent; nut 20–35 mm, conic-oblong to
ovoid, tip acute, shell glabrous inside; mature yr 1. Slopes, canyons,
woodland; < 600 m. ChI (Santa Cruz, Santa Rosa, Santa Catalina
islands). Considered a sp. by Greene but derived from hybrids involving *Q. pacifica*, *Q. lobata*, perhaps others; needs study. Mar–May

Q. pacifica Nixon & C.H. Mull. (p. 811) ISLAND SCRUB OAK
Shrub to 2 m (small tree to 5 m), gen evergreen; twigs finely hairy,
± red or ± brown, glabrous in age, gray. **LF:** 1.5–4 cm; petiole 2–5
mm; blade obovate or oblong, adaxially green, abaxially light green
with minute appressed stellate hairs, in age glabrous, base gradually
tapered, wedge-shaped, or rounded, tip gen rounded, margin entire,
wavy, or ± toothed, teeth gen mucronate. **FR:** cup 8–20 mm wide,
5–15 mm deep, hemispheric to top-shaped, scales moderately to
strongly tubercled; nut 20–30 mm, ovoid to cylindric, shell glabrous
inside; mature yr 1. Slopes, ridges, canyons, chaparral, coastal scrub,
oak woodland, pine forest; < 610 m. ChI (Santa Cruz, Santa Rosa,
Santa Catalina islands). Hybridizes with *Q. lobata*. Mar–Apr ★

Q. palmeri (Engelm.) Engelm. (p. 811) PALMER'S OAK Shrub
2–6 m, evergreen; twigs spreading, rigid. **LF:** 1–3 cm, stiff; petiole
2–5 mm; blade elliptic to round-ovate, adaxially glabrous to sparsely
puberulent, ± shiny, olive-green, abaxially densely glandular-puberulent when young, glabrous, pale gray-green in age, tip gen spine-
toothed, margin wavy, strongly spine-toothed. **FR:** cup 10–25 mm
wide, 6–12 mm deep, gen bowl-shaped, rim ± spreading, scales not
tubercled, densely hairy; nut 20–30 mm, ± ovoid, tip gen obtuse,

shell densely woolly inside; mature yr 2. Uncommon. Rocky slopes,
flats; 300–1600 m. e NCoRI (Colusa Co.), s SNH (e slope), nw SnJV
(Alameda, Contra Costa cos.), SnFrB (Alameda Co.), SCoR, WTR,
SnGb (n slope), SnBr (n slope), e PR, s edge DMoj, DMtns (Little
San Bernardino Mtns); AZ, Baja CA. [*Q. dunnii* Kellogg, illeg.]
Hybridizes with *Q. chrysolepis*. Apr–May

Q. parvula Greene Shrub 1–6 m or tree < 17 m, evergreen. **LF:**
3–9(14) cm; petiole 2–10(15) mm; blade oblong, lanceolate, or ovate
to obovate, adaxially glabrous, olive-green to dark green, abaxially
glabrous, gen ± dull, light olive-green, tip obtuse to acute or acuminate, margin spine-toothed (or long-tapered-dentate) to gen entire.
FR: cup 12–15 mm wide, 6–10 mm deep, gen bowl-shaped, scales
± thin, not tubercled; nut (15)30–45 mm, barrel-shaped to ovoid,
abruptly tapered from above middle to tip, puberulent, shell woolly
inside; mature yr 2.

var. ***parvula*** (p. 811) SANTA CRUZ ISLAND OAK Shrub 1–2
m. **LF:** gen entire. Canyons, slopes, chaparral, woodland; < 500 m.
SCoRO (Santa Barbara Co.), n ChI (Santa Cruz Island), e WTR.
Mar–May ★

var. ***shrevei*** (C.H. Mull.) Nixon (p. 811) SHREVE OAK Tree <
17 m. ± moist woodland; < 1190 m. s NCo, NCoRI, CW (exc SCoRI),
WTR. Hybridizes with *Q. agrifolia*, *Q. kelloggii*. Treated as a synonym of *Q. wislizeni* by Jensen in FNANM 3:452 (1997). Mar–May

var. ***tamalpaisensis*** S.K. Langer TAMALPAIS OAK Shrub 1–6
m. **LF:** 8–14 cm; petiole 5–15 mm; margin long-tapered-dentate.
Understory conifer woodland; 100–750 m. SnFrB (Mount Tamalpais). Mar–Apr ★

Q. sadleriana R. Br. ter (p. 811) DEER OAK Shrub 1–3 m,
evergreen; twigs glabrous. **LF:** 7–11(18) cm; petiole 10–20 mm;
blade elliptic to oblong-obovate, adaxially green, ± shiny, abaxially
sparsely fine-appressed-hairy, pale green, lateral veins (of larger lvs)
gen 20–28, gen straight and prominent, tip ± acute, margin serrate,
teeth 20–32. **FR:** cup 10–18 mm wide, 7–9 mm deep, thin, cup-
shaped to obconic, scales ± tubercled to not; nut 15–20 mm, elliptic
to ± spheric, tip rounded, shell glabrous inside; mature yr 1. Open, rocky
slopes, ridges, conifer forest; 600–2200 m. KR; sw OR. Hybridizes
with *Q. garryana* var. *breweri*. Apr–Jun

Q. tomentella Engelm. (p. 811) ISLAND OAK Tree < 20 m, evergreen; trunk bark in age furrowed, scaly, gray or red-brown; young
twigs tomentose. **LF:** 5–8 cm; petiole 5–18 mm; blade oblong to
oblong-ovate, adaxially ± finely tomentose, glabrous in age, dark
green, abaxially densely tomentose, sparsely tomentose in age, dull,
gray-green, tip acute to obtuse, margin entire to crenate or mucro-
toothed. **FR:** cup 20–30 mm wide, 6–8 mm deep, saucer- to bowl-
shaped, scales thick, tubercled; nut 20–35 mm, widely ovoid, tip
rounded, shell ± woolly inside; mature yr 2. Canyons, slopes, woodland; < 600 m. ChI; Baja CA (Guadalupe Island). Hybridizes with *Q.
chrysolepis*. Apr–May ★

Q. turbinella Greene (p. 811) SHRUB LIVE OAK Shrub 2–5 m
or ± tree < 7 m, evergreen; twigs densely fine-tomentose. **LF:** 1.5–3
cm; petiole 1–3 mm; blade oblong to elliptic, adaxially dull, gray-
green, abaxially with appressed-stellate and glandular, ± yellow hairs,
base rounded to subcordate, tip acute to obtuse, margin gen regularly
spine-toothed. **FR:** stalk 10–15 mm; cup 9–12 mm wide, 4–6 mm
deep, ± hemispheric, scales ± tubercled to not, thin; nut 12–23 mm,
cylindric-ovoid to elliptic, abruptly tapered to tip, gen yellow-brown,
shell glabrous inside; mature yr 1. Pinyon/juniper woodland; 1200–
2000 m. e DMoj (exc DMtns other than New York Mtns); to CO, TX,
Baja CA. Apr–Jun ★

Q. vacciniifolia Kellogg (p. 811) HUCKLEBERRY OAK Shrub <
1.5 m, prostrate to spreading, evergreen; twigs slender, pliable, glabrous. **LF:** 1.5–4 cm; petiole 3–6 mm; blade ± oblong, adaxially glabrous, green, abaxially glabrous, dull, pale green, tip obtuse to acute,
margin entire to mucro-toothed. **FR:** cup 10–12 mm wide, 4–6 mm
deep, thin, gen cup-shaped, scales ± tubercled to not; nut 10–15 mm,
ovoid to ± spheric, tip rounded; shell thin, subglabrous to sparsely
tomentose inside; mature yr 2. Steep slopes, ridges, conifer forest,
subalpine; 150–2930 m. KR, NCoRH, NCoRI, CaRH, SNH, MP;
OR, NV. Hybridizes with *Q. chrysolepis*. May–Jul

Q. wislizeni A. DC. INTERIOR LIVE OAK Shrub 2–4(6) m or tree gen 10–22 m, evergreen; trunk bark in age furrowed, ± checkered, ± gray. **LF**: 2–5 cm; petiole 3–15 mm; blade gen oblong to elliptic or lanceolate, adaxially glabrous, shiny, gen dark green, abaxially glabrous, ± shiny, yellow-green, tip gen acute, abruptly pointed, margin entire to spine-toothed, rarely wavy. **FR**: cup 12–18 mm wide, 12–16 mm deep, cup-shaped to hemispheric, scales not tubercled, ± thin; nut 20–40 mm, cylindric-ovoid, ovoid, or ± obconic, gradually tapered from below middle to tip, shell woolly inside; mature yr 2.

var. ***frutescens*** Engelm. (p. 811) Shrub. **LF**: blade 1.8–4 cm. Valleys, chaparral; 90–2000 m. KR, NCoR, CaRH, SNH, Teh, ScV (Sutter Buttes), SnJV, SnFrB, SCoR, SW (exc ChI); Baja CA. Mar–May

var. ***wislizeni*** (p. 811) Tree. **LF**: blade 2–5 cm. Interior canyons, slopes, pine/oak woodland; < 1600 m. KR, NCoR, CaRF, SNF, s SNH, Teh, ScV (Sutter Buttes), SnFrB, SCoR, TR, PR, SNE; Baja CA. Hybridizes with *Q. agrifolia*, *Q. kelloggii*. Mar–May

FOUQUIERIACEAE OCOTILLO FAMILY

Lisa M. Schultheis & William J. Stone

Shrub, tree, spiny. **ST**: branched near base [or trunk 1, thick, fleshy]. **LF**: simple, alternate, small, ± fleshy, glabrous, gen produced after rains, of 2 kinds: 1° on long-shoots, subtended by decurrent ridge on st, blade gen early-deciduous, petiole persisting as spine; 2° clustered on short-shoot in axils of 1°, ± persistent, not spine-forming. **INFL**: spike, raceme, or panicle, axillary or terminal; fls many. **FL**: sepals 5, unequal, overlapping, ± scarious, persistent; corolla bright red [yellow], tube cylindric, lobes 5, overlapping, spreading; stamens 10–20+, in 1 or ± 2 whorls, filaments free; pistil 1, ovary superior, 3(4)-chambered at base, placenta axile at base, parietal above, ovules ± 3–6 per chamber, style 3(4)-lobed. **FR**: capsule, loculicidal. **SEED**: elliptic, angled or winged. 1 genus (incl *Idria*), 11 spp.: sw US, Mex. [Schultheis & Baldwin 1999 Amer J Bot 86:578–589] Scientific Editor: Thomas J. Rosatti.

FOUQUIERIA

(P.E. Fouquier, Parisian medical professor, 1776–1850)

F. splendens Engelm. subsp. ***splendens*** (p. 811) OCOTILLO **STS**: erect to outwardly arching or ascending, 6–100, 2–10 m, gen < 6 cm diam, cane-like, lfless most of yr; bark gray with darker furrows; spines 1–4 cm. **LF**: 1° 1–5 cm, petiole 1–2.5 cm; 2° 2–6 per cluster, 1–2 cm, 4–9 mm wide, petiole 2–8 mm, blade spoon-shaped to obovate, tip rounded to notched. **INFL**: panicle, gen 10–20 cm, widely to narrowly conic. **FL**: corolla 1.8–2.5 cm; filaments each arising from hairy, ± 5 mm sheath. **FR**: ± 2 cm. $2n=24$. Dry, gen rocky soils; < 700 m. DSon; to TX, c Mex, Baja CA. Sts used for fences, huts; bark for waxes, gums. Mar–Jul

FRANKENIACEAE FRANKENIA FAMILY

Carrie Kiel & R. John Little

1 genus. Scientific Editor: Bruce G. Baldwin.

FRANKENIA FRANKENIA

(Ann) [per] to shrub, gen with rhizome; glands salt-secreting. **ST**: prostrate to erect, nodes swollen, gen rooting. **LF**: opposite, 4-ranked, united by a single sheath, ± clustered, simple, 1–15 mm, salt-encrusted, gen persistent; blade obovate to oblong-oblanceolate or elliptic, gen leathery or fleshy, glabrous to hairy, margin entire, rolled under; stipules 0. **INFL**: gen cyme, axillary. **FL**: gen bisexual, radial; calyx persistent, sepals 4–7, fused, ribbed; petals 4–7, free, white to pink or blue-purple, overlapping, clawed, with adaxial scale-like appendage near base; stamens 3–12(25), 2-whorled, outer whorl shorter; ovary superior, chambers 1–4, styles 1–4, exserted, ovules 1–many. **FR**: loculicidal capsule within persistent calyx. **SEED**: 1–many, ivory to gold-brown. 90 spp.: esp coastal and semi-arid regions; soils high in salt, gypsum, or calcium. (Johann Frankenius, colleague of Linnaeus) [Whalen 1987 Syst Bot Monogr 17:1–93]

1. Ann; st 1–8 cm . . . . [*F. pulverulenta*]
1′ Subshrub or shrub; st 10–60 cm
  2. Lf blade 2–7 mm, gen < 1 mm wide, margins strongly thickened, tightly rolled under, lower surface obscured; petals 3–7 mm . . . . *F. palmeri*
  2′ Lf blade 4–15 mm, 1–6 mm wide, margins not strongly thickened, weakly rolled under, lower surface ± exposed; petals 6–14 mm . . . . *F. salina*

F. palmeri S. Watson PALMER'S FRANKENIA Shrub 10–100 cm diam, rhizome 0. **ST**: prostrate, 10–30 cm, gen with scattered hairs. **LF**: blade 2–7 mm, gen < 1 mm wide, ± fleshy, margins strongly thickened, tightly rolled under, abaxial surface densely hairy. **INFL**: few-fld in distal axils. **FL**: calyx tube 3–5 mm, petals 3–7 mm, white, lower 1/2 gen ± pink; stamens gen 4, 4–9 mm; styles (1)2. **FR**: 2–2.5 mm. **SEED**: 1, 1.5–2 mm, ovoid-conic. Alkali flats, coastal marshes, dunes; < 450 m. SCo; n Mex. Apr–Sep ★

F. salina (Molina) I.M. Johnst. (p. 811) ALKALI HEATH Matted subshrub < 3 m diam. **ST**: ± prostrate, 1–6 dm, glabrous to hairy. **LF**: blade 4–15 mm, 1–6 mm wide, margins not strongly thickened, weakly rolled under, leathery, abaxially glabrous to sparsely hairy. **INFL**: 1 to cymose in upper axils. **FL**: calyx tube 4–9 mm; petals 6–14 mm, white to pink or blue-purple; stamens gen 6, 5–12 mm; styles 3(4). **FR**: 3–5 mm. **SEED**: 1–20, 1–1.5 mm, ± ellipsoid. Salt marshes, alkali flats; < 750 m. GV, CCo, SCo, ChI, SNE, DMoj; to NV, Mex, S.Am. Apr–Sep

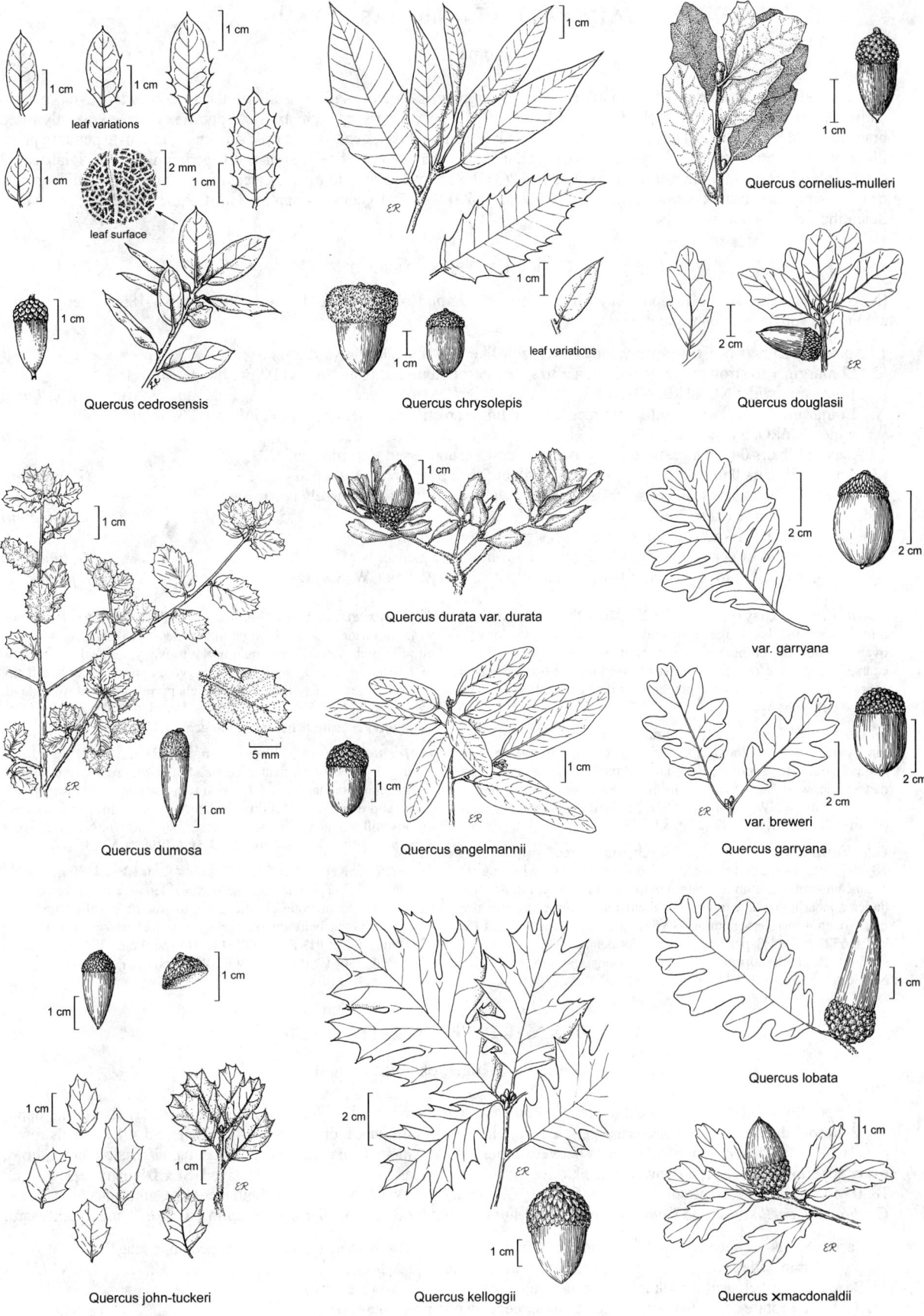

leaf variations

1 cm

2 mm

leaf surface

1 cm

Quercus cedrosensis

ER

1 cm

leaf variations

1 cm

1 cm

Quercus chrysolepis

1 cm

Quercus cornelius-mulleri

2 cm

ER

Quercus douglasii

1 cm

ER

5 mm

1 cm

Quercus dumosa

1 cm

1 cm

Quercus durata var. durata

1 cm

ER

Quercus engelmannii

2 cm

2 cm

var. garryana

2 cm

2 cm

ER

var. breweri

Quercus garryana

1 cm

1 cm

1 cm

1 cm

1 cm

ER

Quercus john-tuckeri

2 cm

ER

1 cm

Quercus kelloggii

1 cm

Quercus lobata

1 cm

ER

Quercus ✕macdonaldii

GARRYACEAE SILK TASSEL FAMILY

Thomas F. Daniel

Shrub, small tree; dioecious. **LF**: simple, opposite, evergreen, petioled; blade ± leathery, flat to concave-convex, margin entire, flat, rolled under, or strongly wavy. **INFL**: catkin-like, pendent; fls small, in axils of opposite, 4-ranked, basally fused bracts. **STAMINATE FL**: (1)3(4) per bract, pedicelled; perianth parts 4, gen fused at tips; stamens 4, alternate perianth parts, filaments free, anthers 2-chambered. **PISTILLATE FL**: 1–3 per bract; pedicel ± 0 or short; perianth 0 or vestigial with 2 small appendages; ovary inferior, chamber 1, styles 2(3). **FR**: berry, spheric to ovoid, green, fleshy, turning dark blue, black, or white-gray, dry, brittle, not or irregularly dehiscent. **SEED**: gen 2. 1 genus, 14 spp.: w US, C.Am, Caribbean; some cult. Scientific Editor: Thomas J. Rosatti.

GARRYA SILK TASSEL BUSH

(N. Garry, 1st secretary of Hudson's Bay Co., explorer of Pacific Northwest with David Douglas, 1782–1856) Intergradation among CA spp. raises doubt about taxonomic status.

1. Abaxial lf hairs dense, curly-wavy, interwoven, felt-like, not appressed toward lf tip
  2. Lf margin ± to strongly wavy and often ± to strongly rolled under; pistillate fls (1)3 per bract; infl in fr 18–28 mm wide; NCo, NCoRO, n SNF, ScV, CCo, SnFrB, SCoRO . . . . *G. elliptica*
  2′ Lf margin flat to ± wavy and sometimes ± rolled under; pistillate fls 1(3) per bract; infl in fr 13–18 mm wide; SCoRO, SW . . . . *G. veatchii*
1′ Abaxial lf hairs 0 to dense, straight to wavy, not felt-like, appressed toward lf tip
  3. Abaxial lf hairs 0 (or sparse on young lvs); fr glabrous or near tip sparsely hairy . . . . *G. fremontii*
  3′ Abaxial lf hairs sparse to dense; fr dense-hairy, glabrous, or near tip sparsely hairy
    4. Abaxial lf hairs dense, fine, wavy . . . . *G. congdonii*
    4′ Abaxial lf hairs sparse to dense, ± coarse, straight to ± curved
     5. Fr glabrous or near tip sparsely hairy; abaxial lf hairs gen dense; NW . . . . *G. buxifolia*
     5′ Fr gen dense-hairy; abaxial lf hairs sparse to ± dense; NW, SN, CW, SW, DMtns . . . . *G. flavescens*

G. buxifolia A. Gray (p. 815)   Shrub < 3 m. **LF**: 14–66 mm, 9–33 mm wide, 1.3–2.3 × longer than wide, flat to ± concave-convex, ovate- to obovate-elliptic; margin flat; abaxial hairs gen dense, ± coarse, straight to ± curved, appressed toward lf tip. **FR**: glabrous or near tip sparsely hairy. Chaparral to yellow-pine forest; 150–2100 m. NW; w OR. Feb–Apr

G. congdonii Eastw. (p. 815)   Shrub < 3 m. **LF**: 15–70 mm, 7–35 mm wide, 1.3–2.6 × longer than wide, flat to ± concave-convex, ovate- to obovate-elliptic; margin flat to strongly wavy; abaxial hairs dense, fine, wavy, appressed toward lf tip. **FR**: hairs dense. Chaparral; 180–750 m. NW, CaRH, n&c SNF, SnFrB. Seems to intergrade equally with *G. elliptica, G. flavescens*. Feb–Apr

G. elliptica Lindl. (p. 815)   Shrub, small tree, < 8 m. **LF**: 18–107 mm, 14–72 mm wide, 1.2–2.6 × longer than wide, flat to ± concave-convex, elliptic; margin ± to strongly wavy, often rolled under, appearing crenate or dentate; abaxial hairs dense, curly-wavy, interwoven, sparse in age or not. **INFL**: in fr 18–28 mm wide. **PISTILLATE FL**: (1)3 per bract. **FR**: hairs dense. 2*n*=22. Seacliffs, sand dunes, chaparral, foothill-pine woodland; < 800 m. NCo, KR, NCoRO, n SNF, ScV (cult?), CW; w OR. Jan–Mar

G. flavescens S. Watson (p. 815)   Shrub < 3 m. **LF**: 19–75 mm, 9–45 mm wide, 1.3–3.3 × longer than wide, flat to ± concave-convex, elliptic to obovate-elliptic; margin flat to wavy; abaxial hairs sparse to ± dense, ± coarse, straight to ± curved, appressed toward lf tip. **FR**: hairs gen dense. Desert slopes, chaparral, pine/oak woodland; 650–2350 m. NW, SN, SnJV, CW, SW, DMoj; to UT, AZ, n Baja CA. Variation in hairiness not related to geog in CA. Feb–Apr

G. fremontii Torr.   Shrub < 3 m. **LF**: 21–120 mm, 11–70 mm wide, 1.3–2.4 × longer than wide, gen flat, ovate- to obovate-elliptic; margin flat; abaxial hairs 0 (or sparse on young lvs), straight to ± wavy, appressed toward lf tip. **FR**: hairs 0 or near tip sparse. Chaparral, foothill woodland, montane forest; 300–2300 m. NW, CaR, n&s SNF, n&c SNH, ScV, SnFrB, PR, Wrn; to WA, w NV. Jan–Apr

G. veatchii Kellogg (p. 815)   Shrub < 2 m. **LF**: 24–90 mm, 8–50 mm wide, 1.3–3.9 × longer than wide, flat to ± concave-convex, lance-ovate to obovate-elliptic; margin flat to ± rolled under or ± wavy; abaxial hairs dense, curly-wavy, interwoven. **INFL**: in fr 13–18 mm wide. **PISTILLATE FL**: 1(3) per bract. **FR**: hairs dense. Chaparral; 250–1750 m. SCoRO, SW; n Baja CA. Feb–Apr

GENTIANACEAE GENTIAN FAMILY

James S. Pringle, except as noted

Ann to per [to trees]. **ST**: decumbent to erect, < 2 m, simple or branched. **LF**: simple, cauline, sometimes also basal, opposite or whorled, entire, sessile or basal ± petioled; stipules 0. **FL**: bisexual, radial, parts in 4s or 5s exc pistil 1; sepals fused, persistent; petals fused, ± persistent, sinus between lobes often unappendaged; stamens epipetalous, alternate corolla lobes; ovary superior, chamber 1, placentas parietal, often intruding, stigmas 1–2. **FR**: capsule, 2-valved. **SEED**: many. ± 90 genera, 1800 spp.: worldwide; some cult (*Eustoma, Exacum, Gentiana*). [Struwe & Albert 2002 Gentianaceae. Cambridge Univ Press] *Gentianella tenella* moved to *Comastoma*. Key to genera revised by Bruce G. Baldwin. Scientific Editor: Thomas J. Rosatti.

1. Base of sinus between corolla lobes truncate or gen with a variously shaped, sometimes fringed appendage, gen > 1 mm wide . . . . **GENTIANA**
1′ Base of sinus between corolla lobes unappendaged, acute, << 1 mm wide
  2. Corolla ± rotate (bell-shaped), lobed ± to base, nectary pits prominent, fringed
    3. Calyx, corolla 4-lobed; cauline lvs opposite or whorled; nectary pits gen 1 per corolla lobe (if 2, st gen > 1 m, > 1.5 cm diam) . . . . **FRASERA**

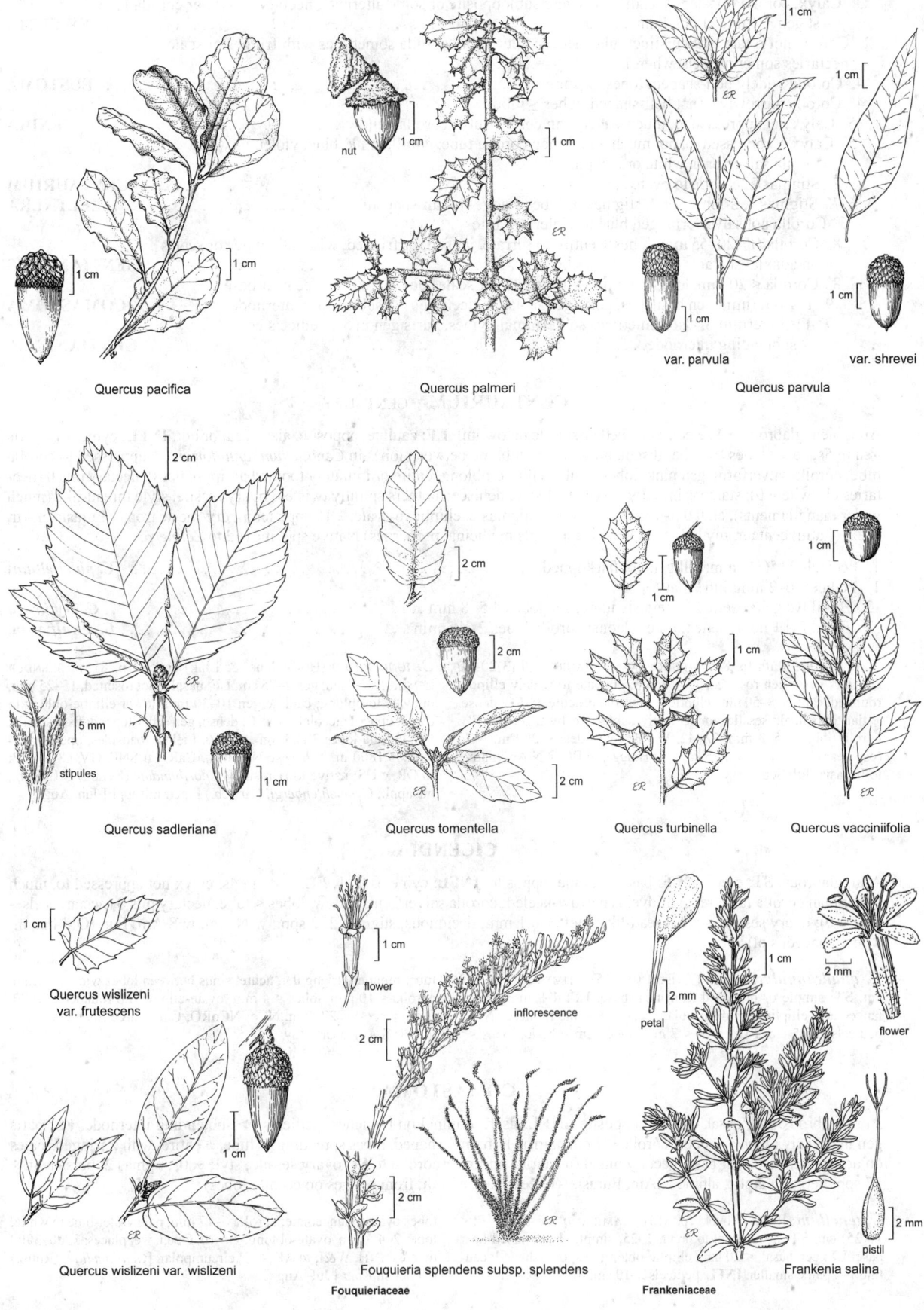

811

Quercus pacifica

nut

Quercus palmeri

var. parvula

var. shrevei

Quercus parvula

stipules

Quercus sadleriana

Quercus tomentella

Quercus turbinella

Quercus vacciniifolia

Quercus wislizeni
var. frutescens

Quercus wislizeni var. wislizeni

flower

inflorescence

Fouquieria splendens subsp. splendens
Fouquieriaceae

petal

flower

pistil

Frankenia salina
Frankeniaceae

3′ Calyx, corolla (4)5-lobed; cauline lvs gen (sub)opposite or some alternate; nectary pits 2 per corolla lobe
— st gen < 1 m, < 1 cm diam . . . . **SWERTIA**
2′ Corolla not rotate, with distinct tube, nectary pits 0 (but corolla sometimes with fringes or scales,
nectaries sometimes elsewhere)
  4. Corolla widely bell-shaped, lobes >> tube . . . . **EUSTOMA**
  4′ Corolla narrowly or not bell-shaped, lobes ≤ tube
    5. Calyx not appressed to, much wider than corolla tube; corolla yellow . . . . **CICENDIA**
    5′ Calyx ± appressed to, not much wider than corolla tube; corolla pink, blue, violet, or white
      6. Corolla salverform, white or gen pink
        7. Stigmas 2, elliptic to ovate . . . . **CENTAURIUM**
        7′ Stigmas 1, 2-lobed, or 2, stigmas or lobes wedge- to fan-shaped . . . . **ZELTNERA**
      6′ Corolla not salverform, gen blue to violet or white
        8. Corolla (17)20–55 mm, lobes ± entire to serrate, jagged, or fringed, without fringes or scales
on adaxial surface. . . . . **GENTIANOPSIS**
        8′ Corolla < 20 mm, lobes ± entire, with fringes or scales on adaxial surface near base
          9. Fls 1, terminal on branches, parts gen in 4s; pedicels gen >> subtending internode . . . . **COMASTOMA**
          9′ Fls in terminal, also sometimes subterminal cymes, parts gen in 5s; pedicels gen
< subtending internode. . . . . **GENTIANELLA**

CENTAURIUM CENTAURY

Ann, bien, glabrous. **ST**: erect, branched or simple below infl. **LF**: cauline, opposite, also basal or not. **INFL**: cyme. **FL**: parts gen in 5s; calyx lobes >> tube (discounting thin membrane between lobes in *Centaurium tenuiflorum*), ± appressed to corolla tube; corolla salverform, gen pink, lobes < tube, elliptic-oblong, entire or minutely toothed at tip, scales 0, nectary pits 0 (nectaries elsewhere 0); stamens initially curved to 1 side, dehisced anthers spirally twisted; ovary sessile, style thread-like (much wider than filaments), cleft 0.5–1 mm, deciduous, stigmas 2, elliptic to ovate. ± 15 spp.: temp, dry-mesic trop, ± Eurasia, n Afr, Mex. (Latin: centaur, mythological discoverer of pls medicinal properties) Native spp. moved to *Zeltnera*.

1. Pedicels 1–5(11) mm; infl gen not flat-topped. . . . . *[C. pulchellum]*
1′ Pedicels 0–2 mm; infl ± flat-topped
  2. Basal lvs gen rosetted, ± persistent; corolla lobes 4.5–8 mm . . . . ***C. erythraea***
  2′ Basal lvs ± not rosetted, ± deciduous; corolla lobes 2–4.5 mm . . . . ***C. tenuiflorum***

C. erythraea Rafn (p. 815) EUROPEAN CENTAURY Pl (3)20–60 cm. **LF**: basal gen rosetted, 15–70 mm, obovate to widely elliptic, rounded; cauline 8–50 mm, elliptic to lanceolate, acute. **INFL**: dense, ± flat-topped; fls sessile, immediately subtended by 2 bracts. **FL**: corolla lobes 4.5–8 mm. 2*n*=42. Fields, roadsides; < 200 m. NCo (esp near Crescent City, Eureka, Fort Bragg); to BC, e N.Am; native to Eurasia. Jul–Sep

C. tenuiflorum (Hoffmanns. & Link) Janch. (p. 815) SLENDER CENTAURY Ann, gen 2–75 cm. **LF**: basal ± not rosetted, 15–25 mm, obovate to oblong; cauline gen 10–30 mm, lower elliptic-oblong to ovate, upper lanceolate. **INFL**: dense, gen ± flat-topped; fls ± sessile. **FL**: corolla lobes 2–4.5 mm. 2*n*=40. Fields, roadsides, open woodland; < 1800 m. NW (exc NCoRH), CaRF, n SNF, GV, CCo, MP; to OR, c US; native to Eurasia. [*C. floribundum* (Benth.) B.L. Rob., misappl.; *C. muehlenbergii* (Griseb.) Piper, misappl.] Jun–Aug

CICENDIA

Ann, glabrous. **ST**: erect, 1. **LF**: basal, cauline, opposite. **INFL**: cyme or fls 1. **FL**: parts in 4s; calyx not appressed to, much wider than corolla tube, sepal midveins narrow-keeled; corolla salverform, yellow, lobes < tube, nectary pits 0 (nectaries elsewhere 0); ovary sessile, style thread-like, cleft ± 0.1 mm, deciduous, stigmas 2. 2 spp.: w N.Am, w S.Am, Eur. (Old Italian: a gentianaceous pl)

C. quadrangularis (Lam.) Griseb. (p. 815) TIMWORT Pl < 9 cm. **ST**: simple or branched from near base. **LF**: 4–9 mm, ovate, lanceolate, elliptic, or oblanceolate; cauline < internodes. **INFL**: pedicels < 5.5 cm. **FL**: calyx < 7 mm, tube hemispheric, lobes << tube, minutely triangular, acute, sinus between lobes wide, truncate; corolla < 10 mm, lobes < 4 mm, ovate-elliptic, tip rounded. 2*n*=26. Open places; < 2700 m. NCo, NCoRO, CaRF, n&c SNF, GV, CCo; to OR; S.Am. Mar–May

COMASTOMA

Ann, glabrous. **LF**: basal, cauline, opposite. **INFL**: fls 1, terminal on branches; pedicels >> subtending internode. **FL**: parts gen in 4s; calyx tube << lobes; corolla < 2 cm, narrowly funnel-shaped, lobes spreading, < tube, ± entire, with 2 fringed scales on adaxial surface near base, nectary pits 0 (nectaries on lower corolla tube); ovary sessile, style ± 0, stigmas 2, persistent. ± 17 spp.: temp to arctic, alpine N.Am, Eurasia. (Greek: hair, mouth, from fringes on corolla lobes)

C. tenellum (Rottb.) Toyok. (p. 815) SAMILAND GENTIAN Pl < 15(25) cm. **ST**: decumbent to erect, 1–25, simple or branched near base. **LF**: gen basal, < 20 mm, elliptic-oblong to spoon-shaped; cauline 1–4 pairs, smaller. **INFL**: pedicels 2–10 cm. **FL**: calyx 4–11 mm, lobes ovate to lanceolate; corolla 6–17 mm, pale violet-blue to white, lobes 2–4.5 mm, ovate-oblong. 2*n*=10. Open, wet places; 3200–3900 m. c&s SNH, W&I; to AK, CO; circumpolar. [*Gentianella t.* (Rottb.) Börner subsp. *t.*] Jul–Aug

EUSTOMA CATCHFLY GENTIAN

Ann, short-lived per, glabrous, ± glaucous. **ST**: erect, branched. **LF**: basal, cauline, opposite. **INFL**: cyme or fls 1. **FL**: parts in 5s; calyx lobed ± to base, lobes linear, narrow-keeled, acuminate; corolla > 2 cm, wide-bell-shaped, lobes >> tube, nectary pit 0 (nectaries at base of ovary); ovary sessile, style slender, entire, deciduous, stigma 2-lobed. 1 sp.: N.Am, C.Am, West Indies. (Greek: beautiful mouth, from corolla tube)

E. exaltatum (L.) G. Don subsp. ***exaltatum*** (p. 815) Pl 1.5–10 dm. **LF**: basal 2–10 cm, spoon-shaped-obovate to elliptic-oblong, obtuse; cauline 1.5–9 cm, elliptic to lanceolate, ± obtuse, upper acute. **INFL**: pedicels 2–14 cm. **FL**: calyx 10–21 mm; corolla 2–4.5 cm, pale to deep violet-blue (rose-violet, white), throat ± white with dark purple blotches, lobes elliptic to obovate, rounded to subacute. 2*n*=±72. Roadsides, alkaline marshes, other open, wet places; 100–600 m. SCo, PR, DSon; to se US, C.Am, West Indies. ± all yr

FRASERA
Bruce G. Baldwin

Per (non-fl rosettes preceding fl-sts, pls dying after fl in *Frasera albomarginata, Frasera parryi, Frasera puberulenta, Frasera speciosa,* and *Frasera umpquaensis*; non-fl rosettes appearing with fl-sts in others). **LF**: basal ± petioled; cauline opposite or whorled, < basal, base often fused-sheathing. **INFL**: cyme or panicle of dense clusters. **FL**: parts in 4s; calyx fused near base, lobes lanceolate; corolla rotate (bell-shaped), lobes >> tube, ridge between stamens fringed or scaled or 0, nectary pits prominent, 1(2) per lobe, margins of openings variously fringed; ovary sessile, style long and well differentiated or short and poorly differentiated, persistent, entire, stigmas 2. ± 15 spp.: temp N.Am. (J. Fraser, Scottish collector of N.Am pls, 1750–1811) [von Hagen & Kadereit 2002 Syst Bot 27:548–572]

1. Nectary pits 2 per corolla lobe (appearing as 1 due to overlapping fringes or not). . . . . **F. speciosa**
1′ Nectary pit 1 per corolla lobe
  2. Proximal to mid-st lvs wide-elliptic, not white-margined; infl dense, < 0.5 dm wide. . . . . **F. umpquaensis**
  2′ Proximal to mid-st lvs narrower than wide-elliptic, white-margined (narrowly so or not); infl dense or ± open, > 0.5 dm wide
    3. Infl ± open, gen > 1 dm wide or proximal-most branches < 1 dm from base and widely divergent, or both; distal pedicels gen within small cyme, not arising directly from main st
      4. Cauline lvs whorled, distal-most opposite or not . . . . **F. albomarginata**
        5. Infl branches, calyx lobes glabrous; nectary pit 2-lobed at tip . . . . var. **albomarginata**
        5′ Infl branches, calyx lobes puberulent; nectary pit ± truncate or ± notched at tip . . . . var. **induta**
      4′ Cauline lvs opposite
        6. Sts, abaxial lf faces glabrous; basal lvs 8–40 mm wide; nectary pit U-shaped . . . . **F. parryi**
        6′ Sts, abaxial lf faces puberulent; basal lvs 6–17 mm wide; nectary pit oblong-obovate. . . . . **F. puberulenta**
    3′ Infl dense, interrupted or not, < 1 dm wide, proximal-most branches > 1 dm from base, strongly ascending; distal pedicels gen arising directly from main st
      7. Nectary pit margins with vertical, oblong, basally tubular, distally fringed extensions . . . . **F. tubulosa**
      7′ Nectary pit margins fringed, without vertical extensions
        8. Nectary pit round to ± square; ridge between stamens low, fringed. . . . . **F. neglecta**
        8′ Nectary pit oblong; ridge between stamens with ± 2 mm scales . . . . **F. albicaulis**
          9. Sts, abaxial lf faces puberulent; larger lvs obtuse to rounded at tip . . . . var. **modocensis**
          9′ Sts, lvs glabrous; larger lvs gen acute at tip. . . . . var. **nitida**

F. albicaulis Griseb. (p. 815) Pl 1–6.5 dm. **ST**: 1–few; rosettes several. **LF**: narrowly white-margined; basal 4–23 cm, 3–12(20) mm wide, oblanceolate; cauline opposite, distal linear-oblong. **INFL**: dense, interrupted proximally; pedicels 2–8(30) mm. **FL**: calyx 3–7(12) mm; corolla 6–12 mm, green-white to pale blue, dark-blue-dotted or not, lobes elliptic-oblong, acute to short-acuminate, ridge between stamens with ± 2 mm, triangular to ovate-oblong, entire or ± coarsely jagged scales, nectary pit 1 per lobe, oblong. KR, NCoR, CaR, n SN, MP; to WA, MT, nw NV. [*Swertia a.* (Griseb.) Kuntze] Vars. intergrade; 4 other vars.

var. ***modocensis*** (H. St. John) N.H. Holmgren **ST**: puberulent. **LF**: puberulent abaxially; larger not arching, tips obtuse to rounded. 2*n*=52. Dry, brushy places; 900–1600 m. CaRH, MP; to WA, MT, nw NV. [*Swertia albicaulis* var. *a.*, as in TJM (1993), in part] May–Jul

var. ***nitida*** (Benth.) C.L. Hitchc. **ST**: glabrous. **LF**: glabrous; larger arching, tips gen acute. Dry, open woodland, chaparral; 150–1900 m. KR, NCoR, CaR, n SN, MP; to OR. [*Swertia a.* var. *n.* (Benth.) Jeps.] May–Jul

F. albomarginata S. Watson Pl 3–6 dm, glabrous or st puberulent. **ST**: 1–few. **LF**: white-margined; basal 2–9 cm, 5–10 mm wide, oblanceolate, tips acute; cauline whorled (distal-most opposite or not), distal lance-linear, tips acuminate. **INFL**: open; pedicels 5–50 mm. **FL**: calyx 5–12 mm; corolla 8–14 mm, green-white, often purple-dotted, lobes lance-oblong, mucronate, ridge between stamens fringed, nectary pit 1 per lobe, oblong, ± truncate to 2-lobed at tip. [*Swertia a.* (S. Watson) Kuntze]

var. ***albomarginata*** (p. 815) DESERT GREEN-GENTIAN **INFL**: branches glabrous. **FL**: calyx lobes glabrous; nectary pit 2-lobed at tip. Dry, open woodland; 1500–2200 m. DMtns; to CO. May–Aug ★

var. ***induta*** (Tidestr.) Card CLARK MOUNTAIN GREEN-GENTIAN **INFL**: branches puberulent. **FL**: calyx lobes puberulent; nectary pit ± truncate or ± notched at tip. Dry, open woodland; 1500–2200 m. DMtns (Clark Mtn Range); s NV. May–Jul ★

F. neglecta H.M. Hall PINE GREEN-GENTIAN Pl 2–5.5 dm, glabrous. **ST**: 1–several; rosettes several. **LF**: narrowly white-margined, linear to narrowly oblanceolate, tips acute; basal 2.5–20 cm, 3–9 mm wide; cauline opposite. **INFL**: dense, interrupted; pedicels 2–25 mm. **FL**: calyx 5–8 mm; corolla 7–15 mm, green-white, purple-streaked, lobes oblong-obovate, abruptly acuminate, ridge between stamens low, fringed, nectary pit 1 per lobe, round to ± square. Dry, open woodland; 1400–2500 m. SCoRO, TR. [*Swertia n.* (H.M. Hall) Jeps.] May–Jul ★

F. parryi Torr. (p. 815) Pl 6–16 dm, glabrous. **ST**: 1 or 2. **LF**: white-margined; basal 5–25 cm, 8–40 mm wide, strap-shaped to elliptic-oblanceolate, tips acute; cauline opposite, ovate to lance-

oblong, tips acute to acuminate. **INFL:** open; pedicels 6–32 mm. **FL:** calyx 8–17 mm; corolla 9–20 mm, light green, purple-dotted, lobes obovate, short-acuminate, ridge between stamens low, fringed, nectary pit 1 per lobe, U-shaped. Dry, open woodland, chaparral; 100–1900 m. SCo, SnGb, SnBr, PR; AZ, Baja CA. [*Swertia p.* (Torr.) Kuntze] Apr–Jul

F. puberulenta Davidson (p. 815)   Pl 1–3 dm; sts, abaxial lf faces puberulent. **ST:** 1–several. **LF:** narrowly white-margined; basal 2–12 cm, 6–17 mm wide, narrowly obovate to elliptic-oblong, tips obtuse or mucronate; cauline opposite, distal oblong to lanceolate, tips acute. **INFL:** open; pedicels 5–35 mm. **FL:** calyx 6–11 mm; corolla 7–13 mm, green-white, purple-dotted or -streaked, lobes oblong-obovate, abruptly acuminate, ridge between stamens low, ± fringed, nectary pit 1 per lobe, oblong-obovate. Dry, open conifer woodland; 1700–3400 m. c&s SNH, W&I; NV. [*Swertia p.* (Davidson) Jeps.] Jun–Aug

F. speciosa Griseb. (p. 815) MONUMENT PLANT   Pl 7–20 dm, glabrous or st, lvs puberulent. **ST:** 1. **LF:** basal 7–50 cm, 1–15 cm wide, oblanceolate to elliptic-obovate, tips rounded to acute; cauline whorled, lance-oblong, tips acute. **INFL:** elongate, open proximally, ± dense distally; pedicels 1–10 cm. **FL:** calyx 12–25 mm; corolla 12–20 mm, light yellow-green, purple-dotted or -streaked, lobes elliptic-oblong, acute to short-acuminate, ridge between stamens with deeply divided scales, nectary pits 2 per lobe, narrowly elliptic. 2*n*=78. Montane meadows, open woodland; 1500–3000 m. KR,

NCoRH, SNH, Wrn; to WA, SD, NM. [*Swertia radiata* (Kellogg) Kuntze var. *r.*] Jul–Aug

F. tubulosa Coville (p. 815)   Pl (0.6)2–11 dm, glabrous. **ST:** gen 1; rosettes 0–few. **LF:** white-margined, curving down near tip; basal 2–9 cm, 3–10(15) mm wide, oblanceolate, tips mucronate; cauline whorled, distal linear-oblong, tips acute. **INFL:** elongate, ± continuous above; pedicels 2–25 mm. **FL:** calyx 6–11 mm; corolla 8–13 mm, bell-shaped, white to pale blue, veins darker blue, lobes elliptic-oblong to obovate, short-acuminate, ridge between stamens 0, nectary pit 1 per lobe, ovate, margins with vertical, oblong, basally tubular, distally fringed extensions. Open pine woodland; 1800–2700 m. s SN. [*Swertia t.* (Coville) Jeps.] Jul–Aug

F. umpquaensis M. Peck & Applegate (p. 815) UMPQUA GREEN-GENTIAN   Pl 3–14 dm, glabrous. **ST:** 1. **LF:** basal 15–30 cm, 3–10 cm wide, spoon-shaped to obovate, tips acute; cauline wide-elliptic, tips acuminate, proximal whorled, distal often opposite, proximal to mid-st lvs not white-margined. **INFL:** dense, interrupted below or not; pedicels 2–10 mm. **FL:** calyx 8–12 mm; corolla 8–12 mm, pale yellow-green, blue-tinged or not, lobes elliptic-ovate, obtuse to acute, ridge between stamens long-fringed, nectary pit 1 per lobe, round. Mtn meadows; 1700–1900 m. KR; sw OR. [*Swertia u.* (M. Peck & Applegate) H. St. John] Treated as a synonym of *Swertia fastigiata* Pursh [*F. fastigiata* (Pursh) A. Heller] in TJM (1993). Jul–Aug ★

GENTIANA GENTIAN

Ann to per, gen glabrous. **ST:** gen simple below infl. **LF:** cauline, opposite (also basal or not). **INFL:** compact cyme or fls 1. **FL:** parts gen in 5s (gen in 4s in *Gentiana prostrata*, 4s or 5s in *Gentiana fremontii*); calyx tube gen > lobes; corolla tube narrow-bell-shaped, lobes spreading, < tube, base of sinus between lobes truncate or gen with a variously shaped, sometimes fringed appendage, nectary pits 0 (nectaries on ovary stalk); ovary stalked, style ± 0 or short, entire, persistent, stigmas 2. ± 400 spp.: temp to subarctic, alpine Am, Eurasia. (Gentius, king of Illyria, who used roots to treat malaria)

1. Ann, bien; st < 1 dm, lf < 15 mm, fl < 2 cm (sect. *Chondrophyllae*)
  2. Basal, cauline lvs not alike, upper cauline strongly ascending, much narrower than basal, margins of all
     lvs white; corolla white to pale blue, abaxially often dark blue . . . . *G. fremontii*
  2′ Basal, cauline lvs ± alike, ± spreading, margins not or ± white; corolla medium to deep blue . . . . *G. prostrata*
1′ Per; st gen > 1 dm; lf > 15 mm; fl > 2 cm (sect. *Pneumonanthe*)
  3. Sts arising laterally from caudex, below rosettes
    4. Sts 20–45 cm; internodes > middle, upper cauline lvs; corolla sinus appendages divided into 2–3 thread-
       like, entire segments . . . . *G. setigera*
    4′ Sts 0.5–10(35) cm; internodes gen < cauline lvs; corolla sinus appendages divided into 2 triangular,
       ± serrate to jagged-cut parts tapered to thread-like tips . . . . *G. newberryi*
      5. Corolla gen medium to deep blue . . . . var. *newberryi*
      5′ Corolla white to pale blue, abaxially dark brown-purple on, below lobes . . . . var. *tiogana*
  3′ Sts arising terminally from caudex, without rosettes
    6. Sinus between corolla lobes ± unappendaged . . . . *G. sceptrum*
    6′ Sinus between corolla lobes with a conspicuous toothed or fringed appendage
      7. Corolla sinus appendages divided ± to base into 2–several thread-like parts; lf-base sheaths at mid-st
         gen > 5 mm . . . . *G. plurisetosa*
      7′ Corolla sinus appendages divided into gen 2 parts, each thread-like only toward tip; lf-base sheaths
         at mid-st gen < 5 mm
        8. Lf, calyx-lobe margins ± entire, minutely ciliate; calyx lobes gen < 1.5 mm wide; corolla lobes gen
           < 7 mm; infls terminal, gen also at 1–6 upper nodes, fls 1–7 . . . . *G. affinis* var. *ovata*
        8′ Lf, calyx-lobe margins gen obscurely, minutely dentate, not ciliate; calyx lobes gen > 2 mm wide;
           corolla lobes 5.5–13 mm; infls gen terminal, also at 1–3 upper nodes or not, fls 1–few . . . . *G. calycosa*

G. affinis Griseb. var. *ovata* A. Gray (p. 819) OREGON GENTIAN   Per. **ST:** from caudex, decumbent to erect, several to many, 5–70 cm, puberulent. **LF:** basal 0; cauline many, 10–45 mm, 4–20 mm wide, minutely ciliate, lower < internodes, elliptic to ovate, upper > internodes, ovate to lanceolate, ± acute. **INFL:** terminal; fls 1–7, also 1–5 at 1–6 upper nodes or not. **FL:** calyx 5–23 mm, lobes linear to narrow-elliptic, gen < 1.5 mm wide, acute, minutely ciliate, some vestigial or not; corolla (12)25–45 mm, blue (white), lobes gen < 7 mm, oblong-ovate, mucronate, sinus appendages divided ± to base into 2 ± triangular parts, these thread-like only toward tip. **SEED:** winged. 2*n*=26. Meadows, shrubby places; < 2300 m. NCo, KR, CaRH, CCo, SnFrB, Wrn; to WA, ID. [*G. a.* var. *parvidentata* Kusn.] Some pls in n CA approach *G. affinis* var. *a.* of Rocky Mtns. Jul–Aug

G. calycosa Griseb. (p. 819) EXPLORERS' GENTIAN   Per. **ST:** from caudex, ± decumbent, 2–many, 5–45 cm. **LF:** basal 0; cauline many, ± even-spaced, gen > internodes, 8–50 mm, 6–30 mm wide, ovate to ± round, obtuse to acute, obscurely, minutely dentate, not ciliate. **INFL:** gen terminal; fls 1–few, also 1–2 at 1–3 upper nodes or not. **FL:** calyx 10–20 mm, lobes narrow- to wide-ovate, gen > 2 mm wide, much reduced or not, obtuse to acute, gen obscurely, minutely dentate, not ciliate; corolla 25–50 mm, deep blue (violet), lobes 5.5–13 mm, oblong-ovate to round, acute, sinus appendages divided < 1/2 way to base into gen 2–3 triangular segments, thread-like only toward tip. **SEED:** wingless. 2*n*=26. Wet mtn meadows, slopes; 1300–3900 m. KR, CaRH, SNH; to BC, MT. Has hybridized with *G. newberryi* var. *tiogana* in s SNH (Tulare Co.). Jul–Sep

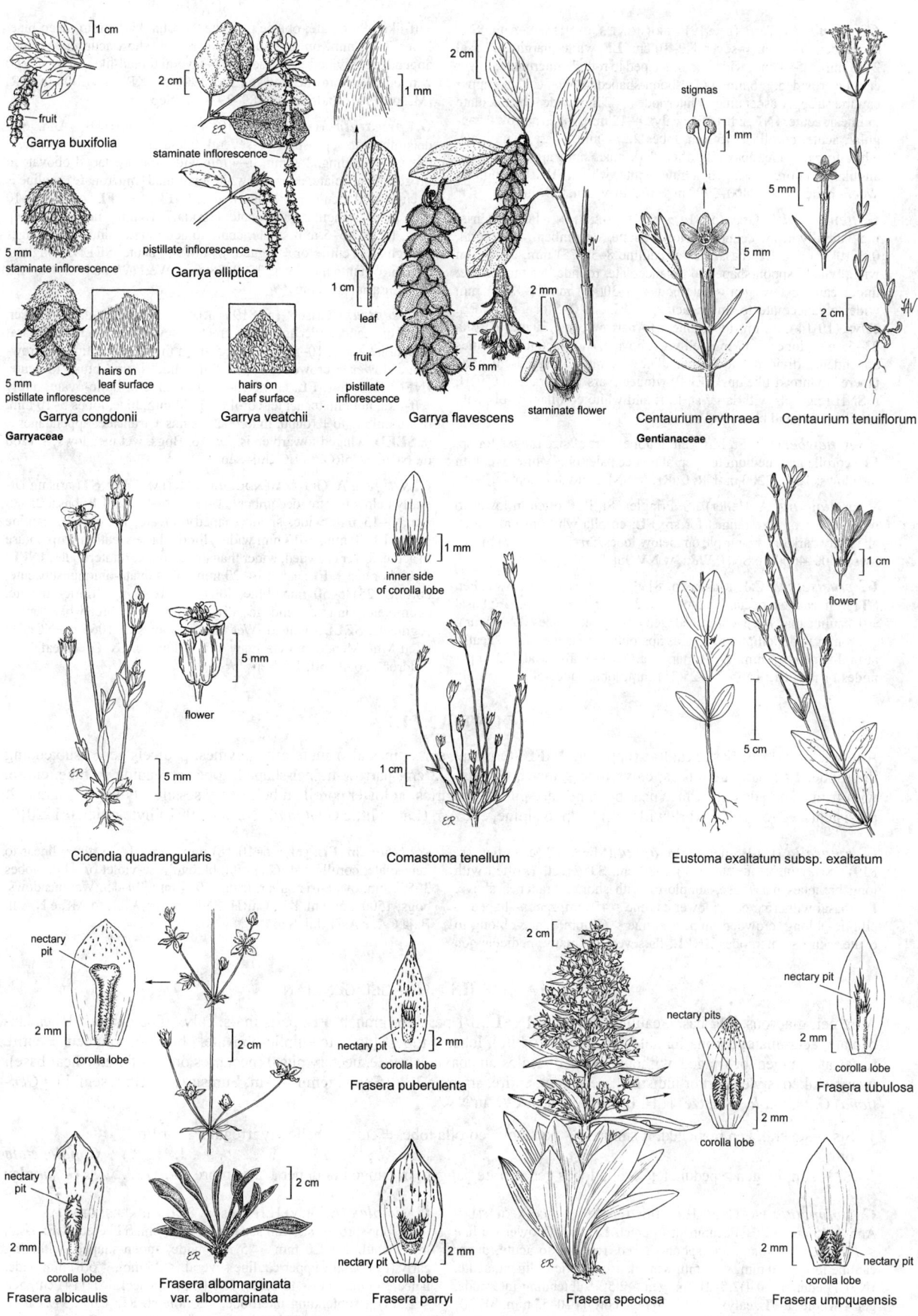

1 cm

Garrya buxifolia

fruit

5 mm
staminate inflorescence

hairs on
leaf surface

5 mm
pistillate inflorescence

Garrya congdonii

Garryaceae

2 cm

ER

staminate inflorescence

pistillate inflorescence

Garrya elliptica

hairs on
leaf surface

Garrya veatchii

1 mm

2 cm

1 cm

leaf

fruit

pistillate
inflorescence

2 mm

5 mm

staminate flower

Garrya flavescens

stigmas

1 mm

5 mm

Centaurium erythraea

Gentianaceae

5 mm

2 cm

HV

Centaurium tenuiflorum

5 mm

flower

Cicendia quadrangularis

ER

5 mm

1 mm
inner side
of corolla lobe

1 cm

ER

Comastoma tenellum

1 cm

flower

5 cm

Eustoma exaltatum subsp. exaltatum

nectary
pit

2 mm

corolla lobe

2 cm

nectary
pit

2 mm

corolla lobe

Frasera albicaulis

Frasera albomarginata
var. albomarginata

2 cm

ER

nectary pit

2 mm
corolla lobe

Frasera puberulenta

nectary pit

2 mm
corolla lobe

Frasera parryi

2 cm

ER

Frasera speciosa

nectary pits

2 mm

corolla lobe

nectary pit

2 mm

corolla lobe

Frasera tubulosa

2 mm

corolla lobe

nectary pit

Frasera umpquaensis

G. fremontii Torr. (p. 819) FREMONT'S GENTIAN Ann. **ST**: decumbent to erect, 1–several, 2–10 cm. **LF**: white-margined; basal 2–13 mm, 1.5–8 mm, wide-spoon-shaped to round, mucronate; lower cauline crowded, < 6 mm wide, spoon-shaped to oblanceolate, upper cauline strongly ascending, < internodes, < 2 mm wide, oblanceolate to linear, acute. **INFL**: fl 1. **FL**: calyx 4–12 mm, lobes narrow-triangular, acute; corolla 7–15 mm, lobes 2.2–4 mm, ovate, acuminate, white to pale blue, abaxially often dark blue, sinus appendages triangular, ± entire or jagged-serrate, acute, white. **SEED**: wingless. Wet mtn meadows; 2400–2700 m. SnBr; to w Can, CO. Jul–Aug ★

G. newberryi A. Gray (p. 819) ALPINE GENTIAN Per. **ST**: arising laterally from caudex, below rosette, decumbent, 1–several, 0.5–10(35) cm. **LF**: basal, lower cauline 8–50(75) mm, 2–25 mm wide, widely spoon-shaped to oblanceolate, rounded or mucronate; upper cauline few, gen > internodes, 6–30(45) mm, 1–8(13) mm wide, oblanceolate to linear, acute. **INFL**: fls 2–5 or gen 1. **FL**: calyx (10)14–30 mm, lobes linear to narrow-ovate, acute; corolla 23–55 mm, lobes 7–17 mm, elliptic-obovate, short-acuminate, sinus appendages divided into 2 triangular, ± serrate to jagged-cut parts tapered to thread-like tips. **SEED**: winged. Vars. intergrade in CaRH, n SNH (e.g., pls with large st, lf, fl and white corollas, or pls with small st, lf, fl and blue corollas).

var. ***newberryi*** St, lf, fl gen in upper part of size ranges for sp. **FL**: corolla gen medium to deep blue, occ paler or ± white. Wet mtn meadows; 1200–2200 m. KR, CaRH, n SNH; OR. Jul–Sep

var. ***tiogana*** (A. Heller) J.S. Pringle St, lf, fl often in lower to middle part of size ranges for sp. **FL**: corolla white to pale blue, abaxially dark brown-purple on, below lobes. 2n=26. Wet mtn meadows; 1500–4000 m. SNH, W&I; w NV. Jul–Sep

G. plurisetosa C.T. Mason (p. 819) KLAMATH GENTIAN Per. **ST**: from caudex, decumbent to erect, 2–several, 5–40 cm. **LF**: basal ± 0; cauline many, ± even-spaced, gen > 0.8 × internodes, 12–60 mm, 7–38 mm wide, elliptic to round, tips obtuse to acute, base sheaths at mid-st gen > 5 mm. **INFL**: terminal; fls 1–5, also 1 at 1–3 upper nodes or not. **FL**: calyx 17–25(35) mm, lobes lance-elliptic, rarely

± lf-like, not ciliate, obtuse to acute; corolla 35–50 mm, deep blue, lobes 7–14 mm, oblong-obovate to round, short-acuminate, sinus appendages divided ± to base into 2–several thread-like parts. **SEED**: winged. Wet mtn meadows; 1200–1900 m. KR, NCoRO; sw OR. [Mason 1991 Madroño 37:289–292] Jul–Sep ★

G. prostrata Haenke (p. 819) PYGMY GENTIAN Ann. **ST**: decumbent to ± prostrate, 1–several, 0.5–8 cm. **LF**: basal, cauline ± alike, ± spreading, 2–6 mm, 1–4 mm wide, spoon-shaped-obovate to broad-oblanceolate, ± or not white-margined, mucronate; cauline ± even-spaced, gen 0.5–2 × internodes. **INFL**: fl 1. **FL**: calyx 3.5–10 mm, lobes triangular, mucronate; corolla 7–18 mm, medium to deep blue, lobes 2.5–5 mm, ovate, acute to acuminate, sinus appendages triangular, ± entire or ± jagged toward tip, acute. **SEED**: wingless. 2n=36. Wet mtn meadows; 3500–3800 m. W&I (White Mtns); to AK, CO, Eurasia. Jul–Aug ★

G. sceptrum Griseb. (p. 819) KING'S SCEPTER GENTIAN Per. **ST**: from caudex, decumbent to erect, 1–several, 5–90 cm. **LF**: basal 0; cauline many, 10–55(70) mm, 5–18 mm wide, elliptic to narrow-ovate, lower < crowded, upper < internodes, tip obtuse to acute. **INFL**: fls 1–few. **FL**: calyx 15–27 mm, lobes elliptic-ovate, rarely ± lf-like, not ciliate, acute; corolla 35–50 mm, blue, lobes 5–10 mm, ovate-oblong to ± round, mucronate, sinus ± truncate, appendages ± 0. **SEED**: winged toward ends. 2n=26. Bogs, wet meadows; < 1200 m. NCo, NCoRO; to BC. Jul–Sep

G. setigera A. Gray MENDOCINO GENTIAN Per. **ST**: arising laterally below rosette, decumbent, 1–few, 20–45 cm. **LF**: basal 25–85 mm, 5–15 mm wide, spoon-shaped-obovate, tip obtuse; cauline many, 10–30 mm, 5–17 mm wide, elliptic, base sheathing, tip obtuse to acute, lower crowded, wider than upper, upper < internodes. **INFL**: fls 2–4 or gen 1. **FL**: calyx 14–23 mm, lobes ovate-oblong, subacute; corolla (25)35–50 mm, blue, lobes 10–16 mm, elliptic-obovate, acuminate, sinus appendages divided into 2–3 thread-like, entire segments. **SEED**: winged. Wet mtn meadows; ± 1065 m. NCoRO (Red Mtn, Mendocino Co.); sw OR. [Chambers & Greenleaf 1989 Madroño 36:49–50] Jul–Sep ★

GENTIANELLA

Ann, glabrous. **LF**: basal, cauline, opposite. **INFL**: terminal, sometimes also subterminal cymes; pedicels gen < subtending internode. **FL**: parts gen in 5s; calyx tube < lobes; corolla < 2 cm, narrow-funnel-shaped, lobes spreading, < tube, entire, with fringes on adaxial surface near base, nectary pits 0 (nectaries on lower corolla tube); ovary sessile; style ± 0, stigmas 2, persistent. ± 275 spp.: ± worldwide, esp temp to alpine, exc Afr. (Latin: little *Gentiana*) [Nesom 1991 Phytologia 70:1–20]

G. amarella (L.) Börner subsp. ***acuta*** (Michx.) J.M. Gillett (p. 819) NORTHERN GENTIAN Pl 5–80 cm. **ST**: erect, 1, often with long branches near base, simple or with shorter branches above. **LF**: basal withering early; lower cauline < 45 mm, spoon-shaped to elliptic-oblong, crowded; upper cauline < 60 mm, lance-oblong to ovate, gen << internodes. **INFL**: fls several to many; pedicels gen

< 2.5(50) cm. **FL**: calyx 5–10(18) mm, lobes 1–6 × tube, linear to lanceolate; corolla 7–15(21) mm, blue to rose-violet or white, lobes 3–5.5 mm, ovate-triangular, fringes 0–4 mm. 2n=18. Wet meadows, bogs; 1500–3500 m. KR, CaRH, SNH, SnBr, W&I; to AK, e N.Am, Baja CA, e Asia. Jul–Sep

GENTIANOPSIS FRINGED GENTIAN

Ann, per, glabrous. **LF**: basal, cauline, opposite. **INFL**: fl 1 per st or branch. **FL**: parts in 4s; calyx tube distinct, lobes lanceolate, acuminate; corolla funnel-shaped, blue (white), lobes ≤ tube, oblong to elliptic-obovate, obtuse or rounded, ± entire to serrate, jagged, or fringed, without fringes or scales on adaxial surface, nectary pits 0 (nectaries on corolla tube near base); ovary stalked, style short or indistinct, persistent, entire, stigmas 2. ± 15 spp.: temp N.Am, Eurasia. (Greek: resembling *Gentiana*) *G. thermalis* (Kuntze) H.H. Iltis occurs near CA, in NV.

1. Sts 1–several, gen branched, if simple, << peduncle; corolla lobes ± entire, shallowly irregular near tip or not . . . . ***G. holopetala***
1′ St 1, simple, gen > peduncle; corolla lobes gen serrate, jagged, or fringed in upper 1/2 or more . . . . ***G. simplex***

G. holopetala (A. Gray) H.H. Iltis (p. 819) SIERRA GENTIAN Ann 3–45 cm. **ST**: ± decumbent to erect. **LF**: basal, lower cauline < 70 mm, < 15 mm wide, spoon-shaped, rounded to acute, upper cauline few, < 60 mm, < 10 mm wide, lance-elliptic to linear, acute. **INFL**: peduncle (0.4)2.5–21 cm, gen > 2.5 × subtending internode, gen > whole st. **FL**: calyx (10)14–36 mm; corolla 20–55 mm. **SEED**: papillate, obtuse. 2n=78. Wet meadows; 1800–4000 m. c&s SNH, W&I; w NV. Jul–Sep

G. simplex (A. Gray) H.H. Iltis (p. 819) HIKERS' GENTIAN Per from root-sprouts, sts well separated, 4–40 cm. **ST**: erect. **LF**: basal, lower cauline < 20 mm, < 5 mm wide, spoon-shaped, withering early (esp basal), upper cauline several, < 25 mm, < 6(8) mm wide, lance-elliptic, obtuse to acute. **INFL**: peduncle 1.3–11(14) cm, gen < 2.5(4) × subtending internode, < whole st. **FL**: calyx 9–27 mm; corolla 17–45 mm. **SEED**: striate-ridged, pointed. Wet meadows; 1200–3400 m. KR, NCoRH, CaRH, SNH, SnBr; to OR, MT. Jul–Sep

SWERTIA

Bruce G. Baldwin

Per; non-fl rosettes appearing with fl-sts. **LF**: basal ± petioled; cauline gen (sub)opposite or some alternate, < basal, bases not fused-sheathing. **INFL**: cyme or panicle of dense clusters. **FL**: parts gen in 5s, sometimes in 4s; calyx fused near base, lobes lanceolate; corolla rotate (bell-shaped), lobes >> tube, ridge between stamens fringed or scaled or neither, nectary pits prominent, 2 per lobe, margins of openings ± fringed; ovary sessile, style short, poorly differentiated, persistent, entire, stigmas 2. ± 135 spp.: temp N.Am, Eurasia, Afr. (E. Sweert, Dutch herbalist, 1552–1612) [von Hagen & Kadereit 2002 Syst Bot 27:548–572] Other taxa in TJM (1993) moved to *Frasera*.

S. perennis L. (p. 819) STAR SWERTIA Pl 1–5 dm, glabrous. **ST**: gen 1; rosettes 1–few. **LF**: basal 3–16 cm, 12–30 mm wide, spoon-shaped to obovate, tips obtuse; cauline elliptic-oblanceolate, tips subacute. **INFL**: open; pedicels 2–30(60) mm. **FL**: calyx 4–8 mm; corolla 7–13 mm, blue-white to -violet, veins darker, lobes lance-oblong, acute, ridge between stamens low, nectary pits round. 2*n*=28. Wet meadows, bogs; 2300–3200 m. KR, n&s SNH; OR (Wallowa Mtns), also AK to CO; Eurasia. Jul–Sep

ZELTNERA CENTAURY

Ann, glabrous. **ST**: erect, gen branched. **LF**: cauline, opposite (sometimes also basal). **INFL**: cyme. **FL**: parts in 4s or gen 5s; calyx lobes >> tube (discounting thin membrane between lobes in *Zeltnera davyi*), appressed to corolla tube, gen not keeled; corolla salverform, white or gen pink, lobes ≤ tube, ± entire, scales 0, nectaries pits 0 (nectaries elsewhere 0); stamens often initially curved to 1 side, dehisced anthers spirally twisted; ovary sessile, style gen thread-like, deciduous, entire or cleft < 1 mm, stigmas 1, 2-lobed, or 2, stigmas or lobes wedge- to fan-shaped. ± 25 spp.: temp, dry-mesic trops, N.Am, S.Am. (L. Zeltner, N. Zeltner, Swiss botanists, b. 1938, 1934) [Mansion 2004 Taxon 53:719–740] Variable, difficult; further study needed, esp of chromosome numbers, hybridization. *Z. calycosa* (Buckley) G. Mans. [*Centaurium calycosum* (Buckley) Fernald] not in CA.

1. Pedicels gen 10–70 mm, gen > closed corollas; fl parts gen in 4s. . . . . ***Z. exaltata***
1′ Pedicels 0–40(60) mm, gen < closed corollas (exc in some robust pls of *Zeltnera venusta*); fl parts in 5s
  2. Lvs linear to thread-like; stigma 1, ± 2-lobed . . . . ***Z. namophila***
  2′ Lvs ovate or lanceolate to elliptic to oblanceolate or obovate; stigmas 1–2
    3. Pedicels of proximal, central fls 0–10 mm, of other fls gen 0–5 mm, of distal often 0
      4. Corolla lobes 2–7 mm, tips rounded; stigmas 2, wide-fan-shaped . . . . ***Z. muehlenbergii***
      4′ Corolla lobes (3)5–10 mm, tips acute to acuminate; stigma 1, lobes ± 2, wedge-shaped . . . . ***Z. trichantha***
    3′ Pedicels 2–40(60) mm
      5. Calyx lobes keeled; corolla lobes 3–7 mm . . . . ***Z. davyi***
      5′ Calyx lobes not keeled; corolla lobes (2)5–20 mm.
        6. Main sts from base gen several; basal lvs often rosetted, gen oblanceolate . . . . ***Z. arizonica***
        6′ Main st from base gen 1; basal lvs 0 or gen not rosetted, narrow-ovate-oblong to lanceolate. . . . . ***Z. venusta***

Z. arizonica (A. Gray) G. Mans. ARIZONA CENTAURY Pl (10)20–60 cm. **LF**: basal often rosetted, gen 15–70 mm, gen oblanceolate; cauline gen 25–70 mm, gen lanceolate. **INFL**: open; pedicels 4–40(60) mm. **FL**: corolla (14)18–25 mm, lobes (7)9–12 mm, lanceolate, acute; stigmas 2. 2*n*=40. Open, damp places, esp streambanks; 50–100 m. Formerly e DSon; to CO, TX, Mex. [*Centaurium a.* (A. Gray) A. Heller] Range barely into CA along Colorado River; no records more recent than ± 1955. Apr–May

Z. davyi (Jeps.) G. Mans. (p. 819) DAVY'S CENTAURY Pl 2–30 cm. **LF**: cauline, gen 8–25 mm, elliptic-oblong to ovate. **INFL**: open; pedicels 2–25(55) mm. **FL**: calyx lobes keeled; corolla 12–17 mm, lobes 3–7 mm, ovate-oblong to elliptic, tip rounded; style wider than thread-like, stigmas 2. Moist coastal bluffs, dunes, open forest; < 1000 m. NCo, CCo, SnFrB, SCoRO, ChI. [*Centaurium d.* (Jeps.) Abrams; *C. muehlenbergii* (Griseb.) Piper, misappl.] Keeled calyx lobes unique in genus. May–Aug

Z. exaltata (Griseb.) G. Mans. TALL CENTAURY, DESERT CENTAURY Pl (3)10–60 cm. **LF**: all ± alike, 10–30(50) mm, oblong-elliptic to linear; basal often rosetted. **INFL**: open; pedicels gen 10–70 mm. **FL**: parts gen in 4s; corolla 9–20 mm, lobes 2.5–7 mm, lance-oblong, tip subacute; stigmas 2. 2*n*=40, 74. Moist, gen alkaline scrub; 900–2200 m. CaR, SCoR, SnGb, PR, GB, w edge DSon; to s BC, NB, Mex. [*Centaurium e.* (Griseb.) Piper] Pls in SCoR, SW, Baja CA disjunct, with smaller fls, pollen grains, needing taxonomic re-evaluation. May–Aug

Z. muehlenbergii (Griseb.) G. Mans. (p. 819) MONTEREY CENTAURY Pl 3–30 cm. **LF**: cauline, 15–25 mm, lower obovate to oblong, upper lance-elliptic. **INFL**: open; pedicels 0–10 mm. **FL**: corolla 12–19 mm, lobes 2–7 mm, lance-oblong, tip rounded; stigmas 2. Moist coastal bluffs, forest openings; < 1600 m. SNF, CW; to sw BC, ID. [*Centaurium m.* (Griseb.) Piper] Jun–Aug

Z. namophila (Reveal et al.) G. Mans. SPRING-LOVING CENTAURY Pl (5)15–60 cm. **LF**: cauline, 10–50 mm, linear to thread-like. **INFL**: open, ± elongate, panicle-like; pedicels gen 0–20 mm. **FL**: corolla 13–18 mm, lobes 5–8 mm, lanceolate, tip obtuse to acute; stigma 1, ± 2-lobed. 2*n*=34. Wet alkaline meadows, gen near springs or lakes or in alkaline marshes; < 700 m. SNE, DMoj; to NV. [*Centaurium exaltatum* (Griseb.) Piper, misappl.] Jul–Sep

Z. trichantha (Griseb.) G. Mans. (p. 819) ALKALI CENTAURY Pl 5–45 cm. **LF**: cauline, 10–40 mm, ovate to lanceolate. **INFL**: ± dense, units ± flat-topped; pedicels 0–3(6) mm. **FL**: corolla 12–22 mm, lobes (3)5–10 mm, linear to lance-oblong, tip acute to acuminate; stigma 1, ± 2-lobed. 2*n*=34. Alkaline and saline flats, dry to moist openings in chaparral, around hot springs, vernal pools; < 800 m. NCoRI, SnFrB. [*Centaurium t.* (Griseb.) B.L. Rob.] Apparently hybridizes with *Z. venusta*, esp in Sonoma Co. May–Aug

Z. venusta (A. Gray) G. Mans. (p. 819) CALIFORNIA CENTAURY, CHARMING CENTAURY Pl 3–50 cm. **LF**: gen cauline, 5–25 mm, narrow-ovate-oblong to lanceolate. **INFL**: ± open; pedicels gen 4–25 mm. **FL**: corolla (7)16–30 mm, lobes (2)5–20 mm, lanceolate to spoon-shaped or elliptic-obovate, tip obtuse to ± acute; stigmas 2. 2*n*=34. Dry grassland, scrub, chaparral, openings in woodland; < 1800 m. KR, s NCoRO, NCoRI, CaRF, SNF, e SnJV, SW (exc n ChI), SNE, DMoj; to Baja CA. [*Centaurium v.* (A. Gray) B.L. Rob.] May–Aug

GERANIACEAE GERANIUM FAMILY

Carlos Aedo, except as noted

Ann, per, or ± woody, gen glandular-hairy. **LF**: simple to compound, basal and cauline; cauline alternate or opposite, stipules 2, ± on st. **INFL**: cyme or pseudo-umbel or 1–2-fld. **FL**: bisexual [unisexual], radial or ± bilateral; sepals 5, free, overlapping in bud; petals gen 5, free, gen with nectar glands at base; stamens gen 5,10[15]; staminodes scale-like or 0; ovary gen 5-lobed, upper part elongating into beak in fr, chambers 5, placentas axile, style 1, stigmas 5, free, persistent in fr. **FR**: septicidal [loculicidal], mericarps 5, dry, gen 1-seeded, each persistent on 1 of 5 linear segments of beak that separate from central column by curving or coiling upward. 6 genera, ± 750 spp.: temp, ± trop. Some cult for orn, perfume oils. [Bakker et al. 2006 Taxon 55:887–896] Scientific Editor: Thomas J. Rosatti.

1. Fl ± bilateral; nectary 1, deeply embedded into receptacle; fertile stamens 1–7, united at base . . . . **PELARGONIUM**
1′ Fl radial (± bilateral); nectaries 5, on petal bases; fertile stamens 5 or 10, free
  2. Fertile stamens gen 10; lf palmately lobed to divided . . . . **GERANIUM**
  2′ Fertile stamens gen 5; lf crenate to shallowly ± palmately lobed or gen pinnately lobed to compound
    3. Staminodes 0; lf simple, ± palmately lobed or not, veins ± palmate . . . . **CALIFORNIA**
    3′ Staminodes 5, alternate stamens; lf simple to pinnately compound, veins pinnate. . . . . **ERODIUM**

CALIFORNIA CALIFORNIA FILAREE

Marisa Alarcón, Carlos Aedo & Carmen Navarro

1 sp.: CA. (Named for the California Floristic Province) [Aldasoro et al. 2002 Anales Jard Bot Madrid 59:209–216] Segregated from *Erodium* on morphological, molecular data.

C. macrophylla (Hook. & Arn.) J.J. Aldasoro et al. (p. 819) ROUND-LEAVED FILAREE Ann, bien, gen scapose. **ST**: < 5 cm, glandular-puberulent. **LF**: simple, 10–15 cm, cauline opposite; blade << petiole, reniform, cordate, or ovate, crenate to shallowly ± palmately lobed, puberulent. **INFL**: umbel. **FL**: radial; sepals 8–10 mm, tips glabrous to puberulent; petals > sepals, gen white, tinged red to purple; stamens 5, free, staminodes 0. **FR**: mericarp body 8–10 mm, indehiscent, fusiform, 1-seeded, glandular-puberulent, base sharply pointed, top truncate, gen with 1 ± round pit on each side of beak segment, pits not subtended by ridges; style column 3–5 cm, persistent on fr; beak segments stiffly hairy adaxially, gen twisted. 2*n*=64. Open sites, grassland, scrub, vertic clay, occ serpentine; < 1200 m. NCoRI, s SNF, GV, CW, SCo, ChI (Santa Cruz, Santa Catalina islands), WTR, PR; to n Mex. [*Erodium m.* Hook. & Arn.; *E. m.* Hook. & Arn. var. *californicum* (Greene) Jeps.] Mar–Jul ★

ERODIUM STORKSBILL, FILAREE

Carlos Aedo & Carmen Navarro

Ann, per. **LF**: simple to pinnately compound, cauline opposite; blade lanceolate to reniform in outline, puberulent or short-hairy, base cordate to truncate. **INFL**: umbel. **FL**: radial; stamens 5, free, alternate 5 scale-like staminodes. **FR**: mericarp body indehiscent, fusiform, 1-seeded, base sharply pointed, top gen with 1 pit on each side of beak segment, pits subtended by 1–4 ridges or not; beak segments stiffly hairy adaxially, gen twisted. ± 74 spp.: temp Am, Eurasia, n Afr, Australia. (Greek: heron, from bill-like fr) [Fiz et al. 2006 Syst Bot 31:739–763] Some cult for forage, dyes; "beak segments" sometimes called "awns" elsewhere. *E. macrophyllum* moved to genus *California*.

1. Basal, lower cauline lvs pinnately compound, lflets lobed to deeply dissected
  2. Pits at top of mericarp without glands . . . . ***E. cicutarium***
  2′ Pits at top of mericarp with glands . . . . ***E. moschatum***
1′ Basal, lower cauline lvs crenate to pinnately lobed to dissected, lobes or segments lobed to dissected, esp basal pair
  3. Pits at top of mericarp with glands. . . . . ***E. malacoides***
  3′ Pits at top of mericarp 0 or without glands
    4. Sepals with glands, spreading hairs
     5. Mericarp top with pits 0 or transversely narrow and inconspicuous, subtended by (2)3–4 glabrous ridges . . . . ***E. botrys***
     5′ Mericarp top with pits ± round, subtended by 1–2 ± hairy ridges . . . . ***E. brachycarpum***
    4′ Sepal without glands, with appressed hairs
     6. Sepal hairs directed basally; petals < 0.8 cm . . . . [***E. cygnorum***]
     6′ Sepal hairs directed apically; petals gen 0.7–1.5 cm . . . . ***E. texanum***

E. botrys (Cav.) Bertol. (p. 825) Ann. **ST**: prostrate to ascending, 1–9 dm, short-hairy. **LF**: lobed to dissected; lobes ± 8–10 mm wide; lower lvs 3–15 cm; blade ± = petiole, ovate to oblong in outline, glabrous to sparsely puberulent, veins short-appressed-hairy. **FL**: sepals 10–13 mm, tip bristly; petals ± > sepals, pink. **FR**: body 8–11 mm, pits 0 or transversely narrow and inconspicuous, subtended by (2)3–4 glabrous ridges; style column 5–12 cm. 2*n*=40, 60. Dry, open or disturbed sites; < 1000 m. CA-FP; native to s Eur. Mar–Jul

E. brachycarpum (Godr.) Thell. (p. 825) Ann. **ST**: ascending, 1–6 dm, ± glandular-hairy. **LF**: lobed; lobes 7–11 mm wide; lower lvs 5–10 cm; blade ± = petiole, gen ovate in outline, glabrous to sparsely puberulent, veins short-appressed-hairy. **FL**: sepals 7–10 mm, tip bristly; petals ± > sepals, pink, veins gen darker. **FR**: body 6–8 mm, pits ± round, subtended by 1–2 ± hairy ridges; style column 5–8 cm. 2*n*=40. Dry, open or disturbed sites; < 1000 m. CA-FP; OR; native to s Eur. Mar–Jul

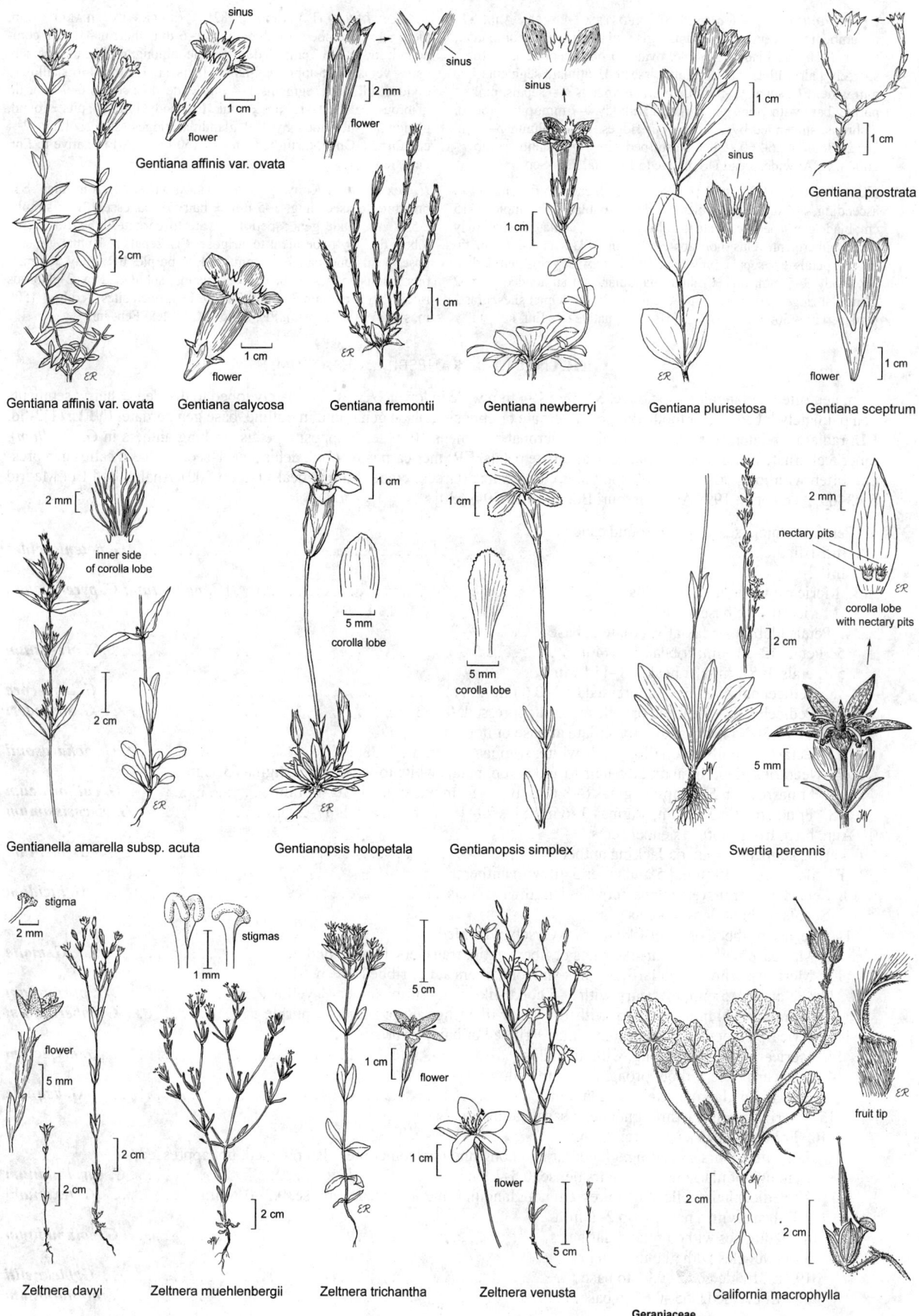

Gentiana affinis var. ovata

Gentiana affinis var. ovata   Gentiana calycosa   Gentiana fremontii   Gentiana newberryi   Gentiana plurisetosa   Gentiana sceptrum

Gentianella amarella subsp. acuta   Gentianopsis holopetala   Gentianopsis simplex   Swertia perennis

Zeltnera davyi   Zeltnera muehlenbergii   Zeltnera trichantha   Zeltnera venusta   California macrophylla

Geraniaceae

E. cicutarium (L.) Aiton (p. 825)   REDSTEM FILAREE   Ann. **ST**: decumbent to ascending, 1–5 dm, ± glandular-hairy. **LF**: compound; lower 3–10 cm, blade > petiole, ovate to oblanceolate in outline, sparsely hairy; lflets 9–13, deeply dissected, ultimate segments 1–2 mm wide. **FL**: sepals 3–5 mm, tip bristly; petals ± = sepals, pink to purple, base with veins gen darker. **FR**: body 4–7 mm, pits ± round, glabrous, subtended by 1–2 glabrous ridges; style column 2–5 cm. 2*n*=20,36,40,42,54,60. Open, disturbed sites, grassland, scrub; < 2000 m. CA; widespread US; native to Eurasia. Feb–Sep ❖

E. malacoides (L.) Aiton (p. 825)   Ann, bien. **ST**: decumbent to ascending, 1–6 dm, ± puberulent; nodes glandular. **LF**: simple, 4–15 cm; blade gen < petiole, ovate to oblong, cordate, crenate to shallowly lobed, puberulent, veins short-appressed-hairy. **FL**: sepals 4–6 mm, tip bristly; petals ± = sepals, pink to purple, base with veins gen darker. **FR**: body 3–5 mm, pits ± round, glandular, gen subtended by 1–2 glandular ridges; style column 2–3 cm. 2*n*=20, 40. Open sites, grassland, scrub; < 500 m. n SnJV, SnFrB, SCo; native to s Eur. Feb–Jul

E. moschatum (L.) Aiton (p. 825)   GREENSTEM FILAREE   Ann, bien. **ST**: decumbent to ascending, 1–6 dm, short hairy. **LF**: compound; lower 5–15 cm; blade > petiole, oblong to obovate, sparsely hairy, veins short-appressed-hairy; lflets 11–15, lobed to shallowly divided, ultimate segments 1–4 mm wide. **FL**: sepals 6–9 mm, tip glabrous; petals 10–15 mm, ± pink. **FR**: body 4–6 mm, pits ± round, glandular, subtended by 1–2 glandular ridges; style column 2–4 cm. 2*n*=20. Open, disturbed sites; < 1500 m. CA-FP; native to Eur. Feb–Sep

E. texanum A. Gray (p. 825)   TEXAS FILAREE   Ann, bien. **ST**: prostrate to ascending, 1–5 dm, ± hairy or canescent. **LF**: simple, 1.5–4 cm; blade gen < petiole, ovate to cordate, lobed to deeply lobed, densely puberulent to strigose. **FL**: sepals 5–10 mm, tip strigose; petals unequal, 7–15 mm, pink to purple. **FR**: body 5–8 mm, pits inconspicuous or 0, without glands, subtended by 2 glabrous ridges; style column 3–7 cm. 2*n*=20. Dry, open sites, scrub; < 1500 m. s SnJV, s SCo (rare), PR, D; to TX, n Mex. Feb–Jun

GERANIUM   CRANESBILL, GERANIUM

Ann, per, often ± glandular, esp above. **ST**: gen 1–4 mm wide in lower 1/2. **LF**: palmately lobed to divided, ± hairy; segments gen palmately lobed and/or toothed; upper alternate or opposite; blade gen round in outline, base gen cordate. **INFL**: (1)2-fld. **FL**: radial (± bilateral); sepals ± awned to mucronate; stamens 10, outer 5 opposite petals (lacking anthers in *G. pusillum*), inner 5 alternate petals, nectary glands 5, alternate petals. **FR**: mericarp gen ovoid, dehiscent, 1-seeded; basal callus gen present, often with long bristles. ± 400 spp.: temp, trop mtns. (Greek: crane, from fr beak) [Aedo 2001 Anales Jard Bot Madrid 59:3–65; Aedo et al. 1998 Ann Missouri Bot Gard 85:594–630]

1. Per, from thick, scaly, underground caudex
  2. Infl 1-fld. . . . . ***G. potentilloides***
  2′ Infl 2-fld
    3. Mericarp without basal callus . . . . [***G. palmatum; G. pyrenaicum***]
    3′ Mericarp with basal callus
      4. Petals glabrous adaxially, ciliate at base
        5. Petals 12–21 mm; fr beak 27–44 mm . . . . ***G. oreganum***
        5′ Petals 3–8.1 mm; fr beak 8.8–15.1 mm
          6. Pedicel hairs reflexed, appressed, 0.2–0.6 mm . . . . ***G. core-core***
          6′ Pedicel hairs spreading or reflexed, not appressed, 0.2–2 mm . . . . ***G. solanderi***
      4′ Petals 1/3–3/4 hairy adaxially, ciliate at base or not
        7. Nectaries woolly abaxially; petals white; stigmas 3–4 mm . . . . ***G. richardsonii***
        7′ Nectaries glabrous abaxially, hair-tufted at top; petals white to red-purple; stigmas 3–8 mm
          8. Fr narrow tip 5–6 mm, stigmas 6–8 mm; lvs 3–8 cm wide; petals 10–15 mm . . . . ***G. californicum***
          8′ Fr narrow tip 3–4 mm, stigmas 3–6 mm; lvs 7.1–16 cm wide; petals 17–20 mm . . . . ***G. viscosissimum***
1′ Ann, bien, from short, ± slender roots
  9. Fertile stamens 5 (outer 5 lacking anthers) . . . . ***G. pusillum***
  9′ Fertile stamens 10 (inner 5 and outer 5 all with anthers)
    10. Sepals with transverse flaps between lengthwise keels . . . . ***G. lucidum***
    10′ Sepals without flaps or keels
      11. Mericarp ribbed or ± net-like or transversely wrinkled
        12. Mericarp without collar-like rings or fibers at tip, transversely wrinkled . . . . ***G. molle***
        12′ Mericarp with 1–5 collar-like rings and long fibers at tip, ribbed or ± net-like
          13. Petals 5–9.5 mm; mericarp with 3–5 collar-like rings at tip; anthers ± yellow. . . . . ***G. purpureum***
          13′ Petals 10–14 mm; mericarp with 1–3 collar-like rings at tip; anthers ± purple . . . . ***G. robertianum***
      11′ Mericarp smooth (although sometimes ± ribbed or hairy or bristly)
        14. Mericarp base with prong, without callus . . . . ***G. dissectum***
        14′ Mericarp base without prong, gen with callus
          15. Mericarp glabrous exc base gen ciliate . . . . ***G. aequale***
          15′ Mericarp ± hairy throughout exc basal callus bristly
            16. Fr beak with a narrow tip < 2 mm
              17. Pedicel hairs spreading, glandular and nonglandular, sometimes also reflexed, not appressed, nonglandular; mericarp hairs dense, 0.5–1.8 mm. . . . . ***G. carolinianum***
              17′ Pedicel hairs reflexed, appressed, nonglandular; mericarp hairs sparse, 0.2–0.7 mm . . . . [***G. texanum***]
            16′ Fr beak with a narrow tip 2–6 mm
               18. Peduncles without glandular hairs . . . . ***G. columbinum***
              18′ Peduncles with glandular hairs
                19. Lf divided 0.85–0.95 to base. . . . . ***G. bicknellii***
                19′ Lf divided 0.55–0.65 to base. . . . . ***G. rotundifolium***

G. aequale (Bab.) Aedo Ann. **ST**: decumbent to erect, 1–4 dm; hairs ± sparse, soft. **LF**: blade 1.5–5.8 cm wide, divided 0.6–0.75 to base, segments 7–9, obtriangular. **FL**: pedicel 10–22 mm; sepals 3–5 mm, smooth, short-awned; petals 3.5–4.5 mm, notched, red-purple. **FR**: mericarp 1.4–1.5 mm, smooth, glabrous exc base gen ciliate, basal callus 0; beak 7–10.5 mm, narrow tip 1–1.5 mm; stigmas 1–2 mm. **SEED**: smooth. 2*n*=26. Open to shaded sites, disturbed ground; < 200 m. s SCo (San Diego Co.); widespread Am, New Zealand; native to Eur. May–Aug

G. bicknellii Britton (p. 825) Ann. **ST**: decumbent to erect, 1–6 dm; hairs short, stiff. **LF**: blade 2–7 cm wide, divided 0.85–0.95 to base, segments 5(7), wedge-shaped. **FL**: pedicel 10–25 mm; sepals 4–6 mm, smooth, short-awned; petals 4–6 mm, ± notched, pale purple. **FR**: mericarp 3–3.5 mm, smooth, minutely bristly; beak 1.5–2.1 cm, narrow tip 3.5–4.5 mm; stigmas ± 1 mm. **SEED**: finely pitted. 2*n*=52. Open, sunny to shaded sites, woodland, conifer forest; 600–1500 m. KR/CaRH, n SCo; to BC, ne N.Am; alien in Venezuela. Jan–Aug

G. californicum G.N. Jones & F.L. Jones (p. 825) Per. **ST**: ascending to erect, 2–7 dm; hairs soft. **LF**: blade 3–8 cm wide, divided 0.7–0.8 to base, segments 5–7, wedge-shaped. **FL**: pedicel 28–80 mm; sepals 8–11 mm, smooth, short-awned; petals 10–15 mm, obtuse to ± notched, white to rose, veins lavender to purple, basal adaxial surface soft-hairy; nectaries hair-tufted at top. **FR**: mericarp 4.5–5 mm, smooth, sparsely glandular; beak 21–23 mm, narrow tip 5–6 mm; stigmas 6–8 mm. **SEED**: faintly pitted. Moist sites, streambanks, meadows, woodland; 1000–3000 m. SNH, SCo, TR, PR, n SNE. [*G. concinnum* G.N. Jones & F.L. Jones] Apr–Aug

G. carolinianum L. (p. 825) Ann. **ST**: gen erect, 1–7 dm; hairs dense, short, spreading or reflexed. **LF**: blade 2.5–8.5 cm wide, divided 0.7–0.9 to base, segments 5(7), oblong to wedge-shaped. **FL**: pedicel 3–11 mm; sepals 5–6.5 mm, smooth, short-awned; petals 5.5–6 mm, ± notched, white to rose-pink. **FR**: mericarp 3–4.5 mm, hairs dense, 0.5–1.8 mm; beak 15–19 mm, narrow tip 1–2 mm; stigmas 0.7–1.4 mm. **SEED**: faintly pitted. 2*n*=52. Open to shaded sites, grassland, scrub, forest; < 1700 m. NW (exc NCoRH), CaR, SNF, c SNH, GV, CW (exc SCoRI), SW (exc WTR, SnJt), MP (exc Wrn); to e N.Am; alien in S.Am, Asia, Eur, Réunion, some Caribbean islands. [*G. sphaerospermum* Fernald] Feb–Aug

G. columbinum L. Ann. **ST**: erect, 0.9–6 dm; hairs reflexed, appressed, nonglandular. **LF**: blade 3–5 cm wide, divided > 0.9 to base, segments 5–7, rhombic. **FL**: pedicel 20–60 mm; sepals 6–11, smooth, awned; petals 8–10 mm, ± notched, purple. **FR**: mericarp 2.2–2.8 mm, smooth, minutely bristly; beak 18–19 mm, narrow tip ± 4 mm; stigmas ± 2 mm. **SEED**: pitted. 2*n*=18. Open, disturbed sites; < 450 m. NCo, NCoRI/ScV; to e N.Am; native to Eur, n Afr, w Asia. May–Sep

G. core-core Steud. (p. 825) Per. **ST**: ascending to erect, 1.5–7.5 dm; hairs sparse, stiff, reflexed, appressed. **LF**: blade 1.5–5.6 cm wide, divided 0.55–0.8 to base, segments 5–7, broadly wedge-shaped. **FL**: pedicel 11–31 mm; sepals 3–5.5 mm, smooth, short-awned; petals 3.1–7.1 mm, rounded or ± notched, purple. **FR**: mericarp 2.6–3.2 mm, smooth, minutely strigose; beak 9.1–14.5 mm, narrow tip 0; stigmas 0.9–1.5 mm. **SEED**: faintly pitted. Disturbed sites; < 200 m. NCo, NCoRO, n SNF, CCo, SnFrB, w SCo, WTR, c PR; native to S.Am. Previously misidentified as *G. retrorsum* DC., native to Australia. Mar–Oct

G. dissectum L. (p. 825) Ann. **ST**: ascending to erect, 0.7–7 dm; hairs rough, spreading to reflexed. **LF**: blade 2.5–5.4 cm wide, divided 0.75–0.95 to base, segments 5–7, rhombic. **FL**: pedicel 6–13 mm; sepals 4.2–7.1 mm, smooth, short-awned; petals 2.9–5.8 mm, ± notched, rose-purple. **FR**: mericarp 2.3–2.9 mm, smooth, minutely bristly, base with prong, callus 0; beak 11.1–13.5 mm, narrow tip 1.6–2.7 mm; stigmas 0.6–1 mm. **SEED**: deeply pitted. 2*n*=22. Open, disturbed sites; < 1300 m. CA-FP; to e N.Am; native to Eur, n Afr, w Asia; alien in some Caribbean islands, S.Am, New Zealand. Mar–Jul ❖

G. lucidum L. Ann. **ST**: erect, 0.5–4.5 dm; hairs ± sparse, soft. **LF**: blade 2–4.2 cm wide, divided 0.55–0.7 to base, segments ± 5, obtriangular. **FL**: pedicel 6–13 mm; sepals 4.5–6.5 mm, with transverse flaps between lengthwise keels, awned; petals 5–9 mm, rounded, ± purple. **FR**: mericarp 2.5–2.8 mm, net-like, glabrous with glandular

cilia on margin, basal callus 0; beak 12–13 mm, narrow tip 5 mm; stigmas ± 1 mm. **SEED**: smooth. 2*n*=20,40–44,60. Open to shaded sites, disturbed ground; < 100 m. CCo (Berkeley, Alameda Co.); OR; native to Eur, n Afr, w Asia. May

G. molle L. (p. 825) Ann. **ST**: decumbent to erect, 1–4.5 dm; hairs ± sparse, soft. **LF**: blade 0.9–5.2 cm wide, divided 0.5–0.75 to base, segments 7–9, obtriangular. **FL**: pedicel 5–15 mm; sepals 1–6 mm, smooth, mucronate; petals 3–10.5 mm, notched, red-purple. **FR**: mericarp 1.8–2.1 mm, transversely wrinkled, glabrous, basal callus 0; beak 6–11 mm, narrow tip 1–3 mm; stigmas 1–2 mm. **SEED**: smooth. 2*n*=26. Open to shaded sites, disturbed ground; < 500 m. CA-FP; widespread N.Am; native to Eur, n Afr, w Asia; alien in S.Am, s Afr, Australia, New Zealand. Feb–Aug

G. oreganum Howell (p. 825) Per. **ST**: erect, 1.5–8 dm; hairs soft. **LF**: blade 7–14 cm wide, divided 0.75–0.9 to base, segments 7, rhombic. **FL**: pedicel 10–35 mm; sepals 9–11 mm, smooth, awned; petals 12–21 mm, rounded or ± notched, red-purple. **FR**: mericarp 4.5–5 mm, smooth, minutely bristly, glandular; beak 2.7–4.4 cm, narrow tip 4–6 mm; stigmas 3–6 mm. **SEED**: pitted. Moist, gen shaded sites, meadows, forest; 250–800 m. NCo, KR, NCoRO, CaRH; to WA. [*G. viscosissimum* var. *incisum* (Torr. & A. Gray) N.H. Holmgren] Mar–Jul

G. potentilloides DC. (p. 825) Per. **ST**: decumbent or ascending, rooting at nodes or not, 1.5–6 dm; hairs short, stiff, spreading to reflexed. **LF**: blade 1–4 cm wide, divided 0.7–0.75 to base, segments 5–7, wedge-shaped. **INFL**: 1-fld. **FL**: pedicel 12–22 mm; sepals 4.7–5.6 mm, smooth, awned; petals 6–9 mm, rounded or ± notched, white to ± pink. **FR**: mericarp 2–3 mm, smooth, hairs sparse; beak 10–11 mm, narrow tip 0.5 mm; stigmas 1.5–1.6 mm. **SEED**: pitted. 2*n*=36,56. Moist, shaded sites, conifer forest; < 500 m. CCo, SnFrB; native to Australia. [*G. microphyllum* Hook. f.] May–Aug

G. purpureum Vill. Ann. **ST**: decumbent to ascending, 1–3.5 dm; hairs soft. **LF**: blade 3–8.7 cm wide; divided > 0.9 to base, segments 5, ± sessile to short-stalked, pinnately lobed to dissected. **FL**: pedicel 5–10 mm; sepals 4.5–7 mm, smooth, soft-hairy, awned; petals 5–9.5 mm, rounded, pink to red-purple; anthers ± yellow. **FR**: mericarp 2.3–3 mm, ribbed or ± net-like, with 3–5 collar-like rings and long fibers at tip, glabrous or hairy, basal callus 0; beak 13–16 mm, narrow tip 4–4.5 mm; stigmas ± 1 mm. **SEED**: smooth. 2*n*=32. Open to shaded sites; < 100 m. NCoRO, NCoRI, CCo, SnFrB; native to Eur, n Afr, w Asia; alien in S.Am, New Zealand, s Afr. Mar–May

G. pusillum L. Ann. **ST**: decumbent to erect, 1–5 dm, often branched; hairs soft. **LF**: blade 1.5–4.8 cm wide, divided 0.3–0.75 to base, segments 7, wedge-shaped. **FL**: pedicel 6–16 mm; sepals 3–4.5 mm, smooth, mucronate to short-awned; petals 2–3 mm, ± notched, pink to violet; fertile stamens 5. **FR**: mericarp 1.7–1.9 mm, smooth, minutely strigose, basal callus 0; beak 7–9 mm, narrow tip 0; stigmas 0.5–0.7 mm. **SEED**: finely net-like. Disturbed open places, dry grassland; < 500 m. NCo, s SNF, n CCo, SnFrB, s SNE; to BC, e N.Am; native to Eur, n Afr, w Asia; alien in S.Am, New Zealand. Mar–Sep

G. richardsonii Fisch. & Trautv. (p. 825) Per. **ST**: erect, 1.5–9 dm; hairs ± soft. **LF**: blade 4.4–11.3 cm wide, divided 0.75–0.9 to base, segments 5–7, rhombic. **FL**: pedicel 15–30 mm; sepals 7–8 mm, smooth, awned; petals 14–20 mm, rounded, white, gen purple-veined, 1/3–1/2 hairy adaxially; nectaries woolly abaxially. **FR**: mericarp 3.5–4 mm, smooth, hairs sparse; beak 13–17 mm, narrow tip 1–2 mm; stigmas 3–4 mm. **SEED**: coarsely pitted. 2*n*=52,56. Moist sites, meadows, conifer forest; 1200–2700 m. SNH, TR, SnJt, Wrn; to Can, Rocky Mtns. May–Sep

G. robertianum L. (p. 825) Ann, bien. **ST**: decumbent to ascending, 1–5 dm, 1–3 mm wide in lower 1/2, hairs ± sparse. **LF**: blade 3.5–10 cm wide; divided > 0.9 to base or midrib, segments 5, ± sessile to short-stalked, pinnately lobed to dissected. **FL**: pedicel 5–20 mm; sepals 6–8 mm, smooth, soft-hairy, awned; petals 10–14 mm, rounded, pink to red-purple; anthers ± purple. **FR**: mericarp 2.5–3.1 mm, net-like, with 1–3 collar-like rings and long fibers at tip, glabrous or hairy, basal callus 0; beak 17–19 mm, narrow tip 4–5 mm; stigmas ± 1 mm. **SEED**: smooth. 2*n*=64. Open to shaded sites; < 500 m. NCo, CCo, SnFrB; widespread N.Am; native to Eur, n Afr, w Asia; alien in S.Am, China, Madagascar. Apr–Sep

G. rotundifolium L.   Ann. **ST**: ascending to erect, 1–4 dm; hairs rough, spreading. **LF**: blade 1.9–4.4 cm wide, divided 0.55–0.65 to base, segments 5–7, obtriangular. **FL**: pedicel 7–20 mm; sepals 4.5–6 mm, smooth, short-awned; petals 6–7 mm, rounded, rose-purple. **FR**: mericarp 2.5–3 mm, smooth, minutely bristly; beak 12–13 mm, narrow tip 2–3 mm; stigmas ± 1 mm. **SEED**: deeply pitted. 2*n*=26. Open, disturbed sites; < 200 m. CCo, SCoR, SnGb; to e N.Am; native to Eur, n Afr, w Asia. May

G. solanderi Carolin   Per. **ST**: ascending to erect, 1.2–10 dm; hairs spreading to reflexed, not appressed, nonglandular. **LF**: blade 1.4–5.7 cm wide, divided 0.45–0.85 to base, segments 5–7, obtriangular. **FL**: pedicel 8–34 mm; sepals 3.2–6.3 mm, smooth, mucronate to short-awned; petals 3–8.1 mm, rounded, ± purple. **FR**: mericarp 2.2–3.2 mm, smooth, minutely strigose; beak 8.8–15.1 mm, narrow tip 0–1 mm; stigmas 0.6–1.6 mm. Disturbed sites; < 100 m. NCo; native to Australia, New Zealand. [*G. pilosum* Sol., nom. nud.; *G. p.* Willd., illeg.] May–Jun

G. viscosissimum Fisch. & C.A. Mey. (p. 825)   Per. **ST**: erect, 3–9 dm; hairs sparse. **LF**: blade 7.1–16 cm wide, divided 0.75–0.85 to base, segments 5–9, rhombic. **FL**: pedicel 6–34 mm; sepals 8–11 mm, smooth, awned; petals 17–20 mm, pink to red-purple, veins red to purple, basal adaxial surface soft-hairy; nectaries hair-tufted at top. **FR**: mericarp 4.5–5.5 mm, smooth, ± glandular puberulent; beak 25–29 mm, narrow tip 3–4 mm; stigmas 3–6 mm. **SEED**: pitted. Meadows, openings in sagebrush scrub, conifer forest; 1000–2500 m. e KR (Quartz Valley), CaRH, MP; to BC, Rocky Mtns. Apr–Sep

PELARGONIUM   GARDEN GERANIUM

Piet Vorster

Ann, gen per, shrub, aromatic or strong-smelling. **ST**: gen erect. **LF**: alternate to ± opposite above; blade lobed to dissected, lobes gen crenate or serrate. **INFL**: umbel, dense to open; fls 3–many. **FL**: ± bilateral [(radial)], nectary 1, deeply embedded into receptacle to form spur fused with pedicel; petals ± equal to strongly unequal, gen striped or marked, upper 2 gen > lower 3, different in shape, position, gen with marks of different and/or more intense color; fertile stamens 1–7, united at base. **FR**: body dehiscent, gen oblong, base acute, 1-seeded; beak segments stiff-hairy adaxially. ± 300 spp.: s Afr, St. Helena, Asia Minor, Madagascar, Australia. (Greek: stork, from beaked fr) [Bakker et al. 2005 *in* Bakker et al. (eds.) Pl Species-Level Syst:75–100. A.R.G. Gantner] *Pelargonium* ×*domesticum*, cult but evidently not escaped in CA; *P. quercifolium* (L.) L'Hér., urban weed.

1. Lf peltate, glabrous . . . . **[*P. peltatum*]**
1′ Lf not peltate, glabrous or not
  2. Lf blade ± as long as wide
    3. Lf membranous, hairs 0 or sparse; petals pink-purple . . . . **[*P. zonale*]**
    3′ Lf ± succulent, hairs dense; petals red to white
      4. Lf blade 15–20 cm wide; petals red to white . . . . **P. ×hortorum**
      4′ Lf blade gen 4–8 cm wide; petals red . . . . **P. inquinans**
  2′ Lf blade longer than wide or wider than long
    5. Fl < 10 mm wide; st slender, < 3 mm diam . . . . **P. grossularioides**
    5′ Fl 15–40 mm wide; st stout, > 3 mm diam
      6. Infl ± lax; fl 30–40 mm wide; lf blade abaxially hairier, of different color than adaxially . . . . **P. panduriforme**
      6′ Infl dense; fl 15–18 mm wide; lf blade abaxially, adaxially equally hairy, of same color
        7. Pl prostrate to sprawling; margins of lower lvs irregularly curled; infl 8–20-fld . . . . **[*P. capitatum*]**
        7′ Pl erect; margins of lower lvs not curled; infl 3–12-fld . . . . **[*P. vitifolium*]**

P. grossularioides (L.) L'Hér.   Ann, occ per; hairs sparse, short. **ST**: prostrate or sprawling, 2–5 dm, < 3 mm diam. **LF**: blade < 6 cm wide, longer than wide, ± round to broadly ovate, ± lobed, lobes coarsely toothed. **INFL**: open; fls 3–50. **FL**: < 1 cm wide; sepals < 5 mm; petals < 6 mm, pink to rose-purple. Disturbed sites; < 300 m. NCo, CCo, SnFrB, SCo; native to s Afr. Apr–Jul

P. ×hortorum L.H. Bailey   FISH GERANIUM   Subshrub, soft-woody. **ST**: branched, 3–6 dm; hairs 0 or ± sparse. **LF**: blade 15–20 cm wide, round, gen with a color band or variegated, ± succulent, deeply notched at base, crenate; hairs dense. **INFL**: dense to open; fls 20–40. **FL**: petals 1–2.5 cm, ± equal, red to white. Disturbed sites; < 300 m. SCo, ChI. Hybrid between *P. inquinans*, *P. zonale*; after a few generations from seed, similarities to *P. inquinans* increase. All yr

P. inquinans (L.) L'Hér.   Subshrub, soft-woody. **ST**: branched, 1–2 m; hairs soft. **LF**: blade gen 4–8 cm wide, round, ± succulent, green, deeply notched at base, gen irregularly 5–7-lobed, lobes crenate; hairs dense. **INFL**: open; fls 5–30. **FL**: sepals 5–7 mm; petals 15–20 mm, ± equal, red. Disturbed sites; < 300 m. SnFrB; native to s Afr. All yr

P. panduriforme Eckl. & Zeyher   OAK-LEAVED GERANIUM   Shrub. **ST**: branched, > 1 m. **LF**: blade < 12 cm, < 10 cm wide, longer than wide, soft, occ ± sticky, gen pinnately divided to near midrib; divisions 3–7, unequal, serrate. **INFL**: ± lax; fls 3–10. **FL**: bilateral, 30–40 mm wide; sepals 8–14 mm; petals 20–35 mm, pink. Disturbed sites; < 300 m. CCo, SCo; native to s Afr. ± all yr

GROSSULARIACEAE   GOOSEBERRY FAMILY

Michael R. Mesler & John O. Sawyer, Jr.

Shrub, gen < 4 m. **ST**: gen erect; nodal spines 0–9; internodal bristles gen 0; twigs gen hairy, gen glandular. **LF**: simple, alternate, gen clustered on short, lateral branchlets, petioled, gen deciduous; blade gen palmately 3–5-lobed, gen thin, gen dentate or serrate, base gen cordate. **INFL**: raceme, axillary, gen pendent, 1–25-fld; pedicel gen not jointed to ovary, gen hairy or glandular; bract gen green. **FL**: bisexual, radial; hypanthium tube exceeding ovary; sepals gen 5, gen spreading; petals gen 5, gen < sepals, gen flat; stamens gen 5, alternate petals, gen inserted at level of petals (hypanthium top), anthers gen free, gen ± not exceeding petals, gen glabrous, tips gen rounded; ovary inferior, chamber 1, ovules many, styles gen 2, gen fused exc at tip, gen glabrous. **FR**: berry. 1 genus, 120 spp.: n hemisphere, temp S.Am. Some cult as food, orn. Hypanthium data refer to part above ovary; statements about ovary hairs actually refer to the hypanthium around the ovary. At one time incl in Saxifragaceae. Scientific Editor: Thomas J. Rosatti.

RIBES CURRANT, GOOSEBERRY

(Arabic: for pls of this genus) [Schultheis & Donoghue 2004 Syst Bot 29:77–96; Senters & Soltis 2003 Taxon 52:51–66]

1. Nodal spines 0
  2. Hypanthium disk- or saucer-shaped, barely exceeding ovary
    3. Lvs evergreen, lobes 0 or very shallow; sw PR, s ChI (Santa Catalina Island) . . . . **R. viburnifolium**
    3′ Lvs deciduous, lobes deep; mainland
      4. Ovary, fr, lf blade abaxially with stalked glands; st spreading or decumbent . . . . **R. laxiflorum**
      4′ Ovary, fr, lf blade abaxially with sessile glands; st ± erect
        5. Sepals green; NCo, w KR, NCoRO . . . . **R. bracteosum**
        5′ Sepals white; n MP. . . . . **R. hudsonianum** var. **petiolare**
  2′ Hypanthium cup- to tube-shaped, clearly exceeding ovary
    6. Sepals yellow . . . . **R. aureum**
      7. Fl odor spicy; petals yellow turning orange; sepals 5–8 mm . . . . var. **aureum**
      7′ Fl odor 0; petals yellow turning deep red; sepals 3–4 mm. . . . . var. **gracillimum**
    6′ Sepals white or white-green to pink, red, or purple
      8. Anther tip rounded or blunt, with cup-like depression
        9. Hypanthium < 2 × longer than wide; stamens inserted at level of petals; fr black to glaucous. . . . **R. viscosissimum**
        9′ Hypanthium > 2 × longer than wide; stamens inserted below level of petals; fr red. . . . . **R. cereum**
          10. Bract tip ± truncate, with several prominent teeth, sides entire; styles gen hairy . . . . var. **cereum**
          10′ Bract tip acute, each side with 1–3 shallow teeth; styles glabrous . . . . var. **inebrians**
      8′ Anther tip rounded, blunt, or shallowly notched, without cup-like depression
        11. Styles glabrous at base
          12. Sepals erect; hypanthium ± as long as wide . . . . **R. nevadense**
          12′ Sepals spreading; hypanthium longer than wide. **R. sanguineum**
            13. Infl pendent; sepals pink to white. . . . . var. **glutinosum**
            13′ Infl erect to spreading; sepals red. . . . . var. **sanguineum**
        11′ Styles hairy, at least at base
          14. Hypanthium wider than long; styles free; infl spike- or head-like, dense. . . . . **R. canthariforme**
          14′ Hypanthium ± longer than wide; styles fused at least in basal 1/2; infl a raceme, open
            15. Hypanthium white, barely longer than wide . . . . **R. indecorum**
            15′ Hypanthium pink, ± 2 × longer than wide. . . . . **R. malvaceum**
              16. Lf blade adaxially dull olive-green. . . . . var. **malvaceum**
              16′ Lf blade adaxially bright green. . . . . var. **viridifolium**
1′ Nodal spines present (sometimes 0 on some shoots)
  17. Infl gen > 5-fld; hypanthium disk- or saucer-shaped; pedicel jointed to ovary
    18. Lf hairs 0 or sparse, nonglandular; fr black. . . . . **R. lacustre**
    18′ Lf glandular-hairy; fr orange-red . . . . **R. montigenum**
  17′ Infl < 5-fld; hypanthium cup- to tube-shaped; pedicel not jointed to ovary
    19. Sepals 4, erect. . . . . **R. speciosum**
    19′ Sepals 5, spreading or reflexed
      20. Anthers not exceeding petals
        21. St low, spreading; internode bristles glandular . . . . **R. tularense**
        21′ St erect or arched; internode bristles 0
          22. Ovary hairs conspicuous, short and long, glandular and not . . . . [2]**R. velutinum**
          22′ Ovary hairs gen 0 or inconspicuous, gen short, gen nonglandular
            23. Hypanthium ± as long as wide — dry interior mtn slopes, 700–2500 m . . . . [2]**R. velutinum**
            23′ Hypanthium longer than wide
              24. Hypanthium > 3 mm; fr red; subalpine, alpine mtn tops, 2100–3100 m . . . . **R. lasianthum**
              24′ Hypanthium < 3 mm; fr black; rocky slopes, oak-covered foothills, chaparral, < 1350 m . . . . **R. quercetorum**
      20′ Anthers exceeding petals
        25. Styles hairy at least at base
          26. Calyx purple to green; filaments exceeding petals by > 3 mm. . . . . **R. divaricatum**
            27. Petals pink or red; styles 8–11 mm . . . . var. **parishii**
            27′ Petals white; styles 5–8 mm . . . . var. **pubiflorum**
          26′ Calyx green-white, purple at base or not; filaments exceeding petals by < 2 mm . . . . **R. inerme**
            28. Lf hairs 0 to sparse, short, soft; sepal hairs 0. . . . . var. **inerme**
            28′ Lf hairs sparse to dense, long, soft to stiff; sepal hairs soft . . . . var. **klamathense**
        25′ Styles glabrous at base
          29. Petals bright yellow, abaxially deeply concave, tip hooded. . . . . **R. marshallii**
          29′ Petals white to pink, flat or abaxially shallowly concave or margins curled inward, nearly touching
            or not, tip not hooded
            30. Petals flat or abaxially shallowly concave; st low, spreading. . . . . **R. binominatum**
            30′ Petal margins curled inward, nearly touching or not; st ± erect

31. Anthers oblong, tips blunt or rounded; styles ± not exceeding anthers
  32. Anthers slightly longer than wide when dehisced, with sessile glands abaxially; internodal
    bristles 0. . . . . *R. lobbii*
  32′ Anthers much longer than wide when dehisced, glabrous; internodal bristles present . . . . *R. sericeum*
31′ Anthers lanceolate to ovate, tips acute or mucronate; styles exceeding anthers
  33. Lf blade abaxially glandular
    34. Internodal bristles 0; hypanthium longer than wide. . . . . *R. amarum*
    34′ Internodal bristles present; hypanthium gen ± as long as wide
      35. Sepals green-white; fr yellow . . . . *R. victoris*
      35′ Sepals purple; fr purple . . . . *R. menziesii*
        36. Lvs strongly aromatic — s SNF . . . . var. *ixoderme*
        36′ Lvs not strongly aromatic . . . . var. *menziesii*
  33′ Lf blade abaxially nonglandular
    37. Hypanthium ± as long as wide . . . . *R. californicum*
      38. Lf hairs 0 or sparse; filaments > 2 × petals . . . . var. *californicum*
      38′ Lf hairy; filaments ± = to barely > petals . . . . var. *hesperium*
    37′ Hypanthium longer than wide
      39. Sepals ± pink, spreading; n ChI (Santa Cruz Island) . . . . *R. thacherianum*
      39′ Sepals purple, reflexed; mainland . . . . *R. roezlii*
        40. Hypanthium, lf blade abaxially glabrous . . . . var. *cruentum*
        40′ Hypanthium, lf blade abaxially hairy or densely white-hairy
          41. Hypanthium, lf blade abaxially densely white-hairy . . . . var. *amictum*
          41′ Hypanthium, lf blade abaxially hairy . . . . var. *roezlii*

R. amarum McClatchie (p. 825) BITTER GOOSEBERRY Pl < 2 m. **ST:** nodal spines 3. **LF:** blade 2–4 cm, glandular-hairy. **INFL:** 1–3-fld. **FL:** hypanthium 5–6 mm, longer than wide; sepals reflexed, 2–4 mm, purple; petals 2 mm, white, margins curled inward; anthers exceeding petals, exceeded by styles, tips mucronate. **FR:** 15–20 mm, purple; bristles stiff, glandular and not. Chaparral; 15–1910 m. SNF, Teh, SnFrB, SCoRO, SW. [*R. a.* var. *hoffmannii* Munz] Feb–Apr

R. aureum Pursh GOLDEN CURRANT Pl < 3 m. **ST:** nodal spines 0; internodes glabrous or puberulent. **LF:** blade firm, 1.5–5 cm, toothed or not, light green, gen glandular in youth, gen glabrous in age, base wedge-shaped to ± cordate. **INFL:** 5–15-fld. **FL:** hypanthium 6–10 mm, longer than wide; sepals 3–8 mm, yellow; petals 2–3 mm; styles fused. **FR:** 6–8 mm, red, orange, or black, glabrous. 2*n*=16.

var. ***aureum*** (p. 825) **FL:** odor spicy; hypanthium 1.5–2 × sepals; sepals 5–8 mm; petals yellow aging orange. Many habitats; 125–1880 m. KR, CaR, SNH, SnJV, GB; to BC, SD, NM. Apr–May

var. ***gracillimum*** (Coville & Britton) Jeps. **FL:** odor 0; hypanthium 2–3 × sepals; sepals 3–4 mm; petals yellow aging deep red. Alluvial areas, forest edges; 105–910 m. NCoRI, SnFrB, SCoR, SW. Feb–May

R. binominatum A. Heller (p. 825) TRAILING GOOSEBERRY Pl gen < 2.5 dm. **ST:** low, spreading, rooting; nodal spines 3; internodes not bristly. **LF:** blade 2–5 cm, glandular-hairy. **INFL:** 1–4-fld. **FL:** hypanthium 2–4 mm, ± as long as wide; sepals reflexed, 4–6 mm, green-white to green with red margins; petals 2–3 mm, white to pink, flat or abaxially shallowly concave; anthers exceeding petals. **FR:** 8–10 mm, yellow-green; prickles stout, nonglandular, yellow, hairs glandular or not. Montane, subalpine forest, meadows; 1000–2300 m. KR, NCoRH; OR. May–Jun

R. bracteosum Douglas (p. 825) STINK CURRANT Pl < 4 m. **ST:** nodal spines 0; internodes sparsely hairy. **LF:** blade deeply 5–7-lobed, 4–20 cm, adaxially shiny, glabrous, abaxially dull, hairs sparse, glands sessile, yellow. **INFL:** erect, 20–50-fld; lower bracts lf-like. **FL:** hypanthium 3–4 mm, saucer-shaped; sepals 3–5 mm, green; petals < 1 mm, white. **FR:** 8–10 mm, black-glaucous; glands sessile. 2*n*=16. Moist forest; 15–1665 m. NCo, w KR, NCoRO; to AK. Mar–Jun

R. californicum Hook. & Arn. HILLSIDE GOOSEBERRY Pl < 1.5 m. **ST:** nodal spines 3; internodes gen glabrous. **LF:** blade 1–3 cm, nonglandular. **INFL:** 1–3-fld. **FL:** hypanthium 2 mm, ± as long as wide; sepals reflexed, 6–8 mm, green to red; petals 3 mm, white, margins curled inward; anthers exceeding petals, exceeded by styles, tips mucronate. **FR:** 9–10 mm, red; shorter bristles glandular.

var. ***californicum*** **LF:** hairs 0 or sparse. **FL:** sepals green or red-tinged; filaments > 2 × petals. Forest openings, woodland; < 720 m. NCoR, SnFrB, SCoR. Feb–Mar

var. ***hesperium*** (McClatchie) Jeps. **LF:** hairy. **FL:** sepals red; filaments ± = to barely > petals. Chaparral, woodland; 150–1670 m. WTR, SnGb, PR (Santa Ana Mtns). Jan–Mar

R. canthariforme Wiggins (p. 831) MORENO CURRANT Pl < 2.5 m. **ST:** nodal spines 0. **LF:** blade 4–6 cm, thick, crenate, glandular, adaxially green, hairs long, soft, wavy, abaxially gray-green, hairs dense. **INFL:** many-fld, dense, spike- or head-like. **FL:** hypanthium 1 mm, wider than long; sepals 2 mm, spoon-shaped, purple with darker veins; petals 1 mm, purple; styles free, bases sparsely hairy. **FR:** 5–6 mm, purple; hairs long, soft, wavy or glandular, 0 in age. Chaparral; 500–1200 m. w PR. Feb–Apr ★

R. cereum Douglas WAX CURRANT Pl < 1.5 m. **ST:** nodal spines 0. **LF:** odor spicy; blade 1–4 cm, round, shallowly lobed, finely toothed, adaxially glossy. **INFL:** 3–7-fld. **FL:** hypanthium 6–8 mm, > 2 × longer than wide; sepals 1–2 mm, reflexed, white to pink; petals < 1 mm, white to pink; stamens inserted below level of petals, anther tips with cup-like depression; styles fused ± to tip. **FR:** 6–7 mm, red, glabrous to sparsely glandular. 2*n*=16.

var. ***cereum*** (p. 831) **LF:** hairs 0 to dense, glandular. **INFL:** bract tip wide, with several prominent teeth. **FL:** styles gen hairy. Dry montane to alpine slopes, among rocks, forest edges; 875–2365 m. KR, CaRH, SNH, Teh, TR, SnJt, GB, DMtns; to BC. Jun–Jul

var. ***inebrians*** (Lindl.) C.L. Hitchc. **LF:** hairs dense, nonglandular. **INFL:** bract tip acute, each side with 1–3 shallow teeth. **FL:** styles glabrous. Open rocky areas; 2100–3850 m. s SNH, W&I; to ID, NE, NV, AZ. Jun–Jul

R. divaricatum Douglas Pl < 3.5 m. **ST:** arched; nodal spines 0–3; internodes bristly or not. **LF:** blade 2–6 cm, coarsely toothed, gen hairy, not glandular. **INFL:** < 5-fld. **FL:** hypanthium 1–4 mm; sepals reflexed, 5–7 mm, obovate, purple to green; petals 1–3 mm, white, pink, or red; filaments exceeding petals by > 3 mm; styles hairy at base. **FR:** 6–10 mm, black, glabrous. 2*n*=16. Var. *divaricatum* OR to BC.

var. ***parishii*** (A. Heller) Jeps. PARISH'S GOOSEBERRY Pl < 3 m. **FL:** hypanthium 3–4 mm; petals 2–3 mm, pink or red; filaments 3–5 mm; styles 8–11 mm. Moist woodland; 60–310 m. e SCo, SnGb. Possibly extinct. Mar–Apr ★

var. ***pubiflorum*** Koehne (p. 831) STRAGGLE BUSH Pl < 3 m. **FL:** hypanthium 1–2 mm; petals 1–2 mm, white; filaments > 5 mm;

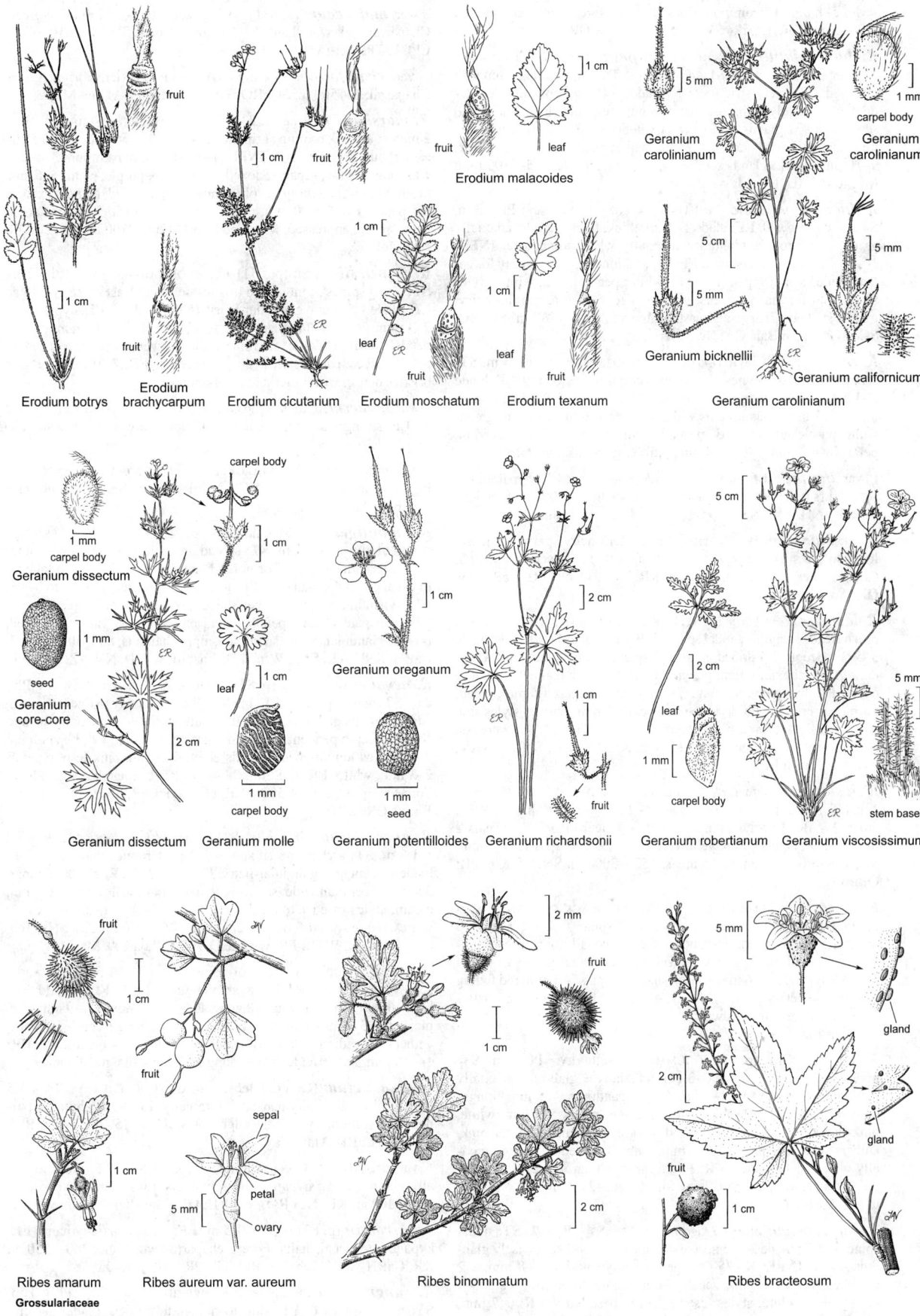

Erodium botrys

Erodium
brachycarpum

fruit

Erodium cicutarium

Erodium malacoides

leaf

fruit

1 cm

leaf

fruit

Erodium moschatum

leaf

fruit

Erodium texanum

Geranium
carolinianum

carpel body

Geranium
carolinianum

5 cm

5 mm

5 mm

Geranium bicknellii

Geranium californicum

Geranium carolinianum

carpel body

1 mm

carpel body

Geranium dissectum

seed

Geranium
core-core

Geranium dissectum

leaf

carpel body

Geranium molle

Geranium oreganum

seed

Geranium potentilloides

fruit

Geranium richardsonii

5 cm

leaf

carpel body

Geranium robertianum

stem base

Geranium viscosissimum

fruit

fruit

Ribes amarum

Grossulariaceae

sepal

petal

ovary

Ribes aureum var. aureum

fruit

fruit

Ribes binominatum

gland

gland

fruit

Ribes bracteosum

styles 5–8 mm. Uncommon. Coastal bluffs, forest edges; 5–1485 m. NW (exc NCoRI), CaR, CW (exc SCoRI); sw OR. Mar–May

R. hudsonianum Richardson var. **petiolare** (Douglas) Jancz. WESTERN BLACK CURRENT Pl < 2 m. **ST:** ± erect; nodal spines 0; internodes sparsely strigose. **LF:** blade 3–10 cm, coarsely double-dentate, adaxially glabrous, abaxially with few soft, shaggy hairs and sessile, yellow glands. **INFL:** erect, 20–50-fld. **FL:** hypanthium 4–5 mm, saucer-shaped; sepals 4–7 mm, white; petals 2 mm, white. **FR:** 9–10 mm, black; glands sessile, yellow. Streamsides; 15–2100 m. n MP; to BC, ID, NV. May–Jul ★

R. indecorum Eastw. WHITE-FLOWERING CURRANT   Pl < 3 m. **ST:** nodal spines 0. **LF:** blade 1–4 cm, thick, finely serrate, adaxially green, hairy, rough, glandular, abaxially white, tomentose. **INFL:** 10–25-fld, open; bracts pink. **FL:** hypanthium 4–5 mm, barely longer than wide, white; sepals 1–2 mm, white; petals 1 mm, white; styles fused ± to tip, hairy at least at base. **FR:** 6–7 mm, purple, hairy; glands stalked. Chaparral, coastal-sage scrub; 20–1760 m. SCoRO, SCo, TR, PR; n Baja CA. Dec–Mar

R. inerme Rydb. WHITE-STEMMED GOOSEBERRY   Pl < 3 m. **ST:** scrambling; nodal spines 0–3; tip internodes white to gray. **LF:** blade 2–3 cm, coarsely toothed, not glandular. **INFL:** 1–5-fld. **FL:** hypanthium 2–3 mm, ± as long as wide; sepals reflexed, 3–4 mm, green-white, purple at base or not; petals 1–2 mm, white; anthers exceeding petals by < 2 mm. **FR:** 7–11 mm, purple, glabrous. 2*n*=16.

var. **inerme** **LF:** hairs 0 to sparse, short, soft. **FL:** hypanthium, sepal hairs 0. Forest, streamsides, meadow edges; 1200–3300 m. KR, CaRH, SNH, Wrn, SNE; to BC, Rocky Mtns. May–Jun

var. **klamathense** (Coville) Jeps. **LF:** hairs sparse to dense, long, soft to stiff. **FL:** hypanthium hairs ± 0 to soft; sepal hairs soft. Conifer forest edges; 150–1200 m. KR, NCoRO, NCoRI, CaRH; sw OR. Apr–May

R. lacustre (Pers.) Poir. (p. 831) SWAMP CURRANT   Pl < 1 m. **ST:** prostrate to ascending; nodal spines 3–9; internodes bristly. **LF:** blade 3–5 cm, deeply 3–7-lobed, hairs 0 or sparse, nonglandular, adaxially dark green, abaxially light green. **INFL:** 5–15-fld; pedicel jointed to ovary. **FL:** hypanthium 1 mm, saucer-shaped; sepals 1.5 mm, green, purple, or cream with darker veins; petals 1 mm, purple; styles free to base. **FR:** 4–6 mm, black; hairs glandular. 2*n*=16. Along creeks, seeps, meadow margins; 1065–2525 m. KR, NCoRH, CaRH; to AK, e N.Am. Jun–Jul

R. lasianthum Greene WOOLLY-FLOWERED GOOSEBERRY   Pl < 1 m. **ST:** nodal spines 1–3. **LF:** blade 1–2 cm, glandular-hairy. **INFL:** erect, 2–4-fld. **FL:** hypanthium > 3 mm, longer than wide; sepals 2 mm, yellow; petals 1 mm, yellow; styles fused ± to tip. **FR:** 6–7 mm, red, ± glabrous. Open, rocky areas; 1520–3000 m. SNH, SnGb, PR; OR, NV. Jun–Jul

R. laxiflorum Pursh (p. 831) TRAILING BLACK CURRANT   Pl < 2.5 m. **ST:** spreading or decumbent; nodal spines 0. **LF:** blade 5–10 cm, deeply 5–7-lobed, ± round-toothed, adaxially dark green, glabrous, abaxially light green, sparsely glandular-hairy. **INFL:** 6–12-fld. **FL:** hypanthium 6 mm, saucer-shaped; sepals 3–4 mm, red fading pale-green; petals 1 mm, red. **FR:** 4–6 mm black-glaucous; bristles glandular. Forest, tree crowns; < 300 m. n NCo; to AK, ID; also Siberia. Mar–May ★

R. lobbii A. Gray (p. 831) GUMMY GOOSEBERRY   Pl < 2 m. **ST:** nodal spines 3. **LF:** blade 2–3 cm, adaxially ± glabrous, abaxially glandular-hairy. **INFL:** 1–3-fld. **FL:** hypanthium 3–5 mm, longer than wide; sepals reflexed, 10–12 mm, red; petals 5–6 mm, white, pink-tinged or not, margins curled inward; filaments white or pink, anthers exceeding petals by > 3 mm, ± not exceeded by styles, abaxially with sessile glands. **FR:** 10–15 mm, oblong, red; bristles glandular, dense. Montane, subalpine forests; 665–2130 m. KR, NCoRH; to BC. May–Jul

R. malvaceum Sm. CHAPARRAL CURRANT   Pl < 2. **ST:** nodal spines 0. **LF:** blade 2–5 cm, coarsely to finely toothed, densely glandular-hairy. **INFL:** 10–25-fld, open. **FL:** hypanthium 5–8 mm, ± 2 × longer than wide, pink; sepals 4–6 mm, pink to purple; petals 2–3 mm, pink to white; styles fused ± to tip, base hairy. **FR:** 6–7 mm, purple-glaucous; hairs white, glandular.

var. **malvaceum** (p. 831) **LF:** blade adaxially dull olive-green. Chaparral, oak woodland; 5–1220 m. NCoRI, SNF, SnFrB, SCoR, ChI, WTR. Oct–Apr

var. **viridifolium** Abrams **LF:** blade adaxially bright green. Chaparral; 5–1500 m. SCoRO, TR, PR; n Baja CA. Jan–May

R. marshallii Greene (p. 831) MARSHALL'S GOOSEBERRY   Pl < 2 m. **ST:** arched, rooting at tip; nodal spines 3. **LF:** blade 2.5–3.5 cm, ± glabrous, nonglandular. **INFL:** 1–3-fld. **FL:** hypanthium 3–4 mm, ± as long as wide; sepals reflexed, 10–15 mm, purple; petals 5–6 mm, bright yellow, abaxially deeply concave, tip hooded; anthers exceeding petals. **FR:** 10–20 mm, oblong, dark red; prickles nonglandular; hairs sparse, appressed. Montane forest; 1200–2100 m. KR; sw OR. May–Jul ★

R. menziesii Pursh (p. 831) CANYON GOOSEBERRY   Pl < 3 m. **ST:** nodal spines 3; internode bristles dense, at least some glandular. **LF:** blade 1.5–4 cm, glandular-hairy. **INFL:** 1–3-fld. **FL:** hypanthium 2–3 mm, gen ± as long as wide; sepals reflexed, 5–10 mm, purple; petals 2–5 mm, white, margins curled inward; anthers exceeding petals, exceeded by styles, tips mucronate. **FR:** 8–10 mm, purple; bristles dense, stiff, at least some glandular.

var. **ixoderme** Quick AROMATIC CANYON GOOSEBERRY   Pl < 2 m. **LF:** strongly aromatic. Chaparral, montane woodland; 900–1100 m. s SNF. Apr ★

var. **menziesii** Pl < 3 m. **LF:** not strongly aromatic. Forest openings, chaparral; < 1820 m. NCo, NCoRO, CCo, SnFrB, SCoRO; sw OR. Feb–Apr

R. montigenum McClatchie (p. 831) WESTERN PRICKLY GOOSEBERRY   Pl < 1 m. **ST:** spreading or decumbent; nodal spines 1–5; internodes ± bristly or not. **LF:** blade 1.5–2.5 cm, gen lobed ± to base, glandular-hairy. **INFL:** gen > 5-fld. **FL:** hypanthium 1 mm, saucer-shaped; sepals 3–4 mm, green, green-white, or pale yellow; petals 1 mm, red. **FR:** 4–5 mm, orange-red; bristles glandular. Many subalpine, alpine habitats; 800–4000 m. KR, CaRH, SNH, TR, SnJt, Wrn, n DMtns; to BC, ID, NV, AZ. Jun–Jul

R. nevadense Kellogg (p. 831) MOUNTAIN PINK CURRANT   Pl < 2 m. **ST:** nodal spines 0. **LF:** blade 3–8 cm, coarsely to finely toothed, adaxially gen glabrous, abaxially with sessile glands, hairy or not. **INFL:** erect to pendent, 8–20-fld, dense; bract pink. **FL:** hypanthium 2 mm, ± as long as wide; sepals erect, 4–5 mm, pink to red; petals 2–3 mm, white. **FR:** 6–8 mm, blue-black-glaucous; hairs glandular. Forest margins; 600–3050 m. KR, NCoRH, CaR, SNH, Teh, TR, PR, Wrn; s OR, w NV. May–Jun

R. quercetorum Greene (p. 831) OAKWOODS GOOSEBERRY   Pl < 1.5 m. **ST:** arched; nodal spines 1(3); internodes puberulent. **LF:** blade 1–2 cm, gen glandular-hairy. **INFL:** 2–3-fld. **FL:** hypanthium < 3 mm, longer than wide; sepals reflexed, 3 mm, yellow; petals 1 mm, cream; styles fused ± to tip. **FR:** 7–8 mm, black, ± glabrous. Rocky slopes, oak-covered foothills, chaparral; 40–1970 m. c&s SNF, Teh, SnFrB, SCoR, WTR, PR, w edge D; AZ, n Baja CA. Mar–May

R. roezlii Regel SIERRAN GOOSEBERRY   Pl < 1.5 m. **ST:** nodal spines 1–3. **LF:** blade 1.2–2.5 cm, nonglandular. **INFL:** 1–3-fld. **FL:** hypanthium 5–7 mm, longer than wide; sepals reflexed, 7–9 mm, purple; petals 3–4 mm, white, pink-tinged or not, margins curled inward; anthers exceeding petals, exceeded by styles, tips mucronate. **FR:** 14–16 mm, red; prickles stout, nonglandular; hairs glandular.

var. **amictum** (Greene) Jeps. HOARY GOOSEBERRY   Pl < 1.5 m. **LF:** blade abaxially densely white-hairy. **FL:** hypanthium, sepals densely white-hairy. Forest edges, woodland; 15–2000 m. KR, c NCoRO, CaRH. Mar–May ★

var. **cruentum** (Greene) Rehder Pl < 1.5 m. **LF:** blade abaxially glabrous. **FL:** hypanthium, sepals glabrous. Forest, woodland; 150–2300 m. KR, NCoRO, NCoRH; OR. Mar–Jun

var. **roezlii** (p. 831) Pl < 1.5 m. **LF:** blade abaxially hairy. **FL:** hypanthium, sepals hairy. Forest, chaparral, woodland; 200–2850 m. KR, CaRH, SNH, Teh, SCoRO, TR, PR, MP. May–Jun

R. sanguineum Pursh RED-FLOWERING CURRANT   Pl < 4 m. **ST:** nodal spines 0. **LF:** thin to moderately thick; blade 2–7 cm,

coarsely to finely toothed. **INFL:** 10–20-fld; bracts white to red. **FL:** hypanthium 2–7 mm, longer than wide; sepals 4–5 mm, white, pink, or red; petals 2–3 mm, white to red; styles fused ± to tip. **FR:** 4–8 mm, blue-black-glaucous; hairs glandular. $2n=16$.

var. ***glutinosum*** (Benth.) Loudon (p. 831) **LF:** blade abaxially sparsely hairy, with short glands on veins or not. **INFL:** pendent. **FL:** sepals pink to white. Many habitats; < 2320 m. NCo, NCoRO, CW (exc SCoRI); OR. Feb–Apr

var. ***sanguineum*** Pl < 3 m. **LF:** blade abaxially white-hairy to finely tomentose, glandular or not. **INFL:** erect to spreading. **FL:** sepals red. Montane forests; 1300–2320 m. KR, NCoRH; to BC. Apr–Jul

R. sericeum Eastw. SANTA LUCIA GOOSEBERRY Pl < 2 m. **ST:** nodal spines 3; internodes densely hairy, bristles gen glandular. **LF:** blade 2–4 cm, adaxially ± glabrous, abaxially glandular-hairy. **INFL:** 1–3-fld. **FL:** hypanthium 3–4 mm, ± as long as wide; sepals reflexed, 6–8 mm, green or red; petals 3–4 mm, white, margins curled inward; anthers exceeding petals, ± not exceeded by styles, much longer than wide when dehisced. **FR:** 10–20 mm, purple; bristles glandular. Forest openings, coastal scrub, streamside thickets; 180–800 m. SCoRO. Dec–Apr ★

R. speciosum Pursh (p. 831) FUCHSIA-FLOWERED GOOSEBERRY Pl < 2 m. **ST:** nodal spines 3; internodes ± bristly. **LF:** semi-deciduous; blade 1–3.5 cm, leathery, sparsely glandular or gen glabrous, adaxially shiny, dark green, abaxially light green, base entire, wedge-shaped to truncate, lobes crenate. **INFL:** 1–4-fld; pedicel bristly. **FL:** hypanthium 2–3 mm, wider than long; sepals 4, erect, 4–5 mm, red; petals 4, 4–5 mm, red, margins curled inward; filaments red, exceeding petals by > 3 mm. **FR:** 10–12 mm, bristles glandular, dense. Coastal-sage scrub, chaparral; 5–2125 m. CCo, s SnFrB, SCoRO, SCo, WTR, PR; n Baja CA. Jan–May

R. thacherianum (Jeps.) Munz SANTA CRUZ ISLAND GOOSEBERRY Pl < 3 m. **ST:** nodal spines 0–3; internodes hairy, bristly. **LF:** blade 2–3 cm, shallowly lobed, adaxially ± glabrous, abaxially hairy. **INFL:** 1–2-fld; pedicel soft-white-hairy. **FL:** hypanthium 4–5 mm, longer than wide; sepals 9–10 mm, ± pink; petals 6 mm, white, margins curled inward; anthers exceeding petals, exceeded by styles, tips mucronate. **FR:** 7–10 mm, purple; bristles dense, hairs soft. Ravines; 30–175 m. n ChI (Santa Cruz Island). Mar–Apr ★

R. tularense (Coville) Fedde SEQUOIA GOOSEBERRY Pl < 0.5 m. **ST:** low, spreading; nodal spines 3; internode hairs long, bristles

glandular. **LF:** blade 2–5 cm, glandular-hairy. **INFL:** 1–3-fld. **FL:** hypanthium 2–4 mm, ± as long as wide; sepals reflexed, 6 mm, green to white; petals 2–3 mm, white, abaxially shallowly concave. **FR:** 8–10 mm, light yellow, prickles nonglandular; bristles glandular. Montane forest; 1660–1740 m. s SNH (Tulare Co.). Closely related to *R. binominatum.* May ★

R. velutinum Greene (p. 831) Pl > 2 m. **ST:** stout, arched; nodal spines 1(3). **LF:** blade 0.5–2 cm, crenate, glabrous, glandular, or gen densely pubescent. **INFL:** 1–4-fld. **FL:** hypanthium 2–3 mm, ± as long as wide; sepals 3 mm, white to yellow; petals 2 mm, white to yellow; ovary hairs 0 or gen conspicuous, short and long, glandular and not. **FR:** 6–7 mm, yellow becoming purple, glabrous, hairy, or glandular-hairy. Sagebrush steppe, juniper woodland, pine forest; 700–3500 m. KR, CaRH, SNH, Teh, TR, GB, DMtns; to UT, AZ. [*R. v.* var. *gooddingii* (M. Peck) C.L. Hitchc.] Apr–Jun

R. viburnifolium A. Gray (p. 831) SANTA CATALINA ISLAND CURRANT Pl < 1 m. **ST:** erect or arched; nodal spines 0; internode glands white, sessile. **LF:** evergreen; blade 2–4 cm, (ob)ovate, leathery, adaxially dark green, glabrous, abaxially with sessile, yellow glands, base acute, tip wide, lobes 0 or sparse, shallow, teeth 0 or sparse. **INFL:** erect, branched at base, 6–15-fld. **FL:** hypanthium 4–5 mm, saucer-shaped; sepals 2–3 mm, brown; petals 2 mm, red. **FR:** 5–6 mm, red, glabrous. $2n=16$. Chaparral, forest openings; 30–600 m. s ChI (Santa Catalina Island), sw PR; Baja CA. Feb–Apr ★

R. victoris Greene VICTOR'S GOOSEBERRY Pl < 2 m. **ST:** nodal spines 1–3; internodes puberulent, sparsely bristly, sticky. **LF:** blade 1.5–5 cm, sparsely glandular. **INFL:** 1–2-fld. **FL:** hypanthium 3 mm, ± as long as wide; sepals reflexed, 6–10 mm, green-white or white, purple-tinged at base; petals 3–5 mm, white, margins curled inward; anthers exceeding petals, exceeded by styles, tips mucronate. **FR:** 8–10 mm, yellow; bristles ± glandular. Canyon forests, chaparral; 130–750 m. NCoRO, SnFrB. Mar–Apr ★

R. viscosissimum Pursh (p. 831) STICKY CURRANT Pl < 1.5 m. **ST:** nodal spines 0. **LF:** fragrant; blade 3–8 cm, thick, crenate, gray-green, glandular. **INFL:** 4–15-fld. **FL:** hypanthium 5 mm, < 2 × longer than wide; sepals 6–7 mm, white-green to pink; petals 3 mm; anther tip with cup-like depression; styles fused ± to tip. **FR:** 10–12 mm, black to glaucous. Sagebrush scrub, forest; 960–3030 m. KR, NCoRH, CaRH, SNH, MP; to BC, MT, CO. [*R. v.* var. *hallii* (Jancz.) Jancz.] May–Jul

GUNNERACEAE GUNNERA FAMILY

Livia Wanntorp & Elizabeth McClintock

Per, often large, terrestrial [semi-aquatic]; rhizome often with large scales. **ST:** ± 0. **LF:** simple, gen alternate, often large; stipules 0; blades < to > petioles, ± round to reniform or ovate, gen toothed or lobed. **INFL:** panicle (gen spikes along 1° axis), terminal or from upper axils, large; fls many, often pistillate below, staminate above, bisexual between. **FL:** very small, unisexual or bisexual; sepals gen 2 [± 0]; petals 2 [0]; stamens 1–2; ovary inferior, chamber 1, styles 2. **FR:** drupe. 1 genus, ± 35 spp.: Mex, s hemisphere, HI; several cult. Rhizome scales previously interpreted as stipules. Scientific Editor: Thomas J. Rosatti.

GUNNERA

(J.E. Gunner, Norwegian bishop, botanist, 1718–1773) [Wanntorp & De Craene 2005 Int J Pl Sci 166:945–953] Pls obtain fixed nitrogen from cyanobacteria (*Nostoc*) living just under surface of rhizomes, roots, in the only such symbiosis in angiosperms.

G. tinctoria (Molina) Mirb. CHILEAN-RHUBARB Rhizome scales 10–20 cm, deeply cut. **LF:** covered with stiff, hard prickles; petiole 1–1.5 m; blade 1–2 m, ± round, ± concave abaxially, thick, rough, main lobes palmate, teeth irregular, veins prominent, esp abaxially. **INFL:** arising ± from ground, 50–75[100] cm, < 10 cm wide, ± conic; spikes many, dense, 2–5 cm, stout; peduncle gen 2–20 cm. **FL:** style < 1.2 mm, ≥ sepals. **FR:** many, 1–2 mm, ± red, ovate to oblong. Uncommon. Disturbed, shaded, damp areas; < 100 m. CCo (Marin, San Francisco cos.); native to Chile. May spread aggressively by seeds, rhizomes; potentially problematic weed. Jul–Oct

HALORAGACEAE WATER-MILFOIL FAMILY

Adolf Ceska & Oldriska Ceska

[Ann, shrub] per, gen aquatic or semiterrestrial, dioecious or monoecious. **LF**: cauline, opposite, alternate or whorled; submersed blades pinnately divided, segments thread-like; emergent lvs simple, entire to divided. **INFL**: raceme, spike, or panicle; fls 1 or clustered, short-pedicelled to ± sessile. **FL**: gen unisexual, small; calyx tube short, fused to ovary, lobes 2–4; petals gen 2–4; stamens 4 or 8, filaments gen short; ovary inferior, chambers 1–4, styles 2–4, separate, stigmas gen plumose. **FR**: fleshy or of nut-like mericarps, dehiscent or not. **SEED**: gen 1 per chamber. 6–8 genera, ± 100 spp.: esp s hemisphere, some cult. [Aiken & McNeill 1980 J Linn Soc Bot 80:213–222] *Haloragis erecta* (Murray) Eichler not naturalized in CA. Scientific Editor: Bruce G. Baldwin.

MYRIOPHYLLUM WATER-MILFOIL

Pl from rhizomes, occ with overwintering buds (late in growing season); occ terrestrial. **ST**: simple or branched, gen green. **LF**: submersed lvs gen whorled, 3–6 per node; emergent lvs entire to pinnately divided, occ bract-like. **INFL**: gen emergent, spike-like, simple or branched, terminal, fls in whorls. **FL**: proximal pistillate, middle occ bisexual, distal staminate; calyx lobes 4; petals gen 4, ephemeral on staminate fls, minute or 0 on pistillate fls; stamens gen 8; ovary 4-chambered. **FR**: mericarps 4, nut-like. ± 60 spp.: worldwide. (Greek: many lvs, from lf segments) [Ceska et al. 1986 Brittonia 38:73–81] *Myriophyllum* specimens best collected in fl or fr.

1. Lvs alternate, or whorled with some scattered single lvs . . . . *M. hippuroides*
1′ Lvs whorled (opposite)
  2. Pls dioecious; lvs narrow, length 3–4 × width; bracts ± = st lvs . . . . *M. aquaticum*
  2′ Pls monoecious; lvs gen wider; bracts < st lvs.
    3. Bracts pinnately dissected; sts green, dark green to ± black-green when dry . . . . *M. verticillatum*
    3′ Bracts entire or dentate; sts dark green, occ red-tinged; gen ± white to pink or dark gray when dry.
      4. Lvs dark green when fresh, darker when dry, glaucous; young shoots with ≥ 1 pair of entire, opposite lvs at base; pls conspicuously rhizomed; fl bracts triangular, 4–10 mm, toothed, glaucous . . . . *M. quitense*
      4′ Lvs dark green when fresh, gen red-tinged, not glaucous when dry, occ waxy from diatoms; young shoots lacking pairs of entire, opposite lvs at base; pls not conspicuously rhizomed; fl bracts lanceolate to ovate or oblanceolate, 1–3 mm, entire to toothed, not glaucous
        5. Lf segments 4–14 pairs, angles, spacing varying throughout lf . . . . *M. sibiricum*
        5′ Lf segments 14–24 pairs, angles, spacing uniform throughout lf. . . . . *M. spicatum*

M. aquaticum (Vell.) Verdc. PARROT'S FEATHER Dioecious; winter buds 0. **ST**: < 1 m, bright green to ± gray-green, lfy throughout. **LF**: submersed lvs 1.5–3.5 cm, length 3–4 × width, segments 20–30 per lf, ± 7 mm. **INFL**: terminal, gen emergent. **STAMINATE FL**: stamens 8; ovary vestigial. **PISTILLATE FL**: stamens 0. **FR**: gen 0 (in CA). Common. Ponds, streams, lakes; < 500 m. NCo, NCoR, CaRF, SNF, GV, CCo, SnFrB, SCo, DMoj; s&e US, warm temp, trop worldwide; native to S.Am. Pls gen pistillate (staminate unknown in CA), clonal; forms dense colonies with emergent branches; escaped from original plantings as orn. Jul–Sep ❖

M. hippuroides Torr. & A. Gray (p. 831) WESTERN WATER-MILFOIL Monoecious; winter buds 0. **ST**: > 1 m. **LF**: alternate, or whorls of 4–5 with scattered single lvs; submersed lvs > 2 cm, midrib and segments linear, segments < 10 mm, 14–20(40) per lf. **INFL**: spike, 3–12 cm, emergent; bracts 5–15 mm, 1–2 mm wide, >> fls, gen entire to ± toothed. Ponds, small streams; < 1100 m. NCoRO, NCoRI, n SNH, SnJV, n CCo, SnFrB; to s BC, introduced in OK, TX. Easily recognized by single lvs originating between whorls. Jul–Sep

M. quitense Kunth ANDEAN WATER-MILFOIL Monoecious; winter buds 0. **ST**: > 1 m, dark gray when dry; from strong ± white rhizomes. **LF**: in whorls of 4–5, 1–2.5 cm, comb-like, with 8–20 thread-like segments; lowermost st lvs on young shoots bract-like, entire, opposite. **INFL**: simple to conspicuously branched spike, 4–8 cm, emergent; bracts 4–10 mm, triangular, minutely serrate. **FL**: stamens 8. Uncommon. In streams, thermal creeks, shallow water along lake-

shores with strong wave action; < 2000 m. CaRH, n SNH, e SnBr; to BC, NB, PE, WY, UT, AZ, Mex, S.Am. Jul–Sep

M. sibiricum Kom. (p. 831) SIBERIAN WATER-MILFOIL Monoecious; winter buds cylindrical, at ends of non-fl branches, gen in fall. **ST**: > 1 m, ± white or pink when dry. **LF**: submersed lvs 1–3 cm, segments linear, < 20 mm, ≤ 28 per lf, angles, spacing varying throughout lf. **INFL**: spike, 3–8 cm, emergent; bracts 1–3 mm, < fls, oblanceolate to ovate, entire to coarsely toothed. Ponds, streams, lakes; < 2600 m. NCo, NCoRI, CaRH, n SNH, SnFrB, SCoRO, SCo, SnGb, SnBr, PR, MP, w DMoj; to AK, Can, e US, Eurasia. [*M. exalbescens* Fernald.] Jun–Sep

M. spicatum L. EURASIAN WATER-MILFOIL Monoecious; winter buds 0. **ST**: > 1 m, ± red to light green when dry. **LF**: submersed lvs < 3 cm, segments ≥ 28, < 10 mm, linear, angles, spacing uniform throughout lf. **INFL**: spike, 4–8 cm, emergent; bracts 1–3 mm, < fls, lanceolate, entire to toothed. Uncommon. Ditches, lake margins; < 2080 m. NCoRI, SNF, n SNH, GV, SnFrB, SCo, SnBr, PR, MP; to BC, e Can, e US; native to Eurasia, n Afr. Jul–Sep ◆

M. verticillatum L. (p. 831) WHORLED WATER-MILFOIL Monoecious; winter buds club-shaped at ends of non-fl branches, gen in fall. **ST**: > 1 m, dark green to ± black when dry. **LF**: submersed lvs < 5 cm, segments 18–34, < 10 mm, linear. **INFL**: spike, 5–12 cm, emergent; bracts 3–8(10) mm, pinnately dissected, segments linear. Lakes, marshes; < 2072 m. NCo, n&s SNH, GB; to AK, Can, e US, Eurasia. Jul–Sep

HYDRANGEACEAE HYDRANGEA FAMILY

Robert E. Preston & Charles F. Quibell

Per to small tree or vine. **ST**: < 3 m, gen erect; bark gen peeling as thin sheets or narrow strips. **LF**: gen simple, opposite, deciduous or not, ± hairy; stipules 0; blade ± round to narrowly elliptic, entire or toothed. **INFL**: cyme, raceme, panicle, or fl

1, terminal or axillary, gen bracted. **FL:** bisexual, radial, fls on infl margins occ sterile and enlarged; sepals 4–10, free or fused at base, spreading or erect; petals 4–7, free, ± round to narrowly elliptic; stamens 8–12 in 2 whorls or many and clustered, filament base linear or wide and flat; pistil 1, ovary superior to inferior, chambers 2–8, ovules 1–2 or many per chamber, placentas axile or parietal, styles 1–8, free or fused at base. **FR:** capsule, loculicidal or septicidal; styles persistent or not. **SEED:** gen many, small to minute, oblong to fusiform, winged or not. 18 genera, ± 250 spp.: gen temp, subtrop n hemisphere; some cult for orn (*Carpenteria, Hydrangea, Philadelphus*). [Hufford et al. 2001 Int J Pl Sci 162:835–846] Philadelphaceae in TJM (1993). Scientific Editors: Douglas H. Goldman, Bruce G. Baldwin.

1. Fls < 8 mm wide, odorless; sepals 1.5–2 mm, ± deciduous; petals 3–6 mm
  2. Lf blade 8–16 mm, 3–6 mm wide; fr ± cylindric, styles persistent; trunk or main st erect;
    W&I, DMtns . . . . **FENDLERELLA**
  2′ Lf blade 15–40 mm, 10–30 mm wide; fr spheric, styles deciduous; main st prostrate; coastal mtns . . . . . **WHIPPLEA**
1′ Fls ≥ 10 mm wide, fragrant; sepals 3–10 mm, persistent; petals 5–25 mm
  3. Stamens 10, alternating long and short; filament base wide, flat; stigma terminal, < 0.5 mm . . . . **JAMESIA**
  3′ Stamens > 20, ± equal, filament base ± linear; stigma linear along style branches, > 1 mm
    4. Fls 3–6 cm wide; lvs 1-veined from base, persistent, leathery; ovary ± superior; petals ± round;
    c&s SNF (Fresno Co.) . . . . **CARPENTERIA**
    4′ Fls 1–3 cm wide; lvs 3-veined from base, deciduous; ovary 1/2 to completely inferior; petals
    oblong to elliptic . . . . **PHILADELPHUS**

CARPENTERIA

1 sp. (William M. Carpenter, LA botanist, 1811–1848)

C. californica Torr. (p. 831)   TREE-ANEMONE   Shrub < 3 m. **ST:** bark ± gray, peeling as thin, wide sheets. **LF:** persistent, leathery; petiole 0.5–1 cm; blade 4–10 cm, 1–2.5 cm wide, narrowly elliptic, 1-veined from base, margin gen rolled under, entire or ± toothed, green and glabrous above, pale and short-tomentose below. **INFL:** cyme or short raceme, terminal, open; fls 3–13; bracts < lvs, sessile. **FL:** 3–6 cm wide, fragrant; sepals 5–7, 8–10 mm, gen glabrous, white-tomentose at inside tip; petals 5–8, 15–25 mm, ± round, white; stamens many, clustered, filaments linear; ovary ± superior, chambers 5–8; placentas axile, ovules many; style 1, branches 5–7, 3–6 mm; stigmas linear along style branches. **FR:** 0.6–1.2 cm wide, conical, weakly septicidal. **SEED:** fusiform, red-brown. 2*n*=20. Streambanks, chaparral, oak woodland; 340–1340 m. c&s SNF (between Kings, San Joaquin rivers, Fresno Co.). Post-fire stump-sprouting, seedling establishment follows burns. May–Jul ★

FENDLERELLA

Shrub < 8 dm. **ST:** bark ± white, peeling as thin sheets or strips; twigs strigose. **LF:** deciduous, leathery, ± sessile. **INFL:** cymes clustered, terminal, dense to open; fls (1)3–11. **FL:** odorless; sepals 5; petals 5, white; stamens 10, alternating long and short, filament base wide, flat; ovary 1/2 inferior, chambers 3, placentas axile, ovule 1 per chamber, styles 3, persistent, spreading in fr, stigma terminal. **FR:** ± cylindric, septicidal. **SEED:** fusiform, red-brown. 3 spp.: sw US, n Mex. (Latin: small *Fendlera*)

F. utahensis (S. Watson) A. Heller (p. 835)   YERBA DESIERTO   **LF:** blade 8–16 mm, 3–6 mm wide, ovate to elliptic, strigose, 3-veined from base, margin entire, ± rolled under. **INFL:** 12–18 mm, short-peduncled. **FL:** sepals ± 1.5 mm, lance-linear, sparsely strigose; petals 3–4 mm, oblong-obovate. **FR:** ± 4 mm. **SEED:** ± 2 mm. Limestone soils, cliffs, rock crevices, slopes, pinyon/juniper woodland; 1300–2800 m. W&I (Inyo Mtns), DMtns; to CO, n Mex. May–Aug ★

JAMESIA   CLIFFBUSH

Shrub < 2 m; herbage gen densely hairy. **ST:** bark gray to ± red-brown, peeling as narrow strips. **LF:** deciduous, widely ovate to ± round, pinnately veined, margin toothed, sparsely hairy to ± glabrous above, canescent below. **INFL:** terminal, fls 1 or in cymes. **FL:** sepals 4 or 5, lanceolate to narrowly ovate; petals 4 or 5, obovate or oblanceolate; stamens [8]10, alternating long and short, filament base wide and flat; chambers 3–5, 1 in fr, placentas parietal, ovules many, styles 3–5, free, stigma terminal. **FR:** conic to ovoid, septicidal, styles persistent, spreading in fr. **SEED:** many, with net-like ridges. 2 spp.: sw US, n Mex. (Edwin P. James, naturalist, 1797–1861) [Holmgren & Holmgren 1989 Brittonia 41:335–350]

J. americana Torr. & A. Gray var. ***rosea*** C.K. Schneid. (p. 835)   ROSY-PETALLED CLIFFBUSH   Shrub < 1 m. **ST:** bark gen gray. **LF:** petiole 2–6 mm; blade 1.5–4 cm, 1–2 cm wide. **INFL:** terminal cyme; fls (1)3–11. **FL:** 1.2–1.5 cm wide, ± fragrant; sepals 5, 3–4 mm, gray-strigose; petals 5, 5–8 mm, elliptic to obovate, gen pink; stamens 10; ovary 1/2-inferior, styles > sepals. **FR:** 1–1.3 cm, conic. **SEED:** fusiform, brown. Rocky slopes, cliffs; 2070–3700 m. c&s SNH, W&I, n DMtns (Panamint Range, Grapevine Mtns); w NV (Sheep Range, Spring Mtns). Other vars. in GB outside CA, Rocky Mtns. Jul–Aug ★

PHILADELPHUS   MOCK ORANGE

Shrub < 3 m. **ST:** bark red-brown, aging gray, peeling as narrow rectangles or strips; twigs glabrous to hairy. **LF:** deciduous, petioled; blade 3-veined from base, ± glabrous to hairy, margin entire to toothed. **INFL:** fl 1, or raceme to panicle, terminal, ± open. **FL:** fragrant; sepals 4–5, glabrous to hairy; petals 4–5, white; stamens gen many, clustered, filaments linear, fused at base; ovary 1/2 to completely inferior, chambers 4–5, placentas axile, ovules many, style 1, branches gen 4, stigmas linear along style branches. **FR:** becoming woody, gen loculicidal. **SEED:** many, gen fusiform, gen brown. ± 65 spp.: temp Am, Eurasia. (Greek: for Ptolemy Philadelphus, Greek king of Egypt, 309–247 BC) [Frazier 1999 New Mexico Botanist 13:1–6] Many intergrading infraspecific taxa described; more study needed to determine which warrant recognition.

1. Fl 2–3 cm wide; infl raceme, fls 6+; lf blade 25–75 mm, sparsely strigose beneath. . . . . *P. lewisii*
1′ Fl 1–1.5 cm wide; infl not raceme, fls 1–3; lf blade 8–25 mm, densely strigose beneath. . . . . *P. microphyllus*

P. lewisii Pursh (p. 835) WILD MOCK ORANGE **LF:** petiole 3–8 mm; blade 20–40 mm wide, ovate, margin entire to toothed, flat. **FL:** sepals 4–7 mm; petals 8–12 mm, obovate to oblong. 2*n*=26. Slopes, canyons in conifer and hardwood forest, montane chaparral; 30–2440 m. NW, CaR, n&c SN, s SNH, ScV (Sutter Buttes), SnGb, PR (Santa Ana Mtns); to BC, MT. May–Jul

P. microphyllus A. Gray (p. 835) LITTLELEAF MOCK ORANGE **LF:** petiole 5–18 mm; blade 3–8 mm wide, narrowly ovate to elliptic, margin entire, ± rolled under. **FL:** sepals 3–5 mm, petals 6–8 mm, widely elliptic. Rocky slopes, cliffs, pinyon/juniper woodland; 1340–2840 m. e PR (SnJt, Santa Rosa Mtns), W&I, e DMtns; to TX, n Mex. [*P. m.* var. *stramineus* (Rydb.) Henrickson] May–Jul

WHIPPLEA MODESTY

1 sp. (Lieutenant A.W. Whipple, 1818–1863, Commander, Pacific Railroad Expedition of 1853–1854)

W. modesta Torr. (p. 835) Per, subshrub, decumbent. **ST:** bark gray-brown, peeling as narrow strips. **LF:** persistent, ± sessile; blade 1.5–4 cm, 1–3 cm wide, ovate to elliptic, coarsely strigose, upper surface hairs ascending, bulbous-based. **INFL:** cyme or raceme, dense; fls 4–12. **FL:** 4–6 mm wide, odorless; sepals 4–6, 1.5–2 mm, narrowly oblong, ± glabrous; petals 4–6, 3–6 mm, obovate, white; stamens 8–12, ± equal, filament base wide, flat; ovary 1/2-inferior, chambers 4–5, placentas ± axile, ovule 1 per chamber, styles 4–5, stigma terminal. **FR:** spheric, segmented in age. **SEED:** 1 per chamber, plump, honeycomb-pitted. Coastal scrub, chaparral, forest, open areas or on streambanks; 45–1525 m. NW, SnFrB, SCoRO (Santa Lucia Range); to WA. Mar–Jul

HYPERICACEAE ST. JOHN'S WORT FAMILY

Robert E. Preston & Jennifer Talbot

Ann to shrub [tree]. **LF:** cauline, simple, opposite or whorled, often gland-dotted; stipules 0. **INFL:** cyme, panicle, or fl 1, terminal or axillary. **FL:** bisexual, radial; sepals persistent, gen 5, often fused at base, overlapping; petals gen 5, free; stamens gen many, free or ± fused into 3–5 clusters; pistil 1, ovary superior, chambers 1–3[5], placentas gen axile, style branches 3. **FR:** capsule, gen septicidal. **SEED:** many, small. 37 genera, 1610 spp.: worldwide, largely trop. [Gustaffson et al. 2002 Int J Pl Sci 163:1045–1054] Sometimes incl in Clusiaceae. Scientific Editors: Douglas H. Goldman, Bruce G. Baldwin.

HYPERICUM

Ann to shrub, glabrous. **LF:** sessile **INFL:** gen terminal cymes, bracted. **FL:** sepals [4]5; petals [4]5, deciduous or persistent, gen ± yellow; anthers occ black-dotted; ovary chambers 1 or 3(5), placentas 3(5), axile or parietal, projecting into chamber. ± 450 spp.: worldwide. (Greek name) [Robson 2002 Bull Nat Hist Mus London, Bot 32:61–123]

1. Petals < 5 mm; styles 1–1.5 mm; fr chamber 1, placenta parietal
  2. Pl prostrate to decumbent, 3–30 cm; fl branches near st tips; lvs below infl elliptic to ± round, 4–15 mm, abruptly reduced above lowest fl branches . . . . **H. anagalloides**
  2′ Pl ± erect, 20–60 cm; fl branches from mid-st or lower; lvs below infl elliptic to ovate, 10–25 mm, slightly reduced above lowest fl branches . . . . **H. mutilum** subsp. **mutilum**
1′ Petals > 5 mm; styles 2–20 mm; fr chambers 3, placentas axile
  3. Per from taproot, rhizome, or woody caudex
    4. Lf linear to lanceolate, gen folded, tip acute; st from woody caudex . . . . **H. concinnum**
    4′ Lf ± elliptic to ovate or oblong, not folded, tip obtuse; st from taproot or rhizome
      5. Lf narrowly oblong; sterile axillary branches gen 2–10 cm; fr unlobed . . . . **H. perforatum** subsp. **perforatum**
      5′ Lf ovate to elliptic; sterile axillary branches gen < 2 cm; fr 3-lobed . . . . **H. scouleri**
  3′ Evergreen shrub or subshrub
    6. Petals ≤ 10 mm, ≤ sepals . . . . [*H. androsaemum*]
    6′ Petals > 10 mm, 2–4 × sepals
      7. St 1 or few-branched; fls gen 1(3) per st. . . . . **H. calycinum**
      7′ St many-branched; fls gen > 1 per branch
        8. Lvs sessile, narrowly lance-oblong; petals 12–17 mm . . . . **H. canariense**
        8′ Lvs petioled, petiole 1–4 mm; lvs ovate to broadly lanceolate; petals 15–30 mm . . . . [*H. hookerianum*]

H. anagalloides Cham. & Schltdl. (p. 835) TINKER'S PENNY Ann or per 3–30 cm, from matted stolons. **ST:** prostrate to decumbent, slender; lower nodes rooting. **LF:** below infl 4–15 mm, elliptic to ± round, gland-dots clear to green, base ± clasping; above lowest fl branch abruptly reduced, linear. **INFL:** fl branches near st tips; fls 1–15 (many) per st. **FL:** sepals 2–4 mm, unequal, lanceolate to ovate, obtuse or acute; petals 2–4 mm, ± = sepals, gold to salmon; stamens gen 15–25; styles 1–1.5 mm. **FR:** ± 3 mm, spheric to oblong. **SEED:** < 1 mm, yellow-brown. Meadows, marshes, seeps, springs, streambanks, lake margins; < 3220 m. NW, CaR, SN (exc Teh), ne ScV, e SnJV, CCo, SnFrB, SnGb, SnBr, PR, MP; to BC, MT, AZ, Baja CA. Large, low elevation pls approach *H. mutilum.* May–Sep

H. calycinum L. AARON'S BEARD Evergreen shrub, 20–60 cm, from creeping stolons. **ST:** erect, 1 or few-branched. **LF:** petiole 0–2 mm; blade 4.5–10 cm, oblong to narrow-ovate, gland-dotted or -streaked. **INFL:** fls 1(3) per st. **FL:** sepals 10–20 mm, unequal, elliptic to obovate, tip hooded; petals 25–40 mm, obovate to oblanceolate, yellow; stamens 90–120 per cluster, anthers red; styles 12–20 mm. **FR:** 10–20 mm, ovoid. **SEED:** 1.5–2 mm, red-brown, surface finely netted. Shaded wildland-urban interface; < 1040 m. NCoRO, c SNH, CCo, SnFrB; native to Turkey. Orn, occ escape from cult. Jun–Nov

H. canariense L. Evergreen shrub, gen 2–5 m. **ST:** erect or ascending. **LF:** 2–7 cm, narrowly lance-oblong, densely glandular; base tapered. **INFL:** fls in upper st nodes, ≤ 30 per st. **FL:** sepals

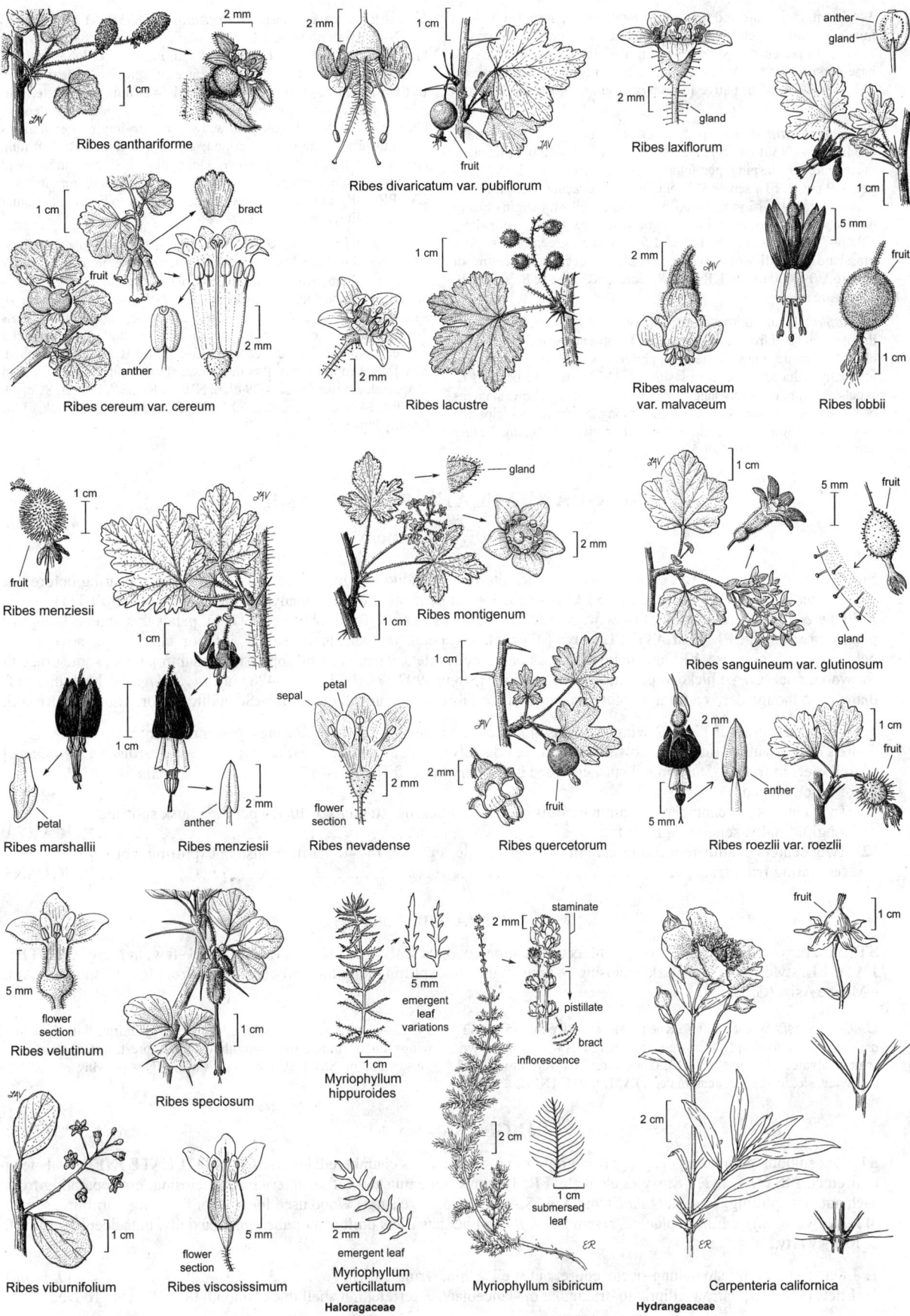

Ribes canthariforme

Ribes divaricatum var. pubiflorum

fruit

Ribes laxiflorum

gland

anther
gland

Ribes cereum var. cereum

bract

fruit

anther

Ribes lacustre

Ribes malvaceum
var. malvaceum

Ribes lobbii

fruit

Ribes menziesii

fruit

gland

Ribes montigenum

fruit

Ribes sanguineum var. glutinosum

gland

Ribes marshallii

petal

Ribes menziesii

anther

sepal
petal

flower
section

Ribes nevadense

Ribes quercetorum

fruit

anther

Ribes roezlii var. roezlii

fruit

flower
section

Ribes velutinum

Ribes speciosum

Myriophyllum
hippuroides

emergent
leaf
variations

staminate

pistillate

bract

inflorescence

fruit

Ribes viburnifolium

flower
section

Ribes viscosissimum

emergent leaf

Myriophyllum
verticillatum

Haloragaceae

submersed
leaf

Myriophyllum sibiricum

Carpenteria californica

Hydrangeaceae

3–4.5 mm, lanceolate and tip acute to ovate and tip rounded, margin minutely toothed; petals 12–17 mm, oblanceolate, bright yellow; stamens 8–10 per cluster; styles 3–6 mm. **FR**: 10–12 mm, ovoid, style base persistent. **SEED**: 1.5–2 mm, yellow-brown. Disturbed areas; < 230 m. CCo, SCo; native to Canary Islands. Orn, escaping from cult. May–Jun ◆

H. concinnum Benth. (p. 835) GOLD-WIRE Per from woody caudex, 15–20(30) cm. **ST**: many, slender. **LF**: 15–40 mm, linear to lanceolate, not clasping, gen folded, sparsely black-dotted. **INFL**: fls gen 3–9 per st. **FL**: sepals 5–9 mm, lanceolate, acute, margin black-dotted; petals 10–15 mm, ± obovate, golden-yellow, margins black-dotted; stamens free or in 3 clusters of 4; styles ± 10 mm, spreading. **FR**: 6–7 mm, 3-lobed. **SEED**: ± 1.5 mm, dark green-brown. Ann grassland, foothill woodland, conifer forest, occ on serpentine or gabbro; 30–1600 m. sw KR, NCoR, CaRF, n&c SN, se ScV, SnFrB. May–Aug

H. mutilum L. subsp. *mutilum* Ann or per, 20–60 cm. **ST**: ± erect, diffuse-branched from mid-st or lower. **LF**: lvs below infl 10–25 mm, elliptic to ovate, gland-dotted; lvs above lowest fl branches not or gradually reduced, uppermost lvs linear. **INFL**: fls 1–15 per st. **FL**: sepals ± 3 mm, narrowly lanceolate; petals ± 2 mm, yellow; stamens 6–12; styles ± 1 mm. **FR**: ± 4 mm, oblong. Streambanks, riparian woodland; < 300 m. ScV, adjacent n SNF; native to e N.Am. 2 other subspp. of e N.Am. Jul–Oct

H. perforatum L. subsp. *perforatum* (p. 835) KLAMATHWEED Per 3–12 dm, taprooted. **ST**: erect, many from base; sterile axillary branches gen 2–10 cm. **LF**: 1.5–2.5 cm, narrowly oblong; margins rolled under, black-dotted; lower surface conspicuously clear-dotted. **INFL**: fls gen 25–100 per st. **FL**: sepals 4–5 mm, lanceolate, acuminate, black- and clear-dotted; petals 8–12 mm, ± oblong, densely black-gland-dotted, bright yellow, twisting, drying in age; stamens in 3 clusters, anthers black-dotted; styles 4–6 mm. **FR**: 7–8 mm, unlobed. **SEED**: ± 1 mm, brown. Open, disturbed areas in many pl communities; 6–1980 m. NW, CaR, n&c SN, ScV (uncommon s SN, CW, PR, MP); to e N.Am; native to Eur. TOXIC to livestock. 3 other subspp. in Eur, Asia. May–Aug ◆

H. scouleri Hook. (p. 835) Per 2–7 dm, taprooted or rhizomed. **ST**: erect, few from base, slender; sterile axillary branches gen < 2 cm. **LF**: 1–3 cm, ovate to elliptic, flat, base ± clasping, margin black-dotted, lower surface obscurely dotted. **INFL**: fls gen 3–25 per st. **FL**: sepals 3–4 mm, oblong to ovate, gen obtuse, black-dotted; petals 7–12 mm, obovate, pale to bright yellow, black-dotted; stamens in 3 clusters, anthers black-dotted; styles 3–5 mm. **FR**: 6–7 mm, 3-lobed. **SEED**: < 1 mm, brown. Wet meadows, streambanks, mesic areas in chaparral, conifer forest; < 2940 m. NW, CaRH, SN, SnFrB, SCoRO, TR, PR, MP; to w Can, MT, NM. [*H. formosum* var. *s.* (Hook.) J.M. Coult.] Jun–Sep

JUGLANDACEAE WALNUT FAMILY

Alan T. Whittemore

Shrub, tree; monoecious. **LF**: odd-(even-)1-pinnate, alternate, deciduous; stipules 0. **INFL**: catkin, gen appearing before lvs, 1 or clustered; fls gen 1 in bract axils. **STAMINATE INFL**: pendent, elongate, many-fld, on last yr's twigs. **PISTILLATE INFL**: erect; fls 1–many, at tip of new twigs. **STAMINATE FL**: calyx [0 or] lobes gen 4 [2, 6]; petals 0; stamens 3–40[50]; pistil 0 or vestigial. **PISTILLATE FL**: calyx [0] or lobes 4; petals 0; stamens 0; ovary 1, inferior, chamber 1 above, gen 2 below, styles 2, plumose. **FR**: nut enclosed in husk or winged nutlets. 9 genera, > 60 spp.: n temp, subtrop mtns; some orn, cult for wood, nuts (*Carya* hickory, pecan; *Juglans* walnut). [Stone 1997 FNANM 3:416–428] In *Carya*, *Juglans*, husk incl in fr diam, even though derived from involucre, calyx, and therefore not technically part of fr. Scientific Editor: Thomas J. Rosatti.

1. Infl pendent in fr, of 15–50 2-winged nutlets, not enclosed in husk; fr body (exc wings) 6–9 mm diam; main
   axis of lf ± winged (at least below lflet attachment distally). . . . . [*Pterocarya stenoptera*]
1′ Infl erect in fr, of 1–3 unwinged nuts, enclosed in leathery husk, fr (incl husk) 2–5 cm diam; main axis
   of lf not winged
   2. Twig centers not chambered; staminate infls clustered, ± sessile, stamens 3–10(15) per fl; fr husk splitting
      longitudinally, separating from fr . . . . **CARYA**
   2′ Twig centers chambered; staminate infl solitary, sessile, stamens 15–40 per fl; fr husk not splitting, not
      separating from fr. . . . . **JUGLANS**

CARYA HICKORY

ST: Bark furrowed in age, gray; twig centers not chambered; buds scaly. **PISTILLATE INFL**: fls 1–few; in fr erect. **PISTILLATE FL**: styles short. **FR**: husk enclosing 1–3 unwinged nuts, splitting longitudinally, separating from fr. ± 17 spp.: e N.Am, e Mex, e Asia. (Greek: nut)

C. illinoinensis (Wangenh.) K. Koch PECAN Tree to 45 m. **ST**: bark shallowly to deeply furrowed; terminal buds 6–7 mm, brown, scales valvate. **LF**: lflets (7)9–13, lanceolate, 110–160 mm, 28–43 mm wide, sickle-shaped, acuminate. **STAMINATE INFL**: ± sessile. **FR**: 3–4 cm, 1.5–2 cm diam, elliptic to ± cylindric, with four low wings along sutures; shell smooth. Riparian forest, watercourse margins; < 600 m. SnJV, SCo; e TX, LA to IL, IN. Apr–May

JUGLANS WALNUT

ST: bark smooth to furrowed in age, gray to gray-brown; twig centers chambered; buds scaly. **PISTILLATE INFL**: fls 1–few; in fr erect. **PISTILLATE FL**: styles elongate. **FR**: 1–3 unwinged nuts per infl; leathery husk enclosing, not separating from each nut, not splitting. 21 spp.: N.Am, temp Asia, S.Am. (Latin: Jove's nut) Wood used for interior finishing, furniture; source of nuts, dye. *J. nigra* L., occ planted, resembles *J. hindsii* but has lflets uniformly pubescent abaxially, nuts deeply grooved, coarsely warty.

1. Lflets 5–11, elliptic to oblong-ovate, entire; nut shell ± thin, wrinkled. . . . . [*J. regia*]
1′ Lflets (9)11–21, narrow-elliptic to -triangular or -lanceolate, ± serrate; nut shell thick, smooth to shallowly grooved.

2. Lflets gen narrow-elliptic to -lance-elliptic (less often lanceolate), rounded to acute; abaxial vein axils glabrous . . . . *J. californica*
2′ Lflets narrow-triangular to -lanceolate, acuminate; abaxial vein axils with tufts of hairs . . . . *J. hindsii*

J. californica S. Watson (p. 835) SOUTHERN CALIFORNIA BLACK WALNUT **ST:** trunk 6–9 m. **LF:** petiole 2–5 cm; lflets (9)11–15(17), 4–9.5 cm, gen narrow-elliptic to -lance-elliptic (less often lanceolate), ± serrate, rounded to acute, abaxial vein axils glabrous. **FR:** 2–3.5 cm diam, nut shell thick, shallowly grooved. Hillsides and canyons; 30–900 m. SCoRO (Santa Lucia Range where cult), SW (exc ChI, SnBr). Mar–May ★

J. hindsii R.E. Sm. NORTHERN CALIFORNIA BLACK WALNUT **ST:** trunk 6–23 m. **LF:** petiole 3–8 cm; lflets 13–21, (6)7–13 cm, narrow-triangular to -lanceolate, ± serrate, acuminate, abaxial vein axils with tufts of hairs. **FR:** 3.5–5 cm diam, nut shell thick, smooth to shallowly grooved. Along streams, disturbed slopes; < 300 m. s NCoRI, s ScV, n SnJV, SnFrB; sw OR. [*J. californica* S. Watson var. *h.* Jeps.] Native to 3 sites, widely cult as rootstock for *J. regia*, escaping sporadically in CW, GV. Apr–May ★

KOEBERLINIACEAE JUNCO FAMILY

Steve Boyd & Staria S. Vanderpool

Shrub or small tree, gen ± lfless, thorny, unscented. **ST:** branches many, rigid, interlocking, glabrous to puberulent, smooth-papillate at 10×. **LF:** simple, alternate, gen scale-like, ephemeral. **INFL:** raceme or ± umbel-like, axillary, bracted. **FL:** radial, bisexual, small; sepals 4, free; petals 4, free; stamens 8, filaments thick, 4-angled in ×-section; ovary ± stalked, spheric, style beak-like, straight, curved, or ± coiled when dry, stigma entire, head-like, or ± 3-lobed. **FR:** berry spheric, fleshy; chambers 2, each 1–2-seeded; style base ± persistent. 1 genus, 2 spp.: sw N.Am, Mex, Bolivia. [Holmes et al. 2008 Brittonia 60:171–184] Often incl in Capparaceae. Scientific Editor: Bruce G. Baldwin.

KOEBERLINIA ALL THORN

(C.L. Koeberlin, German clergyman, botanist, b. 1794)

K. spinosa Zucc. var. *tenuispina* Kearney & Peebles (p. 835) SLENDER-SPINED ALL THORN Pl 3–10 m, puberulent. **ST:** branchlets (25)33–70(82) mm, pale green, tips narrow, sharp-pointed, 3–6 mm, black or dark brown. **LF:** < 2 mm. **INFL:** 1–7 cm, occ thorn-tipped; bracts ± 1.5 mm, linear, ± succulent; pedicels 3–6 mm. **FL:** sepals ± 1.3 mm, ovate-deltate, entire or ± irregularly minutely toothed, puberulent, pale green; petals 4–4.8 mm, short-clawed, blade obovate or oblanceolate, glabrous, white (yellow); stamens 2–4 mm, glabrous; ovary 1.8–2 mm diam, stalk 0.5–0.8 mm, style 1–1.2 mm, straight to ± S-curved when dry, stigma entire to minute-lobed. **FR:** 4–5 mm, glabrous, black. Creosote-bush scrub; 400 m. DSon (Chocolate Mtns); sw AZ, Mex (e Baja CA, nw SON). [*K. s.* subsp. *t.* (Kearney & Peebles) A.E. Murray] Mar–Jul ★

KRAMERIACEAE RHATANY FAMILY

Beryl B. Simpson

Shrub [per], green root-parasite. **ST:** prostrate to erect, much-branched. **LF:** gen simple [1-ternate], alternate, crowded on short-shoots or not, sessile [petioled]; stipules 0; blade linear to ovate, hairy, glandular or not, tip abruptly pointed. **INFL:** fls 1 in axils [terminal racemes]; pedicel jointed, bractlets 2. **FL:** bisexual, bilateral; sepals 4–5, free, petal-like, magenta, red-purple [yellow]; petals gen 5, 3 upper linear to clawed, free or fused basally, held in ± upright flag, 2 lower modified as glands; stamens gen 4, fused to flag base, upcurved, anthers opening by pores; ovary superior, hairy, style slender, upcurved. **FR:** nut-like, with smooth or barbed prickles; seed 1. 1 genus, 18 spp.: Am, arid semitrop, trop. [Simpson et al. 2003 Syst Bot 29:97–108] Pollinating bees collect oils secreted by glandular petals. Scientific Editor: Thomas J. Rosatti.

KRAMERIA RHATANY

(J.G.H. Kramer, Austrian Army physician, botanist, 1684–1744)

1. Flag petals not or ± clawed, free; fr prickles barbed at tip; sepals ± spreading to reflexed . . . . *K. bicolor*
1′ Flag petals clawed, fused at claw bases; fr prickles smooth or barbed along shaft, although sometimes only distally; sepals spreading to ascending . . . . *K. erecta*

K. bicolor S. Watson (p. 841) WHITE RHATANY Densely canescent or silky-hairy. **ST:** < 1 m; branches ± spreading, tips thorn-like. **LF:** narrowly lanceolate. **FL:** buds curved upward; sepals deep purple-red; petal blades oblanceolate, green at base, magenta or purple above; glandular petals purple, outer face glandular covered with equal-sized blisters. **FR:** ± spheric. Dry, rocky or sandy places, esp on lime soils; < 1400 m. D; to NV, TX, n Mex. [*K. grayi* Rose & Painter] Apr–May

K. erecta Schult. (p. 841) PIMA RHATANY, PURPLE HEATHER, LITTLE-LEAVED RHATANY Pl ± strigose to canescent or ± silky-hairy. **ST:** < 2 m; branches often ascending, tips not thorn-like. **LF:** ± linear to lance-linear. **FL:** buds ovate, ± curved; sepals magenta; flag petals ± yellow to green at base, triangular, green and magenta, with ± purple marks above or not; glandular petals magenta with unequal glandular blisters on distal edge abaxially. **FR:** cordate, ± flat. Dry, rocky ridges, slopes; < 1200 m. e PR (Santa Rosa Mtns), D; to NV, TX, n Mex. Mar–May

LAMIACEAE (Labiatae) MINT FAMILY

Dieter H. Wilken & Margriet Wetherwax, family description, key to genera

Ann to shrub [tree, vine], glabrous to hairy, gen aromatic. **ST**: gen erect, gen 4-angled. **LF**: gen simple to deeply lobed, gen opposite, gen gland-dotted. **INFL**: gen cymes, gen many in dense axillary clusters surrounding st, gen separated by evident internodes or collectively crowded, spike- or panicle-like, occ head-like or raceme, subtended by lvs or bracts; fls sessile or pedicelled. **FL**: gen bisexual; calyx gen 5-lobed, radial to bilateral; corolla gen bilateral, 1–2-lipped, upper lip entire or 2-lobed, ± flat to hood-like, occ 0, lower lip gen 3-lobed; stamens gen 4, epipetalous, gen exserted, paired, pairs gen unequal, occ 2, staminodes 2 or 0; ovary superior, gen 4-lobed to base chambers 2, ovules 2 per chamber, style 1, gen arising from center at junction of lobes, stigmas gen 2. **FR**: gen 4 nutlets, gen ovoid to oblong, smooth. ± 230 genera, 7200 spp.: worldwide. Many cult for herbs, oils (*Lavandula*, lavender; *Mentha*, mint; *Rosmarinus*, rosemary; *Thymus*, thyme), some cult as orn (in CA *Cedronella*, *Leonotis*, *Monarda*, *Phlomis*). [Harley et al. 2004 Fam Gen Vasc Pl 7:167–275] *Moluccella laevis* L., shell flower, historical waif in CA. *Satureja calamintha* (L.) Scheele subsp. *ascendens* (Jordan) Briq. reported as alien but not naturalized. *Salazaria* moved to *Scutellaria*; CA *Satureja* moved to *Clinopodium*. Scientific Editors: Douglas H. Goldman, Bruce G. Baldwin.

1. Fertile stamens 2; staminodes 0 or 2
  2. Corolla gen 4-lobed, 2–5 mm, white, not 2-lipped; fr ± compressed, edge corky . . . . **LYCOPUS**
  2′ Corolla 2-lipped, 2.5–35 mm, white or variously colored; fr not compressed, edge not corky
    3. Fertile anther sacs 1 or 2, on thread-like appendage hinged or joined to filament tip; sterile anther sac reduced, modified, or 0 . . . . **SALVIA**
    3′ Fertile anther sacs 2, attached to filament tip side-by-side
      4. Shrub, rounded to mound-like, gen 5–10 dm . . . . **POLIOMINTHA**
      4′ Ann, per, or subshrub, gen < 5 dm
        5. Calyx ± radial, lobes ± equal — e DMoj (New York Mtns) . . . . **MONARDA**
        5′ Calyx ± 2-lipped, lobes of upper versus lower lips unequal
          6. Infl bract margin spiny, veins ± raised . . . . ²**ACANTHOMINTHA**
          6′ Infl bract margin entire or hairy, not spiny, veins not raised
            7. Per; infls ± evenly spaced; calyx tube swollen below middle — DMtns . . . . **HEDEOMA**
            7′ Ann; infls densely clustered, ± terminal, head-like; calyx tube not swollen . . . . ²**POGOGYNE**
1′ Fertile stamens 4; staminodes 0
8. Calyx 2-lipped, entire . . . . **SCUTELLARIA**
8′ Calyx gen 5- or 10-lobed, radial to 2-lipped with lips 2–3-lobed
  9. Calyx 10-lobed, lobe tips recurved or hooked; stamens incl in corolla tube . . . . **MARRUBIUM**
  9′ Calyx gen 5-lobed, lobe tips straight; stamens gen exserted, hidden under upper lip to > corolla lobes
    10. Stamens reclining on lower corolla lip; corolla violet, lower lip pouched; at least younger sts densely stellate-hairy — s DMoj, DSon . . . . **HYPTIS**
    10′ Stamens ascending under, or ± parallel to and exceeding upper corolla lip, if present; corolla violet or not, lower lip ± flat or convex, not pouched; sts glabrous to hairy, not stellate-hairy
      11. Infl bract margin spiny . . . . ²**ACANTHOMINTHA**
      11′ Infl bract margin glabrous to hairy, not spiny
        12. Nutlets, ovary lobes fused laterally below middle; style sunken but not to base of ovary lobes
          13. Corolla lower lip 5-lobed, upper lip 0; lvs lobed or 0 . . . . **TEUCRIUM**
          13′ Corolla not lipped, 5-lobed, lower lobe gen reflexed; lvs entire . . . . **TRICHOSTEMA**
        12′ Nutlets, ovary lobes separate to base; style arising between ovary lobes at their base
          14. Corolla ± radial to ± bilateral, lobes 4 or 5, length ± equal
            15. Fl clusters gen axillary, evenly spaced or together cylindric, gen spike-like; subtended by 2+ lvs . . . . **MENTHA**
            15′ Fl clusters gen terminal, gen 1 per st, head-like; subtended by involucre-like whorl of bracts . . . . **MONARDELLA**
          14′ Corolla clearly bilateral, gen 2-lipped, lobes and lips unequal
            16. Subshrub or shrub
              17. Corolla < 8 mm; calyx not inflated in fr . . . . *Clinopodium chandleri*
              17′ Corolla gen > 20 mm; calyx ± inflated in fr . . . . **LEPECHINIA**
            16′ Ann or per (to subshrub, the corolla then salmon-orange)
              18. Infls gen terminal, collectively ± dense, gen head- to spike- or panicle-like, occ interrupted by 1–3 internodes proximally
                19. 2 or 4 stamens clearly exceeding corolla
                  20. Bracts at base of infl, gen not overlapping, green . . . . **AGASTACHE**
                  20′ Bracts throughout infl, overlapping, pale green or cream (± purple) . . . . ²**ORIGANUM**
                19′ Stamens not clearly exceeding lower corolla lobes
                  21. Ann; infl bracts often ≥ fls, linear to narrowly oblanceolate . . . . ²**POGOGYNE**
                  21′ Per; infl bracts gen < fls, often lanceolate or wider
                    22. Calyx 2-lipped, upper lip 3-toothed, lower lip 2-lobed — lobes acuminate . . . . **PRUNELLA**
                    22′ Calyx ± radial, 5-lobed
                      23. Infl bracts pale green or cream (± purple), persistent; calyx 2–3.5 mm . . . . ²**ORIGANUM**

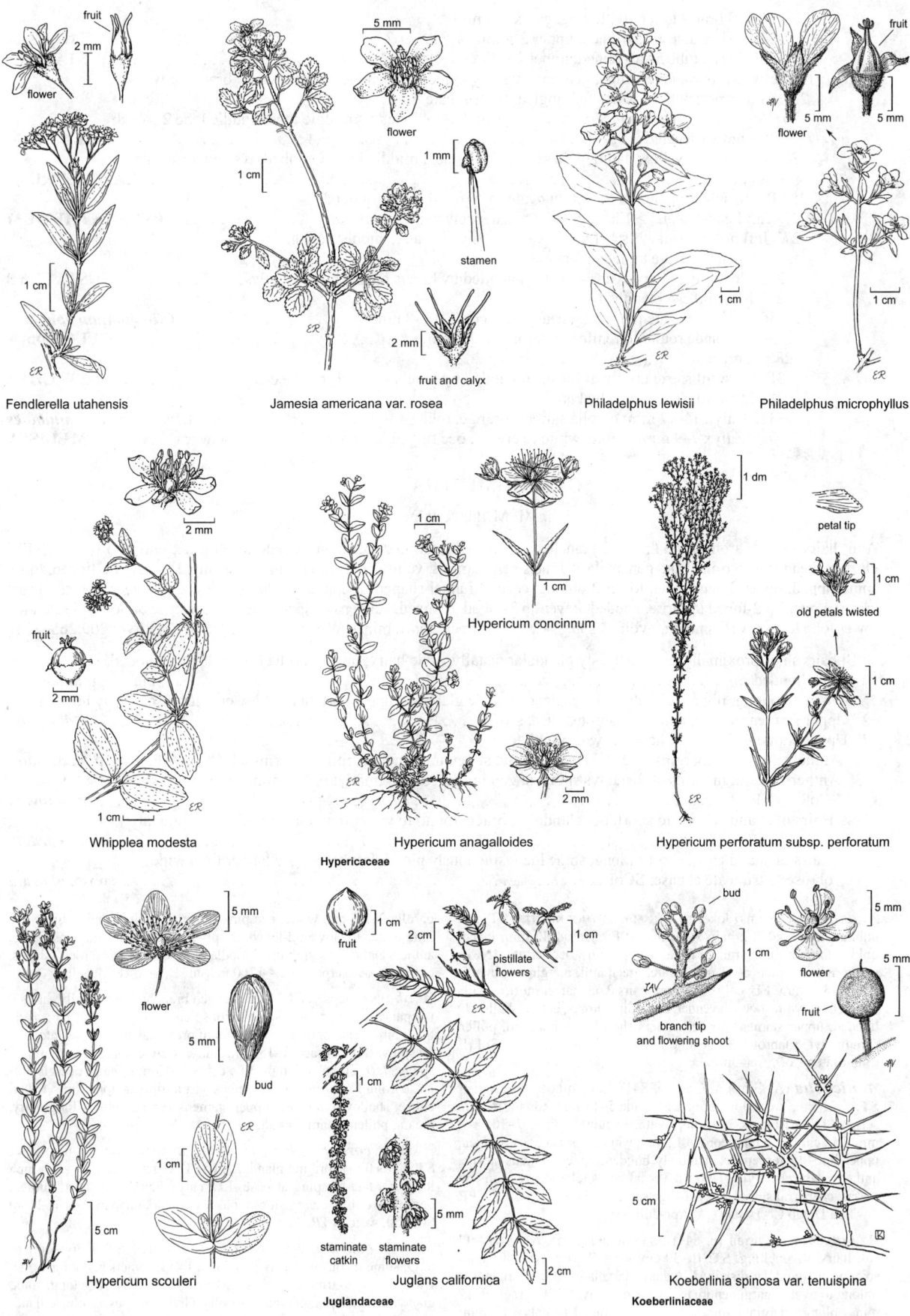

Fendlerella utahensis

fruit
2 mm
flower
1 cm
ER

Jamesia americana var. rosea

5 mm
flower
1 mm
stamen
1 cm
ER
2 mm
fruit and calyx

Philadelphus lewisii

1 cm
ER

Philadelphus microphyllus

fruit
5 mm
5 mm
flower
1 cm
ER

Whipplea modesta

2 mm
fruit
2 mm
1 cm
ER

Hypericum anagalloides

Hypericaceae

1 cm
1 cm
Hypericum concinnum
2 mm

Hypericum perforatum subsp. perforatum

1 dm
petal tip
1 cm
old petals twisted
1 cm
ER

Hypericum scouleri

5 mm
flower
5 mm
bud
1 cm
ER
5 cm

Juglans californica

fruit
1 cm
2 cm
pistillate flowers
1 cm
ER
1 cm
5 mm
staminate catkin   staminate flowers
2 cm

Juglandaceae

bud
1 cm
branch tip and flowering shoot
IAV
5 mm
flower
5 mm
fruit
5 cm

Koeberlinia spinosa var. tenuispina

Koeberliniaceae

23′ Infl bracts green, withering; calyx > 4 mm
  24. Calyx tube ± 15-veined; upper 2 stamens > lower 2 . . . . **NEPETA**
  24′ Calyx tube 5–10-veined; upper 2 stamens < lower 2 . . . . ³**STACHYS**
18′ Infls axillary and terminal, gen separated by evident internodes subtended by bracts or lvs
  25. Ann to per; nutlet ×-section 3-angled, tip truncate
    26. Ann; corolla pink- to red-purple, lower lip lateral lobes << middle lobe, middle lobe 2-lobed;
      lvs not variegated . . . . **LAMIUM**
    26′ Per; corolla yellow, lateral lobes ± < middle lobe, middle lobe not lobed; lvs gen variegated
      . . . . **[LAMIASTRUM]**
  25′ Per to subshrub; nutlet ×-section ovate to ± round, tip ± rounded
    27. Infl head-like, fls > 12; calyx lobe tips densely white-short-hairy . . . . **PYCNANTHEMUM**
    27′ Infl not head-like, fls 1–12; calyx lobe tips glabrous to moderately hairy
      28. St gen prostrate to decumbent
        29. Most infls, exc proximal-most, subtended by bracts or much-reduced lvs . . . . ³**STACHYS**
        29′ Most infls subtended by lvs
          30. Lf blade ovate to broadly triangular; corolla 3–8 mm . . . . *Clinopodium douglasii*
          30′ Lf blade round to reniform; corolla 8–25 mm . . . . **GLECHOMA**
      28′ St erect
        31. Most infls, exc proximal-most, subtended by bracts or much-reduced lvs . . . . ³**STACHYS**
        31′ Most infls subtended by lvs
          32. Calyx 14–17 mm; corolla salmon-orange, tube straight . . . . *Clinopodium mimuloides*
          32′ Calyx 7–9 mm; corolla white to cream, occ tinged lavender, tube curved upward . . . . **MELISSA**

ACANTHOMINTHA　THORNMINT

John M. Miller & James D. Jokerst

Ann, hairy or not, aromatic. **ST:** erect, branched or not. **LF:** petioled; blade lanceolate to obovate, entire to spiny. **INFL:** clusters, head-like, terminal and gen axillary; bracts gen scarious, veins conspicuous, margins spiny. **FL:** calyx 2-lipped, lobes spine-tipped, upper 3 acuminate, lower 2 oblong; corolla funnel-shaped, 2-lipped, white, occ tinged lavender or rose, throat cream, upper lip 2-lobed to entire, hooded, lower lip 3-lobed, reflexed; stamens 4, upper 2 reduced, sterile or not; style slender, lower lobe longer. **FR:** smooth, ovoid. 4 spp.: s CA-FP. (Greek: thorn mint) [Wagstaff et al. 1998 Pl Syst Evol 209:265–274]

1. St hairs short proximally, conspicuously glandular distally; style hairy; upper corolla lip ± = lower, 2-lobed,
   deeply hooded . . . . ***A. lanceolata***
1′ St hairs 0 to long, if glandular then inconspicuous; style glabrous; upper corolla lip < lower, entire, shallowly hooded
  2. Upper stamens sterile; anthers glabrous; SCo, sw PR . . . . ***A. ilicifolia***
  2′ Upper stamens fertile; anthers hairy; CW, WTR
   3. Anthers pink-red; margins of distal lvs not spiny; st gen unbranched; infl gen terminal; CCo, SnFrB . . . . ***A. duttonii***
   3′ Anthers cream; margins of distal lvs spiny; st gen branched proximally; infl terminal and axillary;
    SCoR, WTR . . . . ***A. obovata***
    4. Hairs of st and calyx 0 to short, not glandular; bracts round to wider than long, cordate-clasping at
      base; SCoRO, w&c WTR . . . . subsp. ***cordata***
    4′ Hairs of st and calyx short or long, some inconspicuously glandular; bracts gen longer than wide,
      obtuse to truncate at base; SCoRI . . . . subsp. ***obovata***

A. duttonii (Abrams) Jokerst　SAN MATEO THORNMINT　**ST:** gen unbranched, < 20 cm; hairs 0 to short. **LF:** blade 8–12 mm, lance-oblong to obovate, margin of distal lvs not spiny, occ serrate. **INFL:** gen terminal; bracts 5–11 mm, ovate, green at fl, marginal spines 5, 7, or 9, 3–7 mm. **FL:** calyx 5–8 mm, hairs 0 or sparse, short; corolla 12–16 mm, white, occ ± lavender, upper lip < lower, entire, shallowly hooded; upper stamens fertile, anthers short-hairy, pink-red, pollen cream; style glabrous. Serpentine grassland; < 300 m. CCo, SnFrB (San Mateo Co.). Apr–Jun　★

A. ilicifolia (A. Gray) A. Gray (p. 841)　SAN DIEGO THORNMINT　**ST:** 5–15 cm; hairs 0 to short. **LF:** blade 5–15 mm, round, margin serrate. **INFL:** bracts 7–9 mm, ovate, marginal spines 7–10, 4–8 mm. **FL:** calyx ± 5 mm; corolla 12 mm, white, lobes occ rose-tinged, upper lip < lower, entire, shallowly hooded; upper stamens sterile, anthers glabrous; style glabrous. Vernal pools, clay depressions on mesas, chaparral slopes, coastal-sage scrub; < 1000 m. s SCo, sw PR (sw San Diego Co.); Baja CA. Apr–Jun　★

A. lanceolata Curran (p. 841)　SANTA CLARA THORNMINT　Pl soft-hairy, ill-smelling. **ST:** 10–30 cm; hairs short proximally, conspicuously glandular distally. **LF:** glandular; blade 10–20 mm, lance-oblong to ovate, margin entire to serrate or spiny. **INFL:** bracts 9–12 mm, oblong, marginal spines 7–9, 10–12 mm. **FL:** calyx 12 mm;

corolla 2–2.5 cm, white, occ pink-tipped, glandular-hairy, lips 8–10 mm, upper ± = lower, 2-lobed, deeply hooded; upper stamens fertile, anthers glabrous; style hairy. Woodland, chaparral, talus, rocky slopes, outcrops, occ serpentine; < 1200 m. SnFrB, SCoRI. Mar–Jun　★

A. obovata Jeps.　**ST:** 4–25 cm, gen branched proximally; hairs 0 to sparse, short or long, some inconspicuously glandular. **LF:** blade 8–12 mm, ovate or obovate; margin of proximal entire to serrate, distal spiny. **INFL:** bracts 7–15 mm, shiny, straw-colored at fl, marginal spines 7, 9, or 11, 5–8 mm. **FL:** calyx 7–13 mm, hairs occ glandular; corolla 12–27 mm, white, lobes purple-tipped, upper lip < lower, entire, shallowly hooded; upper stamens fertile, anthers long-hairy, cream, pollen cream. 2n=38.

  subsp. ***cordata*** Jokerst (p. 841)　HEART-LEAVED THORNMINT　**ST:** hairs 0 to short, not glandular. **INFL:** bracts round or wider than long, cordate-clasping at base. **FL:** calyx hairs 0 to short; anthers densely woolly. Grassy slopes, oak woodland, chaparral; < 1600 m. SCoRO, w&c WTR. Apr–Jul　★

  subsp. ***obovata***　SAN BENITO THORNMINT　**ST:** hairs short to long, some inconspicuously glandular. **INFL:** bracts gen longer than wide, obtuse to truncate at base. **FL:** calyx hairs short or long, some glandular; anthers moderately woolly. Grassy slopes, oak woodland, chaparral, vertic clay, occ serpentine; < 1600 m. SCoRI. Apr–Jul　★

AGASTACHE HORSEMINT

Deborah Engle Averett

Per, erect, gen < 1 m, aromatic. **LF**: petioled; blade ± lanceolate to triangular, margin crenate to coarsely serrate. **INFL**: spike of sessile clusters, dense; bracts 1–several at base, lf-like, lanceolate. **FL**: calyx 5-lobed, 2-lipped, pink in age, lobes acuminate; corolla 5-lobed, rose to rose-purple, 2-lipped, lower longer, broader, upper 2-lobed; stamens exserted, 4, in 2 pairs, 1 pair longer, anther sacs spreading; style 2-lobed, exserted. **FR**: ± 2 mm, oblong, brown, smooth, tip with small hairs. 2*n*=18. 22 spp.: N.Am, Mex, Asia. (Greek: many spikes)

1. Lf ± triangular, gen < 3 cm, gen < 2 cm wide . . . . *A. parvifolia*
1′ Lf ± broadly lanceolate, 3–8 cm, 1.5–7 cm wide . . . . *A. urticifolia*

A. parvifolia Eastw. (p. 841) **FL**: corolla rose. Uncommon. Woodland; 1400–2200 m. CaRH, MP. Jun–Aug

A. urticifolia (Benth.) Kuntze (p. 841) **FL**: corolla rose to rose-purple. Common. Gen woodland, but many habitats; 400–3000 m. NCoR, CaR, n&c SNF, SNH, SCoRO, SnBr, Wrn, n SNE; to BC, Rocky Mtns. Coastal pls gen hairier than those inland. Jun–Aug

CLINOPODIUM

Margriet Wetherwax & John M. Miller

Per to shrub, decumbent to erect, < 2 m, aromatic. **LF**: petioled; blade gen ovate-deltate, base rounded to tapered, margin entire to shallowly crenate-dentate, lower surface gen gland-dotted. **INFL**: fls 1–several in lf axils. **FL**: calyx 5-lobed, 2-lipped; corolla 5-lobed, white to lavender or salmon, 2-lipped, lower lip spreading, upper erect, 2-lobed; stamens 4, in 2 pairs, anther sacs spreading; style 2-lobed, exserted. **FR**: 1–2 mm; surface smooth to net-like. 150 spp.: gen Medit, to Japan, Australia, also N.Am, S.Am. (Latin: savory) [Cantino & Wagstaff 1998 Brittonia 50:63–70] *C. hortensis* Kuntze cult as herb (summer savory).

1. Per, decumbent; calyx 4–5 mm . . . . *C. douglasii*
1′ Per to shrub, erect; calyx 6–17 mm
  2. Lf 5–15 mm; calyx 6–8 mm; shrub . . . . *C. chandleri*
  2′ Lf 35–80 mm; calyx 14–17 mm; per to subshrub . . . . *C. mimuloides*

C. chandleri (Brandegee) P.D. Cantino & Wagstaff (p. 841) SAN MIGUEL SAVORY Shrub, erect, < 0.5 m. **ST**: woody; bark red-brown; hairs recurved, white. **LF**: 5–15 mm, 4–16 mm wide, deltate to ovate-deltate, shallowly crenate-dentate; hairs short, white. **INFL**: fls 1–6 per lf axil; pedicel 1–3 mm. **FL**: calyx 6–8 mm, bell-shaped, tinged purple, lobes ± 1 mm; corolla 4–7 mm, white to lavender. **FR**: ± 1.5 mm, shiny dark brown; surface ± net-like. Rocky slopes, chaparral; < 1100 m. PR; n Baja CA. [*Satureja c.* (Brandegee) Druce] Mar–Jul ★

C. douglasii (Benth.) Kuntze (p. 841) YERBA BUENA Per, decumbent, forming mats < 1 m wide. **ST**: ± woody, occ rooting; hairs sparse, minute, recurved. **LF**: 10–35 mm, 5–25 mm wide, ovate to ovate-triangular, shallowly crenate-dentate; hairs sparse, minute. **INFL**: fls 1(3) per lf axil; pedicel 5–20 mm. **FL**: calyx 4–5 mm, tubular, purple in age, lobes ± 0.5 mm; corolla 3–8 mm, white to lavender. **FR**: ± 1 mm, shiny brown, smooth. Shady places, chaparral, woodland; < 900 m. NW, n SNH, CW, SW; to BC, ID. [*Satureja d.* (Benth.) Briq.] Apr–Sep

C. mimuloides (Benth.) Kuntze (p. 841) MONKEY-FLOWER SAVORY Per to subshrub, erect, < 2 m. **ST**: non-woody exc occ at base; hairs long, ± spreading. **LF**: 35–80 mm, 15–55 mm wide, ± ovate to triangular, shallowly, irregularly crenate-dentate; hairs appressed, white. **INFL**: fls 1–6 per lf axil; pedicel 2–3 cm. **FL**: calyx 14–17 mm, tubular-bell-shaped, veins purple, lobes 3–5 mm; corolla 15–35 mm, salmon-orange, gen faded when dry. **FR**: ± 2 mm, brown, surface smooth. Moist places, streambanks, chaparral, woodland; 400–1800 m. CCo, SCoRO, WTR, SnGb. [*Satureja m.* (Benth.) Briq.] Jun–Oct ★

GLECHOMA GROUND IVY

John M. Miller & Dieter H. Wilken

Per, glabrous to sparsely hairy, fls only bisexual or only pistillate. **ST**: prostrate to decumbent, occ erect, gen rooting at lower nodes. **LF**: petioled; blade round to reniform, crenate to toothed. **INFL**: fls 2–5 in lf axils; bracts minute or 0. **FL**: calyx 5-lobed, tube 15-veined, lobes unequal, upper >> lower; corolla 2-lipped, upper lip 2-lobed, ± flat, lower lip 3-lobed, central lobe > lateral lobes; stamens 4, fertile, upper pair >> lower; style lobes ± equal. 10 spp.: temp Eurasia. (Greek: ancient name) [Wagstaff et al. 1998 Pl Syst Evol 209:265–274]

G. hederacea L. **ST**: 1–5 dm; hairs spreading to reflexed. **LF**: petiole gen 1–2 cm; blade 1–2 cm, ± round to reniform. **INFL**: bracts bristle-like. **FL**: calyx 3–7 mm, fine bristly to puberulent; corolla violet, purple-spotted, in bisexual fl gen 15–25 mm, in pistillate fl 8–15 mm. 2*n*=18,36. Disturbed, gen moist, shaded places; < 800 m. NW, SNF, CW, SCo; widespread N.Am; native to Eur. Mar–Jun

HEDEOMA MOCK PENNYROYAL

John M. Miller & Dieter H. Wilken

[Ann] per [subshrub], aromatic; hairs short, spreading to recurved. **ST**: decumbent to erect, branched at base. **LF**: short-petioled to sessile; blade round to linear, entire or toothed. **INFL**: head-like in distal lf axils; bracts minute. **FL**: calyx 2-lipped, upper lip 3-lobed, lower 2-lobed, lobes acuminate, sharp-pointed, tube swollen or pouched proximally; corolla 2-lipped, upper lip > lower, entire to 2-lobed, lower lip 3-lobed; stamens 2, under upper lip or exserted, staminodes minute or 0; style unequally 2-lobed. **FR**: nutlets pitted, glaucous, gelatinous when wet. 38 spp.: N.Am, S.Am. (Greek: ancient name for strongly aromatic mint) [Wagstaff et al. 1998 Pl Syst Evol 209:265–274]

1. Lf blades linear to ± oblong; in fr calyx lobes converging, throat ± closed; sts proximally
glabrous in age . . . . *H. drummondii*
1′ Lf blades ovate to ± round; in fr calyx lobes spreading to ± reflexed, throat ± open;
sts puberulent throughout . . . . *H. nana* subsp. *californica*

H. drummondii Benth. (p. 841) DRUMMOND'S FALSE
PENNYROYAL Pl 15–45 cm. **ST:** puberulent. **LF:** blade 5–11 mm,
1–4 mm wide, tip gen obtuse. **INFL:** fls 3–7; pedicels 2–3.5 mm. **FL:**
calyx 5–6 mm; corolla 7–9 mm, lower lip length = width. 2*n*=34,36.
Rocky, gravelly soils; 1400–1700 m. e DMtns (New York Mtns); to
MT, NE, Mex. May–Jun ★

H. nana (Torr.) Briq. subsp. **californica** W.S. Stewart (p. 841)
CALIFORNIA MOCK PENNYROYAL Pl 10–25 cm. **ST:** puberulent. **LF:**
blade 3–8.5 mm, 2–4.5 mm wide, tip acute. **INFL:** fls 3–5; pedicel
3–4.5 mm. **FL:** calyx 4.5–5.5 mm; corolla 8–9 mm, lower lip length <
width. 2*n*=36. Rocky outcrops, gen limestone; 900–2100 m. DMtns;
NV, AZ, Mex. [*Hedeoma nanum* (Torr.) Briq. subsp. *californicum* W.S.
Stewart, orth. var.] Other subspp. to UT, TX, c Mex. May–Jun ★

HYPTIS

Raymond M. Harley & Dieter H. Wilken

Ann to tree, glabrous to densely hairy. **ST:** erect to spreading, branched. **LF:** gen petioled. **INFL:** axillary; fl clusters loose to
dense, occ in heads, involucre of bracts 0 (present). **FL:** calyx gen 5-lobed, lobes gen equal; corolla 2-lipped, upper lip 2-lobed,
flat, lower lip 3-lobed, central lobe reflexed in age, ± pouched; stamens 4, fertile, curved, reclining on lower lip, exserted; style
lobes ± equal. **FR:** nutlets angled, rough to smooth. ± 280 spp.: warm temp, trop Am, some Old World trop. (Greek: turned
back, from lower lip position) Some spp. used for flavoring (oils), food (seeds), medicinal.

H. emoryi Torr. (p. 841) DESERT LAVENDER Shrub 1–4 m,
densely hairy, hairs branched, white. **ST:** branches erect to spreading; twigs densely hairy, glabrous in age. **LF:** petiole 3–7 mm; blade
1–3 cm, gen ovate to ± round, crenate to dentate. **INFL:** ± spheric;
cymes loose, fls gen 10–15, subtended by small bracts, lower larger,
lf-like; pedicel 2–3.5 mm. **FL:** calyx 4–5 mm, tubular, densely white-
hairy, lobes ± filamentous; corolla 4–6 mm, violet. Gravelly, sandy
washes, canyons, desert scrub; < 1000 m. s DMoj, DSon; AZ, nw
Mex. Jan–May

LAMIUM DEAD NETTLE

John M. Miller & Dieter H. Wilken

Ann [per], glabrous to hairy. **ST:** decumbent to erect; base gen branched. **LF:** petioled to sessile; blade gen ovate to round
or reniform, entire to toothed. **INFL:** terminal and axillary, each head-like, subtended by lvs. **FL:** calyx 5-lobed, lobes ±
equal, gen acuminate; corolla 2-lipped, upper lip hood-like, lower lip ± 3-lobed, lateral lobes < central; stamens 4, fertile, gen
enclosed by upper lip, anthers gen hairy; style ± equally 2-lobed. **FR:** nutlets triangular in ×-section, truncate distally. 40 spp.:
temp Eurasia, n Afr. (Latin: ancient name) [Wagstaff et al. 1998 Pl Syst Evol 209:265–274]

1. Distal cauline lvs sessile, clasping; corolla tube glabrous inside . . . . *L. amplexicaule*
1′ Distal cauline lvs petioled; corolla tube puberulent below filaments . . . . *L. purpureum*

L. amplexicaule L. (p. 841) HENBIT **ST:** 1–4 dm. **LF:** blade
1–2.5 cm, wide-ovate to ± round, base truncate to lobed, margin
crenate to ± lobed. **FL:** calyx 4–7 mm; corolla gen 10–18 mm, red-
purple. 2*n*=18. Disturbed sites, cult or abandoned fields; < 800 m.
CA-FP; widespread N.Am; native to Eurasia. Cleistogamous fl corollas < 8 mm, gen not opening. Apr–Sep

L. purpureum L. (p. 841) **ST:** 1–6 dm. **LF:** petiole 1–2 cm; blade
1–3(4) cm, ovate to ± round, base ± lobed, margin crenate to serrate.
FL: calyx 5–7 mm; corolla 10–20 mm, pink-purple. 2*n*=18. Uncommon. Disturbed sites, meadows; < 300 m. NCo, CCo; widespread
N.Am; native to Eur. Apr–Sep

LEPECHINIA PITCHER SAGE

Deborah Engle Averett

[Per] shrub [tree], < 2 m, aromatic. **LF:** petioled, gen reduced to sessile bracts distally; blade gen 4–12 cm, lanceolate to
ovate, rounded to cordate at base, crenate or serrate to entire. **INFL:** raceme, open; bracts 1 per fl, lf-like, lanceolate; pedicels
persistent after fr-fall. **FL:** calyx 5-lobed, ± 2-lipped, scarlet-purple in age, enlarging in fr, gen falling with fr; corolla 5-lobed,
2-lipped, white to lavender tinged, lower lip longer, upper lip 4-lobed; stamens 4, in 2 pairs, incl in throat, anther sacs spreading; style 2-lobed, incl in throat. **FR:** 2–4 mm, round to oblong, black to dark brown, glabrous or minutely hairy. 2*n*=32. ± 55
spp.: CA, HI, Mex, S.Am. (I.I. Lepechin, Russian botanist, 1737–1802) [Boyd & Mistretta 2006 Madroño 53:77–84]

1. Bracts >> to ± = subtended fl, gradually reduced distally on st, ± overlapping at infl tip; infl axis abruptly
bent 60–90° relative to st . . . . *L. rossii*
1′ Bracts < to << subtended fl, much reduced distally on st, not overlapping at infl tip; infl axis straight to
curved or bent < 60° relative to st
   2. Pedicel 0–1 cm; calyx spheric in fr; fr minutely hairy . . . . *L. calycina*
   2′ Pedicel gen 1–4 cm; calyx bell-shaped in fr; fr glabrous
      3. Calyx lobes clearly < tube . . . . *L. cardiophylla*
      3′ Calyx lobes gen ± ≥ tube
         4. Calyx hairs dense, gen 0.5–2 mm, lobes gen > (<) tube . . . . *L. fragrans*
         4′ Calyx hairs sparse, << 0.5 mm, lobes gen ± = tube . . . . *L. ganderi*

L. calycina (Benth.) Munz (p. 841) Pl with long, branched, nonglandular hairs, occ with sessile to short-stalked glands. **LF:** lanceolate to narrowly ovate, ± entire to crenate-serrate. **FL:** pedicel 0–1 cm; calyx inflated, spheric in fr, gen persistent, lobes < tube. **FR:** hairs minute. Common. Rocky slopes; chaparral, woodland; 150–900 m. NCoR, CaRF, n&c SNF, CCo, SnFrB, SCoRO, WTR. Apr–Jun

L. cardiophylla Epling (p. 841) HEART-LEAVED PITCHER SAGE **ST:** gen with short-stalked glands. **LF:** cordate to ovate, irregularly serrate to ± entire, gen with branched, nonglandular hairs and sessile to short-stalked glands. **FL:** pedicel gen 1–3 cm; calyx spreading at mouth, short-stalked-glandular, lobes < tube. **FR:** glabrous. Chaparral; 600–1200 m. PR. Apr–Jul ★

L. fragrans (Greene) Epling (p. 841) FRAGRANT PITCHER SAGE Pl gen with long, branched, nonglandular hairs, occ with sessile to short-stalked glands. **LF:** lance-deltate to lance-ovate, serrate to entire. **FL:** pedicel gen 1–4 cm; calyx inflated at base, gen persistent, lobes gen > (<) tube. **FR:** glabrous. Chaparral; < 1300 m. SCo, n ChI, WTR, SnGb. Mar–Oct ★

L. ganderi Epling (p. 841) GANDER'S PITCHER SAGE Pl with short, branched, nonglandular hairs and sessile to short-stalked glands. **LF:** lanceolate, serrate to entire. **FL:** pedicel gen 1–2 cm; calyx spreading, lobes gen ± = tube. **FR:** glabrous. Chaparral; 300–1100 m. PR (Otay Mtn); Baja CA. Jun–Jul ★

L. rossii S. Boyd & Mistretta ROSS' PITCHER SAGE Pl with scattered long, branched, nonglandular hairs and ± dense sessile to short-stalked glands. **LF:** 3–13 cm, ovate to deltate-ovate, base truncate to cordate, irregularly and shallowly serrate to dentate. **INFL:** axis bent 60–90° relative to st; bracts ascending. **FL:** pedicel 1.2–1.3 cm; calyx bell-shaped, short-stalked-glandular, lobes < tube, gen ≥ 1 with marginal tooth, 0.5–1.5 mm. **FR:** glabrous. Chaparral; 470–1200 m. WTR (Liebre, Topatopa mtns). May–Sep ★

LYCOPUS BUGLEWEED
John M. Miller & Dieter H. Wilken

Per from rhizome, glabrous or hairy. **ST:** erect, branched or not. **LF:** short-petioled to sessile; blade gen ovate to lanceolate, margin toothed to deeply lobed or cut. **INFL:** head-like in lf axils. **FL:** calyx gen 5-lobed, lobes ± equal, obtuse to short-awned; corolla ± bilateral, not 2-lipped, gen 4-lobed, lobes ± unequal; stamens 2, exserted, staminodes 2, minute, club-shaped; style exserted. **FR:** nutlets ± compressed, truncate to rounded, edges corky-thickened. 14 spp.: temp N.Am, Eurasia, Australia. (Greek: wolf foot, from French common name) [Moon et al. 2006 J Pl Res 119:633–644]

1. Calyx lobes ovate, tip obtuse to acute, ± = fr; lvs gen short-petioled, gen serrate . . . . ***L. uniflorus***
1′ Calyx lobes awl-like, tip acuminate to short-awned, gen > fr; lvs ± sessile to short-petioled, serrate to deeply lobed or cut
2. Lvs gen short-petioled, deeply lobed to cut, esp in lower 1/2; fr 1–1.5 mm, tip rounded . . . . ***L. americanus***
2′ Lvs gen ± sessile, serrate; fr 1.5–2 mm, tip ± truncate . . . . ***L. asper***

L. americanus W.P.C. Barton (p. 841) Rhizomes ± slender, not thickened at tip. **ST:** erect, 2–8 dm, gen glabrous; nodes short-hairy. **LF:** gen short-petioled, 2–8(10) cm; blade oblong to lanceolate, deeply lobed to cut esp in lower 1/2, glabrous to puberulent on veins. **FL:** calyx lobes awl-like, short-awned; corolla 2–3 mm, ± = calyx, white. **FR:** nutlet 1–1.5 mm, tip rounded, smooth. 2n=22. Moist areas, marshes, streambanks; < 1000 m. CA-FP, SNE; to BC, e N.Am. Aug–Sep

L. asper Greene (p. 841) Rhizome tips thicker, tuber-like. **ST:** erect, 3–8(10) dm, puberulent to short-hairy. **LF:** ± sessile, 2.5–7(9) cm, lanceolate to narrowly elliptic, serrate, glabrous to puberulent. **FL:** calyx lobes awl-like, acuminate to short-awned; corolla 3–5 mm, ± > calyx, white. **FR:** nutlet 1.5–2 mm; tip ± truncate, occ minutely toothed. 2n=22. Moist areas, marshes, streambanks; < 1400 m. CA-FP, GB; to w Can, Great Plains. Jun–Oct

L. uniflorus Michx. (p. 847) NORTHERN BUGLEWEED Rhizomes slender, tips abruptly thicker, tuber-like. **ST:** ascending to erect, 1–5 dm, puberulent to finely strigose. **LF:** gen short-petioled, 2–6(8) cm; blade elliptic to lanceolate, gen serrate, glabrous to sparsely puberulent. **FL:** calyx lobes ovate, obtuse to acute; corolla 2.5–4 mm, > calyx, white. **FR:** nutlet 1–2 mm, tip truncate, ± finely toothed. Moist areas, marshes, near springs; < 2000 m. NW, CaR, SN, MP; to BC, e US. Jul–Sep ★

MARRUBIUM HOREHOUND
John M. Miller & Dieter H. Wilken

Per. **ST:** gen erect, gen branched, tomentose. **LF:** petioled to ± sessile; blade gen ovate to round, crenate or toothed. **INFL:** head-like, in lf axils. **FL:** calyx 10-lobed in CA, lobes spreading or recurved, sharp-pointed; corolla 2-lipped, upper lip entire to 2-lobed, lower lip 3-lobed; stamens 4, fertile, lower pair gen > upper pair, incl in tube; style incl, lobes ± equal. **FR:** nutlet tip truncate. 30 spp.: Eur. (Latin: from ancient Hebrew word for bitter juice) [Ryding 1998 Syst Bot 23:235–247] Some spp. cult for folk medicine, flavorings, some TOXIC.

M. vulgare L. (p. 847) **ST:** ascending to erect, 1–6 dm. **LF:** petiole < blade; blade 1.5–5.5 cm, wide-ovate to ± round, base rounded to ± lobed, margin crenate. **FL:** calyx 4–6 mm, lobes short-soft-hairy; corolla > calyx, lips ± equal. 2n=34,36. Disturbed sites, gen overgrazed pastures; < 600 m. CA (exc W&I); widespread worldwide. Formerly cult for tea, flavoring. Mar–Nov ❖

MELISSA
John M. Miller & Dieter H. Wilken

Per. **ST:** erect, simple to branched. **LF:** petioled; blade oblong to ovate, crenate to serrate. **INFL:** ± open, in axils of distal lvs, short-bracted. **FL:** calyx 2-lipped, upper lip > lower, 3-lobed, lower 2-lobed; corolla 2-lipped, upper lip ± entire, hood-like, erect or reflexed, lower lip 3-lobed, tube > calyx, curved upward; stamens 4, fertile, pairs ± equal, incl under upper lip; style lobes unequal. **FR:** ovoid, smooth. 3 spp.: Eur, w Asia. (Greek: bee) [Wagstaff et al. 1998 Pl Syst Evol 209:265–274]

M. officinalis L. LEMON BALM **ST:** branched, 2–15 dm, finely glandular hairy. **LF:** blade 2–14 cm, 1.5–7 cm wide, ovate, crenate. **INFL:** fls 4–12; subtending lvs reduced distally on st; pedicels 2–5 mm. **FL:** calyx 7–9 mm, tube ribbed, long-soft-hairy; corolla 8–15 mm, white to cream, occ tinged lavender. Moist sites, meadows, fields; < 800 m. NW, CaRF, n SNF, CW; native to s Eur. Jun–Sep

MENTHA MINT

Arthur O. Tucker

Per from rhizomes, glabrous to hairy. **ST**: gen ascending to erect, gen branched. **LF**: petioled to sessile; blade elliptic to ovate or lanceolate, toothed to lobed. **INFL**: head-like in lf axils or collectively spike- or panicle-like and subtended by bracts. **FL**: calyx ± radial, gen 10-veined, lobes 4–5, equal or not; corolla ± 2-lipped, lips gen equal, upper lip notched, lower lip 3-lobed; stamens 4, ± equal, gen exserted, filaments glabrous, anthers segments parallel, distinct. **FR**: nutlets ± ellipsoid, tip rounded. 18 spp.: temp. N.Am, Eurasia. (Latin: ancient name for mint) [Tucker et al. 1980 Taxon 29:233–255] Cult for oils, flavoring, herbs. Hybrids in CA gen sterile, spreading from rhizomes.

1. Calyx interior with ring of hairs; vegetative sts decumbent, fl sts upright . . . . ***M. pulegium***
1′ Calyx interior glabrous or hairs few, scattered; both vegetative and fl sts upright
  2. Infl gen axillary
    3. All lvs ± equal in size, blade gen ovate, gen spicy-scented . . . . ***M. arvensis***
    3′ Lvs reduced in size distally, blade gen linear to lanceolate, gen pennyroyal- or spearmint-scented
      4. Lvs gradually reduced distally on st, gen pennyroyal-scented . . . . ***M. canadensis***
      4′ Lvs greatly reduced distally on st, gen spearmint-scented (fruit-scented)
        5. Calyx 1.5–3 mm, lobes triangular . . . . ***M. ×gracilis***
        5′ Calyx 2.5–4 mm, lobes sharply tapering . . . . ***M. ×smithiana***
  2′ Infl terminal
    6. Infl gen head-like, clustered at distal 3–5 nodes
      7. Blade gen ovate, gen lavender- and citrus-scented . . . . ***M. aquatica***
      7′ Blade ovate to lanceolate, gen peppermint- or spearmint-scented (lavender and citrus) . . . . ***M. ×piperita***
    6′ Infl spike-like
      8. Lvs wrinkled
        9. Blade lance-oblong, glabrous (branched-hairy abaxially), spearmint-scented . . . . [2]***M. spicata***
        9′ Blade ovate, with scattered many-branched hairs abaxially, fruit-scented . . . . ***M. suaveolens***
      8′ Lvs not wrinkled, occ deeply veined
        10. Blade lanceolate to lance-oblong
          11. Blade widest near middle, hairs unbranched; fertile anthers ± 0.3–0.4 mm; gen musty-scented . . . . ***M. longifolia***
          11′ Blade widest near base, occ branched-hairy abaxially; fertile anthers ± 0.4–0.5 mm; gen spearmint-scented . . . . [2]***M. spicata***
        10′ Blade oblong to ovate
          12. Calyx ciliate; fertile anthers 4; lf blade base tapered to ± lobed; corolla 2.5–3.5 mm; GV, CCo, TR . . . . ***M. ×rotundifolia***
          12′ Calyx gen canescent; fertile anthers gen 0; lf blade base tapered; corolla 3–4 mm; TR . . . . ***M. ×villosa***

M. aquatica L. WATERMINT **ST**: 3–14 dm, glabrous to hairy. **LF**: 2–5(9) cm; petiole 3–8(25) mm; blade gen ovate, gen serrate, base tapered to ± lobed, tip acute. **INFL**: head-like, clustered at distal 3–5 nodes, subtended by ovate to lance-linear bracts. **FL**: calyx 2.5–4 mm, glabrous or lobes ciliate; corolla 3.5–6 mm, white to pink or violet. 2*n*=96. Moist places, fields; < 950 m. NCo, KR, NCoRO, SNF, SnJV, CCo, SCoRO, SCo, TR, PR, W&I, DMoj, cult elsewhere; to Can, e US; native to Eur, naturalized from cult. Jul–Oct

M. arvensis L. FIELD MINT **ST**: 1–5 dm. **LF**: 2–4(8) cm; petiole 3–8 mm; blade gen ovate, gen serrate, tip acute. **INFL**: axillary, subtending lvs spreading; bracts minute or 0. **FL**: calyx 1.5–2.5 mm, gen glabrous or lobes ciliate; corolla 2.5–3.5 mm, white to pink or violet. 2*n*=72. Moist places, fields; < 550 m. SnJV, cult elsewhere; to Can, e US; native to Eur, naturalized from cult. Jul–Oct

M. canadensis L. (p. 847) AMERICAN CORNMINT, JAPANESE PEPPERMINT **ST**: 1–5(8) dm, puberulent to short-hairy. **LF**: 1.5–5(8) cm; reduced distally on st; petiole 5–25 mm; blade linear to lanceolate, crenate to serrate, base tapered, tip gen acute, lower surface, esp veins, short-hairy. **INFL**: axillary, subtending lvs spreading; bracts minute or 0. **FL**: calyx 1.5–3 mm, short-hairy outside; corolla 4–7 mm, white to pink or violet. 2*n*=96. Moist areas, streambanks, lakeshores, fields; < 2400 m. NW, CaR, SNF, SNH, GV, CW, SCo, TR, PR, MP, DMtns; e Asia. [*M. arvensis* var. *c.* (L.) Kuntze; *M. a.* var. *villosa* (Benth.) S.R. Stewart] Jul–Oct

M. ×gracilis Sole SCOTCH SPEARMINT **ST**: 3–10 dm, gen glabrous. **LF**: 1.5–6(10) cm; petiole 2–8 mm; blade ovate to lanceolate, gen serrate, base tapered, tip acute. **INFL**: gen axillary, occ spike-like and terminal, subtending bracts ovate to lance-linear. **FL**: calyx 1.5–3 mm, gen glabrous to ciliate; corolla 2.5–6 mm, white to pink or violet; stamens gen incl. 2*n*=54,60,61,72,84,96,108,120. Moist places, fields; < 1500 m. NCoRO, NCoRI, c SNH, GV, CCo, SCo, cult else-

where; to Can, e US; native to Eur, naturalized from cult. Hybrid of *M. arvensis* and *M. spicata*. Widely cult. Jul–Oct

M. longifolia (L.) L. **ST**: 3–10 dm, canescent. **LF**: 1.5–4(7) cm; petiole 0–2 mm; blade lance-oblong, gen serrate, base tapered, tip acute. **INFL**: spike-like, bracts subtending fl clusters ovate to lance-linear. **FL**: calyx 1–2 mm, gen canescent; corolla 2–3 mm, white to pink or violet. 2*n*=24. Moist places, fields; < 300 m. SCo, TR, cult elsewhere; to Can, e US; native to Eurasia, Afr, naturalized from cult. Jul–Oct

M. ×piperita L. PEPPERMINT **ST**: 3–5 dm, gen glabrous. **LF**: 2.5–5(7) cm; petiole 3–10 mm; blade ovate to lanceolate, gen serrate, base tapered to ± lobed, tip acute, abaxially gen glabrous. **INFL**: head- to spike-like, clustered at upper 3–5 nodes, subtending bracts ovate to lance-linear. **FL**: calyx 2.5–4 mm, gen glabrous or lobes ciliate; corolla 3.5–6 mm, white to pink or violet; stamens gen incl. 2*n*=72, 84, 108. Moist places, fields; < 1700 m. NCo, NCoRO, NCoRI, CaRH, SNF, n SNH, GV, CW, SCo, TR, PR, MP, DMoj, cult elsewhere; to Can, e US; native to Eur, naturalized from cult. Hybrid of *M. aquatica* × *M. spicata*. Widely cult. Jul–Oct

M. pulegium L. PENNYROYAL **ST**: decumbent to ascending, 1–3 dm, short-hairy. **LF**: 0.5–2.5 cm, reduced distally on st; lower short-petioled, cauline gen ± sessile; blade narrowly ovate to elliptic, entire to finely serrate, base tapered to obtuse, tip gen rounded, abaxially short-hairy. **INFL**: axillary, head-like, subtending lvs or lf-like bracts reflexed. **FL**: calyx 2.5–4 mm, short-hairy; corolla 5–8 mm, violet to lavender; stamens > corolla lobes. 2*n*=20,30,40. Moist places, fields; < 1400 m. NCo, NCoRO, NCoRI, CaR, SN, GV, CW, PR; to OR, e US; native to Eur. Oil TOXIC, fatal when extract ingested by humans; used as insect repellant. Jul–Oct ❖

M. ×rotundifolia (L.) Huds. APPLE MINT **ST**: 3–10 dm, gen short-hairy. **LF**: 2.6–6(8) cm; petiole 0–2 mm; blade ovate, gen ser-

841

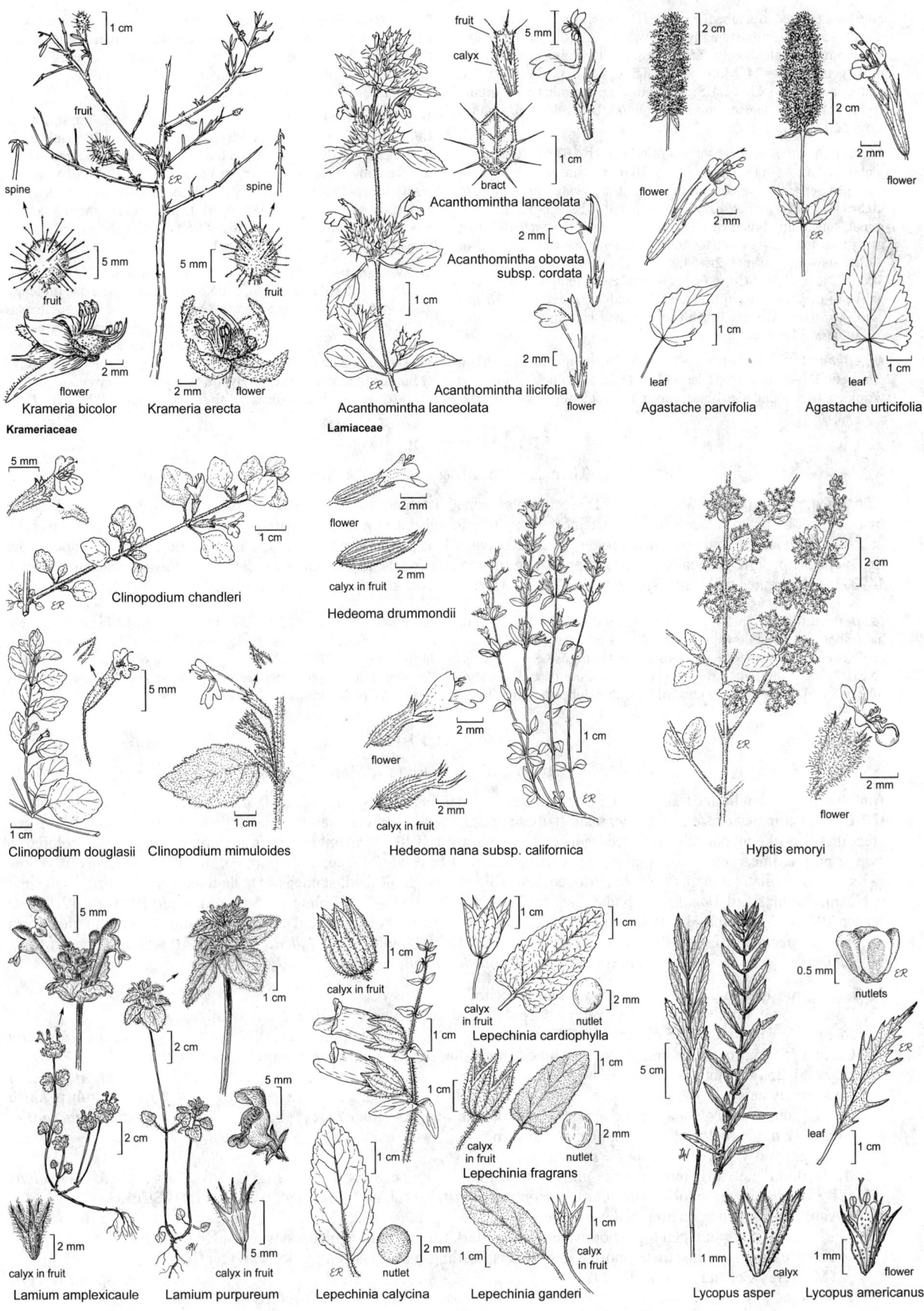

spine

fruit

spine

ER

fruit

5 mm

5 mm

fruit

fruit

flower

2 mm

2 mm

flower

Krameria bicolor

Krameria erecta

Krameriaceae

1 cm

fruit

calyx

5 mm

bract

1 cm

Acanthomintha lanceolata

2 mm

Acanthomintha obovata
subsp. cordata

2 mm

2 mm

Acanthomintha ilicifolia

flower

ER

Acanthomintha lanceolata

Lamiaceae

2 cm

2 cm

2 mm

flower

2 mm

flower

ER

1 cm

leaf

Agastache parvifolia

1 cm

leaf

Agastache urticifolia

5 mm

1 cm

ER

Clinopodium chandleri

2 mm

flower

2 mm

calyx in fruit

Hedeoma drummondii

2 cm

5 mm

2 mm

flower

2 mm

flower

2 mm

calyx in fruit

1 cm

ER

Hedeoma nana subsp. californica

ER

2 mm

flower

Hyptis emoryi

1 cm

1 cm

1 cm

Clinopodium douglasii

Clinopodium mimuloides

5 mm

1 cm

2 cm

2 cm

2 mm

calyx in fruit

Lamium amplexicaule

5 mm

5 mm

calyx in fruit

Lamium purpureum

1 cm

calyx in fruit

1 cm

1 cm

1 cm

ER

nutlet

2 mm

Lepechinia calycina

1 cm

calyx
in fruit

1 cm

nutlet

2 mm

Lepechinia cardiophylla

1 cm

calyx
in fruit

nutlet

2 mm

Lepechinia fragrans

1 cm

1 cm

calyx
in fruit

Lepechinia ganderi

5 cm

1 mm

calyx

Lycopus asper

0.5 mm

ER

nutlets

ER

leaf

1 cm

1 mm

flower

Lycopus americanus

rate, base tapered to ± lobed, tip acute. **INFL:** spike-like, clustered at distal 3–5 nodes, subtending bracts ovate to lance-linear. **FL:** calyx 1.5–2.5 mm, ciliate; corolla 2.5–3.5 mm, white to pink or violet; stamens gen incl. $2n=24$. Moist places, fields; < 800 m. GV, CCo, TR, cult elsewhere; to Can, e US; native to Eur, naturalized from cult. Hybrid of *M. suaveolens* and *M. longifolia*. Difficult to distinguish from *M. ×villosa*. Jul–Oct

M. ×smithiana R.A. Graham   REDSTEM MINT   **ST:** 3–10 dm, gen glabrous. **LF:** 2.5–6(8) cm; petiole 0–10 mm; blade ovate to lanceolate, gen serrate, base tapered to ± lobed, tip acute, abaxially gen glabrous. **INFL:** gen axillary, gen axillary, occ spike-like and terminal, subtending bracts ovate to lance-linear. **FL:** calyx 2.5–4 mm, gen glabrous or lobes ciliate; corolla 3.5–6 mm, white to pink or violet; stamens gen incl. $2n=54,98,108,120$. Moist places, fields; < 300 m. NCo, CCo, SCo, n ChI, TR, cult elsewhere; to Can, e US; native to Eur, naturalized from cult. Hybrid of *M. aquatica*, *M. spicata*, *M. arvensis*. Occ difficult to distinguish from *M. ×gracilis*, *M. ×piperita*. Jul–Oct

M. spicata L.   SPEARMINT   **ST:** 3–10(12) dm, glabrous. **LF:** 1–6 cm, petiole 0–2 mm; blade lanceolate to lance-oblong, smooth to wrinkled, base rounded to obtuse, tip acute to acuminate, gen serrate.

INFL: spike-like, bracts subtending fl clusters ovate to lance-linear. **FL:** calyx 1.5–2.5 mm, gen glabrous; corolla 3–4 mm, white to pink or lavender. $2n=48$. Moist places, marshes, lakeshores, fields; < 2350 m. CA; to e US; native to Eur. Male-sterile pls indistinguishable from *M. ×villoso-nervata* Opiz. Jul–Oct

M. suaveolens Ehrh.   PINEAPPLE MINT   **ST:** 5–10 dm, soft-hairy. **LF:** 1–4 cm, petiole 0–2 mm; blade ovate to oblong or broadly elliptic, wrinkled, base ± lobed, tip gen rounded, crenate to serrate, lower surface tomentose with many-branched hairs. **INFL:** spike-like, bracts subtending fl clusters ovate to lance-linear. **FL:** calyx 1–1.5 mm, short-hairy; corolla 2–3 mm, white to ± pink; stamens > corolla lobes. $2n=24$. Moist places, fields; < 950 m. NCo, NCoR, SNF, CW, SCo, TR, W&I, DMtns, cult elsewhere; native to s Eur. Jul–Oct

M. ×villosa Huds.   APPLE MINT   Per from rhizomes. **ST:** 3–10 dm, gen canescent. **LF:** 2–6(8) cm; petiole 0–2 mm; blade ovate (oblong), base tapered, tip acute, gen serrate. **INFL:** spike-like, bracts subtending fl clusters ovate to lance-linear. **FL:** calyx 1.5–2.5 mm, gen canescent; corolla 3–4 mm, white to pink or violet; stamens gen incl. $2n=36$. Moist places, fields; < 900 m. TR, cult elsewhere; to Can, e US; native to Eur, naturalized from cult. Hybrid between *M. suaveolens* and *M. spicata*. Very difficult to distinguish from *M. ×rotundifolia*. Jul–Oct

MONARDA   BEE BALM

Thomas J. Rosatti & Dieter H. Wilken

Ann, [bien, per, shrub], gen short-hairy. **ST:** erect, gen branched. **LF:** petioled or not. **INFL:** axillary, head-like, lower subtended by lvs, upper by bracts; bracts bristle-tipped [not]. **FL:** ± sessile; calyx ± radial, lobes 5, ± equal; corolla 2-lipped, upper lip entire or ± 2-lobed, hood-like, arched, lower lip gen 2–3-lobed, central lobe gen > lateral lobes; stamens 2, fertile, ascending under upper lip, ≥ upper lip; style unequally lobed. 16 spp.: N.Am; some cult for fls, tea. (Nicolas Monardes, Spanish physician, botanist, 1493–1588) [Prather et al. 2002 Syst Bot 27:127–137] *M. citriodora* Lag. (infl. bracts gen recurved near middle), not persisting.

M. pectinata Nutt. (p. 847)   PLAINS BEE BALM   **ST:** 1.5–3.5 dm; hairs short, ± curled down. **LF:** blade 1.5–4 cm, gen oblong to lanceolate, entire to serrate, ± glabrous to finely strigose esp on veins. **INFL:** bracts ± straight, rarely recurved, then from base. **FL:** calyx tube 6–8 mm, throat densely puberulent within, lobes 2–4 mm, long-

acuminate; corolla 12–25 mm, white to pink, lower lip purple-spotted or not. $2n=18,36$. Washes, rocky slopes, pinyon/juniper woodland; 1150–1500 m. e DMtns (New York Mtns); to w Great Plains, n Mex. Possible allopolyploid between *M. citriodora*, *M. clinopodioides* A. Gray. Collections needed. May–Aug   ★

MONARDELLA

Andrew C. Sanders, Mark A. Elvin & Mark S. Brunell

Ann to shrub, ± gland-dotted, scented. **LF:** entire to serrate, margin flat or wavy; petioles 0 or present, often grading into blade. **INFL:** fls in compact clusters of ≥ 1 per main st, these occ arrayed in panicles (rarely spikes); fls 3–100 per cluster; bracts gen erect in a cup-like involucre or reflexed, reduced in size inward, lf-like to membranous in texture, green or straw-colored to rose or purple, linear to ovate, acuminate to acute or obtuse. **FL:** calyx 5-lobed; 4–25 mm; corolla white to purple or yellow to red, weakly bilateral, upper lip erect, 2-lobed, lower lip recurved, 3-lobed; stamens 4; style unequally 2-lobed. > 30 spp.: w N.Am. (Latin: small *Monarda*) [Elvin & Sanders 2009 Novon 19:315–343; Epling 1925 Ann Missouri Bot Gard 12:1–106; Jepson 1943 Fl California 431–444] Complex; study needed; many taxa intergrade; fl cluster width and bract orientation given for pressed specimens. Lf length incl petiole, if present. M. Brunell authored *M. follettii* (in part), *M. odoratissima* (in part), *M. palmeri*, *M. purpurea*, *M. sheltonii* (in part), *M. stebbinsii*, *M. villosa* (in part).

1. Calyx gen ≥ 12 mm; corolla lobes 1/5–1/2 tube; subshrub
  2. Corolla 15–30 mm, white to cream-yellow or rose-tinged, tube cylindrical, 0.5–2 mm diam at tip, lobes spreading; calyx (10)12–15 mm; anthers < 1 mm . . . . ***M. nana***
  2′ Corolla 35–45 mm, red-orange to yellow, tube funnel-shaped, 1.5–3 mm diam at tip, lobes ascending; calyx 20–25 mm; anthers 1.2–1.5 mm . . . . ***M. macrantha***
    3. St sparsely hairy; lf triangular-ovate, sparsely to densely hairy, ciliate . . . . subsp. ***hallii***
    3′ St glabrous to sparsely hairy; lf elliptic to ovate, ± glabrous, not ciliate . . . . subsp. ***macrantha***
1′ Calyx ≤ 12 mm; corolla lobes > 1/2 tube; ann to shrub
  4. Lf margin wavy, gen ± entire (occ ± serrate)
    5. Lf < 10 mm, narrowly ovate to triangular-ovate; subalpine forest; s SNH . . . . ***M. beneolens***
    5′ Lf ≥ 10 mm, linear to (ob)ovate; dunes, openings in coastal scrub and woodland; CCo, SnFrB, s SCoRO
      6. Ann; main st gen branching below middle . . . . ***M. sinuata***
        7. Pl compact; st stout, gen highly branched; bracts dark-tipped, dark-veined, conspicuously spreading-hairy, esp along veins; distal-most st internode spreading-hairy; lvs gen weakly wavy; n CCo (Monterey Co. and n), SnFrB . . . . subsp. ***nigrescens***
        7′ Pl erect; st slender, sparingly branched; bracts gen not dark-tipped or dark-veined (occ purple), sparsely short-hairy; distal-most st internode ± glabrous; lvs weakly to strongly wavy; s CCo (San Luis Obispo Co. and s), s SCoRO, extirpated in w SCo . . . . subsp. ***sinuata***

6′ Subshrub to shrub; main st branching above middle . . . . **M. undulata**
  8. Pl prostrate to mounded, moderately to densely hairy; lvs appearing pale green, oblanceolate to (ob)
    ovate, obtuse, largest 5–15 mm wide; main st decumbent; lateral and terminal branches erect, short,
    > 1.5 mm diam; pl of active dunes . . . . subsp. *crispa*
  8′ Pl ± erect, sparsely to moderately hairy; lvs green, linear to narrowly (ob)lanceolate, acute to obtuse,
    largest 2–7(8) mm wide; main st ± erect; lateral and terminal branches ascending, < 1.5 mm diam;
    pl of ± stabilized to semi-stabilized dunes
    9. St, lvs sparsely to moderately hairy; fl cluster 18–30 mm wide; bracts lanceolate to ovate, acute
      to obtuse-acuminate, gen 5–9 mm wide; lateral, terminal branches gen 1–1.5 mm diam; Santa
      Barbara Co. . . . . subsp. *arguelloensis*
    9′ St, lvs sparsely hairy; fl cluster 10–20 mm wide; bracts lanceolate to narrowly ovate, acuminate,
      gen 3–5 mm wide; lateral, terminal branches gen < 1 mm diam; San Luis Obispo Co. . . . . subsp. *undulata*
4′ Lf margin slightly or not wavy, entire to serrate
  10. Ann; main st 1; fl clusters several to many per main st, in panicle-like array
    11. Bract lateral veins ± perpendicular to midvein, areas between veins transparent to translucent
      12. St slender; fl cluster 10–15 mm wide; bracts 10–15 mm, narrowly ovate, veins stout; narrow opaque
        band present beside bract veins; NCoR, SnFrB, SCoRI . . . . *M. douglasii*
      12′ St stout; fl cluster 20–30 mm; bracts 15–30 mm, widely ovate; bract veins slender; all of bract
        exc veins transparent to translucent; n&c SNF . . . . *M. venosa*
    11′ Bract lateral veins ± parallel to midvein, areas between veins ± opaque, membranous, or scarious
      13. Bracts not white; calyx lobes not white; corolla rose or purple (white)
        14. Bracts spreading-hairy, hairs > 1 mm . . . . *M. pringlei*
        14′ Bracts appressed-hairy, hairs < 1 mm . . . . *M. breweri*
          15. Bracts gen broadly ovate, strongly acuminate, not or only sparsely cross-veined, gen purple-tinted;
            SnFrB, SCoR, WTR . . . . subsp. *breweri*
          15′ Bracts ± lance-ovate or narrower, acute, weakly to strongly cross-veined, occ purple tinted; KR,
            CaRF, s CaRH, SN, s SCoRO, SW
            16. Terminal fl clusters gen 18–25 mm wide; bracts gen 8–10 per fl cluster, > calyx; calyx gen
              6.5–7.5 mm; mid-cauline lvs gen 40–60 mm; fls 15–75 per cluster; pl 9–65 cm . . . . subsp. *lanceolata*
            16′ Terminal fl clusters gen 5–15 mm wide; bracts gen 5–7 per fl cluster, ≤ calyx; calyx 6–7 mm;
              mid-cauline lvs gen 10–30 mm; fls 3–25 per cluster; pl 9–30 cm
              17. Conical glands abundant on st; calyx sinuses hairy; s SnGb . . . . subsp. *glandulifera*
              17′ Conical glands 0 on st; calyx sinuses gen not hairy; s PR . . . . subsp. *microcephala*
      13′ Bracts at least partly white; calyx lobes, lobe tips, or margins white; corolla white
        18. Calyx lobe tips awl-like, recurved; bracts white throughout (outermost rarely green-tinted);
          stamens incl. . . . . *M. leucocephala*
        18′ Calyx lobe tips acute, erect; bracts partly white; stamens slightly exserted
          19. Bracts conspicuously cross-veined between major parallel veins; fl clusters gen 12–25 mm wide,
            gen closely subtended by lvs, peduncle inconspicuous; main st 1 or gen branching above middle;
            SNF, e SnJV (extirpated in n part of range) . . . . *M. candicans*
          19′ Bracts not or inconspicuously cross-veined; fl clusters 10–20 mm wide, peduncle 1–3 cm; main st
            1 or gen branching below middle (pls in se SNF gen branching distally); se SN, n TR, nw DMoj . . . . . *M. exilis*
10′ Subshrub to shrub; main st 1–many; fl clusters 1–few (several) per main st
  20. Abaxial lf face ± white to ash-gray, short- to long-hairy, felt-like, or woolly, considerably whiter
    than adaxial face
    21. St and lf hairs short, < 0.3 mm, dense, straight; corolla lavender to purple or rose
      22. Fl clusters gen ≥ 20 mm wide; calyx 8–12 mm; lvs 20–40 mm, green adaxially; corolla lavender to
        purple; e SnGb . . . . *M. saxicola*
      22′ Fl clusters gen ≤ 20 mm wide; calyx 7–8 mm; lvs 5–20(30) mm, many ± white (pl appearing pale);
        corolla rose to purple; s NCoRI . . . . *M. viridis*
    21′ St and lf hairs gen > 0.5 mm, gen woolly, curved to curled on abaxial lf face; corolla white to lavender
      23. Branched hairs abundant on st — s CCo, s SCoRO (Santa Barbara to s Monterey cos.)
      . . . . [2]*M. villosa* subsp. *obispoensis*
      23′ Branched hairs 0–few on st
        24. Lf ovate to widely triangular-ovate, base gen truncate . . . . [2]*M. villosa* subsp. *franciscana*
        24′ Lf lance-linear to narrowly ovate, base tapered . . . . *M. hypoleuca*
          25. Lf ± flat or margins weakly rolled-under, 7–19 mm wide, narrowly ovate; adaxial lf face flat;
            corolla lavender; s SCoRO, WTR . . . . subsp. *hypoleuca*
          25′ Lf margins weakly to strongly rolled-under, 3–9(12) mm wide, lance-linear to lanceolate; adaxial
            lf face gen convex; corolla white; PR
            26. Hairs on adaxial lf face sparse to 0, on st sparse, short, or inconspicuous; nw PR . . . . subsp. *intermedia*
            26′ Hairs on adaxial lf face ± dense, on st long, spreading, conspicuous; sw PR (San Diego Co.)
            . . . . subsp. *lanata*
  20′ Abaxial lf face ± green, ± gray, silvery, or purple, not considerably whiter than adaxial face
    27. Pl of DMoj with main infl axis gen 3–several-branched and fl clusters small, 7–20(25) mm wide
      28. St with conical glands . . . . *M. eremicola*

28′ St lacking conical glands
  29. St with nonglandular hairs > 0.25 mm and glandular hairs 0.01–0.03 mm . . . . *M. robisonii*
  29′ St with nonglandular hairs < 0.25 mm, glandular hairs 0
    30. Bracts 8–9 mm; calyx hairs glandular (0.01–0.02 mm) and nonglandular (0.2–0.3 mm); st 12–40
      cm; 1400–1650 m; s-c DMoj (Ord and Rodman mtns) . . . . *M. boydii*
    30′ Bracts 10–11 mm; calyx hairs glandular (0.01–0.1 mm) only; st 30–60 cm; 800–1500 m;
      e DMtns . . . . *M. mojavensis*
27′ Pl of CA-FP or GB, or, if of DMoj, then main infl axis not both branched and with small fl clusters
  31. St hairs short (< 0.25 mm), very dense, uniform, and recurved . . . . *M. linoides*
    32. Lvs ≤ 4 mm wide
      33. Bracts 10–15 mm, ≥ 5 mm wide; PR, sw DMtns (Little San Bernardino Mtns), s DMoj . . . . . . subsp. *linoides*
      33′ Bracts 6–11 mm, < 5 mm wide; n&e SnBr . . . . subsp. *erecta*
    32′ Lvs ≥ 4 mm wide
      34. St 15–25 cm; bracts elliptic to lanceolate, acuminate; Teh, n WTR . . . . subsp. *oblonga*
      34′ St 20–60 cm; bracts broadly lanceolate to ovate; s SNF, c&s SNH, SNE, n DMoj
        35. Bracts 15–25 mm, 10–14 mm wide; peduncle stout, gen 5–15 cm (from distal-most
          lvs to fl cluster), 1–1.5 mm diam just proximal to fl cluster; fl cluster 20–30 mm wide;
          s SNF, s SNH (Kern, Tulare cos.) . . . . subsp. *anemonoides*
        35′ Bracts 7–15 mm, 6–9 mm wide; peduncle slender, gen 1–7 cm (from distal-most lvs to fl cluster),
          ≤ 1 mm diam just proximal to fl cluster; fl clusters 15–26 mm wide; c&s SNH, SNE, n DMoj . . . subsp. *sierrae*
31′ St hairs 0 or long and/or short, but not short (< 0.25 mm) and very dense and uniform and recurved
  36. St hairs relatively long (> 0.3 mm)
    37. Lvs thick; distal-st hairs dense, felt-like . . . . *M. stebbinsii*
    37′ Lvs thin; distal-st hairs sparse to dense, but not felt-like
      38. Pls n of TR . . . . *M. villosa*
        39. Branched hairs abundant and obvious. . . . . [2]subsp. *obispoensis*
        39′ Branched hairs 0 or inconspicuous
          40. Lf ovate to widely triangular-ovate, base gen truncate . . . . [2]subsp. *franciscana*
          40′ Lf ± narrowly ovate, base tapered to obtuse. . . . . subsp. *villosa*
      38′ Pls of TR and s . . . . *M. australis*
        41. St 3–15 cm; lf 5–10 mm; corolla 9–11 mm; SnGb . . . . subsp. *cinerea*
        41′ St 10–35 cm; lf 6–30 mm; corolla 10–13 mm; SnGb, SnBr, and/or SnJt
          42. Bracts lanceolate, exceeding calyx; lvs entire to weakly serrate; gen 1 fl cluster per main st;
            SnGb, SnBr, SnJt . . . . subsp. *australis*
          42′ Bracts narrowly lanceolate to lanceolate, exceeded by or at same level as calyx; lvs dentate to
            serrate (entire); fl clusters 1–several per main st; e SnGb . . . . subsp. *jokerstii*
  36′ St lacking longer hairs (> 0.3 mm)
    43. Pls s of TR
      44. Lvs ≥ 4 mm wide; bracts 7–10 mm, 2.5–4 mm wide; fl clusters 10–15 mm wide. . . . . *M. stoneana*
      44′ Lvs ≤ 4 mm wide; bracts 10–15 mm, 5–9 mm wide; fl clusters 20–30 mm wide . . . . *M. viminea*
    43′ Pls n of TR
      45. Abaxial lf hairs minute (< 0.05 mm), remarkably uniform in length and orientation, velvety
        in appearance. . . . . *M. sheltonii*
      45′ Abaxial lf hairs 0 or long and/or short, but not remarkably uniform in length and orientation
        and velvety in appearance
        46. St sparsely hairy (± glabrous) . . . . *M. odoratissima*
          47. Fl cluster subtended by unmodified lvs or not, if present 8–10 mm, reflexed or not; corolla
            lavender to (± red) purple. . . . . subsp. *glauca*
          47′ Fl cluster gen subtended by reflexed, unmodified lvs 10–20 mm; corolla gen white
            (occ lavender- to purple-tinged) . . . . subsp. *pallida*
      46′ St ± glabrous
        48. Hairs on calyx spreading-glandular; st 30–60 cm . . . . *M. follettii*
        48′ Hairs on calyx not spreading-glandular; st 10–40 cm
          49. Corolla 15–20 mm; calyx 9–12 mm . . . . *M. palmeri*
          49′ Corolla 12–16 mm; calyx 6–8 mm . . . . *M. purpurea*

M. australis Abrams   Subshrub, matted or tufted, rhizomed. **ST:** decumbent to erect, 3–35 cm, sparsely to densely hairy. **LF:** sessile to petioled, 5–30 mm, 2–7 mm wide, lanceolate to triangular-ovate, entire to serrate, green to ash-gray, margin occ faintly wavy, teeth 0–several. **INFL:** fl clusters 1–several per main st, 8–23 mm wide; bracts erect, in cup-like involucre, narrowly lanceolate to ovate, acute to long-acuminate, membranous or not, hairy, green to red or purple. **FL:** calyx lobes hairy; corolla 9–20 mm, white to rose or purple.

  subsp. ***australis*** (p. 847)   Pl matted. **ST:** decumbent to ascending, 10–20 cm. **LF:** ± sessile, 6–17 mm, 2.5–5 mm wide, lanceolate to

narrowly ovate, entire to weakly serrate. **INFL:** fl cluster 1 per main st, 12–20 mm wide; outer bracts ± lf-like, lanceolate, acute to acuminate, hairy, rose-tinged. **FL:** corolla 10–13 mm, rose. Mid-montane to subalpine forest, chaparral, rocky openings; 1450–3300 m. SnGb, SnBr, SnJt. Intergrades with *M. australis* subsp. *cinerea*, *M. linoides* subsp. *erecta*. Some pls in SnBr and SnJt with ± small lvs approach *M. australis* subsp. *cinerea*; more study needed. Jun–Sep

  subsp. ***cinerea*** (Abrams) A.C. Sanders & Elvin (p. 847)   GRAY MONARDELLA   Pl matted; hairs ± long-spreading and short-glandular. **ST:** decumbent, 3–15 cm. **LF:** ± sessile, 5–10 mm, 2–5 mm

wide, narrowly triangular-ovate, margin occ faintly wavy, teeth (0) few to several. **INFL:** fl cluster 1 per main st, 10–18 mm wide; bracts narrowly ovate to ovate, acuminate, hairy, ± purple. **FL:** fragrant; calyx 5–7(8) mm, lobes acuminate, purple, hairs ± long-nongland-ular and short-glandular; corolla 9–11 mm, rose-purple; stamens ± exserted. Mid-montane to subalpine forest; 1800–3100 m. SnGb. [*M. c.* Abrams] Pls approaching *M. australis* subsp. *cinerea* occur in SnBr, SnJt. Jun–Aug ★

subsp. ***jokerstii*** Elvin & A.C. Sanders JOKERST'S MONARDELLA Pl tufted; hairs long-spreading and short-glandular. **ST:** decumbent to erect, 5–35 cm. **LF:** 14–30 mm, 2–7 mm wide, lanceolate, dentate to serrate (entire), teeth (0)few to several, rarely 0; abaxially ± gold-gland-dotted. **INFL:** fl clusters 1–several per main st, 8–23 mm wide, gen subtended by unmodified lvs; bracts gen inconspicuous, membranous, narrowly lanceolate to lanceolate. **FL:** calyx 6–8 mm, hairy, short-glandular, lobes acuminate; corolla 10–11 mm, white to cream with purple markings, appearing pale lavender; stamens exserted. Steep scree or talus, stony benches on canyon bottoms in montane forest (or chaparral); (160)1350–1750 m. e SnGb, (n SCo). Intergrades with *M. australis* subsp. *cinerea*, *M. australis* subsp. *a.* Jul–Sep ★

M. beneolens Shevock et al. SWEET-SMELLING MONARDELLA Subshrub, matted to tufted, rhizomed. **ST:** decumbent, 10–30 cm; hairs spreading, long, soft, some glandular. **LF:** < 10 mm, narrowly ovate to triangular-ovate, margin wavy, entire to ± serrate, hairs dense, long, wavy, and glandular abaxially. **INFL:** fl cluster 10–20 mm wide; outer bracts lf-like or not; middle bracts ± erect, ± in cup-like involucre, short, ovate, green to straw-colored, rose-tinged or not. **FL:** calyx 6–8 mm, hairs dense; corolla 9–11 mm, lavender to pale rose. Rocky granitic or metamorphic slopes in open conifer forest; 2500–3600 m. s SNH. Intergrades with *M. odoratissima* subsp. *pallida.* Apr–Sep ★

M. boydii A.C. Sanders & Elvin Subshrub to shrub. **ST:** erect, 12–40 cm, many-branched distally, appearing gray to silvery, finely densely puberulent, not completely obscuring surface. **LF:** 7–15 mm, 1–3(5) mm wide, narrowly elliptic, acute, gray-green or silvery-green. **INFL:** fl clusters (1)3–5 per st, 10–20 mm wide; bracts deciduous, 8–9 mm, 2–3 mm wide, exceeded by calyx, narrowly elliptic, acuminate, green- to purple-tinged. **FL:** calyx 6–8 mm, lobes acute, puberulent and glandular-puberulent; corolla 10–11 mm, white, purple-marked, appearing lavender, puberulent, tube slightly exserted beyond calyx. Desert scrub, juniper woodland, on canyon bottoms and rocky slopes; 1400–1650 m. s-c DMoj (Ord, Rodman mtns). Aug–Oct ★

M. breweri A. Gray Ann, erect, branched. **ST:** 9–65 cm, short gray-hairy (puberulent) or glandular-hairy. **LF:** petioled, 15–60 mm, lanceolate to narrowly ovate, hairs short. **INFL:** fl cluster 5–30 mm wide; bracts 5–15 mm, narrowly to widely ovate, acute to acuminate, occ stiff-pointed, short-hairy, green to scarious between veins, veins converging at tip, cross-veins present or 0. **FL:** calyx 5–7.5 mm, 12–16-veined, ± glabrous to scabrous externally, exc gen with cluster of spreading hairs at base of sinus, densely ± appressed-hairy on inside of lobes; corolla 12–15 mm, rose to purple, hairy, tube exserted, tips of upper lobes with prominent gland; stamens exserted.

subsp. ***breweri*** (p. 847) **ST:** 15–65 cm, gray-hairy. **LF:** 15–45 mm, narrowly ovate. **INFL:** fl cluster 20–30 mm wide; bracts 10–15 mm, widely ovate, acuminate, stiff-pointed, ± purple, scarious between veins, cross-veins few or 0. **FL:** calyx 6.5–7.5 mm, 12–16-veined; corolla 12–15 mm, rose. Oak woodland, chaparral, pinyon/juniper woodland; < 1500 m. SnFrB, SCoR, WTR. Intergrades extensively with *M. breweri* subsp. *lanceolata* in TR and s SCoR. May–Aug

subsp. ***glandulifera*** (I.M. Johnst.) Elvin **ST:** 9–30 cm, purple, glandular- or short-hairy distally; glands many, stout, conical. **LF:** 15–40 mm, ± lanceolate. **INFL:** fl clusters 5–20(30) mm wide; bracts 5–7 mm, acute, cross-veined, outer green or purple-tipped, inner weakly scarious. **FL:** calyx 6–7 mm, 13–14-veined, sinuses hairy; corolla 12–14 mm, purple. Open, rocky, occ disturbed sites; grassy openings in chaparral, woodland; 500–2000 m. s SnGb (Browns Flat vicinity). [*M. lanceolata* A. Gray, in part in TJM (1993); *M. l.* var. *g.* I.M. Johnst.] Intergrades with *M. breweri* subsp. *lanceolata* in s TR. May–Aug ★

subsp. ***lanceolata*** (A. Gray) A.C. Sanders & Elvin MUSTANG MINT **ST:** 9–65 cm, purple, glandular- or short-hairy distally. **LF:** 15–60 mm, 4–12 mm wide, narrowly lanceolate. **INFL:** fl cluster 5–25(30) mm wide; bracts 10–15 mm, 8–10 mm wide, acute, cross-veined, outer green or purple-tipped, inner weakly scarious. **FL:** calyx gen 6.5–7.5 mm, 13–14-veined; corolla 12–15 mm, purple; stamens exserted. Open, rocky, occ disturbed sites; grassy openings in chaparral, oak woodland, conifer forest; < 3400 m. KR, CaRF, s CaRH, SN, s SCoRO, SW; NV, n Baja CA. [*M. l.* A. Gray] Intergrades extensively with *M. breweri* and with subsp. *microcephala* in s PR. May–Oct

subsp. ***microcephala*** (A. Gray) Elvin & A.C. Sanders **ST:** 9–30 cm, purple, glandular- or short-hairy distally. **LF:** 15–40 mm, ± lanceolate. **INFL:** fl cluster 5–20(30) mm wide; bracts 5–7 mm, acute, cross-veined, outer short-hairy, green or purple-tipped, inner weakly scarious. **FL:** calyx 6–7 mm, 13–14-veined; corolla 12–14 mm, purple. Open, rocky, occ disturbed sites; grassy openings in chaparral, woodland; < 2000 m. s PR; n Baja CA. [*M. lanceolata* A. Gray, in part in TJM (1993); *M. l.* var. *m.* A. Gray; *M. peninsularis* Greene] Intergrades with *M. breweri* subsp. *lanceolata* in PR. May–Aug ★

M. candicans Benth. SIERRA MONARDELLA Ann; branched gen distally or st simple. **ST:** 15–55 cm, ± purple, puberulent. **LF:** conspicuously long-petioled, 20–50 mm, 3–12 mm wide, lanceolate, puberulent. **INFL:** fl cluster 12–25 mm wide; bracts scarious, ovate, acute or obtuse, cross-veined, margin white, veins green. **FL:** calyx lobes erect, acute, margins white-scarious; corolla 8–11 mm, white, purple-spotted or not, tube exserted, lobes each with conspicuous gland at tip; stamens slightly exserted, tissue between pollen chambers entire. Sandy or gravelly soils, oak woodland, chaparral, lower yellow-pine forest; 40–1250 m. SNF, e SnJV (extirpated n SnJV). May–Jul ★

M. douglasii Benth. (p. 847) Ann, branched or not. **ST:** 10–30 cm, slender. **LF:** 15–50 mm, linear-oblong to (ob)lanceolate, hairy. **INFL:** fl cluster 10–15 mm wide; bracts 10–15 mm, narrowly ovate, veins stout, translucent between veins exc narrow opaque band beside vein, hairs between veins present. **FL:** corolla 11–12 mm, deep purple, hairy; stamens exserted. Grassland, openings in woodland and chaparral, serpentine, occ vertic clay; 50–1100 m. NCoR, SnFrB, SCoRI. May–Nov

M. eremicola A.C. Sanders & Elvin Subshrub, erect. **ST:** several to many, 15–55 cm, gray, finely densely glandular-puberulent; glands many, stout, conical. **LF:** ± sessile, 12–27 mm, 3–10 mm wide, narrowly elliptic, acute to obtuse, sparsely puberulent, pale ± gray. **INFL:** fl clusters (1)3–several per main st, 7–20 mm wide; bracts 4.5–9 mm, 2–4.5 mm wide, ± = calyx. **FL:** calyx 5–7 mm, puberulent, spreading gland-tipped hairs 0; corolla 8–11 mm, white with purple markings, appearing lavender. Limestone (occ granite) outcrops in pinyon/juniper woodland, canyons, slopes, wash margins; 1500–2100 m. ne DMtns (Clark, New York mtns, Kingston Range). Intergrades extensively with *M. linoides* subsp. *sierrae*, *M. odoratissima* subsp. *glauca* in n DMtns. Jun–Aug ★

M. exilis (A. Gray) Greene Ann; gen branched from base, occ st simple. **ST:** 6–30 cm, erect, puberulent. **LF:** conspicuously petioled, 15–37 mm, 3–8 mm wide, lanceolate to narrowly ovate. **INFL:** fl cluster 10–20 mm wide; bracts ovate, abruptly acuminate, long-ciliate, puberulent, gen green with purple tinge; margins, tips white-scarious; veins ± palmate, lateral veins 3–4 per side of midrib, cross-veins 0 or inconspicuous. **FL:** calyx lobe tips erect, acute, white, tube ± purple distally, hairy; corolla 10 mm, white, lobes gland-tipped; stamens slightly exserted, anther sacs purple, tissue between pollen chambers notched. Desert scrub, washes, pinyon woodland; 600–2100 m. se SN, n TR, nw DMoj. Apr–Sep

M. follettii (Jeps.) Jokerst (p. 847) FOLLETT'S MONARDELLA Subshrub, erect, open. **ST:** 30–60 cm, glabrous (puberulent or glandular-hairy distally), purple or green; internodes gen 10–75 mm. **LF:** 15–45 mm, 4–12 mm wide, lanceolate to elliptic, entire, ± glabrous (unicellular hairs occ occur), both faces abaxially purple-tinged or not. **INFL:** peduncle ≤ 75 mm; fl cluster 1(several) per main st, gen 12–20 mm wide; bracts 6–10 mm, 4–5 mm wide, leathery, lanceolate to narrowly ovate, hairy, conspicuously spreading-stalked-glandular (hairs > 0.1 mm), ± green or straw-colored. **FL:** calyx 7–8 mm, tube

gen short-hairy proximally, tube and lobes conspicuously spreading-stalked-glandular (hairs > 0.1 mm, equal on sinus area, lobes, tube); corolla 13–15 mm, pink. Forest, open, rocky slopes, serpentine, roadcuts; 700–2000 m. n SNH (Plumas Co.). Jun–Sep ★

M. hypoleuca A. Gray Subshrub, matted or tufted, cespitose to rhizomed. **ST:** 10–55 cm, ascending to erect, glabrous or short- to long-hairy, purple. **LF:** 20–50 mm, 3–19 mm wide, lance-linear to narrowly ovate, ± flat to margins rolled-under; adaxially ± glabrous to densely hairy, green; abaxially white-tomentose, felt-like or woolly. **INFL:** fl clusters 1–several per main st, 10–35 mm wide, occ subtended by unmodified lvs; bracts lanceolate to ovate. **FL:** calyx 6–9 mm; corolla 15–17 mm, white to lavender, lobes obtuse; stamens exserted.

subsp. *hypoleuca* (p. 847) Matted to tufted, weakly rhizomed, fragrant. **ST:** 25–50 cm, glabrous to sparsely hairy. **LF:** 30–50 mm (incl petiole, 3–10 mm), 7–19 mm wide, narrowly ovate, ± flat or margins slightly rolled-under, adaxially glabrous or sparsely hairy, abaxially densely white-tomentose, veins prominent. **INFL:** fl clusters 1–several per main st, 11–35 mm wide; bracts narrowly ovate. **FL:** calyx 6–8 mm; corolla pale lavender to lavender. Oak woodland, chaparral; < 1500 m. s SCoRO, WTR. Intergrades with *M. villosa* subsp. *obispoensis* in areas of overlap. May–Oct

subsp. *intermedia* A.C. Sanders & Elvin Matted, rhizomed, strongly sweet-scented. **ST:** 10–55 cm, ± sparsely short-hairy. **LF:** 20–50 mm, 3–9(12) mm wide, lance-linear to lanceolate, ± arched between rolled-under margins, adaxially ± glabrous to sparsely hairy, abaxially tomentose. **INFL:** fl clusters 1–few per st, 20–35 mm wide, gen subtended by unmodified lvs; bracts 9–13 mm, 4–7 mm wide, lanceolate to ovate. **FL:** calyx 7.5–8.5 mm; corolla 15–17 mm, white to lavender, lobes obtuse. Chaparral, oak woodland, occ conifer forest, dry slopes; 200–1250 m. nw PR (Orange, w Riverside, n San Diego cos.). Intergrades with *M. hypoleuca* subsp. *lanata* in small zone where adjacent. Jun–Sep ★

subsp. *lanata* (Abrams) Munz (p. 847) FELT-LEAVED MONARDELLA Cespitose, strongly sweet-scented. **ST:** 10–50 cm, ± densely hairy. **LF:** 20–40 mm (incl petiole, (0)2–5 mm), 3–9(12) mm wide, lance-linear, ± arched between rolled-under margins, adaxially ± densely hairy. **INFL:** fl clusters 1–few per main st, 11–25 mm wide, gen subtended by unmodified lvs; bracts narrowly ovate to ovate. **FL:** calyx (6)8–9 mm; corolla (10)12–16 mm, white. Chaparral, rocky, granitic slopes or hilltops; 300–1500 m. sw PR (San Diego Co.); n Baja CA. Intergrades with *M. hypoleuca* subsp. *intermedia* in small zone where adjacent. May–Oct ★

M. leucocephala A. Gray (p. 847) MERCED MONARDELLA Ann, branched, hairy, strongly scented. **ST:** 15–26 cm, gray-green, glands in epidermis gen obscured by hairs. **LF:** 10–35 mm, 4–7 mm wide, lanceolate or oblong, hairy, lateral veins obscure. **INFL:** fl cluster 10–15 mm wide; bracts ovate to obovate, gen scarious to membranous, short-acuminate, becoming ± glabrous, white (outermost bracts rarely green-tinted, but still white-tipped), main veins conspicuous, cross-veins few. **FL:** calyx stiff-hairy, lobes white, tips awl-like, recurved, conspicuously spreading-hairy, tube paler at base, dark toward top; corolla 5–5.5 mm, white, tube barely exserted, lobes short and broad, ± ovate; stamens incl, upper 2 sessile. Sandy soil in grassland, interior sand dunes (Delhi sands); 50–100 m. n SnJV (Stanislaus, Merced cos.). Last documented in 1941 (near Delhi, Merced Co.). May–Jul ★

M. linoides A. Gray Subshrub, erect, silvery to ash-gray; hairs dense, uniform, recurved. **ST:** 15–60 cm, few to many from base. **LF:** decurrent or sessile to petioled, 10–35 mm, 2–10 mm wide, linear to narrowly (ob)ovate, entire, ± green, silvery, or ash-gray. **INFL:** fl clusters 1–several per main st 7–30 mm wide; bracts 6–25 mm, 2–14 mm wide, papery, narrowly elliptic to ovate, obtuse-acuminate to acute, ciliate, hairy, ± white or straw-colored to pink, rose or purple. **FL:** calyx lobes stiff-hairy; corolla 10–15 mm, ± white or light blue to purple; stamens exserted.

subsp. *anemonoides* (Greene) Elvin & A.C. Sanders **ST:** few from base, 35–60 cm. **LF:** 15–35 mm (incl petiole, 0–4 mm), 4–8 mm wide, narrowly lanceolate to lanceolate, silvery. **INFL:** fl cluster 1 per main st, 20–30 mm wide; bracts 15–25 mm, 10–14 mm wide, exceeding calyx, narrowly ovate to ovate-acuminate, white to rose. **FL:** calyx (9)10–11 mm; corolla 10–15 mm, light purple to purple. Chaparral, montane forest; 750–2000 m. s SNF, s SNH (Kern, Tulare cos.). [*M. a.* Greene; *M. l.* var. *a.* (Greene) Jeps.; *M. l.* subsp. *l.*, in part in TJM (1993).] Intergrades with *M. linoides* subsp. *oblonga*, *M. linoides* subsp. *sierrae*, *M. odoratissima* subsp. *pallida*. Jun–Aug ★

subsp. *erecta* (Abrams) Elvin & A.C. Sanders Silvery, finely densely puberulent. **ST:** few from base, 15–30 cm. **LF:** decurrent or sessile, 12–19 mm, 2–4 mm wide, narrowly elliptic, obtuse to acute, silvery. **INFL:** fl cluster 1 per main st, 7–18 mm wide; bracts 6–11 mm, 2–4 mm wide, ± = calyx. **FL:** calyx 6–9 mm, puberulent and glandular-puberulent; corolla 10–11 mm, white with purple markings, appearing lavender. Conifer woodland, forest, rocky soils, outcrops; 1800–2600 m. n&e SnBr. Pls in SnBr resembling both *M. linoides* subsp. *erecta* and *M. australis* subsp. *a.*, with different microhabitats, may warrant taxonomic recognition; intermediates also occur among these potential taxa. Intergrades with *M. australis* subsp. *a.* were treated in TJM (1993) as *M. linoides* subsp. *stricta* (the type specimen is an intergrade). Jun–Aug

subsp. *linoides* (p. 853) Silvery. **ST:** few to several, 18–50 cm, densely puberulent. **LF:** 10–25 mm, 2–4 mm wide, linear to narrowly lanceolate. **INFL:** fl cluster 1(3) per main st, 10–22 mm wide; bracts 10–15 mm, 5–12 mm wide, exceeding calyx, broadly lanceolate to ovate, white to rose, sparsely short-ciliate. **FL:** calyx 8–9 mm, puberulent; corolla 10–14 mm, lavender to light blue. Chaparral, montane woodland; coarse soils; 900–2000 m. PR, sw DMtns (Little San Bernardino Mtns), s DMoj; n Baja CA. Pls in Pioneertown and Morongo/Yucca Valley have ± longer hairs, incl here, may be undescribed taxon; warrants further study. Jun–Sep

subsp. *oblonga* (Greene) Abrams (p. 853) TEHACHAPI MONARDELLA Silvery or ash-gray, finely densely puberulent. **ST:** 15–25 cm, several to many. **LF:** 10–20 mm, 4–7 mm wide, narrowly (ob)ovate. **INFL:** fl cluster 1 per main st, 10–20 mm wide; bracts 9–15 mm, 4–7 mm wide, exceeding calyx, lanceolate to elliptic, acuminate, gen rose-purple. **FL:** calyx 9–11 mm; corolla 12–15 mm, pale lavender to violet, tube exceeding calyx by 2–3 mm. Chaparral, conifer woodland to forest, gravelly, dry slopes, flats; 1500–2600 m. Teh, n WTR. Intergrades with *M. linoides* subsp. *anemonoides*, *M. linoides* subsp. *sierrae*. Jun–Aug ★

subsp. *sierrae* Elvin & A.C. Sanders Silvery, finely densely puberulent. **ST:** several to many, 20–50 cm. **LF:** 20–30 mm, (4)5–10 mm wide, lanceolate to broadly (ob)lanceolate, faces pale green, acute at base, adaxially ± glabrous to sparsely puberulent, abaxially sparsely puberulent (esp veins). **INFL:** peduncle ≤ 1 mm diam just proximal to fl cluster, gen 1–7 cm; fl cluster 1 per main st, 15–26 mm wide; bracts 7–15 mm, 6–9 mm wide, = or exceeding calyx, broadly lanceolate to narrowly ovate, acuminate, pale to rose. **FL:** corolla lavender to pink. Chaparral, conifer woodland to forest, coarse, granitic soils; 1000–3500 m. c&s SNH, SNE, n DMoj; NV. Intergrades with *M. linoides* subsp. *anemonoides*, *M. linoides* subsp. *oblonga*, *M. odoratissima* subsp. *glauca*, *M. eremicola*. Jun–Sep

M. macrantha A. Gray Subshrub, low, tufted, rhizomed. **ST:** 5–30 cm. **LF:** 5–30 mm, elliptic to triangular-ovate, entire to minutely serrate, leathery, deep green, shiny. **INFL:** fl cluster 20–40 mm wide; bracts oblong-elliptic; outer bracts like lvs; inner bracts scarious, hairy, ciliate, gen ± red. **FL:** calyx 20–25 mm, 2-lipped, bent, hairy, ± glandular; corolla 35–45 mm, red-orange to yellow, tube 1.5–3 mm diam at tip, funnel-shaped, exserted from calyx, lobes ascending; anthers 1.2–1.5 mm, exserted.

subsp. *hallii* (Abrams) Abrams (p. 853) HALL'S MONARDELLA **ST:** sparsely hairy. **LF:** triangular-ovate, sparsely to densely hairy, ciliate, base truncate. Chaparral, woodland; 600–2000 m. s SnBr, PR. Intermediates to *M. macrantha* subsp. *m.* common. May–Aug ★

subsp. *macrantha* (p. 853) **ST:** glabrous to sparsely hairy. **LF:** elliptic to ovate, ± glabrous, not ciliate, base obtuse. Chaparral, woodland; 600–2000 m. SCoR, TR. May–Aug

M. mojavensis Elvin & A.C. Sanders Subshrub to shrub. **ST:** 30–60 cm, erect, many-branched distally, gray to silvery, finely

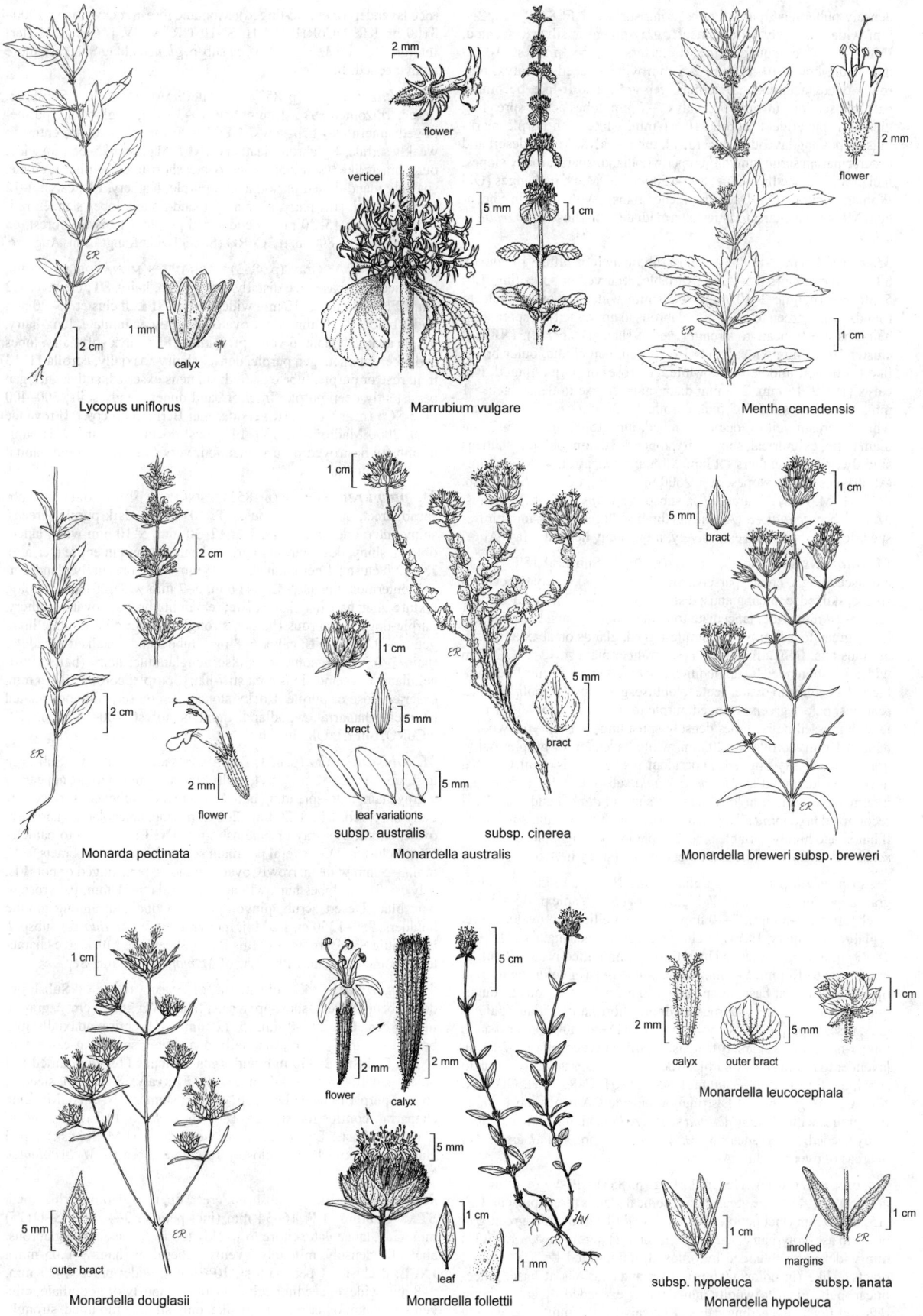

Lycopus uniflorus

2 mm

flower

verticel

5 mm

1 cm

2 cm

1 mm

calyx

Marrubium vulgare

2 mm

flower

Mentha canadensis

ER

1 cm

2 cm

2 cm

2 mm

flower

Monarda pectinata

1 cm

ER

1 cm

bract

5 mm

5 mm

leaf variations

subsp. australis

bract

5 mm

subsp. cinerea

Monardella australis

1 cm

5 mm

bract

ER

Monardella breweri subsp. breweri

1 cm

5 mm

outer bract

ER

Monardella douglasii

flower

2 mm

calyx

2 mm

5 mm

leaf

1 cm

5 cm

1 mm

Monardella follettii

calyx

2 mm

outer bract

5 mm

1 cm

Monardella leucocephala

subsp. hypoleuca

1 cm

subsp. lanata

1 cm

inrolled margins

ER

Monardella hypoleuca

densely puberulent, partially obscuring surface. **LF:** 8–20 mm, 2–4 mm wide, narrowly elliptic, acute, gray-green or silvery, scented. **INFL:** open compound cyme; fl clusters 3–7 per main st, 10–20 mm wide; bracts 10–11 mm, 2–5 mm wide, exceeding calyx, narrowly ovate, acuminate, outermost gen green and lf-like, becoming pale and scarious inward. **FL:** calyx 5–7 mm, lobes acute, spreading glandular-puberulent; corolla 10–11 mm, white with purple markings, appearing lavender, puberulent, tube > calyx. Mixed desert and desert riparian scrub, pinyon/juniper woodland, coarse, rocky slopes, incl granitics, occ limestone; 800–1500 m. e DMtns, incl edges (Old Woman, Granite, Old Dad, Providence mtns); sw NV. Pls in Sheephole Mtns have characteristics intermediate with those of *M. robisonii.* May–Aug

M. nana A. Gray (p. 853) Subshrub, matted or tufted, rhizomed. **ST:** 5–30 cm; hairs 0.5–2 mm, variable, recurved or spreading. **LF:** 5–30 mm (petiole 3–10 mm), 4–20 mm wide, narrowly ovate to round, entire, green to ash-gray, glabrous to appressed- or spreading-hairy, base ± truncate to strongly wedge-shaped (tapered). **INFL:** fl cluster 10–35 mm wide; bracts 15–20 mm, gen ciliate; outer bracts like lvs or not; middle bracts white, gen rose- or purple-tinged. **FL:** calyx (10)12–15 mm, 2–3 mm diam, hairs sparse to dense, gen > 1 mm, wavy, lobes 2.5–3.5 mm; corolla 15–30 mm, 6–15 mm diam, white to cream-yellow, occ rose-tinged, tube 15–25 mm, 0.5–2 mm diam at tip, cylindrical, short-hairy, lobes 4–10 mm, oblong; stamens slightly exserted, anthers < 1 mm. Montane chaparral, woodland, forest, dry desert-like slopes; 800–2600 m. PR; Baja CA. [*M. n.* subsp. *arida* (H.M. Hall) Abrams; *M. n.* subsp. *leptosiphon* (Torr.) Abrams; *M. n.* subsp. *tenuiflora* (A. Gray) Abrams] Previously named infraspecific taxa intergrade extensively; more study needed. May–Aug

M. odoratissima Benth. COYOTE-MINT Subshrub. **ST:** 10–45 cm, ascending to erect, sparsely hairy, (± glabrous), rarely to occ with sparse, stalked, conical glands; distal st diam 0.5–1.2 mm. **LF:** 15–50 mm, 5–18 mm wide, lanceolate to ovate, entire, sparsely to densely hairy, green to ash-gray, gen purple-tinged, glands on abaxial face in obvious pits. **INFL:** fl clusters 1–several per main st, 12–28(37) mm wide; gen subtended by unmodified lvs; bracts erect, in cup-like involucre, lanceolate to ovate, acute to obtuse, glabrous to woolly, ciliate, scarious, pale ± green or tinged purple to rose. **FL:** calyx 6–11 mm, tube appressed-hairy, lobes densely spreading-stiff-hairy or woolly, hairs ≤ 1 mm; corolla 10–20 mm, white, lavender, or purple. Ashy-gray, pale, chaffy-bracted, puberulent pls of ne Siskiyou Co. that have been identified as *M. odoratissima* subsp. *o.* intergrade or are intermediate in morphology between subspp. *glauca* and *pallida*; if recognized taxonomically, pls in NCoRH with spreading, soft, wavy lf hairs (and highly variable bract morphology) assignable to *M. odoratissima* subsp. *pinetorum* (Heller) Epling (study needed).

   subsp. ***glauca*** (Greene) Epling **ST:** 10–30 cm, green to dark gray, appearing glaucous. **LF:** highly variable on same pl, 15–25 mm (incl petiole, 0–8 mm), 5–10 mm wide, gen elliptic to ovate, entire, ± glabrous to hairy. **INFL:** fl cluster 1–several per main st, (13)16–28(33) mm wide; subtended by reflexed, unmodified lvs 8–10 mm or not, bracts 9–15 mm, 5–9 mm wide, occ lf-like (in color, texture) at tip and scarious at base, or spreading to erect and scarious throughout, elliptic to ovate, glabrous to finely short-hairy, ciliate, pale, ± gray, rose, or purple. **FL:** calyx (6.5)8–10(11) mm, tube puberulent to longer-hairy, lobes sparsely to densely stiff-hairy; corolla 10–20 mm, lavender to red-purple or purple. Rocky openings, sagebrush scrub to subalpine forest; 1000–3500 m. KR, NCoRH, CaR, SNH, GB; OR, NV, AZ. [*M. g.* Greene] Distribution outside CA needs more study. Intergrades with *M. odoratissima* subsp. *pallida* along SN crest; more study needed. Intergrades with *M. linoides* subspp., and *M. eremicola* in areas of overlap. Jun–Aug

   subsp. ***pallida*** (A. Heller) Epling (p. 853) Pl ± green or ash-gray. **ST:** 15–45 cm, ± green, hairy, conical glands rarely present. **LF:** (15)25–43 mm (incl petiole, 3–8 mm), (7)8–10 mm wide, green-glabrous to ash-gray-hairy. **INFL:** fl cluster 1(3) per main st, 13–25(37) mm wide; gen subtended by reflexed, unmodified lvs 10–20 mm, bracts lf-like (in color, texture) at tip and scarious at base, lance-linear, puberulent to woolly, inner bracts exceeded by calyx, outer reflexed bracts exceeding calyx. **FL:** calyx 6–10 mm, tube woolly (sparse-hairy), lobes woolly (spreading-stiff-hairy); corolla gen white

(occ lavender- or purple-tinged). Montane forest, rocky slopes; 1000–3100 m. KR, NCoRH, CaRH, SNH; OR, w NV. [*M. p.* A. Heller] Intergrades with *M. odoratissima* subsp. *glauca* along SN crest; more study needed. Jun–Sep

M. palmeri A. Gray (p. 853) PALMER'S MONARDELLA Subshrub, tufted, rhizomed. **ST:** decumbent, 10–30 cm, ± glabrous, purple-tinged; internodes gen < lvs. **LF:** 10–20 mm, lanceolate, entire to weakly serrate, ± glabrous, leathery. **INFL:** fl cluster 25–35 mm wide; outer bracts like lvs; middle bracts ovate, short-branched-hairy, ciliate, sparsely gland-dotted or not, acute, purple, leathery. **FL:** calyx 9–12 mm, 2-lipped, stiff-hairy, obscurely gland-dotted, lobes slender, red-purple; corolla 15–20 mm, slender, red-purple. Chaparral, forest, on serpentine; 200–800 m. n SCoRO (Santa Lucia Range).Jun–Aug ★

M. pringlei A. Gray (p. 853) PRINGLE'S MONARDELLA Ann, branched near base, occ distally, gray-green, hairy. **ST:** erect, 11–32 cm. **LF:** 8–48 mm, 2–12 mm wide, hairy. **INFL:** fl cluster 15–20 mm wide; bracts 6–11 mm wide, ovate, abruptly acuminate, long-hairy, purple adaxially, lateral veins prominent. **FL:** calyx soft-hairy, lobes awl-like, flattened, gen purple, densely hairy adaxially; corolla 11–13 mm, rose or purple, tube exserted; stamens exserted, anther sacs pen persistently deep-purple. Interior sand dunes, sandy soils; 300–400 m. e SCo (near Colton, Riverside, San Bernardino cos.). [Provance et al. 2000 Madroño 47:139–141] Last documented in 1941; habitat mostly destroyed by urbanization, very limited potential habitat remains. Apr–Jun ★

M. purpurea Howell (p. 853) SISKIYOU MONARDELLA Subshrub, erect, open, ± glabrous. **ST:** 10–40 cm, dark purple (green), shiny, internodes gen ≥ lvs. **LF:** 12–30 mm, 5–10 mm wide, lance-oblong, shiny, deep purple to green, leathery, margin entire (serrate). **INFL:** fl cluster 1 per main st, 15–25 mm wide, obscurely gland-dotted; outermost bracts 9–13(14) mm, 5–7 mm wide, lf-like (in color, texture), gen narrowly lanceolate; remaining bracts ± ovate, leathery, purple-tinged, glabrous (hairy), strongly ciliate; cilia multicellular, gen 0.3–0.8 mm. **FL:** calyx 6–8 mm, tube slender-stalked-glandular (hairs ≤ 0.7 mm), sometimes also nonglandular-hairy (hairs multicellular), lobes and sinus area stiff-hairy, purple; corolla 12–16 mm, exserted, rose or purple. Rocky slopes, gen on serpentine or related bedrock, chaparral, woodland, montane forest; 400–1400 m. KR, NCoRO, SnFrB; OR. Jun–Jul

M. robisonii Munz (p. 853) ROBISON'S MONARDELLA Subshrub to shrub. **ST:** 15–50 cm, erect, multi-branched mid-st to tip, appearing ± gray, hairs long-spreading and short-erect or -reflexed, surface partially obscured. **LF:** 8–20 mm, 2–6 mm wide, lanceolate to narrowly ovate, entire, ash-gray or pale ash-gray. **INFL:** ± cyme- to panicle-like; fl cluster (1)3–several per main st, 10–25 mm wide; bracts 9–12 mm, 3–5 mm wide, narrowly ovate, scarious, pink-tinged or not. **FL:** calyx 6–9 mm, lobes narrowly acute; corolla 9–11 mm, pale rose to pale blue. Desert scrub, pinyon/juniper woodland, among granite boulders; 950–1350 m. s DMoj. Intergrades with *M. linoides* subsp. *l.* in w Little San Bernardino Mtns Pls in Sheephole Mtns have characteristics intermediate with those of *M. mojavensis.* Jun–Sep ★

M. saxicola I.M. Johnst. ROCK MONARDELLA Subshrub, decumbent to erect, short-appressed-hairy. **ST:** 20–50 cm, ± gray to light green. **LF:** 20–40 mm, 5–12 mm wide, entire, adaxially glabrous to sparsely hairy, abaxially coarsely ± spreading-canescent. **INFL:** fl cluster 20–30 mm wide, gen subtended by unmodified lvs; bracts gen 10–15 mm, 4–7 mm wide. **FL:** calyx 8–12 mm, slender; corolla purple to lavender, tube > calyx; stamens exserted. Montane chaparral, conifer forest; 425–1800 m. e SnGb. [*M. viridis* subsp. *s.* (I.M. Johnst.) Ewan; *M. hypoleuca* var. *s.* (I.M. Johnst.) Jeps.] Hairs on abaxial lf face closely resemble those of *M. hypoleuca.* May–Aug ★

M. sheltonii Torr. Subshrub, erect, open, ± glabrous, rhizomed. **ST:** ± glabrous. **LF:** 16–34 mm (incl petiole, 3–4 mm), 8–11(15) mm wide, lanceolate, entire to weakly toothed, adaxially ± glabrous, abaxially densely, minutely, evenly puberulent (hairs < 0.05 mm). **INFL:** fl cluster 1 per main st, 10–29 mm wide; bracts 7–17 mm, 4–8 mm wide, exceeding calyx, inconspicuously or not ciliate, cilia gen < 0.5 mm; outer bracts ± lf-like, only slightly modified, strongly reflexed; innermost bracts linear to ovate. **FL:** calyx 6–8 mm, lobes

not or slightly hairier than tube; corolla (12)15–20 mm, purple. Rocky openings, montane forest, oak woodland, chaparral, often serpentine; 425–1600 m. KR, s CaR, n SN. Difficult to distinguish. Intergrades extensively with *M. villosa* subsp. *v.* (into KR, possibly OR) and *M. odoratissima* subsp. *pallida* (in SNH). Likely best treated as an infraspecific taxon but relationships unclear. Study needed. Jun–Aug

M. sinuata Elvin & A.C. Sanders Ann, simple or gen branched proximal to middle; hairs 0 to sparse. **ST:** erect, 8–65 cm, slender to stout, straw-colored to dark red-brown. **LF:** in axillary clusters, short-petioled to decurrent, 10–55 mm, (3)4–10 mm wide, linear to oblanceolate, ± fleshy, subglabrous to sparsely hairy, margin weakly to strongly wavy. **INFL:** fl cluster 10–35 mm wide; bracts 7–16 mm, 3–12 mm wide, elliptic to ovate, obtuse to acute, scarious or green, purple-tinged or not, black-tipped or not, black-veined or not, veins spreading-hairy. **FL:** calyx 5–8 mm, dark-tipped or not, hairy, lobes obtuse; corolla 14–20 mm, lavender to purple, upper 2 lobes gland-tipped. [*M. undulata* Benth., misappl.] Erroneously treated in *M. undulata* since 1868; type of *M. undulata* is a subshrub or shrub.

subsp. ***nigrescens*** Elvin & A.C. Sanders Pl compact, gen highly branched. **ST:** 8–45 cm, stout, dark red-brown, distal-most internode spreading-hairy. **LF:** 10–30 mm, 4–10 mm wide, lanceolate to narrowly oblanceolate, margin gen weakly wavy. **INFL:** fl clusters 1–many per main st, 10–35 mm wide; bracts 9–16 mm, 6–12 mm wide, elliptic to ovate, tips dark (black), veins dark (black), conspicuously spreading-hairy, esp along veins. **FL:** calyx 5–8 mm, dark-tipped, hairy; corolla 14–20 mm, lavender to purple. Dunes, openings in coastal scrub; < 300 m. n CCo (Monterey Co. and n), SnFrB. May–Jul ★

subsp. ***sinuata*** (p. 853) Pl erect, sparingly branched. **ST:** 8–65 cm, slender, straw-colored to red-brown, distal-most st internode glabrous. **LF:** 10–55 mm, (3)4–10 mm wide, linear to oblanceolate, ± glabrous to sparsely hairy, margins weakly to strongly wavy. **INFL:** fl clusters 1–many per main st, occ in spikes, 10–25 mm wide; bracts 7–16 mm, 3–12 mm wide, elliptic to narrowly ovate, sparsely puberulent. **FL:** calyx (4)5–6(7) mm, lobes acute, densely hairy, tube less hairy than lobes; corolla 14–20 mm, purple. Sandy soils, coastal strand, dune and sagebrush scrub, coastal chaparral and oak woodland; < 300 m. s CCo (San Luis Obispo Co. and s), s SCoRO, extirpated w SCo (Ventura Co.). Apr–Sep

M. stebbinsii Hardham & Bartel (p. 853) STEBBINS' MONARDELLA Subshrub, matted or clumped, < 1 m diam; hairs dense, felt-like. **ST:** 15–40 cm. **LF:** 12–22(28) mm (incl petiole, 2–4 mm), 5–12(14) mm wide, thick, narrowly ovate, entire, hairs unicellular and multicellular (latter gen 0.2 mm), abaxially also short-stalked-glandular, ash- to dark gray, purple-blotched. **INFL:** fl cluster 1 per main st (or several in ± terminal spikes or racemes), 8–20(25) mm wide; bracts lanceolate; outer bracts leathery, purple-tinged, soft-hairy and slender-stalked-glandular (gland-tipped hairs gen > 0.1 mm), hairs longer toward margin. **FL:** calyx 8–9 mm, green, purple, or straw-colored; tube slender-stalked-glandular (hairs > 0.1 mm) and nonglandular-hairy (hairs multicellular), lobes narrowly acute, hairs on lobes and sinus area = hairs on tube; corolla 12–18 mm, pink. Rocky serpentine slopes; 800–1100 m. n SNH (Serpentine Canyon, Plumas Co.). Infl, bract, and calyx characters overlap with *M. follettii*, which has more conspicuous stalked glands (because multicellular hairs on bracts and calyx are lacking); more study needed. Jul–Sep ★

M. stoneana Elvin & A.C. Sanders JENNIFER'S MONARDELLA Subshrub, low, compact, strongly pungent-scented. **ST:** several to many, 20–50(60) cm, (± glabrous) sparsely puberulent. **LF:** 17–35 mm (incl petiole, 2–4 mm), 4–10 mm wide, lanceolate to narrowly ovate, ± glabrous to sparsely puberulent, green exc gen purple-tinged near st tip, base acute. **INFL:** fl cluster 1(3) per main st (or several in ± terminal spikes or racemes), 10–15 mm wide; bracts 7–10 mm, 2.5–4 mm wide, lanceolate to narrowly lance-ovate. **FL:** calyx 9–10 mm; corolla pale pink, occ tinged light blue, lobes 2.8–3.7 mm. Rocky streambeds, banks of intermittent streams; < 660 m. sw PR (San Diego Co.); nw Baja CA. May–Sep ★

M. undulata Benth. Subshrub to shrub, erect, tufted, or mounded, < 2.5 m diam, strongly scented. **ST:** few to many, 10–100 cm, branching distal to middle, sparsely hairy to densely white-tomentose. **LF:**

clustered at nodes, 10–40 mm, linear to (ob)ovate, fleshy or not, green to appearing glaucous, sparsely hairy to densely tomentose, margin wavy. **INFL:** fl clusters 1–several in terminal spikes, 10–30 mm wide, lvs subtending fl clusters present or 0; bracts 7–15 mm, 3–10 mm wide, papery, lanceolate to ovate, soft-hairy, straw-colored or purple. **FL:** calyx hairy to woolly; corolla straw-colored or lavender to rose-purple.

subsp. ***arguelloensis*** Elvin & A.C. Sanders Shrub, erect. **ST:** 30–100 cm, main st stout, woody, to 3.5 cm diam, many-branched, sparsely to moderately long-hairy; lateral, terminal branches gen 1–1.5 mm diam. **LF:** 10–40 mm, 3–7 mm wide, ± fleshy, narrowly (ob)lanceolate, obtuse, green, sparsely to moderately hairy, margins strongly wavy. **INFL:** fl cluster 1–5 per main st in open compound cyme, 18–30 mm wide; bracts 9–13 mm, 5–9 mm wide, lanceolate to ovate, acute to obtuse-acuminate, straw-colored or purple-tinged. **FL:** calyx sparsely to moderately hairy, hairs denser at tip; corolla 10–12 mm, rose to purple. Stabilized coastal dunes, sandy soils; 50–150 m. s CCo (Point Arguello, Santa Barbara Co.). Threatened by military activities. May–Sep ★

subsp. ***crispa*** (Elmer) Elvin & A.C. Sanders (p. 853) CRISP MONARDELLA Subshrub to shrub, tufted or in mounds < 2.5 m wide. **ST:** few to many, 10–50 cm, densely white-tomentose, hairs gen wavy. **LF:** 10–40 mm (incl petiole, (3)5–7(8) mm), 5–15 mm wide, oblanceolate to (ob)ovate, fleshy, obtuse, sparsely to densely tomentose, appearing glaucous, margin wavy. **INFL:** fl clusters 1–5 in terminal spike, 15–30 mm wide; gen subtended by lvs; bracts 7–15 mm, 3–8 mm wide, broadly lanceolate to (ob)ovate, soft-hairy, straw-colored or purple. **FL:** calyx 6–7 mm, densely tomentose; corolla 10–14(16) mm, rose-purple. Active dunes; < 100 m. s CCo (San Luis Obispo, Santa Barbara cos.). [*M. c.* Elmer] Intergrades with *Monardella undulata* subsp. *u.* in areas of overlap and transitional habitats. Threatened by coastal development. Apr–Nov ★

subsp. ***undulata*** (p. 853) SAN LUIS OBISPO MONARDELLA Subshrub, erect, open. **ST:** 30–70 cm, several, ± sparsely hairy, purple; lateral sts gen < 1 mm diam. **LF:** 10–30 mm (incl petiole, 0–7 mm), 2–4 mm wide, linear to narrowly (ob)lanceolate, thin, sparsely hairy, green, margin wavy. **INFL:** fl cluster gen 1 per main st, 10–20 mm wide; bracts 7–10 mm, 3–5 mm wide, lanceolate to narrowly ovate, acuminate, straw-colored or purple. **FL:** calyx 4–6(7) mm, hairy, esp lobe tips; corolla rose-purple or purple. Stabilized dunes, coastal scrub, stabilized sandy soils; < 200 m. s CCo (San Luis Obispo Co.). [*M. frutescens* (Hoover) Jokerst;] Intergrades with *M. undulata* subsp. *crispa* in areas of overlap. Threatened by coastal development. Apr–Sep ★

M. venosa (Torr.) A.C. Sanders & Elvin (p. 853) VEINY MONARDELLA Ann, branched or not. **ST:** 15–40 cm, stout. **LF:** 15–50 mm, linear-oblong to (ob)lanceolate, hairy. **INFL:** fl cluster 20–30 mm wide; bracts 15–30 mm, widely ovate, purple, veins slender, area between veins transparent to translucent, hairs between veins sparse to 0. **FL:** corolla 11–12 mm, hairy, purple; stamens exserted. Grassland; 50–400 m. n&c SNF (Butte, Tuolumne cos.). [*M. douglasii* subsp. *v.* (Torr.) Jokerst] Jun–Jul ★

M. villosa Benth. COYOTE-MINT Subshrub, matted to erect, open, rhizomed; hairs gen > 0.5 mm, appressed to spreading, sparse to dense, soft, wavy, or woolly, glandular and not, branched or not. **LF:** 10–30 mm (incl petiole, 5–10 mm), length < 1–3.3 × width, lanceolate to ovate or widely triangular-ovate, serrate or crenate (entire), base truncate to obtuse or tapered; abaxially glandular-hairy, woolly or not. **INFL:** fl clusters gen 1–6(10) per main st, 10–40 mm wide; bracts 10–30 mm, reflexed; bracts gen lf-like in texture, color, hairiness (or innermost of middle series ± scarious proximally), innermost linear to elliptic-ovate. **FL:** calyx tube hairy, glandular, hairs on lobes gen similar to hairs on tube (or stiff-spreading); corolla 10–20 mm, white or pink to purple, lobes obtuse.

subsp. ***franciscana*** (Elmer) Jokerst Pl matted to erect, open; gen densely matted-white-woolly or hairs ± 0. **LF:** thick, ovate to widely triangular-ovate, entire to serrate, base gen truncate; abaxially hairy to white-woolly (appearing white or green), veins sunken. **INFL:** fl cluster 15–40 mm wide, many-fld. Coastal scrub, woodland; < 400 m. s NCo, n CCo, SnFrB (San Mateo, Santa Clara cos.). May–Aug

subsp. ***obispoensis*** (Jeps.) Jokerst   Pl ascending to erect, 15–30 cm, tufted to matted in age; hairs simple and branched, branched hairs many on st, lvs. **LF:** 15–25 mm, 8–15 mm wide, ± narrowly ovate, hairy, base tapered to obtuse; adaxially green; abaxially white-tomentose, often obscuring lf surface. **INFL:** fl cluster 1, 15–25 mm wide, gen subtended by unmodified lvs; bracts ovate. **FL:** corolla white to lavender, tube exserted. Chaparral, oak woodland; < 1500 m. s CCo, s SCoRO. Intergrades with *M. hypoleuca* subsp. *h.* May–Aug

subsp. ***villosa*** (p. 853)   Erect, < 50 cm; hairs simple or inconspicuously branched, ± dense, wavy, soft. **LF:** 10–22 mm, abaxially sparsely hairy to woolly, base tapered to obtuse. **INFL:** fl cluster 10–30 mm wide; outer bracts 8–20 mm. **FL:** corolla pink to purple. Rocky slopes, ephemeral drainages, oak woodland, chaparral, montane forest; < 1300 m. NW, n CW. [*M. antonina* Hardham subsp. *a.*; *M. a.* subsp. *benitensis* (Hardham) Jokerst; *M. siskiyouensis* Hardham; *M. v.* subsp. *globosa* (Greene) Jokerst] May–Aug

M. viminea Greene (p. 853)   WILLOWY MONARDELLA   Subshrub, cespitose. **ST:** erect, 25–50 cm, densely short-hairy, appearing glaucous-green. **LF:** 20–40 mm, 2–4 mm wide, linear to narrowly lanceolate, minutely puberulent, appearing glaucous. **INFL:** fl cluster 1 per main st (or several in ± panicle), 20–30 mm wide; bracts 10–15 mm, 5–9 mm wide, exceeding calyx, lanceolate, acuminate, conspicuously gland-dotted, green-white, occ rose-tipped. **FL:** corolla white to rose. Rocky washes with cobbles, 2° alluvial benches; < 400 m. s SCo (San Diego Co.). [*M. linoides* subsp. *v.* (Greene) Abrams] Threatened by development. Jun–Aug ★

M. viridis Jeps. (p. 853)   GREEN MONARDELLA   Subshrub, cespitose. **ST:** ascending to erect, 10–50 cm. **LF:** petioled, 5–20(30) mm, 4–7 mm wide, adaxially ± glabrous, abaxially finely appressed-canescent. **INFL:** fl cluster 10–20 mm wide; bracts gen 6–10 mm, lanceolate to ovate. **FL:** calyx 7–8 mm; corolla rose to purple, tube incl; stamens incl. Chaparral, oak woodland, conifer forest, also serpentine; 150–800 m. s NCoRI (Napa, Lake, Sonoma cos.). Jun–Aug ★

NEPETA

John M. Miller & Dieter H. Wilken

[Ann], per, glabrous to hairy. **ST:** erect, gen branched. **LF:** gen petioled. **INFL:** axillary or terminal, head- or spike-like, occ panicle, fl clusters subtended by lvs or bracts. **FL:** calyx ± radial, ± 15-veined, lobes gen ± equal; corolla 2-lipped, upper lip ± 2-lobed, < lower, hood-like, lower lip 3-lobed, central lobe > lateral lobes; stamens 4, enclosed by or exceeding upper lip. **FR:** nutlets smooth to rough. ± 250 spp.: Eurasia, Afr. (Latin: ancient name for catnip) [Jamzad et al. 2003 Taxon 52:21–32]

N. cataria L.   CATNIP   Pl < 1.5 m. **ST:** short-hairy to canescent. **LF:** 1.5–7.5 cm; petiole < blade; blade wide-lanceolate to ovate, base lobed, margin crenate to serrate, adaxially short-hairy, abaxially densely short-appressed-hairy. **INFL:** lower short-peduncled, subtended by lvs, upper subtended by short linear bracts. **FL:** calyx 5–6 mm, short-hairy, lobes stiff, acuminate; corolla 6–10 mm, puberulent, white, lower lip purple-spotted, central lobe minutely crenate; upper stamens > lower. 2*n*=36. Moist, gen shaded areas; < 1300 m. CA-FP; to BC, e US; native to Eurasia. Jul–Sep

ORIGANUM

Arthur O. Tucker

Per [subshrub] glabrous to short-hairy. **ST:** decumbent to erect, gen branched. **LF:** petioled to sessile; blade gen ovate, entire to toothed. **INFL:** axillary, sessile or peduncled, collectively spike- or panicle-like; wide bract subtending fl, bracts gen overlapping. **FL:** calyx radial, 5-lobed; corolla 2-lipped, upper lip ± entire, lower 3-lobed; stamens 4, enclosed by upper lip or exserted; style lobes ± unequal. 45 spp.: Medit, w Asia. (Greek: ancient common name, mountain delight) [Tucker & Rollins 1989 Baileya 23:14–27] Cult for tea, cooking herbs, essential oils (*O. dictamnus* L., dittany; *O. majorana* L., sweet marjoram).

1. Calyx funnel-shaped, 2-lipped, 3 upper lobes fused to form a single upper lip, lower lobes
   well developed . . . . ***O. ×majoricum***
1′ Calyx tube-shaped, 5-lobed, lobes equal . . . . ***O. vulgare*** subsp. ***hirtum***

O. ×majoricum Cambess.   ITALIAN OREGANO, HARDY SWEET MARJORAM   Pl from rhizomes. **ST:** 1–6 dm; short-hairy. **LF:** petiole 1–3 mm; blade 0.5–1.2 cm, entire, glabrous to puberulent, minutely gland-dotted; margin ciliate. **INFL:** gen peduncled, collectively panicle-like; bracts 4–6 mm, pale green. **FL:** calyx 2.5–3.5 mm, sparsely puberulent, gland-dotted; corolla 2-lipped for ± 1/3, 5–6 mm, white; stamens poorly developed or to 2.5 mm. Disturbed areas; < 600 m. n SNF, ScV; native to Balearic Islands, Spain, Portugal. Hybrid between *O. vulgare*, *O. majorana* L.; pls sterile or fertile, cult, spreading from rhizomes or seeds. ± summer

O. vulgare L. subsp. ***hirtum*** (Link) Ietsw.   OREGANO   Pl from rhizomes. **ST:** 3–10 dm; short-hairy. **LF:** petiole 3–12 mm; blade 0.2–1.5(3.3) cm, entire to ± serrate, puberulent, minutely gland-dotted, margin ciliate. **INFL:** gen peduncled, collectively panicle-like; bracts 1.5–3(5) mm, pale green or cream. **FL:** calyx 2–2.5(3.5) mm, sparsely puberulent, gland-dotted, short-bristly within; corolla 3–6(7.5) mm, white; stamens poorly developed or 4–5 mm. Disturbed areas; < 1300 m. NCo, CCo, SnFrB, SCo, MP; native to e Medit. ± summer

POGOGYNE   BEARDSTYLE, MESA MINT

Michael Silveira, Michael G. Simpson & James D. Jokerst

Ann, hairy or not, gland-dotted, aromatic. **ST:** decumbent to erect, branched or not, 2.5–45 cm. **LF:** linear to round, entire to toothed, bristly-ciliate or not; short-petioled. **INFL:** clusters of opposite cymes, head-like, spike-like, or interrupted, terminal and axillary, or fls 1 in axils; bracts 2 or more per node, bristly-ciliate or not. **FL:** ± sessile, calyx 2-lipped, lobes 5, deep, ciliate or not, tip extensions 0 to long, outer surface glabrous to coarsely hairy; corolla 2-lipped, lavender to purple or white, raised area on lower lip occ spotted; stamens 2–4, upper 2 sterile and vestigial or 0 in some taxa; style hairy below stigma lobes. **FR:** hairy. 8 spp.: CA, OR, Baja CA. (Greek: bearded style) [Howell 1931 Proc Calif Acad Sci 20:105–128]

1. Corolla inconspicuous, 2.5–8 mm, ± incl in bracts, calyces; fertile stamens 2, lower pair only, < 2 mm, upper pair of stamens sterile and vestigial or 0
2. Infl elongate, spike-like, gen ± from pl base . . . . ***P. floribunda***
2′ Infl gen short, head-like, ± from pl top
  3. Pl inconspicuous; st slender, ± 0.5 mm diam, gen prostrate to decumbent; fr 1–1.3 mm; grassy areas
  . . . . ***P. serpylloides***
  3′ Pl conspicuous; st stout, ± 1 mm diam, gen ascending to erect; fr 1.5–2.5 mm; vernal pools, swales, drainages . . . . ***P. zizyphoroides***
1′ Corolla conspicuous, 9–20 mm, exserted from bracts, calyces; fertile stamens 4, upper and lower pairs, lower pair > 3.5 mm
4. Infl 1–3 cm wide; lower calyx lobes > 1.5 × upper; vernal pools, swales, n of Point Conception (Santa Barbara Co.) . . . . ***P. douglasii***
4′ Infl ≤ 2 cm wide; lower calyx lobes < 1.5 × upper; coastal mesas, terraces (San Diego Co.), creek beds (Monterey Co.)
  5. Infl gen < 10 mm wide; calyx lobe tip extension 0 or short; bract tips gen obtuse to rounded; creek beds, swales (Monterey Co.) . . . . ***P. clareana***
  5′ Infl 8–20 mm wide; calyx lobe tip extension elongate, pointed; bract tips narrowly acute to acuminate; vernal pools (San Diego Co.)
    6. Calyx outer surface with dense, gen curved hairs; infl 8–10 mm wide . . . . ***P. abramsii***
    6′ Calyx outer surface glabrous or hairs sparse, gen straight; infl 10–20 mm wide — Otay Mesa . . . . . . ***P. nudiuscula***

P. abramsii J.T. Howell SAN DIEGO MESA MINT Hairs dense, long, coarse, gen curved. **ST:** spreading to erect, 5–20 cm, gen 0.5–0.8 mm diam at infl base. **INFL:** 8–10 mm wide; bracts gen purple-tinged, tips acuminate. **FL:** calyx densely white-hairy esp on veins, tube 2–2.5 mm, lobes 2–5 mm, tips not flat; corolla 10–12 mm, bell-shaped, hairs sparse; style hairy 2–4 mm below stigma lobes. **FR:** 1–1.5 mm. Coastal terrace vernal pools; 100–200 m. s SCo (San Diego Co.). Threatened by urbanization. Mar–Jun ★

P. clareana J.T. Howell SANTA LUCIA MINT Hairs sparse, short. **ST:** erect, 15–25 cm, slender, few-branched. **INFL:** gen < 10 mm wide; bract tips gen obtuse to rounded. **FL:** calyx 4–4.5 mm, coarsely ciliate, tube 2 mm, lobe tips flat; corolla 10–15 mm, funnel-shaped, hairy outside; style hairy 2 mm below unequal stigma lobes. **FR:** 1.5 mm. Creek beds, swales, vernal pools; 300–400 m. SCoRO (Fort Hunter Liggett, Monterey Co.). Threatened by human disturbance. Mar–Jun ★

P. douglasii Benth. (p. 853) DOUGLAS' BEARDSTYLE Robust. **ST:** erect, < 45 cm. **INFL:** 1–3 cm wide. **FL:** calyx tube 2–4 mm, upper lobes 2–5 mm, tips acute, lower lobes 3–8 mm, 0.75–2.5 × tube, tips lanceolate to acuminate; corolla 9–20 mm, lavender, hairy, raised area on lower lip gen yellow-spotted; style hairy 2–6 mm below stigma lobes. **FR:** 1–2 mm. Vernal pools, swales; < 900 m. NCoRO, NCoRI, SNF, GV, CW. Highly variable, study needed. Mar–Jul

P. floribunda Jokerst PROFUSE-FLOWERED POGOGYNE Hairs sparse to dense, short. **ST:** ascending to erect, 4–10 cm, branched at base. **INFL:** elongate, spike-like, gen ± from pl base, 10–25 mm wide. **FL:** calyx hairy, tube 2–3 mm, lobes 1.5–5 mm; corolla 4.5–6 mm, white, purple-spotted; stamens 2–4, upper 2 vestigial or 0; style ± hairy 1–2 mm below stigma lobes. **FR:** 2–2.5 mm. Vernal pools, seasonal lakes; 1000–1500 m. MP. Resembles *P. zizyphoroides*. Jun–Aug ★

P. nudiuscula A. Gray OTAY MESA MINT Hairs 0 to sparse, short, gen straight, bristle-like. **ST:** erect, stout, 9–30 cm, gen 0.8–1 mm diam at infl base. **INFL:** 10–20 mm wide; bracts green, tips acute. **FL:** calyx glabrous to sparsely hairy, tube 3–5 mm, lobes 2–5 mm, tips not flat; corolla 11–14 mm, bell-shaped, sparsely hairy, purple; style hairy 1.5–4 mm below stigma lobes. **FR:** ± 1.5 mm. Coastal mesa vernal pools; 100–250 m. s SCo (Otay Mesa, San Diego Co.); Baja CA. Threatened by urbanization. Mar–Jun ★

P. serpylloides (Torr.) A. Gray THYMELEAF BEARDSTYLE Pl inconspicuous. **ST:** gen prostrate to decumbent, 2.5–20 cm, ± 0.5 mm diam, gen branched, slender. **INFL:** clusters, head-like, small, dense, terminal and axillary, or fls 1 in distal axils. **FL:** calyx tube 1–3.5 mm, lobes 1.5–4 mm; corolla 2.5–5 mm, lavender; stamens 2–4, upper 2 vestigial or 0; style sparsely hairy just below stigma lobes. **FR:** 1–1.3 mm. 2*n*=38. Grassy, brushy areas; < 1200 m. NCoR, n&c SNF, SnFrB, SCoRO. Mar–Jun

P. zizyphoroides Benth. (p. 853) SACRAMENTO BEARDSTYLE Pl conspicuous. **ST:** gen ascending to erect, 5–16 cm, ± 1 mm diam, simple or branched distally, stout. **INFL:** clusters head-like, dense, terminal and axillary. **FL:** calyx tube 2.5–5 mm, lobes 1.5–6 mm; corolla 4–8 mm, lavender to purple; stamens 2–4, upper 2 vestigial; style sparsely hairy just below stigma lobes. **FR:** 1.5–2.5 mm. 2*n*=38. Vernal pools, depressions; < 400 m. NCoRO, n&c SNF, GV, SnFrB; s OR. Mar–Jun

POLIOMINTHA

John M. Miller & Dieter H. Wilken

Shrub. **ST:** spreading to erect, branched throughout, densely strigose, gen ± gray. **LF:** short-petioled to ± sessile; blade gen narrow, entire. **INFL:** axillary, gen subtended by lvs. **FL:** calyx ± radial, 15-veined, lobes ± equal; corolla 2-lipped, lips ± equal, upper lip ± flat, lower lip 3-lobed, central lobe notched; stamens 2, ± exserted, staminodes short; style unequally lobed. 4 spp.: sw US, n Mex. (Greek: hoary white mint) [Wagstaff et al. 1998 Pl Syst Evol 209:265–274]

P. incana (Torr.) A. Gray (p. 857) FROSTED MINT Pl 5–10 dm, rounded or mound-like. **LF:** ± sessile, gen reduced distally on st; blade 5–18 mm, 2–4 mm wide, oblong-elliptic to narrowly linear. **INFL:** fls 2–6; pedicel 1–2 mm. **FL:** calyx densely short-hairy, tube 3–5 mm, puberulent within, lobes 1–2 mm; corolla 8–10 mm, light blue to lavender, upper lip 2–3.5 mm, lower 3–4 mm, minutely purple-dotted. Sandy soils, rocky slopes; < 1700 m. s DMoj (Cushenbury Springs); to CO, TX, n Mex. Known from 1 documented locality, extirpated by mining. Jun–Jul ★

PRUNELLA SELF-HEAL

John M. Miller & Dieter H. Wilken

Per, glabrous to hairy. **ST**: prostrate to erect, proximal nodes occ rooting. **LF**: basal and cauline, gen petioled; blade gen entire. **INFL**: densely clustered, ± spike-like, terminal; bract gen wide, abruptly acuminate. **FL**: gen bisexual, occ only pistillate; calyx 2-lipped, upper lip = lower, upper lip 3-toothed, lower 2-lobed; corolla finely hairy inside, 2-lipped, lower lip 3-lobed, upper lip ± entire, hood-like, ± enclosing stamens; stamens 4, lower pair > upper, filaments minutely toothed below anthers. **FR**: nutlets obovoid. 4 spp.: temp, esp Eurasia. (Latin: from early German name for pl used to treat chest pains) [Trusty et al. 2004 Syst Bot 29:702–715]

P. vulgaris L. **ST**: 1–5 dm, glabrous to short-hairy. **LF**: lower petioled, petiole 5–30 mm; upper ± sessile; blade 2–7 cm, gen 1–4 cm wide, ovate to elliptic or lanceolate, base gen wedge-shaped. **INFL**: 2–6.5 cm; bract margins ciliate, ± red. **FL**: calyx 7–11 mm, dark green to ± purple; corolla 12–15 mm in bisexual fls, 8–11 mm in pistillate, ± blue-violet, occ pink or white. 2*n*=28,32.

1. Cauline blade length gen 3 × width; st erect, occ decumbent . . . . var. ***lanceolata***

1′ Cauline blade length gen 2 × width; st prostrate, occ decumbent to erect . . . . var. ***vulgaris***

var. ***lanceolata*** (W.P.C. Barton) Fernald (p. 857) Moist areas, gen conifer forest, woodland; < 2500 m. NW, CaR, SN, CW, WTR, PR, MP; N.Am, e Asia. May–Sep

var. ***vulgaris*** Moist, disturbed sites; < 2500 m. NCo, CaR, SN, CCo, SCo, MP; e US; native to Eur. Jun–Sep

PYCNANTHEMUM

Henrietta L. Chambers & Dieter H. Wilken

Per, rhizomatous, glabrous to hairy. **ST**: erect, simple or branched above middle. **LF**: petioled to ± sessile. **INFL**: head-like, axillary at distal nodes and terminal, subtended by lvs or bracts; fls ± sessile. **FL**: calyx 5-lobed, lobes ± equal, deltate, acute to awned; corolla 2-lipped, upper lip gen 2-lobed, ± flat, lower lip 3-lobed, lobes ± equal; stamens 4. **FR**: smooth to rough; glabrous to hairy. 19 spp.: N.Am. (Greek: densely fld)

P. californicum Durand (p. 857) Pl 5–13 dm. **ST**: glabrous to tomentose. **LF**: > internodes, ± overlapping; blade 2.5–7.5 cm, gen ovate to widely lanceolate, entire to finely toothed, gen tomentose, base ± rounded, tip gen acute; petiole < 3 mm. **INFL**: 2–5, head-like; proximal subtended by ± elliptic lvs, distal subtended by lance-linear bracts. **FL**: calyx 4–5.5 mm, puberulent, lobes densely short-hairy; corolla 5–6.5 mm, white, upper lip 2.5–3 mm, lower lip ± 2–3 mm, purple- or violet-spotted. 2*n*=40. Moist sites, chaparral, oak woodland, conifer forest; 500–1900 m. KR, n NCoR, CaR, SN, n ScV, TR, PR, MP. Jun–Sep

SALVIA

Deborah Engle Averett

Ann to shrub. **LF**: entire to lobed or toothed, gen not spine-tipped. **INFL**: clusters gen many-fld, gen head-like, gen ± spheric, gen involucred, gen surrounding nodes in gen ± spike-like, gen interrupted panicles, or fls 1–several per lf axil. **FL**: calyx gen 2-lipped, upper lip entire or of 3 gen shallow, occ spine-tipped lobes, lower lip gen of 2 gen spine-tipped lobes; corolla 2-lipped, upper lip 2-lobed to entire, lower lip with 3 spreading lobes, middle lobe gen expanded; fertile stamens 2, attached in throat, anther sacs 1–2 per stamen, if 2 then separate on thread-like structure with 1 fertile, > other; style forked at tip. ± 900 spp.: ± worldwide, esp trop, subtrop Am. (Latin: to save, from medicinal use) [Walker & Sytsma 2007 Ann Bot 100:375–391] Polyphyletic (taxonomic revision needed); CA natives in monophyletic sect. *Audibertia*. All spp. good bee fodder; seeds edible, a traditional food of native Californians. Historical waifs, *S. microphylla* Kunth last collected in CA in 1943, *S. verbenacea* L. in 1936.

1. Ann
  2. Corolla tube 15–25 mm; bract 2–5 cm; pl white-woolly . . . . ***S. carduacea***
  2′ Corolla tube 6–8 mm; bract < ± 1 cm; pl short-hairy . . . . ***S. columbariae***
1′ Per to shrub
  3. Lf spine-tipped
    4. Lf white-woolly, short-petioled, gen deciduous, spines in 1–2 pairs or 0 on margins; calyx lobes
      triangular, white-woolly . . . . ***S. funerea***
    4′ Lf green-tomentose, sessile or short-petioled, persistent, spines in 2–7 pairs on margins; calyx lobes
      lanceolate, not white-woolly . . . . ***S. greatae***
  3′ Lf not spine-tipped
    5. Per; lf ± green
      6. Corolla tube red to salmon, 25–35 mm, upper lip straight; calyx hairs short, white . . . . ***S. spathacea***
      6′ Corolla tube pale yellow or violet-blue (white), 10–20 mm, upper lip arched; calyx soft-wavy-hairy
        and glandular, to white-woolly
        7. Corolla tube pale yellow; calyx white-woolly . . . . ***S. aethiopis***
        7′ Corolla tube violet-blue (white); calyx soft-wavy-hairy and glandular . . . . ***S. virgata***
    5′ Subshrub or shrub (per); lf ± green to ± gray-white
      8. Stamens incl to ± exserted
        9. Lf linear to linear-elliptic, margins rolled under, abaxially with branched hairs; heterostylous . . . . . . ***S. brandegeei***

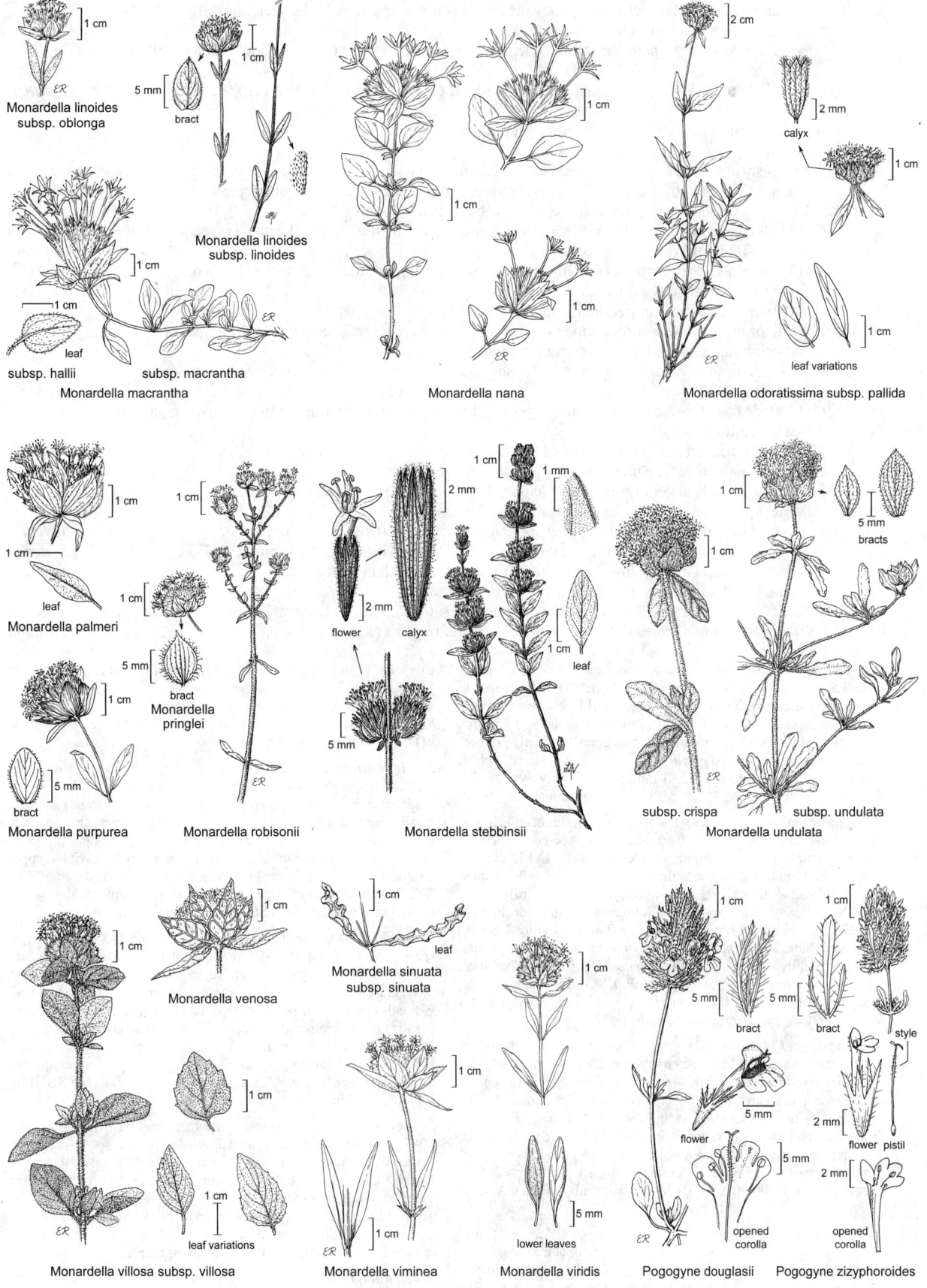

Monardella linoides subsp. oblonga

Monardella linoides subsp. linoides

subsp. hallii   subsp. macrantha
Monardella macrantha

Monardella nana

calyx

leaf variations
Monardella odoratissima subsp. pallida

leaf
Monardella palmeri

bract
Monardella pringlei

bract
Monardella purpurea

Monardella robisonii

flower   calyx

leaf

Monardella stebbinsii

bracts

subsp. crispa   subsp. undulata
Monardella undulata

Monardella venosa

leaf
Monardella sinuata subsp. sinuata

leaf variations
Monardella villosa subsp. villosa

Monardella viminea

lower leaves
Monardella viridis

bract

flower

opened corolla
Pogogyne douglasii

bract   style

flower pistil

opened corolla
Pogogyne zizyphoroides

9′ Lf oblanceolate to oblong-elliptic or obovate, margins not or only ± rolled under, abaxially with
    simple hairs; not heterostylous
  10. Corolla tube white to pale blue or lavender (pale rose); fl clusters 1.6–4 cm wide; lf oblong-elliptic
      to obovate . . . . *S. mellifera*
  10′ Corolla tube dark blue; fl clusters 1–1.5 cm wide; lf oblanceolate to obovate . . . . *S. munzii*
8′ Stamens gen exserted
  11. Lower corolla lip gen > 2 × upper; hairs simple
    12. Lf gen ± green, or ± gray from ± bristly hairs; infl gen < 5 dm
      13. Erect shrub; lf linear; corolla tube white to pale lilac . . . . *S. eremostachya*
      13′ Prostrate subshrub; lf lance-elliptic to obovate; corolla tube blue to lilac or purple . . . . *S. sonomensis*
    12′ Lf ± gray-velvety from minute appressed hairs; infl gen > 10 dm
      14. Lf blade base tapered; calyx lobes ± or not spine-tipped; fls in ± spike-like clusters, arranged in ±
         raceme-like, interrupted panicles . . . . *S. apiana*
      14′ Lf blade base ± truncate to tapered; calyx lobes spine-tipped; fl clusters in ± spike-like,
         interrupted panicles . . . . *S. vaseyi*
  11′ Lower corolla lip < 2 × upper; hairs simple or branched
    15. Width of middle lower corolla lobe < 1/2 length; lf puckered, teeth rounded, hairs simple or
      branched, moderately dense to sparse, spreading
      16. Lf blade base ± truncate to ± cordate; lf ashy-gray, hairs branched; tissue between anther sacs
         glabrous . . . . *S. leucophylla*
      16′ Lf blade base tapered to ± truncate; lf yellow- to gray-green, hairs simple; tissue between anther
         sacs glandular-hairy
        17. Bracts dark, firm, < calyx; spikes of 1–3 fl clusters . . . . *S. clevelandii*
        17′ Bracts pale to dark, papery, >> calyx; fl clusters gen 1 per fl st . . . . *S. mohavensis*
    15′ Width of middle lower corolla lobe ≥ length; lf not puckered, ± entire, scaly or hairs simple, very
      dense, appressed
      18. Corolla tube blue-violet to rose, gen 13–23 mm; bract 10–20 mm; lf 2–5 cm . . . . *S. pachyphylla*
      18′ Corolla tube blue (purple to rose, or white), gen 6–13 mm; bract 5–12(14) mm; lf 0.4–3 cm . . . . *S. dorrii*
      19. Bract, calyx abaxially soft-shaggy-hairy or scaly, margin hairs long . . . . var. *pilosa*
      19′ Bract, calyx glabrous to scaly, margin hairs gen short
      20. Lf abruptly narrowed to petiole, 0.6–2 cm, widest 0.2–1.3 cm from base . . . . var. *dorrii*
      20′ Lf tapered to petiole, 1–4 cm, widest 0.6–2.8 cm from base . . . . var. *incana*

S. aethiopis L. MEDITERRANEAN SAGE Per 30–60 cm, tomentose. **LF:** most basal, 5–30 cm, tomentose; blade widely lanceolate to
triangular, teeth irregular, acute to rounded. **INFL:** clusters 5–10-fld,
in ± open panicle; bracts ± 1 cm, firm, round, acuminate. **FL:** calyx
8–15 mm, white-woolly, lobes spine-tipped, upper lip deeply 3-lobed;
corolla tube 10–20 mm, pale yellow, upper lip ± 2 × lower, arched;
stamens, style exserted. **FR:** 2.5–3 mm, shiny, dark brown. 2*n*=24.
Fields, roadsides; 1250–1550 m. MP; native to Eur. Jun–Jul ◆

S. apiana Jeps. (p. 857) WHITE SAGE Per, subshrub, (< 1)1–2(3)
m. **LF:** 4–8 cm; blade widely lanceolate, base tapered; teeth minute, rounded; hairs dense, minute, simple, appressed. **INFL:** clusters
few-fld, in ± spike-like clusters, these in ± raceme-like, interrupted
panicles; bracts < to > calyx, lance-linear, recurved. **FL:** calyx 8–10
mm, lobes barely or not spine-tipped, upper lip entire; corolla tube
12–22 mm, white and lavender, upper lip < 2 mm, lower lip 4–5 mm,
upcurved, blocking throat; stamens, style exserted. **FR:** 2.5–3 mm,
shiny, light brown. 2*n*=30. Dry slopes, coastal-sage scrub, chaparral,
yellow-pine forest; gen < 1500 m. s SCoRO, SCo, TR, PR, w edge D;
Baja CA. If recognized taxonomically, pls with condensed panicles
assignable to *S. apiana* var. *compacta* Munz. Apr–Aug

S. brandegeei Munz (p. 857) BRANDEGEE'S SAGE Shrub > 1 m
or prostrate; hairs branched; heterostylous. **LF:** 2–6 cm; blade linear to linear-elliptic, puckered, adaxially glabrous, abaxially densely
white-hairy; margins rolled under; teeth small, rounded. **INFL:**
clusters 1.5–2 cm wide; bracts ovate, colored, tips sharp. **FL:** calyx
7–8 mm, hairs long, wavy, upper lip minutely 3-lobed; corolla tube
7–8 mm, pale blue to lavender, upper lip 3–3.5 mm, lower lip 3–4
mm; stamens ± incl. **FR:** 1.5–2.5 mm, oblong, rough, brown. 2*n*=30.
Coastal scrub; < 200 m. n ChI (Santa Rosa Island); n Baja CA. [*Salvia brandegei*, orth. var.] Feb–Aug ★

S. carduacea Benth. (p. 857) THISTLE SAGE Ann 1–10 dm,
white-woolly. **LF:** basal, ± sessile, 3–10(30) cm, oblanceolate, 1-pin-
nately dissected; margin wavy, short-spiny. **INFL:** scapose; clusters
1–4 per fl st, 1.5–3 cm wide; bracts 2–5 cm, lanceolate, spiny. **FL:**

calyx 10–17 mm, lobes spine-tipped, upper lip 3-lobed; corolla tube
15–25 mm, lavender (blue to white), upper lip 2-lobed, lower lip > 2
× upper; stamens exserted. **FR:** 2.5 mm, tan to gray, flecked. 2*n*=32.
Sandy or gravelly soils; < 1400 m. Teh, SnJV, e SnFrB, SCoR, SW,
w D; n Baja CA. Mar–May

S. clevelandii (A. Gray) Greene (p. 857) Shrub < 1.5 m, gray-
green; hairs ± throughout, simple, bent downward. **LF:** 2–4 cm;
blade elliptic-ovate, puckered, base tapered to ± truncate; teeth
small, rounded. **INFL:** spike of 1–3 fl clusters; bracts firm, dark, <
calyx. **FL:** calyx 8–10 mm, distal 1/3 to all purple, upper lip entire
to minutely 3-lobed; corolla tube 12–18 mm, dark blue-violet, upper
lip 2-lobed, 6–8 mm, lower lip 4–5 mm; stamens, style exserted. **FR:**
2–2.5 mm, brown to gray, gen mottled. 2*n*=30. Chaparral, dry slopes,
coastal scrub; < 1350 m. s SCo, PR; n Baja CA. Apr–Jul

S. columbariae Benth. (p. 857) CHIA Ann 1–5 dm; hairs gen
sparse, short. **LF:** basal, 2–10 cm; blade oblong-ovate, 1–2-pinnately
dissected; lobes irregularly rounded, minutely bristly. **INFL:** ± scapose; clusters gen 1–2 per fl st; bracts < ± 1 cm, ± round, awn-tipped.
FL: calyx 8–10 mm, purple-tipped, upper lip unlobed, 2(3)-awned;
corolla tube 6–8 mm, pale- to deep-blue, upper lip entire to shallowly
2-lobed, 2–3 mm, lower lip ± 2 × upper; stamens, style exserted.
FR: 1.5–2 mm, tan to gray. 2*n*=26. Dry, disturbed sites, chaparral,
coastal-sage scrub; gen < 2500 m. CA (exc KR, CaRH, n SNH); to
UT, AZ. Mar–Jun

S. dorrii (Kellogg) Abrams Shrub, spreading to mat-forming,
10–70 cm, densely white-scaly. **LF:** blade linear to spoon-shaped, ±
entire. **INFL:** clusters gen 12–30 mm wide; bracts 5–12(14) mm, ±
round. **FL:** calyx gen 6–11 mm, blue, purple, or rose, upper lip gen
entire, rounded, lower lip lobes acute, not spine-tipped; corolla tube
gen 6–13 mm, blue (purple to rose, or white), upper lip 2-lobed, 2–3
mm, < lower; stamens, style exserted. **FR:** 1.8–3.5 mm, gray to red-
brown. 2*n*=30. Highly variable; vars. intergrade.

  var. *dorrii* (p. 857) Pl spreading to erect. **LF:** 0.6–2 cm, widest
0.2–1.3 cm from base, abruptly narrowed to petiole. **INFL:** bract,

calyx glabrous to scaly, margin hairs gen short. Dry flats, slopes; 450–3200 m. GB, DMoj; to OR, ID, UT, AZ. Apr–Jul

var. ***incana*** (Benth.) Strachan FLESHY SAGE Pl spreading. **LF:** 1–4 cm, widest 0.6–2.8 cm from base, tapered to petiole. **INFL:** bract, calyx glabrous to scaly, margin hairs gen short. Silty to rocky soils; 1250–1550 m. nw CaRH (near Hornbrook, Siskiyou Co.); to WA, ID. CA pls may be intermediate with *S. dorrii* var. *d.* May–Jul ★

var. ***pilosa*** (A. Gray) Strachan & Reveal **LF:** 0.4–3.2 cm, widest 0.1–1.5 cm from base, abruptly narrowed to petiole. **INFL:** bract, calyx abaxially soft-shaggy-hairy or scaly, adaxially glabrous to minutely hairy, margin hairs long. Desert slopes, washes; 300–1900 m. s SNH (e slope), Teh, ne WTR, se PR, GB, n DMoj; NV, AZ. May–Jul

S. eremostachya Jeps. (p. 857) DESERT SAGE Shrub, erect, 60–80 cm, finely branched. **LF:** 1.5–3.3 cm; blade linear, puckered, base truncate; teeth minute, rounded. **INFL:** clusters 1–3 per fl st; bracts 5–12 mm, lanceolate to ovate, rose to purple, papery. **FL:** calyx 6–9 mm, upper lip minutely 3-lobed; corolla tube white to pale lilac, 10–17 mm, upper lip 2-lobed, 1–7 mm, lower lip 4–12 mm; stamens, style exserted. **FR:** 3 mm, yellow-brown. 2*n*=30. Dry, rocky, gravelly places, lower pinyon/juniper woodland; 450–1400 m. w edge DSon; n Baja CA. Mar–May ★

S. funerea M.E. Jones (p. 857) DEATH VALLEY SAGE Shrub 5–12 dm, densely branched, densely white-woolly. **LF:** 9–20 mm, short-petioled, gen deciduous; blade ± ovate; spines 1 at tip, 1–2 pairs or 0 on margins. **INFL:** fls gen 3 per lf axil. **FL:** calyx 4.5–6 mm, lobes 5, ± equal, triangular, spine-tipped; corolla tube 12–16 mm, violet (blue), upper lip 2-lobed, 2–2.5 mm, lower lip ± 2 × upper; stamens, style incl. **FR:** ± 3 mm, smooth, brown. 2*n*=64. Dry washes and canyons; < 1700 m. ne DMoj (Death Valley, Amargosa and Panamint ranges). Mar–Jun ★

S. greatae Brandegee (p. 857) OROCOPIA SAGE Shrub < 1 m; hairs glandular, tangled. **LF:** 9–20 mm, sessile to short-petioled, green-tomentose, persistent; blade ± ovate; spines 1 at tip, 2–7 pairs on margins. **INFL:** clusters many-fld. **FL:** calyx 9–11 mm, upper lip of 3 shallow, spine-tipped lobes; corolla tube 9–11 mm, lavender to rose, upper lip 2-lobed, 2–2.5 mm, lower lip 4–5 mm; stamens, style exserted. **FR:** 2–3 mm, flat, keeled, gray to brown. 2*n*=±30. Alluvial slopes; 30–450 m. DSon (Orocopia, Chocolate mtns). Mar–Apr ★

S. leucophylla Greene (p. 857) PURPLE SAGE Shrub, prostrate to erect, < 1.5 m; hairs dense, branched. **LF:** 2–8 cm; blade lance-oblong, puckered, base ± truncate to ± cordate; margin occ rolled under; teeth small, rounded. **INFL:** clusters 1.5–4 cm wide; bracts ovate, < calyx. **FL:** calyx 8–12 mm, upper lip entire, acute, not spine-tipped, lower lip gen 0; corolla tube 6–13 mm, rose-lavender, upper lip 6–8 mm, ± < lower lip; stamens, style exserted. **FR:** 2–3 mm, brown or dark gray. 2*n*=30. Dry, open hills, sea bluffs; 30–1250 m. s CCo, SCoR, SCo, WTR, SnGb; Baja CA. Apr–Jun

S. mellifera Greene (p. 857) BLACK SAGE Shrub, erect (prostrate), 1–2 m; hairs simple, some glandular. **LF:** 2.5–7 cm; blade oblong-elliptic to obovate, puckered, adaxially ± glabrous, abaxially hairy. **INFL:** clusters 1.6–4 cm wide; bracts gen = calyx, oblong-elliptic to widely ovate, acute to spine-tipped. **FL:** calyx 6–8 mm, hairs wavy, upper lip minutely 3-lobed; corolla tube 5.5–9 mm, white to pale blue or lavender (pale rose), upper lip 2-lobed; stamens, style ± exserted. **FR:** 2–3 mm, gen brown. 2*n*=30. Coastal-sage scrub, lower chaparral; < 1350 m. CW, SW; n Baja CA. Mar–Jun

S. mohavensis Greene (p. 857) Shrub gen < 1 m; hairs minute, simple. **LF:** 1.5–2 cm; blade lance-oblong to ovate, puckered, base tapered to ± truncate; teeth small, rounded; some hairs glandular. **INFL:** ± scapose; fl clusters gen 1 per fl st; bracts gen 1–1.5 cm, >> calyx, papery, ovate to ± round, gen showy, white to dark blue. **FL:** calyx 7–12 mm, minutely glandular-hairy, green, occ tinged blue, upper lip entire to minutely 3-lobed; corolla tube 18–20 mm, sky-

blue to dark blue, upper lip entire, 4–6 mm, ± < lower; stamens, style exserted. **FR:** 2–3 mm, tan to brown. 2*n*=30. Locally common, dry, rocky slopes, blackbrush scrub, pinyon/juniper woodland; 300–1500 m. DMoj (exc n&w), DSon (Whipple Mtns); AZ. Mar–Oct

S. munzii Epling (p. 857) MUNZ'S SAGE Shrub < 2.5 m. **ST:** hairs simple, nonglandular, appressed. **LF:** 1.3–5 cm; blade oblanceolate to obovate, puckered, adaxially ± glabrous, abaxially densely hairy. **INFL:** clusters 1–1.5 cm wide; bracts 7–30 mm, > calyx, lanceolate, lf-like. **FL:** calyx 3–6 mm, upper lip entire or minutely 3-lobed; corolla tube 7–15 mm, dark blue, upper lip 2-lobed, ± 2 mm, lower lip ± 2 × upper; stamens incl; style exserted. **FR:** ± 1 mm, dark brown. 2*n*=30. Coastal-sage scrub, lower chaparral; < 800 m. s PR (San Miguel Mtns, San Diego Co.); n Baja CA. Jan–May ★

S. pachyphylla Munz (p. 857) Shrub, prostrate, 20–80 cm, rooting at nodes, scaly. **LF:** 2–5 cm; blade obovate to spoon-shaped; margins wavy, ± entire. **INFL:** clusters 2–5 per fl st, 20–50 mm wide; fl st ± hidden between clusters; bracts 1–2 cm, papery, ovate to oblong-elliptic, ± green to purple or rose. **FL:** 8–13 mm; calyx like bracts in color, upper lip 3–6 mm, entire, obtuse to abruptly soft-pointed, lower lip 1–3 mm, lobes acuminate, not spine-tipped; corolla tube gen 13–23 mm, blue-violet to rose, upper lip entire, 4–6 mm, < lower lip; stamens, style exserted. **FR:** 2.5–3 mm, tan to brown. 2*n*=30. Dry slopes, pinyon/juniper to yellow-pine forest; 1200–3050 m. s SNH, Teh, SnBr, PR, DMtns; NV, AZ, n Baja CA. Feb–Oct

S. sonomensis Greene (p. 857) Subshrub, prostrate, mat-forming. **ST:** < 4 dm. **LF:** gen 3–6 cm, 5–15 mm wide; blade lance-elliptic to obovate, puckered, adaxially minutely hairy, abaxially white, densely recurved-hairy; teeth minute, rounded. **INFL:** scapose; clusters 1–1.5 cm wide; bracts < 1 cm, lanceolate. **FL:** calyx 5–10 mm, upper lip lobes 3, minute, not spine-tipped; corolla tube 5–15 mm, blue to lilac or purple, upper lip 2-lobed, 1–3 mm, lower lip 3–7 mm; stamens, style exserted. **FR:** 2.5 mm, oblong, brown. 2*n*=30. Locally common. Chaparral, oak woodland, yellow-pine forest, dry slopes; < 2000 m. KR, NCoR, CaRF, n&c SNF, s SnFrB, SCoR, SCo. Mar–Jul

S. spathacea Greene (p. 857) CALIFORNIA HUMMINGBIRD SAGE Per, mat-like, rhizomed; hairs wavy. **LF:** 8–20 cm; blade oblong-hastate, puckered, adaxially sparsely long-hairy, abaxially tomentose; teeth rounded. **INFL:** clusters < 6 cm wide; bracts 1.5–4 cm, green to purple. **FL:** calyx 1.5–3 cm, upper lip gen entire; corolla tube 25–35 mm, red to salmon, upper lip shallowly 2-lobed, 7–8 mm, straight, lower lip 10–12 mm; stamens, style exserted. **FR:** 3.5–6.5 mm, brown. 2*n*=30. Common, oak woodland, chaparral, coastal-sage scrub, open or shady slopes; < 800 m. s ScV (Solano Co.), CW, SCo, TR. Mar–May

S. vaseyi (Porter) Parish (p. 857) Subshrub < 1.5 m. **LF:** basal > cauline, 2–6 cm; blade oblong-ovate, base ± truncate to tapered; teeth minute, rounded; hairs dense, minute, appressed. **INFL:** clusters 1.5–3 cm wide; bracts < 2 cm, lanceolate. **FL:** calyx 8–14 mm, lobes spine-tipped; corolla tube 13–20 mm, white, upper lip shallowly 2-lobed, 2–4 mm, lower lip 6–9 mm; stamens, style exserted. **FR:** 2.5–3 mm, light brown. 2*n*=30. Dry, rocky desert slopes, canyons, creosote-bush scrub; < 1100 m. w edge DSon; n Baja CA. Apr–Jun

S. virgata Jacq. MEADOW SAGE Per ≤ 100 cm. **LF:** 5–27 cm, ± glabrous to sparsely hairy adaxially, soft-hairy abaxially; blade oblong to ovate-oblong, often ± cordate; teeth irregular, acute to rounded. **INFL:** spike, clusters gen 4–6 fld; bracts gen ≤ 1 cm, ovate, acuminate. **FL:** calyx 5–10 mm, tinged blue to purple, soft-wavy-hairy and glandular, upper lip minutely 3-lobed, lower lip lobes acuminate; corolla tube 7–11 mm, upper lip 2+ × > lower, arched; stamens incl (exserted); style exserted. **FR:** ± 2.5 mm, brown. 2*n*=18. Disturbed areas; 270–830 m. e KR, CaR, n SN; native to Eur. Often confused with *S. pratensis* L. Jun–Aug ◆

SCUTELLARIA SKULLCAP
Richard G. Olmstead

Per or shrub, gen hairy, occ glandular, from rhizomes or tubers. **ST**: erect, branched or not. **LF**: basal and cauline; proximal gen petioled; distal cauline ± sessile. **INFL**: fl 1–2 per lf axil, or appearing as a bracted raceme. **FL**: calyx 2-lipped, lips ± equal, enclosing nutlets, back of upper lip dome-like or transversely ridged, gen concave-depressed behind ridge; corolla 2-lipped, white to violet-blue, upper lip < lower, ± entire, hood-like, lower lip 3-lobed; stamens 4, pairs ± equal, enclosed by upper corolla lip, anthers ciliate, lower 2 1-chambered; disk below ovary gen green-yellow. **FR**: gen ovoid, gen minutely papillate, brown or black. ± 300 spp.: gen temp worldwide. (Latin: tray, from calyx dome or ridge) [Olmstead 1990 Contr Univ Michigan Herb 17:223–265] *Salazaria* occ treated as separate genus.

1. Shrub; calyx lobes bladder-like in fr; s SNE, D — lateral branches rigid, tips spine-like in age . . . . ***S. mexicana***
1′ Per; calyx lobes not bladder-like in fr; widespread
  2. Corolla white to pale yellow, lower lip occ blue- to violet-spotted
    3. Distal cauline lf blade crenate (entire), base truncate to ± lobed; st hairs long, spreading; lower corolla
      lip blue-spotted. . . . . ***S. bolanderi***
      4. Corolla 12–14 mm; lf length > 2 × width; SnBr, PR, s DMoj . . . . subsp. ***austromontana***
      4′ Corolla 15–19 mm; lf length 1–2 × width; SN, SCoRI . . . . subsp. ***bolanderi***
    3′ Distal cauline lf blade entire, base obtuse to wedge-shaped; st hairs short, ascending or pointed down;
      lower corolla lip violet-spotted or not
      5. St 1.5–4 dm, hairs appressed-ascending; ovary disk green-yellow; lf blade ovate to oblong . . . . ***S. californica***
      5′ St < 2 dm, hairs pointed or curled down; ovary disk orange-red; lf blade obovate to ± diamond-shaped . . . . ***S. nana***
  2′ Corolla blue to violet-blue, lower lip gen white-patched
    6. Fls gen > 4 in axillary or terminal racemes, bracted . . . . ***S. lateriflora***
    6′ Fls 1, axillary, subtended by lvs
      7. Upper calyx lip abaxially dome-like; fr brown; gen wet habitats . . . . ***S. galericulata***
      7′ Upper calyx lip abaxially transversely ridged, ± concave behind ridge; fr black; gen dry habitats
        8. St, lf hairs 1–3 mm, spreading; pls with tubers . . . . ***S. tuberosa***
        8′ St, lf hairs << 0.5 mm, curled or appressed; rhizomes slender, tubers ± 0
          9. Corolla 13–21 mm, hairs not gland-tipped; lf length 2–3 × width . . . . ***S. antirrhinoides***
          9′ Corolla 25–35 mm, hairs gland-tipped; lf length 3–7 × width . . . . ***S. siphocampyloides***

S. antirrhinoides Benth. (p. 861) Pl 10–35 cm; rhizomes slender, tips ± swollen. **ST**: hairs << 0.5 mm, appressed-ascending or upcurled, occ gland-tipped. **LF**: basal petioles 5–10 mm; distal cauline blades ovate to oblong, entire, bases rounded to tapered, tips rounded. **FL**: pedicel 3–4.5 mm; calyx 3–4 mm, ridged; corolla 13–21 mm, violet-blue, lower lip white-patched or -mottled, inner surface long-hairy. **FR**: black. Dry, rocky, often serpentine slopes, ridges, oak woodland, conifer forest; < 2300 m. NW, CaR, MP; to WA. Hybridizes with *S. californica* in NCoR. Jun–Sep

S. bolanderi A. Gray Pl 30–100 cm; rhizomes slender, tips ± swollen. **ST**: hairs 1–2 mm, spreading, gen gland-tipped. **LF**: basal petioles 2–10 mm; distal cauline blades ovate to cordate, crenate (entire), bases truncate to ± lobed, tips rounded. **FL**: pedicel 2–3 mm; calyx 3–5 mm, ridged; corolla 13–19 mm, white, lower lip blue-spotted, inner surface long-soft-hairy. **FR**: brown to black.

  subsp. ***austromontana*** Epling (p. 861) SOUTHERN MOUNTAINS SKULLCAP **LF**: length > 2 × width. **FL**: corolla 12–14 mm. Gravelly soils, stream banks, oak or pine woodland; 600–2000 m. SnBr, PR, s DMoj. Jun–Jul ★

  subsp. ***bolanderi*** (p. 861) **LF**: length 1–2 × width. **FL**: corolla 15–19 mm. Gravelly soils, stream banks, oak or pine woodland; 100–1500 m. SN, SCoRI (Santa Clara Co.). Jun–Jul

S. californica A. Gray (p. 861) Pl 15–40 cm; rhizomes slender, tip ± swollen. **ST**: hairs << 0.5 mm, appressed-ascending, occ gland-tipped. **LF**: basal petioles 5–10 mm; distal cauline blades ovate to oblong, entire, bases obtuse to wedge-shaped, tips rounded. **FL**: pedicel 3.5–4.5 mm; calyx 3.5–5 mm, ridged; corolla 16–19 mm, white to light yellow, occ tinged pink or blue, lower lip not spotted. **FR**: black. Open sites, scrub, woodland; 50–2200 m. KR, NCoR, CaRF, n SN, SnFrB. Hybridizes with *S. antirrhinoides* in NCoR. Jun–Jul

S. galericulata L. (p. 861) MARSH SKULLCAP Pl 20–80 cm; rhizomes slender. **ST**: glabrous or hairs << 0.5 mm, ± descending, occ gland-tipped. **LF**: basal gen 0; proximal cauline petioles 1–5 mm; distal cauline blades lanceolate to narrowly oblong-ovate, entire to crenate, bases truncate to ± lobed, tips ± acute. **FL**: pedicel 2–4 mm;

calyx 3.5–4.5 mm, upper lip dome-like; corolla 15–20 mm, violet-blue to blue, lower lip white-mottled, inner surface papillate. **FR**: ± spheric, brown. Wet sites, meadows, streambanks, conifer forest; 1000–2100 m. n SNH (Tahoe Basin), MP; to AK, e US; circumboreal. Jun–Sep ★

S. lateriflora L. (p. 861) SIDE-FLOWERING SKULLCAP Pl 20–60 cm; rhizomes slender. **ST**: glabrous or hairs sparse, hairs << 0.5 mm, ascending to upcurled, gen not glandular. **LF**: basal gen 0; proximal cauline petioles 10–20 mm; distal cauline blades ovate to lanceolate, ± dentate, bases rounded to truncate, tips acute. **INFL**: raceme or spike, bracted; bracts < 8 mm. **FL**: pedicel 1–3 mm; calyx 1.5–3 mm, upper lip back dome-like; corolla 6–8 mm, blue, lower lip blue, inner surface glabrous to sparsely soft-hairy. **FR**: ± spheric, brown. Marshes, wet meadows; < 500 m. Deltaic GV, SNE (Saline Valley); to BC, e US. May–Jul ★

S. mexicana (Torr.) A.J. Paton (p. 861) BLADDER-SAGE Shrub, 5–10(15) dm, ± rounded, branched. **ST**: lateral branches spreading, rigid, tips spine-like in age; twigs ± canescent. **LF**: short-petioled to ± sessile; blade 3–15(20) mm, 2–8 mm wide, gen ovate to elliptic, base rounded, margin entire, glabrous to puberulent. **INFL**: axillary at distal 3–10 nodes; fls 2; axis finely glandular-puberulent; bracts 0. **FL**: 2-lipped; calyx lobes ± equal, entire, ± purple, 1–2 cm in age, bladder-like in fr; corolla 15–25 mm, upper lip ± entire, white to light violet, lower lip ± 3 lobed, violet to purple; stamens 4, gen enclosed by upper lip, lower stamen pair < upper pair, anthers ciliate. **FR**: nutlets, widely ovoid, short-stalked, tubercled. Sandy to gravelly slopes, washes, scrub, woodland; < 1900 m. s SNE, D; to UT, TX, n Mex. [*Salazaria m.* Torr.] Mar–Jun

S. nana A. Gray (p. 861) Pl < 2 dm; rhizomes thick. **ST**: hairs << 0.5 mm, ± appressed, pointed down, occ gland-tipped. **LF**: basal petioles 2–5 mm; distal cauline blades obovate to diamond-shaped, entire, bases obtuse to wedge-shaped, tips ± rounded. **FL**: pedicel 1–3 mm; calyx 3–4 mm, ridged; corolla 15–20 mm, white to pale yellow, lower lip purple-spotted, inner surface glabrous to sparsely long-hairy; ovary disk orange-red. **FR**: black. Dry, volcanic soils, scrub; 1000–1900+ m. CaR, MP; to OR, ID, NV. [*S. holmgreniorum* Cronquist, Holmgren's skullcap] Jun–Aug ★

857

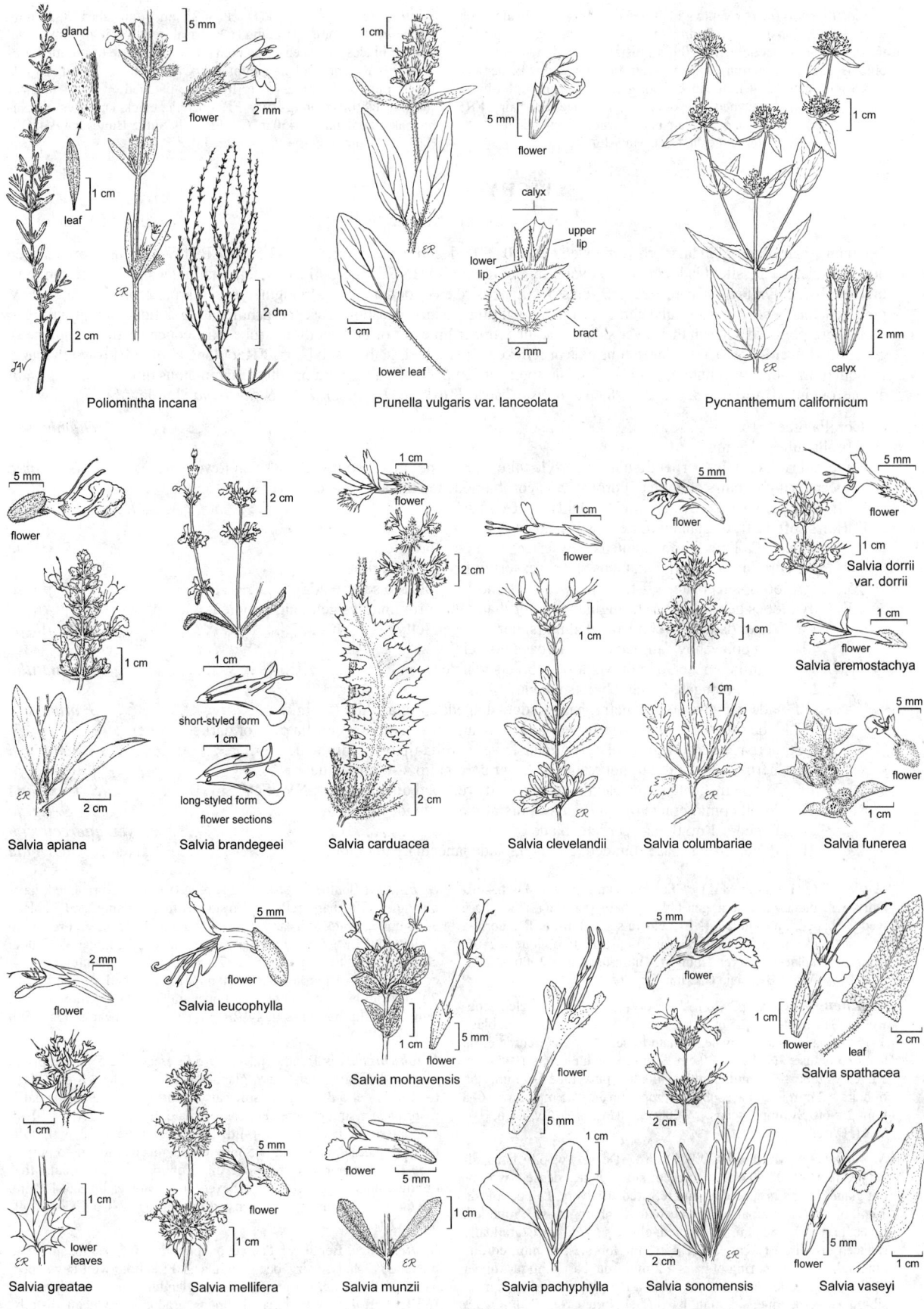

Poliomintha incana

Prunella vulgaris var. lanceolata

Pycnanthemum californicum

Salvia apiana

Salvia brandegeei

Salvia carduacea

Salvia clevelandii

Salvia columbariae

Salvia funerea

Salvia dorrii var. dorrii

Salvia eremostachya

Salvia greatae

Salvia leucophylla

Salvia mellifera

Salvia mohavensis

Salvia munzii

Salvia pachyphylla

Salvia sonomensis

Salvia spathacea

Salvia vaseyi

S. siphocampyloides Vatke   Pl 20–55 cm; rhizomes slender, tips ± swollen. **ST**: hairs << 0.5 mm, appressed-ascending, gen gland-tipped. **LF**: basal petioles 10–20 mm; distal cauline blades ovate to oblong, entire, bases rounded to tapered, tips rounded. **FL**: pedicel 4–5.5 mm; calyx 3–4 mm, ridged; corolla 25–35 mm, violet-blue, lower lip gen white-patched or -spotted, inner surface long-hairy. **FR**: black. Open sites, seeps, dry stream beds, scrub, woodland; 70–2500 m. CA-FP (only Sutter Buttes in GV). May–Jul

S. tuberosa Benth. (p. 861)   Pl < 25 cm, from tubers 5–20 mm. **ST**: hairs 1–3 mm, spreading. **LF**: basal petioles 5–20 mm; distal cauline blades ovate, entire to crenate, bases obtuse to rounded, tips rounded. **FL**: pedicel 2–4 mm; calyx 4–5.5 mm, ridged; corolla 13–20 mm, violet-blue, lower lip white-patched or -spotted, inner surface glabrous to long-hairy. **FR**: obconic, black. Dry sites, chaparral, oak woodland; < 1450 m. CA-FP (only Sutter Buttes in GV); OR, Baja CA. Common after fires. Mar–Jul

STACHYS   HEDGE-NETTLE

John B. Nelson

Per [ann], hairy, gen glandular; rhizome slender or 0. **ST**: decumbent to erect, 0.1–2.5 m. **LF**: 1.5–18 cm, proximal gen petioled, distal ± sessile; blade oblong to ovate, serrate to crenate. **INFL**: spike-like, gen terminal, interrupted or continuous, bracted. **FL**: calyx bell-shaped, ± radial, veins 5–10, lobes 5, erect or spreading, triangular, tips sharp; corolla white, yellow, pink, red, magenta, or purple, tube narrow, with internal ring of hairs gen above base, perpendicular to oblique to tube axis, gen with short, pouched spur on the lower side of the tube, upper lip erect or gen parallel to tube axis, concave, entire (notched), gen hairy, lower lip perpendicular to tube axis or reflexed, 3(2)-lobed, glabrous to hairy. **FR**: oblong to ovoid, brown to black, smooth or irregularly, minutely roughened. ± 300 spp.: gen temp; some cult for orn or edible rhizomatous tubers. (Greek: ear of corn, from infl) [Mulligan & Munro 1989 Naturaliste Canad 116:35–51] *S. arvensis* L., *S. floridana* Shuttlew. historical waifs.

1. Corolla tube > 15 mm . . . . ***S. chamissonis***
1′ Corolla tube < 15 mm
  2. Internal ring of hairs perpendicular to corolla tube; corolla not pouched toward base on lower side . . . . ***S. bullata***
  2′ Internal ring of hairs oblique to corolla tube; corolla pouched toward base on lower side
    3. Hairs densely silky-silvery, woolly; Del Norte Co. . . . . ***S. bergii***
    3′ Hairs soft, stiff, or glandular, never silky-silvery; widespread
      4. Upper corolla lip < 2 mm; stamens ± exserted from corolla tube . . . . ***S. stricta***
      4′ Upper corolla lip > 2 mm; stamens clearly exserted from corolla tube
        5. Calyx lobes lanceolate, ± = tube; proximal lvs short-petioled to sessile; MP . . . . ***S. pilosa***
        5′ Calyx lobes gen deltoid or triangular, shorter than tube; proximal lvs gen long-petioled; widespread
          6. Pl with abundant cobwebby, tangled hairs, herbage gen felty. . . . . ***S. albens***
          6′ Pl lacking cobwebby, tangled hairs, herbage not felty
            7. Mature infl compact, gen only 2 lowest bracts visible. . . . . ***S. pycnantha***
            7′ Mature infl elongated, many bracts visible
              8. Lf blade narrowed, silky hairy, base wedge-shaped; corolla white (± pink). . . . . ***S. ajugoides***
              8′ Lf blade broader, never silky, base rounded, truncate, or cordate; corolla pink or darker (white)
                9. St, herbage densely glandular, resinous; lower corolla lip 6–10 mm wide . . . . ***S. stebbinsii***
                9′ St, herbage ± glandular, not resinous; lower corolla lip 4–6 mm wide
                  10. Corolla magenta to purple; bracts gradually reduced distally on st; NW, ScV, SnFrB . . . . ***S. mexicana***
                  10′ Corolla pink; bracts rapidly reduced distally on st; widespread . . . . ***S. rigida***
                    11. Pl gen < 1 m; lf base cordate, blade ovate . . . . var. ***quercetorum***
                    11′ Pl gen 1 m; lf base rounded (cordate), blade lanceolate to oblong . . . . var. ***rigida***

S. ajugoides Benth. (p. 861)   **ST**: gen decumbent, 0.1–0.6 m, soft-hairy. **LF**: blade 1.5–7 cm, gen oblong, hairs gen long, silky, base wedge-shaped, tip rounded. **INFL**: clusters gen 3–6-fld. **FL**: corolla white (± pink), tube 6–8 mm, ring of hairs strongly oblique. 2*n*=66. Moist, open places, gen remaining wet into summer; < 1000 m. NW, ScV, CW, SW; to BC, Baja CA. Jun–Sep

S. albens A. Gray (p. 861)   Densely hairy, hairs tangled, cobwebby. **ST**: erect, 0.5–2.5 m, gen branched. **LF**: petiole < 5 cm; blade 3–15 cm, lanceolate to ovate, crenate to serrate, base ± cordate, tip acute to obtuse. **INFL**: 10–30 cm, ± interrupted, clusters 10–12-fld. **FL**: calyx tube 4–5.5 mm; corolla white to ± pink, tube 6–9 mm, ring of hairs > 2 mm from base, oblique, upper lip 3.5–5.5 mm, lower 6–8 mm. 2*n*=66. Swamps, seeps; < 3000 m. NCoR, SN, GV, CCo, SW, W&I, D (rare). May–Oct

S. bergii G.A. Mulligan & D.B. Munro   Densely woolly, hairs silvery. **ST**: erect, 0.3–0.6 m, branched; sides, angles densely woolly, not glandular, hairs spreading and reflexed, ± silvery. **LF**: petioles < 1 cm, blade 2–5 cm, narrowly lanceolate to oblong, base ± rounded, tip acute. **INFL**: > 5 cm, interrupted, clusters 6-fld, bracts gradually reduced distally. **FL**: calyx tube 4–5 mm, lobes 2.5–3 mm; corolla pink, tube 7–10 mm, ring of hairs > 2 mm from base, oblique, upper lip 3.5–5.5 mm, lower 6–10 mm. 2*n*=34. Dry slopes near coast, possibly near serpentine; < 900 m. NCo (Del Norte Co.). [*S. ajugoides* Benth. var. *rigida* (Benth.) Jeps. & Hoover, in part] Jun–Sep

S. bullata Benth. (p. 861)   **ST**: erect, 0.4–0.8 m, branched; hairs on angles stiff, sharp, reflexed, on sides soft, gen glandular. **LF**: blade 3–18 cm, ± ovate, soft- and stiff-hairy, glandular, base ± cordate, tip obtuse. **INFL**: > 5 cm, interrupted, internodes gen much elongated; clusters 6-fld. **FL**: calyx tube 6–7.5 mm, densely glandular, hairy; corolla pink to ± purple, tube 7–10 mm, not pouched, ring of hairs < 2 mm from base, perpendicular, upper lip 3.5–5.5 mm, lower 6–10 mm, lower lip hairy abaxially. 2*n*=66. Dry slopes near coast; < 500 m. CW, SW. Mar–Sep

S. chamissonis Benth. (p. 861)   **ST**: erect, 1–2.5 m, simple to branched distally, stiff-hairy, glandular. **LF**: petiole < 8 cm; blade 6–18 cm, deltoid to ovate, soft-hairy, densely glandular abaxially, margins gen straight, crenate, base cordate, tip acute. **INFL**: 10–40 cm, interrupted; clusters 2–6-fld. **FL**: calyx tube 7–12 mm, soft-hairy, glandular, lobes 3.5–4.5 mm; corolla deep magenta to purple, tube 18–24 mm, ring of hairs > 2 mm from base, perpendicular to ± oblique, upper lip 6–10 mm, reflexed in age, lower lip 8.5–15 mm. 2*n*=64. Wet, swampy places, gen coastal; < 150 m. NCo, CCo, SCoR. May–Oct

S. mexicana Benth.   **ST**: erect or sprawling, 0.5–1.5 m, simple to branched distally, hairy, occ glandular. **LF**: petiole well developed, hairy, blade ovate, serrate, hairy and glandular, cordate, tip acute. **INFL**: soft-downy, elongated, bracts gradually reduced distally, appearing lfy; clusters 6–8-fld. **FL**: calyx tube hairy, 3–5 mm, lobes

2–2.5 mm, gen reflexed or spreading in age; corolla magenta to purple, tube 7–8 mm, lower lip 6–7 mm. 2*n*=66. Thickets, damp places, gen coastal; < 500 m. NW, ScV, SnFrB (rare); OR, WA, AK. [*S. ajugoides* Benth. var. *rigida* (Benth.) Jeps. & Hoover, in part] Apr–Sep

S. pilosa Nutt. (p. 861) PRAIRIE WOUNDWORT Hairs of herbage gen hairs soft, spreading, ± glandular. **ST:** erect, 0.3–0.9 m, gen branched. **LF:** petiole 0–1 cm; blade 3.5–9 cm, ovate, oblong to elliptic, crenate, base rounded to ± cordate, tip acute to obtuse. **INFL:** > 5 cm, interrupted; clusters 6-fld. **FL:** calyx tube 3–4.5 mm, soft- to stiff-hairy, gen glandular, lobes ± = tube; corolla pink, tube 6–9 mm, ring of hairs > 2 mm from base, oblique, upper lip 3–5 mm, lower 5–8 mm. 2*n*=68. Moist places; 1200–1500 m. MP; to WA, w N.Am, e Can. [*S. palustris* subsp. *p.* (Nutt.) Epling] Most wide-ranging *Stachys* sp. in N.Am. Jun–Sep ★

S. pycnantha Benth. (p. 861) SHORT-SPIKED HEDGE NETTLE **ST:** decumbent to erect, 0.3–1 m, gen > 0.6 m, gen branched, soft- to stiff-hairy, glandular. **LF:** strongly aromatic; petiole < 5 cm; blade 5–12 cm, ovate or lanceolate to oblong, crenate to serrate, soft- to stiff-hairy, glandular, base rounded to cordate, tip obtuse. **INFL:** gen < 5 cm, ± continuous, occ interrupted proximally; clusters 8–12-fld. **FL:** calyx tube 4–6 mm, densely glandular, soft-hairy; corolla ± white to ± pink, tube 6.5–8.5 mm, ring of hairs > 2 mm from base, oblique, upper lip 3–4 mm, lower 5–7 mm. 2*n*=66. Streambanks, springs, pine/oak forest; < 1100 m. NW, SN (uncommon), CW, PR. May be ± associated with serpentine; more study needed. Jun–Oct

S. rigida Benth. **ST:** gen erect to ± decumbent, 0.6–1 m, ± glabrous to soft- or stiff-hairy, ± glandular. **LF:** blade 5–9 cm, ovate to lanceolate, glabrous to soft-hairy, occ glandular, base rounded to cor-

date, tip acute to obtuse. **INFL:** clusters 6–10-fld. **FL:** corolla pink, tube 6–10 mm, ring of hairs strongly oblique.

var. ***quercetorum*** (A. Heller) G.A. Mulligan & D.B. Munro **LF:** blade ovate, glabrous to soft-hairy, base cordate, tip acute to obtuse, margins gen and prominently crenate. 2*n*=66. Moist to ± dry places; < 2500 m. NW, GV (rare), CW, SW; to OR, Baja CA. [*S. ajugoides* Benth. var. *r.* (Benth.) Jeps. & Hoover, in part] Mar–Oct

var. ***rigida*** (p. 861) **LF:** blade lanceolate to oblong, soft-hairy, base rounded (cordate), tip acute to obtuse, margins minutely serrate to crenate. 2*n*=66. Moist to ± dry places; < 2500 m. CA (± rare D); to WA. [*S. ajugoides* Benth. var. *r.* (Benth.) Jeps. & Hoover] Jul–Aug

S. stebbinsii G.A. Mulligan & D.B. Munro (p. 861) Hairy, gen densely glandular. **ST:** gen erect, gen robust, to 1.5 m, resinous. **LF:** musky-aromatic, longest petioles 5–6 cm, blades 10–12 cm, 4–5 cm wide, base truncate to strongly cordate, margins prominently crenate, tip acute. **INFL:** clusters 6-fld. **FL:** calyx tube 4–5 mm, lobes 2.5–3 mm; corolla pink, tube 8–9 mm, lower lip 7–8.5 mm, 6–10 mm wide. 2*n*=66. Moist to ± dry places; < 2500 m. CW, SW; Baja CA. Summer

S. stricta Greene (p. 861) Herbage soft- to stiff-hairy. **ST:** erect, 0.3–0.8 m, branched, glandular. **LF:** petiole 1–4 cm; blade 5–15 cm, oblong to widely lanceolate, crenate, base cordate to ± truncate, tip obtuse; glands small, many, esp abaxially. **INFL:** 5–10 cm, gen continuous distally and interrupted proximally; clusters 8–12-fld. **FL:** calyx tube 4–5 mm, soft-hairy, glandular; corolla white, tube 5–7 mm, ring of hairs > 2 mm from base, oblique, upper lip < 2 mm, lower 3.5–7 mm. 2*n*=66. Moist, open or shady places; < 600 m. c&s NCoRI, n&c SNF, adjacent GV. Jun–Oct

TEUCRIUM

John M. Miller & Dieter H. Wilken

Ann, per, glabrous to short-hairy. **ST:** ascending to erect, branched or not. **LF:** petioled, crenate to deeply lobed, lobes oblong. **INFL:** gen spike-like, occ few-fld, fls subtended by lvs or bracts. **FL:** calyx ± radial, 10-veined, 5-lobed, lobes ± equal; corolla 1-lipped, tube split above, lip 5-lobed, ± flat, distal lobe > lateral lobes, tip rounded, lateral lobe tips acute to obtuse; stamens 4, lower pair gen > upper; style lobes gen equal. ± 100 spp.: worldwide, esp Medit. (Teucer, a Trojan monarch) [Wagstaff et al. 1998 Pl Syst Evol 209:265–274] *T. canadense* L. var. *occidentale* (A. Gray) E.M. McClintock & Epling, not in CA.

1. Ann, 1–3 dm, simple to branched at base; pedicels ≤ 5 mm; corolla lip 4–8 mm, white to
   ± blue, purple-spotted . . . . ***T. cubense*** subsp. ***depressum***
1′ Per, 5–10 dm, gen branched throughout; pedicels 8–25 mm; corolla lip 10–17 mm, white,
   gen violet-streaked . . . . ***T. glandulosum***

T. cubense Jacq. subsp. ***depressum*** E.M. McClint. & Epling (p. 861) DWARF GERMANDER **LF:** gen withering in fr; lower 2–4 cm, blade ovate to obovate, crenate to lobed; upper 0.5–1.5 cm, gen deeply 3-lobed;. **FL:** calyx tube 1–3 mm, lobes 3–6 mm, bristle-tipped; corolla 7–15 mm, ± puberulent inside; filaments glabrous. Sandy soils, washes, fields, alkaline flats; < 400 m. DSon; to TX, n Mex. Sp. to se US, Mex, Caribbean, S.Am. Mar–May ★

T. glandulosum Kellogg (p. 861) DESERT GERMANDER **LF:** 1–4 cm, gen deeply 3-lobed, ± persistent. **FL:** calyx tube 2–4 mm, lobes 4–8 mm, gen acute; corolla 15–21 mm, densely puberulent inside; filaments short-hairy below middle. Rocky slopes, canyons; 400–500 m. ne DSon (Whipple Mtns); AZ, Baja CA. Apr–May ★

TRICHOSTEMA BLUE CURLS

Harlan Lewis

Ann, shrub, strong-scented. **ST:** hairy, gen glandular. **LF:** simple; blade linear to ovate, entire. **INFL:** cyme (raceme), axillary. **FL:** calyx lobes 5, equal or uppermost 1 narrower; corolla blue to lavender, tube straight or curved upward, occ sharply recurved near throat, incl to much-exserted from calyx, lobes 5, lowest a gen reflexed lip; stamens 4, attached near throat, gen much-exserted, ascending between upper corolla lobes, gen arched. **FR:** nutlets 4, joined in basal ± 1/3, puberulent to hairy, irregularly ridged. ± 17 spp.: N.Am. (Greek: hair, stamen) [Lewis 1945 Brittonia 5:276–303] Ann spp. gen fl late summer, fall.

1. Shrub, < 1.5 m
   2. Corolla tube 9–14 mm; infl hairs 2–3 mm, obscuring pedicel . . . . ***T. lanatum***
   2′ Corolla tube 4–7 mm; infl hairs 1–2 mm, not obscuring pedicel . . . . ***T. parishii***
1′ Ann, < 1 m
   3. Petiole distinct, 5–15 mm
      4. Corolla tube 4–8 mm, exserted, lower lip 4–7 mm . . . . ***T. laxum***
      4′ Corolla tube 1.5–3 mm, incl, lower lip 2–3 mm . . . . ***T. simulatum***
   3′ Petiole indistinct or < 5 mm

5. Corolla tube curved abruptly upward near throat; stamens 5–20 mm
  6. Stamens 5–9 mm . . . . **T. ruygtii**
  6′ Stamens 11–20 mm
    7. Lf blade 2–7 cm, lanceolate to narrowly ovate, length > 3 × width; corolla tube 5–10 mm . . . . **T. lanceolatum**
    7′ Lf blade 1–2 cm, ovate, length < 2 × width; corolla tube 2.5–5.5 mm . . . . **T. ovatum**
5′ Corolla tube curved gradually upward; stamens 2–6 mm
  8. St, lf hairs curled to appressed; stamens 2–3 mm, ± incl to ± exserted . . . . **T. micranthum**
  8′ St, lf hairs straight, gen spreading; stamens 3–6 mm, exserted
    9. Calyx lobes < 2 × tube; calyx tube > mature nutlets; mature calyx gen red-tinged . . . . **T. rubisepalum**
    9′ Calyx lobes > 2 × tube; calyx tube < mature nutlets; mature calyx green
      10. Lf blade widely elliptic, length gen < 4 × width; calyx lobes gen widest above base . . . . **T. oblongum**
      10′ Lf blade elliptic, length gen > 4 × width; calyx lobes widest at base . . . . **T. austromontanum**
        11. Lf ≤ internode above . . . . subsp. **austromontanum**
        11′ Lf >> internode above . . . . subsp. **compactum**

T. austromontanum H. Lewis (p. 861) Ann. **ST**: short hairs appressed, long hairs spreading, some hairs glandular. **LF**: petiole indistinct or < 5 mm; blade 2–5 cm, elliptic, length gen > 4 × width. **FL**: calyx lobes > 2 × tube, widest at base, acute, ± equal; corolla tube 1.5–3 mm, curved gradually upward, ± = calyx, lower lip 1.8–3 mm; stamens 3–5.5 mm.

  subsp. ***austromontanum*** **ST**: < 5 dm. **LF**: < 5 cm, ≤ internode above. 2*n*=28. Drying lake margins, meadows, streambanks; 500–2600 m. c SNH, Teh, TR, PR, SNE; Baja CA. Jul–Oct

  subsp. ***compactum*** H. Lewis HIDDEN LAKE BLUECURLS **ST**: < 1 dm. **LF**: < 3 cm, >> internode above. 2*n*=28. Montane vernal pools; 2650 m. SnJt (Hidden Lake). Jul–Aug ★

T. lanatum Benth. (p. 861) WOOLLY BLUE CURLS Shrub < 15 dm. **ST**: hairs short, appressed on proximal st, occ woolly near infl. **LF**: petiole indistinct or < 3 mm; blade 3.5–7.5 cm, linear, adaxially green, glabrous, abaxially gray-hairy, margin rolled under; smaller lvs gen clustered in axils. **INFL**: hairs densely woolly, 2–3 mm, obscuring pedicel, fine, blue to pink or white. **FL**: calyx lobes ≥ tube, ± equal; corolla tube 9–14 mm, ± straight, exserted, lower lip 7–12 mm; stamens 2.5–4 cm. 2*n*=20. Coastal scrub, chaparral; < 1250 m. SCoR, SCo, WTR, SnGb, PR; Baja CA. Apr–Jul

T. lanceolatum Benth. (p. 861) VINEGAR WEED Ann < 1 m. **ST**: hairs short, appressed, or long-soft-spreading, some glandular. **LF**: petiole indistinct or < 4 mm; blade 2–7 cm, lanceolate to narrowly ovate, length > 3 × width, lateral veins prominent near base. **FL**: calyx lobes 1–3 × tube, uppermost gen narrower; corolla tube 5–10 mm, curved upward and sharply recurved near throat, exserted, lower lip 4–8 mm; stamens 13–20 mm. 2*n*=14. Dry, open, gen disturbed habitats; < 2200 m. CA-FP; OR, Baja CA. Lvs have strong vinegar odor. Jun–Nov

T. laxum A. Gray (p. 861) TURPENTINE WEED Ann < 5 dm. **ST**: hairs short, appressed, or long, spreading, some glandular. **LF**: petiole distinct, 5–15 mm; blade 3–7 cm, lanceolate to narrowly ovate. **FL**: calyx lobes ± = tube, gen red-tinged, uppermost narrower; corolla tube 4–8 mm, abruptly upward near throat, exserted, lower lip 4–7 mm; stamens 7–16 mm. 2*n*=14. Gravelly streambanks or sandy soil; < 1700 m. NW, MP. Lvs have strong turpentine odor. Jun–Oct

T. micranthum A. Gray SMALL-FLOWERED BLUECURLS Ann < 3 dm. **ST**: hairs short, curled to appressed, some glandular. **LF**: petiole indistinct or < 5 mm; blade 2–4.5 cm, narrowly elliptic. **FL**: calyx lobes < 2 × tube, widest at base, acute, ± equal; corolla tube 1–2 mm, curved gradually upward, incl, lower lip 1.5–2 mm; stamens 2–3 mm, ± incl to ± exserted, ± arched to ± straight. 2*n*=14. Dry margins of lakes, meadows, streams; 1500–2500 m. WTR, SnBr; AZ, Baja CA. Jul–Sep ★

T. oblongum Benth. (p. 867) Ann < 5 dm. **ST**: hairs short, appressed to spreading, or hairs, spreading, some glandular. **LF**: petiole indistinct or < 5 mm; blade 2–4 cm, widely elliptic, length gen < 4 × width. **FL**: calyx lobes > 2 × tube, gen widest above base, ± acuminate, ± equal; corolla tube 2–3.5 mm, curved gradually upward, ± = calyx, lower lip 2–3.5 mm; stamens 3–6 mm, exserted, arched. 2*n*=14. Dry margins of meadows, streambanks; 100–3000 m. KR, NCoR, CaR, SN, MP; to WA, ID. Jun–Sep

T. ovatum Curran SAN JOAQUIN BLUECURLS Ann < 8 dm. **ST**: woolly; hairs soft, curled or spreading, some glandular. **LF**: petiole ± indistinct; blade 1–2 cm, ovate, length < 2 × width, lateral veins prominent near base. **FL**: calyx lobes 2–3 × tube, ± equal; corolla tube 2.5–5.5 mm, curved upward and sharply recurved near throat, exserted, lower lip 3.5–5 mm; stamens 11–16 mm. 2*n*=14. Grassland, disturbed sites; < 300 m. s SnJV, w WTR (Ventura Co.). Jul–Oct ★

T. parishii Vasey (p. 867) Shrub < 12 dm. **ST**: hairs short, appressed. **LF**: petiole indistinct or < 5 mm; blade 2–6 cm, linear, adaxially green, puberulent, abaxially gray-hairy, margin rolled under; smaller lvs gen clustered in axils. **INFL**: hairs ± densely woolly, 1–2 mm, not obscuring pedicel, fine, blue to pink or white, occ ± 0. **FL**: calyx lobes ± = tube, ± equal; corolla tube 4–7 mm, ± straight, exserted, lower lip 5–9 mm; stamens 15–25 mm. 2*n*=20. Coastal scrub, chaparral; 600–2000 m. TR, PR; Baja CA. Mar–Jul

T. rubisepalum Elmer (p. 867) HERNANDEZ BLUECURLS Ann < 5 dm. **ST**: hairs long, spreading, some short-glandular. **LF**: petiole indistinct or < 5 mm; blade lanceolate to narrowly ovate. **FL**: calyx lobes < 2 × tube, widest at base, acute, gen red-tinged, uppermost narrower; corolla tube 3.5–4.5 mm, curved gradually upward, exserted, lower lip 2.5–3 mm; stamens 4–6 mm. 2*n*=14. Seasonal seeps and gravelly streambeds, serpentine; 250–1400 m. c SNF (Tuolumne, Mariposa cos.), SCoRI (San Benito Co.). Jul–Sep ★

T. ruygtii H. Lewis NAPA BLUECURLS Ann < 5 dm. **ST**: hairs short, curved, or long, spreading, some glandular. **LF**: petiole 0 or indistinct; blade 1–4 cm, length 3 × width, lanceolate, lateral veins prominent near base. **FL**: calyx lobes ≤ tube, uppermost gen narrower; corolla tube 5–7.5 mm, curved upward and sharply recurved near throat, exserted, lower lip 2–4.5 mm; stamens 5–9 mm. Open areas, gen thin clay soils, possibly seasonally saturated; 30–600 m. s NCoRO/NCoRI (Napa Co.). Jun–Oct ★

T. simulatum Jeps. Ann < 4 dm. **ST**: hairs short, appressed, or long, spreading, some glandular. **LF**: petiole distinct, 7–15 mm; blade 2–5 cm, lanceolate to narrowly ovate. **FL**: calyx lobes ± = tube, triangular, ± equal; corolla tube 1.5–3 mm, ± curved upward, incl, lower lip 2–3 mm; stamens 2.5–5 mm. 2*n*=14. Open, gen sandy or gravelly sites; 500–1600 m. KR, CaR, n SN, MP; s OR. Jul–Sep

LENTIBULARIACEAE   BLADDERWORT FAMILY

Barry A. Rice

Ann, per, carnivorous, of moist or aquatic habitats. **ST**: caudex or stolon, then often with thread-like branches. **LF**: simple, in rosette, or simple or dissected, emerging from caudex or stolon, with minute, carnivorous bladders ± throughout. **INFL**: raceme or 1-fld, scapose. **FL**: bisexual; calyx lips 2[4], upper 3-lobed, lower 2-lobed, or lips unlobed; corolla 2-lipped, spurred

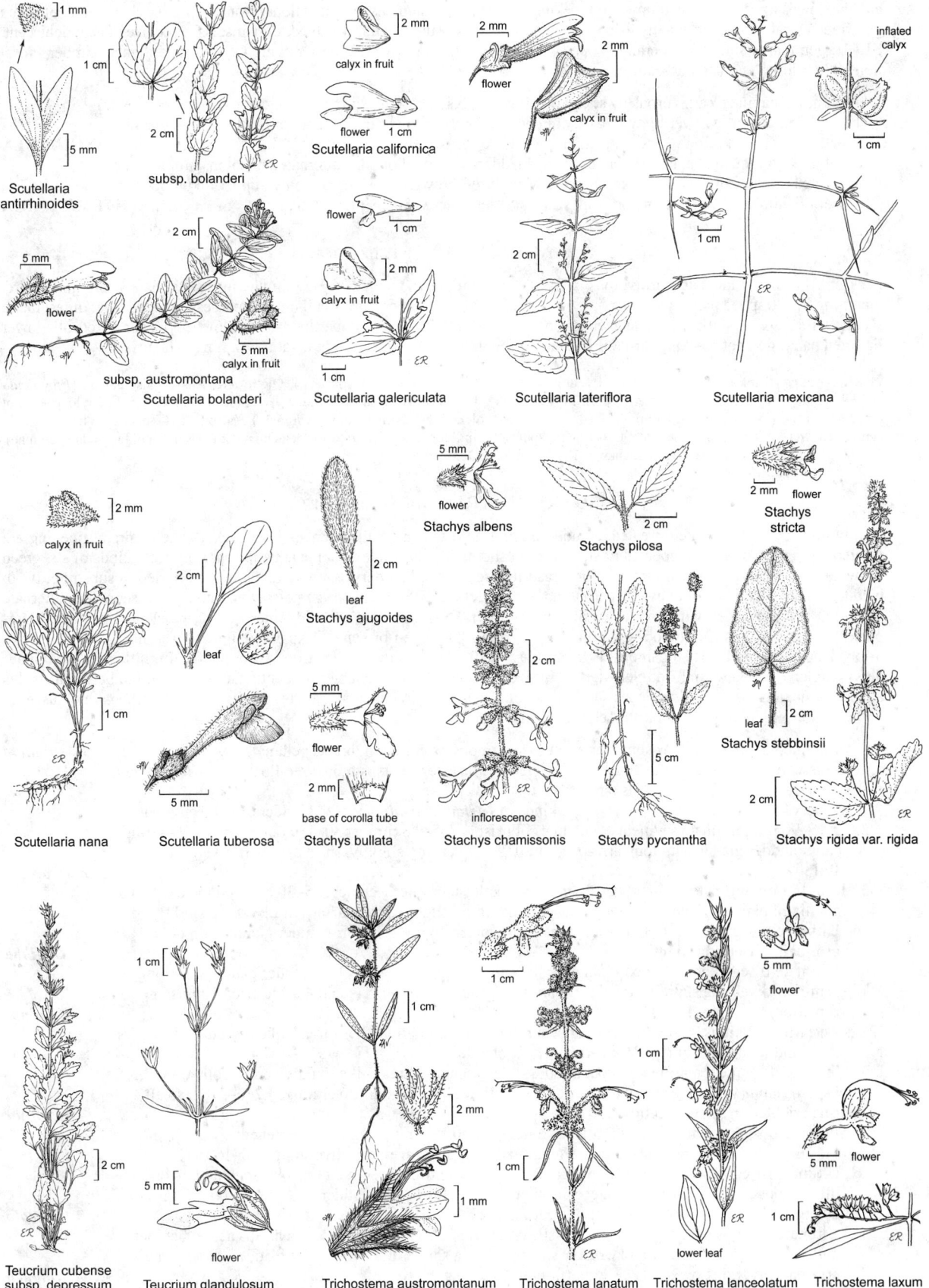

Scutellaria antirrhinoides

subsp. bolanderi

calyx in fruit

flower

Scutellaria californica

flower

calyx in fruit

inflated calyx

5 mm
flower

subsp. austromontana
Scutellaria bolanderi

calyx in fruit

flower

calyx in fruit

Scutellaria galericulata

Scutellaria lateriflora

Scutellaria mexicana

calyx in fruit

leaf

Scutellaria nana

leaf

flower

Scutellaria tuberosa

leaf
Stachys ajugoides

flower

base of corolla tube

Stachys bullata

flower
Stachys albens

inflorescence

Stachys chamissonis

Stachys pilosa

Stachys pycnantha

flower
Stachys stricta

leaf
Stachys stebbinsii

Stachys rigida var. rigida

Teucrium cubense subsp. depressum

flower

Teucrium glandulosum

Trichostema austromontanum

Trichostema lanatum

lower leaf

Trichostema lanceolatum

flower

flower

Trichostema laxum

at base, lower lip flat or arched upward, blocking throat or not; stamens 2, epipetalous; ovary superior, chamber 1, placenta gen free-central; stigma unequally 2-lobed, ± sessile. **FR**: capsule, round, 2-valved, circumscissile, or irregularly dehiscent. **SEED**: gen many, small. 3 genera, 330 spp.: worldwide, esp trop. [Rice 2006 Growing Carnivorous Plants. Timber Press] Scientific Editor: Thomas J. Rosatti.

1. Pl of moist habitats, carnivorous by sticky or slimy lvs; lvs in rosette, simple, elliptic to ovate; corolla (pale) blue-violet, center of lower lip white; calyx upper lip 3-lobed, lower lip 2-lobed; infl 1-fld, scape succulent, bracts 0. . . . . **PINGUICULA**
1′ Pl of moist or aquatic habitats, carnivorous by bladders ± throughout; lvs alternate on stolon, simple or gen dissected into narrow segments; corolla yellow, with red-brown streaks or not; calyx upper lip unlobed, lower lip unlobed; infl a raceme or 1-fld, scape slender or wiry, bracts present . . . . **UTRICULARIA**

PINGUICULA BUTTERWORT

Per [(ann)] of moist habitats; carnivorous by sticky or slimy lvs. **ST**: caudex. **LF**: simple, in rosette, fleshy; margins entire, incurving over trapped prey or not; adaxially with stalked glands that capture small organisms, sessile glands that digest them. **INFL**: 1-fld [rarely 2-fld]; scape succulent, glandular, bracts 0. **FL**: calyx upper lip 3-lobed, lower lip 2-lobed; corolla lower lip often hairy, gen not blocking throat. **FR**: 2-valved. ± 92 spp.: Am, Eur, Medit. (Latin: ± fat, from lf surface)

P. macroceras Link (p. 867) HORNED BUTTERWORT **LF**: 2–5 cm, elliptic to ovate, green to dark brown. **INFL**: 1–5 per rosette, 1–2 dm. **FL**: corolla (incl spur) 13–21 mm, 5-lobed, (pale) blue-violet, center of lower lip white; lobes obovate, throat hairy, spur 6–9 mm. 2*n*=64. Moist slopes, cliffs, serpentine banks; < 1830 m. n KR (Del Norte, Siskiyou cos.); OR to AK, Russia; also Japan. [*Pinguicula vulgaris* subsp. *macroceras* (Link) Calder & Roy L. Taylor] May not be distinct from *P. vulgaris* L. (also 2*n*=64); pls overwinter as dense bud, proliferate by detachment of basal daughter buds (gemmae). Apr–Jul ★

UTRICULARIA BLADDERWORT

Carnivorous by bladders (here treated as modified lvs), into which small organisms are sucked when hairs at opening are triggered [epiphytic]. **ST**: submersed or subterranean shoots [rarely caudex]; some aquatic spp. produce 2 kinds of sts, green (in water or at surface; lvs with flattened or thread-like segments; bladders 0–few) and white (gen buried in substrate; lvs 0; bladders many), the latter not always present in poor collections. **LF**: simple or gen dissected into narrow segments, alternate on stolon, margins often with bristles (visible at 10–30×). **INFL**: raceme or 1-fld, emergent; scape slender or wiry, bracts present. **FL**: calyx lips 2[4], unlobed; corolla yellow [or not], with red-brown streaks or not, upper lip ± entire, lower entire or 3-lobed, spurred; rarely cleistogamous. **FR**: capsule. ± 220 spp.: worldwide, esp trop, Australia. (Latin: little bag, from bladders) [Taylor 1989 Kew Bull Add Ser 14:1–724] Size variable, often unreliable for identification; distinction between sts, lvs uncertain. Glands inside bladders consist of 2 pairs of oppositely directed arms, angles of divergence between which used (at 150× or more) to identify fresh or pressed specimens.

1. Lvs simple, flattened; stolons subterranean; cleistogamous fls common; bracts peltate. . . . . ***U. subulata***
1′ Lvs dissected into 2–many thread-like to flattened ultimate segments; stolons free-floating, loosely rooted, or subterranean; cleistogamous fls rare; bracts attached at base
  2. Sts gen of 1 kind (sometimes weakly of 2 kinds in *Utricularia minor*), with bladders, lvs throughout
    3. Lf 15–90 mm, ultimate segments > 30, margin bristles 0.1–0.3 mm; st 30–150 cm; peduncle 1–3 mm diam; bladder glands with long arm-pair parallel or diverging slightly, short arm-pair diverging by 90–180°. . . . . ***U. macrorhiza***
    3′ Lf 2–15 mm, ultimate segments 2–22, margin bristles 0 or microscopic; st 5–30 cm; peduncle 0.4–0.8 mm diam; bladder glands with long arm-pair diverging slightly, short arm-pair diverging slightly or by 270–300°
      4. Ultimate lf segments 2–6(8); corolla upper lip ≥ lower, spur ± = lower; bladder glands with both arm-pairs diverging slightly. . . . . ***U. gibba***
      4′ Ultimate lf segments 7–22; corolla upper lip, spur both ≤ 1/2 × lower lip; bladder glands with long arm-pair diverging slightly, short arm-pair diverging by 270–300° (so all 4 arms oriented in same direction) . . . . ²***U. minor***
  2′ Sts clearly of 2 kinds, green (in water or at surface; lvs with flattened or thread-like segments; bladders 0–few) and white (gen buried in substrate; lvs 0; bladders many)
    5. Ultimate lf segment margins without bristles; lf segments ± thread-like to flattened; corolla 6–8 mm; bladder glands with long arm-pair diverging slightly, short arm-pair diverging by 270–300° (so all 4 arms oriented in same direction) . . . . ²***U. minor***
    5′ Ultimate lf segment margins with bristles 2–12, 0.08–0.16 mm; lf segments flattened; corolla 1–1.6 cm; bladder glands with long arm-pair diverging by 0–45°, short arm-pair diverging by 0–160°
      6. Ultimate lf segment margins with 5–12(20) bristles, ± entire; lf segment tips obtuse; bladders on lfless shoots, rarely on lfy shoots; corolla spur < or ± = lower lip; bladder glands with both arm-pairs diverging by gen 0–30°. . . . . ***U. intermedia***
      6′ Ultimate lf segments margins with 1 bristle on each of 2–7(10) teeth; lf segment tips acute; bladders on lfless shoots, also often ± few on lfy shoots; corolla spur ± 1/2 × lower lip; bladder glands long arm-pair diverging by 20–45°, short arm-pair diverging by 40–160° . . . . ***U. ochroleuca***

U. gibba L. (p. 867)   Loosely rooted (rarely free-floating) aquatic or creeping in mud. **ST:** gen of 1 kind, thread-like, with lvs, few bladders throughout; winter buds 0. **LF:** 2–15 mm, gen 2-parted at base, each part forked again or not, bladders 0(2); ultimate segments 2–6 (8), thread-like. **INFL:** 1–4-fld; peduncle < 20 cm, < 0.8 mm diam; pedicel erect in fr. **FL:** corolla 6–15(25) mm, lower lip ≤ upper, spur ± = lower lip, conic. **FR:** laterally 2-valved. **SEED:** winged. 2*n*=28. Shallow water, mud, mat-forming at surface of deep waters or not; 10–2300 m. n&c CA-FP. Pls in NCo, GV possibly are not native there, and may have been spread there by horticulturists or agriculture. Jul–Sep

U. intermedia Hayne (p. 867)   FLAT-LEAVED BLADDERWORT **ST:** of 2 kinds, some freely floating, green, with lvs, rarely with bladders, others rooted in mud, white, without lvs, with bladders; winter buds bristly. **LF:** 3–15 mm, 3-parted at base, variously dissected above; ultimate segments < 20, ± linear, flattened, margins with 5–12(20) bristles, ± entire, tip obtuse. **INFL:** 1–5-fld; peduncle < 25 cm, < 1 mm diam; pedicel erect in fr. **FL:** corolla 10–16 mm; lower lip ± 2 × upper, ± > or ± = cylindric, acute spur. **FR:** circumscissile. **SEED:** not seen. 2*n*=44. Shallow (< 1 m) water; 1200–2700 m. CaR, SNH, MP; n US, Can; circumboreal. In some populations, divergence of bladder gland arm-pairs resembling that of *U. ochroleuca*. Jul–Sep ★

U. macrorhiza Leconte (p. 867)   COMMON BLADDERWORT Floating aquatic. **ST:** well defined, weakly branched central stolon with lvs, bladders; winter buds 1–2 cm, bristly. **LF:** 15–90 mm, 1–2-parted at base, each part unequally pinnately dissected above; ultimate segments dense, 30–150, thread-like, margin bristles 0.1–0.3 mm; bladders near base larger than those near tip. **INFL:** 5–20-fld; peduncle 1–4 dm, stout, 1–3 mm diam; pedicel recurved in fr. **FL:** corolla 1–2 cm; lower lip > upper; cylindric spur ± hooked upward near tip. **FR:** circumscissile. **SEED:** 4–6-sided, -winged. 2*n*=±40. Quiet, shallow or deep, rarely flowing, acidic waters; < 2700 m. NW, CaR, SN, SnFrB, SnBr, GB, w DMoj; N.Am exc extreme se, ne Mex, e Asia. *U. vulgaris* L. excluded by recircumscription. Jun–Sep

U. minor L. (p. 867)   LESSER BLADDERWORT Rooted or floating aquatic. **ST:** slender, weakly to strongly of 2 kinds, some with many lvs, some bladders, others with 0–few lvs, ± many bladders; winter buds glabrous. **LF:** 2–15 mm, 3-parted at base, variously dissected above; ultimate segments 7–22, ± thread-like to flattened, margin bristles 0 or microscopic. **INFL:** 2–6-fld; peduncle 3–25 cm, wiry, 0.4–0.8 mm diam; pedicel recurved in fr. **FL:** corolla 6–8 mm; lower lip > 2 × upper, 2 × sac-like spur. **FR:** circumscissile. **SEED:** 4–6-sided, ± unwinged. 2*n*=±40. Shallow (gen < 30 cm) acidic waters; 800–2900 m. CaRH, SNH, MP; AK, n US, Can; circumboreal. Jun–Sep ★

U. ochroleuca R.W. Hartm.   CREAM-FLOWERED BLADDERWORT Rooted aquatic. **ST:** of 2 kinds, some freely floating, green, with lvs, ± few bladders, others rooted in mud, white, without lvs, with bladders; winter buds bristly. **LF:** 5–15 mm, 3-parted at base, variously dissected above; ultimate segments < 20, ± linear, flattened, margins with 1 bristle on each of 2–7(10) teeth, tip acute. **INFL:** 2–5-fld; peduncle < 15 cm, < 1 mm diam. **FL:** corolla 10–15 mm; lower lip ± 2 × upper, ± 2 × conical (sometimes cylindrical-tipped) spur. **FR:** not seen. **SEED:** not seen. 2*n*=±40–48. Shallow (gen < 30 cm) acidic waters; 1300–2400 m. CaRH (Plumas Co.), n SNH (El Dorado Co.), MP (Modoc Co.); n US, Can; circumboreal. [*U. stygia* G. Thor] Probably of hybrid origin (*U. minor* × *U. intermedia*). Known from only 3–4 sites. Jun–Sep ★

U. subulata L.   Creeping in wet substrates. **ST:** creeping in wet substrates, thread-like, with lvs, bladders; winter buds 0. **LF:** lvs at or above surface, simple, < 20 mm, 1 mm wide, entire, flattened, with bladders or not, tip rounded; marginal bristles 0; petiole narrow. **INFL:** 1–many-fld; peduncle 5–25 cm, wiry, 0.3–0.6 mm diam; pedicel 2–10 mm, erect in fr. **FL:** corolla 5–10 mm; lower lip ± 2 × upper, ± = spur; spur narrowly elongate, straight; cleistogamous fls much reduced, < 2 mm. **FR:** dehiscent by ventral valve. **SEED:** unwinged. 2*n*=30. Moist acid wetlands, introduced with other carnivorous pls by horticulturists; < 1200 m. NCo (Mendocino Co.), possibly elsewhere; e US, Mex, ± pantrop. [*U. cleistogama* (A. Gray) Britton] Jun–Sep

LIMNANTHACEAE   MEADOWFOAM FAMILY

Robert Ornduff & Nancy R. Morin

Ann, glabrous to hairy. **ST:** gen branched. **LF:** alternate, deeply pinnately lobed to 1–2-compound; stipules 0. **INFL:** fls solitary in axils, pedicelled. **FL:** gen bisexual, radial; sepals 3–5, free; petals 3–5, free, white to pink or yellow; stamens 3, 8, or 10, free, gen in 2 whorls; nectary glands at bases of outer stamens; pistil 1, ovary deeply 2–5-lobed, style 1, from base of ovary. **FR:** mericarps 1–5, 1-seeded, ovoid to spheric, gen tubercled. 2*n*=10. 2 genera, 10 spp.: temp N.Am. Scientific Editors: Douglas H. Goldman, Bruce G. Baldwin.

1. Sepals gen 3; petals gen 3, << sepals, tips ± entire . . . . **FLOERKEA**
1′ Sepals 4–5; petals 4–5, gen > sepals, tips toothed or jagged . . . . **LIMNANTHES**

FLOERKEA   FALSE MERMAID

1 sp.: N.Am. (H.G. Floerke, German botanist, 1764–1835)

F. proserpinacoides Willd. (p. 867)   Ann, decumbent to erect, fleshy, glabrous. **ST:** < 20 cm. **LF:** 1-ternate or 1-odd-pinnate, ≤ 6 cm; lflets 3–5, ≤ 2 cm, ± oblong. **FL:** sepals gen 3, 2–4 mm; petals gen 3, spoon-shaped, white, tips ± entire; stamens 3–6; ovary lobes 2–3. **FR:** mericarps 2–3, 2–3.5 mm, ± spheric, wrinkled or tubercled above. Moist open places in conifer forest or sagebrush scrub; 1500–3200 m. KR, NCoRH, CaRH, SNH, MP; to BC, e N.Am. May–Jul

LIMNANTHES   MEADOWFOAM

Ann, decumbent to erect. **LF:** gen 1-odd-pinnately lobed or compound; lobes or lflets entire to deeply lobed. **FL:** sepals (4)5; petals (4)5, gen > sepals, tips toothed or jagged; stamens 8 or 10; ovary lobes 4–5. **FR:** mericarps 1–5, 2.5–4 mm, obovoid to nearly spheric, smooth, ridged, or tubercled. 7 spp.: ± coastal w N.Am. (Greek: marsh fl, from habitat) [Mason 1952 Univ Calif Publ Bot 25:455–512] Pls reported as *L. macounii* from CCo are variants of *L. douglasii* (Meyers et al. 2010 Syst Bot 35:552–558).

1. Petals curving over fr in age
  2. Petals < to ± > sepals . . . . *L. floccosa*
    3. Fl cup-shaped; fr papillate, gen with widely conic tubercles . . . . subsp. *californica*
    3′ Fl bell- to urn-shaped; fr with awl-shaped or plate-like tubercles
      4. Sepals glabrous to sparsely hairy; fr with plate-like tubercles, bristles 0 . . . . subsp. *bellingeriana*
      4′ Sepals densely hairy; fr with awl-shaped tubercles, bristles gen present. subsp. *floccosa*
  2′ Petals > sepals
    5. Filaments 2.5–4 mm
      6. Petals aging pink; anthers ± 1 mm; PR . . . . *L. alba* subsp. *parishii*
      6′ Petals aging white; anthers 0.5–1 mm; c&s SNF . . . . *L. montana*
    5′ Filaments 5–6 mm
      7. Sepals densely hairy; fr wide-ridged; s NCoR, n&c SNF, GV. . . . . *L. alba* subsp. *alba*
      7′ Sepals glabrous or sparsely hairy; fr smooth to sharply tubercled; KR, CaRF,
        n&c SN. . . . . *L. alba* subsp. *versicolor*
1′ Petals gen reflexing in age
  8. Lflets gen entire (lower rarely 2–3-lobed); sepals 5–7 mm
    9. Petals 7–9 mm; style 2–3 mm; c NCoRO (near Willits, Mendocino Co.) . . . . *L. bakeri*
    9′ Petals 12–18 mm; style 4.5–6.5 mm; s NCoR (s Sonoma Co.). . . . . *L. vinculans*
  8′ Lflets entire to 2–3-lobed; sepals 4–17 mm. . . . . *L. douglasii*
    10. Fl funnel-shaped; petals white, veins gen dark; anthers cream . . . . *L. douglasii* subsp. *striata*
    10′ Fl cup- or bell-shaped; petals white to yellow, veins occ colored but not dark;
      anthers cream to yellow, dark pink, orange-red, or ± black
      11. Petals yellow, occ white-tipped
        12. Petals yellow, white-tipped; anthers yellow, cream, or red-brown . . . . *L. douglasii* subsp. *douglasii*
        12′ Petals yellow throughout; anthers yellow . . . . *L. douglasii* subsp. *sulphurea*
      11′ Petals ± white, without yellow exc occ when dry
        13. Petals veins white or ± purple; anthers cream to yellow; mericarps ± tubercled above or smooth;
          NCoR, SnFrB, SCoR . . . . *L. douglasii* subsp. *nivea*
        13′ Petals veins rose; anthers cream, dark pink, orange-red, or ± black; mericarps ridged;
          NCoRI, CaRF, c SNF, GV . . . . *L. douglasii* subsp. *rosea*

L. alba Benth. Erect, herbage glabrous to hairy. **ST:** 8–40 cm. **LF:** 2–10 cm; lflets 5–11, linear to ovate or oblong, entire to deeply 3-lobed. **FL:** bowl- to bell-shaped; sepals 4–8 mm; petals 8–15 mm, curving over fr in age, white to cream, aging pink or violet, base occ ± yellow; filaments 2–6 mm, anthers 0.75–2 mm; style 3–6 mm. **FR:** mericarps smooth, wrinkled, wide-ridged, or sharply tubercled.

subsp. ***alba*** Herbage sparsely to densely hairy when young. **FL:** sepals densely hairy; petals 10.5–16 mm, white, drying pink; anthers ± 2 mm, cream. **FR:** ± 4 mm, wide-ridged. Winter-wet grassland, woodland; < 1400 m. s NCoR, n&c SNF, GV; OR. Apr–Jun

subsp. ***parishii*** (Jeps.) Morin Herbage glabrous. **FL:** sepals glabrous; petals 8–10 mm, white, base cream or not, aging pink, tips recurved; anthers ± 1 mm, cream to yellow; style 2–3 mm. **FR:** ± 3 mm, tubercles short, wide. Wet meadows, edges of ephemeral streams; 600–2000 m. PR. [*L. gracilis* subsp. *p.* (Jeps.) Beauch.] Apr–May ★

subsp. ***versicolor*** (Greene) C.T. Mason (p. 867) Herbage glabrous. **FL:** sepals glabrous to sparsely hairy; petals 12–15 mm, cream, tips recurved, drying lilac; anthers ± 1 mm, cream or ± pink. **FR:** 3.5 mm, smooth or sharply tubercled. Winter-wet grassland, woodland, edges of vernal pools, ephemeral streams; 150–2100 m. KR, CaRF, n&c SN. May–Jun

L. bakeri J.T. Howell (p. 867) BAKER'S MEADOWFOAM Erect or ascending, herbage glabrous. **ST:** 10–40 cm. **LF:** < 10 cm. **LF:** lflets 3–9, elliptic to ovate, entire or lower occ 2–3-lobed. **FL:** funnel- to bell-shaped; sepals 5–7 mm; petals 7–9 mm, pale yellow with white tips, reflexed outward in fr; filaments 2.5–4 mm, anthers 0.5 mm, cream; style 2–3 mm. **FR:** mericarp tubercles dense, short, wide. Vernal pools, marshy lake and stream margins; < 500 m. c NCoRO (near Willits, Mendocino Co.). Threatened by development, grazing. Apr–May ★

L. douglasii R. Br. Erect or ascending; herbage glabrous. **ST:** 3–35(100) cm. **LF:** 3–7(25 cm); lflets 5–13, linear to lanceolate to widely ovate, entire or occ toothed or lobed. **FL:** cup-, bell-, or funnel-shaped; sepals 4–17 mm, glabrous or margins hairy; petals 10–18 mm, white to yellow or yellow with white tips, veins purple to pink or cream, occ drying ± green-yellow, reflexed in fr; filaments 3.5–8 mm, anthers 0.8–2 mm, cream to yellow, dark pink, orange-red, or ± black; style 3–8 mm. **FR:** mericarps smooth, ridged, or at least tip tubercled.

subsp. ***douglasii*** (p. 867) **LF:** lflets ovate (to linear), entire or irregularly toothed or lobed. **FL:** cup- or bell-shaped; petals yellow, white-tipped; anthers yellow, cream, or red-brown. **FR:** mericarps tubercles ridge-like. Wet meadows; < 700 m. NCo, NCoRO, CCo, SnFrB; sw OR. Mar–May

subsp. ***nivea*** (C.T. Mason) C.T. Mason **LF:** lflets linear to ovate, entire to deeply lobed. **FL:** cup- or bell-shaped; petals white, drying pale yellow at base or not, veins white or ± purple; anthers cream to yellow. **FR:** mericarps smooth or ± tubercled above. Wet meadows, edges of vernal pools, ephemeral streams; < 1000 m. NCoR, ScV (Butte Co.), SnFrB, SCoR. Mar–May

subsp. ***rosea*** (Benth.) C.T. Mason **LF:** lflets linear to widely ovate, entire to deeply, irregularly lobed. **FL:** cup- or bell-shaped; petals white, occ drying pale pink or yellow, veins rose; anthers cream, dark-pink, orange-red, or ± black. **FR:** mericarps ridged. Wet meadows, edges of vernal pools; < 800 m. NCoRI, CaRF, c SNF, GV. Mar–May

subsp. ***striata*** (Jeps.) Morin **LF:** lflets linear to ovate, entire to toothed or 2–3-lobed. **FL:** funnel-shaped; petals white, bases ± green-yellow, veins gen dark; stamens 2–4 mm, anthers cream. **FR:** mericarps smooth, wrinkled, or tubercled. Vernal pools, stream edges; < 800 m. KR, n&c SNF. [*L. s.* Jeps.] Some KR pls smaller. Mar–May

subsp. ***sulphurea*** (C.T. Mason) C.T. Mason POINT REYES MEADOWFOAM **LF:** lflets ovate, irregularly toothed or lobed. **FL:** petals, anthers yellow. **FR:** mericarp ridges plate-like and appearing slick. Wet meadows of coastal prairie; < 100 m. CCo (Marin Co.). Mar–May ★

L. floccosa Howell (p. 867) Erect to ascending or decumbent; herbage glabrous to sparsely or densely hairy. **ST:** 3–25 cm. **LF:** 1–8 cm; lflets 5–11, linear to ovate-elliptic, entire, toothed, or lobed. **FL:** urn-, cup-, or bell-shaped; sepals 4–10 mm; petals 4.5–10 mm, < to ± > sepals, curving over fr, white; filaments 2–7 mm, anthers 0.4–1.5 mm; style 1.5–4 mm. **FR:** mericarps tubercles low. In CA mericarps fall together with calyx.

subsp. ***bellingeriana*** (M. Peck) Arroyo BELLINGER'S MEADOWFOAM Herbage glabrous. **FL**: urn-shaped; sepals 5.5–8 mm, glabrous to sparsely hairy; petals 5.5–7.5 mm, hairs at base 0; filaments 2–3.5 mm, anthers ± 0.5 mm, dehiscing inward; style 2.3–2.5 mm. **FR**: mericarps 1–2(3), tubercles plate-like, pointed. Vernal pool edges; 300–1100 m. CaR (Shasta Co.); OR. Mar–May ★

subsp. ***californica*** Arroyo BUTTE COUNTY MEADOWFOAM Herbage densely hairy. **FL**: cup-shaped; sepals 7.5–10 mm, densely hairy; petals 8–10 mm, hairs at base present; filaments 3–7 mm, anthers 1–1.5 mm, dehiscing outward; style 3.5–4 mm. **FR**: mericarps, tubercles gen widely conic. Vernal pool edges; < 100 m. ScV (Butte Co.). Mar–May ★

subsp. ***floccosa*** WOOLLY MEADOWFOAM Herbage densely hairy. **FL**: bell- to urn-shaped; sepals 4–9 mm; petals 4.5–8.5 mm, hairs at base 0; anthers 0.4–± 1 mm, gen dehiscing inward; style 1.5–3 mm. **FR**: mericarps, 1–3, tubercles awl-shaped. Vernal pool edges; < 600 m. KR, NCoRI, CaR; OR. Mar–May ★

L. montana Jeps. Ascending to erect; herbage glabrous to sparsely long-hairy. **ST**: 10–40 cm. **LF**: 3–15 cm; lflets 7–11, linear to ovate, entire to deeply 3-lobed, lobes entire to lobed. **FL**: funnel-shaped; sepals 3–6 mm; petals 7–12 mm, curving over fr in age, white, base occ ± yellow, veins ± purple, aging white; filaments 2.5–4 mm, anthers 0.5–0.75(1) mm, cream; style 2.5–4 mm. **FR**: mericarps, tubercles low, conic. Wet meadows, stream edges; 200–1800 m. c&s SNF. Pls in Madera, Mariposa cos. have larger anthers. Mar–May

L. vinculans Ornduff SEBASTOPOL MEADOWFOAM Erect to ascending, glabrous. **ST**: ≤ 30 cm. **LF**: ≤ 10 cm; lflets 3–5, 5–20 mm, entire, narrowly obovate. **FL**: bell-shaped to rotate; sepals 6–7 mm; petals 12–18 mm, white, base drying ± yellow, reflexed in age; filaments 5–7 mm, anthers 1.5–2 mm; style 4.5–6.5 mm. **FR**: mericarps, tubercles dense, short, wide, rounded. Wet meadows; < 300 m. s NCoRO (s Sonoma Co.). Threatened by urbanization, agriculture. Apr–May ★

LINACEAE FLAX FAMILY

Joshua R. McDill

Ann, per [shrub, tree, vine]. **ST**: gen erect [climbing], branched, glabrous to hairy. **LF**: cauline, alternate to opposite or whorled, simple, gen sessile, linear to obovate, entire to minutely toothed or ciliate, teeth occ gland-tipped; stipules small, dark-colored, spheric glands, or 0. **INFL**: raceme, panicle, or cyme [spike]. **FL**: bisexual, radial; sepals [4]5, free; petals = sepals in number, free to adherent; stamens 5[4 or 10], alternate petals, filaments fused basally into a cup-like structure surrounding ovary base; staminodes present, alt stamens at cup rim, or 0; ovary superior, carpels 2–5, fused, styles 2–5, = carpel number, free or partly fused. **FR**: capsule, gen dehiscent [drupe in some trop spp.], gen 10-seeded. 13 genera, ± 250 spp.: cosmopolitan, most temp, some cult. [McDill et al. 2009 Syst Bot 34:386–405] *Hesperolinon, Sclerolinon* are evolutionary lineages within *Linum*. Scientific Editors: Douglas H. Goldman, Robert Patterson, Bruce G. Baldwin.

1. Styles, stigmas 5; fr dehiscing into 5 or 10 segments . . . . **LINUM**
1′ Styles, stigmas 2 or 3; fr dehiscing into 4 or 6 segments, or indehiscent
  2. Lvs gen whorled proximally, alternate or opposite distally; styles gen 3, stigmas ± = styles in width; stipule glands gen present. . . . . **HESPEROLINON**
  2′ Lvs opposite, distal lvs or bracts alternate; styles 2, stigmas > styles in width; stipule glands 0 . . . . **SCLEROLINON**

HESPEROLINON WESTERN FLAX

Ann, erect, 3–50 cm, gen branched (at least in infl). **LF**: gen alternate (occ opposite or whorled), gen sessile, 4–35 mm, 1–3(6) mm wide, thread-like or linear to ovate or obovate, margin gland-toothed or entire; stipular glands present or 0. **INFL**: gen open or dense cyme, bracted; pedicel thread-like, gen ascending to erect. **FL**: sepals lanceolate to ovate, acute or acuminate, glabrous or hairy, or margins glandular-ciliate; petals free, 1–12 mm, yellow, white, or pink, gen with small appendages near base; staminodes 0; carpels 2–3, stigmas linear (± = style in width). **FR**: smooth. **SEED**: ± triangular in ×-section, acute. 12–13 spp.: esp CA, often on serpentine. (Greek, Latin: western flax) [Sharsmith 1961 Univ Calif Publ Bot 32:235–314] *H. serpentinum*, used in TJM (1993), is not validly published.

1. Petals yellow, often veined or tinged orange-red; anthers yellow
  2. Lf base clasping, margin serrate with 1–3 rows of prominent gland-tipped teeth . . . . *H. adenophyllum*
  2′ Lf base not clasping, margin neither serrate nor glandular
    3. Styles gen 2 (occ some pls with both 2- and 3-carpellate fls). . . . . [2]*H. bicarpellatum*
    3′ Styles gen 3 (occ 2-carpellate fls in 3-carpellate pls)
      4. Styles gen 0.5–1 mm, incl; corolla not spreading widely . . . . *H. clevelandii*
      4′ Styles gen ≥ 2 mm, exserted; corolla spreading widely
        5. Fls gen clustered at infl tips; stipular glands gen present throughout pl; petals 4–10 mm; filaments 4–8 mm. . . . . *H. breweri*
        5′ Fls scattered, widely separated throughout infl; stipular glands 0 or present only on proximal st; petals ≤ 5.5 mm; filaments ≤ 4 mm
          6. St, lvs gen glabrous, or st puberulent only above nodes; petals 2–4 mm; styles gen 2–3.5 mm . . . . [2]*H. bicarpellatum*
          6′ St, lvs gen puberulent; petals 3.5–5.5 mm; styles gen 3–10 mm. . . . . *H. tehamense*
1′ Petals white to pink, often veined or tinged darker pink to purple; anthers white to pink or purple
  7. Lvs whorled (gen in 4s) on main st, gen 3–6 mm wide, ovate, margin glandular. . . . . *H. drymarioides*
  7′ Lvs gen alternate on main st, ≤ 3 mm wide, thread-like to linear, margins not glandular
    8. Styles 2. . . . . *H. didymocarpum*
    8′ Styles gen 3 (occ some 2-carpellate fls on 3-carpellate pls)

9. Infl gen dense; pedicels gen ≤ 2 mm at fl (scattered fls at lower nodes occ with pedicels to 5(10) mm),
elongating in fr; stipular glands gen present at lvs, bracts

   10. Sepals glabrous; petals white to white flushed or veined with pink . . . . *H. californicum*

   10′ Sepals hairy; petals light to dark pink . . . . *H. congestum*

9′ Infl gen open; pedicels gen 1–25 mm at fl, to 45 mm in fr; stipular glands 0 or minute, or present only
at proximal lvs

   11. Styles incl; petals only slightly spreading, 0.5–3.5 mm, often ≤ sepals . . . . *H. micranthum*

   11′ Styles exserted; petals widely spreading to reflexed, 3–7 mm, > sepals

      12. Pedicels straight in bud, not pendent . . . . *H. disjunctum*

      12′ Pedicels pendent in bud . . . . *H. spergulinum*

H. adenophyllum (A. Gray) Small (p. 867) GLANDULAR WESTERN FLAX Pl 10–50 cm. **LF:** opposite or whorled proximally, lanceolate, keeled, clasping, margins with stalked, gland-tipped teeth in 1–3 rows. **INFL:** bract margins with stalked, gland-tipped teeth in 1–3 rows; pedicels gen 5–15 mm. **FL:** sepals 2–3 mm; petals 3–5 mm, yellow, often veined or tinged orange, fading white; stamens 4–5.5 mm, anthers yellow; ovary chambers 6, styles 3, 2.5–4 mm, yellow. Serpentine, chaparral; 150–1000 m. n&c NCoR (esp Lake, Mendocino cos.). May–Aug ★

H. bicarpellatum (H. Sharsm.) H. Sharsm. TWO-CARPELLATE WESTERN FLAX Pl 10–30 cm, gen ± glabrous, exc just distal to nodes. **LF:** alternate, thread-like to linear, flat, not clasping. **INFL:** ± equally forked; pedicels 2–12 mm. **FL:** sepals 1.5–3 mm; petals 2–4 mm, yellow; stamens 3–5 mm, anthers yellow; ovary chambers 4, styles 2(3), 2–3.5 mm, yellow. n=17. Serpentine, chaparral; 60–1000 m. s NCoRI. If recognized taxonomically, pls with conspicuous stipular glands and 3-carpellate fls in Lake, Napa cos. assignable to *H. sharsmithiae* R. O'Donnell [O'Donnell 2006 Madroño 53:404–408]. May–Jul ★

H. breweri (A. Gray) Small BREWER'S WESTERN FLAX Pl 5–20 cm, ± glabrous. **LF:** alternate, linear. **INFL:** dense; pedicels gen < 3 mm. **FL:** sepals 3–4 mm, margins gland-toothed; petals 4–10 mm, yellow; stamens 5.5–10 mm, anthers yellow; chambers 6, styles 3, 3–8 mm, yellow. n=18. Chaparral or grassland, occ on serpentine; 30–700 m. s NCoRI (Napa, Solano cos.), nw SnJV, ne SnFrB (Mount Diablo, Contra Costa Co.). May–Jun ★

H. californicum (Benth.) Small (p. 867) Pl 10–25 cm. **LF:** alternate, thread-like to linear; stipule glands well developed, red. **INFL:** pedicels 1–5(10) mm. **FL:** sepals 3.5–4 mm, glabrous, exc margin minutely gland-toothed; petals 4–12 mm, white or partly pink; stamens 6–11 mm, anthers white to rose; ovary chambers 6, styles 3, 4–10 mm, ± white. n=18. Rocky areas, chaparral, grassland, occ on serpentine; 30–1300 m. c&s NCoRI, s CaRF, n SNF, ScV, CCo (around San Francisco Bay), e SnFrB, ne SCoRI. Apr–Jul

H. clevelandii (Greene) Small Pl 5–20 cm, ± glabrous. **LF:** alternate, linear. **INFL:** unequally forked (main axis obvious); pedicels 5–25 mm. **FL:** sepals 1.5–2.5 mm; petals 0.5–1.5 mm, yellow; stamens 1.5–3 mm, anthers yellow; ovary chambers 6, styles 3, gen 0.5–1 mm, yellow. n=18. Chaparral margins, oak woodland, often on serpentine; 150–1400 m. NCoRO, c&s NCoRI, s NCoRH, e SnFrB. Like *H. micranthum* exc fl color. May–Jul

H. congestum (A. Gray) Small (p. 867) MARIN WESTERN FLAX Pl gen 5–15 cm. **LF:** alternate, linear, ± glabrous; stipule glands well developed, exudate red. **INFL:** dense; pedicels 0.5–5(8) mm. **FL:** sepals 3–4 mm, hairy, margins minutely glandular; petals 3–8 mm, pink to rose; stamens 5.5–7 mm, anthers rose to purple; ovary chambers 6, styles 3, 3–5 mm, ± white. n=18. Serpentine, grassland; < 200 m. n CCo, nw SnFrB. Apr–Aug ★

H. didymocarpum H. Sharsm. (p. 867) LAKE COUNTY WESTERN FLAX Pl 10–30 cm, glabrous exc just distal to nodes. **LF:** alternate, 10–25 mm, thread-like to linear, flat, not clasping. **INFL:** ± equally forked; pedicels 2–12 mm. **FL:** sepals 2–3 mm; petals 2.5–4 mm, white or light pink; stamens 3.5–5 mm, anthers white to ± purple; ovary chambers 4, styles 2, 2.5–4 mm, white. n=17. Serpentine, chaparral, grassland; 100–200 m. NCoRI (Big Canyon Creek, Lake Co.). Like *H. bicarpellatum* exc fl color. May–Jun ★

H. disjunctum H. Sharsm. Pl 3–30 cm, stout. **LF:** alternate, linear; stipule glands minute. **INFL:** pedicels gen 1–5 mm. **FL:** sepals 2–3 mm, glabrous exc margins minutely gland-toothed; petals 3–6 mm, white, veins gen pink; stamens gen 4–6 mm, anthers pink (rose, white-margined); ovary chambers 6, styles 3, 2.5–5 mm, ± white. Openings in chaparral, serpentine, vertic clay; 100–1000 m. NCoRI, e SnFrB, SCoRI. Apr–Jul

H. drymarioides (Curran) Small (p. 867) DRYMARIA-LIKE WESTERN FLAX Pl 5–25 cm. **LF:** proximal whorled (gen in 4s), distal alternate or ± opposite; ovate, flat, entire; surfaces, margins stalked-glandular. **INFL:** few-branched; pedicels 1–5(10) mm. **FL:** sepals 2.5–4 mm; petals 3–6 mm, white to pink, darker-veined, reflexed; stamens 3.5–5 mm, anthers white (purple, white-margined); ovary chambers 6, styles 3, 2–3.5 mm, ± white. Serpentine, chaparral or woodland; 100–1000 m. c NCoRI (Colusa, Glenn, Lake cos.), NCoRH. May–Jul ★

H. micranthum (A. Gray) Small Pl 5–20 cm. **LF:** alternate, linear. **INFL:** pedicels gen 5–15(25) mm, ascending to ± pendent. **FL:** sepals 1–4 mm, occ unequal, glabrous; petals 0.5–3.5 mm, white to pink or streaked rose; stamens 2–3 mm, anthers white to deep purple; ovary chambers gen 6, styles (2)3, 0.5–2 mm. n=18. Open areas, woodland margins, occ on serpentine; 50–2000 m. CA-FP, MP; OR, n Baja CA. Mar–Aug

H. spergulinum (A. Gray) Small (p. 867) Pl 10–30 cm, ± glabrous. **LF:** alternate, linear. **INFL:** ± raceme-like, widely spreading; pedicels 6–25 mm, gen pendent or ± reflexed. **FL:** sepals 1.5–3.5 mm, glabrous exc margins minutely gland-toothed; petals 4–7 mm, white or pale pink, gen darker-veined; stamens 4–9 mm, anthers pink to red-purple, margined white; ovary chambers 6, styles 3, 3.5–7 mm, ± white. n=18. Chaparral or woodland margins, incl serpentine; 100–1000 m. c&s NCoR, w SnFrB. May–Aug

H. tehamense H. Sharsm. TEHAMA COUNTY WESTERN FLAX Pl 2–50 cm, hairy. **LF:** alternate, ± linear. **INFL:** pedicels 0.5–3 mm. **FL:** sepals 2–3 mm; petals 3.5–5.5 mm, ± 1/2 recurved, light or bright yellow, veins sometimes ± red, tip notched; stamens 4–5.5 mm, anthers yellow; ovary chambers 6, styles 3, 3–10 mm, yellow. n=18. Serpentine, chaparral; 100–1000 m. n&c NCoRI (Tehama, Glenn cos.), e NCoRH. May–Jul ★

LINUM FLAX

Ann, per. **LF:** gen alternate, occ partially opposite [whorled], ± sessile, gen glabrous; stipule glands present or 0. **INFL:** raceme or cyme. **FL:** inner 2 sepals overlapped by outer 3, all margins gen translucent, gen ciliate or toothed, glandular or not; petals 5–25 mm, gen ephemeral; staminodes 0 or 5; carpels 5, ovary chambers 10, styles free or fused, stigmas 5, ≥ style width. **FR:** 3–10 mm diam, gen spheric, dehiscent, gen 5 or 10 segmented. **SEED:** 5 or 10, lens-shaped, rounded, brown to black, gen glossy. ± 180 spp.: temp & subtrop, esp Eurasia and N.Am. (Latin: flax) [Rogers 1984 North Amer Flora Ser II 12:1–56] *L. usitatissimum* cult for fiber (linen) and seed for oil and food; *L. perenne, L. grandiflorum,* orn; some Eurasian spp. used in cancer treatment.

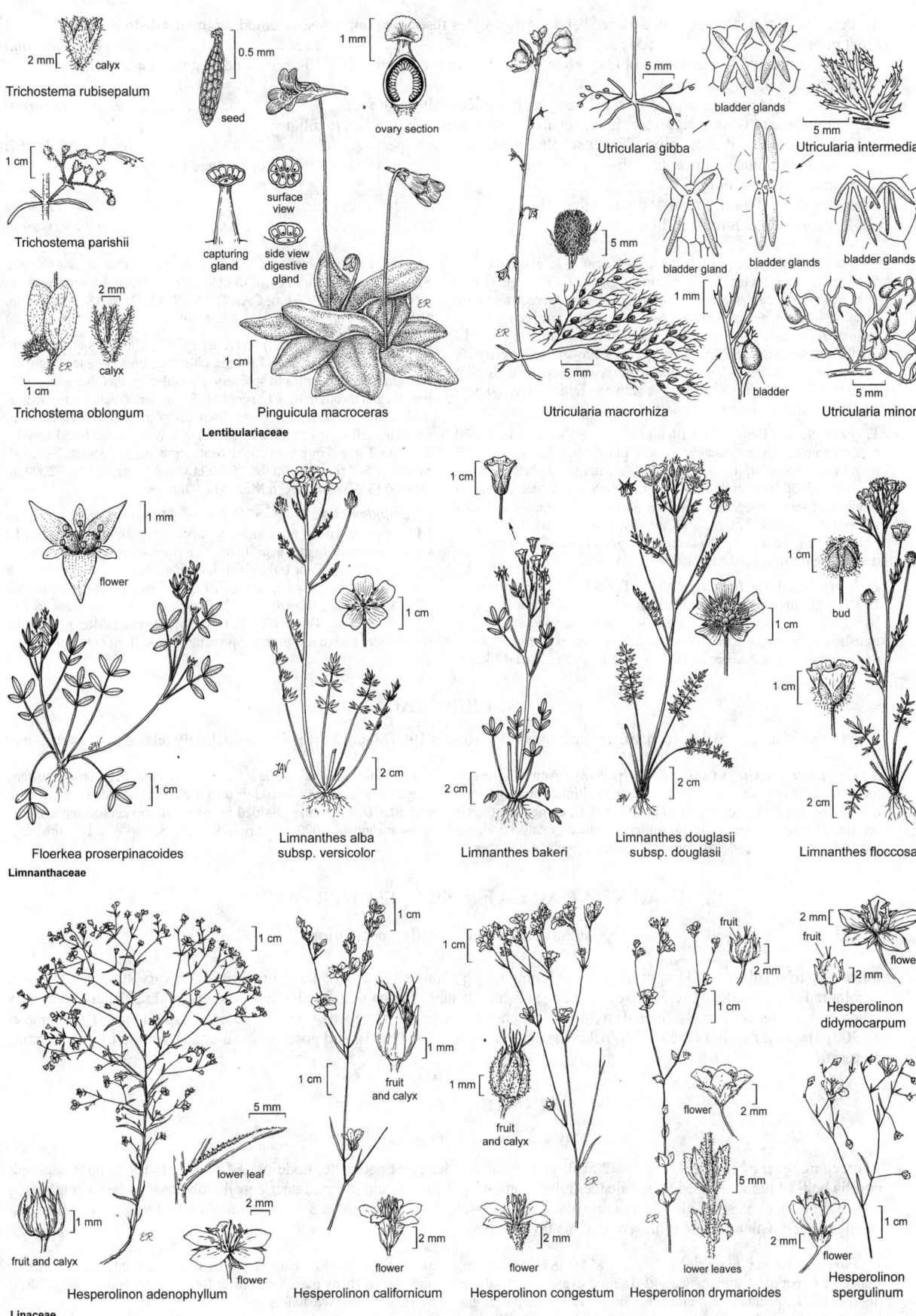

2 mm [calyx

Trichostema rubisepalum

0.5 mm

seed

1 mm

ovary section

surface view

capturing gland

side view digestive gland

1 cm

Trichostema parishii

1 cm

2 mm

Pinguicula macroceras

5 mm

bladder glands

Utricularia gibba

5 mm

Utricularia intermedia

bladder gland

bladder glands

bladder glands

5 mm

1 mm

ER

5 mm

bladder

Utricularia macrorhiza

5 mm

Utricularia minor

ER

calyx

1 cm

Trichostema oblongum

Lentibulariaceae

1 mm

flower

1 cm

1 cm

1 cm

Floerkea proserpinacoides

Limnanthaceae

1 cm

2 cm

Limnanthes alba subsp. versicolor

1 cm

2 cm

Limnanthes bakeri

1 cm

2 cm

Limnanthes douglasii subsp. douglasii

1 cm

bud

1 cm

2 cm

Limnanthes floccosa

1 cm

5 mm

lower leaf

2 mm

1 mm

fruit and calyx

Hesperolinon adenophyllum

Linaceae

1 cm

1 mm

fruit and calyx

1 cm

2 mm

flower

Hesperolinon californicum

1 cm

1 mm

fruit and calyx

2 mm

flower

Hesperolinon congestum

fruit

2 mm

1 cm

flower

2 mm

5 mm

ER

lower leaves

Hesperolinon drymarioides

2 mm [

fruit

2 mm

flower

Hesperolinon didymocarpum

2 mm [

1 cm

flower

Hesperolinon spergulinum

1. Petals yellow to orange, with dark red band at base; styles fused ± to tip; fr 5-segmented, segments distinct; st puberulent. . . . . ***L. puberulum***
1′ Petals bright red to maroon, or blue to white; styles free or bases ± fused; fr 10-segmented, segments ± adherent; st glabrous
  2. Petals bright red to maroon; all sepal margins conspicuously ciliate . . . . ***L. grandiflorum***
  2′ Petals blue to white; margins of inner sepals entire or minutely toothed to ciliate
    3. Stigmas head-like; margins of inner sepals entire, not ciliate; per. . . . . ***L. lewisii*** var. *lewisii*
    3′ Stigmas elongate, linear to club-shaped; margins of inner sepals minutely ciliate or toothed (entire); ann, weak per, or bien
      4. Petals 6–10 mm; fr gen 4–6 mm wide. . . . . ***L. bienne***
      4′ Petals 10–15 mm; fr 5–10 mm wide . . . . ***L. usitatissimum***

L. bienne Mill.   Bien to weak per, 6–60 cm. **ST:** glabrous. **LF:** 5–25 mm, linear to lance-linear; stipule glands 0. **FL:** sepals 4–5 mm, ovate to lanceolate, margins translucent, inner sepals minutely ciliate or toothed; petals light blue, 6–10 mm; styles free, gen ≤ 2 mm, stigmas linear to club-shaped. **FR:** 4–6 mm wide, dehiscing into 10 ± adherent segments. **SEED:** 2–3 mm, dark brown to black. *n*=15. Garden escape; grassland, woodland, disturbed places, esp coastal; < 1000 m. NW, CaRF, n SNF, GV, CW; native to Eurasia. Gen considered progenitor of cult flax, *L. usitatissimum*. Mar–Jun

L. grandiflorum Desf.   Ann 10–60 cm. **ST:** glabrous. **LF:** 10–20 mm, lanceolate, gen glaucous; stipule glands 0. **FL:** sepals 7–10 mm, lanceolate, margins conspicuously ciliate; petals bright red to maroon, 15–30 mm; styles free, stigmas linear. **FR:** dehiscent, 10 ± adherent segments. **SEED:** 2–3 mm, brown. *n*=8. Grassland, disturbed places; < 1000 m. NCo, NCoR, CW (exc SCoRI), SCo, WTR, SnBr, PR; native to n Afr. Uncommon beyond cult; occ incl in commercial wildflower seed packets in US. Apr–Jul

L. lewisii Pursh var. *lewisii* (p. 873)   Per. **ST:** 5–80 cm, glabrous. **LF:** 10–20 mm, linear to lanceolate or narrowly oblanceolate, glabrous; stipular glands 0. **FL:** sepals 4–6 mm, margins entire, gen translucent; petals 6–15 mm, obovate, blue, occ white or mottled white; styles separate, gen ≥ stamen length, stigmas head-like, ± spheric. **FR:** 5–6 mm wide, dehiscent, segments 10, ± adherent. **SEED:** 3–5 mm, dark brown to black. 2*n*=18. Gen dry open areas in mtns, foothills; < 3660 m. CA-FP, MP, W&I, D; to AK, Mississippi Valley. [*L. l.* var. *alpicola* Jeps.] Apr–Jul

L. puberulum (Engelm.) A. Heller (p. 873)   PLAINS FLAX   Ann. **ST:** 5–25 cm, puberulent. **LF:** gen alternate, basal lvs occ opposite; 5–20 mm, linear, entire to sparsely glandular-toothed; stipule glands present gen throughout. **FL:** sepals 4–7 mm, lanceolate, acute-acuminate, gen puberulent, margins glandular-toothed; petals 10–15 mm, obovate, yellow to orange with red band at base; styles fused ± to tip, 3–7 mm. **FR:** ≤ 5 mm wide, dehiscent, segments 5, distinct, 2-seeded. **SEED:** 1.5–3 mm, brown. *n*=15. Rocky, sandy areas; 1000–2000 m. DMtns; to WY, NE, TX, n Mex. May–Jul   ★

L. usitatissimum L. (p. 873)   Ann. **ST:** 20–100 cm, glabrous. **LF:** alternate, 10–40 mm, linear to lanceolate; stipule glands 0. **FL:** sepals 6–9 mm, ovate, acuminate, margin minutely ciliate (entire); petals 10–15 mm, blue (white); styles separate or ± fused at base, 3–6 mm. **FR:** 5–10 mm wide, gen dehiscent, of 10 ± adherent segments. **SEED:** 4–6 mm, brown. 2*n*=30. Disturbed areas, abandoned cult sites; < 1000 m. NW, SNF, GV, CW, SW, D (exc DMtns); Sporadic worldwide; native to Eurasia. Sporadic escape from cult. Apr–May

SCLEROLINON

1 sp. (Greek, Latin: hard flax, for hard, gen indehiscent fr) [Rogers 1966 Madroño 18:181–184] Closely related to *Hesperolinon*.

S. digynum (A. Gray) C.M. Rogers (p. 873)   Ann. **ST:** erect, glabrous. **LF:** opposite, 5–16 mm, oblong to elliptic, proximal gen entire, distal gen serrate; stipule glands 0. **INFL:** cyme; bracts gen serrate. **FL:** sepals unequal, margins gland-toothed; petals 5, yellow, appendages 0; staminodes 0; carpels 2, ovary chambers 4, styles 2, free, ± 1 mm; stigma > style in width, head-like. **FR:** capsule surface rough; indehiscent or breaking into 4 closed, 1-seeded segments in age. **SEED:** ± 1 mm, ± 3-sided in ×-section, tip acute. Gen vernally moist meadows; 1000–1800 m. KR, CaR, SN; to WA, ID. Jun–Aug

LINNAEACEAE   TWIN FLOWER FAMILY

Charles D. Bell, family description

Subshrub [to small tree]. **LF:** opposite, simple [compound]; stipules gen 0. **FL:** calyx tube fused to ovary, limb gen 5-lobed; corolla radial [or bilateral], rotate to cylindric, gen 5-lobed; stamens gen 4 inserted at 2 levels, epipetalous, alternate corolla lobes; ovary inferior, 1–5-chambered, style 1. **FR:** achene or capsule, 1-seeded. 4 genera, 35 spp.: esp n temp. [Donoghue et al. 2001 Harvard Pap Bot 6:459–479] Incl in Caprifoliaceae in TJM (1993), and possibly in future. Scientific Editor: Thomas J. Rosatti.

LINNAEA   TWIN FLOWER

Charles D. Bell & Lauramay T. Dempster

Pl creeping, evergreen. **INFL:** fls gen 2 at forked tip of slender erect peduncle, nodding. **FL:** calyx lobes slender, tapered; corolla bell- to funnel-shaped above slender tube; stamens 4, 2 shorter and inserted near corolla tube base; ovary chambers 3, 2 with aborting ovules, 1 with 1 developing ovule. **FR:** capsule, 3-valved. 1 sp., ± 3 vars.: circumboreal. (Named for Linnaeus, often depicted with a sample of this, one of his favorite pls)

L. borealis L. var. *longiflora* Torr. (p. 873)   **ST:** 15–20 cm, slender. **LF:** petiole ± 2 mm; blade 12–18 mm, ovate, serrate above middle. **INFL:** peduncle ± 6 cm; inner bracts 2, < outer, straight-hairy; outer bracts 2, ± 1 mm, ± round, surrounding ovary, densely glandular-hairy. **FL:** corolla 10–13 mm, hairy adaxially, white to pink. Moist shady places in conifer forest; 200–2600 m. NW, CaR, n SNH, MP; to n AK. Jun–Aug

LOASACEAE LOASA FAMILY

Larry Hufford & Barry Prigge, except as noted

Ann to subshrub; hairs needle-like, barbed, occ stinging. **LF**: alternate [opposite], gen ± pinnate-lobed; stipules 0. **INFL**: cyme, raceme. **FL**: bisexual, radial; sepals gen 5, gen persistent; petals gen 5, free or fused to each other or filament tube; stamens 5–many, filaments thread-like to flat, occ fused at base or in clusters; petal-like staminodes occ present; pistil 1, ovary inferior, chamber gen 1, placentas gen 3, parietal, style 1. **FR**: capsule or achene. **SEED**: 1–many. 18+ genera, 250 spp.: esp Am (Afr, Pacific). [Ernst & Thompson 1963 J Arnold Arbor 44:138–142] Scientific Editors: Douglas H. Goldman, Bruce G. Baldwin.

1. Subshrub; stamens gen 5; stigma ± unlobed; fr achene, seed 1 . . . . **PETALONYX**
1′ Ann to subshrub; stamens gen many; stigma lobes 3 or 5; fr capsule, seeds gen many
  2. Petals fused at least at base; stamens fused to corolla; stigma lobes and placentas 5 . . . . **EUCNIDE**
  2′ Petals free; stamens inserted on hypanthium, not fused to corolla; stigma lobes and placentas gen 3 . . . **MENTZELIA**

EUCNIDE ROCK-NETTLE

Ann or subshrub; hairs needle-like, stinging, barbed. **LF**: widely ovate to ± round, toothed to ± lobed; base widely tapered to cordate. **INFL**: cyme, bracted. **FL**: petals fused at least at base; stamens epipetalous or fused at base into a short tube; ovary funnel-shaped to hemispheric, placentas 5, stigma lobes 5, gen appressed. **FR**: obovoid to spheric, nodding or reflexed, dehiscent at top by 5 valves. **SEED**: many, < 1 mm, ± oblong, grooved or ribbed. 14 spp.: sw US, n Mex. (Greek: strongly nettle-like)

1. Ann; petals fused, tube 8–15 mm, lobes 2–5 mm, ± green; stamen filaments 0–1.5 mm . . . . *E. rupestris*
1′ Subshrub; petals fused at base, 30–50 mm, ± white to pale yellow; stamen filaments 10–20 mm. . . . . *E. urens*

E. rupestris (Baill.) H.J. Thomps. & W.R. Ernst ANNUAL ROCK-NETTLE Pl < 30 cm. **LF**: petiole < 6 cm; blade 1–8 cm, gen round, widely tapered to cordate, toothed to weakly lobed, shiny-green above. **FL**: sepals 3.5–7 mm, 2–3 mm wide, tip obtuse; corolla tube narrow-cylindric with ring of hairs at base, lobes erect; stigma below most anthers. **FR**: 7–15 mm, cylindric; pedicel < 2 cm, reflexed. Crevices, cliffs; 500–600 m. s DSon (Painted Gorge, Imperial Co.); AZ, n Mex (incl islands). Dec–Apr ★

E. urens (A. Gray) Parry (p. 873) Pl 30–100 cm. **LF**: petiole < 5 cm; blade < 10 cm, gen ovate, truncate to cordate, ± irregular-toothed, dull gray-green. **FL**: sepals 15–22 mm, 5–7 mm wide, tip acuminate; petals spreading; stamen tube 2–5 mm; stigma ± above anthers. **FR**: 10–20 mm, club-shaped to obconic; pedicel ascending to erect, < 1.5 cm. $2n=42$. Cliffs, rocky slopes, washes; < 1400 m. D; to sw UT, AZ, n Mex. Apr–Jun

MENTZELIA BLAZING STAR

Joshua M. Brokaw, John J. Schenk & Barry Prigge

Ann to per; hairs barbed to needle-like, not stinging; sts pale pink or gen ± white, branched or not. **LF**: linear to ovate, entire to pinnate-lobed; basal in rosettes, gen petioled; cauline gen sessile, ± reduced distally on st. **INFL**: gen cyme (or fl 1); bracts green to white, margin green. **FL**: sepals lanceolate to deltate, persistent; petals gen 5, free, white to yellow or orange; stamens gen many, ± free, gen unequal, inner filaments gen thread-like; outermost stamens opposite sepal lobes gen modified, ± widened, or petal-like with anther or not; ovary gen cylindric, placentas gen 3, style thread-like, stigma 3-furrowed or -lobed. **FR**: capsule, cup-, barrel-, or urn-shaped to narrowly cylindric, occ curved. **SEED**: gen many, shape variable. ± 100 spp.: w N.Am, ± trop Am. [Darlington 1934 Ann Missouri Bot Gard 21:103–226].

1. Fr > 4 mm wide, cup-shaped to cylindric or barrel-shaped, or fr 3–6 mm wide and urn-shaped; seeds
   transverse-folded or ridged at middle with beak-like end, fusiform and ribbed lengthwise, or lenticular and
   winged; ann, bien, or per
  2. Filament distal lobes 2, anther gen on thin stalk between lobes
    3. Fl bract white-scarious, margin ± green . . . . *M. involucrata*
    3′ Fl bract ± all green
      4. Anther stalk between filament lobes gen < lobes; seeds narrowed at middle . . . . *M. tricuspis*
      4′ Anther stalk gen ≥ filament lobes; seeds widest at middle
        5. Upper lvs sessile; fr erect, 13–25 mm; w DSon. . . . . *M. hirsutissima*
        5′ Upper lvs petioled; fr erect to reflexed, 9–18 mm; c DMoj. . . . . *M. tridentata*
  2′ Filament distal lobes 0
    6. Petals 8; ann . . . . *M. reflexa*
    6′ Petals 5 or petals and petal-like staminodes 10; bien or per
      7. Fr urn-shaped and gradually narrowing; seeds not winged, all filaments thread-like. . . . . *M. torreyi*
      7′ Fr cup-shaped to cylindric; seeds winged; 5 outer stamens with filaments wider than thread-like
        8. 5 outermost stamens with linear filaments, ≤ 2 mm wide, with anthers; petal tips acute; fr 11–44 mm
          9. Petals < 3 cm. . . . . *M. inyoensis*
          9′ Petals > 3 cm. . . . . *M. laevicaulis*
        8′ 5 outermost stamens with spoon-shaped to oblanceolate filaments, ≥ 1 mm wide, with anthers or not;
          petal tips acute to round or obtuse; fr 8–18 mm
          10. 5 outermost stamens without anther; petal tips acute to long-tapered . . . . *M. pterosperma*

10′ 5 outermost stamens with or without anthers; petal tips rounded to obtuse
  11. Fr gen cylindric, ≥ 1 cm; anthers not spiral-twisted in age . . . . *M. longiloba*
  11′ Fr cup-shaped, gen ≤ 1 cm; anthers spiral-twisted in age
    12. Lower lvs gen entire, linear to narrowly elliptic or oblanceolate . . . . *M. polita*
    12′ Lower lvs lobed to toothed, obovate to ovate, elliptic, or oblanceolate
      13. Upper lvs deltate to cordate, base clasping st . . . . *M. oreophila*
      13′ Upper lvs ovate to broadly elliptic, base not clasping st . . . . *M. puberula*
1′ Fr < 4(5) mm wide, cylindric to obconic, if 4–5 mm wide, then obconic only; seeds irregular-rounded to irregular-angular or prism-shaped; ann
14. Seed rows 1 above mid-fr; seeds ± prism-shaped, ×-section gen triangular with grooves along longitudinal edges
  15. Outer 5 filaments widened and distally 2-lobed . . . . *M. micrantha*
  15′ Outer filaments ± thread-like, not lobed
    16. Lower lvs gen lobed (toothed); style 3–6.5 mm; < 1200 m. . . . . *M. affinis*
    16′ Lower lvs gen entire or toothed; style gen 1–3.5 mm; > 900 m . . . . *M. dispersa*
14′ Seed rows ± 3 above mid-fr; seeds irregular-rounded to -angular (winged) above mid-fr, below mid-fr ×-section occ triangular with grooves along longitudinal edges
17. Fl bracts gen entire, green only
  18. Petals gen < 8 mm
    19. Seeds tan and moderately to densely mottled brown to black in age; seed coat cells pointed or domed, in age > 1/2 tall as wide on seed surface edges . . . . [4]*M. albicaulis*
    19′ Seeds tan, not or sparsely mottled in age; seed coat cells flat-surfaced to domed, in age < 1/2 tall as wide on seed surface edges
      20. Seed coat cells gen flat-surfaced in age . . . . *M. desertorum*
      20′ Seed coat cells domed in age . . . . *M. obscura*
  18′ Petals gen ≥ 8 mm
    21. Sepals gen > 8 mm; style 7–15 mm . . . . [2]*M. eremophila*
    21′ Sepals gen ≤ 8 mm; style 4–10 mm
      22. Seeds without conspicuous recurved flap over attachment scar; seed coat cells pointed or domed, in age > 1/2 tall as wide on seed surface edges . . . . *M. jonesii*
      22′ Seeds with conspicuous recurved flap over attachment scar; seed coat cells domed, in age < 1/2 tall as wide on seed surface edges . . . . *M. nitens*
17′ Fl bracts toothed to lobed or fl bracts entire with ± white base and green margin
23. Fl bracts green only
  24. Styles > 15 mm
    25. Petals gen 8–17 mm wide; c&s SNF, s SNH . . . . *M. crocea*
    25′ Petals gen 16–33 mm wide; SnJV, e SnFrB, n SCoRI . . . . *M. lindleyi*
  24′ Styles < 15 mm
    26. Petals gen < 8 mm
      27. Petals orange to orange-yellow, base gen red to orange; fl bracts 3–7-toothed to lobed (entire); fr gen curved < 40° . . . . [3]*M. veatchiana*
      27′ Petals yellow, base gen orange; fl bracts 3-toothed to entire; fr gen curved < 180° . . . . [4]*M. albicaulis*
    26′ Petals gen ≥ 8 mm
      28. Petals yellow only; w DMoj . . . . [2]*M. eremophila*
      28′ Petals orange to yellow, base red to orange; s SNF, s SnJV, s SCoRO, n WTR . . . . *M. pectinata*
23′ Fl bracts ± white at base, margin green
29. Fl bracts ± concealing frs, mostly white-scarious; infl dense . . . . *M. congesta*
29′ Fl bracts not concealing frs, ± white below middle; infl open to ± dense
  30. Petals gen > 8 mm, yellow, base gen orange; seed coat cells domed, in age ± 1/2 tall as wide on seed surface edges
    31. Pine/oak woodland; s SnFrB, SCoR, WTR, w SnGb . . . . *M. gracilenta*
    31′ Desert scrub, Joshua-tree woodland; SnGb, e PR, sw DMoj, w DSon . . . . [2]*M. ravenii*
  30′ Petals gen < 8 mm, orange to yellow, base gen red to orange; seed coat cells gen pointed or domed, in age > 1/2 tall as wide on seed surface edges
    32. Styles gen < 3.5 mm; longest mature fr gen > 10 mm
      33. Fl bracts 3-toothed to entire; ± white base of bracts faint, small; longest mature fr gen > 15 mm, gen curved < 180°; below 2300 m. . . . . [4]*M. albicaulis*
      33′ Fl bracts 3–7-toothed; ± white base of bracts prominent, gen conspicuously extending outwards from midvein; longest mature fr gen < 17 mm, curved < 45°; 600–3400 m. . . . . *M. montana*
    32′ Styles gen ≥ 3.5 mm; longest mature fr gen > 15 mm
      34. Petals orange to orange-yellow, base gen red to orange; fl bracts 3–7-toothed to lobed (entire)
      . . . . [3]*M. veatchiana*
      34′ Petals yellow, base gen orange; fl bracts 7-toothed to entire
        35. Fl bracts 3-toothed to entire, width < 2/3 length; ± white base of bracts faint, small; fr gen curved < 180°. . . . . [4]*M. albicaulis*

35′ Fl bracts 3–7-toothed, width > 1/3 length; ± white base of bracts prominent, gen conspicuously
  extending outwards from midvein; fr gen curved < 40°
  36. St gen spreading; desert scrub, Joshua-tree woodland . . . .[2]*M. ravenii*
  36′ St gen erect; pine/oak woodland, grassland . . . .[3]*M. veatchiana*

M. affinis Greene (p. 873)   Ann 5–47 cm. **ST:** erect, hairy. **LF:** 1–17 cm, lower lobed (toothed), upper entire to lobed. **INFL:** bracts lanceolate to obovate, entire to toothed, green. **FL:** sepals 2–7 mm; petals 3–10 mm, gen ovate, yellow, base yellow to orange; stamens < 6.5 mm; style 3–6.5 mm. **FR:** erect to curved < 90°, 12–28 mm, 1–3 mm wide, narrowly cylindric. **SEED:** 1–2 mm, ± 1 mm wide, ± prism-shaped, ×-section triangular with longitudinal edges grooved, tan, dark-mottled or not; seed coat cells minute. 2*n*=18. Rocky or gray-white soils in grassland, woodland, creosote-bush scrub; < 1200 m. s SN, SnJV, se SnFrB (Mount Hamilton), SCoRI, s ChI, TR, SnJt, D; to AZ, Baja CA. Intergrades with *M. dispersa*. Apr–May

M. albicaulis (Hook.) Torr. & A. Gray (p. 873)   Ann 5–42 cm. **ST:** erect to decumbent, glabrous to hairy. **LF:** 1–11 cm, lower lobed, upper entire to lobed. **INFL:** bracts lanceolate to ovate, 3-toothed to entire, green, occ ± white base, faint, small. **FL:** sepals 1–5 mm; petals 2–7(8) mm, ovate to obovate, yellow, base gen orange; stamens 3–5 mm; style 2–5 mm. **FR:** gen curved < 180°, 8–28(34) mm, 1.5–3.5 mm wide, obconic. **SEED:** 1–1.5 mm, irregular-rounded to -angular above mid-fr, ×-section occ triangular with grooves along longitudinal edges below mid-fr; tan, dark-mottled, to black; seed coat cells pointed or domed, in age > 1/2 tall as wide on seed surface edges. 2*n*=54,72. Sand dunes, gravel fans, washes, creosote-bush scrub, pinyon/juniper woodland; < 2300 m. Teh, SnGb, SnBr, GB, D; to BC, NE, Baja CA. Intergrades with *M. montana*, *M. obscura*, *M. veatchiana*. If recognized taxonomically, pls with *n*=27 assignable to *M. californica* H.J. Thomps. & J.E. Roberts or *M. mojavensis* H.J. Thomps. & J.E. Roberts. Mar–Jul

M. congesta Torr. & A. Gray (p. 873)   Ann 7–40 cm. **ST:** erect, glabrous to hairy. **LF:** 1–9 cm, entire to lobed. **INFL:** dense; outer bract widely ovate, toothed to lobed, ± concealing fr; inner bract ± fused to ovary, obovate, ± toothed to lobed; lower 3/4 of bracts white-scarious, margin green. **FL:** sepals 1–4 mm; petals 3–7(9) mm, obovate, pale yellow to yellow, base orange; stamens ± = style; style 1.5–5 mm. **FR:** gen erect, 5–12 mm, 2–3 mm wide, cylindric. **SEED:** ± 1 mm, irregular-angular with concave sides, tan, dark-mottled or not; seed coat cells gen domed, < 1/2 tall as wide in age. 2*n*=18. Disturbed slopes, pine forest, sagebrush scrub, pinyon/juniper woodland; 1200–2700 m. SNH (e slope), Teh, WTR, SnGb, PR, SNE, DMtns; to ID. May–Jul

M. crocea Kellogg (p. 873)   Ann 34–100 cm. **ST:** erect, hairy. **LF:** 1–40 cm, lower lobed, upper toothed to lobed. **INFL:** bracts ovate to obovate, 6–10-toothed, green. **FL:** sepals 7–20 mm; petals 21–36 mm, 8–17(21) mm wide, elliptic to obovate, yellow, base gen orange; stamens 11–28 mm, outer > inner; style 20–35 mm. **FR:** erect to curved < 45°, 20–35 mm, 3–5 mm wide, obconic. **SEED:** 1.5–2 mm, irregular-rounded to -angular, tan, dark-mottled; seed coat cells minute, domed to pointed. 2*n*=36. Rocky slopes, roadsides, grassland, oak/pine woodland; < 1700 m. c&s SNF, s SNH. Intergrades with *M. lindleyi*. May–Jun

M. desertorum (Davidson) H.J. Thomps. & J.E. Roberts (p. 873)   Ann 5–41 cm. **ST:** erect, glabrous to hairy. **LF:** 1–12 cm, lower toothed to lobed, upper entire to lobed. **INFL:** bracts lanceolate to ovate, gen entire, green. **FL:** sepals 2–4 mm; petals 2.5–6 mm, ovate to obovate, yellow, base yellow to orange; stamens = style; style 2–4 mm. **FR:** gen curved < 180°, 12–27 mm, 1–2.5 mm wide, obconic. **SEED:** ± 1 mm, irregular-rounded to -angular above mid-fr, ×-section occ triangular with grooves along longitudinal edges below mid-fr, tan, not or sparsely dark-mottled; seed coat cells gen flat-surfaced in age. 2*n*=18. Sandy flats in creosote-bush scrub; < 950 m. D; Baja CA. Intergrades with *M. obscura*. Feb–Mar

M. dispersa S. Watson (p. 873)   Ann 7–48 cm. **ST:** erect, hairy. **LF:** < 10 cm, lower entire to toothed (lobed), upper entire to toothed. **INFL:** bracts ovate to round, entire to lobed, green. **FL:** sepals 1–3.5 mm; petals 2–6(8) mm, ovate to obovate, yellow throughout (base orange); stamens < 4.5 mm; style 1–3.5(5) mm. **FR:** erect to curved < 30°, 7–25 mm, 1–2.5 mm wide, narrowly cylindric. **SEED:** 0.5–1.5 mm, 0.5–1 mm wide, ± prism-shaped, ×-section triangular, longitudinal edges grooved, tan, dark-mottled or not; seed coat cells minute. 2*n*=18,36,72. Loamy to sandy or rocky slopes, roadsides; 900–3100 m. CA-FP, GB; to BC, n-c US, NM. Intergrades with *M. affinis*. May–Aug

M. eremophila (Jeps.) H.J. Thomps. & J.E. Roberts (p. 873)   SOLITARY BLAZING STAR   Ann 8–43 cm. **ST:** erect, glabrous to hairy. **LF:** 1–15 cm, lower lobed, upper entire to lobed. **INFL:** bracts lanceolate to ovate, entire to 2-toothed at base, green. **FL:** sepals (7)9–16 mm; petals 12–24 mm, obovate, yellow; stamens 3–10 mm; style 7–15 mm. **FR:** gen curved < 270°, 19–40 mm, 2–3.5 mm wide, obconic. **SEED:** ± 1 mm, irregular-rounded above mid-fr, ×-section occ triangular with grooves along longitudinal edges below mid-fr, tan, ± dark mottled; recurved flap over attachment scar conspicuous; seed coat cells domed, < 1/2 tall as wide in age. 2*n*=18. Canyons, rocky slopes and washes; roadsides, creosote-bush scrub; 600–1250 m. w DMoj. Intergrades with *M. nitens*. Mar–May   ★

M. gracilenta Torr. & A. Gray   Ann 3–60 cm. **ST:** erect, hairy. **LF:** 2–13 cm, lower lobed, upper entire to lobed. **INFL:** bracts ovate to obovate, 3–12-toothed, base ± white. **FL:** sepals 3–8 mm; petals (7)8–18 mm, obovate, yellow, base gen orange; stamens ±, = style; style 5–11 mm. **FR:** erect to curved < 20°, gen 9–15(23) mm, 3–5 mm wide, cylindric to obconic. **SEED:** 1.5–2.5 mm, irregular-angular, tan, dark mottled or not; seed coat cells gen domed, ± 1/2 tall as wide in age. 2*n*=36. Gen unproductive substrates, incl serpentine and calcium-rich soils, grassland to pine/oak woodland; 200–1700 m. s SnFrB, SCoR, WTR, w SnGb. Intergrades with *M. ravenii*, *M. montana*, *M. veatchiana*. Apr–May

M. hirsutissima S. Watson   HAIRY STICKLEAF   Ann 5–31 cm. **ST:** erect, hairy, branched at base. **LF:** 1–11 cm, lanceolate to oblanceolate, toothed to lobed, upper sessile with lower lobes gen clasping. **INFL:** bracts ovate to triangular, 4–10-toothed, ± clasping, green, base occ ± white. **FL:** sepals 9–18 mm; petals 12–31 mm, obovate, pale yellow to orange, tip mucronate; stamens 4–12 mm, filaments ± broadened, gen distally 2-lobed, anther stalk between lobes thin, gen ≥ lobes; style 6–13 mm. **FR:** erect, 13–25 mm, 5–8 mm wide, cylindric. **SEED:** ± 2.5 mm, ± 1.5 mm wide, ± compressed, widest at middle, beaked, ashy-white. Washes, fans, slopes, creosote-bush scrub; < 720 m. w DSon; Baja CA. [*M. h.* var. *stenophylla* (Urb. & Gilg) I.M. Johnst.] Apr–May   ★

M. involucrata S. Watson (p. 873)   Ann 7–35 cm. **ST:** erect, hairy. **LF:** 2–18 cm, lanceolate to ovate, gen irregular-toothed. **INFL:** bracts 4–5 per fl, ovate to obovate, 6–20-toothed, white-scarious, margin ± green. **FL:** sepals 7–23 mm; petals 13–62 mm, obovate, ± white to pale yellow, veins gen orange; stamens 4–26 mm, filaments broadened, distally 2-lobed; style 8–30 mm. **FR:** erect, 14–25 mm, 5–10 mm wide, cylindric. **SEED:** 2–3 mm, 2–2.5 mm wide, ± compressed, middle transverse-ridged, beaked, ashy-white. 2*n*=18. Washes, fans, steep slopes, creosote-bush scrub; < 900 m. D; AZ, n Mex. [*M. i.* var. *megalantha* I.M. Johnst.] Jan–May

M. inyoensis H.J. Thomps. & Prigge   INYO BLAZING STAR   Per 15–40 cm. **ST:** erect, hairy. **LF:** 3–11 cm, 4–20 mm wide, toothed to lobed, lower oblanceolate to elliptic, upper lanceolate. **INFL:** bracts gen linear, entire, occ 2-lobed at base, green. **FL:** sepals 4–12 mm; petals 11–18 mm, 2–4 mm wide, yellow, tip acute; stamens 5–15 mm, filaments of 5 outermost 10–15 mm, 1–1.5 mm wide, linear, with anthers; style 10–13 mm. **FR:** erect, 11–25 mm, 6–8 mm wide, cylindric. **SEED:** 2–3 mm, ± 2 mm wide, lenticular, winged, white to ± gray-white. 2*n*=22. Rocky slopes, canyons, washes, clay hills; 1100–2000 m. W&I (White Mtns); w NV. May–Aug   ★

M. jonesii (Urb. & Gilg) H.J. Thomps. & J.E. Roberts   Ann 5–40 cm. **ST:** erect to decumbent, glabrous to hairy. **LF:** < 14 cm, entire to lobed. **INFL:** bracts lance-ovate, entire, green. **FL:** sepals 2–8(10)

mm; petals (6)8–22 mm, ovate to obovate, yellow, base yellow to orange; stamens 3–10 mm; style 4–10 mm. **FR:** curved < 180° or S-shaped, 15–38 mm, 2–4 mm wide, obconic. **SEED:** ± 1 mm, irregular-rounded to irregular-angular above mid-fr, ×-section occ triangular with longitudinal edges grooved below mid-fr, tan, dark-mottled or not; seed coat cells gen pointed or domed, in age > 1/2 tall as wide on seed surface edges. 2*n*=36,54. Sandy to rocky washes, fans, flats, roadsides; 200–1450 m. W&I, D; NV, AZ. Intergrades with *M. nitens*, *M. albicaulis*. If recognized taxonomically, pls with *n*=27 assignable to *M. californica* H.J. Thomps. & J.E. Roberts (in part). Mar–May

M. laevicaulis (Hook.) Torr. & A. Gray (p. 873) Per 22–100 cm. **ST:** erect, glabrous to hairy. **LF:** lower < 24 cm, oblanceolate, gen pinnately lobed; upper 2–10 cm, lanceolate, gen dentate to serrate. **INFL:** bracts linear to lanceolate, entire to deeply 4–5-lobed, green. **FL:** sepals 15–46 mm; petals 40–80 mm, 8–18 mm wide, adaxially yellow, tip acute; stamens 15–55 mm, filaments of 5 outermost 40–55 mm, 1–2 mm wide, linear, with anthers; style 2.5–7 cm. **FR:** erect, 15–44 mm, 8–13 mm wide, cylindric. **SEED:** 3–4 mm, 2–2.5 mm wide, lenticular, winged, gray. 2*n*=22. Sandy to rocky slopes, washes, roadcuts; < 2900 m. CA (exc DSon); to BC, MT, WY, NV. May–Oct

M. lindleyi Torr. & A. Gray (p. 873) Ann 10–70 cm. **ST:** erect, hairy. **LF:** 2–17 cm, gen deeply lobed. **INFL:** bracts lanceolate to ovate, toothed or lobed, green. **FL:** sepals 9–19 mm; petals 20–40 mm, (12)16–33 mm wide, yellow, base orange; stamens 20–30 mm, outer > inner; style 15–24 mm. **FR:** gen erect to curved < 90°, 25–40 mm, 4–5 mm wide, obconic. **SEED:** 1–2.5 mm, irregularly rounded to winged, tan, dark-mottled or not, seed coat cells minute, domed. 2*n*=36. Rocky, open slopes, coastal-sage scrub, oak/pine woodland; < 1350 m. SnJV, e SnFrB, n SCoRI. Waif populations occur outside native range. Intergrades with *M. crocea*. May–Jun

M. longiloba J. Darl. Bien or Per 15–100 cm. **ST:** erect, glabrous to hairy. **LF:** 1–15 cm, 1–2 cm wide, lower oblanceolate to elliptic, crenate to pinnate-lobed, upper lanceolate to elliptic, toothed. **INFL:** bracts 0 or linear to lanceolate, gen entire, green. **FL:** sepals 6–12 mm; petals 12–20 mm, 4–6 mm wide, yellow, tip obtuse to round; stamens 3–17 mm, filaments of 5 outermost 10–20 mm, 3–6 mm wide, spoon-shaped, tip round, gen without anthers; style 8–13 mm. **FR:** erect, 10–18 mm, 7–10 mm wide, cylindric to occ cup-shaped. **SEED:** 3–4 mm, 2.5–3.5 mm wide, lenticular, winged, ± white. Sandy flats, creosote-bush scrub; < 800 m. D; AZ, n Mex. [*M. multiflora* (Nutt.) A. Gray subsp. *l.* (J. Darl.) Felger] Mar–Jun

M. micrantha (Hook. & Arn.) Torr. & A. Gray (p. 873) Ann 10–80 cm. **ST:** erect, hairy. **LF:** 1–18 cm, irregularly dentate, wavy or lobed. **INFL:** bracts ovate to round, entire to crenate, gen concealing fl, green. **FL:** sepals 1.5–2 mm; petals 2.5–4.5 mm, ovate, yellow; stamens 1.5–4 mm, filaments of 5 outermost widened, distally 2-lobed; style 2–3(5) mm. **FR:** erect to curved < 20°, 6–13 mm, 1.5–2.5 mm wide, cylindric. **SEED:** 5–10, 1.5–2.5 mm, ± 1 mm wide, ± prism-shaped, ×-section triangular, longitudinal edges grooved, tan and dark-mottled to black; seed coat cells minute. 2*n*=18. Open, gen recent-burned or disturbed chaparral and oak woodland; < 2256 m. KR, NCoR, SnFrB, SCoR, SW; Baja CA. Apr–Jun

M. montana (Davidson) Davidson & Moxley Ann 4–48 cm. **ST:** erect, glabrous to hairy. **LF:** < 13 cm, lower lobed, upper entire to lobed. **INFL:** dense to open; bracts lanceolate to obovate, 3–7-toothed, base ± white. **FL:** sepals 1–4 mm; petals 2–6(8) mm, obovate, yellow, base gen orange; stamens 2–7 mm; style 1.5–3.5(6) mm. **FR:** erect to curved < 45°, 6–17(20) mm, 2–3 mm wide, cylindric to obconic. **SEED:** 1–1.5 mm, irregular-rounded to -angular (edges gen acute), tan, dark-mottled; seed coat cells gen pointed or domed, > 1/2 tall as wide on seed surface edges in age. 2*n*=36. Open, gen disturbed slopes, flats, grassland, sagebrush scrub, conifer forest; 600–3400 m. CaR, SN, TR, PR, SNE, n DMtns; to OR, CO, TX, n Mex. Intermediate to *M. congesta*, *M. veatchiana*; intergrades with *M. ravenii*, *M. gracilenta*, *M. veatchiana*, *M. albicaulis*. Apr–Jul

M. nitens Greene (p. 873) Ann 4–34 cm. **ST:** erect to decumbent, glabrous to hairy. **LF:** < 15 cm, lower toothed to lobed, upper entire to lobed. **INFL:** bracts lanceolate, entire, green. **FL:** sepals 3–8 mm; petals (7)8–18 mm, gen obovate, yellow, base yellow to orange; stamens ≤ style; style 4–8 mm. **FR:** gen curved ≤ 180°, 13–26 mm,

2–3.5 mm wide, obconic. **SEED:** ± 1 mm, irregularly rounded above mid-fr, ×-section occ triangular with grooves along longitudinal edges below mid-fr, tan, ± dark-mottled; recurved flap over attachment scar conspicuous; seed coat cells domed, < 1/2 tall as wide in age. 2*n*=18. Sandy washes to rocky slopes; 400–2000 m. SNE, n DMoj; w NV. Intergrades with *M. eremophila*, *M. jonesii*. Apr–Jun

M. obscura H.J. Thomps. & J.E. Roberts (p. 873) Ann 8–45 cm. **ST:** erect to decumbent, glabrous to hairy. **LF:** 1–22 cm; entire to lobed. **INFL:** bracts gen ovate, entire, green. **FL:** sepals 2–6 mm; petals 3–7(8) mm, ovate to occ obovate, yellow, base yellow to orange; stamens 2–7 mm; style 2–6 mm. **FR:** curved < 250°, 11–31 mm, 1.5–3 mm wide, obconic. **SEED:** ± 1 mm, irregularly rounded above mid-fr, ×-section occ triangular with grooves along longitudinal edges below mid-fr, tan, not or sparsely dark-mottled; seed coat cells domed, < 1/2 tall as wide in age. 2*n*=36. Sandy to rocky washes, slopes, roadsides, creosote-bush scrub, blackbush scrub, Joshua-tree woodland; 200–1616 m. D; to UT, AZ, n Mex. Intergrades with *M. desertorum*, *M. albicaulis*. Feb–Mar

M. oreophila J. Darl. (p. 873) Per 15–53 cm. **ST:** erect, hairy. **LF:** lower 2–9 cm, 6–40 mm wide, oblanceolate to elliptic, ± wavy-dentate, upper shorter, deltate to cordate, base clasping. **INFL:** bracts linear to lanceolate, entire, green. **FL:** sepals 4–11 mm; petals 7–17 mm, 3–5 mm wide, yellow, tip obtuse to round; stamens 3–12 mm, filaments of 5 outermost gen 5–10(14) mm, 1.6–3 mm wide, oblanceolate, tip cut to notched or acute, with anther; style 5–9 mm. **FR:** erect, 5–10 mm, 6–11 mm wide, cup-shaped. **SEED:** ± 3.5 mm, ± 3 mm wide, lenticular, winged, white. 2*n*=22. Washes, roadcuts, talus slopes of limestone soils; 150–2200 m. W&I, DMoj, e DSon; s NV. May–Jun

M. pectinata Kellogg (p. 879) Ann 8–54 cm. **ST:** erect to decumbent, hairy. **LF:** 2–12 cm; gen toothed to lobed. **INFL:** bracts lanceolate to ovate, gen toothed to lobed, green. **FL:** sepals 3–13 mm; petals 8–22 mm, obovate, orange to yellow, base red to orange; stamens 4–11 mm; style 5–13 mm. **FR:** erect to curved < 90°, 12–35 mm, 2–4 mm wide, obconic. **SEED:** 1–2 mm, irregularly angular above mid-fr, sides concave, ×-section occ triangular with grooves along longitudinal edges below mid-fr, tan, dark-mottled or not; seed coat cells gen pointed or domed, in age > 1/2 tall as wide on seed surface edges. 2*n*=18. Slopes of sandy to gray-white, calcium-rich soils, grassland to oak woodland; < 1400 m. s SNF, s SnJV, s SCoRO, n WTR. Mar–May

M. polita A. Nelson (p. 879) POLISHED BLAZING STAR Per 12–33 cm. **ST:** erect to decumbent, glabrous to puberulent. **LF:** < 7 cm, 2–10 mm wide, linear to narrowly elliptic or oblanceolate, gen entire. **INFL:** bracts narrowly lanceolate, entire, green. **FL:** sepals 5–7 mm; petals 8–14 mm, 2–4 mm wide, ± white to pale-yellow, tip round; stamens 3–10 mm, filaments of 5 outermost 6–10 mm, 1–2.7 mm wide, oblanceolate, tip gen notched, anther attached gen without thread-like stalk; style 5–7 mm. **FR:** erect, 5–9 mm, 5–9 mm wide, cup-shaped. **SEED:** ± 3 mm, ± 2 mm wide, lenticular, winged, white. 2*n*=22. Washes, limestone, ± white gypsum-rich soils; 1200–1500 m. e DMoj (ne Clark Mtn Range); s NV. May–Jun ★

M. pterosperma Eastw. WING-SEED BLAZING STAR Bien, per 6–42 cm. **ST:** erect, hairy. **LF:** 1–8(13) cm, 6–22 mm wide, lower elliptic to oblanceolate, entire to short-toothed, upper elliptic to lanceolate, toothed (lobed near base). **INFL:** bracts narrowly lanceolate, entire, green. **FL:** sepals 6–10 mm; petals 9–12(24) mm, 2–5 mm wide, yellow, tip acute; stamens 4–15 mm; filaments of 5 outermost 5–15 mm, 1–3 mm wide, narrowly oblanceolate, without anther; style 6–11(17) mm. **FR:** erect, 8–14 mm, 5–10 mm wide, cup-shaped to cylindric. **SEED:** ± 2.5–4 mm, ± 3 mm wide, lenticular, winged, white. 2*n*=22. ± white gypsum-rich soils; 1000–1200 m. e-c DMoj; to CO, nw AZ. Apr–Jun ★

M. puberula J. Darl. DARLINGTON'S BLAZING STAR Per 15–54 cm. **ST:** erect, hairy. **LF:** 2–6.5 cm, 10–40 mm wide, lower widely obovate to elliptic or widely elliptic, toothed, upper sessile, ovate to broadly elliptic, base not clasping. **INFL:** bracts linear, entire, green. **FL:** sepals 5–9 mm; petals 7–14 mm, 2–6 mm wide, yellow, tip obtuse; stamens 3–12 mm, filaments of 5 outermost 5–12 mm, 1.5–3 mm wide, oblanceolate, tip irregularly cut, round to acute,

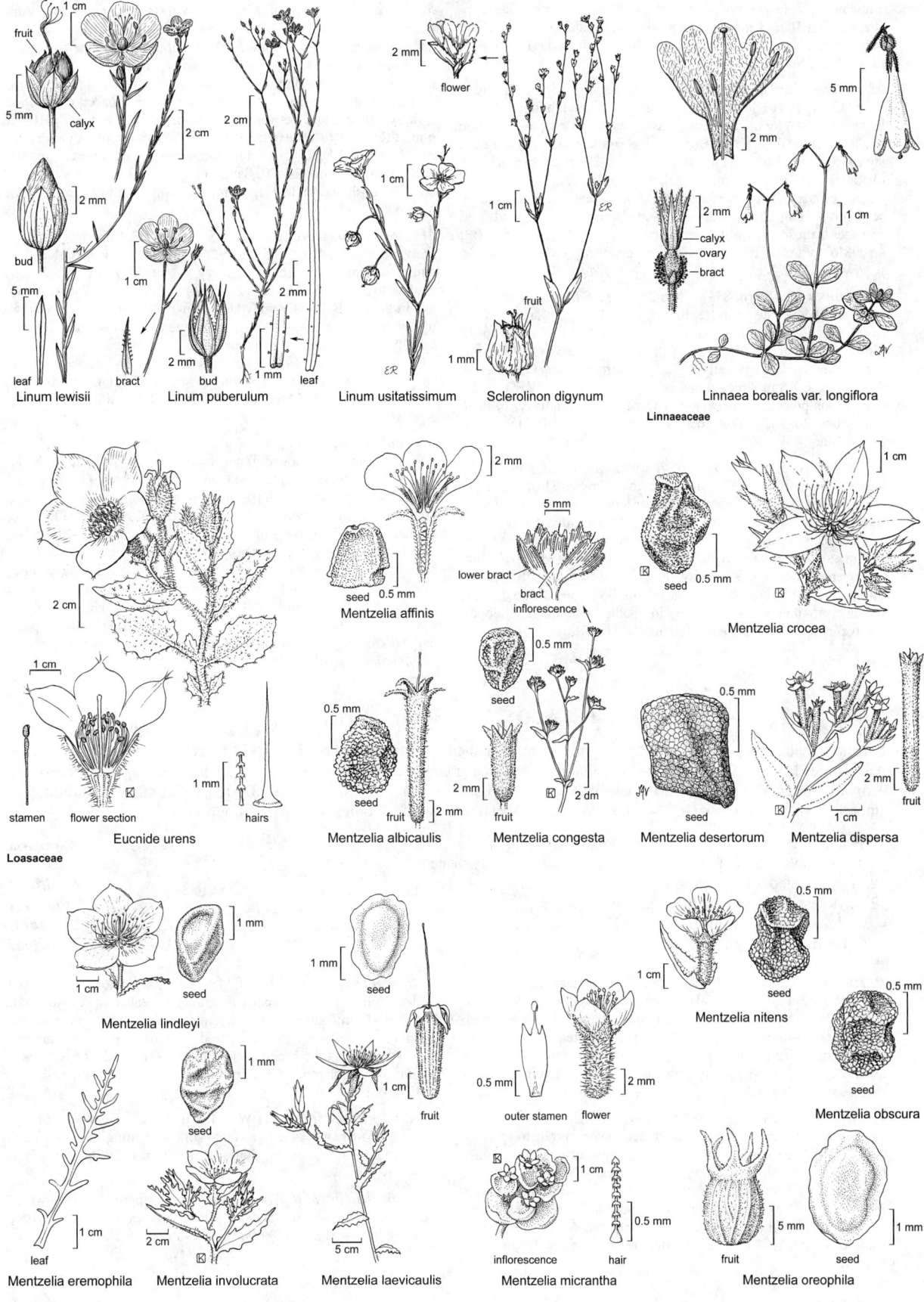

Linum lewisii

fruit
calyx
bud
leaf
bract
bud
leaf

Linum puberulum

Linum usitatissimum

flower
fruit

Sclerolinon digynum

calyx
ovary
bract

Linnaea borealis var. longiflora

Linnaeaceae

stamen   flower section   hairs

Eucnide urens

Loasaceae

seed

Mentzelia affinis

lower bract
bract
inflorescence

seed
fruit

Mentzelia albicaulis

seed

seed
fruit

Mentzelia congesta

seed

Mentzelia crocea

seed

seed
fruit

Mentzelia desertorum   Mentzelia dispersa

seed

Mentzelia lindleyi

seed

fruit

seed

Mentzelia nitens

seed

Mentzelia obscura

leaf

Mentzelia eremophila

seed

Mentzelia involucrata

Mentzelia laevicaulis

outer stamen   flower

inflorescence   hair

Mentzelia micrantha

fruit   seed

Mentzelia oreophila

with anther; style 4–7 mm. **FR**: erect, 5–10 mm, 5–7 mm wide, cup-shaped. **SEED**: ± 3 mm, ± 2 mm wide, lenticular, winged, white. Sandy crevices in cliffs or rocky slopes; 90–1280 m. se DMoj, e DSon; to AZ, Mex. Mar–May ★

M. ravenii H.J. Thomps. & J.E. Roberts Ann 5–45 cm. **ST**: erect, hairy. **LF**: 1–18 cm, lower lobed, upper gen toothed to lobed. **INFL**: bract gen ovate, toothed to lobed, base ± white. **FL**: sepals 2–6 mm; petals 5–11(13) mm, obovate, yellow, base orange; stamens 3–7 mm; style 3.5–8 mm. **FR**: erect to curved < 45°, 8–23 mm, 2–3 mm wide, obconic. **SEED**: 1–2 mm, irregularly angular, tan, dark-mottled, edges gen acute; seed coat cells gen domed, ± 1/2 tall as wide on seed surface edges in age. 2*n*=36. Sandy desert foothills, roadsides, creosote-bush scrub, Joshua-tree woodland; < 1200 m. SnGb, e PR, sw DMoj, w DSon. Intergrades with *M. gracilenta, M. montana, M. veatchiana*. Mar–May

M. reflexa Coville (p. 879) Ann 2–20 cm. **ST**: erect, hairy. **LF**: 1–10 cm, lanceolate to ovate, toothed. **INFL**: bracts lanceolate to ovate, lobed, green. **FL**: sepals 5–7 mm; petals 8 in 2 series, 6–12 mm, oblanceolate, ± white to pale yellow; stamens 3–8 mm, filaments ± broadened, unlobed distally; style 5–6.5 mm. **FR**: erect to reflexed, 9–13 mm, 5–7 mm wide, cylindric to barrel-shaped. **SEED**: 2–2.5 mm, ± compressed, narrowed and folded near middle, ± beaked, ashy-white. 2*n*=20. Washes, rocky slopes, roadsides; < 1524 m. W&I, DMoj. Mar–May

M. torreyi A. Gray (p. 879) TORREY'S BLAZING STAR Per 10–20 cm. **ST**: many from thick caudex, erect to decumbent, hairy. **LF**: 2–4 cm, gen deeply pinnate-lobed, 0–7 linear lobes, margins rolled under. **INFL**: bracts linear, base occ 2-lobed, margins rolled under, green. **FL**: sepals 3–6 mm; petals 9–15 mm, oblanceolate, orange to pale yellow, abaxially puberulent; stamens 7–10 mm; style 8–12 mm. **FR**: ± erect, 4–8 mm, 3–6 mm wide, urn-shaped, gradually narrowing. **SEED**: 2–2.5 mm, ± fusiform, ± spiral-3-ribbed, acute distally, truncate proximally. 2*n*=28. Sandy to alkaline fine-textured, slopes, scrub, pinyon woodland; 900–2100 m. SNE; to OR, ID. Jun–Aug ★

M. tricuspis A. Gray (p. 879) SPINY-HAIR BLAZING STAR Ann 5–27 cm. **ST**: erect, hairy. **LF**: sessile to petioled, 2–12 cm, lanceolate to ovate, wavy to toothed. **INFL**: bracts lanceolate to widely ovate, sessile, 2–8-toothed, green, base ± green. **FL**: sepals 7–17 mm; petals 11–30(50) mm, widely obovate, ± white to pale yellow, tip mucronate; stamens 7–17 mm, filaments ± broadened, 2-lobed distally, anther stalk between lobes thin, gen < lobes; style 10–12 mm. **FR**: erect to reflexed, 9–18 mm, 5–8 mm wide, cylindric to barrel-shaped. **SEED**: 2–2.5 mm, ± compressed, narrowed at middle, beaked, ashy-white. 2*n*=20. Sandy or gravelly slopes or washes in creosote-bush scrub; 150–1280 m. s DMoj, sw DSon; to UT, AZ. Mar–May ★

M. tridentata (Davidson) H.J. Thomps. & J.E. Roberts CREAMY BLAZING STAR Ann 5–25 cm. **ST**: erect, hairy. **LF**: petioled, 1–9 cm, lanceolate to oblanceolate, wavy to toothed. **INFL**: bracts lanceolate to widely ovate, sessile, 2–10-toothed, ± green. **FL**: sepals 5–13 mm; petals 10–40 mm, widely obovate, ± white to pale-yellow, tip mucronate; stamens 6–15 mm, filaments ± broadened, 2-lobed distally, anther stalk between lobes thin, gen > lobes; style 9–12 mm. **FR**: erect to reflexed, 9–18 mm, 5–8 mm wide, cylindric to barrel-shaped. **SEED**: 1.5–2.5 mm, ± compressed, widest at middle, beaked, ashy-white. 2*n*=20. Creosote-bush scrub; 700–1300 m. c DMoj. Apr–May ★

M. veatchiana Kellogg (p. 879) Ann 3–45 cm. **ST**: erect, hairy. **LF**: 1–18 cm, lower lobed, upper toothed to lobed (entire). **INFL**: bracts gen ovate, toothed to lobed (entire), base green to ± white. **FL**: sepals 1–5 mm; petals 4–7(10) mm, obovate, orange to yellow, base gen red to orange; stamens 3–7 mm; style (3)3.5–6 mm. **FR**: erect to curved 40° (70°), 8–28 mm, 2–4 mm wide, obconic. **SEED**: 1–2 mm, irregularly angular, tan, dark-mottled, edges gen acute; seed coat cells gen pointed or domed, > 1/2 tall as wide on seed surface edges in age. 2*n*=54. Loamy to sandy desert, grassland, scrub, oak/pine woodland; 200–2500 m. n SNH, Teh, SnJV, SCoRI, TR, PR, GB, D; OR, NV, AZ, Baja CA. Yellow-petaled forms gen restricted to pine/oak woodland, grassland. Intergrades with *M. gracilenta, M. ravenii, M. montana, M. albicaulis*. Mar–Jun

PETALONYX SANDPAPER-PLANT

Subshrub; hairs barbed. **LF**: linear to ± round, entire to toothed; base tapered to cordate. **INFL**: raceme, gen terminal; bracts 3 per fl, outer 1 > inner 2. **FL**: sepals ± deciduous; petals free or claws adherent, fused below blades; stamens gen 5, free; ovary ± ovoid, placenta 1, stigma 1. **FR**: achene, ± ovoid, gen 5-veined or -ribbed, erect. **SEED**: 1, 1.5–2.5 mm, ± fusiform, gen smooth. 2*n*=46. 5 spp.: sw US, nw Mex. (Greek: petal claw) [Davis & Thompson 1967 Madroño 19:1–32]

1. Petals free; lvs linear to narrowly (ob)lanceolate . . . . ***P. linearis***
1′ Petal claws adherent, ± tubular; lvs lanceolate to widely ovate
  2. Lvs petioled, cauline ± equal . . . . ***P. nitidus***
  2′ Lvs sessile, cauline reduced upwards. . . . . ***P. thurberi***
    3. Hairs soft, spreading; stamens 4–7.5 mm . . . . subsp. ***gilmanii***
    3′ Hairs rough, appressed downward; stamens 6–10 mm . . . . subsp. ***thurberi***

P. linearis Greene Pl 15–100 cm. **LF**: gen sessile, 10–25 mm, linear to narrowly (ob)lanceolate, obtuse to acute, entire to irregularly toothed. **INFL**: 4–10 cm; outer bract 5–8 mm, ovate to ± round; inner bracts 3–4 mm, ovate, ± cordate, acute to notched, lobed; pedicels 1–2 mm. **FL**: petals 2–5.5 mm, free, white; stamens 3–7 mm, ± exserted; style ± 3–6 mm. Sandy or rocky canyons, gen in creosote-bush scrub; < 1000 m. se DMoj, DSon; sw AZ, nw Mex. Mar–May

P. nitidus S. Watson (p. 879) Pl 15–45 cm. **LF**: 15–40 mm, gen ovate, widely tapered to base, acute to acuminate, serrate to sparsely dentate. **INFL**: 3–4.5 cm; outer bract 5–13 mm, narrowly ovate; inner bracts 1–5 mm, elliptic to ovate, truncate, crenate; pedicels 1–2 mm. **FL**: petals 5–11 mm, cream, upper 1/4 of claws adherent; stamens 7–14 mm, well exserted; style 8–15 mm. Sandy washes or rocky canyons, creosote-bush scrub, Joshua-tree woodland, pinyon/juniper woodland; 725–2100 m. W&I, DMtns, w DMoj; to sw UT, nw AZ. May–Jul

P. thurberi A. Gray Pl < 100 cm. **LF**: clasping, 4–45 mm, lanceolate to deltate-ovate, acute to acuminate, entire to few-toothed. **INFL**: 1–4 cm; outer bract 4–7.5 mm, deltate-ovate; inner bracts 2–3 mm, lanceolate to ovate, tapered to cordate, acute to ± acuminate, crenate or lobed; pedicels < 1 mm. **FL**: petals 2.5–6.5 mm, cream, claws adherent, fused in upper 1/5; stamens well exserted; style 3–11 mm.

  subsp. ***gilmanii*** (Munz) W.S. Davis & H.J. Thomps. DEATH VALLEY SANDPAPER-PLANT Hairs soft, spreading. **FL**: stamens 4–7.5 mm. Sandy washes, dunes; < 1200 m. n DMoj. May–Jun, Sep–Nov ★

  subsp. ***thurberi*** (p. 879) Hairs rough, appressed downward. **FL**: stamens 6–10 mm. Sandy or gravelly dunes, washes, canyons, creosote-bush scrub; < 1200 m. D. May–Jul

LYTHRACEAE LOOSESTRIFE FAMILY

Shirley A. Graham

Ann, per, shrub, tree. **ST:** 4-angled or cylindric. **LF:** simple, entire, gen opposite, 4-ranked (alternate, whorled). **INFL:** fls terminal or in axils of upper lvs or lf-like bracts, 1 or in ± dense cymes or along short shoots, sessile or not, subtended by [0]2 bractlets. **FL:** bisexual, gen radial; hypanthium bell-shaped to cylindric, membranous or leathery, persistent in fr; sepals appearing as hypanthium lobes, 4–9, epicalyx lobes alternate sepals or 0; petals, stamens inserted on inner hypanthium; petals 4–6 or 0, alternate sepals, crinkled, deciduous; stamens gen = or 2 × sepals, incl or exserted; ovary gen superior, chambers 2–6[many], style gen slender, stigma head-like. **FR:** dry capsule or leathery berry, dehiscent into 2–4 valves or irregularly. **SEED:** 3–many. ± 28 genera, 600 spp.: temp, trop, gen in wet habitats. Some orn or cult for medicine, dyes. [Graham et al. 2005 Int J Pl Sci 166:995–1017] "Epicalyx lobes" (lobes on calyx) formerly called "appendages"; "hypanthium" in Lythraceae (and Onagraceae) incl receptacle, sometimes called "fl cup" or "fl tube". Punicaceae (*Punica*) incl here. Scientific Editor: Thomas J. Rosatti.

1. Hypanthium leathery, 2–5 cm, wall smooth; ovary inferior; sepals 5–9 . . . . **PUNICA**
1′ Hypanthium membranous, 2–8 mm, wall 4–12-ribbed; ovary superior; sepals 4–6
  2. Hypanthium cylindric, > 2 × longer than wide . . . . **LYTHRUM**
  2′ Hypanthium bell- to urn-shaped, < 2 × longer than wide
    3. Lf with basal ear-like lobes; fls in dense axillary cymes . . . . **AMMANNIA**
    3′ Lf narrowly tapered to base; fls 1 per axil on main shoots and on short axillary shoots
      4. Sepals, stamens gen 6; hypanthium broadly bell-shaped . . . . *Lythrum portula*
      4′ Sepals, stamens gen 4; hypanthium bell- to ± urn-shaped . . . . **ROTALA**

AMMANNIA

Ann. **ST:** prostrate to erect, branched or not, glabrous. **LF:** opposite, sessile, linear to lanceolate [oblanceolate], with basal ear-like lobes. **INFL:** axillary 1–5(14)-fld cymes. **FL:** radial, hypanthium bell- to urn-shaped, 4–8-ribbed; sepals 4, epicalyx lobes ≤ sepals, horn-like, thick; petals (0)4, obovate; stamens 4(5–12), exserted (incl); style long, slender [short, stocky]. **FR:** capsule, ± spheric, wall smooth, not striate at 10×, thin, dry, splitting irregularly. **SEED:** many, ± 1 mm. ± 25 spp.: temp, trop. (Paul Ammann, German botanist, 1634–1691) [Graham et al. 2011 Bot J Linn Soc 166:1–19]

1. Longest peduncles 3–5(9) mm; petals deep rose-purple; fr 3–5 mm wide . . . . *A. coccinea*
1′ Longest peduncles 0–3 mm; petals pale lavender; fr 4–6 mm wide . . . . *A. robusta*

A. coccinea Rottb. (p. 879) **ST:** 1–10 dm. **LF:** 2–8 cm, 2–15 mm wide. **INFL:** (1)3–5(14)-fld. **FL:** hypanthium urn-shaped; petals 2–4(5) mm; stamens 4(7), anthers deep yellow. *n*=33. Wet places, drying ponds, lake, creek margins; < 300 m. CaRF, SNF, GV, CCo, SnFrB, SCoRO, SW, DSon; to e US, S.Am; also s Eur, e Afr as weed in rice fields. Jun–Aug

A. robusta Heer & Regel (p. 879) **ST:** 1–10 dm. **LF:** 1.5–8 cm, 5–15 mm wide. **INFL:** 1–3(5)-fld. **FL:** hypanthium urn-shaped; petals 2–4 mm; stamens 4(5–12), anthers pale yellow. *n*=17. Wet places, drying ponds, ditch margins; < 500 m. NCoR, s SNF, GV, CW, SCo, s ChI (Santa Catalina Island), D; to Great Lakes, LA, S.Am. Jun–Aug

LYTHRUM LOOSESTRIFE

Ann, per. **ST:** prostrate to erect, often 4-angled. **LF:** opposite, alternate, or whorled, linear to ovate or obovate, petiole 0 to short. **INFL:** fls gen 1–2 per axil, sessile or not. **FL:** radial to ± bilateral, of 1–3 style forms (heterostylous); hypanthium cylindric or bell-shaped, ribs gen conspicuous; sepals 4–6, deltate, epicalyx lobes < to > sepals; petals 4–6 or 0; stamens 4–6 or 12, incl or exserted; styles < to > stamens. **FR:** capsule, gen cylindric, rarely spheric, valves 2. **SEED:** many, < 1 mm. ± 35 spp.: temp. (Greek: clotted blood) [Houghton-Thompson et al. 2005 Ann Bot (London) 96:877–885]

1. Hypanthium broadly bell-shaped, 1–2 mm, ribs conspicuous or not; st extensively creeping or decumbent, rooting at nodes . . . . ***L. portula***
1′ Hypanthium cylindric, 4–7 mm, ribs conspicuous; st prostrate or ascending to erect, rooting at base
  2. Fls > 2 per axil, in dense, ± sessile cymes, not on short axillary shoots . . . . ***L. salicaria***
  2′ Fls 1–2 per axil, on short axillary shoots or not
    3. St erect, branches many, above; epicalyx lobes < to > sepals, horn-like; per . . . . ***L. californicum***
    3′ St decumbent to ascending or erect, branches at base, gen 0–few above; epicalyx lobes >> to = sepals; ann or short-lived per
      4. Fls 1 per axil, not on short axillary shoots; hypanthium 4–5 × longer than wide; epicalyx lobes ± 2 × > sepals . . . . ***L. hyssopifolia***
      4′ Fls 1 per axil on main shoots and on short axillary shoots; hypanthium 8–10 × longer than wide; epicalyx lobes ± = sepals . . . . ***L. tribracteatum***

L. californicum Torr. & A. Gray (p. 879) CALIFORNIA LOOSESTRIFE Per, heterostylous. **ST:** erect, 2–6 dm, gray-glaucous; branches many, above. **LF:** 1–7 cm, ± gray-glaucous, lower opposite, linear to lance-linear, upper alternate, linear. **INFL:** fls 1–2 per axil, pedicel 1–3 mm. **FL:** of 2 style forms; hypanthium 4–7 mm, cylindric, 2–3 × longer than wide; sepals < 1 mm, epicalyx lobes < to > sepals, horn-like; petals 4–8 mm, purple; stamens gen 6, incl or exserted; style incl or exserted. **FR:** ovoid, ± = hypanthium. 2*n*=20. Marshes, pond and stream margins; < 2200 m. s NCoRI, SNF, s SNH, GV, CW, SW, W&I, D; to c US, n Mex. Apr–Sep

L. hyssopifolia L. (p. 879) Ann or short-lived per, not heterostylous. **ST:** 1–6 dm; lower branches decumbent to ascending or erect, glabrous, upper few. **LF:** 0.5–3 cm, ± erect, ± glaucous, lower opposite, oblong to elliptic, upper gen alternate, linear. **INFL:** fls 1 per axil, sessile. **FL:** hypanthium 4–6 mm, cylindric, 4–5 × longer than wide; epicalyx lobes awl-like, ± 2 × > sepals; petals 2–5 mm, pink to rose; stamens 4–6, incl; style ± exserted. **FR:** ovoid-oblong, ≥ hypanthium. 2*n*=20. Marshes, drying pond margins, disturbed ground; < 1600 m. CA-FP; to Can, e US; native to Eur. Apr–Oct ❖

L. portula (L.) D.A. Webb (p. 879) Ann, not heterostylous. **ST:** extensively creeping or decumbent, rooting at nodes, 5–25 cm, ± red, glabrous; branched at base. **LF:** 0.5–1.5 cm, opposite, oblong to obovate, fleshy. **INFL:** fl 1 per axil, ± sessile. **FL:** hypanthium 1–2 mm, broadly bell-shaped; epicalyx lobes awl-shaped, 0.5–1 mm, 1–2 × sepals; petals ± 1 mm or 0, white to rose-pink; stamens gen 6. **FR:** ± spheric, > hypanthium. 2*n*=10. Drying ponds, lake margins; 1000–2200 m. CaR, n SN, ScV; native to Eur. Apr–Oct

L. salicaria L. (p. 879) PURPLE LOOSESTRIFE Per, heterostylous. **ST:** erect, 5–15 dm, gray-puberulent; branches few. **LF:** 5–14 cm, sessile, truncate at base, lanceolate to ± ovate. **INFL:** fls > 2 per axil, in dense, ± sessile cymes; pedicel 0–2 mm. **FL:** of 3 style forms; hypanthium 4–6 mm, cylindric, 2+ × longer than wide; sepals < 1 mm, epicalyx lobes linear, ≥ sepals; petals 7–14 mm, red-purple; stamens 12, incl or exserted; style incl or exserted. **FR:** ovoid, < hypanthium. 2*n*=30,50,60. Marshes, ponds, streambanks; < 1000 m. s NCo, NCoRO, n SNF, ScV, CCo, nw MP; worldwide; native to Eur; cult for orn. May–Oct ◆

L. tribracteatum Spreng. (p. 879) Ann, not heterostylous. **ST:** 5–30 cm, glabrous to ± scabrous; branches often at base, decumbent to weakly ascending or erect. **LF:** 0.3–2.5 cm, lower opposite; upper alternate, oblong to oblanceolate. **INFL:** fls 1 per axil on main shoots and on short axillary shoots, ± sessile; ± glandular. **FL:** hypanthium 5–6 mm, narrowly cylindric, 8–10 × longer than wide; sepals, epicalyx lobes ± 0.5 mm, thick; petals < 3 mm, lavender; stamens 4–6, incl; style incl. **FR:** cylindric, ± = hypanthium. *n*=5. Wet areas, drying ponds; < 1500 m. NCoRI, GV, CCo, SnFrB, n MP; native to s Eur. May–Jun

PUNICA POMEGRANATE

Shrub, small tree, glabrous. **ST:** branches gen many, from near base, often a thorn at tip. **LF:** simple, ± opposite, entire, deciduous. **INFL:** fls 1, terminal, 0–4 subterminal, clustered. **FL:** bisexual, radial, hypanthium bell-shaped to cylindric, leathery; sepals 5–9, epicalyx lobes 0; petals 5–9, crumpled; stamens many, inserted at many levels in hypanthium; ovary inferior [partly superior], chambers gen 5 or more, irregular [or regular]. **FR:** berry, ± spheric, leathery, crowned by calyx, splitting irregularly. **SEED:** many, outer seed coat fleshy, juicy, inner hard. 2 spp.: Medit, ne Afr to Himalayas. (Latin: from early name malus Punicus, "apple of Carthage") [Lersten & Horner 2005 Amer J Bot 92:1935–1941]

P. granatum L. **ST:** < 5 m. **LF:** blade 1–9 cm, >> petiole, oblong- or lance-ovate, shiny adaxially. **FL:** 2–3 cm; hypanthium, petals bright orange-red to pale yellow. **FR:** 5–12 cm, red-brown. **SEED:** red (± white). *n*=8,9. Uncommon. Disturbed ground; < 500 m. s SnJV, s CCo, SnGb, n SCo; native se Eur, Asia, to Himalayas. Widely cult in warm areas for fr (seeds edible, source of grenadine; juice popular antioxidant), orn (dwarf, doubled-fld forms); often naturalized. Apr–Jul

ROTALA

Ann, per, glabrous. **ST:** spreading to erect, gen branched, 4-angled to rounded. **LF:** opposite, 4-ranked (whorled), gen sessile. **INFL:** fls 1 per axil on main shoots and on short axillary shoots, sessile. **FL:** radial; hypanthium bell- to ± urn-shaped; sepals gen 4, epicalyx lobes 4 or 0; petals 4, minute, < sepals; stamens gen 4, ± incl (exserted). **FR:** capsule, oblong to ± spheric, golden, wall finely transversely striate at 10× thin, dry, valves 2–4. **SEED:** many, < 1 mm. ± 48 spp.: temp, trop. (Latin: wheel-like) [Baskin et al. 2002 Wetlands (Wilmington) 22:661–668]

1. Lf oblong to obovate or ± round, 0.5–2 cm, paler at margin; epicalyx lobes 0; fr valves gen 2 . . . . ***R. indica***
1′ Lf linear-oblong to elliptic or oblanceolate, 1–5 cm, not paler at margin; epicalyx lobes awl-like,
  to 2 × > sepals; fr valves 3–4 . . . . ***R. ramosior***

R. indica (Willd.) Koehne **ST:** 1–4 dm. **INFL:** fls 1 per axil on main shoots and on short axillary shoots, bractlets lance-linear, ± = hypanthium. **FL:** hypanthium 1–2.5 mm; sepals acutely deltate; petals pink; stamens gen 4, ± incl. **FR:** oblong. Marshes, ponds; < 100 m. ScV; LA, rice-field weed in Eur; native to se Asia. Jun–Sep

R. ramosior (L.) Koehne (p. 879) **ST:** 1–5 dm. **INFL:** fls 1 per axil, bractlets lanceolate, < (to >) hypanthium. **FL:** hypanthium 2–5 mm; sepals broadly deltate; petals white to pink; stamens gen 4, ± incl. **FR:** ovoid to ± spheric. 2*n*=16,32. Lake and pond margins, streams; < 1900 m. s NCoRO, NCoRI, CaR, n&c SN, GV; to WA, e US, S.Am. Jun–Aug

MALVACEAE MALLOW FAMILY

Steven R. Hill, except as noted

Ann to tree; gen with stellate hairs, often with bristles or peltate scales; juice gen mucilage-like; bark fibrous. **LF:** gen cauline, alternate, petioled, simple [palmate-compound], gen palmate-lobed and/or veined, gen toothed, evergreen or not; stipules persistent or not. **INFL:** head, spike, raceme, or panicle, in panicle or not (a compound panicle), or fls ≥ 1 in lf axils, or fls gen 1 opposite a lf or on a spur; bracts lf-like or not; bractlets 0 or on fl stalks, often closely subtending calyx, gen in involucel. **FL:** gen bisexual, radial; sepals 5, gen fused at base, abutting in bud, larger in fr or not, nectaries as tufts of glandular hairs at base; petals (0)5, free from each other but gen fused at base to, falling with filament tube, clawed or not; stamens 5–many, filaments fused for most of length into tube around style, staminodes 5, alternate stamens, or gen 0; pistil 1, ovary superior, stalked or gen not, chambers gen ≥ 5, styles or style branches, stigmas gen 1 or 1–2 × chamber number. **FR:** loculicidal capsule, [berry], or 5–many, disk- or wedge-shaped segments (= mericarps). 266 genera, 4025 spp.: worldwide, esp warm regions; some cult (e.g., *Abelmoschus* okra; *Alcea* hollyhock; *Gossypium* cotton; *Hibiscus* hibiscus). [Angiosperm Phylogeny Group 1998 Ann Missouri Bot Gard 85:531–553] Recently treated to incl Bombacaceae, Sterculiaceae, Tiliaceae. Mature fr needed for identification; "outer edges" are surfaces between sides and back (abaxial surface) of segment. "Fl stalk" used instead of "pedicel", "peduncle", esp where both needed (i.e., when fls both 1 in lf axils and otherwise). Scientific Editors: Steven R. Hill, Thomas J. Rosatti.

1. Fl gen < 3 mm wide; petals with thread-like claws; ovary, fr stalked . . . . **AYENIA**
1′ Fl gen > 3 mm wide; petals 0 or without thread-like claws; ovary, fr not stalked
  2. Petals 0; calyx showy, gen > 25 mm wide; fls 1 in lf axils; shrub, small tree . . . . **FREMONTODENDRON**
  2′ Petals 5, often showy; calyx not showy, gen < 25 mm wide; fls 1 in lf axils or not; ann, per, to small tree
    3. Fr a capsule; seeds 2–many per chamber; fls showy, gen 1 in lf axils or on axillary short-shoots
      4. Fr chambers 3–5; style not branched, gen shallowly lobed at tip; involucel 3(5)-lobed, lf-like, gen
        persistent; seeds obovoid or angular, gen long-white-hairy; fr gen gland-dotted . . . . [*Gossypium hirsutum*]
      4′ Fr chambers 5; style branched or deeply lobed at tip; involucel 3–12-lobed, not lf-like, shed in age or
        not; seeds reniform, short-hairy or glabrous; fr not gland-dotted
        5. Style 5-branched at tip; petals gen white, yellow, lavender, or red; involucel 5–12(20)-lobed,
          persistent . . . . **HIBISCUS**
        5′ Style 5-lobed at tip; petals white, pink, blue, or purple; involucel 3–6-lobed, persistent or not
          6. Per to shrub; petals gen white, yellow, blue, or lilac-blue to purple, membranous; involucel lobes 6,
            linear, persistent . . . . [*Alyogyne huegelii*]
          6′ Small tree; petals white to rose-pink, fleshy; involucel lobes 3–5, gen ovate, gen deciduous . . . . . . **LAGUNARIA**
    3′ Fr of 5–40 segments that gen separate from axis and each other when mature; seeds gen 1–4(15) per
      chamber; fls showy or not, 1–10 in lf axils or > 1 in spike, raceme, or panicle
      7. Stigmas linear, on inner side of style branches; fr segments indehiscent
        8. Anthers near top of filament tube, in 2 concentric series; bractlets subtending calyx 0(1–3), gen not
          forming involucel; fr segments 5–10, beak short or 0 . . . . **SIDALCEA**
        8′ Anthers near top of filament tube, below, not in 2 series; bractlets subtending calyx 3–11, gen forming
          involucel; fr segments 5–40, beak 0
          9. Bractlets 5–11, fused; fls gen > 6 cm wide, in (spike-) or raceme-like cymes; petal veins gen
            indistinct; st erect . . . . **ALCEA**
          9′ Bractlets 3, free or ± fused; fls gen < 6 cm wide, 1–10 in lf axils, gen not in spike- or raceme-like
            cymes (occ in raceme-like cymes in *Malva*); petal veins often distinct; st prostrate to erect
            10. Fr segments gen lacking edges, seed gen not firmly enclosed by, readily separating from fr wall;
              bractlets fused ± 1/2 . . . . **LAVATERA**
            10′ Fr segments gen with edges, seed firmly enclosed by, not readily separating from fr wall; bractlets
              free or fused ± 1/2 . . . . **MALVA**
      7′ Stigmas head-like or obliquely squared, at tip of style branches; fr segments indehiscent or ± dehiscent
        11. Fr segment below indehiscent, firm, net-veined, above dehiscent, with 2 spreading, scarious wings;
          bractlets 0; shrub; PR, DSon . . . . **HORSFORDIA**
        11′ Fr segment indehiscent or not, net-veined or not, wings 0; bractlets 0–3; ann to shrub; widespread
          incl e PR, DSon
          12. Fr segment gen 1–6(15)-seeded
            13. Petals salmon-orange; fr segment 2-chambered, 2-seeded; fls 1 in lf axils; bractlets 3, ± persistent;
              st decumbent to prostrate, rooting . . . . **MODIOLA**
            13′ Petals occ salmon-orange; fr segment 1–±2-chambered, 1–6(15)-seeded; fls 1 in lf axils, or in
              cyme, raceme, or panicle; bractlets 0–3, persistent or not; st erect or prostrate, gen not rooting
              14. Petals gen pale yellow, orange, white, pink, or ± red; bractlets 0; seeds 3–9(15) per fr segment
                15. St gen erect (decumbent); petals yellow, yellow-orange, orange-pink, or ± red; fr segment not
                  inflated, 3–6(15)-seeded, walls firm to woody . . . . **ABUTILON**
                15′ St gen decumbent or trailing; petals white to pale yellow; fr segment inflated, 3-seeded, walls
                  flexible, papery . . . . **HERISSANTIA**
            14′ Petals white, ± rose, pink, ± lavender, or red-orange; bractlets gen 1–3; seeds 1–3(4) per fr segment
              16. Fr segment bristly, ± dehiscent to base, gen not net-veined; bractlets 3, ± persistent; stigmas gen
                oblique-squared, often ± elongate; NCo, NCoRH, CaR, MP (Shasta Co.), KR . . . . **ILIAMNA**
              16′ Fr segment gen not bristly, below indehiscent, strongly net-veined, above dehiscent, smooth;
                bractlets (0)1–3, gen deciduous; stigmas gen symmetric, head-like; esp D or disturbed areas
                . . . . [3]**SPHAERALCEA**
          12′ Fr segment gen 1-seeded
          17. Bractlets 0
            18. Gen subshrub (ann); fr segment below indehiscent, strongly net-veined, above dehiscent, smooth;
              native, esp D . . . . [3]**SPHAERALCEA**
            18′ Gen ann (per); fr segment ± uniform; naturalized, widespread, incl D
              19. Petals ± purple, white, or yellow; lvs gen ± lobed to hastate; fr segment side walls disintegrating . . . **ANODA**
              19′ Petals pale- to orange-yellow; lvs unlobed; fr segment side walls not disintegrating . . . . . . [*Sida rhombifolia*]
          17′ Bractlets gen 1–3, occ deciduous in fl or fr
            20. Fr segment dehiscent; fls gen > 1 in lf axils; widespread, gen chaparral . . . . **MALACOTHAMNUS**
            20′ Fr segment partly or fully indehiscent; fls 1 in lf axils or in gen raceme-like cymes; gen D or
             disturbed sites
             21. Fr segment below indehiscent, sides strongly net-veined, above dehiscent, sides smooth;
              petals gen rose-pink to salmon-orange; infl raceme-like, gen in panicle, fls facing out . . . [3]**SPHAERALCEA**

21′ Fr segment gen indehiscent, sides gen ± weakly net-veined; petals gen white, pale ± yellow, or ± purple; infl raceme or fls gen 1 in lf axils, facing up

22. Ann, prostrate to erect, not weedy, green, glabrous or gen sparsely stellate-hairy; lf base symmetric; petals white to ± purple . . . . **EREMALCHE**

22′ Per, gen prostrate, gen weedy, ± gray, densely stellate-canescent or hairs scaly; lf base asymmetric; petals gen pale ± yellow . . . . **MALVELLA**

ABUTILON

Paul A. Fryxell & Steven R. Hill

Ann, per to shrub, stellate-canescent, tomentose, or bristly. **ST**: gen erect (decumbent). **LF**: cordate to ovate, lobes gen 0(3), crenate or toothed. **INFL**: panicle or fls 1 in lf axils; bracts lf-like or not; involucel 0. **FL**: petals yellow, yellow-orange, orange-pink, to ± red; anthers at top of filament tube; stigmas head-like. **FR**: capsule-like, ± cylindric to ± spheric, segments gen not separating fully from each other or from pl, smooth-sided, dehiscent at top, gen with beak splitting in 2, walls firm to woody. **SEED**: 3–6(15) per segment. 200 spp.: warm regions. (Arabic: possibly father of mallow) [Fryxell 1988 Syst Bot Monogr 25:24–68]

1. Ann; lf gen 10–20 cm wide; fr segments > 10, beaks 3–5 mm . . . . *A. theophrasti*
1′ Per to shrub; lf 1–8 cm wide; fr segments gen ≤ 10, beaks 1–2 mm
  2. Petals (10)20–25 mm, orange; fr segments gen (7)10, ≤ calyx; st erect . . . . *A. palmeri*
  2′ Petals 3–6 mm, orange-pink to ± red; fr segments gen 5(6), >> calyx; st ± decumbent . . . . *A. parvulum*

A. palmeri A. Gray (p. 879) Subshrub, shrub, densely branched, rounded. **ST**: erect, 15–20 dm; hairs stellate, sometimes also simple. **LF**: blade (2)4–8 cm, dentate, acuminate, faces densely velvety; lobes 3, obscure. **INFL**: panicles or fls 1 in lf axils. **FL**: calyx 9–15 mm, ≥ fr; petals (10)20–25 mm, orange. **FR**: 15–16 mm diam; segments gen (7)10, ± 10 mm, 3-seeded, bristly or densely soft-hairy, beaks 1.5–2 mm, ± erect. Uncommon. Dry, gen e-facing mtn slopes, creosote-bush scrub; 600–800 m. DSon, adjacent PR, possibly introduced SCo; AZ, n Mex. Occ cult. Mar–May

A. parvulum A. Gray (p. 879) DWARF ABUTILON Per, from woody root. **ST**: ± decumbent, 1–4 dm, much-branched, stellate-canescent, gen with few simple hairs. **LF**: blade 1–5 cm, dentate, acute to acuminate; hairs scattered, stellate. **INFL**: fls 1 in lf axils. **FL**: calyx 3–5 mm, << fr, lobes reflexed in fr; petals 3–6 mm, orange-pink to ± red. **FR**: segments gen 5(6), 3-seeded, stellate-puberulent, beaks 1–2 mm, ± erect. 2*n*=14. Arid, rocky slopes, shadscale scrub; 900–1300 m. DMtns (Providence Mtns); to s CO, w TX, n Mex. Apr–May ★

A. theophrasti Medik. (p. 883) VELVET-LEAF Ann, gen scented when bruised. **ST**: erect, (5)10–20 dm, gen few-branched, glandular. **LF**: blade gen 10–20 cm wide, shallowly crenate, velvety, densely stellate-hairy to tomentose. **INFL**: panicle or fls 1 in lf axils. **FL**: calyx 6.5–10 mm, < fr; petals 6–8(14) mm, yellow to orange. **FR**: segments > 10, 5–15-seeded, long-soft-hairy, beaks 3–5 mm, backpointing. 2*n*=42. Uncommon. Disturbed places; gen < 100 m. NCoRI, GV (esp ScV), SCoRO, SW, D; widespread worldwide; native to s Asia. Gen weedy; reported from 42 CA cos. (Holt & Boose 2000 Weed Science 48:43–52); seeds long-lived. Jul–Sep

ALCEA

Bien, per. **ST**: erect. **LF**: petiole long; blade ± palmate-lobed. **INFL**: spike, raceme, in panicle or not; bractlets 5–11, fused at base, ≤ sepals. **FL**: showy, > 3 cm, pink, white, red, purple or yellow; petals notched distally; anthers near top of filament tube, below; stigmas thread-like. **FR**: segments 16–40, falling from axis, indehiscent, each with an upper, empty chamber and a lower, 1-seeded chamber, beak 0. 60 spp.: e Medit to c Asia. (Greek: remedy, strength; ancient name) [Fryxell 1988 Syst Bot Monogr 25:68–70]

A. rosea L. (p. 883) HOLLYHOCK **ST**: gen 1–2.5 m, gen unbranched, pithy; stellate-hairy, ± glabrous in age. **LF**: blade gen 7–15(20) cm, cordate or ovate to ± round, weakly palmate-5-7-lobed, stellate-hairy. **INFL**: gen dense, gen branched in age; bracts gen not lf-like; bractlets 6–9, ≤ sepals, widely triangular, fused at base. **FL**: ± sessile, 8–10(12) cm diam; calyx tomentose; stigmas gen 20–40. **FR**: disk-like; segments 20–40, ± 6 mm, horseshoe-shaped, thin, flat, back channeled, winged, hairy. 2*n*=42. Uncommon. Disturbed places; esp < 1100 m. SnFrB, SW; perhaps native to Asia Minor. Occ escaped from cult. Jun–Aug

ANODA

Paul A. Fryxell & Steven R. Hill

Ann (per) [subshrub]. **ST**: decumbent [ascending] to erect. **LF**: gen ± lobed, ± entire to crenate or dentate, reduced distally on st; stipules inconspicuous, deciduous. **INFL**: raceme or panicle, open, or fls 1 in lf axils; fl stalks not jointed, >> fl; bractlets 0. **FL**: calyx gen larger in fr; petals ± purple, white, or yellow; anthers at top of filament tube; styles 5–20, stigmas head-like. **FR**: hemispheric to disk-like; segments 5–20, indehiscent, puberulent to bristly, side walls disintegrating, spur or spine 1 at distal angle or 0. **SEED**: enclosed in net-veined envelope or not, glabrous or puberulent. 25 spp.: esp Mex to S.Am. (Ceylonese or Greek: without knot, from fl stalk not jointed) *A. pentaschista* A. Gray (petals yellow, gen fading ± red, fr 4–5 mm diam, segments 5–8) a weed of citrus groves.

A. cristata (L.) Schltdl. (p. 883) VIOLETTAS **ST**: decumbent to erect, 0.5–1 m, sparsely bristly. **LF**: not leathery; blade 2–9 cm, often with central purple blotch, gen triangular to hastate, ± crenate or dentate; hairs gen simple. **INFL**: gen fls 1 in lf axils, occ raceme-like in age. **FL**: calyx ± 10 mm diam, 20–25 mm diam in fr, lobes acuminate, often ± red, spreading in fr; petals 8–30 mm; styles 10–20. **FR**: bristly. 2*n*=30,60,90. Uncommon in disturbed places; < 800 m. SNF, GV, SCo, D; to se and c US, S.Am, Middle East, Australia; perhaps native to Mex. Aug–Nov

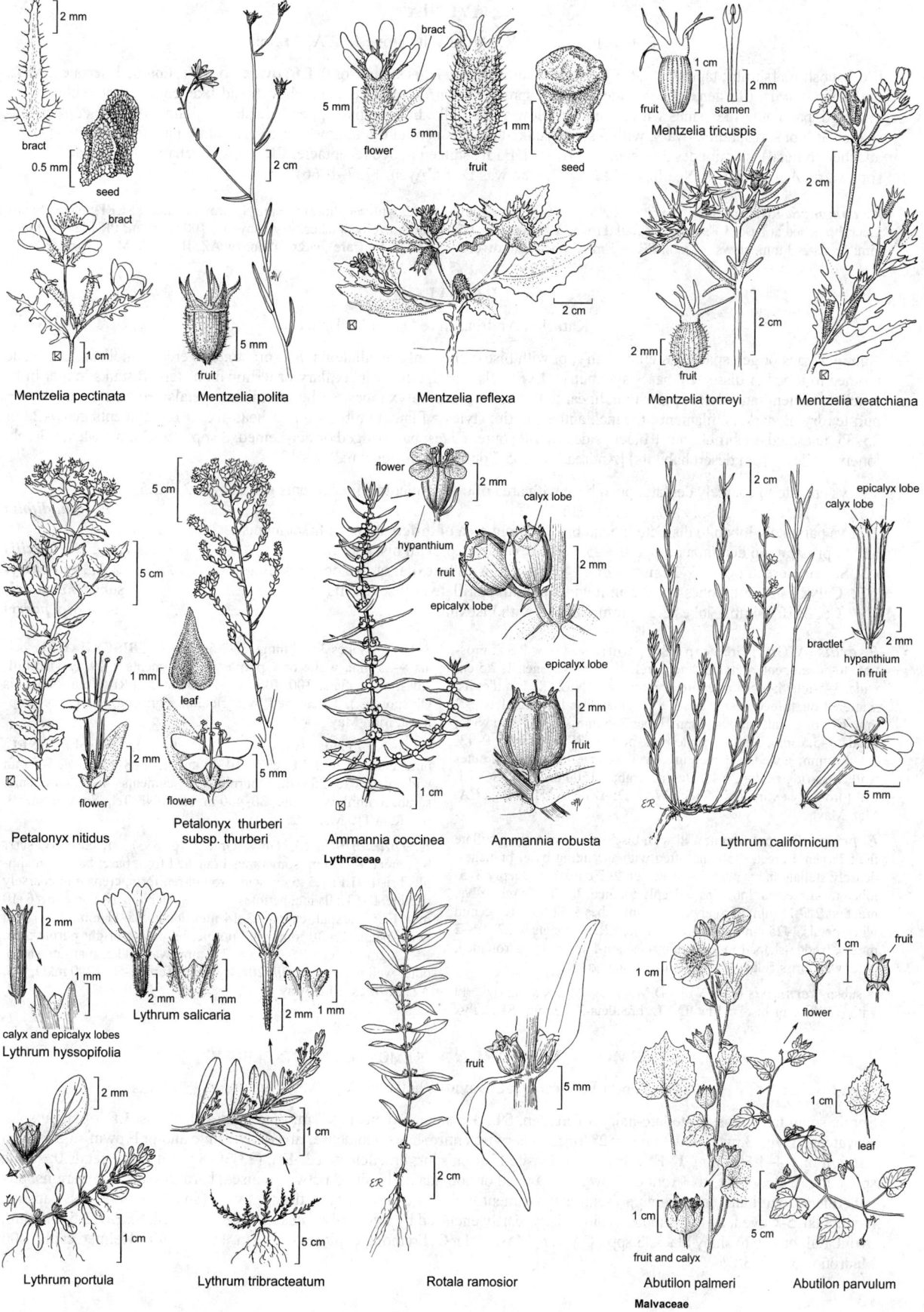

Mentzelia pectinata

Mentzelia polita

Mentzelia reflexa

Mentzelia tricuspis

Mentzelia torreyi

Mentzelia veatchiana

Petalonyx nitidus

Petalonyx thurberi subsp. thurberi

Ammannia coccinea

Lythraceae

Ammannia robusta

Lythrum californicum

Lythrum hyssopifolia

Lythrum salicaria

Lythrum portula

Lythrum tribracteatum

Rotala ramosior

Abutilon palmeri

Abutilon parvulum

Malvaceae

AYENIA

Robert E. Preston, R. David Whetstone & T. A. Atkinson

[Per] subshrub [shrub]; taproot stout. **ST**: erect to trailing, twig hairs stellate or 0. **LF**: ovate-obovate, unlobed, serrate. **INFL**: fls 1(2) in lf axils. **FL**: gen < 3 mm wide; sepals ± spreading, narrowly ovate; petal claw thread-like, coiled, limb ± obcordate, incurved, parachute-like, sinus with an anther below and a stalked, gland-like appendage above; filament tube ± cup- or urn-shaped at top, surrounding ovary, with 5 short, stalked anthers bent out- and downward (each stalk inserted in sinus, thereby attached to petal), staminodes 5, < stamens; ovary (and fr) stalked above receptacle. **FR**: capsule, chambers 1-seeded. 75–80 spp.: warm Am. (Louis de Noailles, 1713–1793, who was Duc d'Ayen, 1737–1766)

A. compacta Rose (p. 883) CALIFORNIA AYENIA Pl 1–4 dm, much-branched at base. **LF**: stipules small, brown. **FL**: sepals ± 1.5 mm; petals 2–3 mm, claws ± 2 mm. **FR**: ± 5 mm, spheric, straw-colored, with ± cylindric, ± purple protuberances. **SEED**: black. Sandy and gravelly washes, dry canyons; 100–1160 m. PR (desert slope), e&s DMoj (rare), w&c DSon; sw AZ; Baja CA. Mar–Apr ★

EREMALCHE

Katarina Andreasen & David M. Bates

Ann, glabrous or gen sparsely stellate-hairy; pl with bisexual or only pistillate fls. **ST**: prostrate to erect, ± hollow. **LF**: blade toothed to lobed or dissected, bases symmetric. **INFL**: fls 1 in lf axils or in axillary or terminal clusters; fl stalks longer in fr; bractlets subtending calyx 3, persistent, linear to thread-like. **FL**: calyx lobes > tube, acuminate; petals gen > calyx, white to ± purple (drying darker); filament tube incl, anthers at tip; styles > filament tube, stigmas head-like. **FR**: segments gen 9–22 or 25–35, unarmed, glabrous, gen ± black, sides fragile, outer edges, back ridged or net-veined. 3 spp.: sw US, nw Mex. (Greek: lonely mallow, from desert habitats) [Andreasen 2005 Conservation Genet 6:399–412]

1. Lvs crenate to coarsely dentate; petal base with area of different color; fr segments gen 25–35, 2.8–3.5 mm
.... ***E. rotundifolia***
1′ Lvs palmately lobed to dissected; petal base without area of different color; fr segments 9–22, 1.4–2 mm
  2. St prostrate to decumbent; petals 4–5.5 mm, ± = calyx; pl with bisexual fls .... ***E. exilis***
  2′ St gen ± erect; petals 3–25 mm, often >> calyx; pl with bisexual or occ only pistillate fls .... ***E. parryi***
   3. Calyx 3–10 mm, lobes 1.5–3.5 mm wide; pl with pistillate or bisexual fls .... subsp. ***kernensis***
   3′ Calyx 10–14 mm, lobes 2.5–5 mm wide; pl with bisexual fls .... subsp. ***parryi***

E. exilis (A. Gray) Greene (p. 883) WHITE MALLOW **ST**: prostrate to decumbent, < 50 cm, finely stellate-hairy. **LF**: gen 1–2.5 cm wide, 3–5-lobed; lobe tips entire or 3-toothed. **INFL**: fls 1 in lf axils, scattered on st, occ near st base, ± = subtending lvs; bractlets 3–7 mm. **FL**: bisexual; calyx 4–7 mm, lobes 3–5 mm, 1.5–2.5 mm wide; petals 4–5.5 mm, white or pale pink-purple. **FR**: segments 9–13, 1.4–1.8 mm, ± wedge-shaped in ×-section, margins rounded, outer wall cross-ridged. 2*n*=20,40. Desert scrub; < 1500 m. SnJV, SCoRI, ChI (probably extirpated), SnGb, W&I, D; AZ, NV, UT, n Baja CA. Mar–May

E. parryi (Greene) Greene Pl with bisexual or occ only pistillate fls. **ST**: gen ± erect, < 50 cm, often with ascending basal branches, densely stellate-hairy near tips. **LF**: gen 2–5 cm wide, deeply 3–5-lobed to -dissected; lobe tips ± deeply toothed. **INFL**: fls gen > lvs; bractlets 2.5–15 mm. **FL**: calyx 3–14 mm, lobes 3–11 mm, 1.5–5 mm wide; petals 3–25 mm, white to ± purple. **FR**: segments 9–22, 1.5–2 mm, ± wedge-shaped in ×-section, cushion-like, margins rounded, outer wall cross-ridged. 2*n*=20. Subspp. intergrade.

  subsp. ***kernensis*** (C.B. Wolf) D.M. Bates KERN MALLOW Pl with pistillate or bisexual fls. **INFL**: bractlets 3–10 mm. **FL**: calyx 3–10 mm, lobes 3.5–8 mm, 1.5–3.5 mm wide. **BISEXUAL FL**: petals 8–25 mm, white or ± purple. **FR**: segments 9–19. Eroded hillsides, alkali flats; 100–1000 m. s SnJV, s SCoRI (Kern, San Luis Obispo cos.). Threatened by agriculture, grazing, energy development. Mar–May ★

  subsp. ***parryi*** PARRY'S MALLOW Pl with bisexual fls. **INFL**: bractlets 7–15 mm. **FL**: calyx 10–14 mm, lobes 8–11 mm, 2.5–5 mm wide; petals 15–25 mm, ± purple. **FR**: segments 14–22. Grassland, scrub, foothill woodland; 30–400 m. c&s SNF, Teh, SnJV, e SnFrB, SCoR, WTR. Mar–May

E. rotundifolia (A. Gray) Greene (p. 883) DESERT FIVESPOT **ST**: erect, 8–60 cm, sometimes branched from base; hairs gen simple, bristly. **LF**: 1.5–6 cm wide, round-reniform, crenate to coarsely dentate. **INFL**: fls incl petioles gen >> subtending lvs; bractlets 6–10 mm. **FL**: bisexual; calyx 9.5–14 mm, lobes 5.5–11 mm, 3.5–7 mm wide; petals 15–30 mm, pink-purple, base with bright purple area. **FR**: segments gen 25–35, 2.8–3.5 mm, wafer-like, margins sharp, outer wall net-veined. 2*n*=20. Dry desert scrub; -50–1200 m. D; NV, AZ, nw Mex. Mar–May

FREMONTODENDRON FREMONTIA, FLANNELBUSH

Robert E. Preston, R. David Whetstone & T. A. Atkinson

Shrub, small tree; densely stellate-hairy; evergreen. **ST**: decumbent to erect, < 7 m; inner bark gelatinous. **LF**: often on spur, ± ovate, gen with 3 main, few to many 2° lobes, otherwise entire; hairs gen denser abaxially, white and/or brown; stipules gen ± 2 mm or ± 4–4.5(9) mm. **INFL**: fls gen 1 opposite lf or on spur; bractlets gen 3. **FL**: (23)30–84 mm wide; petals 0; sepals spreading, widely ovate to ± round, showy, tips awned or not, adaxially pitted between raised, hard, fused basal margins, pits puberulent, long hairs on margins present or 0; filament tube ± = ovary, < style, fleshy; ovary (and fr) sessile. **FR**: capsule, loculicidal, 5-valved, 2–4 cm, acute-ovoid, bristly, partly enclosed by dried calyx, chambers 2–3-seeded. **SEED**: 3.5–5.5 mm, ovoid, dull brown to shiny black. 3 spp.: CA, AZ, Mex. (John C. Frémont, explorer in w Am, 1813–1890) [Kelman et al. 2006 Madroño 53:380–387]

1. Sepal pit margins lacking ± 1 mm silky hairs; pl erect, unbranched near ground; stipules ± 4–4.5(9) mm . . . *F. mexicanum*
1′ Sepal pit margins (and occ surface) with ± 1 mm silky hairs; pl erect, branched near ground, or decumbent;
    stipules ± 2 mm
  2. Pl erect, taller than wide; fl (23)35–60(76) mm wide; sepals yellow, margins ± red or not . . . . *F. californicum*
  2′ Pl decumbent, much wider than tall; fl 30–50 mm wide; sepals orange, coppery, or ± red . . . . *F. decumbens*

F. californicum (Torr.) Coville (p. 883) Pl erect, gen 2–5 m, taller than wide, branched near ground. **LF:** petiole 0.4–3 cm; blade 1–7 cm, palmately lobed (unlobed), soft to leathery, base truncate to shallowly cordate; stipules ± 2 mm. **INFL:** fl stalk 4–18 mm. **FL:** (23)35–60(76) mm wide; sepals yellow, sometimes ± red towards base, keeled on back, pit margins silky-hairy. **SEED:** dull brown to black, stellate-hairy, aril-like structure present. Chaparral, oak/pine woodland; 180–2320 m. NCoRO, NCoRI, s CaR, SN, SnFrB, SCoR, TR, PR; sw OR, AZ, n Baja CA. Highly variable; some variation induced by habitat. Apr–Jul

F. decumbens R.M. Lloyd PINE HILL FLANNELBUSH Pl decumbent, < 1 m, much wider than tall. **LF:** petiole 0.6–3.4 cm; blade 1.5–5.2 cm, palmate-lobed, soft to leathery, base shallowly to deeply cordate; stipules ± 2 mm. **INFL:** fl stalk 10–27 mm. **FL:** 30–50 mm wide; sepals orange, coppery, or ± red, pit margins (and occ surface) with ± 1 mm silky hairs. **FR:** gen not opening without fire. **SEED:** dull brown, stellate-hairy, aril-like structure present. Gabbro outcrops in chaparral/pine woodland; 425–760 m. n SNF (Pine Hill, w El Dorado Co.). [*F. californicum* subsp. *d.* (R.M. Lloyd) Munz] Decumbent pls in Yuba, Nevada cos. morphologically, genetically distinct from *F. decumbens*, needs study. Apr–Jul ★

F. mexicanum Davidson MEXICAN FLANNELBUSH Pl erect, gen 1.5–7 m, unbranched near ground. **LF:** petiole 0.8–4.5 cm; blade 2.5–8 cm, palmately lobed, thick-leathery, base deeply cordate; stipules ± 4–4.5(9) mm. **INFL:** fl stalk 5–15 mm. **FL:** 45–84 mm wide; sepals gen orange, ± red toward bases, pit margins lacking ± 1 mm silky hairs. **SEED:** black, glabrous, shiny, aril-like structure 0. Chaparral; 10–716 m. PR (exc SnJt); n Baja CA. Mar–Jun ★

HERISSANTIA BLADDER MALLOW

John C. La Duke

Ann, per. **ST:** gen decumbent or trailing. **LF:** stipules inconspicuous, deciduous; lower petioled, on fl branches often ± sessile; blade cordate, ovate, dentate. **INFL:** fls 1–few in lf axils; fl stalks slender, jointed; bractlets 0. **FL:** petals white to pale yellow; anthers at tip of filament tube; styles, stigmas (8)10–15, stigmas head-like. **FR:** nodding; segments (8)10–15, dehiscent, bladdery, inflated, rounded at top, side walls thin, flexible, smooth. **SEED:** gen glabrous, 2, 3, 6 per segment. 5 spp.: warm, trop Am. (L.A.P. Herissant, French physician, naturalist, poet, 1745–1769)

H. crispa (L.) Brizicky (p. 883) CURLY HERISSANTIA **ST:** < 10 dm. **LF:** blade 1–7 cm, gen 2 × petiole, reduced upwards. **INFL:** fl stalk gen reflexed at joint. **FL:** calyx 3–7 mm, lobes ± lanceolate; petals 6–11 mm; anthers golden-yellow (orange). **FR:** 10–20 mm diam, spheric, inflated; segments gen taller than wide, short-bristly, papery, silvery-shiny inside. **SEED:** gen 3. 2n=14. Desert scrub; < 800 m. w DSon (c San Diego Co.); AZ, NM, TX, FL; trop. Aug–Sep ★

HIBISCUS HIBISCUS

Ann, per, subshrub [shrub, tree]. **ST:** gen erect, bristly or stellate-hairy to ± glabrous. **LF:** gen simple, cordate, palmate-lobed or -divided or not, ± entire to dentate or crenate-dentate, palmate-veined, tip acute or acuminate; stipules gen persistent. **INFL:** raceme, open, or fls 1 in lf axils; fl stalks often jointed in upper 1/3; bractlets 8–10(20), free or basally fused, gen narrow, persistent. **FL:** gen showy, open ≤ 1 day; calyx 5-lobed; petals white, yellow, lavender, red, or other colors, often with dark basal spot; filament tube 5-toothed at tip, anthers scattered on upper 1/2 below tip; style distally 5-branched, stigmas head-like. **FR:** capsule loculicidal, 5-chambered, ovoid or oblong, glabrous or hairy. **SEED:** several per chamber, hairy or ± glabrous. 200 spp.: esp Am, Afr, Asia, Australia. (Greek: mallow) Other taxa, such as *H. syriacus* L., rose-of-Sharon, cult, possibly escaped.

1. Ann; petals yellow with dark basal spot; calyx inflated, bladder-like in fr . . . . *H. trionum*
1′ Per or subshrub; petals white, lavender, to rose; calyx not inflated in fr
  2. Lf blade 1–3 cm, ovate; fls in lf axils; fl stalk 0.5–1.5 cm; sepals fused only at base; petals 2–2.7 cm;
    seed densely silky; dry habitats . . . . *H. denudatus*
  2′ Lf blade 6–10 cm, cordate; fls near st tip; fl stalk 1–8 cm; sepals fused 1/2; petals 6–10 cm;
    seed glabrous; wet habitats . . . . *H. lasiocarpos* var. *occidentalis*

H. denudatus Benth. (p. 883) PALE FACE Subshrub, densely stellate-hairy. **ST:** 0.5–1 m, gen tangled-branched. **LF:** blade 1–3 cm, ovate, truncate, finely toothed, densely stellate-hairy abaxially, adaxially; stipules 2–3 mm. **INFL:** fls 1 in lf axils; fl stalks 0.5–1.5 cm; bractlets 0.5–4 mm, persistent [(0)]. **FL:** sepals 11–15 mm, fused at base; petals 2–2.7 cm, white or lavender, base gen purple; stamen column 8–9 mm. **FR:** 7–8 mm, ± ≤ calyx, spheric, spread wide open at maturity, glabrous or at tip hairy. **SEED:** ± 2.5 mm, reniform, densely silky, hairs 3–4 mm. 2n=22. Desert scrub of mesas, canyons, creosote-bush scrub; < 800 m. DSon; to NV, w TX, n Mex. Feb–May

H. lasiocarpos Cav. var. ***occidentalis*** (Torr.) A. Gray (p. 883) WOOLLY ROSE-MALLOW Per, subshrub from caudex, gen rhizomed, hairy. **ST:** many from base, prostrate to erect, 1–2 m. **LF:** petiole to 10 cm; blade 6–10 cm, cordate, shallowly 3–5-lobed or entire, toothed, acuminate; ± densely stellate-hairy abaxially, adaxially; stipules 1–4 mm. **INFL:** fls 1 in lf axils; fl stalks 1–8 cm; bractlets 10, ± 2.5 cm, free ± to base, ≥ calyx in bud, fl, forming beak in bud; > calyx in fr. **FL:** calyx 2.5–3 cm, bell-shaped, sepals fused 1/2, veiny in fr; petals 6–10 cm, white with rose-red center. **FR:** 2.5–3 cm, filling calyx, ± spheric to short-cylindric, stellate-hairy. **SEED:** ± 3 mm, glabrous. 2n=38. Freshwater wetlands, wet banks, marshes; < 100 m. CaRF, c&s ScV, deltaic GV. [*H. californicus* Kellogg; *Hibiscus lasiocarpus* Cav., orth. var.] Threatened by riverbank alteration; rest of sp. (typical var.) no nearer CA than nw Mex, NM. Jul–Nov ★

H. trionum L. (p. 883) FLOWER-OF-AN-HOUR Ann, bristly. **ST:** erect, 0.3–0.6 m, branches prostrate or not. **LF:** blade ± 5 cm, 2–3 cm wide, 3–5-lobed or -divided, coarsely toothed, stellate-hairy abaxially, glabrous adaxially. **INFL:** fls 1 in lf axils; bractlets 10, free ± to base, 7–8(10) mm, << calyx. **FL:** calyx fused ± to tip, 5-ridged or -winged, veiny, inflated, papery in fr; petals ± 2 cm, yellow, base purple-black; filament tube ± 11 mm. **FR:** 1.5–2 cm, enclosed within, ≤ calyx. **SEED:** 2–2.5 mm, reniform, sparsely hairy. 2n=28,56. Uncommon. Disturbed places; gen < 350(1450) m. CA-FP; native to Eur, c Afr. Apr–Nov

HORSFORDIA

John C. La Duke

[Subshrub], shrub, densely stellate-tomentose or -scabrous. **ST:** erect. **LF:** petiole stout; blade lanceolate to ± round or cordate, ± entire or irregularly fine-toothed; stipules 2–5 mm. **INFL:** axillary racemes clustered terminally or fls 1 in lf axils [panicle]; bractlets 0. **FL:** petals white, yellow, blue-lavender, rose, or pink; anthers at tip of filament tube; styles 6–11, > anthers, stigmas head-like. **FR:** ± glabrous; segments 6–12, below indehiscent, firm, net-veined, above dehiscent, with 2 spreading, scarious wings. **SEED:** 1–3 per segment, puberulent or minutely ridged. 4 spp.: sw US, nw Mex. (F.H. Horsford, VT botanist, collector, 1855–1923)

1. Petals blue-lavender, rose, or pink, 12–21 mm, exceeding styles; fr segment wings ± lanceolate; seed 1.9 mm, minutely roughened . . . . ***H. alata***
1′ Petals white or yellow, 6–9 mm, extending to style tips; fr segment wings gen ± widely ovate; seed 2.2 mm, densely puberulent . . . . ***H. newberryi***

H. alata (S. Watson) A. Gray (p. 889) PINK VELVET-MALLOW **ST:** 1–2.5(4) m, gray-yellow. **LF:** petioles << blade; blade 4–11 cm, ovate, finely toothed, base ± cordate, tip acute. **INFL:** fl stalks 2–17 mm. **FL:** calyx 5–7 mm, lobed 1/2; styles 9–10. **FR:** 7–8 mm diam, gen ± purple on top, segments 9–10, lower part 1-seeded, upper gen with 2 aborted ovules; wings ± 6 mm, 3 mm wide. **SEED:** 1 per segment, dark brown. 2*n*=30. Rocky canyons, washes; gen 100–600 m. DSon; sw AZ, nw Mex. Dec–Apr ★

H. newberryi (S. Watson) A. Gray (p. 889) NEWBERRY'S VELVET-MALLOW **ST:** 2–3 m, ± yellow. **LF:** petioles << blade; blade 4–10 cm, narrowly ovate, finely toothed, base truncate to ± cordate, tip obtuse to acute. **INFL:** fl stalks 5–12(16) mm. **FL:** calyx 5–6 mm, lobed 1/2; styles ± 10. **FR:** 7–9 mm diam, gen ± purple on top, segments ± 10, lower part 1-seeded, upper gen 2-seeded; wings ± 5 mm, 4 mm wide. **SEED:** 3 per segment, ± white. 2*n*=30. Creosote-bush scrub; gen 100–800 m. PR, w DSon; to sw AZ, n Mex. Mar–Apr, Nov–Dec ★

ILIAMNA GLOBE MALLOW

Tracey Slotta

Per [subshrub], stellate-hairy; caudex woody. **ST:** gen > 1, 0.6–2 m, branched. **LF:** stipules inconspicuous, persistent or not; blade gen 3–7 lobed, base wedge-shaped, truncate, or cordate. **INFL:** spike- or raceme-like, fls 1 in lower lf axils, 2–5 in upper lf axils, lvs reduced above; bractlets 3, free, linear to lanceolate, ± persistent. **FL:** showy; calyx 5-lobed, divided 1/2, stellate-hairy to ± glabrous; petals 5, pale lavender to rose-purple, pink, or white; filament tube with anthers in upper 1/3–1/2; styles gen 6–15, stigmas gen oblique-squared, often ± elongate. **FR:** gen > calyx; oblong; segments gen 6–15, gen 10–15 mm, 6–10 mm wide, beaked or not, attached to fr axis by strong fiber, ± dehiscent to base, bristly, sides thin, smooth, glabrous. **SEED:** gen 2–4 per segment, reniform, glabrous to puberulent. 8 spp.: N.Am. (derivation uncertain)

1. Upper lvs 3-lobed; stipules persistent; petiole 2–5 cm . . . . ***I. bakeri***
1′ Upper lvs 5–7-lobed; stipules persistent or not; petiole 2–7(10) cm
  2. Bractlets 10–14 mm, ± 10 mm wide, 2/3 or > sepals; lf hairs denser abaxially . . . . ***I. latibracteata***
  2′ Bractlets gen 4.5–6 mm, ± 2 mm wide, 1/2 sepals; lf hairs sparse, not denser abaxially . . . . ***I. rivularis***

I. bakeri (Jeps.) Wiggins (p. 889) BAKER'S GLOBE MALLOW Pl 0.3–1.2 m. **LF:** petiole 2–5 cm; blade 1.5–4.5 cm, 2–5 cm wide, 3–5-lobed, crenate, ± densely stellate-hairy, base tapered to truncate. **INFL:** bractlets 5–8 mm; fl stalk ± 5 mm. **FL:** sepals 9–12 mm, puberulent; corolla 3–6 cm wide, rose-purple. **FR:** 7.5–15 mm. **SEED:** 3–4 per segment. Mtn slopes, juniper woodland, lava beds; 1000–2500 m. KR, NCoRH, CaR, MP (Shasta Co.); s-c OR. Jun–Sep ★

I. latibracteata Wiggins (p. 889) CALIFORNIA GLOBE MALLOW Pl 1–2 m. **LF:** petiole 5–14 cm; blade 10–15 cm, 10–15 cm wide, 5–7-lobed, coarsely serrate, abaxially densely short-stellate-hairy, adaxially ± glabrous, base truncate to cordate. **INFL:** bractlets 10–14 mm; fl stalks < 5 mm. **FL:** sepals 8–10 mm, coarsely hairy; corolla 4–6 cm wide, rose-purple. **FR:** 10–20 mm. **SEED:** 2–4 per segment. Conifer forest, streamsides; 500–2000 m. NCo, KR (Humboldt, Siskiyou cos.), CaRH (Siskiyou Co.); sw OR. Jun–Jul ★

I. rivularis (Hook.) Greene STREAMBANK WILD HOLLYHOCK Pl 0.5–2 m. **LF:** petiole 2–7(10) cm; blade 5–14 cm, 5–16 cm wide, 3–7-lobed, coarsely serrate, sparsely hairy, base cordate to reniform. **INFL:** bractlets gen 4.5–6 mm; fl stalks 3–10(15) mm. **FL:** sepals 6–11 mm, ± coarsely hairy; corolla 3–8 cm wide, rose-purple to light pink to white. **FR:** gen ≤ 10 mm. **SEED:** 1–3(4) per segment. 2*n*=66. Mtn streamsides; 500–2000 m. NCo (Humboldt Co.); to BC, WY, MT, CO. Inclusion in flora in error; occurrence in CA based on mis-identification. Jun–Aug

LAGUNARIA

1 sp.: trop, subtrop, e Australia, Norfolk Island, Lord Howe Island. (A. de Laguna, Spanish botanist, physician to Pope Julius III) [Craven et al. 2006 Blumea 51:345–353] Cult widely as orn, rarely escaped; pinnate-veined lvs notable in family; sterile material may be hard to place.

L. patersonia (Andrews) G. Don NORFOLK ISLAND HIBISCUS, COW ITCH TREE Shrub to small tree 2–10(15) m, evergreen. **ST:** bark smooth, gray; twigs stellate-hairy. **LF:** petiole ± 1 cm; blade 5–10 cm, ovate, entire, pinnately veined, blunt, ± thick, leathery, olive-green, densely felty-stellate-hairy abaxially. **INFL:** fls gen 1 in lf axils; fl stalks stout; involucel lobes 3–5, gen ovate, enclosing fl bud, gen deciduous. **FL:** 3.5–8 cm diam; sepals fused, lobes short; petals showy, reflexed, waxy, pink to rose-pink, fading white; anthers throughout filament tube; stigma club-shaped at top, lobes 5, radiating. **FR:** capsule, loculicidal, ± 4 cm, 2 cm diam before opening, filled with irritant hairs; chambers 5, walls firm, persistent, brown, fuzzy-hairy. **SEED:** reniform, smooth, orange to red when fresh. Coastal riparian forest; < 10 m. SCo; trop, subtrop, e Australia, Norfolk Island, Lord Howe Island. Producing seed and naturalized recently, only along Santa Margarita River (Hill 2009); salt tolerant. Jun–Aug

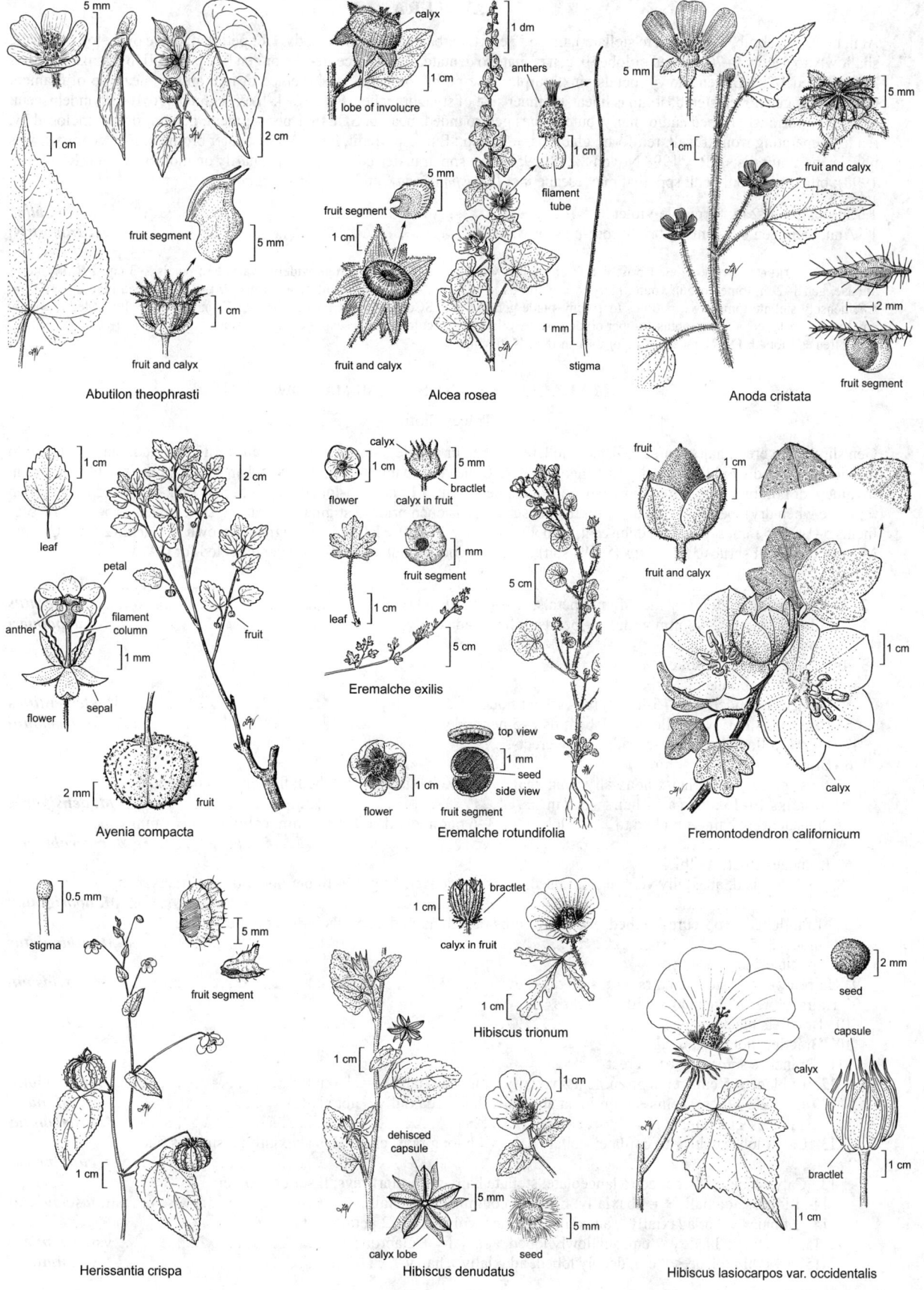

Abutilon theophrasti

Alcea rosea

Anoda cristata

Ayenia compacta

Eremalche exilis

Eremalche rotundifolia

Fremontodendron californicum

Herissantia crispa

Hibiscus trionum

Hibiscus denudatus

Hibiscus lasiocarpos var. occidentalis

LAVATERA

Ann, bien, per, shrub, ± glabrous to stellate-hairy. **ST**: erect, herbaceous to soft-woody. **LF**: petioled; blade ovate to lanceolate, shallowly to deeply 5–7(9)-lobed or lobes 0, gen crenate to dentate. **INFL**: raceme-like or gen fls 1[several] in lf axils; bractlets 3, fused gen 1/2. **FL**: gen showy; petals pink, purple, rose-purple or white, dark-veined or not; anthers near top of filament tube, below, gen not clustered; stigmas linear, on inner side of style branches. **FR**: ± disk-like; segments (6)10–19, indehiscent, smooth or variously ribbed and/or hairy, outer edges gen rounded, beak 0. **SEED**: 1 per segment, gen not firmly enclosed by, readily separating from fr wall, reniform, glabrous. ± 12 spp.: Eur, esp Medit, Asia, Afr. (Lavater brothers, 17th century Swiss physicians, naturalists) [Ray 1998 Novon 8:288–295] Most spp. transferred to *Malva*, primarily on molecular data (Ray 1995, 1998); other commonly cult spp. that may escape incl *L. cashmiriana* Camb., *L. thuringiaca* L.

1. Shrub, gen 1–2 m; fls purple-violet . . . . ***L. olbia***
1′ Ann, bien, gen 0.5–1 m; fls bright rose-pink or white . . . . [***L. trimestris***]

L. olbia L. TREE LAVATERA Gen bristly. **ST**: erect, gen woody at base, gen 1–2 m, tomentose in youth, hairy to ± glabrous in age. **LF**: densely stellate-tomentose; petioles to 10 cm; blade gen to 15 cm, lower 3–5-lobed, acute to obtuse, upper oblong-ovate to lanceolate, often ± 3-lobed. **INFL**: raceme-like or fls 1 in lf axils; fl stalks 2–7 mm; bractlets widely ovate. **FL**: petals 2–3 cm. **FR**: segments 17–19, puberulent or glabrous. 2*n*=42. Disturbed places; < 400 m. SCo (Laguna Canyon, Orange Co.); native sw Eur, Medit; widely cult. Rarely escapes cult. May–Jun

MALACOTHAMNUS BUSH-MALLOW

Tracey Slotta

Gen shrub, gen erect; hairs sparse to dense, stellate (stalked or sessile), simple, and glandular. **LF**: petioled; blade unlobed to 3–7-lobed, coarsely or shaggy-hairy to tomentose. **INFL**: head- to panicle-like, fls gen > 1 in lf axils; bracts lf-like or not, in involucre or not; bractlets 3, free, linear to lanceolate, persistent. **FL**: calyx 5-lobed; petals 5, pale pink-purple or white (often ± purple when dry); petals gen > calyx, fused at base; stamens open prior to stigma maturation; style branches 5–14. **FR**: segments 5–14, 2–5 mm, separating, dehiscent, smooth, top hairy, beak 0. **SEED**: 1 per segment, brown to black. 12 spp.: CA, nw Mex. (Greek: soft shrub) *M. foliosus* (S. Watson) Kearney reported but evidently not documented for CA.

1. Infl head-like . . . . ***M. palmeri***
  2. Involucre bracts 20 mm wide, often > petals; lf ± glabrous adaxially . . . . var. ***involucratus***
  2′ Involucre bracts 12–13 mm wide, ± = petals; lf hairy adaxially . . . . var. ***palmeri***
1′ Infl spike- or panicle-like
  3. Infl spike- or panicle-like
    4. Hairs white; branches ± erect
      5. Branches slender; lvs 3–5-lobed; fls < 10 per node . . . . ***M. clementinus***
      5′ Branches stout; lvs shallowly 5-lobed; fls 3–5 per node . . . . ***M. fremontii***
    4′ Hairs ± yellow; branches spreading or ± erect
      6. Branches slender; lvs thin
        7. Lvs sparsely short-hairy adaxially, long-shaggy-hairy abaxially, 3–5-lobed; fls gen ≥ 10 per node;
          fl stalks 10–15 mm; calyx hairs ≤ 3 mm . . . . ***M. densiflorus***
        7′ Lvs densely hairy, unlobed to ± 3–5-lobed; fls 3–6 per node; fl stalks < 4 mm; calyx hairs < 1 mm
          . . . . ***M. marrubioides***
      6′ Branches stout; lvs thick
        8. Lf 3–5-lobed, abaxially veins ± not raised, hairs not denser; fls gen ± 10 per node; lf-like bracts many
          . . . . ***M. aboriginum***
        8′ Lf unlobed to obscure-3-lobed, abaxially veins raised, hairs denser; fls 3–6 per node; lf-like
          bracts 3–5 . . . . ***M. orbiculatus***
  3′ Infl panicle-like
    9. Hairs ± yellow; stellate hairs with < 10 rays . . . . ***M. davidsonii***
    9′ Hairs white; stellate hairs with > 25 rays
      10. Bractlets black in age . . . . ***M. jonesii***
      10′ Bractlets not black in age
        11. Branches spreading or ± erect
          12. Lf blade < 5 cm, veins prominent abaxially, hairs adaxially ± = abaxially . . . . ***M. abbottii***
          12′ Lf blade ≤ 7 cm, veins ± prominent abaxially, hairs adaxially < abaxially . . . . ***M. hallii***
        11′ Branches spreading . . . . ***M. fasciculatus***
          13. Calyx lobes deltate to cordate; stellate hairs with long rays, esp on lvs abaxially; fl stalks stout
            . . . . var. ***catalinensis***
          13′ Calyx lobes awl-shaped to lanceolate; stellate hairs with short rays; fl stalks slender
            14. Lf hairs adaxially ± = abaxially; filament column ± 1/2 petals . . . . var. ***fasciculatus***
            14′ Lf hairs ± 0 or adaxially > abaxially; filament column > 1/2 petals
              15. Lvs thick, blade < 7 cm, shallowly lobed, adaxially ± glabrous . . . . var. ***nesioticus***
              15′ Lvs thin, blade < 8 cm, deeply lobed, adaxially ± hairy . . . . var. ***nuttallii***

M. abbottii (Eastw.) Kearney (p. 889) ABBOTT'S BUSH-MALLOW
Pl < 1.5 m; hairs dense, white, sessile. **LF:** blade < 5 cm, round, shallowly 3-lobed, hairs adaxially ± = abaxially; veins prominent abaxially. **INFL:** panicle-like; fls 3–4 per node; bracts few; fl stalks < 1
cm. **FL:** buds acuminate; calyx ± 1 cm, ± 0.4 cm wide; petals 1–2
cm; filament column < petals. **FR:** segments dark brown, reniform.
Sandy soils, streambanks, chaparral; < 400 m. SCoR (Monterey Co.).
May–Jul ★

M. aboriginum (B.L. Rob.) Greene (p. 889) INDIAN VALLEY
BUSH-MALLOW Pl 2–3 m, branches stout; hairs dense, coarse, ±
yellow. **LF:** blade < 7 cm, round to ovate, 3–5-lobed, short-stellate-
hairy abaxially, adaxially. **INFL:** spike- or panicle-like; fls gen ± 10
per node; lf-like bracts many; fl stalks < 0.5 cm. **FL:** buds acuminate;
calyx ± 1 cm, ± 0.5 cm wide; petals 1–2 cm; filament column < petals. **FR:** segments 2.5–3 mm, almost as wide, shallowly to deeply
notched, brown. 2*n*=34. Open rocky slopes; 150–700 m. c SNF, SnJV
(San Benito, Monterey, Fresno cos.), s SnFrB, SCoR. May–Jul ★

M. clementinus (Munz & I.M. Johnst.) Kearney (p. 889) SAN
CLEMENTE ISLAND BUSH-MALLOW Shrub, < 1 m, branches slender;
hairs shaggy, white, woolly. **LF:** blade < 6 cm, round to ovate, 3–5-
lobed. **INFL:** spike-like; fls < 10 per node; lf-like bracts few; fl stalks
< 0.5 cm; bractlets > calyx. **FL:** calyx ± 1 cm, 1/2 as wide; petals
1–2 cm; filament column < petals. **FR:** segments 2.5–3 mm, brown.
2*n*=34. Rocky canyon walls; < 310 m. s ChI (San Clemente Island).
Mar–May ★

M. davidsonii (B.L. Rob.) Greene (p. 889) DAVIDSON'S
BUSH-MALLOW Pl 3–5 m, branches stout; hairs dense, ± yellow, stellate with < 10 rays. **LF:** blade 5–11(20) cm, round to ovate, sharp-3–5-
lobed, hairs few, short-rayed-stellate abaxially, adaxially. **INFL:** panicle-like; fls 3–7 per node; lf-like bracts few; fl stalks 1 cm; bractlets
0.5 cm. **FL:** buds acuminate; calyx to 1 cm, 1/2 as wide; petals 1–2 cm;
filament column < petals. **FR:** segments 2.5–3.5 mm, ovate, notched,
brown. 2*n*=34. Slopes, washes; 500–700 m. s SnFrB, SCoRO, SCo,
WTR, SnGb. Intergrades with *M. fasciculatus.* May–Jul ★

M. densiflorus (S. Watson) Greene (p. 889) Pl 2 m, branches slender; hairs dense, ± yellow, woolly, occ glandular. **LF:** blade 3–6 cm,
round to ovate, thin, 3–5-lobed, long-shaggy-hairy abaxially, sparsely
short-hairy adaxially. **INFL:** spike-like; fls gen ≥ 10 per node; lf-like
bracts < 3; fl stalks 1–1.5 cm; bractlets ≥ calyx. **FL:** calyx ≤ 1 cm, ±
0.4 cm wide, hairs ≤ 3 mm; petals 1–2 cm; filament column < petals.
FR: segments 2.2–3.8 mm, ovate, shallowly notched, brown. 2*n*=34.
Chaparral slopes; 100–1100 m. SCo, SnBr, PR, DSon; n Baja CA.
[*M. d.* var. *viscidus* (Abrams) Kearney] May–Jul, Sep–Oct

M. fasciculatus (Torr. & A. Gray) Greene CHAPARRAL MALLOW
Pl 1–5 m, branches slender; hairs sessile (stalked), white, sparse to
velvety, woolly. **LF:** blade 2–8 cm, lobes 0 or 3–5, shallow to deep,
stellate hairs ± equally dense abaxially, adaxially. **INFL:** panicle-like;
fls gen 3–7 per node; lf-like bracts few; fl stalks < 1.5 cm, bractlets to
1 cm, 1/2 as wide, 1/2 calyx. **FL:** buds acuminate; calyx 4–7(11) mm,
1/2 as wide, lobes acute, < 2 × tube; petals 1–3 cm; filament column
< petals. **FR:** segments < 5 mm, deeply notched, brown. 2*n*=34. Vars.
may be untenable, needs study.

var. **catalinensis** (Eastw.) Kearney SANTA CATALINA ISLAND
BUSH-MALLOW Pl 1–3 m, branches ± erect; hairs stalked. **LF:** blade
< 7 cm, < 8 cm wide, thin, 3–5-lobed. **INFL:** fl stalks < 1 cm. **FL:**
calyx < 0.7 cm, lobes deltate to cordate; petals 1.3–2.6 cm; filament
column 1/2 petals. **FR:** segments 3.2–3.8 mm. Open chaparral; < 500
m. s ChI (Santa Catalina Island). May–Jul

var. **fasciculatus** (p. 889) **LF:** blade 2–6 cm, round, unlobed to
shallowly 3–5-lobed. **INFL:** fl stalks 1–1.3 cm; bractlets 0.5–1 cm.
FL: calyx to 1 cm; petals 1–1.5 cm; filament column ± 1/2 petals. **FR:**
segments 2.5–3.2 mm. Coastal scrub, chaparral; < 600 m. SCoRO,
SCo, TR, PR, DSon; Mex. May–Jul

var. **nesioticus** (B.L. Rob.) Kearney SANTA CRUZ ISLAND
BUSH-MALLOW Pl 1–3 m; hairs sparse to short-woolly. **LF:** blade
< 7 cm, thick, shallowly 3–5-lobed, adaxially ± glabrous. **INFL:** fls
< 3 per node. **FL:** calyx < 0.6–0.8 cm, lobes lanceolate; petals < 2
cm; filament column > 1/2 petals. Slopes, chaparral; < 220 m. n ChI
(Santa Cruz Island). May–Jul ★

var. **nuttallii** (Abrams) Kearney Pl 2–3 m. **LF:** blade < 8 cm,
thin, deeply 3–5-lobed, adaxially ± hairy. **INFL:** fls 3–5 per node.
FL: calyx < 1 cm, < 0.5 cm wide; petals 2.5 cm; filament column >
1/2 petals. **FR:** segments 3–5 mm. Open chaparral; SnFrB, SCoRO,
WTR. May–Jul

M. fremontii (A. Gray) Greene (p. 889) Pl < 3 m, branches ±
erect, stout; hairs stalked, white, woolly to long-shaggy-hairy. **LF:**
blade < 9 cm, round, thick, shallowly 5-lobed. **INFL:** spike- to
panicle-like; fls 3–5 per node; lf-like bracts few; fl stalks < 1.5 cm;
bractlets < 1/2 calyx, 1/2 as wide (as long). **FL:** buds rounded; calyx
< 1.2 cm, 1/3 as wide, lobes awl-shaped to lanceolate; petals < 2 cm;
filament column to 1 cm. **FR:** segments 2–4 mm, obovate, shallowly
notched, brown. 2*n*=34. Open chaparral to pine woodland; 600–1300
m. NCoRI, SNF, n SNH, GV, CW (exc SCoRO), SnGb. May–Jul

M. hallii (Eastw.) Kearney HALL'S BUSH-MALLOW Pl ± 3 m,
spreading, branches stout; hairs sessile, white, densely shaggy-hairy.
LF: blade ≤ 7 cm, wider than long, cordate, shallowly 3–5-lobed,
hairs adaxially < abaxially; veins ± prominent abaxially. **INFL:**
panicle-like; fls 3–7 per node; lf-like bracts 3–4; fl stalks < 0.5 cm;
bractlets 1/2 calyx. **FL:** buds rounded; calyx < 1 cm, 0.4 cm wide,
lobes awl-shaped to lanceolate; petals < 1.7 cm; filament column 1/2
petals. **FR:** segments (2)3–5 mm, almost as wide, shallowly notched,
brown. Open chaparral; < 760 m. NCoRO, NCoRI, SNF, c SNH,
SnJV, SnFrB. Sometimes incl in *M. fasciculatus.* May–Jul ★

M. jonesii (Munz) Kearney (p. 889) JONES' BUSH-MALLOW Pl
< 3 m, branches slender, in age occ black; close-dense-woolly, hairs
sessile, white. **LF:** blade < 4 cm, < 4 cm wide, diamond-shaped,
thick, unlobed to shallowly 3-lobed, hairs dense abaxially, adaxially;
veins ± prominent abaxially. **INFL:** panicle-like; fls 3–5 per node; lf-
like bracts < 3; fl stalks slender, < 0.7 cm; bractlets > 1/2 calyx, black
in age. **FL:** calyx < 1 cm, 1/2 as wide, lobes awl-shaped to lanceolate;
petals < 2 cm, filament column to 1 cm. **FR:** segments 3 mm, ± 3 mm
wide, deeply notched, dark brown. 2*n*=34. Open chaparral in foothill
woodland; 250–830 m. NCoRI, SCoRO. May–Jul ★

M. marrubioides (Durand & Hilg.) Greene (p. 889) Pl < 2 m,
branches slender; hairs sessile, ± yellow, close-dense-woolly. **LF:**
blade < 6 cm, cordate, thin, unlobed to ± 3–5-lobed, densely hairy;
veins prominent abaxially. **INFL:** spike- or panicle-like; fls 3–6 per
node; lf-like bracts 3–6, ± = lvs; fl stalks < 4 mm, stout; bractlets > 1/2
calyx. **FL:** calyx 0.7–1.5 cm, 0.4 cm wide, lobes lanceolate to deltate,
acuminate, 2–4 × tube, hairs < 1 mm; petals < 2 cm, pale to deep pink;
filament column < 1 cm. **FR:** segments 3 mm, ± 3 mm wide, deeply
notched, dark brown. 2*n*=34. Chaparral, washes, hillsides; 450–1100
m. SNF, SnJV, s SnFrB, WTR, SnGb, PR; nw Baja CA. May–Jul

M. orbiculatus (Greene) Greene Pl < 2 m, branches stout; hairs
± stalked, ± yellow, shaggy-woolly to velvety. **LF:** blade < 7 cm,
cordate, thick, unlobed to obscure-3-lobed, hairs denser, veins raised
abaxially. **INFL:** spike-like; fls 3–6 per node; lf-like bracts 3–5, at
infl base ± = lvs; fl stalks < 0.6 cm, stout; bractlets 3/4 calyx. **FL:**
calyx < 1.2 cm, 1/3 as wide, lobes awl-shaped to lanceolate; petals
< 2 cm, filament column > 1/2 petals. **FR:** segments 2.2–3.2 mm,
ovate, ± notched, brown. Dry slopes above pine belt; 1300–2800 m.
SNF, s SNH, TR, DMtns. Sometimes incl in *M. fremontii.* May–Jul,
Oct–Nov

M. palmeri (S. Watson) Greene Pl < 2.5 m, branches stout; hairs
stalked, ± yellow, occ glandular and bristly, shaggy woolly to short-
hairy. **LF:** petioles < 6 cm; blade < 7 cm, < 8 cm wide, thin, 5-lobed;
hairs similar abaxially, adaxially. **INFL:** head-like; fls > 10; lf-like
bracts > 10; fl stalks ± 0; bractlets 1.5–2.5 cm, 1/2 as wide. **FL:** calyx
≤ 1.5 cm, lobes awl-shaped to lanceolate; petals < 4 cm, filament
column < 2 cm. **FR:** segments < 4 mm, ovate, deeply or narrowly
notched, brown. 2*n*=34.

var. **involucratus** (B.L. Rob.) Kearney CARMEL VALLEY
BUSH-MALLOW **LF:** ± glabrous adaxially. **INFL:** involucre bracts
20 mm wide, often > petals. **FL:** calyx ≤ 1.5 cm, 1/3 as wide; petals
1–3 cm. Valleys, chaparral; 30–800 m. CCo, SCoRO. May–Jul ★

var. **palmeri** (p. 889) SANTA LUCIA BUSH-MALLOW **LF:** hairy
adaxially. **INFL:** involucre bracts 12–13 mm wide, ± = petals. **FL:**
calyx ≤ 1.5 cm, 1/2 as wide; petals 2–3.5 cm. Interior valleys foothills; 30–800 m. CCo, SCoRO. May–Jul ★

MALVA MALLOW

Ann to shrub, gen taprooted; hairs stellate, simple, or 0. **ST**: prostrate to erect, gen not rooting, herbaceous to soft-woody. **LF**: stipules persistent; petioled; blade round to reniform, shallowly to deeply palmate-5–7(9)-lobed or lobes 0, gen crenate to dentate. **INFL**: raceme-like or gen fls 1–10 in lf axils; fl stalks often jointed above middle; bractlets 3, free or fused ± 1/2. **FL**: showy or not; calyx lobes ± = tube; petals ± 0.4–4.5 cm, gen shallowly notched at tip, pink, purple, rose-purple or white, dark-veined or not; anthers gen on upper 1/3–1/2 of filament tube; stigmas linear, on inner side of style branches. **FR**: ± disk-like; segments 6–15[20], indehiscent, gen edged; walls smooth or ribbed, puberulent or not; beak 0. **SEED**: firmly enclosed by, not readily separating from fr wall, reniform, gen glabrous. ± 30–40 spp.: Eur, esp Medit, Asia, Afr, few Australia, Am. (Greek: mallow, tender) Some spp. reportedly TOXIC to livestock from selenium or nitrate concentration. [Ray 1998 Novon 8:288–295] Incl 3 CA spp. formerly placed in *Lavatera* (Hill 2009).

1. Shrub, st gen 1–4 m; fls showy, petals 2.5–4.5 cm; calyx > 10 mm; segments 6–8 . . . . *M. assurgentiflora*
1′ Ann to per (subshrub), st 0.1–3 m; fls showy or not; petals 0.4–3 cm; calyx < 10 mm; segments 6–15
  2. Petals gen > 15 mm, showy, dark-veined; st erect; bractlets free; calyx 6–8 mm; pl sparsely hairy . . . . . . *M. sylvestris*
  2′ Petals gen < 15 mm, showy or not, dark-veined or not; st prostrate to erect; bractlets free or ± fused;
    calyx 3–6 mm; pl gen sparsely hairy to glabrous
   3. Bractlets linear to thread-like; calyx much larger in fr, lobes spreading, not enclosing fr; petals ± =
    calyx, ≤ 5 mm, ± pink to gen white; fr segment outer edges thin-winged, toothed. . . . . *M. parviflora*
   3′ Bractlets widely linear, lanceolate, ovate, to rounded; calyx not much larger in fr or if so, lobes gen
    enclosing fr; petals ± > or >> calyx, > 5 mm, white to gen ± pink or ± purple; fr segment outer edges not winged
    4. Bractlets widely linear to narrowly lanceolate, free from each other and from calyx; fr segments gen
     smooth or weakly ridged on back, outer edges ± rounded; petals pale lilac-pink to white, veins not dark
    . . . . *M. neglecta*
    4′ Bractlets widely lanceolate to ovate or ± round, bases fused to each other or to calyx; fr segments gen
     net-veined or cross-ridged on back, outer edges sharp (rounded); petals rose or pale lilac-pink to white,
     veins gen dark
     5. Bractlets gen fused to calyx, upper 1/2 free, widely lanceolate to ovate; st gen decumbent or
      ascending, < 1 m; fr segments strongly net-veined-pitted on back, outer edges sharp . . . . *M. nicaeensis*
     5′ Bractlets gen fused to each other, upper 2/3 free, ovate to round; st erect, gen > 1 m; fr segments
      faint-net-veined or cross-ridged on back, outer edges ± sharp (rounded)
      6. Involucel > calyx, lvs densely soft-stellate-hairy; petals rose to lavender with 5 dark veins; st base
       gen woody. . . . . *M. arborea*
      6′ Involucel < calyx; lvs sparsely stellate hairy; petals pale pink to white gen with 3 dark veins;
       st base not woody . . . . *M. pseudolavatera*

M. arborea (L.) Webb & Berthel. TREE MALLOW Bien to subshrub, stellate-tomentose. **ST**: 1–3 m; base gen woody. **LF**: blade 5–20 cm, cordate, unequally shallowly 5–7(9)-lobed, crenate, densely soft-stellate-hairy esp abaxially. **INFL**: fls clustered in lf axils; involucel > calyx, bractlet free parts ± 8 mm, 5–6 mm wide, ovate to round. **FL**: calyx ± 4 mm, not esp larger in fr, stellate-canescent; petals 1.5–2 cm, rose to lavender with 5 dark veins, base dark purple. **FR**: 8–10 mm diam; segments (6)8(9), ± glabrous to puberulent, ridged on sides, back, outer edges sharp. **SEED**: ± 3 mm. 2*n*=36,40,42,44. Uncommon. Disturbed places on coastal bluffs, dunes; < 200 m. NCo, n CCo, SCo; native to Eur, esp Medit. [*Lavatera a.* L.; *M. dendromorpha* M.F. Ray, nom. superfl.] Apr–May

M. assurgentiflora (Kellogg) M.F. Ray (p. 889) ISLAND MALLOW Per, shrub, stellate-hairy to glabrous. **ST**: decumbent to erect, gen 1–4 m. **LF**: blade 5–15 cm, 5–7-lobed, lobes triangular-ovate, ± acute or obtuse, toothed, hairy to glabrous. **INFL**: fls gen 1–2 in lf axils; fl stalks gen ± S-shaped; involucel < calyx, bractlet free parts lanceolate. **FL**: showy; calyx 12–15 mm, not esp larger in fr, gen stellate-puberulent; petals 2.5–4.5 cm, reflexed or not in age, ± rose, or ± purple, veins dark, pale at base; filament tube 1.5–2 cm, sparsely puberulent or stellate-hairy to glabrous, anthers on upper 1/2. **FR**: 12–16 mm diam; segments (6)8, glabrous to hairy, smooth on back, outer edges ± sharp. **SEED**: 1, ± 4 mm. 2*n*=±40. Coastal bluffs; < 350 m. ChI, naturalized CCo, SCo; naturalized Mex. [*Lavatera a.* Kellogg; *L. a.* subsp. *glabra* Philbrick] Cult as orn or windbreak. Feb–Jul ★

M. neglecta Wallr. (p. 889) COMMON MALLOW Ann, bien. **ST**: gen prostrate to ascending, 2–4(6) dm, gen hairy, hair branches 0–few. **LF**: stipules gen ± papery, 3–6 mm, 2.5 mm wide; petioles gradually shorter distally; blade ± 1–6 cm, gen shallowly 5–7 lobed (unlobed), dentate-crenate. **INFL**: fls gen 2–6 in lf axils; fl stalk in fr >> calyx; bractlets free, 3–5(6) mm, ± widely linear. **FL**: calyx gen

4–6 mm, not esp larger in fr, lobes acuminate; petals gen 2 × calyx, white to pale lilac; filament tube puberulent. **FR**: segments 12–15, smooth or weakly ridged, puberulent on back, outer edges ± rounded. 2*n*=42. Disturbed places; < 3000 m. n&c CA; native to Eurasia. Probably also in s CA. May–Oct

M. nicaeensis All. (p. 889) BULL MALLOW Ann, bien. **ST**: gen decumbent or ascending, 2–6 dm, sparsely stellate hairy. **LF**: stipules 4–6 mm, 3–5 mm wide; blade 3–12 cm wide, ± round to reniform, crenate, wavy, lobes 5–7, shallow, ± acute. **INFL**: fls 1–4 in lf axils; fl stalks ± = calyx; bractlets 4–5 mm, 1–2.5 mm wide, gen fused to calyx in lower 1/2, upper free, widely lanceolate to ovate. **FL**: calyx 4–6 mm, larger in fr, ± enclosing fr, veiny; petals 5–12 mm, ± pink to blue-violet, blue when dry, veins gen dark; filament tube hairy. **FR**: segments 7–10, glabrous to puberulent, strongly net-veined-pitted on back, outer edges sharp. 2*n*=42. Disturbed places; < 1200 m. CA-FP; native to Eur, Asia Minor, Medit; naturalized elsewhere, esp Mex. Mar–Jun

M. parviflora L. (p. 889) CHEESEWEED, LITTLE MALLOW Ann. **ST**: prostrate, decumbent, ascending, to gen erect, 2–8 dm, widely branched, ± stellate-hairy near tips, glabrous below. **LF**: stipules 4–5 mm, 2–3 mm wide; blade 2–8 cm wide, ± round to reniform, 5–7-angled to -lobed, teeth ± rounded. **INFL**: fls 2–4 in lf axils; fl stalks in fr gen < calyx; bractlets 1–2 mm, linear to thread-like. **FL**: calyx ± 3 mm, much larger in fr, lobes spreading, widely ovate, net-veined; petals 3–5 mm, ± pink to gen white. **FR**: 6–7 mm diam; segments ± 10–11, glabrous to hairy, wrinkled, net-veined-pitted, ribbed on back, outer edges thin-winged, toothed. 2*n*=42. Common. Disturbed places; < 2500 m. CA-FP, D; native to Eur, Medit, India; widespread weed. Mar–May

M. pseudolavatera Webb & Berthel. CRETAN MALLOW Ann, bien, subshrub, sparsely stellate hairy. **ST**: 10–30 dm; base woody or not. **LF**: blade 4–10 cm, shallowly 5-lobed above, crenate, base truncate to

± cordate. **INFL**: fls in lf axils; involucel < calyx, bractlet free parts ovate to round. **FL**: calyx ± 4 mm, greatly larger in fr; petals 1–1.6 cm, white to pale pink, gen with 3 dark veins. **FR**: segments 7–10, glabrous to puberulent, gen ± cross-ridged on back. 2*n*=44,112. Uncommon. Disturbed places on coastal bluffs, dunes, occ inland; < 50 m. CCo, SCo; native to s Eur. [*Lavatera cretica* L.; *M. linnaei* M.F. Ray, nom. superfl.] Apr–Jun

M. sylvestris L. (p. 889)   HIGH MALLOW   Ann to per. **ST**: erect, 5–30 dm; hairs sparse, simple and stellate. **LF**: stipules 3–5 mm, 3

mm wide; blade gen 5–10 cm wide, ± round to reniform, crenate, distal occ shallowly 3–7-lobed, lobes ± rounded. **INFL**: fls 1–4 in lf axils; fl stalks >> calyx; bractlets free, 3–7 mm, 2–4 mm wide, lanceolate to ovate. **FL**: calyx 6–8 mm, stellate-hairy, not larger in fr; petals gen 15–30 mm, gen bright purple or pink, veins dark; filament tube hairy. **FR**: segments 10–12, ± puberulent, shallowly net-veined, wrinkled on sides, back, outer edges sharp, not winged. 2*n*=42. Uncommon. Disturbed places; gen < 300 m. NW, CW, SCo; native to Eur, Medit, Asia Minor; widely cult as orn. Apr–Jun

MALVELLA

(Ann) per; hairs stellate or scale-like. **ST**: prostrate to decumbent. **LF**: blade gen asymmetric, gen silvery-stellate-hairy. **INFL**: fls 1 in lf axils; fl stalk ± jointed at tip, gen recurved in fr. **FL**: calyx lobes ± = tube, ovate or cordate; petals stellate-hairy in bud, cream-white to yellow gen fading to rose-pink; filament tube glabrous, anthers at tip; styles 7–10, stigmas head-like. **FR**: segments gen 7–10, indehiscent, beak 0. **SEED**: 1 per segment, glabrous. 4 spp.: Am, Medit, introduced elsewhere. (Greek, Latin: small mallow)

M. leprosa (Ortega) Krapov. (p. 895)   ALKALI-MALLOW, WHITE-WEED   Ann, per. **ST**: decumbent, 1–4 dm, densely white-stellate-hairy; some hairs bristly, some scale-like. **LF**: blade 1–3.5 cm, reniform to triangular, densely white-stellate-hairy, base asymmetric, margin toothed, wavy. **INFL**: fl stalk ± = subtending petiole; bractlets (0)3, thread-like, gen deciduous. **FL**: calyx 6–10 mm, divided 1/2, gen pink or pink-dotted abaxially, hairs as on st; petals

10–15 mm, cream-white to yellow, occ with rose tint; styles 7–10, > filament tube. **FR**: ± 7 mm diam; segments 7–10, 3 mm, minutely puberulent on back, gen net-veined on sides. 2*n*=22,32. Valleys, gen saline; < 1500(2500) m. CA (esp GV); to WA, ID, TX, Mex, s S.Am; Australia (introduced). Reported to be TOXIC to sheep, perhaps other livestock. Agricultural weed, esp in orchards. Apr–Nov

MODIOLA

1 sp.: Am; introduced to Old World. (Latin: wheel hub, for fr)

M. caroliniana (L.) G. Don (p. 895)   MODIOLA   (Ann) per, sparsely stellate-hairy exc fr, calyx. **ST**: prostrate to decumbent or creeping, 1.5–5 dm, rooting near base, nodes. **LF**: stipules 3–5 mm, lanceolate, green, persistent; blade (2)5–8 cm, 2–5 cm wide, gen ≤ petiole, reniform, round, or triangular, esp upper deeply toothed to (3)5–7-lobed, lobes ± lobed or toothed. **INFL**: fls 1 in lf axils; fl stalks 5–25 mm; bractlets 3, free, gen narrowly lanceolate. **FL**: calyx 5–6 mm, ± larger in fr, divided ± 1/2, gen sparsely bristly, hairs 1–2 mm; petals 3–8 mm, gen > calyx, red-orange, base gen dark; anthers

at stamen-tube tip; styles 15–25, stigmas head-like. **FR**: segments 15–25, 5–6 mm, reniform, black, bristly on top, chambers 2, lower wrinkly-sided, indehiscent, upper smooth-sided, dehiscent, beaks 2, 1.5–3 mm. **SEED**: 2, ± 1.5 mm, glabrous to minutely puberulent. 2*n*=18. Grassland, disturbed areas; gen < 400 m. CA-FP; to se US; probably native to s S.Am; widely naturalized. Agricultural, lawn weed; tolerant of salt, drought, may accumulate nitrates, then TOXIC to some livestock. Mar–Nov

SIDALCEA   CHECKERBLOOM

Ann, per; with taproot, clustered fleshy roots, caudex, adventitious roots, or occ shallow rhizome. **ST**: ± decumbent or gen erect, some occ stolon-like; erect st, branches terminating in infl. **LF**: gen fewer above, occ ± rosetted; petioles below gen >> petioles above; blades below gen crenate to shallowly lobed, blades above often deeply palmate-lobed or -divided; stipules gen persistent. **INFL**: head, spike, or raceme, in panicle or not, gen more open in fr; bracts 2, gen stipule-like, occ involucre-like, united at base to ± entirely; bractlets 0(3), gen not in involucel. **FL**: fls gen bisexual, protandrous, occ functionally unisexual (occ, pls with either bisexual or pistillate fls in a given sp.); calyx lobes ≥ tube; petals spreading or erect, purple or rose-pink to white, gen with some pale veins, base gen also paler than tips (occ darker), tip ± notched or fringed, petals on pistillate fls shorter, darker, often ≤ 10 mm; filament tube gen stellate-puberulent, anthers near top, in gen 2 concentric series, gen pink, ± purple, or white; stigmas linear, on inner side of style branches, conspicuous in pistillate fls. **FR**: segments gen 5–10, indehiscent, puberulent, glandular, or glabrous, beaked or not, side walls gen ± thin. **SEED**: 1, gen filling chamber, reniform, glabrous. ± 27 spp.: w N.Am: AK, Can, to Mex. (Greek: combination of *Sida*, *Alcea*, 2 other names for mallows) [Andreasen & Baldwin 2003 Amer J Bot 90:436–444; Hill 2008 J Bot Res Inst Texas 2:783–791] Some spp. highly variable, esp in lvs, growth stage; mature pls with fr minimize considerable problems in identification, as does knowledge of pl base, underground parts; needs study.

1. Ann (weak per); gen fl in spring (occ later)
   2. Lobes of mid-, upper-st lvs gen dentate (entire), gen wider than linear; st bristly; fr segments gen glabrous, occ ± pink when fresh; beak 0
      3. Upper paired stipules, bracts divided to base into ≥ 2 linear lobes ≥ calyx; middle lobe of upper st lvs entire or with middle tooth >> lateral 2; infl, calyx gen not glandular; infl in fr congested, axis not visible ..... *S. diploscypha*
      3′ Upper paired stipules, bracts simple, linear, undivided (few divided in robust pls), gen < calyx; middle lobe of upper st lvs with middle tooth ± = lateral 2; infl, calyx gen densely minutely glandular; infl in fr not congested, axis visible .... *S. keckii*
   2′ Lobes of mid-, upper-st lvs ± entire, gen linear; st gen ± glabrous at least below; fr segment glabrous or sparsely puberulent, ± pink or not when fresh; beak present
      4. Outer filaments not fused to tip; fr segment evenly net-veined on back, deeply pitted esp on top, glabrous to puberulent; calyx gen 8–10 mm .... *S. hartwegii*

4′ Outer filaments fused ± to tip; fr segment longitudinally grooved or net-veined on back, gen puberulent (glabrous in *Sidalcea calycosa*); calyx 4–12 mm

    5. Pl gen densely bristly above or st glabrous in age; bracts linear; fr segments ± net-veined-pitted, gen puberulent on back; calyx or fl stalk ± densely stellate-puberulent . . . . **S. hirsuta**

    5′ Pl glabrous or sparsely hairy; bracts wider than linear; fr segments longitudinally grooved, glabrous on back; calyx or fl stalk sparsely to densely silky-bristly . . . . **S. calycosa**

        6. Pl ann, taprooted; bracts 2–6 mm, not silky-hairy, not obscuring calyx; calyx 4–7 mm; petals (9)10–20 mm; NCoRO, CaRF, SNF, n SnFrB . . . . subsp. **calycosa**

        6′ Pl weak per, rhizomed; bracts 8–12 mm, silky-hairy, ± obscuring calyx; calyx 6–12 mm; petals 20–25 mm; c&s NCo, n CCo . . . . subsp. **rhizomata**

1′ Per; gen fl in late spring, summer

  7. Fr segment beak 0; bractlets (0)1–3; lvs gen all along st early to late in season, lobed (unlobed)

    8. Bractlets gen 0(1–2); petals gen white; lvs gen evenly shallowly lobed; pls often > 8 dm . . . . **S. malachroides**

    8′ Bractlets 3; petals pale pink to rose-purple; lvs lobed or not; pls gen < 8 dm

      9. Infl ± head; bracts subtending lower fls involucre-like; stipules widest at base, much wider than st; lf blades ovate, unlobed; n SNF (Nevada Co.) . . . . **S. stipularis**

      9′ Infl spike or raceme; bracts subtending lower fls gen stipule-like; stipules widest above base, occ wider than st; lf blades ± round or fan-like, lobed or not; NCoR, n CCo, SCoRO, WTR . . . . **S. hickmanii**

        10. Lower lvs gen lobed ± to base

          11. Bracts (8)10–12 mm, ± = calyx; SCoRO (c San Luis Obispo Co.). . . . . subsp. **anomala**

          11′ Bracts 5.5–7 mm, < calyx; NCoRI (Napa Co.) . . . . subsp. **napensis**

        10′ Lower lvs unlobed, or deeply crenate to lobed < 1/4 to base

          12. Bracts lanceolate, gen 6–10 mm, 2.5–4 mm wide, ± < calyx; bractlets ± ≤ calyx; upper lvs unlobed or lobed < 1/4 to base; SCoRO, WTR (Santa Barbara Co.), SnBr. . . . . subsp. **parishii**

          12′ Bracts linear, lance-linear, or oblong, 2–7 mm, 0.5–2 mm wide, < calyx; bractlets < calyx; upper lvs unlobed

            13. Pl gen 4–8 dm, ± gray-canescent; calyx densely stellate-puberulent, marginal hairs longest; bractlets 2–7 mm; largest lvs deeply cordate, closely crenate, often > 30 mm wide; infl ± dense — SCoRO. . . . . subsp. **hickmanii**

            13′ Pl gen < 4 dm, ± green, less hairy; calyx moderate- to densely stellate-puberulent, hairs ± equal in length; bractlets 2–5 mm; largest lvs truncate to widely wedge-shaped, coarsely crenate, gen < 25 mm wide, wider in robust pls; infl gen ± open

              14. Bracts of upper fls 1, cupped; lf blades 0.6–1.2 cm, 0.7–2.1 cm wide; upper st hairs appressed, gen 0.2–0.5 mm; st gen < 30 cm; fls gen < 10, infl not a spike; calyx 4–5.5 mm; NCoRH (n Lake Co.) . . . . subsp. **pillsburiensis**

              14′ Bracts of upper fls gen 2, flat or cupped; lf blades 1–2.5 cm, 2–4 cm wide; upper st hairs not appressed, gen 0.5–1.2 mm; sts gen > 30 cm; fls gen > 10, infl a ± spike in age; calyx 4–7 mm; s NCo, n CCo. . . . . subsp. **viridis**

  7′ Fr segment beak 1, occ minute; bractlets 0; lvs only basal or on lower st at least early in season (or not), occ rosette-like, basal gen not lobed, lvs above gen deeply lobed (shallowly lobed or unlobed)

    15. Pl gen (1)1.5–2(2.5) m; rhizomes ≤ (0.5)1 cm diam, bristly; st base hollow, bristly. . . . . **S. gigantea**

    15′ Pl gen < or << 1(1.8) m; rhizomes 0 or gen < to << 1 cm diam; st base solid (hollow), occ bristly

      16. Infl dense, not 1-sided; calyces densely overlapped in fl, occ in fr

        17. Fr segments rough at least on outer edges and/or back, net-veined-pitted; infl in fr ± open and/or interrupted; rhizomes or rooting sts occ present (see also *Sidalcea malviflora* subspp. at 37′)

          18. St base decumbent to ascending; larger petals on bisexual fls 1–2.5 cm, gen white-veined; fr gen not overlapped; NCo, CCo

            19. Sts often freely rooting near base; lvs at mid-st lobed, bristles coarse; n NCo . . . . . . **S. malviflora** subsp. **patula**

            19′ Sts gen not rooting near base; lvs at mid-st unlobed, bristles fine; c&s NCo, n CCo . . . . [3]**S. malviflora** subsp. **rostrata**

          18′ St erect (decumbent); larger petals in bisexual fls 0.5–1.5(2.5) cm, white-veined or not; fr gen overlapped; NW, CaR, SNF, CCo, MP

            20. Petals white-veined, 1–2.5 cm; st gen 3–5 dm; lvs unlobed; pl ± densely finely bristly; coastal, c&s NCo, n CCo . . . . [3]**S. malviflora** subsp. **rostrata**

            20′ Petals gen not white-veined, 0.5–1.5 cm; st gen > 4 dm; lvs lobed or not; pl not densely finely bristly; inland, NW, CaR, SNF, MP

              21. Infl axis below gen visible between fl clusters; pl with ± elongate caudex; sts gen not clustered; s SNF. . . . . **S. ranunculacea**

              21′ Infl axis below ± visible or not between equally spaced individual fls; pl with short caudex; sts clustered; NW, CaR, MP

                22. Calyx ± 6 mm, gen stellate-puberulent and occ with few bristles; infl gen 10–30 cm . . . . [4]**S. oregana** subsp. **oregana**

                22′ Calyx 5–8 mm, stellate-puberulent and with bristles; infl 3–10 cm . . . . **S. setosa**

        17′ Fr segments smooth, gen not net-veined or pitted; infl in fr gen ± dense, not interrupted; rhizomes or rooting sts present or 0

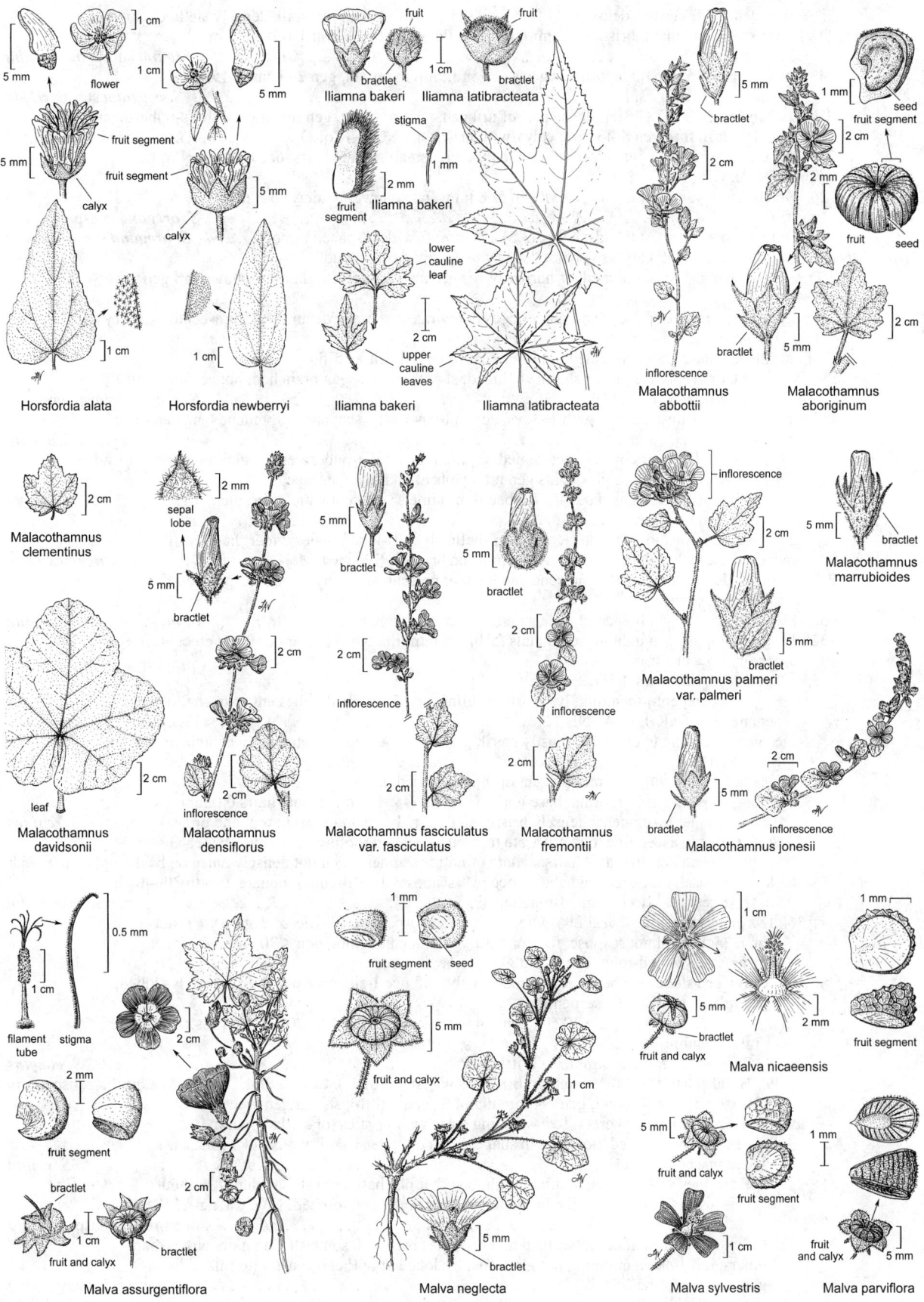

Horsfordia alata

flower

fruit segment

fruit segment

calyx

calyx

Horsfordia newberryi

fruit

bractlet

Iliamna bakeri

fruit

bractlet

Iliamna latibracteata

stigma

fruit segment

Iliamna bakeri

lower cauline leaf

upper cauline leaves

Iliamna bakeri

Iliamna latibracteata

bractlet

inflorescence

Malacothamnus abbottii

seed

fruit segment

fruit

seed

bractlet

Malacothamnus aboriginum

Malacothamnus clementinus

sepal lobe

bractlet

inflorescence

Malacothamnus densiflorus

leaf

Malacothamnus davidsonii

bractlet

inflorescence

Malacothamnus fasciculatus var. fasciculatus

bractlet

inflorescence

Malacothamnus fremontii

inflorescence

bractlet

Malacothamnus palmeri var. palmeri

bractlet

Malacothamnus marrubioides

bractlet

inflorescence

Malacothamnus jonesii

filament tube

stigma

fruit segment

bractlet

fruit and calyx

bractlet

Malva assurgentiflora

fruit segment

seed

fruit and calyx

bractlet

Malva neglecta

bractlet

fruit and calyx

Malva nicaeensis

fruit segment

fruit and calyx

fruit segment

fruit and calyx

Malva sylvestris

Malva parviflora

23. 1° peduncle >> fr cluster; dense part of infl gen < 5 cm; calyx, gen infl stalk densely stellate-canescent
24. Lower st < 5 mm diam, bristles ± 2 mm; infl few-fld, gen 1.5–2.5 cm; 1100–2300 m
. . . . ***S. oregana*** subsp. ***hydrophila***
24′ Lower st gen > 5 mm diam, bristles gen 1–1.5 mm; infl many-fld, gen 2–5 cm; ± 150 m
. . . . ***S. oregana*** subsp. ***valida***
23′ 1° peduncle gen ± ≤ fr cluster; dense part of infl gen > 3 cm; calyx, gen infl stalk stellate-puberulent
25. Calyx larger in fr to gen 8–13 mm; calyx with bristles; n NCo, NCoRO . . . . ***S. oregana*** subsp. ***eximia***
25′ Calyx ± larger in fr to gen 3.5–7(10) mm; calyx with small stellate hairs, occ bristles; NW, CaR,
n&c SNH, GB
26. Bracts gen < youngest fl buds; calyx gen 5–8 mm; fr, esp lower, widely spaced
. . . . ⁴***S. oregana*** subsp. ***oregana***
26′ Bracts gen > youngest fl buds; calyx (3.5)5(7) mm; fr widely spaced or not . . . . ***S. oregana*** subsp. ***spicata***
16′ Infl ± open, occ 1-sided; calyces gen not densely overlapped exc in bud
27. Fr segment gen glabrous, outer edges narrowly winged above; st erect, glabrous above; infl gen
30–40 cm . . . . ***S. robusta***
27′ Fr segment glabrous or puberulent, outer edges not winged; st erect, decumbent, or ascending, hairy
to glabrous above; infl gen < to << 40 cm
28. Fr segment smooth, not net-veined or pitted, glabrous; infl gen > 15-fld
29. Basal lvs 0 or deeply crenate or shallowly lobed; st with > 3 lvs, gen branched, not scapose; infl ±
elongated in fr; NW, CaR, MP . . . . ⁴***S. oregana*** subsp. ***oregana***
29′ Basal lvs deeply dissected; st with 1–3 lvs, gen unbranched, ± scapose; infl much elongated in fr;
SnBr . . . . ***S. pedata***
28′ Fr segment gen finely or coarsely net-veined or pitted, glandular-puberulent or glabrous; infl < 15-fld or not
30. Roots fleshy; rhizome 0, caudex 0; sts gen many; pls of alkaline flats, seeps
31. Lvs gen basal, glaucous, gen deeply 7-lobed, lobes linear; calyx stellate-puberulent;
SNE (Inyo Co.) . . . . ***S. covillei***
31′ Lvs basal, cauline, not glaucous, crenate to shallowly 5(7)-lobed, lobes wider than linear;
calyx stellate-puberulent and with swollen-based bristles; SW, sw DMoj . . . . ***S. neomexicana***
30′ Roots not fleshy; rhizome 0 or present, caudex 0 or present; sts many or not; pls gen not of
alkaline flats, seeps
32. Fls gen > 15, infl gen branched; petals gen 5–15 mm; rhizome 0; sts erect . . . . ⁴***S. oregana*** subsp. ***oregana***
32′ Fls gen < 15, infl gen unbranched; petals (5)10–35 mm; rhizomes present or 0; sts erect or
ascending to ± prostrate
33. Pl glaucous; rhizomes 0; infl ± 1-sided
34. St gen decumbent-ascending; lvs mostly cauline, gen 5(7)-lobed, lobes entire to shallowly
dentate; KR, CaRH, n&c SNH, ... ***S. glaucescens***
34′ St gen ascending to erect; lvs mostly basal, gen 7(9)-lobed, lobes gen deeply dentate or ternate;
SNH, SNE . . . . ***S. multifida***
33′ Pl ± glaucous or not; rhizomes present or 0; infl 1-sided or not
35. St long-prostrate, free-rooting, base with bristly hairs 2–3 mm, stellate hairs 0 (glabrous);
lf hairs simple; fr segments densely bristly-stellate-puberulent on back, top, and/or beak. . . . . ***S. reptans***
35′ St gen erect to ascending, occ prostrate to decumbent and rooting, hairs on st base < 3 mm,
simple, stellate, or mixed; lf hairs stellate or not; fr segments gen not densely hairy on back, top, and/or beak
36. Lvs gen basal, much reduced above, occ pl ± scapose; fls > 10; infl elongate, gen to 30–40 cm
in fr; fr gen weakly net-veined-roughened . . . . ***S. sparsifolia***
36′ Lvs gen cauline (gen basal in *Sidalcea asprella* subsp. *nana*, *Sidalcea elegans*), ± reduced
above (or not), pl not scapose; fls < or > 10; infl elongate or not, gen < 30 cm in fr; fr gen
strongly net-veined-roughened, -pitted, or honeycombed
37. Calyx gen stellate-puberulent, gen not bristly; infl occ 1-sided; pl occ ± glaucous; lvs lobed;
gen inland, mtns; often serpentine
38. Rhizomes gen present, > 10 cm, gen < 4 mm diam; fls gen < 10(20); st occ prostrate at base,
infl ascending
39. Basal st hairs ± 0 or simple; st brittle; KR . . . . ***S. elegans***
39′ Basal st hairs stellate; st not esp brittle; NW (esp NCoRH), CaR, n SNH . . . . . . ***S. asprella*** subsp. ***nana***
38′ Rhizomes 0 or < 10 cm, gen > 4 mm diam; fls gen > (8)10; st, infl gen erect
40. Rhizomes 0; st gen erect, free-standing, base with bristles (or stellate hairs); lvs
gen crenate to lobed, not alike in shape, lobes of upper ± entire; n NCoRI (Shasta,
Tehama cos.), n ScV . . . . ***S. celata***
40′ Rhizomes < 10(30) cm; st often weak, on other pls, base with stellate hairs, without
bristles; lvs ± lobed, ± alike in shape, lobes of upper gen toothed; NW, CaR, n&c SNH
. . . . ***S. asprella*** subsp. ***asprella***
37′ Calyx gen both stellate-puberulent and bristly, if bristles 0 some stellate hairs larger than
others; infl 1-sided or not; pl not glaucous; lvs lobed or not; gen coastal (to inland in SW),
gen not serpentine . . . . ***S. malviflora*** subspp. (in part)
41. Lobes of mid-st lvs deep, ternate with linear segments; st base gen ± rooting. . . . . subsp. ***laciniata***

41′ Lobes of mid-st lvs 0 or not ternate with linear segments; st base rooting or not

42. St base hairs gen ± 2 mm, not stellate; fr segment weakly net-veined; rhizome gen > 10 cm, slender, free-rooting; st lvs lobed; calyx sparsely stellate-puberulent, bristly; infl gen open; SnBr . . . . subsp. *dolosa*

42′ St base hairs < 2 mm, stellate and not; fr segment strongly net-veined-pitted; rhizome < 10 cm, not slender, free-rooting or not; st lvs lobed or not; calyx gen densely stellate-puberulent or not, bristly or not; infl open or dense; not SnBr

   43. Basal lf blade gen < 2(2.5) cm; upper lvs gen unlobed; st base, stipules, calyx gen ± purple; fl stalk gen hair-like; coastal, c&s NCo, n CCo. . . . . subsp. *purpurea*

   43′ Basal lf blade gen > 2 cm; upper lvs lobed or unlobed; st base, stipules, calyx gen not ± purple; fl stalk stouter than hair-like; coastal or not, incl c&s NCo, n CCo, or not

     44. Pl stellate-canescent; upper lvs gen deeply 7-lobed; calyx densely stellate-puberulent; fls gen not dense; petals of bisexual fls ± 25–35 mm; s SCoRO, WTR . . . . subsp. *californica*

     44′ Pl gen densely coarsely stellate-hairy or densely finely bristly; upper lvs unlobed to deeply 7–9-lobed; calyx gen sparsely bristly; fls occ dense, less so in age; petals of bisexual fls often < 20 mm; c&s NCo, NCoRO, CCo, SnFrB, SCo, n ChI

       45. Mid, upper lvs 7–9-lobed; pl not bristly; s NCo, NCoRO, CCo, SnFrB, SCo, n ChI . . . . subsp. *malviflora*

       45′ Mid, upper lvs unlobed; pl finely bristly; c&s NCo (Sonoma, s Mendocino cos.), n CCo (Marin, San Mateo cos.) . . . . [3]subsp. *rostrata*

S. asprella Greene   Per, gen ± stellate-hairy, occ ± glaucous above. **ST:** erect or on other pls, base prostrate, decumbent-ascending to erect, hairs stellate, gen 4-rayed. **LF:** gen lobed, abaxially stellate-puberulent; lobes narrowest at base, tips gen toothed, occ entire; stipules 2–3 mm, ± 1.1 mm wide. **INFL:** open, elongate, fls gen 2–15(30); bracts lf-like to linear, stellate-puberulent; fl stalks 2–5(10) mm, stout. **FL:** calyx 7–10 mm, densely stellate-puberulent, lobes 1–7-veined; petals 1–2.5 cm, pink, pale-veined, shorter, darker in pistillate. **FR:** segments (6)7–8, 3–4 mm, glandular-puberulent to ± stellate-puberulent (glabrous), net-veined, sides, back pitted, beak 0.5–1 mm.

   subsp. **asprella**   HARSH CHECKERBLOOM, SIERRA FOOTHILLS CHECKERBLOOM   Pl 3–10(12) dm, caudex present or 0, rhizomes < 10(30) cm. **ST:** often weak, on other pls, base stellate-hairy. **LF:** basal, cauline, all ± alike, gradually reduced upwards, ± lobed, lobe tips gen toothed; adaxially with gen simple hairs to stellate-puberulent. **INFL:** erect, open, elongate, occ 1-sided, gen 8–15(30)-fld; lowest bract gen lf-like, ± 15 mm, 12 mm wide, bracts above linear, ± 3 mm, gen forked, < fl stalk; fl stalks 2–5(8) mm, stout. **FL:** calyx 7–10 mm, densely stellate-puberulent, lobes gen 6 mm, 3.5 mm wide, veins 5–7, 3 prominent; petals of bisexual fls 2–2.5 cm, of pistillate gen 1–2 cm. **FR:** segments (6)7–8, beak ± 0.5–0.8 mm. 2*n*=20,40,60. Open woodland, dry rocky slopes, foothill woodland, lower conifer forest; (100)200–900(1800) m. NW, CaR, n&c SNH; sw OR. [*S. malviflora* subsp. *a.* (Greene) C.L. Hitchc.] May–Jun

   subsp. **nana** (Jeps.) S.R. Hill   DWARF SIERRAN CHECKERBLOOM   Pl 1–5(8) dm, caudex erect or 0, rhizomes > 10 cm, ± 2(3) mm diam, free-rooting. **ST:** base prostrate, decumbent-ascending, to erect, densely stellate-puberulent, ± sparser above. **LF:** gen basal, 20–23 mm, 22–27 mm wide, cordate, gen 7-lobed; basal (deeply crenate to) lobed 1/2 to base; cauline gen 1–3, deeply lobed, lobes gen 3-toothed at tip; adaxially with hairs gen 2–4-rayed (simple), bristly. **INFL:** ± scapose, ascending, 6–11(30) cm, gen 1-sided, open, stellate-puberulent, axis ± < 1 mm diam; bracts linear, 2–2.5 mm; fl stalks gen 3–4 mm, stout; bisexual 2–9(19)-fld; pistillate ± 9–14-fld. **FL:** calyx ± 7 mm, densely stellate-puberulent, lobes 5 mm, 2.5 mm wide, 1–3-veined; petals gen 1.7–2(2.5) cm in bisexual, gen 0.9–1.1 cm, darker in pistillate. **FR:** segments ± 8, beak ± 1 mm. Open woodland, grassy margins, yellow-pine/Douglas-fir forest, gen serpentine; 460–1725 m. NW (esp NCoRH), CaR, n SNH; sw OR. [*S. malviflora* (DC.) A. Gray subsp. *n.* (Jeps.) C.L. Hitchc.] Jun–Aug

S. calycosa M.E. Jones   VERNAL POOL CHECKERBLOOM   Ann or ± weak per from rhizome. **ST:** 3–5(9) dm, gen ± rooting near base, glabrous or sparsely hairy. **LF:** basal blades gen crenate, cauline with 5–11 deep, linear to oblanceolate lobes **INFL:** dense; bracts entire or gen 2-lobed, gen ciliate, at least some > 2 mm wide, ovate to widely elliptic. **FL:** bisexual or pistillate; calyx 4–12 mm, often ± purple or scarious, ± bristly, stellate-puberulent, lobes narrowly ovate, ±

abruptly pointed; petals (9)10–25 mm, pale purple (white). **FR:** segments ± 9, 2.5–4.5 mm, glabrous, back deeply longitudinal-grooved, beak ± 0.5 mm, often appressed.

   subsp. **calycosa** (p. 895)   VERNAL POOL CHECKERBLOOM, HOGWALLOW CHECKERBLOOM   Ann 3–5(9) dm, taprooted, branched or not, gen glabrous, occ fleshy. **LF:** basal gen unlobed, persistent; blade gen 2–5 cm wide; stipule 2–5 mm. **INFL:** bracts 2–6 mm, ± glabrous to ciliate. **FL:** calyx 4–7 mm; petals (9)10–20 mm, shorter in pistillate pls. **FR:** segment ± 2.5 mm, sides net-veined. Wet places, esp vernal pools, hog wallows, swales, foothill woodland, chaparral openings; < 1200 m. NCoRO, CaRF, SNF, ScV, n SnFrB. Extremely variable in size, dwarfed or ± lfless pls often found. Mar–Jun

   subsp. **rhizomata** (Jeps.) Munz (p. 895)   POINT REYES CHECKERBLOOM   Weak per, rhizomed. **LF:** sparse on rhizomes, dense on fl sts; blade 2.5–10 cm wide; stipule 12–18 mm. **INFL:** bracts 8–12 mm, ± hiding calyx, silky-hairy. **FL:** calyx 6–12 mm; petals 20–25 mm. **FR:** segment ± 4.5 mm, sides weakly net-veined. Marshes; < 30 m. c&s NCo (Mendocino, Sonoma cos.), n CCo (Marin Co.). May–Jul ★

S. celata (Jeps.) S.R. Hill   REDDING CHECKERBLOOM   Pl 4–8(10) dm, caudex present or 0, rhizomes 0. **ST:** gen erect, occ in age glaucous, base with bristles (or stellate hairs). **LF:** petiole 15–18 cm; basal gen crenate to shallowly 7-lobed, upper few, smaller, 5-lobed, lobes ± 3.5 cm, 4 mm wide, ± entire; hairs ± 1.5 mm adaxially, stellate, 6-rayed, 1.5 mm abaxially; stipules ± 3 mm, gen deciduous. **INFL:** open, gen branched, elongate, ± 1-sided, (5)10–12(23)-fld; axis sparsely stellate-puberulent; bracts ± 3 mm, < fl stalks, calyx, puberulent, deciduous. **FL:** calyx (7)9–12 mm, ± uniformly densely stellate-puberulent, lobes 6–8(9) mm, 3–6 mm wide, abruptly acute (acuminate); petals pale pink-lavender, in bisexual fls 2–2.5(3.1) cm, in pistillate gen 1–2 cm; filament tube gen 8–10 mm. **FR:** segments ± 7, 3–4 mm, minutely glandular-puberulent esp in youth, net-veined, sides, back deeply pitted, honeycombed, outer edges occ sharp, beak ± 1 mm. 2*n*=60. Open oak woodland, serpentine or not; 150–370 m. NCoRI (Shasta, Tehama cos.), CaR, n ScV. [*S. malviflora* subsp. *c.* (Jeps.) C.L. Hitchc.] May–Jun

S. covillei Greene (p. 895)   OWENS VALLEY CHECKERBLOOM   Per; roots ± fleshy; caudex 0, rhizome 0. **ST:** many, 2–6 dm, base stellate-hairy to coarsely bristly, hairs smaller above. **LF:** clustered at st base; gen deeply 7-lobed, lobes linear; blade fleshy, glaucous, densely stellate-hairy. **INFL:** open, branched or not, gen > 20-fld, stellate-puberulent; fl stalks 2–8(10) mm. **FL:** calyx 5–8 mm, stellate-puberulent; petals 10–15 mm, pink-lavender. **FR:** segment ± 2.5 mm, sparsely glandular-puberulent, back net-veined, sides strongly so, beak short. 2*n*=20. Alkaline flats; 1100–1300 m. SNE (Owens Valley, Inyo Co.). Threatened by lowering of water table, grazing. May–Jun ★

S. diploscypha (Torr. & A. Gray) A. Gray (p. 895) FRINGED CHECKERBLOOM Ann 4–6 dm. **ST:** stellate-puberulent, long-fine-bristly. **LF:** bristly-puberulent; stipules of mid- to upper-st divided into ≥ 2 linear lobes, gen > 10 mm; lobes of upper lvs narrow, middle lobe entire or with middle tooth >> lateral 2. **INFL:** fl clusters crowded; axis gen not glandular, multicellular hairs gen few or 0; bracts 8–12 mm, divided into 2–4 linear lobes, finely bristly, gen involucre-like in age, ≥ calyx. **FL:** calyx 8–12 mm, lobes lanceolate, long-acuminate, adaxially often with narrow purple line or spot at base, abaxially bristly and stellate-puberulent, gen not glandular, multicellular hairs gen few or 0; petals 20–35 mm, ± 19 mm wide, dark pink, dark spot occ at base, tip minutely fringed; filaments fused to tip. **FR:** segment ± 2.5 mm, glabrous, occ ± pink when fresh, back net-veined, beak 0. Grassland, open woodland, valleys, occ near vernal pools, gen serpentine; < 840 m. NCoRO, CaRF, n&c SNF, ScV, CCo, SnFrB; w OR. Some pls ± intermediate to *S. keckii*, but spp. distinct by molecular evidence (Hill 2009); study needed. Apr–May

S. elegans Greene DEL NORTE CHECKERBLOOM Per 2–6(8) dm; taproot woody, rhizomes gen 2–3 dm, freely rooting. **ST:** < 4 mm diam, brittle, base decumbent to erect, hairs ± 0 to fine, simple, above glabrous-glaucous. **LF:** mostly basal, reduced above, sparsely stellate-hairy (hairs simple), (1)2–10 cm wide, (3)5–7-lobed, basal shallowly lobed, lvs above gen deeply 3–5-lobed ± to base. **INFL:** open, 1-sided, (3)5–10(20)-fld; fl stalks 3–10 mm. **FL:** calyx 7–10 mm, lobes ± 4.5 mm, 2.5 mm wide, acuminate, stellate-puberulent (some rays occ longer than others); bisexual fl petals gen 20–25(33) mm, pistillate fl petals gen 10–15(20) mm. **FR:** segment glandular- and ± stellate-puberulent, beak short. $2n$=40,60. Open dry woodland, gen serpentine; 100–200(900) m. KR; sw OR. Jun–Jul

S. gigantea G.L. Clifton et al. GIANT CHECKERBLOOM Per matted in patches; rhizomes to 40–60 cm, (0.5)1 cm diam; bristles 2.5 mm, reflexed, in age 0. **ST:** erect, (1)1.5–2(2.5) m, diam 10–14 mm, occ glaucous, gen ± purple, hollow below, densely coarsely bristly, hairs 1.5–2 mm, reflexed, stellate hairs few above. **LF:** basal gen 0, cauline gradually reduced upward, stellate-puberulent; petioles below gen 10–14 cm, tip curved with swollen part 5–6 mm; blades below (6.5)11–12 cm, 10–13 cm wide, lowest rounded, 4–5-lobed, upper deeply 5-lobed; stipules deciduous (3.5)5 mm, ± 0.7 mm wide. **INFL:** spike-like raceme, in panicle or not, 14–18 cm, glaucous, sparsely stellate-hairy, fls 10–20, spaced; bracts ± ≥ fl stalks; fl stalks 2(5) mm, stout, stellate-canescent. **FL:** pistillate or bisexual, on separate pls; calyx 5–6 mm, to 8 mm in fr, densely stellate-puberulent, faint-veined; petals pale pink, pistillate 7–9 mm, bisexual 10–17(25) mm; filament tube 5.5–8 mm. **FR:** segments (6)7–8, ± 3 mm, back, outer edges net-veined-honeycombed, back grooved, top sparsely glandular-stellate-puberulent, beak ± 1 mm, tip puberulent. **SEED:** ± 1.5 mm, dark brown, smooth, glabrous. Moist to wet forested slopes, seeps, stream margins, meadows, mid to upper conifer forest; (640)900–1650 m. CaRH, n SNH. Jul–Sep

S. glaucescens Greene (p. 895) WAXY CHECKERBLOOM Per 2–5(7) dm, glaucous; taproot thick, woody, caudex branched, rhizome 0. **ST:** many, gen decumbent-ascending, gen not rooting, below stellate-puberulent (glabrous), above ± glabrous. **LF:** blade 1.5–4(7) cm wide, glabrous or stellate-puberulent, 1° lobes 5(7); entire; uppermost occ linear, unlobed. **INFL:** open, gen unbranched, ± 1-sided, axis ± curved between fls, fls few–15; fl stalks 3–10 mm. **FL:** calyx 6–10 mm, stellate-puberulent, gen larger in fr; petals 10–20 mm, pink to pink-purple. **FR:** segments 6–8, 3–3.5 mm, ± inflated, deeply pitted, sides, back net-veined, back glandular-puberulent, beak very short. $2n$=40. Dry grassy meadows, open gen red-fir forest, often serpentine; (900)1500–3000 m. KR, CaRH, n&c SNH; NV. Intergrades with *S. multifida*. Jun–Aug

S. hartwegii A. Gray (p. 895) HARTWEG'S CHECKERBLOOM Ann. **ST:** 1.5–4 dm, base ± glabrous to sparsely stellate-hairy. **LF:** lower deciduous; lobes gen 5–7, linear; stipules inconspicuous to deciduous. **INFL:** fls gen 4–6, overlapped; bracts inconspicuous or deciduous. **FL:** calyx gen 8–10 mm, ± not larger in fr, stellate-canescent; petals 18–20 mm, rose-purple; outer filaments free above. **FR:** segment 2.5–4 mm, sides smooth, back net-veined-pitted esp on top, glabrous to puberulent, beak 0.5–0.8(1) mm. Dry grassy hillsides, foothill woodland, often serpentine; < 600 m. NCoRI, CaRF, SNF,

ScV, SnJV (e Merced Co.). Beak sometimes appressed, so evidently ± 0. Apr–Jun

S. hickmanii Greene Per from woody caudex, rhizome 0. **ST:** 1–8 dm, coarsely, gen densely stellate-canescent. **LF:** gen all along st, gen alike in size, shape; blade coarsely crenate to gen shallowly or deeply lobed, width 1–6 cm, gen > length. **INFL:** bracts gen stipule-like; fl stalks gen 1–4 mm; bractlets 3, 2–12 mm. **FL:** gen bisexual; calyx lobes 3–10 mm, stellate-puberulent, bristly or not; petals 5–18 mm, pale pink to pink-lavender. **FR:** segments gen 6–10, (1)2–2.5 mm, glabrous, sides gen smooth, thin, outer edges, back gen net-veined, transversely ridged, back gen with medial ridge, beak 0.

subsp. ***anomala*** C.L. Hitchc. (p. 895) CUESTA PASS CHECKERBLOOM Often mounded, 3–6 dm, 5–7 dm wide. **ST:** many, gen crowded in older pls, gen ± red; stellate-hairy, longest hairs ± 1.5 mm, tufted above. **LF:** blades above 2–6 cm wide, deeply 5–7-lobed, lobes crenate-dentate, narrowed to base, sinuses wide. **INFL:** open, branches gen many; bracts (8)10–12 mm, 4–4.5 mm wide, ± = calyx, widely lanceolate; bractlets ± = calyx, lance-linear. **FL:** calyx 10–12 mm; petals 12–15 mm, ± 1 cm wide, pink-lavender. Closed-cone-conifer forest, gen serpentine; 600–800 m. SCoRO (near Cuesta Pass, San Luis Obispo Co.). Threatened by military. May–Jun ★

subsp. ***hickmanii*** (p. 895) HICKMAN'S CHECKERBLOOM Grayish-canescent, hairs coarsely stellate. **ST:** gen 4–8 dm, often brick-red below. **LF:** closely crenate, lobes < 1/4 to base or gen 0, gen rounded with narrow sinus. **INFL:** ± dense; bracts 2–5(7) mm, ± 0.5 mm wide, << calyx, ± linear; bractlets 2–7 mm, < calyx. **FL:** calyx lobes acute, ± 8 mm, 6 mm wide, hairs longest on margin; petals ± 16–17 mm. Chaparral; 330–1200 m. SCoRO (Santa Lucia Range, Monterey Co.). May–Jul ★

subsp. ***napensis*** S.R. Hill NAPA CHECKERBLOOM Pl ± 3 dm, caudex ± elongate. **ST:** few to several, ± stellate-canescent to glabrous, gen ± maroon. **LF:** gen < 5 per st; blade 14–18 mm, 20–25 mm wide, stellate-hairy; mid-st lvs 3–5(7)-lobed ± to base, middle lobe largest, lobes narrowed to base, tips toothed; lower lvs occ coarsely crenate. **INFL:** gen unbranched, fls overlapped above; bracts like stipules, < calyx; bractlets 5–6 mm, 0.7–1 mm wide, < calyx. **FL:** calyx 7–10 mm, not larger in fr, lobes 5–6 mm, 2.5–3 mm wide, acuminate; petals 9–11 mm, ± 6 mm wide, pale pink. Chamise chaparral, rocky rhyolitic volcanic soil; 450–500 m. NCoRI (Napa Co.). Known from 1 historic, 2 recent collections; threatened by agriculture. May ★

subsp. ***parishii*** (B.L. Rob.) C.L. Hitchc. (p. 895) PARISH'S CHECKERBLOOM Coarsely ± gray-stellate-hairy. **ST:** few, 4–8 dm. **LF:** crenate to shallowly lobed, stellate-hairy, hairs gen overlapped. **INFL:** bracts gen 6–10 mm, ≤ calyx, lanceolate; bractlets 6–7 mm, ≤ calyx, 0.8–1.1 mm wide. **FL:** petals 1.1–1.3 cm, ± 0.5 cm wide. $2n$=20. Chaparral, woodland, open conifer forest; 1000–2200 m. SCoRO, WTR (Santa Barbara Co.), SnBr. Threatened by grazing, urbanization, road maintenance. Jun–Aug ★

subsp. ***pillsburiensis*** S.R. Hill LAKE PILLSBURY CHECKERBLOOM Pl (5)10–30(40) cm, 10–20 cm wide. **ST:** several, green, ± densely stellate-canescent, upper st hairs appressed, gen 0.2–0.5 mm. **LF:** blade 6–12 mm, 12–22 mm wide, coarsely crenate. **INFL:** open, unbranched, gen < 10-fld; bracts ± 3 mm, << calyx, ± 1.2 mm wide, oblong, cupped, narrowest at base; bractlets 3 mm, < calyx, 0.8–1 mm wide. **FL:** calyx 4–5 mm, to 5.5 mm in fr, lobes 3–4 mm, 1.8–2.3 mm wide, gen with 3 faint veins, hairs ± equal in length; petals 8–9 mm, ± 4 mm wide, pale pink. Chaparral, ephemeral drainage; ± 700–750 m. NCoRH (n Lake Co.). 1 population known. Jul ★

subsp. ***viridis*** C.L. Hitchc. (p. 895) MARIN CHECKERBLOOM Gen < 3 dm; ± sparsely stellate-hairy. **ST:** several, ± red below, upper st hairs not appressed, gen 0.5–1.2 mm. **LF:** 1–2.5 cm, 2–4 cm wide, coarsely crenate to shallowly lobed. **INFL:** branched or not; fls gen > 10; bracts 2–5 mm, < calyx, linear to narrowly oblong, gen ± 2-lobed; bractlets ± 3–4 mm, < calyx, ± 0.5 mm wide. **FL:** calyx lobe veins faint, hairs ± equal in length; petals 1.1–1.4 cm. Dry ridges near coast, serpentine; 50–400 m. n SnFrB (Carson Ridge, Marin Co.). May–Jun ★

S. hirsuta A. Gray (p. 895) HAIRY CHECKERBLOOM Ann. **ST:** 3–8 dm, stout, ± glabrous below, ± densely finely bristly (glabrous) above. **LF:** blade ± bristly, lobes of upper narrowly linear, acute; stipules inconspicuous or deciduous. **INFL:** dense; bracts 4–8 mm, linear, inconspicuous or deciduous. **FL:** calyx 8–10 mm, ± densely stellate-puberulent, bristly; petals 13–25 mm, dark rose-pink. **FR:** segment 3–4 mm, back wrinkled, net-veined, top, back gen puberulent, beak ± 1 mm, tip bristly. Vernally wet places, pools, grassland; < 1100 m. NCoR, CaRF, n&c SNF, GV. Apr–Jun

S. keckii Wiggins (p. 895) KECK'S CHECKERBLOOM Ann. **ST:** 1.5–4 dm, occ branched, soft stellate-bristly. **LF:** stipules gen < 10 mm, unlobed; blades below shallowly 7–9-lobed or deeply crenate, above with lobes tapered to base, 2–5-toothed, teeth at tip ± equal. **INFL:** ± elongate, few-fld, fr gen not overlapped; axis stellate-puberulent, also with many soft glandular, multicellular hairs, gen bristly; bracts gen not divided, (3)7–11 mm, gen < calyx. **FL:** calyx 8–12 mm, ± divided 4/5 length, lobes lance-linear, gen 6–7-veined, long acuminate, base within with ± purple spot gen 1–2 mm wide, hairs like infl axis; petals 10–20 mm, dark pink, with dark basal spot or not. **FR:** segment 3–4 mm, back net-veined-pitted, gen ± pink, gen glabrous exc 1–5 minute bristles on top, beak 0. Grassy slopes; 75–650 m. s NCoRI (Colusa, Napa, Solano, Yolo cos.), c&s SNF (Fresno, Merced, Tulare cos.). Pls in s NCoRI need study (see note under *S. diploscypha*); vulnerable to development, incl orchards, housing. Apr–May ★

S. malachroides (Hook. & Arn.) A. Gray (p. 895) MAPLE-LEAVED CHECKERBLOOM Per, subshrub from woody caudex, harsh-bristly, stellate-hairy. **ST:** several, 4–15 dm, solid. **LF:** gen all along st, alike in size, shape; blade gen shallowly 5–7-lobed, ± coarsely crenate. **INFL:** head-like spikes, in panicle, < 5 cm; bracts > fl buds, stipule-like; bractlets gen 0(1–2), < sepals. **FL:** bisexual, unisexual, or both; calyx 6–9 mm, larger in fr, densely bristly, often ± purple; petals 7–15 mm, white or pale purple-white. **FR:** segments 5–9, ± 2.5 mm, hairs 0 or few, stellate, sides ± smooth, back ridged, outer edges rounded, beak 0. 2*n*=20. Woodland, clearings near coast; < 700 m. NCo, NCoRO, n&c CCo, SnFrB, n SCoRO; w OR. Apr–Aug ★

S. malviflora (DC.) A. Gray CHECKERBLOOM, CHECKERMALLOW Per 1.5–10 dm, gen rhizomed, caudex woody. **ST:** gen hairy. **LF:** gen dentate or lobed (entire), upper gen much reduced. **INFL:** dense to open; lowest bracts often lf-like, gen divided to base; bractlets 0. **FL:** calyx 5–12 mm, ± larger in fr, gen densely stellate-puberulent, bristly, bristles often on swollen pad; petals 10–20(35) mm, bright to dark pink, gen white-veined. **FR:** segment 2.5–4 mm, ± deeply pitted, net-veined, gen more so on sides than back, beak gen short. 2*n*=20,40,60. Highly variable intergrading complex, many local variants; inland upland taxa probably do not belong in this sp., study needed.

  subsp. ***californica*** (Torr. & A. Gray) C.L. Hitchc. (p. 895) CHAPARRAL CHECKERBLOOM, CALIFORNIA CHECKERBLOOM Pl 4–10 dm, gray, densely soft-stellate-hairy. **ST:** base decumbent. **LF:** blade densely soft-stellate-hairy abaxially, adaxially with some hairs forked, lowest gen shallowly lobed, upper gen deeply 7-lobed. **INFL:** open, gen unbranched. **FL:** calyx 9–13 mm. **FR:** segment 3–3.5 mm, glandular-puberulent, coarsely net-veined-pitted. Coastal scrub, chaparral; < 1000 m. s SCoRO, WTR. Intergrades with *S. malviflora* subsp. *laciniata*, *S. sparsifolia*. Mar–Jun

  subsp. ***dolosa*** C.L. Hitchc. (p. 895) BEAR VALLEY CHECKERBLOOM Pl 2–10 dm, with ± elongate caudex or matted rhizomes. **ST:** base decumbent, rooting; hairs gen ± 2 mm below, occ 0 above. **LF:** blade (2)4–8(12) cm wide, basal unlobed, crenate, upper ± deeply lobed, coarsely stellate-hairy abaxially, stellate-puberulent adaxially; lowest stipules ± 4–5 mm, 2 mm wide. **INFL:** gen open, gen unbranched, elongate, 5–14-fld. **FL:** calyx 6.5–10 mm, sparsely stellate-puberulent, longest hairs simple. **FR:** segment ± 3 mm, weakly net-wrinkled, back glabrous or sparsely glandular-puberulent. Open pine forest; 1500–2300 m. SnBr. Jun–Jul ★

  subsp. ***laciniata*** C.L. Hitchc. GERANIUM-LEAVED CHECKERBLOOM Pl 2–8 dm, from taproot, caudex short. **ST:** gen decumbent, gen ± rooting at base, ± glabrous to bristly, gen soft-

stellate-hairy esp above. **LF:** blade 2–6 cm wide, gen sparsely hairy, abaxially more densely stellate-hairy; basal shallowly 7-lobed, above with deep, linear lobes, gen > 13 segments. **INFL:** open, gen unbranched, gen > 15 cm. **FL:** calyx 7–11 mm, sparsely stellate-puberulent, bristly, marginal, vein hairs longer; petals 1–2.5 cm, pink-lavender (white), gen white-veined. **FR:** segment ± 3.5 mm, ± glandular-puberulent, coarsely net-veined-pitted. 2*n*=40. Grassland, open woodland; gen < 700 m. s NCoR, CW. Intergrades with *S. malviflora* subsp. *californica*, *S. malviflora* subsp. *m.*, *S. sparsifolia*. Mar–Jun

  subsp. ***malviflora*** (p. 895) Pl 1.5–6 dm. **ST:** decumbent at base, rooting, gen densely coarsely stellate-hairy, spreading-bristly esp at base. **LF:** blade 2–6 cm wide, 7–9-lobed, gen ± coarsely hairy, ± fleshy, upper more deeply lobed or not. **INFL:** gen unbranched; fls close but ± evenly spaced. **FL:** calyx densely stellate-puberulent, coarsely bristly, bristles often on swollen pad, hairs at base shorter, denser, marginal longer; petals 1–2.5 cm, pink to rose, white-veined. **FR:** segment 3.5–4 mm, ± sparsely glandular-puberulent, ± net-veined-pitted, beak short. 2*n*=40,60. Coastal prairie, scrub, open forest; gen < 500 m. s NCo, NCoRO, CCo, SnFrB, SCo, n ChI. Intergrades with most other subspp. Once common in, now gen extirpated from Los Angeles area. Mar–Jul

  subsp. ***patula*** C.L. Hitchc. (p. 895) SISKIYOU CHECKERBLOOM Pl 2–6(8) dm. **ST:** decumbent at base, often rooting, bristly below, ± stellate-hairy above. **LF:** blade 3–12 cm wide, 7–9-lobed, more shallowly below, dentate, coarsely bristly and ± stellate-hairy. **INFL:** head- to spike-like, gen unbranched, elongating in fr; fls stiffly erect, overlapped, in pistillate pls gen smaller, more separate; fl stalks gen 2–3(10) mm. **FL:** calyx 8–13 mm, lobes long, narrow, with stellate and gen forked hairs; petals 1–2.5 cm, bright ± pink to rose-pink. **FR:** segment 3.5–4 mm, coarsely net-veined-pitted, ± glandular-puberulent, beak short. 2*n*=40. Open coastal forests, bluffs; gen < 700 m. n NCo; sw OR. Occ intergrades with *S. malviflora* subsp. *rostrata*. May–Aug ★

  subsp. ***purpurea*** C.L. Hitchc. (p. 895) PURPLE-STEMMED CHECKERBLOOM Pl 2–6 dm, gen ± purple, esp st base, stipules, calyx. **ST:** decumbent, glabrous to sparsely bristly. **LF:** blade 1.5–3(4) cm wide, short-bristly, lowest gen < 2(2.5) cm, gen coarsely crenate, unlobed, upper gen unlobed. **INFL:** branched or not, gen few-fld; fl stalks 5–30 mm. **FL:** calyx 7–11 mm, ± purple, sparsely stellate-puberulent, bristly; petals 1–1.5(2.5) cm, occ smaller in pistillate, bright pink-rose, white-veined. **FR:** segment 3–3.5 mm, coarsely net-veined-pitted. 2*n*=20. Meadows, open coastal forest, prairie; gen 0–30 m. c&s NCo (n Sonoma, s Mendocino cos.), n CCo (San Mateo Co.). Occ mistaken for *S. hickmanii* (similar unlobed cauline lvs); occ intermediates to *S. malviflora* subsp. *rostrata*. May–Jun ★

  subsp. ***rostrata*** (Eastw.) Wiggins SEACLIFF CHECKERBLOOM Pl 1.5–6 dm; ± densely stellate-fine-bristly. **ST:** decumbent at base, gen not rooting. **LF:** blade 2–6 cm wide, unlobed. **INFL:** dense, unbranched, gen open in fr. **FL:** calyx densely stellate-puberulent, coarsely bristly, bristles often on swollen pad, hairs at base shorter, denser, tangled, marginal hairs longer; petals 1–2.5 cm, pink to rose, gen white-veined. **FR:** segment 3.5–4 mm, ± sparsely glandular-puberulent, net-veined-pitted, beak short. Open coastal bluffs; < 200 m. c&s NCo (Sonoma, s Mendocino cos.), n CCo (Marin, San Mateo cos.). Intergrades with *S. malviflora* subsp. *purpurea*, *S. malviflora* subsp. *patula*; confused with *S. hickmanii* but lacks bractlets. Unlobed lvs, fine bristles ± distinctive. Apr–May

S. multifida Greene (p. 899) CUT-LEAF CHECKERBLOOM Per 1–4(6) dm, from woody caudex, gray-glaucous; rhizomes 0. **ST:** many, gen ascending to erect, sparsely or densely appressed-stellate-hairy. **LF:** mostly basal; blade 1.5–4 cm wide, fleshy, glaucous, deeply 7(9)-lobed, lobes gen deeply dentate or ternate-divided, segments linear to oblong, narrowest on lvs above. **INFL:** open, gen unbranched, glabrous to sparsely stellate-hairy; fls gen 3–9; fl stalk 3–10 mm. **FL:** calyx 7–10 mm, stellate-puberulent; petals 10–25 mm, rose-pink. **FR:** segments 6–7, 3.5–4 mm, sides, back net-veined-pitted, back glandular-puberulent, beak short. 2*n*=20. Dry places, sagebrush scrub, pine forest; 2000–2800 m. c&s SNH, SNE; NV. Intergrades with, very close to *S. glaucescens*. May–Jul ★

S. neomexicana A. Gray (p. 899)   SALT SPRING CHECKERBLOOM
Per 2–5(9) dm, from clustered, fleshy roots or gen fleshy taproot,
rhizomes 0. **ST:** 1–several, below bristly- to stellate-hairy (glabrous), above gen stellate-hairy. **LF:** blade 2–5(8) cm wide, fleshy,
hairs appressed; basal crenate to shallowly 5(7)-lobed, upper 5-lobed.
INFL: open, branched or not, many-fld, glabrous to sparsely hairy;
fl stalk > calyx; bractlets fused at base. **FL:** calyx gen 5–8 mm, with
few small stellate and longer hairs on swollen pads, lobes acuminate,
veins ± prominent; petals 6–18 mm, pale pink-rose, with paler (white)
veins. **FR:** segment ± 2(3) mm, sides smooth to weakly net-veined, ±
glabrous, beak 0.5–0.8 mm. 2*n*=20. Alkaline springs, marshes; gen <
1500 m. SCo, WTR (extirpated?), SnGb, SnBr, PR, sw DMoj; to NM,
n Mex. Variable; taxonomy below sp. level needs study. Occ like *S.
sparsifolia*. Apr–Jun ★

S. oregana (Torr. & A. Gray) A. Gray   Per 3–15 dm, taproot
woody, crown branched, caudex 0. **ST:** gen clustered (rooting near
base); base glabrous to coarsely stellate-hairy to long-bristly. **LF:**
basal and cauline; blade 3–10(15) cm wide, glabrous to hairy, lower
crenate to deeply lobed, upper deeply (3)5–7-lobed, uppermost simple to 2–3 lobed, lobes narrow, entire to deeply lobed. **INFL:** dense
to open, gen ± spike-like, in panicle or not; fl stalks gen 1–3 mm. **FL:**
calyx 3.5–9 mm, gen ± 5 mm, lobes lanceolate, glabrous to densely
stellate-puberulent or bristly; petals (7)10–20 mm, pink to dark rose-
pink. **FR:** segment 2–3 mm, smooth to weakly net-veined-pitted,
sparsely glandular-puberulent, not stellate-hairy, beak 0.3–0.7 mm.

subsp. ***eximia*** (Greene) C.L. Hitchc. (p. 899)   COAST
CHECKERBLOOM   Pl 9–12 dm. **ST:** base occ rooting, densely bristly,
hairs simple. **INFL:** dense, in panicle, 3–6 cm. **FL:** calyx to ± 10
mm in fr, sparsely to densely stellate-puberulent, bristles 1–2.5 mm.
FR: segment ± 3 mm, smooth, beak 0.5–0.7 mm. Meadows; < 1200
m. n NCo, KR, NCoRO. Intergrades with *Sidalcea oregana* subsp.
spicata, *S. setosa*. Jun–Aug ★

subsp. ***hydrophila*** (A. Heller) C.L. Hitchc. (p. 899)   MARSH
CHECKERBLOOM   Pl 3–9 dm. **ST:** decumbent, occ rooting near
base, lower gen < 5 mm diam, bristles ± 2 mm. **INFL:** head-like, in
panicle, gen 1.5–2.5 cm, few-fld. **FL:** calyx ± 6 mm in fr, stellate-
puberulent, not bristly; petals 8–12 mm, bright pink. **FR:** segment
± 2.2 mm, smooth, gen glabrous or sparsely glandular-puberulent,
beak ≤ 0.3 mm. 2*n*=20. Wet soil of streambanks, meadows; 440–
2300 m. NCoRH, NCoRI. Intergrades with *S. oregana* subsp. *valida*.
Jul–Sep ★

subsp. ***oregana*** (p. 899)   OREGON CHECKERBLOOM   Pl 3–15
dm. **ST:** many, base stellate-hairy, occ bristly, above less hairy, occ
glabrous. **INFL:** in panicle or not, to 3 dm, fls gen not overlapped,
esp in fr; fl stalks 2–10 mm, gen 3 mm in fl, 5 mm in fr; bracts gen <
youngest fl buds. **FL:** calyx gen 5–8 mm, to 10 mm in fr, gen densely
stellate-puberulent to long-bristly, marginal hairs occ longer; petals gen 5–15 mm. **FR:** segment 2.5–3 mm, gen glabrous, back gen
smooth, sides occ net-veined, beak 0.3–0.7 mm. 2*n*=20,40,60. Moist
meadows; 500–2500 m. NW, CaR, MP; to WA, WY, NV. Variable;
intergrades with *S. oregana* subsp. *spicata*, *S. setosa* C.L. Hitchc.
May–Sep

subsp. ***spicata*** (Regel) C.L. Hitchc. (p. 899)   Pl 3–8 dm. **ST:**
gen several, base soft stellate-hairy, occ long-bristly, hairs 1–2 mm,
toward tip occ glabrous. **INFL:** often in panicle, gen dense in fl, open
in fr; bracts gen > youngest fl buds, fl stalks ± 2 mm. **FL:** calyx
(3.5)5(7) mm, gen densely bristly, stellate-puberulent (or stellate-
puberulent only), hairs to 2.5 mm; petals gen 10–15 mm, pink to
rose-pink or magenta. **FR:** segment 2.5–3 mm, gen smooth, sides occ
weakly net-veined, back sparsely glandular-puberulent, beak 0.1–0.3
mm. 2*n*=20,40. Meadows, streamsides; 975–3000 m. KR, n NCoRH,
CaRH, n&c SNH, GB; to OR, w NV. May be confused with other subspp., as well as with *S. setosa*. Jun–Aug

subsp. ***valida*** (Greene) C.L. Hitchc. (p. 899)   KENWOOD MARSH
CHECKERBLOOM   Pl 9–12 dm, roots thick, fleshy. **ST:** many, decumbent, often rooting near base, lower gen > 5 mm diam, bristles
1–1.5 mm, above ± glabrous in age. **INFL:** dense, in panicle, gen
2–5 cm, many-fld. **FL:** calyx < ± 7.5 mm, ± 7.5 mm in fr, stellate-
puberulent, sparsely bristly; petals 10–15 mm, pink-lavender. **FR:**
segment 3 mm, smooth, sparsely glandular-puberulent, beak ± 0.5

mm. 2*n*=20. Marsh; ± 150 m. s NCoRO (near Kenwood, Sonoma
Co.). Threatened by grazing, marsh alteration. Jun–Sep ★

S. pedata A. Gray (p. 899)   BIRD-FOOT CHECKERBLOOM   Per 2–4
dm, from fleshy taproot, caudex 0; pls with either bisexual or pistillate fls. **ST:** many, long-bristly, near base ± stellate-hairy. **LF:** basal,
1–3 cauline, blades 2–5(6) cm wide, lobes deep, ± ternate-dissected,
segments linear to elliptic, narrower in lvs above. **INFL:** below open,
above dense, spike-like, not in panicle, to 25 cm. **FL:** calyx 4–5(7)
mm, ± not larger in fr, stellate-puberulent, marginal hairs longer; petals 9–12 mm, dark rose-pink, veins dark. **FR:** segments 5–6, ± 2.5
mm, smooth, beak ± 0. 2*n*=20. Moist meadows in open woodland;
1520–2500 m. SnBr (Bear Valley, Bluff Lake). Threatened by development, vehicles, grazing. May–Aug ★

S. ranunculacea Greene (p. 899)   Per 2–5 dm, caudex ± elongate. **ST:** ascending, base bristly, above stellate-hairy. **LF:** blade
2.5–6 cm wide, stellate-hairy to sparsely bristly, fleshy; basal shallowly 5-lobed, lobes, teeth rounded, upper deeply 5(7)-lobed, lobes,
teeth acute. **INFL:** spike-like, in panicle or not, axis below gen
visible between fl clusters; fl stalks 1–3 mm. **FL:** calyx 5–9 mm,
stellate-puberulent with marginal bristles ± 1.5 mm; petals 5–15 mm,
magenta-pink, occ white-veined, drying dark purple. **FR:** segment ±
2.5 mm, sides ± weakly net-veined, back rougher, gen sparsely stellate-puberulent. 2*n*=20. Uncommon. Moist meadows, streambanks,
often near *Sequoiadendron* groves; 1820–3050 m. s SNH (Greenhorn
Mtns, Tulare, Kern cos.). Jun–Aug

S. reptans Greene (p. 899)   Per 2–5 dm, from caudex. **ST:** prostrate, free-rooting; lower mid-st densely long-bristly, hairs 2–3 mm,
often on swollen pads. **LF:** lowest long-petioled, unlobed, deeply crenate, blades 2.5–5 cm wide, bristly to ± glabrous, not stellate-hairy,
blades above 5–7-lobed; stipules basally gen ≤ 5 mm wide, above gen
± 6 mm, ± 2 mm wide, sparsely ciliate. **INFL:** ± open, unbranched,
to 20 cm, short-stellate-puberulent, ± glandular, fls gen few, several
cm apart; bracts gen enrolled, tips 2-forked. **FL:** calyx 8–10 mm,
stellate-puberulent, marginal hairs longer (glandular, multicellular);
petals 12–20 mm, dark pink to lavender. **FR:** segment ± 3 mm, deeply
net-veined-pitted; top, back, beak densely stellate-puberulent, beak ±
1 mm. 2*n*=20. Moist meadows, dry places in pine forest; 1120–2000
m. SN (esp c SNH). Jun–Aug

S. robusta Roush (p. 899)   BUTTE COUNTY CHECKERBLOOM   Per
(5)8–12(18) dm, rhizomed, glaucous. **ST:** gen unbranched, clustered,
base densely finely stellate-hairy, hairs spreading, 0 in age above.
LF: blade 5–7-lobed, lobes deeply toothed to ternate-lobed, abaxially sparsely stellate-hairy, glaucous, adaxially bristly. **INFL:** open,
gen unbranched, gen 30–40 cm, glabrous; fls > 10. **FL:** calyx 10–15
mm, densely stellate-puberulent; petals (1.5)2–3.5 cm, pale pink, gen
drying ± yellow; stamens in 3–4 series, 2–4 stamens in each. **FR:**
segment 3–3.5 mm, gen glabrous, sides weakly net-veined-pitted,
back even less so, outer edges narrowly winged above, beak < 0.5
mm. 2*n*=20. Dry banks in chaparral/blue-oak woodland transition;
100–400 m. s CaRF, n SNF (near Chico, Butte Co.). Threatened by
development. Apr–Jun ★

S. setosa C.L. Hitchc.   BRISTLY CHECKERBLOOM   Per 5–10 dm,
from thick taproot, caudex short. **ST:** base soft-appressed-stellate-
hairy and/or long-bristly, above stellate-puberulent. **LF:** gen basal,
long-petioled; lower 5–10 cm wide, shallowly 5–9-lobed, coarsely
crenate, upper 5–9-lobed, lobes coarsely dentate or entire. **INFL:**
spike-like, in panicle, 3–10 cm; bracts ± = calyx; fl stalks 1–2 mm,
to 4 mm in fr. **FL:** calyx 5–8 mm, to 10 mm in fr, stellate-puberulent, bristly with longer coarser hairs gen on pads; petals 8–15 mm,
pink-lavender. **FR:** segment ± 2.5 mm, sides, outer edges ± coarsely
net-veined-pitted, sparsely glandular-puberulent, back weakly net-
veined, beak ± 0.5 mm. 2*n*=40,60. Uncommon. Meadows; < 1200 m.
KR; sw OR. Intergrades with, often combined with *S. oregana* subsp.
o., *S. oregana* subsp. *spicata*. Jun–Jul

S. sparsifolia (C.L. Hitchc.) S.R. Hill   SOUTHERN CHECKERBLOOM
Per 2–8 dm, from thick fibrous crown, caudex short, rhizomes gen 0.
ST: hairs stellate or bristles. **LF:** crowded near st base; blades 2–6(8)
cm wide, stellate-hairy, lower crenate to shallowly 7-lobed, upper
few, lobed ± to base. **INFL:** exceeding lvs, gen branched, (15)30–45
cm, fls > 10. **FL:** calyx 6–10 mm, densely stellate-puberulent, many

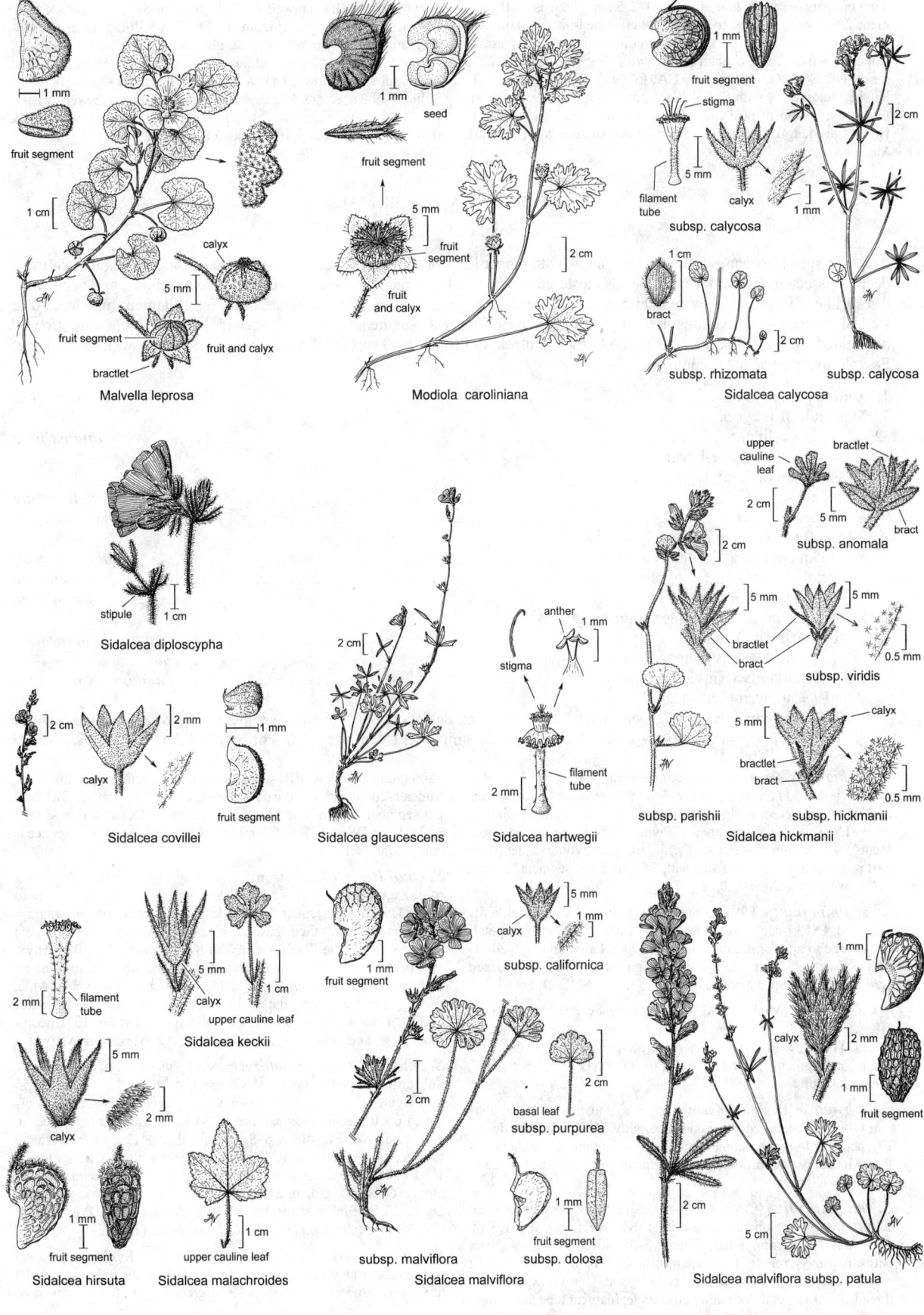

Malvella leprosa

Modiola caroliniana

subsp. calycosa
subsp. rhizomata   subsp. calycosa
Sidalcea calycosa

subsp. anomala
subsp. viridis
subsp. parishii   subsp. hickmanii
Sidalcea hickmanii

Sidalcea diploscypha

Sidalcea covillei

Sidalcea glaucescens

Sidalcea hartwegii

Sidalcea keckii

Sidalcea hirsuta

Sidalcea malachroides

subsp. californica
subsp. purpurea
subsp. malviflora   subsp. dolosa
Sidalcea malviflora

Sidalcea malviflora subsp. patula

hairs on margins, veins longer; petals 1–2.5 cm, pink-rose. **FR**: segment 2.5–3 mm, weakly to moderate-net-veined-pitted, glandular-puberulent, beak short. 2*n*=20,40. Periodically moist to dry grassy slopes or ± flat places, chaparral often with *Artemisia*, oaks; < 2200 m. s SNF, SCoRO, SW; n Baja CA. [*S. malviflora* subsp. *s.* C.L. Hitchc.] Intergrades with *S. malviflora* subsp. *californica*, *S. neomexicana*; evaluation of possible infraspecific taxa, incl those of C.L. Hitchc., needed. With *S. neomexicana*, among southernmost in genus. Mar–Jun

S. stipularis J.T. Howell & G.H. True (p. 899) SCADDEN FLAT CHECKERBLOOM   Per, rhizomed. **ST**: 3–6.5 dm; bristles spreading, simple. **LF**: gen all along st, alike in size, shape; blades ovate, unlobed; stipules < 2 cm, ± oblique-ovate or cordate. **INFL**: ± heads; bracts subtending lower fls involucre-like; fl stalks 1–2 mm; bractlets 3, linear-oblanceolate. **FL**: calyx ± 8 mm, sparsely to densely bristly; petals ± 15 mm, pink. **FR**: segment ± 2 mm, glabrous, smooth, beak 0. Marsh; ± 700 m. n SNF (Scadden Flat, Grass Valley, Nevada Co.). Jun–Aug ★

SPHAERALCEA GLOBEMALLOW

John C. La Duke

Ann, per, subshrub, canescent to stellate-hairy, with longer hairs or not. **LF**: petioled; blade lance-linear to triangular, entire to deeply dissected. **INFL**: raceme-like (fls clustered in bract axils) or panicle; bractlets (0)1–3, inconspicuous, gen deciduous, thread-like. **FL**: petals obovate, white, lavender, pink, rose-pink, salmon- or red-orange, or apricot; filament tube hairs 0 or stellate, anthers gen ± yellow or ± purple; stigmas head-like. **FR**: segments 9–17, 1–2-seeded, below indehiscent, strongly net-veined, above dehiscent, smooth. **SEED**: gray, black, or brown. ± 50 spp.: arid Am. (Greek: globe mallow, from fr shape) Polyploidy, intermediates common.

1. Ann. . . . . ***S. coulteri***
1′ Subshrub (fl 1st yr or not)
  2. Lvs ± lance-linear . . . . ***S. angustifolia***
  2′ Lvs wider than lance-linear
    3. Lvs deeply divided
      4. Sepals < 11 mm . . . . ***S. grossulariifolia***
      4′ Sepals 13–15 mm . . . . ***S. rusbyi*** var. ***eremicola***
    3′ Lvs often lobed but not deeply divided
      5. Infl an open panicle, occ dense in bud. . . . . ***S. ambigua*** (in part)
        6. Petals red-orange to apricot . . . . var. ***ambigua***
        6′ Petals lavender to pink . . . . var. ***rosacea***
      5′ Infl raceme-like or a dense panicle
        7. Lvs coarsely dentate . . . . ***S. munroana***
        7′ Lvs entire, ± wavy, crenate, or finely dentate
          8. Lf markedly wrinkled abaxially. . . . . ***S. ambigua*** var. ***rugosa***
          8′ Lf ± not wrinkled abaxially
            9. Hairs coarse; lvs gray-green; fr segment 4.5–5 mm, dehiscent part ± 60% of segment. . . . . ***S. emoryi*** var. ***emoryi***
            9′ Hairs fine; lvs gen yellow-green; fr segment 2.5–3 mm, dehiscent part < 20% of segment . . . . ***S. orcuttii***

S. ambigua A. Gray   Subshrub, canescent. **ST**: erect, 5–10 dm. **LF**: blade 15–50 mm, ± triangular, 3-veined, green or yellow-green, crenate, wavy, base wedge-shaped, truncate, or cordate; lobes 3, weak. **INFL**: panicle, gen open, tip without lf-like bracts. **FR**: segments 9–13, < 6 mm, < 3.5 mm wide, truncate-cylindric, dehiscent part < 3.5 mm, 60–75% of segment. **SEED**: 2 per segment, brown, glabrous to hairy. 2*n*=10,20,30.

var. ***ambigua*** (p. 899) APRICOT MALLOW   **LF**: not markedly wrinkled. **INFL**: occ dense in bud; fls ≥ 1 per bract axil. **FL**: petals red-orange to apricot; filament tube 3–9 mm, hairy, anthers yellow-purple. **FR**: dehiscent part ± 60% of segment. 2*n*=10,20,30. Desert scrub; pinyon/juniper woodland; 150–2500 m. s SNE, D. Feb–Jul

var. ***rosacea*** (Munz & I.M. Johnst.) Kearney PARISH MALLOW   **LF**: not markedly wrinkled. **FL**: petals lavender to pink; filament tube ± 11 mm, ± glabrous, anthers purple-gray. **FR**: dehiscent part ± 60% of segment. Desert scrub; 150–800 m. D (esp DSon); AZ, Mex. Mar–Jul

var. ***rugosa*** (Kearney) Kearney (p. 899) ROUGHLEAF APRICOT MALLOW   **LF**: markedly wrinkled abaxially. **INFL**: panicle, dense. **FL**: petals red-orange; filament tube ± 5 mm, ± glabrous, anthers yellow. **FR**: dehiscent portion 75% of segment. Desert scrub; 150–2500 m. D. Mar–Sep

S. angustifolia (Cav.) G. Don (p. 899) NARROW-LEAVED GLOBEMALLOW   Subshrub, canescent. **ST**: erect, gen 0.5–1.5 m. **LF**: blade 15–48 mm, ± lance-linear, 3–5-veined, light gray-green, entire to wavy-crenate, base tapered; lobes 0 or lf hastate. **INFL**: raceme-like, tip gen with lf-like bracts; fl stalk ± = calyx; bractlets thread-like. **FL**: petals 7–9 mm, red-orange; filament tube 5.5–7 mm,

hairy, anthers yellow. **FR**: segments 9–13, 4–7 mm, 1.5–2 mm wide, truncate-conic, dehiscent part 3–4 mm, ± 75% of segment. **SEED**: 2 per segment, brown-black, hairy. 2*n*=10,20,30. Desert scrub; -6–500 m. DMoj, n DSon; to KS, TX, n Mex. Relationship to *S. emoryi* needs study. Mar–Oct

S. coulteri (S. Watson) A. Gray (p. 899) COULTER'S GLOBEMALLOW   Ann, with few long, soft hairs. **ST**: sprawling to erect, 1.5–15 dm, slender. **LF**: blade 15–45 mm, ± triangular, gray-green, coarsely toothed, base cordate or ± truncate; lobes 3 or 5. **INFL**: gen raceme-like, tip gen with lf-like bracts; fl stalk > calyx. **FL**: petals < 11 mm, salmon-orange; filament tube ± 5 mm, hairy, anthers yellow. **FR**: segments ± 15, 1.5–2 mm, 2–2.5 mm wide, dehiscent part ± 1 mm, flat, ± 30% of segment, projecting inward. **SEED**: 1 per segment, brown, glabrous or ± hairy. 2*n*=10. Uncommon. Dry, sandy places; < 300 m. s DSon; AZ, Mex. Mar–May

S. emoryi A. Gray var. ***emoryi*** (p. 899) EMORY'S GLOBEMALLOW   Subshrub, coarsely canescent. **ST**: erect, < 21 dm. **LF**: blade 25–55 mm, ovate-triangular, 3- or 5-veined, gray-green, crenate, base cordate, tip ± truncate to acute; lobes 3. **INFL**: raceme-like below, dense panicle above. **FL**: calyx 6–8 mm; petals 10–12 mm, red-orange to lavender; filament tube ± 6 mm, anthers yellow. **FR**: segments 10–16, 4.5–5 mm, 2.5 mm wide, truncate-conic, dehiscent part < 3 mm, ± 60% of segment. **SEED**: 1–2 per segment, brown or black. 2*n*=20,30,50. Fields, roadsides; < 600 m. SCo, s PR, s DMoj, DSon; NV, AZ, Mex. Intergrades with *S. angustifolia*. Feb–Jul

S. grossulariifolia (Hook. & Arn.) Rydb. (p. 899) GOOSEBERRY-LEAVED GLOBEMALLOW   Subshrub. **ST**: erect, 6–10 dm, white-canescent (± glabrous and green or purple); base woody.

LF: blade 17–35 mm, 5-veined, green to gray-green, base cordate; lobes 3, deep, rounded or pointed, each ± lobed or divided. **INFL**: raceme-like, fls rarely not in clusters; tip without lf-like bracts; fl stalks < or > calyx. **FL**: sepals < 11 mm; petals ± 11 mm, red-orange; filament tube ± 6 mm, ± hairy, anthers yellow. **FR**: segments 10–12, ± 2.5 mm, ± 2.5 mm wide, truncate-conic to spheric, dehiscent part < 1.5 mm, ± 60% of segment. **SEED**: 1 per segment, gray, glabrous to ± hairy. 2*n*=20. Dry, volcanic soils; < 1700 m. se MP (Lassen Co.); to WA, ID, UT. Var. *fumariensis* S.L. Welsh & N.D. Atwood, a Utah endemic, recognized by some. May–Jun ★

S. munroana (Lindl.) A. Gray (p. 899)   MUNRO'S DESERT-MALLOW Subshrub, canescent. **ST**: erect, ± 7.5 dm. **LF**: blade < 4.5 cm, triangular, 5-veined, green to gray-green, coarsely dentate, base truncate to tapered; lobes 0 or 5, shallow. **INFL**: raceme-like, tip without lf-like bracts; fl stalks < calyx. **FL**: petals 11–14 mm, red-orange; filament tube 7–8 mm, ± hairy, anthers yellow. **FR**: segments ± 12, 3.5–4 mm, 2.5–3 mm wide, spheric, with sharp reflexed tip, dehiscent part < 2 mm, ± 55% of segment. **SEED**: 1 per segment, brown, ± hairy. Dry, open places; ± 2000 m. n SNH (Squaw Creek, Placer Co.); to WA, MT, WY, UT. Evidently only 1 specimen from CA. May–Jun ★

S. orcuttii Rose (p. 899)   CARRIZO MALLOW   [Ann], subshrub, ± yellow-canescent, hairs fine. **ST**: erect, 5–12 dm. **LF**: blade 30–50 mm, rounded to triangular, prominently 3-veined, gen yellow-green, entire to ± wavy, base tapered to truncate; lobes ± 3. **INFL**: raceme-like or dense panicle, tip without lf-like bracts; fl stalk < calyx. **FL**: petals 10–12 mm, red-orange; filament tube 5–6 mm, hairy, anthers yellow. **FR**: segments 12–17, 2.5–3 mm, 2–3 mm wide, ± hemispheric, dehiscent part < 1 mm, < 20% of segment. **SEED**: 1 per segment, brown, glabrous or ± hairy. 2*n*=10. Dry, sandy, ± alkaline desert scrub; -20–900 m. s DSon; AZ, Mex. Feb–Sep

S. rusbyi A. Gray var. *eremicola* (Jeps.) Kearney (p. 899) RUSBY'S DESERT-MALLOW   Subshrub. **ST**: erect, ± 3 dm. **LF**: blade 15–20 mm, widely ovate, 5-veined, light green, entire, base truncate to cordate; lobes 3, deep, rounded or pointed, each ± lobed or divided. **INFL**: panicle, tip gen with lf-like bracts; fl stalks < calyx. **FL**: sepals 13–15 mm; petals < 20 mm, red-orange; filament tube ± 9 mm, hairy, anthers yellow. **FR**: segments ± 13, ± 5 mm, 2 mm wide, ± truncate-spheric, dehiscent part ± 3 mm, ± 60% of segment. **SEED**: 1–2 per segment, black-gray, glabrous to ± hairy. 2*n*=10,20. Desert scrub; 1000–1500 m. n DMtns (Death Valley region, e Inyo Co.; Clark Mtn Range, ne San Bernardino Co.), e DMoj exc DMtns. [*S. r.* subsp. *e.* (Jeps.) Kearney] May ★

MARTYNIACEAE   UNICORN-PLANT FAMILY

Margriet Wetherwax & Lawrence R. Heckard

Ann, per, glandular-hairy, gen strongly scented. **LF**: simple, opposite or alternate; stipules 0; petiole long. **INFL**: raceme, terminal, bracted; bractlets 2, just below fl. **FL**: bisexual; sepals 5, ± unequal; corolla 2-lipped, gen 5-lobed; stamens epipetalous, gen 2 long, 2 short, 1 vestigial; ovary superior, 1-chambered, placentas 2, parietal, 2-lobed, style > ovary, curved, stigma 2-lobed, flat, gen closing when touched. **FR**: capsule, drupe-like; outer layer fleshy, deciduous; inner layer ultimately exposed, woody, tip incurved, splitting to form 2 claws. 4 genera, 16 spp.: gen ± trop Am; some cult. [Wagstaff & Olmstead 1997 Syst Bot 22:165–179] No evidence that insects stuck to glands are digested. Scientific Editor: Thomas J. Rosatti.

PROBOSCIDEA   UNICORN-PLANT

Taproot branched or tuberous. **ST**: prostrate to spreading, gen < 1 m. **LF**: blade broadly ovate to round or triangular, palmately veined (gen palmately lobed), base cordate. **INFL**: bractlets < calyx. **FL**: calyx 1–2 cm, gen 5-lobed, split to base on lower side, or sepals free; corolla 2–5 cm, bell- to funnel-shaped, showy, tube cylindric, gen < 1 cm, bent downward, throat 10–30 mm, limb with 5 flared lobes, throat and lower limb with colored lines (nectar guides). **FR**: body 5–10 cm, fusiform, smooth, rough, or spiny, crested with branched projections gen only along upper suture; claws 1.5–3 × body. **SEED**: 8–13 mm, angled, gen black, corky. 8 spp.: Am. (Greek: elephant's trunk) [Bretting 1982 Amer J Bot 69:1531–1537] Fr dispersed by attachment of claw to animals. *P. parviflora* subsp. *p.* not in CA according to some reports, naturalized in CA according to others.

1. Sepals free; woody fr-body short-spiny (sect. *Ibicella*) . . . . *P. lutea*
1′ Sepals ± fused; woody fr-body smooth or rough but not spiny (sect. *Proboscidea*)
  2. Per; largest lvs ≤ 7 cm wide; corolla yellow to orange with various darker markings; fr body ± 1 cm wide, lanceolate; taproot in width > st base . . . . *P. althaeifolia*
  2′ Ann; largest lvs gen > 7 cm wide; corolla white with ± purple tinge, various darker markings; fr body 1.5–3 cm wide, narrow-ovate; taproot in width ± = st base . . . . *P. louisianica* subsp. *louisianica*

P. althaeifolia (Benth.) Decne. (p. 903)   DESERT UNICORN-PLANT Per; taproot fusiform, tuber-like, in width > st base, yellow. **ST**: decumbent. **LF**: blade 3–7 cm wide, broad-ovate to round or deltate, gen palmately 3–5-lobed, crenate. **INFL**: 5–50-fld, gen exceeding lvs. **FL**: calyx 5-lobed, lower 3 > upper; corolla yellow to orange with various darker markings. **FR**: body ± 1 cm wide, lanceolate, spines 0. **SEED**: 6–7 mm. Sandy places; < 1000 m. se DMoj, DSon; to TX, n Mex, Peru. May–Aug ★

P. louisianica (Mill.) Thell. subsp. *louisianica*   Ann; taproot in width ± = st base. **LF**: blade gen 5–20 cm wide, broad-triangular to round, entire to shallowly indented. **INFL**: 20–40-fld, exceeding lvs. **FL**: calyx 5-lobed, lower 3 > upper; corolla white with ± purple tinge,

darker markings. **FR**: body 1.5–3 cm wide, narrowly ovate, spines 0. **SEED**: 7–9 mm. 2*n*=30. Uncommon. Open, disturbed areas; < 500 m. NCoRO, NCoRI, GV, SCoRO, SCo, PR, cult elsewhere; to s-c US (where perhaps native), e US. Widely cult as novelty. May–Aug

P. lutea (Lindl.) Stapf (p. 903)   Ann. **LF**: blade 10–20 cm wide, blade ± rounded, angled or not, entire to dentate. **INFL**: few–many-fld, gen exceeded by lvs. **FL**: sepals free, upper 3 narrower; corolla yellow, often with orange tinge, gen with darker markings. **FR**: body 3–4 cm wide, ovate, short-spiny. 2*n*=30,32. Uncommon. Open, disturbed places; < 500 m. GV, SnFrB, expected elsewhere; native to S.Am. Jul–Aug

MELIACEAE MAHOGANY FAMILY

Robert E. Preston & Elizabeth McClintock

[(Per), shrub] or tree; wood hard, often aromatic. **LF:** gen alternate, often clustered near st tips, gen odd-2-pinnate, bases gen swollen; stipules 0. **INFL:** panicle [raceme or umbel]. **FL:** gen bisexual, radial; sepals gen 3–5, fused at base or not; petals gen 3–5, free [or ± fused at base or to filament tube]; stamens gen 8–12, filaments gen fused; disk gen between stamens and ovary; ovary superior, chambers gen 2–5, placentas axile, style gen 1, stigma gen head-like, lobed. **FR:** gen drupe. **SEED:** many, often winged or with an aril. ± 50 genera, ± 550 spp.: trop, subtrop (some temp). Timber, incl mahogany (*Swietenia*). [Muellner et al. 2006 Molec Phylogen Evol 40:236–250] Scientific Editor: Thomas J. Rosatti.

MELIA BEAD TREE

LF: deciduous, petioled. **INFL:** panicle; fls many. **FL:** white or purple; sepals gen 5; petals gen 5; filament tube 10–12-lobed at tip (lobes sometimes further divided), anthers 10–12; pistil surrounded by, ± = filament tube, ovary chambers 5–8, style ± as wide as ovary, stigma. ± 10 spp.: trop Asia, Australia. (Greek: ash tree, from lf shape)

M. azedarach L. CHINA BERRY, PERSIAN LILAC Tree, < 10 m. **ST:** branches broadly spreading; bark furrowed. **LF:** ± 20–40 cm; 1° lflets ± 5–9; 2° lflets gen 5–7 per 1° lflet, 2.5–5 cm, ovate to lanceolate, toothed. **FL:** white to lilac, fragrant; sepals ± 2–3 mm; petals ± 5 mm, oblong; filament tube ± 5 mm, purple, aging black. **FR:** 10–15 mm, spheric, yellow. **SEED:** 1, bony. Washes, riparian areas, coastal scrub, or persisting near old habitations; < 1280 m. SnFrB, SCo, SnGb, SnBr, PR; native to se Asia, n Australia. Fast-growing, used in reforestation; fr pulp mildly toxic; seeds used for beads. Mar–Jul

MENYANTHACEAE BUCKBEAN FAMILY

Robert F. Thorne, C. Barre Hellquist & William J. Stone

Per [ann], ± aquatic, herbage glabrous; rhizomes thick. **LF:** simple or of 3 lflets, gen alternate; stipules gen 0; petiole bases sheathing. **INFL:** various. **FL:** bisexual [unisexual], radial; sepals 5, united or not; corolla rotate to funnel-shaped, lobes 5, fringed or coarse-hairy marginally or adaxially or not; stamens 5, epipetalous, alternate corolla lobes, anthers sagittate; nectaries gen 5, at ovary base; pistil 1, ovary gen ± superior, chamber 1, stigma 2-lobed. **FR:** capsule [± fleshy], valves 2–4. **SEED:** few to many, smooth, shiny. 5 genera, ± 70 spp.: worldwide. Scientific Editor: Thomas J. Rosatti.

1. Lvs of 3 lflets; fls white or pink . . . . **MENYANTHES**
1′ Lvs simple; fls yellow . . . . **NYMPHOIDES**

MENYANTHES BUCKBEAN, BOGBEAN

1 sp. (Greek: disclosing, fl, from fls opening in succession in infl)

M. trifoliata L. (p. 903) Rhizome covered with old lf bases. **ST:** prostrate or fl branches ascending. **LF:** basal, alternate, emergent; lflets 3, 2–12 cm, 1–5 cm wide, oblong-obovate, ± entire; petioles 5–30 cm; stipules as wing-margins of petiole. **INFL:** raceme; peduncle 20–40 dm; pedicel 5–25 mm. **FL:** calyx persistent, tube short-conic, lobes 5, 2–5 mm, oblong; corolla funnel-shaped, white to pink, tube 5–8 mm, lobes 5–8 mm, spreading, gen ± pink at tip, coarse-hairy adaxially; filaments thread-like, anthers sagittate; style persistent. **FR:** 2-valved, ± ellipsoid. **SEED:** ± elliptic, ± compressed. 2*n*=54,108. Ponds, bogs, swamps, wet meadows, seeps, margins of shallow lakes; 900–3200 m. c NCo, KR, CaR, SN (exc Teh), n CCo; to AK, CO, MI, ME, circumboreal. Lvs sometimes used in beer-making as hops substitute. May–Aug

NYMPHOIDES FLOATING-HEART

LF: simple, submersed or floating, deeply cordate, gen alternate, those on fl sts gen opposite; petioles > blades. **INFL:** ± umbel, in axils of petioles near top of long, petiole-like fl st, often with clusters of tuberous roots. **FL:** corolla rotate, white or yellow, lobes with a glandular appendage near base; filaments < to > style; stigma persistent, 2-lobed. **FR:** irregularly dehiscent. ± 50 spp.: N.Am, S.Am, Eurasia, Afr, Australia. (Greek: Nymphaea-like, from resemblance of lvs to those of water lily) Sometimes in Gentianaceae, but pls differ in their ± aquatic habitats, gen alternate, petioled lvs, and several other ways.

N. peltata (S.G. Gmel.) Kuntze WATER FRINGE Heterostylous. **ST:** < 160 cm, creeping; fl sts floating. **LF:** 3–10 cm wide; margin wavy or entire. **INFL:** fls 2–5. **FL:** pedicel 3–10 cm, ± thick; calyx lobes lance-oblong; corolla 30–40 mm diam, yellow, lobe margins wide, thin, wavy, fringed-ciliate; styles < filaments on some pls, > filaments on others. **SEED:** ovate, flat, coarsely ciliate. 2*n*=54. Still water, slow-moving rivers; ± 1100 m. n SNH (Trout Lake, El Dorado Co.), CCo (reported); WA, ne US, AR, AZ; native to Eurasia, Medit. Cult in aquatic gardens; listed as invasive in many states. Jun–Sep

MOLLUGINACEAE CARPET-WEED FAMILY

Michael A. Vincent & Wayne R. Ferren, Jr.

Ann [per, shrub], glabrous or hairy. **ST:** prostrate to erect. **LF:** simple, gen basal and cauline, alternate, opposite, or whorled, entire, (fleshy), petioled [not]; stipules 0 or small, deciduous. **INFL:** axillary or terminal cyme [umbel] or fls 1; fls pedicelled

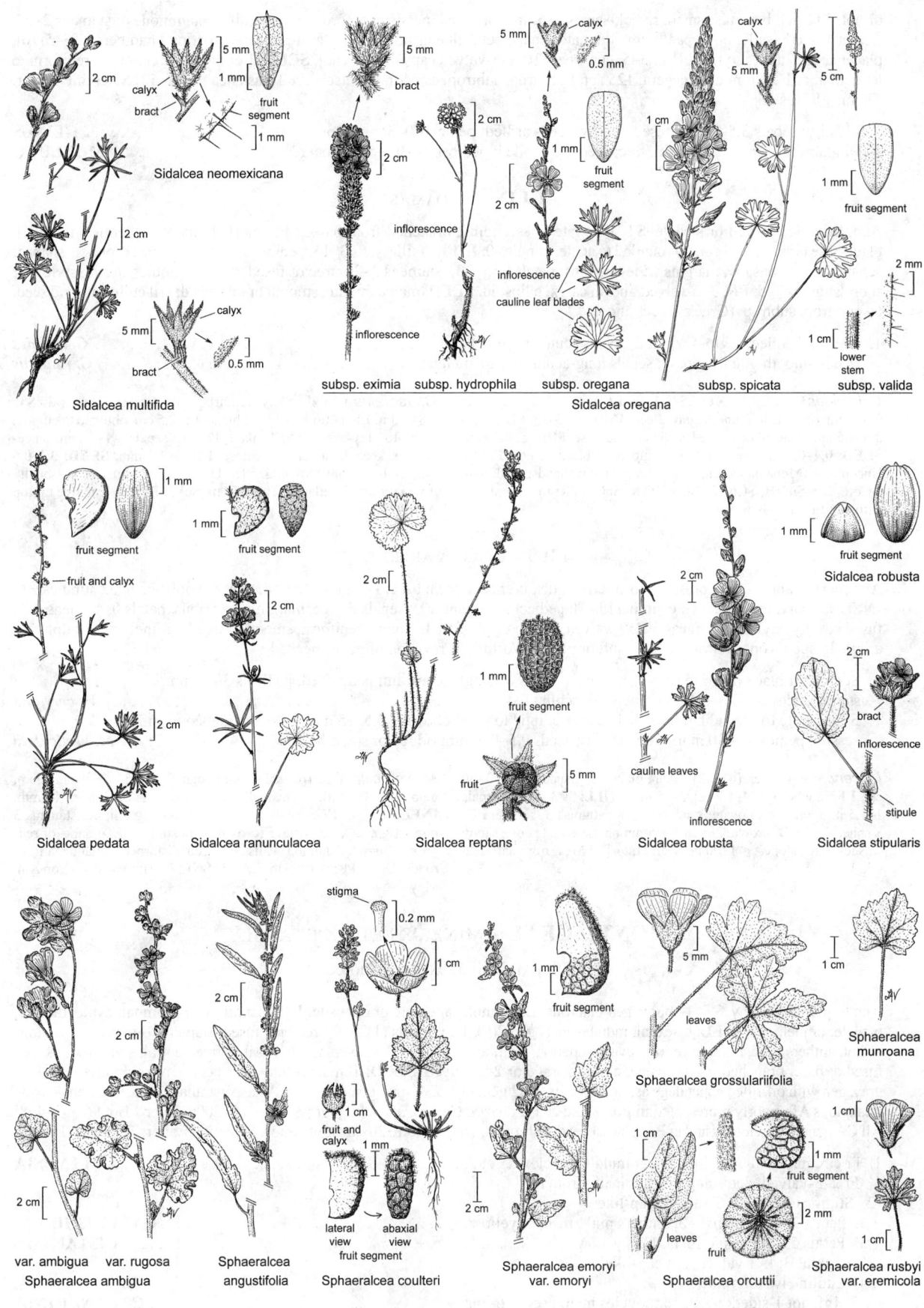

Sidalcea neomexicana

Sidalcea multifida

5 mm bract

5 mm calyx 0.5 mm

5 mm calyx 5 mm

calyx 5 mm 5 cm

1 mm fruit segment

2 cm

inflorescence

1 mm fruit segment

2 cm

inflorescence

cauline leaf blades

1 cm

2 mm

1 cm lower stem

subsp. eximia   subsp. hydrophila   subsp. oregana   subsp. spicata   subsp. valida

Sidalcea oregana

1 mm fruit segment

1 mm fruit segment

fruit segment

fruit and calyx

2 cm

2 cm

Sidalcea pedata

Sidalcea ranunculacea

2 cm

1 mm fruit segment

fruit

5 mm

Sidalcea reptans

2 cm

cauline leaves

inflorescence

Sidalcea robusta

1 mm fruit segment

Sidalcea robusta

2 cm

bract

inflorescence

stipule

Sidalcea stipularis

var. ambigua   var. rugosa

Sphaeralcea ambigua

2 cm

2 cm

2 cm

Sphaeralcea angustifolia

stigma 0.2 mm

1 cm

fruit and calyx 1 cm

1 mm

lateral view   abaxial view
fruit segment

2 cm

Sphaeralcea coulteri

1 mm fruit segment

2 cm

Sphaeralcea emoryi var. emoryi

5 mm

leaves

Sphaeralcea grossulariifolia

1 cm

leaves

fruit

1 cm

1 cm

fruit segment

2 mm

Sphaeralcea orcuttii

1 cm

1 mm

1 cm

Sphaeralcea munroana

Sphaeralcea rusbyi var. eremicola

or not. **FL:** gen bisexual, small, radial; sepals 4–5, free or fused, persistent; corolla 0 or small (= staminodes); stamens 2–25, free or fused basally, on hypanthium, alternate sepals, petal-like or not; nectary a ring; ovary superior, chambers [1]3–5[10], placentas gen axile, styles [0]1 or 3–5, gen free. **FR:** 3–5-valved capsule [achene]. **SEED:** 1 or more per chamber, reniform to lenticular, arilled or not. 14 genera, 125 spp.: gen trop, subtrop, esp Afr. [Boetsch 2002 Castanea 67:42–53] Scientific Editor: Thomas J. Rosatti.

1. Pl hairy; sepals 3.5–7 mm, fused basally; seeds arilled; pedicels 0–3 mm, stout . . . . **GLINUS**
1′ Pl glabrous; sepals 1–2.5 mm, free; seeds not arilled; pedicels 3–30 mm, slender . . . . **MOLLUGO**

GLINUS DAMASCISA

Ann, hairs gen forked or stellate. **ST:** prostrate to ascending, branched from base. **LF:** whorled [alternate], entire [toothed]; blade base tapered, tip broadly rounded to acute; stipules 0. **INFL:** axillary, fls 3–15, pedicel 0–3 mm. **FL:** flask- or bell-shaped; sepals 5, fused basally, margins wide-scarious; petals 0 [5–20]; stamens 3–20, free or fused in groups, outer sterile; style 0 or short, stigmas 3–5. **FR:** 3–5-valved, incl in calyx, ellipsoid. **SEED:** many, minute, smooth or tubercled; aril coiled around seed. 6 spp.: trop, subtrop. (Greek: sweet juice)

1. Seeds tubercled, ± glossy or dull; sepals acute or ± mucronate. . . . . *G. lotoides*
1′ Seeds smooth, highly glossy; sepals long-acuminate to attenuate . . . . *G. radiatus*

G. lotoides L. (p. 903)   **ST:** 0.5–3.5 dm. **LF:** petiole 1–7 mm; blade 0.5–3 cm, obovate to round, ± gray-green. **INFL:** fls 3–15. **FL:** sepals 3.5–4.5 mm, lanceolate, keeled, stellate-tomentose. **FR:** 3.5–4.5 mm. **SEED:** 0.4–0.6 mm, orange-brown, tubercles black or not. 2*n*=36. Uncommon. Moist or seasonally dry margins of wetlands; < 1000 m. NCoR, GV, SnFrB, SCoRO, SCo, WTR, SnGb, PR; to AK, se US; native to Eur. Jun–Nov

G. radiatus (Ruiz & Pav.) Rohrb. SHINING DAMASCISA **ST:** 0.8–5 dm. **LF:** petiole 1–7 mm; blade 0.5–2.5 cm, obovate to elliptic, green to gray-green. **INFL:** fls 3–11. **FL:** sepals 4.1–7 mm, lance-oblong, keeled, stellate-tomentose. **FR:** 3–3.5 mm. **SEED:** 0.4–0.5 mm, red- to gold-brown. 2*n*=18. Uncommon. Moist or seasonally dry margins of wetlands; 100–450 m. ScV, PR; to LA; native to trop Am. Jun–Sep

MOLLUGO CARPET-WEED

Ann [per], glabrous. **ST:** prostrate to erect, slender, branched from base. **LF:** gen whorled, linear to oblanceolate; stipules ± 0. **INFL:** axillary, terminal, cyme, umbel [fls 1]; pedicel 3–30 mm. **FL:** sepals 5, free, margins ± scarious; petals 0; stamens 3–5, fused basally; styles 3–5, linear. **FR:** 3-valved, incl in calyx. **SEED:** many, reniform, smooth, ridged, or finely net-sculptured; aril 0. 35 spp.: trop, subtrop. (Greek: soft or pliant) Worldwide revision of genus needed.

1. St erect; lf blade linear to spoon-shaped, 1–5 mm wide, glaucous; infl peduncled; pedicels 3–11 mm; stamens 5; seeds 0.3–0.4 mm, finely net-sculptured . . . . *M. cerviana*
1′ St prostrate to ascending; lf blade linear or elliptic to wide-obovate, 0.5–15 mm wide, not glaucous; infl ± sessile; pedicels 3–30 mm; stamens 3[4]; seeds 0.5–0.6 mm, ridged or smooth . . . . *M. verticillata*

M. cerviana (L.) Ser. (p. 903)   Ann, glabrous, glaucous. **ST:** 3–20 cm. **LF:** in whorls of 4–10[12], 3–15 mm. **INFL:** fls 3–4. **FL:** sepals 1–1.5 mm, elliptic to ovate; stamens 1 mm; stigmas 3, sessile. **FR:** ± spheric. **SEED:** brown. 2*n*=18. Uncommon. Seasonal pools, sandy washes, flats, slopes; < 1700 m. SnJt, D; to TX, Mex, trop; native to Old World. Sep–Mar

M. verticillata L. (p. 903)   Ann, mat-forming, < 50 cm diam, glabrous. **ST:** forked unequally. **LF:** in whorls of 3–8, 5–40 mm. **INFL:** fls 2–6. **FL:** sepals 1.5–2.5 mm, oblong-elliptic; stamens 3 mm; stigmas 3, ± sessile. **FR:** ovoid-ellipsoid. **SEED:** dark- or red-brown. 2*n*=64. Common. Moist, exposed, disturbed wetland margins, roadsides, fields; < 1000 m. CA-FP, SNE; N.Am; native to trop Am. May–Nov

MONTIACEAE MINER'S LETTUCE FAMILY

John M. Miller, except as noted

Ann to per; gen fleshy. **ST:** 1–many, gen glabrous. **LF:** simple, alternate or opposite. **INFL:** axillary or terminal; cyme, raceme, panicle, umbel, or fl 1. **FL:** bisexual, radial; sepals gen 2(9), free; petals (1)2–19, free or ± fused; stamens 1–many, epipetalous or not, anthers pink, rose, or yellow; ovary superior, chamber 1, ovules 1–many, placenta basal or free-central; styles (0)1–8, gen fused at base, branched. **FR:** capsule, circumscissile or 2–3-valved. **SEED:** 1–many, shiny or ± pebbly or sculptured, black or gray, gen with oil-filled appendage as food for ants. ± 22 genera, ± 230 spp.: gen temp Am, Asia, Australia, Eur, Kerguelen Is, New Zealand, s Afr, poorly represented in Eur; some cult (*Lewisia, Calandrinia*). [Ogburn & Edwards 2009 Amer J Bot 96:391–408] All CA genera previously incl in Portulacaceae; details of fls, seeds require 20× magnification. Scientific Editor: Thomas J. Rosatti.

1. Fr circumscissile, dehiscing near middle or below . . . . **LEWISIA**
1′ Fr 2–3-valved, dehiscing longitudinally from top
   2. Stigmas 2–3; fr 2–3-valved, cap-like
      3. Petals 2–4; stamens 1–3, anthers pink, rose, or yellow . . . . **CALYPTRIDIUM**
      3′ Petals 5; stamens 5–10, anthers yellow . . . . **CISTANTHE**
   2′ Stigmas 3; fr 3-valved, not cap-like
      4. Cauline lvs alternate
         5. Infl not 1-sided; petals red; ovules many; seeds 6–many . . . . **CALANDRINIA**
         5′ Infl gen 1-sided; petals white to pink; ovules 3; seeds 1–3 . . . . ²**MONTIA**

4′ Cauline lvs ± opposite
  6. Cauline lvs gen 2, free, ± fused on 1 side, or fused into ± disk; ovules, seeds 3–6 . . . . **CLAYTONIA**
  6′ Cauline lvs > 2, free; ovules 3, seeds 1–3 . . . . ²**MONTIA**

CALANDRINIA

Ann [per], ± fleshy, ± glabrous or glaucous. **ST:** several to many, prostrate to erect, 3–45 cm. **LF:** simple, alternate; blade linear to spoon-shaped, flat [cylindric]. **INFL:** raceme or panicle; bracts lf-like [or scarious]. **FL:** sepals 2, overlapped, persistent in fr; petals (3)5(7), red [(white)]; stamens 3–15; style 3-branched. **FR:** 3-valved. **SEED:** 6–many, ovate to ± elliptic, gen black, smooth, finely tubercled, or with fine, net-like pattern. 150 spp.: w Am, Australia. (J.L. Calandrini, Swiss scientist, 1703–1758) [Hershkovitz 1993 Ann Missouri Bot Gard 80:333–365, 366–396] Other taxa in TJM (1993) moved to *Cistanthe*.

1. Fr gen > calyx by 3+ mm; seed finely tubercled . . . . *C. breweri*
1′ Fr gen not > calyx by 3+ mm; seed with fine, net-like pattern . . . . *C. ciliata*

C. breweri S. Watson (p. 903) BREWER'S CALANDRINIA Ann, ± glabrous. **ST:** prostrate to ascending. **LF:** 2–8 cm, ± ovate to spoon-shaped. **INFL:** raceme, elongate; pedicel 6–20 cm, gen curved in fr. **FL:** sepals 4–6 mm, glabrous to ± ciliate; petals gen 5, 3–5 mm; stamens 3–6. **SEED:** 10–15, 1–2 mm wide, ± elliptic. Sandy to loamy soil, disturbed sites, burns; < 1200 m. NCoR, c SNF, SnFrB, SCoRO, SCo, WTR; n Baja CA. Feb–May ★

C. ciliata (Ruiz & Pav.) DC. (p. 903) RED MAIDS Ann, ± glabrous. **ST:** spreading. **LF:** 1–10 cm, linear to oblanceolate. **INFL:** raceme, elongate; pedicel 4–25 mm, gen straight in fr. **FL:** sepals 2.5–8 mm, glabrous to ciliate; petals gen 5, 4–15 mm; stamens 3–15. **SEED:** 10–20(–30), 1–2.5 mm wide, elliptic. 2*n*=24. Common. Sandy to loamy soil, grassy areas, cult fields; < 2200 m. CA-FP, w MP, s SNE, n DMtns (Coso Range); to NM, C.Am; nw S.Am. Feb–May

CALYPTRIDIUM PUSSYPAWS

John M. Miller & C. Matt Guilliams

Ann, per, ± fleshy, from taproot or fibrous roots, glabrous. **ST:** 1–several, gen spreading to ascending. **LF:** basal or basal and cauline, simple, oblanceolate to spoon-shaped; basal rosetted. **INFL:** raceme, panicle, or umbel, scapose, bracts gen < sepals, lf-like or not; fls gen on 1 side of axis, persistent in fr or not; pedicels ± jointed with a transverse groove or constriction. **FL:** sepals 2, ovate to reniform, gen scarious or scarious-margined, persistent in fr; petals 2–4, < sepals, tips adherent, forming cap in fr (fr cap), falling as 1 unit; stamens 1–3, anthers pink, rose, or yellow; style incl to exserted, stigmas 2. **FR:** 2-valved, gen compressed, narrowly oblong to ± round, gen translucent, deciduous or not. **SEED:** 1–many, black, dull, fine- to coarse-papillate to shiny, smooth. 8 spp.: w N.Am. (Greek: cap, for petal tips in fr) [Hershkovitz 2006 Gayana Bot 63:13–74]

1. Per; style thread-like, 2–4 mm, exserted
  2. Infl ≥ 2 per rosette; cauline lvs gen present; petals rose to white . . . . *C. monospermum*
  2′ Infl gen 1 per rosette; cauline lvs gen 0; petals white . . . . *C. umbellatum*
1′ Ann; style thread-like or not, gen < 2 mm, gen incl
  3. Style 1–2 mm; seeds 1–3, 0.7–0.9 mm; c SN . . . . *C. pulchellum*
  3′ Style < 0.5 mm; seeds ≥ 4, 0.4–0.8 mm; widespread, incl e slope SNH (for *C. roseum*) but otherwise exc c SN
    4. Seed < 0.5 mm, papillae gen 0 or marginal; pedicel 1–3 mm, slender
      5. Petals 4; stamens gen 3 . . . . *C. pygmaeum*
      5′ Petals 2; stamen 1 . . . . *C. roseum*
    4′ Seed ≥ 0.5 mm, papillae marginal or throughout (exc *Calyptridium parryi* var. *arizonicum*); pedicel < 2
      mm, thickened (slender, 1–2 mm in *Calyptridium quadripetalum*)
      6. Petals gen 3; stamen 1; fr oblong to linear; seeds 4–10, gen from base to above middle of fr, often
        in ± 1 row . . . . *C. monandrum*
      6′ Petals 4; stamens 1–3; fr widely ovate or lanceolate to elliptic or oblong; seeds 5–many, gen from
        base to middle of fr, not in ± 1 row
        7. Seed coarsely papillate gen throughout; basal lvs gen persistent in fr; fls deciduous in fr; sepals
          scarious throughout . . . . *C. quadripetalum*
        7′ Seed smooth or ± finely papillate throughout or on margin; basal lvs gen withering in fr; fls persistent
          in fr to not; sepals gen scarious on margin or becoming scarious throughout. . . . . *C. parryi*
          8. Seed smooth, 0.7–0.8 mm; fr 6–7 mm; w DSon . . . . var. *arizonicum*
          8′ Seed papillate, at least on margin, gen 0.5–0.7 mm; fr < 6 mm; sw SnFrB, n SCoR, s SCoRO, TR, SnJt, SNE, DMtns
            9. Seed papillae throughout; s SCoRO, TR, SnJt . . . . var. *parryi*
            9′ Seed papillae marginal; sw SnFrB, n SCoR, SNE, DMtns
              10. Seed 0.5–0.6 mm; sepals ovate, with gen narrowly scarious margin; 600–1050 m, sw SnFrB, n
                SCoR . . . . var. *hesseae*
              10′ Seed 0.6–0.7 mm; sepals ovate to reniform, with widely scarious margin or scarious throughout;
                1500–2990 m, SNE, DMtns . . . . var. *nevadense*

C. monandrum Nutt. (p. 903) Ann 1.5–18 cm; taproot slender. **ST:** spreading to decumbent, lfy. **LF:** basal 1–5 cm, narrow-oblanceolate to -spoon-shaped, withering in fr; cauline 1–2(4) cm. **INFL:** axillary, raceme or panicle, 1–4 cm, gen open; bracts narrowly ovate; fls ± sessile, gen persistent in fr. **FL:** sepals 1–2 mm, ovate to deltate, fleshy, narrowly white on margin in fl, scarious throughout in fr; petals gen 3, 1–3 mm, pink to ± red; stamen 1; stigma sessile. **FR:** 3–6 mm, oblong to linear, in width gen ± > seed. **SEED:** 4–10, 0.5–0.8 mm. 2*n*=44. Widespread in desert and scrub, open areas, sandy soils, burns; gen < 2000 m. s SN, Teh, SnJV, SnFrB, SCoR, SW, SNE, D; NV, AZ, nw Mex. [*Cistanthe m.* (Nutt.) Hershk.] Jan–Jul

C. monospermum Greene (p. 903) Per < 50 cm; caudex short, thick; taproot slender to thick. **ST:** gen spreading to ascending, lfy to not. **LF:** basal 1.5–6 cm, rosette 1; cauline gen present, 0.8–3 cm. **INFL:** axillary, umbel, ± open (dense in small pls), ≥ 2 per rosette, 1–10 cm diam; pedicel ± 0 to short. **FL:** sepals 3–8 mm, ± round, scarious; petals 4, 3–7 mm, rose to white; stamens 3; style 2–4 mm, thread-like, exserted, falling with fr cap. **FR:** 2–3.5 mm, widely ovate to ± round; deciduous or not. **SEED:** 1–4(8). 2*n*=44. Open areas, sandy or gravelly soils, conifer forest; 300–3970 m. KR, NCoRH, CaR, SN, SnFrB, TR, SnJt, GB; s OR, NV, n Baja CA. [*Cistanthe m.* (Greene) Hershk.] Hybridizes with *C. umbellatum* in SN. Apr–Sep

C. parryi A. Gray Ann 2–11 cm; taproot slender. **ST:** spreading to ascending, lfy. **LF:** basal and cauline, 1–3 cm, basal gen withering in fr. **INFL:** axillary, raceme or panicle, open to dense, 1–3.5 cm; bracts ovate to elliptic; pedicel stout, ≤ 1 mm, gen jointed at base; fls persistent in fr to not. **FL:** sepals 2–5 mm, gen unequal (outer >, wider than inner), round, ovate, or reniform, gen scarious on margin or becoming scarious throughout; petals 4, 1.5–3 mm, gen white; stamens (1)3; style 0.5–1 mm, stigma ± sessile. **FR:** 3–7 mm, ovate to oblong, deciduous or not. **SEED:** 5–20, 0.5–0.8 mm. 2*n*=44. [*Cistanthe p.* (A. Gray) Hershk.] Unnamed var. in Baja CA.

var. ***arizonicum*** J.T. Howell **FL:** sepals widely reniform, margin scarious. **FR:** 6–7 mm, exserted from calyx. **SEED:** 0.7–0.8 mm, smooth, shiny, papillae 0. Coarse, well-drained soils in desert scrub, wash; 600–800 m. w DSon; s AZ, nw Mex. [*Cistanthe p.* var. *a.* (J.T. Howell) Kartesz & Gandhi] 2 populations known in CA, in n Pinyon Mtn area in Anza-Borrego Desert State Park. Mar–Apr ★

var. ***hesseae*** J.H. Thomas (p. 903) SANTA CRUZ MOUNTAINS PUSSYPAWS **FL:** sepals ovate, margin gen narrowly scarious. **FR:** readily deciduous. **SEED:** 0.5–0.6 mm, papillae marginal. Sandy soils in chaparral, oak woodland, conifer forest; 600–1050 m. sw SnFrB, n SCoR. [*Cistanthe p.* var. *h.* (J.H. Thomas) Kartesz & Gandhi] Apr–Jul ★

var. ***nevadense*** J.T. Howell **FL:** sepals ovate to reniform, with widely scarious margin or scarious throughout. **FR:** deciduous or not. **SEED:** 0.6–0.7 mm, papillae marginal. Mixed desert scrub, pinyon/juniper woodland; 1500–2990 m. SNE, DMtns; w NV, Baja CA. [*Cistanthe p.* var. *n.* (J.T. Howell) Kartesz & Gandhi] Apr–Jul

var. ***parryi*** (p. 903) **FL:** sepals ovate to reniform, margin often scarious. **FR:** gen not deciduous. **SEED:** 0.6–0.7 mm, papillate throughout. Open, sandy areas in chaparral, conifer forest; 1400–3500 m. s SCoRO, TR, SnJt. May–Jul

C. pulchellum (Eastw.) Hoover MARIPOSA PUSSYPAWS Ann 2–7 cm; roots fibrous. **ST:** spreading to ascending, lfy. **LF:** basal 0.5–2 cm, rosette 1; cauline 0.2–1.5 cm. **INFL:** axillary, panicle of head-like clusters of fls, open, 1–4 cm, gen 2+ per rosette; bracts ovate to deltate; fls ± sessile, gen persistent in fr. **FL:** sepals 3–4 mm,

± reniform to round, scarious; petals 4, ± 3 mm, rose; stamens 3; style 1–2 mm, thread-like, gen incl. **FR:** 1.5–2.5 mm, ovoid or widely elliptic to ± round. **SEED:** 1–3, 0.7–0.9 mm, shiny. Sandy soils, decomposed granite or metamorphic rocks, chaparral, gray pine, oak woodland; 400–1100 m. c SN (s Mariposa, Madera, n Fresno cos.). [*Cistanthe p.* (Eastw.) Hershk.] Apr–Jul ★

C. pygmaeum Rydb. PYGMY PUSSYPAWS Ann 0.5–8 cm; taproot slender or roots fibrous. **ST:** spreading to erect, lfy. **LF:** basal and cauline, 5–1.5 cm, gen persistent in fr. **INFL:** axillary, raceme or panicle, ± dense, 0.5–3 cm; bracts ovate to ± round; pedicel 1–3 mm; fls persistent in fr. **FL:** sepals 2–4 mm, ovate, fleshy, membranous in age, margin white or not; petals 4, 2–3 mm, white; stamens gen 3; stigma ± sessile. **FR:** 3–5 mm, ± ovate to elliptic. **SEED:** 15–30, 0.4–0.5 mm, smooth, shiny. 2*n*=44. Sandy to gravelly soils, conifer forest; 2100–3200 m. c&s SNH, SnBr. [*Cistanthe p.* (Rydb.) Hershk.] Jun–Jul ★

C. quadripetalum S. Watson (p. 903) FOUR-PETALED PUSSYPAWS Ann 1.5–13 cm; taproot slender or roots fibrous. **ST:** spreading to erect, lfy. **LF:** basal and cauline, 0.5–6 cm, gen persistent in fr. **INFL:** axillary, raceme or panicle, ± dense, 0.5–4 cm; bracts ovate to ± round; pedicel 1–2 mm, slender; fls deciduous in fr. **FL:** sepals 4–6 mm, reniform to round, scarious; petals 4, 2–3 mm, white to pink; stamens 1–3; stigma sessile. **FR:** 3–4 mm, lanceolate to ovate. **SEED:** 10–20, 0.5–0.6 mm, dull, coarsely papillate gen throughout. 2*n*=44. Sandy or gravelly areas, gen serpentine; 400–2000 m. NCoRH, NCoRI. [*Cistanthe q.* (S. Watson) Hershk.] Apr–Jun ★

C. roseum S. Watson (p. 903) ROSY CALYPTRIDIUM Ann 1.5–10 cm; taproot slender or fibrous. **ST:** spreading to ascending, lfy. **LF:** basal and cauline, 0.5–4 cm, gen persistent in fr. **INFL:** axillary, raceme or panicle, open, 1–4 cm; bracts ovate to elliptic; pedicel slender, 1–3 mm; fl persistent in fr to ± not. **FL:** sepals 2–3 mm, ovate, gen scarious-margined; petals 2, ± 1 mm, white; stamen 1; stigmas ± sessile. **FR:** 2–3 mm, ovate to oblong. **SEED:** 10–25, 0.4–0.5 mm, shiny, papillae marginal. 2*n*=44. Gravelly soils, conifer forest, sagebrush scrub; 1500–3800 m. e slope SNH, n SNE, W&I; to OR, WY, UT. [*Cistanthe r.* (S. Watson) Hershk.] May–Aug

C. umbellatum (Torr.) Greene (p. 903) Per < 6 dm; caudex short, thick; taproot slender to thick. **ST:** gen spreading to ascending, not lfy. **LF:** basal 1.5–7 cm, rosettes gen ≥ 2; cauline gen 0. **INFL:** terminal, umbel, compound (simple in small pls), 1–7 cm diam, gen 1 per rosette, dense; pedicel ± 0. **FL:** sepals 3–8 mm, ± round, scarious; petals 4, 3–8 mm, white; stamens 3; style 2–4 mm, thread-like, exserted. **FR:** 2–3 mm, widely ovate to ± round, deciduous or not at maturity. **SEED:** 1–8. 2*n*=44. Open, sandy to rocky soils, conifer forest, alpine; 240–4300 m. KR, NCoRH, CaR, SN, GB; to BC, MT, WY, UT. [*C. u.* var. *caudiciferum* (A. Gray) Jeps.; *Cistanthe u.* (Torr.) Hershk.] May–Oct

CISTANTHE

John M. Miller & C. Matt Guilliams

Ann [per], ± fleshy, from taproot or fibrous roots, gen glabrous. **ST:** gen several, gen spreading to ascending. **LF:** cauline [basal and cauline], simple, linear to spoon-shaped, fleshy; basal rosetted. **INFL:** scapose, raceme, panicle, or umbel, bracts lf-like or not; bracts gen < sepals, scarious; fls gen on 1 side of axis, deciduous or not in fr. **FL:** sepals 2, ovate, green, scarious-margined or not, persistent in fr; petals 5, > sepals; stamens 5–10, anthers yellow; style 1, stigmas 3. **FR:** 3-valved, ovoid. **SEED:** 6–many, dull to shiny, black. 25–35 spp.: Am. (Greek: rockrose-fl) [Hershkovitz 2006 Gayana Bot 63:13–74]

1. Lvs not glaucous; petals white . . . . ***C. ambigua***
1′ Lvs glaucous; petals rose-red to purple . . . . ***C. maritima***

C. ambigua (S. Watson) Hershk. (p. 903) DESERT CISTANTHE **ST:** spreading to erect. **LF:** 1.5–6 cm, linear to ± spoon-shaped, ± cylindric. **INFL:** panicle of umbel-like clusters, compact; bracts scarious; pedicel 1–3 mm, straight in fr. **FL:** sepals 2–5 mm, glabrous, margins white-scarious; petals 3–5, 2–5 mm; stamens 5–7(10). **FR:** < calyx. **SEED:** 6–15, 1–2 mm wide, ± elliptic to ovate, smooth, shiny, black. Desert scrub, sandy to silty soil, often alkaline; < 1000 m. D; sw AZ, nw Mex. [*Calandrinia a.* (S. Watson) Howell] Feb–May

C. maritima (Nutt.) Hershk. SEASIDE CISTANTHE **ST:** spreading or ascending. **LF:** 1–6 cm, obovate to spoon-shaped, flat. **INFL:** panicle, gen above lvs; bracts scarious; pedicel 5–15 mm, ± straight in fr. **FL:** sepals 3–5 mm, glaucous, gen purple-veined; petals gen 5, 3–6 mm; stamens gen 5. **FR:** gen > calyx. **SEED:** 20–40, 0.5–1 mm wide, ± elliptic, smooth, dull gray, with short, white hairs, black. Coastal scrub, sandy soil, sea bluffs; < 300 m. SCo, ChI; n Baja CA. [*Calandrinia m.* Nutt.] Feb–May ★

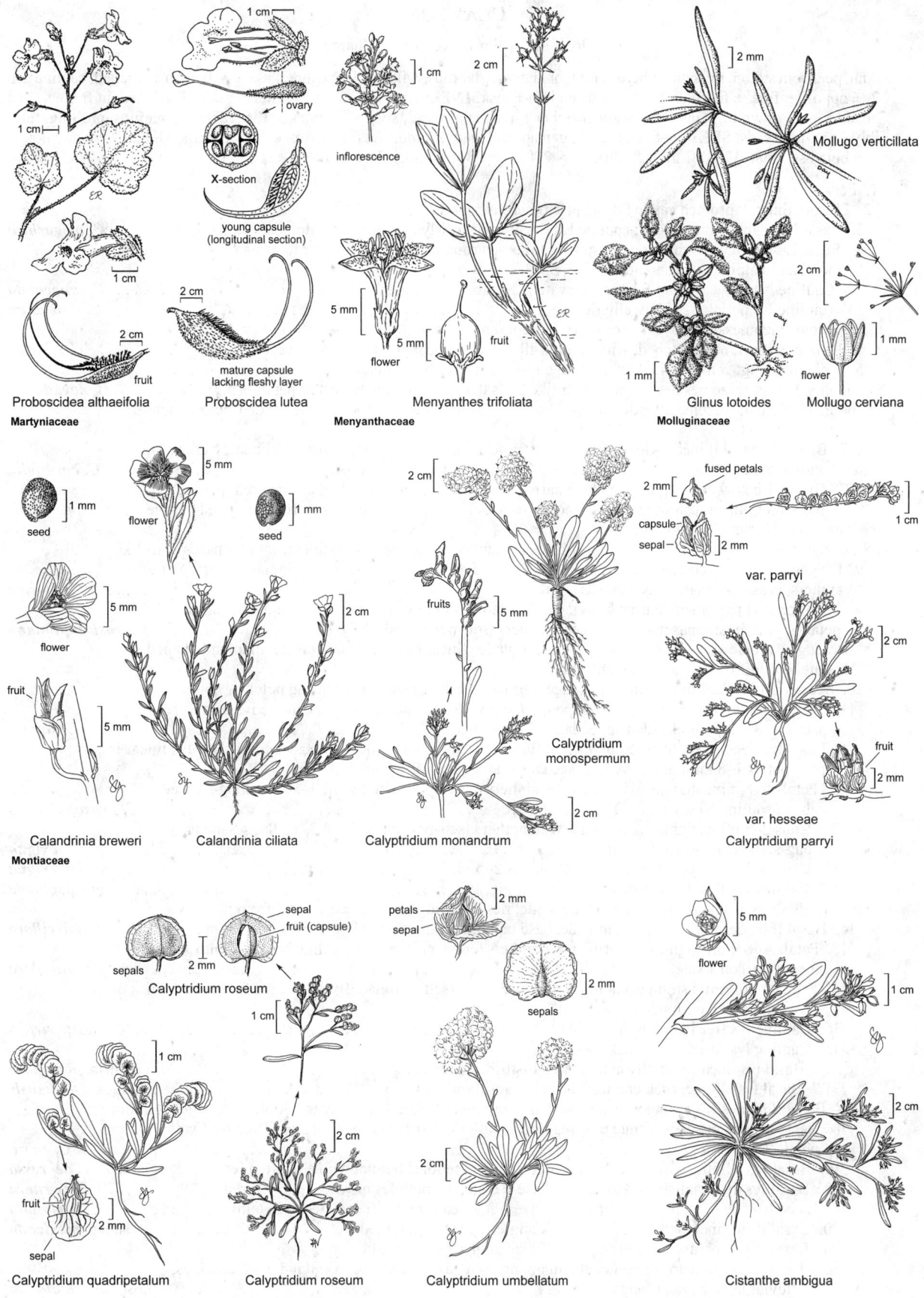

Proboscidea althaeifolia
Martyniaceae

Proboscidea lutea

Menyanthes trifoliata
Menyanthaceae

Mollugo verticillata

Glinus lotoides
Molluginaceae

Mollugo cerviana

1 cm

ovary

X-section

young capsule
(longitudinal section)

mature capsule
lacking fleshy layer

inflorescence

flower

fruit

fruit

2 mm

2 cm

1 mm

flower

seed

flower

seed

flower

fruit

Calandrinia breweri
Montiaceae

Calandrinia ciliata

fruits

Calyptridium
monospermum

Calyptridium monandrum

fused petals

capsule

sepal

var. parryi

var. hesseae

Calyptridium parryi

fruit

sepals

fruit (capsule)

sepal

Calyptridium roseum

petals

sepal

sepals

flower

fruit

sepal

Calyptridium quadripetalum

Calyptridium roseum

Calyptridium umbellatum

Cistanthe ambigua

CLAYTONIA

John M. Miller & Kenton L. Chambers

Ann, per, from stolon, rhizome, tuberous root, or taproot, glabrous, ± fleshy. **LF**: entire; basal gen 0–many, rosetted; cauline gen 2, ± opposite, free, ± fused on 1 side, or fused into ± disk. **INFL**: terminal, raceme, 1-sided; pedicel reflexed, in fr erect. **FL**: petals 5, pink or white; stamens 5; ovary chamber 1, placentas basal, style 1, stigmas 3. **FR**: valves 3, margins inrolling, forcibly expelling seeds. **SEED**: 3–6, gen black, gen appendaged. 27 spp.: C.Am, N.Am, e Asia, Siberia. (John Clayton, colonial Am botanist, 1694–1773) [Miller & Chambers 2006 Syst Bot Monogr 78:1–236]

1. Per
  2. Pl from thick, taprooted caudex or spheric tuberous root
    3. Basal lvs many, from caudex; cauline lvs linear to narrowly oblanceolate; alpine; n&c SNH, Wrn .... ***C. megarhiza***
    3′ Basal lvs 0 to few, from oblong to spheric tuberous root; cauline lvs ± linear to ovate or widely elliptic; montane to alpine; widespread, not incl SNH, incl Wrn
      4. Cauline lvs gen ± sessile, ± linear to ovate .... ***C. lanceolata***
      4′ Cauline lvs petioled, widely elliptic .... ***C. umbellata***
  2′ Pl from rhizomes, caudex gen 0 or short, taproot 0 or slender
    5. Rhizome short, not branched, without bulb-like winter buds; petals ± = to >> sepals .... ²***C. sibirica***
    5′ Rhizome long, gen branched, often with bulb-like winter buds; petals >> sepals
      6. Pl < 15 cm; rhizomes fleshy, pale, bulb-like buds 0; bract 1, below lowest fl; petals ± pink .... ***C. nevadensis***
      6′ Pl 15–60 cm; rhizomes slender, often ± brown, bulb-like buds at nodes; bracts 0 or 1 below each fl; petals gen white
        7. Bracts 0; basal lf blade widely ovate, base truncate to cordate, petiole distinct, linear; cauline lvs widely ovate, sessile .... ***C. cordifolia***
        7′ Bracts 1 below each fl; basal lf blade narrowly elliptic to oblanceolate, base tapered, petiole ± indistinct, linear; cauline lvs oblanceolate to widely elliptic, sessile or tapered to winged petiole ..... ***C. palustris***
1′ Ann (rarely per in *Claytonia rubra* subsp. *rubra*)
  8. Bracts 1 per 1–4 fls; petals 4–12 mm; lf blades oblanceolate to ovate, if ± linear, some lf bases bulb-like
    9. Fls gen 1–2 per bract; cauline lvs lanceolate to ovate, free; petals 6–12 mm; stigmas maturing after anthers (cross-pollinated); NW, CaRH, SnFrB .... ²***C. sibirica***
    9′ Fls often 2–4 per bract; cauline lvs widely ovate to ± round, free or fused < 5 mm on 1 side; petals 4–5 mm; stigmas maturing ± with anthers (self-pollinated); NCo .... ***C. washingtoniana***
  8′ Bract gen 1, below lowest fl; petals 1–8 mm; lf blades linear elliptic, oblanceolate, diamond-shaped, deltate, or reniform, lf bases not bulb-like
    10. Seeds dull black, ± tubercled, appendage ± brown, fleshy or shrunken, filling notch or not
      11. Basal lvs ± prostrate; pl cushion-shaped, < 6 cm wide, gen not glaucous; cauline lvs ovate, free; petals 5–8 mm, ± pink; seed appendage filling notch .... ***C. saxosa***
      11′ Basal lvs decumbent to erect; pl open or tufted, often > 6 cm wide, gen glaucous; cauline lvs linear if free; petals 1–8 mm, pink or white; seed appendage not filling notch
        12. Petals 5–8 mm; stigmas maturing after anthers (cross-pollinated); infl 3–30-fld, long-stalked above cauline lvs .... ***C. gypsophiloides***
        12′ Petals 2–5 mm; stigmas maturing ± with anthers (self-pollinated); infl 3–15-fld, sessile to long-stalked above cauline lvs .... ***C. exigua***
          13. Cauline lvs free or ± fused on 1 side; petals 2–5 mm .... subsp. ***exigua***
          13′ Cauline lvs fused into ± disk; petals ± 2 mm .... subsp. ***glauca***
    10′ Seeds shiny black, smooth, appendage white, fleshy, filling notch or ± exserted from it
      14. Basal lf blades > 3 × longer than wide, base tapered gradually to petiole; petals 1–6 mm .... ***C. parviflora***
        15. Petals 4–6 mm; stigmas maturing after anthers (cross-pollinated); cauline lvs fused into ± disk (or ± fused on 1 side) — SNF, Teh .... subsp. ***grandiflora***
        15′ Petals < 4(6) mm; stigmas maturing ± with anthers (self-pollinated); cauline lvs fused into ± disk or free (or ± fused on 1 side)
          16. Cauline lvs free (or ± fused on 1 side) .... subsp. ***viridis***
          16′ Cauline lvs fused into ± disk
            17. Basal lvs linear; vernally moist, often disturbed sites .... subsp. ***parviflora***
            17′ Basal lvs elliptic; rock crevices, alluvial fans, boulder fields .... subsp. ***utahensis***
      14′ Basal lf blades < 3 × longer than wide, base cordate to tapered abruptly to petiole; petals gen < 5 mm
        18. Basal lvs gen many, prostrate to spreading, smaller inward; cauline fused or ± free on 1 side; petals 2–3.5 mm .... ***C. rubra***
          19. Basal lvs elliptic to obovate, base wedge-shaped; petioles but not whole pl often red .... subsp. ***depressa***
          19′ Basal lvs diamond-shaped to deltate, base ± truncate; petioles or whole pl often red .... subsp. ***rubra***
        18′ Basal lvs few to many, spreading to erect, ± equal; cauline lvs fused; petals 2–6 mm .... ***C. perfoliata***
          20. Basal lf tips mucronate; cauline lf disk angles gen short-pointed .... subsp. ***mexicana***
          20′ Basal lf tips obtuse to acute; cauline lf disk angles gen 0 or ± obtuse
            21. Basal lvs elliptic to round-deltate (diamond-shaped); shrubland, woodland, rock crevices, alluvial fans, boulder fields .... subsp. ***intermontana***
            21′ Basal lf blades elliptic to deltate; vernally moist, often shady or disturbed sites .... subsp. ***perfoliata***

C. cordifolia S. Watson   Per; caudex short, 2–10 mm diam, horizontal, ± brown; rhizome long, with bulb-like buds; stolons 0. **ST**: 10–40 cm, erect. **LF**: basal 2–25 cm, blade 1–9 cm, widely ovate, base truncate or cordate, tip obtuse to acute, petiole linear; cauline 1–4 cm, free, widely ovate, sessile, obtuse to acute. **INFL**: stalked, open; bracts 0; fls 3–12. **FL**: sepals 3–5 mm; petals 8–13 mm, white. **FR**: 4–5 mm. **SEED**: 1.5–2 mm, round, shiny. 2*n*=10,20. Gen shaded swamps, streambanks, seeps, wet meadows; 1200–2300 m. KR, NCoRH, n SNH; to BC, MT, UT. Jun–Aug

C. exigua Torr. & A. Gray   Ann, ± glaucous. **ST**: 1–15 cm, spreading or erect. **LF**: basal < 12 cm, ± linear; cauline < 8 cm. **INFL**: sessile to long-stalked; fls 3–15, lowest bracted. **FL**: sepals 1.5–3 mm; petals 2–5 mm, white or ± pink; anthers maturing ± with stigmas. **FR**: 1.5–2.5 mm. **SEED**: 1–1.5 mm, elliptic, dull; appendage minute. Self-pollinated.

  subsp. ***exigua*** (p. 909)   **LF**: cauline linear and free or crescent-shaped and ± fused on 1 side. **FL**: petals 2–5 mm, white or ± pink. 2*n*=16,32,48. Dry or moist, disturbed bare clay to sandy soils, often serpentine; < 1000 m. NW, SNF, GV, CW, TR, PR; to BC. Apr–Jul

  subsp. ***glauca*** (Torr. & A. Gray) John M. Mill. & K.L. Chambers (p. 909)   **LF**: cauline fused into ± disk. **FL**: petals 2 mm, white. 2*n*=16. Dry or moist, disturbed bare clay to sandy soils, often serpentine; < 1000 m. KR, NCoR, SnFrB, SCoRO; to BC. Apr–Jul

C. gypsophiloides Fisch. & C.A. Mey. (p. 909)   Ann, ± glaucous. **ST**: 3–25 cm, spreading to erect. **LF**: basal 2–15 cm, linear; cauline < 6 cm, linear and free, fused on 1 side, or fully fused into 2-toothed disk. **INFL**: 2–15 cm, long-stalked; fls 3–30, lowest bracted. **FL**: sepals 2–3 mm; petals 5–8 mm, gen ± pink; anthers maturing before stigmas. **FR**: 1.5–2 mm. **SEED**: 1–1.5 mm, elliptic, dull; appendage minute. 2*n*=16. Moist, bare, often stony sites, in sun or shade, often serpentine; < 1300 m. NCoR, SnFrB, SCoR (Santa Lucia Range, Figueroa Mtn). Cross-pollinated. Mar–May

C. lanceolata Pursh (p. 909)   WESTERN SPRING BEAUTY   Per; caudex 0; tuberous root 1–3 cm wide, spheric, ± brown; rhizomes, stolons 0. **ST**: 5–15 cm, erect. **LF**: basal 0 or shriveled, 5–8 cm, elliptic, base wedge-shaped, tip acute, petiole thread-like; cauline 1–7 cm, free, ± linear to ovate, gen ± sessile. **INFL**: gen short-stalked or sessile, 1-bracted at base; fls 3–15. **FL**: sepals 3–7 mm; petals 5–12 mm, white or ± pink, base yellow or not. **FR**: 3.5–4.5 mm. **SEED**: 2–2.5 mm, round, shiny. Gravelly woodland, meadows; 1500–2600 m. KR, NCoRH, CaRH, n&c SNH, SnGb, SnBr, MP, DMtns (Panamint Range); to w Can, MT, CO, NM. [*C. l.* var. *peirsonii* Munz & I.M. Johnst.] May–Jul

C. megarhiza (A. Gray) S. Watson (p. 909)   FELL-FIELDS CLAYTONIA   Per; caudex long, 5–30 mm diam, vertical, ± brown, scaly with withered lf bases; rhizomes, stolons 0. **ST**: 1–6 cm, spreading. **LF**: basal 2–10 cm, crowded, blade 1–3 cm, widely elliptic to obovate, base tapered, tip obtuse; cauline 1–3 cm, free, ± linear, acute. **INFL**: ± sessile, dense, bracted throughout; fls 2–6. **FL**: sepals 4–6 mm; petals 5–9 mm, white or ± pink. **FR**: 3.5–4.5 mm. **SEED**: 2–2.5 mm, round, shiny. 2*n*=12,16,24,32,34,36. Subalpine, alpine gravel, talus, crevices; 2600–3300 m. n&c SNH, Wrn; to w Can, MT, CO, NM. May–Aug   ★

C. nevadensis S. Watson (p. 909)   Per; caudex long, 1–3 mm diam, horizontal, white or ± yellow, continuous with fleshy, much-branched rhizomes; bulb-like buds, stolons 0. **ST**: 2–10 cm, spreading or erect. **LF**: basal 2–15 cm, blade 1–5 cm, elliptic to widely ovate, base wedge-shaped, tip gen obtuse, petiole linear, often buried; cauline < 2 cm, free, ovate, sessile, tip obtuse. **INFL**: short-stalked or sessile, gen dense; bract 1, below lowest fl; fls 2–8. **FL**: sepals 3–8 mm; petals 6–10 mm, ± pink. **FR**: 3–4 mm. **SEED**: 1.5–2 mm, round, shiny, smooth. 2*n*=14. Subalpine streams, springs, melting snow beds, gravel or sand; 2200–3500 m. KR, CaRH, SNH, SNE; OR. Jul–Sep

C. palustris Swanson & Kelley (p. 909)   MARSH CLAYTONIA   Per; caudex short, 2–5 mm diam, horizontal, white; rhizomes, stolons slender, with bulb-like buds. **ST**: 10–60 cm, spreading to erect. **LF**: basal 8–30 cm, blade 2–8 cm, narrowly elliptic to oblanceolate, base tapered to linear, ± indistinct petiole, tip obtuse to acute; cauline 2–9

cm, free, oblanceolate to widely elliptic, sessile or tapered to winged petiole, obtuse to acute. **INFL**: stalked, open, bracts 1 below each fl; fls 5–18. **FL**: sepals 3–4 mm; petals 5–9 mm, gen white (± pink). **FR**: 2–3 mm. **SEED**: 1.5–1.8 mm, round, dull. 2*n*=12. Marshy meadows, springs, streambanks; 1000–2500 m. KR, CaRH, SNH. May–Aug   ★

C. parviflora Hook.   Ann. **ST**: 1–30 cm, spreading to erect. **LF**: basal 1–18 cm, linear to narrowly oblanceolate, blade gradually tapered to petiole, tip obtuse to acute; cauline free (or ± fused on 1 side), < 6 cm, linear (elliptic to diamond-shaped), or fused ± into < 5 cm diam, round or ± square disk. **INFL**: stalked or not, open or dense, 1-bracted at base; fls 3–40. **FL**: sepals 1.5–4 mm; petals 1–6 mm, white or ± pink. **FR**: 1.5–4 mm. **SEED**: 1.2–2.3 mm, ovate to round, shiny, smooth.

  subsp. ***grandiflora*** John M. Mill. & K.L. Chambers (p. 909)   STREAMBANK SPRING BEAUTY   **LF**: cauline fused ± into disk. **INFL**: stalked, open. **FL**: sepals 1.5–2.5 mm; petals 4–6 mm; anthers maturing before stigmas. **SEED**: 1.2–1.5 mm. 2*n*=12. Vernally moist, often disturbed sites; 150–1200 m. SNF, Teh. Cross-pollinated. May–Jul   ★

  subsp. ***parviflora***   **LF**: basal linear; cauline fused ± into disk. **INFL**: stalked or not, open or dense. **FL**: sepals 1.5–4 mm; petals 2–6 mm; anthers maturing ± with stigmas. **SEED**: 1.2–2.3 mm. 2*n*=24,36,48. Vernally moist, often disturbed sites; < 2300 m. CA-FP; to BC, MT, AZ, Baja CA. Variable; intergrades with other members of *C. perfoliata* complex. Self-pollinated. Mar–Jun

  subsp. ***utahensis*** (Rydb.) John M. Mill. & K.L. Chambers (p. 909)   **LF**: basal elliptic; cauline fused ± into disk. **INFL**: stalked or not, open or dense. **FL**: sepals 1.5–4 mm; petals 2–6 mm; anthers maturing ± with stigmas. **SEED**: 1.2–2.3 mm. 2*n*=24,36. Rock crevices, alluvial fans, boulder fields; 1000–1500 m. Teh, DMoj; to UT, AZ. [*C. p.* Hook. subsp. *p.*, sensu TJM (1993), in part] Variable; intergrades with other members of *C. perfoliata* complex. Self-pollinated. Apr–Jul

  subsp. ***viridis*** (Davidson) John M. Mill. & K.L. Chambers (p. 909)   **LF**: cauline free (or ± fused on 1 side), linear (elliptic to diamond-shaped), often curved, spreading or erect. **FL**: sepals 1.5–2 mm; petals 2–3.5 mm; anthers maturing ± with stigmas. **SEED**: 1.2–1.5 mm. 2*n*=24,36. Shrub- or woodland, dry or not; decomposed granite, sandstone rock crevices, boulder fields; 100–2000 m. s SN, SCoR, TR, PR, SNE, DMtns; to NV, AZ, n Baja CA. Intergrades with *C. parviflora* subsp. *utahensis*, *C. rubra*. Self-pollinated. Apr–Jun

C. perfoliata Willd.   MINER'S LETTUCE   Ann. **ST**: 1–40 cm, spreading to erect. **LF**: basal 1–25 cm, blade < 4 cm, < 3 × longer than wide, elliptic to reniform, tip rounded to acute, petiole linear; cauline fused into < 10 cm diam, round, ± obtuse-angled, or ± square disk (or free on 1 side). **INFL**: stalked or not, open or dense, 1-bracted at base; fls 5–40. **FL**: sepals 1.5–5 mm; petals 2–6 mm, white or ± pink. **FR**: 1.5–4 mm. **SEED**: 1.2–2.7 mm, ovate to round, shiny, smooth. Highly variable; subspp. difficult because of environmental modification of character states, genetic mixing among polyploids, and geog overlap of distinct, self-pollinated forms.

  subsp. ***intermontana*** John M. Mill. & K.L. Chambers (p. 909)   **LF**: basal elliptic to round-deltate (diamond-shaped), tip obtuse to acute; cauline gen round or ± obtuse-angled. 2*n*=24,36,48. Shrubland, woodland, rock crevices, alluvial fans, boulder fields; 500–2000 m. KR, SN, GB, DMoj; to BC, WY, CO, AZ. [*C. p.* Willd. subsp. *p.*, sensu TJM (1993), in part] Polyploids from hybrids with *C. parviflora*, *C. rubra*. Apr–Jun

  subsp. ***mexicana*** (Rydb.) John M. Mill. & K.L. Chambers (p. 909)   **LF**: basal deltate to reniform, mucronate; cauline disk angles gen 2, gen short-pointed. 2*n*=12,24. Shrubland, woodland, rock crevices, rockslides; 500–2000 m. NCoR, SnFrB, SCoR, TR, PR; AZ, Mex, C.Am. Intergrades with *C. perfoliata* subsp. *p.*, *C. parviflora*, *C. rubra*. Feb–Apr

  subsp. ***perfoliata***   **LF**: basal elliptic to deltate, tip obtuse to acute (mucronate); cauline gen round or ± obtuse-angled. 2*n*=24,36,48,60. Vernally moist, often shady or disturbed sites; < 1000 m. CA-FP; to BC, MT, UT, c AZ. Polyploids from hybrids with *C. perfoliata* subsp. *mexicana*, *C. parviflora*, *C. rubra*. Jan–May

C. rubra (Howell) Tidestr.   Ann (per). **ST**: 1–15 cm. **LF**: basal 1–8 cm, blade < 2 cm, elliptic to widely deltate, base truncate to wedge-shaped, tip gen obtuse, petiole linear; cauline fused or ± free on 1 side, < 4 cm wide, gen round or with 2 ± square corners. **INFL**: ± dense, 1-bracted at base; fls 3–30. **FL**: sepals 1.5–3 mm; petals 2–3.5 mm, white to pink-white. **FR**: 2–3 mm. **SEED**: 1–2 mm, elliptic, shiny, smooth. Intergrades with *C. parviflora*, *C. perfoliata*.

subsp. ***depressa*** (A. Gray) John M. Mill. & K.L. Chambers (p. 909)   **LF**: basal elliptic to obovate, base wedge-shaped; cauline fused. 2*n*=24,36. Open, often disturbed sites, scrub; < 2500 m. NCo, KR, n CCo, SnFrB; to BC, MT. Feb–Apr

subsp. ***rubra*** (p. 909)   **LF**: basal diamond-shaped to deltate, base ± truncate; cauline fused or partly free on 1 side. 2*n*=12. Vernally moist dunes, conifer forest, woodland, scrub; < 2500 m. NW (exc NCo), CaRH, SNH, Teh, w CW (exc n CCo), TR, PR, GB; to BC, SD, CO. Apr–Jul

C. saxosa Brandegee (p. 909)   Ann. **ST**: 1–3 cm, spreading to erect. **LF**: basal crowded, < 2 cm, oblong to obovate, sessile, tip obtuse; cauline < 1.5 cm, free, ovate, sessile, obtuse. **INFL**: sessile, dense; bracts 0; fls 2–10. **FL**: sepals 2–3 mm; petals 5–8 mm, ± pink. **FR**: 2–3.5 mm. **SEED**: 1.3–2 mm, elliptic, dull. 2*n*=16. Open, rocky sites, gen serpentine; 800–2150 m. KR, NCoRO, NCoRH; s OR. Mar–May

C. sibirica L. (p. 909)   CANDY FLOWER   (Ann) per; caudex short, < 1 cm diam, vertical, ± brown; rhizomes, stolons short, forming off-set rosettes or not; taproot slender. **ST**: 5–60 cm, spreading to erect. **LF**: basal 3–30 cm, blade 1–8 cm, oblanceolate to deltate, base trun-cate to long-tapered, tip obtuse to long-tapered, often some petioles with bulb-like base; cauline 1–8 cm, free, lanceolate to ovate, sessile, gen acute. **INFL**: branched or not, stalked, open, bracted throughout; fls gen 10–20. **FL**: sepals 3–6 mm; petals 6–12 mm, gen ± pink; anthers maturing before stigmas. **FR**: 2.5–3.5 mm. **SEED**: 1.5–2 mm, round to elliptic, shiny or dull. 2*n*=12,24,36. Gen shady moist wood-land, streambanks, marshes; < 1300 m. NW, CaRH, SnFrB; to AK, MT; Siberia (Komandorski Islands). Feb–Aug

C. umbellata S. Watson   GREAT BASIN CLAYTONIA   Per from oblong to spheric, ± brown tuberous root 1–5 cm diam; caudex, rhi-zomes, stolons gen 0; taproot gen 2–5 mm diam. **ST**: 5–25 cm, erect, mostly underground. **LF**: basal 0–3, 5–25 cm, blade 1–3 cm, widely elliptic to ovate, base acute or truncate, tip obtuse, petiole mostly underground; cauline 1.5–5 cm, blade 1–3 cm, widely elliptic, obtuse, abruptly tapered to linear petiole. **INFL**: ± sessile, dense, 1–2-bracted at base; fls 2–12. **FL**: sepals 4–7 mm; petals 6–12 mm, ± pink. **FR**: 5–6 mm. **SEED**: 2.5–3 mm, round, shiny. 2*n*=16. Talus slopes, stony flats, rock crevices; 1900–3500 m. KR, GB; OR, NV. May–Aug ★

C. washingtoniana (Suksd.) Suksd. (p. 909)   Ann. **ST**: 6–40 cm, spreading to erect. **LF**: basal 4–30 cm, blade 1–5 cm, gen widely diamond-shaped to ovate or deltate, base wedge-shaped to truncate, tip gen acute; cauline 1–4 cm, free or fused < 5 mm on 1 side, widely ovate to ± round, sessile, obtuse. **INFL**: stalked or not, open, bracted throughout; fls 5–25, clustered. **FL**: sepals 2.5–4 mm; petals 4–5 mm, ± pink [or not]; anthers maturing ± with stigmas. **FR**: 2.5–3 mm. **SEED**: 1.5–2 mm, elliptic, shiny or dull. 2*n*=24. Gen shady moist woodland, streambanks, seeps; < 150 m. NCo; to BC. Fertile deriva-tive of *C. sibirica* × *C. perfoliata*. Self-pollinated. Jan–Jun

LEWISIA

John M. Miller & Lauramay T. Dempster

Per gen from short, thick, ± branched taproot; tuberous root gen 0 (or spheric). **ST**: prostrate to erect, scape-like or branched. **LF**: gen in basal rosette and cauline, simple, entire or not; base wide; margin gen ± translucent. **INFL**: ± scapose; cyme, panicle, raceme, or ± umbel; sts 1–many, gen lfless but bracted, disjointing in age or not, 1–many fld; pedicel 0–30 mm. **FL**: sepals 2–8, free, persistent; petals 4–19, white, cream, yellow, orange, pink, rose, purple, overlapped in bud, often with pink or dark purple veins; stamens 1–50; styles 2–8, fused at base, stigmas 2–8, thread-like. **FR**: 6–9 mm, spheric or ovoid, circum-scissile near middle or below, translucent. **SEED**: 1–50, dark, gen shiny, smooth or finely tubercled, 1–4 mm in size. 18 spp.: w N.Am, 16 in flora. (Captain Meriwether Lewis, of Lewis & Clark Expedition, 1774–1809) [Wilson et al. 2005 W N Amer Naturalist 65:345–358] Many hybrids, cult, incl *Lewisia* ×*whiteae* Purdy in CA; *L. columbiana* (A. Gray) B.L. Rob. not in CA.

1. Cauline lvs 3–5, whorled, less often 2, opposite; taproot 0; tuberous root spheric, with many rootlets . . . . ***L. triphylla***
1′ Cauline lvs 0–many, opposite or alternate; taproot present; tuberous root 0
  2. Basal lvs green after fl, not shriveled
    3. Basal lf blades linear to narrowly oblanceolate; lf margin entire
      4. Lvs flat or adaxialy grooved . . . . ***L. ×whiteae***
      4′ Lvs ± cylindric, adaxially not grooved . . . . ***L. leeana***
    3′ Basal lf blades ovate, oblanceolate, ± spoon-shaped, or obovate to round (± linear); lf margin entire or toothed
      5. Petals 5–9 mm, white to pale pink with darker veins; infl panicle, open
        6. Pedicel 0.3–4 mm; petals 6–9 mm . . . . ***L. cantelovii***
        6′ Pedicel 3–8 mm; petals 5–6 mm . . . . ***L. serrata***
      5′ Petals (8)10–20 mm, gen pink-purple with pale and darker stripes, less often white, cream with
          pink-orange stripes, yellow, or orange; infl cyme, ± umbel, or panicle, ± dense . . . . ***L. cotyledon***
        7. Lf margin ± entire; petals (8)12–14 . . . . var. ***cotyledon***
        7′ Lf margin dentate or wavy; petals 12–20 mm
          8. Lf margin dentate; 225–2200 m . . . . var. ***heckneri***
          8′ Lf margin ± wavy; 100–400 m . . . . var. ***howellii***
  2′ Basal lvs shriveled after fl, turning brown
    9. Fls sessile; bract, sepal pairs resembling 4-parted calyx
      10. Sepal margin entire, not glandular; petals 12–26 mm . . . . ***L. brachycalyx***
      10′ Sepal margin toothed, glandular or not; petals 10–30 mm . . . . ***L. kelloggii***
        11. Lf blade > 4.5 cm, > 1 cm wide; petals ≥ 20 mm . . . . subsp. ***hutchisonii***
        11′ Lf blade < 4.5 cm, < 1 cm wide; petals ≤ 20 mm . . . . subsp. ***kelloggii***
    9′ Fls pedicelled; bract, sepal pairs not resembling 4-parted calyx
      12. Fls breaking apart in fr; sepals 2–9, scarious, margins entire to ± jagged, not toothed
        13. Sepals 2; petals 5–9; proximal bracts 2–4 per node . . . . ***L. disepala***
        13′ Sepals (4)6–9; petals 10–19; proximal bracts 4–7(8) per node . . . . ***L. rediviva***
          14. Sepals 10–12(15) mm; petals ± 15 mm; stamens 20–30 . . . . var. ***minor***
          14′ Sepals 15–25 mm; petals 18–35 mm; stamens 30–50 . . . . var. ***rediviva***

12′ Fls not breaking apart in fr; sepals 2, not scarious, margins entire, toothed, or gland-toothed
  15. Cauline lvs opposite, not markedly smaller than basal . . . . ***L. oppositifolia***
  15′ Cauline lvs 0 or alternate, markedly smaller than basal
    16. Basal lvs oblanceolate, spoon-shaped, or obovate; infl 3–100-fld
      17. Cymes panicle-like, 20–100-fld; petals 6–7, pale pink, base yellow-green; stamens 4–5 . . . . ***L. congdonii***
      17′ Cymes ± umbel-like, 3–11-fld; petals 7–10, magenta or red, bases white or ± pink;
        stamens 10–13. . . . . ***L. stebbinsii***
    16′ Basal lvs thread-like to linear to narrow-lanceolate or -oblanceolate; infl < 16-fld
      18. Fls 2.5–4 cm diam; petals ≥ 15 mm . . . . ***L. longipetala***
      18′ Fls 0.5–2 cm diam; petals ≤ 20 mm
        19. Petals 9–20 mm; sepal margins entire to ± jagged, not gland-toothed (obscurely or irregularly
          toothed), tips ± acute . . . . ***L. nevadensis***
        19′ Petals 4–10 mm; sepal margins ± jagged, toothed, or gland-toothed (entire), tips ± rounded or truncate
          20. Bract, sepal glands ± red or dark; stamens 6; stigmas 4 . . . . ***L. glandulosa***
          20′ Bract, sepal glands 0 or pale; stamens 4–8; stigmas 3–6 . . . . ***L. pygmaea***

L. brachycalyx A. Gray (p. 909) SHORT-SEPALED LEWISIA **LF:** basal several to many, in spreading rosette, 2–8 cm, oblanceolate, ± fleshy, entire, tapered to petiole, tip blunt. **INFL:** sts 1–many, 1–3.5 cm, 1-fld; fls incl in lvs; bracts 2, closely subtending, ± like sepals; pedicel 0. **FL:** sepals 2 (seemingly 4 from 2 bracts), 1/3–1/2 × corolla, obovate-obtuse to round, entire; petals 5–9, 12–26 mm, oblong, white or ± pink, veins ± pink, tip blunt or notched; stamens 9–15; stigmas 5–8. **FR:** 6–9 mm. **SEED:** 40–50, 1.5 mm. *n*=10. Sandy, wet meadows, seeps, open conifer forest; 1370–2450 m. SnBr, PR; to UT, AZ, n Baja CA. Feb–Jun ★

L. cantelovii J.T. Howell (p. 909) CANTELOW'S LEWISIA **LF:** many, in spreading rosette, (2)2.5–8(14) cm, spoon-shaped, finely dentate, tapered to wide petiole, tip truncate or notched (or rounded). **INFL:** sts several, 15–30 cm, each with a wide, open, gen few-fld panicle; fls well-exserted from lvs; bracts among fls, few below, minute, lance-ovate, gland-toothed; pedicel 0.3–4 mm. **FL:** sepals 2, ± 1/2 × corolla, ± round, 3–12 mm, margin gland-toothed; petals 5–7, 6–9 mm, obovate, white to pale pink, veins 5–7, tip rounded; stamens 5–6; stigmas 3. **FR:** ± 3 mm. **SEED:** 1–3, 1.2–1.5 mm. *n*=14. Granite cliff faces, rocky outcrops, ravines, serpentine seeps, chaparral, woodland, conifer forest; 385–1370 m. KR, n&s SNH. May–Oct ★

L. congdonii (Rydb.) S. Clay (p. 909) CONGDON'S LEWISIA **LF:** few to many in rosette; cauline few, 5–30 cm, narrowly oblanceolate to obovate, entire, tapered to long, slender petiole, tip acute or obtuse. **INFL:** sts 1–7, 2–6 dm, each with open, 20–100-fld, panicle-like cyme; fls well-exserted from lvs; bracts among fls and below, awl-like, gland-toothed; pedicel 5–10 mm. **FL:** sepals 2, 1/4–1/3 × corolla, ± round, margin gland-toothed; petals 6–7, ± 8 mm, lanceolate, pale pink, base yellow-green, veins magenta, tip jagged; stamens 4–5; stigmas 3. **FR:** 3–4 mm. **SEED:** 1–few, 2 mm. *n*=±12. Granite, metamorphic outcrops, crevices, rock slides, chaparral, woodland, conifer forest; 500–2800 m. c SN. Apr–Jun ★

L. cotyledon (S. Watson) B.L. Rob. CLIFF MAIDS **LF:** many, rosetted, 3–14 cm, ovate or spoon-shaped, fleshy, entire or not, tapered to base, tip rounded. **INFL:** sts gen 1–6, 10–30 cm, each with a ± flat-topped, 12–many-fld cyme, ± umbel, or panicle; fls exserted from lvs; bracts at each fl node, gland-toothed; pedicels gen < fls. **FL:** sepals 2, ± 1/3 × corolla, round or truncate, margin gland-toothed; petals 5–13, (8)10–20 mm, oblanceolate or obovate, gen pink-purple with pale and darker stripes, less often white, cream with pink-orange stripes, yellow, or orange, tip ± notched; stamens 5–12; stigmas 2–4. **FR:** 3–5 mm. **SEED:** 4–15, 1.5 mm. *n*=14. Hybrids with *L. leeana* are *L.* ×*whiteae.*

var. ***cotyledon*** (p. 909) **LF:** margin ± entire. **FL:** petals (8)12–14 mm. Crevices in cliffs or pavement of granite, serpentine, metamorphics, alpine slopes, subalpine forest; 300–2300 m. KR; sw OR. May–Jul

var. ***heckneri*** (C.V. Morton) Munz (p. 909) HECKNER'S LEWISIA **LF:** margin dentate. **FL:** petals 12–20 mm. Crevices in cliffs, rocky slopes of granite or basalt, conifer forest; 225–2200 m. KR. May–Jul ★

var. ***howellii*** (S. Watson) Jeps. (p. 909) HOWELL'S LEWISIA **LF:** margin ± wavy. **FL:** petals 12–20 mm. Rock outcrops, crevices on canyon walls, open woodland; chaparral, conifer forest; 100–400 m. KR; sw OR. Apr–Jun ★

L. disepala Rydb. (p. 909) YOSEMITE LEWISIA **LF:** many, rosetted, 0.5–2 cm, linear or club-shaped, fleshy, entire, tapered to base, tip blunt. **INFL:** sts few, < 1 cm, 1-fld, disjointing near middle, leaving proximal ring of 2–4 scarious, ovate, bracts; fls exserted from lvs; pedicel 1–2 mm. **FL:** sepals 2, 0.7 × corolla, petal-like, scarious, ± round, entire to ± jagged; petals 5–9, 10–18 mm, obovate, pale rose-pink, tip round; stamens 1–15; stigmas 2–4. **FR:** < 10 mm. **SEED:** 11–15, 2.8–4 mm. Sand, granite of exposed mtn summits, knobs, subalpine conifer forest, alpine fell-fields; 1340–3500 m. c&s SNH. Feb–Jun ★

L. glandulosa (Rydb.) S. Clay (p. 909) GLANDULAR LEWISIA **LF:** many, dense-rosetted, gen 2–10 cm, thread-like to narrowly lanceolate, entire, gen persistent after withering, tapered to fleshy, expanded base, tip obtuse. **INFL:** sts many, 1–3.5 cm, (1)2-fld; fls ± exserted or not from lvs; bracts in 1–2 irregular pairs, above st middle, ± ovate, marginal teeth with ± red or dark glands; pedicel 2–5 mm. **FL:** sepals 2, ± 1/2 × corolla, ± rounded or truncate, marginal teeth with ± red or dark glands; petals 6–8, ± 8 mm, obovate, white, pink, or ± red, tip irregular, gen acuminate; stamens 6; stigmas 4. **FR:** 4–15 mm. **SEED:** ≤ 24, 1–2 mm. Granite sand, rock cracks, wet meadows; 3000–4000 m. c&s SNH, SNE. L. glandulosa (Rydb.) Dempster, illeg. Jul–Sep

L. kelloggii K. Brandegee KELLOGG'S LEWISIA **LF:** many, dense-rosetted, 1.5–9 cm, spoon-shaped or obovate, leathery, entire, abruptly narrowed to long petiole, tip obtuse. **INFL:** sts several, 1–4 cm, 1-fld; fls often exserted from lvs; bracts 2, closely subtending, ± like sepals; pedicel 0. **FL:** sepals 2 (seemingly 4 from 2 bracts), ± 1/2 × corolla, lance-oblong, margins toothed, glandular or not; petals 5–12, 10–30 mm, obovate, white or pink, tip blunt, ± notched; stamens 8–26; stigmas 3–6. **FR:** 8 mm. **SEED:** 12–15, 2 mm. Pls in n part of CA range (e.g., Sierra Co.) larger, with fls more deeply colored.

subsp. ***hutchisonii*** Dempster HUTCHISON'S LEWISIA **LF:** blade > 4.5 cm, > 1 cm wide. **FL:** petals ≥ 20 mm. Decomposed granite, slate, volcanic rubble, conifer forest; 1800–2135 m. KR, n SN. Pls in c SNH conform to this subsp. exc lvs too short; study needed. Jul–Aug ★

subsp. ***kelloggii*** (p. 915) **LF:** blade < 4.5 cm, < 1 cm wide. **FL:** petals ≤ 20 mm. Decomposed granite, volcanic ash, rubble, conifer forest; 1370–2360 m. c&s SN. May–Jul ★

L. leeana (Porter) B.L. Rob. (p. 915) QUILL-LEAVED LEWISIA **LF:** many, dense-rosetted, 1.5–6 cm, ± linear, ± cylindric, entire, ± fleshy, tapered to narrow base, tip blunt. **INFL:** sts 1–several, 8–24 cm, each with open, rounded, many-fld, panicle-like cyme; fls well exserted from lvs; bracts among fls and below, lanceolate, entire or increasingly gland-toothed distally; pedicel 3–15 mm. **FL:** sepals 2, ± 1/3 × corolla, ovate or ± round, ± gland-toothed; petals 6–8, 5–6.5 mm, obovate, white, pink, purple, or magenta, tip gen rounded; stamens 4–8; stigmas 2. **FR:** 4–5 mm. **SEED:** 1–2, 2–2.5 mm. *n*=14.

Granite, serpentine cliffs, rocky slopes, conifer forest; 1300–3350 m. KR, NCoRH, CaR, c&s SNH; sw OR. [*Lewisia leana*, orth. var.] *L. leeana* hybridizes with *L. cotyledon* (see *L. ×whiteae*). Jun–Aug

L. longipetala (Piper) S. Clay LONG-PETALED LEWISIA **LF:** many, open-rosetted, 3–6 cm, linear to narrowly oblanceolate, thin, entire, tapered to base, tip blunt. **INFL:** sts 3–8, each 1–2-fld; fls not or ± exserted from lvs; bracts in 1–2 pairs, ± near middle of st, ± ovate, gland-toothed; pedicel 10–25 mm. **FL:** sepals 2, < 1/3 × corolla, fan-shaped, margin red-gland-toothed; petals 5–10, ≥ 15 mm, oblanceolate, white to rose, often red-gland-toothed; stamens 5–9; stigmas 5–6. **FR:** ± 8 mm. **SEED:** 20–50, 1.5 mm. *n*=11. Boulder, rock fields, crevices, scree fed by snow-melt, subalpine forest; 2500–2925 m. n&c SNH. Jul–Aug ★

L. nevadensis (A. Gray) B.L. Rob. (p. 915) NEVADA LEWISIA **LF:** few to many, ± open-rosetted, 3–13 cm, thread-like to narrowly lanceolate, entire, tapered to base, tip obtuse. **INFL:** 1-fld (or 2–3-fld cyme); fls ± exserted from lvs; bracts 2, below st middle, 6–18 mm, erect, lanceolate, margins entire, translucent; pedicel 10–40 mm. **FL:** sepals 2, 1/2–2/3 × corolla, widely ovate, entire to ± jagged (obscurely or irregularly toothed), tips ± acute, spreading; petals 5–10, 9–20 mm, ± obovate, white or pale-pink, tips blunt or ± pointed; stamens 6–15; stigmas 3–6. **FR:** 5–10 mm. **SEED:** 20–50, 1.3 mm. *n*=28. Grassy meadows, moist gravel flats, open forest; 1300–3000 m. KR, NCoRH, CaRH, SNH, TR, PR, Wrn, SNE; to WA, Rocky Mtns. Hybridizes with *L. triphylla*. May–Aug

L. oppositifolia (S. Watson) B.L. Rob. (p. 915) OPPOSITE-LEAVED LEWISIA **LF:** few to many, reduced on lower part of st or not, 4–11 cm, linear-oblanceolate, entire, tapered to slender base, tip blunt. **INFL:** sts 1–3, 6–20 cm, each with 2–5-fld, ± umbel-like cluster; fls rarely incl in lvs; bracts among fls and below, lanceolate, ± toothed, glands 0; pedicel 20–75 mm. **FL:** sepals 2, ± 1/3 × corolla, round or truncate, jagged-dentate, glands 0; petals 8–11, 12–17 mm, ovate, pink fading to white, tips blunt or ± jagged; stamens 8–21; stigmas 3–5. **FR:** 5–6 mm. **SEED:** 5–15, 1–1.8 mm. Moist sites, conifer forest; 300–1300 m. KR; sw OR. Mar–May ★

L. pygmaea (A. Gray) B.L. Rob. DWARF LEWISIA **LF:** gen several, rosetted, 2–9 cm, thread-like to lance-linear, fleshy, entire, tapered to expanded base, tip blunt. **INFL:** sts several to many, 1–5 cm, each 1-(several-)fld, fls gen incl in lvs; bracts 2, at or below st middle, ± widely lanceolate, entire or dentate, glands 0 or pale; pedicel 2–5(10) mm. **FL:** sepals 2, ± 1/2 × corolla, ± ovate, rounded or truncate, margin ± jagged or toothed, glands 0 or pale; petals 5–9, 4–10 mm, obovate, white, pink, or magenta (base green), ± striped, tip ± jagged; stamens 4–8; stigmas 3–6. **FR:** 4–5 mm. **SEED:** 15–24, 1–2 mm. *n*=±33. Rocky slopes, wet granite sand, gravel, moist meadows, along streams; 1700–4020 m. KR, NCoRH, CaRH, SNH, WTR, SnBr, Wrn, SNE; to AK, Rocky Mtns. May–Aug

L. rediviva Pursh **LF:** many, rosetted, 0.5–5 cm, linear, thick, entire, tapered at base, tip blunt. **INFL:** sts several to many, 2–6 cm, each 1-fld, disjointing near middle, leaving proximal ring of 4–7(8) scarious, awl-like bracts; fls exserted from lvs; pedicel 1–15(30). **FL:** sepals (4)6–9, ± 3/4 × corolla, petal-like, scarious, widely obovate, entire to ± jagged; petals 10–19, 12–35 mm, obovate-oblong, white, pink, rose, lavender, base ± white, tip obtuse-notched; stamens 20–50; stigmas 4–9. **FR:** 5–6 mm. **SEED:** 6–25, 2–2.5 mm. *n*=13,14.

var. ***minor*** (Rydb.) Munz (p. 915) SMALL BITTERROOT **LF:** blades linear, club-shaped to narrowly oblanceolate, grooved adaxially. **INFL:** pedicel (1)3–8 mm. **FL:** sepals 10–12(15) mm; petals ± 15 mm; stamens 20–30. Rocky open conifer woodland, scrub; 1900–2800 m. SCoRI, TR, SnJt, W&I, DMtns; NV, UT. May–Jun

var. ***rediviva*** BITTERROOT **LF:** blades linear, not grooved adaxially. **INFL:** pedicel 10–15(30) mm. **FL:** sepals 15–25 mm; petals 18–35 mm; stamens 30–50. Rocky, sandy ground, open conifer woodland, scrub; 60–1900 m. KR, NCoRH, NCoRI, SN, SnFrB, SCoRI, SNE; to BC, Rocky Mtns. Mar–Jun

L. serrata Heckard & Stebbins SAW-TOOTHED LEWISIA **LF:** many, in spreading rosette, < 8 cm, narrowly spoon-shaped, coarsely dentate, tapered to wide petiole, tip rounded **INFL:** sts several, 10–20 cm, each with wide, open, gen many-fld panicle; fls well exserted from lvs; bracts among fls, few below, minute, lance-ovate, gland-toothed; pedicel 3–8 mm. **FL:** sepals 2, ± 1/2 × corolla, ± round, margin gland-toothed; petals 5–6, 5–6 mm, obovate, pale pink with 3(5) longitudinal veins, tip round; stamens 5–6; stigmas 3. **FR:** 2.5–3 mm. **SEED:** 1–3. ± 1 mm. *n*=14. Metamorphic cliff faces, rocky outcrops, riparian scrub, woodland, conifer forest; 900–1435 m. n SNH. May–Jul ★

L. stebbinsii Gankin & W.R. Hildreth STEBBINS' LEWISIA **LF:** 5–15, rosetted, < 9 cm, oblanceolate to obovate, fleshy, entire or ± wavy, tapered to long petiole, tip obtuse. **INFL:** sts 1–9, ± prostrate, 1–14 cm, each with 3–11-fld, ± umbel-like cyme; fls gen exserted from lvs; bracts among fls, several below, often paired, gland-toothed; pedicel 8–25 mm. **FL:** sepals 2, 1/3–1/2 × corolla, widely ovate, margins red-gland-toothed; petals 7–10, 8–15 mm, oblanceolate-obovate, magenta or red, bases white or ± pink, veins purple, tips blunt, ± jagged; stamens 10–13; stigmas 3–4. **FR:** 5–7 mm. **SEED:** 20, 1.5–2 mm. Open, gravelly sites, serpentine or not, conifer forest; 1600–2050 m. NCoRH. May–Jul ★

L. triphylla (S. Watson) B.L. Rob. (p. 915) THREE-LEAVED LEWISIA Taproot 0; tuberous root spheric, with many rootlets. **LF:** basal withering before fl, cauline lvs 3–5, whorled, less often 2, opposite, 1–6 cm, linear or thread-like, entire, tapered to base, tip obtuse. **INFL:** sts 1–several, gen 2–7 cm above ground, each with open, 1–25-fld cluster; fls incl in or exserted from lvs; bracts among fls, ± lanceolate, entire; pedicels 5–15(25) mm. **FL:** sepals 2, ± 1/2 × corolla, ovate, entire; petals 5–9, 4–7 mm, oblong-oblanceolate, white or ± pink, veins purple, tips rounded; stamens 3–5; stigmas 3–5. **FR:** 3–4 mm. **SEED:** 8–25, 1 mm. Moist sandy or gravelly slopes, grassy meadows, open conifer forest; 1300–3400 m. KR, NCoRH, CaRH, SN, Wrn; to BC, Rocky Mtns. Hybridizes with *L. nevadensis*. May–Aug

L. ×whiteae Purdy **LF:** many, dense-rosetted, 2–10 cm, narrowly oblanceolate, flat or abaxially grooved, fleshy, entire, tapered to base, tip blunt. **INFL:** sts several, 12–30 cm, each with open, many-fld panicle; fls exserted from lvs; bracts among fls, few below, gland-toothed; pedicel 5–13 mm. **FL:** sepals 2, ± 1/4 × corolla, ± round, gland-toothed; petals 4–11, 5–13 mm, ovate, rose, pink, or white with magenta veins, tip notched; stamens 5–6; stigmas 2–3. **FR:** 3–6 mm. **SEED:** 1–7, 1.5–2 mm. *n*=15. Granite, sandstone, serpentine slopes, cliffs, conifer forest; 2100–2300 m. KR; sw OR. Hybrid of *L. cotyledon*, *L. leeana*. May–Aug

MONTIA

John M. Miller & Kenton L. Chambers

Ann, per, glabrous, ± fleshy, ± aquatic or not, matted or not. **LF:** cauline > 2, alternate or opposite, free, entire. **INFL:** raceme, 1-sided; lowest fl gen bracted; pedicel recurved, erect in fr. **FL:** petals (3)5, equal or 2 larger, white to pink; stamens (3)5, epipetalous; ovary chamber 1, placentas basal, style 1, stigmas 3. **FR:** valves 3, margins rolled in, forcibly expelling seeds. **SEED:** 1–3, gen black, smooth to tubercled, appendaged or not. 12 spp.: Worldwide. (Giuseppe Monti, Italian botanist, 1682–1760) [Heenan 2007 New Zealand J Bot 45:437–439] Sometimes divided into 9 genera.

1. Lvs opposite, not much reduced upwards
  2. Per, from lfy stolons with pink overwintering bulblets; petals 5–9 mm, >> sepals . . . . ***M. chamissoi***
  2′ Ann, without stolons or overwintering bulblets; petals 1–2 mm, ± = sepals . . . . ***M. fontana***
1′ Lvs alternate, reduced upwards or not
  3. Per; lvs basal and on stolons; petals 7–15 mm . . . . ***M. parvifolia***

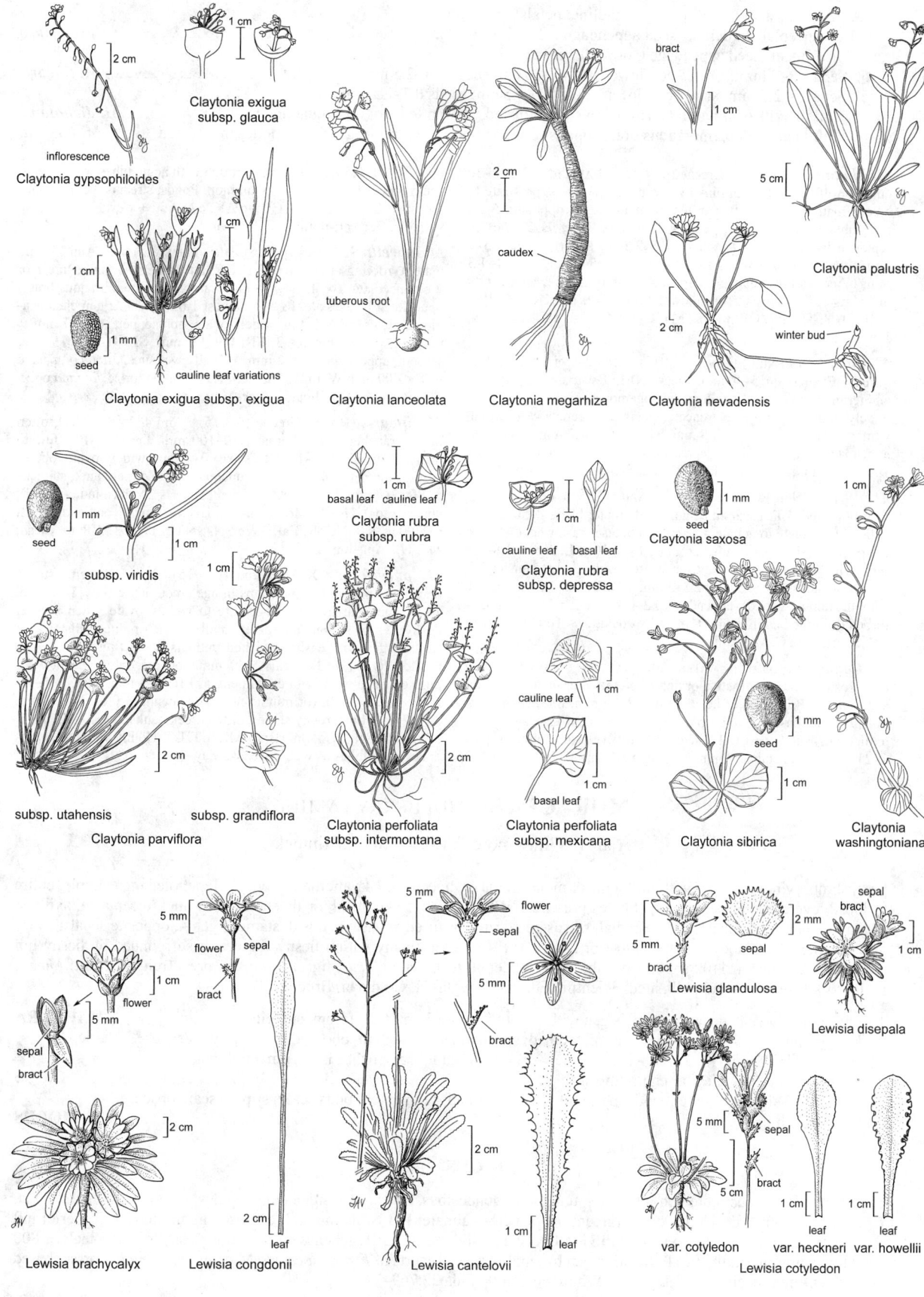

Claytonia gypsophiloides

inflorescence

Claytonia exigua
subsp. glauca

seed

Claytonia exigua subsp. exigua

cauline leaf variations

Claytonia lanceolata

tuberous root

Claytonia megarhiza

caudex

bract

Claytonia palustris

Claytonia nevadensis

winter bud

seed

subsp. viridis

subsp. utahensis

subsp. grandiflora

Claytonia parviflora

basal leaf   cauline leaf

Claytonia rubra
subsp. rubra

Claytonia perfoliata
subsp. intermontana

cauline leaf   basal leaf

Claytonia rubra
subsp. depressa

cauline leaf

basal leaf

Claytonia perfoliata
subsp. mexicana

seed

Claytonia saxosa

seed

Claytonia sibirica

Claytonia
washingtoniana

flower   sepal

bract

flower

sepal

bract

Lewisia brachycalyx

leaf

Lewisia congdonii

flower
sepal

bract

leaf

Lewisia cantelovii

sepal

bract

Lewisia glandulosa

sepal
bract

Lewisia disepala

sepal

bract

var. cotyledon   var. heckneri   var. howellii

leaf   leaf

Lewisia cotyledon

3′ Ann; lvs basal and cauline or gen cauline; petals < 6.5 mm
  4. Lvs lanceolate to deltate; seed appendage 0.2–0.4 mm . . . . *M. diffusa*
  4′ Lvs ± linear; seed appendage 0 or << 0.2 mm
    5. Sepals 3–7 mm; st 4–25 cm; lower st lvs 2–10 cm; seeds 1.4–2.5 mm . . . . *M. linearis*
    5′ Sepals < 2.5 mm; st 1–9 cm; lower st lvs 0.5–4.5 cm; seeds 0.7–1.2 mm
      6. Infl in axils of st lvs and terminal; pls often tufted, not matted, not rooting at nodes . . . . *M. dichotoma*
      6′ Infl in axils of membranous bracts opposite st lvs; pls matted, rooting at lower nodes, not tufted . . . . . . *M. howellii*

M. chamissoi (Spreng.) Greene (p. 915) TOAD LILY Per 2–30 cm, from lfy stolons with pink overwintering bulblets, prostrate to erect, floating or not, often tufted or matted. **LF:** opposite, 5–50 mm, oblanceolate. **INFL:** gen few, some axillary; fls 2–8, often replaced by bulblets. **FL:** sepals 1.5–2 mm, spheric; petals 5–9 mm, equal, pink or white; stamens 5. **FR:** 1–1.5 mm. **SEED:** 1–1.3 mm; tubercles low, rounded; appendage ± flat. 2*n*=22. Wet, sandy or loamy soil, seeps, wet meadows; 1100–3700 m. KR, NCoRH, CaR, SNH, TR, PR, GB; to AK, MT, CO, NM; also c Can, n-c US. Jun–Aug

M. dichotoma (Nutt.) Howell (p. 915) Ann 2–9 cm, erect, often tufted. **LF:** alternate, 5–40 mm, linear. **INFL:** few to many, in lf axils and terminal; fls 4–16. **FL:** often cleistogamous; sepals 1.5–2.5 mm, widely obovate, obtuse to truncate; petals ± ≥ sepals, ± unequal, white; stamens 3. **FR:** 1.4–1.8 mm. **SEED:** 0.8–1.2 mm, ± smooth; appendage 0. 2*n*=14. Moist depressions, grassland, open woodland; 1000–1600 m. KR, CaR, MP; to BC, ID. Self-pollinated. Apr–Jun

M. diffusa (Nutt.) Greene (p. 915) Ann 5–20 cm, erect, gen diffuse-branched. **LF:** basal and cauline, alternate; blade 8–50 mm, ≤ petiole, lanceolate to deltate, tip acute to obtuse, base wedge-shaped to cordate. **INFL:** gen many, terminal; fls 3–10. **FL:** sepals 1.8–3.5 mm, widely obovate, obtuse to truncate; petals 3–4.5 mm, equal, pink or white; stamens 5. **FR:** 2.3–3 mm. **SEED:** 1.3–1.5 mm, gen ± smooth (minute-netted); appendage 0.2–0.4 mm. 2*n*=16. Often disturbed or burned conifer forest, mixed woodland; < 1000 m. NW, n SN, SnFrB; to BC. May–Jul

M. fontana L. (p. 915) WATER CHICKWEED, BLINKS Ann 1–30 cm, prostrate to erect, floating or not, often tufted or matted, rooting at lower nodes. **LF:** opposite, ± sessile, 3–20 mm, linear to widely oblanceolate, base tapered, tip acute to obtuse. **INFL:** many, some axillary; fls 1–8, lowest 1–2-bracted. **FL:** often cleistogamous; sepals 1–2 mm, round; petals 1–2 mm, ± unequal, white; stamens 3. **FR:** 1–2

mm. **SEED:** 0.5–1.2 mm, gen ± rough with acute tubercles; appendage round or flat. 2*n*=20,40. Common. Ponds, streams, vernal pools, seeps; < 3200 m. CA-FP, GB; to AK, e N.Am; ± worldwide. Highly variable. Gen self-pollinated. Feb–Jul

M. howellii S. Watson (p. 915) HOWELL'S MONTIA Ann 1–9 cm, matted, rooting at lower nodes. **LF:** alternate, 5–25 mm, linear or ± oblanceolate. **INFL:** few to many, in axils of membranous bracts opposite st lvs, sessile; fls 2–6, lowest 1 bracted. **FL:** gen cleistogamous; sepals 0.8–1.5 mm, widely ovate, obtuse; petals 0–0.7 mm, ± unequal, white; stamens 3. **FR:** 0.9–1.2 mm. **SEED:** 0.7–0.9 mm, smooth; appendage << 0.2 mm. Vernally wet sites, often compacted soil; < 400 m. NW; to BC. Sometimes mistaken for *M. fontana* or *M. dichotoma*. Self-pollinated. Mar–May ★

M. linearis (Hook.) Greene (p. 915) Ann 4–25 cm, erect, often much-branched. **LF:** alternate, 10–100 mm, linear. **INFL:** few to many, terminal; fls 3–14. **FL:** sepals 3–7 mm, widely ovate, obtuse to truncate; petals 4–6.5 mm, ± unequal, white or pale pink; stamens 3. **FR:** 3–4 mm. **SEED:** 1.4–2.5 mm, gen ± smooth (minute-netted); appendage 0. 2*n*=28. Moist grassland, scrub, open woodland, fields; < 2300 m. KR, NCoR, CaR, n&c SN, SnFrB, TR, PR, MP; to w Can, MT, UT. Apr–Jun

M. parvifolia (DC.) Greene (p. 915) Per 5–40 cm; stolon branched, matted, creeping, tips in age erect, densely lfy. **LF:** basal 15–60 mm, narrowly oblanceolate to widely ovate, often sessile, fleshy; lvs of stolons alternate, smaller, much reduced distally on st, axils gen with easily detached bulblets. **INFL:** terminal; 1+ fls bracted; fls 2–12. **FL:** sepals 2–3.5 mm, obovate, obtuse; petals 7–15 mm, equal, pink or white; stamens 5. **FR:** 2–3 mm. **SEED:** 1–1.5 mm, gen ± smooth (or minute-netted); appendage 0 to << 0.2 mm. 2*n*=22,44. Moist, rocky slopes, cliffs, creek banks, forest; < 2600 m. NW, CaR, SN, CCo, SnFrB; to AK, MT, UT. Highly variable; reproduces by bulblets as well as seeds. May–Jul

MORACEAE MULBERRY FAMILY

Alan T. Whittemore & Elizabeth McClintock

[Per] shrub, [vine] tree, gen with milky juice; monoecious or dioecious. **LF:** alternate [opposite], petioled, gen simple, entire to lobed, evergreen or deciduous; stipules present. **INFL:** raceme, spike, head, or fls enclosed in thick receptacle, axillary. **FL:** unisexual or bisexual, small, ± radial; sepals gen 4, free or fused at base; petals 0; stamens gen 4, opposite sepals; ovary gen superior, 1-chambered, style simple or 2-parted. **FR:** achenes many within fleshy calyces or surrounded by fleshy infl receptacle. 37 genera, 1100 spp.: trop, subtrop, some temp; many cult (*Ficus*, fig; *Artocarpus*, breadfruit, jackfruit; *Morus*, mulberry). Insect- or wind-pollinated. Scientific Editors: Douglas H. Goldman, Bruce G. Baldwin.

1. Lvs pinnately veined, entire; st axils gen with 1 stout thorn to 3 cm; fr 9–15 cm, spheric . . . . **MACLURA**
1′ Lvs palmately 3–5-veined, lobed or not, toothed; sts unarmed; fr 1–8 cm, obovoid to short-cylindric or ± spheric.
  2. Lf blade 11–32 cm, 13–37 cm wide, petiole 4–12 cm; fr not lobed, surface not segmented; buds inside
    conic stipules; stipule scar encircling st . . . . **FICUS**
  2′ Lf blade 5–16 cm, 3–16 cm wide, petiole 1–4 cm; fr lobed, segmented; buds scaly; stipule scars obscure,
    not encircling st . . . . **MORUS**

FICUS FIG

Shrub, tree, [vine, occ rooted on other pls], unarmed; monoecious. **ST:** buds inside conic stipules; stipule scar encircling st. **LF:** entire or lobed, deciduous or evergreen; major veins palmate. **INFL:** fls internal, enclosed in an obovoid [spheric] infl receptacle with a small, scaly opening. **PISTILLATE FL:** style simple. **FR:** achenes many within fleshy infl receptacle. ± 800 spp.: trop, subtrop. (Latin: fig) Pollination gen by small wasps. Reports of *F. pseudocarica* Miq. and *F. palmata* Forssk. based on misidentified specimens of *F. carica* (Whittemore 2006 Sida 22:769–775).

F. carica L.   EDIBLE FIG   Pl < 10 m. **LF**: petiole 4–12 cm; blade 11–32 cm, widely ovate, ± scabrous adaxially, hairy abaxially, palmately lobed, major lobes gen 3–7, gen > 1/2 to base. **FR**: 5–8 cm, obovoid, green, yellow, or red to purple. Creeks, riverbanks, flood- plains, seeps, disturbed areas; < 800 m. SNF, GV, SnFrB, SCo; native to sw Asia, naturalized in c Am, Australia. Commonly cult. Mar–Apr ❖

MACLURA   OSAGE ORANGE

Tree, thorny; dioecious. **ST**: buds scaly; stipule scars obscure, not encircling st. **LF**: alternate, or clustered with infls, entire, deciduous; major veins pinnate. **INFL**: ± erect, spheric; staminate an umbel or umbel-like raceme, > 1 per axil, peduncled; pistillate a head, 1 per axil, sessile. **PISTILLATE FL**: style simple. **FR**: spheric, bumpy, of many achenes within fleshy calyces, yellow-green [red]. ± 12 spp.: warm parts of Am, Afr, Asia, Australia. (William McClure, Am geologist, 1760–1840) Wind-pollinated.

M. pomifera (Raf.) C.K. Schneid.   Pl to 20 m; thorns to 3 cm. **LF**: petiole 1–4 cm; blade 3–14 cm, ovate to lance-oblong, dark green, sparsely soft-hairy. **FR**: 9–15 cm diam, yellow-green, densely irregu- larly warty. Streambanks, disturbed areas; < 440 m. GV, SCo, WTR; native to s-c US. Widely planted; fr inedible. Much less thorny with age. Apr–Jun

MORUS   MULBERRY

Tree, unarmed; monoecious or dioecious. **ST**: buds scaly; stipule scars obscure, not encircling st. **LF**: alternate, occ clustered with infls, unlobed or 3–5-lobed, toothed, 3–5-veined from base, deciduous. **INFL**: catkins, ± pendent, peduncled. **PISTIL- LATE FL**: style deeply 2-parted. **FR**: of many achenes within fleshy calyces, resembling blackberries. ± 20 spp.: temp, warm temp n hemisphere. (Latin: mulberry) [Whittemore 2006 Sida 22:769–775] Wind-pollinated; *M. nigra*, black mulberry, waif in urban areas.

M. alba L.   WHITE MULBERRY   Pl 10–15 m. **LF**: petiole 5–35 mm; blade 5–12 cm, ovate, coarsely toothed, abaxially glabrous or hairy only in axils of and on major veins, largest lobes gen 0–3, occ on 1 side, shallow to deep. **FR**: 1–2.5 cm, fleshy, white to ± pink or red- black. Disturbed areas, moist soil, streambanks; < 1300 m. SNF, GV, WTR; native to China. Widely cult; fr edible; lvs food of silkworm larva. Mar–May

MYRICACEAE   WAX MYRTLE FAMILY

Allan J. Bornstein

Shrub, small tree, gen aromatic, evergreen or deciduous; gen monoecious or dioecious; roots gen with nitrogen-fixing bacteria; hairs gen both nonglandular, thread-like and glandular, peltate. **LF**: simple, alternate, entire to pinnately lobed, resin-dotted; stipules gen 0. **INFL**: spike, catkin-like, axillary; staminate, pistillate separate. **FL**: gen unisexual, small; perianth gen 0. **STAMINATE FL**: subtended by 1 bract, 0 [1–4] bractlets. **PISTILLATE FL**: subtended by 1 bract, 2 or 4–6[3(8)] bractlets; ovary superior [to ± inferior], ovule 1, style 1, stigmas 2, short. **FR**: gen drupe or nut-like, small, rough or smooth, waxy or not, sometimes winged or bur-like from fused bractlets. 4 genera, ± 45 spp.: gen temp, subtrop; fr of some *Morella* boiled to make fragrant wax. [Carlquist 2002 Aliso 21:7–29] *Myrica* as treated in TJM (1993) split into 2 genera based on Wilbur 1994 Sida 16:93–107. Scientific Editor: Thomas J. Rosatti.

1. Terminal buds present; lvs evergreen, shiny, hairs 0 or soft, shaggy adaxially along midrib; fls unisexual
   (pls monoecious) or bisexual; staminate bract < stamens; pistillate bractlets 4–6, not enlarging, not spongy
   in fr, not adherent to fr; fr with fleshy papillae, pubescent, ± spheric, gen with waxy coating drying white
   to gray . . . . **MORELLA**
1′ Terminal buds 0; lvs deciduous, ± dull, hairs soft, shaggy abaxially, adaxially; fls unisexual (pls dioecious);
   staminate bract > stamens; pistillate bractlets 2, enlarging, spongy in fr, adherent to fr; fr not papillate,
   glabrous but gen dotted with minute, ± orange resin glands, gen ovoid, ± flat, without waxy coating . . . . **MYRICA**

MORELLA

Shrub, small tree. **LF**: ± spicy-scented; blade unlobed, ± serrate, esp in upper 1/2. **STAMINATE FL**: stamens gen 6–12(22). ± 40 spp.: Afr, e Asia, Philippines, Malaysia, N.Am, S.Am, temp, subtrop.

M. californica (Cham. & Schltdl.) Wilbur (p. 915)   WAX MYRTLE Pl 2–10 m. **LF**: gen ± narrowly lance-elliptic; tip gen sharply acute. **STAMINATE FL**: stamens 7–16. **FR**: few per pistillate infl, 4–6.5(8) mm diam. Coastal dunes and scrub, closed-cone pine, redwood for- est; < 150 m near coast, < 500+ m inland. NCo, w KR, NCoRO, CCo, w SnFrB, SCo; to BC. [*Myrica c.* Cham. & Schltdl.] Mar–Apr

MYRICA

Shrub. **LF**: aromatic; blade unlobed, ± serrate, esp in upper 1/2. **STAMINATE FL**: stamens gen 3–5. 2 spp. in N.Am, Eurasia, temp. (Greek: old name for fragrant shrub) *M. californica* moved to *Morella*.

M. hartwegii S. Watson (p. 915)   SIERRA SWEET BAY   Pl 1–2 m. **LF**: gen ± widely oblanceolate; tip gen obtuse to rounded. **STAMI- NATE FL**: stamens 3–6. **FR**: many per pistillate infl, gen 1.5–2.5 mm diam. Streambanks, moist places in foothills or low montane yellow-pine forest; 300–1800 m. n&c SN; OR. May–Jun ★

MYRSINACEAE MYRSINE FAMILY

Anita F. Cholewa

Ann, per, [shrub, tree], gen hairy, often glandular, resin canals appearing as dark dots or streaks on sts, lvs, or fls. **LF**: simple, alternate, opposite, subopposite, or whorled, petioled or not; stipules 0. **INFL**: terminal or axillary, fls 1 or not. **FL**: bisexual, radial; parts in [4s] 5s to 7s; calyx deeply lobed, often persistent; corolla lobes gen spreading; stamens epipetalous, opposite corolla lobes; ovary superior, 1-chambered, placenta free-central, style 1, stigma head-like. **FR**: capsule, circumscissile or 5–6-valved [drupe, drupe-like]. **SEED**: [1]–many. ± 40 genera, 800 spp.: esp trop, subtrop; some orn (*Anagallis*, *Lysimachia*). [Lens et al. 2005 Syst Bot 30:163–183] Based on molecular evidence, non-rosette terrestrial members of Primulaceae as treated in TJM (1993) removed to Myrsinaceae. Based on phylogenetic research, all CA members of Myrsinaceae have been or need to be transferred to *Lysimachia* (see 2009 Willdenowia 39:49–54): *Anagallis arvensis* L. is now *Lysimachia arvensis* (L.) U. Manns & Anderb., *A. minima* (L.) E.H.L. Krause is now *L. minima* (L.) U. Manns & Anderb., *A. monelli* L. is now *L. monelli* (L.) U. Manns & Anderb., *Glaux maritima* L. is now *L. maritima* (L.) Galasso et al.; and *Trientalis europaea* L. is now *L. europaea* (L.) U. Manns & Anderb.; unfortunately, a combination for *T. latifolia* Hook. in *Lysimachia* had not yet been published by the time of this writing. Scientific Editor: Thomas J. Rosatti.

1. Lvs mostly in ± 1 whorl near st tip, a few also distributed along st below or not . . . . **TRIENTALIS**
1′ Lvs distributed along st
  2. Corolla < calyx . . . . *Anagallis minima*
  2′ Corolla 0 or ≥ calyx
    3. Corolla 0, calyx corolla-like; fls ± sessile, 1 in lf axils . . . . **GLAUX**
    3′ Corolla ≥ calyx, calyx not corolla-like; fls pedicelled, 1 or in racemes in lf axils
      4. Corolla salmon, blue or blue-white; fr circumscissile; ann . . . . **ANAGALLIS**
      4′ Corolla yellow; fr 5–6-valved; per . . . . **LYSIMACHIA**

ANAGALLIS PIMPERNEL

Pl ± erect to spreading. **ST**: simple to diffusely branched. **LF**: cauline, alternate, opposite or whorled, ± sessile, gen entire. **INFL**: fls 1 in axils of upper lvs, gen pedicelled. **FL**: calyx persistent; filaments free or fused at base. **FR**: circumscissile, spheric. 20 spp.: gen Eurasia. (Greek: pimpernel) [Manns & Anderberg 2005 Int J Pl Sci 166:1019–1028] Inclusion of *A. minima*, *Centunculus minimus* in TJM (1993), controversial, based primarily on morphology; reports of *A. foemina* from CA based on blue-fld forms of *A. arvensis*.

1. Per; calyx 4–6 mm . . . . [*A. monelli*]
1′ Ann; calyx < 5 mm
  2. Calyx ≤ corolla; fl distinctly pedicelled . . . . *A. arvensis*
  2′ Calyx > corolla; fl sessile or nearly so . . . . *A. minima*

A. arvensis L. (p. 915) SCARLET PIMPERNEL Ann, spreading. **ST**: 5–40 cm, freely branched. **LF**: opposite or whorled; blade 5–20 mm, ovate to elliptic, upper lanceolate to ovate. **FL**: pedicel 1–3 cm, gen > subtending lf, recurved in fr; parts in 5s; calyx 3–5 mm, ≤ corolla, divided nearly to base; corolla hidden by calyx in bud, 4–7 mm, salmon, blue or blue-white, marginal hairs on petals many, terminal cell spheric. 2*n*=40. Common. Disturbed places, ocean beaches; gen < 1000 m. CA-FP, D; to e N.Am; native to Eur. TOXIC to livestock, humans. Mar–May

A. minima (L.) E.H.L. Krause (p. 915) CHAFFWEED Ann, ± erect. **ST**: 3–10 cm, branched or not. **LF**: alternate or lowest opposite, blade 2–5 mm, oblanceolate to widely obovate. **FL**: sessile or nearly so; parts gen in 4s; calyx ± 3 mm, > corolla, divided about 3/4 to base; corolla pink; filaments wider at base. 2*n*=22. Vernal pools, moist places; < 950 m. NW, CaRF, n SNF, GV, CW, SCo, n ChI, PR; to BC, e N.Am, Eur, S.Am. [*Centunculus m.* L.] Mar–May

GLAUX SEA MILKWORT

1 sp.: n temp, arctic. (Greek: blue-green) [Hao et al. 2004 Molec Phylogen Evol 31:323–339]

G. maritima L. (p. 915) Per, fleshy, erect, tufted. **ST**: 5–40 cm. **LF**: cauline, opposite to subopposite or alternate near st tip, sessile, 4–20 mm, linear to oblong, entire. **INFL**: fls 1 in lf axils, ± sessile. **FL**: 3–4 mm; parts in 5s; calyx corolla-like, lobes ± 2 × tube, ovate or oblong, white to ± red or lavender; corolla 0; stamens 5, alter-nate calyx lobes, filaments awl-like, anthers cordate. **FR**: 5-valved, < calyx, ovoid. **SEED**: few, elliptic in outline, ± flat. 2*n*=30. Coastal salt marshes, saline meadows; < 2100 m. NCo, deltaic GV, CCo, SnFrB, SNE, MP (exc Wrn); to BC and e to MN, e coast US, Eurasia. May–Jul

LYSIMACHIA LOOSESTRIFE

Per. **LF**: cauline, opposite or whorled, petioled or not, lanceolate to widely ovate, some gland-dotted, entire. **INFL**: raceme, panicle, or fls 1. **FL**: parts gen in 5s or 6s; corolla > calyx, yellow; filaments free or fused at base, tooth-like staminodes present or 0, united to corolla; ovary superior. **FR**: 5–6-valved, ovate. 150 spp.: gen n temp. (Greek: loose dagger) [Hao et al. 2004 Molec Phylogen Evol 31:323–339]

1. St creeping; petiole 2–4 mm; fls 1 in lf axils; corolla lobes 7–13 mm . . . . *L. nummularia*
1′ St erect; petiole 0; fls in racemes in lf axils; corolla lobes 3–5 mm . . . . *L. thyrsiflora*

L. nummularia L. (p. 915) CREEPING-JENNY **ST**: often rooted at nodes, branched or not, 20–50 cm, glabrous. **LF**: 1.5–2.5 cm, ovate to widely so, base truncate to rounded. **FL**: calyx lobes 5–13 mm; corolla lobes cordate-ovate; filaments free or ± fused at base, wider at base, glandular. **FR**: < calyx, black-dotted, at least near tip. 2*n*=36. Moist meadows; ± 1000 m. n SN (Folsom, Sacramento Co.; near Quincy, Plumas Co.); e US; native to Eur. Jun–Aug

L. thyrsiflora L. (p. 915) TUFTED LOOSESTRIFE **ST**: unbranched, glabrous. **LF**: 5–12 cm, lanceolate, base long-tapered. **FL**: calyx lobes 1–4 mm; corolla lobes linear; filaments free, not wider at base, glabrous. **FR**: ± > calyx, black-dotted. 2*n*=40,54. Wet places; 800–1300 m. KR, CaR, n SNF (Calaveras Co.), n SNH (Plumas Co.), MP; to AK, e N.Am, Eurasia. Jun–Aug ★

TRIENTALIS STARFLOWER

Per, low, glabrous; roots tuber-like. **ST**: erect, simple. **LF**: at base of st scale-like, others (cauline) well developed, mostly in ± 1 whorl near st tip. **INFL**: fls 1 in axils of few uppermost lvs; pedicels slender. **FL**: parts gen in 5s to 7s; sepals ± free, persistent; corolla spreading, ± flat, = or > calyx, gen white or ± pink to rose, lobes free ± to base; filaments fused at base, slender, anthers oblong. **FR**: 5-valved, spheric. **SEED**: few. 3 spp.: N.Am, n Eurasia. (Latin: 1/3 foot, from height of pl)

1. Corolla gen white; pedicels > lvs; cauline lvs mostly in ± 1 whorl near st tip, a few alternate, distributed along st below . . . . ***T. europaea***
1′ Corolla gen ± pink to rose; pedicels gen < lvs; cauline lvs all in ± 1 whorl near st tip. . . . . ***T. latifolia***

T. europaea L. (p. 915) ARCTIC STARFLOWER **ST**: 5–20 cm. **LF**: 15–50 mm, 12–16 mm wide, elliptic to obovate. **FL**: corolla 12–16 mm wide. 2*n*=±170. Bogs, wet areas; gen < 10 m. NCo (near Crescent City, Del Norte Co.); to AK, w Can, ID. [*T. arctica* Hook.; *T. e.* subsp. *a.* (Hook.) Hultén] May–Jun ★

T. latifolia Hook. (p. 915) **ST**: 5–30 cm. **LF**: 25–90 mm, 10–50 mm wide, ovate to obovate. **FL**: corolla 8–15 mm wide. Shaded places, esp woodland; < 1400 m. NW, CaR, n&c SN, ScV, CW; to BC. [*T. borealis* Raf. subsp. *l.* (Hook.) Hultén] Apr–Jul

MYRTACEAE MYRTLE FAMILY

Leslie R. Landrum, except as noted

[Subshrub] shrub, tree, trunk bark smooth or scaly; glands 0 or embedded in epidermis. **LF**: opposite or alternate, persistent, gen glandular when young. **INFL**: gen axillary, raceme, panicle, cyme, or fls 1. **FL**: gen bisexual, parts in 4s, 5s, gen ± white; hypanthium exceeding ovary or not; stamens gen many; ovary [rarely superior to] inferior, 2–5(18)-chambered; placentas axillary, just below top, or basal, ovules few to many, gen in 2–many series. **FR**: berry, capsule, nut. **SEED**: 1–many; coat membranous to ± leathery or hard, bony; embryo starchy or oily (of great taxonomic importance). 100 genera, ± 3500 spp.: many spp. trop Am, Australasia, fewer Afr, s Asia; economically important for timber (*Eucalyptus*), spices (*Syzygium aromaticum* (L.) Merr. & L.M. Perry, cloves; *Pimenta dioica* (L.) Merr., allspice), edible frs (*Psidium guajava* L., guava; *Acca sellowiana* (O. Berg) Burret, pineapple guava), many orns (*Eucalyptus*, *Melaleuca*, several other genera). [McVaugh 1968 Taxon 17:354–418] Apparently of Gondwanan origins; trop, subtrop, Medit climates. *Chamelaucium uncinatum* Schauer, *Luma apiculata* (DC.) Burret, *Melaleuca citrina* (Curtis) Dum.Cours., *Myrtus communis* L., *Syzygium australe* (Link) B. Hyland are waifs. Scientific Editor: Thomas J. Rosatti.

1. Fr a berry; seeds often > 5 mm; lvs gen opposite (whorled).
2. Calyx lobes, petals 5; seed coat shiny, hard. . . . . [**MYRTUS**]
2′ Calyx lobes, petals 4; seed coat membranous
3. Ripe fr dark purple; seeds gen > 1, lenticular; cotyledons ± = embryonic st . . . . [**LUMA**]
3′ Ripe fr red or purple; seed gen 1, spheric; cotyledons >> embryonic st . . . . [**SYZYGIUM**]
1′ Fr a capsule or nut; seeds gen 1–3 mm; mature lvs alternate or opposite
4. Perianth parts fused into bud cap in bud, shed at fl as a unit . . . . **EUCALYPTUS**
4′ Perianth parts free in bud, each segment persistent or falling separately
5. Fls in a dense cylindrical cluster surrounding st at tip or just below; stamens several × > perianth . . . . **MELALEUCA**
5′ Fls 1 or in small clusters, not surrounding st; stamens ≤ perianth
6. Lvs opposite, linear or awl-shaped; fr a 1-chambered nut; style with ring of hairs just below stigma. . . . . [**CHAMELAUCIUM**]
6′ Lvs alternate, oblanceolate to obovate-oblong; fr a many-chambered capsule; style glabrous . . . . **LEPTOSPERMUM**

EUCALYPTUS EUCALYPTUS, GUM TREE

Matt Ritter

Tree, shrub. **ST**: gen erect; bark shedding, smooth, or persistent near base (occ) or throughout, rough; twigs gen round. **LF**: juvenile gen opposite, horizontal, sessile, ± cordate, entire, glaucous; adult gen alternate, vertical, petioled, ± lanceolate, entire, glandular, glabrous, gen same color on both sides. **INFL**: axillary, (1)3–many-fld, stalked umbel or panicle-like cluster of such umbels. **FL**: perianth (gen, entirety of calyx lobes, petals) fused into bud cap in bud, bud cap shed at fl; stamens many, in several series, gen all fertile, white (yellow, red, pink); ovary chambers 3–6, fused to hypanthium. **FR**: capsule, thick-walled, woody, gen smooth, gen dehiscing at top. **SEED**: gen 1–3 mm, wind-dispersed. ± 700 spp.: most endemic to Australia; > 250 spp. Cult in CA; important for oils, tannins, timber, orns. (Greek: true cap, for bud cap) [Brooker 2000 Austral Syst Bot 13:79–148] World's largest fl pls, some > 100 m; *E. pulverulenta* Sims excluded, cult only.

1. Infl a panicle of 3–many-fld umbels
    2. Lvs lanceolate, lemon-scented; bark smooth . . . . *E. citriodora*
    2′ Lvs ovate, elliptic, or round, not lemon-scented; bark rough . . . . *E. polyanthemos*
1′ Infl a 3–many-fld umbel or fls 1 in axils
    3. Fls 1 in lf axils, ± sessile; fr glaucous, > 1.5 cm wide . . . . *E. globulus*
    3′ Fls 3–many in stalked umbels; fr not glaucous, < 1 cm wide
        4. Bark deeply furrowed, dark brown to ± black; anthers 0 on outer stamens. . . . . *E. sideroxylon*
        4′ Bark smooth, shedding, occ rough near base, mottled, gray, white, orange, or tan; anthers on all stamens
            5. Fls, fr fused at base into cluster 3–6 cm wide . . . . *E. conferruminata*
            5′ Fls, fr all free at base
                6. Hypanthium cylindric or urn-shaped; lvs lighter abaxially . . . . *E. cladocalyx*
                6′ Hypanthium obconic, ovoid, or hemispheric; lvs same color on both sides
                    7. Infl a 3-fld umbel. . . . . *E. viminalis*
                    7′ Infl a 5–15-fld umbel
                        8. Lvs linear, gen < 0.5 cm wide; fr valves not exserted . . . . *E. pulchella*
                        8′ Lvs lanceolate, gen > 1 cm wide; fr valves exserted
                            9. Bud cap hemispheric, beaked, ± = hypanthium . . . . *E. camaldulensis*
                            9′ Bud cap horn-shaped to conic, not beaked, ± 2 × hypanthium . . . . *E. tereticornis*

E. camaldulensis Dehnh. (p. 921)   RED GUM, RIVER RED GUM
ST: < 25 m, ± straight; branches often hanging in clumps; bark persistent near base, gen shed in irregular strips, smooth, gray or tan. **LF:** 6–20 cm, 1.5–2.5 cm wide, lanceolate. **INFL:** umbel, 7–11-fld. **FL:** hypanthium 2–3 mm, hemispheric, ± = bud cap; bud cap hemispheric, beaked (bluntly conic); stamens white. **FR:** 0.5–1 cm, ± hemispheric; valves exserted. Common. Disturbed areas; < 300 m. NCoRO, GV, CW (exc SCoRI), SCo, n ChI (Santa Cruz Island), TR, PR; native to Australia; naturalized in AZ. As cult sp., ranks 1 outside CA, 2 in CA. Apr–Jul ❖

E. citriodora Hook.   LEMON-SCENTED GUM   **ST:** 20–35 m, straight, slender; bark smooth, shed in irregular pieces, occ spotted, white or golden when first exposed, tan in age. **LF:** 10–20 cm, 1–2 cm wide, lanceolate, lemon-scented. **INFL:** panicle of 3–5-fld umbels. **FL:** hypanthium 5–6 mm, hemispheric, > bud cap; bud cap gen beaked; stamens white. **FR:** < 15 mm, urn-shaped; valves incl. Uncommon. Disturbed coastal areas; gen < 200 m. SCo; native to ne Australia. Commonly cult in CA; treated by some as *Corymbia citriodora* (Hook.) Hill & Johnson (Hill & Johnson 1995 Telopea 6:185–504). Dec–May

E. cladocalyx F. Muell.   SUGAR GUM   **ST:** 10–20 m, gen straight; bark shed in large irregular patches, ± smooth, white, often mottled gray, orange, or tan. **LF:** 8–15 cm, 2–3 cm wide, ± widely lanceolate, lighter abaxially. **INFL:** umbel, 7–11-fld, gen on lfless branches. **FL:** hypanthium < 1 cm, cylindric or urn-shaped, ± ridged; bud cap hemispheric to conic, < hypanthium, > hypanthium in width; stamens white. **FR:** 1–1.5 cm, ± urn-shaped, ribbed; valves incl. Uncommon. Disturbed coastal areas; gen < 200 m. CCo, SCo, PR; native to s Australia. TOXIC to livestock in Australia. Commonly cult in s CA. Apr–Jul

E. conferruminata D.J. Carr & S.G.M. Carr   SPIDER GUM, BUSHY YATE   **ST:** 1–5 m, irregularly branched; bark smooth, shed in strips, short ribbons, light gray or tan. **LF:** 5–9 cm, 1–4 cm wide, elliptic to elongate-elliptic, glossy, light green. **INFL:** umbel, fls 7–19, fused at base; peduncle flat, 1–2 cm wide. **FL:** bud cap finger-shaped; stamens yellow-green. **FR:** sessile, fused at base into cluster 3–6 cm wide; valves 3, exserted, style remnants persistent. Uncommon. Disturbed coastal areas; gen < 200 m. CCo, SCo; native to sw Australia. Commonly cult as screen, SnFrB south. Apr–Jul

E. globulus Labill. (p. 921)   BLUE GUM   **ST:** < 60 m, straight; bark sometimes persistent near base, otherwise shed in irregular strips, smooth, blue-gray; twigs ± square or winged. **LF:** 10–30 cm, 2.5–4 cm wide, gen narrowly lanceolate, often sickle-shaped, gen aromatic. **INFL:** fls 1 in axils, ± sessile. **FL:** hypanthium < 2 cm, ± 4-ribbed, obconic, glaucous; bud cap flat-hemispheric, with central knob, gen < hypanthium, warty, glaucous; stamens white. **FR:** > 2 cm, ± 4-ribbed, warty, glaucous, rim wide, thickened; valves ± not exserted. Common. Disturbed areas; < 300 m. NCoRO, GV, CW, SW; native to

se Australia. Most cult, most naturalized, tallest fl pls in CA; easily recognized by large, solitary fls, frs; growth rapid. Oct–Jan ❖

E. polyanthemos Schauer   SILVER DOLLAR GUM, RED BOX   **ST:** < 25 m; bark highly varied, rough, fibrous, persistent or smooth, shed in flakes or irregular strips, gray or tan. **LF:** 5–10 cm, 1.5–5 cm wide, round, elliptic, or ovate, gray-green, silver, or blue-green, occ glaucous. **INFL:** terminal or axillary panicle of 5–7 fld umbels. **FL:** hypanthium ± 4 mm, ovoid to obconic, ± 2 × bud cap; bud cap conic to hemispheric; stamens white. **FR:** 5–6 mm, pear- to bowl-shaped; valves incl. Uncommon. Disturbed coastal areas; < 200 m. SnJV, CCo, SnFrB, SCoRO, SCo, s ChI (Santa Catalina Island), WTR; native to se Australia. Commonly cult in CA; often juvenile, adult, and transitional lvs in mature crowns of naturalized pls; juvenile lvs used in fl arrangements. Dec–Feb

E. pulchella Desf.   WHITE PEPPERMINT   **ST:** < 20 m, branches erect; bark shedding in long strips, smooth, occ shaggy or rough near base, white to blue-gray. **LF:** 5–10 cm, gen < 0.5 cm wide, linear, dark green, ± peppermint odor when crushed. **INFL:** umbel, 9–15-fld. **FL:** hypanthium ± 2 mm, obconic, ± = bud cap; bud cap hemispheric; stamens white. **FR:** 4–6 mm, ovoid to cup-shaped; valves not exserted. Uncommon. Disturbed areas; < 200 m. SnFrB; native to se Australia, rarely cult in CA. Dec–Feb

E. sideroxylon Woolls   RED IRON BARK   **ST:** 7–25 m; bark persistent, deeply furrowed, hard, dark brown to ± black. **LF:** 6–14 cm, 1–2 cm wide, lanceolate, dull gray-green. **INFL:** umbel, 5–7-fld, pendent. **FL:** hypanthium 4–6 mm, gen glaucous, ovoid to hemispheric, > bud cap; bud cap conic; stamens gen pink to red (white), anthers 0 on outer. **FR:** ± 1 cm, ovoid; valves incl. Uncommon. Coastal, disturbed areas; < 200 m. CCo, SCoRO, SCo, n ChI, WTR; native to se Australia, commonly cult in CA. Dec–Feb

E. tereticornis Sm.   FOREST RED GUM   **ST:** 10–25(50) m, straight; bark persistent near base, gen shed in irregular strips, smooth, gray or tan. **LF:** 8–20 cm, 1–2.5 cm wide, lanceolate. **INFL:** umbel, 7–11-fld. **FL:** hypanthium 2–3 mm, obconic to hemispheric, ± 1/2 × bud cap; bud cap horn-shaped to conic, smooth; stamens white. **FR:** 7–9 mm, hemispheric; valves exserted. Uncommon. Disturbed coastal areas; < 200 m. SnJV, SCoRO, SCo, n ChI, TR, PR; native to e Australia. Commonly cult in CA. May–Aug

E. viminalis Labill.   MANNA GUM, RIBBON GUM   **ST:** 25–50 m, straight; bark gen shed in long ribbons, persistent near base or not, smooth, white, gray, or tan. **LF:** 10–15 cm, 1–2.5 cm wide, lanceolate. **INFL:** umbel, 3-fld. **FL:** hypanthium 2–3 mm, ovoid to obconic, ± = bud cap; buds ± sessile; bud cap conic, smooth; stamens white. **FR:** 5–7 mm, ± hemispheric; valves exserted. Uncommon. Disturbed areas; < 100 m. NCoRO, CCo, SCoRO, SCo, PR; native to se Australia. Commonly cult in CA. Jul–Sep

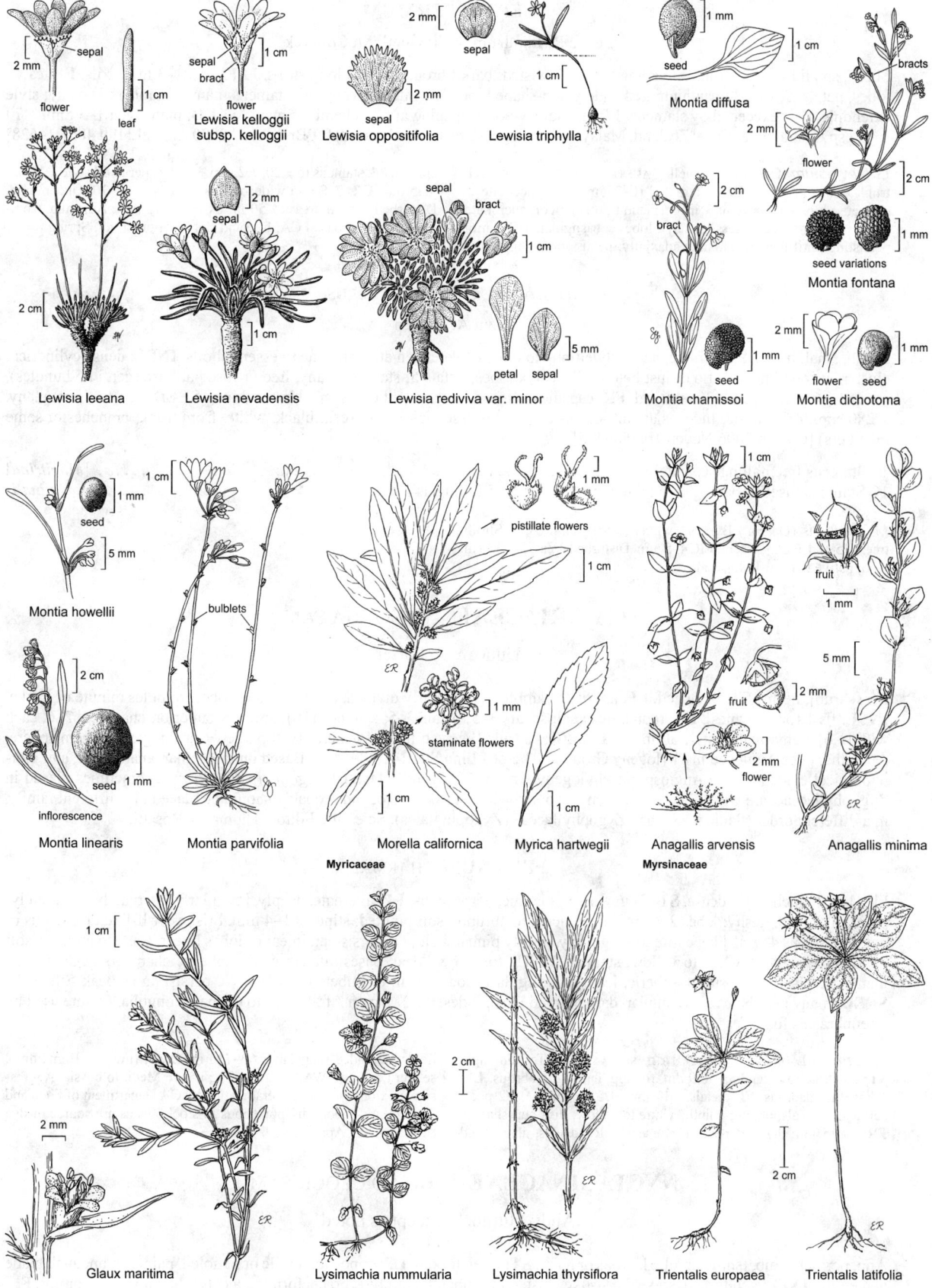

915

LEPTOSPERMUM

Leslie R. Landrum & Elizabeth McClintock

Shrub, small tree. **ST**: trunk spreading to erect, twisted; bark fibrous, shed in long strips. **INFL**: fls 1–3 in lf axils. **FL**: calyx lobes, petals, 5, free; hypanthium gen widely cup-shaped; petals white, pink, or red; stamens many; chambers 5–many; style extending to or exceeded by stamens. **FR**: capsule, woody, opening at top; chambers many. **SEED**: many, gen 1–3 mm. ± 70 spp.: esp Australia, few New Zealand, Malay Archipelago. (Greek: slender seed) [O'Brien et al. 2000 Austral J Bot 48:621–628]

L. laevigatum (Gaertn.) F. Muell. AUSTRALIAN TEA TREE **ST**: trunk gen spreading. **LF**: alternate, 10–25 mm, oblanceolate to obovate-oblong; main veins 3, inconspicuous; tip blunt or mucronate. **FL**: 1.5–2 cm wide, ± sessile; calyx lobes ± triangular, 1–1.5 mm, persisting until fr matures, silky adaxially; petals spreading, white, ephemeral; stamens to ± 20, ± 2 mm, < (or =) perianth; style ± 1 mm, glabrous. **FR**: 7–8 mm wide; valves ± exserted, 7–10. Uncommon. Disturbed coastal areas; < 50 m. CCo; native to se Australia. Commonly cult in coastal CA, for sand stabilization or not. Spring

MELALEUCA BOTTLEBRUSH

Lyn A. Craven

Shrub, small tree. **LF**: alternate, narrowly elliptic to oblanceolate, gen aromatic due to essential oils. **INFL**: dense cylindrical cluster surrounding st at tip or just below. **FL**: calyx lobes, petals 5; stamens many, free or fused at base (gen in 5 bundles), several × > perianth, 13–30 mm, red. **FR**: capsule enclosed in persistent cup-shaped hypanthium. **SEED**: gen 1–3 mm, many. ± 280 spp.: Australasia, Indonesia; cult, gen as orn, or for essential oils. (Greek: black, white, from trunk, branches of some members) [Craven 2006 Novon 16:468–475]

1. Stamens free, falling separately . . . . *[M. citrina]*
1′ Stamens fused at base, gen in 5 bundles of 9–14, all 5 falling as 1 unit . . . . **M. viminalis**

M. viminalis (Gaertn.) Byrnes WEEPING BOTTLEBRUSH Shrub, tree < 15 m. **LF**: 2.5–14 cm. **FR**: 4–5 mm. Disturbed areas; 30–330 m. SCo; native to e Australia. ± all yr

NITRARIACEAE NITRARIA FAMILY

Duncan M. Porter

Per [shrub], fleshy. **ST**: branched. **LF**: alternate, simple to irregularly divided, occ toothed or lobed; stipules minute or lf-like. **INFL**: fls 1 [or in cymes]. **FL**: radial, bisexual; sepals 3–5; petals 3–5; stamens (10)15; ovary superior, chambers 2–6, each with [1,6] many ovules, style at tip or basal. **FR**: capsule [berry, drupe]. 4 genera, 16 spp.: gen ± arid regions, n hemisphere, Australia. [The Angiosperm Phylogeny Group 2003 Bot J Linn Soc 141:399–436] Based on molecular, anatomical, and morphological evidence, The Angiosperm Phylogeny Group (2003) indicates that *Peganum* and *Nitraria*, sometimes placed in Zygophyllaceae, are widely separated from Zygophyllaceae *sensu stricto*, and probably should be placed in a different family, in a different order (Sapindales) than Zygophyllaceae (Zygophyllales). Scientific Editor: Thomas J. Rosatti.

PEGANUM HARMAL

Pl < 1 m; branches, lvs dense. **ST**: from woody rhizome, herbaceous. **LF**: alternate, deeply, irregularly 2-ternately or ternately-pinnately lobed, fleshy; lobes 2–6 cm, linear, acute to abruptly soft pointed; stipules 1–4 mm, bristle- or lf-like, deciduous or not. **FL**: sepals 4–5, lf-like, entire to irregularly, deeply pinnately lobed, persistent, linear or lobes linear, acute to abruptly soft pointed; petals 4–5, white to yellow; stamens 12–15, filaments linear, bases dilated; ovary spheric, chambers (2)3(4), each many-ovuled. **FR**: capsule ± spheric, leathery, irregularly loculicidal; chambers (2)3(4); style persisting as beak > fr body. **SEED**: many per chamber, triangular, dark, curved. 3 spp.: deserts, N.Am, n Afr, Spain; to China; Mongolia. (Name used by Theophrastus for "rues")

P. harmala L. (p. 921) AFRICAN RUE, SYRIAN RUE To 3 dm high. **ST**: prostrate to ascending, to 9 dm, rooting, hairy to glabrous. **LF**: ± hairy to glabrous. **FL**: petals < 17 mm, 1/2 as wide, ± 1/2 pedicel, gen < sepals, oblong-elliptic, white to yellow; disk cup-shaped. **FR**: to 15 mm diam, < pedicel, << sepals. Roadsides, abandoned fields, degraded rangeland; 60–730 m. DMoj (nw San Bernardino, se Kern cos.); to WA, MT, NV, TX; native Medit to c Asia. Aggressive, noxious weed under eradication by CA Department of Food and Agriculture; reportedly poisonous to stock. Seeds are source of dye "Turkey-red". Apr–Nov ◆

NYCTAGINACEAE FOUR O'CLOCK FAMILY

Andy Murdock, except as noted

Ann, per, subshrub, [shrub, tree], glabrous or hairy. **ST**: often forked. **LF**: opposite, sessile or petioled, pairs gen unequal; blade gen entire. **INFL**: gen forked; spike, head-like cluster, or umbel, fls rarely 1, bracts forming a calyx-like involucre or not. **FL**: bisexual, gen ± radial (bilateral), sometimes cleistogamous in some genera; perianth of 1 whorl, gen petal-like, bell- to trumpet-shaped, base hardened, tightly surrounding ovary in fr, lobes 3–5, gen notched to ± lobed; stamens 1–many; ovary superior

(appearing inferior due to hardened perianth base), style 1. **FR**: achene in hardened perianth base; round to ± flat; smooth, angled, ribbed, or winged; glabrous, hairy, or glandular. 30 genera, 350 spp.: warm regions, esp Am; some orn (*Bougainvillea*; *Mirabilis*, four o'clock). [Spellenberg 2003 FNANM 4:14–17] Scientific Editor: Thomas J. Rosatti.

1. Stigma linear, incl; infl a head or umbel; fr gen winged
  2. Fr wings thick or 0, not surrounding fr body; fls maturing outer before inner or ± simultaneously;
    receptacle conic, ± smooth . . . . **ABRONIA**
  2′ Fr wings translucent, surrounding fr body; fls maturing 1 side of infl before the other; receptacle flat
    or rounded, studded with peg-shaped pedicel-like projections . . . . **TRIPTEROCALYX**
1′ Stigma ± spheric, gen exserted; infl a head or umbel or not; fr gen unwinged
  3. Bracts 3 or 5(9), gen at least partially fused, forming a calyx-like involucre
    4. Fls strongly bilateral, in 1 involucre 3, blooming ± together; bracts 3; fr bilateral . . . . **ALLIONIA**
    4′ Fls radial or ± bilateral, in 1 involucre 1–16, gen not blooming together; bracts 5(9); fr radial . . . . **MIRABILIS**
  3′ Bracts 1–3, free, not forming an involucre
    5. St erect, > 7 mm diam; fr glabrous, with 10 inconspicuous ribs . . . . **ANULOCAULIS**
    5′ St decumbent to erect, gen << 5 mm diam; fr hairy or not, with (4)5 prominent ribs or wings
      6. Perianth tube 30–170 mm, trumpet-shaped; fr oblong, glabrous, 5–10 mm . . . . **ACLEISANTHES**
      6′ Perianth tube < 5 mm, bell-shaped; fr club-shaped, glabrous or hairy, < 4 mm . . . . **BOERHAVIA**

ABRONIA SAND-VERBENA

Ann, per, gen glandular. **ST**: prostrate to ascending, gen ± red. **LF**: gen fleshy, petioled. **INFL**: head or umbel; fls maturing outer before inner or ± simultaneously; receptacle conic, ± smooth; bracts 5–10. **FL**: perianth salverform to trumpet-shaped, gen fragrant, lobes 4–5; stamens 4–5, incl; stigma linear, incl. **FR**: body fusiform; wings (0)2–5, lobe-like, prominent, opaque, thick, not continuous above fr body. 25 spp.: w N.Am. (Greek: graceful) [Galloway 2003 FNANM 4:61–69] Closely related to *Tripterocalyx*; relationships among spp. *A. gracilis* Benth. (Mex), *A. maritima*, *A. umbellata*, and *A. villosa* need study; hybrids involve *A. latifolia*, *A. maritima*, *A. umbellata*; *A. gracilis* incl in TJM (1993) based on misidentifications.

1. Sts densely cespitose or in mats < 20 cm wide; montane; per
  2. Infl 1–5-fld; pl in mats; lvs cauline, blade ± round; fr wing 0 . . . . *A. alpina*
  2′ Infl ≥ 6-fld; pl densely tufted; lvs basal, blade oblong to ± round; fr winged . . . . *A. nana* var. *covillei*
1′ Sts elongate, not cespitose, when prostrate in mats >> 25 cm; coastal, desert, or montane; ann or per
  3. Per; lf fleshy; perianth limb yellow or wine-red; coastal
    4. Lf blade ± as long as wide; perianth limb yellow . . . . *A. latifolia*
    4′ Lf blade longer than wide; perianth limb wine-red . . . . *A. maritima*
  3′ Ann (per); lf ± not fleshy; perianth limb white, pink, red, or magenta; coastal or inland
    5. Fr wings thick, hollow, inflated or ± collapsed and folded together; lvs ± glabrous (sparsely puberulent);
      inland, mostly desert . . . . *A. turbinata*
    5′ Fr wings thin, solid, rarely 0; lvs hairy to glandular-hairy; coastal or inland, mostly desert
      6. Fr wings 2(3); fr broadly obcordate, with sinus at top between wings . . . . *A. pogonantha*
      6′ Fr wings gen 3–5 or 0; fr cordate, obconic, or ovoid, with no sinus at top between wings
        7. Fr body ± smooth; immediate coast . . . . *A. umbellata*
          8. Perianth limb light to bright magenta, throat with a cream to light yellow spot; infl sparse, fls 10–18;
            fr wings angled near tip . . . . var. *breviflora*
          8′ Perianth limb light to dark magenta, throat with white spot; infl ± dense, fls 10–27; fr wings broadly
            rounded near tip . . . . var. *umbellata*
        7′ Fr body with gen raised veins (most evident near tip); ± inland . . . . *A. villosa*
          9. Perianth tube 2–3.5 cm, limb (1)1.5–1.8 cm wide; fr wings each with 1 prolonged, acute lobe distally . . . . . var. *aurita*
          9′ Perianth tube 1.3–2 cm, limb 0.6–1.2(1.5) cm wide; fr wings each with 0 or 1 shallow, rounded lobe
            distally . . . . var. *villosa*

A. alpina Brandegee (p. 921) RAMSHAW MEADOWS ABRONIA Per, in mats < 25 cm, glandular-puberulent. **LF**: petiole 1–2 cm; blade 4–9 mm, ± round. **INFL**: peduncle < petioles; bracts 2–3 mm, narrowly ovate; fls 1–5. **FL**: perianth tube 10–15 mm, ± white, limb 6–9 mm wide, white to lavender-pink. **FR**: 3–5 mm, 5-angled, net-veined; wing 0. Dry, open, granitic meadows; 2400–2700 m. s SNH (Ramshaw, Templeton meadows, Tulare Co.). May–Sep ★

A. latifolia Eschsch. (p. 921) Per, fleshy, finely glandular-hairy; root deep, thick. **ST**: prostrate, < 2 m; branches short, forming a dense mat. **LF**: petiole 1–6 cm; blade 2–5 cm, broadly ovate to reniform, thick. **INFL**: peduncle 1–6 cm; bracts 5–9 mm, ovate; fls 17–34. **FL**: perianth tube 9–12 mm, yellow to ± green, limb 8–13 mm wide, yellow. **FR**: 8–15 mm, net-veined, glabrous or ± hairy, tapered to base, tip; wings 0 or 5. Coastal dunes, scrub; < 100 m. NCo, CCo, n SCo (Santa Barbara Co.); to BC. May–Oct

A. maritima S. Watson RED SAND-VERBENA Per, densely glandular-hairy; roots spreading. **ST**: prostrate, < 2 m; branches short, erect, forming thick horizontal mat. **LF**: petiole 5–30 mm; blade 5–7 cm, fleshy, broadly elliptic to oblong. **INFL**: peduncle 3–10 cm; bracts 7–11 mm, lanceolate or narrowly ovate; fls 10–18. **FL**: perianth tube 6–10 mm, ± green to red, limb 7–10 mm wide, wine-red. **FR**: 10–14 mm, glandular-hairy near top, wings 3–5, thick, truncate, widest at base, coarsely net-veined. Coastal dunes; < 100 m. s CCo, SCo, ChI; Baja CA. Feb–Oct ★

A. nana S. Watson var. *covillei* (Heimerl) Munz COVILLE'S DWARF ABRONIA Per, densely tufted. **ST**: < 6 cm. **LF**: basal, glaucous; petiole 1–4 cm; blade 5–20 mm, oblong to ± round. **INFL**: scapose; peduncle < 10 cm; bracts 6–8 mm, lanceolate; fls > 6. **FL**: perianth tube 11–15 mm, white to light pink, limb 6–8 mm wide, white to pink. **FR**: 7–8 mm; wings 5, thin, rounded at top. Dry sandy places; 1600–2800 m. SnBr, W&I, DMtns; sw NV. [*A. n.* S. Watson subsp. *c.* (Heimerl) Munz] Pls from e DMoj have wider bracts, longer fls, approach *A. nana* var. *n.* of NV. Jun–Aug ★

A. pogonantha Heimerl (p. 921) Ann, glandular-hairy. **ST**: decumbent to ascending, 10–55 cm. **LF**: petiole 0.8–4 cm; blade 1.5–5.5 cm, 1–3 cm wide, ovate to oblong-ovate. **INFL**: peduncle

2–7 cm; bracts 4–9 mm, lanceolate to broadly ovate; fls 12–24. **FL:** perianth tube 1–2 cm, red or basally ± green, limb 6–8 mm wide, white to pink. **FR:** 4–6 mm, broadly obcordate, with sinus at top between wings; wings 2(3), interior spongy. Sand, desert communities; < 2000 m. s SnJV, SnGb, DMoj, DMoj/SN; w NV. Apr–Jul

A. turbinata S. Watson (p. 921)   Ann (per), glabrous or sparsely glandular-hairy. **ST:** ascending to erect, < 50 cm. **LF:** petiole 1–5 cm, slender; blade 1–5 cm, broadly ovate or cordate to round. **INFL:** peduncle 3–9 cm; bracts 3–10 mm, lanceolate to ovate; fls 15–35. **FL:** perianth tube 6–18 mm, pink to ± green, limb 5–8 mm wide, white to pale magenta. **FR:** 3–8 mm, wings gen 5, thick, hollow, inflated or ± collapsed and folded together. Dry, sandy soil, desert scrub; 900–2500 m. s MP (Sierra Valley), s SNF, WTR, SNE, DMoj; se OR, w NV. May–Jul

A. umbellata Lam.   Ann, ± glabrous to glandular-hairy. **ST:** prostrate, > 1 m. **LF:** petiole 1–6 cm, slender; blade 15–70 mm, 8–50 mm wide, ovate to ± diamond-shaped. **INFL:** peduncle 2.5–15 cm, slender, gen erect in fr; bracts 5–7 mm, lanceolate to ovate; fls [8]10–27. **FL:** perianth tube 6–16(20) mm, throat with white to cream spot, limb 5–12(17) mm wide, light to dark magenta. **FR:** 6–13 mm, glandular-hairy at base, glabrous above, wings 4(5), with no sinus at top between. 1 other var., in BC, WA.

var. **breviflora** (Standl.) L.A. Galloway   PINK SAND-VERBENA **INFL:** fls 10–18. **FL:** perianth tube 6–10 mm, ± green to magenta, throat with a cream to light yellow spot, limb 5–11 mm wide, light to bright magenta. **FR:** wings poorly to moderately developed. Dis-

turbed sandy areas, coastal dunes and scrub; < 100 m. NCo, CCo (Marin Co.); s OR. [*A. u.* subsp. *b.* (Standl.) Munz] Populations gen small. Jun–Sep ★

var. **umbellata**   **INFL:** fls 10–27. **FL:** perianth tube 12–16(20) mm, ± green to maroon, throat with white spot, limb 7–12(17) mm wide, light to dark magenta. **FR:** wings poorly to well developed, broadly rounded near tip. Disturbed sandy areas, coastal dunes and scrub; < 100 m. NCo (Sonoma Co.), CCo, SCo, ChI, DSon; Baja CA. [*A. u.* subsp. *u.*] All yr

A. villosa S. Watson   Ann, glandular-hairy. **ST:** prostrate to ascending, < 80 cm. **LF:** petiole 0.5–5 cm; blade 1–5 cm, 1–4.5 cm wide, triangular-ovate to ± round. **INFL:** peduncle 2–10 cm; bracts 3–11 mm, lanceolate to narrowly ovate; fls 15–35. **FL:** perianth tube 1.3–3.5 cm, ± pink, throat with white spot, limb 6–18 mm wide, pale to bright magenta. **FR:** 5–10 mm, body with prominent raised veins, top with hardened beak (base of fl tube); wings 3–5, thin, rounded or angled, or 0.

var. **aurita** (Abrams) Jeps.   CHAPARRAL SAND-VERBENA   **FL:** perianth tube 2–3.5 cm, limb (1)1.5–1.8 cm wide. **FR:** body nearly smooth; wings exceeding body. Sandy places in coastal-sage scrub, chaparral; < 1600 m. c&s SCo, w DSon. Mar–Aug   ★

var. **villosa** (p. 921)   **FL:** perianth tube 1.3–2 cm, limb 0.6–1.2(1.5) cm wide. **FR:** body prominently wrinkled, appearing pitted; wings 0 or not or barely exceeding body. Sandy places in creosote-bush scrub; < 1000 m. D; s NV, sw AZ, nw Mex. Feb–Jul

ACLEISANTHES   TRUMPETS

Per from thick taproot, minutely hairy. **ST:** prostrate to ascending. **LF:** petioled; blade < 8 cm, ± firm to fleshy, paired lvs ± unequal. **INFL:** bracts 1–3, free, not forming involucre; fl gen 1, or paired. **FL:** nocturnal or cleistogamous; perianth trumpet-shaped, white, tube slender; stamens 5, ± exserted; stigma ± spheric, ± exceeding anthers. **FR:** oblong, glabrous; wings translucent or 0. 17 spp.: esp sw US, ne Mex. (Greek: without closure, from absence of involucre) [Poole 2003 FNANM 4:33–39] Pollinated by hawkmoths. *Selinocarpus* moved to *Acleisanthes*.

1. Perianth 7–17 cm, limb 1–3 cm wide; fr with 5 ribs; e DSon . . . . ***A. longiflora***
1′ Perianth 3–4 cm, limb ± 1 cm wide; fr with 5 wings; DMtns . . . . ***A. nevadensis***

A. longiflora A. Gray (p. 921)   ANGEL TRUMPETS   Ann; puberulent to ± glabrous. **ST:** < 1 m. **LF:** petiole 1–15 mm; blade < 4 cm, lanceolate to triangular, acute to acuminate, margins ± wavy. **FL:** perianth tube ± green or pale purple, limb white. **FR:** 6–10 mm, oblong, with 2 parallel grooves between adjacent ribs. Dry places, gen on limestone; 10–2500 m. e DSon (Santa Maria Mtns, e Riverside Co.); to TX, n Mex. May   ★

A. nevadensis (Standl.) B.L. Turner (p. 921)   DESERT WING-FRUIT Ann; hairs appressed and divergent, white, also glandular-puberulent. **ST:** prostrate to erect, < 3 dm. **LF:** petiole 5–20(26) mm; blade ovate to round. **FL:** perianth tube ± green, limb white. **FR:** 5–7 mm; wings 2 mm. Dry, rocky areas; 1250 m. DMtns (ne Kingston Range, se Inyo Co.); to sw UT, nw AZ. [*Selinocarpus n.* (Standl.) B.A. Fowler & B.L. Turner] Jun–Sep   ★

ALLIONIA   WINDMILLS

Ann, short-lived per, glabrous to densely glandular-hairy. **ST:** trailing, < 1 m. **LF:** petioled; blade < 4 cm, ovate to oblong, paler below. **INFL:** head, 3-fld but resembling 1 fl; bracts 3, free to ± 1/2 fused, forming involucre resembling 1 calyx, hairy; fls 3 in 1 involucre, blooming ± together. **FL:** strongly bilateral (entire 3-fld head ± radial); stamens gen 4(7), exserted; stigma ± spheric, exserted. **FR:** bilateral. 2 spp.: Am. (C. Allioni, Italian botanist, 1725–1804) [Spellenberg 2003 FNANM 4:58–60]

A. incarnata L.   **FL:** perianth tube funnel-shaped, limb 3–15 mm, lobes 3, notched, oblique, red-purple. **FR:** 2.5–6 mm, compressed, bilateral, adaxially convex, abaxially ± flat, with 2 rows of sticky glands; 2 lateral wings strongly incurved abaxially, entire or with 3–5 irregular teeth. $2n=40$. Vars. ± intergrade where geog overlaps.

1. Perianth limb 3–10 mm, head 5–12 mm wide
   at maturity, peduncle gen < 15 mm; stamens
   gen ± incl . . . . var. ***incarnata***
1′ Perianth limb 8–15 mm, head 20–25 mm wide
   at maturity, peduncle > 15 mm; stamens
   gen exserted . . . . var. ***villosa***

var. **incarnata** (p. 921)   **ST:** glandular-puberulent, sparsely long-hairy. **LF:** tomentose, upper lvs often reduced. **FR:** gen 2–5 mm. Creosote-bush scrub; < 1600 m. D; to CO, TX, S.Am. [*A. i.* var. *nudata* (Standl.) Munz] Mar–Sep

var. **villosa** (Standl.) Munz   **ST:** glandular-puberulent, gen densely long-hairy. **LF:** long-hairy, upper lvs ± not reduced. **FR:** gen 4–6 mm. Creosote-bush scrub; 100–1200 m. D; NV, UT, AZ, NM, Mex. Mar–Sep

ANULOCAULIS RINGSTEM

Andy Murdock & Richard Spellenberg

Per from thick caudex. **ST**: little-branched, erect, > 7 mm diam; internodes with sticky brown ring. **LF**: few, ± in lower 1/2, petioled; blade oblong to round, thick. **INFL**: openly branched; fls in head, raceme, or umbel-like cluster; bracts 1–3, free, not forming involucre. **FL**: perianth funnel-shaped; stamens 3, exserted; stigma ± spheric, exserted. **FR**: inconspicuously 10-ribbed, glabrous. 5 spp.: esp Chihuahuan Desert, ne Mex. (Latin: ring st, from sticky ring on internode) [Spellenberg 2003 FNANM 4:28–30] Closely related to *Boerhavia*.

A. annulatus (Coville) Standl. (p. 921) **ST**: < 1.5 m. **LF**: blade 3–10 cm, oblong to round-reniform, hairs stiff, with enlarged, dark, glandular base. **INFL**: terminal, panicle-like, 1–2 branches per node, each ending in head-like cluster of fls; bracts lf-like, reduced; peduncle 15–40 mm. **FL**: perianth ± 8 mm, tube ± green, hairy, limb pale pink. **FR**: 4–5 mm, spindle-shaped, gray-brown. 2*n*=48. Rocky slopes, canyons; < 1200 m. ne DMoj (Death Valley region). Apr–May

BOERHAVIA SPIDERLING

Ann, per. **ST**: prostrate to erect; internode often with sticky region. **LF**: petioled; blade 1–6 cm, paler beneath, often brown-dotted. **INFL**: panicle-like, branches ending in umbel, spike-like raceme, or paired or solitary fls; bracts 1–3, free, not forming involucre. **FL**: closing by evening; perianth bell-shaped, tube < 5 mm, limb < 3 mm; stamens 1–5; stigma ± spheric, gen exserted. **FR**: < 4 mm, club-shaped; ribs 4–5; wings 0. ± 40 spp.: warm regions worldwide. (H. Boerhaave, Dutch botanist, 1668–1738) [Spellenberg 2003 FNANM 4:17–28]

1. Perianth red to purple (pale red); per; fr glandular hairy
　2. Branches hairy, gen glandular; lvs throughout pl . . . . ***B. coccinea***
　2′ Branches ± glabrous; lvs ± in lower 1/2 of pl. . . . . ***B. diffusa***
1′ Perianth pale pink to white; ann; fr glabrous
　3. Infl gen ending in umbels of 3–6(8) fls or fls 1 or paired . . . . ***B. triquetra*** var. ***intermedia***
　3′ Infl ending in spike-like raceme
　　4. Bracts deciduous, << fr; fr ribs gen 5; st short-hairy . . . . ***B. coulteri*** var. ***palmeri***
　　4′ Bracts ± persistent, ± = fr; fr ribs gen 4; st glandular-hairy. . . . . ***B. wrightii***

B. coccinea Mill. (p. 921) SCARLET SPIDERLING Per. **ST**: prostrate or sprawling, < 15 dm; hairy, gen glandular. **LF**: blade broadly ovate, blunt, slightly wavy. **INFL**: umbel. **FL**: perianth limb 1–2 mm, red-violet. **FR**: 2.5–3.5 mm, glandular-hairy, wide, smooth between ribs; tip rounded. 2*n*=24. Dry disturbed places; < 1000 m. s SnJV, SCo, PR, DSon; to se US, nw S.Am. Apr–Jul

B. coulteri (Hook. f.) S. Watson var. ***palmeri*** (S. Watson) Spellenb. (p. 921) Ann. **ST**: decumbent to erect, < 8 dm, puberulent. **LF**: blade lanceolate to ovate-triangular, ± acute. **INFL**: spike-like raceme; bracts << fr, deciduous. **FL**: < 1.5 mm; perianth pale pink to white. **FR**: 2.5 mm, glabrous; narrow, wrinkled between gen 5 wide ribs; tip rounded to ± truncate. Gravelly hillsides, washes; < 1450 m. ChI (possibly introduced), se DMoj (desert mtns and lower montane slopes), DSon; to AZ, Mex. *B. coulteri* var. *c.* expected as agricultural weed in s CA. Sep–Nov

B. diffusa L. Per. **ST**: decumbent to erect, < 1 m, ± glabrous. **LF**: blade broadly lanceolate to round; glabrous to minutely puberulent; **INFL**: umbel of (2)3–5 fls, bracts gen strongly reduced, not lf-like. **FL**: 1–1.5 mm; perianth red-purple (pale red). **FR**: 3–4.5 mm; stalked glands on ribs, less dense between; tip broadly conic. Disturbed places, often in sand, gravel along rivers; < 50 m. ScV (Yolo Co.); trop, subtrop worldwide. ± all yr

B. triquetra S. Watson var. ***intermedia*** (M.E. Jones) Spellenb. (p. 921) Ann. **ST**: ascending or erect, 2–5 dm; hairs fine, sparse. **LF**: blade broadly lanceolate or ovate, acute to obtuse. **INFL**: gen umbel of 3–6(8) fls or fls 1 or paired. **FL**: 1.5–2 mm; perianth pale pink to white. **FR**: 2–2.7 mm, glabrous, length 3 × width; ribs (3)5, rounded to angled; tip ± truncate to abruptly rounded. Gravelly washes, flats; < 1300 m. e PR, e DMtns (Clark, Ivanpah mtns, Kingston Range), se DMoj exc DMtns, e DSon; to TX, Mex. [*B. i.* M.E. Jones] In n Mex, less so in CA, intergrading with *B. triquetra* S. Watson var. *t.* of Mex; easily confused with *B. erecta* L. (fr 3–4.5 mm), widespread in trop Am, expected as agricultural weed in CA near Mex. Aug–Oct

B. wrightii A. Gray (p. 921) Ann. **ST**: branched from base, erect, < 7 dm, glandular-hairy. **LF**: blade lanceolate to oblong-ovate, acute. **INFL**: spike-like raceme; bracts wide, ± red, ± persistent, ± = fr. **FL**: 1–2 mm, 2–3 mm in fr; perianth pale pink to white. **FR**: 2–2.5 mm, nearly as wide, wrinkled between gen 4 wide ribs. Dry, sandy places; < 1400 m. D; to NV, TX, Mex. Aug–Dec

MIRABILIS FOUR O'CLOCK

Per, subshrub. **ST**: repeatedly forked, decumbent to erect. **LF**: gen petioled. **INFL**: branches ending in umbel-like cluster or solitary fls; bracts 5(9), ± fused (or not) into calyx-like, bell- to saucer-shaped involucre; fls in 1 involucre 1–16, gen not blooming together; fls cleistogamous or not. **FL**: radial or ± bilateral; perianth funnel- to bell-shaped, lobes 5; stamens 3–5, gen exserted; stigma ± spheric, gen exserted. **FR**: ± round to club-shaped; ribs or angles 0, 5, 10; wings 0. ± 60 spp.: Am, Himalayas. (Latin: wonderful) [Spellenberg 2003 FNANM 4:40–57] Fls open in evening, close in morning; spp. intergrade, taxonomy unsettled.

1. Fr < 3 mm wide, ribs 5, strong, gen with prominent warts or coarse wrinkles between; involucre enlarged,
　　brown, papery in fr
　2. Lf petioled, lanceolate or wider
　　3. Involucre in fl hairy throughout on outside, in fr gen < 8 mm. . . . . ***M. albida***
　　3′ Involucre in fl short-hairy at base, glabrous on lobes, in fr 10–15 mm . . . . [***M. nyctaginea***]
　2′ Lf ± sessile, linear to narrowly lanceolate
　　4. Perianth bright red, 3–4 × length of involucre in fl. . . . . ***M. coccinea***
　　4′ Perianth pale pink to magenta, ± 2 × length of involucre in fl . . . . ***M. linearis*** var. ***linearis***
1′ Fr often > 3 mm wide, ribs or angles 0, 5, 10, moderate to inconspicuous, sometimes with low wrinkles
　　or warts between; involucre little changed in fr

5. Involucre 3–16-fld, bracts > 15 mm, free to fused
  6. Perianth ± 15 mm, bell-shaped; bracts 15–30 mm, free to ± 1/2 fused . . . . ***M. alipes***
  6′ Perianth 40–60 mm, funnel-shaped; bracts > 22 mm, > 1/2 fused
    7. Fr bluntly 5-angled . . . . ***M. greenei***
    7′ Fr not angled, often with 10 lines or low ribs . . . . ***M. multiflora***
      8. Fr faintly warty, ribs gen inconspicuous, gelatinous when wet . . . . var. ***glandulosa***
      8′ Fr ± smooth, with 10 slender, tan, sometimes raised ribs, gen alternating with 10 brown, often
        interrupted lines, not gelatinous when wet . . . . var. ***pubescens***
5′ Involucre 1(2)-fld, bracts < 15 mm, fused
  9. Perianth 30–50 mm; fr with 5 blunt ribs, ± wrinkled or warty between . . . . ***M. jalapa*** var. ***jalapa***
  9′ Perianth < 15 mm; fr unmarked or with gen 10 obscure lines, ± wrinkled or warty between or not
    10. Involucre 8–13 mm, lobes > tube, narrowly lanceolate; perianth ± white; lvs ascending . . . . ***M. tenuiloba***
    10′ Involucre 3–7 mm, lobes < tube, ± ovate; perianth white to magenta; lvs widely spreading . . . . ***M. laevis***
      11. Perianth pink to purple-red (white); lf puberulent or glandular-hairy . . . . var. ***crassifolia***
      11′ Perianth white (pale pink); lf gen ± glandular-hairy
        12. Fr ± spheric, gen with 5–10 visible lines; sts, lvs with short, reflexed hairs, gen also ± glandular-
          hairy . . . . var. ***retrorsa***
        12′ Fr gen ± ovoid, gen with 0 visible lines, rarely 5–10 barely visible lines; sts, lvs glandular-hairy . . . . var. ***villosa***

M. albida (Walter) Heimerl (p. 929)   St, lf hairs short to long, glandular or not. **ST:** decumbent to erect, 10–50 cm (in CA). **LF:** petiole < 4 cm, blade 2–6 cm lanceolate or wider. **INFL:** axillary involucres gen 1, with cleistogamous fls; involucre cup-shaped, green or ± purple, hairy, (1)3(5)-fld, enlarged, < 8 mm, brown, papery in fr; bracts 5, ≥ 1/2 fused. **FL:** perianth 8–12 mm, broadly funnel-shaped, light pink to magenta. **FR:** 3–5 mm, tapered at both ends; ribs 5, wide, with warts or wrinkles between. Dry, rocky areas; 1400–2500 m. SnBr, SnJt, W&I, DMtns; N.Am. [*M. oblongifolia* (A. Gray) Heimerl; *M. pumila* (Standl.) Standl.] Complex incl variable, poorly defined, intergrading taxa and many named forms, some of which in CA merit further study. May–Aug

M. alipes (S. Watson) Pilz (p. 929)   **ST:** decumbent to erect, 2–4 dm; glaucous, hairs 0 or sparse upward. **LF:** blade 2–7 cm, broadly ovate to round, fleshy, hairs 0 or sparse, short. **INFL:** involucre 1 per upper axil, peduncled, ± cup-shaped, glabrous; bracts 5–7(9), free to ± 1/2 fused, 15–30 mm, broadly ovate, each subtending, fused to 1 pedicel. **FL:** perianth 15–19 mm, funnel-shaped, magenta (creamy white). **FR:** 5.5–7 mm, elliptic, glabrous; ribs 10, slender, tan. Dry slopes, flats; 1200–2000 m. W&I, DMtns (Panamint Range); to w CO. May–Jun

M. coccinea (Torr.) Benth. & Hook. f.   RED FOUR O'CLOCK   **ST:** ascending to erect, < 6 dm, glabrous, glaucous. **LF:** ± sessile; blade 2–12 cm, linear, fleshy, glabrous. **INFL:** involucre 1 per upper axil, peduncled, bell-shaped, short-hairy, 1–3-fld, enlarged, 5–8 mm, papery in fr; bracts 5, ± 1/2 fused. **FL:** perianth 15–20 mm, ± funnel-shaped, bright red. **FR:** ± 5 mm, club-shaped; ribs 5, coarsely wrinkled between; hairs fine. Dry, rocky slopes, washes; 1300–1800 m. DMtns; to w NM, nw Mex. Easily confused with *M. linearis* var. *l.* without mature fls. May–Jul ★

M. greenei S. Watson   GREENE'S FOUR O'CLOCK   **ST:** ascending to erect, 4–8 dm, hairs 0 or sparse upward. **LF:** blade 3–10 cm, round to ovate, fleshy, hairs 0 or sparse, short. **INFL:** involucre 1 per upper axil, peduncled, bell-shaped, ± glabrous, 6–16-fld; bracts gen 5–6, 25–40 mm, > 1/2 fused. **FL:** perianth 4–5 cm, narrowly funnel-shaped, magenta. **FR:** ± 7 mm, broadly club-shaped or elliptic, bluntly 5-angled, sparsely short-hairy. Dry slopes, flats; < 1000 m. e KR, n&c NCoRI. May–Jun ★

M. jalapa L. var. ***jalapa***   **ST:** erect, < 1 m, hairs 0 or sparse, short. **LF:** blade 5–14 cm, ovate to cordate. **INFL:** ± terminal, short-peduncled, umbel-like clusters; involucre bell-shaped, 1-fld; bracts 5, 5–15 mm, 1/2–3/4 fused. **FL:** perianth 30–50 mm, narrowly funnel-shaped, gen bright magenta (yellow, white, orange, or variegated). **FR:** 8–10 mm, ovoid, hairs 0 to minute; ribs 5, blunt, ± wrinkled or warty between. 2*n*=54. Disturbed places with mild winters; < 300 m. CA-FP (esp SnFrB); native to trop Am. TOXIC: intestinal irritant in fr, roots. Cult as orn with many perianth colors, some escaping, possibly hybridizing. Gen May–Oct

M. laevis (Benth.) Curran   **ST:** ascending to erect, < 8 dm, broadly forked, clambering or forming clumps, basally woody, [glabrous] scabrous, or glandular-hairy. **LF:** fleshy, hairs 0 or various, often glandular. **INFL:** involucres in terminal, umbel-like cluster or gen 1 in axils, 3–7 mm, 1(2)-fld, bracts 5, lobes < tube, ± ovate. **FL:** perianth white to magenta, 5–14 mm. **FR:** ovoid, gen dark colored, often spotted, with 10 obscure vertical lines. Var. *laevis* restricted to Mex.

var. ***crassifolia*** (Choisy) Spellenb. (p. 929)   WISHBONE BUSH   **ST:** trailing to ascending, ± woody, ± gray in age, scabrous or ± glandular-hairy towards tips; sts with infls gen unbranched above. **LF:** blade 1–4.5 cm, ovate, puberulent or glandular-hairy (hairs of youngest lvs with conic base). **INFL:** involucre bell-shaped, 5–8 mm. **FL:** perianth 5–14 mm, broadly funnel-shaped, pink to purple-red (white). **FR:** ± 5 mm, ovoid, gen lightly dotted or wrinkled (smooth), glabrous. Grassy areas, chaparral, dunes, dry rocky areas, washes; < 2500 m. c&s SNF, Teh, CCo, SCoR, SW, W&I, w edge D; Baja CA. [*M. californica* A. Gray] Mostly Dec–Jun

var. ***retrorsa*** (A. Heller) Jeps.   **ST:** (and lf) with short, reflexed hairs, gen also ± glandular-hairy; sts with infls gen branched throughout. **LF:** blade 0.5–3.5 cm. **INFL:** involucre bell-shaped, 5–6 mm. **FL:** perianth 8–12 mm, widely funnel-shaped, white (pale pink). **FR:** ± spheric, gen with 5–10 visible lines. Rocky places; < 2300 m. W&I, D; OR; to UT, AZ, nw Mex. [*M. bigelovii* var. *r.* (A. Heller) Munz] Feb–Jun

var. ***villosa*** (Kellogg) Spellenb.   **ST:** (and lf) glandular-hairy; sts with infls gen branched throughout. **LF:** blade 0.5–4 cm. **INFL:** involucre bell-shaped, 5–6 mm. **FL:** perianth 8–12 mm, widely funnel-shaped, white (pale pink). **FR:** gen ± ovoid, lightly dotted, often mottled, gen with 0 visible lines, rarely 5–10 barely visible lines. Rocky places; < 2000 m. W&I, D (esp e D); to UT, AZ, nw Mex. [*M. bigelovii* A. Gray] ± all yr

M. linearis (Pursh) Heimerl var. ***linearis***   **ST:** ascending to erect, < 10 dm, glabrous to finely strigose. **LF:** ± sessile; blade 3–10 cm, linear to narrowly lanceolate, ± fleshy, gray-hairy. **INFL:** broad cyme; involucre cup-shaped, densely glandular-hairy, gen 3-fld, enlarged, 6–10 mm, papery in fr; bracts 5, 1/2–2/3 fused. **FL:** perianth ± 10 mm, ± funnel-shaped, pale pink to magenta. **FR:** ± 5 mm, club-shaped, hairy, ribs 5, strong, wrinkled between. Riverbeds, railroads, roadsides, gravelly places; < 300 m. SCo (Orange, Riverside cos.); native AZ to c US, Mex. Cult as orn; easily confused with *M. coccinea* without mature fls. Jul–Oct

M. multiflora (Torr.) A. Gray (p. 929)   **ST:** ascending to erect, 3–8 dm. **LF:** blade 3–12 cm, round to ovate, fleshy, glandular-hairy or glabrous in age. **INFL:** involucre 1 per upper axil, peduncled, bell-shaped, ± glabrous to minutely glandular-hairy, 6-fld; bracts 5, 22–35 mm, 1/2–3/4 fused. **FL:** perianth 40–60 mm, narrowly funnel-shaped, magenta. **FR:** 6–11 mm, elliptic, often with 10 lines or low ribs. Vars. difficult to distinguish where they overlap in CA; 1 other var., AZ to CO, TX.

var. ***glandulosa*** (Standl.) J.F. Macbr.   **INFL:** bracts gen acute [± obtuse], tip often short-acuminate. **FR:** faintly warty, with low ribs, gelatinous when wet. Dry, rocky or sandy places; 900–2500 m. W&I, n DMoj (Inyo Co.); to w CO. May–Aug

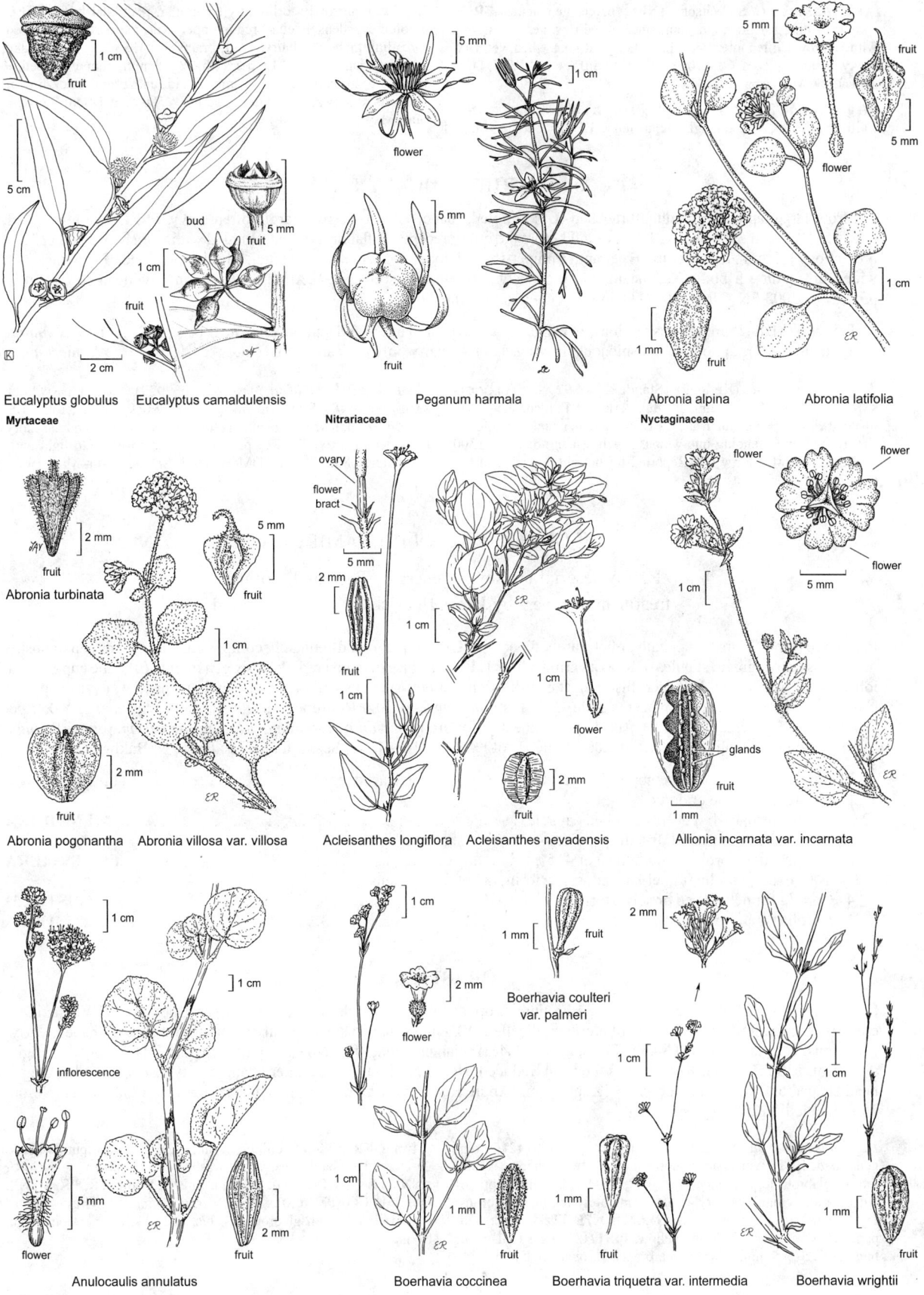

Eucalyptus globulus   Eucalyptus camaldulensis
Myrtaceae

Peganum harmala
Nitrariaceae

Abronia alpina   Abronia latifolia
Nyctaginaceae

Abronia turbinata

Abronia pogonantha   Abronia villosa var. villosa

Acleisanthes longiflora   Acleisanthes nevadensis

Allionia incarnata var. incarnata

Anulocaulis annulatus

Boerhavia coccinea

Boerhavia coulteri
var. palmeri

Boerhavia triquetra var. intermedia

Boerhavia wrightii

var. *pubescens* S. Watson **INFL**: bracts gen acute. **FR**: ±
smooth, with 10 slender, tan, sometimes raised ribs, gen alternating
with 10 brown, often interrupted lines, not gelatinous when wet. Dry,
rocky or sandy places; 50–2100 m. s SNF, s SnJV, e PR, W&I, D; to
sw UT, nw Mex. Apr–Aug

M. tenuiloba S. Watson SLENDER-LOBED FOUR O'CLOCK **ST**:
trailing to erect, < 5 dm, glandular-hairy. **LF**: ascending; blade

2.5–5 cm, narrowly to broadly triangular, glandular-hairy. **INFL**:
involucres ± densely clustered in upper axils, narrowly bell-shaped,
spreading glandular-hairy, 1-fld; bracts 5, 8–13 mm, lobes > tube,
narrowly lanceolate. **FL**: perianth 12–15 mm, funnel-shaped, ±
white, lightly hairy. **FR**: ± 5 mm, ovoid, unmarked, black-brown,
glabrous. Rocky slopes in desert scrub; < 500 m. w DSon; Baja CA.
Mar–May ★

TRIPTEROCALYX

Per from large taproot. **ST**: much-branched. **LF**: petioled; blade < 8 cm, fleshy, margin often wavy. **INFL**: head or umbel;
bracts 5–10, green; fls maturing 1 side of infl before the other; receptacle flat or rounded, studded with peg-shaped pedicel-like
projections 1–3 mm. **FL**: nocturnal, perianth trumpet-shaped, lobes 4–5; stamens 3–5, incl; stigma linear, incl. **FR**: wings 2–4,
wide, surrounding fr body, translucent, conspicuously net-veined. 4 spp.: arid N.Am. (Greek, Latin: 3-winged cup, from fr)
[Galloway 2003 FNANM 4:70–71] Closely related to *Abronia*.

1. Perianth (8)18–25 mm, lobes obvious; fr wings 2(3), hairy on margins and gen on veins near base; MP . . . *T. crux-maltae*
1′ Perianth 6–18 mm, lobes inconspicuous; fr wings (2)3(4), hairy only on margins; DMoj . . . . *T. micranthus*

T. crux-maltae (Kellogg) Standl. (p. 929) KELLOGG'S
SAND-VERBENA **ST**: < 3 dm, glandular-sticky. **LF**: blade 2.5–7 cm,
ovate to elliptic, ± glandular-hairy. **FL**: perianth gen pink to magenta.
FR: raised ribs extending into wings. Sagebrush scrub; 1200–1500
m. MP (Lassen Co.); nw NV. Reported for but not collected in CA
(Lassen Co.). May–Jun ★

T. micranthus (Torr.) Hook. (p. 929) SMALL-FLOWERED
SAND-VERBENA **ST**: < 6 dm, glandular-sticky or scabrous. **LF**:
blade 1–6 cm, narrowly ovate to elliptic, glabrous to glandular-hairy,
hairs denser abaxially. **FL**: perianth gen green-white to light pink.
Sand dunes; 800–2450 m. DMoj; to MT, SD, NM. Apr–May ★

OLEACEAE   OLIVE FAMILY

Family description, key to genera by Thomas J. Rosatti;
treatment of genera by James Henrickson, except as noted

Per to tree [vine], hairs 0 or peltate or not; rarely dioecious. **LF**: simple to odd-pinnately compound, alternate or gen opposite,
deciduous or evergreen; stipules 0. **INFL**: various; fls ≥ 1. **FL**: gen bisexual, gen radial; calyx gen minute (0), tube cup-shaped,
lobes 4–15; petals (0)4–6(8), gen fused; nectar disk often present; stamens (0)2(4–5), epipetalous; pistil 1, ovary superior,
chambers 2, each 2–4 ovuled, placenta axile, style 1, stigma gen 2-lobed. **FR**: drupe, capsule, or winged achene. **SEED**: 1 per
chamber. ± 25 genera, 900 spp.: ± worldwide; some cult for orn (*Forsythia*; *Jasminum*, jasmine; *Ligustrum*, privet; *Syringa*,
lilac) or food (*Olea*, olive). [Lee et al. 2007 Molec Biol Evol 24:1161–1180] Scientific Editor: Bruce G. Baldwin.

1. Lf simple or gen pinnately compound, lflets 0 or (1)3–9; fr winged achene . . . . **FRAXINUS**
1′ Lf simple; fr capsule or drupe
 2. Lvs gen alternate; fr a deeply 2-lobed capsule . . . . **MENODORA**
 2′ Lvs opposite or clustered; fr a drupe
  3. Fls unisexual; corolla 0; stamens 0 or 4–5 . . . . **FORESTIERA**
  3′ Fls bisexual; corolla funnel-shaped, salverform, or rotate; stamens 2
   4. Corolla funnel-shaped or salverform; fr 1–4-seeded . . . . **LIGUSTRUM**
   4′ Corolla rotate; fr 1-seeded . . . . **OLEA**

FORESTIERA

Shrub, erect to mounded; gen dioecious. **LF**: simple, opposite or clustered, gen deciduous, short-petioled. **INFL**: axillary
clusters; staminate fls ± sessile; pistillate fls pedicelled. **FL**: unisexual [bisexual]; calyx minute, lobes awl-shaped, early-
deciduous; corolla 0. **STAMINATE FL**: stamens (1)4(6); filaments long, free (in CA); anthers white [black]; pistil vestigial
if present. **PISTILLATE FL**: stamens 0 or 1–4, reduced, sterile (in CA); ovules 2 per chamber; style short, slender; stigma
head-shaped or 2-lobed. **FR**: drupe. ± 12 spp.: Am. (André Robert Forestier, French physician, teacher, 1736–1812) [Nesom
2009 Lundellia 12:8–14]

F. pubescens Nutt. (p. 929) DESERT OLIVE Shrub, 12–50 dm,
gen rounded. **ST**: many, bark smooth, tan-gray; twigs short, ± rigidly
thorny, glabrous to puberulent. **LF**: blade 15–40(50) mm, elliptic or
lanceolate to ovate, leathery, entire to minutely toothed, glabrous.
INFL: gen appearing before lvs. **STAMINATE FL**: 1–5 per bud;
pedicel ± 0; filaments 3–5 mm, yellow. **PISTILLATE FL**: 1–8 per
bud; pedicel 2–5 mm, green, some branched; ovary > 1 mm, style

± 1 mm. **FR**: 5–8 mm, elliptic in outline, ± glaucous, purple-black.
$2n=46$. Streambanks, canyons, washes; 100–1800 m. se NCoRI,
s SNF, c&s SNH (e slope), Teh, e SnFrB, SCoR, TR, PR, s SNE,
DMoj; to CO, TX, n Mex. [*F. p.* var. *parviflora* (A. Gray) Nesom]
CA specimens with glabrous sts, lvs, pedicels, exc some in e San
Bernardino Co. Mar–Apr

FRAXINUS ASH

Shrub or tree; gen dioecious, often bisexual (in CA). **ST**: older bark smooth or becoming furrowed, gen gray; lenticels broadly elliptic; twigs cylindric to 4-angled, glabrous to hairy; developing short-shoot spurs. **LF**: simple or gen odd-pinnate, opposite, deciduous; petioles channeled, occ winged, hairy or not; if compound, lflets (1)3–9, lanceolate to ovate or obovate, gen acute to acuminate at tip, entire or ± crenate-serrate, gen dark green adaxially, pale abaxially, thin to ± leathery in drier habitats, gen glabrous or with simple hairs abaxially or throughout, proximal opposite on rachis, stalked or not, terminal gen largest, stalk longer. **INFL**: axillary, of clusters or long-branched panicles; fls pedicelled. **FL**: unisexual or bisexual; calyx 1–2 mm, shallowly ± 4-lobed to cut, persistent on fr; petals 0, 2, or 4, free or fused to basal filaments. **STAMINATE FL**: stamens 2(3); pistil vestigial. **PISTILLATE FL**: stamens 0; style slender; ovules 2 per chamber. **FR**: achenes, winged, wings gen flat, extending to tip or base of seed-containing chamber. **SEED**: gen 1. ± 65 spp.: temp. N.Am, Eurasia, trop Asia. (Latin: ancient name) [Little 1952 J Washington Acad Sci 42:369–380; Miller 1955 Cornell Univ Agric Exp Sta Mem 335:1–64] *F. uhdei* (Wenzig) Lingelsheim, Mexican ash, cult in w US; similar to *F. velutina*, with ± larger lvs and lflets, gen with stiff hairs to 0.5 mm bordering abaxial midvein and occ 2° veins abaxially (as occ in *F. velutina*), and ± larger fr, but margins tapered to near base of fr body; native n Mex to Honduras.

1. Small to large tree; lvs compound, lflets (3)4–11(14) cm; fr pedicel tip expanded; pls with either staminate
   or pistillate fls
   2. Lflets ovate, or oblong-(ob)ovate; lateral lflets ± sessile; fr body 15–18 mm, 25–50 mm incl wing, tapered
      wing extending proximally to or below mid-body . . . . *F. latifolia*
   2′ Lflets lanceolate to lance-ovate or lance-obovate; lateral lflets gen stalked; fr body 12–14 mm, 15–38 mm
      incl wing, tapered wing extending proximally to distal 1/4 of body . . . . *F. velutina*
1′ Shrub or small tree, sts many; lvs simple or compound, lflets 2–5(6) cm; fr pedicel tip ± slender; fls gen bisexual.
   3. Petals gen 0; lvs simple or lflets 3(5); calyx irregularly cut; n&e DMtns . . . . *F. anomala*
   3′ Petals 2, white or cream, 3.5–6.5 mm; lvs simple or lflets 3–9; calyx gen 4-toothed; CA-FP
      4. Petals and filaments free exc at base; lvs, incl petioles, 7–18 cm; lflets (3)5–7(9), gen serrate-crenate . . . . *F. dipetala*
      4′ Petals fused with proximal filaments for 0.5–1.5 mm at base; lvs, incl petioles, 2–5(6) cm, simple and/or
         lflets 3; simple lvs and/or lflets gen entire . . . . *F. parryi*

F. anomala S. Watson (p. 929) SINGLE-LEAF ASH Shrub to tree, 1.5–5 m. **ST**: many; bark gray; twigs gen 4-angled, tan; buds glandular-puberulent. **LF**: simple and/or compound; petiole 1.2–3 cm; blade narrowly ovate to ± rounded, cordate to tapered at base, acute to rounded at tip, gen crenate-serrate to ± entire, thick, glabrous, yellow-green; if compound, lflets 3–5, 2–10 cm, 1.5–6 cm wide. **INFL**: 3–9 cm; bracts glandular-puberulent. **FL**: gen bisexual; calyx 1.2–1.5 mm, irregularly cut, thin, green to ± purple; petals gen 0; anthers 1.5–2.5 mm, filaments 2–2.5 mm; stigma ≤ style. **FR**: 13–24 mm, 5–11 mm wide; body oblong-oblanceolate, broadly winged to near base; fr pedicel tip slender. 2*n*=46. Washes, rocky slopes, shrubland, pinyon/juniper woodland; 1100–2400 m. n&e DMtns; to CO, NM. Fls bisexual but only 1/2 of pls produce fr; pls rarely with petals, source of erroneous reports of *F. dipetala* in D. Apr–May

F. dipetala Hook. & Arn. (p. 929) CALIFORNIA ASH Shrub to small tree, 1.5–3(6) m. **ST**: older st bark gray, smooth; twigs cylindric to 4-angled, gray; buds ± glandular-puberulent. **LF**: compound, 7–18 cm, 5–9(11) cm wide; petiole 2.4–4.5 cm; lflets (3)5–7(9), 2–4.5 cm, 1–2.5 cm wide, ovate to ± rounded, tapered at base, obtuse to ± rounded at tip, gen serrate-crenate, gen thin, glabrous, dark green adaxially, pale abaxially. **INFL**: 8–15 cm. **FL**: gen bisexual; calyx 1.2–2 mm, ± toothed, thin, green; petals 2, 3.5–6.5 mm, 1.3–3 mm wide, oblong-ovate, cream-white, ± cupped, narrowed and fused with filaments at base (rarely forming basal tube); anthers 2.4–4.2 mm; filaments 0.8–2.8 mm; stigmas ± < style. **FR**: 20–32 mm, 5–9 mm wide; body broadly oblanceolate, flat, broadly winged to near base; fr pedicel tip slender. 2*n*=46. Canyons, slopes, chaparral, oak/pine woodland; 100–1300 m. KR, NCoR, CaR, SNF, c&s SN, CW, TR, PR. Apr–Jun

F. latifolia Benth. (p. 929) OREGON ASH Tree < 25 m, trunk < 1.5 m diam; dioecious. **ST**: bark gray-brown, furrowed; twigs cylindric, brown-gray, ± long-shaggy-hairy or glabrous. **LF**: compound, 12–33(50) cm, ± long-shaggy-hairy or glabrous; petioles 3–7(9) cm, channeled; lflets (3)5–7, 4–11(14) cm, 2.4–7.5 cm wide, ovate or oblong-(ob)ovate, broadly wedge-shaped to ± rounded at base, acuminate at tip, entire to ± serrate, lateral lflets ± sessile, terminal lflet ± larger, stalk to 10–35 mm. **FL**: petals 0. **STAMINATE FL**: calyx > 0.5 mm, 4-toothed; anthers 2, 2–3.5 mm. **PISTILLATE FL**: calyx ± 1 mm, finely irregularly cut; style ± 3 mm; stigma ± 1.3 mm. **FR**: 25–50 mm, 5–9 mm wide; body 15–18 mm, ± cylindric, wing flat, extending proximally as tapering margin down 1/2–3/4 body; fr pedicel tip much expanded. 2*n*=46. Canyons, streambanks, woodland; < 1700 m. NW, CaR, SN, GV, SnFrB, MP; to BC. Pure in n CA; mostly introgressed with *F. velutina* in s CA. Mar–May

F. parryi Moran CHAPARRAL ASH Shrub to tree, 1.5–3(5) m. **ST**: older st bark gray, smooth; twigs cylindric to 4-angled; buds glandular-puberulent. **LF**: simple and/or compound, 2–5(6) cm; petiole 0.6–1.5 cm; if compound, lflets 3, 2–5(6) cm, 1.5–3.5 cm wide, ovate to rounded, 1–4 cm, 1–2.5 cm wide, broadly tapered at base, obtuse to rounded-notched at tip, gen entire, green adaxially, pale abaxially. **INFL**: 3–10 cm; bracts ± glandular-puberulent. **FL**: bisexual; calyx 1.2–2 mm, short-toothed, thin, green, persisting on fr; petals 2, 4.5–6.5 mm, 2.2–4.3 mm wide, oblong ovate, cream-white, united with filament base for 0.5–1.5 mm; anthers 2, 2.3–4 mm, free filaments 1.5–2 mm; stigma > narrow style. **FR**: 22–30 mm, 7–9 mm wide, body broadly oblong-oblanceolate, flat, broadly winged to near base; fr pedicel tip slender. Canyons, slopes, margins of mixed chaparral; 600 m. PR (s San Diego Co.); Baja CA. [*F. dipetala* var. *trifoliolata* Torr.; *F. t.* (Torr.) H. Lewis & Epling, illeg.] Feb–Mar ★

F. velutina Torr. (p. 929) VELVET ASH Tree < 15 m, trunk to 3 dm diam; dioecious. **ST**: bark gray, furrowed; twigs cylindric, gray-brown, minutely coarse-hairy to velvety or becoming glabrous. **LF**: compound, 9–20(30) cm, occ stiff-leathery, minutely coarse-hairy to velvety throughout or gen adaxially, often becoming glabrous, hairs gen erect, straight, to 0.5 mm; petiole 2–8 cm, channeled; lflets (3)5–7, 3–10 cm, 1.5–3.5 cm wide, lanceolate to lance-ovate or lance-obovate, tapered to base, long tapered at tip, entire to serrate, lateral lflets gen smaller, with stalk 4–6(10) mm, terminal lflet more tapered at base, with stalk 10–27 mm. **FL**: petals 0. **STAMINATE FL**: calyx < 1 mm, anthers 2(3), 2–3 mm. **PISTILLATE FL**: calyx 1–2 mm, green, ± unequally cut; style 0.5, stigma 2–3.5 mm. **FR**: 15–38 mm, 3–6(8) wide; body 12–14 mm, ± cylindric, wing flat, extending proximally as tapering margin onto distal 1/4 of fr body; fr pedicel tip much expanded. 2*n*=46,92. Canyons, streambanks, woodland; 200–1600 m. s SN, SCo, TR, PR, s SNE, DMoj; to sw UT, TX, n Mex. Many s CA specimens show introgression with *F. latifolia*. Mar–Apr

LIGUSTRUM PRIVET

Thomas J. Rosatti

Shrub, small tree, gen evergreen [deciduous], gen glabrous [hairy]. **LF**: simple, opposite, blade >> petiole, entire, thick. **INFL**: terminal panicle, axes ± puberulent or not. **FL**: bisexual; calyx minute, bell-shaped, 4-toothed, deciduous; corolla funnel-shaped or salverform, 4-lobed, white; stamens 2. **FR**: gen berry-like drupe, ± 8–10 mm wide, gen purple-black. **SEED**: 1–4. ± 45 spp.: e Asia, Malaysia to Australia, Eur, n Afr; many cult for orn lvs. (Classical name of *Ligustrum vulgare* L.)

1. Anthers ± acute at tip, 2–4 mm; style not exserted from tube; corolla tube 2–3 × (± >) lobes;
   petiole 3–10 mm. . . . . ***L. ovalifolium***
1′ Anthers ± obtuse at tip, 1–2 mm; style exserted from tube; corolla tube ± < to ± > lobes; petiole 5–20 mm
  2. Pl < 3 m; lf 4–8(10) cm; petiole 5–12 mm; blade ovate-elliptic to obovate; corolla tube ± ≥ lobes . . . . . . ***L. japonicum***
  2′ Pl < 10 m; lf 6–10(12) cm, petiole 10–20 mm, blade ± widely ovate; corolla tube ± ≤ lobes . . . . ***L. lucidum***

L. japonicum Thunb. JAPANESE PRIVET Shrub, < 3 m. **LF**: 5–8(10) cm; petiole 5–12 mm; blade ovate-elliptic to obovate, base acute to rounded, tip acute. **INFL**: 5–12 cm. **FL**: corolla tube ± ≥ lobes; anthers 1–2 mm, tip ± obtuse; style exserted from tube. $2n=44$. Canyon slopes; < 45 m. SCo; native to Japan, Korea. May–Sep

L. lucidum W.T. Aiton CHINESE PRIVET Shrub, small tree, < 10 m. **LF**: 6–10(12) cm; petiole 10–20 mm; blade ± widely ovate, base subacute to rounded, tip acute. **INFL**: 10–20 cm. **FL**: corolla tube ± ≤ lobes; anthers 1–2 mm, tip ± obtuse; style exserted from tube. $2n=46$.

Riparian zones; < 1100 m. NCo (Mendocino Co.), ScV, SCo, SnBr; native to China, Japan, Korea. Potentially problematic weed. Jun–Sep

L. ovalifolium Hassk. CALIFORNIA PRIVET Shrub, < 4 m, ± evergreen. **LF**: 4–10 cm, petiole 3–10 mm; blade base, tip acute to obtuse. **INFL**: 5–10 cm. **FL**: corolla tube 2–3 × (± >) lobes; anthers 2–4 mm, tip ± acute; style not exserted from tube. $2n=46$. Alder woodland, thickets, disturbed areas; < 300 m. NCo, ScV, CCo; native to Japan. Potentially problematic weed. Jul–Aug

MENODORA

Timothy Chumley

Per to shrub. **LF**: simple, opposite or alternate, sessile or short-petioled, entire to lobed, prominently 1-veined below. **INFL**: axillary cluster or terminal compound cyme. **FL**: bisexual; calyx persistent, lobes (4)5–10(12), ± linear; corolla ± rotate or funnel-shaped, lobes (4)5(8); style slender, stigma head-like, ± 2-lobed. **FR**: papery capsule, deeply 2-lobed to near base. **SEED**: 2–4[8]. ± 23 spp.: Am, s Afr. (Greek: gift of force, for sustenance provided to horses of Humboldt and Bonpland in Mexico) [Chumley 2007 Ph.D. Dissertation, Univ Texas, Austin]

1. Shrub, densely mounded; branches many, spreading to ascending, twigs short, stout, becoming thorny;
   corolla white, tinged red or purple; fr ± indehiscent . . . . ***M. spinescens***
  2. Calyx tube in fl gen > 1.5 mm; corolla tube > 7 mm . . . . var. ***mohavensis***
  2′ Calyx tube in fl gen < 1.5 mm; corolla tube < 5 mm . . . . var. ***spinescens***
1′ Per to subshrub, broom-like; branches 0–few, ascending to erect, slender; corolla yellow; fr circumscissile . . . ***M. scabra***
  3. Pls gen > 30 cm, ± glabrous; lvs proximal to infl > 5 × width, gen linear or lance-linear; internodes
    proximal to infl gen > 1.5 × subtending lvs; calyx lobes 5–7(10) . . . . var. ***glabrescens***
  3′ Pls gen < 28 cm, ± scabrous, often densely so; lvs proximal to infl < 4 × width, gen narrowly lanceolate
    or oblanceolate; internodes proximal to infl gen < 1.5 × subtending lvs; calyx lobes (8)9–10(12) . . . . var. ***scabra***

M. scabra A. Gray ROUGH MENODORA Per to subshrub, broom-like, (1.5)2–4(5) dm. **ST**: 2–many from woody base, ascending to erect, angled in ×-section, ± glabrous to densely scabrous, unarmed. **LF**: basal-most 2–4 nodes opposite, becoming alternate distally, grading into alternate lf-like bracts in infl, sessile or ± petioled, (3)10–45(60) mm, linear to ovate or obovate, ± glabrous to scabrous. **INFL**: terminal, compound cyme. **FL**: calyx gen densely scabrous, lobes (4)5–10(12), (1)2.6–5(10) mm, linear; corolla opening in evening, often scented, yellow, gen red in bud, tube 3–6 mm, lobes (4)5–6, (4)5–8(10) mm, oblong or obovate; anthers yellow; stigma exserted. **FR**: circumscissile, (4)5–7(8) mm, glabrous. **SEED**: gen 4, 4–6 mm, obovate, gen 3-sided in ×-section, spongy; seed coat deeply netted. $2n=22,44$.

var. ***glabrescens*** A. Gray Gen > 30 cm. **ST**: 3–5 dm, straight, erect, ± glabrous exc axils gen ± bristly; branching distally gen only in infl. **LF**: linear to lance-linear, > 5 × longer than wide, gen < 2/3 internode, reduced in size approaching infl, ± glabrous. **FL**: calyx lobes 5–7(10). Rocky or sandy soils, desert scrub, woodland; (400)700–2400 m. e PR, e&s DMoj, w DSon; AZ, Mex. [*M. scoparia* A. Gray] May–Jun

var. ***scabra*** (p. 929) **ST**: 1–2.8 dm, ascending to erect, ± scabrous; gen branching throughout. **LF**: narrowly lanceolate or oblanceolate (linear), < 4 × longer than wide, gen > 2/3 internode, ± uniform in size proximal to infl, gen scabrous. **FL**: calyx lobes (8)9–10(12). Rocky or sandy soils, desert scrub, woodland; 1000–1800 m. e DMtns (Clark, Eagle, New York mtns); to NM, n Mex. May–Jun ★

M. spinescens A. Gray Shrub, dense, mounded, 2–5(6) dm. **ST**: many from gen short, stout trunk, spreading to ascending, round

in ×-section, densely puberulent, terminating in thorns, these often appearing forked at ends of major branches. **LF**: alternate (lf-like bracts appearing clustered on short fl shoots), sessile, (2)4–10(16) mm, linear-oblanceolate to oblanceolate, 3-cleft to middle, densely puberulent, gen only present at fl. **INFL**: axillary, compact cluster. **FL**: calyx puberulent, lobes (4)5–6(8), (1.8)2–4(5) mm, linear; corolla white tinged red or purple, red in bud, lobes 4–5(6), obovate. **FR**: each lobe ± indehiscent, opening irregularly, (5)6.3–7.9(9) mm, ± glabrous to slightly scabrous. **SEED**: gen 2; (5)5.7–7.2(8) mm, elliptic to obovate in outline, planoconvex to lenticular in ×-section, seed coat appearing smooth, shiny.

var. ***mohavensis*** Steyerm. **FL**: bisexual; calyx tube 1.5–2.1(2.4) mm; corolla tube (7.5)7.9–10.2(11) mm, lobes 2.9–4.9(5.8) mm; filaments 0.5–2 mm, anthers yellow, 2–4 mm, incl, or rudimentary; stigma well exserted, head-like or 2-lobed. Rocky desert hillsides, canyons; 690–2000(2300) m. SnBr (n slope), DMoj. Apr–May ★

var. ***spinescens*** **FL**: unisexual or bisexual, calyx tube in fl gen < 1.4 mm; filaments < 1 mm, anthers gen appearing sessile; bisexual fls in showier, denser infl, fls larger. **BISEXUAL FL**: corolla tube (4)4.3–5.1(5.2) mm, lobes (1.9)2–2.9(3) mm; filaments 0.5–2 mm, anthers 2 mm, yellow, 1/2-exserted; stigma exserted, exceeding anthers, 2-forked. **PISTILLATE FL**: corolla tube (2.5)2.6–3.7(3.8) mm, lobes (1)1.2–2.2; anthers green, < 1 mm, incl, separate; stigma well exserted, head-like or 2-lobed. Rocky desert hillsides, canyons; 690–2000(2300) m. SnBr (n slope), s SNE, n&e DMoj exc DMtns; s NV, sw UT, nw AZ. Apr–May

OLEA OLIVE

Tree [shrub]. **LF**: simple, opposite, short-petioled, entire, leathery. **INFL**: raceme or panicle, axillary or terminal. **FL**: gen bisexual; calyx 4-toothed to -lobed; corolla rotate, lobes 4; stamens 2; ovary 1. **FR**: drupe. 20 spp.: trop, warm temp Eurasia, Afr. (Greek: ancient name)

O. europaea L.   Tree gen < 10 m. **ST**: trunks becoming gnarled; bark gray, with age furrowed. **LF**: petiole 2–7 mm; blade 20–70 mm, 6–16 mm wide, narrowly elliptical to oblanceolate, green adaxially, closely silver-scaly abaxially. **INFL**: axillary, narrow raceme-like panicle, 15–40 mm, branches opposite; bract 1–3 mm; pedicel 1–3 mm; fls in 1–3s. **FL**: calyx cup-like, 4-toothed, ± 1 mm; corolla 2.5–4 mm, white, tube < 1 mm, lobes 4, 2–3.5 mm, 1.5–2 mm wide, ovate-elliptical, margins inrolled; stamens 2, filaments < 1 mm; anthers yellow, 1.5–2.5 mm; ovary superior, ± 1 mm, style < 1 mm, stigma ± head-shaped, ± 1 mm; some ovaries not developing. **FR**: 9–20 mm, ovoid, oily, green, becoming black. 2*n*=46. Gen waif, persisting from cult; < 200 m. s NCoR, GV, CCo, SnFrB, w SCoRO, SCo, n ChI (Santa Cruz Island); native to w Asia. [*O. e.* subsp. *africana* (Mill.) P.S. Green] Introduced widely; cult for food and cooking oil in Medit for ± 6000 yrs. Feb–Jun ❖

ONAGRACEAE EVENING-PRIMROSE FAMILY

Warren L. Wagner & Peter C. Hoch, family description, key to genera;

treatment of genera by Warren L. Wagner, except as noted

Ann to per (to tree). **LF**: cauline or basal, alternate, opposite, or whorled, gen simple and toothed (to pinnately compound); stipules 0 or gen deciduous. **INFL**: spike, raceme, panicle, or fls 1 in axils; bracted. **FL**: gen bisexual, gen radial, often opening at either dawn or dusk; hypanthium gen prolonged beyond ovary (measured from ovary tip to sepal base); sepals 4(2–7); petals 4(2–7, rarely 0), often fading darker; stamens 2 × or = sepals in number, anthers 2-chambered, opening lengthwise, pollen interconnected by threads; ovary inferior, chambers gen as many as sepals (sometimes becoming 1), placentas axile or parietal, ovules 1–many per chamber, style 1, stigma 4-lobed (or lobes as many as sepals), club-shaped, spheric, or hemispheric. **FR**: capsule, loculicidal (sometimes berry or indehiscent and nut-like). **SEED**: sometimes winged or hair-tufted. 22 genera, ± 657 spp.: worldwide, esp w N.Am; many cult (*Clarkia, Epilobium, Fuchsia, Oenothera*). [Wagner et al. 2007 Syst Bot Monogr 83:1–240] *Gaura* moved to *Oenothera. Fuchsia magellanica* Lam. naturalized in n CA. Scientific Editors: Robert Patterson, Bruce G. Baldwin.

1. Sepals persistent after fl; fl parts in (3)4–5(7)s; hypanthium 0 . . . . **LUDWIGIA**
1′ Sepals deciduous after fl (along with other fl parts); fl parts in (2)4s; hypanthium present, often elongate, or if 0 then petals rose-purple to pink, rarely white, or fls bilateral
  2. Stipules present, occ deciduous; fr indehiscent, bur-like with hooked hairs; fl parts in 2s; stigmas maturing before anthers . . . . **CIRCAEA**
  2′ Stipules 0; fr gen dehiscent capsule, rarely indehiscent but not bur-like; fl parts in (3)4s; anthers gen maturing before stigmas
    3. Seeds hair-tufted (tufts rarely secondarily lost); sepals erect; stigmas dry with multicellular papillae
      4. Lvs gen alternate; hypanthium 0; petals entire; stamens ± equal, in 1 whorl; pollen blue-gray, grains shed singly . . . . **CHAMERION**
      4′ Lvs opposite at least at base; hypanthium 0.3–34 mm; petals notched; stamens in 2 unequal whorls; pollen cream, grains gen shed in 4s. . . . . **EPILOBIUM**
    3′ Seeds not hair-tufted; sepals reflexed; stigmas mainly wet and nonpapillate or with dry unicellular papillae
      5. Stigma 4-lobed, dry, papillae unicellular; anthers gen maturing before stigmas. . . . . **CLARKIA**
      5′ Stigma hemispheric, peltate or 4-lobed, wet, papillae 0; anthers not maturing before stigmas
        6. Fr 2-chambered; sts slender, hair-like . . . . **GAYOPHYTUM**
        6′ Fr (3)4-chambered; sts gen not slender
          7. Seeds with concave and convex sides, concave side with thick wing, both sides densely covered with club-shaped hairs; petals white, base yellow . . . . **CHYLISMIELLA**
          7′ Seeds gen not concave and convex, wing 0, club-shaped hairs 0; petals yellow, purple, red, or white, if white, gen lacking yellow base
            8. Ovary tip projection 0.4–18 cm; st 0
              9. Per; fr cylindric-ovoid, ± straight or slightly curved, ± angled to nearly smooth, lacking pointed wing near center top of each valve, walls gen thin, distended by seeds; sterile projection of ovary persistent with fertile part in fr, without abscission line at juncture between hypanthium and fertile part of ovary . . . . **TARAXIA**
              9′ Ann; fr irregularly obovoid, sharply 4-angled with pointed wing near center top of each valve, thick walled, not distended by seeds; sterile projection of ovary with clear abscission line at juncture between short hypanthium and fertile part of ovary. . . . . **TETRAPTERON**
            8′ Ovary tip projection 0, occ ± tapering; st gen present
              10. Stigma 4-lobed . . . . **OENOTHERA**
              10′ Stigma hemispheric to spheric or cylindric
                11. Seeds in 2 rows per chamber; fr pedicelled; lvs often ± basal, pinnately lobed with large terminal lobe, occ either lateral lobes ± reduced or lf ovate-cordate and long-petioled; abaxial lf surface or margin with conspicuous (brown) oil cells. . . . . **CHYLISMIA**

11′ Seeds in 1 row per chamber; fr sessile or occ short-pedicelled *(Camissonia kernensis)*; lvs gen not predominately basal, entire to pinnately lobed and if so then terminal lobe not conspicuously larger than lateral; lf oil cells 0

 12. Petals white, pink, or rarely red; fls night-blooming . . . . **EREMOTHERA**

 12′ Petals yellow, often with red flecks; fls day-blooming

  13. Fl sts wand-like; lvs pinnately lobed to occ entire; petals yellow with red flecks near base; seeds gen with purple spots; hypanthium with lobed disk . . . . **EULOBUS**

  13′ Fl sts not wand-like; lvs entire to toothed; petals yellow, occ with 2 red spots at base, or white; seeds without purple spots; hypanthium without lobed disk

   14. Distal st densely lfy, proximal st nearly lfless; branches many, slender, ascending from base of pl; fr strongly flattened, straight, 0.5–1 cm . . . . **NEOHOLMGRENIA**

   14′ St gen lfy throughout, branched throughout or with a few basal branches; fr cylindric to quadrangular, not flattened, often wavy or curled, (0.8)1–6 cm

    15. Fr ± cylindric; pls fl from distal nodes; lvs gen linear to narrowly elliptic; seeds glossy, triangular in ×-section, often < 1 mm . . . . **CAMISSONIA**

    15′ Fr 4-angled, at least when dry; pls gen fl from basal nodes; lvs gen narrowly lanceolate, narrowly elliptic or ovate; seeds dull, often > 1 mm, flattened. . . . . **CAMISSONIOPSIS**

CAMISSONIA SUN CUP

Ann, from taproot; rosette gen ± 0. **LF:** cauline, alternate, simple, gen linear to narrowly elliptic. **INFL:** bracted; spike or raceme, nodding in bud, erect in fr, fl only at distal nodes. **FL:** radial, gen opening at dawn; sepals 4, reflexed singly or in pairs; petals 4, yellow, gen fading red, often with red basal spots; stamens 8, longer ones opposite sepals, anthers attached at middle, pollen grains 3-angled exc in polyploid taxa, at 20×; ovary chambers 4, stigma hemispheric, gen > anthers and cross-pollinated, or ± = anthers and self-pollinated. **FR:** ± cylindric, straight to wavy, distorted by seeds at maturity, dehiscent throughout most of its length; pedicel ± 0 or ≤ 2(15) mm, 0 or shorter in fl. **SEED:** in 1 row per chamber, narrowly obovoid, smooth (minutely pitted), glossy. 12 spp.: w N.Am (esp CA-FP), 1 S.Am. (L.A. von Chamisso, French-born German botanist, 1781–1838) [Wagner et al. 2007 Syst Bot Monogr 83:1–240] Polyploidy and self-pollination have predominated in evolution of genus. Not monophyletic as treated in TJM (1993); segregates moved to *Camissoniopsis, Chylismia, Chylismiella, Eremothera, Eulobus, Neoholmgrenia, Taraxia, Tetrapteron* (Wagner et al. 2007).

1. Infl spike or raceme; pedicel in fr ± 0–15 mm; sepals all separating when fl opens
 2. Petals 8–18 mm; sepals 5–11 mm; stigma exceeding anthers. . . . . *C. kernensis*
  3. Pl ± open, sparsely spreading-hairy, sparsely glandular-hairy; lvs not clustered at base; pedicel in fr ± 0–5 mm. . . . . subsp. *gilmanii*
  3′ Pl compact, densely spreading-hairy, sparsely glandular-hairy; lvs clustered at base; pedicel in fr 3–15 mm. . . . . subsp. *kernensis*
 2′ Petals 1.8–4 mm; sepals ≤ 3.8 mm; stigma surrounded by anthers
  4. Lf ± entire; pl ± glabrous or minutely strigose, gen sparsely glandular-hairy, rarely sparsely spreading-nonglandular-hairy. . . . . *C. parvula*
  4′ Lf serrate or minutely so; pl hairs gen spreading, some glandular
   5. Hypanthium 1.3–3 mm; sepals 2.2–3.8 mm; fr (18)26–50 mm . . . . *C. pubens*
   5′ Hypanthium 0.8–1.6 mm; sepals 1.2–2 mm; fr 18–32 mm. . . . . *C. pusilla*
1′ Infl spike; pedicel in fr ± 0; sepals gen remaining fused in pairs when fl opens
 6. Stigma exceeding anthers; sepals 3–8(12) mm; petals (3.5)5–15 mm
  7. Lf lanceolate to narrowly ovate or elliptic, ± entire; sepals < 4.2 mm . . . . ²*C. sierrae* subsp. *sierrae*
  7′ Lf linear to narrow-elliptic or -oblanceolate, minutely to coarsely serrate; sepals 3–8(12) mm. . . . . *C. campestris*
   8. St gen erect; lf minutely serrate. . . . . subsp. *campestris*
   8′ St gen decumbent; lf coarsely serrate . . . . subsp. *obispoensis*
 6′ Stigma gen surrounded by anthers; sepals 1.2–4(5.5) mm; petals 2–7 mm
  9. Lf entire (teeth 1–2, small). . . . . *C. integrifolia*
  9′ Lf gen minutely serrate *(Camissonia sierrae* occ with only several small teeth)
   10. Lf lanceolate to narrowly ovate or elliptic, base obtuse or rounded . . . . *C. sierrae*
    11. Sepals 1.2–3 mm; petal base without red dots; style 2.8–5 mm . . . . subsp. *alticola*
    11′ Sepals 3–4.2 mm; petal base with 2 red dots; style 4.5–7 mm . . . . ²subsp. *sierrae*
   10′ Lf linear to narrowly elliptic, base acute or long-tapered
    12. Pl minutely strigose, or some hairs spreading and glandular, or occ coarsely spreading and nonglandular toward base; < 10% of pollen grains 4-angled . . . . *C. strigulosa*
    12′ Pl hairs spreading, gen fine (coarse in *Camissonia contorta)*, often glandular in infl; < 10% or > 30% of pollen grains 4-angled
     13. Sepals (3)3.8–5.5 mm; petals (4)4.5–7 mm . . . . *C. lacustris*
     13′ Sepals 2.5–4 mm; petals 3–5 mm
      14. Lf not (or slightly) blue-green; hypanthium ± 1.2 mm; pl nonglandular hairs of 2 types (linear, white; rod-shaped, transparent); < 10% of pollen grains 4-angled; SCoRI. . . . . *C. benitensis*
      14′ Lf gen blue-green; hypanthium 1.2–2.3 mm; pl nonglandular hairs of 1 type (rod-shaped, transparent); gen > 30% of pollen grains 4-angled; widespread . . . . *C. contorta*

C. benitensis P.H. Raven (p. 929) SAN BENITO EVENING-PRIMROSE
Slender; hairs spreading, of 2 types (linear, white; rod-shaped, transparent), in infl also glandular. **ST**: erect or decumbent, 3–20 cm, peeling; branches widely spreading, wiry. **LF**: 7–20 mm, narrowly elliptic, minutely serrate. **FL**: hypanthium ± 1.2 mm; sepals 3.2–3.5 mm, remaining adherent in pairs; petals 3.5–4 mm, yellow fading ± red, basal spots 2; < 10% of pollen grains 4-angled. **FR**: 15–40 mm, 0.8–1.3 mm wide, slightly swollen by seeds, straight or wavy, ± sessile. **SEED**: 0.6–0.8 mm. 2*n*=28. Sandy or gravelly serpentine soil; ± 600–1200 m. SCoRI (Fresno, Monterey, San Benito cos.). Self-pollinated. Threatened by off-road vehicles. Apr–Jun ★

C. campestris (Greene) P.H. Raven Slender; hairs 0, coarse, or glandular. **ST**: decumbent or erect, 5–25(50) cm, peeling. **LF**: 5–30 mm, linear to narrowly elliptic or narrowly oblanceolate, minutely to coarsely serrate. **FL**: hypanthium 1.5–5 mm; sepals 3–8(12) mm, remaining adherent in pairs; petals (3.5)5–15 mm, yellow fading ± red, basal spots (1)2; stigma exceeding anthers. **FR**: 20–43 mm, 0.7–2 mm wide, alternately narrow and swollen by seeds, straight or wavy, ± sessile. **SEED**: 0.8–1.6 mm. 2*n*=14. Cross-pollinated. Subspp. intergrade extensively.

subsp. ***campestris*** **ST**: gen erect. **LF**: linear to narrowly elliptic or narrowly oblanceolate, minutely serrate. Open sandy flats, desert scrub, inland grassland; < 2000 m. SNF, GV, CW, e SW, DMoj. Sometimes hybridizes with *C. contorta*, hybrids seldom fertile. Mar–May

subsp. ***obispoensis*** P.H. Raven **ST**: gen decumbent. **LF**: narrowly elliptic, coarsely serrate. Marine deposits in openings in chaparral and oak woodland; 100–500 m. SCoRO. Hybridizes with *C. contorta, C. strigulosa*. Mar–May

C. contorta (Douglas) Kearney Slender; hairs spreading, gen of 1 type (rod-shaped, transparent) only at pl base or throughout, in infl also sparsely glandular. **ST**: decumbent or erect, gen 3–30 cm, wiry, peeling. **LF**: gen 10–35 mm, linear to narrowly elliptic, minutely serrate, gen blue-green. **FL**: hypanthium 1.2–2.3 mm; sepals 2.5–4 mm, remaining adherent in pairs; petals 3–5 mm, yellow fading ± red, basal spots 0 or 2; gen > 30% of pollen grains 4-angled. **FR**: gen 25–35 mm, 0.7–1.2 mm wide, ± swollen by seeds, straight or wavy, ± sessile. **SEED**: 0.7–0.9 mm. 2*n*=42. Sandy soil, slopes, flats, often disturbed, grassland, chaparral, pinyon/juniper woodland; < 2300 m. NW, CaR, SNF, GV, CW, MP; to WA, ID, w NV. Self-pollinated. Probably derived from *C. strigulosa* × *C. campestris* subsp. *c.*; sterile hybrids formed where both occur. May–Jun

C. integrifolia P.H. Raven KERN RIVER EVENING-PRIMROSE
Slender, ± glabrous or sparsely minutely strigose (exc infl densely strigose, ± gray). **ST**: decumbent or erect, gen < 30 cm, wiry, peeling. **LF**: gen 10–30 mm, linear, entire; teeth 0–2, small. **FL**: hypanthium 1.5–2.5 mm; sepals 1.6–4 mm, remaining adherent in pairs; petals gen 2–4 mm, yellow fading ± red, basal spots 0–2. **FR**: 45–60 mm, 0.8–1.3 mm wide, ± swollen by seeds, straight or wavy, ± sessile. **SEED**: 1–1.2 mm. 2*n*=28. Gen sagebrush slopes; 700–1000 m. s SNF (Kern River area, Kern Co.). Self-pollinated. Probably derived from *C. strigulosa*; forms sterile hybrids where both occur. May ★

C. kernensis (Munz) P.H. Raven Robust; hairs dense, spreading, some glandular, or ± 0, esp in infl. **ST**: erect, 5–30 cm. **LF**: 10–38(55) mm, gen narrowly elliptic, sparsely serrate. **FL**: hypanthium 2.2–3.8(5.5) mm; sepals 5–11 mm, free; petals 8–18 mm, yellow fading ± red, basal spots 2, large. **FR**: 22–37 mm, 1.5–1.7 mm wide, ± swollen by seeds, straight or wavy; pedicel ± 0–15 mm. **SEED**: 1.1–1.2 mm. 2*n*=14. Cross-pollinated. Related to *C. parvula, C. pubens, C. pusilla*. Subspp. intergrade extensively.

subsp. ***gilmanii*** (Munz) P.H. Raven (p. 929) Pl ± open; hairs ± 0–few, glandular and spreading. **ST**: < 30 cm. **LF**: not clustered at base. **FR**: pedicel ± 0–5 mm. Washes, slopes; 760–1800 m. se SNH, s SNE (esp Inyo Co.), n&w DMoj; s NV. May

subsp. ***kernensis*** (p. 929) KERN COUNTY EVENING-PRIMROSE
Pl compact; hairs dense, spreading, few glandular. **ST**: 5–15(22) cm. **LF**: clustered at base. **FR**: pedicel 3–15 mm. Sandy slopes, flats, gen in sagebrush scrub or Joshua-tree woodland; 850–1800 m. se SNH, w DMoj (Piute, El Paso mtns, Grapevine Canyon, Kern Co.). Often locally abundant. May ★

C. lacustris P.H. Raven Slender; hairs ± dense, spreading, glandular in infl. **ST**: decumbent or erect, < 50 cm, wiry, peeling. **LF**: 8–35 mm, linear to narrowly elliptic, minutely serrate. **FL**: hypanthium 1.7–2.7 mm; sepals (3)3.8–5.5 mm, remaining adherent in pairs; petals (4)4.5–7 mm, yellow fading ± red, basal spots 0 or 2. **FR**: < 45 mm, 0.8–1.3 mm wide, ± swollen by seeds, straight or wavy, ± sessile. **SEED**: 0.6–0.8 mm. 2*n*=28. Open grassland; on serpentine, 400–600 m; not on serpentine, 200–1600 m. s NCoRI (c Lake Co.), SNF. Self-pollinated. Probably derived from *C. strigulosa*. Apr–Jun

C. parvula (Torr. & A. Gray) P.H. Raven (p. 929) Slender, ± glabrous or minutely strigose, gen sparsely glandular-hairy (sparsely spreading-nonglandular-hairy). **ST**: erect, < 30 cm, wiry. **LF**: 10–30 mm, linear, ± entire. **FL**: hypanthium < 2 mm; sepals < 2.2 mm, free; petals 1.8–4 mm, yellow fading ± red. **FR**: ± 20–30 mm, 0.6–0.9 mm wide, ± swollen by seeds, straight or wavy; pedicel ± 0–2 mm. **SEED**: 0.7–0.8 mm. 2*n*=28. Sandy soils, gen sagebrush scrub; 1000–2000 m. GB; to WA, WY, CO. Self-pollinated. Related to *C. kernensis, C. pubens, C. pusilla*. May–Jun

C. pubens (S. Watson) P.H. Raven Slender; hairs gen glandular, some spreading. **ST**: erect, < 38 cm. **LF**: 15–45 mm, narrowly lanceolate, wavy-serrate. **FL**: hypanthium 1.3–3 mm; sepals 2.2–3.8 mm, free; petals 2.2–4 mm, yellow fading ± red, basal spots 1–few. **FR**: (18)26–50 mm, 0.8–1.2 mm wide, ± swollen by seeds, straight or wavy; pedicel ± 0–2 mm. **SEED**: 0.7–0.8 mm. 2*n*=28. Sandy soils, gen sagebrush scrub or pinyon/juniper woodland; 1000–3000 m. GB, n DMoj (scattered); w NV. Self-pollinated. Related to *C. kernensis, C. parvula, C. pusilla*. May–Jun

C. pusilla P.H. Raven (p. 929) Slender; hairs glandular, gen also spreading. **ST**: erect, 2–22 cm. **LF**: 10–30 mm, linear, minutely serrate. **FL**: hypanthium 0.8–1.6 mm; sepals 1.2–2 mm, free; petals 1.8–3.1 mm, yellow fading ± red, basal spots 2. **FR**: 18–32 mm, 0.6–0.9 mm wide, ± swollen by seeds, straight or wavy; pedicel ± 0–2 mm. **SEED**: 0.7–0.8 mm. 2*n*=14. Sandy soils, gen sagebrush scrub; 760–3000 m. n slope SnBr (Cactus Flats), GB, DMoj (scattered); to WA, ID, UT. Self-pollinated. Related to *C. kernensis, C. parvula, C. pubens*. May–Jun

C. sierrae P.H. Raven Slender; hairs spreading, coarse, in infl also glandular. **ST**: decumbent to ascending, < 15 cm, peeling. **LF**: 5–18 mm, lanceolate to narrowly ovate or elliptic; teeth few or minute. **FL**: hypanthium 1–2.2 mm; sepals 1.2–4.2 mm, remaining adherent in pairs; petals 2.2–7 mm, yellow fading ± red, basal spots 0 or 2. **FR**: 20–30 mm, 0.5–0.7 mm wide, ± swollen by seeds, straight or wavy, ± sessile. **SEED**: 0.8–1.6 mm. 2*n*=14. Related to *C. campestris*.

subsp. ***alticola*** P.H. Raven (p. 929) MONO HOT SPRINGS EVENING-PRIMROSE **LF**: lanceolate; teeth several, small. **FL**: hypanthium 1–2.2 mm; sepals 1.2–3 mm; petals 2.2–4 mm, not red-dotted; style 2.8–5 mm. Shallow soil on granite outcrops, ponderosa-pine forest; 2000–2350 m. c SNH (Mariposa, ne Fresno cos.). Self-pollinated. Pls from Merced Lake (± 2200 m, e Mariposa Co.) may be evolutionarily distinct. May–Jul ★

subsp. ***sierrae*** (p. 929) YOSEMITE EVENING-PRIMROSE **LF**: narrowly ovate or elliptic, minutely serrate. **FL**: hypanthium 1.3–2.2 mm; sepals 3–4.2 mm; petals 4–7 mm, basal spots 2; style 4.5–7 mm. Ponderosa-pine or foothill-pine/blue-oak forest; 500–1300 m. c SN (Madera, Mariposa, Tuolumne cos.). Cross- or self-pollinated. May be locally abundant. Apr–Jun ★

C. strigulosa (Fisch. & C.A. Mey.) P.H. Raven (p. 929) Slender, minutely strigose (hairs glandular or not, toward base also coarse, spreading). **ST**: decumbent or erect, < 50 cm, wiry, peeling. **LF**: 8–35 mm, linear to narrowly elliptic, minutely serrate. **FL**: hypanthium 1.6–2.7 mm; sepals 1.6–4 mm, remaining adherent in pairs; petals 2.1–4.5 mm, yellow fading ± red, basal spots 0–2. **FR**: 15–45 mm, 0.8–1.3 mm wide, ± swollen by seeds, straight or wavy, ± sessile. **SEED**: 0.6–0.8 mm. 2*n*=28. Open sandy soils of dunes, grassland, desert scrub; < 2100 m. s edge s SNH, Teh, CW, SW (exc s ChI), w DMoj; n Baja CA. Self-pollinated. Related to S.Am *C. dentata* Reiche; hybridizes with *C. campestris* subsp. *obispoensis, C. contorta, C. integrifolia, C. kernensis* subsp. *k.* Mar–May

CAMISSONIOPSIS

Ann to subshrub, from taproot. **LF**: basal and cauline, alternate, simple, gen narrowly lanceolate or narrowly elliptic to ovate. **INFL**: spike, nodding in bud, gen fl from basal-most to distal nodes. **FL**: opening at dawn; sepals 4, reflexed singly or in fused pairs; petals 4, yellow, fading red, gen with 1+ red basal spots, with no ultraviolet pattern; stamens 8, longer opposite sepals, anthers attached at middle, pollen grains 3-angled exc in polyploid taxa at 20×; ovary chambers 4, stigma ± spheric or hemispheric, exceeding anthers and cross-pollinated or ± = anthers and self-pollinated. **FR**: 4-angled at least when dry, gen proximally thick, contorted or curled 1–5 times, or straight, not swollen by seeds, sessile. **SEED**: in 1 row per chamber, narrowly obovoid, flattened, dull brown-black. 14 spp.: CA, AZ, OR, Baja CA. (Greek: like *Camissonia*) [Wagner et al. 2007 Syst Bot Monogr 83:1–240] Polyploidy and self-pollination have predominated in evolution of genus. Incl in *Camissonia* in TJM (1993).

1. Per or subshrub; ± coastal . . . . *C. cheiranthifolia*
  2. Per; hairs rarely dense and silvery; style 6–9 mm; petals 6–11 mm; stigma surrounded
     by anthers . . . . subsp. *cheiranthifolia*
  2′ Subshrub; hairs gen dense and silvery; style 13–23 mm; petals (10)12–20 mm; stigma
     exceeding anthers . . . . subsp. *suffruticosa*
1′ Ann or short-lived per; inland or coastal
  3. Stigma exceeding anthers. . . . . *C. bistorta*
  3′ Stigma gen surrounded by anthers
    4. Fr stout, 2.8–3.5 mm wide at base, deeply grooved . . . . *C. guadalupensis* subsp. *clementina*
    4′ Fr not esp stout, 0.7–2.2 mm wide at base, not deeply grooved
      5. > 25% of pollen grains 4–5-angled (visible with light microscopy)
        6. Infl hairs nonglandular; 25–60% of pollen grains 4–5-angled — SCoRO (Monterey Co.). . . . . *C. luciae*
        6′ Infl hairs both glandular and other types nonglandular; 70–100% of pollen grains 4–5-angled
          7. Fr 1.3–1.6 mm wide, ± cylindric, drying ± 4-angled; SCoRO (Monterey, San Luis Obispo cos.). . . *C. hardhamiae*
          7′ Fr 1.5–2 mm wide, 4-angled; ChI, PR. . . . . *C. robusta*
      5′ < 5% of pollen grains 4–5-angled (visible with light microscopy)
        8. Fr conspicuously 4-angled, 1.8–2.2 mm wide — SCo, w PR . . . . *C. lewisii*
        8′ Fr cylindric or weakly 4-angled when fresh, 0.7–1.8 mm wide
          9. Petioles of distal lvs < 25 mm; fr < 5-coiled . . . . *C. ignota*
          9′ Petioles of distal lvs 0 or < 2 mm; fr straight to 3-coiled
            10. Pl strigose, ± gray — s SnJV, n slope SnBr, D. . . . . *C. pallida*
              11. Petals 6.5–13 mm; hypanthium 3.8–4.2 mm; style 6.5–10.5 mm . . . . subsp. *hallii*
              11′ Petals 2–6.5 mm; hypanthium 1–3 mm; style 2.1–6.5 mm . . . . subsp. *pallida*
            10′ Pl not strigose, ± gray or not
              12. Fr 0.7–0.9 mm wide; distal lvs narrowly ovate to ovate; st ascending. . . . . *C. hirtella*
              12′ Fr 0.8–1.8 mm wide; distal lvs narrowly lanceolate to narrowly ovate; st ± erect or decumbent
                13. Distal lvs gen narrowly lanceolate; infl hairs gen nonglandular . . . . *C. micrantha*
                13′ Distal lvs lanceolate to narrowly ovate; infl gen with glandular hairs
                  14. Pl ± gray with spreading hairs; hypanthium 2–3.8 mm; sepals 3.2–8 mm . . . . *C. confusa*
                  14′ Pl with spreading hairs, but not conspicuously ± gray; hypanthium 1.2–2 mm;
                    sepals 1–2.5 mm . . . . *C. intermedia*

C. bistorta (Torr. & A. Gray) W.L. Wagner & Hoch (p. 935) CALIFORNIA SUN CUP Ann or short-lived per, rosetted; strigose or hairs spreading, in infl short, erect. **ST**: decumbent to ± ascending, 50–80 cm, peeling. **LF**: 12–120 mm, petioled or distal gen ± sessile; cauline gen lanceolate (linear), minutely dentate to ± entire. **FL**: hypanthium 2–5(7.5) mm; sepals (2.3)5–8(11) mm; petals (4.2)7–15 mm, basal spots gen 1–2; stigma exceeding anthers. **FR**: 12–40 mm, 1.5–2.5 mm wide, ± 4-angled, gen straight or slightly wavy and twisted. **SEED**: 0.9–1 mm. 2*n*=14. Sandy fields near coast or clay soils in grassland to openings in coastal-sage scrub or chaparral; < 600 m. SW; n Baja CA. [*Camissonia b.* (Torr. & A. Gray) P.H. Raven] Cross-pollinated. Intergrades with *C. cheiranthifolia* subsp. *suffruticosa*. Mar–Jun

C. cheiranthifolia (Spreng.) W.L. Wagner & Hoch (p. 935) BEACH EVENING-PRIMROSE Per or subshrub, short-lived, rosetted, densely strigose (glabrous); hairs of infl gen erect, short. **ST**: prostrate to ± ascending, < 60(130) cm, peeling. **LF**: 5–50 mm, narrowly ovate to obovate, minutely serrate; cauline petioles 0–10 mm. **FL**: hypanthium 2.1–8.5 mm; sepals 4–11.5 mm; petals 6–20 mm, basal spots 0–2. **FR**: 10–25 mm, 2–2.5 mm wide, 4-angled, gen 1–2-coiled. **SEED**: 1.2–1.3 mm. 2*n*=14. [*Camissonia c.* (Spreng.) Raim.] Subspp. intergrade on ChI.

subsp. ***cheiranthifolia*** Per; hairs rarely dense and silvery. **FL**: hypanthium 2.1–4.2(4.8) mm; petals 6–11 mm, basal spots 0(2);

anthers 1–1.5 mm; style 6–9 mm. Sandy slopes, flats, coastal dunes; < 100 m. NCo, CCo, ChI; sw OR. Gen self-pollinated. Apr–Aug

subsp. ***suffruticosa*** (S. Watson) W.L. Wagner & Hoch Subshrub; hairs gen dense and silvery. **FL**: hypanthium 5–8.5 mm; petals (10)12–20 mm, basal spots 1–2; anthers 2.2–3 mm; style 13–23 mm. Sandy slopes, flats, coastal dunes; < 100 m. SCo; n Baja CA. [*Camissonia c.* subsp. *s.* (S. Watson) P.H. Raven] Gen cross-pollinated, ± self-incompatible; hybridizes widely with *C. bistorta*. Apr–Aug

C. confusa (P.H. Raven) W.L. Wagner & Hoch Ann, robust, rosetted, ± gray; hairs dense, spreading, infl also glandular or ± glabrous. **ST**: ± erect, < 60 cm. **LF**: 10–50 mm; cauline lanceolate to narrowly ovate, minutely dentate, ± sessile. **FL**: hypanthium 2–3.8 mm; sepals 3.2–8 mm; petals 5–10 mm, basal spots 1–2. **FR**: 13–23 mm, 0.8–1.1 mm wide, cylindric, drying 4-angled, gen 1–2-coiled. **SEED**: 0.7–1.1 mm. 2*n*=28. Dry inland slopes, gen chaparral; 300–2000 m. SW. [*Camissonia c.* P.H. Raven] Self-pollinated. Probably derived from *C. hirtella* × *C. pallida* subsp. pallida. Mar–May

C. guadalupensis (S. Watson) W.L. Wagner & Hoch subsp. ***clementina*** (P.H. Raven) W.L. Wagner & Hoch SAN CLEMENTE ISLAND EVENING-PRIMROSE Ann, rosetted; hairs spreading, in infl also glandular. **ST**: erect, 2–18(35) cm, ± fleshy, peeling. **LF**: 10–38(95) mm; cauline narrowly ovate, minutely dentate or ± entire, ± sessile. **FL**: hypanthium 1.6–2.4 mm; sepals 1.9–3.2 mm; petals

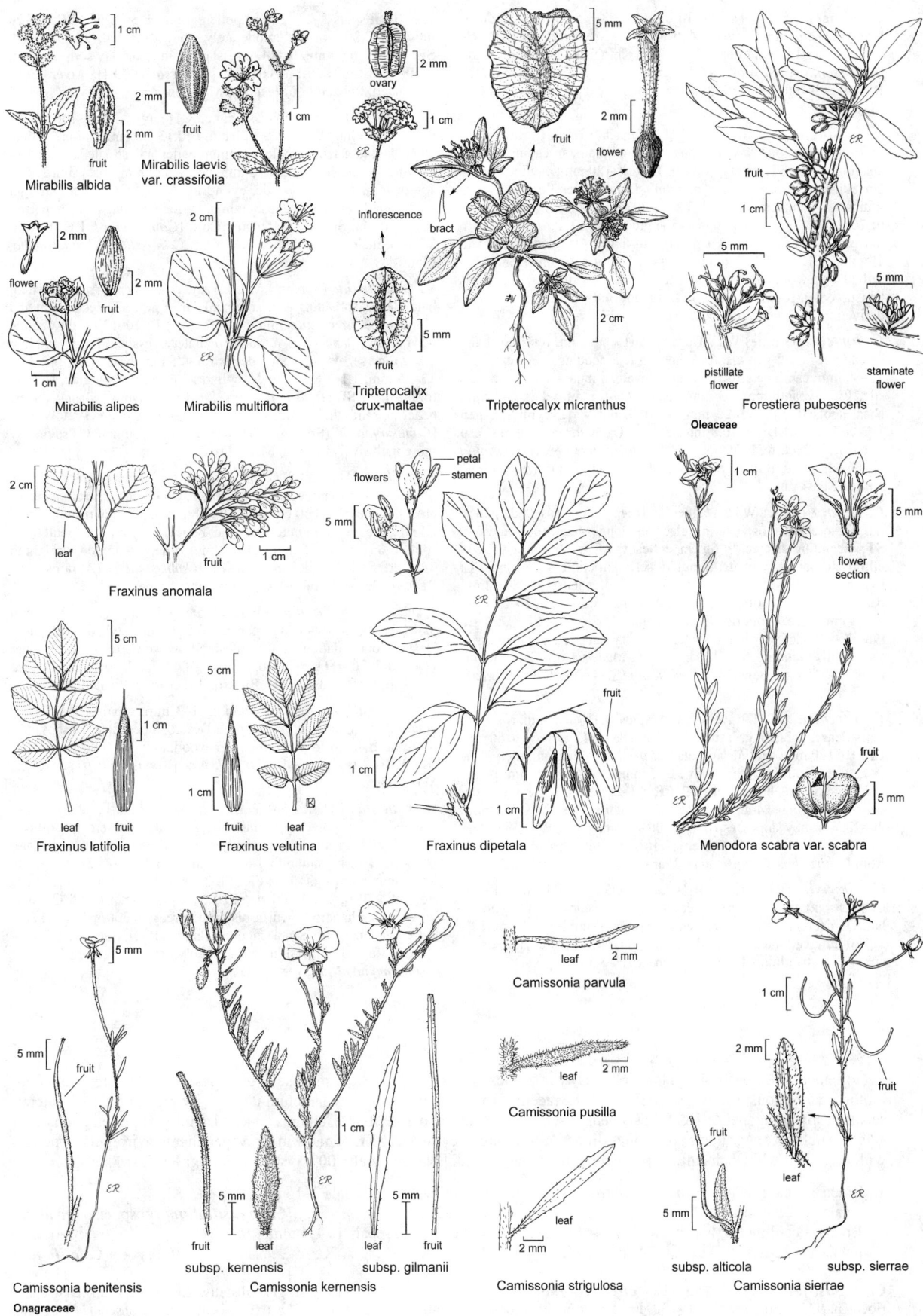

929

Mirabilis albida

1 cm

2 mm

2 mm

fruit

Mirabilis laevis var. crassifolia

1 cm

ovary

2 mm

1 cm

inflorescence

bract

ER

5 mm

fruit

Tripterocalyx crux-maltae

5 mm

fruit

2 mm

flower

2 cm

Tripterocalyx micranthus

ER

fruit

1 cm

5 mm

5 mm

pistillate flower

staminate flower

Forestiera pubescens

Oleaceae

flower

2 mm

2 mm

fruit

Mirabilis alipes

2 cm

ER

Mirabilis multiflora

2 cm

leaf

fruit

1 cm

Fraxinus anomala

5 cm

1 cm

leaf   fruit

Fraxinus latifolia

5 cm

1 cm

fruit   leaf

Fraxinus velutina

flowers

petal

stamen

5 mm

ER

fruit

1 cm

1 cm

Fraxinus dipetala

1 cm

5 mm

flower section

ER

fruit

5 mm

Menodora scabra var. scabra

5 mm

5 mm

fruit

ER

Camissonia benitensis

Onagraceae

1 cm

fruit   leaf

subsp. kernensis

5 mm

leaf   fruit

subsp. gilmanii

5 mm

Camissonia kernensis

leaf

2 mm

Camissonia parvula

leaf

2 mm

Camissonia pusilla

leaf

2 mm

Camissonia strigulosa

1 cm

2 mm

fruit

fruit

leaf

subsp. alticola

5 mm

ER

subsp. sierrae

Camissonia sierrae

2.8–4.2 mm, basal spot 1. **FR:** 10–18 mm, 2.8–3.5 mm wide, stout, ± 4-angled, straight or ± curved outward, deeply grooved. **SEED:** 0.7–0.9 mm. 2*n*=14. Sand dunes; < 30 m. s ChI (San Clemente Island). [*Camissonia g.* (S. Watson) P.H. Raven subsp. *c.* (P.H. Raven) P.H. Raven] Self-pollinated. Other subsp. on Guadalupe Island, Baja CA. May–Jun ★

C. hardhamiae (P.H. Raven) W.L. Wagner & Hoch HARDHAM'S EVENING-PRIMROSE Ann, robust, rosetted; hairs spreading, in infl also glandular. **ST:** erect, < 60 cm. **LF:** 10–120 mm; cauline widely lanceolate to narrowly ovate, minutely dentate, ± sessile. **FL:** hypanthium 1.7–2 mm; sepals 1.8–3.2 mm; petals 2–4 mm, basal spots 0; 70–100% of pollen grains 4–5-angled. **FR:** 13–25 mm, 1.3–1.6 mm wide, not esp stout, ± cylindric, drying ± 4-angled, straight to 1-coiled. **SEED:** 0.7–1.1 mm. 2*n*=42. Sandy soil, limestone, disturbed oak woodland; 240–600 m. SCoRO (Monterey, San Luis Obispo cos.). [*Camissonia h.* P.H. Raven] Self-pollinated. Probably derived from *C. micrantha* × *C. intermedia.* Mar–May ★

C. hirtella (Greene) W.L. Wagner & Hoch Ann, rosetted; hairs spreading, in infl gen also glandular. **ST:** ascending, < 60 cm. **LF:** 10–80 mm; cauline narrowly ovate to ovate, minutely dentate, ± sessile. **FL:** hypanthium 1–3 mm; sepals 2.5–6 mm; petals 2–9 mm, basal spots 0. **FR:** 13–20 mm, 0.7–0.9 mm wide, cylindric, gen 1–2-coiled. **SEED:** 0.7–0.8 mm. 2*n*=14. Open shrubby slopes, esp in burns; < 2300 m. NW, SN, CW, SW; n&c Baja CA. [*Camissonia h.* (Greene) P.H. Raven] Self-pollinated. Rarely hybridizes with *C. ignota.* Mar–Aug

C. ignota (Jeps.) W.L. Wagner & Hoch Ann, rosetted, gen ± red, minutely strigose exc infl glabrous, longer-hairy, or glandular. **ST:** ascending or erect, < 55 cm, ± fleshy. **LF:** < 60 mm, narrowly elliptic to narrowly ovate, minutely serrate; petiole < 25 mm. **FL:** hypanthium 1.1–3 mm; sepals 2.5–5.5 mm; petals (3)4.8–8 mm, basal spots 0; < 5% of pollen grains 4–5-angled. **FR:** gen 20–30 mm, 0.8–1 mm wide, cylindric, drying 4-angled, < 5-coiled. **SEED:** ± 1.2 mm. 2*n*=14. Gen clay fields, slopes in coastal-sage scrub or chaparral, sandy soils in mtns; 100–1100 m. c SNF (Madera Co.), GV (uncommon), CW, SW; n Baja CA. [*Camissonia i.* (Jeps.) P.H. Raven] Self-pollinated. Mar–Aug

C. intermedia (P.H. Raven) W.L. Wagner & Hoch Ann, rosetted; hairs dense, spreading, in infl also glandular. **ST:** ± erect, < 60 cm. **LF:** 10–120 mm; cauline lanceolate to narrowly ovate, minutely dentate, ± sessile. **FL:** hypanthium 1.2–2 mm; sepals 1–2.5 mm; petals 1.5–3.5(4.5) mm, basal spots 1–2. **FR:** 13–25 mm, 1.1–1.2 mm wide, cylindric, drying ± 4-angled, straight to 1-coiled. **SEED:** 0.7–1 mm. 2*n*=28. Shrubby slopes, esp burns; 300–800 m. NW, CW, SW; n Baja CA. [*Camissonia i.* P.H. Raven] Self-pollinated. Probably derived from *C. hirtella* × *C. micrantha.* Mar–May

C. lewisii (P.H. Raven) W.L. Wagner & Hoch LEWIS' EVENING-PRIMROSE Ann, rosetted; hairs spreading, in infl glandular. **ST:** decumbent and branched or erect and simple, < 60 cm. **LF:** 10–80 mm; cauline narrowly lance-elliptic, minutely dentate, ± sessile. **FL:** hypanthium 1.5–4 mm; sepals 1.7–3.5 mm; petals 2.5–5.5

mm, basal spots 1–2; < 5% of pollen grains 4–5-angled. **FR:** 13–20 mm, 1.8–2.2 mm wide, moderately stout, 4-angled, gen 1-coiled. **SEED:** 0.7–0.8 mm. 2*n*=14. Grassland, sandy or clay soils, coastal; < 300 m. SCo, w PR; n Baja CA. [*Camissonia l.* P.H. Raven] Self-pollinated. Related to *C. bistorta.* Mar–Jun ★

C. luciae (P.H. Raven) W.L. Wagner & Hoch Ann, rosetted; hairs dense, spreading. **ST:** erect, < 60 cm. **LF:** 13–55 mm; cauline widely lanceolate to narrowly oblong, minutely dentate, sessile. **FL:** hypanthium 2–3 mm; sepals 2.5–4.5 mm; petals 4–7 mm, basal spot 1. **FR:** 15–20 mm, 1.3–2 mm wide, cylindric, drying ± 4-angled, straight to 2-coiled. **SEED:** 1.3–1.5 mm. 2*n*=42. Openings in chaparral; 300–1400 m. SCoRO (Monterey Co.). [*Camissonia l.* P.H. Raven] Self-pollinated. Probably derived from *C. hirtella* × *C. intermedia.* Mar–May

C. micrantha (Spreng.) W.L. Wagner & Hoch Ann, rosetted; hairs dense, spreading, gen nonglandular in infl. **ST:** decumbent and branched or erect and simple, < 60 cm. **LF:** 10–120 mm; cauline gen narrowly lanceolate, minutely dentate, ± sessile. **FL:** hypanthium 1.2–2 mm; sepals 1–2.5 mm; petals 1.5–4.5 mm, basal spots 1–2. **FR:** 13–25 mm, 1.1–1.8 mm wide, cylindric, drying ± 4-angled, straight to 1-coiled. **SEED:** 0.7–1.1 mm. 2*n*=14. Canyons, coastal plains, beaches, sandy fields, washes; < 300(800) m. w edge GV, CW, SW. [*Camissonia m.* (Spreng.) P.H. Raven] Self-pollinated. Previously more inclusively defined. Mar–May

C. pallida (Abrams) W.L. Wagner & Hoch Ann, rosetted, ± gray, densely strigose, infl also glandular. **ST:** decumbent and branched or erect and simple, < 60 cm. **LF:** 10–30 mm; cauline narrowly lance-elliptic, ± entire to minutely dentate; petiole < 2 mm. **FL:** petal basal spots 1–3. **FR:** 10–25 mm, 1.1–1.2 mm wide, ± 4-angled, straight to 3-coiled. **SEED:** 1–1.5 mm. 2*n*=14. [*Camissonia p.* (Abrams) P.H. Raven] Gen self-pollinated. Subspp. intergrade.

subsp. ***hallii*** (Davidson) W.L. Wagner & Hoch **FL:** hypanthium 3.8–4.2 mm; sepals 4.8–8 mm; petals 6.5–13 mm; style 6.5–10.5 mm. Desert slopes, flats, washes, creosote-bush scrub to pinyon/juniper woodland; 30–1800 m. n slope SnBr, s DMoj, n DSon. [*Camissonia p.* subsp. *h.* (Davidson) P.H. Raven] Occ cross-pollinated. Mar–Aug

subsp. ***pallida*** **FL:** hypanthium 1–3 mm; sepals 1.5–5.5 mm; petals 2–6.5 mm; style 2.1–6.5 mm. Desert slopes, flats, washes, creosote-bush scrub to pinyon/juniper woodland; 30–1800 m. s SnJV (Kern Co.), n slope SnBr, D; NV, AZ. [*Camissonia p.* subsp. *p.*] Mar–Aug

C. robusta (P.H. Raven) W.L. Wagner & Hoch (p. 935) Ann, rosetted; hairs spreading, in infl also glandular. **ST:** erect, < 60 cm. **LF:** 10–80 mm; cauline narrowly lance-elliptic, minutely dentate, ± sessile. **FL:** hypanthium 1.8–3.7 mm; sepals 2.6–4.2 mm; petals 3.2–7 mm, basal spots 1–2. **FR:** 14–25 mm, 1.5–2 mm wide, not esp stout, 4-angled, gen 1-coiled. **SEED:** 0.9–1.2 mm. 2*n*=42. Coastal-sage scrub, chaparral, gen in disturbed areas; < 300 m. ChI (Santa Cruz, San Clemente, Santa Catalina islands), PR; n Baja CA. [*Camissonia r.* Raven] Self-pollinated. Probably derived from *C. lewisii* × *C. intermedia.* Mar–Aug

CHAMERION

Peter C. Hoch

Per, often clumped or forming large colonies by rhizomes. **ST:** gen unbranched, strigose or glabrous. **LF:** alternate, gen ± fine-toothed; veins conspicuous or obscure. **INFL:** raceme. **FL:** nodding in bud; hypanthium 0 (exc as ± green disk); ± bilateral; sepals 4, spreading; petals 4, entire; stamens 8, subequal, maturing before stigma, anthers attached at middle, pollen grains shed singly, gen blue-gray; stigma spreading-4-lobed. **FR:** straight, cylindric. **SEED:** many in 1 row per chamber, irregularly netted, with persistent white hair-tuft. 8 spp.: n temp montane, boreal. [Wagner et al. 2007 Syst Bot Monogr 83:78–81]

1. Bracts << cauline lvs, ± linear; lf veins conspicuous; petioles 2–7 mm; sepals 7–16 mm;
   petals 10–25 mm . . . . ***C. angustifolium*** subsp. ***circumvagum***
1′ Bracts = cauline lvs, lanceolate; lf veins obscure; petioles 0–3 mm; sepals 10–24 mm;
   petals 13–32 mm . . . . ***C. latifolium***

C. angustifolium (L.) Holub subsp. ***circumvagum*** (Mosquin) Hoch (p. 935) FIREWEED Pl < 30 dm, gen strongly colonial, ±

glabrous to densely strigose distally. **LF:** 1.5–20 cm, lanceolate; midrib strigose abaxially. **INFL:** dense, gen canescent. **FL:** petals

gen deep pink to magenta; stamens < pistil, pollen blue-gray. **FR**: 4–10 cm, gray-hairy; pedicel 7–20 mm. **SEED**: 1–1.3 mm. 2*n*=72. Common. Open places, gravel bars, roadsides, esp after fires; < 3300 m. NCo, KR, NCoRO, CaRH, SNH, SnBr, W&I, ne DMtns; circumboreal. [*Epilobium a*. L. subsp. *c*. Mosquin] *C. angustifolium* subsp. *a*. (2*n*=36), farther n and at higher elevations, might be expected in CA. Jul–Sep

C. latifolium (L.) Holub (p. 935)   Pl < 7 dm, clumped from woody caudex, ± glaucous to densely strigose. **LF**: 1–10 cm, elliptic to widely lanceolate, glabrous or hairy on margins and veins, ± obtuse to acute. **INFL**: gen ± strigose. **FL**: petals (white) pink to rose-purple; stamens ± = pistil; pollen ± gray. **FR**: 3–10.5 cm, ± glabrous to gray-hairy; pedicel 5–18 mm. **SEED**: 1.3–2.4 mm. 2*n*=36(CA),72. Gravel bars, talus slopes, glacial outwashes; 2500–3100 m. n&c SNH; ± circumboreal. [*Epilobium l*. L.] Jul–Aug

CHYLISMIA

Ann, occ per, from taproot. **LF**: basal and cauline, alternate, lanceolate to ovate, pinnately lobed with large terminal lobe (simple), margin dentate to entire, abaxial face and/or margin with ± conspicuous brown oil cells, long-petioled. **INFL**: erect or nodding raceme. **FL**: gen opening at dawn (occ at dusk); sepals 4, reflexed; petals 4, yellow or white (often fading orange-red) or lavender, gen fading red, if yellow gen strongly ultraviolet reflective, often with 1+ red spots near base, occ non-reflective near base or throughout; stamens (4)8, longer opposite sepals, anthers attached at middle, filaments long-ciliate or glabrous, pollen grains 3-angled at 20×; stigma entire and spheric or rarely conic-peltate, gen > anthers and cross-pollinated or ± = anthers and self-pollinated. **FR**: straight to curved, not twisted or coiled, valves with obvious midrib, pedicelled. **SEED**: in 2 rows per chamber, lenticular to narrowly ovoid, with ± pronounced membranous margin when immature. 16 spp.: w N.Am. (Greek: juice) [Wagner et al. 2007 Syst Bot Monogr 83:1–240] Incl in *Camissonia* in TJM (1993).

1. Hypanthium 4.5–40 mm; pl with cauline, simple lvs; rosette 0 or poorly developed (sect. *Lignothera*)
  2. Hypanthium 18–40 mm; infl open; sepals 8–15 mm. . . . . ***C. arenaria***
  2′ Hypanthium 4.5–14 mm; infl dense; sepals 3–9 mm . . . . ***C. cardiophylla***
    3. Hypanthium 4.5–12 mm; hairs gen spreading, occ some glandular; c&s DMoj (San Bernardino Co.), DSon. . . . . subsp. ***cardiophylla***
    3′ Hypanthium 9–14 mm; hairs gen glandular, some spreading, nonglandular; n DMoj (Inyo Co.). . . . . . . subsp. ***robusta***
1′ Hypanthium 0.4–8 mm; pl gen with well-developed basal rosette, occ cauline lvs also present; lvs gen 1-pinnate (sect. *Chylismia*)
  4. Petals lavender, base gen yellow, lavender-dotted; rosette poorly developed. . . . . ***C. heterochroma***
  4′ Petals yellow or white, occ fading purple or red or base purple- or red-dotted; rosette well developed
    5. Fr gen < 2 mm wide, ± same width throughout
      6. Sepals 1.5–4 mm; hypanthium 1–1.5 mm; stigma ± = anthers; infl erect; lateral lflets < 30 mm . . . . ***C. walkeri*** subsp. ***tortilis***
      6′ Sepals 5–9 mm; hypanthium 3–8 mm; stigma > anthers; infl nodding; lateral lflets gen < 10 mm or 0 . . . ***C. brevipes***
        7. Fl bud reflexed; petals gen fading red . . . . subsp. ***arizonica***
        7′ Fl bud not reflexed; petals not fading red
          8. Pl hairs spreading; sepals in bud with free subterminal tips; petals not red-dotted. . . . . subsp. ***brevipes***
          8′ Pl strigose (spreading-hairy proximally); sepals in bud without free tips; petals gen red-dotted. . . . subsp. ***pallidula***
    5′ Fr > 2 mm wide, wider toward tip
      9. Pedicel and fr becoming reflexed; petals yellow. . . . . ***C. munzii***
      9′ Pedicel and fr spreading or ascending; petals white or yellow. . . . . ***C. claviformis***
        10. Pl spreading-hairy proximally; free tips of sepals in bud conspicuous; petals gen yellow . . . . subsp. ***peirsonii***
        10′ Pl strigose or glabrous proximally (exc *Chylismia claviformis* subsp. *cruciformis*); free tips of sepals in bud 0 or inconspicuous (exc *Chylismia claviformis* subsp. *funerea*); petals white or yellow
          11. Petals yellow
            12. Pl strigose distally, also glandular or not; free tips of sepals in bud 0 or short. . . . . subsp. ***yumae***
            12′ Pl glabrous distally or glandular-hairy; free tips of sepals in bud 0
              13. Terminal lflet ovate to ± cordate, lateral lflets few to many; fl gen opening at dawn . . . . subsp. ***cruciformis***
              13′ Terminal lflet lanceolate, lateral lflets 0–few; fl opening at dusk . . . . subsp. ***lancifolia***
          11′ Petals white (pale yellow in *Chylismia claviformis* subsp. *claviformis*)
            14. Lateral lflets gen well developed
              15. Pl strigose distally (glandular-hairy) . . . . subsp. ***aurantiaca***
              15′ Pl ± glabrous distally or with few glandular hairs. . . . . subsp. ***claviformis***
            14′ Lateral lflets gen 0 or poorly developed
              16. Free tips of sepals in bud conspicuous; pl gen strigose distally. . . . . subsp. ***funerea***
              16′ Free tips of sepals in bud 0 or minute; pl ± glabrous distally or gen with some glandular hairs . . . . subsp. ***integrior***

C. arenaria A. Nelson (p. 935)   Ann or bushy per, erect; hairs spreading, in infl a few glandular. **ST**: < 180 cm. **LF**: petiole < 60 mm; blade < 60 mm, cordate-deltate, teeth coarse or larger and smaller. **INFL**: open, nodding. **FL**: opening at dusk; hypanthium 18–40 mm; sepals 8–15 mm; petals 8–20 mm, yellow. **FR**: 30–44 mm, ascending, cylindric, ± straight; pedicel 2–5 mm. **SEED**: 0.5–0.7 mm. 2*n*=14. Sandy washes, rocky slopes, desert scrub; < 430 m. DSon; sw AZ, n Mex (SON). [*Camissonia a*. (A. Nelson) P.H. Raven] Gen cross-pollinated. Mar–Apr   ★

C. brevipes (A. Gray) Small   Ann, strigose or hairs spreading. **ST**: 3–75 cm. **LF**: gen basal, simple to 1-pinnate; terminal lflet < 65 mm, lateral lflets gen < 10 mm or 0. **INFL**: nodding. **FL**: opening at dawn; hypanthium 3–8 mm; sepals 5–9 mm, tips in bud gen free and ± terminal; petals 3–18 mm, yellow, rarely bases red-dotted; stamens ± equal. **FR**: 18–92 mm, ascending to spreading, cylindric, straight or curved; pedicel 2–20 mm. **SEED**: 1–1.5 mm. 2*n*=14. [*Camissonia b*. (A. Gray) P.H. Raven]

subsp. ***arizonica*** (P.H. Raven) W.L. Wagner & Hoch (p. 935)
Hairs spreading. **FL:** bud reflexed; hypanthium 3–5 mm; sepal tips in
bud free or not; petals 3–8 mm, gen red-dotted, gen fading red. **FR:**
18–60 mm; pedicel 2.5–5 mm. Rocky slopes and flats; 70–300 m. se
DSon (Imperial Co.); sw AZ. [*Camissonia b.* subsp. *a.* (P.H. Raven)
P.H. Raven] Hybridizes with *C. claviformis* subsp. *yumae.* Mar–Apr

subsp. ***brevipes*** (p. 935) Hairs spreading. **FL:** bud not reflexed;
hypanthium 4–8 mm; petals 6–18 mm, not red-dotted, not fading red.
FR: 20–92 mm; pedicel 5–20 mm. Sandy slopes, washes, alluvial
fans, Joshua-tree woodland (moister areas than other subspp.); -70–
1800 m. D; to sw UT. Intergrades with *C. brevipes* subsp. *pallidula;*
hybridizes with *C. claviformis, C. munzii.* Mar–May

subsp. ***pallidula*** (Munz) W.L. Wagner & Hoch (p. 935) Pl
strigose (hairs rarely spreading proximally). **FL:** bud not reflexed;
hypanthium 4–5 mm; sepal tips in bud not free; petals 7–12 mm, gen
red-dotted, not fading red. **FR:** 20–42 mm; pedicel 2–10 mm. Dry
flats, desert pavement, gen with *Larrea, Ambrosia;* 70–1100 m. D (se
Inyo, ne Imperial cos.); to sw UT, nw AZ. [*Camissonia b.* subsp. *p.*
(Munz) P.H. Raven] Mar–May

C. cardiophylla (Torr.) Small   Ann, per; rosette 0; hairs spreading, glandular or not. **ST:** < 100 cm. **LF:** blade < 55 mm, ovate to
rounded-cordate, irregularly dentate; petiole < 75 mm. **INFL:** dense,
nodding. **FL:** opening at dusk; hypanthium 4.5–14 mm; sepals 3–9
mm; petals 3–12 mm, yellow or cream. **FR:** 20–55 mm, ascending,
cylindric, ± straight; pedicel 1–18 mm. **SEED:** 0.5–0.7 mm. 2*n*=14.
[*Camissonia c.* (Torr.) P.H. Raven] Gen cross-pollinated. 3 subspp.,
2 in CA.

subsp. ***cardiophylla*** (p. 935) Occ some hairs glandular. **LF:**
cordate. **FL:** hypanthium 4.5–12 mm; petals 3–12 mm. Sandy washes,
slopes, rocky walls, creosote-bush scrub; < 600 m. c&s DMoj (San
Bernardino Co.), DSon; s AZ, nw Mex. Mar–May

subsp. ***robusta*** (P.H. Raven) W.L. Wagner & Hoch (p. 935)
Hairs both glandular and not. **LF:** cordate-round. **FL:** hypanthium
9–14 mm; petals 7–11 mm. Sandy washes, slopes, rocky walls, creosote-bush scrub; 600–1400 m. n DMoj (Inyo Co.). [*Camissonia c.*
subsp. *r.* (P.H. Raven) P.H. Raven] Mar–May

C. claviformis (Torr. & Frém.) A. Heller   Ann. **ST:** 3–70 cm.
LF: gen basal, gen 1-pinnate; terminal lflet 8–90 mm, lanceolate to
cordate; lateral lflets < 25 mm or 0. **INFL:** nodding. **FL:** gen opening near dusk; hypanthium 1–6.5 mm; sepals 2–8 mm, free tips in
bud present or not; petals 1.5–8 mm, yellow or white; stamens ±
equal. **FR:** ascending or spreading, 8–38 mm, wider to tip, straight
or curved; pedicel 4–40 mm. **SEED:** 0.6–1.5 mm. 2*n*=14. [*Camissonia c.* (Torr. & Frém.) P.H. Raven] Cross-pollinated; most complex,
widespread sp. in genus; 11 subspp., 8 in CA.

subsp. ***aurantiaca*** (Munz) W.L. Wagner & Hoch (p. 935) Pl
strigose (glandular-hairy distally). **LF:** terminal lflet < 30 mm, narrowly ovate; lateral lflets well developed, like terminal. **FL:** hypanthium 3–5 mm; sepal tips not free in bud (inconspicuously so); petals
2.5–8 mm, white gen fading purple, bases rarely purple-dotted. Sandy
flats, washes, creosote-bush scrub; -70–900 m. D; s NV, w AZ, n Baja
CA. [*Camissonia c.* subsp. *a.* (Munz) P.H. Raven] Intergrades extensively with *C. claviformis* subsp. *peirsonii, C. claviformis* subsp.
yumae, and those with white petals; occ hybridizes with *C. brevipes,
C. munzii.* Feb–May

subsp. ***claviformis*** Pl glabrous or strigose proximally, ± glabrous or glandular-hairy distally. **LF:** terminal lflet < 60 mm, narrowly ovate, purple-dotted or not; lateral lflets gen large. **FL:** hypanthium 3–5.5 mm; sepal tips free in bud, inconspicuous; petals 3.5–8
mm, white (pale yellow) gen fading purple, bases purple-dotted or
not. Alluvial slopes, flats, creosote-bush scrub; 850–1700 m. DMoj
and edges. Intergrades widely and gradually with *C. claviformis*
subsp. *aurantiaca, C. claviformis* subsp. *funerea;* hybridizes with *C.
brevipes* subsp. *b.* Mar–May

subsp. ***cruciformis*** (Kellogg) W.L. Wagner & Hoch (p. 935) Pl
strigose or glandular-hairy proximally, glabrous or glandular-hairy
distally. **LF:** terminal lflet < 80 mm, ovate to ± cordate; lateral lflets
few to many. **FL:** gen opening at dawn; hypanthium 2–6.5 mm; sepal

tips not free in bud; petals 2.5–8 mm, yellow gen fading purple, bases
gen red-dotted. Sagebrush scrub; 1200–1600 m. MP; to c OR, ID,
nw NV. [*Camissonia claviformis* subsp. *cruciformis* (Kellogg) P.H.
Raven] Outside CA, intergrades with *C. claviformis* subsp. *integrior.*
Mar–May ★

subsp. ***funerea*** (P.H. Raven) W.L. Wagner & Hoch (p. 935) Pl
gen strigose, densely so at least proximally. **LF:** terminal lflet < 80
mm, ovate, gen cordate; lateral lflets gen 0. **FL:** hypanthium 3–5.5
mm; sepal tips free in bud, conspicuous, subterminal; petals 3.5–7.5
mm, white gen fading purple. Dry slopes, flats, creosote-bush scrub;
-70–900 m. n DMoj (Eureka, Saline, Death valleys). [*Camissonia
c.* subsp. *f.* (P.H. Raven) P.H. Raven] Intergrades with *C. claviformis* subsp. *aurantiaca, C. claviformis* subsp. *c.;* occ hybridizes with
C. munzii, C. brevipes subsp. *b.* Mar–May

subsp. ***integrior*** (P.H. Raven) W.L. Wagner & Hoch   Pl strigose
proximally, gen glandular-hairy or ± glabrous distally. **LF:** terminal
lflet < 70 mm, ± ovate, base gen ± cordate; lateral lflets 0–few, small.
FL: hypanthium 3–6 mm; sepal tips not free in bud, or inconspicuously so; petals 4.5–8 mm, white fading purple, bases purple-dotted or not. Dry flats, desert scrub; 1200–2000 m. SNE; c OR, NV.
[*Camissonia c.* subsp. *i.* (P.H. Raven) P.H. Raven] Intergrades with
Camissonia claviformis subsp. *aurantiaca, C. claviformis* subsp. *cruciformis;* hybridizes with *C. brevipes.* Apr–Jul

subsp. ***lancifolia*** (A. Heller) W.L. Wagner & Hoch (p. 935) Pl
strigose proximally, glabrous and glaucous distally. **LF:** terminal
lflet < 50 mm, lanceolate; lateral lflets 0–few, small. **FL:** hypanthium
3.5–6 mm; sepal tips not free in bud; petals 3.5–7 mm, yellow, bases
gen red-dotted. Sandy soils, sagebrush scrub; 1200–1700 m. SNE,
DMtns. [*Camissonia c.* subsp. *l.* (A. Heller) P.H. Raven] Apr–Jul

subsp. ***peirsonii*** (Munz) W.L. Wagner & Hoch (p. 935) Pl
spreading-hairy, rarely strigose or glandular. **LF:** terminal lflet < 90
mm, narrowly ovate; lateral lflets gen large. **FL:** hypanthium 2.5–4.5
mm; sepal tips in bud free, conspicuous; petals 4.5–7 mm, gen yellow. Sandy flats, creosote-bush scrub; -70–300 m. PR, DSon (Imperial Co.); n Baja CA. [*Camissonia c.* subsp. *p.* (Munz) P.H. Raven]
Intergrades with *C. claviformis* subsp. *aurantiaca.* Feb–Apr

subsp. ***yumae*** (P.H. Raven) W.L. Wagner & Hoch (p. 935) Pl strigose
(gen densely so), occ also glandular-hairy distally. **LF:** terminal lflet
< 65 mm, lanceolate; lateral lflets reduced or not. **FL:** hypanthium
2.5–4 mm; sepal tips not free in bud, or inconspicuously so; petals
4–5.5 mm, yellow, occ fading ± red. Dunes or sandy flats, creosote-
bush scrub; < 300 m. se DSon (se Imperial Co.); sw AZ, nw Mex.
[*Camissonia c.* subsp. *y.* (P.H. Raven) P.H. Raven] Probably derived
from *C. claviformis* subsp. *aurantiaca* × *C. claviformis* subsp. *peeble-
sii* (Munz) P.H. Raven in AZ; intergrades with *C. claviformis* subsp.
aurantiaca. Feb–May

C. heterochroma (S. Watson) Small (p. 935) SHOCKLEY'S
EVENING-PRIMROSE   Ann, glandular-hairy, or ± glabrous and glaucous distally. **ST:** 10–100 cm. **LF:** gen basal, < 70 mm, ovate; base gen
cordate. **INFL:** erect, longer in fr. **FL:** opening at dawn; hypanthium
2–5 mm; sepals 1.5–3.5 mm, tips not free in bud; petals 2–6 mm, lavender, bases gen yellow, lavender-dotted. **FR:** 7–13 mm, erect, club-
shaped, straight; pedicel 2–5 mm. **SEED:** 1–1.2 mm. 2*n*=14. Alluvial
slopes, rock slides, creosote-bush scrub to pinyon/juniper woodland;
600–2100 m. SNE, n DMtns (Grapevine Mtns); NV. [*Camissonia h.*
(S. Watson) P.H. Raven] Cross-pollinated. May–Jun

C. munzii (P.H. Raven) W.L. Wagner & Hoch   Ann, strigose. **ST:**
8–50 cm. **LF:** gen basal, 1-pinnate; terminal lflet < 60 mm, ovate; lateral lflets well developed, like terminal. **INFL:** nodding. **FL:** opening
at dawn; hypanthium 2–3 mm; sepals 4–7 mm, tips not free in bud;
petals 3–10 mm, yellow, bases red-dotted; stamens ± equal. **FR:** 8–24
mm, wider toward tip, ± straight; pedicel, fr reflexed in age, 8–28
mm. **SEED:** 0.8–1.6 mm. 2*n*=14. Slopes, washes, mtns; 600–1600
m. ne DMoj; s NV. [*Camissonia m.* (P.H. Raven) P.H. Raven] Cross-
pollinated. Occ hybridizes with *C. brevipes* subsp. *b.* and *C. claviformis* subsp. *aurantiaca.* Mar–Jun

C. walkeri A. Nelson subsp. ***tortilis*** (Jeps.) W.L. Wagner & Hoch
Ann, short-lived per; hairs spreading, in infl also glandular or bristly.
ST: 10–60 cm. **LF:** gen basal, simple to 2-pinnate, 30–220 mm, gen

purple-dotted; lateral lflets < 30 mm. **INFL**: erect; buds drooping. **FL**: opening at dawn; hypanthium 1–1.5 mm; sepals 1.5–4 mm, gen purple-dotted, in bud tips free, ± terminal; petals 3–6 mm, yellow. **FR**: ascending or spreading, 10–45 mm, 1.2–1.8 mm wide, cylindric, straight or curved; valves gen twisted; pedicel 5–30 mm. **SEED**: 0.6–1.2 mm. 2*n*=14. Rocky places near cliffs, along ephemeral streams, creosote-bush scrub to pinyon/juniper woodland; 600–1800 m. W&I, n DMoj (Inyo, ne San Bernardino cos.); to sw UT. [*Camissonia w.* (A. Nelson) P.H. Raven subsp. *t.* (Jeps.) P.H. Raven] Self-pollinated. Other var. outside CA. Apr–Jun

CHYLISMIELLA

1 sp.: w US. (Latin: diminutive of *Chylismia*) [Wagner et al. 2007 Syst Bot Monogr 83:1–240] Incl in *Camissonia* in TJM (1993).

C. pterosperma (S. Watson) W.L. Wagner & Hoch (p. 935)   Ann, from taproot; hairs bristly, in infl also glandular. **ST**: erect, branched, 2–14 cm, slender, peeling near base. **LF**: cauline, alternate, ± sessile; blade 3–30 mm, narrowly lanceolate to oblanceolate, entire. **INFL**: raceme, nodding in bud. **FL**: opening at dawn; hypanthium 1–2 mm; sepals 4, 1.5–2.5 mm, reflexed singly or occ in pairs; petals 4, 1.5–2.5 mm, notched at tip, white, with yellow area at base, fading purple; longer stamens opposite sepals, anthers attached at base, pollen grains 3-angled; stigma hemispheric, ± = anthers, self-pollinated. **FR**: (3)4-chambered, ascending or spreading, 12–28 mm, 0.6–0.8 mm wide, straight or slightly curved, cylindric, slightly swollen by seeds; pedicel 4–8 mm. **SEED**: in 2 rows per chamber, 1–1.5 mm, with concave and convex sides, concave side with a thick wing, both sides densely covered with glassy club-shaped hairs. 2*n*=14. Uncommon. Well-drained, gen volcanic slopes, pinyon/juniper woodland, sagebrush scrub; 1400–2400 m. W&I (Inyo Mtns), DMtns (Panamint Range); to OR, w UT, nw AZ. [*Camissonia p.* (S. Watson) P.H. Raven] Self-pollinated. May–Jun

CIRCAEA ENCHANTER'S NIGHTSHADE

Peter C. Hoch

Per from tuber-tipped rhizomes or stolons. **LF**: opposite, petioled, entire to toothed; stipules present, occ deciduous. **INFL**: raceme or panicle. **FL**: hypanthium present; biradial; sepals 2, often reflexed, deciduous after fl (along with other fl parts); petals 2, erect; stamens 2, pollen yellow, grains shed singly; ovary chambers 1–2, stigmas maturing before anthers. **FR**: indehiscent, gen club-shaped, bur-like with hooked hairs. **SEED**: 1 per chamber, adhering to inner fr wall. 8 spp.: n hemisphere. (Greek: Circe, the enchantress) [Wagner et al. 2007 Syst Bot Monogr 83:78–81] Often self-pollinated.

C. alpina L. subsp. ***pacifica*** (Asch. & Magnus) P.H. Raven (p. 935)   **ST**: 1–5 dm, simple, erect, slender, cylindric, gen densely strigose. **LF**: 3–11 cm, ovate to ± round, glabrous or ± hairy; base round to ± cordate; tip acute; petiole 1.5–5 cm. **INFL**: raceme, erect, densely strigose and glandular; pedicel in fr 2–5 mm. **FL**: hypanthium 0.3–0.5 mm; sepals 1–2 mm, white, reflexed; petals 1–1.5 mm, white; stamens ± = pistil; stigma 2-lobed. **FR**: ± 2 mm, 1-chambered. 2*n*=22. Cool, moist, conifer forest; < 2700 m. NW, CaR, n SNF, SNH, SnFrB, SnBr, Wrn; to w Can, MT, NM. May–Aug

CLARKIA

Harlan Lewis

Ann < 1.5 m. **ST**: prostrate to erect, glabrous, often glaucous, or puberulent (hairs long, spreading). **LF**: pinnately veined; petiole < 4 cm or 0; blade 1–10 cm, linear to elliptic or ovate, entire or shallow-toothed, glabrous or sparsely puberulent. **INFL**: spike, raceme; bracts lf-like; axis in bud straight or recurved at tip, in fl ± straight; buds erect or not. **FL**: hypanthium obconic to cup-shaped, or long, slender, gen with ring of hairs within; sepals 4, gen fused to tip in bud, reflexed at least at base, staying fused at least at tip, in 4s or 2s, or all coming free; corolla bowl-shaped to rotate, petals 5–60 mm, often lobed or clawed, lavender or pink to dark red, pale yellow, or white, often spotted, flecked, or streaked with red, purple, or white; stamens 8, in 2 like or unlike series, or 4, filaments cylindric to wider distally, subtended by ciliate scales or gen not, anthers attached at base, pollen white or yellow to blue-gray, lavender, or ± red; ovary 4-chambered, glabrous or not, cylindric, fusiform, or wider distally, gen shallowly to deeply 4- or 8-grooved, stigma lobes 4, gen prominent. **FR**: gen capsule, elongate (short, indehiscent, nut-like). **SEED**: gen many, rarely 1–2, 0.5–2 mm, angled, crested or not, brown, gray, or mottled. ± 41 spp.: w N.Am, 1 S.Am. (Captain William Clark, 1770–1838, of Lewis & Clark Expedition) [Lewis & Lewis 1955 Univ Calif Publ Bot 20:241–392] Self-fertile; self-pollinated or outcrossed; on herbarium specimens, curvature of infl axis in bud gen reliable, pollen color gen not.

1. Fr 2–3 mm, nut-like, indehiscent, 1–2-seeded . . . . ***C. heterandra***
1′ Fr > 5 mm, not nut-like, dehiscent, many-seeded
  2. Stamens 4; petals 3-lobed
    3. Petal length ± = width, middle lobe narrower . . . . ***C. breweri***
    3′ Petal length ± 2 × width, lobes ± equally wide . . . . ***C. concinna***
      4. Stigma exserted beyond anthers; petals 15–30 mm . . . . subsp. ***concinna***
      4′ Stigma not exserted beyond anthers; petals 10–20 mm
        5. Petal lobes prominent, separated by deep sinuses . . . . subsp. ***automixa***
        5′ Petal lobes short, not separated by deep sinuses . . . . subsp. ***raichei***
  2′ Stamens 8; petals not 3-lobed
    6. Axis of infl in bud straight; buds erect or reflexed
      7. Buds reflexed; corolla rotate or bowl-shaped; inner anthers < outer, paler
        8. Petal lobes 2, with slender tooth between . . . . ***C. xantiana***
          9. Petals 6–12 mm; stigma not exserted beyond anthers . . . . subsp. ***parviflora***
          9′ Petals 12–20 mm; stigma exserted beyond anthers . . . . subsp. ***xantiana***

8′ Petals entire
  10. Corolla bowl-shaped; petals not clawed
    11. Seeds brown; s SCoRO (San Luis Obispo Co. and s), SCo, WTR, nw PR . . . . *C. bottae*
    11′ Seeds gray; n SCoRO (Monterey Co.) . . . . *C. jolonensis*
  10′ Corolla rotate; petals clawed
    12. Petal claw < blade . . . . *C. delicata*
    12′ Petal claw ≥ blade
      13. Calyx and ovary puberulent and with longer, straight, spreading hairs < 3 mm . . . . *C. unguiculata*
      13′ Calyx and ovary sparsely to densely puberulent, longer straight spreading hairs 0
        14. Lvs bright green, not glaucous . . . . *C. exilis*
        14′ Lvs gray-green or ± red, glaucous
          15. Sepals gen dark red-purple — stigma exserted beyond anthers . . . . *C. springvillensis*
          15′ Sepals green, red-tinged or not . . . . *C. tembloriensis*
            16. Petals ± 10 mm wide; stigma not exserted beyond anthers . . . . subsp. *calientensis*
            16′ Petals gen < 10 mm wide; stigma exserted beyond anthers or not . . . . subsp. *tembloriensis*
7′ Buds erect; corolla bowl-shaped; anthers alike
  17. Sepals staying fused in 4s
    18. Ovary and immature fr 8-grooved
      19. Petals 5–15 mm; immature fr cylindric . . . . *C. affinis*
      19′ Petals 30–60 mm; immature fr fusiform . . . . *C. amoena* subsp. *whitneyi*
    18′ Ovary and immature fr 4-grooved
      20. Petals with red spot or marks near middle . . . . *C. amoena*
        21. St decumbent; infl dense; internode gen < subtending fl . . . . subsp. *amoena*
        21′ St erect; infl open; internode gen > subtending fl . . . . subsp. *huntiana*
      20′ Petals with ± red spot or zone at base
        22. Petals 5–13 mm; stigma not exserted beyond anthers . . . . *C. franciscana*
        22′ Petals 10–30 mm; stigma exserted beyond anthers . . . . *C. rubicunda*
  17′ Sepals all coming free or staying fused in 2s
    23. Stigma not exserted beyond anthers; petals 5–15 mm
      24. St gen erect; lvs gen linear to lanceolate, acute . . . . *C. purpurea* subsp. *quadrivulnera*
      24′ St prostrate to decumbent; lvs elliptic to obovate, tip gen obtuse
        25. Petals 5–11 mm, spot 0 . . . . *C. davyi*
        25′ Petals 10–15 mm, ± red spot above base . . . . *C. prostrata*
    23′ Stigma exserted beyond anthers; petals gen 15–30 mm
      26. Sepals free at tip in bud . . . . *C. williamsonii*
      26′ Sepals fused to tip in bud
        27. Infl dense; ovary gen > distal internode
          28. Hypanthium conspicuously veined; petals with a large wedge-shaped purple-red spot distally . . . *C. imbricata*
          28′ Hypanthium not conspicuously veined; petal spot near middle
            29. Lvs broadly lanceolate to elliptic or ovate, glabrous to sparsely puberulent . . . *C. purpurea* subsp. *purpurea*
            29′ Lvs lanceolate, puberulent . . . . *C. speciosa* subsp. *nitens*
        27′ Infl open; ovary gen < distal internode
          30. Petals lacking red or red-purple spot . . . . *C. speciosa* subsp. *immaculata*
          30′ Petals with red or red-purple spot
            31. Petal spot distal (darker proximally or not) . . . . *C. purpurea* subsp. *viminea*
            31′ Petal spot near middle
              32. Fls many on well-developed branches; s SNF, Teh. . . . . *C. speciosa* subsp. *polyantha*
              32′ Fls few on well-developed branches; SCoR . . . . *C. speciosa* subsp. *speciosa*
6′ Axis of infl in bud recurved at tip; buds pendent
  33. Petal claw broad, 2-lobed
    34. Stigma not exserted beyond anthers; petals 6–12 mm
      35. Petals gen spotted; pollen blue-gray . . . . *C. rhomboidea*
      35′ Petals unspotted; pollen yellow . . . . *C. stellata*
    34′ Stigma exserted beyond anthers; petals gen > 12 mm
      36. Axis of infl in bud recurved within 3 nodes of open fl; petal length 1.4–1.6 × width . . . . *C. mildrediae*
        37. Anthers yellow to orange-red or red; fresh pollen yellow to tan . . . . subsp. *lutescens*
        37′ Anthers magenta; fresh pollen blue-gray . . . . subsp. *mildrediae*
      36′ Axis of infl straight 4 or more nodes distal to open fl; petal length 1.6–3 × width
        38. Fl buds fusiform, tip acute . . . . *C. borealis*
          39. Seed 1.8–2.5 mm; se KR, CaRF . . . . subsp. *arida*
          39′ Seed 1.5–1.8 mm; KR . . . . subsp. *borealis*
        38′ Fl buds obovate, tip obtuse
          40. Petal length 1.6–2 × width . . . . *C. mosquinii*
          40′ Petal length 1.9–3 × width
            41. Lvs lance-linear . . . . *C. australis*
            41′ Lvs elliptic to ovate . . . . *C. virgata*

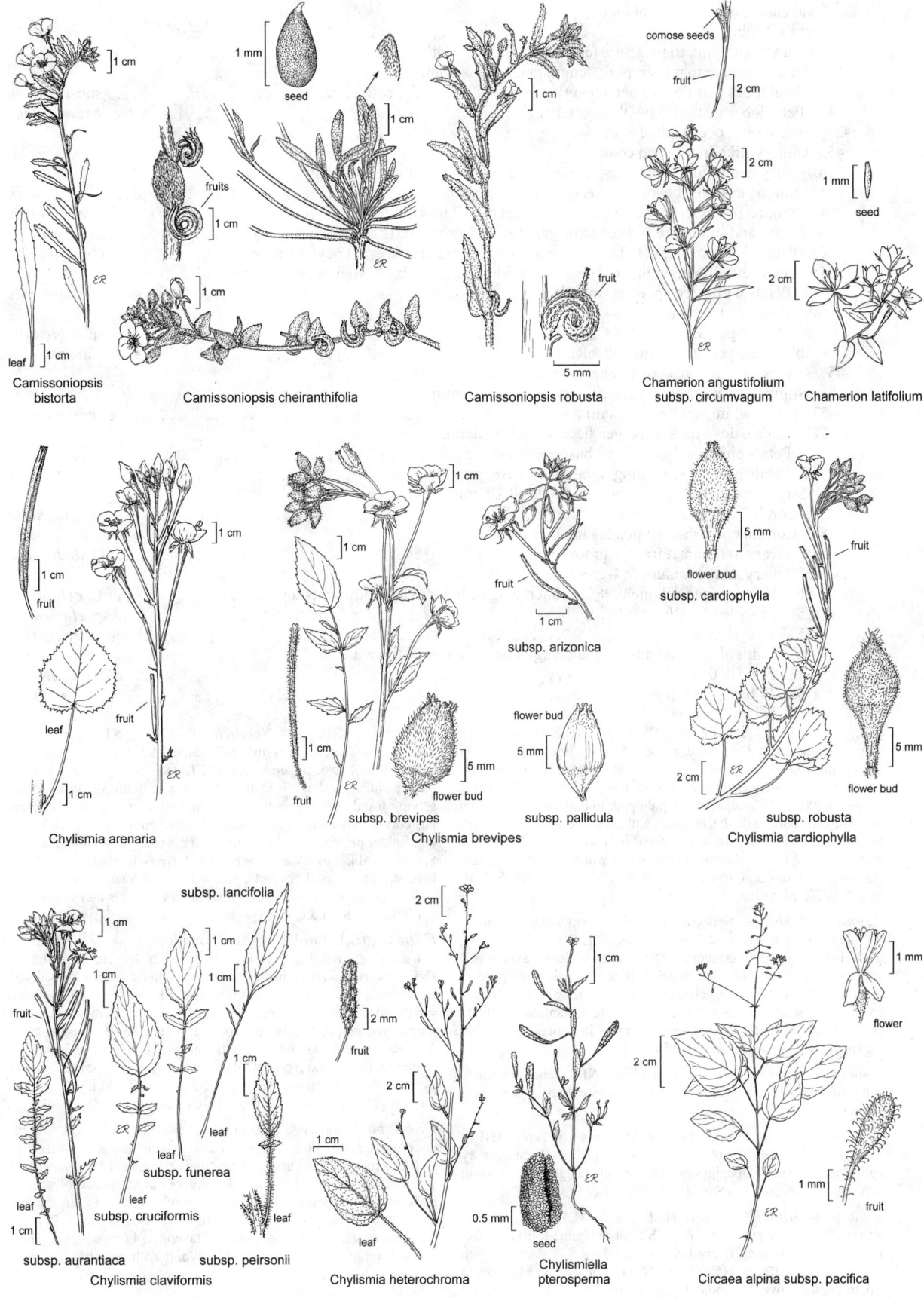

Camissoniopsis
bistorta

Camissoniopsis cheiranthifolia

Camissoniopsis robusta

Chamerion angustifolium
subsp. circumvagum

Chamerion latifolium

seed

fruits

fruit

comose seeds

fruit

seed

Chylismia arenaria

leaf

fruit

fruit

subsp. arizonica

fruit

subsp. brevipes

Chylismia brevipes

flower bud

subsp. pallidula

flower bud

subsp. cardiophylla

fruit

flower bud

subsp. robusta

Chylismia cardiophylla

subsp. lancifolia

subsp. funerea

subsp. cruciformis

subsp. peirsonii

leaf

leaf

leaf

leaf

leaf

fruit

subsp. aurantiaca

Chylismia claviformis

fruit

leaf

Chylismia heterochroma

seed

Chylismiella
pterosperma

flower

fruit

Circaea alpina subsp. pacifica

33′ Petal claw 0 or < 2 mm, not lobed
  42. Petals 2-lobed . . . . *C. biloba*
    43. Petals bright pink to magenta, length gen > 1.5 × width . . . . subsp. *australis*
    43′ Petals lavender to purple-pink, length gen < 1.5 × width
      44. Petal lobes gen 1/5–1/2 petal length . . . . subsp. *biloba*
      44′ Petal lobes gen < 1/5 petal length . . . . subsp. *brandegeeae*
  42′ Petals entire, occ notched at tip
    45. Anthers alike in size and color
      46. Ovary and immature fr conspicuously 8-grooved; fr not wider distally
        47. Stigma exserted beyond anthers; pedicel 5–15 mm . . . . *C. arcuata*
        47′ Stigma not exserted beyond anthers; pedicel 0–3 mm . . . . *C. lassenensis*
      46′ Ovary and immature fr 4- or rarely shallowly 8-grooved; fr gen wider distally . . . . *C. gracilis*
        48. Petals 6–23 mm, spot 0, base darker or not; stigma not exserted beyond anthers . . . . subsp. *gracilis*
        48′ Petals 20–40 mm with a red spot near middle or red at base; stigma exserted beyond anthers
          49. Petals with a red spot near middle . . . . subsp. *sonomensis*
          49′ Petals red at base
            50. Petals gen 30–40 mm; s CaRF . . . . subsp. *albicaulis*
            50′ Petals gen 20–30 mm; NCoRI . . . . subsp. *tracyi*
    45′ Anthers in 2 unlike series, inner smaller, paler
      51. Stigma not exserted beyond anthers; petals 5–12 mm
        52. Petals white, not flecked, fading pink . . . . *C. epilobioides*
        52′ Petals pale to dark pink, gen flecked, fading darker
          53. Petals pink, not lighter near base, gen darker flecked . . . . *C. modesta*
          53′ Petals pale pink, shading lighter near base, purple flecked . . . . *C. similis*
      51′ Stigma exserted beyond anthers; petals 10–35 mm
        54. Corolla rotate; petals oblanceolate . . . . *C. lingulata*
        54′ Corolla bowl-shaped; petals fan-shaped
          55. Ovary and immature fr 8-grooved . . . . *C. dudleyana*
          55′ Ovary and immature fr 4-grooved
            56. Width of outer filaments ± 2 × inner; ring of hairs in hypanthium below rim . . . . *C. cylindrica*
              57. Fr wider distally . . . . subsp. *clavicarpa*
              57′ Fr cylindric . . . . subsp. *cylindrica*
            56′ Width of all filaments ± equal; ring of hairs in hypanthium at rim
              58. Fr beak 0–3 mm . . . . *C. lewisii*
              58′ Fr beak 7–15 mm . . . . *C. rostrata*

C. affinis H. Lewis & M. Lewis (p. 939) **ST**: erect, < 8 dm, puberulent. **LF**: petiole < 3 mm or 0; blade 1.5–7 cm, linear to narrowly lanceolate. **INFL**: axis in bud straight; buds erect. **FL**: hypanthium 1.5–4 mm, not slender; sepals staying fused in 4s; corolla bowl-shaped, petals 5–15 mm, obovate, pale pink to dark wine-red, often purple-flecked or -marked; stamens 8, anthers alike; ovary 8-grooved, length > 9 × width, stigma not exserted beyond anthers. **FR**: 1.5–3 cm; beak 3–7 mm, cylindric and grooved when immature. *n*=26. Openings in woodland, chaparral; < 500 m. NCoRI, c SNF, SnFrB, SCoR, WTR. May–Jun

C. amoena (Lehm.) A. Nelson & J.F. Macbr. FAREWELL-TO-SPRING **ST**: decumbent to erect, < 1 m, puberulent. **LF**: petiole < 1 cm; blade 1–6 cm, linear to lanceolate. **INFL**: open or dense; axis in bud straight; buds erect. **FL**: hypanthium 3–10 mm; sepals staying fused in 4s; corolla bowl-shaped, petals obovate to fan-shaped, pink to lavender, rarely white, red spot or marks near middle; stamens 8, anthers alike; stigma exserted beyond anthers. *n*=7. Intermediates among subspp. frequent. Other subspp. to BC.

  subsp. ***amoena*** (p. 939) **ST**: decumbent. **INFL**: dense; internode gen < subtending fl. **FL**: petals 20–35 mm; ovary cylindric, grooves 4. Open slopes, bluffs on or near coast; < 100 m. NCo, n CCo. Jun–Aug

  subsp. ***huntiana*** (Jeps.) H. Lewis & M. Lewis **ST**: erect. **INFL**: open; internode gen > subtending fl. **FL**: petals 15–30 mm; ovary cylindric, grooves 4. Openings in forest, woodland, gen not on coast; < 700 m. KR, NCoRO, n SnFrB; OR. Jun–Aug

  subsp. ***whitneyi*** (A. Gray) H. Lewis & M. Lewis (p. 939) WHITNEY'S FAREWELL-TO-SPRING **ST**: decumbent. **INFL**: dense. **FL**: petals 30–60 mm; ovary fusiform, grooves 8, 4 deeper. Open coastal scrub; < 100 m. NCo (formerly n of Fort Bragg, Mendocino Co. to Shelter Cove, Humboldt Co.). Reportedly only 1 wild population remains. Jul–Aug ★

C. arcuata (Kellogg) A. Nelson & J.F. Macbr. **ST**: erect, < 8 dm; hairs 0 or scattered, spreading. **LF**: sessile, 1.5–6 cm, linear to narrowly lanceolate or oblanceolate. **INFL**: axis in bud recurved at tip; buds pendent; pedicel 5–15 mm. **FL**: hypanthium 3–7 mm; sepals staying fused in 4s, puberulent, hairs spreading, some short, glandular; corolla bowl-shaped, petals 10–30 mm, fan-shaped, pink-lavender, lighter proximally, often with a dark ± red spot at base; stamens 8, anthers alike; ovary 8-grooved, gen ± sparsely glandular-puberulent, stigma exserted beyond anthers. **FR**: 1–3.5 cm; beak < 7 mm. *n*=7. Openings in woodland, chaparral, drying soils incl serpentine; < 1700 m. CaRF, n&c SN. Apr–Jun

C. australis E. Small SMALL'S SOUTHERN CLARKIA **ST**: erect, < 1 m, puberulent. **LF**: petiole 1–3 cm; blade 2–5 cm, lance-linear. **INFL**: axis in bud straight 4 or more nodes distal to open fl, recurved at tip; buds pendent, obovate, tip obtuse. **FL**: hypanthium 2–4 mm; sepals all coming free; corolla rotate, petals 8–15 mm, lavender-purple, red-purple-mottled or -spotted, length 1.9–3 × width, claw 2-lobed, blade ± diamond-shaped; stamens 8, subtended by ciliate scales, anthers alike, pollen blue-gray; ovary 4-grooved, stigma exserted beyond anthers. **SEED**: 1–1.5 mm. *n*=5. Yellow-pine forest; 800–1500 m. c SNF, n&c SNH. Jun–Jul ★

C. biloba (Durand) A. Nelson & J.F. Macbr. **ST**: erect, < 1 m, puberulent. **LF**: petiole < 1.5 cm; blade 2–6 cm, linear to lanceolate. **INFL**: axis in bud recurved at tip; buds pendent. **FL**: hypanthium 1–4 mm; sepals staying fused in 4s, pink or purple-red; corolla rotate to bowl-shaped, petals 10–25 mm, narrowly wedge- to fan-shaped, lavender to bright pink or magenta, often red-flecked, lobes 2, tooth between 0; stamens 8, outer anthers lavender, inner smaller, paler; ovary 8-grooved, stigma exserted beyond anthers. *n*=8. Subspp. intergrade.

subsp. ***australis*** H. Lewis & M. Lewis (p. 939) MARIPOSA CLARKIA **LF:** linear to narrowly lanceolate. **FL:** petals bright pink to magenta, length gen > 1.5 × width, lobes gen 1/5–1/3 petal. Chaparral, woodland; 300–500 m. c SNF (Merced River drainage, Tuolumne, Mariposa cos.). May–Jun ★

subsp. ***biloba*** (p. 939) **LF:** narrowly lanceolate. **FL:** petals purple-pink, length gen < 1.5 × width, lobes gen 1/5–1/2 petal. Foothill woodland, serpentine or not; < 1000 m. n&c SNF, e SnFrB. May–Aug

subsp. ***brandegeeae*** (Jeps.) H. Lewis & M. Lewis (p. 939) BRANDEGEE'S CLARKIA **LF:** lanceolate. **FL:** petals lavender, length gen < 1.5 × width, lobes gen < 1/5 petal, occ obscure. Foothill woodland; < 500 m. n SNF. [*Clarkia biloba* subsp. *brandegeae*, orth. var.] Jun–Jul ★

C. borealis E. Small **ST:** erect, < 1 m, puberulent. **LF:** petiole 1.5–5 cm; blade 2–5 cm, elliptic to ovate. **INFL:** axis in bud straight 4 or more nodes distal to open fls, recurved at tip; buds pendent, fusiform, tip acute. **FL:** hypanthium 2–4 mm; sepals all coming free; corolla rotate, petals 13–19 mm, lavender-pink, often dark-flecked, length 1.6–2 × width, claw 2-lobed, blade triangular to semicircular; stamens 8, subtended by ciliate scales, anthers alike, pollen blue-gray; ovary 4-grooved, stigma exserted beyond anthers. *n*=7.

subsp. ***arida*** E. Small (p. 939) SHASTA CLARKIA **SEED:** 1.8–2.5 mm. Foothill woodland; 500 m. se KR, CaRF. Jun–Jul ★

subsp. ***borealis*** NORTHERN CLARKIA **SEED:** 1.5–1.8 mm. Foothill woodland, forest margin; 400–800 m. KR (e Trinity, w Shasta cos.). Jun–Jul ★

C. bottae (Spach) H. Lewis & M. Lewis (p. 939) PUNCHBOWL GODETIA **ST:** erect, < 1 m, glabrous (puberulent), glaucous. **LF:** petiole < 5 mm; blade 3–10 cm, narrowly lanceolate to lanceolate. **INFL:** axis in bud straight; buds reflexed. **FL:** hypanthium 2–3 mm; sepals staying fused in 4s; corolla bowl-shaped, petals 15–30 mm, fan-shaped, pale lavender to pink-lavender, often white toward base, gen red-flecked; stamens 8, outer anthers lavender, inner << outer, paler; stigma exserted beyond anthers. **SEED:** brown. *n*=9. Openings in chaparral, woodland, coastal scrub; < 1000 m. s SCoRO, SCo, WTR, nw PR. Apr–Jul

C. breweri (A. Gray) Greene BREWER'S CLARKIA **ST:** decumbent or erect, < 2 dm, glabrous or sparsely puberulent. **LF:** petiole < 2 cm; blade 2–5 cm, linear to lanceolate. **FL:** hypanthium 20–35 mm, slender, ring of hairs within 0; sepals staying fused in 4s, not petal-like, green to magenta; corolla rotate, petals 15–25 mm, pink, length ± = width, 3-lobed, middle lobe longer, narrower, linear to oblanceolate; stamens 4, filaments wider distally; stigma exserted beyond anthers. *n*=7. Openings in woodland, chaparral; < 1000 m. se SnFrB (Mount Hamilton Range), SCoRI. Apr–Jun ★

C. concinna (Fisch. & C.A. Mey.) Greene RED RIBBONS **ST:** erect, < 4 dm, glabrous or puberulent. **LF:** petiole 5–25 mm; blade 1–4.5 cm, lanceolate to elliptic or ovate. **FL:** hypanthium 10–25 mm, slender, ring of hairs within 0; sepals staying fused in 4s, petal-like proximally, thin distally, red; corolla rotate, petals bright pink, gen white-streaked, length ± 2 × width, lobes 3, ± equally wide, middle ≥ others, gen oblanceolate; stamens 4, filaments cylindric. *n*=7.

subsp. ***automixa*** R.N. Bowman SANTA CLARA RED RIBBONS **FL:** sepals staying fused only near tip; petals 10–20 mm, lobes prominent, separated by deep sinuses; stigma not exserted beyond anthers. Woodland; < 1500 m. SnFrB. Apr–Jun ★

subsp. ***concinna*** (p. 939) **FL:** sepals staying fused only near tip; petals 15–30 mm, lobes prominent, separated by deep sinuses; stigma exserted beyond anthers. Mixed-evergreen forest, woodland, coastal scrub; < 1500 m. NW, n SNF, e SnFrB (Oakland Hills, Mount Diablo). Apr–Jul

subsp. ***raichei*** G.A. Allen et al. RAICHE'S RED RIBBONS **FL:** sepals staying fused in distal 2/3; petals 10–20 mm, lobes short, not separated by deep sinuses; stigma not exserted beyond anthers. Exposed sites; < 100 m. n CCo (only at type locality near Tomales). Apr–May ★

C. cylindrica (Jeps.) H. Lewis & M. Lewis **ST:** erect, < 6 dm, puberulent or glabrous. **LF:** petiole < 5 mm; blade 1–6 cm, linear to narrowly lanceolate. **INFL:** axis in bud recurved at tip; buds pendent. **FL:** hypanthium 2–7 mm, ring of hairs below rim; sepals staying fused in 4s, red-purple; corolla bowl-shaped, petals 10–35 mm, fan-shaped, base bright purple-red, middle ± white, distally purple to pink-lavender, middle and distally often red-purple-flecked; stamens 8, width of outer filaments ± 2 × inner, outer anthers lavender, inner smaller, paler; ovary 4-grooved, stigma exserted beyond anthers. **FR:** 2–5 cm, cylindric or wider distally; beak 3–5 mm. *n*=9. Subspp. intergrade.

subsp. ***clavicarpa*** W.S. Davis (p. 939) **FL:** hypanthium length > 2 × ovary width; fr wider distally. Open grassland, woodland, chaparral; < 1000 m. SNF, Teh. Apr–Jun

subsp. ***cylindrica*** (p. 939) **FL:** hypanthium length ± = ovary width; fr cylindric. Open grassland, woodland, chaparral; < 1000 m. SCoR, WTR. Apr–Jun

C. davyi (Jeps.) H. Lewis & M. Lewis **ST:** prostrate to decumbent, < 9 dm, sparsely puberulent. **LF:** ± sessile; blade 1–2.5 cm, oblanceolate to elliptic or obovate, tip gen obtuse. **INFL:** axis in bud straight; buds erect. **FL:** hypanthium 2–5 mm; sepals all coming free or staying fused in 2s; corolla bowl-shaped, petals 5–11 mm, fan-shaped to obovate, lavender-pink shading in middle to white or pale yellow, spot 0; stamens 8, anthers alike; ovary 8-grooved, stigma not exserted beyond anthers. *n*=17. Coastal grassland, bluffs; < 100 m. NCo, n CCo, n ChI (Santa Rosa Island). Apr–Jun

C. delicata (Abrams) A. Nelson & J.F. Macbr. (p. 939) DELICATE CLARKIA **ST:** erect, < 7 dm; glabrous, ± glaucous distally, gen puberulent proximally. **LF:** petiole < 1 cm; blade 1.5–4 cm, lanceolate to elliptic or ovate. **INFL:** axis in bud straight; buds reflexed. **FL:** hypanthium ± 2 mm; sepals staying fused in 4s; corolla rotate, petals 8–12 mm, oblanceolate to obovate, rose-lavender to pale pink, claw < blade, tapered, entire; stamens 8, outer anthers orange-red, inner smaller, paler; ovary 8-grooved, stigma not exserted beyond anthers. *n*=18. Oak woodland, chaparral; < 1000 m. s PR (San Diego Co.); n Baja CA. Probably from hybrids between *C. unguiculata*, *C. epilobioides*. Apr–May ★

C. dudleyana (Abrams) J.F. Macbr. (p. 939) **ST:** erect, < 7 dm, puberulent. **LF:** petiole < 1 cm; blade 1.5–7 cm, narrowly lanceolate to lanceolate. **INFL:** axis in bud recurved at tip; buds pendent. **FL:** hypanthium 1–3 mm; sepals staying fused in 4s, gen pink or purple-red; corolla bowl-shaped, petals 10–30 mm, fan-shaped, lavender-pink, gen white-streaked, often red-flecked; stamens 8, outer anthers lavender, inner smaller, paler; ovary 8-grooved, stigma exserted beyond anthers. *n*=9. Openings in woodland, chaparral, yellow-pine forest; < 1500 m. c&s SNF, Teh, TR. May–Jul

C. epilobioides (Torr. & A. Gray) A. Nelson & J.F. Macbr. **ST:** erect, < 7 dm, sparsely puberulent. **LF:** petiole < 7 mm; blade 15–25 mm, linear to narrowly lanceolate or oblanceolate. **INFL:** axis in bud recurved at tip; buds pendent. **FL:** hypanthium 0.5–1.5 mm; sepals staying fused in 4s or 2s, often red; corolla bowl-shaped, petals 5–12 mm, obovate, entire, white, fading pink, flecks 0; stamens 8, inner anthers smaller, paler; ovary cylindric, not obviously grooved, stigma not exserted beyond anthers. *n*=9. Gen shady sites in woodland, chaparral; < 1000 m. CW, SW; to AZ, n Baja CA. Apr–May

C. exilis H. Lewis & Vasek SLENDER CLARKIA **ST:** erect, < 1 m, glabrous, glaucous. **LF:** petiole < 5 mm or 0; blade 1–6 cm, lanceolate to narrowly elliptic, glabrous, not glaucous, bright green. **INFL:** axis in bud straight; buds reflexed. **FL:** hypanthium 1–3 mm; sepals staying fused in 4s, gen green, sparsely to densely puberulent; corolla rotate, petals 5–15 mm, lavender, pink, or white, often with dark ± purple spot, claw ≥ blade, slender, blade gen diamond-shaped; stamens 8, outer anthers gen ± red, inner smaller, paler; ovary 8-grooved, hairs as on sepals, stigma ± not beyond anthers. **FR:** 1.5–3 cm, ± 2 mm wide. *n*=9. Woodland; < 1000 m. s SNF, Teh. Apr–May ★

C. franciscana H. Lewis & P.H. Raven (p. 939) PRESIDIO CLARKIA **ST:** erect, < 4 dm, puberulent. **LF:** petiole < 5 mm; blade 1–3.5 cm, narrowly lanceolate. **INFL:** axis in bud straight; buds erect. **FL:** hypanthium 1–3 mm; sepals staying fused in 4s; corolla

bowl-shaped, petals 5–13 mm, wedge-shaped, ± truncate at tip, lavender-pink shading to white near middle, base ± red; stamens 8, anthers alike; ovary 4-grooved, stigma not exserted beyond anthers. $n=7$. Serpentine soil; ± 50 m. CCo (Presidio, San Francisco), SnFrB (Oakland hills). May–Jun ★

C. gracilis (Piper) A. Nelson & J.F. Macbr. **ST:** erect, < 9 dm; hairs 0 to dense. **LF:** petiole < 1 cm; blade 2–7 cm, linear to narrowly lanceolate. **INFL:** axis in bud recurved at tip; buds pendent. **FL:** hypanthium 1.5–10 mm; sepals staying fused in 4s; corolla bowl-shaped, petals obovate to fan-shaped, pink-lavender, often lighter toward base, spot 0 or 1 near middle or base, red; stamens 8, anthers alike; ovary 4- or rarely shallowly 8-grooved, puberulent. **FR:** 2.5–5 cm, gen wider distally; beak slender, < 10 mm. $n=14$. From hybrids between *C. amoena* and a probably extinct relative of *C. lassenensis*.

subsp. *albicaulis* (Jeps.) H. Lewis & M. Lewis (p. 939) WHITE-STEMMED CLARKIA **FL:** petals gen 30–40 mm, pink-lavender to light purple shading to white near middle, base red; stigma exserted beyond anthers. Foothill woodland (known only from 3–4 small populations); ± 500 m. s CaRF (Butte Co.). Local. May–Jul ★

subsp. *gracilis* (p. 939) **FL:** petals 6–23 mm, ± pink to lavender, spot 0, base darker or not; stigma not exserted beyond anthers. Common. Openings in woodland, forest; < 1500 m. NW, n SNF, ScV, SnFrB, MP; to WA. Apr–Jul

subsp. *sonomensis* (C.L. Hitchc.) H. Lewis & M. Lewis (p. 939) **FL:** petals 20–40 mm, pink-lavender shading to white proximally, with red spot near middle; stigma exserted beyond anthers. Openings in woodland, forest; < 500 m. KR, NCoR, n SnFrB; s OR. May–Jun

subsp. *tracyi* (Jeps.) Abdel-Hameed & R. Snow TRACY'S CLARKIA **FL:** petals gen 20–30 mm, pink-lavender, base red; stigma exserted beyond anthers. Gen serpentine soil; 100–500 m. NCoRI. May–Jul ★

C. heterandra (Torr.) H. Lewis & P.H. Raven (p. 939) **ST:** erect, < 6 dm, glandular-puberulent. **LF:** petiole 5–20 mm; blade 2–8 cm, lanceolate to ovate. **INFL:** axis in bud straight. **FL:** hypanthium 1–2 mm; sepals staying fused in 4s; corolla rotate, petals 5–6 mm, elliptic, pink; stamens 8, inner anthers smaller, sterile; stigma not exserted beyond anthers. **FR:** 2–3 mm, nut-like, indehiscent. **SEED:** 1–2. $n=9$. Shady sites, woodland, yellow-pine forest; 500–1700 m. KR, NCoRI, SN, s SCoRO, TR. May–Jul

C. imbricata H. Lewis & M. Lewis (p. 939) VINE HILL CLARKIA **ST:** erect, < 6 dm, glabrous or sparsely puberulent. **LF:** petiole 0–2 mm; blade 2–2.5 cm, lanceolate. **INFL:** dense; axis in bud straight; buds erect. **FL:** hypanthium 10–15 mm, conspicuously veined; sepals fused to tip in bud, all coming free; corolla bowl-shaped, petals 20–25 mm, fan-shaped, lavender shading paler near middle, with large wedge-shaped purple-red spot distally; stamens 8, anthers alike; ovary > distal internode, 8-grooved, stigma exserted beyond anthers. $n=8$. Clearings, roadsides, perhaps chaparral; ± 50 m. NCoRO (Sonoma Co.). Jun–Jul ★

C. jolonensis D.R. Parn. JOLON CLARKIA **ST:** erect, < 6 dm, glabrous, glaucous. **LF:** petiole < 5 mm; blade 3–5 cm, narrowly lanceolate to lanceolate. **INFL:** axis in bud straight; buds reflexed. **FL:** hypanthium 2–3 mm; sepals staying fused in 4s; corolla bowl-shaped, petals 1–2 cm, fan-shaped, pale lavender to pink-lavender, gen red-flecked; stamens 8, outer anthers lavender, inner smaller, paler; stigma exserted beyond anthers. **SEED:** gray. $n=9$. Dry woodland; ± 500 m. n SCoRO (near Jolon, Monterey Co.). Apr–Jun ★

C. lassenensis (Eastw.) H. Lewis & M. Lewis (p. 939) **ST:** erect, < 9 dm, puberulent. **LF:** petiole < 1 cm; blade 2–5 cm, linear to narrowly lanceolate. **INFL:** axis in bud recurved at tip; buds pendent; pedicel 0–3 mm. **FL:** hypanthium 3–5 mm; sepals staying fused in 4s; corolla bowl-shaped, petals 8–16 mm, obovate to fan-shaped, pink-lavender shading lighter toward red-purple base; stamens 8, anthers alike; ovary 8-grooved, hairs dense, stigma not exserted beyond anthers. **FR:** 2.5–4 cm, not wider distally, beak ± 2 mm. $n=7$. Woodland, conifer forest; 500–2000 m. KR, NCoRI, CaR, n SN, MP; OR, NV. May–Jun

C. lewisii P.H. Raven & D.R. Parn. (p. 939) LEWIS' CLARKIA **ST:** erect, < 5 dm, puberulent or glabrous. **LF:** petiole < 7 mm;

blade 2–5 cm, narrowly lanceolate to lanceolate. **INFL:** axis in bud recurved at tip; buds pendent. **FL:** hypanthium 1.5–4 mm, ring of hairs at rim; sepals staying fused in 4s, pink to red-purple; corolla bowl-shaped, petals 10–30 mm, fan-shaped, pink-lavender shading to white near middle, base purple-red or with a red line, often red-purple-flecked; stamens 8, filaments alike, outer anthers lavender, inner smaller, paler; ovary 4-grooved, stigma exserted beyond anthers. **FR:** 1.5–7 cm, cylindric; beak 0–3 mm. $n=9$. Coastal scrub, woodland, chaparral; < 300 m. CCo, n SCoR (Monterey, San Benito cos.). May–Jun ★

C. lingulata H. Lewis & M. Lewis (p. 939) MERCED CLARKIA **ST:** erect, < 6 dm, puberulent. **LF:** petiole < 1.5 cm; blade 2–6 cm, linear to narrowly lanceolate. **INFL:** axis in bud recurved at tip; buds pendent. **FL:** hypanthium 1–4 mm; sepals staying fused in 4s, pink or purple-red; corolla rotate, petals 1–2 cm, oblanceolate, bright pink, red-flecked or not; stamens 8, outer anthers lavender, inner << outer, paler; ovary 8-grooved, stigma exserted beyond anthers. $n=9$. Open chaparral, steep n-facing slopes; 400–450 m. c SNF (2 sites, Merced River Canyon, Mariposa Co.). Derived from *C. biloba*. May–Jun ★

C. mildrediae (A. Heller) H. Lewis & M. Lewis (p. 939) **ST:** erect, < 1 m, puberulent. **LF:** petiole 1.5–4 cm; blade 3–6 cm, elliptic to ovate. **INFL:** axis in bud recurved within 3 nodes of open fl; buds pendent, tips acute. **FL:** hypanthium 2–3 mm; sepals gen all coming free; corolla rotate, petals 1.5–2 cm, lavender to purple-red, often darker-flecked or -spotted, length 1.4–1.6 × width, claw broad, 2-lobed, blade triangular to semicircular; stamens 8, subtended by ciliate scales, anthers alike; ovary 4-grooved, stigma exserted beyond anthers. $n=7$.

subsp. *lutescens* Gottlieb & Janeway GOLDEN-ANTHERED CLARKIA **FL:** petal blade 7–16 mm wide; anthers yellow to orange-red or red; fresh pollen of yellow anthers bright yellow, for anthers of other colors, pollen yellow to tan. Yellow-pine forest; 450–1700 m. n SN (Feather River drainage, se of subsp. *mildrediae*). Jun–Jul ★

subsp. *mildrediae* MILDRED'S CLARKIA **FL:** petal blade 12–18 mm wide; anthers magenta; fresh pollen blue-gray. Yellow-pine forest; 450–1700 m. s CaR, n SN (Feather River drainage, nw of subsp. *lutescens*). Jun–Jul ★

C. modesta Jeps. **ST:** erect, < 7 dm, puberulent. **LF:** petiole < 15 mm; blade 2–4 cm, linear to narrowly lanceolate or elliptic. **INFL:** axis in bud recurved at tip; buds pendent. **FL:** hypanthium 1–3 mm; sepals staying fused in 4s; corolla gen rotate, petals 8–12 mm, oblanceolate to diamond-shaped, pink, not lighter near base, gen darker flecked, fading darker; stamens 8, inner anthers << outer, paler; ovary 8-grooved, stigma not exserted beyond anthers. $n=8$. Shady places in woodland; < 1000 m. NCoRI, c&s SNF, SnFrB, SCoR. Apr–May

C. mosquinii E. Small (p. 939) MOSQUIN'S CLARKIA **ST:** erect, < 1 m, puberulent. **LF:** petiole 1–3 cm; blade 2–5 cm, lance-linear to ovate or elliptic. **INFL:** axis in bud straight 4 or more nodes above open fls, recurved at tip; buds pendent, obovate, tip obtuse. **FL:** hypanthium 2–4 mm; sepals all coming free; corolla rotate, petals 10–20 mm, lavender-purple, often red-purple-spotted, length 1.6–2 × width, claw-lobed, blade ± diamond-shaped; stamens 8, subtended by ciliate scales, anthers alike, pollen blue-gray; ovary 4-grooved, stigma exserted beyond anthers. **SEED:** < 1 mm. $n=6$. Dry, rocky places, probably foothill woodland; 180–1200 m. n SNF (ne Butte Co.). [*C. m.* subsp. *xerophila* E. Small; *Clarkia mosquinii* subsp. *xerophylla*, orth. var.] Jun–Jul ★

C. prostrata H. Lewis & M. Lewis (p. 945) **ST:** prostrate to decumbent, < 5 dm, sparsely puberulent. **LF:** ± sessile; blade 1–2.5 cm, elliptic to oblanceolate, tip gen obtuse. **INFL:** axis in bud straight; buds erect. **FL:** hypanthium 3–7 mm; sepals gen staying fused in 2s; corolla bowl-shaped, petals 10–15 mm, fan-shaped to obovate, lavender-pink shading to pale yellow proximally, ± red spot distal to base; stamens 8, anthers alike; ovary 8-grooved, stigma not exserted beyond anthers. $n=26$. Coastal bluffs in grassland, closed-cone-pine forest; < 100 m. s CCo. Possibly from hybrids between *C. davyi*, *C. speciosa*. May–Jul

C. purpurea (Curtis) A. Nelson & J.F. Macbr. **ST:** decumbent to erect, < 1 m, glabrous, to densely puberulent (some hairs longer), glaucous or not. **LF:** petiole 0–2 mm; blade 1.5–7 cm, linear or

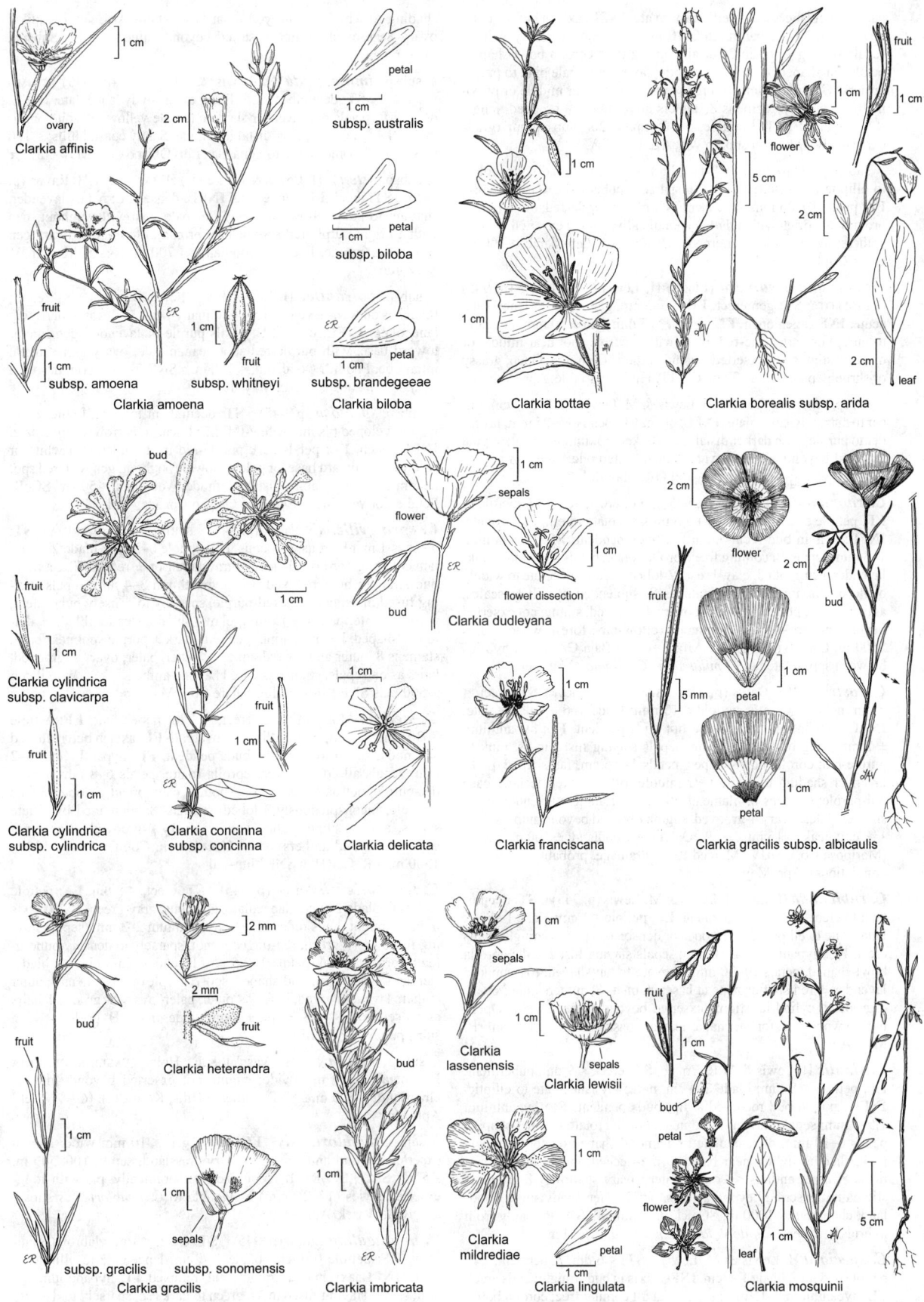

Clarkia affinis

ovary

1 cm

petal
1 cm
subsp. australis

petal
1 cm
subsp. biloba

petal
1 cm
subsp. brandegeeae

Clarkia biloba

1 cm

Clarkia bottae

fruit
1 cm

flower
1 cm

5 cm

2 cm

leaf
2 cm

Clarkia borealis subsp. arida

fruit

ER

1 cm

fruit
1 cm
subsp. amoena

fruit
1 cm
subsp. whitneyi

Clarkia amoena

bud

fruit
1 cm

Clarkia cylindrica
subsp. clavicarpa

fruit
1 cm

Clarkia cylindrica
subsp. cylindrica

1 cm

fruit
1 cm

ER

Clarkia concinna
subsp. concinna

flower

bud

1 cm
sepals

1 cm

flower dissection

Clarkia dudleyana

1 cm

Clarkia delicata

1 cm

Clarkia franciscana

2 cm

flower

fruit

5 mm

petal
1 cm

petal
1 cm

2 cm

bud

ER

Clarkia gracilis subsp. albicaulis

fruit

bud

1 cm

subsp. gracilis

ER

2 mm

2 mm

fruit

Clarkia heterandra

sepals
1 cm

subsp. sonomensis

Clarkia gracilis

bud

1 cm

ER

Clarkia imbricata

sepals
1 cm

Clarkia
lassenensis

1 cm

sepals

Clarkia lewisii

Clarkia
mildrediae

petal
1 cm

Clarkia lingulata

fruit
1 cm

bud

petal

flower

flower

leaf
1 cm

5 cm

Clarkia mosquinii

narrowly oblanceolate to elliptic or ovate. **INFL**: axis in bud straight; buds erect. **FL**: hypanthium 2–10 mm, not conspicuously veined; sepals staying fused in 2s or all coming free; corolla bowl-shaped, petals fan-shaped, obovate or elliptic, lavender or pale pink to purple or dark wine-red, often with red or purple spot near middle or proximally or distally; stamens 8, anthers alike; ovary 8-grooved, length < 8 × width. **FR**: 1–3 cm; beak 0–2 mm. *n*=26. Subspp. intergrade extensively. Other subspp. to WA, AZ, Baja CA.

subsp. ***purpurea*** (p. 945) **LF**: 1.5–4.5 cm, broadly lanceolate to elliptic or ovate, glabrous to sparsely puberulent. **INFL**: dense. **FL**: petals 10–25 mm, lavender to purple or purple-red, often lighter proximally, often with a darker spot distally; stigma exserted beyond anthers; ovary > distal internode. Uncommon. Grassland; < 100 m. GV, CCo; OR. Apr–Jun

subsp. ***quadrivulnera*** (Lindl.) H. Lewis & M. Lewis (p. 945) FOUR-SPOT **ST**: gen erect. **LF**: 1.5–5 cm, gen linear to lanceolate, acute. **INFL**: gen open. **FL**: petals ≤ 15 mm, gen ± 10 mm, lavender to purple or dark wine-red, often with a purple spot near middle or distally; stigma not exserted beyond anthers. Common. Open, grassy or shrubby places; < 1500 m. CA-FP. Highly variable. Apr–Aug

subsp. ***viminea*** (Hook.) H. Lewis & M. Lewis **LF**: 3–7 cm, linear to narrowly lanceolate. **INFL**: open. **FL**: petals 15–25 mm, lavender to purple, with darker distal spot, darker proximally or not; stigma exserted beyond anthers, ovary < distal internode. Open, grassy or shrubby places; < 1500 m. CA-FP; OR. May–Jul

C. rhomboidea Douglas (p. 945) **ST**: erect, < 1 m, puberulent. **LF**: petiole 5–25 mm; blade 1–6 cm, lanceolate to elliptic or ovate. **INFL**: axis in bud recurved at tip; buds pendent. **FL**: hypanthium 1–3 mm; sepals all coming free; corolla rotate, petals 6–12 mm, pink-lavender, gen spotted, claw broad, 2-lobed, blade lanceolate to widely ovate or diamond-shaped; stamens 8, subtended by ciliate scales, anthers alike, pollen blue-gray; ovary 4-grooved, stigma not exserted beyond anthers. *n*=12. Common. Yellow-pine forest, woodland; < 2500 m. CA-FP (exc SnJV), MP; ± w US, Baja CA. From hybrids between spp. related to *C. mildrediae, C. virgata.* Mar–Sep

C. rostrata W.S. Davis (p. 945) BEAKED CLARKIA **ST**: erect, < 6 dm, puberulent. **LF**: petiole < 10 mm; blade 1–6 cm, lanceolate. **INFL**: axis in bud recurved at tip; buds pendent. **FL**: hypanthium ± 2 mm, ring of hairs ± at rim; sepals staying fused in 4s, pink to purple-red; corolla bowl-shaped, petals 1–2.5 cm, fan-shaped, pink-lavender shading to white near middle, often darker-flecked, base red-purple; stamens 8, filaments alike, outer anthers lavender, inner smaller, paler; ovary 4-grooved, stigma exserted beyond anthers. **FR**: 1–3 cm; beak 7–15 mm. *n*=9. Oak/pine woodland; ± 500 m. c SNF (Mariposa Co.), SnJV (Merced River drainage, probably ephemeral populations). Apr–May ★

C. rubicunda (Lindl.) H. Lewis & M. Lewis (p. 945) **ST**: decumbent to erect, < 1.5 m, puberulent. **LF**: petiole < 1 cm; blade 1–4 cm, lanceolate to elliptic. **INFL**: open or dense; axis in bud straight; buds erect. **FL**: hypanthium 4–10 mm; sepals staying fused in 4s; corolla bowl-shaped, petals 10–30 mm, obovate to fan-shaped, rosy-pink to lavender, ± red spot or zone at base; stamens 8, anthers alike; ovary 4-grooved, cylindric, stigma exserted beyond anthers. *n*=7. Openings in woodland, forest, chaparral near coast; < 500 m. CCo, SnFrB. May–Aug

C. similis H. Lewis & W.R. Ernst **ST**: erect, < 9 dm, puberulent. **LF**: petiole < 8 mm; blade 2–4 cm, narrowly lanceolate to elliptic. **INFL**: axis in bud recurved at tip; buds pendent. **FL**: hypanthium 1.5–2 mm; sepals staying fused in 4s; corolla rotate to bowl-shaped, petals 6–10 mm, oblanceolate to diamond-shaped or obovate, pale pink, shading lighter near base, purple-flecked, fading darker; stamens 8, inner anthers << outer, paler; ovary shallowly 8-grooved, stigma not exserted beyond anthers. *n*=17. Gen shady sites, woodland, chaparral; < 1000 m. SCoRI (uncommon), SW. Probably from hybrids between *C. epilobioides, C. modesta.* Apr–Jun

C. speciosa H. Lewis & M. Lewis **ST**: < 6 dm, puberulent. **LF**: petiole 0–5 mm; blade 1–6 cm. **INFL**: axis in bud straight; buds erect. **FL**: hypanthium 5–15 mm; sepals gen all coming free; corolla bowl-shaped, petals 1–2.5 cm, fan-shaped, lavender to deep wine-red often

shading to white or pale yellow at base; stamens 8, anthers alike; ovary 8-grooved, stigma exserted beyond anthers. *n*=9. Subspp. intergrade.

subsp. ***immaculata*** H. Lewis & M. Lewis (p. 945) PISMO CLARKIA **ST**: decumbent. **LF**: linear to narrowly lanceolate. **INFL**: open. **FL**: petals red-lavender shading to pale yellow or white proximally, spot 0; ovary gen < distal internode. Sandy coastal hills; < 100 m. s CCo (± Pismo Beach to Edna, San Luis Obispo Co.). May–Jul ★

subsp. ***nitens*** (H. Lewis & M. Lewis) H. Lewis & P.H. Raven (p. 945) **ST**: erect. **LF**: lanceolate. **INFL**: dense. **FL**: petals lavender, shading to pale yellow proximally or pale yellow throughout, gen with a bright purple-red spot near or proximal to middle; ovary gen > distal internode. Foothill woodland; < 700 m. c&s SNF, e SnJV. May–Jun

subsp. ***polyantha*** H. Lewis & M. Lewis (p. 945) **ST**: erect; branches on well-developed pls few, many-fld. **LF**: linear to narrowly lanceolate. **INFL**: open. **FL**: petals purple or lavender, gen lighter toward base, with purple-red spot near middle; ovary gen < distal internode. Foothill woodland; < 700 m. s SNF, Teh. Intergrades with *C. speciosa* subsp. *nitens.* May–Jun

subsp. ***speciosa*** (p. 945) **ST**: decumbent to erect; branches on well-developed pls many, few-fld. **LF**: linear to narrowly lanceolate. **INFL**: open. **FL**: petals dark purple-red to lavender, gen white or pale yellow toward base, or pale yellow throughout, gen with red spot near middle; ovary gen < distal internode. Woodland; < 500 m. SCoR. Petal color variable in San Luis Obispo Co. May–Jul

C. springvillensis Vasek (p. 945) SPRINGVILLE CLARKIA **ST**: erect, < 1 m, glabrous, glaucous. **LF**: petiole 0–5 mm; blade 2–9 cm, lanceolate, glabrous, glaucous, gray-green or ± red. **INFL**: axis in bud straight; buds reflexed. **FL**: hypanthium 3–4 mm; sepals staying fused in 4s, gen dark red-purple, sparsely to densely puberulent; corolla rotate, petals ± 15 mm, claw ≥ blade, slender, blade ± diamond-shaped, lavender-pink, gen with dark ± purple spot near base; stamens 8, outer anthers red, inner << outer, paler; ovary 8-grooved, hairs as on sepals, stigma exserted beyond anthers. *n*=9. Woodland; ± 500 m. s SNF (Springville, Tulare Co.). May ★

C. stellata Mosquin **ST**: erect, < 1 m, puberulent. **LF**: petiole 5–30 mm; blade 1–5 cm, elliptic to ovate. **INFL**: axis in bud recurved < 3 nodes distal to open fls; buds pendent. **FL**: hypanthium 1.5–2 mm; sepals all coming free; corolla rotate, petals 6–8 mm, lavender-purple, not dark-flecked or -spotted, claw broad, 2-lobed, blade obovate, inconspicuously 3-lobed; stamens 8, subtended by ciliate scales, anthers alike, pollen yellow; ovary 4-grooved, stigma not exserted beyond anthers. *n*=7. Uncommon. Conifer forest; 1000–1500 m. KR, CaRH, n SNH. Jun–Jul

C. tembloriensis Vasek (p. 945) **ST**: erect, < 8 dm. **LF**: petiole 0–5 mm; blade 2–7 cm, lanceolate, glaucous, gray-green. **INFL**: axis in bud straight; buds reflexed. **FL**: hypanthium 2–3 mm; sepals staying fused in 4s, green, red-tinged or not, sparsely to densely puberulent; corolla rotate, petals 1–2.5 cm, lavender-pink, claw ≥ blade, slender, blade ± diamond-shaped, spot ± purple or 0; stamens 8, outer anthers lavender to red, inner << outer, paler; ovary 8- grooved, hairs as on sepals. **FR**: 1.5–3 cm, ± 3 mm wide. *n*=9. Hybrids between subspp. have low fertility.

subsp. ***calientensis*** (Vasek) K.E. Hols. VASEK'S CLARKIA **FL**: petals ± 10 mm wide; stigma not exserted beyond anthers. Grassland; ± 500 m. s SNF (Caliente Hills, Kern Co.). [*C. c.* Vasek] Apr–May ★

subsp. ***tembloriensis*** **FL**: petals gen < 10 mm wide; stigma exserted beyond anthers or not. Dry grassland, scrub; 100–500 m. s SNF, SnJV, SCoRI. If recognized taxonomically, pls with long-exserted styles (17–25 mm) assignable to *C. tembloriensis* subsp. *longistyla* Vasek. Apr–May

C. unguiculata Lindl. (p. 945) **ST**: erect, < 1 m, glabrous, glaucous. **LF**: petiole 0–1 cm; blade 1–6 cm, lanceolate to elliptic or ovate. **INFL**: axis in bud straight; buds reflexed. **FL**: hypanthium 2–5 mm; sepals staying fused in 4s, green to dark red, sparsely to densely puberulent and with longer, straight, spreading hairs < 3 mm; corolla

rotate, petals 1–2.5 cm, lavender-pink to salmon or dark red-purple, claw ≥ blade, slender, blade triangular or diamond-shaped to ± round; stamens 8, outer anthers red, inner << outer, paler; ovary 8-grooved, hairs as on sepals, stigma exserted beyond anthers. *n*=9. Common. Woodland; < 1500 m. NCoR, SNF, Teh, ScV (Sutter Buttes), SnFrB, SCoR, SCo, WTR, PR. Apr–Sep

C. virgata Greene (p. 945) SIERRA CLARKIA **ST:** erect, < 1 m, puberulent. **LF:** petiole 1.5–5 cm; blade 2–5 cm, elliptic to ovate. **INFL:** axis in bud straight 4 or more nodes distal to open fls, recurved at tip; buds pendent, obovate, tip obtuse. **FL:** hypanthium 2–4 mm; sepals all coming free; corolla rotate, petals 8–15 mm, lavender-purple, red-purple-mottled or -spotted, length 1.9–2.7 × width, claw broad, 2-lobed, blade ± diamond-shaped; stamens 8, subtended by ciliate scales, anthers alike, pollen blue-gray; ovary 4-grooved, stigma exserted beyond anthers. **SEED:** 1–1.5 mm. *n*=5. Yellow-pine forest, foothill woodland; 400–1100 m. n&c SN (El Dorado to Tuolumne cos.). Forms sterile hybrids with *C. australis*. Jun–Jul ★

C. williamsonii (Durand & Hilg.) H. Lewis & M. Lewis (p. 945) **ST:** erect, < 1 m, puberulent. **LF:** petiole 0–10 mm; blade 2–7 cm, linear to narrowly lanceolate. **INFL:** open; axis in bud straight; buds

erect. **FL:** hypanthium 7–13 mm; sepals all coming free or staying fused in 2s, free at tip in bud; corolla bowl-shaped, petals 10–30 mm, fan-shaped, lavender, white near middle, with purple spot distally, rarely uniformly wine-red; stamens 8, anthers alike; ovary 8-grooved, stigma exserted beyond anthers. *n*=9. Foothill woodland, yellow-pine forest; 400–2000 m. n&c SN, s SNF, Teh. Apr–Sep

C. xantiana A. Gray **ST:** erect, < 8 dm, glabrous, glaucous. **LF:** petiole 0–2 mm; blade 2–6 cm, linear to lanceolate. **INFL:** axis in bud straight; buds reflexed. **FL:** hypanthium 2–5 mm; sepals staying fused in 4s; corolla rotate, petals lavender to red-purple, clawed, blade lobes 2, with slender, 1–3 mm tooth between, upper petals gen with white-surrounded spot; stamens 8, outer anthers lavender to purple, inner anthers << outer, paler; ovary 8-grooved. *n*=9.

subsp. ***parviflora*** (Eastw.) H. Lewis & P.H. Raven KERN CANYON CLARKIA **FL:** petals 6–12 mm, lavender-pink or white; stigma not exserted beyond anthers. Dry slopes; 1000–1500 m. s SN (esp Kern River drainage). May–Jun ★

subsp. ***xantiana*** (p. 945) **FL:** petals 12–20 mm, lavender to red-purple; stigma exserted beyond anthers. Foothill woodland; 500–2000 m. s SN (esp Kern River drainage), Teh, WTR. May–Aug

EPILOBIUM WILLOWHERB

Peter C. Hoch

Ann to subshrub. **LF:** gen opposite proximally (or clustered in axils), gen ± fine-toothed; veins gen obscure. **INFL:** gen raceme, bracted. **FL:** radial or rarely ± bilateral; sepals 4, erect; petals 4, notched; stamens 8, anthers attached at middle, pollen grains gen shed in 4s, cream-yellow; ovary chambers 4, stigma gen club-like, occ 4-lobed. **FR:** straight, cylindric to club-like. **SEED:** gen in 1 row per chamber, gen with white, deciduous hair-tuft. 165 spp.: worldwide exc trop. (Greek: upon pod, from inferior ovary) [Raven 1976 Ann Missouri Bot Gard 63:326–340; Wagner et al. 2007 Syst Bot Monogr 83:81–95] Incl *Boisduvalia*, *Zauschneria*. Most taxa polyploid; many with anthers ± = stigma self-pollinated; many hybrids. Taxa with alternate lvs moved to *Chamerion*.

1. Ann from taproot; lvs opposite only near base, alternate or clustered distally, ± narrowly lanceolate; proximal st gen peeling
   2. Seed hair-tuft 0
      3. Fr wall tough, gen splitting only on distal 1/3; seeds in 2 rows per chamber; basal branches gen prostrate or decumbent
         4. Fr 8–12 mm, sharply 4-angled. . . . . *E. cleistogamum*
         4′ Fr 3.5–8 mm, cylindric . . . . *E. campestre*
      3′ Fr wall flexible, splitting to base; seeds in 1 row per chamber or compressed into 1 staggered row per capsule; basal branches ascending or 0
         5. Fr not beaked, central axis persistent . . . . *E. densiflorum*
         5′ Fr beaked, central axis disintegrating
            6. Hypanthium 1.5–3 mm, sepals 2–6 mm, petals 3–10 mm; fr 10–21 mm; seeds compressed into 1 staggered row per capsule, 1.8–2.3 mm. . . . . *E. pallidum*
            6′ Hypanthium 0.4–1 mm, sepals 0.7–2 mm, petals 1–3.2 mm; fr 8–14 mm; seeds in 1 row per chamber, 0.9–1.6 mm. . . . . *E. torreyi*
   2′ Seed hair-tuft present
      7. Pl 2–20 dm, ± glandular; hypanthium 1.5–8(16) mm, sepals 1.2–8 mm, petals 2–15(20) mm; seed 1.5–2.7 mm. . . . . *E. brachycarpum*
      7′ Pl 0.5–4.5 dm, nonglandular; hypanthium 0.4–1.5 mm, sepals 1.3–2.5 mm, petals 1.8–5 mm; seed 0.6–1.2 mm
         8. Lvs gen folded along midrib, distal clustered; infl dense; petals 1.8–3 mm; seeds 0.6–0.9 mm, papillate . . . . *E. foliosum*
         8′ Lvs flat, not clustered; infl open; petals 2–5 mm; seeds 0.9–1.2 mm, netted. . . . . *E. minutum*
1′ Per to subshrub from caudex, taproot 0; lvs gen opposite up to infl, gen narrowly lanceolate to ovate; proximal st not peeling (exc *Epilobium rigidum*)
   9. Hypanthium 17–34 mm; corolla red-orange, ± bilateral
      10. Matted per 1–2 dm; proximal lvs white-canescent, nonglandular, distal lvs glandular; NCo, NCoRO . . . . *E. septentrionale*
      10′ Clumped per to subshrub 1–12 dm; all lvs alike, gen not white-canescent, glandular or not; widespread CA-FP. . . . . *E. canum*
         11. Lvs linear to lanceolate, gen < 6 mm wide, often clustered; per to subshrub 2–12 dm . . . . subsp. *canum*
         11′ Lvs lanceolate to widely ovate, gen > 6 mm wide, not clustered; per 1–5 dm
            12. Lvs widely ovate, prominently toothed, lateral veins conspicuous. . . . . subsp. *garrettii*
            12′ Lvs lanceolate to ovate, ± entire to gen toothed, lateral veins obscure. . . . . subsp. *latifolium*
   9′ Hypanthium gen 0.4–8(16) mm; corolla white or cream-yellow to purple

13. Stigma 4-lobed; sepals (3)6–15 mm; petals (6)10–24 mm, pink to rose-purple or cream
  14. Petals cream-yellow; hair-tufts ± red, persistent; fr 35–75 mm . . . . *E. luteum*
  14′ Petals pink to rose-purple; hair-tufts white, deciduous; fr 8–45 mm
    15. St 4–10 dm; lf 30–90 mm, veins ± red, conspicuous; seed 0.9–1.3 mm, ridged. . . . . *E. oreganum*
    15′ St < 4 dm; lf 6–45 mm, veins obscure; seed 1.4–3.4 mm, papillate
      16. Petals 6–10 mm; seeds 1–2 per chamber (8 per fr); fr 8–16 mm . . . . *E. nivium*
      16′ Petals 11–24 mm; seeds > 8 per chamber; fr 20–45 mm
        17. Pl 1–4 dm, nonglandular; lf 20–45 mm, petiole 2–6 mm; hypanthium 1–1.8 mm; seed
          2.5–3.4 mm . . . . *E. rigidum*
        17′ Pl < 2.5 dm, sometimes glandular (esp ovaries); lf 6–26 mm, petiole 0–3 mm; hypanthium 2–6
          mm; seed 1.4–2.1 mm
          18. Pedicel in fr 2–3 mm; lf 6–20 mm, tip round to obtuse, petiole 0–3 mm; pl ± glaucous . . . . *E. obcordatum*
          18′ Pedicel in fr 6–25 mm; lf 13–26 mm, tip ± acute; petiole 0; pl not glaucous . . . . *E. siskiyouense*
13′ Stigma entire, club-like (occ cylindric); sepals 1–7.5 mm; petals 1.6–14 mm, white to rose-purple
  19. St ± erect, with offset rosettes or fleshy underground shoots
    20. Lf veins ± obscure; st with fleshy bulb-like shoot; seed papillate
      21. Pedicel in fr 10–40 mm; infl ± nodding; lf 5–45 mm . . . . *E. hallianum*
      21′ Pedicel in fr 0–5 mm; infl erect; lf 10–65 mm . . . . *E. saximontanum*
    20′ Lf veins conspicuous; st with rosette or fleshy shoot; seed ridged . . . . *E. ciliatum*
      22. Petals 2–6(9) mm, white to pink; lf lanceolate, reduced in open infl — basal rosette present . . . . subsp. *ciliatum*
      22′ Petals 4–14 mm, pink to rose-purple; lf lanceolate to ovate, little reduced on crowded infl
        23. St with fleshy shoot; infl long . . . . subsp. *glandulosum*
        23′ St with lfy rosette or shoot; infl flat-topped. . . . . subsp. *watsonii*
  19′ St ± ascending, clumped or cespitose; lfy or thread-like stolons sometimes tipped by fleshy bulblets
    24. Pl ± glabrous (exc scattered infl hairs)
      25. Pl 0.8–3(4) dm, matted; stolons thread-like; pedicels 20–65 mm; infl open (bracts << internodes);
        lf tip obtuse. . . . . *E. oregonense*
      25′ Pl < 8.5 dm, clumped; stolons short, scaly; pedicels 5–25 mm; infl crowded (bracts ± = internodes);
        lf tip subacute. . . . . *E. glaberrimum*
        26. St < 3.5 dm; lf 10–35 mm, lanceolate to narrowly ovate; fr 20–55 mm . . . . subsp. *fastigiatum*
        26′ St 2–8.5 dm; lf 20–70 mm, ± narrowly lanceolate; fr 45–75 mm . . . . subsp. *glaberrimum*
    24′ Pl variously hairy, often in decurrent lines on st
      27. St densely glandular-hairy; petals 2–3 mm, white. . . . . *E. howellii*
      27′ St hairy, nonglandular; petals 2–9 mm, white or pink to rose-purple
        28. Stolons thread-like, tipped by fleshy bulblets; lf ± narrowly lanceolate; seed 1.4–2.2 mm, hair-tuft persistent
          29. Lf densely minutely strigose; infl minutely strigose and glandular. . . . . *E. leptophyllum*
          29′ Lf minutely strigose only on margins and midrib; infl minutely strigose, nonglandular. . . . . *E. palustre*
        28′ Stolons short, lfy; lf gen lanceolate or wider; seeds 0.7–2.1 mm, hair-tuft gen deciduous
          30. Pl cespitose, < 2.2 dm; lf 8–28 mm; fr 17–40 mm
            31. Infl ± nodding; pedicel in fr 10–55 mm; seed 0.7–1.4 mm, netted. . . . . *E. anagallidifolium*
            31′ Infl erect; pedicel in fr 2–18 mm; seed 1.5–2.1 mm, papillate. . . . . *E. clavatum*
          30′ Pl loosely clumped, 1–5 dm; lf 15–55 mm; fr 40–100 mm
            32. Pedicel in fr 5–15 mm; fr 40–65 mm; seed 0.9–1.2 mm, papillate; petals gen pink to
              rose-purple. . . . . *E. hornemannii* subsp. *hornemannii*
            32′ Pedicel in fr 20–45 mm; fr 50–100 mm; seed 1.1–1.6 mm, netted; petals gen white . . . . *E. lactiflorum*

E. anagallidifolium Lam. (p. 945) Per < 2 dm, cespitose, ascending to erect, ± strigose in decurrent lines; stolons short, lfy. **LF:** 8–25 mm; basal spoon-shaped to oblong, glabrous; distal elliptic to lanceolate, sparsely strigose, obtuse, ± entire; petiole 1–6 mm. **INFL:** ± nodding, sometimes glandular. **FL:** hypanthium 0.6–1.2 mm; sepals 1.5–5 mm; petals 2.5–8 mm, pink to rose-purple; stigma club-like. **FR:** 17–40 mm, ± glabrous; pedicel 10–55 mm. **SEED:** 0.7–1.4 mm, netted, hair-tuft 0. 2*n*=36. Moist alpine slopes, meadows, streambanks; 1500–4500 m. KR, CaRH, SNH, SnBr, Wrn, n W&I; circumboreal. Jun–Sep

E. brachycarpum C. Presl (p. 945) Ann 2–20 dm, glabrous and peeling proximally, strigose and gen glandular-hairy distally. **LF:** gen early-deciduous, 10–55 mm, linear to narrowly elliptic, acuminate, gen folded along midrib, ± glabrous; petiole 0–4 mm. **INFL:** panicle or raceme. **FL:** hypanthium 1.5–8(16) mm; sepals 1.2–8 mm; petals 2–15(20) mm, white to rose-purple; stamens ≤ pistil; stigma sometimes 4-lobed. **FR:** 15–32 mm, glabrous or glandular; pedicel 1–17 mm. **SEED:** 1.5–2.7 mm, papillate, hair-tuft readily deciduous. 2*n*=24. Common. Dry open or disturbed woodland, grassland, roadsides; < 3300 m. CA-FP (exc ChI), MP, W&I; to BC, MN, NM, also e Can; introduced in s S.Am, Eur. Highly variable, esp fl size. Jun–Sep

E. campestre (Jeps.) Hoch & W.L. Wagner (p. 945) Ann 1–5.5 dm, taprooted, basal branches gen decumbent, glabrous and peeling proximally, minutely strigose or spreading-hairy distally. **LF:** opposite only near base, ± sessile, 8–35 mm, lanceolate (narrowly), gen hairy. **INFL:** crowded; bracts narrowly ovate or wider. **FL:** gen cleistogamous; hypanthium 0.3–1.1 mm; sepals 0.7–2 mm; petals 1–3.2 mm, pink; stigma club-like, sometimes barely 4-lobed. **FR:** sessile, 3.5–8 mm, cylindric, tough, proximal 1/2 ± indehiscent; axis disintegrating. **SEED:** in 2 rows per chamber, 0.9–1.3 mm, netted, hair-tuft 0. 2*n*=30. Vernal pools, clay mud flats; 30–3000 m. CA-FP (exc ChI, e TR), MP; to w Can, ND, UT, n Baja CA; s S.Am. [*Boisduvalia glabella* (Nutt.) Walp.; *B. g.* var. *c.* (Jeps.) Jeps.; *E. pygmaeum* (Speg.) Hoch & P.H. Raven, illeg.] May–Sep

E. canum (Greene) P.H. Raven CALIFORNIA FUCHSIA, ZAUSCHNERIA Per (clumped with basal scaly shoots) to subshrub, strigose to ± densely spreading-hairy and gen glandular. **LF:** ± sessile, 8–70 mm, linear to widely ovate, green to ± gray, ± entire to strongly toothed. **FL:** red-orange; hypanthium 17–34 mm; sepals 4–10 mm; petals 7–15 mm; stamens << pistil; stigma 4-lobed. **FR:** 15–30 mm, ± beaked, hairy; pedicel 0–3 mm. **SEED:** 1.2–2.3 mm, low-papillate. Hummingbird-pollinated. Subspp. intergrade, esp in s CA.

subsp. ***canum*** (p. 945) Per or subshrub, 2–12 dm, strigose to spreading hairy, occ glandular. **LF:** often clustered, linear to (narrowly) lanceolate, ± entire to barely toothed, gen ± gray. 2*n*=30,60. Dry slopes, ridges; < 2100 m. CA-FP (exc n SNH); to nw Mex. Jun–Dec

subsp. ***garrettii*** (A. Nelson) P.H. Raven Per 1.5–4 dm, gen spreading-hairy and glandular. **LF:** opposite, widely ovate, prominently toothed, gen green. 2*n*=30. Dry slopes, ridges; 1200–3400 m. s SNH, DMtns; to se ID, nw WY, UT. Jun–Oct

subsp. ***latifolium*** (Hook.) P.H. Raven (p. 945) Per 1–5 dm, gen spreading-hairy and glandular. **LF:** opposite, lanceolate to ovate, ± entire to gen toothed, gen green. 2*n*=60. Dry slopes, ridges; 50–3300 m. KR, NCoRH, CaR, SN, TR, PR, DMtns; to sw OR, w NM, nw Mex. Jun–Dec

E. ciliatum Raf. Per, erect, ± loosely clumped, with basal rosettes or fleshy shoot, gen strigose in lines, glandular distally, occ spreading-hairy. **LF:** 1–12 cm, narrowly lanceolate to ovate, fine-toothed; veins conspicuous; petiole 0–5(8) mm. **INFL:** densely strigose, ± spreading-hairy, gen glandular. **FL:** hypanthium 0.5–2.6 mm; sepals 2–7.5 mm; petals white to rose-purple; stamens ≤ pistil; stigma club- or head-like. **FR:** 15–100 mm, hairy; pedicel 0–15(40) mm. **SEED:** 0.8–1.6 mm, ridged, hair-tuft deciduous. 2*n*=36.

subsp. ***ciliatum*** (p. 945) Pl 5–12(19) dm; rosettes well developed. **LF:** narrowly lanceolate, distal reduced. **INFL:** openly branched, not lfy. **FL:** petals 2–6(9) mm, white to pink. Common. Disturbed places, moist meadows, streambanks, roadsides; < 4100 m. ± CA. Jun–Oct

subsp. ***glandulosum*** (Lehm.) Hoch & P.H. Raven Pl 2–11(17) dm; shoot fleshy, scaly, rarely lfy, gen below ground; strigose in lines, glandular distally. **LF:** narrowly ovate to ovate (lanceolate), little reduced upward. **INFL:** dense, lfy racemes. **FL:** petals 4–14 mm, pink to rose-purple. Often ± shaded streambanks, seeps, meadows; < 3500 m. KR, NCoR, CaR, SNH, TR, Wrn; to AK, Rocky Mtns, ne N.Am; introduced to n Eur. [*E. g.* Lehm.; *E. brevistylum* var. *b.*] Variable; larger-fld pls cross-pollinated. Jun–Oct

subsp. ***watsonii*** (Barbey) Hoch & P.H. Raven (p. 945) Robust, 3–15(18) dm; rosettes well developed. **LF:** lanceolate to subovate, distal not much reduced. **INFL:** dense, lfy, flat-topped. **FL:** petals 5–14 mm, rose-purple to pink. Moist coastal bluffs, streamsides, ± disturbed sites; < 200 m. NCo, CCo; to BC. [*E. w.* Barbey; *E. w.* var. *franciscanum* (Barbey) Jeps.] Distinctive. May–Oct

E. clavatum Trel. (p. 945) Per < 2.2 dm, cespitose, ascending, ± strigose (esp distally); stolons wiry, lfy. **LF:** 8–28 mm, ovate to ± elliptic, obtuse to subacute; petiole 0–3 mm. **INFL:** erect, sparsely glandular. **FL:** hypanthium 0.6–2 mm; sepals 2.5–4.2 mm; petals 3.5–6 mm, rose-purple to pink; stigma (widely) club-like. **FR:** 17–40 mm, ± thick, gen sparsely hairy; pedicel 2–18 mm. **SEED:** 1.5–2.1 mm, papillate. 2*n*=36. Subalpine scree, rocky streambanks; 1200–4200 m. KR, CaRH, n&c SNH, Wrn; to AK, MT, CO. Jun–Sep

E. cleistogamum (Curran) Hoch & P.H. Raven (p. 945) Ann 1.5–3.2 dm, peeling proximally, occ with prostrate basal branches, ± glabrous to spreading-hairy. **LF:** opposite only near base, ± sessile, 15–55 mm, linear to narrowly elliptic; distal lvs hairy, often folded along midrib. **INFL:** lfy, spreading and glandular hairy. **FL:** gen cleistogamous; hypanthium 0.5–1 mm; sepals 1–3 mm; petals 2–5.5 mm, white to pale pink; stigma head-like or occ barely 4-lobed. **FR:** 8–12 mm, sharply 4-sided, tough, sessile, splitting only on distal third; central axis disintegrating. **SEED:** in 2 rows per chamber, 1–1.5 mm, netted, glabrous, hair-tuft 0. 2*n*=30. Vernal pools, clay flats; < 300 m. SNF, GV, SCoRO. May–Jul

E. densiflorum (Lindl.) Hoch & P.H. Raven (p. 945) Ann 0.5–15 dm, simple or branched distally, peeling proximally, spreading hairy or strigose. **LF:** opposite only near base, ± sessile, 14–85 mm, (narrowly) lanceolate to sublinear; distal lvs hairy. **INFL:** lfy, ± dense spike, sometimes glandular. **FL:** hypanthium 1.3–4 mm; sepals 2–8 mm; petals 3–10 mm, rose-purple to white; stigma head-like to irregularly 4-lobed. **FR:** 4–11 mm, cylindric, pliable, not beaked, dehiscent to base; axis persistent; pedicel 0–2.5 mm. **SEED:** 1.2–1.6(1.9) mm, netted, glabrous, hair-tuft 0. 2*n*=20. Streambanks, outwashes, seasonal moist flats; < 2600 m. CA-FP (exc SCo, ChI), MP; to BC, MT, AZ, n Baja CA. Highly variable. May–Oct

E. foliosum (Torr. & A. Gray) Suksd. (p. 945) Ann ≤ 4.5 dm, taprooted, ± peeling proximally, ± strigose. **LF:** basal pairs opposite, gen alternate, distal clustered, 5–30 mm, sublinear to lanceolate, gen folded along midrib; petiole 1–12 mm. **INFL:** crowded, densely strigose. **FL:** hypanthium 0.4–0.8 mm; sepals 1.3–2.5 mm; petals 1.8–3 mm, white. **FR:** 12–20 mm, sparsely hairy; pedicel 2–5 mm. **SEED:** 0.6–0.9 mm, papillate. 2*n*=32. Dry, open, disturbed areas, roadsides; 20–1900 m. KR, NCoR, CaR, n SNH, SNH, ScV, CW (exc SCoRI), WTR, SnGb; to BC, ID, AZ, nw Mex. Apr–Aug

E. glaberrimum Barbey Per, clumped, gen ascending, ± glabrous, glaucous; stolons short, scaly. **LF:** ± sessile, narrowly lanceolate to narrowly ovate, clasping. **FL:** hypanthium 0.7–2.3 mm; petals gen pink to rose-purple (white); stigma broadly club-like. **FR:** occ sparsely hairy. **SEED:** papillate. 2*n*=36.

subsp. ***fastigiatum*** (Nutt.) Hoch & P.H. Raven (p. 945) **ST:** < 3.5 dm, gen simple, gen glaucous. **LF:** 10–35 mm, lanceolate to narrowly ovate. **FL:** sepals 2.5–5.2 mm; petals 3.4–8 mm, rose-purple to pink. **FR:** 20–55 mm; pedicel 5–17 mm. **SEED:** 0.8–1(1.3) mm. Gravel bars, scree, roadsides, moist rocky areas; 1200–3800 m. KR, NCoRH, CaRH, SNH, Wrn; to w Can, MT, WY. [*E. g.* var. *f.* (Nutt.) Trel., nom. nud.] Jun–Sep

subsp. ***glaberrimum*** (p. 945) **ST:** 2–8.5 dm, gen branched distally. **LF:** 20–70 mm, ± narrowly lanceolate. **FL:** sepals 3–7.5 mm; petals 5–10(12) mm, rose-purple to pink or white. **FR:** 45–75 mm; pedicel 5–25 mm. **SEED:** 0.7–1 mm. Well-drained, gravelly soils, streambanks, roadsides; 150–3000 m. KR, NCoRH, CaRH, SNH, TR, PR, Wrn, n W&I; to WA, ID, UT, n Mex. Jun–Sep

E. hallianum Hausskn. (p. 945) Per 1–6 dm, erect, gen sparsely strigose with minutely strigose lines; bulb-like shoot small, underground. **LF:** 5–45 mm, proximal ovate, distal lanceolate, ciliate and low-dentate in distal lvs, obtuse to subacute; petiole 0–1.5 mm. **INFL:** ± nodding, gen ± glandular. **FL:** hypanthium 0.5–1.7 mm; sepals 1.2–2.8 mm; petals 1.6–5.5 mm, white to pink, stigma club-like. **FR:** 24–60 mm, gen hairy; pedicel 10–40 mm. **SEED:** 1.1–1.6 mm, papillate. 2*n*=36. Moist meadows, streambanks; 100–3700 m. KR, NCoR, CaR, SNH, SnFrB, TR, PR, MP; to w Can, SD, CO, AZ. [*Epilobium halleanum* Hausskn., orth. var.] Jun–Sep

E. hornemannii Reichb. subsp. ***hornemannii*** (p. 945) Per 1–4.5 dm, loosely clumped, ascending, ± strigose in lines (esp distally); stolons short, lfy. **LF:** 15–55 mm, lanceolate to ovate (distal narrower), gen ± glabrous, obtuse to acute; petiole 0–8 mm. **INFL:** glandular. **FL:** hypanthium 1–2.2 mm; sepals 3–6.6 mm; petals 4–11 mm, gen pink to rose-purple (white). **FR:** 40–65 mm, hairy; pedicel 5–15 mm. **SEED:** 0.9–1.2 mm, papillate. 2*n*=36. Moist meadows, streambanks, subalpine ridges; 1200–3900 m. KR, NCoRH, CaRH, SNH, SnBr, SnJt, Wrn, n W&I; ± circumboreal. Jun–Aug

E. howellii Hoch SUBALPINE FIREWEED Per < 2 dm, loosely clumped; stolons short, thread-like, minutely lfy. **ST:** densely glandular-hairy. **LF:** 4–20 mm, round or ovate on proximal st to lanceolate distally, sparsely strigose; tip round to obtuse; sessile. **INFL:** glandular (ovaries sparsely so). **FL:** bud gen nodding; hypanthium 0.4–0.8 mm; sepals 1.5–2 mm; petals 2–3 mm, white; stigma head-like. **FR:** 35–45 mm, sparsely hairy; pedicel 25–40 mm. **SEED:** 0.8–1.1 mm, low-papillate. 2*n*=36. Wet meadows, mossy seeps; 1950–2700 m. n&c SNH. Jul–Aug ★

E. lactiflorum Hausskn. (p. 945) Per 1.5–5 dm, gen clumped, ascending, minutely strigose in lines (esp distally); stolons short, lfy. **LF:** 20–55 mm, narrowly ovate to narrowly lanceolate (distal narrower), ± ciliate, low-dentate, obtuse to acute; petiole ± winged, 3–12 mm. **INFL:** nodding to erect, strigose, glandular. **FL:** hypanthium 1–2.2 mm; sepals 3–5.5 mm; petals 4–8.8 mm, gen white; stigma club-like. **FR:** 50–100 mm, scattered hairy; pedicel 20–45 mm. **SEED:** 1.1–1.6 mm, netted. 2*n*=36. Damp meadows, streambanks, talus; 1400–3350 m. KR, CaRH, SNH, n W&I; to AK, MT, NM, also e N.Am, n Eurasia. Jun–Sep

E. leptophyllum Raf. Per 1–10 dm, erect, ± densely strigose; stolons thread-like, tipped with fleshy bulblets. **LF:** ± sessile, 20–75 mm, linear to narrowly elliptic. **INFL:** minutely strigose and glandular. **FL:** hypanthium 0.8–1.5 mm; sepals 2.5–4.5 mm; petals 3.5–7 mm, white or pink. **FR:** 35–80 mm, gray-hairy; pedicel 5–35 mm.

SEED: 1.5–2.2 mm, papillate; hair-tuft persistent. 2*n*=36. Uncommon. Boggy meadows, seeps, damp places; 800–1630 m. NCoRI, CaRH, c&s SNE; to BC, MT, CO; native to e N.Am. Jun–Sep

E. luteum Pursh (p. 945) YELLOW WILLOWHERB Per 1–8 dm, loosely clumped, ± strigose in decurrent lines; stolons short, lfy or scaly. **LF**: 25–80 mm, spoon-shaped on proximal st to ovate or elliptic, acute to acuminate; veins conspicuous; petiole 0–3 mm. **INFL**: nodding in bud, densely glandular, fine-strigose. **FL**: hypanthium 1.2–3 mm; sepals 10–12 mm; petals 12–22 mm, cream-yellow; stamens < pistil; stigma 4-lobed. **FR**: 35–75 mm, ± hairy; pedicel 10–22 mm. **SEED**: 1–1.2 mm, netted; hair-tufts ± red, persistent. 2*n*=36. Moist streambanks, montane meadows; ± 1200–2500 m. n KR, n SNH; to AK, w Can. Jul–Sep ★

E. minutum Lindl. (p. 945) Ann < 4 dm, taprooted, ± peeling proximally, ± glabrous to strigose. **LF**: opposite near base, alternate distally, 9–25 mm, gen lanceolate to narrowly elliptic, ± glabrous; petiole 0–2 mm. **INFL**: open, strigose. **FL**: hypanthium 1.1–1.5 mm; sepals 1.3–2.5 mm; petals 2–5 mm, white or pink; stigma club-like to barely lobed. **FR**: 9–28 mm, ± glabrous; pedicel 3–10 mm. **SEED**: 0.9–1.2 mm, netted; hair-tuft readily deciduous. 2*n*=26. Dry, open, disturbed areas, vernal pools, often after fire; 15–2320 m. NW, CaR, n&c SN, ScV, CW, SnBr, PR, GB, n DMoj; to BC, MT, w NV. Apr–Sep

E. nivium Brandegee (p. 945) SNOW MOUNTAIN WILLOWHERB Subshrub 1–2.5 dm, sts many from woody caudex, ± gray strigose. **LF**: occ clustered, 8–18 mm, lance-elliptic, densely spreading-hairy; petiole 0–3 mm. **INFL**: erect, densely spreading-hairy. **FL**: hypanthium 5–10 mm; sepals 3–4.5 mm; petals 6–10 mm, rose-purple; gen stamens < pistil; stigma 4-lobed. **FR**: 8–16 mm, fusiform, glandular; pedicel 2–5 mm. **SEED**: 1–2 per chamber, 1.5–2.4 mm, ± obovoid, low-papillate. 2*n*=30. Dry talus, shaly slopes; 1500–2400 m. s NCoRH. Jul–Sep ★

E. obcordatum A. Gray (p. 945) Per < 1.5 dm, clumped from woody caudex, ± subglabrous, ± glaucous; many wiry, scaly basal shoots. **LF**: 6–20 mm, widely lance-elliptic to nearly round; tip round to obtuse; petiole 0–3 mm. **INFL**: gen densely hairy (glandular). **FL**: hypanthium 3–5.5 mm; sepals (3)6–15 mm; petals 11–24 mm, pink to rose-purple; stamens < pistil; stigma 4-lobed. **FR**: 20–45 mm, widely club-like, densely glandular; pedicel 2–3 mm. **SEED**: 1.4–2.1 mm, low-papillate; hair-tuft white, deciduous. 2*n*=36. Rocky ridges, dry talus; 1700–4000 m. CaRH, SNH, MP; to e OR, c ID, ne NV. Jul–Sep

E. oreganum Greene (p. 951) OREGON FIREWEED Per 4–10 dm from lfy, basal shoots, ± glabrous, ± glaucous. **LF**: 30–90 mm, narrow-lanceolate to -ovate; veins ± red, conspicuous; petiole 1–3 mm. **INFL**: open, sparsely glandular. **FL**: hypanthium 2–3 mm; sepals 6–10 mm; petals (6)10–15 mm, pink to rose-purple; stamens < pistil; stigma 4-lobed. **FR**: 25–45 mm, hairy and glandular; pedicel 3–6 mm. **SEED**: 0.9–1.3 mm, ridged, hair-tuft deciduous. 2*n*=36. Bogs, small streams; 550–1800 m. KR, NCoRO; sw OR. [*E. exaltatum* Drew] Often confused with *E. ciliatum*, esp *E. ciliatum* subsp. *glandulosum*. Jul–Aug ★

E. oregonense Hausskn. (p. 951) Per 0.8–3(4) dm, matted, delicate, ± glabrous, ± purple distally; stolons sprawling, thread-like, with minute rounded lvs. **LF**: sessile, clasping, 5–25 mm, broadly elliptic near base to linear and widely spaced distally; tip obtuse. **INFL**: gen erect, open (bracts << internodes), sparsely glandular. **FL**: hypanthium 0.8–1.8 mm; sepals 2.5–4.5 mm; petals 5–8 mm, white to pink; stigma broadly club-like. **FR**: 20–50 mm, ± glabrous; pedicel 20–65 mm, gen > fr. **SEED**: 1–1.4 mm, low-papillate. 2*n*=36. ± boggy areas, shaded streambanks; 1200–3500 m. KR, CaRH, SNH, SnGb, SnBr, SnJt, Wrn, SNE; to BC, w MT, WY. Jul–Aug

E. pallidum (Eastw.) Hoch & P.H. Raven (p. 951) Ann 0.5–6 dm, st peeling proximally, densely spreading-hairy or strigose. **LF**: opposite only at base, 12–50 mm, narrowly lance-elliptic, glabrous (distal hairy), ± sessile. **INFL**: crowded spike, sometimes glandular; bracts = lvs. **FL**: hypanthium 1.5–3 mm; sepals 2–6 mm; petals 3–10 mm, rose-purple; stamens ≤ pistil; stigma 4-lobed. **FR**: 10–21 mm, beaked, ± spreading-hairy, pliable, sessile, dehiscent to base; septa

and axis disintegrating. **SEED**: compressed into 1 staggered row per capsule, 1.8–2.3 mm, netted, glabrous, hair-tuft 0. 2*n*=20. Edges of seasonally moist ponds, streams; 60–2050 m. KR, NCoRI, CaR, n SN, ScV, MP; to sw OR, ID. May–Aug

E. palustre L. MARSH WILLOWHERB Per 1–8 dm, ± fine-strigose, esp in lines from lf base and distally; stolons thread-like, sparsely lfy, tipped with compact, fleshy bulblets. **LF**: ± sessile, 15–70 mm, sublinear to elliptic. **INFL**: ± nodding to erect, densely strigose, nonglandular. **FL**: hypanthium 0.6–1.8 mm; sepals 1.4–4.5 mm; petals 2–9 mm, white (pink); stigma occ cylindric. **FR**: 30–90 mm, hairy; pedicel 15–35(60) mm. **SEED**: 1.4–2.2 mm, finely papillate; hair-tufts persistent. 2*n*=36. Wet meadows, seeps, bogs, disturbed wet areas; 1950–2400 m. n SNH (Grass Lake, El Dorado Co.), Wrn; circumboreal. Jul–Sep ★

E. rigidum Hausskn. (p. 951) SISKIYOU MOUNTAINS WILLOWHERB Per 1–4 dm, clumped from woody caudex, ascending, wiry shoots, glabrous and glaucous proximally, fine-strigose (densely) distally, nonglandular. **ST**: peeling proximally. **LF**: 20–45 mm, gen ± (ob)ovate, ± glabrous or hairy, ± entire; petiole 2–6 mm. **FL**: hypanthium 1–1.8 mm; sepals 9–15 mm; petals 16–20 mm, pink to rose-purple; stamens < pistil; stigma 4-lobed. **FR**: 20–35 mm, hairy, club-like, fine-strigose; pedicel 9–13 mm. **SEED**: 2.5–3.4 mm, papillate, hair-tuft deciduous. 2*n*=36. Dry, open places, dry streambeds, sometimes on serpentine-like soils; 100–1200 m. nw KR (Del Norte Co.); sw OR. Jul–Sep ★

E. saximontanum Hausskn. (p. 951) Per < 6 dm, erect, from fleshy, scaly, underground shoots/bulblets, ± strigose in decurrent lines, strigose and glandular distally. **LF**: ± sessile, gen clasping, 10–65 mm, narrowly elliptic to (ob)ovate. **INFL**: erect, fine-strigose and glandular. **FL**: hypanthium 0.8–1.4 mm; sepals 1.2–3.5 mm; petals 2.2–7 mm, white (pink); stigma club-like. **FR**: 20–55(70) mm, hairy; pedicel 0–5 mm. **SEED**: 1–1.8 mm, low-papillate (netted). 2*n*=36. Moist montane meadows, streambanks, ± disturbed roadsides; 1400–3500 m. c&s SNH, SNE; to se OR, sw Can (also e Can), MT, NM. [*E. drummondii* Hausskn.] Jul–Sep

E. septentrionale (D.D. Keck) R.N. Bowman & Hoch (p. 951) HUMBOLDT COUNTY FUCHSIA Per 0.5–2(3) dm, matted, wiry, scaly proximally, from ± woody caudex, occ decumbent, densely white-canescent and nonglandular below infl. **LF**: gen opposite, ± sessile, 10–25 mm, lanceolate to ovate, white-canescent; distal bract-like, ± green, glandular and spreading-hairy. **INFL**: densely glandular. **FL**: red-orange; hypanthium 17–23 mm; sepals 7–12 mm; petals 8–14 mm; stamens << pistil; stigma 4-lobed. **FR**: 20–26 mm, short-beaked, hairy; pedicel 0–2 mm. **SEED**: 1.8–2.4 mm, low-papillate, hair-tuft deciduous. 2*n*=30. Dry, sandy or rocky ledges (serpentine slopes); 20–1900 m. NCo, NCoRO. Bird-pollinated. Jul–Sep ★

E. siskiyouense (Munz) Hoch & P.H. Raven (p. 951) SISKIYOU FIREWEED Per or subshrub, 1–2.5 dm, clumped, with scaly shoots from woody caudex, ± densely fine-strigose and glandular, rarely ± glabrous. **LF**: sessile, 13–26 mm, lanceolate to widely ovate, ± sparsely strigose. **INFL**: ± densely glandular. **FL**: hypanthium 2–4 mm; sepals 5–10 mm; petals 11–24 mm, pink to rose-purple; stamens < pistil; stigma 4-lobed. **FR**: 25–45 mm, hairy; pedicel 6–25 mm. **SEED**: 1.4–1.9 mm, papillate, hair-tuft white, deciduous. 2*n*=36. Scree, moist ledges, serpentine ridges; 1700–2500 m. n KR; sw OR. [*E. obcordatum* A. Gray subsp. *s.* Munz; *E. o.* A. Gray var. *laxum* (Hausskn.) Dempster] Jul–Sep ★

E. torreyi (S. Watson) Hoch & P.H. Raven (p. 951) Ann 0.4–6.5 dm, taprooted, ± gray spreading-hairy, peeling proximally. **LF**: opposite only near base (where glabrous), ± sessile, 5–45 mm, ± lance-linear, hairy. **INFL**: glandular, bracts lf-like. **FL**: gen cleistogamous; hypanthium 0.4–1 mm; sepals 0.7–2 mm; petals 1–3.2 mm, pink or white; stigma club-like or rarely 4-lobed. **FR**: sessile, 8–14 mm, cylindric, beaked, flexible, dehiscent to base; axis disintegrating. **SEED**: 0.9–1.6 mm, netted; hair-tuft 0. 2*n*=18. Seasonally moist streambanks, seeps, roadside ditches; 50–2600 m. NW (exc NCo), CaR, SN, GV, SnFrB, SCoRO, MP, sw DMoj; to BC, ID, n NV. [*Boisduvalia stricta* (A. Gray) Greene; *B. t.* (S. Watson) S. Watson] May–Aug

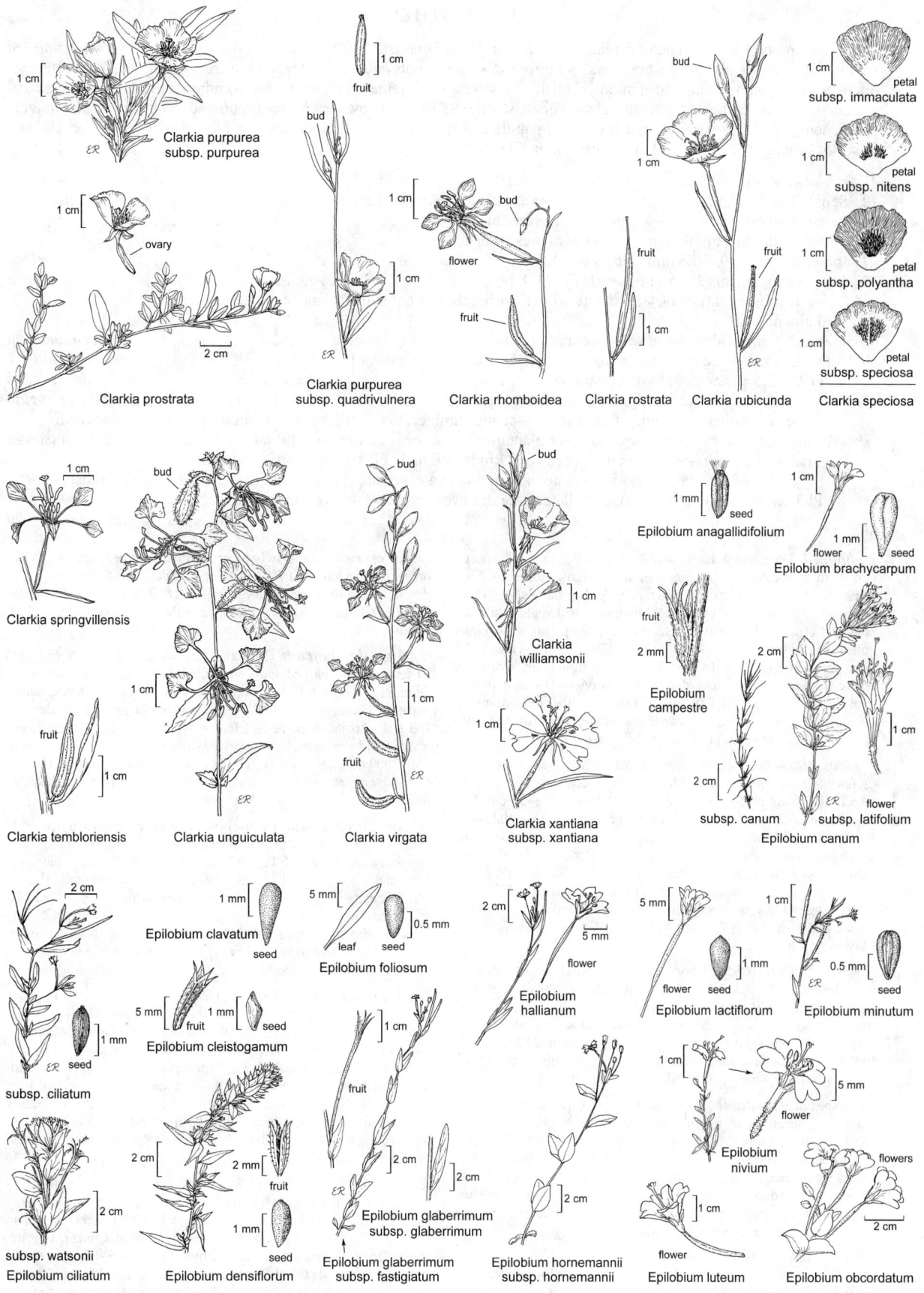

Clarkia purpurea subsp. purpurea

Clarkia prostrata

Clarkia purpurea subsp. quadrivulnera

Clarkia rhomboidea

Clarkia rostrata

Clarkia rubicunda

subsp. immaculata
subsp. nitens
subsp. polyantha
subsp. speciosa
Clarkia speciosa

Clarkia springvillensis

Clarkia tembloriensis

Clarkia unguiculata

Clarkia virgata

Clarkia williamsonii

Clarkia xantiana subsp. xantiana

Epilobium anagallidifolium

Epilobium brachycarpum

Epilobium campestre

subsp. canum   subsp. latifolium
Epilobium canum

Epilobium clavatum

Epilobium cleistogamum

Epilobium foliosum

Epilobium hallianum

Epilobium lactiflorum

Epilobium minutum

subsp. ciliatum

subsp. watsonii
Epilobium ciliatum

Epilobium densiflorum

Epilobium glaberrimum subsp. glaberrimum

Epilobium glaberrimum subsp. fastigiatum

Epilobium hornemannii subsp. hornemannii

Epilobium nivium

Epilobium luteum

Epilobium obcordatum

EREMOTHERA

Ann from taproot. **LF**: basal and/or cauline, alternate, simple to 2-pinnate. **INFL**: spike. **FL**: opening at dusk; sepals 4, reflexed singly or in pairs; petals 4, gen white, pink, or rarely red, without spots or ultraviolet reflective area, fading red; longer stamens opposite sepals, anthers attached at middle, pollen grains 3-angled; stigma hemispheric, gen > anthers and cross-pollinated or ± = anthers and self-pollinated. **FR**: straight to coiled, sessile. **SEED**: in 1 row per chamber, obovoid to oblanceoloid, minutely pitted, sometimes those near base of fr coarsely papillate. 7 spp.: w N.Am. (Greek: desert + *Oenothera*) [Wagner et al. 2007 Syst Bot Monogr 83:1–240] Incl in *Camissonia* in TJM (1993).

1. Fr ± same width throughout; fls at proximal nodes 0
  2. Sepals 1.5–2.5 mm; petals 1.8–3 mm; stigma ± surrounded by anthers . . . . ***E. chamaenerioides***
  2′ Sepals 4–6 mm; petals 3.5–7 mm; stigma exceeding anthers . . . . ***E. refracta***
1′ Fr base wider than tip; fls at proximal nodes present or 0
  3. Infl erect; petals 0.8–1.3 mm; sepals 0.8–1.8 mm . . . . ***E. minor***
  3′ Infl nodding; petals 3–7.5 mm; sepals (2.7)4–8 mm . . . . ***E. boothii***
    4. Rosette present at time of 1st fl in late winter, spring; bracts not lf-like, ± inconspicuous; pl ± glabrous
      or hairs not spreading
      5. Fr 2–3.8 mm wide at base, woody, curved outward but not downward . . . . subsp. ***condensata***
      5′ Fr 1–2.3 mm wide at base, not woody, curved outward or downward
        6. Fr 1.7–2.3 mm wide at base, curved outward . . . . subsp. ***decorticans***
        6′ Fr 1–1.6 mm wide at base, gen curved downward. . . . . subsp. ***desertorum***
    4′ Rosette gen withered by time of 1st fl in late spring, summer; bracts lf-like, conspicuous; pl hairs often spreading
      7. Pl strigose, also rarely with spreading or glandular hairs; seeds all minutely pitted . . . . subsp. ***alyssoides***
      7′ Pl hairs spreading, some glandular; seeds of 2 kinds, minutely pitted and coarsely papillate
        8. Pl gen 15–40 cm; cauline lvs lanceolate to narrowly ovate, serrate . . . . subsp. ***boothii***
        8′ Pl 5–20 cm; cauline lvs ± narrowly lanceolate or lower oblanceolate, ± entire to minutely serrate
        . . . . subsp. ***intermedia***

E. boothii (Douglas) W.L. Wagner & Hoch Pl gen ± red; rosette gen 0 (to well-developed); hairs minutely strigose and spreading, some glandular, esp in infl. **ST**: erect, peeling. **LF**: lanceolate to narrowly elliptic or narrowly ovate, sparsely minutely dentate or serrate; proximal oblanceolate or 0. **INFL**: nodding; fls gen 0 at proximal nodes. **FL**: hypanthium 4–8 mm; sepals (2.7)4–8 mm; petals 3–7.5 mm, gen white (red). **FR**: 8–35 mm, 1–3.8 mm wide, cylindric, tapered to tip, ± curved outward to strongly wavy and twisted, persistent, tardily dehiscent. **SEED**: 1.4–2.1 mm, gen of 2 kinds, minutely pitted in rows, pale brown and coarsely papillate, dark brown. 2*n*=14. [*Camissonia b.* (Douglas) P.H. Raven] Cross-pollinated.

subsp. ***alyssoides*** (Hook. & Arn.) W.L. Wagner & Hoch PINE CREEK EVENING-PRIMROSE Pl hairs gen densely, minutely strigose (esp in infl, hairs rarely spreading or glandular). **ST**: 3–35 cm. **LF**: 10–40 mm, lanceolate to narrowly ovate; proximal oblanceolate or 0. **INFL**: bracts lf-like; fls sometimes present at proximal nodes. **FR**: 1–1.4 mm wide, gen strongly wavy and twisted. **SEED**: all alike. Sandy slopes, flats, gen sagebrush scrub; 600–1700 m. MP, W&I; to s ID, w UT. [*Camissonia b.* subsp. *a.* (Hook. & Arn.) P.H. Raven] Intergrades widely with subspp. *E. boothii* subsp. *b.*, *E. boothii* subsp. *intermedia* in NV; much like *E. boothii* subsp. *desertorum*. May–Aug ★

subsp. ***boothii*** (p. 951) BOOTH'S EVENING-PRIMROSE Pl hairs spreading and glandular. **ST**: gen 15–40 cm. **LF**: 30–80 mm, lanceolate to narrowly ovate, serrate. **INFL**: bracts lf-like. **FR**: 1.4–2 mm wide, gen strongly wavy and twisted. Sandy flats, steep loose slopes, Joshua-tree and pinyon/juniper woodland; 900–2400 m. SNE; to WA, nw AZ. Intergrades widely with *E. boothii* subsp. *alyssoides*, *E. boothii* subsp. *intermedia* in NV. Jun–Aug ★

subsp. ***condensata*** (Munz) W.L. Wagner & Hoch (p. 951) Pl stout, ± glabrous, exc infl minutely strigose or glandular-hairy; rosette well developed. **ST**: 5–20(30) cm. **LF**: 25–100(130) mm, gen lanceolate to oblanceolate, ± entire to minutely dentate. **INFL**: bracts inconspicuous. **FR**: 2–3.8 mm wide, curved outward. Sandy slopes, washes, desert scrub; -70–1200 m. D; to s UT, w AZ, nw Mex. [*Camissonia b.* subsp. *c.* (Munz) P.H. Raven] Intergrades widely with *E. boothii* subsp. *desertorum*. Feb–May

subsp. ***decorticans*** (Hook. & Arn.) W.L. Wagner & Hoch SHREDDING EVENING-PRIMROSE Pl stout, ± glabrous exc infl; rosette well developed. **ST**: 12–65 cm. **LF**: gen 20–80 mm, gen lan-ceolate or proximal narrowly ovate, entire to minutely dentate. **INFL**: bracts inconspicuous. **FR**: 1.7–2.3 mm wide, curved outward. Open, gen steep and rocky, esp shale slopes; < 1850 m. s SNF, Teh, s SnJV, SnFrB, SCoRI, WTR. [*Camissonia b.* subsp. *d.* (Hook. & Arn.) P.H. Raven] Mar–Jun

subsp. ***desertorum*** (Munz) W.L. Wagner & Hoch (p. 951) Rosette well developed; hairs sparse, minutely strigose (or also glandular, esp in infl). **ST**: 10–35 cm. **LF**: gen 10–40 mm, lanceolate to narrowly ovate, or proximal oblanceolate, entire to minutely dentate. **INFL**: bracts inconspicuous. **FR**: 1–1.6 mm wide, gen curved downward. Sandy or gravelly slopes, washes, gen creosote-bush scrub; 450–2000 m. s SNH, s SNE, DMoj. [*Camissonia b.* subsp. *d.* (Munz) P.H. Raven] Intermediate between *E. boothii* subsp. *condensata*, *E. boothii* subsp. *decorticans*. Mar–Jul

subsp. ***intermedia*** (Munz) W.L. Wagner & Hoch (p. 951) BOOTH'S HAIRY EVENING-PRIMROSE Hairs dense, spreading (and glandular, esp in infl). **ST**: 5–20 cm. **LF**: gen < 25 mm, ± narrowly lanceolate, or proximal oblanceolate, ± entire to minutely serrate. **INFL**: bracts lf-like; fls sometimes present at proximal nodes. **FR**: 1–1.4 mm wide, gen curved outward or ± wavy and twisted. Sandy soils, sagebrush scrub; 1500–2150 m. SNE, DMtns; NV. [*Camissonia b.* subsp. *i.* (Munz) P.H. Raven] Intermediate between *E. boothii* subsp. *alyssoides*, *E. boothii* subsp. *b.*; ± uniform. May–Jul(Oct) ★

E. chamaenerioides (A. Gray) W.L. Wagner & Hoch Pl gen ± red, glandular-hairy, infl also minutely strigose; rosette gen ± 0. **ST**: erect, 8–50 cm, peeling. **LF**: < 80 mm, narrowly elliptic to narrowly lanceolate, sparsely minutely dentate. **INFL**: nodding. **FL**: hypanthium 1.6–2.3 mm; sepals 1.5–2.5 mm; petals 1.8–3 mm, white fading ± red. **FR**: 35–55 mm, 0.8–0.9 mm wide, spreading, cylindric, ± straight. **SEED**: 0.9–1 mm, minutely pitted in rows. 2*n*=14. Sandy slopes, flats, desert scrub; ± -50–1300 m. W&I, D; to UT, TX, nw Mex. [*Camissonia c.* (A. Gray) P.H. Raven] Self-pollinated. Related to *E. refracta.* Mar–Jun

E. minor (A. Nelson) W.L. Wagner & Hoch NELSON'S EVENING-PRIMROSE Pl gen ± red, gen densely and minutely ± gray strigose, in infl also glandular; rosette ± 0. **ST**: ± erect, 3–30 cm, peeling. **LF**: 5–45 mm, ± oblanceolate; distal-most ± linear, ± entire. **INFL**: erect; fls from pl base. **FL**: hypanthium 0.5–1.9 mm; sepals 0.8–1.8 mm; petals 0.8–1.3 mm, white fading ± red. **FR**: 15–25 mm, 0.8–1.2 mm wide, spreading, ± cylindric, tapered to tip, twisted and

wavy. **SEED**: 1.1–1.2 mm, minutely pitted in rows. 2*n*=14. Sandy slopes, flats, sagebrush scrub; ± 1200 m. MP (Wendel, Surprise Valley); to WA, WY, CO. [*Camissonia m.* (A. Nelson) P.H. Raven] Self-pollinated. Apr–Jul ★

E. refracta (S. Watson) W.L. Wagner & Hoch (p. 951)   Pl gen ± red, sparsely minutely strigose, esp in infl also glandular; rosette ± 0. **ST**: erect, 6–45 cm, peeling. **LF**: < 60 mm, narrowly elliptic to narrowly lanceolate, proximal-most gen oblanceolate, sparsely minutely dentate. **INFL**: nodding. **FL**: hypanthium 4–6 mm; sepals 4–6 mm; petals 3.5–7 mm, white fading ± red; 70–100% of pollen grains 4–5-angled. **FR**: 20–50 mm, 0.7–1 mm wide, spreading or reflexed, cylindric, straight or wavy. **SEED**: 0.9–1.5 mm, minutely pitted in rows. 2*n*=14. Sandy slopes, flats, desert scrub; -30–1300 m. D; to s NV, sw UT, w AZ. [*Camissonia r.* (S. Watson) P.H. Raven] Cross-pollinated. Mar–May

EULOBUS

Ann to subshrub, from taproot; st often thick and fleshy. **LF**: basal and/or cauline, alternate, simple, narrowly elliptic to narrowly lanceolate, pinnately lobed to occ entire. **INFL**: loose wand-like spike. **FL**: opening at dawn; hypanthium short, ± with a lobed, red-brown, fleshy disk within; sepals 4, reflexed (occ 2–3 remaining adherent); petals 4, yellow, finely flecked with red near base, this area not ultraviolet reflective, remainder of petals strongly so, fading orange-red; longer stamens opposite sepals, anthers attached at middle, pollen grains 3-angled; stigma ± hemispheric or cylindric, gen > anthers and cross-pollinated, or ± = anthers and self-pollinated. **FR**: straight to curved, sessile. **SEED**: in 1 row per chamber, narrowly obovoid, finely papillate, gen purple-spotted. 4 spp.: 3 in Baja CA, 1 from w-c CA to w&s AZ, nw Mex. (Greek: well lobed) [Wagner et al. 2007 Syst Bot Monogr 83:1–240] Incl in *Camissonia* in TJM (1993).

E. californicus Torr. & A. Gray (p. 951)   Ann, ± glabrous or lvs minutely strigose, infl sparsely glandular-hairy; rosette well developed. **ST**: erect, straight, 2–180 cm, slender; ± glaucous or green. **LF**: < 30 cm, narrowly elliptic, irregularly and sharply pinnately lobed, much reduced distally on st. **INFL**: fls widely spaced. **FL**: hypanthium 0.6–1.5 mm; sepals 3.9–8 mm; petals 6–14 mm. **FR**: reflexed, 45–110 mm, cylindric early, drying 4-angled, ± straight. **SEED**: 1.3–1.6 mm, olive-green, purple-spotted. 2*n*=14,28. Open places in coastal-sage scrub, chaparral, desert scrub; < 1300 m. NCoRO (Sonoma Co.), SnJV (Fresno Co.), SCoR, SW, D; s&w AZ, nw Mex. [*Camissonia c.* (Torr. & A. Gray) P.H. Raven] Self-pollinated. Apr–Jun

GAYOPHYTUM

Harlan Lewis

Ann. **ST**: gen erect, < 1 m, slender; hairs 0 to dense, rarely glandular. **LF**: cauline, alternate (or ± opposite near base), petioled or not, narrow-lanceolate, entire. **INFL**: fls axillary, pedicelled or not, opening at dawn. **FL**: hypanthium inconspicuous; sepals 4, staying fused in 2s or all coming free; petals 4, 0.5–8 mm, white, with 1–2 yellow or ± green spots at base, fading pink or red; stamens 8, those opposite sepals longer, pollen ± yellow; ovary chambers 2, stigma gen not exserted beyond anthers, gen touching them, gen ± spheric. **FR**: capsule, ± cylindric or flat; valves 4, gen all coming free, gen equal. **SEED**: few to many, gen all maturing, gen appressed to septum, alternate or ± opposite between chambers, in each chamber gen in 1 row and gen not overlapped, 0.5–2.3 mm, ovoid, glabrous or hairy, brown or gray mottled with brown; appendages 0. ± 9 spp.: w N.Am, 2 S.Am. (C. Gay, French author of Flora of Chile, 1800–1873) Self-fertile; taxa with petals < 3 mm self-pollinated.

1. Petals 3–8 mm; stigma exserted beyond anthers (see also *Gayophytum heterozygum*)
  2. Petals 3–7 mm; CaR, n SNH (Plumas Co.), SnBr, Wrn . . . . ***G. diffusum*** subsp. ***diffusum***
  2′ Petals 4–8 mm; SN (s of Placer Co.) . . . . ***G. eriospermum***
1′ Petals 0.5–3 mm; stigma gen not exserted beyond anthers
  3. Seeds ± 50% aborted; fr very irregular, lumpy . . . . ***G. heterozygum***
  3′ Seeds all maturing; fr not irregular, smooth to very knobby
    4. Fr flat; pl gen branched only proximally, 2° branches 0–many
      5. Lateral fr valves staying attached; petals 0.8–1.5 mm; larger stamens ± 0.7 mm . . . . ***G. humile***
      5′ All fr valves coming free; petals 1.3–1.8 mm; larger stamens 0.8–1.5 mm . . . . ***G. racemosum***
    4′ Fr ± cylindric; pl branched only distally to throughout, 2° branches gen many
      6. Seeds in each chamber overlapped, gen in 2 rows
        7. Petals 1.2–3 mm; pedicel gen < fr . . . . [3]***G. diffusum*** subsp. ***parviflorum***
        7′ Petals 0.7–1.5 mm; pedicel gen > fr . . . . ***G. ramosissimum***
      6′ Seeds in each chamber not overlapped, in 1 row
        8. Seeds > 9, ± opposite; fr slightly knobby
          9. Branches ± throughout, gen 2–8 nodes between . . . . ***G. decipiens***
          9′ Branches at base or not, many distally, gen 0–1 node between . . . . [3]***G. diffusum*** subsp. ***parviflorum***
        8′ Seeds (1)3–9, gen alternate; fr very knobby
          10. Seeds 6–9; fr gen > pedicel . . . . [3]***G. diffusum*** subsp. ***parviflorum***
          10′ Seeds (1)3–5; fr ± = pedicel . . . . ***G. oligospermum***

G. decipiens H. Lewis & Szweyk. **ST**: < 50 cm; branches ± throughout, gen 2–8 nodes between. **LF**: 1–3 cm, gen ± reduced distally on st. **INFL**: 1st fl gen 1–5 nodes distal to base. **FL**: petals 1.1–1.8 mm; larger stamens 0.8–1.5 mm; ovary hairy. **FR**: 6–8 mm, > pedicel, ± not flat, slightly knobby. **SEED**: 10–25, ± opposite, glabrous to dense-puberulent. 2*n*=14. Pinyon/juniper woodland, pine, fir forest; 1800–4200 m. SNH, SnGb, SnBr, MP, W&I, DMtns (Panamint Range); ± w US. May–Sep

G. diffusum Torr. & A. Gray   **ST**: < 60 cm; branches at base or not, gen forked distally. **LF**: 1–6 cm, gen reduced distally on st. **INFL**: 1st fl 1–20 nodes distal to base. **FL**: ovary hairy. **FR**: 3–15 mm, sessile or gen > pedicel, cylindric, slightly to very knobby. **SEED**: 3–18, alternate or ± opposite, in each chamber occ in 2 rows and overlapped, glabrous to densely puberulent. 2*n*=28. Complex from several 2*n*=14 spp.; subspp. may intergrade locally.

subsp. *diffusum* **FL:** petals 3–7 mm; larger stamens 2–6 mm; stigma exserted beyond anthers, hemispheric. Uncommon in CA. Open montane forest, sagebrush scrub; 800–3700 m. CaR, n SNH (Plumas Co.), SnBr, Wrn; w US to BC, Baja CA. May–Sep

subsp. *parviflorum* H. Lewis & Szweyk. (p. 951) **FL:** petals 1.2–3 mm; larger stamens 0.9–2 mm; stigma not exserted beyond anthers, ± spheric. **FR:** seeds in each chamber occ in 2 rows and overlapped. Common. Open montane forest, sagebrush scrub; 800–3700 m. NW, CaR, SN, TR, PR, GB. Variable; most small-fld pls assigned by Munz, others to *G. nuttallii* belong here; may occur with any taxon of genus. May–Sep

G. eriospermum Coville **ST:** < 1 m; branches 0 near base, forked distally. **LF:** 2–7.5 cm, much reduced distally on st. **INFL:** 1st fl gen 10–20 nodes distal to base. **FL:** petals 4–8 mm; larger stamens 3–7 mm; ovary hairy, stigma exserted beyond anthers, hemispheric. **FR:** 4–16 mm, > pedicel, slightly knobby. **SEED:** 4–12, alternate, glabrous to densely puberulent. 2*n*=14. Open montane forest; 100–3200 m. SN (s of Placer Co.). Jun–Oct

G. heterozygum H. Lewis & Szweyk. (p. 951) **ST:** < 80 cm; branches gen 0 at base, forked distally. **LF:** 1.5–6 cm, much reduced distally on st. **INFL:** 1st fl gen 10–20 nodes distal to base. **FL:** petals gen 2–3 mm; larger stamens 2–4 mm, pollen ± 50% aborted; ovary hairy, stigma exserted beyond anthers, hemispheric or not. **FR:** 6–15 mm, > pedicel, very irregular, lumpy. **SEED:** 2–10 maturing, ± 50% aborted, irregularly alternate, glabrous to densely puberulent. 2*n*=14. Open montane forest; 500–3000 m. KR, NCoRH, CaRH, SNH, Teh, SCoRO, TR, PR, Wrn; to WA, NV. Gen self-pollinated; stable hybrid, probably between *G. eriospermum*, *G. oligospermum*. Jun–Oct

G. humile A. Juss. (p. 951) Herbage gen glabrous. **ST:** < 30 cm; branches 0 or gen only proximal, not forked, tips ± glandular-puberulent or not. **LF:** 1–2.5 cm, little reduced distally. **INFL:** 1st fl 1–3

nodes distal to base. **FL:** petals 0.8–1.5 mm; larger stamens ± 0.7 mm; ovary glabrous or glandular-puberulent. **FR:** 8–17 mm, sessile, ± like lvs, flat, grooved along midline, slightly to very knobby; lateral 2 valves staying attached, wider. **SEED:** 24–50, ascending, ± opposite, glabrous. 2*n*=14. Drying margins of wet sites, snowbeds; 750–3200 m. NW, CaR, SN, SCoR, PR, TR, Wrn; ± w US; S.Am. [*G. nuttallii* Torr. & A. Gray] May–Sep

G. oligospermum H. Lewis & Szweyk. Pl herbage gen glabrous. **ST:** < 70 cm; branches gen 0 at base, forked distally, tips occ barely hairy, 2° branches gen many. **LF:** 1.5–6 cm, much reduced distally on st. **INFL:** 1st fl gen 10–20 nodes distal to base. **FL:** petals 1.5–2.5 mm; larger stamens 1.2–1.5 mm; ovary puberulent. **FR:** 4–9 mm, ± = pedicel, very knobby. **SEED:** (1)3–5, alternate, glabrous. 2*n*=14. Open pine forest; 1200–2800 m. TR, PR. Jun–Oct

G. racemosum Torr. & A. Gray **ST:** < 40 cm; branches gen proximal, not forked, 2° branches gen many. **LF:** 1–2.5 cm, not much reduced distally on st. **INFL:** 1st fl 1–3 nodes distal to base. **FL:** petals 1.3–1.8 mm; larger stamens 0.8–1.5 mm; ovary hairy or not. **FR:** 10–15 mm, > pedicel, flat, ± grooved along midline, smooth or slightly knobby; lateral 2 valves not attached, wider. **SEED:** 10–35, appressed or above ascending, ± opposite, glabrous or densely puberulent. 2*n*=28. Wet meadows, drying margins; 1000–4000 m. CaRH, SNH, Teh, WTR, Wrn; to WA, AB, CO. Probably from hybrids between *G. decipiens*, *G. humile*. May–Oct

G. ramosissimum Torr. & A. Gray Herbage ± glabrous. **ST:** < 50 cm; branches throughout, forked exc at base, 2° branches gen many. **LF:** 1–4 cm, much reduced distally on st. **INFL:** 1st fl 5–15 nodes distal to base. **FL:** petals 0.7–1.5 mm; larger stamens ± 0.5 mm; ovary glabrous or puberulent. **FR:** 3–9 mm, gen < pedicel, cylindric, slightly knobby. **SEED:** 10–30, in each chamber in 2 rows, overlapped, glabrous. 2*n*=14. Sagebrush scrub; 500–3500 m. CaRH, n SN, GB; ± w US. May–Sep

LUDWIGIA   FALSE LOOSESTRIFE, WATER PRIMROSE

Peter C. Hoch & Brenda J. Grewell

Ann to subshrub or emergent aquatic, often floating, rooting at nodes. **LF:** alternate to opposite. **INFL:** spike; fls 1 per bract. **FL:** radial; hypanthium 0; sepals 4–5(7), persistent; petals (0)4–5(7), white to yellow; stamens 4 or 10(12), pollen gen shed singly (in CA); stigma club-shaped to spheric. **FR:** irregularly dehiscing; wall thick or thin. **SEED:** free or embedded in woody piece of fr wall. 82 spp.: ± worldwide. (C.G. Ludwig, German botanist, physician, 1709–1773) [Raven 1963 Reinwardtia 6:327–427] Many polyploids.

1. Lvs opposite; sepals 4; petals 0 or 4, 1–3 mm; stamens 4; fr 1–10 mm, erect; seed free from fr wall (sect. *Isnardia*)
2. Petals 0; sepals 1.1–2 mm; fr 1.5–5 mm; pedicels 0–0.5 mm; ovary with 4 green stripes. . . . . **L. palustris**
2′ Petals 4; sepals 1.8–5 mm; fr 4–10 mm; pedicels 0–3 mm; ovary not striped . . . . **L. repens**
1′ Lvs alternate; sepals (4)5(6); petals (4)5(6), 9–29 mm; stamens (8)10(12) in 2 unequal sets; fr 10–32 mm, gen reflexed; seeds embedded in woody inner fr wall (sect. *Oligospermum*)
3. Bracts lanceolate to lance-ovate; sepals 8–19 mm; petals 15–29 mm
4. St gen ± glabrous; pedicels 13–25(47) mm; sepals 8–12(15) mm; petals 15–18(20) mm; fr 13–25(27) mm, ± glabrous . . . . ***L. grandiflora***
4′ St glabrous to spreading-hairy; pedicels (9)13–25(85) mm; sepals 12–19 mm; petals 18–29 mm; fr 14–26 mm, ± spreading-hairy . . . . ***L. hexapetala***
3′ Bracts deltate; sepals 6–14 mm; petals 9–22 mm . . . . ***L. peploides***
5. Pl ± spreading-hairy (exc st base); lf tip gen mucronate; sepals spreading-hairy; petals 10–16(22) mm; fr (20)24–32 mm . . . . subsp. ***montevidensis***
5′ Pl gen glabrous to ± glabrous; lf tip gen not mucronate; sepals ± glabrous; petals 9–13 mm; fr 10–17(20) mm. . . . . subsp. ***peploides***

L. grandiflora (Michx.) Greuter & Burdet LARGE-FLOWERED PRIMROSE-WILLOW Per or subshrub, 14–31(45) dm, occ with thick white spongy roots at floating nodes. **ST:** prostrate to erect, fl shoot erect, ± branched distally, gen ± glabrous. **LF:** alternate, 3–10(13) cm, petiole (0)1–20(33) mm; blade narrowly elliptic or oblanceolate to widely obovate, entire, ± glabrous with hairy midrib to spreading-hairy, tip gen not mucronate. **INFL:** bracts lanceolate to lance-ovate; pedicel 13–25(47) mm. **FL:** sepals 5(6), 8–12(15) mm; petals 5, 15–18(20) mm; stamens 10(12) in 2 unequal sets, anthers 1.7–2.3 mm. **FR:** gen reflexed, falling with pedicel; 13–25(27) mm, gen

cylindric, tapered to pedicel, ± glabrous. **SEED:** 0.8–1 mm, embedded in woody piece of inner fr wall. 2*n*=48. Streambanks, shores; < 300 m. SCo, PR; se and s US, C.Am (native?), s S.Am. [*Jussiaea uruguayensis* Cambess., in part] Invasive weed of emergent wetlands, ponds. May–Nov

L. hexapetala (Hook. & Arn.) Zardini et al. (p. 951) URUGUAYAN PRIMROSE-WILLOW Per, subshrub, emergent aquatic, 3–20(40) dm, matted, with thick white spongy roots at floating nodes, creeping to erect on land, often climbing on other pls. **ST:** prostrate to erect, fl shoot erect, simple or branched distally, glabrous (floating)

to spreading-hairy (erect). **LF:** alternate, 3–11(15) cm; petiole (0)5–30(56) mm; blade narrowly elliptic to oblanceolate or widely obovate, entire, ± glabrous, tip gen mucronate. **INFL:** bracts lanceolate to lance-ovate; pedicel (9)13–25(85) mm. **FL:** sepals 5(6), 12–19 mm; petals 5(6), 18–29 mm; stamens 10(12) in 2 unequal sets, anthers 1.5–4.5 mm. **FR:** reflexed, falling with pedicel; 14–26 mm, cylindric, tapered to pedicel, ± spreading-hairy. **SEED:** 1.2–1.5 mm, embedded in woody inner fr wall. 2*n*=80. Lake margins, wetlands; < 300 m. NCo, s NCoRO, GV, CCo, SnFrB, SCo; OR, WA, se US (native?), C.Am, s S.Am, Eur. [*Jussiaea uruguayensis* Cambess., in part; *L. grandiflora* subsp. *h.* (Hook. & Arn.) G.L. Nesom & Kartesz] Invasive weed. May–Dec ❖

L. palustris (L.) Elliott (p. 951)   Per 1–5 dm, matted. **ST:** prostrate or ascending, rooting at nodes, well branched, ± glabrous. **LF:** opposite, < 5 cm; blade narrowly elliptic to subovate, entire, ± glabrous. **INFL:** pedicel 0–0.5 mm. **FL:** sepals 4, 1.1–2 mm; petals 0; stamens 4, anthers 0.2–0.4 mm; ovary stripes 4, green. **FR:** erect, 1.5–5 mm, ± oblong, minutely strigose. **SEED:** 0.5–0.7 mm, free from fr wall. 2*n*=16. Roadside ditches, wet meadows, pond margins; < 1000 m. w NW, n SN, c SNH, SnJV, CCo, SnFrB, SCo; to BC, e US, C.Am, n S.Am; introduced ± worldwide. Highly variable, often weedy. Jun–Sep

L. peploides (Kunth) P.H. Raven   Per, matted, floating, or creeping. **ST:** prostrate to erect, simple or branched. **LF:** alternate, ± clustered, (2)3–8(11) cm; blade oblong to round, ± entire, glabrous to spreading-hairy distally on st. **INFL:** bracts deltate. **FL:** sepals (4)5(6); stamens (8)10(12) in 2 unequal sets, anthers 0.7–1.5 mm. **FR:** gen reflexed; body hard, ± glabrous to spreading-hairy. **SEED:**

1–1.5 mm, embedded in woody inner fr wall. 2*n*=16. May be serious wetland weed.

subsp. *montevidensis* (Spreng.) P.H. Raven   Pl (5)9–23(32) dm, ± spreading-hairy (exc st base), ± sticky when fresh. **ST:** fl shoot floating, creeping, or erect. **LF:** tip gen mucronate; petioles (1)5–32(57) mm. **FL:** sepals 6–14 mm, spreading-hairy; petals 4–6, 10–16(22) mm. **FR:** pedicels 7–38 mm; body (20)24–32 mm, cylindric, tapered to pedicel. Lakeshores, streambanks, floodplains, seasonal wetlands; < 500 m. s NCo, n&c SNF, GV, SCo; se US (native?), Eur, Australia; native to s S.Am. [*Jussiaea repens* L. var. *m.* (Spreng.) Munz] May–Nov ❖

subsp. *peploides* (p. 951)   Pl (4)6–14(23) dm, (sub)glabrous, not sticky when fresh. **ST:** fl shoot floating or creeping. **LF:** tip gen not mucronate; petioles 3–20(35) mm. **FL:** sepals 6–10 mm, ± glabrous; petals 5, 9–13 mm. **FR:** pedicel 5–20 mm; body 10–17(20) mm, cylindric or 5-angled. Lakeshores, streambanks, seasonal wetlands; < 900 m. NCo, NCoRO, SNF, GV, CCo, SnFrB, SCo, WTR, sw DMoj; OR, TX (native?), C.Am, S.Am, Australia. [*Jussiaea repens* L. var. *p.* (Kunth) Griseb.] May–Oct

L. repens J.R. Forst.   Per 1–3 dm, matted. **ST:** decumbent, rooting at nodes, ± branched, ± glabrous. **LF:** opposite, < 5 cm; blade narrowly elliptic to ± round, entire, ± glabrous to densely and minutely strigose. **INFL:** pedicel 0–3 mm. **FL:** sepals 4, 1.8–5 mm; petals 4, 1–3 mm, yellow; stamens 4, anthers 0.4–0.9 mm, pollen shed ± in 4s; ovary stripes 0. **FR:** erect, 4–10 mm, oblong to narrowly obconic, sometimes ± hairy. **SEED:** 0.6–0.8 mm, free from fr wall. 2*n*=32. Muddy or sandy streambanks, ponds; < 900 m. SnBr, PR, sw DMoj; to se US, Caribbean, n Mex, s Asia, Japan. Jul–Sep

NEOHOLMGRENIA

Ann from taproot; branches many, slender, ascending from base. **ST:** nearly naked proximally, densely lfy distally. **LF:** cauline, alternate, mostly closely spaced at st tips, entire. **INFL:** fls 1 in axils. **FL:** opening at dawn; sepals (3)4, reflexed separately or in pairs; petals (3)4, yellow, fading yellow; stamens (4)8, longer ones opposite sepals, anthers attached at middle, pollen grains 3-angled exc in polyploid taxa (visible with hand lens); stigma hemispheric, > anthers and cross-pollinated or ± = anthers and self-pollinated. **FR:** sessile, 0.5–1 cm, ± straight, strongly flattened, walls ± swollen by seeds. **SEED:** in 1 row per chamber, narrowly obovoid, smooth, shiny, dots or blotches 0. 2 spp.: ne CA, NV, c UT, BC, AB. (Arthur H. Holmgren (1912–1992), Noel H. Holmgren (b. 1937), and Patricia K. Holmgren (b. 1940)) [Wagner & Hoch 2009 Novon 19:130–132] Incl in *Camissonia* in TJM (1993).

N. andina (Nutt.) W.L. Wagner & Hoch (p. 951)   Pl minutely strigose, infl gen more densely so. **ST:** 1–15 cm. **LF:** many, 10–30 mm, narrowly oblanceolate. **INFL:** erect. **FL:** hypanthium 0.8–2 mm; sepals 0.8–2 mm; petals 0.8–2.3 mm; stamens opposite petals, sometimes reduced or 0. **FR:** ascending, 5–10 mm, 1–1.3 mm wide.

SEED: 0.7–1.3 mm. 2*n*=28,42. Seasonally moist flats, gen clay soil, sagebrush scrub to pinyon/juniper woodland; 1200–2000 m. CaRH, MP; to BC, AB, MT, UT. [*Camissonia a.* (Nutt.) P.H. Raven; *Holmgrenia a.* (Nutt.) W.L. Wagner & Hoch] Self-pollinated, rarely cleistogamous. May–Jul

OENOTHERA   EVENING-PRIMROSE

Ann to per, gen from taproot, occ rhizomed. **LF:** basal or cauline, alternate, gen pinnately toothed to lobed, gen sessile. **INFL:** spike, raceme-like, or fls in axils of distal, reduced lvs. **FL:** radial or (sect. *Gaura*) bilateral, gen opening at dusk; sepals 4, reflexed in fl (sometimes 2–3 remaining adherent); petals 4, yellow, white, rose, or ± purple, gen fading ± orange to ± purple, tip notched or toothed; stamens 8, filaments sometimes (sect. *Gaura*) with paired teeth at base, anthers attached at middle; ovary chambers 4, stigma gen deeply lobed, gen > anthers and cross-pollinated (or ± = anthers and self-pollinated). **FR:** gen dehiscent, cylindric to ovoid or obovoid, cylindric to 4-winged or -angled, straight to curved, gen sessile (base sometimes seedless, stalk-like). **SEED:** in gen 2(1–3) rows per chamber, or clustered or reduced to 1–4 per fr. 145 spp.: Am, some widely naturalized. (Greek: wine-scented) [Wagner et al. 2007 Syst Bot Monogr 83:1–240]

1. Fr indehiscent, short
  2. Fr ± linear; stalk-like base 2–8 mm, slender . . . . ***O. sinuosa***
  2′ Fr fusiform to ovoid; stalk-like base 0–3 mm, thick if present
    3. Ann; sepals 2–3.5 mm, barely opening; stalk-like base of fr 0 . . . . ***O. curtiflora***
    3′ Per; sepals 5–14 mm, widely opening; stalk-like base of fr short, thick
      4. Fr base > 1/2 diam of widest part, fr becoming wider ± gradually; caudex woody, branched below
        ground. . . . . ***O. suffrutescens***
      4′ Fr base ± 1/4 diam of widest part, fr with a conspicuous, abrupt bulge near middle; rhizomed . . . . ***O. xenogaura***
1′ Fr dehiscent, gen elongate
  5. Fr winged, angled or ribbed, basal part sterile
    6. Petals yellow; fr winged in distal 2/3, not narrower toward base . . . . ***O. flava***

6′ Petals white, rose, or rose-purple; basal part of fr narrower, sterile; seeds clustered

   7. Hypanthium 4–8 mm; petals 5–10 mm; fr base tapered . . . . **[*O. rosea*]**

   7′ Hypanthium 10–23 mm; petals 25–40 mm; fr base cylindric . . . . **O. speciosa**

5′ Fr not winged, often angled or tubercled, fertile throughout

  8. Fr tubercled; petals white

    9. Ann; hypanthium 20–37(47) mm; sepals 4.5–12 mm; petals 6.5–20(25) mm . . . . **O. cavernae**

    9′ Per; hypanthium 30–165 mm; sepals 16–50 mm, tips in bud not free; petals 16–56 mm . . . . **O. cespitosa**

      10. Fr 10–34 mm, lanceolate to elliptic-ovate, gen S-shaped, stalk-like base 0–1 mm; petals fading rose
         to purple; margin of seed-cavity lobed . . . . subsp. **crinita**

      10′ Fr 25–68 mm, cylindric, ± straight, stalk-like base 0–55 mm; petals fading lavender to pink;
         margin of seed-cavity entire . .: . . . . subsp. **marginata**

  8′ Fr not tubercled; petals yellow or white

    11. Petals white; st peeling; bud nodding; seeds in 1 row per chamber

      12. Per; new shoots gen from lateral roots; free sepal tips in bud 0 or < 1 mm . . . . **O. californica**

        13. St hairs dense, short, appressed and long, spreading, wavy; roots fleshy; new rosettes at st tips
           . . . . subsp. **eurekensis**

        13′ St hairs 0 (to dense, short, appressed and long, spreading); roots not fleshy; new rosettes not
           forming at st tips

          14. Cauline lvs gen ± pinnately lobed; pl ± gray-green . . . . subsp. **avita**

          14′ Cauline lvs ± entire to deeply wavy-dentate; pl green to slightly ± gray . . . . subsp. **californica**

      12′ Ann or per; new shoots from lateral roots 0; free sepal tips in bud 0–1.2 mm (pl ann or short-lived
         per) or 1–9 mm (pl per) . . . . **O. deltoides**

        15. Per, gray-green; free sepal tips in bud 1–9 mm . . . . subsp. **howellii**

        15′ Ann or short-lived per, green; free sepal tips in bud 0–1 mm

          16. Distal lvs pinnately lobed; st gen < 1 dm; petals 15–25(30) mm . . . . subsp. **piperi**

          16′ Distal lvs ± entire to ± dentate, rarely pinnately lobed; st gen 2–10 dm; petals 15–43 mm

            17. Spreading hairs ± 2 mm distally; fr base 3–5 mm wide . . . . subsp. **cognata**

            17′ Spreading hairs 1–1.5 mm distally; fr base 2–3 mm wide . . . . subsp. **deltoides**

11′ Petals yellow; st gen not peeling; buds gen erect(nodding in *Oenothera pubescens*); seeds gen in 2
   rows per chamber

  18. Fr lanceolate to ovate; seed 2.4–3.5 mm, ± wrinkled

    19. Petals fading deep salmon-red; fr twisted; 1 side of seed with 2 small ridges; c&s SNH . . . . **O. xylocarpa**

    19′ Petals fading red-purple to orange; fr not twisted; 1 side of seed with thick, U-shaped area
      surrounding a groove; D. . . . . **O. primiveris**

      20. Petals (22)29–40 mm; lf gray-green. . . . . subsp. **bufonis**

      20′ Petals 6–25(28) mm; lf gen green . . . . subsp. **primiveris**

  18′ Fr ± cylindric (sometimes slightly wider toward tip); seed 1–2 mm, pitted

    21. St decumbent to erect; ann or per (exc *Oenothera stricta*, *Oenothera pubescens* bien); fls few, in distal axils

      22. Bud erect; petal base gen with a red spot. . . . . **[*O. stricta* subsp. *stricta*]**

      22′ Bud nodding or curved upward; petal base unspotted

        23. Bud curved upward; sepal free tips in bud 0.3–3 mm . . . . **O. laciniata**

        23′ Bud nodding; sepal free tips in bud 0.1–1 mm. . . . . **O. pubescens**

    21′ St erect; bien; fls in gen dense spikes; seeds angled, irregularly pitted

      24. Hypanthium 60–135 mm . . . . **O. longissima**

      24′ Hypanthium 20–55 mm

        25. Petals 25–52 mm; stigma well exceeding anthers; pollen ± all fertile or ± 50% sterile (sterile
          grains small and shriveled at 10×)

          26. Lf crinkled; seeds and pollen ± 50% sterile. . . . . **O. glazioviana**

          26′ Lf ± flat, ± not crinkled; seeds and pollen ± all fertile. . . . . **O. elata**

            27. Sepals green or red-flushed; red blister-like hair bases on sepals ± 0; anthers 8–15 mm
              . . . . subsp. **hirsutissima**

            27′ Sepals red-flushed; red blister-like hair bases on sepals conspicuous; anthers 12–23 mm. . . . subsp. **hookeri**

        25′ Petals 7–25(30) mm; stigma ± not exceeding anthers; pollen ± 50% sterile

          28. Petals 13–23 mm, gen < sepals; gen coastal. . . . . **O. wolfii**

          28′ Petals 7–25(30) mm, ± = sepals; gen not near coast

            29. Spike dense (internodes in fr < fr); longest hairs gen without red blister-like bases; petals
              10–25(30) mm; sepals green or ± yellow (± red) . . . . **O. biennis**

            29′ Spike ± open (internodes in fr gen > fr); longest hairs gen with red blister-like bases; petals
              7–20 mm; sepals often marked ± red . . . . **O. villosa** subsp. **strigosa**

O. biennis L. COMMON EVENING-PRIMROSE Bien, rosetted; hairs glandular (esp infl), spreading (blister-like bases gen 0), and minutely strigose. **ST:** erect, 3–20 dm. **LF:** cauline 5–20 cm, oblanceolate to elliptic, ± entire to dentate, ± lobed toward base. **INFL:** spike, dense. **FL:** hypanthium (20)25–40 mm; sepals 12–20(28) mm, green or ± yellow, rarely ± red, free tips in bud 1.5–3 mm; petals 10–25(30) mm, yellow fading duller or orange. **FR:** 20–40 mm, 4–6 mm wide, cylindric, ± straight. **SEED:** 1–2 mm, angled, irregularly pitted. 2*n*=14. Disturbed places; gen < 300 m. NW, CW, SW; native to c&e N.Am; ± weedy worldwide. Gen cross-pollinated; permanent translocation heterozygote. Jun–Sep

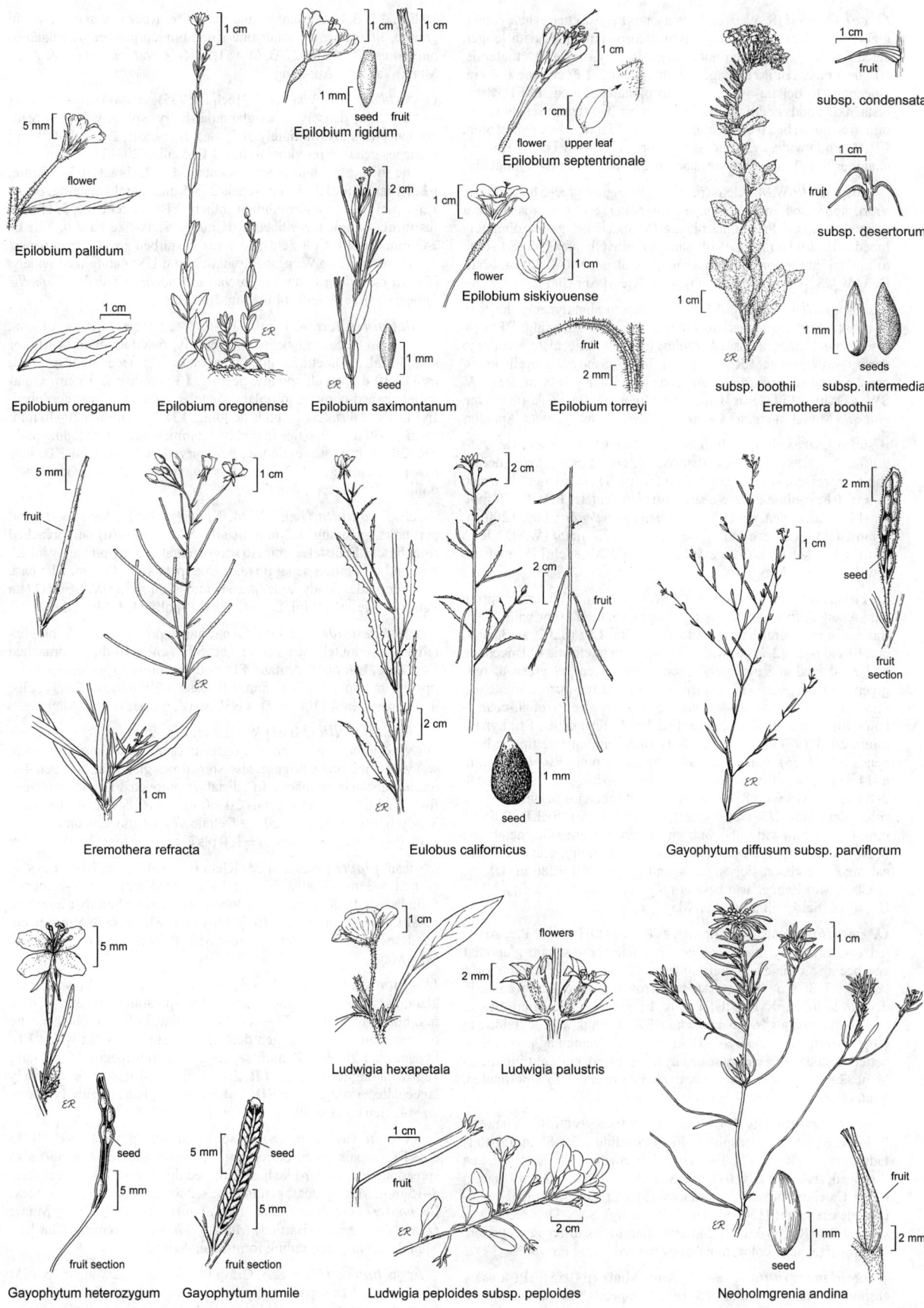

Epilobium pallidum

flower

Epilobium oreganum

Epilobium oregonense

Epilobium rigidum

seed   fruit

Epilobium saximontanum

seed

Epilobium septentrionale

flower   upper leaf

Epilobium siskiyouense

flower

Epilobium torreyi

fruit

subsp. condensata

fruit

subsp. desertorum

fruit

seeds

subsp. boothii   subsp. intermedia

Eremothera boothii

Eremothera refracta

fruit

Eulobus californicus

fruit

seed

Gayophytum diffusum subsp. parviflorum

seed

fruit section

Gayophytum heterozygum

seed

fruit section

Gayophytum humile

seed

fruit section

Ludwigia hexapetala

flowers

Ludwigia palustris

fruit

Ludwigia peploides subsp. peploides

Neoholmgrenia andina

seed

fruit

O. californica (S. Watson) S. Watson Per, rosetted when young, glabrous to densely minutely strigose, also sometimes with longer, spreading hairs; roots gen not fleshy, gen new shoots from laterals. **ST:** decumbent or ascending, 1–8 dm, peeling. **LF:** cauline 1–6 cm, lanceolate to deltate-ovate, entire to pinnately lobed. **INFL:** fls in distal axils; buds nodding. **FL:** hypanthium 20–40 mm; sepals 15–30 mm, free tips in bud 0 or < 1 mm; petals 15–35 mm, white fading pink. **FR:** 2–3 mm wide, cylindric, straight or curved. **SEED:** in 1 row per chamber, 1.4–3 mm, obovate, smooth. $2n$=14,28. Cross-pollinated.

subsp. ***avita*** W.M. Klein (p. 955) Pl ± gray-green; hairs dense, short, appressed and also longer, spreading. **ST:** new rosettes not formed at tips. **LF:** cauline oblong to lanceolate, gen ± pinnately lobed. **FR:** 20–80 mm. $2n$=14. Sandy or gravelly areas, desert scrub to pinyon/juniper or ponderosa-pine woodland; 800–2500 m. SNE, D; ± sw US. [*O. a.* (W.M. Klein) W.M. Klein] Apr–Jun

subsp. ***californica*** (p. 955) Pl green to slightly gray; hairs 0, or dense, short, appressed, also sometimes long, spreading. **ST:** new rosettes not formed at tips. **LF:** cauline oblong to lanceolate, ± entire to deeply wavy-dentate. **FR:** 30–55 mm. $2n$=28. Sandy or gravelly areas, open, coastal-sage scrub, chaparral, oak woodland; < 1900 m. c&s CW, SW, s DMtns (Little San Bernardino Mtns); nw Baja CA. [*O. c.* var. *glabrata* Munz] Intergrades with *O. californica* subsp. *avita*. Apr–Jun

subsp. ***eurekensis*** (Munz & J.C. Roos) W.M. Klein (p. 955) EUREKA DUNES EVENING-PRIMROSE Roots fleshy; hairs dense, short, appressed, also long, spreading, wavy. **ST:** new rosettes formed at tips. **LF:** cauline deltate-ovate, entire to dentate. **FR:** 30–70 mm. $2n$=14. Dunes, gen with *Psorothamnus polydenius*; 900–1200 m. n-most DMoj (Eureka Valley, ne Inyo Co.). [*O. avita* (W.M. Klein) W.M. Klein subsp. *e.* (Munz & J.C. Roos) W.M. Klein] Populations few, large. Mar–May ★

O. cavernae Munz CAVE EVENING-PRIMROSE Winter or spring ann, rosetted, from a ± fleshy tapering taproot. **ST:** 0 or with a central and gen several lateral sts to ± 3–4 cm. **LF:** (0.5)2.5–13(19.5) cm, (0.2)0.6–2.3(2.7) cm wide, oblanceolate to elliptic-oblanceolate or lyre-shaped and pinnately lobed to nearly entire, green to red-green, lustrous, gen with numerous scattered red-purple splotches, minutely glandular-hairy, sometimes also sparsely rough-coarse-hairy, finely crenate-dentate to entire. **INFL:** fls in axils. **FL:** hypanthium 20–37(47) mm; sepals 4.5–12 mm, without free tips in bud; petals 6.5–20(25) mm, white, fading to a pale pink. **FR:** 12–38 mm, 6–14 mm wide, ellipsoid-ovoid to ovoid, tapering to a sterile beak 2–8 mm, valves with 8–20 nearly distinct tubercles or with a wavy ridge, dehiscing 1/3–1/2; sessile (pedicel ≤ 1 cm). **SEED:** 2.5–3.1 mm, 1.1–1.4 mm wide, obovoid, with a cavity on one side, membrane 0. $2n$=14. Joshua-tree woodland, desert scrub in dry, gravelly (often calcareous) soils on slopes, cliffs, and ridges; 760–1280 m. DMtns (e Clark Mtn Range, also base of range in Ivanpah Valley); s NV, s UT, n AZ. Self-pollinated. Apr–May ★

O. cespitosa Nutt. FRAGRANT EVENING-PRIMROSE Per, rosetted; caudex woody, new shoots gen from lateral roots; hairs glandular and occ also coarse and nonglandular. **ST:** sprawling, < 2 dm, or ± 0. **LF:** 1.7–36 cm, oblanceolate to narrowly elliptic, gen irregularly dentate to lobed. **INFL:** fls in axils. **FL:** sepals 16–50 mm, tips in bud not free; petals 16–56 mm, white. **FR:** 4–9 mm wide, cylindric to elliptic-ovate, tubercled. **SEED:** obovate to ± triangular, papillate or netted, 1 side with a cavity sealed by a depressed, gen splitting membrane. $2n$=14,28. [*Oenothera caespitosa*, orth. var.] Cross-pollinated. 5 intergrading subspp., 2 in CA.

subsp. ***crinita*** (Rydb.) Munz CESPITOSE EVENING-PRIMROSE Pl loosely to densely cespitose. **FL:** hypanthium 30–85 mm; petals fading rose to purple. **FR:** 10–34 mm, lanceolate to elliptic-ovate, gen S-shaped; stalk-like base 0–1 mm. **SEED:** 2.9–3.5 mm; cavity margin lobed. Calcium soils in bristlecone-pine forest, pinyon/juniper woodland, desert scrub; 1150–3370 m. MP (Likely), SNE, D; w US. [*O. c.* var. *crinita* (Rydb.) Munz] 2 intergrading forms differ in elevation, habit, lf size, petal color; more study needed. Jun–Sep ★

subsp. ***marginata*** (Hook. & Arn.) Munz (p. 955) Pl loosely cespitose. **FL:** hypanthium 65–165 mm; petals fading lavender to pink. **FR:** 25–68 mm, cylindric, ± straight; stalk-like base 0–55 mm.

SEED: 2.2–3.4 mm; cavity margin entire. Rocky or sandy sites in granite, limestone, or sandstone soils, pinyon/juniper woodland to pine forest; < 2400 m. GB, D; w US. [*O. c.* var. *m.* (Hook. & Arn.) Munz] Variable. Apr–Aug

O. curtiflora W.L. Wagner & Hoch (p. 955) LIZARD-TAIL, VELVET WEED Ann, densely short-glandular-hairy, sparsely long-spreading-hairy (lvs also minutely strigose). **ST:** erect, 20–200(300) cm; branches gen 0 or proximal to infl. **LF:** cauline 20–125 mm, narrow-elliptic to -ovate, slightly wavy-dentate. **INFL:** bracts 1.5–5.5 mm. **FL:** hypanthium 1.5–5 mm; sepals 2–3.5 mm, barely opening; petals 1.5–3 mm, stigma ± surrounding anthers. **FR:** gen reflexed, 5–11 mm, fusiform, 4-angled, 8-ribbed proximally; stalk-like base 0. **SEED:** 2–3 mm. $2n$=14. Cult fields, pastures, disturbed areas, streambanks; < 400 m. SnFrB, SW; probably native to c US; naturalized widely. [*Gaura parviflora* Lehm.; *G. p.* var. *lachnocarpa* Weath.; *G. mollis* James, nom. rej.] Self-pollinating. Jul–Aug

O. deltoides Torr. & Frém. DEVIL'S LANTERN, LION-IN-A-CAGE, BASKET EVENING-PRIMROSE Loosely rosetted; hairs curly or straight, also sometimes glandular, or 0. **ST:** decumbent or erect, (< 1)2–10 dm, stout, spongy, peeling. **LF:** cauline 2–15 cm, ± diamond-shaped-obovate to oblanceolate, ± entire to pinnately lobed. **INFL:** fls in distal axils; buds nodding. **FL:** hypanthium 20–40 mm; sepals 8–30 mm, free tips in bud 0–1.2 mm; petals white fading pink. **FR:** 20–60(80) mm, cylindric, gen curved, ± twisted. **SEED:** in 1 row per chamber, 1.5–2 mm, obovate, smooth. $2n$=14. Gen cross-pollinated. 5 subspp., 4 in CA.

subsp. ***cognata*** (Jeps.) W.M. Klein (p. 955) Ann, short-lived per; hairs spreading, ± 2 mm distally. **ST:** gen 2–4(6) dm, branched from base. **LF:** distal ± entire to wavy-dentate, rarely pinnately lobed. **FL:** bud tip obtuse; sepal tips not free; petals 25–40 mm. **FR:** base 3–5 mm wide. Sandy soils, grassland; < 700 m. SnJV, SCoRO (La Panza Range), w DMoj. Sometimes self-pollinated. Mar–May

subsp. ***deltoides*** (p. 955) Ann; hairs spreading, 1–1.5 mm distally, also minutely strigose or not. **ST:** gen 2–10 dm, ± branched from base. **LF:** distal dentate. **FL:** bud tip obtuse to acute; sepal free tips 0–1 mm; petals 15–43 mm. **FR:** base 2–3 mm wide. Sandy soils, incl dunes; gen < 1100 m. D; s NV, w AZ, nw Mex. Mar–May

subsp. ***howellii*** (Munz) W.M. Klein (p. 955) ANTIOCH DUNES EVENING-PRIMROSE Per, gray-green; hairs spreading, 1–3 mm, wavy, also minutely strigose, also sometimes glandular. **ST:** gen 4–8 dm, ± branched throughout. **LF:** distal wavy-lobed. **FL:** bud tip obtuse; free sepal tips 1–9 mm; petals 20–40 mm. **FR:** base 3–4 mm wide. Sandy bluffs, dunes; < 100 m. Deltaic GV (Antioch, Contra Costa Co.). Populations with few to ± 150 pls at any one time. Mar–Sep ★

subsp. ***piperi*** (Munz) W.M. Klein (p. 955) Ann; hairs wavy or curly, 1.5–3 mm distally. **ST:** gen < 1 dm, gen simple or few-branched from base. **LF:** distal pinnately lobed. **FL:** bud tip bluntly acute; free sepal tips 0–1 mm; petals 15–25(30) mm. **FR:** base 3–5 mm wide. Sand, incl dunes, sagebrush scrub; 850–1800 m. MP; c OR, w NV. Mar–May

O. elata Kunth Bien, densely minutely strigose and (esp in infl) glandular; hairs also long, appressed to spreading, sometimes with red, blister-like base. **ST:** erect. **LF:** cauline 4–25 cm, oblanceolate to lanceolate or elliptic, gen dentate to ± entire. **INFL:** spike. **FL:** hypanthium 20–48(55) mm; sepals 27–48 mm; petals 25–52 mm, yellow fading red-orange. **FR:** 20–65 mm, 4–7 mm wide, narrowly lanceolate, ± straight. **SEED:** 1–1.8 mm, angled, irregularly pitted. $2n$=14. Gen cross-pollinated. 3 subspp., 2 in CA.

subsp. ***hirsutissima*** (S. Watson) W. Dietr. (p. 955) **ST:** 10–25 dm. **FL:** sepals green or red-flushed, hairs glandular or not, also spreading, red, blister-like base ± 0, free tips in bud 3–6 mm; anthers 8–15 mm. Moist places, gen inland; < 2800 m. CA; w US, nw Mex. [*O. hookeri* Torr. & A. Gray subsp. *angustifolia* (R.R. Gates) Munz; *O. h.* subsp. *grisea* (Bartlett) Munz; *O. h.* subsp. *venusta* (Bartlett) Munz] Several intergrading forms. Jun–Sep

subsp. ***hookeri*** (Torr. & A. Gray) W. Dietr. & W.L. Wagner (p. 955) **ST:** 4–8 dm. **FL:** sepals red-flushed, hairs glandular and also spreading with conspicuous, red, blister-like base, free tips in bud 1–5 mm;

anthers 12–23 mm. Moist, coastal, slightly inland, sandy bluffs; < 200 m. CW, SW. [*O. h.* subsp. *montereyensis* Munz] Jun–Sep

O. flava (A. Nelson) Garrett (p. 955) Per, ± rosetted, green; hairs glandular, also gen minutely strigose; taproot fleshy. **ST**: ± 0. **LF**: 3–36 cm, oblanceolate to oblong, irregularly pinnately lobed, ± fleshy. **INFL**: fls in axils. **FL**: hypanthium 24–80 (150) mm; sepals 10–34 mm, free tips in bud 1–2(5) mm; petals 10–30(38) mm, yellow fading pale orange. **FR**: 10–40 mm, 4–7 mm wide, narrowly ovate to elliptic; wings 2–6 mm wide. **SEED**: 1.8–2.6 mm, obliquely wedge-shaped, minutely beaded, narrow-winged distally and along 1 margin. 2*n*=14. Drying depressions, streambanks, gen clay soils, sagebrush scrub to pinyon/juniper woodland; 900–1600 m. CaR, n SNH, MP; to s Can, CO, Mex. Gen self-pollinated. May–Jul

O. glazioviana Micheli (p. 955) Bien, rosetted, densely minutely strigose; hairs also long, spreading, gen with red blister-like base, in infl also glandular. **ST**: erect, 5–15 dm. **LF**: cauline 5–15 cm, elliptic to lanceolate, crinkled, dentate to ± entire. **INFL**: spikes. **FL**: hypanthium 35–50 mm; sepals 28–45 mm, free tips in bud 5–8 mm; petals 35–50 mm, yellow fading red-orange. **FR**: 20–35 mm, cylindric, ± straight. **SEED**: 1.3–2 mm, angled, irregularly pitted, ± 50% sterile. 2*n*=14. Disturbed places; < 500 m. NW, CW, SW; ± worldwide. [*O.* ×*erythrosepala* Borbás] Commonly cult; naturalized occurrences scattered. Gen cross-pollinated. Permanent translocation heterozygote. Possibly derived in Eur from hybrids between two N.Am spp. Jun–Sep

O. laciniata Hill Ann, short-lived per, rosetted, minutely strigose; hairs distally gen also long, spreading, and glandular. **ST**: decumbent to erect, 0.5–5 dm. **LF**: cauline 2–10 cm, narrowly oblanceolate to ± elliptic or oblong, ± entire to pinnately lobed. **INFL**: fls in distal axils; buds curved upward. **FL**: hypanthium 12–35 mm; sepals 5–15 mm, free tips in bud 0.3–3 mm; petals 5–22 mm, yellow fading orange. **FR**: 20–50 mm, 2–4 mm wide, cylindric. **SEED**: 1–1.8 mm, ± spheric or widely elliptic, pitted. 2*n*=14. Open, gen sandy, disturbed places; gen < 500 m. CW, SW, DMoj; native to e US; widely naturalized, but occurrences scattered. Self-pollinated. Permanent translocation heterozygote. May–Jul

O. longissima Rydb. LONG-STEM EVENING-PRIMROSE Bien, rosetted, minutely strigose; hairs also gen long, spreading, with red, bristle-like base, sometimes some glandular. **ST**: erect, 6–30 dm. **LF**: cauline 5–22 cm, narrowly oblanceolate to ± elliptic, ± entire to dentate. **INFL**: spike. **FL**: hypanthium 60–135 mm; sepals 23–47 mm, free tips in bud 2–6 mm; petals 28–65 mm, yellow fading red-orange. **FR**: 25–55 mm, 4–9 mm wide, ± cylindric, ± straight. **SEED**: 1–2 mm, angled, irregularly pitted. 2*n*=14. Seasonally moist places in creosote-bush scrub, pinyon/juniper woodland; 1000–1700 m. e DMtns (New York Mtns); to CO, AZ. [*O. l.* subsp. *clutei* (A. Nelson) Munz] Gen cross-pollinated. Jul–Sep ★

O. primiveris A. Gray Ann, rosetted, minutely strigose, in infl gen glandular; hairs also coarse, with red, blister-like base or not. **ST**: gen 0 (sometimes erect or ascending, < 3.5 dm). **LF**: 4–28 cm, oblanceolate, wavy-dentate to 1–2-pinnately lobed. **INFL**: fls in axils. **FL**: hypanthium 20–72 mm; sepals 7–30 mm, free tips in bud 0; petals yellow fading red-purple to orange. **FR**: 10–60 mm, 4–8 mm wide, lanceolate to ovate, straight, curved, or S-shaped. **SEED**: 3–3.5 mm, irregularly obovate to oblanceolate, papillate, 1 side coarsely wrinkled in distal 1/2, other side with thick, U-shaped area forming groove and small cavity at tip. 2*n*=14.

subsp. ***bufonis*** (M.E. Jones) Munz (p. 955) **LF**: gray-green. **FL**: petals (22)29–40 mm. Sandy flats, low hills, dune margins, arroyos; 30–1400 m. D; to UT, w AZ, nw Mex. Gen cross-pollinated. Mar–May

subsp. ***primiveris*** **LF**: gen green. **FL**: petals 6–25(28) mm. Uncommon. Sandy flats, low hills, dune margins, arroyos; 30–1400 m. D; to TX, nw Mex. [*O. p.* subsp. *caulescens* (Munz) Munz] Self-pollinated. Mar–May

O. pubescens Spreng. (p. 955) Ann, bien, rosetted, minutely strigose; hairs also long, spreading, and gen some glandular. **ST**: decumbent to erect, 0.5–8 dm. **LF**: cauline 2–8 cm, narrow-oblanceolate to lanceolate or elliptic, pinnately lobed to ± entire. **INFL**: fls in distal axils; buds nodding. **FL**: hypanthium 15–50 mm; sepals 5–25 mm, free tips in bud 0.1–1 mm; petals 5–35 mm, yellow fading

orange. **FR**: 20–45 mm, 2–4 mm wide, cylindric. **SEED**: 1–1.5 mm, ± spheric, pitted. 2*n*=14. Open places; ± 600 m. c DMoj (Newberry Springs, San Bernardino Co.); to NM, S.Am. [*O. laciniata* subsp. *p.* (Spreng.) Munz] Permanent translocation heterozygote. CA pls possibly introduced from AZ. May–Jul

O. sinuosa W.L. Wagner & Hoch (p. 955) WAVY-LEAVED GAURA Per, forming large mats, rhizomed; hairs gen sparse or 0. **ST**: 20–60 cm, branched, sparsely minutely strigose and with long, spreading hairs. **LF**: 10–110 mm, linear to narrow-oblanceolate, slightly wavy-dentate. **INFL**: bracts 1–5 mm. **FL**: hypanthium 2.5–5 mm; sepals 7–14 mm; petals 7–14.5 mm. **FR**: erect, 4–16 mm, ± linear, narrowly 4-winged; stalk-like base 2–8 mm, slender, tapered. **SEED**: 1–4, 2–3 mm. 2*n*=28. Light sandy loam of cult fields; < 1000 m. GV, CW, SW, DMoj; native to OK, TX; widely naturalized, esp se US. [*Gaura sinuata* Ser.; *G. villosa* Torr. var. *v.*, misappl.; *G. v.* var. *mckelveyae* Munz, misappl.] Limited by self-sterility. Jun–Sep ◆

O. speciosa Nutt. (p. 955) Per, rosetted when young, forming large patches from woody caudices and rhizomes, minutely strigose (some hairs also longer). **ST**: weakly ascending to erect, 1–5 dm. **LF**: cauline 2.5–8 cm, oblanceolate to ± elliptic, ± entire to wavy-lobed. **INFL**: fls in distal axils; st tip nodding. **FL**: hypanthium 10–23 mm; sepals 15–30 mm, free tips in bud 1–4 mm; petals 25–40 mm, white fading pink or rose-purple. **FR**: 10–25 mm, widening upward (to 3–5 mm), 8-ribbed; stalk-like base (4)8–15 mm, 1.5–2 mm wide, cylindric. **SEED**: clustered in each chamber, 1–1.5 mm, obliquely oblanceolate, finely granular-papillate. 2*n*=14,28,42. Disturbed places; gen < 500 m. SCo; native NM to c US, c Mex. [*O. s.* var. *childsii* (L.H. Bailey) Munz] Commonly cult, doubtfully naturalized; cross-pollinated. May–Sep

O. suffrutescens (Ser.) W.L. Wagner & Hoch (p. 955) WILD HONEYSUCKLE, LINDA TARDE Per, gen minutely strigose and long-spreading-hairy, or ± glabrous; caudex woody, branched below ground. **ST**: 10–120 cm. **LF**: 10–70 mm, linear to narrow-elliptic, entire to coarsely wavy-serrate. **INFL**: bracts 2–5 mm. **FL**: hypanthium 4–13 mm; sepals 5–10 mm; petals 3–8 mm. **FR**: erect or spreading, 4–9 mm, 4-angled; stalk-like base short, thick, > 1/2 diam of widest part. **SEED**: 1.5–3 mm. 2*n*=14,42,56. Dry slopes, gen limestone, Joshua-tree or pinyon/juniper woodland; 900–1600 m. DMtns (naturalized Teh, SW); to w Can, c US, Mex. [*Gaura coccinea* Pursh; *G. c.* var. *glabra* (Lehm.) Munz] Apr–Jun

O. villosa Thunb. subsp. ***strigosa*** (Rydb.) W. Dietr. & P.H. Raven (p. 955) Bien, rosetted, minutely strigose, esp in infl also glandular; hairs also long, spreading, gen with red, blister-like base. **ST**: erect, 5–20 dm. **LF**: cauline 10–30 cm, lanceolate or elliptic, entire to minutely dentate. **INFL**: spike, ± open, few-fld; internodes in fr gen > fr. **FL**: hypanthium 25–40 mm; sepals 9–18 mm, often marked ± red, free tips in bud 0.5–2.5 mm; petals 7–20 mm, yellow fading duller to pale orange. **FR**: 20–35 mm, 4–7 mm wide, ± cylindric, ± straight. **SEED**: 1–2 mm, angled, irregularly pitted. 2*n*=14. Moist openings in forest; esp 500–2000 m. KR, NCoRI (Tehama Co.), CaRH (Plumas Co.), MP; to sw Can, c US. [*O. s.* (Rydb.) Mack. & Bush; *O. biennis* L., misappl.; *O. hookeri*, misappl.] Permanent translocation heterozygote. Jul–Aug

O. wolfii (Munz) P.H. Raven et al. (p. 955) WOLF'S EVENING-PRIMROSE Bien, rosetted, densely minutely strigose; many hairs also with red, blister-like base, some glandular. **ST**: erect, 5–10 dm. **LF**: cauline 5–18 cm, narrowly lanceolate to elliptic, wavy-dentate, distal dentate. **INFL**: spike. **FL**: hypanthium 30–46 mm; sepals 17–28 mm, free tips in bud erect, 1–3 mm; petals 13–23 mm, yellow fading red-orange. **FR**: 30–48 mm, 5–8 mm wide, ± cylindric, ± straight. **SEED**: 1–2 mm, angled, irregularly pitted. 2*n*=14. Coastal sand, incl dunes, bluffs, roadsides, gen moist places (perhaps also inland); < 100 m (± 800 m, Carrville, Trinity Co.). n NCo, KR; OR. [*O. hookeri* subsp. *w.* Munz] Permanent translocation heterozygote. May–Oct ★

O. xenogaura W.L. Wagner & Hoch (p. 955) Per, forming large mats, rhizomed; minutely strigose and gen long-spreading-hairy. **ST**: 1–several, 2–6(12) cm. **LF**: 5–95 mm, narrowly lanceolate to elliptic, ± entire to shallowly and coarsely wavy-dentate. **INFL**: bracts 2–8 mm. **FL**: hypanthium 4–14 mm; sepals 7–14 mm; petals 6–10 mm. **FR**: erect, 7–13 mm, ovate, 4-angled, with conspicuous, abrupt bulge near middle; stalk-like base short and thick, ± 1/4 diam of widest

part. **SEED**: 2–8, 2–2.5 mm. 2*n*=28. Cult fields; < 300 m. CW, SW; native c TX to c Mex. [*Gaura drummondii* (Spach) Torr. & A. Gray; *G. odorata* Lag., misappl.] May–Sep ◆

O. xylocarpa Coville (p. 955)   Per, rosetted, gray-green; hairs short (on fl long, coarse), erect; taproot fleshy. **ST**: 0. **LF**: main segment ± round to oblanceolate, 26–62 mm; lobes few, gen small. **INFL**: fls in axils. **FL**: hypanthium 27–45(55) mm; sepals 25–30 mm, tips not free

in bud; petals 25–38 mm, yellow fading deep salmon-red. **FR**: 35–90 mm, 7–11 mm wide, narrowly lanceolate, curved, twisted, wrinkled. **SEED**: in 1 row (or near base 2 rows) per chamber, 2.4–3.2 mm, obovate; tip gen truncate, coarsely wrinkled, papillate; 1 side with 2 small ridges. 2*n*=14. Gravelly to pumice meadows, Jeffrey-pine or lodgepole-pine/fir forests; 2200–3100 m. c&s SNH; w NV. Gen cross-pollinated. Locally common. Jul–Aug

TARAXIA

Per, from thick taproot that may branch in age producing multiple rosettes, st 0. **LF**: basal, simple, ovate or lanceolate to narrowly elliptic, nearly entire to deeply pinnately lobed. **INFL**: fls 1 in axils, erect in bud. **FL**: radial, opening at dawn; sepals 4, reflexed separately; petals 4, yellow (or white), unspotted, strongly ultraviolet reflective (non-reflective at base), fading ± purple or ± red; stamens 8, longer ones opposite sepals, anthers attached at base, pollen grains 3-angled exc in polyploid taxa (visible with hand lens); ovary chambers 4, stigma hemispheric, gen > anthers and cross-pollinated (or ± = anthers and self-pollinated), sterile projection of ovary persistent with fertile part in fr, without abscission lines at juncture between hypanthium and fertile part of ovary. **FR**: cylindric-ovoid, ± angled to nearly cylindric, walls gen thin, distended by seeds, tip gradually attenuate into a slender sterile portion (0.4)1.5–18 cm, occ persistent for > 1 yr, sessile. **SEED**: in 2 rows per chamber, elongate-ovoid, cylindric to oblong-ellipsoid, pitted or coarsely papillate, giving a shaggy appearance. 4 spp.: CA to WA & sw Can, NV to CO (esp CA-FP). (Lvs similar to *Leontodon taraxacoides*) [Wagner et al. 2007 Syst Bot Monogr 83:1–240] Incl in *Camissonia* in TJM (1993).

1. Lf pinnately lobed; pl hairs ± dense, spreading or appressed; hypanthium 4–8.5 mm; stigma exceeding anthers . . . . ***T. tanacetifolia***
1′ Lf not pinnately lobed (or pl ± glabrous); hypanthium 1.5–3 mm; stigma at same level as or slightly exceeding anthers
    2. Lf hairs short, erect; fr ± cylindric (swollen by seeds), papery . . . . ***T. ovata***
    2′ Lf ± glabrous; fr 4-angled, barely swollen by seeds, leathery . . . . ***T. subacaulis***

T. ovata (Torr. & A. Gray) Small (p. 955)   Pl ± fleshy. **LF**: blade 30–150 mm, narrowly elliptic to ovate, ± entire to wavy, veins and margin with short, erect hairs; petiole 8–150 mm. **FL**: hypanthium 2–3 mm; sepals 11–19 mm; petals 8–23 mm; sterile tip of ovary 25–180 mm. **FR**: 11–30 mm, linear-oblanceolate, ± straight or slightly curved, swollen by seeds, papery, tardily dehiscent, ± sessile. **SEED**: 1.8–2.2 mm, papillate, brown. 2*n*=14. Grassy fields, gen clay soil; < 500 m. NW, CW; sw OR. [*Oenothera o.* Torr. & A. Gray; *Camissonia o.* (Torr. & A. Gray) P.H. Raven] Mar–Jun

T. subacaulis (Pursh) Rydb. (p. 955)   Pl ± fleshy, ± glabrous. **LF**: blade 20–220 mm, lanceolate to narrowly elliptic, ± entire to irregularly pinnately lobed, rarely sparsely and minutely strigose; petiole 10–120 mm. **FL**: erect; hypanthium 1.5–3 mm; sepals 4.1–13 mm; petals 5–16 mm; sterile tip of ovary 15–80 mm. **FR**: 11–28 mm, linear-ovoid, 4-angled, barely swollen by seeds, ± straight or slightly

curved, leathery; pedicel 0–10 mm. **SEED**: 1.3–1.9 mm, pitted, pale brown. 2*n*=14. Moist meadows, gen clay soils; 450–2600 m. SN, GB; to WA, MT, CO. [*Camissonia s.* (Pursh) P.H. Raven; *Oenothera s.* (Pursh) Garrett; *O. s.* var. *taraxacifolia* (S. Watson) Jeps.] May–Jul

T. tanacetifolia (Torr. & A. Gray) Piper (p. 955)   Hairs ± dense (sparse), short, spreading or appressed. **LF**: 65–320 mm, narrowly elliptic, deeply and irregularly pinnately lobed; petiole 10–80 mm. **FL**: hypanthium 4–6.5(8.5) mm; sepals 5.5–13 mm; petals 8–23 mm; sterile tip of ovary 14–55 mm. **FR**: 7–25 mm, long-tapered, swollen by seeds, leathery, ± straight or slightly curved, disintegrating irregularly. **SEED**: 1.5–2 mm, pitted in rows, pale brown. 2*n*=14,28,42. Open fields, moist slopes, clay soils; 700–2500 m. CaR, SN, GB; to WA, ID, NV. [*Oenothera t.* Torr. & A. Gray; *Camissonia t.* (Torr. & A. Gray) P.H. Raven; *C. t.* subsp. *quadriperforata* P.H. Raven] May–Jul ★

TETRAPTERON

Ann, from taproot, sts rarely present, to 25 mm. **LF**: mostly basal, or cauline on short sts, alternate, broadly sessile or clasping. **INFL**: fls 1 in axils of rosette, nodding in bud. **FL**: opening at dawn; sepals 4, reflexed in pairs; petals 4, yellow, unspotted, strongly ultraviolet-reflective, fading red-orange; longer stamens opposite sepals, anthers attached near base, pollen grains 3-angled; ovary tip with sterile 0.4–18 cm projection that often breaks off in fr, with clear abscission line at juncture between short hypanthium and fertile part of ovary, stigma hemispheric, ± = anthers and gen self-pollinated (or > anthers and cross-pollinating). **FR**: irregularly obovoid, sharply 4-angled with pointed wing near center top of each valve, ± sessile, very tardily dehiscent in distal 1/2 only, thick-walled (not distended by seeds), persistent for > 1 yr, sessile. **SEED**: in 2 crowded rows per chamber, obovoid to narrowly obovoid, finely papillate or deeply pitted, tan or brown with dark patches. 2 spp.: CA, disjunct in nw NV, s OR, n Baja CA. (Greek: four winged) [Wagner et al. 2007 Syst Bot Monogr 83:1–240] Incl in *Camissonia* in TJM (1993).

1. Lf linear to narrowly lanceolate or oblanceolate; petals 5–18 mm; hypanthium 1.6–3.2 mm; pl densely spreading-hairy . . . . ***T. graciliflorum***
1′ Lf narrowly oblanceolate; petals 2–3.5 mm; hypanthium 0.8–1.3 mm; pl strigose and sparsely spreading-hairy . . . ***T. palmeri***

T. graciliflorum (Hook. & Arn.) W.L. Wagner & Hoch (p. 961) HILL SUN CUP   **ST**: to 25 mm, not swollen, not peeling. **LF**: 10–98 mm, entire or weakly and minutely serrate. **FL**: sepals 4.5–8 mm; petals ± = anthers and gen self-pollinated or exceeding anthers and cross-pollinated; sterile tip of ovary 6–45 mm. **FR**: 4–8 mm, 4-angled (4-winged near tip), leathery, tardily dehiscent. **SEED**: 1.2–2 mm. 2*n*=14. Open or shrubby slopes, gen clay soils, grassland, oak and Joshua-tree woodland; < 800 m. NW, CaR, SNF, GV, CW, WTR; sw OR, Baja CA. [*Oenothera g.* Hook. & Arn.; *Camissonia g.* (Hook. & Arn.) P.H. Raven] Gen self-pollinated. Mar–Apr

T. palmeri (S. Watson) W.L. Wagner & Hoch (p. 961)   **ST**: to 20 mm, swollen, peeling. **LF**: 15–55 mm, minutely serrate. **FL**: sepals 1.6–2.3 mm; petals ± = anthers; sterile tip of ovary 5–12 mm. **FR**: 5–7 mm, 4-angled, 4-winged near tip, ± straight, leathery, tardily dehiscent. **SEED**: 1.2–2 mm. 2*n*=14. Desert flats, sagebrush scrub; 600–1400 m. s GV, SCoRI, WTR (Tejon Pass), PR (Jacumba), W&I, DMoj; OR, NV. [*Oenothera p.* S. Watson; *Camissonia p.* (S. Watson) P.H. Raven] Self-pollinated. Apr–May

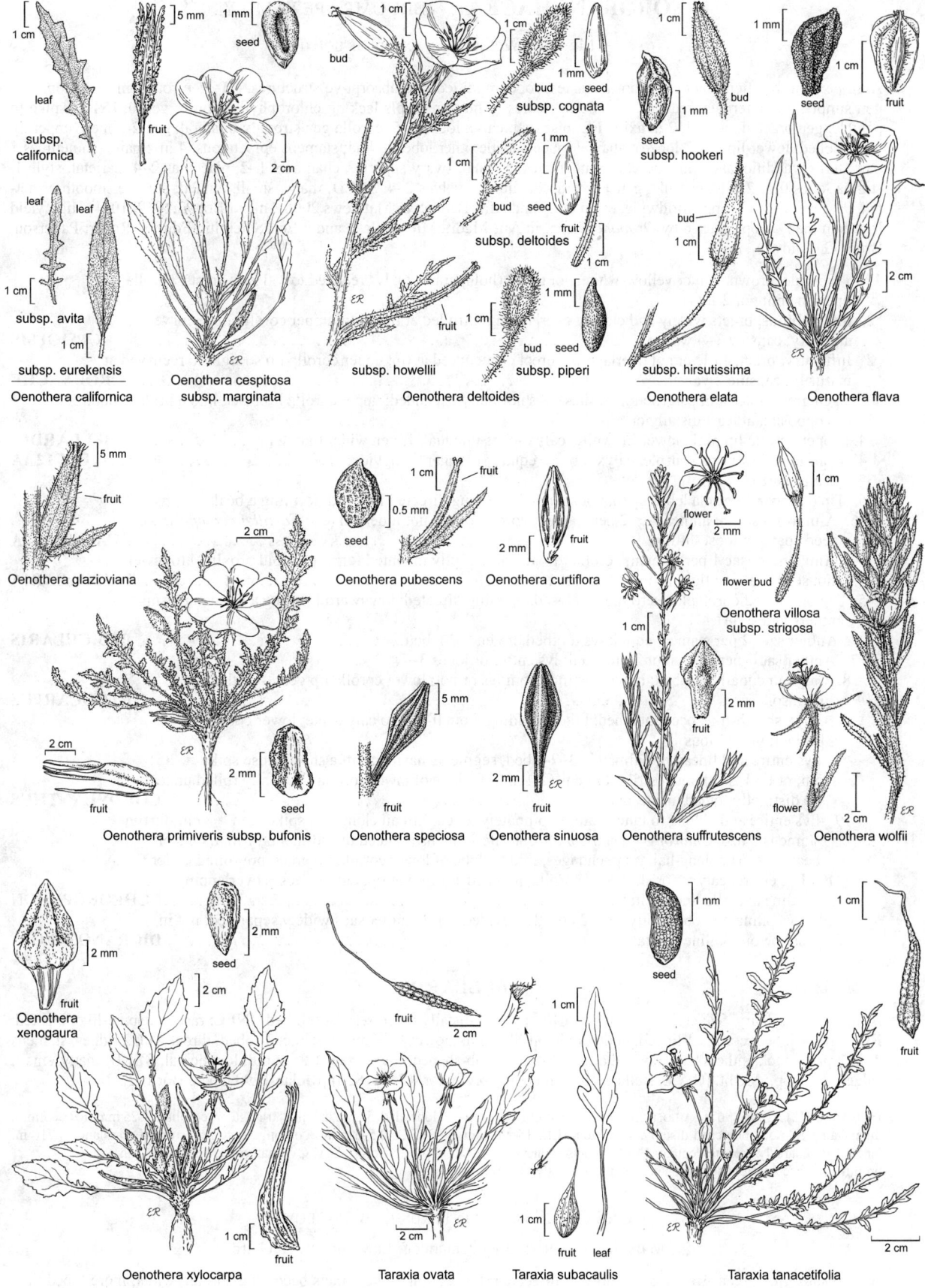

subsp. californica

leaf

1 cm

leaf

leaf

1 cm

subsp. avita

1 cm

subsp. eurekensis

Oenothera californica

5 mm

fruit

1 mm

seed

2 cm

Oenothera cespitosa
subsp. marginata

1 cm

bud

1 cm

ER

1 cm

bud

seed

subsp. cognata

1 mm

1 cm

bud

seed

1 mm

fruit

subsp. deltoides

1 cm

1 cm

bud

seed

1 mm

fruit

subsp. howellii

subsp. piperi

Oenothera deltoides

1 cm

bud

seed

1 mm

subsp. hookeri

bud

1 cm

subsp. hirsutissima

Oenothera elata

1 mm

seed

1 cm

fruit

2 cm

ER

Oenothera flava

5 mm

fruit

Oenothera glazioviana

2 cm

2 cm

fruit

2 mm

seed

ER

Oenothera primiveris subsp. bufonis

1 cm

0.5 mm

seed

fruit

Oenothera pubescens

2 mm

fruit

Oenothera curtiflora

5 mm

fruit

2 mm

ER

fruit

Oenothera speciosa

Oenothera sinuosa

flower

2 mm

1 cm

1 cm

flower bud

Oenothera villosa
subsp. strigosa

2 mm

fruit

flower

2 cm

ER

Oenothera wolfii

Oenothera suffrutescens

2 mm

fruit

Oenothera
xenogaura

2 mm

seed

2 cm

1 cm

fruit

ER

Oenothera xylocarpa

fruit

2 cm

1 cm

ER

Taraxia ovata

1 cm

1 cm

fruit   leaf

Taraxia subacaulis

1 mm

seed

1 cm

fruit

ER

2 cm

Taraxia tanacetifolia

OROBANCHACEAE BROOMRAPE FAMILY

Margriet Wetherwax, except as noted

Ann, per, shrub; often glandular; root-parasites, roots modified into absorptive structures. **ST**: gen round in ×-section. **LF**: gen simple, gen alternate, reduced to ± fleshy scales in non-green pls lacking chlorophyll; stipules gen 0. **INFL**: spike to panicle, gen bracted, or fls 1–2 in axils. **FL**: bisexual; calyx lobes 0–5; corolla gen strongly bilateral, gen 2-lipped (upper lip gen 2-lobed, lower lip gen 3-lobed), abaxial lobes outside other lobes in bud; stamens epipetalous, 4 in 2 pairs (sometimes 1 pair sterile), additional staminode 0(1), anther sacs unequal; ovary superior, chambers 1–2, placentas 2–4, parietal, style 1, stigma lobes 0 or 2. **FR**: capsule, gen ± ovoid, loculicidal, valves 2–4. **SEED**: many, small, angled; surface smooth or netted. 99 genera, 2060 spp.: worldwide, esp n temp and Afr. [Bennett & Matthews 2006 Amer J Bot 93:1039–1051] High yield losses in many crops caused by *Orobanche* spp. in Afr, Medit, Middle East, and e Eur. Scientific Editors: Robert Patterson, Bruce G. Baldwin.

1. Pl ± purple, brown, tan, or yellow-white, not green (holoparasites); lvs reduced to ± fleshy bracts; corolla upper lip entire or 2-lobed
   2. Infl glabrous; bracts widely and closely overlapping, inrolled at tip; tip of upper corolla lip incurved at maturity; capsule 3–4 valved . . . . **KOPSIOPSIS**
   2′ Infl hairy; bracts little or not overlapping, erect or recurved at tip; upper corolla lip straight to recurved at maturity; capsule 2-valved . . . . **OROBANCHE**
1′ Pl gen green at least in part (hemiparasites); lvs normally expanded; upper corolla lip fused into a beak or hood
   3. Lvs opposite, at least distally on st
      4. Upper corolla lip pink, lower lip white; calyx lobes unequal; fr gen wider than long. . . . . **BELLARDIA**
      4′ Corolla yellow or red-purple; calyx lobes ± equal; fr longer than wide. . . . . **PARENTUCELLIA**
   3′ Lvs alternate or ± basal
      5. Tip of upper 2 corolla lobes open, not rounded, opening directed forward, forming a beak
         6. Ann to subshrub; anther sacs 2 per stamen; corolla throat not indented (exc *Castilleja campestris*); seed coat gen loose-fitting . . . . **CASTILLEJA**
         6′ Ann; anther sac 1 per stamen; corolla throat gen abruptly indented forming a fold proximal to lower lip; seed coat gen tight-fitting . . . . **TRIPHYSARIA**
      5′ Tip of upper 2 corolla lobes rounded, closed, opening directed downward forming a hood enclosing anthers and style
         7. Anther sacs 2 per stamen, equal; lvs toothed or gen > 7-lobed. . . . . **PEDICULARIS**
         7′ Anther sac 1 per stamen or 2, unequal; lvs entire or lobes 3–7
            8. Calyx unequally 4-lobed; bracts distinct from lvs or not; lower corolla lip gen 3-toothed; infl a spike. . . . . **ORTHOCARPUS**
            8′ Calyx sheath-like, occ ± notched; bracts grading from lf-like to calyx-like; lower corolla lip 3-lobed; infl various
               9. Lvs entire and linear, or palmately 3–7-lobed, segments narrow; infl a short dense spike, < 2(5) cm, or fls 1, scattered; fertile stamens (2)4; middle lobe of lower corolla lip tightly rolled under, tip distinctly folded inside-out . . . . **CORDYLANTHUS**
               9′ Lvs entire and oblong to lanceolate, or pinnately lobed; infl an elongated spike, gen > 3 cm; fertile stamens 2 (exc *Chloropyron maritimum*), sterile stamens reduced to rudimentary filaments or bearing sterile, long-hairy, appendages; middle lobe of lower corolla lip erect, not rolled under
                  10. Lvs entire; calyx = or slightly < corolla, notched ± 1 mm at tip; anther sacs ± overlapping; saline and alkaline habitats. . . . . **CHLOROPYRON**
                  10′ Lvs pinnately lobed; calyx ± 1/2 corolla, divided > 1/2; anther sacs widely separated; not in saline or alkaline habitats. . . . . **DICRANOSTEGIA**

BELLARDIA

Ann, glandular-hairy. **ST**: erect, gen simple. **LF**: opposite distally on st, sessile, toothed. **INFL**: raceme, spike-like, bracted. **FL**: calyx 4-lobed; corolla 2-lipped, upper lip hood-like, pink, lower lip > upper, white, 3-lobed, throat with 2 ridges; stamens 4, in 2 pairs, incl, anthers hairy, awned at base; stigma club-shaped, ± exserted. **FR**: ovoid, loculicidal. **SEED**: many, small, ridged. 1–2 spp.: Medit. (C.A.L. Bellardi, Italian botany professor, 1740–1826) [Molau 1990 Opera Bot 102:27]

B. trixago (L.) All. (p. 961) MEDITERRANEAN LINSEED Pl glandular-hairy. **ST**: occ branched distally, 15–80 cm. **LF**: 15–90 mm, lanceolate, coarsely crenate-dentate. **INFL**: bracts reduced distally on infl axis, becoming ovate, cordate, ± entire. **FL**: calyx 8–10 mm, lobes 1–1.5 mm, unequal, triangular; corolla 20–25 mm. **FR**: ± 7 mm. **SEED**: 0.5–1 mm, ± oblong. 2*n*=24. Disturbed grassland; < 710 m. NCo, NCoRO, ScV, CW; native to Medit. Apr–Jun ❖

CASTILLEJA PAINTBRUSH, OWL'S-CLOVER

Margriet Wetherwax, T.I. Chuang & Lawrence R. Heckard

Ann to subshrub, green. **LF**: sessile, entire to dissected. **INFL**: spike-like; bracts becoming shorter, wider, more lobed than lvs, mature tips gen cream to red or green. **FL**: calyx unequally 4-lobed, colored like bract tips; corolla upper 2 lip lobes fused,

beak-like, tip open, lower lip reduced, 3-toothed to -pouched; stamens 4, anther sacs 2, unequal; stigma entire to 2-lobed, gen exserted. **FR**: ± asymmetric. **SEED**: gen ± brown, attached at base; coat netted, net-like walls ladder-like or not. ± 200 spp.: esp w N.Am. (Domingo Castillejo, Spanish botanist, 1744–1793) [Tank & Olmstead 2008 Amer J Bot 95:608–625] Hybridization and polyploidy common. Biologically consistent taxa difficult to define. *C. chrymactis* Pennell not in CA, sole (1947) record a misidentified, incomplete specimen.

1. Infl gen green, white, yellow, ± red-gray, bright pink, or ± purple, but not bright red; corolla 1–3 cm,
   beak gen ≤ tube; gen bee-pollinated; lower corolla lip gen 3-pouched (exc *Castilleja grisea*, *Castilleja plagiotoma*), teeth gen ± erect, white or green to purple-red
   2. Per to subshrub
      3. Herbage gray-green, exc bract lobes and calyx white-woolly; calyx divided deeper on sides than
         abaxially and adaxially . . . . ²*C. plagiotoma*
      3′ Herbage, incl bract lobes and calyx, gray-green to densely white-woolly; calyx divided deeper abaxially
         and adaxially than on sides, or divided ± equally
         4. Calyx divided much deeper abaxially and adaxially than on sides
            5. Hairs stellate; San Clemente Island . . . . ²*C. grisea*
            5′ Hairs simple or branched; KR, CaRH, SNH, SnBr, GB
               6. Herbage glandular; corolla 16–20 mm, lower lip 5–7 mm, teeth ± green-white or ± purple, ±
                  triangular; st gen unbranched; bract lobes ± acute . . . . *C. lemmonii*
               6′ Herbage stiff-hairy, nonglandular; corolla 13–16 mm, lower lip 2 mm, teeth dark green, incurved;
                  st much-branched; bract lobes truncate or rounded . . . . *C. praeterita*
         4′ Calyx divided ± equally
            7. Herbage ± white-woolly
               8. Pl 8–30 cm; lvs 20–60 mm; hairs simple; KR, NCoRH, CaRH, n SNH, MP . . . . *C. arachnoidea*
               8′ Pl 8–15 cm; lvs 5–20 mm; hairs branched; KR . . . . *C. schizotricha*
            7′ Herbage puberulent to coarsely hairy
               9. Herbage ashy-puberulent; bract tips dusty red, ± purple, or green-yellow; hairs dense, partitioned,
                  much-branched; seed coat walls irregularly netted — seed coat tight-fitting; SnBr . . . . *C. cinerea*
               9′ Herbage not ashy-puberulent; bract tips gen yellow-green or ± purple; hairs ± stiff, ± unbranched;
                  seed coat walls with ladder-like thickenings
                  10. Bract lobes acuminate, green-margined; seed coat loose-fitting. *C. nana*
                  10′ Bract lobes gen truncate or rounded (acute), white-margined; seed coat tight-fitting . . . . *C. pilosa*
   2′ Ann
      11. Bracts green throughout
         12. Lvs and bracts entire . . . . *C. campestris*
            13. Lvs and bracts linear, thin, flexible; bracts ≤ fls; corolla light to bright yellow; lower anther sac
               1/4–1/3 upper . . . . subsp. *campestris*
            13′ Lvs and bracts lanceolate, thick, ± brittle; bracts > fls; corolla gen yellow to orange; lower anther sac
               ± 1/2 upper . . . . subsp. *succulenta*
         12′ Lvs entire or lobed; bracts lobed
            14. Corolla beak densely white-hairy — SnBr, n&c PR . . . . *C. lasiorhyncha*
            14′ Corolla beak puberulent
               15. Lower corolla lip pouches 2–4 mm wide, ± 2 mm deep; stigma incl . . . . *C. tenuis*
               15′ Lower corolla lip pouches 4–10 mm wide, 3–6 mm deep; stigma gen ± exserted
                  16. Corolla 13–22 mm, pouches 4–8 mm wide; KR, NCoRO, NCoRI, CaR, SN, MP. . . . . *C. lacera*
                  16′ Corolla 20–28 mm, pouches 8–10 mm wide; NCoR, CaRF, ScV, CCo, SnFrB . . . . *C. rubicundula*
                     17. Corolla yellow . . . . subsp. *lithospermoides*
                     17′ Corolla white, turning ± pink. . . . . subsp. *rubicundula*
      11′ Bracts and calyx lobes tipped ± purple, ± yellow, or ± white
         18. Corolla beak tip hooked, densely shaggy-hairy; filaments puberulent. *C. exserta*
            19. Distal bracts gen 5–7 mm wide, lobes < 2 mm, tipped pale lavender; infl appearing alternately light-
               and dark-banded — NCo, n&c CCo. . . . . subsp. *latifolia*
            19′ Distal bracts gen < 5 mm wide, lobes > 2 mm; infl not appearing alternately light- and dark-banded
               20. Corolla white, pale yellow or rose to ± purple; widespread . . . . subsp. *exserta*
               20′ Corolla bright rose-red exc yellow- to orange-tipped pouch; w DMoj. . . . . subsp. *venusta*
         18′ Corolla beak straight, puberulent; filaments glabrous
            21. Infl 10–20 mm wide; corolla linear, pouches ± 2 mm wide; stigma incl
               22. Bract lobes 3, tipped white or pale yellow; seed < 1 mm, coat shallowly netted . . . . *C. attenuata*
               22′ Bract lobes 3–5, tipped pink or purple-red; seed 1–1.5 mm, coat ± deeply netted. . . . . *C. brevistyla*
            21′ Infl 20–40 mm wide; corolla wider distally, pouches gen 3–7 mm wide; stigma gen ± exserted
               23. St ± stiffly spreading-hairy, glandular; bract tips cream to pale yellow — SNF . . . . *C. lineariloba*
               23′ St glabrous to puberulent, not glandular; bract tips yellow (cream) to pink- or rose-purple
                  24. Branches gen many; lf lanceolate to oblong, > 3 mm wide; seed coat tight-fitting . . . . *C. ambigua*
                     25. Bracts tipped yellow . . . . subsp. *ambigua*

25′ Bracts tipped ± yellow-pink to rose-purple
  26. Pl ± fleshy, branches 0–few from mid-st; salt marshes . . . . subsp. *humboldtiensis*
  26′ Pl not fleshy, much-branched from base; grassy coastal bluffs . . . . subsp. *insalutata*
24′ Branches 0 or few from mid-st; lf lance-linear, < 3 mm wide; seed coat loose-fitting . . . . *C. densiflora*
  27. Calyx 5–10 mm; lower corolla lip widened abruptly, pouches gen shorter than deep . . . . subsp. *gracilis*
  27′ Calyx 8–20 mm; lower corolla lip widened gradually, pouches longer than deep
    28. Infl gen rose-purple (cream) . . . . subsp. *densiflora*
    28′ Infl cream to pale yellow. . . . . subsp. *obispoensis*
1′ Infl gen red, orange, or yellow; corolla 1.2–5 cm, beak often ≥ tube; gen hummingbird-pollinated;
  lower corolla lip of small, incurved, gen dark green, red, black, ± yellow, or ± purple teeth
29. Ann; st ± simple; wet places — lvs and bracts entire, lance-linear. . . . . *C. minor*
  30. Corolla 15–20(30) mm; pl shaggy-hairy, most hairs glandular . . . . subsp. *minor*
  30′ Corolla 25–35 mm; pl puberulent, many hairs nonglandular. . . . . subsp. *spiralis*
29′ Per to subshrub; sts gen branched; gen drier places (exc *Castilleja miniata*)
31. Herbage gray-green to densely white-woolly, hairs branched
  32. Herbage ± gray-green; calyx white-woolly, divided deeper on sides than abaxially and adaxially
  . . . . ²*C. plagiotoma*
  32′ Entire pl densely white-woolly or ash-gray; calyx sinuses deeper abaxially and adaxially than on sides
    33. Lvs oblong to obovate, entire, tip rounded; bract tips rounded or 0–3-lobed, yellow(-green) — n ChI . . . *C. mollis*
    33′ Lvs linear to lanceolate, distal sometimes 3-lobed, tip acute to obtuse; infl red, orange, or yellow;
      bracts 0–7-lobed, tips colored like calyx
      34. Herbage white-felty, with long, intertwined, ± branched hairs; lvs entire — n ChI . . . . *C. hololeuca*
      34′ Herbage white-woolly or ash-gray with shorter, much-branched or stellate hairs; upper lvs
        sometimes 3-lobed
        35. Infl pale yellow-green; pl ash-gray; hairs stellate; bracts 5–7-lobed — s ChI . . . . ²*C. grisea*
        35′ Infl bright red to bright yellow (yellow-green); pl white-woolly or ash-gray; hairs much-branched;
          bracts 0–5-lobed
          36. Herbage ± white to ± gray; calyx entire or barely notched on sides . . . . *C. foliolosa*
          36′ Herbage ash-gray or green; calyx divided 1/5–1/4 on sides
            37. Calyx lobes ovate, obtuse; lower corolla lip dark green to ± black; lf lobes 0–3; SnGb. . . . . *C. gleasoni*
            37′ Calyx lobes lanceolate, acute; lower corolla lip green; lf lobes 0–5; NW, CaR, n&c SN, MP. . . . ²*C. pruinosa*
31′ Herbage not both with branched hairs and gray-green to white-woolly
38. Calyx divided ± 2/3 abaxially, 1/6–1/3 adaxially, lobes gen curved upward; corolla (incl lower lip) gen
  curved out through abaxial calyx sinus
  39. Herbage glabrous to slightly puberulent; lvs linear, margins rolled upward; bracts gen narrowly
    3-lobed, < calyx; corolla beak sparsely puberulent; CaR, e slope SNH, TR, GB, DMtns. . . . . *C. linariifolia*
  39′ Herbage puberulent to hairy; lvs lanceolate, not rolled upward; bracts gen entire, ≥ calyx; corolla
    beak densely shaggy-puberulent; mostly c&s CA-FP. . . . . *C. subinclusa*
    40. Infl 2-colored, red and yellow; corolla yellow-orange; calyx lobes clearly curved upward, sides ±
      entire to divided < 1/5. . . . . subsp. *franciscana*
    40′ Infl ± red throughout; corolla ± red; calyx lobes barely curved upward, sides divided 1/5–1/3 . . . . subsp. *subinclusa*
38′ Calyx divided gen 1/4–1/2 (exc *Castilleja miniata*, *Castilleja peirsonii*) abaxially and adaxially, lobes
  not curved upward; corolla not curved out through abaxial side of calyx sinus
41. Hairs dense, branched (puberulent) . . . . ²*C. pruinosa*
41′ Hairs gen unbranched (or ± 0), pl puberulent to long-soft-hairy or bristly (*Castilleja affinis*
  subsp. *affinis* occ with some branched hairs)
  42. Pl glandular-puberulent to -sticky-hairy proximal to infl
    43. Calyx divided ≥ 1/2 abaxially; corolla beak ± 1/2 tube; stigma clearly 2-lobed. . . . . *C. peirsonii*
    43′ Calyx gen divided 1/4–1/2 abaxially; corolla beak 1–2 × tube; stigma gen entire or slightly 2-lobed
      44. Branches gen crowded proximal to infl; lvs gen crowded; seed coat deeply netted; coastal . . . . *C. wightii*
      44′ Branches few proximal to infl; lvs gen ± well spaced; seed coat shallowly netted; inland
        45. Pl glandular but not sticky; lvs entire or shallowly lobed; calyx lobes obtuse to rounded or acute
          46. Lf lobes 0–7; calyx lobes obtuse; nw KR. . . . . *C. brevilobata*
          46′ Lf lobes 0–3; calyx lobes acute; e SnBr . . . . *C. montigena*
        45′ Pl densely glandular-sticky; lvs 0–3-lobed; calyx lobes gen acute . . . . *C. applegatei*
          47. St 10–25 cm; lf lobes gen 3; calyx 13–15 mm, divided ± 1/4 on sides; red fir forest to alpine
            barrens — SNH, GB . . . . subsp. *pallida*
          47′ St 30–80 cm; most lvs entire; calyx 12–25 mm, gen divided ≤ 1/5 on sides; gen below subalpine
            48. Calyx lobes obtuse. . . . . subsp. *martinii*
            48′ Calyx lobes acute
              49. Calyx 12–18 mm . . . . subsp. *disticha*
              49′ Calyx 16–25 mm . . . . subsp. *pinetorum*
  42′ Pl gen not glandular proximal to infl (exc occ *Castilleja affinis* or *Castilleja latifolia*)

50. Lvs ± fleshy, oblong to rounded, lobes 0–3, truncate-rounded; bract lobes 0–3, truncate-rounded; hairs gen bristly or shaggy
    51. Corolla 20–30 mm, beak 10–15 mm; c&s CCo . . . . *C. latifolia*
    51′ Corolla 30–45 mm, beak 15–25 mm; c NCo . . . . *C. mendocinensis*
50′ Lvs gen not fleshy, linear to lanceolate; lvs and bracts entire or lobed, tips acute to obtuse; gen puberulent or glabrous (bristly)
    52. Calyx divided 1/4–1/3 abaxially and adaxially — dry sagebrush scrub, pinyon/juniper woodland
    . . . . *C. chromosa*
    52′ Calyx divided 1/3–2/3 or more abaxially and adaxially
        53. Herbage gen ± bristly-puberulent; lvs 0–5-lobed . . . . *C. affinis* (in part)
        54. Infl gen bright red to orange-red (yellow), 30–50 mm wide . . . . subsp. *affinis*
        54′ Infl gen yellow (pink or ± red-orange), 15–25 mm wide — s NCoRI, SnFrB, open serpentine slopes . . . . subsp. *neglecta*
        53′ Herbage glabrous to sparsely puberulent, to long-soft-hairy distally; lvs gen entire
            55. Bract and calyx lobes obtuse; seed coat deeply netted, tight-fitting; gen dry sea bluffs — n&c NCo . . . . *C. affinis* subsp. *litoralis*
            55′ Bract and calyx lobes acute; seed coat shallowly netted, loose-fitting; moist places . . . . *C. miniata*
                56. St slender; infl ± pink to yellow-orange; serpentine bogs; nw KR . . . . subsp. *elata*
                56′ St stout; infl gen bright red to orange-red; widespread . . . . subsp. *miniata*

C. affinis Hook. & Arn. Per 15–60 cm, few-branched, green becoming ± purple, ± glabrous to bristly-puberulent, gen nonglandular. **LF**: linear to lanceolate; lobes 0–5, tips ± rounded. **INFL**: 5–30 cm; bracts 17–25 mm, lobes 0–5. **FL**: calyx divided ± 1/3–1/2 abaxially and adaxially, ± 1/4–1/3 on sides, long-nonglandular- and short-glandular-hairy, lobes acute to obtuse, gen not curved upward; corolla beak ± 1–1.5 × tube, adaxially gen shaggy-hairy, margins ± red or ± yellow, lower lip 2–3 mm, green to dark purple; stigma slightly notched. **FR**: 10–15 mm. **SEED**: 1.5–2 mm; coat deeply netted, tight-fitting, most walls ladder-like. Many forms, some geog isolated; hybridizes with other spp.

subsp. ***affinis*** (p. 961) Pl ± bristly, occ some hairs branched. **LF**: 30–80 mm, ± lanceolate; lobes 0–5. **INFL**: 30–50 mm wide, gen bright red to orange-red (yellow). **FL**: calyx 20–35 mm; corolla 25–40 mm. $2n$=48,72,96. Chaparral, coastal scrub; < 1800 m. c NCo (Mendocino Co.), KR, NCoRO, NCoRI, n CaRF, SNF, CW, SW; Baja CA. Coastal pls tend to have ± fleshy lvs, inflated calyx. Dune pls of s CCo, n SCo with some branched hairs may represent past hybridization with diploid *C. mollis* of n ChI. Mar–Jun

subsp. ***litoralis*** (Pennell) T.I. Chuang & Heckard (p. 961) OREGON COAST PAINTBRUSH Pl glabrous to sparsely puberulent. **LF**: 30–80 mm, ± oblong, gen entire. **INFL**: gen 30–50 mm wide, bright red to orange-red; bract lobes obtuse. **FL**: calyx 20–25 mm, lobes obtuse; corolla 25–40 mm. $2n$=120,144. Gen dry sea bluffs; < 160 m. n&c NCo; OR. May–Aug ★

subsp. ***neglecta*** (Zeile) T.I. Chuang & Heckard TIBURON PAINTBRUSH Pl ± bristly. **LF**: 20–40 mm, ± lanceolate; lobes 0–5. **INFL**: 15–25 mm wide, gen yellow (pink or ± red-orange). **FL**: calyx 15–20 mm; corolla 18–22 mm. $2n$=72. Open serpentine slopes; < 300 m. s NCoRI (Napa Co.), SnFrB (Marin, Santa Clara cos.). Apr–Jun ★

C. ambigua Hook. & Arn. JOHNNY-NIP Ann 10–30 cm, gen much-branched, decumbent, puberulent. **LF**: 10–50 mm, > 3 mm wide, lanceolate to oblong; lobes 0–5. **INFL**: 3–12 cm, 3–4 cm wide, often dense; bracts 15–25 mm, oblong to ovate, tipped yellow (cream) to rose-purple, lobes 3–9, central lobe gen rounded. **FL**: calyx 12–20 mm, divided 1/2 abaxially and on sides, 2/3 adaxially, lobes linear; corolla 14–25 mm, pale yellow or rose-purple, beak 4–5 mm, straight, puberulent, lower lip 3–4 mm, pouches 3–7 mm wide, 1–2 mm deep, gen purple-dotted at base; filaments glabrous; stigma gen ± exserted. **FR**: 8–12 mm. **SEED**: coat shallowly netted, tight-fitting. $2n$=24. Highly variable and difficult; many local, ± ecological forms; needs study.

subsp. ***ambigua*** (p. 961) Pl not fleshy. **INFL**: bract tips acute or rounded, yellow. **FL**: corolla yellow, lower lip teeth ± 1 mm. **SEED**: ± 1 mm. Coastal bluffs, grassland; < 500 m. NCo, s NCoR, n&c CCo. May–Aug ★

subsp. ***humboldtiensis*** (D.D. Keck) T.I. Chuang & Heckard (p. 961) HUMBOLDT BAY OWL'S-CLOVER Pl ± fleshy, ascending; branches 0–few from mid st. **INFL**: bract tips rounded to truncate, ± yellow-pink to rose-purple. **FL**: corolla pink or rose-purple (yellow), lower lip teeth 2–3 mm. **SEED**: 2–2.5 mm. Salt marshes; ± 0 m. NCo (Humboldt, Mendocino cos.), n CCo (Marin Co.). Threatened by coastal development. Yellow-fld pls from Marin Co. are possible hybrids with *C. ambigua* subsp. *a.* May–Jun ★

subsp. ***insalutata*** (Jeps.) T.I. Chuang & Heckard (p. 961) Pl not fleshy, much-branched from base. **INFL**: bract tips acute or rounded, ± rose-purple. **FL**: corolla ± rose-purple, lower lip teeth 1.5–2 mm. **SEED**: ± 1 mm. Grassy coastal bluffs; < 100 m. c CCo (Monterey, San Luis Obispo cos.). May–Jul ★

C. applegatei Fernald Per, few-branched, green, gen short-glandular-sticky-hairy and long-nonglandular-hairy. **ST**: gen few. **LF**: 20–70 mm, ± lanceolate, gen wavy-margined. **INFL**: 5–20 cm; bracts 15–25 mm, lobes 0–7, bright red to ± yellow. **FL**: calyx divided 1/3–1/2 abaxially and adaxially, ± 1/8 on sides, lobes acute to obtuse; corolla beak ± = tube, adaxially puberulent, margins ± red, lower lip 1–3 mm, dark green, incl or exserted; stigma slightly 2-lobed, exserted. **FR**: 8–15 mm. **SEED**: 1–1.5 mm; coat shallowly netted, loose-fitting, side walls ladder-like. Highly variable complex (other subsp. outside CA), unique in combination of glandular-sticky herbage, wavy lf margins.

subsp. ***disticha*** (Eastw.) T.I. Chuang & Heckard (p. 961) **ST**: 30–80 cm. **LF**: lobes 0(3). **FL**: calyx 12–18 mm, divided > 1/3 abaxially and adaxially, ± 1/6 on sides, lobes lanceolate, acute; corolla 25–35 mm, beak 14–18 mm. $2n$=24. Open conifer forest; 2000–3000 m. c&s SNH. Intergrades with *C. applegatei* subsp. *pinetorum*. Jun–Aug

subsp. ***martinii*** (Abrams) T.I. Chuang & Heckard (p. 961) **ST**: 30–80 cm. **LF**: lobes 0(3). **FL**: calyx 15–25 mm, divided ± 1/3 abaxially and adaxially, 1/5–1/6 on sides, lobes ovate, obtuse; corolla 25–40 mm, beak 12–18 mm. $2n$=24,48,72. Dry chaparral, open yellow-pine forest, sagebrush scrub; 300–2800 m. KR, NCoRO, NCoRI, e SnFrB, SCoR, SW (exc ChI), SNE, DMtns; s NV, n Baja CA. May–Sep

subsp. ***pallida*** (Eastw.) T.I. Chuang & Heckard (p. 961) **ST**: 10–25 cm. **LF**: lobes gen 3. **FL**: calyx 13–15 mm, divided > 1/3 abaxially and adaxially, ± 1/4 on sides, lobes lanceolate, acute; corolla 16–22 mm, beak 7–10 mm. $2n$=48. Dry rocky slopes and flats, red-fir forest to alpine barrens; 1900–3600 m. SNH, GB. Jun–Aug

subsp. ***pinetorum*** (Fernald) T.I. Chuang & Heckard (p. 961) **ST**: 30–60 cm. **LF**: lobes 0(3). **FL**: calyx 16–25 mm, divided 1/3–1/2 abaxially and adaxially, 1/8–1/5 on sides, lobes lanceolate, acute; corolla 25–35 mm, beak 11–16 mm. $2n$=24,48. Open conifer forest, dry sagebrush scrub; 300–3600 m. KR, NCoRH, CaR, SN, GB, n DMtns; to OR, ID. Jun–Aug

C. arachnoidea Greenm. (p. 961) Per 8–30 cm, white-woolly; hairs simple. **LF:** 20–60 mm, lance-linear, lobes 0–5. **INFL:** 3–12 cm; bracts 10–25 mm, white-woolly, lobes gen 3–5, obtuse, dull yellow to rusty red. **FL:** calyx 12–18 mm, ± equally divided ± 1/2, lobes lance-linear; corolla 12–20 mm, beak 3–5 mm, lower lip 2–4 mm, pouches shallow, pale green or purple-red, teeth pale yellow; stigma ± exserted, notched, dark. **FR:** 8–12 mm. **SEED:** ± 1 mm; coat shallowly netted, loose-fitting. 2*n*=24. Open summits, dry rocks; 1300–3300 m. KR, NCoRH, CaRH, n SNH, MP; sw OR. [*C. a.* subsp. *shastensis* Pennell] Jun–Aug

C. attenuata (A. Gray) T.I. Chuang & Heckard (p. 961) VALLEY TASSELS Ann 10–50 cm, nonglandular, spreading-hairy. **LF:** 20–80 mm, ± linear; lobes 0–3. **INFL:** 3–30 cm, 1–2 cm wide; bracts 15–35 mm, lobes 3, ± lanceolate, tips white or pale yellow. **FL:** calyx 10–20 mm, divided ± 1/2 abaxially, 3/4 adaxially, 1/3 on sides; corolla 10–25 mm, linear, beak 4–5 mm, straight, puberulent, lower lip 3–4 mm, pouches ± 2 mm wide, 1–1.5 mm deep, ± white, purple-dotted; filaments glabrous; stigma incl. **FR:** 7–11 mm. **SEED:** < 1 mm; coat shallowly netted, loose-fitting. 2*n*=24. Grassland; gen < 1600 m. CA-FP; to BC, n Baja CA; also c Chile. Mar–May

C. brevilobata Piper (p. 961) SHORT-LOBED PAINTBRUSH Per 10–50 cm, green or ± yellow, glandular-puberulent, or hairs glandular and nonglandular, but not sticky. **LF:** 15–60 mm, lanceolate to ovate; lobes 0–7, tips rounded to acute, margin not wavy. **INFL:** 5–20 cm, open proximally; bracts 15–45 mm, lobes gen 3–5, shallow, bright red to yellow. **FL:** calyx 15–30 mm, divided ± 1/3 abaxially and adaxially, 1/5–1/4 on sides, soft-hairy, lobes obtuse; corolla 20–35 mm, beak ± = tube, ± yellow-green, adaxially puberulent, margins ± red, lower lip 1–2 mm, dark green, incl; stigma slightly 2-lobed. **FR:** 8–13 mm. **SEED:** 1–2 mm; coat shallowly netted, loose-fitting. 2*n*=24. Dry, open serpentine, forest edges; 200–1850 m. nw KR; sw OR. [*C. hispida* Benth. subsp. *b.* (Piper) T.I. Chuang & Heckard] May–Aug ★

C. brevistyla (Hoover) T.I. Chuang & Heckard (p. 961) Ann 6–40 cm, nonglandular, sparsely hairy. **LF:** 20–60 mm, ± linear; lobes 0–3. **INFL:** 5–25 cm, 1–2 cm wide; bracts 10–20 mm, tipped pink or purple-red, lobes 3–5, ± linear. **FL:** calyx 15–30 mm, divided < 1/3 abaxially, 2/3 adaxially, 1/3 on sides; corolla 15–28 mm, linear, beak 4–6 mm, straight, puberulent, lower lip 3–5 mm, pouches ± 2 mm wide, 1–1.5 mm deep, purple-red or blotched pink; filaments glabrous; stigma incl. **FR:** 8–12 mm. **SEED:** 1–1.5 mm; coat ± deeply netted, loose-fitting. 2*n*=48. Grassland; < 1200 m. s SNF, SnJV, s SCoRI. Mar–Jun

C. campestris (Benth.) T.I. Chuang & Heckard Ann 10–30 cm, ± glabrous. **LF:** 15–40 mm, linear to lanceolate, entire. **INFL:** 3–15 cm, 2–3 cm wide; bracts 10–25 mm, lance-linear, entire, green. **FL:** calyx ± 7 mm, divided 1/3 abaxially, 1/2 adaxially, 1/4 on sides, ± hairy; corolla 15–25 mm, beak 5–6 mm, straight, lower lip 4–5 mm, pouches 6–10 mm wide, 3–4 mm deep; stigma ± exserted. **FR:** 5–7 mm. **SEED:** ± 0.7 mm; coat shallowly netted, tight-fitting. 2*n*=24. Occ confused with *Triphysaria eriantha*, which has 1 anther sac per stamen.

subsp. ***campestris*** (p. 961) **LF:** narrowly linear, thin, flexible. **INFL:** bracts ≤ fls. **FL:** corolla light to bright yellow; lower anther sac 1/4–1/3 upper. Vernal pools, moist places; < 2100 m. s NCoRI, CaR, n SN, GV, MP; to c OR. Apr–Jul

subsp. ***succulenta*** (Hoover) T.I. Chuang & Heckard (p. 961) SUCCULENT OWL'S-CLOVER **LF:** lanceolate, thick, ± brittle. **INFL:** bracts > fls. **FL:** corolla gen yellow to orange; lower anther sac ± 1/2 upper. Vernal pools, moist places; < 750 m. s SNF, se ScV, e SnJV. Threatened by urbanization, agriculture. Apr–Jul ★

C. chromosa A. Nelson (p. 961) DESERT PAINTBRUSH Per 15–45 cm, few-branched, gray-green, hairs ± bristly, unbranched, nonglandular. **LF:** 20–70 mm, lance-linear, lobes (0)3–5, widely spreading. **INFL:** 4–15 cm; bracts 20–30 mm, lobes 3–5, bright red to ± yellow-orange. **FL:** calyx 15–25 mm, divided 1/4–1/3 abaxially and adaxially, ± 1/7 on sides, long-nonglandular- and short-glandular-hairy, lobes obtuse to rounded; corolla 20–35 mm, beak ± = tube, ± yellow-green, adaxially puberulent, margins ± red, lower lip 2–3 mm,

dark green, incl; stigma 2-lobed. **FR:** 10–15 mm. **SEED:** 1.5–2 mm; coat deeply netted, most walls ladder-like. 2*n*=24,48. Dry sagebrush scrub, pinyon/juniper woodland; < 3000 m. CaRH, s SNH, ne SnBr, GB, DMoj; to OR, MT, WY, CO, NM. [*C. angustifolia* (Nutt.) G. Don, misappl.] May–Sep

C. cinerea A. Gray (p. 961) ASH-GRAY PAINTBRUSH Per 5–15 cm, ascending to erect, densely ashy-puberulent. **LF:** 10–20 cm, lance-linear; lobes gen 0(3). **INFL:** 3–6 cm; bracts 12–20 mm, densely branched-hairy, lobes 3–5, truncate or rounded, tips dusty red, ± purple, or green-yellow. **FL:** calyx 15–20 mm, ± equally divided 1/3–1/2, lobes ± linear, densely branched-hairy; corolla 15–18 mm, incl, beak 4–5 mm, pale ± yellow, lower lip ± 2 mm, ± green, pouches shallow, teeth minute, incurved; stigma ± exserted. **FR:** 6–10 mm. **SEED:** ± 1 mm; coat ± deeply netted, tight-fitting, walls irregularly net-thickened. Dry sagebrush scrub; 1800–3300 m. SnBr. Threatened by grazing, development, vehicles. May–Aug ★

C. densiflora (Benth.) T.I. Chuang & Heckard Ann 10–40 cm, branches 0 or few from mid-st, ± glabrous. **LF:** 20–80 mm, < 3 mm wide, lance-linear, lobes 0–3. **INFL:** 3–25 cm, 2.5–4 cm wide; bracts 10–25 mm, tipped white or ± purple, lobes 3–5, linear. **FL:** calyx divided 1/2 abaxially and on sides, 2/3 adaxially, lobes ± linear, tip wider; corolla 10–25 mm, yellow (cream), pink, or rose-purple, beak 5–6 mm, straight, puberulent, lower lip 4–5 mm, pouches 4–6 mm wide, 2–3 mm deep; filaments glabrous; stigma ± exserted, ± 2-lobed. **FR:** 7–10 mm. **SEED:** ± 0.5 mm; coat shallowly netted, loose-fitting. 2*n*=24. Highly variable; many local forms; more study needed.

subsp. ***densiflora*** (p. 961) **INFL:** gen rose-purple (cream); bracts = fls. **FL:** calyx 8–20 mm; corolla ± incl, lower lip widened gradually, pouches longer than deep, teeth ± 2 mm. Grassland; < 1600 m. NCoR, SNF, SnFrB, SCoR. Mar–May

subsp. ***gracilis*** (Benth.) T.I. Chuang & Heckard (p. 961) **INFL:** rose-purple; bracts < fls. **FL:** calyx 5–10 mm; corolla exserted, lower lip widened abruptly, pouches gen shorter than deep, teeth gen < 1 mm. Grassland; < 1550 m. SCoR, SW; n Baja CA. Mar–May

subsp. ***obispoensis*** (D.D. Keck) T.I. Chuang & Heckard SAN LUIS OBISPO OWL'S-CLOVER **INFL:** cream to pale yellow; bracts = fls. **FL:** calyx 8–20; corolla ± incl, lower lip widened gradually, pouches longer than deep, teeth ± 2 mm. Coastal grassland; < 400 m. s CCo (San Luis Obispo Co.). Mar–Jun ★

C. exserta (A. Heller) T.I. Chuang & Heckard PURPLE OWL'S-CLOVER Ann 10–45 cm, glandular-puberulent, stiff-hairy. **LF:** 10–50 mm; lobes 5–9, ± thread-like. **INFL:** 2–20 cm, 2–4 cm wide; bracts 10–25 mm, white to purple-red, lobes 5–9 (lowest pair often again 2–4-lobed), ± linear, tips wider. **FL:** calyx 10–22 mm, divided 1/2 abaxially, 2/3 adaxially, 1/3 on sides, colored like bracts; corolla 12–30 mm, beak 6–7 mm, densely shaggy-hairy, tip hooked, lower lip 4–6 mm, pouches 3–8 mm wide, 3–4 mm deep; filaments puberulent; stigma ± incl. **FR:** 10–15 mm. **SEED:** 1–2 mm; coat deeply netted, loose-fitting. Highly variable; hybridizes with *C. attenuata*, *C. densiflora*, *C. lineariloba*.

subsp. ***exserta*** (p. 965) **INFL:** distal bracts gen < 5 mm wide, lobes > 2 mm, tipped white, pale yellow, or rose to ± purple. **FL:** corolla colored like bracts. 2*n*=24. Open fields, grassland; < 1600 m. NW, SNF, GV, CW, SW, w DMoj; AZ. Mar–May

subsp. ***latifolia*** (S. Watson) T.I. Chuang & Heckard (p. 965) **ST:** branches many, decumbent. **INFL:** banded alternately light and dark; distal bracts gen 5–7 mm wide, lobes < 2 mm, tipped pale lavender. **FL:** corolla colored like bracts. 2*n*=24. Coastal grassland, dunes; < 500 m. NCo, n&c CCo. Mar–May

subsp. ***venusta*** (A. Heller) T.I. Chuang & Heckard **INFL:** distal bracts gen < 5 mm wide, lobes > 2 mm, dark red, tipped deep pink or purple. **FL:** corolla bright rose-red exc pouch yellow- to orange-tipped. Dry sand, washes; 600–1500 m. w DMoj. Mar–May

C. foliolosa Hook. & Arn. (p. 965) WOOLLY PAINTBRUSH Per to subshrub, 30–60 cm; hairs woolly, felt-like, much-branched, ± white to ± gray. **ST:** much-branched; short, axillary shoots present. **LF:** 10–50 mm, ± linear; lobes 0–3, tips obtuse. **INFL:** 3–20 cm; bracts 15–25 mm, lobes 0–5, orange-red (yellow-green). **FL:** calyx

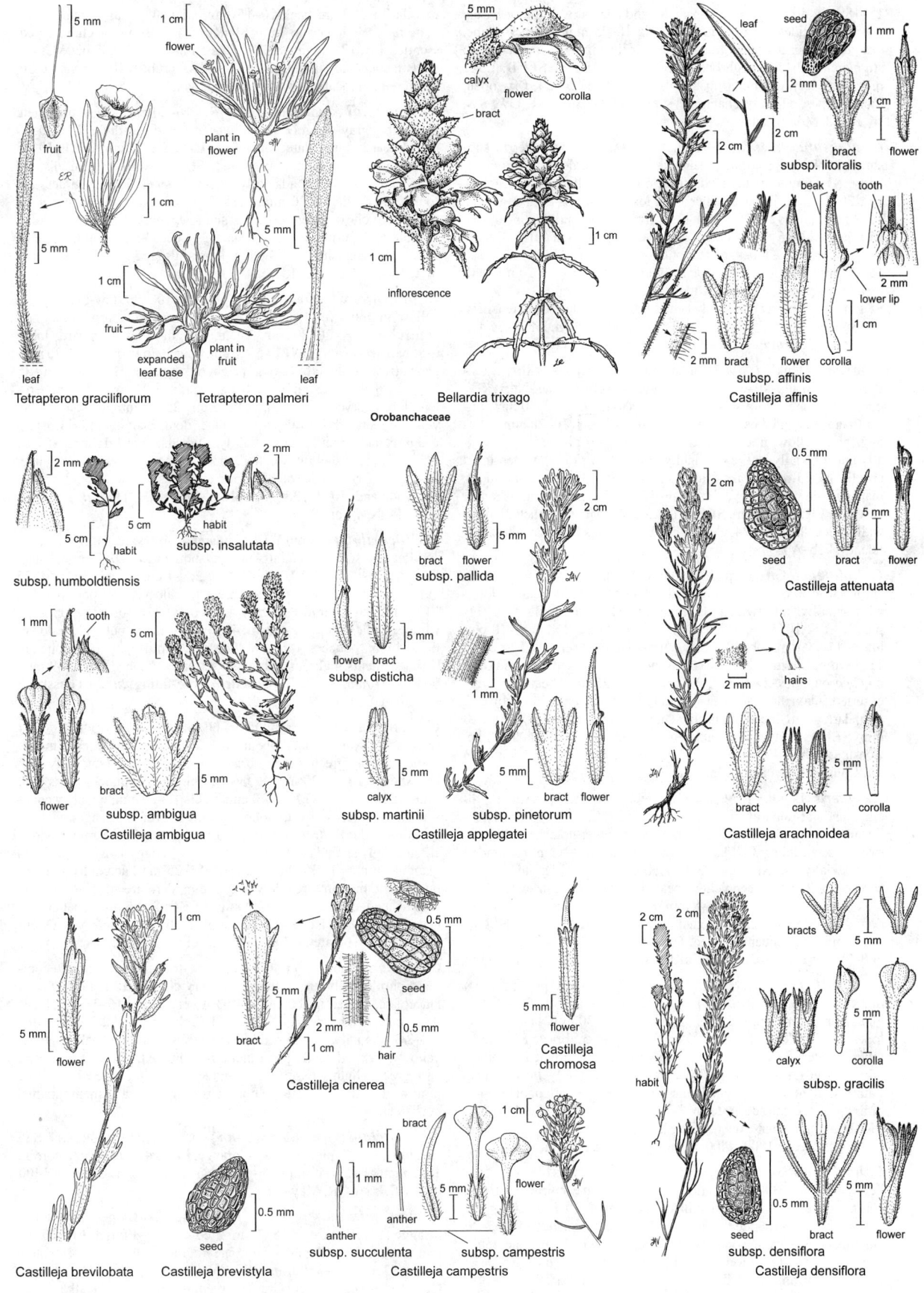

Tetrapteron graciliflorum

Tetrapteron palmeri

Bellardia trixago

Orobanchaceae

Castilleja affinis
subsp. litoralis
subsp. affinis

subsp. humboldtiensis
subsp. insalutata

Castilleja ambigua
subsp. ambigua

Castilleja applegatei
subsp. pallida
subsp. disticha
subsp. martinii
subsp. pinetorum

Castilleja attenuata

Castilleja arachnoidea

Castilleja brevilobata

Castilleja brevistyla

Castilleja cinerea

Castilleja campestris
subsp. succulenta
subsp. campestris

Castilleja chromosa

Castilleja densiflora
subsp. gracilis
subsp. densiflora

15–18 mm, divided 1/3–2/5 abaxially and adaxially, entire or barely notched on sides, swollen in fr; corolla 18–25 mm, beak ± = tube, exserted, puberulent, margins pale, lower lip 2 mm, dark green, incl; stigma club-shaped, slightly 2-lobed. **FR**: 10–15 mm. **SEED**: 1.5–2 mm; coat deeply netted, most walls ladder-like. $2n$=24. Dry, open, rocky slopes, edges of chaparral; < 1800 m. NCoR, SNF, CW, SW, sw edge DMoj; n Baja CA. Mar–Jun

C. gleasoni Elmer MOUNT GLEASON PAINTBRUSH Per to subshrub, 30–80 cm, ash-gray, glandular-puberulent, densely branched-hairy. **ST**: simple or branched from middle, axillary shoots gen 0. **LF**: 20–60 mm, linear to lanceolate, lobes 0–3, tips obtuse. **INFL**: 10–30 cm, bracts 15–20 mm, lobes 3 from below middle, linear, red. **FL**: calyx 13–15 mm, divided 2/5 abaxially and adaxially, divided 1/5–1/4, on sides, lobes ovate, obtuse, red; corolla 25–30 mm, beak 15–20 mm, ± yellow with wide red margins, lower lip ± 2 mm, dark green to ± black, spreading; stigma 2-lobed, exserted. **FR**: ± 15 mm. **SEED**: 1–1.5 mm; coat deeply netted, most walls ladder-like. Cliffs, rocky slopes in open yellow-pine forest; 1100–2200 m. SnGb (Mount Gleason). [*C. pruinosa* Fernald, in part, misappl.] May–Jun ★

C. grisea Dunkle (p. 965) SAN CLEMENTE ISLAND PAINTBRUSH Per to subshrub, 40–60 cm, ash-gray, densely stellate-hairy. **ST**: openly branched, with short lfy axillary shoots. **LF**: 10–50 mm, linear to lanceolate; lobes 0–3. **INFL**: 3–10 cm; bracts 10–20 mm, lobes 5–7, pale yellow-green. **FL**: calyx 10–20 mm, divided 1/3 abaxially and adaxially, entire or slightly notched on sides, swollen in fr; corolla 15–25 mm, beak ≤ tube, dull yellow, adaxially puberulent, margins pale yellow, lower lip 2 mm, dark green; stigma club-shaped, exserted. **FR**: 10–12 mm. **SEED**: 1–1.5 mm; coat deeply netted, most walls ladder-like. $2n$=24. Coastal bluffs; 400 m. s ChI (San Clemente Island). Feb–Apr ★

C. hololeuca Greene (p. 965) ISLAND PAINTBRUSH Per or subshrub, 30–100 cm, much-branched, white-felty; hairs densely long-white-woolly, intertwined, ± branched. **ST**: short axillary shoots present. **LF**: 10–50 mm, linear, entire; tip obtuse. **INFL**: 3–10 cm; bracts 15–20 mm, lobes 3, deep, gen orange-red (yellow). **FL**: calyx 15–18 mm, divided 2/5 abaxially and adaxially, entire or slightly notched on sides, swollen in fr; corolla 20–25 mm, beak ± = tube, exserted, adaxially glandular-puberulent, margins pale, lower lip 2–3 mm, dark green, incl; stigma club-shaped, slightly 2-lobed. **FR**: ± 10 mm. **SEED**: 1–1.5 mm; coat deeply netted, most walls ladder-like. Coastal scrub; < 400 m. n ChI. [*C. lanata* A. Gray subsp. *h.* (Greene) T.I. Chuang & Heckard] Mar–Aug ★

C. lacera (Benth.) T.I. Chuang & Heckard (p. 965) Ann 10–40 cm, glandular-puberulent and spreading-hairy. **LF**: 10–50 mm, lance-linear; lobes 0–5. **INFL**: 3–15 cm, 2–3 cm wide; bracts 10–20 mm, ovate, green, lobes 3–7, lance-linear. **FL**: calyx 8–13 mm, divided 1/2 abaxially and on sides, 2/3 adaxially, stiff-short-glandular-hairy; corolla 13–22 mm, deep yellow, beak 4–6 mm, straight, puberulent, lower lip 3–5 mm, pouches 4–8 mm wide, 3–6 mm deep, gen purple-dotted at base; stigma ± exserted, ± 2-lobed. **FR**: 5–8 mm. **SEED**: < 1 mm; coat ± deeply netted, loose-fitting. $2n$=22,24. Grassland; 200–2800 m. KR, NCoRO, NCoRI, CaR, SN, MP; s OR. Apr–Jul

C. lasiorhyncha (A. Gray) T.I. Chuang & Heckard (p. 965) SAN BERNARDINO MOUNTAINS OWL'S-CLOVER Ann gen 10–20(40) cm, glandular-puberulent, spreading-hairy. **LF**: 10–30 mm, lance-linear; lobes 0–3. **INFL**: 2–15 cm, 2–3 cm wide; ± open; bracts 8–20 mm, lobes 3–5, lance-linear, green. **FL**: calyx 8–12 mm, divided 1/3 abaxially and on sides, 1/2 adaxially; corolla 14–22 mm, yellow, beak ± 7 mm, straight, densely white-hairy, lower lip ± 5 mm, pouches 4–8 mm wide, 3–4 mm deep, teeth ± 2 mm; stigma incl. **FR**: 6–9 mm. **SEED**: ± 1 mm; coat ± deeply netted, loose-fitting. $2n$=24. Meadows, flats, open forest; 1000–2400 m. SnBr, n&c PR. Jun–Jul ★

C. latifolia Hook. & Arn. (p. 965) MONTEREY COAST PAINTBRUSH Per or subshrub, 30–60 cm, gray-green becoming ± purple, ± bristly, gen nonglandular. **ST**: short axillary shoots many. **LF**: 5–20 mm, ± fleshy, oblong to rounded; lobes 0–3, truncate-rounded. **INFL**: 5–20 cm; bracts 15–20 mm, widely wedge-shaped to widely obovate, lobes 0–3, bright red to yellow, central lobe tip truncate to obtuse. **FL**: calyx 15–25 mm, divided ± 1/3–1/2 abaxially and adaxially, < 1/8 on sides, long-nonglandular- and short-glandular-hairy, lobes obtuse;

corolla 20–30 mm, beak 10–15 mm, adaxially shaggy-puberulent, margins ± red, lower lip 2 mm, dark green, incl; stigma club-shaped, ± entire. **FR**: 12–20 mm. **SEED**: 2–2.5 mm; coat deeply netted, most walls ladder-like. $2n$=24. Coastal dunes, scrub; < 100 m. c&s CCo. Scattered but locally abundant in good yrs. Mar–Sep ★

C. lemmonii A. Gray (p. 965) Per 10–20 cm, gen unbranched, green or ± gray-green, ± spreading-hairy and glandular. **LF**: 20–40 mm, linear to lanceolate; lobes 0–3. **INFL**: 3–12 cm; bracts 10–15 mm, lobes 3–5, ± acute, purple-red. **FL**: calyx 16–18 mm, divided 1/2–2/3 abaxially and adaxially, ± 1/8 on sides, lobes gen acute to rounded; corolla 16–20 mm, beak 7–9 mm, pale yellow, lower lip 5–7 mm, yellow-green, pouches shallow, short, teeth ± triangular, ± white or ± purple, erect; stigma ± 2-lobed. **FR**: 7–9 mm. **SEED**: 1–1.5 mm; coat shallowly netted, loose-fitting. $2n$=24. Moist meadows; 1550–3700 m. CaRH, SNH. Jul–Aug

C. linariifolia Benth. (p. 965) Per 30–100 cm, few-branched, ± yellow to gray-green, gen becoming ± purple, glabrous to slightly puberulent. **LF**: 20–80 mm, linear, margins folded upward; lobes 0–3, narrow, < calyx. **INFL**: 5–20 cm, open below; bracts 15–30 mm, gen narrowly 3-lobed, < calyx, bright red to yellow. **FL**: calyx 20–35 mm, divided 2/3 abaxially, ± 1/3 adaxially, ± 1/8 on sides, puberulent, lobes curved upward, acute; corolla 25–45 mm, beak ± = tube, yellow-green, adaxially sparsely puberulent, margins red, lower lip 2–3 mm, dark green; stigma slightly 2-lobed. **FR**: 10–15 mm. **SEED**: 1.5–2 mm; coat shallowly netted, loose-fitting, side walls ladder-like. $2n$=24,48. Dry plains, rocky slopes, sagebrush scrub or pinyon/juniper woodland; 1000–3350 m. CaR, e slope SNH, TR, GB, DMoj; to OR, MT, NM. Jun–Sep

C. lineariloba (Benth.) T.I. Chuang & Heckard (p. 965) Ann 15–45 cm, ± stiffly spreading-hairy, glandular. **LF**: 20–75 mm, lance-linear; lobes 0–7. **INFL**: 20–50 cm, 2.5–4 cm wide; bracts 12–25 mm, lobes 5–7, linear, tipped cream to pale yellow (pale purple). **FL**: calyx 15–25 mm, ± equally divided ± 2/3; corolla 15–30 mm, ± white, ± yellow, or ± rose, beak 4–5 mm, straight, puberulent, lower lip 3–4 mm, pouches 4–5 mm wide, ± 2 mm deep, purple-dotted at base; filaments glabrous; stigma exserted, ± 2-lobed. **FR**: 7–9 mm. **SEED**: ± 1 mm; coat ± deeply netted, loose-fitting. $2n$=20. Grassland; < 1800 m. SNF, c SNH. Apr–Jun

C. mendocinensis (Eastw.) Pennell (p. 965) MENDOCINO COAST PAINTBRUSH Per, decumbent to ascending, 40–60 cm, much-branched, gray-green, shaggy-bristly, nonglandular. **ST**: with lfy axillary shoots. **LF**: 5–20 mm, ± fleshy, oblong to rounded; lobes 0–3, truncate-rounded. **INFL**: 5–20 cm; bracts 15–20 mm, widely wedge-shaped to widely obovate, lobes 0–3, bright red to orange-red, central lobe wide; tip truncate-rounded. **FL**: calyx 20–25 mm, divided 1/2 abaxially and adaxially, < 1/8 on sides, shaggy-hairy (some hairs glandular); corolla 30–45 mm, beak 15–25 mm, adaxially shaggy-puberulent, margin ± red, lower lip 2 mm, dark green, ± incl; stigma club-shaped, entire. **FR**: 15–20 mm. **SEED**: 2–2.5 mm; coat deeply netted, most walls ladder-like. Coastal scrub; < 100 m. c NCo (Mendocino Co.). Threatened by coastal development. May–Aug ★

C. miniata Hook. Per 40–80 cm, few-branched, green or becoming ± purple, glabrous, to long-soft-hairy distally. **LF**: 30–60 mm, ± lanceolate, entire; tip acute. **INFL**: 3–15 cm; bracts 15–35 mm, lobes 0–5, acute, bright red to ± yellow. **FL**: calyx divided 2/3+ abaxially, 1/2–2/3 adaxially, ± 1/4 on sides, lobes acute; corolla beak ± = tube, yellow-green, adaxially puberulent, margins red, lower lip 1–2 mm, dark green, slightly exserted; stigma slightly 2-lobed. **SEED**: 1.5–2 mm; coat shallowly netted, loose-fitting, inner walls membranous, persistent.

subsp. ***elata*** (Piper) Munz (p. 965) SISKIYOU PAINTBRUSH **ST**: slender. **INFL**: ± pink to yellow-orange. **FL**: calyx 9–17 mm; corolla 15–25 mm. **FR**: 6–10 mm. $2n$=24. Bogs, often on serpentine; < 1400 m. nw KR; sw OR. May–Aug ★

subsp. ***miniata*** (p. 965) **ST**: stout. **INFL**: gen bright red (orange-red). **FL**: calyx 15–30 mm; corolla 20–40 mm. **FR**: 10–12 mm. $2n$=24,48,72,96,120. Common. Wet montane meadows, streambanks; (350)1500–3500 m. NW, CaR, SNH, c CCo, SCoRO, SW, GB. [*C. oblongifolia* A. Gray] If recognized taxonomically, pls in

lowland s NCoRO (Pitkin Marsh, Sonoma Co., ± 60 m) with yellow infl assignable to *C. uliginosa* Eastw., Pitkin Marsh paintbrush. May–Sep

C. minor (A. Gray) A. Gray   Ann, ± simple, ± slender, 30–150 cm, green to ± gray, variously hairy. **LF:** 40–100 mm, lance-linear, entire. **INFL:** 10–40 cm, narrow, open proximally; bracts 20–50 mm, like lvs, entire, tips red, ± long-tapered; proximal pedicels < 10 mm (distal 0). **FL:** calyx 14–28 mm, divided 2/3–3/4 abaxially and adaxially, ± 1/8 on sides, soft-hairy, lobes narrow, acute; corolla beak < tube, adaxially ± puberulent, margins pale, lower lip 2–3 mm, ± yellow, exserted from calyx, spreading; stigma slightly 2-lobed. **FR:** 10–15 mm. **SEED:** 1–1.5 mm; coat ± deeply netted, most walls ladder-like. 2*n*=24.

subsp. ***minor*** (p. 965)   Shaggy-hairy, most hairs glandular. **FL:** corolla 15–20(30) mm. Alkaline marshes; 850–2300 m. GB; to WA, NM. Jul–Sep

subsp. ***spiralis*** (Jeps.) T.I. Chuang & Heckard   Puberulent, many hairs nonglandular. **FL:** corolla 25–35 mm. Wet places; < 2600 m. NCoR, c&s SNF, s SNH, Teh, SnJV, CCo, n SnFrB, SW. Jun–Oct

C. mollis Pennell (p. 965)   SOFT-LEAVED PAINTBRUSH   Per or subshrub, ± prostrate, 30–40 cm, openly branched, white-woolly, hairs branched, some glandular. **ST:** with lfy axillary shoots. **LF:** 10–30 mm, oblong to obovate, entire, tip rounded. **INFL:** 3–8 cm; bracts 15–20 mm, fleshy, obovate, tip rounded or 3-toothed, yellow(-green). **FL:** calyx 16–18 mm, divided ± 1/2 abaxially and adaxially, ± 1/8 on sides, lobes ovate, acute; corolla 17–18 mm, beak gen < tube, ± yellow-green, slightly exserted, adaxially densely puberulent, margins pale, lower lip 2 mm, green; stigma slightly 2-lobed, dark green, slightly exserted. **FR:** 16 mm. **SEED:** 2 mm; coat deeply netted, most walls ladder-like. 2*n*=24. Coastal dunes; < 20 m. n ChI (Santa Rosa, e San Miguel islands). [Heckard et al. 1991 Madroño 38:141–142] Apr–Aug ★

C. montigena Heckard   HECKARD'S PAINTBRUSH   Per to subshrub, 15–45 cm, branching distally, gray-green, glandular-puberulent throughout. **LF:** 2–5.5 cm, lanceolate to lance-linear, margin occ wavy, lobes 0–3, narrow-triangular, ± 1 cm. **INFL:** 3–4 cm wide, shaggy-hairy; bracts < lvs, 3–5-lobed, tips gen acute, red to orange. **FL:** calyx 15–20 mm, divided ± 1/4 abaxially and adaxially, ± 1/8 on sides, lobes acute, red; corolla 20–40 mm, beak 1–2 × tube, tube 12–23 mm, pale green-yellow; lower lip exserted; stigma slightly 2-lobed, exserted. **FR:** 10–15 mm. **SEED:** ± 1.5 mm, coat shallowly netted, walls irregularly ladder-like. *n*=24,36. Dry, rocky, open slopes and flats in open forest, pinyon/juniper woodland; 1800–2900 m. e SnBr. [Heckard et al. 1980 Syst Bot 5:71–85] May–Aug ★

C. nana Eastw. (p. 965)   Per 5–25 cm, green or ± purple, spreading-hairy, gen nonglandular. **LF:** 10–35 mm, lance-linear; lobes 0–5. **INFL:** 3–13 cm; bracts 15–30 mm, lobes 3–5, acuminate, yellow-green or ± purple, green-margined. **FL:** calyx 12–20 mm, subequally divided ± 1/2, lobes lance-linear; corolla 15–20 mm, beak 4–6 mm, pale yellow blotched ± purple, lower lip 3–5 mm, pouches shallow; stigma ± exserted, dark. **FR:** 8–12 mm. **SEED:** 1–1.5 mm; coat shallowly netted, loose-fitting, with ladder-like thickenings. Dry, ± alpine barrens; 2400–4200 m. SNH, W&I. Highly variable, intergrades with *C. pilosa*; needs study. Pls on e slope SNH and W&I ± purple. Jul–Aug

C. peirsonii Eastw. (p. 965)   Per 15–40 cm, ± glabrous to long-nonglandular- and short-glandular-hairy. **LF:** 15–50 mm, ± lanceolate to oblong; lobes 0–5. **INFL:** 3–15 cm; bracts 15–25 mm, lobes 3–5, gen bright red. **FL:** calyx 12–20 mm, gen divided ≥ 1/2 abaxially, 1/2 adaxially, 1/4–1/3 on sides, lobes acute to notched; corolla 15–28 mm, beak ± 1/2 tube, ± yellow-green, adaxially sparsely puberulent, margins ± red, lower lip 1–2 mm, dark green, incl; stigma exserted, 2-lobed, ± black. **FR:** 6–9 mm. **SEED:** 1–1.5 mm; coat shallowly netted, loose-fitting. 2*n*=24. Montane to alpine meadows; 1500–3400 m. CaRH, SNH; to AK. [*C. parviflora* Bong., misappl.] Jul–Aug

C. pilosa (S. Watson) Rydb. (p. 965)   Per 8–35 cm, often decumbent, spreading-stiff-hairy, nonglandular. **LF:** 10–50 mm, lance-linear; lobes 0–3. **INFL:** 3–20 cm; bracts 10–30 mm, lobes 3–5, pale green or ± purple, white-margined, central lobe truncate or rounded

(acute). **FL:** calyx 10–20, subequally divided ± 1/2, lobes lance-linear; corolla 13–22 mm, beak 4–6 mm, pale yellow-green, lower lip 3–5 mm, pouches shallow; stigma ± exserted, dark. **FR:** 8–12 mm. **SEED:** ± 1 mm; coat deeply netted, tight-fitting, walls ladder-like. 2*n*=24. Dry sagebrush scrub to alpine barrens; 1200–3400 m. CaRH, SNH, GB; to e OR, c ID, w WY. Highly variable, intergrades with *C. nana*; needs further study. Jun–Aug

C. plagiotoma A. Gray (p. 965)   MOJAVE PAINTBRUSH   Per 30–60 cm, gray-green, becoming ± dark red, puberulent (esp lvs); hairs branched. **LF:** 20–50 mm, ± linear; lobes 3–5. **INFL:** 3–20 cm; bracts 13–20 mm, white-woolly lobes 3–5, central lobe wide, truncate, green. **FL:** calyx 12–18 mm, pale yellow, white-woolly, divided 1/4 abaxially, 1/8 adaxially, ± 1/2 on sides; corolla 12–20 mm, beak ± = tube, ± yellow, adaxially puberulent, margins pale, lower lip 1 mm, pale green; stigma barely exserted, head-like. **FR:** ± 10 mm. **SEED:** 1–1.5 mm; coat deeply netted, tight-fitting, most walls ladder-like. Dry sagebrush scrub, pinyon woodland; 300–2500 m. s SN, Teh, SCoRI, TR, DMoj. Apr–Jun ★

C. praeterita Heckard & Bacig. (p. 965)   Per 10–45 cm, much-branched, stiff-hairy, nonglandular. **LF:** 30–50 mm, ± linear; lobes 0–3. **INFL:** 8–14 cm, 1.5–2 cm wide; bracts 15–25 mm, pale green, tipped lemon yellow or pale red, lobes 3(5), truncate or rounded. **FL:** calyx 14–18 mm, divided ± 1/2 abaxially and adaxially, 1/8 on sides, lobes acute to obtuse; corolla 13–16 mm, incl or exserted, beak ± 5 mm, puberulent, margins ± yellow or dark purple, lower lip 2 mm, pouches narrow, teeth dark green, incurved; stigma ± exserted, 2-lobed. **FR:** 8–10 mm. **SEED:** 1–1.5 mm; coat shallowly netted, loose-fitting. 2*n*=24. Dry places, esp in *Artemisia rothrockii* meadows; 2200–3400 m. s SNH. Jul–Aug

C. pruinosa Fernald (p. 965)   Per to subshrub, 30–80 cm, few-branched with short axillary shoots, densely ± gray pubescent or green, and branched-hairy. **LF:** 20–80 mm, ± lanceolate; lobes 0–5, tips obtuse. **INFL:** 3–20 cm; bracts 10–25 mm, lobes 0–5, gen bright red to orange-red. **FL:** calyx 13–20 mm, divided 1/3–1/2 abaxially and adaxially, 1/5–1/4 on sides, glandular, lobes lanceolate, acute; corolla 25–32 mm, beak 1–2 × tube, adaxially puberulent, margins ± red, lower lip 1–2 mm, green, appressed to upper lip; stigma unlobed, club-shaped. **FR:** 8–15 mm. **SEED:** 1.5–2 mm; coat deeply netted, loose-fitting, side walls ladder-like, inner walls membranous, splitting. 2*n*=48. Dry, open serpentine or forest edge; < 2600 m. NW, CaR, n&c SN, MP; s OR. Highly variable and confusing complex; needs further study. Apr–Aug

C. rubicundula (Jeps.) T.I. Chuang & Heckard   CREAM SACS   Ann 20–70 cm, spreading-hairy, glandular. **LF:** 20–80 mm, lanceolate; lobes 0–7. **INFL:** 5–15 cm, 3–4 cm wide, dense; bracts 15–30 mm, ovate, green, lobes 5–9, ± lanceolate. **FL:** calyx 8–10 mm, subequally divided 1/2; corolla 20–28 mm, beak 5–7 mm, straight, puberulent, lower lip 4–6 mm, pouches 8–10 mm wide, 4–6 mm deep, gen purple-dotted at base; stigma ± 2-lobed, exserted. **FR:** 7–10 mm. **SEED:** < 1 mm; coat ± deeply netted, loose-fitting. 2*n*=24.

subsp. ***lithospermoides*** (Benth.) T.I. Chuang & Heckard (p. 965)   **FL:** corolla yellow. Open grassland; < 900 m. NCoR, CaRF, CCo, SnFrB; sw OR. Apr–Jun

subsp. ***rubicundula***   PINK CREAMSACS   **FL:** corolla white, turning ± pink. Open grassland; < 900 m. s NCoRI, ScV, s SnFrB. Apr–Jun ★

C. schizotricha Greenm. (p. 969)   SPLIT-HAIR PAINTBRUSH   Per 8–15 cm, ± white-woolly, hairs branched. **LF:** 5–20 mm, lance-linear; lobes 0–3. **INFL:** 3–8 cm; bracts 10–20 mm, ± pink to dusty red, lobes 3, obtuse. **FL:** calyx 13–18 mm, divided subequally ± 1/2; corolla 15–20 mm, tomentose, distal 1/2 purple-red, beak 4–5 mm, lower lip ± 4 mm, pouches shallow; stigma ± exserted, ± notched, dark. **FR:** 8–10 mm. **SEED:** ± 1 mm; coat shallowly netted, loose-fitting. Decomposed granite or marble; 1500–2300 m. KR (Siskiyou Co.); sw OR. Jul–Aug ★

C. subinclusa Greene   Per 30–120 cm, (gray-)green, becoming ± purple, ± puberulent to hairs long- and short-glandular. **ST:** with short axillary shoots. **LF:** 30–80 mm, lanceolate; lobes 0–3. **INFL:** 6–40 cm, open proximally; bracts 20–60 mm, gen entire, tips bright

red to orange-red; pedicels < 10 mm (distal 0). **FL:** calyx 20–32 mm, divided 2/3 abaxially, 1/6–1/3 adaxially, glandular-puberulent and long-hairy, lobes acute to obtuse, ± curved upward; corolla (incl lower lip) gen curved out through abaxial calyx sinus, 30–50 mm, beak ± = tube, yellow-green, margins rose-pink, beak densely shaggy-puberulent, lower lip 2–3 mm, dark green or ± purple; stigma unlobed. **FR:** 10–15 mm. **SEED:** ± 2 mm; coat deeply netted, tight-fitting, most walls ladder-like. Range is discontinuous.

subsp. ***franciscana*** (Pennell) T.I. Chuang & Heckard (p. 969) **INFL:** strongly 2-colored, red and yellow. **FL:** calyx sides ± entire to divided < 1/5, lobes linear to narrowly deltate, obviously curved upward; corolla yellow-orange. 2*n*=24,48,72,96. Coastal scrub; < 100 m. s NCo (s Mendocino, Sonoma cos.), n CCo (to Santa Cruz Co.), w SnFrB. Mar–Jul

subsp. ***subinclusa*** (p. 969) **INFL:** ± red throughout. **FL:** calyx sides divided 1/5–1/3, lobes long-tapered, barely curved upward; corolla ± red. 2*n*=±72. Open chaparral; < 2200 m. SNF, Teh, SCoRI, WTR, SnGb, s PR; n Baja CA. Apr–Jul

C. tenuis (A. Heller) T.I. Chuang & Heckard (p. 969) Ann 10–45 cm, spreading-hairy, glandular distally. **LF:** 10–40 mm, lance-linear,

lobes 0–3(5). **INFL:** 5–25 cm, 1–3 cm wide; bracts 10–30 mm, ovate, green, lobes 3–7, ± lanceolate. **FL:** calyx 8–12 mm, divided 1/3 abaxially and on sides, 1/2 adaxially; corolla 12–20 mm, white or yellow, beak 4–5 mm, straight, puberulent, lower lip 3–4 mm, pouches 2–4 mm wide, ± 2 mm deep; stigma incl. **FR:** 6–9 mm. **SEED:** ± 1 mm; coat ± deeply netted, loose-fitting. 2*n*=24,48. Moist flats, meadows; 1000–2800 m. KR, NCoR, CaR, SNH, SnBr (rare), PR (rare), MP, n SNE; to AK. May–Aug

C. wightii Elmer (p. 969) Per 30–80 cm, much-branched proximal to infl, (yellow-)green tinged ± purple, densely long-bristly and glandular-sticky, hairs simple. **ST:** short, lfy axillary shoots present, many, proximal to infl. **LF:** 20–60 mm, gen crowded, lanceolate to ± ovate; lobes 0–3. **INFL:** 5–20 cm; bracts 10–25 mm, lobes 3, bright red to yellow. **FL:** calyx 15–28 mm, divided 1/3–1/2 abaxially and adaxially, 1/8–1/4 on sides, lobes acute or rounded; corolla 20–30 mm, beak 1–2 × tube, adaxially shaggy-hairy, margins ± red to ± yellow, lower lip 2 mm, dark green, incl; stigma slightly 2-lobed. **FR:** 10–15 mm. **SEED:** 2 mm; coat deeply netted, most walls ladder-like. 2*n*=24,48. Coastal scrub; < 300 m. c&s NCo, n CCo, SnFrB. Mar–Aug

CHLOROPYRON SALTY BIRD'S-BEAK

Margriet Wetherwax & David C. Tank

Ann 10–60 cm, green, branches few to many; roots ± yellow. **LF:** alternate, sessile, 5–35 mm, entire, lanceolate to oblong. **INFL:** spike, loose to dense, subtended by outer, lf-like bracts; inner bract 1 per fl, ± lf-like, entire or pinnately lobed. **FL:** calyx = or slightly < corolla, sheath-like, gen cut completely to base abaxially, partially surrounding corolla tube laterally, notched ± 1 mm at tip; corolla 2-lipped, club-shaped, tubular below, expanded laterally, upper lip folded lengthwise, tip rounded, closed, opening directed downward forming a hood enclosing anthers and style; lower lip ≤ upper lip, obscurely 3-lobed; fertile stamens 4 and staminodes 0, or 2 with adaxial pair of staminodes (attached deeper in corolla tube), anther sacs gen 2 per stamen, ± overlapping, tufted-hairy at base, unequal in size and placement; ovary 2-chambered, glabrous, ovules many, style bent near tip, stigma barely exserted. **SEED:** attached at side; seed coat tight-fitting, netted. *n*=14,15,21. 4 spp.: saline and alkaline habitats, w N.Am. (Greek: salt pl) [Tank et al. 2009 Syst Bot 34:182–197] Formerly incl in *Cordylanthus*. Close to *Dicranostegia*, together forming the *Pseudocordylanthus* clade (see Tank et al. 2009); distinguished by salt-tolerant ecology, infl, calyx, stamens; fls May–Nov.

1. Fertile stamens 4; inner bracts entire or slightly notched . . . . ***C. maritimum***
2. Inner bracts gen entire; seeds 25–40, 1–1.5 mm; inland . . . . subsp. ***canescens***
2′ Inner bracts slightly notched; seeds 10–20, 2–3 mm; gen coastal
   3. St gen much-branched, distal branches gen > central spike; seeds ± 2 mm; s CCo, SCo . . . . subsp. ***maritimum***
   3′ St 0–few-branched, branches ≤ central spike; seeds 2–3 mm; NCo, n CCo . . . . subsp. ***palustre***
1′ Fertile stamens 2, staminodes 2; inner bracts 3–7-lobed
   4. Lf 1–2 mm wide; inner bracts 3-lobed; style puberulent . . . . ***C. tecopense***
   4′ Lf 3–8 mm wide; inner bracts 3–7-lobed; style glabrous
     5. Pl soft-hairy to becoming glabrous, longest hairs < 1 mm; seed coat deeply netted, wavy-crested . . . . . . ***C. palmatum***
     5′ Pl, esp infl, glandular-puberulent and stiff-long-nonglandular-hairy, longest hairs > 1 mm; seed coat
       deeply netted, not wavy-crested . . . . ***C. molle***
      6. St much-branched from near base; infl gen 2–6 cm; corolla pouch and tube sparsely tomentose;
        seed 1–1.5 mm; GV . . . . subsp. ***hispidum***
      6′ St gen few-branched from middle; infl 5–15 cm; corolla pouch and tube densely tomentose; seed 2–3
        mm; n CCo, deltaic GV . . . . subsp. ***molle***

C. maritimum (Benth.) A. Heller Pl 10–40 cm, gray-green, glaucous, often tinged purple and salt-encrusted, gen ± short-hairy. **LF:** 5–25 mm, lance-linear, entire. **INFL:** spike, 20–90 mm, many-fld; inner bract 15–30 mm. **FL:** calyx 15–25 mm; corolla 15–25 mm, white to cream, puberulent, lips pale to ± brown or purple-red; fertile stamens 4, anther sacs 2 (proximal pair) or 1 (distal pair). **SEED:** ± reniform, deeply netted, dark brown. 2*n*=30. [*Cordylanthus m.* Benth.]

subsp. ***canescens*** (A. Gray) Tank & J.M. Egger (p. 969) **ST:** branches gen many, ± erect, distal gen > central spike. **INFL:** dense; inner bracts gen entire. **SEED:** 25–40, 1–1.5 mm. Inland alkaline flats; 600–1900 m. GB, n DMoj; to s OR, UT. [*Cordylanthus m.* subsp. *c.* (A. Gray) T.I. Chuang & Heckard] Jun–Sep

subsp. ***maritimum*** (p. 969) SALT MARSH BIRD'S-BEAK **ST:** branches gen many, decumbent to ascending, distal gen > central

spike. **INFL:** loose or dense; inner bracts slightly notched. **SEED:** 15–20, ± 2 mm. Coastal salt marsh; < 10 m. s CCo (Morro Bay), SCo; n Baja CA. [*Cordylanthus m.* Nutt. ex Benth. subsp. *m.*] May–Oct ★

subsp. ***palustre*** (Behr) Tank & J.M. Egger (p. 969) POINT REYES SALTY BIRD'S-BEAK **ST:** branches 0–few, ascending, ≤ central spike. **INFL:** dense; inner bracts slightly notched. **SEED:** 10–20, 2–3 mm. Coastal salt marsh; < 10 m. n NCo (Humboldt Co.), n CCo (Marin, Sonoma cos.); sw OR. [*Cordylanthus m.* subsp. *p.* (Behr) T.I. Chuang & Heckard] May–Oct ★

C. molle (A. Gray) A. Heller Pl 10–40 cm, gray-green, often tinged purple, glandular-puberulent and stiffly long-nonglandular-hairy. **LF:** 10–25 mm, ± oblong, entire to 7-lobed. **INFL:** spike, 20–150 mm; outer bracts lf-like; inner bracts 15–25 mm, 3–7-lobed. **FL:** calyx 15–20 mm; corolla 15–20 mm, ± white, ± densely tomentose, middle

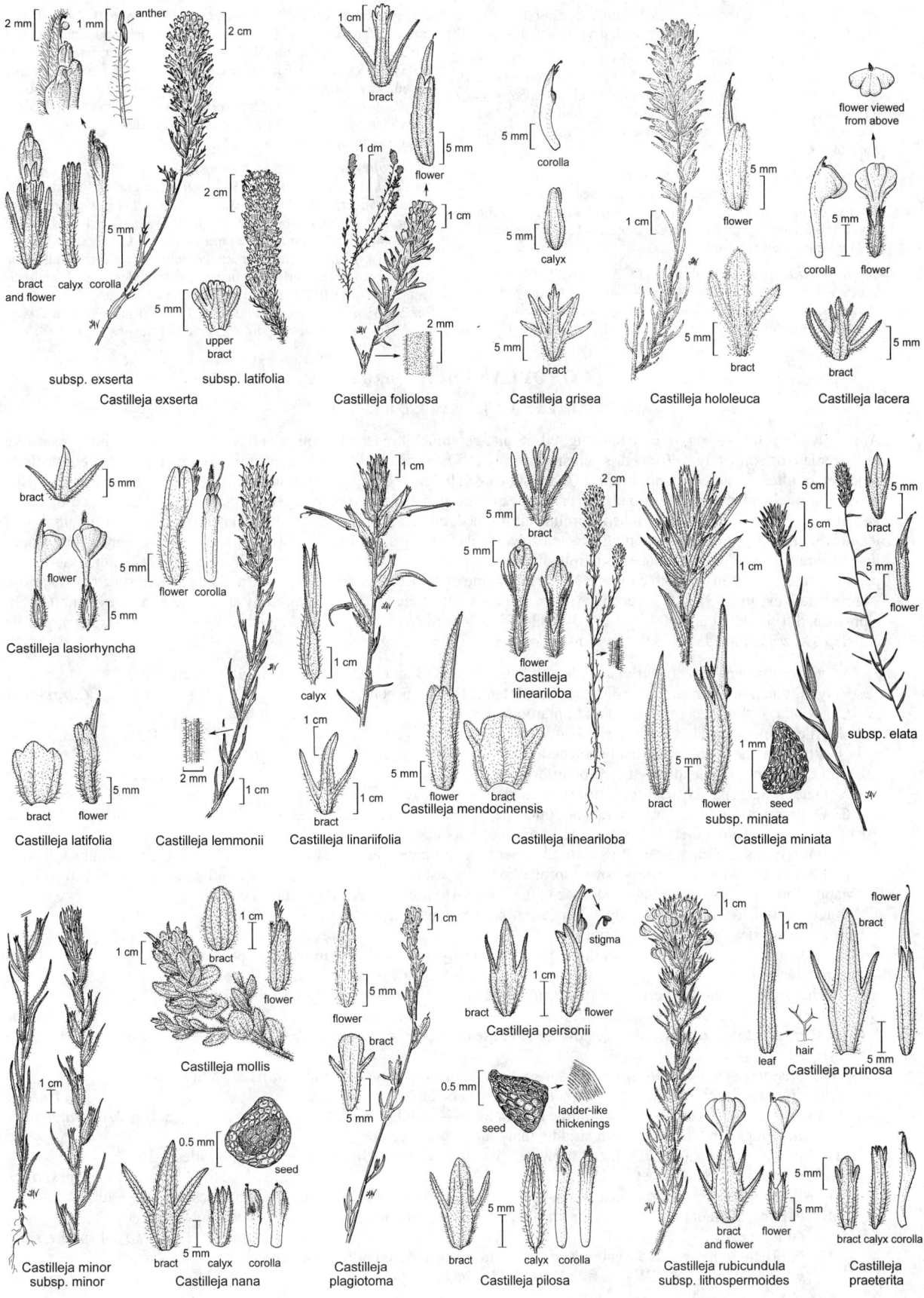

subsp. exserta   subsp. latifolia
Castilleja exserta

Castilleja foliolosa

Castilleja grisea

Castilleja hololeuca

Castilleja lacera

Castilleja lasiorhyncha

Castilleja lineariloba

subsp. elata

Castilleja latifolia

Castilleja lemmonii

Castilleja linariifolia

Castilleja mendocinensis

subsp. miniata

Castilleja lineariloba

Castilleja miniata

Castilleja mollis

Castilleja peirsonii

Castilleja pruinosa

Castilleja minor
subsp. minor

Castilleja nana

Castilleja plagiotoma

Castilleja pilosa

Castilleja rubicundula
subsp. lithospermoides

Castilleja praeterita

lobe of lower lip erect; fertile stamens 2, staminodes 2; style glabrous. **SEED:** 1–3 mm, ± reniform, deeply netted, dark brown. 2*n*=28.

subsp. ***hispidum*** (Pennell) Tank & J.M. Egger (p. 969) HISPID SALTY BIRD'S-BEAK Pl bristly. **ST:** branches many, from near base, spreading. **INFL:** gen 2–6 cm. **FL:** corolla pouch and tube sparsely tomentose. **SEED:** 1–1.5 mm. Saline marshes and flats; < 130 m. GV. [*Cordylanthus m.* subsp. *h.* (Pennell) T.I. Chuang & Heckard] Jun–Jul ★

subsp. ***molle*** (p. 969) SOFT SALTY BIRD'S-BEAK Pl gen ± soft-hairy. **ST:** branches gen few, from middle, ± ascending. **INFL:** 5–15 cm. **FL:** corolla pouch and tube densely tomentose. **SEED:** 2–3 mm. Coastal salt marshes; < 10 m. n CCo, deltaic GV. [*Cordylanthus m.* A. Gray subsp. *m.*] Jul–Nov ★

C. palmatum (Ferris) Tank & J.M. Egger (p. 969) PALMATE SALTY BIRD'S-BEAK Pl 10–30 cm, 3–8 mm wide, gray-green, ± glandular, soft-hairy, longest hairs < 1 mm, becoming glabrous. **LF:** 7–20 mm, ± oblong, entire to 5-lobed. **INFL:** spike, 50–150 mm, dense; outer bracts lf-like; inner bracts 15–20 mm, 3–7-lobed. **FL:** calyx ± 15 mm; corolla 15–20 mm, ± white, finely puberulent, sides often ± pale lavender, middle lobe of lower lip erect; fertile stamens 2, staminodes 2; style glabrous. **SEED:** 2.5–3 mm, ± reniform, deeply netted, wavy-crested, ± dark-brown. 2*n*=42. Alkaline flats; < 60 m. GV (Colusa, Yolo, Alameda, San Joaquin, Madera, Fresno cos.). [*Cordylanthus p.* (Ferris) J.F. Macbr.] Threatened by agriculture, urbanization. Jun–Aug ★

C. tecopense (Munz & J.C. Roos) Tank & J.M. Egger (p. 969) TECOPA SALTY BIRD'S-BEAK Pl 10–60 cm, ± gray or tinged purple, sparsely puberulent, glaucous. **LF:** 5–15 mm, 1–2 mm wide, lance-linear, entire. **INFL:** spike, 20–150 mm, loose; outer bracts lf-like; inner bracts 10–15 mm, 3-lobed near middle. **FL:** calyx 10–13 mm; corolla 10–15 mm, pale lavender, densely puberulent, middle lobe of lower lip erect; fertile stamens 2, staminodes 2; style puberulent. **SEED:** 8–10, 2–3 mm, ± reniform, deeply netted, light brown. 2*n*=28. Alkaline meadows and flats; 100–900 m. s SNE, n DMoj; w NV. [*Cordylanthus t.* Munz & J.C. Roos] Aug–Oct ★

CORDYLANTHUS BIRD'S-BEAK

Margriet Wetherwax & David C. Tank

Ann, gray- or yellow-green, often becoming red-purple, gen much-branched; roots ± yellow. **LF:** sessile, entire, thread-like to lanceolate or palmately 3–7-lobed, segments narrow. **INFL:** short dense spike (subtended by bracts), < 2(5) cm or fls 1, scattered or often clustered but not in spikes (each subtended by outer bract); outer bracts ± lf-like; inner bract calyx-like (formerly confused with calyx), 0–7-lobed. **FL:** calyx sheath-like, gen divided to base abaxially, partially surrounding corolla tube laterally, tip entire or shallowly notched; corolla club-shaped, tubular proximally, expanded laterally; upper corolla lip folded lengthwise, tip rounded, closed, opening directed downwards forming a hood enclosing anthers, style; lower corolla lip ≤ upper lip, obscurely 3-lobed, middle lobe tightly rolled under, tip distinctly folded inside-out; fertile stamens (2)4, anther sacs gen 2 per stamen, densely hairy at both ends and ciliate along line of dehiscence, unequal in size and placement; style bent near tip, stigma barely exserted. **SEED:** attached at side; seed coat tight-fitting, netted or irregularly striate. 13 spp.: w N.Am. (Greek: club-shaped fl) [Tank et al. 2009 Syst Bot 34:182–197] Close to *Orthocarpus*, distinguished by infl, calyx, stamens; gen fls Jul–Sep. Other taxa in TJM (1993) moved to *Chloropyron*, *Dicranostegia*.

1. Mature corolla gen > 3 × longer than wide; seed netted; esp GB and D in sagebrush or juniper scrub
   2. Calyx divided to base; corolla ± dusty yellow, often marked maroon . . . . **C. ramosus**
   2′ Calyx tube 1–4 mm; corolla gen pink to maroon (± yellow)
      3. Fertile stamens 2; calyx divided ± 1/3 — Wrn . . . . **C. capitatus**
      3′ Fertile stamens 4; calyx shallowly notched
         4. Pl canescent, not glandular-sticky; outer bracts 3–7-lobed . . . . **C. eremicus**
            5. Outer bracts sparsely scabrous, tips thickened, maroon . . . . subsp. **eremicus**
            5′ Outer bracts distinctly bristly, tips not thickened, green . . . . subsp. **kernensis**
         4′ Pl densely glandular-sticky; outer bracts gen 3- or 5-lobed
            6. Infl a spike, 2–3 cm; corolla lips ± equal; inner bract pinnately lobed; s MP, SNE . . . . **C. kingii** subsp. **helleri**
            6′ Fls 1 or 2–4 in loose clusters; lower corolla lip << upper; inner bract gen entire; e DMtns . . . . **C. parviflorus**
1′ Mature corolla 2–3 × longer than wide; seed finely wavy-striate; esp CA-FP in foothill woodland
   7. Outer bracts fan-shaped, shallowly 3–7-lobed; inner bract and calyx tips rough-papillate; corolla 8–9 mm, pouch sides yellow — serpentine, NCoR. . . . . **C. pringlei**
   7′ Outer bracts entire to deeply 3-lobed; inner bract and calyx tips scabrous to puberulent; corolla 10–20 mm, pouch sides ± white
      8. Anther sac 1 (+ bearded vestige) per stamen; calyx tube ± 1 mm; outer bracts 3-lobed, < fl — s SNH, TR, PR . . . . **C. nevinii**
      8′ Anther sacs 2 per stamen; calyx gen divided to base (exc *Cordylanthus rigidus*); outer bracts entire or 3-lobed, gen ≥ fl
         9. Pl ± bristly (exc *Cordylanthus rigidus* subsp. *littoralis*), hairs gen not glandular; fls 5–15 in ± dense clusters; calyx tube 1–2 mm; corolla maroon in U-shape on lower side. . . . . **C. rigidus**
            10. Outer bract < fl, gen 5–10 mm, middle lobe oblong — s SNH . . . . subsp. **brevibracteatus**
            10′ Outer bract gen > fl, 10–20 mm, middle lobe linear to lanceolate
               11. Middle lobe of outer bract linear, tip wider, notched, thickened, maroon, marginal bristles > 2 × lobe width — SCoRO, SW . . . . subsp. **setiger**
               11′ Middle lobe of outer bract gen lanceolate or oblong, tip gen not wider, marginal bristles ± = lobe width
                  12. Pl puberulent or soft-hairy; outer bract weakly ciliate and puberulent, tip of middle lobe tapered; dunes, c&s CCo . . . . subsp. **littoralis**
                  12′ Pl bristly; outer bracts ± scabrous, tip of middle lobe tapered or blunt and notched; esp chaparral, woodland, c&s SN, CW (exc n SnFrB), WTR, SnGb . . . . subsp. **rigidus**

9′ Pl soft-hairy, becoming glabrous, hairs glandular or not; fls 1–7 in loose clusters; calyx divided to base; corolla veined, streaked or blotched maroon

13. St decumbent, branches ascending < 15 cm; outer bracts 3-lobed near base — serpentine, ne SnFrB . . . ***C. nidularius***

13′ St erect, branches ascending gen > 20 cm; outer bracts entire or 3-lobed in distal 2/3

14. Main st densely puberulent (partly glandular) and long-soft-hairy; tips of outer bracts (or middle lobes) > 1.5 mm wide, longest hairs 2–3 mm; corolla barely marked maroon . . . . ***C. pilosus***

15. Outer bracts entire (tip wider, obtuse to 2-notched); NCoR, n SnFrB. . . . . subsp. ***pilosus***

15′ Outer bracts 3-lobed; CaRF, n&c SNF

16. Outer bracts lobed in distal 1/3, lateral lobes < 3 mm; corolla ≥ calyx and inner bract . . . . subsp. ***hansenii***

16′ Outer bracts lobed in proximal 1/2; lateral lobes > 3 mm; corolla < calyx and inner bract . . . . . . subsp. ***trifidus***

14′ Main st ± glabrous to sparsely long-hairy or ± glandular; tips of outer bracts (or middle lobes) < 1.5 mm wide, longest hairs < 2 mm; corolla heavily blotched maroon. . . . . ***C. tenuis***

17. St glabrous or minutely puberulent; lvs (or lobes) thread-like, tightly folded up, gen ± glabrous; serpentine outcrops — s NCoR

18. Outer bracts entire; sts minutely and sparsely glandular, esp proximal to fls . . . . subsp. ***brunneus***

18′ Outer bracts 3-lobed; sts ± glabrous (not glandular proximal to fls) — sw NCoRO. . . . . subsp. ***capillaris***

17′ St gen puberulent and long-hairy, sometimes glandular; lvs (or lobes) linear to ± oblong, flat to channeled, puberulent and gen glandular; gen not on serpentine

19. Fls 3–7 in dense clusters; outer bracts gen 3-lobed, densely soft-hairy; longest hairs of inner bract > 2 mm; s SNH. . . . . subsp. ***barbatus***

19′ Fls 1–6 in loose clusters; outer bracts entire or 3-lobed, ± stiff-hairy; longest hairs of inner bract < 1 mm

20. Outer bracts entire. . . . . subsp. ***tenuis***

20′ Outer bracts 3-lobed

21. Bracts and calyx yellow-green, puberulent, weakly glandular; fls gen 3–6 per cluster — CaRH . . . . subsp. ***pallescens***

21′ Bracts and calyx green or maroon-tinged, glandular, puberulent and with long, stiff hairs; fls 1–3 per cluster . . . . subsp. ***viscidus***

C. capitatus Benth. (p. 969) YAKIMA BIRD'S-BEAK Pl 10–50 cm, glaucous-green or gray-purple, densely glandular- and nonglandular-hairy. **LF**: 20–40 mm, linear, entire. **INFL**: head-like spike, 15–20 mm, 2–5-fld; outer bracts 4–7, 10–20 mm, lobes 3 in lower 1/2, lance-linear; inner bract 12–18 mm. **FL**: calyx 10–15 mm, tube 2–4 mm, divided ± 1/3; corolla 10–20 mm, maroon, yellow-tipped, throat 4–6 mm wide; fertile stamens 2, anther sac 1. **SEED**: 4–6, 2–2.5 mm, ± reniform, shallowly netted, ± smooth between nets. 2*n*=26. Open conifer forest; juniper woodland; 1800–2150 m. Wrn; to c WA, ID. Aug–Oct ★

C. eremicus (Coville & C.V. Morton) Munz Pl 10–80 cm, (yellow-)green, often tinged red, canescent, not glandular-sticky. **LF**: 10–40 mm, thread-like or linear, 0–7-lobed. **INFL**: spike, 15–25 mm, 3–14-fld; outer bracts 3–7-lobed at base of spike, 5–20 mm, 3–7-lobed; inner bract 10–18 mm. **FL**: calyx 10–18 mm, slightly notched, tube 1–3 mm; corolla 10–20 mm, ± purple, ± pink, or yellow-green, often yellow-tipped, throat base blotched maroon, lower lip 4–6 mm wide; soft-white-hairy; anther sacs overlapping. **SEED**: 1.5–2 mm, ± ovoid, deeply netted, pale brown. 2*n*=26.

subsp. ***eremicus*** (p. 969) DESERT BIRD'S-BEAK **INFL**: outer bracts sparsely scabrous, tips wider, thickened, maroon. **FL**: calyx tube 2–3 mm; corolla lavender to ± pink, blotched purple. **SEED**: papillate between nets. Sagebrush scrub, pinyon/juniper woodland; 1000–2800 m. n SnBr (near Cushenbury), n DMtns. Aug–Oct ★

subsp. ***kernensis*** T.I. Chuang & Heckard (p. 969) KERN PLATEAU BIRD'S-BEAK **INFL**: outer bracts distinctly bristly, tips acute, unthickened, unmarked. **FL**: calyx tube 1 mm; corolla gen ± pink. **SEED**: smooth between nets. Open Jeffrey-pine or juniper forest; 2100–3000 m. s SNH (Kern Plateau, Inyo, Kern, Tulare cos.). Jun–Aug ★

C. kingii S. Watson subsp. ***helleri*** (Ferris) T.I. Chuang & Heckard (p. 969) Pl 10–60 cm, gray-green, tinged red, densely glandular-sticky. **LF**: gen 10–25 mm, linear. **INFL**: spike, 20–30 mm, 1–4-fld; outer bracts 3–6, 10–15 mm, gen 3-lobed; inner bract 10–15 mm, 3–5-lobed, glandular. **FL**: calyx 15–20 mm, notched ± 2 mm, tube 2.5 mm; corolla 15–25 mm, rosy-lavender to dull purple-red or -yellow, hairy. **SEED**: 2–2.5 mm, ± ovoid, deeply netted, pale brown, papillate between nets. 2*n*=26. Open pinyon/juniper woodland, sagebrush scrub; 1200–3200 m. s MP, SNE; NV. Other subsp. in NV, UT. Jul–Sep

C. nevinii A. Gray (p. 969) Pl 20–80 cm, gen gray-green, tinged red-purple, densely glandular-puberulent and long-soft-hairy. **LF**: 5–30 mm, linear, lobes 0–3. **INFL**: fls 1 or 2–3 in loose clusters, often inverted; outer bracts 1–3, 5–10 mm, 3-lobed, tips wider, sometimes notched, cream; inner bract 10–15 mm, tip scabrous. **FL**: calyx 10–15 mm, tube ± 1 mm; corolla 12–18 mm, white, yellow-tipped, pouch 5–8 mm wide; anther sac 1 per stamen, 2nd sac vestigial, bearded. **SEED**: 2 mm, finely wavy-striate, dark brown. 2*n*=28. Dry, open Jeffrey-pine/oak forest; 1400–2600 m. s SNH (Piute Mtns), TR, PR; w AZ, n Baja CA. Jul–Sep

C. nidularius J.T. Howell (p. 969) MOUNT DIABLO BIRD'S-BEAK Pl < 15 cm, decumbent, gray-green, maroon-tinged, glandular-puberulent and long-nonglandular-hairy. **ST**: decumbent, openly branched. **LF**: 10–30 mm, narrowly linear, lobes 0–3. **INFL**: fls 1 or 2–3 in loose clusters; outer bracts 2–3, 10–15 mm, 3-lobed near base, lobes linear; inner bract ± 15 mm. **FL**: calyx 13–16 mm; corolla ± 15 mm, white, lower lip and throat maroon-lined. **SEED**: 7–10, 1.5–2 mm, ± ovoid, finely wavy-striate, dark brown. 2*n*=28. Dry, open serpentine in chaparral; 600–800 m. ne SnFrB (e slope Mount Diablo). Jul–Aug ★

C. parviflorus (Ferris) Wiggins (p. 969) SMALL-FLOWERED BIRD'S-BEAK Pl 20–60 cm, gray-green, tinged red, glandular-sticky and long-hairy. **LF**: 5–30 mm, linear, lobes 0–3. **INFL**: fls 1 or 2–4 in loose clusters; outer bract gen 1, 5–15 mm, 3–5-lobed, tips obtuse, ciliate, densely glandular; inner bract 10–12 mm. **FL**: calyx 10–15 mm, tube 1–1.5 mm, tip slightly notched; corolla 15–20 mm, pink to lavender, pouch 5–7 mm wide, often dark-veined. **SEED**: 1.5–2 mm, ± ovoid, shallowly netted, dark brown, densely papillate between nets. 2*n*=26. Dry sagebrush scrub, pinyon/juniper and Joshua-tree woodland; 700–2200 m. e DMtns (New York, Providence mtns); to se NV, sw UT, nw AZ; also s-c ID. Aug–Oct ★

C. pilosus A. Gray (p. 969) Pl 20–120 cm, (gray-)green, gen tinged purple, densely puberulent (partly glandular) and long-soft-hairy. **LF**: 10–40 mm, linear to lanceolate. **INFL**: fls 1 or 2–3 in loose clusters; outer bracts 1–4, 15–20 mm, linear, entire or 3-lobed in distal 2/3, tips > 1.5 mm wide, angled to slightly 3-lobed, often ivory-thickened; inner bract 15–22 mm, tip abruptly pointed to notched. **FL**: calyx 15–20 mm; corolla 15–20 mm, ± white, yellow-green-tipped, pouch 5–8 mm wide, ± lightly marked maroon. **SEED**: 1.5–2.5 mm, ± ovoid, finely wavy-striate, dark brown. 2*n*=28. Variable; close to *C. tenuis*.

subsp. *hansenii* (Ferris) T.I. Chuang & Heckard (p. 969) **INFL:** outer bracts ± linear, 3-lobed in distal 1/3, lobes < 3 mm, tips ± truncate. **FL:** corolla ≥ calyx and inner bract. Open foothill woodland; < 1000 m. s KR, s CaR, n&c SNF. Jul–Aug

subsp. *pilosus* (p. 969) **INFL:** outer bracts entire, tip wider, obtuse to 2-notched. **FL:** corolla ≤ calyx and inner bract. Open foothill woodland, chaparral, on serpentine; gen < 1500 m. NCoR, n SnFrB. Jul–Sep

subsp. *trifidus* (B.L. Rob. & Greenm.) T.I. Chuang & Heckard (p. 969) **INFL:** outer bracts ± lanceolate, 3-lobed in proximal 1/2, lobes > 3 mm, tips truncate to obtuse. **FL:** corolla < calyx and inner bract. Foothill woodland; gen < 1100 m. n&c SNF. Jun–Aug

C. pringlei A. Gray (p. 973) Pl 30–150 cm, green, lightly maroontinged, glabrous or puberulent. **LF:** 10–40 mm, thread-like, lobes 0–3. **INFL:** fls 2–4 in head-like clusters; outer bracts 1–3, 5–8 mm, fan-shaped, shallowly 3–7-lobed; inner bract 8–10 mm, tip roughpapillate. **FL:** calyx 8–10 mm; corolla 8–9 mm, ± yellow, mottled maroon on top. **SEED:** 4–6, 2.5–3 mm, ± ovoid, finely wavy-striate, dark brown. 2*n*=28. Open dry serpentine in chaparral and mixedevergreen forest; 300–1850 m. NCoRH, NCoRI. Jul–Sep

C. ramosus Benth. (p. 973) Pl 10–90 cm, gray-green or tinged red, ± canescent. **LF:** 10–40 mm, ± thread-like, entire to 5-lobed. **INFL:** fls 3–7 in spike-like clusters, clusters 15–25 mm; outer bract 1 per fl, 10–20 mm, entire to 7-lobed, sometimes bristly, segments ± thread-like; inner bract 10–20 mm, entire. **FL:** calyx 10–15 mm, divided to base, tip slightly notched; corolla 10–20 mm, dusty yellow, often marked maroon, ± puberulent. **SEED:** ± 2 mm, ± ovoid, deeply netted, light brown. 2*n*=24. Rocky to alkaline soils in sagebrush scrub; 1050–2850 m. n CaRH (Shasta Valley), GB, DMtns (Panamint Range); to OR, WY, CO. Jul–Aug

C. rigidus (Benth.) Jeps. Pl 30–150 cm, yellow-green or tinged red, gen ± bristly, gen not glandular. **LF:** 10–40 mm, ± linear, often inrolled, lobes 0–3. **INFL:** fls 5–15 in ± compact, head-like clusters; outer bract gen 1 per fl, 5–20 mm, gen 3-lobed; inner bract 14–20 mm. **FL:** calyx 10–20 mm, tube 1–2 mm; corolla 12–20 mm, ± yellow, lower side marked maroon in U-shape, pouch 3–10 mm wide, white. **SEED:** 1.5–2 mm, ± ovoid, wavy-striate, dark brown. 2*n*=28. Highly variable, intergrading. Many geog races, variably distinct; 5 most discrete (1 in Baja CA) considered subspp.

subsp. *brevibracteatus* (A. Gray) Munz SHORT-BRACTED BIRD'S-BEAK **INFL:** 2–12-fld, 10–20 mm wide; outer bract gen 5–10 mm (< fl), lobed in distal 1/2, ± long-bristly, middle lobe oblong, tip slightly wider, unthickened, ± green. **FL:** corolla 8–17 mm. Granitic openings in Jeffrey-pine and pinyon/juniper forest, sagebrush scrub; 850–2560 m. s SNH. Jul–Sep ★

subsp. *littoralis* (Ferris) T.I. Chuang & Heckard (p. 973) SEASIDE BIRD'S-BEAK Pl puberulent to soft-hairy. **INFL:** 5–8-fld, 10–20 mm wide; outer bract 15–20 mm, lobed in distal 1/2, middle lobe ± lanceolate, tip tapered. **FL:** corolla 15–20 mm. Dunes; < 200 m. c CCo (s Monterey Bay and Peninsula), s CCo (Santa Barbara Co.). Jul–Aug ★

subsp. *rigidus* (p. 973) **INFL:** 5–15-fld, 20–40 mm wide; outer bract 15–20 mm, lobed in lower 1/2, ± scabrous, middle lobe linear to oblong, tip linear or wider and notched. **FL:** corolla 12–20 mm. Open

foothill woodland, chaparral margins, conifer forest; < 2700 m. c&s SN, CW (exc n SnFrB), WTR, SnGb. Jul–Sep

subsp. *setiger* T.I. Chuang & Heckard (p. 973) **INFL:** 5–13-fld, 20–30 mm wide; outer bract 10–20 mm, lobed in lower 1/2, densely long-bristly, hairs purple-based, middle lobe linear, tip wider, notched, thickened, maroon. **FL:** corolla 13–17 mm. Open coastal-sage scrub, chaparral, oak woodland, conifer forest; < 2200 m. SCoRO, SW; n Baja CA. Jun–Sep

C. tenuis A. Gray Pl 20–120 cm, (gray- or yellow-)green or tinged maroon, ± glabrous to sparsely long-hairy or glandular-sticky. **ST:** ± wiry. **LF:** 10–60 mm, ± linear. **INFL:** fls 1 or 2–7 in loose clusters; outer bracts 1–4, 5–20 mm, linear, entire or 3-lobed, tips gen ± wider (< 1.5 mm) and thickened, longest hairs < 2 mm; inner bract 10–20 mm. **FL:** calyx 10–20 mm, tip notched; corolla 10–20 mm, 4–8 mm wide, ± white, ± yellow-tipped, heavily blotched maroon. **SEED:** 6–16, 1.5–2.5 mm, ± ovoid or angled, finely wavy-striate, dark brown. 2*n*=28. Highly variable. Close to *C. pilosus*.

subsp. *barbatus* T.I. Chuang & Heckard (p. 973) FRESNO COUNTY BIRD'S-BEAK **ST:** gen puberulent and long-hairy, sometimes glandular. **LF:** linear to ± oblong. **INFL:** gen 3–7 in dense clusters; outer bracts gen 3-lobed; inner bract bearded, hairs opaque, longest > 2 mm. **FL:** corolla 15–18 mm. Open mixed forest; 1300–2400 m. s SNH (Madera, Fresno cos.). Jul–Aug ★

subsp. *brunneus* (Jeps.) Munz (p. 973) SERPENTINE BIRD'S-BEAK **ST:** sparsely glandular-puberulent, esp proximal to fls. **LF:** thread-like, gen folded or channeled. **INFL:** clusters loosely 1–4-fld; outer bracts entire, thread-like; inner bract sparsely glandular-puberulent. **FL:** corolla 12–15 mm. Serpentine in mixed-evergreen forest, chaparral; gen < 1350 m. s NCoR (Lake, Sonoma, Napa cos.). Jun–Jul ★

subsp. *capillaris* (Pennell) T.I. Chuang & Heckard (p. 973) PENNELL'S BIRD'S-BEAK **ST:** ± glabrous. **LF:** thread-like, gen folded. **INFL:** clusters loosely 1–4-fld; outer bracts 3-lobed, thread-like; inner bract tip gen puberulent. **FL:** corolla 14–16 mm. Serpentine in chaparral; ± 200 m. sw NCoRO (near Occidental, Sonoma Co.). Jun–Jul ★

subsp. *pallescens* (Pennell) T.I. Chuang & Heckard (p. 973) PALLID BIRD'S-BEAK **ST:** gen puberulent and long-hairy, weakly glandular. **LF:** linear. **INFL:** clusters loosely 3–6-fld; outer bracts 3-lobed; inner bract hairs < 1 mm. **FL:** corolla 10–15 mm. Open volcanic alluvium; 900–1200 m. CaRH (near Black Butte, Siskiyou Co.). 1 large population. Jun–Sep ★

subsp. *tenuis* **ST:** gen puberulent (esp > 1600 m) to glandular-sticky (esp < 900 m). **LF:** linear, entire. **INFL:** clusters loosely 1–3-fld; outer bracts entire; inner bract hairs < 1 mm. **FL:** corolla 10–20 mm. Open conifer forests, foothill woodland; 300–2600 m. c KR, SN. Pls from c KR (Scott Mtns) that key here may be derived from *C. tenuis* subsp. *viscidus*. Jul–Sep

subsp. *viscidus* (Howell) T.I. Chuang & Heckard **ST:** densely glandular-sticky, puberulent, and long-hairy. **LF:** linear. **INFL:** clusters loosely 1–3-fld; outer bracts 3-lobed; inner bract hairs < 1 mm. **FL:** corolla 10–18 mm. Open yellow-pine forest on serpentine; 200–2000 m. KR, NCoRH, CaR, n SN, MP; OR. Jul–Aug

DICRANOSTEGIA

Margriet Wetherwax & David C. Tank

1 sp.: n Baja CA, adjacent CA. (Latin: 2-forked calyx) [Tank et al. 2009 Syst Bot 34:182–197] Formerly incl in *Cordylanthus*. Close to *Chloropyron*, together forming the *Pseudocordylanthus* clade (see Tank et al. 2009), distinguished by infl, calyx, stamens, and ecology.

D. orcuttiana (A. Gray) Pennell (p. 973) ORCUTT'S BIRD'S-BEAK, BAJA BIRD'S-BEAK Ann, green, often red-tinged, 10–50 cm, gen much-branched, ± stiff-hairy; roots ± yellow. **LF:** alternate, sessile, (2)3–6(8) cm, pinnately 8–11 lobed. **INFL:** elongated spike, 2–10 cm, dense, subtended by ± lf-like outer bracts; inner bracts 15–25 mm, 3–7-lobed, lance-oblong. **FL:** calyx sheath-like, ± 1/2 corolla, divided > 1/2, cut abaxially to base; corolla 18–25 mm, club-shaped, white, yellow-tipped, tubular proximally, expanded laterally, upper lip folded lengthwise, tip rounded, closed, opening directed downward forming a hood enclosing anthers and style, lower lip < upper,

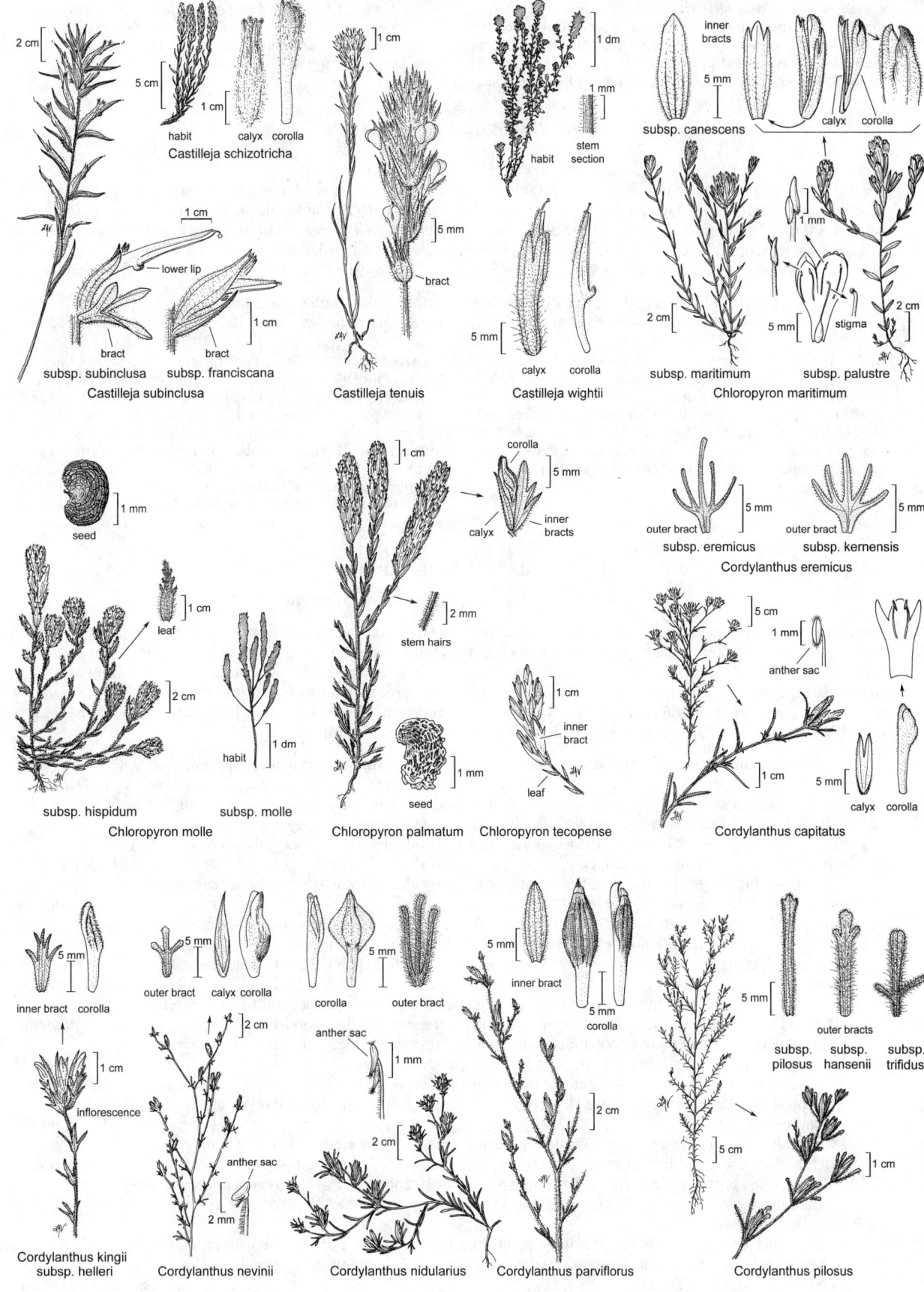

Castilleja schizotricha
habit — calyx — corolla

Castilleja subinclusa
subsp. subinclusa — lower lip — bract
subsp. franciscana — bract

Castilleja tenuis
bract

Castilleja wightii
calyx — corolla
habit — stem section

Chloropyron maritimum
subsp. canescens — inner bracts — calyx — corolla
subsp. maritimum — subsp. palustre — stigma

Chloropyron molle
seed — subsp. hispidum — leaf — subsp. molle — habit

Chloropyron palmatum
stem hairs — seed

Chloropyron tecopense
corolla — calyx — inner bracts — inner bract — leaf

Cordylanthus eremicus
subsp. eremicus — outer bract — subsp. kernensis — outer bract

Cordylanthus capitatus
anther sac — calyx — corolla

Cordylanthus kingii
subsp. helleri
inner bract — corolla — inflorescence

Cordylanthus nevinii
outer bract — calyx — corolla — anther sac

Cordylanthus nidularius
corolla — outer bract — anther sac

Cordylanthus parviflorus
inner bract — corolla

Cordylanthus pilosus
outer bracts
subsp. pilosus — subsp. hansenii — subsp. trifidus

densely white-bearded, obscurely 3-lobed, middle lobe erect, not rolled under; fertile stamens 2, anther sacs 2 per stamen, widely separated, sterile stamens reduced to rudimentary filaments or bearing sterile, long-hairy, appendages; ovary 2-chambered, glabrous, style elongate, bent near tip, stigma barely exserted. **SEED**: attached at side, 30–40, 1–1.5 mm, seed coat netted, tight-fitting, wavy-crested, dark brown. 2n=32. Coastal scrub; < 350 m. s SCo (sw San Diego Co.); n Baja CA. [*Cordylanthus o.* A. Gray] Threatened by urbanization. Mar–Aug ★

KOPSIOPSIS GROUND-CONE

Alison E.L. Colwell

Per 7–30 cm, not green (holoparasite); roots 0. **ST**: erect, simple, 1–many from spheric, tuber-like base attached to host root; tuber surface covered with polygonal plates 2–3 mm diam. **LF**: true lvs 0. **INFL**: spike-like; bracts densely overlapping; pedicels gen < 2 mm; bractlets on pedicel 0–2, linear. **FL**: calyx cup-shaped, lobes tapered; corolla with ring of hairs in distal tube at base of stamens, upper lip entire or indented, margin incurved, lower lip 3-lobed; distal end of filament and anther hairy. **FR**: placenta 1 per valve. 2 spp.: w N.Am. (Jan Kops, Dutch botanist, 1765–1849) [Gilkey 1945 Oregon State Monogr Stud Bot 9]

1. Infl 1.5–2.5 cm diam; largest bracts lanceolate to obovate, 6–8 mm wide; corolla 10–15 mm . . . . ***K. hookeri***
1′ Infl 2.5–5.5 cm diam; largest bracts gen ovate to spoon-shaped, 8–15(19) mm wide; corolla 15–20 mm. . . ***K. strobilacea***

K. hookeri (Walp.) Govaerts (p. 973) SMALL GROUND-CONE Pl 7–16(20) cm. **INFL**: 3–7 cm, pale yellow to ± purple throughout; largest bracts 7–11(14) mm. **FL**: calyx lobes 2–4, 2–4 mm, tapered; corolla pale, lips 3–4 mm, erect, upper > lower, margins sparsely ciliate. **SEED**: ± 1.5 mm. 2n=82. Open woodland, mixed conifer forest, gen on *Gaultheria shallon*, occ on *Arbutus menziesii, Arctostaphylos uva-ursi*; < 700 m. NCoR, SnFrB (Marin Co.); to BC. [*Boschniakia h.* Walp.] Apr ★

K. strobilacea (A. Gray) Beck (p. 973) CALIFORNIA GROUND-CONE Pl (6)10–30 cm. **INFL**: 4–15 cm, gen red-brown to dark ± purple (pale); largest bracts 15–20 mm. **FL**: calyx lobes 0(2–4), 3–7 mm, abruptly tapered; corolla gen purple (lobe margins pale), lips 5–6 mm, ± equal, lower lip gen spreading, margins glabrous. **SEED**: ± 2 mm. 2n=82. Open woodland, chaparral, gen on *Arctostaphylos*; < 3000 m. NW, CaR, n SN, SnFrB, TR, PR; s OR, Baja CA. [*Boschniakia s.* A. Gray] Widely scattered, most common in NW, TR. A few populations intermediate to *K. hookeri* in corolla pubescence and bract shape occur in Siskiyou, Sonoma, Shasta cos. Apr–Jun

OROBANCHE BROOMRAPE

Alison E.L. Colwell & Lawrence R. Heckard

Ann, per, not green (holoparasites), gen glandular-puberulent distally; root attachment occ tuber-like. **ST**: simple or branched. **INFL**: gen ± spike-like (proximal fls often short-pedicelled or on short branches), gen dense; fls gen > 20; bracts gen lanceolate to deltate (wider on peduncle); bractlets 0 or 2. **FL**: calyx lobes 4–5; corolla glandular-puberulent, gen lacking ring of hairs at stamen bases, upper lip erect to reflexed, 2-lobed, lower lip 3-lobed, spreading, throat floor with yellow folds; anthers glabrous to hairy; stigma lobes gen 2. **FR**: 2-valved; placentas gen 2, often lobed. **SEED**: < 0.7 mm. 140 spp.: worldwide, esp Medit. (Greek: vetch strangler, from parasitic habit) [Heckard 1973 Madroño 22:41–70]

1. Calyx 4-lobed, cut deeper on upper side; infl open, fls sessile, becoming well separated — pest on vegetable
    crops. . . . . ***O. ramosa***
1′ Calyx 5-lobed, cut equally or deeper on lower side; infl dense, gen remaining so; at least proximal fls ± pedicelled
  2. Pedicels 3–15 cm, bractlets on pedicel 0; fls 1–20; nectary gland at base of ovary evident
    3. Fls gen 5–20; bracts > 6; distal pedicels < st; corolla gen ± pink or yellow, lobes glabrous to soft-hairy;
        gen on *Artemisia, Eriodictyon, Eriogonum, Galium* . . . . ***O. fasciculata***
    3′ Fls gen 1–3; bracts gen < 6; distal pedicels > st; corolla gen pale ± purple to ± yellow, lobes minutely
        ciliate; gen on *Sedum*, Saxifragaceae, Asteraceae . . . . ***O. uniflora***
  2′ Pedicels < 2(5) cm, bractlets on pedicel 2; fls gen > 20; nectary gland at base of ovary obscure
    4. Infl and fls ± yellow to dark purple (may be ± red in *Orobanche pinorum*); calyx gen 5–12 mm
      5. Calyx divided nearly to base on lower side; hairs of infl and corolla papillate, not glandular; on
          Adenostoma . . . . ***O. bulbosa***
      5′ Calyx divided ± equally; most hairs of infl and corolla glandular, not papillate; on various genera
        6. Corolla with ring of hairs at base of filaments; root attachment rounded, coral-like; on *Holodiscus* . . . . ***O. pinorum***
        6′ Corolla lacking ring of hairs; root attachment branched (rarely rounded, coral-like in *Orobanche cooperi*)
          7. Infl 4–5 cm wide; corolla 18–32 mm, lips 5–10 mm; on Asteraceae . . . . ***O. cooperi***
          7′ Infl 2–3 cm wide; corolla 12–18 mm, lips 3–5 mm; on *Garrya* . . . . ***O. valida***
            8. Corolla 14–18 mm, hairy outside, densely so at sinuses, hairs 0.4–0.7 mm; filament base and
                anther hairy. . . . . subsp. ***howellii***
            8′ Corolla 12–14 mm, puberulent outside, sparsely so to glabrous at sinuses, hairs ± 0.1 mm;
                filament base and anther glabrous . . . . subsp. ***valida***
    4′ Infl and fls gen buff to ± pink, corolla lips white to pink or lavender, occ with darker veins (some
        populations of *Orobanche californica, Orobanche parishii* subsp. *parishii* darker-fld); calyx gen 10–20 mm
      9. Corolla 13–30 mm, lips 4–10 mm
        10. Infl branched, forming convex or ± flat-topped cluster; fls crowded; corolla tube with prominent
            hump on upper side, lobes erect; anthers densely hairy throughout; gen on *Artemisia tridentata*;
            n CaR, SNH, GB, n DMtns . . . . ***O. corymbosa***

10′ Infl of elongate, gen unbranched, ± raceme-like units; fls less crowded and more regularly-spaced; corolla tube without prominent hump on upper side, lobes various; anthers glabrous or hairy along dehisced margin; on *Artemisia*, other Asteraceae

 11. Fls pedicelled throughout; bracts lanceolate; mature fl corolla lobes strongly recurved, narrowly acute . . . . *O. vallicola*

 11′ Only proximal fls pedicelled; bracts ± ovate; mature fl upper corolla lobes slightly recurved at tip, rounded or obtuse . . . . *O. parishii*

  12. Corolla 15–24 mm, lips 4–6 mm; calyx lobes gen 7–10 mm; sandy soil near ocean . . . . subsp. *brachyloba*

  12′ Corolla 20–25 mm, lips 6–8 mm; calyx lobes 10–16 mm; dry openings in forest, chaparral, scrub . . . . subsp. *parishii*

9′ Corolla 20–50 mm, lips 10–14 mm, (see also *Orobanche corymbosa* for pls of SNH, GB) . . . . *O. californica*

 13. Corolla lavender to purple; calyx lobes, pedicels, and bracts similarly colored, drying dark purple; NCo, n&c CCo . . . . subsp. *californica*

 13′ Corolla white to ± yellow to ± pink or ± red-tinged; calyx lobes, pedicels, and bracts pale to pink-tinged, drying brown; gen CA-FP (inland exc *Orobanche californica* subsp. *grandis*), MP, s SNE

  14. Corolla tube ± stout, > 4 mm wide at constriction, abruptly expanded to form hump-back throat

   15. St delicate; ± 500 m; s CCo, SCoRI, SCoRO, w WTR . . . . subsp. *condensa*

   15′ St stout; 700–2500 m; s SNH, Teh, TR, PR, s SNE, DMoj . . . . subsp. *feudgei*

  14′ Corolla tube slender below, < 4 mm wide at constriction, gradually expanded

   16. Corolla 35–50 mm, lower lobes narrowly ovate, > 5 mm wide; gen on *Corethrogyne filaginifolia*; coastal dunes, marine terraces; s CCo, s SCoRO (Santa Barbara Co.), c SCo (Los Angeles Co.), n ChI (Santa Rosa Island) . . . . subsp. *grandis*

   16′ Corolla 20–40 mm, lower lobes narrowly triangular to oblong, < 5 mm wide

    17. Pl 4–10 cm; st proximal to infl 1–5 cm; infl a convex to ± flat-topped cluster < 5 cm; on Asteraceae; moist or vernally wet meadows, stream margins . . . . subsp. *grayana*

    17′ Pl 10–35 cm; st proximal to infl gen > 6 cm; infl elongate, 5–20 cm; gen dry places; on various per and shrub Asteraceae . . . . subsp. *jepsonii*

O. bulbosa Beck (p. 973) Pl 8–30 cm, dark ± purple above ground, white-papillate, not glandular. **ST:** arising from round, coral-like root attachment, stout, thickened at base, bulb-like, with overlapping scales. **INFL:** occ pyramid-shaped; pedicel < 2(5) cm. **FL:** calyx 6–10 mm, tube cut nearly to base on lower side, lobes 5, narrowly triangular, gen ≤ tube; corolla 10–18 mm, ± yellow to ± purple, lobes 2–4 mm, obtuse to acute; anthers glabrous to sparsely hairy; stigma 2-lobed, margins recurved. 2*n*=48. Openings in chaparral, gen on *Adenostoma fasciculatum*; < 1700 m. s NCoRO, NCoRI, CaR, SN, SnFrB, SCoRO, SW; Baja CA. Apr–Jun

O. californica Cham. & Schltdl. Pl 4–35 cm, pale to pink-purple above ground, glandular-puberulent. **ST:** 1 or clustered, slender to stout, branched proximally or throughout. **INFL:** long or branched and ± flat- to convex-topped; pedicels 0–4 cm, distal shorter; fls gen crowded. **FL:** calyx 12–20 mm, gen pale to ± pink, lobes 5, linear-triangular, >> tube; corolla ± purple or white to pink with darker veins, ± glandular-puberulent, lips 10–14 mm, gen widely flaring; upper lip obtuse to rounded; anthers woolly; placentas 4, stigma lobes 2, triangular, margins recurved.

 subsp. ***californica*** Pl 5–27 cm, gen pale to pink-purple above-ground. **ST:** gen branched. **INFL:** < 8 cm, ± head-like or round-topped. **FL:** corolla 22–45 mm, throat abruptly wider than tube, 8–10 mm wide, lips ± purple, paler than throat. 2*n*=48. Uncommon. Sandy or heavy soils of coastal bluffs, gen on *Grindelia*; < 150 m. NCo, n&c CCo. Pale form with corolla 10–15 mm, erect corolla lobes on *Eriophyllum staechadifolium* is undescribed sp. Jun–Aug

 subsp. ***condensa*** Heckard Pl 5–15 cm. **ST:** gen branched distally, delicate. **INFL:** round-topped; branches gen < 9 cm. **FL:** corolla 25–35 mm, tube ± stout, abruptly expanded above sinus, > 4 mm wide at constriction, forming hump on throat 8–10 mm wide, lips moderately recurved, buff to ± yellow, purple-tinged and/or ± red veined. 2*n*=48. Dry washes, flats, on *Heterotheca sessiliflora* subsp. *echioides*; ± 500 m. s CCo, SCoRI, SCoRO, w WTR. May–Jul

 subsp. ***feudgei*** (Munz) Heckard (p. 973) Pl 10–30 cm. **ST:** gen branched distally, stout. **INFL:** ± flat- or round-topped; branches gen < 9 cm. **FL:** corolla 25–35 mm, tube ± stout, abruptly expanded above sinus, > 4 mm wide at constriction, forming hump on throat 8–10 mm wide, lips moderately recurved, ± white to ± yellow, purple-tinged and/or ± red veined. 2*n*=48. Dry washes, slopes, flats, primarily on *Artemisia tridentata*; 700–2500 m. s SNH, Teh, TR, PR, s SNE,

DMoj; to Baja CA. Pls with elongate infl from SW chaparral/oak woodland on *Isocoma*, or from SNE, DMoj on *Chrysothamnus* that key here are undescribed subspp. May–Jul

 subsp. ***grandis*** Heckard Pl 8–30 cm. **ST:** gen branched distally, stout. **INFL:** gen < 5 cm, round-topped or not. **FL:** corolla 35–50 mm, ± pink or pale brown-red, veins dark, tube < 4 mm wide at constriction, throat 9–10 mm wide, lips widely recurved, lower lobes narrowly ovate. Uncommon. Coastal dunes, marine terraces; < 250 m. s CCo, s SCoRO (Santa Barbara Co.), c SCo (Los Angeles Co.), n ChI (Santa Rosa Island). May–Jun

 subsp. ***grayana*** (Beck) Heckard Pl 4–10 cm. **ST:** branched or not, 1–5 cm. **INFL:** < 5 cm, convex to ± flat-topped, few-fld. **FL:** corolla 20–30 mm, tube gradually widening to 5–8 mm, lips ± white to ± yellow to pale ± purple with lavender veins. 2*n*=48. Uncommon. Moderately moist meadows, stream margins, on *Symphyotrichum* and *Erigeron*; 1200–1800 m. NCoRH, CaRH, n&c SNH, SnFrB (s Sonoma Co., ± 60 m), MP; to WA. Pls with Asteraceae hosts other than *Erigeron* in drier habitats are undescribed taxa, possibly of hybrid origin. May–Aug

 subsp. ***jepsonii*** (Munz) Heckard Pl 10–35 cm. **ST:** gen branched, gen > 6 cm. **INFL:** 5–20 cm. **FL:** corolla 25–40 mm, ± white or ± pink to pale yellow-brown, tube gradually widened to 5–8 mm, veins gen ± purple. 2*n*=48. Uncommon. Gen dry flats, slopes, gen on per and shrub Asteraceae; < 2200 m. KR, NCoR, CaR, SNF, GV, SnFrB, SCoR. Variation suggests an unnatural group; may incl hybrids of other *O. californica* subspp. Jul–Sep

O. cooperi (A. Gray) A. Heller (p. 973) Pl 10–40 cm, gen dark ± purple above ground, glandular-puberulent; root attachment occ a coral-like thickening. **ST:** simple or branched, often forming large clumps, stout, little enlarged at base. **INFL:** 4–5 cm wide; proximal pedicels < 5 cm, distal 0. **FL:** calyx 8–12 mm, lobes 5, > tube, triangular, acuminate; corolla 18–32 mm, ± purple, hairs long-stalked, gen glandular, tube lacking ring of hairs, lips 5–10 mm, upper lobes 6–10 mm, > lower, obtuse; anthers gen hairy; stigma lobes 2, thin, recurved. 2*n*=24,48,72. Sandy flats, washes, on Asteraceae (gen *Ambrosia*, *Encelia farinosa*); -30–1500 m. D; to UT, AZ, Baja CA. An undescribed form (probably best a subsp.), 2*n*=96, with smaller, shorter-lobed corolla and peltate, bowl-shaped stigma occurs on same hosts over range of sp. Jan–May

O. corymbosa (Rydb.) Ferris (p. 973) Pl 3–17 cm, pale to purple-tinged aboveground, glandular-puberulent. **ST:** branches clustered, stout, gen thickened at base. **INFL:** 3–4 cm, branched, gen round-topped, few-fld; proximal pedicels 5–30 mm, distal 0. **FL:** calyx 12–18 mm, lobes 5, > tube, linear-triangular; corolla 20–30 mm, sparsely short-glandular, lips 5–8 mm, ± purple to pink, veins darker, upper lobes rounded; anthers densely hairy; stigma 2-lobed, peltate. $2n=48,96$. Openings in sagebrush scrub, gen on *Artemisia tridentata*; 1200–2800 m. n CaR, SNH, GB, n DMtns (Panamint Range); to BC, MT, UT. Apparently intergrading with *O. californica* throughout its CA range. Jun–Aug

O. fasciculata Nutt. (p. 973) CLUSTERED BROOMRAPE **ST:** 1 or clustered, 5–20 cm, branched or not. **INFL:** raceme, ± flat-topped, gen 5–20-fld; bracts > 6, glandular-puberulent; pedicels 3–15 cm, distal shorter, bractlets 0. **FL:** calyx lobes 5, 3–7 mm, gen < tube, deltate, gen ± acuminate; corolla 15–30 mm, curved, becoming erect, gen ± pink or yellow, lobes rounded to narrowly acute; anthers gen hairy; stigma 2-lobed, recurved; orange nectary gland at base of ovary. $2n=48$. Dry, gen ± bare places, gen on shrubs (gen *Artemisia, Eriodictyon, Eriogonum, Galium*); < 3300 m. CA-FP, GB, DMtns; to YT, c N.Am, n Mex. Undescribed pls intermediate to *O. uniflora*, with prominent red nectary gland at base of ovary are scattered in CA-FP, on *Galium*. Apr–Jul

O. parishii (Jeps.) Heckard Pl ± yellow-white. **ST:** gen simple, stout, glandular-puberulent. **INFL:** bracts narrowly ovate, with > 5 conspicuous, parallel veins. **FL:** calyx 10–20 mm, lobes 5, ± narrowly triangular, pale; corolla buff to ± pink, lobes rounded or obtuse, veins ± red.

   subsp. ***brachyloba*** Heckard (p. 973) SHORT-LOBED BROOMRAPE Pl 5–18 cm. **INFL:** 3–8 cm, elongate, fls arranged in regular spiral along axis. **FL:** calyx lobes gen 7–10 mm; corolla 15–24 mm, lips 4–6 mm, erect or slightly spreading; anthers gen glabrous; stigma lobes gen narrow, recurved. $2n=96$. Sandy soil near ocean, gen on *Isocoma menziesii*; < 300 m. SCo, ChI; Baja CA. May–Aug ★

   subsp. ***parishii*** Pl 15–26 cm. **INFL:** 5–14 cm, elongate, fl arrangement asymmetrical. **FL:** calyx lobes 10–16 mm; corolla 20–25 mm, lips 6–8 mm, spreading; anthers glabrous or hairy; stigma lobes wide, spreading. $2n=48$. Uncommon. Conifer forest, openings in chaparral, scrub, gen on shrubs; 300–3000 m. s SNH, Teh, SW, W&I, DMtns; Baja CA. Comprises 2 or more undescribed taxa. One has narrow ± purple spikes, in conifer forest at 2000–3000 m on *Artemisia dracunculus* in SnJt and WTR (Mount Pinos), represented by specimen from Bear Valley (S.B. Parrish, s.n. June 1894). Another has narrow buff to yellow spikes, often with ± red markings on corolla lobes, in oak-chaparral at 300–1500 m on *Isocoma* and other Asteraceae in San Diego, Orange, Los Angeles, Riverside cos. May–Jul

O. pinorum Hook. (p. 973) Pl 10–30 cm, glandular-puberulent aboveground. **ST:** slender distally; base gen enlarged, with many overlapping bracts; root attachment rounded, coral-like. **INFL:** at first dense, gen becoming open; proximal pedicels 2–6 mm, distal 0, deep purple (golden brown or ± red). **FL:** calyx 5–8 mm, lobes 5, ± = tube, triangular-acuminate; corolla 12–20, ± yellow, hairy in a ring at filament bases, lips erect, lobes tinged pale purple (brick-red);

anthers glabrous or sparsely hairy; stigma lobes 2, recurved. $2n=48$. Uncommon. Rocky, open forest slopes, on *Holodiscus*; < 2100 m. NW, CaRH, n SNH, SnFrB; to WA, ID, NM. Not known on conifers. Jul–Sep

O. ramosa L. Pl 10–60 cm, ± yellow, glandular-puberulent. **ST:** branches many from near base, slender. **INFL:** open; pedicels 0, bractlets subtending fl 2. **FL:** calyx divided more deeply on upper side, lobes 4; corolla 10–15 mm, tube ± white, throat, lobes pale blue or lavender; anthers gen glabrous; stigma 2-lobed. $2n=24$. Persisting in and near cult (esp tomato) fields; < 50 m. n SnJV, SnFrB, SCoRI, SCo; worldwide; native to Eur. Jul–Sep ◆

O. uniflora L. (p. 973) NAKED BROOMRAPE **ST:** 0.5–5 cm. **INFL:** gen raceme; fls gen 1–3; bracts gen < 6, gen glabrous; pedicels 3–12 cm, scapose, bractlets 0. **FL:** calyx lobes 5, gen 4–8 mm, > tube, narrowly triangular; corolla 12–35 mm, ± horizontal, gen pale ± purple to ± yellow, lobes gen rounded, minutely ciliate; anthers gen hairy; stigma lobes 2, margins recurved; orange nectary gland at base of ovary. $2n=36,48,±70$. Gen moist places, on herbs, esp *Sedum*, Saxifragaceae, Asteraceae; < 3100 m. NW, CaR, SN, ScV (Sutter Buttes), CW, SCo, n ChI, Wrn; to YT, e N.Am. If recognized taxonomically, pls < 20 cm of n&c CA to Can with corollas deep violet, 25–35 mm, assignable to *O. uniflora* var. *purpurea* (A. Heller) Achey. Apr–Jul

O. valida Jeps. Pl 6–35 cm, dark ± purple, glandular-puberulent. **ST:** gen 1; branches (if any) slender to stout, base enlarged or not. **INFL:** 2–3 cm wide. **FL:** calyx 5–11 mm, lobes 5, subequal, linear-triangular; corolla 12–18 mm, lips 3–5 mm, acute, purple; anthers glabrous or hairy; stigma bowl-shaped, slightly 2-lobed.

   subsp. ***howellii*** Heckard & L.T. Collins HOWELL'S BROOMRAPE Pl 6–20 cm; glandular hair stalks 0.2–0.4 mm, gen 3-celled. **ST:** slender, gen enlarged at base. **FL:** corolla 14–18 mm, densely hairy outside at sinuses, hairs 0.4–0.7 mm; folds of throat floor puberulent; filament base and anther hairy. $2n=48$. Volcanic and serpentine slopes, open chaparral, on *Garrya*; 200–1700 m. s NCoRH, c&s NCoRI. Jun–Sep ★

   subsp. ***valida*** (p. 973) ROCK CREEK BROOMRAPE Pl 10–35 cm; glandular hair stalks ± 0.1 mm, gen 2-celled. **ST:** gen stout, not enlarged at base. **FL:** corolla 12–14 mm, sparsely hairy outside at sinuses, hairs ± 0.1 mm; folds of lower throat glabrous; filament base and anther glabrous. $2n=48$. Decomposed granite, on *Garrya fremontii*; 1250–2000 m. c WTR (Topatopa Mtns), SnGb. May–Sep ★

O. vallicola (Jeps.) Heckard (p. 973) Ann or per, 8–40 cm. **ST:** stout; yellow with rosy tinge, glandular-puberulent; base gen thickened, single or multiple-branched from base. **INFL:** > 4 cm, occ of > 1 raceme; bracts lanceolate, veins 3–5, inconspicuous; pedicels < 2 cm. **FL:** calyx 8–15 mm, pale or ± pink, lobes 5; corolla 17–30 mm, ± yellow to pink, with pink veins (pale in contrast to st); lips 5–9 mm, lobes narrowly acute, strongly recurved in mature fls; anthers glabrous to hairy; stigma 2-lobed, margins recurved. $2n=48$. Riparian woodland, on *Sambucus nigra*; < 300 m. NCoR, CaRF, GV, CW, PR. St base may be large, woody in per pls, with past yrs' black stalks evident. Pls with hosts other than *Sambucus nigra* that key here are undescribed taxa (e.g., SNE on *Ericameria*; PR on *Solidago*). Jul–Sep

ORTHOCARPUS

Margriet Wetherwax, T. I. Chuang & Lawrence R. Heckard

Ann, green. **LF:** alternate, sessile, entire to 3-lobed. **INFL:** spike; bracts gen distinct from lvs, 1 per fl, entire to 5-lobed, tips gen colored. **FL:** calyx unequally 4-lobed, deepest sinus adaxial; corolla club-shaped, upper lip folded lengthwise, tip rounded, closed, opening directed downward forming a hood enclosing anthers and style, lower lip shorter, ± 3-pouched, (0)3-toothed; stamens 4, anther sacs 2, unequal; style, stigma slender. **FR:** gen ± notched. **SEED:** gen 8–15, often ± curved, ± keeled, attached at side; coat netted or ridged, tight-fitting. 9 spp.: w N.Am. (Greek: straight fr) [Chuang & Heckard 1992 Syst Bot 17:560–582] Close to *Cordylanthus*; other spp. formerly placed here are *Castilleja* (owl's-clovers) or *Triphysaria* (Johnny-tuck).

1. Bracts grading into distal lvs, uniformly green or occ distal-most purple-tinged, lobes 3, triangular to lanceolate
  2. Corolla white to ± purple, gen 12–20 mm; beak tip strongly hooked; lower corolla lip deeply pouched, teeth 0 . . . . ***O. bracteosus***
  2′ Corolla golden-yellow, 10–15 mm; beak tip obscurely hooked; lower corolla lip moderately pouched, teeth 3, incurved . . . . ***O. luteus***

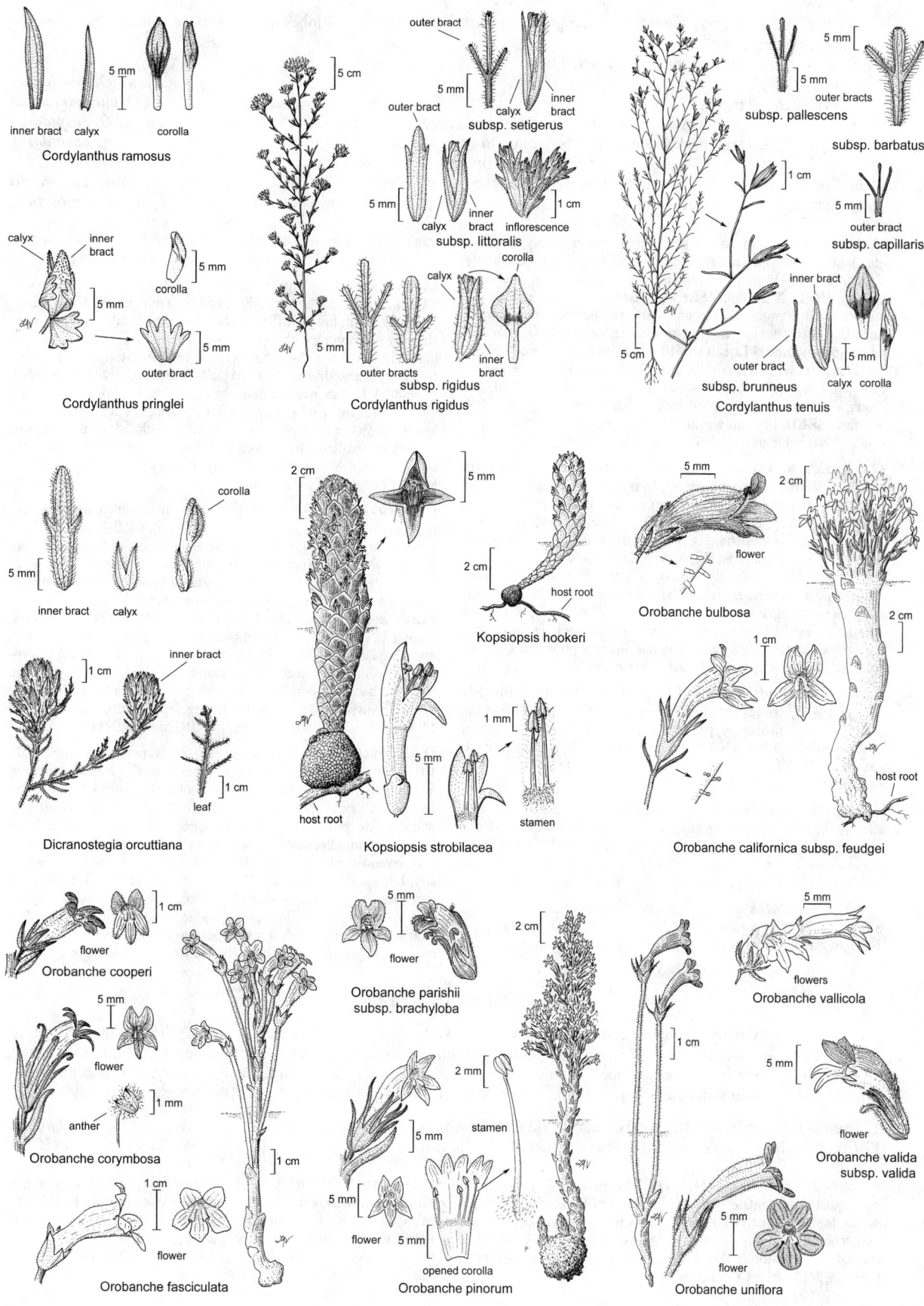

973

1′ Bracts differing abruptly from distal lvs, ± purple or white at tips, lateral 2–4 lobes much narrower than ovate or oblong central lobe
  3. Beak with conspicuous, cylindric, hooked tip
    4. Corolla 10–15 mm . . . . *O. imbricatus*
    4′ Corolla 25–30 mm . . . . *O. pachystachyus*
  3′ Beak straight, without cylindric, hooked tip . . . . *O. cuspidatus*
    5. Corolla 16–25 mm; beak 7–8 mm, gen 1–4 mm > lower lip; pouches 4–5 mm deep . . . . subsp. *cuspidatus*
    5′ Corolla 9–18 mm; beak 3–6 mm, ≤ 1 mm > lower lip; pouches 1–3 mm deep
      6. Corolla 14–18 mm, exserted, beak 5–6 mm, pouches ± 3 mm deep . . . . subsp. *copelandii*
      6′ Corolla 9–14 mm, incl, ± hidden by bract, beak 3–4 mm, pouches 1–2 mm deep . . . . subsp. *cryptanthus*

O. bracteosus Benth. (p. 979) Pl 10–40 cm, minutely scabrous, glandular-puberulent, gen becoming ± purple-tinged. **ST:** gen simple, slender. **LF:** 15–35 mm, ± linear; proximal entire; distal 3-lobed to middle. **INFL:** 3–20 cm, densely puberulent; bracts grading into distal lvs, 10–20 mm, ± ovate, green or distal occ becoming purple-tinged, 2 lateral lobes proximal to middle, central lobe ± lanceolate, 3–5 mm wide. **FL:** calyx 6–10 mm, divided 1/2 abaxially and on sides, 2/3 adaxially; corolla gen 12–20 mm, white to ± purple, exserted, lips ± equal, beak 4–6 mm, beak tip strongly hooked, glabrous, lower lip deeply pouched, teeth 0; stigma barely exserted. **FR:** 5–7 mm. **SEED:** light brown. 2*n*=30. Moist meadows; 500–2000 m. CaR, n SNH (Plumas Co.), MP; to BC. Jun–Aug

O. cuspidatus Greene Pl 10–40 cm, puberulent to scabrous, sparsely glandular, becoming ± purple-tinged. **ST:** simple to much-branched, gen slender. **LF:** 10–50 mm, ± lanceolate; proximal entire, distal with 3 deep, linear lobes. **INFL:** 2–10 cm, dense; bracts differing abruptly from distal lvs, 10–20 mm, ± ovate, ± purple-pink on distal 1/3 of distal-most bracts, with 2 narrow, ± basal lobes, central lobe 7–15 mm wide, tip abruptly pointed. **FL:** calyx 7–10 mm, divided 1/2 abaxially, 2/3 adaxially, 1/4 on sides; corolla 9–25 mm, exserted or not, lips ± purple-pink, densely puberulent, beak straight, without cylindric, hooked tip, 3–8 mm, 0–4 mm > lower lip, lower lip ± pouched, teeth 1–2 mm, triangular, densely puberulent; stigma well incl. **FR:** 6–8 mm. **SEED:** dark brown. 2*n*=28.

  subsp. ***copelandii*** (Eastw.) T.I. Chuang & Heckard (p. 979) **FL:** corolla 14–18 mm, exserted, beak 5–6 mm, 0–1 mm > lower lip, pouches ± 3 mm deep; upper anther sac ± 1.3 mm. Open, grassy to rocky slopes; 1200–2300 m. KR, NCoRH, CaR, n SNH (Plumas Co.); sw OR. [*Orthocarpus copelandii* Eastw.] Intergrades with *O. cuspidatus* subsp. *c.* Jun–Aug

  subsp. ***cryptanthus*** (Piper) T.I. Chuang & Heckard (p. 979) **FL:** corolla 9–14 mm, incl, ± hidden by bract, beak 3–4 mm, 0–1 mm > lower lip, pouches 1–2 mm deep; upper anther sac 0.7–1 mm. Drying meadows, open sagebrush; 1500–3200 m. CaRH, n&c SNH, GB; to OR, n-c US. [*O. copelandii* var. *c.* (Piper) D.D. Keck] Jun–Aug

  subsp. ***cuspidatus*** (p. 979) SISKIYOU MOUNTAINS ORTHOCARPUS **ST:** gen stout. **FL:** corolla 16–25 mm, exserted, beak 7–8 mm, gen 1–4 mm > lower lip, pouches 4–5 mm deep; upper anther sac 1.5–2

mm. Open, grassy to rocky slopes; 700–2200 m. NW, CaR; sw OR. Intergrades with *O. cuspidatus* subsp. *copelandii*. Jun–Aug ★

O. imbricatus S. Watson (p. 979) Pl 10–35 cm, ± puberulent. **ST:** simple or branched distally. **LF:** 20–50 mm, entire, ± lanceolate. **INFL:** 2–8 cm; bracts differing abruptly from distal lvs, 10–15 mm, strongly net-veined, with 2 small, lateral lobes from near base, central lobe widely oblong, 5–10 mm wide, ± rose on distal 1/3. **FL:** calyx 5–7 mm, hairy, divided 3/4 abaxially and adaxially, 1/4 on sides; corolla 10–15 mm, nearly hidden by bract, ± rose, beak ± 4 mm, ± 0.5 mm > lower lip, with conspicuous, cylindric, hooked tip, cylindric, lower lip moderately pouched, ± yellow, teeth 0.5 mm, triangular, densely puberulent; stigma well incl. **FR:** ± 5 mm. **SEED:** 3–5, dark brown. 2*n*=28. Mtn meadows, grassy or rocky slopes; 1600–2400 m. KR, NCoRO, CaR; to BC. Jun–Aug

O. luteus Nutt. (p. 979) Pl 10–40 cm, yellow-green, often becoming ± purple-tinged, glandular- and longer-nonglandular-hairy. **ST:** gen simple, slender. **LF:** 15–50 mm, ± linear, entire or upper deeply 3-lobed. **INFL:** 5–20 cm, densely glandular-puberulent; bracts grading into distal lvs, 10–20 mm, ± green, 2 lateral lobes proximal to middle, narrowly triangular, central lobe ± lanceolate, 2–5 mm wide. **FL:** calyx 5–8 mm, divided 1/3 abaxially, 3/4 adaxially, 1/4 on sides; corolla 10–15 mm, golden-yellow, exserted, puberulent (esp beak), lips ± equal, beak 2–4 mm, tip obscurely hooked, downward-projecting, cylindric, lower lip moderately pouched, teeth 3, 0.5 mm, blunt, incurved; stigma incl. **FR:** 5–7 mm. **SEED:** 20–35, ± yellow to dark brown. 2*n*=28. Moist fields, sagebrush scrub, mtn meadows; 1300–3000 m. GB, adjacent edges; to BC, n-c US, NM. Jul–Aug

O. pachystachyus A. Gray (p. 979) SHASTA ORTHOCARPUS Pl 10–20 cm, scabrous, sparsely soft-hairy, and (partly glandular-)puberulent. **ST:** simple, stout. **LF:** 30–50 mm; lobes of distal lvs 3–5, ± linear. **INFL:** 5–10 cm; bracts differing abruptly from distal lvs, 20–30 mm, widely ovate, prominently net-veined, gen with 4 narrow lateral lobes proximal to middle, central lobe ± ovate, 7–10 mm wide, distal margin pink-lavender. **FL:** calyx 15–20 mm, divided 1/2 abaxially, 2/3 adaxially, 1/4 on sides; corolla 25–30 mm, ± rose above, beak ± 10 mm, ± 3 mm > lower lip, glabrous, strongly curved, tip with conspicuous, cylindric, hooked tip, lower lip slightly pouched, puberulent, teeth rounded; stigma barely incl. **FR:** 5–7 mm. Openings in sagebrush scrub; < 1000 m. e KR, adjacent w CaR (n-c Siskiyou Co.). May–Jun ★

PARENTUCELLIA

Ann, green, sticky-hairy, gen erect and unbranched in CA. **LF:** gen opposite, sessile, toothed **INFL:** raceme, spike-like, bracted. **FL:** calyx 4-lobed, lobes ± equal, ± lanceolate; corolla 2-lipped, yellow or red-purple, upper lip entire or notched, forming hood, lower lip > upper, 3-lobed; stamens 4, in 2 pairs, incl in upper corolla lip, anthers hairy, awned at base. **FR:** cylindric. **SEED:** many, minute, smooth. 2 spp.: w Eur, Medit, c Asia. (Tomaso Parentucelli [Pope Nicholas V], 1397–1455)

1. Corolla red-purple, persistent; calyx lobes ± 1/2 tube length . . . . *P. latifolia*
1′ Corolla yellow, deciduous; calyx lobes ± = tube. . . . . *P. viscosa*

P. latifolia (L.) Caruel (p. 979) **ST:** < 30 cm. **LF:** 4–12 mm, triangular-lanceolate, dentate, margins ± purple. **INFL:** interrupted; bracts deeply lobed. **FL:** calyx 6–12 mm, lobes triangular-lanceolate, margins ± purple; corolla 8–10 mm, with 2 yellow, partly free ridge-like appendages near lower throat. **FR:** = calyx, glabrous. Uncommon. Pastures; ± 100 m. n CCo; native to Eur. May

P. viscosa (L.) Caruel (p. 979) **ST:** < 50 cm. **LF:** 20–40 mm, lanceolate, serrate, green throughout. **INFL:** dense; bracts lf-like. **FL:** calyx 10–16 mm, lobes lanceolate; corolla ± 20 mm, throat 2-ridged. **FR:** slightly > calyx tube, hairy. Damp, grassy places; gen < 700 m. NCo, NCoRO, NCoRI, CaRF, n&c SNF, ScV, CCo; OR; native to Eur. Apr–Jun ❖

PEDICULARIS LOUSEWORT

Linda Ann Vorobik

Per, ± green. **ST:** decumbent to erect, gen 1–several from gen short caudex. **LF:** alternate, gen ± basal, gen < infl, toothed or gen > 7-lobed, gen reduced distally on st; petiole gen < blade. **INFL:** raceme, spike-like; bracts (at least proximal) gen ± like distal lvs; pedicels 1–6 mm. **FL:** calyx lobes (2,4)5, distal-most gen shortest (all gen < tube), lateral fused in pairs; corolla white or yellow to red or purple, upper lip hood- or beak-like, curved or not, lower lip 3-lobed, narrow to fan-shaped, central lobe gen smallest; fertile stamens 4, gen glabrous, anthers gen incl, sacs 2 per stamen, equal; stigma head-like, gen exserted. **FR:** gen ± ovate or lanceolate in outline, asymmetric, opening mostly on upper side. **SEED:** smooth or netted. ± 500 spp.: cool wet n temp, circumboreal, S.Am. (Latin: lice, from belief that ingestion by stock promoted lice infestation) [Ree 2005 Int J Pl Sci 166:595–613]

1. Lf simple, ± toothed; calyx lobes gen 2 (4 in some pls); fr ± lanceolate in outline
 2. Upper corolla lip hooded, not beaked; lf ± doubly crenate; st gen simple . . . . **P. crenulata**
 2′ Upper corolla lip extended in a long, down-curved beak; lf serrate to dentate; st gen branched distally . . . **P. racemosa**
1′ Lf (at least some) deeply lobed to compound; calyx lobes 5; fr ovoid
 3. Proximal cauline lf segments gen 3–5(11), ovate, entire to toothed, terminal >> lateral; basal lvs 0 at fl . . . . **P. howellii**
 3′ Proximal cauline lf segments > 5, serrate to divided, terminal segment not largest; basal lvs gen present
  at fl (exc *Pedicularis bracteosa*)
 4. Corolla not club-like, upper lip with a long, narrow beak
  5. Corolla white to ± yellow, upper lip 4–5.5 mm, beak curved downward, not like an elephant's trunk . . . **P. contorta**
  5′ Corolla pink to purple, upper lip 3.5–18 mm, beak curved upward, ± like an elephant's trunk
   6. Infl densely hairy; beak 3–5 mm; lower lip ± fan-like . . . . **P. attollens**
   6′ Infl ± glabrous; beak 6–13 mm; lateral lobes of lower lip resembling ears . . . . **P. groenlandica**
 4′ Corolla ± club-like, upper lip hooded, not beaked
  7. Corolla deep red to red-purple (yellow to orange), lower lip ± 1/4 upper . . . . **P. densiflora**
  7′ Corolla ± yellow or ± purple, lower lip gen >> 1/4 upper
   8. St 30–110 cm; infl > lvs; basal lvs gen 0 at fl; corolla ± yellow . . . . **P. bracteosa** var. **flavida**
   8′ St ≤ 30 cm; infl gen ≤ lvs; basal lvs present at fl; corolla ± yellow or ± purple
    9. Corolla ± yellow, gen tinged red or tipped purple, hairy; filaments hairy . . . . **P. semibarbata**
    9′ Corolla pink to ± purple, glabrous; filaments glabrous
     10. Corolla 30–42 mm, pale purple, darker-tipped, lower lip > 1/2 upper, its lobes rounded,
      wavy-margined; lf margin wavy, white, thickened . . . . **P. centranthera**
     10′ Corolla 17–24 mm, pink or purple to purple-red, lower lip < upper, its lobes pointed, not
      wavy-margined; lf margin not wavy, white, or thickened . . . . **P. dudleyi**

P. attollens A. Gray (p. 979) LITTLE ELEPHANT'S HEAD **ST:** 6–60 cm, tomentose distally. **LF:** basal 3–20 cm, ± linear, segments 17–41, linear, toothed. **INFL:** 2–30 cm; bracts densely hairy, < fls. **FL:** calyx 4–6 mm, tomentose; corolla 4.5–7 mm, light pink to purple, marked darker, glabrous, upper lip 3.5–6 mm, curved upward, beak 3–5 mm, lower lip 3–5 mm, ± fan-like; anthers ± 1 mm, base obtuse. **FR:** 6–10 mm. **SEED:** 2.5–4 mm, surface finely netted. Wet meadows, streamsides, bogs; 1200–4000 m. KR, CaR, SNH, MP, W&I; OR. Jun–Sep

P. bracteosa Benth. var. ***flavida*** (Pennell) Cronquist (p. 979) YELLOWISH LOUSEWORT **ST:** 30–110 cm, sparsely hairy distally. **LF:** ± ovate; segments singly to doubly toothed; basal lvs 8–25 cm, 5–21-segmented, petiole ± = blade, gen 0 at fl; cauline lvs 3–15 cm, proximal 13–25-segmented, distal dentate. **INFL:** 4–17 cm, > lvs, glabrous; proximal bracts > fls. **FL:** calyx 7–13 mm, lobes minutely ciliate; corolla 15–24 mm, ± club-like, ± yellow, upper lip 8–10 mm, deeply hooded, lower lip 4–6 mm; anthers 1.5–3 mm, base acute to obtuse. **FR:** 9–10 mm. **SEED:** 2.5–3.5 mm, surface netted. Moist upper montane conifer forest; 1200–2400 m. KR; OR. Jul–Aug ★

P. centranthera A. Gray (p. 979) GREAT BASIN LOUSEWORT **ST:** 4–10 cm, ± glabrous; caudex long. **LF:** basal 6–20 cm, > infl, ± lanceolate, basal lvs present at fl; segments 13–25, oblong, crenate, dentate, or lobed, margin wavy, white, thickened. **INFL:** 3–10 cm; proximal bracts ≤ fls. **FL:** calyx 15–22 mm, ciliate or long-hairy; corolla 30–42 mm, ± club-like, pale purple, darker-tipped, glabrous, upper lip 10–18 mm, hooded, lower lip 9–15 mm, lobes ± equal, widely rounded, wavy-margined; anthers 3.5–5 mm, exserted, base acuminate, filaments glabrous. **FR:** 10–13 mm. **SEED:** 3–4.5 mm, surface shallowly netted. Sagebrush scrub, alluvial fans; 1300–1500 m. se MP (e Lassen Co.); to OR, CO, NM. May ★

P. contorta Benth. (p. 979) CURVED-BEAK LOUSEWORT **ST:** 15–40 cm, subglabrous. **LF:** basal 3–18 cm, lance-oblong; segments 25–41, linear, entire to serrate. **INFL:** 5–28 cm; bracts gen < fls, ciliate. **FL:** calyx 6–10 mm, ciliate, sometimes hairy; corolla 10–13 mm, white to ± yellow, upper lip 4–5.5 mm, beak curved downward, not like an elephant's trunk, lower lip 5–8 mm, ± fan-like; anthers 2.5–3 mm, base acute. **FR:** 6–10 mm. **SEED:** 2–2.5 mm, surface finely netted. Bogs, meadows, streamsides, moist, open montane conifer forest; 1600–2400 m. KR; to w Can, MT, WY, UT. Jul–Aug ★

P. crenulata Benth. (p. 979) SCALLOPED-LEAVED LOUSEWORT **ST:** 12–40 cm, gen simple, distally tomentose. **LF:** basal and cauline, 3–11 cm, ± linear; margins ± doubly crenate, thick, rolled under, wavy, white. **INFL:** 2–12 cm, gen simple; bracts gen < fls. **FL:** calyx 8–12 mm, tomentose, lobes gen 2(4 in some pls); corolla 20–26 mm, ± club-like, white in CA, glabrous, upper lip 11–15 mm, hooded, lower lip 7–12 mm, ± fan-like, margins ± wavy, lobes ± equal; anthers 2–3 mm, base acute. **FR:** 10–20 mm, ± lanceolate in outline. **SEED:** 1.5–2 mm, smooth. Wet meadows, streambanks; 2100–2300 m. SNE (Convict Creek, Mono Co.); to NV, WY, CO. Jun–Jul ★

P. densiflora Hook. (p. 979) WARRIOR'S PLUME Pl soft- to coarse-brown-hairy. **ST:** 6–55 cm. **LF:** basal 5–28 cm, lance-oblong, segments 13–41, ± linear to ovate, doubly toothed to lobed. **INFL:** 4–12 cm; lower bracts > fls. **FL:** calyx 8–15 mm, gen hairy, lobes ± equal; corolla 23–36 mm, straight, club-like, deep red to red-purple (yellow to orange), gen minutely hairy, upper lip 8–17 mm, hooded, lower lip 2–4 mm, lobes ± equal; anthers 2–3 mm, base acute. **FR:** 8–13 mm. **SEED:** 2.5–4.5 mm, surface netted. Dry chaparral, oak/pine or yellow-pine forest; < 2100 m. NW, CaR, SNF, Teh, CW, SW; s OR. Mar–May

P. dudleyi Elmer (p. 979) DUDLEY'S LOUSEWORT Pl gen ± hairy. **ST:** 10–30 cm. **LF:** basal 3–26 cm, > infl, lance-oblong, segments

15–25, oblong to ovate, doubly toothed to lobed, present at fl. **INFL:** 2–6 cm, gen ≤ lvs; proximal bracts > fls. **FL:** calyx 10–14 mm, tomentose; corolla 17–24 mm, ± club-like, pink or purple to purple-red, marked darker, glabrous, upper lip 9–11 mm, hooded, lower lip 4–7 mm, lobes pointed, ± equal, not wavy-margined; anthers 1.5–2.5 mm, gen incl, base acute, filaments glabrous. **FR:** 8–13 mm. **SEED:** 3.5–5 mm, surface netted. Coastal chaparral and forest; < 350 m. CCo, SnFrB (Santa Cruz Mtns), SCoRO. Widely scattered. Pls from c CCo (Arroyo de la Cruz, San Luis Obispo Co.) warrant further study (smaller, lvs < infl, anthers often exserted with base ± acuminate). Mar–Jun ★

P. groenlandica Retz. (p. 979) ELEPHANT'S HEAD Pl ± glabrous. **ST:** 8–80 cm. **LF:** basal 3–25 cm, ± lanceolate; segments 25–51, linear to oblong, toothed. **INFL:** 1–30 cm, gen glabrous; proximal bracts ± = fl. **FL:** calyx 3.5–6 mm, lobes ± equal, densely short-ciliate; corolla 8–15 mm, light pink to red-purple, glabrous, tube 4–10 mm, upper lip 7–18 mm, hood curved downward, beak 6–13 mm, narrow, curved upward, lower lip 3–7 mm, lateral lobes resembling ears; anthers 1.5–2.5 mm, base acute. **FR:** 6–9 mm. **SEED:** 2.5–4 mm, surface netted. Wet meadows, streamsides, bogs; 1000–3600 m. KR, CaR, SNH; to AK, e N.Am, NM. Jun–Sep

P. howellii A. Gray (p. 979) HOWELL'S LOUSEWORT **ST:** 6–45 cm, glabrous to puberulent; caudex long. **LF:** basal 0 at fl (4–20 cm, widely lanceolate, segments 7–17, ovate, entire-dentate to lobed); cauline lvs 2–9 cm, ovate, lobes or segments 3–5(11), ovate, entire to toothed, terminal >> lateral. **INFL:** closely subtended by 2–3 lvs, 1.5–4 cm; bracts < fls, ± deltate, hairy. **FL:** calyx 6–7 mm, hairy;

corolla 7–9 mm, sickle-shaped, ± white to ± purple, marked darker, glabrous, upper lip 5–6 mm curved, beak ± 4 mm, tapered, lower lip 2–3 mm, ± incl in calyx; anthers 1–2 mm, base acute. **FR:** 7–9 mm. **SEED:** 1.5–2 mm, surface finely netted. Dry ridges, open red-fir forest; 1500–1900 m. n KR; s OR. Jul–Aug ★

P. racemosa Benth. (p. 979) LEAFY LOUSEWORT **ST:** few to many, ± decumbent, gen branched distally, 12–80 cm, subglabrous. **LF:** cauline, 2–10 cm, ± narrowly lanceolate, serrate to dentate. **INFL:** 1–5 cm; proximal bracts ≥ fls. **FL:** calyx 4.5–8 mm, glabrous, lobes gen 2 (4 in some pls); corolla 10–16 mm, ± white, ± yellow, or ± purple, glabrous, upper lip 5–7.5 mm, extended in a long, down-curved beak, lower lip 5–9 mm, fan-like; anthers 1.5–2.5 mm, base acute. **FR:** 10–16 mm, ± lanceolate in outline. **SEED:** 1.5–3 mm, smooth. Open conifer forest; 900–2300 m. KR, CaR, n SNH, SNE; to w Can, WY, NM. If recognized taxonomically, pls e of CA-FP with linear lvs, ± white to ± yellow corollas assignable to *P. racemosa* subsp. *alba* Pennell. Jun–Aug

P. semibarbata A. Gray (p. 979) **ST:** mostly underground, 3–20 cm, sparsely tomentose; caudex long. **LF:** basal 5–22 cm, gen > infl, ± lanceolate, segments 11–25, lanceolate to ovate, toothed to lobed, present at fl. **INFL:** 3–12 cm; bracts hairy, proximal >> fls. **FL:** calyx 8–10 mm, ciliate; corolla 15–24 mm, ± club-like, ± yellow, gen tinged red or tipped purple, hairy, upper lip 7–10 mm, hooded, lower lip 5–7 mm, lobes ± equal, rounded; anthers 2–2.5 mm, exserted, base acuminate, filaments hairy. **FR:** 6.5–10 mm. **SEED:** 3.5–4.5 mm, smooth. Dry ridgetops, conifer forest, often with red fir; 1500–3500 m. NCoRH, CaRH, SN, TR, PR; NV. May–Jul

TRIPHYSARIA

Margriet Wetherwax, T I. Chuang & Lawrence R. Heckard

Ann, green to yellow-brown or ± purple; gen ascending-branched. **LF:** alternate or ± basal, sessile, ± linear, distal finely divided. **INFL:** spike and/or fls solitary in axils; bracts gen lf-like. **FL:** calyx ± equally 4-lobed; corolla tube gen slender, upper lip folded lengthwise, tip of upper 2 corolla lobes open, opening directed forward, forming a beak, lower lip deeply 3-pouched, 3-toothed, throat abruptly gen indented forming a fold proximal to lower lip; stamens 4, anther sac 1 per stamen, gen incl; stigma gen slightly enlarged, entire to slightly 2-lobed. **SEED:** 20–80, 0.8–1.2 mm, ovoid, attached at base, dark brown; coat netted, gen tight-fitting. 2*n*=22. 6 spp.: w N.Am, China. (Greek: 3 bladders, from lower lip pouches) [Chuang & Heckard 1991 Syst Bot 16:644–666] Hybrids common. Related to *Castilleja*, *Orthocarpus*; isolated reproductively.

1. Pl gen ≤ 20 cm, decumbent-branched from base; corolla 4–7 mm, beak hooked, pouches ± 1 mm deep . . . . ***T. pusilla***
1′ Pl gen ≥ 10 cm, ± ascending-branched; corolla 8–25 mm, beak not hooked, pouches gen 1.5–4 mm deep
 2. Stamens exserted; pouch ± 2 mm deep . . . . ***T. floribunda***
 2′ Stamens ± incl; pouch depth gen > 2 mm deep (exc *Triphysaria micrantha*)
  3. Pl green- or yellow-brown, gen glabrous; corolla yellow or white, beak ± yellow . . . . ***T. versicolor***
   4. Corolla yellow; coastal to inland . . . . subsp. ***faucibarbata***
   4′ Corolla white fading rose-pink; coastal . . . . subsp. ***versicolor***
  3′ Pl ± purple, gen puberulent at least distally; corolla sulphur-yellow or white to rose, beak dark purple
   5. Corolla 8–15 mm, ± = bract, pouches 1–2 mm deep . . . . ***T. micrantha***
   5′ Corolla 10–25 mm, > bract, pouches 3–4 mm deep . . . . ***T. eriantha***
    6. Corolla yellow . . . . subsp. ***eriantha***
    6′ Corolla white, fading rose-pink . . . . subsp. ***rosea***

T. eriantha (Benth.) T.I. Chuang & Heckard BUTTER-AND-EGGS, JOHNNY-TUCK Pl gen 10–35 cm, ± purple, gen puberulent, glandular- and nonglandular-hairy. **LF:** 10–50 mm, 3–7-lobed. **INFL:** 2–15 cm, dense distally; bracts 5–18 mm, 3–5-lobed. **FL:** calyx 8–13 mm, divided 1/4–1/2; corolla 10–25 mm, yellow, or white fading rose-pink, beak dark purple, tube densely puberulent, lower lip ± = beak, pouches 3–4 mm deep. **FR:** 5–8 mm, oblong. **SEED:** 30–50, dark brown.

subsp. ***eriantha*** (p. 979) **FL:** corolla yellow. Grassland, foothills; < 1600 m. NCo, NCoRO, NCoRI, CaRF, SNF, GV, CW, PR, MP; sw OR. Mar–May

subsp. ***rosea*** (A. Gray) T.I. Chuang & Heckard **FL:** corolla white, fading rose-pink. Coastal fields, bluffs; < 300 m. NCo, CCo, SnFrB, SCo. Mar–May

T. floribunda (Benth.) T.I. Chuang & Heckard (p. 979) SAN FRANCISCO OWL'S-CLOVER Pl 10–30 cm, yellow-brown, ± glabrous or sparsely stiff-hairy. **LF:** 10–40 mm, 5–9-lobed. **INFL:** gen 1–5 cm,

dense; bracts 5–12 mm, 3–7-lobed, ± glabrous. **FL:** calyx 4–6 mm, divided 1/4–1/2; corolla 10–14 mm, creamy white, tube glabrous, lower lip ± = beak, pouches ± 2 mm deep; stamens exserted. **FR:** 4–5 mm. **SEED:** 20–30, dark brown. Coastal grassland, serpentine slopes; < 200 m. n CCo, w SnFrB (Marin, San Mateo cos.). Apr–May ★

T. micrantha (A. Heller) T.I. Chuang & Heckard (p. 979) Pl 5–15 cm, gen ± dark purple, glabrous proximally, puberulent, glandular- and nonglandular-hairy distally. **LF:** 7–25 mm, 0–5-lobed. **INFL:** 3–6 cm, slender, ± open; bracts 8–15 mm, 3–5-lobed, long-hairy. **FL:** calyx 6–7 mm, divided 1/3–2/3, hairy; corolla 8–15 mm, beak dark purple, 1 mm > lower lip, marked dark purple, pouches 1–2 mm deep; stigma slightly 2-lobed. **FR:** 4–5 mm. **SEED:** 50–80, dark brown; coat loose-fitting. Grassland; < 1000 m. c&s SNF, SnJV, c CCo, SCoR. Mar–May

T. pusilla (Benth.) T.I. Chuang & Heckard (p. 979) Pl 5–20 cm, yellow-brown or ± purple, finely spreading-hairy. **ST:** decumbent-branched from base; main axis obscure. **LF:** 5–30 mm, 3–9-lobed.

INFL: 4–15 cm, ± from st base; bracts 5–12 mm, 3–11-lobed. **FL**: calyx 5–7 mm, divided 1/3–2/3, hairy; corolla 4–7 mm, ± purple (yellow), beak hooked, 1 mm > lower lip, pouches ± 1 mm deep. **FR**: 4–6 mm, round, compressed. **SEED**: 20–30, dark brown. Grassland; < 1200 m. NW, GV, SNF, CW; to BC. Self-pollinating. Apr–Jun

T. versicolor Fisch. & C.A. Mey. Pl 10–60 cm, green- to yellow-brown, gen glabrous. **LF**: 20–80 mm, 5–9-lobed. **INFL**: 5–20 cm, dense distally; bracts 8–18 mm, 3–5-lobed, lobes lanceolate. **FL**: calyx 5–10 mm, divided ± 1/3; corolla 12–22 mm, yellow or white,

tube >> calyx, densely puberulent, beak ± yellow, ± 1 mm > lower lip, pouches 2–4 mm deep, margins purple-dotted. **FR**: 6–9 mm. **SEED**: 30–50, dark brown.

subsp. ***faucibarbata*** (A. Gray) T.I. Chuang & Heckard **FL**: corolla yellow. Grassland; < 500 m. NCoR, ScV, CCo, SnFrB; sw OR, introduced in BC. Apr–Jun

subsp. ***versicolor*** (p. 979) **FL**: corolla white fading rose-pink. Grassland; < 100 m. NCo, CCo; s OR. Apr–Jun

OXALIDACEAE OXALIS FAMILY

Robert E. Preston & Robert Ornduff

Ann, per [vine, shrub, tree]. **LF**: compound (palmate [pinnate, or lflet 1]), alternate, often ± basal in rosettes or at st or rhizome tips in clusters, gen petioled, stipules gen 0, lflets gen sessile. **INFL**: cyme, umbel- or raceme-like or not, or fls 1, gen in axils; peduncle bracted. **FL**: gen bisexual, radial; sepals 5, free or fused at base; petals 5, free or fused above base; stamens 10[15], fused below, of 2 lengths; pistil 1, ovary superior, chambers [3]5, placentas axile, styles [1]5, gen ± free. **FR**: gen capsule, loculicidal. **SEED**: gen arilled. 5 genera, number of spp. uncertain: esp temp. [Matthews & Endress 2003 Bot J Linn Soc 140:321–381] Often heterostylous. Scientific Editor: Thomas J. Rosatti.

OXALIS

Roots fibrous or woody; bulbs, tubers, or rhizomes often present. **ST**: ± 0 or not. **LF**: stipules 0 or small; lflets 3, gen ± obcordate [not], gen entire, gen green. **FL**: petals clawed; styles erect or curved. **FR**: cylindric to spheric, explosively dehiscent. **SEED**: flat, often ridged; aril translucent. 500–950 spp.: esp temp. (Greek: sour, from acidic taste) [Lourteig 2000 Bradea 7:201–629] Taxonomy difficult, needs study; gen heterostylous; many (esp aliens in CA exc *O. micrantha*) orn; some noxious weeds; oxalates may be TOXIC to livestock; *O. latifolia* Kunth possibly naturalized in CCo (Keil 30389, just n of San Simeon), differs from *O. purpurea* in having fls in umbel-like cyme; *O. hirta* L. an historical waif (no recent collections), excluded.

1. Petals yellow to yellow-orange, drying ± red or not
  2. Ann; fr < 5 mm.. ***O. micrantha***
  2′ Per; fr ≥ 6 mm or 0
    3. Lvs in loose, ± basal rosette; bulbs present .... ***O. pes-caprae***
    3′ Lvs cauline; bulbs 0
      4. Petals 12–20 mm; NCo .... ***O. suksdorfii***
      4′ Petals 5–12 mm; NCo, elsewhere
        5. Petals gen < 8 mm; taproot fibrous; st rooting at nodes .... ***O. corniculata***
        5′ Petals gen 8–12 mm; taproot ± woody; st not rooting at nodes (sometimes rooting at nodes in *O. pilosa*)
          6. Hairs gen sparse, appressed or arched upward; pedicels > 2 × fr .... ***O. californica***
          6′ Hairs dense, gen spreading; pedicels < 2 × fr .... ***O. pilosa***
1′ Petals white to pink, red, purple-rose, or purple
  7. Bulbs 0
    8. Rhizome woody, not creeping; disturbed places .... ***O. articulata*** subsp. ***rubra***
    8′ Rhizome fleshy, creeping; ± undisturbed forests
      9. Fls 1 .... ***O. oregana***
      9′ Fls 3–9 in umbel-like cyme .... ***O. trilliifolia***
  7′ Bulbs present
    10. St elongate .... ***O. incarnata***
    10′ St ± 0
      11. Fls in umbel-like cyme .... [***O. bowiei***]
      11′ Fls 1 .... [***O. purpurea***]

O. articulata Savigny subsp. ***rubra*** (A. St.-Hil.) Lourteig WINDOWBOX WOOD-SORREL Per; rhizome woody, not creeping, with bulb-like tubercles; bulbs 0. **ST**: ± 0. **LF**: basal; petiole < 30 cm; lflets < 2 cm. **INFL**: umbel-like, < 10-fld; peduncle > petiole. **FL**: sepals < 5 mm, widely lanceolate, tips with 2 orange tubercles; petals white to purple-rose. **FR**: < 8 mm, ovoid. Disturbed places; < 800 m. NCo, ScV, CCo, SnFrB, SCo, n ChI, PR; native to Brazil. [*O. r.* A. St.-Hil.] Orn, escaping cult. Apr–Oct

O. californica (Abrams) R. Knuth CALIFORNIA WOOD-SORREL Per; taproot ± woody, gen > 4 mm diam; bulbs 0. **ST**: < 40 cm, decumbent to erect, not rooting at nodes, hairs gen sparse, appressed or arched upward. **LF**: cauline; petiole < 7 cm; lflets < 1.5 cm, appressed-hairy. **INFL**: 1–3-fld; pedicel 2–3 cm, sparsely appressed-

hairy. **FL**: sepals 3–7 mm, ovate, purple-tinged, tip ciliate; petals gen 8–12 mm, yellow-orange, drying ± red. **FR**: 10–15 mm, < 2 × pedicel, hairs simple, short, appressed, rarely multicellular. Coastal scrub, chaparral; < 2035 m. SW (exc SnBr, SnJt); Baja CA. [*O. albicans* Kunth subsp. *c.* (Abrams) G. Eiten] Probably closely related to *O. pilosa*. Feb–Jun

O. corniculata L. (p. 983) Per; taproot fibrous, gen < 4 mm diam, bulbs 0. **ST**: rooting at nodes, < 50 cm, hairs appressed to spreading. **LF**: cauline; petiole < 7 cm; lflets < 2 cm, ciliate, hairs gen 0 adaxially, sparse, appressed abaxially. **INFL**: cyme, ± umbel-like or not, 2–7-fld; pedicel < 1 cm, hairs appressed. **FL**: sepals < 4.5 mm, linear to narrowly ovate; petals gen < 8 mm, oblong to spoon-shaped, yellow, often with red spots below middle. **FR**: 6–25 mm, cylindric,

± angled, hairs simple, short, gen appressed. Disturbed areas; gen < 1900 m. NCoRO, n&c SNF, GV, CCo, SnFrB, SCo, PR, expected elsewhere; worldwide; native to Medit Eur. Possibly TOXIC in quantity to sheep. ± all yr

O. incarnata L. CRIMSON WOOD-SORREL Per; rhizome slender; bulbs < 2 cm, pale brown, beaked. **ST:** erect, branched, < 30 cm, glabrous; axils often with bulblets. **LF:** cauline, in whorl-like clusters of < 10, lower opposite or not; petiole 2–7 cm; lflets < 1.5 cm. **INFL:** fls 1; peduncle < 7 cm; bracts 2, near middle. **FL:** sepals < 6 mm, oblong, acute, tips with 2 orange tubercles; petals < 2 cm, white to pale pink. **FR:** 0 in CA. Shady woodland; < 400 m. NCo, SnFrB; native to s Afr. Orn, escaping cult. Mar–Jun

O. micrantha Colla DWARF WOOD-SORREL Ann; roots fibrous; bulbs 0. **ST:** erect, < 20 cm, hairy. **LF:** cauline; petiole < 5 cm; lflets < 13 mm, hairs sparse. **INFL:** gen raceme-like, 6–14-fld; peduncle < 15 cm; pedicel recurved in fr. **FL:** sepals < 4 mm, linear to lanceolate; petals < 12 mm, oblong, yellow. **FR:** < 5 mm, ovoid to ± spheric. Foothill woodland, rock outcrops, disturbed sites; < 1000 m. SNF, ScV, CCo; S.Am. [*O. laxa* Hook. & Arn., misappl.] Apr–May

O. oregana Nutt. (p. 983) REDWOOD SORREL Per; rhizome creeping, fleshy, scaly; bulbs 0. **ST:** ± 0. **LF:** < 10, clustered at rhizome tips; petiole < 20 cm, hairy; lflets < 4 cm, often ± purple abaxially, midribs pale. **INFL:** fls 1; peduncle < 15 cm, bracts 2, near tip. **FL:** sepals < 1 cm, lanceolate; petals < 2.5 cm, white to deep pink. **FR:** < 12 mm, ± ovoid. Moist conifer forest; < 1000 m. NCo, w KR, NCoRO, CCo, SnFrB; to WA. Feb–Aug

O. pes-caprae L. (p. 983) BERMUDA BUTTERCUP Per; bulbs many on rhizomes, root tips, < 2.5 cm, white to brown. **ST:** gen underground, vertical, short. **LF:** < 40, in loose ± basal rosette at enlarged st tip; petiole < 12 cm; lflets < 3.5 cm, often purple-spotted,

abaxially hairy. **INFL:** umbel-like, < 20-fld; peduncle < 30 cm. **FL:** sepals < 7 mm, lanceolate to oblong, tips often with 2 orange tubercles; petals < 2.5 cm, yellow; filaments < 7 mm, hairy. **FR:** 0 in CA. Disturbed areas, roadsides, grassland, dunes; < 820 m. NCo, NCoRO, n SNF, ScV, CW (exc SCoRI), SW (exc SnJt); worldwide alien; native to s Afr. Possibly TOXIC in quantity to sheep. Cult as orn; common garden weed. Jan–May ❖

O. pilosa Nutt. (p. 983) Taproot ± woody, gen > 4 mm diam. **ST:** erect to decumbent, < 40 cm, rooting at nodes or not, hairs 0.7–1.2 mm, dense, gen spreading. **LF:** lflets < 1.5 cm, appressed-hairy. **INFL:** 1–3-fld; pedicel < 2 cm, hairs dense, gen spreading, occ appressed. **FL:** sepals 3.5–6 mm, lanceolate, tips ciliate; petals gen 8–12 mm, yellow. **FR:** 12–18 mm, hairs simple, short, appressed, and multicellular, spreading. Coastal grassland, scrub, chaparral; < 1800 m. NCo, NCoRO, n SNF, CW (exc SCoRI), SCo, WTR, PR (exc SnJt); to BC, Mex. [*O. corniculata* L. subsp. *p.* (Nutt.) Lourteig; *O. albicans* subsp. *p.* (Nutt.) G. Eiten; *O. corniculata* var. *wrightii* (A. Gray) B.L. Turner] Feb–Sep

O. suksdorfii Trel. (p. 983) SUKSDORF'S WOOD-SORREL Per, taprooted, often stoloned; bulbs 0. **ST:** < 25 cm, ± trailing to erect, ± hairy. **LF:** cauline; petiole < 5 cm; lflets < 2 cm. **INFL:** umbel-like or not, 1–3-fld; peduncle < 10 cm. **FL:** sepals < 8 mm, obtuse; petals 12–20 mm, yellow. **FR:** 1–1.5 cm, oblong. Dry, shrubby or wooded areas; < 700 m. NCo; to WA. Jun–Jul ★

O. trilliifolia Hook. (p. 983) Per; rhizome creeping, fleshy, scaly; bulbs 0. **ST:** ± 0. **LF:** clustered at rhizome tips; petiole < 30 cm, hairy; lflets < 4 cm. **INFL:** umbel-like, 3–9-fld; peduncle < 25 cm. **FL:** sepals < 5 mm, lanceolate; petals < 1.3 cm, white or ± pink. **FR:** < 3 cm, linear. Dense conifer forest; < 1900 m. NCo, KR, NCoRO; to WA. May–Aug

PAEONIACEAE PEONY FAMILY

Fosiée Tahbaz

Per, subshrub; roots gen in fleshy clusters. **ST:** 1–several from root crown. **LF:** basal and cauline, alternate, large, deeply, gen ternately, dissected to compound, ± fleshy; stipules 0. **INFL:** terminal, fls 1 or few in cluster. **FL:** sepals gen 5, free, leathery, persistent in fr; petals gen 5–10; stamens many, maturing outward; pistils 2–5, simple, thick-walled, surrounded at base by lobed disk, style ± 0. **FR:** follicles, 2–5. **SEED:** large, gen several per follicle, gen black, arilled. 1 genus: w US, Eurasia, esp temp e Asia.; many cult. [Sang et al. 1997 Amer J Bot 84:1120–1136] Disk evidently not nectar-secreting. Scientific Editor: Thomas J. Rosatti.

PAEONIA PEONY

± 30 spp.: Eurasia, 2 spp. in w US. (Greek: Paeon, physician to the gods)

1. Petals ± round, < inner sepals, maroon to bronze, margins ± yellow or ± green; lf segment tips gen obtuse or
   rounded; st 20–40 cm; lvs gen 5–8 per st, glaucous, esp abaxially . . . . *P. brownii*
1′ Petals elliptic, > inner sepals, black-red, margins pink to red; lf segment tips gen long-acute; st 35–75 cm;
   lvs gen 7–12 per st, paler green but not strongly glaucous abaxially . . . . *P. californica*

P. brownii Douglas (p. 983) **ST:** gen simple. **LF:** 1° divisions 3–6 cm, 2–5 cm wide; ultimate segments ± elliptic, bases tapered. **FL:** petals 8–13 mm; filaments 3–5 mm, anthers 2–4 mm. **FR:** 2–4 cm. **SEED:** ± 11 mm. *n*=5. Open, dry pine forest, scrub; 200–3000 m. KR, NCoR, CaR, n&n-c SN, se SnFrB (Mount Hamilton), MP, n SNE; to WA, WY, NV. Apr–Jun

P. californica Nutt. (p. 983) **ST:** gen branched. **LF:** 1° divisions 3–9 cm, 1–6 cm wide; ultimate segments ± wide-linear or oblong, bases ± or not tapered. **FL:** petals 15–25 mm; filaments 5–8 mm, anthers 3–7 mm. **FR:** 3–4 cm. **SEED:** ± 16 mm. *n*=5. Chaparral, coastal scrub; < 1500 m. s CCo, SCoR, SCo, TR, PR; Baja CA. Jan–May

PAPAVERACEAE POPPY FAMILY

Gary L. Hannan & Curtis Clark, except as noted

Ann to small tree; sap colorless, yellow, orange, red, or white. **LF:** basal, cauline, or both, simple and entire, toothed, or lobed, or 1–3-pinnate-dissected or compound; cauline gen alternate; stipules 0. **INFL:** terminal, 1-fld or cyme, raceme, or panicle; bracts gen present. **FL:** bisexual, radial, bilateral, or biradial; sepals 2–3, shed after fl; petals gen 2 × sepals in number; stamens gen many; ovary 1, superior, chamber 1, style 0 or 1, stigmas or lobes 2–many, ovules few to many. **FR:** capsule, dehiscent by valves or pores, ± nut, or breaking transversely into 1-seeded, indehiscent units. **SEED:** fleshy appendage gen 0. 25–30

979

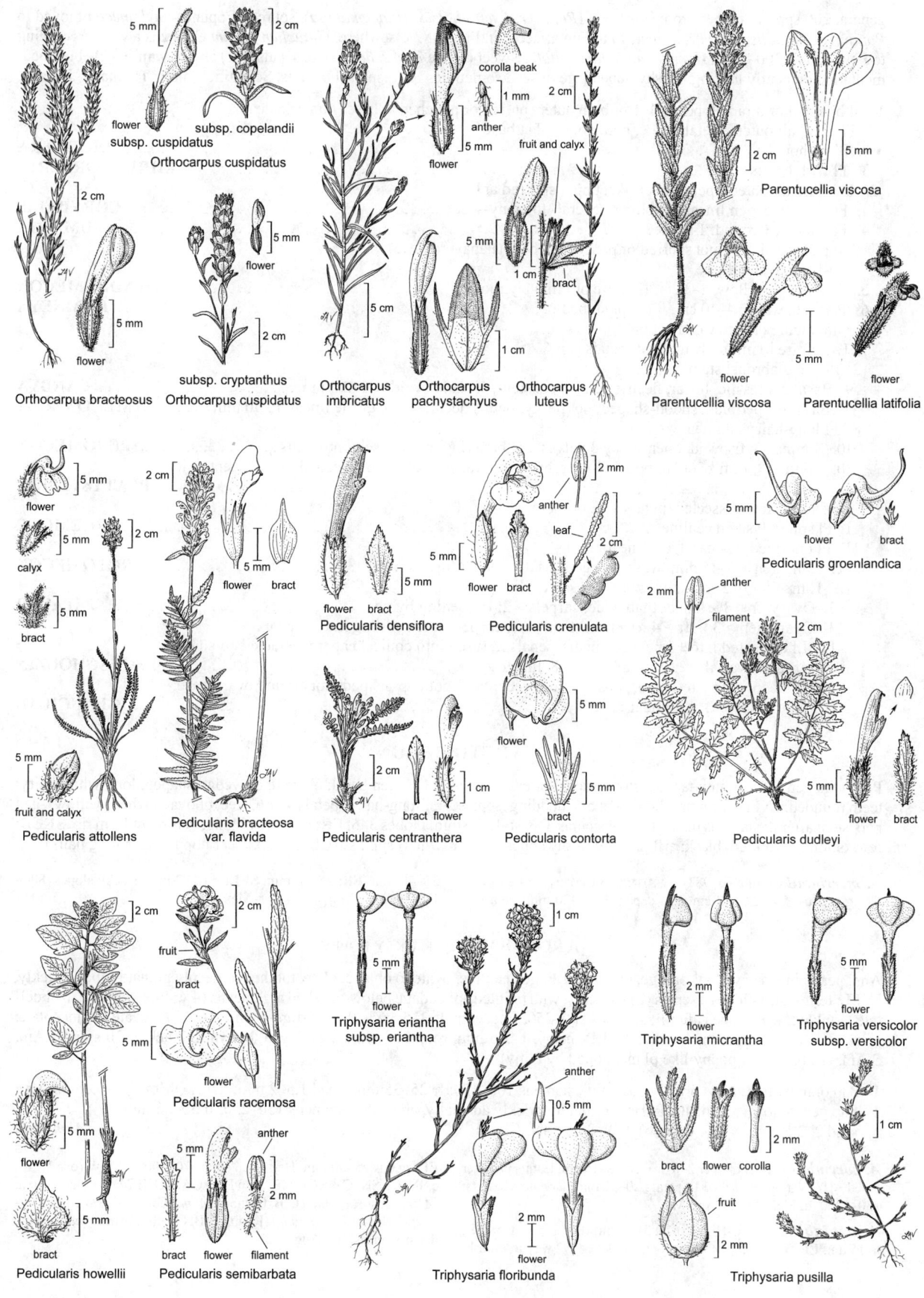

flower
subsp. cuspidatus
subsp. copelandii
Orthocarpus cuspidatus

flower
Orthocarpus bracteosus
subsp. cryptanthus
Orthocarpus cuspidatus

Orthocarpus imbricatus
Orthocarpus pachystachyus
Orthocarpus luteus

corolla beak
anther
flower
fruit and calyx
bract

Parentucellia viscosa
flower
Parentucellia viscosa
flower
Parentucellia latifolia

flower
calyx
bract
fruit and calyx
Pedicularis attollens

Pedicularis bracteosa var. flavida
flower
bract

flower
bract
Pedicularis densiflora

anther
leaf
flower
bract
Pedicularis crenulata

flower
bract
Pedicularis groenlandica

bract flower
Pedicularis centranthera

flower
bract
Pedicularis contorta

anther
filament
flower
bract
Pedicularis dudleyi

fruit
bract
flower
Pedicularis racemosa

flower
bract
Pedicularis howellii

bract flower filament
anther
Pedicularis semibarbata

flower
Triphysaria eriantha subsp. eriantha

anther
flower
Triphysaria floribunda

flower
Triphysaria micrantha

flower
Triphysaria versicolor subsp. versicolor

bract flower corolla
fruit
Triphysaria pusilla

genera, 200 spp.: n temp, n trop; some cult (*Papaver, Eschscholzia, Hunnemannia*), source of opiates. *Stylomecon* moved to *Papaver. Corydalis, Dicentra, Fumaria* in Fumariaceae in FNANM, elsewhere. *Glaucium flavum* Crantz is a waif. According to FNANM (3:300–301), *Hunnemannia fumariifolia* Sweet (± like *Eschscholzia* exc sepals free) an occ waif in CA, but documentation evidently lacking. Fleshy appendage of seed sometimes for dispersal by ants. Scientific Editor: Thomas J. Rosatti.

1. Fl bilateral or biradial; petals 4, 1 or both outer spurred or pouched at base (Fumarioideae)
 2. Fl biradial; outer 2 petals alike, both pouched at base
  3. Pl without lfy sts . . . . **DICENTRA**
  3′ Pl with lfy sts . . . . **EHRENDORFERIA**
 2′ Fl bilateral; outer 2 petals not alike, upper spurred at base
  4. Fr a capsule, gen linear to oblong, several- to many-seeded . . . . **CORYDALIS**
  4′ Fr ± a nut, ± round, 1-seeded . . . . **FUMARIA**
1′ Fl radial; petals ≥ 4, not spurred or pouched at base (Papaveroideae)
 5. Subshrub to small tree
  6. Petals 4(6), yellow, 2–3 cm; lf entire or minutely toothed . . . . **DENDROMECON**
  6′ Petals 6, white, 4–10 cm; lf deeply lobed . . . . **ROMNEYA**
 5′ Ann, bien, per, occ woody at base
  7. Lvs entire or minutely toothed; petals gen 6
   8. Pl gen glabrous; stamens 4–15
    9. Basal lvs sessile, linear, fleshy; petals persistent; capsule ovoid, 1.5–2.5 mm . . . . **CANBYA**
    9′ Basal lvs petioled, spoon-shaped, not fleshy; petals deciduous; capsule linear, to 50 mm . . . . **MECONELLA**
   8′ Pl long-hairy; stamens > 12
    10. Stigmas 3; fr ovoid, opening by 3 valves, not breaking transversely into units . . . . **HESPEROMECON**
    10′ Stigmas ≥ 6; fr ovoid to widely linear, breaking transversely into 1-seeded, indehiscent units
     . . . . **PLATYSTEMON**
  7′ Lvs toothed to dissected; petals 4, 6, or more
   11. Pl spiny; lvs gen cauline . . . . **ARGEMONE**
   11′ Pl unarmed; lvs basal, cauline, or both
    12. Hairs gen ± 5–15 mm, wavy; lvs gen basal — ne DMoj . . . . **ARCTOMECON**
    12′ Hairs 0 or to 3 mm, straight; lvs basal, cauline, or both
     13. Ovary, fr < 3 × longer than wide; carpels > 2; fr opening by pores . . . . **PAPAVER**
     13′ Ovary gen > 3 × (fr > 4 ×) longer than wide; carpels 2; fr splitting into 2 valves
      14. Lf dissected into ± linear segments; sepals 2, fused into conical cap; receptacle cup-shaped
       around ovary base; stigma lobes 4–8, linear . . . . **ESCHSCHOLZIA**
      14′ Lf deeply pinnate-lobed; sepals 2, free; receptacle not cup-shaped, not around ovary base;
       stigma lobes 2, not linear . . . . **[GLAUCIUM]**

ARCTOMECON

Per; hairs gen ± 5–15 mm; taproot stout; sap colorless or yellow. **LF**: gen basal, obovate or wedge-shaped, long white-hairy; teeth rounded. **INFL**: terminal, 1-fld. **FL**: bud nodding; sepals 2–3, long-hairy; petals 4 or 6, free, obovate, white [yellow], gen persistent after fl; stamens many, free; placentas 3–6, style 1, stigma lobes 3–6. **FR**: ovate to oblong, dehiscent from tip. **SEED**: few, oblong, wrinkled, black; aril present; fleshy appendage present. 3 spp.: sw US. (Greek: bear poppy, from long hairs)

A. merriamii Coville (p. 983) WHITE BEAR POPPY Pl 2–5 dm. **ST**: glaucous. **LF**: 2.5–12 cm. **FL**: sepals 3, 15–20 mm; petals 6, 25–40 mm. **FR**: 25–40 mm. **SEED**: 1.5–2 mm. Rocky slopes; 800–1600 m. ne DMoj; s NV. Apr–May ★

ARGEMONE   PRICKLY POPPY

Ann, per, spiny; sap yellow, orange, or white. **LF**: gen cauline, ovate to oblanceolate, toothed or deeply pinnate-lobed, prickly. **INFL**: terminal, 1-fld. **FL**: sepals (2)3, prickly, with pointed appendage below tip, shed at fl; petals (4)6, free, obovate to obdeltate, crinkled, white, shed after fl; stamens 100–250, free; carpels 3–5, style 0 or 1, stigma lobes 3–5. **FR**: ovate to lanceolate, prickly, dehiscent from top by slits. **SEED**: many, 1–2.5 mm, round to ovate, net-ridged, brown or black. ± 30 spp.: N.Am, S.Am, HI. (Greek: a poppy-like pl mentioned by Pliny)

1. Sap orange; lf gen less prickly adaxially; stamens 100–120; fr 25–35 mm; seed 1.5–2 mm; SNE, DMoj . . . *A. corymbosa*
1′ Sap gen yellow (red in NCoRH); lf ± equally prickly adaxially, abaxially; stamens 150–250; fr 35–55 mm;
  seed 2–2.5 mm; NW (exc NCo), CW, SW, s SN, GB, D . . . . *A. munita*

A. corymbosa Greene   Per 2–8 dm. **LF**: 8–15 cm, leathery; lower lobed < 1/2 way to midrib. **FL**: petals 20–35 mm. Dry slopes, flats; 230–1200 m. SNE, DMoj. Apr–May

A. munita Durand & Hilg. (p. 983) CHICALOTE Ann or per, 6–15 dm. **LF**: 5–15 cm, not leathery; lower lobed 1/2 way to midrib.

FL: petals 25–50 mm. Dry, open areas; 70–3000 m. NW (exc NCo), c SNH, s SN, CW, SW, GB, D; n Baja CA. TOXIC but not gen eaten. [*A. m.* subsp. *argentea* G.B. Ownbey.; *A. m.* subsp. *robusta* G.B. Ownbey; *A. m.* subsp. *rotundata* (Rydb.) G.B. Ownbey] Variable in prickle density on sts, lvs. Aug

CANBYA

Ann; sap colorless. **LF**: ± basal, linear-oblong, entire. **INFL**: axillary, 1-fld; peduncles > lvs. **FL**: sepals 3; petals 5–7, free, elliptic, white, persistent after fl; stamens 6–15, free; carpels 3(4), ovary ± spheric, style 0, stigma lobes 3(4), linear, radiating from ovary top. **FR**: ovate, dehiscent from tip. **SEED**: many, shiny, brown. 2 spp.: CA, OR, NV. (W.M. Canby, DE philanthropist, botanist, 1831–1904)

C. candida A. Gray (p. 983)   WHITE PYGMY-POPPY   Pl 10–30 mm, tufted, ± glabrous. **LF**: 5–9 mm, fleshy. **INFL**: peduncle 10–20 mm. **FL**: petals 3–4 mm, closing over fr. **FR**: 1.5–2.5 mm. **SEED**: 0.6 mm. Sandy places; 600–1350 m. w DMoj, adjacent SN. Apr–May ★

CORYDALIS

Ann to per, glabrous, glaucous; sap colorless. **LF**: pinnately dissected to compound. **INFL**: raceme or panicle. **FL**: bilateral; sepals 2, shed at fl or not; petals 4, yellow or white to pink, persistent after fl, outer 2 free, not alike, keeled, upper spurred at base, inner 2 adherent at tips, oblanceolate, crested on back; stamens 6, ± fused in 2 sets, opposite outer petals; ovary obovoid, placentas 2, style 1, stigma lobes 4–8. **FR**: gen linear to oblong, dehiscent from tip. **SEED**: several to many, 2–2.5 mm, round-reniform, smooth or rough, black; fleshy appendage gen present. ± 100 spp.: n hemisphere, s Afr (some orn). (Greek: crested lark)

1. Petals yellow, spur 4–5 mm; ann, bien; fr cylindric . . . . **_C. aurea_**
1′ Petals white to pink, purple-tipped in age, spur 9–16 mm; per; fr elliptic . . . . **_C. caseana_** subsp. **_caseana_**

C. aurea Willd. (p. 983)   GOLDEN CORYDALIS   Pl 10–40 cm. **LF**: several to many, 3–18 cm. **INFL**: raceme. **FL**: spurred petal 13–16 mm. **FR**: 18–25 mm. Loose soil in open areas; 1500–2800 m. GB; to AK, e US, n Mex. May–Aug

C. caseana A. Gray subsp. **_caseana_** (p. 983)   SIERRA CORYDALIS, FITWEED   Pl 50–100 cm. **LF**: several, 15–35 cm. **INFL**: raceme or panicle. **FL**: spurred petal 16–25 mm. **FR**: 10–15 mm. Moist sites in forest and along streams; 1100–2800 m. CaRH, n SNH, s SNH (Tulare Co.), s MP. TOXIC, eaten by naive livestock. Other subspp. in OR, Rocky Mtns. Jun–Aug

DENDROMECON   TREE POPPY

Shrub to small tree, evergreen; sap colorless. **LF**: simple, entire or minutely toothed. **INFL**: terminal, 1-fld. **FL**: receptacle funnel-shaped around ovary base; sepals 2, shed at fl; petals 4(6), free, 2–3 cm, obovate or wedge-shaped, yellow, satiny, shed after fl; stamens many, free; placentas 2, style 0, stigma lobes 2, flat. **FR**: linear, dehiscent from base. **SEED**: many, 2–3 mm, obovate, smooth, brown or black; fleshy appendage present. 2 spp.: CA, Baja CA. (Greek: tree poppy)

1. Lf 1.5–3 × longer than wide, entire, tip gen ± rounded-mucronate; ChI . . . . **_D. harfordii_**
1′ Lf 3–8 × longer than wide, minutely toothed, tip acute-mucronate; not ChI . . . . **_D. rigida_**

D. harfordii Kellogg   CHANNEL ISLAND TREE POPPY   Pl 2–6 m. **LF**: 3–8 cm, 1.5–4.5 cm wide. **FL**: petals 2–3 cm. **FR**: 7–10 cm. **SEED**: 2.5–3 mm. Shrubby slopes; < 600 m. ChI. [*D. h.* var. *rhamnoides* (Greene) Munz] Apr–Jul ★

D. rigida Benth. (p. 983)   BUSH POPPY   Pl 1–3 m. **LF**: 2.5–10 cm, 0.7–2.5 cm wide. **FL**: petals 2–3 cm. **FR**: 5–10 cm. **SEED**: 2–2.5 mm. Dry slopes, washes, esp recent burns; < 1900 m. s NW, CaRF, SNF, CW (except ChI), SW; n Baja CA. Apr–Jun

DICENTRA

Gary L. Hannan

Per, glabrous, glaucous or not; sap colorless. **LF**: basal, deeply dissected, segments often lobed. **INFL**: raceme, panicle, or 1-fld. **FL**: biradial, nodding to erect; sepals 2, shed after fl; petals 4, white or cream to pale yellow to pink or pink-tipped, persistent or not, outer 2 free, lanceolate, alike, both pouched at base, inner 2 adherent at tips, oblanceolate, ± crested on back; stamens 6, ± fused in 2 sets opposite outer petals; ovary cylindric to long-conic, placentas 2, style 1, stigma lobes 2. **FR**: oblong, fusiform to ovate, or conic, dehiscent from tip. **SEED**: few, 1–2 mm, oblong to reniform, smooth to finely netted, black; fleshy appendage present. 18 spp.: N.Am, Asia; some orn. (Greek: twice spurred, from outer petals) Other spp. in TJM (1993) moved to *Ehrendorferia* (Liden et al. 1997 Pl Syst Evol 206:411–420).

1. Infl 2–30-fld; crest on inner petals conspicuous, extending 1–2 mm beyond tip; pl rhizomed, tubers 0, bulblets 0
   2. Base of central stamen of each set aligned with lateral 2; NW, CaR, SNH, n CW; < 2300 m . . . . **_D. formosa_**
   2′ Base of central stamen of each set looped into base of outer petal; s SNH; 2200–3100 m . . . . **_D. nevadensis_**
1′ Infl 1–3-fld; crest on inner petals 0 or obscure, not extending beyond tip; pl with tubers or bulblets
   3. Outer petals recurved < 1/2 length; infl 1–3-fld . . . . **_D. pauciflora_**
   3′ Outer petals recurved > 1/2 length; infl 1-fld . . . . **_D. uniflora_**

D. formosa (Haw.) Walp. (p. 987)   PACIFIC BLEEDING HEART   Pl 20–45 cm, rhizomed. **LF**: 2-ternate-dissected, 20–50 cm. **INFL**: 2–30-fld. **FL**: nodding; outer petals 14–19 mm, rose-purple to cream or pale yellow, not drying black; base of central stamen of each set aligned with lateral 2. **FR**: 14–20 mm, oblong. Damp, shaded areas; < 2400 m. NW, CaR, SNH, n CW; to BC. Further study needed to determine if small pls from nw KR (lvs blue-glaucous adaxially; petals

cream to pale yellow with rose tips), described as *D. formosa* subsp. *oregana* (Eastw.) Munz, Oregon bleeding heart, deserve taxonomic recognition. Mar–Jul

D. nevadensis Eastw.   TULARE COUNTY BLEEDING HEART   Pl 20–45 cm, rhizomed. **LF**: 2-ternate-dissected, 15–25 cm. **INFL**: 2–20-fld. **FL**: nodding; outer petals 12–18 mm, white to pale yellow

or ± pink, drying black; base of central stamen of each set looped into base of outer petal. **FR**: 13–20 mm, oblong. Meadows, in gravelly soils; 2200–3100 m. s SNH (Sequoia, Kings Canyon national parks, Tulare Co.). Jul ★

D. pauciflora S. Watson Pl 3–7 cm, tubered. **LF**: 1–3, 2–3-ternate-dissected, 4–7 cm. **INFL**: 1–3-fld. **FL**: nodding to erect; petals 18–22 mm, white to pink or lavender, recurved tips of outer petals < 1/2 length of petal; inner petals ± purple. **FR**: 10–15 mm, fusiform

to ovate. Open, gravelly areas; 1200–3000 m. KR, CaRH, n&s SNH, WTR; to OR. Jun–Jul

D. uniflora Kellogg (p. 987) STEER'S HEAD Pl 3–7 cm, tubered. **LF**: 1–3, 2–3-ternate-dissected, 4–7 cm. **INFL**: 1-fld. **FL**: nodding to erect; petals 12–16 mm, white to pink or lavender, recurved tips of outer petals > 1/2 length petal; inner petals purple-tipped. **FR**: 9–14 mm, conic. Gravelly or rocky areas; 1000–3300 m. KR, NCoRH, CaRH, SNH, MP; to BC, MT, WY, UT. May–Jul

EHRENDORFERIA

Gary L. Hannan

Per, erect, glabrous, glaucous, taprooted; st lfy, hollow; sap colorless. **LF**: basal and cauline, 2- to 3-pinnate-dissected, segments lobed. **INFL**: panicle, 20–50 cm, 5–many-fld, pedicel ascending to erect. **FL**: biradial; sepals 2, shed after fl; petals 4, yellow or cream with purple tips, outer 2 free, lanceolate, alike, both pouched at base, not spurred, inner 2 adherent at tips, oblanceolate, crested on back; stamens 6, ± fused in 2 sets opposite outer petals. **FR**: erect, long-conic, dehiscent from tip. **SEED**: reniform; seed coat densely papillate. 2 spp.: N.Am. (Friedrich Ehrendorfer, University of Vienna, b. 1927)

1. Petals golden yellow, 12–16(22) mm, outer recurved ± 1/2 length; crest of inner petal exceeding tip by
   2–4 mm . . . . ***E. chrysantha***
1′ Petals ± white to cream, purple-tipped, 20–25 mm, outer recurved < 1/2 length; crest of inner petal
   exceeding tip by < 2 mm . . . . ***E. ochroleuca***

E. chrysantha (Hook. & Arn.) Rylander (p. 987) GOLDEN EARDROPS Pl 50–160 cm. **LF**: 2- to 3-pinnate-dissected, 15–30 cm. **INFL**: many-fld. **FR**: 15–25 mm. Dry slopes, burns, disturbed areas; < 2300 m. NW, SN, CW, SW; n Baja CA. TOXIC to livestock, esp when abundant after burns. [*Dicentra c.* (Hook. & Arn.) Walp.] Apr–Sep

E. ochroleuca (Engelm.) Fukuhara WHITE EARDROPS Pl 50–200 cm. **LF**: 3-pinnate-dissected, 15–35 cm. **INFL**: 5- many-fld. **FR**: 15–30 mm. Dry slopes, burns, disturbed areas; < 1100 m. s CW, SW. [*Dicentra o.* Engelm.] May–Jul

ESCHSCHOLZIA

Ann, per; sap colorless or orange. **LF**: basal or basal and cauline, 1–4-pinnate-dissected, segments narrow. **INFL**: cyme, 1–many-fld. **FL**: receptacle funnel-shaped, tip cupped around ovary base, outer receptacle rim occ spreading; sepals 2, fused, shed as unit at fl; petals gen 4 (exc doubled fls), free, obovate or wedge-shaped, gen yellow to orange (white or pink), shed after fl leaving crown-like membrane (inner receptacle rim); stamens 12–many, free; carpels 2, style 0, stigma lobes 4–8, spreading, linear. **FR**: oblong, dehiscent from base. **SEED**: many, 1–2 mm, round to ovate, net-ridged, prominent-discontinuous-ridged, or minutely pitted, tan, brown, or black. 12 spp.: w N.Am. (J.F.G. von Eschscholtz, Russian surgeon, botanist, 1793–1831)

1. Outer receptacle rim 0.5–5 mm . . . . ***E. californica***
1′ Outer receptacle rim 0–0.3 mm
  2. Lvs basal
    3. Fl bud gen nodding; lf 3 × ternate-divided, segments many, short-linear, condensed at end of long
       petiole, so lf broom-like; petals (10)12–25 mm; seed minutely pitted, tan to ashy-gray; D . . . . ***E. glyptosperma***
    3′ Fl bud erect; lf 2 × ternate-divided, segments fewer than many, long-linear, sparsely distributed along lf,
       so lf not broom-like; petals 7–12 mm; seed prominent-discontinuous-ridged, brown; SNF, GV . . . . ***E. lobbii***
  2′ Lvs basal and cauline
    4. Receptacle tapered at base, but with parallel sides distally
      5. Peduncle = basal lvs; petals yellow, 3–15 mm . . . . ***E. rhombipetala***
      5′ Peduncle 2 × basal lvs; petals orange or deep yellow, 15–40 mm . . . . ***E. lemmonii***
        6. Bud gen erect, glabrous . . . . subsp. ***kernensis***
        6′ Bud nodding, gen hairy . . . . subsp. ***lemmonii***
    4′ Receptacle obconic
      7. Fl bud gen hairy, nodding; petiole gen hairy . . . . ***E. hypecoides***
      7′ Fl bud glabrous, nodding or erect; petiole gen glabrous
        8. Fl bud gen nodding; SNE, D
          9. Lf segments widened to tip, short, tip gen obtuse, notched; lvs gray- or blue-green; seeds gen oblong
             to elliptic . . . . ***E. minutiflora***
          9′ Lf segments not widened to tip, short or not, tip acute, ± notched or not; lvs bright- or yellow-green;
             seeds gen round . . . . ***E. parishii***
        8′ Fl bud gen erect; not SNE, D
          10. Fl bud long-pointed, tip gen > 1/4 bud; pl denser below than above; mainland . . . . ***E. caespitosa***
          10′ Fl bud blunt or short-pointed, tip < 1/4 bud; pl denser above than below; ChI . . . . ***E. ramosa***

E. caespitosa Benth. (p. 987) Ann, erect, tufted, 5–30 cm, glabrous, occ ± glaucous. **LF**: segments obtuse or acute. **FL**: bud gen erect, glabrous, long-pointed, tip gen > 1/4 bud; receptacle obconic; petals 10–25 mm, yellow, bases occ orange-spotted. **FR**: 4–8 cm. **SEED**: 1.5–2.4 mm wide, elliptic to obovate, net-ridged, brown to

black. 2*n*=12. Open chaparral; < 1800 m. CA-FP (exc KR, CaR, ChI); sw OR. Mar–Jun

E. californica Cham. (p. 987) CALIFORNIA POPPY Ann (or per from heavy taproot), erect or spreading, 5–60 cm, glabrous, occ glaucous. **LF**: segments obtuse or acute. **FL**: bud erect, acute

983

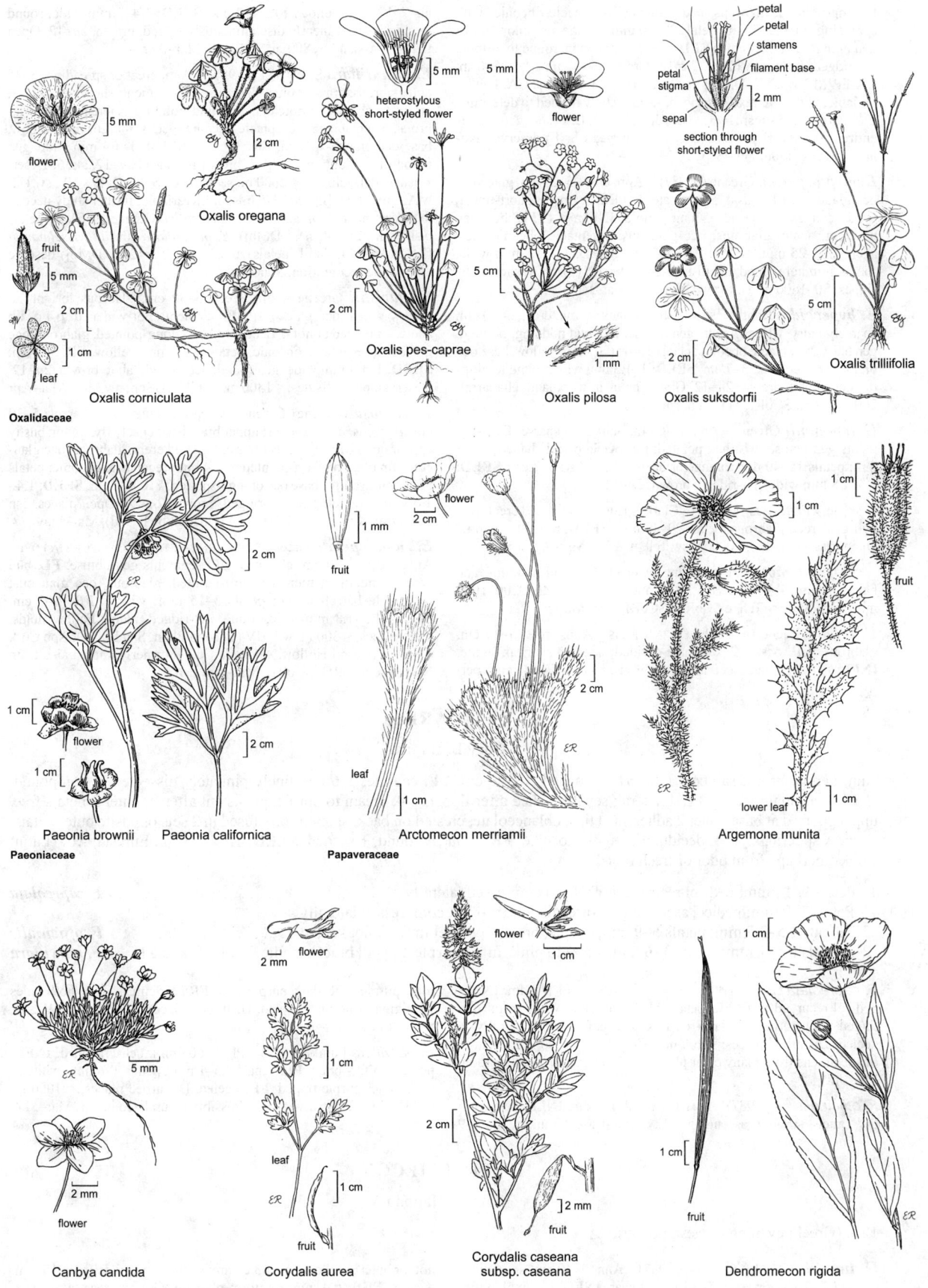

flower

flower

heterostylous
short-styled flower

flower

petal
petal
stamens
filament base
petal
stigma
sepal
section through
short-styled flower

Oxalis oregana

fruit

leaf

Oxalis corniculata

Oxalidaceae

Oxalis pes-caprae

Oxalis pilosa

Oxalis suksdorfii

Oxalis trilliifolia

flower

flower

fruit

leaf

Paeonia brownii   Paeonia californica

Paeoniaceae

Arctomecon merriamii

Papaveraceae

fruit

lower leaf

Argemone munita

flower

leaf

fruit

Canbya candida

flower

leaf

fruit

Corydalis aurea

flower

fruit

Corydalis caseana
subsp. caseana

fruit

Dendromecon rigida

to long-pointed, glabrous, occ glaucous; receptacle obconic, with spreading rim 0.5–5 mm; petals 20–60 mm, orange, or yellow, bases gen orange. **FR:** 3–9 cm. **SEED:** 1.5–1.8 mm wide, round to elliptic, net-ridged, brown to black. 2*n*=12. Grassy, open areas; < 2500 m. CA-FP, MP, w SNE, DMoj; to s WA, NV, NM, nw Baja CA. Highly variable, with > 90 taxa described; further study needed to determine if *E. californica* subsp. *mexicana* (Greene) C. Clark (ann; cotyledons entire; DMtns), *E. procera* Greene (pl large; s SNF) deserve taxonomic recognition. Feb–Sep

E. glyptosperma Greene (p. 987) Ann, erect, 5–25 cm, glabrous, occ glaucous. **LF:** basal; 3 × ternate-divided; segments short-linear, acute, condensed at end of long petiole. **INFL:** fl 1. **FL:** bud gen nodding, acute, glabrous, occ glaucous; receptacle obconic; petals (10)12–25 mm, yellow. **FR:** 4–7 cm. **SEED:** 1.2–1.8 mm wide, round, minutely pitted, tan to ashy-gray. 2*n*=14. Desert washes, flats, slopes; 30–1600 m. D; to sw UT, w AZ. Mar–May

E. hypecoides Benth. SAN BENITO POPPY Ann, erect, 5–30 cm, sparsely hairy. **LF:** segments gen obtuse. **FL:** bud nodding, pointed, gen hairy; receptacle obconic; petals 10–15(20) mm, yellow, base occ orange-spotted. **FR:** 3–7 cm. **SEED:** 1–1.3 mm wide, round to elliptic, net-ridged, brown. 2*n*=12. Grassy areas in woodland, chaparral; 200–1600 m. SCoRI. Mar–Jun ★

E. lemmonii Greene Ann 5–30 cm, hairs 0 to sparse. **LF:** segments gen obtuse. **FL:** bud pointed; receptacle barrel-shaped, > 1.5 mm; petals 15–40 mm, orange or deep yellow. **FR:** 3–7 cm. **SEED:** 1.3–1.8 mm wide, net-ridged, brown. 2*n*=12.

subsp. ***kernensis*** (Munz) C. Clark TEJON POPPY Pl erect. **FL:** bud gen erect, glabrous; petals 20–40 mm. **SEED:** round to elliptic. Open grassland; 200–1000 m. sw Teh, n WTR. Mar–Apr ★

subsp. ***lemmonii*** Pl spreading or erect. **FL:** bud nodding, gen hairy; petals 15–40 mm. **SEED:** elliptic. Open grassland; 200–1000 m. s-most SNF, w Teh, e SCoRO, SCoRI. Mar–May

E. lobbii Greene (p. 987) FRYING PANS Ann, erect, 5–15 cm, glabrous. **LF:** basal; 2 × ternate-divided; segments linear, acute. **INFL:** fl 1. **FL:** bud erect, acute, glabrous; receptacle obconic; pet-

als 7–12 mm, yellow. **FR:** 4–7 cm. **SEED:** 1.4–2 mm wide, round to elliptic, prominently discontinuously ridged, brown. 2*n*=12. Open fields, grassland; < 800 m. SNF, GV. Mar–Apr

E. minutiflora S. Watson (p. 987) Ann, erect or spreading, 5–35 cm, glabrous, gray- or blue-glaucous. **LF:** segments short, gen obtuse, entire to shallowly notched. **FL:** bud nodding, short-pointed, glabrous, occ glaucous; receptacle obconic; petals 3–6(26) mm, yellow, base occ orange-spotted. **FR:** 3–6 cm. **SEED:** 1–1.4 mm wide, gen oblong to elliptic, net-ridged, brown to black. 2*n*=12,24,36. Desert washes, flats, slopes; < 2600 m. se SCoRO, SNE, D; to s NV, sw UT, w AZ, nw Mex. [*E. parishii*, misappl.] Variable; further study needed to determine if *E. minutiflora* subsp. *covillei* (Greene) C. Clark (petals 6–18 mm; 2*n*=24; n&c DMoj), *E. minutiflora* subsp. *twisselmannii* C. Clark & M. Faull (petals 10–26 mm; 2*n*=12; w DMoj) deserve taxonomic recognition. Mar–May

E. parishii Greene Ann, erect, 5–30 cm, glabrous, bright- to yellow-green, occ glaucous. **LF:** segments not widened to tip, tip acute, ± notched or not. **FL:** bud nodding, long-pointed, glabrous, occ glaucous; receptacle obconic; petals 8–30 mm, yellow. **FR:** 5–7 cm. **SEED:** 1–1.4 mm wide, gen round, net-ridged, tan to brown. 2*n*=12. Desert slopes, hillsides; < 1300 m. s DMoj, DSon; nw Mex. Mar–Apr

E. ramosa (Greene) Greene ISLAND POPPY Ann, erect, 5–30 cm, glabrous, occ glaucous; upper branches densely lfy, giving bushy appearance. **LF:** segments obtuse. **FL:** bud erect, glabrous, occ glaucous, tip blunt or short-pointed, < 1/4 bud; receptacle obconic; petals 5–20 mm, yellow, base occ orange-spotted. **FR:** 4–7 cm. **SEED:** 1.4–1.6 mm wide, elliptic, net-ridged, brown. 2*n*=24. Open places, esp chaparral; < 500 m. ChI; Baja CA (Guadalupe Island). Mar–May ★

E. rhombipetala Greene DIAMOND-PETALED CALIFORNIA POPPY Ann, erect, 5–30 cm, glabrous. **LF:** segments gen obtuse. **FL:** bud erect or nodding, blunt or short-pointed, glabrous, occ glaucous; receptacle barrel-shaped; petals 3–15 mm, yellow. **FR:** 4–7 cm. **SEED:** 1.3–1.8 mm wide, round, net-ridged, black. Fallow fields, open places; < 300 m. w SnJV (Carrizo Plain, San Luis Obispo Co.), e SnFrB (Corral Hollow, Alameda Co.) (Formerly known also from NCoRI, e SCoRO, SCoRI). Mar–Apr ★

FUMARIA

Gary L. Hannan

Ann, gen glabrous; sap colorless. **ST:** branched, 10–70 cm. **LF:** cauline, 2–6 cm, finely pinnately dissected or compound. **INFL:** terminal raceme. **FL:** bilateral; sepals 2, shed after fl; petals 4, cream to purple, persistent after fl, outer 2 petals free, upper spurred at base, inner 2 adherent at tips, oblanceolate, crested on back; stamens 6, ± fused in 2 sets opposite outer petals; ovary ± spheric, style 1, deciduous, stigma dot-like. **FR:** ± nut, ± round, 1-seeded. **SEED:** 1. ± 50 spp.: Eurasia, Afr. (Latin: smoky, perhaps from odor of fresh roots)

1. Petals 9–14 mm; pedicel recurved in fr; fr ± compressed laterally . . . . ***F. capreolata***
1′ Petals 3–9 mm; pedicel ascending or spreading in fr; fr not compressed laterally
  2. Sepals 1.5–3.5 mm; petals 5–9 mm, purple; bract < pedicel in fr; fr not keeled . . . . ***F. officinalis***
  2′ Sepals 0.5–1 mm; petals 3–6 mm, white to pink, inner purple-tipped; bract > pedicel in fr; fr keeled . . . ***F. parviflora***

F. capreolata L. WHITE RAMPING FUMITORY Pl 1–8 dm. **INFL:** pedicel recurved in fr. **FL:** petals 9–14 mm, outer white to purple-tinged with purple or black-red tips. **FR:** compressed laterally, ± keeled. Disturbed places; < 100 m. CCo; FL; native to Eur. Identity of specimens from many other places in CA in need of confirmation. Mar–Jul

F. officinalis L. (p. 987) FUMITORY Pl 1–7 dm. **INFL:** pedicel in fr > bracts, straight, ascending. **FL:** sepals 1.5–3.5 mm; petals 5–9

mm, purple with dark purple tips. **FR:** ± depressed on both sides of stigma or not, not keeled. Disturbed places; < 1000 m. CW, SW; native to Eur. Apr–Jul

F. parviflora Lam. (p. 987) Pl 10–60 cm, much-branched. **INFL:** pedicel in fr ≤ bract. **FL:** sepals 0.5–1 mm; petals 3–6 mm, white to pink, inner purple-tipped. **FR:** keeled. Disturbed places; < 1000 m. CCo, SCo; c TX; native to Eur. Possibly naturalized in CA. Mar–May

HESPEROMECON

Gary L. Hannan

1 sp. (Greek: evening or west, poppy)

H. linearis (Benth.) Greene (p. 987) Ann, 3–40 cm, spreading-hairy. **LF:** basal or ± so, 5–85 mm, linear. **INFL:** terminal, 1-fld; peduncle 2.5–38 cm, > lvs, spreading-hairy. **FL:** buds nodding; petals 3–20 mm, 2–10 mm wide, ovate to obovate, cream, base yellow or

not, or outer 3 yellow, inner 3 cream; stamens many. **FR:** 10–15 mm, ovate. **SEED:** 0.4 mm, reniform-obovate. Grassy areas, washes; < 1000 m. c&s SNF, Teh, SnJV, CW; s OR. [*Meconella l.* (Benth.) A. Nelson & J.F. Macbr.] Mar–Jun

MECONELLA
Gary L. Hannan

Ann, glabrous; sap colorless. **LF**: basal and a few cauline, gen entire, glabrous; basal spoon-shaped, cauline linear to linear-spoon-shaped. **INFL**: gen terminal and axillary, 1-fld; peduncle > lvs. **FL**: sepals 3; petals 6, free, obovate to oblong, white (yellow), shed after fl; stamens 4–6 or 12, free; placentas 3, style 0, stigma lobes 3, ± erect, ovate. **FR**: linear, dehiscent from tip, twisted. **SEED**: many, 0.5–1 mm, ovate to reniform, smooth, shiny, brown or black. 3 spp.: w N.Am. (Greek, Latin: little poppy) *M. linearis* moved to *Hesperomecon*.

1. Receptacle without expanded ring beneath sepals; anthers linear-oblong, ≥ 1/2 filaments; stamens 6, in 1 series
   .... ***M. denticulata***
1′ Receptacle with expanded ring beneath sepals; anthers widely elliptic to round, < 1/2 filaments; stamens (6)12(6), in 2 series, or 4–6, in 1 series
   2. Stamens gen 12 (fewer in smallest pls), in 2 series, outer filaments < inner...................... ***M. californica***
   2′ Stamens 4–6, in 1 series, outer filaments = inner .... ***M. oregana***

M. californica Torr. & Frém. (p. 987) Pl 3–20 cm. **LF**: 3–25 mm. **FL**: petals 2–7 mm; stamens (6)12(16), in 2 series, anthers widely elliptic to round; receptacle with expanded ring beneath sepals. **FR**: 15–50 mm. **SEED**: 0.6 mm, obovate. Open, rocky areas; < 1000 m. CaRF, SNF, n CCo, SnFrB. Mar–May

M. denticulata Greene Pl 3–21 cm. **LF**: 2–40 mm, entire to minutely toothed. **FL**: petals 2–6 mm; stamens 6, in 1 series, anthers linear-oblong, 1 mm, ≥ 1/2 filaments. **FR**: 20–30 mm. **SEED**: 0.5 mm, obovate-reniform. Shaded canyons; < 1250 m. SCoRO, SW. Mar–Jun

M. oregana Nutt. OREGON MECONELLA Pl 2–16 cm. **LF**: 3–18 mm, entire to minutely toothed. **FL**: petals 1–5 mm; stamens 4–6, in 1 series, anthers widely elliptic to round; receptacle with expanded ring beneath sepals. **FR**: 20–30 mm. **SEED**: 0.5 mm, obovate-reniform. Shaded canyons; < 1000 m. CCo (Fort Ord), SnFrB (Berkeley Hills, Mount Hamilton); s OR, WA. Mar–May ★

PAPAVER
Gary L. Hannan

Ann [per]; sap white, orange, or red. **LF**: basal and cauline [or basal], deeply pinnate-lobed, glabrous or hairy, glaucous or not. **INFL**: axillary or terminal, 1-fld. **FL**: bud nodding; sepals 2, shed at fl; petals gen 4, free, obovate to wedge-shaped, white to red or purple; stamens many, free; placentas 4–20, style 0(1), stigma disk- or head-like, lobes 4–20. **FR**: dehiscent by pores below stigma. **SEED**: many, < 0.7 mm, reniform, net-ridged, brown or black. 70+ spp.: esp Eurasia. (Latin: poppy) *P. dubium* L. (fr ± oblong, glabrous; peduncle hairs near fl appressed, stiff) perhaps in CA as waif.

1. Style present; stigma head-like, lobes 4–11... ***P. heterophyllum***
1′ Style 0; stigma disk-like, lobes 4–20
   2. Cauline lf base clasping; fr 3–5 cm, round .... [***P. somniferum***]
   2′ Cauline lf base not clasping; fr 1–2 cm, narrowly obovate to round
      3. Peduncle hairs appressed; lf segments linear to lanceolate; fr bristly
         4. Fr oblong to club-shaped, hairs sparse, soft .... [***P. argemone***]
         4′ Fr obovate to round, hairs dense, stiff.................................... ***P. hybridum***
      3′ Peduncle hairs 0 or spreading; lf segments oblong to lanceolate; fr glabrous
         5. Petals 10–20 mm; peduncle glabrous or soft-hairy; fr narrowly obovate; native, after fire or disturbance
            .... ***P. californicum***
         5′ Petals 20–40 mm; peduncle bristly; fr widely obovate to round; alien, from cult.................. ***P. rhoeas***

P. californicum A. Gray (p. 987) FIRE POPPY Pl 30–60 cm, glabrous to shaggy-hairy. **LF**: 3–9 cm. **FL**: petals 10–20 mm, gen brick-red, base spotted ± green; filaments yellow-green, anthers yellow; carpels 4–8(11). **FR**: 1–1.6 cm, narrowly obovate. **SEED**: many, 0.4–0.5 mm, roughly net-ridged. 2n=28. Burns, disturbed areas, open woodland; < 1200 m. CW, SW. Apr–May

P. heterophyllum (Benth.) Greene (p. 987) WIND POPPY Pl 30–60 cm; sap yellow. **LF**: basal and cauline, 2–12 cm, deeply pinnate-lobed, glabrous to sparsely hairy, petioled; 1° lobes toothed or lobed. **INFL**: axillary, 1-fld; peduncle 5–20 cm. **FL**: sepals 4–10 mm; petals 10–20 mm, wedge-shaped, orange-red, purple-spotted at base, shed after fl; filaments purple, anthers yellow; carpels 4–11, style 1, slender, stigma head-like, lobes 4–11. **FR**: 1–2 cm, obconic to obovate. **SEED**: many, 0.4 mm, reniform, rough-net-ridged, brown or black. 2n=56.

Grassy areas, openings in chaparral; < 1500 m. s NW, c&s SNF, SnJV, CW, SW; n Baja CA. [*Stylomecon h.* (Benth.) G. Taylor] Apr–May

P. hybridum L. ROUGH POPPY To 50 cm, stiff-hairy. **LF**: 3–10 cm. **INFL**: peduncle appressed-hispid. **FL**: petals to 25 mm, red to purple-red with dark basal spot; filaments purple, anthers blue. **FR**: to 1.5 cm, widely obovate to round; densely stiff-bristly. Disturbed areas, cult fields; < 700 m. SnJV, CW, w DMoj, expected elsewhere; e US; native to Eurasia. Cult in CA. Apr

P. rhoeas L. CORN POPPY Pl 30–80 cm, hairy. **LF**: 3–15 cm. **FL**: petals 20–40 mm, white, with red-marks or not, or red or purple; filaments purple, anthers blue. **FR**: 1–2 cm, widely obovate to round. Disturbed areas, fallow fields; < 1800 m. CA-FP; N.Am; native to Eurasia. Cult, perhaps only a waif in CA. Jun–Jul

PLATYSTEMON
Gary L. Hannan

1 sp. (Greek: wide stamen)

P. californicus Benth. (p. 993) CREAM CUPS Ann 3–30 cm, shaggy-hairy; sap colorless to orange. **LF**: basal and cauline, alternate and whorled, 1–9 cm, linear to lanceolate or narrowly oblong, entire. **INFL**: axillary and terminal, 1-fld; peduncle 3.4–26 cm, > lvs. **FL**: buds nodding; sepals 3, hairy; petals 6, free, 6–19 mm, narrowly ovate to obovate, cream with yellow base, tip, or both (or all yellow), often persistent after fl; stamens > 12, free, filaments flat; carpels gen 9–18, fused, glabrous to densely long-hairy, separating

in fr. **FR**: 10–16 mm, ovoid to widely linear, gen narrowed between seeds, breaking transversely into 1-seeded, indehiscent units. **SEED**: 1 mm, elliptic to reniform, smooth, black. Open grassland, sandy soil, burns; < 1000 m. CA-FP, w D; to OR, UT, AZ, Baja CA. Highly vari-

able; further study needed to determine if pls of s ChI (Santa Barbara Island), described as *P. californicus* var. *ciliatus* Dunkle, Santa Barbara Island cream cups, deserve taxonomic recognition. Mar–May

ROMNEYA MATILIJA POPPY

Subshrub or shrub, 100–250 cm; sap colorless; rhizomes creeping. **LF**: cauline, gray-green-glaucous; lobes 3–5, deep, lanceolate to ovate. **INFL**: terminal, 1-fld. **FL**: buds erect; sepals 3; petals 6, free, obovate, crinkled, white, shed after fl; stamens many, free; ovary chambers 1–12, style 0, stigma lobes 7–12, flat. **FR**: oblong to ovate, dehiscent from tip, bristly. **SEED**: many, 1.3–1.5 mm, ovate. 2 spp.: CA, Baja CA. (T. Romney Robinson, Irish astronomer, 1792–1882)

1. Sepals glabrous; peduncle glabrous; petals 60–100 mm; lf 5–20 cm . . . . ***R. coulteri***
1′ Sepals appressed-hairy; peduncle ± bristly at top; petals 40–80 mm; lf 3–10 cm . . . . ***R. trichocalyx***

R. coulteri Harv. (p. 993) COULTER'S MATILIJA POPPY **FR**: 3–4 cm. **SEED**: papillate, dark brown. Dry washes, canyons; < 1200 m. SCo, w WTR, PR. Largest fls of any pl native to CA. Mar–Jul ★

R. trichocalyx Eastw. HAIRY MATILIJA POPPY **FR**: 2.5–3.5 cm. **SEED**: smooth, ± light brown. Dry washes, canyons; < 1500 m. w SCo, WTR, PR; n Baja CA. Incl in *R. coulteri* by some. Apr–Jun

PARNASSIACEAE GRASS-OF-PARNASSUS FAMILY

Peter W. Ball

Per from caudex or rhizome [ann], gen glabrous, often with ± red marks on lvs, fls when dry. **ST**: scape with gen 1 lf-like bract [scape 0]. **LF**: simple, basal [cauline], alternate [subopposite], gen petioled, often with ± red marks when dry; veins ± palmate. **INFL**: 1-fld [cyme]. **FL**: gen bisexual, ± radial; hypanthium minute, free from ovary; calyx lobes gen 5; petals gen 5 [0], free, gen white; stamens gen 5, opposite sepals; staminodes gen 5, alternate stamens, lobes gen present, thread-like to oblong, gland-tipped; pistil 1, ovary superior, chamber ± 1, placentas 4 [3], axile below, parietal above, styles very short, stigmas [3]4. **FR**: [3]4-valved capsule. **SEED**: many, winged, netted. 2 genera, ± 70 spp.: n temp, low arctic, alpine, temp S.Am (*Lepuropetalon, Parnassia*). [Zhang & Simmons 2006 Syst Bot 31:122–137] Formerly incl in Saxifragaceae. Scientific Editor: Thomas J. Rosatti.

PARNASSIA GRASS-OF-PARNASSUS

LF: basal, blade ovate to reniform, entire, base tapered to cordate. **FL**: petals white, with yellow, green or gray-brown lines. ± 70 spp. (Mount Parnassus, Greece)

1. Petals entire
   2. Petals 1.5–2 × sepals; staminode lobes (7)9–27. . . . . ***P. palustris***
   2′ Petals 1–1.5 × sepals; staminode lobes < 7(9) . . . . ***P. parviflora***
1′ Petals fringed basally
   3. Staminode lobes oblong, tips flat, round; calyx lobes irregularly small-dentate to short-fringed . . . . ***P. fimbriata***
   3′ Staminode lobes thread-like to linear, tips spheric; calyx lobes entire to minute-dentate. . . . . ***P. cirrata***
      4. Larger petals 3.3–5.2(7) mm wide, longer fringe segments (3.3)3.5–6.5 mm . . . . var. ***cirrata***
      4′ Larger petals (4)5–10 mm wide, longer fringe segments 1–3(3.5) mm . . . . var. ***intermedia***

P. cirrata Piper **LF**: 3–18 cm; blade 1–6 cm, round-ovate, base tapered to weakly cordate. **INFL**: 17–43 cm; bract above scape middle, ovate, base cordate, clasping. **FL**: calyx lobes 3–7 mm, elliptic, entire to minute-dentate, reflexed in fr; petals 8–15 mm, ovate to ± elliptic, fringed basally; staminodes 3–6 mm, lobes < 15, ± equal, thread-like to linear, tips spheric; anthers 1.5–2.2 mm. **FR**: 5–13 mm.

var. ***cirrata*** SAN BERNARDINO GRASS-OF-PARNASSUS **LF**: larger blades 0.7–2.5(3.5) cm wide. **FL**: larger petals 3.3–5.2(7) mm wide; longer fringe filaments (3.3)3.5–6.5 mm. Wet places; 700–2500 m. SnGb, SnBr; Mex (Durango). Jul–Oct ★

var. ***intermedia*** (Rydb.) P.K. Holmgren & N.H. Holmgren CASCADE GRASS-OF-PARNASSUS **LF**: larger blades (1)1.3–5 cm wide. **FL**: larger petals (4)5–10 mm wide; longer fringe filaments 1–3(3.5) mm. Wet places; 700–2900 m. KR; to WA, ID, NV. Aug–Sep ★

P. fimbriata K.D. Koenig (p. 993) **LF**: 8–16 cm; blade 1–5 cm, round-reniform to reniform, base cordate. **INFL**: 10–40 cm; bract gen above scape middle, ovate, base cordate, clasping. **FL**: calyx lobes 3–6 mm, elliptic, irregularly small-dentate to short-fringed, gen spreading to ascending; petals 7–14 mm, obovate, fringed basally; staminodes 3–5 mm, lobes < 9, oblong, often unequal, tips flat, round; anthers 1.5–2.5 mm. **FR**: 8–12 mm. 2*n*=36. Uncommon. Wet banks,

meadows, rocky seeps; 1500–2700 m. KR, SNH, Wrn, W&I; to AK, AB, CO. Jul–Sep

P. palustris L. (p. 993) **LF**: 4–14 cm; blade 2–5 cm, lance-ovate, base tapered to cordate. **INFL**: 15–47 cm; bract gen above scape middle (gen at or below middle in SN) [rarely above middle in N.Am outside CA], elliptic. **FL**: calyx lobes 3–11 mm, elliptic, entire, spreading in fr; petals 8–20 mm, 1.5–2 × sepals, round-ovate, entire; staminodes 5–9 mm, lobes (7)9–27, thread-like, tips spheric; anthers 1.5–2.8 mm. **FR**: 8–12 mm. 2*n*=18,36[27,54]. Wet banks, meadows; < 3600 m. KR, NCoRO, NCoRH, CaRH, SNH, SnFrB, SCoR, SnBr, SNE; to AK, e Can; Eurasia. [*P. californica* (A. Gray) Greene] Jul–Oct

P. parviflora DC. (p. 993) SMALL-FLOWERED GRASS-OF-PARNASSUS **LF**: 2–7 cm; blade 1–3.5 cm, ovate to oblong, base gen truncate. **INFL**: 10–35 cm; bract at or below scape middle, ± elliptic. **FL**: calyx lobes 4–7 mm, elliptic, entire, spreading in fr; petals 3.5–10 mm, 1–1.5 × sepals, ovate-elliptic, entire; staminodes 3.5–5 mm, lobes < 7(9), tips spheric; anthers 1–1.6 mm. **FR**: 7–13 mm. Rocky seeps; 2000–2800 m. c SNH (Convict Lake, Mono Co.); WA to MI, Can. [*P. palustris* var. *p.* (DC.) B. Boivin] Probably hybridizes with *P. palustris*. Some reports of *P. parviflora* refer to pls of *P. palustris* that have small fls and the bract gen at or below middle of scape. Aug–Sep ★

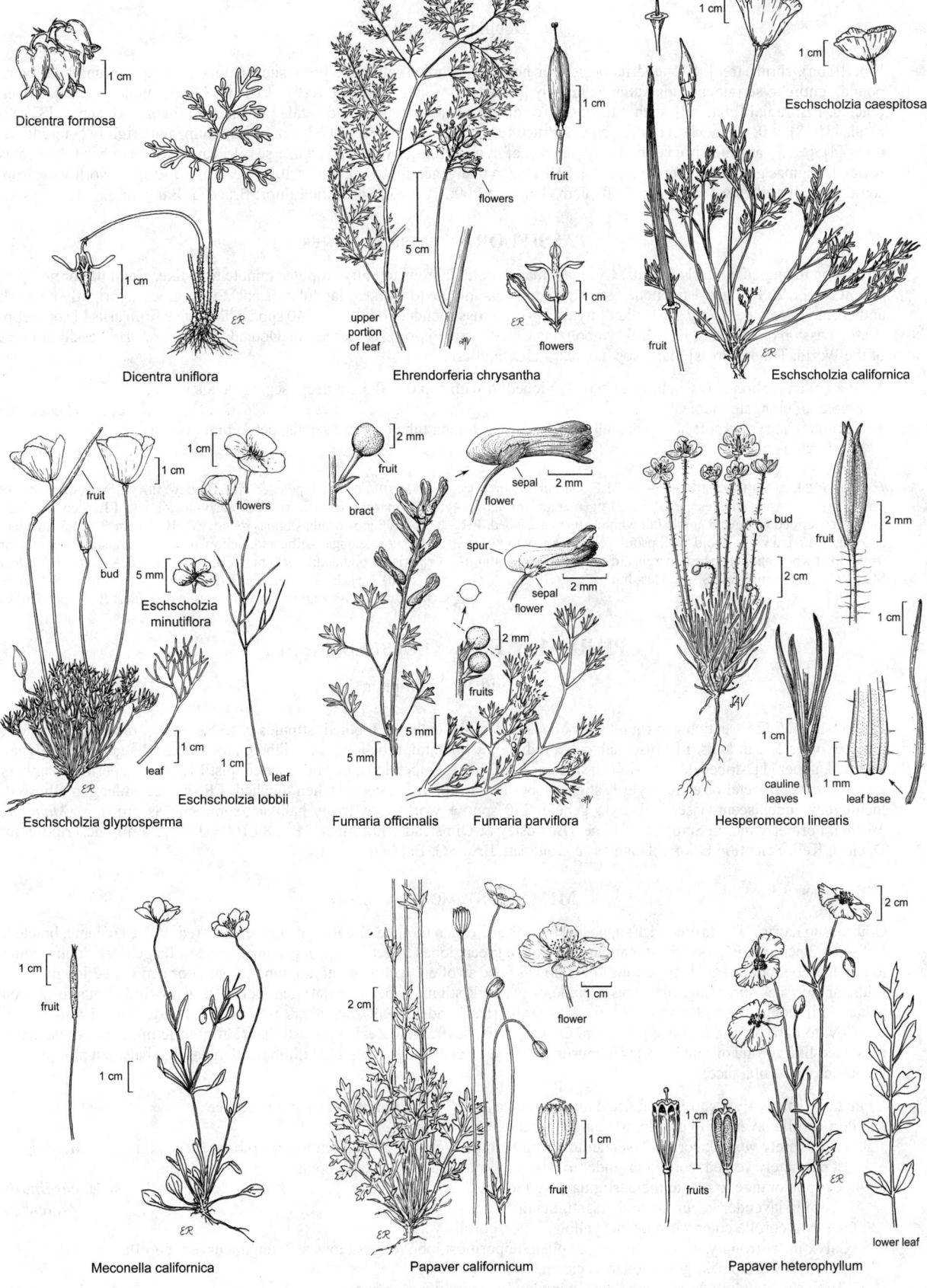

Dicentra formosa

Dicentra uniflora

Ehrendorferia chrysantha

Eschscholzia caespitosa

Eschscholzia californica

Eschscholzia glyptosperma

Eschscholzia minutiflora

Eschscholzia lobbii

Fumaria officinalis

Fumaria parviflora

Hesperomecon linearis

Meconella californica

Papaver californicum

Papaver heterophyllum

PASSIFLORACEAE   PASSION FLOWER FAMILY

Douglas H. Goldman

Vine [(ann), shrub, tree]. **ST**: tendrils present or not. **LF**: petioled, alternate [opposite], palmately lobed to unlobed [compound], entire to serrate, gen glandular, palmately [pinnately] veined; stipuled. **INFL**: axillary, 1–2 per node [cymes]; fls gen bracted. **FL**: radial [bilateral], with a tube, cup, or disk from fused sepals and petals [and stamen filaments]; sepals [3]5[8], petals [3]5[8] or 0; gen a whorl of filamentous structures or knobs ("corona") at edge of hypanthium, gen brightly colored; stamens [4]5[± 25], attached just below ovary [or edge of hypanthium]; ovary stalked to ± sessile, carpels 3, chamber 1, placentas parietal, stigmas gen 3. **FR**: berry [capsule]. **SEED**: gen many, gen flattened, with aril. 17 genera, ± 750 spp.: worldwide trop, some temp. [Feuillet & MacDougal 2007 Fam Gen Vasc Pl 9:270–281] Scientific Editor: Bruce G. Baldwin.

PASSIFLORA   PASSION FLOWER

ST: round to angled, tendrils in axils. **LF**: glandular or not, glabrous to hairy; stipules minute to lf-like, glandular or not, persistent or not. **INFL**: gen 1 per node; gen bracted, bracts minute to lf-like, glandular or not. **FL**: ± green to brightly colored; anthers easily rotated; ovary gen stalked, styles 3 [4], stigmas rounded to lobed. ± 540 spp.: edible juice from arils of some spp. (Latin: passion or suffering + fl, for fl symbolizing Christ's crucifixion) [Ulmer & MacDougal 2004 *Passiflora*: Passionflowers of the World. Timber Press] Many spp. popular in horticulture.

1. St angular, glabrous; lvs glabrous, (3)5–7(9)-lobed; fl with cup 0.4–0.5 cm deep; sepals, petals white; corona of elongate filaments . . . . *P. caerulea*
1′ St round, hairy; lvs soft-hairy adaxially, 3-lobed; fl with long tube 6–8 cm; sepals, petals pink; corona of small knobs . . . . *P. tarminiana*

P. caerulea L.   BLUE PASSION FLOWER   **LF**: margin ± entire, occ serrate at lobe bases; stipules persistent. **FL**: gen erect; corona filaments gen striped white and purple (all white); stigmas ± lobed. **FR**: 3–5 cm, 3–3.5 cm wide, ovoid to ellipsoid, yellow-orange to orange. *n*=9. Open woodland, chaparral margins, disturbed areas; < 400 m. SCo, WTR, PR; native to c S.Am. Mar–Jun

P. tarminiana Coppens & V.E. Barney   BANANA PASSION FRUIT   **LF**: margin serrate; stipules early-deciduous. **FL**: gen pendent; corona knobs white; stigmas round. **FR**: 10–14 cm, 3.5–4.5 cm wide, oblong to elongate-ellipsoid, yellow to yellow-orange. Pine, oak, or riparian woodland; < 100 m. CCo; native to n S.Am. [*P. mollissima* (Kunth) L.H. Bailey, misappl.] *P. mixta* L. f., a waif in SnFrB, is similar but with angular sts and persistent stipules. Jun–Dec

PHRYMACEAE   LOPSEED FAMILY

David M. Thompson

Ann to shrub. **ST**: ×-section gen round. **LF**: opposite, simple, entire or toothed; stipules 0. **INFL**: spike, raceme, or panicle, bracted, or fls 1–2 in axils. **FL**: bisexual; calyx radial or ± bilateral, tube long, gen ribbed, lobes 5; corolla gen bilateral, gen 2-lipped, upper [1]2-lobed, lower 3-lobed; stamens 4 in 2 pairs, epipetalous, incl or exserted; pistil 1, ovary superior, chambers 1–2, placentas parietal or axile, style 1, stigma lobes 2, flat, folding together when touched. **FR**: gen capsule, gen ellipsoid, loculicidal [indehiscent, 1-seeded]. ± 15 genera, 230 spp.: ± worldwide, many habitats; some cult as orn (e.g., *Mimulus*, *Mazus*). Formerly incl in Scrophulariaceae. [Beardsley & Olmstead 2002 Amer J Bot 89:1093–1102] Family description by David J. Keil. Scientific Editors: Douglas H. Goldman, Bruce G. Baldwin.

MIMULUS   MONKEYFLOWER

Glabrous to hairy. **ST**: gen erect. **LF**: opposite, gen ± sessile, gen toothed or entire, gen green or ± red. **INFL**: raceme, bracted, or fls gen 2 per axil. **FL**: occ cleistogamous; calyx gen green, lobes 5, gen << tube, gen unequal; corolla gen deciduous, white to red, maroon, purple, gold, or yellow, lower lip base occ swollen, ± closing mouth, tube-throat floor gen with 2 longitudinal folds; anther sacs spreading; placentas 2, axile or parietal; stigma lobes gen flat, gen incl. **FR**: gen ovoid to fusiform, gen upcurved if elongate, gen ± fragile, loculicidal near tip (hard, indehiscent), chambers 1–2. **SEED**: many, gen < 1 mm, ovoid, ± yellow to dark brown. ± 100 spp.: N.Am, Chile, e Asia, s Afr, New Zealand, Australia. (Latin: little mime or comic actor, from face-like corolla of some spp.) [Thompson 2005 Syst Bot Monogr 75:1–213] Limb width measured between most distant points across corolla face.

1. Pedicel gen > calyx; corolla gen deciduous; placenta axile, fused, not parted by fr dehiscence
   2. Per; corolla lavender to purple, or orange to red — tube-throat > 2 cm
      3. Lf pinnately veined; corolla lavender to purple, mouth ± closed by swollen lower lip base . . . . *M. ringens*
      3′ Lf palmately veined; corolla lavender to purple or orange to red, mouth open
         4. Corolla orange to red; anthers, stigma exserted . . . . *M. cardinalis*
         4′ Corolla lavender to purple; anthers, stigma incl . . . . *M. lewisii*
   2′ Gen ann; corolla color various, incl yellow; if per, corolla yellow
      5. Calyx in fr strongly, asymmetrically swollen, uppermost lobe longest, lowest 2 gen upcurved; corolla yellow, lower lip base gen swollen, ± closing mouth
         6. Bracts subtending fls a ± round disk, completely encircling st, glaucous . . . . *M. glaucescens*

6′ Bracts or lvs subtending fls not a round disk, petioled or only base encircling st, not glaucous

  7. At least some lvs ± pinnately and narrowly lobed or dissected . . . . *M. laciniatus*

  7′ Lvs not pinnately lobed or dissected, but base occ irregularly dissected or lobed

    8. Bracts or lvs subtending fls linear to lanceolate, not fused at base around st . . . . *M. nudatus*

    8′ Bracts or lvs subtending fls ovate to cordate or round, occ fused at base around st

      9. Fls gen > 5 per st, in a bracted raceme, cleistogamous or not, corolla tube-throat 2–40 mm; ann or

        per . . . . *M. guttatus*

      9′ Fls 1–5 per st, 1 in axils of upper lvs, not in bracted raceme, fls not cleistogamous, corolla

        tube-throat 17–45 mm; per . . . . *M. tilingii*

5′ Calyx in fr not or symmetrically swollen, lobes gen ± equal or lowest 2 longest, not upcurved; corolla
yellow or not, lower lip base not swollen, not closing mouth

10. Calyx lobes ± = tube; pl hairs dense, long, soft, wavy . . . . *M. pilosus*

10′ Calyx lobes << tube; pl glabrous, puberulent, or hairs ± straight

  11. Per; stolons or rhizomes present

    12. Calyx glabrous; lf palmately 3-veined . . . . *M. primuloides*

      13. Lvs linear to oblanceolate, ± erect, not in rosettes . . . . var. *linearifolius*

      13′ Lvs oblong to obovate, gen ± spreading in ± distinct rosettes . . . . var. *primuloides*

    12′ Calyx ± glabrous to hairy; lf pinnately ≥ 5-veined

      14. Corolla tube-throat 28–40 mm, tube funnel-shaped, gen > 4 mm wide . . . . *M. dentatus*

      14′ Corolla tube-throat 15–26 mm, tube ± cylindric, 2–4 mm wide exc at mouth . . . . *M. moschatus*

  11′ Ann; stolons and rhizomes 0

    15. Calyx tube ± hairy

      16. Lf palmately veined; calyx 8–10(13 in fr) mm; corolla ± pink to white . . . . *M. parishii*

      16′ Lf ± pinnately veined; calyx 3–8 mm; corolla yellow

        17. Corolla lobe base without large blotches; calyx lobes ± acute; pedicel in fr not growing away

          from light . . . . *M. floribundus*

        17′ Corolla lobe base each with 1 large, red-purple dot; calyx lobes rounded; pedicel in fr growing

          away from light . . . . *M. norrisii*

    15′ Calyx tube glabrous to puberulent

      18. Calyx very swollen in fr, membranous; corolla lavender, rose to ± white, or yellow

        19. Upper lvs sessile; corolla lavender, pink to ± white, or yellow

          20. Basal lvs not in rosette; corolla yellow . . . . *M. evanescens*

          20′ Basal lvs in rosette; corolla pink to lavender or white (± yellow)

            21. Corolla limb < 14 mm wide, tube-throat 7–16 mm; drained upland areas . . . . *M. inconspicuus*

            21′ Corolla limb 4–6 mm wide, tube-throat 7–11 mm; vernally wet depressions . . . . *M. latidens*

        19′ Upper lvs petioled; corolla gen yellow

          22. Corolla tube-throat 3–6.5 mm . . . . *M. breviflorus*

          22′ Corolla tube-throat 7–13 mm . . . . *M. pulsiferae*

      18′ Calyx ± enlarged in fr, ± corky or membranous; corolla maroon, purple, pink, lavender, yellow,
or partly white

        23. All corolla lobes yellow or white

          24. Lower calyx lobes > upper . . . . *M. alsinoides*

          24′ Lower calyx lobes ± = upper

            25. Corolla tube-throat 4–10 mm

              26. Corolla tube-throat 6–10 mm; calyx lobes ciliate . . . . [2]*M. rubellus*

              26′ Corolla tube-throat 4–6 mm; calyx lobes glabrous . . . . *M. suksdorfii*

            25′ Corolla tube-throat 10–19 mm

              27. White corolla lobes 0–4; calyx ± corky in fr; pedicel thickened, ± rigid in fr . . . . *M. bicolor*

              27′ White corolla lobes 0; calyx membranous in fr; pedicel thin, flexible in fr . . . . [2]*M. montioides*

        23′ Some or all corolla lobes maroon, purple, pink, or lavender

          28. Corolla tube-throat 2–10 mm

            29. Calyx 2–3.5 mm; SnBr . . . . *M. exiguus*

            29′ Calyx 3.5–9 mm; widespread

              30. Corolla persistent; pedicel 2–15 mm . . . . *M. breweri*

              30′ Corolla deciduous; pedicel 5–27 mm

                31. Calyx lobes ± glabrous . . . . *M. androsaceus*

                31′ Calyx lobes ciliate

                  32. Corolla clearly 2-lipped, upper 2 lobes << lower 3; limb 5–9 mm wide . . . . *M. gracilipes*

                  32′ Corolla not clearly 2-lipped, upper 2 lobes ± = lower 3; limb 3–4 mm wide . . . . [2]*M. rubellus*

         28′ Corolla tube-throat 8–22 mm

           33. Upper corolla lip 4-lobed, lower large, 2-lobed, yellow . . . . *M. shevockii*

          33′ Corolla not lipped, or upper corolla lip 2-lobed, lower of 3 ± equal, partly yellow or all rose
to purple lobes

            34. Calyx gen with red-brown spots, widely corky-ribbed in fr, lobes acute to long-tapered . . . . . . *M. filicaulis*

  34′ Calyx gen without spots, membranous in fr, lobes gen obtuse to rounded or truncate

   35. All lvs linear, gen 1–3 mm wide; pedicel 3–25 mm . . . . [2]***M. montioides***

   35′ At least lower lvs ovate, > 3 mm wide; pedicel 10–65 mm

    36. Corolla not lipped, tube-throat 10–22 mm, lobes equally colored; upper lvs ± linear. . . . . ***M. palmeri***

    36′ Corolla strongly 2-lipped, tube-throat 8–13 mm, upper lip darker; upper lvs elliptic to ovate

    . . . . ***M. purpureus***

1′ Pedicel gen < calyx; corolla gen persistent; placenta parietal, at least tip parted by fr dehiscence (or fr hard, indehiscent)

 37. Rhizomatous per to shrub; calyx 1.7–4.5 cm; lf edges flat to ± rolled under; main lf axils gen with clusters of smaller lvs

  38. Rhizomatous per, 1 infl per st; lf adaxially hairy; calyx base gen swollen; corolla bright yellow; fr splitting 4-parted at tip . . . . ***M. clevelandii***

  38′ Subshrub or shrub, > 1 infl per st; lf adaxially glabrous; calyx base not swollen; corolla pale yellow to ± yellow or yellow-brown, yellow-orange, orange, or red; fr splitting only along upper suture. . . . . . ***M. aurantiacus***

   39. Lower lf surface puberulent to densely hairy, gen paler than upper

    40. Calyx tube glabrous . . . . var. ***aurantiacus***

    40′ Calyx tube puberulent to hairy . . . . var. ***pubescens***

   39′ Lower lf surface glabrous, not paler than upper

    41. Corolla tube-throat 26–37 mm; corolla brightly dark yellow, orange, or red

     42. Lvs lance-elliptic to ovate or obovate, gen flat or ± rolled under at edges . . . . var. ***parviflorus***

     42′ Lvs linear, edges gen tightly rolled under . . . . var. ***puniceus***

    41′ Corolla tube-throat 37–57 mm; corolla pale yellow or pale orange

     43. Corolla lobes entire; calyx tube flared distally . . . . var. ***aridus***

     43′ Corolla lobes 2-lobed; calyx tube appressed to corolla tube distally . . . . var. ***grandiflorus***

37′ Ann (per); calyx gen < 2 cm; lf edges ± flat; main lf axils not with clusters of smaller lvs

 44. Corolla radial, ± salverform, purple-brown and ± white, throat floor without longitudinal folds

  45. Corolla limb ± purple-brown at base, veins << 0.5 mm wide, unpatterned, purple-brown, fading into white border . . . . ***M. mohavensis***

  45′ Corolla limb ± all white, veins > 0.5 mm wide, bold, intricately patterned, purple-brown . . . . ***M. pictus***

 44′ Corolla bilateral, gen not purple-brown and white, gen not salverform, throat floor gen with 2 longitudinal folds — fls occ cleistogamous in some spp.

  46. Fr hard, gen oblique at base, tardily dehiscent; fl gen Mar–Apr

   47. St < 1 cm; lf gen ± linear, glabrous to puberulent, at least basal 1/2 ± ciliate; vernally wet depressions, seepage areas

    48. Corolla tube-throat 5–10 mm. . . . . ***M. pygmaeus***

    48′ Corolla tube-throat 15–60 mm

     49. Corolla tube-throat glabrous outside; lowest corolla lobe magenta, base with a large, dark purple spot . . . . ***M. angustatus***

     49′ Corolla tube-throat puberulent outside; lowest corolla lobe gold, base with scattered, small, dark dots . . . . ***M. pulchellus***

   47′ St gen > 1 cm; lf gen not linear, gen not ciliate, or both ciliate and hairy; vernally wet depressions or not

    50. Corolla lobes with large dark spot at center of base; stigma lobes ± equal; fls not cleistogamous

     51. Corolla pink to white; fls 2 per node; limestone crevices . . . . ***M. rupicola***

     51′ Corolla magenta to purple; fls 1 per node; vernally wet depressions . . . . ***M. tricolor***

    50′ Corolla lobes not all with large dark spots; lower stigma lobe >> upper; fls occ cleistogamous

     52. Calyx 18–21 mm. . . . . ***M. traskiae***

     52′ Calyx 5–17 mm

      53. Lvs gen ciliate at base; corolla limb 3–11 mm wide; if fls cleistogamous then lvs ± purple abaxially

       54. Corolla limb widely spreading; tube-throat not marked laterally with yellow, throat and lower lip gen without folds and rarely yellow . . . . ***M. congdonii***

       54′ Corolla limb slightly spreading; tube-throat yellow near stamen attachment, floor and lower lip base with 2 prominent yellow folds . . . . ***M. latifolius***

      53′ Lvs gen hairy, not ciliate; corolla limb 10–18 mm wide; if fls cleistogamous then lvs green abaxially

       55. Corolla lower lip ± absent; fr 2.5–6.5 mm; asymmetric-ovoid . . . . ***M. douglasii***

       55′ Both corolla lips well developed; fr 6–12 mm, nearly cylindrical. . . . . ***M. kelloggii***

  46′ Fr gen fragile, symmetric at base, gen rapidly dehiscent; fl gen Apr–Sep

   56. All fl nodes with 1 fl or fr, calyx gen inflated or swollen

    57. Longitudinal folds of corolla throat floor densely hairy, white or yellow proximally, white distally; anthers ciliate . . . . ***M. viscidus***

     58. Corolla lobes dark at base, without red-purple midveins; throat ceiling hairy; limb face glabrous; lower stigma lobe 3–4 × upper. . . . . var. ***compactus***

     58′ Corolla lobes with dark red-purple midveins; throat ceiling glabrous, limb face gen hairy; lower stigma lobe 1.5 × upper . . . . var. ***viscidus***

    57′ Longitudinal folds of corolla throat floor glabrous or minutely puberulent, all yellow; anthers glabrous

     59. Corolla tube-throat 7–10 mm, limb 4–7 mm wide; lower stigma lobe 5–7 × upper . . . . ***M. rattanii***

     59′ Corolla tube-throat 9–23 mm, limb 8–26 mm wide; stigma lobes ± equal . . : . . . . ***M. fremontii***

 60. Corolla magenta to dark red-purple . . . . var. *fremontii*
 60′ Corolla yellow. . . . . var. *vandenbergensis*
56′ Most fl nodes with 2 fls or frs; calyx occ inflated
 61. Stigma exserted or at end of corolla tube-throat
  62. Corolla lobes evenly spread; anthers glabrous
   63. Corolla throat floor white; lower stigma lobe 2–3 × upper; calyx in fr prominently ribbed and
    inflated, hairy, mouth slightly oblique . . . . **M. bolanderi**
   63′ Corolla throat floor yellow; stigma lobes equal; calyx in fr not ribbed or inflated, minutely
    puberulent, mouth strongly oblique . . . . **M. johnstonii**
  62′ Upper 2 corolla lobes divergent from lower 3; anthers ciliate
   64. Tip of distal lvs sharply acuminate or sharp-tipped. . . . . **M. cusickii**
   64′ Tip of distal lvs acute to rounded . . . . **M. nanus**
    65. Lvs and calyces hairy, hairs ≥ 0.7 mm; corolla tube gen puberulent externally, corolla magenta,
     dark red-purple, or yellow. . . . . var. *mephiticus*
    65′ Lvs and calyces minutely puberulent, hairs ≤ 0.3 mm; corolla tube glabrous externally, corolla
     magenta or lavender-purple
     66. Corolla tube-throat 5.5–12 mm; fr gen 3.5–6 mm . . . . var. *jepsonii*
     66′ Corolla tube-throat 11–19 mm; fr gen 6–12 mm . . . . var. *nanus*
 61′ Stigma incl
  67. Corolla tube-throat 6–8 mm, limb 3–5 mm wide . . . . **M. leptaleus**
  67′ Corolla tube-throat 8–34 mm, limb 8–30 mm wide
   68. Lower stigma lobe 1.5–2 × upper
    69. Corolla yellow; tube-throat glabrous externally. . . . . **M. brevipes**
    69′ Corolla rose-pink to magenta; tube-throat puberulent externally
     70. Corolla throat near mouth gen white with red-purple dots (throat all yellow), lower lip bent
      downward; calyx with wide, dark ribs, occ inflated, lobes acute to acuminate . . . . **M. layneae**
     70′ Lower lip of corolla with 2 distinct conspicuous yellow stripes extending onto base from
      throat folds, lower lip projecting forward; calyx all pale green, ribs of veins only, not inflated,
      lobes obtuse. . . . . **M. torreyi**
   68′ Stigma lobes equal
    71. Corolla throat floor white at mouth; if corolla mostly yellow, then 3–7 longitudinal maroon-black
     stripes extend from throat onto lobe base; w slope of s SN, Teh, n WTR
     72. Calyx inflated, hairy, many hairs > 1 mm; corolla ± pink to red-purple; largest lvs
      round-toothed distally. . . . . **M. constrictus**
     72′ Calyx not inflated, puberulent, hairs < 0.5 mm; corolla magenta or yellow; lvs all entire . . . . **M. whitneyi**
    71′ Corolla throat floor yellow at mouth; if corolla yellow overall, then lower lip with 6–8 ± red
     spots in arc around mouth; high desert
     73. Calyx puberulent, hairs < 0.5 mm; upper calyx lobe largest, widely rounded to acute; corolla
      magenta or yellow . . . . **M. parryi**
     73′ Calyx hairy, many hairs > 1 mm; calyx lobes ± equal, acuminate to attenuate; corolla
      magenta . . . . **M. bigelovii**
      74. Distal lf width gen < basal lvs, distal longer tapered than basal; lower st internodes
       gen >> others . . . . var. *bigelovii*
      74′ Distal lf width gen ≥ basal lvs, distal gen acute; st internodes gen ± equal. . . . . var. *cuspidatus*

M. alsinoides Benth. (p. 993) Ann 0.5–15 cm, puberulent. **LF:** petiole 0.5–20 mm; blade 3–18 mm, ovate to round. **FL:** pedicel 7–27 mm; calyx 4–8 mm, minutely puberulent, upper 3 lobes ± 0.5 mm, acute to acuminate, lower 2 lobes ± 1 mm, widely rounded; corolla yellow, lower lip base gen with large, ± red spot, tube-throat 7–11 mm; placentas axile. **FR:** 5–7 mm. Moist to wet places, mossy rock crevices; < 860 m. KR; to BC. Mar–Jun

M. androsaceus Greene (p. 993) Ann 0.5–9 cm, gen minutely puberulent. **LF:** 3–13 mm, ± lanceolate to oblong or ovate. **FL:** pedicel 7–27 mm, ± spreading, ascending in fr; calyx 3.5–6.5 mm, tube gen minutely puberulent, lobes equal, 0.2–1 mm, ± glabrous; corolla red-purple, tube-throat 5–8 mm, limb 4–6 mm wide; placentas axile. **FR:** 4–5 mm. Uncommon. Moist runoff areas on gentle slopes; < 2100 m. NCoRI, Teh, CW, WTR, SnBr, e PR (Santa Rosa Mtns, Riverside Co.), w edge DMoj. Mar–Jun

M. angustatus (A. Gray) A. Gray (p. 993) Ann, compact, glabrous exc calyx. **ST:** < 1 cm. **LF:** 5–36 mm, ± linear, at least basal 1/2 ciliate. **FL:** pedicel 0–1 mm; calyx 6–14 mm, hairy, lobes unequal, 1–4 mm, obtuse, gen ciliate; corolla persistent, magenta-purple, lowest lobe base with large, dark purple spot, tube-throat 20–60 mm, glabrous outside; placentas parietal. **FR:** 2–4 mm, ± oblique-ovoid,

hard, indehiscent. 2n=18. Vernally wet depressions; 250–1200 m. c NCoRO (Longvale, Mendocino Co.), s NCoRI, s CaRF, n SNF, s SNF (Pinehurst, Fresno Co.). Mar–Jun

M. aurantiacus Curtis (p. 993) Subshrub, shrub, glabrous to hairy. **ST:** 10–150 cm; main lf axils gen with small lf clusters. **LF:** 10–88 mm; upper surface glabrous, lower gen sticky. **INFL:** pedicel 3–30 mm. **FL:** calyx 17–45 mm, not swollen at base, glabrous to hairy, lobes unequal, 2–11 mm, acute to acuminate; corolla persistent, ± yellow, orange, or red, tube-throat 25–57 mm; placentas parietal. **FR:** 13–25 mm, splitting at upper suture. 2n=20. Complex; vars. intergrading, hybridizing. *M. aurantiacus* subsp. *australis* Munz, *M. aurantiacus* subsp. *lompocensis* (McMinn) Munz are intervarietal hybrids.

  var. **aridus** (Abrams) D.M. Thomps. **LF:** ovate to lanceolate or oblanceolate, dentate to serrate, edges gen rolled under, uniformly green, glabrous. **INFL:** fl 2 per node; pedicel 3–10 mm. **FL:** calyx 23–40 mm, tube glabrous, flared distally; corolla pale yellow, tube-throat 37–52 mm, lobes entire; anthers incl or ± exserted. Crevices in granite boulders or outcrops; 660–1180 m. se PR; c Baja CA. [*M. aridus* (Abrams) A.L. Grant] Mar–Jun ★

  var. **aurantiacus** **LF:** narrowly elliptic to linear, entire to serrate, edges gen rolled under; lower surface pale and puberulent to

densely hairy, at least some hairs branched to stellate. **INFL:** fl 2–4 per node; pedicel 5–27 mm. **FL:** calyx 18–36 mm, tube glabrous, distally appressed to corolla tube; corolla ± yellow-orange to orange (cream), tube-throat 26–38 mm, lobes entire to 2-lobed distally; anthers incl. Disturbed areas, coastal cliffs, canyon sides; < 800 m. NW, CW, n&c SNF; Baja CA (Cedros Island). Mar–Jun

var. ***grandiflorus*** (Lindl. & Paxton) D.M. Thomps. **LF:** narrowly elliptic to linear, ± entire, edges ± flat, rolled under in drought, uniformly green, glabrous. **INFL:** fl 2 per node; pedicel 4–25 mm. **FL:** calyx 23–33 mm, tube glabrous, distally appressed to corolla tube; corolla pale yellow or orange, tube-throat 38–57 mm, lobes each 2-lobed; anthers incl. Steep banks of canyons, rocky hillsides; 90–1530(1830) m. n SN; n SCoR. Mar–Jun

var. ***parviflorus*** (Greene) D.M. Thomps. **LF:** lance-elliptic to ovate or obovate, entire to crenate or dentate, edges gen flat or ± rolled under, more so in drought, uniformly green, glabrous. **INFL:** fl gen 2(4) per node; pedicel 8–24 mm. **FL:** calyx 17–26 mm, tube glabrous, appressed to corolla tube distally; corolla dark yellow-orange to orange or red, tube-throat 32–37 mm, lobes entire to 2-lobed distally; anthers incl or ± exserted. Rocky hillsides, canyon walls, cliffs; < 600 m. ChI (Anacapa, San Clemente, Santa Cruz, Santa Rosa islands). [*M. flemingii* Munz] Mar–Jun ★

var. ***pubescens*** (Torr.) D.M. Thomps. **LF:** narrowly elliptic, serrate, edges gen ± flat, rolled under in drought; lower surface hairy. **INFL:** fl 2 per node; pedicel 3–12 mm. **FL:** calyx 25–45 mm, tube puberulent to hairy, distally appressed to corolla tube; corolla pale yellow, tube-throat 37–52 mm, lobes entire to 2-lobed; anthers incl. Well-drained, exposed sites; crevices in boulders or rock outcrops in desert areas; 7–2440 m. s SN, s CW, SW (incl Santa Cruz, Santa Rosa islands), s DMtns; n Baja CA. Mar–Jun

var. ***puniceus*** (Nutt.) D.M. Thomps. **LF:** gen linear, ± entire; edges gen tightly rolled under, uniformly green, glabrous. **INFL:** fl 2–4 per node; pedicel 9–22 mm. **FL:** calyx 19–27 mm, tube glabrous, distally appressed to corolla tube; corolla dark yellow-orange to orange or red, tube-throat 27–35 mm, lobes entire to 2-lobed; anthers incl or ± exserted. Scrub communities, disturbed areas; < 1220 m. SW; n Baja CA. Mar–Jun

M. bicolor Benth. (p. 993) Ann 4–27 cm, densely puberulent. **LF:** 4–30 mm, linear to oblanceolate. **INFL:** pedicel 6–30 mm. **FL:** calyx 5.5–12 mm, ± corky in fr, gen red-dotted, puberulent, lobes equal, 1–3.5 mm, acute; corolla tube-throat 10–19 mm, lobes each 2-lobed, upper lip white or yellow, lower lip yellow; placentas axile. **FR:** 4–8 mm. Moist places, gen on clay soils along drainages; 360–2100 m. se KR, s CaR, SN. Apr–Jun

M. bigelovii (A. Gray) A. Gray   Ann 2–25 cm, densely hairy. **LF:** 7–35 mm, elliptic to round, gen abruptly acuminate. **FL:** pedicel 1–4 mm; calyx 6–13 mm, occ ± purple on ribs, hairy, lobes spreading, ± equal, 1.5–6 mm, ± acuminate; corolla persistent, magenta, tube-throat 12–22 mm, throat floor yellow, mouth gen with 2 lateral, dark maroon patches; placentas parietal. **FR:** 7–13 mm, exserted. 2*n*=16. Variable; vars. intergrade.

var. ***bigelovii*** (p. 993) **ST:** lower internodes gen >> upper. **LF:** lower elliptic, occ acuminate but not abruptly, upper narrower, longer-tapered than lower. Rocky desert slopes, margins of washes; 76–1980 m. D; s NV, w AZ. Feb–Jun

var. ***cuspidatus*** A.L. Grant   **ST:** lower internodes ± = upper. **LF:** lower gen widely ovate to ± round, abruptly sharp-pointed, distal lvs as wide or wider, gen acute. Rocky desert slopes, margins of washes; 15–3350 m. s SNE, n D; to sw UT, s NV, nw AZ. Feb–Jun

M. bolanderi A. Gray (p. 993) Ann 2–90 cm, gen densely hairy. **LF:** 5–60 mm, oblanceolate to ovate, lowest lvs glabrous, upper densely hairy. **FL:** pedicel 2–5 mm; calyx 7–27 mm, widely ribbed, swollen esp in fr, densely hairy, lobes gen ± spreading in fr, unequal, 1–9 mm, acute to acuminate; corolla persistent, magenta, throat floor white, tube-throat 10–30 mm, throat floor with small magenta hairs; placentas parietal. **FR:** 8–20 mm. 2*n*=16. Granite outcrops, burns, openings in chaparral, disturbed areas; gen 300–1700 m. KR, NCoR, SN (common c SNF), CW, WTR. Very small pls much like *M. rattanii*. Apr–Jul

M. breviflorus Piper (p. 993) Ann 2–17 cm, puberulent. **LF:** petioled, 3–28 mm, elliptic. **FL:** pedicel 3–16 mm, ascending in fr; calyx 2–6 mm in fl, 3.5–9.5 mm and swollen in fr, gen glabrous, lobes equal, < 1 mm, acute, gen ciliate; corolla yellow, tube-throat 3–6.5 mm; placentas axile. **FR:** 3–6.5 mm. Uncommon. Moist places; 1500–2200 m. n SNH, MP; to WA, ID, NV. May–Jun

M. brevipes Benth. (p. 993) Ann 5–80 cm, gen densely hairy. **LF:** 7–90 mm, lanceolate to obovate; lowest glabrous. **FL:** pedicel 2–10 mm; calyx 10–25 mm, hairy, lobes acuminate, strongly unequal, upper lobes 4–13 mm, lower 1–5 mm; corolla yellow, persistent, tube-throat 15–30 mm; placentas parietal. **FR:** 8–14 mm. 2*n*=16. Bare, disturbed or burned areas on slopes in chaparral; 30–2200 m. s SCoRO, SW (incl Santa Catalina Island); Baja CA. Apr–Jul

M. breweri (Greene) Coville (p. 993) Ann 3–21 cm, ± hairy. **LF:** 3–35 mm, ± linear. **FL:** pedicel 2–15 mm, ascending in fr; calyx 3.5–9 mm, ridges puberulent, lobes equal, 1–2 mm, puberulent; corolla pink to lavender, persistent, tube-throat 4.5–8 mm; placentas axile. **FR:** 3.5–6.5 mm. Common. Moist places along streams, seepage areas; gen 1200–3400 m. NW, CaR, SNH, SCoRO, TR, SnJt, MP; to BC, MT, WY, UT. Jun–Aug

M. cardinalis Benth. (p. 997) Per, rhizomed, hairy. **ST:** 25–80 cm, gen decumbent to ascending. **LF:** 20–80 mm, oblong to obovate, palmately 3–5-veined, upper clasping st. **FL:** pedicel 50–80 mm; calyx 20–30 mm, hairy, lobes equal, 4–5 mm, acute to acuminate; corolla orange to red, tube-throat 40–50 mm, upper lip arched forward, lower lip reflexed; anthers, stigma exserted; placentas axile. **FR:** 16–18 mm. 2*n*=16. Moist to wet places along streams, seepage areas; < 2400 m. CA-FP, W&I, DMtns (Panamint Range); to OR, NV, UT, NM; Baja CA. May–Sep

M. clevelandii Brandegee (p. 997) CLEVELAND'S BUSH MONKEYFLOWER Per, rhizomed. **ST:** 30–95 cm, hairy; main lf axils gen with clusters of smaller lvs. **LF:** 20–112 mm, lanceolate, edges gen rolled under; hairy. **INFL:** 1 per st. **FL:** pedicel 2–7 mm; calyx 20–35 mm, gen swollen at base, narrowed above ovary, hairy, lobes ± equal, 5–11 mm, acute; corolla yellow, persistent, tube-throat 27–40 mm; placentas parietal. **FR:** 8–15 mm, splitting into 4 parts at tip. 2*n*=20. Disturbed areas, open borders of woodland, chaparral; 915–1465 m. PR; Baja CA. Rarely hybridizes with *M. aurantiacus*. Apr–Jun ★

M. congdonii B.L. Rob. (p. 997) Ann < 12 cm, puberulent to hairy. **LF:** blade 8–32 mm, oblanceolate to elliptic, upper surface ± green, lower ± purple; base, petiole ± ciliate. **FL:** cleistogamous or open; pedicel 2–5 mm; calyx 5–14 mm, ribs hairy, lobes unequal, 0.2–1 mm, obtuse; corolla magenta, persistent, tube-throat 8–30 mm; placentas parietal. **FR:** 4–8.5 mm, angled, laterally compressed, oblique, hard, indehiscent. 2*n*=18. Disturbed areas, sloped runoff areas, gen granitic or serpentine soils; 120–1100(1700) m. NCoR, SNF, SnFrB, n SCoRO (Monterey Co.), WTR (Ventura Co.), PR (San Diego Co.). Mar–May

M. constrictus (A.L. Grant) Pennell (p. 997) Ann 2–24 cm, hairy. **LF:** 5–32 mm, narrowly elliptic to obovate. **FL:** pedicel 0.5–3 mm; calyx 7–12 mm, ribbed, swollen esp in fr, coarsely hairy, ± oblique at mouth, lobes erect, ± equal, 1–4 mm, acuminate; corolla ± pink to red-purple, persistent, tube-throat 13–22 mm, throat ± white with radiating, dark maroon marks, throat floor and limb long-hairy; placentas parietal. **FR:** 8–12 mm. 2*n*=16. Slopes, road banks near runoff areas; 790–2090(2380) m. s SN, Teh, n WTR. Intergrades with *M. whitneyi* in s SNH, *M. johnstonii* in s WTR and s SCoRO. May–Aug

M. cusickii (Greene) Rattan (p. 997) CUSICK'S MONKEYFLOWER Ann 3–24 cm, densely puberulent to hairy. **LF:** 15–45 mm, ovate to obovate, at least upper lvs sharply acuminate or sharp-tipped. **FL:** pedicel 1–3 mm; calyx 10–17 mm, densely puberulent, lobes ± unequal, 1.5–5 mm, acuminate; corolla magenta, throat floor folds yellow at mouth, persistent, tube-throat 20–28 mm; stigma and gen lower anthers exserted; placentas parietal. **FR:** 10–17 mm. 2*n*=16. Steep canyon slopes, scree; gen 600–1600 m. s Wrn; to WA, ID. May–Aug ★

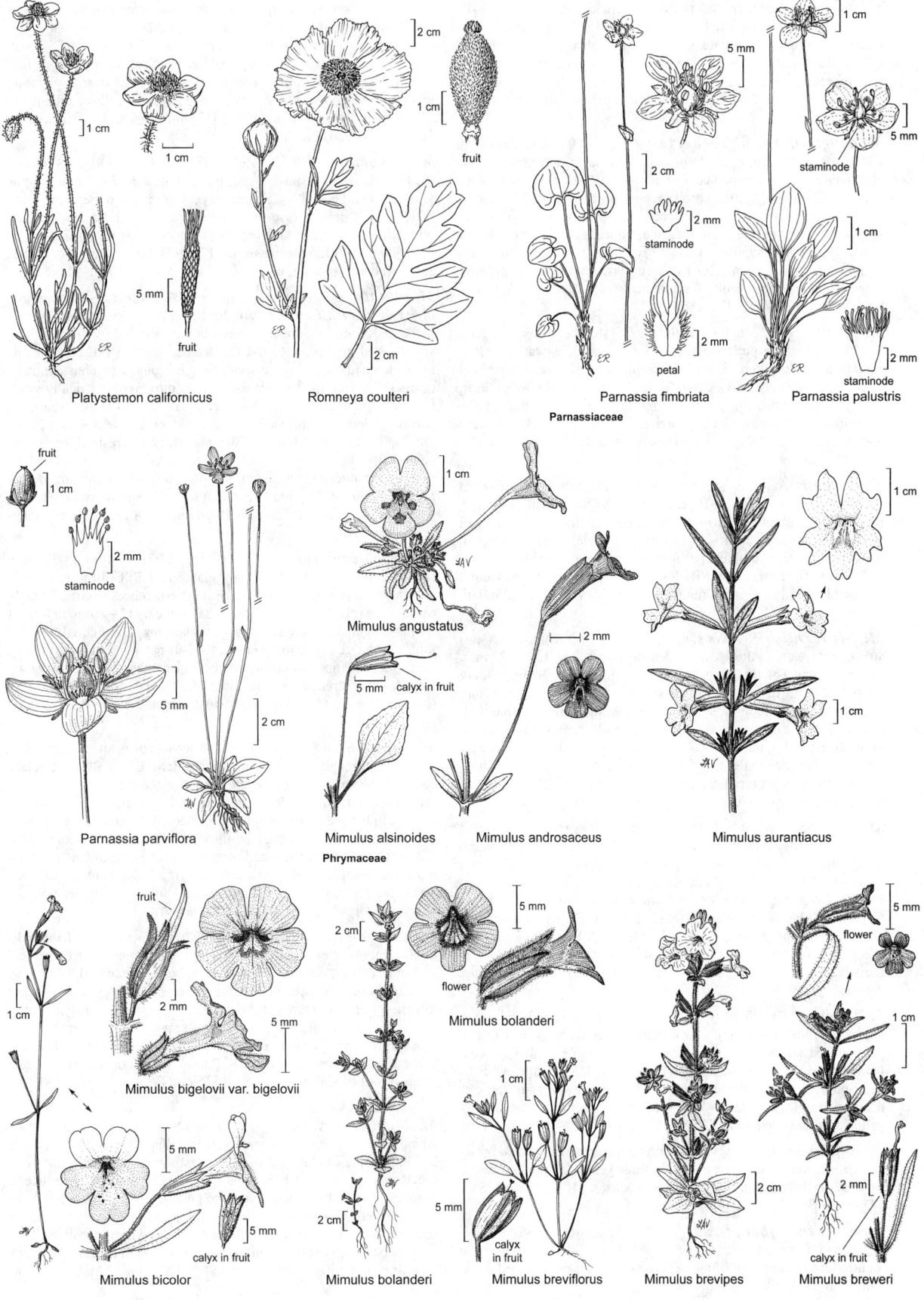

Platystemon californicus

Romneya coulteri

Parnassia fimbriata

Parnassia palustris

staminode

staminode

2 mm
staminode

2 mm
petal

2 mm
staminode

Parnassiaceae

fruit

staminode

Parnassia parviflora

Mimulus angustatus

calyx in fruit

Mimulus alsinoides

Mimulus androsaceus

Mimulus aurantiacus

Phrymaceae

fruit

Mimulus bigelovii var. bigelovii

Mimulus bicolor

calyx in fruit

flower

Mimulus bolanderi

Mimulus bolanderi

Mimulus breviflorus

calyx
in fruit

Mimulus brevipes

flower

calyx in fruit

Mimulus breweri

M. dentatus Benth. (p. 997)   Per, rhizomed, ± hairy. **ST:** 15–40 cm, ascending. **LF:** petiole 0–12 mm; blade 15–70 mm, ovate, pinnately ± 5-veined. **FL:** pedicel 20–50 mm; calyx 9–18 mm, ± glabrous, ribs ± hairy, lobes 2–7 mm, ± equal, ± ciliate; corolla yellow, tube-throat 28–40 mm, tube funnel-shaped, gen > 4 mm wide; placentas axile. **FR:** 8–9 mm. Coastal streambanks, gen in partial shade; < 400 m. n NCo; to WA. May–Aug

M. douglasii (Benth.) A. Gray (p. 997)   Ann 0.3–4 cm, ± hairy. **LF:** blade 5–28 mm, ovate to obovate, adaxially shiny green. **FL:** cleistogamous or open; pedicel 2–4 mm; calyx 8–14 mm, hairy, lobes unequal, 0.5–2 mm, obtuse; corolla persistent, tube-throat 20–41 mm, tube gradually to ± abruptly widened, throat long, boldly mottled, striped gold and purple, limb magenta, upper lobes 4–5 mm, lower lobes < 1 mm; placentas parietal. **FR:** 2.5–6.5 mm, asymmetric-ovoid, hard, indehiscent. 2*n*=18. Bare clay, serpentine or granitic soils; gen along upper banks of small creeks; 45–1200 m. NW, CaRF, SNF, CW, Wrn; sw OR. Feb–Apr

M. evanescens Meinke   EPHEMERAL MONKEYFLOWER   Ann 6–25 cm, glandular-puberulent. **LF:** 1–4 cm, broadly ovate to ± lanceolate, lowest not in basal rosette, sessile exc at pl base. **FL:** pedicel 8–18 mm, gen < lvs, calyx 3.5–6.5 mm, 7–11 mm and swollen in fr, tube glabrous, lobes equal, 0.8–1.6 mm in fr, ciliate; corolla yellow, tube-throat 4–9.5 mm; placentas axile. **FR:** 5–9 mm. Among rocks and boulders on moist gravel, previously flooded; 1200–1700 m. MP; to sw ID. Jul   ★

M. exiguus A. Gray (p. 997)   SAN BERNARDINO MOUNTAINS MONKEYFLOWER   Ann 2–10 cm, minutely puberulent. **LF:** 2–8 mm, narrowly elliptic to narrowly ovate, not sheathing st. **FL:** pedicel 6–20 mm, ascending in fr; calyx 2–3.5 mm, ± glabrous, lobes equal, 0.5–1 mm, acute, ± minutely puberulent; corolla lavender, tube-throat 2–3.5 mm; placentas axile. **FR:** 3–4 mm, ± spheric, > calyx. Gentle slopes, along small streams, runoff areas in clay soils; 1800–2300 m. SnBr; n Baja CA. Jun–Jul   ★

M. filicaulis S. Watson (p. 997)   SLENDER-STEMMED MONKEYFLOWER   Ann 4–30 cm, densely puberulent. **LF:** 8–23 mm, linear to oblanceolate. **FL:** pedicel 9–25 mm; calyx 5–9 mm, widely corky-ribbed in fr, gen red-brown-spotted, puberulent, lobes equal, 1–2 mm, acute to long-tapered; corolla not lipped, rose-purple with dark, red-purple throat marks ± extending onto upper 4 lobes, yellow on folds extending onto lower lobe, tube-throat 13–17 mm, lobes all notched; placentas axile. **FR:** 5.5–6.5 mm. 2*n*=16. Disturbed, moist loamy soil, gen in partial shade; 1200–1750 m. c SNH (Tuolumne, Mariposa cos.). May–Jul   ★

M. floribundus Lindl. (p. 997)   Ann, hairy, gen ± slimy. **ST:** 3–50 cm, gen decumbent or ± climbing. **LF:** petiole 0–20 mm, blade 5–45 mm, lanceolate to ovate, base gen rounded to cordate, veins ± pinnate. **FL:** pedicel 5–30 mm, not reflexed in fr; calyx 3–8 mm, ± hairy, lobes equal, 1–2 mm, ± acute; corolla yellow, tube-throat 6–15 mm; placentas axile. **FR:** 4–7 mm. 2*n*=32. Crevices, granite outcrop seeps, near streams; < 2500 m. CA-FP (esp c&s SNF), W&I, DMtns (Panamint Range); to BC, SD, n Mex. Many minor, ± indistinct forms. If corolla tube-throat > 15 mm, see *M. moschatus.* Apr–Jul

M. fremontii (Benth.) A. Gray   Ann 1–20 cm. **LF:** 2–30 mm, gen narrowly elliptic (obovate); lowest glabrous, upper hairy. **INFL:** fl 1 per node. **FL:** pedicel 1–4 mm; calyx 5–14 mm, widely ribbed, swollen, gen ± red, gen ± white-hairy, lobes ± equal, 0.5–3.5 mm, acute to short-pointed; corolla magenta to red-purple (yellow), persistent, gen darker at mouth, tube-throat 9–23 mm, throat floor ± glabrous, folds yellow; placentas parietal. **FR:** 6.5–13 mm. *n*=8.

var. ***fremontii*** (p. 997)   **FL:** corolla magenta to dark red-purple, gen darker in throat near mouth, folds yellow at mouth. Sandy, shrubby, disturbed areas, gen on streambanks; < 2100 m. SCoR, SW, s DMoj; n Baja CA. [*M. subsecundus* A. Gray] Mar–Jun

var. ***vandenbergensis*** D.M. Thomps.   VANDENBERG MONKEYFLOWER   **FL:** corolla yellow with red-brown spots near mouth. Open, sandy sites among shrubs; 75–120 m. s SCoRO (sw Santa Barbara Co.). May–Jun   ★

M. glaucescens Greene (p. 997)   SHIELD-BRACTED MONKEYFLOWER   Ann 6–80 cm, ± glabrous, glaucous. **LF:** petiole 0–60 mm; blade 5–70 mm, ovate to ± round. **INFL:** raceme, gen > 5-fld; bracts 5–45 mm wide, ± round disks completely encircling st, glaucous. **FL:** pedicel 5–35 mm; calyx 7–25 mm, asymmetrically swollen in fr, ± glabrous, lobes unequal, lowest 2 upcurved in fr; corolla yellow, tube-throat 15–35 mm; placentas axile. **FR:** 6–12 mm. 2*n*=28. Seepage areas; < 600 m. s CaRF, n SNF. Mar–May   ★

M. gracilipes B.L. Rob. (p. 997)   SLENDER-STALKED MONKEYFLOWER   Ann 2–8 cm, puberulent. **LF:** 3–13 mm, elliptic to ovate. **FL:** pedicel 8–25 mm; calyx 4–6 mm, ± puberulent, lobes equal, 0.5–2 mm, obtuse to rounded, densely ciliate, tip ± small-pointed; corolla pink to lavender, 2-lipped, tube-throat 7–10 mm, limb 5–9 mm wide; placentas axile. **FR:** 4–5 mm. Disturbed or burned areas on decomposed granite; 500–1300 m. c SNF. Apr–May   ★

M. guttatus DC. (p. 997)   Ann or rhizomed per, 2–150 cm, glabrous to hairy. **LF:** abruptly reduced upwards; petiole 0–95 mm; blade 4–125 mm, ovate to round, gen crenate, base gen irregularly small-lobed or dissected. **INFL:** raceme, gen > 5-fld; bracts ovate to cordate, fused at base or not, not glaucous. **FL:** cleistogamous or open; pedicel 10–80 mm; calyx 6–30 mm, asymmetrically swollen in fr, glabrous to hairy, lobes unequal, lowest 2 upcurved in fr; corolla yellow, tube-throat 2–40 mm; placentas axile. **FR:** 5–12 mm. 2*n*=28,30,32,48,56. Common. Wet places, gen terrestrial, occ emergent or floating in mats; < 2500 m. CA; to AK, w Can, Rocky Mtns, n Mex; introduced in ne US, e Can. [*M. cupriphilus* Macnair; *M. glabratus* Kunth subsp. *utahensis* Pennell; *M. microphyllus* Benth.] Complex: local populations possibly unique but forms readily intergrade. Mar–Aug

M. inconspicuus A. Gray (p. 997)   SMALL-FLOWERED MONKEYFLOWER   Ann 3–30 cm, glabrous. **LF:** 6–40 mm, broadly elliptic, lowest in basal rosette, ± petioled, others sessile, broadly ovate, base rounded. **FL:** pedicel 5–25 mm; calyx 5–9 mm, in fr 6–11 mm and symmetrically swollen, gen glabrous, lobes equal, 0.5–1.5 mm; corolla gen rose to lavender, tube-throat 7–16 mm, limb < 14 mm wide, lobe tips gen deeply notched; placentas axile. **FR:** 4–9 mm. Near hillside streams or seeps, in partial shade; 160–2000 m. SNF (s of El Dorado Co.). [*M. acutidens* Greene; *M. grayi* A.L. Grant] Apr–Jun   ★

M. johnstonii A.L. Grant (p. 997)   JOHNSTON'S MONKEYFLOWER   Ann 3–20 cm, gen finely ± white-puberulent. **LF:** 7–30 mm, oblanceolate to obovate, tip gen acute. **FL:** pedicel 1–5 mm; calyx 6–11 mm, membranous, ± red, minutely puberulent, mouth strongly oblique in fr, lobes erect to spreading, unequal, 1–3 mm, ± acuminate; corolla magenta, persistent, tube-throat 9–15 mm, throat floor yellow, mouth gen with 2 lateral, dark maroon patches; stigma ± exserted; placentas parietal. **FR:** 7–12 mm. 2*n*=16. Road banks, disturbed areas, esp scree; gen 975–2920 m. SnGb, SnBr. Intergrades with *M. constrictus* in s WTR and s SCoRO. May–Aug   ★

M. kelloggii (Greene) A. Gray (p. 997)   Ann 1–31 cm, hairy. **LF:** 6–40 mm, elliptic to obovate; abaxially gen ± purple. **FL:** pedicel 2–6 mm; calyx 8–16 mm, puberulent, lobes ± unequal, 0.5–1.5 mm, obtuse; corolla persistent, magenta to red-purple, becoming darker toward spotted-gold throat, tube-throat 20–45 mm, ± 2× calyx, gradually wider and funnel-like toward tip, upper lobes 4–5 mm, lower lobes 2–3 mm; placentas parietal. **FR:** 6–12 mm, ± cylindrical, ± curved, hard, indehiscent. 2*n*=18. Bare, unstable or disturbed areas; steep slopes of soil or scree; 50–1525 m. NW, CaRF, n&c SNF; sw OR. Mar–Jun

M. laciniatus A. Gray (p. 997)   CUT-LEAVED MONKEYFLOWER   Ann 3–38 cm, glabrous to sparsely hairy. **LF:** petiole 0–35 mm; blade 3–55 mm, oblanceolate to ± ovate, at least some ± pinnately, narrowly lobed or dissected. **INFL:** raceme, gen > 5-fld; bracts ± lanceolate to pinnately lobed, base long-tapered to petioled. **FL:** cleistogamous or open; pedicel 10–45 mm; calyx 3–15 mm, asymmetrically swollen in fr, ± glabrous, lobes unequal, lowest 2 upcurved in fr; corolla yellow, tube-throat 4–15 mm; placentas axile. **FR:** 3–8 mm. 2*n*=28. Seeps on granite outcrops; > 900 m. n&c SNF, SNH, SnJV. May–Aug   ★

M. latidens (A. Gray) Greene (p. 997) Ann 10–27 cm, branched only near base. **LF:** 4–25 mm, gen lanceolate to ± ovate; lowest ± petioled, in basal rosette; others sessile, ovate, base rounded. **FL:** pedicel 6–30 mm; calyx 6–7 mm, ± glabrous; in fr 9–12 mm, strongly swollen, pleated, membranous; lobes equal, 1–2 mm; corolla pale pink to white (± yellow), tube-throat 7–11 mm, limb 4–6 mm wide, lobes entire; placentas axile. **FR:** 7–8 mm. Vernally wet depressions; < 900 m. NCoRI, GV, CW, e SCo (Menifee Valley, w Riverside Co.); Baja CA. Scattered. Apr–Jun

M. latifolius A. Gray Ann 1–10 cm, puberulent to hairy. **LF:** petiole 1–4 mm; blade 6–39 mm, elliptic to ovate, gen ciliate in basal 1/2, lower surface ± purple. **FL:** pedicel 1.5–4 mm; calyx 7.5–13 mm, hairy, lobes unequal, 0.5–3 mm, obtuse; corolla magenta, persistent, tube-throat 12–19 mm, yellow near stamen attachment, with 2 prominent yellow folds on floor and lower lip base; placentas parietal. **FR:** 6–11 mm, oblique, angled, hard, indehiscent. Presumed extirpated in CA. Rocky places; < 150 m. n ChI (Santa Cruz Island), s ChI (Santa Catalina Island); Baja CA (Guadalupe Island). [*M. brandegeei* Pennell] Mar–Apr ★

M. layneae (Greene) Jeps. (p. 997) Ann 2–28 cm, ± hairy. **LF:** 6–25 mm, narrowly elliptic to obovate. **FL:** pedicel 1–3 mm; calyx 5–11 mm, dark-ribbed, ± hairy, lobes ± equal, 1–3 mm, acute to acuminate; corolla pink to magenta (± white), darker central marks on lobes, persistent, tube-throat 13–21 mm, throat near mouth gen white (yellow) with red-purple dots, lips ± equal, lower bent downward; placentas parietal. **FR:** 6–10 mm. 2*n*=16. Bare or disturbed areas, gen serpentine or granitic; gen 200–2000 m. NW, CaR, SN, SW, MP. Intermediates with *M. nanus* (NW, s SN) have yellow at corolla mouth. May–Aug

M. leptaleus A. Gray (p. 997) Ann 0.5–14 cm, puberulent. **LF:** 5–25 mm, linear to oblanceolate. **FL:** pedicel gen 0.5–1 mm; calyx 3–6 mm, puberulent, lobes ± equal, 0.8–2 mm, acute; corolla magenta or white, persistent, tube-throat 6–8 mm, throat ± white, dark-dotted, limb 3–5 mm wide, gen not spreading; placentas parietal. **FR:** 4–6 mm. 2*n*=16. Granitic soils or sand near outcrops, disturbed areas; 2000–3400 m. SNH; w NV (Mount Rose, Washoe Co.). Intermediates with *M. whitneyi* in s SNH gen have ± yellow fls. Jun–Aug

M. lewisii Pursh (p. 997) Per, rhizomed, hairy. **ST:** 25–80 cm. **LF:** 20–70 mm, gen clasping st, oblong to elliptic, palmately 3–5-veined. **FL:** pedicel gen 30–70 mm; calyx 15–25 mm, hairy, lobes equal, 4–6 mm, acute; corolla lavender to purple, tube-throat 30–50 mm, lips spreading; placentas axile. **FR:** 13–14 mm. 2*n*=16. Streambanks, seeps; 1200–3100 m. KR, CaRH, SN; to MT, WY, UT, w Can. Jun–Aug

M. mohavensis Lemmon (p. 997) MOJAVE MONKEYFLOWER Ann 2–10 cm, ± puberulent. **ST:** cylindric. **LF:** 7–27 mm, narrowly elliptic, red-purple. **FL:** pedicel 2–5 mm; calyx 7–15 mm, enlarged in fr, red-purple, ± puberulent along veins, lobes spreading, ± unequal, 1.5–4.5 mm, acuminate; corolla radial, ± salverform, gen deciduous, tube-throat 9–15 mm; tube-throat, limb at base ± purple-brown, veins << 0.5 mm wide, unpatterned, purple-brown, fading into white border; placentas parietal. **FR:** 8–13 mm. 2*n*=16. Gravelly banks of desert washes; 600–1000 m. DMoj (w-c San Bernardino Co.). Apr–May ★

M. montioides A. Gray (p. 999) Ann 1–18 cm, puberulent. **LF:** 4–32 mm, 1–3 (> 3) mm wide, linear. **FL:** pedicel 3–25 mm; calyx 3.5–8.5 mm, minutely puberulent, membranous in fr, lobes equal, 0.5–2 mm, obtuse to rounded, ± glabrous to ciliate, tip small-pointed; corolla yellow, purple, or yellow with purple upper lip, tube-throat 10–16 mm, lobes gen each notched; placentas axile. **FR:** 3–6 mm. Disturbed areas along small streams, gen in granitic soils; > 1800 m. c&s SN (s of Madera Co.), W&I; near Carson City, NV. Corolla colors mixed in some areas, uniform in others; pls with purple corollas at lower elevation resemble *M. palmeri.* Mar–Jun

M. moschatus Lindl. (p. 999) MUSK MONKEYFLOWER Per, rhizomed, ± glabrous to densely slimy-hairy, gen musk-scented. **ST:** 5–30 cm, prostrate to ascending. **LF:** petiole 0–15 mm; blade 10–60 mm, oblong to ovate, pinnately ± 5-veined. **FL:** pedicel 10–50 mm; calyx 8–12 mm, ± glabrous to hairy, lobes equal, 2–5 mm; corolla yellow, tube ± cylindric, 2–4 mm wide exc mouth, tube-throat 15–26 mm, throat floor deeply grooved; placentas axile. **FR:** 4–9 mm. 2*n*=32. Common. Seeps, streambanks, gen partial shade; < 2900 m. CA-FP, GB; to BC, Rocky Mtns; naturalized in ne US, Chile, Eur. Jun–Aug

M. nanus Hook. & Arn. Ann 1–15 cm, ± puberulent to hairy. **LF:** 3–30 mm, oblanceolate to linear; lower surface ± purple; tip acute to rounded. **FL:** pedicel 0.5–7 mm; calyx lobes ± unequal, acute to ± long-tapered, mouth in fr ± oblique; corolla persistent, throat floor gen with 2 yellow folds surrounded by and gen dotted deeper magenta; stigma ± exserted; placentas parietal. **FR:** 3.5–12 mm. 2*n*=16. Intermediates with *M. layneae* occur in NW and s SN.

var. ***jepsonii*** (A.L. Grant) D.M. Thomps. (p. 999) **ST:** lowest internodes >> others, primary axis dominant. **LF:** minutely puberulent. **INFL:** fr nodes gen congested at st tips. **FL:** calyx 3–6 mm, minutely puberulent, lobes 0.5–1.7 mm; corolla lavender to rose-purple, tube-throat 5.5–12 mm, glabrous externally, limb 5–10 mm wide. **FR:** 3.5–6 mm. Shallow drainage areas in pine forest openings; 1220–2380 m. e KR, CaRH, n SNH, w MP; s OR, w NV (Mount Rose, Washoe Co.). [*M. j.* A.L. Grant] May–Aug

var. ***mephiticus*** (Greene) D.M. Thomps. (p. 999) **ST:** lowest internodes ± longer than others, primary axis gen dominant, tufted at highest elevations. **LF:** dull olive-green, at least margins hairy. **INFL:** fr nodes ± evenly spaced. **FL:** calyx 5–11 mm, at least veins hairy, lobes 1–3 mm; corolla magenta, dark red-purple, or yellow, tube-throat 11–19 mm, gen puberulent externally, limb 9–17 mm wide. **FR:** gen 6–12 mm. Openings in sagebrush or on disturbed slopes, gen around granite; 1520–3445 m. SNH, s MP, SNE; NV. [*M. m.* Greene] May–Aug

var. ***nanus*** (p. 999) **ST:** lowest internodes ± longer than others, primary axis not dominant; tufted or mound-like. **LF:** minutely puberulent, green, basal gen ± purple below. **INFL:** fr nodes ± evenly spaced. **FL:** calyx 6–10 mm, minutely puberulent, lobes 1–4 mm; corolla magenta or lavender-purple, tube-throat 11–19 mm, glabrous externally, limb 8–14 mm wide. **FR:** 6–12 mm. Openings in sagebrush or on disturbed slopes, gen around granite; gen 1000–2300 m. NW, CaRH, GB; to WA, MT, WY. May–Aug

M. norrisii Heckard & Shevock (p. 999) KAWEAH MONKEYFLOWER Ann gen 3–15 cm, ascending, hairy. **LF:** 15–45 mm; blade ovate, ± tapered to petiole, veins pinnate. **FL:** pedicel 10–50 mm; calyx 3.5–6 mm, hairy, ribs wide, ± purple, lobes equal, 1–2 mm, rounded; corolla yellow, tube-throat 15–30 mm, lobe base with 1 large, central, red-purple dot, 2 ± white dots occ present between lobe sinuses; placentas axile. **FR:** 4–6 mm, growing away from light, late-dehiscent. 2*n*=32. Marble crevices; 600–1300 m. s SNF (Kaweah River drainage, Tulare Co.). Mar–May ★

M. nudatus Greene (p. 999) BARE MONKEYFLOWER Ann 4–30 cm, ± glabrous. **LF:** petiole 0–30 mm, abruptly reduced upward; blade 3–25 mm, proximal lvs lanceolate to ovate, distal narrower; pairs at nodes not fused. **INFL:** fls 1 in upper axils, gen > 5 per st. **FL:** pedicel 10–50 mm; calyx 4–15 mm, asymmetrically swollen in fr, glabrous, lobes unequal, lowest 2 upcurved in fr; corolla yellow, tube-throat 10–20 mm; placentas axile. **FR:** 5.5–9.5 mm. Seeps on serpentine outcrops; 250–700 m. NCoR (Lake, Napa cos.). May–Jun ★

M. palmeri A. Gray (p. 999) Ann 1–28 cm, ± puberulent. **LF:** 3–28 mm, ovate, upper ± linear. **FL:** pedicel 10–65 mm; calyx 4–10 mm, glabrous to puberulent, lobes equal, 0.5–1.5 mm, rounded to truncate, glabrous to densely ciliate, tip small-pointed; corolla magenta to purple with variable darker marks, not lipped, tube-throat 10–22 mm, throat floor folds yellow; placentas axile. **FR:** 3.5–9 mm. Sandy washes, disturbed areas; < 2100 m. s SNF, Teh, n SCoRO, TR, PR, w DMoj; n Baja CA. [*M. diffusus* A.L. Grant] Intergrades with *M. montioides* in SN, TR. Mar–Jun

M. parishii Greene (p. 999) Ann 3–85 cm, hairy. **LF:** 8–75 mm, oblanceolate to ovate, palmately 3-veined. **FL:** pedicel 15–18 mm; calyx 8–10 mm, to 13 mm in fr, hairy, lobes equal, 1–2 mm; corolla ± pink to white, tube-throat 9–14 mm; placentas axile. **FR:** 6–10 mm. Uncommon. Wet, sandy streambanks; < 2100 m. s SN, SW, w D, n&e DMtns (Granite, New York mtns, Panamint Range); n Baja CA. May–Aug

M. parryi A. Gray (p. 999) PARRY'S MONKEYFLOWER Ann 1–17 cm, densely puberulent. **LF:** 3–26 mm, ± linear to oblanceolate, tip rounded to acute. **FL:** pedicel 1–5 mm; calyx 6–12 mm, membranous, gen ± purple throughout, puberulent, lobes spreading, upper lobe 2–3 mm, rounded to acute, other lobes gen 0.5–2 mm, acute or short-pointed; corolla persistent, magenta or yellow, tube-throat 11–18 mm; placentas parietal. **FR:** 5.5–10 mm, incl. $2n$=16. Steep hillsides, along washes; (770)1200–1830 m. W&I (esp Inyo Mtns); to sw UT, nw AZ. ± common outside CA. Apr–Jul ★

M. pictus (Greene) A. Gray (p. 999) CALICO MONKEYFLOWER Ann 2–38 cm, hairy. **ST:** 4-sided. **LF:** 7–45 mm, obovate. **FL:** pedicel 2–6 mm; calyx 6–13 mm, to 18 mm in fr, hairy, lobes ± unequal, 1–4 mm, obtuse; corolla radial, salverform, gen deciduous, tube dark, purple-brown, tube-throat 6.5–18 mm, limb ± white, veins > 0.5 mm wide, bold, intricately patterned, purple-brown; placentas parietal. **FR:** 7–17 mm. $2n$=16. Bare, sunny, shrubby areas, around granite outcrops; 135–1250 m. s SNF, Teh. Mar–May ★

M. pilosus (Benth.) S. Watson (p. 999) Ann 2–35 cm, densely long-soft-wavy-hairy. **LF:** 10–30 mm, lanceolate to oblong. **FL:** pedicel gen 10–15 mm; calyx 6–7 mm, densely hairy, lobes unequal, 3–4 mm, ± = tube, obtuse; corolla yellow, tube-throat 7–8 mm; placentas axile. **FR:** 4–7 mm. Moist, sandy areas, esp by small streams, disturbed areas; < 2600 m. CA-FP, SNE, DMtns; to WA, UT, AZ, Baja CA. Apr–Aug

M. primuloides Benth. Per; rhizomes or stolons forming mats of ± distinct rosettes or tufted pls; forming bulblets in fall. **ST:** glabrous. **LF:** adaxially glabrous to densely long-hairy; palmately 3-veined. **FL:** pedicel 10–120 mm, stiffly erect; calyx glabrous, lobes equal, 0.5–1.5 mm; corolla yellow, tube-throat 8–20 mm, base of lower lip lobes gen red-spotted; placentas axile. **FR:** 6–7 mm.

var. ***linearifolius*** A.L. Grant (p. 999) **ST:** 4–12 cm, tufted. **LF:** ± erect, some ± basal but not in rosettes, 15–50 mm, linear to oblanceolate. **FL:** calyx 7–12 mm. Wet meadows, seeps, streambanks; 600–2200 m. KR. [*M. p.* subsp. *l.* (A.L. Grant) Munz] Jul–Aug

var. ***primuloides*** (p. 999) **ST:** 0.5–4 cm. **LF:** gen ± spreading in ± distinct rosettes, 7–35 mm, oblong to obovate. **FL:** calyx 5–9 mm. $2n$=34. Wet meadows, seeps, streambanks; 600–3400 m. NW, CaR, SN, WTR, SnBr, SnJt, GB; to WA, NV. If recognized taxonomically, occ small, hairy pls (of e.g., Echo Summit, El Dorado Co.) assignable to *M. primuloides* var. *pilosellus* (Greene) Smiley but variation ± continuous. Jun–Aug

M. pulchellus (Greene) A.L. Grant (p. 999) YELLOW-LIP PANSY MONKEYFLOWER Ann < 1 cm, compact, puberulent. **LF:** 8–35 mm, ± linear, at least the basal 1/2 ciliate. **FL:** pedicel 0–2 mm; calyx 7–15 mm, hairy, lobes unequal, 1–3 mm, obtuse, ± ciliate; corolla persistent, lavender to purple exc lower lip, at least lowest lobe yellow, lowest lobe base with scattered, small, dark dots, tube-throat 20–40 mm, puberulent outside; placentas parietal. **FR:** 3–5.5 mm, oblique-ovoid, hard, indehiscent. n=9. Vernally wet depressions or seepage areas; 600–2000 m. n&c SNF (Calaveras, Tuolumne, Mariposa cos.). Apr–Jul ★

M. pulsiferae A. Gray (p. 999) Ann 2–21 cm, puberulent. **LF:** petiole 1–6 mm, blade 3–20 mm; basal lvs in ± rosette, ovate, cauline elliptic. **FL:** pedicel 6–24 mm, ascending; calyx 3.5–7.5 mm, 5–12 mm in fr, minutely puberulent, lobes ± equal, 0.5–1.2 mm, acute; corolla gen yellow, tube-throat 7–13 mm; placentas axile. **FR:** 5–10 mm. Runoff areas, moist places along streams; 500–1900 m. NW, CaR, n&c SNH, MP; to WA. Limb gen ± white with ± pink border in n NW. Apr–Jun

M. purpureus A.L. Grant (p. 999) LITTLE PURPLE MONKEYFLOWER Ann 0.5–7 cm, puberulent. **LF:** 3–16 mm, elliptic to ovate. **FL:** pedicel 15–56 mm; calyx 4–8 mm, tube puberulent, lobes equal, 0.5–1 mm, rounded to truncate, glabrous, tip small-pointed; corolla tube-throat 8–13 mm, 2-lipped, upper lip red-purple, lower lip rose, limb 8–10 mm wide; placentas axile. **FR:** 4–7 mm. Along small streams on open, gentle slopes; 1900–2300 m. SnBr; Baja CA. Vars. not distinct. Threatened by development, vehicles. Jun–Jul ★

M. pygmaeus A.L. Grant (p. 999) EGG LAKE MONKEYFLOWER Ann < 1 cm, compact, puberulent. **LF:** 2–15 mm, oblanceolate to narrowly elliptic, basal 1/2 ciliate. **FL:** pedicel 0–0.5 mm; calyx 4–8 mm, lobes unequal, 0.5–1.5 mm, rounded, occ ± ciliate; corolla yellow, persistent, tube-throat 5–10 mm; placentas parietal. **FR:** 2–4 mm, ovoid, hard, indehiscent. $2n$=20. Vernally wet depressions or along stream channels, on clay soils; 1095–1790 m. s CaRH, n SNH (Lake Almanor region, Plumas Co.), MP (Egg Lake, Modoc Co., w of Eagle Lake, Lassen Co.); s OR. May–Jun ★

M. rattanii A. Gray (p. 999) Ann 1–18 cm. **LF:** 3–46 mm, narrowly elliptic to obovate, proximal glabrous, distal hairy. **INFL:** fl 1 per node. **FL:** pedicel 1–3 mm; calyx 5–9 mm, widely ribbed, swollen, hairy, lobes ± equal, 0.7–2 mm, obtuse or short-pointed; corolla pink to magenta, persistent, tube-throat 7–10 mm, throat floor ± glabrous, folds yellow at mouth; placentas parietal. **FR:** 7–11 mm. $2n$=16. Sandy, open places, esp around sandstone outcrops or burns, other disturbed areas; 90–1220 m. NCoRI, CW, PR. [*M. r.* subsp. *decurtatus* (A.L. Grant) Pennell] Apr–Jul

M. ringens L. (p. 999) Per, rhizomed, glabrous. **ST:** 20–130 cm, gen ascending, 4-angled. **LF:** 25–80 mm, lanceolate to narrowly oblong, gen ± clasping st; veins pinnate. **FL:** pedicel 20–35 mm; calyx 10–16 mm, ± puberulent, lobes equal, 2–6 mm, acute, minutely ciliate at base; corolla lavender to purple, tube-throat 20–30 mm, lower lip base swollen, ± closing mouth; placentas axile. **FR:** 10–12 mm. $2n$=16,22,24. Wet places; < 200 m. ne SnJV (near La Grange, Stanislaus Co.); to Can to e US; also ID, CO. Probably introduced in CA. Apr–Jun

M. rubellus A. Gray (p. 999) Ann 2–32 cm, minutely puberulent. **LF:** 3–31 mm, lanceolate to ovate, proximal ± petioled, distal sessile. **FL:** pedicel 5–18 mm, ascending in fr; calyx 4.5–9 mm, gen glabrous, ribs ± brown, lobes equal, 0.5–1.5 mm, rounded, ciliate, tip gen small-pointed; corolla yellow or purple-magenta, tube-throat 6–10 mm, limb 3–4 mm wide; placentas axile. **FR:** 3.5–7 mm. Gen around washes; 800–2600 m. TR, GB, DMtns; to WY, NM, n Baja CA. Apr–Jun

M. rupicola Coville & A.L. Grant (p. 999) DEATH VALLEY MONKEYFLOWER Per 0–17 cm, puberulent. **LF:** 18–60 mm, oblanceolate, not ciliate. **INFL:** fls 2 per node. **FL:** pedicel 1–3 mm; calyx 8–17.5 mm, puberulent, lobes unequal, 1.5–7 mm, long-tapered; corolla persistent, ± pink to white with large, magenta-purple spot at center of lobe base, tube-throat 17–35 mm; placentas parietal; not cleistogamous. **FR:** 3–8 mm, ovoid-oblong, ± curved, hard, indehiscent. **SEED:** few, 1–2 mm. $2n$=16. Limestone cliff crevices; 310–1830 m. n DMtns (Cottonwood, Funeral, Grapevine mtns, Last Chance, n Panamint ranges). Feb–Jun ★

M. shevockii Heckard & Bacig. (p. 999) KELSO CREEK MONKEYFLOWER Ann 2–12 cm, minutely puberulent. **LF:** 3–10 mm, lanceolate to ovate, clasping st; pairs at nodes fused or not. **FL:** pedicel 10–22 mm; calyx 4–7 mm, with ± red spots or all red, ± puberulent, lobes equal, 0.5–1 mm, gen rounded; corolla tube-throat 8–12 mm, upper lip 4-lobed, maroon-purple, lower lip 2-lobed, yellow, maroon-dotted at base; anthers and stigma exserted; placentas axile. **FR:** 5–6 mm. $2n$=32. Alluvial fans, dry streamlets, gen granitic soils; 900–1300 m. s SNF (Cortez, Cyrus canyons, Kern Co.), w edge DMoj (Kelso Creek). Apr–May ★

M. suksdorfii A. Gray (p. 1005) Ann 0.5–10 cm, minutely puberulent. **LF:** 4–23 mm, ± linear to ovate; distal sessile. **FL:** pedicel 2–8 mm, spreading and ± S-curved in fr; calyx 3–5.5 mm, glabrous, lobes equal, 0.3–1 mm, tip rounded or small-pointed; corolla yellow, tube-throat 4–6 mm; placentas axile. **FR:** 2.5–5 mm. Moist, gen clay soils in ± full sun; 1100–4000 m. CaRH, SNH, WTR, SnBr, SnJt, n SNE (Sweetwater Mtns), W&I, n DMtns (Grapevine Mtns, Inyo Co.); to WA, MT, WY, CO, AZ. May–Aug

M. tilingii Regel (p. 1005) Per 2–35 cm, rhizomed, glabrous to ± hairy. **LF:** petiole 0–25 mm; blade 5–30 mm, elliptic to ± round; pairs at nodes not fused. **INFL:** fls 1–5 per st, 1 in axils of distal lvs, not in bracted raceme. **FL:** opening; pedicel 10–90 mm; calyx 7–25 mm, asymmetrically swollen in fr, glabrous to puberulent, lobes unequal,

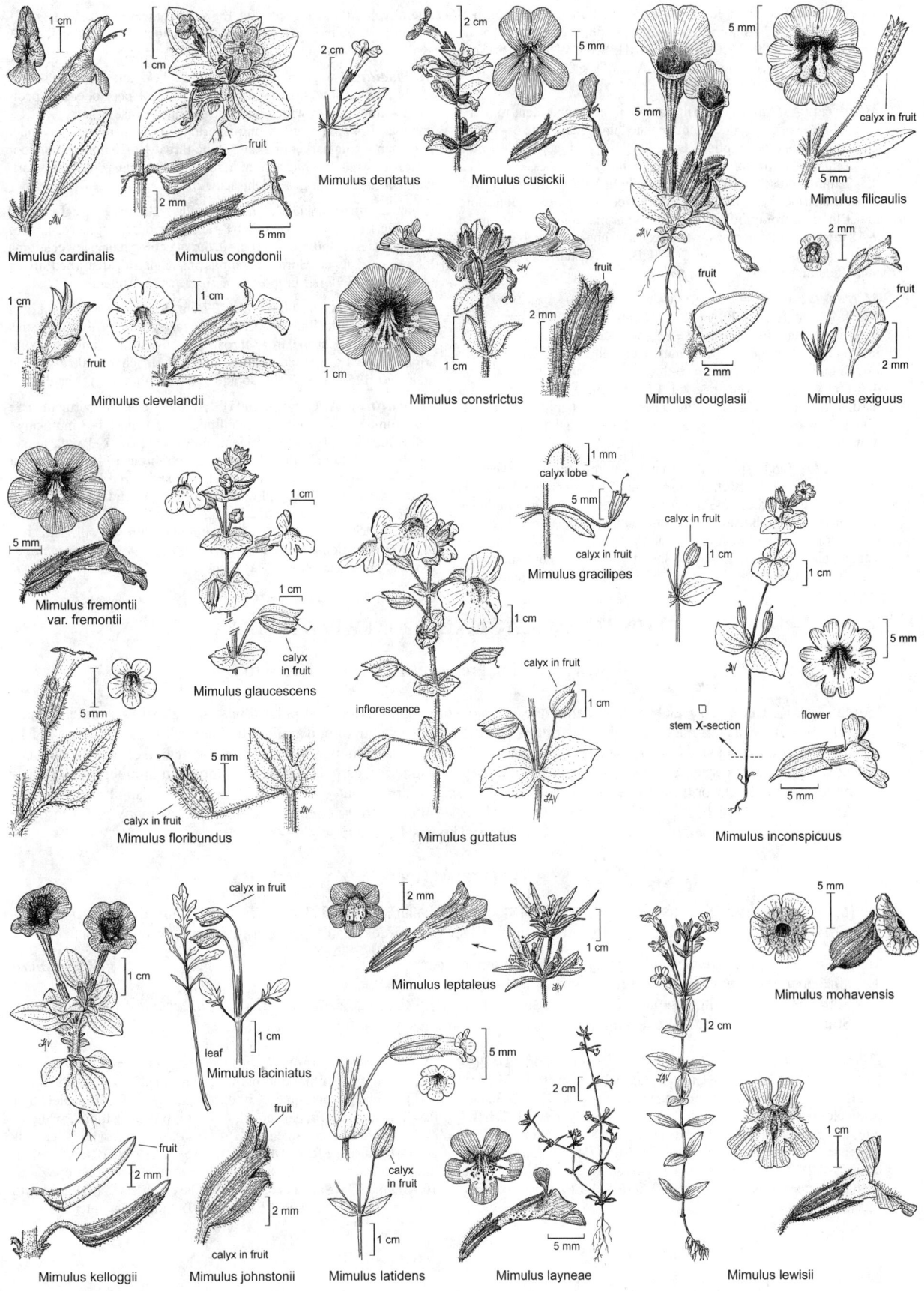

Mimulus cardinalis

Mimulus congdonii

fruit

Mimulus dentatus

Mimulus cusickii

Mimulus filicaulis

calyx in fruit

Mimulus clevelandii

fruit

Mimulus constrictus

fruit

Mimulus douglasii

fruit

Mimulus exiguus

fruit

Mimulus fremontii var. fremontii

Mimulus glaucescens

calyx in fruit

Mimulus gracilipes

calyx lobe

calyx in fruit

Mimulus inconspicuus

calyx in fruit

stem X-section

flower

Mimulus floribundus

calyx in fruit

inflorescence

Mimulus guttatus

calyx in fruit

Mimulus kelloggii

fruit

Mimulus johnstonii

calyx in fruit

fruit

Mimulus laciniatus

calyx in fruit

leaf

Mimulus leptaleus

Mimulus mohavensis

Mimulus latidens

calyx in fruit

Mimulus layneae

Mimulus lewisii

lowest 2 upcurved in fr; corolla yellow, tube-throat 17–45 mm; placentas axile. **FR:** 5–10 mm. $2n$=28,30,48,50,56. Seeps, streambanks, wet meadows; 1400–3400 m. NW, CaRH, SNH, WTR, SnBr, SnJt, n SNE (Sweetwater Mtns), W&I; to AK, MT, CO. Occ intergrades with *M. guttatus.* Jul–Sep

M. torreyi A. Gray (p. 1005) Ann 4–38 cm, puberulent to hairy. **LF:** petiole ± obscure, 0–10 mm, reduced upwards; blade 6–40 mm, gen oblanceolate or elliptic. **FL:** pedicel 1–3 mm; calyx 5–10 mm, membranous, uniformly pale green, puberulent, lobes ± unequal, 0.5–2 mm, obtuse; corolla persistent, light rose to magenta, with 2 yellow stripes with dark magenta lower margins extending onto lower lip base from throat floor folds, other marks gen 0, tube-throat 9–18 mm, lower lip gen projecting forward, upper lip ± reflexed; placentas parietal. **FR:** 6–11 mm. $2n$=20. Bare or disturbed, open areas; 600–2000 m. s CaRH, SN. May–Aug

M. traskiae A.L. Grant SANTA CATALINA ISLAND MONKEYFLOWER Ann 8–12 cm, hairy. **LF:** petiole 1–12 mm; blade 12–41 mm, ovate to obovate, gen ciliate near base **FL:** pedicel 3–5 mm; calyx 18–21 mm, hairy, lobes unequal, 2–4 mm, acute; corolla persistent, upper lip white, lower red-purple, tube-throat 20–23 mm, ± = calyx, limb 4–5 mm wide; placentas parietal. **FR:** ± 8 mm, ± oblique-oblong, hard, probably indehiscent. Presumed extinct; habitat and elevation unknown; s ChI (Avalon, Santa Catalina Island). Probably Mar–Apr ★

M. tricolor Lindl. (p. 1005) Ann 1–14 cm, densely puberulent. **LF:** 8–45 mm, narrowly elliptic to oblanceolate, not ciliate. **INFL:** fls 2 per node. **FL:** pedicel 1–3 mm; calyx 10–23 mm, puberulent, lobes unequal, 1–5 mm, obtuse; corolla persistent, magenta to purple with ± white throat with yellow folds, mottled with dark dots, lobe base with 1 large, maroon-purple spot, tube-throat 15–50 mm, puberulent outside; placentas parietal. **FR:** 3–8 mm, oblique-ovoid, ± compressed

laterally, hard, indehiscent. $2n$=18. Vernally wet depressions; gen < 600 m (< 1500 m in MP). s NCoRO (s Sonoma Co.), NCoRI, e GV, sw MP; OR. Mar–Jun

M. viscidus Congdon Ann, hairy. **LF:** 4–45 mm, obovate to narrowly elliptic; lowest gen glabrous. **INFL:** fl 1 per node. **FL:** pedicel 1–4 mm; calyx widely ribbed, swollen, densely hairy, lobes ± unequal, 1–4 mm, acute to long-tapering; corolla persistent, lavender to magenta with darker markings, tube-throat 10–20 mm, throat floor white or ± yellow, white at mouth, folds densely hairy; placentas parietal. **FR:** 6–11 mm, gen indehiscent. $2n$=16.

var. ***compactus*** D.M. Thomps. Pl 2–28 cm. **FL:** calyx 6–10 mm, gen ± red in fr; corolla limb 8–12 mm wide, face glabrous, lobes dark at base, red-purple midveins 0, throat ceiling hairy; lower stigma lobe 3–4 × upper. Burns, openings in chaparral, disturbed areas; 90–1250 m. SNF (s Mariposa to n Tulare cos.). Apr–Jul

var. ***viscidus*** (p. 1005) Pl 6–37 cm. **FL:** calyx 8–15 mm, gen green in fr; corolla limb 8–20 mm wide, face gen hairy, lobe midveins with red-purple stripe from throat, throat ceiling glabrous, lower stigma lobe 1.5 × upper. Burns, openings in chaparral, disturbed areas; 90–1250 m. SNF (El Dorado to n Mariposa cos.). Apr–Jul

M. whitneyi A. Gray (p. 1005) Ann 1–14 cm, puberulent. **LF:** 7–23 mm, ± linear to narrowly elliptic. **FL:** pedicel 1–3 mm; calyx 4–8 mm, dark-ribbed, puberulent, lobes ± equal, 1–3 mm, acute; corolla magenta or yellow, persistent, tube-throat 13–18 mm, throat with 3–5 red-brown longitudinal stripes and 2 red-brown broader blotches most visible in yellow fls; placentas parietal. **FR:** 6–10 mm. $2n$=16. Disturbed areas, road banks, exposed runoff areas, granitic soils; gen 1500–3270 m. s SNH. Intergrades with *M. constrictus* to s at lower elevations; with *M. leptaleus* and *M. nanus* var. *mephiticus* to ne at higher elevations. May–Sep

PHYTOLACCACEAE POKEWEED FAMILY

John W. Thieret, final revision by Thomas J. Rosatti

[Ann to] per [to tree], gen ± glabrous. **LF:** alternate, simple, entire, gen petioled; stipules 0 or vestigial. **INFL:** spike, raceme [panicle], [axillary or] terminal (then often lateral, ± opposite a lf, by growth of bud axillary to that lf); bract 1; bractlets 2. **FL:** gen bisexual, radial; sepals [4]5[10], fused at base, gen persistent; petals 0; stamens 4–many, gen on disk, free or fused at base, in 1–2 whorls or not; carpels gen 1–12, [free] to fused, ovary gen superior, carpels = styles, chambers in number, ovules 1 per chamber, stigmas linear or thread-like. **FR:** berry [achene, capsule, drupe, nut, samara]. ± 18 genera, 130 spp.: ± worldwide, esp Am trop, subtrop. [Rogers 1985 J Arnold Arbor 66:1–37] A broad family concept adopted here, as many as 6 segregate families sometimes recognized by others. Scientific Editor: Thomas J. Rosatti.

PHYTOLACCA

Per [to tree]; axes often ± red to purple. **FL:** sepals 5[8], petal-like; stamens gen 5–30; carpels 5–12, ± fused. **FR:** berry, chambers 5–12. **SEED:** 1 per chamber. ± 25 spp.: trop, subtrop, esp Am. (Greek, Latin: plant, crimson dye, from fr color)

1. Infl spike or spike-like, pedicel 0 or 0.5–2(4) mm; stamens gen in 2 whorls . . . . ***P. icosandra***
1′ Infl ± open raceme; pedicel 2–13 mm; stamens in 1–2 whorls
  2. Stamens gen in 1 whorl; sepals ± equal . . . . ***P. americana*** var. ***americana***
  2′ Stamens gen in 2 whorls; sepals unequal (largest ± 2 × smallest in width) . . . . [***P. heterotepala***]

P. americana L. var. *americana* (p. 1005) POKEWEED Pl to 3(7) m. **LF:** blade 8–35 cm, 3–18 cm wide, lanceolate to ovate; petiole 1–6 cm. **INFL:** ± open raceme, spreading to gen drooping, (10)12–30 cm; peduncle to 15 cm; pedicel 3–13 mm. **FL:** sepals ± equal, 2.5–3.3 mm, ovate to ± round, ± white to ± pink; stamens 10, gen in 1 whorl; carpels 6–12, fused at least in basal 1/2. **FR:** 6–11 mm diam, purple-black. **SEED:** 2.5–3 mm, lenticular, black, shiny. $2n$=36. Uncommon. Disturbed areas, roadsides; gen < 1000 m. CA-FP; native to e US. Summer–fall ❖

P. icosandra L. TROPICAL POKEWEED Pl to 3 m. **LF:** blade < 30 cm, 15 cm wide, elliptic to obovate (lanceolate); petiole 0.5–6 cm. **INFL:** spike or spike-like, < 30 cm; peduncle < 10 cm; pedicel 0 or 0.5–2(4) mm. **FL:** sepals equal, < 3 mm, broadly elliptic, white or pink to pale ± red; stamens (8)10–22(30), gen in 2 whorls; carpels gen 6–10, fused. **FR:** 7–8 mm diam, purple-black. **SEED:** 2.5–3 mm, thickly lenticular, black, shiny. Disturbed areas, chrome ore piles; 10–400 m. SCo (San Diego Co.), PR (Otay Mtn); AZ, MD; n Mex to n S.Am. Some authors incl *P. octandra* L. Summer–winter

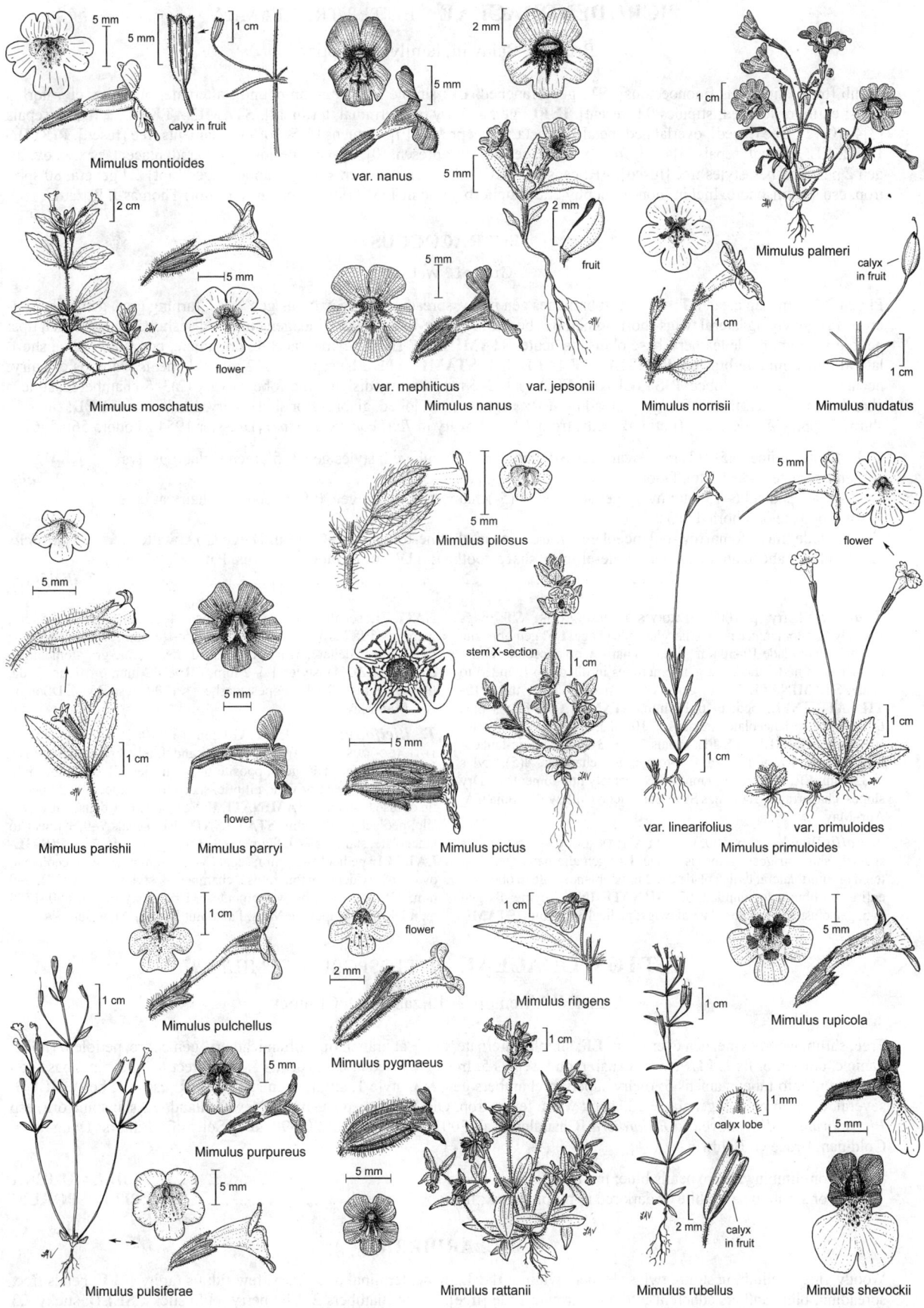

Mimulus montioides

calyx in fruit

var. nanus

2 mm

fruit

Mimulus palmeri

calyx in fruit

Mimulus moschatus

flower

var. mephiticus   var. jepsonii

Mimulus nanus

Mimulus norrisii

Mimulus nudatus

Mimulus pilosus

flower

Mimulus parishii   Mimulus parryi

flower

stem X-section

Mimulus pictus

var. linearifolius   var. primuloides

Mimulus primuloides

Mimulus pulchellus

Mimulus purpureus

Mimulus pulsiferae

flower

Mimulus pygmaeus

Mimulus ringens

Mimulus rupicola

calyx lobe

calyx in fruit

Mimulus rattanii

Mimulus rubellus   Mimulus shevockii

PICRODENDRACEAE BITTER-TREE FAMILY

Bruce G. Baldwin, family description

Shrub [tree]; dioecious [monoecious]. **ST**: gen branched. **LF**: simple [palmate-compound], alternate, opposite, or whorled in 3s; entire or toothed; stipules 0 [present]. **INFL**: gen axillary [subterminal, terminal]. **STAMINATE FL**: ± radial; sepals [3]4–10[12], free [fused], overlapped; petals 0; nectary disk present [0]; stamens [2]5–10[55], filaments free [fused]. **PISTILLATE FL**: ± radial; sepals [3]4–13, free; petals 0; nectary disk present [0]; ovary superior, compound, placentas axile, ovules gen 2 per chamber, styles free [fused]. **FR**: capsule. **SEED**: 1–2 per chamber, scar gen appendaged [not]. 24 genera, 80 spp.: trop, esp s hemisphere. Incl in (non-monophyletic) Euphorbiaceae in TJM (1993). Scientific Editor: Thomas J. Rosatti.

TETRACOCCUS

Grady L. Webster

Pl gen 0.5–2 m; sap clear. **ST**: axis erect; branches gen many, spreading to erect; twigs gen ± red, turning gray, gen hairy, glabrous in age; young lateral twigs short, sometimes becoming spine-like. **LF**: cauline, gen clustered at short, lateral branch tips; petiole < 2 mm; blade leathery, base obtuse to acute. **STAMINATE INFL**: cyme, raceme, or panicle, occ clustered on short, lateral twigs, minute-bracted. **PISTILLATE INFL**: fl 1. **STAMINATE FL**: sepals 0.5–2 mm; filaments glabrous or hairy; nectary disk ± minute-lobed. **PISTILLATE FL**: sepals 2–5 mm; nectary disk minute-lobed; ovary (2)3–5-chambered, style = chamber in number, free, ± flat, gen spreading. **FR**: ± spheric, gen lobed, glabrous or short-hairy, gen brown. **SEED**: smooth, shiny. 5 spp.: CA, AZ, Mex. (Latin: 4 seeds, from 4-lobed ovary in *Tetracoccus dioicus*) [Dressler 1954 Rhodora 56:45–61]

1. Pistillate pedicel 0.5–1(3) mm; ovary gen 3-lobed, gen 3-chambered; styles gen 3; filaments glabrous; lvs gen alternate — se DMoj, DSon . . . . ***T. hallii***
1′ Pistillate pedicel 6–15 mm; ovary gen 4-lobed, 4-(5-)chambered; styles gen 4; filaments soft-hairy at base; lvs opposite or whorled in 3s
  2. Lf blade linear to narrow-oblanceolate, entire or sparsely fine-toothed; s SCo (San Diego Co.), w PR . . . . . . ***T. dioicus***
  2′ Lf blade lanceolate to ovate or wide-elliptic, sharp-toothed; n DMtns (Grapevine Mtns, Panamint Range)
. . . . ***T. ilicifolius***

T. dioicus Parry (p. 1005) PARRY'S TETRACOCCUS **ST**: twigs sparsely fine-tomentose near axils, glabrous in age. **LF**: gen opposite or 3-whorled; blade 10–30 mm, linear to narrow-oblanceolate, entire or sparsely fine-toothed, margin sometimes inrolled, tip rounded to acute. **STAMINATE INFL**: gen raceme; pedicel 3–10 mm. **PISTILLATE INFL**: pedicel 6–15 mm. **STAMINATE FL**: sepals 6–10, ovate to lanceolate; stamens 5–10, filaments 2.5–4 mm, base soft-hairy. **PISTILLATE FL**: sepals 7–13, 3–5 mm, wide-lanceolate to ovate; ovary 4-lobed, fine-tomentose, chambers 4(5), styles 3–3.5 mm. **FR**: ± 6 mm, 7–9 mm wide, sparsely fine-tomentose. Dry slopes, chaparral; < 1000 m. s SCo (San Diego Co.), w PR; Baja CA. Apr–May ★

T. hallii Brandegee (p. 1005) HALL'S TETRACOCCUS **ST**: twigs sparsely short-strigose, glabrous in age. **LF**: gen alternate, gen clustered on short, lateral twigs; blade 2–12 mm, oblanceolate to obovate, entire, tip obtuse to rounded. **STAMINATE INFL**: cyme; fls gen 1–5, gen clustered on short, lateral twigs; pedicel 3–5.5 mm. **STAMI-**NATE FL: sepals 4–6, ± round; stamens 4–8, filaments 1.5–2.5 mm, glabrous. **PISTILLATE FL**: pedicel 0.5–1(3) mm; sepals gen 5, 2–5 mm, ovate to deltate; ovary gen 3-lobed, dense-, fine-gray-tomentose, chambers (2)3(4), styles 1.5–2 mm. **FR**: 8–12 mm, 6–10 mm wide, fine-tomentose. Rocky slopes, washes; < 1200 m. se DMoj, DSon; w AZ. Mar–May ★

T. ilicifolius Coville & Gilman (p. 1005) HOLLY-LEAVED TETRACOCCUS **ST**: twigs sparsely and finely brown-tomentose, glabrous in age. **LF**: gen opposite or 3-whorled; blade 15–30 mm, lanceolate to ovate or wide-elliptic, sharp-toothed, teeth 8–20 per lf, tip obtuse to acute. **STAMINATE INFL**: panicle, ± dense; fls ± sessile; pedicel << 0.5 mm. **STAMINATE FL**: sepals 7–9, ± linear to lanceolate; stamens 7–9, filaments 2–3 mm, base soft-hairy. **PISTILLATE FL**: pedicel 8–15 mm; sepals 5–8, 2–4 mm, wide-lanceolate to ovate; ovary densely tomentose, chambers 4, styles ± 3 mm. **FR**: 8–9 mm, 6–8 mm wide, brown-tomentose. Dry, rocky slopes; 600–1700 m. n DMtns (Grapevine Mtns, Panamint Range). May–Jun ★

PITTOSPORACEAE PITTOSPORUM FAMILY

Alan T. Whittemore & Elizabeth McClintock

Tree, shrub, woody vine, gen evergreen. **LF**: simple, alternate, occ ± at branch tips, often leathery, gen entire, petioled. **INFL**: panicle, cluster, or fls 1. **FL**: gen bisexual, radial; sepals 5, free or ± fused at base; petals 5, free, erect or spreading, base gen ± adherent into tube; stamens 5; ovary superior, chambers gen 2–3, style 1, stigma gen spheric. **FR**: capsule, berry. **SEED**: several, often in pulp. 9 genera, ± 200 spp.: warm temp, trop, Old World, esp Australia, New Zealand, e Asia; some orn, esp *Pittosporum*. *Sollya* moved to *Billardiera* [Chandler et al. 2007 Aust Syst Bot 20:390–401] Scientific Editors: Douglas H. Goldman, Bruce G. Baldwin.

1. Vine or climbing shrub; petals blue; fr a berry . . . . **BILLARDIERA**
1′ Tree or shrub; petals white or dark red to purple-black; fr a woody capsule . . . . : . . **PITTOSPORUM**

BILLARDIERA

Woody vine or climbing shrub; twigs slender, twining. **INFL**: cyme, terminal or axillary, few-fld; fls rarely 1. **FL**: petals free, spreading, blue; anthers adherent into cone around style [free]; ovary chambers 2. **FR**: berry; pulp sticky. **SEED**: sticky. 23 spp.: w Australia. (Jacques Julien Houtou de Labillardière, French botanist, early student of w Australia flora, 1755–1834)

B. heterophylla (Lindl.) L.W. Cayzer & Crisp AUSTRALIAN BLUEBELL **LF**: 3–6 cm, lanceolate to oblanceolate or narrow-ovate; petiole 0–0.5 cm. **INFL**: pendent. **FL**: petals 8–10 mm, ovate; style < 5 mm. **FR**: cylindric, 1.3–2 cm, ± black. Uncommon. Disturbed areas; < 200 m. CCo, SCo; native to w Australia. [*Sollya h.* Lindl.] Some in cult with white petals. CA reports of *Sollya fusiformis* (Labill.) Briq. based on misidentified *B. heterophylla*. Apr–Nov

PITTOSPORUM

Tree, shrub. **LF**: tip acute or rounded. **INFL**: panicle, umbel-like cluster, or fls 1, terminal or axillary. **FL**: gen functionally unisexual; petals gen adherent proximally, spreading distally; anthers free; ovary chambers 2–3. **FR**: capsule, woody, 2–3 valved; pulp resinous. **SEED**: sticky. ± 150 spp.: warm parts of Australia, New Zealand, Pacific islands, e Asia, Afr. (Greek: pitch, seed, from resinous seed coating) Some spp. medicinal and poisonous; saponins in *P. crassifolium*.

1. Lf abaxially white, densely hairy . . . . **P. crassifolium**
1′ Lf abaxially green, glabrous or subglabrous
  2. Lf margin ± turned under, stiff, leathery . . . . **P. tobira**
  2′ Lf flat, ± wavy, flexible, not leathery
    3. Lf pale gray-green, no odor when crushed; petals dark red to purple-black . . . . [**P. tenuifolium**]
    3′ Lf bright green, strong odor when crushed; petals white . . . . **P. undulatum**

P. crassifolium Banks & Sol. Shrub, small tree. **ST**: < 9 m; twigs densely hairy. **LF**: gen 4.5–7 cm, oblong to obovate, leathery, margin gen ± turned under; abaxially white, densely hairy; petiole 8–16 mm. **INFL**: umbel, terminal. **FL**: not fragrant; petals ± 1 cm, dark red to purple-black. **FR**: 1.5–3 cm, spheric or ovoid, dense-hairy, gen 3-valved. **SEED**: purple-black. Uncommon but aggressively spreading. Disturbed areas, coastal sage scrub; < 200 m. CCo; native to New Zealand. Nov–May

P. tobira (Thunb.) W.T. Aiton JAPANESE PITTOSPORUM, MOCK ORANGE Shrub, small tree. **ST**: < 8 m; twigs hairy early. **LF**: 3–14 cm, obovate, glabrous, leathery, margin turned under; petiole 5–15 mm. **INFL**: umbel-like cluster, terminal. **FL**: fragrant; petals 10–12 mm, white. **FR**: 10–18 mm, ovoid, dense-hairy, 3-valved. **SEED**: dark red. Disturbed areas; < 200 m. SCo; native to e Asia. Nov–May

P. undulatum Vent. VICTORIAN BOX, MOCK ORANGE Tree. **ST**: < 15 m; twigs glabrous or sparsely hairy. **LF**: 7–15 cm, oblong or lance-elliptic, thin, glabrous, margin ± wavy, tip acuminate; petiole 5–25 mm. **INFL**: umbel-like cluster, terminal. **FL**: fragrant; petals 10–15 mm, white. **FR**: 10–15 mm, ± round, glabrous, 2-valved. **SEED**: ± red. Uncommon but aggressively spreading. Disturbed areas, coastal scrub; < 200 m. CCo, SCo, s ChI; native to se Australia. Nov–Jun

PLANTAGINACEAE PLANTAIN FAMILY

Margriet Wetherwax, except as noted

Ann to shrub, some aquatic. **LF**: basal or cauline, alternate or opposite (whorled), simple, entire to dentate or lobed, venation gen pinnate; stipules 0. **INFL**: raceme, spike, or fls axillary in 1–few-fld clusters; fls few to many, each subtended by 1 bract. **FL**: unisexual or bisexual, radial or bilateral; sepals 4–5, gen fused at base; corolla 4–5 lobed, scarious or not, persistent or not, gen 2-lipped, upper lip gen 2-lobed, lower gen 3-lobed, spur present or not, tube sac-like at base or not; stamens 2 or 4, alternate corolla lobes, epipetalous, staminode 0 or 1–2, anthers opening by 2 slits; ovary superior, [1]2–4-chambered, style 1, stigma lobes 0 or 2. **FR**: gen a capsule, septicidal, loculicidal, circumscissile, or dehiscing by terminal slits or pores. 110 genera, ± 2000 spp.: worldwide, esp temp. [Angiosperm Phylogeny Group 1998 Ann Missouri Bot Gard 85:531–553; Olmstead et al. 2001 Molec Phylogen Evol 16:96–112] Veronicaceae sensu Olmstead et al. Recently treated to incl Callitrichaceae, Hippuridaceae, and most non-parasitic CA genera of Scrophulariaceae (exc *Buddleja*, *Limosella*, *Mimulus*, *Myoporum*, *Scrophularia*, *Verbascum*). CA *Maurandya* moved to *Holmgrenanthe* and *Maurandella*. *Limnophila* ×*ludoviciana* Thieret an occ agricultural weed in rice fields. *Hebe* ×*franciscana* (Eastw.) Souster, *H. speciosa* (R. Cunn.) Andersen only cult. Scientific Editors: Robert Patterson, Bruce G. Baldwin.

1. Pl gen aquatic, or pls in wet soil or mud
  2. Corolla and calyx clearly present; pls in wet soil or mud
    3. Stamens 4, staminode 0 . . . . ²**BACOPA**
    3′ Stamens 2, staminodes 2 (forming corolla throat ridges and with free, forked, filament-like tips in *Lindernia*)
      4. Distal lvs scale-like; corolla 4–7 mm . . . . ²**DOPATRIUM**
      4′ Distal lvs lanceolate to ovate; corolla 7–10 mm . . . . ²**LINDERNIA**
  2′ Corolla 0; calyx gen 0, exc occ present as minute rim on ovary; pls aquatic
    5. Lvs gen opposite; ovary superior, styles 2 . . . . CALLITRICHE
    5′ Lvs in whorls of 6–12; ovary inferior, style 1 . . . . HIPPURIS
1′ Pl gen terrestrial, gen of drier habitats
  6. Corolla dry, scarious, translucent, persistent in fr; lvs gen basal, veins gen ± parallel; fr circumscissile. . . . **PLANTAGO**
  6′ Corolla of petal-like texture, variously colored, but not scarious or translucent, gen not persistent in fr;
    lvs basal or basal and cauline, veins gen pinnate or palmate; fr gen not circumscissile
    7. Lvs gen alternate (exc proximal gen opposite in *Antirrhinum*, *Cymbalaria*, *Mohavea*, *Pseudorontium*;
      proximal-most opposite on seedlings in *Maurandella*), not all basal
      8. Corolla base neither spurred nor sac-like, throat open, floor without longitudinal folds, or throat ±
        closed; fr gen loculicidal or opening by irregular bursting near tip
        9. Corolla 40–60 mm, lower lip not swollen, throat open; fr gen loculicidal . . . . **DIGITALIS**
        9′ Corolla 8–9 mm, lower lip swollen, ± closing throat; fr opening by irregular bursting near tip
        . . . . **PSEUDORONTIUM**

8′ Corolla base spurred or sac-like (exc *Holmgrenanthe*, corolla throat open, floor with 2 longitudinal folds); fr gen dehiscent by terminal slits or pores (circumscissile)

  10. Lvs palmately veined, entire to lobed; petioles 5–33 mm

    11. Corolla base spurred . . . . **CYMBALARIA**

    11′ Corolla base not spurred

      12. Lvs ± round to reniform; neither sts or petioles twining; corolla white to pale yellow; pedicels 1–4 mm . . . . **HOLMGRENANTHE**

      12′ Lvs hastate to sagitate; corolla pink, red, to violet distally; sts or petioles twining; pedicels 12–47 mm . . . . **MAURANDELLA**

  10′ Lvs not palmately veined, gen entire, gen ± sessile (or petioles ≤ 30 mm)

    13. St twining or decumbent

      14. Pl glabrous (exc st base, infl); pedicels twining or U-shaped, or bractlets subtended by twining branchlets, these reduced at distal nodes; lvs linear to ovate . . . . ²**ANTIRRHINUM**

      14′ Pl hairy; pedicels straight; branchlets subtending bracts 0; lvs ovate to hastate or sagittate — fr circumscissile . . . . ²**KICKXIA**

    13′ St erect

      15. Corolla base spurred; pl gen glabrous

        16. Fls yellow, red-violet, purple, or occ 2-colored; corolla lower lip ≥ upper lip . . . . ²**LINARIA**

        16′ Fls blue to violet; corolla lower lip >> upper lip . . . . ²**NUTTALLANTHUS**

      15′ Corolla base sac-like; pl gen hairy (at least infl), rarely glabrous throughout

        17. Stamens 4, anther sacs 2 per stamen; fr asymmetric, dehiscing by 1–2 pores near tip . . . . . ²**ANTIRRHINUM**

        17′ Stamens 2, anther sacs 1 per stamen; fr symmetric, dehiscent by irregular slit at tip . . . . **MOHAVEA**

7′ Lvs gen opposite, sometimes whorled or all basal

  18. Stamens 2, staminodes 0 or 2

    19. Lvs basal, long-petioled; infl scapose . . . . **SYNTHYRIS**

    19′ Lvs cauline, sessile to short-petioled; infl not scapose

      20. Corolla ± radial, gen 4-lobed, tube << lobes; sepals gen 4(5) . . . . **VERONICA**

      20′ Corolla 2-lipped, 5-lobed, tube > lobes; sepals gen 5

        21. Anther sacs of each stamen separated, parallel; corolla tube 4-angled . . . . **GRATIOLA**

        21′ Anther sacs of each stamen touching, not parallel; corolla tube cylindric

          22. Distal lvs scale-like; corolla 4–7 mm . . . . ²**DOPATRIUM**

          22′ Distal lvs lanceolate to ovate; corolla 7–10 mm . . . . ²**LINDERNIA**

  18′ Stamens 4, staminode 0 or 1

    23. Calyx tube well developed

      24. Stamens incl in keeled lower central corolla lobe . . . . **COLLINSIA**

      24′ Stamens exserted, lower central corolla lobe not keeled . . . . **TONELLA**

    23′ Calyx segments ± free

      25. Corolla spurred at base of tube

        26. St decumbent; fl 1 in axils . . . . ²**KICKXIA**

        26′ St erect; infl a terminal raceme

          27. Fls yellow, red-violet, purple, or occ 2-colored; corolla throat swelling obvious, not white-ridged . . . ²**LINARIA**

          27′ Fls blue; corolla throat swelling ± obscure, white-ridged . . . . ²**NUTTALLANTHUS**

      25′ Corolla not spurred, occ sac-like at base

        28. Staminode 0

          29. Lvs in whorls of 3; fls > 2 cm, bright red . . . . **GAMBELIA**

          29′ Lvs opposite; fls < 2 cm, not red

            30. Lvs gen narrowly ± ovate to ± round, ± entire; corolla gen white to pink . . . . ²**BACOPA**

            30′ Lvs ± lanceolate, toothed; corolla violet to purple . . . . **STEMODIA**

        28′ Staminode 1

          31. Fertile filament bases glabrous, attached to corolla at different levels . . . . **PENSTEMON**

          31′ Fertile filament bases densely hairy, attached to corolla at 1 level

            32. Subshrub or shrub; anthers glabrous; seeds angled . . . . **KECKIELLA**

            32′ Per; anthers woolly; seeds winged . . . . **NOTHOCHELONE**

ANTIRRHINUM SNAPDRAGON

Margriet Wetherwax & David M. Thompson

Ann, per, glabrous to hairy. **ST:** ascending, erect, or vine-like, often clinging by twining pedicels or branchlets. **LF:** proximal gen opposite, distal alternate, gen reduced distally on st; veins pinnate. **INFL:** cleistogamous or opening; raceme or fl 1 in axils. **FL:** uppermost calyx lobe gen largest; corolla tube of opening fls truncate or with rounded sac-like extension at base, lower lip base gen swollen, closing throat; stamens 4, gen incl, staminode 0; style incl, straight or curved, glabrous or glandular-puberulent to near tip, stigma inconspicuous. **FR:** ovoid to spheric; chambers 2, gen dehiscent by 1–2 pores near tip, lower chamber gen larger, upper occ indehiscent. **SEED:** many, gen with tubercles or netted ridges, winged or not. 35 spp.: w N.Am, w Medit. (Greek: nose-like, from corolla shape) [Oyama & Baum 2004 Amer J Bot 91:918–925; Vargas et al. 2004 Pl Syst Evol 249:151–172] N.Am taxa more closely related to *Mohavea* than to Medit taxa; revision needed. *A. cyathiferum* moved to *Pseudorontium*.

1. Pedicels 3–10 cm, twining or with U-shaped hook
  2. Corolla of opening fls yellow and gold with maroon flecks on lower lip; fr fragile, opening by irregular
    bursting on sides . . . . *A. filipes*
  2′ Corolla of opening fls lavender to deep blue-purple, veins darker; fr firm, dehiscent by 2 slits at tip . . . . . *A. kelloggii*
1′ Pedicels 0.1–2.5 cm, not twining, not hooked
  3. Proximal-most fls not subtended by branchlets; fls all opening; ann to per; pls self-supporting
    4. Lower calyx lobes ≥ corolla tube; seed gen smooth, 1 side flattened, with raised, rough, irregular border
      . . . . [*A. orontium*]
    4′ Lower calyx lobes < to << corolla tube; seed netted, ridged, or winged
      5. Corolla 25–45 mm . . . . [*A. majus*]
      5′ Corolla 7–18 mm
        6. Fls in terminal racemes; corolla 13–18 mm, pale pink to red; often per from caudex
          7. Pl sticky-glandular-hairy . . . . *A. multiflorum*
          7′ Pl glabrous . . . . *A. virga*
        6′ Fls 1 in lf axils; corolla gen white, cream, or pale lavender, often with violet or purple veins or marks,
          7–14 mm; ann
          8. Corolla veins violet; swollen lower lip base with dense, cylindric hairs to 1 mm; KR, NCoRI, w CaR,
            n ScV . . . . *A. cornutum*
          8′ Corolla veins not contrasting in color; swollen lower lip base with sparse, sphere-tipped hairs;
            SN (esp SNF) . . . . *A. leptaleum*
  3′ Proximal-most fls and often others gen subtended by branchlets; fls all opening or some cleistogamous;
    ann, rarely bien; pls often not self-supporting, but climbing or sprawling
    9. Calyx lobes ± equal
      10. Only st base and infl hairy; upper fr chamber indehiscent . . . . *A. coulterianum*
      10′ Pl hairy throughout; both fr chambers dehiscent . . . . *A. nuttallianum*
        11. Hairs very dense, very fine, of mixed lengths, tips ± not enlarged . . . . subsp. *nuttallianum*
        11′ Hairs sparse to moderately dense, coarse, of ± = length, tips much-enlarged . . . . subsp. *subsessile*
    9′ Calyx lobes unequal
      12. Fls cleistogamous and opening; corolla of opening fls 5–7 mm, white, veins violet . . . . *A. kingii*
      12′ Fls all opening; corolla 8–20 mm, white to tan, cream tinged pink, or lavender
        13. Corolla lower lip base not swollen, not closing throat, throat floor with 2 longitudinal folds . . . . *A. ovatum*
        13′ Corolla lower lip base swollen, closing throat, throat floor without folds
          14. Corolla white to tan, throat floor expanded at mouth, lower lip lobes reflexed, poorly developed
            . . . . *A. subcordatum*
          14′ Corolla lavender, throat uniformly narrowed, lower lip lobes thrust forward, erect or
            spreading conspicuous . . . . *A. vexillocalyculatum*
            15. Proximal st glabrous or with nonglandular hairs to 3 mm; corolla 11–17 mm —
              s NCoR, ScV (Sutter Buttes), SnFrB, n SCoRI . . . . subsp. *vexillocalyculatum*
            15′ Proximal st glandular-hairy, with or without nonglandular hairs; corolla 8–14 mm
              16. Branchlets subtending fls with 1 lf at proximal-most node; KR, NCoR,
                CaRF . . . . subsp. *breweri*
              16′ Branchlets subtending fls gen with 2 lvs at proximal-most node; n&c SN . . . . subsp. *intermedium*

A. cornutum Benth. (p. 1005) Ann, glandular-hairy throughout. **ST:** erect, self-supporting, 6–40 cm. **LF:** petiole 0–12 mm; blade 7–43 mm, linear to oblanceolate, not reduced distally on st, tip obtuse to rounded. **INFL:** fls 1 in axils, all opening; pedicels gen 1–2(4) mm, not changing orientation in fr, subtending branchlets 0. **FL:** calyx densely short-glandular and long-nonglandular hairy, lobes 4–5.5 mm equal, lower < corolla tube; corolla 9–11 mm, white, veins violet, swollen lower lip base rounded, with dense, cylindric hairs to 1 mm. **FR:** 5–6 mm; chambers unequal, upper chamber indehiscent. **SEED:** 0.7–0.9 mm, ovoid, black, ridged, wing 0. *n*=16. Uncommon. Dry stream margins, disturbed areas, often on serpentine; < 1220 m. KR, NCoRI, w CaR, n ScV. Jun–Aug

A. coulterianum A. DC. (p. 1005) Ann, glabrous exc st base, infl hairy. **ST:** erect but weak, often clinging to other pls or debris, 12–150 cm. **LF:** basal rosette occ present; petiole 0–3 cm, winged; blade < 11 cm, linear to lanceolate, tip acute to rounded. **INFL:** raceme, terminal, fls cleistogamous and opening; pedicels 1–5 mm, proximal-most gen to 13 mm, subtended by twining branchlets. **FL:** calyx densely glandular hairy, hairs to 2.5 mm, lobes 3–6.4 mm, ± equal; corolla 9–12 mm, white to lavender; lower stamens becoming exserted. **FR:** 5–10 mm; upper chamber indehiscent. **SEED:** ± 1 mm, ovoid, black, netted and ridged. *n*=15. Among shrubs in desert, gen on burns elsewhere; < 2700 m. s SCoRO, SW (exc ChI), nw edge DSon; n Baja CA. Apr–Jul

A. filipes A. Gray (p. 1005) Ann, glabrous. **ST:** vine-like, climbing, 9–100 cm, exc st wooly at base. **LF:** petiole 0–5 mm; blade 6–50 mm, linear to ovate; distal linear, sessile. **INFL:** fls 1 in axils, opening and cleistogamous; pedicels 3–10 cm, thread-like, twining, subtending branchlets gen 0. **FL:** calyx ± glabrous, lobes 3–4.4 mm, equal; cleistogamous fls minute, white; corolla of opening fls 10–13 mm, yellow and gold with maroon flecks on lower lip. **FR:** 3–5 mm, spheric; chambers equal, fragile, opening by irregular bursting on sides. **SEED:** ± 1 mm, ovoid to spheric, black; ridges 4–6, thick, wing-like. *n*=15. On shrubs, debris, gen in washes; < 1650 m. D; to sw UT, w AZ, nw Mex. Mar–May

A. kelloggii Greene (p. 1005) Ann, ± glabrous. **ST:** ascending to vine-like, gen clinging to other pls or debris, 7–80 cm. **LF:** petiole 0–7 mm; blade 10–58 mm, linear to ovate, distal sessile. **INFL:** fls 1 in axils, opening and cleistogamous; pedicels 3–9 cm, thread-like, twining or with U-shaped hook, subtending branchlets gen 0. **FL:** calyx ± glabrous, lobes 5–8 mm, equal; cleistogamous fls minute, white; corolla of opening fls 10–14 mm, lavender to deep blue-purple, veins darker. **FR:** 5–7 mm, ovoid to spheric; chambers equal, firm, dehiscent by 2 slits at tip. **SEED:** < 1 mm, ovoid, black; tubercles scattered, block-like. *n*=15. Disturbed areas, esp burns; < 1300 m. s NCoRO, CW, SW; nw Baja CA. Apr–Jun

A. kingii S. Watson (p. 1005) Ann, gen glabrous proximal to infl. **ST:** erect but weak, often clinging to other pls or debris, 3–45 cm. **LF:** sessile, 2–35 mm, linear to elliptic. **INFL:** ± sparsely glandular-puberulent; fls 1 in axils, opening and cleistogamous; distal pedicels gen 1–4 mm, proximal-most gen to 12 mm, subtended by twining branchlets. **FL:** calyx ± glandular-puberulent, lobes unequal, upper 4–5.5 mm; cleistogamous fls minute, white; corolla of opening fls 5–7 mm, white, veins violet. **FR:** 3–4.5 mm, oblique-ovoid to spheric, chambers unequal, dehiscent by 1 or 2 pores at tip. **SEED:** < 1 mm, ovoid, ± black, netted, ridged, or tubercled. *n*=15. Uncommon. Washes, scree; 500–2300 m. e slope SNH (Kern Co.), SNE, DMtns (esp ne San Bernardino Co.); to se OR, w UT, nw AZ. Apr–Jul

A. leptaleum A. Gray (p. 1005) Ann, glandular-hairy. **ST:** erect, self-supporting, 8–60 cm. **LF:** petiole 1–7 mm; blade 7–40 mm, oblanceolate to ovate, obtuse to truncate at tip, not reduced distally on st. **INFL:** fls 1 in axils, all opening; pedicels 1–2 mm, subtending branchlets 0. **FL:** calyx glandular- and nonglandular-hairy, < corolla tube, lobes 3–7 mm, ± equal; corolla 7–14 mm, white to pale lavender, veins not contrasting. **FR:** 5–8 mm; dehiscent by 2 pores at tip of lower chamber, upper chamber indehiscent. **SEED:** ± 1 mm, ovoid, black, netted and ridged. *n*=16. Uncommon. Small washes, disturbed areas; 200–2100 m. SN (esp SNF). Jun–Aug

A. multiflorum Pennell, illeg. (p. 1011) Ann, per, densely glandular-hairy, sticky. **ST:** erect, self-supporting, 3–15 dm, often from woody caudex. **LF:** sessile, 8–63 mm, linear to lanceolate, tip acute to obtuse. **INFL:** raceme, terminal, fls all opening; pedicels 2–4(10) mm, subtending branchlets 0. **FL:** calyx lobes 5–13.5 mm unequal, < corolla tube; corolla 13–18 mm, pale pink to red with tan-brown withered area on lower lip. **FR:** 7–11 mm, oblique-ovoid; dehiscent by 3 pores at tip, 1 in upper chamber, 2 in lower. **SEED:** 0.8–1.3 mm, ovoid, black, ± ridged. *n*=16. Rocky or disturbed areas, burns; < 2200 m. n SNF (Calaveras Co.), s SnFrB, SCoR, n ChI, TR, PR. *A. m.* a later homonym; legitimate name in *A.* evidently not available. Apr–Aug

A. nuttallianum A. DC. Ann, rarely bien, glandular-hairy. **ST:** erect but weak, often clinging to other pls or debris. **LF:** petiole < 25 mm; blade < 60 mm, ovate, acute, mucronate. **INFL:** fl 1 in axils, opening and cleistogamous; pedicels gen 2–18(25) mm, proximal-most gen subtended by twining branchlets, fl branchlets, or both. **FL:** calyx lobes ± equal; cleistogamous fls minute, white; corolla of opening fls 7–12 mm, lavender to blue-purple with 1–2 blue-veined white blotches on lower lip base and gold hairs in mouth (hair color unique in CA pls). **FR:** oblique-ovoid; chambers unequal, dehiscent by 3 pores at tip, 1 in upper chamber, 2 in lower. **SEED:** < 1 mm, ovoid, brown. *n*=16. Subspp. hybridize near coast.

subsp. ***nuttallianum*** Pl 6–200 cm; hairs very dense, very fine, of mixed lengths, tips ± not enlarged. **LF:** opposite at proximal-most 5–8 nodes of main st; petiole < 25 mm; blade < 60 mm. **INFL:** fls not present at proximal-most 1-lvd nodes; distal pedicels gen > 6 mm. **FL:** calyx lobes 1–3.4 mm; ± white blotch on lower corolla lip uninterrupted. **FR:** 3.5–8 mm. **SEED:** ridges 9–10, entire, longitudinal. Rocky areas, gen inland areas, esp burns; < 1300 m. s SCo, SnBr, PR; n Baja CA. May–Jul

subsp. ***subsessile*** (A. Gray) D.M. Thomps. (p. 1011) Pl 6–90 cm; hairs sparse to moderately dense, coarse, of ± = length, tips much enlarged. **LF:** opposite at proximal-most 2–5 nodes of main st; petiole < 14 mm; blade < 40 mm. **INFL:** fls at all 1-lvd nodes; distal pedicels gen < 6 mm. **FL:** calyx lobes 0.7–2.5 mm; ± white blotch on lower corolla lip gen interrupted by lavender. **FR:** 3–11 mm. **SEED:** ridges 9–10, broken, fragments longitudinal or unpatterned. Stabilized coastal dunes, rocky or disturbed areas; < 1300 m. s CCo, SCo, ChI; AZ, s Baja CA. Mar–Aug

A. ovatum Eastw. (p. 1011) OVAL-LEAVED SNAPDRAGON Ann, glandular-hairy. **ST:** erect but weak, often clinging to other pls or debris, 8–60 cm. **LF:** petiole 0–15 mm; blade 4–49 mm, lanceolate to

obovate. **INFL:** raceme-like or fls 1 in axils, fls all opening; pedicels 2–5 mm, proximal-most gen subtended by twining branchlets. **FL:** calyx lobes unequal, 4–5 mm, upper 11–12 mm; corolla 17–20 mm, cream, pink-tinged, lower lip base not swollen, not closing throat, throat floor with 2 longitudinal folds (unique in genus). **FR:** 7–10 mm, oblique-ovoid, chambers unequal; style projecting forward. **SEED:** ± 1 mm, ovate, black, ridged. *n*=16. Heavy, adobe-clay soils on gentle, open slopes, also disturbed areas; 200–1400 m. s SnJV (esp w Kern, e San Luis Obispo cos.), s SCoRI, w WTR. Abundant every 20–50 yrs. May–Jul ★

A. subcordatum A. Gray (p. 1011) DIMORPHIC SNAPDRAGON Ann, nonglandular-hairy proximal to infl. **ST:** erect but weak, often clinging to other pls or debris, 8–90 cm. **LF:** petiole 0–20 mm; blade 5–60 mm, ovate, rounded at tip. **INFL:** raceme-like or fls 1 in axils, glandular-hairy, fls all opening; pedicels 1–3 mm, proximal-most subtended by twining branchlets. **FL:** calyx lobes unequal, 7–9 mm, upper 8–14 mm; corolla 13–17 mm, white to tan, throat floor expanded at mouth, lower lip lobes reflexed, poorly developed. **FR:** ± 6 mm, dehiscent by 3 pores at tip, 1 in upper chamber, 2 in lower. **SEED:** ± 1 mm, ovoid, dark brown to black, netted, ridged. *n*=15. Gentle, open slopes on serpentine, often under shrubs; 150–800 m. n&c NCoRI. May–Jun ★

A. vexillocalyculatum Kellogg Ann, branched or not, glabrous to hairy proximal to infl. **ST:** erect but weak, often clinging to other pls or debris. **LF:** petiole 0–27 mm; blade 1.5–60 mm, elliptic to ovate or oblong, tip obtuse to rounded. **INFL:** raceme-like or fls 1 in axils, glandular-hairy, fls all opening; pedicels 1–4 mm, proximal-most subtended by twining branchlets, fl branchlets, or both. **FL:** calyx lobes unequal; corolla lavender, veins often vaguely darker, throat uniformly narrowed, curved upward at mouth, lower lip lobes conspicuous, thrust forward, erect or spreading. **FR:** 4–8 mm; dehiscent by 3 pores at tip, 1 in upper chamber, 2 in lower. **SEED:** 0.7–1.3 mm, ovoid, dark brown to black, tubercled, ridged. *n*=15. [*Antirrhinum vexillo-calyculatum*, orth. var.] Subspp. intergrade in NCoR (exc *A. vexillocalyculatum* subsp. *v.*).

subsp. ***breweri*** (A. Gray) D.M. Thomps. **ST:** 7–76 cm, proximal st hairs < 3 mm, all glandular. **INFL:** branchlets subtending fls with 1 lf at proximal-most node. **FL:** upper calyx lobe 3.3–5.3 mm, lower 2.4–3.6 mm; corolla 8–12 mm, some veins dark. Gravelly lower slopes of rockslides, disturbed areas, often on serpentine; < 2000 m. KR, NCoR, CaRF; sw OR. [*A. b.* A. Gray] Jun–Sep

subsp. ***intermedium*** D.M. Thomps. **ST:** 10–100 cm, proximal st hairs glandular, often also longer (< 3 mm), nonglandular. **INFL:** branchlets subtending fls gen with 2 lvs at proximal-most node. **FL:** upper calyx lobe 4.4–10 mm, lower 3–6 mm; corolla 10–14 mm, some veins dark. Gravelly lower slopes of rockslides, disturbed areas, often on serpentine; 100–1400 m. n&c SN. Jun–Sep

subsp. ***vexillocalyculatum*** (p. 1011) **ST:** 7–170 cm, proximal st glabrous or hairs sparse, ≤ 3 mm, nonglandular. **INFL:** branchlets subtending fls with 2 lvs at proximal-most node. **FL:** upper calyx lobe 6.8–14 mm, lower 4–8.6 mm; corolla 11–17 mm, darker veins 0. Gravelly lower slopes of rockslides, disturbed areas, often on serpentine; < 1200 m. n NCoR, ScV (Sutter Buttes), SnFrB, n SCoRI. Jun–Aug

A. virga A. Gray (p. 1011) TWIG-LIKE SNAPDRAGON Per from woody caudex, glabrous. **ST:** erect, self-supporting, 40–220 cm, lfy, wand-like. **LF:** spirally arranged, sessile, 5–12 cm, linear, acute, reduced basally to persistent scales, distal reduced to bracts in infl, 4–11 mm. **INFL:** raceme, terminal, open, fls all opening; pedicels 2–6 mm, subtending branchlets 0. **FL:** calyx lobes 6–8 mm, < corolla tube, equal; corolla 13–18 mm, pink, lips withering, turning brown. **FR:** 6–9 mm, oblique-ovoid; chambers unequal, ± truncate distally, dehiscent by 2 pores each. **SEED:** 1–1.3 mm, ovoid, black, ± ridged. *n*=16. Openings in chaparral, rocky areas, often on serpentine; 200–2000 m. s NCoRH, s NCoRI. Jun–Jul ★

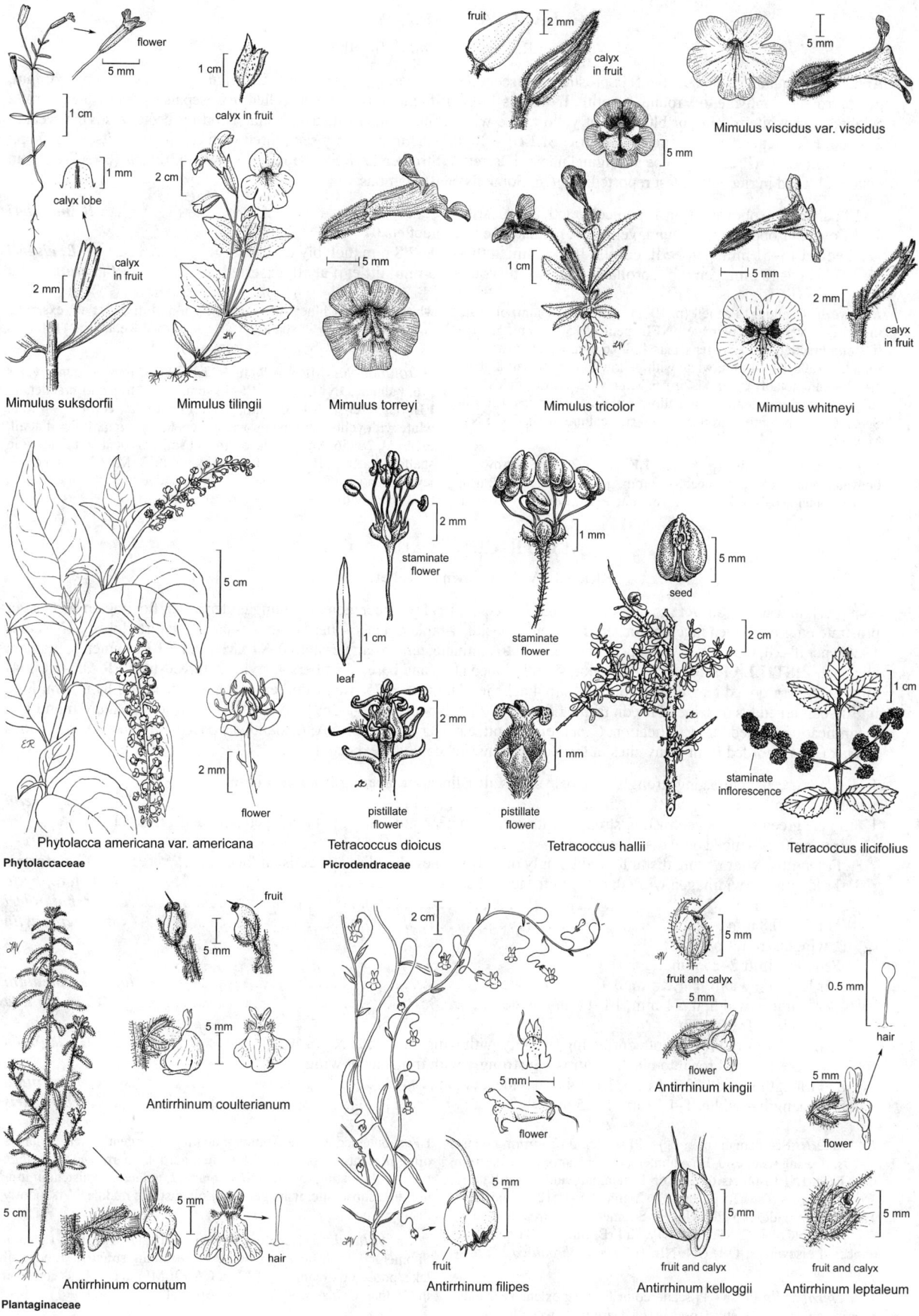

flower
5 mm
1 cm
1 mm
calyx lobe
2 mm
calyx in fruit
Mimulus suksdorfii

1 cm
calyx in fruit
2 cm
Mimulus tilingii

5 mm
Mimulus torreyi

1 cm
Mimulus tricolor

fruit
2 mm
calyx in fruit
5 mm
5 mm
Mimulus viscidus var. viscidus

5 mm
2 mm
calyx in fruit
Mimulus whitneyi

5 cm
2 mm
flower
Phytolacca americana var. americana
Phytolaccaceae

2 mm
staminate flower
1 cm
leaf
2 mm
pistillate flower
Tetracoccus dioicus
Picrodendraceae

1 mm
staminate flower
1 mm
pistillate flower
Tetracoccus hallii

5 mm
seed
2 cm
1 cm
staminate inflorescence
Tetracoccus ilicifolius

fruit
5 mm
5 mm
Antirrhinum coulterianum

5 cm
5 mm
hair
Antirrhinum cornutum
Plantaginaceae

2 cm
5 mm
flower
5 mm
fruit
Antirrhinum filipes

5 mm
fruit and calyx
5 mm
flower
Antirrhinum kingii

5 mm
fruit and calyx
Antirrhinum kelloggii

0.5 mm
hair
5 mm
flower
5 mm
fruit and calyx
Antirrhinum leptaleum

BACOPA WATER-HYSSOP

Kim R. Kersh & John L. Strother

Ann, ± per, aquatic or subaquatic. **ST:** prostrate to ascending, or floating, gen < 6 dm. **LF:** cauline, opposite, gen < 40 mm, gen narrowly ± obovate to ± round, ± entire. **INFL:** fls 1–3 per lf axil, sessile or pedicelled. **FL:** sepals (4)5, unequal; corolla 5-lobed, gen white to pink or blue, throat yellow, occ with white; stamens 4; ovary subtended by nectary, stigma weakly 2-lobed. **FR:** loculicidal or septicidal, spheric. **SEED:** > 30, 0.1–0.3 mm, evenly sculptured with rectangular pits. ± 100 spp.: trop, warm temp. (Presumed to be aboriginal name) [Barrett & Strother 1978 Syst Bot 3:408–419] *B. repens* (Sw.) Wettst. an historical weed in rice fields, last reported in 1976. Some fls cleistogamous.

1. Lf narrowly ± obovate, vein 1; pedicel bractlets 2, near tip . . . . ***B. monnieri***
1′ Lf ovate or obovate to ± round, veins > 6, palmate; pedicel bractlets 0
  2. Pedicel 15–51 mm, gen >> lf; corolla 10–14 mm; anthers 1.8–2.8 mm, notably exserted . . . . ***B. eisenii***
  2′ Pedicel 5–18 mm, gen < lf; corolla 5–8 mm; anthers 0.6–1.3 mm, little, if at all, exserted . . . . ***B. rotundifolia***

B. eisenii (Kellogg) Pennell (p. 1011) **LF:** 12–34 mm, obovate to ± round; veins > 6, palmate. **INFL:** pedicel 15–51 mm, gen >> lf, stout; bractlets 0. **FL:** outer sepals 4.3–6.6 mm, widely ovate to ± round; corolla 10–14 mm, white with yellow in tube/throat; anthers 1.8–2.8 mm, notably exserted. 2*n*=56. Muddy places, rice fields; wet soil or rooted and floating in shallow water; < 100 m (< 1200 in SNE). GV (apparently introduced with rice culture ScV), SNE; NV. May–Oct

B. monnieri (L.) Wettst. (p. 1011) **LF:** gen 5–25 mm, narrowly ± obovate; vein 1. **INFL:** pedicel 8–40 mm, gen > lf; bractlets 2, near tip. **FL:** outer sepals 5–7 mm, lanceolate to ovate; corolla 8–10 mm, white to pink or blue; anthers 0.6–1.1 mm, little, if at all, exserted. 2*n*=64. Wet soil or in shallow water, disturbed wetlands; < 180 m. s SCo, DSon; to s US; native to trop. All yr ◆

B. rotundifolia (Michx.) Wettst. **LF:** 12–36 mm, ± round; veins > 6, palmate. **INFL:** pedicel 5–18 mm, gen < lf, stout; bractlets 0. **FL:** outer sepals 3.1–5.3 mm, ovate to ± round; corolla 5–8 mm, white with yellow in tube/throat; anthers 0.6–1.3 mm, little, if at all, exserted. 2*n*=56. Muddy places, in wet soil or rooted and floating in shallow water; < 100 m (< 1100 m in MP). GV, MP (Fall River Valley, Shasta Co.); to ID, e US, AZ, Mex; native to c US. In CA often associated with rice cult. May–Aug

CALLITRICHE WATER-STARWORT

Richard V. Lansdown & Robert E. Preston

Ann [per], in water or on wet ground; monoecious [dioecious]. **ST:** slender, gen ascending under water, floating on surface, or prostrate on ground, gen much-branched. **LF:** gen opposite, 4-ranked, lance-linear to spoon-shaped, entire [lobed]. **INFL:** fls 1–2(4) per lf axil, bracts 2, ± white, gen inflated, or 0. **FL:** minute, unisexual; perianth 0. **STAMINATE FL:** stamen 1, filament elongate. **PISTILLATE FL:** ovary superior, slightly lobed at tip and base, chambers 4, styles 2, thread-like. **FR:** 0.6–1.6[2.4] mm, ± dry, ± grooved lengthwise, splitting into 4 achene-like units. ± 75 spp.: trop, temp. (Greek: beautiful hair, from long linear submerged lvs of some of Medit taxa) [Lansdown 2009 Novon 19:364–369] Taxonomically difficult; mature fr and 10× magnification needed for identification. *C. peploides* Nutt. a nursery weed in s CA; *C. stenocarpa* Hegelm., misappl. to various CA taxa, but recorded from many sites in the Rocky Mtns and could be expected in CA.

1. Lf dark green, contrasting strongly with pale st; lvs all ± linear or tapering from near base; no lvs with > 1 central vein . . . . ***C. fassettii***
1′ Lf pale green and not contrasting strongly with st; lvs ± linear and vein 1, to broad, petioled, and veins > 1
  2. Fr wing 0 or winged only distal to middle
    3. Fr length ≥ width, wing distal to middle only or 0; fr with evident lines of cells on faces . . . . ***C. palustris***
    3′ Fr length ± = width, gen wing 0; cells on fr faces obscure . . . . ***C. heterophylla***
      4. Fr 0.8–1.4 mm . . . . var. ***bolanderi***
      4′ Fr 0.6–0.8 mm . . . . var. ***heterophylla***
  2′ Fr winged from tip to base
    5. Peduncle in fr 2–25 mm
      6. Fr length ≥ width, 1–1.2 mm, 0.9–1.2 mm wide . . . . ***C. longipedunculata***
      6′ Fr length < width, 0.7–1 mm, 1.1–1.4 mm wide . . . . ²***C. marginata***
    5′ Peduncle in fr 0–1 mm
      7. Mature fr gray-brown, not contrasting strongly with wing . . . . ***C. stagnalis***
      7′ Mature fr dark brown to black, contrasting strongly with translucent wing
        8. Fr length < width, 0.7–1 mm, 1.1–1.4 mm wide . . . . ²***C. marginata***
        8′ Fr length = width, 1–1.5 mm, 1–1.5 mm wide . . . . ***C. trochlearis***

C. fassettii Schotsman (p. 1011) Pl scattered or forming dense stands, floating rosettes 0. **LF:** submersed, ± linear or tapering from near base. **INFL:** bracts 0; fls 1 or 1 staminate and 1 pistillate in 1 of a pair of axils. **FR:** peduncle < 2 mm; 1.5–1.8(2.5) mm, 1.1–1.5(1.8) mm wide, broadly winged. Submerged in streams, ponds; < 2100 m. NCoRI, CaRF, c SN, GV, SnFrB, SnBr, MP (exc Wrn), expected elsewhere; OR, ON, NE. [*C. hermaphroditica* L., misappl.] Apr–Jul

C. heterophylla Pursh Pl scattered or forming extensive stands, submerged or with floating rosettes, or dense mats when terrestrial. **LF:** submersed lvs linear-oblong; floating, emergent, or terrestrial lvs often present, spoon-shaped lvs often abruptly narrowed to petiole; often with floating rosette. **INFL:** bracts 2, inflated, persistent in fr or not. **FR:** subsessile; wing gen 0 or only distal to middle. Vars. poorly distinguished.

  var. ***bolanderi*** (Hegelm.) Fassett (p. 1011) **FR:** 0.8–1.4 mm. Submerged, with floating rosettes or becoming stranded at edge of lake, pools, or streams; < 2000 m. CA-FP, MP; to BC. Less common in CA than *C. heterophylla* var. *h.* Apr–Jul

var. **heterophylla** (p. 1011) **FR:** 0.6–0.8 mm. Submerged, with floating rosettes or becoming stranded at edge of lake, pools, or streams; < 2000 m. NW, GV, CW, SW; Can, e US, n Mex; naturalized in New Zealand. Apr–Jul

C. longipedunculata Morong **LF:** spoon-shaped, gradually to abruptly narrowed to petiole; blade 2.8–5 mm, 1.8–3.3 mm wide; petiole 1–5.7 mm. **INFL:** bracts 0.4–0.8 mm, early-deciduous; fls gen 1 staminate, 1(2) pistillate in each axil. **FR:** peduncles 2–25 mm; 1–1.2 mm, 0.9–1.2 mm wide, black; wing 0.03–0.1, narrowly winged from tip to base. Becoming stranded or submersed in vernal pools; < 625 m. NCoRI, CaRF, GV, SnFrB, w PR. Mar–May

C. marginata Torr. (p. 1011) **LF:** oblong to spoon-shaped, 4.8–7.1(7.2) mm, 0.7–1.4 mm wide. **INFL:** bracts 0; fls 1–2 pistillate and 1 staminate in each axil. **FR:** 0.7–1 mm, 1.1–1.4 mm wide, length < width, black when mature; wing 0.03–0.2 mm, slightly wider at base than at tip. Becoming stranded (often in vernal pools); < ± 1500 m. CA-FP; to BC, Baja CA. Mar–Jun

C. palustris L. (p. 1011) **LF:** floating rosette gen present; submersed lvs linear-oblong; floating, emergent, or terrestrial lvs often present, spoon-shaped lvs gen gradually narrowed to petiole; petiole 1–3.5 mm; terrestrial lvs 3–4.5 mm, 0.6–1 mm wide; blade 2.5–5.5 mm, 1–5 mm wide. **INFL:** bracts 2, 0.3–1 mm, persistent in fr or early-deciduous. **FR:** subsessile, 0.7–1 mm, 0.7–1 mm wide (wider distal to middle); wing distal to middle only or 0. Submerged, with floating rosettes or becoming stranded at edge of lake, pools, or streams; < 4000 m. CA-FP, GB; n&w US, Can, Eur, Asia; naturalized in Australia. [*C. verna* L.] May–Aug

C. stagnalis Scop. **LF:** petiole 3.5–5.5 mm, variable, from broadly parallel-sided to ± round in ×-section; blade 4–9 mm, 1–4 mm wide. **INFL:** bracts translucent, ± white, sickle-shaped, persistent; staminate bracts 0.2–1.1 mm, pistillate bracts 0.5–1.1 mm. **FR:** subsessile, 1.2–1.7 mm, 1.2–1.8 mm wide, light brown to gray; wing gen uniformly wide from base to tip, transition to fr wall gradual. Floating rosettes in permanent water, terrestrial on mud or becoming stranded by streams, or ponds; < 800 m. NCo, NCoR, n SNF; w&e N.Am; native to Eur. Apr–Jun

C. trochlearis Fassett (p. 1011) **LF:** submersed lvs lance-linear; floating, emergent, or terrestrial lvs spoon-shaped. **INFL:** bracts 2, persistent in young fr. **FR:** 1–1.5 mm, 1–1.5 mm wide, dark brown to black, groove shallow; wing uniformly wide from base to tip, transition to fr wall abrupt; pedicel gen 0–0.5 mm. Becoming stranded or submersed < ± 6 dm; < ± 500 m. NW, CW; w OR. Mar–Jun

COLLINSIA CHINESE-HOUSES

Michael S. Park & Elizabeth Chase Neese

Ann, often glandular, sometimes brown-staining. **LF:** opposite; proximal petioled. **INFL:** bracted, often interrupted; fls 1–many in lf axils. **FL:** calyx lobes 5, gen glabrous on inner surface; corolla ± pea-like, uniformly pale, or gen with pale regions, esp throat and base of upper lip (± uniformly dark in *Collinsia greenei*), gen glabrous outside, tube short, throat ± angled to tube, ± pouched on upper side, lips gen ± = throat, upper lobes 2, ± reflexed, lower lobes 3, lateral spreading, central lobe keeled, enclosing stamens and style; stamens 4, attached unequally near throat base, spur at base of upper filaments > 1 mm, vestigial, or 0; staminode gland-like; style > 2 mm, stigma minutely 2-lobed. **FR:** septicidal and loculicidal (valves 2-lobed). **SEED:** gen few, ± oblong, gen plump; inner surface ± concave. ± 20 spp.: N.Am, esp CA. (Zaccheus Collins, Philadelphia botanist, 1764–1831) Late-season fls gen atypically small.

1. Corolla ± uniformly dark purple; base of each upper lobe with 2 crested projections bulging > 1 mm away
   from throat opening . . . . **C. greenei**
1′ Corolla uniformly pale, or with pale regions, esp throat and base of upper lip; base of each upper lobe with
   projection 0 or bulging away < 1 mm away from throat opening
  2. Fl whorls dense; proximal pedicels < calyx
    3. Upper filaments with curved basal spur 1–2 mm, curved into pouch; corolla throat as wide as long
       . . . . **C. heterophylla**
      4. Corolla 10–15 mm, upper lip < lower lip; lf (sparsely) hairy abaxially . . . . var. **austromontana**
      4′ Corolla (13)15–20 mm, lips ± equal; lf glabrous abaxially (at most ciliate on midvein) . . . . var. **heterophylla**
    3′ Filaments ± unspurred or spur ≤ 0.5 mm (exc 0.8 mm in *Collinsia tinctoria*); corolla throat longer than wide
      5. Upper corolla lip ≤ 2/3 lower lip
        6. Upper lip 1/2 to 2/3 lower lip; corolla 8–12 mm — lf glabrous abaxially; c SNH . . . . **C. bartsiifolia** var. **stricta**
        6′ Upper lip < 1/2 lower lip; corolla 12–22 mm
          7. Lf glabrous abaxially; infl ± head-like; corolla upper lobes ± 1 mm, dry, ± brown, lateral lobes
             glabrous, throat barely angled to tube, pouch low; infl sparsely and finely glandular; NCo . . . . **C. corymbosa**
          7′ Lf hairy abaxially; infl gen of several whorls; corolla upper lobes >> 1 mm, red-banded, lateral
             lobes gen hairy, throat strongly angled to tube, pouch prominent; infl densely glandular, staining;
             widespread . . . . **C. tinctoria**
      5′ Corolla lips ± = in fully open fls
        8. Corolla blue to blue-purple, not drying veiny, upper lobes notched, lateral lobes and tip of lowest lobe
           gen hairy; lvs thin, entire to toothed; SCo (Riverside Co.), PR . . . . **C. concolor**
        8′ Corolla gen white to ± purple, gen drying veiny, upper lobes ± toothed (backs often touching or
           parallel); lvs ± thick, crenate; widespread . . . . **C. bartsiifolia** (in part)
          9. Corolla 15–20 mm . . . . var. **bartsiifolia**
          9′ Corolla 9–14 mm . . . . var. **davidsonii**
  2′ Fls 1 to loosely whorled; proximal pedicels > calyx
    10. Lvs gen ± lance-deltate, gen coarsely toothed; proximal-most pedicels >> calyx, distal pedicels ± =
        calyx — n&c CCo, SnFrB (San Mateo Co.) . . . . **C. multicolor**
    10′ Lvs linear to ovate or oblong, gen entire, crenate, or finely toothed; all pedicels ± equal
      11. Inner face of calyx lobes white-hairy — pls on white shale scree . . . . **C. antonina**
      11′ Inner face of calyx lobes not hairy

12. Calyx lobe tips sharp to long-tapered
  13. Upper filaments, middle lower corolla lobe glabrous; calyx ± = fr, lobes not widely spreading; seeds
    plump, oblong, unwinged
    14. Corolla 8–15 mm, main lobes gen 2–6 mm wide, obovate, throat strongly angled to tube . . . . *C. grandiflora*
    14′ Corolla 4–8 mm, main lobes ± 1 mm wide, oblong, throat barely angled to tube . . . . *C. parviflora*
  13′ Upper filaments, middle lower corolla lobe sparsely long-hairy near tip; calyx gen >> fr, lobes
    widely spreading in fr; seeds flat, ± round, ± winged — fr gen red-blotched . . . . *C. sparsiflora*
    15. Corolla 5–8 mm, throat barely angled to tube, pouch hidden by calyx. . . . . var. *collina*
    15′ Corolla 7–20 mm, throat strongly angled to tube, pouch gen evident to prominent. . . . . var. *sparsiflora*
12′ Calyx lobe tips ± blunt
  16. Infl ± glandless . . . . *C. parryi*
  16′ Infl conspicuously glandular
    17. Pedicels spreading to ascending in fr; distal-most bracts > 2 mm
      18. Middle and distal lvs oblong to ovate to (ob)lanceolate, gen < 6 × longer than wide
        19. Pl fleshy; distal lvs clasping; seeds 6–8, ± 2 mm; calyx ± = fr . . . . *C. callosa*
        19′ Pl not fleshy; distal lf bases tapered; seeds 2, ± 3 mm; calyx > fr. . . . . *C. childii*
      18′ Middle and distal lvs ± linear, gen > 6 × longer than wide
        20. Corolla gen 8–12(15) mm, >> 2 × calyx, throat strongly angled to tube, pouch prominent . . . . *C. linearis*
        20′ Corolla 4–8 mm, gen < 2 × calyx, throat barely angled to tube, pouch ± hidden by calyx . . . . *C. rattanii*
    17′ Some pedicels strongly reflexed in fr; distal-most bracts 0–2 mm . . . . *C. torreyi*
      21. Lateral corolla lobes exceeding lowest by 1–2 mm
        22. Corolla gen 6–9 mm . . . . var. *brevicarinata*
        22′ Corolla gen 4–6 mm . . . . var. *wrightii*
      21′ Lateral corolla lobes exceeding lowest by << 1 mm, or not exceeding it
        23. Lvs elliptic to ovate, gen 2–5 × longer than wide. . . . . var. *latifolia*
        23′ Lvs ± linear; gen > 6 × longer than wide . . . . var. *torreyi*

C. antonina Hardham SAN ANTONIO COLLINSIA Pl 4–15 cm.
LF: oblong, crenate **INFL:** open, ± finely scaly, gen sparsely and finely glandular; pedicels 1–3 per node, >> calyx. **FL:** calyx lobe tips ± blunt; corolla 4.5–8 mm, throat white with red-purple spots below upper lobes, lobes purple (white); upper filaments sparsely hairy, basal spur 0. **SEED:** 6–8. *n*=7. Margins of oak scrub on white shale scree; 250–400 m. n SCoRO (Monterey Co.). [*C. parryi* A. Gray, misappl.] Mar–Apr ★

C. bartsiifolia Benth. **LF:** gen 1–4 cm, ± thick, ± oblong, obtuse, crenate, rolled under, gen finely hairy. **INFL:** interrupted, ± glandular or shaggy-hairy; whorls dense; pedicel < calyx. **FL:** calyx lobe tips ± blunt; corolla drying veiny, throat longer than wide, hairy inside, lips ± equal or upper lip 1/2–2/3 lower lip, upper lobes ± oblong, ± toothed, backs often touching or parallel, lateral lobes obovate, notched or ± entire; upper filaments hairy, basal spur 0–0.5 mm. **SEED:** many, ± plump, 1–1.5 mm. *n*=7. Corolla size in populations varies with climate across yrs. Vars. not distinct in SCoRO and n WTR; study needed.

var. ***bartsiifolia*** (p. 1011) Pl (10)20–35 cm. **FL:** corolla 15–20 mm, gen white to pale lavender, lips ± equal, lateral lobes notched. Open, sandy places; gen < 600 m. NCoRO/KR (Humboldt Co.), NCoRI, CaRF, c&s SNF, Teh, GV, s SnFrB (Santa Cruz Co.), SCoRO, n WTR. Pls called *C. bartsiifolia* var. *hirsuta* (Kellogg) Pennell with upper corolla lip intermediate in size to short lip of *C. corymbosa*, historically known from n CCo (San Francisco, Marin cos.), apparently extirpated. Mar–Jun

var. ***davidsonii*** (Parish) Newsom Pl 5–20(25) cm. **FL:** corolla 9–14 mm, gen pale lavender to pink-purple, lips ± equal, upper lobes often paler than lower lobes, lateral lobes notched. Open, sandy places; 500–1300 m. SCoR, n TR, w DMoj. Apr–Jun

var. ***stricta*** (Greene) Newsom Pl 12–35 cm. **FL:** corolla 8–12 mm, fleshy to pink-lavender, upper lip 1/2–2/3 lower lip, lobes sometimes purple tipped with purple spots at base, lateral lobes ± entire. *n*=7. Chaparral, open oak woodland, dry mixed woodland; 700–1300 m. SNF/SNH, c SNH (Wawona, likely extirpated). [*C. b.* var. *davidsonii*, in part, misappl.] Last collected in 1975. May–Jul

C. callosa Parish (p. 1011) Pl 4–25 cm, stout, fleshy. **LF:** gen < 3 cm, oblong to ovate, obtuse, gen entire, ± rolled under; middle and distal lvs clasping. **INFL:** open, conspicuously glandular; bracts > 2 mm; fls 1–3 per node, pedicel > calyx, spreading to ascending.

FL: calyx ± = fr, in fr gen > 5 mm wide, base ± truncate, lobe tips ± blunt; corolla gen 7–9 mm, lavender-blue; filaments gen glabrous, upper pair rarely sparsely hairy, basal spur 0–0.5 mm. **SEED:** 6–8, ± 2 mm. Disturbed rocky slopes, open chaparral, sagebrush scrub, pinyon/juniper and pine woodland; 1000–2300 m. s SN, Teh, s SCoRO, e WTR, SnGb, DMoj/SnGb, DMoj/SNE, DMtns (Panamint, Argus ranges). Apr–Jun

C. childii A. Gray (p. 1011) Pl 8–35 cm, not fleshy. **LF:** oblong to (ob)lanceolate, gen flat, entire or finely toothed; base tapered. **INFL:** open, densely glandular; bracts > 2 mm; pedicels 2–5 per node, > calyx, spreading to ascending in fr. **FL:** calyx > fr, in fr < 5 mm wide, base rounded, lobes spreading, tips ± blunt; corolla 6–9(11) mm, ± white or pale lavender, upper lip ± = lower; filaments glabrous, basal spur 0. **SEED:** 2, ± 3 mm. Shaded slopes, open oak and mixed-conifer woodland; 1000–2200 m. c&s SNH, Teh, SCoRO, TR, PR. (Apr)May–Jul

C. concolor Greene (p. 1011) Pl 15–40 cm. **LF:** thin, narrowly oblong to widely lanceolate, entire (to toothed), subglabrous. **INFL:** interrupted, finely- to shaggy-hairy, gen finely glandular; whorls dense; pedicel gen < calyx. **FL:** calyx gen long-hairy, lobe tips widely acute, ciliate; corolla 11–15 mm, blue to blue-purple, not drying veiny, throat longer than wide, hairy inside, lips ± equal in fully open fls, upper lip evenly purple-dotted in triangular white area near base, upper and lateral lobes notched, lateral lobes obovate, gen sparsely hairy, lowest lobe gen sparsely hairy at tip; upper filaments hairy, basal spur 0–0.5 mm. **SEED:** many, ± round, flat. *n*=7. Openings and margins of chaparral, oak and pinyon/juniper woodland; 300–1700 m. SCo (Riverside Co.), PR; Baja CA. Apr–Jun

C. corymbosa Herder (p. 1011) ROUND-HEADED CHINESE-HOUSES Pl 5–25 cm. **ST:** branches gen long, decumbent, gen ± red, densely scaly. **LF:** lanceolate to ovate, ± thick, crenate, obtuse at tip, subglabrous to finely gray-hairy adaxially, glabrous abaxially. **INFL:** ± head-like, sparsely and finely glandular; pedicel < calyx. **FL:** calyx ± hairy, lobe tips blunt; corolla 14–22 mm, gen ± white, throat longer than wide, barely angled to tube, pale lavender, gen undotted, hairy inside, pouch low, smoothly arched, upper lobes reflexed, ± 1 mm, ± brown, dry, papery, lateral lobes spoon-shaped, glabrous; filaments hairy, basal spur 0. **SEED:** many, ± concave. *n*=7. Coastal sand dunes; < 20 m. NCo (Mendocino Co.). See notes under *C. bartsiifolia* var. *b.* for pls from n CCo with longer upper lip. Apr–Jun ★

C. grandiflora Lindl. (p. 1011) Pl 6–35 cm. **LF:** narrowly oblong to lanceolate, subentire. **INFL:** open, glabrous to finely glandular or scaly-hairy; pedicels 1–4 per node, gen > calyx, ascending to reflexed. **FL:** calyx ± = fr, lobes not widely spreading, tips sharp; corolla 8–15 mm, ± blue, throat strongly angled to tube, longer than wide, pouch prominent, angular, upper lip pale at center, main lobes gen 2–6 mm wide, obovate, notched, middle lobe glabrous; filaments glabrous, basal spur 0. **SEED:** gen 4, plump, oblong, unwinged. *n*=7. Gravelly or grassy margins of conifer woodland; 400–1600 m. KR, NCoRO, NCoRH; to BC. Apr–Jul

C. greenei A. Gray (p. 1011) Pl 10–30 cm, gen ± purple, ± candelabra-like. **LF:** narrowly (ob)lanceolate to ovate, entire to toothed, obtuse at tip. **INFL:** interrupted, glandular; pedicels 1–5 per node, ± = calyx. **FL:** calyx > fr, lobe tips blunt; corolla gen 10–15 mm, ± uniformly dark purple, upper lip ± 1/2 lower, base of each upper lobe with 2 crested projections bulging > 1 mm away from throat opening; filaments glabrous, occ hairy at base of upper pair, basal spur 0. **SEED:** 2–4, wafer-like, ± concave. *n*=7. Open chaparral or conifer forest, serpentine slopes; 300–2500 m. KR, NCoRH, NCoRI. Apr–Aug

C. heterophylla Graham Pl 10–50 cm. **LF:** lance-deltate, toothed, often deeply lobed in seedlings. **INFL:** interrupted, glabrous to hairy, ± glandular; whorls dense; pedicel < calyx. **FL:** calyx lobe tips gen acute, glabrous to shaggy-hairy; corolla throat hairy inside, strongly angled to tube, as wide as long, pouch prominent, ± square, upper lip white to lavender or tipped dark violet, wine-spotted and gen ± red lined near base, lower lip ± white to rose-purple, lowest lobe gen with darker red tip; upper filaments hairy, basal spur 1–2 mm, curved into pouch. **SEED:** many, ovate, ± flat. *n*=7.

var. ***austromontana*** (Newsom) Munz DOWNY CHINESE-HOUSES **LF:** (sparsely) hairy abaxially **FL:** corolla 10–15 mm, upper lip < lower lip. Shady places in chaparral, mixed woodland; 300–1700 m. SnGb, SnBr, PR (Santa Ana Mtns). May–Aug

var. ***heterophylla*** (p. 1011) CHINESE-HOUSES **LF:** glabrous abaxially (at most ciliate on midvein) **FL:** corolla (13)15–20 mm, upper lip ± = lower lip. Shady places in chaparral, open mixed woodland, oak woodland; < 1300 m. NCoRO, NCoRI, CaRF, SNF, Teh, ScV (Sutter Buttes), CW, SCo, ChI, WTR, PR (exc SnJt); n Baja CA. Mar–Jun

C. linearis A. Gray (p. 1011) Pl 10–40 cm. **LF:** gen > 6 × longer than wide, ± linear, margins rolled under. **INFL:** open, finely scaly and spreading-glandular-hairy; bracts > 2 mm; pedicels gen 1–3(5) per node, > calyx, ascending. **FL:** calyx ± = fr, lobe tips blunt; corolla gen 8–12(15) mm, >> 2 × calyx, white to blue-purple, throat strongly angled to tube, as wide as long, pouch ± square, prominent, upper lip with projections bulging ± 0.5 mm away from throat opening at base of each lobe; upper filaments hairy at base, basal spur 0–0.5 mm. **SEED:** 3–4(+), ± plump, gen unwinged. *n*=7. Open conifer forest; 200–2000 m. KR, n NCoRO; s OR. Pls on metamorphic rocky slopes in c SNH with 2 seeds/fr treated as *C. linearis* in TJM (1993) not yet described. May–Jul

C. multicolor Lindl. & Paxton (p. 1011) SAN FRANCISCO COLLINSIA Pl gen 30–60 cm. **ST:** loosely branched, weak. **LF:** middle and distal clasping, ± lance-deltate, gen coarsely toothed. **INFL:** ± glandular-clammy; proximal-most pedicels 1–2 per node, >> calyx, distal pedicels 3+ per node, ± crowded, ± = calyx. **FL:** calyx lobe tips acute; corolla 12–18 mm, throat longer than wide, pouch rounded, not prominent, upper lip ± white, not or faintly dotted and lined, lower lip lavender to blue-purple, lateral lobes obovate, notched, lowest lobe sometimes sparsely hairy; upper filaments hairy, basal spur 0–0.5 mm. **SEED:** 8+, ± plump. Moist, ± shady scrub, forest; < 300 m. n&c CCo, SnFrB (San Mateo Co.). Mar–May ★

C. parryi A. Gray (p. 1011) Pl 10–40 cm. **ST:** finely hairy. **LF:** ± lanceolate, obtuse, entire to crenate; distal clasping. **INFL:** ± glandless; pedicels gen 1–3(5) per node, gen >> calyx. **FL:** calyx ± = fr, lobe tips ± blunt, gen ciliate; corolla 4–10 mm, blue-violet to lavender (white), glabrous; upper filaments sparsely short-spreading-hairy, basal spur 0. **SEED:** 8–12, sometimes slightly winged. *n*=7. Open chaparral, sagebrush scrub, mixed woodland; 500–1600 m. TR, PR (Santa Ana Mtns), SnJt/SnBr, DMoj/SnBr. Apr–May(Jun)

C. parviflora Lindl. (p. 1011) BLUE-EYED MARY Pl 3–40 cm. **LF:** ± lance-linear, obtuse, margin gen rolled under. **INFL:** open, lfy, glabrous to sparsely and finely glandular; pedicels > calyx, gen 1 per node proximally, 3–5 per node distally, often reflexed. **FL:** calyx > 2/3 corolla, ± = fr, lobe tips sharp; corolla 4–8 mm, throat barely angled to tube, tube and throat white, narrowed to lips, pouch angular, ± hidden by calyx, upper lip ± white or blue-tipped, main lobes ± 1 mm wide, oblong, gen blue (± purple); filaments glabrous, basal spur 0. **SEED:** gen 4. *n*=7. Common. Moist, ± shady places, montane; 800–3500 m. KR, n NCoRO, NCoRH, CaRH, SNH, SnFrB (Mount Hamilton Range), TR, PR, GB; to YT, e Can, NM. Mar–Jul

C. rattanii A. Gray (p. 1011) Pl 8–40 cm. **LF:** gen > 6 × longer than wide, margin rolled under, ± linear, gen entire (crenate), gen finely hairy and gray-green adaxially, ± purple and subglabrous abaxially. **INFL:** open, scaly and spreading-glandular-hairy; bracts > 2 mm; pedicels 1–3(5) per node, ascending, > calyx. **FL:** calyx lobe tips ± blunt; corolla 4–8 mm, gen < 2 × calyx, gen purple-lavender (white), glabrous or keel finely glandular, throat barely angled to tube, pouch ± hidden by calyx, upper lip with projections bulging < 0.5 mm away from throat opening at base of each lobe; upper filaments hairy at base, basal spur 0–0.5 mm. **SEED:** 2–6, plump to ± flat and narrowly winged. *n*=7. Open conifer forest; 100–1500 m. KR, NCoR, w CaRH, n SNH; to WA. May–Aug

C. sparsiflora Fisch. & C.A. Mey. **LF:** gen linear to oblong, entire. **INFL:** open, glabrous to finely hairy, nonglandular; pedicels 1–2(3) per node, ± equal, gen >> calyx. **FL:** calyx gen >> fr, lobe tips sharp to long-tapered, gen ciliate; corolla gen lavender to purple (white), lowest lobe sparsely long-hairy near tip; upper filaments (sparsely) short-spreading-hairy; basal spur 0. **FR:** spheric; top gen red-blotched, revealed by widely spreading calyx lobes. **SEED:** 4–12, flat, ± round, gen concave, ± winged. Large- and small-fld vars. often co-occur regionally, sometimes locally; fl sizes of vars. do not overlap within a region.

var. ***collina*** (Jeps.) Newsom Pl 5–20 cm. **FL:** corolla 5–8 mm, throat barely angled to tube, pouch hidden by calyx. Disturbed grassy fields, roadbanks, open chaparral, open oak and dry mixed woodland; 100–1200 m. s NCoRO, s NCoRI, CaRF, SNF, c SNH, GV, n CW. Mar–Apr

var. ***sparsiflora*** (p. 1015) Pl 5–30 cm. **FL:** corolla 7–20 mm, throat strongly angled to tube, pouch gen evident to prominent. *n*=7. Grassy, sometimes disturbed or rocky places, chaparral, oak woodland, dry mixed woodland; < 1000 m. NCoRO, NCoRI, CaRF, n&c SNF, GV, n CW; OR. If recognized taxonomically, pls with esp large corollas (12–20 mm) in s NCoRO, s NCoRI, n SnFrB assignable to *C. sparsiflora* var. *arvensis* (Greene) Jeps. Mar–May

C. tinctoria Benth. (p. 1015) Pl 20–60 cm, often robust. **LF:** gen lance-deltate, ± entire to serrate, (sub)glabrous adaxially, densely hairy abaxially, gen strongly mottled. **INFL:** interrupted, densely glandular, staining; whorls dense; pedicels gen crowded, < calyx. **FL:** calyx >> fr, lobes deep, tips blunt; corolla 12–22 mm, gen white to ± yellow or pale lavender (purple), throat strongly angled to tube, pouch prominent, projecting backward 2–4 mm from tube, upper lip < 1/2 lower, reflexed portion gen 2–5 mm, red-banded, lateral lobes gen red-dotted, gen hairy adaxially, lowest lobe glandular, hairy; upper filaments hairy, basal spur 0–0.8 mm. **SEED:** gen 4, flat, gen slightly winged. *n*=7. Rocky, dry mixed woodland, conifer forest; 100–2500 m. NCoRO, NCoRI, CaRF, SN (exc Teh), SnFrB. May–Aug

C. torreyi A. Gray Pl 5–25 cm. **INFL:** open, conspicuously densely glandular, staining; distal-most bracts 0–2 mm; pedicels gen >> calyx, some strongly reflexed in fr, S-shaped. **FL:** calyx ± = fr, lobe tips ± blunt; corolla 4–10 mm, throat angled to tube, as wide as long, throat and lower lip gen blue (lavender ± purple), upper lip ± white or pale lavender, with projections bulging < 1 mm away from throat opening at base of each lobe; filaments glabrous, upper pair sometimes hairy at base, basal spur 0. **SEED:** 2. Var. *wrightii*, var. *brevicarinata* perhaps best treated as *C. wrightii* S. Watson.

var. ***brevicarinata*** Newsom **LF:** > 5 × longer than wide, gen linear to elliptic. **FL:** corolla gen 6–9 mm. Conifer forest, often in sandy, granitic soil; 1800–3000 m. s SNH. May–Jul

var. *latifolia* Newsom **LF:** gen 2–5 × longer than wide, elliptic to ovate. **FL:** corolla gen 6–9 mm, upper lip with projections bulging ± 0.5 mm away from throat opening at base of each lobe. Conifer forest, often in sandy soil; 1000–2500 m. KR, NCoRH, CaRH, MP; s OR. Jun–Aug

var. *torreyi* (p. 1015) **LF:** gen > 6 × longer than wide, ± linear. **FL:** corolla gen 6–9 mm, upper lip with projections bulging ± 0.5

mm away from throat opening at base of each lobe. *n*=21. Conifer forest, often in sandy, granitic soil; 1000–3000 m. SNH; w NV (SN). May–Aug

var. *wrightii* (S. Watson) I.M. Johnst. **LF:** gen > 5 × longer than wide, linear to elliptic. **FL:** corolla gen 4–6 mm. Conifer forest, often in sandy, granitic soil; 800–4000 m. NCoRH, CaRH, SNH, Teh, TR, Wrn. May–Aug

CYMBALARIA

Margriet Wetherwax & David M. Thompson

Ann, per, glabrous to hairy. **ST:** decumbent or vine-like. **LF:** proximal gen opposite, distal alternate, long-petioled; blade round to reniform, entire to palmately lobed; veins palmate. **INFL:** fls 1 in lf axils. **FL:** calyx lobes 5, deep, ± unequal; corolla tube with conic or cylindric spur at base, lower lip base swollen, closing mouth. **FR:** spheric; chambers dehiscent by several slits radiating from tip. **SEED:** many, gen ridged or tubercled. *n*=7. ± 8 spp.: esp Medit. (Latin: round lvs)

C. muralis G. Gaertn. et al. (p. 1015) KENILWORTH IVY **LF:** 1–3 cm wide, lobes 5–9, moderately shallow, rounded to triangular, often abruptly pointed. **FL:** calyx 2–2.5 mm; corolla 9–15 mm, pale lilac to violet, spur 1–3 mm, cylindric, lower lip base ± yellow. **FR:** ± 4 mm, glabrous, pedicel growing away from light. *n*=7. Rock walls, shady, disturbed areas; < 1000 m. NCo, CCo, SCo; native to Medit. Cult as orn. May–Sep

DIGITALIS FOXGLOVE

Robert E. Preston & Margriet Wetherwax

Bien, per. **ST:** erect. **LF:** basal, in rosette, and cauline, alternate. **INFL:** raceme, 1-sided, bracted. **FL:** nodding; calyx deeply 5-lobed; corolla ± bilateral, long-bell-shaped, gen pink or white, lower 3 lobes forming prominent lip; stamens 4, in 2 pairs, incl; stigma lobes 2, flat. **FR:** gen loculicidal. ± 25 spp.: Eur (esp Medit), w&c Asia. (Latin: finger, from corolla shape) Some cult as orn or as source of the cardiac glycoside digitalis, a medically important heart stimulant.

D. purpurea L. (p. 1015) Gen bien. **ST:** < 18 dm, simple, gray-tomentose and glandular, esp distally. **LF:** 10–30 cm; petiole winged; blade lanceolate to ovate, crenate to dentate, adaxially green and soft-hairy, abaxially gray-tomentose. **INFL:** fls many; pedicel 6–25 mm, tomentose. **FL:** calyx lobes < 18 mm, lanceolate to ovate; corolla 40–60 mm, white to pink-purple with darker spots on lower inside surface, lobes ciliate, sparsely long-hairy inside; stamens, style incl. **FR:** ± 12 mm, ovoid. **SEED:** many, ± 0.5 mm. 2*n*=56. Acid soil in open woodland, disturbed areas; < 1700 m. NCo, KR, NCoRO, n&c SN, s SNH, CCo, SnFrB, WTR, PR; to BC; native to w Eur, nw Afr. All parts TOXIC to humans and livestock. May–Jul ❖

DOPATRIUM

Robert E. Preston & Margriet Wetherwax

Ann, fleshy. **ST:** erect. **LF:** opposite, sessile to short-petioled, fleshy. **INFL:** fl 1 in axils; proximal bracts lf-like, distal scale-like. **FL:** calyx 5-lobed; corolla 2-lipped, yellow or pale blue to lavender, upper lip erect, 2-lobed, lower lip 3-lobed, middle lobe > lateral, tube cylindric, > lobes; fertile stamens 2, anther sacs of each stamen touching, not parallel, staminodes 2. **FR:** loculicidal, spheric. ± 14 spp.: Afr, Asia, Australia. (Latin: aboriginal name)

D. junceum (Roxb.) Benth. (p. 1015) Glabrous exc pedicels. **ST:** 10–30 cm. **LF:** basal few, 10–25 mm, 2–10 mm wide, oblong, tip obtuse; cauline lvs and bracts lance-linear, 2–7 mm, 0.5–2 mm wide. **INFL:** proximal fls cleistogamous, sessile; distal fls open; pedicels 4–11 mm, sparsely glandular, spreading in fr. **FL:** calyx 1–2 mm, lobes fused to middle, tips obtuse; corolla 4–7 mm, pale blue to lavender, tube slightly > calyx. **FR:** ± 2 mm. **SEED:** net-like. Uncommon. Emergent in drying wetlands; < 100 m. ScV; native to Afr, Asia, Australia. Introduced with rice cult. Jun–Sep

GAMBELIA

Per, shrub, glabrous to hairy. **ST:** erect, arching, or pendent, often much-branched. **LF:** opposite or whorled in 3s, entire; veins pinnate. **INFL:** raceme. **FL:** calyx lobes 5, entire, ± equal; corolla gen red, tube with sac-like extension at base, lower lip base ± swollen, gen not fully closing mouth, not spurred; stamens 4, often exserted, staminode 0; stigma exserted. **FR:** ovoid to spheric; chambers dehiscent by 1–2 pores near tip. **SEED:** many; ridges thin, netted. 2 spp.: CA, nw Mex. (William Gambel, Am naturalist, collector, 1823–1849) [Elisens & Nelson 1993 Syst Bot 18:454–468] S.Am spp. remain in *Galvezia*.

G. speciosa Nutt. (p. 1015) SHOWY ISLAND SNAPDRAGON, SHOWY GREENBRIGHT Pl spreading, < 1 m, < 2 m wide. **ST:** arching, often ± pendent. **LF:** whorled (opposite on seedlings), ± thick. **INFL:** pedicel 1–2 cm. **FL:** calyx 7–10 mm; corolla 2–2.5 cm, bright red. **FR:** 6–7 mm, ovoid, asymmetric; lower chamber larger. *n*=15. Rocky cliffs, canyons; 3–500 m. s ChI; Baja CA (Guadalupe Island). [*Galvezia s.* (Nutt.) A. Gray] Mar–Jun ★

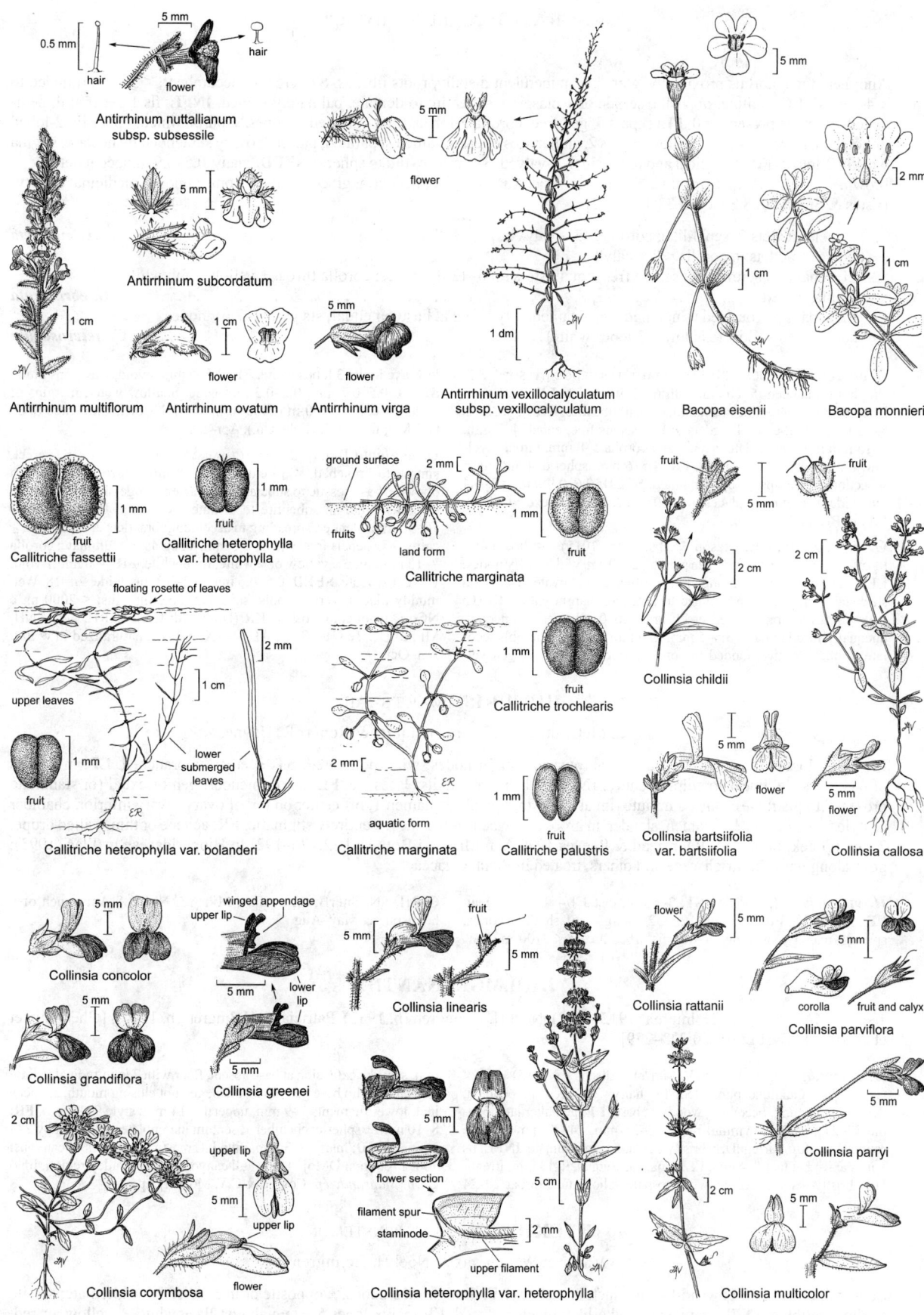

GRATIOLA HEDGE-HYSSOP

Dwayne Estes

Ann; herbage glabrous proximally, glandular-puberulent distally; roots fibrous. **ST**: erect to decumbent, < 30 cm, rounded to ± 4-angled. **LF**: cauline, opposite, sessile or ± clasping, subentire to dentate, palmately veined. **INFL**: fls 1 per lf axil, pedicelled, bractlets present or 0. **FL**: sepals 5, gen free; corolla 5-lobed, tube 4-angled, > lobes, ± purple-lined, upper lip 2-lobed or notched, lower 3-lobed; fertile stamens 2, anther sacs of each stamen separated, parallel; ovary subtended by nectary, stigma weakly 2-lobed. **FR**: septicidal and loculicidal, 4-valved, 3–6 mm, ovoid to spheric. **SEED**: many, 0.5–0.9 mm, coat net-like, ± brown. ± 25 spp.: temp, trop mtns, N.Am, Eur, Asia, Australia, S.Am. (Latin: grace or favor, from supposed medicinal quality) [Estes & Small 2008 Syst Bot 33:176–182]

1. Pedicel bractlets 2, sepal-like; corolla 2–3 × calyx .... ***G. neglecta***
1′ Pedicel bractlets 0; corolla < 2 × calyx
  2. Lf, sepal tips acuminate; sepals free, equal; sts with 6–12 lfy nodes; corolla throat ± yellow, limb white, ± pink-tinged .... ***G. ebracteata***
  2′ Lf, sepal tips rounded to notched; sepals unequally fused at base, unequal; sts with 4–7 lfy nodes; corolla gen yellow (exc lower 3 lobes white) .... ***G. heterosepala***

G. ebracteata A. DC. (p. 1015) BRACTLESS HEDGE-HYSSOP **ST**: simple to branched, 5–20 cm, with 6–12 lfy nodes. **LF**: 7–25 mm, lance-ovate to lance-linear, acuminate, gen subentire. **INFL**: pedicels 5–25 mm, stout, erect; bractlets 0. **FL**: sepals free, equal, 4–7 mm, 6–15 mm in fr, lanceolate, acuminate; corolla 5–8 mm, throat ± yellow, limb white ± pink-tinged. **FR**: 3.5–6 mm, spheric, 4-angled in ×-section, valve tips abruptly pointed. **SEED**: 0.6–0.9 mm, 0.3–0.4 mm wide. Wet, muddy places; < 2400 m. n&c CA-FP (exc NCoRH), MP; to BC, MT. Apr–Jun

G. heterosepala H. Mason & Bacig. (p. 1015) BOGGS LAKE HEDGE-HYSSOP **ST**: erect, gen simple, 2–10 cm, with 4–7 lfy nodes. **LF**: 2–20 mm, proximal lance-linear, oblong to obovate distally on st, distal ± clasping, tip rounded to notched, margin entire. **INFL**: pedicels 5–25 mm, slender, erect; bractlets 0. **FL**: sepals unequal, unequally fused at base, 3.5–5 mm, to 6 mm in fr, elliptic-oblanceolate to obovate, tip rounded to notched; corolla 5–8 mm, gen yel-

low, exc lower 3 lobes white. **FR**: 3.5–5 mm, ovoid, valve tips blunt. **SEED**: 0.5–0.7 mm, 0.2–0.3 mm wide. Shallow water, margins of vernal pools; < 1600 m (2400 m in Wrn). NCoRI, CaR, n&c SNF, GV, MP; to s-c OR (Lake Co.). Apr–Sep ★

G. neglecta Torr. (p. 1015) CLAMMY HEDGE-HYSSOP **ST**: erect, simple to branched, < 30 cm. **LF**: 5–50 mm, narrowly elliptic to obovate, base sessile to subclasping and gen wedge-shaped, tip acute to obtuse, margin subentire to dentate toward tip. **INFL**: pedicels 10–30 mm, slender, spreading to ascending; bractlets 2, sepal-like, ≥ calyx. **FL**: sepals free, equal, 4–6 mm, lanceolate, acuminate; corolla 7–12 mm, tube ± yellow or ± white, limb white. **FR**: 4–6 mm, ovoid, valve tips acute. **SEED**: 0.5–0.6 mm, 0.2–0.3 mm wide. $2n=18$. Wet, muddy places, vernal pools, sandbars, lake margins; < 2000 m. c NCo (Mendocino Co.), w KR (Humboldt Co.), CaRH, n&c SNH, MP; to BC, se Can, SC, c AL, se TX, n-c AZ; naturalized in w Eur. Jun–Oct

HIPPURIS MARE'S-TAIL

Elizabeth McClintock, C. Barre Hellquist & Robert R. Haynes

Per from rhizome, emergent aquatic, glabrous; rooting at nodes; wind-pollinated. **ST**: ± erect, unbranched. **LF**: in whorls of 6–12, sessile, linear to elliptic, entire. **INFL**: fl 1 in upper axils, ± sessile. **FL**: inconspicuous, gen bisexual (or staminate proximal to pistillate); calyx a minute rim at ovary top; petals 0; stamen 1, off center on top of ovary; ovary inferior, chamber 1, style 1, off-center, ± = stamen, slender, in groove between anther sacs, ± entirely stigmatic. **FR**: achene or thin-walled drupe. 2 spp. (Greek: horse tail) [Olmstead & Reeves 1995 Ann Missouri Bot Gard 82:176–192] In Hippuridaceae in TJM (1993); now, along with Callitrichaceae and others, treated in Plantaginaceae.

H. vulgaris L. (p. 1015) **ST**: 3–6 dm; distal 1/4–1/2 emergent. **LF**: 1–3.5 cm. **FR**: 2–3 mm. $2n=32$. Margins of shallow ponds, springs, marshy, swampy, or wet disturbed areas; < 2700 m. NW, CaRH, SN, SnFrB, SnBr, SnJt, MP; AZ, NM, s S.Am; much of n hemisphere. May–Aug

HOLMGRENANTHE

1 sp.: se CA. (Arthur H. Holmgren (1912–1992), Noel H. Holmgren (b. 1937), Patricia K. Holmgren (b. 1940)) [Ghebrehiwet et al. 2000 Pl Syst Evol 220:223–239]

H. petrophila (Coville & C.V. Morton) Elisens (p. 1015) ROCK LADY Per, glandular-puberulent to -hairy. **ST**: erect to pendent, densely branching, becoming woody at base. **LF**: gen alternate; petiole 12–27 mm, not twining; blade 12–35 mm, 14–27 mm wide, ± round to reniform, irregularly bristly-dentate; veins palmate. **INFL**: fls 1 in lf axils; pedicel 1–4 mm. **FL**: sepals 5, ± equal, 9–13 mm, irregularly bristly-dentate; corolla white to pale yellow, tube-throat 20–24

mm, sac-like extension at base absent, floor with 2 longitudinal yellow folds, lower lip base gen not swollen, gen not closing mouth; stamens incl, lower filaments 7–9 mm, upper 12–14 mm; style 9–10 mm. **FR**: 8–10 mm, ± spheric; chamber 1, septum incomplete, dehiscent by 2–3 pores. **SEED**: many, 2–3 mm, pitted. Limestone crevices of canyons; 700–1700 m. n DMoj (Titus, Fall canyons, Death Valley region, Inyo Co.). [*Maurandya p.* Coville & C.V. Morton] Apr–Jun ★

KECKIELLA BUSH PENSTEMON

Margriet Wetherwax & Noel H. Holmgren

Subshrub or shrub. **ST**: wand-like to much-branched. **LF**: drought-deciduous, ± opposite or in 3s, distal occ alternate; sessile to short-petioled. **INFL**: panicle or spike-like; bracts reduced. **FL**: calyx lobes 5, ± equal; corolla ± white, ± yellow, or red,

short-glandular outside, upper lip ± hooded, 2-lobed, lobes rounded, external in bud, lower lip rounded, lobes often reflexed; stamens 4, filaments attached at 1 level, densely nonglandular-hairy at base, anthers small, glabrous, anther sacs gen spreading flat at dehiscence; staminode well developed, glabrous to densely bearded; nectary a disk; stigma unlobed. **FR**: septicidal and sometimes also loculicidal at tip, ovoid. **SEED**: many, irregularly angled. 7 spp.: esp CA, w NV, w AZ, Baja CA. (David D. Keck, CA botanist, 1903–1995; *Keckia* used earlier, for genus of fossil algae) [Freeman et al. 2003 Syst Bot 28:782–790] Red-fld spp. are hummingbird-pollinated.

1. Corolla bright pink to red or red-orange, tube 16–25 mm, throat indistinct
  2. Lvs ± opposite
    3. Lf base rounded to cordate; anther sacs 1.1–1.5 mm . . . . ***K. cordifolia***
    3′ Lf base wedge-shaped; anther sacs 0.9–1 mm . . . . ***K. corymbosa***
  2′ Lvs gen whorled in 3s (occ few opposite) . . . . ***K. ternata***
    4. Calyx glandular, 3.8–5.5 mm . . . . var. ***septentrionalis***
    4′ Calyx glabrous, 4.5–7.2 mm . . . . var. ***ternata***
1′ Corolla white, yellow, or brown-purple, short tube and distinct throat together 4–11 mm
  5. Corolla 10–15 mm, tube + throat > upper lip
    6. Infl a panicle; fls pedicelled; anther sac valves spreading widely; staminode densely hairy; young sts glabrous, glaucous . . . . ***K. lemmonii***
    6′ Infl spike-like; fls ± sessile; anther sac valves barely spreading; staminode glabrous near tip; young sts densely short-hairy, green . . . . ***K. rothrockii***
      7. Corolla 13–15 mm, sparsely long-hairy; lvs green, becoming glabrous . . . . var. ***jacintensis***
      7′ Corolla 10–12 mm, becoming glabrous; lvs canescent . . . . var. ***rothrockii***
  5′ Corolla 12–23 mm, tube + throat < upper lip
    8. Staminode densely yellow-hairy; corolla yellow; anther sacs 1.1–1.8 mm . . . . ***K. antirrhinoides***
      9. Calyx 3–6 mm, lobes widely ovate, tips obtuse to acute; pl finely hairy; anther sacs 1.4–1.8 mm
        . . . . var. ***antirrhinoides***
      9′ Calyx 5.5–9 mm, lobes lanceolate, tips acute to acuminate; pl canescent; anther sacs 1.1–1.5 mm
        . . . . var. ***microphylla***
    8′ Staminode glabrous near tip; corolla white, cream, or rose-tinged; anther sacs 0.6–0.8 mm . . . . ***K. breviflora***
      10. Calyx glandular . . . . var. ***breviflora***
      10′ Calyx glabrous . . . . var. ***glabrisepala***

K. antirrhinoides (Benth.) Straw **ST**: spreading to erect, 6–25 dm; young sts canescent (glabrous). **LF**: ± opposite, in axillary clusters on older sts; blade 5–20 mm, (ob)lanceolate to narrowly (ob) ovate, base tapered, margin gen entire. **INFL**: finely short-hairy and sparsely glandular. **FL**: corolla 15–23 mm, yellow (drying ± black), tube + widely expanded throat 6–10 mm, upper lip 8–15 mm; anther sacs 1.1–1.8 mm; staminode densely yellow-hairy, exserted.

var. ***antirrhinoides*** (p. 1015) Pl finely hairy. **FL**: calyx 3–6 mm, lobes widely ovate, tips obtuse to acute; anther sacs 1.4–1.8 mm. Chaparral, oak forest; 100–1100 m. SCo, TR, PR; Mex. Apr–May

var. ***microphylla*** (A. Gray) N.H. Holmgren (p. 1015) Pl canescent. **FL**: calyx 5.5–9 mm, lobes lanceolate, tips acute to acuminate; anther sacs 1.1–1.5 mm. Pinyon/juniper woodland, Joshua-tree scrub; 300–1800 m. e PR, D; AZ. Apr–Jun

K. breviflora (Lindl.) Straw **ST**: 5–20 dm; young sts glabrous, glaucous. **LF**: opposite, subsessile; main lvs 10–40 mm, ± (ob)lanceolate, gen serrate. **INFL**: glabrous to sticky-hairy. **FL**: calyx 4–8 mm, lobes lanceolate to ovate, acuminate; corolla 12–18 mm, white, cream, or rose-tinged, lobes sometimes rose-tinged, lined ± purple or ± pink, tube + throat 4–8 mm, upper lip 8–12 mm; anther sacs 0.6–0.8 mm; staminode glabrous near tip, slightly exserted. 2*n*=16.

var. ***breviflora*** (p. 1015) **FL**: calyx glandular. Rocky slopes, forest, chaparral; 200–2000 m. SN, SnFrB, SCoR, WTR, SN/SNE. May–Jul

var. ***glabrisepala*** (D.D. Keck) N.H. Holmgren **FL**: calyx glabrous. Rocky slopes, forest, chaparral; < 2800 m. s CaR, NCoRI, SN, ScV, SCoRI, n SNE exc W&I; w NV. Jun–Aug

K. cordifolia (Benth.) Straw (p. 1015) **ST**: often spreading, < 30 dm; young sts glabrous to short-hairy. **LF**: ± opposite; blade 20–65 mm, ovate, base rounded to cordate, margin gen shortly 3–11-toothed. **INFL**: glandular and stiffly hairy, gen more densely hairy than herbage. **FL**: calyx 7–13 mm, lobes ± lanceolate; corolla 31–43 mm, red to red-orange, tube 18–25 mm, throat indistinct, upper lip 11–21 mm; anther sacs 1.1–1.5 mm; staminode densely yellow-hairy, incl. 2*n*=16. Chaparral, forest; < 1600 m. s SCoRO, SW; n Baja CA. May–Jul

K. corymbosa (A. DC.) Straw (p. 1015) Shrub 3–6 dm, < 1 m wide. **ST**: young sts glabrous or hairy. **LF**: opposite; blade 10–35 mm, (ob)lanceolate to narrowly (ob)ovate, base wedge-shaped, margin entire to 3–5-toothed. **INFL**: pedicels (and calyces) glandular-hairy and densely coarse-hairy. **FL**: calyx 6.4–11 mm, lobes ± lanceolate; corolla 22–40 mm, bright pink to red, tube 17–22 mm, throat indistinct, upper lip 9–15 mm; anther sacs 0.9–1 mm; staminode densely yellow-hairy, incl. Rocky slopes in conifer or hardwood forests, (chaparral); 100–2000 m. NW, CaRF, n SNF, ScV (Sutter Buttes), n&c CW. Jun–Oct

K. lemmonii (A. Gray) Straw (p. 1015) **ST**: 5–15 dm; young sts glabrous, glaucous, wand-like. **LF**: opposite; blade 10–65 mm, lanceolate to ovate, base rounded to widely wedge-shaped, margin gen 2–12-toothed. **INFL**: panicle, glandular-hairy; fls pedicelled. **FL**: calyx 3.5–6.7 mm, lobes ± lanceolate; corolla 11–15 mm, purple-brown, lower lip pale yellow, brown-purple-lined, tube + expanded throat 5.5–9.5 mm, upper lip 2.6–6 mm; anther sacs 0.6–0.9 mm, valves spreading widely; staminode densely yellow-hairy, exserted. Rocky slopes, conifer and mixed forests, chaparral; < 2100 m. NW, n SN; OR, NV. Jun–Aug

K. rothrockii (A. Gray) Straw Pl wide, low, 3–6 dm. **ST**: young sts green, densely short-hairy. **LF**: subopposite or in 3s, subsessile; main lvs 5–16 mm, (ob)lanceolate to widely obovate, entire or finely serrate towards tip. **INFL**: spike-like, short-hairy; fls ± sessile. **FL**: calyx 4–7 mm, sometimes glandular, lobes lanceolate; corolla brown-yellow, pale yellow, or cream, purple- or red-brown-lined, tube + expanded throat 7–11 mm, upper lip 4–5 mm; anther sacs 0.8–1.1 mm, valves barely spreading at dehiscence; staminode glabrous near tip, ± incl.

var. ***jacintensis*** (Abrams) N.H. Holmgren **LF**: green, becoming glabrous. **FL**: corolla 13–15 mm, sparsely long-hairy. Conifer forest, pinyon/juniper woodland; 2200–2800 m. SnJt. Jun–Aug

var. ***rothrockii*** (p. 1015) **LF**: canescent. **FL**: corolla 10–12 mm, becoming glabrous. Conifer forest, pinyon/juniper woodland; 1600–3100 m. c&s SNH, SNE, DMtns; NV. Jun–Aug

K. ternata (A. Gray) Straw **ST:** spreading to erect, < 25 dm; young sts glabrous, glaucous, wand-like. **LF:** gen whorled in 3s (occ few opposite); blade 15–60 mm, linear to narrowly (ob)lanceolate, acutely tapered at both ends, often folded and curved, margin gen 5–11-toothed. **FL:** calyx lobes lanceolate to ovate; corolla 21–31 mm, red, tube 16–24 mm, throat indistinct, upper lip 5.5–9.5 mm; anther sacs 0.8–1.1 mm; staminode densely yellow-hairy, incl.

var. ***septentrionalis*** (Munz & I.M. Johnst.) N.H. Holmgren (p. 1021) **INFL:** glandular-hairy. **FL:** calyx 3.8–5.5 mm, glandular. Mixed-hardwood forest, chaparral; 800–1900 m. Teh, WTR. Jun–Sep

var. ***ternata*** (p. 1021) **INFL:** glabrous. **FL:** calyx 4.5–7.2 mm, glabrous. Pinyon/juniper woodland, chaparral, forest; 100–2100 m. SnGb, SnBr, PR; n Baja CA. Jun–Sep

KICKXIA

Margriet Wetherwax & David M. Thompson

Ann [per, shrub], hairy, glandular. **ST:** gen decumbent, much-branched. **LF:** alternate (proximal-most sometimes opposite), gen entire, ± sessile; veins pinnate. **INFL:** fl 1 in axils; pedicel straight. **FL:** calyx segments 5, ± free, equal; corolla tube with conic or cylindric spur at base, lower lip base swollen, closing mouth; stamens 4, staminode minute. **FR:** ± spheric; chambers equal, circumscissile. **SEED:** many, 0.3–1.3 mm, ± netted or ridged. ± 9 spp.: Eur, Medit., w Asia. (Jean Kickx father and/or son, 17th, 18th century Belgian professors) [Ghebrehiwet 2001 Nord J Bot 20:655–690]

1. Proximal lvs ovate to hastate, distal lvs hastate to sagittate . . . . ***K. elatine***
1′ Lvs narrowly to widely ovate or subcordate throughout. . . . . ***K. spuria***

K. elatine (L.) Dumort. (p. 1021) **FL:** calyx lobes lanceolate, not enlarging in fr; corolla 7–15 mm, ± yellow or ± blue, upper lip violet. **FR:** 4–4.5 mm. 2*n*=18. Disturbed, open places; < 1000 m. CA-FP (exc SNH); native to Eur. Apr–Oct

K. spuria (L.) Dumort. (p. 1021) **FL:** calyx lobes ovate, enlarging in fr, base in fr ± cordate; corolla 10–15 mm, yellow, upper lip deep purple. **FR:** 3–5 mm. 2*n*=18. Disturbed, open places; < 600 m. NCoRO, NCoRI, n&c SNF, GV, SnFrB, SCoRI, SCo, TR, PR; native to Eur. Jul–Dec

LINARIA TOADFLAX

Robert E. Preston & Margriet Wetherwax

Ann to per, gen glabrous. **ST:** erect, simple or branched at base. **LF:** gen opposite or whorled (or distal alternate), sessile, linear to ovate, gen wider on non-fl shoots, entire to dentate, pinnately veined. **INFL:** spike or raceme, terminal; bracts reduced, alternate. **FL:** sepals 5, free to near base, lobes ± equal; corolla 5-lobed, 2-lipped, lower lip ≥ upper, lower side of tube spurred at base, lower side of throat swollen, ± hairy, ± closing corolla below lips; stamens 4, in 2 pairs, incl; stigma small, lobes 0 or 2, flat. **FR:** opening by slits into chambers near tip, ± spheric. **SEED:** many, flat and winged or pyramid-shaped and 0–3-ridged. ± 100 spp.: Eur, Asia, n Afr; many cult. (Latin: flax, from flax-like lvs of some) [Sutton 1988 Revision tribe Antirrhineae. Oxford Univ Press] Corolla length incls spur. *L. supina* (L.) Chaz. mistakenly reported for CA in TJM (1993). *L. canadensis* moved to *Nuttallanthus.*

1. Corolla yellow
   2. Lf lanceolate to ovate; corolla 20–50 mm; seed ± pyramid-shaped, ridged . . . . ***L. dalmatica*** subsp. ***dalmatica***
   2′ Lf ± linear; corolla 18–32 mm; seed flat, winged . . . . ***L. vulgaris***
1′ Corolla gen not yellow (exc sometimes throat)
   3. Per from woody base; lf linear to oblanceolate . . . . **[*L. purpurea*]**
   3′ Ann; lf linear to elliptic
      4. Infl glabrous; corolla violet, throat swelling white with orange patch at base . . . . ***L. bipartita***
      4′ Infl ± glandular-puberulent; corolla red-violet to magenta, throat swelling red, yellow, or white
         5. Calyx lobes lanceolate; infl open in fr; corolla 20–38 mm . . . . ***L. maroccana***
         5′ Calyx lobes ovate; infl dense in fr; corolla 15–18 mm . . . . ***L. pinifolia***

L. bipartita (Vent.) Willd. Ann, glabrous, glaucous. **ST:** 10–30 cm. **LF:** 30–50 mm, linear. **FL:** calyx 2.5–6 mm; corolla 18–20 mm, violet, throat swelling white with orange at base; stigma lobes 2, flat. **FR:** ± 2 mm. **SEED:** minute, pyramid-shaped, ridged. Disturbed areas; < 550 m. CCo, SCoRO, SCo, s ChI, PR; native to Medit. Dec–Jul

L. dalmatica (L.) Mill. subsp. *dalmatica* (p. 1021) DALMATIAN TOADFLAX Per, glabrous. **ST:** erect, 20–120 cm, branched. **LF:** crowded, < 60 mm, lanceolate to ovate, rigid, ± clasping, tip acute to long-tapered. **INFL:** dense to open; pedicels 1–13 mm, ± = bracts. **FL:** calyx 2–12 mm, lobes linear to triangular-ovate; corolla 20–50 mm, yellow, lower lip closing throat, densely white- to orange-hairy; stigma lobes 0. **FR:** 3–7 mm. **SEED:** ± 1.2 mm, ± pyramid-shaped, ridged. 2*n*=12. Roadsides, fields, open areas in yellow-pine forest, pinyon/juniper woodland, sagebrush scrub; < 2080 m. NW (exc NCoRH), CaR, n SN, GV, CW (exc SCoRI), SW (exc ChI, SnJt), GB (exc Wrn, W&I); widely naturalized; native to Medit. [*L. genistifolia* (L.) Mill. subsp. *d.* (L.) Maire & Petitm.] Other subsp. OR, WA, e US. Apr–Sep ❖

L. maroccana Hook. f. Ann. **ST:** < 50 cm. **LF:** 20–40 mm, linear. **INFL:** dense, becoming open in fr, ± glandular-puberulent. **FL:** calyx 3–6 mm, lobes lanceolate, pubescent at base and on midrib; corolla 20–38 mm, red-violet to magenta, throat swelling red, yellow, or white; spur gen ± straight; stigma lobes 2, flat. **FR:** 3–5 mm. **SEED:** ± 1 mm, pyramid-shaped, ridged, black. 2*n*=12. Disturbed areas; < 830 m. NCoRO, GV, CW (exc SCoRI), SW (exc ChI, SnJt), w DMoj (Joshua Tree National Park), w DSon (Coachella Valley); native to Medit. Garden escape and incl in wildflower seed mixes (variably colored). Mar–Jun

L. pinifolia (Poir.) Thell. Ann. **ST:** 80–100 cm, with decumbent non-fl shoots. **LF:** 20–40 mm, linear, elliptic on non-fl shoots. **INFL:** dense, ± glandular-puberulent. **FL:** calyx 3–5 mm, sepals ovate, ciliate; corolla 15–18 mm, red-violet, throat swelling yellow, veins purple; stigma lobes 2, flat. **FR:** ± 4 mm. **SEED:** ± 1 mm, pyramid-shaped, ridged. 2*n*=12. Disturbed areas; < 305 m. SCoRO; native to Medit. [*L. reticulata* (Sm.) Desf.] Garden escape. May–Jun

L. vulgaris Mill. (p. 1021) BUTTER-AND-EGGS Per. **ST:** 30–100 cm, ascending to erect, simple or not. **LF:** crowded, 25–50 mm, ± linear, glabrous to sparsely soft-shaggy-hairy. **INFL:** raceme, dense, ± glandular-hairy; pedicel 1–4 mm. **FL:** calyx lobes ± 3 mm, unequal; corolla 18–32 mm, yellow, lower lip orange-hairy; stigma lobes 0. **FR:** 9–12 mm. **SEED:** 1.5 mm, flat, winged. 2*n*=12. Disturbed areas; < 2150 m. NCo, KR, NCoRO, CaRH, n SN, ScV, CW (exc SCoRI), SCo, MP; widely naturalized; native to Medit. Jun–Aug ❖

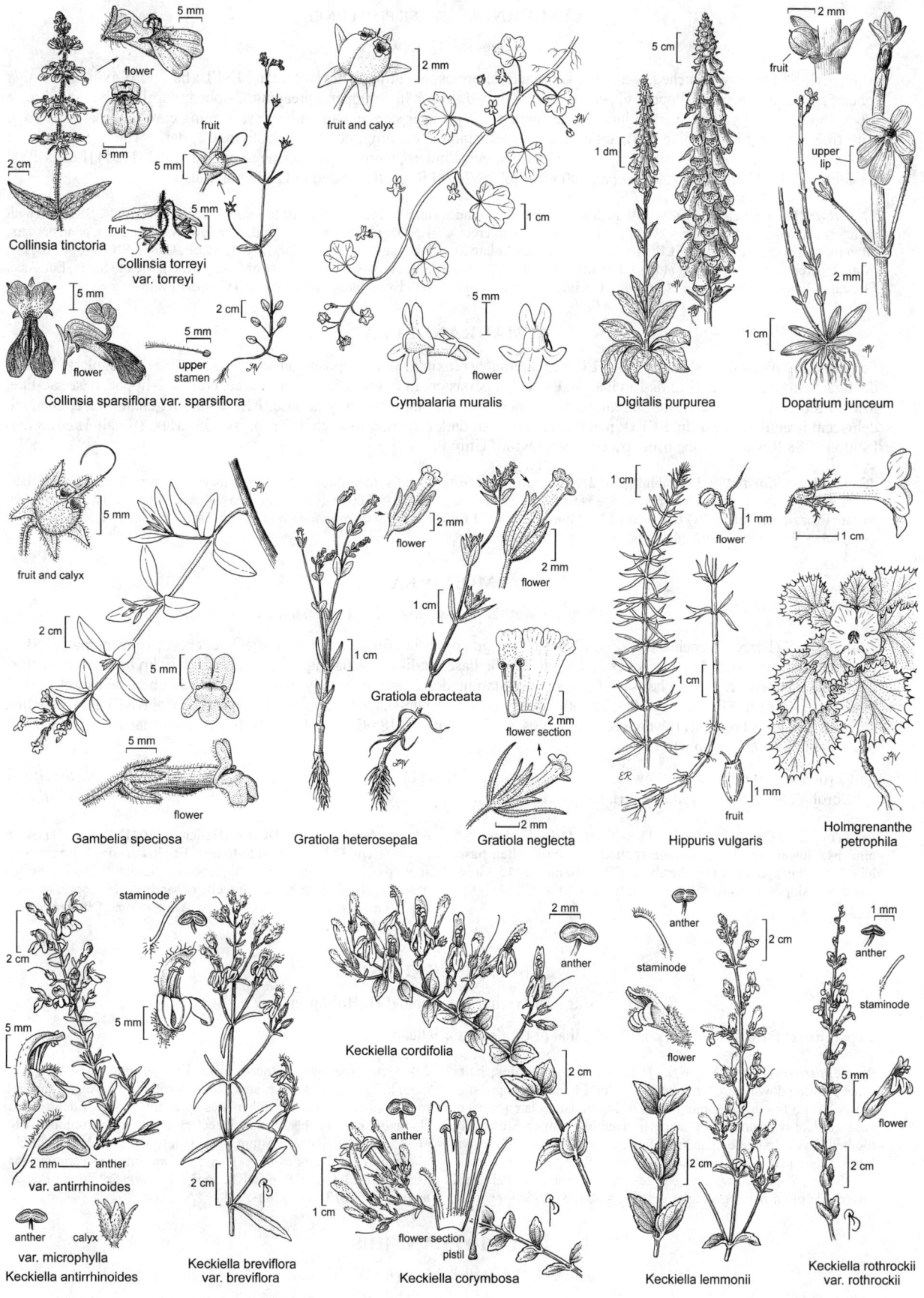

Collinsia tinctoria

Collinsia torreyi
var. torreyi

Collinsia sparsiflora var. sparsiflora

flower

fruit

fruit

upper
stamen

Cymbalaria muralis

fruit and calyx

flower

Digitalis purpurea

Dopatrium junceum

fruit

upper
lip

Gambelia speciosa

fruit and calyx

flower

Gratiola heterosepala

Gratiola ebracteata

flower

flower

flower section

Gratiola neglecta

Hippuris vulgaris

flower

fruit

Holmgrenanthe
petrophila

Keckiella antirrhinoides

var. antirrhinoides

anther

var. microphylla

anther   calyx

staminode

Keckiella breviflora
var. breviflora

Keckiella cordifolia

anther

anther

flower section
pistil

Keckiella corymbosa

anther

staminode

flower

Keckiella lemmonii

anther

staminode

flower

Keckiella rothrockii
var. rothrockii

LINDERNIA FALSE PIMPERNEL

Deborah Q. Lewis

Ann [per]. **ST**: openly branched, 4-angled. **LF**: cauline, opposite, entire to finely dentate. **INFL**: fls 1 in lf axils. **FL**: calyx lobes 5, ± equal; corolla 2-lipped, upper lip erect, 2-lobed, lower lip > upper, spreading, 3-lobed, corolla tube cylindric, > lobes, throat with 2 yellow, hairy ridges; fertile stamens 2[4] (lower stamen pair antherless, forming corolla throat ridges and with free, forked, filament-like tips), anther sacs of each stamen touching, not parallel; stigmas 2, flat. ± 100 spp.: temp and trop N.Am to Eur, Asia, Australia, Afr, S.Am. (Franz B. von Lindern, German physician & botanist, 1682–1755) [Lewis 2000 Castanea 65:93–122; Rahmanzadeh et al. 2005 Pl Biol 7:67–78] Recently treated in Linderniaceae.

L. dubia (L.) Pennell (p. 1021) Pl glabrous or sparsely and minutely stalked-glandular. **ST**: < 27 cm [38 cm], spreading to erect, often rooting at proximal nodes. **LF**: 1–37 mm, sessile, lanceolate to ovate, tapered to round-clasping at base. **INFL**: pedicel 0.5–28 mm. **FL**: calyx lobed nearly to base, < 7 mm, lobes linear; corolla 7–10 mm, white to ± blue or lavender. **FR**: < or > sepals. **SEED**: length 1.5–3 × width, yellow to golden-yellow. Wet meadows, pond borders, lakes and streams; < 1700 m. KR, NCoR, CaRH, SN, GV, SnFrB, PR, SNE; to WA, s Can, e US (e of Rocky Mtns); Mex to S.Am, Eur, Asia. [*L. d.* var. *anagallidea* (Michx.) Cooperr.] Jun–Aug

MAURANDELLA

Per, climbing by twining sts or petioles. **LF**: gen alternate (proximal-most opposite on seedlings), entire to irregularly bristly-dentate; veins palmate. **INFL**: fls 1 in lf axils. **FL**: calyx persistent in fr, lobes 5, ± equal, ± free; corolla 2-lipped, base sac-like, lower lip base swollen, ± closing throat, hairy; stamens 4, incl, filaments hairy at base. **FR**: ± spheric; chambers 2, unequal, dehiscent irregularly near tip. **SEED**: many, ± rectangular, dark brown, ± tubercled. 2 spp.: sw US, Mex. (Small *Maurandya*) [Sutton 1988 Revision of the tribe Antirrhineae. Oxford Univ Press]

M. antirrhiniflora (Willd.) Rothm. (p. 1021) VIOLET TWINING SNAPDRAGON Pl ± glabrous. **LF**: petiole 5–33 mm; blade 5–30 mm, hastate to sagittate, entire. **INFL**: pedicel 12–47 mm, glabrous. **FL**: calyx lobes 9–14 mm, entire, glabrous; corolla pink, red, to violet distally, tube-throat 13–17 mm, lobes 4–8 mm. 2*n*=24. Desert flats, washes; < 2600 m. e DMtns (Providence Mtns); to TX, Mex. [*Maurandya antirrhiniflora* Willd.] Feb–Dec ★

MOHAVEA

Margriet Wetherwax & David M. Thompson

Ann, hairy. **ST**: erect; branches 0 or few. **LF**: proximal gen opposite, distal alternate; lanceolate, entire; veins pinnate. **INFL**: crowded; fls 1 in axils. **FL**: calyx lobes 5, ± equal; corolla base sac-like, lips flaring, ± fan-shaped, lower lip base swollen, closing throat; stamens 2, incl, anther sacs 1 per stamen, staminodes 2. **FR**: obliquely ovoid, fragile; chambers dehiscent by 1–2 large pores near tip. **SEED**: ovate, flat, smooth; wing incurved, ± cup-shaped. 2 spp.: sw US, n Mex. (Mojave River, where first collected by John Frémont) [Oyama & Baum 2004 Amer J Bot 91:918–925] Best incl in *Antirrhinum* (name in *Antirrhinum* unavailable for *M. breviflora*).

1. Corolla 15–20 mm, dark yellow . . . . ***M. breviflora***
1′ Corolla 25–35 mm, ± yellow to white. . . . . ***M. confertiflora***

M. breviflora Coville (p. 1021) Pl 5–20 cm. **INFL**: pedicel 2–5 mm. **FL**: lower corolla lip maroon-spotted only on swollen base, lobed to within 2–3 mm of swollen base. **FR**: 8–10 mm. *n*=15. Gravelly desert slopes, washes; < 1850 m. DMoj; s NV, nw AZ. Mar–Apr

M. confertiflora (A. DC.) A. Heller (p. 1021) GHOST FLOWER Pl 10–40 cm. **INFL**: pedicel 5–10 mm. **FL**: lower corolla lip maroon-spotted on swollen base and limb, lobed to within 6–8 mm of swollen base. **FR**: 10–12 mm. *n*=15. Gravelly or sandy desert slopes, washes; < 1250 m. D (exc Inyo Co.); s NV, w AZ, nw Mex. [*Antirrhinum c.* A. DC.] Mar–Apr

NOTHOCHELONE

Margriet Wetherwax & Noel H. Holmgren

1 sp. (Greek: false turtle, from corolla like that of *Chelone*, turtlehead)

N. nemorosa (Lindl.) Straw (p. 1021) Per from caudex; hairs often pointing downward. **ST**: erect, 3–10 dm. **LF**: cauline, opposite, sessile to short-petioled; larger blades 4–14 cm, lanceolate to ovate, base cordate or rounded, tip acute to acuminate, margin coarsely toothed. **INFL**: panicle, glandular, few-fld; bracts small. **FL**: calyx lobes 5, ± equal, 5–13 mm, narrowly lanceolate to ovate; corolla 26–33 mm, 8–11 mm wide when pressed, rose-purple to dark red, lower side often paler, glandular outside, glabrous inside, upper lip 2.4–5 mm, shallowly 2-lobed, lower lip 5–9 mm; stamens 4, filament hairs longest at base, attached to corolla at 1 level, anthers ± 1 mm, long-woolly-hairy, anther sacs spreading ± flat at dehiscence; staminode coarsely bearded, incl; nectary a disk; stigma unlobed. **FR**: 10–15 mm. **SEED**: 2–3.5 mm, flat, widely winged. 2*n*=30. Rocky places in open Douglas-fir, yellow-pine, and mixed-evergreen forests; 1200–2300 m. KR, n CaRH; to BC. [*Chelone n.* Lindl.; *Penstemon n.* (Lindl.) Trautv.] Jun–Aug

NUTTALLANTHUS

Robert E. Preston & Margriet Wetherwax

Ann to bien. **ST**: erect, simple or branched at base. **LF**: on fl st gen alternate, sessile, linear, entire to dentate, pinnately veined; on non-fl sts whorled, gen wider. **INFL**: raceme, terminal; bracts reduced, alternate. **FL**: calyx lobes 5, deep, ± equal; corolla

5-lobed, unequally 2-lipped, lower lip >> upper, lower side of tube spurred at base or spur 0, lower side of throat swollen, ± hairy, ± closing corolla proximal to lips; stamens 4, in 2 pairs, incl; stigma small, unlobed. **FR**: dehiscent by slits into chambers near tip, ± spheric. **SEED**: many, prismatic, 4–7-angled, faces smooth or tubercled. 4 spp.: N.Am, S.Am, introduced elsewhere. (Thomas Nuttall, English naturalist, 1786–1859) [Crawford & Elisens 2006 Amer J Bot 93:582–591]

N. texanus (Scheele) D.A. Sutton (p. 1021) BLUE TOADFLAX
Gen glabrous. **ST**: 10–60 cm, slender, with decumbent non-fl shoots. **LF**: 5–25 mm, narrowly linear, obtuse. **INFL**: raceme, dense in fl, open in fr, ± glandular-puberulent; pedicels 1.5–6 mm, > bracts; fls opening or cleistogamous. **FL**: calyx ± 3 mm, lobes lance-linear, tips acute; corolla 10–24 mm (incl spur), violet to blue, lips spread-ing, lower lip 6–11 mm, >> upper, throat swelling ± obscure, white-ridged, spur 6–11 mm, straight or curved, slender. **FR**: ± 3 mm. **SEED**: 0.5 mm, faces ± tubercled. *n*=12. Sand or gravel; < 1800 m. NW (exc NCoRH), SNF, GV, CW, SW (exc SnBr, SnJt), DMtns (Granite Mtns, w San Bernardino Co.); sw&s-c US, Mex, temp S.Am. [*Linaria canadensis* (L.) Chaz., in part, misappl.] Mar–May

PENSTEMON BEARDTONGUE
Margriet Wetherwax & Noel H. Holmgren

Per to shrub. **LF**: gen opposite, entire to toothed; distal sessile. **INFL**: panicle, raceme, cyme, or fls in whorls; bracts gen small. **FL**: calyx lobes 5, ± equal; corolla tube ± cylindric or lower side expanded, ± 2-lipped, gen pink or blue to purple (some red, yellow, or white), upper lip 2-lobed, external in bud; stamens 4, filament bases glabrous, attached to corolla at different levels, anther sacs 2, valves gen spreading ± flat at dehiscence; staminode attached near base of corolla tube, well developed, gen hairy adaxially; nectaries 2, at bases of upper stamens; stigma unlobed. **FR**: septicidal and sometimes also loculicidal at tip. **SEED**: gen many, irregularly angled. 250 spp.: N.Am, esp w US. (Latin & Greek: almost thread, from stamen-like staminode) [Wolfe et al. 2006 Amer J Bot 93:1699–1713] Largest genus of fl pls endemic to N.Am. *P. subglaber*, *P. strictus* may persist in SNH, from commercial wildflower seed mixes or plantings; both native to Rocky Mtns.

Key to Groups

1. Anther sacs not dehiscing full length (opening only partly)
  2. Anther sacs dehiscing at distal end (proximal portion indehiscent) (subg. *Habroanthus*). . . . . **Group 1**
  2′ Anther sacs dehiscing at proximal end (leaving distal portion indehiscent) (subg. *Saccanthera*). . . . . **Group 2**
1′ Anther sacs dehiscing full length, valves gen spreading widely apart
  3. Corolla mouth closed by arched floor (subg. *Cryptostemon*). . . . . ***P. personatus***
  3′ Corolla mouth open
  4. Anthers densely woolly (subg. *Dasanthera*). . . . . **Group 3**
  4′ Anthers glabrous (subg. *Penstemon*). . . . . **Group 4**

Group 1

1. Corolla blue or blue-violet, tube ± abruptly expanded into throat, 6–13 mm wide at widest point
  2. Calyx 3–4(5) mm; corolla throat 6–8 mm wide when pressed, floor hairy; anther sacs 1.6–2.2 mm . . . . . ***P. pahutensis***
  2′ Calyx (4)6–13 mm; corolla throat 7–13 mm wide when pressed, floor glabrous; anther sacs 1.8–3 mm . . . ***P. speciosus***
1′ Corolla (orange-)red (yellow), tube gradually expanding into throat, 4–9 mm wide at widest point
  3. Corolla 30–40 mm, distinctly 2-lipped, upper lip projecting, hood-like, lobes short, lower lip reflexed,
     lobes longer. . . . . ***P. labrosus***
  3′ Corolla 24–33 mm, nearly radial, lobes ± equal, projecting. . . . . ***P. eatonii***
  4. Pl glabrous throughout. . . . . var. ***eatonii***
  4′ Pl short-hairy . . . . var. ***undosus***

Group 2

1. Corolla red to orange, lips long, lower lip strongly reflexed. . . . . ***P. rostriflorus***
1′ Corolla lavender, ± blue, or ± purple, lips short, lower lip projecting or spreading, not reflexed
  2. Lvs sharply serrate; filaments white-hairy distally; corolla lobes finely ciliate . . . . ***P. venustus***
  2′ Lvs ± entire (weakly serrate in *Penstemon parvulus*, *Penstemon purpusii*); filaments glabrous distally;
     corolla lobes not finely ciliate
  3. Staminode hairy; anthers glabrous on inner margins
    4. Corolla 15–20 mm; lvs glabrous; anther sacs 1–1.3 mm. . . . . ***P. gracilentus***
    4′ Corolla 24–34 mm; lvs hairy; anther sacs 1.4–1.9 mm
      5. Lvs well-distributed on sts; anther sacs 1.7–1.9 mm, dehiscing < 1/3 length . . . . ***P. papillatus***
      5′ Lvs gen basal; anther sacs 1.4–1.7 mm, dehiscing for ± 1/2 length . . . . ***P. scapoides***
  3′ Staminode glabrous; anthers hairy on inner margins near filament attachment
    6. Lvs gen basal; corolla floor hairy. . . . . ***P. caesius***
    6′ Lvs well-distributed on sts; corolla floor glabrous
      7. Herbage densely canescent-hairy; pl 0.5–2 dm . . . . ***P. purpusii***
      7′ Herbage glabrous to short-hairy; pl gen > 2 dm

8. Infl glandular; peduncles widely spreading
  9. Corolla 13–22 mm; anther sacs 1.2–2.2 mm, dehiscing for 2/5–3/5 length
    10. Lvs thread-like, gen ± 0.5 mm wide; anther sacs 1.2–1.4 mm . . . . ***P. filiformis***
    10′ Lvs linear to (ob)lanceolate, 2–9 mm wide; anther sacs 1.5–2.2 mm . . . . ***P. roezlii***
  9′ Corolla 21–38 mm; anther sacs 1.7–3.2 mm, dehiscing for 3/5–4/5 length
    11. Herbage glaucous, glabrous . . . . ***P. neotericus***
    11′ Herbage green, hairy, lvs sometimes glabrous. . . . . ***P. laetus***
      12. Calyx 8–14 mm, lobes narrowly lanceolate . . . . var. ***leptosepalus***
      12′ Calyx 4.7–9 mm, lobes gen lanceolate to ovate
        13. Corolla throat open, lips ± spreading; anther sacs dehiscing for 3/5–2/3 length. . . . . var. ***laetus***
        13′ Corolla throat ± narrowed at mouth, lips projecting; anther sacs dehiscing for 3/5–4/5 length
        . . . . var. ***sagittatus***
8′ Infl ± glabrous; peduncles appressed-ascending
  14. Distal lvs widest at base, gen cordate-clasping; herbage gen glabrous, glaucous
    15. Corolla 14–20 mm; anther sacs 1.4–2 mm . . . . ***P. parvulus***
    15′ Corolla 20–35 mm; anther sacs 1.8–3.5 mm . . . . ***P. azureus***
      16. Calyx 5–7.5 mm, lobes ± ovate, long-tapered tips 1.5–3.2 mm; lvs linear, 4–9 cm,
        2–7.5 mm wide. . . . . var. ***angustissimus***
      16′ Calyx 3–6.5 mm, lobes (ob)lanceolate to round, pointed tip 0.2–1.5 mm; lvs lanceolate,
        1–6.5 cm, 5–17 mm wide . . . . var. ***azureus***
  14′ Distal lvs tapered to base; herbage glabrous or hairy, rarely glaucous . . . . ***P. heterophyllus***
    17. Lvs narrowly (ob)lanceolate, 2–7.5 mm wide, rarely with axillary lf clusters. . . . . var. ***purdyi***
    17′ Lvs linear to narrowly (ob)lanceolate, 0.5–5 mm wide, gen with axillary lf clusters
      18. Pl short-hairy. . . . . var. ***australis***
      18′ Pl glabrous, sometimes proximal sts short-hairy. . . . . var. ***heterophyllus***

Group 3

1. Fl st gen < 1.2 dm
  2. Corolla blue-violet or blue-purple; staminode densely hairy; lvs green, glabrous . . . . ***P. davidsonii*** var. ***davidsonii***
  2′ Corolla rose, lavender, or violet; staminode glabrous to sparsely hairy; lvs glaucous, gen puberulent . . . . . ***P. rupicola***
1′ Fl st 1.2–3 dm — corolla rose-red to dark rose-purple. . . . . ***P. newberryi***
  3. Corolla 27–35 mm, throat well-expanded, 7.5–12 mm wide when pressed, floor long-curly-hairy;
    stamens incl. . . . . var. ***berryi***
  3′ Corolla 20–30 mm, throat moderately expanded, 4.7–7.5 mm wide when pressed, floor short-hairy;
    stamens, at least longer pair, exserted
    4. Corolla rose-red; lvs of fl st << basal lvs. . . . . var. ***newberryi***
    4′ Corolla dark rose-purple; lvs of fl st barely reduced. . . . . var. ***sonomensis***

Group 4

1. Herbage hairy
  2. Herbage glandular-hairy; corolla lobes glandular-hairy inside
    3. Herbage densely glandular-hairy; fl st erect, gen 4–7 dm, stout. . . . . ***P. sudans***
    3′ Herbage sparsely glandular-hairy; fl st gen decumbent at base, gen < 4 dm . . . . ***P. deustus***
      4. Corolla 12–18 mm, upper lip white; lvs 8–18 mm wide . . . . var. ***deustus***
      4′ Corolla 9–15 mm, upper lip ± brown; lvs 3–28 mm wide
        5. Lvs lanceolate, 2.5–8(14) mm wide; calyx 2.5–5 mm, lobes lanceolate, acute; corolla 10–12(15) mm
        . . . . var. ***pedicellatus***
        5′ Lvs ovate to ± round, 7–28 mm wide; calyx 3.5–6.5 mm, lobes narrowly lanceolate, sharply acute or
        long-tapered; corolla 9–11 mm . . . . var. ***suffrutescens***
  2′ Herbage nonglandular-hairy (distal st, infl occ also glandular); corolla lobes glabrous inside or corolla
    floor nonglandular-hairy
    6. Infl glabrous; calyx 1–4.5 mm; corolla 6–10 mm
      7. Lvs gen all cauline, ± linear, folded lengthwise, arching-recurved . . . . [2]***P. cinicola***
      7′ Lvs best developed on basal sterile sts, these (ob)lanceolate to obovate, cauline lvs ascending (folded
        lengthwise), not arching-recurved (see 29 for vars.). . . . . [2]***P. procerus***
    6′ Infl hairy, gen glandular; calyx 3–13 mm; corolla 10–30 mm
      8. Calyx 3–8.5 mm
        9. Basal lvs (15)20–75 mm, (2)4–32 mm wide, petioled, cauline lvs sessile . . . . ***P. humilis*** var. ***humilis***
        9′ Lvs similar throughout, 5–20 mm, 0.8–6 mm wide, petioled
          10. Pl spreading to ascending, not mat-forming; st 1–3 dm; calyx lobes ovate . . . . ***P. californicus***
          10′ Pl low, gen mat-forming; st 0.5–1.5 dm; calyx lobes lanceolate. . . . . ***P. thompsoniae***
      8′ Calyx (4.5)6–13 mm
        11. Corolla gradually expanded into throat, throat 4–6 mm wide when pressed; staminode ± incl, not coiled at tip
          12. Anther sacs 0.6–0.7 mm, valves spreading widely; corolla 13–17 mm . . . . ***P. calcareus***
          12′ Anther sacs 1.1–1.4 mm, valves barely spreading; corolla 15–20 mm . . . . ***P. monoensis***

    11′ Corolla abruptly and widely expanded into throat, throat 7–12 mm wide (4–5 mm in *Penstemon*
        barnebyi) when pressed; staminode exserted, coiled at tip
      13. Corolla 10–14 mm, throat 4–5 mm wide when pressed. . . . . ***P. barnebyi***
      13′ Corolla 18–28 mm, throat 7–12 mm wide when pressed. . . . . ***P. janishiae***
1′ Herbage (proximal to infl) glabrous
  14. Infl glabrous; lf margin gen entire (serrate)
    15. Corolla 17–34 mm
      16. Anther sacs 0.6–1.2 mm; corolla floor glabrous
        17. Corolla gen 13–20 mm, moderately expanded, lavender, magenta, or violet . . . . [2]***P. patens***
        17′ Corolla 17–30 mm, cylindric or narrowly funnel-shaped, pink to red (± purple)
          18. Corolla lobes projecting, not spreading, glabrous; lvs cordate-clasping. . . . . ***P. centranthifolius***
          18′ Corolla lobes spreading, glandular; lvs not cordate-clasping . . . . ***P. utahensis***
      16′ Anther sacs 1.2–2.4 mm; corolla floor glabrous or hairy
        19. Distal lf pairs fused around st, sharply serrate
          20. Corolla 19–22 mm, bright pink; anther-sacs 1.3–1.6 mm . . . . ***P. clevelandii*** var. ***connatus***
          20′ Corolla 24–34 mm, blue-purple or blue; anther-sacs 1.8–2.4 mm . . . . ***P. spectabilis*** var. ***spectabilis***
        19′ Lf pairs free (or barely fused in some hybrids), entire or weakly serrate
          21. Lvs lanceolate to ovate, 10–35 mm wide; staminode incl, glabrous — common hybrid
          . . . . (see *Penstemon centranthifolius* description)
          21′ Lvs linear to narrowly lanceolate, 3–9 mm wide; staminode exserted, densely hairy . . . . ***P. fruticiformis***
            22. Corolla 22–24 mm, throat 8–10 mm wide when pressed, glandular outside; calyx lobes ovate,
              gen 5–7.5 mm . . . . var. ***amargosae***
            22′ Corolla 24–28 mm, throat 10–14 mm wide when pressed, glabrous outside; calyx lobes widely
              ovate to ± round, gen 4.5–6 mm . . . . var. ***fruticiformis***
    15′ Corolla 6–17 mm
      23. Openly branched shrub; lvs narrowly linear, gen rolled upward, gen < 1.4 mm wide. . . . . ***P. thurberi***
      23′ Per, fl sts gen little-branched; lvs gen lanceolate to oblanceolate to ovate (± linear in *Penstemon*
        cinicola), flat to folded, 1–22 mm wide
        24. Lvs thick; anther sacs 1–1.2 mm; corolla floor glabrous . . . . [2]***P. patens***
        24′ Lvs ± thin; anther sacs 0.3–1 mm; corolla floor hairy
          25. Corolla 9–17 mm (if < 10 mm anther sacs 0.5–0.8 mm)
            26. Pl 8–12 cm; distal cauline lvs ovate to ± round, 10–20 mm; anther sacs ± 0.4 mm; calyx 2.5–3 mm. . . ***P. tracyi***
            26′ Pl 15–60 cm; distal cauline lvs (ob)lanceolate, 15–70 mm; anther sacs 0.5–1 mm; calyx 3–5 mm
              27. St base gen buried in sand; proximal-most cauline lvs scale-like . . . . ***P. albomarginatus***
              27′ St base above ground; proximal cauline lvs well developed . . . . ***P. rydbergii*** var. ***oreocharis***
          25′ Corolla 6–10 mm; anther sacs 0.3–0.5 mm
            28. Lvs gen all cauline, ± linear, folded lengthwise and arching-recurved. . . . . [2]***P. cinicola***
            28′ Lvs best developed on basal, sterile sts, (ob)lanceolate to obovate, cauline lvs ascending, not
              arching-recurved. . . . . [2]***P. procerus***
              29. Pl 10–40 cm; distal cauline lvs 10–45 mm, 3–10 mm wide. . . . . var. ***brachyanthus***
              29′ Pl 5–12 cm; distal cauline lvs gen 5–20 mm, 1–5.5 mm wide. var. ***formosus***
14′ Infl gen glandular or glandular-hairy (exc occ *Penstemon stephensii* infl glabrous); lf margin ± serrate or
  dentate (exc *Penstemon heterodoxus*, *Penstemon incertus*, occ *Penstemon clevelandii*)
  30. Distal lf pairs fused around st
    31. Anther sacs 0.8–1.3 mm, valves spreading flat — desert areas
      32. Corolla gen 17–25 mm, throat 5–9 mm wide when pressed; lower lip strongly lined; anther sacs
        1.1–1.3 mm . . . . ***P. pseudospectabilis***
      32′ Corolla gen < 20 mm, throat 4–6 mm wide when pressed, lower lip lacking prominent guide-lines;
        anther sacs 0.8–1.1 mm . . . . ***P. stephensii***
    31′ Anther sacs 1.3–2.4 mm, valves hardly spreading (exc *Penstemon bicolor*)
      33. Corolla 18–24(27) mm; anther sacs 1.3–2 mm . . . . ***P. bicolor***
      33′ Corolla 24–34 mm; anther sacs 1.8–2.4 mm
        34. Corolla pale pink to lavender-pink; staminode densely hairy, exserted . . . . ***P. palmeri*** var. ***palmeri***
        34′ Corolla blue-purple or blue; staminode glabrous, incl . . . . ***P. spectabilis*** var. ***subviscosus***
  30′ Lvs free, not fused around st
    35. Corolla 10–18 mm
      36. Lvs ± serrate, distal cordate-clasping; upper side of corolla abruptly expanded . . . . ***P. anguineus***
      36′ Lvs entire, distal not cordate-clasping; upper side of corolla cylindric to moderately expanded . . . ***P. heterodoxus***
        37. Pl 5–20 cm; infl of 1(2) fl clusters. . . . . var. ***heterodoxus***
        37′ Pl gen > 20 cm; infl gen of 2–6 fl clusters
          38. Infl strongly glandular . . . . var. ***cephalophorus***
          38′ Infl moderately glandular . . . . var. ***shastensis***
    35′ Corolla 18–35 mm
      39. Lvs thin; upper side of corolla tube abruptly expanded; calyx 6–12 mm. . . . . ***P. rattanii***
        40. Calyx lobes ovate, obtuse or rounded at tip, calyx 6–7 mm . . . . var. ***kleei***
        40′ Calyx lobes lanceolate, acute or long-tapered at tip, calyx 6–12 mm. var. ***rattanii***

39′ Lvs thick; corolla gradually expanded or lower side ± abruptly expanded; calyx 3–8.5 mm

  41. Lvs linear to narrowly lanceolate, 1.5–7 mm wide . . . . ***P. incertus***

  41′ Lvs lanceolate to widely ovate, 10–60 mm wide

    42. Corolla gradually expanded, throat 4.5–9 mm wide when pressed

      43. Corolla 24–30 mm; anther sacs 1.2–1.6 mm; main upper lvs 45–90 mm — common hybrid

      . . . . (see *Penstemon centranthifolius* description)

      43′ Corolla 18–24 mm; anther sacs 0.8–1.3 mm; main distal lvs 20–60 mm . . . . ***P. clevelandii*** (in part)

        44. Lvs entire to moderately serrate; staminode glabrous or sparsely hairy . . . . var. ***clevelandii***

        44′ Lvs sharply serrate; staminode densely hairy . . . . var. ***mohavensis***

    42′ Corolla abruptly expanded (exc *Penstemon floridus* var. *austinii*), throat 6–20 mm wide when pressed

      45. Staminode incl, glabrous; anther-sacs 1.3–1.9 mm . . . . ***P. floridus***

        46. Corolla 21–27 mm, throat 6–11 mm wide when pressed, mouth perpendicular to tube,

        tube 6–8 mm . . . . var. ***austinii***

        46′ Corolla 24–30 mm, ± abruptly expanded to throat, 10–16 mm wide when pressed, mouth

        oblique to tube, tube 7–12 mm . . . . var. ***floridus***

      45′ Staminode exserted, densely hairy; anther-sacs 1.6–2.2 mm . . . . ***P. grinnellii***

        47. Pl 10–60 cm; young branches light green; corolla 22–30 mm, ± white, tinged pink or lavender

        . . . . var. ***grinnellii***

        47′ Pl 45–85 cm; young branches glaucous; corolla 26–35 mm, blue-violet. . . . . var. ***scrophularioides***

P. albomarginatus M.E. Jones (p. 1021) WHITE-MARGINED BEARDTONGUE (Group 4) Per 15–35 cm, glabrous; st base gen buried in sand. **LF:** proximal-most scale-like, distal 15–50 mm, oblanceolate, white-margined, entire or weakly dentate. **INFL:** glabrous. **FL:** calyx 3–5 mm, lobes ovate-elliptic, white-margined, finely serrate; corolla 13–17 mm, pink to purple, glabrous exc floor hairy; anther sacs ± 0.9–1 mm, spreading flat; staminode glabrous. 2*n*=16. Loose desert sand, gen on stabilized dunes; 700–900 m. DMoj; NV, AZ. Mar–May ★

P. anguineus Eastw. (p. 1021) (Group 4) Per 20–90 cm, glabrous proximal to infl. **LF:** thin, ± serrate; basal lvs lanceolate to narrowly ovate; middle cauline lvs narrowly elliptic, slightly narrowed in middle, widest at sessile, rounded or truncate base; distal cauline lvs 10–70 mm, lanceolate to ovate, cordate-clasping. **INFL:** glandular. **FL:** calyx 5–8 mm, lobes narrowly lanceolate; corolla 13–18 mm, abruptly expanded throat on upper side, blue to purple-violet, glandular outside, floor gen long-hairy; anther sacs 1–1.3 mm, dehiscing full length, valves barely spreading; staminode sparsely hairy. 2*n*=16. Open, gen logged areas of ± conifer forest; 600–2100 m. KR, NCoRH; OR. Jun–Aug

P. azureus Benth. (Group 2) Per 20–70 cm, with woody branches proximally, glabrous. **LF:** gen cauline; distal lvs widest at gen cordate-clasping base, entire. **INFL:** glabrous. **FL:** corolla 20–35 mm, (lavender-)blue, glabrous; anther sacs 1.8–3.5 mm, dehiscing across common end 1/2–2/3 their length, inner margins hairy; staminode glabrous.

  var. ***angustissimus*** A. Gray (p. 1023) **LF:** 4–9 cm, linear. **FL:** calyx 5–7.5 mm, lobes ± ovate, long-tapered tip 1.5–3.2 mm. 2*n*=48. Gen ± moist woodland, open forest; 300–800 m. KR, NCoRH, CaR, n&c SN. May–Aug

  var. ***azureus*** (p. 1023) **LF:** 1–6.5 cm, lanceolate. **FL:** calyx 3–6.5 mm, lobes (ob)lanceolate to round, pointed tip 0.2–1.5 mm. 2*n*=48. Gen ± moist woodland, open forest; 500–2500 m. NW, CaR, n&c SNH; OR. May–Aug

P. barnebyi N.H. Holmgren (p. 1023) BARNEBY'S BEARDTONGUE (Group 4) Per 6–30 cm; hairs short, backward-pointing. **LF:** 20–75 mm; basal lvs well developed; distal cauline lvs lanceolate, (sub)entire. **INFL:** glandular. **FL:** calyx 4.5–12 mm, lobes ± lanceolate; corolla 10–14 mm, abruptly expanded to throat on lower side, throat 4–5 mm wide when pressed, violet, blue distally, glandular outside, throat white, dark-lined, floor ± yellow-hairy; anther sacs 0.7–0.9 mm, spreading flat; staminode much exserted, coiled at tip, densely orange-yellow-hairy. Limestone gravel or silt in sagebrush scrub or pinyon/juniper woodland; 1500–2500 m. W&I; NV. May ★

P. bicolor (Brandegee) Clokey & D.D. Keck (Group 4) Per < 150 cm; herbage glabrous. **LF:** thick; distal cauline 4–11 cm, ovate, sharply serrate, base fused around st. **INFL:** glandular. **FL:** calyx 4–6 mm, lobes ± ovate; corolla 18–24(27) mm, throat 6–11 mm wide

when pressed, cream to magenta, strongly lined, glandular outside, floor long- ± white-hairy; anther sacs 1.3–2 mm, spreading flat; staminode incl, densely yellow-hairy. Gravelly or rocky soils, creosote-bush or blackbush scrub, Joshua-tree woodland; 700–1500 m. DMtns (Castle Mtns); NV. May

P. caesius A. Gray (p. 1023) (Group 2) Subshrub 20–80 cm, ± scapose; herbage glabrous, glaucous. **LF:** gen basal, petioled; blade 15–43 mm, widely obovate to round, entire. **INFL:** glandular. **FL:** calyx 3.7–7.5 mm, lobes lanceolate or ovate; corolla 15–23 mm, purple-blue, glandular outside, floor hairy; anther sacs 1.3–1.6 mm, dehiscing across common end slightly > 1/2 their length, inner margins hairy; staminode glabrous. Rocky ridges and slopes in open conifer forest and alpine communities; 1800–3400 m. s SNH, SnGb, SnBr. Jun–Aug

P. calcareus Brandegee (p. 1023) LIMESTONE BEARDTONGUE (Group 4) Per 7–25 cm; hairs fine, backward-pointing, ashy, short; distal st densely glandular. **LF:** distal cauline 20–60 mm, widely lanceolate, entire to shallowly dentate. **INFL:** densely glandular. **FL:** calyx 6.5–8.5 mm, lobes ± lanceolate; corolla 13–17 mm, cylindric to funnel shaped, bright pink to rose-purple, glandular outside, throat 4–6 mm wide when pressed, floor nearly glabrous; anther sacs 0.6–0.7 mm, spreading flat; staminode ± incl, densely yellow-hairy. Limestone crevices, rocky slopes in pinyon/juniper woodland, Joshua-tree scrub; 1200–1600 m. DMtns. Apr–May ★

P. californicus (Munz & I.M. Johnst.) D.D. Keck (p. 1023) CALIFORNIA BEARDTONGUE (Group 4) Per 10–30 cm, spreading to ascending; hairs appressed, flat, scale-like, backward-pointing. **LF:** petioled; 7–15 mm, linear to narrowly oblanceolate, entire. **FL:** calyx 4–8.5 mm, lobes ovate, hairy as on lvs (or also sparsely glandular); corolla 14–18 mm, (blue-)purple, white inside, dark-lined, sparsely glandular outside, floor sparsely hairy; anther sacs 1–1.3 mm, dehiscing full length; staminode hairy. 2*n*=16. Sandy soils, yellow-pine forest or pinyon/juniper woodland; 1200–2300 m. PR; Mex. May–Jun ★

P. centranthifolius Benth. (p. 1023) SCARLET BUGLER (Group 4) Per 30–120 cm, glabrous, glaucous. **LF:** thick; middle cauline gen largest, 40–100 mm, lanceolate to ovate, cordate-clasping, entire. **FL:** calyx 3.5–6.5 mm, lobes ovate to ± round; corolla 20–30 mm, cylindric, lobes projecting, not spreading, red, glabrous (incl floor); anther sacs 0.8–1.2 mm, spreading flat; staminode glabrous. 2*n*=16. Dry, open chaparral or oak woodland; < 1800 m. NCoR, n SNF, margins of GV, SnFrB, SCoR, SW (exc SCo), sw edge DMoj; Mex. If recognized taxonomically, common hybrids with *P. spectabilis* assignable to *P.* ×*parishii* A. Gray. Apr–Jul

P. cinicola D.D. Keck (p. 1023) ASH BEARDTONGUE (Group 4) Per 15–40 cm, glabrous or short-hairy proximally. **LF:** gen all cauline, 30–60 mm, ± linear, entire, folded lengthwise, arching-recurved. **INFL:** glabrous. **FL:** calyx 1–2.5 mm, lobes widely obovate, tips gen jagged-toothed; corolla 6–9 mm, blue-purple, glabrous exc hairy

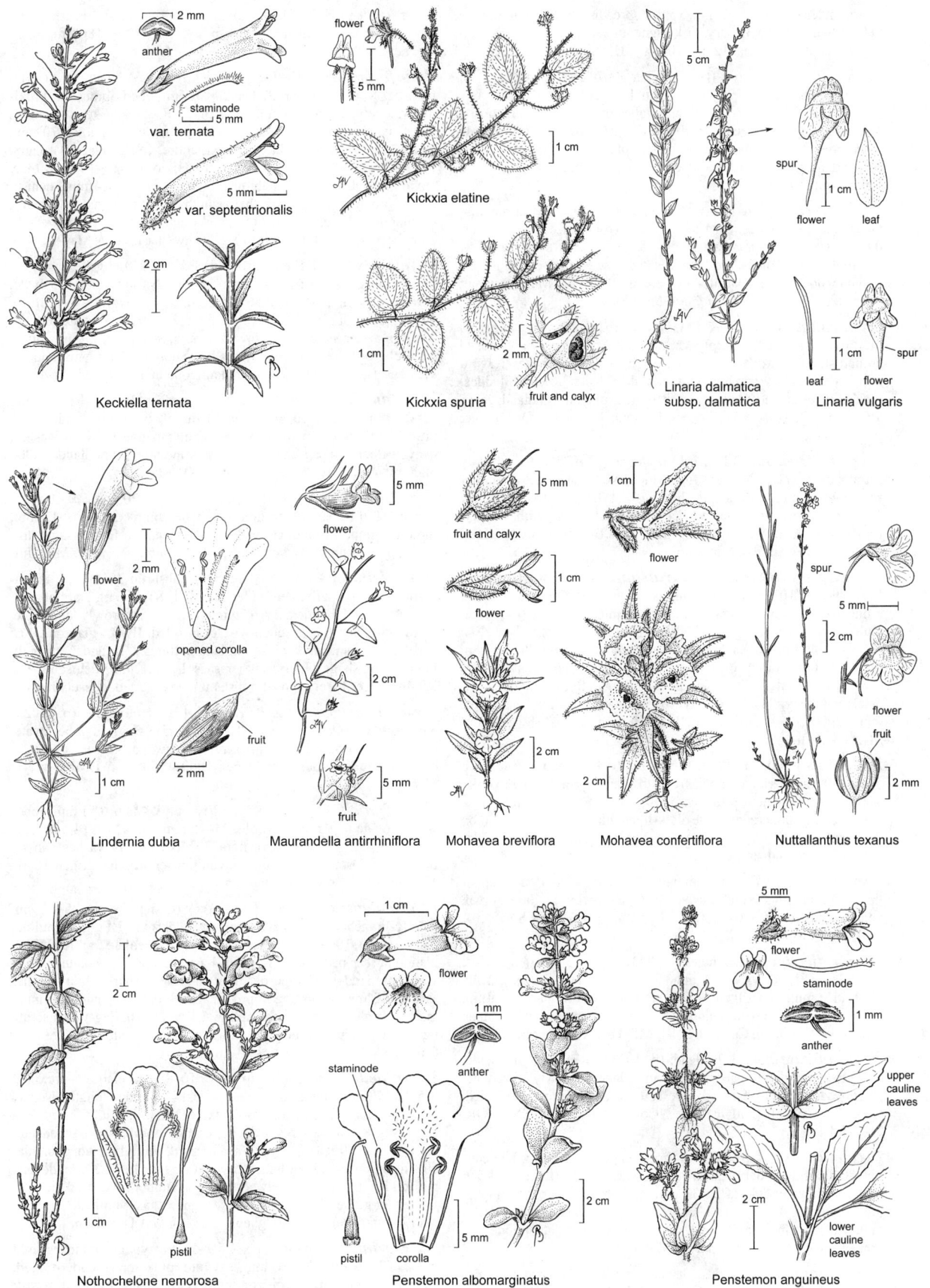

2 mm
anther
staminode
5 mm
var. ternata

5 mm
var. septentrionalis

2 cm

Keckiella ternata

flower
5 mm

Kickxia elatine

1 cm

1 cm
2 mm
Kickxia spuria
fruit and calyx

5 cm

spur
1 cm
flower
leaf

Linaria dalmatica
subsp. dalmatica

leaf
1 cm
flower
spur
Linaria vulgaris

flower
2 mm
opened corolla
fruit
2 mm

Lindernia dubia

flower
5 mm

2 cm

fruit
5 mm

Maurandella antirrhiniflora

fruit and calyx
5 mm

flower
1 cm

2 cm

Mohavea breviflora

flower
1 cm

2 cm

Mohavea confertiflora

1 cm

spur
5 mm
2 cm

flower
fruit
2 mm

Nuttallanthus texanus

2 cm

pistil
1 cm

Nothochelone nemorosa

1 cm
flower

staminode
anther
1 mm

staminode
pistil
corolla
5 mm

2 cm

Penstemon albomarginatus

5 mm
flower
staminode
anther
1 mm

upper
cauline
leaves

2 cm

lower
cauline
leaves

Penstemon anguineus

floor; anther sacs 0.3–0.5 mm, valves spreading widely; staminode yellow-hairy. 2*n*=16,32. Dry, rocky, igneous soils, in sagebrush openings of montane forests; 1250–2700 m. GB; OR. Jun–Aug ★

P. clevelandii A. Gray (Group 4) Per 30–70 cm, much-branched; herbage glabrous. **LF:** 20–60 mm, thick, ± ovate. **INFL:** glabrous or glandular. **FL:** calyx 3.5–6.5 mm, lobes ovate to ± round; corolla 18–24 mm, gradually expanded into throat, throat 4.5–8 mm wide when pressed, pink, magenta, or red-purple, unlined, glandular outside and inside (floor lacks nonglandular hairs); anther sacs dehiscing full length; staminode incl.

var. ***clevelandii*** (p. 1023) **LF:** entire to moderately serrate; distal cauline lf bases free, cordate-clasping to rounded. **FL:** anther sacs 0.8–1.3 mm, spreading flat; staminode glabrous or sparsely hairy. 2*n*=16. Rocky hillsides, rock crevices in creosote-bush scrub, juniper/pinyon woodland, chaparral; 400–1800 m. PR/DSon; Mex. Variable; hybridizes with *P. centranthifolius*. Mar–May

var. ***connatus*** Munz & I.M. Johnst. (p. 1023) SAN JACINTO BEARDTONGUE **LF:** sharply serrate; distal cauline lf bases fused around st. **FL:** corolla 19–22 mm, bright pink; anther sacs 1.3–1.6 mm, valves barely spreading; staminode ± hairy. Rocky hillsides, rock crevices in creosote-bush scrub, juniper/pinyon woodland, chaparral; 400–1700 m. e PR (e slope SnJt, Santa Rosa mtns). Distinctive; perhaps should be recognized as sp. Mar–May ★

var. ***mohavensis*** (D.D. Keck) McMinn (p. 1023) **LF:** sharply serrate; distal cauline lf bases free, widely wedge-shaped. **FL:** anther sacs 0.9–1.1 mm, valves spreading but not flat; staminode densely hairy. Rocky hillsides, rock crevices in creosote-bush scrub, juniper/pinyon woodland, chaparral; 400–1500 m. s DMtns (Little San Bernardino, Granite mtns). Mar–May

P. davidsonii Greene var. ***davidsonii*** (p. 1023) (Group 3) Low shrub < 10 cm, mat-forming, short-hairy. **LF:** ± basal (much reduced distally on st); main blades 5–30 mm, elliptic to (ob)ovate, (sub)entire, glabrous, green. **INFL:** glandular; pedicels short. **FL:** calyx 7–13 mm, lobes linear to lanceolate; corolla 20–36 mm, blue-violet or blue-purple, floor ± white-shaggy-hairy; anther sacs 0.9–1.3 mm, valves spreading flat, white-woolly; staminode incl, densely pale yellow-hairy. 2*n*=16. Montane to alpine outcrops, talus; 2000–3750 m. KR, CaRH, SNH, Wrn, n SNE (Sweetwater Mtns); OR. Jul–Aug

P. deustus Lindl. HOT-ROCK PENSTEMON (Group 4) Subshrub gen < 40 cm; herbage subglabrous to glandular-hairy. **LF:** opposite (in CA), dentate; cauline lvs 8–50 mm. **INFL:** ± glandular. **FL:** corolla 8–15 mm, cream-white, dark-lined, glandular outside and on floor; anther sacs 0.5–0.7 mm, valves spreading flat; staminode glabrous or bearded distally. 2*n*=16.

var. ***deustus*** **LF:** 8–18 mm wide. **FL:** corolla 12–18 mm, upper lip white; staminode glabrous at tip. Open, rocky, granitic gravel, volcanic substrates; 600–2200 m. KR, NCoR, CaRH, n&c SNH, GB; e OR, e WA to w WY. May–Jul

var. ***pedicellatus*** M.E. Jones (p. 1023) **LF:** 2.5–8(14) mm wide, lanceolate. **FL:** calyx 2.5–5 mm, lobes lanceolate, acute; corolla 10–12(15) mm, upper lip ± brown; staminode short, glabrous at tip. Sagebrush scrub, pinyon/juniper woodland, yellow-pine and montane forests; 900–3000 m. CaRH, n SNH, MP; OR, NV, UT. May–Jul

var. ***suffrutescens*** L.F. Hend. (p. 1023) **LF:** 7–28 mm wide, ovate to ± round. **FL:** calyx 3.5–6.5 mm, lobes narrowly lanceolate, sharply acute or long-tapered; corolla 9–11 mm; staminode bearded. Open forest; 600–2200 m. KR, NCoRH, s CaRH, n SNH; OR. May–Jul

P. eatonii A. Gray (p. 1023) (Group 1) Per 40–100 cm. **LF:** cauline 30–90 mm, widely lanceolate to ovate, entire. **FL:** calyx 3.5–6 mm, lobes ovate; corolla 24–33 mm, cylindric, obscurely 2-lipped, lobes subequal, barely spreading, red, glabrous; anther sacs 1.4–2.8 mm, dehiscing in distal 1/2–3/4, short-hairy on sides; staminode glabrous to sparsely hairy at tip. 2*n*=16.

var. ***eatonii*** Pl glabrous. Dry sagebrush scrub, pinyon/juniper woodland; 1400–1800 m. DMtns; to ID, WY, NM. Mar–Jul

var. ***undosus*** M.E. Jones Pl short-hairy. Dry sagebrush scrub, pinyon/juniper woodland, yellow-pine forest; 1200–1800 m. SnBr, D; NV, CO, NM. Mar–Jul

P. filiformis (D.D. Keck) D.D. Keck (p. 1023) THREAD-LEAVED BEARDTONGUE (Group 2) Per 20–50 cm, woody-branched proximally, finely downward-pointing-hairy. **LF:** thread-like, sometimes glabrous; short basal lvs densely clustered; cauline lvs 20–70 mm, gen ± 0.5 mm wide, tightly inrolled, entire. **INFL:** glandular-hairy. **FL:** calyx 3.4–5.7 mm, lobes lanceolate; corolla 13–16 mm, blue, glandular outside, floor glabrous; anther sacs 1.2–1.4 mm, dehiscing at proximal end to 1/2 their length, inner margins short-hairy; staminode glabrous. 2*n*=16. Open, rocky places among shrubs, yellow-pine forest; 400–1700 m. KR (n Trinity, nw Shasta cos.). May–Jul ★

P. floridus Brandegee (Group 4) Per 50–120 cm; herbage glabrous, glaucous. **LF:** thick; distal cauline lvs lanceolate or ovate, cordate-clasping, gen dentate (distal-most sometimes subentire). **INFL:** glandular-hairy. **FL:** calyx 4.2–6.2 mm, lobes ovate; corolla 21–30 mm, throat narrowed toward mouth, rose-pink, strongly lined, glandular outside and inside (floor lacks nonglandular hairs); anther sacs 1.3–1.9 mm, spreading flat; staminode incl, glabrous. 2*n*=16.

var. ***austinii*** (Eastw.) N.H. Holmgren (p. 1023) **FL:** corolla 21–27 mm, expanded gradually to throat, 6–11 mm wide when pressed, tube 6–8 mm, mouth perpendicular to tube. Gravelly washes, canyon floors, in sagebrush scrub, pinyon/juniper woodland; 1400–3000 m. W&I (Inyo Mtns), n DMtns; NV. May–Jun

var. ***floridus*** (p. 1023) **FL:** corolla 24–30 mm, ± abruptly expanded to throat, 10–16 mm wide when pressed, tube 7–12 mm, mouth oblique to tube. Gravelly washes, canyon floors, sagebrush scrub, pinyon/juniper woodland; 1600–2700 m. SNE; NV. May–Jul

P. fruticiformis Coville (Group 4) Subshrub 30–60 cm, much-branched proximally, gen wider than tall. **ST:** young sts glabrous, gen glaucous. **LF:** thick, 25–65 mm; distal lvs ± narrowly lanceolate, (sub)entire, gen folded lengthwise or inrolled. **INFL:** glabrous. **FL:** corolla pale pink to ± white, limb sometimes ± lavender, strongly lined, floor shaggy-hairy; anther sacs 1.6–2.1 mm, dehiscing full length, valves barely spreading; staminode exserted, densely hairy.

var. ***amargosae*** (D.D. Keck) N.H. Holmgren (p. 1023) AMARGOSA BEARDTONGUE **FL:** calyx (4.5)5–7.5 mm, lobes ovate; corolla 22–24 mm, throat 8–10 mm wide when pressed, glandular outside. Creosote-bush scrub; gen 1000–1750 m. n DMoj; w NV. May–Jun ★

var. ***fruticiformis*** (p. 1023) **FL:** calyx 4.5–6(6.5) mm, lobes widely ovate to ± round; corolla 24–28 mm, throat 10–14 mm wide when pressed, glabrous exc on floor. 2*n*=16. Gravelly washes, canyon floors in creosote-bush scrub, pinyon/juniper woodland; gen 1000–1700 m. s SNE, n DMoj. May–Jun

P. gracilentus A. Gray (p. 1023) (Group 2) Per 25–65 cm, woody-branched proximally; herbage glabrous. **LF:** gen cauline; distal lvs 40–100 mm, linear to narrowly lanceolate, entire. **INFL:** glandular (esp pedicels). **FL:** calyx 4–6 mm, lobes lanceolate; corolla 15–20 mm, red- to blue-purple, floor glabrous to shaggy-hairy; anther sacs 1–1.3 mm, dehiscing at proximal end to 1/2 their length, inner margins glabrous; staminode yellow-hairy. 2*n*=16. Sagebrush scrub, juniper woodland, yellow-pine to subalpine forests; 900–2800 m. CaR, n SNH, MP; OR, NV. Jun–Aug

P. grinnellii Eastw. (Group 4) Per 10–85 cm, glabrous proximal to infl. **LF:** 50–90 mm, lanceolate, thick, gen folded lengthwise and arching-recurved, subentire to dentate. **INFL:** glandular-hairy. **FL:** calyx 4.7–8.5 mm, lobes ± ovate; corolla abruptly expanded into throat, throat 10–18 mm wide when pressed, strongly lined, glandular outside and inside, floor long-hairy; anther sacs 1.6–2.2 mm, dehiscing full length, valves barely spreading; staminode exserted, densely yellow- or golden-hairy. 2*n*=16. If recognized taxonomically, hybrids with *P. centranthifolius* assignable to *P.* ×*dubius* Davidson.

var. ***grinnellii*** (p. 1027) Pl 10–60 cm; young branches light green. **FL:** corolla 22–30 mm, ± white, pink- or lavender-tinged. Chaparral, foothill and pinyon/juniper woodland, montane forest; 300–2700 m. s SN, w SnJV, TR, PR, DMoj (Victorville). May–Aug

1023

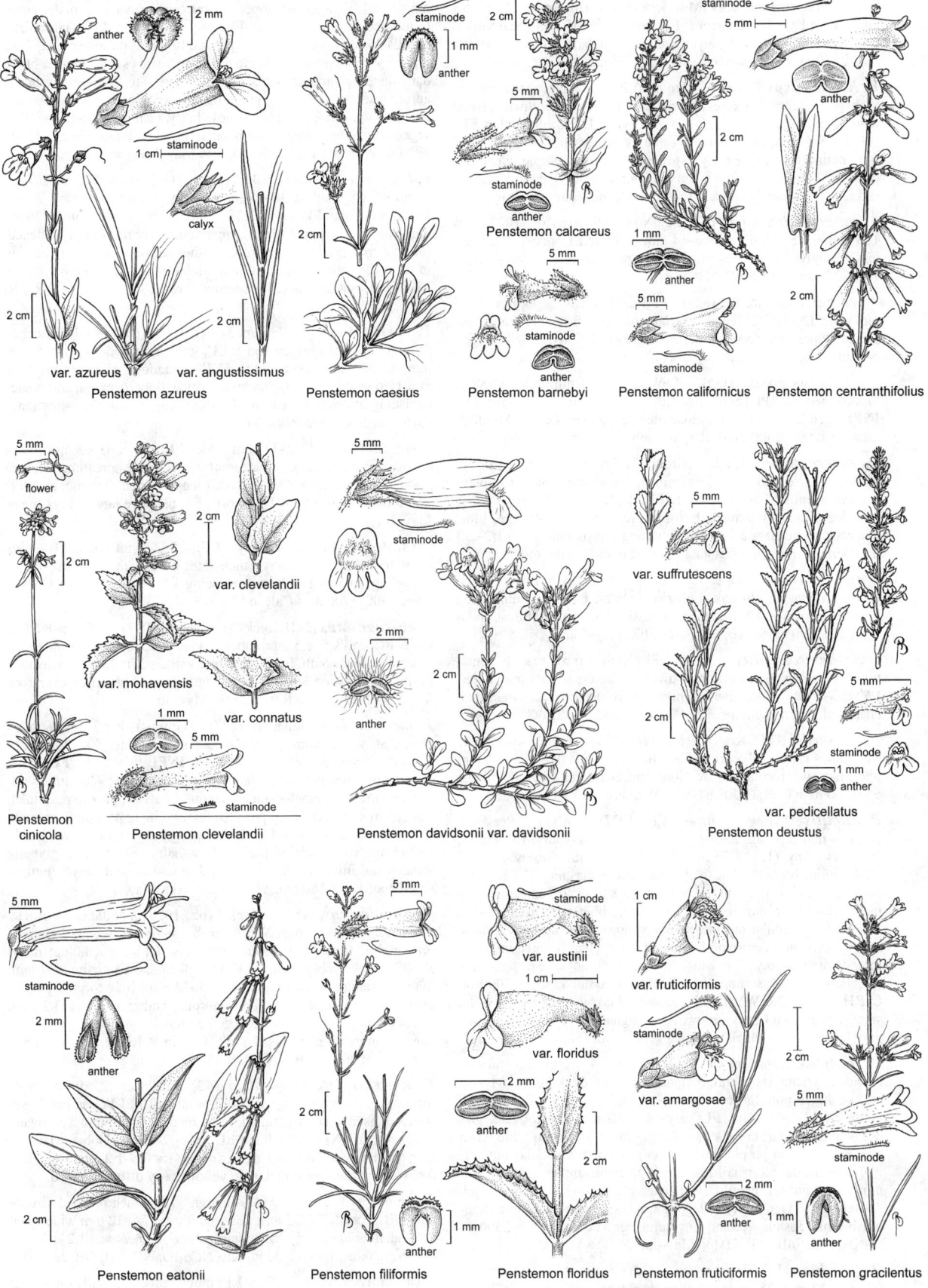

var. azureus   var. angustissimus

Penstemon azureus

anther

staminode

calyx

staminode
anther

Penstemon caesius

staminode
anther
Penstemon calcareus

staminode
anther

staminode
anther

staminode

Penstemon barnebyi

staminode

anther

staminode

Penstemon californicus

staminode

anther

Penstemon centranthifolius

flower

var. clevelandii

var. mohavensis

var. connatus

staminode

Penstemon
cinicola

Penstemon clevelandii

staminode

anther

Penstemon davidsonii var. davidsonii

var. suffrutescens

staminode
anther

var. pedicellatus

Penstemon deustus

staminode

anther

Penstemon eatonii

anther

Penstemon filiformis

var. austinii

var. floridus

anther

Penstemon floridus

var. fruticiformis

staminode

var. amargosae

anther

Penstemon fruticiformis

staminode

anther

Penstemon gracilentus

var. *scrophularioides* (M.E. Jones) N.H. Holmgren Pl 45–85 cm; young branches glaucous. **FL:** corolla 26–35 mm, blue-violet. Chaparral, foothill and pinyon/juniper woodland, montane forest; 500–2850 m. s SN, SnFrB (Santa Clara Co.), SCoR, TR. Apr–Jul

P. heterodoxus A. Gray   (Group 4) Per 5–65 cm, mat-forming; herbage ± glabrous. **LF:** entire, sometimes folded lengthwise; basal many; cauline narrowly lanceolate to ovate. **INFL:** glandular. **FL:** calyx 2.5–6 mm, lobes narrowly oblong to obovate; corolla 10–16 mm, cylindric to moderately expanded, deep blue-purple, glandular outside, floor yellow-brown-hairy; anther sacs 0.5–1 mm, dehiscing full length but barely spreading; staminode moderately yellow-hairy.

var. *cephalophorus* (Greene) N.H. Holmgren (p. 1027) Pl 15–35 cm. **LF:** 20–70 mm, 6–14 mm wide. **INFL:** gen 2–6 fl clusters, strongly glandular. Subalpine and alpine slopes, meadows, scree; 1900–3200 m. n&s SNH. Jul–Aug

var. *heterodoxus* (p. 1027) Pl 5–20 cm. **LF:** 5–50 mm, 2–10 mm wide. **INFL:** 1(2) fl clusters, glandular. 2*n*=16. Montane and subalpine slopes, meadows, scree; 2700–3900 m. SNH, n SNE, W&I; NV. Jul–Aug

var. *shastensis* (D.D. Keck) N.H. Holmgren SHASTA BEARDTONGUE Pl 15–65 cm. **LF:** 30–140 mm, 6–20 mm wide. **INFL:** gen 2–6 fl clusters, moderately glandular. 2*n*=32. Montane meadows; 900–2400 m. CaRH, MP. Jun–Aug   ★

P. heterophyllus Lindl.   (Group 2) Per 25–150 cm, woody-branched proximally. **LF:** gen cauline, 20–95 mm, linear to oblanceolate, ± entire; base tapered. **INFL:** ± glabrous. **FL:** calyx 4.2–8 mm, lobes lance-acuminate to (ob)ovate; corolla 23–40 mm, blue, glabrous; anther sacs 2.2–3 mm, dehiscing at proximal end to 1/2–2/3 their length, inner margins long-hairy; staminode glabrous.

var. *australis* Munz & I.M. Johnst. (p. 1027) Pl short-hairy, bearing axillary lf clusters at proximal nodes. **LF:** 0.5–4 mm wide, linear to narrowly (ob)lanceolate. Grassland, chaparral, forest openings; 50–1700 m. SnFrB, SCoR, TR, PR. May–Jun

var. *heterophyllus* (p. 1027) Pl glabrous (proximal sts sometimes short-hairy), gen bearing axillary lf clusters at proximal nodes. **LF:** 0.5–5 mm wide, linear to narrowly (ob)lanceolate. Grassland, chaparral, forest openings; 50–1900 m. NCoR, SCoR, WTR. Apr–Jul

var. *purdyi* (D.D. Keck) McMinn (p. 1027) Pl short-hairy (exc sometimes lf tips, calyces); axillary lf clusters gen 0. **LF:** 2–7.5 mm wide, narrowly (ob)lanceolate. Grassland, chaparral, forest openings; 50–1900 m. NCoR, CaRF, n SN, ScV, SnFrB, SCoRI. May–Jun

P. humilis A. Gray var. **humilis** (p. 1027)   (Group 4) Per 5–35 cm, gen mat-forming; herbage ± short-(occ ashy-)hairy. **LF:** entire; basal lvs many, (15)20–75 mm, (2)4–32 mm wide, (ob)ovate, petioled; cauline lvs lanceolate to obovate, sessile, clasping. **INFL:** glandular. **FL:** calyx 3–6 mm, lobes lanceolate to ovate; corolla 11–15 mm, cylindric to narrowly funnel shaped, blue with lighter floor, dark-lined, glandular outside, floor ± yellow- or white-hairy; anther sacs 0.5–0.8 mm, dehiscing full length, valves barely spreading; staminode orange- to yellow-hairy. 2*n*=16. Open montane to subalpine forests, sagebrush scrub, pinyon/juniper woodland; 1500–3000 m. CaRH, GB; to OR, WY, CO. If recognized taxonomically, small-fld, small-lvd pls in n CA, s OR, nw NV assignable to *P. cinereus* Piper, gray beardtongue, which intergrades fully with *P. humilis*. May–Jul

P. incertus Brandegee (p. 1027)   (Group 4) Shrub 20–100 cm, rounded; young sts glabrous, glaucous. **LF:** thick; largest lvs at mid st, 40–70 mm, linear to narrowly lanceolate, gen rolled inward, entire. **INFL:** glandular. **FL:** calyx 4.5–7.3 mm, lobes ovate, glabrous to glandular; corolla 23–32 mm, throat 8–12 mm wide when pressed, violet to purple (limb ± blue), unlined, glandular outside, glabrous inside exc small hair patch on floor; anther sacs 1.8–2.4 mm, dehiscing full length, valves barely spreading; staminode incl, densely hairy. Gen sandy soil along washes, canyon slopes, in sagebrush scrub, Joshua-tree and pinyon/juniper woodland; 900–2300 m. s SNH, Teh, SnBr, PR, DMoj. May–Jun

P. janishiae N.H. Holmgren (p. 1027)   JANISH'S BEARDTONGUE (Group 4) Per 8–25 cm; hairs short, backward-pointing. **LF:** 20–60 mm; basal lvs well developed; distal cauline lvs lanceolate, entire to toothed. **INFL:** glandular. **FL:** calyx 6–13 mm, lobes lanceolate; corolla 18–28 mm, abruptly expanded to throat on lower side, 7–12 mm wide when pressed, pink to dull purple (lobes sometimes ± blue), dark-lined, glandular outside, floor white- to pale-yellow-hairy; anther sacs 0.8–1.2 mm, spreading flat; staminode much exserted, coiled at tip, densely orange-yellow-hairy. Gen igneous-clay soils in sagebrush scrub, juniper/shrub savanna, yellow-pine forest; 1350–2350 m. n SNH, MP; to OR, ID, NV. May–Jul   ★

P. labrosus (A. Gray) Hook. f. (p. 1027)   (Group 1) Per 30–70 cm, glabrous. **LF:** cauline 30–85 mm, linear, gen rolled inward, entire. **INFL:** glabrous. **FL:** calyx 3.5–5.5 mm, lobes ovate; corolla 30–40 mm, ± cylindric, upper lip forming hood, lower lip strongly reflexed, (orange-)red (yellow), glabrous; anther-sacs 1.5–2 mm, dehiscing in distal 2/3, glabrous; staminode glabrous. 2*n*=16. Pinyon/juniper woodland, pine and mixed-hardwood forests; 1200–3100 m. TR, PR; Mex. Jul–Aug

P. laetus A. Gray   (Group 2) Per 15–75 cm, woody-branched proximally; herbage gen hairy. **LF:** gen cauline; distal lvs 15–100 mm, linear to lanceolate, entire. **INFL:** glandular. **FL:** corolla 21–35 mm, (violet-)blue, short-glandular outside, floor glabrous; anther sacs dehiscing at proximal end to 4/5 their length, inner margins long-hairy; staminode glabrous. 2*n*=16.

var. *laetus* **LF:** 2–11 mm wide. **FL:** calyx 4.7–9 mm, lobes ± widely lanceolate; corolla throat open, lips ± spreading; anthers 1.8–2.3 mm, dehiscing 3/5–2/3 their length. Foothill to montane forest; 400–2500 m. SN, WTR. Most s SN pls have larger calyx lobes. May–Jul

var. *leptosepalus* A. Gray **LF:** 4.5–22 mm wide. **FL:** calyx 8–14 mm, lobes narrowly lanceolate; corolla throat open, lips ± spreading; anthers 2–2.7 mm, dehiscing 3/5–2/3 their length. Conifer forest; 300–1700 m. s CaR, n SN, MP. May–Jul

var. *sagittatus* (D.D. Keck) McMinn (p. 1027) **LF:** 1.5–8.5 mm wide. **FL:** calyx 5–8.5 mm, lobes ± widely lanceolate; corolla throat ± narrowed at mouth, lips projecting; anthers 2.1–2.8 mm, narrowly sagittate, dehiscing 3/5–4/5 their length. Sagebrush scrub, evergreen forest; 400–2300 m. KR, CaR; OR. May–Jul

P. monoensis A. Heller (p. 1027)   (Group 4) Per 7–30 cm; hairs dense, ashy, backward-pointing. **LF:** gen cauline, 50–120 mm, ± (widely) lanceolate, entire to toothed. **INFL:** glandular. **FL:** calyx 8–11 mm, lobes gen lanceolate (ovate); corolla 15–20 mm, cylindric to narrowly funnel-shaped, glandular outside, pink to red-violet, throat 4–6 mm wide when pressed, floor white to pale pink, sparsely hairy; anther sacs 1.1–1.4 mm, dehiscing full length, valves barely spreading; staminode ± incl, yellow-hairy. Sandy and gravelly washes and hills, sagebrush scrub, Joshua-tree and pinyon/juniper woodland; 1200–1800 m. SNE, DMoj. Apr–May

P. neotericus D.D. Keck (p. 1027)   PLUMAS COUNTY BEARDTONGUE   (Group 2) Subshrub 25–80 cm, woody-branched proximally; herbage glabrous, glaucous. **LF:** gen cauline; distal 30–85 mm, lanceolate, entire. **INFL:** glandular. **FL:** calyx 4–7 mm, lobes lanceolate to ovate; corolla 23–38 mm, blue to pink-purple, glandular outside, floor white, glabrous; anther sacs 2.4–3.2 mm, dehiscing at proximal end 2/3–4/5 their length, inner margins white-hairy; staminode glabrous. 2*n*=32. Gen in volcanic soils of scrub, open forest; 500–2500 m. CaR, n SN. May–Aug

P. newberryi A. Gray   (Group 3) Subshrub 12–30 cm, mat-forming; hairs gen short, backward-pointing (0). **LF:** gen basal, gen reduced distally, short-petioled; main blades, 10–40 mm, (ob)ovate, finely serrate. **INFL:** glandular. **FL:** calyx 7–12 mm, lobes ± lanceolate; corolla floor hairy on ridges; anther sacs 0.8–1.2 mm, spreading flat, woolly; staminode incl, pale yellow-hairy distally. 2*n*=16.

var. *berryi* (Eastw.) N.H. Holmgren **LF:** much reduced distally on st. **FL:** corolla 27–35 mm, rose-red, throat 7.5–12 mm wide when pressed, floor long-curly-hairy; stamens incl, anthers sacs 1.2–1.7 mm. Outcrops, talus; 1000–2400 m. KR, NCoRO, NCoRH; OR. Jun–Aug

var. *newberryi* (p. 1027) **LF:** much reduced distally on st. **FL:** corolla 20–30 mm, rose-red, throat 4.7–7.5 mm wide when pressed,

floor short-hairy; 2–4 stamens exserted, anther sacs 0.8–1.3 mm. Outcrops, talus; 1000–3700 m. KR, CaRH, SNH, n Teh; NV. Jun–Aug

var. ***sonomensis*** (Greene) Jeps. SONOMA BEARDTONGUE **LF:** little reduced distally on st. **FL:** corolla 22–30 mm, dark rose-purple, throat 5.5–6.5 mm wide when pressed, floor short-hairy; 2–4 stamens exserted, anther sacs ± 1.2 mm. Outcrops, talus; 500–2400 m. NCoR (peaks of Lake, Napa, Sonoma cos.). Jun–Aug ★

P. pahutensis N.H. Holmgren (p. 1027) PAHUTE BEARDTONGUE (Group 1) Per 15–35 cm, glabrous, sometimes glaucous. **LF:** cauline 30–100 mm, linear to narrowly lanceolate, entire. **INFL:** glabrous. **FL:** calyx 3–4(5) mm, lobes widely ovate; corolla 17–30 mm, blue-violet, throat 6–8 mm wide when pressed, floor yellow- or white-hairy; anther sacs 1.6–2.2 mm, ± S-shaped, dehiscing in distal 2/3, glabrous; staminode densely golden-yellow-hairy. 2*n*=16. Sagebrush scrub, pinyon/juniper woodland; 1900–2300 m. ne DMtns (Grapevine Mtns); NV. Jun ★

P. palmeri A. Gray var. ***palmeri*** (p. 1027) (Group 4) Per 50–200 cm; herbage glabrous, glaucous. **LF:** thick; distal cauline lvs 40–120 mm, triangular-clasping, fused around st, ± dentate. **INFL:** glandular. **FL:** fragrant; calyx 4–7 mm, lobes ± ovate; corolla 25–32 mm, abruptly expanded into throat, 12–20 mm wide when pressed, pale (lavender-)pink, strongly lined, glandular outside and inside, floor sparsely long-± white hairy; anther sacs 1.8–2.2 mm, dehiscing full length, valves barely spreading; staminode exserted, densely long-spreading-yellow-hairy. 2*n*=16,32. Washes, roadsides, canyon floors, in creosote-bush scrub to pinyon/juniper woodland; 300–2600 m. DMoj; to UT, AZ. May–Jun

P. papillatus J.T. Howell (p. 1027) INYO BEARDTONGUE (Group 2) Subshrub 20–40 cm, woody-branched proximally; herbage ashy-hairy. **LF:** gen cauline, 25–40 mm, 7–25 mm wide, narrowly elliptic to oblanceolate, gen widest at base, cordate-clasping, entire. **INFL:** glandular. **FL:** calyx 7–12 mm, lobes narrowly lanceolate; corolla 24–30 mm, blue-violet, glandular outside, floor glabrous; anthers sacs 1.7–1.9 mm, dehiscing only at proximal end, inner margins glabrous; staminode pale yellow-hairy. Rocky openings of pinyon/juniper woodland and montane forest communities; 2100–2900 m. c&s SNH, SNE. Jun–Jul ★

P. parvulus (A. Gray) Krautter (p. 1027) (Group 2) Per 15–30 cm, loosely matted, glabrous, glaucous, woody-branched proximally. **LF:** gen cauline, 10–45 mm, lanceolate to ovate, clasping, weakly serrate. **INFL:** glabrous. **FL:** calyx 3–5 mm, lobes lanceolate to widely obovate; corolla 14–20 mm, blue(-violet), glabrous; anther sacs 1.4–2 mm, dehiscing at proximal end 1/2–2/3 their length, inner margins hairy; staminode glabrous. 2*n*=32 (s SNH). Rocky, open foothill and montane forests; 400–3100 m. KR, s SNH; OR. Some n CA and OR pls intergrade with *P. azureus* var. *a.* Jun–Aug

P. patens (M.E. Jones) N.H. Holmgren (p. 1027) (Group 4) Per 15–40 cm, glabrous, glaucous. **LF:** thick; basal lvs well developed; cauline lvs 25–90 mm, 5–20 mm wide, lanceolate, entire. **INFL:** glabrous. **FL:** calyx 3–7 mm, lobes widely ovate; corolla 13–20 mm, lavender, magenta, or violet, glabrous; anther sacs 1–1.2 mm, dehiscing full length, valves barely spreading; staminode minutely orange- to yellow-hairy. 2*n*=16. Sagebrush scrub, pinyon/juniper woodland, yellow-pine forest; 1600–3000 m. c&s SNH, s SNE; s NV. May–Jun

P. personatus D.D. Keck (p. 1027) CLOSED-THROATED BEARDTONGUE Per 30–50 mm, gen glabrous (exc infl), glaucous. **LF:** cauline; proximal-most scale-like; largest (at mid st) short-petioled, blade 30–50 mm, ovate, (sub)entire. **INFL:** hairy, glandular. **FL:** calyx 5–6 mm, lobes widely lanceolate; corolla 20–25 mm, blue-purple, glabrous to glandular outside, densely hairy throughout inside, arched floor closing throat; anther sacs 1.2–1.4 mm, dehiscing full length, valves barely spreading; staminode incl, coarsely hairy. Yellow-pine, montane forests; 1050–1800 m. n SN (Butte, Plumas cos.). Jul ★

P. procerus Graham (Group 4) Per 5–40 cm, gen matted, glabrous (exc some proximal sts). **LF:** entire; basal lvs many, (ob)lanceolate to obovate; distal cauline lvs < basal, narrowly lanceolate to narrowly ovate. **INFL:** glabrous. **FL:** calyx 1.5–4.5 mm, lobes (ob)ovate; corolla 6–10 mm, blue-purple, ± glabrous, floor ± white,

white- to yellow-hairy; anther sacs 0.3–0.5 mm, widely spreading; staminode glabrous or sparsely orange-yellow-hairy. Other vars. outside CA to AK, MT, WY, CO; vars. separated by elevation in KR.

var. ***brachyanthus*** (Pennell) Cronquist Pl 10–40 cm. **LF:** distal 10–45 mm, 3–10 mm wide. Montane meadows; 1300–2400 m. KR, NCoRH; OR. Jul–Aug

var. ***formosus*** (A. Nelson) Cronquist (p. 1029) Pl 5–12 cm. **LF:** upper 5–20 mm, 1–5.5 mm wide. 2*n*=32. Alpine barrens; 2100–3600 m. KR, SNH; OR, NV. SN pls have lvs folded lengthwise. Jul–Aug

P. pseudospectabilis M.E. Jones (p. 1029) (Group 4) Shrub 30–100 cm. **ST:** young sts glabrous, glaucous. **LF:** ± thin; distal cauline lvs 30–90 mm, widely triangular-ovate, ± serrate, bases fused around st; distal lf pairs ± disk-like. **INFL:** gen glandular. **FL:** calyx 3.5–7.5 mm, lobes ovate; corolla (17)20–25 mm, gradually expanded to throat, (5)6–9 mm wide when pressed, red-pink, strongly lined, glandular outside and inside (floor lacks nonglandular hairs); anther sacs 1.1–1.3 mm, spreading flat; staminode incl, glabrous. 2*n*=16. Gravelly or rocky desert washes, canyon floors, creosote-bush scrub, pinyon/juniper woodland; gen 100–1750 m. D; AZ. Mar–May ★

P. purpusii Brandegee (p. 1029) SNOW MOUNTAIN BEARDTONGUE (Group 2) Subshrub 5–20 cm, woody-branched proximally; herbage densely canescent-hairy. **LF:** cauline 10–30 mm, ± (ob)lanceolate, gen folded lengthwise, weakly serrate. **INFL:** glandular. **FL:** calyx 6–10 mm, lobes ± lanceolate; corolla 27–31 mm, blue-violet, glandular outside, floor glabrous; anther sacs 2.4–2.6 mm, dehiscing at proximal end 2/3–3/4 their length, inner margins hairy; staminode glabrous. Rocky ridges, peaks, open slopes in montane forest; 800–2500 m. w KR, NCoRH. Jun–Aug

P. rattanii A. Gray (Group 4) Per 25–120 cm; herbage ± glabrous. **LF:** main cauline lvs thin, 25–140 mm, narrowly elliptic, slightly narrowed just distal to rounded to truncate base (where widest), shallowly toothed, ± = basal. **INFL:** glandular. **FL:** corolla 20–30 mm, abruptly expanded to throat, gen on upper side, blue-violet, ± white within, glandular outside, floor ± hairy; anther sacs 1.3–1.7 mm, dehiscing full length, valves barely spreading; staminode yellow-hairy. 2*n*=16.

var. ***kleei*** (Greene) A. Gray SANTA CRUZ MOUNTAINS BEARDTONGUE **FL:** calyx 6–7 mm, lobes ovate. Redwood, hardwood forests; 400–600 m. SnFrB. May–Jun ★

var. ***rattanii*** (p. 1029) **FL:** calyx 6–12 mm, lobes lanceolate. Grassy slopes in grand-fir/sitka-spruce, redwood, mixed-evergreen forests; 10–1200 m. n NCo, w KR, n NCoRO; OR. May–Aug

P. roezlii Regel (p. 1029) (Group 2) Subshrub 15–55 cm, woody-branched proximally; herbage hairy. **LF:** gen cauline; distal lvs 15–70 mm, linear to (ob)lanceolate, gen folded lengthwise, entire. **INFL:** glandular. **FL:** calyx 3.5–6 mm, lobes ± linear to narrowly ovate; corolla 14–22 mm, ± purple-blue, glandular outside, floor glabrous; anther sacs 1.5–2.2 mm, dehiscing at proximal end 1/2–3/5 their length, inner margins long-hairy; staminode glabrous. 2*n*=16. ± Dry sagebrush or juniper scrub, conifer forest; 300–3500 m. KR, CaR, SNH, MP; OR, NV. May–Jul

P. rostriflorus Kellogg (p. 1029) (Group 2) Subshrub 30–100 cm, glabrous or st finely hairy proximally; clumps large, woody-branched. **LF:** gen cauline, 20–70 mm, linear to lanceolate, entire. **INFL:** glandular. **FL:** calyx 4–8 mm, lobes lanceolate to narrowly ovate; corolla 22–33 mm, upper lip forming hood over anthers, lower lip strongly reflexed, red to orange, sparsely glandular outside, floor glabrous; anther sacs 1.8–2.5 mm, dehiscing at proximal end 1/4–1/3 their length, glabrous on sides; staminode glabrous. 2*n*=16. Dry sagebrush or Joshua-tree scrub, pinyon/juniper woodland, montane forest; 500–3500 m. c&s SNH, TR, PR, SNE, DMtns; to CO, NM. Jun–Aug

P. rupicola (Piper) Howell (p. 1029) (Group 3) Subshrub 5–14 cm, matted; herbage spreading-hairy (exc some lvs). **LF:** gen basal (reduced distally on st); main blades 7–20 mm, widely (ob)ovate to ± round, (sub)entire, glaucous; petiole 1.5–6 mm. **INFL:** glandular. **FL:** calyx 4–10 mm, lobes ± lanceolate; corolla 25–38 mm, rose, lavender or violet, floor glabrous or sparsely and finely hairy; stamens incl, anther sacs 1–1.4 mm, spreading flat, woolly; staminode

glabrous to gen sparsely long-hairy distally. $2n=16$. Outcrops, rocky slopes in Douglas-fir forest; 1300–2300 m. KR, CaRH; to WA. In CA, hybridizes readily with *P. davidsonii, P. newberryi.* Jun–Aug

P. rydbergii A. Nelson var. ***oreocharis*** (Greene) N.H. Holmgren (p. 1029) (Group 4) Per 20–60 cm, gen glabrous, occ finely hairy on st proximal to infl; st base above ground; fl sts few-branched. **LF:** entire; basal and proximal cauline lvs well developed, oblanceolate; distal cauline lvs 25–70 mm, lanceolate. **INFL:** nonglandular. **FL:** calyx 3–5 mm, lobes narrowly oblong to (ob)ovate; corolla 9–14 mm, blue to purple, glabrous (exc white- to yellow-hairy floor); anther sacs 0.5–0.8 mm, dehiscing full length; staminode densely golden-yellow-hairy. $2n=16$. Moist meadows, streambanks, gen in montane to subalpine forests; 1000–3600 m. KR, NCoRH, CaR, SNH, GB; OR, NV. May–Aug

P. scapoides D.D. Keck (p. 1029) PINYON BEARDTONGUE (Group 2) Subshrub 15–60 cm, ± scapose, woody-branched proximally. **ST:** glabrous, glaucous. **LF:** gen dense, in basal mat, entire, densely hairy; basal lvs 1.5–6 mm, ovate to ± round; cauline lvs few, narrowly linear to narrowly oblanceolate. **INFL:** glandular. **FL:** calyx 3–5 mm, lobes oblong to widely ovate; corolla 25–34 mm, pale lavender to purple or blue, glandular outside, floor yellow-hairy; anther sacs 1.4–1.7 mm, dehiscing at proximal end ± 1/2 length, inner margins glabrous; staminode pale yellow-hairy. $2n=16$. Sagebrush scrub, pinyon/juniper, and bristlecone-pine woodland; 2000–3200 m. W&I, n DMtns (Last Chance Range). Jun–Jul ★

P. speciosus Lindl. (p. 1029) (Group 1) Per 5–60 cm, short-hairy on sts. **LF:** distal cauline 35–90 mm, lanceolate, clasping, sometimes folded lengthwise, entire, gen glabrous. **INFL:** glabrous to short-hairy, rarely glandular. **FL:** calyx (4)6–13 mm, lobes lanceolate to ovate; corolla 25–37 mm, ± abruptly expanded into throat, 7–13 mm wide when pressed, blue, gen white inside, glabrous; anther sacs 1.8–3 mm, ± S-shaped, dehiscing in distal 2/3, pubescent on sides; staminode glabrous or tip hairy. $2n=16$. Open sagebrush scrub to subalpine forest; 850–3800 m. KR, CaR, SNH, TR, GB, n DMtns; to WA, ID, UT. May–Aug

P. spectabilis A. Gray (p. 1029) (Group 4) Per 80–120 cm; herbage glabrous, green or glaucous. **LF:** ± thin; distal cauline lvs 35–100 mm, lanceolate to widely ovate, pairs fused at base, gen folded lengthwise and recurved at tip, ± serrate. **FL:** calyx 3.5–6.3 mm, lobes ovate to ± round; corolla 24–34 mm, blue(-purple), throat 8–14 mm wide when pressed, pale violet to lavender, ± white within, glandular outside and on throat roof, floor gen glabrous; anther sacs 1.8–2.4 mm, dehiscing full length but barely spreading; staminode incl, glabrous. Vars. hybridize in e TR.

var. ***spectabilis*** **INFL:** glabrous. $2n=16$. Gravelly and sandy slopes, banks of washes, coastal-sage scrub, chaparral, oak woodland; 100–1800 m. SnGb, SnBr, PR; Mex. Apr–Jun

var. ***subviscosus*** (D.D. Keck) McMinn **INFL:** glandular. Gravelly and sandy slopes, banks of washes, coastal-sage scrub, chaparral, oak woodland; 500–1400 m. TR. Apr–Jun

P. stephensii Brandegee (p. 1029) STEPHENS' BEARDTONGUE (Group 4) Shrub 30–150 cm, herbage glabrous. **LF:** ± thin; distal cauline lvs 25–50 mm, 1.8–5(6) mm wide, triangular-ovate, pairs fused at base, finely and sharply serrate. **INFL:** glabrous to glandular. **FL:** calyx 3–5.2 mm, lobes ovate to ± round; corolla 15–20 mm, rose to magenta, unlined, throat 4–6 mm wide when pressed, 2-ridged, glan-

dular outside and inside (floor lacks nonglandular hairs); anther sacs 0.8–1.1 mm, spreading flat; staminode incl, glabrous. Rocky slopes, washes, rock crevices in creosote-bush scrub, pinyon/juniper woodland; 1000–2200 m. DMtns. Apr–Jun ★

P. sudans M.E. Jones (p. 1029) SUSANVILLE BEARDTONGUE (Group 4) Per 4–7 dm, glandular-hairy, sticky throughout. **LF:** gen cauline, 30–60 mm, ovate, serrate. **FL:** calyx 3.5–5.5 mm, lobes lanceolate to ovate; corolla 9–11 mm, cream-white, dark-lined; anther sacs 0.5–0.7 mm, spreading widely; staminode gen glabrous. Open, rocky, igneous soils in sagebrush scrub, yellow-pine and montane forests; 1200–2200 m. s CaRH, MP; NV. May–Jul ★

P. thompsoniae (A. Gray) Rydb. (p. 1029) THOMPSON'S BEARDTONGUE (Group 4) Per 5–15 cm, ± prostrate, gen matted; hairs appressed, backward-pointing, scale-like, ashy. **LF:** 5–20 mm, ± obovate, entire, petioled. **FL:** calyx 4–7.5 mm, lobes lanceolate, some hairs gen glandular; corolla 10–18 mm, violet to blue, glandular outside, floor pale yellow-hairy; anther sacs 0.8–1.3 mm, dehiscing full length but barely spreading; staminode densely orange- to golden-yellow-hairy. White calcareous soil in pinyon/juniper woodland; 1700–1900 m. e DMtns (New York Mtns, Clark Mtn Range); to UT, AZ. May–Jun ★

P. thurberi Torr. (p. 1029) THURBER'S BEARDTONGUE (Group 4) Shrub 20–80 cm, ± round, glabrous. **LF:** cauline, narrowly linear; main blades 10–45 mm, gen rolled upward, nearly round in ×-section, entire. **INFL:** glabrous. **FL:** calyx 2.4–4.5 mm, lobes ovate; corolla 8–15 mm, ± funnel-shaped (lips obliquely spreading), lavender, rose, or blue-purple, glabrous outside, floor hairy; anther sacs 0.7–0.9 mm, spreading flat; staminode glabrous. $2n=16$. Sandy and gravelly slopes and mesas, in chaparral, creosote-bush or Joshua-tree scrub, pinyon/juniper woodland; 200–1600 m. PR, DMoj; to NV, NM, Mex. May–Jun ★

P. tracyi D.D. Keck (p. 1029) TRACY'S BEARDTONGUE (Group 4) Per 8–12 cm, glabrous. **LF:** gen basal, obovate, (sub)entire; cauline lvs 10–20 mm, ovate to ± round, base widely wedge-shaped. **INFL:** glabrous. **FL:** calyx 2.5–3 mm, lobes ovate, glabrous or glandular-ciliate; corolla 11–13 mm, glabrous (exc densely long-hairy floor); anther sacs ± 0.4 mm, spreading flat; staminode tip hairy-tufted. Exposed outcrops; 2000–2250 m. s KR (n Trinity Co.). Jul–Aug ★

P. utahensis Eastw. (p. 1033) UTAH BEARDTONGUE (Group 4) Per 15–50 cm, glabrous, glaucous. **LF:** thick, gen folded lengthwise; basal lvs well developed; distal cauline lvs 15–55 mm, 5–20 mm wide, ± lanceolate, entire. **INFL:** glabrous. **FL:** calyx 2.5–5 mm, lobes widely ovate; corolla 17–25 mm, cylindric to funnel-shaped (limb slightly oblique, lobes spreading), pink to red (± purple), glandular outside and inside; anther sacs 0.6–1.1 mm, spreading flat; staminode glabrous or tip minutely hairy. Sagebrush scrub, pinyon/juniper woodland; 1200–1800 m. e DMtns (Kingston, New York mtns); to UT, AZ. Apr–May ★

P. venustus Lindl. (p. 1033) (Group 2) Shrub 30–80 cm, spreading, gen glabrous (or sts short-hairy). **LF:** stiff, cauline; largest (at mid-st) 50–120 mm, lanceolate, sharply serrate. **INFL:** glabrous. **FL:** calyx 3.4–6.2 mm, lobes ovate to lanceolate; corolla 22–38 mm, lavender to purple, glabrous exc lobes finely ciliate; filaments and inner margins of anthers long-white-hairy, anther sacs 1.5–2 mm, dehiscing at proximal end; staminode long-white-hairy. $2n=64$. Dry, rocky, exposed places; 900–2200 m. s CaRH (sw Lassen Co.); native to OR, WA, ID. Naturalized from cult. Jul–Aug

PLANTAGO PLANTAIN

Thomas J. Rosatti

Ann to per, gen scapose. **ST:** decumbent to erect. **LF:** gen basal, veins gen ± parallel. **INFL:** spike, gen dense; fls few to many, cleistogamous or opening (both). **FL:** gen bisexual; calyx deeply 4-lobed, lobes gen overlapped, persistent, margin gen scarious; corolla radial or bilateral, salverform or cylindric, scarious, persistent in fr, colorless exc for lobe midribs or not, lobes 4, spreading to erect; stamens gen 4; ovules several per chamber, stigma long, hairy. **FR:** circumscissile ± at or proximal to middle. **SEED:** 2–many, gelatinous when wetted. ± 250 spp.: worldwide, esp temp; some weedy, some (esp *Plantago afra* L., psyllium) cult for laxative. (Latin: sole of foot) [Meyers & Liston 2008 Int J Pl Sci 169:954–962] *P. sempervirens* Crantz, *P. heterophylla* Nutt., reported but not documented, possibly naturalized in CA.

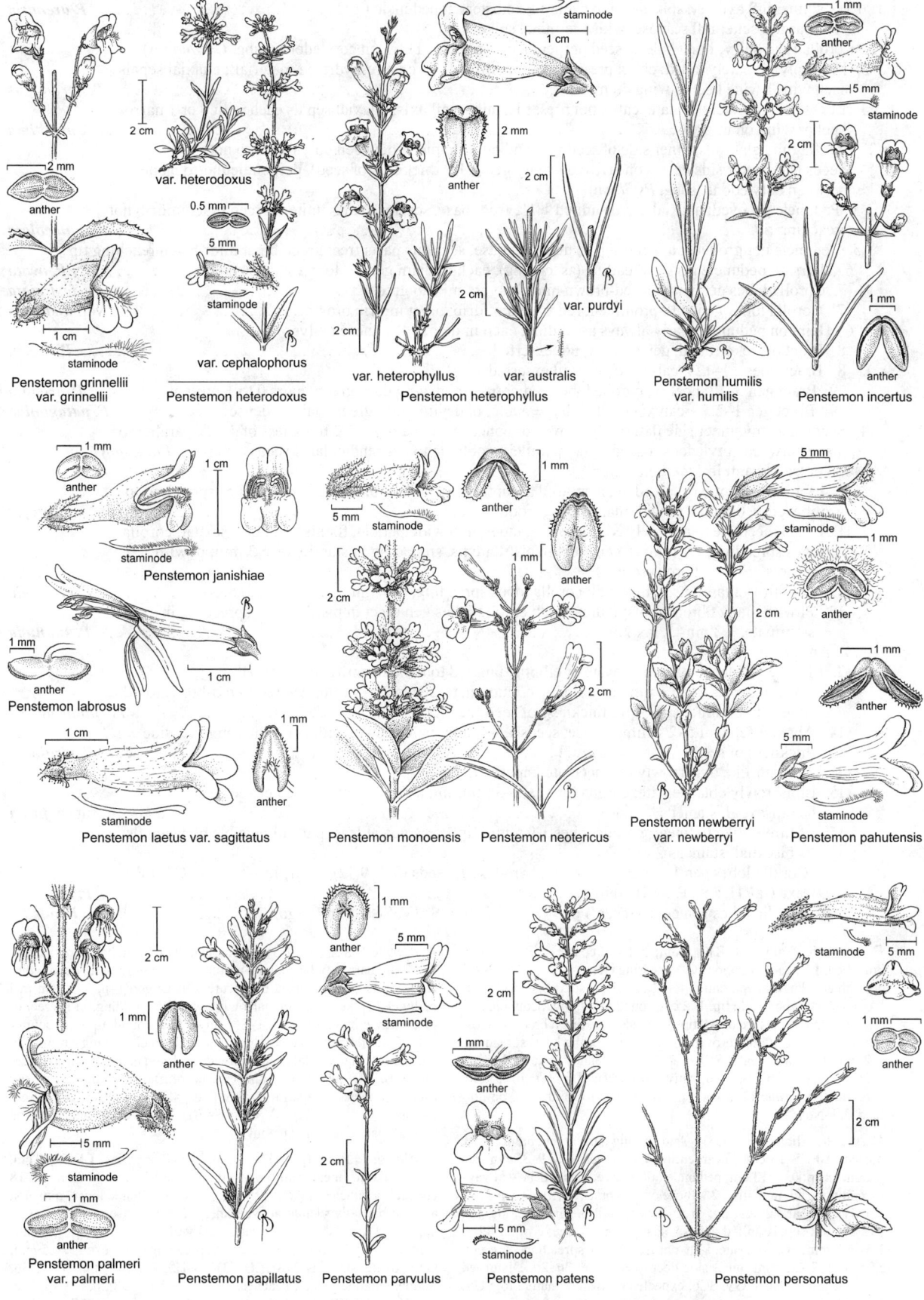

Penstemon grinnellii var. grinnellii

Penstemon heterodoxus — var. heterodoxus, var. cephalophorus

Penstemon heterophyllus — var. heterophyllus, var. australis, var. purdyi

Penstemon humilis var. humilis

Penstemon incertus

Penstemon janishiae

Penstemon labrosus

Penstemon laetus var. sagittatus

Penstemon monoensis

Penstemon neotericus

Penstemon newberryi var. newberryi

Penstemon pahutensis

Penstemon palmeri var. palmeri

Penstemon papillatus

Penstemon parvulus

Penstemon patens

Penstemon personatus

1. Lvs cauline; infl axillary, spikes gen in cyme-like clusters at peduncle tip. . . . . ***P. arenaria***
1′ Lvs in basal rosette; infl scapose, with 1 terminal spike
  2. Corolla tube hairy; inner side of seed perpendicular to plane between cotyledons (subg. *Coronopus*)
    3. Lf deeply pinnately lobed; calyx pressed against infl axis (exc on re-hydrated material); adaxial sepals
      each with 1 wide keel or wing on midrib . . . . ***P. coronopus***
    3′ Lf entire to sparsely dentate; calyx not pressed against infl axis; adaxial sepals each with 0 or 1 narrow
      keel or wing on midrib. . . . . ***P. maritima***
  2′ Corolla tube glabrous; inner side of seed perpendicular to plane between cotyledons to not
    4. Seeds (1)2, inner side deep-concave (concavity gen ≥ 1/2 thickness of seed), perpendicular to plane
      between cotyledons (subg. *Psyllium*)
      5. Peduncle grooved; abaxial sepals united > 1/2 from base, scarious parts united; green bract midrib not
        reaching tip. . . . . ***P. lanceolata***
      5′ Peduncle not grooved; abaxial sepals united at base, scarious parts free; green bract midrib gen reaching tip
        6. Hairs on peduncle gen ± spreading (ascending), each in diam gen << to = 2 × those on lvs. . . . . ***P. ovata***
        7. Corolla lobes gen without red-brown midrib; bract midrib green. . . . . var. ***fastigiata***
        7′ Corolla lobes gen with prominent red-brown midrib; bract midrib brown . . . . var. ***insularis***
        6′ Hairs on peduncle nearly always ascending, each in diam ± 2 × those on lvs
          8. Bract ovate to round, gen ≤ calyx, not exserted. . . . . ***P. erecta***
          8′ Bract linear, 1–12 × calyx, some or all exserted
            9. Bract gen 3–12 × calyx; corolla lobes gen obtuse; pl drying dark green, hairs 0 to ± present. . . . . ***P. aristata***
            9′ Bract gen 1–2.5 × calyx; corolla lobes ± acute; pl drying light green, hairs ± dense. . . . . ***P. patagonica***
    4′ Seeds 1–many, inner side flat, rarely convex or concave (concavity ≤ 1/2 thickness of seed), parallel to
      plane between cotyledons if seeds 1–2, parallel to oblique to perpendicular if seeds > 2 (subg. *Plantago*)
      10. Per (rarely ann in *Plantago major*)
        11. Lf widely elliptic to ± cordate, narrowed abruptly to petiole; fl bisexual; corolla lobes spreading to
          reflexed, ± 0.5 mm; pl with many fibrous roots . . . . ***P. major***
        11′ Lf oblanceolate to narrowly elliptic-ovate, tapered to wide petiole; fl unisexual or bisexual; corolla
          lobes spreading, ± 1 mm, or gen erect in pistillate fls, spreading in staminate, ± 3 mm; pl with 1–2
          taproots or many fibrous roots
            12. Pl with 1–2 taproots; fl bisexual; corolla lobes spreading, ± 1 mm. . . . . ***P. eriopoda***
            12′ Pl with many fibrous roots; fl unisexual; corolla lobes gen erect in pistillate fls, spreading in
              staminate, ± 3 mm. . . . . ***P. subnuda***
      10′ Ann, bien
        13. Lf widely oblanceolate to obovate or elliptic, tapered to petiole, entire to ± toothed
          14. Abaxial sepals 2.7–3.6 mm, acuminate to mucronate; seeds 2.5–3 mm, gen red to red-brown, with
            concavity on inner side < 1/2 thickness of seed. . . . . ***P. rhodosperma***
          14′ Abaxial sepals 1.5–2.4 mm, obtuse; seeds 1.5–2 mm, pale brown, with concavity on inner side ± 1/2
            thickness of seed. . . . . ***P. virginica***
        13′ Lf thread-like to narrowly oblanceolate, tapered to base, gen entire
          15. Lf narrowly oblanceolate; corolla lobes ± 2–3 mm; infl axis gen not visible between fls;
            fls unisexual, stamens 0 or 4. . . . . ***P. truncata*** subsp. ***firma***
          15′ Lf thread-like to linear; corolla lobes ± 0.5 mm; infl axis gen at least partially visible between fls;
            fls bisexual, stamens 2
            16. Corolla lobes gen 1 erect, 3 spreading or reflexed; seeds (3)4–9(12) per fr, 1.5–2.5 mm; CA-FP
              (exc CaRH, s SNF, SNH, Teh) . . . . ***P. elongata***
            16′ Corolla lobes gen erect; seeds 4 per fr, 0.7–1.8 mm; SnJV, SCo (San Diego Co.) . . . . ***P. pusilla***

P. arenaria Waldst. & Kit. Ann; hairs ± coarse. **ST:** 2–6 cm, branched. **LF:** cauline, opposite, appearing clustered at nodes, 2–4.5 cm, thread-like or linear, entire. **INFL:** axillary; spikes gen in cyme-like clusters at peduncle tip, 1–2 cm, round to elliptic; peduncle 3–7 cm; proximal bracts exserted, tips long, slender, distal bracts enclosing fls, gen = calyx, round-ovate, obtuse. **FL:** corolla lobes spreading, ± 2 mm, narrowly ovate. **SEED:** 2, 2.5 mm. Sandy disturbed areas; < 200 m. CCo, SnFrB, SCo; native to Eurasia. [*P. indica* L., nom. superfl.; *P. psyllium* L., nom. rej.] Often used for bird seed, poultry feed. Jul–Nov

P. aristata Michx. (p. 1033) Ann; drying dark green, hairs 0 to ± present. **LF:** 5–15 cm, linear, entire. **INFL:** gen 1–6, 9–35 cm incl peduncle; spike 1–13 cm; peduncle not grooved, hairs nearly always ascending, each in diam ± 2 × those on lvs; bract exserted, gen 3–12 × calyx, linear, ascending, green bract midrib gen reaching tip. **FL:** abaxial sepals united at base, scarious parts free; corolla lobes 1.5–2.3 mm, round-ovate, gen obtuse, 3 gen spreading, 1 erect. **SEED:** 2, 2–2.5 mm, inner side deeply concave. 2*n*=20. Disturbed areas; < 200 m. NCo, GV, SCo, expected elsewhere; native to e US. Jun–Jul

P. coronopus L. (p. 1033) Ann, bien, taprooted; hairs coarse. **LF:** 4–25 cm, narrowly lanceolate, gradually tapered to base, deeply pinnately lobed; lobes ascending, acute. **INFL:** gen 1–6, 5–50 cm incl peduncle; spike 2–22 cm, narrowly cylindric, nodding, in fr erect or often wavy; bract exserted, 1–2 × calyx, base round, tip long-acuminate. **FL:** adaxial sepals each with 1 wide keel or wing on midrib; corolla tube hairy, 3 lobes gen spreading, 1 gen erect, ± 1 mm, lance-ovate. **SEED:** gen 3, ± 1 mm. Coastal bluffs, salt marshes, weedy trampled places, chaparral, grassy flats; < 300 m. NCo, KR, n SNF (Amador, Nevada cos.), GV, CCo, SnFrB, SCo, s ChI (Santa Catalina Island); native to Eur. Apr–Jul

P. elongata Pursh (p. 1033) Ann; hairs 0 or sparse. **LF:** 3–10 cm, thread-like to linear, entire or teeth few. **INFL:** few to many, 2–18 cm incl peduncle; spike 0.5–5 cm, dense or often ± loose, axis gen at least partially visible between fls; bract not exserted, ± = calyx, appressed to calyx, round-ovate, keel wide, fleshy, margin translucent. **FL:** corolla lobes gen 1 erect, 3 spreading or reflexed, ± 0.5 mm, lance-ovate; stamens 2. **SEED:** (3)4–9(12), 1.5–2.5 mm. 2*n*=12,36. Saline, alkaline places, beaches, vernal pools; < 200 m. CA-FP (exc CaRH, s SNF, SNH, Teh); to BC. [*P. bigelovii* A. Gray; *P. e.* subsp. *pentasperma* Bassett] Apr–Jun

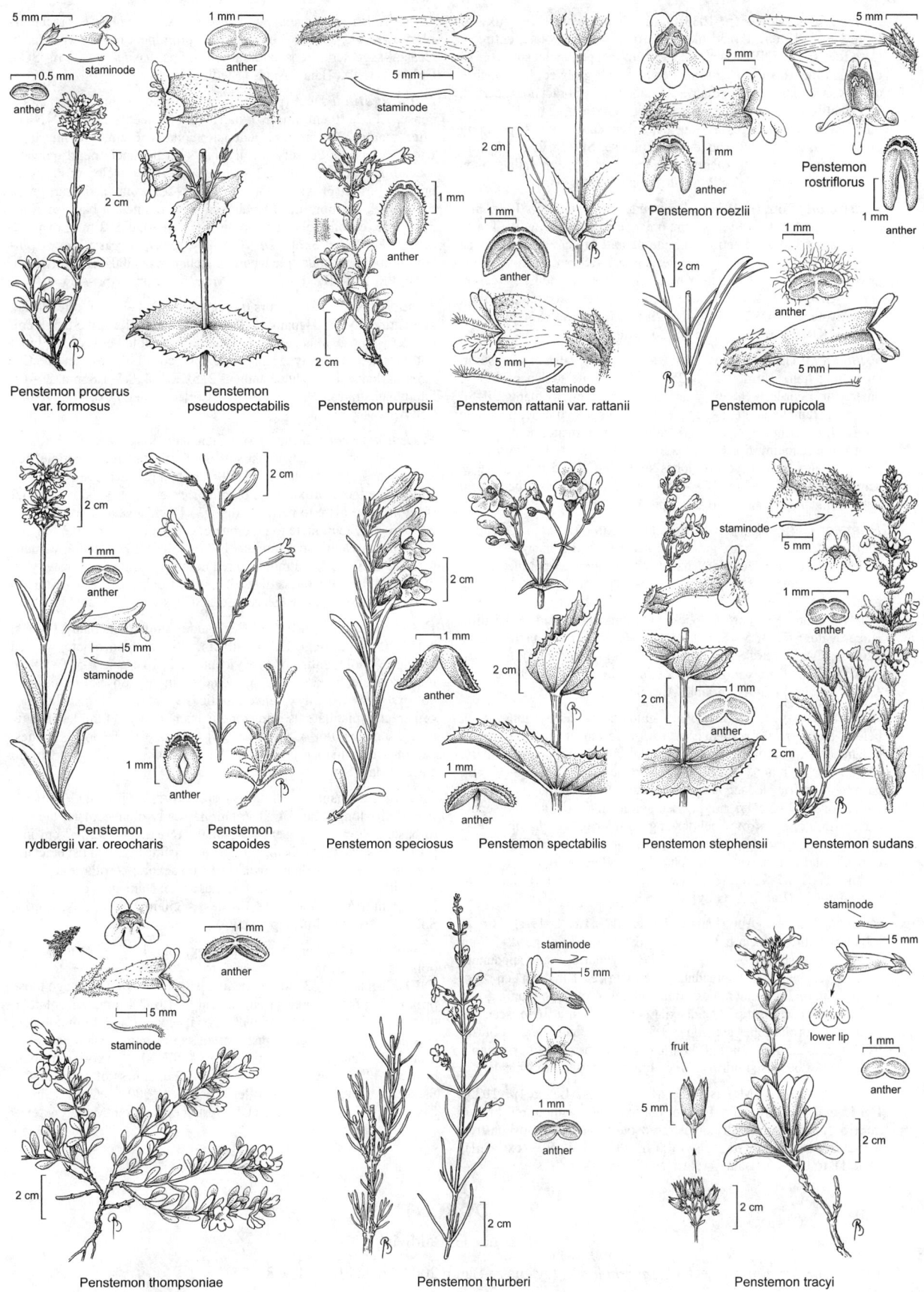

Penstemon procerus
var. formosus

Penstemon
pseudospectabilis

Penstemon purpusii

Penstemon rattanii var. rattanii

Penstemon roezlii

Penstemon
rostriflorus

Penstemon rupicola

Penstemon
rydbergii var. oreocharis

Penstemon
scapoides

Penstemon speciosus

Penstemon spectabilis

Penstemon stephensii

Penstemon sudans

Penstemon thompsoniae

Penstemon thurberi

Penstemon tracyi

P. erecta E. Morris (p. 1033)   Ann; hairs scattered, ± silky or long. **LF**: 3–13 cm, thread-like to narrowly oblanceolate, entire or teeth few, small. **INFL**: 1–many, 3–30 cm incl peduncle; spike 0.5–3 cm, head-like to short-cylindric; peduncle not grooved, hairs nearly always ascending, each in diam ± 2 × those on lvs; bract not exserted, gen ≤ calyx, ovate to round, green midrib gen reaching tip. **FL**: abaxial sepals united at base, scarious parts free; corolla lobes spreading or reflexed, 2–2.7 mm, round-ovate, obtuse. **SEED**: 2, 2–2.5 mm. 2*n*=20. Sandy, clay, or serpentine substrates, grassy slopes, flats, open woodland; < 700 m. CA-FP; to OR, Baja CA. Mar–May

P. eriopoda Torr. (p. 1033)   Per, caudex stout, taproots 1–2; hairs gen 0 exc peduncle. **LF**: 5–25 cm, oblanceolate to narrowly elliptic-ovate, tapered to wide petiole, minute-dentate or gen ± wavy. **INFL**: gen 1–6, 10–55 cm incl peduncle; spike 8–18 cm, narrow-cylindric, arched in age, loose near base; bract not exserted, ± = calyx, appressed to calyx, wide-ovate. **FL**: corolla lobes spreading, ± 1 mm, lance-ovate. **SEED**: 2–4, ± 2 mm. 2*n*=24. Moist, alkaline places; < 1000 m. KR, n CaR; to Can, c US. Jul–Aug

P. lanceolata L. (p. 1033)   ENGLISH PLANTAIN   Per, caudex stout, taprooted; hairs short. **LF**: 5–25 cm, lanceolate to lance-oblong, tapered to petiole, gen finely dentate. **INFL**: few to many, 20–80 cm incl peduncle; spike 2–8 cm, ± ovoid, in age cylindric; peduncle grooved; bract not exserted, ± = calyx, widely ovate, acute, green midrib not reaching tip. **FL**: abaxial sepals united > 1/2 from base, scarious parts united; corolla lobes spreading, 2–2.5 mm, ovate, ± acute. **SEED**: 1–2, ± 3 mm. 2*n*=12,24,96. Common. Disturbed areas; < 1600 m. CA-FP; native to Eur. Apr–Aug ❖

P. major L. (p. 1033)   COMMON PLANTAIN   (Ann) per, caudex short, roots fibrous; hairs gen 0 exc peduncle. **LF**: blade 5–18 cm, widely elliptic to ± cordate, narrowed abruptly to petiole, entire or ± finely dentate. **INFL**: gen 3–7, 5–60 cm incl peduncle; spike gen 3–20 cm, linear-cylindric, in age ± loose; bract not exserted, ± = calyx, ovate. **FL**: corolla lobes spreading to reflexed, ± 0.5 mm, lance-ovate. **SEED**: 5–16, < 1 mm. 2*n*=6,12,24. Disturbed areas; < 2200 m. CA-FP (exc SNF, Teh), GB, DMoj (uncommon); to e US; native to Eur. [*P. m.* var. *pilgeri* Domin; *P. m.* var. *scopulorum* Fries & S.P. Broberg] Highly variable. Apr–Sep

P. maritima L. (p. 1033)   Per, caudex stout, taproot long, thick below; hairs gen 0 exc peduncle. **LF**: 3–15 cm, linear to narrowly oblanceolate, gradually tapered to base, entire to sparsely dentate. **INFL**: few to many, gen 8–30 cm incl peduncle; spike 2–10 cm, narrow-cylindric, looser near base; bract not exserted, ± = calyx, widely ovate. **FL**: calyx not pressed against infl axis; adaxial sepals each with 0 or 1 narrow keel or wing on midrib; corolla tube hairy, 3 lobes gen spreading, 1 erect, ± 1 mm, lance-ovate. **SEED**: 2–3, 2 mm. Coastal bluffs, wet, saline places; < 150 m. NCo, CCo, SnFrB, n ChI; to AK, ne N.Am. [*P. m.* var. *californica* (Fernold) Pilg; *P. m.* var. *juncoides* (Lam.) A. Gray] May–Sep

P. ovata Forssk.   Ann; hairs ± dense, silky. **LF**: 2–17 cm, linear (oblong), entire or teeth few, minute. **INFL**: few–many, gen 2–27 cm incl peduncle; spike 0.5–3.5 cm, gen short-cylindric, woolly; peduncle not grooved, hairs gen ± spreading (ascending), each in diam gen << to = 2 × those on lvs; bract not exserted, ± = calyx, ovate to round, green midrib gen reaching tip. **FL**: abaxial sepals united at base, scarious parts free; corolla lobes spreading, 1.3–2.8 mm, round-ovate, obtuse. **SEED**: 2, 2–2.5 mm. 2*n*=8. Herbarium specimens gen not assignable to var. in CA because midrib color on bracts, petals not reliable.

var. ***fastigiata*** (Morris) S.C. Meyers & A. Liston (p. 1033) **INFL**: bract midrib green. **FL**: corolla lobes gen without red-brown midrib. Sandy or gravelly areas, creosote-bush scrub, Joshua-tree woodland, sagebrush scrub; < 1400 m. SnJV, CW, SW (exc ChI), SNE, D; to UT, TX, Baja CA; also Medit. Jan–Apr

var. ***insularis*** (Eastw.) S.C. Meyers & A. Liston **INFL**: bract midrib brown. **FL**: corolla lobes gen with prominent red-brown midrib. Coastal-sage scrub, coastal strand, hilltops, basins; < 300 m. SCo, ChI; to UT, TX, Baja CA; also Medit. Feb–Apr

P. patagonica Jacq. (p. 1033)   Ann; drying light green, hairs ± dense. **LF**: 2–10 cm, linear or narrowly oblanceolate, entire. **INFL**: often many, 2–18 cm incl peduncle; spike 1–6 cm, ± cylindric, ± wider at base, ± densely woolly, hairs on peduncle nearly always ascending, each in diam ± 2 × those on lvs; peduncle not grooved; bract exserted (proximal) or not, gen 1–2.5 × calyx, linear, green midrib gen reaching tip. **FL**: abaxial sepals united at base, scarious parts free; corolla lobes spreading (or 1 erect), 1.5–2 mm, ovate, ± acute. **SEED**: 2, ± 3 mm. 2*n*=20. Sandy, rocky, or grassy slopes, pinyon/juniper or Joshua-tree woodland, chaparral; 500–2200 m. SnGb, SnBr, PR, D; to BC, n-c US, TX, Mex; Argentina. Apr–Aug

P. pusilla Nutt.   Ann; hairs 0 or sparse. **LF**: 1–10 cm, thread-like, gen entire. **INFL**: 1–many, 2–19 cm incl peduncle; spike 0.5–8 cm, slender, ± loose, axis gen at least partially visible between fls; bract not exserted, ± = calyx, lanceolate. **FL**: corolla lobes gen erect, ± 0.5 mm, lanceolate, acute; stamens 2. **SEED**: 4, 0.7–1.8 mm. 2*n*=12. Sandy substrates; < 100 m. SnJV, SCo (San Diego Co.); native to e US. Apr–May

P. rhodosperma Decne.   Ann, bien; hairs long, fine. **LF**: 5.5–29.5 cm, widely oblanceolate to obovate or elliptic, tapered to petiole, gen ± toothed. **INFL**: 1–many, 3.5–29.5 cm incl peduncle; spike 1.5–17 cm, in age loose proximally; bract not exserted, ± ≤ calyx, 2.5–3.2 [4] mm, lanceolate to narrow-elliptic. **FL**: unisexual; abaxial sepals 2.7–3.6 mm, acuminate to mucronate; corolla lobes upcurved in pistillate fls, spreading in staminate, 2–3 mm, lanceolate, acute; stamens 0 or 4. **SEED**: 2, 2.5–3 mm, gen red to red-brown, with concavity on inner side < 1/2 thickness of seed. 2*n*=24. Rocky, sandy, or disturbed places; < 1000 m. CaRF (Shasta Co.), SCo; s-c&sw US, ne Mex. May

P. subnuda Pilg. (p. 1033)   Per, caudex stout, roots many, fibrous; hairs ± 0 to ± present. **LF**: 12–40 cm, elliptic-oblanceolate, tapered to wide petiole, entire or finely dentate. **INFL**: gen 1–5, 9–50 cm incl peduncle; spike 8–30 cm, cylindric, in age loose at base; bract not exserted, ± = calyx, lance-elliptic. **FL**: unisexual; corolla lobes gen erect in pistillate fls, spreading in staminate, ± 3 mm, lanceolate, acute; stamens 0 or 4. **SEED**: 3, ± 1.6 mm. Wet meadows, ditches, coastal bluffs, marshes; < 300 m. NCo, CCo, SnFrB, SCo, n ChI; to c Mex. May–Sep

P. truncata Cham. & Schltdl. subsp. *firma* (Walp.) Pilg. (p. 1033) Ann; hairs long, fine. **LF**: 1–6 cm, narrowly oblanceolate, tapered to base, gen entire or teeth few, small. **INFL**: 1–few, 3–7 cm incl peduncle; spike 0.5–2.5 cm, axis gen not visible between fls; bract not exserted, ± = calyx, lance-ovate. **FL**: unisexual; corolla lobes erect in pistillate fls, ± 2–3 mm, lanceolate, acute; stamens 0 or 4. **SEED**: 2, ± 1.8 mm. Moist meadows, ditches; < 350 m. n SNF, ScV, n SnJV, SnFrB; native to Chile. Apr–May

P. virginica L.   Ann, bien; hairs long, fine. **LF**: 2–12 cm, widely oblanceolate to obovate, tapered to petiole, entire to ± toothed. **INFL**: 1–many, 3–33 cm incl peduncle; spike 1–14 cm, in age loose proximally; bract not exserted, ± = calyx, 1.6–2.4 mm, lanceolate to narrowly elliptic. **FL**: unisexual; abaxial sepals 1.5–2.4 mm, obtuse; corolla lobes upcurved in pistillate fls, spreading in staminate, 2–3 mm, lanceolate, acute; stamens 0 or 4. **SEED**: 2, 1.5–2 mm, pale brown, with concavity on inner side ± 1/2 thickness of seed. 2*n*=24. Sandy or disturbed places, often weedy; < 1000 m. CaRF (Shasta Co.), GV, SCo; native to c&e US. Spring ephemeral. May

PSEUDORONTIUM

Bruce G. Baldwin

1 sp. (false *Orontium*, a sect. of *Antirrhinum*). [Oyama & Baum 2004 Amer J Bot 91:918–925]

P. cyathiferum (Benth.) Rothm. (p. 1033)   DEEP CANYON SNAPDRAGON   Ann, herbage glandular-sticky-hairy. **ST**: erect, self-supporting, 1–70 cm, coarsely much-branched throughout. **LF**: proximal-most opposite, most alternate; petiole 2–21 mm; blade 2.5–40

mm, gradually reduced distally on st, ovate, tip obtuse to acute; veins pinnate. **INFL:** fls 1 per lf axil, all opening; pedicels 1–3 mm in fl, 2–6 mm and bending down in fr, subtending branchlets 0. **FL:** calyx lobes 4–5 mm, equal; corolla 8–9 mm, cream and purple, veins purple, tube ± not extended at base, lower lip swollen, ± closing throat; stamens 4, incl, staminode 0; style incl, slightly curved upward,

sparsely glandular-puberulent proximally, stigma inconspicuous. **FR:** spreading or pendant at maturity, ± spheric, sparsely glandular-puberulent distally; chambers equal, opening by irregular bursting near tip. **SEED:** many; wing large, cup-shaped. n=13. Washes, rocky slopes; < 800 m. w DSon (Deep Canyon, Riverside Co.); s AZ, nw Mex. [*Antirrhinum c.* Benth.] Jan–Apr ★

STEMODIA

Robert E. Preston & Margriet Wetherwax

[Ann] per [shrub]. **ST:** 1–many, ascending to erect. **LF:** gen opposite, clasping, ± dentate to pinnately-lobed. **INFL:** fls in terminal interrupted spikes, 2–4 per node; bracts 2 proximal to calyx. **FL:** sepals 5, free; corolla 5-lobed, 2-lipped, white to violet or purple, tube 4-angled, upper lip 2-lobed, arched, lower lip reflexed; stamens 4, in 2 pairs, anther sacs well separated by thick tissue; stigma lobes 2, flat. **FR:** loculicidal, ± ovoid. **SEED:** many, surface net-like. ± 50 spp.: trop Am, Afr, Asia, Australia. (Latin: from stemodiacra, stamens with 2 tips)

S. durantifolia (L.) Sw. (p. 1033) PURPLE STEMODIA Pl glandular-hairy. **ST:** 10–100 cm. **LF:** 2–3 per node, ± sessile, 2–4 cm, 5–20 mm wide, ± lanceolate, ± dentate; reduced distally on st. **INFL:** pedicels 2–12 mm; bracts ± = sepals. **FL:** sepals 5–7 mm, lanceolate,

long-tapered; corolla 7–10 mm, violet to purple. **FR:** 4–5 mm, ovoid-cylindric. Riparian habitats, on wet sand or rocks, drying streambeds; < 400 m. PR, w DSon; AZ, TX, to S.Am; introduced elsewhere. Var. *chilensis* (Benth.) C.C. Cowan in Chile. All yr ★

SYNTHYRIS

Larry Hufford

Per, rhizomed. **LF:** basal, deciduous to persistent, long-petioled; blade ovate to round, base acute to cordate, tip acute to rounded, margin serrate, some deeply cut to pinnately lobed. **INFL:** scapose; axillary raceme fusiform to conical, dense to open; fls 5–100, bracts subtending fls or not. **FL:** bilateral; sepals 2–4, fused laterally; petals 3–4 or 0, fused basally, bell-shaped [2-lipped], blue to purple-blue or white; stamens 2, epipetalous, exserted; stigmas 2, fused, minutely capped, ovules 2–16. **FR:** loculicidal, laterally flattened, notched at tip. **SEED:** flattened front-to-back or ± cupped, oval; coat net-like, brown. 19 spp.: N.Am. (Greek: united door, from basally united valves of fr)

1. Infl erect in fr; fls ≥ 15 per infl; ovules 10–16 per fl — Wrn . . . . ***S. missurica* subsp. *missurica***
1′ Infl curved to reclining on soil in fr; fls 5–10 per infl; ovules 4 per fl
  2. Lf blade gen longer than wide, base lobed to cordate, tip acute to obtuse; corolla tube glabrous to minutely hairy; NW . . . . ***S. cordata***
  2′ Lf blade wider than long, base lobed, tip obtuse to rounded; corolla tube long-soft-wavy-hairy; NW, SnFrB . . . . ***S. reniformis***

S. cordata (A. Gray) A. Heller SERPENTINE SNOW QUEEN **LF:** petiole ≤ 13 cm; blade ≤ 7 cm, ≤ 6.5 cm wide, ovate to widely ovate, base lobed to cordate, tip acute to obtuse. **INFL:** curved to reclining on soil in fr, ≤ 14 cm; fls 5–10. **FL:** corolla ≤ 9 mm, blue, tube glabrous to minutely hairy; filaments ≤ 7.5 mm; style ≤ 9 mm; ovules 4. **FR:** ≤ 3.5 mm, ≤ 9 mm wide; margins sparsely hairy. **SEED:** slightly cupped. 2n=24. Moist forest, serpentine; 100–1000 m. NW; to OR. Feb–Apr

S. missurica (Raf.) Pennell subsp. ***missurica*** (p. 1033) KITTENTAIL **LF:** petiole ≤ 27 cm; blade ≤ 11 cm, ≤ 15 cm wide, widely ovate, base lobed, tip rounded. **INFL:** erect in fr, axis thick, ≤ 33 cm; fls 15–100, bracted, sterile bracts ≤ 3. **FL:** corolla ≤ 10.5

mm, blue, tube glabrous to long-soft-wavy-hairy; filaments ≤ 9 mm; style ≤ 8 mm; ovules 10–16. **FR:** ≤ 8 mm, ≤ 10 mm wide, glabrous to sparsely hairy. **SEED:** round, flat. 2n=24,48. Moist forest; 30–2900 m. Wrn; to WA, MT. [*S. major* (Hook.) A. Heller; *S. stellata* Pennell] Mar–Jul ★

S. reniformis (Benth.) Benth. (p. 1033) SNOW QUEEN **LF:** petiole ≤ 13 cm; blade ≤ 7.5 cm, ≤ 8.5 cm wide, widely ovate, base lobed, tip obtuse to rounded. **INFL:** raceme, curved to reclining on soil in fr, ≤ 18 cm; fls < 15. **FL:** corolla ≤ 10 mm, blue, tube long-soft-wavy-hairy; filaments ≤ 9 mm; style ≤ 11 mm; ovules 4. **FR:** ≤ 4 mm, ≤ 9 mm wide; sparsely hairy. **SEED:** slightly cupped. 2n=24. Moist forest; < 1500 m. NW, SnFrB; to WA. Feb–Jun

TONELLA

Michael S. Park, Elizabeth Chase Neese & Margriet Wetherwax

Ann. **ST:** erect, slender, branched. **LF:** opposite, entire to 3-lobed. **INFL:** raceme, bracted. **FL:** calyx well developed, deeply 5-lobed; corolla 2-lipped, upper lip 2-lobed, lower lip 3-lobed (middle lobe wider than lateral); stamens 4, equal, exserted, filaments hairy; stigmas fused. **FR:** loculicidal, spheric to ovoid. **SEED:** large, wingless. 2 spp.: N.Am. (Derivation unknown)

T. tenella (Benth.) A. Heller (p. 1033) **ST:** ascending, 5–30 cm. **LF:** 10–15 mm; adaxial face soft-shaggy-hairy; proximal lvs petioled, ovate to round, becoming sessile distally on st, entire to deeply 3-lobed or ternate. **INFL:** minutely glandular-hairy distally; bracts lanceolate, subtending 1–3 fls; pedicels 8–15 mm. **FL:** calyx < 3 mm,

lobes < 2 mm, tips acute to obtuse, minutely ciliate; corolla 2–2.5 mm, upper lobes < lower, white proximally, blue or violet distally, often with purple spots. **FR:** 2–2.5 mm. **SEED:** 1 per chamber, < 1.5 mm. Moist, shaded places in chaparral, oak and mixed woodland; < 1600 m. KR, NCoR, CaR, n SNH, SnFrB; to WA. Mar–Jun

VERONICA SPEEDWELL, BROOKLIME

Ann, per. **ST**: erect or prostrate. **LF**: cauline, opposite, sessile to short-petioled. **INFL**: raceme, terminal or axillary, or fls 1 in axils; bracts small, alternate. **FL**: sepals gen 4(5), ± free, gen unequal; corolla ± radial, ± rotate, gen 4-lobed, tube << lobes, upper lobe wide (perhaps formed by fusion of upper pair), blue or violet to white; stamens 2, exserted; stigma unlobed. **FR**: flattened perpendicular to septum, gen obcordate, loculicidal and septicidal. ± 250 spp.: n temp, esp Eurasia. (Named for Saint Veronica) *V. beccabunga* L., *V. chamaedrys* L. not in CA; *V. filiformis* Sm., *V. hederifolia* L. occ as lawn weeds. *V. biloba* L., native to e Eur, Asia, a waif in s SNF, MP, differs from *V. persica* in ways incl shorter styles (< 1 mm).

1. Racemes axillary; herbage gen glabrous; per
  2. Lvs petioled. . . . . *V. americana*
  2′ Lvs sessile (proximal rarely short-petioled)
    3. Racemes alternate; lf lance-linear, length gen >> 5 × width; fr deeply notched . . . . *V. scutellata*
    3′ Racemes opposite; lf elliptic or lanceolate to ovate, length < 5 × width; fr entire or slightly notched
      4. Lvs elliptic to ovate, length 1.5–3 × width; corolla 5–10 mm, lavender to blue; fr rounded to obcordate
         . . . . *V. anagallis-aquatica*
      4′ Lvs lanceolate, length 3–5 × width; corolla 3–5 mm, pink; fr notched 0.1–0.3 mm . . . . *V. catenata*
1′ Racemes mostly terminal, or fls 1 in axils; herbage gen hairy; ann or per
  5. Per, rhizomed
    6. Style 6–9 mm, > fr; corolla 8–13 mm
      7. Lvs hairy; corolla 8–10 mm, pale blue to lavender-rose; e KR . . . . *V. copelandii*
      7′ Lvs gen glabrous; corolla 10–13 mm, deep blue; CaR, n&c SNH . . . . *V. cusickii*
    6′ Style 0.8–3 mm, gen < fr; corolla 6–10 mm
      8. Style 2–3 mm; fr gen < calyx, wider than long; st decumbent . . . . *V. serpyllifolia* subsp. *humifusa*
      8′ Style 0.8–1.3 mm; fr > calyx, longer than wide; st erect nearly from base . . . . *V. wormskjoldii*
  5′ Ann, fibrous-rooted or taprooted
    9. Pedicel 0.5–2 mm, < calyx; seeds gen many per chamber, flat, smooth
      10. Lf triangular to ovate, crenate to serrate; sepals unequal (outer pair larger); corolla blue to violet . . . . . *V. arvensis*
      10′ Lf oblong to spoon-shaped, entire to ± serrate; sepals ± equal; corolla ± white . . . . *V. peregrina* subsp. *xalapensis*
    9′ Pedicel 4–30 mm, gen > calyx; seeds 5–12 per chamber, concave, outer surface rough
      11. Pedicel 15–30 mm; lf crenate to serrate; style 2–3 mm; fr lobes spreading . . . . *V. persica*
      11′ Pedicel 4–15 mm; lf palmately 3–9 lobed; style 1.2–1.6 mm; fr lobes ± parallel (or fr ± round) . . . . . [*V. triphyllos*]

V. americana Benth. (p. 1039) AMERICAN BROOKLIME Per, rhizomed, gen glabrous. **ST**: gen decumbent, rooting at proximal nodes, branched, 5–60(100) cm. **LF**: petioled; blade 5–50 mm, lanceolate to ovate, ± serrate, tip acute to obtuse. **INFL**: racemes axillary; fls 10–25; bracts linear; pedicels 5–10(13) mm. **FL**: sepals ± 3 mm, lanceolate, acute; corolla 7–10 mm, violet-blue, dark-lined; style 1.5–3(4) mm. **FR**: 3–4 mm, ± round, entire to barely notched. **SEED**: 0.5 mm, flat. 2*n*=36. Common. Moist to wet soil, springs, slow streams, meadows, lakeshores; < 3300 m. CA-FP, GB, DMtns (uncommon); US, Can, temp & arctic Asia. May–Aug

V. anagallis-aquatica L. WATER SPEEDWELL Per, rhizomed, glabrous. **ST**: gen decumbent, rooting at proximal nodes, simple to many-branched from base, 10–60(100) cm. **LF**: sessile (proximal rarely short-petioled); 20–80 mm, elliptic to ovate, length 1.5–3 × width, clasping to cordate, entire to serrate, light green. **INFL**: racemes axillary, opposite, glabrous to ± glandular-puberulent; fls gen > 30; bracts lance-linear; pedicels 4–8 mm, upcurved. **FL**: sepals 3–5.5 mm, elliptic to ovate; corolla 5–10 mm, lavender to blue, violet-lined; style 1.5–3 mm. **FR**: 2.5–4 mm, at least as wide, rounded to obcordate. **SEED**: 0.5 mm, flat. 2*n*=18,36. Wet meadows, streambanks, slow streams; < 3000 m. CA (uncommon D); widely naturalized in N.Am, S.Am; native to Eur. May–Sep

V. arvensis L. Ann, hairy; roots fibrous. **ST**: prostrate to erect, ± branched, 5–40 cm. **LF**: proximal ± short-petioled; blade 2–15 mm, triangular to ovate, cordate to truncate at base, crenate to serrate. **INFL**: racemes terminal; bracts > pedicels; pedicels 0.5–2 mm. **FL**: sepals 3–4 mm, unequal (outer pair larger), lanceolate; corolla 2–3 mm, blue to violet; style 0.4–1 mm. **FR**: 3–4 mm, flat, ciliate; notch 0.5–0.8 mm. **SEED**: gen many per chamber, 1 mm, flat, smooth. 2*n*=14,16,18. Meadows; < 1500 m. KR, NCoR, SNF, GV, SnFrB, SCoRO, TR, PR, MP, D; widely naturalized; native to Eurasia. Apr–Jul

V. catenata Pennell CHAIN SPEEDWELL Per, rhizomed, glabrous. **ST**: ascending to erect, rooting at proximal nodes, branched, 10–60(100) cm. **LF**: sessile; 25–90 mm, lanceolate, length 3–5 × width, ± entire, dark green, base clasping, tip acute. **INFL**: racemes axillary, opposite; fls 15–25; bracts lance-linear; pedicels 3–7 mm, ± straight, spreading. **FL**: sepals ± 3 mm, lanceolate to ovate, acute to obtuse; corolla 3–5 mm, pink; style < 2 mm. **FR**: 3–4 mm, obcordate to rounded, notched 0.1–0.3 mm. **SEED**: 0.5 mm, flat. 2*n*=36. Wet meadows, slow streams; < 2500 m. CA-FP; widespread US, adjacent Can; native to Eur. Jul–Sep

V. copelandii Eastw. (p. 1039) COPELAND'S SPEEDWELL Per, rhizomed, shaggy-hairy, ± glandular. **ST**: ascending, branched, 5–12 cm. **LF**: sessile; 5–35 mm, oblong to elliptic, entire, acute to obtuse at tip, hairy. **INFL**: racemes terminal; pedicels 6–8 mm. **FL**: sepals gen 5, 2–3 mm, unequal, elliptic; corolla 8–10 mm, pale blue to lavender-rose; style ± 7 mm, > fr. **FR**: longer than wide, barely notched. Subalpine meadows, slopes; < 2600 m. e KR (Trinity, Siskiyou cos.). Aug ★

V. cusickii A. Gray (p. 1039) CUSICK'S SPEEDWELL Per, rhizomed. **ST**: ascending, branched, 10–15(20) cm, glandular-hairy. **LF**: sessile; 5–25 mm, elliptic to ovate, entire, acute to obtuse at tip, gen glabrous. **INFL**: terminal, glandular-hairy; pedicels 3–9 mm. **FL**: sepals 2–3 mm, ± unequal, lanceolate; corolla 10–13 mm, deep blue; style 6–9 mm, > fr. **FR**: 5–6 mm, longer than wide, deeply notched, glandular-hairy. Forest openings, meadows; 1800–3000 m. CaR, n&c SNH; to WA. Jul–Aug ★

V. peregrina L. subsp. *xalapensis* (Kunth) Pennell (p. 1039) PURSLANE SPEEDWELL Ann, taprooted, gen glandular-hairy. **ST**: erect, gen branched, 5–30 cm. **LF**: proximal ± petioled; 5–25 mm, oblong to spoon-shaped, entire to ± serrate. **INFL**: racemes terminal, open; bracts lanceolate, > pedicels; pedicels 0.5–2 mm. **FL**: sepals 3–6 mm, ± equal, lanceolate; corolla 2–3 mm, ± white; style 0.1–0.4 mm. **FR**: 3–4 mm, obovate; notch 0.2–0.5 mm. **SEED**: gen many, 0.5 mm, flat, smooth. 2*n*=52. Moist places; < 3100 m. CA-FP, SNE, D (uncommon); to w Can, Mex, S.Am. *V. peregrina* subsp. *p.* (glabrous) is widespread in e N.Am, Eur. Apr–Aug

V. persica Poir. (p. 1039) PERSIAN SPEEDWELL Ann, taprooted, hairy. **ST**: prostrate, simple or branched, 5–60 cm. **LF**: short-

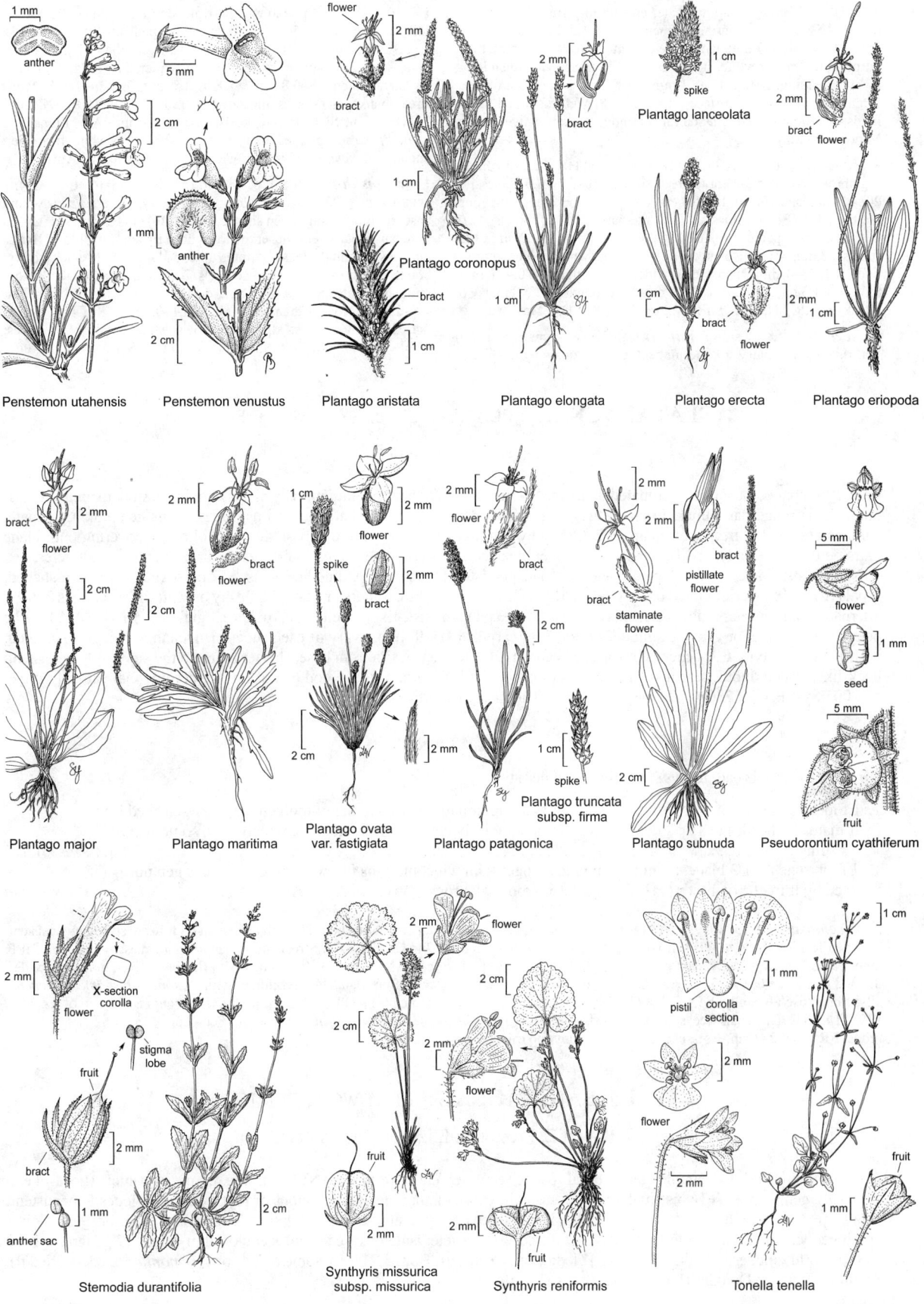

1 mm
anther

5 mm

2 cm

1 mm

anther

1 mm

2 cm

Penstemon utahensis

Penstemon venustus

flower
2 mm
bract

1 cm

Plantago coronopus

bract

1 cm

Plantago aristata

2 mm
bract

1 cm

Plantago elongata

spike
1 cm

Plantago lanceolata

1 cm
2 mm
bract
flower

Plantago erecta

2 mm
bract
flower

1 cm

Plantago eriopoda

bract
flower
2 mm

2 cm

Plantago major

2 mm
flower
bract
flower
2 cm

Plantago maritima

1 cm
spike
2 mm
flower
2 mm
bract

2 cm

Plantago ovata
var. fastigiata

2 mm
flower
bract

2 cm

1 cm

spike

Plantago truncata
subsp. firma

2 mm

2 mm
bract
bract
pistillate
flower
staminate
flower

2 cm

Plantago subnuda

5 mm

flower

1 mm
seed

5 mm

fruit

Pseudorontium cyathiferum

2 mm
X-section
corolla
flower

stigma
lobe

fruit

2 mm

bract

1 mm

anther sac

2 cm

2 cm

Stemodia durantifolia

2 mm
flower

2 cm

2 cm

fruit

2 mm

Synthyris missurica
subsp. missurica

2 cm
2 mm
flower

2 mm
fruit

Synthyris reniformis

pistil
corolla
section

1 mm

flower

2 mm

2 mm

1 cm

1 mm
fruit

Tonella tenella

petioled; 5–25 mm, ovate, truncate, crenate to serrate, tip acute to obtuse. **INFL**: racemes terminal, open; pedicels 15–30 mm, recurved in fr. **FL**: sepals 4–7 mm, lanceolate to ovate; corolla 8–12 mm, blue, purple-lined, center white; style 2–3 mm. **FR**: much wider than long; lobes spreading; notch 0.7–1.2 mm. **SEED**: 5–11 per chamber, ± 2 mm, concave, outer surface rough. $2n=28$. Wet, disturbed areas, fields; < 1100 m. CA-FP, DSon (uncommon); N.Am; native to Asia Minor. Feb–May

V. scutellata L. (p. 1039) MARSH SPEEDWELL Per, rhizomed, ± glabrous. **ST**: decumbent to erect, slender, 10–60 cm. **LF**: sessile; 20–40 mm, lance-linear, length gen >> 5 × width, ± entire, often purple-tinged. **INFL**: racemes axillary, alternate; fls 5–20; bracts < 1/2 pedicels, linear; pedicels 6–16 mm, spreading at right angles in fr. **FL**: sepals ± 3 mm, ± equal, ovate; corolla 5–7 mm, white or blue, purple-lined. **FR**: 3–4 mm, longer than wide; deeply notched 0.4–1 mm. **SEED**: ± 1.5 mm, flat. $2n=18$. Wet meadows, ponds; < 2300 m. NCo, NCoRO, CaR, SN, CCo, SnBr, MP; to BC, e US; Eur. May–Aug

V. serpyllifolia L. subsp. ***humifusa*** (Dicks.) Syme (p. 1039) Per, rhizomed, ± hairy. **ST**: decumbent, erect at tips only, 5–30 cm.

LF: 10–25 mm, elliptic to widely ovate, entire to crenate, tip obtuse. **INFL**: racemes terminal, glandular-hairy; pedicels 2.5–7 mm. **FL**: sepals 2.5–4 mm, ± equal, oblong to ovate; corolla 6–7 mm, bright blue; style 2–3 mm. **FR**: 2.8–3.7 mm, wider than long, ± glandular-hairy; notch 0.3–0.8 mm. **SEED**: 0.5 mm, flat. $2n=14,28$. Moist streambanks, lakeshores, meadows; < 3200 m. NW, CaR, SN, CCo, SnFrB, SnBr, SNE; to AK, ne US, NM; also in S.Am, Eurasia. *V. serpyllifolia* subsp. *s.* less hairy, pedicels shorter, fls white, fr smaller; uncommon lawn weed in ScV, CCo, SCo; native to Eur. Apr–Aug

V. wormskjoldii Roem. & Schult. (p. 1039) AMERICAN ALPINE SPEEDWELL Per, rhizomed, long-wavy-hairy. **ST**: erect nearly from base, rooting at nodes, gen simple, 10–25(40) cm. **LF**: sessile; 20–40 mm, lanceolate to elliptic, entire to crenate, tip acute to obtuse. **INFL**: racemes terminal, dense to interrupted, glandular- to sticky-hairy; bracts linear to lanceolate; pedicels 2–6 mm. **FL**: sepals 3.5–5.5 mm, lanceolate; corolla 6–10 mm, deep blue; style 0.8–1.3 mm. **FR**: 4.5–6.5 mm, longer than wide; notch 0.1–0.5 mm. **SEED**: ± 1 mm, flat. $2n=18,36$. Moist alpine meadows, streambanks, lakeshores; 1500–3500 m. KR, CaR, SN, MP; to AK, ne US, n Mex. Jun–Aug

PLATANACEAE PLANE-TREE or SYCAMORE FAMILY

Robert Lee Allen

Tree, gen monoecious, wind-pollinated; hairs many-branched. **ST**: bark peeling in scaly plates, leaving ± smooth areas of various colors, in age dark, thick, fissured; twig hairs dense. **LF**: simple, alternate, deciduous; lobes, veins gen 3,5(7), palmate; stipules gen lf-like, free or fused around st, shed by lf maturity or not; petiole at base dilated, hollow, ± covering bud; blade hairs dense, ± 0 in age. **INFL**: heads 1–7, ± evenly spaced on axis, spheric, many-fld, sessile or on pendent peduncles, gen unisexual; staminate breaking apart in age; pistillate persistent; bracts subtending heads, fls. **FL**: unisexual; calyx cup-shaped, sepals (0)3–6(8), free or united basally. **STAMINATE FL**: petals 3–6, minute or vestigial, fleshy or scale-like; stamens 3–6(8), alternate petals, anthers subsessile, axis above anther expanded, disk-like, ± peltate; carpels vestigial or 0. **PISTILLATE FL**: petals 3–6, minute, or gen 0; staminodes often 3–4; carpels (3)5–9, free, ovary of each superior, 1-chambered, gen 1-ovuled, with 1 ± linear style. **FR**: achenes in spheric head, small, each with hairs from base, shorter hairs up the side; style persistent, beak-like, or deciduous. 1 genus, ± 8 spp.: n temp; some cult for orn, shade; wood gen of limited commercial value. [Feng et al. 2005 Syst Bot 30:786–799] Scientific Editor: Thomas J. Rosatti.

PLATANUS

(Greek: probably broad, for lvs) Fr length excludes style.

1. Lf lobes gen < 1/3 blade, terminal gen as long as wide, margins few- to many-toothed; stipules gen shed by lf maturity; heads (1)2(3); stigma brown-maroon, densely covered with tan hairs; widely cult, sometimes escaping . . . . *[P. ×hispanica]*
1' Lf lobes gen ≥ 1/3 blade, terminal gen much longer than wide, margins 0(few)-toothed; stipules gen not shed by lf maturity; heads (1)2–7; stigma ± maroon, glabrous; native . . . . *P. racemosa*

P. racemosa Nutt. (p. 1039) WESTERN SYCAMORE **ST**: 10–35 m, often leaning; base < 1(2+) m wide; outer bark light gray, tan, inner paler. **LF**: stipules 2–3 cm, gen persistent after maturity; petiole 3–8 cm; blade ± 10–25 cm, ± round, glabrous to ± hairy adaxially, tomentose abaxially. **STAMINATE FL**: sepals 0; petals free. **PISTILLATE FL**: sepals free; style red-tipped, stigma maroon, glabrous. **FR**: head 2–3 cm, ± sessile; achene 7–10 mm, top truncate or tapered, basal hairs ± 2/3 fr, persistent on fr head; style gen persistent. $2n=42$. Common. Streamsides, canyons, arroyos; < 2000 m. CaRF, c&s SNF, Teh, GV, CW, SW, DMoj (Mojave River, n of Victorville), nw DSon; Baja CA. Hybridizes with *Platanus ×hispanica* Muenchh; some pls in e PR with lvs similar to *P. wrightii* S. Watson of AZ, NM, Mex, but with fr of *P. racemosa*. Feb–Apr

PLUMBAGINACEAE LEADWORT FAMILY

Robert E. Preston & Elizabeth McClintock

Ann to shrub [vine]. **LF**: simple, gen in basal rosette (cauline), entire or lobed. **INFL**: raceme, cyme, or panicle (head-like in *Armeria*), gen scapose. **FL**: bisexual, radial, gen small; calyx tubular, gen membranous or partly scarious, lobes 5, persistent; petals 5, ± free to ± fused, clawed, ± intertwined; stamens 5, opposite petals, occ epipetalous; ovary superior, gen 5-lobed or -ribbed, chamber and ovule 1, styles 5, occ fused. **FR**: utricle, achene, or capsule, ± enclosed in calyx. 27 genera, ± 1000 spp.: ± worldwide, esp Medit, w&c Asia. [Lledo et al. 1998 Syst Bot 23:21–29] Some cult as orn (*Limonium* used as dried fl). Scientific Editors: Douglas H. Goldman, Bruce G. Baldwin.

1. Infl head-like, on unbranched, scapose peduncle; lf linear . . . . **ARMERIA**
1′ Infl a raceme or panicle, branched; lf oblanceolate to obovate
  2. Lvs appearing all basal, cauline lvs gen reduced to bract-like scales; corolla > calyx by 0–4 mm . . . . **LIMONIUM**
  2′ Lvs cauline; corolla > calyx by 25–50 mm . . . . [**PLUMBAGO**]

ARMERIA THRIFT, SEA-PINK

Per, rhizomed. **LF**: many, sessile, linear [to lanceolate], entire; veins 1–7, parallel. **INFL**: head-like, on unbranched, scapose peduncle; involucre sheathing, recurved; bracts subtending individual fls. **FL**: calyx 10-ribbed; styles fused at base, hairy. ± 50 spp.: N.Am, Eur, Medit, s S.Am. (Latin, from Old French: name for a dianthus with spheric heads) [Fuertes Aguilar & Nieto Feliner 2003 Molec Phylogen Evol 28:430–447]

A. maritima (Mill.) Willd. subsp. ***californica*** (Boiss.) A.E. Porsild (p. 1039) **LF**: 6–12 cm, 1–2.5 mm wide, flat, glabrous. **INFL**: < 2 cm wide; peduncle < 60 cm. **FL**: calyx < corolla, both gen pink; calyx tube ± hairy on ribs, lobe tips short, sharp. Ocean bluffs, ridges, coastal strand, sand, exposed grassland; < 245 m. NCo, CCo, n ChI; to BC. Other subspp. in boreal N.Am, nw Eur. Feb–Sep

LIMONIUM SEA-LAVENDER, MARSH-ROSEMARY

(Ann) per, scapose; rhizome ± woody. **LF**: few to many; basal gen in rosette, oblanceolate to obovate, entire or lobed, gen petioled; cauline gen small, scale-like. **INFL**: panicle, branched ± at pl base, fls in sessile, 1–3(5) fld spike-like clusters, each 3-bracted at base, clusters evenly spaced along branches to crowded at branch tips. **FL**: calyx funnel-shaped, base tubular, lobes gen fused, gen 5–10-ribbed, gen pink to blue; petals free, white to yellow or pale violet; styles 5, ± free. ± 300 spp.: ± worldwide, gen in saline soils. (Greek: meadow, from habitat of many spp.) [Karis 2004 Biol J Linn Soc 144:461–482] Many cult spp.; those of Medit origin potentially could escape and naturalize. *L. arborescens* Kuntze, illeg. may have been misappl. to the waif *L. brassicifolium* (Webb & Berthel.) Kuntze.

1. Infl branch wings 0
  2. Infl with some non-fl branches; proximal infl axis bracts round, clasping. . . . . [***L. otolepis***]
  2′ Infl non-fl branches 0; proximal infl axis bracts deltate to linear, not clasping
    3. Basal lvs obovate to oblong, 1.5–6 cm wide . . . . ***L. californicum***
    3′ Basal lvs oblanceolate to spoon-shaped, 0.5–2 cm wide
      4. Basal lvs 1–4 cm, 5–9 mm wide, oblanceolate to spoon-shaped, tip obtuse; infl clusters 1–3 fld, ±
        curved, evenly distributed on terminal 2–3(6) cm of branch tips, 2–4 per cm . . . . ***L. duriusculum***
      4′ Basal lvs 3–10 cm, 7–20 mm wide, obovate to oblanceolate or spoon-shaped, tip acute to rounded;
        infl clusters 2–5 fld, crowded in terminal 1–4 cm of branch tips, 4–8 per cm. . . . . ***L. ramosissimum***
1′ Terminal infl branches winged
  5. Lf pinnately lobed at least at base
    6. Wings of infl axes 5–20 mm wide; lf-like bracts 0 at st branch points. . . . . [***L. brassicifolium***]
    6′ Wings of infl axes ≤ 5 mm wide; ± linear lf-like bracts gen present at st branch points. . . . . ***L. sinuatum***
  5′ Lf ± entire
    7. Lf margins ciliate with simple to 3-branched hairs; lf tip long-acuminate, ≥ 4 mm; proximal fl bract
      awned, distal rounded . . . . ***L. perezii***
    7′ Lf margins glabrous or with few, simple hairs; lf tip acuminate, < 3 mm; fl bract tips mucronate to rounded
      8. Lf blade base gen cordate; fl clusters 2–4, in terminal 0.5 cm of branch tips; calyx sparsely hairy . . . . . [***L. preauxii***]
      8′ Lf blade base long-tapered; fl clusters 1–4(5), in terminal 1.5 cm of branch tips; calyx glabrous . . . . . [***L. sventenii***]

L. californicum (Boiss.) A. Heller (p. 1039) WESTERN MARSH-ROSEMARY Per, erect, < 35 cm. **LF**: petiole gen < blade, stout, narrow-winged; blade 5–15 cm, 1.5–6 cm wide, obovate to oblong, ± entire or margin wavy, ± thick, leathery, base tapered. **INFL**: wings 0; proximal infl bracts deltate, not clasping; branches all fl; clusters 1–2-fld, crowded in terminal 1–3 cm of branch tips; fl bracts 2.5–4.5 mm. **FL**: calyx tube 4–5 mm, ribs gen ± hairy, lobes ± 0.5 mm, white; corolla blue, ± > calyx. Common; coastal dunes, salt marshes; < 50 m. NCo, CCo, SCo; OR, s NV, n Baja CA. Jul–Dec

L. duriusculum (Girard) Fourr. Erect, 20–30 cm. **LF**: basal lvs 1–4 cm, 5–9 mm wide, oblanceolate to spoon-shaped, tip obtuse. **INFL**: wings 0; proximal infl axis bracts deltate to linear, not clasping; fls on proximal 1–5 branches gen sterile; clusters 1–3-fld, ± curved, evenly distributed on terminal 2–3(6) cm of branch tips, 2–4 per cm; fl bracts 4.5–5 mm. **FL**: calyx (4.5)5.5–6 mm, limb 1.5 × tube, midrib extended to tip, persisting and spreading to recurved in fr; corolla ± 8 mm, pale pink. Coastal salt marsh, coastal scrub, riparian scrub, disturbed areas; < 445 m. SCo, SnGb (Verdugo Mtns), PR; native to w Medit. Sep–Jun

L. perezii (Stapf) F.T. Hubb. Erect, 15–45 cm. **LF**: petiole ≥ blade; blade 4–15 cm, 2.5–7 cm wide, round to wide-ovate, ± entire, margin ciliate with simple to 3-branched hairs, base ± truncate, tip long-acuminate. **INFL**: winged; clusters 2-fld, crowded in terminal 1–3 cm of branch tips; fl bracts 2–5 mm, proximal awned, distal rounded, margins ciliate. **FL**: calyx ± 10 mm, blue-purple, tube short-stiff-hairy; corolla > calyx, white. Disturbed coastal areas, cliffs, sand dunes, roadsides; < 100 m. SW (exc SnJt); native to Canary Islands. Mar–Sep

L. ramosissimum (Poir.) Maire Erect, 20–50 cm. **LF**: basal lvs 3–10 cm, 7–20 mm wide, obovate to oblanceolate or spoon-shaped, tip acute to rounded. **INFL**: wings 0; proximal infl axis bracts deltate to linear, not clasping; all branches fl, clusters 2–5 fld, crowded in terminal 1–4 cm of branch tips, 4–8 per cm; fl bracts 4–5.5 mm. **FL**: calyx 4–6 mm, white; corolla 5–7 mm, pale pink. Coastal salt marshes, coastal or riparian scrub, grassland, disturbed areas; < 500 m. CCo (Morro Bay, San Francisco Bay), SCo, PR; native to Medit. Highly variable; our material may represent > 1 subsp. or sp. Aug–Jun ❖

L. sinuatum (L.) Mill. Erect, 20–50 cm. **ST**: rough-hairy, narrowly 3–5-winged; ± linear lf-like bracts gen present at st branch points. **LF**: petiole < to > blade, winged; blade 3–12 cm, ± obovate, pinnately lobed, rough-hairy, base tapered. **INFL**: wings ≤ 5 mm wide; clusters 2-fld, crowded in terminal 1–3 cm at branch tips; fl bracts 5–7 mm, proximal awned, distal irregularly lobed. **FL**: calyx 10–12 mm, blue, tube short-stiff-hairy; corolla > calyx, white to pale yellow. Disturbed coastal areas, roadsides; < 300 m. CCo, SnFrB, SCoRO, SCo; native to Medit, w Asia. All yr

POLEMONIACEAE PHLOX FAMILY

Robert Patterson, family description, key to genera

Ann, per, shrub, vine. **LF**: simple or compound, cauline (or most basal), alternate or opposite; stipules 0. **INFL**: cymes, heads, clusters, or fl 1; bracts in involucres or not. **FL**: sepals gen 5, fused at base, translucent membrane gen connecting lobes, torn by fr; corolla gen 5-lobed, radial or bilateral, salverform to bell-shaped, throat often well defined; stamens gen 5, epipetalous, attached at ≥ 1 level, filaments of ≥ 1 length, pollen white, yellow, blue, or red; ovary superior, chambers gen 3, style 1, stigmas gen 3. **FR**: capsule. **SEED**: 1–many, when wetted swelling or not, gelatinous or not. 26 genera, 314 spp.: Am, n Eur, n Asia; some cult (*Cantua*, *Cobaea* (cup-and-saucer vine), *Collomia*, *Gilia*, *Ipomopsis*, *Linanthus*, *Phlox*). [Porter & Johnson 2000 Aliso 19:55–91] *Leptodactylon* moved to *Linanthus*. Scientific Editors: Robert Patterson, Thomas J. Rosatti.

1. Lvs pinnate-compound . . . . **POLEMONIUM**
1′ Lvs 0 or simple, entire to deeply pinnate- or palmate-lobed
  2. Lvs 0; filaments < 0.5 mm . . . . **GYMNOSTERIS**
  2′ Lvs present; filaments > 0.5 mm
    3. Per, subshrub, woody above ground, at least below (see also per *Aliciella*, *Collomia*, *Phlox*)
      4. Calyx lobes unequal; infl, bracts, calyx with woolly tufted hairs . . . . ³**ERIASTRUM**
      4′ Calyx lobes equal; infl, bracts, calyx with hairs or not, but without woolly tufted hairs
        5. Lvs alternate
          6. Infl head-like; lf lobes not rigid, not sharp-tipped; stamens exserted. . . . . *Ipomopsis congesta*
          6′ Infl not head-like; lf lobes rigid, sharp-tipped; stamens incl . . . . ⁶**LINANTHUS**
        5′ Lvs opposite, at least on lower st
          7. Stamens attached at > 1 level; lvs not lobed. . . . . ²**PHLOX**
          7′ Stamens attached at 1 level; lvs lobed
            8. Calyx 5-parted; corolla glandular abaxially, open in day; lvs palmate-lobed, lobes pliable . . . ³**LEPTOSIPHON**
            8′ Calyx gen 6-parted; corolla glabrous abaxially, open gen at night; lvs pinnate-lobed, lobes rigid —
              SnJt . . . . *Linanthus jaegeri*
    3′ Ann or rhizomed, short-lived per, gen not woody above ground exc sometimes for parts of woody caudex
      9. Calyx membrane narrow, spout-like distally. . . . . **COLLOMIA**
      9′ Calyx membrane narrow to wide, not spout-like distally
        10. Lvs cauline, opposite below, alternate or opposite above
          11. Stamens attached at > 1 level; lvs not lobed
            12. Ann; seeds gelatinous when wet . . . . **MICROSTERIS**
            12′ Per; seeds not gelatinous when wet . . . . ²**PHLOX**
          11′ Stamens attached at 1 level; lvs palmate-lobed
            13. Fls in head subtended by bracts or in small clusters
              14. Corolla salverform to funnel-shaped, tube long-exserted . . . . ³**LEPTOSIPHON**
              14′ Corolla bell- or short-funnel-shaped, tube gen incl . . . . ⁶**LINANTHUS**
            13′ Fls in small clusters
              15. Pedicel > 2 mm, if 0 then corolla with hairy ring inside tube; corolla gen open daytime, without
                red marks at throat. . . . . ³**LEPTOSIPHON**
              15′ Pedicel 0–2 mm; corolla without hairy ring inside tube; corolla gen open at night, if open
                daytime then with red marks at throat (see also *Linanthus dichotomus* subsp. *meridianus*) . . . . ⁶**LINANTHUS**
        10′ Lvs in rosette and/or cauline, alternate or ± opposite below, alternate above
          16. Calyx lobes unequal
            17. Infl open, fls in groups of 1–3 . . . . ³**ERIASTRUM**
            17′ Infl dense, fls in groups of gen > 3
              18. Infl woolly . . . . ³**ERIASTRUM**
              18′ Infl hairy or not, but not woolly . . . . ²**NAVARRETIA**
          16′ Calyx lobes equal
            19. Lower cauline lvs palmate-lobed
              20. Lvs petioled, sinuses between lobes not to lf midrib. . . . . ³**IPOMOPSIS**
              20′ Lvs sessile, sinuses between lobes ± to lf midrib. . . . . ⁶**LINANTHUS**
            19′ Lower cauline lvs simple to pinnate-lobed
              21. Lf teeth, calyx lobes bristle-tipped
                22. Corolla radial; lf hairs branched; filaments equal . . . . **LANGLOISIA**
                22′ Corolla bilateral; lf hairs simple; filaments unequal . . . . **LOESELIASTRUM**
              21′ Lf teeth, calyx lobes with short or long point or not, but not bristle-tipped
                23. Corolla rotate, bell-shaped, or short-funnel-shaped, lobes > tube
                  24. Calyx lobes >> tube, margins scarious, rarely ciliate . . . . ⁶**LINANTHUS**
                  24′ Calyx lobes < tube, margins scarious, not ciliate
                    25. Lobes of basal lvs >> 2, pinnately arranged. . . . . ²**ALICIELLA**
                    25′ Lobes of basal lvs 0(1–2). . . . . ²**LATHROCASIS**
                23′ Corolla funnel-shaped or salverform, lobes < tube

26. Calyx lobes >> tube (± free to base), margins not scarious, rarely ciliate . . . . ⁶**LINANTHUS**
26′ Calyx lobes ≤ tube, margins scarious, not ciliate
  27. Seeds 1 per chamber, upper lvs linear, simple, entire; pls without skunky odor . . . . ²**LATHROCASIS**
  27′ Seeds gen > 1 per chamber, if 1 then upper lvs ± palmate-compound, pls with skunky odor
    28. Seeds not gelatinous when wet . . . . ²**ALICIELLA**
    28′ Seeds gelatinous when wet
      29. Infl dense, fls gen sessile
        30. Lvs 2-pinnate-lobed or -dissected, margins gen toothed; corolla gen funnel-shaped, throat
           well developed . . . . ²**GILIA**
        30′ Lvs entire to pinnate-lobed, margins not toothed; corolla narrowly funnel-shaped to
           salverform, throat 0 or not well developed
          31. Lvs pinnate-lobed or linear, entire; corolla radial to bilateral . . . . ³**IPOMOPSIS**
          31′ Lvs lance-elliptic, entire or dentate; corolla ± bilateral . . . . *Loeseliastrum depressum*
      29′ Infl gen open, fls pedicelled, if subsessile then infl not dense
        32. Seeds medium to dark brown or black; lvs basal and cauline, upper cauline ± palmate-
           lobed with terminal lobe width > lateral, or entire, similar to lower cauline (or 2-pinnate-lobed)
          33. Hairs, or at least some, > 0.5 mm, often longer, glandular or not, gen > 1 type or length
             throughout; cotyledons spoon-shaped . . . . **ALLOPHYLLUM**
          33′ Hairs < 0.3 mm, glandular, ± 1 type or length throughout; cotyledons linear . . . . ²**NAVARRETIA**
        32′ Seeds ± green, tan, light brown, or light gray; lvs gen in basal rosette, upper cauline 0 or
           simple, entire, different from basal, or pinnate-lobed with terminal lobe width ≤ lateral
          34. Corolla salverform, radial, or funnel-shaped, bilateral . . . . ³**IPOMOPSIS**
          34′ Corolla bell- or funnel-shaped, ± radial
            35. Hairs on basal lvs nonglandular, white cobwebby (at least in axils), or white and sharply
               bent, or translucent, then infl glands long-stalked, diam < stalk (hair) . . . . ²**GILIA**
            35′ Hairs on basal lvs minutely glandular, translucent; infl glands subsessile,
               diam > stalk (hair) . . . . **SALTUGILIA**

ALICIELLA

J. Mark Porter

ST: erect, ascending or decumbent, glabrous, hairy, or glandular. **LF**: simple, gen alternate, tips acute, acuminate, or mucro-nate; basal gen in rosette, entire, toothed, or 1–2-pinnate-lobed; cauline gen reduced. **INFL**: fls 1–3 in bract axils. **FL**: calyx membranous between lobes, lobes < tube, membranes glandular, splitting or expanding in fr; corolla > calyx, lobes gen < tube, gen ovate, acute, acuminate. **FR**: spheric to ovoid; chambers 3; valves separating from top. **SEED**: 3–many, yellow to brown, not gelatinous when wet. ± 25 spp.: w N.Am. (Alice Eastwood, w Am botanist, 1859–1953) [Porter 1998 Aliso 17:23–46]

1. Basal lf blades 1–7 cm wide, obovate, coarsely dentate, teeth needle-like; filaments (at least longest) papillate
  2. Stamens attached in lower throat; corolla lobes pink or magenta adaxially, pale pink abaxially; ann
     . . . . *A. latifolia* subsp. *latifolia*
  2′ Stamens attached near base of tube, filaments 4 short, 1 long; corolla lobes magenta; per . . . . *A. ripleyi*
1′ Basal lf blades gen ≤ 2 cm wide, lanceolate (oblanceolate to obovate for *Aliciella triodon*), pinnate-divided,
   entire, or if dentate then teeth mucronate but not needle-like; filaments smooth
  3. Corolla throat ± constricted near mouth; corolla lobes 3-toothed, central tooth gen longest . . . . *A. triodon*
  3′ Corolla throat widened near mouth; corolla lobes lanceolate- to ovate- or ± truncate-acuminate
    4. Corolla lobes truncate-acuminate
      5. Pedicels >> calyx; branches ± spreading; fr spheric . . . . *A. humillima*
      5′ Pedicels of 2 lengths, terminal < calyx, axillary > calyx; branches ascending; fr narrowly ellipsoid . . *A. leptomeria*
    4′ Corolla lobes lanceolate- to ovate-acuminate
      6. Corolla gen 2–3.5 mm, pedicels >> calyx, spreading or recurved; fr ± spheric . . . . *A. micromeria*
      6′ Corolla gen 4.5–14 mm; pedicels of 2 lengths, terminal < calyx, axillary > calyx, all straight,
         ascending to erect (nodding in bud in *A. monoensis*); fr ellipsoid or ovoid
        7. Basal lvs 1–2 × pinnate-divided (in depauperate pls dentate but then midrib narrow) glandular hairs
           on basal lvs long
          8. Corolla glandular-puberulent abaxially . . . . *A. hutchinsifolia*
          8′ Corolla glabrous abaxially . . . . ²*A. monoensis*
        7′ Basal lvs dentate, serrate, or ± 1-pinnate-divided (if dentate then midrib wide); glandular hairs on
           basal lvs short, at least abaxially
          9. Basal lvs adaxially glabrous . . . . *A. lottiae*
          9′ Basal lvs adaxially glandular-puberulent . . . . ²*A. monoensis*

A. humillima (Brand) J.M. Porter SMALLEST ALICIELLA Ann. **ST**: 4–20 cm, branches many, ± spreading, thread-like, ± glandular-puberulent. **LF**: basal 1–7.5 cm, 1-pinnate-lobed, lobes spreading, length 0.5–1 × width of lf axis; upper lvs linear, entire. **FL**: calyx ± 2 mm; corolla 4–7 mm, bell- to funnel-shaped, tube 1.5–3 × calyx, white, throat yellow, lobes truncate-acuminate, white adaxially, purple abaxially; stamens, style ± exserted, pollen white. **FR**: ± 3 mm, ± = calyx, spheric. **SEED**: 6–24, 0.6–0.9 mm. $2n$=18. Uncommon. Rocky or sandy soil; 1200–1670 m. GB (Modoc, Inyo cos.); to OR, NV. Previously incl in *Gilia leptomeria* A. Gray. Apr–May

A. hutchinsifolia (Rydb.) J.M. Porter DESERT PALE ALICIELLA
Ann; densely glandular-puberulent, glandular-hairy below, odor
skunk-like. **ST:** 5–40 cm, branched from base or above. **LF:** basal
2–8 cm, ± erect, 1–2-pinnate, axis linear, lobes > 2 × width of lf axis;
upper lvs linear, entire. **FL:** calyx 3–4 mm, lobes linear; corolla 7–14
mm, tube well exserted, white, glandular-puberulent abaxially, throat
green-spotted, yellow above, lobes 2–4 mm, gen wavy-margined,
white to lavender; stamens, style ± exserted, pollen white. **FR:** 3–6
mm, ≥ calyx. **SEED:** many. 2*n*=18. Sandy or gravelly flats, slopes,
dunes; 400–1800 m. GB, DMoj; to UT, AZ. [*Gilia h.* Rydb.] Sand
grains often stuck to glands. Mar–May(Jun)

A. latifolia (S. Watson) J.M. Porter subsp. ***latifolia*** BROAD-
LEAVED ALICIELLA Ann, odor skunk-like. **ST:** 10–30 cm, branches
spreading, glandular-hairy below. **LF:** basal petioled, occ not rosetted, blade obovate, 1–7 cm wide, coarsely dentate (teeth needle-like),
glandular-hairy, hairs often appressed; upper lvs reduced, needle-like.
FL: calyx fused in lower 1/2, lobes fine-pointed; corolla 7–11 mm,
tube white, lobes pink or magenta adaxially, pale pink abaxially; stamens attached in lower throat, unequal, longest ± exserted, filaments
(at least longest) papillate, pollen white. **FR:** 5–7 mm, ≥ calyx, ovoid.
SEED: many, deep red-brown. 2*n*=36. Common. Rocky slopes,
washes; < 1800 m. W&I, D; to UT, AZ. [*Gilia l.* S. Watson] Other
subsp. in UT. (Jan)Apr–May(Jul)

A. leptomeria (A. Gray) J.M. Porter (p. 1039) SAND ALICIELLA
Ann. **ST:** 7–23 cm, branches ascending, 1–several, thread-like,
glandular-puberulent. **LF:** basal 1–6 cm, lanceolate or widely linear, toothed or lobed, lobes rounded, mucronate, gen < lf axis width;
cauline linear, entire. **FL:** calyx 2–3 mm, lobe tips thickened; corolla
4–7 mm, tube 1.5–3 × calyx, thread-like, purple, throat yellow, lobes
truncate-acuminate, white adaxially, purple abaxially; stamens, style
± exserted, pollen white. **FR:** 3–5 mm, = calyx, narrowly ellipsoid.
SEED: many. 2*n*=34,36. Common. Open, sandy or rocky areas;
800–2100 m. GB, DMtns (uncommon); to OR, ID, CO, NM. [*Gilia
l.* A. Gray] Apr–Jun

A. lottiae (A.G. Day) J.M. Porter (p. 1039) LOTT'S ALICIELLA
Ann. **ST:** 5–43 cm, branches gen spreading from base, glandular-
puberulent. **LF:** basal 2–11 cm, lanceolate, dentate, serrate, or ±
1-pinnate-divided, fleshy, adaxially glabrous, lobes spreading, midrib wide; cauline linear, entire. **FL:** calyx 2–3 mm, lobe tips long-
tapered; corolla 6–7 mm, tube ± exserted, stout, white, throat > tube,
white, yellow-spotted, lobes lanceolate to ovate, pink, lavender, or
white; stamens, styles ± exserted, pollen white. **FR:** 3–5 mm, < 1.5
× calyx, ovoid, tip pointed. **SEED:** many. 2*n*=50. Sandy soils, sagebrush scrub; 400–2100 m. GB, w DMoj, e DMtns (Clark Mtn Range);
to WA, ID, UT. [*Gilia l.* A.G. Day] Apr–Jul

A. micromeria (A. Gray) J.M. Porter (p. 1045) DAINTY
ALICIELLA Ann. **ST:** 4–14 cm, branches many, ± spreading, often
at ± 75–90°, thread-like, ± glandular-puberulent. **LF:** basal 1–6 cm,
1-pinnate-lobed, lobes spreading, length 0.5–1 × lf axis width; upper
cauline linear, entire. **FL:** calyx ± 2 mm; corolla gen 2–3.5 mm, tube
gen incl, white, throat pale yellow, lobes ovate, entire, white or lavender; stamens 3–5, stamens, style ± exserted, pollen white. **FR:** ± 3
mm, ± = calyx, ± spheric. **SEED:** 6–24. 2*n*=18. Uncommon. Rocky
or sandy, gen saline soil; 1200–1670 m. GB (Modoc, Inyo cos.); to
OR, CO. [*Gilia m.* A. Gray] Mar–Jun

A. monoensis J.M. Porter (p. 1045) MONO LAKE ALICIELLA
Ann; densely glandular-puberulent, glandular-hairy below, odor
skunk-like. **ST:** 5–30 cm, glandular-puberulent; branches gen
spreading from base. **LF:** basal in ± erect cluster, 1–7 cm, glandular-
puberulent, 1-pinnate-lobed, midrib narrow, lobes widely obtuse to
pointed, mucronate; cauline entire, linear. **FL:** calyx 2–3.5 mm, lobe
0.5–1.2 mm; corolla 4–7 mm, glabrous abaxially, tube + throat 3–6.2
mm, white, stout, throat > tube, well-expanded, yellow-spotted, lobes
ovate, white, lavender-streaked abaxially; stamens exserted, pollen
white; mature stigma gen exserted. **FR:** 3–5 mm, 1.5–2 × calyx,
ovoid; tip pointed. **SEED:** 21–36. 2*n*=16. Sandy soils, sagebrush
scrub, pinyon/juniper woodland; 500–2500 m. SNE, DMoj; to OR,
NV. [*A. subacaulis* (Rydb.) J.M. Porter & L.A. Johnson, misappl.;
Gilia s. Rydb., misappl.] *Gilia subacaulis* [*A. subacaulis*], the name
formerly applied to this taxon in CA, applies instead to a narrow
endemic of WY. Sand grains often stuck to glands. Apr–Jul

A. ripleyi (Barneby) J.M. Porter (p. 1045) RIPLEY'S ALICIELLA
Per. **ST:** 10–30 cm, densely glandular-hairy. **LF:** basal petioled,
lower in clusters, 3–6 cm, blades 1–7 cm wide, obovate, holly-like,
coarsely dentate, tips of teeth needle-like, glandular-hairy, veins
raised abaxially; upper cauline < 1 cm, gland-dotted, needle-like. **FL:**
calyx 3.5–6 mm, tube 1.9–3 mm, lobes acuminate, margins membranous ± to tip; corolla 7–10 mm, tube 4–5.7 mm, white, throat white,
gen not distinguishable from tube, lobes > tube, ovate, magenta;
stamens attached near base of tube, incl or longest ± exserted, filaments (at least longest) papillate, pollen white; style ± exserted. **FR:**
3–5 mm, ovoid, ± = calyx. **SEED:** 57–72, deep red-brown. 2*n*=18.
Limestone cliffs; 65–1400 m. n DMtns (Inyo Co.); NV (rare). [*Gilia
r.* Barneby] Apr–Jun ★

A. triodon (Eastw.) Brand (p. 1045) COYOTE GILIA Ann, glandular-puberulent. **ST:** gen 1, branches above, spreading, 5–13 cm,
thread-like. **LF:** basal, 5–20 mm, oblanceolate to obovate, entire,
toothed, or lobed, lobes spreading, short-pointed, length < lf axis
width; upper lvs linear, entire. **FL:** calyx 2–3 mm, lobes short-pointed,
tips thick; corolla 5–7 mm, tube + throat 3–5.8 mm, thread-like,
proximally purple, distally yellow, throat narrower than tube, yellow,
lobes 3-toothed, central tooth gen longest, base yellow, tips white;
stamens, stigma at mouth or ± exserted, pollen white. **FR:** 3–4 mm,
narrowly ovoid. **SEED:** many. 2*n*=18. Open, sandy or rocky areas,
sagebrush scrub, juniper woodland; 1200–1700 m. e DMtns (Clark
Mtn Range, uncommon); to CO, NM. [*Gilia t.* Eastw.] Apr–May ★

ALLOPHYLLUM

Leigh A. Johnson & Alva G. Day

Ann, hairy, ± minutely glandular. **ST:** erect, gen branched. **LF:** alternate, dark green, gen deeply pinnate-lobed, ± palmate-
lobed upward; lobes linear to lanceolate, blunt-tipped, central lobe widest. **INFL:** fls gen in clusters. **FL:** calyx tube narrowly
membranous between lobes, lobes blunt-tipped, translucent below in fr, membrane splitting; corolla radial or bilateral, funnel-
shaped, throat narrow, tapered, lobes narrowly obovate; stamens attached in tube. **FR:** < calyx, spheric; valves gen falling.
SEED: 1–3 per chamber, concave, black or brown, gelatinous when wet, ends rounded. 4 spp.: w N.Am. (Greek: other lf)
[Grant & Grant 1955 Aliso 3:93–110]

1. Widest lvs or lf lobes 2–4 mm wide, linear to narrowly lanceolate; corolla dark blue-purple; st glands
   short-stalked . . . . ***A. gilioides***
   2. Fls 4–8 in ± dense clusters . . . . subsp. ***gilioides***
   2′ Fls 1 or loosely paired or in 3s, not clustered . . . . subsp. ***violaceum***
1′ Widest lvs or lf lobes 3–15 mm wide, lanceolate to elliptic; corolla not dark blue-purple; st glands short- or long-stalked
   3. Corolla ± 2-lipped, lobes 1/2 to = tube + throat; longest stamens well exserted . . . . ***A. glutinosum***
   3′ Corolla not 2-lipped; lobes < 1/3 tube + throat; longest stamens exserted or incl
      4. Lower lvs 3–13-lobed; corolla 8–22 mm, tube red-purple, lobes pink . . . . ***A. divaricatum***
      4′ Lower lvs entire or coarsely toothed; corolla ≤ 11 mm, white to pale blue . . . . ***A. integrifolium***

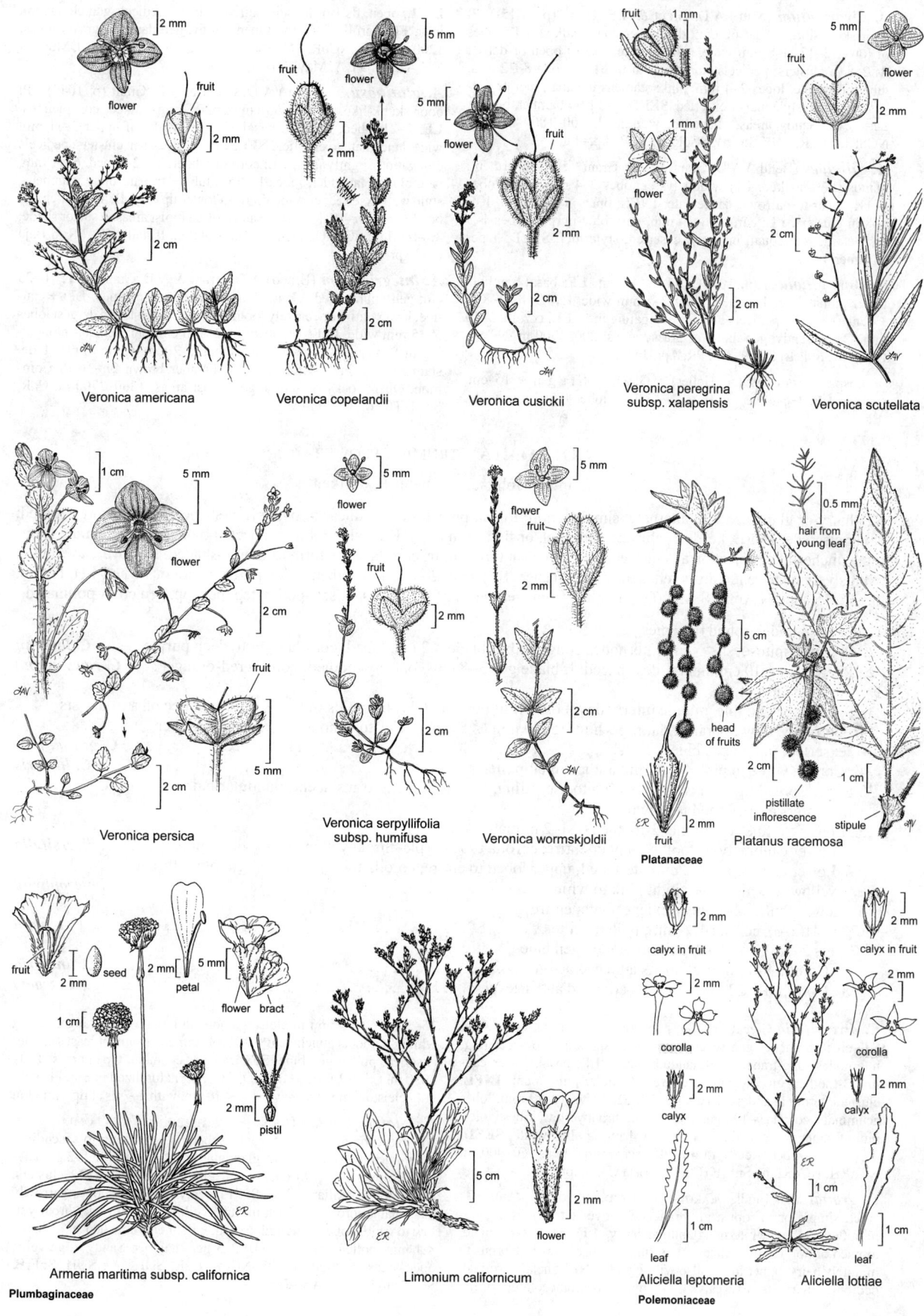

Veronica americana

Veronica copelandii

Veronica cusickii

Veronica peregrina subsp. xalapensis

Veronica scutellata

Veronica persica

Veronica serpyllifolia subsp. humifusa

Veronica wormskjoldii

Platanaceae

Platanus racemosa

Armeria maritima subsp. californica

Plumbaginaceae

Limonium californicum

Aliciella leptomeria

Aliciella lottiae

Polemoniaceae

A. divaricatum (Nutt.) A.D. Grant & V.E. Grant (p. 1045) Pl odor skunk-like. **ST**: < 60 cm; glands gen long-stalked. **LF**: lobes of lower 3–13, 4–8 mm wide, ± lanceolate. **INFL**: open or dense; fls 2–8 in clusters; pedicels elongating in fr. **FL**: corolla 8–22 mm, tube red-purple, lobes 2–4 mm, pink; stamens unequal, incl or longest exserted; style incl or exserted. **SEED**: 1–3 per chamber, brown. 2*n*=16,18. Sandy areas, chaparral, woodland; 300–1800 m. KR, NCoRI, s CaR, SNF, SnFrB, SCoR, TR, PR. Apr–Jun

A. gilioides (Benth.) A.D. Grant & V.E. Grant **ST**: Puberulent; glands short-stalked. **LF**: widest lvs or lobes 2–4 mm wide, lobes 0–11, linear to narrowly lanceolate. **INFL**: open or dense; fls 1 or 2–8 in clusters. **FL**: corolla 6–10 mm, dark blue-purple, lobes 1–3 mm; stamens ± equal, incl to ± exserted; style incl. **SEED**: 1 per chamber, black.

subsp. **gilioides** (p. 1045) **ST**: < 40 cm. **LF**: basal, cauline, lower pinnately 5–11-lobed, lobes 2–4 mm wide. **INFL**: fls 4–8 in ± dense clusters; pedicels gen not elongating in fr. **FL**: corolla 6–10 mm. Open, sandy, gen damp or grassy areas; 200–1900 m. NCoR, CaR, SN, SnFrB, SCoR, TR, PR. Apr–Jun

subsp. **violaceum** (A. Heller) A.G. Day **ST**: gen ± 15 cm, slender. **LF**: ± cauline, entire or 3–7-lobed, lobes 0.5–3 mm wide.

INFL: open; fls 1 or loosely paired or in 3s; pedicels gen elongating in fr. **FL**: corolla 5–8 mm. Open, sandy, gen damp or grassy areas; 1200–2900 m. NCoR, SN, SnFrB, SCoR, TR, PR, n SNE, DMtns; w NV, AZ, Baja CA. May–Jul

A. glutinosum (Benth.) A.D. Grant & V.E. Grant (p. 1045) Pl odor skunk-like. **ST**: 10–60 cm; glands long-stalked, conspicuous. **LF**: 7–21-lobed, lobes of basal 1–2 mm wide, of upper 3–11 mm wide, lanceolate to elliptic. **INFL**: ± open; fls 2–8 in clusters; pedicels elongating slightly in fr. **FL**: corolla bilateral, ± 2-lipped, 6–11 mm, pale blue to lavender, lobes 1/2 to = tube + throat; longest stamens, style well exserted, curving up from lower lip. **SEED**: 1–3 per chamber, brown. 2*n*=18. Rocky or sandy, often moist areas in sun or shade; 30–1600 m. SCoRO, s ChI (Santa Catalina Island), TR, PR; n Baja CA. Apr–Jun

A. integrifolium (Brand) A.D. Grant & V.E. Grant **ST**: 10–25 cm, gen unbranched; glands long-stalked, conspicuous. **LF**: ± cauline, lower entire or coarsely toothed, upper gen 3-lobed, largest lobes 5–15 mm wide. **INFL**: ± open; fls 2–4 in clusters; pedicels elongating in fr. **FL**: corolla 6–11 mm, lobes 1–2 mm, white to pale blue; stamens, style incl. **SEED**: 1–2 per chamber, brown. 2*n*=16,18. Common. Open, rocky or sandy, gen moist areas; 1300–2700 m. CaR, SNH, TR; w NV. Jun–Aug

COLLOMIA TRUMPET FLOWER

Leigh A. Johnson & Dieter H. Wilken

ST: hairy or glandular. **LF**: alternate, simple, entire to gen pinnate-lobed, linear to ovate [fan-shaped]; basal short-petioled; cauline sessile. **INFL**: heads or clusters, terminal, or fls 1–3 in axils. **FL**: calyx lobes connected by narrow membrane forming a pitcher-like projection at sinus, enlarging, not rupturing in fr; corolla salverform to funnel-shaped. **FR**: ovate to elliptic, explosively dehiscent, valves reflexed on dehiscence. **SEED**: 1(2–3) per chamber, oblong, gen gelatinous when wet, brown. 2*n*=16. 15 spp.: N.Am, s S.Am. (Greek: glue, from wet seed surface) Ann spp. self-pollinated; per spp. gen cross-pollinated.

1. Per from underground rhizomes
  2. Pl gen cespitose; st < 5 cm, internodes congested; lf blade < 2 cm, lobed; corolla light to deep purple . . . . . **C. larsenii**
  2′ Pl erect; st > 10 cm, internodes spaced; lf blade gen 3–8 cm, coarsely toothed; corolla red-orange . . . . . **C. rawsoniana**
1′ Ann from taproot
  3. St gen erect, simple, middle internodes short-white-nonglandular-hairy; fls in heads terminal, occ on axillary sts
    4. Corolla yellow to orange, fading white, 15–30 mm (< 5 mm in cleistogamous fls); at least 1 anther
      exserted; pollen gen blue . . . . **C. grandiflora**
    4′ Corolla white to pink, ≤ 15 mm; anthers incl; pollen gen white . . . . **C. linearis**
  3′ St gen spreading, branches spreading to ascending, internodes long-translucent-glandular-hairy; fls 1 or
    in clusters axillary and terminal
    5. Lower cauline lvs elliptic, 3-toothed or 1–2 pinnate-lobed
      6. Lower cauline lvs 3-toothed, upper entire; corolla tube purple, throat yellow, lobes violet-pink . . . . . . **C. diversifolia**
      6′ Lower cauline lvs 1–2 pinnate-lobed, upper lobed to entire; corolla tube yellow to light pink, throat
        yellow to white, lobes light pink to white . . . . **C. heterophylla**
    5′ Lower cauline lvs linear to lanceolate, entire
      7. Fls 1(2–3), corolla 4–6 mm; pollen white . . . . **C. tenella**
      7′ Fls 2–3(5), corolla 8–22 mm; pollen gen blue
        8. Corolla 8–14 mm, stamens attached at 1 level . . . . **C. tinctoria**
        8′ Corolla 12–22 mm, stamens attached at > 1 level . . . . **C. tracyi**

C. diversifolia Greene (p. 1045) SERPENTINE COLLOMIA Ann. **ST**: erect, branches gen several, subequal, spreading to ascending; internodes long-translucent-glandular-hairy. **LF**: basal, lower cauline 3-toothed; upper cauline entire, glandular-puberulent. **INFL**: clusters, terminal and axillary; fls 7–15. **FL**: calyx 10–12 mm, lobes acuminate; corolla 9–12 mm, tube purple, throat yellow, lobes violet-pink; filaments attached at > 1 level, < 1 mm, pollen white. **SEED**: 2–3 per chamber. Rocky to gravelly serpentine areas; 60–900 m. NCoRH, NCoRI, ne SnFrB (Contra Costa Co.). Apr–Jul ★

C. grandiflora Lindl. LARGE-FLOWERED COLLOMIA Ann. **ST**: erect, simple or in robust pls branched above; internodes at mid-pl reflexed-short-white-nonglandular-hairy. **LF**: basal lanceolate, toothed or not; cauline lanceolate to linear, entire, lower glabrous to minutely hairy, upper in age also with glands. **INFL**: head, terminal, occ on axillary sts. **FL**: calyx 7–10 mm, lobes lanceolate; corolla

15–30 mm (< 5 mm in cleistogamous fls), yellow to orange, fading white; filaments attached at > 1 level, some > 1 mm, at least 1 anther exserted; pollen gen blue. **SEED**: 1 per chamber. Open areas; 120–3050 m. CA-FP, GB; to BC, ID, CO, AZ; naturalized in Eur. Pls with only cleistogamous fls have st 1, < 10 cm, with 3–7 fls at tip. Apr–Jun

C. heterophylla Hook. VARIABLE-LEAF COLLOMIA Ann. **ST**: erect, branches gen several, subequal, spreading to ascending; internodes long-translucent-glandular-hairy. **LF**: lower 1–2 pinnate-lobed; upper lobed to entire, gen glandular-hairy. **INFL**: clusters, terminal and axillary; fls 7–25. **FL**: calyx 5–8 mm, lobes narrowly lanceolate; corolla 10–14 mm, tube yellow to light pink, throat yellow to white, lobes light pink to white; filaments attached at > 1 level, < 1 mm, pollen white. **SEED**: 2–3 per chamber. Sandy to gravelly, open areas; < 2000 m. KR, NCoR, CaR, SNF, n&s SNH, SnFrB, SCoR; to BC, ID. Apr–Jun

C. larsenii (A. Gray) Payson (p. 1045) TALUS COLLOMIA Per, gen cespitose, from slender rhizomes. **ST**: gen branched, < 5 cm; internodes congested, glandular-hairy. **LF**: blade < 2 cm, 1–2-pinnate- or -palmate-lobed from elongate, clasping petiole; lobes linear-oblong, narrowed at base, glandular-hairy. **INFL**: clusters, terminal; fls 6–9. **FL**: calyx 5–9 mm; corolla 10–15 mm, light to deep purple; filaments attached at 1 level, > 3 mm, pollen blue. **SEED**: 1 per chamber. Volcanic talus; 2225–3500 m. CaRH (Lassen, Magee peaks, Little Mount Hoffman); to WA. Jun–Sep ★

C. linearis Nutt. (p. 1045) NARROW-LEAF MOUNTAIN TRUMPET Ann. **ST**: erect, simple or in robust pls branched above, or occ branches many, internodes congested, pls then gen as wide as tall; internodes at mid-pl reflexed-short-white-nonglandular-hairy. **LF**: basal lanceolate, toothed or not; cauline lanceolate to linear, entire, lower ± glandular, upper gen glabrous. **INFL**: bracted head, terminal, occ on axillary sts; fls 7–20. **FL**: calyx 4–7 mm; corolla 8–15 mm, white to pink; filaments attached at > 1 level, < 1 mm, anthers incl, pollen gen white. **SEED**: 1 per chamber. Open areas; 600–3650 m. KR, CaR, SN, SnBr, GB; to AK, e N.Am, AZ; naturalized elsewhere. May–Aug

C. rawsoniana Greene (p. 1045) RAWSON'S FLAMING-TRUMPET Per, erect, from slender rhizomes. **ST**: > 10 cm, internodes spaced, glandular-hairy. **LF**: cauline, gen 3–8 cm, ovate to elliptic, coarsely toothed, glandular-hairy. **INFL**: terminal heads or clusters; fls 3–7. **FL**: corolla 8–12 mm; corolla 25–40 mm, red-orange; filaments attached at 1 level, > 1 cm, stamens, style ± exserted; pollen gen blue. **SEED**: 1 per chamber. Shaded areas near streams in woodland; 1000–2200 m. c SN (Mariposa, Madera cos.). Jun–Sep ★

C. tenella A. Gray SLENDER COLLOMIA Ann. **ST**: erect, branches many, from near base to tip, spreading; internodes long-translucent-glandular-hairy. **LF**: linear to lanceolate, entire, glandular. **INFL**: axillary and terminal clusters; fls 1(2–3). **FL**: calyx 3–4 mm, lobes short-triangular; corolla 4–6 mm, tube white, lobes white to lavender; filaments attached at 1 level, ± 1 mm, anthers, stigma exserted; pollen white. **SEED**: 1 per chamber. Dry open areas, often in sagebrush; 1700–2600 m. Wrn; to BC, WY, UT. May–Jul ★

C. tinctoria Kellogg (p. 1045) STAINING COLLOMIA Ann. **ST**: 2–8 cm, branches gen several, subequal, spreading to ascending; internodes long-translucent-glandular-hairy. **LF**: lower cauline linear to lanceolate, entire, glandular. **INFL**: gen axillary clusters; fls gen 2–3. **FL**: calyx 5–7 mm, lobes long-tapered, awned; corolla 8–14 mm, slender, tube maroon to violet, lobes pink; filaments attached at 1 level, some > 1 mm, 1–2 anthers, stigma exserted; pollen blue. **SEED**: 1 per chamber. Gravelly to rocky, open areas; 600–3000 m. KR, NCoRH, CaRH, n&c SNH, SCoRO (San Rafael Mtns), WTR, GB; to WA, ID, NV. Jun–Sep

C. tracyi H. Mason (p. 1045) TRACY'S COLLOMIA Ann. **ST**: 2–8 cm, erect, branches gen several, subequal, spreading to ascending; internodes long-translucent-glandular-hairy. **LF**: lower cauline linear to lanceolate, entire, glandular. **INFL**: gen axillary clusters; fls gen 2–3. **FL**: calyx 6–8 mm; corolla 12–22 mm, gen white to lavender; filaments attached at > 1 level, some > 1 mm, upper anthers gen in throat, less than exserted; stigmas incl, below anthers, or exserted; pollen blue. **SEED**: 1 per chamber. Rocky, gravelly, or sandy areas; 30–2100 m. KR, n NCoRH. Similar to but not intergrading with *C. tinctoria*. Jun–Sep ★

ERIASTRUM

Sarah De Groot, David Gowen & Robert Patterson

Glabrous to woolly, glandular or not. **ST**: gen erect. **LF**: cauline, alternate, gen pinnate-lobed or simple; lobes gen linear or lanceolate. **INFL**: head-like, bracted, gen densely woolly; bracts lf-like; fls sessile. **FL**: calyx lobes unequal, gen woolly; corolla radial or bilateral, funnel-shaped to salverform; stamens equal or not, anthers gen sagittate, pollen white or blue; style incl or exserted. **SEED**: 1–several per chamber. 16 spp.: w N.Am. (Greek: woolly star) [Harrison 1972 Brigham Young Univ Sci Bull, Biol Ser 16:1–26] Apparently much undescribed variation; genus being revised. Key to spp. by David Gowen, Sarah De Groot.

1. Per .... ***E. densifolium***
  2. Corolla 20–37 mm
    3. Corolla 20–25 mm, lobes 1/3–1/2 corolla; lvs glabrous to subglabrous; coastal dunes, San Luis Obispo,
      Santa Barbara cos. .... subsp. ***densifolium***
    3′ Corolla 25–37 mm, lobes ± 1/3 corolla; lvs woolly; Santa Ana River drainage, sw San Bernardino Co
      .... subsp. ***sanctorum***
  2′ Corolla < 20 mm
    4. Lvs strongly recurved, blade wider at base than tip, lobes spine-tipped — SNE, DMoj. .... subsp. ***mohavense***
    4′ Lvs not recurved, blade gen equally wide at base, tip, gen not spine-tipped
      5. Bracts gen 5–9 lobed; infl gen terminal, 10–20 fld; lvs subglabrous to ± canescent .... subsp. ***austromontanum***
      5′ Bracts gen (0)3–5(7)-lobed; infls gen terminal and axillary, 1–10(15)-fld; lvs ± canescent to woolly
        .... subsp. ***elongatum***
1′ Ann (occ short-lived per)
  6. Stamens attached at or just below corolla sinus, corolla 11–24 mm .... ***E. pluriflorum***
    7. Pl ± open; lvs thread-like; corolla gen salverform, tube + throat gen > 2 × lobes; chaparral, grassland,
      savanna, woodland, pine forest .... subsp. ***pluriflorum***
    7′ Pl gen dense, occ cespitose; lvs linear; corolla narrowly funnel-shaped, tube + throat gen ≤ 2 × lobes;
      open sandy flats, deserts .... subsp. ***sherman-hoytiae***
  6′ Stamens attached well below corolla sinus, if higher, corolla < 15 mm
    8. Stamens exserted > 1/2 length of corolla lobes
      9. Corolla lobes bright yellow .... ***E. luteum***
      9′ Corolla lobes white, lavender, or variously blue
        10. Corolla gen bilateral; stamens gen unequal .... ***E. eremicum*** subsp. *eremicum*
        10′ Corolla ± radial (or bilateral from unequal sinuses in *Eriastrum sapphirinum*), stamens gen equal
          11. Stamens exceeded by tips of corolla lobes; corolla lobes < 4 mm. .... ***E. filifolium***
          11′ Stamens equalling to exceeding tips of corolla lobes (almost equalling to exceeding in *Eriastrum*
            sapphirinum subsp. *dasyanthum*); corolla lobes ≥ 4 mm
            12. Corolla tube > 3 × throat, tube + throat gen ≥ 1.5 × lobes; pls ± not glandular; CCo, n SCoR
              (Monterey, San Benito cos.) .... ***E. virgatum***

12′ Corolla tube < 3 × throat, tube + throat gen ≤ 1.5 × lobes; pls often glandular; s SNF, SW
(exc ChI), w D.. ***E. sapphirinum***
   13. Fls 3+ per cluster; corolla lobes dark to royal blue, pale blue, or white; calyx, bracts occ ±
      glandular, woolly .... subsp. ***dasyanthum***
   13′ Fls 1–3 per cluster; corolla lobes bright blue to lavender; calyx, bracts glandular, ± woolly
      to subglabrous.. subsp. ***sapphirinum***
8′ Stamens incl or exserted < 1/2 length of corolla lobes
  14. Anthers incl
    15. Upper lvs, bracts gen pinnate-3–7-lobed; infl many-fld, densely bracted, densely woolly .... ***E. abramsii***
    15′ Upper lvs, bracts pinnate-1–4-lobed or palmate-5-lobed; infl few-fld, few-bracted, ± woolly
      16. Corolla 5–7 mm, white; seeds (1)2–4 per chamber .... ***E. hooveri***
      16′ Corolla ≥ 7 mm, white, pink, lavender, or light blue; seeds 1(2) per chamber
        17. Stamens 1.5–2 mm, filaments ± 2 × anther; corolla lobes narrowly elliptic, ± 3 × longer than wide
        .... ***E. brandegeeae***
        17′ Stamens 0.5–1.5 mm, filaments ± = anther; corolla lobes widely elliptic, ± 2 × longer than wide ..... ***E. tracyi***
  14′ Anthers exserted
    18. Stamens unequal; corolla lobes gen > 1.25 mm wide
      19. Corolla gen 6–9 mm; st wiry; pls erect or spreading, gen subglabrous; lvs gen 1–3-lobed near base;
      SNE, D .... ***E. diffusum***
      19′ Corolla 9–14 mm; st gen not wiry; pls erect, woolly-hairy to occ subglabrous; lvs entire or
      1–7-lobed near base; SNH, GB, n DMtns .... ***E. wilcoxii***
    18′ Stamens equal; corolla lobes gen ≤ 1.25 mm wide
      20. Pl minutely glandular-hairy .... ***E. sparsiflorum***
      20′ Pl woolly-hairy to woolly, gen ± not glandular
        21. Corolla lobes pale yellow to white, without spots; sand dunes in creosote-bush scrub .... ***E. harwoodii***
        21′ Corolla lobes pale blue to pink, base gen with maroon spot that is occ faint; pinyon/juniper
        woodland, sagebrush scrub, pine/oak woodland .... ***E. signatum***

E. abramsii (Elmer) H. Mason Ann. **ST**: 5–15 cm. **LF**: 10–45 mm; gen pinnate-3–7-lobed, lobes thread-like, woolly. **INFL**: many-fld, densely bracted, densely woolly. **FL**: corolla 5–8 mm, salverform, tube 3–4 mm, white to ± yellow, throat 1–1.25 mm, lobes 2–3 mm, white or pale blue; stamens attached in upper tube or lower throat, incl. **SEED**: 1 per chamber. Dry gentle slopes, flats, gen chaparral; < 1200 m. NCoRI, SnFrB, SCoRI. Apr–Jun

E. brandegeeae H. Mason (p. 1045) BRANDEGEE'S ERIASTRUM Ann. **ST**: 5–30 cm. **LF**: 10–25 mm; pinnate-1–3-lobed or palmate-5-lobed, lobes thread-like, woolly. **FL**: corolla 7–11 mm, salverform or ± bilateral, white, pale blue, or lavender, tube 4–5 mm, throat 1.5–2 mm, lobes 3–4 mm; stamens attached in upper tube or lower throat, 1.5–2 mm, equal, incl, filaments ± 2 × anthers. **SEED**: 1(2) per chamber. Open flats of volcanic soils, shales; 400–1000 m. n&c NCoRI. [*Eriastrum brandegeae*, orth. var.] Threatened by development, grazing, vehicles. Probably a sp. complex. May–Aug ★

E. densifolium (Benth.) H. Mason Per from woody caudex. **ST**: erect or spreading, ± glabrous to woolly. **LF**: 10–50 mm; lobes 0–15, glabrous to woolly. **FL**: corolla 16–37 mm, funnel-shaped, blue, lavender, or white, lobes 5–12 mm; stamens attached in sinus or upper throat, equal, exserted. 2*n*=14.

subsp. ***austromontanum*** (T.T. Craig) H. Mason **LF**: lobes 2–15, subglabrous to ± canescent. **INFL**: gen terminal; bracts gen 5–9-lobed; fls 10–20. **FL**: corolla gen 16–18 mm, narrowly funnel-shaped, blue, tube puberulent, lobes 5–7 mm; stamens 5–6 mm. Dry, open slopes, open woodland; 1300–2900 m. s SN, SCo, TR, PR. Jun–Aug

subsp. ***densifolium*** **LF**: gen entire or 2–4-lobed, glabrous to subglabrous. **INFL**: heads terminal or in compact clusters. **FL**: corolla 20–25 mm, widely funnel-shaped, bright blue, lobes ≤ 11 mm; stamens ± 8 mm. **SEED**: ± 6 per chamber. Coastal dunes; gen < 15 m. s CCo. Jun–Jul

subsp. ***elongatum*** (Benth.) H. Mason (p. 1045) **LF**: lobes gen 2–8, 5 mm, ± canescent to woolly. **INFL**: terminal, axillary; bracts gen (0)3–5(7)-lobed; fls gen 1–10(15). **FL**: corolla to 20 mm, bright blue or lavender, veins occ darker, tube + throat 8–13 mm, white or yellow, lobes 7 mm; stamens 5–7 mm. Dry places; < 2200 m. s SN, SCoR, SW, w edge DMoj; Baja CA. Jun–Sep

subsp. ***mohavense*** (T.T. Craig) H. Mason **LF**: axis 1–4 mm wide, strongly recurved; blade base wider than tip; lobes 2–8, spine-tipped, densely woolly. **INFL**: heads several per branch. **FL**: corolla 16 mm, light blue to lavender-blue, throat 2 mm, tube 9 mm, puberulent, lobes gen 5 mm; stamens 4 mm. Sandy slopes, flats; 600–2800 m. SNE, w DMoj. Jun–Oct

subsp. ***sanctorum*** (Milliken) H. Mason (p. 1045) SANTA ANA RIVER WOOLLYSTAR **LF**: lobes gen 2–6, woolly, occ densely so. **INFL**: heads terminal or upper branches ± raceme-like. **FL**: corolla 25–37 mm, lavender-blue, veins occ darker, tube + throat to 25 mm, yellow, ± puberulent, lobes 8–12 mm; stamens 6.5–8 mm. **SEED**: ± 5 per chamber. Washes, floodplains, dry riverbeds; < 500 m. e SCo (Santa Ana River drainage, sw San Bernardino Co.). Threatened by habitat alteration. May–Sep ★

E. diffusum (A. Gray) H. Mason Ann. **ST**: erect or spreading, 3–20 cm, thin, wiry, gen subglabrous, often brown. **LF**: 10–25 mm; lobes gen 1–3 near base, ± glabrous to woolly. **FL**: corolla gen 6–9 mm, gen funnel-shaped, tube 3–4(5) mm, white, yellow, or pale blue, throat ± 1 mm, lobes 2–3(4) mm, gen > 1 mm wide, light blue or cream; stamens attached in upper throat, 2–3 mm, unequal, exserted. 2*n*=14. Open areas, sandy soil; < 2000 m. SNE, D; to CO, TX; n Mex. Pl not always open or diffuse. Mar–Jun

E. eremicum (Jeps.) H. Mason subsp. ***eremicum*** (p. 1045) Ann. **ST**: 5–30 cm. **LF**: 5–55 mm; lobes 2–7, ± glabrous to woolly. **FL**: corolla 11–18 mm, gen bilateral, tube 4–9 mm, blue or yellow, throat 3–3.5 mm, gen yellow, lobes 4–8 mm, light to dark blue, sinuses unequal; stamens attached in upper tube or lower throat, 2–9 mm, gen unequal, bent toward lower corolla lip, exserted. 2*n*=14. Open areas in sandy soils; < 1800 m. SNE, D; to UT, AZ. Other subsp. in AZ. Apr–Jun

E. filifolium (Nutt.) Wooton & Standl. Ann. **ST**: 4–40 cm. **LF**: 20–40 mm, ± glabrous to woolly, gen red-brown in age, entire or 3-lobed at base, lobes thread-like. **FL**: corolla 7–9 mm, salverform to narrowly funnel-shaped, tube 4 mm, yellow, throat 1–1.5 mm, yellow, lobes 3 mm, blue; stamens attached in lower throat, 3.5 mm, equal, exserted. **SEED**: 3–5 per chamber. Dry slopes; < 1700 m. SCoRO (nw Santa Barbara Co.), SW; Baja CA. Apr–Jul

E. harwoodii (T.T. Craig) D. Gowen Ann. **ST**: to 20 cm. **LF**: 10–35 mm, thread-like, entire or 3-lobed near base, yellow-green,

densely woolly. **FL:** corolla 6–7.5 mm, narrowly funnel-shaped, pale yellow or cream to white, tube 4 mm, throat yellow, lobes elliptic, ± 2 mm; stamens attached ± 0.2 mm below sinus, equal. **SEED:** gen 2 per chamber. Sand dunes in creosote-bush scrub; < 1000 m. DMoj, n DSon. [*E. sparsiflorum* (Eastw.) H. Mason subsp. *h.* (T.T. Craig) H.K. Harrison] Mar–May ★

E. hooveri (Jeps.) H. Mason (p. 1045) HOOVER'S ERIASTRUM Ann. **ST:** 3–15 cm. **LF:** 5–25 mm, linear, entire or 3-lobed at base, ± glabrous to woolly. **INFL:** few-fld, -bracted. **FL:** corolla 5–7 mm, ± salverform, white, tube 2–2.5 mm, throat 1–1.5 mm, lobes 2–3 mm; stamens attached 1 mm below sinus, 1–1.25 mm, incl. **SEED:** (1)2–4 per chamber. Alkaline flats, above dry streambeds; < 900 m. s SNF (ne Kern Co.), SnJV, WTR (ne Santa Barbara Co.). Pls with pale blue to white fls in w DMoj attributed to *E. hooveri* probably an undescribed taxon; study needed. Mar–Jul ★

E. luteum (Benth.) H. Mason YELLOW-FLOWERED ERIASTRUM Ann. **ST:** 2–25 cm. **LF:** 20–40 mm, thread-like, entire or 2-lobed at base, woolly. **FL:** corolla 7.5–9 mm, funnel-shaped or ± bilateral, bright yellow, tube 3 mm, throat 1.5–2 mm, lobes 3–4 mm; stamens attached at base of throat, 5–5.5 mm, equal, exserted. **SEED:** 1–2 per chamber. Drying slopes; < 1000 m. SCoR (Monterey, San Luis Obispo cos.). May–Jun ★

E. pluriflorum (A. Heller) H. Mason Ann, gen not glandular. **ST:** 2–32 cm. **LF:** 20–50 mm, thread-like, entire or 2–10-lobed, glabrous to woolly. **FL:** corolla 11–24 mm, salverform to narrowly funnel-shaped, tube gen 8–14 mm, white, yellow, or blue, throat 2.5–3 mm, white, yellow, or blue, lobes 4–7 mm, pale to dark or bright blue, or lavender; stamens attached at or just below sinuses, exserted. 2*n*=14.

subsp. *pluriflorum* (p. 1045) MANY-FLOWERED ERIASTRUM Ann, ± open, occ resprouting after summer rain. **ST:** 2–32 cm. **LF:** 20–50 mm, thread-like, glabrous to woolly. **INFL:** gen prominently woolly. **FL:** corolla 16–24 mm, gen salverform, tube 8–14 mm, white, yellow, or blue, throat ± 3 mm, yellow or blue, lobes 5–6(7) mm, to 3 mm wide, dark or bright blue, or lavender, veins occ darker; stamens attached at or just below sinuses, 3–5 mm, gen equal (occ 1 short), exserted. **SEED:** gen 1–3 per chamber. Chaparral, grassland, woodland, pine forest; < 2000 m. c&s SNF, s SNH, SnJV, SnFrB, e SCoR, WTR. Variable; needs study. May–Jul

subsp. *sherman-hoytiae* (T.T. Craig) H. Mason Ann, occ cespitose, gen dense, occ resprouting after summer rain. **ST:** 2–15 cm. **LF:** 20–40 mm, linear, woolly. **FL:** corolla (11)12–20 mm, narrowly funnel-shaped, tube 8–10 mm, white, yellow, or blue, throat 2.5–3 mm, yellow or white, lobes 4–6 mm, to 4 mm wide, bright to pale blue; stamens attached just below sinuses, 2–4 mm, gen equal (occ 1 short), exserted. Open sandy flats, deserts; < 2000 m. w DMoj. May–Jun

E. sapphirinum (Eastw.) H. Mason Ann. **ST:** 5–40 cm; hairs minute, often glandular. **LF:** 5–55 mm, thread-like, entire or 2-lobed near base, ± glabrous to woolly, often glandular, lobes linear. **FL:** calyx often glandular; corolla 10–15(17.5) mm, funnel-shaped, ± radial to bilateral from unequal sinuses, tube 2–4.5 mm, blue, white, or yellow, throat 2–4.5 mm, yellow and/or white, lobes bright to pale blue to lavender, occ with yellow or white blotch or dark purple streaks at base; stamens attached in upper tube or lower throat, to 10 mm, equal, exserted, gen equalling or exceeding corolla lobe tips; anthers yellow to white.

subsp. *dasyanthum* (Brand) H. Mason **INFL:** bracts, calyx occ ± glandular, woolly, fls 3+ per cluster. **FL:** corolla 10–15 mm, funnel-shaped, ± radial, tube 3–4.5 mm, blue or yellow, throat 2–3 mm, yellow, white, or both, lobes 5–7.5 mm, dark to royal blue, pale blue, or white; stamens exserted, ± equalling to exceeding corolla lobe tips. Woodland, chaparral, sandy washes; 700–2700 m. s SNF, SW (exc ChI), w D; n Baja CA. May–Aug

subsp. *sapphirinum* (p. 1045) **INFL:** bracts, calyx glandular, ± woolly to subglabrous, fls 1–3 per cluster. **FL:** corolla 10–15(17.5) mm, funnel-shaped, bilateral from unequal sinuses, tube 2–3 mm, white or yellow, throat 3–4.5 mm, yellow, lobes 5–7.5(10) mm, bright blue to lavender, often with dark streaks; stamens attached in upper tube or lower throat, equal, exserted, gen exceeding corolla lobe tips. 2*n*=14. Woodland, chaparral; 700–2700 m. SW (exc ChI), w D; n Baja CA. [*E. s.* subsp. *ambiguum* (M.E. Jones) H. Mason; *E. s.* subsp. *gymnocephalum* (Brand) H. Mason] May–Aug

E. signatum D. Gowen Ann. **ST:** to 35 cm. **LF:** 5–30 mm, thread-like, entire or 2-lobed near base, woolly. **FL:** corolla 7–11 mm, ± salverform, radial, tube 3.5–6 mm, throat ± 1 mm, yellow to white, lobes 2–4 mm, pale blue to pink, gen with maroon spot at base (occ faint); stamens 1.5–2 mm, equal, ± exserted. **SEED:** 2 per chamber. Pinyon/juniper woodland, sagebrush scrub, pine/oak woodland; 1000–2500 m. CaRH (Shasta Valley), s SNF, SNH, TR, SNE (exc W&I), MP; OR, NV. Pls in SCoRI with 1 seed per chamber probably represent an undescribed taxon; study needed. May–Aug

E. sparsiflorum (Eastw.) H. Mason Ann, minutely glandular-hairy or tufted-woolly-hairy to glabrous. **ST:** 10–30(35) cm. **LF:** (5)10–30(35) mm, thread-like, entire or 2-lobed near base, woolly. **FL:** corolla 7–10 mm, salverform or narrowly funnel-shaped, tube 3–6 mm, white or yellow, throat 1–1.5 mm, white or yellow, lobes 2–3 mm, ≤ 1.2 mm wide, white, cream, or pale blue-lavender; stamens attached in upper tube or lower throat, 2–2.5 mm, equal, ± exserted. **SEED:** gen 2 per chamber. Open areas of granitic sand, desert slopes, pinyon/juniper woodland, yellow-pine forest, sagebrush scrub; < 2000 m. SNH, s MP, SNE (exc W&I); NV. May–Jul

E. tracyi H. Mason TRACY'S ERIASTRUM Ann. **ST:** to 22 cm, ± tufted-woolly-hairy. **LF:** 15–25 mm, awl-shaped, entire or 2–4-lobed near base, awn-tipped. **FL:** corolla 9–10 mm, ± salverform, radial, tube 5.5 mm, white to purple, throat ± 1 mm, white to yellow, lobes 2.5 mm, widely elliptic, ± 2 × longer than wide, pale blue, pink, or white; stamens 0.5–1.5 mm, equal, filaments ± = anther. **SEED:** 1 per chamber. Open areas on shale or alluvium, open woodland, chaparral; 400–1000 m. KR, n NCoRI, s SNF, SnFrB, MP (ne Shasta Co.). Threatened by habitat loss due to agriculture and development. May–Aug ★

E. virgatum (Benth.) H. Mason VIRGATE ERIASTRUM Ann. **ST:** 3–40 cm. **LF:** 20–50 mm, thread-like, entire or 2-lobed near base, woolly. **FL:** corolla gen 15–22 mm, funnel-shaped, tube 8–12 mm, yellow, throat ± 2 mm, yellow, lobes 7–10 mm, bright blue; stamens attached in base of throat, 6–11 mm, gen equal, gen equalling corolla lobe tips. **SEED:** 1(2) per chamber. Sandy soils; < 600 m. CCo, n SCoR (Monterey, San Benito cos.). May–Jul ★

E. wilcoxii (A. Nelson) H. Mason Ann, erect, gen robust, woolly-hairy to occ subglabrous. **ST:** 3–30 cm. **LF:** 15–30 mm, thread-like, entire or 1–7-lobed near base, densely woolly. **FL:** corolla 9–14 mm, gen funnel-shaped, tube 5–8 mm, yellow or white, throat ± 2 mm, yellow or white, lobes 4–5 mm, blue; stamens attached in upper tube or lower throat, 1.5–4.5 mm, unequal, exserted. **SEED:** gen 3 per chamber. 2*n*=14. Desert flats, slopes; gen < 2900 m. SNH, GB, n DMtns; to ID, WY. Pls in SCoRO with corollas light blue with a dark spot at lobe bases probably an undescribed taxon; study needed. Jun–Aug

GILIA

J. Mark Porter

Ann. **ST:** decumbent to erect, glabrous, hairy, glandular, or tufted-woolly-hairy. **LF:** simple, 1–3-pinnate-lobed or -dissected, gen alternate, margins entire, toothed, or lobed, tips acute, acuminate, or mucronate; basal gen in rosette; cauline gen reduced. **INFL:** fls 1–many in bract axils. **FL:** calyx membranous between lobes, membranes splitting or expanding in fr; corolla > calyx, lobes gen ovate, acute or acuminate. **FR:** spheric to ovoid; chambers 3; valves separating from top, to base and detaching or not to base and staying attached to receptacle. **SEED:** 3–many, yellow to brown, gelatinous when wet. ± 40 spp.:

w N.Am, S.Am. (Filippo L. Gilii, Italian naturalist, 1756–1821) Stamens, styles said to be exserted protrude beyond fused part of corolla, that is, beyond corolla throat. Other taxa in TJM (1993) moved to *Aliciella, Lathrocasis, Linanthus, Navarretia, Saltugilia. G. mexicana* A.D. Grant & V.E. Grant recently found in San Diego Co.

1. Fls gen in heads or clusters of > 8
  2. Heads hemispheric; corolla throat > tube; stamens exceeded by to reaching corolla lobes
    . . . . **G. achilleifolia** subsp. **achilleifolia**
  2′ Heads spheric; corolla throat ≤ tube; stamens reaching or exceeding corolla lobes . . . . **G. capitata**
   3. Corolla 5–8 mm (7–11 in subsp. tomentosa), lobes linear or narrowly oblong; calyx lobes acute, erect or ± recurved
    4. Base of heads tomentose
     5. Corolla pale blue-violet to white; heads 12–25 mm wide; n&c SN . . . . subsp. **mediomontana**
     5′ Corolla bright blue-violet; heads 20–35 mm wide; NCo, se NCoRI, n CCo . . . . subsp. **tomentosa**
    4′ Base of heads glabrous, glandular, or sparsely tufted-woolly-hairy, but not tomentose
     6. Calyx membranes white; corolla lobes < 1 mm wide; seeds gen 1–6. subsp. **capitata**
     6′ Calyx membranes blue-violet; corolla lobes 1–2 mm wide; seeds (3)6–25 . . . . subsp. **pacifica**
   3′ Corolla 7–13 mm, lobes oblong; calyx lobes acuminate, recurved
    7. Base of heads glabrous, sparsely tufted-woolly-hairy, not tomentose; pedicels 1–2 mm; esp s CA
    . . . . subsp. **abrotanifolia**
    7′ Base of heads densely tomentose; pedicels 0; esp n&c CA
     8. Corolla bright blue-violet; pl with skunk-like odor. . . . . subsp. **chamissonis**
     8′ Corolla pale blue-violet; pl lacking skunk-like odor
      9. Infl 10–20 mm wide; corolla lobes ± 2 mm wide. . . . . subsp. **pedemontana**
      9′ Infl 20–30 mm wide; corolla lobes ± 3 mm wide . . . . subsp. **staminea**
1′ Fls not in heads, if clustered at ends of branches then gen ≤ 8
  10. Basal lvs, lower st with white, sharply bent, acute-tipped hairs . . . . **G. stellata**
  10′ Basal lvs, lower st glabrous or hairy, but not with white, sharply bent, acute-tipped hairs
   11. Cauline lvs (particularly mid- to upper lvs) clasping or expanded at base
    12. Corolla 9–35 mm, 2–7 × calyx
     13. St tufted-woolly-hairy below middle
      14. Corolla 10–20 mm, throat widely tapered (widely V-shaped); fr ± = calyx . . . . . . ²**G. brecciarum** subsp. **neglecta**
      14′ Corolla 18–35 mm, throat narrowly tapered; fr > calyx
       15. Corolla tube stout, throat > 7 mm wide, corolla gen 3 × lobes . . . . ²**G. latiflora** subsp. **davyi**
       15′ Corolla tube slender, throat < 7 mm wide, corolla 4–6 × lobes. . . . . **G. latiflora** subsp. **elongata**
     13′ St glabrous, glaucous below middle
      16. Pedicels elongate in fr but not in fl; corolla purple from base to mid-throat or higher
      . . . . ²**G. latiflora** subsp. **davyi**
      16′ Pedicels elongate in fl; corolla purple from base to base of throat
       17. Calyx 2–4 mm; corolla 9–16 mm . . . . **G. latiflora** subsp. **cuyamensis**
       17′ Calyx 4–5 mm; corolla 15–22 mm. . . . . **G. latiflora** subsp. **latiflora**
    12′ Corolla 5–12 mm, 1–2 × calyx
     18. St below middle glabrous, glaucous; upper lvs gen not deeply toothed
      19. Stamens, style reaching corolla lobe middle; pollen blue . . . . **G. diegensis**
      19′ Stamens, style reaching corolla lobe base; pollen white . . . . **G. sinuata**
     18′ St below middle green, not glaucous, ± to densely tufted-woolly-hairy; upper lvs gen deeply toothed
      20. Basal lvs strap-shaped, 1-pinnate-lobed, lobes < lf axis; corolla throat yellow, midveins occ purple
      . . . . **G. modocensis**
      20′ Basal lvs not strap-shaped, 2-pinnate-lobed, lobes > lf axis; corolla throat purple or distally yellow,
       purple-veined
       21. Branches decumbent; throat gen purple. . . . . ²**G. jacens**
       21′ Branches spreading to ± erect; throat yellow with purple veins . . . . ²**G. brecciarum** subsp. **brecciarum**
   11′ Cauline lvs not clasping or expanded at base
    22. Calyx glabrous or hairy, nonglandular
     23. Stamens reaching or exceeding corolla lobes
      24. Basal lvs densely tufted-woolly-hairy; calyx membranes gen dull purple-spotted (colorless), flat;
       SNH, w MP, nw SNE (Mono Co.) . . . . **G. salticola**
      24′ Basal lvs glabrous or ± tufted-woolly-hairy; calyx membranes bright purple or purple-spotted,
       inflated-puckered or keeled; s SN, n SnGb, n SnBr, DMoj . . . . **G. aliquanta**
       25. Corolla lobes = tube + throat . . . . subsp. **aliquanta**
       25′ Corolla lobes gen 1/2 tube + throat . . . . subsp. **breviloba**
     23′ Stamens exceeded by corolla lobes
      26. Corolla throat with purple spots
       27. Corolla throat funnel-shaped; lobes 3–5 mm wide; style incl or ± exserted. . . . . ²**G. clivorum**
       27′ Corolla throat widely bell-shaped; lobes 5–14 mm wide; style exserted . . . . **G. tricolor**
        28. Fl gen in open clusters; pedicel 15–35 mm; corolla 7–13 mm. . . . . ²subsp. **diffusa**
        28′ Fls gen in dense clusters; pedicel 1–5 mm; corolla 10–19 mm . . . . subsp. **tricolor**
      26′ Corolla throat without purple spots

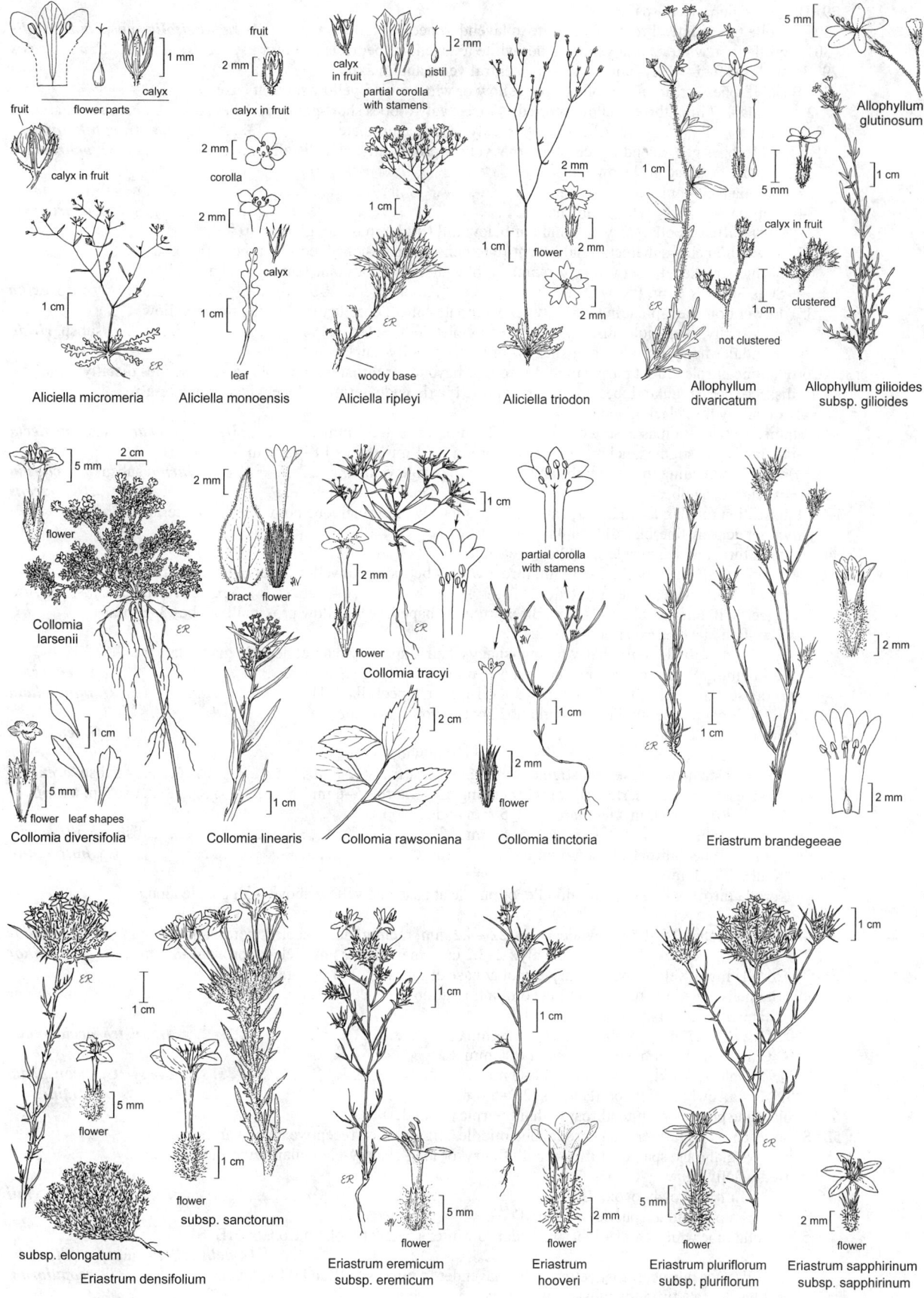

flower parts — fruit, calyx

fruit

calyx in fruit

1 mm

Aliciella micromeria

fruit

calyx in fruit 2 mm

corolla 2 mm

calyx 2 mm

leaf 1 cm

Aliciella monoensis

calyx in fruit

partial corolla with stamens

pistil 2 mm

woody base

Aliciella ripleyi

flower 2 mm

2 mm

1 cm

Aliciella triodon

Allophyllum glutinosum 5 mm

calyx in fruit

not clustered

clustered

Allophyllum divaricatum 1 cm 5 mm

Allophyllum gilioides subsp. gilioides 1 cm

flower 5 mm

bract flower 2 mm

2 cm

Collomia larsenii

flower 5 mm

flower leaf shapes 1 cm

Collomia diversifolia

Collomia linearis 1 cm

flower 2 mm

1 cm

Collomia tracyi

2 cm

Collomia rawsoniana

partial corolla with stamens

flower 2 mm

1 cm

Collomia tinctoria

2 mm

2 mm

1 cm

Eriastrum brandegeeae 2 mm

subsp. elongatum

flower 5 mm

flower 1 cm

subsp. sanctorum 1 cm

Eriastrum densifolium

flower 5 mm

Eriastrum eremicum subsp. eremicum 1 cm

flower 2 mm

Eriastrum hooveri 1 cm

flower 5 mm

Eriastrum pluriflorum subsp. pluriflorum

flower 2 mm

Eriastrum sapphirinum subsp. sapphirinum 1 cm

29. Basal lvs glabrous to sparsely short-hairy
   30. Corolla tube white, throat, lobes white to lavender; pedicel 1–30 mm . . . . . [2]***G. achilleifolia*** subsp. ***multicaulis***
   30′ Corolla tube white, throat yellow, lobes white to lavender; pedicels 2–3 mm . . . . **G. angelensis**
29′ Basal lvs tufted-woolly-hairy at least on central veins and in axils
   31. Basal lf lobes spreading; corolla throat yellow or white with 5 yellow spots at base
      32. Corolla 4–7 mm; throat yellow-spotted, ≤ lobes; calyx lobes short-pointed . . . . **G. clokeyi**
      32′ Corolla 7–12 mm; throat yellow, ≥ lobes; calyx lobes acuminate . . . . **G. ophthalmoides**
   31′ Basal lf lobes gen ascending; corolla throat yellow proximally, blue distally. . . . . **G. ochroleuca**
      33. St glabrous, glaucous below infl
         34. Corolla 8–14 mm; style > stamens . . . . subsp. **bizonata**
         34′ Corolla 4–6 mm; style ≤ stamens . . . . subsp. **ochroleuca**
      33′ St gen tufted-woolly-hairy or glandular below infl (if glabrous, not glaucous)
         35. Lower branches suberect; basal lvs entire to gen 1-pinnate-lobed; corolla tube white, gen
            purple-streaked, throat yellow proximally, blue distally, lobes blue (entire corolla ± yellow);
            SCo, SnGb, s SnBr, PR, w D . . . . subsp. **exilis**
         35′ Lower branches spreading; basal lvs 2–3-pinnate-lobed; corolla tube purple, throat yellow
            proximally, deep violet distally, lobes deep violet; Teh, SnGb . . . . subsp. **vivida**
22′ Calyx glandular-hairy, occ also long-shaggy- or tufted-woolly-hairy
  36. Corolla tube purple, throat purple to mid-throat or beyond, or throat with purple spots or stripes distally
    37. Lf dissection ± irregular, lobes not linear; cauline lf axis wide, terminal lobe wider than lateral;
       calyx densely fine-black-glandular
      38. Stamens, style, stigmas exserted; corolla 9–20 mm, lobes 3–5.7 mm . . . . [2]***G. brecciarum*** subsp. ***neglecta***
      38′ Stamens, style, stigmas incl or ± exserted; corolla 5–11 mm, lobes 1.8–3.5 mm
         39. Branches spreading to ± erect; throat yellow with purple veins . . . . [2]***G. brecciarum*** subsp. ***brecciarum***
         39′ Branches decumbent; throat gen purple. . . . . [2]***G. jacens***
    37′ Lf dissection regular, lobes linear; cauline lf axis narrow, terminal lobe not wider than lateral;
       calyx not densely fine-black-glandular
      40. Corolla throat with 2 purple spots below each lobe
         41. Pedicel 10–40 mm; corolla 7–13 mm, throat widely bell-shaped, yellow or orange proximally,
            lobes 3–6 mm wide; style exserted. . . . . [2]***G. tricolor*** subsp. ***diffusa***
         41′ Pedicel 1–10 mm; corolla 6–11 mm, throat funnel-shaped, pale yellow proximally, lobes 3–5 mm
            wide; style incl or ± exserted
            42. St sparsely tufted-woolly-hairy to long-shaggy-hairy, sparsely but not densely glandular; fr
               3–4.8 mm; calyx 5–6 mm in fr; corolla 6–8 mm. . . . . [2]***G. clivorum***
            42′ St densely glandular; fr 7–9 mm; calyx 8–11 mm in fr; corolla 8–11 mm . . . . **G. millefoliata**
      40′ Corolla throat purple-marked but without 2 spots below each lobe
         43. Corolla 11.5–22 mm . . . . **G. tenuiflora**
           44. Corolla lobes 2.1–4.2 mm wide; throat 2.2–3.8 mm wide at top
              45. Longest stamens ± exserted; stigmas among anthers; fr 5–6.2 mm . . . . subsp. **arenaria**
              45′ Longest stamens exserted; stigmas exceeding anthers; fr 3.5–6 mm . . . . subsp. **tenuiflora**
           44′ Corolla lobes 4–6 mm wide; throat 3.7–5 mm wide at top
              46. Corolla tube + throat 6.7–11 mm; fr 5–8 mm. . . . . subsp. **amplifaucalis**
              46′ Corolla tube + throat 13–14.5 mm; fr 6.5–7.5 mm. . . . . subsp. **hoffmannii**
        43′ Corolla 4–11.3 mm
         47. Corolla throat purple to above middle, or purple at base and yellow above with purple along
            veins; fls 1–2 per bract
           48. Basal lf lobes 0.9–1.5 mm wide; calyx 2.5–4.2 mm, membranes ± keeled; corolla 6–10.7 mm. . . . [2]***G. malior***
           48′ Basal lf lobes 0.6–0.9 mm wide; calyx 2–3.2 mm, membranes not keeled; corolla 4–8 mm. . . . . . . [2]***G. minor***
         47′ Corolla throat yellow, with 1 purple spot at base of lobes; fls 1–5 per bract
           49. Stamens equal in length, reaching corolla throat top. . . . . [2]***G. minor***
           49′ Stamens unequal in length, exserted
             50. Calyx 2–2.5 mm; corolla lobes 1.4–2.8 mm. . . . . [2]***G. austro-occidentalis***
             50′ Calyx 2.6–4.6 mm; corolla lobes 1.6–5 mm
               51. Fr oblong-ovoid; corolla lobes 1.6–2 mm . . . . [2]***G. inconspicua***
               51′ Fr narrowly ovoid; corolla lobes (2)3–5 mm. . . . . [2]***G. interior***
36′ Corolla tube purple or white, throat without purple (lavender)
  52. Style reaching no farther than corolla lobe middle, stigma when receptive among anthers
    53. Basal lvs glabrous, sparsely tufted-woolly-hairy, or long-shaggy-glandular-hairy, but not densely
       tufted-woolly-hairy
      54. Fr 6–8 mm, narrowly ovoid; ChI . . . . **G. nevinii**
      54′ Fr 3–6 mm, ovoid to spheric; mainland
         55. Corolla tube white, throat, lobes lavender to white; s NCoRI (Solano Co.), SnFrB, SCoR
         . . . . [2]***G. achilleifolia*** subsp. ***multicaulis***
         55′ Corolla tube purple, throat yellow, lobes lavender to pink; s SNE, e DMoj, DSon. . . . . **G. scopulorum**
    53′ Basal lvs densely tufted-woolly-hairy, at least on midvein

56. Longest stamens exceeding corolla lobe middle
  57. Lf lobes ascending; corolla lobes 1.6–2 mm; GB . . . . ²*G. inconspicua*
  57′ Lf lobes spreading; corolla lobes (2)3–5.5 mm; s SNF, Teh, WTR . . . . ²*G. interior*
56′ Longest stamens exceeded by to reaching corolla lobe middle
  58. Branches decumbent; corolla without spots, lobe tips obtuse . . . . ²*G. malior*
  58′ Branches spreading to erect; corolla with purple spots below lobes, lobe tips acute
    59. Basal lf lobes 1–6 mm; corolla tube purple; fr 3–4 mm. . . . . ²*G. austro-occidentalis*
    59′ Basal lf lobes 3–11 mm; corolla tube red- or maroon-streaked; fr 5–6 mm . . . . *G. transmontana*
52′ Style exceeding corolla lobes, stigma when receptive above anthers
  60. Corolla 7–8.5 mm, white to pale lavender or blue . . . . *G. yorkii*
  60′ Corolla 9–32 mm, tube, throat, lobes of different colors
    61. Longest stamens gen exceeding corolla lobes; corolla throat yellow or white . . . . *G. leptantha*
     62. Corolla tube 2–4 × calyx
      63. Corolla lobes 1–3 mm wide; SnBr. . . . . subsp. *leptantha*
      63′ Corolla lobes 3–5 mm wide; s SN. . . . . subsp. *purpusii*
     62′ Corolla tube < 2 × calyx
      64. Corolla throat obconic; longest stamens exceeding corolla lobes . . . . subsp. *pinetorum*
      64′ Corolla throat cup-shaped; longest stamens exceeded by corolla lobes . . . . subsp. *transversa*
    61′ Longest stamens exceeded by corolla lobes; corolla throat blue distally, yellow proximally. . . . . *G. cana*
     65. Adjacent pedicels spreading, shorter pedicel > 1/2 longer
      66. Corolla throat conic, tube 2–4 × calyx . . . . subsp. *speciformis*
      66′ Corolla throat cup-shaped, tube gen < 2 × calyx . . . . subsp. *triceps*
     65′ Adjacent pedicels ascending, shorter pedicel < 1/2 longer
      67. Fr narrowly ovoid, gen < 5 mm; basal lvs gen tufted-woolly-hairy-matted. . . . . subsp. *cana*
      67′ Fr widely ovoid, 5–9 mm; basal lvs tufted-woolly-hairy, not matted
       68. Corolla tube = throat; stamens unequal, longest exserted . . . . subsp. *bernardina*
       68′ Corolla tube 3–6 × throat; stamens equal, ± exserted . . . . subsp. *speciosa*

G. achilleifolia Benth. CALIFORNIA GILIA **ST**: 6–70 cm, glabrous or ± hairy below, ± glandular in infl or throughout. **LF**: basal in rosette, 1–2-pinnate-lobed, 5–10 cm, axis linear, lobes 3–25 mm, ± 2 mm wide, linear, spreading, gen curved, glabrous or sparsely tufted-woolly-hairy in axils or at base of lobes. **INFL**: head, cluster, or fls 1. **FL**: calyx 3.5–7 mm, lobes acute; corolla 5–21 mm, lavender or white, lobes spreading or suberect; stamens, style gen exserted; stamens exceeded by to reaching corolla lobes. **SEED**: (6)9–21. 2*n*=18.

subsp. ***achilleifolia*** (p. 1051) **ST**: branches spreading to erect. **INFL**: ± head, hemispheric; fls 8–25; pedicel 1–2 mm. **FL**: calyx 4.5–7 mm, tufted-woolly-hairy or glandular; corolla 10–21 mm, 2–3 × calyx, lavender, throat > tube; stamens exceeded by to reaching corolla lobes; stigmas among to exceeding anthers. **FR**: 3–6.2 mm, ≤(>) calyx, ovoid. **SEED**: 1.5–2.3 mm, 0.8–1.2 mm wide, ovoid or angular, red-brown. Open or shaded, gen grassy places, sandy or rocky soil; 60–1200 m. SnFrB, SCoR. Occ garden escape elsewhere. Mar–Jun

subsp. ***multicaulis*** (Benth.) V.E. Grant & A.D. Grant (p. 1051) **ST**: trailing to erect. **INFL**: open cymes, fls 1–7; pedicel 1–30 mm. **FL**: calyx 3.5–6 mm, tufted-woolly-hairy or glandular; corolla 5–10 mm, 1–2 × calyx, lobes 2.2–3.8 mm, 1–1.5 mm wide, tube white, throat, lobes white to lavender; stamens, style ± exserted; stigmas among anthers. **FR**: 3–5.5 mm, < to > calyx, ovoid to spheric. **SEED**: 0.9–2.2 mm, 0.6–1 mm wide, ovoid, angular, or reniform, red-brown. Open or shaded, gen grassy places, sandy or rocky soil; 60–1200 m. s NCoRI (Solano Co.), SnFrB, SCo, SCoR, s ChI, WTR. Often grows with *G. achilleifolia* subsp. *a.* and may ultimately be better treated as a separate sp. Feb–Jun

G. aliquanta A.D. Grant & V.E. Grant PUFFCALYX GILIA, WESTERN GILIA **ST**: 8–16 cm, branches spreading from base, glabrous or ± tufted-woolly-hairy below, glandular in infl. **LF**: basal in suberect rosette, 1–3 cm, 1-pinnate-lobed; axis 1–3 mm wide; lobes 2–7 mm, gen > axis width. **INFL**: open; pedicels 1–11 mm, in unequal pairs. **FL**: calyx 3–5 mm, glabrous or sparsely tufted-woolly-hairy, bright purple or purple-spotted, membranes inflated-puckered or keeled; corolla 6–12 mm, tube purple, throat yellow to cream, often purple-streaked, lobes 0.9–7.5 mm, 1.2–4.4 mm wide, obovate, lavender, tips rounded; stamens unequal, exserted; style exserted. **FR**: 3–5.5 mm, ≥ calyx, widely ovoid; valves detaching. **SEED**: (9)12–15, 1.2–1.9 mm, 0.8–1 mm wide, ovoid, angular, or reniform, yellow-brown.

subsp. ***aliquanta*** (p. 1051) **FL**: corolla tube incl, throat spreading, lobes = tube + throat, widely obovate; longest stamens gen exceeding corolla lobes; stigmas among or exceeding anthers. 2*n*=18. Rocky slopes, washes; 700–1300 m. s SN, n SnGb, n SnBr, DMoj. Mar–Jun

subsp. ***breviloba*** A.D. Grant & V.E. Grant (p. 1051) **FL**: corolla tube exserted, throat narrow, tapered, lobes gen 1/2 tube + throat, narrowly obovate; stamens exceeded by to reaching corolla lobes; stigmas among anthers. Uncommon. Rocky slopes, washes; 700–1600 m. DMoj; s NV. Mar–May

G. angelensis V.E. Grant CHAPARRAL GILIA **ST**: branches ± erect from base, 7–70 cm, hairs below, short, translucent. **LF**: basal in ± erect cluster, 1–2-pinnate-lobed, glabrous or short-hairy; axis, lobes linear. **INFL**: open cluster; fls 1–10; pedicels yellow-glandular. **FL**: calyx 3–5 mm, gen short-hairy, lobes 1–2 mm, acute, wider than membranes, membranes gen blue; corolla 6–11 mm, tube incl, white, throat cup-shaped, yellow, not spotted, lobes 2.3–5 mm, 1.5–4 mm wide, white to lavender; stamens, style exserted. **FR**: 4–5 mm, < calyx, ovoid. **SEED**: 18–30, 0.8–1.4 mm, 0.5–0.8 mm wide, ovoid, reniform, or angular, light brown. 2*n*=18. Open, sandy or rocky, gen grassy areas; (5)200–1900 m. CCo, e SnFrB (Mount Hamilton Range), SCoR, SW; Baja CA. Feb–Jun

G. austro-occidentalis (A.D. Grant & V.E. Grant) A.D. Grant & V.E. Grant SOUTHWESTERN GILIA **ST**: 10–30 cm, branches spreading from base, tufted-woolly-hairy near base, glandular above. **LF**: basal in rosette, 1-pinnate-lobed, glabrous or tufted-woolly-hairy; axis linear; lobes spreading, 1–6 mm, linear, entire or toothed. **INFL**: cluster, open in fr; fls 1–5; pedicels unequal. **FL**: calyx 2–2.5 mm, glandular or tufted-woolly-hairy, green or ± purple, lobes acuminate, as wide as membranes; corolla 6–8 mm, tube purple, throat yellow, lobes 1.4–2.8 mm, lavender, with 1 purple spot at base; stamens unequal, longest reaching corolla lobe middle. **FR**: 3–4 mm, ± = calyx, widely ovoid or spheric; valves detaching. **SEED**: 24–30. 2*n*=18. Sandy flats, interior, coastal valleys; 600–1300 m. SCoR, s SnJV, n WTR. Similar to *G. interior*, *G. minor*, *G. inconspicua*. Mar–May

G. brecciarum M.E. Jones BREAK GILIA, NEVADA GILIA Occ with skunk-like odor. **ST**: 8–35 cm, densely tufted-woolly-hairy near base; branches spreading, decumbent, or erect. **LF**: basal in suberect

cluster, 2–5 cm, 2-pinnate-lobed, lobes ± irregular, lanceolate, > lf axis width, axis not strap-shaped; cauline axis wide, upper cauline lf lobes gen finger-like, terminal lobe wider than lateral. **INFL:** clusters; pedicel glands dense, minute, black, stalked. **FL:** calyx 2.5–6 mm, lobes spreading, acute, thick, densely fine-black-glandular, wider than membranes, membranes gen purple-tinged; corolla 8–20 mm; stamens exserted or ± so. **FR:** 4–7 mm, ± = calyx, widely ovoid. **SEED:** 18–33. 2*n*=18. Subspp. intergrade ± in s CA.

subsp. ***brecciarum*** (p. 1051)   **ST:** branches spreading to ± erect. **FL:** calyx 2.5–4 mm; corolla 8–11 mm, tube incl, purple, throat exserted, narrowly V-shaped, purple below, yellow, purple-veined above, lobes 1.8–3.5 mm, pink-lavender, base white; stamens, stigmas at corolla throat top or ± exserted. Sandy flats in open woodland, scrub; 915–2560 m. s SNH, Teh, n edge SnGb, SnBr, GB, DMtns (Cottonwood Mtns); to se OR, UT. Apr–Jun

subsp. ***neglecta*** A.D. Grant & V.E. Grant (p. 1051)   **ST:** spreading to erect. **FL:** calyx 3–6 mm; corolla (9)10–20 mm, tube gen exserted, purple, throat widely V-shaped, lower 1/2 purple, white above, yellow-spotted, lobes 3–5.7 mm, white, tips lavender; stamens, style exserted. Sandy desert flats; 650–2100 m. c&s SNH, Teh, DMoj. Mar–Jun

G. cana (M.E. Jones) A. Heller   DESERT GILIA, SHOWY GILIA
Pl occ with skunk-like odor. **ST:** 9–32 cm, branches 1–several from base, tufted-woolly-hairy near base, gen glandular above. **LF:** basal in rosette, 1–2-pinnate-lobed, tufted-woolly-hairy; axis < 3 mm wide; lobes gen ascending, toothed on both sides, teeth short-pointed or acuminate. **INFL:** spreading, showy; fls few to many. **FL:** calyx 2–5.7 mm, gland-dotted, lobes acute to acuminate; corolla 8–32 mm, tube purple, throat yellow proximally, blue distally, lobes pink; stamens reaching below corolla lobe middle; style > stamens. **FR:** 3–9 mm, < calyx, ovoid to spheric; valves detaching. **SEED:** 12–60.

subsp. ***bernardina*** A.D. Grant & V.E. Grant   **LF:** basal densely tufted-woolly-hairy, not matted. **INFL:** fls 2–3; adjacent pedicels ascending, shorter < 1/2 longer. **FL:** calyx 2.8–4 mm, tufted-woolly-hairy to ± glandular; corolla 17–23 mm, tube exserted, throat = tube, widely conic, lobes 4–6 mm wide; stamens unequal, longest exserted. **FR:** 5–9 mm, widely ovoid. **SEED:** 12–18. 2*n*=18. Open, gravelly or sandy flats, washes; 800–1890 m. n SnBr, sw edge DMoj. Some populations intergrade with *G. leptantha* subsp. *transversa.* Apr–May

subsp. ***cana*** (p. 1051)   **LF:** basal gen tufted-woolly-hairy, matted. **INFL:** pedicels ascending, shorter < 1/2 longer. **FL:** calyx 2.5–4 mm, tufted-woolly-hairy or glandular; corolla 14–26 mm, tube exserted, throat narrowly conic, lobes 2–3 mm wide; stamens gen unequal, longest reaching above corolla lobe middle. **FR:** 3–5(6) mm, narrowly ovoid. **SEED:** 6–12. 2*n*=18. Rocky, sandy, or granitic slopes; 1800–3100 m. c&s SNH (e slope), n SNE, DMoj. Jun–Aug

subsp. ***speciformis*** A.D. Grant & V.E. Grant (p. 1051)   **LF:** basal 2–5 cm, densely tufted-woolly-hairy. **INFL:** fls 4–8; pedicels slender, in spreading, subequal pairs. **FL:** calyx 2.6–4.2 mm; corolla 15–29 mm, tube 2–4 × calyx, throat narrowly conic, lobes 3–9 mm wide; stamens equal, reaching below corolla lobe middle. **FR:** 5–8 mm, widely ovoid to spheric. **SEED:** 12–18. Gen basalt gravel, sand; 90–2290 m. s SNH, DMoj; sw NV, AZ. Mar–May

subsp. ***speciosa*** (Jeps.) A.D. Grant & V.E. Grant (p. 1051)   **LF:** basal 2–4 cm, densely tufted-woolly-hairy. **INFL:** fls 4–8; pedicels slender, ascending, unequal, ± glabrous. **FL:** calyx 3.7–5.7 mm, glabrous or glandular; corolla 20–32 mm, tube 3–6 × throat, throat narrow, lobes 4–9 mm wide; stamens equal, ± reaching corolla lobe bases. **FR:** 5–9 mm, widely ovoid. **SEED:** 30–60. 2*n*=18. Open, gravelly or sandy flats, washes; 200–2900 m. s SNH (e slope), DMoj. Locally abundant. Intergrades with *G. cana* subsp. *c.* Mar–May

subsp. ***triceps*** (Brand) A.D. Grant & V.E. Grant (p. 1051)   **LF:** basal 1–5 cm, ± tufted-woolly-hairy. **INFL:** fls many; pedicels slender, spreading, pairs ± equal. **FL:** calyx 2–4 mm; corolla 8–23 mm, tube gen < 2 × calyx, throat cup-shaped, lobes ± 3 mm wide; stamens reaching below corolla lobe middle. **FR:** 4–7 mm, widely ovoid to spheric. **SEED:** 12–18. 2*n*=18. Sandy flats, gen limestone; 90–2400 m. s SNH, W&I, WTR, DMoj; sw NV, w AZ. Mar–May

G. capitata Sims   BLUEHEAD GILIA   **ST:** branched, 10–90 cm, lfy below infl, glabrous, glandular, or with white hairs. **LF:** lower 1–2-pinnate-lobed, lobes 5–20 mm, axis, lobes ± 1 mm wide, linear, end lobe gen wider, axils, midribs gen with white hairs; upper cauline reduced or lobes finger-like. **INFL:** head, terminal, spheric; fls 25–100. **FL:** calyx 3–5.2 mm, expanded in fr, lobes acute to acuminate, narrower than membranes; corolla tube incl, throat ≤ tube, gen exserted, lobes linear to oblong; stamens, style exserted; stigmas < 2 mm. **FR:** spheric, ovoid, or obovoid. **SEED:** 1–25. 2*n*=18.

subsp. ***abrotanifolia*** (Greene) V.E. Grant (p. 1051)   **ST:** 20–80 cm, glabrous to glandular. **LF:** 1–2-pinnate-lobed. **INFL:** 17–35 mm wide, base glabrous, glandular, or sparsely tufted-woolly-hairy; pedicels 1–2 mm; fls 25–50. **FL:** calyx hairy, green or brown, lobes recurved, acuminate, membranes colorless or ± blue; corolla (7)8–12(13) mm, pale blue-violet or white, lobes 2–3 mm wide, oblong. **FR:** 2.5–4.7 mm, ≤ calyx, ovoid to spheric; valves detaching. **SEED:** 9–18. Sandy, loamy slopes; gen < 1900(3018) m. s SN, c SNF, Teh, SCoR, ChI, TR, PR; Baja CA. Mar–Jul

subsp. ***capitata*** (p. 1051)   **ST:** 20–80 cm, glabrous, glandular, or hairy. **LF:** 2-pinnate-lobed. **INFL:** 14–40 mm wide; pedicels < 1 mm. **FL:** calyx gen glabrous, lobes gen erect, acute, membranes white; corolla 6–8 mm, pale blue-violet, lobes < 1 mm wide, linear; stigmas < 0.5 mm. **FR:** 2.5–3.7 mm, ≥ calyx, spheric; valves detaching. **SEED:** 1–6(9). Rocky slopes; 30–2400 m. NW, CaR, SnFrB; to BC, ID. Apr–Aug(Sep–Nov)

subsp. ***chamissonis*** (Greene) V.E. Grant   BLUE COAST GILIA
Pl with skunk-like odor. **ST:** branches spreading, 15–70 cm, stout, glandular. **LF:** basal in rosette, fleshy, glandular, 2-pinnate-lobed. **INFL:** 25–35 mm wide, base densely tomentose; pedicels 0. **FL:** calyx tomentose, lobes acuminate, green, tips recurved, membranes ± purple; corolla 9–10 mm, bright blue-violet, lobes ± 3 mm wide, oblong. **FR:** 4–4.7 mm, < calyx, ovoid; valves not detaching. **SEED:** 10–25. Coastal sandhills; < 185 m. n CCo. Apr–Jul   ★

subsp. ***mediomontana*** V.E. Grant   **ST:** simple or branches ± erect, 15–40 cm, lfy below middle, glabrous or glandular. **LF:** in open basal rosette, 1–2-pinnate-lobed. **INFL:** 12–25 mm wide, base tomentose; pedicels 0. **FL:** calyx hairy, lobes erect to recurved, acute, membranes colorless; corolla 5–8 mm, pale blue-violet to white, lobes ± 1 mm wide, linear. **FR:** 3–4.5 mm, ≤ calyx, ovoid to spheric; valves detaching. **SEED:** 3–10. Open, rocky slopes; 900–2100 m. n&c SN. May–Aug

subsp. ***pacifica*** V.E. Grant   PACIFIC GILIA   **ST:** 25–50 cm. **LF:** 2-pinnate-lobed; axis, lobes ± 1 mm wide. **INFL:** 12–40 mm wide; pedicels 0–3 mm. **FL:** calyx glabrous to ± hairy or glandular, lobes erect, green, acute, membranes blue-violet; corolla 6–8 mm, pale to bright blue-violet, lobes 1–2 mm wide, linear. **FR:** 3–5.2 mm, ≤ calyx, spheric, ovoid, or obovoid; valves detaching. **SEED:** (3)6–25. Steep slopes, ravines, open flats, or coastal bluffs, grassland, dunes; gen < 400(761) m. n&c NCo, NCoRO, w KR; OR. May–Aug   ★

subsp. ***pedemontana*** V.E. Grant   **ST:** 30–90 cm, glabrous or glandular. **LF:** 1–2-pinnate-lobed. **INFL:** 10–20 mm wide, base tomentose; pedicels 0. **FL:** calyx densely hairy, lobes recurved, acuminate, membranes colorless; corolla 7–11 mm, pale blue-violet, lobes ± 2 mm wide, oblong. **FR:** 3–4.5 mm, ≤ calyx, spheric to ovoid; valves not detaching. **SEED:** 6–15(18). Open, rocky slopes; 60–2000 m. CaRF, SNF, n&c SNH, ScV (Sutter Buttes), n SnJV. Apr–Jun

subsp. ***staminea*** (Greene) V.E. Grant (p. 1051)   **ST:** 10–60 cm, glabrous or glandular. **LF:** in open basal rosette, gen 1-pinnate-lobed. **INFL:** 20–30 mm wide, base tomentose; pedicels 0. **FL:** calyx hairy, lobes acuminate, tips recurved, membranes ± blue or colorless; corolla 7–13 mm, pale blue-violet, lobes ± 3 mm wide, oblong. **FR:** 3.7–5 mm, ≤ calyx, ovoid; valves gen not detaching. **SEED:** 10–25. Sandhills, flats; 100–1300 m. s NCoRI, GV, SnFrB, SCo; AZ. Intergrades with *G. capitata* subsp. *pedemontana.* Mar–May

subsp. ***tomentosa*** (Brand) V.E. Grant   WOOLLY-HEADED GILIA
ST: 10–70 cm, stout, gen hairy. **LF:** 2-pinnate-lobed. **INFL:** 20–35 mm wide, base tomentose; pedicels 0. **FL:** calyx densely hairy, lobes acute, tips ± recurved, membranes colorless; corolla 7–11 mm, bright blue-violet, lobes 1–2 mm wide, linear. **FR:** 3–3.8 mm, ≤ calyx,

spheric to ovoid; valves detaching. **SEED**: 3–10(24). Sea bluffs, outcrops (serpentine); < 30 m. NCo, se NCoRI, n CCo. May–Jul ★

G. clivorum (Jeps.) V.E. Grant (p. 1051) PURPLESPOT GILIA **ST**: branches spreading or erect, gen many, 6–30 cm, lfy, sparsely tufted-woolly-hairy to long-shaggy-hairy, occ also sparsely glandular in infl. **LF**: basal 2–6 cm, 1–2-pinnate-lobed, axis, lobes narrowly linear, ascending. **INFL**: open clusters; fls 2–5. **FL**: calyx 4–5 mm, 5–6 mm in fr, hairy or glandular, lobes wider than membranes, membranes gen purple; corolla 6–8 mm, tube, throat gen incl, funnel-shaped, pale yellow, purple-spotted, lobes 3–5 mm wide, light blue or white; stamens, style incl or ± exserted. **FR**: 3–4.8 mm, < calyx, ovoid, valves separating only in upper 1/3, not detaching. **SEED**: (15)24–36. 2*n*=36. Common. Open, grassy areas; < 1500 m. s NCoR, CW, ChI, WTR, SCo, PR; AZ (alien), Baja CA. Feb–Jun

G. clokeyi H. Mason (p. 1051) CLOKEY'S GILIA **ST**: 8–17 cm, tufted-woolly-hairy below middle, glandular above. **LF**: 1–3 cm, gray-green, tufted-woolly-hairy; basal in rosette, 1–2-pinnate-lobed, lobes short-pointed, spreading; upper cauline palmate, middle lobe widest. **INFL**: open; pedicels paired, spreading, unequal, thread-like. **FL**: calyx 2–4 mm, glabrous (or tufted-woolly-hairy in early fls), lobes short-pointed, tip thick; corolla 4–7 mm, tube gen incl, throat exserted, ≤ lobes, white with 5 yellow spots at base, lobes 1–3 mm, acute, white, gen blue-streaked; stamens, style ± exserted. **FR**: 3–6 mm, ± = calyx, spheric or widely ovoid, tip pointed; valves detaching at base. **SEED**: 9–24. 2*n*=18. Open, rocky slopes, sandy washes, gen limestone; 400–2500 m. SNE, e DMoj, DMtns; to CO, NM. In fr calyx translucent, papery, lobes red, triangular, pedicel white, ± inflated below fr. Mar–Jun

G. diegensis (Munz) A.D. Grant & V.E. Grant COASTAL GILIA **ST**: 10–40 cm, glabrous, glaucous below middle, glandular above. **LF**: basal in prostrate rosette, 1–7 cm, strap-shaped, toothed or pinnate-lobed, lobes spreading, tufted-woolly-hairy; cauline shorter, ± clasping, lanceolate, serrate. **INFL**: cluster, open in fr; pedicels unequal. **FL**: calyx 3–5 mm; corolla (6)8–12 mm, tube incl or ± exserted, purple, lower throat purple, upper yellow, lobes 2–5 mm, lavender or white; stamens unequal, reaching corolla lobe middle, pollen blue; style gen reaching to exceeding stamens. **FR**: 3.7–7 mm, < calyx, ovoid. **SEED**: 24–30(39). 2*n*=18. Sandy areas, open forest, scrub; 220–2200 m. SnGb, SnBr, PR, s edge DMoj, w edge DSon; Baja CA. Apr–Jun

G. inconspicua (Sm.) Sweet (p. 1051) SHY GILIA **ST**: 8–32 cm, branches ascending or spreading, tufted-woolly-hairy below infl, black-glandular above. **LF**: basal in rosette, 1-pinnate-lobed, tufted-woolly-hairy, lobes 2–10 mm, ≥ axis width, linear (or rounded, short-pointed), entire or toothed, ascending. **INFL**: cluster, open in fr; fls 2–4; pedicels unequal. **FL**: calyx 2.6–4.6 mm, glabrous or gland-dotted (or tufted-woolly-hairy in early fls); corolla 6–11 mm, 3–4 × lobes, tube, throat gen yellow, lobes 1.6–2 mm, lavender with purple spot at base or in throat; stamens, style exserted, longest stamens exceeding corolla lobe middle. **FR**: 5–8 mm, < calyx, oblong-ovoid; valves detaching. **SEED**: 12–18. 2*n*=36. Rocky or sandy sagebrush slopes, washes; 1200–2300 m. GB; to WA, ID, NV. Apr–Jun

G. interior (H. Mason & A.D. Grant) A.D. Grant INLAND GILIA **ST**: branches spreading, 6–13 cm, tufted-woolly-hairy below infl, glandular above. **LF**: basal in rosette, 1-pinnate-lobed, tufted-woolly-hairy; axis 1–2 mm wide; lobes entire or toothed, 1–2 × axis width, spreading. **INFL**: fls 1–2; pedicels 2–20 mm, unequal. **FL**: calyx 2.6–3.6 mm, glandular; corolla 6–11 mm, 2–3 × lobes, tube purple, throat yellow, purple-spotted or with 1 spot at lobe base, lobes (2)3–5 mm, lavender; stamens, style exserted, longest stamens exceeding corolla lobe middle. **FR**: 3–6 mm, ≤ calyx, narrowly ovoid. **SEED**: 9–27. 2*n*=18. Rocky slopes, open woodland, scrub; 700–1700 m. s SNF, Teh, WTR. Apr–May ★

G. jacens A.D. Grant & V.E. Grant **ST**: branches long-decumbent, densely tufted-woolly-hairy. **LF**: basal in suberect rosette, 2–5 cm, 1–2-pinnate-lobed, tufted-woolly-hairy; axis 0.5–1.5 mm wide, lobes > axis width; upper cauline lobes gen lanceolate, 1-pinnate-lobed. **INFL**: cluster; pedicel glands dense, minute, black, stalked. **FL**: calyx 3–4 mm; corolla 5–7 mm, tube incl, purple, throat gen incl, purple; stamens, style ± exserted. **FR**: 4–5 mm, ± < calyx, ovoid. **SEED**: 24–36. 2*n*=18. Low, sandy or clay flats; 500–2500 m. Teh,

SCoRI, WTR, w DMoj. [*G. brecciarum* subsp. *j.* (A.D. Grant & V.E. Grant) A.G. Day] A cryptic but reproductively isolated sp. confused with *G. brecciarum*. Apr–Jun

G. latiflora (A. Gray) A. Gray BROAD-FLOWERED GILIA Pl ± scapose. **LF**: basal in prostrate rosette, 2–7 cm, tufted-woolly-hairy, strap-shaped, toothed or lobed, lobes spreading; cauline shorter, clasping, entire or lobed at base, tapered. **INFL**: gen branches ascending-spreading; pedicels unequal. **FL**: showy, fragrant; calyx 2–7 mm, ± glandular, or early fls tufted-woolly-hairy; corolla 9–35 mm, tube purple, upper throat, lobe bases white, tips lavender; stamens exserted, exceeded by style, corolla lobes. **FR**: 3–9 mm, gen > calyx, ovoid to obovoid. **SEED**: 11–60. Makes showy displays. Subspp. variable within populations, intergrading where ranges overlap.

subsp. ***cuyamensis*** A.D. Grant & V.E. Grant CUYAMA GILIA **ST**: 6–25 cm, glabrous, glaucous below middle. **INFL**: pedicels elongate. **FL**: calyx 2–4 mm; corolla 9–16 mm, tube exserted, throat tapered or widely expanded, purple at base. **FR**: 3–6 mm. 2*n*=18. Sandy flats, pinyon/juniper woodland, lower river valleys; 600–2100 m. s SCoRI, n WTR, s SNF (near Lake Isabella). Mar–May ★

subsp. ***davyi*** (Milliken) A.D. Grant & V.E. Grant (p. 1051) DAVY'S BROAD-FLOWERED GILIA **ST**: 10–30 cm, glabrous, glaucous below middle, or tufted-woolly-hairy. **INFL**: pedicels elongate in fr. **FL**: calyx 4–7 mm; corolla 18–24 mm, gen 3 × lobes, tube exserted, throat gen long-tapered, purple to mid-throat or above. **FR**: 6–9 mm. Common. Open, sandy flats; 400–1700 m. SCoRI, WTR, s&w DMoj. Mar–May

subsp. ***elongata*** A.D. Grant & V.E. Grant (p. 1051) **ST**: 13–32 cm, tufted-woolly-hairy below middle. **INFL**: pedicels elongate. **FL**: calyx 4–6 mm; corolla 21–35 mm, 4–6 × lobes, tube > 3 × calyx, slender, white-veined, throat < 7 mm wide, narrowly tapered, yellow, lobes white. **FR**: 6–9 mm. Open, sandy slopes; 700–1900 m. w DMoj (Rand, El Paso mtns, Black Rock Hills). Apr–May

subsp. ***latiflora*** (p. 1051) BROAD-LEAVED GILIA **ST**: 10–33 cm, glabrous, glaucous below middle. **INFL**: pedicels elongate. **FL**: calyx 4–5 mm; corolla 15–22 mm, tube ± exserted, white-veined, throat widely expanded, purple at base, otherwise white with yellow spots near base. **FR**: 6–9 mm. 2*n*=18. Common. Open, sandy flats; 120–2500 m. Teh, WTR, SnGb, SnBr (n base), sw DMoj. Apr–May

G. leptantha Parish FINEFLOWERED GILIA **ST**: spreading or erect, tufted-woolly-hairy below middle, glandular above. **LF**: basal in rosette, 1-pinnate-lobed, tufted-woolly-hairy, axis 1–2 mm wide, linear, lobes toothed on both sides, teeth short-pointed or acuminate. **INFL**: clusters, open or not; fls 2–6; pedicels ± spreading, unequal. **FL**: calyx 2.3–5.2 mm, ± glandular (or tufted-woolly-hairy in early fls); corolla 9–23 mm, tube yellow or purple with yellow veins, throat yellow or white, lobes pink to lavender; stamens unequal, longest gen exceeding corolla lobes. **FR**: 3–8 mm, < to > calyx, spheric to obovoid. **SEED**: 2–63. 2*n*=18.

subsp. ***leptantha*** (p. 1051) SAN BERNARDINO GILIA **ST**: spreading or erect, 15–45 cm. **FL**: calyx 3.5–4 mm; corolla 13–23 mm, tube 2–4 × calyx, narrow, flared to narrow throat, lobes 1–3 mm wide, narrowly oblong. **FR**: 3–4.5 mm, < calyx, ovoid. **SEED**: 2–12. Open, rocky soil, forest, streambanks; 1500–2300 m. SnBr. May–Jul ★

subsp. ***pinetorum*** A.D. Grant & V.E. Grant (p. 1051) PINE GILIA **ST**: lower branches long-decumbent, 5–25 cm. **LF**: basal in compact rosette, densely tufted-woolly-hairy. **FL**: calyx 2.8–4.2 mm; corolla 9–14 mm, tube 1–1.5 × calyx, throat obconic, > tube, lobes 3–4 mm wide, obovate. **FR**: 4–5 mm, < calyx, spheric. **SEED**: 2–12. Bare summits, open, rocky or sandy, with pines; 900–2900 m. Teh, SCoR, WTR (Mount Pinos). May–Jul ★

subsp. ***purpusii*** (Milliken) A.D. Grant & V.E. Grant PURPUS' FINEFLOWERED GILIA **ST**: < 60 cm. **FL**: calyx 3–5.2 mm; corolla 13–23 mm, tube 2–4 × calyx, slender, throat flared, lobes 3–5 mm wide, gen narrowly oblong; longest stamens exserted, exceeded by corolla lobes. **FR**: 3–8 mm, < to > calyx, ovoid to obovoid. **SEED**: 9–63. Common. Open, rocky or sandy soil in forests, chaparral; 700–2900 m. s SN. Fls smaller to s, grading into *G. leptantha* subsp. *pinetorum* in Teh. May–Jul

subsp. *transversa* A.D. Grant & V.E. Grant **ST**: 16–40 cm, densely glandular above rosette. **LF**: basal, lower cauline 2–8 cm. **FL**: calyx 2.3–4.5 mm, glandular or tufted-woolly-hairy; corolla 13–17 mm, tube 1–1.5 × calyx, throat cup-shaped, lobes 3–6 mm wide, obovate; stamens unequal, longest exserted, exceeded by style, corolla lobes. **FR**: 3–8 mm, < to > calyx, ovoid to obovoid. **SEED**: 9–24. Rocky or sandy soil, gen near streams; 900–2450 m. SnGb (n slope), SnBr, sw edge DMoj. Apr–Aug

G. malior A.G. Day & V.E. Grant SCRUB GILIA, GREAT BASIN GILIA **ST**: 5–20 cm; lower branches decumbent, spreading, tufted-woolly-hairy below infl, gland-dotted above. **LF**: lower in basal rosette, 1-pinnate-lobed, axis, lobes 1–2 mm wide, linear, tufted-woolly-hairy, lobes spreading, occ toothed; upper palmate to entire above. **INFL**: clusters, open in fr; fls 1–2; pedicels unequal. **FL**: calyx 2.5–4.2 mm, gen ± glandular, membranes ± keeled, wrinkled, puffed out; corolla 6–10.7 mm, tube, throat purple to above middle or lacking purple in throat, yellow above, throat tapered, lobes obovate, lavender, base often with white spot adaxially, tip obtuse; stamens, style ± exserted. **FR**: 3.4–7 mm, = calyx, ovoid, tip rounded; valves detaching. **SEED**: 10–18. 2*n*=36. Open, sandy, rocky flats; 300–2600 m. Teh, s SnJV, SCoRI, WTR (Mount Pinos, Liebre Mtns), SNE, DMoj; s OR, w NV. Intermediate between *G. minor*, *G. aliquanta*. Mar–Jun

G. millefoliata Fisch. & C.A. Mey. (p. 1051) DARK-EYED GILIA Pl densely glandular, odor skunk-like. **ST**: 8–30 cm; main st short; branches gen long-decumbent. **LF**: ± fleshy; basal in rosette, 1–2-pinnate-lobed, lobes 2–5 mm; upper shorter, palmate. **INFL**: clusters; fls 2–6; pedicels 2–5 mm. **FL**: calyx 4–6 mm, 8–11 mm in fr, glandular, membranes narrower than lobes, purple or colorless; corolla 8–11 mm, tube yellow, throat incl, funnel-shaped, pale yellow with paired purple spots, lobes 3–5 mm wide, obovate, blue-white; stamens, style ± reaching corolla lobe bases; pollen white. **FR**: 7–9 mm, < calyx, narrowly ovoid to ellipsoid. **SEED**: 25–50. 2*n*=18. Stabilized coastal dunes; < 10 m. NCo, n CCo; s OR. Formerly in San Francisco. Mar–Jul ★

G. minor A.D. Grant & V.E. Grant (p. 1051) LITTLE GILIA **ST**: lower branches decumbent, 6–20 cm, tufted-woolly-hairy below infl. **LF**: basal in rosette, 1-pinnate-lobed, axis and lobes 0.6–0.9 mm wide, lobes linear, tufted-woolly-hairy, ascending; upper glandular. **INFL**: clusters, open in fr; pedicels unequal. **FL**: calyx 2–3.2 mm, glandular (or tufted-woolly-hairy in early fls), membranes not keeled; corolla 4–8 mm, ± 2 × calyx, tube incl, purple, throat purple or yellow with purple veins, sometimes yellow with a purple spot at lobe bases, tapered, lobes ovate, acute, lavender; longest stamens, style reaching corolla throat top. **FR**: 5–7 mm, 2 × calyx, narrowly ovoid; valves not detaching completely. **SEED**: 18–27. 2*n*=18. Firm sand, gen at base of shrubs; 300–1100 m. SCoR, SnJV, WTR, SnGb, DMoj; AZ. Mar–Apr

G. modocensis Eastw. MODOC GILIA **ST**: 10–34 cm, branches gen several, spreading from below, ± tufted-woolly-hairy sometimes also glandular near base. **LF**: basal in rosette, strap-shaped, pinnate-lobed, lobes short, spreading, toothed; cauline clasping, gen deeply lobed. **INFL**: clusters, open in fr; pedicels glandular, glands black. **FL**: calyx 3.5–6 mm, gland-dotted, lobes wider than membranes, erect or spreading in fr; corolla 7–11 mm, tube incl, purple, throat yellow, midveins occ purple, lobes 2–3 mm, lavender; stamens, style ± exserted. **FR**: 4–6.5 mm, ≤ calyx, widely ovoid. **SEED**: 15–30. 2*n*=36. Open, rocky areas, pinyon/juniper woodland; 400–2450 m. Teh, s SCoRO, TR, GB, DMtns; OR, ID, NV. If recognized taxonomically, decumbent pls in n WTR (Mount Pinos) with lower lvs 2-pinnate-lobed, densely tufted-woolly-hairy would be assignable to *G. tetrabreccia* A.D. Grant & V.E. Grant. Apr–Jun

G. nevinii A. Gray (p. 1051) NEVIN'S GILIA **ST**: branches ± erect, 10–40 cm, lfy, white- or translucent-hairy. **LF**: basal hairy, but not densely tufted-woolly-hairy, 2–3-pinnate-lobed, axis, lobes < 1 mm wide. **INFL**: glandular. **FL**: calyx 4–6 mm, longer in fr, lobes wider than membranes, acute; corolla 8–14 mm, lavender with narrow, yellow throat, lobes 2–3 mm, short-pointed; stamens ± exserted, pollen blue. **FR**: 6–8 mm, < calyx, narrowly ovoid. **SEED**: 24–36. 2*n*=36. Rocky, grassy slopes, coastal canyons; 7–600 m. ChI; Baja CA (Guadalupe Island). Mar–Jul ★

G. ochroleuca M.E. Jones VOLCANIC GILIA **ST**: branches 5–30 cm, glabrous or tufted-woolly-hairy, gen glandular in infl. **LF**: lower in rosette, 1–2-pinnate-lobed or entire, axis, lobes gen linear, gen angled toward tip; cauline palmate. **INFL**: pedicels spreading, gen in subequal pairs, thread-like, glandular; lf-like bracts gen entire. **FL**: calyx 2–4 mm, lobes linear, membranes colorless; corolla 4–14 mm, tube purple, throat expanded, yellow proximally, blue distally, lobes pink to white; stamens ± exserted. **FR**: 2.4–5 mm, spheric; valves detaching. **SEED**: 3–15. 2*n*=18.

subsp. *bizonata* A.D. Grant & V.E. Grant Pl gray-green. **ST**: lower branches spreading, gen glabrous, glaucous below infl. **LF**: tufted-woolly-hairy or glabrous, 1–2-pinnate-lobed; lobes 1–15 mm. **FL**: calyx 2.1–3.7 mm, glabrous (tufted-woolly-hairy), lobes acute; corolla 8–14 mm, tube gen incl, < throat; style exceeding stamens. **FR**: 3.5–4.5 mm, spheric. Common. Flats; 240–2600 m. s SNF, SCoRI, s SCoRO, PR, TR (n slope), w DMoj. Mar–Jul

subsp. *exilis* (A. Gray) A.D. Grant & V.E. Grant (p. 1051) Pl yellow-green. **ST**: branches suberect, gen tufted-woolly-hairy (glandular). **LF**: tufted-woolly-hairy, entire to gen 1-pinnate-lobed. **INFL**: glabrous or glandular. **FL**: calyx 2.3–3.2 mm, ± tufted-woolly-hairy, lobes acute to acuminate; corolla 8–11 mm, tube ± exserted, > throat, tube white, gen purple-streaked, throat yellow proximally, blue distally, lobes blue (entire corolla ± yellow); style ± exserted; stigmas near or just exceeding anthers. **FR**: 3.8–4.5 mm, spheric to ovoid. Common. Flats; 100–2500 m. SCo, SnGb, s SnBr, PR, w D; Baja CA. Small-fld race in se PR. Mar–Aug

subsp. *ochroleuca* (p. 1051) Pl gray-green. **ST**: branches spreading, < 30 cm, glabrous, glaucous below infl. **LF**: ± tufted-woolly-hairy, 1–2-pinnate-lobed, lobes 3–10 mm, ascending; infl lvs palmate. **FL**: calyx 2.4–3.2 mm, glabrous; corolla 4–6 mm, tube incl, throat = tube; style exceeded by to equalling stamens, stigmas near anthers. **FR**: 2–5 mm, spheric. Desert sand; 600–2050 m. s SNH, W&I, w DMoj. Apr–Jun

subsp. *vivida* (A.D. Grant & V.E. Grant) A.D. Grant & V.E. Grant (p. 1051) Pl low, spreading, dark gray-green. **ST**: gen glabrous. **LF**: densely woolly-hairy-matted or tufted, 2–3-pinnate-lobed; axis 2–4 mm wide; lobes linear or reduced to teeth. **INFL**: longer pedicel of pair 3–6 × shorter. **FL**: calyx 2.8–4 mm, glabrous, tufted-woolly-hairy, or fine-glandular, lobes acute to acuminate; corolla 9–14 mm, tube 1–1.5 × calyx, throat ± > tube, tube purple, throat yellow proximally, deep violet distally, lobes deep violet; stamens exceeded by style. **FR**: 2.5–3.7 mm, spheric. Common. Rocky or sandy soil, pine forest; 1500–2700 m. Teh, SnGb. May–Aug

G. ophthalmoides Brand (p. 1055) EYED GILIA, YELLOW-EYED GILIA **ST**: branches suberect, 15–30 cm, densely tufted-woolly-hairy below middle, glandular above. **LF**: lower in rosette, densely tufted-woolly-hairy, 1–2-pinnate-lobed, axis linear, lobes spreading, gen toothed (or lobed) on both sides. **INFL**: pedicels unequal, thread-like. **FL**: calyx 2.5–4.5 mm, glabrous, lobes acuminate, ± red; corolla 7–12 mm, tube 1.5–2 × calyx, slender, purple, throat ≥ lobes, narrow, long-tapered, bright yellow, lobes 1–3 mm, pink; stamens, style ± exserted. **FR**: 4–6.5 mm, gen < calyx, ovoid; valves detaching. **SEED**: 6–12. 2*n*=36. Open, rocky soil, gen pinyon/juniper woodland; 1100–2600 m. n SNE, W&I, DMtns; to CO, AZ. May–Jun

G. salticola Eastw. (p. 1055) SALT GILIA **ST**: branches spreading, 4–20 cm, densely tufted-woolly-hairy. **LF**: basal in rosette, densely tufted-woolly-hairy, 1–2-pinnate-lobed, axis, lobes 2–3 mm wide, entire or toothed. **INFL**: open clusters, black-gland-dotted. **FL**: calyx 2.2–3.8 mm, glabrous or tufted-woolly-hairy, lobes acuminate, outcurved in fr, membranes dull purple-spotted or colorless; corolla 6–15 mm, tube incl, yellow, throat yellow, lobes 3–6 mm, gen > throat, bright pink-lavender; stamens, style equalling to exceeding corolla lobes. **FR**: 4–6 mm, < calyx, ovoid. **SEED**: 10–15. 2*n*=18. Open, volcanic or granitic areas; 1350–2700 m. SNH, w MP, nw SNE (Mono Co.); w NV. May–Jun

G. scopulorum M.E. Jones (p. 1055) ROCK GILIA **ST**: branches ± erect, 10–30 cm, glandular-hairy. **LF**: glandular-hairy; hairs translucent; lower lvs in suberect rosette, 1–2-pinnate-lobed, axis linear below, winged above, merging with lobes, lobes toothed, teeth short-

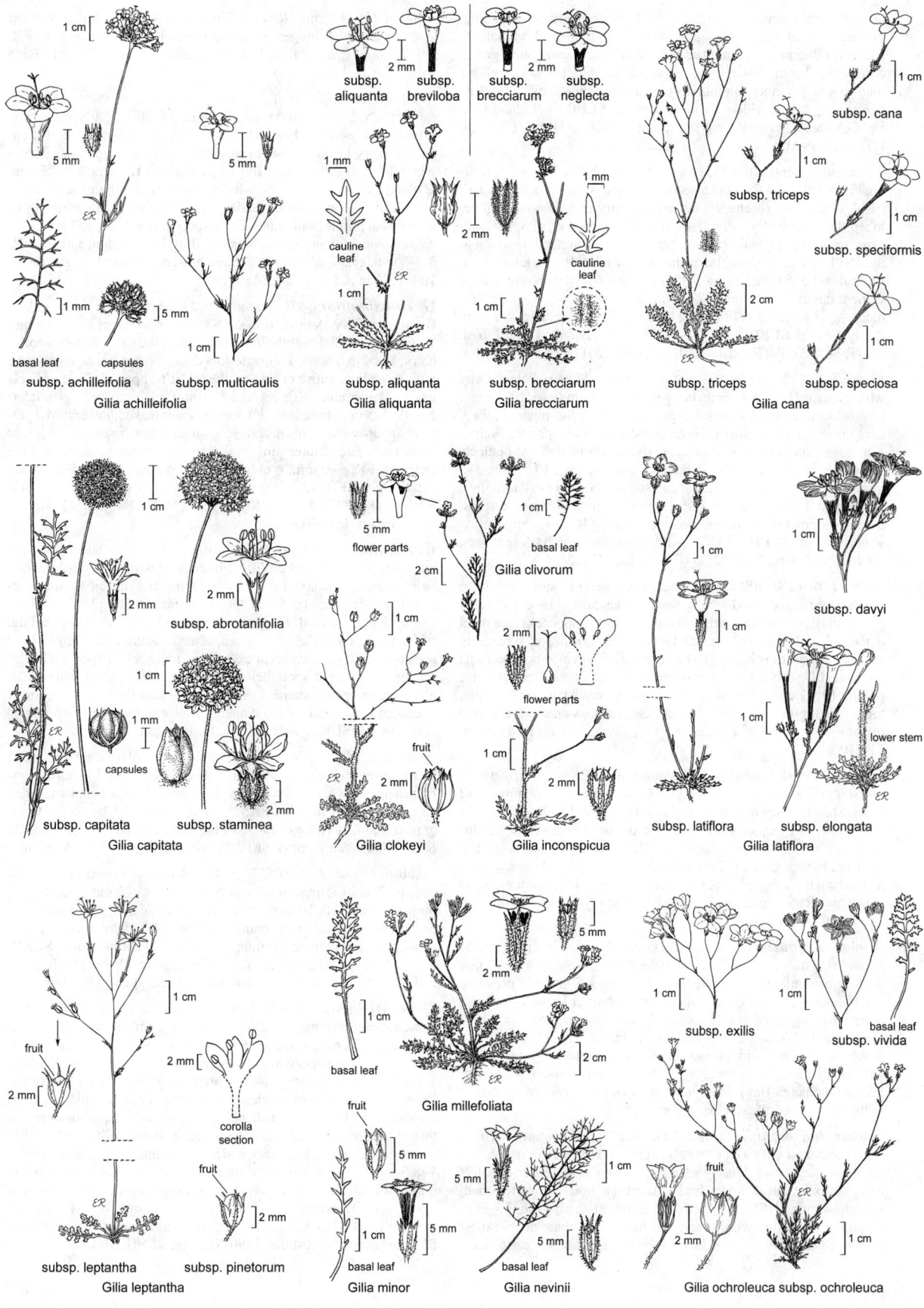

1 cm

subsp.
aliquanta subsp.
breviloba subsp.
brecciarum subsp.
neglecta

2 mm 2 mm

5 mm 5 mm

ER

1 mm

basal leaf capsules

subsp. achilleifolia subsp. multicaulis

Gilia achilleifolia

1 mm

cauline
leaf

ER

1 cm

2 mm

subsp. aliquanta

Gilia aliquanta

1 mm

cauline
leaf

1 cm

ER

subsp. brecciarum

Gilia brecciarum

1 cm

subsp. cana

1 cm

subsp. triceps

1 cm

subsp. speciformis

2 cm

ER

subsp. triceps

1 cm

subsp. speciosa

Gilia cana

1 cm

2 mm

subsp. abrotanifolia

1 cm

1 mm

capsules

subsp. capitata

ER

1 mm

2 mm

subsp. staminea

Gilia capitata

5 mm

flower parts

2 cm

1 cm

basal leaf

Gilia clivorum

1 cm

ER

fruit

2 mm

Gilia clokeyi

2 mm

flower parts

1 cm

2 mm

Gilia inconspicua

1 cm

1 cm

subsp. latiflora

1 cm

subsp. davyi

1 cm

lower stem

ER

subsp. elongata

Gilia latiflora

1 cm

fruit

2 mm

2 mm

corolla
section

fruit

2 mm

subsp. leptantha subsp. pinetorum

Gilia leptantha

1 cm

basal leaf

5 mm

2 mm

2 cm

ER

Gilia millefoliata

fruit

5 mm

1 cm

basal leaf

5 mm

Gilia minor

5 mm

5 mm

basal leaf

Gilia nevinii

1 cm

1 cm

subsp. exilis

1 cm

basal leaf
subsp. vivida

fruit

2 mm

2 mm

ER

1 cm

Gilia ochroleuca subsp. ochroleuca

pointed; upper minute, 2-toothed. **INFL:** open; pedicels spreading, in unequal pairs, thread-like, glands black, flat-topped, short-stalked. **FL:** calyx 3–5 mm, ± glandular, lobes wider than membranes, green; corolla 9–17 mm, 2–4 × calyx, tube purple, throat < tube, yellow, lobes ovate, lavender to pink, acute; stamens, style ± exserted. **FR:** 5–6 mm, ≤ calyx, widely ovoid to spheric. **SEED:** 6–9. 2*n*=18,36. Semi-shaded, rocky ravines; < 1200 m. s SNE, e DMoj, DSon; to UT, AZ. Apr–May

G. sinuata Benth. (p. 1055) ROSY GILIA   **ST:** 1–several spreading from base, 13–34 cm, glabrous, glaucous below middle. **LF:** basal in prostrate rosette, tufted-woolly-hairy or glabrous, pinnate-lobed, lobes spreading, gen toothed; cauline reduced, clasping with expanded base, toothed or entire, gen not deeply toothed, tapering to tip. **INFL:** clusters, glandular; pedicels unequal, 1–9 mm, longer in fr. **FL:** calyx 3.5–5 mm; corolla 7–12 mm, tube exserted, purple, white-veined, throat yellow, lobes 2–3 mm, lavender, pink, or white; stamens, style reaching corolla lobe bases; pollen white. **FR:** 4–7 mm, ≤ calyx, ovoid. **SEED:** 9–27. 2*n*=36. Open, sandy flats; 150–2750 m. s SNF, SCoRO, WTR, SnJt, GB, DMoj; to WA, ID, CO, AZ. Apr–Jun

G. stellata A. Heller (p. 1055) STAR GILIA   Pl hairy below; hairs white, sharply bent. **ST:** branches gen several from base, 10–40 cm, gland-dotted above. **LF:** basal in rosette, ± gray, 1–2-pinnate-lobed, axis linear, lobes toothed on both sides, teeth short-pointed; upper cauline minute, entire or 2-toothed. **INFL:** open; fls 4–8; pedicels spreading, in unequal pairs, glabrous or gland-dotted. **FL:** calyx 3–5 mm, hairy or glandular, glands black, stalked; corolla 6–10 mm, funnel-shaped, tube incl or ± exserted, throat yellow with purple spots, lobes pink or white; stamens, style ± exserted. **FR:** 5–7 mm, ≤ calyx, widely ovoid. **SEED:** 9–18. 2*n*=18. Common. Sandy desert flats, washes; < 1770 m. W&I, D; to UT, AZ, Baja CA. Mar–May

G. tenuiflora Benth.   **ST:** branches gen several, spreading from below, glabrous or tufted-woolly-hairy or glandular at base. **LF:** basal gen in cluster or rosette, tufted-woolly-hairy on upper surface or in axils or ± glabrous, 1–2-pinnate-lobed, lobes gen linear, spreading, axis linear or strap-shaped. **INFL:** ± clusters, open or not; pedicels unequal. **FL:** calyx 3.3–5.8 mm, glandular or tufted-woolly-hairy; corolla 10–22 mm, tube purple, throat gen tapered, part or all purple, lobes bright pink-lavender, white at base; stamens unequal, exserted or longest ± exserted. **FR:** 3.5–8 mm, ≤ calyx, ovoid to obovoid. **SEED:** 9–42.

subsp. ***amplifaucalis*** A.D. Grant & V.E. Grant (p. 1055) TRUMPET-THROATED GILIA   **ST:** 16–24 cm, stout, glabrous or tufted-woolly-hairy near base, ± glandular above. **INFL:** dense clusters, showy. **FL:** calyx 4–5.8 mm, tufted-woolly-hairy or ± glandular, lobes in width ≤ membranes; corolla 13–22 mm, tube stout, 1–2 × calyx, tube + throat 6.7–11 mm, throat wide, 3.7–5 mm, purple in proximal 1/2, lobes 4–6.3 mm, 4–5.5 mm wide; stamens, style exserted. **FR:** 5–8 mm, < calyx. Sandy soil of dry creeks, floodplains, slopes; 39–900 m. SCoRI. Mar–Apr ★

subsp. ***arenaria*** (Benth.) A.D. Grant & V.E. Grant (p. 1055) MONTEREY GILIA   **ST:** lower branches decumbent, 6–17 cm, gen densely glandular or base tufted-woolly-hairy. **LF:** basal in prostrate rosette, serrate or 1-pinnate-lobed, lobes short. **FL:** calyx 3.4–4.2 mm, densely glandular, lobes wider than membranes; corolla 10–14 mm, tube 1–2 × calyx, slender, tube = throat, 8.3–9.6 mm, lobes 3–4.6 mm, 3–4 mm wide; longest stamens ± exserted, stigmas among anthers. **FR:** 5–6.2 mm, ≤ calyx. 2*n*=18. Coastal sand dunes; < 30 m. c CCo (Monterey Bay). Intergrades with *G. tenuiflora* subsp. *t.* near Salinas River mouth. Mar–May ★

subsp. ***hoffmannii*** (Eastw.) A.D. Grant & V.E. Grant (p. 1055) HOFFMANN'S SLENDER-FLOWERED GILIA   **ST:** 6–12 cm, stout, lfy, glabrous near base, densely glandular above. **LF:** tufted-woolly-hairy in axils; basal, lower lvs in rosette or not, 1-pinnate-lobed, lobes 1–4 mm. **INFL:** clusters (open in fr). **FL:** calyx 4.6–5.7 mm, glandular or tufted-woolly-hairy, lobes wider than membranes; corolla 18–20 mm, throat < lobes, tube stout, 1.5–2.5 × calyx, tube

+ throat 13–14.5 mm, lobes 4.5–6.6 mm, 4–6 mm wide, obovate; stamens exserted, longest exceeding stigmas, style ± exserted. **FR:** 6.5–7.5 mm, gen < calyx. Coastal sandhills; < 30 m. n ChI (Santa Rosa Island). Apr–Jun ★

subsp. ***tenuiflora*** (p. 1055) SLENDER-FLOWERED GILIA, GREATER YELLOWTHROAT GILIA   **ST:** 15–40 cm, glabrous or ± tufted-woolly-hairy; branches spreading from base. **LF:** tufted-woolly-hairy in axils; basal in suberect cluster, 2–6 cm, 1–2-pinnate-lobed, lobes 1–2 mm wide. **INFL:** glandular. **FL:** calyx 3.3–4.3 mm, ± glandular or tufted-woolly-hairy, lobes wider than membranes; corolla 11.5–22 mm, tube slender, 1.5–3 × calyx, tube + throat 8–12 mm, throat purple, with pale yellow spots or not, lobes 4–5 mm, 2.1–4.2 mm wide; longest stamens exserted, style exceeding anthers. **FR:** 3.5–6 mm, gen = calyx. 2*n*=18. Sand hills, floodplains, dry riverbeds; 100–1530 m. CCo, SCoR. Mar–Jun

G. transmontana (H. Mason & A.D. Grant) A.D. Grant & V.E. Grant   TRANSMONTANE GILIA   **ST:** branches erect to spreading, 10–32 cm, ± tufted-woolly-hairy below middle. **LF:** ± tufted-woolly-hairy; basal in rosette, 1-pinnate-lobed, axis, lobes linear, lobes 3–11 mm, spreading, entire or few-toothed. **INFL:** open; fls 2–3; pedicels unequal, glandular. **FL:** calyx 2.5–4 mm, glabrous or ± glandular; corolla 5–8 mm, tube incl, red- or maroon-streaked, lower throat yellow, upper white, with purple spots near lobes bases or not, lobes > throat, ovate, acute, pink, lavender, or white; stamens, style ± exserted. **FR:** 5–6 mm, ≥ calyx, ovoid, tip pointed or round; valves detaching. **SEED:** 12–21(36). 2*n*=36. Rocky or sandy desert slopes, washes; 480–2250 m. s SCoRO, n WTR, SNE, DMoj; to UT. Similar to *G. clokeyi*. Mar–May

G. tricolor Benth.   BIRD'S-EYE GILIA   **ST:** branches many from base, spreading, 10–38 cm, lfy, glabrous or white-hairy below. **LF:** hairy in axils, on adaxial surfaces; basal in open cluster, 1–2-pinnate-lobed; upper palmate, lobes linear, entire. **INFL:** glands minute, black, slender-stalked. **FL:** calyx 3–7 mm, lobes erect or spreading, wider than membranes, flat, green, acute or acuminate; corolla 7–19 mm, tube incl, yellow, throat exserted, widely bell-shaped, yellow or orange with purple spots below lobes, lobes spreading, blue-violet at tip, paler below; stamens attached near corolla sinuses, unequal, exceeded by corolla lobes; stamens, style exserted. **FR:** 3–6 mm, ≤ calyx, ovoid. **SEED:** 9–36. 2*n*=18. Subspp. intergrade.

subsp. ***diffusa*** (Congdon) H. Mason & A.D. Grant (p. 1055) **INFL:** open clusters; pedicels 10–40 mm. **FL:** calyx 3–4 mm, membranes purple or colorless; corolla 7–13 mm, lobes 3–6 mm wide, gen pale, throat spots in pairs. **FR:** 3–6 mm, ovoid. **SEED:** 12–36. Open grassland, hills, valleys; 90–1530 m. NCo, NCoRO, NCoRI, CaRF, SNF, ScV (Sutter Buttes), SnJV, SnFrB, SCoRI, w DMoj. Mar–Jul

subsp. ***tricolor*** (p. 1055) **INFL:** dense, occ open clusters; fls 2–5; pedicels 1–5 mm, longer in fr. **FL:** calyx 4–7 mm, membranes purple; corolla 10–19 mm, lobes 4–8 mm wide, bright, distal tube, proximal throat yellow to orange, throat with 5 pairs of dark purple spots, the pairs merged, forming a ring or not. **FR:** 4–6 mm. **SEED:** 9–36. Open, grassland, hills, valleys; < 1200 m. CaRF, NCoR, SNF, SnJV, SnFrB, SCoR, ScV (Sutter Buttes). (Jan–Feb)Mar–May

G. yorkii Shevock & A.G. Day (p. 1055) MONARCH GILIA   **ST:** branches ± spreading, 10–25 cm, with long, soft, wavy hairs below, glandular puberulent above. **LF:** sparsely hairy; hairs translucent; lower lvs in non-persistent, suberect rosette, 1–2-pinnate-lobed, 1.5–2.5 cm, axis linear below, narrowly winged above, lobes distinct, lobes entire to toothed, teeth sharp-pointed; upper minute, entire to 2-toothed. **INFL:** open; pedicels spreading, in unequal pairs or 3s, thread-like, glands colorless, flat-topped, short-stalked. **FL:** calyx 3–3.6 mm, ± glandular, lobes wider than membranes, green; corolla 7–8.5 mm, 1.5–3 × calyx, white to pale lavender or blue, throat = tube, lobes ovate, obtuse; stamens reaching corolla throat top or ± exserted, style exceeding anthers. **FR:** 3–4.5 mm, ≥ calyx, widely ovoid. **SEED:** (1)6–12. 2*n*=18. Sunny to semi-shaded, sand-, gravel-filled terraces of limestone; 1290–1830 m. s SNH. May–Jul ★

GYMNOSTERIS

Dieter H. Wilken

Ann. **ST**: erect, gen 1, glabrous. **LF**: 0. **INFL**: head; bracts lf-like, in involucre; fls 1–5. **FL**: calyx urn-shaped, scarious, abruptly awned, puberulent; corolla salverform, white to lavender, throat gen yellow; stamens attached in throat, filaments < 0.5 mm. **FR**: ovoid. **SEED**: 1–5 per chamber, angled. 2 spp.: w US. (Greek: naked st) Self-pollinated.

G. parvula A. Heller (p. 1055)   **ST**: < 7 cm. **INFL**: bracts 2–13 mm, lanceolate to ovate, glabrous. **FL**: calyx < 4 mm; corolla < 7 mm, tube = calyx, lobes gen oblong, acuminate, white or tinged pink. Gravelly, sandy areas, gen in meadows or wet depressions in scrub; 2400–3700 m. n&c SNH, GB; to OR, WY, CO. May–Jun

IPOMOPSIS   SCARLET GILIA

Dieter H. Wilken

Ann, per, [± subshrub]. **ST**: gen branched at base. **LF**: alternate, simple, smaller upward, entire to pinnate- or palmate-lobed; lobes gen small-pointed at tip. **INFL**: clusters, lateral or open to head-like, terminal. **FL**: calyx gen bell-shaped, tube, sinuses membranous, glabrous to hairy, lobes gen small-pointed at tip; corolla gen salverform, radial or bilateral, white to red or lavender. **SEED**: slender, angled, ± winged, white to light brown. 30 spp.: w N.Am, se US, s S.Am. (Greek: like *Ipomoea*) Per cross-, ann gen self-pollinated. *I. depressa* moved to *Loeseliastrum*.

1. Corolla tube gen < 10 mm; infl open to head-like, terminal
  2. Ann
    3. Corolla ± bilateral, deep pink to white mottled pink; stamens exserted . . . . **[*I. effusa*]**
    3′ Corolla radial, gen white; stamens gen incl . . . . ***I. polycladon***
  2′ Per . . . . ***I. congesta***
    4. Lf pinnate-lobed . . . . subsp. ***congesta***
    4′ Lf palmate-lobed . . . . subsp. ***montana***
1′ Corolla tube 10–45 mm; infl clusters, lateral
  5. Corolla white to pink or lavender, ± speckled at lobe bases, tube 25–45 mm . . . . ***I. tenuituba***
  5′ Corolla gen red, mottled white on lobes, throat or not, tube 5–30 mm
    6. Corolla ± bilateral, lobes notched; stamens strongly exserted . . . . ***I. tenuifolia***
    6′ Corolla radial, lobes acute to acuminate; stamens incl or exserted
      7. Corolla tube 10–20 mm; anthers incl; e DMtns . . . . ***I. arizonica***
      7′ Corolla tube 20–30 mm; anthers exserted; KR, NCoRH, CaRH, SNH, GB . . . . ***I. aggregata***
        8. Pollen white, light yellow, or ± blue; cauline lf lobes acute . . . . subsp. ***aggregata***
        8′ Pollen gen blue; cauline lf lobes blunt to rounded. . . . . subsp. ***bridgesii***

I. aggregata (Pursh) V.E. Grant   Short-lived per. **ST**: erect, glabrous or glandular, ± hairy. **LF**: basal 3–5 cm, pinnate-9–11-lobed, withered at fl; cauline 5–7-lobed, glabrous to puberulent. **INFL**: clusters lateral, compact. **FL**: corolla radial, tube 20–30 mm, gen red with yellow mottling on throat, lobe bases, lobes acute to acuminate; stamens attached at > 1 level, exserted; style exserted. $2n$=14. 8 subspp., others in Rocky Mtns. Hybridizes with *I. tenuituba*.

  subsp. ***aggregata*** (p. 1055)   Fl gen once. **LF**: lobes acute. **INFL**: fls 3–7 per cluster. **FL**: calyx lobes 3–4 mm, acuminate; stamens exserted, pollen white, light yellow, or ± blue. Openings in forest, woodland; 1100–2500 m. KR, NCoRH, CaRH, n&c SNH, GB; to WA, CO. [*I. a.* subsp. *formosissima* (Greene) Wherry, misappl.] Jun–Sep

  subsp. ***bridgesii*** (A. Gray) V.E. Grant & A.D. Grant (p. 1055)   Fl gen > 1 once. **LF**: lobes blunt to rounded. **INFL**: fls 1–5 per cluster. **FL**: calyx lobes 1–2 mm, acute; 1–4 stamens exserted, pollen gen blue. Openings in forests, woodland; 1800–3300 m. SNH. Jun–Sep

I. arizonica (Greene) Wherry (p. 1055)   Per, fl gen once. **ST**: erect, glabrous or glandular, ± hairy. **LF**: basal 3–5 cm, pinnate-7–11-lobed; cauline hairy. **INFL**: fls 5–13 per cluster, on upper 1/3 of axis. **FL**: calyx lobes 1–3 mm; corolla radial, tube 10–20 mm, red, lobes 5–10 mm, narrowly acute; stamens attached at 1 level, incl; style incl. $2n$=14. Open, sandy to rocky areas in canyons; 1500–3100 m. e DMtns; s NV, n AZ. [*I. aggregata* subsp. *a.* (Greene) V.E. Grant & A.D. Grant] May–Oct

I. congesta (Hook.) V.E. Grant   Per. **ST**: decumbent to erect, 1–3 dm, glabrous to densely puberulent. **LF**: 1–4 cm, gen hairy, entire or pinnate- to palmate-lobed. **INFL**: head-like, terminal. **FL**: calyx 3–5 mm; corolla gen > calyx, tube gen < 10 mm, yellow, lobes gen oblong, white; stamens attached at 1 level, exserted; style incl. $2n$=14. 7 subspp. total, esp GB.

  subsp. ***congesta***   **ST**: 1–3 dm, branched above. **LF**: pinnate-3–5-lobed. Valleys, basins; 1200–2450 m. KR, CaRH, n&c SNH, MP, n NE; to OR, w Great Plains. May–Sep

  subsp. ***montana*** (A. Nelson & P.B. Kenn.) V.E. Grant (p. 1055)   Per, cespitose. **ST**: 1–2 dm. **LF**: palmate-3–5-lobed. Montane to alpine slopes; 1500–3700 m. CaRH, n&c SNH, GB; to OR, ID. [*I. c.* subsp. *palmifrons* (Brand) A.G. Day] May–Sep

I. polycladon (Torr.) V.E. Grant (p. 1055)   Ann. **ST**: decumbent to prostrate, < 1 dm, glandular-puberulent. **LF**: < 2 cm, entire to toothed or pinnate-5–7-lobed; basal, cauline crowded below infl, terminal lobe, axis equal in width, glabrous adaxially, puberulent abaxially. **INFL**: head-like, terminal. **FL**: corolla radial, 3–6 mm, gen white, lobes < 2 mm; stamens attached at 1 level, gen incl, filaments = anthers; style incl. **FR**: < 5 mm, ellipsoid. $2n$=14. Sandy, gravelly soils; 750–2150 m. GB, DMoj; to TX, n Mex. Mar–Jul

I. tenuifolia (A. Gray) V.E. Grant (p. 1055)   SLENDER-LEAVED IPOMOPSIS   Per. **ST**: 1–4 dm, sparsely puberulent. **LF**: basal 5–35 mm, pinnate-3–5-lobed, lobes remote, linear; cauline entire. **INFL**: open, in upper axils; fls 1–7. **FL**: calyx lobes < 3 mm; corolla ± bilateral, 15–28 mm, tube 5–10 mm, throat 6–11 mm, lobes 5–7 mm, oblong, notched, red mottled white on lobes, throat; stamens strongly exserted; style exserted. $2n$=14. Gravelly to rocky slopes, canyons; 100–1200 m. se PR (s San Diego Co.), DSon; Baja CA. Apr–Jun ★

I. tenuituba (Rydb.) V.E. Grant (p. 1055)   Per, fl gen once. **ST**: erect, glabrous or glandular, ± hairy. **LF**: basal 3–6 cm, pinnate-9–17-lobed, withered at fl; cauline gen puberulent. **INFL**: fls 3–7 per cluster, lower clusters spaced more than upper. **FL**: calyx lobes tapered; corolla radial, tube 25–45 mm, lobes white to pink or lavender, ± speckled at bases; stamens attached at > 1 level, incl, pollen white to yellow (blue); style incl to ± exserted. $2n$=14. Gravelly to rocky slopes; 2400–3050 m. s CaRH, SNH, MP, n SNE; to Rocky Mtns. [*I. aggregata* subsp. *attenuata* (A. Gray) V.E. Grant & A.D. Grant, misappl.] Hybridizes with *I. aggregata*. Jun–Sep

LANGLOISIA

Dieter H. Wilken & Steven L. Timbrook

1 sp. (Rev. A.B. Langlois, LA botanist, 1832–1990)

L. setosissima (Torr. & A. Gray) Greene   Ann, bristly, gen hairy; hairs branched, nonglandular. **ST**: erect. **LF**: gen 0 below, alternate, simple, linear or oblanceolate, teeth 3–5, at tip, each with 1 bristle, basal teeth of upper lvs reduced to clusters of 2–3 bristles. **INFL**: terminal, head-like; bracts lf-like; pedicels 0–short. **FL**: calyx lobes equal, bristle-tipped; corolla radial, funnel-shaped; stamens attached at or below sinuses, equal, exserted, pollen white to blue; style exserted. **FR**: lance-oblong, triangular in ×-section; outer wall of valve flat. **SEED**: gelatinous when wet. 2*n*=14. Self-compatible; gen cross-pollinated. Subspp. intergrade.

1. Corolla white to light blue, lobes 1/2 to ± = tube, dotted purple, gen with 2 yellow dots in middle of each; filaments > 3 mm . . . . .subsp. ***punctata***
1′ Corolla lavender to blue, lobes 1/3–1/2 tube, unmarked or streaked purple; filaments < 3 mm. . . . . . subsp. ***setosissima***

subsp. ***punctata*** (Coville) Timbrook (p. 1055)   LILAC SUNBONNET   Common. Washes, flats, slopes, gravelly to sandy soils; < 1800 m. SNE, DMoj; to NV. Smaller-fld populations in w ID, e OR. Feb–Jun

subsp. ***setosissima*** (p. 1055)   BRISTLY LANGLOISIA   Common. Washes, flats, slopes, gravelly to sandy soils; gen < 1800 m. SNE, D; to NV, AZ, n Mex. Jan–Jun

LATHROCASIS

Leigh A. Johnson

1 sp. (Latin: hidden sister) [Johnson & Weese 2000 W N Amer Naturalist 60:355–373]

L. tenerrima (A. Gray) L.A. Johnson (p. 1055)   DELICATE GILIA   Ann, open, stalked-glandular. **ST**: 3–35 cm, erect, thread-like, branched. **LF**: alternate, 0.5–4 cm, lanceolate to oblanceolate, entire (1–2 lobed), reduced upward. **INFL**: loose, fls 1, bract-opposed; pedicels spreading or recurved in fr. **FL**: calyx 1–2 mm, lobe margins membranous to tip, fused 0.6–0.8 × length; corolla 1.5–3.5 mm, tube, throat white, yellow spotted, incl, lobes white to lavender or ± blue; stamens attached at mid corolla; stamens, style ± exserted; pollen white. **FR**: 1–3 mm, ± spheric, valves spreading, detaching. **SEED**: 1 per chamber, gelatinous when wet. 2*n*=36. Sandy, gravelly slopes; 1100–3000 m. SNE; OR to MT, CO. [*Gilia t.* A. Gray] May–Sep

LEPTOSIPHON

Robert Patterson & Robyn Battaglia

Ann, per. **ST**: gen erect, gen branched from base. **LF**: cauline, opposite, entire or lobes 3–9, palmate, linear to narrowly lanceolate or spoon-shaped, gen not fused by membrane. **INFL**: head, open clusters, few-fld cyme, or fl 1; bracts ± lf-like, gen palmate-lobed, lobes gen not connected by translucent membrane; fls sessile or not. **FL**: sepals gen equal; corolla funnel-shaped, salverform, or bell-shaped, with hairy ring inside tube or gen not (determined at 10×); stamens attached at 1 level, pollen yellow. 30 spp.: w N.Am, Chile. (Greek: narrow tube, for corollas of some spp.) [Battaglia & Patterson 2001 Madroño 48:62–78] Calyx lobe membrane gen expressed as length relative to calyx or lobe length, or as width relative to calyx lobe.

1. Per
  2. Lf lobes 0 . . . . ***L. floribundus*** subsp. ***hallii***
  2′ Lf lobes palmate
    3. Lf lobes gen 3; pedicel gen 1–6 mm . . . . ***L. floribundus***
    4. Lf, calyx hairy . . . . subsp. ***floribundus***
    4′ Lf, calyx glabrous . . . . subsp. ***glaber***
    3′ Lf lobes gen 5; pedicel 0 or << 1 mm
      5. Corolla tube > calyx; lf lobes lance-linear; seeds > 2 mm . . . . ***L. pachyphyllus***
      5′ Corolla tube ≤ calyx; lf lobes gen linear; seeds < 2 mm. . . . . ***L. nuttallii***
      6. Lf, calyx sparsely hairy . . . . subsp. ***nuttallii***
      6′ Lf, calyx densely hairy
        7. Lf lobes 3–7 mm; NCoRH . . . . subsp. ***howellii***
        7′ Lf lobes 5–10 mm; SNE. . . . . subsp. ***pubescens***
1′ Ann
  8. Corolla not both salverform and in dense bracted heads
    9. Fls sessile
      10. Corolla 15–30 mm; c CW. . . . .***L. grandiflorus***
      10′ Corolla < 15 mm; s SW, DSon. . . . . ***L. lemmonii***
    9′ Fls pedicelled
      11. Hairy ring inside corolla 0, filaments glabrous; corolla gen < calyx
        12. St branches gen above base or 0; corolla tube << lobes; stamens exserted; seeds 1 per chamber. . . . ***L. harknessii***
        12′ St branches at base; corolla tube ≥ lobes; stamens incl; seeds > 1 per chamber . . . . ***L. pygmaeus***
        13. Corolla white; SNF, GV, CW, SW (exc ChI) . . . . subsp. ***continentalis***
        13′ Corolla blue; s ChI (San Clemente) . . . . subsp. ***pygmaeus***
      11′ Hairy ring inside corolla present or, if 0, then filaments hairy; corolla > calyx

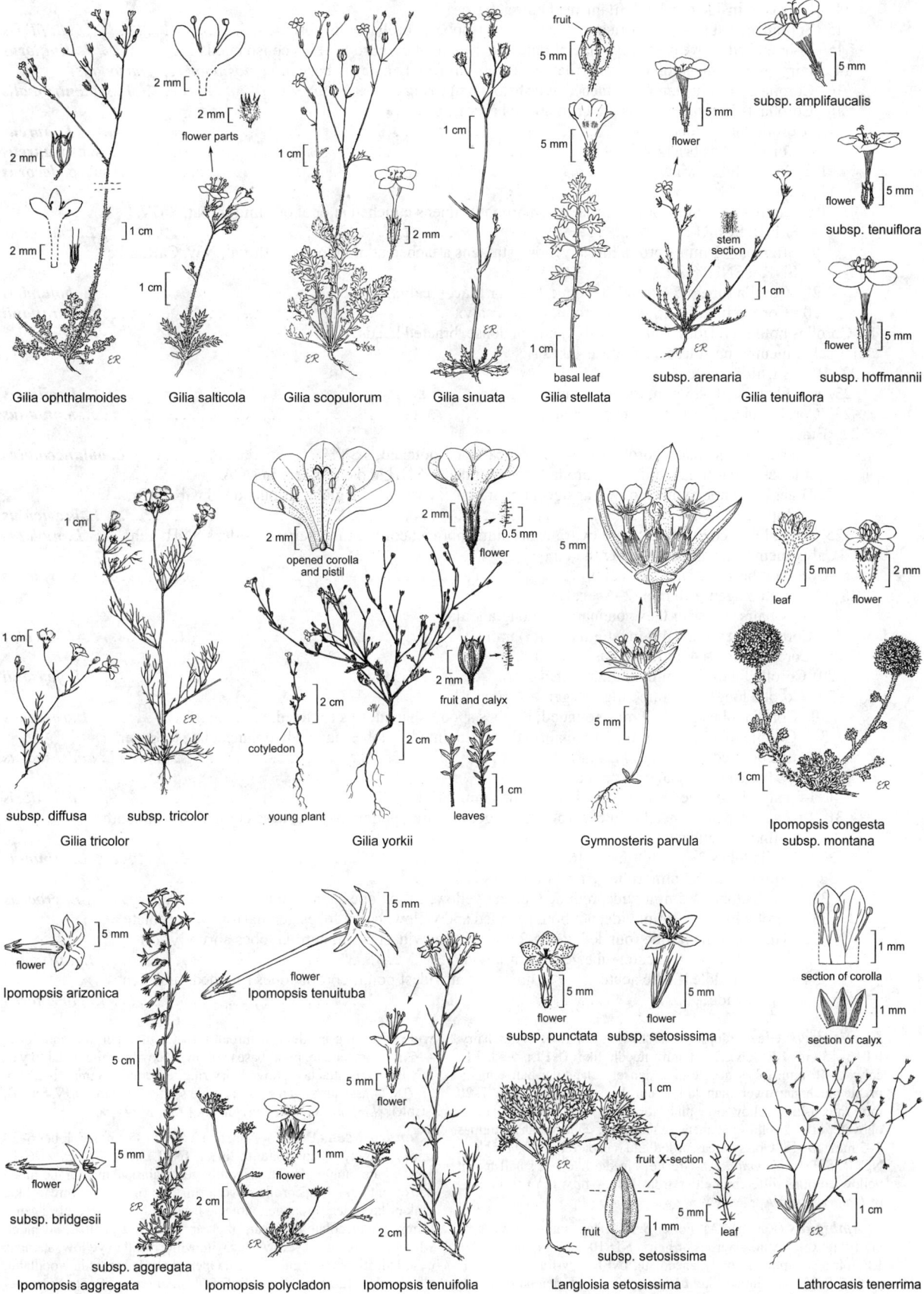

Gilia ophthalmoides

Gilia salticola

Gilia scopulorum

Gilia sinuata

Gilia stellata

subsp. arenaria

Gilia tenuiflora

subsp. amplifaucalis

subsp. tenuiflora

subsp. hoffmannii

flower parts

fruit

basal leaf

stem section

subsp. diffusa

subsp. tricolor

Gilia tricolor

opened corolla and pistil

flower

fruit and calyx

cotyledon

young plant

Gilia yorkii

leaves

Gymnosteris parvula

leaf

flower

Ipomopsis congesta subsp. montana

Ipomopsis arizonica

flower

Ipomopsis tenuituba

flower

subsp. bridgesii

flower

subsp. aggregata

Ipomopsis aggregata

Ipomopsis polycladon

flower

Ipomopsis tenuifolia

flower

subsp. punctata

flower

subsp. setosissima

subsp. setosissima

Langloisia setosissima

fruit X-section

fruit

leaf

section of corolla

section of calyx

Lathrocasis tenerrima

14. Hairy ring inside corolla 0; filaments hairy at base
  15. St branched at base; calyx membrane as wide as lobes; corolla veins obscure. . . . . ***L. filipes***
  15′ St branched above base; calyx membrane wider than lobes; corolla veins conspicuous . . . . ***L. liniflorus***
14′ Hairy ring inside corolla present; filaments glabrous (rarely hairy at base in *Leptosiphon septentrionalis*)
  16. Corolla 2–4 mm; stamens attached at or below hairy ring . . . . ***L. septentrionalis***
  16′ Corolla 4–15 mm; stamens attached above hairy ring
    17. Corolla tube < lobes . . . . ***L. aureus***
      18. Corolla lobes bright yellow. . . . . subsp. ***aureus***
      18′ Corolla lobes white . . . . subsp. ***decorus***
    17′ Corolla tube > lobes
      19. Stigmas 1–2 mm; corolla throat gen maroon; stamens attached at or above mid throat; SnJV, SnFrB, SCoRI . . . . ***L. ambiguus***
      19′ Stigmas ≤ 1 mm; corolla throat yellow; stamens attached at or below mid throat; NW, CaR, n&c SNF, n SNH, SnFrB, MP
        20. Corolla tube < 5 mm, white, pink, blue, or lilac, ≤ calyx. . . . . ***L. bolanderi***
        20′ Corolla tube 10–15 mm, maroon, > 1.5 × calyx . . . . ***L. rattanii***
8′ Corollas both salverform (tube >> calyx) and in dense, bracted heads
  21. Calyx membrane at least as wide as lobes
    22. Bracts white-ciliate
      23. Corolla tube 10–25 mm, lobes 2–4 mm. . . . . ***L. ciliatus***
      23′ Corolla tube 25–30 mm, lobes 5–8 mm . . . . ***L. montanus***
    22′ Bracts not white-ciliate
      24. Lf lobes oblanceolate; corolla lobes gen truncate or ± notched; s SNH . . . . ***L. oblanceolatus***
      24′ Lf lobes linear to lance-linear; corolla lobes rounded; s SNH and elsewhere in s CA
        25. Bract lobes not connected by translucent membrane; corolla tube glabrous outside; SnGb, SnBr, DMoj . . . . ***L. breviculus***
        25′ Bract lobes connected at base by translucent membrane; corolla tube hairy outside; s SNH, Teh . . . . ***L. nudatus***
  21′ Calyx membrane much narrower than lobes
    26. Corolla tube 7–8 mm, < 2 × calyx. . . . . ***L. serrulatus***
    26′ Corolla tube gen > 10 mm, 2–6 × calyx
      27. Calyx hairs sparse or 0 exc on margins, nonglandular
        28. Corolla lobes gen < 4 mm; stigmas ≤ 1(3) mm . . . . ***L. bicolor***
        28′ Corolla lobes > 4 mm; stigmas < or > 1 mm
          29. Corolla lobes 4–6 mm; stigmas gen < 2 mm . . . . ***L. jepsonii***
          29′ Corolla lobes 6–14 mm; stigmas gen > 2 mm
            30. Corolla lobes gen 6–9 mm, rounded; lf lobes spoon-shaped, tips rounded. . . . . ***L. rosaceus***
            30′ Corolla lobes gen > 8 mm, often abruptly soft-pointed; lf lobes narrowly oblanceolate to linear, tips pointed . . . . ***L. androsaceus***
      27′ Calyx hairs dense, glandular or not
        31. Lf, calyx lobes needle-like; corolla lobes 3(5) mm, bright yellow . . . . ***L. acicularis***
        31′ Lf, calyx lobes not needle-like; corolla lobes gen > 5 mm, yellow or otherwise colored, but not both < 5 mm and yellow
          32. Corolla lobes 2–4 mm, tube 5–16 mm . . . . ***L. minimus***
          32′ Corolla lobes ≥ 4 mm, tube gen > 14 mm
            33. Corolla lobes > 5 mm wide, rounded, bright yellow; calyx lobes gen deltate. . . . . ***L. croceus***
            33′ Corolla lobes < 5 mm wide, not both rounded and yellow; calyx lobes gen narrowly lanceolate
              34. Tip of middle lf lobe rounded, gen > 2 mm wide at widest point; corolla lobes abruptly soft pointed or gen truncate, tube gen > 1 mm wide. . . . . ***L. latisectus***
              34′ Tip of middle lf lobe acute, gen < 2 mm wide at widest point; corolla lobes rounded, tube gen < 1 mm wide. . . . . ***L. parviflorus***

L. acicularis (Greene) Jeps. BRISTLY LEPTOSIPHON Ann, hairy. **ST:** 3–15 cm. **LF:** lobes 3–11 mm, needle-like. **INFL:** head. **FL:** calyx 5–10 mm, lobes needle-like, densely glandular-hairy, membrane much narrower than lobes; corolla tube 7–20 mm, thread-like, yellow or ± pink, lobes 3(5) mm, oblanceolate to obovate, bright yellow; stamens exserted (or ± at mouth); stigmas 2–5 mm. 2*n*=18. Grassy areas, woodland, chaparral; < 700 m. NCo, NCoR, SnFrB. [*Linanthus a.* Greene] Needle-like lvs, small bright yellow corollas differentiate this from yellow-fld s CA populations of *L. parviflorus*. Apr–May ★

L. ambiguus (Rattan) J.M. Porter & L.A. Johnson SERPENTINE LEPTOSIPHON Ann, glabrous or hairy. **ST:** 10–20 mm, thread-like. **LF:** lobes 3–5 mm, narrowly lanceolate. **INFL:** few-fld cyme or fl 1; pedicel 5–25 mm, thread-like. **FL:** calyx 3–6 mm, gen hairy, membrane gen = calyx; corolla funnel-shaped, tube 3–6 mm, purple, with wide hairy ring inside near throat, throat 3 mm, gen maroon, lobes 4–6 mm, lanceolate, pink, bases yellow, occ with white distal to yellow; stamens attached above hairy ring, exserted; stigmas 1–2 mm. 2*n*=18. Grassy areas gen on serpentine soil; < 1000 m. SnJV, SnFrB, SCoRI. [*Linanthus a.* (Rattan) Greene] Apr–May ★

L. androsaceus Benth. Ann, hairy. **ST:** 5–45 cm. **LF:** lobes 6–30 mm, narrowly oblanceolate to linear. **INFL:** head. **FL:** calyx 6–13 mm, ciliate, nonglandular, abaxially gen glabrous, membrane much narrower than lobes; corolla salverform, tube 10–33 mm, thread-like, pink or lavender, throat gen violet at base, yellow above, lobes gen > 8 mm, oblanceolate to linear, often abruptly soft-pointed, lavender, white, or pink, bases yellow, occ with white distal to yellow; stamens exserted; stigma 2–6 mm. 2*n*=18. Open or shaded areas in woodland, chaparral; < 1200 m. NCoR, SnFrB. [*Linanthus a.* (Benth.) Greene] Similar to *L. rosaceus, L. latisectus*. Apr–Jun ★

L. aureus (Nutt.) E. Vilm. (p. 1061)   Ann. **ST**: thread-like, glabrous or hairy, glandular or not. **LF**: lobes 3–6 mm, linear. **INFL**: fl 1; peduncle 5–15 mm, thread-like. **FL**: calyx 4–6 mm, membrane wider than lobes; corolla funnel-shaped, tube 3–5 mm, with hairy ring inside and out, lobes 5–7 mm, oblanceolate; filaments glabrous, attached above hairy ring, in throat; stigmas 3–4 mm, exserted. New name possibly needed because *L. aureus* might have been used earlier for different taxon. Following subspp. overlap in geog but do not occur together.

　subsp. ***aureus***   **FL**: corolla tube, lobes bright yellow, throat gen brighter yellow or maroon. $2n=18$. Pinyon-pine/oak/juniper woodland, desert flats; < 2000 m. WTR (n slope), D; to UT, NM, Baja CA. [*Linanthus a.* (Nutt.) Greene subsp. *a.*] Mar–Jun

　subsp. ***decorus*** (A. Gray) J.M. Porter & L.A. Johnson   **FL**: corolla tube, lobes white, throat maroon. $2n=18$. Pinyon-pine/oak/juniper woodland, desert flats; < 2000 m. DMoj; to NV, AZ. [*Linanthus a.* subsp. *d.* (A. Gray) H. Mason] Mar–May

L. bicolor Nutt.   Ann, hairy. **ST**: 2–21 cm. **LF**: lobes 3–13 mm, narrowly obovate to linear. **INFL**: head, gen 1 fl open at a time in each infl, late morning, closed by early evening. **FL**: calyx 6–11 mm, ciliate, nonglandular, membrane obscure; corolla salverform, tube 12–32 mm, thread-like, ± red, throat yellow, lobes gen < 4 mm, obovate, rounded or truncate, pink, white, or pale yellow; stamens exserted; stigmas ≤ 1(3) mm. $2n=18$. Common. Open, grassy areas, chaparral, woodland; gen < 1700 m. NCoR, SNF, SnFrB, SCoR, s ChI, WTR. [*Linanthus b.* (Nutt.) Greene] Mar–Jun

L. bolanderi (A. Gray) J.M. Porter & L.A. Johnson   Ann. **ST**: 5–20 cm, thread-like, glabrous or hairy, glandular or not. **LF**: lobes 2–5 mm, linear. **INFL**: fl 1; peduncle 5–25 mm, thread-like. **FL**: calyx 4–6 mm, membrane 1/2 to ± = calyx, gen hairy, glandular; corolla funnel-shaped, tube < 5 mm, ≤ calyx, white, pink, blue, or lilac, with hairy ring inside, throat < 2 mm, gen yellow, lobes 2–3 mm, white or pink; stamens attached above hairy ring, exserted; stigmas ≤ 1 mm. $2n=18$. Drying areas in woodland, chaparral; 200–1700 m. NW, CaR, n&c SNF, SNH, SnFrB, MP; OR. [*Linanthus b.* (A. Gray) Greene] Some populations difficult to distinguish from *L. ambiguus* or *L. rattanii.* Mar–Jul

L. breviculus (A. Gray) J.M. Porter & L.A. Johnson   Ann, hairy. **ST**: 10–25 cm. **LF**: lobes 3–10 mm, linear to lance-linear. **INFL**: head; bract lobes not connected at base. **FL**: calyx 7–8 mm, membranes as wide as lobes; corolla salverform, tube 15–25 mm, maroon, glabrous, throat purple, lobes 4–6 mm, rounded, white, pink, or blue; stamens gen incl. $2n=18$. Deserts, dry montane areas; < 2400 m. SnGb, SnBr, DMoj. [*Linanthus b.* (A. Gray) Greene] May–Aug

L. ciliatus (Benth.) Jeps. (p. 1061)   WHISKER BRUSH   Ann, hairy. **ST**: 2–30 cm. **LF**: lobes 5–20 mm, linear. **INFL**: head; bracts white-ciliate. **FL**: calyx 7–10 mm, membranes ± 1/2 calyx, wider than lobes; corolla salverform, tube 10–25 mm, white or pink, hairy outside, throat yellow, lobes 2–4 mm, rounded or truncate, light to deep pink, base gen with darker pink or red spot; stamens exserted. $2n=18$. Common. Open or wooded areas; < 3000 m. CA-FP, MP. [*Linanthus c.* (Benth.) Greene] Mar–Jul

L. croceus (Eastw.) J.M. Porter & L.A. Johnson   COAST YELLOW LEPTOSIPHON   Ann, hairy. **ST**: much-branched from base, 2 cm, internodes short. **LF**: lobes 4–7 mm, narrowly obovate, ± thickened, middle lobe rounded. **INFL**: heads. **FL**: calyx 6–11 mm, lobes gen deltate, 1 mm wide at middle, densely glandular-hairy, membrane obscure; corolla salverform, tube 26–39 mm, yellow to ± red, throat yellow-orange, lobes 6–8 mm, obovate, rounded, bright yellow, gen with 2 red spots at base; stamens exserted; stigmas 2–5 mm. $2n=18$. Local, open, grassy areas, coastal bluffs; ± 0 m. n CCo (Montara, San Mateo Co.). [*Linanthus c.* Eastw.] Apr–May　★

L. filipes (Benth.) J.M. Porter & L.A. Johnson   Ann. **ST**: diffuse-branched at base, 5–20 cm, thread-like, hairy. **LF**: lobes 3–7 mm, linear. **INFL**: fl 1; peduncle 4–12 mm, thread-like. **FL**: calyx 2–3 mm, membrane 2/3 calyx, as wide as lobes, lobes hairy; corolla funnel-shaped, tube 1–2 mm, white or pink, throat yellow, lobes 2–3 mm, pink or white; stamens ± exserted, hairy at base; stigmas exserted. $2n=18$. Openings in woodland; < 1300 m. NCoR, CaRF, SN. [*Linanthus f.* (Benth.) Greene] Apr–Jul

L. floribundus (A. Gray) J.M. Porter & L.A. Johnson   Per, hairy or glabrous. **LF**: lobes gen 3, palmate, 8–20 mm. **INFL**: clusters; pedicels gen 1–6 mm. **FL**: calyx 7–9 mm, membrane much narrower than lobes; corolla funnel-shaped, tube 5–9 mm, white, throat yellow, lobes 5–9 mm, white; stamens incl or ± exserted. Gen self-pollinated.

　subsp. ***floribundus***   St, lf, calyx hairy. $2n=18$. Open, wooded areas; < 2300 m. SnGb, SnBr, PR. [*Linanthus f.* (A. Gray) Milliken subsp. *f.*] Mar–Aug

　subsp. ***glaber*** (R. Patt.) J.M. Porter & L.A. Johnson   St, lf, calyx glabrous. $2n=18$. Open, wooded areas; < 2100 m. SnBr, PR; Baja CA. [*Linanthus f.* subsp. *g.* R. Patt.] Mar–Aug

　subsp. ***hallii*** (Jeps.) J.M. Porter & L.A. Johnson   SANTA ROSA MOUNTAINS LEPTOSIPHON   St, lf, calyx glabrous. **LF**: lobes 0. Desert canyons; 1000–2000 m. e PR (Santa Rosa Mtns). [*Linanthus f.* subsp. *h.* (Jeps.) H. Mason] Apr–May　★

L. grandiflorus Benth. (p. 1061)   LARGE-FLOWERED LEPTOSIPHON   Ann. **ST**: branched above, hairy. **LF**: ascending, lobes 10–30 mm, linear, hairy or glabrous. **INFL**: head. **FL**: calyx 10–14 mm, hairy, membrane 2/3 calyx, wider than lobes; corolla funnel-shaped, 15–30 mm, with hairy ring inside, throat yellow, lobes white or pink; stamens incl. $2n=18$. Open, grassy flats, gen sandy soil; < 1200 m. NCo, CCo, SnFrB, n SCoR. [*Linanthus g.* (Benth.) Greene] Apr–Jul　★

L. harknessii (Curran) J.M. Porter & L.A. Johnson (p. 1061)   Ann. **ST**: 5–15 cm, thread-like, gen glabrous, branches gen above or 0. **LF**: lobes 5–15 mm, thread-like, glabrous or hairy. **INFL**: few-fld cyme or fl 1; pedicels 5–20 mm, thread-like. **FL**: calyx 1–3 mm, glabrous, membrane 1/2 calyx; corolla funnel-shaped, white or pale blue, gen incl in calyx, glabrous, tube << lobes; stamens exserted, glabrous. $2n=18$. Open flats; 1000–3200 m. KR, NCoR, CaR, SNH, GB; to WA. [*Linanthus h.* (Curran) Greene] Jun–Aug

L. jepsonii (Schemske & Goodwillie) J.M. Porter & L.A. Johnson   JEPSON'S LEPTOSIPHON   Ann, hairy. **ST**: 4–12 cm. **LF**: lobes 8–13 mm, narrowly obovate to linear. **INFL**: head. **FL**: calyx 6–9 mm, ciliate, nonglandular, membrane obscure; corolla salverform, tube 20–36 mm, thread-like, pink, throat yellow, lobes 4–6 mm, elliptic to obovate, white or pink; stamens exserted; stigmas gen < 2 mm. $2n=18$. Open or partially shaded grassy slopes; < 500 m. s NCoR. [*Linanthus j.* Schemske & Goodwillie] Apr–May　★

L. latisectus (E.G. Buxton) J.M. Porter & L.A. Johnson   BROAD-LOBED LEPTOSIPHON   Ann, hairy. **ST**: 3–20 cm. **LF**: lobes 3–10 mm, narrowly obovate to linear, tip of middle lobe rounded. **INFL**: head; fls staying open at night. **FL**: calyx 5–9 mm, lobes narrowly acute, < 1 mm wide at mid lobe, densely hairy, glandular, membrane obscure, corolla salverform, tube, 15–27 mm, red to purple, lobes 5–7 mm, < 5 mm wide, abruptly soft-pointed or gen truncate, red-pink to white; stamens exserted; stigmas 2–6 mm. $2n=18$. Open or partially shaded grassy slopes; < 1500 m. NCoR. [*Linanthus l.* E.G. Buxton] Corolla rarely closing after blooming. Similar to *L. rosaceus, L. androsaceus.* Mar–Jun　★

L. lemmonii (A. Gray) J.M. Porter & L.A. Johnson   Ann, hairy. **ST**: 5–15 cm, gen glandular. **LF**: lobes 2–5 mm, linear. **INFL**: head. **FL**: calyx 4–5 mm, membrane 1/2 calyx; corolla funnel-shaped, tube 1–3 mm, light yellow, with hairy ring inside, throat yellow or maroon, lobes 2–3 mm, yellow or white; stamens exserted. $2n=18$. Dry openings gen inland, chaparral, woodland, desert; < 1900 m. SW, w DSon; Baja CA. [*Linanthus l.* (A. Gray) Greene] Apr–Jun

L. liniflorus (Benth.) J.M. Porter & L.A. Johnson   Ann. **ST**: branched above base, 10–50 cm, glabrous or hairy. **LF**: lobes 1–3 cm, linear. **INFL**: few-fld cyme or fl 1; pedicels 1–3 cm. **FL**: calyx 3–5 mm, hairy, membrane wider than lobes; corolla funnel-shaped, white, tube 1–2 mm, throat yellow, 2–4 × tube, lobes 8–10 mm, veins gen purple; stamens exserted, hairy at base. $2n=18$. Woodland, openings, common on serpentine, desert; < 1700 m. CA-FP, w DMoj; to WA. [*Linanthus l.* (Benth.) Greene] Apr–Jun

L. minimus (H. Mason) R. Battaglia   Ann, hairy. **ST**: 2–10 cm. **LF**: lobes 3–13 mm, lobes narrowly obovate to linear. **INFL**: head, gen 1 fl open at a time in each infl, late morning, closed by early evening. **FL**: calyx 4–8 mm, densely hairy, membrane obscure;

corolla salverform, tube 5–16 mm, thread-like, ± red, throat yellow, lobes 2–4 mm, obovate, rounded or truncate, pink or white; stamen exserted; stigmas gen ≤ 1(3) mm. Grassy areas; < 500 m. NW; to WA. [*Linanthus bicolor* subsp. *m.* H. Mason] May

L. montanus (Greene) J.M. Porter & L.A. Johnson MUSTANG CLOVER Ann, hairy. **ST:** 10–60 cm. **LF:** lobes 20–30 mm, linear. **INFL:** head; bracts white-ciliate. **FL:** calyx ± 10 mm, hairy-ciliate, membranes 1/2 calyx, wider than lobes; corolla salverform, tube 25–30 mm, maroon, hairy outside, throat yellow, lobes 5–8 mm, white or pink with purple spot at base flanked by white; stamens at mouth or exserted. 2*n*=18. Dry openings in woodland; 300–1700 m. SNF. [*Linanthus m.* (Greene) Greene] Apr–Jul

L. nudatus (Greene) J.M. Porter & L.A. Johnson (p. 1061) Ann, hairy. **ST:** 5–30 cm. **LF:** lobes 3–12 mm, linear, gen fused by membrane for 1/3–1/2 their length, often glandular. **INFL:** head; bract lobes connected at base by translucent membrane. **FL:** calyx 5 mm, membrane > 1/2 calyx; corolla salverform, tube 10–12 mm, hairy outside, throat yellow, lobes 3–4 mm, white or pink; stamens exserted. 2*n*=18. Openings in woodland, chaparral; 600–2400 m. s SN, Teh. [*Linanthus n.* Greene] Apr–Aug

L. nuttallii (A. Gray) J.M. Porter & L.A. Johnson (p. 1061) Per, hairy. **ST:** many, 10–20 cm. **LF:** lobes gen 5, gen linear. **INFL:** clusters. **FL:** calyx 8–9 mm, membrane gen obscure, gen much narrower than lobes; corolla funnel-shaped, tube ≤ calyx, white or light yellow, throat yellow, lobes 4–5 mm, white; stamens incl or ± exserted; stigmas ± exserted. **SEED:** < 2 mm. 2*n*=18.

subsp. ***howellii*** (T.W. Nelson & R. Patt.) J.M. Porter & L.A. Johnson MOUNT TEDOC LEPTOSIPHON **LF:** lobes 3–7 mm, ± gray, densely hairy. Open Jeffrey pine forest, Douglas-fir/white-fir/ponderosa-pine forest, serpentine or not; 1400–1800 m. KR, NCoRH. [*Linanthus n.* (A. Gray) Milliken subsp. *h.* T.W. Nelson & R. Patt.] Jun–Jul ★

subsp. ***nuttallii*** **LF:** lobes 5–10 mm, green, sparsely hairy. Openings in forest; 500–2200 m. KR, NCoRO, CaRH, Wrn; to BC, Rocky Mtns, n Mex. [*Linanthus n.* subsp. *n.*] Jun–Aug

subsp. ***pubescens*** (R. Patt.) J.M. Porter & L.A. Johnson **LF:** lobes 5–10 mm, gray-green, densely hairy. Dry flats, openings in forest; 2800–3500 m. SNE; to NV. [*Linanthus n.* subsp. *p.* R. Patt.] Jun–Aug

L. oblanceolatus (Brand) J.M. Porter & L.A. Johnson SIERRA NEVADA LEPTOSIPHON Ann, hairy. **ST:** 2–12 cm. **LF:** lobes 5–15 mm, oblanceolate. **INFL:** head. **FL:** calyx ± 5 mm, membrane 1/2 calyx; corolla salverform, tube 8–12 mm, white to pink, throat yellow, hairy outside, lobes ± 2 mm, lobes gen truncate or ± notched, white; stamens incl. Open flats near meadows; 2800–3700 m. c&s SNH. [*Linanthus o.* (Brand) Jeps.] Jul–Aug ★

L. pachyphyllus (R. Patt.) J.M. Porter & L.A. Johnson Per, hairy. **ST:** 10–20 cm. **LF:** lobes gen 5, 10–16 mm, lance-linear. **INFL:** clusters. **FL:** calyx 9–10 mm, densely hairy, membrane much narrower than lobes; corolla funnel-shaped, tube 12–15 mm, white or pale yellow, throat yellow, lobes 6–9 mm, white; stamens gen exserted; stigmas ± exserted. **SEED:** > 2 mm. 2*n*=36. Open, wooded areas; 1700–2500 m. SNH, SNE. [*Linanthus p.* R. Patt.] Jun–Sep

L. parviflorus Benth. (p. 1061) Ann, hairy. **ST:** 4–40 cm. **LF:** lobes 2–18 mm, narrowly obovate to linear, tip of middle lobe acute. **INFL:** head; fls closing at night. **FL:** calyx 4–10 mm, densely hairy, glandular, membrane obscure; corolla salverform, tube 11–46 mm, thread-like, maroon, pink, or yellow, throat yellow, purple, or orange, lobes 4–8 mm, elliptic to oblanceolate, pink, white, yellow, or purple, often with red marks at lobe bases; stamens exserted; stigmas 1–7 mm. 2*n*=18. Abundant. Open or wooded areas; < 1200 m. CA-FP. [*Linanthus p.* (Benth.) Greene] Mar–Jun

L. pygmaeus (Brand) J.M. Porter & L.A. Johnson Ann. **ST:** 2–10 cm, puberulent, branches at base. **LF:** lobe 2–6 mm, thread-like. **INFL:** fl 1; pedicels 4–15 mm, thread-like. **FL:** calyx 3–5 mm, membrane 2/3 calyx; corolla funnel-shaped, white or blue, tube 2–3 mm, glabrous, lobes 1–2 mm; stamens exserted, glabrous.

subsp. ***continentalis*** (P.H. Raven) J.M. Porter & L.A. Johnson **FL:** corolla white. 2*n*=18. Dry openings; < 1700 m. SNF, GV, CW, SW (exc ChI). [*Linanthus p.* (Brand) J.T. Howell subsp. *c.* P.H. Raven] Mar–Jun

subsp. ***pygmaeus*** PYGMY LEPTOSIPHON **FL:** corolla blue. Dry openings; < 500 m. s ChI (San Clemente Island); Baja CA (Guadalupe Island). [*Linanthus p.* subsp. *p.*] Apr ★

L. rattanii (A. Gray) J.M. Porter & L.A. Johnson (p. 1061) RATTAN'S LEPTOSIPHON Ann. **ST:** 5–20 cm, thread-like, gen hairy, glandular. **LF:** lobes 2–5 mm, linear. **INFL:** fl 1; peduncle 5–25 mm, thread-like. **FL:** calyx 4–6 mm, membrane extending to calyx middle or tip, gen hairy, glandular; corolla funnel-shaped, tube 10–15 mm, exserted from calyx, maroon, with hairy ring inside, throat < 2 mm, yellow, lobes 2–3 mm, white; stamens attached above hairy ring, exserted; stigmas ≤ 1 mm. 2*n*=18. Dry openings in conifer forest; 1700–2000 m. NCoRH, NCoRO. [*Linanthus r.* (A. Gray) Greene] May–Jul ★

L. rosaceus (Hook. f.) R. Battaglia ROSE LEPTOSIPHON Ann, low, hairy. **ST:** 3–18 cm. **LF:** lobes 5–12 mm, narrowly obovate to linear, rounded. **INFL:** head. **FL:** calyx 7–9 mm, lobes deltate, ± glabrous, nonglandular, membrane much narrower than lobes; corolla salverform, tube 19–28 mm, pink or yellow, throat yellow, lobes gen 6–9 mm, obovate, rounded, white or pink; stamens exserted; stigmas 3–6 mm. 2*n*=18. Open, grassy slopes, coastal bluffs; ± 0 m. n CCo. [*Linanthus r.* (Hook. f.) Greene] Similar to *L. latisectus*, *L. androsaceus*. Apr–Jun ★

L. septentrionalis (H. Mason) J.M. Porter & L.A. Johnson Ann. **ST:** 5–30 cm, thread-like, glabrous or hairy. **LF:** lobes 5–20 mm, thread-like. **INFL:** fl 1; peduncle 10–25 mm, thread-like. **FL:** calyx 1–3 mm, membrane 2/3 calyx; corolla funnel-shaped, tube 1–2 mm, white, throat yellow, lobes 1–2 mm, white or pale blue; stamens attached at or below hairy ring, exserted. Common. Sagebrush scrub, pinyon/juniper woodland; 2000–3000 m. GB; to w Can, CO. [*Linanthus s.* H. Mason] May–Jul

L. serrulatus (Greene) J.M. Porter & L.A. Johnson MADERA LEPTOSIPHON Ann. **ST:** 5–18 cm, puberulent. **LF:** lobes 4–10 mm, linear, hairy. **INFL:** head. **FL:** calyx 5–7 mm, sparsely hairy, membrane < 1/2 calyx, much narrower than lobes; corolla funnel-shaped, tube 7–8 mm, white below, dark purple at transition to throat, throat 4–6 mm, yellow, lobes 5–7 mm, white; stamens incl. Openings in woodland, chaparral; 300–1300 m. s SN. [*Linanthus s.* Greene] Apr–May ★

LINANTHUS

Robert Patterson & J. Mark Porter

Ann, per, subshrub. **ST:** gen erect, gen branched from base. **LF:** cauline, alternate or opposite, entire or lobes 3–9, pinnate or palmate, linear to narrow-lanceolate or spoon-shaped. **INFL:** open or dense clusters or cyme or fl 1; bracts lf-like; fls sessile or not. **FL:** corolla funnel-shaped, salverform, or bell-shaped; stamens attached at 1 level, incl or exserted, pollen yellow. **FR:** capsule, valves 3(4). **SEED:** gen many, when wet gelatinous to not. 24 spp.: w N.Am. (Greek: flax fl) [Porter & Johnson 2000 Aliso 19:55–91] Other taxa in TJM (1993) moved to *Leptosiphon*.

1. Per, subshrub; lf lobes sharp-tipped
2. Lvs opposite, pinnate-3-lobed; corolla lobes 5–6; capsules gen 4-valved — SnJt . . . . ***L. jaegeri***

2′ Lvs gen alternate, palmate- or pinnate-3–many-lobed; corolla lobes 5; capsules gen 3-valved

  3. Corolla salverform, lobes 9–18 mm, gen pink; 0–1500 m . . . . *L. californicus*

  3′ Corolla funnel-shaped, lobes 7–10 mm, white or pink; 1700–4000 m . . . . *L. pungens*

1′ Ann; lf lobes not sharp-tipped

  4. Corolla gen open evening (if open daytime then limb > 10 mm wide); calyx membrane wider than lobes; corolla lacking red marks near throat

    5. Calyx glandular-hairy . . . . *L. jonesii*

    5′ Calyx hairy or not, not glandular

      6. Calyx 4–5 mm, hairy adaxially . . . . *L. arenicola*

      6′ Calyx 8–14 mm, glabrous adaxially

        7. Corolla tube without hairy pads where stamens attached . . . . *L. bigelovii*

        7′ Corolla tube with hairy pads where stamens attached . . . . *L. dichotomus*

          8. Corolla closed daytime, open evening; CA exc MP . . . . subsp. *dichotomus*

          8′ Corolla open daytime, evening; n SnFrB, NCoR . . . . subsp. *meridianus*

4′ Corolla open daytime, limb gen < 10 mm wide; calyx membrane gen narrower than lobes (if wider then corolla with red marks near throat)

  9. Lvs alternate (opposite or ± alternate in *Linanthus concinnus*)

    10. Lf linear to thread-like; corolla lobes yellow . . . . *L. filiformis*

    10′ Lf oblong, lanceolate, or narrowly oblanceolate; corolla lobes white

      11. Pl < 3 cm, st not glandular; calyx membrane ciliate; DMtns (Little San Bernardino Mtns), w DSon . . . . *L. maculatus*

      11′ Pl > 3 cm, st glandular; calyx membrane not ciliate; s SNH, GB, DMoj

        12. Corolla throat with 2 purple marks below lobes, tube and throat exserted from calyx . . . . *L. campanulatus*

        12′ Corolla throat lacking purple marks, tube and throat incl in calyx . . . . *L. inyoensis*

  9′ Lvs opposite (or ± alternate in *Linanthus concinnus*)

    13. Membrane ± not connecting calyx lobes; corolla tube inconspicuous

      14. St thread-like, openly branched at base; calyx lobes with purple marks at base . . . . *L. bellus*

      14′ St not thread-like, compactly branched; calyx lobes without purple marks at base

        15. Corolla bell-shaped, lobes white, base with 2 red marks . . . . *L. demissus*

        15′ Corolla funnel-shaped, lobes blue-purple or white, base with 1–2 purple marks . . . . *L. parryae*

    13′ Membrane connecting calyx lobes; corolla tube conspicuous

      16. Lf lobes 0; corolla lobe tips fine-toothed . . . . *L. dianthiflorus*

      16′ Lf lobes deep; corolla lobe tips ± entire

        17. Calyx membrane much wider than lobes — SnGb . . . . *L. concinnus*

        17′ Calyx membrane as wide as lobes

          18. Corolla lobes gen white, with 1 red mark at base, 3–4 mm, tube and throat ± exserted from calyx — c SnBr . . . . *L. killipii*

          18′ Corolla lobes gen pink, with 2 purple marks at base, 1 at each sinus, 5–8 mm, tube and throat long-exserted from calyx . . . . *L. orcuttii*

L. arenicola (M.E. Jones) Jeps. & V.L. Bailey SAND LINANTHUS Ann. **ST**: 1–8 cm, glabrous to ± hairy. **LF**: lobes 3–12 mm, adaxially hairy, abaxially glabrous. **INFL**: cyme. **FL**: open evening; calyx 4–5 mm, hairy adaxially, lobes unequal, membrane 2/3 calyx; corolla 5–7 mm, funnel-shaped, light yellow, throat purple; stamens attached in lower throat. **FR**: ≤ calyx, ellipsoid. **SEED**: 15–27, not gelatinous when wet. Dunes, sandy flats; 800–1400 m. s SNE, DMoj; NV. Mar–Apr

L. bellus (A. Gray) Greene (p. 1061) DESERT BEAUTY Ann. **ST**: openly branched at base, long, thread-like, gen glabrous. **LF**: lobes 2–4 mm, linear. **INFL**: fl 1, sessile. **FL**: calyx lobes 3–4 mm, with purple marks at base, membrane ± = but not connecting lobes; corolla widely funnel-shaped, tube 2 mm, purple, throat yellow, lobes 5–10 mm, widely lanceolate, pink with 2 purple marks at base; stamens attached in throat base. **FR**: < calyx, obovoid. **SEED**: 6–18, not gelatinous when wet. 2*n*=18. Desert chaparral areas in sandy soils; 1000–1400 m. se PR (se San Diego Co.); Baja CA. Mar–May ★

L. bigelovii (A. Gray) Greene Ann, glabrous. **ST**: erect, 5–20 cm. **LF**: simple, 10–30 mm, linear. **INFL**: cyme. **FL**: open evening; calyx 8–12 mm, glabrous adaxially, membrane much wider than lobes; corolla funnel-shaped, tube 4–5 mm, lobes 7–8 mm, cream or white with light purple shading on abaxial margins; stamens attached in upper tube. **FR**: < calyx, cylindric. **SEED**: 9–28, swelling, ± gelatinous when wet. 2*n*=18. Deserts, dry areas; gen < 1700 m. se SnFrB, SCoRI, WTR, SnGb (e edge), SnBr (e edge), SNE exc W&I, PR, D. Mar–May

L. californicus (Hook. & Arn.) J.M. Porter & L.A. Johnson (p. 1061) PRICKLY PHLOX Subshrub. **ST**: 3–10 dm. **LF**: lobes palmate, 3–12 mm, sharp-tipped. **FL**: open during day; calyx lobes equal, membrane much wider than lobes; corolla salverform, lobes 9–18 mm, elliptic, obovate, or round, gen pink; stamens attached in mid tube. **FR**: < calyx, obovoid. **SEED**: 12–30, not gelatinous when wet. Scrub, forest, coastal strand; < 1500 m. s CCo, s SCoRO, n&c SW. [*Leptodactylon c.* Hook. & Arn.; *L. c.* Hook. & Arn. subsp. *brevitrichomum* Gordon-Reedy; *L. c.* Hook. & Arn. subsp. *c.*; *L. c.* Hook. & Arn. subsp. *glandulosum* (Eastw.) H. Mason; *L. c.* Hook. & Arn. subsp. *leptotrichomum* Gordon-Reedy; *L. c.* Hook. & Arn. subsp. *tomentosum* Gordon-Reedy★] Jan–Jul

L. campanulatus (A. Gray) J.M. Porter & L.A. Johnson **ST**: 3–12 cm; branches spreading, glandular-hairy. **LF**: basal few, rosette 0; lower cauline narrowly oblanceolate, entire or 3–7-toothed, hairs white, jointed; upper cauline spreading or recurved, entire, gland-dotted. **INFL**: fls 2 per st; pedicel 3–11 mm, thread-like, glandular. **FL**: calyx lobes acuminate, membrane ± = but not connecting lobes; corolla 7–9 mm, 2–3 × calyx, bell-shaped, yellow with white lobes, 2 brown-purple or yellow marks ± below lobes, tube = 1 mm, throat > lobes; stamens attached near base of tube, unequal in length, incl; style incl. **FR**: 2–3 mm, < calyx. **SEED**: 7–8 per chamber, gelatinous when wet. 2*n*=18. Open, sandy flats; 900–2100 m. SNE; NV. [*Gilia c.* A. Gray] May–Jun

L. concinnus Milliken (p. 1061) SAN GABRIEL LINANTHUS Ann, glandular-hairy. **ST**: 1–12 cm. **LF**: opposite or ± alternate; lobes 8–15 mm, linear. **INFL**: dense; fls 3–7. **FL**: calyx 10 mm, membrane much wider than lobes; corolla funnel-shaped, white, tube 1–2 mm, throat 6–8 mm, yellow, lobes 5–10 mm, widely lanceolate, ± entire, base with 2 dark purple marks; stamens incl. **FR**: < calyx, obovoid. **SEED**: 6–12, not gelatinous when wet. Dry rocky slopes; 1700–2800 m. SnGb. May–Jun ★

L. demissus (A. Gray) Greene (p. 1061)　Ann. **ST:** decumbent, 2–10 cm, hairy or glandular. **LF:** lobes 6–10 mm, hairy or glabrous. **INFL:** fls gen sessile. **FL:** calyx 3–4 mm, tube obscure, membrane ± = but not connecting lobes; corolla bell-shaped, tube 1–2 mm, white or yellow-green, throat 4–6 mm, white or light yellow, lobes 2–3 mm, white, base with 2 red marks; stamens incl to ± exserted. **FR:** < calyx, obovoid. **SEED:** 18–32, not gelatinous when wet. $2n=18$. Limestone soils, desert pavement, sandy areas; < 1700 m. SNE, DMoj, DSon (ne corner); to UT, AZ. Mar–May

L. dianthiflorus (Benth.) Greene (p. 1061)　Ann. **ST:** erect or spreading, 5–12 cm, hairy. **LF:** 5–20 mm, thread-like or linear-oblong, entire, glabrous. **INFL:** open to dense, few-fld clusters, cyme, or fl 1, gen sessile. **FL:** calyx 10–16 mm, gen hairy, membrane wider than lobes; corolla funnel-shaped, tube 4–6 mm, light pink, throat 5–10 mm, yellow, lobes 8–12 mm, toothed at tip, pink or white, base with purple mark; stamens incl. **FR:** < calyx, obovoid. **SEED:** 6–18, not gelatinous when wet. Openings; < 1300 m. SCo, ChI, WTR, s SnBr, PR; to Baja CA. Feb–Jun

L. dichotomus Benth.　Ann, glabrous. **ST:** 5–20 cm, glaucous. **LF:** lobes 3–7, 10–22 mm, linear. **INFL:** cyme. **FL:** calyx 8–14 mm, glabrous adaxially, membrane much wider than lobes; corolla funnel-shaped, tube 7–10 mm, purple, throat white or cream, lobes 10–16 mm, white, gen with light purple shading on abaxial margins; stamens attached in lower tube, incl. **FR:** < calyx, ellipsoid to cylindric. **SEED:** 15–28, not gelatinous when wet. $2n=18$.

subsp. ***dichotomus*** (p. 1061)　EVENING SNOW　**FL:** gen open evening, closed daytime. $2n=18$. Common. Drying openings, esp serpentine; < 1800 m. CA (exc MP); NV, AZ. Apr–Jun

subsp. ***meridianus*** (Eastw.) H. Mason　**FL:** open daytime. Drying openings, esp serpentine; < 1700 m. n SnFrB, NCoR. Apr–Jun

L. filiformis (A. Gray) J.M. Porter & L.A. Johnson　Pl glabrous or sparsely glandular-hairy. **ST:** 5–15 cm, branches few to many, thread-like, glaucous. **LF:** basal rosette 0 or ephemeral; cauline spreading, 1–3 cm, thread-like, entire. **INFL:** fls gen 2 per st; pedicel 2–16 mm. **FL:** calyx 2–4 mm, tube < 1 mm, lobes acuminate, spreading in fr; corolla 4–7 mm, yellow, lobes 3–5 mm, oblong; stamens attached near base; stamens, style well exserted, gen exceeded by corolla lobes. **FR:** 2–4 mm. **SEED:** many, gelatinous when wet. $2n=18$. Open, rocky canyon slopes, washes; 300–1800 m. SNE, D; to UT, AZ. [*Gilia f.* A. Gray] Mar–May

L. inyoensis (I.M. Johnst.) J.M. Porter & L.A. Johnson (p. 1061)　**ST:** 3–10 cm, branches spreading, fine-glandular, white-hairy below. **LF:** white-jointed-hairy; basal few, 4–8 mm, oblanceolate to obovate, entire or toothed; cauline 3–6 mm, spreading or recurved, lower entire or toothed, upper entire. **INFL:** fls 2 per st; pedicels spreading, 4–8 mm, thread-like. **FL:** calyx 2–4 mm, tube < 1 mm; corolla 4–7 mm, tube, throat yellow, incl in calyx, lobes white, obovate, short-pointed; stamens attached near tube base, ± equal, exserted, style ± exserted. **FR:** 2–4 mm. **SEED:** 10–15 per chamber, gen not gelatinous when wet. Common. Open, sandy flats in pine forest or sagebrush scrub; 1900–2600 m. s SNH, SNE; NV. [*Gilia i.* I.M. Johnst.] Apr–Jul

L. jaegeri (Munz) J.M. Porter & L.A. Johnson (p. 1061)　SAN JACINTO LINANTHUS　Per, subshrub, cespitose, 2–10 cm; st, lvs, calyx lobes glandular-hairy. **LF:** opposite; lobes 3, pinnate, 10–15 mm, linear, sharp-tipped. **INFL:** crowded in upper axils. **FL:** calyx lobes gen 6, gen unequal, membrane much wider than lobes; corolla funnel-shaped, white, tube and throat 17–30 mm, lobes 5–6, 7–9 mm, oblanceolate; stamens gen 6, attached at throat. **FR:** < calyx, obovoid.

valves gen 4. **SEED:** 12–30, not gelatinous when wet. Dry rocky areas; 2900–3000 m. SnJt. [*Leptodactylon j.* (Munz) Wherry] May–Jul ★

L. jonesii (A. Gray) Greene　Ann, glandular-hairy. **ST:** 3–15 cm. **LF:** 10–20 mm, linear, entire. **INFL:** cyme. **FL:** open evening; calyx 7–8 mm, membrane wider than lobes; corolla funnel-shaped, tube 5–6 mm, white, throat yellow, lobes 3–5 mm, white or cream-yellow, with light purple shading on abaxial margins; stamens incl. **FR:** < calyx, cylindric. **SEED:** 9–28, not gelatinous when wet. Sandy flats, washes; < 900 m. D; AZ, Baja CA. Mar–May

L. killipii H. Mason (p. 1061)　BALDWIN LAKE LINANTHUS　Ann, hairy. **ST:** 5–15 cm. **LF:** lobes 3–10 mm, linear. **INFL:** fls gen sessile. **FL:** calyx 6–7 mm, membrane as wide as lobes; corolla funnel-shaped, tube 4–5 mm, white, throat 3–4 mm, yellow, lobes 3–4 mm, gen white, base with 1 red mark; stamens exserted. **FR:** < calyx, obovoid. **SEED:** 6–12, not gelatinous when wet. $2n=18$. Dry openings in pinyon/juniper woodland; 1700–2400 m. c SnBr (Baldwin Lake). Threatened by development, vehicles. May–Jun ★

L. maculatus (Parish) Milliken (p. 1061)　LITTLE SAN BERNARDINO MOUNTAINS LINANTHUS　Pl hairy. **ST:** 1–3 cm, not glandular. **LF:** 2–5 mm, oblong, lanceolate, or narrowly oblanceolate, entire. **INFL:** cluster, dense; fls ± sessile. **FL:** calyx tube < 1 mm, membrane ± = but not connecting lobes, ciliate; corolla 3–5 mm, white, lobes 1–2 mm, wavy, recurved, base with 1(2) red spots; stamens attached at tube base, equal, exserted, pollen yellow; style ± exserted. **FR:** < calyx, widely ovoid. **SEED:** 6–16, not gelatinous when wet. $2n=18$. Sandy washes, flats; 900–1100 m. DMtns (Little San Bernardino Mtns), DMoj exc DMtns (sw corner), w DSon. [*Gilia m.* Parish] Threatened by development. Apr–May ★

L. orcuttii (Parry & A. Gray) Jeps. (p. 1061)　ORCUTT'S LINANTHUS　Ann. **ST:** 5–10 cm, sparsely puberulent. **LF:** lobes 5–12 mm, linear, hairy. **INFL:** cluster or fl 1; fl sessile. **FL:** calyx 6–10 mm, membrane as wide as lobes; corolla funnel-shaped, tube 5–15 mm, stout, white grading to purple (on any given tube), throat 3–5 mm, white or yellow, lobes 5–8 mm, gen pink, with 2 purple marks at base, 1 at each sinus; stamens incl or at mouth. **FR:** < calyx, obovoid. **SEED:** 15–30, not gelatinous when wet. Chaparral, pine forest, desert scrub; 1100–2150 m. SnBr, PR, D; Baja CA. Apr–Jun ★

L. parryae (A. Gray) Greene　Ann. **ST:** decumbent or short-erect, hidden by lvs, 2–10 cm, glandular-hairy. **LF:** crowded, lobes 5–15 mm, linear, hairy. **INFL:** bracts crowded; fls sessile. **FL:** calyx 6–8 mm, tube obscure, membrane ± = but not connecting lobes; corolla funnel-shaped, blue-purple or white, tube 1 mm, throat 1–2 mm, lobes 8–12 mm, base with 1–2 purple marks, tip gen jagged; stamens incl. **FR:** < calyx, obovoid, angled. **SEED:** 18–36, swelling, ± gelatinous when wet. $2n=18$. Sandy, open, flat areas; < 2000 m. s SN, Teh, s SnJV, SCoRI, WTR, SNE exc W&I, DMoj. Some populations mixed in having blue- and white-fld pls. Mar–May

L. pungens (Torr.) J.M. Porter & L.A. Johnson (p. 1067)　GRANITE GILIA　Per, subshrub, gen hairy, glandular or not. **ST:** 1–3 dm. **LF:** alternate, lobes 3–7, palmate, sharp-tipped, middle gen longest. **FL:** gen open evening; calyx lobes gen unequal, membrane much wider than lobes; corolla funnel-shaped, tube and throat 7–15 mm, white or light pink, lobes 7–10 mm, obovate, white or pink with darker shading on abaxial margins; stamens attached at throat. **FR:** < calyx, narrowly ovoid. **SEED:** 15–30, not gelatinous when wet. $2n=18$. Open, rocky areas in montane, subalpine forest, alpine fell-fields; 1700–4000 m. CA; to BC, Rocky Mtns. [*Leptodactylon p.* (Torr.) Rydb.] Some proposed subspp. sort well elsewhere, but not in CA; further study needed. May–Aug

LOESELIASTRUM

Dieter H. Wilken & Steven L. Timbrook

Ann, bristly, gen soft; hairs simple. **ST:** erect. **LF:** gen 0 below, alternate, simple, 1–4 cm, linear to oblanceolate, entire to toothed; teeth acute to pointed or bristle-tipped. **INFL:** bracts lf-like; pedicels glandular. **FL:** calyx lobes equal, pointed to bristle-tipped; corolla bilateral, 2-lipped, white to deep pink, upper lip gen 3-lobed, lower 2-lobed; stamens attached at or below sinuses, unequal, gen curved, incl to exserted, pollen yellow; style gen exserted. **FR:** 2–5 mm, ovoid, 3-lobed in ×-section; outer wall of valve rounded or indented between walls separating chambers. **SEED:** gelatinous when wet. $2n=14$. 3 spp.: sw US, n Mex. (Latin: like *Loeselia*) [Timbrook 1986 Madroño 33:157–174] Self-compatible; self- to cross-pollinated.

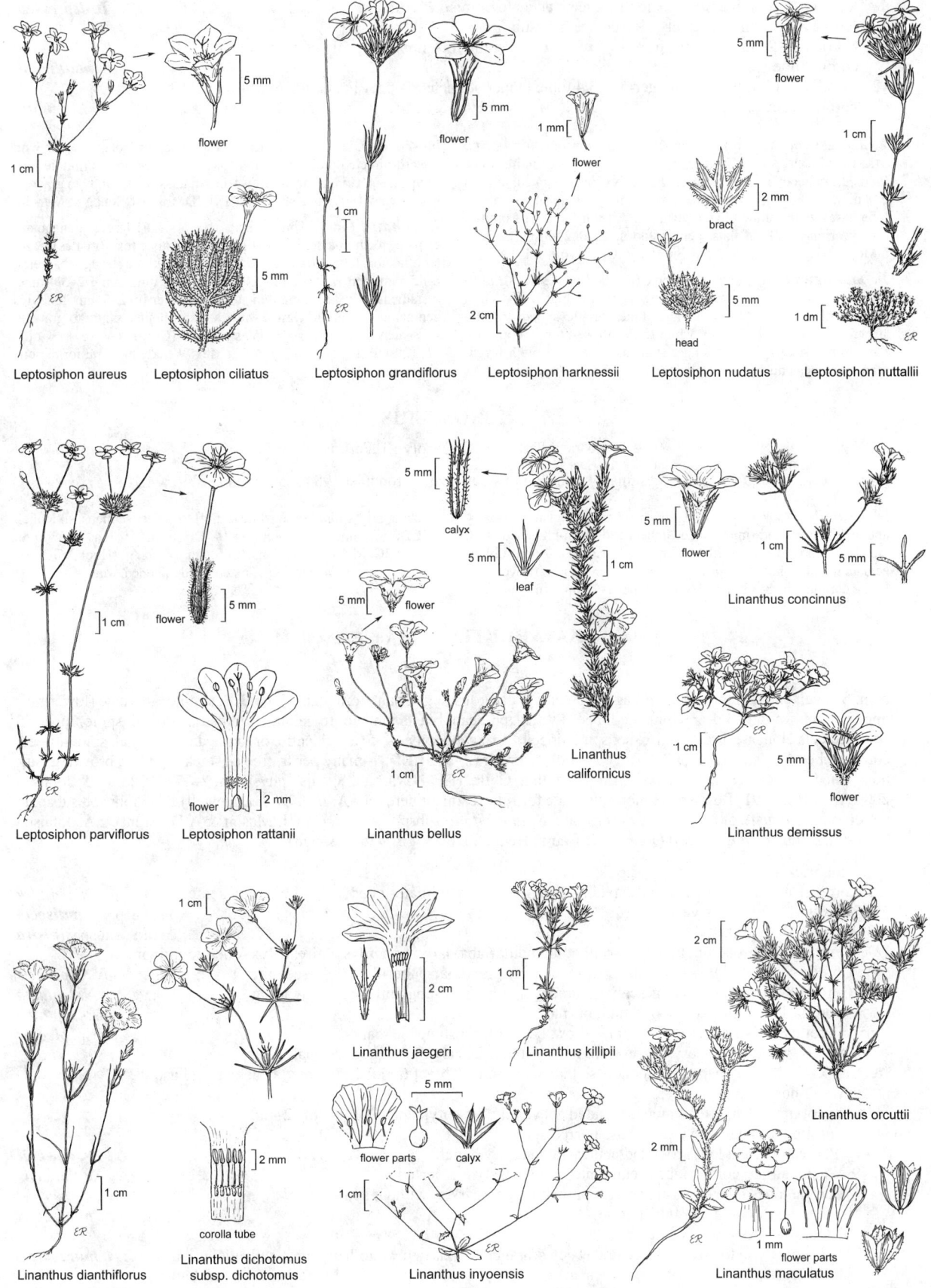

Leptosiphon aureus

Leptosiphon ciliatus

Leptosiphon grandiflorus

Leptosiphon harknessii

Leptosiphon nudatus

Leptosiphon nuttallii

Leptosiphon parviflorus

Leptosiphon rattanii

Linanthus bellus

Linanthus californicus

Linanthus concinnus

Linanthus demissus

Linanthus dianthiflorus

Linanthus dichotomus subsp. dichotomus

Linanthus jaegeri

Linanthus killipii

Linanthus inyoensis

Linanthus orcuttii

Linanthus maculatus

1. Corolla 5–8 mm; lf entire to ± toothed, teeth acute to pointed . . . . ***L. depressum***
1′ Corolla 8–21 mm; lf coarsely toothed, teeth bristle-tipped
  2. Corolla 11–21 mm, upper lip 3/4 to > tube; longer stamens > upper lip; calyx (exc bristles) 1/2–3/4
    corolla tube . . . . ***L. matthewsii***
  2′ Corolla 8–15 mm, upper lip gen 1/2–3/4 tube; longer stamens < upper lip; calyx (exc bristles) > 3/4
    corolla tube . . . . ***L. schottii***

L. depressum (A. Gray) J.M. Porter & L.A. Johnson DEPRESSED STANDING-CYPRESS **LF:** largest narrowly elliptic, entire to ± toothed, tip, teeth acute to pointed. **FL:** calyx lobes pointed; corolla 5–8 mm, ± 2-lipped, white, upper lip 1–2 mm, < 1/4 tube, > stamens, lobe bases with yellow blotch, lower lip ± 1 mm, lobe tips obtuse. Sandy or clay soils of flats, gentle slopes; 1000–2100 m. SNE, n DMoj; to UT. [*Ipomopsis d.* (A. Gray) V.E. Grant] Apr–Jul ★

L. matthewsii (A. Gray) Timbrook (p. 1067) DESERT CALICO **LF:** largest gen oblanceolate, teeth coarse, bristle-tipped. **FL:** calyx (exc bristles) 1/2–3/4 corolla tube, lobes bristle-tipped; corolla strongly bilateral, 11–21 mm, white to deep rose-purple, upper lip 5–11 mm, 3/4 to > tube, < longer stamens, lobe bases with bright maroon arches and white blotch, rarely suffused with yellow, lower

lip 4–7 mm, lobe tips truncate, 3-toothed, or notched, gen with inward-directed projections in sinuses on either side of middle lobe of upper lip. Common. Desert washes, flats, slopes, sandy to gravelly soils; gen < 1800 m. n TR, n PR, SNE, DMoj, w DSon; NV. Mar–Jul

L. schottii (Torr.) Timbrook (p. 1067) **LF:** largest gen oblanceolate, teeth coarse, bristle-tipped. **FL:** calyx (exc bristles) > 3/4 corolla tube, lobes bristle-tipped; corolla weakly to strongly bilateral, 8–15 mm, gen white to pink, upper lip 3–7 mm, gen 1/2–3/4 tube, > stamens, lobe bases maroon-streaked, lower lip 2–5 mm, lobe tips gen acute. Common. Desert washes, flats, slopes, sandy to gravelly soils; gen < 1800 m. sw SnJV, s SCoRO (Cuyama Valley), SCoRI, WTR, SNE, D; to sw UT, w AZ, n Mex. Weakly bilateral forms self-pollinated. Mar–Jun

MICROSTERIS

Robert Patterson & Carolyn J. Ferguson

1 sp. (Greek: small, to support, presumably referring to small size) [Cronquist 1984 Intermountain Flora 4:108]

M. gracilis (Hook.) Greene (p. 1067) Ann; gen glandular-hairy above. **ST:** erect and simple to decumbent and much-branched, < 20 cm. **LF:** opposite below, often alternate above, 10–30 mm, oblanceolate to lanceolate. **FL:** calyx 5–10 mm, membrane narrower than lobes; corolla tube 8–12 mm, ± yellow, lobes 1–2 mm, gen truncate or

notched, bright pink (base white or not) to white; stamens, style incl. **SEED:** gelatinous when wet. 2*n*=14. Dry to moist areas; < 3300 m. CA; to BC, MT, CO, Mex; also in S.Am. [*Phlox g.* (Hook.) Greene] Subspp., vars. have been recognized; study needed. Mar–Aug

NAVARRETIA NAVARRETIA

Leigh A. Johnson

Ann. **ST:** gen erect; branches ascending or spreading; glabrous to gen hairy, often glandular. **LF:** simple, alternate (lowermost opposite), entire to gen deeply pinnate-lobed. **INFL:** gen head, bracts pinnate- to palmate-toothed or -lobed, spine-tipped (fl 1–2, pedicels elongate, bracts entire, not spine-tipped). **FL:** calyx lobes 4–5, equal, entire or toothed, or unequal, spine-tipped; corolla lobes 4–5; stigmas 2 or 3. **FR:** gen ovoid, chambers 1–3. **SEED:** 1–many per fr, free or stuck together, brown, gelatinous when wet. 2*n*=18. 35 spp.: w N.Am, Argentina, Chile. (F.F. Navarrete, Spanish physician, ?–1742) [Porter & Johnson 2000 Aliso 19:55–91] Revised taxonomy, too late for full treatment here, incl *N. linearifolia* (Howell) L.A. Johnson subsp. *l.*, a ± cryptic segregate of *Navarretia sinistra*, and *N. linearifolia* subsp. *pinnatisecta* (H. Mason & A.D. Grant) L.A. Johnson [*N. sinistra* subsp. *pinnatisecta*] (Johnson & Cairns-Heath 2010 Syst Bot 35:618–628).

1. Pl not prickly; fls 1–2 on long pedicels; calyx lobes equal
  2. Corolla tube, throat ± glabrous abaxially; upper lvs often palmate-lobed; calyx lobes gen acute . . . . ***N. sinistra***
    3. Corolla 10–20 mm, style exserted . . . . subsp. ***pinnatisecta***
    3′ Corolla 5–10 mm, style incl. . . . . subsp. ***sinistra***
  2′ Corolla tube, throat minutely long-stalked-glandular abaxially; upper lvs entire; calyx lobes acuminate
    4. Corolla 6–8 mm, lobes blue-white; calyx densely glandular-puberulent; lvs ascending . . . . ***N. capillaris***
    4′ Corolla 8–21 mm, lobes pink; calyx sparsely glandular to subglabrous; lvs spreading . . . . ***N. leptalea***
      5. Corolla 13–21 mm, throat 6–8 mm, purple . . . . subsp. ***leptalea***
      5′ Corolla 8–15 mm, throat 2–5 mm, yellow, occ with short purple marks . . . . subsp. ***bicolor***
1′ Pl prickly, at least when dry; fls several to many in heads; calyx lobes gen unequal
  6. Calyx lobes strap-shaped, wider at base than membranes; bract tip lobes 7–many; fr dehiscing base
    upward, not to tip
    7. Axis of upper lvs, often bracts expanded above middle; cotyledons 2–3, deeply 2-lobed
      8. Corolla lobes white to cream, base purple-spotted
        9. Pl erect; corolla lobes rounded-acute; stamens, style incl. . . . . ***N. gowenii***
        9′ Pl spreading; corolla lobes acuminate; stamens, style exserted . . . . ***N. ojaiensis***
      8′ Corolla lobes blue to purple or light pink, base gen darker, without spots
        10. Corolla throat, upper tube purple . . . . ***N. pubescens***
        10′ Corolla throat, tube white
          11. Pl spreading to ascending, central st gen < branches; axis gen white-hairy, bracts glandular-hairy . . . ***N. mitracarpa***
          11′ Pl erect, central st gen > branches; axis, bracts puberulent or occ mid bract with narrow hairy
            band . . . . ***N. setiloba***

7′ Axis of upper lvs, bracts not expanded above middle; cotyledons 2, entire

  12. Corolla lobes purple, base spotted; fr 8-valved . . . . ***N. jepsonii***

  12′ Corolla lobes yellow, cream, or white, base spotted or not, tip occ light blue; fr 4-valved

    13. Infl puberulent, with longer hairs or not; corolla throat not spotted, lobes 4; style often long-exserted

      from fl bud . . . . ***N. cotulifolia***

    13′ Infl white-hairy; corolla throat gen spotted, lobes 4–5; style incl in fl bud

     14. Corolla white (occ lobe tips blue), lobes 4 or 5; fr 1–2 seeded

      15. Corolla 8–12 mm, style exserted . . . . ***N. eriocephala***

      15′ Corolla 6–8 mm, style incl . . . . ***N. heterandra***

     14′ Corolla yellow, lobes 5; fr 2–9 seeded . . . . ***N. nigelliformis***

      16. Corolla 12–16 mm; herbage dark green . . . . subsp. ***nigelliformis***

      16′ Corolla 9–11 mm; herbage gray green . . . . subsp. ***radians***

6′ Calyx lobes tapered, narrower at base than membranes; bract tip lobes gen 0–few; fr dehiscing tip

  downward (if dehiscing upward, then only when wet)

  17. Pl gen not glandular, st hairs gen recurved; seeds stuck together until wet; fr translucent, stuck to

    seeds, dehiscing at base when wet

    18. Axis of inner bracts concave-expanded at base; corolla lobes ovate

     19. Bract tips with 3 spreading, needle-like lobes; corolla lobes with 1 vein entering base; stigmas 2. . . . ***N. intertexta***

      20. Pl erect, open; corolla 7–11 mm, gen > calyx; filaments 1.5–3.5 mm. . . . . subsp. ***intertexta***

      20′ Pl ascending to spreading, dense; corolla 4–7 mm, gen ≤ calyx; filaments 0.5–2 mm . . . . subsp. ***propinqua***

     19′ Bract tips with few to several fine spines; corolla lobes with 3 veins entering base; stigmas 3

      21. Calyx hairs fine, obscure; calyx lobes entire . . . . ***N. subuligera***

      21′ Calyx coarsely appressed-hairy; calyx lobes toothed . . . . ***N. tagetina***

    18′ Axis of inner bracts widely membranous-winged at base, not concave-expanded; corolla lobes linear

      or narrowly ovate

     22. Mature infl head (fls sessile); pl prostrate; calyx tube long-soft-wavy-hairy

      23. Corolla 7–9 mm, tube ± exserted; lvs at base of head lobed . . . . ***N. prostrata***

      23′ Corolla 12–21 mm, tube long-exserted; lvs at base of head entire or lobed in proximal 1/2 . . . . ***N. myersii***

       24. Corolla blue, tube 1–1.2 × calyx; outer bract lobes many . . . . subsp. ***deminuta***

       24′ Corolla white, tube 2–4 × calyx; outer bracts lobes few . . . . subsp. ***myersii***

     22′ Mature infl dense cyme, fls subsessile or short-pedicelled; pl gen not prostrate (exc *Navarretia*

      leucocephala); calyx tube gen not long-soft-wavy-hairy (exc *Navarretia fossalis*)

      25. Calyx lobes hairy; filament ≤ 1 mm; s CA . . . . ***N. fossalis***

      25′ Calyx lobes glabrous or ± hairy; filaments 1–3 mm; n&c CA . . . . ***N. leucocephala***

       26. Stamens attached at corolla throat

        27. Corolla 7–9 mm, tube exserted, throat 2–3 mm wide . . . . subsp. ***leucocephala***

        27′ Corolla 4–6 mm, tube incl, throat 1–2 mm wide . . . . subsp. ***minima***

       26′ Stamens attached at or just below corolla sinuses

        28. Infl 2–20-fld; outer bracts gen with 3–5 entire or forked lobes . . . . subsp. ***pauciflora***

        28′ Infl 10–60-fld; outer bracts gen with ≥ 7 lobes, each 3–4-branched

         29. St gen elongate, branches ascending; corolla white . . . . subsp. ***bakeri***

         29′ St short, branches spreading; corolla blue . . . . subsp. ***plieantha***

  17′ Pl gen glandular, st nonglandular hairs 0 or not recurved; seeds not stuck together; fr opaque, not stuck

    to seeds, dehiscing tip to base

    30. Lf, bract axis ± linear (tapered in *Navarretia filicaulis*); terminal lobe >> to = lobed part of lf

     31. Branches ascending, sparse, along st; outer bracts pinnate-lobed, lowermost lobes departing above

      base, or base strongly concave

      32. Lf lobed near base; bract base concave; corolla bright purple. . . . . ***N. filicaulis***

      32′ Lf lobed near, above base; bract base flat; corolla lavender-white or yellow

       33. Corolla yellow . . . . ***N. breweri***

       33′ Corolla lavender-white . . . . ***N. peninsularis***

     31′ Branches gen spreading, dense, just below terminal head; outer bracts palmate-lobed at base, not concave

      34. Corolla 8–10 mm; stamens, style exserted . . . . ***N. prolifera***

      35. Corolla lobes yellow . . . . subsp. ***lutea***

      35′ Corolla lobes blue to purple . . . . subsp. ***prolifera***

      34′ Corolla 4–7 mm; stamens, style incl . . . . ***N. divaricata***

      36. Corolla tube white, lobes pink-tipped; longest bract 5–10 mm. . . . . subsp. ***divaricata***

      36′ Corolla purple; longest bract 7–20 mm . . . . subsp. ***vividior***

    30′ Lf (at least upper), bract axis gen wide; terminal lobe < lobed part of lf

     37. Outer bract tip lobes gen 3, ± separated from non-tip lobes

      38. Lf axis 2–6 mm wide, gen ≥ lobes; upper lf, bract lobes ascending. . . . . ***N. atractyloides***

      38′ Lf axis gen 1–2 mm wide (wider in *Navarretia hamata* subsp. *parviloba*), < lobes; lf, bract lobes

       spreading . . . . ***N. hamata***

      39. Corolla pale blue or lavender; bracts long-hairy, not glandular, 3 lobes at tip ± equal . . . . subsp. ***parviloba***

39′ Corolla bright pink or purple; bracts gland-dotted or glandular-hairy, 3 lobes at tip gen unequal
  40. Corolla tube incl, throat wide, lobes ± 0.3 × corolla . . . . subsp. *hamata*
  40′ Corolla tube exserted, throat narrow, lobes ± 0.25 × corolla . . . . subsp. *leptantha*
37′ Outer bract tip lobes 0, non-tip lobes ± adjacent to each other
  41. Bract bases not concave, lobes in 2 planes; anthers incl; stigma below anthers
    42. Corolla 5–7 mm, = calyx, lobes light blue . . . . *N. mellita*
    42′ Corolla 9–12 mm, > calyx, lobes dark blue . . . . *N. squarrosa*
  41′ Bract bases concave, lobes in 1 plane; anthers exserted or at mouth of throat, stigma at level of anthers
    43. Bract axis strap-shaped, much longer than wide . . . . *N. viscidula*
    43′ Bract axis widely ovate, not much longer than wide
      44. Corolla > calyx, purple; stamens, style exserted . . . . *N. heterodoxa*
      44′ Corolla = calyx, lavender to white; stamens, style ± at mouth . . . . *N. rosulata*

N. atractyloides (Benth.) Hook. & Arn. (p. 1067) Pl erect, 1° axis gen 1. **ST:** 5–29 cm, branches gen above, ascending; glandular-hairy. **LF:** strap-shaped to lanceolate, pinnate-lobed; cauline axis 2–6 mm wide, gen ≥ lobes; basal lobes spreading, upper unequal, ascending. **INFL:** bracts gland-dotted or glandular-hairy, recurved, base widely clasping, lobes at tip gen 3, unequal, ascending, ± separated from other lobes. **FL:** calyx lobes entire or toothed; corolla ± incl, 8–9 mm, gen purple (white), tube red-veined, lobes < 2 mm; stamens, style ± exserted; stigmas 3. **FR:** < calyx, dehiscing from tip. 2*n*=18. Open, rocky or sandy areas; < 2500 m. KR, NCoR, CaRH, c SNF, s SNH, CW, ChI, TR, PR, MP; OR, Baja CA. Like *N. hamata*, but odor not skunk-like, bract tips not hooked. May–Jul

N. breweri (A. Gray) Greene (p. 1067) Pl erect to spreading, gen dense, 1° axes 1–several. **ST:** 3–8 cm, gen as wide as high, glandular-puberulent, brown or red. **LF:** green, pinnate-lobed; axis < 0.5 mm wide; lobes above base, needle-like, entire or forked. **INFL:** outer bracts like lvs but axis wider, shorter, bract base flat. **FL:** calyx 6–9 mm, sinuses U-shaped, lobes needle-like, hairy adaxially; corolla 6–7 mm, yellow, tube minutely glandular abaxially, lobes ± 1 mm; stamens, style exserted; stigmas 3, minute. **FR:** obovoid, dehiscing from tip. Open, wet areas, meadows, streamsides; 1000–3300 m. SNH, SnBr, GB, DMtns; to WA, ID, CO, AZ. Jun–Aug

N. capillaris (Kellogg) Kuntze (p. 1067) Pl erect, 1° axes 1–several. **ST:** 4–20 cm, branches ascending; glandular-puberulent. **LF:** cauline, ascending, 1–2 mm wide, linear to narrowly lanceolate, entire. **INFL:** fls 1–3 per st, not in heads. **FL:** calyx densely glandular-puberulent, united to middle, membranes splitting to base in fr, lobes linear, acuminate, translucent at tip; corolla 6–8 mm, 2 × calyx, tube, yellow, minutely long-stalked-glandular, throat, lobes blue-white; stamens unequal, attached in upper throat, ± exserted; style < stamens, ± exserted. **FR:** < calyx, splitting to base, detaching. Open, wet, gravelly areas, meadows, streamsides, snow pockets; 200–3100 m. KR, NCoRH, CaRH, NCoRI, CaRH, SNH, WTR, PR, Wrn; to OR, ID, UT. [*Gilia c.* Kellogg] See note under *N. leptalea* subsp. *bicolor*. Jun–Aug

N. cotulifolia (Benth.) Hook. & Arn. COTULA NAVARRETIA Pl erect, 1° axes 1–several. **ST:** 3–31 cm, internodes < lvs, branches from base or above, ascending; glandular-hairy. **LF:** 2-pinnate-lobed; base hairy; axis, lobes < 0.5 mm wide, linear; lobes clustered, ascending; cotyledons 2, entire. **INFL:** heads lfy at base; bracts pinnate-lobed, lower lobes forked; bracts, calyces hairy at middle. **FL:** calyx lobes 4, strap-shaped, lobes entire; corolla exserted, 8–11 mm, pale yellow, tube thread-like, throat not spotted, lobes 4, ± 3 mm; stamens, style often long-exserted from fl bud; stigmas 2. **FR:** chamber 1; valves 4, dehiscing in lower 1/2. Heavy soils; < 500 m. NCoRI, ScV, SnFrB, SCoRI. May–Jun ★

N. divaricata (A. Gray) Greene MOUNTAIN NAVARRETIA Pl erect to spreading, 1° axis short, 2° axes gen several. **ST:** 1–10 cm; branches 2–5, just below terminal head, spreading, slender, brown or purple; glandular-puberulent or hairy. **LF:** basal thread-like, entire; cauline pinnate-lobed, lobes thread-like, mucronate. **INFL:** bracts, calyces white-hairy at middle, outer bracts palmate-lobed at base, middle lobe > lateral. **FL:** corolla incl or exserted, 4–7 mm, lobes < 2 mm; stamens, style incl; stigmas 3, minute. **FR:** 2–4 mm, dehiscing tip to base.

  subsp. ***divaricata*** (p. 1067) **INFL:** longest bracts 5–10 mm,

bract tips gen glabrous. **FL:** corolla 4–5 mm, ≤ calyx, white, throat yellow, lobe tips pink. Open, gravelly, often volcanic areas; 100–2600 m. NW, CaR, SN, s SCoRO (Santa Barbara Co.), GB; to WA, ID. Jun–Aug

  subsp. ***vividior*** (Jeps. & V.L. Bailey) H. Mason (p. 1067) **INFL:** longest bracts 7–20 mm, bract lobes glandular-puberulent to tip. **FL:** corolla exserted, 5–7 mm, purple. Clay or volcanic soil; 60–2000 m. NW, CaR, ScV, SN, MP; OR. Jun–Aug

N. eriocephala H. Mason (p. 1067) HOARY NAVARRETIA Pl erect, 1° axis gen 1. **ST:** 5–25 cm, often branched above, white-hairy, hairs gen recurved. **LF:** 2-pinnate-lobed; axis, lobes thread-like; upper lobes clustered, ascending; cotyledons 2, entire. **INFL:** dense, white-woolly; bract lobes puberulent, tips exposed, spreading. **FL:** calyx lobes 4–5, strap-shaped, wider at base than membranes, hairy at middle, longest toothed; corolla ± exserted, 8–12 mm, white, throat spots gen 0, lobes 4–5, 2–3 mm, tips occ blue; stamens unequal, longest exserted; style exserted, stigmas 2. **FR:** obovoid; chamber 1; valves 4, dehiscing in lower 1/2. Heavy soil of seasonally wet flats; < 400 m. n&c SNF, ScV, e SnFrB. Intergrades with *N. heterandra*. May–Jun ★

N. filicaulis (A. Gray) Greene (p. 1067) Pl erect, 1° axis gen 1. **ST:** 7–18 cm, branches along st, ascending, slender; glandular-puberulent. **LF:** 1–3 cm, 1-pinnate-lobed near base; axis, lobes narrow, tapered; lobes 0 or 2–6, terminal ± = to >> lobed part of lf. **INFL:** bracts gland-dotted, pinnate-lobed near base, base wide, concave; outer bracts lf-like. **FL:** calyx 3–4 mm; corolla 5–7 mm, bright purple, tube thread-like, exserted, lobes 1–2 mm, narrowly ovate; stamens attached in lower throat; stamens, style exserted; stigmas 2, minute. **FR:** 2–4-valved, dehiscing tip to base. Open areas, chaparral, woodland, gravel, clay; 300–1200 m. CaR, SN, MP. Jun–Jul

N. fossalis Moran SPREADING NAVARRETIA Pl spreading, gen not prostrate, 1° axis gen 1, short. **ST:** 1–15 cm, branches from base or below heads, spreading; sparsely hairy, hairs recurved. **LF:** subglabrous, pinnate-lobed; axis, lobes linear; lobes few, entire or forked. **INFL:** 1–2 cm wide; fls clustered; longest bracts gen < 2 × head, hairy below middle, membranous-winged at base, gen ciliate; bract lobes few, 2–4-branched. **FL:** calyx tube hairy, ± glandular at base, narrower at base than membranes, lobes glabrous, membranes truncate, ciliate; corolla exserted, 4–7 mm, white, lobes linear, < 1 mm wide; stamens, style ± exserted, filament ≤ 1 mm; stigmas 2, minute. **FR:** chambers 2, translucent, stuck to seeds until wet. **SEED:** separating when wet. 2*n*=18. Vernal pools, ditches; 30–1300 m. s SCoRO (San Luis Obispo Co.), SW, DMoj (Los Angeles Co.); Baja CA. Apr–Jun ★

N. gowenii L.A. Johnson LIME RIDGE NAVARRETIA Pl erect, 1° axis gen 1. **ST:** 5–30 cm, tan to red-brown; branches above, ascending, glandular. **LF:** 2-pinnate-lobed; axis, lobes linear, tip linear to lanceolate, many-toothed; glandular-hairy; cotyledons 2–3, deeply lobed. **INFL:** bracts, calyces glandular hairy at middle; outer bracts 1.5–2 × head, expanded at base, gen narrower above; inner bracts shorter, base clasping, tip narrow, toothed. **FL:** calyx lobes strap-shaped, entire or longest toothed; corolla 8–10 mm, > calyx, white to cream with purple spot at lobe bases, lobes 1.2–2 mm, tips rounded-acute, entire, glandular abaxially; stamens attached at ± 1 level in throat, incl to ± exserted; style incl, stigmas 2, minute. **FR:** chamber 1; valves 4, dehiscing in lower 1/2. Clay, serpentine soils; 200–300 m. SnJV (Stanislaus Co.), SnFrB (Contra Costa Co.). May–Jun ★

N. hamata Greene   Pl erect, 1° axes gen 1–5; odor gen skunk-like. **ST**: 8–30 cm, branches ascending; glandular-puberulent. **LF**: pinnate-lobed; axis linear to widely lanceolate; lobes spreading, not hooked; tip 3-lobed, ± hooked. **INFL**: bracts widely clasping, outer lanceolate, recurved, hooked at tip, pinnate-lobed, lobes spreading, at tip gen 3, ± separated from others. **FL**: calyx lobes entire or toothed; corolla gen purple or ± pink; stigmas 3. **FR**: < calyx, dehiscing tip to base. Subspp. intergrade.

subsp. *hamata* (p. 1067)   **LF**: cauline axis linear or lanceolate. **INFL**: gen terminal; bracts gland-dotted or glandular-hairy, lobes gen unequal. **FL**: calyx lobes entire or toothed; corolla tube incl, lobes 3–5 mm. 2*n*=18. Dry, sandy, rocky places in coastal, inland chaparral; < 1200 m. TR, PR. Pls in TR with wider lvs, axillary infl much like *N. hamata* subsp. *parviloba* but with larger, brighter fls. Apr–Jun

subsp. *leptantha* (Greene) H. Mason (p. 1067)   **LF**: axis, lobes linear. **INFL**: terminal; bracts gland-dotted or glandular-hairy, lobes of hook gen unequal. **FL**: calyx lobes gen entire; corolla tube exserted, lobes 2–3 mm. Dry, sandy, rocky places in coastal chaparral; < 700 m. SCo, ChI, PR; Baja CA. Apr–Jun

subsp. *parviloba* A.G. Day (p. 1067)   **LF**: axis linear to widely lanceolate, narrowest below hook, lobes of hook equal, spreading, middle recurved. **INFL**: axillary, terminal; bracts long-hairy, not glandular, hooks like those of lvs, 3 lobes at tip ± equal. **FL**: calyx lobes entire or largest 2-toothed; corolla tube incl, lobes 2–3 mm. Open, sandy areas, often sand hills; < 1000 m. CCo, SnFrB, SCoRO. Apr–Jun

N. heterandra H. Mason   TEHAMA NAVARRETIA   Pl erect to ± spreading, 1° axes 1–several. **ST**: 3–11 cm, branches ascending, white-hairy, hairs gen recurved. **LF**: 2-pinnate-lobed; axis linear; lobes linear or needle-like, spreading; cotyledons 2, entire. **INFL**: dense, densely white-hairy in center; bract tips spreading, often red, glabrous. **FL**: calyx lobes 4–5, strap-shaped, hairy above middle, longest toothed; corolla 6–8 mm, white, throat with purple spots below lobes, lobes 4–5, tips occ blue; stamens exserted, unequal; style incl; stigmas 2. **FR**: obovoid; chamber 1; valves 4, dehiscing in lower 1/2. Heavy soil, vernal pools, wet or drying flats; < 1100 m. NCoRI, CaR, ScV, e SnFrB, SCoRI, MP; OR. May–Jun ★

N. heterodoxa (Greene) Greene (p. 1067)   Pl erect, 1° axes 1–several; odor skunk-like. **ST**: 8–24 cm, branches ascending; glandular-hairy. **LF**: pinnate-lobed; basal axis, lobes thread-like; cauline recurved at tip, axis wider than thread-like, lobes acuminate. **INFL**: bract axis widely ovate, ± longer than wide, bract palmate-lobed, glandular, lobe length < axis width, middle lobe acuminate, tip recurved; fls many. **FL**: calyx 5–8 mm; corolla 6–11 mm, > calyx, purple, lobes ± 2 mm, narrowly oblong; stamens, style exserted, stigmas 3, minute. **FR**: dehiscing tip to base. Open, rocky, gen serpentine slopes; 100–900 m. s NCoRI, s NCoRO, SnFrB. May–Jun

N. intertexta (Benth.) Hook.   Pl erect, narrow to as wide as tall, 1° axes 1–several. **ST**: 5–25 cm, branches ascending or spreading; white-hairy, hairs gen reflexed. **LF**: 1–2-pinnate-lobed, glabrous or white-hairy near base; axis, lobes needle-like, spreading at tip. **INFL**: bracts, calyces white-hairy about middle; bracts pinnate-lobed, lower lobes needle-like, forked, 3 lobes at tip spreading; inner bracts clasping, base expanded-concave, narrowly membrane-margined. **FL**: calyx membranes V-shaped, hairy adaxially, lobes toothed or not; corolla gen white, lobes ovate, white to light blue, 1 vein entering base; stamens, style exserted, stigmas 2. **FR**: chambers 1–2, translucent below middle, stuck to seeds until wet. **SEED**: dark brown, pitted.

subsp. *intertexta* (p. 1067)   Pl erect, open. **ST**: 8–28 cm, branched. **INFL**: densely white-hairy; bracts 10–20 mm, gen exceeding fls by 1.5–2 times. **FL**: corolla 7–11 mm, gen > calyx, blue or white; filaments 1.5–3.5 mm. Open, seasonally wet areas, meadows, vernal pools; < 2100 m. KR, NCoR, CaR, SNF, GV, SnFrB, SCoR, SW, MP; to BC, ID, NV, Baja CA. May–Jul

subsp. *propinqua* (Suksd.) A.G. Day (p. 1067)   Pl ascending to spreading, dense. **ST**: 3–10 cm, gen wider than high, branched. **INFL**: sparsely hairy; bracts 5–10 mm, gen exceeding fls by < 1.5

times. **FL**: corolla 4–7 mm, gen ≤ calyx, white, lobes white to light blue; filaments 0.5–2 mm. Open, seasonally wet areas, meadows, disturbed sites, vernal pools; 800–2500 m. NCoRH, CaR, SNH, PR, MP; to BC, ID, UT. Jun–Aug

N. jepsonii Jeps. (p. 1069)   JEPSON'S NAVARRETIA   Pl erect to ascending, 1° axes 1–several. **ST**: 5–15 cm, branches ascending, ± red; densely white-puberulent. **LF**: glabrous or ± puberulent, 1–2-pinnate-lobed; axis, lobes needle-like; cotyledons 2, entire. **INFL**: bracts dark red, coarsely white-hairy below, lobes gland-dotted. **FL**: calyx coarsely white-hairy below, lobes strap-shaped, wider at base than membranes, toothed or not; corolla 9–11 mm, tube white, lobes 2–3 mm, purple with darker spot at base; stamens, style exserted; stigmas 2. **FR**: chamber 1; valves 8, dehiscing in lower 1/2. Open, grassy or clay flats, serpentine areas; 150–800 m. NCoRI. Apr–Jun ★

N. leptalea (A. Gray) L.A. Johnson   Pl erect, 1° axis gen 1. **ST**: 4–33 cm, branches ascending to spreading; subglabrous to glandular, glands black. **LF**: cauline, spreading, linear to narrowly elliptic, gen entire (lower pinnate-lobed). **INFL**: not heads; pedicels elongate, thread-like. **FL**: calyx sparsely glandular to subglabrous, lobes tapered, fine-pointed, membranes in fr splitting; corolla tube gen exserted, minutely glandular abaxially, lobes pink; stamens unequally attached in throat; style, longest stamens well exserted. **FR**: < calyx; valves detaching from tip. [*Gilia l.* (A. Gray) Greene] Forms large showy populations. Subspp. intergrade.

subsp. *bicolor* (H. Mason & A.D. Grant) L.A. Johnson (p. 1069)   **ST**: 4–26 cm, branches ± spreading. **FL**: corolla 8–15 mm, throat 2–5 mm, < 2 × lobes, tube, throat yellow, occ with short purple lines below lobes. Open rocky areas in forest, meadows; 1500–3100 m. SNH. [*Gilia l.* (A. Gray) Greene subsp. *b.* H. Mason & A.D. Grant] Small pls (1–4 cm, corolla ± 6 mm) in rocky meadows, c&s SNH, 2200–3100 m, differ from *N. capillaris* in having pink corolla lobes, exserted stamens, more spreading habit. Jun–Sep

subsp. *leptalea* (p. 1069)   **ST**: 8–33 cm. **FL**: corolla 13–21 mm, throat 6–8 mm, 2–3 × lobes, tube yellow, throat purple with yellow veins. 2*n*=18. Open rocky areas in forest, meadows; 900–2100 m. CaR, SN, MP; OR. [*Gilia l.* (A. Gray) Greene subsp. *l.*] Jun–Aug

N. leucocephala Benth.   Pl erect to prostrate, 1° axes 1–several. **ST**: 3–15 cm, branches ascending to prostrate; puberulent or hairy below heads, hairs recurved. **LF**: 1–2-pinnate-lobed; lobes linear; lower lvs glabrous, upper hairy at base. **INFL**: outer bracts lf-like, lobes needle-like, 2–4-branched at base, lobes near tip 2, shorter, entire, ascending; inner bracts simpler, base membranous-winged, ciliate, wing width > midrib. **FL**: calyx lobes glabrous or ± hairy, tapered, gland-dotted below, membranes truncate, ciliate; corolla 4–10 mm, lobes with 1 vein entering base; stamens exserted; stigmas 2, minute. **FR**: translucent, stuck to seeds until wet. **SEED**: separating when wet. Some subspp. intergrade. Variable with water level, duration.

subsp. *bakeri* (H. Mason) A.G. Day   BAKER'S NAVARRETIA   **ST**: 2–10 cm, gen erect; branches ascending. **INFL**: longer bracts < 2 × head, lobes 2–4-branched abaxially. **FL**: calyx lobes ± hairy or not; corolla white, tube incl, throat 1–2 mm wide, lobes linear, exserted; style exserted. Vernal pools; < 1700 m. KR, NCoR, CaRH, w ScV, n SnFrB. Intermediate between *N. leucocephala* subsp. *l.*, *N. leucocephala* subsp. *plieantha*. Apr–Jul ★

subsp. *leucocephala* (p. 1069)   **ST**: 2–22 cm. **INFL**: longer bracts > 2 × head, lobes 3–4-branched abaxially. **FL**: calyx lobes gen glabrous; corolla white, tube exserted, lobes narrowly ovate; style exserted. 2*n*=18. Vernal pools; < 2100 m. NCoRO, CaR, SN, GV, MP; OR. Prostrate pls with spreading branches in Butte Co; small pls with few-fld infl in MP. Intergrades with *N. leucocephala* subsp. *minima* in CaR, n SNF. Apr–May

subsp. *minima* (Nutt.) A.G. Day   **ST**: 2–8 cm. **INFL**: 1–2 cm wide, bracts < 2 × head, lobes 3–4-branched abaxially. **FL**: calyx lobes gen toothed; corolla white, tube incl, throat 1–2 mm wide, lobes linear; style incl or ± exserted. 2*n*=18. Vernal pools; < 2200 m. CaR, n&c SNH, MP; to WA, UT. Jun–Aug

subsp. ***pauciflora*** (H. Mason) A.G. Day FEW-FLOWERED NAVARRETIA **ST:** 1–4 cm. **INFL:** ± 1 cm wide; longest bracts 1.5–2 × head, bract lobes few, entire or forked; fls 2–20, in ≤ 4 clusters. **FL:** corolla white to blue, tube incl, lobes linear. Vernal pools; 400–900 m. s NCoRI (Lake, Napa cos.). Intergrades rarely with *N. leucocephala* subsp. *plieantha*. May–Jun ★

subsp. ***plieantha*** (H. Mason) A.G. Day (p. 1069) MANY-FLOWERED NAVARRETIA **ST:** 1–3 cm, spreading, branches many. **INFL:** bracts < 2 × head, lobes 2–4-branched abaxially; fls 10–60, short-pedicelled, in clusters. **FL:** calyx lobes ± glandular at base, puberulent at middle; corolla blue, tube incl, lobes linear. Vernal pools; 800–1100 m. s NCoR (Lake, Sonoma cos.). May–Jun ★

N. mellita Greene (p. 1069) Pl erect, 1° axes 1–several; odor strong but not skunk-like. **ST:** 5–20 cm, branches gen many, ascending; glandular-hairy. **LF:** basal pinnate-lobed, axis, lobes linear; cauline axis shorter, wider. **INFL:** outer bracts palmate, base expanded, middle lobe ovate, acuminate, > head, lateral lobes linear to lanceolate, forked abaxially. **FL:** corolla 5–7 mm, = calyx, throat white, lobes light blue; stamens attached in tube, incl; style incl, stigmas 3. **FR:** dehiscing tip to base. Open, wet, sandy or gravelly areas, chaparral; 90–1200 m. NCoR, c SNF (Tuolumne Co.), n SNF (Calaveras Co.), SnFrB, SCoR, WTR. May–Jul

N. mitracarpa Greene (p. 1069) Pl ascending, 1° axes gen several. **ST:** 3–20 cm, branches spreading to ascending; gen white-hairy. **LF:** lower 1–2-pinnate-lobed, glabrous, axis linear, lobes needle-like; upper hairy below middle, tip toothed along wide axis; cotyledons 2–3, deeply lobed. **INFL:** bracts hairy, gland-dotted, tip of outer toothed along wide axis. **FL:** calyx lobes strap-shaped, toothed or not; corolla 7–11 mm, exserted, tube, throat white, glandular-puberulent, lobes 2–4 mm, blue; stamens, style exserted; stigmas 2. **FR:** chamber 1; valves 4, dehiscing in lower 1/2. Open, grassy, serpentine areas; 200–500 m. SCoR, n SW. [*N. jaredii* Eastw.] May–Jul ★

N. myersii P.S. Allen & A.G. Day Pl prostrate, 1° axis 1, minute. **ST:** < 20 mm, branches 0–few; glabrous or puberulent, hairs recurved. **LF:** gen spreading from base of central head, 4–8 cm, pinnate-lobed, axis linear, base membrane-winged, lobes few, gen in lower 1/2, linear. **INFL:** gen 1, sessile or on prostrate peduncle, 1–2 cm wide; outer bracts lf-like, >> head, basal wings hairy; inner bracts < head, basal wings ciliate; fls 5–60, sessile. **FL:** calyx lobes long-hairy, gland-dotted below, tapered, glabrous above, membranes truncate, ciliate; corolla 12–21 mm, 2–4 × calyx, white or blue, tube thread-like, throat cup-shaped, lobes 2.5–3 mm, linear or narrowly ovate; stamens, style exserted, stigmas 2, minute. **FR:** translucent, stuck to seeds until wet. **SEED:** separating when wet.

subsp. ***deminuta*** A.G. Day SMALL PINCUSHION NAVARRETIA **INFL:** outer bract lobes near base, 0–few above middle. **FL:** calyx tube = lobes; corolla 12–14 mm, blue, tube 1–1.2 × calyx. Vernal pools; 20–90 m. NCoRI (1 site, Lake Co.). Apr–May ★

subsp. ***myersii*** (p. 1069) PINCUSHION NAVARRETIA **INFL:** outer bract lobes gen few at base, 0 above middle. **FL:** calyx tube > lobes; corolla 17–21 mm, white, tube 2–4 × calyx. Vernal pools; 20–90 m. n&c SNF, c GV (few sites, Sacramento, Amador, Merced cos.). May ★

N. nigelliformis Greene Pl erect to spreading, 1° axes 1–several. **ST:** 9–32 cm, branched gen from base; puberulent, hairs white, recurved. **LF:** 2-pinnate-lobed; axis, lobes linear, < 1 mm wide, spreading (or upper clustered, ascending); cotyledons 2, entire. **INFL:** bracts pinnate-lobed, axis, lobes linear, hairy below middle. **FL:** calyx lobes strap-shaped, hairy near middle, longest toothed; corolla yellow, throat with paired, purple or brown spots below lobes; stigmas 2, minute. **FR:** chamber 1; valves 4, dehiscing in lower 1/2. Subspp. intergrade; further study needed.

subsp. ***nigelliformis*** (p. 1069) ADOBE NAVARRETIA Herbage dark green. **ST:** branches spreading-ascending. **INFL:** ± white-hairy in center; bract tips glandular-puberulent. **FL:** corolla exserted, 12–16 mm, lobes 3–4 mm; stamens, style > corolla lobes. Vernal pools, clay depressions; 10–1000 m. NCoRI, SNF, Teh, GV, SCoR. Apr–Jun ★

subsp. ***radians*** (J.T. Howell) A.G. Day (p. 1069) SHINING NAVARRETIA Pl wider than high; herbage light gray-green. **ST:** branches decumbent. **INFL:** densely white-hairy in center; bract tips gen glabrous. **FL:** corolla incl, 9–11 mm, lobes 1–2 mm; stamens exserted; style incl. Vernal pools, clay depressions; 150–1000 m. SCoR. May–Jul ★

N. ojaiensis Elvin et al. OJAI NAVARRETIA Pl 1° axes gen several. **ST:** 4–33 cm, branches spreading to ascending. **LF:** 2-pinnate-lobed, lower axis, lobes linear, tip linear to enlarged-lanceolate, many-toothed; glandular-hairy; cotyledons 2–3, deeply lobed. **INFL:** bracts, calyces glandular hairy at middle; outer bracts 1.5–2 × head; inner bracts shorter, base clasping, tip narrow, toothed. **FL:** calyx lobes strap-shaped, entire or longest toothed; corolla 6–11 mm, > calyx, white with purple spot at lobe bases, glandular abaxially, lobes 1.4–2.2 mm, acuminate, ± mucronate; stamens attached at ± 1 level in throat, exserted; style exserted, stigmas 2, minute. **FR:** chamber 1; valves 4, dehiscing in lower 1/2. Clay soils; 300–1000 m. WTR. May–Jul ★

N. peninsularis Greene (p. 1069) BAJA NAVARRETIA Pl gen as wide as high, 1° axes 1–several. **ST:** 3–25 cm, branches along st, ascending; glandular-hairy. **LF:** 1–3 cm, glandular hairy near base, pinnate-lobed; axis, lobes linear, < 1 mm wide; lobes 3–5 mm, needle-like, simple or forked, middle longest. **INFL:** outer bracts 10–15 mm, lf-like exc axis wider, shorter than on lvs. **FL:** calyx lobes needle-like; corolla 6–9 mm, ± > calyx, lavender-white, lobes 1–2 mm; stamens unequal, longest exserted; style incl, stigmas 3. **FR:** dehiscing tip to base. Wet areas in open forest; 1400–2300 m. Teh, TR, PR; AZ, Baja CA. Like *N. breweri*. Jun–Aug ★

N. prolifera Greene Pl erect, 1° axis gen 1. **ST:** 6–16 cm, branches 2–4, just below terminal head, occ along st, ascending, slender, brown, glabrous to minutely glandular. **LF:** basal entire, thread-like; cauline pinnate-lobed, axis, lobes thread-like, glandular-hairy at base. **INFL:** bracts, calyces densely white-hairy at middle, glandular-puberulent above; bract lobe tips glabrous, needle-pointed; outer bracts palmate-lobed at base, middle lobe < 2 × longest lateral. **FL:** calyx lobes entire; corolla 8–10 mm, gen > calyx, lobes ovate, 1–3 mm; stamens, style exserted; stigmas 3. **FR:** dehiscing tip to base.

subsp. ***lutea*** (Brand) H. Mason YELLOW BUR NAVARRETIA **FL:** corolla yellow. Dry, rocky flats near drainage channels; 700–2000 m. n SNF, n SNH (El Dorado Co.). May–Jul ★

subsp. ***prolifera*** **FL:** corolla tube, throat yellow, lobes blue or purple. Dry, rocky flats near drainage channels; 600–1500 m. SNF, n SNH. May–Jun

N. prostrata (A. Gray) Greene PROSTRATE VERNAL POOL NAVARRETIA Pl prostrate, with central head, 1° axis gen ± 0, 2° axes several, each with head. **ST:** sparsely hairy (densely hairy below heads), hairs recurved. **LF:** clustered just below heads, > 2 × head diam, subglabrous, sparsely gland-dotted, 1–2-pinnate-lobed, axis 1–4 mm wide, lobes linear, ascending. **INFL:** bracts like lvs but shorter, hairy below, basal wing wide, membranous, ciliate, lower lobes 2–3-branched abaxially. **FL:** calyx lobes tapered, hairy, tube ± glandular at base, glabrous, often 3-toothed; corolla 7–9 mm, blue to white, lobes linear, < 1 mm wide; stamens, style exserted, stigmas 2, minute. **FR:** translucent, stuck to seeds until wet. **SEED:** separating when wet. Alkaline floodplains, vernal pools; < 700 m. w SnJV (Merced Co.), CCo (w Alameda Co.), SnFrB (Alameda Co.), SCoR, c SCo (Los Angeles Co.), PR (Santa Rosa Plateau). Apr–Jul ★

N. pubescens (Benth.) Hook. & Arn. (p. 1069) Pl erect, 1° axis gen 1. **ST:** 14–33 cm, tan to red-brown, branches ascending, internodes gen < lvs. **LF:** 2-pinnate-lobed, axis, lobes linear; lower lvs glabrous, lobes clustered; upper lvs puberulent or hairy, tip linear or wider, many-toothed. **INFL:** bracts, calyces gen glandular hairy at middle, puberulent above; bracts pinnate-lobed; outer bracts 1.5–2 × head, axis narrow or wider above, toothed; inner bracts shorter, base clasping, tip narrow, toothed. **FL:** calyx lobes strap-shaped, gen 2 toothed; corolla 10–16 mm, > calyx, tube, throat red-purple, lobes bright blue-purple, glandular abaxially; stamens attached in mid-throat, exserted; style exserted, stigmas 2, minute. **FR:** chamber 1; valves 4, dehiscing in lower 1/2. Open, slopes, gravel, clay; < 1850 m. NCoR, CaR, SNF, c SNH, ScV, SnFrB, SCoR; s OR. May–Jul

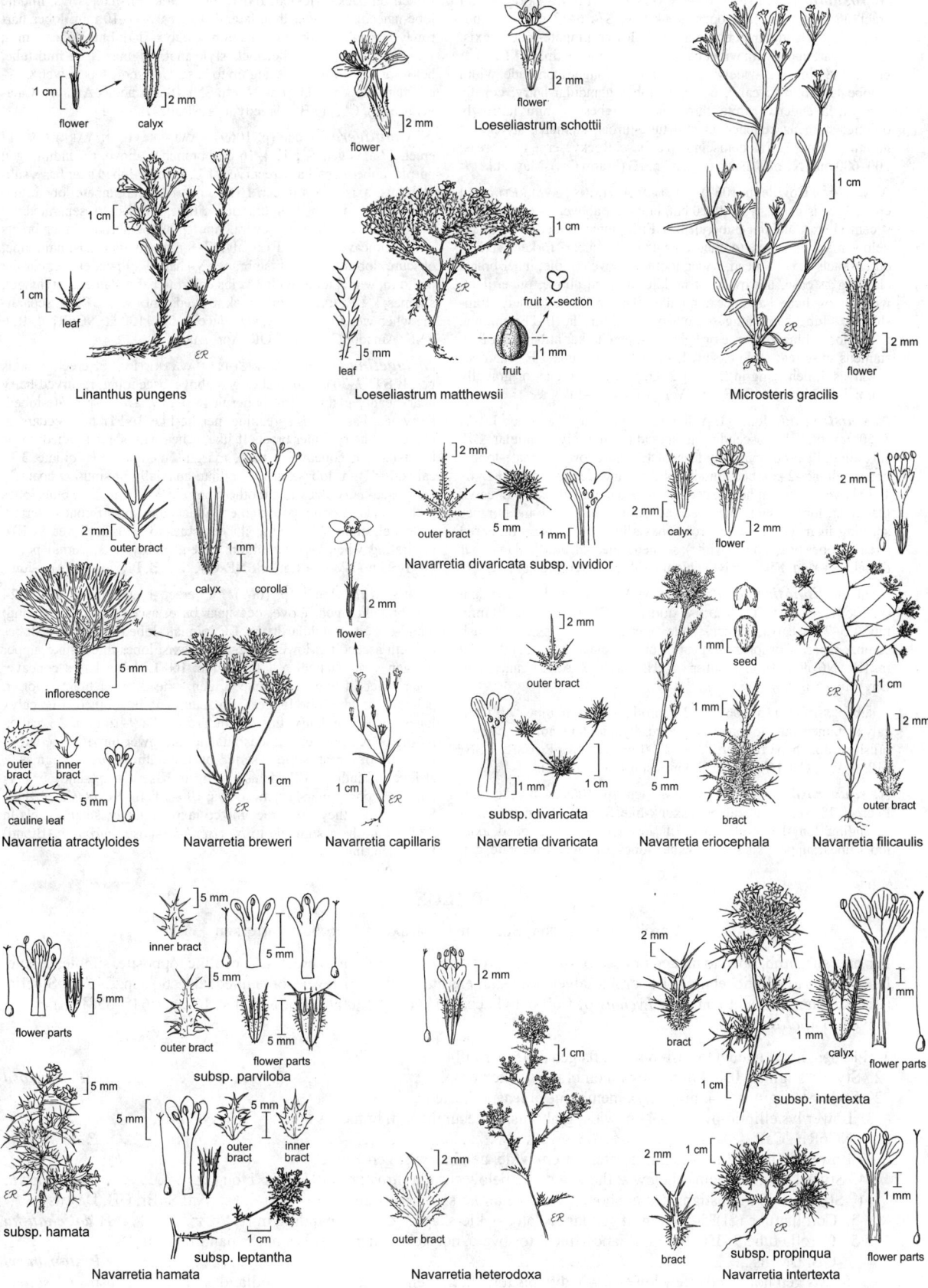

flower — calyx
Linanthus pungens
leaf

flower
Loeseliastrum matthewsii
leaf — fruit — fruit X-section

Loeseliastrum schottii
flower

Microsteris gracilis
flower

outer bract
inflorescence
outer bract — inner bract
cauline leaf
Navarretia atractyloides
calyx — corolla
Navarretia breweri
flower
Navarretia capillaris
Navarretia divaricata subsp. vividior
outer bract
outer bract
subsp. divaricata
Navarretia divaricata
calyx — flower
seed
bract
Navarretia eriocephala
outer bract
Navarretia filicaulis

inner bract
flower parts
outer bract
flower parts
subsp. parviloba
subsp. hamata
outer bract — inner bract
subsp. leptantha
Navarretia hamata
outer bract
Navarretia heterodoxa
bract — calyx — flower parts
subsp. intertexta
bract — subsp. propinqua — flower parts
Navarretia intertexta

N. rosulata Brand (p. 1069) MARIN COUNTY NAVARRETIA Pl erect, 1° axis 1–several; odor skunk-like. **ST:** 6–13 cm, branches ascending; glandular-puberulent. **LF:** lower pinnate-lobed, axis, lobes linear; upper with wider base, ovate-acuminate tip. **INFL:** outer bracts 7–8 mm, axis widely ovate, not much longer than wide, width > lobe length. **FL:** calyx 6–7 mm, lobes glandular-hairy; corolla = calyx, lavender to white, throat narrow, lobes ± 1 mm, narrowly oblong; stamens attached in mid tube-throat, stamens, style ± at mouth; stigmas 3. **FR:** dehiscing tip to base. Rocky, serpentine areas; 200–600 m. s NCoRI (Napa Co.), n SnFrB (Marin Co.). May–Jul ★

N. setiloba Coville (p. 1069) PIUTE MOUNTAINS NAVARRETIA Pl erect, 1° axis gen 1. **ST:** 10–20 cm, branches above, ascending, gen < central st; glandular-puberulent. **LF:** 2-pinnate-lobed; axis linear below middle, wider, toothed above; lobes linear, forked. **INFL:** outer bracts 1.5–2 × head, wider, toothed above middle; inner bracts clasping at base, narrowest at middle, toothed above, puberulent, with narrow hairy band occ at middle. **FL:** calyx lobes widely strap-shaped, < tube, hairy, longest toothed or not; corolla 10–11 mm, tube white, lobes blue to purple or light pink, gen darker at base, spots 0; stamens exserted; style exserted, stigmas 2, minute. **FR:** chamber 1; valves 4, dehiscing in lower 1/2. Depressions in clay or gravelly loam; 500–2100 m. s SNF, s SnJV, n WTR. Apr–Jul ★

N. sinistra (M.E. Jones) L.A. Johnson Pl erect, 1° axis gen 1. **ST:** 5–50 cm, branches ascending to spreading; densely glandular. **LF:** spreading, linear or narrowly lanceolate, entire lower pinnate-lobed, upper palmate-2–5 lobed, middle lobe much wider, longer. **INFL:** fls 1–2 above a lf, not in heads; pedicels unequal, thread-like. **FL:** calyx glandular, lobes equal, linear, gen acute, margins ± red, membranes splitting in fr; corolla ± glabrous abaxially, tube, throat with ± red streaks, lobes bright pink (white); stamens unequal, attached in upper throat, exserted. **FR:** ≤ calyx, oblong. Subspp. intergrade in KR.

  subsp. ***pinnatisecta*** (H. Mason & A.D. Grant) L.A. Johnson (p. 1069) PINNATE-LEAVED NAVARRETIA **FL:** corolla 10–20 mm, tube 1.5–2.5 × calyx; stamens, style exserted. Openings in sagebrush scrub, chaparral, or forest, serpentine or red volcanic soils; 300–2200 m. KR, NCoRI. [*Gilia s.* subsp. *p.* (H. Mason & A.D. Grant) A.G. Day] Jun–Aug ★

  subsp. ***sinistra*** (p. 1069) **FL:** corolla 5–10 mm, tube < to 1.5 × calyx; stamens incl to longest exserted, style incl. Openings in sagebrush scrub, chaparral, forest; 50–2700 m. KR, NCoR, CaR, n SN, MP; to WA, CO. [*Gilia s.* M.E. Jones subsp. *s.*] Jun–Aug

N. squarrosa (Eschsch.) Hook. & Arn. (p. 1069) SKUNKWEED Pl erect, 1° axes 1–several; odor skunk-like. **ST:** 10–60 cm, branches ascending; long-hairy, glandular. **LF:** lower 1–2-pinnate-lobed, axis, lobes linear; upper axis shorter, wider, lobes narrowly lanceolate, gen

forked on back, prickled. **INFL:** outer bracts lf-like, < lvs, middle lobe lanceolate, wider than lateral. **FL:** calyx 7–10 mm, lobes narrowly lanceolate; corolla 9–12 mm, > calyx, dark blue, lobes 2 mm; stamens attached in tube, incl; style incl, stigmas 3, in mid tube, below anthers. **FR:** dehiscing tip to base. Common. Open, wet, gravelly flats, slopes; < 1100 m. NW, n SNF (Sacramento, Amador, Calaveras cos.), CW; to BC. Weedy in Australia. Jun–Aug

N. subuligera Greene (p. 1069) AWL-LEAVED NAVARRETIA Pl erect, 1° axis gen 1. **ST:** 4–16 cm, branches above, ascending; gen purple, puberulent, hairs recurved. **LF:** pinnate-lobed near base, subglabrous; axis, lobes linear. **INFL:** outer bracts pinnate-lobed, glabrous, basal margins membranous, ciliate, lobes short, separated, tip awl-shaped with few to several fine spines, glabrous; inner bracts wider, concave at base. **FL:** calyx 7–8 mm, subglabrous, hairs fine, obscure, lobes tapered, membranes V-shaped, ciliate, entire; corolla 6–7 mm, white, lobes with 3 veins entering base; stamens, style incl, stigmas 3. **FR:** translucent, stuck to seeds until wet. **SEED:** separating when wet. Open, rocky, wet places; 150–1100 m. NCoRI, CaR, n SNF (Amador Co.), ScV; OR. Apr–Aug ★

N. tagetina Greene MARIGOLD NAVARRETIA Pl erect, 1° axis gen 1. **ST:** 7–30 cm, branches gen above, ascending; recurved-hairy near heads, glabrous or puberulent below. **LF:** 2-pinnate-lobed, hairy near base; lobes spreading, needle-like. **INFL:** bracts coarsely appressed-hairy; outer bracts lf-like, > head, tip short-toothed; inner bracts wider, concave at base, margins membranous, ciliate. **FL:** calyx 7–8 mm, lobes tapered, ciliate adaxially at sinuses, coarsely appressed-hairy abaxially, toothed; corolla 9–11 mm, pale blue, lobes with 3 veins entering base; stamens attached in upper throat, exserted; style incl, stigmas 3. **FR:** translucent, stuck to seeds until wet. **SEED:** separating when wet. Common. Open, grassy flats, vernal pools; 10–1600 m. NW, CaR, n&c SNF, GV, SnFrB, PR; to WA. Apr–Jun

N. viscidula Benth. (p. 1069) Pl erect, 1° axis gen 1. **ST:** 3–24 cm; branches gen above, occ near base, ascending to spreading; gland-dotted, glandular-hairy. **LF:** pinnate-lobed or -toothed; lobes or teeth ascending; lower lvs 2–5 cm, axis, lobes thread-like; upper lvs gen strap-shaped, > 1 mm wide. **INFL:** bract bases concave; outer bracts lf-like, strap-shaped, gen > head; inner bracts shorter, palmate, middle lobe longest, tip acuminate or toothed. **FL:** calyx lobes lanceolate, hairy, tips glandular; corolla 9–16 mm, 2 × calyx, purple or red-purple; stamens attached in lower throat, exserted or at mouth of throat; style exserted or at mouth of throat, stigmas 3, at level of anthers. **FR:** dehiscing tip to base. Open, sandy or clay flats, near pools, marshes, meadows; 100–900 m. NCoR, CaRF, SNF, SnFrB. Were they to be recognized taxonomically, small-fld pls in SNF would be assignable to *N. viscidula* subsp. *purpurea* (Brand) H. Mason. Jun–Jul

PHLOX

Carolyn J. Ferguson, Suzanne C. Strakosh & Robert Patterson

Per or ± subshrub, open to matted or cushion-like. **ST:** prostrate or decumbent to erect. **LF:** cauline, opposite, simple, sessile, lance-linear to elliptic, entire. **FL:** corolla salverform; stamens attached at > 1 level, some unequal. ± 60 spp.: N.Am, Siberia. (Greek: flame, ancient name for *Lychnis* of Caryophyllaceae) [Locklear 2009 J Bot Res Inst Texas 3:645–658] *P. gracilis* moved to *Microsteris*.

1. Pl open, not ± matted to cushion-like; fls gen 3–15 per infl
  2. Style < stigmas, 0.4–2 mm; calyx membrane not keeled . . . . **P. speciosa**
  2′ Style >> stigmas, ≥ 4 mm; calyx membrane at least ± keeled
    3. Lower lvs elliptic-ovate, 1–4 cm wide, glabrous; st decumbent, fl branches erect — KR, NCoRO, NCoRH, CaR . . . . **P. adsurgens**
    3′ Lower lvs lance-linear to lanceolate, < 1 cm wide, gen hairy; st gen ± erect
      4. Style 4–7 mm; stigmas below anthers, anthers below corolla throat; n CaRH (Siskiyou Co.) . . . . **P. hirsuta**
      4′ Style 11–40 mm; stigmas at or above 1 or more anthers, at least 1 anther near corolla throat; SnBr, GB, DMtns
        5. Corolla tube (21)35–50 mm; lf gen lanceolate, sickle-shaped; calyx expanding in fr; SnBr . . . . **P. dolichantha**
        5′ Corolla tube 9–35(39) mm; lf lance-linear to -ovate, not sickle-shaped; calyx not expanding in fr; GB, DMtns . . . . **P. stansburyi**
          6. Sts gen not woody (few lower ± woody); pls together; open pinyon/juniper woodland . . . . subsp. **stansburyi**

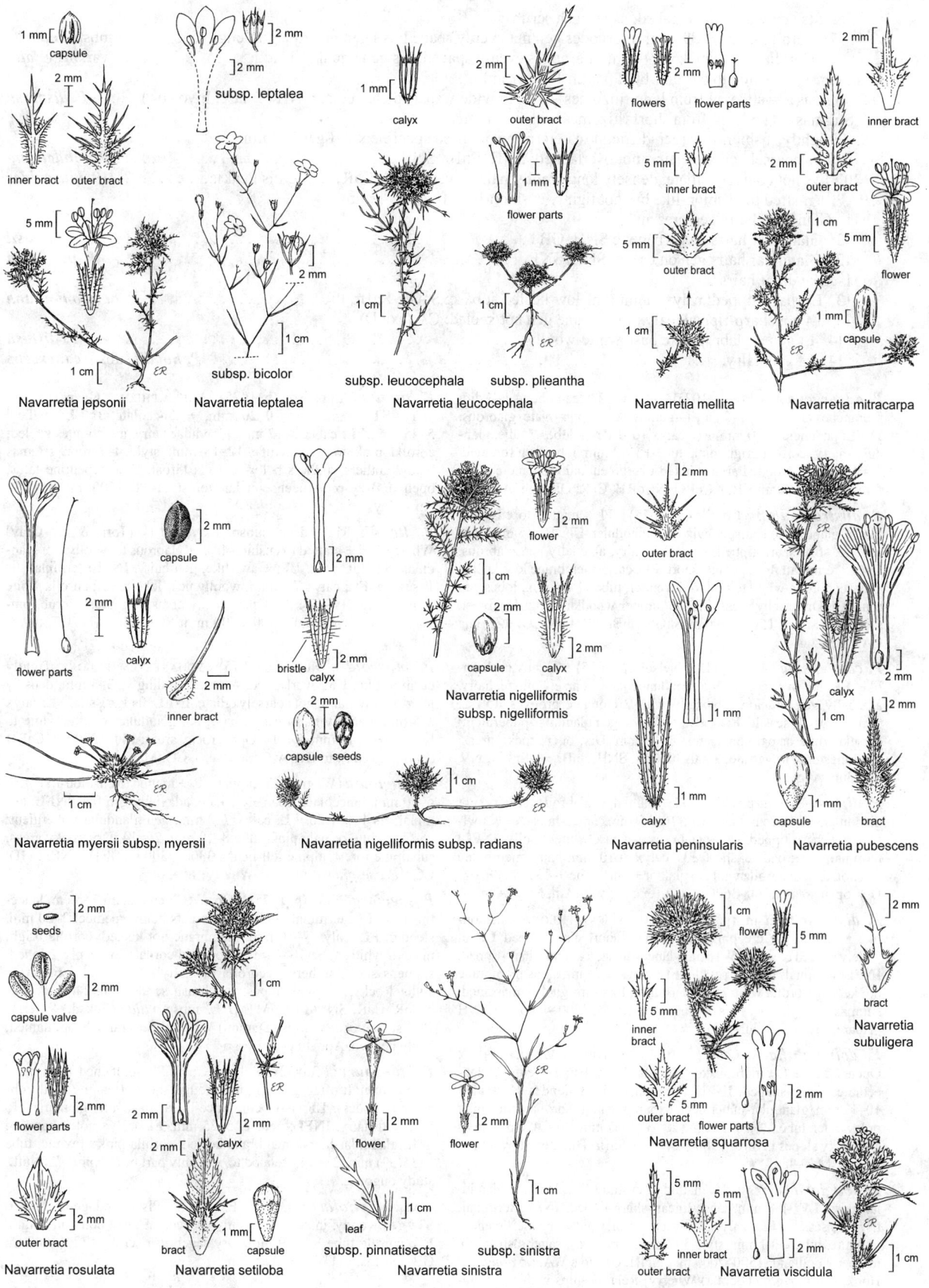

Navarretia jepsonii

subsp. leptalea
subsp. bicolor
Navarretia leptalea

calyx
outer bract
flower parts
subsp. leucocephala
subsp. plieantha
Navarretia leucocephala

flowers
flower parts
inner bract
outer bract
Navarretia mellita

inner bract
outer bract
flower
capsule
Navarretia mitracarpa

flower parts
capsule
calyx
inner bract
Navarretia myersii subsp. myersii

bristle
calyx
capsule
seeds
flower
capsule
calyx
Navarretia nigelliformis subsp. nigelliformis

Navarretia nigelliformis subsp. radians

outer bract
calyx
calyx
Navarretia peninsularis

calyx
capsule
bract
Navarretia pubescens

seeds
capsule valve
flower parts
outer bract
Navarretia rosulata

calyx
bract
capsule
Navarretia setiloba

flower
leaf
subsp. pinnatisecta
flower
subsp. sinistra
Navarretia sinistra

flower
inner bract
outer bract
flower parts
Navarretia squarrosa
bract
Navarretia subuligera

outer bract
inner bract
flower parts
Navarretia viscidula

6′ Sts gen woody; pls spaced; sagebrush scrub
  7. Corolla tube 18–39 mm; internodes gen not evenly spaced; lvs gen not equal in length . . . . subsp. ***superba***
  7′ Corolla tube gen 8.5–19 mm; internodes evenly spaced; lvs gen equal in length . . . . var. ***brevifolia***
1′ Pl dense, ± matted to cushion-like; fls gen 1–5 per infl
  8. Stigmas ± = style; pl from long rhizomes — lvs in widely spaced clusters; s SNH (Tulare, Inyo cos.) . . . . . ***P. dispersa***
  8′ Stigmas << style; pl from short rhizomes with sts clumped
    9. Pl tightly cushion-like (gen difficult to determine on dried specimens); lf gen < 4 mm
      10. Lvs coarsely ciliate, hairy, not overlapped; SNH, SnBr, SNE . . . . ***P. condensata***
      10′ Lvs not coarsely ciliate, densely long-woolly-hairy, overlapped; CaRH (Loomis Peak). . . . . ***P. muscoides***
    9′ Pl ± matted to cushion-like but not tightly cushion-like; lf gen > 4 mm
      11. Glandular hairs gen present
        12. Glandular hairs on herbage; c SNH, GB . . . . ***P. douglasii***
        12′ Glandular hairs gen on calyx; SNH, SNE. . . . . ***P. pulvinata***
      11′ Glandular hairs 0
        13. Lf sharp-tipped; calyx membrane low-keeled at base; SnGb, SnBr, PR. . . . . ***P. austromontana***
        13′ Lf not sharp-tipped; calyx membrane gen not keeled; CA (exc D)
          14. Lf gen ± glabrous exc base white-woolly. . . . . ***P. diffusa***
          14′ Lf gen hairy. . . . . ***P. hoodii*** subsp. ***canescens***

P. adsurgens A. Gray (p. 1075)   Open. **ST**: decumbent, 1–3 dm; fl branches erect. **LF**: 1–3 cm, 1–4 cm wide, elliptic-ovate, glabrous. **INFL**: peduncle 5–20 mm. **FL**: calyx 10–13 mm, lobes > tube, glandular-hairy; corolla bright pink, tube 12–20 mm, lobes gen rounded. $2n=14$. Open, wooded areas, mixed-evergreen and montane conifer forest; 500–2000 m. KR, NCoRO, NCoRH, CaR; OR. Jun–Aug

P. austromontana Coville (p. 1075)   Pl ± matted to cushion-like, ± glabrous to sparsely hairy, not glandular. **LF**: 10–15 mm, lanceolate, stiff, sharp-tipped, adaxially hairy, abaxially gen glabrous. **INFL**: terminal; fls 1; pedicel short. **FL**: calyx membrane low-keeled at base; corolla white to pink or lavender, tube 11–14 mm, lobes gen rounded. Dry, rocky areas, pinyon/juniper woodland, conifer forest, sagebrush scrub; 1500–2700 m. SnGb, SnBr, PR; to w CO, AZ, Baja CA. May–Jun

P. condensata (A. Gray) E.E. Nelson (p. 1075)   Tightly cushion-like. **LF**: not overlapped, 3–5 mm, lanceolate, coarsely ciliate, hairy, adaxially gen concave, abaxially with 2 elongate grooves. **INFL**: terminal; fls 1, sessile. **FL**: calyx 5–6 mm, gen glandular-puberulent; corolla white or pale pink, tube 8–10 mm. Dry, open, rocky areas, esp limestone, travertine; 2000–4000 m. SNH, SnBr, SNE; OR, NV, CO. Jun–Aug

P. diffusa Benth. (p. 1075)   Pl ± matted, ± glabrous to hairy, not glandular. **ST**: decumbent. **LF**: 10–15 mm, lance-linear or ± awl-like, not sharp-tipped, gen ± glabrous exc base white-woolly. **INFL**: terminal; fls 1; pedicel short. **FL**: calyx 8–10 mm, hairy, membrane gen not keeled; corolla white to pink or ± blue, tube 9–13 mm. $2n=28$. Dry, open areas; 1100–3600 m. CA (exc D); w N.Am. May–Aug

P. dispersa Sharsm. (p. 1075)   HIGH SIERRA PHLOX   Cushion-like; rhizome long, creeping, slender, hairs short, gland-tipped. **LF**: in widely spaced clusters, 5–10 mm, lance-linear, leathery, sharp-tipped. **INFL**: terminal; fls few, sessile. **FL**: calyx ± 7 mm, sinus membrane not keeled; corolla white, tube ± 10 mm, lobes irregularly margined; stigmas ± = style. Dry flats of loose granite; gen 3600–4200 m. s SNH (Tulare, Inyo cos.). Jul–Aug   ★

P. dolichantha A. Gray (p. 1075)   BIG BEAR VALLEY PHLOX   Open. **ST**: erect, ± glabrous to glandular-hairy. **LF**: 2–6 cm, gen lanceolate, sickle-shaped. **INFL**: terminal; pedicel slender. **FL**: calyx 10–12 mm, glandular-puberulent, expanding in fr; corolla white, pink or lavender, tube (21)35–50 mm; style 12–37 mm. $2n=14$. Open areas and rocky slopes in forest; 2000–2700 m. SnBr. Threatened by urbanization. May–Jun   ★

P. douglasii Hook.   Pl ± matted to cushion-like, herbage glandular-hairy. **LF**: 4–8 mm, lance-linear, sharp-tipped. **INFL**: terminal; fls 1–3, sessile. **FL**: calyx 7–9 mm; corolla pink or pale lavender to white, tube ± 10 mm; style 3–7 mm. Dry areas, sagebrush scrub, juniper woodland; 1500–2000 m. c SNH, GB; to s WA. [*P. douglasii* Hook. subsp. *rigida* (Benth.) Wherry] Relationships with *P. caespitosa* Nutt., *P. diffusa* need study. Apr–May

P. hirsuta E.E. Nelson (p. 1075)   YREKA PHLOX   Open, coarsely hairy. **ST**: erect. **LF**: 10–20 mm, ± lanceolate. **INFL**: pedicel 5–25 mm. **FL**: calyx 8–12 mm, glandular-hairy, membranes keeled; corolla pink to white, tube 12–15 mm; style 4–7 mm, stigmas below anthers, anthers below corolla throat. Dry serpentine talus, open Jeffrey-pine/incense-cedar forest; 1000–1500 m. ne KR. Apr–May   ★

P. hoodii Richardson subsp. ***canescens*** (Torr. & A. Gray) Wherry   Pl ± matted to cushion-like, ± glabrous to woolly, not glandular. **ST**: glabrous. **LF**: ± awl-like, gen hairy. **INFL**: terminal; fls 1, sessile. **FL**: calyx 7–8 mm, woolly near lobe bases; corolla white to lilac, tube 10–12 mm. Open, rocky areas, sagebrush scrub, pinyon/juniper woodland; 1500–2700 m. n SNH, GB; MT, UT, n AZ. May–Jul

P. muscoides Nutt. (p. 1075)   SQUARESTEM PHLOX   Tightly cushion-like. **LF**: overlapped, 4-ranked, hiding st, 3–5 mm, densely long-woolly-hairy, not coarsely ciliate. **INFL**: fls 1, sessile. **FL**: calyx ± 5 mm, densely hairy esp at base, not glandular; corolla white to lilac, tube 7–10 mm. $2n=14$. Open, rocky areas; 1400–2700 m. CaRH (Loomis Peak); to OR, w Great Plains, CO. May–Jun   ★

P. pulvinata (Wherry) Cronquist   Pl ± matted to cushion-like. **LF**: 4–10 mm, lanceolate, coarsely ciliate, adaxially gen flat. **INFL**: terminal; fls 1, sessile. **FL**: calyx 7–8 mm, gen glandular-puberulent; corolla white or pale pink, tube 8–10 mm. $2n=14$. Dry, rocky areas, subalpine forest, alpine fell-fields; 3300–4300 m. SNH, SNE; to ID, CO. [*P. caespitosa* subsp. *p.* Wherry] Jul–Aug

P. speciosa Pursh (p. 1075)   Open. **ST**: erect. **LF**: 1–5 cm, lance-linear. **INFL**: terminal, with lf-like bracts below; pedicel 3–20 mm, slender. **FL**: calyx 7–10 mm, membrane not keeled; corolla bright pink to white, tube 10–15 mm, lobes obcordate to deeply 2-lobed; stamens short, anthers in corolla tube; style 0.4–2 mm, stigmas > style. Rocky, wooded slopes, sagebrush scrub; 500–2400 m. KR, NCoR, CaR, SN; to BC, MT. [*P. s.* subsp. *nitida* (Suksd.) Wherry; *P. s.* subsp. *occidentalis* (Durand) Wherry] Several subspp. named; study needed. Apr–Jun

P. stansburyi (Torr.) A. Heller   Open. **ST**: branched from base, woody or not; often growing up through shrubs (leaning, with longer internodes). **LF**: 1–11 cm, lance-linear to -ovate, glandular- to long-soft-hairy. **INFL**: pedicel 5–25 mm. **FL**: calyx not expanding in fr, glandular-hairy, membrane keeled; corolla pink to white, tube 9–35(39) mm. $2n=14$. Related to, possibly part of *P. longifolia* Nutt.; study ongoing.

  var. ***brevifolia*** (A. Gray) E.E. Nelson   Pls spaced, gen compact. **ST**: gen woody, internodes evenly spaced. **LF**: gen equal in length. **FL**: corolla tube 8.5–19 mm. Dry sagebrush scrub; 1700–3000 m. GB, DMtns; to UT, AZ. Apr–Jun

subsp. ***stansburyi*** (p. 1075) Pls close together, gen compact. **ST**: gen not woody. **LF**: short, wide or long, linear. **FL**: corolla tube 17–25 mm. Open areas of pinyon/juniper woodland, disturbed roadsides; 1700–3000 m. GB, DMtns; to UT, NM. Apr–Jun

subsp. ***superba*** (Brand) Wherry Pls spaced, gen elongate. **ST**: woody, internodes gen not evenly spaced. **LF**: gen not equal in length. **INFL**: fls clustered. **FL**: corolla tube 18–39 mm. Open sagebrush scrub; 1700–3000 m. W&I; to NV, AZ. Apr–Jun

POLEMONIUM

Ruth E. Timme & Dieter H. Wilken

Ann, per. **ST**: decumbent to erect, 10–100 cm, glandular-hairy, hairy, or glabrous. **LF**: pinnate-compound, alternate; basal petiole base membranous or not, sheathing or not; cauline sessile above; lflets entire to divided, glabrous to glandular-hairy. **INFL**: cyme or head. **FL**: calyx bell-shaped, membranous in age but not separated into membrane and lobes, glandular-hairy; corolla rotate to funnel- or bell-shaped, tube << throat, lobes white to blue or purple; stamens attached at 1 level, filaments hairy at base; ovary gen ± 1 mm, ± 1 mm wide. **FR**: ovoid to spheric. **SEED**: ≤ 10, gen 1–3 mm, elliptic to ovate, ± gelatinous when wet, brown to black. ± 30 spp.: Am, Eurasia. (Greek: perhaps from Polemon, Athenian philosopher, or polemos, strife or war) Per gen cross-pollinated, ann self-pollinated.

1. Ann; fls 1–2 in axils . . . . *P. micranthum*
1′ Per; fls gen clustered at st tips
  2. Lflets deeply 3–5-lobed; infl a dense, rounded head
    3. Stamens, style ± exserted; petiole base membranous . . . . *P. chartaceum*
    3′ Stamens, style incl; petiole base green, gen not membranous . . . . *P. eximium*
  2′ Lflets entire; infl various but not head-like
    4. Pl erect, gen < 10 dm; lvs cauline
      5. Corolla lobes gen pale pink to purple; lflets elliptic . . . . *P. carneum*
      5′ Corolla lobes purple to gen blue; lflets lanceolate. . . . . *P. occidentale* subsp. *occidentale*
    4′ Pl cespitose, < 3 dm; lvs ± basal
      6. Terminal lflet ± fused with adjacent pair; corolla limb 7–15 mm diam; dry, open to shaded areas
        in montane forest . . . . *P. californicum*
      6′ Terminal lflet gen free from adjacent pair; corolla limb 5–11 mm diam; subalpine to alpine talus
         . . . . *P. pulcherrimum*
        7. Corolla lobes white; herbage soft-shaggy-hairy . . . . var. *pilosum*
        7′ Corolla lobes blue to purple; herbage sparsely glandular-hairy . . . . var. *pulcherrimum*

P. californicum Eastw. (p. 1075) Per, cespitose, soft-hairy; rhizomed. **ST**: decumbent to erect, 12–25 cm, glandular-hairy. **LF**: ± basal, < 15 cm, 1–5 cm wide, cauline reduced or not, glandular-hairy; petioles 2–6 cm, bases not membranous, ± sheathing; lflets 9–25, 10–25 mm, 5–15 mm wide, elliptic to lanceolate, entire, terminal ± fused with adjacent pair. **INFL**: open cyme, gen same level as highest lvs, 5–25-fld; pedicel 1–7 mm. **FL**: calyx 6–8 mm, lobes > tube, acute, glandular-hairy; corolla bell-shaped, limb 7–15 mm diam, throat 2–3 mm, lobes 3–4 mm, light to dark blue or purple; stamens 5 mm, incl; pistil 4–7 mm, exserted or not. **FR**: 2–4 mm, 2–3 mm wide. **SEED**: 6–10, brown. 2*n*=36. Dry, open to shaded areas in montane forest; 1600–3100 m. KR, CaRH, n&c SNH; to WA. Self-incompatible. Intergrades with *P. pulcherrimum* at high elevations in SNH. Jun–Aug

P. carneum A. Gray OREGON POLEMONIUM Per, ± glabrous; rhizome long, slender. **ST**: erect, 40–100 cm, glabrous or ± hairy. **LF**: cauline, < 20 cm, 3–8 cm wide, not reduced, glabrous; petioles 2–7 cm, bases not membranous, not sheathing; lflets 9–21, 2–4 cm, 4–15 mm wide, elliptic, entire, terminal gen free. **INFL**: open cyme, 4–11-fld; pedicel 4–13 mm. **FL**: calyx 8–10 mm, soft-hairy, lobes < tube, acute; corolla rotate to bell-shaped, limb 12–20 mm diam, throat 5–7 mm, lobes 8–10 mm, gen pale pink to purple; stamens 5–10 mm, incl; pistil 7–12 mm, incl, ovary ± 2 mm, ± 1 mm wide. **FR**: 4–8 mm, 3–6 mm wide. **SEED**: ≤ 6, dark brown to black. Moist to dry, open areas; < 1800 m. NW, CCo, SnFrB; to WA. Apr–Jun ★

P. chartaceum H. Mason (p. 1075) MASON'S SKY PILOT Per, cespitose, hairy; rhizome short. **ST**: erect, < 20 cm, hairy, ± purple. **LF**: basal 3–7 cm, 3–6 mm wide, hairy, cauline reduced; petioles 1–5 cm, bases membranous, sheathing; lflets 15–25, < 4 mm, deeply 3–5-lobed. **INFL**: head, many-fld; pedicel 1–6 mm. **FL**: calyx 5–7 mm, hairy, often purple, lobes < tube; corolla funnel-shaped, limb 5–12 mm diam, throat 4–7 mm, lobes 3–5 mm, blue to purple; stamens 6–9 mm, ± exserted; pistil 5–10 mm, ± exserted. **FR**: ± 4 mm, ± 3 mm wide. **SEED**: ≤6, brown. Rocky slopes, talus; 2600–4200 m. KR (Mount Eddy), n SNE, W&I. Morphology, molecular data suggest KR pls a separate sp. Jul–Aug ★

P. eximium Greene (p. 1075) Per, cespitose, hairy; rhizomed. **ST**: erect, 10–40 cm, glandular-hairy to hairy. **LF**: basal 4–13 cm, 4–9 mm wide, glandular-hairy, cauline reduced; petioles 2–6 cm, bases gen not membranous, sheathing; lflets 20–35, 3–6 mm, deeply 3–5-lobed. **INFL**: head, many-fld; pedicel 1–3 mm. **FL**: calyx 6–8 mm, glandular-hairy, ± purple, lobes < tube; corolla funnel-shaped, limb 9–15 mm diam, throat 6–11 mm, lobes ± 5 mm, blue to purple; stamens ± 5 mm, incl; pistil 4–7 mm, incl. **FR**: ± 5 mm, 3 mm wide. **SEED**: ≤ 6, brown. Rocky outcrops, talus; 3000–4200 m. c&s SNH. Jul–Aug

P. micranthum Benth. (p. 1075) Ann, soft-hairy. **ST**: decumbent to erect, 5–25 cm, glandular-hairy. **LF**: basal and cauline, < 5 cm, 5–11 mm wide, cauline not reduced, soft-glandular-hairy; petioles 5–13 mm, bases not membranous, not sheathing; lflets 5–15, 2–7 mm, 1–3 mm wide, entire, lanceolate, entire, terminal fused to adjacent pair. **INFL**: fls 1–2 in axils; pedicel 3–25 mm. **FL**: calyx 3–9 mm, glandular-hairy, lobes > tube, acute; corolla bell-shaped, limb 3–5 mm diam, throat ± 1 mm, lobes 1–2 mm, light blue to white; stamens ± 1 mm, incl; pistil 1.5 mm, incl. **FR**: 3–4 mm, 3 mm wide. **SEED**: ≤ 6, dark brown. Open, seasonally wet areas, gen among shrubs; 600–1800 m. KR, CaRH, n SNH, s SnJV, se SCoRO, n WTR, MP; to BC, MT; S.Am. Apr–Jun

P. occidentale Greene subsp. ***occidentale*** (p. 1075) Per, gen glabrous; rhizome short, thin. **ST**: erect, 40–100 cm, glabrous at base, glandular-hairy above. **LF**: cauline, 6–40 cm, 1–9 cm wide, reduced upward, glabrous; petioles 2–4 cm, bases not membranous, not sheathing; lflets 15–23, 8–45 mm, 4–15 mm wide, lanceolate, entire, terminal fused to adjacent pair or not. **INFL**: open to dense cyme, 10–35 fld; pedicel 1–3 mm. **FL**: calyx 4–5 mm, lobes > tube, glandular-hairy, acute; corolla bell-shaped, limb 10–17 mm diam, throat 3–6 mm, lobes 5–8 mm, purple to gen blue; stamens 7–9 mm, incl; pistil 12–15 mm, exserted, ovary < 1 mm wide. **FR**: 3–5 mm, 2–4 mm wide. **SEED**: ≤ 10, dark brown. Moist areas, meadows, streambanks; 900–3300 m. KR, CaRH, SN, SnBr, SNE, Wrn; to BC, CO. Jun–Aug

P. pulcherrimum Hook. Per, cespitose, hairy; rhizome short. **ST**: erect, 5–20 cm. **LF**: basal < 6 cm, 5–10 mm wide, cauline much reduced, glandular-hairy; petioles 5–13 mm, bases not membranous, sheathing; lflets 9–22, 3–4 mm, 2–4 mm wide, ovate to round, entire, terminal gen free from adjacent pair. **INFL**: dense, 4–5 fld, gen above lvs; pedicel 2–5 mm. **FL**: calyx 4–5 mm, lobes ± = tube, glandular-hairy, acute; corolla rotate to bell-shaped, limb 5–11 mm diam, throat ± 3 mm, lobes ± 4 mm; stamens 3–5 mm, incl; pistil 5–6 mm, incl, ovary < 1.5 mm, < 1 mm wide. **FR**: 3–4 mm, ± 2 mm wide. **SEED**: ≤3, dark brown to black.

var. ***pilosum*** (Greenm.) Brand Pl gen < 10 cm; herbage soft-shaggy-hairy. **FL**: corolla lobes white. Volcanic talus; 2700–3000 m. CaRH, MP; to WA. Jul–Aug

var. ***pulcherrimum*** Pl gen 10–20 cm; herbage sparsely glandular-hairy. **FL**: corolla lobes blue to purple. Talus; 2400–3700 m. KR, NCoRH, CaRH, n&c SNH, MP, n SNE; to AK. Jun–Aug

SALTUGILIA

Leigh A. Johnson

Ann, not cobwebby. **ST**: erect, branched above; glabrous to glandular-hairy below, glandular but with 0 nonglandular hairs in infl. **LF**: simple, alternate; basal gen suberect in rosette, 2–15 cm, gen 2–3-pinnate, axis linear, lobes spaced, ascending, segments narrower than or equalling lf axis, tips acute; hairs long, shiny, translucent, minutely gland-tipped; cauline lvs reduced. **INFL**: open; bracts linear or lobed at base; pedicels 1 or in unequal pairs, glandular; glands flat-topped, wider than stalk. **FL**: calyx membranes wider than lobes, gen purple-spotted; corolla > calyx, funnel-shaped, throat yellow-spotted; shortest anther gen attached perpendicular to corolla tube (exc in *Saltugilia caruifolia*); pollen blue. **FR**: narrowly ovoid, valves separating from top. **SEED**: 4–many per chamber, tan to golden, gelatinous when wet. 2*n*=18. 4 spp.: s CA, Baja CA. (Latin: woodland + gilia, after Filippo Luigi Gilii, Italian naturalist, 1756–1821) [Johnson 2007 Novon 17:193–197] Seed germination stimulated by charcoal.

1. Corolla tube, throat white; tube incl . . . . ***S. australis***
1′ Corolla tube, gen throat pink, purple or lavender-blue (throat white or not in *Saltugilia caruifolia*); tube exserted
  2. Corolla lobe base with purple marks; stamens attached at mid-throat, gen ≥ corolla lobes. . . . . ***S. caruifolia***
  2′ Corolla lobe base lacking purple marks; stamens attached at corolla sinuses, ≤ corolla lobes
    3. Corolla 7.5–11 mm, tube not glandular. . . . . ***S. latimeri***
    3′ Corolla gen ≥ 11 mm, tube minutely glandular . . . . ***S. splendens***
      4. Corolla tube 7–18 mm, 2–5 × calyx. . . . . subsp. ***grantii***
      4′ Corolla tube 4–10 mm, 1–2 × calyx . . . . subsp. ***splendens***

S. australis (H. Mason & A.D. Grant) L.A. Johnson **ST**: gen 1, 5–50 cm. **INFL**: distal branches ascending; pedicel 4–25 mm. **FL**: calyx 2–4 mm, gen glabrous or in age so; corolla 5–10 mm, tube incl, glabrous, white, throat white, lobes white to light pink or blue, tips short-pointed; stamens attached at corolla sinuses; style ± exserted. **FR**: 4–7 mm, 1–2 × calyx. **SEED**: 7–13 per chamber. Chaparral, sandy-gravelly soil; 600–1200 m. TR, PR, D; Baja CA. [*Gilia a.* (H. Mason & A.D. Grant) V.E. Grant & A.D. Grant] Mar–Jun

S. caruifolia (Abrams) L.A. Johnson CARAWAY-LEAVED WOODLAND-GILIA **ST**: gen 1, 10–100 cm. **INFL**: distal branches spreading; pedicel 1–11 mm. **FL**: calyx 3–4 mm, gen glabrous or in age so; corolla 7–13 mm, tube exserted, glabrous, purple, throat white or not, lobes lavender to blue, base purple-marked, tips round; stamens attached at mid-throat, exserted, gen ≥ corolla lobes; style > stamens. **FR**: 3–5 mm, 1–1.5 × calyx. **SEED**: 4–7 per chamber. Openings in chaparral or forest, rocky soil; 1400–2300 m. PR; Baja CA. [*Gilia c.* Abrams] May–Aug ★

S. latimeri T.L. Weese & L.A. Johnson LATIMER'S WOODLAND-GILIA **ST**: gen 1°, several 2° from base, 5–30 cm. **INFL**: distal branches ascending; pedicel 2–16 mm. **FL**: calyx 2–4 mm, glandular, lobes each with 6–35 glands; corolla 7.5–11 mm, tube exserted, glabrous, purple, throat, lobes pink, lobe tips acute; stamens attached at corolla sinuses; style ± exserted. **FR**: 3.5–5 mm, ± = calyx. **SEED**: 6–9 per chamber. Dry desert slopes, coarse sand to rocky soils; 400–1900 m. TR, PR, D. [*Gilia l.* (T.L. Weese & L.A. Johnson) V.E. Grant] Mar–Jun ★

S. splendens (H. Mason & A.D. Grant) L.A. Johnson SPLENDID WOODLAND-GILIA **ST**: gen 1, 10–100 cm. **INFL**: distal branches ascending; pedicel 3–20 mm. **FL**: calyx 3–4 mm, glabrous or in age so; corolla tube exserted, minutely glandular, red-purple, throat, lobes pink, tips obtuse; stamens attached at corolla sinuses; style exserted gen beyond stamens. **FR**: 4–8 mm, 2 × calyx. **SEED**: 10–20 per chamber. [*Gilia s.* H. Mason & A.D. Grant; *S. grinnellii* (Brand) L.A. Johnson, illeg.]

subsp. ***grantii*** (Brand) L.A. Johnson (p. 1075) **FL**: corolla 20–36 mm, tube 7–18 mm, 2–5 × calyx. Openings in chaparral or forest, rocky soil; 800–2400 m. SnGb, SnBr. [*Gilia s.* subsp. *g.* (Brand) V.E. Grant & A.D. Grant] May–Aug

subsp. ***splendens*** (p. 1075) **FL**: corolla 10–23 mm, tube 4–10 mm, 1–2 × calyx. Openings in chaparral or forest, rocky soil; 300–2200 m. SCoR, TR, SnJt. [*Gilia s.* subsp. *s.*] May–Aug

POLYGALACEAE MILKWORT FAMILY

Robert E. Preston & Thomas L. Wendt

[Ann] per, subshrub, shrub [tree, vine]; hairs unbranched. **LF**: simple, gen alternate (opposite or whorled); veins pinnate; margin gen ± entire; stipules gen 0. **INFL**: raceme, spike, or panicle. **FL**: bisexual, gen bilateral and ± pea-fl-like [or ± radial]; sepals 5, fused or not, lateral or inner pair gen larger and petal-like (called wings); petals 5[3], fused to stamen tube, [± similar or] different with 1 lower keel petal, 2 strap-like upper petals, and 2[0] lateral petals; stamens 3–10, ± fused, tube open at top; ovary chambers 1–8 with 1 ovule each, style 1 or 0. **FR**: capsule [drupe or nut; occ winged]. **SEED**: often with aril. 20 genera, 1000 spp.: esp trop, subtrop; few cult. [Persson 2001 Taxon 50:763–779] Scientific Editor: Bruce G. Baldwin.

POLYGALA MILKWORT

Root odor gen wintergreen. **INFL**: raceme or spike, occ grouped and panicle-like; cleistogamous fls occ solitary. **FL**: bilateral; lateral 2 sepals enlarged; petals 3 or 5, keel petal gen with cylindric beak or fringed crest at tip; stamens 6–8, anthers dehiscent at tip, appearing 1-chambered; with nectary disk or gland; ovary chambers 2, stigma 2-lobed. **FR**: capsule. **SEED**: fusiform or ovoid, black, gen hairy, gen with prominent white aril on 1 end. ± 500 spp.: trop, temp. (Greek: much milk, some Eur spp. said to increase milk in cows) [Wendt 1979 J Arnold Arbor 60:504–514]

1. Infl thornless
   2. Fls 2.5–5 mm, ± cleistogamous, keel petal beakless . . . . *²P. californica*
   2′ At least some fls > 7 mm, opening, keel petal beaked
      3. Fl 9–14.5 mm (occ 2.5–5 mm cleistogamous fls near pl base), keel petal beak oblong, gen notched or contorted, ± 0.7–1 mm diam near tip; aril glabrous . . . . *²P. californica*
      3′ Fl 7–14 mm, of 1 kind, keel petal beak linear, entire, ± 0.2 mm diam near tip; aril hairy . . . . *P. cornuta*
         4. Fl 8.5–14 mm, wings dense-puberulent; upper sepal gen acute to acuminate . . . . var. *cornuta*
         4′ Fl 7–11.2 mm, wings gen ciliate, surface glabrous or puberulent only near tip; upper sepal gen rounded to obtuse . . . . var. *fishiae*
1′ Infl thorn-tipped
   5. Fl 2.5–5.3 mm, wings cream or ± green; twigs densely hairy
      6. Pedicels, outer sepals, and lvs spreading-hairy . . . . *P. acanthoclada*
      6′ Pedicels glabrous; outer sepals glabrous or ciliate, occ sparsely hairy near tip; lf hairs incurved or appressed . . . . *P. intermontana*
   5′ Fl 6–13.5 mm, wings pink; twigs glabrous to short-stiff-hairy
      7. Keel petal beak prominently notched below; seeds hairier near aril . . . . *P. heterorhyncha*
      7′ Keel petal beak entire or ± jagged; seeds gen evenly hairy . . . . *P. subspinosa*

P. acanthoclada A. Gray (p. 1075) THORNY MILKWORT Subshrub, shrub; hairs spreading. **ST**: sprawling to erect, < 10 dm, twigs densely hairy, white. **LF**: 5–25 mm, oblanceolate to narrowly elliptic or obovate, hairy. **INFL**: thorn-tipped; fls 1–15; pedicel 1.5–5.8 mm, hairy. **FL**: 3–5.3 mm; outer sepals hairy, wings cream; keel petal beak 0 or minute. **SEED**: 2–2.5 mm incl hairs; aril glabrous. 2*n*=18. Desert scrub, Joshua-tree or pinyon/juniper woodland, gen in loose, sandy or gravelly soil; 945–1830 m. s DMoj (Lucerne Valley), DMtns (Eagle, New York mtns); to s UT, AZ. May–Aug ★

P. californica Nutt. (p. 1075) CALIFORNIA MILKWORT Per, gen from rhizome. **ST**: gen decumbent, 0.5–3.5 dm. **LF**: 7–60 mm, lanceolate or elliptic to (ob)ovate. **INFL**: thornless. **FL**: open fls 9–14.5 mm, or ± cleistogamous fls 2.5–5 mm and occ present in separate, gen basal raceme; sepal wings ciliate, tip occ puberulent, pink (white); keel petal beak 1.2–3 mm, notched or contorted (entire) below, ± 0.7–1 mm diam near tip, yellow or white when pollen shed. **FR**: 4.5–10.5 mm incl stalk, green. **SEED**: 3.5–6 mm incl hairs; aril glabrous. 2*n*=18. Coastal prairie and forest, chaparral, occ on serpentine; 10–1400 m. NW, CW (exc SCoRI), n ChI; sw OR. Apr–Jul

P. cornuta Kellogg Gen from rhizome. **LF**: 10–65 mm, linear to ovate. **FL**: wings ciliate, glabrous to puberulent, cream or ± green to pink; keel petal beak 0.5–2.5 mm, entire, ± 0.2 mm diam near tip, dull rose to green when pollen shed. **FR**: 5.9–10 mm incl stalk, dark yellow-brown. **SEED**: 4.5–7.3 mm incl hairs; aril hairy. 2*n*=18.

var. ***cornuta*** (p. 1075) SIERRA MILKWORT Subshrub. **ST**: prostrate to erect, 1–10 dm. **LF**: length gen ± 2 × width. **FL**: 8.5–14 mm; upper sepal gen acute or acuminate, outer sepals and wings cream or ± green to pink in bud, wings densely puberulent. Moist to dry, open areas in chaparral, woodland, conifer forest, occ on serpentine; 150–1830 m. KR, CaR, n&c SN, s SNH (uncommon), w MP. A population in Mariposa Co. has glabrous fls. May–Sep

var. ***fishiae*** (Parry) Jeps. (p. 1075) FISH'S MILKWORT Shrub, gen forming thickets < 2 m diam. **ST**: decumbent to erect, 6–25 dm. **LF**: length > 2 × width. **FL**: 7–11.2 mm; upper sepal gen rounded or obtuse, wings and at least outer sepals tips dark pink in bud, wings ciliate, glabrous or puberulent only near tip. Chaparral, oak woodland; 90–1270 m. s SCoRO, WTR, SnGb, PR; n Baja CA. Pls in s SCoRO, WTR (Santa Ynez Mtns) with green-white fls, wings puberulent, named *P. cornuta* var. *pollardii* Munz, Pollard's milkwort. May–Aug ★

P. heterorhyncha (Barneby) T. Wendt (p. 1075) NOTCH-BEAKED MILKWORT Per, subshrub, forming thorny mats 1–2 dm, < 7 dm diam. **ST**: glabrous to short-stiff-hairy, ± glaucous. **LF**: 4–20 mm, ovate to elliptic or obovate, scattered short-sharp-hairy; base tapered to rounded. **INFL**: prominently thorn-tipped. **FL**: 7.5–13.5 mm; sepal wings pink; keel petal beak 1.4–4 mm, prominently notched below, yellow. **FR**: 4.2–7.8 mm incl stalk. **SEED**: 3–4.4 mm incl hairs; seed body widely elliptic to round in ×-section, hairier near aril; aril glabrous. 2*n*=36. Rocky areas in desert scrub; 900–1600 m. DMtns (Funeral Mtns, Inyo Co.); s NV. Apr–May ★

P. intermontana T. Wendt (p. 1075) INTERMOUNTAIN MILKWORT Subshrub, shrub, stiff-branched, ± open, < 10 dm, occ mat-like. **ST**: twig hairs dense, white, appressed to irregularly ascending. **LF**: 3–25 mm, linear to obovate, hairs incurved or appressed. **INFL**: thorn-tipped; fls 1–7; pedicel 2.5–9 mm, glabrous. **FL**: 2.5–5.2 mm; outer sepals glabrous or ciliate, occ sparsely hairy near tip, wings cream or ± green; keel petal beak 0 or minute. **FR**: 3.5–5.8 mm incl stalk. **SEED**: 2.8–4.2 mm, incl hairs; sparsely pubescent to glabrous; aril glabrous. 2*n*=18. Pinyon/juniper woodland; 2600–3080 m. SNE (Mono Co.); to UT, n AZ. May–Aug ★

P. subspinosa S. Watson (p. 1075) SPINY MILKWORT Per, shrub. **ST**: gen < 2.5 dm, glabrous to short-stiff-hairy, occ glaucous. **LF**: 4–31 mm, obovate or elliptic, ± glabrous or with short, stiff hairs; base acuminate. **INFL**: gen weakly thorn-tipped. **FL**: 6–13 mm; sepal wings pink; keel petal beak 1–3 mm, entire or ± jagged, yellow or ± green. **FR**: 5.5–10 mm incl stalk. **SEED**: 3.3–4.9 mm incl hairs; seed body elliptic in ×-section, gen evenly hairy throughout; aril glabrous. 2*n*=18,36. Desert scrub, volcanic mesas; 1350–2285 m. MP (Lassen Co.), SNE (Sweetwater Mtns); to UT, sw CO, nw NM, n AZ. Jun–Jul ★

POLYGONACEAE BUCKWHEAT FAMILY

Mihai Costea & James L. Reveal, family description, key to genera;
treatment of Polygonoideae genera revised by Mihai Costea; treatment of Eriogoneae
revised by James L. Reveal & Thomas J. Rosatti, except as noted

Ann to shrub [tree]. **ST**: nodes swollen or not. **LF**: simple, basal or cauline, gen alternate; ocreae present or 0, gen scarious, persistent or not. **INFL**: fl clusters in axillary to terminal cyme-, panicle-, raceme-, spike-, umbel- or head-like arrangements, entire infl or main infl branches gen subtended by bracts ("infl bracts"); peduncles present or 0; fl clusters in Eriogoneae–Eriogonoideae subtended by involucre of ≥ 1 free or ± fused, sometimes awn-tipped bracts ("involucre bracts") or, in Polygonoideae and rarely in Eriogonoideae, not (if bracts completely fused, involucre "tubular"); pedicels in Eriogoneae each often subtended by 2 free, transparent, linear bractlets or in Polygonoideae all subtended by 2+ fused, membranous, wide bractlets. **FL**: gen bisexual, small, 1–200 per node; perianth parts 2–6, gen in 2 whorls, free or basally fused, gen petal-like, often ± concave adaxially, often darker at midvein, often turning ± red or ± brown in age; stamens [1]3 or 6–9 in 2 whorls; ovary superior, 1-chambered, ovule 1, styles 1–3. **FR**: achenes, incl in or exserted from perianth, gen 3-angled, ovoid or elliptic, gen glabrous. 48 genera, ± 1200 spp.: worldwide, esp n temp; some cult for food (*Coccoloba*, sea-grape; *Fagopyrum*, *Rheum*, *Rumex*) or orn (*Antigonon*, lovechain; *Coccoloba*; *Muehlenbeckia*; *Persicaria*; *Polygonum*), a few timbered (*Coccoloba*; *Triplaris*). Several (*Emex*; *Fallopia*; *Persicaria*; *Polygonum*; *Rumex*) are weeds. [Freeman & Reveal 2005 FNANM 5:216–601] Treatment of genera in Eriogonoideae based on monographic work of James L. Reveal. Involucre number throughout is number (1–many) per ultimate grouping, at tips of ultimate branches; fl number is per fl cluster or involucre, unless otherwise stated. *Fagopyrum esculentum* Moench not naturalized, considered an historical waif (or garden weed ± presently), therefore not treated. Scientific Editors: Thomas J. Rosatti, Bruce G. Baldwin.

1. Ocreae present, persistent or not; nodes gen swollen; fl cluster not subtended by involucre; pedicels all
   subtended by 2+ fused, membranous, wide bractlets (Polygonoideae)
  2. Perianth parts (5)6 (Rumiceae)
    3. Fls unisexual, outer perianth parts of pistillate fl spine-tipped in fr . . . . **EMEX**
    3′ Fls gen bisexual, outer perianth parts not spine-tipped in fr . . . . **RUMEX**
  2′ Perianth parts 4–5
    4. Perianth parts 4; fr elliptic, 2-winged; lvs ± basal . . . . **OXYRIA**
    4′ Perianth parts 4–5; fr gen ovoid; lvs gen cauline
    5. Vine-like shrub; perianth in fr fleshy (Coccolobeae) . . . . **MUEHLENBECKIA**
    5′ Herb, or if shrub then not vine-like; perianth in fr gen not fleshy (Polygoneae)
      6. Outer perianth parts winged or keeled, or ± so
        7. Fl base stalk-like; outer perianth parts winged (keeled in *Fallopia convolvulus*); lf blade base
           sagittate, hastate, cordate, truncate, or ovate; ocrea papery, not 2-lobed distally . . . . **FALLOPIA**
        7′ Fl base not stalk-like; outer perianth parts keeled; lf blade base tapered (ovate); ocrea gen
           translucent, 2-lobed distally . . . . **²POLYGONUM**
      6′ Outer perianth parts not winged or keeled
        8. Lvs gen basal; infl terminal, spike-like; sts not branched . . . . **BISTORTA**
        8′ Lvs cauline; infl terminal, axillary, spike-like or not; sts gen branched
          9. Ocreae gen translucent, glabrous, 2-lobed, fibrous in age . . . . **²POLYGONUM**
          9′ Ocreae opaque, glabrous to scabrous, not 2-lobed
            10. Infl raceme- or panicle-like; perianth parts fused ± 1/4; stamens 8 . . . . **ACONOGONON**
            10′ Infl ± head-, spike-, or panicle-like; perianth parts fused 1/4–2/3; stamens 5–8 . . . . **PERSICARIA**
1′ Ocreae 0; nodes not swollen; fl clusters subtended by involucre of 1+ free or fused bracts (or not, in
   Gilmania); pedicels each often subtended by 2 free, transparent, linear bractlets (Eriogoneae–Eriogonoideae)
  11. Involucre bract 1, inflated, awns 0; lvs gen lobed; stamens 6 (Pterostegiinae) . . . . **PTEROSTEGIA**
  11′ Involucre 0 or tubular or of 2–5 ± free bracts, not inflated, awned or not; lvs entire; stamens 3–9
     (Eriogoninae)
    12. Ann to shrub, if ann then involucre tubular, awnless
      13. Involucres of 2–5 awnless bracts . . . . **DEDECKERA**
      13′ Involucres tubular . . . . **ERIOGONUM**
    12′ Ann; involucres tubular and awned or of free or basally fused bracts, awned or not, 0 in *Gilmania*
      14. Lvs basal and cauline, opposite . . . . **GOODMANIA**
      14′ Lvs basal or if cauline then alternate or whorled
        15. Involucre 0 . . . . **GILMANIA**
        15′ Involucre present
          16. Involucres of free or basally fused bracts
            17. Stamens 3
              18. Perianth lobes leathery; involucre bracts 3, awned . . . . **LASTARRIAEA**
              18′ Perianth lobes petal-like; involucre bracts many, awnless . . . . **NEMACAULIS**
            17′ Stamens 6 or 9
              19. Involucre bracts 3(4), awned; pl tomentose . . . . **HOLLISTERIA**
              19′ Involucre bracts 4–7, awnless; pl silky-puberulent . . . . **JOHANNESHOWELLIA**
          16′ Involucres tubular

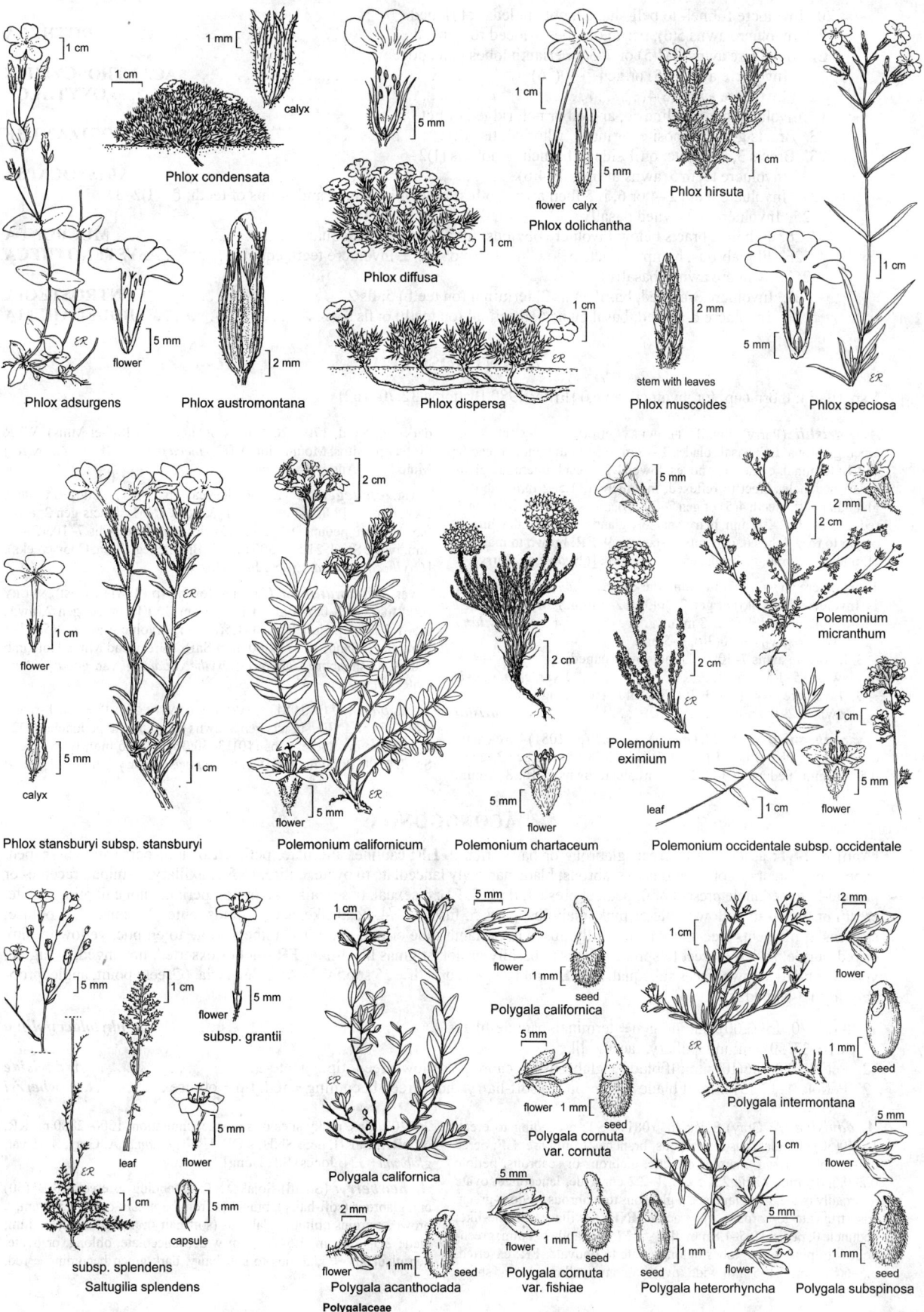

Phlox condensata

calyx

Phlox diffusa

Phlox dolichantha

flower   calyx

Phlox hirsuta

Phlox adsurgens

flower

Phlox austromontana

Phlox dispersa

Phlox muscoides

stem with leaves

Phlox speciosa

Phlox stansburyi subsp. stansburyi

flower

calyx

Polemonium californicum

flower

Polemonium chartaceum

flower

Polemonium micranthum

Polemonium eximium

leaf

flower

Polemonium occidentale subsp. occidentale

subsp. grantii

flower

leaf   flower

capsule

subsp. splendens

Saltugilia splendens

Polygala californica

flower

Polygala acanthoclada

flower   seed

Polygala californica

flower

seed

flower   seed

Polygala cornuta var. cornuta

flower   seed

Polygala cornuta var. fishiae

Polygala intermontana

flower

seed

seed   flower

Polygala heterorhyncha

flower

seed

Polygala subspinosa

Polygalaceae

20. Involucre funnel- to bell-shaped, not angled, not ridged
  21. Involucre awns 5(6); perianth lobes 3-lobed to fringed . . . . **SIDOTHECA**
  21′ Involucre awns (3)4(5) or 7–36; perianth lobes gen entire
    22. Involucre awns 4(5) or gen 7–30(36) . . . . **ACANTHOSCYPHUS**
    22′ Involucre awns (3)4 . . . . **OXYTHECA**
20′ Involucre gen ± cylindric, angled or not, ridged or not
  23. Bracts gen 2, opposite, or many, whorled; fls 1(2) . . . . **CHORIZANTHE**
  23′ Bracts 3, alternate, on 1 side of branch or not; fls (1)2–6
    24. Involucre teeth 5, awns at tips; fls 4(6) . . . . **ARISTOCAPSA**
    24′ Involucre teeth 2–4 or 6, if 5 then awns both at base of involucre and at tips of teeth; fls (1)2–3
      25. Involucre not awned basally
        26. Fl hairy; bracts below involucre obvious; involucre teeth unequal . . . . **MUCRONEA**
        26′ Fl glabrous, bumpy; bracts below involucre obscure; involucre teeth equal . . . . **SYSTENOTHECA**
      25′ Involucre awned basally
        27. Involucre 3-angled, basal awns 3, terminal (on teeth) 5; fls 2 . . . . **CENTROSTEGIA**
        27′ Involucre 6-angled, basal awns 6, terminal (on teeth) 6; fls 3 . . . . **DODECAHEMA**

ACANTHOSCYPHUS

1 sp. (Latin: thorn cup, for involucre awns) [Ertter 1980 Brittonia 32:70–102]

A. parishii (Parry) Small FLOWERY PUNCTUREBRACT Ann, erect, glabrous. **LF**: basal; blades 1–7 cm, 0.5–2 cm wide; ocreae 0. **INFL**: terminal, cyme-like, nodes, lower 1/2 of each internode glandular; peduncles erect to reflexed; involucre 1, 1.5–2 mm, tubular, glandular at tip, teeth 4(5) or gen 7–30(36), awns 2–5 mm, ivory or ± red. **FL**: 3–20, 2–2.5 mm, hairy, sparsely glandular; perianth white or cream to rose, lobes 6, gen entire; stamens 9. **FR**: brown to maroon, elliptic, 1.7–2 mm, glabrous; embryo curved. [*Oxytheca p.* Parry]

1. Involucre awns 7–16, 3–4 mm, dark red . . . . var. *abramsii*
1′ Involucre awns 4(5) or gen 7–30(36), 2–5 mm, ivory
 2. Involucre awns 4(5), 2–3 mm. . . . . var. *goodmaniana*
 2′ Involucre awns 7–30(36), 3–5 mm
  3. Involucre awns 7–10; infl bracts awl-shaped,
    awns 0.5–1.5 mm . . . . var. *cienegensis*
  3′ Involucre awns (10)13–30(36); infl bracts triangular,
    awns 0.2–0.5 mm . . . . var. *parishii*

var. **abramsii** (E.A. McGregor) Reveal (p. 1081) ABRAMS' OXYTHECA Pl 1–3 dm. **LF**: 1–3(5) cm. **INFL**: bracts gen 3, awns 0.5–1.5 mm; peduncles 0.8–2 cm; involucre awns 7–16, 3–4 mm, dark red. Sand; 1700–2000 m. s SCoRO (San Rafael Mtns), WTR (Topatopa Mtns, Mount Pinos). [*Oxytheca p.* var. *a.* (E.A. McGregor) Munz] Jun–Aug ★

var. **cienegensis** (Ertter) Reveal (p. 1081) CIENEGA SECA OXYTHECA Pl 0.5–4 dm. **LF**: 1–3.5 cm. **INFL**: bracts gen 2, awns 0.5–1.5 mm; peduncles 0.5–2(3.5) cm; involucre awns 7–10, (3)4–5 mm, ivory. Sand; 2100–2500 m. e SnBr (Cienega Seca, Coon creeks). [*Oxytheca p.* var. *c.* Ertter] Jun–Sep ★

var. **goodmaniana** (Ertter) Reveal (p. 1081) CUSHENBURY OXYTHECA Pl 0.5–3 dm. **LF**: 1–3 cm. **INFL**: bracts gen 2, awns 0.5–1.5 mm; peduncles 0.3–1.5(2) cm; involucre awns 4(5), 2–3 mm, ivory. Sand; 1300–2300 m. n SnBr (Greenlead Mine, Holcomb Valley, Cushenbury Canyon). [*Oxytheca parishii* var. *goodmaniana* Ertter] May–Sep ★

var. **parishii** (p. 1081) PARISH'S OXYTHECA Pl erect, 1–6 dm. **LF**: 1–7 cm. **INFL**: bracts gen 3, awns 0.2–0.5 mm; peduncles (1)2–5(7.5) cm; involucre awns (10)13–30(36), (3)4–5 mm, ivory. *n*=20. Sand; 1900–2600 m. TR. [*Oxytheca parishii* var. *p.*] Jun–Oct ★

ACONOGONON

[Ann] per. **ST**: ascending to erect, glabrous or hairy, ribs 0. **LF**: cauline, alternate, petioled or not; ocrea funnel-shaped, papery, persistent or not, glabrous to scabrous; blade narrowly lanceolate to ovate, entire. **INFL**: axillary, terminal, raceme- or panicle-like; peduncle present or 0; pedicel present; fls 1–5. **FL**: bisexual, base stalk-like or not; perianth not enlarging, rotate, cream or green- to yellow-white or pink, glabrous, parts 5, fused ± 1/4, petal-like, ± of 2 kinds, outer 2 < inner 3, ± obtuse; stamens 8, filaments free, not wider basally, fused to perianth tube or not, glabrous, anthers ovate to elliptic, yellow to pink or red-purple; styles 3, erect to spreading, fused basally or not, stigmas head-like. **FR**: incl or exserted, unwinged, 3-angled, yellow- or dark-brown, faces subequal. **SEED**: embryo gen curved. ± 25 spp.: w N.Am, Eur, Asia. (Greek: point, angle, probably from 3-angled fr)

1. St (50)70–150(200) cm; infl gen ± terminal, panicle-like . . . . ***A. phytolaccifolium***
1′ St 10–35(50) cm; infl axillary, raceme-like
 2. Petiole 0.2–0.6(10) mm; lf blade ± glabrous to scabrous, yellow-green, tip ± acute . . . . ***A. davisiae***
 2′ Petiole 4–12(15) mm; lf blade ± glabrous to soft-hairy, dark-green becoming ± red, tip ± obtuse . . . . ***A. newberryi***

A. davisiae (A. Gray) Soják (p. 1081) **ST**: ascending to erect, 15–40(50) cm, glabrous to scabrous, branches from base. **LF**: ocrea 0.5–3 cm, ± brown, margins oblique, glabrous or scabrous; petiole 0.2–0.6(10) mm; blade 2.2–5 cm, 1–2.2 cm wide, lanceolate to broadly ovate, gen glaucous, ± glabrous to scabrous, yellow-green, base truncate to cordate, tip ± acute. **INFL**: axillary, raceme-like; peduncle 0; pedicel 0.5–1.8 mm; fls 1–5. **FL**: perianth 2–4 mm, green-yellow to pink-white, lobes oblong-ovate to obovate. **FR**: exserted, 3.5–6(8) mm, 2.1–5 mm wide, ovate-oblong, yellow-brown, shiny.

$2n$=20. Talus, rocky sites of snow accumulation; 1500–2800 m. KR, NCoRH, CaRH, n&c SNH; s OR. [*Polygonum d.* A. Gray; *A. d.* var. *glabrum* (G.N. Jones) S.P. Hong] Jun–Aug

A. newberryi (Small) Soják **ST**: ascending to erect, 10–35(50) cm, glabrous (soft-hairy), branches from base. **LF**: ocrea 0.5–2 cm, ± brown, margins oblique, glabrous (soft-hairy); petiole 4–12(15) mm; blade 2.5–7(9) cm, 1.5–3(5) cm wide, lanceolate, oblong, or ovate, gen glaucous, ± glabrous to soft-hairy, dark-green becoming ± red,

base acute, cordate to semi-sagittate, tip ± obtuse. **INFL**: axillary; peduncle 0; pedicel 0.5–2 mm; fls 3–6. **FL**: perianth 2.5–4(4.5) mm, green-white to ± pink, lobes oblong-ovate to obovate. **FR**: exserted, 3.5–5(7.5) mm, 2–4.5 mm wide, ovate-oblong, yellow-brown, shiny. Volcanic slopes, esp pumice; 1500–2400 m. KR, CaRH (esp Mount Shasta), n&c SNH, MP; to WA. [*Polygonum n.* Small] Further study needed to determine if this and *A. davisiae* belong to 1 sp. Jun–Aug

A. phytolaccifolium (Small) Rydb. (p. 1081) **ST**: erect, (50)70–150(200) cm, glabrous or hairy. **LF**: ocrea 1–3 cm, ± brown, margins oblique, glabrous or hairy; petiole 5–20 mm; blade 5–15(19) cm, 1.4–7.5 cm wide, lanceolate to lance-ovate, not glaucous, glabrous or densely hairy, base rounded, margins gen scabrous to ciliate (glabrous), tip obtuse to acuminate. **INFL**: gen ± terminal, panicle-like; peduncle 0.5–10 cm, glabrous or hairy; pedicels 1–4 mm; fls 1–3. **FL**: perianth 2.5–3.5 mm, white to ± green, lobes ovate to obovate, tip obtuse. **FR**: gen exserted, (3)4–6.5 mm, 2.5–4.5 mm wide, ovoid, yellow-brown, smooth, shiny. Meadows, moist rocky places; 1500–2800 m. KR, NCoRH, n&c SNH, MP; to WA, ID, NV. [*Polygonum p.* Small; *A. p.* var. *glabrum* S.P. Hong] Jun–Aug

ARISTOCAPSA

1 sp. (Latin: awned box, for awned involucre)

A. insignis (Curran) Reveal & Hardham (p. 1081) INDIAN VALLEY SPINEFLOWER Ann, erect, 0.2–1 dm, glandular. **LF**: basal; blades (0.3)0.5–1.5 cm, (0.1)0.2–0.4 cm wide, glabrous; ocreae 0. **INFL**: terminal, cyme-like; peduncles erect, 1–2 mm; involucre 1, 3–5 mm, tubular, teeth 5, awns (1)2–3 mm. **FL**: (4)6, 1.5 mm, hairy; perianth white to pink or rose, lobes 6, entire; stamens 9. **FR**: 1.5 mm, light green-brown to tan, obconic, glabrous; embryo curved. *n*=14. Sand; 300–600 m. SCoRI (Monterey, San Luis Obispo cos.). Extirpated from Monterey Co. May–Jun ★

BISTORTA

1 sp. (Latin: twice twisted, from contorted rhizomes)

B. bistortoides (Pursh) Small (p. 1081) WESTERN BISTORT Per, rhizomes contorted. **ST**: (10)20–70 cm, branches 0. **LF**: gen basal; ocrea cylindric, 10–25(30) mm, brown, margins oblique, glabrous; petiole 10–30(50) mm; blade 4–30 cm, 1–4.8 cm wide, elliptic to lance-oblong or oblanceolate, base tapered to rounded, gen oblique, abaxially hairs 0 or ± white or ± brown, glaucous, adaxially hairs 0, margins entire, tip gen acute to acuminate; cauline lvs 2–4, petioled proximally, sessile distally, gradually reduced upward; blade elliptic or lanceolate to lance-linear. **INFL**: terminal, spike-like, (10)20–40(50) mm, 10–25 mm wide, short-cylindric to ovoid; peduncle 1–10 cm; pedicel ascending or spreading, 2–10 mm; fls 1–2. **FL**: bisexual; perianth 4–5 mm, not enlarging, bell-shaped, glabrous, white or pale pink, parts 5, fused ± 1/5, oblong, tips obtuse to acute; stamens 5–8, exserted, anthers yellow, elliptic. **FR**: 3–4 mm, 1.2–2 mm wide, obovoid, light- or olive-brown, shiny, smooth. 2*n*=24. Wet meadows, streambanks, alpine slopes; 1500–3000 m. CA-FP (uncommon in coastal freshwater marshes, 0–20 m, NCo, n CCo), n SNE; to AK, e N.Am. [*Polygonum b.* Pursh] Jul–Sep

CENTROSTEGIA

1 sp. (Greek: spurred roof, for basal awns of involucre)

C. thurberi A. Gray (p. 1081) RED TRIANGLES Ann, erect, 0.3–2(3) dm, sparsely glandular. **LF**: basal; blades (0.5)1–3.5(4) cm, 0.3–0.8(1) cm wide, glabrous; ocreae 0. **INFL**: terminal, cyme-like; peduncles 0; involucre 1, (2)3–6(8) mm, tubular; basal awns 3, 0.2–2 mm, teeth 5, each with 0.3–1 mm awn. **FL**: 2, 2–3(3.5) mm, hairy; perianth white to pink, lobes 6, 2-lobed; stamens 9. **FR**: brown, obconic, 2–2.5 mm, glabrous; embryo curved. *n*=19. Common. Sand or gravel; 300–2400 m. s SnJV, e SCoRI, TR, SNE, D; to UT, AZ, nw Mex. Mar–Jul

CHORIZANTHE SPINEFLOWER

Ann [per], prostrate to erect, gen hairy. **LF**: basal or cauline, alternate, linear to narrow, awns 0; ocreae 0. **INFL**: terminal, gen cyme- or head-like, hairy [(glabrous)]; peduncles 0; involucres gen 1–6, tubular, 3–6-ribbed, teeth 3, 5, or 6, awned. **FL**: 1(2), glabrous or hairy; perianth white, yellow, red, maroon, or purple, lobes (5)6, entire, notched or lobed to fringed; stamens gen 3, 6, or 9. **FR**: gen brown or black, gen elliptic, glabrous; embryo straight or curved. ± 50 spp.: temp w N.Am; sw S.Am. (Greek: divided fl, for perianth lobes) Involucre length measurements incl teeth and awns unless otherwise specified.

1. Involucre 3–5-toothed or -ribbed
  2. Involucre teeth (4)5
   3. Involucre awns straight; fl glabrous; stamens fused to tube base — w DMoj . . . . **C. spinosa**
   3′ Involucre awns hooked; fl hairy; stamens fused to near tube top
    4. Involucre bell-shaped, tube 2–2.5 mm, abaxial tooth not longer, not wider than others; fl white to rose, 1.5–1.8(2) mm; c&n CA . . . . ²**C. polygonoides** var. **polygonoides**
    4′ Involucre cylindric, tube 3–4.5 mm, abaxial tooth longer, wider than others; fl yellow, 1.5–2.5 mm; sw SnJV, n edge TR, GB, w DMoj . . . . **C. watsonii**
  2′ Involucre teeth 3
   5. Involucre cylindric, markedly transversely ridged; stamens 6 — D . . . . **C. corrugata**
   5′ Involucre urn- to bell-shaped, not markedly transversely ridged; stamens 9
    6. Pl prostrate; involucre bell-shaped; abaxial tooth 1.8–2 mm, awn hooked; s SCo . . . . **C. orcuttiana**
    6′ Pl erect; involucre urn-shaped; abaxial tooth 5–10 mm, awn straight; SNE, D . . . . **C. rigida**

1′ Involucre 6–toothed or -ribbed
  7. Involucre teeth with scarious margins
    8. Involucre teeth scarious margins continuous across sinuses
      9. Perianth lobes obcordate to 2-lobed — NCoRI, CaRF, SNF, e SnFrB . . . . ²*C. stellulata*
     9′ Perianth lobes entire
      10. Involucre hairy, 3–5 mm; fl 3.5–4(4.5) mm; SCoR . . . . **C. douglasii**
      10′ Involucre tomentose to ± glabrous, 3–4 mm; fl (1.5)2.5–3 mm; more widespread in CA-FP . . . . **C. membranacea**
    8′ Involucre teeth scarious margins not continuous across sinuses
     11. Involucre awns straight
      12. Involucres 3–4 in small bracted clusters; perianth lobes alike — NCoRI, CaRF, SNF, e SnFrB . . . . . ²*C. stellulata*
      12′ Involucres 1; perianth lobes not alike
        13. Fl (3)3.5–4.5 mm, hairy gen throughout; pl gen spreading to decumbent; c NCo . . . . **C. howellii**
        13′ Fl (4)5–6 mm, hairy on lower 1/2; pl erect to spreading; n CCo. . . . . ²*C. valida*
     11′ Involucre awns hooked
      14. Fl glabrous; perianth lobes entire, tube yellow — CCo, w SnFrB, w SCoRO . . . . **C. diffusa**
      14′ Fl hairy; perianth lobes 2-lobed, jagged, or short-pointed, tube white
        15. Involucre tube 2.5–4 mm, thinly hairy; fl 2.5–4 mm . . . . **C. robusta**
        16. Involucre lobes rose-pink; pl erect; s SnFrB . . . . var. **hartwegii**
        16′ Involucre lobes white; pl spreading or decumbent; n-c CCo, sw SnFrB. . . . . var. **robusta**
       15′ Involucre tube 1–2.5(3) mm, densely hairy; fl 2–3.5 mm
        17. Perianth lobes awned. . . . . ²*C. cuspidata* var. *cuspidata*
        17′ Perianth lobes minutely jagged
          18. Involucre teeth margins ± pink; involucre tube 1.5–2(2.5) mm; fl 2–3 mm; stamens 3 or 6–9
           . . . . ²*C. angustifolia*
          18′ Involucre teeth margins white (± pink) or dark ± pink to purple; involucre tube 2–2.5(3) mm;
           fl 2–3.5 mm; stamens 9 . . . . **C. pungens**
           19. Involucre teeth margins dark ± pink to purple; pl ± ascending to erect; 90–500 m . . . . var. **hartwegiana**
           19′ Involucre teeth margins white (± pink); pl prostrate to ± ascending; 0–65 m. . . . . var. **pungens**
  7′ Involucre teeth without scarious margins (exc *Chorizanthe fimbriata*, *Chorizanthe procumbens*)
    20. Involucre teeth of unequal, not of equal or alternating, lengths, abaxial tooth >> others
     21. Abaxial awn hooked. . . . . **C. clevelandii**
     21′ Abaxial awn straight
      22. Outer perianth lobes entire or ± 2-lobed, inner jagged; stamens 9; SCoRO . . . . **C. rectispina**
      22′ Outer perianth lobes with 1 or 3 teeth, inner entire; stamens 3; s SN, e CW, n WTR. . . . . **C. uniaristata**
    20′ Involucre teeth of equal or alternating lengths, abaxial tooth not >> others
     23. Perianth lobes all or at least inner fringed or 2-lobed
      24. Outer and inner perianth lobes fringed — SW . . . . **C. fimbriata**
        25. Perianth lobes with terminal segment linear to lanceolate, wider than lateral; fl 6–7(8) mm . . . . . . var. **fimbriata**
        25′ Perianth lobes with terminal segment linear, barely wider than lateral; fl 8–9(10) mm . . . . var. **laciniata**
      24′ Outer and/or inner perianth lobes not fringed
        26. Perianth lobes white to ± pink
          27. Inner perianth lobes 2-lobed. . . . . **C. blakleyi**
          27′ Inner perianth lobes fringed. . . . . **C. obovata**
        26′ Perianth lobes red, rose, maroon, or dark purple
         28. Outer perianth lobes entire; involucre tube 3.5–4 mm, awns 0.5–1 mm . . . . **C. palmeri**
         28′ Outer perianth lobes 2-lobed or minutely jagged; involucre tube 4–6 mm, awns 0.5–2 mm
          29. Outer perianth lobes minutely jagged or wavy-margined, inner fringed or ± 2-lobed; fl 4–4.5 mm;
           involucre tube 4–4.5 mm, swollen; SCoRI . . . . **C. ventricosa**
          29′ Outer perianth lobes 2-lobed or notched, inner fringed; fl (4.5)5–6 mm; involucre tube 4–6 mm,
           ± swollen; SCoR . . . . **C. biloba**
           30. Outer perianth lobes deeply 2-lobed. . . . . var. **biloba**
           30′ Outer perianth lobes notched or ± cordate at tip . . . . var. **immemora**
  23′ Perianth lobes entire, awned, or minutely jagged
    31. Perianth lobes alike, outer and inner similar in width and length
     32. Pl gen erect; branches breaking at nodes; stamens 3, fused to tube top . . . . **C. brevicornu**
      33. Lf blades 1–3(5) mm wide; involucres ribbed . . . . var. **brevicornu**
      33′ Lf blades 5–10 mm wide; involucres obscurely ribbed . . . . var. **spathulata**
     32′ Pl prostrate to erect; branches not breaking at nodes; stamens 3–9 and fused to tube base or (6)9
      and fused to tube top
      34. Stamens fused to tube top; perianth lobes entire or minutely notched; CA-FP . . . . **C. polygonoides**
        35. Involucre tube 1.5–2 mm, awns of prominent teeth 2–3 mm; PR . . . . var. **longispina**
        35′ Involucre tube 2–2.5 mm, awns of prominent teeth 1.5–2 mm; NW, CaR, n SNF, ScV, CW
        . . . . ²var. **polygonoides**
      34′ Stamens fused to tube base; perianth lobes entire, awned, or minutely jagged
        36. Perianth lobes entire; stamens fused basally — c&s SCo, s TR, w PR. . . . . **C. procumbens**

36′ Perianth lobes awned or minutely jagged; stamens free
  37. Perianth lobes minutely jagged; CCo . . . . ²*C. angustifolia*
  37′ Perianth lobes awned; n CCo, SnFrB
    38. Involucre awns hooked, teeth with or without scarious margins; n CCo, SnFrB . . . ²*C. cuspidata* var. *cuspidata*
    38′ Involucre awns straight, teeth without scarious margins; n CCo . . . . *C. cuspidata* var. *villosa*
31′ Perianth lobes not alike, inner narrower and shorter
  39. Involucre awns straight
    40. Involucre tube 1.5–2 mm; WTR . . . . *C. parryi* var. *fernandina*
    40′ Involucre tube 3–4(4.5) mm; n CCo . . . . ²*C. valida*
  39′ Involucre awns hooked
    41. Involucre tube 1.5–2 mm — TR . . . . *C. parryi* var. *parryi*
    41′ Involucre tube 2.5–6 mm
      42. Proximal lf-like bracts of st early-deciduous or 0
        43. Fl 4.5–6 mm; e PR . . . . *C. leptotheca*
        43′ Fl 3–4(5) mm; c&s CCo, SCoRO, SW (exc e PR) . . . . *C. staticoides*
      42′ Proximal lf-like bracts of st persistent
        44. Fl 4.5–6 mm, long exserted from involucre . . . . *C. xanti*
          45. Involucre densely hairy; e SnBr, n SnJt . . . . var. *leucotheca*
          45′ Involucre thinly hairy; s SN, Teh, s SCoRI, n&e TR (exc e SnBr), SNE . . . . var. *xanti*
        44′ Fl 2.5–3.5 mm, ± exserted from involucre
          46. Fl 3–3.5 mm; involucre tube 2.5–3 mm; stamens 9; SCoRO . . . . *C. breweri*
          46′ Fl 2.5–3 mm; involucre tube 2–2.5 mm; stamens 6; n ChI . . . . *C. wheeleri*

C. angustifolia Nutt. NARROW-LEAF SPINEFLOWER Pl gen decumbent, 0.3–1 dm, 0.5–10(13) dm diam, hairy. **LF:** blades (0.5)1–4(5) cm, (0.2)0.3–0.6 cm wide. **INFL:** bracts 2, awns 0; involucre 3-angled, 6-ribbed, tube 1.5–2(2.5) mm, teeth 6, 1.5–2.5 mm, gen without scarious margins, abaxial longest, awns 1–1.5 mm, hooked. **FL:** 1, 2–3 mm, hairy; perianth 2-colored, fl tube white, lobes white to rose, gen jagged at tip; stamens 3 or 6–9. **FR:** 2–2.5 mm. n=19,20(21–23). Uncommon. Sand; 10–500 m. s CCo (San Luis Obispo, Santa Barbara cos.). Apr–Jul

C. biloba Goodman Pl gen erect, (0.5)1–3(4) dm, hairy. **LF:** blades 1–3(5) cm, 0.4–1(1.3) cm wide. **INFL:** bracts gen 2 proximally, a whorl of 3–5 at middle, awns straight, 1–3 mm; involucre 4–6 mm, 3-angled, 6-ribbed, ± swollen basally, ± transversely ridged, teeth 6, 1–2 mm, awns 0.5–2 mm, gen hooked. **FL:** 1, (4.5)5–6 mm, sparsely hairy; perianth 2-colored, fl tube white to yellow, lobes red, maroon or dark purple, outer 2-lobed or notched, inner fringed; stamens 9. **FR:** 4–4.5 mm.

  var. ***biloba*** TWO-LOBE SPINEFLOWER **FL:** outer perianth lobes deeply 2-lobed, occ jagged. n=20. Uncommon. Sand, gravel, clay; 200–700 m. e SCoRO, w SCoRI, e SCoRI (near Parkfield Grade, w Fresno Co.). May–Aug

  var. ***immemora*** Reveal & Hardham HERNANDEZ SPINEFLOWER **FL:** outer perianth lobes notched or ± cordate at tip. n=17–23. Serpentine, gravel, vertic clay; 600–800 m. e SCoRI (se San Benito, e Monterey cos.). May–Sep ★

C. blakleyi Hardham (p. 1081) BLAKLEY'S SPINEFLOWER Pl gen ascending, 0.5–1.5 dm, 0.5–3 dm diam, thinly hairy. **LF:** blades 0.5–2.5 cm, 0.3–0.8 cm wide. **INFL:** bracts 2, awns 1–2.5 mm, straight; involucre 3–4.5 mm, 3-angled, 6-ribbed, ± swollen basally, ± transversely ridged, teeth 6, 1–3 mm, abaxial longest, awns 0.5–2 mm, straight or hooked. **FL:** 1, 5–6 mm, sparsely hairy; perianth 2-colored, fl tube green-white to white, lobes white to ± pink, outer 2-lobed, inner 2-lobed, jagged; stamens 9. **FR:** 3–3.5 mm. n=±19. Sand or gravel; 600–1600 m. s SCoRO (n slope, Sierra Madre Mtns, Santa Barbara Co.). May–Jul ★

C. brevicornu Torr. Pl gen erect, thinly hairy. **INFL:** bracts 2, awns straight, 0.2–0.5 mm; involucre 3–5 mm, 3-angled, 6-ribbed, teeth 6, 0.4–1.2 mm, awns 0.2–0.5 mm, hooked. **FL:** 1, 2–4 mm, glabrous; perianth green-white to white or pale yellow-white, lobes entire; stamens 3, fused to tube top. **FR:** 3–4 mm.

  var. ***brevicornu*** (p. 1081) BRITTLE SPINEFLOWER Pl 0.5–3 dm. **LF:** blades 1.5–3(4) cm, 0.1–0.3(0.5) cm wide, tips acute. **INFL:** involucre ribbed. n=19–21(23). Sand, gravel; 60–2300 m. s SNE, D; to UT, AZ, nw Mex. Feb–Jul

  var. ***spathulata*** (Rydb.) C.L. Hitchc. GREAT BASIN BRITTLE SPINEFLOWER Pl 0.5–2(3) dm. **LF:** blades 1–2 cm, 0.5–1 cm wide, tips round. **INFL:** involucre obscurely ribbed. n=19. Sand, gravel; 700–2900 m. SNE; to n&c NV, also se OR, s ID. Apr–Jul

C. breweri S. Watson (p. 1081) BREWER'S SPINEFLOWER Pl gen decumbent, (0.3)0.5–1.5(2) dm, 1–5(7) dm diam, thinly hairy. **LF:** blades 0.5–2 cm, 0.3–1.2(1.5) cm wide, occ tomentose. **INFL:** bracts 2, awns 0.5–1 mm, straight; involucre 3-angled, 6-ribbed, tube 2.5–3 mm, transversely ridged, teeth 6, 0.4–1.2 mm, awns 0.3–0.6 mm, hooked. **FL:** 1, 3–3.5 mm, hairy; perianth white to rose or red, lobes entire; stamens 9. **FR:** 2.5–3 mm. n=19. Gravel or rocks; 60–800 m. SCoRO (sw San Luis Obispo Co.). Mar–Jul ★

C. clevelandii Parry (p. 1081) CLEVELAND'S SPINEFLOWER Pl gen decumbent, 0.2–0.8(1) dm, 0.5–5(7) dm diam, hairy. **LF:** blades 0.5–1.5(2) cm, 0.3–0.6(0.8) cm wide. **INFL:** bracts 2, awns 1–3 mm, straight; involucre 3–3.5 mm, 3-angled, 6-ribbed, ± transversely ridged, teeth 6, 0.3–6 mm, abaxial longest, awns 0.3–0.6 mm, hooked. **FL:** 1, 2.5–3 mm, sparsely hairy; perianth 2-colored, fl tube green-white, lobes white, outer entire or ± 2-lobed, inner entire to jagged, ± fringed or 2-lobed; stamens 3. **FR:** 2.5–3 mm. n=21. Common. Sand or gravel; 400–2000 m. NW, s SN, Teh, CW, n WTR. May–Sep

C. corrugata (Torr.) Torr. & A. Gray (p. 1081) WRINKLED SPINEFLOWER Pl erect, 0.3–1.5 dm, sparsely tomentose. **LF:** blades ± round, (0.3)0.5–1.5(2) cm, gen tomentose. **INFL:** bracts 2, awns 0.5–1 mm, ± curved; involucre 1.8 mm, 3-angled, 3-ribbed, ± glabrous, transversely ridged, teeth 3, 2–4.5 mm, awns 0.6–1 mm, hooked. **FL:** 1, 2–2.5 mm, hairy; perianth white, lobes entire; stamens 6. **FR:** 2.5–3 mm. n=19. Common. Sand or gravel; 70–1000 m. D; to s NV, w AZ, nw Mex. Feb–May

C. cuspidata S. Watson Pl decumbent to ascending, 0.5–10 dm diam, hairy. **INFL:** bracts 2, awns 0.5–1.2 mm; involucre 3-angled, 6-ribbed, tube 1–3 mm, gen swollen, transversely ridged, teeth 6, 0.5–2 mm, gen without scarious margins, abaxial longest, awns 1–3 mm, hooked or straight. **FL:** 1, 2–3 mm, hairy; perianth 2-colored, fl tube white, lobes white to rose, entire or 3-lobed and awned; stamens 9. **FR:** 2–3 mm.

  var. ***cuspidata*** (p. 1081) SAN FRANCISCO BAY SPINEFLOWER Pl decumbent to prostrate, 0.5–1.5 dm. **LF:** blades 0.5–2(2.5) cm, 0.3–0.7 cm wide. **INFL:** involucre tube 1–2 mm, teeth with or without scarious, ± pink margins, awns hooked. **FL:** 2–2.5 mm. **FR:** 2–2.5 mm. Sand; < 300 m. n CCo, SnFrB. Apr–Jul ★

  var. ***villosa*** (Eastw.) Munz WOOLLY-HEADED SPINEFLOWER Pl gen ascending, 0.5–2 dm. **LF:** blades 1–5 cm, 0.4–1 cm wide. **INFL:** involucre tube (2)2.5–3 mm, awns straight. **FL:** 2.5–3 mm. **FR:** 2.5–3 mm. n=19–21. Sand; < 60 m. n CCo. May–Aug ★

C. diffusa Benth. DIFFUSE SPINEFLOWER Pl gen spreading to prostrate, 0.3–1(1.5) dm, 0.5–2(10) dm diam, hairy. **LF:** blades 0.3–2 cm, 0.1–0.4 cm wide. **INFL:** bracts 2, awns 0; involucre 2–2.5 mm, 3-angled, 6-ribbed, teeth 6, 0.5–1 mm, margins white or ± pink to purple, abaxial longest, awns 1–2 mm, hooked. **FL:** 1, 2.5–3 mm, glabrous; perianth 2-colored, fl tube lemon-yellow, lobes white, entire; stamens 3–9. **FR:** 2–2.5 mm. *n*=19–21. Common. Sand or gravel; 30–800 m. CCo, w SnFrB, w SCoRO. Apr–Jul

C. douglasii Benth. (p. 1081) DOUGLAS' SPINEFLOWER Pl erect, 1–4(5) dm, hairy. **LF:** blades 0.5–2(4) cm, 0.1–0.4(1) cm wide. **INFL:** bracts gen 2 proximally, a whorl of 3–5 at middle, awns 0; involucre 3–5 mm, 3-angled, 6-ribbed, ± swollen, finely transversely ridged, teeth 6, (0.7)1–1.5 mm, margins purple, awns 0.5–1 mm, straight. **FL:** 1, 3.5–4(4.5) mm, hairy; perianth white to rose, lobes gen 2-lobed or toothed; stamens 9. **FR:** 3.5–4 mm. *n*=20. Sand or gravel; (200)300–1600 m. e SCoRO, SCoRI. Apr–Jul ★

C. fimbriata Nutt. Pl gen erect, hairy, glandular. **LF:** blades tomentose abaxially. **INFL:** bracts gen 2, awns 1–2 mm, straight; involucres 4–6(7) mm, 3-angled, 6-ribbed, finely transversely ridged, teeth 6, 0.3–3 mm, margins scarious, awns straight, longer three 1–2.5(3) mm, shorter three (0.3)0.5–1.5 mm. **FL:** 1, 6–9(10) mm, glabrous; perianth 2-colored, fl tube yellow to yellow-white, lobes white to rose or red, oblong, fringed; stamens 9. **FR:** 3–4 mm.

var. ***fimbriata*** (p. 1081) FRINGED SPINEFLOWER Pl 1–3(3.5) dm. **LF:** blades 1–3.5 cm, 0.2–1(2.5) cm wide. **FL:** 6–7(8) mm; perianth lobes fringed, distal segments wider than lateral. *n*=20,22. Sand, gravel, rocks; 30–1700 m. w PR; n Baja CA. Mar–Jul

var. ***laciniata*** (Torr.) Jeps. (p. 1081) LACINIATE SPINEFLOWER Pl 1–2(2.5) dm. **LF:** blades 1–2(3) cm, 0.2–1(1.5) cm wide. **FL:** 8–9(10) mm; perianth lobes finely fringed, distal segments barely wider than lateral. *n*=20. Sand, gravel, rocks; 400–1500 m. w PR; n Baja CA. Mar–Jul

C. howellii Goodman (p. 1081) HOWELL'S SPINEFLOWER Pl gen spreading or decumbent, 0.3–1 dm, 1–5 dm diam, hairy. **LF:** blades 1–3 cm, 0.5–1.5(1.8) cm wide. **INFL:** bracts 2, awns 0; involucre 3–4 mm, 3-angled, 6-ribbed, teeth 6, 0.5–1 mm, margins white, awns 0.5–2 mm, straight, abaxial longest. **FL:** 1, (3)3.5–4.5 mm, hairy; perianth 2-colored, fl tube white, lobes white to rose, jagged; stamens 9. **FR:** 3–4.5 mm. *n*=(36–38)40(41–45). Coastal dunes; < 20 m. NCo (c Mendocino Co.). May–Jul ★

C. leptotheca Goodman PENINSULAR SPINEFLOWER Pl gen spreading, 0.5–3(3.5) dm, 0.5–3(5) dm diam, thinly hairy. **LF:** blades 0.5–2(3) cm, 0.3–0.5(0.7) cm wide, gen tomentose abaxially. **INFL:** bracts 2, awns 0.5–1 mm, straight; involucre 3–4 mm, 3-angled, 6-ribbed, teeth 6, 0.7–1.5 mm, awns 0.5–1 mm, hooked. **FL:** 1, 4.5–6 mm, hairy; perianth rose to red, lobes entire; stamens 9. **FR:** 3–4 mm. *n*=19. Sand or gravel; (300)600–1600 m. e PR; n Baja CA. May–Aug ★

C. membranacea Benth. (p. 1081) PINK SPINEFLOWER Pl erect, 1–6(10) dm, hairy. **LF:** blades (1)1.5–5 cm, 0.1–0.3 cm wide, gen linear, tomentose abaxially. **INFL:** bracts gen 2 (or a whorl of 3–5), awns 0.5–1 mm, straight; involucre 3–4 mm, 3-angled, 6-ribbed, tomentose to ± glabrous, ± swollen, teeth 5, 0.5–1 mm, margins scarious, awns 0.7–1.5 mm, hooked. **FL:** 1(2), (1.5)2.5–3 mm, hairy; perianth white to rose, lobes entire; stamens 9. **FR:** 2.5–3 mm. *n*=19–20,(22),40–41. Common. Sand, gravel or rocks; 40–1400 m. NW, CaR, SNF, Teh, GV, CW, n WTR; s-c OR. Apr–Jul

C. obovata Goodman SPOON-SEPAL SPINEFLOWER Pl erect to prostrate, (0.5)1–3(4) dm, 1–4(5) dm diam, hairy. **LF:** blades 0.5–2.5 cm, 0.3–1 cm wide. **INFL:** bracts gen 2, awns 1–2 mm, straight; involucre 3–4 mm, 3-angled, 6-ribbed, ± swollen, ± transversely ridged, teeth 6, 1–2 mm, awns straight or hooked, abaxial longest. **FL:** 1, 4–4.5(5) mm, sparsely hairy; perianth 2-colored, fl tube green-white to white, lobes white to pink, outer gen entire, inner shorter, fringed; stamens (6)9. **FR:** 3–3.5 mm. *n*=19–21. Sand or gravel; 10–1300 m. s SCoRO. Abaxial tooth of involucre reported by some to be >> others, contrary to key, further study needed. May–Jul

C. orcuttiana Parry (p. 1081) ORCUTT'S SPINEFLOWER Pl prostrate, 0.1–0.5 dm, 0.3–2(2.5) dm diam, hairy. **LF:** blades 0.5–1.5

cm, 0.2–0.35(0.5) cm wide. **INFL:** bracts 2, awns 0 or 0.6–1 mm, straight; involucre 0.8–2 mm, 3-angled, 3-ribbed, faintly transversely ridged, teeth 3, 1.8–2 mm, awns 0.6–1 mm, hooked. **FL:** 1, 1.5–1.8 mm, densely hairy; perianth yellow, lobes entire; stamens 9. **FR:** 2–2.2 mm. *n*=(18–19)20(21). Sand; 60–200 m. s SCo (Del Mar to Point Loma, San Diego Co., extirpated elsewhere). Mar–May ★

C. palmeri S. Watson (p. 1093) PALMER'S SPINEFLOWER Pl gen erect, (0.5)1–3(4) dm, hairy. **LF:** blades 1–3 cm, 0.4–0.8 cm wide. **INFL:** bracts gen 2, gen a whorl of 3–5 at middle, awns 1–3 mm, straight; involucre 3-angled, 6-ribbed, tube 3.5–4 mm, ± swollen, ± transversely ridged, teeth 6, 1–2 mm, abaxial longest, awns 0.5–1 mm. **FL:** 1, 4–5 mm, glabrous or hairy; perianth 2-colored, fl tube white to yellow, lobes red, maroon or dark purple, outer entire, inner fringed; stamens 9. **FR:** 3–3.5 mm. *n*=19,20(24). Serpentine; 60–700 m. SCoRO (w Monterey, w San Luis Obispo cos.). May–Aug ★

C. parryi S. Watson Pl spreading to erect, 0.2–1(1.5) dm, 0.5–4(6) dm diam, hairy. **LF:** blades 0.5–2.5(4) cm, 0.2–0.6(1.3) cm wide. **INFL:** bracts mostly 2, awns 0.4–1 mm, straight; involucre 3-angled, 6-ribbed, tube (1.5)1.7–2 mm, ± swollen, transversely ridged, teeth 6, 0.5–3 mm, awns 0.2–1.5 mm. **FL:** 1, 2.5–3 mm, hairy; perianth 2-colored, fl tube green-white, lobes white, outer gen fringed, inner narrower, entire or toothed; stamens 9. **FR:** 2.5–3 mm.

var. ***fernandina*** (S. Watson) Jeps. (p. 1093) SAN FERNANDO VALLEY SPINEFLOWER **INFL:** involucre awns straight. Sand; 90–500 m. WTR (Laskey Mesa, Ventura Co.; n Santa Susana Mtns, Los Angeles Co.). Apr–Jun ★

var. ***parryi*** PARRY'S SPINEFLOWER **INFL:** involucre awns hooked. Sand; 90–800 m. c&e SCo, e TR, nw edge DSon. May–Jun ★

C. polygonoides Torr. & A. Gray Pl prostrate, 0.1–0.5 dm, 0.3–2(2.5) dm diam, hairy. **LF:** blades 0.3–1 cm, 0.2–0.3 cm wide. **INFL:** bracts 2, awns 0; involucre 3-angled, 5–6-ribbed, tube transversely ridged, hairy, teeth 5–6, awns 1–3 mm, hooked or straight. **FL:** 1, 1.5–1.8(2) mm, hairy; perianth white to rose, lobes entire or minutely notched; stamens (6)9, fused to tube top. **FR:** 2–2.5 mm.

var. ***longispina*** (Goodman) Munz (p. 1093) LONG-SPINED SPINEFLOWER Pl gen ± red. **INFL:** involucre tube 1.5–2 mm, awns of prominent teeth 2–3 mm. *n*=20. Sand; 30–1500 m. PR; n Baja CA. Apr–Jun ★

var. ***polygonoides*** (p. 1093) KNOTWEED SPINEFLOWER Pl gen ± green. **INFL:** involucre tube 2–2.5 mm, awns of prominent teeth 1.5–2 mm. *n*=20. Sand or gravel; 100–1500 m. NW, CaR, n SNF, ScV, CW. Apr–Jun

C. procumbens Nutt. (p. 1093) PROSTRATE SPINEFLOWER Pl gen decumbent, 0.2–0.8 dm, 0.5–4(5) dm diam, hairy. **LF:** blades (0.5)1–3(4) cm, 0.1–0.7(1.2) cm wide. **INFL:** bracts 2, awns 0.2–1 mm, straight; involucre 1.5–3 mm, 3-angled, 6-ribbed, faintly transversely ridged, teeth 6, 1–5 mm, margins scarious, awns 0.2–0.5 mm, hooked. **FL:** 1, (1.7)2–3 mm, hairy; perianth white or yellow, lobes entire; stamens 9, fused and hairy at base. **FR:** 1.5–2.5 mm. *n*=(19)20(21–23). Common. Sand or gravel; (0)10–1300 m. c&s SCo, s TR, w PR; n Baja CA. Apr–Jun

C. pungens Benth. Pl prostrate to erect, 0.5–2.5 dm diam, hairy. **LF:** blades (0.5)1–5(7) cm, (0.3)0.4–0.7(1) cm wide. **INFL:** bracts 2, awns 0.5–1.2 mm; involucre 3-angled, 6-ribbed, tube 2–2.5(3) mm, ± swollen, transversely ridged, teeth 6, 0.5–1 mm, margins white or pink to purple, awns 1–3 mm, hooked. **FL:** 1, 2–3.5 mm, hairy; perianth 2-colored, fl tube white, lobes white to rose, jagged; stamens 9. **FR:** 2–2.5 mm.

var. ***hartwegiana*** Reveal & Hardham BEN LOMOND SPINEFLOWER Pl ascending to erect, 0.5–2.5 dm. **INFL:** involucre teeth margins dark pink to purple. *n*=20. Sand; 90–500 m. sw SnFrB (Santa Cruz Sandhills, Santa Cruz Co.). Apr–Jul ★

var. ***pungens*** (p. 1093) MONTEREY SPINEFLOWER Pl prostrate to ascending, 0.5–1.5 dm. **INFL:** involucre teeth margins white (± pink). *n*=20. Sand; 0–65 m. CCo (s Santa Cruz, n Monterey cos.; extirpated San Luis Obispo Co.), SnFrB (s Santa Cruz Co.). Apr–Jul ★

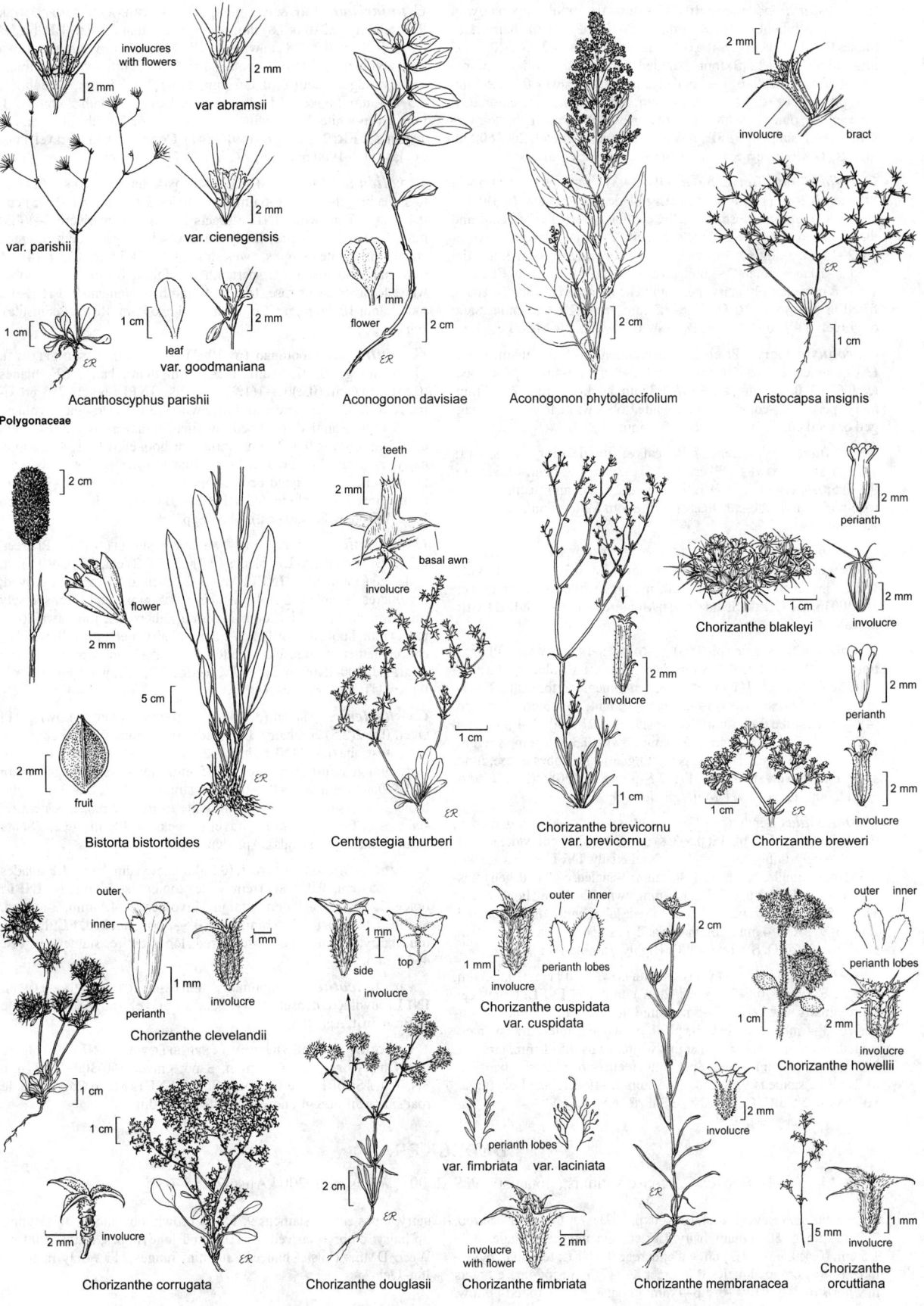

Polygonaceae

involucres with flowers

var abramsii

2 mm

var. parishii

var. cienegensis

2 mm

1 cm

1 cm

leaf

var. goodmaniana

2 mm

ER

Acanthoscyphus parishii

flower

1 mm

2 cm

ER

Aconogonon davisiae

2 cm

Aconogonon phytolaccifolium

2 mm

involucre   bract

ER

1 cm

Aristocapsa insignis

2 cm

flower

2 mm

5 cm

2 mm

fruit

ER

Bistorta bistortoides

teeth

2 mm

basal awn

involucre

1 cm

ER

Centrostegia thurberi

2 mm

involucre

1 cm

Chorizanthe brevicornu
var. brevicornu

perianth

2 mm

2 mm

involucre

Chorizanthe blakleyi

perianth

2 mm

2 mm

involucre

1 cm

ER

Chorizanthe breweri

outer

inner

perianth

1 mm

1 mm

involucre

Chorizanthe clevelandii

side   top

1 mm

involucre

var. fimbriata   var. laciniata

perianth lobes

involucre
with flower

2 mm

Chorizanthe fimbriata

outer   inner

1 mm

perianth lobes

involucre

Chorizanthe cuspidata
var. cuspidata

2 cm

1 cm

ER

Chorizanthe membranacea

2 mm

involucre

outer   inner

perianth lobes

1 cm

2 mm

involucre

Chorizanthe howellii

5 mm

1 mm

involucre

Chorizanthe
orcuttiana

outer
inner

perianth

1 cm

involucre

1 cm

1 cm

2 mm   involucre

ER

Chorizanthe corrugata

2 cm

ER

Chorizanthe douglasii

C. rectispina Goodman (p. 1093) STRAIGHT-AWNED SPINEFLOWER
Pl gen decumbent, 0.3–0.8(1) dm, 0.5–4(5) dm diam, hairy. **LF:**
blades 0.5–1.5(2) cm 0.2–0.6 cm wide. **INFL:** bracts 2, awns 0.5–1.5
mm; involucre 2–2.5(3) mm, 3-angled, 6-ribbed, ± swollen, ± transversely ridged, teeth 6, 1–2 mm, abaxial longest, awns 0.3–2.5 mm,
straight or hooked. **FL:** 1, 3.5–4 mm, hairy; perianth 2-colored, fl
tube yellow, outer lobes white, entire or ± 2-lobed, inner shorter, yellow, jagged; stamens 9. **FR:** 3–3.5 mm. Sand or gravel; 200–600 m.
SCoRO (s Monterey, San Luis Obispo cos.). May–Jul ★

C. rigida (Torr.) Torr. & A. Gray (p. 1093) DEVIL'S SPINEFLOWER
Pl erect, 0.2–0.8(1.5) dm, hairy. **LF:** blades 0.5–2.5 cm, (0.3)0.5–2
cm wide, occ tomentose abaxially, cauline blades awned, hard and
thorn-like in age. **INFL:** bracts 2, awns 2–4 mm, straight; involucre 2–3 mm, 3-angled, 3-ribbed, transversely ridged, teeth 3, abaxial
tooth 5–10 mm, others 2–5 mm, awns straight, 0.5–2.5 mm. **FL:** 1–2,
1.5–1.8 mm, densely hairy; perianth yellow, lobes entire; stamens 9,
fused to tube top. **FR:** (1.5)1.8–2.2 mm. *n*=19,20. Common. Sand
or gravel; 60–1900 m. SNE, D; to sw UT, w AZ, nw Mex. Feb–Jun

C. robusta Parry Pl erect to decumbent, 1–6 dm diam, hairy.
INFL: bracts 2, awns 0; involucre 2.5–4 mm, 3-angled, 6-ribbed,
teeth 6, 0.3–0.8(1) mm, awns 0.7–1.3 mm, hooked. **FL:** 1, 2.5–4 mm,
hairy; perianth 2-colored, fl tube white, lobes white to rose, gen jagged or toothed; stamens 9. **FR:** 3.5–4 mm.

var. ***hartwegii*** (Benth.) Reveal & Rand. Morgan SCOTTS
VALLEY SPINEFLOWER Pl erect, (0.5)1–3 dm. **LF:** blades 1.5–5 cm,
0.2–0.5(0.7) cm wide. **INFL:** involucre 2.5–3.5 mm, teeth margins
rose-pink. Sand; 200–300 m. s SnFrB (Scotts Valley, Santa Cruz Co.).
Apr–Jul ★

var. ***robusta*** (p. 1093) ROBUST SPINEFLOWER Pl spreading
or decumbent, 1–2 dm. **LF:** blades 1.5–5 cm, 0.3–0.7(1) cm wide.
INFL: involucre 2.5–4 mm, teeth margins white. Sand or gravel;
10–300 m. n-c CCo (s Santa Cruz, n Monterey cos.), sw SnFrB (extirpated). May–Sep ★

C. spinosa S. Watson (p. 1093) MOJAVE SPINEFLOWER Pl prostrate, 0.3–0.8(1) dm, 0.5–8 dm diam, hairy. **LF:** blades (0.3)0.5–
1.5(2) cm, (0.3)0.5–1(1.2) cm wide, gen tomentose abaxially. **INFL:**
bracts 3, whorled, awns 1–3.5 mm, straight; involucre 3-angled,
(4)5-ribbed, tube 2–2.5 mm, teeth (4)5, abaxial tooth 2–4 mm, awn
1–2.5 mm, straight, others 0.5–1 mm, awns 0.3–0.8 mm, straight.
FL: 1, 2.5–3.5 mm, glabrous; perianth gen white, lobes entire, inner
shorter, narrower; stamens 9. **FR:** 2.5–3 mm. *n*=(20)22(23). Sand or
gravel; 600–1300 m. w DMoj. Apr–Jul ★

C. staticoides Benth. (p. 1093) TURKISH RUGGING Pl gen erect,
0.5–6 dm, hairy. **LF:** blades 0.5–3(8) cm, 0.3–1(2.5) cm wide, tomentose abaxially, thinly hairy or glabrous adaxially. **INFL:** bracts 2, awns
0.5–2 mm, straight; involucre 3–4(5) mm, 3-angled, 6-ribbed, gen transversely ridged, teeth 6, 0.7–1.3(1.5) mm, awns 0.5–1 mm, hooked. **FL:**
1, 3–4(5) mm, hairy; perianth rose to red, lobes entire or occ toothed;
stamens 9. **FR:** 3–4 mm. Common. Sand, gravel or rocks; 300–1700 m.
c&s CCo, SCoRO, SW (exc e PR). Highly variable. Apr–Jul

C. stellulata Benth. STARLITE SPINEFLOWER Pl erect, 0.5–3 dm,
hairy. **LF:** blades 0.5–2 cm, 0.8–2(2.2) cm wide. **INFL:** bracts gen
2 proximally, a whorl of 3–5 near middle, awns 0; involucres congested, 3–4 mm, 3-angled, 6-ribbed, ± swollen, finely transversely
ridged, teeth 6, 1–1.5 mm, margins white, awns 0.5–1 mm, straight.
FL: 1, 4–4.5(5) mm, ± hairy; perianth cream or rose, lobes obcordate
to 2-lobed; stamens 9. **FR:** 3.5–4.5 mm. *n*=19–21. Sand or gravel;
30–900 m. NCoRI, CaRF, SNF, e SnFrB. Apr–Jul

C. uniaristata Torr. & A. Gray (p. 1093) ONE-AWN SPINEFLOWER
Pl spreading, 0.2–0.6(0.8) dm, 0.5–4(5) dm diam, hairy. **LF:** blades
0.5–1.5(2) cm, 0.2–0.8 cm wide. **INFL:** bracts 2, awns straight, 1.5–4
mm; involucre 2–3 mm, 3-angled, 6-ribbed, slightly swollen, ± transversely ridged, teeth 6, 0.3–6 mm, awns 2.5–5.5 mm, straight, or
0.3–0.5 mm, hooked. **FL:** 1, 2–3 mm, hairy; perianth 2-colored, fl
tube green-white, lobes white, outer with 1 or 3 teeth, inner entire;
stamens 3. **FR:** 2–3 mm. *n*=(39)40(41). Common. Sand, gravel, talus
or clay; 800–1900 m. s SN, e CW, n WTR. Apr–Jul

C. valida S. Watson (p. 1093) SONOMA SPINEFLOWER Pl erect
to spreading, 1–3 dm, 1–6 dm diam, hairy. **LF:** blades 1–2.5(5) cm,
0.4–0.8(1.2) cm wide. **INFL:** bracts 2, awns 0; involucre 3–4(4.5)
mm, 3-angled, finely transversely ridged, teeth 6, erect,
equal, with white margins, awns straight, 0.5–1.3 mm. **FL:** 1, (4)5–6
mm, hairy on lower 1/2; perianth 2-colored, fl tube white, lobes
white to lavender or rose, fringed to toothed; stamens 9. **FR:** 3–4.5
mm. Sand; 10–90 m. n CCo (extirpated exc Point Reyes Peninsula).
Jun–Aug ★

C. ventricosa Goodman (p. 1093) POTBELLIED SPINEFLOWER
Pl spreading, (0.5)1–5 dm, 1–5(7) dm diam, hairy. **LF:** blades
(0.5)1–3(4) cm, (0.2)0.4–1(1.2) cm wide. **INFL:** bracts 2–3 proximally, a whorl of 3–5 near middle, awns 1–3 mm, straight; involucre
4–4.5 mm, 3-angled, 6-ribbed, swollen, transversely ridged, teeth
6, 1–3 mm, awns 0.5–2 mm, straight or hooked. **FL:** 1, 4–4.5 mm,
hairy; perianth 2-colored, fl tube white to green-yellow, lobes red
to maroon, inner fringed or ± 2-lobed; stamens 9. **FR:** 3–3.5 mm.
Serpentine; 500–1000 m. SCoRI (s San Benito, se Monterey, ne San
Luis Obispo, w Fresno cos.). May–Sep ★

C. watsonii Torr. & A. Gray WATSON'S SPINEFLOWER Pl erect,
0.2–1(1.5) dm, hairy. **LF:** blades (0.3)0.5–1.5(2) cm, 0.2–0.4(0.5) cm
wide, gen tomentose. **INFL:** bracts 2, awns 0.5–1 mm, ± curved;
involucres 3-angled, 5-ribbed, tube 3–4.5 mm, finely transversely
ridged, teeth 5, abaxial tooth 2–6 mm, others 1–2 mm, awns 0.4–
0.8(1) mm, hooked. **FL:** 1, 1.5–2.5 mm, hairy; perianth yellow, lobes
entire; stamens 9, fused to tube top. **FR:** 2.5–3 mm. Common. Sand
or gravel; 300–2400 m. sw SnJV, n edge TR, GB, w DMoj; e WA, s
ID, sw UT, nw AZ. Apr–Aug

C. wheeleri S. Watson (p. 1093) WHEELER'S SPINEFLOWER Pl
erect, 0.5–2(2.5) cm, hairy. **LF:** blades 0.5–2 mm, 0.2–0.6 cm wide,
tomentose abaxially. **INFL:** bracts 2, awns 0.5–1 mm, straight; involucre 3-angled, 6-ribbed, tube 2–2.5 mm, transversely ridged, teeth
6, 0.3–0.8(1) mm, awns 0.3–0.5 mm, straight. **FL:** 1, 2.5–3 mm, glabrous or sparsely hairy; perianth white to rose or red, lobes entire;
stamens 6. **FR:** 2.5–3 mm. Gravel to rocks; < 400 m. n ChI (Santa
Cruz, Santa Rosa islands). Apr–Jun ★

C. xanti S. Watson Pl erect, (0.3)0.5–2.5(3) dm, hairy. **LF:** blades
0.3–1(1.5) cm, 0.3–0.8(1) cm wide, tomentose abaxially. **INFL:**
bracts 2, awns 0.5–1 mm, straight; involucre 3–4.5 mm, 3-angled,
6-ribbed, teeth 6, 0.7–1.5 mm, awns 0.5–1 mm, hooked. **FL:** 1, 4.5–6
mm, hairy; perianth gen rose to red, lobes entire; stamens 9. **FR:**
4–4.5 mm.

var. ***leucotheca*** Goodman WHITE-BRACTED SPINEFLOWER
INFL: involucre densely hairy. Sand or gravel; (60)400–1300 m. e
SnBr, n SnJt. Apr–Jun ★

var. ***xanti*** (p. 1093) PINYON SPINEFLOWER **INFL:** involucre
thinly hairy. *n*=19,21. Common. Sand, gravel; (60)300–1600 m. s
SN, Teh, s SCoRI, n&e TR (exc e SnBr), SNE (s Mono Co.). Possible
roadside waif in SNE, now extirpated. Apr–Jul

DEDECKERA

1 sp. (Mary C. DeDecker, CA conservationist, botanist, 1909–2000) [Wiens et al. 2004 Aliso 21:55–63]

D. eurekensis Reveal & J.T. Howell (p. 1093) JULY GOLD Shrub,
2–7(10) dm, 5–20 m diam, hairy. **LF:** cauline, alternate; blades 0.7–
1.5 cm, (0.4)0.5–0.8(1.3) cm wide; ocreae 0. **INFL:** terminal, cyme-
like, 1–4(6) cm, peduncles 0 or erect, 2–6(7) mm; involucre bracts
in whorls of 2–5. **FL:** 4–6, 1.8–4 mm; perianth yellow to red-yellow,
hairy, lobes entire; stamens 9. **FR:** ± brown, obconic, 2–3(3.5) mm,
tip hairy; embryo curved. *n*=20. Limestone slopes; 1200–2200 m.
W&I, DMtns (Last Chance, Panamint ranges). Fr rarely matures.
Jun–Oct ★

DODECAHEMA

1 sp. (Greek: 12 darts or javelins, for involucre awns)

D. leptoceras (A. Gray) Reveal & Hardham (p. 1093) SLENDER-
HORNED SPINEFLOWER Ann, gen erect, 0.3–0.8(1) dm, sparsely
glandular. **LF**: basal; blades (1)1.5–4(6) cm, 0.2–0.4(0.7) cm wide,
glabrous; ocreae 0. **INFL**: terminal, cyme-like; peduncles erect,
0.5–1 mm; involucre 1, cylindric, 6-angled, basal awns 6, teeth 6,
each with awn. **FL**: 3, 1.2–1.8(2) mm, hairy; perianth white to pink,
lobes entire; stamens 9. **FR**: dark brown to black, obconic, 1.7–2 mm,
glabrous; embryo curved. *n*=±17. Sand or gravel; 200–700 m. c&e
SCo, adjacent foothills of TR, PR. May–Jun ★

EMEX

Ann, glabrous; monoecious. **ST**: decumbent to erect, 1–8 dm, base often ± red. **LF**: cauline, alternate, petioled; ocrea soon torn,
falling; blade ± hastate to cordate. **STAMINATE INFL**: axillary, terminal, raceme-like; fls 1–8. **PISTILLATE INFL**: axillary,
spike-like; fls 1–8. **STAMINATE FL**: perianth parts 5–6, free; stamens 4–6. **PISTILLATE FL**: perianth parts 6, fused; outer
3 spine-tipped in fr, inner 3 tubercled; styles 3, free. **FR**: incl, 2–4.5 mm, ovate, shiny, spines 3; perianth hardened. 2*n*=20. 2
spp.: native to Medit, S.Afr, spread worldwide. (Latin: removal from *Rumex*) Hybrids occur readily when spp. together; highly
sterile when self-pollinated but can backcross with either parent to form viable seeds.

1. Perianth in fr (incl spines) 9–13 mm; inner perianth parts of pistillate fls widely triangular-ovate,
   tips mucronate .... ***E. australis***
1′ Perianth in fr (incl spines) 3–7 mm; inner perianth parts of pistillate fls lance-linear, tips acute .... ***E. spinosa***

E. australis Steinh. (p. 1093) SOUTHERN THREE-CORNERED JACK
ST: 1–5(6) dm. **LF**: petiole (0.5)1–6(15) cm; blade 1–5(9) cm, 0.5–
3(6) cm wide. **FR**: perianth 3-angled, each face with 3–4 longitudinally elongated depressions, spines 3.5–6 mm. Coastal, sandy areas;
< 200 m. CCo; introduced in HI, West Indies, Afr, Asia, Australia;
native to s Afr. Possibly an historical waif. May–Dec ◆

E. spinosa (L.) Campd. (p. 1093) DEVIL'S THORN **ST**: 3–6(8)
dm. **LF**: petiole 1.5–4 cm, blade 3–13 cm, 1–12 cm wide. **FR**: perianth 3-angled, each face with 8–12 pit-like depressions, spines 1.5–3
mm. Dry, sandy, disturbed places; < 500 m. CCo, SCo; MA, NJ, FL,
TX; also HI, S.Am, Eurasia, Afr, Australia; native to Medit. Cleistogamous fls produced underground. May–Dec ◆

ERIOGONUM WILD BUCKWHEAT

Ann to shrub, matted or not. **ST**: prostrate to erect, occ 0, glabrous or hairy, occ glandular. **LF**: basal, sheathing (on st above
basal, beneath other, ± non-sheathing cauline), or cauline, alternate, opposite, or whorled; blades linear to ± round, gen longer
than wide, margins flat, wavy, or rolled under; ocreae 0. **INFL**: variable, glabrous or hairy, occ glandular; peduncles 0 or erect
to reflexed (pointed down); involucres 1 or in clusters, tubular, glabrous or hairy, teeth gen 4–10, awns 0. **FL**: (2)6–100(200),
with a stalk-like base ("fl stipe" or "stipe") or not; perianth gen white to red or yellow, glabrous or hairy, occ glandular, lobes
6, gen entire; stamens 9. **FR**: gen brown or black, gen obconic, glabrous or hairy; embryo curved or straight. ± 250 spp.: temp
N.Am. (Greek: woolly knees, for hairy nodes of 1st sp. named) One of largest genera in CA; st in descriptions refers to the
main st(s), not branches of infl. A per allied to *E. austrinum* (S. Stokes) Reveal and *E. moranii* Reveal of e-c Baja CA, with
spreading (rather than reflexed or erect) involucres on peduncles < 2 mm, occurring in DMtns (Bristol, Granite mtns), has been
known for nearly 25 yrs yet remains undescribed. *E. puberulum* moved to *Johanneshowellia*.

Key to Groups

1. Ann (exc *Eriogonum inflatum*, per, keyed in Group 2 when fl 1st yr)
   2. Involucres angled to strongly ribbed, sessile and strongly appressed or at tips of short bracted branchlets
      (or peduncled) (subg. *Oregonium*) .... **Group 1**
   2′ Involucres smooth, not ribbed or angled, gen peduncled, if sessile then not appressed (subg. *Ganysma*) ..... **Group 2**
1′ Per
   3. Fl stipe present, sometimes weakly so, not winged; infl bracts 2–13+ (subg. *Oligogonum*) .... **Group 3**
   3′ Fl stipe 0 or ± winged; infl bracts gen 3 (mainly subg. *Eucycla*) .... **Group 4**

Group 1

Ann; involucres angled to strongly ribbed, sessile and strongly appressed or at tips
of short bracted branchlets (or peduncled) (subg. *Oregonium*)

1. Involucres at tips of slender branchlets, not appressed; peduncles present or 0
   2. Fls 0.7–1.5 mm, yellow or cream; n SNE, w DMoj
      3. Fls cream to ± red, 1–1.5 mm; infl 5–10(15) cm wide; n SNE .... ***E. ampullaceum***
      3′ Fls yellow, 0.7–1 mm; infl 5–20 cm wide; w DMoj .... ***E. mohavense***
   2′ Fls 1–2.5(3) mm, yellow or white to pink, rose, or red
      4. St glabrous; involucres sessile
         5. Basal lf blades oblong-ovate; involucres 3–4 mm, 5-toothed; c NCoRI, n CCo, n SnFrB
            .... ***E. luteolum*** var. ***caninum***

5′ Basal lf blades gen round to reniform; involucres 2–4 mm, 5–8-toothed; c&s CW, WTR

  6. Involucres 2–2.5 mm, 5-toothed; fls 2–2.5(3) mm, minutely puberulent, white to rose or yellow in fr;

    c&s CW, WTR . . . . ***E. covilleanum***

  6′ Involucres 3–4 mm, 6–8-toothed; fls 1–2 mm, glabrous, white to rose; n SCoR . . . . ***E. nortonii***

4′ St tomentose, if glabrous then involucres penduncled at lower nodes

  7. Involucres glabrous; s SnFrB, n SCoRI . . . . ***E. argillosum***

  7′ Involucres tomentose; ScV, ne SnFrB, ne&s SCoRI

    8. Lvs basal, blades ± round; styles 0.1–0.3 mm; s SCoRI . . . . ***E. eastwoodianum***

    8′ Lvs basal or basal and cauline; ± round or narrowly oblong to ovate; styles 0.2–1 mm; ne SnFrB, SCoRI

      9. Involucres 2.5–3.5(4) mm; styles 0.2–0.3 mm; ScV (extirpated), ne SnFrB . . . . ***E. truncatum***

      9′ Involucres (1.5)1.8–2.5 mm; styles 0.2–1 mm; SCoRI

        10. Fls smooth; involucres 2–2.5 mm; lvs basal and subbasal, rarely cauline; s SCoRI . . . . ***E. temblorense***

        10′ Fls papillate; involucres (1.5)1.8–2 mm; lvs basal and cauline; ne SCoRI . . . . ***E. vestitum***

1′ Involucres not at tips of slender branchlets, gen appressed; peduncles 0

11. Fls hairy, white to rose — c&s NCoR . . . . ***E. dasyanthemum***

11′ Fls glabrous or sparsely glandular, white, cream, yellow, rose, or red

  12. Involucres 1–2 mm, if 2 mm then outer perianth lobes fan-shaped or ± hastate (exc oblong to oblong-

    ovate in *Eriogonum baileyi*)

    13. Sts glabrous, if tomentose then fls glandular and pl of e SN, GB; outer perianth lobes oblong to oblong-obovate

      14. Fls yellow or pale ± yellow (or ± white), 0.6–0.8(1) mm; SNE, DMoj . . . . ***E. brachyanthum***

      14′ Fls white to rose, 1–2 mm; SNF, Teh, SnFrB, SCoR, TR, GB, w DMoj

        15. Fls 1–1.5 mm, glabrous or puberulent; st, infl glabrous — SnFrB, SCoR, WTR . . . . ***E. elegans***

        15′ Fls 1.5–3 mm, gen minutely glandular, occ glabrous in SNF; st, infl glabrous or tomentose . . . . ***E. baileyi***

        16. Infl glabrous; SNF, Teh, s SCoR, TR, GB, w DMoj. . . . . var. ***baileyi***

        16′ Infl tomentose; GB . . . . var. ***praebens***

    13′ Sts tomentose to sparsely so; outer perianth lobes narrowly to widely fan-shaped or ± hastate

      17. Outer perianth lobes ± hastate in fr; fls 0.8–1.2 mm; involucres bell-shaped, 1–2 mm; SnBr, PR . . . . ***E. evanidum***

      17′ Outer perianth lobes fan-shaped; fls 1–3.5 mm; involucres bell-shaped to obconic, 1–2.5 mm; GB, D

        18. Lvs cauline; lf blades narrowly oblanceolate to widely elliptic; pl narrowly erect; fls white to pink or

        red; se DMoj . . . . [***E. polycladon***]

        18′ Lvs basal; lf blades ± round to cordate; pl gen spreading; fls white to pink or pale yellow to yellow; GB, D

          19. Fls pale yellow or yellow to red; outer perianth lobes widely fan-shaped; infl gen densely

          branched, branches incurved; fr 1–1.3 mm; GB, D . . . . ***E. nidularium***

          19′ Fls white to pink or pale yellow, outer perianth lobes narrowly fan-shaped; infl gen sparsely

          branched, branches not incurved; fr 1.5–1.8 mm; SNE, D . . . . ***E. palmerianum***

  12′ Involucres (1.8)2–5(7) mm; outer perianth lobes not fan-shaped or hastate (see also *Eriogonum baileyi* at 12.)

    20. Lf blades oblong-obovate to oblanceolate or oblong (basal occ ± rounded in *E. cithariforme*)

      21. Involucres (3.5)4–5 mm; outer perianth lobes narrowly obovate to oblong; fr 1.8–2(2.2) mm;

      NW, SNF, Teh, CW, TR . . . . ***E. roseum***

      21′ Involucres (1.8)2–3 mm; outer perianth lobes lanceolate to oblong or oblong-obovate; fr 1–2 mm; CA-FP

        22. Infl branches upcurved; basal lf blades oblanceolate or elliptic to ovate or ± rounded, (0.3)1–2 cm;

        petiole gen winged; involucres 2.5–3 mm; fls 1.5–2 mm, white to rose; SCoR, TR . . . . ***E. cithariforme***

        23. St, infl glabrous; SCoR, WTR . . . . var. ***agninum***

        23′ St, infl tomentose; SCoR, TR. . . . . var. ***cithariforme***

        22′ Infl branches straight; basal lf blades oblanceolate to oblong, (0.8)1–4(6) cm; petiole not winged;

        involucres (1.8)2–3 mm; fls 1.5–3 mm, white to pink or yellow; CA-FP . . . . ***E. gracile***

        24. Infl densely to sparsely tomentose; widespread. . . . . var. ***gracile***

        24′ Infl glabrous; PR. . . . . var. ***incultum***

    20′ Lf blades round-ovate to rounded or reniform, if not then st and infl branches glabrous and pl of SNH;

    st, infl branches glabrous to tomentose

      25. Infl sparsely tomentose, esp basally, or glabrous; lvs basal; fls white to rose or pale yellow —

      ne KR, CaR, n SNH, MP . . . . ***E. vimineum***

      25′ Infl tomentose or gen glabrous, if sparsely tomentose then lvs basal and cauline; fls white to pink,

      red or yellow

        26. Involucres (3.5)4–5(7) mm; pl 4–10 dm; n WTR, SnBr, PR. . . . . ***E. molestum***

        26′ Involucres 2–4 mm; pl 1–5 dm; widespread

          27. Lvs basal; involucres 3–4 mm; fls 1.5–2 mm; st glabrous; SW, SNE, DMoj . . . . ***E. davidsonii***

          27′ Lvs basal and cauline, involucres 3–3.5 mm and fls 1.5–2 mm with sparsely tomentose st and infl,

          if lvs only basal and st glabrous then involucres 2–3.5 mm and pl of n&c SN; n CA-FP. . . ***E. luteolum*** (in part)

            28. Involucres 2–3 mm; fls 2–2.5 mm; pl of granitic sand; n&c SNH . . . . var. ***saltuarium***

            28′ Involucres 3–3.5 mm; fls 1–2 mm; pl of serpentine; NW, CaR, SNF, SnFrB

              29. Fls 1.8–2 mm; NW, CaR, SnFrB. . . . . var. ***luteolum***

              29′ Fls 1–1.5(1.8) mm; SNF . . . . var. ***pedunculatum***

Group 2

Ann (exc 1st-yr fl *Eriogonum inflatum*); involucres smooth, not ribbed or angled, gen peduncled,
if sessile then not appressed (subg. *Ganysma*)

1. Per, fl 1st yr or not — GB, D . . . . ²**E. inflatum**
1′ Ann (per *Eriogonum inflatum* may key here if fl 1st yr)
 2. Lvs basal and cauline, not basal only or basal and sheathing
  3. Involucres glandular-hairy or puberulent
   4. Perianth lobes alike, oblong to elliptic, not inflated
    5. Involucre glandular-hairy, densely tomentose adaxially, 2.7–3 mm; fl 1.5–1.7 mm; s SNF, sw SnJV
    . . . . **E. gossypinum**
    5′ Involucre not densely tomentose, 1.8–2 mm; fl 2–2.5 mm; s SN, SCoR, TR . . . . **E. gracillimum**
   4′ Perianth lobes not alike, outer elliptic to obovate or ± round, inflated or not, inner narrower
    6. Outer perianth lobes gen elliptic to obovate, not obviously inflated, if so then only basally; stamens
     exserted; c&s SNF, Teh, SnJV, CW, TR, s PR, w DMoj . . . . **E. angulosum**
    6′ Outer perianth lobes elliptic to ± round or obovate, obviously inflated basally or at tip; stamens incl;
     SnJV, SCoR, s SN, TR, GB, w DMoj
     7. Outer perianth lobes inflated base to middle; involucres glandular-hairy; s SN, TR, GB . . . . **E. maculatum**
     7′ Outer perianth lobes inflated above middle; involucres glandular; SnJV, SCoR, TR, w DMoj . . . . . **E. viridescens**
  3′ Involucres glabrous, long-stiff or long-soft hairy, not glandular-hairy or puberulent
   8. Fls hairy, hairs hooked
    9. Fls 2, 0.8–1.1 mm; fr tip exserted; NW, SN, n CW, TR . . . . **E. hirtiflorum**
    9′ Fls 4–6, 1.2–1.8 mm; fr tip not exserted; s NCoR, c&s SN, Teh, CW, WTR . . . . **E. inerme**
     10. Involucres short-stiff hairy; c&s SN, Teh . . . . var. **hispidulum**
     10′ Involucres glabrous; s NCoR, CW, WTR . . . . var. **inerme**
   8′ Fls glabrous, if hairy then hairs not hooked
    11. Lvs sparsely tomentose or glabrous, oblong-oblanceolate to obovate; fls densely short-hairy; pl
     glabrous or sparsely tomentose; s SNF, Teh, SCoRI, n WTR . . . . ²**E. ordii**
    11′ Lvs short-hairy, linear; fls glabrous or sparsely hairy; pl glabrous or glandular and short-hairy;
     NCoRH, CaRH, SN, n SCoRO, TR . . . . **E. spergulinum**
     12. Infl internodes not glandular; pl prostrate to ascending — s SNH . . . . var. **pratense**
     12′ Infl internodes glandular; pl erect
      13. Fl 1.5–2.5 mm; NCoRH, CaRH, SN, n SCoRO, TR . . . . var. **reddingianum**
      13′ Fl 2.5–3.5 mm; SN . . . . var. **spergulinum**
 2′ Lvs basal and sometimes sheathing, not both basal and cauline
  14. Lf blades glabrous, or hairy on one or both surfaces, or if sparsely tomentose then fls densely short-soft-hairy
   15. Fls gen glabrous — c SNH, SNE . . . . **E. esmeraldense** var. **esmeraldense**
   15′ Fls hairy
    16. Fls densely short-soft hairy or puberulent, white to pink or red (or ± yellow)
     17. Lf blades oblong-oblanceolate to obovate, sparsely tomentose or glabrous; st glabrous to sparsely
      tomentose; s SNF, Teh, SCoRI, n WTR . . . . ²**E. ordii**
     17′ Lf blades oblanceolate to obovate, sparsely soft-shaggy hairy and glandular or coarse-hairy;
      st glabrous and sparsely glandular; s SN, TR, PR, n W&I, sw DMtns
      18. Fls pink to red, aging to white, 0.5–0.9 mm; lf blades coarse-hairy; involucres 0.5–0.9 mm;
       s SN, TR, PR, n W&I . . . . **E. parishii**
      18′ Fls white to ± red (or yellow), 1.5–2.5 mm; lf blades soft-shaggy hairy, some glandular;
       involucres 1.2–1.5 mm; e SnBr, c&s PR, sw DMtns . . . . **E. apiculatum**
    16′ Fls soft-shaggy- or coarse-hairy, white to ± pink or gen yellow
     19. Pl glandular throughout; fls white to ± pink — n&ne DMtns . . . . **E. glandulosum**
     19′ Pl glabrous, exc sometimes glandular or hairy near base or at nodes; fls yellow to green-yellow or ± red
      20. St glandular near base
       21. Infl glandular at lower nodes; involucre teeth (4)5; pl (0.3)0.5–3 dm; lf blades gen ± round,
        (0.3)0.5–1(1.4) cm; fr 1.5–1.8(2) mm; ne DMoj . . . . **E. contiguum**
       21′ Infl glabrous; involucre teeth 4; pl (2)4–18(22) dm; lf blades round to reniform, (0.5)1–2.5(4) cm,
        (0.5)1–2(3) cm wide; fr 2–2.5 mm; CW, SW, s SNE, D . . . . ²**E. clavatum**
      20′ St glabrous or basally coarse- or long-soft-hairy
       22. St, infl ± gray; per but appearing ann when fl 1st yr — GB, D . . . . ²**E. inflatum**
       22′ St, infl branches ± green or yellow-green; ann
        23. Involucres 1–1.5(1.8) mm; pl (2)4–18(22) dm; infl 30–150(170) cm; fr 2–2.5 mm; CW, SW,
         s SNE, D . . . . ²**E. clavatum**
        23′ Involucres 0.7–1 mm; pl 1–4.5(6) dm; infl 5–30 cm; fr 1–1.5 mm; D . . . . **E. trichopes**
  14′ Lf blades tomentose to sparsely so on one or both surfaces; fls glabrous, glandular, or variously hairy,
   but not short-soft-hairy

24. Outer perianth lobes gen oblong, obovate or round, cordate basally
  25. Involucres sessile and erect or on erect peduncles ≤ 0.5 cm — DMoj
    26. St 0.5–3 cm; pl 1–4 dm; infl branches spreading. . . . . ***E. bifurcatum***
    26′ St (5)10–20 cm; pl (0.5)3–6(10) dm; infl branches ± erect . . . . ***E. exaltatum***
  25′ Involucres sessile and reflexed or on reflexed peduncles ≤ 1.5 cm
    27. St, infl glandular; SNE, DMoj . . . . ***E. brachypodum***
    27′ St, infl glabrous; s SNH, Teh, s SnJV, TR, GB, D
      28. Involucres widely bell-shaped; peduncles 0; fls yellow to red-yellow; SNE. . . . . ***E. hookeri***
      28′ Involucres narrowly obconic or narrowly bell-shaped; peduncles ≤ 1.5 cm; fls white to pink;
        s SNH, Teh, s SnJV, TR, D
        29. Involucres 1–1.5 mm; infl of many horizontal tiers; ne DMoj. ***E. rixfordii***
        29′ Involucres 1.5–2.5 mm; infl not of many horizontal tiers; s SNH, Teh, s SnJV, TR, D . . . ***E. deflexum*** (in part)
          30. Involucres (2)2.5–3 mm; peduncles (0.3)0.5–1.5 cm; st inflated; s SNH, Teh, s SnJV, TR,
            n&w DMoj. . . . . var. ***baratum***
          30′ Involucres 1.5–2.5 mm; peduncles 0 or gen ≤ 0.5 cm; st not inflated; D. . . . . var. ***deflexum***
24′ Outer perianth lobes oblong to oblanceolate or ovate, truncate to obtuse basally
  31. Fls papillate basally — peduncles curving, ascending, 1–5 cm; MP . . . . ***E. collinum***
  31′ Fls smooth, not papillate
    32. Perianth lobes alike
      33. Sts, infl glandular; peduncles 0 or reflexed; fls 2–2.5 mm. . . . . ***E. eremicola***
      33′ Sts, infl glabrous; peduncles erect; fls 1.5–1.8 mm . . . . ***E. hoffmannii***
        34. Lf margins flat, blades 2–4 cm wide; petioles 1–5 cm; pl 0.5–5 dm . . . . var. ***hoffmannii***
        34′ Lf margins wavy, blades 3–8 cm wide; petioles (3)5–10 cm; pl 4–10 dm. . . . . var. ***robustius***
    32′ Perianth lobes not alike, if alike then glandular-hairy
      35. Fls glandular
        36. Fls white to red, glandular-hairy with a tuft of long white hairs adaxially; e PR, s DMoj, DSon;
          outer perianth lobes gen fan-shaped. . . . . ***E. thurberi***
        36′ Fls yellow, if in age ± white then outer perianth lobes sac-like basally, glandular or short-stiff
          hairy, without a tuft of long white hairs adaxially; SCoRI, WTR, GB, D
          37. Outer perianth lobes cordate, in age sac-like basally, initially yellow then white to rose;
            involucres 0.6–1.2 mm, glabrous; D . . . . ***E. thomasii***
          37′ Outer perianth lobes oblong-elliptic to obovate or ovate, not sac-like basally, yellow; involucres
            1–2 mm, glabrous or glandular-hairy; SCoRI, WTR, GB, D
            38. Involucres glandular-hairy; SCoRI, WTR, GB, D . . . . ***E. pusillum***
            38′ Involucres glabrous; D . . . . ***E. reniforme***
    35′ Fls glabrous
      39. Peduncles 0 or reflexed; outer perianth lobes oblong; SNE. . . . . ***E. deflexum*** var. ***nevadense***
      39′ Peduncles gen curving downward; outer perianth lobes ± fiddle-shaped or oblong to oval; n&c SNH, GB
        40. Outer perianth lobes ± fiddle-shaped; peduncles glabrous (or 0); c SNH, GB, probably
          introduced in n SNH (Nevada Co.). . . . . ***E. cernuum***
        40′ Outer perianth lobes oblong to oval; peduncles glandular (glabrous); n SNH, n SNE. . . . . ***E. nutans***
          41. Peduncles glandular; n SNE . . . . var. ***nutans***
          41′ Peduncles glabrous; n SNH (introduced but now extirpated) . . . . [var. ***glabratum***]

Group 3

Per; fl stipe present, sometimes weakly so, not winged; infl bracts 2–13+ (subg. *Oligogonum*)

1. Involucre teeth not lobe-like, << tube, erect or ± so
 2. Fls hairy
  3. Lf blades tomentose to densely so abaxially, gen glabrous adaxially; nw KR. . . . . . ***E. pyrolifolium*** var. ***coryphaeum***
  3′ Lf blades glabrous or short hairy on both surfaces; n KR, CaRH, c&s SNH, W&I, DMtns
    4. Lf blades glabrous; fls white to rose; st gen ascending; CaRH. . . . . ***E. pyrolifolium*** var. ***pyrolifolium***
    4′ Lf blades short hairy; fls cream to yellow; st erect; n KR, W&I, DMtns
      5. Fls bright yellow, 3–3.5(4) mm; n KR . . . . ***E. hirtellum***
      5′ Fls cream to pale yellow, 3–6 mm; c&s SNH, W&I, n DMtns. . . . . ***E. latens***
 2′ Fls glabrous
  6. Infl not immediately subtended by a whorl of bracts; st with whorl of bracts near middle
    7. Fls ± white to pink or rose-red; lf blades silky-tomentose abaxially, 0.1–0.3(0.4) cm wide; c NCoRO . . . ***E. kelloggii***
    7′ Fls yellow; lf blades tomentose abaxially, 0.2–2(3) cm wide; not c NCoRO
      8. Sts (0.3)0.4–0.6 dm; lf blades 1–2(3) cm wide — e KR . . . . ***E. alpinum***
      8′ Sts 0.5–3 dm; lf blades 0.2–0.7(0.8) cm wide
        9. Involucre teeth 5–8, 0.5–1.5 mm; s KR, n NCoRH, n NCoRI . . . . ***E. libertini***
        9′ Involucre teeth 8–10, 1–3 mm; SN (see also 29′) . . . . [2]***E. prattenianum***
  6′ Infl immediately subtended by whorl of bracts; st without a whorl of bracts near middle

10. Fls chalky-white, white to rose or ± pale yellow; fl stipe 0.1–0.4 mm
    11. Fls white to rose, 5–7 mm; involucres 5–10(12) mm; NW, CaR, SN, GB . . . . ²*E. lobbii*
    11′ Fls chalky-white, 2.5–3.5 mm; involucres (2)2.5–3.5(4) mm; s SNH . . . . *E. polypodum*
10′ Fls yellow or ± brown to sulphur yellow or cream, occ with blush of pink-red to maroon (or white);
    fl stipe (0.1)0.3–1.5 mm
    12. Pl unisexual, pistillate and staminate pls morphologically different
        13. Lf blades ± glabrous, bright green to olive-green adaxially; e KR, CaRH, n&c SNH . . . . *E. marifolium*
            14. Involucres 4–5 mm, 5–7 mm wide; lf-blades narrow; CaRH . . . . var. *cupulatum*
            14′ Involucres 2–3 mm, 1.5–2 mm wide; lf-blades wide to round; e KR, CaRH, n&c SNH . . . . . . var. *marifolium*
        13′ Lf blades tomentose to densely so, gen not green adaxially; KR, SNH
            15. Pistillate infl gen persistently head-like (or umbel-like after fertilization); lf blades (0.5)1–2 cm,
                0.5–1.5 cm wide, densely tomentose; petioles 0.7–3 cm; KR . . . . *E. diclinum*
            15′ Pistillate infl initially head-like, umbel-like after fertilization; lf blades 0.5–1.5 cm, 0.3–0.7 cm
                wide, gen tomentose; petioles (0.3)0.5–1 cm; SNH . . . . *E. incanum*
    12′ Pl bisexual (but gen with some aborted male or female parts), otherwise not morphologically different
        16. Fls sulphur yellow; involucres long-soft-wavy-hairy; infl often compound umbel-like
            17. Lf blades narrow-elliptic to -oblong, 0.3–0.6(0.8) cm wide; subshrub; KR . . . . *E. congdonii*
            17′ Lf blades oblong to obovate, (0.4)0.8–1.3 cm wide; per mat; n KR, s NCoRO, n NCoRH . . . . *E. ternatum*
        16′ Fls white, cream, pale yellow (yellow); involucres tomentose; infl umbel-like
            18. Sts 0.2–0.6(1) dm; infl ± head-like; fls yellow to red or maroon, or white to pink-rose or deep red;
                s NCoRH, c&s NCoRI
                19. Involucres 4–6.5 mm; stipe 0.1–0.3 mm; fls yellow to red or maroon; s NCoRO . . . . *E. cedrorum*
                19′ Involucres 3–4 mm; stipe 0.5–0.8 mm; fls white to pink-rose or deep red; s NCoRH,
                    c&s NCoRI . . . . *E. nervulosum*
            18′ Sts 0.4–4 dm; infl compound-umbel- or umbel-like; fls cream or ± yellow, blushed or not with
                pink-red to maroon; KR, s CaR, n SN . . . . *E. ursinum*
                20. Fls 5–9 mm, cream (or yellow); fr (5)5.5–8 mm; KR . . . . var. *erubescens*
                20′ Fls 4–6 mm, pale yellow (or yellow); fr 3–3.5 mm; KR, s CaR, n SN . . . . var. *ursinum*
1′ Involucre teeth lobe-like, at least 1/2 tube, gen reflexed or spreading
  21. Fls hairy
    22. St bractless — GB . . . . *E. caespitosum*
    22′ St with whorl of subtending bracts at base of umbel, middle of st, or middle of branches
      23. Infl scapose, involucres 1
        24. Lf blades densely tomentose on both surfaces; e KR, e CaRH, n SNH, MP . . . . *E. douglasii* var. *meridionale*
        24′ Lf blades sparsely tomentose to ± glabrous adaxially; CaRH, n&s SNH, MP
          25. Lf blades gen narrowly oblanceolate; CaRH, n SNH, MP . . . . ²*E. sphaerocephalum* var. *halimioides*
          25′ Lf blades oblanceolate to elliptic; s SNH . . . . *E. twisselmannii*
      23′ Infl not scapose, involucres ≥ 1
        26. Fls 4–5 mm, yellow; lf blades densely tomentose; pl 2.5–5 dm; NCoRI, n&c SNF . . . . *E. tripodum*
        26′ Fls (5)6–9 mm, cream or pale to bright yellow; lf blades densely tomentose to glabrous; pl 0.5–4 dm;
            CaRH, n SNH, MP . . . . *E. sphaerocephalum*
            27. Fls pale yellow to cream; CaRH, n SNH, MP . . . . ²var. *halimioides*
            27′ Fls bright yellow; n SNH, MP . . . . var. *sphaerocephalum*
  21′ Fls glabrous
    28. Infl head-like, involucres not immediately subtended by lfy bracts; st with a whorl of lfy bracts near
        middle; fls yellow
        29. Lf blades (0.3)0.5–0.8 cm, sparsely tomentose to ± glabrous, bright green or olive-green adaxially;
            gen matted per; e KR . . . . *E. siskiyouense*
        29′ Lf blades 0.5–1.5(2) cm, densely tomentose or ± glabrous adaxially, gen pale green adaxially; pl erect
            or rounded per or subshrub, occ mat; n&c SN, s SNH . . . . ²*E. prattenianum*
            30. Lf blades tomentose abaxially, subglabrous adaxially; c&s SNH . . . . var. *avium*
            30′ Lf blades tomentose on both surfaces; n&c SN . . . . var. *prattenianum*
    28′ Infl umbel-like or compound-umbel-like, involucres immediately subtended by lfy bracts; st with or
        without an additional whorl of bracts near middle; fls white, yellow, or rose
        31. St with whorl of lf-like bracts near middle — lf blades gen linear-oblanceolate to oblanceolate; Wrn
            . . . . *E. heracleoides* var. *heracleoides*
        31′ Sts without a whorl of lf-like bracts near middle
            32. Fl stipe 0.1–0.4 mm — st prostrate to decumbent or weakly erect; infl ± head- to umbel-like —
                NW, CaR, SN, GB . . . . ²*E. lobbii*
            32′ Fl stipe 0.7–2 mm
                33. Lf blades (2)7–25 cm; st gen ± inflated; NW, CaR . . . . *E. compositum* var. *compositum*
                33′ Lf blades 0.3–3(4) cm; st not inflated; most of CA . . . . *E. umbellatum*

34. Main infl branches of 1 order, none with a whorl of bracts near middle
  35. Fls gen ± white or cream to red, occ ± yellow
    36. Lf blades densely tomentose on both surfaces — fls gen lemon yellow to yellow-red; SnGb, SnBr. . . . . var. *minus*
    36′ Lf blades tomentose to glabrous adaxially
      37. Fls pale yellow to cream or ± white (or green-white); lf blades ± green adaxially; Wrn, W&I
      . . . . var. *dichrocephalum*
      37′ Fls yellow, in age red-brown to rose or pink, with large ± red spot on each lobe; lf blades gen ± red adaxially; W&I (Inyo Mtns), DMtns . . . . ²var. *versicolor*
  35′ Fls bright yellow
    38. Fl clusters with branches gen > 2.5 cm
      39. Lf blades sparsely tomentose abaxially; subshrub — SN, GB, nw DMoj . . . . ²var. *nevadense*
      39′ Lf blades densely white-tomentose or tomentose abaxially; subshrub or spreading to prostrate mat
        40. Shrub 3–5 dm, on non-serpentine; KR, NCoRH, NCoRI, CaR, n SN, MP. . . . . var. *dumosum*
        40′ Per mat (0.7)1–4.5(5) dm, on serpentine; KR, n NCoRH, n CaRH
          41. Sts 1–2.5(4) dm; lf blades 0.5–2(3.5) cm; fls 6–8(9) mm . . . . var. *goodmanii*
          41′ Sts 0.5–1.5 dm; lf blades 0.5–1(1.5) cm; fls 3–6 mm . . . . var. *humistratum*
    38′ Fl clusters with branches ≤ 2.5 cm
      42. Pl prostrate, subalpine and alpine — c&s SNH, n W&I. . . . . var. *covillei*
      42′ Pl erect to ± spreading, gen not subalpine or alpine
        43. Lf blades sparsely tomentose abaxially; SN, GB, nw DMoj. . . . . ²var. *nevadense*
        43′ Lf blades tomentose to densely so abaxially; KR, CaR, SN, MP, DMtns
          44. Infl branches (2.5)3–10(15) cm; KR, CaR, n SN, MP . . . . var. *modocense*
          44′ Infl branches 1–2.5(4) cm; KR, s SNH, DMtns
            45. Involucre tubes 3–4 mm; pl on serpentine; KR. . . . . var. *nelsoniorum*
            45′ Involucre tubes 2–3 mm; pl not on serpentine; s SNH, n DMtns . . . . var. *canifolium*
34′ Main infl branches of ≥ 2 orders, or at least some with a whorl of bracts near middle
  46. Infl with a whorl of bracts at middle (see also *Eriogonum umbellatum* var. *polyanthum*)
    47. Fls cream or ± white, 4–7 mm; Wrn . . . . var. *glaberrimum*
    47′ Fls yellow, 7–10 mm; n SNH . . . . var. *torreyanum*
  46′ Infl without a whorl of bracts at middle, bracts only at base of infl or involucres
    48. Fls 7–10(12) mm; shrub; infl 2–4-branched; KR . . . . var. *speciosum*
    48′ Fls 3–8 mm; per mat or subshrub, if shrub then infl 3-branched; not of KR
      49. Fls cream, ± white, pale yellow, or green-yellow, or yellow, in age red-brown to rose or pink
        50. Fls cream, ± white, pale yellow, or green-yellow, without large ± red spot on each lobe; subshrub or shrub; e DMtns . . . . var. *juniporinum*
        50′ Fls yellow, in age red-brown to rose or pink, with large ± red spot on each lobe; spreading to ± prostrate mat; W&I (Inyo Mtns), DMtns . . . . ²var. *versicolor*
      49′ Fls yellow, in age not red-brown to rose or pink
        51. Lf blades glabrous to densely tomentose (see also *Eriogonum umbellatum* var. *munzii* of TR, SnJt)
          52. Lf blades densely tomentose — NCoRI, CW . . . . var. *bahiiforme*
          52′ Lf blades densely tomentose to glabrous
            53. Lf blades tomentose to densely so abaxially, tomentose to sparsely so adaxially; NW
              54. Lf blades (0.5)0.8–1.8(2) cm wide; involucre lobes (3)4–6 mm; KR . . . . ²var. *lautum*
              54′ Lf blades 0.3–0.7 cm wide; involucre lobes 1–3 mm; NCoR . . . . ²var. *smallianum*
            53′ Lf blades gen sparsely tomentose to glabrous; not NW
              55. Lf blades glabrous; s SNH, SNE, nw DMoj. . . . . var. *chlorothamnus*
              55′ Lf blades sparsely tomentose to glabrous adaxially; se SN, TR, SNE, DMtns . . . . var. *subaridum*
      51′ Lf blades sparsely to densely tomentose, or densely tomentose abaxially, less so to glabrous adaxially (rarely both surfaces tomentose in *Eriogonum umbellatum* var. *munzii* of TR, SnJt)
        56. Shrub, densely branched — KR, n SNF, n SNH/CaRH
          57. Lf blades rusty-woolly to tomentose abaxially; infl 3–4-branched, branches tomentose to woolly, central branch without a whorl of bracts near middle . . . . var. *ahartii*
          57′ Lf blades white-tomentose abaxially; infl 1–2(3)-branched, branches sparsely tomentose to glabrous, central branch occ with a whorl of bracts near middle. . . . . var. *polyanthum*
        56′ Per mat or more openly and sparsely branched subshrub or shrub
          58. Sts gen sparsely tomentose or glabrous
            59. Lf margins finely wavy; sts often with lf-like bract near middle; KR, c NCoRH . . . . var. *argus*
            59′ Lf margins flat; sts without lf-like bract near middle; SNH . . . . var. *furcosum*
          58′ Sts gen tomentose or sparsely tomentose (or ± glabrous in age)
            60. Involucre lobes (3)4–6 mm; lf blades gen widely elliptic — KR . . . . ²var. *lautum*
            60′ Involucre lobes 1–3 mm; lf blades elliptic
              61. Lf blades 0.5–1 cm wide; TR, SnJt. . . . . var. *munzii*
              61′ Lf blades 0.3–0.7 cm wide; NCoR . . . . ²var. *smallianum*

Group 4

Per; fl stipe 0 or ± winged; infl bracts gen 3 (mainly subg. *Eucycla*)

1. St basally jointed, breaking into short-cylindric sections; fls 150–200 — n DMtns (subg. *Clastomyelon*)
.... ***E. intrafractum***
1′ St not jointed, not breaking into sections; fls 7–100
  2. Fls with dense coarse, curved, ± white hairs; lvs gen coarse-hairy on both surfaces;
     GB, D (subg *Ganysma*).. ***E. inflatum***
  2′ Fls glabrous or not as above; lvs tomentose to sparsely so at least abaxially, not coarse-hairy
    3. Fl stipe ± winged
      4. Fls bright yellow; infl cyme- or umbel-like, 0.5–3 cm; fr 2.5–3 mm; s WTR .... ***E. crocatum***
      4′ Fls white to rose or ± yellow; infl cyme-like, 10–25 cm; fr 3.5–4 mm; s SN, CW, SW, SNE, DMtns .... ***E. saxatile***
    3′ Fl stipe 0
     5. Pl forming cespitose or cushion-like mat; infl head-like, if not then pl 0.1–1 dm
      6. Infl branched
        7. Fls glandular-hairy, ± white to ± red; st gen glandular-hairy; s SNH .... ***E. breedlovei***
         8. St erect; infl umbel-like (head-like) .... var. ***breedlovei***
         8′ St suberect to prostrate; infl cyme-like .... var. ***shevockii***
        7′ Fls glabrous, white to pink or rose or ± yellow with ± red spot; st not glandular; s SNH, n DMtns
         9. Fls 3.5–5 mm, lobes not alike, outer round; involucres 1.5–2 mm; n DMtns .... ***E. gilmanii***
         9′ Fls (1.5)2–2.5 mm, lobes alike; involucres 0.8–1.7(2) mm; s SNH .... [2]***E. wrightii*** var. ***olanchense***
      6′ Infl head-like
        10. Perianth lobes not alike, outer gen 2 × as wide as inner; sts sparsely to densely tomentose or
           ± glabrous — KR, CaR, SNH, SnBr, GB .... ***E. ovalifolium***
         11. Lf blades gen 1–6 cm (or shorter); st (1)5–30(40) cm; involucres (3.5)4–6.5(8) mm
          12. St gen 1–5(7.5) cm; — lf blades densely tomentose; n SNH (Job's Peak, Alpine Co.) .... [2]var. ***eximium***
          12′ St gen (4)7–30(40) cm (or shorter)
           13. Fls 5–7 mm; involucres 5–7 mm; rare, ne SnBr.. [2]var. ***vineum***
           13′ Fls 4–5 mm; involucres 4–6.5 mm; common, not SnBr
            14. Fls yellow; SNH, GB.. var. ***ovalifolium***
            14′ Fls white to rose or purple; KR, CaRH, SNH, GB................................... var. ***purpureum***
         11′ Lf blades 0.2–1.2 (2) cm; st gen 0.3–5(9) cm (or longer); involucres gen 2–4.5(8) mm
          15. Fls yellow — c SNH .... var. ***caelestinum***
          15′ Fls white, rose, purple or red
           16. Lf blades densely tomentose, margins ± brown — n SNH (Job's Peak, Alpine Co.) .... [2]var. ***eximium***
           16′ Lf blades tomentose to densely so, margins not ± brown
            17. Involucres 3–4.5 mm
              18. Lf blades tomentose, to sparsely so adaxially; rare, MP .... var. ***depressum***
              18′ Lf blades tomentose to densely so on both surfaces; common, CaRH, SNH, W&I .... var. ***nivale***
            17′ Involucres 5–8 mm — s SNH or ne SnBr
             19. Lf blades tomentose to sparsely so; s SNH .... var. ***monarchense***
             19′ Lf blades densely tomentose; ne SnBr .... [2]var. ***vineum***
        10′ Perianth lobes alike, if not then sts glandular-hairy
         20. Fls yellow or red-yellow
          21. Involucres glandular and sparsely hairy; st glandular-hairy; n&c SNH, n SNE....... ***E. rosense*** var. ***rosense***
         21′ Involucres glabrous or sparsely tomentose, occ glandular; st glabrous or glandular; MP
          22. Lf blades obovate or not, 1–2 cm, (0.3)0.5–1(1.5) cm wide; st 0.5–1.5(2) dm, glandular below
           infl; involucres sparsely tomentose, occ sparsely glandular .... ***E. ochrocephalum*** var. ***ochrocephalum***
          22′ Lf blades not obovate, 0.3–1(1.4) cm, 0.15–0.4(0.6) cm wide; st 0.2–0.6(0.8) dm, glabrous;
           involucres glabrous or sparsely tomentose... ***E. prociduum***
        20′ Fls white to rose or pink
          23. Involucres sparsely glandular to glandular-hairy or hairy, pliable; c&s SNH, n W&I .... ***E. gracilipes***
         23′ Involucres glabrous or tomentose, rigid; c&s SN, Teh, TR, SNE, n DMtns
          24. Fr tomentose; SNE, n DMtns (Last Chance Range)....................... ***E. shockleyi*** var. ***shockleyi***
          24′ Fr glabrous; c&s SN, Teh, TR, SNE, nw DMtns
           25. Pedicels hairy; lvs (1)2–5 cm, 0.8–2 cm wide; pl 3–10(11) dm diam; Teh........... [2]***E. callistum***
           25′ Pedicels glabrous; lvs 0.2–1(1.2) cm, 0.05–0.4 cm wide; pl 1–4 dm diam; c&s SN, TR, SNE,
           nw DMtns .... ***E. kennedyi***
            26. Lf hairs white; pl forming dense mat; c&s SN, SNE, nw DMtns.................... var. ***purpusii***
            26′ Lf hairs gray- or red- to brown-white; pl gen forming loose, open mat; se SNF, TR
             27. Lf blades oblong, 0.3–0.5 cm, 0.1–0.4 cm wide; se SNF .... var. ***pinicola***
             27′ Lf blades oblanceolate to elliptic, 0.2–1.2 cm, 0.05–0.2 cm wide; TR
              28. St 0.5–2(3) cm; SnGb, SnBr... var. ***alpigenum***
              28′ St 4–15 cm; n WTR, SnBr
               29. St sparsely tomentose; involucres 2.5–4 mm........................... var. ***austromontanum***
               29′ St glabrous; involucres 1.5–2.5 mm.. var. ***kennedyi***

5′ Shrub to subshrub or spreading to erect herb; infl branched, if head-like then pl 1–5 dm
  30. Involucres clustered 2–10+ per node
    31. Perianth lobes not alike . . . . **E. strictum**
      32. Fls yellow — CaR, ne SN, MP . . . . var. **anserinum**
      32′ Fls white to rose or purple
        33. Lf hairs white, densely tomentose to tomentose; KR . . . . var. **greenei**
        33′ Lf hairs ± gray or ± green, tomentose to sparsely so; KR, n NCoRO, NCoRH, CaR, ne SNH,
          w MP . . . . var. **proliferum**
    31′ Perianth lobes alike
      34. Herb
        35. Lf blades lanceolate to narrowly ovate, 4–15(25) cm, long-soft-hairy . . . . **E. elatum**
          36. St, infl glabrous; n CA . . . . var. **elatum**
          36′ St, infl long-soft-hairy; KR, CaR, n SN, MP . . . . var. **villosum**
        35′ Lf blades oblanceolate to ovate or elliptic, gen 2–6 cm, tomentose abaxially, tomentose to
          glabrous adaxially
          37. Involucres 3–6 mm wide; st, infl glabrous (to sparsely tomentose); infl ± head- to cyme-like;
            lf blades densely tomentose abaxially, tomentose to ± glabrous adaxially; ChI . . . . **E. grande**
            38. Fls pink to red or rose; pl 2–5 dm; involucres 5–7 mm; Santa Rosa, San Miguel, Santa Cruz
              islands, occ CW . . . . var. **rubescens**
            38′ Fls white; pl 1–15 dm; involucres 4–6 mm; various islands exc Santa Rosa Island
              39. Pl 5–15 dm; involucres 5–6 mm; San Miguel, Santa Cruz, Anacapa, Santa Catalina,
                San Clemente islands, occ CW . . . . var. **grande**
              39′ Pl 1–2(2.5) dm; involucres 4–5 mm; San Nicolas Island . . . . var. **timorum**
        37′ Involucres (1.5)2–4 mm wide; st, infl glabrous to tomentose; infl head- to umbel- or cyme-like;
          lf blades tomentose to densely so on both surfaces, or tomentose to glabrous adaxially; mainland
          40. Pedicels hairy; fls densely hairy — Teh . . . . [2]**E. callistum**
          40′ Pedicels glabrous; fls glabrous or hairy
            41. Involucres, sts tomentose to (glabrous); lf blades densely white- to yellow-brown-tomentose
              on both surfaces, or tomentose to glabrous adaxially; subshrub or mat — NCo, n&c CCo . . . . **E. latifolium**
            41′ Involucres, sts glabrous, if tomentose then pl not in NCo, n&c CCo; lf blades tomentose to
              densely so on both surfaces or tomentose to glabrous adaxially; pl gen erect . . . . [2]**E. nudum**
              42. Involucres, infl branches tomentose to sparsely so; lvs gen basal
                43. Fls white to rose or yellow; st tomentose to hairy; lf blades 2–4 cm; NW, CaR, n SN . . . . var. **oblongifolium**
                43′ Fls white; st densely tomentose (to tomentose); lf blades 2–3.5 cm; s SNF . . . . [2]var. **regirivum**
            42′ Involucres, infl branches glabrous, if not then lvs sheathing
              44. Lvs sheathing, margins gen strongly wavy
                45. Sts tomentose
                  46. Fls hairy, 1.5–2 mm; involucres 3–4 mm; s SNF . . . . [2]var. **regirivum**
                  46′ Fls glabrous, 2–4 mm; involucres 3–6 mm; SnFrB
                    47. Lf blades (1)3–8 cm, (1)2–4 cm wide, sheathing 1–5(10) cm; c SnFrB . . . . [2]var. **auriculatum**
                    47′ Lf blades 1–3 cm, 1–1.5 cm wide, sheathing 5–20 cm; s SnFrB . . . . var. **decurrens**
                45′ Sts glabrous
                  48. Lf blades densely tomentose abaxially, tomentose adaxially; involucres 5–10; s SNF . . . var. **murinum**
                48′ Lf blades tomentose abaxially, less so to glabrous adaxially; involucres 1–5; NW, deltaic
                  GV (Contra Costa Co.), CW
                  49. Involucres solitary; st inflated; fls white to yellow, glabrous; w edge SnJV, e SCoRI . . . var. **indictum**
                  49′ Involucres 1–5 on same pl; st not or ± inflated; fls white to pink (or ± yellow), glabrous or
                    hairy; NW, deltaic GV (Contra Costa Co.), CW
                    50. Involucres (2)3–5 per cluster; st not or ± inflated; fls white to pink (or ± yellow),
                      glabrous; NW, CW . . . . [2]var. **auriculatum**
                    50′ Involucres 1–3; st not inflated; fls white to pink, hairy; deltaic GV (Contra Costa Co.)
                    . . . . var. **psychicola**
            44′ Lvs basal, margins not or ± wavy
              51. Involucres 1
                52. St inflated; fls yellow (or white) — s SN, Teh, s CW, TR, s SNE, w DMoj . . . . [2]var. **westonii**
                52′ St not inflated, if so then pl of PR; fls white (or yellow)
                  53. Fls hairy (or glabrous); TR, PR . . . . [2]var. **pauciflorum**
                  53′ Fls glabrous; NW, CaR, SN, SnFrB, GB
                    54. Lf blades 1–2 cm; SNH, GB . . . . var. **deductum**
                    54′ Lf blades 1–5 cm; NW, CaR, SN, SnFrB . . . . [2]var. **nudum**
             51′ Involucres 2–10
              55. Fls hairy, gen yellow
                56. St not inflated; lf blades glabrous or woolly-tufted adaxially, margins flat; NW, CaR, SN,
                ScV, SnFrB . . . . var. **pubiflorum**
                56′ St ± to distinctly inflated; lf blades tomentose to sparsely so adaxially, margins wavy;
                s SNE, w DMoj . . . . [2]var. **westonii**

55′ Fls gen glabrous, white (or yellow)
  57. Infl gen head-like or ± so; alpine, c&s SNH. . . . . var. *scapigerum*
  57′ Infl cyme-like, if head-like or ± so then not alpine
    58. Involucres 5–7 mm, 1(2); TR, PR . . . . [2]var. *pauciflorum*
    58′ Involucres 3–5 mm, 2–10; NW, CaR, SN, SnFrB
      59. Infl cyme-like, branched 2+ times; involucres 2–5; NW, CaR, SN, SnFrB . . . . [2]var. *nudum*
      59′ Infl head- or cyme-like, branched 1–2 times; involucres 5–10; n NCo . . . . var. *paralinum*
34′ Shrub
  60. Lf blades narrowly linear to oblanceolate, ≤ 2 cm, gen strongly clustered; widespread . . . . . . **E. fasciculatum**
    61. Lvs thinly white-tomentose abaxially, glabrous adaxially; fls, involucres gen glabrous; pl gen
      decumbent; CCo . . . . var. *fasciculatum*
    61′ Lvs, fls, involucres hairy, if ± glabrous then pl of deserts; pl erect to rounded; CA-FP, SNE, D
      62. Lvs light yellow-green, glabrous or ± so adaxially; involucres, fls glabrous or occ hairy; D . . . . var. *flavoviride*
      62′ Lvs dark green or ± gray, gen hairy adaxially; involucres, fls hairy; CA-FP, SNE, D
        63. Infl open, mostly cyme-like; lvs densely white-tomentose abaxially, less so to green, sparsely
          tomentose adaxially, margins gen tightly rolled under; CW, SW, occ NW, D. . . . . var. *foliolosum*
        63′ Infl head- to umbel-like (or cyme-like); lvs gray-hairy on both surfaces or densely
          gray-tomentose abaxially, gray-hairy adaxially, margins not or occ rolled under; s SN, SCoRI,
          e SCo, TR, e PR, SNE, D . . . . var. *polifolium*
  60′ Lf blades linear-oblong to round, if linear then > 2 cm and not clustered; CW, SW
    64. Fls glabrous; CCo, SCo. . . . . **E. parvifolium**
    64′ Fls long-soft-hairy; CCo, SCo, ChI, occ established elsewhere
      65. Lf blades linear to narrowly oblong, 2–4(5) cm, white-tomentose abaxially, gray-hairy to
        ± glabrous adaxially, margins rolled under; n ChI, occ established elsewhere . . . . **E. arborescens**
      65′ Lf blades ovate or lanceolate to oblong- or narrow-ovate, 1.5–10 cm, white- or gray-tomentose
        gen on both surfaces, margins flat; CW, SW
        66. Lf blades ovate, 1.5–3 cm, 1–2.5(3) cm wide; infl gen head-like; s CCo, w SCo, n ChI (Santa
          Rosa Island), occ established CW, SW . . . . **E. cinereum**
        66′ Lf blades oblong-ovate to ovate or lanceolate to narrowly oblong, 2–7(10) cm, 1–5 cm wide;
          infl gen open, cyme-like; s ChI, occ established CW, SW. . . . . **E. giganteum**
          67. Lf blades lance-oblong to lanceolate; San Clemente Island . . . . var. *formosum*
          67′ Lf blades oblong-ovate to ovate; Santa Catalina, Santa Barbara islands
            68. Lf blades 2.5–3.5(6) cm, 1.5–2(4) cm wide; infl rarely open; Santa Barbara Island . . . . var. *compactum*
            68′ Lf blades 3–7(10) cm, 2–5 cm wide; infl open; Santa Catalina Island, occ CW, SW. . . . . var. *giganteum*
30′ Involucres 1 per node
69. Involucres in raceme-like arrangements at tips of infl branches
  70. Involucres (4)6–7 mm; per, 6–12(18) dm — c&s CW, SW . . . . **E. elongatum** var. *elongatum*
  70′ Involucres (0.8)1–6 mm; herb to shrub, (1)1.5–5 dm
    71. Shrub — s MP, SNE . . . . **E. nummulare**
    71′ Subshrub or herb
      72. Pl basally ± woody, much-branched; lf blades oblanceolate to elliptic, 0.1–1 cm wide, margins
        often rolled under
        73. Involucres (4)5–6 mm; fls (3)4–5 mm, cream to rose; n SCoRO . . . . **E. butterworthianum**
        73′ Involucres 0.7–4 mm; fls 1.5–4 mm, white to pink or rose; widespread . . . . **E. wrightii**
          74. Herb, loosely to compactly matted
          75. Pl 0.1–0.3(0.6) dm; lf blades 0.1–0.25 cm; involucres 0.8–1.7(2) mm; s SNH . . . . [2]var. *olanchense*
          75′ Pl 0.5–2.5(3) dm; lf blades 0.5–1(1.2) cm; involucres 1.5–4 mm; SN, CW, TR, e PR . . . . var. *subscaposum*
          74′ Shrub or subshrub
            76. Sts, infl ± gray, tomentose to densely so — e PR, s DMoj, w DSon. . . . . var. *nodosum*
            76′ Sts, infl ± white, ± red, or ± green, tomentose to woolly-tufted
              77. Petiole bases forming distinct ring around st; lf blades 0.2–0.6(1) cm, 0.1–0.3(0.4) cm wide;
                SW. . . . . var. *membranaceum*
             77′ Petiole bases not forming ring around st; lf blades 0.5–3 cm, 0.2–1 cm wide; NW,
                n SNF, DMoj
                78. Lf blades 1.5–3 cm, 0.5–1 cm wide; involucres 2–4 mm; fls 2.5–4 mm; NW, n SNF
                  . . . . var. *trachygonum*
                78′ Lf blades 0.5–1.5 cm, 0.2–0.5(0.7) cm wide; involucres 2–2.5 mm; fls 2.5–3.5 mm; DMoj . . . var. *wrightii*
      72′ Pl basally not woody, little-branched; lf blades round to widely ovate, (0.3)1–3.5(4) cm wide,
        margins flat — W&I, DMtns
        79. Lf blades ± round, (0.3)0.5–1.5 cm, petioles 1–2 cm; involucres bell-shaped . . . . **E. mensicola**
        79′ Lf blades ovate to elliptic, oblong, or obovate, 1.5–4.5(5) cm, petioles 1–7 cm; involucres gen
          narrower than bell-shaped
          80. Lf blades ovate to elliptic or obovate, 1.5–4 cm, 1–2.5 cm wide, petioles 1–5 cm; involucres
            3–5 mm, 2–4 mm wide . . . . **E. panamintense**
          80′ Lf blades elliptic to oblong, (2)2.5–4.5(5) cm, 1.5–3.5(4) cm wide, petioles (2)3–7 cm;
            involucres 3–4 mm, 3–4 mm wide . . . . **E. rupinum**

69′ Involucres in forks of branches
 81. Herb
  82. Lf blades ovate to elliptic or oblanceolate, (0.3)1–4 cm wide; CA (see 41' above) . . . . ²*E. nudum*
  82′ Lf blades round-ovate, 0.3–0.5(1) cm wide; n SNF . . . . *E. apricum*
   83. Sts erect to ± spreading . . . . var. *apricum*
   83′ Sts prostrate . . . . var. *prostratum*
 81′ Subshrub to shrub
  84. Fls hairy
   85. Fls green-yellow to yellow; involucres 4-toothed — s DSon . . . . *E. deserticola*
   85′ Fls white; involucres 5–8-toothed
    86. Involucres 3.5–5 mm; nw KR . . . . *E. pendulum*
    86′ Involucres 2–3 mm; CaRH . . . . *E. spectabile*
  84′ Fls glabrous
   87. Infl panicle- or cyme-like, gen extensively branched, gen open, not terminal; involucres glabrous
    88. Outer perianth lobes obovate; infl branches ± gray, gen spreading, tiered; D . . . . *E. plumatella*
    88′ Outer perianth lobes obovate to round; infl branches green, ascending, not tiered; c&s SNH,
     Teh, s SCoR, TR, SNE, DMoj. . . . . *E. heermannii*
     89. Sts, infl branches round or angled, minutely scabrous or papillate-scabrous; pl gen densely
      branched; DMtns
      90. Sts, infl branches round, glabrous or minutely scabrous . . . . var. *argense*
      90′ Sts, infl branches sharply angled, deeply grooved, minutely scabrous . . . . var. *sulcatum*
     89′ Sts, infl branches round, not angled, gen smooth, glabrous or sparsely tomentose to woolly-
      tufted; pl sparsely branched; widespread
      91. Involucres not in raceme-like arrangement, or only uppermost 2–3 so disposed; infl branches
       glabrous — c&s SNH (e slope), SNE, n DMtns . . . . var. *humilius*
      91′ Involucres in raceme-like arrangement, at least at tips of branches; infl branches glabrous or
       sparsely tomentose to woolly-tufted
       92. Infl sparsely tomentose to woolly-tufted; s CA . . . . var. *floccosum*
       92′ Infl branches glabrous; s-c, sw CA
        93. Lf blades 0.5–1.5 cm, glabrous to sparsely tomentose abaxially; infl branches stout; s SNH
         (Kern Co.), Teh, se SCoRO (se San Luis Obispo Co.), n TR . . . . var. *heermannii*
        93′ Lf blades 1.5–3(4) cm, gen tomentose to sparsely so abaxially (at least in early fl);
         infl branches slender; c SCoRI . . . . var. *occidentale*
   87′ Infl cyme-like, gen moderately branched, gen compact, terminal; involucres tomentose to glabrous
    94. Pl matted subshrub, 0.5–1.5 dm; lf blades 0.4–0.6 cm, margins rolled under; e DMtns . . . . *E. thornei*
    94′ Pl erect to spreading subshrub or shrub, 0.2–15 dm; lf blades (0.3)1–3.5 cm, margins flat,
     or if rolled under; widespread . . . . *E. microthecum*
     95. Fls yellow
      96. Lf blades (0.2)0.3–0.6(0.8) cm wide; fls (1.5)2–2.5(3) mm; involucres 2–2.5 mm;
       fr 1.5–2 mm; SNH (e slope), GB . . . . var. *ambiguum*
      96′ Lf blades 0.5–1.2 cm wide; fls 2.5–3 mm; involucres 2.5–4 mm; fr 2.5–3 mm; s MP . . . var. *schoolcraftii*
     95′ Fls white, cream, orange, pink, or red
      97. Hairs ± white (see also *Eriogonum microthecum* var. *alpinum*) or st, infl rarely glabrous
       98. Lf margins flat; infl sparsely tomentose or glabrous; SN, GB . . . . var. *laxiflorum*
       98′ Lf margins gen rolled under; infl tomentose to densely so or rarely subglabrous; s SNE,
        DMoj. . . . . var. *simpsonii*
      97′ Hairs ± brown or ± red (gen white in *Eriogonum microthecum* var. *alpinum*) or sts, infl gen glabrous
       99. Shrub 3–6 dm
        100. Sts, infl tomentose to densely so; fls 2–2.5(3) mm; fr 2.5–3 mm; e SnBr . . . . var. *corymbosoides*
        100′ Sts, infl tomentose in youth, sparsely tomentose in age; fls 1.5–2(2.5) mm; fr 1.8–2 mm;
         s W&I, n DMtns . . . . var. *panamintense*
       99′ Subshrub 0.2–1.5(2) dm
        101. Lf blades elliptic or ovate, margins flat; fls (1.5)2–3.5(4) mm
         102. Lf blades 0.5–1 cm, (0.2)0.3–0.5(0.6) cm wide; involucres (2)2.5–3 mm; fls (2.5)3–3.5(4)
          mm; e SnGb, w SnBr . . . . var. *johnstonii*
         102′ Lf blades 0.3–0.7(0.8) cm, 0.1–0.4 cm wide; involucres 2.5–3.5 mm; fls (1.5)2–3.5 mm;
          s W&I . . . . var. *lapidicola*
        101′ Lf blades linear or linear-oblanceolate to narrowly elliptic, margins often rolled under; fls 1.5–2.5 mm
         103. Involucres (1.5)2–3 mm; fls white to pink or rose; n&c SNH, n SNE . . . . var. *alpinum*
         103′ Involucres 3–4 mm; fls cream; SnBr . . . . var. *lacus-ursi*

E. alpinum Engelm. (p. 1093) TRINITY BUCKWHEAT (Group 3)
Per mat 0.2–0.6(0.8) dm, (0.3)0.4–1 dm diam. **ST**: (0.3)0.4–0.6 dm, sparsely tomentose, bracts 3–5, lf-like, near middle. **LF**: basal; blades ± round, 1–2(3) cm, tomentose. **INFL**: head-like, 1–2(2.5) cm wide;

involucres 3–5(6) mm, 4–7(10) mm wide, tomentose, teeth 6–12. **FL**: (3)4–8 mm, stipe 0.5–0.8 mm, glabrous; perianth bright yellow, lobes oblong-obovate. **FR**: 4–5 mm, glabrous. Serpentine; 2000–2800 m. e KR (Mount Eddy area, s Siskiyou, ne Trinity cos.). Jul–Sep ★

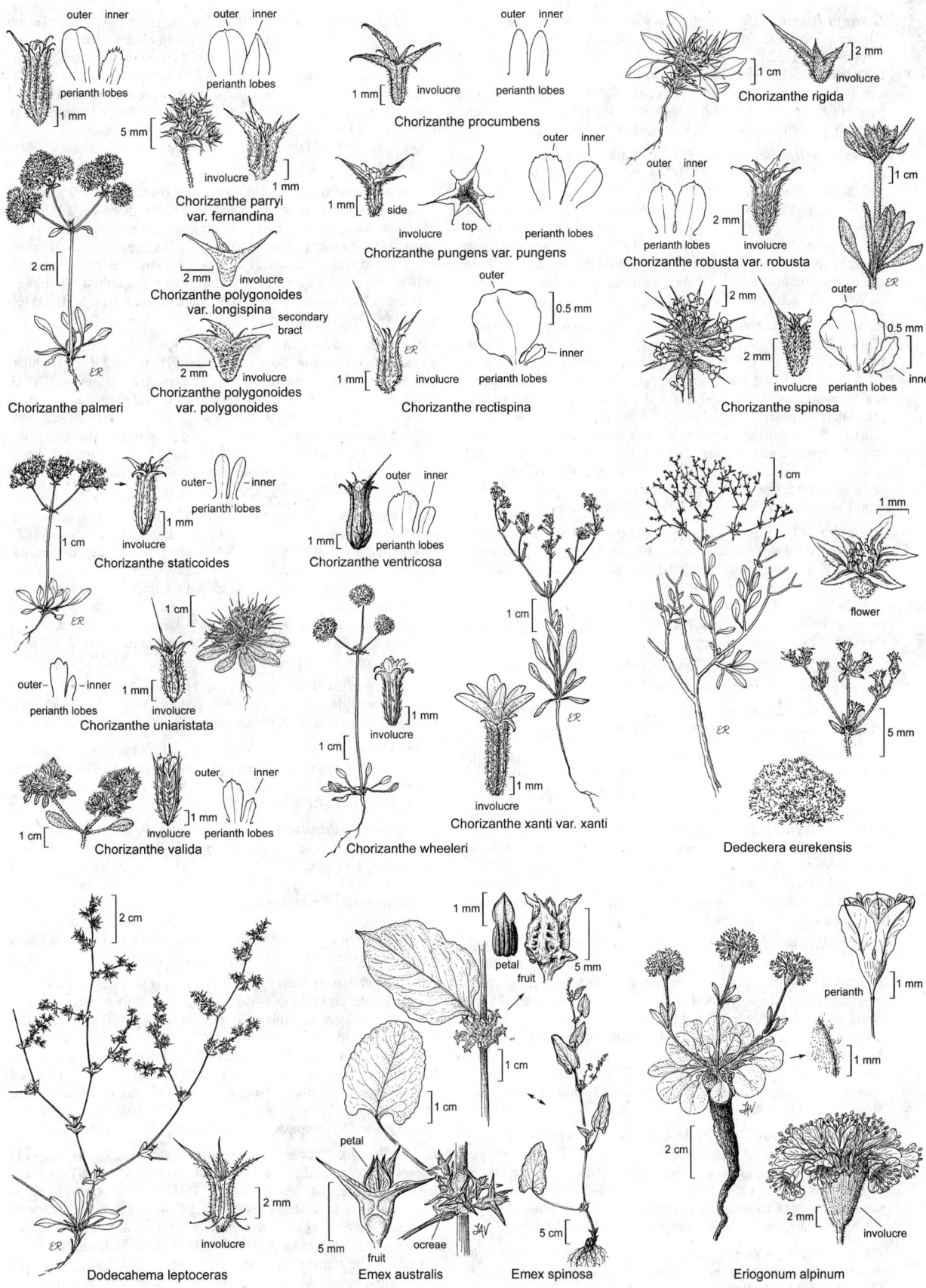

Chorizanthe palmeri

Chorizanthe parryi var. fernandina

Chorizanthe polygonoides var. longispina

Chorizanthe polygonoides var. polygonoides

Chorizanthe procumbens

Chorizanthe pungens var. pungens

Chorizanthe rectispina

Chorizanthe rigida

Chorizanthe robusta var. robusta

Chorizanthe spinosa

Chorizanthe staticoides

Chorizanthe uniaristata

Chorizanthe valida

Chorizanthe ventricosa

Chorizanthe wheeleri

Chorizanthe xanti var. xanti

Dedeckera eurekensis

Dodecahema leptoceras

Emex australis

Emex spinosa

Eriogonum alpinum

E. ampullaceum J.T. Howell (p. 1097) MONO WILD BUCKWHEAT (Group 1) Ann 1–3 dm. **ST:** 0.2–1 dm, glabrous. **LF:** basal, blades ± round, 0.5–2(2.5) cm, gen tomentose. **INFL:** 5–25 cm, 5–10(15) cm wide; branches glabrous; peduncles 0; involucres gen at tips of slender branchlets, 1.5–2 mm, glabrous. **FL:** 1–1.5 mm, glabrous; perianth cream to ± red, lobes obovate. **FR:** 1–1.3 mm, glabrous. Uncommon. Sand; 1700–2200 m. n SNE; w NV. Jul–Sep

E. angulosum Benth. (p. 1097) ANGLE-STEM WILD BUCKWHEAT (Group 2) Ann 1–5(10) dm. **ST:** 0.5–1 dm, gen tomentose. **LF:** basal and cauline; blades (0.5)1–4(4.5) cm, (0.2)0.5–1(1.3) cm wide, gen tomentose. **INFL:** 5–80 cm, 10–60 cm wide; branches gen tomentose to ± glabrous; peduncles erect, 1–2 cm, straight, slender, tomentose or glabrous; involucres 1.5–2.5(3) mm, hairy. **FL:** 1.5–1.8 mm, glandular-hairy; perianth white to rose, lobes gen elliptic to obovate, outer occ inflated basally. **FR:** 1–1.5 mm, glabrous. Common. Clay; < 800 m. c&s SNF, Teh, SnJV, CW, TR, s PR, w DMoj. All yr

E. apiculatum S. Watson (p. 1097) SAN JACINTO WILD BUCKWHEAT (Group 2) Ann 2–9 dm. **ST:** 0.5–1.5 dm, gen glabrous. **LF:** basal; blades (0.5)1–4 cm, 0.5–1.5 cm wide, hairy, some glandular. **INFL:** 30–80 cm, 10–50 cm wide; branches glabrous, sparsely glandular at nodes; peduncles 0 or reflexed, (0.1)0.2–0.35 cm, straight, slender, glandular; involucres 1.2–1.5 mm, 1–1.3 mm wide, glabrous; teeth 4. **FL:** 1.5–2.5 mm, hairy; perianth white to ± red (or yellow), lobes oblong-obovate, with a sharp, 1–2 mm tip or not. **FR:** 1.3–1.7(2.2) mm, glabrous. *n*=20. Gen (200)700–2700 m. e SnBr (Pipes Spring area, San Bernardino Co.), c&s PR (Santa Rosa, Palomar, Cuyamaca mtns), sw DMtns (Little San Bernardino Mtns). May–Nov

E. apricum J.T. Howell (Group 4) Per mat 0.08–0.2 dm, 0.1–0.25 dm diam. **ST:** 0.4–0.8 dm, glabrous. **LF:** basal; blades round-ovate, 0.3–0.5(1) cm, tomentose abaxially, gen glabrous adaxially. **INFL:** 0.5–1.5 cm, 1–2 cm wide; branches glabrous; peduncles 0; involucres 2–2.5 mm, 2–2.5 mm wide, glabrous. **FL:** 2–3 mm, glabrous; perianth white, lobes oblong. **FR:** 2.5–3 mm, glabrous. Both vars. threatened by vehicles, mining.

var. ***apricum*** (p. 1097) IONE BUCKWHEAT **ST:** erect to ± spreading. *n*=40. Clay; 80–200 m. n SNF (w Amador Co.). Jun–Oct ★

var. ***prostratum*** Myatt IRISH HILL BUCKWHEAT **ST:** prostrate. Clay; 90–200 m. n SNF (w Amador Co.). Jun–Sep ★

E. arborescens Greene (p. 1097) SANTA CRUZ ISLAND WILD BUCKWHEAT (Group 4) Shrub 6–15(20) dm, 5–30 dm diam. **ST:** 0.5–1(1.5) dm, glabrous. **LF:** cauline; blades 2–4(5) cm, 0.1–0.4(0.6) cm wide, tomentose abaxially, gray-hairy to ± glabrous adaxially. **INFL:** 5–10 cm, 5–15(20) cm wide; branches glabrous; peduncles 0 or erect, 0.1–0.5 cm, stout, hairy; involucres 2–3 mm, 2.5–4 mm wide, hairy, teeth 5–7. **FL:** 2–3.5(4) mm, hairy; perianth white to ± pink, lobes oblanceolate to narrowly obovate. **FR:** 2.5–3.5 mm, glabrous. *n*=20. Gravel; 10–600 m. n ChI (exc San Miguel Island); widely cult, occ established CW, SW, elsewhere. Apr–Oct

E. argillosum J.T. Howell (p. 1097) CLAY BUCKWHEAT (Group 1) Ann 1–3 dm. **ST:** (0.3)0.5–1.5(2.5) dm, gen glabrous. **LF:** basal and cauline; blades (0.5)1–3(5) cm, 0.2–1.2(1.6) cm wide, tomentose abaxially, hairy adaxially. **INFL:** (2)5–20(25) cm; branches glabrous; peduncles 0 or erect, 0.5–4.5(6) cm, straight, slender, glabrous; involucres gen terminal at tips of slender branchlets, (2)2.5–3 mm, 2–3 mm wide, glabrous. **FL:** 1.5–2 mm, glabrous; perianth white to rose, lobes oblong; style 0.2–0.3 mm. **FR:** 2–2.5 mm, glabrous; tip smooth. Steep clay slopes, occ serpentine; 150–600 m. s SnFrB, n SCoRI. Mar–Oct ★

E. baileyi S. Watson (Group 1) Ann 1–4(5) dm. **ST:** 0.5–1 dm, glabrous or tomentose. **LF:** basal; blades ± round, 0.5–2 cm, tomentose. **INFL:** 5–35(45) cm, glabrous or tomentose; peduncles 0; involucres appressed, 1–2 mm, 0.5–1 mm wide. **FL:** 1.5–3 mm, gen minutely glandular, occ glabrous in SNF; perianth white to rose, lobes oblong to oblong-obovate, gen constricted near middle. **FR:** 1–1.5 mm.

var. ***baileyi*** (p. 1097) BAILEY'S WILD BUCKWHEAT **INFL:** glabrous. Common. Sand, gravel; (100)500–2900 m. SNF, Teh, s SCoR, TR, GB, w DMoj; to WA, ID, w UT. May–Oct

var. ***praebens*** (Gand.) Reveal BAILEY'S WOOLLY BUCKWHEAT **INFL:** tomentose. Sand; 1300–2500 m. GB; to c ID, NV. Jun–Oct ★

E. bifurcatum Reveal (p. 1097) FORKED BUCKWHEAT (Group 2) Ann 1–4 dm. **ST:** 0.05–0.3 dm. **LF:** basal; blades ± round, 1–3 cm, gen tomentose. **INFL:** 5–30 cm, 10–50 cm wide; branches spreading, glabrous; peduncles 0 or erect, 0.1–0.5 cm, straight, slender, glabrous; involucres, 2–2.5 mm, 1.3–2 mm wide, glabrous. **FL:** 1.5–2 mm, glabrous; perianth white to faintly pink; outer lobes obovate to cordate, inner lanceolate. **FR:** 2–2.3 mm, glabrous. *n*=20. Sand; 600–800 m. DMoj (s Inyo, ne San Bernardino cos.); s NV. May–Jun ★

E. brachyanthum Coville SHORT-FLOWER WILD BUCKWHEAT (Group 1) Ann (0.5)1–3 dm. **ST:** 0.3–0.8 dm, glabrous. **LF:** basal; blades ± round, 0.5–2 cm, gen tomentose. **INFL:** gen 5–30 cm; branches glabrous; peduncles 0; involucres appressed, 1–1.2 mm, 0.4–0.6 mm wide, glabrous. **FL:** 0.6–0.8(1) mm, glabrous; perianth yellow or pale ± yellow (or ± white), lobes gen oblong to oblong-obovate. **FR:** 0.8–1 mm. Common. Sand; 600–2300 m. SNE, DMoj; w NV, s-c OR. Apr–Nov

E. brachypodum Torr. & A. Gray (p. 1097) PARRY'S WILD BUCKWHEAT (Group 2) Ann 0.5–4 dm. **ST:** 0.2–0.7 dm, glandular. **LF:** basal; blades ± round, gen 1–3(5) cm, gen tomentose. **INFL:** 3–40 cm, 3–100 cm wide; branches glandular; peduncles 0 or reflexed, ≤ 1.5 cm,, straight, ± stout glandular; involucres 1–2.5 mm, 1.5–2.5 mm wide, glandular. **FL:** 1–2.5 mm, glabrous; perianth white to ± red; outer lobes ovate to oblong, inner gen lanceolate. **FR:** 1.5–2 mm, glabrous. *n*=20. Common. Sand; 100–2300 m. SNE, DMoj; to nw AZ, sw UT. All yr

E. breedlovei (J.T. Howell) Reveal (Group 4) Per mat 0.2–1 dm, 0.8–1.5(2) dm diam. **ST:** 0.15–0.6(0.8) dm, gen glandular-hairy. **LF:** basal; blades 0.2–0.8(1) cm, 0.2–0.4(0.6) cm wide, gen tomentose. **INFL:** 0.5–3 cm, 1–2.5 cm wide, rarely head-like; branches gen glandular; peduncles gen erect to spreading, 0.05–0.5(1) cm, slender, gen glandular; involucres 2–4, (2.5)3.5–4 mm, 3.5–4 mm wide, rigid, gen glandular-hairy, teeth 7–9. **FL:** 2.5–3.5(4) mm, glandular-hairy; perianth ± white to ± red, lobes gen obovate, obtuse. **FR:** 2.5–3 mm, tip sparsely glandular.

var. ***breedlovei*** (p. 1097) BREEDLOVE'S BUCKWHEAT **ST:** erect, densely glandular-hairy. **INFL:** peduncles 0.05–0.2 cm. Quartzite; 2300–2500 m. s SNH (Piute Mtns, Kern Co.). Jun–Sep ★

var. ***shevockii*** J.T. Howell (p. 1097) THE NEEDLES BUCKWHEAT **ST:** suberect to prostrate, thinly glandular-hairy. **INFL:** peduncles 0.3–0.5(1) cm. Granite; 1800–2300 m. s SNH (The Needles, Baker Point, Little Kern River Gorge, Kern, Tulare cos.). Jun–Oct ★

E. butterworthianum J.T. Howell (p. 1097) BUTTERWORTH'S BUCKWHEAT (Group 4) Subshrub 1–3 dm, 1–4 dm diam. **ST:** tomentose. **LF:** sheathing; blades (0.5)1–2 cm, 0.1–0.4 cm wide, tomentose. **INFL:** gen 1–2 cm; branches tomentose; peduncles 0; involucres (4)5–6 mm 3–4 mm wide, tomentose. **FL:** (3)4–5 mm, glabrous; perianth cream to rose, lobes obovate. **FR:** 3–3.5 mm, glabrous. Sandstone; 650–700 m. n SCoRO (The Indians, Santa Lucia Range, c Monterey Co.). Jun–Sep ★

E. caespitosum Nutt. (p. 1097) MATTED WILD BUCKWHEAT (Group 3) Per mat (0.1)0.3–1 dm, 1–5(12) dm diam. **ST:** (0.1)0.3–0.8(1) dm, hairy or glabrous. **LF:** basal; blades 0.2–1(1.5) cm, 0.15–0.4(0.5) cm wide, gen tomentose. **INFL:** head-like, 0.5–2 cm wide; branches 0; involucres 2–3.5 mm, 2–4 mm wide, teeth 6–9, reflexed. **FL:** 2.5–10 mm, stipe 0.5–1 mm, hairy; perianth yellow to ± red or rose, lobes gen oblanceolate. **FR:** (3.5)4–5 mm, glabrous or beak sparsely hairy. Common. Sand; (1300) 1500–3000 m. GB; to OR, MT, UT, nw AZ. [*Eriogonum cespitosum*, orth. var.] Apr–Jul

E. callistum Reveal TEHACHAPI BUCKWHEAT (Group 4) Per mounded mat (0.5)1–3.5 dm, 3–10(11) dm diam. **ST:** densely white-tomentose. **LF:** basal; blades (1)2–5 cm, 0.8–2 cm wide, elliptic, gray-white-tomentose. **INFL:** (1.5)2–4(4.5) dm; peduncles 0; involucres (5)6–8(9), 2–4 mm wide, tomentose abaxially, glabrous adaxially. **FL:** 2–5 mm; perianth pink-white in bud, white in age, densely white tomentose abaxially, glabrous adaxially exc glands on midribs; outer lobes ± wider than inner. **FR:** 3–4 mm, glabrous. Open limestone outcrops, ridges in chaparral; 1400–1500 m. Teh. Spring–summer ★

E. cedrorum Reveal & Raiche THE CEDARS BUCKWHEAT (Group 3) Per mat 1–3(5) dm diam. **ST:** 0.2–0.8 dm, hairs long, soft, wavy. **LF:** basal; blades 0.7–1.5 cm, 0.4–1 cm wide, tomentose abaxially, sparsely hairy or glabrous adaxially, margin ± flat. **INFL:** 1–2 cm, 1–3 cm wide; branch hairs long, soft, wavy; involucres 4–6.5 mm, 3.5–5.5(6) mm wide hairs long, soft, wavy; teeth 6–8. **FL:** 2–6 mm, stipe 0.1–0.3 mm, glabrous; perianth yellow to red or red-maroon, lobes obovate. **FR:** 4.5–6 mm, glabrous. Serpentine; 300–600 m. NCoRO. May–Oct ★

E. cernuum Nutt. (p. 1097) NODDING WILD BUCKWHEAT (Group 2) Ann 0.5–6 dm. **ST:** 0.3–2 dm, glabrous. **LF:** basal; blades (0.5)1–2(2.5) cm, gen ± round, gen tomentose. **INFL:** gen 5–50 cm; branches glabrous; peduncles 0 or spreading to reflexed, 0.1–2.5 cm, slender, glabrous; involucres (1)1.5–2 mm, 1–1.5 mm wide, glabrous. **FL:** 1–2 mm, glabrous; perianth white to ± pink, rose or red; outer lobes ± fiddle-shaped, inner obovate. **FR:** 1.5–2 mm, glabrous. Uncommon. Sand; 600–3100 m. n SNH (probably introduced, Nevada Co.), c SNH (e slope), GB; to WA, w&c Can, SD, NM. [*E. c.* var. *viminale* (S. Stokes) Reveal] Apr–Oct

E. cinereum Benth. (p. 1097) COASTAL WILD BUCKWHEAT (Group 4) Shrub 6–15 dm, 10–20(25) dm diam. **ST:** 1–4 dm, hairy. **LF:** cauline; blades 1.5–3 cm, 1–2.5(3) cm wide, white-tomentose. **INFL:** gen head-like 1–2.5 cm wide; branches hairy; peduncles 0; involucres 3–10, 3–5 mm, 2–3 mm wide, hairy. **FL:** 2.5–3 mm, hairy; perianth white to ± pink, lobes spoon-shaped to narrowly obovate. **FR:** 2–2.5 mm, glabrous. *n*=40. Uncommon. Sand; < 400 m. s CCo, w SCo, n ChI (Santa Rosa Island); widely cult, occ established CW, SW. All yr

E. cithariforme S. Watson (Group 1) Ann 2–3(5) dm. **ST:** 0.5–1 dm, glabrous or tomentose. **LF:** basal, sometimes cauline; blades (0.3)1–2 cm, narrow to ± round, gen tomentose. **INFL:** 5–25 cm; branches glabrous or tomentose; peduncles 0; involucres gen appressed, 2.5–3 mm, 1.5–2 mm wide, glabrous. **FL:** 1.5–2 mm, glabrous; perianth white to rose, lobes oblong-obovate. **FR:** 1–1.5 mm, glabrous.

var. *agninum* (Greene) Reveal SANTA YNEZ WILD BUCKWHEAT **LF:** petiole not winged; basal blades elliptic to ovate or ± rounded. **INFL:** glabrous. Uncommon. Sand; 500–1800 m. SCoR, WTR. May–Oct

var. *cithariforme* (p. 1097) CITHARA WILD BUCKWHEAT **LF:** petiole gen winged; basal blades oblanceolate. **INFL:** tomentose. Uncommon. Sand; (100)500–2200 m. SCoR, TR. May–Oct

E. clavatum Small HOOVER'S DESERT TRUMPET (Group 2) Ann (2)4–18(22) dm. **ST:** 0.5–3.5(5) dm, inflated, glabrous, glandular and gen hairy basally. **LF:** basal; blades (0.5)1–2.5(4) cm, (0.5)1–2(3) cm wide, round to reniform, coarse-hairy. **INFL:** 30–150(170) cm, 10–80 cm wide; branches gen inflated, glabrous; peduncles gen spreading, 1–4 cm, ± bent distally, slender, glabrous; involucres 1–1.5(1.8) mm, 0.6–0.9(1.2) mm wide, glabrous, teeth 4. **FL:** 1.5–2.5 mm, hairy; perianth yellow, lobes narrowly ovate. **FR:** 2–2.5 mm, glabrous. *n*=16. Common. Clay; 400–1100 m. CW, SW, s SNE, D; to s NV, n Baja CA. [*E. trichopes* var. *hooveri* Reveal] All yr

E. collinum M.E. Jones HILL BUCKWHEAT (Group 2) Ann 1–5(7) dm. **ST:** 0.3–1 dm, gen glabrous. **LF:** basal; blades gen (0.5)1–2.5(3.5) cm, ± round, gen tomentose abaxially, gen glabrous adaxially. **INFL:** 5–60 cm, 5–45 cm wide; branches glabrous; peduncles ascending, 1–5 cm, curving, slender, glabrous; involucres (1.5)2–3 mm, (1)1.5–2.5 mm wide, glabrous. **FL:** 1–2.5 mm, glabrous, bumpy basally; perianth white to pale yellow, lobes lanceolate to spoon-shaped or ovate. **FR:** 2–2.5 mm, glabrous. *n*=18. Sand; 1300–2000 m. MP; to nw NV, s-c OR. May–Nov ★

E. compositum Benth. var. *compositum* (p. 1097) ARROW-LEAF WILD BUCKWHEAT (Group 3) Per 4–7 dm, 2–5 dm diam. **ST:** 2–5 dm, glabrous, gen ± inflated. **LF:** basal; blades (2)7–25 cm, 2–8 cm wide, gen tomentose. **INFL:** 3–20 cm; branches glabrous; involucres 6–10 mm, 4–10 mm wide, densely tomentose, teeth (5)7–10, gen weakly reflexed. **FL:** 5–6 mm, stipe 0.7–1.5 mm, glabrous; perianth pale to bright yellow, lobes oblong to oblong-ovate. **FR:** 5–6 mm, glabrous exc tip hairy. *n*=20. Uncommon. Serpentine; 30–2500 m. NW, CaR; to WA, ID. Other vars. in OR, WA, ID. Apr–Jul

E. congdonii (S. Stokes) Reveal (p. 1097) CONGDON'S BUCKWHEAT (Group 3) Subshrub 1.5–5 dm, 1–3 dm diam. **ST:** 1–2(2.5) dm, gen glabrous. **LF:** basal; blades 0.5–2 cm, 0.3–0.6(0.8) cm wide, tomentose abaxially, gen glabrous adaxially. **INFL:** 0.3–3 cm 1–3.5 cm wide; branches gen glabrous; involucres 5–6 mm, 3–4 mm wide, tomentose, teeth 6–8. **FL:** 4–6 mm, stipe 0.4–0.6 mm, glabrous; perianth yellow, lobes obovate. **FR:** 4–5.5 mm, glabrous exc beak hairy. Serpentine; (1000)1500–2300 m. KR. Jul–Sep ★

E. contiguum (Reveal) Reveal (p. 1097) REVEAL'S BUCKWHEAT (Group 2) Ann (0.3)0.5–3 dm. **ST:** 0.1–0.5 dm, glabrous, glandular basally. **LF:** basal; blades (0.3)0.5–1(1.4) cm, ± round, coarse-hairy. **INFL:** gen (2)5–18 cm; branches glabrous exc glandular at lower nodes; peduncles gen erect, 0.3–1.2(2) cm, slender, glabrous, glandular basally; involucres 1–1.3 mm, 0.6–1 mm wide, glabrous, teeth (4)5. **FL:** 1–2.5 mm, coarse-hairy; perianth yellow to ± red, lobes lanceolate. **FR:** 1.5–1.8(2) mm, glabrous. *n*=16. Sand; 20–900 m. ne DMoj (se Inyo, ne San Bernardino cos.); sw NV. Apr–Jun ★

E. covilleanum Eastw. (p. 1097) COVILLE'S WILD BUCKWHEAT (Group 1) Ann 1–4 dm. **ST:** 0.3–1 dm, glabrous. **LF:** basal; blades (0.3)0.5–1.5(1.8) cm, ± round, tomentose abaxially, glabrous adaxially. **INFL:** 5–35 cm; branches glabrous; peduncles 0; involucres gen at tips of slender branchlets, not appressed, 2–2.5 mm, 1–1.5 mm wide, glabrous. **FL:** 2–2.5(3) mm, minutely puberulent; perianth pink, to white, rose or yellow in fr, lobes narrowly elliptic. **FR:** 1.8–2 mm. *n*=17. Uncommon. Shale or serpentine; 200–1400 m. c&s CW, WTR. Apr–Aug

E. crocatum Davidson (p. 1097) CONEJO BUCKWHEAT (Group 4) Subshrub 3–5 dm, 5–10 dm diam. **ST:** 0.2–1 dm, tomentose. **LF:** cauline, blades 1–3(3.5) cm, 0.8–2.5(3) cm wide, tomentose. **INFL:** 0.5–3 cm, 3–8 cm wide; branches tomentose; peduncles 0; involucres 3–4 mm, 3–5 mm wide, tomentose, teeth 5–6. **FL:** 5–6 mm, incl ± winged stipe, glabrous; perianth bright yellow; outer lobes 3–4 mm, 0.7–0.9 mm wide, narrowly oblong, inner 3.5–5 mm, 1–1.5 mm wide, oblong to spoon-shaped. **FR:** 2.5–3 mm, glabrous. *n*=20. Volcanics; 60–150 m. s WTR (nw Santa Monica Mtns, Ventura Co.). Apr–Jul ★

E. dasyanthemum Torr. & A. Gray (p. 1097) CHAPARRAL WILD BUCKWHEAT (Group 1) Ann 2–6 dm. **ST:** 0.5–2 dm, tomentose to sparsely so (or glabrous). **LF:** basal and occ cauline, blades (0.3)1–2 cm, (0.3)1–2 cm wide, ± round, tomentose abaxially, sparsely tomentose to ± glabrous or glabrous adaxially. **INFL:** 10–50 cm; branches sparsely tomentose to ± glabrous (or glabrous); peduncles 0; involucres appressed, (3.8)4–4.5 mm, 2–3 mm wide, ± glabrous, teeth 5. **FL:** (1.8)2–2.5 mm, hairy; perianth white to rose, lobes oblong-obovate. **FR:** 1.5–2 mm, glabrous; tip scabrous. *n*=12. Uncommon. Sand; 50–1400 m. c&s NCoR. May–Oct

E. davidsonii J.A. Clark DAVIDSON'S WILD BUCKWHEAT (Group 1) Ann 1–5 dm. **ST:** 0.5–1.5(2) dm, glabrous. **LF:** basal; blades (0.3)1–2(4) cm, ± round, tomentose abaxially, sparsely tomentose to ± glabrous adaxially. **INFL:** 5–40 cm; branches glabrous; peduncles 0; involucres appressed, 3–4 mm, 2–2.5 mm wide, glabrous, teeth 5. **FL:** 1.5–2 mm, glabrous; perianth white to pink or red (or yellow), lobes oblong-obovate. **FR:** 2 mm. *n*=20. Common. Sand; (400)900–2600 m. SW, SNE, DMoj; to n AZ, sw UT, n Baja CA. May–Sep

E. deflexum Torr. SKELETON WEED (Group 2) Ann (0.5)1–10(20) dm. **ST:** 0.3–3(4) dm, glabrous, occ inflated. **LF:** basal; blades 1–2.5(4) cm, 2–4(5) cm wide, gen ± round, tomentose abaxially, sparsely tomentose or ± glabrous adaxially. **INFL:** 10–90(180) cm, 5–50 cm wide; branches occ inflated, glabrous; peduncles 0 or reflexed (or some ± erect), ≤ 1.5 cm, straight, slender to stout, glabrous; involucres 1.5–2.5(3) mm, glabrous, teeth 5. **FL:** 1–2.5 mm, glabrous; perianth white to ± pink to ± red; outer lobes oblong or cordate to ovate, inner lanceolate to narrowly ovate. **FR:** (1.5)2–3 mm, glabrous.

var. *baratum* (Elmer) Reveal (p. 1097) TALL SKELETON WEED Pl 3–10 dm. **ST:** inflated. **INFL:** gen narrowly erect; branches gen inflated; peduncles reflexed, (0.3)0.5–1.5 cm; involucres (2)2.5–3 mm. **FL:** 2–2.5 mm; outer lobes cordate. *n*=20. Locally common. Sand; 700–2900 m. s SNH (e slope), Teh, s SnJV, TR, n&w DMoj; s NV. Jul–Oct

var. ***deflexum*** (p. 1097) FLAT-TOPPED SKELETON WEED Pl (0.5)1–5(20) dm. **ST:** not inflated. **INFL:** flat-topped, spreading, hemispheric, rarely narrowly erect with whip-like branches; branches not inflated; peduncles 0 or reflexed (or some ± erect), gen ≤ 0.5 cm; involucres 1.5–2.5 mm. **FL:** 1–2 mm; outer lobes cordate to ovate. *n*=20. Common. Sand; -50–1900 m. D; to w&s UT, sw NM, nw Mex. [*E. d.* var. *rectum* Reveal; *E. insigne* S. Watson] All yr

var. ***nevadense*** Reveal NEVADA SKELETON WEED Pl 0.5–3(5) dm; st not inflated. **INFL:** flat-topped, spreading; branches not inflated; peduncles 0 or reflexed, ≤ 0.5 cm; involucres 1.5–2 mm. **FL:** 1–1.5(2) mm, lobes oblong. *n*=20. Locally common. Sand; 1000–2200 m. SNE; NV. Jun–Oct

E. deserticola S. Watson (p. 1097) COLORADO DESERT WILD BUCKWHEAT (Group 4) Shrub 6–15(18) dm, 10–20(35) dm diam. **ST:** 1.5–4 dm, tomentose to sparsely so, occ ± glabrous. **LF:** cauline, blades (0.3)0.5–1.5(3.5) cm, (0.5)0.7–1.7(2) cm wide, gen ± round, tomentose. **INFL:** 15–100 cm; branches tomentose to sparsely so or ± glabrous; peduncles 0 or erect, ≤ 0.5 cm, slender, sparsely tomentose to ± glabrous; involucres 1.5–2.5(3) mm, 1.3–1.6(2) mm wide, tomentose; teeth 4. **FL:** (2.5)3–4(4.5) mm, hairy; perianth green-yellow to yellow, lobes oblong to oblong-obovate. **FR:** 3–4(4.5) mm, glabrous exc minutely scabrous beak. Sand; 65–200 m. s DSon (Imperial Co.); sw AZ, ne Baja CA, nw SON. Jul–Jan

E. diclinum Reveal (p. 1097) JAYNES CANYON BUCKWHEAT (Group 3) Per mat 3–8 dm diam; dioecious. **ST:** (0.5)1–2 dm, erect, sparsely tomentose. **LF:** basal, blades (0.5)1–2 cm, 0.5–1.5 wide ± densely gray-tomentose. **INFL:** head-like (or umbel-like); branches tomentose to sparsely so; involucres 1–3, staminate 2.5–3 mm, teeth 5–6, pistillate 3–4 mm, teeth 4–5. **FL:** glabrous; staminate 2–3 mm incl 0.5–0.7 mm stipe, pistillate (3.5)5–8 mm incl 0.5–0.8 mm stipe; perianth yellow to brown-yellow, lobes oblanceolate to obovate. **FR:** 3–4 mm, glabrous exc hairy beak. Serpentine; 1700–2400 m. KR (Siskiyou, Trinity cos.); sw OR. Jun–Sep ★

E. douglasii Benth. var. ***meridionale*** Reveal (p. 1097) SOUTHERN WILD BUCKWHEAT (Group 3) Per mat 0.5–6 dm diam. **ST:** 0.4–1.2 dm, tomentose; bracts 4–8, lf-like, near middle. **LF:** basal, blades 0.4–1.5(1.9) cm, 0.1–0.5 cm wide, densely tomentose. **INFL:** head-like, 0.8–1.5 cm wide; branches 0; involucres 2.5–3.5 mm, 2–2.5 mm wide, teeth 6–14, lobe-like, strongly reflexed. **FL:** 4–5(6) mm, stipe 0.7–1.3 mm, hairy; perianth yellow to rose-red, lobes obovate. **FR:** 3–4 mm, glabrous exc hairy tip. Sand or gravel; 1200–2500 m. e KR, e CaRH, n SNH, MP; s-c OR, w-c NV. [*E. d.* var. *d.*, misappl.] Other vars. in n OR, e WA, e NV. Apr–Jun

E. eastwoodianum J.T. Howell EASTWOOD'S BUCKWHEAT (Group 1) Ann 2–5 dm. **ST:** (0.4)0.8–1.5 dm, tomentose. **LF:** basal, blades 1–3 cm, ± round, tomentose abaxially, ± glabrous adaxially. **INFL:** 10–40 cm; branches tomentose; peduncles erect, 1–3.5 cm, straight, slender, tomentose; involucres gen at tips of slender branchlets, not appressed, 2–2.5 mm, 1.5–2 mm wide, tomentose. **FL:** 1.5–2.5(3) mm, glabrous; perianth white, lobes elliptic to oblong-obovate; style 0.1–0.3 mm. **FR:** 1.6–2(2.5) mm, glabrous exc granular tip. *n*=17. Shale; (200)500–1000 m. s SCoRI (Fresno, Monterey cos.). May–Sep ★

E. elatum Benth. (Group 4) Per 4–8(15) dm. **ST:** 1.5–4(8) dm, occ inflated, glabrous or hairy. **LF:** basal, blades 4–15(25) cm, 1.5–6 cm wide, hairy. **INFL:** 15–50 cm; branches glabrous or hairy; peduncles 0 or erect, 0.5–4 cm, slender, glabrous or hairy; involucres 1 or 2–5, (2.5)3–4 mm, 2.5–3 mm wide, glabrous or tomentose. **FL:** 2.5–4 mm, glabrous; perianth white, lobes obovate. **FR:** 3.5–4 mm, glabrous.

var. ***elatum*** (p. 1101) TALL WILD BUCKWHEAT St, infl glabrous. *n*=20. Common. Sand or gravel; 600–3000 m. n CA; to WA, ID, c NV. May–Oct

var. ***villosum*** Jeps. TALL WOOLLY WILD BUCKWHEAT St, infl hairy. Uncommon. Sand or gravel; 500–2000 m. KR, CaR, n SN, MP; s OR, w NV. Jun–Sep

E. elegans Greene (p. 1101) ELEGANT WILD BUCKWHEAT (Group 1) Ann 1–4 dm. **ST:** 0.5–1 dm, glabrous. **LF:** basal; blades 0.3–1.5 cm, gen ± round, gen tomentose. **INFL:** 5–35 cm, glabrous;

peduncles 0; involucres gen appressed, 1–1.5 mm, 0.6–0.8 mm wide. **FL:** 1–1.5 mm; perianth white to rose or ± yellow, lobes gen oblong-obovate. **FR:** 1–1.5 mm, glabrous. Uncommon. Sand or gravel; 200–1200 m. SnFrB, SCoR, WTR. May–Nov

E. elongatum Benth. var. ***elongatum*** (p. 1101) LONG-STEM WILD BUCKWHEAT (Group 4) Per 6–12(18) dm. **ST:** 1–4 dm, gen tomentose. **LF:** cauline; blades 1–3 cm, 0.5–2 cm wide, tomentose. **INFL:** 15–120 cm; branches gen tomentose; peduncles 0; involucres (4)6–7 mm, 2–4 mm wide, gen tomentose. **FL:** 2.5–3 mm, glabrous; perianth white to rose, lobes oblong to obovate. **FR:** 2–3 mm, glabrous. *n*=34. Common. Sand or clay; 60–1900 m. c&s CW, SW; Baja CA. Other vars. in Baja CA. Jul–Nov

E. eremicola J.T. Howell & Reveal (p. 1101) WILDROSE CANYON BUCKWHEAT (Group 2) Ann 0.8–2.5 dm. **ST:** 0.3–1 dm, glandular. **LF:** basal; lf blades 1–2.5 cm, ± round, tomentose abaxially, sparsely tomentose or glabrous adaxially. **INFL:** 5–20 cm; branches glandular; peduncles 0 or reflexed, gen 0.5–1 cm, straight, slender, glandular; involucres 1.8–2 mm, 1–1.5 mm wide, glandular. **FL:** 2–2.5 mm, glabrous; perianth white to red, lobes oblong-ovate. **FR:** 2–2.5 mm, glabrous. Sand or gravel; 2200–3100 m. s W&I (Inyo Mtns), n DMtns (Panamint Range). Jun–Sep ★

E. esmeraldense S. Watson var. ***esmeraldense*** ESMERALDA WILD BUCKWHEAT (Group 2) Ann (0.5)1–5(10) dm. **ST:** 0.5–1(3) dm, glabrous. **LF:** basal; blades 0.5–2.5 cm, 0.4–2 cm wide, gen ± round, hairy, often ± glandular. **INFL:** 5–40(70) cm, 5–50 cm wide; branches glabrous; peduncles spreading to reflexed, 0.2–1.5 cm, straight, gen slender, glabrous; involucres 0.8–1.8 mm, 0.5–1.2 mm wide, glabrous; teeth 4–5. **FL:** 1–2 mm, gen glabrous; perianth white to ± pink or ± red, lobes oblanceolate to oblong. **FR:** 1.4–1.8 mm, glabrous. Locally common. Sand; 1700–3200 m. c SNH (e slope), SNE; NV, c UT. Other var. in c NV. May–Oct

E. evanidum Reveal (p. 1101) VANISHING WILD BUCKWHEAT (Group 1) Ann 1–2 dm. **ST:** 0.3–0.6 dm, gen tomentose. **LF:** basal; blades 0.7–1.2 cm, ± round, tomentose abaxially, hairy adaxially. **INFL:** 5–15 cm; branches gen tomentose; peduncles 0; involucres ± appressed, 1–2 mm, glabrous. **FL:** 0.8–1.2 mm, glabrous; perianth cream, outer lobes ovate, ± hastate in fr, inner lanceolate to elliptic. **FR:** 1.3–1.5 mm, glabrous. Sand; 1100–2100 m. SnBr (Bear Valley), PR (scattered); n Baja CA. [*E. foliosum* S. Watson, misappl.] Possibly extirpated. Jul–Oct ★

E. exaltatum M.E. Jones LADDER WILD BUCKWHEAT (Group 2) Ann (0.5)3–6(10) dm. **ST:** (0.5)1–2 dm, glabrous. **LF:** basal; blades (1.5)2–5(8) cm, ± round, tomentose abaxially, sparsely tomentose adaxially. **INFL:** (5)10–50(80) cm, 10–50 cm wide; branches ± erect, whip-like, glabrous; peduncles 0 or erect, ≤ 0.2 cm, straight, slender, glabrous; involucres 2–2.5(3) mm, 1.5–2.5 mm wide, glabrous. **FL:** 1.5–2 mm, glabrous; perianth white to ± pink, outer lobes oblong, inner lanceolate. **FR:** 2–2.5 mm, glabrous. *n*=40. Sand; 500–1400 m. DMoj (se Inyo Co.); nw AZ, s NV, sw UT. [*E. insigne* S. Watson, misappl.] May–Oct

E. fasciculatum Benth. CALIFORNIA BUCKWHEAT (Group 4) Per mat to shrub, (1)2–15 dm, 2–25(30) dm diam. **ST:** 0.3–2.5(3) dm, tomentose, hairy, or glabrous. **LF:** cauline; blades 0.6–1.5(1.8) cm, 0.05–0.4(0.6) cm wide, gen linear, tomentose, hairy or glabrous, margin gen rolled under. **INFL:** 0.2–20 cm, 0.2–15 cm wide; branches tomentose, hairy, or glabrous; peduncles 0; involucres (1)3–8, 2–4 mm, 1.5–3 mm wide, tomentose, hairy, or glabrous. **FL:** 2.5–3 mm, glabrous or hairy; perianth white to ± pink, lobes gen elliptic to obovate. **FR:** 1.8–2.5 mm, glabrous. Other var. in c Baja CA.

var. ***fasciculatum*** COASTAL CALIFORNIA BUCKWHEAT Pl gen decumbent mat, 1–5 dm, 5–30 dm diam. **ST:** glabrous, ± gray. **LF:** blades 0.6–1(1.2) cm, 0.05–0.2(0.4) cm wide, thinly tomentose abaxially, glabrous adaxially, margin rolled under. **INFL:** gen head-like; branches gen glabrous; involucres 3–4 mm, 1.5–2 mm wide, gen glabrous. **FL:** gen glabrous. *n*=40. Sand; < 300 m. CCo; nw Baja CA. All yr

var. ***flavoviride*** Munz & I.M. Johnst. SONORAN DESERT CALIFORNIA BUCKWHEAT Pl gen rounded shrub, 2–5 dm, 3–6(10) dm diam, thinly hairy or glabrous, yellow-green. **ST:** gen glabrous.

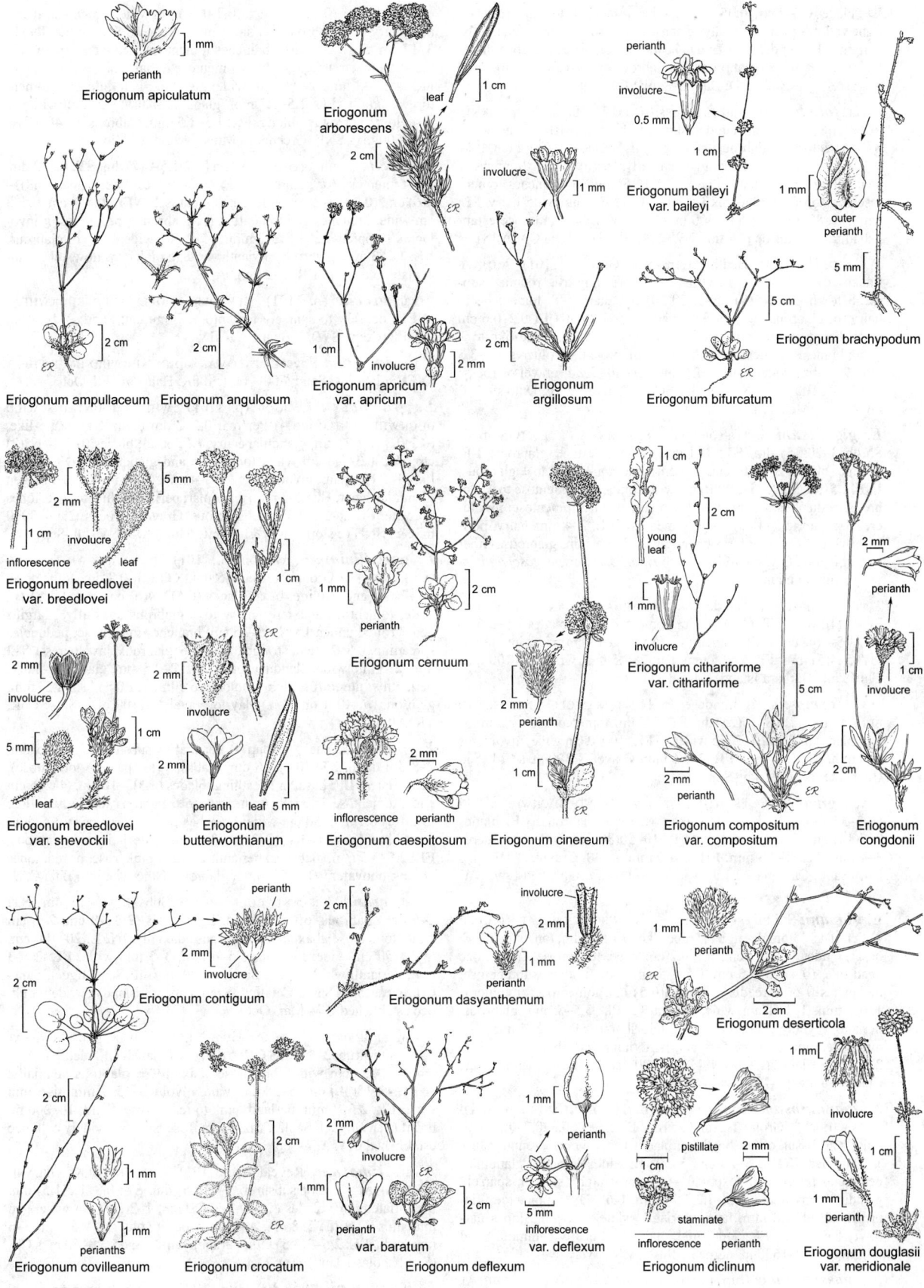

Eriogonum apiculatum
perianth
1 mm

Eriogonum
arborescens
leaf
1 cm
2 cm
ER

Eriogonum baileyi
var. baileyi
perianth
involucre
0.5 mm
1 cm

involucre
1 mm

Eriogonum brachypodum
1 mm
outer
perianth
5 mm
5 cm

Eriogonum ampullaceum
2 cm
ER

Eriogonum angulosum
2 cm

Eriogonum apricum
var. apricum
1 cm
involucre
2 mm

Eriogonum
argillosum
2 cm

Eriogonum bifurcatum
ER

Eriogonum breedlovei
var. breedlovei
2 mm
involucre
5 mm
inflorescence
1 cm
leaf

Eriogonum breedlovei
var. shevockii
2 mm
involucre
5 mm
leaf
1 cm

Eriogonum
butterworthianum
2 mm
involucre
2 mm
perianth
leaf 5 mm
1 cm
ER

Eriogonum cernuum
1 mm
perianth
2 cm

Eriogonum caespitosum
2 mm
inflorescence
2 mm
perianth

Eriogonum cinereum
2 mm
perianth
1 cm
ER

Eriogonum cithariforme
var. cithariforme
1 cm
young
leaf
2 cm
1 mm
involucre
5 cm
2 mm
perianth
ER

Eriogonum compositum
var. compositum

Eriogonum
congdonii
2 mm
perianth
involucre
1 cm
2 cm

Eriogonum contiguum
perianth
2 mm
involucre
2 cm

Eriogonum dasyanthemum
2 cm
involucre
2 mm
1 mm
perianth

Eriogonum deserticola
1 mm
perianth
2 cm
ER

Eriogonum covilleanum
2 cm
1 mm
1 mm
perianths

Eriogonum crocatum
2 cm
ER

var. baratum
2 mm
involucre
1 mm
perianth
2 cm
ER

var. deflexum
1 mm
perianth
5 mm
inflorescence

Eriogonum deflexum

Eriogonum diclinum
pistillate
1 cm
2 mm
staminate
inflorescence
perianth

Eriogonum douglasii
var. meridionale
1 mm
involucre
1 cm
1 mm
perianth

LF: blades 0.6–1 cm, 0.05–0.2 cm wide, tomentose to ± glabrous and light yellow-green abaxially, glabrous or ± so and green adaxially, margin rolled under. **INFL**: gen head-like; branches glabrous; involucres 2–3 mm, gen glabrous. **FL**: glabrous or occ hairy. n=40. Sand or gravel; 50–1300 m. D; Baja CA, nw SON. All yr

var. *foliolosum* (Nutt.) Abrams (p. 1101) LEAFY CALIFORNIA BUCKWHEAT Pl gen rounded shrub, 6–15 dm, (8)10–25 dm wide. **ST**: tomentose or glabrous, gen ± gray. **LF**: blades 0.6–1.2 cm, 0.1–0.4 cm wide, densely tomentose abaxially, less to sparsely tomentose adaxially, margins gen rolled under. **INFL**: open; branches tomentose or glabrous; involucres 3–4 mm, (1.5)2–2.5 mm wide, hairy. **FL**: hairy. n=80. Gravel; 60–1300 m. CW, SW; widely cult, occ established from roadside planting NW, CW, SW, D; n Baja CA. All yr

var. *polifolium* (Benth.) Torr. & A. Gray (p. 1101) MOJAVE DESERT CALIFORNIA BUCKWHEAT Pl spreading to ± rounded subshrub to shrub, 2–5(8) dm 2–20(30) dm diam. **ST**: tomentose to hairy (or glabrous), ± gray. **LF**: blades 0.6–1.8 cm, (0.1)0.2–0.6 cm wide, hairy, margin not or occ rolled under. **INFL**: head-like to ± open; branches tomentose to hairy (or glabrous); involucres 2.5–3.5 mm, 2–3 mm wide, hairy. **FL**: hairy. n=40. Sand, gravel or rocks; (60)300–2500 m. s SN, SCoRI, e SCo, TR, e PR, SNE, D; to sw UT, nw AZ, nw Mex. All yr

E. giganteum S. Watson ST. CATHERINE'S-LACE (Group 4) Shrub 3–20(35) dm. **ST**: 1–4 dm, tomentose to ± glabrous. **LF**: cauline; blades 2–7(10) cm, 1–5 cm wide, tomentose to ± glabrous. **INFL**: 5–30 cm, (2)5–50(80) cm wide; branches tomentose to ± glabrous; peduncles 0 or erect, 0.1–0.5 cm, slender, tomentose; involucres 3–5 mm, 2.5–4 mm wide, tomentose. **FL**: 2–4 mm, hairy; perianth white to rose, lobes obovate. **FR**: 2–3.5 mm, glabrous. Other var. in n Baja CA. Hybridizes with *E. arborescens, E. fasciculatum* where cult on mainland.

var. *compactum* Dunkle SANTA BARBARA ISLAND BUCKWHEAT Pl 4–6(10) dm. **ST**: (1.5)2–3 dm, tomentose. **LF**: blades 2.5–3.5(6) cm, 1.5–2(4) cm wide, tomentose. **INFL**: 2–15 cm wide; involucres (3)3.5–4(5) mm. **FL**: 2–2.5 mm. **FR**: 2–2.5 mm. Gravel; < 100 m. s ChI (Santa Barbara Island). May–Oct ★

var. *formosum* K. Brandegee (p. 1101) SAN CLEMENTE ISLAND BUCKWHEAT Pl 3–15(25) dm. **ST**: 2–4 dm, tomentose to ± glabrous. **LF**: blades 5–8 cm, 1–2 cm wide. **INFL**: 10–50 cm wide; involucres 4–5 mm. **FL**: 3–4 mm. **FR**: 3–3.5 mm. Gravel; < 300 m. s ChI (San Clemente Island). May–Sep ★

var. *giganteum* SANTA CATALINA ISLAND BUCKWHEAT Pl (3)5–35 dm. **ST**: 1–3 dm, tomentose to ± glabrous. **LF**: blades 3–7(10) cm, 2–5 cm wide. **INFL**: 10–50(80) cm wide; involucres 3–4 mm. **FL**: 2–2.5 mm. **FR**: 2–2.5 mm. n=40. Gravel; < 450 m. s ChI (Santa Catalina Island); widely cult, occ established CW, SW. All yr ★

E. gilmanii S. Stokes (p. 1101) GILMAN'S BUCKWHEAT (Group 4) Per mat 1–3 dm diam. **ST**: scape-like, 0.1–0.2 dm, tomentose. **LF**: basal; blades 0.2–0.4 cm, 0.1–0.2 cm wide, tomentose. **INFL**: occ head-like, (0.4)0.8–1.5 cm, 0.5–1.5 cm wide; branches tomentose; peduncles 0 or spreading, 0.1–0.3(0.5) cm, tomentose; involucres 1.5–2 mm, 1–1.5 mm wide, tomentose. **FL**: 3.5–5 mm, glabrous; perianth ± yellow, with dark ± red midrib, outer lobes 3–4 mm, 3–4 mm wide, round, inner 3.5–5 mm, 1–1.5 mm wide, oblanceolate. **FR**: 2.5–3 mm, glabrous. Gravel; 1500–2000 m. n DMtns (Panamint, Last Chance ranges). May–Sep ★

E. glandulosum (Nutt.) Benth. (p. 1101) GLANDULAR WILD BUCKWHEAT (Group 2) Ann (0.5)1–2.5 dm. **ST**: 0.3–0.7 dm, glandular. **LF**: basal, often sheathing; blades 0.5–1.5 cm, ± round, hairy, ± glandular. **INFL**: 5–20 cm, 5–30 cm wide; branches glandular; peduncles reflexed or ± so, 0.01–0.05 cm, straight, slender, sparsely glandular; involucres 0.8–1.2(1.5) mm, 0.6–1(1.3) mm wide, glabrous. **FL**: 1–1.8 mm, hairy; perianth white to ± pink, lobes narrowly lanceolate. **FR**: 1–1.3 mm, glabrous. Locally common. Sand or gravel; 900–1600 m. n&ne DMtns; sw NV. May–Nov

E. gossypinum Curran (p. 1101) COTTONY BUCKWHEAT (Group 2) Ann 0.5–2 dm. **ST**: 0.2–0.5 dm, tomentose. **LF**: basal and

cauline; blades (0.3)0.5–4 cm, (0.2)0.5–1 cm wide, tomentose abaxially, hairy to ± glabrous adaxially, margin occ ± rolled under. **INFL**: 3–17 cm, 5–25 cm wide; branches tomentose; peduncles spreading, 0.2–1.5 cm, straight, hair-like, tomentose to hairy or glabrous; involucres 2.7–3 mm, 2–2.5 mm wide, glandular-hairy, densely tomentose adaxially. **FL**: 1.5–1.7 mm, glandular-hairy; perianth white to rose, lobes narrowly oblong. **FR**: 1.3–1.5 mm, glabrous. n=40. Clay; 100–500 m. s SNF (Greenhorn Mtns), sw SnJV. All yr ★

E. gracile Benth. (Group 1) Ann (0.7)1.5–5(7) dm. **ST**: 0.5–2 dm, gen tomentose (or glabrous). **LF**: basal and cauline; blades (0.8)1–4(6) cm, (0.3)0.5–2 cm wide, gen tomentose. **INFL**: 0.5–4 cm, 0.5–3 cm wide; branches gen tomentose (or glabrous); peduncles 0; involucres ± appressed, (1.8)2–3 mm, 1.5–2 mm wide, hairy or glabrous. **FL**: 1.5–3 mm, glabrous; perianth white to pink or yellow, lobes lanceolate to oblong. **FR**: 1–2 mm. n=22.

var. *gracile* (p. 1101) SLENDER WOOLLY WILD BUCKWHEAT **INFL**: densely to sparsely tomentose. Common. Sand; < 1400 m. CA-FP (exc w SW); n Baja CA. All yr

var. *incultum* Reveal PALOMAR MOUNTAIN WILD BUCKWHEAT **INFL**: glabrous. Sand; 700–1600 m. PR; Baja CA. Jul–Oct

E. gracilipes S. Watson (p. 1101) WHITE MOUNTAINS WILD BUCKWHEAT (Group 4) Per mat 0.5–2 dm diam. **ST**: scape-like, (0.2)0.3–1(1.2) dm, glandular-hairy. **LF**: basal; blades (0.3)1–1.5(2) cm, (0.2)0.3–0.6 cm wide, tomentose and glandular. **INFL**: head-like, 1–2 cm diam; involucres 5–7, 2–4 mm, 2–3 mm wide, thin, gen glandular-hairy. **FL**: 2–3 mm, glandular; perianth white to rose, lobes obovate. **FR**: 2–2.3(2.5) mm, glabrous. Gravel or rocks; 2900–3900 m. c&s SNH (e slope), n W&I (White Mtns); w-c NV. Jul–Sep

E. gracillimum S. Watson (p. 1101) ROSE-AND-WHITE WILD BUCKWHEAT (Group 2) Ann 1–5 dm. **ST**: 0.1–0.8 dm, tomentose. **LF**: basal and cauline; basal blades (0.5)2–4(6) cm, 0.2–1(1.5) cm wide, tomentose abaxially, hairy to ± glabrous adaxially, margins gen ± rolled under. **INFL**: 5–35 cm; branches tomentose; peduncles spreading, 0.8–2.5 cm, straight, hair-like, glabrous; involucres 1.8–2 mm, 2–3 mm wide, glandular-hairy. **FL**: 2–2.5 mm, glandular-hairy; perianth white to rose, lobes oblong to elliptic. **FR**: 1–1.2(1.5) mm, glabrous. n=40. Common. Clay to gravel; < 1100 m. s SN, SCoR, TR. All yr

E. grande Greene (Group 4) Per mat or subshrub, (1)5–15 dm, 2–8 dm diam. **ST**: (0.8)2–6 dm, glabrous (to sparsely tomentose), occ inflated. **LF**: basal or sheathing; blades (1.5)3–10 cm, (1)2–6 cm wide, tomentose abaxially, tomentose to ± glabrous adaxially. **INFL**: (5)20–100 cm, 4–50 cm wide; branches glabrous (or hairy); peduncles 0; involucres 1–3, (4)5–7 mm, 3–6 mm wide, hairy to ± glabrous. **FL**: 2.5–3.5 mm, glabrous; perianth white to pink, rose, or red, lobes oblong-obovate. **FR**: 2.5–3 mm, glabrous. Other var. in n Baja CA.

var. *grande* ISLAND BUCKWHEAT Subshrub 5–15 dm. **ST**: 2–6 dm, glabrous, often inflated. **LF**: blades (2)3–10 cm, 2–6 cm wide, tomentose abaxially, ± glabrous adaxially. **INFL**: 20–100 cm, (5)10–50 cm wide; involucres 5–6 mm, 3–5 mm wide. **FL**: 2.5–3 mm; perianth white. n=20. Sand; < 300 m. ChI (San Miguel, Santa Cruz, Anacapa, Santa Catalina, San Clemente islands); widely cult, occ established CW. Mar–Oct ★

var. *rubescens* (Greene) Munz RED-FLOWERED BUCKWHEAT Per or subshrub, 2–5 dm, occ inflated. **LF**: blades 2–5(6) cm 1–3(4.5) cm wide, tomentose abaxially, ± glabrous adaxially. **INFL**: 5–15(30) cm, 5–20 cm wide; involucres 5–7 mm, 4–6 mm wide. **FL**: 2.5–3 mm; perianth pink to red or rose. Sand; 10–200 m. n ChI (San Miguel, Santa Cruz, Santa Rosa islands); widely cult, occ established CW. Apr–Sep ★

var. *timorum* Reveal (p. 1101) SAN NICOLAS ISLAND BUCKWHEAT Per or subshrub, 1–2(2.5) dm. **ST**: 0.8–1.2(1.5) dm, not inflated. **LF**: blades (1.5)2–3.5(4.5) cm, 1–2(2.5) cm wide, gen tomentose. **INFL**: 5–8 cm, 4–8 cm wide; involucres 4–5 mm, 3–6 mm wide. **FL**: 2.5–3(3.5) mm; perianth white. Sand; 20–60 m. s ChI (San Nicolas Island). Apr–Oct ★

E. heermannii Durand & Hilg. (Group 4) Subshrub or shrub, (0.5)1–20 dm, 2–25 dm diam. **ST**: 0.02–0.5 dm, tomentose or gla-

brous. **LF**: cauline; blades (0.4)1–2(4) cm, 0.1–0.8 cm wide, linear to narrow, gen tomentose or glabrous. **INFL**: 1–25(30) cm, 1–30(35) cm wide; branches smooth or angled to ridged and grooved, glabrous or hairy, occ scabrous; peduncles 0; involucres 0.7–3 mm, 0.7–4 mm wide, glabrous or hairy. **FL**: (1.5)2–4 mm, glabrous; perianth white, yellow-white, pink, or ± red, outer lobes obovate to round, inner narrowly lanceolate to oblong. **FR**: 2–5 mm, glabrous. Other vars. in NV, UT, AZ.

var. ***argense*** (M.E. Jones) Munz HEERMANN'S ROUGH WILD BUCKWHEAT Shrub 1–10 dm, 2–10(12) dm diam. **LF**: 0.5–1.2(1.5) cm, 0.1–0.6 cm wide, tomentose or glabrous. **INFL**: (1)3–20 cm, (1)3–25 cm wide; branches smooth, glabrous or minutely scabrous; involucres 0.9–1.8 mm, 0.7–1.3 mm wide. **FL**: 1.5–4 mm; perianth white to yellow-white or ± red. Gravel or rocks; 800–2800 m. DMtns; AZ, s NV. Apr–Nov

var. ***floccosum*** Munz CLARK MOUNTAIN BUCKWHEAT Shrub 3–6 dm, 4–8 dm diam. **LF**: blades (0.5)0.8–1.5 cm, 0.2–0.5 cm wide, tomentose to hairy. **INFL**: 5–12 cm, 5–20 cm wide; branches smooth, sparsely tomentose to woolly-tufted; involucres 1–1.5(2) mm, 1–1.5 mm wide. **FL**: 2–3 mm; perianth yellow-white. Sand or gravel; 900–2300 m. DMtns; nw AZ, s NV. May–Oct ★

var. ***heermannii*** (p. 1101) HEERMANN'S WILD BUCKWHEAT Shrub 5–15 dm, 6–15(20) dm diam. **LF**: blades 0.5–1.5 cm, 0.2–0.6 cm wide, glabrous to slightly tomentose. **INFL**: (5)10–20 cm, (5)8–25 cm wide; branches smooth, glabrous; involucres 2–2.5 mm. **FL**: (2.5)3–4 mm; perianth yellow-white. Gravel; 900–1700 m. s SNH (Kern Co.), Teh, se SCoRO (se San Luis Obispo Co.), n TR. May–Nov

var. ***humilius*** (S. Stokes) Reveal HEERMANN'S GREAT BASIN WILD BUCKWHEAT Shrub or subshrub, 3–7 dm, 5–12(15) dm diam. **LF**: blades 0.8–1.5 cm, 0.4–0.8 cm wide, hairy or glabrous. **INFL**: 3–15(23) cm, 5–20 cm wide; branches round, smooth, glabrous; involucres 1–1.5 mm, 1.5–3 mm wide. **FL**: 2.5–3 mm; perianth white. Common. Gravel; 1100–2500 m. c&s SNH (e slope), SNE, n DMtns; NV. Jun–Oct

var. ***occidentale*** S. Stokes WESTERN HEERMANN'S BUCKWHEAT Shrub 10–20 dm, 10–25 dm diam. **LF**: blades 1.5–3(4) cm, 0.5–0.8 cm wide, tomentose to hairy, occ glabrous in age. **INFL**: 10–30 cm, 10–35 cm wide; branches round, smooth, glabrous; involucres 2.5–3 mm, 2.5–3.5 mm wide. **FL**: 3–4 mm; perianth white to pink. Gravel bars, steep, clay slopes, often serpentine; (100)400–1000 m. c SCoRI (San Benito, s Monterey cos.). Jul–Oct ★

var. ***sulcatum*** (S. Watson) Munz & Reveal (p. 1101) HEERMANN'S GROOVED WILD BUCKWHEAT Subshrub (0.5)1–8 dm, 2–8 dm diam. **LF**: blades 0.4–1.2(1.5) cm, 0.2–0.8 cm wide, tomentose. **INFL**: 1–5(8) cm, 3–10 cm wide; branches sharply ridged, deeply grooved; involucres 0.7–1.5(2) mm, 0.7–1.5(2) mm wide. **FL**: 1.5–2.5 mm; perianth yellow-white. Gravel or rocks; 700–2700 m. DMtns; to nw AZ, sw UT. Apr–Oct

E. heracleoides Nutt. var. ***heracleoides*** (p. 1101) PARSNIP-FLOWERED BUCKWHEAT (Group 3) Per mat 1–6 dm, 2–10 dm diam. **ST**: (0.5)1–3(4) dm, gen tomentose; bracts (2)5–10, lf-like, near middle. **LF**: basal; blades (1.5)2–5 cm, 0.2–1(1.5) cm wide, gen tomentose. **INFL**: 1–10 cm; branches tomentose to hairy; involucres 3–4.5 mm, 2.5–5(6) mm wide, gen tomentose; teeth 6–12+, reflexed. **FL**: 4–9 mm, stipe 1.5–3 mm, glabrous; perianth white to cream or yellow-white, lobes spoon-shaped to oblong-ovate. **FR**: (2)3.5–5 mm, glabrous exc sparsely hairy tip. Locally common. Sand to gravel; (300)600–3100 m. Wrn; to BC, MT, CO. Other var. in e WA, w-c ID. May–Sep ★

E. hirtellum J.T. Howell & Bacig. (p. 1101) KLAMATH MOUNTAIN BUCKWHEAT (Group 3) Per 1–3.5 dm, 2–4(6) dm diam. **ST**: 0.8–2.5(3) dm, gen glabrous. **LF**: basal; blades 0.5–2(2.5) cm, 0.3–0.8(1.2) cm wide, short hairy. **INFL**: (1.5)2–5 cm wide; branches ± glabrous; involucres 5–6 mm, 2–3 mm wide, hairy; teeth 5–6. **FL**: 3–3.5(4) mm, stipe 0.1–0.2 mm, sparsely hairy; perianth bright yellow, outer lobes 2.5–3 mm wide, spoon-shaped, inner 1 mm wide, oblanceolate. **FR**: 3–3.5 mm, glabrous exc hairy tip. Serpentine; (1100)1300–1700 m. n KR (e Del Norte, Siskiyou cos.). Jul–Sep ★

E. hirtiflorum S. Watson (p. 1101) HAIRY-FLOWER WILD BUCKWHEAT (Group 2) Ann 0.5–1.5 dm. **ST**: 0.2–0.5 dm, glandular. **LF**: basal and cauline; blades (0.1)1–2.5 cm, (0.1)0.4–0.8 cm wide, glabrous. **INFL**: 5–13 cm, 5–25 cm wide; branches glandular; peduncles 0 or erect, 0.1–0.5 cm, straight, thread-like, glandular; involucres 0.8–1 mm, 0.7–0.9 mm wide, sparsely hairy; teeth 4. **FL**: 2, 0.8–1.1 mm, hairy with hooked hairs; perianth pink to ± red, lobes oblong. **FR**: elliptic, 1–1.3 mm, glabrous; tip exserted. Sand or gravel; 200–2000 m. NW, SN, n CW, TR. May–Oct

E. hoffmannii S. Stokes (Group 2) Ann 0.5–10 dm. **ST**: 0.4–4 dm, glabrous. **LF**: basal or sheathing; blades 1–5 cm, 2–8 cm wide, ± round, tomentose abaxially, hairy or glabrous adaxially. **INFL**: 5–70 cm, 10–60 cm wide, glabrous; peduncles 0 or erect, 0.01–0.1 cm, straight, stout; involucres 1–2 mm, 1–1.8 mm wide. **FL**: 1.5–1.8 mm; perianth white to ± red, lobes lanceolate to spoon-shaped or ovate. **FR**: 2 mm, glabrous. *n*=20.

var. ***hoffmannii*** HOFFMANN'S BUCKWHEAT Pl 0.5–5 dm. **LF**: basal; blades 1–4 cm, 2–4 cm wide, margin flat. Gravel; 1000–1700 m. n DMtns (w slope Panamint Range). Jul–Sep ★

var. ***robustius*** S. Stokes ROBUST HOFFMANN'S BUCKWHEAT Pl 4–10 dm. **LF**: basal or sheathing; blades 2–5 cm, 3–8 cm wide, margin wavy. Sand; 100–1700 m. ne DMtns (Black, Funeral mtns). Aug–Nov ★

E. hookeri S. Watson (p. 1101) HOOKER'S WILD BUCKWHEAT (Group 2) Ann 1–6 dm. **ST**: 0.1–0.4 dm, glabrous. **LF**: basal; blades (1)2–6 cm, ± round, tomentose. **INFL**: 5–35 cm, 5–50 cm wide; branches glabrous; peduncles 0; involucres reflexed, 1–2 mm, 1.5–3(3.5) mm wide, glabrous. **FL**: 1.5–2 mm, glabrous; perianth yellow to red-yellow, outer lobes round, inner narrowly ovate. **FR**: 2–2.5 mm, glabrous. *n*=20. Uncommon. Sand or gravel; 1300–2500 m. SNE; to ID, WY, CO, NM. Jun–Oct

E. incanum Torr. & A. Gray FROSTED WILD BUCKWHEAT (Group 3) Per mat 1–4 dm diam; dioecious. **ST**: 0.1–2(2.5) dm, gen tomentose. **LF**: basal; blades 0.5–1.5 cm, 0.3–0.7 cm wide, gen tomentose. **INFL**: head-like, 0.5–2 cm wide, mature pistillate pl branched, 1–3 cm, 1–4 cm wide; branches gen tomentose; involucres 2.5–3 mm, 2–2.5 mm wide, tomentose, teeth 5–8. **FL**: glabrous, stipe 0.5–1 mm; perianth yellow; staminate fl 2–3 mm, lobes ovate; pistillate fl 4–6 mm, lobes oblanceolate, often ± red in fr. **FR**: 3–3.5 mm, glabrous exc sparsely hairy tip. Common. Sand; (1900)2100–4000 m. SNH; w-c NV. Jun–Sep

E. inerme (S. Watson) Jeps. (Group 2) Ann 0.5–3 dm. **ST**: 0.1–0.5 dm, glandular. **LF**: basal and cauline; blades (0.3)1–4(4.5) cm, (0.1)0.4–1.5(1.8) cm wide, glabrous. **INFL**: 5–30 cm; branches glandular; peduncles 0 or erect, 0.1–1 cm, straight, slender, glandular; involucres 1.5–1.9 mm, 1–1.2 mm wide, glabrous or hairy; teeth 4. **FL**: 4–6, 1.2–1.8 mm, hairy with hooked hairs; perianth pink to ± red, lobes oblong. **FR**: elliptic, 1.5–1.9 mm, glabrous; tip not exserted.

var. ***hispidulum*** Goodman GOODMAN'S UNARMED WILD BUCKWHEAT **INFL**: involucres short-stiff hairy. Sand; 800–2100 m. c&s SN, Teh. May–Sep

var. ***inerme*** (p. 1100) UNARMED WILD BUCKWHEAT **INFL**: involucres glabrous. Sand; 600–2200 m. s NCoR, CW, WTR; c ID, disjunct as waif. May–Aug

E. inflatum Torr. & Frém. (p. 1101) DESERT TRUMPET (Groups 2, 4) Per, occ fl 1st yr, 1–10(15) dm. **ST**: (0.2)2–5 dm, glabrous, occ hairy at base, gen inflated. **LF**: basal; blades (0.5)1–2.5(3) cm, ± round, (0.5)1–2(2.5) cm wide, short-hairy. **INFL**: 5–70 cm, 5–50 cm wide; branches occ inflated, glabrous; peduncles erect, 0.5–2(3.5) cm, straight, gen thread-like, glabrous; involucres 1–1.5 mm, 1–1.8 mm, glabrous. **FL**: (1)2–3(4) mm, hairy; perianth yellow, lobes gen ovate. **FR**: light 2–2.5 mm, glabrous. *n*=16. Common. Sand or gravel; -30–1800 m. GB, D; to CO, NM, nw Mex. [*E. i.* var. *deflatum* I.M. Johnst.] All yr

E. intrafractum Coville & C.V. Morton (p. 1105) JOINTED BUCKWHEAT (Group 4) Per 6–15 dm, 0.3–0.9 dm diam. **ST**: 5–12 dm, glabrous, partially hollow, ± inflated, basally jointed, breaking into short-cylindric sections. **LF**: basal; blades 2.5–7 cm, 0.7–2(3)

cm wide, hairy. **INFL:** 1–3 cm, 2–6 cm wide; branches occ breaking, hairy or glabrous; involucres 2.5–3.5 mm, 3–5 mm wide, hairy. **FL:** 1.5–3 mm, hairy; perianth yellow to red-yellow, outer lobes 0.9–1.2 mm wide, oblanceolate, inner 1–1.5 mm wide, widely oblanceolate to fan-shaped. **FR:** 2–2.5 mm, glabrous exc hairy tip. *n*=20. Gravel or rocks; (600)800–1600 m. n DMtns (Grapevine Mtns, Panamint Range). May–Oct ★

E. kelloggii A. Gray (p. 1105) KELLOGG'S BUCKWHEAT (Group 3) Per, 0.5–1 dm, 1–5 dm diam. **ST:** (0.4)0.5–0.8 dm, gen tomentose; bracts 2–4, lf-like, near middle. **LF:** basal; blades 0.4–1 cm, 0.1–0.3(0.4) cm wide, silky-tomentose abaxially. **INFL:** head-like, 0.5–1.5 cm wide; involucres (4)5–6 mm, 2.5–4 mm wide; teeth 6–8. **FL:** 5–7 mm, stipe 0.5–0.8(1) mm, glabrous; perianth ± white to ± pink or rose-red, lobes obovate. **FR:** 4–5 mm, glabrous exc sparsely hairy tip. Serpentine; 1000–1200 m. c NCoRO (Red Mtn, Mendocino Co.). May–Aug ★

E. kennedyi S. Watson (Group 4) Per mat. **ST:** scape-like, (0.5)1.5–4.5(5) dm, glabrous or gen tomentose. **LF:** basal or sheathing; blades 0.2–1(1.2) cm, 0.05–0.4 cm wide, tomentose, margins occ rolled under. **INFL:** head-like, 0.4–1 cm diam; peduncles 0; involucres 3–7, 1.5–4 mm, 1–3.5 mm wide, rigid, glabrous or tomentose. **FL:** 1.5–4 mm, glabrous; perianth white to pink or rose, lobes oblanceolate to elliptic or obovate to ± oval. **FR:** 1.8–4 mm, glabrous.

var. *alpigenum* (Munz & I.M. Johnst.) Munz & I.M. Johnst. (p. 1105) SOUTHERN ALPINE BUCKWHEAT Pl 0.2–0.5 dm, 1–4 dm diam. **ST:** 0.5–2(3) cm, tomentose. **LF:** blades 0.2–0.4 cm, 0.07–0.15 cm wide, red- or brown-white-tomentose, margin gen rolled under. **INFL:** involucres 1.5–2 mm, glabrous or tomentose. **FL:** (1.5)2–2.5 mm. **FR:** 1.8–2 mm. Gravel; 2500–3500 m. SnGb, SnBr. Jul–Aug ★

var. *austromontanum* Munz & I.M. Johnst. (p. 1105) SOUTHERN MOUNTAIN BUCKWHEAT Pl 0.8–1.5 dm, 1.5–3.5 dm diam. **ST:** 8–15 cm, sparsely tomentose. **LF:** blades (0.4)0.6–1(1.2) cm, 0.1–0.2 cm wide, gray-white-tomentose, margin flat. **INFL:** involucres 2.5–4 mm, tomentose. **FL:** 2–3 mm. **FR:** 3.5–4 mm. Gravel; 2000–2200 m. n WTR (Mount Pinos), SnBr (Bear Valley). Jun–Aug ★

var. *kennedyi* KENNEDY'S WILD BUCKWHEAT Pl 0.4–1.2 dm, 1–3 dm diam. **ST:** 4–12 cm, glabrous. **LF:** blades 0.2–0.4(0.5) cm, 0.05–0.15(0.2) cm wide, gray- to brown-white-tomentose, margin occ rolled under. **INFL:** involucres 1.5–2.5 mm, glabrous or sparsely tomentose. **FL:** 1.5–2.5 mm. **FR:** 2–2.5 mm. Gravel; 1700–2700 m. n WTR (Mount Pinos), SnBr. Apr–Jul

var. *pinicola* Reveal KERN BUCKWHEAT Pl 0.5–1.3 dm, 1–3 dm diam. **ST:** 5–13 cm, glabrous. **LF:** blades 0.3–0.5 cm, 0.1–0.4 cm wide, gray- or red-white-tomentose, margin flat. **INFL:** involucres 2.5–3.5 mm, sparsely tomentose. **FL:** 2–4 mm. **FR:** 2.5–3 mm. Gravel; 1700–1800 m. se SNF (s of Cache Peak, se Kern Co.). May–Jun ★

var. *purpusii* (Brandegee) Reveal PURPUS' WILD BUCKWHEAT Pl 0.4–1.2 dm, 1–3 dm diam. **ST:** 0.4–1 cm, glabrous (or hairy). **LF:** blades (0.25)0.3–0.6 cm, 0.15–0.35 cm wide, white-tomentose, margin flat. **INFL:** involucres 1.5–2 mm, glabrous or sparsely tomentose. **FL:** 2–2.5 mm. **FR:** 2.5–3 mm. Gravel; 1500–2500 m. c&s SN (e slope), SNE, nw DMtns (Argus, Coso ranges); w-c NV. May–Jul

E. latens Jeps. INYO WILD BUCKWHEAT (Group 3) Per (1.5)2.5–4.5(5) dm, 1–2 dm diam. **ST:** 1–4 dm, gen glabrous. **LF:** basal; blades 1–3(3.5) cm, 0.8–2.5 cm wide, short hairy. **INFL:** head-like, 2–3.5(4) cm diam; branches 0; involucres gen 2–5, 6–8 mm, hairy; teeth 5–8. **FL:** 3–6 mm, stipe 0.1–0.2 mm, sparsely hairy; perianth cream to pale yellow, lobes obovate to spoon-shaped. **FR:** 3–5 mm, glabrous. Gravel; 2600–3400 m. c&s SNH (e slope), W&I, n DMtns (Panamint Range); w-c NV. Jun–Aug

E. latifolium Sm. (p. 1105) SEASIDE WILD BUCKWHEAT (Group 4) Subshrub or per mat, 2–7 dm, 5–20 dm diam. **ST:** 2–6 dm, tomentose to (glabrous). **LF:** cauline; blades (1.5)2.5–5 cm, 1.5–4 cm wide, gen tomentose, occ glabrous adaxially. **INFL:** head-like or branched, 3–40 cm, 2–20 cm wide, tomentose to (glabrous); peduncles 0; involucres (3)5–20, 3.5–5(6) mm, 2–4 mm wide. **FL:** 3–3.5

mm, glabrous; perianth white to pink or rose, lobes obovate. **FR:** 3.5–4 mm, glabrous. *n*=20. Common. Sand; < 80 m. NCo, n&c CCo; to OR. All yr

E. libertini Reveal (p. 1105) DUBAKELLA MOUNTAIN BUCKWHEAT (Group 3) Per mat 0.5–2 dm, 1–4 dm diam. **ST:** 0.5–1.5(1.8) dm, hairy; bracts 3, lf-like, near middle. **LF:** basal; blades 0.5–1.5(2) cm, 0.3–0.5(0.8) cm wide, gen tomentose. **INFL:** head-like, 1–1.5 cm diam; involucres 4–8 mm, 5–8 mm wide, tomentose; teeth 5–8. **FL:** 5–8 mm, stipe 1–1.5 mm, sparsely hairy; perianth sulphur yellow, lobes spoon-shaped to oblong. **FR:** 4–5 mm, glabrous exc sparsely hairy tip. Serpentine; 1100–1600 m. s KR, n NCoRH, n NCoRI (Trinity, Shasta, Tehama cos.). Jun–Aug ★

E. lobbii Torr. & A. Gray (p. 1105) LOBB'S WILD BUCKWHEAT (Group 3) Per, occ mat, 0.3–3 dm, 1–2.5 dm diam. **ST:** gen decumbent, 0.5–1.5(2) dm, gen tomentose. **LF:** basal; blades 1–4(5) cm, gen ± round, tomentose abaxially, occ glabrous adaxially. **INFL:** 1–4 cm; branches gen tomentose; involucres 5–10(12) mm, gen tomentose; teeth 6–10, gen reflexed. **FL:** 5–7 mm, stipe 0.1–0.4 mm, glabrous; perianth white to rose, lobes oblong-obovate. **FR:** 4.5–6 mm, glabrous. *n*=20. Common. Sand or gravel; (1000)1600–3800 m. NW, CaR, SN, GB; sw OR, w NV. Jun–Aug

E. luteolum Greene (Group 1) Ann 0.5–6 dm. **ST:** 0.2–2 dm, glabrous or occ tomentose. **LF:** basal or basal and cauline; blades 0.5–5 cm, 0.5–3.5 cm wide, tomentose abaxially, hairy or glabrous adaxially. **INFL:** 2–50 cm, 3–40 cm wide; branches glabrous or occ tomentose; peduncles 0 (or erect), 0.1–0.5 cm, straight, slender, glabrous; involucres 2–4(4.5) mm, (0.8)1.5–2.5 mm wide, sessile or at tips of slender branchlets, glabrous (or hairy). **FL:** 1–2.5 mm, glabrous; perianth white to rose or yellow, lobes obovate. **FR:** 1–2 mm, glabrous.

var. *caninum* (Greene) Reveal (p. 1105) TIBURON BUCKWHEAT Pl 0.5–3 dm, glabrous. **ST:** gen prostrate to weakly erect, 0.2–1 dm. **LF:** basal and cauline; blades wide. **INFL:** 3–30 cm; involucres 3–4 mm, sessile. **FL:** 1.5–2.5 mm; perianth white to rose. **FR:** 1.4–1.6 mm. *n*=12. Serpentine; < 700 m. c NCoRI (Colusa Co.), n CCo, n SnFrB (Marin, Alameda cos.). May–Oct ★

var. *luteolum* (p. 1105) GOLDEN-CARPET WILD BUCKWHEAT Pl (2)3–5(6) dm, glabrous or occ tomentose. **ST:** erect, 0.5–1 dm. **LF:** basal and cauline; blades gen ± round. **INFL:** 10–45 cm; involucres 3–3.5 mm, sessile. **FL:** 1.8–2 mm; perianth white to rose or yellow. **FR:** 1.8–2 mm. *n*=12. Common. Serpentine; 50–1600 m. NW, CaR, SnFrB. Jul–Nov

var. *pedunculatum* (S. Stokes) Reveal MOKELUMNE HILL WILD BUCKWHEAT Pl 3–6 dm, glabrous. **ST:** erect, 0.5–1 dm. **LF:** basal or basal and cauline; blades wide. **INFL:** 10–50 cm; involucres 3–3.5 mm, sessile (or on erect peduncles 1–5 mm). **FL:** 1–1.5(1.8) mm; perianth white. **FR:** 1–1.4 mm. Serpentine; 100–1500 m. SNF. Jun–Oct

var. *saltuarium* Reveal JACK'S WILD BUCKWHEAT Pl 2–4 dm, glabrous. **ST:** erect, 0.2–2 dm. **LF:** basal or basal and cauline; blades ± round. **INFL:** 2–35 cm; involucres 2–3 mm, sessile. **FL:** 2–2.5 mm; perianth white to rose. **FR:** 1.8–2 mm. Granitic sand; 1700–2400 m. n&c SNH (Alpine, Tuolumne cos.). Jul–Sep ★

E. maculatum A. Heller (p. 1105) SPOTTED WILD BUCKWHEAT (Group 2) Ann 1–2(3) dm. **ST:** 0.1–0.5 dm, tomentose. **LF:** basal and cauline; blades 0.5–3(4) cm, 0.3–1.5(2) cm wide, gen tomentose. **INFL:** 5–25 cm, 10–30 cm wide; branches tomentose; peduncles spreading, (0.5)1–3 cm, straight or ± so, thread-like, glandular-hairy; involucres 1–1.5(2) mm, 1.5–3(3.5) mm wide, glandular-hairy. **FL:** 1–2.5 mm, glandular-hairy; perianth white to yellow, pink or red, with a rose-purple spot, outer lobes elliptic to ± round, inflated base to middle, inner lanceolate. **FR:** 1–1.5 mm, glabrous. *n*=20. Common. Sand to gravel; 100–2500 m. s SN, TR, GB, D; to e WA, UT, AZ, n Baja CA. Apr–Nov

E. marifolium Torr. & A. Gray (Group 3) Per mat 2–8 dm diam; dioecious. **ST:** (0.1)0.5–4 dm, hairy or glabrous. **LF:** basal; blades 0.3–3 cm, 0.3–1 cm wide, tomentose abaxially, ± glabrous adaxially. **INFL:** head-like, 0.5–2 cm wide, mature pistillate pl 1–5 cm, 1–7 cm wide; branches gen glabrous; involucres 2–5 mm, 1.5–7 mm

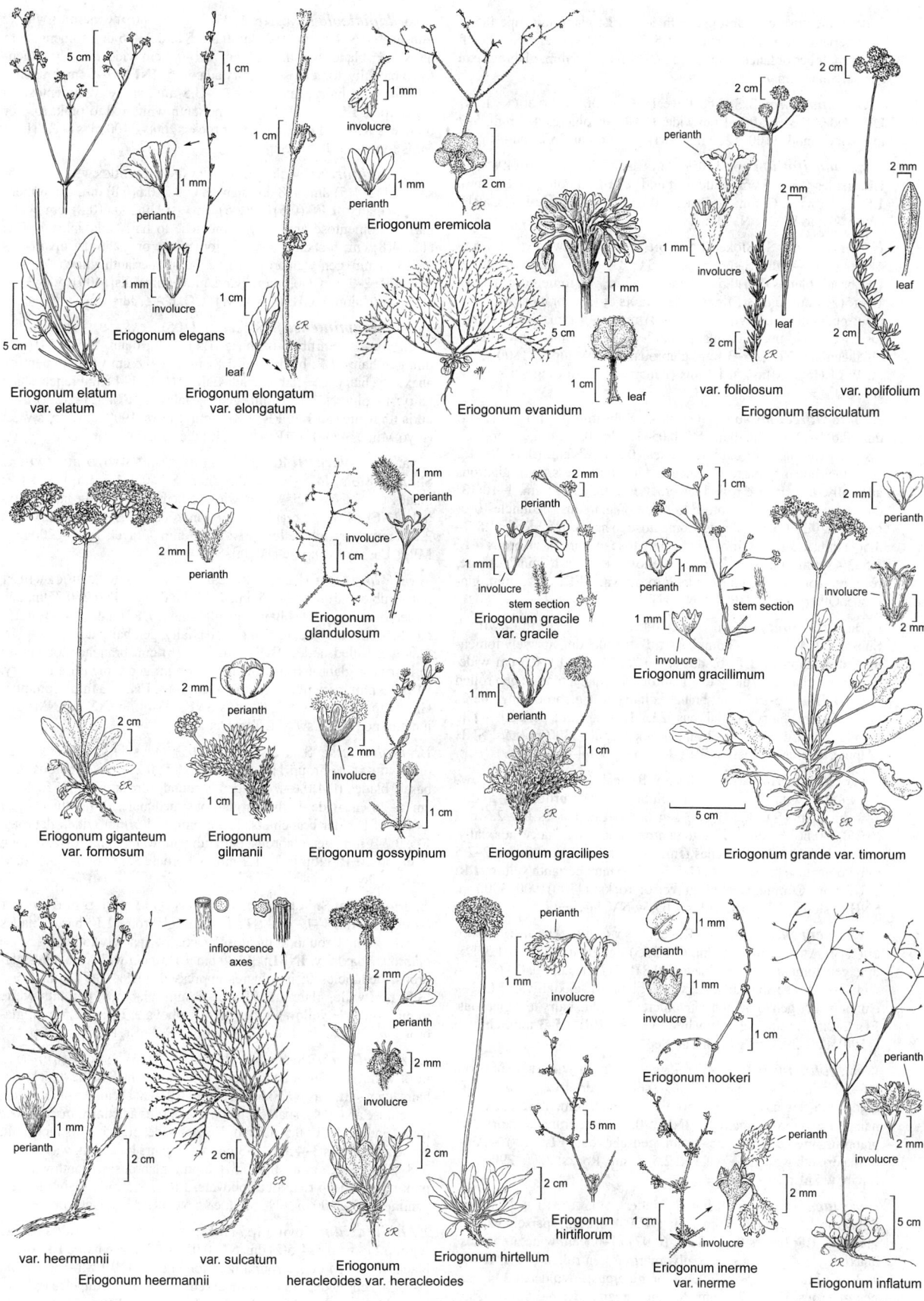

wide, tomentose or glabrous; teeth 5–6. **FL:** glabrous, stipe 0.5–1 mm; perianth yellow; staminate fl 1.5–3 mm, lobes ovate; pistillate fl 4–7 mm, lobes oblanceolate, ± red in fr. **FR:** 3.5–5 mm, glabrous exc sparsely hairy tip. *n*=32.

var. ***cupulatum*** (S. Stokes) Reveal   MCCLOUD WILD BUCKWHEAT **LF:** blades 1–3 cm, 0.5–1 cm wide. **INFL:** involucres 4–5 mm, 5–7 mm wide. Sand; 1100–1200 m. CaRH (s Siskiyou Co.). Jun–Aug

var. ***marifolium*** (p. 1105)   MARUM-LEAF WILD BUCKWHEAT **LF:** blades 0.3–1.5 cm, wide to round. **INFL:** involucres 2–3 mm, 1.5–2 mm wide. Common. Sand; (900)1100–3100 m. e KR, CaRH, n&c SNH; to WA, w NV. Jun–Aug

E. mensicola S. Stokes   PINYON MESA BUCKWHEAT   (Group 4) Per (1)1.5–3 dm, 1–2 dm diam. **ST:** 0.5–1(1.3) dm, tomentose. **LF:** basal; blades (0.3)0.5–1.5 cm, ± round, gen tomentose. **INFL:** 10–15(20) cm, 0.6–3(5) cm wide, bracts lf-like; branches tomentose; peduncles 0; involucres 2–3(4) mm, 2–4 mm wide, tomentose. **FL:** 3–4 mm, glabrous; perianth white to white-brown, lobes oblanceolate. **FR:** 2.5–3 mm, glabrous. Rocky slopes; 1800–2700 m. W&I (Inyo Mtns), n DMtns (Panamint, Coso ranges); sw NV. Jul–Oct ★

E. microthecum Nutt.   (Group 4) Subshrub or shrub, 0.2–1.5 dm, (0.6)1–13(16) dm diam. **ST:** 0.05–1.5 dm, 0.1–1 dm, gen tomentose or glabrous. **LF:** cauline, blades (0.3)1–3.5 cm, (0.07)0.1–1.2 cm wide, linear to narrow, tomentose abaxially, less so or glabrous adaxially, margin occ rolled under. **INFL:** 0.5–6(12) cm, 1–10(13) cm wide; branches tomentose to hairy or glabrous; peduncles 0 or gen erect, 0.3–1.5 cm, slender, tomentose to hairy; involucres (1.5)2–3.5(4) mm, 1.3–2.5(3) mm wide, tomentose to hairy or glabrous. **FL:** 1.5–3(4) mm, glabrous; perianth yellow or white to pink, orange, rose, red (or cream), lobes oblong to obovate. **FR:** 1.5–3 mm, glabrous. Other vars. in OR, ID, NV, UT.

var. ***alpinum*** Reveal   NORTHERN LIMESTONE BUCKWHEAT Subshrub 0.4–1 cm, 1–3 dm diam. **ST:** 0.1–0.3 dm, sparsely tomentose to ± glabrous. **LF:** blades 0.3–0.7(0.9) cm, 0.1–0.3 cm wide, tomentose abaxially, hairy to ± glabrous adaxially, margin gen rolled under. **INFL:** 0.5–2(3) cm; branches hairy to ± glabrous; involucres (1.5)2–3 mm, hairy or glabrous. **FL:** 1.5–2.5 mm; perianth white to pink or rose. **FR:** 1.5–2 mm. Rocks; 2500–3300 m. n&c SNH (Alpine, Tuolumne cos.), n SNE (Mono Co.). Jul–Sep ★

var. ***ambiguum*** (M.E. Jones) Reveal (p. 1105)   YELLOW-FLOWERED WILD BUCKWHEAT   Subshrub or shrub, 0.5–5 dm, 1–8 dm diam. **ST:** 0.2–1 dm, gen hairy. **LF:** blades 0.8–2.5 cm, (0.2)0.3–0.6(0.8) cm wide, tomentose abaxially, hairy adaxially. **INFL:** 1–5(12) cm; branches tomentose to hairy; involucres 2–2.5 mm, tomentose to hairy. **FL:** (1.5)2–2.5(3) mm; perianth yellow. **FR:** 1.5–2 mm. Common. Sand, gravel or rocks; (1100)1900–3300 m. SNH (e slope), GB; to se OR, sw ID, w NV. Jul–Sep

var. ***corymbosoides*** Reveal   SAN BERNARDINO WILD BUCKWHEAT   Shrub 3–6 dm, 6–12(15) dm diam. **ST:** 0.5–1.3(1.5) dm, gen tomentose. **LF:** blades (0.8)1–2(2.5) cm, (0.4)0.6–1 cm wide, tomentose abaxially, hairy to ± glabrous adaxially. **INFL:** 1–4 cm; branches gen tomentose; involucres 2–3 mm, hairy to ± glabrous. **FL:** 2–2.5(3) mm; perianth white to ± red. **FR:** 2.5–3 mm. Gravel; (1500)1800–2900 m. e SnBr. Jul–Sep

var. ***johnstonii*** Reveal   JOHNSTON'S BUCKWHEAT   Subshrub 0.6–1.3 dm, 2–5 dm diam. **ST:** 0.3–0.6 dm, hairy to ± glabrous. **LF:** blades 0.5–1 cm, (0.2)0.3–0.5(0.6) cm wide, tomentose abaxially, hairy to ± glabrous adaxially. **INFL:** 0.5–3 cm; branches hairy to ± glabrous; involucres (2)2.5–3 mm, gen glabrous. **FL:** (2.5)3–3.5(4) mm; perianth white to ± red. **FR:** 2.5–3 mm. Rocks; 2600–2900 m. e SnGb, w SnBr. Jul–Sep ★

var. ***lacus-ursi*** Reveal & A.C. Sanders   BEAR LAKE BUCKWHEAT Subshrub 1.5–2 dm, 4–6 dm diam. **ST:** 0.4–0.8 dm, sparsely hairy or glabrous. **LF:** blades 0.7–1.5 cm, (0.07)0.1–0.3 cm wide, tomentose abaxially, gen glabrous adaxially, margins gen rolled under. **INFL:** 1–3 cm; branches sparsely hairy or glabrous; involucres 3–4 mm, gen glabrous. **FL:** 2–2.5 mm; perianth cream. **FR:** 2–2.5 mm. Clay; 2000–2100 m. SnBr (Bear Valley). Jul–Aug ★

var. ***lapidicola*** Reveal (p. 1105)   INYO MOUNTAINS BUCKWHEAT Subshrub 0.5–1.5 dm, 1–3 dm diam. **ST:** 0.2–0.6 dm, tomentose to hairy. **LF:** blades 0.3–0.7(0.8) cm, 0.1–0.4 cm wide, densely tomentose abaxially, tomentose to hairy adaxially. **INFL:** 2–6 cm; branches tomentose to hairy; involucres 2.5–3.5 mm, sparsely tomentose to ± glabrous. **FL:** (1.5)2–3.5 mm; perianth white-red to pink, rose or orange. **FR:** 2.5–3 mm. Gravel or rocks; 2600–3100 m. s W&I (Inyo Mtns); to w-c UT. Jul–Sep ★

var. ***laxiflorum*** Hook.   GREAT BASIN WILD BUCKWHEAT   Subshrub (1)2–4(5) dm, 3–8 dm diam. **ST:** 0.2–0.6(0.8) dm, gen tomentose to hairy. **LF:** (0.5)1–2(2.5) cm, (0.1)0.2–0.6(0.8) cm wide, narrow, tomentose abaxially, tomentose to hairy adaxially. **INFL:** (1)2–4(8) cm; branches sparsely tomentose or glabrous; involucres 2–3(3.5) mm, gen ± glabrous. **FL:** 2–3 mm; perianth white to pink or rose. **FR:** 2–3 mm. Common. Sand, gravel or rocks; (400)1500–3200 m. SNH (e slope), GB; to WA, MT, CO, n AZ. Jun–Oct

var. ***panamintense*** S. Stokes (p. 1105)   PANAMINT MOUNTAINS BUCKWHEAT   Shrub 3–6 dm, 5–12(15) dm diam. **ST:** 0.5–1(1.5) dm, gen hairy. **LF:** blades 0.6–1.8 cm, 0.3–0.8 cm wide, tomentose abaxially, hairy to ± glabrous adaxially. **INFL:** 1–4 cm; branches gen hairy; involucres 2–2.5 mm, gen ± glabrous. **FL:** 1.5–2(2.5) mm; perianth white to ± red. **FR:** 1.8–2 mm. Rocks; 1900–2800 m. s W&I (Inyo Mtns), n DMtns (Panamint Range). Jul–Oct ★

var. ***schoolcraftii*** Reveal   SCHOOLCRAFT'S WILD BUCKWHEAT Shrub 3–8 dm, (4)5–10(12) dm diam. **ST:** 0.5–1.5 dm, hairy. **LF:** blades 1.5–3.5 cm, 0.5–1.2 cm wide, tomentose abaxially, hairy adaxially. **INFL:** 1–8 cm; branches hairy; involucres 2.5–4 mm, hairy. **FL:** 2.5–3 mm; perianth yellow. **FR:** 2.5–3 mm. Gravel; 1400–2200 m. s MP (s Lassen Co.); w-c NV. Jul–Sep ★

var. ***simpsonii*** (Benth.) Reveal   SIMPSON'S WILD BUCKWHEAT Subshrub or shrub, (1)2–15 dm, 4–16 dm diam. **ST:** 0.2–0.7 dm, gen tomentose, sometimes hairy (or glabrous). **LF:** blades 0.5–1.8(2.5) cm, 0.1–0.2 cm wide, tomentose abaxially, gen hairy adaxially, margins gen rolled under. **INFL:** (1.5)2–4(6) cm; branches tomentose to hairy (or glabrous); involucres 2–3 mm, gen tomentose to hairy. **FL:** 2–3 mm; perianth white to pink or rose. **FR:** 2–3 mm. Common. Gravel to rocks; 1400–2300 m. s SNE, DMoj; to CO, nw NM; disjunct in ne ID and sw MT. Jun–Oct

E. mohavense S. Watson (p. 1105)   WESTERN MOJAVE WILD BUCKWHEAT   (Group 1) Ann 1–3 dm. **ST:** 0.2–1 dm, glabrous. **LF:** basal; blades (0.4)0.6–2 cm, gen ± round, tomentose. **INFL:** 5–25 cm, 5–20 cm wide; branches glabrous; peduncles 0; involucres gen at tips of slender branchlets, 1.7–2 mm, 1–1.5 mm wide, glabrous. **FL:** 0.7–1 mm, gen glabrous; perianth yellow, lobes narrowly oblong to elliptic. **FR:** elliptic, 1–1.2 mm, glabrous. Sand; 600–1200 m. w DMoj. May–Sep

E. molestum S. Watson (p. 1105)   PINELAND WILD BUCKWHEAT (Group 1) Ann 4–10 dm. **ST:** 1–4 dm, glabrous. **LF:** basal; blades (0.5)1–4 cm, ± round, tomentose abaxially, sparsely tomentose to ± glabrous adaxially. **INFL:** 30–80 cm 10–50 cm wide; branches glabrous; peduncles 0; involucres appressed, (3.5)4–5(7) mm, 2.5–3 (3.5) mm wide, glabrous. **FL:** 1.5–3 mm, glabrous; perianth white to pink (or pale yellow), lobes oblong-obovate. **FR:** 2–2.5 mm, glabrous. *n*=20. Sand; 1100–2200 m. n WTR, SnBr, PR. May–Sep

E. nervulosum (S. Stokes) Reveal (p. 1105)   SNOW MOUNTAIN BUCKWHEAT   (Group 3) Per mat 1–3(5) dm diam. **ST:** 0.2–0.6(1) dm, hairs long, soft, wavy. **LF:** basal; blades 0.4–0.8(1) cm, 0.5–1 cm wide, tomentose abaxially, sparsely hairy or glabrous adaxially, margins occ rolled under. **INFL:** 0.5–1.5 cm, 1–2 cm wide; branch hairs long, soft, wavy; involucres 3–4 mm, 2–3 mm wide, hairs long, soft, wavy; teeth 6–8. **FL:** 3.5–5.5 mm, stipe 0.5–0.8 mm, glabrous; perianth white to pink-rose or deep red, lobes obovate. **FR:** 4.5–5 mm, glabrous. Serpentine; 400–2100 m. s NCoRH, c&s NCoRI. May–Oct ★

E. nidularium Coville (p. 1105)   BIRDNEST WILD BUCKWHEAT (Group 1) Ann 0.5–1.5(2) dm. **ST:** 0.05–0.3 dm, gen hairy. **LF:** basal; blades 1–2(2.5) cm, ± round, gen tomentose. **INFL:** 5–15(20) cm, 5–20(30) cm wide; branches incurved, gen hairy; peduncles 0; involucres appressed, 1–1.3 mm, 0.5–0.7 mm wide, hairy. **FL:** 1–3(3.5) mm, glabrous; perianth pale yellow or yellow to red, outer lobes

widely fan-shaped, inner oblanceolate. **FR**: 1–1.3 mm, glabrous. Common. Sand or gravel; (300)500–2300 m. GB, D; to se OR, sw ID, sw UT, nw AZ. Mar–Oct

E. nortonii Greene (p. 1105) PINNACLES BUCKWHEAT (Group 1) Ann 0.5–2 dm. **ST**: gen prostrate, 0.1–0.3 dm, glabrous. **LF**: basal and cauline; blades 0.5–1.5 cm, ± round, tomentose abaxially, gen glabrous. **INFL**: 4–18 cm, 10–30 cm wide; branches glabrous; peduncles 0; involucres gen at tips of slender branchlets, 3–4 mm, 2.5–3.5 mm wide, glabrous; teeth 6–8. **FL**: 1–2 mm, glabrous; perianth white to rose, lobes obovate. **FR**: 1–1.3 mm. Sand; 300–1200 m. n SCoR (Gabilan, e Santa Lucia ranges, San Benito, Monterey cos.). May–Aug ★

E. nudum Benth. (Group 4) Per (0.5)1–15(20) cm, 0.5–3 dm diam. **ST**: 0.3–4(10) dm, glabrous or tomentose to hairy, occ inflated. **LF**: basal or sheathing up st; blades 1–6 cm, (0.3)1–4 cm wide, tomentose abaxially, tomentose to hairy or glabrous adaxially. **INFL**: rarely head-like, 2–100(150) cm, 2–40(80) cm diam; branches glabrous or tomentose to hairy; peduncles 0; involucres 1–10, (2.5)3–5(7) mm, (1.5)2–4 mm wide, glabrous or tomentose to hairy; teeth 5–8. **FL**: (1.5)2–4 mm, glabrous or hairy; perianth white or yellow to pink or rose, lobes oblong to obovate. **FR**: 1.5–3.5 mm, glabrous.

var. **auriculatum** (Benth.) Jeps. (p. 1105) EAR-SHAPED WILD BUCKWHEAT Pl 5–15(20) dm. **ST**: 2–5(10) dm, glabrous (or tomentose), occ inflated. **LF**: sheathing; blades (1)3–8 cm, (1)2–4 cm wide, tomentose abaxially, gen ± glabrous adaxially. **INFL**: 30–100(150) cm, 10–80 cm wide; branches glabrous (or tomentose); involucres (2)3–5, 3–4 mm, glabrous or sparsely hairy. **FL**: 2.5–3 mm, glabrous; perianth white to pink (or ± yellow). *n*=40. Common. Sand or gravel; < 1200 m. NW, CW. May–Oct

var. **decurrens** (S. Stokes) M.L. Bowerman (p. 1105) BEN LOMOND BUCKWHEAT Pl 5–12(15) dm. **ST**: 3–6 dm, tomentose, gen inflated. **LF**: sheathing; blades 1–3 cm, 1–1.5 cm wide, tomentose abaxially, thinly hairy or glabrous adaxially. **INFL**: 50–100 cm, 30–60 cm wide; branches tomentose; involucres 1–2, 4–6 mm, tomentose. **FL**: 3–4 mm, glabrous; perianth white. Sand; 90–200 m. s SnFrB (Santa Cruz Sandhills, Santa Cruz Co.). Jul–Oct ★

var. **deductum** (Greene) Jeps. (p. 1105) REDUCED WILD BUCKWHEAT Pl 2–3(5) dm. **ST**: 0.5–1.5 dm, glabrous. **LF**: basal; blades 1–2 cm, 0.3–0.8(1) cm wide, tomentose abaxially, hairy or glabrous adaxially. **INFL**: 5–15 cm, 5–15(20) cm wide; branches glabrous; involucres 1(2), 2.5–3.5 mm, gen glabrous. **FL**: 2–3 mm, glabrous; perianth white. *n*=20. Common. Sand or gravel; (1100)1500–3000 m. SNH, GB; w NV. Intergrades with *E. nudum* var. *n*. Jun–Sep

var. **indictum** (Jeps.) Reveal PROTRUDING BUCKWHEAT Pl 5–8(10) dm. **ST**: 2.5–3.5 dm, glabrous, inflated. **LF**: sheathing; blades 1–6 cm, 1–2(3) cm wide, tomentose abaxially, tomentose to sparsely so adaxially. **INFL**: 25–50 cm, 15–25 cm wide; branches glabrous; involucres 4–5 mm, glabrous. **FL**: 2.5–3 mm, glabrous; perianth white to yellow. *n*=40. Clay; 100–1100 m. w edge SnJV, e SCoRI. May–Oct ★

var. **murinum** Reveal (p. 1105) MOUSE BUCKWHEAT Pl 3–6 dm. **ST**: 2–3 dm, glabrous. **LF**: sheathing; blades 1.5–3.5 cm, 1–2 cm wide, tomentose. **INFL**: 15–30 cm, 10–20 cm wide; branches glabrous; involucres 5–10, (4)5–6 mm, glabrous. **FL**: 3–4 mm, hairy; perianth white. Sand; 400–700 m. s SNF (Kaweah River drainage, Tulare Co.). May–Oct ★

var. **nudum** (p. 1105) NAKED WILD BUCKWHEAT Pl 3–10 dm. **ST**: (1.5)2–4 dm, glabrous. **LF**: basal; blades 1–5 cm, 0.5–2 cm wide, tomentose abaxially, thinly hairy or glabrous adaxially. **INFL**: 20–50 cm; branches glabrous; involucres 1–5, 3–5 mm, glabrous or sparsely hairy. **FL**: 2–3 mm, glabrous; perianth white (or yellow). *n*=20. Common. Sand or gravel; 10–2100 m. NW, CaR, SN, SnFrB; to s WA. Not on immediate coast. Jun–Sep

var. **oblongifolium** S. Watson HARFORD'S WILD BUCKWHEAT Pl 5–10(18) dm. **ST**: 2–5(8) dm, tomentose to hairy. **LF**: basal; blades 2–4 cm, 1.5–2 cm wide, tomentose abaxially, thinly hairy adaxially. **INFL**: 20–50(100) cm, 10–50 cm wide; branches tomentose to hairy; involucres 3–6, 3–5 mm, tomentose. **FL**: 3–4 mm, hairy; perianth white to rose or yellow. *n*=20. Common. Sand or gravel; 20–1900 m. NW, CaR, n SN; s OR, w NV. May–Oct

var. **paralinum** Reveal (p. 1105) DEL NORTE BUCKWHEAT Pl 0.5–2 dm. **ST**: 0.3–1.2 dm, glabrous. **LF**: basal; blades 1–2 cm, 0.5–1.5 cm wide, tomentose abaxially, glabrous adaxially. **INFL**: 5–20 cm; branches glabrous; involucres 5–10, 3.5–5 mm, glabrous. **FL**: 3–3.5 mm, glabrous; perianth white. Sand; < 80 m. n NCo (Del Norte Co.); sw OR. Jun–Oct ★

var. **pauciflorum** S. Watson LITTLE-FLOWER WILD BUCKWHEAT Pl 3–8 dm. **ST**: 1.5–5 dm, glabrous, rarely inflated. **LF**: basal; blades 1.5–3 cm, 0.8–1.8 cm wide, tomentose, glabrous adaxially. **INFL**: 20–50 cm, 10–30 cm wide; branches glabrous; involucres 1(2), 5–7 mm, glabrous or sparsely hairy. **FL**: 2–2.5 mm, hairy (or glabrous); perianth white or yellow. *n*=20. Sand; 1100–2800 m. TR, PR; n Baja CA. Jun–Oct

var. **psychicola** Reveal ANTIOCH DUNES BUCKWHEAT Pl 5–15(20) dm. **ST**: 2–5(10) dm, glabrous. **LF**: sheathing; blades 3–7 cm, 2–4 cm wide, tomentose abaxially, hairy to glabrous adaxially. **INFL**: 30–100(150) cm, 10–80 cm wide; involucres 1–3, 3–4 mm, glabrous. **FL**: 2.5–3 mm, hairy; perianth white to pink. Sand; 3–20 m. Deltaic GV (Contra Costa Co.). Jun–Oct ★

var. **pubiflorum** Benth. FREMONT'S WILD BUCKWHEAT Pl 3–8(10) dm. **ST**: 1–3 dm, glabrous. **LF**: basal; blades 1–4 cm, 0.5–2 cm wide, tomentose abaxially, glabrous or woolly-tufted adaxially. **INFL**: 20–70 cm, 20–40 cm wide; branches glabrous; involucres 2–7, 3–5 mm, glabrous or sparsely hairy. **FL**: 2.5–3 mm, hairy; perianth white or yellow. *n*=20. Common. Sand or gravel; 50–2200 m. NW, CaR, SN, ScV, SnFrB, SCoRO; s OR, w NV. Intergrades with *E. nudum* var. *oblongifolium, E. nudum* var. *westonii*. Jun–Oct

var. **regirivum** Reveal & J.C. Stebbins (p. 1105) KINGS RIVER BUCKWHEAT Pl 5–10 dm. **ST**: 1–3(4) dm, tomentose. **LF**: gen basal; blades 2–3.5 cm, 1.5–2.5 cm wide, tomentose. **INFL**: 30–100(140) cm, 20–40 cm wide; branches tomentose; involucres 1, 3–4 mm, tomentose. **FL**: 1.5–2 mm, hairy; perianth white. *n*=20. Gravel; 200–600 m. s SNF (Kings River near Pine Flat Reservoir, Fresno Co.). Aug–Nov ★

var. **scapigerum** (Eastw.) Jeps. (p. 1105) SIERRAN CREST WILD BUCKWHEAT Pl 1–2 dm. **ST**: 1–1.8 dm, glabrous. **LF**: basal; blades 1–2 cm, 0.3–0.8(1) cm wide, tomentose abaxially, gen glabrous adaxially. **INFL**: gen head-like, 2–5 cm wide; branches glabrous; involucres 3–6, 2–3 mm, glabrous or sparsely hairy. **FL**: 2–3 mm, gen glabrous; perianth white. Sand, gravel or rocks; 2800–3800 m. c&s SNH. Perhaps only a depauperate phase of *E. nudum* var. *deductum*. Jul–Sep

var. **westonii** (S. Stokes) J.T. Howell WESTON'S WILD BUCKWHEAT Pl 3–6 dm. **ST**: 1–3 dm, glabrous, inflated. **LF**: basal; blades 1–3 cm, 0.3–0.8 cm wide, tomentose abaxially, tomentose to hairy adaxially. **INFL**: 10–30 cm, 5–20 cm wide; branches glabrous; involucres 1(2), 3–5 mm, glabrous. **FL**: 2.5–3 mm, hairy; perianth yellow or white. *n*=20. Common. Sand or gravel; 700–1900 m. s SN, Teh, s CW, TR, s SNE, w DMoj. [*E. n.* var. *gramineum* (S. Stokes) Reveal] May–Sep

E. nummulare M.E. Jones (p. 1109) MONEY WILD BUCKWHEAT (Group 4) Shrub (1.5)3–8(10) dm, 3–12(15) dm diam. **ST**: 0.5–1 dm, gen tomentose. **LF**: cauline; blades (0.5)1–3 cm, 0.4–1.2(1.5) cm wide, narrow [or ± round], tomentose. **INFL**: 5–50 cm, 5–80 cm wide; branches tomentose to hairy (or ± glabrous); peduncles 0 or erect, 0.5–3 cm, tomentose; involucres 2–3 mm, 1.5–2.7 mm wide, gen tomentose. **FL**: 1.5–3 mm, glabrous; perianth white, lobes obovate. **FR**: 2–3 mm, glabrous. *n*=40. Common. Sand; 800–2600 m. s MP, SNE; to w UT. Jul–Oct

E. nutans Torr. & A. Gray (Group 2) Ann 0.5–3 dm. **ST**: 0.3–1.5 dm, glandular or glabrous. **LF**: basal; blades 0.5–2.5 cm, gen ± round, tomentose abaxially, hairy to ± glabrous adaxially. **INFL**: 5–20 cm; branches glandular or glabrous; peduncles recurved, curved or straight, slender, 0.3–1 cm, glandular (glabrous); involucres 2–3 mm, 2–3.5 mm wide, glandular or glabrous. **FL**: 2–3 mm, glabrous; perianth white to rose or red, outer lobes oblong to oval, inner oblanceolate. **FR**: 1.7–2 mm, glabrous.

var. **nutans** (p. 1109) DUGWAY WILD BUCKWHEAT **INFL**: peduncles, involucres glandular. *n*=40. Sand; 1200–2300 m. n SNH (e Mono Co.), s MP (ne Lassen Co.); to se OR, w UT, nw AZ. May–Sep ★

E. ochrocephalum S. Watson var. ***ochrocephalum*** (p. 1109)
OCHRE-FLOWERED BUCKWHEAT (Group 4) Per mat, scapose, 0.5–3 dm diam. **ST:** 0.5–1.5(2) dm, glandular below infl. **LF:** basal and sheathing; blades 1–2 cm, (0.3)0.5–1(1.5) cm wide (in CA), tomentose. **INFL:** head-like, 1.5–2.5 cm wide; peduncles 0 (in CA); involucres 5–8, 3–5(6) mm, 3–4.5(5) mm wide, hairy, occ sparsely glandular (in CA); teeth 6–8. **FL:** 2–2.5(3) mm (in CA), glabrous or weakly glandular; perianth yellow, lobes widely oblong. **FR:** 1.5–2.5 mm (in CA), glabrous. Clay; 1300–1700 m. MP; se OR, sw ID, NV. Other var. in se OR, sw ID, NV. *E. ochrocephalum* var. *alexanderae* Reveal, now treated as *E. alexanderae* (Reveal) Grady & Reveal, in NV near Mono Co. line, unknown in CA. May–Jun ★

E. ordii S. Watson (p. 1109) FORT MOHAVE WILD BUCKWHEAT (Group 2) Ann (0.5)1–7 dm. **ST:** (0.3)0.7–3 dm, sparsely tomentose to glabrous. **LF:** gen basal; blades (0.7)2–8 cm, (0.2)1–3 cm wide, sparsely tomentose or glabrous. **INFL:** (5)10–50 cm, 5–50 cm wide; branches glabrous, hairy nodes; peduncles erect, straight, thread-like, 0.5–2 cm, glabrous or thinly hairy; involucres 1–1.5(1.8) mm, 0.6–1.2 mm wide, glabrous; teeth 4. **FL:** 1–2.5(3) mm, densely short-hairy; perianth white or pale yellow to ± red, lobes oblong to narrowly ovate, densely covered with short-soft hairs. **FR:** 1.8–2 mm, glabrous. Clay; 200–1400 m. s SNF, Teh, SCoRI, n WTR. Mar–Jul

E. ovalifolium Nutt. (Group 4) Per mat, scapose (in CA), 0.5–4 dm diam. **ST:** 0.03–2 dm, sparsely to densely tomentose or ± glabrous. **LF:** basal; blades 0.2–6 cm, (0.1)0.2–1.5 cm wide, narrow to ± round, gen tomentose. **INFL:** head-like (in CA); peduncles 0; involucres 1–15, (2)3.5–5(8) mm, 2–4 mm wide, tomentose to hairy. **FL:** (2.5)3–6(7) mm, glabrous; perianth yellow or white to cream, rose, red or purple, outer lobes 2–4 mm, 2–4 mm wide, gen oval to round, inner 3–7 mm, 0.8–1.5 mm wide, oblanceolate to elliptic. **FR:** 2–3 mm, glabrous. Additional vars. elsewhere.

var. ***caelestinum*** Reveal HEAVENLY WILD BUCKWHEAT Pl 0.5–1(4) dm diam. **ST:** erect, 1–6 cm, hairy. **LF:** blades 0.2–0.5 cm, tomentose, margins not ± brown. **INFL:** 1–2 cm wide; involucres 1–2, 2–2.5 mm. **FL:** perianth yellow. *n*=20. Rocks; 3000–3600 m. c SNH (Bloody Canyon, Mono Co.); c NV. Jul–Aug

var. ***depressum*** Blank. DEPRESSED WILD BUCKWHEAT Pl 1–2.5 dm diam. **ST:** gen suberect to decumbent, 1–4(8) cm, hairy. **LF:** blades 0.4–0.8 cm, tomentose to sparsely so, margins not ± brown. **INFL:** 1–1.5(2) cm wide; involucres 2–4, 3–3.5 mm. **FL:** 4–5 mm; perianth white to rose. Dry playas; 1700 m. MP (e Lassen Co.); to w MT, w UT, w WY. Jun–Aug

var. ***eximium*** (Tidestr.) J.T. Howell (p. 1109) BROWN-MARGINED BUCKWHEAT Pl 1–3 dm diam. **ST:** erect, 1–5(7.5) cm, tomentose. **LF:** blades (0.3)0.5–1.5(2) cm, densely tomentose, margins ± brown. **INFL:** 1–2.5 cm wide; involucres 3–10, 4–5(6.5) mm. **FL:** 4–5 mm; perianth white. Granite; 1700–2500 m. n SNH (Job's Peak, Alpine Co.); w-c NV. Jun–Sep ★

var. ***monarchense*** D.A. York MONARCH BUCKWHEAT Pl 1.5–3 dm diam. **ST:** decumbent to ascending, 2–6(9) cm, tomentose to hairy. **LF:** blades 0.3–1.2 cm, gen tomentose, margins not ± brown. **INFL:** 1.5–4 cm wide; involucres 4–6, 5–8 mm. **FL:** 4–6 mm; perianth white to cream. Rocks; 1800 m. s SNH (Kings River Canyon area, Fresno Co.). Jun–Aug ★

var. ***nivale*** (Coville) M.E. Jones (p. 1109) SIERRAN CUSHION WILD BUCKWHEAT Pl 0.5–3 dm diam. **ST:** erect, 0.3–5(13) cm, gen tomentose. **LF:** blades 0.2–0.8 cm, tomentose, margins occ ± brown. **INFL:** 1–2.5 cm wide; involucres 2–4, 3–4.5 mm. **FL:** 4–5 mm; perianth white to rose or red. *n*=20. Common. Sand or gravel; 1700–4200 m. CaRH, SNH, W&I; to BC, w UT. Jun–Sep

var. ***ovalifolium*** (p. 1109) CUSHION WILD BUCKWHEAT Pl 2.5–4 dm diam. **ST:** erect, (4)5–20 cm, tomentose. **LF:** blades (1)3–6 cm, tomentose to hairy, margins rarely ± brown. **INFL:** 1–3.5 cm wide; involucres 3–15, 4–5(6.5) mm. **FL:** 4–5 mm; perianth yellow. *n*=20. Common. Sand or gravel; 600–2600 m. SNH (e slope), GB; to e WA, w MT, nw CO. Apr–Aug

var. ***purpureum*** (Nutt.) Durand PURPLE CUSHION WILD BUCKWHEAT Pl 2.5–4 dm diam. **ST:** erect, (4)5–20 cm, tomen-

tose. **LF:** blades 0.5–2 cm, tomentose to hairy, margins not ± brown. **INFL:** 1.5–3.5 cm wide; involucres 3–15, 4–5 mm. **FL:** 4–5 mm; perianth white to rose or purple. *n*=20. Common. Sand or gravel; 700–3100 m. KR, CaRH, SNH (e slope), GB; to sw Can, w CO, nw NM. Apr–Aug

var. ***vineum*** (Small) A. Nelson CUSHENBURY BUCKWHEAT Pl 1.5–2.5 dm diam. **ST:** erect or ± so, 3–6 cm, hairy to ± glabrous. **LF:** blades 0.7–1.2 cm, densely tomentose, margins not ± brown. **INFL:** 1.5–3.5 cm wide; involucres 2–4(5), 5–7 mm. **FL:** 5–7 mm; perianth white to cream. Gravel or rocks; 1500–2100 m. ne SnBr (Cushenbury Canyon). May–Jun ★

E. palmerianum Reveal (p. 1109) PALMER'S WILD BUCKWHEAT (Group 1) Ann (0.5)1–3 dm. **ST:** 0.3–0.8 dm, hairy to tomentose. **LF:** basal; blades 0.5–2 cm, tomentose abaxially, less so to ± glabrous adaxially. **INFL:** 5–2.5 cm, 10–30 cm wide; branches not incurved, hairy to tomentose; peduncles 0; involucres appressed, 1.5–2 mm, hairy to tomentose. **FL:** 1.5–2 mm, glabrous; perianth white to pink or pale ± yellow, in age pink to red, outer lobes narrowly fan-shaped, inner oblanceolate. **FR:** 1.5–1.8 mm. *n*=20. Common. Sand or gravel; (300)600–2300 m. SNE, D; to sw ID, CO, NM. Mar–Oct

E. panamintense C.V. Morton (p. 1109) PANAMINT MOUNTAIN WILD BUCKWHEAT (Group 4) Per 1.5–3 dm, 1–4 dm diam. **ST:** 0.5–1.5 dm, tomentose. **LF:** basal; blades 1.5–4 cm, 1–2.5 cm wide, gen tomentose. **INFL:** (10)12–20 cm, 1–5 cm wide; branches occ with lf-like bracts; peduncles 0; involucres 3–5 mm, 2–4 mm wide, tomentose. **FL:** 3.5–5 mm, glabrous; perianth white to white-brown, lobes oblanceolate. **FR:** 2.5–3 mm, glabrous. Gravel or rocks; (1600)1800–2900 m. W&I, DMtns; s NV, nw AZ. May–Oct

E. parishii S. Watson (p. 1109) PARISH'S WILD BUCKWHEAT (Group 2) Ann 1–4(5) dm. **ST:** 0.3–1 dm, glabrous, glandular distally. **LF:** basal; blades 2–6 cm, 0.5–2 cm wide, coarse-hairy. **INFL:** 10–35 cm, 10–50 cm wide; branches glabrous, glandular at nodes; peduncles spreading, 0.4–1.5(2.5) cm, straight, thread-like, glabrous or sparsely glandular; involucres 0.5–0.9 mm, 0.5–0.7 mm wide, glabrous; teeth 4. **FL:** 0.5–0.9 mm, hairy; perianth pink to red, aging to white, outer lobes ovate, inner oblong. **FR:** 1–1.3 mm, glabrous. *n*=20. Common. Sand; (1000)1300–3200 m. s SN, TR, PR, n W&I (White Mtns); n Baja CA. Jun–Oct

E. parvifolium Sm. (p. 1109) SEACLIFF WILD BUCKWHEAT (Group 4) Shrub 3–10 dm, 5–20(25) dm diam. **ST:** 0.2–1 dm, tomentose or glabrous. **LF:** cauline; blades 0.5–3 cm, 0.3–0.8(1.2) cm wide, narrow to ± round, tomentose abaxially, gen glabrous adaxially. **INFL:** rarely head-like, 20–30 cm, 10–20 cm wide; branches tomentose or glabrous; peduncles 0; involucres 2–7, (2.5)3–4 mm, 2–3.5 mm wide, hairy to ± glabrous. **FL:** 2.5–3 mm, glabrous; perianth white to pink- or green-yellow, lobes obovate. **FR:** 2.5–3 mm, glabrous. *n*=20. Common. Sand; < 410 m. CCo, SCo. Cult, occ established SnFrB. All yr

E. pendulum S. Watson (p. 1109) WALDO WILD BUCKWHEAT (Group 4) Shrub 2–5 dm, 2–8 dm diam. **ST:** 2–4 dm, tomentose. **LF:** sheathing; blades (1.5)2–4(5) cm, 1–2.5(3) cm wide, narrow to wide, tomentose abaxially, hairy adaxially. **INFL:** 15–40 cm; branches tomentose; peduncles 0 or erect, (1)3–10 cm, slender, tomentose; involucres 3.5–5 mm, 2.5–4 mm wide, tomentose. **FL:** 3–6(7) mm, hairy; perianth white, lobes narrowly oblong. **FR:** 3–5 mm, hairy. Serpentine; 200–800 m. nw KR (n Del Norte Co.); sw OR. Jul–Sep ★

E. plumatella Durand & Hilg. (p. 1109) YUCCA WILD BUCKWHEAT (Group 4) Shrub 3–10(12) dm, 3–6(8) dm diam. **ST:** 0.5–2 dm, tomentose or glabrous. **LF:** basal and cauline; blades 0.6–1.5 cm, 0.2–0.3 cm wide, gen tomentose. **INFL:** 15–40 cm; branches zigzag, gen spreading, tiered, tomentose or glabrous; peduncles 0; involucres 2–2.5 mm, 1.5–2 mm wide, glabrous. **FL:** 2–2.5 mm, glabrous; perianth white to pale yellow, outer lobes obovate, inner oblong. **FR:** 2.5–3 mm, glabrous. Common. Sand or gravel; 400–1700 m. D; to s NV, nw AZ. Apr–Oct

E. polypodum Small (p. 1109) TULARE COUNTY BUCKWHEAT (Group 3) Per mat 1–4(5) dm diam; gen dioecious. **ST:** 0.5–1.5 dm, gen tomentose. **LF:** basal; blades 0.2–1 cm, 0.1–0.6(0.8) cm wide, tomentose on both surfaces, soon glabrous adaxially. **INFL:**

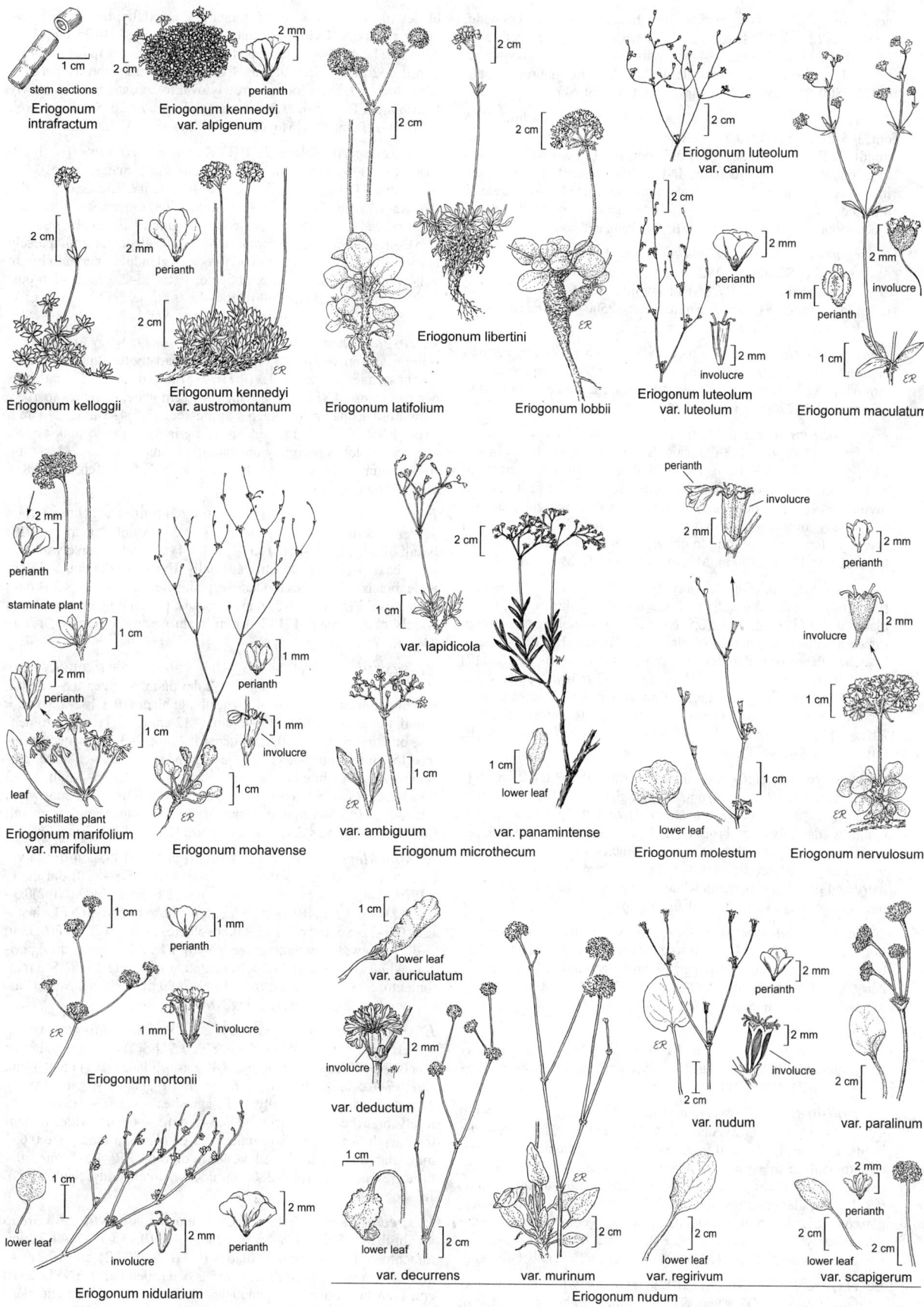

stem sections
Eriogonum
intrafractum

1 cm

2 cm

2 mm

perianth

Eriogonum kennedyi
var. alpigenum

2 cm

2 cm

Eriogonum luteolum
var. caninum

2 cm

2 mm

perianth

Eriogonum kelloggii

2 cm

Eriogonum kennedyi
var. austromontanum

2 cm

Eriogonum libertini

ER

Eriogonum latifolium

Eriogonum lobbii

2 cm

2 mm

perianth

2 mm

involucre

Eriogonum luteolum
var. luteolum

2 mm

involucre

2 mm

perianth

1 mm

perianth

1 cm

ER

Eriogonum maculatum

2 mm

perianth

staminate plant

1 cm

2 mm

perianth

leaf

pistillate plant

Eriogonum marifolium
var. marifolium

1 cm

1 mm

perianth

1 mm

involucre

1 cm

ER

Eriogonum mohavense

2 cm

var. lapidicola

1 cm

var. ambiguum

1 cm

lower leaf

var. panamintense

Eriogonum microthecum

ER

1 cm

perianth

involucre

2 mm

lower leaf

Eriogonum molestum

1 cm

perianth

2 mm

involucre

2 mm

1 cm

ER

Eriogonum nervulosum

1 cm

perianth

1 mm

1 mm

involucre

ER

Eriogonum nortonii

1 cm

lower leaf
var. auriculatum

2 mm

involucre

var. deductum

2 cm

lower leaf

perianth

2 mm

involucre

2 mm

ER

2 cm

var. nudum

perianth

2 mm

2 cm

var. paralinum

1 cm

lower leaf

2 mm

involucre

2 mm

perianth

Eriogonum nidularium

1 cm

lower leaf

2 cm

var. decurrens

ER

2 cm

var. murinum

2 cm

lower leaf
var. regirivum

2 mm

perianth

2 cm

lower leaf

2 cm

var. scapigerum

Eriogonum nudum

head-like, 0.5–2 cm wide; pistillate ones branched in fr, 2–3 cm wide; involucres (2)2.5–3.5(4) mm, 2–3 mm wide, tomentose; teeth 5–7. **FL:** 2.5–3.5 mm, stipe 0.2–0.4 mm, glabrous; perianth chalky white, lobes oblong to narrowly obovate. **FR:** 3–3.5 mm, glabrous. Sand; (2400)2800–3500 m. s SNH (esp Tulare Co.). Jul–Sep ★

E. prattenianum Durand (Group 3) Per mat or subshrub, 1–5 dm, 3–5 dm diam. **ST:** (0.5)1–3 dm, glabrous; bracts 3–6, lf-like, near middle. **LF:** basal; blades 0.5–1.5(2) cm, 0.2–0.7 cm wide, tomentose, occ ± glabrous adaxially. **INFL:** head-like, 0.8–2 cm wide; involucres 3–4 mm, 3–8 mm wide, tomentose; teeth 8–10, erect to reflexed. **FL:** 3–8 mm, stipe 1–1.5 mm, glabrous; perianth yellow, lobes spoon-shaped to obovate. **FR:** 4–5 mm, glabrous.

var. ***avium*** Reveal & Shevock (p. 1109) KETTLE DOME BUCKWHEAT Subshrub, 1–2 dm. **LF:** blades tomentose abaxially, subglabrous adaxially. **INFL:** involucres 3–5 mm wide, teeth reflexed, 2–3 mm. **FL:** gen 3–6 mm. Rocks; 2500–2900 m. c&s SNH (Madera, Fresno cos.). Jul–Aug ★

var. ***prattenianum*** NEVADA CITY WILD BUCKWHEAT Per mat or subshrub, 3–5 dm. **LF:** blades tomentose. **INFL:** involucres 5–8 mm wide; teeth erect to ± reflexed, 1–2 mm. **FL:** gen 5–8 mm. Sand or gravel; 800–2200 m. n&c SN. May–Jul

E. prociduum Reveal (p. 1109) PROSTRATE BUCKWHEAT (Group 4) Per mat, scapose, 1–3 dm diam. **ST:** 0.2–0.6(0.8) dm, glabrous. **LF:** basal; blades 0.3–1(1.4) cm, 0.15–0.4(0.6) cm wide, narrow to wide, tomentose. **INFL:** head-like, 0.8–1.2 cm wide; peduncles 0; involucres 4–7, 2–3 mm, (2.5)3–4 mm wide, weakly rigid, glabrous or sparsely tomentose; teeth 5–6. **FL:** 2–3 mm, glabrous; perianth bright yellow, lobes oblong to oblong-obovate. **FR:** 2–3 mm, glabrous. Clay; 1400–2700 m. MP; nw NV, s-c OR. May–Jul ★

E. pusillum Torr. & A. Gray (p. 1109) YELLOW TURBANS (Group 2) Ann 0.5–3 dm. **ST:** 0.2–0.8 dm, gen glabrous. **LF:** basal; blades 0.5–2.5(3) cm, 0.4–2(2.5) cm wide, ovate-oblong to ± round, tomentose abaxially, hairy to ± glabrous adaxially. **INFL:** 5–25 cm; branches glabrous; peduncles erect, mostly straight, slender, 0.1–5(7) cm, glabrous; involucres 1–1.5(1.7) mm, 1.5–3 mm wide, glandular-hairy. **FL:** 1–3 mm, glandular; yellow to red-yellow or red, outer lobes oblong-elliptic to obovate, inner oblong. **FR:** gen elliptic, 0.6–0.8 mm, glabrous. *n*=16. Common. Sand; -70–2600 m. SCoRI, WTR, GB, D; to se OR, w UT, nw AZ. Feb–Aug

E. pyrolifolium Hook. (Group 3) Per mat 0.5–3 dm diam. **ST:** gen decumbent, 0.3–1.5(1.8) dm, hairy or glabrous. **LF:** basal; blades 1–2.5(4) cm, 0.8–2 cm wide, ovate to ± round, tomentose abaxially, glabrous adaxially or glabrous on both surfaces. **INFL:** head-like or branched, 1–3(5) cm, 1–4 cm wide; involucres 4–6 mm, (3)4–8 mm wide, hairy or glabrous; teeth 4–5. **FL:** 4–6 mm, stipe 0.1–0.2 mm, hairy and glandular; perianth white to rose, lobes obovate. **FR:** 4–5 mm, glabrous exc soft, shaggy hairs on tip.

var. ***coryphaeum*** Torr. & A. Gray HAIRY SHASTA WILD BUCKWHEAT **ST:** prostrate to weakly erect, 0.4–1.5(1.8) dm. **LF:** blades tomentose abaxially, gen glabrous adaxially. **FL:** densely hairy, glandular hairs obscure. *n*=20. Sand; 1600–3100 m. nw KR; to s BC, w MT. Jun–Sep

var. ***pyrolifolium*** (p. 1109) PYROLA-LEAVED BUCKWHEAT **ST:** gen ascending, 0.3–1 dm. **LF:** blades glabrous. **FL:** sparsely hairy, glandular hairs obvious. Locally common. Sand; (800)1600–3300 m. CaRH; WA, ID. Jul–Sep ★

E. reniforme Torr. & Frém. (p. 1109) KIDNEY-LEAF WILD BUCKWHEAT (Group 2) Ann, 0.5–4 dm. **ST:** 0.2–0.8 dm, gen glabrous. **LF:** basal; blades 0.5–2(2.5) cm, ± round, tomentose abaxially, tomentose to ± glabrous adaxially. **INFL:** 5–35 cm; branches glabrous; peduncles erect, mostly curved, slender to thread-like, 0.3–1.5 cm, glabrous; involucres 1.5–2(2.2) mm, 1.5–2.5 mm wide, glabrous. **FL:** 1–2 mm, glandular; perianth ± yellow to ± red, outer lobes widely ovate, inner oblong. **FR:** elliptic, 0.8–1 mm, glabrous. *n*=16. Common. Sand; < 1700 m. D; NV, AZ, n Baja CA. Feb–Aug

E. rixfordii S. Stokes (p. 1109) PAGODA WILD BUCKWHEAT (Group 2) Ann, (1.5)2–4 dm. **ST:** 0.3–2 dm, glabrous. **LF:** basal;

blades (0.5)1–3 cm, ± round, tomentose abaxially, less so to ± glabrous adaxially. **INFL:** of many horizontal tiers, 10–35 cm, 5–35 cm wide; branches glabrous; peduncles 0; involucres reflexed, 1–1.5 mm, 1.5–2 mm wide, glabrous. **FL:** 1.3–1.5 mm, glabrous; perianth white to ± red, outer lobes narrowly oval to cordate, inner narrowly lanceolate. **FR:** elliptic, 1.3–1.5 mm, glabrous. *n*=20. Sand or gravel; (30)100–1600 m. ne DMoj (Death Valley); sw NV. Jun–Dec

E. rosense A. Nelson & P.B. Kenn. var. ***rosense*** (p. 1109) MOUNT ROSE WILD BUCKWHEAT (Group 4) Per mat, scapose, 0.5–5 dm diam. **ST:** 0.1–0.9(1.1) dm, glandular-hairy. **LF:** basal; blades 0.4–1.5(1.7) cm, (0.15)0.25–0.6(1) cm wide, tomentose, glandular. **INFL:** head-like, 0.6–1.5 cm wide; peduncles 0; involucres 3–6, (2.5)3–4(4.5) mm, 2.5–3.5 mm wide, rigid, glandular and sparsely hairy; teeth 5–8. **FL:** 2–3 mm, glabrous or glandular; perianth bright yellow to red-yellow, lobes obovate. **FR:** 1.5–2.5 mm, glabrous. *n*=20. Sand or gravel; (2300)2500–4000 m. n&c SNH, n SNE; w NV. [*E. beatleyae* Reveal, misappl.] Other var. in NV. Jul–Sep

E. roseum Durand & Hilg. (p. 1109) WAND WILD BUCKWHEAT (Group 1) Ann 1–8 dm. **ST:** 0.5–3 dm, sparsely tomentose to hairy. **LF:** basal and cauline; blades (0.5)1–3(5) cm, (0.3)0.5–1(2) cm wide, gen tomentose. **INFL:** 10–70 cm, 10–45 cm wide; branches sparsely tomentose to hairy; involucres appressed, (3.5)4–5 mm, 2–3 mm wide. **FL:** 1.5–2(2.5) mm, glabrous; perianth white to pink or red, occ yellow, lobes narrowly obovate to oblong. **FR:** 1.8–2(2.2) mm. *n*=9. Common. Sand or gravel; < 2200 m. NW, SNF, Teh, CW, TR; s OR, n Baja CA. May–Nov

E. rupinum Reveal WYMAN CREEK WILD BUCKWHEAT (Group 4) Per 3–5 dm, 0.5–1 dm diam. **ST:** 1.5–2.5 dm, tomentose. **LF:** basal; blades (2)2.5–4.5(5) cm, 1.5–3.5(4) cm wide, sparsely tomentose abaxially, less so to hairy adaxially. **INFL:** 15–25 cm, 5–15 cm wide, bracts 0; branches tomentose; peduncles 0; involucres 3–4 mm, tomentose. **FL:** 2.5–3.5 mm, glabrous; perianth gen cream, lobes widely oblanceolate. **FR:** 2–4 mm, glabrous. *n*=20. Gravel; 1700–3100 m. W&I; NV. Jul–Sep

E. saxatile S. Watson (p. 1109) HOARY WILD BUCKWHEAT (Group 4) Per mat (1)2–4 dm, 0.5–2 dm diam. **ST:** gen 0.5–1.5 dm, tomentose or hairy. **LF:** basal or sheathing; blades (0.3)1–2(2.5) cm, ± round, tomentose. **INFL:** 10–25 cm, 5–12 cm wide; branches tomentose or hairy; peduncles 0; involucres 3–4 mm, 2–3 mm wide, tomentose to hairy; teeth 5–6. **FL:** (3)5–7 mm, incl ± winged stipe, glabrous; perianth white to rose or ± yellow, outer lobes 3–5 mm, 1.5–2 mm wide, oblanceolate to lanceolate, inner 4–6 mm, 2–3 mm wide, obovate. **FR:** 3.5–4 mm, glabrous. *n*=20. Common. Sand or gravel; (300)800–3400 m. s SN, CW, SW, SNE, DMtns; s NV. May–Oct

E. shockleyi S. Watson var ***shockleyi*** (p. 1109) SHOCKLEY'S BUCKWHEAT (Group 4) Per mat, scapose, (0.5)1–4(20) dm diam. **ST:** (0.05)0.1–0.3 dm, hairy to tomentose. **LF:** basal; blades (0.2)0.3–0.8(1.2) cm, 0.2–0.4(0.6) cm wide, tomentose to hairy. **INFL:** head-like, 0.8–2 cm wide; peduncles 0; involucres 2–4(6), (2)2.5–5(6) mm, 3–6(7) mm wide, tomentose; teeth 5–10. **FL:** 2.5–4 mm, hairy; perianth white to rose or yellow, lobes oblong to obovate. **FR:** 2.5–3 mm, tomentose. Gravel or clay; (800)1200–2600 m. SNE, n DMtns (Last Chance Range); to s ID, CO, nw NM. May–Aug ★

E. siskiyouense Small (p. 1109) SISKIYOU BUCKWHEAT (Group 3) Per, gen matted, 1–5 dm diam. **ST:** 0.5–1.5(2) dm, gen glabrous; bracts 2–4, lf-like, near middle. **LF:** basal; blades (0.3)0.5–0.8 cm, (0.2)0.3–0.5(0.7) cm wide, gen tomentose abaxially, gen sparsely hairy to ± glabrous adaxially. **INFL:** gen head-like, 0.8–1.5 cm wide, rarely branched; involucres (3)3.5–4 mm, 4–6 mm wide, tomentose; teeth 6–10, lobe-like, reflexed. **FL:** (4)4.5–6 mm, stipe 0.6–1 mm, glabrous; perianth yellow, lobes oblong. **FR:** 4.5–5 mm, glabrous. Serpentine; 1600–2800 m. e KR (Mount Eddy, Scott Mtns). Jul–Sep ★

E. spectabile B.L. Corbin et al. BARRON'S BUCKWHEAT (Group 4) Shrub 1–1.5 dm, 1.7–2.5 dm diam. **ST:** (0.15)0.6.–1.3(1.7) dm, tomentose. **LF:** gen basal; blades (0.7)1.2–1.7(2.2) cm, (0.2)0.4–0.7(0.9) cm wide, tomentose, margins occ rolled under. **INFL:** 2–10 cm; branches tomentose; peduncles erect, (1.5)2–8.5(9.5) cm, slender, tomentose to ± glabrous; involucres 2–3 mm, (2)3–4(5) mm wide,

tomentose and glandular abaxially; teeth 5–7. **FL:** 4–6 mm, hairy and glandular; perianth white, lobes obovate. **FR:** 3–4 mm, hairy. Gravel; 2000 m. CaRH (near Chester, Lassen Co.). Jul–Sep ★

E. spergulinum A. Gray (Group 2) Ann 0.5–4 dm. **ST:** 0.1–0.5 dm, glabrous or glandular and short-hairy. **LF:** basal and cauline; blades (0.3)1–3(4) cm, 0.05–0.3 cm wide, linear, short-hairy, margins occ rolled under. **INFL:** 4–25 cm, 5–35 cm wide; branches sparsely hairy, internodes gen glandular; peduncles erect, 0.4–1.5 cm, straight, thread-like, glabrous; involucres 0.5–1 mm, 0.4–0.8 mm wide, glabrous; teeth 4. **FL:** 1.5–3.5 mm, glabrous or sparsely hairy; perianth white to ± pink or rose, lobes oblong. **FR:** elliptic, 1.5–2.3 mm, glabrous.

var. ***pratense*** (S. Stokes) J.T. Howell   MOUNTAIN MEADOW WILD BUCKWHEAT   Pl prostrate to ascending, 0.5–1 dm. **INFL:** internodes not glandular. **FL:** 1.8–2 mm, hairy. Sand; (2500)2800–3500 m. s SNH. Jul–Aug

var. ***reddingianum*** (M.E. Jones) J.T. Howell   REDDING'S WILD BUCKWHEAT   Pl erect, 0.8–4 dm. **INFL:** internodes glandular. **FL:** 1.5–2.5 mm, glabrous or sparsely hairy. Common. Sand or gravel; 1300–3400 m. NCoRH, CaRH, SN, n SCoRO, TR, GB; to OR, ID, NV. Jun–Sep

var. ***spergulinum*** (p. 1109)   SPURRY WILD BUCKWHEAT   Pl erect, 1–4 dm. **INFL:** internodes glandular. **FL:** 2.5–3.5 mm, glabrous. Sand; 1200–3000 m. SN. Jun–Sep

E. sphaerocephalum Benth.   (Group 3) Subshrub, 0.5–4 dm, 3–5(6) dm diam. **ST:** (0.3)0.5–1 dm, thinly hairy or glabrous; bracts 4–8, lf-like, near middle. **LF:** basal; blades 1–3(4) cm, (0.1)0.3–0.6(1) cm wide, tomentose to glabrous, margins occ rolled under. **INFL:** head-like or branched, 1–5 cm; branches thinly hairy or glabrous; involucres 3–4 mm, 2.5–4.5 mm wide; teeth 6–10, ± reflexed. **FL:** (5)6–9 mm, stipe 1–2 mm, hairy; perianth cream or pale to bright yellow, lobes obovate to oblong-ovate. **FR:** 3–4 mm, glabrous exc sparsely hairy tip. Other vars. in OR, WA, ID.

var. ***halimioides*** (Gand.) S. Stokes (p. 1109)   HALIMIUM WILD BUCKWHEAT   **FL:** perianth pale yellow to cream. *n*=20. Common. Sand or gravel; 300–2300 m. CaRH, n SNH, MP; to WA, ID, NV. May–Jul

var. ***sphaerocephalum*** (p. 1109)   ROCK WILD BUCKWHEAT   **FL:** perianth bright yellow. Sand or gravel; (100)300–2000 m. n SNH, MP; to WA, ID, NV. May–Jul

E. strictum Benth.   (Group 4) Per 1–5 dm, 1–10 dm diam. **ST:** 1–3 dm. **LF:** basal; blades 0.5–2.5(4) cm, (0.3)0.5–1.5 cm wide, gen tomentose to sparsely so. **INFL:** 1–20 cm, 3–25 cm wide; branches gen tomentose to hairy; peduncles 0; involucres 4–6 mm, 1.5–5 mm wide, tomentose or glabrous. **FL:** 3–5(6) mm, glabrous; perianth yellow or white to rose or purple, outer lobes 2–3 mm, 2–3 mm wide, elliptic to ± round, inner 3–4 mm, 1–2 mm wide, oblanceolate to oblong. **FR:** 3–3.5 mm, glabrous. KR, n NCoRO, NCoRH, CaR, ne SNH, MP; to WA, MT, NV. Other var. in OR, WA, ID.

var. ***anserinum*** (Greene) S. Stokes   GOOSE LAKE WILD BUCKWHEAT   Pl 2–4 dm diam. **LF:** blades 0.5–2 cm, tomentose to hairy. **INFL:** 1–3(5) cm; branches tomentose to hairy; involucres 4–5.5 mm, 4–5 mm wide, tomentose. **FL:** 3–4.5 mm; perianth yellow. Common. Sand or gravel; (100)400–2600 m. CaR, ne SN, MP; to WA, ID, NV. May–Aug

var. ***greenei*** (A. Gray) Reveal (p. 1109)   GREENE'S BUCKWHEAT   Pl 2–10 dm diam. **LF:** blades 0.5–1 cm, tomentose. **INFL:** 3–5 cm; branches tomentose; involucres 4–6 mm, 2–3 mm wide, tomentose. **FL:** 4–5 mm; perianth white. Serpentine; 1500–2400 m. KR. Jun–Sep ★

var. ***proliferum*** (Torr. & A. Gray) C.L. Hitchc. (p. 1109)   PROLIFEROUS WILD BUCKWHEAT   Pl 2–4 dm diam. **LF:** blades 1–3 cm, tomentose to hairy. **INFL:** 5–15 cm; branches tomentose to hairy; involucres 4–6 mm, 2–4 mm wide, tomentose. **FL:** 3–5 mm; perianth white to rose or purple. Common. Sand or gravel; (100)400–2700 m. KR, n NCoRO, NCoRH, CaR, ne SNH, w MP; to ne WA, w MT, n NV. Jun–Sep

E. temblorense J.T. Howell & Twisselm. (p. 1109)   TEMBLOR BUCKWHEAT   (Group 1) Ann 1–8 dm. **ST:** 0.5–1(1.5) dm, tomentose. **LF:** ± basal, rarely cauline; basal blades 1.5–4 cm, 1–1.5 cm wide (or ± round, 2–4 cm), tomentose. **INFL:** 8–70 cm, 10–50 cm wide; peduncles erect, 1–4 cm, straight, slender, tomentose; involucres terminal on branchlets, 2–2.5 mm, 1.5–2 mm wide, tomentose. **FL:** 1.5–2.5 mm, glabrous, smooth; perianth white, lobes oblong; styles 0.2–1 mm. **FR:** 2–2.8 mm; tip granular. *n*=17. Sand; 300–900 m. s SCoRI (s Monterey, e San Luis Obispo, w Kern cos.). May–Sep ★

E. ternatum Howell (p. 1109)   TERNATE BUCKWHEAT   (Group 3) Per mat 3–13 dm diam. **ST:** 1–2.5(3) dm, tomentose. **LF:** basal; blades (0.7)1–1.5 cm, (0.4)0.8–1.3 cm wide, tomentose abaxially, tomentose to ± glabrous adaxially. **INFL:** 0.8–3 cm, 1–3.5 cm wide; branches tomentose; involucres 5–8 mm, 3–5 mm wide, tomentose; teeth 5–8. **FL:** 3–5 mm, stipe 0.3–0.6 mm, glabrous; perianth yellow, lobes spoon-shaped to obovate. **FR:** 3.5–5 mm, glabrous exc sparsely hairy tip. Serpentine; 400–1700 m. n KR, s NCoRO (nw Sonoma Co.), n NCoRH; sw OR. Jun–Aug ★

E. thomasii Torr. (p. 1109)   THOMAS' WILD BUCKWHEAT   (Group 2) Ann 0.5–3 dm. **ST:** 0.2–1 dm, glabrous exc for few glands proximally. **LF:** basal; blades 0.5–2 cm, round to round-reniform, densely white-tomentose abaxially, woolly-tufted to ± glabrous adaxially. **INFL:** 5–25 cm; branches glabrous; peduncles spreading to ± recurved, 0.5–3 cm, thread-like, glabrous; involucres 0.6–1.2 mm, 0.7–1.3 mm wide, glabrous. **FL:** 0.8–2 mm, hairy; perianth yellow then white to rose, outer lobes cordate, in age sac-like basally, inner spoon-shaped. **FR:** gen elliptic, 0.8–1 mm, glabrous. *n*=20. Common. Sand; -70–1200 m. D; to sw UT, AZ, nw Mex. All yr

E. thornei (Reveal & Henrickson) L.M. Shultz   THORNE'S BUCKWHEAT   (Group 4) Per mat 0.4–1(2.5) dm diam. **ST:** 0.1–0.2 dm, gen hairy. **LF:** cauline; blades 0.4–0.6 cm, 0.05–0.1 cm wide, linear, tomentose adaxially, hairy adaxially, margins rolled under. **INFL:** 0.5–1 cm; branches gen glabrous; peduncles 0; involucres 1.5–2 mm, 1–1.5 mm wide, hairy. **FL:** 1.5–2 mm, glabrous; perianth white, outer lobes 1–1.5 mm wide, obovate, inner 0.8–1 mm wide, oblanceolate. **FR:** 2–2.5 mm, glabrous exc bumpy tip. Gravel; 1800 m. e DMtns (New York Mtns). [*E. ericifolium* var. *t.* Reveal & Henrickson] May–Jul ★

E. thurberi Torr. (p. 1109)   THURBER'S WILD BUCKWHEAT   (Group 2) Ann 0.5–4 dm. **ST:** 0.3–1 dm, glabrous, sparsely tomentose and glandular proximally. **LF:** basal; blades 0.8–4.5 cm, 0.5–3 cm wide, tomentose abaxially, sparsely tomentose or glabrous adaxially. **INFL:** 5–30 cm, 5–50 cm wide; branches sparsely glandular to glabrous; peduncles erect, 0.5–2.5 cm, straight, thread-like, glabrous and glandular-hairy distally; involucres 1.8–2 mm, minutely glandular-hairy. **FL:** 1–1.7 mm, glandular-hairy with adaxial tuft of long white hairs; perianth white to red, outer lobes gen fan-shaped, inner oblanceolate. **FR:** gen elliptic, 0.6–0.8 mm, glabrous. *n*=20. Common. Sand; 100–1200 m. e PR, s DMoj, DSon; to sw NM, nw Mex. All yr

E. trichopes Torr. (p. 1109)   LITTLE DESERT TRUMPET   (Group 2) Ann 1–4.5(6) dm. **ST:** 0.5–2(3) dm, occ inflated, glabrous, minutely hairy proximally. **LF:** basal; blades (0.5)1–2.5(4) cm, (0.5)1–2(3) cm wide, widely oblong, hairy. **INFL:** 5–30 cm, 5–50 cm wide; branches rarely inflated, glabrous; peduncles mostly erect, 0.5–1.5 cm, straight, thread-like, glabrous; involucres 0.7–1 mm, 0.6–0.9 mm wide, glabrous; teeth 4(5). **FL:** 1–2 mm, hairy; perianth yellow to green-yellow, lobes narrowly ovate. **FR:** occ elliptic, 1–1.5 mm, glabrous. Common. Sand or gravel; -60–1500 m. D; to sw UT, sw NM, nw Mex. All yr

E. tripodum Greene (p. 1113)   TRIPOD BUCKWHEAT   (Group 3) Subshrub 2.5–5 dm, 3–6 dm diam. **ST:** 2–3 dm, ± glabrous. **LF:** basal; blades 1.5–2(2.5) cm, 0.5–1 cm wide, densely tomentose. **INFL:** 0.4–1 cm; branches ± glabrous; involucres 3–4.5 mm, tomentose; teeth 6–10. **FL:** 4–5 mm, stipe 1–1.5 mm, hairy; perianth yellow, lobes obovate. **FR:** 2.5–3 mm, glabrous exc sparsely hairy tip. Serpentine; (100)300–800 m. NCoRI, n&c SNF. May–Jul ★

E. truncatum Torr. & A. Gray   (p. 1113)   MOUNT DIABLO BUCKWHEAT   (Group 1) Ann 1–4.5 dm. **ST:** (0.1)0.2–0.6(0.8) dm,

tomentose. **LF:** basal and cauline; blades (0.8)1–5(7) cm, (0.2)1–2(3) cm wide, tomentose abaxially, hairy to ± glabrous adaxially. **INFL:** (5)10–30(40) cm, 5–30 cm wide; branches tomentose to hairy; peduncles 0; involucres at tips of slender branchlets, 2.5–3.5(4) mm, 2–2.5 mm wide, tomentose. **FL:** (1.5)1.7–2(2.2) mm, glabrous; perianth white to rose, lobes elliptic to oblong or obovate; styles 0.2–0.3 mm. **FR:** 1.7–2 mm; tip smooth. Sand; 200–400 m. ScV (extirpated near Suisun, Solano Co.), ne SnFrB (Mount Diablo, Contra Costa Co.; extirpated Alameda Co.). Apr–Aug　★

E. twisselmannii (J.T. Howell) Reveal TWISSELMANN'S BUCKWHEAT　(Group 3) Per mat 3–4 dm diam. **ST:** (0.2)0.5–1.2 dm, tomentose; bracts 4–8, lf-like, near middle. **LF:** basal; blades 0.5–1 cm, (0.2)0.4–0.6 cm wide, tomentose abaxially, tomentose to ± glabrous adaxially. **INFL:** head-like, 0.7–1.5 cm wide; involucres (2.5)4–5 mm, 4–6 mm wide; teeth 6–9, reflexed. **FL:** (4)5–6 mm, stipe 1–1.3 mm, hairy; perianth pale yellow, outer lobes 3–5 mm, obtuse, inner 5–6 mm, widely oblanceolate. **FR:** 5–5.5 mm, glabrous exc sparsely hairy tip. Rocks; 2300–2600 m. s SNH (The Needles, Slate Mtn, Tulare Co.). Jun–Sep　★

E. umbellatum Torr. SULPHUR FLOWER　(Group 3) Per to shrub, (0.2)1–12(20) dm, (0.5)1–12(20) dm diam. **ST:** (0.1)0.5–3(4) dm, tomentose to hairy or glabrous. **LF:** basal; blades 0.3–3(4) cm, 0.1–2.5 cm wide, tomentose or glabrous. **INFL:** umbel-like (main branches of 1 order) or compound umbel-like (main branches of ≥ 2 orders), rarely head-like, 3–25 cm, 2–18 cm wide; branches tomentose to hairy or glabrous, rarely with bracts near middle of branches; involucres 1–6 mm, (1)1.5–10 mm wide, tomentose to hairy or glabrous; teeth 6–12, reflexed. **FL:** 2–10(12) mm, stipe (0.7)1.3–2 mm, glabrous; perianth white, yellow or red, lobes gen spoon-shaped to obovate. **FR:** 2–7 mm, glabrous exc sparsely hairy tip. Highly variable; many vars. outside CA, to w Can, MT, CO, AZ.

var. ***ahartii*** Reveal AHART'S BUCKWHEAT　Shrub 3–8 dm, 5–13 dm diam. **ST:** 1–2 dm, gen tomentose. **LF:** blades 1–2.5(3) cm, 0.7–1.5 cm wide, rusty-woolly to tomentose abaxially, hairy or glabrous adaxially. **INFL:** main branches of ≥ 2 orders, tomentose to woolly-tufted; involucre tubes 2.5–4 mm, teeth 2–3 mm. **FL:** 5–8 mm; perianth yellow. Serpentine; 400–1000 m. n SNF (near Paradise, Lumpkin ridges, Butte Co.). Jun–Sep　★

var. ***argus*** Reveal ONE-EYED SULPHUR FLOWER　Mat 5–15 dm diam. **ST:** (0.8)1–2 dm, tomentose to hairy or glabrous; often with lf-like bract near middle. **LF:** blades (0.7)1–2(2.5) cm, 0.4–1 cm wide, tomentose abaxially, hairy or glabrous adaxially, margins wavy. **INFL:** main branches of ≥ 2 orders, hairy or glabrous; involucre tubes 2–3 mm, teeth 2–3(4) mm. **FL:** 3–8 mm; perianth yellow. Serpentine; (900)1500–2200 m. KR, c NCoRH; sw OR. Jun–Sep

var. ***bahiiforme*** (Torr. & A. Gray) Jeps. BAY BUCKWHEAT　Mat 3–6 dm diam. **ST:** 0.5–1.5 dm, tomentose. **LF:** blades 0.5–1.5(1.7) cm, 0.3–0.7 cm wide, densely tomentose. **INFL:** main branches of ≥ 2 orders, tomentose; involucre tubes 2–3 mm, teeth 2–3.5(4) mm. **FL:** 5–8 mm; perianth yellow. Serpentine; 700–2000 m. NCoRI, CW. Jul–Sep　★

var. ***canifolium*** Reveal SHERMAN PASS SULPHUR FLOWER　Mat 3–10 dm diam. **ST:** (0.5)1–1.8(2) dm, tomentose. **LF:** blades 0.4–2 cm, 0.3–0.7(0.9) cm wide, gen tomentose. **INFL:** main branches of 1 order, tomentose to hairy; involucre tubes 2–3 mm, teeth 2–3.5 mm. **FL:** (4)5–7(8) mm; perianth yellow. Sand; 2400–2600 m. s SNH (Inyo, Tulare cos.), DMtns (Argus Range). Unnamed var. allied to this taxon in KR (near Gazelle Mtn, Siskiyou Co.) with densely gray-tomentose lf-blades 0.4–0.8 cm, 0.3–0.5 cm wide. Jun–Sep

var. ***chlorothamnus*** Reveal SHERWIN GRADE SULPHUR FLOWER　Subshrub or shrub, (4)5–10(12) dm, 5–12 dm diam. **ST:** 1–2.5 dm, glabrous. **LF:** blades 0.5–2 cm, 0.3–1 cm wide, glabrous. **INFL:** main branches of ≥ 2 orders, glabrous; involucre tubes 2–2.5 mm, teeth 1–1.5 mm. **FL:** 3–6 mm; perianth yellow. Sand; 1600–2600 m. s SNH, SNE, nw DMoj. Jul–Sep

var. ***covillei*** (Small) Munz & Reveal COVILLE'S SULPHUR FLOWER　Mat 1–5 dm diam. **ST:** 0.3–0.9 dm, tomentose to ± glabrous. **LF:** blades 0.3–0.6(1) cm, 0.2–0.4(0.6) cm wide, gen tomentose. **INFL:** main branches of 1 order, tomentose to hairy; involucre

tubes 1.5–2.5 mm, teeth 1–3 mm. **FL:** 2–4(5) mm; perianth yellow. Gravel or rocks; 3000–3600 m. c&s SNH (Inyo, Tulare cos.), W&I (n White Mtns, Mono Co.). Jul–Sep

var. ***dichrocephalum*** Gand. BICOLOR SULPHUR FLOWER　Mat 5–10 dm diam. **ST:** (0.5)1–2.5 dm, tomentose to hairy (or glabrous). **LF:** blades 1–2(2.5) cm, 0.5–1.5(2) cm wide, tomentose to hairy abaxially, tomentose or glabrous adaxially. **INFL:** main branches of 1 order, tomentose; involucre tubes 2–3 mm, teeth 1–2.5 mm. **FL:** 4–8 mm; perianth pale yellow to cream or ± white (or green-white). Sand to gravel; (1200)1400–3100 m. Wrn, W&I; to e OR, sw MT, n UT. Jun–Sep

var. ***dumosum*** (Greene) Reveal AMERICAN VALLEY SULPHUR FLOWER　Shrub 3–5 dm, 3–10 dm diam. **ST:** 0.8–2(2.5) dm, tomentose to hairy. **LF:** blades 1–2.5(3) cm, 0.5–1.2 cm wide, tomentose abaxially, hairy or glabrous adaxially. **INFL:** main branches of 1 order, tomentose to hairy; involucre tubes 2–3(5) mm, teeth 2.5–4 mm. **FL:** 5–9 mm; perianth yellow. Sand or gravel; (300)600–1200 m. KR, NCoRH, NCoRI, CaR, n SN, MP; s-c OR. [*E. u.* var. *polyanthum* (Benth.) M.E. Jones, misappl., in part] Jun–Sep

var. ***furcosum*** Reveal (p. 1113) SIERRA NEVADA SULPHUR FLOWER　Subshrub 3–6 dm, 3–8 dm diam. **ST:** 0.5–2 dm, sparsely hairy or glabrous. **LF:** blades (0.7)1–2.5(3) cm, 0.3–0.8(1.3) cm wide, tomentose abaxially, gen glabrous adaxially. **INFL:** main branches of ≥ 2 orders, gen glabrous; involucre tubes 2–3(4.5) mm, teeth 1.5–3(4) mm. **FL:** (5)6–8 mm; perianth yellow. Common. Sand or gravel; 1200–3000 m. SNH; w-c NV. Jun–Sep

var. ***glaberrimum*** (Gand.) Reveal WARNER MOUNTAINS BUCKWHEAT　Mat 3–8 dm diam. **ST:** 1–2 dm, glabrous. **LF:** blades 1–2 cm, 0.3–1 cm wide, glabrous. **INFL:** main branches of 1 order, glabrous, with a whorl of bracts near middle; involucre tubes 4–5 mm, teeth 1–3.5(4) mm. **FL:** 4–7 mm; perianth cream or ± white. Sand or gravel; 1600–2300 m. Wrn; s-c OR. Jul–Sep　★

var. ***goodmanii*** Reveal GOODMAN'S SULPHUR FLOWER　Mat 4–7 dm diam. **ST:** 1–2.5(4) dm, hairy. **LF:** blades 0.5–2(3.5) cm, 0.5–1(1.5) cm wide, gen tomentose. **INFL:** main branches of 1 order, hairy; involucre tubes 2–3 mm, teeth 2.5–4 mm. **FL:** 6–8(9) mm; perianth yellow. Serpentine; 400–1700 m. n KR; sw OR. May–Sep

var. ***humistratum*** Reveal (p. 1113)　MOUNT EDDY BUCKWHEAT　Mat 1–3 dm diam. **ST:** 0.5–1.5 dm, hairy. **LF:** blades 0.5–1(1.5) cm, 0.5–1 cm wide, gen tomentose. **INFL:** main branches of 1 order, hairy; involucre tubes 1.5–2 mm, teeth 1.5–3 mm. **FL:** 3–6 mm; perianth yellow. Serpentine; 1700–2800 m. KR (White Mtn, Mount Eddy, Scott Mtn, Marble Mtns), n CaRH (Mount Shasta). Jun–Sep　★

var. ***juniporinum*** Reveal JUNIPER SULPHUR-FLOWERED BUCKWHEAT　Subshrub or shrub, 4–8 dm, 5–10 dm diam. **ST:** 1–2.5 dm, hairy or glabrous. **LF:** blades (0.7)1–2 cm, (0.3)0.5–1(1.2) cm wide, hairy or glabrous. **INFL:** main branches of ≥ 2 orders, hairy or glabrous; involucre tubes (2.5)3–3.5 mm, teeth 1–2.5 mm. **FL:** (4)5–6 mm; perianth cream, ± white or pale- to green-yellow, without large ± red spots. Sand or gravel; 1300–2300 m. e DMtns (e San Bernardino Co.); NV, sw UT, nw AZ. Jun–Oct　★

var. ***lautum*** Reveal SCOTT VALLEY BUCKWHEAT　Mat 3–10 dm diam. **ST:** 1–2 dm, tomentose. **LF:** blades (1)1.5–4 cm, (0.5)0.8–1.8(2) cm wide, gen tomentose. **INFL:** main branches of ≥ 2 orders, tomentose; involucre tubes (3)3.5–5 mm, lobes (3)4–6 mm. **FL:** 4–7 mm; perianth bright yellow. Sand or gravel; 800–900 m. KR (Scott Valley, Siskiyou Co.). Jul–Sep　★

var. ***minus*** I.M. Johnst. (p. 1113) ALPINE SULPHUR-FLOWERED BUCKWHEAT　Mat 0.5–2 dm diam. **ST:** 0.2–0.8(1.5) cm, tomentose. **LF:** blades 0.3–0.8(1) cm, tomentose. **INFL:** main branches of 1 order, tomentose; involucre tubes 1.5–2 mm, teeth 1.5–2 mm. **FL:** (2.5)4–6 mm; perianth lemon yellow to yellow-red, becoming red. Gravel; (1800)2400–3100 m. SnGb, SnBr. Jul–Sep　★

var. ***modocense*** (Greene) S. Stokes (p. 1113) MODOC SULPHUR FLOWER　Mat 1–3(5) dm diam. **ST:** 0.7–2.5 dm, gen tomentose. **LF:** blades (0.3)1–1.5(2) cm, 0.1–1(1.2) cm wide, tomentose abaxially, gen glabrous adaxially. **INFL:** main branches of 1 order, gen tomentose; involucre tubes 2–3 mm, teeth 2–3 mm. **FL:** 4–8 mm; perianth

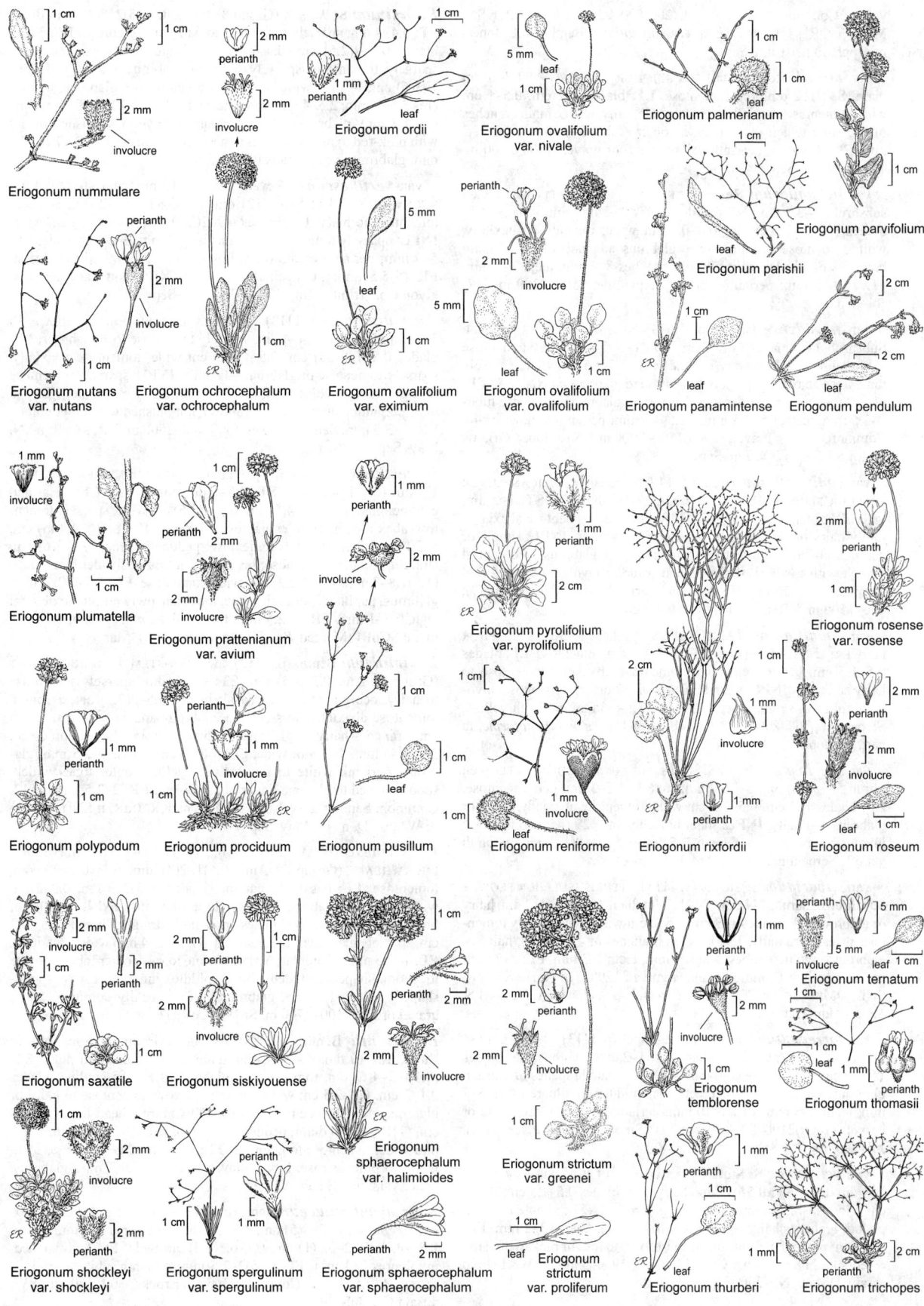

Eriogonum nummulare

Eriogonum ordii

Eriogonum ovalifolium var. nivale

Eriogonum palmerianum

Eriogonum parishii

Eriogonum parvifolium

Eriogonum nutans var. nutans

Eriogonum ochrocephalum var. ochrocephalum

Eriogonum ovalifolium var. eximium

Eriogonum ovalifolium var. ovalifolium

Eriogonum panamintense

Eriogonum pendulum

Eriogonum plumatella

Eriogonum prattenianum var. avium

Eriogonum pyrolifolium var. pyrolifolium

Eriogonum rosense var. rosense

Eriogonum polypodum

Eriogonum prociduum

Eriogonum pusillum

Eriogonum reniforme

Eriogonum rixfordii

Eriogonum roseum

Eriogonum saxatile

Eriogonum siskiyouense

Eriogonum sphaerocephalum var. halimioides

Eriogonum strictum var. greenei

Eriogonum temblorense

Eriogonum ternatum

Eriogonum thomasii

Eriogonum shockleyi var. shockleyi

Eriogonum spergulinum var. spergulinum

Eriogonum sphaerocephalum var. sphaerocephalum

Eriogonum strictum var. proliferum

Eriogonum thurberi

Eriogonum trichopes

yellow. Common. Sand or gravel; (200)600–2300 m. KR, CaR, n SN, MP; to e OR, s ID, n NV. [*E. u.* var. *polyanthum* (Benth.) M.E. Jones, misappl., in part] Jun–Sep

var. ***munzii*** Reveal MUNZ'S SULPHUR FLOWER Mat 3–6 dm diam. **ST**: 1–2 dm, gen tomentose. **LF**: blades 1–2 cm, 0.5–1 cm wide, tomentose abaxially, gen hairy adaxially. **INFL**: main branches of ≥ 2 orders, gen tomentose; involucre tubes 2–3 mm, teeth 1–2 mm. **FL**: 3–8 mm; perianth yellow. Sand or gravel; 1500–2600 m. TR, SnJt. Jun–Sep

var. ***nelsoniorum*** Reveal NELSONS' SULPHUR FLOWER Gen subshrub, 1–2.5 dm, 2–7 dm diam. **ST**: 1–2 dm, thinly hairy. **LF**: blades (0.5)1–1.5(2) cm, 0.4–0.8 cm wide, tomentose abaxially, white-tomentose to hairy (or ± glabrous adaxially). **INFL**: main branches of 1 order, hairy; involucre tubes 3–4 mm, teeth 2–4 mm. **FL**: (5)6–7 mm; perianth yellow. Serpentine; 1500–2300 m. KR. Jul–Sep

var. ***nevadense*** Gand. (p. 1113) NEVADA SULPHUR FLOWER Subshrub 1–5 dm, 2–6 dm diam. **ST**: 1–3 dm, sparsely tomentose to ± glabrous. **LF**: blades 1–2(2.5) cm, 0.5–1.5 cm wide, sparsely tomentose abaxially, sparsely tomentose to glabrous adaxially. **INFL**: main branches of 1 order, gen hairy to ± glabrous; involucre tubes 2–3.5 mm, teeth 1.5–3.5 mm. **FL**: 4–7 mm; perianth yellow. *n*=40. Common. Sand or gravel; (1000)1500–3000 m. SN (e slope), GB, nw DMoj; se OR, w NV. Jun–Sep

var. ***polyanthum*** (Benth.) M.E. Jones AMERICAN RIVER SULPHUR FLOWER Shrub 4–10 dm, 5–10 dm diam. **ST**: 1–2 dm, hairy. **LF**: blades 1–3 cm, 0.3–1(1.3) cm wide, tomentose abaxially, thinly hairy or glabrous and light green adaxially. **INFL**: simple or compound; branches sparsely tomentose or glabrous, occ central branch with a whorl of bracts near middle; involucre tubes 2.5–4 mm, teeth 2–3.5 mm. **FL**: 4–7 mm; perianth yellow. Serpentine(?); 800–1500 m. KR, n SNH/CaRH. Jun–Sep

var. ***smallianum*** (A. Heller) S. Stokes SMALL'S SULPHUR FLOWER Mat 3–5 dm diam. **ST**: 0.5–1.5 dm, tomentose. **LF**: blades 0.5–1.5 cm, 0.3–0.7 cm wide, tomentose abaxially, tomentose to hairy adaxially. **INFL**: main branches of ≥ 2 orders, tomentose; involucre tubes 2–3 mm, lobes 1–3 mm. **FL**: 4–6 mm; perianth yellow. Serpentine; 700–2000 m. NCoRO, NCoRI. [*E. u.* var. *bahiiforme*, in part, misappl.] Jul–Sep ★

var. ***speciosum*** (Drew) S. Stokes BEAUTIFUL SULPHUR FLOWER Shrub 5–15(20) dm, 5–20 dm diam. **ST**: 1.5–3 dm, gen tomentose. **LF**: blades 1–3 cm. 0.5–2.5 cm wide, tomentose abaxially, hairy or glabrous adaxially. **INFL**: main branches of ≥ 2 orders, tomentose; involucre tubes 4–6 mm, teeth 3–5 mm. **FL**: 7–10(12) mm; perianth yellow. Serpentine; 100–800 m. KR. Jun–Sep

var. ***subaridum*** S. Stokes (p. 1113) FERRIS' SULPHUR FLOWER Subshrub or shrub, 2–7 dm, 3–9(12) dm diam. **ST**: (0.5)1–3 dm, hairy or glabrous. **LF**: blades 1–3 cm, 0.5–2 cm wide, gen sparsely tomentose at least abaxially. **INFL**: main branches of ≥ 2 orders, hairy or glabrous; involucre tubes 2–3(3.5) mm, teeth 1–3 mm. **FL**: 3–7 mm; perianth yellow. Common. Sand or gravel; 1200–3100 m. se SN, TR, SNE, DMtns; to sw CO, n AZ. [*E. u.* subsp. *ferrissii* (A. Nelson) S. Stokes] Jun–Oct

var. ***torreyanum*** (A. Gray) M.E. Jones (p. 1113) DONNER PASS BUCKWHEAT Mat 4–8 dm diam. **ST**: 1–2 dm, glabrous. **LF**: blades 1–3(4) cm, 1–2 cm wide, glabrous. **INFL**: main branches of 1 order, glabrous, with a whorl of bracts near middle; involucre tubes 5–7 mm, teeth 2–5 mm. **FL**: 7–10 mm; perianth yellow. *n*=20. Sand or gravel; (1800)2100–2400 m. n SNH (Sierra, Nevada, Placer cos.). Jul–Sep ★

var. ***versicolor*** S. Stokes PANAMINT SULPHUR FLOWER Mat 1–4 dm diam. **ST**: 0.5–1.5 dm, hairy. **LF**: blades 0.5–1.5 cm, 0.3–1 cm wide, sparsely tomentose or ± glabrous. **INFL**: simple or compound; branches hairy; involucre tubes 2–3 mm, teeth 1–2 mm. **FL**: 3–6 mm; perianth yellow, in age red-brown to rose or pink, with large ± red spot on each lobe. Gravel to rock; 1900–3300 m. W&I (Inyo Mtns), DMtns; NV. Jun–Sep

E. ursinum S. Watson (Group 3) Per mat, (1.5)2–8(12) dm diam. **ST**: 0.4–4 dm, sparsely tomentose to hairy or ± glabrous. **LF**: basal; blades 0.7–2(2.5) cm, 0.4–1.2(2) cm wide, tomentose abaxially, sparsely tomentose, sparsely hairy or glabrous adaxially. **INFL**: 1–3(6) cm, 1–5 cm wide, sparsely tomentose to ± glabrous; involucres 3.5–8 mm, 2.5–4 mm wide, hairy; teeth 5–8. **FL**: 4–9 mm, stipe 0.5–1.3 mm, glabrous; perianth cream or ± yellow, blushed or not with pink-red to maroon, lobes gen widely ovate or obovate. **FR**: 3–8 mm, glabrous exc tip sparsely hairy.

var. ***erubescens*** Reveal & J. Knorr BLUSHING WILD BUCKWHEAT Pl (1.5)2–8(12) dm diam. **ST**: 0.4–2 dm, sparsely tomentose to hairy. **LF**: blades 0.7–2(2.2) cm, 0.4–1(1.2) cm wide. **INFL**: sparsely tomentose; involucres 5–8 mm, 3–4 mm wide. **FL**: 5–9 mm; perianth cream (or yellow), blushed pink-red to maroon. **FR**: (5)5.5–8 mm. Gravel; 1600–1900 m. KR (Scott Bar Mtns, Siskiyou Co.; Trinity Mtn, Trinity Co.). Jun–Sep ★

var. ***ursinum*** (p. 1113) BEAR VALLEY WILD BUCKWHEAT Pl 3–6 dm diam. **ST**: 0.4–4 dm, sparsely tomentose to ± glabrous. **LF**: blades 0.8–1.4(2.5) cm, 0.5–1.2(2) cm wide, tomentose abaxially, sparsely tomentose or glabrous adaxially. **INFL**: sparsely tomentose to ± glabrous; involucres 3.5–4.5(5) mm, 2.5–4 mm wide. **FL**: 4–6 mm; perianth pale yellow (or yellow), not blushed with other colors. **FR**: 3–3.5 mm. Sand or gravel; (500)900–2500 m. KR, s CaR, n SN. May–Sep

E. vestitum J.T. Howell (p. 1113) IDRIA BUCKWHEAT (Group 1) Ann (0.5)1–4(5) dm. **ST**: 0.5–1 dm, tomentose. **LF**: basal and cauline; blades (0.7)2–4(5) cm, 0.5–2(3.5) cm wide, sparsely tomentose abaxially, hairy to ± glabrous adaxially. **INFL**: 10–40(45) cm, 5–30 cm wide; branches tomentose; peduncles erect, 0.5–4(6) cm, straight, slender, tomentose; involucres at tips of slender branchlets, (1.5)1.8–2 mm, 1.5–2(2.2) mm wide, tomentose. **FL**: 1.5–2(2.5) mm, glabrous, papillate; perianth white, lobes narrowly elliptic to oblong; style 0.7–1 mm. **FR**: 2–2.5 mm; tip papillate. *n*=17. Clay; 400–700 m. ne SCoRI (Merced, San Benito, Fresno cos.). Mar–Nov ★

E. vimineum Benth. (p. 1113) WICKER-STEM WILD BUCKWHEAT (Group 1) Ann 0.5–3(5) dm. **ST**: 0.5–1 dm, sparsely tomentose to hairy, occ glabrous. **LF**: basal; blades 0.5–2(2.5) cm, ± round, tomentose abaxially, less so to ± glabrous adaxially. **INFL**: 5–25 cm; branches hairy or glabrous; peduncles 0; involucres appressed, 2–3.5(4) mm, 1–2 mm wide, glabrous (hairy). **FL**: 2–2.5 mm, glabrous; perianth white to rose or pale yellow, outer lobes widely spoon-shaped to obovate, inner oblanceolate. **FR**: 2–2.5 mm. *n*=12. Common. Sand or gravel; 50–2400 m. ne KR, CaR, n SNH, MP; to se WA, w ID, n NV. May–Sep

E. viridescens A. Heller (p. 1113) TWO-TOOTHED WILD BUCKWHEAT (Group 2) Ann (0.5)1–2(3) dm. **ST**: 0.2–0.8 dm, tomentose. **LF**: basal and cauline; blades (0.5)2–3 cm, 0.3–2 cm wide, tomentose abaxially, hairy to ± glabrous. **INFL**: 5–30 cm; branches tomentose; peduncles spreading, straight, thread-like, 1–2 cm, sparsely glandular; involucres 2–3 mm, 2–4 mm wide, glandular. **FL**: 1–2.5 mm, glandular; perianth white to rose, outer lobes obovate to spoon-shaped, inflated above middle, inner narrowly spoon-shaped. **FR**: 1–1.5 mm, glabrous. *n*=20. Locally common. Sand, gravel or clay; 100–1700 m. SnJV, SCoR, TR, w DMoj. Apr–Nov

E. wrightii Benth. BASTARD-SAGE (Group 4) Subshrub or shrub, 1.5–10 dm, 1–15(18) dm diam or mat 0.5–4 dm diam. **ST**: (0.1)0.5–4(6) dm, tomentose to glabrous. **LF**: gen cauline; blades 0.1–3 cm, 0.1–2.5 cm wide, linear to narrow, tomentose to hairy or glabrous, margins occ rolled under. **INFL**: rarely head-like, (1)5–20 cm, (1)10–40 cm diam; peduncles 0; involucres (0.7)1–4 mm, 1–2.5 mm wide, tomentose to glabrous. **FL**: 1.5–4 mm, glabrous; perianth white to pink or rose, lobes obovate. **FR**: (1)1.5–3 mm, glabrous. Other vars. in n Baja CA.

var. ***membranaceum*** Jeps. (p. 1113) RINGED-STEM BASTARD-SAGE Subshrub 2–4(5) dm, 3–6 dm diam, gen sparsely tomentose. **LF**: blades 0.2–0.6(1) cm, 0.1–0.3(0.4) cm wide. **INFL**: branched; involucres 2–3 mm. **FL**: (1.5)2–2.5 mm; perianth white to pink or rose. **FR**: 2.5–3 mm. Common. Gravel or rocks; 300–2200 m. SW; n Baja CA. Jul–Oct

var. ***nodosum*** (Small) Reveal (p. 1113) KNOT-STEM BASTARD-SAGE Shrub 3–10 dm, (3)5–15 dm diam, tomentose. **LF**: blades 0.8–1.2 cm, 0.25–0.5(0.9) cm wide. **INFL**: branched; involucres 0.7–2.5 mm. **FL**: 1–4 mm; perianth white to pink or rose. **FR**: 1–3 mm. Uncommon. Sand or gravel; 100–1600 m. e PR, s DMoj, w DSon; s AZ, nw Mex. Aug–Feb

var. ***olanchense*** (J.T. Howell) Reveal (p. 1113) OLANCHA PEAK BUCKWHEAT Mat 0.1–0.3(0.6) dm, 0.5–3 dm diam, sparsely tomentose. **LF**: blades 0.1–0.25 cm, 0.06–0.12 cm wide. **INFL**: gen head-like; involucres (0.8)1–1.7(2) mm. **FL**: (1.5)2–2.5 mm; perianth white to pink. **FR**: 1.5–2 mm. Gravel or rock; 3500–3600 m. s SNH (Olancha Peak, Tulare Co.). Jul–Aug ★

var. ***subscaposum*** S. Watson SHORT-STEMMED BASTARD-SAGE Mat 0.5–2.5(3) dm, 1–3(5) dm diam, tomentose or glabrous. **LF**: blades 0.5–1(1.2) cm, 0.2–0.4(0.5) cm wide. **INFL**: branched;

involucres 1.5–4 mm. **FL**: 2–3 mm; perianth white to pink. **FR**: 2–2.5 mm. *n*=17. Common. Gravel or rocks; 200–3400 m. SN, CW, TR, e PR, SNE, expected n DMtns. Jun–Sep

var. ***trachygonum*** (Benth.) Jeps. (p. 1113) ROUGH-NODE BASTARD-SAGE Subshrub 1.5–4 dm, 1–5 dm diam, tomentose. **LF**: blades 1.5–3 cm, 0.5–1 cm wide. **INFL**: branched; involucres 2–4 mm. **FL**: 2.5–4 mm; perianth white to pink or rose. **FR**: 2.5–3 mm. Common. Gravel; 40–800 m. NW, n SNF. Jul–Oct

var. ***wrightii*** WRIGHT'S BASTARD-SAGE Shrub or subshrub, (1)1.5–5(7.5) dm, 1–12(18) dm diam. **LF**: blades 0.5–1.5 cm, 0.2–0.5(0.7) cm wide. **INFL**: branched; involucres 2–2.5 mm. **FL**: 2.5–3.5 mm; perianth white to pink or rose. **FR**: 2.5–3 mm. Uncommon. Sand, gravel or rocks; (300)900–2200 m. DMoj; to w TX, n Mex. Jul–Oct

FALLOPIA

Ann, per, vine or not. **ST**: erect, trailing, or twining, glabrous or hairy, ribs 0. **LF**: cauline, alternate, petioled; ocrea persistent or not, cylindric, papery, opaque; blade broad-ovate to triangular, entire. **INFL**: axillary, terminal, spike-, panicle-, or raceme-like; fls 3–7; peduncle present or 0. **FL**: bisexual or pistillate (1 kind per pl); perianth gen enlarging, bell-shaped, pale green or white to pink, glabrous, base stalk-like, parts 5, fused basally or ± completely, of 2 kinds, outer 3 gen winged, > inner 2; stamens 6–8, filaments free, wider basally, glabrous or hairy, anthers yellow to pink or red, ovate to elliptic; styles 3, spreading, fused basally or ± completely, stigmas head-like, fringed, or peltate. **FR**: incl or exserted, 3-angled, brown to black. **SEED**: embryo straight. 12 spp.: Am, Eur, Asia, Afr. (Gabriele Fallopio, Italian botanist, 1523–1562)

1. Rhizomes present, per; sts erect; stigmas fringed
  2. St gen 1–2 m; lf blade 6–13 cm, 5–10 cm wide, base truncate . . . . ***F. japonica***
  2′ St 2–4(5) m; lf blade 15–30(40) cm, 7–25 cm wide, base cordate . . . . ***F. sachalinensis***
1′ Rhizomes 0; ann, per; sts trailing, twining, or climbing; stigmas head-like or peltate
  3. St woody; infl panicle-like . . . . ***F. baldschuanica***
  3′ St not woody; infl spike-like . . . . ***F. convolvulus***

F. baldschuanica (Regel) Holub BUKHARA FLEECEFLOWER Per vine, rhizomes 0. **ST**: woody, climbing, 2–10 m, glabrous, branches from near base. **LF**: ocrea 3–8 mm, ± brown, margins truncate to oblique, glabrous; blade 3–10 cm, 1–5 cm wide, narrow- to oblong-ovate, abaxially glabrous or scabrous on midvein, rarely minute-dotted, adaxially glabrous, base cordate to sagittate, tip obtuse to acuminate. **INFL**: axillary, terminal, spreading or drooping, panicle-like, 3–15 cm; peduncle 1–3 cm; pedicels 1.5–4 mm, glabrous or scabrous; fls 3–6. **FL**: bisexual, perianth 5–8 mm, green-white with white wings or mostly pink, bright pink in fr or not, glabrous, lobes elliptic, tips obtuse to rounded; stamens 6–8, filaments hairy basally; styles fused basally, stigmas peltate. **FR**: incl, 2–4 mm, 1.8–2.2 mm wide, dark-brown to black, shiny, smooth. 2*n*=20. Disturbed places; < 500 m. SnFrB, SCo, expected elsewhere; to e N.Am; orn native to c Asia often escaping from cult; introduced to C.Am, Eur. Potentially problematic weed [*Polygonum b.* Regel; *P. aubertii* L. Henry] Aug–Oct

F. convolvulus (L.) A. Löve BLACK BINDWEED Ann, rhizomes 0. **ST**: not woody, trailing or twining, 0.4–1.2 m, puberulent, fine-granular or not, branches proximal. **LF**: ocrea 2–4 mm, tan or green-brown, margin oblique, glabrous or scabrous; blade 2–6(15) cm, 2–5(10) cm wide, ± ovate, abaxially gen fine-granular, rarely minute-dotted, adaxially glabrous, base ovate, hastate, or sagittate, tip acuminate. **INFL**: axillary, spike-like, 2–10(15) cm; peduncle 0–10 cm; pedicels 1–3 mm, glabrous (scabrous); fls 3–6. **FL**: bisexual, perianth not enlarging, 3–5 mm, green-white, often with ± pink base, glabrous or outer 3 lobes short-hairy, lobes elliptic to obovate, tips obtuse to acute, outer 3 ± keeled; stamens 8, filaments glabrous; styles fused distally, stigmas head-like. **FR**: incl, 4–5(6) mm, 1.8–2.3 mm wide, black, dull, minute-granular-tubercled, esp on faces. 2*n*=40. Disturbed places; < 1000 m. CA-FP, MP; to e N.Am; native to Eur. [*Polygonum c.* L.] May–Dec

F. japonica (Houtt.) Ronse Decr. (p. 1113) JAPANESE KNOTWEED Per, rhizomed. **ST**: gen in dense clumps, erect, gen 1–2 m, glaucous,

glabrous, branches few. **LF**: ocrea 5–8 mm, ± brown, margin oblique, glabrous or puberulent; blade 6–13 cm, 5–10 cm wide, widely ovate, abaxially with 1-celled hairs < 0.1 mm on veins, adaxially glabrous, base truncate, tip acuminate. **INFL**: axillary, gen distal, erect or spreading, panicle-like, 3–10 cm; peduncle 0–2.5 cm; pedicels 3–5 mm, glabrous; fls 4–6. **FL**: bisexual or functionally ± unisexual; perianth 4.5–6.5 mm, winged, ± white, glabrous, lobes obovate to elliptic, tips obtuse to acute; stamens 6–8, filaments glabrous; styles fused basally, stigmas fringed. **FR**: incl, 3.8–5 mm, 2–2.2 mm wide, dark-brown, shiny, smooth. 2*n*=44,88. Disturbed places; < 1000 m. n CA-FP (esp s NCoR), SnFrB; to e N.Am; native to Japan. [*Polygonum cuspidatum* Siebold & Zucc.; *Reynoutria j.* Hoult] Introduced as orn, spreading aggressively. Aug–Oct ◆

F. sachalinensis (F. Schmidt) Ronse Decr. (p. 1113) GIANT KNOTWEED Per, rhizomed. **ST**: gen in dense clumps, erect, 2–4(5) m, glabrous, branches few. **LF**: ocrea 6–12 mm, ± brown, margins oblique, glabrous or puberulent; blade 15–30(40) cm, 7–25 cm wide, ovate-oblong, abaxially minute-dotted, with many-celled hairs 0.2–0.6 mm on veins, adaxially glabrous, base cordate, tip acute to acuminate. **INFL**: axillary, gen distal, erect or spreading, panicle-like, 3–8 cm; peduncle 0–4 cm; pedicels 2–4 mm, glabrous; fls 4–7. **FL**: bisexual or functionally ± unisexual; perianth 4.5–6.5 mm, ± green, glabrous, lobes obovate to elliptic, tips obtuse to acute; stamens 6–8, filaments glabrous; styles fused basally, stigmas fringed. **FR**: incl, 2.8–4.5 mm, 1.1–1.8 mm wide, brown, shiny, smooth. 2*n*=44,66,102,132. Disturbed places; < 500 m. NCo, SnFrB, SCo, expected elsewhere; native to c Asia. [*Polygonum s.* F. Schmidt; *Reynoutria s.* (F. Schmidt) Nakai] Introduced as orn, spreading aggressively; hybrids with *F. japonica* (*F.* ×*bohemica* (Chrtek & Chrtková) J.P. Bailey), intermediate in lf shape, size, potentially spreading vegetatively. Aug–Sep ◆

GILMANIA　GOLDEN-CARPET

1 sp. (M. French Gilman, Death Valley naturalist, 1871–1944)

G. luteola (Coville) Coville (p. 1113)　GOLDEN-CARPET GILMANIA
Ann, decumbent, 0.3–1.2(1.5) dm, 0.3–1.5(2) dm diam, glabrous or
sparsely hairy. **LF:** basal and cauline, in whorls of 3; blades 0.5–
1.5 cm, 0.3–0.8(1) cm wide, glabrous or sparsely hairy; ocreae 0.
INFL: terminal, cyme-like; peduncles 0; involucre 0. **FL:** 3–9, 1–2
mm, thinly hairy; perianth yellow, lobes 6, entire; stamens 9. **FR:**
brown, obconic, 1.5–2 mm, glabrous; embryo curved. Clay; 10–500
m. DMoj (Death Valley, Inyo Co.). Feb–May　★

GOODMANIA

1 sp. (G.J. Goodman, *Chorizanthe* authority, 1904–1999)

G. luteola (Parry) Reveal & Ertter (p. 1113)　GOLDEN GOODMANIA
Ann, spreading, (0.3)0.5–1.5 dm, (0.1)0.3–1.5 dm diam, hairy or
glabrous. **LF:** basal and cauline, opposite; basal blades 0.2–0.5(0.7)
cm, gen ± round, cauline 0.3–0.5 cm, 0.05–0.15 cm wide, gen linear,
awned; ocreae 0. **INFL:** terminal, cyme-like; peduncles 0; involucre
bracts in 1 whorl of 5, 2–5 mm, narrow, gen glabrous, awned. **FL:** (6)9–
15, 0.8–1 mm, tomentose; perianth yellow, lobes 6, entire; stamens 9.
FR: light brown, obconic, 1–1.2 mm, glabrous; embryo curved. *n*=21.
Clay; 70–2200 m. s SnJV (extirpated), SNE (s Mono, n Inyo cos.), w
DMoj (s Kern, n Los Angeles cos.); w-c NV. Apr–Aug　★

HOLLISTERIA

1 sp. (W.W. Hollister, CA rancher, 1818–1886)

H. lanata S. Watson (p. 1117)　FALSE SPIKEFLOWER　Ann,
decumbent to prostrate, 0.3–0.8(1) dm, (0.5)0.8–3(5) dm diam,
tomentose. **LF:** basal and cauline, alternate; blades (0.5)1–5(6) cm,
(0.1)0.3–0.7(1) cm wide, tomentose, spiny-tipped; ocreae 0. **INFL:**
terminal, cyme-like; peduncles 0; involucre bracts in 1 whorl of
3(4), narrow, 1–2 mm, awned. **FL:** 1, 1.5–2 mm, densely tomentose;
perianth ± yellow, lobes 6, entire; stamens 6 or 9. **FR:** 1.7–2 mm,
obconic, glabrous, brown to black; embryo curved. *n*=21. Sand, clay
or gravel; 10–1000 m. s SnJV, SCoRI. Mar–Jul

JOHANNESHOWELLIA

Ann, weakly erect to spreading, silky-puberulent. **LF:** basal; blades obovate to ± round [± reniform], awns 0; ocreae 0. **INFL:**
terminal, cyme-like; peduncles 0; involucre bracts obscure, 4–7 in a spiral, narrow, awns 0. **FL:** 4–7, glabrous, smooth or
papillate; perianth white or pale ± yellow to rose or red, lobes 6, entire; stamens 9. **FR:** light brown, obconic, glabrous; embryo
curved. 2 spp.: w N.Am. (J.T. Howell, CA botanist, *Eriogonum* scholar, 1903–1994) [Reveal 2004 Brittonia 56:299–306]

J. puberula (S. Watson) Reveal (p. 1117)　DOWNY BUCKWHEAT
Pl erect, 0.5–3 dm. **LF:** blades 0.5–1.5 cm. **INFL:** involucre bracts
0.5–1.2 mm, 0.1–0.3 mm wide, outermost 2–3-lobed. **FL:** 1–2 mm,
minutely papillate; perianth white or pale yellow to rose or red. **FR:**
1–1.5 mm. Sand; (500)800–2800 m. n DMtns (Cottonwood Mtns,
Death Valley region, Inyo Co.); to sw UT. [*Eriogonum p.* S. Watson]
May–Sep　★

LASTARRIAEA

Ann, prostrate to ascending, hairy. **LF:** basal; blades linear, awns 0; ocreae 0. **INFL:** terminal, cyme-like, peduncles 0; involucre bracts in 1 whorl of 3, linear to narrow, awned. **FL:** 1, hairy; perianth light green to green-white, lobes 5, leathery, awned;
stamens 3. **FR:** elliptic, glabrous, ± brown; embryo straight. 3 spp.: w N.Am, s S.Am. (J.V. Lastarria Santander, founder of
Liberal Party in Chile, 1817–1888)

L. coriacea (Goodman) Hoover (p. 1117)　LEATHER-SPINEFLOWER
Pl 0.2–1.5 dm, 0.5–3(5) dm diam, hairy. **LF:** blades 0.5–3 cm,
0.02–0.08(0.1) cm wide. **INFL:** readily breaking; involucre bracts
0.3–1(1.2) cm, 0.03–0.1 cm wide, linear, awns 0.5–2 mm, hooked.
FL: 2–3.5 mm (incl awns); perianth lobes narrow. **FR:** 2.5–3 mm.
n=21, 30. Common. Sand or gravel; < 800 m. SN, GV, CW, SW; nw
Mex. Feb–Jun

MUCRONEA

Ann, erect to spreading, glandular-hairy. **LF:** basal; blades entire, gen narrow, awns 0; ocreae 0. **INFL:** terminal, cyme-like;
bracts fused, on ± 1 side of or around node, awned; peduncles 0; involucres 1–3, (2)3–4-angled, tubular, occ swollen, teeth
(2)3–4, unequal, awned. **FL:** 1(2), hairy; perianth white to pink, lobes 6, entire or fringed; stamens 6 or 9. **FR:** gen elliptic,
glabrous, brown to black; embryo straight. 2 spp.: CA. (Latin: sharp point, for awns of bracts, involucres)

1. Bracts fused, on ± 1 side of node; involucres 1–3; involucre teeth (2)3(4), 2.5–5(7) mm; involucre awns
　　(0.5)1–2.5(3) mm; fls 1–2; perianth lobes entire . . . . ***M. californica***
1′ Bracts fused, around node; involucres 1; involucre teeth 4, 3–5(6) mm; involucre awns (0.3)0.5–1.2 mm;
　　fls 1; perianth lobes fringed, occ entire or 2-lobed . . . . ***M. perfoliata***

M. californica Benth. (p. 1117)　CALIFORNIA SPINEFLOWER
Pl (0.3)0.5–3(5) dm, 1–6(8) dm diam. **LF:** blades (0.5)1–5 cm,
(0.1)0.2–0.8(1.2) cm wide. **INFL:** lower bracts 3(5), fused 1/2, awns
1–2.5(3) mm; involucres (2)3(4)-angled, obscurely ribbed, not transversely ridged, not swollen, 2.5–5(7) mm. **FL:** 1.5–2.5(3) mm, hairy
basally; stamens 6–9. **FR:** 2–3 mm. *n*=19. Sand; < 1000 m. s CW,
SW. Mar–Aug　★

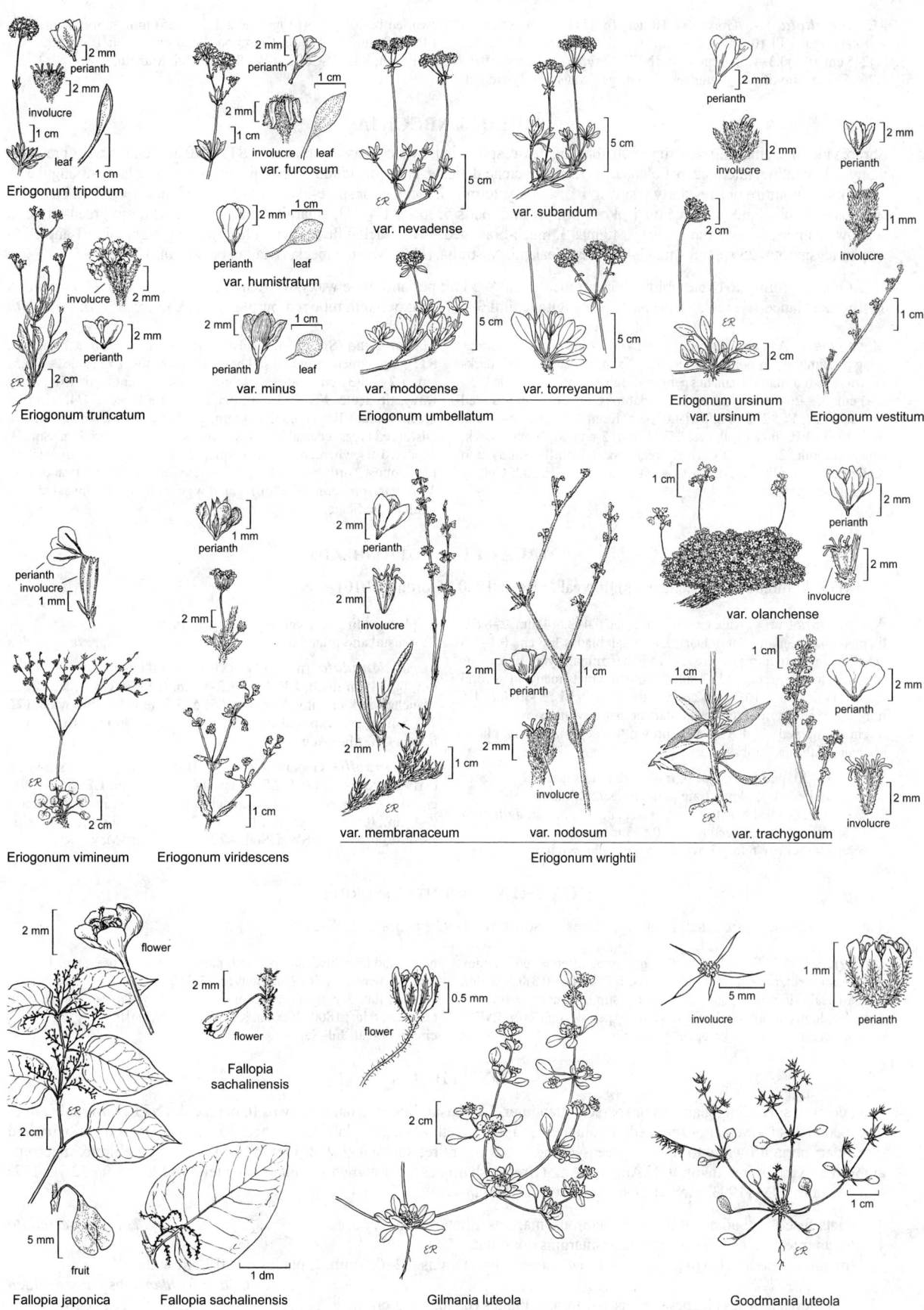

Eriogonum tripodum

Eriogonum truncatum

var. furcosum
var. humistratum
var. minus
Eriogonum umbellatum

var. nevadense

var. subaridum

var. modocense

var. torreyanum

Eriogonum ursinum
var. ursinum

Eriogonum vestitum

Eriogonum vimineum

Eriogonum viridescens

var. membranaceum

var. nodosum

var. olanchense

var. trachygonum

Eriogonum wrightii

Fallopia
sachalinensis

Fallopia japonica

Fallopia sachalinensis

Gilmania luteola

Goodmania luteola

M. perfoliata (A. Gray) A. Heller (p. 1117) PERFOLIATE SPINEFLOWER Pl (0.2)0.3–2(3) dm, 0.5–5 dm diam. **LF**: blades (1)2–5 cm, (0.2)0.3–1.2(2) cm wide. **INFL**: lower bracts 3, awns 0.3–1.2(1.5) mm; involucre 4-angled, ribbed, gen transversely ridged, swollen basally, 3–5(6) mm. **FL**: 1.5–3(3.5) mm, hairy; stamens 9. **FR**: 2–3 mm. n=19, 20. Common. Sand or gravel; 100–1600 m. s SNF, Teh, SnJV, SCoRI, n WTR, w DMoj. Mar–Jul

MUEHLENBECKIA

Shrub, vine-like, climbing or, if support 0, forming dense, sprawling, twisted masses, rhizomed. **ST**: to 10 m, slender, wiry, branches many. **LF**: cauline, alternate, petioled, ± deciduous; ocreae delicate, early-deciduous; blade linear to round or lance-triangular, ± thick, margins entire or irregularly wavy. **INFL**: axillary, terminal, raceme- or spike-like; fls 1–3(5); peduncles ± 0. **FL**: bisexual, staminate, or pistillate (all three on 1 pl or not); perianth parts 5, fused 1/4–1/2, enlarging, fleshy in fr, white to green-white to -yellow; stamens 8; style rudimentary, stigmas 3, much-branched. **FR**: ± incl in fleshy perianth, appearing berry-like; 3-angled to ± spheric, smooth. 23 spp.: S.Am, C.Am, New Zealand, Australia. (H.G. Muehlenbeck, Alsatian physician, 1798–1845)

1. Lf blade ± round to ovate-oblong, base truncate; infl < 3 cm; perianth tube waxy-white in fr . . . . **M. complexa**
1′ Lf blade lance-triangular, base hastate or sagittate; infl 3–5(7) cm; perianth tube red-purple to black in fr . . . . **M. hastulata**

M. complexa (A. Cunn.) Meisn. MAIDENHAIR VINE Shrub, straggling or climbing, in dense tangles. **ST**: ≤ 5 m, wiry, ± twisted, dark-brown or ± red; distal branches ± brown-puberulent. **LF**: blade 0.5–2(4) cm, 0.5–2(4) cm wide, entire, ± glabrous, leathery, tip rounded to mucronate. **FL**: 2–5 mm; perianth parts fused ± 1/2, yellow-green or ± green. **FR**: incl or exserted, 3–4 mm, 3-angled, brown-black, shiny, smooth. $2n$=20. Disturbed areas, coastal bluffs, sandy and rocky places; < 500 m. NCo, CCo; native to New Zealand. Cult as orn. Jul–Sep

M. hastulata (Sm.) I.M. Johnst. WIREVINE Shrub, ± climbing. **ST**: ≤ 3 m, green-brown; distal branches glabrous. **LF**: blade 2–4(5.5) cm, 1.2–2.5(3) cm wide, ± glabrous, margins entire or irregularly wavy, tip acute. **FL**: 1–3 mm; perianth parts fused ± 1/4, white to green-white. **FR**: gen incl, 3–4 mm, ± spheric, black, shiny, smooth. Disturbed areas, coastal bluffs, sandy, rocky places; < 500 m. SnFrB, expected elsewhere; native to S.Am. [*Muehlenbeckia hastatula* (Sm.) I.M. Johnst., orth. var.; *M. h.* var. *fascicularis* (Meisn.) Brandbyge; *M. h.* var. *rotundata* (F. Phil.) Brandbyge] Cult as orn; invasive, difficult to eradicate. Jul–Sep

NEMACAULIS COTTONHEADS

1 sp. (Greek: thread st, for slender sts) [Reveal & Ertter 1980 Madroño 27:101–109]

N. denudata Nutt. Ann, prostrate to erect, 0.4–2.5(4) dm, 2–8 dm diam, glabrous, glandular or hairy. **LF**: basal; blades 1–8 cm, 0.1–1.5 cm wide, linear to narrow; ocreae 0. **INFL**: terminal, cyme-like; awns 0; peduncles 0 or erect, 0.5–3 mm; involucre bracts many, in several whorls, (1)2–4 mm, (0.3)0.5–2 mm wide, awns 0. **FL**: 5–30, 0.8–1.5 mm, glabrous or minutely glandular; perianth white to rose, lobes 6, entire; stamens 3. **FR**: 1 mm, brown to deep maroon or black, obconic, glabrous; embryo curved.

1. Fl (5)12–30; peduncles gen 0; involucre bracts gen red with white hairs; pl prostrate to decumbent; lf blades gen wide; coastal beaches . . . . var. **denudata**
1′ Fl 5(12); peduncles gen present, 0.5–3 mm; involucre bracts light brown to yellow-green with ± brown hairs; pl ascending to erect; lf blades gen linear; coastal and inland deserts . . . . var. **gracilis**

var. **denudata** (p. 1117) COAST WOOLLY-HEADS Pl 0.4–2 dm, 0.2–8 dm diam. **LF**: blades 2–8 cm, 0.3–1.5 cm wide. **INFL**: branches dark red; involucre bracts 1.5–3.5 mm, 1–2 mm wide. **FL**: 1–1.5 mm, gen exposed in hairs. Beaches; < 100 m. s CCo, SCo; n Baja CA. Mar–Aug ★

var. **gracilis** Goodman & L.D. Benson (p. 1117) SLENDER COTTONHEADS Pl 0.4–2.5(4) dm, 0.4–2 dm diam. **LF**: blades 1–7 cm, 0.1–0.6 cm wide. **INFL**: branches light brown; involucre bracts 2–4 mm, 0.5–1 mm wide. **FL**: 0.5–1.2 mm, gen obscured by hairs. Deserts; 10–500 m. SW, DSon; AZ, nw Mex. Jan–May ★

OXYRIA MOUNTAIN SORREL

4 spp. (Greek: sour, from acidic taste) [Chrtek & Sourková 1992 Preslia 64:207–210]

O. digyna (L.) Hill (p. 1117) Per, glabrous, often ± red; caudex thick, scaly; rhizome rare; roots fibrous. **ST**: erect, (0.3)0.5–5 dm. **LF**: ± basal, alternate, petioled, ± fleshy, stipules fused, ± translucent, deciduous or not; blade 0.5–6.5 cm, << petiole, reniform. **INFL**: panicle, erect, 2–20 cm, ± open; peduncle 1–15 cm. **FL**: 2–6 per node, nodding, bisexual, perianth parts 4, outer 2 spreading, 1.2–2.5 mm; stamens 6; styles 2, stigmas red. **FR**: 3–4.5 mm, 2.5–5 mm wide, elliptic, flat, 2-winged, ± red or ± pink, veiny. $2n$=14. Alpine rock crevices, talus; 1800–4000 m. KR, CaRH, SN, SnBr, SnJt, Wrn, SNE; circumboreal. Jul–Sep

OXYTHECA

Ann, erect to spreading, glabrous or sparsely glandular. **LF**: basal, linear to narrow, awns 0; ocreae 0. **INFL**: terminal, cyme-like; peduncles 0 or erect or reflexed; involucres 1, tubular, teeth (3)4, gen glabrous, awned. **FL**: 2–10, glabrous or hairy and glandular; perianth white to rose or green-yellow, lobes 6, entire; stamens 9. **FR**: brown to maroon, gen elliptic, glabrous; embryo curved. 3 spp.: temp w N.Am, s S.Am. (Greek: sharp case, for awned involucre) [Ertter 1980 Brittonia 32:70–102] Other taxa in TJM (1993) moved to *Acanthoscyphus*, *Sidotheca*.

1. Bracts fused around node; lf blades glabrous, margins ciliate . . . . **O. perfoliata**
1′ Bracts free; lf blades hairy, glandular, margins not ciliate
  2. Involucres peduncled (occ sessile) at lower nodes; bract awns 0.2–0.5 mm; lf blades 0.1–0.7 cm wide
  . . . . **O. dendroidea** subsp. **dendroidea**
  2′ Involucres gen sessile; bract awns 1–3 mm; lf blades (0.1)0.5–1.2 cm wide . . . . **O. watsonii**

O. dendroidea Nutt. subsp. *dendroidea* (p. 1117) TREE-LIKE PUNCTUREBRACT Pl erect to spreading, 0.4–4 dm, 0.3–4.5 dm diam. **LF:** blades 1–4.5 cm, 0.1–0.7 cm wide, gen linear, hairy, glandular. **INFL:** bracts 2–3(4), free, awns 0.2–0.5 mm; peduncles erect or reflexed, 5–15 mm, glabrous, occ 0; involucre 1–2 mm, lobes (3)4, awns ± gray, 0.5–3 mm. **FL:** 2–6, 1–2 mm, occ glabrous; perianth white to pink. **FR:** 2–2.5 mm. *n*=20. Common. Sand or gravel; 300–3000 m. s SNH, GB; to e WA, sw WY, e NV. Other subsp. in Andes of Chile, Argentina. Jun–Oct

O. perfoliata Torr. & A. Gray (p. 1117) ROUND-LEAF PUNCTUREBRACT Pl spreading, 0.6–2 dm, 0.5–4 dm diam. **LF:** blades 1–6 cm, 0.3–1.5 cm wide, glabrous, margins ciliate. **INFL:** bracts 4–5, fused around node, 1–2.5 cm diam, glabrous or glandu-lar, awns 1–3 mm; peduncles 0 or erect, 0.3–0.8 mm; involucre 2–5 mm, occ sparsely glandular, lobes 4, awns 2–3 mm, ± red. **FL:** 5–10, 1.5–2.5 mm; perianth white to yellow-green or pink. **FR:** 1.5–2 mm. *n*=20. Common. Sand or gravel; 600–1900 m. s SnJV, GB, DMoj; to sw UT, nw AZ. Apr–Aug

O. watsonii Torr. & A. Gray (p. 1117) WATSON'S OXYTHECA **ST:** erect to spreading, 0.5–2.5 dm, 0.4–4 dm diam. **LF:** blades 0.7–4 cm, (0.1)0.5–1.2 cm wide, hairy and glandular adaxially, less so or glabrous abaxially. **INFL:** bracts gen 3, free, awns 1–3 mm; peduncles reflexed, 2–5 mm, or gen 0; involucre 1.5–2 mm, lobes 4, awns 2.5–3 mm, ± red. **FL:** 2–4(7), 1–1.5 mm; perianth white to pink. **FR:** 1–1.5 mm. *n*=20. Sand; 1200–2000 m. se W&I (Santa Rosa Hills, Inyo Co.); NV. Jun–Oct ★

PERSICARIA SMARTWEED

Ann, per, rhizomed or stoloned. **ST:** prostrate to erect, ribbed or ± so or not, glabrous or hairy, gen with adventitious roots. **LF:** cauline, alternate, petioled or not; ocrea papery, rarely ± lf-like, opaque, persistent or disintegrating, glabrous to variously hairy; blade lanceolate or ovate to hastate ± or sagittate, entire. **INFL:** axillary, terminal, gen spike-like; fls 1–14; peduncle present, pedicels present [or 0]. **FL:** bisexual or functionally ± unisexual, base not stalk-like; perianth not or ± enlarging, bell-shaped (urn-shaped, rotate), glabrous, gland-dotted or not, green-white, white, pink, or red; perianth parts 4–5, fused 1/4–2/3, outer 2 > inner 2 or 3; stamens 5–8, filaments free, cylindric, thread-like, glabrous, outer fused to perianth tube or not, anthers elliptic to ovate, yellow, pink, or red; styles 2–3, erect to reflexed, free or fused, stigmas head-like. **FR:** incl or exserted, brown or dark-brown to black, not winged, discoid, lens-shaped, or 3-angled. **SEED:** embryo curved. ± 100 spp.: ± worldwide. (Latin: peach, pertaining to; from resemblance of lvs of some spp. to those of peach)

1. Infl ± head-like; petioles gen winged, ear-like lobes at base (sect. *Cephalophilon*) . . . . *P. capitata*
1′ Infl spike- or panicle-like; petioles not winged, ear-like lobes 0
2. Infl panicle-like; perianth rotate, parts fused < 1/5 (sect. *Rubrivena*) . . . . *P. wallichii*
2′ Infl spike-like; perianth bell-shaped or ± so, parts fused > 1/3 (sect *Persicaria*)
3. Perianth gland-dotted
4. Perianth lobe margins ± red; fr dull; lvs bitter-tasting . . . . *P. hydropiper*
4′ Perianth lobe margins gen white; fr shiny; lvs not bitter-tasting . . . . *P. punctata*
3′ Perianth not gland-dotted
5. Per, gen rhizomed or stoloned
6. Infl terminal, 1 or paired, 8–20 mm wide; styles 2; fr lens-shaped . . . . *P. amphibia*
6′ Infl axillary, terminal, 2 or more, 2–5 mm wide; styles 3; fr 3-angled . . . . *P. hydropiperoides*
5′ Ann, rhizomes or stolons 0
7. Ocreae lf-like distally . . . . *[P. orientalis]*
7′ Ocreae papery throughout
8. Peduncle glands 0; bristles 2–3.5(4.5) mm on ocrea margins . . . . *P. maculosa*
8′ Peduncle glands ± stalked; bristles 0 or < 1 mm on ocrea margins
9. Perianth lobes 4(5), outer with anchor-shaped veins; infl gen ± nodding . . . . *P. lapathifolia*
9′ Perianth lobes 5, outer without anchor-shaped veins; infl gen erect . . . . *P. pensylvanica*

P. amphibia (L.) Delarbre (p. 1117) WATER SMARTWEED Per, rhizomed or stoloned. **ST:** prostrate to erect, 20–120 (300) cm, glabrous or hairy, ± ribbed. **LF:** ocrea 5–50 mm, ± cylindric, lf-like distally or not, tan to dark-brown, margins truncate to oblique, glabrous or ciliate, bristles 0.5–4.5 mm, surface glabrous or appressed-hairy, not gland-dotted; petiole 0–3(7) cm; blade 2–15(20) cm, 1–6(8) cm wide, ovate- or lance-oblong to elliptic, glabrous or sparsely strigose, gen not gland-dotted, adaxial dark blotch 0, base tapered to acute or rounded (cordate), tip acute to acuminate. **INFL:** 1 or paired, terminal, ascending to erect, gen uninterrupted, 10–150 mm, 8–20 mm wide; peduncle 10–500 mm, glabrous or strigose, often stalked-glandular; bractlets gen overlapped; pedicels ascending, 0.5–1.5 mm; fls 1–4. **FL:** perianth 4–6 mm, bell-shaped, ± enlarging, not gland-dotted, pink to red, parts 5, obovate to elliptic, veins prominent, not anchor-shaped, tip rounded to acute; stamens 5, incl or exserted, anthers pink or red; styles 2, incl or exserted, fused ± 2/3. **FR:** incl, 2–3 mm, 1.6–2.6 mm wide, lens-shaped, dark-brown, smooth or minute-granular, shiny or dull. 2*n*=66,132. Shallow lakes, streams, shores; < 3000 m. CA (exc e DMoj); n hemisphere; naturalized in S.Am, s Afr. [*P. coccinea* (Willd.) Greene; *Polygonum a.* L.; *P. a.* var. *emersum* Michx.; *P. a.* var. *stipulaceum* N. Coleman] High variation ± correlated with ecology, but ± continuous so that vars. not recognized. Jun–Nov

P. capitata (D. Don) H. Gross [Ann] per, rhizomes, stolons 0. **ST:** prostrate, 5–50 cm, glabrous or glandular-hairy. **LF:** ocrea 5–10 mm, cylindric to funnel-shaped, brown or red-brown, margins oblique, glabrous or ciliate, bristles < 1.5 mm, surface woolly-hairy, also glandular-hairy or not; petiole 2–5 mm, winged distally, with ear-like lobes basally; blade 1.5–4(6) cm, 0.6–2.5(3.3) cm wide, ovate to elliptic, abaxially glandular-hairy, gen dark-blotched adaxially, base ± wedge-shaped, tip acute. **INFL:** terminal, ± head-like, 5–20 mm, 6–20 mm wide; peduncle 10–40 mm, glabrous or stalked-glandular; bractlets overlapped; pedicels spreading, 0.5–1 mm; fls 1–5. **FL:** 2–3 mm, perianth urn-shaped, not enlarging, pink-red, lobes 5, elliptic, tips acute to obtuse; stamens 8, anthers pink to red; styles 3, fused > 1/2. **FR:** incl, 1.5–2.2 mm, 1–1.5 mm wide, 3-angled, red- to black-brown, shiny, smooth or minute-dotted. 2*n*=22. Disturbed places; cult; < 500 m. NCo, CCo, SnFrB, SCo, PR; OR, LA, HI; native to Himalayas. [*Polygonum c.* D. Don] Jun–Sep

P. hydropiper (L.) Delarbre WATERPEPPER Ann, rhizomes, stolons 0. **ST:** ascending to erect, 20–60(100) cm, gland-dotted. **LF:** ocrea 8–15 mm, brown, cylindric or funnel-shaped, margins truncate, ciliate, bristles 1–4 mm, surface glabrous or strigose, gen gland-dotted; petiole 0–0.8 cm; blade (1.5)4–10(15) cm, 0.4–2.5 cm wide,

lanceolate to narrowly diamond-shaped, glabrous or scabrous on midveins, ± gland-dotted, adaxial dark blotch 0, base wedge-shaped, tip acute to acuminate. **INFL:** axillary, terminal, erect or nodding, interrupted or not distally, 30–180 mm, 5–9 mm wide; peduncle (0)10–50 mm, glabrous, gland-dotted; bractlets not overlapped; pedicels ascending, 1–3 mm; fls 1–3(5). **FL:** perianth 2–3.5 mm, ± bell-shaped, gland-dotted, green-white or pink, lobes 4–5, obovate, margins ± red, veins ± prominent, not anchor-shaped, tips obtuse to rounded; stamens 6–8, incl, anthers pink or red; styles 2–3, fused basally. **FR:** incl or exserted, 2–3 mm, 1.5–2 mm wide, lens-shaped or 3-angled, brown-black, dull, minutely rough. 2*n*=20. Shores of lakes, ponds, riverbanks, forested wetlands, pastures, disturbed ground; < 1500 m. n&c CA-FP; to n&e N.Am, also Asia, Afr, Australia; native to Eur. [*Polygonum h.* L.] May–Dec

P. hydropiperoides (Michx.) Small (p. 1117)   FALSE WATERPEPPER Per, rhizomed. **ST:** decumbent to ascending, 15–100 cm, glabrous or obscurely strigose distally. **LF:** ocrea 5–20 mm, cylindric, brown, margins truncate, ciliate, bristles (2)4–10 mm, surface glabrous or strigose, not gland-dotted; petiole 0.2–2 cm; blade 5–25 cm, 0.4–3.7 cm wide, linear- to widely lanceolate, glabrous or appressed-hairy on midveins, also on surfaces or not, gen abaxially gland-dotted, adaxial dark blotch 0, base tapered or acute, tip acuminate. **INFL:** axillary, terminal, erect, gen uninterrupted, 30–80 mm, 2–5 mm wide; peduncle 10–30 mm, glabrous or strigose; bractlets overlapped distally, often not proximally; pedicels ascending, 1–1.5 mm; fls 2–6. **FL:** perianth (1.5) 2.5–4 mm, bell-shaped, ± enlarging, not gland-dotted, pink or white-± green, lobes 5, obovate, veins prominent or not, not anchor-shaped, tips obtuse to rounded; stamens 8, incl or exserted, anthers pink or red; styles 3, fused nearly 1/2. **FR:** incl or exserted, 1.5–3 mm, 1–2.3 mm wide, 3-angled, brown to black, shiny, smooth. 2*n*=40. Wet banks, shallow water, marshes, moist prairies; < 1500 m. CA-FP, DSon; to e N.Am, Mex. [*Polygonum h.* Michx.] Confused with *P. maculosa* (see key). Jun–Oct

P. lapathifolia (L.) Delarbre (p. 1123)   WILLOW WEED Ann; rhizomes, stolons 0. **ST:** ascending to erect, 10–100 cm, glabrous to appressed-hairy distally, gland-dotted or not. **LF:** ocrea 4–25(35) mm, cylindric, brown, margins truncate, glabrous or ciliate, bristles 0 or < 1 mm, surface glabrous (strigose), not gland-dotted; petiole 0–1.5 cm; blade 4–12(22) cm, (0.3)0.5–4(6) cm wide, narrowly to widely lanceolate, strigose on main veins, glabrous or tomentose, gland-dotted abaxially, dark-blotched or not adaxially, base ± wedge-shaped, tip acuminate. **INFL:** axillary, terminal, gen ± nodding, uninterrupted, 30–80 mm, 5–12 mm wide; peduncle 2–25 mm, glands ± stalked; bractlets gen overlapped; pedicels ascending, 0.5–2.3 mm; fls 4–13. **FL:** perianth 3–4 mm, bell-shaped, ± enlarging, not gland-dotted, green-white to pink, lobes 4(5), obovate to elliptic, veins prominent, those of outer parts 2-branched distally, anchor-shaped, tips obtuse to rounded; stamens 5–6, incl, anthers pink or red; styles 2(3), fused basally. **FR:** incl or exserted, 1.5–3.2 mm, 1.6–3 mm wide, gen lens-shaped (< 1% 3-angled), brown to black, shiny or dull, smooth. 2*n*=22. Moist places, disturbed areas, roadsides, fields; < 1500 m. CA; N.Am, S.Am, Eur, Asia, Afr. [*Polygonum l.* L.] Highly variable, > 20 infraspecific taxa named but not worthy of taxonomic recognition. Jun–Oct

P. maculosa Gray (p. 1123)   LADY'S THUMB   Ann, rhizomes, stolons 0. **ST:** prostrate to erect, 10–70(130) cm, glabrous or appressed-hairy. **LF:** ocrea 5–10(15) mm, cylindric, light-brown, margin truncate, ciliate, bristles 2–3.5(4.5) mm, surface glabrous or strigose, not gland-dotted; petiole 0–0.8 cm; blade (1)5–10(18) cm, (0.2)1–2.5(4) cm wide, lanceolate to narrowly ovate, glabrous or strigose, gland-dotted abaxially or not, often dark-blotched adaxially, base tapered or wedge-shaped, tip acute to acuminate. **INFL:** axillary, terminal, ± erect, interrupted proximally or not, 10–45(60) mm, 7–12 mm wide;

peduncle 10–50 mm, ± glabrous, glands 0; bractlets gen overlapped; pedicels ascending, 1–2.5 mm; fls 4–13. **FL:** perianth 2–3.5 mm, bell-shaped, ± enlarging, not gland-dotted, pink or rose, lobes 4–5, obovate, veins prominent, not anchor-shaped, tips obtuse to rounded; stamens 4–8, incl, anthers yellow or pink; styles 2–3, fused basally. **FR:** incl or exserted, 2–2.7 mm, (1.5) 1.8–2.2 mm wide, lens-shaped to 3-angled, brown-black to black, shiny, smooth. 2*n*=44. Moist disturbed areas, fields; < 1500 m. CA; N.Am, native to Eurasia. [*Polygonum persicaria* L.] Probably of hybrid origin, with *P. lapathifolia* as one parent. Jun–Nov

P. pensylvanica (L.) M. Gómez   PINKWEED Ann, rhizomes, stolons 0. **ST:** ascending to erect, 10–200 cm; glabrous or appressed-hairy distally, glands 0. **LF:** ocrea 5–20 mm, cylindric, brown, margin truncate, glabrous or ciliate, bristles 0 or < 0.5 mm, surface glabrous or appressed-hairy, not gland-dotted; petiole 0.1–2(3) cm; blade 4–17(20) cm, (0.5)1–4.8 cm wide, narrowly to widely lanceolate, glabrous or appressed-hairy, ± gland-dotted abaxially, occ adaxially, adaxially dark-blotched or not, base tapered to wedge-shaped, tip acuminate. **INFL:** axillary, terminal, erect, dense, not interrupted; peduncle glands ± stalked; bractlets overlapped; pedicels ascending, 1.5–4.5 mm. **FL:** perianth 2.5–5 mm, bell-shaped, ± enlarging, glabrous, not gland-dotted, white to pink, lobes 5, obovate to elliptic, veins prominent, not anchor-shaped, tips obtuse to rounded; stamens 6–8, incl, anthers yellow, pink, or red; styles 2(3), fused basally. **FR:** incl or exserted, 2–3.4 mm, 1.8–3 mm wide, lens-shaped or 3-angled, brown to black, shiny, smooth. 2*n*=88. Moist disturbed places, drying ponds, riverbanks, fields; ± 30 m. e ScV, expected elsewhere; native to e US, introduced otherwise in N.Am, Eur. [*Polygonum p.* L.] Variable, of hybrid origin, possibly with *P. lapathifolia* as one parent. May–Dec

P. punctata (Elliott) Small (p. 1123)   Ann, per, gen rhizomed. **ST:** ascending to erect, 15–100 cm, glabrous, gland-dotted. **LF:** ocrea (5)9–18 mm, cylindric, brown, margins truncate, ciliate, bristles 2–11 mm, surface glabrous or strigose, gland-dotted; petiole 0–1 cm; blade 5–10(15) cm, 0.6–2.4 cm wide, lanceolate to lance-ovate or sub-diamond-shaped, glabrous or scabrous on midveins, gland-dotted, adaxial dark blotch 0, base tapered or wedge-shaped, tip acute to acuminate. **INFL:** axillary, terminal, gen erect, interrupted, 50–200 mm, 4–8 mm wide; peduncle 30–60 mm, glabrous, gland-dotted; bractlets gen not overlapped; pedicels ascending, 1–4 mm; fls 2–6. **FL:** perianth 3–3.5 mm, bell-shaped, ± enlarging, gland-dotted, gen ± green to white, lobes 5, margins gen white, veins prominent or not, not anchor-shaped, tips obtuse to rounded; stamens 6–8, incl, anthers pink or red; styles 2–3, fused basally. **FR:** incl or exserted, 2–3.2 mm, 1.5–2.2 mm wide, gen 3-angled (lens-shaped), brown-black, shiny, smooth. 2*n*=44. Shallow water, shores, marshes, floodplain forest; < 1500 m. CA; N.Am to S.Am. [*Polygonum p.* Elliott] Highly variable but named vars. not distinct. Confused with *P. hydropiper*. Jun–Nov

P. wallichii Greuter & Burdet   HIMALAYAN KNOTWEED Per, gen rhizomed, base woody or not. **ST:** ascending or erect, 7–12(250) cm, glabrous or densely hairy, ± ribbed; adventitious roots 0. **LF:** ocrea 10–40 mm, cylindric, red-brown, surface glabrous or densely hairy; petiole 0.3–2(3) cm; blade (7.5)9–22(27) cm, 2.8–7.8 cm wide, hairy abaxially, glabrous or hairy adaxially, base cordate to truncate, tip acuminate to acuminate-tailed. **INFL:** axillary, terminal, panicle-like, 40–120 mm, 10–55 mm wide; peduncle 10–80 mm, glabrous or hairy; bractlets not overlapped; pedicels ascending to spreading, 1–4 mm, glabrous; fls 3–10. **FL:** perianth 2.5–4 mm, rotate, not enlarging, gland-dotted, white or ± pink, lobes 5, oblong to obovate, tips obtuse to rounded; anthers red to purple; styles free. **FR:** incl, 2.1–2.5 mm, 1.3–1.8 mm wide, 3-angled, brown, dull, minutely rough (sometimes 0 in CA). Moist disturbed places, marshes; < 500 m. NCo (esp Del Norte, Humboldt cos.), n CCo; OR, BC; native to s-c Asia. [*Polygonum polystachyum* Meisn.] Aug–Oct ◆

POLYGONUM KNOTWEED

Ann, per to shrub. **ST:** prostrate to erect, 8–16-ribbed or 4–5-angled with ribs 0 or obscure; glabrous or papillate-scabrous. **LF:** cauline, alternate, petioled or not; ocrea gen jointed to lf, gen cylindric proximally, gen translucent distally, white or silvery, 2-lobed, glabrous, disintegrating to fibers or completely; blade linear, lanceolate, elliptic, ovate, or subround, entire. **INFL:**

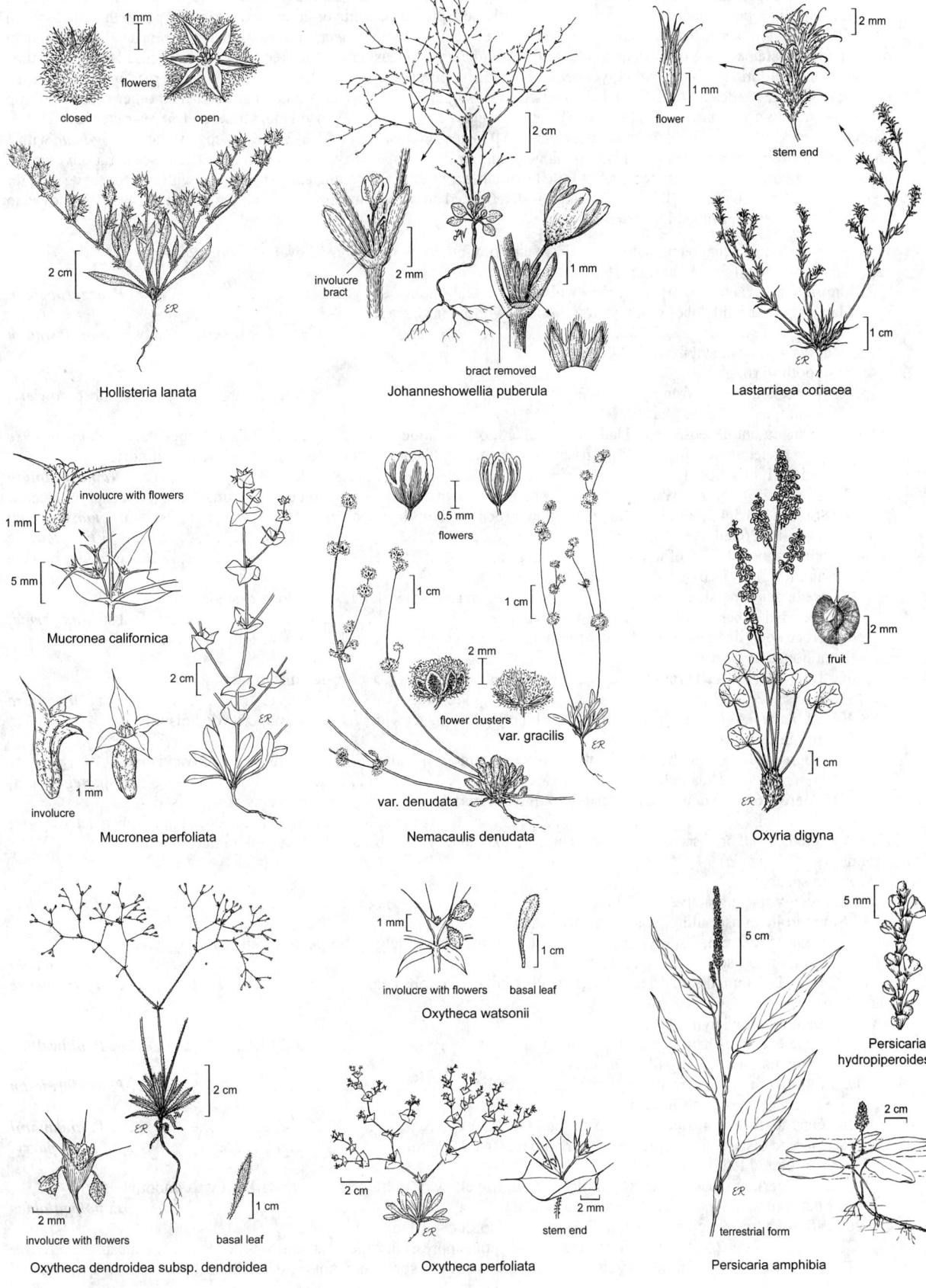

flowers
closed
open

Hollisteria lanata

2 cm

involucre bract

2 mm

bract removed

1 mm

Johanneshowellia puberula

flower

1 mm

2 mm

stem end

1 cm

Lastarriaea coriacea

involucre with flowers

1 mm

5 mm

Mucronea californica

involucre

1 mm

2 cm

Mucronea perfoliata

flowers

0.5 mm

1 cm

1 cm

2 mm

flower clusters
var. gracilis

var. denudata

Nemacaulis denudata

fruit

2 mm

1 cm

Oxyria digyna

2 cm

involucre with flowers

2 mm

basal leaf

1 cm

Oxytheca dendroidea subsp. dendroidea

involucre with flowers

1 mm

basal leaf

1 cm

Oxytheca watsonii

2 cm

2 mm

stem end

Oxytheca perfoliata

5 mm

Persicaria hydropiperoides

5 cm

terrestrial form

2 cm

Persicaria amphibia

axillary, terminal, gen spike-like; peduncle 0; pedicels present or 0, incl in to exserted from bractlets; fls 1–7(10). **FL:** bisexual, base not stalk-like; perianth not enlarging, bell- to urn-shaped, glabrous, white or green-white to pink; perianth parts 5, fused 3–60[70]%, petal- or sepal-like, similar [or not], outer ± keeled or not, < to > inner, midveins gen a different shade or color than rest of perianth; stamens 3–8 (some staminodes or not), filaments or at least innermost free, wider at base, fused to perianth tube or not, anthers elliptic to oblong, white-yellow or gen pink to purple (orange-pink); styles (2)3, gen spreading, free or fused basally, stigmas 2–3, head-like. **FR:** incl or exserted, wings 0, angles (2)3, 1 face much narrower than other (1)2 or not; tip beak-like, yellow-green, brown, or black. **SEED:** embryo curved. ± 65 spp.: ± worldwide; sect. *Duravia* restricted to w N.Am. (Greek: many, knee joint or seed, of uncertain meaning) [Costea 2005 Brittonia 57:1–27] Many spp. of sect. *Polygonum* with 2 kinds of fr, differing in germination and morphology (summer fr brown, ovate, tubercled to smooth; late-season fr olive-green, lanceolate, smooth, 2–5 × > summer), but of little taxonomic importance. Mature, early-season pls with lvs, fls, frs needed for identification. Fls "closed" or "1/2-open" should be determined on herbarium specimens. Other taxa in TJM (1993) moved to *Aconogonon, Bistorta, Fallopia, Persicaria*.

1. St 8–16 ribbed; lf venation pinnate, abaxial 2° veins obvious; anthers white-yellow (sect. *Polygonum*)
  2. Distal lvs ≤ distal fls; infl axillary, terminal
   3. Margins of perianth lobes pink (red or white); fr 1.3–2.3 mm . . . . ***P. argyrocoleon***
   3′ Margins of perianth lobes green-yellow, yellow (white, pink); fr (2.3)2.5–3.5 mm
    . . . . ²***P. ramosissimum*** subsp. ***ramosissimum***
  2′ Distal lvs > to ≫ distal fls; infl axillary
   4. Fr smooth to rough
    5. Fr tip beaked, edges strongly concave . . . . [***P. fowleri*** subsp. ***fowleri***]
    5′ Fr tip not beaked, edges flat
     6. Pl ± succulent; lf red-tinged, blade tips rounded; ocrea funnel-shaped proximally; fls 1/2-open . . . . . ***P. marinense***
     6′ Pl not ± succulent; lf light-yellow-green to -blue-green, blade tips acute to acuminate; ocrea cylindric
       proximally; fls closed . . . . ***P. ramosissimum***
      7. St lvs 1–2.5 × branch lvs, blue-green fresh, dark-brown or black dry; pedicels 1–2 mm . . . . subsp. ***prolificum***
      7′ St lvs 2.1–3.5(4.2) × branch lvs, gen yellow-green fresh or dry; pedicels 2.5–6 mm . . . . ²subsp. ***ramosissimum***
   4′ Fr striate-tubercled . . . . ***P. aviculare***
    8. Perianth tube 40–57% of length . . . . subsp. ***depressum***
    8′ Perianth tube (15)20–40% of length
     9. Ocreae gen ± persistent distally, esp on upper lvs; perianth 0.9–1.3(1.5) × as long as wide,
       ± pouched at base . . . . [subsp. ***buxiforme***]
     9′ Ocreae soon disintegrating distally to fibers or completely; perianth 1.5–2.9 × as long as wide,
       not pouched at base
      10. Lf blade (6)10–20 mm wide, 2–4.5 × as long as wide; fls 3–8, ± crowded distally; fr incl or
        ± exserted . . . . subsp. ***aviculare***
      10′ Lf blade 0.5–6.8(8) mm wide, (3.4)4.2–15(19) × as long as wide; fls 1–3(5), not crowded distally;
        fr gen exserted
       11. Ocrea (3)4–8 mm, distally disintegrating ± completely, veins obscure; lateral veins of lvs visible
         but not raised adaxially . . . . [subsp. ***neglectum***]
       11′ Ocrea (6)8–12 mm, distally disintegrating to fibers, veins obvious; lateral veins of lvs raised
         adaxially . . . . [subsp. ***rurivagum***]
1′ St ± angled, ribs 0 or obscure; lf venation parallel, abaxial 2° veins obscure; anthers pink to purple (orange-pink) (sect. *Duravia*)
 12. Per to shrub
  13. St erect, wiry; perianth 2.6–3.2 mm . . . . ***P. bolanderi***
  13′ St prostrate or ascending, not wiry; perianth (4.5)5–10 mm
   14. Ocreae 15–20 mm, distally ± persistent; lf blade midrib coarsely scabrous to ciliate abaxially;
     coastal dunes, scrub . . . . ***P. paronychia***
   14′ Ocreae 3–5 mm, distally deciduous; lf blade midrib glabrous abaxially; mtns . . . . ***P. shastense***
 12′ Ann
  15. Ocrea not jointed to lf
   16. Ocreae entire or shallowly dentate distally . . . . ***P. bidwelliae***
   16′ Ocreae disintegrating to fibers distally
    17. St 4–40 cm; pl with branches 0 or widely spreading . . . . ***P. californicum***
    17′ St 1.5–5(8) cm; pl cushion-like
     18. Ocreae distally disintegrating to straight fibers; fr 2–2.5 mm . . . . ***P. hickmanii***
     18′ Ocreae distally disintegrating to curly fibers; fr 1.2–1.6 mm . . . . ***P. parryi***
  15′ Ocrea jointed to lf
   19. Tips of perianth lobes acute to acuminate; fr light-yellow or light- or green- to dark-brown, smooth
     or net-like with longitudinal ridges . . . . ***P. polygaloides***
    20. Infl bract margin green, if white then scarious border < 0.2 mm wide
     21. Fr lanceolate, 2–2.5 mm; infl bracts lance-elliptic, appressed, rigid; stamens 5–8 . . . . ²subsp. ***esotericum***
     21′ Fr ovate, 1.3–1.7 mm; infl bracts linear to lance-linear, ± spreading, not rigid; stamens 3 . . . . subsp. ***kelloggii***

20′ Infl bract margin white, scarious border (0.2)0.25–1 mm wide
    22. Infl ovate to cylindric, gen distal, rarely also continuous from st bases; fr 1.3–2.1 mm,
        lance-ovate to ovate . . . . subsp. *confertiflorum*
    22′ Infl narrowly cylindric, continuous from st bases; fr 2–2.5 mm, lanceolate . . . . [2]subsp. *esotericum*
19′ Tips of perianth lobes rounded; fr black, smooth to minute- and/or striate-tubercled
  23. Pedicels reflexed
    24. Ocrea 3–5 mm; fr 2–2.5 mm . . . . *P. austiniae*
    24′ Ocrea 5–12 mm; fr 3–5 mm
      25. Fl closed; pedicel 2–6 mm; perianth tube 20–28% of perianth length . . . . *P. douglasii*
      25′ Fl open or half-open; pedicel 0.5–1 mm; perianth tube 9–17% of perianth length . . . . *P. majus*
23′ Pedicels erect to spreading
    26. Ocrea 1–5 mm, distally entire to dentate-torn; perianth 1.8–2.5 mm; fr 1.8–2.3 mm . . . . *P. minimum*
    26′ Ocrea 4–12 mm, distally disintegrating to fibers; perianth (2.5)3–5 mm; fr 2.5–5 mm
      27. Infl dense, fls crowded distally . . . . *P. spergulariiforme*
      27′ Infl not dense, fls not crowded distally . . . . *P. sawatchense*
        28. St, ocrea papillate-scabrous; lf blade elliptic to oblong-elliptic, margin papillate-fine-dentate;
          fls, or at least some, open . . . . subsp. *oblivium*
        28′ St, ocrea glabrous; lf blade lance-linear to narrowly oblong or oblanceolate, margin smooth,
          entire; fls closed . . . . subsp. *sawatchense*

P. argyrocoleon Kunze (p. 1123) PERSIAN KNOTWEED **ST**: decumbent to erect, 15–100 cm, green, glabrous, ribbed, branches gen from base. **LF**: distal ≤ distal fls; ocreae 4–8 mm, each proximally cylindric, distally disintegrating to curly or straight fibers; petiole 0–1.5 mm; blade 15–50 mm, 2–8 mm wide, lanceolate or lance-linear, blue-green, margin flat, tip acute. **INFL**: axillary, terminal, spike-like; pedicels incl, 1–2 mm; fls 4–6, crowded distally. **FL**: closed; perianth 1.8–2.4 mm, green-white, margin pink (red or white), tube 10–22% of length, lobes overlapped, oblong to obovate, not keeled, tip rounded, hood-shaped, midvein gen unbranched; stamens 7–8. **FR**: gen incl, 1.3–2.3 mm, ovate, brown, shiny, smooth; late-season 0. $2n=40$. Disturbed places, often in saline soils; < 1000 m. CA (esp s); s&w US; native to sw Asia. Jun–Oct

P. austiniae Greene MRS. AUSTIN'S KNOTWEED Ann. **ST**: ascending to erect, 5–10(20) cm, papillate-scabrous, green to ± purple, ± angled, branches many from base. **LF**: ocreae 3–5 mm, papillate-scabrous or glabrous, each proximally funnel-shaped, distally torn; petiole 0.1–2 mm; blade 5–15 mm, 4–7 mm wide, ovate to elliptic or obovate, margin flat or narrowly rolled under, papillate-toothed, tip acute or mucronate. **INFL**: axillary, terminal, spike-like, slender; pedicels exserted, reflexed, 1–2.5 mm; fls 1–4. **FL**: closed; perianth (1.8)2–2.6 mm, green or purple, margins narrow, ± white, tube 20–28% of length, lobes overlapped, oblong, ± keeled, tip rounded, hood-shaped, midvein unbranched; stamens 8. **FR**: incl, 2–2.5 mm, elliptic to obovate, black, shiny, ± smooth. Dry to moist flats on banks, sagebrush plains, ponderosa-pine forest; 1300–1600 m. MP; to BC, MT, WY. [*P. douglasii* subsp. *a.* (Greene) E. Murray] Jun–Sep

P. aviculare L. KNOTWEED, KNOTGRASS Ann. **ST**: prostrate to erect, 5–200 cm, glabrous, green, ribbed, branched. **LF**: distal > distal fls; ocreae 3–15 mm, distally disintegrating to fibers or ± completely persistent; petiole 0.3–9 mm; blade 6–50(60) mm, 0.5–22 mm wide, narrowly elliptic, lanceolate, elliptic, or obovate, green to gray-green, margin flat, tip acute or rounded. **INFL**: axillary; fls 1–6(8), crowded distally or not; pedicels incl to exserted, 1.5–5 mm. **FL**: closed or half-open; perianth 1.8–5.5 mm, gen not pouched at base, ± green, margin white to red, tube 20–57% of length, lobes overlapped or not, not keeled, oblong to obovate, tip rounded, gen hood-shaped, midvein branched or not; stamens 5–8. **FR**: incl or exserted, 1.2–4.2 mm, ovate, light- to dark-brown, dull, (2)3–angled, faces ± unequal, gen coarsely striate-tubercled; late-season common or not, 2–5 mm. $2n=40,60$. Taxonomically controversial, gen self-pollinated, self-fertilized, polyploid complex in which vars. to spp. have been named, but intergradation precludes recognition as spp.

   subsp. *aviculare* **ST**: 1–3, ascending or erect, 10–75(100) cm; basal branches widely spreading. **LF**: st lvs 1.4–4 × > branch lvs (basal lvs often 0 at maturity); ocreae (5.5)6–10(14) mm, proximally cylindric, distally soon disintegrating to fibers; petiole 1–6(8) mm; blade 18–50(60) mm, (6)10–20 mm wide, elliptic to oblanceolate, 2–4.5 × as long as wide, tip acute or obtuse. **INFL**: pedicel gen

exserted, 2–5 mm; fls 3–8, ± crowded distally. **FL**: perianth (2.3)2.8–4.7(5) mm, green, margins white to red, tube (15)20–37% of length, lobes overlapped, oblong, tips hood-shaped; veins branched, thickened; stamens 7–8. **FR**: incl or ± exserted, (2.1)2.7–3.7 mm, ovate, ± dark-brown, 3-angled, faces subequal, tip straight, gen striate-tubercled; late-season uncommon, 3.5–5 mm. $2n=40,60$. Disturbed places, roadsides, cult fields; < 2000 m. CA; worldwide. May–Nov

   subsp. *depressum* (Meisn.) Arcang. (p. 1123) Ann, short-lived per, gen mat-forming. **ST**: 3–15, prostrate to ascending, (5)10–50(100) cm, branches many. **LF**: st lvs 1–2.3(3.4) × > branch lvs; ocreae either 3–5.5 mm, distally soon disintegrating ± completely, or 7–12 mm, distally membranous, ± torn on margins, overlapped on distal sts; petiole 0.5–3 mm; blade (6.2)8–27(35) mm, (1.4)2–7(10) mm wide, elliptic to narrowly elliptic or oblanceolate, 2.8–5.7(6.5) × as long as wide, (longitudinally striate), tip obtuse or acute. **INFL**: pedicels incl, 1–2.5 mm fls; 2–7, crowded distally or not. **FL**: perianth (1.8)2–3.4(4) mm, green or red-brown, margins white, tube 40–57% of length, lobes overlapped, ± spreading in fr, oblong, flat, or obscurely hood-shaped in fr, midveins unbranched, thin to thickened; stamens 5–7. **FR**: ± exserted, 1.5–2.7(3) mm, ovate, dark-brown, (2)3-angled, 1 face ± narrower, tip straight or ± bent toward ± narrow face, gen striate-tubercled; late-season common, 2.5–4.5 mm. $2n=40,60$. Disturbed places; < 2000 m. CA; worldwide. [*P. arenastrum* Boreau] More trample-resistant than *P. aviculare* subsp. *a.*; *P. montereyense* Brenckle ± distinctive (ocreae 8–12 mm, esp silvery, persistent) may deserve infraspecific recognition but further study needed. May–Nov

P. bidwelliae S. Watson (p. 1123) BIDWELL'S KNOTWEED Ann. **ST**: erect, 2–20 cm, ± wiry, minutely papillate-scabrous, green, ± angled, branches 0 or widely spreading. **LF**: lvs crowded distally; ocreae not jointed to lf, 8–13 mm, papillate-scabrous, proximally cylindric, distally overlapped, obscuring lvs, fls, entire or shallowly dentate; petiole 0; blade 5–15(20) mm, 0.5–1.5(2) mm wide, linear, margin rolled under, papillate-fine-dentate, tip spine-tipped. **INFL**: in distal axils; pedicels 0; fls 1. **FL**: closed; perianth 2–3 mm, pink, margin white or pink, tube 10–18% of length, lobes overlapped, elliptic, ± keeled, tips acute to acuminate, midvein unbranched; stamens 8. **FR**: incl, 1.8–2.3 mm, ovate-elliptic, light-brown to brown, shiny, smooth. Thin volcanic soils, chaparral, montane woodland valleys, grassland; 60–1200 m. CaR (Butte, Tehama, Shasta cos.), ne ScV (Butte Co.). May–Jun ★

P. bolanderi W.H. Brewer (p. 1123) BOLANDER'S KNOTWEED Per, shrub. **ST**: erect, gnarled in age, 20–60 cm, wiry, glabrous, brown, ± angled, branches 0. **LF**: lvs crowded distally; ocreae not jointed to lf, 6–10 mm, glabrous or papillate-scabrous, proximally cylindric, distally deeply fringed, disintegrating; petiole 0; blade 3–15(25) mm, 0.4–1.5 mm wide, linear to awl-like, margins flat, smooth, tip acuminate to spine-tipped. **INFL**: in distal axils; pedicels 0; fls 1–2. **FL**: 1/2-open; perianth 2.6–3.2 mm, white or pink,

tube 18–33% of length, lobes overlapped, ± recurved, elliptic-oblong to obovate, ± keeled, tip rounded, midvein unbranched; stamens 8. **FR:** incl, 2.5–3 mm, lanceolate to oblong-ovate, light-brown, shiny, smooth. Open, dry, gravelly, rocky places; 300–1500 m. NW (esp Napa Co.), CaR, n&c SN. Jun–Nov

P. californicum Meisn. (p. 1123) Ann. **ST:** erect, 4–40 cm, ± wiry, papillate-scabrous, green, ± angled, branches 0 or widely spreading. **LF:** basal lvs gen early-deciduous; ocreae not jointed to lf, 5–10 mm, glabrous or papillate-scabrous, proximally cylindric, distally disintegrating to straight bristly-fringed fibers; petiole 0; blade 5–25(30) mm, 0.5–2 mm wide, linear, margin rolled under, papillate-fine-dentate or smooth, tip mucronate or weakly spine-tipped. **INFL:** in distal axils; pedicels 0; fls 1. **FL:** open or closed; perianth 2.5–3.5 mm, tube 10–20% of length, lobes overlapped, elliptic, ± keeled, white to pink, tip acute to acuminate, midvein unbranched; stamens 8. **FR:** incl or ± exserted, 1.8–2.2 mm, narrowly elliptic, brown, shiny, smooth. Open places (incl serpentine); 40–1200 m. NW, CaR, SN, GV, sw MP; to WA. May–Oct

P. douglasii Greene (p. 1123) DOUGLAS' KNOTWEED Ann. **ST:** erect, 5–80 cm, glabrous or sparsely papillate-scabrous, green, ± angled, branches 0 or not. **LF:** basal lvs early-deciduous, distal abruptly reduced to bracts; ocreae 6–12 mm, glabrous or minutely papillate-scabrous, proximally cylindric, distally deeply cut; petiole 0.1–2 mm; blade 15–55 mm, 2–8(12) mm wide, lance-linear, narrowly oblong, or oblanceolate, margin rolled under, smooth or papillate-fine-dentate; tip acute to mucronate. **INFL:** axillary, terminal, spike-like; pedicels gen exserted, reflexed, 2–6 mm; fls 2–4. **FL:** closed; perianth 3–4.5 mm, green to ± tan, margin white or pink, tube 20–28% of length, lobes overlapped, oblong, ± keeled, tip rounded, hood-shaped, midvein gen branched; stamens 8. **FR:** incl, 3–4(4.5) mm, elliptic or oblong to ovate, black, shiny or dull, smooth to minute-striate-tubercled. 2*n*=40. Dry, rocky outcrops, sandy ground, disturbed places; < 3500 m. CA (esp mtns); to Can, e N.Am. Jun–Oct

P. hickmanii H.R. Hinds & Rand. Morgan SCOTTS VALLEY POLYGONUM Ann, compact, cushion-like. **ST:** erect, 2–5 cm, glabrous, ± angled. **LF:** ocreae 4–6 mm, glabrous, proximally cylindric, distally disintegrating to straight fibers; petiole 0; blade 5–35 mm, 1–1.5 mm wide, linear, margin rolled under, smooth, tip acuminate. **INFL:** axillary; pedicels 0; fls 1. **FL:** perianth 2–3 mm, closed, white, margin ± white or pink, tube 6–18% of length, lobes oblong, tip acute, mucronate, midvein unbranched; stamens 8, anthers orange-pink. **FR:** incl, 2–2.5 mm, ovate, olive-brown, shiny, smooth. Open, seasonally dry grassland; 200–300 m. SnFrB (n end of Scotts Valley, Santa Cruz Co.). May–Oct ★

P. majus (Meisn.) Piper (p. 1123) WIRY KNOTWEED Ann. **ST:** erect, 15–60 cm, ± wiry, gen papillate-scabrous, green, ± angled, branches 0 or not. **LF:** basal lvs often early-deciduous; ocreae 5–12 mm, glabrous or papillate-scabrous, proximally cylindric, distally torn or disintegrating to fibers; petiole 0–2 mm; blade 15–70 mm, 2–8 mm wide, lance-linear to narrowly oblong or lanceolate, margin rolled under, papillate-fine-dentate, tip acute or mucronate. **INFL:** axillary, terminal, spike-like; pedicels exserted, reflexed, 0.5–1 mm; fls 2–5. **FL:** open or 1/2-open; perianth (3.5)4–5 mm, white or pink, tube 9–17% of length, lobes overlapped, oblong to oblong-obovate, ± keeled in distal 1/4, tips rounded, hood-shaped, midvein branches 0 or short; stamens 8. **FR:** incl, 3.5–5 mm, elliptic, black, shiny or dull, smooth to striate-tubercled. Dry plains, meadows, serpentine or not; 500–2000 m. KR, NCoR, CaR, SN, SnBr, w MP; to BC, ID. [*P. douglasii* subsp. *m.* (Meisn.) J.C. Hickman] Jun–Sep

P. marinense T.R. Mert. & P.H. Raven (p. 1123) MARIN KNOTWEED Ann, ± succulent. **ST:** prostrate to ascending, 15–40 cm, glabrous, green to red-tinged, ribbed, branches from base. **LF:** distal >> distal fls; ocreae 4–6 mm, proximally funnel-shaped, distally soon disintegrating ± completely; petiole 2–5 mm; blade 20–35 mm, 9–16 mm wide, elliptic to obovate or oblanceolate, gen red-tinged, margin flat, tip rounded; st lvs (1.3)2–2.6(3.5) × branch lvs. **INFL:** in most lf axils; pedicels gen exserted, 2–4 mm; fls 1–4. **FL:** 1/2-open; perianth 3–3.5(4) mm, green, margin white or pink, tube 18–25% of length, lobes overlapped, not keeled, widely rounded, hood-shaped, midvein

unbranched; stamens 8. **FR:** exserted, 2.8–3.4(4) mm, ovate, brown, shiny, rough, faces subequal or 1 ± narrower; late-season uncommon, 4.5–5 mm. 2*n*=60. Coastal salt, brackish marshes, swamps; < 10 m. n CCo (< 15 sites, Marin, Napa, Solano, Sonoma cos.). Origin, phylogeny needing further study. Apr–Aug ★

P. minimum S. Watson (p. 1123) Ann. **ST:** prostrate to erect, 2–30 cm, wiry, often zigzagged, papillate-scabrous, red-brown, angled, branches from base or 0. **LF:** crowded distally or not; ocreae jointed to lf, 1–5 mm, papillate-scabrous, proximally cylindric, distally entire to dentate-torn; petiole 0.1–3 mm; blade 6–27 mm, 3–8 mm wide, narrowly elliptic, ovate, obovate, or ± round, margins flat, smooth, irregularly thickened or papillate-fine-dentate, tip mucronate. **INFL:** pedicel incl in bractlets, erect to spreading, 2–3 mm; fls 1–3, near st base, also crowded distally or not. **FL:** 1/2-open or closed; perianth 1.8–2.5 mm, ± green, margin narrow, white or pink, tube 22–29% of length, lobes overlapped, oblong, hood-shaped, ± keeled, tip rounded, midvein thickened, unbranched; stamens 8. **FR:** incl or exserted, 1.8–2.3 mm, elliptic to ovate, black, shiny, smooth. Meadows, open rocky or ± barren soil; 700–3500 m. NW, CaR, SN; w N.Am. Jul–Sep

P. paronychia Cham. & Schltdl. (p. 1123) Per to shrub. **ST:** prostrate or ascending, 10–100 cm, glabrous, covered with ocreae remains, brown, ± angled, branched, rooting at nodes. **LF:** crowded distally; ocreae 15–20 mm, glabrous, proximally cylindric to funnel-shaped, distally ± persistent, disintegrating to white-gray curly fibers; petiole 0–0.5 mm; blade (5)10–20(33) mm, 3–8 mm wide, linear to oblanceolate, leathery, margin rolled under, smooth, tip acute or mucronate. **INFL:** axillary; pedicels incl or ± at same level as bractlets, erect to spreading, 2–5 mm; fls 2–5, crowded distally. **FL:** 1/2-open or open; perianth (4.5)5–10 mm, pink or white, (red-brown when dry), tube 22–48% of length, lobes partially overlapped, oblong-ovate to ± lanceolate, not keeled, tip rounded, midvein branched; stamens 8. **FR:** incl or ± exserted, 4–5 mm, ovate, black, shiny, smooth. Coastal dunes, scrub; < 50 m. NCo, CCo; to BC. Mar–Sep

P. parryi Greene (p. 1123) PARRY'S OR PRICKLY KNOTWEED Ann, cushion-like. **ST:** erect, 2–5 cm, glabrous, green-brown. **LF:** ocreae not jointed to lf, 2–4(5) mm, proximally cylindric, distally deeply torn, disintegrating to white, curly fibers; petiole 0; blade 5–13(20) mm, 0.4–1 mm wide, lance-linear or awl-shaped, margin rolled under, smooth, tip spine-tipped. **INFL:** pedicels 0; fls 1, in most axils. **FL:** closed; perianth 1.5–2(2.5) mm, gen ± red, margin white to red, tube 6–15% of length, lobes oblong, ± keeled, tip acute, midvein unbranched; stamens 8. **FR:** ± exserted, 1.2–1.6 mm, ovate, dark-brown, shiny, smooth. Vernally moist, open, sandy, rocky places; 500–2000 m. NW, CaR, SN, SnFrB, PR; to WA. May–Jul

P. polygaloides Meisn. POLYGALA KNOTWEED Ann. **ST:** erect, (2)6–20(25) cm, ± wiry, glabrous, green, ± angled, branches gen widely spreading (0). **LF:** basal lvs early-deciduous, distal abruptly reduced to bracts; ocreae 4–8 mm, glabrous, proximally cylindric, distally cut; petiole 0; blade 10–40 mm, 1–2.5 mm wide, narrowly linear, margin rolled under, smooth, tip acute or mucronate. **INFL:** axillary, terminal, spike-like, ± spheric to cylindric; pedicels incl, erect, 0–2 mm; fls 1–3, in most axils or crowded distally. **FL:** gen closed; perianth 1.5–3 mm, white to red, tube 19–40% of length, lobes overlapped, lance-oblong, ± keeled, tip acute to acuminate, midvein thickened, branches 2, proximal, or gen 0; stamens 3–8. **FR:** incl, 1.3–2.5 mm, ovate to lanceolate, light-yellow or light- or green-to dark-brown, shiny or dull, smooth or net-like with longitudinal ridges. Highly variable; intermediates occur between all subspp. exc *P. polygaloides* subsp. *p.*, of nw US (exc CA), AB.

subsp. ***confertiflorum*** (Piper) J.C. Hickman (p. 1123) **ST:** 2–15 mm. **INFL:** gen distal, rarely also continuous from st bases, 7–40 mm, 5–10 mm wide, ovoid to cylindric; bracts ascending to appressed, 4–11 mm, linear to lance-like, ± rigid, margin rolled under, white, scarious border 0.25–0.4 mm wide, midvein ± heavily thickened, branches obscure. **FL:** perianth 1.8–2.2(2.4) mm, white or red, tube 22–30% of length, tube and base of lobes smooth or papillate; stamens 3. **FR:** 1.3–2.1 mm, lance-ovate to ovate, brown to dark-brown, dull, net-like with longitudinal ridges. Vernal pools, wet meadows; 500–1900 m. NW, CaR, n SN, MP; to SK, MT, WY. May–Aug

subsp. *esotericum* (L.C. Wheeler) J.C. Hickman (p. 1123) MODOC COUNTY KNOTWEED **ST:** 5–12 cm. **INFL:** continuous from st bases, 2–10 mm, 5–7 mm wide, narrowly cylindric; bracts appressed, 4–8 mm, lance-elliptic, rigid, margin flat, green, if white then scarious border 0.1–0.2 mm wide; veins prominent adaxially. **FL:** perianth 2.1–2.9 mm, white or pink, tube 29–40% of length, tube and base of lobes smooth; stamens 5–8. **FR:** 2–2.5 mm, lanceolate, dark-brown, dull, net-like to ± smooth. Vernal pools, seasonally wet places, pinyon/juniper woodland; ± 1500 m. MP (near Goose Lake, Modoc Co.; Sierra Valley, s Plumas Co.). Jun–Aug ★

subsp. *kelloggii* (Greene) J.C. Hickman (p. 1123) **ST:** 1–15 cm. **INFL:** gen terminal, 3–15 mm, 5–15 mm wide, ovoid, or continuous from st bases; bracts ± spreading, 7–25 mm, linear to lance-linear, soft, margin flat or rolled under, green, not scarious; veins obscure. **FL:** perianth 1.5–2.3 mm, pink or white, tube 19–34% of length, tube and base of lobes smooth; stamens 3. **FR:** 1.3–1.7 mm, ovate, light-yellow to green-brown, shiny, smooth to obscurely longitudinally net-like. Mtn meadows, seeps, rainpools; 1500–3300 m. NW, CaR, SN, TR, PR; to BC, MT, CO, AZ. Jun–Sep

P. ramosissimum Michx. BUSHY KNOTWEED **Ann. ST:** erect, 10–100(200) cm, glabrous, green, ribbed, branches gen profuse in distal 1/2. **LF:** proximal often early-deciduous, distal < to >> fls; ocreae 6–12(15) mm, proximally cylindric, distally soon disintegrating to brown fibers; petiole 2–4 mm; blade 8–70 mm, 4–18(35) mm wide, narrowly elliptic, lanceolate, or oblanceolate (ovate), light-yellow-green to -blue-green, margins flat, tip acute to acuminate or obtuse. **INFL:** axillary, terminal, spike-like; pedicels incl or exserted, 1–6 mm; fls 2–5, crowded distally or not. **FL:** closed; perianth (2)2.2–3.6(4) mm, green-yellow, margin green-yellow to yellow or white to pink, tube 20–38% of length, lobes overlapped, not keeled, elliptic to oblong, hood-shaped, midvein thickened or not; stamens 3–6(8). **FR:** incl or exserted, 1.6–3.5 mm, ovate, dark-brown, shiny or dull, rough to gen smooth; late-season common, 4–15 mm. $2n=60$. [*P. patulum* M. Bieb., misappl.] An unnamed variant of *P. ramosissimum* in saline marshes in SnFrB mistakenly identified as *P. patulum* M. Bieb., may be taxonomically significant; study needed.

subsp. *prolificum* (Small) Costea & Tardif Pl blue-green fresh, dark-brown or black dry. **ST:** 10–80 cm. **LF:** distal >> fls; blade 8–30(35) mm, 4–6 mm wide, oblanceolate, tip rounded or obtuse; st lvs 1–2.5 × branch lvs. **INFL:** axillary; pedicels incl, 1–2 mm; fls 2–4, crowded distally or not. **FL:** perianth 2–2.8(3) mm, margins white to pink; stamens 3. **FR:** 1.6–2(2.4) mm, shiny or dull, smooth or rough; late-season 4–10 mm. Wet, saline places; 100–2000 m. n SNH (Lake Tahoe area), SnFrB; native to e N.Am. [*P. p.* (Small) B.L. Rob.] Jun–Oct

subsp. *ramosissimum* (p. 1123) Pl gen yellow-green fresh or dry. **ST:** 30–200 cm. **LF:** distal 1–2 mm, < fls, or 5–15 mm, >> fls; blade 35–70 mm, 7–18(35) mm wide, narrowly elliptic to lanceolate (ovate), tip acute to acuminate; st lvs 2.1–3.5(4.2) × branch lvs. **INFL:** axillary, terminal, spike-like; pedicels exserted, 2.5–6 mm; fls 2–5, crowded distally. **FL:** perianth (2.6)3–4 mm, margin green-yellow, yellow (white or pink); stamens 3–6(8). **FR:** (2.3)2.5–3.5 mm, shiny or dull, smooth to rough; late-season common, 5–12 mm. Disturbed places, saline marshes; < 200 m. n CA-FP, SnFrB; N.Am. Jul–Sep

P. sawatchense Small **Ann. ST:** erect, 4–50 cm, glabrous or papillate-scabrous, green to ± brown, ± angled, branches 0 or from base. **LF:** ocreae 4–10 mm, glabrous or papillate-scabrous, proximally cylindric, distally disintegrating to fibers; petiole 0–2 mm;

blade 8–45 mm, 2–8(12) mm wide, linear or narrowly oblong to oblanceolate, margin rolled under or flat, smooth or papillate-fine-dentate, tip acute, mucronate, rarely obtuse. **INFL:** axillary, terminal, spike-like; pedicels incl or exserted, erect, 1–4 mm; fls 2–4. **FL:** closed or at least some open; perianth (2.5)3–4 mm, ± green or ± red, flushed purple or not, margin white or pink, tube 20–40% of length, lobes overlapped, oblong or oblong-elliptic, ± keeled, tip rounded, hood-shaped, midvein branches 0 or few, at base; stamens 3–8. **FR:** incl, 2.5–3.3 mm, elliptic or ovate, black, shiny, smooth.

subsp. *oblivium* Costea & Tardif **ST:** 5–15(25) cm, papillate-scabrous, green, branches 0 or few. **LF:** persistent, gradually reduced distally on st; ocreae papillate-scabrous; blade 8–20(25) mm, 5–10 mm wide, elliptic to oblong-elliptic, margin flat, papillate-fine-dentate. **INFL:** gen axillary, if also terminal then < 5 cm; pedicels 1–3 mm; fls 1–3(4). **FL:** at least some open; perianth 3–4 mm, green-white or -yellow, flushed with purple or not, margin white, tube 20–30% of length; stamens 8. **FR:** 2.6–3.3 mm, elliptic to elliptic-ovate. Dry, moist meadows, pastures, forest, sandy, gravelly, rocky substrates; 1000–3100 m. NW, CaRH, n&c SNH, Wrn; to BC. Jun–Aug

subsp. *sawatchense* (p. 1123) **ST:** 4–50 cm, glabrous, green to ± brown, gen branched. **LF:** persistent or not, abruptly reduced distally to bracts, rarely >> than fl; ocreae glabrous; blade 15–45 mm, 2–8(12) mm wide, lance-linear to narrowly oblong or oblanceolate, margin gen rolled under, smooth, entire. **INFL:** axillary, terminal, open, 5–15 cm; pedicels 1–4 mm; fls (1)2–4. **FL:** closed; perianth (2.5)3–3.5 mm, ± green or ± red, margin white or pink, tube 25–40% of length; stamens 3–8. **FR:** 2.5–3 mm, ovate or elliptic. Dry meadows, pastures, sagebrush, forest, sandy, gravelly, rocky substrates; 800–3000 m. KR, NCoRH, CaRH, SN, SnJV, SW, GB, DMtns; to NM; Can. [*P. douglasii* subsp. *johnstonii* (Munz) J.C. Hickman; *P. triandrum* Coolidge] Jun–Aug

P. shastense W.H. Brewer (p. 1123) Per, shrub. **ST:** prostrate to ascending, gnarled, 5–40 cm, glabrous, brown, ± angled, branched. **LF:** crowded distally or ± not; ocreae jointed to lf, 3–5 mm, glabrous, proximally cylindric, distally membranous, falling with lvs; petiole 0–0.5 mm; blade 5–25 mm, 2–5 mm wide, 1-veined, without pleats, lanceolate to elliptic, leathery, margin rolled under, smooth, tip acute. **INFL:** in distal axils; pedicels incl or ± exserted, erect, 2–4 mm: fls 2–6. **FL:** open or 1/2-open; perianth 5–9 mm, pink or white, tube 7–15% of length, lobes partially overlapped, ovate to ovate-round, tip rounded, midvein branches many; stamens 8. **FR:** incl or ± exserted, 3–4 mm, narrowly ovate, brown, shiny, smooth. Rocky, gravelly slopes; 2100–3400 m. CaRH, SNH; OR, NV. Jul–Sep

P. spergulariiforme Small (p. 1123) **Ann. ST:** erect, 5–50 cm, gen papillate-scabrous (± glabrous in age), green, branches widely spreading. **LF:** basal lvs gen early-deciduous, distal abruptly reduced to bracts; ocreae 8–12 mm, papillate-scabrous, proximally cylindric, distally disintegrating to fibers; petiole 0–2 mm; blade 35–60 mm, 1–3 mm wide, linear to lanceolate, margin flat or narrowly rolled under, smooth or papillate-fine-dentate, tip acute, mucronate. **INFL:** axillary, terminal, spike-like, dense; pedicels ± incl or ± exserted, erect to spreading, 0.5–2 mm; fls 2–5, crowded distally. **FL:** open or 1/2-open; perianth 3–5 mm, pink or white, tube 9–17% of length, lobes overlapped, oblong-obovate, ± keeled in distal 1/4, tip rounded, hood-shaped, midvein branched; stamens 8. **FR:** incl, 3–5 mm, narrow-elliptic or -ovate or lance-elliptic, black, shiny, smooth to striate-tubercled, faces subequal. Open, rocky places (incl serpentine); 10–2000 m. NW, CaR, n SN, SnFrB, w MP; to BC. [*P. douglasii* subsp. *s.* (Small) J.C. Hickman] Jun–Oct

PTEROSTEGIA WOODLAND THREADSTEM

1 sp. (Greek: wing covering, for winged bract)

P. drymarioides Fisch. & C.A. Mey. (p. 1123) Ann, gen sprawling, 0.1–10(12) dm diam, hairy. **LF:** cauline, opposite; blades entire or gen lobed, wide, 0.3–2 cm, 0.5–2.5(3) cm wide, awns 0; ocreae 0. **INFL:** terminal, cyme-like; peduncles 0; involucre bract erect, 2-winged, swollen on 1 side, veined, 1–1.5(3) mm, (1.5)2–3(3.5) mm

wide, awns 0. **FL:** 2–3, 0.9–1.2 mm, hairy; perianth pale yellow to pink or rose, lobes (5)6, entire; stamens 6. **FR:** 1.2–1.5 mm, obconic, winged, glabrous, gen brown; embryo straight. $n=14$. Common. Sand or gravel; < 1600 m. CA; to s OR, sw UT, w AZ, nw Mex. Mar–Jul

RUMEX DOCK
Scott Simono

Ann to per; glabrous or papillate and hairy; rhizomed, stoloned, or gen from taproot with a short caudex; occ dioecious. **ST**: prostrate, decumbent, or gen erect to ascending, often ± ridged, red-brown in fr, nodes ± swollen. **LF**: gen basal and cauline, alternate, petioled exc uppermost cauline; ocreae deciduous to persistent. **INFL**: axillary or gen terminal, gen panicle-like. **FL**: gen bisexual, bell-shaped, glabrous, green, ± pink, or red; perianth lobes 6, persistent, in fr outer 3 ± inconspicuous, inner 3 enlarged, hardened, ± veiny, covering fr, midrib of 1–3 often expanded into tubercle; stamens 6; stigmas 3, fringed. **FR**: achene, glabrous, ± black or dark brown to ± red. 190–200 spp.: ± worldwide, 63 in N.Am. (Latin: sorrel) Mature inner perianth lobes gen needed for identification. Spp. often hybridize.

1. Fls unisexual, on different pls (some fls bisexual); lf blade hastate or sagittate or base tapered; pedicels gen jointed
    2. Pedicels jointed distally; inner perianth lobes gen 1.2–1.7(2) mm, ± = fr; lf blade widely tapered or gen
        hastate or sagittate; from a creeping rhizome and/or vertical taproot; CA-FP, GB . . . . ***R. acetosella***
    2′ Pedicels jointed near middle or proximally; inner perianth lobes 2.8–3.8 mm, > fr; lf blade base tapered;
        from a vertical taproot and densely tufted underground stolons; CaRH, SNH, MP, W&I . . . . ***R. paucifolius***
1′ Fls gen bisexual, sometimes also unisexual in 1 infl; lf blades not hastate or sagittate; pedicels jointed or not
    3. Lvs cauline; sts decumbent to erect, gen with lfy axillary shoots below terminal infl; lf blade base tapered
        to ± rounded; inner perianth margins gen entire (irregularly minutely notched, toothed)
        4. Inner perianth lobes (20)23–30 mm wide; MP . . . . ***R. venosus***
        4′ Inner perianth lobes gen < 15 mm wide; MP, elsewhere
            5. Perianth tubercles gen 0
                6. Inner perianth lobe margin gen minute-toothed; infl ± open; sts ascending (decumbent or ± erect);
                    CA (esp mtns, coast). . . . . ***R. californicus***
                6′ Inner perianth lobe margin gen entire; infl dense; sts ascending or gen erect; c&s SNH, expected
                    elsewhere . . . . ***R. utahensis***
            5′ Perianth tubercles ≥ 1
                7. Perianth tubercle width < 1/3 inner lobe, if tubercle 1 then width ≥ 1/2 lobe
                    8. Pl terrestrial, submersed, or floating; taproot vertical or creeping; inner perianth lobes 2–2.5(3) mm,
                        1–1.5(2) mm wide, gen ovate or elliptic; lf lanceolate to lance-ovate, gen 3.5–5 × longer than wide; GB . . . ***R. lacustris***
                    8′ Pl terrestrial; taproot vertical; inner perianth lobes (2)2.5–3.5(4) mm, (2)2.5–3(3.5) mm wide,
                        widely triangular; lf linear to lanceolate, gen 5–6 × longer than wide; CaRH, SN, ScV (Sutter
                        Buttes), SnGb, GB . . . . ***R. triangulivalvis***
                7′ Perianth tubercle nearly as wide as inner lobe on at least 1 lobe
                    9. Lf blade thick, leathery, lance-ovate to -elliptic or ovate-elliptic, ≤ 3.5 × as long as wide; inner
                        perianth lobes (3)4–5 mm, (2.5)3–4 mm wide; perianth tubercle 1 . . . . ***R. crassus***
                    9′ Lf blade thick or gen thin, occ ± leathery, linear to lanceolate, gen > 3.5 × as long as wide; inner
                        perianth lobes gen 2–4 mm, gen 1.5–3 mm wide; perianth tubercles 1 or 3
                        10. Inner perianth lobes (1.8)2–2.5(3) mm, tubercle 1, smooth to ± warty . . . . ***R. salicifolius***
                        10′ Inner perianth lobes (2.5)3–4 mm, tubercles (1)3, gen smooth . . . . ***R. transitorius***
    3′ Lvs basal and cauline (basal often withered when fr mature); st gen ± erect, gen unbranched below infl;
        lf blade base tapered to cordate; inner perianth margins entire or variously dentate
        11. Perianth tubercles 0 or inconspicuous
            12. Inner perianth lobes 11–16 mm; ocreae persistent; roots tuber-like . . . . ***R. hymenosepalus***
            12′ Inner perianth lobes 5–10(12) mm; ocreae deciduous to partly persistent; roots not tuber-like . . . . . ***R. occidentalis***
        11′ Perianth tubercles 1–3, conspicuous
            13. Inner perianth lobe margin gen entire (if toothed or notched, teeth or notches < 0.3 mm)
                14. Inner perianth lobe 2–3.5(4) mm, tubercle ≥ 1/3 its width; infl ± open
                    15. Per; perianth tubercles ± equal, almost as wide as lobe. . . . . ***R. conglomeratus***
                    15′ Ann; perianth tubercles gen unequal, width ± 1/3 lobe. . . . . ²***R. violascens***
                14′ Inner perianth lobe 3.5–7(7.5) mm, tubercle gen << 1/3 its width; infl gen dense
                    16. Lf blade 20–55(70) cm, margins flat to ± wavy; inner perianth lobe 4–7(7.5) mm; pedicel 5–13 mm,
                        joint ±0; st 8–15(20) dm . . . . ***R. britannica***
                    16′ Lf blade 15–30(35) cm, margin strongly wavy; inner perianth lobe 3.5–6 mm; pedicel (3)4–8 mm,
                        joint swollen; st 4–10(15) dm . . . . ***R. crispus***
            13′ Inner perianth lobe margin variously dentate, at least some teeth ≥ 0.3 mm
                17. Ann, bien
                    18. Inner perianth lobe 3–5.5(6) mm, teeth 1–3(5) mm . . . . ***R. dentatus***
                    18′ Inner perianth lobe gen 1.5–3.5 mm, teeth 0.5–4 mm
                        19. Widest inner perianth lobe narrow-triangular or -kite-shaped, teeth 1–3 mm; infl dense, bracts lf-like
                            20. Perianth tubercle straw-colored, oblong-ovate, nearly as wide as inner lobe, tip obtuse; inner
                                perianth lobe teeth ± = width of lobe . . . . ***R. persicarioides***
                            20′ Perianth tubercle ± brown or ± red, lance-linear to fusiform, ≤ 1/2 width of inner lobe, tip ±
                                acute; inner perianth lobe teeth gen 1.5–2.5(4) × width of lobe . . . . ***R. fueginus***
                        19′ Widest inner perianth lobe widely ovate-triangular to ± deltate, teeth ≤ 0.5 mm; infl ± open, bracts
                            ± not lf-like . . . . ²***R. violascens***

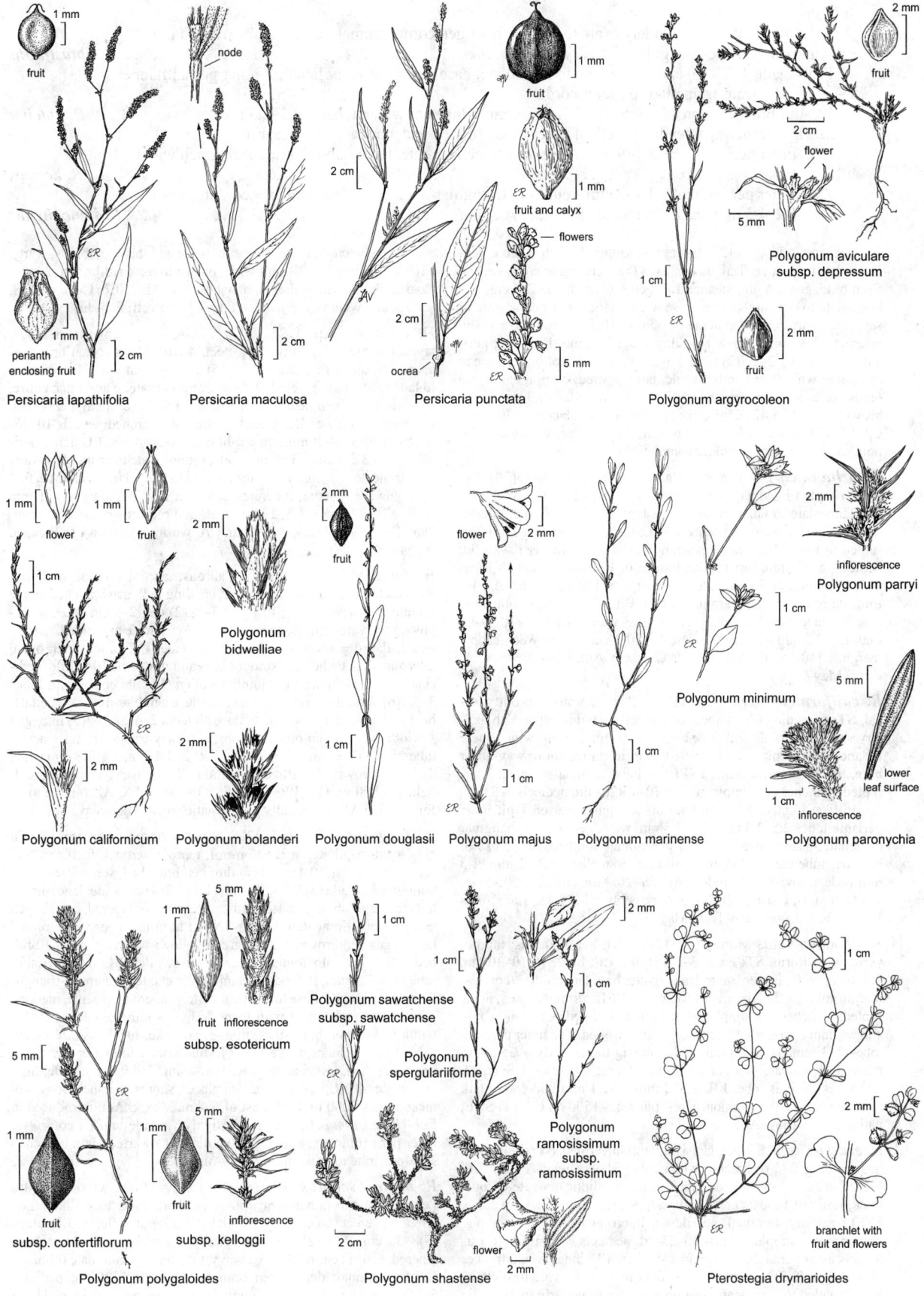

fruit

node

2 cm

ER

perianth
enclosing fruit

2 cm

1 mm

Persicaria lapathifolia

2 cm

2 cm

Persicaria maculosa

2 cm

2 cm

ocrea

fruit

fruit and calyx

flowers

1 mm

1 mm

5 mm

ER

Persicaria punctata

fruit

flower

5 mm

2 cm

1 cm

ER

fruit

2 mm

2 mm

**Polygonum aviculare
subsp. depressum**

Polygonum argyrocoleon

flower

1 mm

fruit

1 mm

1 cm

ER

2 mm

Polygonum californicum

2 mm

2 mm

**Polygonum
bidwelliae**

fruit

2 mm

1 cm

Polygonum bolanderi

Polygonum douglasii

flower

2 mm

1 cm

ER

Polygonum majus

1 cm

ER

1 cm

Polygonum marinense

2 mm

inflorescence

Polygonum parryi

1 cm

Polygonum minimum

5 mm

1 cm

inflorescence

lower
leaf surface

Polygonum paronychia

5 mm

fruit inflorescence

subsp. esotericum

1 mm

5 mm

ER

1 mm

fruit

subsp. confertiflorum

1 mm

fruit

subsp. kelloggii

5 mm

inflorescence

Polygonum polygaloides

1 cm

1 cm

**Polygonum sawatchense
subsp. sawatchense**

**Polygonum
spergulariiforme**

ER

2 cm

flower

2 mm

Polygonum shastense

2 mm

1 cm

**Polygonum
ramosissimum
subsp.
ramosissimum**

1 cm

2 mm

ER

branchlet with
fruit and flowers

Pterostegia drymarioides

17′ Per

  21. Lf blade 20–40 cm, widely ovate to oblong, base gen cordate; inner perianth lobe triangular to
      narrowly ovate-triangular. . . . . ***R. obtusifolius***

  21′ Lf blade 4–25(30) cm, gen oblong to lanceolate, base gen truncate or tapered; inner perianth lobe
      widely ovate-triangular to round-cordate

    22. Infl branches spreading; teeth of inner perianth lobe not wider at base, 0.3–2.5 mm . . . . ***R. pulcher***

    22′ Infl branches ascending; teeth of inner perianth lobe wider at base, 0.2–1.5 mm

      23. Inner perianth lobe 6–8(9) mm, round, base gen cordate, teeth 0.2–0.5 mm, each much wider
        at base . . . . ***R. kerneri***

      23′ Inner perianth lobe 3.5–5 mm, round- to triangular-ovate, base truncate or ± cordate, teeth
        0.2–1.5 mm, each ± wider at base. . . . . ***R. stenophyllus***

R. acetosella L. (p. 1127) SHEEP SORREL Per; rhizome creeping and/or taproot vertical, dioecious. **ST:** ascending or erect, several from base, gen < 4 dm, slender. **LF:** gen ± basal; blade 2–6 cm, gen lanceolate to obovate-oblong, base widely tapered or gen hastate or sagittate, margin flat, tip acute or obtuse. **INFL:** terminal, ± open, interrupted; whorls (3)5–8(10)-fld; pedicel 1–3 mm. **FL:** inner perianth lobes gen 1.2–1.7(2) mm, 0.5–1.3 mm wide, not or ± enlarged in fr, free wing 0 or barely visible, base tapered, margin entire, tip obtuse or ± acute; tubercles 0. **FR:** 0.9–1.5 mm, 0.6–0.9 mm wide, brown. 2*n*=14,28,42. ± disturbed, often acidic places; < 3000 m. CA-FP, GB; worldwide; native to Eur, Asia. Large, variable, taxonomically complex, polyploid complex. Apr–Jul ❖

R. britannica L. Per, gen glabrous. **ST:** erect, 8–15(20) dm, stout, deeply ridged; taproot fusiform, vertical. **LF:** blade 20–55(70) cm, lanceolate to lance-oblong, base tapered, rounded, or ± truncate, margin flat to ± wavy, tip acute. **INFL:** terminal, gen dense to interrupted in lower 1/2, branches straight or arched; whorls 15–25-fld; pedicels 5–13 mm, thread-like, joint ± 0. **FL:** inner perianth lobes 4–7(7.5) mm, 3.5–7 mm wide, round or round-ovate (-deltate), base truncate or ± cordate, margin entire or ± notched, flat, tip obtuse to ± acute; tubercles 3, ± equal, gen << 1/3 width of lobe. **FR:** 3–4.5 mm, 1.5–2.5 mm wide, red-brown. 2*n*=20. Uncommon. Wet meadow margins; 1100 m. n SNH (Plumas Co.); e N.Am. [*R. orbiculatus* A. Gray] May–Aug

R. californicus Rech. f. (p. 1127) Per, glabrous; taproot vertical. **ST:** ascending (decumbent or ± erect), 3–6 dm, often with axillary shoots below 1° infl. **LF:** blade 5–10 cm, 1–3 cm wide, linear to lance-oblanceolate, base tapered, margin entire, flat or wavy near base, tip acute or attenuate. **INFL:** axillary, terminal, ± open, interrupted, branches gen simple; whorls 10–15(20)-fld; pedicels 3–8 mm, thread-like, jointed. **FL:** gen both bi- and unisexual on 1 pl; inner perianth lobes 2.5–3.5 mm, 2.2–3.3 mm wide, gen widely triangular or deltate, base truncate, margin gen minute-toothed, tip obtuse or ± acute; tubercles gen 0, or 1 midvein ± swollen. **FR:** 2 mm, 1.3 mm wide, brown to dark red-brown. 2*n*=20. Moist places; < 3500 m. CA (esp mtns, coast); OR, NV, AZ, possibly n Mex. [*R. salicifolius* Weinm. var. *denticulatus* Torr.] May–Sep

R. conglomeratus Murray (p. 1127) Per, gen glabrous; taproot vertical, fusiform. **ST:** erect, 3–8(15) dm. **LF:** blade (5)10–30 cm, 2.5–6 cm wide, lanceolate to lance-ovate, base tapered to ± cordate, margin entire, ± flat, tip ± acute (obtuse). **INFL:** terminal, open, interrupted, branches gen simple; whorls 10–20-fld; pedicels 1–4(5) mm, gen ± ≥ inner perianth, thread-like, joints swollen. **FL:** inner perianth lobes 2–3 mm, 1–1.6(2) mm wide, oblong to narrowly ovate, base tapered or truncate, margin entire, tip obtuse; tubercles 3, ± equal, almost as wide as lobe. **FR:** 1.5–2 mm, 1–1.4 mm wide, gen dark red-brown. 2*n*=20. Common. Moist places; < 1500 m. CA-FP, SNE; native to Eur. May–Aug

R. crassus Rech. f. Per, glabrous; taproot vertical. **ST:** decumbent to ascending, 2–5(6) dm, gen with axillary shoots below 1° infl. **LF:** blade 3–12 cm, 1–5 cm wide, lance-ovate to -elliptic or ovate-elliptic, thick, leathery, base tapered or truncate, margin entire, flat, tip acute. **INFL:** axillary, terminal, gen dense, interrupted near base or not, branches gen simple; whorls 10–25-fld; pedicels 4–9 mm, ± 2 × as long as inner perianth, thick, joints swollen. **FL:** inner perianth lobes (3)4–5 mm, (2.5)3–4 mm wide, widely to narrow-ovate to -deltate, base rounded to ± truncate, margin entire to irregularly toothed, tip

± acute; tubercle 1, large, almost as wide as lobe, ± glabrous, warty. **FR:** 2–2.5 mm, 1.7–2.1 mm wide, brown to dark red-brown. 2*n*=20. Coastal dunes, sandy shores, marshes; < 100 m. NCo, CCo; OR. [*R. salicifolius* Weinm. var. *c.* (Rech. f.) J.T. Howell] Feb–Jul

R. crispus L. (p. 1127) CURLY DOCK Per (bien), glabrous; taproot vertical, fusiform. **ST:** erect, 4–10(15) dm, often branched above middle. **LF:** blade 15–30(35) cm, 2–6 cm wide, lance-linear to lanceolate, base tapered, truncate, or ± cordate, margin gen entire, strongly wavy, esp near base, tip acute. **INFL:** terminal, dense, occ interrupted in lower 1/2, branches straight or arched; whorls 10–25-fld; pedicels (3)4–8 mm, thread-like, joints swollen. **FL:** inner perianth lobes 3.5–6 mm, 3–5 mm wide, round-ovate to triangular-ovate, base truncate or ± cordate, margin entire to irregularly ± toothed, flat, tip obtuse or ± acute; tubercles gen 3, at least 1(2) larger, > 1 mm wide. **FR:** 2–3 mm, 1.5–2 mm wide, gen red-brown. 2*n*=60. Abundant. Disturbed places; < 2700 m. CA; worldwide; native to Eurasia. All yr ❖

R. dentatus L. Ann (bien), glabrous; taproot vertical, fusiform. **ST:** erect, < 7(8) dm, slender, often bending in fl, gen branched above middle, occ from base. **LF:** blade 3–8(12) cm, 2–5 cm wide, lance-oblong to ovate-elliptic, base ± cordate to truncate or tapered, margin entire, flat to ± wavy, tip obtuse to ± acute. **INFL:** terminal, open, interrupted, branches gen straight, ascending; whorls 10–20-fld; pedicels 2–5 mm, thread-like, joints swollen. **FL:** inner perianth lobes 3–5.5(6) mm, 2–3 mm wide exc teeth, triangular-ovate or deltate, base truncate, margin gen dentate with teeth 2–4(5), 1–3(5) mm, gen ≤ width of lobe, gen on both sides, narrowly triangular, tip ± acute; tubercles (1)3, ± equal, lanceolate. **FR:** 2–2.8 mm, 1.4–1.8 mm wide, dark red-brown. 2*n*=40. Uncommon. Wet, disturbed places, cult fields; < 800 m. GV, PR; OR, AB, ON, MO, TX, AZ; native to se Eur, Asia, n Afr. Potentially problematic weed. May–Sep

R. fueginus Phil. (p. 1127) GOLDEN DOCK Ann (bien), gen pubescent, papillate, at least on infl; taproot vertical, fusiform. **ST:** prostrate to erect, (0.4)1.5–6(7) dm, gen branched, sometimes stout, hollow. **LF:** blade (3)5–25(30) cm, (1)1.5–3(4) cm wide, lance-linear to lanceolate (oblong), base cordate, truncate, or tapered, margin gen entire, wavy, tip acute (obtuse). **INFL:** terminal, dense, interrupted below, gen red-brown, red, or green-yellow; whorls gen 15–30-fld; pedicels 3–7(9) mm, joints weak, gen not swollen. **FL:** inner perianth lobes 1.5–2.5 mm, 0.7–0.9(1.2) mm wide exc teeth, narrow-triangular or -kite-shaped, base truncate or widely tapered, tip acute, margins gen dentate (subentire) with teeth 2–3, 1–3 mm, gen 1.5–2.5(4) × width of lobe, on both sides, thin, bristle-like; tubercles 3, ± equal, ≤ 1/2 width of lobes, lance-linear to fusiform, ± brown or ± red, gen net-like, pitted, tips ± acute. **FR:** 1–1.4 mm, 0.6–0.8 mm wide, light brown. 2*n*=40. Riparian, disturbed places, shores, marshes, bogs, wet meadows; < 2000 m. CA; most of N.Am, Mex, S.Am; introduced in Eur. [*R. maritimus* L., misappl.] Highly variable in size, color, lvs, inner perianth lobes; alluvial pls sometimes dwarfed, with branched sts ± prostrate to ascending. May–Aug

R. hymenosepalus Torr. (p. 1127) CANAIGRE, WILD-RHUBARB Per, glabrous to puberulent; rhizomes short, roots tuber-like, clustered. **ST:** erect (ascending), 2.5–9(10) dm, stout, ± fleshy. **LF:** blade (5)8–30 cm, 2–8(12) cm wide, fleshy, lanceolate to oblong, base tapered, margin entire, flat to ± wavy, tip acute to acuminate (obtuse). **INFL:** terminal, dense, gen branched; whorls 5–20-fld; pedicels 5–15(20) mm, thread-like, ± jointed. **FL:** inner perianth lobes 11–16

mm, 9.5–14 mm wide, oblong- to round-cordate, base wavy, margin entire or at base notched, tip obtuse or ± acute; tubercles 0. **FR:** 4–5(7) mm, 2.5–4.5(5) mm wide, brown to red-brown. $2n=40$. Sandy, rocky places, plains, slopes, streambeds, alkaline soils; < 2000 m. SnJV, SW, DMoj; to MT, TX; Mex (Baja CA, Chihuahua). Reported from CCo, SCoR. Jan–May

R. kerneri Borbás   Per, gen papillate; taproot vertical, fusiform. **ST:** 5–10(15) dm, ± slender. **LF:** blade 15–25 cm, 5–9 cm wide, oblong to lanceolate, base truncate or ± cordate, margin entire, flat to ± wavy, tip acute or long-tapered. **INFL:** terminal, dense, interrupted esp below, branches gen simple, arched or gen straight; whorls 15–20-fld; pedicels 6–12 mm, thread-like, joints swollen. **FL:** inner perianth lobes 6–8(9) mm, 6–7.5(8) mm wide, round, base gen cordate, margin entire or toothed esp near base with teeth 0.2–0.5 mm, tip gen acute; tubercles gen 1 (if 3, then unequal), < 2 × lobe width. **FR:** 2.5–3.3 mm, 1.8–2.3 mm wide, brown. $2n=80$. Disturbed areas, oak woodland, grassland; < 700 m. CaRF, n ScV, sw SnJV (Carrizo Plain), s CCo, SCoR, n SCo; native to se Eur. Apr–Jun

R. lacustris Greene (p. 1127)   Per, terrestrial, submersed, or floating, glabrous or not; occ rhizomed, taproot vertical or creeping. **ST:** decumbent to erect, 4–7(90) dm, with axillary shoots below 1° infl esp if aquatic. **LF:** blade (2)3–7 cm, (0.5)1–3 cm wide, lance-ovate to lanceolate, base tapered, margin entire, flat or wavy, tip acute to ± obtuse; submersed ± glabrous, terrestrial abaxially papillate-hairy. **INFL:** axillary, terminal, dense, interrupted or not; branches short, simple; whorls 10–20-fld; pedicels 2.5–4(6) mm, thread-like, jointed. **FL:** inner perianth lobes 2–2.5(3) mm, 1–1.5(2) mm wide, ovate, elliptic, or elliptic-triangular, base truncate or round-tapered, margin entire or irregularly minute-toothed, tip ± acute to obtuse; tubercles 3, ± equal, gen << 1/3 lobe width, smooth or ± warty. **FR:** 1.5–2.2 mm, 0.8–1.2 mm wide, brown or dark red-brown. Beds, shores of ± salty lakes; 1000–2500 m. GB (esp MP); OR, NV. [*R. salicifolius* Weinm. var. *l.* (Greene) J.C. Hickman] Pl form varies with water level. Mar–Jun

R. obtusifolius L. (p. 1127)   BITTER DOCK   Per, glabrous, papillate or not; taproot vertical. **ST:** 6–12(15) dm, gen branched. **LF:** blade 20–40 cm, 10–15 cm wide, widely ovate to oblong, base gen cordate, margin gen entire, flat, tip obtuse to ± acute. **INFL:** terminal, open, interrupted below, branches gen ascending; whorls 10–25-fld; pedicels 2.5–8.5(10) mm, thread-like, joints swollen. **FL:** inner perianth lobe 3–6 mm, 2–3.5 mm wide exc teeth, triangular to narrowly ovate-triangular, base truncate, margin gen dentate with teeth 2–5, 0.5–1.8 mm, tip obtuse to ± acute; tubercles 3, 1 larger, or gen 1, smooth. **FR:** 2–2.7 mm, 1.2–1.7 mm wide, brown to red-brown. $2n=40$. Moist places; < 2000 m. NCo, NCoRO, n SN, GV, CCo, SnFrB, SCo, expected elsewhere; throughout N.Am; native to Eur, w Asia. Hybridizes with *R. crispus*, other Eur spp. May–Sep

R. occidentalis S. Watson (p. 1127)   WESTERN DOCK   Per, gen glabrous; taproot ± vertical. **ST:** gen erect, 5–25 dm, stout, branched. **LF:** blade 10–35 cm, 5–12 cm wide, lanceolate to ovate-triangular, ± leathery, base truncate to cordate, margin entire, ± wavy, tip acute (obtuse). **INFL:** terminal, dense above, ± interrupted below; branches often branched; whorls 12–25-fld; pedicels 5–13(17) mm, thread-like, ± jointed. **FL:** inner perianth lobes 5–10(12) mm, 5–8(11) mm wide, round to widely ovate-triangular, base truncate to ± cordate, margin entire to ± toothed, tip obtuse to ± acute; tubercles 0. **FR:** 3–4.5(5) mm, 1.5–2.5 mm wide, red-brown. $2n=120$. Uncommon. Wet habitats; < 2500 m. NCo, KR, CaRH, SNH, deltaic GV, CCo, GB; w US, Can. May–Aug

R. paucifolius Nutt.   Per, glabrous; stolons underground, densely tufted, taproot vertical; dioecious (some fls bisexual). **ST:** erect to ascending, clustered, 1–4(6) dm, branched or not, slender. **LF:** blade 3–7(10) cm, (0.6)1–3(4) cm wide, wide- or lance-ovate, base tapered, margin entire, flat, tip obtuse to ± acute. **INFL:** terminal, gen open, interrupted, gen branched; whorls 3–10(12)-fld; pedicels 1–3(5) mm, thread-like, joints ± swollen. **FL:** inner perianth lobes 2.8–3.8 mm, 2.7–3.6 mm wide, widely ovate or ± round, base cordate or rounded-truncate, tip obtuse or ± acute; tubercles 0. **FR:** 1.2–1.8 mm, 0.8–1 mm wide, brown. $2n=14,28$. Moist places; 1500–4000 m. CaRH, SNH, MP, W&I; to BC, AB, UT. May–Aug

R. persicarioides L.   Ann (bien), gen papillate, hairy; taproot vertical, fusiform. **ST:** erect, (0.5)1.5–7(9) dm, branched, slender, or stout, hollow. **LF:** blade 5–25(30) cm, 1.5–5 cm wide, lanceolate to ovate, base truncate or cordate to tapered, margin entire, ± wavy, tip gen acute. **INFL:** terminal, dense, interrupted below; whorls gen 15–25-fld; pedicels 3–7 mm, thread-like, ± jointed. **FL:** inner perianth lobes 2–2.5 mm, 0.75–1(1.5) mm wide exc teeth, narrow-triangular or -kite-shaped, base truncate or widely tapered, margin dentate with teeth 2–3, 1–1.5(1.7) mm, ± = width of lobe, gen on both sides, bristle-like, tip ± acute; tubercles 3, ± equal, oblong-ovate, straw-colored, gen netted, pitted, tip obtuse. **FR:** 0.9–1.5 mm, 0.5–0.8(1) mm wide, brown. $2n=40$. Wet, ± salty places, coastal; < 50 m. NCo, CCo, SCo; to AK, e N.Am, S.Am, Eurasia. Jun–Aug

R. pulcher L. (p. 1127)   FIDDLE DOCK   Per, glabrous, papillate or not; taproot vertical, fusiform. **ST:** erect to arching at tip, 2–6(7) dm, branched, slender. **LF:** blade 4–10(15) cm, (2)3–5 cm wide, lance-oblong to ovate-oblong, narrowed at or below middle or not, base ± cordate, rounded, or gen truncate, margin entire, flat or ± wavy, tip obtuse to ± acute. **INFL:** terminal, open, interrupted, branches spreading; whorls 10–20-fld; pedicels 2–5(6) mm, thick, joints swollen. **FL:** inner perianth lobes 3–6 mm, 2–3 mm wide exc teeth, ovate-triangular to oblong-deltoid, base truncate, margin gen dentate (± entire) with teeth 2–5(9), 0.3–2.5 mm, gen on both sides, tip obtuse to ± acute; tubercles (1)3, equal or not, gen warty. **FR:** 2–2.8 mm, 1.3–2 mm wide, dark red-brown to ± black. $2n=20$. Disturbed places, meadows, moist or dry habitats; gen < 1500 m. CA; OR, NV, OK, AZ, e US; native to Medit. Highly variable. May–Sep

R. salicifolius Weinm. (p. 1127)   WILLOW DOCK   Per, glabrous; taproot vertical. **ST:** ascending to erect, 3–6(9) dm, gen with axillary shoots below 1° infl. **LF:** blade 5–13 cm, (0.5)1–2.5 cm wide, lance-linear to occ ± linear, base tapered, margin entire, flat or ± wavy, tip acute to attenuate. **INFL:** terminal, axillary, ± open, interrupted, branches gen simple; whorls 7–20-fld; pedicels 3–5 mm, ± thread-like, ± thicker distally, joints ± not swollen. **FL:** inner perianth lobes (1.8)2–2.5(3) mm, 1.5–2.1 mm wide, widely triangular, base widely tapered or truncate, margin entire or notched, tip acute; tubercle 1, in width ± ≤ lobe, smooth to ± warty. **FR:** 1.8–2 mm, 1–1.3 mm wide, dark red-brown. Wet places, margins, rocky slopes; < 3500 m. SN, CW, SW, SNE, D; to NV, AZ, n Mex. May–Jul

R. stenophyllus Ledeb. (p. 1127)   Per, gen glabrous; taproot vertical. **ST:** erect, 4–8(13) dm. **LF:** blade 15–25(30) cm, 2–7 cm wide, oblong to narrowly lanceolate, base tapered or truncate, margin entire to irregularly toothed, gen wavy (flat), tip acute to ± obtuse. **INFL:** terminal, dense, interrupted below or not, branches gen straight (arched); whorls 20–25-fld; pedicels 3–8 mm, thread-like, joints swollen. **FL:** inner perianth lobes 3.5–5 mm, 3–5 mm wide, round- to triangular-ovate, base truncate or ± cordate, margin teeth 4–10, 0.2–1.5 mm, on both sides, tip ± acute, tubercles gen 3, ± equal, oblong. **FR:** 2–2.5(3) mm, 1–1.5 mm wide, dark or red-brown. $2n=60$. Uncommon. Wet, disturbed or marginal habitats; ± saline soils; < 150(1250) m. GV, SCo, MP; Can, w US, SC; native to Eurasia. Jun–Aug

R. transitorius Rech. f. (p. 1127)   Per, glabrous; taproot vertical. **ST:** ± decumbent to erect, 2.5–7 dm, simple or with axillary shoots below 1° infl. **LF:** blade 6–15(17) cm, 2–4 cm wide, lance-linear to lanceolate, base tapered, margin entire, ± wavy, tip acute. **INFL:** axillary, terminal, dense, interrupted below or not, branches gen simple; whorls 10–20(25)-fld; pedicels 3–7 mm, thread-like, joints swollen. **FL:** inner perianth lobes (2.5)3–4 mm, 2–2.5 mm wide, ovate, lance-ovate, or ± triangular, base rounded or truncate, margin gen entire, tip obtuse to ± acute; tubercles (1)3, unequal, gen smooth. **FR:** 1.8–2.4 mm, 1–1.5 mm wide, dark red-brown. $2n=20$. Coastal dunes, wet margins of meadows; < 2250 m. NCo, NCoRH, SNH, CCo, reported from ScV; to AK. [*R. salicifolius* var. *t.* (Rech. f.) J.C. Hickman] Apr–Jun

R. triangulivalvis (Danser) Rech. f. (p. 1127)   Per, glabrous; rhizomes 0 or short-creeping, taproot vertical. **ST:** ascending to erect, (3)4–10 dm, with axillary shoots below 1° infl. **LF:** blade 6–17 cm, 1–4(5) cm wide, linear to lanceolate, base tapered, margin entire, flat to wavy, tip acute. **INFL:** axillary, terminal, dense, interrupted or

not, branches branched; whorls 10–25-fld; pedicels 4–8 mm, thread-like, ± thicker distally, joints ± not swollen. **FL:** inner perianth lobes, (2)2.5–3.5(4) mm, (2)2.5–3(3.5) mm wide, widely triangular, base truncate or rounded, margin entire or ± notched near base, tip gen acute; tubercles gen 3, ± equal, width < 1/3 lobe (if 1 then width ≥ 1/2 lobe), glabrous or minute-warty. **FR:** 1.7–2.2 mm, 1–1.5 mm wide, brown to dark red-brown. 2*n*=20. Many habitats, often marginal or disturbed; < 3000 m. CaRH, SN, ScV (Sutter Buttes), SnGb, GB; N.Am, Mex; naturalized in Eur. [*R. salicifolius* var. *t.* (Danser) J.C. Hickman] Apr–Jul

R. utahensis Rech. f.   Per, glabrous; taproot vertical. **ST:** ascending or gen erect, 1.5–4(6) dm, with axillary shoots below 1° infl. **LF:** blade 6–15 cm, 2–3 cm wide, linear to lanceolate, base tapered, margin entire, gen flat, tip acute. **INFL:** axillary, terminal, dense, interrupted below or not, branches gen simple; whorls 10–25-fld; pedicels 4–7 mm, thread-like, thicker distally, joints swollen or not. **FL:** inner perianth lobes 2.5–3 mm, 2.5–3 mm wide, deltate to widely ovate-deltate, base gen ± truncate, margin entire, tip acute; tubercles 0. **FR:** 1.8–2 mm, 1–1.3 mm wide, dark red-brown to black. 2*n*=40. Wet meadows, river and streambanks, rocky slopes; 1000–3500 m. c&s SNH, expected elsewhere; OR to AB, CO, NV. May–Aug

R. venosus Pursh (p. 1127)   WINGED DOCK   Per, gen glabrous; rhizomes creeping. **ST:** ascending (erect), (1)1.5–3(4) dm, gen

with axillary shoots near base. **LF:** blade (2)4–12(15) cm, 1–5(6) cm wide, lance-ovate to ovate-elliptic, ± leathery, base gen tapered, margin entire, flat to ± wavy, tip acute to acuminate. **INFL:** axillary, terminal, dense, interrupted below or not; whorls 5–15-fld; pedicels (8)10–16 mm, thread-like to ± thickened, joints ± swollen. **FL:** inner perianth lobe 13–18(20) mm, (20)23–30 mm wide, round to widely round-cordate, base gen deep-notched, margin entire, tip rounded, obtuse, or widely triangular; tubercles gen 0. **FR:** 5–7 mm, 4–6 mm wide, brown to dark brown. 2*n*=40. Dry, sandy places; 1200–1800 m. MP; w N.Am. Mar–Jun ★

R. violascens Rech. f. (p. 1127)   Ann, bien (short-lived per), glabrous; taproot vertical or subvertical. **ST:** erect, 2.5–7.5 dm, gen stout (appearing per), branched above. **LF:** blade 6–12(15) cm, (2)3–4(5) cm wide, oblanceolate to elliptic, base widely tapered or rounded (truncate), margin entire, gen flat (± wavy), tip gen obtuse. **INFL:** terminal, ± open, interrupted, branches bent at nodes, arched; whorls 10–20-fld; pedicels 3–8 mm, thread-like, joints swollen. **FL:** inner perianth lobe 2.5–3.5(4) mm, 2–2.8(3) mm wide exc teeth, widely ovate-triangular to ± deltate, base truncate, margin gen dentate, teeth ≤ 0.5 mm, 2–3, on both sides, triangular; tubercles 3, ovoid, gen unequal, width ± 1/3 lobe. **FR:** 1.8–2.3 mm, 1.2–1.5 mm wide, brown to red-brown. 2*n*=20. Uncommon. Wet places along streams, rivers; < 1000 m. Deltaic ScV, SnJV, SCo, w D, reported from SCoR (Salinas Valley); to TX, Mex, also MA. Mar–Aug

SIDOTHECA   STARRY PUNCTUREBRACT

Ann, gen spreading, glabrous or glandular. **LF:** basal, linear to narrow, hairy, glandular, awns 0; ocreae 0. **INFL:** terminal, cyme-like; peduncle erect to spreading; involucre 1, tubular, teeth 5(6), awned. **FL:** 2–5(10), hairy, glandular; perianth white to rose or green-yellow to red, lobes 6, 3-lobed to fringed; stamens 9. **FR:** brown, obconic, glabrous; embryo curved. 3 spp.: w N.Am. (Greek: star case, for involucre)

1. Involucre funnel-shaped, white-margined, teeth fused 3/4; perianth lobes with 3–5 fringed lobes . . . . *S. emarginata*
1′ Involucres narrowly to widely obconic, 1-colored, teeth fused ± 1/2; perianth lobes 3-lobed.
  2. Fl 1–2 mm, green-yellow to ± red; perianth lobed 1/5; bract awns 0.2–0.5 mm . . . . *S. caryophylloides*
  2′ Fl 2.5–4 mm, white to pink or ± red; perianth lobed 1/3–1/2; bract awns 0.8–1 mm . . . . *S. trilobata*

S. caryophylloides (Parry) Reveal (p. 1127)   CHICKWEED OXYTHECA   Pl 1–2.5 dm, (0.3)1–4(5) dm diam. **LF:** blades 1–6(8) cm, 0.3–1.2(1.8) cm wide. **INFL:** bract awns 0.2–0.5 mm; peduncle erect, 0–1 cm, glandular or glabrous; involucre 4–7 mm, glandular, teeth 5(6), awns 0.3–1 mm, ± green to ± red. **FL:** 2–3, 1–2 mm; perianth green-yellow to ± red, lobes 3-lobed. **FR:** 1.2–1.5 mm. *n*=20. Sand or gravel; 1300–2600 m. s SNH (Tulare Co.), TR, SnJt. [*Oxytheca c.* Parry] Jun–Sep ★

S. emarginata (H.M. Hall) Reveal (p. 1127)   WHITE-MARGINED OXYTHECA   Pl 0.3–3 dm, 0.3–5 dm diam. **LF:** blades 1.5–7.5 cm, 0.4–1.5 cm wide. **INFL:** bract awns 1–2 mm; peduncles erect, 0.5–3 cm, glandular; involucre 4–8 mm, 9–12 mm wide, glabrous, margin white, teeth 5, awns 1–1.5 mm, ± red. **FL:** 3–6, (2)3–4.5(5) mm;

perianth white to pink, lobes with 3–5 fringed linear lobes. **FR:** 1.8–2 mm. Gravel; 1200–2500 m. e PR (SnJt, Santa Rosa Mtns, Riverside Co.). [*Oxytheca e.* H.M. Hall] Feb–Aug ★

S. trilobata (A. Gray) Reveal (p. 1127)   THREE-LOBED STARRY PUNCTUREBRACT   Pl 0.7–5 dm, 0.7–6(11) dm diam. **LF:** blades 1–5(9) cm, 0.2–0.7(2) cm wide, linear or narrow. **INFL:** bract awns 0.8–1 mm; peduncles erect to spreading, 0.5–1.5 cm, glabrous, glandular proximally; involucre 3–8 mm, glandular, teeth 5(6), awns 0.3–2 mm, ± green to ± red. **FL:** 3–5(10), 2.5–4 mm; perianth white to pink or ± red, lobes 3-lobed. **FR:** 1.2–2 mm. *n*=20. Common. Sand; 700–2100 m. SnGb, SnBr, PR, w DMtns (Little San Bernardino Mtns); Baja CA. [*Oxytheca t.* A. Gray] Apr–Sep

SYSTENOTHECA

1 sp. (Greek: tapered case, for involucre teeth)

S. vortriedei (Brandegee) Reveal & Hardham (p. 1127)   VORTRIEDE'S SPINEFLOWER   Ann, spreading, 0.2–1.5 dm, 0.2–1 dm diam, glandular. **LF:** basal; blades (1)2–5 cm, (0.4)0.5–1 cm wide, awn 0; ocreae 0. **INFL:** terminal, cyme-like; peduncle 0; involucre 1, 2.5–4 mm, 4-angled, swollen, teeth 4, short awned. **FL:** 2, upper

perfect, 1–1.5 mm, lower pistillate, 2–2.5 mm; perianth 2-colored, fl tube yellow, lobes 6, white to pink or rose, glabrous, papillate, 2-lobed; stamens 9. **FR:** 2–2.5 mm, brown to black, obconic, glabrous; embryo curved. *n*=19, 20. Sand; 700–1500 m. SCoRO (Santa Lucia Range, Monterey, Santa Luis Obispo cos.). May–Jul ★

PORTULACACEAE   PURSLANE FAMILY

Gilberto Ocampo

Ann to per, gen fleshy. **ST:** [1]several to many, spreading [to erect], gen glabrous. **LF:** simple, alternate or opposite, linear, obovate, spoon-shaped, [elliptic, ± round], flat or cylindric, hairs in axils, inconspicuous or not [0], upper 2–5 forming involucre. **INFL:** fls 1 or clustered at st tips. **FL:** bisexual, radial; sepals 2, fused at base, lower part fused to ovary and gen persistent

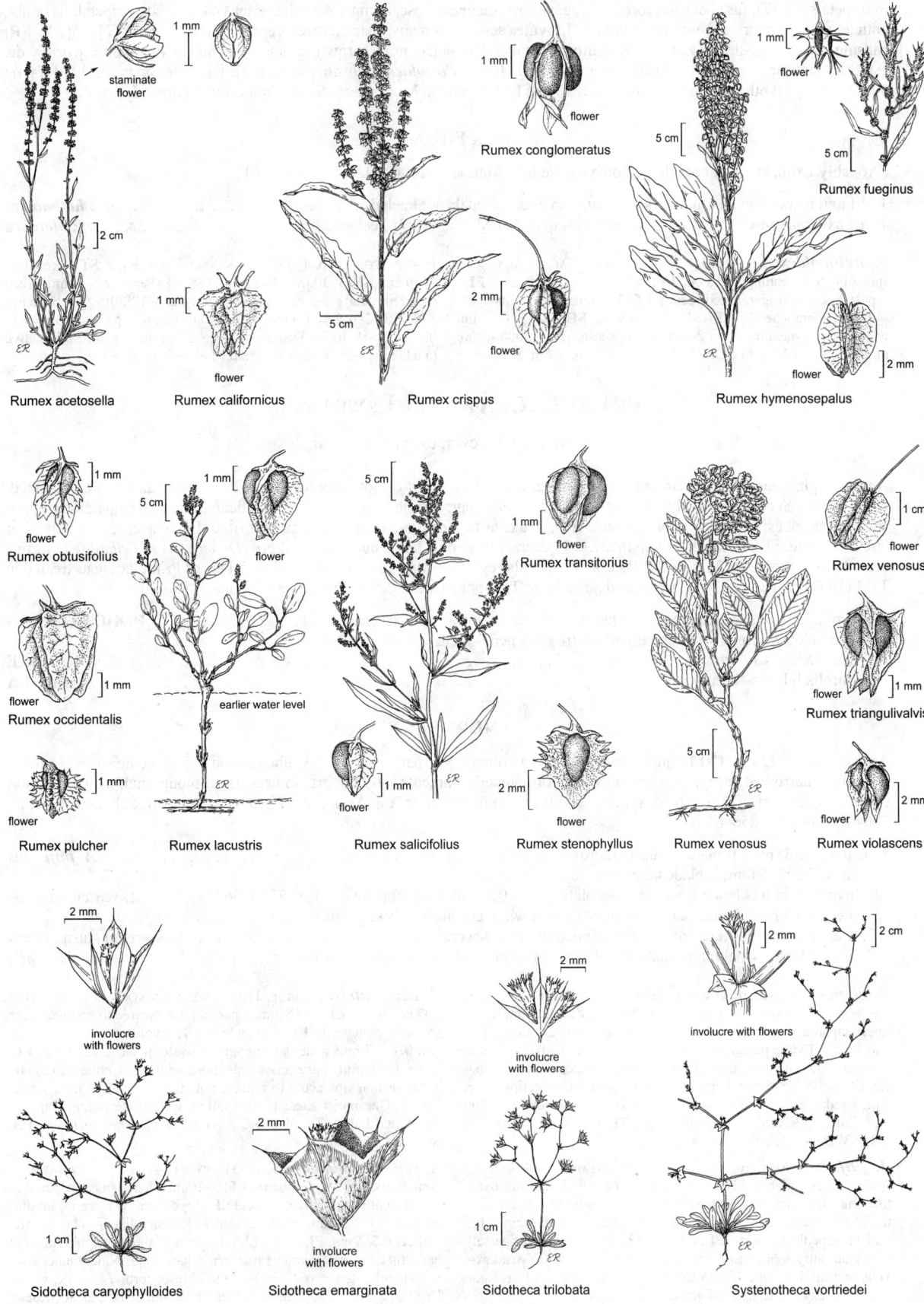

staminate flower

1 mm

2 cm

ER

Rumex acetosella

1 mm

flower

Rumex californicus

5 cm

ER

2 mm

flower

Rumex crispus

1 mm

flower

Rumex conglomeratus

1 mm

flower

5 cm

5 cm

Rumex fueginus

5 cm

ER

2 mm

flower

Rumex hymenosepalus

1 mm

flower

Rumex obtusifolius

1 mm

flower

Rumex occidentalis

1 mm

flower

Rumex pulcher

1 mm

5 cm

flower

earlier water level

ER

Rumex lacustris

5 cm

1 mm

1 mm

flower

Rumex transitorius

ER

1 mm

flower

Rumex salicifolius

2 mm

flower

Rumex stenophyllus

5 cm

ER

Rumex venosus

1 cm

flower

Rumex venosus

1 mm

flower

Rumex triangulivalvis

2 mm

flower

Rumex violascens

2 mm

involucre
with flowers

1 cm

Sidotheca caryophylloides

2 mm

involucre
with flowers

Sidotheca emarginata

2 mm

involucre
with flowers

1 cm

ER

Sidotheca trilobata

2 mm

involucre with flowers

2 cm

ER

Systenotheca vortriedei

in fr; petals [4]5[7], fused at base, forming ring, yellow [magenta, rose, orange, white]; stamens 4–20[> 20], epipetalous or not, anthers yellow; ovary 1/2-inferior, chamber 1, ovules several to many, placenta free-central; style branches [2]3–6[12]. **FR**: capsule, circumscissile. **SEED**: many, reniform, ± tubercled, black, gray, [brown, ± blue]. 1 genus, ± 100 spp.: ± worldwide, esp trop, subtrop; some cult (*Portulaca grandiflora* Hook.; *Portulaca umbraticola* Kunth). [Nyffeler & Eggli 2010 Taxon 59:227–240] All other CA genera previously in this family now in Montiaceae. Scientific Editor: Thomas J. Rosatti.

PORTULACA

(Probably Latin: small gate or door, from capsule lid) [Matthews 2004 FNANM 4:496–501]

1. Lf axil hairs many, > 1 mm; lvs ± linear, ± cylindric; sepals not keeled . . . . ***P. halimoides***
1′ Lf axil hairs few, < 1 mm; lvs obovate to spoon-shaped, flat; sepals keeled . . . . ***P. oleracea***

P. halimoides L. (p. 1135) DESERT PORTULACA Ann. **ST**: spreading to ascending, 2–15 cm. **LF**: 3–15 mm. **INFL**: fls 2–8. **FL**: sepals 1–2.5 mm, gen ± red; petals 1.5–3.5 mm, yellow; stamens 4–18; style branches 3–4. **FR**: 1.5–2 mm wide. **SEED**: 0.3–0.5 mm wide, black or metallic silver. 2*n*=18. Sandy washes, flats; 1000–1200 m. PR, DSon, DMoj; to CO, OK, TX, Mex, C.Am, S.Am. Sep ★

P. oleracea L. (p. 1135) PURSLANE Ann, ± per. **ST**: spreading, 3–40 cm. **LF**: 3–30 mm. **INFL**: fls 1–8. **FL**: sepals 2.5–5 mm, green or ± red; petals 3–5 mm, yellow; stamens 5–12[20]; style branches 3–6. **FR**: 2–4.5 mm wide. **SEED**: 0.6–1 mm wide, dark brown to black. 2*n*=18,36,54. Disturbed soil; < 2100 m. CA; probably native to e hemisphere, widely naturalized elsewhere. Apr–Oct

PRIMULACEAE PRIMROSE FAMILY

Anita F. Cholewa, except as noted

Ann, per, glabrous to glandular-hairy. **LF**: simple, ± basal, petioled or not; stipules 0. **INFL**: scapose umbel. **FL**: bisexual, radial; parts gen in 4s or 5s; calyx deeply lobed, often persistent; corolla lobes spreading to reflexed; stamens epipetalous, opposite corolla lobes; ovary superior, 1-chambered, placenta free-central, style 1, stigma head-like. **FR**: capsule, 2–6-valved or circumscissile. **SEED**: small, few to many. ± 9 genera, 600 spp.: n hemisphere; several orn (*Dodecatheon, Primula*). [Martins et al. 2003 Pl Syst Evol 237:75–85] Based on molecular evidence, non-rosette terrestrial members of Primulaceae as treated in TJM (1993) moved to Myrsinaceae, and *Samolus* to Theophrastaceae. Scientific Editor: Thomas J. Rosatti.

1. Corolla lobes reflexed; anthers ± adherent into a cone; lvs gen 12–30 cm . . . . **DODECATHEON**
1′ Corolla lobes spreading or erect; anthers free; lvs gen < 8 cm
  2. Corolla lobes 1–3 mm . . . . **ANDROSACE**
  2′ Corolla lobes 5–10 mm . . . . **PRIMULA**

ANDROSACE

Ann, per, gen < 12 cm. **INFL**: umbel subtended by involucre. **FL**: parts in 5s; calyx tube scarious, lobes acute; corolla salverform, tube narrowed at top, lobe tips obcordate or notched; filaments ± 0 or short, anthers free, oblong, incl; ovary superior, spheric, style short. **FR**: 5-valved, spheric. ± 100 spp.: n temp, arctic, esp Asia. (Greek: uncertain sea-pl) [Schneeweiss et al. 2004 Syst Biol 53:856–876]

1. Calyx ± 2(3) mm; lf blade abruptly narrowed to petiole . . . . ***A. filiformis***
1′ Calyx (2.5)3–6 mm; lf blade tapered to petiole
  2. Involucre bracts lance-linear to lanceolate, gen < 0.5 mm wide; corolla ≥ calyx; 2700–3600 m . . . . ***A. septentrionalis***
  2′ Involucre bracts lanceolate to ovate, < 1.7 mm wide; corolla < calyx; < 1700 m
    3. Calyx lobes awl-like, tips acute, stiff; peduncles 1–several . . . . ***A. elongata*** subsp. ***acuta***
    3′ Calyx lobes widely lanceolate to deltate, tips acute to obtuse, not stiff; peduncles gen 1 . . . . ***A. occidentalis***

A. elongata L. subsp. **acuta** (Greene) G.T. Robbins (p. 1135) CALIFORNIA ANDROSACE Ann 2–8 cm, hairy. **LF**: 5–20 mm, lance-linear, tapered to petiole, acute to acuminate, gen entire or finely dentate, ciliate. **INFL**: peduncles 1–several; involucre bracts 2.5–5 mm, < 1 mm wide, lanceolate to ovate, acute to acuminate; pedicel 0.6–6.4 cm. **FL**: calyx 3.5–5 mm, hairy, lobes gen = tube, awl-like, tips acute, stiff, ± red; corolla < calyx, white. 2*n*=40. Dry grassy slopes; < 1200 m. NCoRI, CaR, s SNF, GV, SnFrB, SCoRI, SCo, WTR, SnBr, PR; OR, NV, Baja CA. Feb–Apr ★

A. filiformis Retz. (p. 1135) SLENDER-STEMMED ANDROSACE Ann 3–12 cm, glabrous or ± glandular-hairy. **LF**: 3–20 mm, ovate to ± triangular, abruptly narrowed to petiole, finely dentate esp near tip. **INFL**: peduncles gen several; involucre bracts 0.8–1.2 mm, < 0.5 mm wide; pedicel 1–4 cm. **FL**: calyx ± 2(3) mm, glabrous to sparsely glandular-hairy, lobes < to ± = tube; corolla > calyx, white. Meadows; 1800 m. CaRH (s slope Willow Creek Mtn, n Siskiyou Co.); sporadic to WA, Rocky Mtns; also in Eurasia. May?–Jun ★

A. occidentalis Pursh (p. 1135) WESTERN ROCK-JASMINE Ann 3–7 cm, hairy. **LF**: 5–15 mm, lance-elliptic, tapered to petiole, entire to finely dentate. **INFL**: peduncle gen 1; involucre bracts 2–5 mm, gen 0.7–1.7 mm wide, lance-ovate to ovate; pedicel 0.5–3 cm. **FL**: calyx 3.6–6 mm, hairy, lobes gen = tube, widely lanceolate to deltate, ± red or not, tips acute to obtuse, not stiff; corolla < calyx, white. 2*n*=20. Gen moist sites; 1580–1650 m. n SNH (Emigrant Gap); to BC, Great Lakes, TX (exc OR, WA). [*A. o.* var. *simplex* (Rydb.) H. St. John] Apr–Aug ★

A. septentrionalis L. (p. 1135) PYGMY-FLOWER ROCK-JASMINE Ann, weak per, 1–6 cm, hairy. **LF**: 5–20 mm, lance-linear, tapered to petiole, entire to finely dentate. **INFL**: peduncles (1)several; involucre bracts 1.7–3 mm, gen < 0.5 mm wide, lance-linear to lanceolate; pedicel 0.5–5 cm. **FL**: calyx (2.5)3–4 mm, glabrous or puberulent at base, tube > lobes, scarious between ridges, lobes widely lanceolate to triangular, gen ± red, tips acute to obtuse; corolla ≥ calyx, white. Dry, rocky sites; 2700–3600 m. c&s SNH, SnBr, SNE; to BC, Rocky Mtns. [*A. s.* subsp. *subumbellata* (A. Nelson) G.T. Robbins] Jul–Aug

DODECATHEON SHOOTING STAR

Sylvia Kelso

Per, glabrous or glandular-hairy; roots fleshy-fibrous. **INFL**: umbel subtended by involucre. **FL**: nodding; sepals reflexed, later erect, persistent; corolla tube short, gen not covering anther bases, lobes reflexed, often with white and/or yellow and/or dark purple base; filaments very short, wide, often fused, anthers ± lanceolate, exserted, erect, ± adherent into a cone around style; ovary superior, style slender, ± exserted from anthers. **FR**: ± 5-valved or circumscissile, oblong-ovate to cylindric. ± 15 spp.: gen N.Am. (Greek: 12 gods, presumably the Olympians) [Mast & Reveal 2007 Brittonia 59:79–82] Monophyletic genus closest to *Primula* subg. *Auriculastrum* and recently treated in *Primula* (Mast & Reveal 2007); polyploid group; spp. often intergrade; "anther connective" refers to tissue between pollen sacs, esp near base; dehiscence must be determined on fr that has aged and dried naturally, because e.g., green fr of circumscissile taxa (e.g., *D. clevelandii*) sometimes split longitudinally as a result of pressing and thereby may appear valved. For another, in some cases different, treatment of genus, see Reveal 2009 FNANM 8:268–271.

1. Stigma enlarged (>> style in width); anther connective wrinkled, dark maroon to black; filaments < 1.5 mm, free or joined by thin membrane but not fused into tube
    2. Corolla tube gen covering anther bases; fl parts in 5s; anther tips acute; pl densely glandular-hairy . . . . . . ***D. redolens***
    2′ Corolla tube not covering anther bases; fl parts in 4s or 5s; anthers tips truncate to obtuse; pl glabrous to glandular-hairy
        3. Lf 2–20 cm, linear to linear-oblanceolate; st glabrous; infl glabrous to ± glandular-hairy; fl parts in 4s; fr with valves . . . . ***D. alpinum***
        3′ Lf 9–50 cm, linear-oblanceolate to oblanceolate; st glandular-hairy; infl glandular-hairy; fl parts in 4s or 5s; fr circumscissile or with valves, sometimes both on 1 pl . . . . ***D. jeffreyi***
1′ Stigma ± not enlarged (± = style in width); anther connectives smooth or wrinkled, yellow or dark; filaments 1–4 mm, fused into tube, or < 1.5 mm, free
    4. Anther connectives smooth (to longitudinally wrinkled when dry); fr circumscissile or with valves
        5. Corolla lobe gen 9–14 mm, magenta to lavender; roots white, without bulblets; fr with valves . . . . ***D. pulchellum***
        5′ Corolla lobe gen 5–9 mm, magenta to white; roots ± red, with many bulblets; fr circumscissile . . . . ²***D. subalpinum***
    4′ Anther connectives transversely wrinkled; fr circumscissile
        6. Lf blade gen tapered to petiole, length gen > 2.5 × width; filament tube > 2 mm; connective dark, smooth or obscurely wrinkled (if lf narrowed abruptly to petiole, then filament tube < 1.5 mm, connective dark, gen smooth)
            7. Anther 5–9 mm, filament 0.5–1.5 mm, free or fused into tube; corolla lobe gen 7–20 mm . . . . ***D. conjugens***
            7′ Anther 3–4 mm, filament 2–3.5 mm, fused into tube; corolla lobe gen 5–9 mm . . . . ²***D. subalpinum***
        6′ Lf blade gen narrowed abruptly to petiole, length gen < 2.5 × width; filament tube 1–4 mm; connective dark, gen wrinkled
            8. Lf blade length gen < 2 × width; filament tube gen < 3 mm wide; root bulblets present at fl . . . . ***D. hendersonii***
            8′ Lf blade length gen > 2 × width; filament tube gen 3–4 mm wide; root bulblets 0 . . . . ***D. clevelandii***
                9. Anther connectives yellow; filament tube without yellow or white spot below each anther, sometimes all yellow . . . . subsp. *clevelandii*
                9′ Anther connectives maroon to black; filament tube with or without yellow or white spot below each anther
                    10. Filament tube without yellow or white spot below each anther . . . . subsp. ***insulare***
                    10′ Filament tube with yellow or white spot below each anther
                        11. Anther gen dark purple, tip obtuse to notched . . . . subsp. ***patulum***
                        11′ Anther dark red, black, or gen yellow, tip acute to obtuse . . . . subsp. ***sanctarum***

D. alpinum (A. Gray) Greene (p. 1135) ALPINE SHOOTING STAR Pl gen glabrous. **ST**: 4–30 cm. **LF**: 2–20 cm; blade linear to linear-oblanceolate, tapered to petiole. **INFL**: 1–10-fld, glabrous to ± glandular-hairy. **FL**: parts in 4s; corolla lobes 8–16 mm, magenta to lavender; filaments free, 0.5 mm, black, anthers 4.7–8.5 mm, connective wrinkled, purple-black; stigma enlarged. **FR**: 5-valved. *n*=22. Boggy meadows, streambanks; 1700–3400 m. KR, NCoRH, SNH, n WTR (Mount Pinos), SnBr, SnJt, Wrn, SNE; to OR, UT, AZ. [*Primula tetrandra* (Greene) Mast & Reveal] Jun–Aug

D. clevelandii Greene   Pl glandular-hairy. **ST**: 12–45 cm. **LF**: 1–18 cm; blade gen oblanceolate, gen narrowed abruptly to petiole, margin dentate to ± entire. **INFL**: 1–16-fld. **FL**: parts in 5s; corolla lobes 6–25 mm, magenta to white, lobes 6–25 mm; filament tube 2.5–4 mm, anthers 3–5 mm, connective transversely wrinkled, yellow to black; stigma not enlarged. **FR**: circumscissile. [*Primula c.* (Greene) Mast & Reveal] Extremes (subspp. here) gen segregated by geog but intergrade; possibly hybridizes with *D. hendersonii* complex.

subsp. ***clevelandii*** (p. 1135)   **INFL**: 1–16-fld. **FL**: filament tube without yellow or white spot below each anther, sometimes all yellow, anther connective yellow or purple and yellow. *n*=22. Grassy slopes, flats; gen < 600 m. SCo, TR, PR; Baja CA. Feb–Apr

subsp. ***insulare*** H.J. Thomps.   **INFL**: 5–9-fld. **FL**: filament tube without yellow or white spot below each anther, anther connective maroon to black. *n*=22. Grassland, woodland; gen < 600 m. SCoRO, ChI; Baja CA (Guadalupe Island). [*Primula c.* var. *i.* (H.J. Thomps.) Mast & Reveal] Feb–Apr

subsp. ***patulum*** (Kuntze) H.J. Thomps.   **INFL**: 1–6-fld. **FL**: filament tube with yellow or white spot below each anther, anthers gen dark purple, connective maroon or maroon and yellow to black, tip obtuse to notched. *n*=22,44. Moist places, often on serpentine or in ± alkaline sites; gen < 600 m. SNF, c SNH, GV, CW. [*Primula c.* var. *p.* (Kuntze) Mast & Reveal; *D. c.* var. *p.* (Kuntze) Reveal] Mar–May

subsp. ***sanctarum*** (Greene) Abrams   **INFL**: 3–7-fld. **FL**: filament tube with yellow or white spot below each anther, anthers dark red, black, or gen yellow; connectives maroon to black, tip acute to obtuse. *n*=22,33,44. Woodland; gen < 600 m. SnFrB, SCoR, WTR, SnGb. [*D. c.* Greene var. *gracile* (Greene) Reveal; *Primula c.* var. *g.* (Greene) Mast & Reveal] Mar–May

D. conjugens Greene (p. 1135)   Pl glabrous. **ST**: to 16 cm. **LF**: 3–14 cm; blade linear-oblanceolate to obovate, tapered or narrowed abruptly to petiole, entire. **INFL**: 1–7-fld. **FL**: parts in 5s; corolla

lobes gen 7–20 mm, magenta to white; filaments free or fused into tube 0.5–1.5 mm, anthers 5–9 mm, connective transversely wrinkled, dark maroon to black (yellow or yellow and purple); stigma ± not enlarged. **FR:** circumscissile. *n*=22. Moist slopes, meadows, often in sagebrush scrub; 1200–1900 m. CaRH, MP; to w Can, WY. [*Primula c.* (Greene) Mast & Reveal] Apr–Jul

D. hendersonii A. Gray (p. 1135) MOSQUITO BILL(S), SAILOR CAPS Pl glabrous to glandular-hairy; roots with rice-like bulblets at fl. **ST:** 12–48 cm. **LF:** 2–16 cm; blade elliptic to ovate or obovate, gen narrowed abruptly to petiole, entire to ± toothed. **INFL:** 3–17-fld. **FL:** parts in 4s or 5s, even on same pl; corolla lobes 6–23 mm, magenta to deep lavender or white; filament tube 1–3 mm, anthers 3–5 mm, connective transversely wrinkled, dark maroon to black; stigma ± not enlarged. **FR:** circumscissile. *n*=22,33,66. Gen in shady sites; < 1900 m. NW, CaR, SNF, n SNH, GV, SnFrB, CW, SnBr; to s BC, ID. [*Primula h.* (A. Gray) Mast & Reveal] Highly variable; possibly hybridizes with *D. clevelandii.* Mar–Jul

D. jeffreyi Van Houtte SIERRA SHOOTING STAR Pl glabrous exc st, infl glandular-hairy. **ST:** 10–60 cm. **LF:** 9–50 cm; blade linear-oblanceolate to oblanceolate, gen tapered to petiole, entire to crenate. **INFL:** 3–18-fld. **FL:** parts in 4s or 5s; corolla lobes 10–25 mm, magenta to lavender or white; filaments free or partly fused, gen < 1.5 mm, anthers 6.5–11 mm, connective wrinkled, gen dark purple; stigma enlarged. **FR:** circumscissile or with valves, sometimes both on 1 pl. 2*n*=42,43,44, 66,86. Moist to dry meadows, streambanks; 600–3000 m. NW (exc NCo), CaR, SN, MP; to AK, MT. [*Primula j.* (Van Houtte) Mast & Reveal] Highly variable, esp in s SN. Intergrades with *D. alpinum* and *D. redolens.* Jun–Aug

D. pulchellum (Raf.) Merr. (p. 1135) BEAUTIFUL SHOOTING STAR Pl gen glabrous. **ST:** 10–40 cm. **LF:** 4–25 cm; blade oblanceolate to ovate, gen narrowed ± gradually to petiole, gen entire. **INFL:**

2–15-fld. **FL:** parts in 5s; corolla lobes gen 9–14 mm, magenta to lavender; filament tube 1.5–3.5 mm, dark maroon to black, anthers 3–5.5 mm, connective smooth (to longitudinally wrinkled when dry), maroon or maroon and yellow to black; stigma ± not enlarged. **FR:** 5-valved. 2*n*=44,88, 132. Wet meadows; 1200–2200 m. CaRH, GB (expected but evidently not collected W&I), DMtns; to AK, e US, Mex. [*Primula pauciflora* (Greene) Mast & Reveal; *D. p.* var. *monanthum* (Greene) B. Boivin; *D. p.* var. *shoshonense* (A. Nelson) Reveal] Highly variable polyploid complex in need of further study. May–Aug ★

D. redolens (H.M. Hall) H.J. Thomps. (p. 1135) Pl densely glandular-hairy. **ST:** 15–60 cm. **LF:** 20–40 cm; blade oblanceolate, tapered to petiole, entire. **INFL:** 5–10-fld. **FL:** parts in 5s; corolla tube gen covering anther bases, lobes 15–25 mm, gen magenta to lavender; filaments free, gen < 1 mm, anthers 7–11 mm, connective transversely wrinkled, gen dark maroon to black; stigma enlarged. **FR:** 5-valved. Moist sites; 2400–3600 m. c&s SNH, SnGb, SnBr, SnJt, SNE, DMtns; to w UT. [*Primula fragrans* Mast & Reveal] Intergrades with *D. alpinum, D. jeffreyi;* forms with ± white petals, light purple or red connectives, may occur sporadically. Jun–Jul

D. subalpinum Eastw. Pl glabrous, ± red exc lf blades; roots ± red, rice-like bulblets many. **ST:** 10–20 cm. **LF:** 3–10 cm; blade oblanceolate, tapered to petiole, entire. **INFL:** 1–5-fld. **FL:** parts in 5s; corolla lobes gen 5–9 mm, magenta to white; filament tube 2–3.5 mm, anthers 3–4 mm, connective smooth (to longitudinally wrinkled when dry) or transversely wrinkled, maroon to black; stigma ± not enlarged. **FR:** circumscissile. 2*n*=66. Moist places, often shaded; 2100–4000 m. c&s SNH. [*Primula s.* (Eastw.) Mast & Reveal] Evidently a high elevation ecotype, but possibly more appropriately treated as a var. of *D. hendersonii,* for which the name *D. hendersonii* var. *yosemitanum* H. Mason is available. Jun–Aug

PRIMULA PRIMROSE

Sylvia Kelso

Per, subshrub, rhizomed or stoloned. **LF:** basal or crowded near ground, sessile to petioled. **INFL:** umbel 1 per scape, terminal, subtended by bracts. **FL:** parts in 5s; calyx tube angled; corolla funnel-shaped or salverform, lobes spreading or erect, entire or notched at tip; stamens incl, anthers free, oblong, obtuse. **FR:** 5-valved, elliptic to ovoid. **SEED:** often many. 450 spp.: gen n temp. (Latin: diminutive of first, from early fl) Spp. often heterostylous. Forms monophyletic group with *Dodecatheon* and non-CA genera (e.g., *Dionysia*); transfers to *Primula* underway (Mast & Reveal 2007 Brittonia 59:79–82).

P. suffrutescens A. Gray (p. 1135) SIERRA PRIMROSE Subshrub. **ST:** creeping, branched; base woody. **LF:** 15–35 mm, ± spoon-shaped, glabrous, ± fleshy, dentate, base tapered, tip gen rounded. **INFL:** glandular; scape 3.5–12 cm; pedicels 2–several, < 1.5 cm. **FL:** calyx 4–8 mm, lobes lanceolate; corolla magenta, throat yellow, tube 5–10 mm, = lobes. **FR:** ± = calyx, ovoid. 2*n*=44. Gen rock crevices; gen 2000–4200 m. KR, SNH, SNH/SNE. Jul–Aug

PROTEACEAE PROTEA FAMILY

David J. Keil

[(Herb, subshrub) shrub] tree; evergreen. **LF:** alternate [opposite, whorled], [simple, entire to] pinnately divided or compound; stipules 0. **INFL:** raceme [spike, head, umbel, fls 1]. **FL:** bisexual [unisexual], bilateral [radial]; sepals 4, petal-like, fused at base; petals 0; stamens gen 4, filaments fused to sepals [free]; pistil 1; ovary superior, chamber 1, ovules 1–2[many], at tip or marginal, style 1, stigma 1. **FR:** follicle [achene, drupe, nut]. 75 genera, 1050 spp.: Australia, s Afr, S.Am. Scientific Editor: Thomas J. Rosatti.

GREVILLEA

FL: calyx tube divided down 1 side, lobes curled to other side; style in bud strongly curved, in fl to ± straight, exserted. 360 spp.: Australia. (C.F. Greville, British collector, horticulturist, 1749–1809)

G. robusta R. Br. Pl ≤ 30 m. **LF:** petioled, 15–40 cm, ovate to elliptic, compound or divided with pinnately divided lflets, adaxially green, abaxially appressed-silvery-hairy. **INFL:** 5–15 cm, appearing 1-sided; pedicels spreading, turned up, orange. **FL:** calyx ± 10 mm, tube yellow or orange with ± red-brown center, lobes narrowly spoon-shaped, yellow; style 20–25 mm. **FR:** narrowed to stalk-like base, body 12–20 mm, style persistent; seeds 1–2, flat, winged. Escaped from cult, brush-covered hillsides, canyons; < 450 m. s SCoRO, SCo, PR; native to e Australia. Mar–May

RANUNCULACEAE BUTTERCUP FAMILY

Margriet Wetherwax & Dieter H. Wilken, family description, key to genera

Ann, per, woody vine [shrub], occ aquatic. **LF**: gen basal and cauline, alternate or opposite, simple or compound; petioles at base gen flat, occ sheathing or stipule-like. **INFL**: cyme, raceme, panicle, or fls 1. **FL**: gen bisexual, gen radial; sepals 3–6(20), free, early-deciduous or withering in fr, gen green; petals 0–many, gen free; stamens gen 5–many, staminodes gen 0; pistils 1–many, ovary superior, chamber 1, style 0–1, gen ± persistent as beak, ovules 1–many. **FR**: achene, follicle, berry, ± utricle in *Trautvetteria*, in aggregate or not, 1–many-seeded. ± 60 genera, 1700 spp.: worldwide, esp n temp, trop mtns; many orn (*Adonis, Aquilegia, Clematis, Consolida, Delphinium, Helleborus, Nigella*). some highly TOXIC (*Aconitum, Actaea, Delphinium, Ranunculus*). [Whittemore & Parfitt 1997 FNANM 3:85–271] Taxa of *Isopyrum* in TJM (1993) moved to *Enemion*; *Kumlienia* moved to *Ranunculus*. Scientific Editors: Douglas H. Goldman, Bruce G. Baldwin.

1. Fl bilateral; sepals petal-like, not alike, uppermost spurred or hooded, gen > others
  2. Uppermost sepal hooded, not spurred, enclosing petals, stamens . . . . **ACONITUM**
  2′ Uppermost sepal spurred, ± flat or curved but not hooded, gen not enclosing petals, stamens
    3. Ann, gen from taproot; petals 2, fused; pistil 1 . . . . **CONSOLIDA**
    3′ Per, gen from ± fibrous or fleshy roots; petals 4, free; pistils 3(5) . . . . **DELPHINIUM**
1′ Fl radial; sepals petal-like or not, ± alike
  4. Sepal spurs present, petal spurs 0; pl gen tufted; lvs basal, simple, linear . . . . **MYOSURUS**
  4′ Sepal spurs 0, petal spurs occ; pl not tufted; lvs gen basal and cauline, gen dissected or compound (or basal, simple, lanceolate to reniform)
    5. Pistils 1; fr a berry . . . . **ACTAEA**
    5′ Pistils (1)2–many; fr achene or follicle (± utricle in *Trautvetteria*)
      6. Ovules ≥ 2 per ovary; fr follicle
        7. Petals spurred . . . . **AQUILEGIA**
        7′ Petals 0 or not spurred
          8. Lf simple . . . . **CALTHA**
          8′ Lf compound
            9. Pl from slender rhizomes or stolons, scapose; sepals 5–8, linear; petals 5–7 . . . . **COPTIS**
            9′ Pl from clustered, slender to fusiform or ± spheric fleshy roots, not scapose; sepals gen 5, oblong to obovate; petals 0 . . . . **ENEMION**
      6′ Ovule 1 per ovary; fr achene (± utricle in *Trautvetteria*)
        10. Cauline lvs gen opposite or 0 but involucre bracts lf-like, gen in 1–2 whorls of 2–5
          11. Pl per, not vine; basal lvs present, cauline 0 but involucre bracts lf-like, gen in 1–2 whorls of 2–5; infl terminal; sepals 5–10 . . . . **ANEMONE**
          11′ Pl ± woody vine; basal lvs 0, cauline gen many, opposite; infl axillary; sepals gen 4 . . . . **CLEMATIS**
        10′ Cauline lvs alternate or 0
          12. Perianth parts in 2 whorls; sepals gen green to ± yellow, not petal-like; petals 3–many, yellow to orange, occ white or purple
            13. Anthers ± purple; petals yellow to orange, base ± purple, nectaries 0 . . . . **ADONIS**
            13′ Anthers yellow; petals gen yellow, occ white or purple, nectaries near base . . . . **RANUNCULUS**
          12′ Perianth parts in ± 1 whorl (inner << outer, gland-like in *Ranunculus hystriculus*); sepals gen green to white or ± purple, petal-like or not; petals gen 0
            14. Lf compound, lflets entire, crenate, or lobed; fl unisexual or bisexual . . . . **THALICTRUM**
            14′ Lf simple, crenate to deeply lobed; fl bisexual
              15. Lf crenate to palmately lobed, lobes rounded to obtuse, entire to toothed or lobed; infl scapose, fls 1(3); sepals 6–13 mm, blade ± flat; petals 2–4 mm, gland-like . . . . ***Ranunculus hystriculus***
              15′ Lf deeply palmately lobed, lobes ± wedge-shaped, toothed distally; infl not scapose, fls ≥ 5 in terminal panicle; sepals 2.5–6 mm, blade cup-like; petals 0 . . . . **TRAUTVETTERIA**

ACONITUM MONKSHOOD

Petra Foerster

Per from rhizome or tuber; roots fibrous or fleshy. **ST**: 1–few, gen erect, gen simple. **LF**: palmately divided; segments 3–7, toothed to lobed; cauline gradually reduced distally on st. **INFL**: raceme or panicle, terminal, bracted; pedicels ascending. **FL**: bilateral; sepals 5, petal-like, lower 2, < others, pendent, lateral 2, round-reniform, upper 1 > others, hooded, sac-like, crescent-shaped to rounded-conic or cylindric, tip gen rounded to beaked; petals 2, covered by sepal hood, long-clawed, blades gen inflated, spurred; stamens 20–50; pistils gen 3. **FR**: follicle. **SEED**: deltoid, gen with small transverse wings, dark brown to black. > 100 spp.: boreal arctic, temp montane to alpine N.Am, Eurasia. (Greek: aconiton, of unknown origin). Most spp. highly TOXIC, causing death in livestock, humans. [Brink & Woods 1997 FNANM 3:191–195]

A. columbianum Nutt. Pl 3–15(20) dm. **ST**: erect, less gen reclining or twining above; bulblets 0 or in axils of lvs, infl. **LF**: 5–15 cm wide; deeply 3–5 divided, segments wedge- to diamond-shaped, toothed to irregularly cut or lobed distally. **INFL**: open. **FL**: sepals deep ± blue-purple to white or yellow-green, lower 7–15 mm, lanceolate to ovate, lateral 8–18 mm, ± round to reniform, upper 10–22(30) mm, 8–20(25) mm wide; petals blue to ± white. **FR**: glabrous to puberulent, glandular or not.

1. Lf axils, infl with conspicuous, deciduous
   bulblets to 5 mm . . . . .subsp. *viviparum*
1′ Lf axils, infl without bulblets . . . . subsp. *columbianum*

    subsp. ***columbianum*** (p. 1135) $2n$=16,18. Streambanks, moist areas, meadows, conifer forest; 300–3500 m. KR, NCoRH, CaR,

SNH, MP, n SNE; to BC, SD, NM, n Mex, also IA, WI, OH, NY. [*A. geranioides* Greene; *A. leibergii* Greene] Jul–Sep

    subsp. ***viviparum*** (Greene) Brink (p. 1135) $2n$=18,19,20. Streambanks, moist areas, meadows, conifer forest; 900–2500 m. KR, CaR, n SNH (s Lake Tahoe); OR. [*A. c.* var. *howellii* (A. Nelson & J.F. Macbr.) C.L. Hitchc.; *A. hansenii* Greene; *A. vivipara* Greene] Jul–Sep

ACTAEA BANEBERRY

Bruce A. Ford & Dieter H. Wilken

Per from stout, branched caudex. **ST**: 1–few, ascending to erect, branched or not. **LF**: 1–4, gen 1–3-ternate or -pinnate. **INFL**: raceme, axillary or terminal. **FL**: sepals 3–5, petal-like, early-deciduous; petals 4–10, spoon-shaped to obovate, clawed; pistil 1, placentas 2, ovules several. **FR**: berry. ± 8 spp.: temp N.Am, Eurasia. (Greek: ancient name, from wet habitat, similarity to *Sambucus* lvs) Fr TOXIC to humans.

A. rubra (Aiton) Willd. (p. 1135) Pl (2)3–10 dm. **ST**: few-branched distally, sparsely puberulent. **LF**: 10–40 cm, lower 2–3-ternate, upper 1–2-ternate, lflets 2–9 cm, toothed to irregularly cut, lateral lanceolate to ovate, terminal widely ovate to ± round. **INFL**: pedicels spreading to ascending, 5–8 mm, 6–37 mm in fr. **FL**: sepals 2–5 mm, ± white or ± purple-green; petals ± = sepals, white; stamens 3–7 mm. **FR**: 5–10 mm, red or white, shiny. n=8. Deep soils, moist, open to shaded sites, mixed-evergreen or conifer forests; < 2800 m. NW, CaR, SN (exc Teh), SnFrB, SCoRO, SnBr; to AK, ne N.Am. May–Sep

ADONIS PHEASANT'S EYE

Bruce D. Parfitt & Dieter H. Wilken

Ann [per] gen from taproot. **ST**: erect, 1–few, branched or not. **LF**: 2–3[many]-pinnately dissected, cauline, alternate; segments gen linear. **INFL**: raceme or fl 1, terminal or axillary. **FL**: sepals 5, gen ± green; petals 5–20; pistils many. **FR**: achene, rough to wrinkled or ridged, beaked. ± 35 spp.: temp Eurasia, n Afr; some cult as orn. (Greek: Adonis of mythology, from whose blood the pl allegedly grew)

A. aestivalis L. SUMMER PHEASANT'S EYE Pl 3–7 dm, glabrous. **LF**: 2–7 cm. **FL**: sepals 5–8 mm, obovate to oblong; petals (6)8–10, 8–15 mm, oblong to oblanceolate, yellow to orange, bases ± purple; anthers ± purple. **FR**: 4–5 mm, 3.5–5 mm wide, minute-ridged later-ally below middle, abaxial keel base with rounded tooth, beak 1–2 mm, straight, receptacle 1.5–3 cm. Disturbed sites, fields, open pine forest; 1200–1400 m. MP; to WA, MT, UT; native to Eur. May–Jul

ANEMONE ANEMONE

Scott Simono

Per from caudex, rhizome, or tuber. **LF**: basal, gen many, simple to compound, gen petioled; blade or lflets lobed to dissected or not, margins entire or toothed; in fl or fr withered or not. **INFL**: terminal, fls 1 or 2–7[9] in cymes; peduncle erect; pedicel elongated in fr; involucre bracts sessile or stalked, gen in 1–2 whorls of 2–5[9], simple to compound, ± like lvs or lflets in size, shape. **FL**: receptacle elongated in fr; sepals 5–10[27], petal-like; petals gen 0; stamens 10–200; pistils many, styles persistent as beaks. **FR**: achene. ± 150 spp.: arctic, temp worldwide; some cult for orn. (Greek: fl shaken by wind) [Dutton et al. 1997 FNANM 3:139–155] Spp. with long, plumose styles sometimes placed in *Pulsatilla*.

1. Fr beak (18)20–40(50) mm, long-shaggy-hairy, plumose; sepals 10–17(19) mm wide . . . . ***A. occidentalis***
1′ Fr beak ≤ 6 mm, glabrous to puberulent, not plumose; sepals gen 3.5–15(20) mm wide
  2. Lvs gen > 1; lvs, involucre bracts simple or 1–2-ternate, lobed to dissected; pl from tuber or caudex;
     fr body densely woolly or silky; sepals ≤ 10 mm wide
    3. Fr aggregate ellipsoid, (15)20–30 mm; fr beak soft-puberulent; e DMtns . . . . ***A. tuberosa***
    3′ Fr aggregate gen spheric, 10–20 mm; fr beak glabrous; KR, CaR, SN
      4. Stamens 80–100; lf ultimate segments 1–1.5(2) mm wide; involucre bracts ± 2-ternate, in 1 whorl;
        sepals white to ± blue . . . . ***A. drummondii*** var. ***drummondii***
      4′ Stamens 50–80; lf ultimate segments (1.5)2–3.5(5) mm wide; involucre bracts ± 1-ternate, in (1)2
        whorls; sepals green to yellow, blue, purple, or red (white) . . . . ***A. multifida*** var. ***multifida***
  2′ Lvs 0–2; lvs, involucre bracts simple or 1-ternate, crenate or serrate to lobed, not dissected; pl from
    rhizome; fr body glabrous to puberulent or rough-hairy; sepals 1.5–15(20) mm wide
    5. Involucre bracts gen simple, ± sessile; sepals gen 5, (8)10–15(20) mm wide, white; stamens ≥ 100 . . . . . . ***A. deltoidea***
    5′ Involucre bracts gen 1-ternate, stalked; sepals gen 5–7, gen 1.5–8 mm wide, blue to purple, ± red, pink,
      or white; stamens ≤75
      6. Fr beak (0.5)1–1.5 mm; sepals 10–20 mm; stamens 30–60(75) . . . . ***A. oregana*** var. ***oregana***
      6′ Fr beak ± 0.5–1 mm; sepals 3.5–15 mm; stamens ≤ 40
        7. Sepals 7–15 mm, 4–8 mm wide; pedicel gen (0.5)3–10 cm in fr; fr beak 0.6–1 mm; stamens 25–40 . . . . ***A. grayi***
        7′ Sepals 3.5–8(10) mm, 1.5–3(3.5) mm wide; pedicel 1–3(4) cm in fr; fr beak ± 0.5 mm; stamens 10–30(35) . . . ***A. lyallii***

A. deltoidea Hook. (p. 1135) Pl (7.5)10–30 cm; rhizome spreading, slender. **LF**: gen 0–2; petiole 10–15 cm, short-stiff-hairy; blade 1-ternate, lflets ± like involucre bracts in size, shape. **INFL**: fl 1; peduncle sparse-bristly to glabrous; involucre bracts in 1 ± whorl of 3, ± sessile, gen simple, (2.5)4–8 cm, gen ± ovate to diamond-shaped, margins crenate or sharp-toothed to lobed on distal 2/3. **FL**: sepals gen 5, (12)15–25 mm, (8)10–15(20) mm wide, ovate to obovate (oblanceolate), white, glabrous; stamens 100–120. **FR**: body 2.5–4 mm,

ovoid, glabrous to rough-hairy; pedicel (5)7–12 cm; beak ≤ 0.5 mm, glabrous; aggregate 9–12 mm, ± spheric. 2*n*=14. Open to shaded sites, conifer forest; 100–2000 m. KR, n NCoRO, NCoRH; to WA. Apr–Jul

A. drummondii S. Watson var. *drummondii* (p. 1135) Pl (7)10–25(30) cm; caudex branched. **LF**: 5–15; petiole 2–10 cm, soft-hairy; blade 2-ternate, 2–5 cm, ± soft-shaggy-hairy; lflet margins dissected on distal 1/3–1/2; terminal lflet 0.5–3 cm, 0.5–2 cm wide; ultimate segments linear, 1–1.5(2) mm wide. **INFL**: fls 1–2(3); peduncle soft-hairy; involucre bracts in 1 whorl of 3(4), short-stalked, ± 2-ternate. **FL**: sepals (5)6–9, 8–20 mm, 6–10 mm wide, ovate (to narrowly obovate), white to ± blue, abaxially gen soft-hairy, adaxially glabrous; stamens 80–100, ± white; styles white. **FR**: body 2–4 mm, ovoid, woolly; pedicel (2)3–10 cm; beak 2–4(6) mm, straight, glabrous; aggregate 10–15 mm, spheric (cylindric). 2*n*=32. Rocky slopes, conifer forest, alpine; 1200–3350 m. KR, CaR, SN; to AK, ID; Asia. Jun–Aug

A. grayi Behr & Kellogg Pl (3)10–30(40) cm; rhizome spreading (ascending). **LF**: gen 0(1); petiole (1.5)2.5–20(25) cm; lflets ± like involucre bracts in size, shape. **INFL**: fl 1; peduncle proximally glabrous, distally soft-shaggy- to fine-hairy; involucre bracts in 1 whorl of 3, 1-ternate; terminal lflet-like unit diamond-shaped to ovate or oblanceolate, finely soft-hairy to ± glabrous, margins crenate or occ coarsely serrate on distal 1/2–2/3. **FL**: sepals 5–6, 7–15 mm, 4–8 mm wide, gen elliptic to obovate, white or blue, glabrous; stamens 25–40. **FR**: body 3–4 mm, elliptic, flat, finely-puberulent to soft-shaggy-hairy; pedicel (0.5)3–10 cm; beak 0.6–1 mm, curved, glabrous; aggregate ± spheric. 2*n*=16. Moist shaded slopes, redwood and mixed-evergreen forests; 100–900 m. KR, NCoR, SnFrB, n SCoRO; sw OR. Rhizomes gen knobby as compared to *A. lyallii*, *A. oregana*; hybrids occur in populations where these 3 spp. overlap. Feb–Jun

A. lyallii Britton Pl 5–30(40) cm; rhizome spreading. **LF**: 0–1; petiole 5–8 cm; lflets ± like involucre bracts in size, shape. **INFL**: fl 1; peduncle glabrous or ± hairy distally; involucre bracts in 1 whorl of 3, 1-ternate, glabrous (± hairy); terminal lflet-like unit (1)1.5–3(4) cm, (0.4)0.7–1.5(2) cm wide, ovate or oblanceolate, margins crenate to serrate on distal 1/2–2/3. **FL**: sepals 5(7), 3.5–8(10) mm, 1.5–3(3.5) mm wide, oblong (narrowly ovate), white, pink, or blue, glabrous; stamens 10–30(35). **FR**: body 3–4 mm, elliptic, flat, ± white-hairy; pedicel 1–3(4) cm; beak ± 0.5 mm, tapered, occ ± curved, glabrous; aggregate ± spheric. 2*n*=16. Moist shaded slopes, subalpine ridges; 100–1900 m. KR, CaR, n SN; to BC. May intergrade with *A. grayi* or *A. oregana*; study needed. Mar–Jul

A. multifida Poir. var. *multifida* (p. 1135) Pl (30)40–70 cm; caudex branches ascending to erect. **LF**: 3–6(10); petiole (2)4–10(14) cm, silky-hairy; blade 1–2-ternate; lflet margins dissected in distal 1/3; terminal lflet (1.5)2.5–4.5(5.5) cm, (1)3–10 cm wide, broadly, irregularly diamond-shaped to obovate; ultimate segments (1.5)2–

3.5(5) mm wide. **INFL**: fls (2)5–7; peduncle soft-shaggy-hairy; involucre bracts in (1)2 whorls of gen 2–5, gen sessile, ± 1-ternate. **FL**: sepals 5–9, 6–17 mm, ovate or oblong, green to yellow, blue, purple, or red (white), adaxially soft-hairy; stamens 50–80. **FR**: body 3–4 mm, ellipsoid to elliptic, flat, woolly to densely silky-hairy; pedicel 6–15(23) cm; beak 1–2 mm, ± straight, glabrous; aggregate spheric. 2*n*=32. Open, gravelly or rocky slopes, ± subalpine; 1700–2750 m. KR (Marble Mtns), n SNH (The Dardanelles, Alpine Co.); w N.Am, to ne US, e Can; Chile, Argentina. Apr–Jul

A. occidentalis S. Watson (p. 1139) Pl 10–60(75) cm; caudex branches 0–few. **LF**: (2)3–6(8); petiole 6–8(12) cm; blade gen 1-ternate, 1–2-pinnately lobed to dissected, soft-shaggy-hairy; terminal lflet (2.5)3–6(8) cm, ovate; ultimate segments 2–3 mm wide, linear. **INFL**: fls gen 1; peduncle woolly to shaggy-hairy or ± glabrous; involucre bracts in 1 whorl of gen 3. **FL**: sepals 5–7, 15–30 mm, 10–17(19) mm wide, ovate to obovate (elliptic), white to ± purple, abaxially soft-hairy; stamens 150–200. **FR**: body 3–4 mm, ellipsoid, densely woolly to soft-hairy; pedicel 15–20(22) cm; beak (18)20–40(50) mm, curved to reflexed, long-shaggy-hairy, plumose; aggregate spheric (cylindric). 2*n*=16. Open, rocky slopes, alpine; 1200–3200 m. KR, CaR, SN; to BC, MT. May–Sep

A. oregana A. Gray var. *oregana* (p. 1139) Pl 8–30(35) cm; rhizome spreading, thick. **LF**: 0–1, in fl persistent or not; petiole 4–20 cm; lflets ± like involucre bracts in size, shape. **INFL**: fl 1; peduncle glabrous; involucre bracts in 1 whorl of 3, stalked, gen 1-ternate; terminal lflet-like unit 2–8 cm, 1–3(3.5) cm wide, ± diamond-shaped or ovate to oblanceolate, ± glabrous to strigose, margins crenate to sharply serrate on distal 1/2 (2/3). **FL**: sepals 5–7(8), 10–20 mm, 5–8(10) mm wide, ovate to oblong, blue to purple, ± red, or pink (white), glabrous; stamens 30–60(75). **FR**: body 4–5 mm, oblong to elliptic, puberulent (glabrous); pedicel (1.5)2–5(7) cm; beak (0.5)1–1.5 mm, ± straight, glabrous; aggregate 8–15 mm, ± spheric. 2*n*=16. Shaded sites, conifer forest; 100–1900 m. KR, CaRH; to WA. Mar–Jun

A. tuberosa Rydb. (p. 1139) Pl 10–30(40) cm; caudex slender, top with tuber. **LF**: 1–3(5); petiole 5–7 cm; blade 1–2 ternate, gen glabrous; lflet margins ± lobed or dissected in distal 2/3; terminal lflet (1.5)2–3(3.5) cm, 1–2(2.5) cm wide; ultimate segments 4–8(12) mm wide. **INFL**: fls 1–3(5); peduncle ± glabrous proximally, woolly distally; involucre bracts in (1)2 whorls of 3, ± sessile, simple, 1–2(3)-pinnately lobed or dissected, sparsely soft-hairy, margins irregularly minutely serrate; ultimate segments occ 1.5–2.5 mm wide. **FL**: sepals 8–10, 10–14(20) mm, (2)3–5(6) mm wide, linear-oblong, pink to white, sparsely hairy; stamens 50–60. **FR**: body 2–3.5 mm, round, flat, densely woolly; pedicel (5)7–15(22) cm; beak ± 1.5 mm, straight, soft-puberulent; aggregate (15)20–30 mm, ellipsoid. 2*n*=16. Rocky slopes, ledges; 900–1900 m. e DMtns, e DMoj, ne DSon; to UT, TX, n Mex. Apr–May

AQUILEGIA COLUMBINE

Justen Whittall, Scott A. Hodges & Dieter H. Wilken

Per; caudex thick, branched to not. **ST**: 1–few, ascending to erect, branched to not, scapose to not, glabrous to glandular-hairy. **LF**: basal 1–3-ternate, petiole gen long; cauline 0–few, gen much reduced, deeply 3-lobed to 1–2-ternate, petiole short to ± 0; segments gen wedge-shaped to obovate, abaxially pale green to glaucous, adaxially green to gray, glabrous to glandular. **INFL**: few-fld raceme or fl 1, terminal; axis, pedicels glabrous to glandular; fls gen nodding. **FL**: sepals 5, petal-like, spreading [to ± reflexed]; petals 5, spurs between sepals, mouths < to > 90° to exposed filaments; pistils gen 5. **FR**: follicle, glabrous to glandular. **SEED**: smooth, shiny, brown to black. ± 70 spp.: temp N.Am, Eurasia. Many spp., hybrids cult as orn; natural hybrids common; recent adaptive radiation with specialized pollinations syndromes (bee, hummingbird, hawkmoth).

1. Fl spreading to erect; spurs gen > 25 mm; sepals gen cream to yellow or pink . . . . ***A. pubescens***
1′ Fl nodding; spurs gen < 25 mm; sepals red
  2. Petal blade 0, mouth elliptic to triangular, > 90° to exposed filaments; pl densely glandular; fr beak 12–20 mm; gen serpentine seeps . . . . ***A. eximia***
  2′ Petal blade 1–8 mm, mouth ± round, ≤ 90° to exposed filaments; pl gen glabrous, glaucous at least proximally; fr beak 9–12 mm; non-serpentine
    3. Lflets of basal, lower cauline lvs green adaxially; basal lf gen 2-ternate; pl not of desert seeps; CA-FP (exc GV, SCo, ChI), GB . . . . ***A. formosa***
    3′ Lflets of basal, lower cauline lvs gray-green adaxially; basal lf gen 3-ternate; pl of desert seeps, elsewhere; W&I, DMtns . . . . ***A. shockleyi***

A. eximia Planch. (p. 1139)   Pl 20–160 cm, densely glandular. **LF:** basal, lower cauline 2–3-ternate, petioles 4–30 cm, lflets 8–35(50) mm; upper cauline gen simple to deeply 3-lobed. **FL:** sepals 10–28 mm, red; petal blade 0, spur 12–25(35) mm, red, tip 2–4 mm wide, mouth > 90° to exposed filaments, 6–10 mm wide, elliptic to triangular, yellow; stamens 10–25 mm. **FR:** 15–25 mm, beak 12–20 mm. 2*n*=14. Gen serpentine seeps, occ moist ravines, mixed-evergreen or conifer forests; 100–1800 m. NCoR, SnFrB, SCoR, w WTR. Hummingbird-pollinated; fls later than *A. formosa* where ranges overlap. May–Oct

A. formosa DC. (p. 1139)   Pl 20–80(150) cm, glabrous, glaucous at least proximally. **LF:** basal, lower cauline gen 2-ternate, petioles 5–30(40) cm, lflets 7–45(130) mm; upper cauline gen simple to deeply 3-lobed. **FL:** sepals 10–20(25) mm, red; petal blade 0 or 1–7 mm, yellow, spur 12–23 mm, red, tip 1.5–4 mm wide, mouth ≤ 90° to exposed filaments, 4–6 mm wide, ± round; stamens 10–18 mm. **FR:** 15–28 mm, beak 9–12 mm. 2*n*=14. Streambanks, seeps, moist places, chaparral, oak woodland, mixed-evergreen or conifer forests; < 3300 m. CA-FP (exc GV, SCo, ChI), GB; to AK, MT, Baja CA. [*A. f.* var. *hypolasia* (Greene) Munz; *A. f.* var. *truncata* (Fisch. & C.A. Mey.) Baker] Gen hummingbird-pollinated; lf, petal blade variation needs study. Apr–Sep

A. pubescens Coville (p. 1139)   Pl 15–50 cm. **LF:** basal, lower cauline 1–2-ternate, petioles 5–25 cm, lflets 10–25(40) mm; upper cauline gen deeply 3-lobed to 1-ternate. **INFL:** fls spreading to erect. **FL:** sepals 10–24 mm, cream to yellow or pink; petal blade 7–20 mm, cream to yellow, spur (20)25–50 mm, cream to yellow or pink, tip 1–2.5 mm wide, mouth ≤ 90° to exposed filaments, 6–10 mm wide, round, ± cream; stamens 12–21 mm. **FR:** 19–25 mm, beak 10–12 mm. 2*n*=14. Open, gen rocky slopes, scrub, subalpine forest, alpine; 2600–3650 m. SNH. Gen hawkmoth-pollinated; fl color variable; hybridizes with *A. formosa* in SNH. Jul–Aug

A. shockleyi Eastw.   Pl 40–100 cm, gen glabrous, glaucous at least proximally. **LF:** basal, lower cauline gen 3-ternate, petioles 8–40 cm, lflets 11–38 mm; upper cauline gen simple to deeply 3-lobed. **FL:** sepals 10–20(25) mm, red (± yellow or green); petal blade 1–8 mm, yellow, spur 12–23 mm, pink or red, tip 1.5–4 mm wide, mouth ≤ 90° to exposed filaments, 4–8 mm wide, ± round; stamens 10–18 mm. **FR:** 14–23 mm, beak 9–12 mm. 2*n*=14. Seeps, springs, moist places in pinyon/juniper woodland; 1200–2700 m. W&I, DMtns; NV. Hummingbird-pollinated; may occur with *A. formosa* in W&I. May–Aug

CALTHA   MARSH MARIGOLD

Bruce A. Ford

Per from short caudex [long, slender stolons], gen fleshy, glabrous. **ST:** 1–few. **LF:** simple, oblong-ovate to spheric-reniform or cordate, crenate to dentate [entire]; basal petioles > blades. **INFL:** cyme or fls 1, terminal or axillary, bracts lf-like. **FL:** sepals 5–12, petal-like, white to yellow; petals 0; pistils 5–many, ovules. **FR:** follicle, sessile to short-stalked, gen beaked. **SEED:** brown, wrinkled. 10 spp.: worldwide. (Greek: ancient name, from bowl-shaped fl)

1. St lvs 0–1; sepals 8–20 mm, white to yellow . . . . *C. leptosepala*
1′ St lvs > 1; sepals (6)10–25 mm, yellow to orange . . . . *C. palustris*

C. leptosepala DC. (p. 1139)   Pl 8–48 cm. **LF:** petiole 3–25 cm, < to > blade; blade 2–9 cm wide, ± crenate. **INFL:** peduncle gen > lvs, 1–4-fld. **FL:** sepals 5–11, oblong to elliptic. **FR:** 4–15, 7–18 mm; beak straight or ± curved. Marshes, pond margins, streambanks, conifer forest; 900–3300 m. KR, CaR, SNH, MP; to AK, MT, NM. [*C. l.* var. *biflora* (DC.) G. Lawson] May–Jul

C. palustris L.   Pl 10–70(80) cm. **LF:** petiole 3–50 cm, > blade; blade 2–20 cm wide, entire, crenate, or dentate. **INFL:** peduncle gen > lvs, 1–7-fld. **FL:** sepals 4–9, elliptic to obovate. **FR:** 6–16, 12–20 mm; beak straight or ± curved. Roadside bogs, edges of marshes, swamps, lakes, streams; 900–3300 m. NCo, NCoR, CCo; OR to AK, AB to ON, NC. Apr–Jul

CLEMATIS   CLEMATIS, VIRGIN'S BOWER

James S. Pringle & Frederick B. Essig

Pl ± woody vine; occ dioecious. **LF:** gen 1–2-pinnate, cauline, opposite; petiole gen tendril-like; lflets ovate to lanceolate, gen irregularly 2–3-lobed or coarsely toothed, occ entire. **INFL:** 1-fld to panicle, axillary [terminal]. **FL:** unisexual; sepals gen 4, free, petal-like, white to cream [brightly colored]; petals 0; stamens many, free; pistils 5–many. **FR:** achene, each gen with elongate, feathery style. 300 spp.: worldwide; *Clematis terniflora* DC., cult. (Greek: twig) [Pringle 1999 Clematis 1999:12–19] *C. drummondii* Torr. & A. Gray undocumented for CA.

1. Lflets 5–15; infl several- to many-fld; fl Jun–Sep . . . . *C. ligusticifolia*
1′ Lflets gen 3–5; infl 1–3(12)-fld; fl Jan–Jun
  2. Sepals hairy abaxially, adaxially; fr body hairy . . . . *C. lasiantha*
  2′ Sepals hairy abaxially, glabrous adaxially; fr body glabrous . . . . *C. pauciflora*

C. lasiantha Nutt. (p. 1139)   CHAPARRAL CLEMATIS, PIPESTEM CLEMATIS   **LF:** lflets 3–5, ± 3-lobed, toothed, largest gen 1.5–6 cm. **INFL:** gen 1-fld. **FL:** sepals 10–21 mm, hairy abaxially, adaxially; stamens 50–100, 7–13 mm, << sepals; pistils 75–100. **FR:** body hairy. Hillsides, chaparral, open woodland; < 2000 m. KR, NCoRI, CaRF, SNF, ScV (Sutter Buttes), CW, SW; Baja CA. Jan–Jun

C. ligusticifolia Nutt. (p. 1139)   WESTERN VIRGIN'S BOWER   **LF:** lflets 5–15, irregularly lobed or toothed, largest 2–9 cm. **INFL:** several- to many-fld. **FL:** sepals 6–10 mm, hairy abaxially, adaxially; stamens 25–50, 5–9 mm, ± = sepals; pistils 25–65. **FR:** body hairy.

Along streams, wet places; < 2400 m. CA; to BC, SD, NM, nw Mex. Jun–Sep

C. pauciflora Nutt. (p. 1139)   SOUTHERN CALIFORNIA CLEMATIS, FEW-FLOWERED CLEMATIS   **LF:** lflets gen 3–5, ± 3-lobed, gen toothed, largest gen 1–3.5 cm. **INFL:** 1–3(12)-fld. **FL:** sepals 7–12 mm, hairy abaxially, glabrous adaxially; stamens 30–50, 6–12 mm, ± = sepals; pistils 25–50. **FR:** body glabrous. Dry chaparral; < 1300 m. SW, DMtns (Little San Bernardino Mtns); Baja CA. Pls ± intermediate to *C. lasiantha* in SW. Jan–Jun

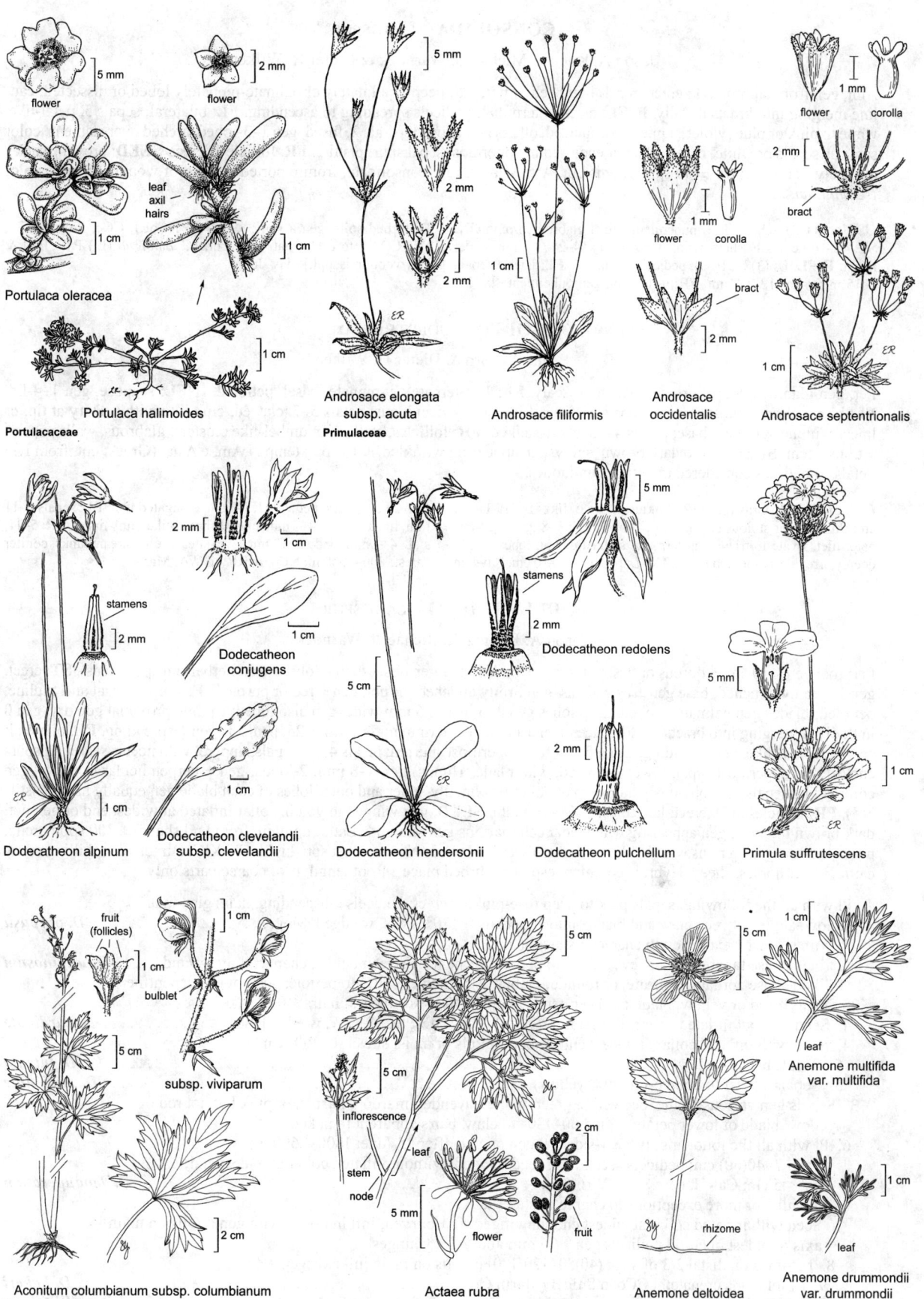

flower

5 mm

flower

2 mm

flower

1 mm

corolla

5 mm

leaf axil hairs

1 cm

1 cm

2 mm

2 mm

2 mm

1 cm

bract

2 mm

1 cm

flower

corolla

bract

2 mm

Portulaca oleracea

Portulaca halimoides

Portulacaceae

Androsace elongata
subsp. acuta

Primulaceae

Androsace filiformis

Androsace
occidentalis

Androsace septentrionalis

2 mm

1 cm

stamens

2 mm

Dodecatheon
conjugens

1 cm

1 cm

5 mm

stamens

2 mm

Dodecatheon redolens

5 mm

2 mm

5 cm

2 mm

1 cm

ER

1 cm

Dodecatheon alpinum

Dodecatheon clevelandii
subsp. clevelandii

Dodecatheon hendersonii

Dodecatheon pulchellum

Primula suffrutescens

fruit
(follicles)

1 cm

1 cm

bulblet

5 cm

subsp. viviparum

5 cm

inflorescence

leaf

stem

node

5 mm

flower

2 cm

2 cm

fruit

5 cm

rhizome

5 cm

leaf

Anemone multifida
var. multifida

1 cm

leaf

Aconitum columbianum subsp. columbianum

Ranunculaceae

Actaea rubra

Anemone deltoidea

Anemone drummondii
var. drummondii

CONSOLIDA LARKSPUR

Jason A. Koontz, Michael J. Warnock & Dieter H. Wilken

Ann, gen from taproot. **ST**: erect, gen 1, branched or not. **LF**: deeply palmately or palmate-pinnately lobed or dissected, cauline merging into bracts distally. **INFL**: raceme, terminal; pedicels spreading to ascending. **FL**: bilateral; sepals 5, petal-like, white to pink or blue [violet], uppermost spurred, others gen clawed; petals 2, fused, gen 1–3-lobed, arched over stamens, color same as sepals or white, base with spur enclosed in uppermost sepal spur; pistil 1. **FR**: follicle 1, erect. **SEED**: brown [black], minutely scaly in rows. ± 40 spp.: Medit Eur, Afr, Asia. (Latin: consolidate, from reported healing of wounds) Segregate of *Delphinium*.

C. ajacis (L.) Schur   Pl ± puberulent exc st glabrous proximally. **ST**: 45–100 cm. **LF**: short-petioled, segments 14–21, < 2 mm wide, linear. **INFL**: fls (3)8–21(60); pedicels 5–16 mm. **FL**: lateral sepals 8–15 mm, spur 12–18 mm. **FR**: 9–18 mm, gen puberulent. 2*n*=16. Disturbed soils, gen a waif; < 300 m. NCoRI, CCo, SnFrB, SCo, WTR, MP; to e US; native to Eur. [*C. ambigua* (L.) P.W. Ball & Heywood, misappl.] May–Jul

COPTIS GOLDTHREAD

Bruce A. Ford & Dieter H. Wilken

Per, gen glabrous. **ST**: short, simple, stout, scaly. **LF**: 1–2-ternate or -pinnate, basal, petioled. **INFL**: scapose, gen 1–4-fld. **FL**: bisexual or some staminate; sepals 5–8, petal-like, early-deciduous; petals 5–7, clawed, club-like with nectary at tip, or linear with nectary near base; pistils 4–15, short-stalked. **FR**: follicles, stalked, in umbel-like clusters, glabrous, walls papery, ± translucent. **SEED**: tan to dark brown, shiny, gen appearing wrinkled. ± 10 spp.: temp N.Am, e Asia. (Greek: cut, from lvs) Petals sometimes considered modified staminodes.

C. laciniata A. Gray (p. 1139)   OREGON GOLDTHREAD   Pl 11–24 cm; rhizomes or stolons slender, pale brown. **LF**: 3–8, gen 1(2)-ternate; lflets ovate to triangular, terminal stalked or not, lobes gen 3, deeply, irregularly toothed to cut. **INFL**: peduncle 5–10 cm, < lvs, in fr to 25 cm, > lvs; pedicels 1.5–3 cm, elongated in fr. **FL**: sepals 6–11 mm, linear; petals 4–7 mm, claw ± thread-like, limb linear. **FR**: 5–11, stalk 4–7 mm, body 8–13 mm. Wet sites, seeps, streambanks, conifer forest; 500–2000 m. NCo, w KR; to WA. Mar–Apr   ★

DELPHINIUM LARKSPUR

Jason A. Koontz & Michael J. Warnock

Per; root gen < 10 cm, ± fibrous or fleshy; rootstock buds in life gen obscure (0 or obscure on herbarium specimens). **ST**: erect, gen 1, gen unbranched; base gen ± as wide as, gen firmly attached to root, gen ± red or purple. **LF**: simple, basal and cauline, petioled; blades gen palmately lobed, deep lobes gen 3–5, gen < 6 mm wide, gen also lobed; cauline proximal gen dry, gen 0 in fl, distal merging into bracts. **INFL**: raceme or ± branched, terminal; fls gen 10–25; pedicels gen ± spreading. **FL**: bilateral; sepals 5, petal-like, gen spreading, gen ± dark blue, uppermost spurred; petals 4, << sepals, upper 2 with nectar-secreting spurs enclosed in uppermost sepal, lower 2 clawed, with blades (limbs) gen 4–8 mm, 2-lobed, gen ± perpendicular to claws, gen colored like sepals, gen obviously hairy esp on lobes proximally, inner and outer lobes of each blade gen equally hairy; pistils 3(5). **FR**: follicles 3(5), erect, length gen 2.5–4 × width. **SEED**: gen winged in youth, collar inflated at widest end or gen not, dark brown to black, gen appearing white; coat cell margins gen straight. (Latin: dolphin, from bud shape) ± 300 spp.: arctic, n temp, subtrop, trop mtns worldwide; 3 commonly cult as orn in N.Am. Most spp. highly TOXIC, attractive to, killing many cattle, fewer horses, sheep. Hybrids common, esp in disturbed places. Root length is of coarse parts only.

1. Pl with all the following: sepals pink to deep rose-pink; fr erect; pedicels ± ascending, hairs glandular, yellow; angle between claw and blade on lower petals > 130°; s SN, w edge DMoj . . . . ***D. purpusii***
1′ Pl with at least 1 exception to character states at 1.
  2. Lf blade base tapered, lobes few, < 50% to petiole; serpentine streambanks, chaparral, grassland . . . . . ***D. uliginosum***
  2′ Lf blade base cordate, truncate, or rounded, lobes (3)5–many, > 50% to petiole; gen not on serpentine
    3. Sepals ± red or yellow; blade of lower petals angled 140–180° to claw, hairs ± 0 to naked eye
      4. Seed without inflated collar; fr ± straight; pls of ± dry sites; SCoR, SW, w edge DSon . . . . ***D. cardinale***
      4′ Seed with inflated collar at widest end: fr curved; pls of moist sites; SCoRO & n
        5. Sepals bright yellow . . . . ***D. luteum***
        5′ Sepals scarlet to orange-red (dull yellow) . . . . ***D. nudicaule***
    3′ Sepals gen variously blue, occ white, green-white, lavender, maroon, purple, or pink but not red or yellow; blade of lower petals angled 60–130° to claw, hairs apparent to naked eye
      6. Pl with all the following: infl ± 1-sided; fl gen on and after 15 June; 1800–2600 m; st gen 7–40(60) cm; pedicels ascending; seed winged, without inflated collar at widest end; root < 5 cm; CaRH, n&c SNH, Wrn . . . . ***D. depauperatum***
      6′ Pl with 2 or more exceptions to character states at 6.
        7. Seed with inflated collar at widest end, unwinged; fr ± curved; infl in mid-fl with gen < 6 fls on main axis or oldest open-fld pedicels gen > 25 mm and > 2 × youngest
          8. Most lvs on distal 2/3 of st; st (40)50–120(180) cm; fls on main infl axis gen > 15
            9. Lf lobe tips crenate; n CCo, n SnFrB (Marin Co.) . . . . ***D. bakeri***
            9′ Lf lobe tips sharply, irregularly cut; NCo, NCoRO (n of Sonoma Co.) . . . . ***D. trolliifolium***

8′ Most lvs on proximal 2/3 of st; st < 50(90) cm; fls on main infl axis gen < 20

  10. Sepals dark blue-purple (gen faded, mottled on herbarium specimens), puberulent abaxially, gen
     not reflexed; proximal st hairy; lower petal blades 6–11 mm . . . . ***D. decorum***

    11. Lf gen with 5 lobes > 50% to petiole; < 200 m; open coastal grassland, chaparral . . . . subsp. ***decorum***

    11′ Lf gen with > 5 lobes > 50% to petiole; 700–2300 m; meadows in montane forest . . . . subsp. ***tracyi***

  10′ Sepals gen bright blue to white or pink (gen neither faded nor mottled on herbarium specimens),
     gen glabrous, gen reflexed; proximal st glabrous to puberulent; lower petal blades 3–11 mm (if
     sepals ± purple, puberulent abaxially, not reflexed, then proximal st ± glabrous or lower petal blade < 6 mm)

    12. Lf ± fleshy; pl restricted to moist talus slopes . . . . ***D. antoninum***

    12′ Lf not fleshy; pl not restricted to moist talus slopes

      13. Lf with ≥ 6 lobes > 50% to petiole, < 7 mm wide; pedicels glabrous or puberulent; SNH and n
       . . . . ***D. nuttallianum***

      13′ Lf with ≤ 5 lobes > 50% to petiole, gen > 7 mm wide; pedicels glabrous to glandular or
        glandular-puberulent (if > 5 lobes > 50% to petiole, then pedicels puberulent); SN, SCoRI and s
        (also s KR, s CaR for *Delphinium gracilentum*)

        14. Terminal lf lobe widest above middle; pedicel at 70–90° angle to axis; lf lobes gen 5 . . . . . . ***D. gracilentum***

        14′ Terminal lf lobe widest near middle; pedicel at gen < 70° angle to axis; lf lobes 3–10 . . . . ***D. patens***

          15. Lobes of proximal lvs gen > 1.5 cm wide; basal, proximal cauline lvs gen cut < 80% to petiole
           — SCoRO, SW (exc ChI). . . . . subsp. ***hepaticoideum***

          15′ Lobes of proximal lvs gen < 1.5 cm wide; basal (if present), proximal cauline lvs gen cut
           > 80% to petiole

            16. Pedicels puberulent . . . . subsp. ***montanum***

            16′ Pedicels gen glabrous. . . . . subsp. ***patens***

7′ Seed gen without inflated collar (exc *Delphinium andersonii*), winged or not; fr ± straight; infl in
  mid-fl with gen > 6 fls on main axis or oldest open-fld pedicel gen < 25 mm or < 2 × youngest

  17. Sts gen > 1 m, gen ≥ 2; enlarged rootstock buds in life gen ± white, brown on herbarium specimens

    18. Fl gen June or earlier; sepals ± lavender or ± green-white, puberulent; < 1100 m . . . . ***D. californicum***

    19. Sepals ± lavender; main infl axis puberulent; upper petals ± hairy. . . . . subsp. ***californicum***

    19′ Sepals ± green-white; main infl axis ± glabrous; upper petals ± glabrous . . . . subsp. ***interius***

    18′ Fl gen July or later; sepals gen ± blue, ± canescent; ≤ 3600 m (if sepals ± lavender or ± green-
      white, then > 1100 m)

      20. Lvs present on proximal 1/5 of st in fl

        21. Fl spur 9–12 mm; sepals white to light blue. . . . . ***D. inopinum***

        21′ Fl spur 11–22 mm; sepals dark blue . . . . ***D. polycladon***

      20′ Lvs gen 0 on proximal 1/5 of st in fl

        22. Sepals ± purple-blue, abaxial midline not lighter; proximal st glaucous; KR, SNH, SnGb, SnBr,
         n SNE. . . . . ***D. glaucum***

        22′ Sepals ± bright blue, abaxial midline lighter due to hairs; proximal st puberulent; Wrn. . . . . . ***D. stachydeum***

17′ St gen < 1 m, gen 1; enlarged rootstock buds gen 0

  23. Proximal petiole hairs > 0.5 mm, straight, spreading, white, and some not

    24. Seeds fine-prickly, fuzzy to naked eye; lateral sepals 7–13 mm; fls gen > 12 on main axis. . . . . ***D. hansenii***

      25. Sepals violet-purple to maroon — SNF, SnJV. . . . . subsp. ***ewanianum***

      25′ Sepals dark blue-purple to white or pink

        26. Basal lvs gen 0 in fl; cauline lvs ≥ 3; NCoRI, CaRF, SNF, c&s SNH, ScV. . . . . subsp. ***hansenii***

        26′ Basal lvs present but gen dry in fl (so 0 on herbarium specimens); cauline lvs gen ≤ 2; s SN,
          Teh, w edge DMoj. . . . . subsp. ***kernense***

    24′ Seeds not prickly, not fuzzy to naked eye; lateral sepals 10–25 mm; fls gen < 12 on main axis

      27. Lower petal blade margin hairs 0; sepal spur tip gen down-curved > 3 mm; st 25–100 cm —
        CCo (c Monterey Co.) . . . . ***D. hutchinsoniae***

      27′ Lower petal blade margin hairs present; sepal spur straight or tip down-curved < 3 mm; st gen
        < 50 cm . . . . ***D. variegatum***

        28. Pl of mainland — sepals gen dark royal blue (white or lavender) . . . . subsp. ***variegatum***

        28′ Pl of ChI

          29. Sepals white to light blue; fl Jan–Apr; pls on 0–15° slopes . . . . subsp. ***kinkiense***

          29′ Sepals bright blue; fl Feb–May; pls on 15–35° slopes . . . . subsp. ***thornei***

  23′ Proximal petiole hairs 0–0.5 mm, or curved, or both (if some > 0.5 mm, straight, then pl of NW or
    SnFrB, seeds winged)

    30. Blade of each lower petal hairier on inner lobe than outer; proximal st ± striate — lateral sepals
      7–16 mm; seeds winged; sepals not reflexed; pedicels ascending at < 45° angle; NW, CaRF, ScV,
      SnFrB, SCoRI, PR. . . . . ***D. hesperium***

      31. Lateral sepals ≤ 4 mm wide; c PR. . . . . subsp. ***cuyamacae***

      31′ Lateral sepals > 4 mm wide; n of TR

        32. Sepals dark blue-purple. . . . . subsp. ***hesperium***

        32′ Sepals white to ± pink or light blue . . . . subsp. ***pallescens***

    30′ Blade of each lower petal equally hairy on inner and outer lobe; proximal st not striate

33. Seed coat cell margins wavy, ± visible at 10× when held to light; fr length gen ≤ 3 × width; sepals light blue to pink or white
  34. Pl of grassland, open woodland; sepals white to ± pink or light blue, rarely reflexed; st 30–150 cm . . . . **D. gypsophilum**
  34′ Pl of gen deserts, scrub, juniper woodland (grassland); sepals white to pink or light, ± sky, or dark blue, esp lateral gen reflexed (spreading to erect in *Delphinium parishii* subsp. *pallidum*, spreading in *Delphinium parishii* subsp. *subglobosum*); st < 78(95) cm
    35. Pl of fine, alkaline soil; sepals gen light blue; lower petals white — GV, s SCoRI (Caliente Range), w DMoj . . . . **D. recurvatum**
    35′ Pl of gen coarse, not very alkaline soil; sepals white or pink to ± sky or dark blue; lower petals same color as sepals. . . . . **D. parishii**
      36. Sepals ± sky blue, lateral reflexed — s SNH, Teh, TR, SNE, D. . . . . subsp. *parishii*
      36′ Sepals white or pink or blue to dark blue, not sky blue, lateral not reflexed
        37. Gen fl after 20 May; sepals white to pink or blue; sw SnJV, SCoRI, WTR . . . . subsp. *pallidum*
        37′ Gen fl before 20 May; sepals dark blue; w DSon . . . . subsp. *subglobosum*
33′ Seed coat cell margins straight, ± visible at 10× when held to light; fr length gen > 3× width; sepals blue to dark blue
  38. Green lvs gen present on proximal 1/5 of st in fl; proximal st, petioles ± glabrous
    39. In fl, basal lvs 4–10, cauline 1–4, lobes 3–5(9), shorter lobes 0.5–6 cm, lobe tips rounded — DSon. . . . . **D. scaposum**
    39′ In fl, basal lvs 0–3, cauline 3–7, lobes 3–30, longer lobes 2–6 cm, lobe tips tapered to point
      40. Lobes of proximal lvs > 4 mm wide; lateral sepals reflexed; SCoRO, WTR . . . . **D. umbraculorum**
      40′ Lobes of proximal lvs < 4 mm wide; lateral sepals rarely reflexed; CaRH, ne SNH, GB. . . . **D. andersonii**
  38′ Green lvs gen 0 on proximal 1/5 of st in fl; proximal st, petioles ± puberulent — Teh, SnJV, CW, SW . . . . **D. parryi**
    41. Basal lvs gen 0 in fl
      42. Lateral sepals gen 16–25 mm . . . . subsp. *blochmaniae*
      42′ Lateral sepals 9–15 mm. . . . . subsp. *parryi*
    41′ Basal lvs gen present in fl
      43. Pl from > 700 m . . . . subsp. *purpureum*
      43′ Pl from < 700 m
        44. Sepals gen reflexed. . . . . subsp. *eastwoodiae*
        44′ Sepals gen spreading . . . . subsp. *maritimum*

D. andersonii A. Gray (p. 1139) ANDERSON'S LARKSPUR Root gen > 10 cm, distally branched. **ST:** (20)30–60(90) cm; base gen narrower than root, firmly attached, ± glabrous. **LF:** gen on proximal 1/2 of st, ± glabrous; lobes 7–30, < 4 mm wide. **INFL:** cylindric; pedicels 8–68 mm, 5–25 cm apart, ± S-shaped, glabrous to puberulent. **FL:** sepals dark blue, lateral 9–16 mm, spur 11–18 mm. **FR:** 17–32 mm, length gen > 4 × width. **SEED:** coat smooth, shiny, ± translucent, collar inflated. 2*n*=16. Talus, dry sandy soils in sagebrush scrub; 1300–2700 m. CaRH, ne SNH, GB; to se OR, c ID, s UT. Hybridizes with *D. nuttallianum*, *D. parishii*. May–Jun

D. antoninum Eastw. (p. 1139) ANTHONY PEAK LARKSPUR Root gen > 15 cm. **ST:** 7–30(60) cm; base narrower than root, not firmly attached, glabrous to puberulent. **LF:** gen on proximal 1/3 of st, ± fleshy, ± glabrous; lobes 3–15. **INFL:** pedicels 6–32 mm, 5–25 mm apart, gen puberulent; fls 3–25. **FL:** sepals blue, lateral spreading to reflexed, 11–13 mm, spur 12–16 mm; lower petal blades 3–5 mm, each hairier on inner lobe than outer. **FR:** 14–22 mm, curved. **SEED:** ± bumpy, collar inflated. Moist talus slopes; 1100–2600 m. KR, NCoRH. Uncommon. Hybridizes with *D. decorum* subsp. *tracyi*, *D. nudicaule*. Jun–Jul

D. bakeri Ewan (p. 1139) BAKER'S LARKSPUR **ST:** (45)50–100 cm, base narrower than root, not firmly attached, glabrous. **LF:** gen on distal 2/3 of st; basal ± glabrous or gen 0; lobes > 1 cm wide, tips crenate. **INFL:** pedicels 8–91 mm, > 10 mm apart, ± glandular. **FL:** sepals dark blue, lateral 9–11 mm, spur 9–13 mm; lower petal blades each hairier on inner lobe than outer. **FR:** 18–20 mm, ± curved. **SEED:** smooth, shiny, collar inflated. Coastal scrub, decomposing shale slopes; 80–305 m. n CCo, n SnFrB (Marin Co.). Mar–May ★

D. californicum Torr. & A. Gray Root gen > 15 cm, distally branched; rootstock buds in life prominent. **ST:** gen ≥ 2, (60)100–160(220) cm, gen puberulent. **LF:** lobes 3–15, 5–60 mm wide, tips ± sharply cut. **INFL:** gen branched; most proximal bracts ± lf-like; pedicels 5–65 mm, 5–25 mm apart, puberulent; fls gen > 50. **FL:** sepals forward-pointing, ± lavender or ± green-white; lower petal blades 3–5 mm. **FR:** 11–16 mm. **SEED:** with ± overlapping scales.

subsp. **californicum** (p. 1139) COAST OR CALIFORNIA LARKSPUR **LF:** abaxial surface, margins puberulent. **INFL:** main axis puberulent. **FL:** sepals ± lavender, densely puberulent, lateral 7–11 mm, spur 7–14 mm; upper petals ± hairy. 2*n*=16. Gen slopes in dense chaparral, w side of coast ranges; < 1000 m. CCo, w SnFrB. Apr–Jun

subsp. **interius** (Eastw.) Ewan (p. 1139) HOSPITAL CANYON LARKSPUR **LF:** glabrous. **INFL:** main axis ± glabrous; pedicel tips gen puberulent. **FL:** sepals ± green-white, puberulent gen near tips, lateral 6–8 mm, spur 7–10 mm; upper petals ± glabrous. Gen slopes in open woodland, e side of coast ranges; 300–1000 m. e SnFrB, SCoRI. Apr–Jun ★

D. cardinale Hook. (p. 1143) CARDINAL OR SCARLET LARKSPUR Root gen > 15 cm, distally branched. **ST:** 30–270 cm, base equal in width or occ narrower than root, firmly attached, gen curly-puberulent. **LF:** basal present in fl or not, ± glabrous; lobes 5–27. **INFL:** pedicels 15–55 mm, (6)15–80 mm apart, puberulent. **FL:** sepals gen ± forward-pointing, red, lateral 11–15 mm, spur 15–24 mm; lower petals flattened, blades 4–5 mm, with hairs few, short, yellow, obscure to naked eye. **FR:** 12–18 mm, ± straight. **SEED:** bumpy. 2*n*=16. Slopes, gen talus, chaparral; 300–1500 m. SCoR, SW, w edge DSon; Baja CA. Hybridizes with *D. parryi* (*D.* ×*inflexum* Davidson). Gen hummingbird-pollinated. Feb–Jul

D. decorum Fisch. & C.A. Mey. Roots clustered, ± spheric or not. **ST:** base narrower than root, not firmly attached, hairy. **LF:** gen basal, ± glabrous adaxially, gen ± puberulent abaxially and on margins; lobes occ > 6 mm wide. **INFL:** pedicels 10–63 mm, 10–25 mm apart; fls 2–20. **FL:** sepals gen not reflexed, dark blue-purple (gen faded, mottled on herbarium specimens), puberulent abaxially; lower petal blades 6–11 mm, each gen hairier on inner lobe than outer. **FR:** 9–20 mm, ± curved. **SEED:** bumpy, collar inflated.

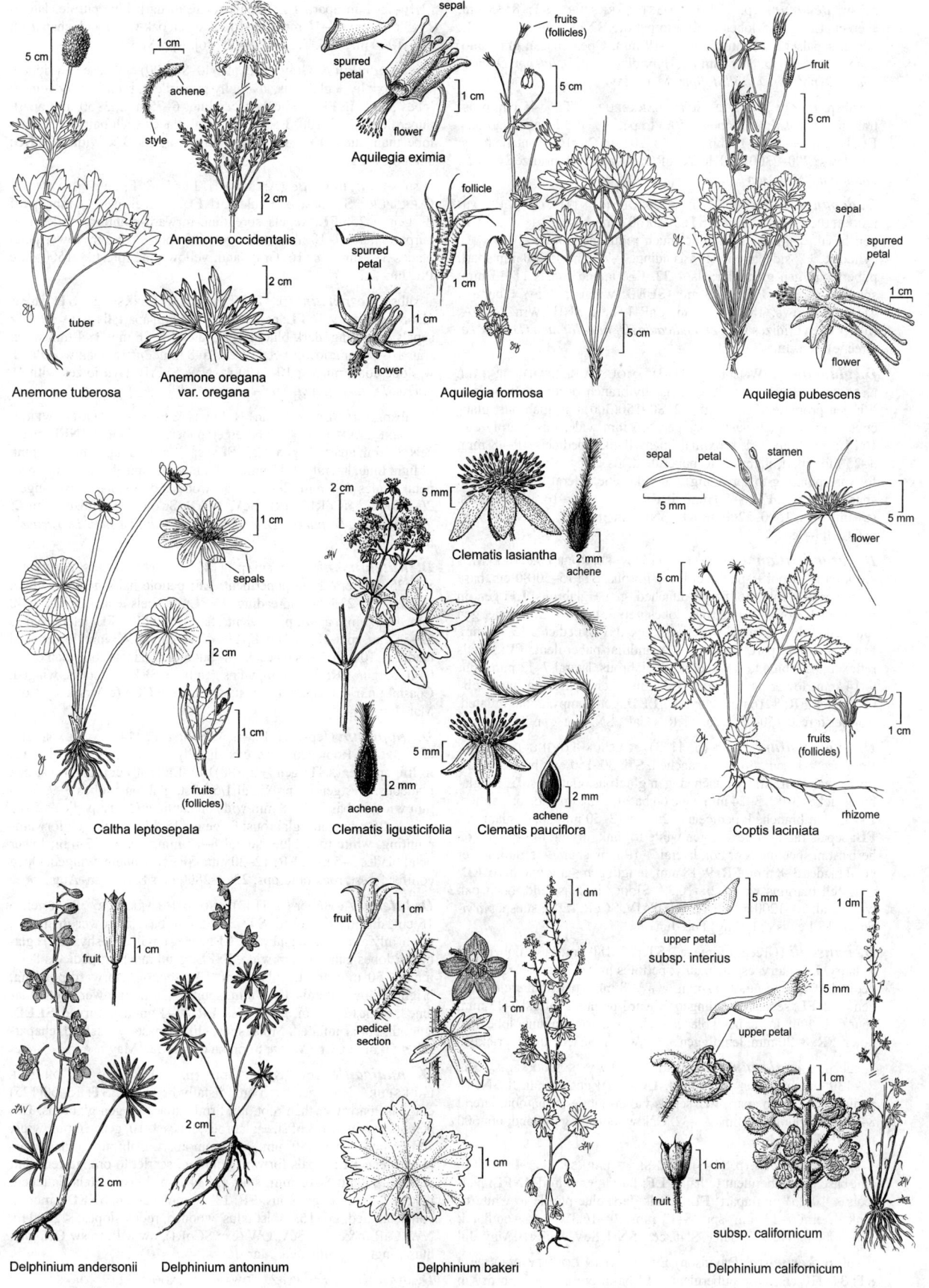

Anemone tuberosa

Anemone oregana var. oregana

Anemone occidentalis

achene

style

tuber

5 cm

1 cm

2 cm

2 cm

Aquilegia eximia

sepal

spurred petal

flower

spurred petal

flower

follicle

fruits (follicles)

Aquilegia formosa

5 cm

1 cm

1 cm

1 cm

5 cm

Aquilegia pubescens

fruit

sepal

spurred petal

flower

5 cm

1 cm

Caltha leptosepala

sepals

fruits (follicles)

1 cm

2 cm

1 cm

Clematis lasiantha

Clematis ligusticifolia

Clematis pauciflora

achene

achene

achene

2 cm

5 mm

2 mm

5 mm

2 mm

5 mm

2 mm

Coptis laciniata

sepal

petal

stamen

flower

fruits (follicles)

rhizome

5 mm

5 mm

5 cm

1 cm

Delphinium andersonii

Delphinium antoninum

fruit

2 cm

1 cm

2 cm

Delphinium bakeri

fruit

pedicel section

1 dm

1 cm

1 cm

1 cm

Delphinium californicum

upper petal

subsp. interius

upper petal

subsp. californicum

fruit

5 mm

1 dm

5 mm

1 cm

1 cm

subsp. ***decorum*** (p. 1143) COAST LARKSPUR **ST:** 8–35 cm, ± erect. **LF:** gen 5 lobes > 50% to petiole. **INFL:** puberulent. **FL:** lateral sepals 12–24 mm, spur 13–19 mm. Open coastal grassland, chaparral; < 200 m. NCo, SnFrB. Hybridizes with *D. luteum, D. nudicaule, D. patens, D. trolliifolium.* Mar–May

subsp. ***tracyi*** Ewan TRACY'S LARKSPUR **ST:** 7–45 cm, prostrate to erect. **LF:** ≥ 6 lobes > 50% to petiole. **INFL:** gen glabrous. **FL:** lateral sepals 11–18 mm, spur 13–20 mm. 2*n*=16. Montane forest meadows; 700–2300 m. KR, NCoRO, NCoRH. Hybridizes with *D. antoninum.* May–Jun

D. depauperatum Nutt. DWARF, BLUE MOUNTAIN, OR MOUNTAIN LARKSPUR Root < 5 cm. **ST:** 7–40(60) cm, base ± glabrous. **LF:** gen basal, ± glabrous, cauline much reduced; lobes 5–10. **INFL:** cylindric, ± 1-sided; pedicels ascending, 7–73 mm, 15–50 mm apart, puberulent, gen glandular; fls 4–22. **FL:** lateral sepals 11–13 mm, spur 12–16 mm. **FR:** 9–16 mm. **SEED:** winged, shiny, ± bumpy. Moist meadows; 1800–2600 m. CaRH, n&c SNH, Wrn; to WA, MT, NV. Hybridizes with *D. nudicaule, D. nuttallianum* (*D. ×burkei* Greene). Jun–Jul

D. glaucum S. Watson (p. 1143) MOUNTAIN, GIANT, OR TALL LARKSPUR Root gen > 15 cm, distally branched; rootstock buds in life gen prominent. **ST:** gen ≥ 2, 80–150(300) cm, glabrous, glaucous. **LF:** gen glabrous; lobes gen > 6 mm wide, tips sharply cut. **INFL:** gen branched; proximal bracts lf-like; pedicels 10–48 mm, 3–25 mm apart, glabrous to puberulent; fls gen > 50. **FL:** sepals ± forward-pointing to spreading, ± purple-blue, lateral 8–14(21) mm, spur 10–19 mm. **FR:** 9–20 mm. **SEED:** bumpy. 2*n*=16. Wet thickets, streambanks; 1500–3200 m. KR, SNH, SnGb, SnBr, n SNE; to AK, w CO. Jul–Sep

D. gracilentum Greene (p. 1143) SLENDER OR GREENE'S LARKSPUR Root ± spheric or loose-fibrous. **ST:** 15–50(80)cm, base narrower than root, not firmly attached, gen ± glabrous. **LF:** gen on proximal 1/3 of st, ± glabrous; lobes gen 5, gen > 6 mm wide, gen rounded at tips, terminal widest distally. **INFL:** pedicels 15–45 mm, gen > 10 mm apart, glabrous to glandular-puberulent. **FL:** sepals reflexed, pale blue to white or pink, glabrous, lateral 7–13 mm, spur 9–14 mm; lower petal blades 3–5 mm, each hairier on inner lobe than outer. **FR:** 8–16 mm, curved. **SEED:** gen bumpy, collar inflated. Conifer forest; 150–2700 m. s KR, s CaR, SN. Mar–Jun

D. gypsophilum Ewan (p. 1143) GYPSUM-LOVING LARKSPUR Root occ > 15 cm, distally branched. **ST:** 30–150 cm; base occ narrower than root, firmly attached, gen glabrous, glaucous. **LF:** ± glabrous, lobes 3–12, 3–24 mm wide on basal, 1–8 mm wide on cauline. **INFL:** gen branched; pedicels 5–25 mm, 3–50 mm apart, glabrous. **FL:** sepals rarely reflexed, gen white to pink, drying to sky blue on herbarium specimens or not, lateral 7–19 mm, spur 7–15 mm; lower petal blades 3–8 mm. **FR:** 9–18 mm, length gen ≤ 3 × width. **SEED:** coat cell margins wavy. 2*n*=16,32. Slopes in grassland, open oak woodland; 90–1200 m. s SNF, Teh, SnJV, SCoR. [*D. g.* subsp. *parviflorum* H. Lewis & Epling] Feb–Jun ★

D. hansenii (Greene) Greene **ST:** 25–180 cm; base puberulent to hairy. **LF:** hairy, esp abaxially; petioles hairy; lobes 3–18. **INFL:** pedicels ± ascending, 3–57 mm, gen < 8 mm apart, puberulent; fls gen > 12. **FL:** sepals spreading to forward-pointing, lateral 7–13 mm, spur 6–16 mm; lower petal blades each gen hairier on inner lobe than outer. **FR:** 8–20 mm, length gen < 3 × width. **SEED:** finely prickly.

subsp. ***ewanianum*** M.J. Warnock EWAN'S LARKSPUR **ST:** 25–130 cm; base gen puberulent. **LF:** basal gen few in fl. **INFL:** pedicels 6–25 mm apart. **FL:** sepals violet-purple to maroon, lateral 8–12 mm, spur 6–16 mm. 2*n*=32. Oak woodland, grassland; 60–600 m. SNF, SnJV. Mar–May ★

subsp. ***hansenii*** (p. 1143) HANSEN'S LARKSPUR **ST:** 40–180 cm; base gen puberulent to hairy. **LF:** basal gen 0 in fl. **INFL:** pedicels < 10(25) mm apart. **FL:** sepals dark blue-purple to white or pink, lateral 7–13 mm, spur 8–13 mm. 2*n*=16,32. Oak woodland; 150–3000 m. NCoRI, CaRF, SNF, c&s SNH, ScV, e SnFrB. May–Jul

subsp. ***kernense*** (Davidson) Ewan KERN COUNTY LARKSPUR **ST:** 34–110 cm, base puberulent. **LF:** basal present but gen dry in fl (so 0 on herbarium specimens); cauline gen ≤ 2. **INFL:** pedicels

(7)10–25 mm apart. **FL:** sepals white to dark blue-purple, lateral 7–13 mm, spur 8–16 mm. 2*n*=16,32. Open oak woodland, chaparral; 800–1900 m. s SN, Teh, w edge DMoj. Apr–May

D. hesperium A. Gray **ST:** (10)40–80(120) cm, base gen ridged. **LF:** adaxially ± glabrous, abaxially puberulent, prominently veined; lobes 3–14. **INFL:** pedicels ascending, 6–75 mm, 3–50 mm apart, puberulent; fls 5–100. **FL:** lower petal blades each hairier on inner lobe than outer. **FR:** 8–18 mm, length occ < 3 × width. **SEED:** smooth, winged.

subsp. ***cuyamacae*** (Abrams) H. Lewis & Epling CUYAMACA LARKSPUR **ST:** base puberulent. **INFL:** pedicels gen < 8 mm apart; fls gen > 25. **FL:** sepals spreading-forward-pointing, dark blue-purple, lateral 7–10 mm, ≤ 4 mm wide, spur 8–12 mm; lower petal blades 3–5 mm. 2*n*=16. Grassland, yellow-pine forest; ± 1500 m. c PR. Jun–Jul ★

subsp. ***hesperium*** (p. 1143) WESTERN LARKSPUR **ST:** base ± glabrous to hairy. **INFL:** pedicels gen > 8 mm apart; fls gen < 30. **FL:** sepals spreading, dark blue-purple, lateral 8–16 mm, > 4 mm wide, spur 10–18 mm; lower petal blades 5–8 mm. 2*n*=16. Oak woodland, w slope coast ranges; 10–1100 m. NW, SnFrB. Hybridizes with *D. parryi, D. variegatum.* Mar–Jun

subsp. ***pallescens*** (Ewan) H. Lewis & Epling PALE-FLOWERED WESTERN LARKSPUR **ST:** base gen puberulent (hairy). **INFL:** pedicels > 8 mm apart; fls gen < 25. **FL:** sepals spreading, white to ± pink or light blue, lateral 10–15 mm, > 4 mm wide, spur 10–17 mm; lower petal blades 4–7 mm. 2*n*=16. Oak woodland, e slope coast ranges; 20–1000 m. NCoRI, CaRF, ScV, SnFrB, SCoRI. Hybridizes with *D. gypsophilum, D. parryi, D. recurvatum, D. uliginosum, D. variegatum.* Mar–May

D. hutchinsoniae Ewan (p. 1143) HUTCHINSON'S LARKSPUR **ST:** 25–100 cm, base puberulent. **LF:** petiole hairs gen spreading; lobes 3–17, 2° lobes spreading. **INFL:** pedicels ± ascending, 8–40 mm, 10–25 mm apart, puberulent; fls (2)7–20(30). **FL:** lateral sepals 12–24 mm, spur 11–19 mm, tip gen down-curved > 3 mm; lower petal blades 5–10 mm, hairs on each few but more on inner lobe than outer, 0 on margin. **FR:** 9–21 mm, veins gen blue. **SEED:** smooth, winged. Coastal prairie, chaparral, forest; < 430 m. c CCo (c Monterey Co.). Mar–Jun ★

D. inopinum (Jeps.) H. Lewis & Epling (p. 1143) UNEXPECTED LARKSPUR Root gen > 15 cm, distally branched; rootstock buds in life apparent. **ST:** gen ≥ 2, (80)100–140(150) cm, glabrous, gen glaucous. **LF:** gen on proximal 1/5 of st, glabrous; lobes 3–9, 5–28 mm wide on basal, 3–18 mm wide on cauline. **INFL:** pedicels 5–25 mm, 6–25 mm apart, glabrous; fls gen > 25. **FL:** sepals gen forward-pointing, white to ± blue, lateral 8–12 mm, spur 9–12 mm; lower petal blades 3–5 mm. **FR:** 12–20 mm. **SEED:** smooth, winged. Open conifer forest, rock outcrops; 2200–2800 m. s SNH. Jun–Aug ★

D. luteum A. Heller (p. 1143) GOLDEN LARKSPUR Root gen > 15 cm, distally branched. **ST:** 20–55 cm, base narrower than root, not firmly attached, ± glabrous. **LF:** gen ± basal, ± fleshy, gen ± glabrous; lobes gen > 6 mm wide. **INFL:** gen branched; pedicels 8–68 mm, 8–50 mm apart, puberulent. **FL:** sepals ± forward-pointing, bright yellow, lateral 11–16 mm, spur 13–20 mm; lower petals flattened, blades 3–4 mm, ± glabrous. **FR:** 11–14 mm, ± curved. **SEED:** smooth, collar inflated. Moist sites, cliffs, coastal grassland, chaparral; < 50 m. n CCo (Marin, Sonoma cos.). Mar–May ★

D. nudicaule Torr. & A. Gray (p. 1143) RED OR ORANGE LARKSPUR Root gen > 15 cm, distally branched. **ST:** 15–50(125) cm, base narrower than root, not firmly attached, gen glabrous. **LF:** gen on proximal 1/3 of st, ± glabrous; lobes 3–10, gen > 6 mm wide. **INFL:** pedicels 15–80 mm, 7–50 mm apart, glabrous to glandular-puberulent. **FL:** sepals forward-pointing, scarlet to orange-red (dull yellow), lateral 8–16 mm, spur 12–34 mm; lower petals flattened, blades 2–3 mm, ± glabrous. **FR:** 13–26 mm, curved. **SEED:** smooth, collar inflated. 2*n*=16. Moist talus, wooded, rocky slopes; < 2600 m. NW, CaR, n&c SN, ScV, CW (exc SCoRI), nw MP; to sw OR. Gen hummingbird-pollinated. Mar–Jun

D. nuttallianum Pritz. DWARF, MEADOW, SLIM, OR SONNE'S LARKSPUR Roots clustered or not, ± spheric. **ST:** 5–50(100) cm,

base narrower than root, not firmly attached, glabrous to puberulent. **LF:** gen on proximal 1/5 of st, ± glabrous; lobes 7–25, < 7 mm wide, ≥ 6 > 50% to petiole. **INFL:** pedicels 7–75 mm, 15–50 mm apart, glabrous or puberulent; fls gen < 12. **FL:** sepals gen ± reflexed, gen ± glabrous, lateral 8–18 mm, spur 8–20 mm. **FR:** 7–17 mm, ± curved. **SEED:** smooth, shiny, collar inflated. 2*n*=16. Open woodland, sagebrush scrub, meadow edges, streambanks; 300–3300 m. NW, SNH, MP; to BC, CO, n AZ. Hybridizes with *D. andersonii*, *D. depauperatum*, *D. nudicaule*, *D. polycladon*; extremely difficult, variable complex. May–Jul

D. parishii A. Gray   Herbage glabrous to puberulent; root gen > 15 cm, branched. **ST:** base gen narrower than root, firmly attached. **LF:** lobes gen < 6 mm wide. **INFL:** pedicels 3–48 mm; fls 6–75. **FR:** length gen ≤ 3 × width. **SEED:** ± winged, coat inflated, cell margins wavy.

subsp. ***pallidum*** (Munz) M.J. Warnock   PALE-FLOWERED PARISH'S LARKSPUR   **ST:** 27–95 cm. **LF:** gen basal, cauline much-reduced; lobes 3–7, gen > 6 mm wide. **INFL:** pedicels 4–17 mm apart. **FL:** sepals spreading to erect, white to pink or blue, lateral 6–11 mm, 2–4 mm wide, spur 7–13 mm; lower petal blades 3–4 mm. **FR:** 11–14 mm. 2*n*=16. Uncommon. Sagebrush scrub, chaparral; 900–1900 m. sw SnJV, SCoRI, WTR. May–Jun

subsp. ***parishii*** PARISH'S, DESERT, SKY BLUE, OR MOJAVE LARKSPUR   **ST:** 17–100 cm. **LF:** basal 3–5-lobed, gen 0 in fl; cauline 3–15-lobed. **INFL:** pedicels 8–25 mm apart. **FL:** sepals ± sky blue, lateral reflexed, 8–12 mm, 3–6 mm wide, spur 8–15 mm; lower petal blades 3–6 mm. **FR:** 9–21 mm. 2*n*=16. Desert scrub, juniper woodland; 300–2500 m. s SNH, Teh, TR, SNE, D; to sw UT, w AZ, n Baja CA. Mar–May

subsp. ***subglobosum*** (Wiggins) H. Lewis & Epling   COLORADO DESERT LARKSPUR   **ST:** 19–78 cm. **LF:** basal gen present in fl; cauline much-reduced; lobes 7–12. **INFL:** pedicels 8–17 mm apart. **FL:** sepals spreading, dark blue, lateral 9–13 mm, 5–7 mm wide, spur 12–14 mm; lower petal blades 4–6 mm. **FR:** 8–11 mm. Chaparral, desert scrub; 600–1300 m. w DSon; n Baja CA. Mar–Apr ★

D. parryi A. Gray   Root occ > 10 cm. **ST:** 15–80(110) cm, base gen curly-puberulent. **LF:** gen curly-puberulent; lobes 5–27, gen < 6 mm wide. **INFL:** pedicels ± ascending, 5–68 mm, 8–50 mm apart, gen puberulent; fls 3–60. **FL:** sepals reflexed or spreading, lateral 9–25 mm, spur 8–21 mm; lower petal blades 3–10 mm. **FR:** 10–19 mm. **SEED:** ± bumpy, winged.

subsp. ***blochmaniae*** (Greene) H. Lewis & Epling (p. 1143)   DUNE LARKSPUR   Root < 10 cm. **LF:** basal gen 0 in fl; lobes 5–15. **FL:** sepals gen reflexed, lateral 16–25 mm, spur 11–16 mm; lower petal blades 7–10 mm, paler than sepals, esp on herbarium specimens. 2*n*=16. Coastal chaparral, sand; < 200 m. s CCo, SCo. Apr–May ★

subsp. ***eastwoodiae*** Ewan   EASTWOOD'S LARKSPUR   Root < 10 cm. **LF:** basal, cauline present in fl, gen on lower 1/3 of st; lobes 5–15. **FL:** sepals gen reflexed, lateral 11–20 mm, spur 11–17 mm; lower petal blades 6–9 mm. Uncommon. Coastal chaparral, grassland, on serpentine; 100–500 m. s CCo, SCoRO (San Luis Obispo Co.). Mar–May ★

subsp. ***maritimum*** (Davidson) M.J. Warnock   MARITIME LARKSPUR   Root < 10 cm. **LF:** basal, cauline present in fl; lobes 5–10, gen > 6 mm wide. **FL:** sepals gen spreading, lateral 9–20 mm, spur 8–21 mm; lower petal blades 4–11 mm. 2*n*=16. Coastal chaparral; < 300 m. CCo, SCo, ChI; n Baja CA. Mar–May

subsp. ***parryi*** PARRY'S LARKSPUR   Root 5–20 cm. **LF:** basal gen 0 in fl; lobes 7–27. **FL:** sepals gen spreading, lateral 9–15 mm, spur 8–15 mm; lower petal blades 3–8 mm. 2*n*=16. Chaparral, oak woodland; 200–1700 m. SnJV, CW, SW; n Baja CA. Hybridizes with *D. cardinale* (*D.* ×*inflexum* Davidson), *D. gypsophilum*, *D. hesperium*, *D. umbraculorum*. Apr–Jun

subsp. ***purpureum*** (H. Lewis & Epling) M.J. Warnock   MOUNT PINOS LARKSPUR   Root gen > 10 cm. **LF:** basal, cauline present in fl, gen on lower 1/3 of st; lobes 3–20. **FL:** sepals gen reflexed, ± purple to dark blue, lateral 7–11 mm, spur 10–13 mm; lower petal blades 3–5 mm. **FR:** 10–15 mm. 2*n*=16. Sagebrush scrub, dry chaparral; 1000–2600 m. Teh, SCoRO, WTR. Apr–Jun ★

D. patens Benth.   SPREADING OR ZIGZAG LARKSPUR   Root ± ellipsoid to diffuse-fibrous. **ST:** base gen narrower than root, not firmly attached, ± glabrous. **LF:** gen on lower 1/3 of st, ± glabrous; terminal lobe widest near middle. **INFL:** pedicels 10–78 mm, gen > 10 mm apart. **FL:** sepals reflexed, bright or dark blue. **FR:** 12–23 mm, ± curved. **SEED:** smooth, shiny, collar inflated.

subsp. ***hepaticoideum*** Ewan (p. 1143)   **ST:** 25–80 cm. **LF:** basal, lower cauline gen divided < 80% to petiole; lobes 3–5, at least those of lower lvs gen > 15 mm wide. **INFL:** pedicels gen glabrous. **FL:** lateral sepal 11–17 mm, spur 10–18 mm; lower petal blades 5–8 mm. Riparian woodland; 300–1300 m. SCoRO, SW (exc ChI). Apr–Jun

subsp. ***montanum*** (Munz) Ewan   **ST:** 30–70 cm. **LF:** basal 0 in fl, cauline gen divided > 80% to petiole; lobes 5–10, < 10 mm wide. **INFL:** pedicels puberulent, gen glandular. **FL:** lateral sepals 7–11 mm, spur 8–14 mm; lower petal blades 3–6 mm. Open conifer forest, drier, e sides of mtn ranges; 1500–2800 m. s SNH, s SCoRI, TR, PR. Apr–Jun

subsp. ***patens*** **ST:** 10–90 cm. **LF:** basal, lower cauline divided > 80% to petiole; lobes 3–5, gen < 15 mm wide. **INFL:** pedicels gen glabrous. **FL:** lateral sepals 9–20 mm, spur 10–15 mm; lower petal blades 4–6 mm, each gen hairier on inner lobe than outer. 2*n*=16. Grassland, open woodland; 80–1100 m. s NCoR, SN, GV, SnFrB, n SCoR. Hybridizes with *D. decorum*, *D. nudicaule*, *D. variegatum*. Mar–Jun

D. polycladon Eastw.   HIGH MOUNTAIN LARKSPUR   Root gen > 15 cm, distally branched; rootstock buds in life prominent. **ST:** gen ≥ 2, (15)80–120(160) cm, glabrous. **LF:** gen on proximal 1/3 of st, glabrous; lobes gen > 6 mm wide. **INFL:** gen ± 1-sided; pedicels 10–150 mm, ± S-shaped, 10–80 mm apart, glabrous to puberulent; fls 3–35. **FL:** lateral sepals 12–18 mm, spur 11–22 mm. **FR:** 13–20 mm. **SEED:** ± striate. 2*n*=16. Stream-banks, wet talus; 2200–3600 m. SNH, W&I. Hybridizes with *D. depauperatum*, *D. nuttallianum*, *D. glaucum*. Jul–Sep

D. purpusii Brandegee (p. 1143)   ROSE-FLOWERED LARKSPUR   Root gen > 15 cm, branched distally. **ST:** 30–120 cm, base gen narrower than root, not firmly attached, ± glabrous. **LF:** gen on proximal 1/2 of st, ± puberulent; lobes 5–30(50) mm wide. **INFL:** narrow; pedicels ± ascending, 5–48 mm, 8–50 mm apart, hairs glandular, yellow. **FL:** sepals gen reflexed, pink to deep rose-pink, lateral 10–16 mm, spur 10–19 mm; lower petal blades 3–4 mm, angled > 130° to claw, ± glabrous. **FR:** 11–29 mm. **SEED:** shiny, winged, coat inflated, ± clear. 2*n*=16. Talus, cliffs; 300–1300 m. s SN, w edge DMoj. Mar–May ★

D. recurvatum Greene   RECURVED LARKSPUR   Herbage ± glabrous; root > 15 cm or not. **ST:** 18–60(85) cm, base gen narrower than root, firmly attached, glabrous. **LF:** basal gen >> cauline; lobes 3–11. **INFL:** pedicels 10–56 mm, 7–25 mm apart. **FL:** sepals gen reflexed, gen light blue, lateral 11–16 mm, spur 10–18 mm; lower petals white. **FR:** 8–21 mm, length gen < 3 × width. **SEED:** winged, coat cell margins wavy. 2*n*=16. Poorly drained, fine, alkaline soils in grassland, *Atriplex* scrub; 30–600 m. ScV (extirpated), SnJV, s SCoRI (Caliente Range), w DMoj. Mar–Jun ★

D. scaposum Greene   BARE-STEM LARKSPUR   **ST:** 40–85 cm, base gen narrower than root, firmly attached, glabrous. **LF:** gen basal, glabrous; lobes 3–5(9), ≥ 4 mm wide. **INFL:** cylindric; pedicels ± ascending, 10–25 mm, 10–60 mm apart, glabrous. **FL:** sepals dark blue but brightly reflective, lateral 7–10 mm, spur 8–12 mm. **FR:** 12–16 mm, length gen 3 × width, ± glabrous to sparsely pubescent. **SEED:** smooth, seed coat cells ± rectangular, cell margins straight. 2*n*=16. Canyons, sandy, gravelly soil, juniper woodland; 269–1055 m. DSon (Whipple Mtns); to CO, NM. Mar–Apr ★

D. stachydeum (A. Gray) Tidestr.   SPIKED LARKSPUR   Root gen > 15 cm, distally branched; rootstock buds in life ± prominent. **ST:** gen > 1, (40)100–180 cm, puberulent. **LF:** gen on proximal 1/2 of st, but gen 0 on proximal 1/5; lobes 3–18. **INFL:** gen branched; pedicels 10–30 mm, 4–25 mm apart, puberulent; fls gen > 30. **FL:** sepals ± bright blue, paler along midline from hairs, lateral 9–13 mm, spur 11–16 mm. **FR:** 10–15 mm. **SEED:** ± striate. Conifer forest edges, sagebrush scrub; 2300–2600 m. Wrn; to se WA, sw ID, nw UT. Jul–Aug ★

D. trolliifolium A. Gray (p. 1143) COW POISON, POISON LARKSPUR **ST:** (40)60–120(180) cm, base gen narrower than root, not firmly attached, glabrous to hairy. **LF:** glabrous; lobes > 1 cm wide, tips sharply, irregularly cut. **INFL:** pedicels 7–96 mm, 4–50 mm apart, glabrous to hairy. **FL:** lateral sepals 8–21 mm, spur 10–23 mm. **FR:** 15–34 mm, length gen > 4 × width, curved. **SEED:** smooth, collar inflated. 2*n*=16. Oak woodland, coastal chaparral; 30–1100 m. NCo, n&c NCoRO; OR. Hybridizes with *D. decorum, D. nudicaule, D. nuttallianum.* Apr–Jun

D. uliginosum Curran (p. 1143) SWAMP LARKSPUR **ST:** 8–70 cm, base narrower than root, firmly attached, ± glabrous. **LF:** basal, ± fleshy, glabrous, blade fan-shaped, gen divided < 50% to petiole, base tapered. **INFL:** pedicels ascending, 3–104 mm, puberulent; fls 5–45. **FL:** lateral sepals 9–14 mm, spur 10–14 mm; lower petal blades each hairier on inner lobe than outer. **FR:** 10–18 mm. **SEED:** bumpy, ± winged. 2*n*=16. Streambanks, chaparral, grassland, on serpentine; 400–600 m. s NCoRI (very local). Hybridizes with *D. hesperium.* May–Jun ★

D. umbraculorum H. Lewis & Epling UMBRELLA LARKSPUR **ST:** 40–85 cm, base gen narrower than root, firmly attached, ± glabrous. **LF:** basal, cauline ± glabrous; lobes 3–10, > 4 mm wide on proximal. **INFL:** pedicels ± ascending, 6–73 mm, 10–50 mm apart, glabrous or puberulent. **FL:** lateral sepals reflexed, 10–16 mm, spur 11–14 mm. **FR:** 15–19 mm. **SEED:** smooth, winged. 2*n*=16. Moist oak forest; 400–1600 m. SCoRO, WTR. Hybridizes with *D. parryi, D. patens* subsp. *montanum.* Apr–Jun ★

D. variegatum Torr. & A. Gray (p. 1143) ROYAL LARKSPUR **ST:** gen < 50 cm, base hairy. **LF:** gen on proximal 1/3 of st; petioles hairy; lobes 3–15, gen overlapping. **INFL:** gen branched; pedicels ± ascending, 6–74 mm, 10–25 mm apart, gen puberulent. **FL:** spur 10–19 mm, straight or down-curved < 3 mm at tip; lower petal blades each hairier on inner lobe than outer, and hairier on margins. **FR:** 9–19 mm; veins gen colored. **SEED:** winged.

subsp. ***kinkiense*** (Munz) M.J. Warnock SAN CLEMENTE ISLAND LARKSPUR **INFL:** fls on main axis gen < 12. **FL:** sepals light blue to white, lateral 11–18 mm; lower petal blades 4–9 mm. Coastal grassland, 0–15° slopes; < 500 m. s ChI (San Clemente Island). Jan–Apr ★

subsp. ***thornei*** Munz THORNE'S ROYAL LARKSPUR **INFL:** fls on main axis gen < 12. **FL:** sepals bright blue, lateral 17–21 mm; lower petal blades 6–11 mm. Grassland, oak woodland, 15–35° slopes; < 500 m. s ChI (San Clemente Island). Feb–May ★

subsp. ***variegatum*** **INFL:** fls on main axis gen < 10. **FL:** sepals gen dark royal-blue (white or lavender), lateral 10–25 mm; lower petal blades 4–11 mm. 2*n*=16,32. Grassland, open oak woodland; 20–800 m. NCo, se KR, NCoR, CaR, SNF, GV, SnFrB, SCoR. Hybridizes with *D. hansenii, D. hesperium, D. parryi, D. recurvatum*; large fls throughout, but esp n from SnFrB. Mar–May

ENEMION FALSE RUE-ANEMONE

Bruce A. Ford

Per from clustered, slender to fusiform or ± spheric fleshy roots, glabrous. **ST:** ascending to erect; branches 0(–few). **LF:** 2-ternate; basal petioles gen > blades, cauline short to ± 0; lflets wide-ovate to wedge-shaped-obovate, margins entire to deeply 2–3 lobed. **INFL:** terminal or axillary, cymes or racemes, 2–10 fld, or fls 1. **FL:** sepals gen 5, petal-like; petals 0; stamens 10–many; pistils [2]3–10, stalk-like base 0 or short. **FR:** follicle, glabrous, veins obvious, stalk-like base curved or not, occ 0, beak straight to recurved. **SEED:** ± red-brown, smooth, wrinkled or minutely pubescent. 6 spp.: temp N.Am, Eurasia.

1. Fr stalk-like base 0; filaments > 20, ± thread-like, not flat; sepals 5–11 mm; lf segment lobes 2–9 mm wide
....*E. occidentale*
1′ Fr stalk-like base 0.5–2 mm; filaments < 15, flat; sepals 3.5–6 mm; lf segment lobes 0.5–4 mm wide......*E. stipitatum*

E. occidentale (Hook. & Arn.) J.R. Drumm. & Hutch. (p. 1143) WESTERN RUE-ANEMONE Pl 8–34 cm. **ST:** 1–3, erect, gen simple. **LF:** 3–12 cm; segment lobes 2–3, gen < 1/2 segment length. **FL:** sepals 3–7 mm wide, oblong to ovate, white, occ tinged pink; stamens 3–6 mm; pistils 5–8. **FR:** 8–10.5 mm. Shaded slopes, chaparral, oak woodland, conifer forest; 200–1500 m. NCoRI, SN, CW (exc CCo), n WTR. [*Isopyrum o.* Hook. & Arn.] Mar–May

E. stipitatum (A. Gray) J.R. Drumm. & Hutch. (p. 1143) SISKIYOU RUE-ANEMONE Pl 4–12 cm. **ST:** (1)3–7, decumbent to erect, gen simple. **LF:** 4–11 cm; segment lobes gen 3, gen > 1/2 segment length. **FL:** sepals 1–2.5 mm wide, white; stamens 2–3 mm; pistils 3–10. **FR:** 4–7 mm. Shaded slopes, chaparral, oak woodland, mixed-evergreen forest; 200–1500 m. NW, CaR, n SN, se SnFrB (Mount Hamilton Range), MP; to OR. [*Isopyrum s.* A. Gray] SnFrB pls grow with, similar to *E. occidentale.* Feb–Apr

MYOSURUS MOUSETAIL

Alan T. Whittemore

Ann, short-lived, gen tufted. **LF:** basal, simple, linear. **INFL:** scapose, 1-fld. **FL:** sepals (3)5(8), 1.5–4 mm, alike, free, green or scarious-margined, spurred; petals 0–5, free, linear to narrowly spoon-shaped, long-clawed, white; stamens 5–25; pistils 10–400, 1–2.5 mm, ovules 1, styles thread-like. **FR:** achene, angled, wall not veined, styles persistent, ± enlarged in fr; receptacle elongate, gen growing and producing ovules after fl. ± 15 spp.: temp worldwide. (Greek: mouse tail) [Whittemore 1997 FNANM 3:135–138]

1. Fr body not compressed, length ± = width, keel in depression; e DMtns ....*M. cupulatus*
1′ Fr body ± compressed laterally, length > width, keel not in depression; CA-FP, GB
    2. Infl in fr < lvs, peduncle 0–0.2 cm...*M. sessilis*
    2′ Infl in fr > lvs, peduncle 0.9–13 cm
        3. Fr beak erect, parallel to fr surface, 0.05–0.4 mm...............................*M. minimus*
        3′ Fr beak ± spreading, diverging from fr surface, 0.6–1.4 mm....................*M. apetalus*
            4. Fr aggregate 4–9 mm; sepal 1-veined, widely scarious-margined.........var. *borealis*
            4′ Fr aggregate 11–26 mm; sepal 3-veined, narrowly scarious-margined.........var. *montanus*

M. apetalus Gay Pl 1.5–12.5 cm. **INFL:** scape 0.9–10.5 cm. **FR:** outer face 1–2.2 mm, 0.4–1 mm wide, not bordered; beak 0.6–1.4 mm, diverging from fr surface; aggregate 1.5–2 mm wide, exserted from lvs.

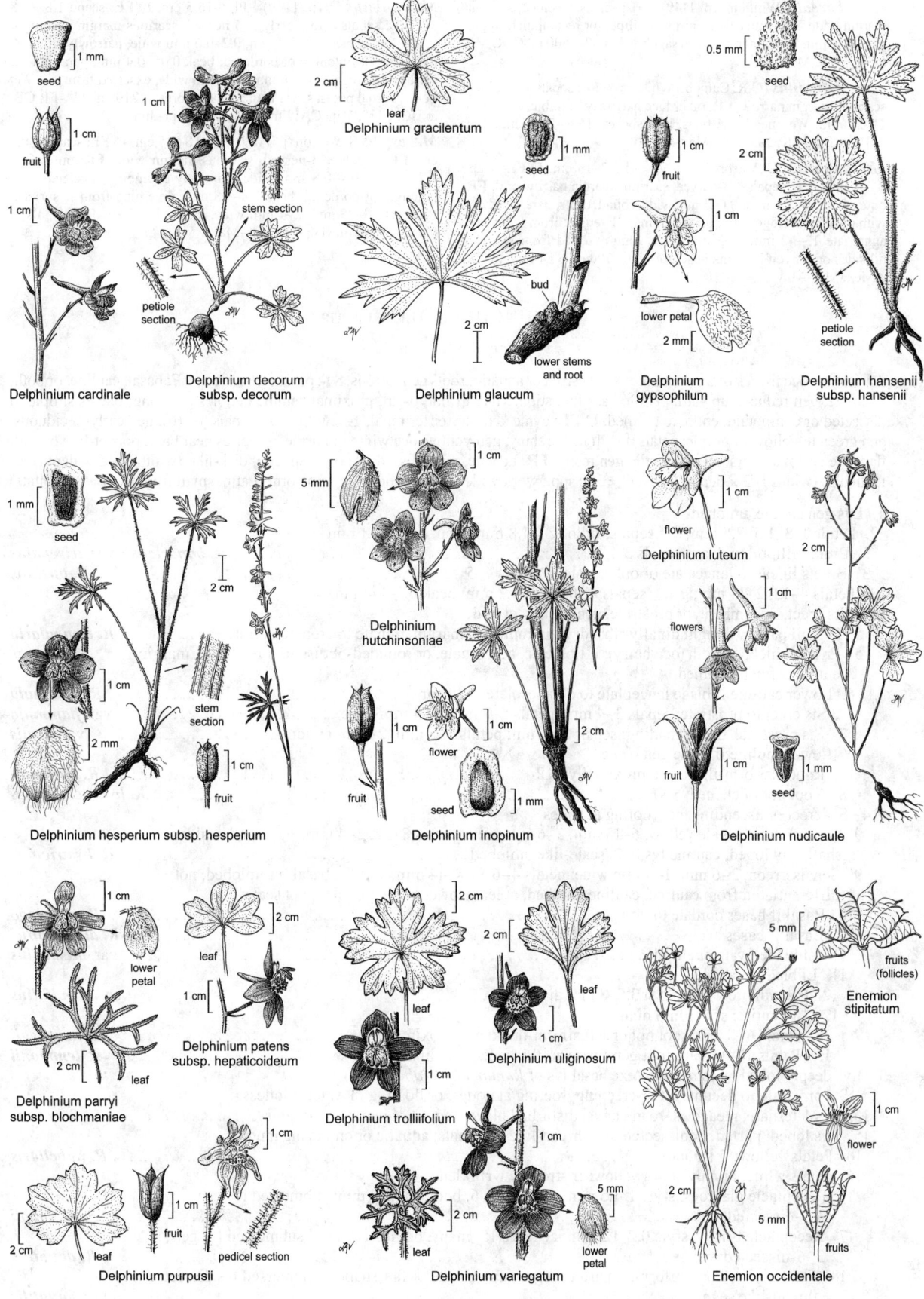

seed
1 mm
fruit
1 cm
1 cm
Delphinium cardinale

1 cm
stem section
petiole section
Delphinium decorum subsp. decorum

2 cm
leaf
Delphinium gracilentum
seed
1 mm
bud
2 cm
lower stems and root
Delphinium glaucum

0.5 mm
seed
2 cm
fruit
1 cm
1 cm
lower petal
2 mm
Delphinium gypsophilum
petiole section
Delphinium hansenii subsp. hansenii

1 mm
seed
2 cm
1 cm
stem section
2 mm
1 cm
fruit
Delphinium hesperium subsp. hesperium

5 mm
1 cm
Delphinium hutchinsoniae
fruit
1 cm
flower
seed
1 mm
2 cm
Delphinium inopinum

1 cm
flower
Delphinium luteum
1 cm
flowers
fruit
1 cm
seed
1 mm
2 cm
Delphinium nudicaule

1 cm
lower petal
2 cm
leaf
Delphinium parryi subsp. blochmaniae
2 cm
leaf
1 cm
Delphinium patens subsp. hepaticoideum

2 cm
leaf
1 cm
Delphinium trolliifolium
2 cm
leaf
1 cm
Delphinium uliginosum

5 mm
fruits (follicles)
Enemion stipitatum
1 cm
flower
2 cm
5 mm
fruits
Enemion occidentale

2 cm
leaf
1 cm
fruit
pedicel section
Delphinium purpusii
1 cm
2 cm
leaf
1 cm
5 mm
lower petal
Delphinium variegatum

var. ***borealis*** Whittem. (p. 1149) **FL:** sepals 1-veined, scarious margin wide. **FR:** outer face narrowly elliptic or rectangular; aggregate 4–9 mm. Flats, marshes, gen sagebrush; 600–1500 m. CaR, MP; to BC, WY. May

var. ***montanus*** (G.R. Campb.) Whittem. **FL:** sepals ± 3-veined, scarious margin narrow. **FR:** outer face narrowly rhombic; aggregate 11–26 mm. Wet meadows, bogs, lake shores; 1500–2500 m. SNH, SnBr, MP; to SK, ND, NM. May–Jul

M. cupulatus S. Watson (p. 1149) Pl 3.3–16 cm. **INFL:** scape 2.2–12 cm. **FL:** sepals ± 3-nerved, scarious margin narrow or 0. **FR:** outer face 0.8–1.2 mm, 0.6–1 mm wide, round, occ square, bordered with prominent ridge; beak 0.6–1.2 mm, ± divergent from fr surface; aggregate 13–42 mm, 2–3 mm wide, long-exserted from lvs. Dry hillsides or canyon bottoms in scrub; 350–1700 m. e DMtns; to TX, n Mex. Mar–May

M. minimus L. (p. 1149) Pl 4–16.5 cm. **INFL:** scape 1.8–12.8 cm. **FL:** sepals ± to clearly 3–5 nerved, scarious margin narrow or 0. **FR:** outer face 0.8–1.4 mm, 0.2–0.6 mm wide, narrowly rhombic to elliptic or oblong, not bordered; beak 0.05–0.4 mm, parallel to fr surface; aggregate 16–50 mm, 1–3 mm wide, exserted from lvs. Wet fields, vernal pools, streambanks, lake shores; < 2100 m. CA-FP, GB; to BC, e US, Baja CA; Eur, Asia, n Afr. Apr–Jun

M. sessilis S. Watson (p. 1149) Pl 0.8–2.5 cm. **INFL:** scape 0–0.2 cm. **FL:** sepals ± 3-nerved, scarious margin wide. **FR:** outer face 1.2–2 mm, 0.4–0.8 mm wide, narrowly rhomboid, occ narrowly oblong, not bordered; beak 0.8–1.8 mm, diverging from fr surface; aggregate 8–25 mm, 2–4 mm wide, incl in lvs. Vernal pools, alkali flats; 3–1600 m. GV; OR. Mar–May

RANUNCULUS BUTTERCUP

Alan T. Whittemore

Ann to per, occ from stolons or caudices, terrestrial or aquatic; roots gen fibrous. **ST:** prostrate to erect. **LF:** basal, cauline, or both, alternate, gen reduced upward; petiole base flat, stipule-like or not; basal, proximal cauline petioles gen long; blades simple to dissected or compound, entire to toothed. **INFL:** cyme, axillary or terminal, 1–few-fld. **FL:** sepals 3–5(6), gen early-deciduous, gen green to yellow or purple; petals 0–17[(150)], shiny, gen yellow, occ white or purple, nectaries near base, pocket-like or with flap-like scale; anthers yellow; pistils gen many. **FR:** achene, compressed or not, ± spheric, disk-like (width 3–15 × depth), or lenticular (width 1–2 × depth), beaked. ± 300 spp.: worldwide exc lowland trop; some orn. (Latin: small frog, from wet habitats)

1. Lvs gen simple, unlobed
  2. Petals 1–3, 1.5–2.5 mm, ± = sepals; fr body 1–1.8 mm, beak 0 or < 0.1 mm
    3. Bracts elliptic to ovate; sepals 3 . . . . ***R. bonariensis*** var. ***trisepalus***
    3′ Bracts linear to lanceolate or oblanceolate; sepals 4–5 . . . . ***R. pusillus***
  2′ Petals 4–12, 2–14 mm, gen ≥ sepals; fr body 1–4.2 mm, beak 0.1–1.4 mm.
    4. Sts erect, decumbent, or prostrate, gen rooting at nodes
      5. Fr wall papery, longitudinally ribbed; lf tip rounded, margin crenate to crenate-serrate . . . . ***R. cymbalaria***
      5′ Fr wall thick, smooth (occ hairy); lf tip acute, acuminate, or rounded-obtuse to thread-like, margin entire or finely toothed
        6. Lower cauline lf blade lanceolate to oblanceolate or linear . . . . ***R. flammula***
          7. Sts erect to prostrate; sepals 3–4 mm; petals 5–7 mm, 3–4 mm wide . . . . var. ***flammula***
          7′ Sts prostrate, occ ascending; sepals 2–3 mm; petals 3–5 mm, 2–3 mm wide . . . . var. ***ovalis***
        6′ Lower cauline lf blade gen ovate
          8. Root fusiform-thickened proximally; KR . . . . ***R. gormanii***
          8′ Root not thickened; SNE . . . . ***R. hydrocharoides***
    4′ Sts erect or ascending, not rooting at nodes
      9. Sepals white or pale yellow, 6–13 mm, 3–6 mm wide; petals 8–12, 2–4 mm, ± green; basal lvs shallowly lobed, cauline lvs 0–2, scale-like, unlobed . . . . ***R. hystriculus***
      9′ Sepals green, 2–6 mm, 1–4 mm wide; petals 4–6(8), 4–14 mm, yellow; basal lvs unlobed, not differentiated from cauline, cauline unlobed, at least lower large, petioled, not scale-like
        10. Basal lf bases cordate to obtuse . . . . ***R. populago***
        10′ Basal lf bases acute . . . . ***R. alismifolius***
          11. Lf blade ovate or elliptic . . . . var. ***alismellus***
          11′ Lf blade lanceolate
            12. Lvs minutely serrate; st 3.5–8 mm diam . . . . var. ***alismifolius***
            12′ Lvs entire; st 1–3 mm diam
              13. Petals 6–8 mm; root not or ± fusiform-thickened proximally . . . . var. ***hartwegii***
              13′ Petals 10–14 mm; root gen fusiform-thickened proximally . . . . var. ***lemmonii***
1′ Lvs deeply lobed or compound (exc basal lvs of *Ranunculus glaberrimus*)
  14. Sts prostrate to decumbent or creeping, rooting at nodes, or floating in water, rootless
    15. Lvs 1-ternate; meadows, road banks, disturbed places, edges of marshes, streams . . . . ***R. repens***
    15′ Lvs lobed, parted, or dissected into thread-like segments; aquatic or on drying mud
      16. Petals yellow; fr smooth . . . . ***R. flabellaris***
      16′ Petals white, or only claws yellow; fr strongly wrinkled
        17. Receptacle glabrous; style 1–1.5 mm; achenes 2–6, bodies 2–2.4 mm; submersed lvs gen 2–3-dissected; ann . . . . ***R. lobbii***
        17′ Receptacle bristly; style 0.2–1.2 mm; achenes 15–many, bodies 1–2 mm; submersed lvs gen 3–6-dissected; per . . . . ***R. aquatilis***
          18. Floating lvs 0 or reniform, 3-parted, segments obovate or fan-shaped, submersed lvs dissected into thread-like segments . . . . var. ***aquatilis***
          18′ Floating and submersed lvs dissected into thread-like segments . . . . var. ***diffusus***

14′ Sts erect to ascending or decumbent (prostrate), gen not rooting at nodes, not floating, occ decumbent, rooting only at lowest nodes

  19. Style 0; fr margins thick, corky; pl emergent aquatic, occ on mud . . . . **R. sceleratus**

    20. Fr faces smooth; lvs parted, gen deeply, segments lobed or parted, deeply crenate or again lobed

      . . . . var. **multifidus**

    20′ Fr faces finely transversely wrinkled; lvs lobed or parted, segments undivided or lobed, crenate

      . . . . var. **sceleratus**

  19′ Style present; fr margins ridged to narrowly winged, or differentiated margins not evident; pl on wet or dry soil

    21. Fr wall papery, inflated; petals ± pink-white . . . . **R. andersonii**

    21′ Fr wall thick, not inflated; petals yellow, occ ± purple abaxially

      22. Cauline lvs 0; fr beak >> body . . . . **R. testiculatus**

      22′ Cauline lvs well developed; fr beak gen < body

        23. Fr lenticular, width 1–2 × depth; st glabrous

          24. Basal lvs entire or with 3 wide, rounded teeth distally . . . . **R. glaberrimus**

            25. Basal lvs elliptic to very narrowly elliptic, entire (or with 3 wide, rounded teeth distally); bract lobes 3, middle >> lateral . . . . var. **ellipticus**

            25′ Basal lvs ovate to obovate or reniform, with 3 wide, rounded, teeth distally (or entire); bract lobes 3, equal . . . . var. **glaberrimus**

        24′ Basal lvs 3–9-lobed or 3-parted . . . . **R. eschscholtzii**

          26. Ultimate lf segments acute to acuminate . . . . var. **suksdorfii**

          26′ Ultimate lf segments rounded, obtuse, or widely rounded-acute

            27. Persistent lf bases on caudex 0–few; basal lvs 3-parted . . . . var. **eschscholtzii**

            27′ Persistent lf bases on caudex many; basal lvs shallowly 5–9-lobed or occ 3-parted . . . . var. **oxynotus**

      23′ Fr disk-like, width 3–15 × depth; st hairy or glabrous

        28. Fr faces papillate to spiny

          29. Petals 1–2 mm; fr faces papillate, each papilla tipped with hooked bristle

            30. Fr ornamented on faces, margins, beaks lanceolate . . . . **R. hebecarpus**

            30′ Fr ornamented on faces, not margins, beaks deltate . . . . **R. parviflorus**

          29′ Petals 4–10 mm; fr faces gen papillate or spiny, hooked bristles 0

            31. Sepals spreading; achenes 5–9 in a single whorl, faces and margins spiny . . . . **R. arvensis**

            31′ Sepals reflexed; achenes 10–60 in ovoid to globose heads, faces papillate to spiny (smooth), margins smooth

              32. Basal lvs simple; fr beak 2–2.5 mm . . . . **R. muricatus**

              32′ Basal lvs 1–2 ternate; fr beak 0.3–0.7 mm . . . . ²**R. sardous**

      28′ Fr faces smooth

        33. Petals 2–6 mm, ± = sepals

          34. Petals 4–6 mm; basal lvs 1-ternate . . . . **R. macounii**

          34′ Petals 2–4(6) mm; basal lvs 3-parted, occ 1-ternate . . . . **R. uncinatus**

        33′ Petals 5–26 mm, if 5–6 mm then >> sepals

          35. Sepals spreading . . . . [**R. acris**]

          35′ Sepals reflexed 1–3 mm from base

            36. Petals 9–17

              37. Fr body 3.4–4.2 mm, beak deltate . . . . **R. canus** var. **ludovicianus**

              37′ Fr body 1.8–3.2 mm, beak lanceolate . . . . **R. californicus**

                38. St erect to decumbent, glabrous or spreading-hairy; ultimate segments of basal lvs acute to rounded-acute . . . . var. **californicus**

                38′ St prostrate, strigose; ultimate segments of basal lvs rounded to obtuse . . . . var. **cuneatus**

            36′ Petals 5–7

              39. Fr beak straight

                40. Receptacle glabrous . . . . ²**R. occidentalis** (in part)

                  41. Ultimate lf segments lanceolate to oblanceolate . . . . var. **dissectus**

                  41′ Ultimate lf segments narrowly elliptic . . . . var. **howellii**

              40′ Receptacle bristly . . . . **R. orthorhynchus**

                42. Basal lvs simple to compound, lflets 0 or 3, undivided, crenate; petals notched . . . . var. **bloomeri**

                42′ Basal lvs compound, lflets 3–5, lobed or parted, entire, dentate, or dentate-lobed; petals rounded

                  43. Fr aggregate hemispheric, occ spheric, 5–7 mm; petals gen red abaxially . . . . var. **orthorhynchus**

                  43′ Fr aggregate spheric or ovoid, 8–13 mm; petals yellow abaxially . . . . var. **platyphyllus**

              39′ Fr beak curved or tip hooked

              44. Base of st bulbous, corm-like . . . . [**R. bulbosus**]

              44′ Base of st not enlarged

                45. Receptacle long-hairy; fr wall coarsely papillate (smooth), beak 0.4–0.7 mm . . . . ²**R. sardous**

                45′ Receptacle glabrous; fr wall gen smooth, beak 0.2–1.4 mm

                  46. Fr beak deltate, body 3.4–4.4 mm . . . . **R. canus** var. **canus**

                  46′ Fr beak lanceolate, body 2.6–3.6(4) mm . . . . ²**R. occidentalis** (in part)

                    47. Petals 3–6 mm wide; 0–1500 m . . . . var. **occidentalis**

                    47′ Petals 1.5–2.5 mm wide; 1300–2500 m . . . . var. **ultramontanus**

R. alismifolius Benth. Per, erect or ascending, not rooting at nodes. **LF**: tip obtuse to acuminate; basal bases acute. **INFL**: bracts lanceolate. **FL**: receptacle glabrous; sepals 5, spreading or reflexed from base, 2–6 mm, 1–4 mm wide, early-deciduous; petals 4–6(8). **FR**: body 1.6–2.8 mm, 1.2–2 mm wide, lenticular, wall thick, smooth, beak 0.4–1.2 mm, straight or ± curved, awl-shaped.

var. ***alismellus*** A. Gray   Pl 6–21 cm; roots not or ± fusiform-thickened proximally. **ST**: 1–3 mm diam. **LF**: blade 2–5 cm, ± 1 cm wide, ovate or elliptic; cauline bases rounded to widely acute, margins entire. **FL**: petals 5–8 mm, 2–6 mm wide. Damp meadows, swamps, streams; 1400–3400 m. KR, NCoR, SNH, SnBr, SnJt, MP; to WA, MT, nw Baja CA. May–Aug

var. ***alismifolius*** (p. 1149)   Pl 20–70 cm; roots not or ± fusiform-thickened proximally. **ST**: 3.5–8 mm diam. **LF**: blade 6–14 cm, 1–3 cm wide, lanceolate; cauline bases acute or acuminate, margins minutely serrate. **FL**: petals 7–11 mm, 4–6 mm wide. Wet meadows, bogs, shallow water of streams, ponds; 1300–2600 m. KR, NCoR, n SNH, MP; to BC, MT. May–Jun

var. ***hartwegii*** (Greene) Jeps.   Pls 13–30 cm; roots not or ± fusiform-thickened proximally. **ST**: 1–3 mm diam. **LF**: blade 3–10 cm, ± 1 cm wide, lanceolate; cauline bases gen acute or acuminate, margins entire. **FL**: petals 6–8 mm, 3–5 mm wide. Meadows; 1200–2100 m. NCoRH, n&c SNH, MP; to WA, WY. Apr–Aug

var. ***lemmonii*** (A. Gray) L.D. Benson (p. 1149)   Pl 7–35 cm; roots gen fusiform-thickened proximally. **ST**: 1–3 mm diam. **LF**: blade 3–11 cm, ± 1 cm wide, narrowly lanceolate; cauline bases acuminate, margins entire. **FL**: petals 10–14 mm, 5–8 mm wide. Meadows; 1400–2900 m. n SNH, MP; to NV. Apr–Jul

R. andersonii A. Gray (p. 1149)   Per 8–18 cm, erect from short caudex, not rooting at nodes. **LF**: basal, 1.5–3.8 cm, 2.1–3.8 cm wide, cordate, 1–2-ternate, lflets 2–3 × parted, ultimate segments elliptic to linear, margins entire or few-toothed, tips obtuse to acuminate; cauline 0 or ± like, ≪ basal, ± sessile. **FL**: receptacle bristly; sepals 5, spreading, 9–15 mm, 5–9 mm wide, persistent in fr; petals 5, 12–18 mm, 9–13 mm wide, ± pink-white. **FR**: body 6–12 mm, 4–6 mm wide, wall inflated, papery, longitudinal veins strong, beak 0.2–0.6 mm, deltate or awl-shaped from a deltate base. Dry rocky slopes; 1000–2500 m. SNH, GB, DMtns; to OR, UT. Mar–May

R. aquatilis L.   Per (5)20–80 cm, aquatic or creeping on mud, rooting at nodes. **LF**: submersed or floating and submersed; submersed gen 3–6-dissected; stipules tapered, united. **FL**: receptacle bristly; sepals 5, spreading or reflexed, 2–4 mm, 1–2 mm wide, early-deciduous; petals 5, 4–7 mm, 1–5 mm wide, white or white with yellow claws; style 0.2–1.2 mm. **FR**: 15–many, body 1–2 mm, 0.8–1.4 mm wide, lenticular, wall thick, transversely wrinkled, beak persistent, thread-like. [*Ranunculus aquatilis*, orth. var.]

var. ***aquatilis*** (p. 1149)   **LF**: floating 0 or 0.4–1.1 cm, 0.7–2.3 cm wide, reniform, 3-parted, segments obovate or fan-shaped, shallowly cleft, crenate; submersed dissected into thread-like segments. **FR**: body 1.6–2 mm, 1–1.4 mm wide, beak 0.1–0.4 mm. Marshes, streams, ponds, lakes; < 2000 m. NW, CaR, SN, GV, CW, PR, GB; N.Am, S.Am, Eurasia. [*R. a.* var. *hispidulus* E. Drew] Feb–Jun

var. ***diffusus*** With. (p. 1149)   **LF**: floating and submersed dissected into thread-like segments. **FR**: body 1–1.8 mm, 0.8–1.2 mm wide, beak 0.2–1.2 mm. Ponds, lakes, streams, river edges; < 3200 m. CA-FP, GB; N.Am, Eurasia, Australia. [*R. a.* var. *capillaceus* (Thuill.) DC.; *R. a.* var. *subrigidus* (W.B. Drew) Breitung] Mar–Sep

R. arvensis L. (p. 1149)   Ann 10–40 cm, erect or ascending, not bulbous-based; roots basal, not tuberous. **LF**: basal, proximal cauline 1.8–5.2 cm, 1.6–4.2 cm wide, obovate to wedge-shaped, deeply lobed or dissected, lflets oblanceolate, divided into entire or distally dentate, oblanceolate or linear segments or not, lflet base narrowly acuminate, tip rounded, dentate or acuminate. **FL**: receptacle sparsely bristly; sepals 5, spreading, 4–7 mm, 1–2 mm wide, early-deciduous; petals 5, 5–8 mm, 2–4 mm wide. **FR**: body 4–6.4 mm, 8–9 mm wide, disk-like, wall spiny; beak 1.6–3.8 mm, straight, lanceolate to awl-shaped. Roadsides, fields, disturbed areas; < 1000 m. KR, NCoRI, CaR, n SNF, ScV, MP; to WA, NY, S.Am, Australia; native to Eurasia. Apr–May

R. bonariensis Poir. var. ***trisepalus*** (Hook. & Arn.) Lourteig (p. 1149) Ann 4–20(30) cm, decumbent, rooting at most proximal nodes, or erect. **LF**: cauline simple, 0.8–2.3 cm, 0.5–1.2 cm wide, elliptic to ovate, base rounded or obtuse, margin entire to finely dentate, tip widely rounded-acute to rounded. **INFL**: bracts elliptic to ovate. **FL**: receptacle glabrous; sepals 3, spreading or reflexed from base, 1.5–3 mm, 0.5–2 mm wide, early-deciduous; petals 1–3, 1.5–2.5 mm, 0.5–1 mm wide. **FR**: body 1.4–1.8 mm, 1–1.2 mm, lenticular, wall thick, smooth, beak 0. Vernal pools, stream edges; 30–1000 m. n SNF, GV; S.Am. Mar–May

R. californicus Benth.   Per (11)18–70 cm, not rooting at nodes. **LF**: basal, proximal cauline widely ovate or cordate, 3-lobed or -parted to 1-ternate; distal much reduced, deeply parted or compound. **FL**: receptacle glabrous (bristly); sepals 5, reflexed 2–3 mm from base, 4–8 mm, 2–4 mm wide, early-deciduous; petals 9–17, (6)7–14 mm, 2–6 mm wide. **FR**: disk-like, wall thick, smooth, beak 0.2–0.8 mm, curved, lanceolate.

var. ***californicus*** (p. 1149)   **ST**: erect to decumbent, glabrous or spreading-hairy. **LF**: 2.8–5.8 cm, 4–6 cm wide, segments undivided or 1–2 × lobed or parted, ultimate segments oblong-elliptic to lanceolate or linear, toothed or entire, tip acute to rounded-acute. **FR**: body 1.8–3.2 mm, 1.4–2.4 mm wide. Grassland, open woodland; < 2000 m. NW, SNF, GV, CW, SW; to BC. Mar–Aug

var. ***cuneatus*** Greene   **ST**: prostrate, strigose. **LF**: 4–5 cm, 4–6 cm wide, segments undivided or 1 × lobed, ultimate segments elliptic to round, toothed or crenate, tip rounded to obtuse. **FR**: body 1.8–2.2 mm, 1.4–1.8 mm wide. Coastal bluffs, hillsides; < 200 m. NCo, CCo, n ChI; OR. Jan–Apr

R. canus Benth.   Per 11–65 cm, erect to decumbent, not rooting at nodes. **LF**: basal, proximal cauline 3.3–9.5 cm, 3.5–9.4 cm wide, ovate to narrowly ovate, 3-parted, segments 1–3 × lobed; distal cauline much reduced, deeply parted or compound. **FL**: receptacle glabrous; sepals 5, reflexed 1–2 mm from base, 3–8 mm, 2–4 mm wide, early-deciduous; petals 6–12 mm, 3–6 mm wide. **FR**: body 3.4–4.4 mm, 2.4–3.6 mm wide, disk-like, wall thick, smooth, beak 0.2–1.2 mm, curved, deltate.

var. ***canus*** (p. 1149)   **LF**: ultimate segments ovate to oblong-ovate or lanceolate, toothed, acute or obtuse. **FL**: petals 5–7. Grassland, open oak woodland; < 1200 m. NCoRI, CaRF, n SNF, GV, e SnFrB. Mar–Jul

var. ***ludovicianus*** (Greene) L.D. Benson (p. 1149)   **LF**: ultimate segments lanceolate to lance-oblong, entire or toothed, acute or rounded-acute. **FL**: petals 13–17. Meadows; 1000–2300 m. s SNH (Kern Co.), Teh, SnBr. Mar–Aug

R. cymbalaria Pursh (p. 1149)   Per 3–20(30) cm, with stolons. **LF**: basal, proximal cauline, and on stolon 0.7–3.8 cm, 0.8–3.2 cm wide, oblong to cordate or round, simple, undivided, base rounded to cordate, margin crenate or crenate-serrate, tip rounded; distal cauline much reduced, simple, undivided. **FL**: receptacle bristly or glabrous; sepals 5, spreading, 2.5–6 mm, 1.5–3 mm wide, early-deciduous; petals 5, 2–7 mm, 1–3 mm wide. **FR**: body 1–1.4(2.2) mm, 0.8–1.2 mm wide, lenticular, wall papery, longitudinally ribbed, beak persistent, 0.1–0.2 mm, straight, conic, straight. Muddy places, gen brackish or alkaline; < 3200 m. CaR, SNH, SW, GB, DMoj; to AK, NL; S.Am, Eurasia. [*R. c.* var. *saximontanus* Fernald] May–Aug

R. eschscholtzii Schltdl.   Per 5–25 cm, erect or decumbent, not bulbous-based; roots basal, not tuberous. **LF**: segment tips rounded; distal cauline much-reduced, deeply parted or compound. **FL**: receptacle glabrous or sparsely long-hairy; sepals 5, spreading, 4–8 mm, 2–6 mm wide, early-deciduous; petals 5–8, 6–13 mm, 4–13 mm wide. **FR**: body 1.4–2 mm, 1–1.6 mm wide, lenticular, wall thick, smooth, beak 0.6–1.8 mm, straight but occ curved in youth, lanceolate or awl-shaped.

var. ***eschscholtzii***   **ST**: caudex 1–3 cm, persistent lf bases 0–few. **LF**: basal 1–2.3 cm, 1.5–3.7 cm wide, reniform or cordate, 3-parted, at least lateral segments again lobed, base truncate or cordate, middle segment 0–1 × lobed, ultimate segments rounded-obtuse or widely rounded-acute. Open, rocky alpine slopes, meadows; 2200–3600 m. KR, SNH; to AK, CO. Jul–Aug

var. ***oxynotus*** (A. Gray) Jeps. (p. 1149) **ST**: caudex 2.5–4.5 cm, persistent lf bases many. **LF**: basal 0.5–1.5 cm, 0.8–2 cm wide, reniform, shallowly 5–9-lobed or occ 3-parted, lateral segment again lobed, base truncate, middle segment unlobed, ultimate segments rounded or obtuse. Open, rocky alpine slopes, meadows; 2700–4300 m. SNH, SnBr, SnJt, Wrn, SNE (Sweetwater, White mtns); NV. Jul–Sep

var. ***suksdorfii*** (A. Gray) L.D. Benson (p. 1149) **ST**: caudex 1–3 cm, persistent lf bases 0–few. **LF**: basal 0.8–2.1 cm, 1.5–3.5 cm wide, reniform, 3-parted, all segments again lobed, base truncate or cordate, middle segment 1-lobed, ultimate segments acute to acuminate. Open, rocky slopes, meadows; 1500–2200 m. KR (Marble Mtns); to BC, WY. Jul–Aug

R. flabellaris Raf. (p. 1149) Per < 70 cm, floating or decumbent, glabrous, rooting at proximal nodes. **LF**: basal gen 0; cauline 1.2–7.3 cm, 1.9–10.8 cm wide, semicircular to reniform, blade 1–6 × lobed, parted or dissected, segments entire or crenate, tip rounded to thread-like. **FL**: receptacle sparsely bristly; sepals 5, spreading or ± reflexed, 5–7 mm, 3–6 mm wide, early-deciduous; petals 5–6(14), 7–12 mm, 5–9 mm wide. **FR**: body 1.8–2.2 mm, 1.6–2.2 mm wide, lenticular, wall thick, smooth, beak 1–1.8 mm, straight, lanceolate. Shallow water, drying mud; 900–2000 m. KR, NCoRH, MP; to BC, NB, NC. Apr–Aug

R. flammula L. Per (3)10–45 cm, gen rooting at nodes. **LF**: base acute to thread-like, margin entire or minutely serrate, tip acute to thread-like. **INFL**: bracts lanceolate to oblanceolate. **FL**: receptacle glabrous; sepals 5, spreading or ± reflexed, 1–2 mm wide, early-deciduous; petals 5–6, 1–4 mm wide. **FR**: body 1.2–1.6 mm, 1–1.4 mm wide, lenticular, wall thick, smooth, beak 0.1–0.6 mm, straight or curved, lanceolate to linear.

var. ***flammula*** **ST**: erect to prostrate, 0.8–3 mm diam. **LF**: 1.8–4.5 cm, 0.3–1 cm wide, lanceolate to oblanceolate. **FL**: sepals 3–4 mm; petals 5–7 mm, 3–4 mm wide. Shallow water or muddy shores of ponds; < 1000 m. NW; to BC; e Can, Eurasia. May–Aug

var. ***ovalis*** (Bigelow) L.D. Benson (p. 1149) **ST**: prostrate, occ ascending, 0.5–2 mm diam. **LF**: 0.8–3.3 cm, 0.2–0.8 cm wide, lance-elliptic to lanceolate or linear. **FL**: sepals 2–3 mm; petals 3–5 mm, 2–3 mm wide. Muddy ground, shallow water; < 2300 m. NW, CaR, SN, CCo, SnBr, MP; to AK, NL, MA. May–Sep

R. glaberrimus Hook. Per 5–26 cm, st prostrate or ascending, not bulbous-based, glabrous; roots basal, not tuberous. **LF**: basal, occ lower cauline persistent. **FL**: receptacle glabrous; sepals 5, spreading, 5–8 mm, 3–7 mm wide, early-deciduous; petals 5–10, 8–13 mm, 5–12 mm wide. **FR**: body 1.4–2.2 mm, 1.1–1.8 mm wide, lenticular, wall thick, smooth, beak 0.4–1 mm, straight or curved, lanceolate to awl-shaped.

var. ***ellipticus*** (Greene) Greene (p. 1149) **LF**: basal 1.5–5.2 cm, 0.7–2 cm wide, elliptic to very narrowly elliptic, entire (or with 3 wide, rounded teeth distally), base obtuse to attenuate, tip acute to rounded. **INFL**: bract lobes 3, middle >> lateral. Moist, seepy slopes; 1200–3600 m. n SNH (e slope), MP, n SNE, W&I; to BC, ND. Mar–Jun

var. ***glaberrimus*** (p. 1149) **LF**: basal 0.7–1.9 cm, 1–1.7 cm wide, ovate to obovate or reniform, with 3 wide, rounded, teeth distally (or entire), base obtuse to truncate, tip rounded. **INFL**: bract lobes 3, equal. Meadows, open woodland, scrub; 900–2000 m. CaR, n SNH, MP; to BC, SD. Mar–May

R. gormanii Greene (p. 1149) Per 5–20 cm, prostrate, occ rooting at nodes; root fusiform-thickened proximally. **LF**: lower cauline 1.2–4 cm, 0.7–2 cm wide, narrowly to widely ovate, base rounded, truncate or occ obtuse, margin entire or minutely dentate, tip obtuse or acute. **INFL**: bracts ovate, occ lanceolate. **FL**: receptacle glabrous; sepals 5, spreading or reflexed from ± base, 2–4 mm, 1–3 mm wide, early-deciduous; petals 5–6, 4–6 mm, 2–4 mm wide. **FR**: body 1.2–2 mm, 1.2–1.4 mm wide, lenticular, wall thick, smooth, beak 0.6–0.8 mm, straight or curved, lanceolate to awl-shaped. Damp meadows, streambanks; 900–1900 m. KR; OR. Jun–Jul

R. hebecarpus Hook. & Arn. (p. 1149) Ann (2)8–30 cm, erect, not bulbous-based; roots basal, not tuberous. **LF**: basal, proximal

cauline 0.6–2.3 cm, 1.2–3.5 cm wide, cordate-reniform, deeply 3-parted, segments entire or 2–4-lobed or -dentate, base shallowly cordate, ultimate segment tip acute. **FL**: receptacle glabrous; sepals 5, spreading, 1.1–1.8 mm, 0.5–1 mm wide, early-deciduous; petals 0–5, 1.3–2 mm, 0.3–0.7 mm wide. **FR**: body 1.7–2.3 mm, ± 3 mm wide, disk-like, wall finely papillate on faces, margins, each papilla tipped with hooked bristle, beak 0.5–0.7 mm, hooked distally, lanceolate. Grassland, open woodland; 50–900 m. NCoRI, CaRF, SNF, ScV, CW, SW; to WA, Baja CA. Mar–May

R. hydrocharoides A. Gray (p. 1151) FROG'S-BIT BUTTERCUP Per 5–25 cm, erect to prostrate, gen rooting at nodes; root not thickened. **LF**: lower cauline 0.8–2.7 cm, 0.8–1.9 cm wide, ovate to widely ovate, base rounded to ± cordate, margin entire or dentate, tip rounded or obtuse. **INFL**: bracts lanceolate to oblanceolate, occ ovate. **FL**: receptacle glabrous; sepals 5, spreading or reflexed from base, 1.5–3 mm, 1–2 mm wide, early-deciduous; petals 5–6, 3–5 mm, 1–2 mm wide. **FR**: body 1.2–1.4 mm, 1–1.2 mm wide, lenticular, wall thick, smooth, beak 0.4–1 mm, straight or curved, lanceolate to thread-like. Wet ground, shallow water, creek edges, lakes; 1200–2800 m. SNE; to NM, C.Am. Jun–Aug ★

R. hystriculus A. Gray (p. 1151) Per 8–25 cm, erect from short caudex. **LF**: basal 1.2–4.6 cm, 1.8–6.6 cm wide, semicircular or reniform, shallowly 5–7-lobed, ultimate segments semicircular, margin crenate, tip rounded or small-pointed; cauline lvs 0–2, scale-like, unlobed. **FL**: receptacle glabrous; sepals 5, spreading, early-deciduous, 6–13 mm, 3–6 mm wide, white or pale yellow; petals 8–12, 2–4 mm, 0.6–1.6 mm wide, ± green. **FR**: body 3.8–4.2 mm, 0.8–1 mm wide, cylindric, wall papery, veins strong, longitudinal, beak 1.2–1.4 mm, thread-like, hooked, persistent. Wet places near streams, esp waterfalls; 300–1800 m. SN. [*Kumlienia h.* (A. Gray) Greene] Feb–Jul

R. lobbii (Hiern) A. Gray (p. 1151) LOBB'S AQUATIC BUTTERCUP Ann 20–80 cm, floating, roots 0, or creeping on mud, rooting at nodes. **LF**: floating lvs 0.5–0.8 cm, 0.9–1.5 cm wide, reniform, deeply 3-parted, segments elliptic or obovate, occ notched; submersed lvs gen 2–3-dissected, gen petioled, stipules gradually reduced distally on, fused to st. **FL**: receptacle glabrous; sepals 5, spreading, 2–3 mm, 1–1.5 mm wide, early-deciduous; petals 5, 4–6 mm, 2–5 mm wide, white or white with yellow claws; style 1–1.5 mm. **FR**: 2–6, body 2–2.4 mm, 1.4–1.8 mm wide, lenticular, wall thick, transversely wrinkled, beak deciduous, occ leaving stub to 0.2 mm. Ponds; < 500 m. NCoR, SnFrB; to BC. Mar–May ★

R. macounii Britton (p. 1151) MACOUN'S BUTTERCUP Per 20–65 cm, erect, or base decumbent, rooting at most proximal nodes. **LF**: basal, proximal cauline 3.7–7.5 cm, 4.5–9.5 cm wide, cordate to reniform, 1-ternate, lflets 3-lobed or -parted, ultimate segments ± elliptic, toothed or again lobed, tip acute to widely acute; distal cauline much-reduced, deeply parted or compound. **FL**: receptacle rough-stiff-hairy; sepals 5, spreading or reflexed ± 1 mm from base, 4–6 mm, 1.5–3 mm wide, early-deciduous; petals 5, 4–6 mm, 3.5–5 mm wide. **FR**: body 2.4–3 mm, 2–2.4 mm wide, disk-like, wall thick, smooth, beak 1–1.2 mm, straight or ± so, lanceolate. Wet meadows, shallow water; 1200–1500 m. MP; to AK, e Can. Occ fls 1st yr. Jun–Jul ★

R. muricatus L. (p. 1151) Ann, bien (5)15–50 cm, decumbent or erect, not bulbous-based; roots basal, not tuberous. **LF**: basal, proximal cauline simple, 2–5 cm, 3–6.5 cm wide, widely cordate to reniform or semicircular, entire or 3-lobed and coarsely crenate, base rounded to cordate, tip rounded. **FL**: receptacle bristly; sepals 5, reflexed, 4–7 mm, 2–3 mm wide, early-deciduous; petals 5, 4–8 mm, 2–4.5 mm. **FR**: body 5–5.5 mm, 3–3.5 mm wide, disk-like, faces spiny, margins smooth, beak 2–2.5 mm, curved, lanceolate. Streambanks, drainages, low meadows; < 700 m. GV, SNF, SnFrB, SCo; worldwide; native to Eurasia. Apr–Jun

R. occidentalis Nutt. Per 10–60 cm, not rooting at nodes. **LF**: basal, proximal cauline 1.5–5.3 cm, 2.2–8 cm wide, widely ovate to semicircular or reniform; distal cauline reduced, deeply parted or compound. **FL**: receptacle glabrous; sepals 5, reflexed 2–3 mm from base, 4–7(9) mm, 2–4 mm wide, early-deciduous; petals 5–6. **FR**: disk-like, wall thick, smooth. 3 other vars., not in CA.

var. *dissectus* L.F. Hend. Erect to spreading. **LF:** basal lvs 3-parted, occ 1-ternate, ultimate segments lanceolate to oblanceolate, entire or sparsely dentate. **FL:** petals 6–10 mm, 3–6 mm wide. **FR:** body 2.6–3.6 mm, 2–2.8 mm wide, glabrous, beak 1.2–2.2 mm, straight, awl-shaped. Wet or dry meadows; 1000–1800 m. KR, CaRH, MP; OR. May–Jul

var. *howellii* Greene Erect to decumbent. **LF:** basal lvs 3-parted, ultimate segments narrowly elliptic, entire or dentate. **FL:** petals 6–10 mm, 3–5 mm wide. **FR:** body 3.4–4.8 mm, 2.6–3.2 mm wide, glabrous, beak 1.6–2.2 mm, straight, lanceolate. Meadows; 900–1400 m. KR; OR. Apr–Jul

var. *occidentalis* (p. 1151) Erect to decumbent. **LF:** basal 3-parted or 1-ternate, ultimate segments oblong to elliptic, lanceolate, or oblanceolate, dentate. **FL:** petals 5–10 mm, 3–6 mm wide. **FR:** body 2.6–3.6(4) mm, 1.8–3(3.2) mm wide, glabrous or bristly, beak (0.6)1–1.4 mm, curved, lanceolate. Grassy slopes in meadows or open woodland; < 1500 m. NW, CaR, SN, CW, MP; to BC, NV. Mar–Jul

var. *ultramontanus* Greene Decumbent. **LF:** basal 1-ternate (3-parted), ultimate segments oblong to elliptic, lanceolate, or oblanceolate, lobed, or dentate. **FL:** petals 6–8 mm, 1.5–2.5 mm wide. **FR:** body 3–3.4 mm, 2–2.6 mm wide, glabrous (bristly), beak 0.4–1.2 mm, curved, lanceolate. Meadows; 1300–2500 m. KR, CaRH, SNH; OR, NV. May–Jul

R. orthorhynchus Hook. Per 15–50(85) cm, ± erect or decumbent, not rooting at nodes. **LF:** basal, proximal cauline 2.8–12.5 mm, 2.5–14 cm wide, ultimate segments round to linear, entire to dentate or crenate, tip rounded to narrowly acute; distal cauline much-reduced, deeply parted or compound. **FL:** receptacle bristly; sepals 5, reflexed 1–2 mm from base, 5–11 mm, 2–4 mm wide, early-deciduous; petals 5–6, 8–18 mm, red abaxially, yellow adaxially. **FR:** body 2.8–4.5 mm, disk-like, wall thick, smooth, beak 1.8–3.8(4.8) mm, straight, narrowly lanceolate to awl-shaped.

var. *bloomeri* (S. Watson) L.D. Benson (p. 1151) **LF:** basal cordate to oblong or round, simple, undivided, or 1-ternate; lflets 0 or undivided, crenate, rounded to obtuse or occ acute. **FL:** petals 6–9 mm wide, yellow abaxially, notched. **FR:** beak 1.8–2.2 mm; aggregate 7–9 mm, hemispheric to ovoid. Meadows, marshy areas; < 100 m. NCoRO, NCoRI, SnFrB; OR. Mar–May

var. *orthorhynchus* (p. 1151) **LF:** basal lvs narrowly ovate to semicircular, 1-ternate or -pinnate, lflets 3–5, 1–2 × lobed or parted, segments elliptic to linear, entire to dentate-lobed, obtuse to narrowly acute. **FL:** petals 4–6 mm wide, rounded, gen red abaxially. **FR:** beak 3–3.8(4.8) mm; aggregate 5–7 mm, hemispheric, occ spheric. Meadows, marshy areas; < 2200 m. NW, CaRH, SNH, MP; to AK. Mar–Aug

var. *platyphyllus* A. Gray **LF:** basal lvs ovate to semicircular, 1-ternate or -pinnate, lflets 3–5, 1 × lobed or parted, segments narrowly elliptic to linear, dentate, obtuse to narrowly acute. **FL:** petals 7–11 mm wide, rounded, yellow abaxially. **FR:** beak 2–3 mm; aggregate 8–13 mm, spheric or ovoid. Meadows, marshy areas; < 2100 m. NW, CCo, SnFrB, CaR, GV; to WA, WY. Mar–Jul

R. parviflorus L. Ann 8–40 cm, ± erect, not bulbous-based; roots basal, not tuberous. **LF:** basal, proximal cauline 1.5–3.2 cm, 1–2.4 cm wide, semicircular or reniform, 3-parted or -divided, again lobed and dentate, base cordate, tip rounded. **FL:** receptacle glabrous; sepals 5, reflexed, 1.5–2 mm, 0.8–1.2 mm wide, early-deciduous; petals 0–5, 1.1–1.8 mm, 0.2–0.7 mm wide. **FR:** body 1.7–2 mm, 3–5 mm wide, disk-like, wall finely papillate on faces, not margins, each papilla tipped with hooked bristle, beak 0.4–0.6 mm, deltate, tip slender, recurved. Roadsides, fields; < 800 m. NW, SnFrB; e US, Australia; native to Eur. Apr–Jun

R. populago Greene (p. 1151) Per 8–30 cm, erect or ascending, not rooting at nodes. **LF:** basal, lower cauline 1.2–5.1 cm, 1.5–2.9 cm wide, semicircular to cordate or ovate, bases cordate to obtuse, margin entire or crenate, tip widely acute to rounded. **INFL:** bracts narrowly elliptic to ovate or lanceolate. **FL:** receptacle glabrous or bristly; sepals 4–5, spreading or reflexed from base, 3–5 mm, 2–4

mm wide, green, early-deciduous; petals 5–6, 4–9 mm, 2–5 mm wide. **FR:** body 1.6–1.8 mm, 1.2 mm wide, lenticular, wall thick, smooth, beak 0.2–1 mm, straight, awl-shaped. Wet ground, shallow water, wet meadows, streams; 1500–2500 m. CaR, n SNH; to WA, MT. Jun–Jul

R. pusillus Poir. (p. 1151) Per 8–50 cm, erect or ascending, rooting at most proximal nodes. **LF:** simple, unlobed, lower cauline 1.2–4.2 cm, 0.5–1.2 cm wide, ovate or lanceolate, base acute to truncate, margin entire or minutely dentate, tip acuminate to rounded. **INFL:** bracts linear to lanceolate or oblanceolate. **FL:** receptacle glabrous; sepals 4–5, spreading or reflexed from base, 1.5–3 mm, 1–1.5 mm wide, early-deciduous; petals 1–3, 1.5–2 mm, 0.5–1 mm wide. **FR:** body 1–1.2 mm, 0.6–0.8 mm wide, lenticular, wall thick, smooth, beak ± rudimentary, < 0.1 mm. Ditches, streams, wet meadows, fields; < 500 m. NCo, NCoRO, ScV, c SNF, CCo; e US. Apr–Jun

R. repens L. Per 10–60 cm, decumbent or creeping, rooting at nodes. **LF:** basal, proximal cauline 1–8.5 cm, 1.5–10 cm wide, ovate to reniform, 1-ternate, lflets lobed, parted, or parted and again lobed, ultimate segments obovate to elliptic or occ narrowly oblong, toothed, tip obtuse to acuminate; distal cauline reduced, deeply parted or compound. **FL:** receptacle bristly (glabrous); sepals 5, spreading or reflexed from base, 4–7(10) mm, 1.5–3(4) mm wide, early-deciduous; petals 5(150), 6–18 mm, 5–12 mm wide. **FR:** body 2.6–3.2 mm, 2–2.8 mm wide, disk-like, wall thick, smooth, beak 0.8–1.2 mm, curved, lanceolate to lanceolate-thread-like. Meadows, road banks, disturbed places, edges of marshes, streams; < 1600 m. NCo, NCoRI, CaRH, n&c SNH, CCo, SnFrB, MP; worldwide; native to Eurasia. All yr ❖

R. sardous Crantz Ann, bien 18–50 cm, ± erect, not bulbous-based; roots basal, not tuberous. **LF:** basal, proximal cauline lvs 2–6 cm, 2–6 cm wide, ovate to cordate, 1–2-ternate, lflets crenate-dentate, base truncate to acute, tip rounded to obtuse. **FL:** receptacle long-hairy; sepals 5, reflexed, 3–8 mm, 1.5–3 mm wide, early-deciduous; petals 5, 7–10 mm, 4–8 mm wide. **FR:** body 2–3 mm, 6–7 mm wide, disk-like, wall coarsely papillate (smooth), each papilla tipped with hooked bristle, beak 0.4–0.7 mm, curved, oblong to deltate. Ponds, wetlands; < 300 m. NCo, CaRF, n&c SNF, ScV; OR, e US, Australia; native to Eur. May–Jul

R. sceleratus L. Ann, emergent aquatic, occ on mud, 15–50 cm, erect, glabrous, gen not rooting at proximal nodes. **LF:** basal, proximal cauline 1–5 cm, 1.6–6.8 cm wide, reniform to semicircular, 3-lobed or -parted, segments gen again lobed or parted, occ undivided, base truncate to cordate, margin crenate, tip rounded or occ obtuse. **FL:** receptacle hairy or glabrous; sepals 3–4, reflexed ± from base, 2–5 mm, 1–3 mm wide, early-deciduous; petals 3–5, 2–5 mm, 1–3 mm wide; style 0. **FR:** body 1–1.2 mm, 0.8–1 mm wide, lenticular, wall thick, smooth or ± wrinkled, margins thick, corky, beak 0.1 mm, gen straight, deltate.

var. *multifidus* Nutt. **LF:** parted, gen deeply, segments lobed or parted, deeply crenate or again lobed. **FR:** faces smooth. Wet ground or shallow water; 1000–2000 m. KR, CaRH, MP; to BC, e Can. Jul–Aug

var. *sceleratus* (p. 1151) **LF:** lobed or parted, segments undivided or lobed, crenate. **FR:** faces finely transversely wrinkled. Ponds, riverbanks; < 1800 m. NCo, n SNF, GV, SnBr, PR; to BC, e US; native to Eur. Apr–Jun, Oct

R. testiculatus Crantz (p. 1151) Ann 1–6(10) cm, erect or ascending, not rooting at nodes. **LF:** basal, 0.9–3.8 cm, 0.5–1.5 cm wide, widely spoon-shaped, 1–2 × dissected, segments linear, margin entire, tip obtuse to acuminate; cauline lvs 0. **FL:** receptacle glabrous; sepals 5, spreading, 3–6 mm, 1–2 mm wide, persistent; petals 5, 3–5 mm, 1–3 mm wide. **FR:** body 1.6–2 mm, 1.8–2 mm, ellipsoid, width ± = depth, wall thick, smooth, beak 3.5–4.5 mm, lanceolate, persistent. Disturbed areas, esp in grassland; 1000–2300 m. s SNF, Teh, SnBr, GB, DMtns; to OH; native to Eurasia. Apr–May

R. uncinatus D. Don (p. 1151) Per (occ fl 1st yr) 15–60 cm, erect, not rooting at nodes. **LF:** basal, proximal cauline 1.8–5.6 cm, 2.8–8.3 cm wide, cordate to reniform, 3-parted, occ 1-ternate, segments again lobed, ultimate segments elliptic to lanceolate, toothed or crenate-toothed, tip acute to rounded-obtuse; distal cauline lvs reduced, deeply parted or compound. **FL:** receptacle glabrous; sepals 5,

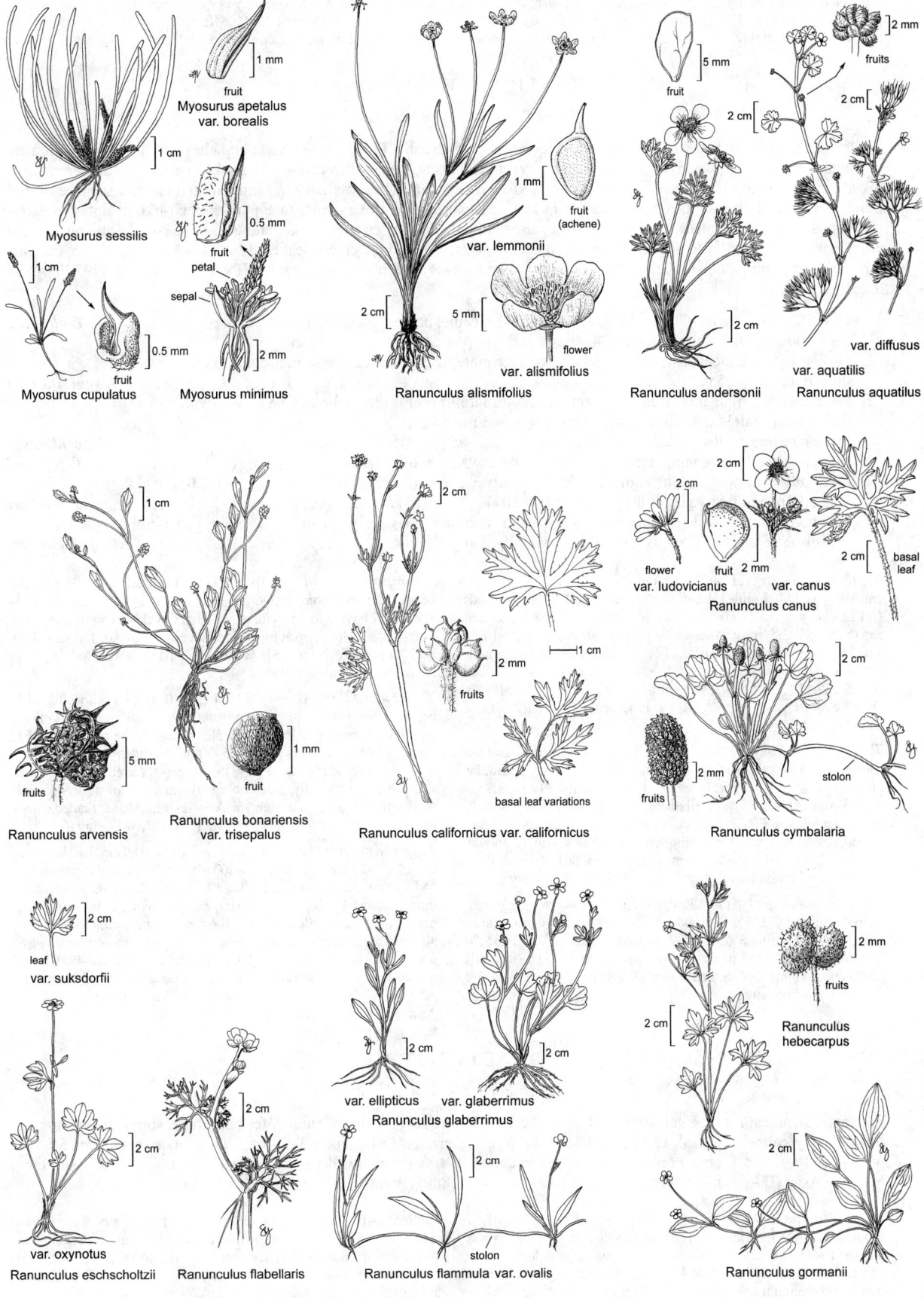

Myosurus sessilis

fruit
Myosurus apetalus
var. borealis

fruit

petal
sepal
fruit
Myosurus cupulatus

Myosurus minimus

var. lemmonii

fruit
(achene)

flower
var. alismifolius
Ranunculus alismifolius

fruit

fruits

var. diffusus
var. aquatilis

Ranunculus andersonii

Ranunculus aquatilus

fruits
Ranunculus arvensis

fruit
Ranunculus bonariensis
var. trisepalus

fruits

basal leaf variations
Ranunculus californicus var. californicus

flower
var. ludovicianus

fruit
var. canus

basal
leaf
Ranunculus canus

fruits

stolon
Ranunculus cymbalaria

fruits
Ranunculus
hebecarpus

leaf
var. suksdorfii

var. oxynotus
Ranunculus eschscholtzii

Ranunculus flabellaris

var. ellipticus   var. glaberrimus
Ranunculus glaberrimus

stolon
Ranunculus flammula var. ovalis

Ranunculus gormanii

reflexed or occ spreading, 2–3.5 mm, 1–2 mm wide, early-deciduous; petals 5, 2–4(6) mm, 1–2(3) mm wide. **FR:** body 2–2.8 mm, 1.6–2 mm, disk-like, wall thick, smooth, beak 1.2–2.5 mm, curved, hooked, lanceolate. Moist meadows or woodland, gen along streams; < 2800 m. NW, CCo, CaR, n&c SNH, MP, SnBr; to AK, NM. [*R. u.* var. *parviflorus* (Torr.) L.D. Benson] Apr–Jul

THALICTRUM MEADOW-RUE

Bruce D. Parfitt & Dieter H. Wilken

Per from caudex or rhizomes, gen glabrous; dioecious or fls bisexual. **ST:** 1–few, gen erect; branches 0 or few. **LF:** 1–4-ternate or pinnate, basal or basal and cauline, alternate, gen reduced distally on st; lflets wedge-shaped to ± round, entire, crenate, or lobed; pale green abaxially, gen green adaxially. **INFL:** raceme or panicle, axillary or terminal, gen erect, ± scapose or not; bracts simple to 1-ternate; pedicels gen erect in fr. **FL:** sepals 4–5, ± green-white to ± purple, petal-like or not, gen early-deciduous; petals 0; stamens 8–many, gen > sepals, filaments flat or gen thread-like, anthers gen narrowly oblong, tip gen pointed; pistils (1)2–22. **FR:** achene, compressed laterally or not, ribbed or veined, beaked. 120–200 spp.: temp N.Am, Eurasia, Afr; some orn, medicinal. (Greek: name given by Dioscorides, Greek physician-botanist) [Park & Festerling 1997 FNANM 3:258–271]

1. Pl 5–15(20) cm; lvs basal (1 cauline), 1.5–6 cm; infl raceme; pedicel recurved in fr . . . . ***T. alpinum***
1′ Pl 40–200 cm; lvs basal and cauline, 4–46 cm; infl gen panicle; pedicel ± erect in fr
  2. Fls bisexual; filaments flat; anthers obtuse to mucronate; fr body sides gen semicircular to crescent-shaped, beak 1–1.5 mm . . . . ***T. sparsiflorum***
  2′ Fls unisexual; filaments thread-like; anthers acuminate; fr body sides obliquely narrowly ovate to fusiform or widely obovate to ± round, beak 1.5–4.5 mm
    3. Fr gen reflexed, ribs ± straight; n NCo, nw KR . . . . ***T. occidentale***
    3′ Fr spreading to ascending, ribs ± curved; n NCo, nw KR, and elsewhere . . . . ***T. fendleri***
      4. Fr body ± compressed throughout, sides obliquely ± ovate to ± obovate, 2–3-ribbed, veins 0; distal lvs abaxially gen finely glandular-puberulent (at 20×) . . . . var. ***fendleri***
      4′ Fr body compressed gen only near margins, sides obliquely ± widely obovate to ± round, gen 1-ribbed, veins several, wavy, net-like; distal lvs abaxially gen glabrous . . . . var. ***polycarpum***

T. alpinum L. (p. 1151) ARCTIC MEADOW-RUE Pl 5–15(20) cm. **LF:** basal (1 cauline), 1.5–6 cm; segments 4–10 mm, glabrous. **INFL:** raceme, ± scapose; pedicel recurved in fr. **FL:** bisexual; sepals 5, 1.5–2.5 mm; stamens 7–12(15), filaments thread-like, anthers linear-oblong, mucronate. **FR:** 1–6; body 2–3.5 mm, ± compressed laterally, sides obliquely ± ovate to ± obovate, 2–3-ribbed, curved. 2*n*=14. Meadows, gen moist, gravelly soils; 2900–3700 m. n&c SNH, n W&I; to AK, NM; Greenland, arctic Eurasia. Jun–Aug ★

T. fendleri A. Gray Pl 60–200 cm; gen dioecious. **LF:** basal and cauline, 7–46 cm; segments 8–20 mm, glabrous to finely glandular-puberulent. **INFL:** panicle, bracts lf-like proximally. **FL:** sepals gen 4, 2–5 mm; stamens 15–28, filaments thread-like, anthers oblong-linear, acuminate. **FR:** 7–20, spreading to ascending; body 4–8 mm, ± compressed laterally, sides obliquely ± ovate or ± widely obovate to ± round, ribs 1–3, ± curved, veins 0–several, beak 1.5–4 mm. Pls in NCoR occ with bisexual fls; vars. in CA difficult, need study.

  var. ***fendleri*** (p. 1151) **LF:** distal abaxially gen finely glandular-puberulent (at 20×). **FR:** body ± compressed throughout, sides obliquely ± ovate to ± obovate, 2–3-ribbed, veins 0. 2*n*=28,56,70. Moist, open to shaded places, woodland, forest; 500–3200 m. KR, CaR (uncommon), SN, SnFrB, SCoRO, TR, PR, GB; to OR, WY, TX, n Mex. May–Aug

var. ***polycarpum*** Torr. (p. 1151) **LF:** distal abaxially gen glabrous. **FR:** body compressed gen only near margins, sides obliquely ± widely obovate to ± round, gen 1-ribbed, veins several, wavy, net-like. 2*n*=28. Moist, open to shaded places, woodland, forest; < 2600 m. NCo, w KR, NCoR, SN (± rare), CW, TR, w PR; to WA. [*T. p.* (Torr.) S. Watson] Mar–Jun

T. occidentale A. Gray (p. 1151) Pl 40–100 cm; dioecious. **LF:** basal and cauline, 6–40 cm; segments 11–35 mm, glabrous to finely glandular-puberulent. **INFL:** panicle, bracts lf-like proximally. **FL:** sepals gen 4, 2–5 mm; stamens 15–30, filaments thread-like, anthers oblong-linear, acuminate. **FR:** (3)6–14, gen reflexed; body 4–6 mm, ± compressed laterally, sides obliquely narrowly ovate to fusiform, ribs gen 3, ± straight, beak 3–4.5 mm. Moist, shaded places, conifer forest; < 200 m. n NCo, nw KR; to BC, WY. May–Jul

T. sparsiflorum Fisch. & C.A. Mey. Pl 60–180 cm. **LF:** basal and cauline, 4–30 cm; segments 12–20 mm, finely glandular-puberulent. **INFL:** panicle, bracts gen lf-like. **FL:** bisexual; sepals gen 5, 2.5–4 mm; stamens 10–20, filaments flat, anthers ovoid, obtuse to mucronate. **FR:** 6–22, ± reflexed; body 4–6 mm, strongly compressed laterally, sides gen semicircular to crescent-shaped, ribs or veins 3–4(5), weakly defined, beak 1–1.5 mm, pedicel ascending. 2*n*=42. Uncommon. Moist places, streambanks, conifer forest; 1400–3500 m. CaR, n&c SN, SnBr, SnJt, SNE (Sweetwater, White mtns); to AK, CO; Asia. Jul–Aug

TRAUTVETTERIA

Bruce D. Parfitt & Dieter H. Wilken

Per from rhizomes; roots gen clustered. **ST:** gen 1, erect, simple. **LF:** basal and cauline, few, alternate, simple, round to reniform, deeply palmately lobed. **INFL:** panicle, ± flat-topped, terminal. **FL:** sepals 3–7, petal-like; petals 0; stamens 50–100; pistils 10–16, ovule 1, style persistent, ± hooked or coiled. **FR:** ± utricle; wall papery, shiny, veined or ribbed. 1 sp.: temp N.Am, e Asia. (E.R. von Trautvetter, Russian botanist, 1809–1889) [Parfitt 1997 FNANM 3: 138–139]

T. caroliniensis (Walter) Vail (p. 1151) Pl 3–1.5 m. **ST:** gen glabrous. **LF:** basal 1–2, lobes 5–11, ± wedge-shaped, toothed distally, petiole 15–45 cm; cauline reduced distally on st, petiole 0–15 cm. **INFL:** bracts < 2 cm; pedicels 4–10 mm; fls ≥ 5. **FL:** sepals 2.5–6 mm, early-deciduous, blade widely ovate, cup-like, ± green-white; stamens 5–10 mm, outer filaments flat, width > anthers, inner thread-like. **FR:** 2.5–4.5 mm, 4-angled. Moist, shaded places, streambanks; 640–2100 m. KR, CaR (uncommon), SNH; to BC, MT, WY, NM; also e US; e Asia. [*T. c.* var. *occidentalis* (A. Gray) C.L. Hitchc.] Jul–Aug

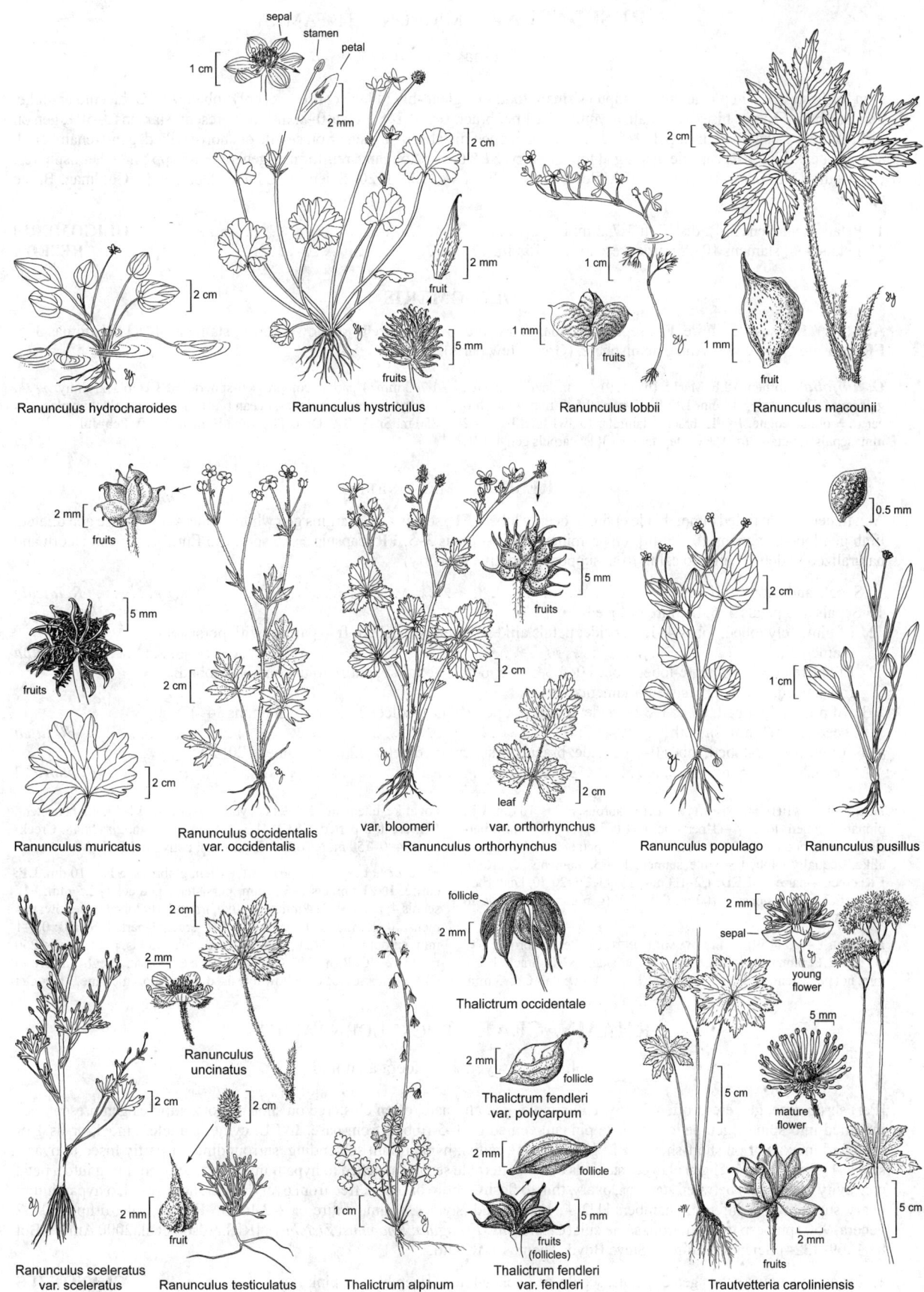

Ranunculus hydrocharoides

Ranunculus hystriculus

Ranunculus lobbii

Ranunculus macounii

Ranunculus muricatus

Ranunculus occidentalis
var. occidentalis

var. bloomeri   var. orthorhynchus
Ranunculus orthorhynchus

Ranunculus populago

Ranunculus pusillus

Ranunculus sceleratus
var. sceleratus

Ranunculus testiculatus

Ranunculus
uncinatus

Thalictrum alpinum

Thalictrum occidentale

Thalictrum fendleri
var. polycarpum

Thalictrum fendleri
var. fendleri

Trautvetteria caroliniensis

RESEDACEAE　MIGNONETTE FAMILY

Thomas F. Daniel

Ann to shrub. **LF**: simple, alternate; stipules small, tooth- or gland-like; blade entire to deeply lobed. **INFL**: raceme or spike, gen terminal. **FL**: gen bisexual, small, asymmetric, 1 per bract; sepals 2–8; petals 0–8; disk occ present; stamens 3–50+, gen on disk, anthers 2-chambered; carpels 2–8, gen fused, gen open distally, ovary superior, sessile or short-stalked, gen 1-chambered, stigmas beak-like. **FR**: capsule, gaping at top, or berry. **SEED**: few to many, reniform. 6 genera, ± 85 spp.: n&e hemispheres, esp Medit. [Martín-Bravo et al. 2007 Molec Phylogen Evol 44:1105–1120] Scientific Editors: Douglas H. Goldman, Bruce G. Baldwin.

1. Petals 2; stamens 3(4); disk 0; fr 1.7–3 mm . . . . **OLIGOMERIS**
1′ Petals 4–6; stamens 10–25; disk present; fr 3–15 mm . . . . **RESEDA**

OLIGOMERIS

Ann (per). **LF**: sessile, entire. **FL**: sepals 2–6, margins white; petals entire to shallowly lobed; stamens 3(4)[12]; stigmas 3–5. **FR**: capsule. 3 spp.: w N.Am, e hemisphere. (Greek: few parts)

O. linifolia (Hornem.) J.F. Macbr. (p. 1159)　Ann (per), ± fleshy, glabrous. **ST**: erect, < 45 cm. **LF**: 8–45 mm, 0.5–2 mm wide, linear to ± oblanceolate. **INFL**: bracts triangular to awl-like. **FL**: 1–2 mm; sepals 4; petals 2(>2), ± white; stamens 3(4); carpels gen 4. **FR**: 1.7–3 mm, 4-parted, compressed-spheric. **SEED**: black, shiny. *n*=24. Rocky slopes, open dunes, ocean bluffs, roadsides, alkaline places; < 850 m. SnJV, SW, D; to TX, Mex; Eurasia, n Afr. Feb–Jul

RESEDA　MIGNONETTE

Ann to per. **LF**: petioled or not; blade entire to deeply lobed. **FL**: sepals 4–8, margins gen white; petals 4–6[8], base gen dilated, limb gen lobed; stamens 10–25[40], on prominent disk; stigmas 3–5. **FR**: capsule. ± 65 spp.: esp Eur, Medit; cult as orn and naturalized widely. (Latin: to calm, from supposed sedative property)

1. Sepals and petals 4; lf entire . . . . ***R. luteola***
1′ Sepals and petals 5–6; lf entire or lobed
　2. Lf pinnately lobed, lobes 4–11 per side; petals alike, ± equally 3-lobed; fr 4-parted, with persistent
　　filaments . . . . ***R. alba***
　2′ Lf entire or 3- or pinnate-lobed, lobes 0–3 per side; petals not alike, entire to unequally 6-lobed;
　　fr 3-parted, without persistent filaments
　　3. Lf pinnately lobed, lobes 1–3 per side; bracts not persistent; pedicel 2–4.5 mm; stamens 14–17;
　　　seed 1.5–1.7 mm, smooth . . . . ***R. lutea***
　　3′ Lf entire to 3-lobed, lobes 0–1 per side; bracts persistent; pedicel 4–12 mm; stamens 20–25;
　　　seed 2–2.2 mm, rough . . . . [***R. odorata***]

R. alba L.　WHITE MIGNONETTE　Per, glabrous. **ST**: < 10 dm. **LF**: pinnately lobed; lobes 4–11 per side. **INFL**: bracts 3.5–9 mm, persistent in fr; pedicel 2–8 mm. **FL**: sepals 5–6; petals 5–6, 4.5–6 mm, alike, ± equally 3-lobed, ± white; stamens 11–13, filaments persistent. **FR**: erect, 4-parted. **SEED**: 1.2–1.3 mm, rough. 2*n*=20,40. Disturbed areas, fields, roadsides; < 100 m. CW, SW (gen coastal); native to Medit. May–Nov

R. lutea L. (p. 1159)　YELLOW MIGNONETTE　Per, glabrous. **ST**: < 7 dm. **LF**: pinnately-lobed; lobes 1–3 per side. **INFL**: bracts 1.5–2 mm, not persistent; pedicel 2–4.5 mm. **FL**: sepals 6; petals 6, 2–3 mm, not alike, unequally 2–3-lobed, yellow; stamens 14–17, not persistent. **FR**: erect, 3-parted. **SEED**: 1.5–1.7 mm, smooth. 2*n*=24,48. Creekbeds; 400–750 m. NCoRI (s Lake Co.); native to Medit. May–Sep

R. luteola L.　DYER'S ROCKET　Bien, glabrous. **ST**: < 10 dm. **LF**: entire. **INFL**: bracts 2–3.5 mm, persistent; pedicel 1–2.5 mm. **FL**: sepals 4; petals 4, 2–4 mm, not alike, irregularly lobed, ± yellow; stamens 20–25, filaments persistent. **FR**: erect, 3-parted. **SEED**: 0.8–1 mm, smooth. 2*n*=24,26,28. Disturbed areas, fields, roadsides; < 250 m. NCo, NCoR, n SNF, CCo, SnFrB, s SCo (San Diego); native to Old World. Source of yellow dye used from Neolithic time. May–Oct

RHAMNACEAE　BUCKTHORN FAMILY

John O. Sawyer, Jr., except as noted

[Per] shrub, tree, gen erect, often thorny. **LF**: simple, gen alternate, often clustered on short-shoots; stipules gen present, occ modified into spines; gen petioled; blade pinnate-veined or 1–5-ribbed from base. **INFL**: cyme, panicle, umbel, or fls 1 or clustered in axils or on short-shoots. **FL**: gen bisexual, radial; hypanthium subtending, surrounding, or partly fused to ovary; sepals 4–5; petals 0, 4–5, gen clawed; stamens 0, 4–5, alternate sepals, attached to hypanthium top, each gen fitting into a petal concavity; disk (0 or) between stamens, ovary, thin to fleshy, entire or lobed, free from ovary, adherent or fused to hypanthium; ovary superior or ± inferior, chambers [1]2–4, 1–2-ovuled, style 1, stigma entire or 2–3-lobed. **FR**: capsule, drupe. 50–52 genera, 950 spp.: esp trop, subtrop; some cult (*Ceanothus*; *Frangula*; *Rhamnus*; *Ziziphus*). [Richardson et al. 2000 Amer J Bot 87:1309–1324] Scientific Editors: Steve Boyd, Thomas J. Rosatti.

1. Fls conspicuous; petals gen 5; petals, sepals, pedicels white to deep blue or pink . . . . **CEANOTHUS**
1′ Fls inconspicuous; petals 0–5, gen green or white; sepals, pedicels white, green, or gray

2. Fr a capsule

  3. Branches opposite, green; twigs jointed at base; SCo, PR . . . . **ADOLPHIA**

  3′ Branches alternate, ± white; twigs not jointed at base; s DMoj, DSon. . . . . **COLUBRINA**

2′ Fr a drupe

  4. Fr of 2–4 separate stones; widespread

    5. Winter bud scales 0; petals 5; sepals fleshy, erect, keeled adaxially. . . . . **FRANGULA**

    5′ Winter bud scales present; petals 0 or 4–5; sepals thin, spreading, not keeled adaxially . . . . **RHAMNUS**

  4′ Fr of 1 stone; D, e PR, SCo

    6. Petals 0; disk 0 or thin in early fl . . . . **CONDALIA**

    6′ Petals 5; disk thick, ± obscuring ovary in early fl . . . . **ZIZIPHUS**

ADOLPHIA

Shrub. **ST**: branches dense, opposite, rigid, green; twigs spreading, jointed at base, thorn-tipped, glabrous to puberulent. **LF**: opposite, clustered in axils or not, early-deciduous; stipules ± triangular, black, shiny; blade entire, 1–3-ribbed from base. **INFL**: small axillary clusters; pedicel gen white to green-white. **FL**: hypanthium bowl-shaped, gen white to green-white; sepals 5, triangular, persistent; petals 5, spoon-shaped, white; stamens 5; disk fused to hypanthium, 5-angled, fleshy, gen white to green-white; ovary superior, chambers 3, 1-ovuled, stigma 3-lobed. **FR**: capsule, spheric, explosively dehiscent. 2 spp.: CA, TX, Mex. (Adolphe T. Brongniart, student of Rhamnaceae, 1801–1876)

A. californica S. Watson (p. 1159) CALIFORNIA ADOLPHIA Pl < 1.5 m. **ST**: twigs stout. **LF**: petiole 2–4 mm; blade, 3–12 mm, elliptic-oblong to obovate, base wedge-shaped, tip obtuse to acute. **FL**: hypanthium 3 mm wide; petals 2 mm. **FR**: ± 5 mm. Coastal scrub, chaparral; < 400 m. SCo, PR; Mex. Dec–Apr ★

CEANOTHUS CALIFORNIA-LILAC

Dieter H. Wilken

Shrub, tree-like or not, gen erect or mat- to mound-like. **ST**: branches gen arranged like lvs; twigs thorn-like or not, gen not angled. **LF**: alternate or opposite, some clustered on short-shoots or not, deciduous or evergreen; stipules scale-like, thin, deciduous, or knob-like, corky, thick, base persistent; blade flat or wavy, tip gen acute to obtuse, margin thick (i.e., thicker than adjacent blade) or not, rolled under or not, wavy or not, entire or gland- or sharp-toothed, glands gen dark, teeth pale, alternate blade 1–3-ribbed from base, gen thin, opposite blade 1-ribbed from base, thick, firm. **INFL**: umbel-, raceme-, or panicle-like aggregations of few-fld clusters, axillary or terminal; pedicels white to deep blue or pink. **FL**: conspicuous, gen < 5 mm; hypanthium surrounding fleshy disk below ovary base, in fr thick, not splitting; sepals gen 5, lance-deltate, incurved, colored like petals, persistent; petals gen 5, blade hood-like, white to deep blue or pink; stamens gen 5, opposite petals; ovary 1/2-inferior, 3-lobed, chambers 3, each 1-ovuled, styles 3. **FR**: capsule, ± spheric, gen ± 3-lobed, gen smooth, 3-ridged or not, horned or not. **SEED**: 3, 2–5 mm. ± 55 spp.: N.Am. (Greek: thorny pl) [Fross & Wilken 2006 *Ceanothus*. Timber Press] Hybrids common (named hybrids not recognized here), discussed in Fross & Wilken; hybrid forms do not key easily. As recircumscribed here, *C. greggii* A. Gray restricted to Mex.

1. Stipules knob-like, corky, thick, bases persistent; lvs gen opposite (gen alternate in *Ceanothus megacarpus* var. *megacarpus*, *Ceanothus verrucosus*), clustered or not; infl gen umbel-like (subg. *Cerastes*)

  2. Lvs gen alternate

    3. Lf blade 10–25 mm, 5–12 mm wide, length gen > 2 × width, elliptic to widely oblanceolate, margin entire; fr 7–12 mm wide. . . . . ***C. megacarpus*** var. ***megacarpus***

    3′ Lf blade 5–14 mm, 3–10 mm wide, length gen < 2 × width, widely obovate to round, margin entire or teeth ± sharp; fr 4–6 mm wide . . . . ***C. verrucosus***

  2′ Lvs gen opposite, occ clustered

    4. Pl mat- to mound-like, gen < 1 m; sts spreading, ± ascending, occ arched, rooting at nodes or not

      5. Lf margin gen entire, occ ± 3–5-toothed distally

        6. Lf blade 8–20 mm, gray-tomentose abaxially, margin thick or ± rolled under; CCo (nw San Luis Obispo Co.), < 60 m . . . . ***C. maritimus***

        6′ Lf blade gen 4–12 mm, ± glabrous to strigose abaxially, margin not thick, ± not rolled under; KR, n NCoR, n&c SN, > 260 m

          7. Lvs gen clustered, gen erect . . . . [2]***C. roderickii***

          7′ Lvs ± evenly spaced, gen not clustered, spreading to ascending

            8. Pl gen mound-like, gen 0.3–1 m, sts not rooting at nodes; twigs brown to gray-brown. . . . . ***C. arcuatus***

            8′ Pl mat-like, < 0.3 m, sts rooting at nodes; twigs gen red-brown . . . . ***C. fresnensis***

      5′ Lf blade margin 2–35-toothed

        9. Lf blade gen wide-elliptic or -obovate or round, length gen < 2 × width;, margin 5–35-toothed

          10. Twigs gen not angled; fls white to pale blue or blue-lavender

            11. Lf blade 10–20 mm; fr 6–9 mm wide; horns 1–2 mm, wrinkled; KR (Trinity Mtns), s SNH (Kern Plateau), 1050–2750 m . . . . ***C. pinetorum***

            11′ Lf blade 4–10 mm; fr 4–6 mm wide; horns < 1 mm, smooth; CCo, < 400 m. . . . . [2]***C. rigidus***

          10′ Twigs gen angled distally; fls blue to blue-purple . . . . ***C. gloriosus***

            12. Pl < 0.3 m; sts gen spreading; lf blade 23–31 mm, margin 13–35-toothed . . . . var. ***gloriosus***

    12′ Pl < 0.5 m; sts spreading to ± ascending; lf blade 10–21 mm, margin 9–19-toothed . . . . var. ***porrectus***

  9′ Lf blade gen elliptic to obovate or oblanceolate, length gen > 2 × width; margin 2–11-toothed

    13. Lf blade 2–6 mm wide, margin 0–5-toothed distally

      14. Lvs not clustered, spreading, tips gen truncate, margin gen 2–3-toothed; KR, NCoR. . . . . ***C. pumilus***

      14′ Lvs gen clustered, gen erect, tips obtuse to ± notched, margin 0–5-toothed; n SNF

        (w El Dorado Co.) . . . . ²***C. roderickii***

    13′ Lf blade 4–16 mm wide, margin gen 2–11-toothed distally

      15. Lf blade gen not ± wavy, gen not ± folded lengthwise; fr 6–9 mm wide, horns 1–2 mm, thick,

        wrinkled (slender in *Ceanothus prostratus* var. *occidentalis*). . . . . ***C. prostratus***

        16. Pl mound-like; lf blade ± folded lengthwise, concave adaxially, teeth 5–9, in distal 1/2;

          fr horns spreading . . . . var. ***occidentalis***

        16′ Pl gen mat-like; lf blade ± not folded lengthwise, not concave adaxially, teeth 3–5(7),

          in distal 2/3; fr horns erect . . . . var. ***prostratus***

      15′ Lf blade gen ± wavy or ± folded; fr 4–6 mm wide, horns 1.5–2.5 mm, ± slender, smooth

        17. Pl mound-like, 0.2–0.5 m, sts spreading to ± ascending, occ rooting at nodes . . . . ***C. confusus***

        17′ Pl erect to mound-like, 0.5–1.5 m, sts ascending to ± erect, gen not rooting at nodes . . . . ²***C. divergens***

  4′ Pl gen erect, 1–4+ m; sts ascending to erect, occ intricately branched, not rooting at nodes

    18. Lf blade ± wavy to ± folded lengthwise, margin teeth spine-like

      19. Lf blade elliptic to narrowly obovate . . . . ²***C. divergens***

      19′ Lf blade widely elliptic to ± round

        20. Lvs ± spreading, margin 3–5-toothed . . . . ***C. sonomensis***

        20′ Lvs gen reflexed, margin 7–15-toothed

          21. Sepals, petals 6–8; fr 5–7 mm, wrinkled, horns thick; fls white to blue; gen serpentine . . . . ***C. jepsonii***

          21′ Sepals, petals 5; fr 4–5 mm, smooth, horns slender; fls dark blue to purple; volcanic substrates. . . . ***C. purpureus***

    18′ Lf blade gen flat or convex to concave adaxially, margin entire or teeth ± sharp, not spine-like

      22. Lf blade densely tomentose abaxially, veins obscure, margin gen rolled under . . . . ***C. crassifolius*** var. ***crassifolius***

      22′ Lf blade gen glabrous to strigose or puberulent abaxially (sparsely tomentose in *Ceanothus*

        crassifolius var. *planus*, ± tomentose in *Ceanothus otayensis*), veins gen evident, margin not rolled

        under (exc *Ceanothus otayensis*)

        23. Lf margin gen entire, or teeth < 5, unevenly distributed

          24. Lf blade 3–7 mm, 1–3 mm wide, narrow-oblanceolate to -obovate; infl gen < 1 cm; fls gen pale

            blue, occ ± pink. . . . . ***C. ophiochilus***

          24′ Lf blade 5–30 mm, 3–19 mm wide, elliptic, oblanceolate, obovate, or ± round; infl gen 1–2 cm;

            fls gen white (pale blue to lavender in some *Ceanothus cuneatus*)

            25. Fl disk dark; fr 5–12 mm wide

              26. Lf tip obtuse to rounded; fr 5–9 mm wide; s SCoRO, WTR (Santa Barbara, Ventura cos.)

                . . . . ***C. crassifolius*** var. ***planus***

             26′ Lf tip truncate to ± notched; fr 7–12 mm wide; ChI . . . . ***C. megacarpus*** var. ***insularis***

            25′ Fl disk gen not dark, white to pale blue or lavender; fr 3–6 mm wide

             27. Twigs gen dense-puberulent to short-tomentose, pale gray to ± white; lf blade short-curly-

               puberulent abaxially . . . . ²***C. vestitus***

             27′ Twigs gen glabrous to ± puberulent, gen brown to gray-brown, lf blade glabrous to ± strigose

               abaxially . . . . ***C. cuneatus***

               28. Fl gen white; twigs gen gray-brown; lf blade tip acute to ± rounded. . . . . var. ***cuneatus***

               28′ Fl gen blue to lavender; twigs gen brown; lf blade tip truncate to notched

                 29. Lvs at nodes and in axillary clusters, node lvs widely obovate, clustered lvs oblanceolate

                 to narrowly obovate . . . . var. ***fascicularis***

               29′ Lvs at nodes, blades widely obovate to ± round . . . . var. ***ramulosus***

      23′ Lf margin gen dentate to ± serrate, teeth gen 5–35 (occ < 5 in *Ceanothus vestitus*), ± evenly

        distributed, or in distal 1/2, occ minute

        30. Twigs gen angled, esp distally

          31. Lf blade 13–45 mm, widely obovate to ± round, margin 13–35-toothed. . . . . ***C. gloriosus*** var. ***exaltatus***

          31′ Lf blade 7–21 mm, obovate to narrowly obovate, margin 9–17-toothed . . . . ***C. masonii***

        30′ Twigs gen not angled

          32. Lf blade gen 10–25+ mm

            33. Lf blade dark green, obovate to widely elliptic; fr 6–9 mm wide, ± wrinkled, horns 1–2 mm;

             se SnFrB (Santa Clara Co.) . . . . ***C. ferrisiae***

            33′ Lf blade green to yellow-green, widely elliptic to ± round; fr 3–5 mm wide, gen smooth or

             ± 3-bulged distally, horns < 1 mm or 0; e TR, PR . . . . ***C. perplexans***

        32′ Lf blade gen ≤ 12 mm (exc some *Ceanothus vestitus* 5–20 mm)

          34. Lf blade ± tomentose abaxially, glabrous in age, veins obscure, margin thick or rolled under —

            s PR (Otay, San Miguel mtns) . . . . ***C. otayensis***

          34′ Lf blade glabrous to short-strigose or short-curly-puberulent abaxially, occ glabrous in age,

            veins evident, margin gen not rolled under, occ thick

            35. Fls gen blue to lavender (white), lf tip rounded to truncate, margin entire or 5–9-toothed. . . . . ²***C. rigidus***

            35′ Fls gen white; lf tip gen acute to obtuse, margin entire or 3–5-toothed . . . . ²***C. vestitus***

1′ Stipules scale-like, thin, deciduous; lvs alternate; infl gen raceme- to panicle-, occ ± umbel-like (subg. *Ceanothus*)
  36. Twigs rigid, gen thorn-like
    37. Lf blade 1-ribbed from base, shiny green adaxially; twigs green. . . . . ²*C. spinosus*
    37′ Lf blade 3-ribbed from base, dull green adaxially; twigs pale gray to gray-green
      38. Pl gen < 1.5 m; sts gen spreading; infl 1–4 cm, gen sessile . . . . *C. cordulatus*
      38′ Pl gen 1.5–4 m; sts ascending to erect; infl 3–15 cm, gen stalked
        39. Lf blade wide-elliptic to -ovate, 20–50+ mm, 10–45 mm wide; fr coarse-wrinkled when dry . . . . ²*C. incanus*
        39′ Lf blade elliptic to ovate, 12–30 mm, 3–15 mm wide; fr smooth to ± wrinkled when dry . . . . *C. leucodermis*
  36′ Twigs flexible, not thorn-like
    40. Lf blade gen 1-ribbed from base
      41. Pl mat- to mound-like, gen < 0.5 m; st spreading to ascending
        42. Lf margin rolled under; blade tomentose abaxially, tip truncate to notched . . . . *C. hearstiorum*
        42′ Lf margin not rolled under; blade glabrous, puberulent, or short-wavy-hairy abaxially, tip acute to ± rounded
          43. Petiole 1–3 mm; lf blade ± wavy, ± folded lengthwise, glabrous abaxially exc veins;
            margin entire to few-toothed distally. . . . . *C. foliosus* var. *vineatus*
          43′ Petiole 3–11 mm; lf blade gen not ± wavy, not ± folded lengthwise, puberulent to tomentose
            abaxially; margin 27–45-toothed
            44. Pl gen mat-like, gen < 0.3 m, rooting at nodes; infls on short, erect twigs; twigs green,
              occ tinged red . . . . ²*C. diversifolius*
            44′ Pl erect to mound-like, 0.5–1 m, not rooting at nodes; infls not on short, erect twigs; twigs
              pale green to gray-green . . . . ³*C. lemmonii*
      41′ Pl gen erect, occ tree-like, > 0.5 m (occ mound-like in *Ceanothus papillosus*); st ascending to erect
        45. Lf blade margin thick or rolled under
          46. Lf blade widely ovate to ± round, furrowed along veins adaxially, tip acute to obtuse. . . . . *C. impressus*
            47. Lf blade 5–14 mm; pl 0.5–1.5 m, gen dense. . . . . var. *impressus*
            47′ Lf blade 11–25 mm; pl 1–3 m, gen open . . . . var. *nipomensis*
          46′ Lf blade ± oblong or narrowly elliptic, not furrowed adaxially, tip truncate to notched
            48. Pl gen 0.5–1.5 m; lf blade 4–16 mm, 2–8 mm wide, not glandular-papillate adaxially, glandular-
             papillate near margin . . . . *C. dentatus*
            48′ Pl 1–3.5 m; lf blade 11–50 mm, 6–15 mm wide, glandular-papillate adaxially and on margin . . . . . *C. papillosus*
        45′ Lf blade margin gen not thick, gen not rolled under (partly rolled under in *Ceanothus parryi*)
          49. Lf margin gen gland-toothed, glands gen dark, ± persistent
            50. Lf margin ± rolled under, esp below middle, occ minutely glandular; blade cobwebby to
             loose-tomentose abaxially . . . . ²*C. parryi*
            50′ Lf margin entire or minutely gland-toothed; blade glabrous to puberulent abaxially
              51. Twigs pale green to gray-green; petiole 3–6 mm; lf blade flat. . . . . ³*C. lemmonii*
              51′ Twigs green to red-green; petiole 1–3 mm; lf blade gen wavy, ± folded lengthwise . . . . ²*C. foliosus*
                52. Lf blade ± glabrous to sparsely puberulent abaxially; pls ± open . . . . ²var. *foliosus*
                52′ Lf blade densely short-wavy-hairy abaxially; pls ± dense. . . . . ²var. *medius*
          49′ Lf margin gen entire (or minutely gland-toothed, glands not dark, not persistent)
           53. Pl gen < 1.5 m, sts ± spreading to gen ascending; lf blade tip obtuse; infl 4–9 cm, raceme- to
            panicle-like; SNH . . . . ²*C. parvifolius*
           53′ Pl gen 1–6 m, occ tree-like, sts ascending to erect; lf blade tip acute to obtuse; infl 4–21 cm,
            gen panicle-like; not SNH
             54. Fls blue to pale blue; lf blade gen dark shiny green adaxially — s SCoRO, WTR, PR . . . . ²*C. spinosus*
             54′ Fls white; lf blade gen dull to light green adaxially, occ shiny
               55. Lf blade elliptic to ± oblong, thin; fr 4–6.5 mm wide; sw SnFrB (Santa Cruz Mtns)
                . . . . *C. integerrimus* var. *integerrimus*
               55′ Lf blade elliptic to oblong-ovate, ± thick; fr 6–9 mm wide; n&c SNF, s SCoR, TR, PR . . . . *C. palmeri*
    40′ Lf blade gen 3-ribbed from base
      56. Lf margin entire (minutely toothed)
        57. Main st gen ascending to ± spreading; lf blade 3–12 mm wide, shiny green adaxially. . . . . ²*C. parvifolius*
        57′ Main st ascending to erect; lf blade 10–45 mm wide, dull green adaxially
          58. Twigs rigid, pale gray to gray-green; lf blade abaxially gray-green, strigose; fr wrinkled . . . . ²*C. incanus*
          58′ Twigs flexible, gen green; lf blade abaxially pale green, glabrous to sparsely short-hairy, esp veins;
            fr smooth to bulged or ± 3-ridged distally. . . . . *C. integerrimus* var. *macrothyrsus*
      56′ Lf margin minutely toothed or serrate, teeth gen gland-tipped
        59. Lf blade gen tomentose abaxially, veins ± obscured by hairs
          60. Lf blade 25–80 mm, thick, leathery; petiole 10–25 mm; ChI. . . . . ²*C. arboreus*
          60′ Lf blade 10–25 mm, gen thin, not leathery; petiole 1–3 mm; mainland. . . . . *C. tomentosus*
        59′ Lf blade ± glabrous to puberulent or short-wavy-hairy abaxially, veins not obscured by hairs
          61. Lf blade gen 25–80 mm, 13–60 mm wide, gen widely elliptic to ovate
           62. Fls gen blue; fr 5–8 mm wide, black; ChI . . . . ²*C. arboreus*
           62′ Fls white; fr 3–4.5 mm wide, brown; mainland
             63. Lf deciduous, not aromatic; blade thin, not leathery, not sticky adaxially . . . . *C. sanguineus*
             63′ Lf evergreen, aromatic; blade leathery, ± sticky adaxially . . . . *C. velutinus*

61′ Lf blade gen 5–45 mm, 3–25 mm wide, elliptic, oblong-elliptic, oblanceolate (ovate)
   64. Pl mat- to mound-like, gen < 1 m; sts ± ascending to spreading
      65. Twigs angled distally; lf blade abaxial veins ± raised . . . . ²*C. thyrsiflorus*
         66. Twigs ± puberulent; lf blade ovate to widely elliptic, margin rolled under. . . . . ²var. *griseus*
         66′ Twigs ± glabrous; lf blade oblong-ovate to elliptic, margin not to partly rolled under. . . . . . . ²var. *thyrsiflorus*
      65′ Twigs not angled distally; lf blade abaxial veins ± not raised
         67. Pl gen mat-like, gen < 0.3 m; st rooting at nodes; twigs green, occ tinged red; fr 4–5 mm wide
          . . . . ²*C. diversifolius*
         67′ Pl ± mound-like, gen > 0.5 m, st not rooting at nodes; twigs pale green to gray-green;
          fr 3–4 mm wide. . . . . ³*C. lemmonii*
   64′ Pl not mat- to mound-like, gen 1–4 m; sts ascending to erect
      68. Twigs gen angled distally, internodes ridged
         69. Lf blade ± oblong to narrowly elliptic, length gen > 2 × width, cobwebby to loose-tomentose
          abaxially, glabrous in age . . . . ²*C. parryi*
         69′ Lf blade gen ovate to elliptic, length gen < 2 × width, glabrous to ± puberulent abaxially, esp on veins
          70. Infl 5–30 cm; twigs papillate; lf blade abaxial veins ± not raised . . . . *C. cyaneus*
          70′ Infl 1.5–7 cm; twigs not papillate; lf blade abaxial veins ± raised. . . . . ²*C. thyrsiflorus*
            71. Twigs ± puberulent; lf blade ovate to widely elliptic, margin rolled under . . . . ²var. *griseus*
            71′ Twigs ± glabrous; lf blade oblong-ovate to elliptic, margin not to partly rolled under . . . . . ²var. *thyrsiflorus*
      68′ Twigs not angled distally, internodes not ridged
         72. Lf blade flat or gen wavy, ± folded lengthwise; petiole 1–3 mm; fr 3–4 mm wide . . . . ²*C. foliosus*
          73. Lf blade glabrous to sparsely puberulent abaxially; pl ± open . . . . ²var. *foliosus*
          73′ Lf blade densely short-wavy-hairy abaxially; pl ± dense. . . . . ²var. *medius*
         72′ Lf blade gen flat; petiole 3–7 mm; fr 4–7 mm wide. . . . . *C. oliganthus*
          74. Fl disk, ovary short-hairy; fr prominently wrinkled . . . . var. *orcuttii*
          74′ Fl disk, ovary glabrous; fr gen smooth or ± ridged to bulged near top
            75. Twigs puberulent; lf blade short-hairy adaxially . . . . var. *oliganthus*
            75′ Twigs ± glabrous; lf blade glabrous to sparsely short-hairy adaxially. . . . . var. *sorediatus*

C. arboreus Greene FELTLEAF CEANOTHUS Pl < 7 m, occ tree-like. **ST**: gen erect; twigs flexible, not thorn-like, brown. **LF**: alternate, evergreen; stipules scale-like; petiole 10–25 mm; blade 25–80 mm, 20–40 mm wide, widely ovate to elliptic, thick, leathery, adaxially dark green, gen glabrous, abaxially pale green, gen short-tomentose, veins ± obscured by hairs, occ ± glabrous, margin serrate, teeth 35–65, gen gland-tipped. **INFL**: raceme- to panicle-like, 5–8 cm. **FL**: gen blue. **FR**: 5–8 mm wide, ± sticky, ± 3-ridged, black; horns 0. Slopes, chaparral, oak woodland; 60–580 m. ChI; Baja CA (Guadalupe Island). Feb–May

C. arcuatus McMinn ARCHING CEANOTHUS Pl gen mound-like, dense, gen 0.3–1 m. **ST**: spreading to ascending, not rooting at nodes; twigs brown to gray-brown, glaucous. **LF**: opposite, gen not clustered, evergreen, spreading to ascending; stipules knob-like; petiole < 3 mm; blade 7–12 mm, 2–5 mm wide, elliptic to obovate, adaxially dull green, glabrous, abaxially pale green, glabrous to strigose, margin gen entire, ± not thick, ± not rolled under. **INFL**: umbel-like, < 1.5 cm. **FL**: gen white, occ light blue. **FR**: 4–6 mm, white, ± 3-ridged; horns 1.5–2 mm, ± slender. Open rocky slopes, montane forest; 580–2140 m. KR, n NCoR, s CaRH, n&c SNH; OR. Previously treated as part of *C. fresnensis* or in error as hybrid. Apr–Jun

C. confusus J.T. Howell (p. 1159) RINCON RIDGE CEANOTHUS Pl mound-like, 0.2–0.5 m. **ST**: spreading to ± ascending, occ rooting at nodes; twigs gen red-brown. **LF**: opposite, evergreen; stipules knob-like; petiole < 2 mm; blade 10–20 mm, 5–14 mm wide, length gen > 2 × width, elliptic to narrowly obovate, ± wavy, ± folded lengthwise, adaxially ± shiny green, glabrous, abaxially paler, ± glabrous exc veins short-strigose, margin ± thick, teeth 3–9 in distal 2/3, sharp. **INFL**: umbel-like, 1–2 cm. **FL**: blue to purple. **FR**: 4–6 mm wide, smooth to ± 3-ridged distally; horns 1.5–2.5 mm, slender. Volcanic slopes, chaparral, pine/oak woodland; 75–1100 m. s NCoR (Lake, Napa, Sonoma cos.). Feb–Apr ★

C. cordulatus Kellogg MOUNTAIN WHITETHORN Pl ± open, gen < 1.5 m. **ST**: gen spreading; twigs rigid, thorn-like, pale gray to gray-green. **LF**: alternate, evergreen; stipules scale-like; petiole 2–8 mm; blade 9–25 mm, 6–18 mm wide, ovate to elliptic, 3-ribbed from base, adaxially dull green, glabrous to puberulent, abaxially paler, glabrous to puberulent, margin gen entire. **INFL**: gen raceme- to panicle-like, gen sessile, 1–4 cm. **FR**: 3.5–5 mm wide, ± 3-ridged distally; horns

0. 2*n*=24. Ridges, slopes, open conifer forests; 365–3355 m. KR, n NCoR, SNH, TR, SnJt, MP, DMtns (Panamint Range); OR, NV, n Baja CA. May–Jul

C. crassifolius Torr. HOARYLEAF CEANOTHUS Pl erect, ± open, < 4 m. **ST**: ascending to erect; twigs white- to rusty-tomentose. **LF**: opposite, evergreen; stipules knob-like; petiole 2–5 mm; blade 10–25 mm, 5–15 mm wide, ± elliptic, adaxially olive-green, glabrous, abaxially paler, short-strigose to densely tomentose, tip obtuse to rounded. **INFL**: umbel-like, 1–2 cm. **FL**: white exc disk, ovary gen dark. **FR**: 5–9 mm wide; horns 1–2 mm. Vars. intergrade in TR.

var. **crassifolius** (p. 1159) **LF**: blade densely tomentose abaxially, veins obscure, margin ± dentate, gen rolled under. 2*n*=24. Ridges, slopes, chaparral; 60–1100 m. TR, PR; n Baja CA. Jan–Apr

var. **planus** Abrams **LF**: blade adaxially pale green to ± white between evident veins, abaxially sparsely tomentose, margin gen thick, not rolled under, gen entire. Ridges, slopes, chaparral; 60–1280 m. s SCoRO, WTR (Santa Barbara, Ventura cos.). Jan–Apr

C. cuneatus (Hook.) Nutt. Pl erect (mound-like), gen open. **ST**: ascending to spreading, occ arched, not rooting at nodes; twigs gen brown to gray-brown, gen glabrous to ± puberulent. **LF**: opposite, some clustered or not, evergreen; stipules knob-like; petiole < 3 mm, gen glabrous to ± puberulent; blade 6–30 mm, 3–18 mm wide, elliptic to ± round, dull green, adaxially glabrous, abaxially paler, glabrous to ± strigose, tip obtuse to notched, margin entire or teeth ± sharp. **INFL**: umbel-like, gen 1–2 cm. **FL**: white, pale blue, blue, or lavender. **FR**: 4–6 mm wide, ± 3-ridged distally; horns 0.5–2 mm. 2*n*=24.

var. **cuneatus** (p. 1159) BUCKBRUSH Pl < 3 m. **ST**: gen ascending to spreading; twigs gen gray-brown. **LF**: some occ clustered; blade elliptic, oblanceolate, or obovate, tip acute to ± rounded, margin gen entire. **FL**: gen white. Sandy to rocky flats, slopes, ridges; < 2133 m. CA-FP (exc ChI), MP; OR, n Baja CA. Intergrades with *C. cuneatus* var. *ramulosus* in s SCoRO, w TR. Were they to be recognized taxonomically, pls from Santa Cruz Mtns with lf blades 15–27 mm, 9–20 mm wide, wide-elliptic to -obovate, would be assignable to *C. cuneatus* var. *dubius* J.T. Howell; study needed. Feb–May

var. **fascicularis** (McMinn) Hoover LOMPOC CEANOTHUS Pl erect, ± open, gen < 2.5 m. **ST**: gen erect, ± arched; twigs gen brown.

LF: some clustered; margin entire; node blade 5–11 mm, 4–7 mm wide, widely obovate, tip truncate to notched, clustered blade 9–15 mm, 3–6 mm wide, oblanceolate to narrowly obovate, tip gen obtuse. **FL:** pale blue to lavender. Sandy substrates, coastal chaparral; < 275 m. s CCo (Santa Barbara, San Luis Obispo cos.). Intergrades with *C. cuneatus* var. *ramulosus*. Feb–May ★

var. **ramulosus** Greene   Pl erect (mound-like), gen dense, gen < 1.5 m. **ST:** erect to spreading, ± arched; twigs gen brown. **LF:** lvs 2 per node; blade 5–15 mm, 3–12 mm wide, widely obovate to ± round, tip truncate to notched, margin entire or teeth few, near tip, sharp. **FL:** pale blue to lavender (white). Sandy substrates, chaparral; < 700 m. s CCo, SnFrB, s SCoRO, w TR. Feb–May

C. cyaneus Eastw.   LAKESIDE CEANOTHUS   Pl open, < 3 m. **ST:** ascending to erect; twigs flexible, not thorn-like, green, ± dark-papillate, angled, internodes ridged. **LF:** alternate, evergreen; stipules scale-like; petiole 2–6 mm; blade 15–45 mm, 15–20 mm wide, gen < 2 × width, ± ovate-elliptic, adaxially green, glabrous, abaxially pale green, ± glaucous, glabrous exc veins puberulent, margin ± serrate, teeth 23–58, minutely gland-tipped or not. **INFL:** panicle-like, 5–30 cm. **FL:** bright blue. **FR:** 3–5 mm wide, ± 3-ridged distally; horns 0. 2n=24. Slopes, ridges, chaparral; 45–1050 m. s PR (San Diego Co.); n Baja CA. Apr–Jun ★

C. dentatus Torr. & A. Gray (p. 1159)   Pl dense, gen 0.5–1.5 m. **ST:** ascending to erect; twigs flexible, not thorn-like, brown to gray-brown. **LF:** alternate, evergreen; stipules scale-like; petiole 1–2 mm; blade 4–16 mm, 2–8 mm wide, flat, elliptic to oblong-oblanceolate, adaxially dark green, not furrowed, not glandular-papillate, puberulent, abaxially pale green, tomentose or short-wavy-hairy, 1-ribbed from base, tip truncate to notched, margin rolled under, ± glandular-papillate. **INFL:** gen raceme-like, 1.5–3 cm. **FL:** deep blue. **FR:** 2.5–4 mm wide, 3-ridged distally; horns 0. Sandy substrates, coastal bluffs, slopes; 5–1500 m. CCo, SnFrB (Santa Cruz Mtns), n SCoRO. Mar–Jun

C. divergens Parry (p. 1159)   CALISTOGA CEANOTHUS   Pl erect or mound-like, 0.5–1.5 m. **ST:** ascending to ± erect, gen not rooting at nodes, red-brown. **LF:** opposite, evergreen; stipules knob-like; petiole < 2 mm; blade 10–19 mm, 5–12 mm wide, length gen > 2 × width, elliptic to narrowly obovate, ± wavy to ± folded lengthwise, adaxially ± shiny green, glabrous, abaxially paler, ± glabrous exc veins short-strigose, margin ± thick or ± rolled under, teeth 3–11 in distal 2/3, spine-like. **INFL:** umbel-like, 1–2 cm. **FL:** blue to purple. **FR:** 4–6 mm wide, smooth to ± 3-ridged distally; horns 1.5–2.5 mm, slender. 2n=24. Volcanic slopes, chaparral, pine/oak woodland; 150–950 m. NCoR (Napa, Sonoma cos.). Feb–Apr ★

C. diversifolius Kellogg   PINE MAT   Pl gen mat-like, dense, gen < 0.3 m. **ST:** spreading or ± ascending, rooting at nodes; twigs not angled, flexible, not thorn-like, green, occ tinged red, puberulent. **LF:** alternate, evergreen; stipules scale-like; petiole 3–11 mm; blade 11–35 mm, 6–20 mm wide, ovate, obovate, or elliptic, gen flat, short-wavy-hairy, adaxially blue-green, abaxially paler, 1(3)-ribbed from base, tip obtuse to ± rounded, margin not thick, not rolled under, glandular-serrate, teeth 27–42, glands narrowly conic. **INFL:** raceme-, occ ± umbel-like, 1–2 cm, on short, erect twigs. **FL:** blue to ± white. **FR:** 4–5 mm wide, smooth, 3-ridged distally; horns 0. Slopes, flats, oak-conifer forest; 630–2300 m. KR, NCoRH, CaRH, SNH. Apr–Jun

C. ferrisiae McMinn (p. 1159)   COYOTE CEANOTHUS   Pl erect, ± open, 1–2 m. **ST:** ascending to erect; twigs brown to gray-brown, not angled, gen ± short-hairy, glabrous in age. **LF:** opposite, evergreen; stipules knob-like; petiole < 2 mm; blade 12–25 mm, 7–18 mm wide, obovate to widely elliptic, adaxially dark green, glabrous, abaxially paler, short-strigose between evident veins, tip obtuse to ± notched, margin entire or teeth ± sharp. **INFL:** umbel-like, 1–1.5 cm. **FL:** white; disk, ovary gen not dark. **FR:** 6–9 mm wide, ± wrinkled; horns 1–2 mm, thick, erect. 2n=24. Rocky, serpentine slopes, chaparral; 120–320 m. se SnFrB (Santa Clara Co.). [*Ceanothus ferrisae*, orth. var.] Jan–May ★

C. foliosus Parry   Pl erect to mat-like, ± open to dense. **ST:** erect to spreading; twigs flexible, not thorn-like, not angled, green to red-green. **LF:** alternate, evergreen; stipules scale-like; petiole 1–3 mm;

blade 5–20 mm, 3–13 mm wide, flat or gen wavy, ± folded length-wise, oblong to widely elliptic or obovate, adaxially dark green, abaxially paler, glabrous to puberulent, 1-ribbed or faintly 3-ribbed from base, tip acute to ± rounded, margin not thick, not rolled under, gland-toothed or not, teeth 31–42, glands gen dark, persistent. **INFL:** raceme-, occ ± umbel-like, 1–4 cm. **FL:** blue to ± purple. **FR:** 3–4 mm wide, ± 3-ridged distally; horns 0.

var. **foliosus**   WAVYLEAF CEANOTHUS   Pl erect, not mat-like, ± open, < 3.5 m. **ST:** ascending to erect. **LF:** blade oblong to widely elliptic, abaxially glabrous to sparsely puberulent, margin wavy, gland-toothed. 2n=24. Rocky slopes, flats, chaparral, woodland, mixed-evergreen forest; < 1500 m. NCoR, SnFrB, SCoRO, PR. Mar–Jun

var. **medius** McMinn   LA CUESTA CEANOTHUS   Pl erect, not mat-like, ± dense, < 2 m. **ST:** ascending to erect. **LF:** blade elliptic, abaxially densely short-wavy-hairy, margin wavy, gland-toothed. Rocky slopes, flats, chaparral, woodland, mixed-evergreen forest; < 650 m. SnFrB, SCoR. Mar–Jun

var. **vineatus** McMinn   VINE HILL CEANOTHUS   Pl mat- to mound-like, ± open, < 0.5(0.8) m. **ST:** spreading. **LF:** blade widely elliptic to obovate, ± wavy, abaxially glabrous exc veins, margin not thick, not rolled under, entire or teeth few. Rocky slopes, flats, chaparral, woodland, mixed-evergreen forest; < 300 m. s NCoR. Mar–Jun ★

C. fresnensis Abrams   FRESNO CEANOTHUS   Pl mat-like, dense, < 0.3 m. **ST:** spreading, rooting at nodes; twigs gen red-brown. **LF:** opposite, gen not clustered, evergreen, spreading to ascending; stipules knob-like; petiole 1–2 mm; blade 4–12 mm, 3–8 mm wide, elliptic to oblanceolate, adaxially ± shiny green, glabrous to strigose, abaxially glabrous to strigose, esp veins, margin entire, ± not thick, ± not rolled under. **INFL:** umbel-like, 1–2 cm. **FL:** blue. **FR:** 4–6 mm wide; horns 1.5–2 mm, thick. Rocky slopes, flats, conifer forest; 900–2200 m. n&c SN. May–Jun ★

C. gloriosus J.T. Howell   Pl mat-like to erect, ± open. **ST:** erect to spreading, occ arched, rooting at nodes or not; twigs gen angled near tips, green, red-brown, or brown. **LF:** opposite, some clustered, evergreen; stipules knob-like; petiole 1–4 mm; blade 10–45 mm, 5–24 mm wide, length gen < 2 × width, widely elliptic, obovate, or round, adaxially dark green, glabrous, abaxially glabrous to puberulent, tip obtuse, truncate, or notched, margin teeth sharp. **INFL:** ± umbel-like, 1–2.5 cm. **FL:** blue to blue-purple. **FR:** 4–6 mm wide, smooth to ± 3-ridged distally; horns < 1.5 mm. 2n=24.

var. **exaltatus** J.T. Howell   GLORY BRUSH   Pl erect, < 2 m. **ST:** spreading, arched, not rooting at nodes. **LF:** blade 13–45 mm, widely obovate to ± round, margin 13–35-toothed. Sandy or rocky substrates; < 500 m. NCo, NCoRO, n SnFrB. Mar–May ★

var. **gloriosus**   POINT REYES CEANOTHUS   Pl mat- to mound-like, < 0.3 m. **ST:** gen spreading, occ rooting at nodes. **LF:** blade 23–31 mm, margin 13–35-toothed. Sandy places, coastal bluffs, closed-cone-pine forest; < 500 m. s NCo, n CCo (Marin Co.). Mar–May ★

var. **porrectus** J.T. Howell   MOUNT VISION CEANOTHUS   Pl mound-like, < 0.5 m. **ST:** spreading to ± ascending, ± arched. **LF:** blade 10–21 mm, margin 9–19-toothed. Coastal bluffs, scrub, closed-cone pine forest; < 300 m. n CCo (Inverness Ridge). Mar–May ★

C. hearstiorum Hoover & Roof (p. 1159)   HEARSTS' CEANOTHUS   Pl mat-like, dense, < 0.3 m. **ST:** spreading; twigs flexible, not thorn-like, ridged, green to ± brown. **LF:** alternate, evergreen; stipules scale-like; petiole < 2 mm; blade 9–17 mm, 2–10 mm wide, flat, ± oblong, leathery, adaxially dark green, glandular-papillate, abaxially tomentose, 1-ribbed from base, tip truncate to notched, margin rolled under, minutely glandular-papillate. **INFL:** umbel- to ± raceme-, occ umbel-like, 1–4 cm. **FL:** gen blue. **FR:** 4–5 mm wide; horns 0. Coastal bluffs; < 200 m. CCo (nw San Luis Obispo Co.). Mar–Apr ★

C. impressus Trel.   Pl open to dense. **ST:** ascending to erect; twigs flexible, not thorn-like, dark brown to gray-brown. **LF:** alternate, evergreen; stipules scale-like; petiole 1–4 mm; blade 5–25 mm, 4–15 mm wide, gen flat, widely ovate to ± round, adaxially dark green,

puberulent, furrowed along veins, abaxially densely short-hairy, 1-ribbed from base, tip acute to obtuse, margin thick to rolled under, gen minutely gland-toothed **INFL:** umbel- to ± raceme-like, 1–3.5 cm. **FL:** blue. **FR:** 3–4 mm wide, ± 3-ridged; horns 0.

var. *impressus* Pl erect, gen dense. **LF:** blade 5–14 mm, 3–12 mm wide. 2*n*=24. Sandy substrates, flats, canyons; < 320 m. s CCo (w Santa Barbara Co.). Feb–Apr

var. *nipomensis* McMinn (p. 1159)   Pl erect, gen open, 1–3 m. **LF:** blade 11–25 mm, 7–20 mm wide, flat. Sandy substrates, flats, canyons; < 200 m. s CCo (Nipomo Mesa, Irish Hills). Feb–Apr

C. incanus Torr. & A. Gray COAST WHITETHORN   Pl erect, open, < 4 m. **ST:** ascending to erect; twigs rigid, thorn-like or not, pale gray to gray-green. **LF:** alternate, evergreen; stipules scale-like; petiole 3–12 mm; blade 20–50+ mm, 10–45 mm wide, ± widely ovate, adaxially dull green, glabrous to sparsely puberulent, abaxially gray-green, strigose, 3-ribbed from base, margin gen entire. **INFL:** gen panicle-like, 3–7 cm, gen stalked. **FL:** white. **FR:** 4–6 mm wide, sticky, coarse-wrinkled when dry; horns minute. 2*n*=24. Flats, slopes, chaparral, mixed-evergreen forest; < 900 m. NW, CCo, SnFrB. Apr–Jun

C. integerrimus Hook. & Arn. (p. 1161)   DEER BRUSH   Pl occ tree-like, open, < 4 m. **ST:** ascending to erect; twigs flexible, not thorn-like, gen green. **LF:** alternate, deciduous; stipules scale-like; petiole 3–12 mm; blade 15–53 mm, 10–45 mm wide, thin, flat, lanceolate to widely ovate, adaxially dull green, glabrous to puberulent, abaxially pale green, glabrous to sparsely short-hairy, 1–3-ribbed from base, margin gen entire. **INFL:** panicle-like, 4–20 cm. **FL:** white to blue (pink). **FR:** 4–6.5 mm wide, sticky, smooth to bulged or ± 3-ridged distally; horns 0.

var. *integerrimus* **LF:** blade elliptic to ± oblong, thin, 1-ribbed from base. **FL:** white. Mixed conifer forest; ± 600 m. sw SnFrB (Santa Cruz Mtns). May–Jul

var. *macrothyrsus* (Torr.) G.T. Benson   **LF:** blade lanceolate to widely ovate (oblong), 3-ribbed from base. **FL:** white to blue (pink). 2*n*=24. Chaparral, oak woodland, conifer forest; 70–2600 m. KR, NCoR, CaR, SN, SnFrB, SCoR, TR, PR, MP; to WA, AZ. [*C. i.* var. *californicus* G.T. Benson; *C. i.* var. *puberulus* (Greene) Abrams] May–Jul

C. jepsonii Greene (p. 1161)   MUSK BRUSH   Pl gen erect, ± open, < 1.5 m. **ST:** ascending to erect, intricately branched; twigs gen brown. **LF:** opposite, evergreen, gen reflexed; stipules knob-like; petiole < 2 mm; blade 10–20 mm, 5–13 mm wide, ovate to elliptic, ± folded lengthwise, adaxially yellow-green, glabrous, abaxially ± short-strigose between veins, margin thick to ± rolled under, wavy, 7–11-spine-toothed. **INFL:** umbel-like, 1–2.5 cm. **FL:** sepals, petals 6–8, blue to white. **FR:** 5–7 mm wide, 3-ridged, wrinkled distally; horns 1–3 mm, thick. 2*n*=24. Rocky, serpentine slopes; 36–1740 m. KR, NCoR, SnFrB. [*C. j.* var. *albiflorus* J.T. Howell] Mar–Apr

C. lemmonii Parry   Pl erect to gen mound-like, ± open, 0.5–1 m. **ST:** ascending to erect, not rooting at nodes; twigs flexible, not thorn-like, pale green to gray-green. **LF:** alternate, evergreen; stipules scale-like; petiole 3–6 mm; blade 12–33 mm, 6–15 mm wide, flat, elliptic to ± oblong, adaxially green, ± glabrous to strigose, abaxially paler, densely puberulent, 1- or ± 3-ribbed from base, tip acute to obtuse, margin not thick, not rolled under, minutely gland-toothed, teeth 34–45, glands ± spheric. **INFL:** raceme-, occ ± umbel-like, 2–7 cm. **FL:** blue to purple-blue. **FR:** 3–4 mm wide, 3-ridged distally; horns 0. 2*n*=24. Open, rocky sites; 150–1650 m. KR, NCoRI, CaR, n&c SN. Apr–May

C. leucodermis Greene (p. 1161)   CHAPARRAL WHITETHORN   Pl ± open, < 4 m. **ST:** erect; twigs rigid, thorn-like, pale gray to gray-green. **LF:** alternate, evergreen; stipules scale-like; petiole 4–7 mm; blade 12–30 mm, 3–15 mm wide, ovate to elliptic, adaxially dull green, glabrous to sparsely puberulent, abaxially ± strigose, esp veins, 3-ribbed from base, margin entire to ± glandular-serrate. **INFL:** raceme- to panicle-like, 3–15 cm, gen stalked. **FL:** pale blue to white. **FR:** 3–5 mm wide, sticky, ± 3-ridged, smooth to ± wrinkled when dry; horns 0. 2*n*=24. Rocky slopes, chaparral; 270–2150 m. SNF, SnFrB, SCoR, TR, PR; n Baja CA. Apr–Jun

C. maritimus Hoover (p. 1161)   MARITIME CEANOTHUS   Pl mat- to mound-like, ± open, gen < 1 m. **ST:** spreading or ± ascending, occ arched, twigs red-brown. **LF:** opposite, evergreen; stipules knob-like; petiole 1–2 mm; blade 8–20 mm, 4–12 mm wide, obovate or oblong-obovate, adaxially shiny green, glabrous, abaxially gray-tomentose, tip truncate or notched, margin thick to ± rolled under, entire or ± 3–5-toothed distally. **INFL:** umbel-like, < 2 cm. **FL:** gen blue. **FR:** 5–8 mm, wrinkled near top; horns 1–2 mm. Coastal hills, bluffs; < 60 m. CCo (nw San Luis Obispo Co.). Feb–May ★

C. masonii McMinn   MASON'S CEANOTHUS   Pl gen erect, < 2 m. **ST:** ascending, twigs gen angled distally, red-brown to dark brown, gen gray-brown in age. **LF:** opposite, evergreen; stipules knob-like; petiole < 3 mm; blade 7–21 mm, 4–13 mm wide, obovate to narrowly obovate, flat, adaxially dark green, glabrous, abaxially short-strigose, tip obtuse to truncate, margin 9–17-toothed. **INFL:** umbel-like, 1–2 cm. **FL:** blue to purple-blue. **FR:** 4–5 mm wide; horns < 1.5 mm. 2*n*=24. Rocky slopes, chaparral; 150–450 m. SnFrB (Bolinas Ridge, sw Marin Co.). Probably better treated as part of *C. gloriosus*. Mar–May ★

C. megacarpus Nutt.   Pl erect, open, < 4 m. **ST:** ascending to erect; twigs brown to gray-brown. **LF:** opposite or alternate, evergreen; stipules knob-like; petiole 1–4 mm; blade 10–25 mm, 5–12 mm wide, length gen > 2 × width, elliptic to widely oblanceolate, adaxially dull green, glabrous, abaxially gray-green, short-strigose, esp veins, tip truncate to ± notched, margin entire, ± thick near middle, not rolled under. **INFL:** umbel-like, 1–2 cm. **FL:** gen white, disk dark. **FR:** 7–12 mm wide, 3-ridged or not; horned or not.

var. *insularis* (Eastw.) Munz (p. 1161)   ISLAND CEANOTHUS   **LF:** gen opposite. **FR:** smooth or ± 3-ridged; horns gen 0. Rocky slopes, canyons, chaparral; < 475 m. ChI. Feb–Mar ★

var. *megacarpus* (p. 1161)   BIGPOD CEANOTHUS   **LF:** gen alternate. **FR:** 3-ridged; horns < 1.5 mm. 2*n*=24. Rocky slopes, canyons, chaparral; < 900 m. n SCo, s ChI, WTR, SnGb, n PR. [*C. m.* var. *pendulus* McMinn] Dec–Mar

C. oliganthus Nutt.   Pl erect, tree-like or not, open, < 3.5 m. **ST:** ascending to erect; twigs flexible, not thorn-like, not angled. **LF:** alternate, evergreen; stipules scale-like; petiole 3–7 mm; blade 11–30 mm, 5–25 mm wide, ovate to elliptic, gen flat, adaxially dark green, ± glabrous to short-hairy, abaxially paler, sparsely to densely short-hairy, esp veins, margin minutely gland-toothed, teeth 20–70, glands ± spheric. **INFL:** gen raceme-like, 1–5 cm. **FL:** blue to purple-blue; disk, ovary glabrous or short-hairy. **FR:** 4–7 mm wide, gen sticky, ± ridged to bulged distally; horns 0.

var. *oliganthus* (p. 1161)   **ST:** twigs green to ± red-brown, puberulent. **LF:** blade adaxially short-hairy. **FL:** disk, ovary glabrous. **FR:** gen smooth. Slopes, ridges, chaparral, conifer forest; 60–1800 m. NCoRO, SnFrB, SCoRO, TR, PR. Dec–Jun

var. *orcuttii* (Parry) Jeps.   **ST:** twigs ± red-brown, puberulent. **LF:** blade adaxially short-hairy. **FL:** disk, ovary short-hairy. **FR:** gen prominent-wrinkled. Slopes, ridges, chaparral, conifer forest; 400–1500 m. PR; n Baja CA. Feb–Jun

var. *sorediatus* (Hook. & Arn.) Hoover   JIM BRUSH   **ST:** twigs ± red-brown, ± glabrous. **LF:** blade adaxially glabrous to sparsely short-hairy. **FL:** disk, ovary glabrous. **FR:** gen smooth. 2*n*=24. Slopes, ridges, chaparral, conifer forest; 60–1220 m. KR, NCoR, n SNH, CW, TR, PR. Jan–May

C. ophiochilus S. Boyd et al.   VAIL LAKE CEANOTHUS   Pl erect, open, < 2 m. **ST:** ascending to erect; twigs gen red-brown. **LF:** opposite, clustered or not, evergreen; stipules knob-like; petiole < 1 mm; blade 3–7 mm, 1–3 mm wide, narrow-oblanceolate to -obovate, dull green, glabrous, veins obscure, tip acute to truncate, margin gen entire, flat. **INFL:** umbel-like, gen < 1 cm. **FL:** gen pale blue, occ ± pink. **FR:** 3–3.5 mm wide; horns ± 0. Rocky slopes, ridges of pyroxenite-rich substrate, chaparral; 600–1100 m. n PR (Vail Lake, Agua Tibia Mtn). Mar–Apr ★

C. otayensis McMinn   OTAY MOUNTAIN CEANOTHUS   Pl erect, ± open, < 2 m. **ST:** ascending to erect; twigs not angled, gen gray-brown. **LF:** opposite, evergreen; stipules knob-like; petiole 1–2 mm; blade 5–12 mm, 4–10 mm wide, obovate to oblong-obovate,

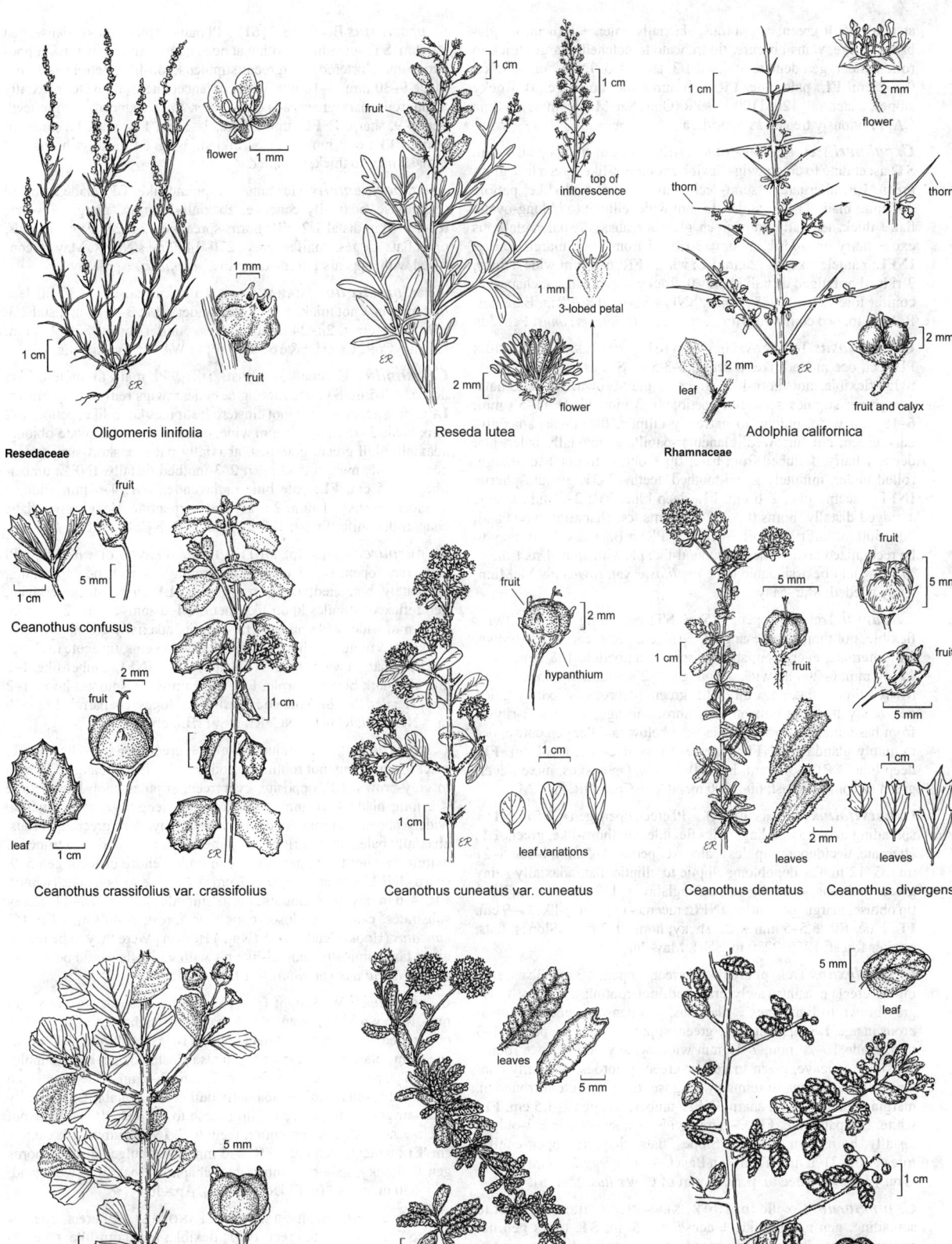

Oligomeris linifolia
Resedaceae

Reseda lutea

Adolphia californica
Rhamnaceae

Ceanothus confusus

Ceanothus crassifolius var. crassifolius

Ceanothus cuneatus var. cuneatus

Ceanothus dentatus   Ceanothus divergens

Ceanothus ferrisiae

Ceanothus hearstiorum

Ceanothus impressus var. nipomensis

adaxially dull green, ± glabrous, abaxially paler, ± tomentose, glabrous in age, veins obscure, tip truncate to notched, margin thick to rolled under, gen dentate in distal 1/3, teeth 3–5. **INFL:** umbel-like, 0.5–2 cm. **FL:** pale blue. **FR:** 4–6 mm wide; horns gen 0. Rocky slopes, chaparral; 120–1100 m. s PR (Otay, San Miguel mtns); n Baja CA. Previously treated as hybrid. Jan–Apr ★

C. palmeri Trel. PALMER CEANOTHUS Pl gen erect, open, < 4 m. **ST:** ascending to erect; twigs flexible, not thorn-like, green to ± gray-green. **LF:** alternate, ± semi-deciduous; stipules scale-like; petiole 2–7 mm; blade 11–35 mm, 9–15 mm wide, elliptic to oblong-ovate, flat, ± thick, adaxially shiny green, glabrous, abaxially paler, glabrous exc ± hairy on midrib, 1- or ± 3-ribbed from base, margin entire. **INFL:** panicle-like, 4–12 cm. **FL:** white. **FR:** 6–9 mm wide, sticky, 3-ridged or bulged distally; horns 0. Rocky slopes, ridges, chaparral, conifer forest; 100–1985 m. n&c SNF, s SCoR, TR, PR; n Baja CA. Related to, occ confused with *C. spinosus, C. integerrimus.* Feb–Jun

C. papillosus Torr. & A. Gray (p. 1161) WARTLEAF CEANOTHUS Pl erect, occ mound-like, ± dense, 1–3.5 m. **ST:** ascending to erect; twigs flexible, not thorn-like, green to gray-brown. **LF:** alternate, evergreen; stipules scale-like; petiole 1–3 mm; blade 11–50 mm, 6–15 mm wide, oblong to narrowly elliptic, flat, thick, adaxially dark green, not furrowed, glandular-papillate, abaxially paler, gen densely hairy, 1-ribbed from base, tip ± obtuse to notched, margin rolled under, minutely gland-toothed, teeth 17–31, glands spheric. **INFL:** raceme-like, 2–6 cm. **FL:** deep blue. **FR:** 2–3 mm, sticky, ± ridged distally; horns 0. 2*n*=24. Open sites, chaparral, woodland; 90–1500 m. SnFrB, SCoR, WTR, nw PR; n Baja CA. Were they to be recognized taxonomically, mound-like pls with arched sts from w WTR would be assignable to *C. papillosus* var. *roweanus* McMinn; study needed. Mar–May

C. parryi Trel. Pl open, < 5 m. **ST:** ascending to erect; twigs flexible, not thorn-like, ± angled, ± ridged, gen green or red-brown. **LF:** alternate, evergreen; stipules scale-like; petiole 1–5 mm; blade 12–40 mm, 6–20 mm wide, length gen > 2 × width, ± oblong to narrowly elliptic, flat, adaxially dark green, glabrous, abaxially paler, cobwebby to loose-tomentose, glabrous in age, 1- or ± 3-ribbed from base, margin ± rolled under, esp below middle, gen entire, occ minutely glandular. **INFL:** raceme- to panicle-like, 6–14 cm. **FL:** deep blue. **FR:** 2–3.5 mm; horns 0. 2*n*=24. Open sites, mixed-evergreen or conifer forest; 60–1200 m. NCo, NCoR; OR. Apr–May

C. parvifolius (S. Watson) Trel. Pl erect, open, gen < 1.5 m. **ST:** ± spreading to gen ascending; twigs flexible, not thorn-like, green. **LF:** alternate, deciduous; stipules scale-like; petiole 1–5 mm; blade 8–21 mm, 3–12 mm wide, oblong-elliptic to elliptic, flat, adaxially shiny green, glabrous, abaxially paler, gen glabrous, 1–3-ribbed from base, tip obtuse, margin gen entire. **INFL:** raceme- to panicle-like, 4–9 cm. **FL:** blue. **FR:** 3.5–5 mm wide, sticky; horns 0. 2*n*=24. Slopes, flats, conifer forest; 1255–2220 m. SNH. May–Jul

C. perplexans Trel. (p. 1161) Pl erect, ± open, < 3 m. **ST:** ascending to erect, occ intricately branched, not rooting at nodes; twigs gray-brown to light gray, ± glaucous, gen densely puberulent, glabrous in age. **LF:** opposite, evergreen; stipules knob-like; petiole 1–3 mm; blade 10–20 mm, 7–17 mm wide, widely elliptic to ± round, adaxially concave, green to yellow-green, glabrous, abaxially convex, paler, glabrous to minutely strigose, tip gen acute to rounded, margin teeth gen 7–11, sharp. **INFL:** umbel-like, gen 1–1.5 cm. **FL:** white, occ pale blue. **FR:** 3–5 mm wide, gen smooth or ± 3-bulged distally; horns 0 or < 1 mm. Slopes, flats, chaparral, open conifer forest; 305–2100 m. e TR, PR; n Baja CA. [*C. greggii* A. Gray var. *p.* (Trel.) Jeps.] Related to, perhaps part of *C. vestitus.* Mar–May

C. pinetorum Coville (p. 1161) KERN CEANOTHUS Pl erect to spreading, gen mound-like, ± dense, < 1.5. m. **ST:** twigs gen not angled, gen red-brown. **LF:** opposite, occ clustered, evergreen; stipules knob-like; petiole 1–3 mm; blade 10–20 mm, 8–19 mm wide, widely elliptic to ± round, adaxially green, gen glabrous, abaxially paler, glabrous to strigose, tip rounded, margin thick, occ ± rolled under, teeth 9–15, sharp. **INFL:** umbel-like, 1–2.5 cm. **FL:** pale blue to blue-lavender. **FR:** 6–9 mm wide, wrinkled, 3-ridged distally; horns 1–2 mm, wrinkled. Slopes, ridges, flats, conifer forest; 1050–2750 m. KR (Trinity Mtns), s SNH (Kern Plateau). May–Jun ★

C. prostratus Benth. (p. 1161) Pl mat- or mound-like, dense, gen < 0.3 m. **ST:** spreading, rooting at nodes; twigs red-brown. **LF:** opposite, evergreen; stipules knob-like; petiole 1–3 mm; blade 9–30 mm, 4–16 mm wide, oblanceolate to obovate, adaxially ± glabrous, dark green, abaxially paler, tip ± rounded, margin teeth gen 3–9, sharp. **INFL:** umbel-like, 1–2 cm. **FL:** blue, lavender, or purple. **FR:** 6–9 mm wide, wrinkled, 3-ridged distally; horns 1–2 mm, slender to thick, wrinkled.

var. *occidentalis* McMinn Pl mound-like. **LF:** blade ± folded lengthwise, adaxially concave, abaxially convex, margin ± wavy, teeth 5–9, in distal 1/2. **FR:** horns spreading. 2*n*=24. Volcanic soils, open flats, ridges, conifer forest; 270–1400 m. s NCoR. May be confused with, appears intermediate to *C. confusus.* Apr–May

var. *prostratus* MAHALA MAT Pl gen mat-like. **LF:** blade ± flat, margin ± not thick, ± not rolled under, teeth 3–5(7), in distal 2/3. **FR:** horns erect. 2*n*=24. Flats, ridges, conifer forest; 800–2700 m. KR, n NCoR, CaRH, n&c SNH, MP; to WA, w NV. Apr–Jun

C. pumilus Greene SISKIYOU MAT Pl mat- to mound-like, dense, < 0.5 m. **ST:** occ rooting at nodes; twigs red- to gray-brown. **LF:** opposite, evergreen, not clustered; stipules knob-like; petiole 1–2 mm; blade 5–15 mm, 3–6 mm wide, oblanceolate to obovate-oblong, adaxially dull green, glabrous, abaxially paler, ± short-strigose, tip gen truncate, margin thick, gen 2–3-toothed distally. **INFL:** umbel-like, 1–1.5 cm. **FL:** pale blue or lavender. **FR:** 4–6 mm wide, ± 3-ridged; horns < 1 mm. 2*n*=24. Gen serpentine, slopes, open flats, chaparral, conifer forest; 180–2290 m. KR, NCoR; sw OR. Apr–Jun

C. purpureus Jeps. (p. 1161) HOLLY-LEAVED CEANOTHUS Pl gen erect, open, < 1.5 m. **ST:** ascending to erect, occ spreading, intricately branched; twigs red-brown. **LF:** opposite, evergreen, gen reflexed; stipules knob-like; petiole 1–2 mm; blade 12–25 mm, 7–20 mm wide, wide-obovate to -elliptic, adaxially green, glabrous, abaxially ± minutely short-strigose, hairy on veins, tip acute to sharp-toothed, margin wavy, teeth 7–15, spine-like. **INFL:** umbel-like, 1–2 cm. **FL:** dark blue to purple. **FR:** 4–5 mm wide, smooth; horns 1–2 mm, slender. 2*n*=24. Volcanic substrates, slopes, chaparral; 145–670 m. s NCoRO/NCoRI, s NCoRI, n SnFrB. Feb–Apr ★

C. rigidus Nutt. Pl mound-like to ± erect, dense, < 1.5 m. **ST:** erect to spreading, not rooting at nodes; twigs gen not angled, brown to gray-brown. **LF:** opposite, evergreen; stipules knob-like; petiole < 2 mm; blade 4–10 mm, 4–6 mm wide, length gen < 2 × width, widely obovate to round, adaxially gen shiny, dark green, glabrous, abaxially paler, short-strigose or puberulent, tip rounded to truncate, margin gen not thick, gen not rolled under, entire or teeth gen 5–9, sharp. **INFL:** umbel-like, 0.5–2 cm. **FL:** blue to lavender (white). **FR:** 4–6 mm wide; ± smooth; horns smooth, < 1 mm. 2*n*=24. Sandy substrates, chaparral, closed-cone-pine forest; < 400 m. CCo. [*C. cuneatus* (Hook.) Nutt. var. *r.* (Nutt.) Hoover] Were they to be recognized taxonomically, mound-like pls with white fls would be assignable to *C. rigidus* var. *albus* Roof. Mar–May ★

C. roderickii W. Knight (p. 1167) PINE HILL CEANOTHUS Pl mat- to mound-like, open, < 0.5 m. **ST:** spreading to ± ascending, occ arched, often rooting at distal nodes; twigs red-brown. **LF:** opposite, some clustered, evergreen, gen also clustered, gen erect; stipules knob-like; petiole 1–2 mm; blade 4–11 mm, 2–6 mm wide, length gen > 2 × width, oblanceolate, adaxially dull green, ± glabrous, abaxially short-strigose between veins, tip obtuse to ± notched, margin ± not thick, ± not rolled under, entire, teeth 0–5. **INFL:** umbel-like, 0.5–1 cm. **FL:** white tinged blue. **FR:** 4–5 mm wide, bulged distally; horns gen 0. Rocky, gabbroic substrates, chaparral, oak/pine woodland; 260–630 m. n SNF (w El Dorado Co.). Apr–Jun ★

C. sanguineus Pursh REDSTEM CEANOTHUS Pl erect, open, < 3 m. **ST:** ascending to erect; twigs flexible, not thorn-like, green to red-brown. **LF:** alternate, deciduous; stipules scale-like; petiole 5–25 mm; blade 25–95 mm, 17–60 mm wide, ovate to widely elliptic, thin, not leathery, adaxially green, ± glabrous, not sticky, abaxially paler, gen puberulent, esp veins, tip obtuse to rounded, margin minutely glandular-serrate, teeth 30–80+. **INFL:** panicle-like, 3–8 cm. **FL:** white. **FR:** 3–4.5 mm wide, brown, sticky, ± 3-ridged distally; horns 0. Flats, canyon, mixed conifer forest; 700–1700 m. KR; to BC, MT. Apr–Jun

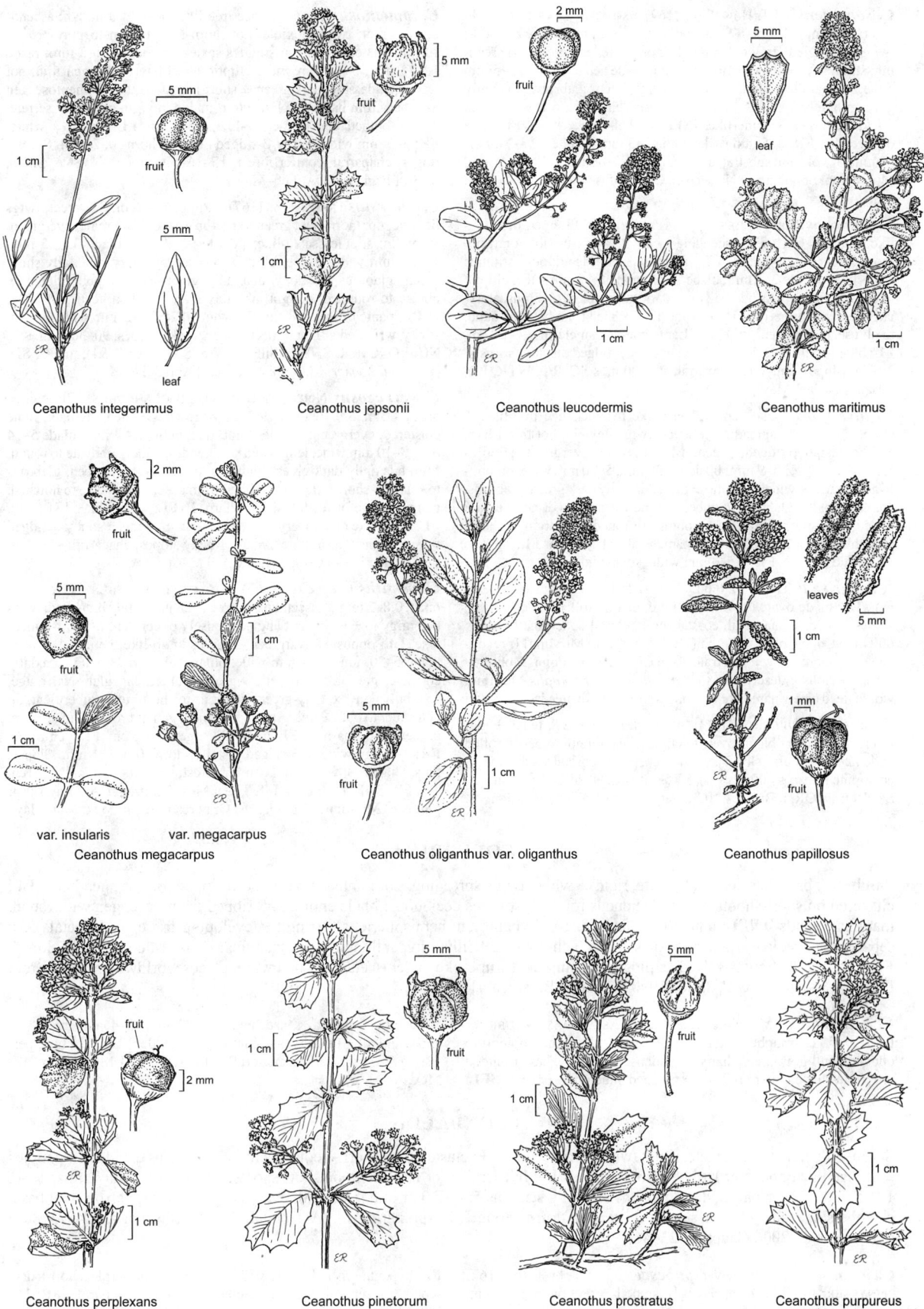

Ceanothus integerrimus

Ceanothus jepsonii

Ceanothus leucodermis

Ceanothus maritimus

var. insularis var. megacarpus
Ceanothus megacarpus

Ceanothus oliganthus var. oliganthus

Ceanothus papillosus

Ceanothus perplexans

Ceanothus pinetorum

Ceanothus prostratus

Ceanothus purpureus

C. sonomensis J.T. Howell (p. 1167) SONOMA CEANOTHUS Pl erect, ± open, < 1.5 m. **ST:** ascending to erect, not rooting at nodes; twigs not angled; gray to brown. **LF:** opposite, evergreen, ± spreading; stipules knob-like; petiole < 1 mm; blade gen 5–11 mm, 2–5 mm wide, obovate to round, adaxially shiny green, glabrous, abaxially gray, minutely appressed-tomentose, esp veins, tip notched, margin thick, teeth 3–5, ± spine-like. **INFL:** umbel-like, ± 1 cm. **FL:** blue or lavender. **FR:** 4–5 mm wide; horns 2–3 mm, slender. 2*n*=24. Serpentine or volcanic substrates, chaparral; 140–600 m. NCoRO (Hood Mtn region, Sonoma and Napa cos.). Occ confused with *C. divergens.* Mar–Apr ★

C. spinosus Nutt. GREENBARK CEANOTHUS Pl erect, occ tree-like, open, < 6 m. **ST:** ascending to erect; twigs flexible or rigid, thorn-like or not, green. **LF:** alternate, semi-deciduous; stipules scale-like; petiole 2–7 mm; blade 11–30 mm, 9–15 mm wide, elliptic to oblong, flat, ± firm, adaxially gen dark shiny green, glabrous, abaxially paler, glabrous or hairy, esp midrib, 1-ribbed from base, tip acute to obtuse, margin gen entire. **INFL:** raceme- to panicle-like, 4–21 cm. **FL:** blue to pale blue. **FR:** 4–7 mm wide, ± 3-ridged distally; horns 0. 2*n*=24. Slopes, canyons, chaparral; < 1200 m. s SCoRO, WTR, PR; n Baja CA. Jan–May

C. thyrsiflorus Eschsch. Pl mat-like to erect, occ tree-like, ± open, < 6 m. **ST:** spreading to erect; twigs flexible, not thorn-like, angled distally, ridged, green. **LF:** alternate, evergreen; stipules scale-like; petiole 3–9 mm; blade 10–39 mm, 5–20 mm wide, oblong-ovate, ovate, or widely elliptic, ± firm, adaxially dark green, glabrous, abaxially paler, glabrous to puberulent between veins or not, veins ± raised, sparsely puberulent, tip obtuse to rounded, margin ± gland-toothed. **INFL:** gen raceme- or panicle-like, 1.5–7 cm. **FL:** light to deep blue (white). **FR:** 2.5–4 mm wide, sticky; horns 0.

var. ***griseus*** Trel. CARMEL CEANOTHUS **ST:** twigs ± puberulent. **LF:** blade ovate to widely elliptic, margin rolled under. Bluffs, slopes, canyons, chaparral, coastal scrub, closed-cone-pine forest; < 600 m. NCo, NCoRO, CCo, SnFrB. [*C. g.* (Trel.) McMinn] If recognized taxonomically, ± prostrate pls in CCo (Yankee Point, Monterey Co.), previously described as *C. griseus* var. *horizontalis* McMinn, would need a new name under *C. thyrsiflorus*; study needed. Feb–Jun

var. ***thyrsiflorus*** (p. 1167) BLUE BLOSSOM **ST:** twigs ± glabrous. **LF:** blade oblong-ovate to elliptic, margin not to partly rolled under. 2*n*=24. Bluffs, slopes, canyons, chaparral, coastal scrub, closed-cone-pine forest; < 800 m. NCo, KR, s NCoRI, NCoRO, CCo, SnFrB, SCoRO; sw OR, n Baja CA. [*C. t.* var. *repens* McMinn] Mar–Jun

C. tomentosus Parry Pl occ tree-like, open, < 3 m. **ST:** ascending to erect; twigs flexible, not thorn-like, brown to gray-brown. **LF:** alternate, evergreen; stipules scale-like; petiole 1–3 mm; blade 10–25 mm, 5–12 mm wide, elliptic to widely ovate, gen thin, not leathery, adaxially dark green, ± short-hairy, abaxially tomentose, gen 3-ribbed from base, tip obtuse to rounded, margin glandular-serrate, 40–60-toothed. **INFL:** raceme-like, 1.5–5 cm. **FL:** blue to ± white. **FR:** 3–5 mm wide, sticky, 3-ridged distally; horns 0. 2*n*=24. Slopes, ridges, chaparral, conifer forest; 15–1675 m. n&c SN, SCo, SnGb, SnBr, PR; n Baja CA. Feb–May

C. velutinus Douglas (p. 1167) Pl open, < 6 m. **ST:** erect; twigs flexible, not thorn-like, green to red-brown. **LF:** alternate, evergreen, aromatic; stipules scale-like; petiole 9–32 mm; blade 33–75 mm, 13–55 mm wide, widely elliptic to ovate, leathery, adaxially shiny green, glabrous, ± sticky, abaxially glabrous to ± puberulent, tip obtuse to rounded, margin minutely gland-toothed, teeth 90–150+. **INFL:** panicle-like, 3–8 cm. **FL:** white. **FR:** 3–4.5 mm wide, brown, sticky, wrinkled or not; horns 0. Open, rocky slopes; 90–3050 m. KR, NCoRO, CaRH, SNH, n SnFrB, Wrn, SNE (exc W&I); to BC, SD. [*C. v.* var. *hookeri* M.C. Johnst., illeg.] Apr–Jul

C. verrucosus Nutt. WART-STEMMED CEANOTHUS Pl erect, ± open, < 3 m. **ST:** twigs angled, gray-brown. **LF:** gen alternate, some clustered, evergreen; stipules knob-like; petiole < 3 mm; blade 5–14 mm, 3–10 mm wide, length gen < 2 × width, widely obovate to round, firm, adaxially dark green, glabrous, abaxially gray-green, glabrous to sparsely short-strigose, 1-ribbed from base, tip truncate to notched, margin entire or teeth 9–12, ± sharp. **INFL:** umbel-like, 1–1.5 cm. **FL:** white exc disk; ovary dark. **FR:** 4–6 mm wide, ± sticky, ± ridged or not; horns 0 or minute. 2*n*=24. Rocky slopes, chaparral; < 350 m. s SCo (San Diego Co.); n Baja CA. Jan–Apr ★

C. vestitus Greene (p. 1167) MOJAVE CEANOTHUS Pl erect, ± open, 0.8–2 m. **ST:** ascending to erect, gen intricately branched; twigs pale gray to ± white, not angled, densely puberulent to short-tomentose. **LF:** opposite, evergreen; stipules knob-like; petiole 1–3 mm; blade 5–20 mm, 3–19 mm wide, oblanceolate to ± round, adaxially concave, gray-green to yellow-green, puberulent, glabrous in age, abaxially convex, gray-green, glabrous to short-curly-puberulent, tip gen acute to obtuse, margin gen not thick, gen not rolled under, entire or teeth 3–5, sharp. **INFL:** umbel-like, gen 1–1.5 cm. **FL:** gen white. **FR:** 3–5 mm wide, ± 3-ridged distally; horns 0 or < 1 mm. Slopes, flats, chaparral, woodland, conifer forest; 550–2600 m. s SNH, Teh, TR, PR, SNE, DMtns; to UT, TX, n Mex. [*C. greggii* A. Gray var. *v.* (Greene) McMinn] *C. greggii* A. Gray restricted to Mex. Mar–May

COLUBRINA

Shrub. **ST:** branches dense, alternate, rigid, ± white; twigs spreading, not jointed at base, thorn-tipped [or not], gen hairy. **LF:** clustered on short-shoots or not, deciduous [or not]; stipules deciduous; blade entire, 3–5-ribbed from base, gen with round, marginal glands. **INFL:** umbel-like, few-fld. **FL:** hypanthium hemispheric, adhering to developing fr; sepals 5; petals 5, = sepals, oblanceolate; stamens 5; disk fleshy, adhering to, ± filling hypanthium; ovary chambers 3, 1-ovuled, stigma 3-lobed. **FR:** capsule, shallow-3-valved, explosively dehiscent (drupe-like, indehiscent). 31 spp.: warm places worldwide. (Latin: from French for serpent tree) [Bastos 1990 Pesquisas Botanica 41:99–122]

C. californica I.M. Johnst. (p. 1167) LAS ANIMAS COLUBRINA Pl < 3 m. **ST:** straight, tomentose. **LF:** blade 12–35 mm, oblong to obovate, dull gray-green, hairs silky, denser abaxially, base rounded or wedge-shaped, tip rounded to ± notched, mucronate or not. **INFL:** 5–10 mm, 3–12-fld, dense; pedicel 1–2 mm, 2–4 mm in fr. **FL:** appearing after rain; hypanthium ± 3 mm wide. **FR:** 8–10 mm, persisting 3–6 months. Desert scrub; 240–920 m. s DMoj, DSon; AZ, Mex. Apr–May ★

CONDALIA

Shrub. **ST:** branches alternate, rigid; twigs thorn-tipped. **LF:** clustered on short-shoots, deciduous; stipules deciduous; petioles ± 0; blade gen obovate, 1-ribbed from base, entire. **INFL:** fls 1 or in clusters on short-shoots. **FL:** hypanthium hemispheric, 1–1.5 mm wide; sepals 5, deciduous; petals 0 [5]; stamens 5; disk 0 or thin in early fl; ovary spheric, strong-narrowed at base, chambers 2, each 1-ovuled, stigma entire. **FR:** drupe, stone 1. 18 spp.: arid Am. (A. Condal, Spanish physician, 1745?–1804) [Christie et al. 2006 Canotia 2:23–46]

C. globosa I.M. Johnst. var. ***pubescens*** I.M. Johnst. (p. 1167) SPINY ABROJO Pl < 4 m. **ST:** bark smooth, gray; twigs 3–13 cm, pale olive-green or purple, short-hairy. **LF:** in clusters of 2–7; stipules brown; blade 3–12 mm, narrowly oblanceolate to obovate, ± thickened, densely to sparsely hairy, base wedge-shaped. **INFL:** 1–8-fld. **FL:** hypanthium 1–1.5 mm wide, olive-green or purple, short-hairy; sepals ± 1 mm, olive-green; stamens < sepals; pistil purple. **FR:** 3–5 mm, black, juicy. Desert scrub; < 1000 m. DSon; AZ, Mex. Other var. (twigs glabrous) in nw Mex. Mar–Apr ★

FRANGULA COFFEE BERRY

Shrub, small tree. **ST**: branches alternate, flexible; winter bud scales 0. **LF**: scattered along branches or clustered on short-shoots, deciduous or not; stipules gen deciduous; petioled; blade veins prominent or not. **INFL**: umbel or fls 1 in axils. **FL**: bisexual; hypanthium 1–3 mm wide, cup-shaped; sepals 5, erect, fleshy, keeled adaxially; petals 5, short-clawed; stamens 5; disk thin, adherent to hypanthium; ovary ± inferior, chambers 2–3, 1–2-ovuled, stigma 2–3-lobed. **FR**: drupe, 2–3[4]-stoned. 50 spp.: temp, w. Med, Eurasia. (Frangible: capable of being broken) [Sawyer & Edwards 2007 Madroño 54:172–174] Often a subg. of *Rhamnus*; some of value in food, medicine.

1. Lvs deciduous, blades thin
  2. Fr 3-stoned; lf blade (50)80–150 mm . . . . *F. purshiana*
   3. Lf blade blue- or green-gray, ± glaucous when fresh, papillate, densely hairy or adaxially velvety,
     light green, abaxially sparsely to densely hairy . . . . ²subsp. *ultramafica*
   3′ Lf blade green, not papillate, glabrous to sparsely hairy
    4. Lf base tapered; KR, CaR, n&c SN . . . . subsp. *annonifolia*
    4′ Lf base rounded or cordate; NCo, KR, NCoRO, NCoRH . . . . subsp. *purshiana*
  2′ Fr 2(3)-stoned; lf blade 15–80 mm . . . . *F. rubra*
   5. Lf blade finely hairy . . . . subsp. *yosemitana*
   5′ Lf blade glabrous or abaxially puberulent on midrib, veins
    6. Twigs gray; lvs clustered on short-shoots . . . . subsp. *modocensis*
    6′ Twigs red to gray; lvs scattered along st
     7. Lf base, tip rounded . . . . subsp. *obtusissima*
     7′ Lf base, tip acute . . . . subsp. *rubra*
1′ Lvs semi-deciduous or evergreen, blades ± leathery or thin
  8. Lvs semi-deciduous, fr 2–3-stoned
   9. Lf blade thin, 20–60 mm, elliptic; fr 2-stoned . . . . ²*F. californica* subsp. *cuspidata*
   9′ Lf blade ± leathery, 50–100 mm, widely oblong or ovate to obovate; fr 3-stoned . . . ²*F. purshiana* subsp. *ultramafica*
  8′ Lvs gen evergreen; fr gen 2-stoned . . . . *F. californica*
   10. Lf blade ± glabrous or abaxially ± puberulent
    11. Lf blade dark green adaxially, bright green or yellow abaxially; fr 2-stoned . . . . subsp. *californica*
    11′ Lf blade ± yellow-green; fr 3-stoned . . . . subsp. *occidentalis*
   10′ Lf blade adaxially glabrous to tomentose, abaxially glabrous to tomentose, velvety, or silvery
    12. Lf blade adaxially green, glabrous or minutely puberulent, abaxially glabrous to white-tomentose
      mixed with long hairs
     13. Lf tip abruptly pointed or not, margin dentate to dentate-serrate, long hairs conspicuous abaxially;
       c&s SN, Teh, TR, nw PR, SnJt, SNE, DMoj . . . . ²subsp. *cuspidata*
     13′ Lf tip acute to rounded, margin entire to serrate, long hairs inconspicuous abaxially; DMtns
       (Clark, New York, Providence mtns) . . . . subsp. *ursina*
    12′ Lf blade white-tomentose, or adaxially dull green, glabrous, abaxially white-tomentose or velvety
      to silvery, long hairs 0.
     14. Lf blade narrowly elliptic, abaxially velvety to silvery . . . . ²subsp. *tomentella*
     14′ Lf blade widely elliptic, abaxially white-tomentose or velvety to silvery
      15. Lf blade adaxially white-tomentose, tip obtuse; s KR, NCoRI . . . . subsp. *crassifolia*
      15′ Lf blade adaxially glabrous (white-tomentose), tip acute; s KR, NCoR, CaRF, SNF, n SNH,
        ScV, SnFrB, SCoR, SW . . . . ²subsp. *tomentella*

F. californica (Eschsch.) A. Gray (p. 1167) CALIFORNIA COFFEE BERRY Shrub < 5 m. **ST**: bark bright gray, brown, or red; twigs brown, gray, or red, glabrous, tomentose, or hairs of 2 lengths; terminal bud glabrous to velvety. **LF**: gen evergreen; petiole 3–10 mm; blade 20–100 mm, elliptic to ovate, ± leathery, ± glabrous to tomentose, base acute to rounded, tip truncate to acute or mucronate, margin entire to toothed, ± rolled under or not, veins gen prominent. **INFL**: 5–60-fld; pedicel < 20 mm. **FL**: hypanthium 1–2 mm wide. **FR**: gen 2-stoned, 10–15 mm, black. [*Rhamnus c.* Eschsch.] Subspp. intergrade in intermediate habitats.

subsp. ***californica*** **ST**: twigs red, glabrous. **LF**: blade 20–80 mm, narrowly to widely elliptic, dark green adaxially, bright green or yellow abaxially, glabrous or ± puberulent, base, tip acute. 2*n*=24. Coastal-sage scrub, desert scrub, chaparral, forest, woodland; < 2800 m. NW, CW, SW, DMtns (Providence Mtns). [*Rhamnus c.* Eschsch. subsp. *c.*] May–Jul

subsp. ***crassifolia*** (Jeps.) Kartesz & Gandhi **ST**: twigs gray, tomentose. **LF**: blade 30–100 mm, widely elliptic, white-tomentose, tip obtuse, margin entire or blunt-toothed. Chaparral, woodland; < 1400 m. s KR, NCoRI. [*Rhamnus tomentella* subsp. *crassifolia* (Jeps.) Sawyer] Morphologically similar pls in PR here considered *F. californica* subsp. *tomentella*. Feb–Apr

subsp. ***cuspidata*** (Greene) Kartesz & Gandhi Pl < 2 m. **ST**: twigs red, hairs of 2 lengths. **LF**: evergreen or semi-deciduous; blade 20–60 mm, elliptic, thin, green, glabrous or minutely puberulent adaxially, white-tomentose mixed with conspicuous long hairs abaxially, tip mucronate or not, margin dentate to dentate-serrate. Chaparral, desert scrub, montane woodland; 400–2300 m. c&s SN, Teh, TR, nw PR, SnJt, SNE, DMoj. [*Rhamnus tomentella* subsp. *cuspidata* (Greene) Sawyer] Pls at high elevations ± winter-deciduous. Apr–Jul

subsp. ***occidentalis*** (Greene) Kartesz & Gandhi Pl < 2 m. **ST**: twigs brown, glabrous. **LF**: blade 20–80 mm, ovate to elliptic, ± yellow-green, glabrous, base rounded, tip acute, rounded, or truncate, veins not prominent. **FR**: 3-stoned. Chaparral, woodland on serpentine; < 2300 m. KR, NCoRO, n SNF; sw OR. [*Rhamnus c.* subsp. *o.* (Greene) C.B. Wolf] Mar–Jun

subsp. ***tomentella*** (Benth.) Kartesz & Gandhi **ST**: twigs gray, tomentose. **LF**: blade 30–70 mm, narrowly (widely) elliptic, dull green, (white-tomentose or) adaxially glabrous, abaxially velvety or silvery, long hairs 0, tip acute, margin entire or blunt-toothed. Chaparral, woodland; < 2200 m. s KR, NCoR, CaRF, SNF, n SNH, ScV, SnFrB, SCoR, SW; Baja CA. [*Rhamnus t.* Benth. subsp. *t.*] Jan–Apr

subsp. ***ursina*** (Greene) Kartesz & Gandhi Pl < 2 m. **ST**: twigs gray, tomentose. **LF**: blade 30–85 mm, elliptic or ovate, green,

glabrous or minutely puberulent adaxially, pale green, glabrous to minutely puberulent mixed with inconspicuous long hairs abaxially, tip acute to rounded, margin entire to serrate. Desert scrub, woodland; 1000–2100 m. SnBr, DMtns (Clark, New York, Providence mtns); to AZ, NV, NM, Baja CA. [*Rhamnus tomentella* subsp. *u.* (Greene) Sawyer; *R. t.* var. *u.*, ined.] May–Jul

F. purshiana (DC.) J.G. Cooper CASCARA **ST:** bark gray; twigs green, gray, red, or dull brown, gen glabrous or densely hairy; terminal bud brown-hairy. **LF:** gen deciduous; petiole 5–25 mm; blade (50)80–150 mm, widely elliptic to obovate, gen thin, gen green, gen not papillate, glabrous to sparsely hairy, or blue- or green-gray, ± glaucous when fresh, papillate, densely hairy or velvety adaxially, light green, sparsely to densely hairy abaxially, base rounded, cordate, or tapered, tip obtuse to truncate or notched, margin entire to toothed, gen not wavy, veins prominent, 1°, 2°, 3° veins gen glabrous or sparsely hairy. **INFL:** < 25-fld; pedicel < 25 mm. **FL:** hypanthium 3 mm wide. **FR:** 3-stoned, 5–10 mm, black. Bark and fr TOXIC in excess, esp to children. [*Rhamnus p.* DC.] Cathartic drugs from bark.

subsp. **annonifolia** (Greene) Sawyer & S.W. Edwards Shrub, < 5 m. **ST:** twigs red to brown. **LF:** petiole 5–25 mm; blade thin, green, glabrous to sparsely hairy, base tapered, tip obtuse to truncate, margin irregularly toothed to entire. Conifer forest edges, streamsides, non-serpentine; < 2000 m. KR, CaR, n&c SN; s OR. [*Rhamnus p.* DC. var. *a.* (Greene) Jeps.] Mar–Jun

subsp. **purshiana** (p. 1167) Tree, shrub, < 12 m. **ST:** bark gray; twigs red to brown. **LF:** petiole 5–25 mm; blade thin, green, glabrous to sparsely hairy, base rounded or cordate, tip obtuse to truncate, margin irregularly toothed to entire. Coastal scrub, conifer forest, forest edges, non-serpentine; < 2000 m. NW (exc NCoRI); to BC, MT. Feb–Jun

subsp. **ultramafica** Sawyer & S.W. Edwards CARIBOU COFFEE BERRY Shrub, gen < 2 m. **ST:** twigs green to gray or dull brown, densely hairy. **LF:** clustered near st tips, deciduous or semi-deciduous; petiole 5–15 mm; blade widely oblong or ovate to obovate, ± leathery, blue- or green-gray, ± glaucous when fresh, papillate, sparsely to densely hairy or velvety, abaxially light green, base obtuse or tapered, tip obtuse, often notched, margin entire to ± minutely serrate, often wavy, 1°, 2° veins densely hairy, 3° veins less so. Open conifer forest, montane chaparral, seeps, serpentine; 820–1950 m. n SN (Plumas Co.). Apr–Jun ★

F. rubra (Greene) Grubov (p. 1167) SIERRA COFFEE BERRY Shrub, < 2 m. **ST:** bark red to bright gray; twigs red to gray; terminal bud hairy. **LF:** gen scattered along st, deciduous; petiole 2–12 mm; blade 15–80 mm, narrowly elliptic to obovate, thin, green or gray, gen glabrous to finely hairy or abaxially puberulent on midrib, veins, acute to rounded at base, tip, margin finely toothed to entire, veins not prominent. **INFL:** 4–15-fld; pedicel 1–12 mm. **FL:** hypanthium 2 mm wide. **FR:** 2(3)-stoned, 12 mm, black. [*Rhamnus r.* Greene] 1 other subsp., in NV.

subsp. **modocensis** (C.B. Wolf) Kartesz & Gandhi **ST:** twigs gray. **LF:** lvs clustered on short-shoots; blade 15–40 mm, narrowly elliptic, glabrous or abaxially puberulent on midrib, veins, acute at base, tip. Montane forest, sagebrush steppe; 1000–2200 m. CaR, MP. [*Rhamnus r.* subsp. *m.* C.B. Wolf] Apr–Jun

subsp. **obtusissima** (Greene) Kartesz & Gandhi **ST:** twigs red to gray. **LF:** blade 25–60 mm, oblong to obovate, green, glabrous or abaxially puberulent on midrib, veins, rounded at base, tip. Chaparral, montane forest; 1000–2000 m. KR, NCoRH, CaR, n&c SNH. [*Rhamnus r.* subsp. *o.* (Greene) C.B. Wolf] Mar–Jun

subsp. **rubra** **ST:** twigs red. **LF:** blade 20–60 mm, narrowly elliptic to oblong, glabrous, base, tip acute. Chaparral, montane forest; 1000–2200 m. KR, NCoRH, CaR, n&c SNH. [*Rhamnus r.* Greene subsp. *r.*] Mar–Jun

subsp. **yosemitana** (C.B. Wolf) Kartesz & Gandhi **ST:** twigs red to gray. **LF:** blade 30–70 mm, narrowly elliptic to oblong, finely hairy, base, tip acute to rounded. Chaparral, montane forest; 1000–2200 m. c SNH. [*Rhamnus r.* Greene subsp. *y.* C.B. Wolf] Apr–Jun

RHAMNUS BUCKTHORN

Shrub, small tree, < 10 m. **ST:** branches alternate, stiff or flexible; twigs gen not thorn-tipped; winter bud scales present, gen ± 3 mm. **LF:** scattered along branches or clustered on short-shoots, deciduous or evergreen; stipules gen deciduous; petioles gen glabrous; blade veins prominent or not. **INFL:** fls 1 or in cyme-like clusters in axils. **FL:** unisexual (bisexual), gen on separate pls, gen < 3 mm; hypanthium bell-shaped to cup-like, 2–3 mm wide; sepals 4–5, thin, spreading, not keeled adaxially; petals 0 or 4–5; disk thin, adhering to hypanthium; ovary appearing superior or partly inferior, chambers 2–4, each 1-ovuled, style 1, stigma 2–4-lobed. **FR:** drupe, 2–3[4]-stoned. 110 spp.: temp, few trop; some of value in medicine or as dyes. (Greek: name for pls of this genus) [Bolmgren & Oxelman 2004 Taxon 53:383–390] W.H. Brewer collected *R. cathartica* L., considered invasive in parts of US, in 1861, but it apparently never naturalized. Other taxa in TJM (1993) moved to *Frangula*.

1. Lvs deciduous; sepals 5; fr black, 3-stoned . . . . *R. alnifolia*
1′ Lvs evergreen; sepals 4–5; fr red, 2-stoned, or black, 3-stoned
　2. Lf blade 20–60 mm, elliptic or ovate to lance-ovate; sepals 5; fr black, 3-stoned — CCo (Monterey Co.),
　　SnFrB, SCo (Newport Back Bay), expected elsewhere. . . . . *[R. alaternus]*
　2′ Lf blade 10–40 mm, elliptic to round; sepals 4; fr red, 2-stoned (*Rhamnus crocea* complex)
　　3. Lf blade ≤ 15 mm, flat abaxially; branches spreading . . . . *R. crocea*
　　3′ Lf blade ≥ 15 mm, flat to concave abaxially; branches gen ascending
　　　4. Lf blade elliptic; small tree; ChI . . . . *R. pirifolia*
　　　4′ Lf blade elliptic to round; shrub; mainland
　　　　5. Lf blade glabrous or hairy abaxially, glabrous adaxially; twigs glabrous to finely hairy;
　　　　　CA-FP, DMtns . . . . *R. ilicifolia*
　　　　5′ Lf blade soft hairy; twigs densely hairy; s PR (San Diego Co.) . . . . *R. pilosa*

R. alnifolia L'Hér. ALDER BUCKTHORN Shrub, < 2 m. **ST:** bark gray; branches puberulent to glabrous; twigs brown; winter bud scales ± 5 mm. **LF:** deciduous; petiole 4–16 mm; blade 45–110 mm, elliptic to ovate, thin enough to transmit light, glabrous to puberulent, base acute to obtuse, tip acute, margin irregularly toothed, veins prominent, arched. **INFL:** 1–3-fld; pedicels 2–10 mm. **FL:** unisexual; hypanthium ± 1 mm wide; sepals 5; petals 0, 4, 5. **FR:** 3-stoned, ± 8 mm, black. Wet meadow edges, seeps, stream sides; 1450–2020 m. n SNH; to Can, n Rocky Mtns, n US. May–Jul ★

R. crocea Nutt. (p. 1167) SPINY REDBERRY Shrub, < 2 m. **ST:** bark gray; branches many, spreading, stiff, rooting; twigs ± thorn-tipped, red or red-purple. **LF:** evergreen; petiole 1–4 mm; blade 10–15 mm, elliptic to obovate, thick, glabrous, abaxially flat, base acute to rounded, tip rounded, margin sharp-toothed or entire, veins not prominent. **INFL:** 1–6-fld, glabrous; pedicel 1–6 mm. **FL:** gen unisexual; hypanthium ± 2 mm wide; sepals 4; petals 0. **FR:** 2-stoned, 6 mm, red. Coastal-sage scrub, chaparral, woodland; < 1150 m. KR, NCoRO, SNH, CW, SW; Baja CA. Jan–Apr

R. ilicifolia Kellogg (p. 1167)   HOLLYLEAF REDBERRY   Shrub <
4 m. **ST:** bark gray; branches stiff, gen ascending; twigs glabrous
to finely hairy. **LF:** evergreen; petiole 2–10 mm; blade 20–40 mm,
ovate to round, thick, glabrous adaxially, glabrous or hairy, flat to
concave abaxially, base rounded, tip obtuse, rounded, or widely
notched, margin entire, irregularly toothed, or prickly, veins prominent or not. **INFL:** 1–6-fld, gen glabrous; pedicel 2–4 mm. **FL:** gen
unisexual; hypanthium ± 2 mm wide; sepals 4; petals 0. **FR:** 2-stoned,
4–8 mm, red. 2*n*=24. Chaparral, desert scrub, montane forest; < 1150
m. CA-FP, DMtns; s OR, AZ, Baja CA. Intermediates between *R.
ilicifolia*, *R. insula* Kellogg (Baja CA) in PR. Mar–Jun

R. pilosa (Curran) Abrams (p. 1167)   Shrub, < 2 m. **ST:** bark gray;
branches few, gen flexible, gen ascending; twigs densely hairy. **LF:**
evergreen; petiole 2–5 mm, hairy; blade 15–20 mm, ovate to round,
thick, soft-hairy, abaxially flat to concave, base tip rounded to acute,
margin finely toothed, flat to rolled under, veins not prominent. **INFL:**
1–6-fld, soft-hairy; pedicel 2–4 mm. **FL:** gen unisexual; hypanthium
± 2 mm wide; sepals 4; petals 0, 4. **FR:** 2-stoned, 6 mm, red. Uncommon. Chaparral; 75–1650 m. s PR (San Diego Co.). Jan–Mar

R. pirifolia Greene (p. 1167)   ISLAND REDBERRY   Small tree,
< 10 m. **ST:** bark gray; branches ascending; twigs purple. **LF:** ever-
green; petiole 5–10 mm; blade 20–50 mm, elliptic, thick, glabrous,
abaxially concave, base rounded, tip acute to rounded, mucronate,
margin entire to toothed, veins prominent. **INFL:** 1–6-fld, glabrous;
pedicels 3–6 mm. **FL:** gen unisexual; hypanthium ± 2 mm wide;
sepals 4; petals 0. **FR:** 2-stoned, 6–8 mm, red. Coastal-sage scrub,
chaparral; 10–520 m. ChI; Mex (Guadalupe Island). Jan–Apr   ★

ZIZIPHUS

Tree, shrub [vine]. **ST:** branches alternate, flexible or stiff, ± 2–3-ranked; twigs ± pendent, zigzag. **LF:** clustered on short-
shoots or not, deciduous or not, petioled; stipules ± spine-like or not, unequal; blade elliptic to obovate, ± entire to serrate,
1–5-ribbed from base. **INFL:** cyme or small panicle. **FL:** hypanthium surrounding base of ovary; sepals 5; petals 5, < to >
sepals; stamens 5; disk thick, ± obscuring ovary in early fl; ovary widely attached at base, chambers 2, each 1-ovuled, stigma
2-lobed. **FR:** drupe, stone 1. 100 spp.: gen trop. (Latin: from Arabic "zizouf", common jujube, *Ziziphus zizyphus*) [Islam &
Simmons 2006 Syst Bot 31:826–842]

1. Tree, lvs ≥ 25 mm, blade shiny adaxially . . . . ***Z. zizyphus***
1′ Shrub; lvs ≤ 25 mm, blade not shiny adaxially
  2. Fr 7–10 mm, blue-black, not beaked . . . . ***Z. obtusifolia*** var. ***canescens***
  2′ Fr 10–25 mm, brown, beaked. . . . . ***Z. parryi*** var. ***parryi***

Z. obtusifolia (Torr. & A. Gray) A. Gray var. ***canescens*** (A. Gray)
M.C. Johnst. (p. 1173)   GRAYTHORN   Shrub, < 3 m. **ST:** bark gray,
smooth; twigs 1–8 cm, stiff, spreading, thorn-tipped, densely short-
white-hairy. **LF:** deciduous; stipules brown; blade 2–20 mm, ovate or
oblong, firm, dull gray, margin entire or teeth 2–10, glandular. **INFL:**
2–30-fld. **FL:** hypanthium 1.5–2 mm wide, olive-green, glabrous to
tomentose; sepals ± yellow to ± orange or ± purple; petals white; pistil olive-green. **FR:** fleshy, 7–10 mm, not beaked, blue-black. Uncommon. Desert scrub; 45–1250 m. D; AZ, NV, UT, Mex. 1 other var., in
NM, OK, TX, Mex. Apr–Jun

Z. parryi Torr. var. ***parryi*** (p. 1173)   PARRY'S JUJUBE   Shrub, <
4 m. **ST:** bark gray to brown, smooth; branches pale green-yellow
to ± purple; twigs 1.3–3 cm, stiff, spreading, thorn-tipped, with 1
node, 1 short-shoot, glabrous. **LF:** deciduous, membranous; stipules brown; blade 10–25 mm, elliptic to obovate, dull olive-green,
margin ± entire. **INFL:** 2–4-fld. **FL:** hypanthium 2–2.2 mm wide,
purple-green, glabrous; sepals green; petals white; pistil green. **FR:**
dry, 10–25 mm, beaked, brown. Uncommon. Chaparral; 15–1220 m.
SCo, SnBr, PR, DMoj, w edge DSon; Mex. 1 other var., in Baja CA
incl Cedros Island. Feb–Apr

Z. zizyphus (L.) H. Karst.   JUJUBE   Tree, < 12 m. **ST:** bark gray
to black, shaggy; twigs pendent, zigzag. **LF:** deciduous; blade 25–60
mm, elliptic to obovate, ± leathery, bright green, shiny adaxially,
paler, dull abaxially, margin serrate. **INFL:** 2–8-fld clusters at nodes
on twigs. **FL:** hypanthium 2–2.2 mm wide, green; sepals < petals,
both yellow; stamens < petals; pistil green. **FR:** Fleshy, red. Occ garden escape; 270 m. n SNF, n ScV; US, Old World. [*Z. jujuba* Mill.]
Correct name could become *Z. jujuba* Mill. May–Jul

ROSACEAE   ROSE FAMILY

Daniel Potter & Barbara Ertter, family description, key to genera;
treatment of genera by Daniel Potter, except as noted

Ann to tree, glandular or not. **LF:** simple to palmately or pinnately compound, gen alternate; stipules free to fused (0), persistent to deciduous. **INFL:** cyme, raceme, panicle, cluster, or fls 1; bractlets on pedicel ("pedicel bractlets") gen 0–3(many),
subtended by bract or gen not. **FL:** gen bisexual, radial; hypanthium free or fused to ovary, saucer- to funnel-shaped, subtending bractlets ("hypanthium bractlets") 0–5, alternate sepals; sepals gen 5; petals gen 5, free; stamens (0,1)5–many, anther
pollen sacs gen 2; pistils (0)1–many, simple or compound, ovary superior to inferior, styles 1–5. **FR:** 1–many per fl, achene
(fleshy-coated or not), follicle, drupe, or pome with gen papery core, occ drupe-like with 1–5 stones. **SEED:** gen 1–5 (per fr,
not per fl). 110 genera, ± 3000 spp.: worldwide, esp temp; many cult for orn, fr, esp *Cotoneaster, Fragaria, Malus, Prunus,
Pyracantha, Rosa, Rubus*. Number of teeth is per lf or lflet, not per side of lf or lflet, exc in *Drymocallis*. [Potter et al. 2007 Pl
Syst Evol 266:5–43] Scientific Editors: Daniel Potter, Thomas J. Rosatti.

1. Ann to per
  2. Pl 1–2 m; lvs 2–3-pinnately compound; fls unisexual; fr follicles . . . . **ARUNCUS**
  2′ Pl gen < 1 m; lvs simple to 1-ternately or 1-palmately or 1-pinnately compound, incl 2-ternately
    dissected; fls gen bisexual; fr achenes or follicles
    3. Petals 0, sepals gen 4; frs 1(3) per fl, inside ± urn-shaped hypanthium
      4. Lf palmately lobed; infl few-fld cluster, axillary, ± hidden by sheathing stipules; hypanthium not
        hardened in fr; ann . . . . **APHANES**

4′ Lf pinnately compound; infl head or spike, axillary or terminal, not hidden; hypanthium ± hardened in fr; gen per

   5. Hypanthium ± prickly, not angled in fr; stamens 2 or 4 . . . . **ACAENA**

   5′ Hypanthium not prickly, 4-angled in fr; stamens 0–many

      6. Lflets lobed > 2/3 to midvein; ann, bien; basal lvs withered at fl . . . . **POTERIDIUM**

      6′ Lflets toothed < 1/3 to midvein; per; basal lvs not withered at fl

         7. Largest lflet blade gen 5–20 mm, on stalk gen 1–4 mm, teeth gen < 15; stamens many, filaments thread-like; open, esp disturbed areas; CA-FP (exc SNH) . . . . **POTERIUM**

         7′ Largest lflet blade 25–50 mm, on stalk 3–25 mm, teeth gen > 15; stamens 2–4, filaments not thread-like; bogs, streams; c NCo, nw KR, n NCoRO . . . . **SANGUISORBA**

3′ Petals, sepals gen 5; frs 1–many per fl, on or outside shallow to ± obconic hypanthium

   8. Lvs ± 2-ternately dissected, segments linear; fr follicles . . . . **²LUETKEA**

   8′ Lvs simple to compound, ternately compound or not but not ternately dissected, segments linear or not or 0; fr achenes, sometimes fleshy-coated (*Rubus*) or on fleshy receptacle (*Duchesnea*, *Fragaria*)

      9. Hypanthium obconic to cup-shaped, rim with hooked bristles; infl ± raceme . . . . **AGRIMONIA**

      9′ Hypanthium gen shallower than obconic to cup-shaped, rim without hooked bristles; infl gen cyme or fls 1

         10. Hypanthium bractlets 0; achenes fleshy-coated, finely hairy . . . . **³RUBUS**

         10′ Hypanthium bractlets gen present; achenes not fleshy-coated, gen glabrous (exc *Geum*)

            11. Style continuous to fr, if not or inconspicuously hooked then plumose — lvs 1-pinnately compound . . . **GEUM**

            11′ Style jointed to fr, not hooked, not hairy

               12. Lf (sub)palmately to ternately compound

                  13. Receptacle in fr strawberry-like (enlarged, red, fleshy); stolons gen present; petals white or yellow; lvs ternately compound

                     14. Petals yellow; fls 1 from axils of lvs on stolons; hypanthium bractlets gen wider than sepals . . . **DUCHESNEA**

                     14′ Petals gen white; fls 1–several from axils of basal lvs, stolons lfless; hypanthium bractlets narrower than sepals . . . . **FRAGARIA**

                  13′ Receptacle in fr not strawberry-like; stolons 0; petals ± yellow; lvs pinnately, palmately, or ternately compound

                     15. Stamens 5; petals ± 1 mm; lflets 3, gen 3-toothed at tip . . . . **SIBBALDIA**

                     15′ Stamens 10–25; petals (2)4–20 mm; lflets 3–7, gen > 3-toothed . . . . **²POTENTILLA**

               12′ Lf ± 1-pinnately to subpalmately compound

                 16. Petals dark red, elliptic to ± ovate, << sepals . . . . **COMARUM**

                 16′ Petals yellow to white, elliptic to obcordate (linear), < to > sepals

                  17. Style fusiform, attached below fr middle . . . . **DRYMOCALLIS**

                  17′ Style slender throughout or widest near base, gen attached below fr tip

                    18. Hypanthium cup-like, ± flat-bottomed; filaments gen ± flat, often forming tube; petals white or ± pink

                     19. Stamens 10; lflets 2–15 per side; CA-FP, GB. . . . . **HORKELIA**

                     19′ Stamens 20; lflets 15–35 per side; c&s SNH, SNE. . . . . **HORKELIELLA**

                  18′ Hypanthium ± shallow, if cup-like, not flat-bottomed; filaments gen thread-like, not flat (exc ± flat in *Ivesia argyrocoma* var. *argyrocoma*), not forming tube; petals white to yellow

                    20. Pl gen hanging clump or rosette in vertical rock crevices, ± resin-scented . . . . **²IVESIA**

                    20′ Pl not hanging clump or rosette, gen not in vertical rock crevices, resin-scented or not

                     21. Lf gen ± cylindric; lflets 4–80 per side; pistils 1–8(20); stamens 5–20(40); petals 1–5(7) mm, linear to obovate or round . . . . **²IVESIA**

                     21′ Lf gen ± flat; lflets 2–8(13) per side; pistils gen > 10; stamens 20–25; petals (2)4–20 mm, gen ± widely obcordate . . . . **²POTENTILLA**

1′ Subshrub to tree

  22. Lvs deeply lobed (gen to midrib) or compound

     23. Ovary gen inferior, chambers 1–5; styles 1–5; fr pome . . . . **SORBUS**

     23′ Ovary superior, chamber 1; style 1; fr achene or follicle

        24. Pistils gen 10–many

          25. Petals yellow; hypanthium bractlets 5; style attached near fr base. . . . . **DASIPHORA**

          25′ Petals white to red; hypanthium bractlets 0 (5 in *Fallugia*); style attached at fr tip

             26. Pl not prickly; lvs pinnately lobed, rusty-scaly abaxially, margins rolled under; style plumose in fr; e DMtns. . . . . **FALLUGIA**

             26′ Pl gen prickly; lvs pinnately to palmately compound, not rusty-scaly, margins not gen rolled under; style short-hairy to ± glabrous in fr; gen CA-FP, GB

                27. Lvs pinnately compound; hypanthium urn-shaped; fr not on spongy receptacle, gen enclosed in fleshy hypanthium. . . . . **ROSA**

                27′ Lvs palmate-lobed or -compound; hypanthium flat to saucer-shaped; fr on spongy receptacle, not enclosed by hypanthium (aggregate raspberry- or blackberry-like). . . . . **³RUBUS**

        24′ Pistils 1–6

          28. Lvs opposite; 1° lflets 5–14 cm; infl gen flat-topped; trunk bark peeling in strips; ChI. . . . . . **²LYONOTHAMNUS**

          28′ Lvs alternate or clustered; 1° lflets or lobes gen < 5 cm; infl not flat-topped; trunk bark gen not peeling in strips; mainland

             29. Lf 3–9-lobed; infl gen 1-fld; styles in fr 2–6 cm, plumose . . . . **⁴PURSHIA**

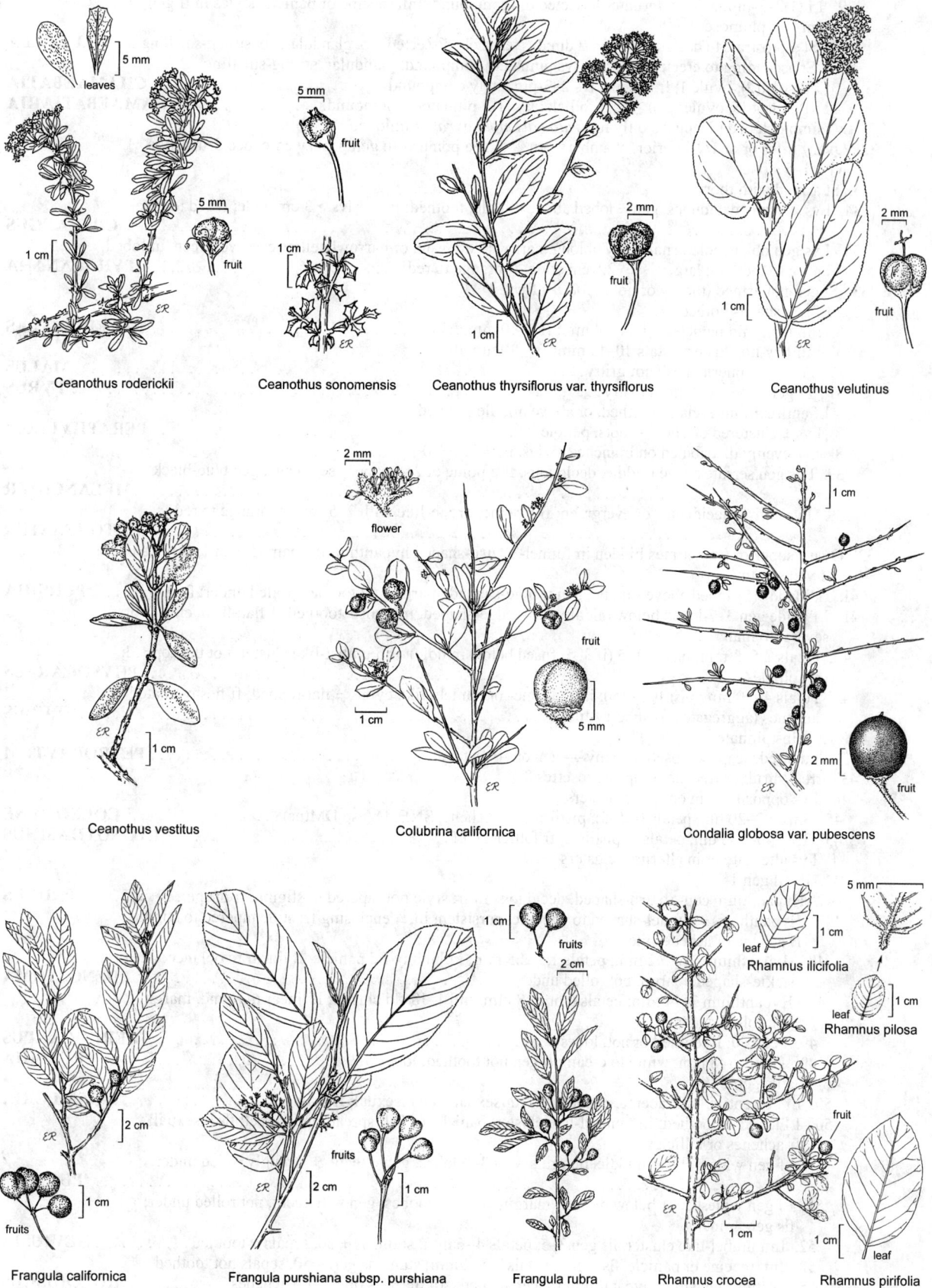

Ceanothus roderickii

Ceanothus sonomensis

Ceanothus thyrsiflorus var. thyrsiflorus

Ceanothus velutinus

Ceanothus vestitus

Colubrina californica

Condalia globosa var. pubescens

Frangula californica

Frangula purshiana subsp. purshiana

Frangula rubra

Rhamnus crocea

Rhamnus pirifolia

Rhamnus ilicifolia

Rhamnus pilosa

29′ Lf (1)2–3-pinnate- or -ternate-dissected or -compound; infl raceme or panicle; styles in fr gen << 1 cm, not plumose

  30. St ± prostrate to decumbent; pl < 2 dm; lvs ternately dissected, not glandular, not strong-smelling .... [2]**LUETKEA**

  30′ St ascending to erect; pl > 2 dm; lvs pinnately compound, glandular, strong-smelling

    31. Pistil 1(2); ovule 1; fr achene; lvs 2–3-pinnately compound .... **CHAMAEBATIA**

    31′ Pistils 4–5; ovules gen ≥ 2; fr follicle; lvs (1)2-pinnately compound............... **CHAMAEBATIARIA**

22′ Lvs simple, lobed (deeply so only in *Crataegus monogyra*) or unlobed

  32. Ovary inferior or 1/2-inferior, chambers gen 2–5; fr a pome with gen papery core, occ drupe-like with 1–5 stones

    33. St armed with thorns

      34. Lvs ± ovate, deciduous, gen ± lobed above middle, toothed; larger lvs > 3 cm wide; fr red to black
.... **CRATAEGUS**

      34′ Lvs gen oblanceolate, narrowly oblong, narrowly obovate, or narrowly elliptic, ± evergreen, unlobed, entire to toothed; larger lvs < 2.5 cm wide; fr orange to red.......................... **PYRACANTHA**

    33′ St gen not armed (thorny or not in *Malus, Pyrus*)

      35. Lf gen sharp-toothed

        36. Infl many-fld panicle; petals < 4 mm; fr 5–10 mm diam............................ **HETEROMELES**

        36′ Infl few-fld cluster; petals 10–15 mm; fr ≥ 10 mm diam

          37. Fr gen ± spheric, flesh not gritty... **MALUS**

          37′ Fr gen ± obovoid, flesh gritty.. **PYRUS**

      35′ Lf entire, minute-gland-toothed, or above middle toothed

        38. Lvs ± clustered on short-shoots, petiole ± 0........................... **PERAPHYLLUM**

        38′ Lvs evenly distributed on branches, twigs, petiole > ± 0

          39. Lvs gen serrate above middle, deciduous; fr a pome of 2–5 papery segments, gen blue-black
.... **AMELANCHIER**

          39′ Lvs entire, deciduous, or evergreen; fr a pome, drupe-like with 2–5 stones, orange to red
.... **COTONEASTER**

  32′ Ovary superior (sometimes hidden in funnel- or urn-shaped hypanthium), chamber gen 1; fr not a pome

    40. Lf veins palmate

      41. Lf blade 3(5)-lobed above middle, gen wedge-shaped, margin not toothed, rolled under; fls 1........ [4]**PURSHIA**

      41′ Lf blade gen 3–7-lobed below middle, not wedge-shaped, margin ± toothed, ± flat; fls in cluster of several to many

        42. Petals 2.5–3 mm; pistils 1–5 (if 3–5, fused below middle); infl umbel-like cluster, not flat-topped; fr follicles .... **PHYSOCARPUS**

        42′ Petals 4–30 mm; pistils 5–150; infl raceme- or panicle-like cyme, ± flat-topped; fr fleshy-coated achenes (aggregate ± raspberry-like).. [3]**RUBUS**

    40′ Lf veins pinnate

      43. Pl matted, scapose; rosettes many — gen on limestone .... **PETROPHYTUM**

      43′ Erect shrub or tree, not scapose; rosettes 0

        44. Lvs opposite or in opposite clusters

          45. Shrub 2–20 dm; petals 0 (1–5); pistil 1(2); fr achene; SNE, D (esp DMtns) .... **COLEOGYNE**

          45′ Tree 40–150 dm; petals 5; pistils 2; fr follicle; ChI .... [2]**LYONOTHAMNUS**

        44′ Lvs alternate or in alternate clusters

          46. Pistil gen 1

            47. Hypanthium cup- to urn-shaped, deciduous in fr; style not tapered to stigma; fr drupe .... **PRUNUS**

            47′ Hypanthium ± funnel-shaped to obconic, persistent in, ± enclosing fr; style tapered to stigma; fr achene or achene-like

              48. Hypanthium 0.6–3.2 mm; petals 1.5–2.5 mm; infl many-fld panicle; lf linear to oblanceolate or sickle-shaped, margin not rolled under .... **ADENOSTOMA**

              48′ Hypanthium 2–14 mm; petals 0 or 6–8 mm; infl 1–18-fld cluster; lf linear to round, margin gen rolled under

                49. Petals 0; lf toothed or not, lobes 0.. **CERCOCARPUS**

                49′ Petals 6–8 mm, white to cream; lf gen not toothed, lobes 3–9 .... [4]**PURSHIA**

          46′ Pistils gen 2–6

            50. Lf gen entire; infl raceme, pendent; fls unisexual, occ bisexual; fr drupes................. **OEMLERIA**

            50′ Lf lobed or toothed; infl umbel-like cluster, panicle, or fl 1, spreading to erect; fls bisexual; fr achenes or follicles

              51. Lf gen widest above middle, margins with 3–9 lobes, gen 0 teeth, ± strongly rolled under; fls 1.. [4]**PURSHIA**

              51′ Lf gen widest at or below middle, margins with ± 0 lobes, gen with teeth, not rolled under; fls gen 3–many

                52. Infl umbel-like cluster, fls gen 3–5; petals 4–6 mm; stamens ± 50, sepals ± toothed .... **NEVIUSIA**

                52′ Infl raceme or panicle, fls many; petals 1.5–2 mm; stamens gen < 50, sepals not toothed

                  53. Petals gen white; hypanthium saucer-shaped; stigma ± 2-lobed; fr 5 achenes .... **HOLODISCUS**

                  53′ Petals pink to rose; hypanthium obconic to bell-shaped; stigma head-like; fr 5 follicles........ **SPIRAEA**

ACAENA

B.H. Macmillan & Barbara Ertter

Per, nonglandular. **LF:** gen ± basal, odd-1-pinnate; lflets ± evenly toothed to lobed. **INFL:** dense spike or head; pedicel bractlets 0. **FL:** hypanthium ± obconic, bractlets lf-like to linear; sepals gen 4; petals 0; stamens 2 or 4, opposite sepals; pistils 1(2), ovary superior, continuous to style at top, stigma many-branched, exserted from hypanthium. **FR:** hypanthium ± hardened, encasing elliptic achenes; prickles gen 4–many, gen barbed. ± 45 spp.: esp s hemisphere. (Greek: thorn, from fr)

1. Infl a spike; hypanthium prickles in fr > 4, longest gen 1–3 mm, ± throughout; pl from ± woody,
    branched caudex; lflets pinnately dissected . . . . *A. pinnatifida* var. *californica*
1' Infl a head; hypanthium prickles in fr 4, longest 7.5–15 mm, ± on top; pl from stolons; lflets toothed
  2. Hypanthium in fr with longest prickle 7.5–12 mm, without vestigial prickles below top; st 1–2 mm diam;
      lflets dull, smooth adaxially, gen ± glaucous abaxially . . . . *A. novae-zelandiae*
  2' Hypanthium in fr with longest prickle 9–15 mm, gen with a few vestigial prickles below top; st 2–3 mm
      diam; lflets shiny, wrinkled adaxially, pale with uneven wax layer abaxially . . . . *A. pallida*

A. novae-zelandiae Kirk (p. 1173) BIDDY-BIDDY **ST:** 10–20 cm, 1–2 mm diam. **LF:** 2–6 cm [or not]; stipules lflet-like; lflets 3–5 per side, 5–15 mm, elliptic-oblanceolate, evenly toothed < 1/4 to midvein, dull-green, smooth adaxially, gen ± glaucous abaxially. **INFL:** head, ± 10 mm diam, 30–35 mm in fr. **FL:** sepals ± 1.5 mm, elliptic-ovate; stamens 2, pale. **FR:** hypanthium body 2.5–4 mm, obconic; prickles 4, ± on top, longest 7.5–12 mm. Gen ± disturbed areas; < 200 m. NCo, CCo, n SCo; OR; native to Australia, New Zealand, New Guinea. [*A. anserinifolia* J.R. Forst. & G. Forst., misappl.] Mar–May ◆

A. pallida (Kirk) Allan **ST:** 12–15 cm, 2–3 mm diam. **LF:** 3–12 cm; stipules lflet-like; lflets 4–7 per side, 6–30 mm, elliptic-oblanceolate, evenly toothed < 1/4 to midvein, shiny, wrinkled adaxially, pale with uneven wax layer abaxially. **INFL:** head, ± 10 mm diam, 20–40

mm in fr. **FL:** sepals ± 2 mm, elliptic-ovate, stamens 2, pale. **FR:** hypanthium body 4–6 mm, obtriangular; prickles 4, ± on top, longest 9–15 mm, gen a few vestigial below. Coastal sand, ± disturbed areas; ± 0 m. CCo (Stinson Beach, Marin Co.); native to Australia, New Zealand. Mar–May ◆

A. pinnatifida Ruiz & Pav. var. *californica* (Bitter) Jeps. (p. 1173) **ST:** 10–60 cm, 3–5 mm diam. **LF:** 3–12 cm; stipules 0; lflets 5–8 per side, 4–15 mm, ± oblanceolate-ovate, pinnately dissected into 3–8 lance-linear segments. **INFL:** spike, 10–20 mm diam, interrupted below. **FL:** sepals 2.5–3.5 mm, lance-elliptic; stamens gen 4, purple-black. **FR:** hypanthium body 4–7 mm, obovate, ± ridged; prickles > 4, ± throughout, longest gen 1–3 mm. Coastal grassland, open, rocky slopes; 50–400 m. s NCo, CCo, w SnFrB, SCoR. Mar–May

ADENOSTOMA

William Jones

Shrub, small tree, ± resinous. **ST:** bark shredding. **LF:** simple, alternate, clustered or not, evergreen, short-petioled, stipuled or not, linear to oblanceolate or sickle-shaped, stiff or flexible. **INFL:** panicle, terminal, 0.5–17 cm, many-fld; pedicel bractlets gen present, lanceolate to elliptic. **FL:** hypanthium 0.6–3.2 mm, throat obconic, glandular, persistent, ± enclosing fr, bractlets 0; petals 1.5–2.5 mm; calyx lobes 1–1.5 mm; corolla lobes ± round, cream to white; stamens 10–15, in 5 groups of 2 or 3, alternate petals; ovary superior, chamber 1, ovules 1–2. **FR:** achene-like. 2*n*=18. 2 spp.: CA, Baja CA. (Greek: glandular mouth, from hypanthium ring gland)

1. Lvs not clustered, flexible, glandular, stipules 0; pedicel bractlets 5, unlobed, translucent; stamens 10–12
    . . . . *A. sparsifolium*
1' Lvs clustered, stiff, glabrous to puberulent, stipules < 1.5 mm; pedicel bractlets 1–3, 3-lobed, not
    translucent; stamens 15 . . . . *A. fasciculatum*
  2. Pls low, mounded; branches decumbent; CCo, ChI . . . . var. *prostratum*
  2' Pls tall, not mounded; branches erect to ascending; NCoR, CaRF, SN, CW, SW
    3. Twigs glabrous (hairy); lvs gen linear to oblanceolate, 5–13 mm, tips gen acute-acuminate . . . . var. *fasciculatum*
    3' Twigs hairy; lvs oblanceolate to club-like, 2–6.5 mm, tips gen obtuse-mucronate . . . . var. *obtusifolium*

A. fasciculatum Hook. & Arn. (p. 1173) CHAMISE, GREASEWOOD Pl < 4 m, burled, much-branched. **ST:** trunk bark gray-brown. **LF:** clustered, sickle-shaped or not, glabrous to puberulent, stiff; stipules < 1.5 mm. **INFL:** dense to open; pedicels 0–1.1 mm, bractlets 1–3, not enclosing buds, 3-lobed, lanceolate to narrow-elliptic, not translucent. **FL:** hypanthium 0.8–3.2 mm, strongly 10-ribbed; calyx lobe width > length; petals round to widely obovate; stamens 15. **FR:** obovoid, tip oblique-truncate, height ≤ hypanthium rim.

  var. *fasciculatum* Pl < 4 m, not mounded. **ST:** branches erect to ascending, twigs glabrous (hairy). **LF:** 5–13 mm, gen linear to oblanceolate, tips gen acute-acuminate. **INFL:** open, 0.5–17.2 cm. Dry slopes, ridges, chaparral; < 1830 m. NCoR, CaRF, SN, CW, SW; sw OR, Baja CA. May–Jun

  var. *obtusifolium* S. Watson Pl < 2 m, not mounded. **ST:** branches erect to ascending, twigs short- to long-soft-wavy-hairy. **LF:** 2–6.5 mm, oblanceolate to club-like, tips gen obtuse-mucronate.

INFL: open, 1.2–9.7 cm. Dry slopes, ridges, chaparral; < 800 m. s SCo, sw PR; Baja CA. May–Jun

  var. *prostratum* Dunkle Pl gen < 0.5(1.5) m, mounded. **ST:** branches decumbent, twigs glabrous (hairy). **LF:** 1.9–6.3 mm, gen linear to oblanceolate, tips acute-acuminate. **INFL:** compact, 0.5–6.5 cm. Dry slopes, ridges, chaparral; < 750 m. CCo, ChI. May–Jun

A. sparsifolium Torr. (p. 1173) RED SHANK, RIBBON WOOD Pl 2–6 m. **ST:** trunk bark red-brown; twigs glabrous. **LF:** not clustered, 3.6–26.3 mm, linear, glandular, short-erect-hairy, flexible; stipules 0. **INFL:** open, 0.8–7.7 cm; pedicels 0–2.8 mm, bractlets 5, enclosing bud, unlobed, narrow-elliptic to ovate, translucent. **FL:** hypanthium 0.6–2 mm, weak-10-ribbed; calyx lobe length > width; petals elliptic to ovate. **FR:** elliptic, tip rounded, exceeding hypanthium. Dry slopes, flats, ravines, chaparral, pinyon woodland; 275–2000 m. s SCoRO, e SCo, s WTR, PR; Baja CA. Jul–Sep

AGRIMONIA AGRIMONY

Genevieve J. Kline

Per, finely glandular. **ST**: 1–several, erect, rhizomed. **LF**: odd–1-pinnate; lflets evenly toothed, gen alternately large, small. **INFL**: spike-like raceme, terminal, often also axillary; pedicel bractlets 2, near tip, fused at base. **FL**: hypanthium stalk 1–2 mm, reflexed in fr, bractlets 0; petals ± elliptic to ± obovate [or otherwise], yellow; stamens 5–15; pistils 2, ovary superior, continuous to style at top. **FR**: hypanthium obconic to cup-shaped, hard, ridged, rim with 3–5 rows of spreading hooked bristles; sepal tips converged inward, with hypanthium, gen encasing 1 achene. ± 20 spp.: gen n temp, S.Afr, Brazil, Argentina. (Greek: eye disease, from former use as cure) [Kline & Sorensen 2008 Brittonia 60:11–33]

1. St, lower infl axis with nonglandular hairs of 1 kind: coarse, straight; stipule outer margins toothed or lobed above middle; fls alternate throughout raceme; hypanthium with short-stalked glands . . . . ***A. gryposepala***
1′ St, lower infl axis with nonglandular hairs of 2 kinds: soft, shaggy and coarse, straight; stipule outer margins entire above middle; fls ± opposite at raceme middle, above; hypanthium with coarse nonglandular hairs . . . . . . ***A. striata***

A. gryposepala Wallr. (p. 1173) COMMON AGRIMONY **ST**: gen 25–130 cm; glands short-stalked, occ also larger, dot-like. **LF**: largest gen 10–25 cm; stipules 0.5–4 cm, gen half-ovate; major lflets 3–11, 1–10 cm, elliptic to ± diamond-shaped to obovate; abaxially with coarse, straight nonglandular hairs, gen with stalked and dot-like glands. **INFL**: 9–43 cm, gen 10–50-fld; pedicels gen 1–12 mm. **FL**: sepals 1.5–3 mm, tips long-tapered; petal 2–4.5 mm. **FR**: hypanthium 2.5–6 mm; bristles 1–4 mm, in 4–5 rows, lowermost reflexed; converged sepal tips hooked. 2*n*=56. Moist places, gen in woodland; 100–1700 m. NW (exc NCo), CaRH, n SNF/n SNH, n SNH, SnBr, PR; to e N.Am, mtns of Mex, Guatemala. Jun–Sep

A. striata Michx. BRITTON'S AGRIMONY, GROOVED AGRIMONY **ST**: gen 20–150 cm; glands short-stalked, also larger, dot-like above. **LF**: largest gen 12–21 cm; stipules 0.5–3 cm, ± sickle-shaped to half-ovate; major lflets 3–11, 1–11 cm, ± diamond-shaped to elliptic; abaxially with soft, shaggy and coarse, straight nonglandular hairs, stalked and dot-like glands. **INFL**: gen 8–60 cm, gen 10–60-fld; pedicels gen 3–6 mm. **FL**: sepals 1–3 mm, tips often long-tapered; petal 2–4 mm. **FR**: hypanthium 2–7 mm; bristles 3–4 mm, in 3 rows, lowermost spread ± 90° (pressed upward on dried specimens); converged sepal tips not hooked. 2*n*=56. Moist places, gen in woodland; 1000–3000 m. SnBr (Oak Glen), W&I (White Mtns); to e N.Am, mtns of Mex. Jun–Aug

AMELANCHIER SERVICE-BERRY

Christopher S. Campbell

Shrub, small tree. **ST**: bark gray- to red-brown; overwintering buds ovate to lanceolate, ± red to ± purple. **LF**: simple, deciduous; stipules deciduous. **INFL**: raceme, cluster (panicle), fls 3–16+; pedicel bractlets gen 1–2. **FL**: hypanthium bell- to urn-shaped, bractlets 0; sepals persistent; petals erect to spreading, white (suffused with red); stamens ± 10–20; ovary inferior, 2–5-chambered, styles 2–5. **FR**: pome of 2–5 papery segments, berry-like, gen spheric, gen blue-black. ± 25 spp.: temp N.Am, Eurasia, n Afr. (Latin: from old French common name) Fr of some spp. used by Native Americans for food.

1. Lf finely hairy abaxially in fr; twigs glabrous (n DMtns) to gen white-hairy; styles 2–4(5) . . . . ***A. utahensis***
1′ Lf glabrous abaxially in fr; twigs glabrous; styles 4–5 . . . . ***A. alnifolia***
   2. Ovary top glabrous; petals 8–12 mm; lf glabrous or occ sparsely hairy abaxially in fr; fr 8–9 mm diam . . . var. ***pumila***
   2′ Ovary top gen hairy; petals 12–15 mm; lf finely or occ densely hairy abaxially in fl; fr 10–13 mm diam
   . . . . var. ***semiintegrifolia***

A. alnifolia (Nutt.) M. Roem. **ST**: twigs glabrous. **LF**: blade elliptic to round, gen serrate above middle, glabrous in fr. **FL**: petal ovate to ± round; styles 4–5.

var. ***pumila*** (Torr. & A. Gray) C.K. Schneid. (p. 1173) Pl 1–3 m. **LF**: blade 10–50 mm, 10–20 mm wide, abaxially glabrous or occ sparsely hairy in fl. **INFL**: 2–4 cm; fls 4–8. **FL**: petal 8–12 mm; ovary top glabrous. **FR**: 8–9 mm diam. Open, often moist scrub, mtn slopes; 1400–2600 m. n&c SNH (e slope); to MT, CO. May–Jun

var. ***semiintegrifolia*** (Hook.) C.L. Hitchc. (p. 1173) Pl 1–12 m. **LF**: blade 30–40 mm, 20–30 mm wide, abaxially finely or occ densely hairy in fl. **INFL**: 4–8 cm; fls 5–15. **FL**: petal 12–15 mm; ovary top gen hairy. **FR**: 10–13 mm diam. 2*n*=68. Open conifer or mixed-evergreen forest, slopes; 50–2500 m. NCo, KR, n NCoRO; to AK. Mar–Jun

A. utahensis Koehne (p. 1173) UTAH SERVICE-BERRY Pl 0.5–5 m. **ST**: twigs glabrous (n DMtns) to gen white-hairy. **LF**: blade 13–45 mm, 10–45 mm wide, gen serrate above middle, abaxially hairy in fl, finely hairy in fr. **INFL**: 2–3 cm; fls 3–6. **FL**: petals 6–11 mm; ovary top hairy; styles 2–4(5). **FR**: 6–10 mm diam. Open, rocky slopes, canyons, banks of creeks, deserts, conifer forest; 200–3400 m. NW, CaR, SN, CW, SW, SNE, DMtns; to OR, MT, TX, Baja CA. Variable; possibly warranting taxonomic status are *A. utahensis* var. *covillei* (Standl.) N.H. Holmgren (n DMtns; gen < 2 m; twigs, lvs glabrous), *A. pallida* Greene (petals > 9 mm, lvs with 7–9 pairs of lateral veins); study needed. Apr–Jun

APHANES

Barbara Ertter

Ann, inconspicuous, soft-hairy, nonglandular. **ST**: spreading to erect. **LF**: palmately lobed. **INFL**: few-fld cluster, axillary, ± hidden by sheathing stipules; pedicel bractlets 0. **FL**: hypanthium ± urn-shaped, bractlets 0 or 4; sepals 4; petals 0; stamen gen 1, pollen sac 1, horseshoe-shaped; pistil gen 1, ovary superior, style attached near base. **FR**: hypanthium encasing achene, not hardened. 10–20 spp.: worldwide, esp Medit. (Greek: unseen, from hidden fls)

A. occidentalis (Nutt.) Rydb. (p. 1173) **ST**: gen 2–10 cm. **LF**: gen 3–12 mm; stipules widely ovate, deeply lobed; petiole gen 1–5 mm; blade gen 2–5 mm, ± round, main lobes 3, > 2/3 to base, again toothed or lobed. **FL**: 0.5–2 mm; hypanthium bractlets 0 or < 0.5 mm; sepals 0.2–0.6 mm. **FR**: achene ± 1 mm, ovoid. Seasonally moist grassland, chaparral, woodland; 30–1200 m. NW, CaRF, SNF, ScV, CW, SW; to WA, Baja CA. Highly variable; several ± separable forms. Mar–May

ARUNCUS GOAT'S BEARD

T. Lawrence Mellichamp & Margriet Wetherwax

1 sp.: n temp N.Am, Eur. (Latin: goat's beard, for infl)

A. dioicus (Walter) Fernald var. **acuminatus** (Rydb.) H. Hara (p. 1173) Per; rhizome stout; dioecious. **ST**: 1–2 m, glabrous. **LF**: < 60 cm, 2–3-pinnately compound, petioled; stipules 0; lflets 3–16 cm, ovate, acuminate, 2-serrate, ± hairy. **INFL**: panicle, < 50 cm; pedicels < 1 mm, bractlet 1, at top. **FL**: hypanthium saucer-shaped, bractlets 0; sepals 5, 1–2 mm; petals 5, 1–2 mm, oblong to obovate, white to pale ± yellow. **STAMINATE FL**: stamens 15–30, exserted. **PISTILLATE FL**: pistils 3(5), styles short. **FR**: follicles 3(5), 3–5 mm, ± cylindric, reflexed, adaxially dehiscent. **SEED**: ± 2 mm. 2*n*=18. Moist streambanks, conifer or mixed-evergreen forest; < 1900 m. NCo, KR, n NCoRO, CaR; to AK. [*A. d.* var. *pubescens* (Rydb.) Fernald, misappl.] Other vars. in e N.Am. Jun–Sep

CERCOCARPUS MOUNTAIN-MAHOGANY

Brian Vanden Heuvel & Richard Lis

Shrub, small tree, evergreen. **ST**: trunk < 80 cm diam; bark gen gray to red-brown; twigs short. **LF**: gen clustered, simple; stipules deltate to lanceolate, gen deciduous; blade ± thin to leathery, entire to toothed. **INFL**: clusters, fls 1–18. **FL**: hypanthium funnel-like, tube persistent in fr, rim cup-like, deciduous, bractlets 0; petals 0; stamens 10–46, in ± 3 rows on hypanthium rim, anthers glabrous or hairy; pistil 1, free from hypanthium tube, ovary superior, 1-ovuled, style terminal, persistent in fr, straight or twisted in age, plumose. **FR**: achene, cylindric, hairy, incl in hypanthium tube. 11 spp.: w N.Am, Mex. (Greek: tailed fr) [Lis 1992 Int J Pl Sci 153:258–272]

1. Lf blade entire; anthers glabrous . . . . ***C. ledifolius***
2. Lf blade gen 3–10 mm, margin inrolled to midrib; pls highly branched . . . . var. ***intricatus***
2′ Lf blade 10–30 mm, margin flat to inrolled but not to midrib; pls moderately branched
 3. Lf oblanceolate to lance-elliptic, (3)9(11) mm wide, abaxially glabrous to sparsely woolly, midrib,
  veins visible . . . . var. ***intermontanus***
 3′ Lf narrow-lanceolate, 2–5(7) mm wide, abaxially densely woolly, midrib, veins obscure . . . . var. ***ledifolius***
1′ Lf blade toothed to crenate; anthers hairy
 4. Lf leathery, abaxially white-woolly — s ChI (Santa Catalina Island) . . . . ***C. traskiae***
 4′ Lf not leathery, abaxially ± sparsely hairy
  5. Lf thin, abaxial areoles glabrous — PR (Riverside, San Diego cos.) . . . . ***C. minutiflorus***
  5′ Lf thin to thick, abaxial areoles sparsely hairy . . . . ***C. betuloides***
   6. Lf blade gen 10–27(40) mm, < 15 mm wide . . . . var. ***betuloides***
   6′ Lf blade gen (20)27–70 mm, > 15 mm wide
    7. Lf wide- to oblong-ovate, base rounded; fls, frs gen 1–6(12) . . . . var. ***blancheae***
    7′ Lf obovate, base tapered; fls, frs 1–3 . . . . var. ***macrourus***

C. betuloides Nutt. (p. 1173) Branches spreading to erect. **LF**: petiole 1–10(16) mm; blade 1–7 cm, widely elliptic to obovate, thin to thick, serrate to dentate or crenate, abaxially sparsely hairy. **FL**: hypanthium glabrous to sparsely hairy; stamens 25–45, anthers hairy. **FR**: 8–12 mm, strigose; style 5–11(12) cm.

 var. **betuloides** (p. 1173) BIRCH-LEAF MOUNTAIN-MAHOGANY Shrub 1–3 m. **LF**: petiole 1–6(8) mm; blade 1–2.7(4) cm, obovate to ± round, finely toothed to serrate. **INFL**: fls 1–6(12). **FL**: hypanthium 5–8 mm, 7–13 mm in fr, rim 4–7 mm diam. **FR**: style 5–9(12) cm. 2*n*=18. Dry, rocky slopes, chaparral; < 2500 m. NW, CaR, SN, CW, SW, MP. Mar–May

 var. **blancheae** (C.K. Schneid.) Little (p. 1173) ISLAND MOUNTAIN-MAHOGANY Shrub, small tree, 2–7 m. **LF**: petiole 5–9(16) mm; blade (2.5)2.7–7 cm, wide- to oblong-ovate, widely dentate to serrate, lateral veins 6–8(9). **INFL**: fls gen 1–6(12). **FL**: hypanthium 7–9 mm, 8–13 mm in fr, rim 4–7 mm diam. **FR**: style 5–8 cm. Chaparral; < 600 m. ChI (exc San Clemente Island), s WTR. [*C. betuloides* subsp. *blancheae* (C.K. Schneid.) Thorne] Mar–Apr ★

 var. **macrourus** (Rydb.) Jeps. (p. 1173) Shrub, small tree, 2–5 m. **LF**: petiole 5–10 mm; blade (2)2.7–7 cm, obovate, dentate or crenate, lateral veins 5–10. **INFL**: fls 1–3. **FL**: hypanthium 7–9 mm, 9–14 mm in fr, rim 4–7 mm diam. **FR**: style 8–11 cm. Conifer forest; 280–1800 m. KR, CaR, MP; sw OR. Jun

C. ledifolius Nutt. CURL-LEAF MOUNTAIN-MAHOGANY **LF**: petiole 0–6 mm; blade 0.3–3 cm, 1–9(11) mm wide, linear to lanceolate, entire, leathery, abaxially glabrous to densely hairy or woolly. **INFL**: fls 1–10. **FL**: hypanthium 2–6 mm, 3–10 mm in fr, rim 1–3.5 mm diam; stamens 10–25, anthers glabrous. **FR**: 6–11 mm; style 3–7 cm.

 var. **intermontanus** N.H. Holmgren (p. 1173) Shrub, small tree 2–7 m. **LF**: petiole 2.5–6 mm; blade 1–3 cm, (3)9(11) mm wide, oblanceolate to lance-elliptic, abaxially glabrous to sparsely woolly, midrib, veins visible. **INFL**: fls 4–10. **FL**: hypanthium 2.5–6 mm, 6–10 mm in fr, rim 2.5–3.5 mm diam; stamens 15–25. **FR**: 6–11 mm; style 5–7 cm. 2*n*=18. Pinyon/juniper woodland, sagebrush scrub; 1000–3256 m. KR, NCoRH, CaRH, SNH, Teh, TR, n PR (incl SnJt), GB, DMtns; to WA, WY, AZ, Baja CA. Putative hybrids with *C. ledifolius* var. *intricatus* (*C. ledifolius* var. *intercedens* C.K. Schneid.) do not merit taxonomic status. Apr–Aug

 var. **intricatus** (S. Watson) M.E. Jones (p. 1173) Shrub 1–3 m, highly branched. **LF**: petiole 0–1 mm; blade gen 0.3–1 cm, linear, thick-leathery, entire, inrolled to midrib, abaxially glabrous to gray-white-hairy. **INFL**: fls 1–5; pedicels 0.3–0.8 mm, 0.5–1 mm in fr. **FL**: hypanthium 2–5 mm, 3–8 mm in fr, rim 1–2 mm diam; stamens 10–18. **FR**: 6–7 mm; style 3–4.5 cm. Dry, rocky outcrops, slopes, pinyon/juniper woodland; 1000–3000 m. s SNH, SNE, DMtns; to CO, AZ. [*C. i.* S. Watson] May

 var. **ledifolius** (p. 1173) Shrub, small tree 1–3 m. **LF**: petiole 2–4 mm; blade 1–3 cm, 2–5(7) mm wide, narrow-lanceolate, leathery, abaxially densely white-woolly, midrib, veins obscure. **INFL**: fls 1–5. **FL**: hypanthium 4–6 mm, 5–7 mm in fr, rim 2–3 mm diam; stamens 15–25. **FR**: 6–9 mm; style 4–6 cm. Uncommon. Steep slopes, open pine forest; 1200–3000 m. s SNH; to e OR, sw MT, n WY, n UT. Apr–May

C. minutiflorus Abrams (p. 1177) Shrub 1–6 m. **LF**: petiole 2–6 mm; blade 1–2.5 cm, ± widely (ob)ovate, thin, ± serrate, abaxial areoles glabrous, lateral veins 3–6. **INFL**: fls gen 2–10. **FL**: hypanthium 5–8 mm, 8–11 mm in fr, glabrous or sparsely hairy, rim 4–7

mm diam; stamens 15–25, anthers hairy. **FR**: 8–12 mm; style 4.5–6 cm. Chaparral; < 1400 m. PR (Riverside, San Diego cos.); n Baja CA. Mar–May

C. traskiae Eastw. (p. 1177) CATALINA ISLAND MOUNTAIN-MAHOGANY Tree 3–8 m. **LF**: petiole 5–8(11) mm; blade 2–6 cm, elliptic to (ob)ovate, leathery, ± serrate or crenate, rolled under, abaxially white-woolly, lateral veins 5–7(8). **INFL**: fls 3–16. **FL**: hypanthium 7–14 mm, white-tomentose, rim 4–7 mm diam; stamens 20–40, anthers hairy. **FR**: 7–11 mm; style 4–6 cm. Dry, rocky soils; 100–250 m. s ChI (Santa Catalina Island). Only 7 pls survived in 1990. Mar ★

CHAMAEBATIA

Brian Vanden Heuvel & Thomas J. Rosatti

Shrub, strong-smelling, evergreen, gen stellate-hairy, glandular. **LF**: odd-2–3-pinnately compound; stipules entire; 1° lflets 8–17 per side. **INFL**: terminal, panicle or raceme, 3–10 cm, fls (1)2–7(10); pedicel bractlets 1–2. **FL**: hypanthium densely glandular and hairy, bractlets 0; sepals 5, deltate; petals 5, spreading, ± round, white; stamens 35–65; pistil 1(2), ovary superior, free, ovule 1, style base hairy. **FR**: achene, obovoid, leathery, dark brown-black. 2 spp.: CA, Baja CA. (Greek: low bramble) [Armstrong 1980 Madroño 27:111]

1. Lf gen elliptic, gen 2-pinnately compound, gland-tip gen sessile or sunken; 300–1230 m, s PR . . . . *C. australis*
1′ Lf gen obovate, gen 3-pinnately compound, gland-tip gen stalked; 600–2350 m, CaR, SN . . . . *C. foliolosa*

C. australis (Brandegee) Abrams (p. 1177) SOUTHERN MOUNTAIN MISERY Pl possibly colonial. **ST**: bark ± gray-black. **LF**: 3–6 cm. **FL**: hypanthium gen ± 3 mm; sepals ± 3 mm; petals 4–6 mm; ovary glabrous. Dry slopes, chaparral; 300–1230 m. s PR; n Baja CA. Some populations threatened by clearing for avocado orchards. Nov–May ★

C. foliolosa Benth. (p. 1177) MOUNTAIN MISERY Pl low, extensively colonial. **ST**: bark dark red-brown. **LF**: 2–10 cm. **FL**: hypanthium ± 5–6 mm; sepals ± 5 mm; petals 6–9 mm; ovary base hairy. Conifer forest; 600–2350 m. CaR, SN. May–Jul

CHAMAEBATIARIA

Brian Vanden Heuvel & Thomas J. Rosatti

1 sp. (Greek: *Chamaebatia*-like)

C. millefolium (Torr.) Maxim. (p. 1177) Shrub 6–20 dm, strong-smelling, evergreen, densely branched, gen stellate-hairy, glandular. **LF**: alternate, odd-(1)2-pinnately compound, 2–8 cm, oblong; stipules entire; 1° lflets 13–25; 2° lflets 6–10, 0.8–2 mm, sessile, entire, lobed, or toothed; petioles, axes hairy adaxially. **INFL**: panicle or raceme, 3–15 cm, fls 20–400; pedicel bractlets 1–21. **FL**: hypanthium bractlets 0; sepals 5, 2.5–4 mm, lanceolate, acute, abaxially glandular-hairy; petals 5, ± 5 mm, ± round, white; pistils 4–5, ovaries superior, ± fused below, ovules gen ≥ 2, styles free. **FR**: follicles, 3–5 mm, red-brown, leathery, dehiscent on inner suture, upper 1/2 of outer. **SEED**: few, 2.5–3.5 mm, narrow-fusiform, ± yellow. *n*=9. Dry, rocky sagebrush scrub, pinyon/juniper woodland, pine forest; 900–3400 m. KR, CaR, SN (e slope), GB, ne DMtns; to OR, s ID, UT, AZ. Jun–Aug

COLEOGYNE BLACKBUSH

Bruce D. Parfitt & Thomas J. Rosatti

1 sp. (Greek: sheath female, from hypanthium enclosing pistil)

C. ramosissima Torr. (p. 1177) Shrub 2–20 dm, much-branched, thorny, ± strigose. **LF**: in opposite clusters, simple, 5–15 mm, linear-oblanceolate, ± thick, entire; stipules persistent. **INFL**: fls 1, terminal; pedicel bractlets paired, closely subtending fl. **FL**: hypanthium bell-shaped, leathery, sheath at top enclosing pistil, 4–9 mm, bractlets 0; sepals 4(5), spreading, 5–8 mm, ovate, gen yellow, often ± red abaxially, persistent, inner 2 widely scarious-margined; petals 0(1–5, 5–8 mm, yellow); stamens 20–25(30); pistil 1(2), ovary superior, style lateral, long-hairy esp below, persistent. **FR**: achene, ± 4–6 mm, ± crescent-shaped, red-brown, glabrous. Dry, open slopes, creosote-bush scrub, pinyon/juniper woodland; 600–2000 m. SNE, D (esp DMtns); to CO, NM. Apr–Jun

COMARUM

Barbara Ertter

1 sp. (Greek: fr of *Arbutus unedo*) [Ertter 2007 J Bot Res Inst Texas 1:31–46]

C. palustre L. (p. 1177) MARSH CINQUEFOIL Per, nonglandular, openly matted. **ST**: non-fl portion creeping, often floating and/ or rooting at nodes; fl-st erect to ascending, 20–50 cm, ± glabrous to ± strigose. **LF**: alternate, odd-1-pinnate to subpalmate, 5–20 cm; lflets 2–3 per side, toothed, terminal gen ± = lateral, 15–70 mm, oblanceolate-elliptic, distal 1/3 toothed ± 1/4 to midvein, ± glabrous to sparsely strigose. **INFL**: cyme, ± open, gen < 10-fld; pedicels straight. **FL**: hypanthium ± shallow, 5–10 mm wide; bractlets 5; sepals ± triangular; petals << sepals, 2–6 mm, ± ovate to elliptic, dark red; stamens 20–25, filaments 1.5–2.5 mm, anthers 1–1.5 mm, pollen sac 1, horseshoe-shaped; pistils many, ovaries superior, styles 0.8–2 mm, slender, jointed laterally to fr. **FR**: achene, 1–1.5 mm, glabrous, glossy, ± brown, sometimes remaining attached to spongy receptacle. 2*n*=28,42–64. Bogs, marshes; < 2400 m. NCo, CaRH, n SNH, ScV, Wrn; circumboreal. [*Potentilla p.* (L.) Scop.] May–Aug

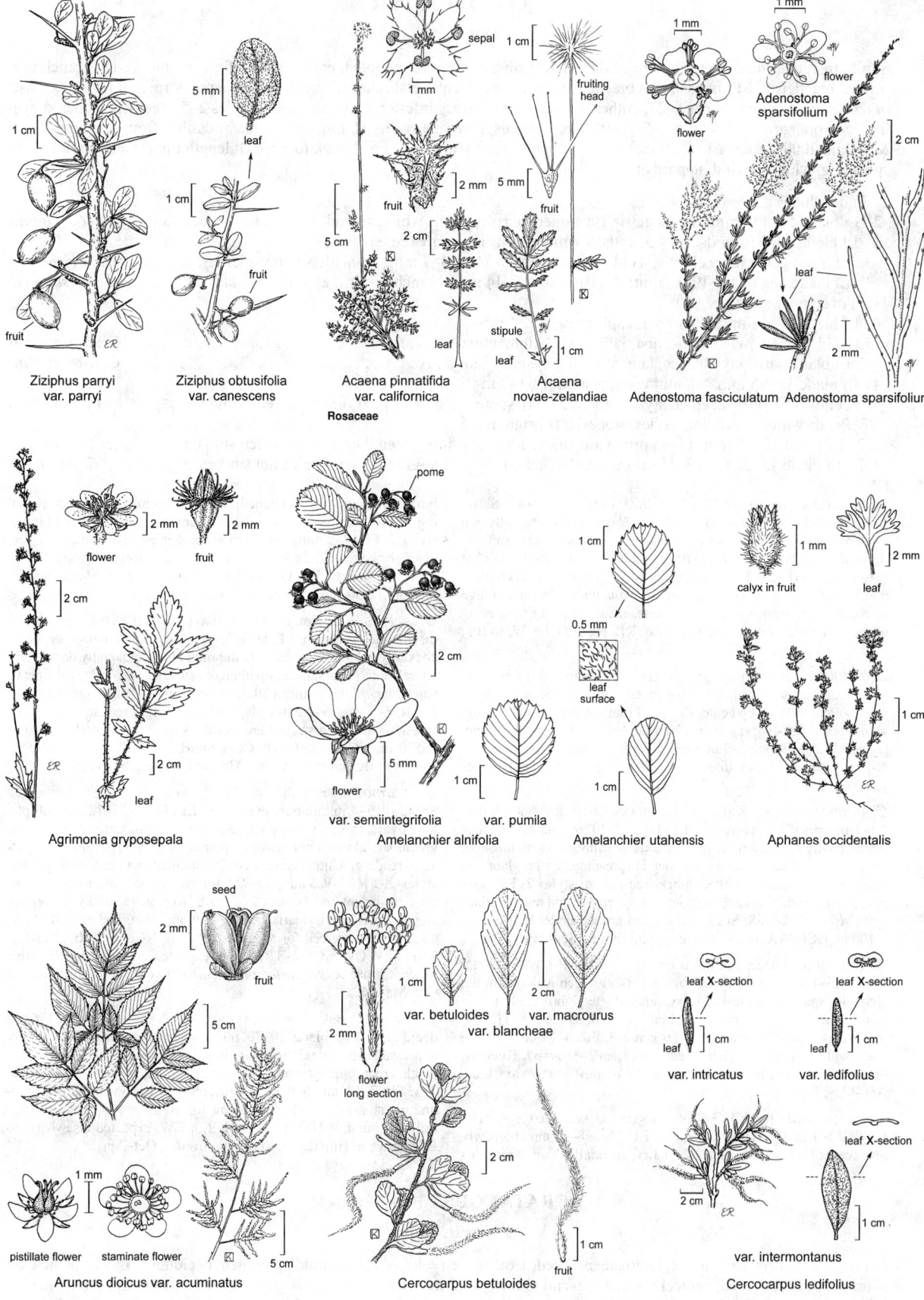

Ziziphus parryi var. parryi

Ziziphus obtusifolia var. canescens

Acaena pinnatifida var. californica

Acaena novae-zelandiae

Adenostoma fasciculatum   Adenostoma sparsifolium

Adenostoma sparsifolium

Rosaceae

Agrimonia gryposepala

var. semiintegrifolia   var. pumila
Amelanchier alnifolia

Amelanchier utahensis

Aphanes occidentalis

Aruncus dioicus var. acuminatus

var. betuloides   var. macrourus
var. blancheae
Cercocarpus betuloides

var. intricatus   var. ledifolius

var. intermontanus
Cercocarpus ledifolius

COTONEASTER

Peter F. Zika

Shrub, tree, unarmed; evergreen or deciduous. **LF**: simple, stipuled, petioled, entire. **INFL**: fls 1 or clustered at branch tips; pedicel bractlets 0. **FL**: hypanthium bractlets 0; sepals persistent; petals clawed, erect, pink to ± red or rose, at least near base, or spreading, white; stamens 8–21, anthers darker after fl; ovary inferior, 2–5-chambered, styles 2–5, free. **FR**: pome, drupe-like, gen orange to red, stones 2–5. ± 400 spp.: e hemisphere; many orn. (Latin: quince-like, possibly from lf shape) [Fryer & Hylmö 2009 Cotoneasters. Timber Press] 2 subgenera, 1 with petals erect, pink to ± red (fl length important), 1 with petals spreading, white (fl width important).

1. Pl deciduous
  2. Lf blade 55–150 mm; fls, frs gen > 20; anthers purple; petals white, spreading . . . . ***C. frigidus***
  2′ Lf blade 4–30 mm; fls, frs < 5; anthers white; petals pink to ± red, erect
    3. Lf blade gen widely elliptic, 4–14 mm; stamens 8–11; fr 4–7 mm; branchlets many, planar . . . . ***C. horizontalis***
    3′ Lf blade gen ovate, 10–30 mm; stamens 20; fr 7–14 mm; branchlets few, ascending . . . . ***C. simonsii***
1′ Pl evergreen
  4. Lf blade 5–15 mm, < 8 mm wide; pl sprawling, ≤ 1 m
    5. Lf blade narrowly elliptic, abaxially densely tomentose. . . . . ***C. hodjingensis***
    5′ Lf blade narrowly oblanceolate, abaxially sparsely hairy . . . . ***C. integrifolius***
  4′ Lf blade 15–95 mm, > 8 mm wide; pl arching 1–9 m
    6. Petals pink or rose, erect; styles, stones 2–3(4), some > 2 on all pls; fr orange to red-orange . . . . ***C. franchetii***
    6′ Petals white, spreading; styles, stones 2; fr bright red
      7. Lf blade 35–95 mm, 17–45 mm wide, thick, adaxially dark green, lateral veins often sunken. . . . . ***C. lacteus***
      7′ Lf blade 15–32 mm, 8–15 mm wide, thin, adaxially ± blue-green, lateral veins not sunken . . . . ***C. pannosus***

C. franchetii Bois (p. 1177) FRANCHET'S COTONEASTER Shrub 1–3 m, arching; evergreen. **LF**: blade 18–62 mm, ovate, abaxially gen ± yellow- or ± gray-tomentose, adaxially shiny, lateral veins sunken, tip acute to acuminate. **FL**: 7–10 mm; petals erect, pink or rose, at least at base; stamens 16–21, filaments pink, tips pale, anthers pink to purple; styles 2–3(4). **FR**: 8–12 mm, 6–9 mm wide, obovate, orange to red-orange; stones 2–3(4). Thickets, meadows, forest, riparian edges, disturbed places; < 500 m. NCo, KR, NCoRO, n CW; to BC; native to China. May–Jul, fruiting Oct–Apr. ❖

C. frigidus Lindl. TREE COTONEASTER Shrub, tree, 5–15 m, arching; deciduous. **LF**: blade 55–150 mm, narrowly elliptic, abaxially ± hairy, adaxially dull, tip acute. **FL**: 8–13 mm wide; petals spreading, white, gen tufted-hairy; stamens 20, filaments white, anthers purple, styles 2. **FR**: 5–7 mm, 5–7 mm wide, ± spheric, red; stones 2. 2*n*=34. N-facing canyon slopes, thickets; < 500 m. SnFrB; native to Himalayan Asia. May–Jun, fruiting Sep–Oct.

C. hodjingensis G. Klotz EARTHQUAKE COTONEASTER Shrub 0.5–1 m, sprawling; evergreen. **LF**: blade 5–10 mm, narrowly elliptic, abaxially densely tomentose, adaxially shiny, margins inrolled, tip gen acute. **FL**: 7–8 mm wide; petals spreading, white, glabrous; stamens 20, filaments white, anthers red-purple; styles 2. **FR**: 6–8 mm, 5–8 mm wide, spheric to depressed-spheric, bright red, not glaucous; stones 2. 2*n*=68. Shaded, disturbed ground under bird roosts; < 100 m. CCo; WA; native to China. Jun, fruiting Sep–Feb.

C. horizontalis Decne. WALL OR HERRINGBONE COTONEASTER Shrub 0.5–1 m; deciduous. **LF**: blade 4–14 mm, gen widely elliptic, abaxially sparsely appressed hairy, adaxially ± glabrous, shiny, tip acute. **FL**: 5–7 mm; petals erect, pink to ± red; stamens 8–11, filaments dark red, anthers white; styles gen 2–3. **FR**: 4–7 mm, 6–7 mm wide, widely ovate to ± spheric, red; stones gen 2–3. 2*n*=68. Thickets, brushy slopes, pasture edges; < 100 m. NCo; to BC; native to China. Apr–May, fruiting Sep–Mar.

C. integrifolius (Roxb.) G. Klotz ENTIRE-LEAVED COTONEASTER Shrub 0.5–1 m, sprawling; evergreen. **LF**: blade 8–15 mm, narrowly oblanceolate, abaxially sparsely hairy, adaxially shiny, margins

inrolled, tip gen blunt (notched). **FL**: 7–15 mm wide; petals spreading, white, glabrous; stamens 20, filaments white, anthers red-purple, styles 2. **FR**: 7–9 mm, 7–10 mm wide, depressed-spheric, deep red, glaucous; stones 2. 2*n*=68. Disturbed conifer forest, brushy slopes; < 500 m. SnFrB, PR; OR; native to Asia. [*C. microphyllus* Lindl., misappl.] Apr–Jun, fruiting Aug–Mar.

C. lacteus W.W. Sm. (p. 1177) LATE COTONEASTER Shrub 1–9 m, arching; evergreen. **LF**: blade 35–95 mm, elliptic to obovate, thick, abaxially pale-tomentose, hairs thinning in age, adaxially dark green, lateral veins often sunken, tip blunt or acute. **FL**: 8–9 mm wide; petals spreading, white; stamens 20, filaments white, anthers purple; styles 2. **FR**: 6–7 mm, 6–7 mm wide, ± spheric, bright red; stones 2. Open forest, meadows, disturbed ground, thickets, creeks, ponds, canyons; < 500 m. NCo, KR, CaRF, CCo, SnFrB, SCo, PR, expected elsewhere; to BC; native to China. May–Jul, fruiting Nov–Apr. ❖

C. pannosus Franch. (p. 1177) SILVERLEAF COTONEASTER Shrub 0.15–4 m, arching; evergreen. **LF**: blade 15–32 mm, elliptic to ± ovate, thin, abaxially white-tomentose, adaxially dull, ± blue-green, lateral veins not sunken, tip acute. **FL**: 8.5–10 mm wide; petals spreading, white; stamens 16–20, filaments white, anthers purple; styles 2. **FR**: 7–9.2 mm, 7.9–9.5 mm wide, obovoid to depressed-spheric, bright red; stones 2. 2*n*=68. Meadows, rocky or brushy slopes, open forest, riparian zones, canyons, disturbed ground, thickets, mixed-evergreen forest; gen < 1000 m. NCo, NCoRO, CaRF, n SNF, ScV, CW, SCo, SnGb, expected elsewhere; OR; native to sw China. [*Cotoneaster pannosa*, orth. var.] May–Jul, fruiting Oct–May. ❖

C. simonsii Baker HIMALAYAN COTONEASTER Shrub 1–6 m; deciduous. **LF**: blade 10–30 mm, gen ovate, abaxially sparsely appressed hairy, adaxially glabrous, shiny, tip acute. **FL**: 7.5–10 mm; petals erect, pink; stamens 20, filaments pink, anthers white; styles 3–5. **FR**: 7–14 mm, 6–9.5 mm wide, obovate to obovate-elliptic, red-orange; stones 3–5. Thickets, open forest, meadows, pastures, disturbed ground; < 1000 m. NCo, KR, n CW, expected elsewhere; to BC; native to Himalayas. May–Jun, fruiting Oct–Apr.

CRATAEGUS HAWTHORN

James B. Phipps

Shrub, tree, thorny. **LF**: simple, alternate, petioled, ± ovate, gen ± lobed above middle, toothed, deciduous. **INFL**: panicle on short-shoot tips, domed; pedicel bractlets several to many, ± 5 mm, narrow, margins glandular. **FL**: hypanthium urn-shaped, bractlets 0; sepals small, margins entire to finely toothed, gen glandular; petals white; stamens ± 10 or 20; ovary inferior, styles 1–5, free. **FR**: pome, drupe-like, red to black, gen lighter in color before fully mature, core of 1–5 laterally pitted [or

not] stones; sepals reflexed [or not]. ± 200 spp.: n temp. (Greek: hard, for wood) [Phipps & O'Kennon 2002 Sida 20:115–144]
Several spp. cult, escaped.

1. Short-shoot lvs deeply lobed, veined to sinuses; fr red to deep red. . . . . ***C. monogyna***
1′ Short-shoot lvs unlobed to shallowly lobed, not veined to sinuses; fr deep purple to black
  2. Stamens ± 20; short-shoot lvs unlobed to shallowly lobed; thorns 8–12 mm. . . . . ***C. gaylussacia***
  2′ Stamens ± 10; short-shoot lvs gen shallowly lobed; thorns 15–23 mm
   3. Infl hairy; thorns 18–23 mm . . . . ***C. castlegarensis***
   3′ Infl glabrous; thorns 15–18 mm. . . . . ***C. douglasii***

C. castlegarensis J.B. Phipps & O'Kennon **ST:** thorns 18–23 mm, in 2s to 4s or not. **LF:** of short-shoots 3.5–6 cm, ± elliptic-diamond-shaped to narrow-obovate, base wedge-shaped, shallowly lobed, tip acute. **INFL:** hairy. **FL:** 12–15 mm diam; sepals distally glandular-serrate; stamens ± 10, anthers pink; styles 3–4. **FR:** 10–12 mm diam, ± spheric, deep purple to black, stones 3–4. 2*n*=68. Streamsides in meadows, scrub, forest; 900–1300 m. CaR, MP; to BC, SK, MT. Described in 2002, found in CA in 2004. May–Aug ★

C. douglasii Lindl. (p. 1177) **ST:** thorns 15–18 mm. **LF:** of short-shoots 3.5–6 cm, ± elliptic-diamond-shaped to narrow-obovate, base wedge-shaped, gen shallowly lobed, tip acute. **INFL:** glabrous. **FL:** 12–15 mm diam; sepals distally glandular-serrate; stamens ± 10, anthers pink; styles 3–4. **FR:** 10–12 mm diam, spheric [to elliptic], deep purple to black, stones 3–4. 2*n*=68. Streamsides in meadows, scrub, forest; 600–2450 m. KR, CaR, n SNH; to BC, MT, SK, ON. [*C. columbiana* Howell] May–Aug

C. gaylussacia A. Heller (p. 1177) **ST:** thorns 8–12 mm. **LF:** of short-shoots 3–8 cm, wide-elliptic to -ovate or narrow-obovate, base wedge-shaped, unlobed to shallowly lobed, thin to ± leathery. **INFL:** glabrous. **FL:** 12–13 mm wide; sepals distally glandular-serrate; stamens ± 20, anthers pink; styles 3–5. **FR:** 10–12 mm diam, ± spheric, deep purple to black, stones 3–5. 2*n*=34,51,68. Streamsides in meadows, scrub, forest; 30–1250 m. KR, NCoR, CaR, SNH, n CCo (Marin Co.); to AK, MT. [*C. suksdorfii* (Sarg.) Kruschke] May–Jun

C. monogyna Jacq. **ST:** thorns of indefinite growth. **LF:** of short-shoots 3–5 cm, widely ovate, deeply sharp-lobed. **FL:** 13–16 diam; sepals triangular; stamens 20, anthers pink-purple; style 1. **FR:** 10 mm diam, elliptic to ± spheric, red to deep red; stone 1. 2*n*=34,51. Naturalized in scattered places; 500 m. NCo, NCoR, SnFrB; temp N.Am; native to temp w Eurasia. May–Aug ❖

DASIPHORA

Barbara Ertter

Shrub, nonglandular. **LF:** alternate, leathery, ± odd-1-pinnately to subpalmately compound; lflets entire. **INFL:** 1 to cyme; pedicels straight, bractlets 0. **FL:** hypanthium ± shallow, bractlets 5; sepals ± triangular; petals > sepals, obovate to round, yellow [white]; stamens gen 20–25, pollen sac 1, horseshoe-shaped; pistils gen many, ovaries superior, styles ± club-shaped, attached near fr base. **FR:** achene, ± hairy. 5–7 spp.: n temp, arctic. (Greek: referring to hairy achenes) [Ertter 2007 J Bot Res Inst Texas 1:31–46]

D. fruticosa (L.) Rydb. (p. 1177) SHRUBBY CINQUEFOIL Pl < 1 m. **LF:** 1–3 cm; lflets 2–3 per side, 5–20 mm, linear to narrow-elliptic, upper pair decurrent. **INFL:** fls 1–6 at end of twigs. **FL:** hypanthium 5–8 mm wide; petals 5–10 mm; filaments 2–3 mm, anthers ± 1 mm; styles ± 2 mm. **FR:** ± 1.5 mm. *n*=7,14. Meadows, rocks; 2000–3600 m. KR, CaRH, SNH, Wrn, W&I; circumboreal. [*Potentilla f.* L.] Jun–Sep

DRYMOCALLIS

Barbara Ertter

Per, hairs short, simple, nonglandular and/or long, cross-walled, glandular. **ST:** ± erect, from ± branched caudex or rhizomes. **LF:** basal and cauline, alternate, odd-1-pinnately compound; lflets toothed, terminal gen ≥ lateral. **INFL:** cyme, ± open; pedicels straight, bractlets 0. **FL:** hypanthium ± shallow, bractlets 5; sepals ± triangular; petals < to > sepals, white to yellow; stamens gen 20–25, pollen sac 1, horseshoe-shaped; pistils many, styles fusiform, attached below fr middle. **FR:** achene, glabrous. *n*=7. 30 spp.: n temp. (Greek: wood beauty) [Ertter 2007 J Bot Res Inst Texas 1:31–46] Recognition based on morphological, molecular evidence. *D. ashlandica* (Green) Rydb. (infl narrow, petioles glandular, petals yellow) in sw OR, possibly nw CA.

1. Fls opening narrowly, petals ± erect; styles 1.5–3 mm, ± slender
  2. Petiole gen densely hairy with nonglandular hairs; terminal lflet gen > 10-toothed, tip obtuse to rounded;
   n CA . . . . ***D. rhomboidea***
  2′ Petiole sparsely hairy with glandular and nonglandular hairs; terminal lflet < 10-toothed, tip ± truncate;
   TR . . . . ***D. cuneifolia***
   3. St 20–45 cm; lvs 5–15 cm; SnBr . . . . var. ***cuneifolia***
   3′ St gen 5–20 cm; lvs 2–10 cm; SnGb . . . . var. ***ewanii***
1′ Fls opening widely, petals spreading; styles gen < 1.5 mm, swollen
  4. Petals ± ≤ sepals, ± elliptic-(ob)ovate; st and pedicel hairs gen glandular . . . . ***D. glandulosa***
   5. Petals ± elliptic-ovate, 3.5–6 mm wide, cream to pale yellow; infl appearing ± lfy (bracts ± 1/2
    subtended branches); pedicels 1–5 mm
    6. Petals gen 3.5–5 mm, gen pale yellow, sometimes cream; inland . . . . var. ***glandulosa***
    6′ Petals gen 4.5–6.5 mm, cream (± yellow); coastal . . . . var. ***wrangelliana***
   5′ Petals narrow-elliptic-obovate, 2–3.5 mm wide, yellow; infl not appearing lfy (bracts < 1/2 subtended
    branches); pedicels 2–10 mm
    7. Infl branches spreading (angle gen 25–55°); lflet teeth mostly double; sepal tip gen obtuse . . . . var. ***reflexa***
    7′ Infl branches ascending (angle gen 10–30°); lflet teeth mostly single; sepal tip acute . . . . var. ***viscida***

4′ Petals > sepals, ± obovate to round; st and pedicel hairs glandular or not — > 900 m
  8. St gen 5–25 cm; sheathing lf base gen appressed-hairy — gen glandular above; rocky areas, 2300–3900 m . . . . ***D. pseudorupestris***
    9. Lflets gen 4 per side; calyx with prominent nonglandular hairs 1–1.5 mm; styles gen ± red . . . . var. ***crumiana***
    9′ Lflets gen 3 per side; calyx with nonglandular hairs gen < 1 mm or 0; styles gen golden-brown . . . . var. ***saxicola***
  8′ St gen 10–90 cm; sheathing lf base gen glabrous
    10. Infl open, branches spreading (angle gen 20–40°); petals ± yellow; pedicel hairs glandular; n CA
       . . . . ***D. lactea*** var. ***austiniae***
    10′ Infl congested or narrow, branches ascending (angle gen 10–30°); petals cream or ± yellow; pedicel hairs nonglandular; CaR, SN, TR, SnJt, SNE
      11. St-base hairs gen 2–3 mm, glandular; st 40–90 cm; lflet teeth gen double; 1200–2200 m . . . . ***D. hansenii***
      11′ St-base hairs < 1 mm, nonglandular; st gen 20–60 cm; lflet teeth gen single; 1800–3700 m . . . ***D. lactea*** var. ***lactea***

D. cuneifolia Rydb. Matted, open. **ST:** glandular hairs gen sparse at base. **LF:** sheathing base gen strigose, lateral lflet pairs 3–5, terminal lflet flattened to rounded, teeth ± single, 2–4 per side. **INFL:** not lfy, spreading, branch angle 20–75°; pedicels 2–15 mm, lowermost to 30 mm, glandular hairs and short nonglandular hairs ± sparse. **FL:** opening narrowly; hypanthium bractlets ± 1–2 mm, 0.5 mm wide, ± linear-elliptic; sepals 2–5 mm, widely obtuse; petals erect, 2–4 mm, ± = sepals, narrow-obovate, yellow; styles 1.5–2.5 mm. **FR:** brown.

  var. ***cuneifolia*** WEDGELEAF WOODBEAUTY **ST:** 20–45 cm. **LF:** basal 5–15 cm, terminal lflet gen 10–25 mm, ± fan-shaped. **FR:** ± 1 mm. Riparian scrub; 1800–2200 m. SnBr (n of Big Bear Lake). [*Potentilla peirsonii* Munz] Jun–Jul ★

  var. ***ewanii*** (D.D. Keck) Ertter (p. 1177) EWAN'S WOODBEAUTY **ST:** gen 5–20 cm. **LF:** basal 2–10 cm, terminal lflet 6–15 mm, wedge-shaped to ± round. **FR:** < 1 mm. Edges of seeps, small waterways; 1900–2450 m (Mount Islip area). [*Potentilla glandulosa* subsp. *e.* D.D. Keck] May–Jul ★

D. glandulosa (Lindl.) Rydb. Tufted. **ST:** glandular hairs abundant at base. **LF:** sheathing base glabrous to glandular. **INFL:** pedicel glandular hairs gen abundant, short nonglandular hairs gen sparse. **FL:** opening widely; hypanthium bractlets ± elliptic; petals spreading; styles gen ± 1 mm. **FR:** 1–1.5 mm, ± red to brown. [*Potentilla g.* Lindl.]

  var. ***glandulosa*** (p. 1177) **ST:** 20–60 cm. **LF:** basal gen 10–25 cm, lateral lflet pairs gen 2–3, terminal lflet gen 25–55 mm, obovate, teeth double, ± 7–15 per side. **INFL:** lfy, spreading, branch angle gen 30–55°; pedicels gen 1–5 mm, lowermost to 15 mm. **FL:** hypanthium bractlets 2.5–6 mm, 0.5–2 mm wide; sepals spreading, gen 4.5–7 mm, widely obtuse; petals gen 3.5–5 mm, 3–4 mm wide, obovate-elliptic to round, pale yellow to cream. Gen ± shady or moist areas; gen 400–2000 m. KR, NCoR, CaR, SN, ScV (Sutter Buttes), TR, PR, MP; to BC, MT, AZ. [*Potentilla g.* subsp. *g.*] May–Jul

  var. ***reflexa*** (Greene) Ertter (p. 1177) **ST:** gen 15–80 cm. **LF:** basal gen 6–15 cm, lateral lflet pairs 3, terminal lflet gen 15–40 mm, obovate, teeth ± double, ± 6–16 per side. **INFL:** not lfy, spreading, branch angle gen 25–55°; pedicels gen 2–10 mm, lowermost to 20 mm. **FL:** hypanthium bractlets 1.5–4 mm, 0.5–1 mm wide; sepals ± reflexed, gen 4–6 mm, ± obtuse; petals gen 3–4 mm, 2–2.5 mm wide, narrow-obovate, yellow. Moist or ± shaded places; 450–2600 m. KR, NCoRH, CaRH, SNH, SnGb, SnBr, PR; s OR, w NV, n Baja CA. [*Potentilla g.* subsp. *r.* (Greene) D.D. Keck] May–Aug

  var. ***viscida*** (Parish) Ertter **ST:** gen 6–20 cm. **LF:** basal gen 10–25 cm, lateral lflet pairs gen 2–3, terminal lflet gen 20–40 mm, obovate to ± diamond-shaped, teeth ± single, 4–9 per side. **INFL:** not lfy, narrow, branch angle gen 15–30°; pedicels gen 2–10 mm, lowermost to 20 mm. **FL:** hypanthium bractlets gen 1.5–3 mm, ± 1 mm wide; sepals ± reflexed, gen 4–6 mm, acute; petals gen 2–4 mm, ± 2 mm wide, narrow-obovate-elliptic, ± yellow. Along streams, open areas under pines; 1100–2500 m. TR, PR. Combines features of *D. glandulosa* var. *reflexa*, *D. lactea* var. *l.* May–Aug

  var. ***wrangelliana*** (Fisch. & Avé-Lall.) Ertter **ST:** 20–70 cm. **LF:** basal gen 10–25 cm, lateral lflet pairs gen 3, terminal lflet gen 30–60 mm, widely obovate, teeth double, ± 9–17 per side. **INFL:** lfy, spreading, branch angle gen 30–55°; pedicels gen 1–5 mm, lowermost to 30 mm. **FL:** hypanthium bractlets gen 4–7 mm, 1.5–2.5 mm wide; sepals spreading, gen 5–11 mm, widely obtuse; petals 4.5–6.5

mm, 3–5.5 mm wide, widely ovate-elliptic, gen cream or pale yellow. Openings in coastal scrub, moist or ± shaded places; < 1000 m. NCo, NCoRO, CW, SCo; sw OR, n Baja CA. Mar–Jun

D. hansenii (Greene) Rydb. YOSEMITE WOODBEAUTY **ST:** gen 1, 30–90 cm, glandular hairs abundant at base. **LF:** basal gen 10–25 cm, sheathing base gen glabrous, lateral lflet pairs 3–4, terminal lflet gen 20–50 mm, widely obovate, rounded, teeth single to double, 6–11 per side. **INFL:** not lfy, narrow, branch angle gen 10–30°; pedicels gen 1–10 mm, lowermost to 20 mm, glandular hairs and short nonglandular hairs abundant. **FL:** opening widely; hypanthium bractlets 2–4 mm, 0.5–1 mm wide, linear to narrow-elliptic; sepals 4–8 mm, acute; petals spreading, 4–7 mm, ≥ sepals, widely obovate, cream; styles ± 1 mm. **FR:** < 1 mm, light brown. Moist meadows; 1200–2200 m. CaR, SN. [*Potentilla glandulosa* subsp. *h.* (Greene) D.D. Keck] Coarser pls from Wrn probably undescribed taxon. Intergrades with *D. lactea* at higher elevation. Jun–Sep

D. lactea (Greene) Rydb. Tufted. **ST:** gen 10–60 cm. **LF:** basal gen 5–20 cm, sheathing base gen glabrous, lateral lflet pairs gen 3–4, terminal lflet gen 10–40 mm, ± obovate, ± obtuse, teeth ± single, gen 4–10 per side. **INFL:** not lfy; pedicels 2–10 mm, lowermost to 30 mm, glandular hairs gen 0–few, short nonglandular hairs gen dense. **FL:** opening widely; hypanthium bractlets lance-linear, 2–5 mm, 0.5–1 mm wide; sepals 3–8 mm, ± acute; petals spreading, gen 4–8 mm, > sepals, widely obovate; style gen ± 1 mm. **FR:** ± 1 mm, brown.

  var. ***austiniae*** (Jeps.) Ertter (p. 1185) AUSTIN'S WOODBEAUTY **ST:** glandular hairs present at base or not. **INFL:** spreading, branch angle gen 20–40°. **FL:** petals gen ± yellow. Gen ± moist, often rocky places; 900–2600 m. NW, CaR, n SNH, MP; sw OR. [*Potentilla glandulosa* subsp. *ashlandica* (Greene) D.D. Keck, misappl.] Incl most yellow-fld pls formerly in *Potentilla glandulosa* subsp. *nevadensis*. May–Sep

  var. ***lactea*** (p. 1185) SIERRAN WOODBEAUTY **ST:** glandular hairs gen 0–sparse at base. **INFL:** narrow, branch angle gen 10–20°. **FL:** petals gen cream. Gen ± moist, often rocky places; 1800–3700 m. SNH, TR, SnJt, SNE; OR, NV. [*Potentilla glandulosa* subsp. *nevadensis* (S. Watson) D.D. Keck] Jun–Sep

D. pseudorupestris (Rydb.) Rydb. Tufted to matted. **ST:** gen 5–25 cm, glandular hairs abundant at base. **LF:** basal gen 6–9 cm, sheathing base gen appressed-hairy, terminal lflet gen 5–20 mm, widely obovate to fan-shaped, ± rounded, teeth gen ± single, 4–10 per side. **INFL:** not lfy, spreading, branch angle gen 20–40°; pedicels gen 3–10 mm, lowermost to 20 mm, glandular hairs gen abundant, short nonglandular hairs 0–many. **FL:** opening widely; hypanthium bractlets 2–5 mm, 1–2 mm wide, lance-linear to elliptic-ovate; sepals gen 4–6 mm, acute to obtuse; petals spreading, gen 4–8 mm, > sepals, ± obovate, cream to pale yellow; styles 1–1.5 mm. **FR:** ± 1 mm, light brown. [*Potentilla glandulosa* subsp. *p.* (Rydb.) D.D. Keck]

  var. ***crumiana*** Ertter CRUM'S WOODBEAUTY **ST:** short nonglandular hairs gen 0. **LF:** lflets gen 4 per side. **FL:** calyx with nonglandular hairs 1–1.5 mm; styles gen dark ± red. Rocky areas; 3200–3900 m. SNH, W&I. Jul–Aug

  var. ***saxicola*** Ertter CLIFF WOODBEAUTY **ST:** short nonglandular hairs gen present. **LF:** lflets gen 3 per side. **FL:** calyx with nonglandular hairs < 1 mm or 0; styles gen golden-brown. Rocky areas; 2300–3500 m. NW, CaRH, SNH, Wrn, W&I; to BC, MT. Jul–Sep

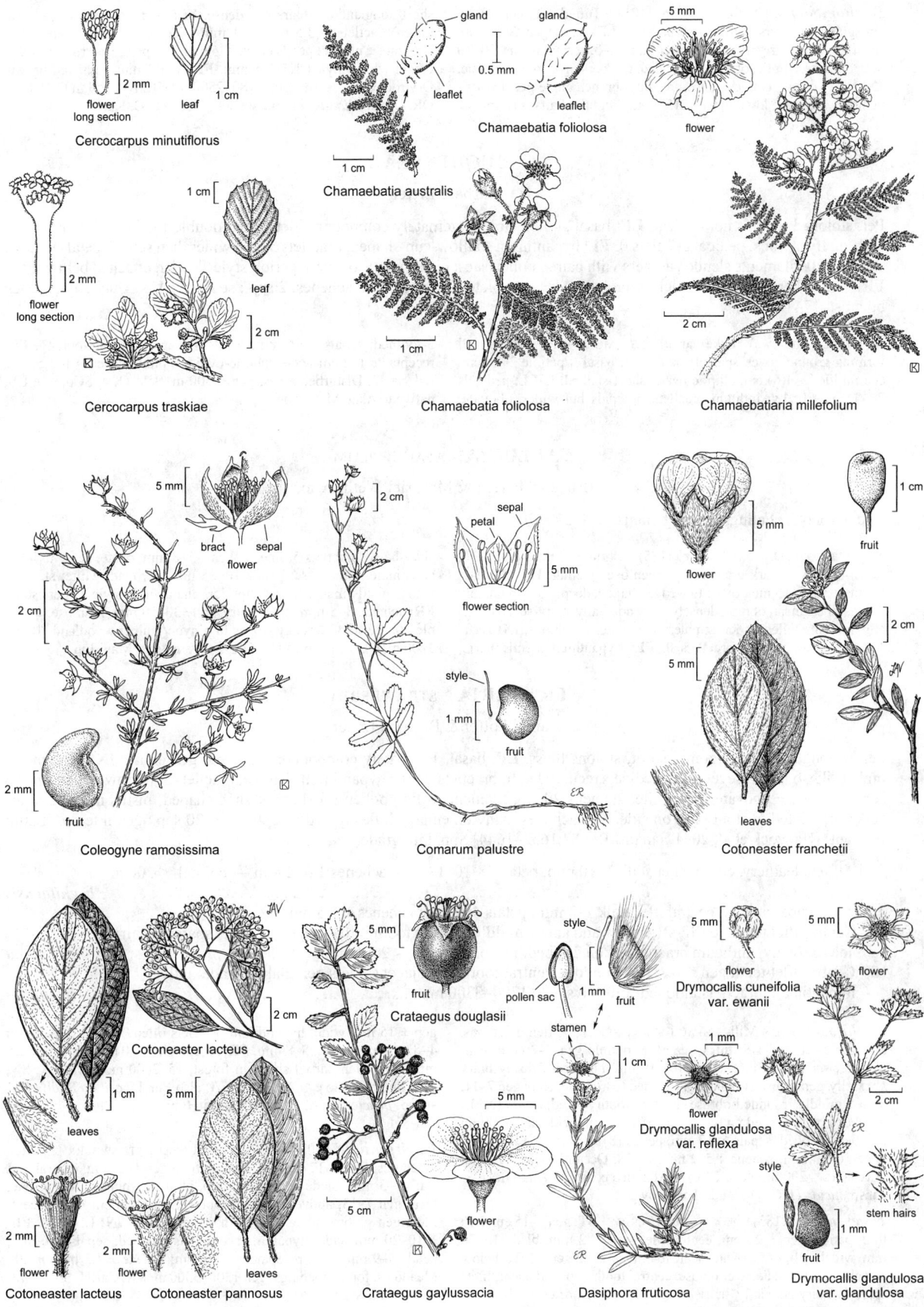

Cercocarpus minutiflorus

flower long section
leaf
2 mm
1 cm

Chamaebatia foliolosa

gland
leaflet
gland
leaflet
0.5 mm

Chamaebatia australis
1 cm
1 cm

flower
5 mm

Cercocarpus traskiae

flower long section
2 mm
leaf
1 cm
2 cm

Chamaebatia foliolosa
1 cm

Chamaebatiaria millefolium
2 cm

Coleogyne ramosissima

flower
bract
sepal
5 mm
2 cm
fruit
2 mm

Comarum palustre

2 cm
petal
sepal
5 mm
flower section
style
1 mm
fruit

Cotoneaster franchetii

flower
5 mm
fruit
1 cm
leaves
5 mm
2 cm

Cotoneaster lacteus

leaves
1 cm
flower
2 mm

Cotoneaster pannosus

flower
2 mm
leaves
5 mm

Crataegus douglasii

fruit
5 mm

Crataegus gaylussacia

flower
5 mm
5 cm

Dasiphora fruticosa

style
pollen sac
stamen
fruit
1 mm
flower
1 cm

Drymocallis cuneifolia var. ewanii

flower
5 mm
flower
5 mm

Drymocallis glandulosa var. reflexa

flower
1 mm
1 mm
2 cm

Drymocallis glandulosa var. glandulosa

style
fruit
1 mm
stem hairs

D. rhomboidea (Rydb.) Rydb. (p. 1185) Tufted. **ST:** gen 20–45 cm, glandular hairs 0 to sparse at base. **LF:** basal gen 5–18 cm, sheathing base strigose, lateral lflet pairs 2–3, terminal lflet 10–30 mm, obovate-elliptic, obtuse to rounded, teeth ± single, 8–12 per side. **INFL:** lfy or not, congested to spreading, branch angle gen 15–40°; pedicels 1–5 mm, lowermost to 15 mm, glandular hairs 0 to sparse, short nonglandular hairs gen dense. **FL:** opening narrowly; hypanthium bractlets 2–3.5 mm, ± 1 mm wide, lance-linear; sepals gen 4–6 mm, ± obtuse; petals erect, 3–5 mm, < sepals, ± narrow-obovate, cream-white; styles 1.5–2.5 mm. **FR:** 1–1.3 mm, light red-brown. Dry rocky slopes, roadcuts; 180–2500 m. KR, NCoRH, n CaRH; sw OR. [*Potentilla glandulosa* subsp. *globosa* D.D. Keck] Jun–Jul

DUCHESNEA

Barbara Ertter

Per, stoloned; glands not evident. **LF:** basal and on stolons, ternately compound; lflets often double-toothed. **INFL:** fls 1 in axils on lfy stolons; pedicel bractlets 0. **FL:** hypanthium shallow-cup-shaped, bractlets 5, gen wider than sepals; petals yellow; stamens 20, filaments slender, anthers with paired pollen sacs; pistils many, ovary superior, style slender, attached below fr tip. **FR:** ± strawberry-like; receptacle enlarged, red, ± completely covered with achenes. 2 spp.: se Asia. (A.N. Duchesne, French botanist, 1747–1827)

D. indica (Andrews) Focke var. ***indica*** MOCK-STRAWBERRY Pl forming ground-cover, sparsely hairy. **LF:** basal lf petiole 2–20 cm, central lflet 1–4(6) cm, elliptic-ovate, short-stalked. **INFL:** pedicels 2–13 cm. **FL:** hypanthium bractlets = sepals but wider, obovate, 3-toothed; sepals 4–10 mm; petals 4–8 mm, obovate-oblong. **FR:** receptacle 1–2 cm wide, spheric-ovoid, ± pithy; achenes 1–1.5 mm, red. *n*=42. Disturbed areas; gen < 500 m. ScV, CCo, SCo; to e US; native to Asia. May–Aug

FALLUGIA APACHE PLUME

Bruce D. Parfitt & Margriet Wetherwax

1 sp. (Abbot V. Fallugi, Italian botanist, ± 1627–1707)

F. paradoxa (D. Don) Torr. (p. 1185) Shrub < 2 m, ± erect. **ST:** much-branched; bark gray-white-tomentose, peeling. **LF:** alternate to clustered, 7–15 mm, ovate to wedge-shaped, deeply 3–7-pinnately lobed, lobes linear, obtuse, densely hairy adaxially, rusty-scaly abaxially, margins rolled under; stipules lanceolate, deciduous. **INFL:** fls 1–3, terminal; pedicel bractlets 0. **FL:** hypanthium hemispheric, silky-hairy, bractlets 5, linear; sepals 5–8 mm, ovate, acute to long-acuminate, tomentose; petals 10–25 mm, ± round, white; stamens many; pistils many, ovary superior, chamber 1, style 1, persistent. **FR:** achene, 3–5 mm, silky-hairy, style 30–50 mm, plumose, ± purple. 2*n*=28. Dry, ± rocky slopes in pinyon/juniper woodland; 1000–2200 m. e DMtns; to CO, w OK, w TX, n Mex. May–Jun

FRAGARIA STRAWBERRY

Daniel Potter & Barbara Ertter

Per, ± nonglandular, rhizomes short, stolons lfless. **LF:** basal, 1-ternately compound; lflet teeth gen entire. **INFL:** cyme, ± umbel-like, open, 1–several-fld; pedicels recurved in fr, bractlets 0. **FL:** hypanthium shallow, bractlets 5, narrower than sepals; sepals, petals ± obovate, gen white; stamens 20–35, filaments ± flat, pollen sac 1, horseshoe-shaped; pistils many, ovaries superior, jointed to stout style on side. **FR:** achenes many on enlarged, fleshy, red receptacle. ± 20 spp.: gen n temp. (Latin: fragrant) [Hancock et al. 2004 Canad J Bot 82:1632–1644] Spp. intergrade.

1. Lf thick, leathery, central lflet stalk 1–10 mm; petals (8)10–18 mm; achenes 1.5–2 mm — coastal, < 200 m
.... ***F. chiloensis***
1′ Lf thin, not leathery, central lflet stalk 1–3 mm; petals 4–9 mm; achenes ± 1.5 mm
  2. Central lflet teeth gen 12–21, below and above middle, central tooth < to > adjacent; lflets sparsely hairy adaxially; hypantheum bractlets often 2-lobed; infl often > lvs; < 2000 m .... ***F. vesca***
  2′ Central lflet teeth gen < 13, above middle, central tooth << adjacent; lflets gen glabrous adaxially; hypantheum bractlets unlobed; infl < lvs; gen 1200–3300 m .... ***F. virginiana***

F. chiloensis (L.) Mill. BEACH STRAWBERRY Often dioecious. **ST:** gen 5–20 cm. **LF:** thick, leathery; petiole gen 2–20 cm; central lflet stalk 1–10 mm, blade 10–60 mm, obovate, densely hairy abaxially, gen glabrous adaxially, rounded to truncate, teeth gen 7–11, above middle, rounded-obtuse, central tooth << adjacent. **INFL:** exceeded by to exceeding lvs. **FL:** gen 20–40 mm wide; hypanthium bractlets unlobed; sepals 6–10 mm; petals (8)10–18 mm. **FR:** receptacle 10–20 mm; achene 1.5–2 mm. *n*=28. Ocean beaches, coastal grassland; < 200 m. NCo, CCo; to AK; also coastal S.Am, HI. [*F. c.* subsp. *lucida* (J. Gay) Staudt] Feb–Nov

F. vesca L. (p. 1185) WOOD STRAWBERRY **ST:** gen 3–15 cm. **LF:** thin; petiole gen 3–25 cm; central lflet stalk < 2 mm, blade 15–70 mm, widely elliptic-obovate, acute to obtuse, teeth gen 12–21, below and above middle, acute or obtuse, central tooth < to > adjacent; lflets sparsely hairy adaxially, hairier abaxially. **INFL:** often >> lvs. **FL:** gen ± 15 mm wide; hypanthium bractlets often 2-lobed; sepals gen 4–8 mm; petals gen 5–8 mm. **FR:** receptacle 5–10 mm; achene ± 1.5 mm. *n*=7. Gen partial shade in forest; 15–2000 m. NW, CaR, SN, CW, SnBr, PR; to e N.Am, Baja CA; also Eur. [*F. crinita* Rydb.; *F. v.* subsp. *bracteata* (A. Heller) Staudt] Hybridizes with *F. chiloensis.* Jan–Jul

F. virginiana Mill. (p. 1185) MOUNTAIN STRAWBERRY Often dioecious. **ST:** 2–12 cm. **LF:** thin; petiole gen 1–25 cm; central lflet stalk 1–3 mm, blade gen 15–60 mm, obovate, rounded to truncate, teeth 7–13, above middle, ± obtuse, central tooth << adjacent; lflets gen glabrous adaxially, ± hairy abaxially. **INFL:** < lvs. **FL:** ± 10–20 mm wide; hypanthium bractlets unlobed; sepals 3–6 mm; petals 4–9 mm. **FR:** receptacle ± 10 mm; achene ± 1.5 mm. *n*=28. Meadows, forest openings; gen 1200–3300 m. KR, CaRH, SNH, MP; to e N.Am. May–Aug

GEUM

Joseph R. Rohrer

Per from thick caudex or elongate rhizome. **ST:** ascending to erect, gen hairy. **LF:** gen basal, gen odd-1-pinnately compound, upper cauline simple or not; lflets lobed, alternately large, small, teeth uneven. **INFL:** gen cyme, open; pedicel bractlets 0. **FL:** hypanthium shallow, bractlets gen 5; stamens > 20; pistils many, ovary superior, continuous to style. **FR:** achene, ovoid to fusiform, ± flat; style elongated, persistent. 40–50 spp.: gen n temp, arctic. (Latin: ancient name)

1. Style in fr 15–40 mm, plumose, not or inconspicuously hooked; fls nodding, ± cup-shaped; petals cream to
    pale yellow, pink-tinged or purple-veined, persistent; hypanthium bractlets 5–15 mm . . . . *G. triflorum* var. *ciliatum*
1′ Style in fr 2–9 mm, not plumose, hooked; fls gen erect, rotate; petals yellow, deciduous; hypanthium bractlets ≤ 4 mm
  2. Hypanthium bractlets 2–4 mm; style in fr with few bristles at base, nonglandular; pl drying black-green
    . . . . *G. aleppicum*
  2′ Hypanthium bractlets gen 0 or ≤ 2 mm; style in fr without bristles, gen glandular; pl drying ± green
    . . . . *G. macrophyllum*
    3. Pedicel nonglandular; upper st lvs simple, divided < 3/4 to base; NCo . . . . var. *macrophyllum*
    3′ Pedicel glandular; upper st lvs compound or simple, divided ± to base; NW (exc NCo), CaRH, SNH,
      SnBr, GB . . . . var. *perincisum*

G. aleppicum Jacq. ALEPPO OR YELLOW AVENS Pl tufted, drying black-green. **ST:** 30–120 cm. **LF:** 10–40 cm; main lflets 1–4 per side, largest nearly = lobes of terminal; terminal lflet 5–10 cm, gen 3-lobed nearly to base, sharp-toothed. **INFL:** 3–7(10)-fld; pedicels hairy, nonglandular. **FL:** rotate, gen erect; hypanthium bractlets 2–4 mm, lance-linear; sepals reflexed, 4–8 mm; petals 4–9 mm, ovate-round, yellow, deciduous. **FR:** body 4–5 mm; style 3–6 mm, hooked below deciduous tip, with few bristles at base, nonglandular. 2*n*=42. Meadows; 1000–1600 m. n CaRH (Mount Shasta), MP; to n&e N.Am, Mex, Eurasia. Jun–Aug ★

G. macrophyllum Willd. LARGE-LEAVED AVENS Pl tufted, drying ± green. **ST:** 20–110 cm. **LF:** 10–45 cm; main lflets gen 2–4 per side, gen << terminal; terminal lflet gen 8–10 cm, ± blunt-cordate-reniform, 3-lobed < 3/4 to base, irregularly toothed. **INFL:** 3–16-fld; pedicels densely hairy. **FL:** rotate, gen erect; hypanthium bractlets gen 0 or ≤ 2 mm, linear; sepals reflexed, 2.5–5.5 mm; petals 3–7 mm, obovate-round, yellow, deciduous. **FR:** body 2.5–3.5 mm; style 2.5–6 mm, hooked below deciduous tip, without bristles, gen glandular. 2*n*=42.

var. ***macrophyllum*** (p. 1185) **LF:** upper st lvs simple, divided < 3/4 to base, segments diamond-shaped to oblong **INFL:** pedicels

nonglandular. Meadows, streambanks; < 300 m. NCo; to n&e N.Am, Eurasia. May–Aug

var. ***perincisum*** (Rydb.) Raup **LF:** upper st lvs compound or simple, divided ± to base, segments oblanceolate to obovate. **INFL:** pedicels with stalked glands. Meadows, streambanks; 1000–3300 m. NW (exc NCo), CaRH, SNH, SnBr, GB; to n&e N.Am. May–Aug

G. triflorum Pursh var. ***ciliatum*** (Pursh) Fassett (p. 1185) PRAIRIE-SMOKE Pl in patches, rhizomed, ± gray-green. **ST:** gen 10–50 cm. **LF:** 4–30 cm; lflets wedge-shaped, gen 2–3-lobed > 1/2 to base, lobes deeply few-toothed, main lflets 3–9 per side, largest 1–3 cm, ± = terminal. **INFL:** (1)3–5(7)-fld; pedicels tomentose, occ glandular. **FL:** ± cup-shaped, nodding; hypanthium bractlets 5–15 mm, linear-oblanceolate, outcurved; sepals erect, 6–14 mm, maroon, purple, or ± green and purple-tinged; petals erect, 7–13 mm, ± elliptic, cream to pale yellow, pink-tinged or purple-veined, persistent. **FR:** body 2.5–5 mm; style 15–40 mm below tardily deciduous tip, not or inconspicuously hooked, plumose. 2*n*=42. Dry meadow edges, sagebrush scrub, open yellow-pine forest; 1300–3200 m. c KR (Marble Mtns), CaRH, n&c SNH, GB; to BC, MT, CO. [*G. t.* var. *canescens* (Greene) Kartesz & Gandhi] May–Jul

HETEROMELES CHRISTMAS BERRY, TOYON

James B. Phipps

1 sp. (Greek: different apple) [Phipps 1992 Canad J Bot 70:2138–2162]

H. arbutifolia (Lindl.) M. Roem. (p. 1185) Shrub, small tree, < 10 m, unarmed. **ST:** trunk bark ± gray; twigs puberulent. **LF:** petioled, simple, evergreen; blade 5–10 cm, ± elliptic, leathery, finely toothed, veined to teeth. **INFL:** panicle, domed, branches white-tomentose; pedicel bractlets several to many, scattered throughout, gland-tipped. **FL:** hypanthium urn-shaped, smooth, bractlets 0; sepals

short, triangular, over hypanthium in fr; petals < 4 mm, white; stamens ± 10, filaments short; ovary inferior, 2–3-chambered, styles 2–3, free. **FR:** pome, 5–10 mm diam, elliptic, bright red (yellow), pulp mealy. **SEED:** 2–3, large, smooth, brown. Chaparral, oak woodland, mixed-evergreen forest; < 1300 m. NW, SNF, n&c SNH, GV, CW, SW; BC, Mex. Cult for foliage, fls, fr. (May)Jun–Aug

HOLODISCUS

Shrub, ± hairy. **LF:** simple, alternate, toothed; stipules 0. **INFL:** raceme or panicle, terminal, many-fld, persistent; pedicels slender, bractlets 1–3, linear. **FL:** hypanthium saucer-shaped, prominent nectary-disk below inner rim, bractlets 0; petals gen white; stamens 15–20; pistils 5, ovaries superior, 2-ovuled, hairs dense, bristle-like, persistent in fr, style persistent, stigma ± 2-lobed. **FR:** achenes 5. 3–5 spp.: w N.Am, C.Am, n S.Am. (Greek: whole disk)

H. discolor (Pursh) Maxim. OCEANSPRAY Pl 0.3–6 m. **ST:** bark ± red, in age gray, shredding; twigs glabrous to hairy, occ glandular. **LF:** 0.3–8 cm, ovate to obovate, strong-veined abaxially, glabrous to hairy, occ glandular, teeth entire to compound; base truncate to wedge-shaped; petiole distinct or not. **INFL:** 2–25 cm, 1.5–25 cm wide. **FL:** hypanthium 3–5 mm wide; sepals 1–2 mm; petals 1.5–2 mm; style 1 mm. **FR:** achenes 1–1.5 mm, often with sessile glands. [*H. saxicola* A. Heller] Highly variable; vars. intergrade.

1. Lf blade 1.5–8 cm, teeth toothed, base gen truncate
    to rounded; infl 5–25 cm, 5–25 cm wide, branches
    gen many. . . . . var. *discolor*
1′ Lf blade 0.3–3 cm, teeth entire, above middle, base gen
    wedge-shaped; infl 2–8 cm, 1.5–5 cm wide, branches 0–few
  2. Lf puberulent to glabrous both surfaces,
      glands visible. . . . . var. *glabrescens*
  2′ Lf ± long-hairy 1 or both surfaces, glands 0
      or gen obscured by hairs . . . . var. *microphyllus*

var. ***discolor*** (p. 1185) Pl 1.5–6 m, ± open. **ST**: twigs hairy, glands 0. **LF**: petiole 2–15 mm; blade 1.5–8 cm, ovate to elliptic, hairs sparse adaxially, dense abaxially, teeth toothed, base gen truncate to rounded. **INFL**: 5–25 cm, 5–25 cm wide; branches gen many. Moist woodland edges, rocky slopes; < 3200 m. NW, SNH, GV (Sutter Buttes), CW, ChI, WTR, SnGb (300–1300 m), PR; to BC, MT, CO, TX, Mex. May–Aug

var. ***glabrescens*** (Greenm.) Jeps. Pl 0.3–1 m, ± dense. **ST**: twigs glabrous to puberulent; glands 0 or gen visible. **LF**: petiole gen indistinct; blade 0.3–3 cm, round to obovate, both surfaces glabrous to puberulent, hairs much longer on margins, veins, glands visible, teeth entire, above middle, base gen wedge-shaped. **INFL**: gen 2–8 cm, 1.5–5 cm wide; branches 0–few. Rocky places, outcrops; 600–3020 m. CaRH, NCoRH (Mendocino Co.), MP; to OR, UT. [*H. microphyllus* var. *g.* (Greenm.) F.A. Ley] Jun–Aug

var. ***microphyllus*** (Rydb.) Jeps. (p. 1185) Pl 0.3–1 m, ± dense. **ST**: twigs hairy to long-hairy; glands 0 or obscured by hairs. **LF**: petiole gen indistinct; blade 0.3–3 cm, ± round to obovate, ± long-hairy on 1 or both surfaces, hairs not greatly longer on margins, veins, glands 0 or gen obscured by hairs, teeth entire, above middle, base gen wedge-shaped. **INFL**: gen 2–8 cm, 1.5–5 cm wide; branches 0–few. Rocky places, outcrops; 1100–4000 m. SNH, SnGb, SnBr, SnJt, SNE, DMtns; to OR, ID, WY, CO, AZ, Baja CA. [*H. m.* Rydb.; *H. m.* Rydb. var. *m.*] Jun–Sep

HORKELIA

Barbara Ertter

Per, gen ± glandular, gen resinous-smelling; caudex gen branched. **ST**: gen ascending to erect. **LF**: gen basal, odd-1-pinnately compound, gen ± flat; cauline alternate, reduced upward; lflets 2–15 per side, uppermost lateral gen ± fused with terminal. **INFL**: cyme; pedicels gen straight, bractlets 0. **FL**: hypanthium cup-like, ± flat-bottomed, width ± 2 × length, bractlets 5, gen 2/3 sepals; sepals often reflexed; petals gen ± = sepal, blunt, white; stamens 10, filaments ± flat, often forming a tube; pistils 2–many, ovary superior, style attached below fr tip, ± thicker at base. **FR**: achene. 20 spp.: w N.Am. (J. Horkel, German pl physiologist, 1769–1846) [Ertter & Reveal 2007 Novon 17:315–325] Many attractive to bees; data apply to basal lvs, pressed hypanthia.

1. Fls gen in head-like clusters; pl gen rosetted or tufted; NW, CaR, SN, MP, n SNE — hypanthium bractlets < 0.5 mm wide; hypanthium gen 2–5 mm wide; lower pedicel often reflexed in fr; > 300 m
  2. Lflets 2–5 per side, entire to ± 3-toothed; petal linear to obovate; fr 1.5–2 mm; 300–2500 m
    3. St densely glandular above, ascending to erect; petal gen 2.5–4 mm, gen widely obovate — nw KR
      . . . . *H. congesta* var. *nemorosa*
    3′ St ± glabrous to sparsely glandular above, often decumbent; petal 1.5–4 mm, linear to widely oblanceolate . . . . *H. tridentata*
      4. Petal ± widely oblanceolate, often < sepal; hypanthium 2.5–5 mm wide, inner wall hairy (exc in NCoRH); KR, NCoR, n SNH . . . . var. *flavescens*
      4′ Petal gen narrow-oblanceolate, gen ≥ sepal; hypanthium gen 2–3.5 mm wide, inner wall gen glabrous; e KR, CaR, SN, MP . . . . var. *tridentata*
  2′ Lflets 3–15 per side, > 3-toothed; petal ± wedge-shaped; fr 1–1.8 mm; > 950 m . . . . *H. fusca*
    5. Lflets gen divided ± 3/4 to base, gen 8–15 per side, ± crowded; filament gen < 0.5 mm, gen wider than long — s CaRH . . . . ²var. *tenella*
    5′ Lflets gen toothed ≤ 1/2 to base, gen 4–8 per side, ± separated; filament often > 0.5 mm, gen longer than wide
      6. Petal 4–6 mm; lflets gen 10–20 mm, toothed < 1/4 to base, gen ± obovate; anther ± 0.6 mm; style 1–1.5 mm; fr ± 1.8 mm; Wrn . . . . var. *pseudocapitata*
      6′ Petal 2–4 mm; lflets gen 5–15 mm, toothed 1/4–1/2 to base, ± round to wedge-shaped; anther ± 0.4 mm; style ± 1 mm; fr 1–1.2 mm; n CA (exc Wrn), SNH
        7. Lflets ± gray-hairy, narrow-wedge-shaped; CaR, MP . . . . var. *brownii*
        7′ Lflets green, wedge-shaped to round; e KR, SNH, n SNE . . . . var. *parviflora*
1′ Fls gen either in ± flat-topped clusters or infl open; pl often matted; CA-FP, W&I
  8. Pedicel ± recurved in fr; infl open, often ± 1-sided
    9. Petal 4–7 mm; lflets toothed 1/4 to midrib; pistils gen 20–50; n&c SNF (esp Ione Formation) . . . . *H. parryi*
    9′ Petal 2–3 mm; lflets lobed 1/2 to 3/4 to base; pistils < 5; SnBr . . . . *H. wilderae*
  8′ Pedicel erect in fr; infl open or dense, not 1-sided
    10. Filament gen < 0.5 mm, gen wider than long; style ± 1 mm — petal ± wedge-shaped; s CaRH . . . . ²*H. fusca* var. *tenella*
    10′ Filament ≥ 0.5 mm, gen longer than wide; style 1–4 mm
      11. Pl gen rosetted from simple or few-branched caudex; resinous odor weak; lflets gen < 10-lobed; fr ≥ 2 mm; often ± serpentine clay
        12. Lflets 5–10 per side, 5–25 mm, lobed > 3/4 to base; st base hairs 2–3 mm; infl open to dense; hypanthium 3.5–5 mm wide; petal ± obovate, cream; pistils 5–15; style 2.5–4 mm . . . . *H. daucifolia*
          13. Lflets ± gray, lobes 2–5, 1–3 mm wide; petal 2–4 mm wide; pedicels nonglandular . . . . var. *daucifolia*
          13′ Lflets green, lobes 5–15, < 1 mm wide; petal 4–8 mm wide; pedicels glandular . . . . var. *indicta*
        12′ Lflets ± 15 per side, 2–10 mm, gen lobed ± 1/2 to base; st base hairs ± 1 mm; infl open; hypanthium 2–3 mm wide; petal obcordate, often pink-tinged; pistils 2–6; style ± 1.5–2 mm
          14. Stipules of basal lvs pinnately divided; lvs ± green, 5–15 cm . . . . *H. howellii*
          14′ Stipules of basal lvs entire or forked; lvs silvery, 3–10 cm . . . . *H. sericata*
      11′ Pl gen tufted to matted from a few- to many-branched caudex; resinous odor strong; lflets often > 10-toothed or -lobed; fr often < 2 mm; gen non-serpentine
        15. Lflets pinnately veined, oblong to obovate, ± evenly > 10-toothed — hypanthium width 4–8 mm, > 2 × length, bractlets ± 2–3 mm wide; CW, SW
          16. Lflets 1–3 per side, < distinct terminal lflet, teeth gen double, gen > 20; petal ± round, ± 5 mm wide; 400–1300 m; PR . . . . *H. truncata*

16′ Lflets 5–12 per side, ± = gen indistinct terminal lflet, teeth gen single, ± 10–15; petal obovate or
    narrower, 1.5–4 mm wide; < 870 m; CCo, SCoRO, SCo. . . . . ***H. cuneata***
  17. Hairs ± all glandular; infl open; hypanthium inner rim ± glabrous; filament base gen 0.5–2 mm
     wide. . . . . var. ***puberula***
  17′ Hairs mostly glandless; infl open to dense; hypanthium inner rim gen hairy; filament base 0.5–1 mm wide
    18. Infl gen ± open; hairs ± spreading, ± dense; glands ± obvious . . . . var. ***cuneata***
    18′ Infl dense to ± open; hairs ascending to appressed, dense; glands hidden, if any . . . . var. ***sericea***
15′ Lflets ± palmately veined at base, wedge-shaped or ± round-ovate, ± unevenly few- to many-toothed or -lobed
  19. Hypanthium width 4–10 mm, 1–2 × length, bractlets ovate, gen ± 2 mm wide, ± = sepal; lflets round
    to ovate, 5–60 mm, gen ± separated, gen > 20-toothed or -lobed; st ± lfy; fr ± 1 mm. . . . . ***H. californica***
    20. Lflets 3–5 per side, gen unlobed, 15–60 mm, elliptic to ovate — hypanthium inner wall glabrous;
      sepal not red-mottled; style 2–3 mm; CW . . . . var. ***frondosa***
    20′ Lflets 4–9 per side, ± lobed, 5–40 mm, ovate to round
      21. Lflets gen few-lobed ± 1/2 to base, 10–40 mm; sepal red-mottled inside; hypanthium bractlets gen
        toothed; hypanthium inner wall ± hairy; filament gen 1.5–3 mm; style gen 3–4 mm; NCo, CCo
        . . . . var. ***californica***
      21′ Lflets lobed 1/2–3/4 to base, 5–25 mm; sepal not red-mottled; hypanthium bractlets gen entire;
        hypanthium inner wall glabrous; filament gen 0.5–1.5 mm; style gen 2–3 mm; NCoR, c&s SN,
        SnFrB, SnBr . . . . var. ***elata***
19′ Hypanthium width gen < 6 mm, gen 2+ × length, bractlets lanceolate to ovate, ≤ 2 mm wide, <
    sepal; lflets round to wedge-shaped, gen < 15 mm, often crowded, < 20-toothed or -lobed; st lvs
    few, ± reduced; fr > 1 mm
  22. St < 25 cm; hypanthium bractlets gen < 0.5 mm wide; lvs 2–10 cm; fr 1.5–2.5 mm; > 2000 m;
    KR, s SNH, W&I
    23. Hypanthium densely hairy on inner rim; sepal 3.5–6 mm, 2+ × hypanthium length; longer hairs
      ± 1 mm, soft — ne KR . . . . ***H. hendersonii***
    23′ Hypanthium ± hairy on inner wall; sepal gen 2–4 mm, < 2 × hypanthium length; longer hairs
      ± 0.5 mm, stiff
      24. Lflets gen 10–14 per side, gen 2.5–4 mm; pistils 10–20; W&I . . . . ***H. hispidula***
      24′ Lflets gen 4–6 per side, gen 3.5–8 mm; pistils 5–12; s SNH. . . . . ***H. tularensis***
22′ St 10–70 cm; hypanthium bractlets gen ≥ 0.5 mm wide; lvs 3–20 cm; fr 1.5–2 mm; < 2800 m;
    NCo, NCoR, CW, SW
  25. St decumbent to ascending, gen 10–30 cm; pl matted, odor strong; infl ± dense; lvs gen 4–10 cm;
    lflets gen 5–10 per side, ± crowded — hairs dense; NCo, CCo . . . . ***H. marinensis***
  25′ St ascending to erect, 10–70 cm; pl tufted or matted, odor strong or not; infl gen ± open; lvs gen
    3–20 cm; lflets 6–16 per side, separated or crowded
    26. Hypanthium bractlets 1–2 mm wide, lanceolate to ovate; anther ± 1 mm — hypanthium gen 3–6
      mm wide; st hairs spreading; SCoRO. . . . . ***H. yadonii***
    26′ Hypanthium bractlets gen 0.5–1 mm wide, linear or elliptic to lanceolate; anther 0.5–1 mm
      27. St and petiole hairs gen spreading; pl often green
        28. Lflets toothed ± 1/3 to base; hypanthium inner wall glabrous; PR . . . . ***H. clevelandii*** var. ***clevelandii***
        28′ Lflets divided > 1/2 to base, lobes linear to oblanceolate; hypanthium inner wall ± hairy;
          NCo, NCoRO, SnFrB. . . . . ***H. tenuiloba***
      27′ St and petiole hairs gen ascending to appressed; pl gen ± gray — hypanthium inner wall ± hairy
        29. Lf 3–8 cm; pistils gen 10–20; style 1–2 mm; NCoRI . . . . ***H. bolanderi***
        29′ Lf 4–20 cm; pistils gen 20–50; style gen 2–3 mm; TR . . . . ***H. rydbergii***

H. bolanderi A. Gray BOLANDER'S HORKELIA Pl ± matted, ± gray. **ST**: 10–30 cm; hairs ascending to appressed. **LF**: 3–8 cm; stipules entire; lflets ± 7 per side, gen ± crowded, 4–10 mm, wedge-shaped to obovate, ± 5-toothed ± 1/3 to base, densely hairy. **INFL**: ± dense to open, several to many-fld; pedicels gen 2–4 mm. **FL**: hypanthium width 2–4 mm, > 2 × length, inner wall ± hairy, bractlets 0.5–1 mm wide, ± elliptic; sepals ± 3–4 mm; petals 3–5.5 mm, oblong to oblanceolate; filaments 0.5–2 mm, base 0.3–0.4 mm wide, anthers ± 0.5 mm; pistils gen 10–20, style 1–2 mm. **FR**: ± 1.2 mm. Edges of vernally wet places in pine forest; 450–1100 m. NCoRI. May–Sep ★

H. californica Cham. & Schltdl. Pl clumped, green. **ST**: 10–120 cm. **LF**: stipules entire; lflets ± separated, ± ovate to round, ± hairy. **INFL**: open, of few to many separate fls and few-fld clusters; pedicels gen 1–20 mm. **FL**: hypanthium width 4–10 mm, 1–2 × length, bractlets ± 4–6 mm, gen ± 2 mm wide, ovate, often toothed; sepals gen 4–6 mm; petals 3–8 mm, gen oblanceolate to elliptic; anthers 0.8–1.8 mm; pistils gen > 50. **FR**: ± 1 mm. Often confused with *Drymocallis* spp. (terminal lflets distinct). Vars. intergrade.

var. ***californica*** (p. 1185) **LF**: 8–40 cm; sheathing base ± hairy; lflets 4–9 per side, 10–40 mm, toothed, gen few-lobed ± 1/2 to base; terminal lflet 10–40 mm. **FL**: hypanthium inner wall ± hairy, bractlets gen toothed; sepals red-mottled inside; filaments gen 1.5–3 mm, base 0.5–1 mm wide; style gen 3–4 mm. *n*=28. Grassy openings, edges of coastal scrub, esp n slopes; < 400 m. NCo, CCo. [*H. c.* subsp. *c.*] Mar–Sep

var. ***elata*** (Greene) Ertter & Reveal **LF**: gen 5–25 cm; sheathing base gen glabrous; lflets 7–9 per side, 5–25 mm, unevenly toothed, lobed 1/2–3/4 to base; terminal lflet 10–40 mm. **FL**: hypanthium inner wall glabrous, bractlets gen entire; sepals not red-mottled inside; filaments gen 0.5–1.5 mm, base gen < 0.5 mm wide; style gen 2–3 mm. 2*n*=28. Shady meadow edges, seasonal streams, open chaparral; 50–1830 m. NCoR, c&s SN, SnFrB, SnBr. [*H. c.* subsp. *dissita* (Crum) Ertter] Jun–Sep

var. ***frondosa*** (Greene) Ertter & Reveal (p. 1185) **LF**: 10–40 cm; sheathing base ± hairy; lflets 3–5 per side, 15–60 mm, ± evenly double-toothed < 1/4 to base, gen unlobed; terminal lflet 20–90 mm. **FL**: hypanthium inner wall glabrous; sepals not red-mottled inside;

filaments 1–2 mm, base 0.5–1.5 mm wide; style 2–3 mm. Coastal scrub, canyons, poison-oak thickets; 10–400 m. CW. [*H. c.* subsp. *f.* (Greene) Ertter] May–Oct

H. clevelandii (Greene) Rydb. var. ***clevelandii*** Pl tufted, green to ± gray. **ST**: 10–50 cm; hairs spreading. **LF**: 5–18 cm; stipules entire; lflets 6–12 per side, gen separated, gen 5–12 mm, wedge-shaped to round, 5–10-toothed ± 1/3 to base, ± densely hairy. **INFL**: gen open, ± 5–30-fld; pedicels gen 1.5–6 mm. **FL**: hypanthium width 2–4 mm, ± 2 × length, inner wall glabrous, bractlets 0.5–1.5 mm wide, gen lanceolate; sepals ± 3–4 mm; petals 3–6 mm, widely oblanceolate; filaments 0.5–1.5 mm, base ± 0.5 mm wide, anthers 0.5–1 mm; pistils gen 10–50, style ± 1.5 mm. **FR**: ± 1.2 mm. Meadows, under pines, on granite; 1200–2500 m. PR. [*H. bolanderi* subsp. *c.* (Greene) D.D. Keck] May–Aug

H. congesta Hook. var. ***nemorosa*** (D.D. Keck) M. Peck JOSEPHINE HORKELIA Pl tufted or rosetted, ± gray-green, ± odorless; caudex 0–few-branched. **ST**: 15–30 cm, densely glandular above. **LF**: 4–8 cm; stipules entire to forked; lflets 2–5 per side, separated, 5–12 mm, ± elliptic, < 5-toothed < 1/4 to base, densely hairy. **INFL**: clusters 1–several, head-like, gen 5–15-fld; pedicels gen ± 2 mm. **FL**: hypanthium width gen 3–4.5 mm, ≤ 2 × length, bractlets < 0.5 mm wide, linear; sepals 2–4.5 mm; petals gen 2.5–4 mm, gen widely obovate; filaments 0.5–2 mm, base 0.2–0.5 mm wide, anthers ± 0.5 mm; pistils ± 10, style 2–3 mm. **FR**: ± 2 mm. Vernally moist, rocky clay, gen serpentine; 300–800 m. nw KR (Del Norte Co.); sw OR. [*H. c.* subsp. *n.* D.D. Keck] Apr–Jun ★

H. cuneata Lindl. Pl matted, green or ± gray. **ST**: gen 20–70 cm. **LF**: gen 10–30 cm; stipules entire or basally lobed; lflets 5–12 per side, ± separated, gen 10–25 mm, ± = gen indistinct terminal lflet, ± elliptic, pinnately veined, teeth gen single, ± 10–15, gen < 1/3 to midvein, ± glabrous to densely hairy. **FL**: hypanthium width 4–7 mm, > 2 × length, bractlets ± 2 mm wide, ovate; sepals 4–6 mm; petals 4–8 mm, 1.5–4 mm wide, obovate or narrower; filaments 1–3 mm, anthers ± 1 mm; pistils gen 30–60, style gen 2–3 mm. **FR**: 1.5–1.8 mm. Vars. intergrade.

var. ***cuneata*** Hairs ± dense, glandular and nonglandular, ± spreading. **INFL**: gen ± open; clusters several to many, few-fld; pedicels gen 1–12 mm. **FL**: hypanthium inner rim gen ± hairy; filament base 0.5–1 mm wide. *n*=14. Old dunes, coastal sandhills; gen < 500 m. CCo, SCoRO, SCo. [*H. c.* subsp. *c.*] Mar–Jul

var. ***puberula*** (Rydb.) Ertter & Reveal (p. 1185) MESA HORKELIA Hairs sparse, ± all glandular. **INFL**: open, of many ± separate fls; pedicels gen 5–30 mm. **FL**: hypanthium inner rim ± glabrous; filament base gen 0.5–2 mm wide. Dry, sandy, coastal chaparral; 70–870 m. SCoRO, SCo (esp foothill edge of Los Angeles Basin), PR. [*H. c.* subsp. *p.* (Rydb.) D.D. Keck] More inland than other vars. Mar–Jul ★

var. ***sericea*** (A. Gray) Ertter & Reveal KELLOGG'S HORKELIA Hairs dense, ± nonglandular, ascending to appressed. **INFL**: dense to ± open; clusters several to many, several-fld; pedicels gen 1–12 mm. **FL**: hypanthium inner rim many wide; filament base 0.5–1 mm wide. *n*=14. Old dunes, coastal sandhills; gen < 200 m. CCo. [*H. c.* subsp. *s.* (A. Gray) D.D. Keck] Remaining pls less distinct from *H. cuneata* var. *c.* than those formerly near San Francisco. Threatened by coastal development. Apr–Aug ★

H. daucifolia (Greene) Rydb. Pl rosetted, ± gray; caudex simple to branched; odor indistinct. **ST**: 15–30 cm; hairs at base 2–3 mm. **LF**: stipules pinnately divided; lflets 5–10 per side, ± crowded, lobed > 3/4 to base, lobes linear to oblanceolate, hairs sparse to many. **INFL**: dense to open, ± flat-topped, 5–25-fld; pedicels gen 3–9 mm. **FL**: hypanthium width 3.5–5 mm, > 2 × length, bractlets ± 0.5 mm wide, linear; sepals 4–6 mm; petals cream drying ± yellow; filament base 0.5–1.5 mm wide, anthers ± 0.7 mm; pistils 5–15, style 2.5–4 mm. **FR**: 2–2.5 mm.

var. ***daucifolia*** (p. 1185) **LF**: < 15 cm, lflets 5–25 mm, lobes 2–5, 1–3 mm wide, gray-hairy. **INFL**: pedicels 2–10 mm, finely short-hairy. **FL**: petals 4–6 mm, 2–4 mm wide, wedge-shaped to

obovate; filaments 1–2.5 mm. Dry open places, often on serpentine clay; 500–1650 m. w KR, CaRH (esp Shasta, Scott valleys); OR (sw Klamath Co.). Apr–Jul

var. ***indicta*** (Jeps.) Ertter & Reveal JEPSON'S HORKELIA **LF**: < 11 cm, lflets 5–15 mm, lobes 5–15, < 1 mm wide, green. **INFL**: pedicels 5–20 mm, short-glandular. **FL**: petals 5–7.5 mm, 4–8 mm wide, obovate-cordate; filaments 2–3 mm. Dry open places, often on serpentine clay; 240–670 m. n NCoRI, CaRF. Apr–Jun ★

H. fusca Lindl. Pl gen tufted (± matted), green to ± gray. **ST**: gen 10–60 cm. **LF**: stipules entire. **INFL**: clusters 1–several, gen ± head-like; pedicels gen 1–3 mm. **FL**: hypanthium width gen 2–3.5 mm, ± 1–2 × length, bractlets < 0.5 mm wide, linear; sepals gen 2–3 mm; petals ± wedge-shaped; filament base 0.2–1 mm wide; pistils gen 10–20.

var. ***brownii*** (Rydb.) Ertter & Reveal **LF**: gen 4–12 cm; lflets 3–7 per side, ± separated, gen 5–15 mm, narrow-wedge-shaped, 4–6-toothed 1/4–1/2 to base, hairs dense, ± gray. **INFL**: clusters gen 5–20-fld. **FL**: petals 2–3 mm; filaments 0.2–1 mm, gen longer than wide, anthers ± 0.4 mm; style ± 1 mm. **FR**: 1–1.2 mm. *n*=14. Dry openings in forest and chaparral, esp on pumice; 950–2000 m. CaR, MP (exc Wrn); NV. [*H. f.* subsp. *pseudocapitata* (Howell) D.D. Keck, misappl.] May–Aug

var. ***parviflora*** (Hook. & Arn.) Wawra (p. 1185) **LF**: gen 4–15 cm; lflets gen 4–8 per side, separated, gen 5–15 mm, wedge-shaped to round, ± 5-toothed 1/4–1/2 to base, hairs sparse to dense, green. **INFL**: clusters gen 5–20-fld. **FL**: petals 2–4 mm; filaments 0.2–1 mm, gen longer than wide, anthers ± 0.4 mm; style ± 1 mm. **FR**: ± 1.2 mm. *n*=14. Dry meadow edges, open forest, volcanic or granitic soils; 1400–3300 m. e KR, SNH, n SNE; to OR, WY. [*H. f.* subsp. *p.* (Hook. & Arn.) D.D. Keck] Pls in KR/OR, ID/WY probably distinct. Jun–Sep

var. ***pseudocapitata*** (Howell) M. Peck **LF**: 10–15 cm; lflets 4–7 per side, separated, gen 10–20 mm, gen ± obovate, ± 10-toothed < 1/4 to base, hairs gen sparse, green. **INFL**: clusters gen 10–30-fld. **FL**: petals 4–6 mm; filaments 0.5–1.5 mm, gen longer than wide, anthers ± 0.6 mm; style 1–1.5 mm. **FR**: ± 1.8 mm. 2*n*=28. Dry meadow edges, open forest, volcanic or granitic soils; 1800–2300 m. Wrn; to OR, NV. [*H. f.* subsp. *p.* (Howell) D.D. Keck; *H. f.* subsp. *capitata* (Lindl.) D.D. Keck, misappl.] Jul–Aug

var. ***tenella*** S. Watson **LF**: gen 4–15 cm; lflets gen 8–15 per side, ± crowded, gen 5–15 mm, widely obovate, gen divided ± 3/4 to base, lobes 5–15, oblanceolate, hairs sparse, green. **INFL**: clusters gen 5–10-fld. **FL**: petals gen 2–4 mm; filaments gen < 0.5 mm, gen wider than long, anthers ± 0.5 mm; style ± 1 mm. **FR**: 1.2–1.5 mm. 2*n*=28. Gen lodgepole-pine forest; 1200–2200 m. s CaRH. [*H. f.* subsp. *t.* (S. Watson) D.D. Keck] Jun–Aug

H. hendersonii Howell HENDERSON'S HORKELIA Pl matted, ± gray. **ST**: < 20 cm. **LF**: 3–8 cm, gen ± cylindric; stipules entire; lflets 5–10 per side, crowded, 4–10 mm, wedge-shaped, ± 5-toothed ± 1/2 to base, hairs dense. **INFL**: ± dense, ± flat-topped, 5–20-fld; pedicels 1–7 mm. **FL**: hypanthium width 2–4 mm, ± 2 × length, inner rim densely hairy, bractlets < 0.5 mm wide, linear; sepals 3.5–6 mm; petals gen 3–4 mm, linear to narrow-oblanceolate; filaments ± 2 mm, base ± 0.5 mm wide, anthers ± 0.5 mm; pistils ± 10–15, style 2–3 mm. **FR**: ± 2 mm. Dry granitic flats; 2000–2300 m. ne KR; sw OR (Mount Ashland). Jul–Aug ★

H. hispidula Rydb. WHITE MOUNTAINS HORKELIA Pl matted, gen ± gray. **ST**: < 25 cm. **LF**: 3–10 cm, gen ± cylindric; stipules entire; lflets gen 10–14 per side, crowded, gen 2.5–4 mm, divided > 3/4 to base, lobes 3–7, oblanceolate, hairs many. **INFL**: ± dense or open, gen 3–15-fld; pedicels gen 1–6 mm. **FL**: hypanthium width 3–4 mm, ± 2 × length, inner wall ± hairy, bractlets < 0.5 mm wide, linear to lanceolate; sepals 2.5–4 mm; petals 3–5 mm, oblong to oblanceolate; filaments 0.5–2 mm, base ± 0.5 mm wide, anthers ± 0.5 mm; pistils 10–20, style ± 2 mm. **FR**: ± 1.5 mm. Dry flats; 3000–3400 m. n W&I (White Mtns); NV. Jun–Aug ★

H. howellii (Greene) Rydb. (p. 1185) HOWELL'S HORKELIA Pl ± tufted, ± green; odor indistinct; caudex few-branched. **ST:** 15–50 cm; hairs at base ± 1 mm. **LF:** 5–15 cm, often ± cylindric; stipules forked to pinnately divided; lflets ± 15 per side, ± crowded, 3–10 mm, gen lobed ± 1/2 to base, lobes < 10, elliptic, hairs few to many. **INFL:** open; fls many, ± separate or in few-fld clusters; pedicels gen 1–4 mm. **FL:** hypanthium 2–3 mm wide, 1–2 × length, bractlets ± 0.5 mm wide, lance-linear; sepals 2.5–4 mm; petals 3–5 mm, narrow-obcordate, often pink-tinged; filaments 1–1.5 mm, base ± 0.5 mm wide, anthers ± 0.5 mm; pistils 2–6, style ± 1.5–2 mm. **FR:** 2–2.7 mm. Dry, rocky serpentine clay, open chaparral or pine forest; 60–1200 m. w KR; sw OR. Previously incl in *H. sericata.* Jun–Aug ★

H. marinensis (Elmer) Crum POINT REYES HORKELIA Pl matted, ± gray; resinous odor strong. **ST:** decumbent to ascending, gen 10–30 cm. **LF:** gen 4–10 cm; stipules entire to basally lobed; lflets gen 5–10 per side, ± crowded, 7–12 mm, gen ± wedge-shaped, ± 5–10-toothed ± 1/3–1/2 to base, hairs dense. **INFL:** ± dense; clusters indistinct, 5–10-fld; pedicels gen 1–6 mm. **FL:** hypanthium width 4–5 mm, > 2 × length, inner wall hairy, bractlets ± 1 mm wide, lanceolate; sepals 3–6 mm; petals gen 4–6 mm, oblong to oblanceolate; filaments 1–2.8 mm, base ± 0.5 mm wide, anthers ± 0.7 mm; pistils gen 20–30, style 2–4 mm. **FR:** 1.5–2 mm. $2n=56$. Sandy coastal flats; ± 15–760 m. c NCo (Fort Bragg), n CCo (Point Reyes to Santa Cruz). Pls in c NCo may be distinct. May–Sep ★

H. parryi Greene PARRY'S HORKELIA Pl openly matted, green. **ST:** gen 10–30 cm. **LF:** 5–10 cm; stipules entire or basally lobed; lflets 3–6 per side, separated, 5–15 mm, ± obovate, 5–10-toothed 1/4 to midrib, hairs dense. **INFL:** open, ± 5–10-fld; pedicels 5–15 mm, ± recurved in fr. **FL:** hypanthium width ± 3–4 mm, > 3 × length, bractlets 0.5–1.5 mm wide, lanceolate; sepals 4–6 mm; petals 4–7 mm, obovate to elliptic; filaments gen 1–3 mm, base ± 1 mm wide, anthers 0.6–1 mm; pistils gen 20–50, style ± 2 mm. **FR:** ± 1.5 mm. $2n=28.$ Open chaparral; 80–900 m. n&c SNF (esp Ione Formation). Reported once from limestone. Apr–Sep ★

H. rydbergii Elmer Pl ± matted, gray-green. **ST:** 10–70 cm; hairs ascending to ± appressed. **LF:** 4–20 cm; stipules entire; lflets 7–14 per side, separated or crowded, 5–15 mm, ± wedge-shaped, ± 5–10-toothed ± 1/3 to base, hairs gen dense. **INFL:** ± open, gen 8–40-fld; pedicels gen 2–8 mm. **FL:** hypanthium width gen 2.5–4 mm, ± > 2 × length, inner wall ± hairy, bractlets 0.5–1 mm wide, ± lanceolate; sepals 3–5 mm; petals 4–5.5 mm, oblong to oblanceolate; filaments 0.5–2 mm, base 0.5–1 mm wide, anthers 0.6–1 mm; pistils gen 20–50, style gen 2–3 mm. **FR:** ± 1.2 mm. $2n=28.$ Meadows, streambanks, under pines; 1200–2800 m. TR. Jun–Aug

H. sericata S. Watson SILKY HORKELIA Pl ± tufted, silvery; caudex few-branched; odor indistinct. **ST:** 15–40 cm; hairs at base ± 1 mm. **LF:** 3–10 cm, often ± cylindric; stipules entire or forked; lflets ± 15 per side, crowded, 2–8 mm, gen divided ± 1/2 or more to base, lobes < 5, elliptic, hairs dense. **INFL:** open; fls many, ± separate or in few-fld clusters; pedicels 1–4 mm. **FL:** hypanthium 2–3 mm wide, 1–2 × length, bractlets ± 0.5 mm wide, lance-linear; sepals 2.5–4 mm; petals 3–7 mm, narrow-obcordate, often pink-tinged; filaments ± 1 mm, base ± 0.5 mm wide, anthers ± 0.5 mm; pistils 2–6, style ± 1.5 mm. **FR:** 2–2.5 mm. $n=14.$ Dry, rocky serpentine clay, open chaparral or pine forest; 180–1200 m. nw KR (Del Norte Co.); sw OR. Jun–Aug ★

H. tenuiloba (Torr.) A. Gray THIN-LOBED HORKELIA Pl loosely matted, gen ± green. **ST:** < 40 cm; hairs ± spreading. **LF:** 5–15 cm, often ± cylindric; stipules entire; lflets gen 8–15 per side, ± crowded, 3–10 mm, divided > 1/2 to base, lobes 3–8, linear to oblanceolate, hairs sparse to dense. **INFL:** dense or ± open, few- to many-fld; pedicels gen 1–6 mm. **FL:** hypanthium width 3–4.5 mm, > 2 × length, inner wall ± hairy, bractlets 0.5–1 mm wide, lance-linear; sepals 3–5 mm; petals ± 4 mm, oblanceolate; filaments ± 1.5 mm, base ± 0.5 mm wide, anthers ± 0.5 mm; pistils 10–25, style ± 2 mm. **FR:** ± 1.5 mm. $2n=28.$ Sandy soils, open chaparral; 50–500 m. c&s NCo, c&s NCoRO, nw SnFrB. Apr–Jul ★

H. tridentata Torr. (p. 1185) Pl rosetted or tufted, ± gray, ± odorless; caudex 0–few-branched. **ST:** < 45 cm, ± glabrous to sparsely glandular above. **LF:** 3–12 cm; stipules entire to forked; lflets 2–5 per side, separated, 5–30 mm, gen elliptic to oblong, ± 3-toothed < 1/4 to base (entire), at least lower surface ± densely hairy. **INFL:** clusters ± head-like, gen 3–40-fld; pedicels gen 1–6 mm. **FL:** hypanthium width 1–2 × length, bractlets < 0.5 mm wide, linear; sepals gen 1.5–3 mm; petals 1.5–4 mm; filament base gen ± 0.2–0.5 mm wide, anthers 0.2–0.5 mm; pistils 5–15, style 1–2.5 mm. Vars. intergrade extensively.

var. ***flavescens*** (Rydb.) Ertter & Reveal **ST:** decumbent. **INFL:** cluster gen 1. **FL:** 6–10 mm wide; hypanthium 2.5–5 mm wide, inner wall hairy (exc in NCoRH); petals often < sepal, ± widely oblanceolate; filaments 1–2 mm. **FR:** 2–2.5 mm. Often on ± serpentine; 750–2000 m. KR, NCoR, n SNH (esp Plumas Co.); sw OR. [*H. t.* subsp. *f.* (Rydb.) D.D. Keck] NCoRH pls may be distinct. Apr–Jul

var. ***tridentata*** **ST:** decumbent to ± erect. **INFL:** clusters 1–many. **FL:** 4–7 mm wide; hypanthium gen 2–3.5 mm wide, inner wall gen glabrous; petals gen ≥ sepals, gen narrow-oblanceolate; filaments gen 0.5–1 mm. **FR:** 1.5–2 mm. $n=14.$ Granitic or volcanic soils; 300–2500 m. e KR, CaR, SN, MP; s OR. [*H. t.* subsp. *t.*] Apr–Aug

H. truncata Rydb. (p. 1185) RAMONA HORKELIA Pl tufted, green. **ST:** 20–60 cm. **LF:** 4–13 cm; stipules entire to lobed; lflets 1–3 per side, separated, 10–30 mm, oblong to obovate, gen > 20-toothed < 1/4 to midvein, ± glabrous; terminal lflet largest, gen unlobed. **INFL:** open, 5–20-fld; pedicels gen 4–20 mm. **FL:** hypanthium width 5–8 mm, > 2 × length, bractlets ± 2–3 mm wide, widely ovate; sepals gen 3.5–5.5 mm; petals 5–7 mm, ± 5 mm wide, ± round; filaments 1–2 mm, base 0.5–2 mm wide, anthers ± 1 mm; pistils 50–80, style ± 2 mm. **FR:** ± 1.3 mm. Dry red clay, open chaparral; 400–1300 m. PR; n Baja CA. Mar–Jun ★

H. tularensis (J.T. Howell) Munz (p. 1185) KERN PLATEAU HORKELIA Pl ± matted, ± gray. **ST:** < 25 cm. **LF:** 2–10 cm, often ± cylindric; stipules entire; lflets gen 4–6 per side, ± crowded, gen 3.5–8 mm, divided > 3/4 to base, lobes 5–8, ± oblanceolate, hairs dense. **INFL:** ± dense to open, gen 3–15-fld; pedicels gen 2–7 mm. **FL:** hypanthium width 2.5–4.5 mm, ± 2 × length, inner wall hairy, bractlets gen < 0.5 mm wide, linear; sepals ± 3–4.5 mm; petals ± 3–4 mm, oblong to oblanceolate; filaments 1–2 mm, base ± 0.5 mm wide, anthers ± 0.8 mm; pistils 5–12, style ± 2 mm. **FR:** ± 2–2.5 mm. Dry, rocky balds, flats; 2350–2850 m. s SNH (Bald Mtn area, Tulare Co.). Jun–Aug ★

H. wilderae Parish (p. 1185) BARTON FLATS HORKELIA Pl ± rosetted, green; caudex gen simple. **ST:** gen decumbent, gen 10–25 cm. **LF:** 4–10 cm; stipules entire to lobed; lflets 4–7 per side, separated, 3–10 mm, ± obovate, lobed 1/2–3/4 to base, lobes 5–15, hairs ± sparse. **INFL:** open, gen 5–15-fld; pedicels 3–15 mm, ± recurved in fr. **FL:** hypanthium width ± 2–3 mm, ± 2 × length, bractlets gen < 0.5 mm wide, lanceolate to ovate; sepals ± 2 mm; petals 2–3 mm, oblanceolate to oblong; filaments 0.5–1 mm, base 0.5–1 mm wide, anthers ± 0.4 mm; pistils < 5, style ± 1–1.5 mm. **FR:** ± 2 mm. Edge of chaparral, under pines; 1880–3000 m. SnBr (Barton Flats area). Jun–Aug ★

H. yadonii Ertter SANTA LUCIA HORKELIA Pl tufted to matted, gray-green. **ST:** gen 20–60 cm; hairs spreading. **LF:** gen 6–20 cm; stipules entire; lflets 7–16 per side, separated to ± crowded, gen 4–15 mm, wedge-shaped to round, ± 3–20-toothed or -lobed 1/5–1/2 to base, often deeply notched, hairs gen dense. **INFL:** ± open, gen 5–10-fld; pedicels gen 1–7 mm. **FL:** hypanthium width gen 3–6 mm, ± > 2 × length, inner wall ± hairy, bractlets 1–2 mm wide, lanceolate to ovate; sepals gen 4–6 mm; petals 3–5 mm, oblanceolate to elliptic; filaments 1–2 mm, base 0.5–1 mm wide, anthers ± 1 mm; pistils gen > 20, style 2.5–3 mm. **FR:** ± 1.5 mm. Sandy meadow edges, seasonal streambeds in chaparral or foothill-pine woodland; 350–1900 m. SCoRO. Jun–Sep ★

HORKELIELLA
Barbara Ertter

Per, tufted, ± glandular; odor ± resinous. **ST**: ascending to erect, 15–50(90) cm. **LF**: gen basal, odd-pinnately compound; lflets 15–35 per side, ± overlapped, toothed to palmately divided, segments ± oblanceolate. **INFL**: ± cyme; pedicels straight, bractlets 0. **FL**: hypanthium cup-like, flat-bottomed, bractlets 5; sepals often reflexed; petals white or ± pink, midvein often ± red; stamens 20, filaments ± flat, often an erect tube; pistils many, ovaries superior, style attached below fr tip, ± rough, thick. **FR**: achene, ± 1.5 mm. 2 spp.: CA. (Latin: small *Horkelia*) Fl ± like *Horkelia*.

1. Filament opposite sepal center > adjacent ones; lflet lobes (or 2° lflets) gen 5–10; nonglandular hairs ± 0
   to sparse, not obscuring glands; c SNH (e slope), w SNE. . . . . ***H. congdonis***
1′ Filament opposite sepal center < adjacent ones; lflet lobes gen < 5; nonglandular hairs ± 0 to dense,
   often obscuring glands; s SNH (w slope) . . . . ***H. purpurascens***

H. congdonis (Rydb.) Rydb. (p. 1189) Pl green, gen glandular-sticky. **LF**: basal gen 8–25 cm, lflets 25–35 per side, gen 2–6 mm; cauline 2–7. **INFL**: ± dense; pedicels gen ± 2 mm. **FL**: ± 15 mm wide; sepals 4–8 mm; petals 3–6 mm, oblong-oblanceolate; anthers ± 1 mm, filaments 1.5–3 mm; style 2–5 mm. Meadows in sagebrush flats; 1500–3100 m. c SNH (e slope), w SNE, DMtns (Coso Range). Jul–Aug

H. purpurascens (S. Watson) Rydb. Pl often ± gray, often not sticky. **LF**: basal gen 7–17 cm, lflets 15–30 per side, gen 3–10 mm; cauline 2–4. **INFL**: open at least in fr; pedicels 1–15 mm. **FL**: 10–15 mm wide; sepals 3–6 mm; petals 3–7 mm, ± oblanceolate; anthers 0.5–1 mm, filaments 1–2.5 mm; style ± 3–4 mm. Partial shade at granitic meadow edges in conifer forest; 1400–2900 m. s SNH (w slope). Jun–Aug

IVESIA
Barbara Ertter

Per, glandular; odor resinous. **LF**: gen basal, odd-1-pinnately compound, gen ± cylindric; cauline gen alternate, reduced; lflets 4–80 per side, gen overlapped, gen divided ± to base. **INFL**: cyme; pedicel bractlets 0. **FL**: receptacle gen not stalked; hypanthium shallow or deep, bractlets (0)5, gen < sepals; petals gen 5, 1–5(7) mm, linear to obovate or round, acute to rounded; stamens 5–20(40), filaments gen thread-like; pistils 1–8(20), ovary superior, style attached below fr tip, base ± rough-thickened. **FR**: achene. 30 spp.: w N.Am. (Eli Ives, Yale University, CT pharmacologist, 1779–1861) [Ertter & Reveal 2007 Novon 17:315–325] Lf, lflet data for basal lvs.

1. Lf ± flat, lflets toothed or lobed gen < 3/4 to base, gen ± separate — pls rosettes or hanging clumps
   2. Stamens 15–40 — s SNH, SnBr, PR, SNE, DMtns. . . . . ***I. saxosa***
   2′ Stamens 5–10
      3. Hypanthium bractlets 5; MP, adjacent SNH . . . . ***I. baileyi***
         4. Petals pale yellow; hypanthium bractlets ± = sepals. . . . . var. ***baileyi***
         4′ Petals white; hypanthium bractlets ± < 1/2 sepals . . . . var. ***beneolens***
      3′ Hypanthium bractlets gen 0; DMtns
         5. Hypanthium width ≤ length, receptacle stalked in pistil-bearing portion; n DMtns . . . . . . ***I. arizonica*** var. ***arizonica***
         5′ Hypanthium width > 2 × length; receptacle not stalked; e DMtns. . . . . ***I. patellifera***
1′ Lf ± cylindric, lflets gen lobed ± to base, overlapped
   6. Stamens 10–20 (if 10, cauline lvs ≥ 3)
      7. Lflets < 8 per side; sts hanging to ± matted; caudex few- to many-branched; rock outcrops
         8. Lflet lobes < 4; petals white; SnJt . . . . ***I. callida***
         8′ Lflet lobes 3–6; petals yellow; e DMtns . . . . ***I. jaegeri***
      7′ Lflets > 10 per side; sts decumbent to erect; caudex 0–few-branched; meadows, sandy flats
         9. Lflets gen 15–20 per side; sheathing lf bases glabrous; stamens 10–16, filament ± < 1 mm; style < 1.5 mm
            10. Petals 4(5), pale yellow; infl not red-tinged; s SNH . . . . ***I. campestris***
            10′ Petals 5, white or pink-tinged; infl often red-tinged; c&s SNH . . . . ***I. unguiculata***
         9′ Lflets ≥ 20 per side; sheathing lf bases gen ± strigose; stamens 15–20; filament 1–4 mm; style ± > 2 mm
            11. Infl of separate fls or gen < 5-fld clusters; pedicels gen > 5 mm; lflets gen 30–80 per side
               12. Lf mousetail-like, lflets indistinct, obscured by dense, silvery hairs, lobes < 1.5 mm; stamens 15;
                  pistil 1; granitic sand — SNH, TR, SnJt . . . . ***I. santolinoides***
               12′ Lf not mousetail-like, lflets overlapped but distinct, not obscured by hairs, lobes ≥ 2 mm; stamens
                  20; pistils ≥ 2; meadows
                  13. Herbage hairs ± 0 or appressed, < 1 mm; petals ± obovate; alkali meadows; n SNE. . . . . . . ***I. kingii*** var. ***kingii***
                  13′ Herbage hairs spreading, < 4 mm; petals ± oblanceolate; gen on serpentine clays; c KR . . . . ***I. pickeringii***
            11′ Infl of gen > 5-fld clusters; pedicels gen < 3 mm (exc lowest); lflets 20–35 per side
               14. St 10–20 cm, ± decumbent; lvs gen 4–8 cm, cauline ± 2; filament ± flat; SnBr
                  . . . . ***I. argyrocoma*** var. ***argyrocoma***
               14′ St gen > 20 cm, decumbent to erect; lvs > 8 cm, cauline ≥ 3; filament thread-like, not flat; n SNH, MP
                  15. Petals white; hypanthium length ≥ width; hairs of st base ± spreading, 2–4 mm. . . . . ***I. sericoleuca***
                  15′ Petals yellow; hypanthium length ≤ width; hairs of st base ascending, < 2 mm . . . . ***I. aperta***
                     16. Petals 2–3 mm, oblanceolate; filament 1–1.5 mm; n SNH (exc Dog Valley), s MP . . . . var. ***aperta***
                     16′ Petals gen 4–7 mm, ± obovate; filament 2–4 mm; n SNH (Dog Valley, e Sierra Co.) . . . . var. ***canina***

style
flower
var. lactea
5 mm
fruit
1 mm
Drymocallis
rhomboidea

var. austiniae
2 cm
stem hairs
var. lactea
2 cm
Drymocallis lactea

5 cm
fruiting head
1 cm
Fallugia paradoxa

5 cm
Fragaria virginiana

5 cm
fruit
1 cm
Fragaria vesca

inflorescence
2 cm
fruiting head
1 cm
fruit
2 mm
5 cm
Geum macrophyllum
var. macrophyllum

fruiting head
2 cm
fruit
2 cm
Geum triflorum
var. ciliatum
5 mm

2 cm
1 cm
fruit
flower
2 mm
Heteromeles arbutifolia

1 cm
2 mm
leaf
short shoot
1 cm
var. discolor
5 cm

short shoot
1 cm
fruit
0.5 mm
leaf
1 cm
var. microphyllus
Holodiscus discolor
2 cm

flower
var. californica
5 mm
leaf
var. frondosa
2 cm
Horkelia californica

Horkelia fusca var. parviflora
1 cm
2 mm

leaf
2 cm
flower
5 mm
Horkelia cuneata var. puberula
5 cm

1 cm
5 cm
Horkelia daucifolia
var. daucifolia
Horkelia howellii

leaf
2 mm
5 mm
leaf
1 cm
Horkelia
truncata
Horkelia
tularensis

leaf
1 cm
Horkelia
tridentata
2 cm
Horkelia wilderae

6′ Stamens 5 or 10; cauline lvs 1–3
   17. Hypanthium bractlets > sepals — KR (Castle Crags) . . . . **I. longibracteata**
   17′ Hypanthium bractlets ± 1/2 sepals
      18. Infl gen ± open between separate fls or of ± loose clusters; pl ± matted from much-branched caudex;
         pedicels gen ± S-shaped in fr; sheathing lf bases strigose
         19. Stamens 10; pistils 10–30 — c&s SNH . . . . [2]**I. pygmaea**
         19′ Stamens 5; pistils gen < 5
            20. Petals white to pale ± yellow, linear, ± 1 mm; pl densely hairy, glands not obvious; 1400–1800 m;
               MP . . . . **I. paniculata**
            20′ Petals yellow, oblanceolate, ± 2 mm; pl moderately hairy, glands many; 2700–4000 m;
               n&c SNH, W&I . . . . **I. shockleyi** var. **shockleyi**
      18′ Infl of 1–few gen head-like, 3–20-fld clusters (becoming more open in *Ivesia webberi*); pl gen
         rosetted from 0–few-branched caudex; pedicels straight in fr; sheathing lf-bases hairy or not
         21. Cauline lvs 2, ± opposite; lflets 4–8 per side, lobes linear to lanceolate; 1500–1900 m — s-most MP,
            adjacent n SNH . . . . **I. webberi**
         21′ Cauline lvs gen 1 or alternate; lflets > 8 per side, lobes oblanceolate to ± round; > 1800 m
            22. Lf mousetail-like, lflets indistinct, < 1 mm, hidden by dense, silvery hairs; petals 1–2 mm —
               c&s SNH . . . . **I. muirii**
            22′ Lf not mousetail-like, lflets overlapped but distinct, gen > 1 mm, ± glabrous to ± hairy; petals ≥ 2 mm
               23. Stamens 10; sheathing lf-bases gen ± strigose; hypanthium length ± < 1/2 width — c&s SNH . . . . . [2]**I. pygmaea**
               23′ Stamens 5; sheathing lf-bases not strigose (± glabrous or glandular); hypanthium length > 1/2 width
                  24. Hypanthium length ≥ width; petals narrow-oblanceolate; pistils gen 2–4; fr ± 2 mm,
                     mottled brown . . . . **I. gordonii**
                     25. Sts ascending to erect, gen not ± red . . . . var. **alpicola**
                     25′ Sts prostrate to ascending, dark ± red . . . . var. **ursinorum**
                  24′ Hypanthium length ± < width; petals ± obovate; pistils gen 5–15; fr 1–1.5 mm, not mottled . . . . **I. lycopodioides**
                     26. Lflet lobes ± 1 mm, ± round, ± glabrous; petals 2–3 mm — rocky areas; n&c SNH, SNE
                        (Sweetwater Mtns) . . . . var. **lycopodioides**
                     26′ Lflet lobes ≥ 1 mm, narrow-oblanceolate to obovate, often ± hairy; petals gen > 3 mm
                        27. Lflet lobes 2–8 mm, ± glabrous to sparsely hairy, bristle-tip 0–0.5 mm; wet meadows;
                           c&s SNH . . . . var. **megalopetala**
                        27′ Lflet lobes 1–3 mm, moderately to densely hairy, bristle-tip gen 0.5–1 mm; vernally moist,
                           rocky areas; c&s SNH, W&I . . . . var. **scandularis**

I. aperta (J.T. Howell) Munz Pl tufted, ± green or white-hairy; caudex 0–few-branched. **ST:** 15–45 cm. **LF:** 10–20 cm; sheathing bases densely strigose; lflets 20–35 per side, lobes < 5, 3–15 mm, elliptic to oblanceolate; cauline lvs 3–8. **INFL:** open; clusters many, ± head-like, 10–20 mm wide; pedicels gen < 3 mm (exc lowest), straight. **FL:** 5–15 mm wide; hypanthium length ≤ width; petals < to > sepals, yellow; stamens gen 20; pistils 2–7. **FR:** 2–3 mm, smooth, brown.

var. ***aperta*** (p. 1189) SIERRA VALLEY IVESIA **ST:** gen ascending to erect. **INFL:** clusters gen > 10-fld. **FL:** hypanthium 2–3 mm wide; petals 2–3 mm, oblanceolate; filaments 1–1.5 mm; style ± 2.5 mm. Dry, rocky meadows, gen volcanic soils; 1500–2300 m. n SNH (exc Dog Valley), s MP; w NV. Jun–Aug ★

var. ***canina*** Ertter (p. 1189) DOG VALLEY IVESIA **ST:** decumbent to ascending. **INFL:** clusters gen < 10-fld. **FL:** hypanthium 4–5 mm wide; petals gen 4–7 mm, ± obovate; filaments 2–4 mm; style 3–4 mm. Dry, rocky meadows, gen volcanic soils; 1600–2000 m. n SNH (Dog Valley, e Sierra Co.). Jun–Aug ★

I. argyrocoma (Rydb.) Rydb. var. ***argyrocoma*** SILVER-HAIRED IVESIA Pl rosetted or tufted, silvery-hairy; caudex gen simple. **ST:** ± decumbent, 10–20 cm. **LF:** gen 4–8 cm; sheathing bases densely strigose; lflets 25–35 per side, lobes ± 3, gen 2–3 mm, elliptic to obovate; cauline lvs ± 2. **INFL:** clusters 1–several, loosely head-like, 10–20 mm wide, < 20-fld; pedicels < 3 mm (exc lowest), ± straight. **FL:** ± 10 mm wide; hypanthium length = width; petals 2–4 mm, > sepals, obovate, white; stamens 20, filaments ± flat (unique in *Ivesia*); pistils 4–8. **FR:** 2–2.5 mm, smooth, brown. Pebble plains; 1450–2300 m. SnBr. Apr–Jul ★

I. arizonica (J.T. Howell) Ertter var. ***arizonica*** (p. 1189) YELLOW PURPUSIA Pls rosettes or hanging clumps, green. **ST:** 5–10 cm. **LF:** gen 5–10 cm, flat; sheathing bases gen glabrous; lflets 2–4 per side, separated, 5–15 mm, ± round, ± evenly toothed or lobed < 3/4 to base; cauline lvs 1–3. **INFL:** open, gen 1–20(60)-fld; pedicels 5–30 mm, often ± S-shaped in fr. **FL:** 4–10 mm wide; receptacle stalked in pistil-bearing portion (unique in *Ivesia*); hypanthium length 1–2 × width, bractlets gen 0; petals 2–3 mm, ± = sepals, oblanceolate to elliptic, yellow; stamens 5; pistils 2–10. **FR:** 1.5–2 mm, ± ridged, pale. Limestone crevices; 1200–3100 m. n DMtns (Inyo Co.); s NV, nw AZ. May–Sep ★

I. baileyi S. Watson Pls rosettes or hanging clumps, green. **ST:** 5–20 cm. **LF:** gen 3–10 cm, ± flat; sheathing bases ± glabrous; lflets 2–6 per side, ± separated, ± 4–10 mm, ± round, toothed ± < 1/2 to base; cauline lvs 1–2. **INFL:** open, (1)5–40-fld; pedicels S-shaped in fr. **FL:** 4–8 mm wide; hypanthium length < 1/2 width; petals 1.5–2 mm, < sepals, ± oblanceolate; stamens 5; pistils 1–6. **FR:** ± 1.5 mm, pale.

var. ***baileyi*** (p. 1189) BAILEY'S IVESIA **INFL:** pedicels 2–12 mm. **FL:** hypanthium bractlets ± = sepals; petals pale yellow. **FR:** smooth. Volcanic crevices; 1700–2200 m. s MP, adjacent SNH; nw NV. Jun–Jul ★

var. ***beneolens*** (A. Nelson & J.F. Macbr.) Ertter OWYHEE IVESIA **INFL:** pedicels 6–12 mm. **FL:** hypanthium bractlets ± < 1/2 sepals; petals white. **FR:** ± ridged. Volcanic crevices; ± 2150 m. Wrn; to se OR, s ID, n NV. Jun–Jul ★

I. callida (H.M. Hall) Rydb. (p. 1189) TAHQUITZ IVESIA Pl hanging to ± matted, green; caudex branched. **ST:** gen 2–15 cm. **LF:** gen 1–7 cm; sheathing bases ± hairy; lflets ± 6 per side, lobes < 4, 2–7 mm, oblanceolate to elliptic; cauline lvs ± 2. **INFL:** open, 1–10(15)-fld; pedicels gen 5–15 mm, ± S-shaped in fr. **FL:** 7–10 mm wide; hypanthium length < 1/2 width; petals ± 3 mm, > sepals, obovate, white; stamens 20; pistils 4–8. **FR:** ± 1.5 mm, smooth, pale. Granite crevices; ± 2500 m. SnJt. Jul–Sep ★

I. campestris (M.E. Jones) Rydb. FIELD IVESIA Pl rosetted, green to white-hairy; caudex simple. **ST:** ascending, ± 10–30 cm. **LF:** 5–15 cm; sheathing bases glabrous; lflets gen 15–20 per side, lobes 3–5, 2–10 mm, oblanceolate; cauline lvs 3–4. **INFL:** clusters 1–few, loosely head-like, 10–20 mm wide, gen < 15-fld; pedicels < 4 mm, straight. **FL:** 7–10 mm wide; hypanthium length ± 1/2 width; petals 4(5) (4 unique in *Ivesia*), 3–4 mm, > sepals, narrow-obovate, pale yellow; stamens gen 12–16; pistils 4–20. **FR:** 1–1.5 mm, smooth, light brown. Meadow edges; 2200–3100 m. s SNH. Jul–Sep ★

I. gordonii (Hook.) Torr. & A. Gray (p. 1189) Pl tufted, green; caudex ± branched. **ST:** 5–20 cm. **LF:** gen 3–8 cm; sheathing bases ± glandular; lflets gen 10–16 per side, overlapped but distinct, lobes 4–8, gen 2–5 mm, oblanceolate to obovate; cauline lf gen 1. **INFL:** cluster gen 1, gen head-like, 10–30 mm wide, 10–20-fld; pedicels gen < 3 mm, straight. **FL:** 5–9 mm wide; hypanthium length ≥ width; petals 2–3 mm, ≤ sepals, narrow-oblanceolate, yellow; stamens 5; pistils gen 2–4. **FR:** ± 2 mm, ± smooth, mottled brown. Vars. tentatively distinct in CA.

var. *alpicola* (Howell) Ertter & Reveal **ST:** ascending to erect, gen not ± red. **FL:** anthers gen not red-margined. Open, dry, rocky ridges, slopes; 2100–3300 m. KR (Marble Mtns), NCoRH, CaRH, n&c SNH; to WA, MT. Jun–Sep

var. *ursinorum* (Jeps.) Ertter & Reveal **ST:** prostrate to ascending, dark ± red. **FL:** anthers red-margined. Open, dry, rocky ridges, slopes; 1800–3500 m. KR (Scott Mtns, Mount Eddy), n&c SNH (Sonora Pass, Alpine, Tuolumne cos.), Wrn, SNE (Sweetwater Mtns); to ID, NV. Jun–Aug

I. jaegeri Munz & I.M. Johnst. JAEGER'S IVESIA Pl hanging to ± matted, green; caudex branched. **ST:** 3–15 cm. **LF:** 2–10 cm; sheathing bases sparsely hairy; lflets 4–6 per side, lobes 3–6, 2–5 mm, oblanceolate to obovate; cauline lvs ± 2. **INFL:** open, gen 3–15-fld; pedicels 6–30 mm, ± S-shaped in fr. **FL:** 5–11 mm wide; hypanthium length < 1/2 width; petals gen 2 mm, ≤ sepals, narrow-oblanceolate, yellow; stamens 20; pistils 3–8. **FR:** 1–2 mm, ± ridged, pale. Limestone crevices; 1890–2300 m. e DMtns (Clark Mtn Range); sw NV. Jun–Jul ★

I. kingii S. Watson var. *kingii* ALKALI IVESIA Pl rosetted, glabrous or short-appressed-hairy, glaucous or not; caudex gen simple. **ST:** decumbent to ascending, 15–40 cm. **LF:** 7–15 cm; sheathing bases gen strigose; lflets 30–50 per side, overlapped but distinct, lobes < 4, 2–6 mm, oblanceolate to obovate; cauline lvs 4–13. **INFL:** open; clusters gen < 10, loosely head-like, 10–20 mm wide, gen < 5-fld; pedicels 3–25 mm, straight. **FL:** 8–12 mm wide; hypanthium length ± 1/2 width; petals 3–5 mm, > sepals, ± obovate, white; stamens 20; pistils 2–6. **FR:** 2–2.5 mm, smooth, light brown. Moist alkaline clay; 1200–2100 m. n SNE; to UT. Jun–Aug ★

I. longibracteata Ertter (p. 1189) CASTLE CRAGS IVESIA Pl tufted, green; caudex 0–few-branched. **ST:** ascending to erect, 3–12 cm. **LF:** gen 2–4 cm, ± flat; sheathing bases ± ciliate; lflets 4–6 per side, lobes 2–7, 2–6 mm, ± oblanceolate; cauline lvs 1–3. **INFL:** loosely head-like, 10–20 mm wide, 3–14-fld; pedicels 2–6 mm, straight. **FL:** 8–10 mm wide; hypanthium length < 1/2 width, bractlets > sepals (unique in *Ivesia*); petals 1.5–2.5 mm, ± = sepals, ± linear, pale ± yellow; stamens 5; pistils 6–11. **FR:** 1–1.5 mm, ± veined, pale. Granite crevices; 1200–1400 m. e KR (Castle Crags). Jun–Jul ★

I. lycopodioides A. Gray CLUB-MOSS IVESIA Pl rosetted, green; caudex gen simple. **ST:** decumbent to erect. **LF:** sheathing bases ciliate; lflets 10–35 per side, overlapped but distinct, lobes 4–10; cauline lf gen 1. **INFL:** cluster gen 1, ± dense or head-like, gen 5–25 mm wide, 3–20-fld; pedicels gen < 5 mm, straight. **FL:** 6–12 mm wide; hypanthium length ± < width; petals > sepals, yellow; stamens 5; pistils gen 5–15. **FR:** 1–1.5 mm, smooth, pale. Vars. intergrade.

var. *lycopodioides* **ST:** 3–15 cm. **LF:** 1–7 cm; lobes ± 1 mm, ± round, ± glabrous, bristle-tip gen 0. **FL:** petals 2–3 mm, gen < 2 mm wide, obovate; filaments ± 1 mm; style 1–2 mm. 2*n*=28. Rocky areas; 3000–4000 m. n&c SNH, n SNE (Sweetwater Mtns). [*I. l.* subsp. *l.*] Jul–Aug

var. *megalopetala* (Rydb.) Ertter & Reveal (p. 1189) **ST:** 10–30 cm. **LF:** 4–15 cm; lobes 2–8 mm, ± oblanceolate, ± glabrous to sparsely hairy, bristle-tip 0–0.5 mm. **FL:** petals 3–5 mm, gen > 2 mm wide, widely obovate; filaments 1–2 mm; style 2.5–3 mm. Wet meadows; 2300–3700 m. c&s SNH. [*I. l.* subsp. *m.* (Rydb.) D.D. Keck] Jun–Sep

var. *scandularis* (Rydb.) Ertter & Reveal **ST:** gen 5–15 cm. **LF:** gen 3–8 cm; lobes 1–3 mm, ± obovate, moderately to densely hairy, bristle-tip gen 0.5–1 mm. **FL:** petals 3–5 mm, gen > 2 mm wide, widely obovate; filaments ± 1.5 mm; style 2–3 mm. Vernally moist, open, rocky areas; 3000–4115 m. c&s SNH, W&I. [*I. l.* subsp. *s.* (Rydb.) D.D. Keck] Jul–Aug

I. muirii A. Gray (p. 1189) Pl tufted, silvery; caudex gen simple. **ST:** ascending to erect, gen 5–15 cm. **LF:** 2–5 cm, mousetail-like; sheathing bases densely strigose; lflets 25–40 per side, indistinct, lobes ± 3, < 1 mm, obovate to round; cauline lf 1, gen ± bract-like. **INFL:** clusters 1–few, head-like, 10–15 mm wide, gen 10–20-fld; pedicels ± < 2 mm, straight. **FL:** 5–6 mm wide; hypanthium length 1/2–1 × width; petals 1–2 mm, ± = sepals, narrow-oblong to -oblanceolate, pale yellow; stamens 5; pistils 1–4. **FR:** ± 2 mm, smooth, gray spotted red. Rocky areas; 2900–4000 m. c&s SNH. Intermediates with *I. pygmaea* in s SNH (Center Basin) may be hybrids. Jul–Aug

I. paniculata T.W. Nelson & J.P. Nelson (p. 1189) ASH CREEK IVESIA Pl matted, gray-green; caudex much-branched. **ST:** prostrate in fr, gen 5–15 cm. **LF:** gen 2–5 cm; sheathing bases densely strigose; lflets gen 8–15 per side, lobes gen > 5, < 2 mm, obovate; cauline lf 1. **INFL:** ± open; clusters 1–many, ± loose, 10–20 mm wide, ± 10–20-fld; pedicels 2–6 mm, ± S-shaped in fr. **FL:** 4–6 mm wide; hypanthium length ± 1/2 width; petals ± 1 mm, < sepals, linear, white to pale ± yellow; stamens 5; pistils 1–3. **FR:** 1–1.5 mm, smooth, brown. Shallow, rocky soil, open sagebrush; 1400–1800 m. MP (Ash Valley, n-c Lassen Co.). May–Jul ★

I. patellifera (J.T. Howell) Ertter (p. 1189) KINGSTON MOUNTAINS IVESIA Pls rosettes or hanging clumps, green. **ST:** 10–20 cm. **LF:** gen 5–12 cm, flat; sheathing bases glabrous; lflets 2–3 per side, separated, 5–20 mm, ± round, evenly toothed or lobed < 1/2 to base; cauline lvs 0–2. **INFL:** open, gen 3–20-fld; pedicels 5–30 mm, gen ± S-shaped in fr. **FL:** 7–10 mm wide; hypanthium length < 1/2 width, bractlets gen 0; petals 2–3 mm, < sepals, narrow-oblanceolate, yellow; stamens 5–10; pistils 4–10. **FR:** 1.5–2 mm, ± ridged, pale. Granite crevices; 1400–2200 m. e DMtns (Kingston Range). May–Jul ★

I. pickeringii A. Gray PICKERING'S IVESIA Pl tufted, ± gray, long-spreading-hairy; caudex 0–few-branched. **ST:** ascending to erect, 30–50 cm. **LF:** 10–20 cm; sheathing bases strigose; lflets 35–50 per side, overlapped but distinct, lobes gen 3–5, 2–5 mm, oblanceolate to obovate; cauline lvs 5–10. **INFL:** open; fls 10–100, gen separate; pedicels 2–10 mm, straight. **FL:** 8–13 mm wide; hypanthium length ± = width; petals 3–5 mm, > sepals, ± oblanceolate, white or pink-tinged; stamens 20; pistils 2–4. **FR:** 2.5–3 mm, smooth, dark brown. 2*n*=28. Wet, rocky meadows, gen on serpentine clay; 800–1500 m. c KR. Jul–Aug ★

I. pygmaea A. Gray Pl ± matted, green; caudex branched. **ST:** decumbent to erect, gen 3–15 cm. **LF:** 1–10 cm; sheathing bases gen ± strigose; lflets gen 10–15 per side, lobes gen 5–8, gen 1–3 mm, ± widely oblanceolate; cauline lf 1. **INFL:** dense to ± open, 8–30 mm wide, gen 5–10-fld; pedicels 2–10 mm, straight. **FL:** 9–11 mm wide; hypanthium length ± < 1/2 width; petals 2–4 mm, ≤ sepals, widely oblanceolate, yellow; stamens 10; pistils 10–30. **FR:** 1–1.5 mm, smooth, pale. Rocky (granitic) places; 2700–4000 m. c&s SNH. Jul–Sep

I. santolinoides A. Gray (p. 1189) MOUSETAIL IVESIA Pl tufted, silvery; caudex 0–several-branched. **ST:** erect, gen 15–40 cm. **LF:** mousetail-like, gen 4–10 cm; sheathing bases densely strigose; lflets 60–80 per side, indistinct, lobes 1–5, < 1.5 mm, obovate to round; cauline lvs 1–3. **INFL:** open; fls 30–200, separate; pedicels 5–30 mm, straight. **FL:** gen 5–8 mm wide; hypanthium length 1/2 width; petals ± 2 mm, 2 × sepals, obovate to round, white; stamens 15; pistil 1. **FR:** ± 2 mm, smooth, mottled gray-brown. *n*=14. Bare places, sandy, granite ledges; 1500–3600 m. SNH, TR, SnJt. Jun–Sep

I. saxosa (Greene) Ertter (p. 1189) Pls hanging clumps, green. **ST:** 5–30 cm. **LF:** gen 5–15 cm, flat; sheathing bases ± glabrous; lflets 2–4 per side, separated, gen 5–15 mm, ± round, evenly shallow-toothed to unevenly lobed ± to base; cauline lvs 2–4. **INFL:** open, few- to many-fld; pedicels 7–30 mm, gen ± S-shaped in fr. **FL:** 7–10 mm wide; hypanthium length < 1/2 width; petals 2–4 mm, < sepals, oblanceolate to obovate, yellow; stamens 15–40; pistils 3–20. **FR:** 1–1.5 mm, ± ridged, pale. Granitic or volcanic crevices; 900–3300 m. s SNH, SnBr, PR, SNE, DMtns; n Baja CA. Apr–Aug

I. sericoleuca (Rydb.) Rydb. (p. 1189) PLUMAS IVESIA Pl tufted, ± green to white-hairy; caudex 0–few-branched. **ST:** gen decumbent to ascending, 15–45 cm. **LF:** 10–20 cm; sheathing bases densely strigose; lflets 20–35 per side, lobes 1–4, 3–15 mm, elliptic

to oblanceolate; cauline lvs 3–8. **INFL**: open; clusters ± 5–20, 10–20 mm wide, ± head-like, ± 5–10-fld; pedicels < 3 mm (exc lowest), straight. **FL**: 10–15 mm wide; hypanthium length ± 1–2 × width; petals 4–7 mm, > sepals, wide-obovate to -obcordate, white; stamens gen 20; pistils 2–7. **FR**: 2–3 mm, smooth, brown. *n*=14. Dry, gen volcanic meadows; 1300–2320 m. n SNH, s MP (Sierra Valley). May–Sep ★

I. shockleyi S. Watson var. ***shockleyi*** Pl matted, green; caudex much-branched. **ST**: spreading, < 15 cm. **LF**: gen 2–8 cm; sheathing bases strigose; lflets gen 5–10 per side, lobes 3–5, 1–5 mm, oblanceolate to obovate; cauline lf 1, gen ± bract-like. **INFL**: open, 2–10-fld; pedicels 4–10 mm, S-shaped in fr. **FL**: 5–10 mm wide; hypanthium length 1/2 width; petals ± 2 mm, ≤ sepals, oblanceolate, yellow; stamens 5; pistils 2–5(6). **FR**: ± 2 mm, smooth, light brown. Rocky areas; 2700–4000 m. n&c SNH, W&I; se OR, n&c NV. Jun–Aug

I. unguiculata A. Gray YOSEMITE IVESIA Pl rosetted, silvery- to gray-hairy; caudex gen simple. **ST**: ascending, gen 10–30 cm. **LF**: 7–15 cm; sheathing bases glabrous; lflets gen 15–25 per side, lobes

3–8, gen 3–5 mm, linear to oblanceolate; cauline lvs 3–6. **INFL**: clusters 1–several, ± 10 mm wide, head-like, gen < 10-fld; pedicels < 3 mm, straight. **FL**: 6–9 mm wide; hypanthium length ± = width; petals 3–4 mm, > sepals, ± obovate, white or pink-tinged; stamens 10–15; pistils 3–9. **FR**: ± 1.5 mm, smooth, light brown. Meadows; 1500–2500 m. c&s SNH. Historical record from SNE (Owens Valley) unconfirmed. Jun–Sep ★

I. webberi A. Gray WEBBER'S IVESIA Pl rosetted, green; caudex gen simple. **ST**: decumbent to ascending, gen 5–15 cm. **LF**: 3–7 cm; sheathing bases strigose; lflets 4–8 per side, lobes 5–12, 3–10 mm, linear to lanceolate; cauline 2, ± opposite (unique in *Ivesia*). **INFL**: cluster 1, 15–50 mm wide, head-like (open in fr), 5–15-fld; pedicels gen 2–5 mm, straight. **FL**: 8–12 mm wide; hypanthium length 1/2 width; petals gen 2–3 mm, < sepals, oblanceolate, yellow; stamens 5; pistils ± 5. **FR**: 2.5 mm, ± smooth, light brown, mottled darker. Rocky clay in sagebrush flats; 1500–1900 m. n SNH (Dog Valley, e Sierra Co.), s MP (Sierra Valley); w NV. May–Jun ★

LUETKEA

Daniel Potter & Thomas J. Rosatti

1 sp. (Count F.P. Lütke, Russian sea captain, 1797–1882)

L. pectinata (Pursh) Kuntze (p. 1189) Subshrub 5–25 cm, ± prostrate or long-trailing to decumbent; not glandular. **LF**: ± 2-ternately dissected, ± sessile, stipules 0; lf 5–15 mm; segments linear, gen acute, ribbed abaxially, grooved adaxially. **INFL**: raceme, terminal, 1–10 cm, narrow; bracts ternately dissected; pedicel bractlets 0. **FL**: hypanthium hemispheric, bractlets 0; sepals ± 2 mm, ovate; petals ± 3 mm, ± obovate, white; stamens ± 20; ovaries 4–6, free, superior. **FR**: follicles 4–6, ± 4 mm, leathery, sparsely hairy, dehiscent along both sutures. **SEED**: > 1, ± fusiform, flat, smooth. Moist slopes, often near snow, conifer forest; 1800–2800 m. KR, CaRH; to AK. Jun–Sep

LYONOTHAMNUS CATALINA IRONWOOD

Steve Junak & Dieter H. Wilken

1 sp. (Greek: Lyon's shrub, for W.S. Lyon, early resident of Los Angeles)

L. floribundus A. Gray Tree, evergreen. **ST**: trunk bark gray to red-brown, peeling in strips. **LF**: opposite, simple or compound; stipules deciduous; petiole 1–3 cm; blade ± gray abaxially, dark shiny green adaxially. **INFL**: panicle, gen flat-topped; pedicel bractlets 1–3, at top. **FL**: hypanthium bell-shaped, tomentose, bractlets 0; sepals persistent; petals ± round, white; stamens ± 15; pistils 2, ovary superior, style stout, stigma ± head-like. **FR**: follicles 2, 3–4 mm, woody. **SEED**: 1–4, ± 2 mm, compressed, ± brown. Subspp. hybridize in cult; endemic to ChI.

1. Lf gen palmately to pinnately compound
. . . . .subsp. ***aspleniifolius***
1′ Lf gen simple. . . . . subsp. ***floribundus***

subsp. ***aspleniifolius*** (Greene) P.H. Raven (p. 1193) SANTA CRUZ ISLAND IRONWOOD Pl 4–12 m. **LF**: 9–21 cm; 1° lflets gen 3–7, 5–14 cm, linear; 2° lflets gen 20–30, 4–12 mm, subopposite. 2*n*=54. Rocky slopes, canyons, oak woodland, chaparral; 20–500 m. ChI (Santa Cruz, Santa Rosa, San Clemente islands). Threatened by grazing. May–Jul ★

subsp. ***floribundus*** (p. 1193) SANTA CATALINA ISLAND IRONWOOD Pl 5–15 m. **LF**: 7–17 cm, linear to oblong, gen entire (crenate to irregularly lobed). 2*n*=±48. Rocky slopes, canyons, oak woodland, chaparral; 100–500 m. s ChI (Santa Catalina Island). Threatened by feral animals. Seedling lvs often compound. May–Jun ★

MALUS

Daniel Potter & Thomas J. Rosatti

Shrub to tree, thorny or not. **LF**: simple, gen toothed (lobed). **INFL**: few-fld, ± umbel-like cluster; pedicel bractlets 0 or 1, deciduous. **FL**: hypanthium bractlets 0; stamens many; ovary inferior, chambers (2)5, 2-ovuled, styles (2)5, ± fused at base. **FR**: pome, ± spheric, flesh not gritty. 25–50 spp.: n temp. (Classical name of apple)

1. Main lvs (at least some) gen lobed; fr 10–15 mm, oblong, yellow to gen purple-red or -black . . . . ***M. fusca***
1′ Main lvs unlobed; fr gen > 30 mm, round, ± red. . . . . ***M. pumila***

M. fusca (Raf.) C.K. Schneid. (p. 1193) OREGON CRAB APPLE Shrub, small tree. **LF**: 3–12 cm, gen widely lanceolate; petiole 15–50 mm. **FL**: petals white. **FR**: pedicel 20–30 mm. *n*=17. Moist, open conifer forest; < 800 m. NCo, NCoRO, CaRF, n CCo; to AK. Apr–May

M. pumila Mill. APPLE Tree. **LF**: 5–11 cm, elliptic to widely ovate; petiole 10–40 mm. **FL**: petals white or pink. **FR**: pedicel 10–25 mm. 2*n*=34,51. Disturbed places; < 2080 m. CA-FP; native to Eurasia. [*M. sylvestris* Mill., misappl.] Apr–May

NEVIUSIA

Alice Long Heikens & Barbara Ertter

Shrub, ± strigose. **LF**: deciduous; shallow-lobed, toothed; stipules linear. **INFL**: umbel-like cluster; pedicel bractlets 0. **FL**: hypanthium flat, bractlets 0; sepals 5–6, ± toothed; petals 0–2; stamens ± 50[100+], filaments showy, white, ± expanded; pistils 2–6, ovary superior, strigose, chamber 1, ovule 1. **FR**: achene, wall ± soft. 2 spp.: CA, se US. (Reverend R.D. Nevius, 1827–1913)

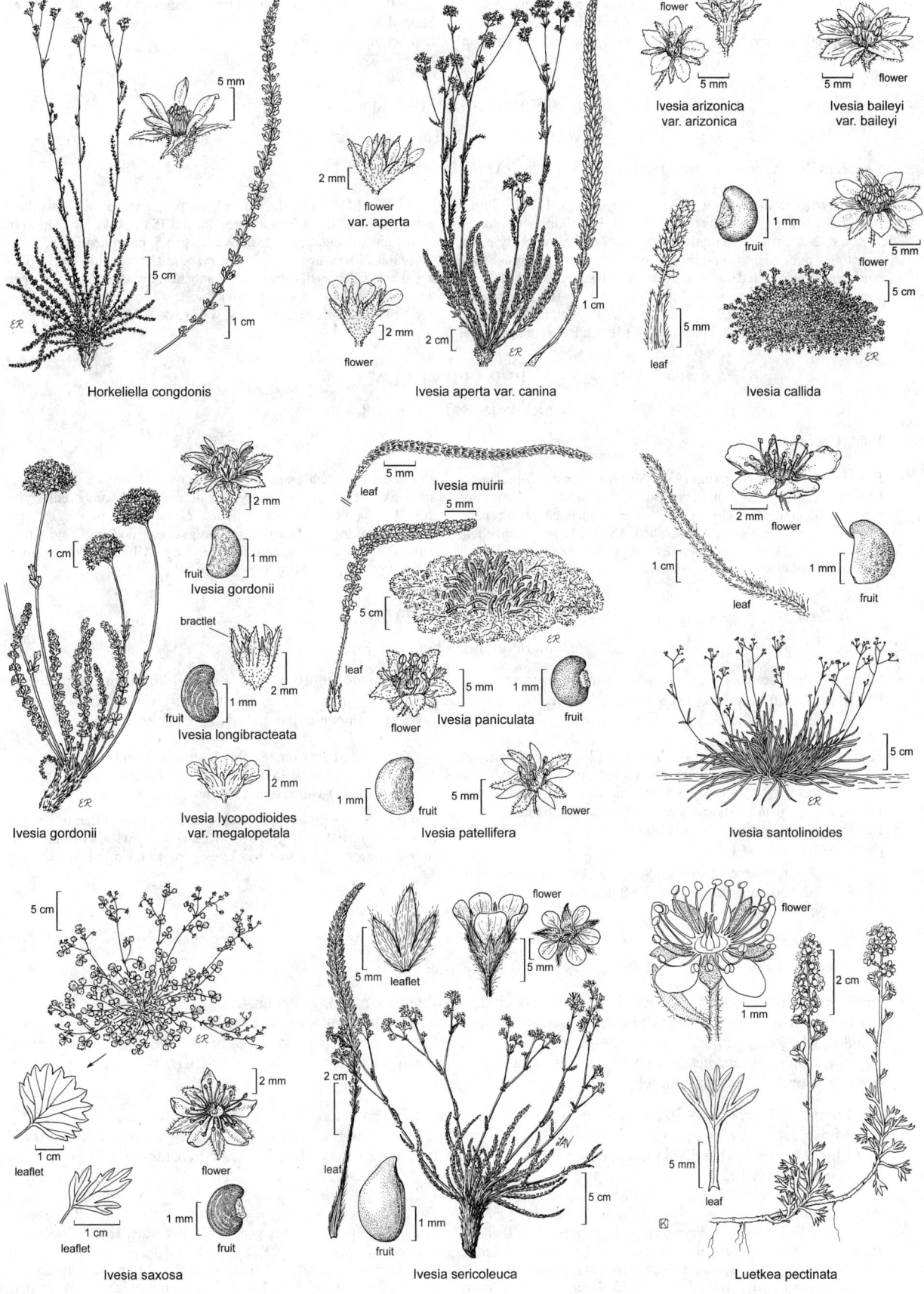

Horkeliella congdonis

Ivesia aperta var. canina

Ivesia arizonica var. arizonica

Ivesia baileyi var. baileyi

Ivesia callida

flower var. aperta

flower

leaf

flower

fruit

flower

leaf

Ivesia gordonii

Ivesia longibracteata

Ivesia lycopodioides var. megalopetala

Ivesia gordonii

bractlet

fruit

Ivesia muirii

leaf

leaf

flower

Ivesia paniculata

fruit

fruit

flower

Ivesia patellifera

flower

leaf

fruit

Ivesia santolinoides

Ivesia saxosa

leaflet

leaflet

flower

fruit

Ivesia sericoleuca

leaflet

flower

leaf

fruit

Luetkea pectinata

flower

leaf

N. cliftonii Shevock et al. (p. 1193) SHASTA SNOW-WREATH Pl 5–25 dm, erect. **ST:** branches slender. **LF:** petiole gen 4–10 mm; blade 20–60 mm, ovate to cordate, sharp-toothed, sparsely strigose. **INFL:** fls gen 3–5; pedicels 1–3 cm. **FL:** sepals 4–6 mm, ± obovate; petals 0–2, 4–6 mm, clawed, white; filaments 4–5 mm. **FR:** 3–4 mm. Shaded, n-facing slopes; 300–600 m. KR (near Lake Shasta). Petal number uncertain. Apr–May ★

OEMLERIA OSO BERRY

William J. Hess & Thomas J. Rosatti

1 sp. (A.G. Oemler, German naturalist at Savannah, GA, 1773–1852)

O. cerasiformis (Hook. & Arn.) J.W. Landon (p. 1193) Shrub, small tree, 1–6 m; dioecious (monoecious). **LF:** simple, deciduous, 5–13 cm, 2–5 cm wide, elliptic to narrow-obovate, abaxially paler, sometimes puberulent becoming glabrous, adaxially glabrous; margins gen entire, ± rolled under; stipules early-deciduous; petiole 5–15 mm. **INFL:** racemes on short lateral branches, 3–10 cm, pendent, bracted; pedicel bractlets 1–2, near top. **FL:** fragrant, unisexual, occ bisexual; hypanthium 3–4.5 mm, bractlets 0; petals 3–6 mm, clawed, white. **STAMINATE FL:** hypanthium persistent; petals > pistillate petals; stamens 15, in 3 series, prominent. **PISTILLATE FL:** most of hypanthium deciduous after fl; ovaries gen 5, prominent, free, styles deciduous. **FR:** drupes 1–5, 5–15 mm, bean-shaped, blue-black, glaucous. *x*=8. Chaparral, canyons, streambanks, lowland wet to dry open woodland, coast to shaded conifer forest; < 1850 m. NW, w CaR, w SN, ScV (Sutter Buttes), s-c SnJV, w CW, sw WTR; to s BC. Feb–Apr

PERAPHYLLUM

Daniel Potter & Thomas J. Rosatti

1 sp. (Greek: very lfy)

P. ramosissimum Nutt. (p. 1193) Shrub 1–3 m, much-branched. **LF:** simple, ± clustered on short-shoots, deciduous, 1–4 cm; stipules early-deciduous; petiole ± 0; blade ± oblanceolate, mucronate, strigose, entire to minute-gland-toothed. **INFL:** fls 1–3, pedicelled, on short-shoots; pedicel bracts 0. **FL:** hypanthium ± funnel-shaped, bractlets 0; sepals spreading to reflexed, 3–5 mm, hairy at least adaxially, persistent; petals spreading, 6–8 mm, ± obovate, white to rose; stamens ± 15–20, free; ovary 1, inferior, 4(6)-chambered, styles 2(3). **FR:** pome, 8–10 mm, spheric, yellow to red, purple, or blue. Dry washes, sagebrush scrub, pinyon/juniper woodland, pine forest; 1010–2500 m. e CaRH, n&c SNH (e slope), GB, DMtns (Panamint Range); to OR, CO, NM. Apr–May

PETROPHYTUM

Daniel Potter & Thomas J. Rosatti

Shrub, matted, scapose. **LF:** crowded, simple, evergreen, gen ± oblanceolate, entire. **INFL:** ± spike-like; pedicel bractlets 0. **FL:** hypanthium bractlets 0; sepals persistent; petals white; stamens 20–40; pistils gen 5, simple, ovary superior, hairy, styles thread-like. **FR:** follicles, dehiscing along both sutures. **SEED:** 1–several, linear. ± 4 spp.: w N.Am. (Greek: rock pl)

P. caespitosum (Nutt.) Rydb. (p. 1193) Pl 3–8 dm wide; rosettes many. **LF:** 1–3-veined abaxially. **INFL:** 1–5 cm, dense; peduncle 3–10 cm, bracted. **FL:** sepals ± 1.5 mm, narrow-ovate, acute; petals ± 1.5 mm, gen obtuse; style ± 3 mm. **FR:** ± 2 mm. **SEED:** 1–2, ± 1.5 mm, linear to obovoid, brown, smooth. [*Petrophyton caespitosum* (Nutt.) Rydb., orth. var.]

1. Lf 10–18 mm, sparsely hairy; s SNH . . . . . subsp. ***acuminatum***
1′ Lf 5–12 mm, densely hairy; W&I, DMtns. . . subsp. ***caespitosum***

subsp. ***acuminatum*** (Rydb.) Munz Limestone cliffs, conifer forest; 900–2350 m. s SNH. [*Petrophyton caespitosum* subsp. *acuminatum* (Rydb.). Munz, orth. var.] Jun–Sep ★

subsp. ***caespitosum*** Limestone ledges, rocks, often in pinyon/juniper woodland; 1350–3050 m. W&I, DMtns; to Rocky Mtns. [*Petrophyton caespitosum* subsp. *caespitosum*, orth. var.] May–Sep

PHYSOCARPUS NINEBARK

Sang-Hun Oh

Shrub, gen ± stellate-hairy. **LF:** petioled, stipuled, deciduous, ovate to ± round, gen palmately 3–7-lobed, crenate to serrate. **INFL:** umbel-like cluster, bracted; pedicel bractlets gen 0. **FL:** hypanthium bell-shaped, bractlets 0; sepals persistent; petals rounded, white or pale pink; stamens 20–40, exserted; pistils 1–5, fused at base (if 3–5, fused below middle), ovary superior, style thread-like, stigma head-like. **FR:** follicles, inflated or ± flat, dehiscent along both sutures. **SEED:** 1–4, ovoid; coat hard, shiny. ± 6 spp.: N.Am, Asia. (Greek: bladdery fr)

1. Lf blade of fl branches 5–20 mm, petiole 3–10 mm; pistil gen 1; fr ± flat laterally, ± = hypanthium; W&I, n DMtns . . . . ***P. alternans***
1′ Lf blade of fl branches 22–85 mm, petiole 7–30 mm; pistils 3–5, fused at base; fr inflated, >> hypanthium; CA-FP (exc GV) . . . . ***P. capitatus***

P. alternans (M.E. Jones) J.T. Howell (p. 1193) NEVADA NINEBARK Pl 5–15 dm. **LF:** gen densely hairy, lobes 3–7, shallow, gen crenate. **INFL:** 10-fld, densely hairy. **FL:** hypanthium 1.7–3 mm, 3.2–5 mm wide at rim; sepals 1.8–3.3 mm; petals 2.5–3 mm; stamens 20–30, longest filament 1.3–2.5 mm. **FR:** 3–4.5 mm, densely hairy. Dry, rocky pinyon/juniper woodland, limestone outcrops; 1800–3100 m. W&I, n DMtns; to ID, CO, UT. Jun–Jul ★

P. capitatus (Pursh) Kuntze (p. 1193) Pl 10–25 dm. **LF:** glabrous to moderately hairy; lobes gen 3–5, gen serrate. **INFL:** >> 10-fld, sparsely to densely hairy. **FL:** hypanthium 1.5–2.5 mm, 4–4.8 mm wide at rim; sepals 2.5–3.5 mm; petals 2.5–3 mm; stamens 25–30, longest filament 4–5 mm. **FR:** 3–5, 8–10 mm, glabrous to ± hairy. *n*=9. Moist banks, n-facing slopes, mixed-conifer forest; < 1400 m. CA-FP (exc GV); to AK, ID. May–Jul

POTENTILLA CINQUEFOIL
Barbara Ertter

Ann to per; odor gen 0. **LF**: gen basal, odd-1-pinnately, 1-palmately, or 1-ternately compound; lflets 1–8(13) per side, ± toothed, gen ± separated, terminal gen ± = lateral; margins gen flat. **INFL**: gen cyme, gen ± open; pedicels gen ± straight, bractlets 0. **FL**: hypanthium ± shallow, bractlets gen 5, gen < sepals, gen flat; sepals ± triangular; petals (2)4–20 mm, ≥ sepals, gen ± widely obcordate, gen yellow; stamens 10–25; pistils gen > 10, ovaries superior, styles slender to ± tapering, gen attached near fr tip. **FR**: achene, gen glabrous. ± 400 spp.: mostly n temp, arctic. (Latin: diminutive of powerful, for reputed medicinal value) Other taxa in TJM (1993) moved to *Comarum, Dasiphora, Drymocallis*.

1. Fls 1, arising from axils of slender stolons
  2. Lf palmate . . . . ***P. anglica***
  2′ Lf pinnate . . . . ***P. anserina***
    3. Lf 3–15 cm, gen densely hairy adaxially; pedicel gen 2–7 cm; 1200–2600 m; SNH, SnBr, GB . . . . . subsp. ***anserina***
    3′ Lf 3–50(75) cm, ± glabrous adaxially; pedicel gen 5–30 cm; gen < 150 m; NCo, CCo, SCo . . . . subsp. ***pacifica***
1′ Fls in cymes; stolons 0
  4. Styles < or ± 1 mm, tapered from rough-thickened base to tip; petals variable
    5. Per; basal lvs gen present in fl; lflet margins ± rolled under; rocky alpine barrens and meadows, > 2700 m
      6. Lf ± palmate; lflets ± 5, toothed > 3/4 to midvein . . . . ***P. pseudosericea***
      6′ Lf ± pinnate; lflets gen 5–13, toothed ± 3/4 to midvein
        7. Lflets gen 2–3 per side; petiole gen ± = blade; hypanthium bractlets < sepals, flat . . . . ***P. jepsonii***
        7′ Lflets 3–6 per side; petiole gen < blade; hypanthium bractlets gen ± = sepals, margins ± rolled-up. . . ***P. pensylvanica***
    5′ Ann or bien (to short-lived per); basal lvs often withered or fallen in fl; lflet margins flat; gen moist or disturbed areas, < 3100 m
      8. Lf pinnate, lflets 9–17; petals > sepals, cream, ± obcordate — lflets toothed > 1/2 to midvein; MP . . . [2]***P. newberryi***
      8′ Lf ternate or ± palmate, lflets 3–7; petals < sepals (exc *Potentilla recta*), yellow (occ fading white), gen elliptic to obovate
        9. St-base hairs gen ± dense, glandular or not, < 1 mm, spreading or not; petals ≤ 2.5 mm; hypanthium gen 2–4 mm wide; fr smooth, < 1 mm, ± white or light brown
          10. St hairs glandular and not; lflets 3, ± obovate; petals oblanceolate-elliptic . . . . ***P. biennis***
          10′ St hairs glandless; lflets 3–5, oblanceolate to elliptic; petals obovate-elliptic . . . . ***P. rivalis***
        9′ St-base hairs sparse, glandless, < 2 mm, spreading; petals ≥ 3 mm; hypanthium 3–10 mm wide; fr veined, ± 1–1.5 mm, ± brown
          11. Lflets gen 3; petals 3–4 mm; anthers ± 0.3 mm; ann to short-lived per . . . . ***P. norvegica***
          11′ Lflets gen 6–7; petals 6–9 mm; anthers ± 1 mm; short-lived per . . . . [2]***P. recta***
  4′ Styles ≥ 1 mm, ± slender (exc *Potentilla recta*); petals ± obcordate, > sepals
    12. Lflets 3; infl gen < 7-fld (see also *Fragaria, Sibbaldia*)
      13. Lflets secondarily lobed as well as toothed ± 1/2 to base, sparsely hairy; styles ± rough-thickened at base; KR . . . . ***P. cristae***
      13′ Lflets only toothed, not secondarily lobed, ± nonglandular; styles slender throughout; KR, CaRH, SNH
        14. Teeth of central lflet gen 7–15, gen uneven, ± 1/4 to midvein; pedicel 10–20 mm; pistils > ± 20; st ± glabrous or sparsely spreading-hairy; KR, CaRH, SNH . . . . ***P. flabellifolia***
        14′ Teeth of central lflet ± 7, even, 1/4–1/2 to midvein; pedicel 10–40 mm; pistils < ± 20; st sparsely strigose; SNH . . . . ***P. grayi***
    12′ Lflets > 3; infl gen > 7-fld (exc *Potentilla hickmanii*)
      15. Lf pinnate to subpalmate (see also *Horkelia, Horkeliella, Ivesia*)
        16. Ann to short-lived per; petals cream; style 1–1.5 mm; fr veined — receding shorelines; MP . . . . [2]***P. newberryi***
        16′ Per; petals yellow; style gen > 1.5 mm; fr gen smooth
          17. Lf densely white-tomentose abaxially, ± green and strigose adaxially; style ± 1.5–2 mm — alpine barrens; c SNH, n W&I . . . . ***P. morefieldii***
          17′ Lf hairs ± similar adaxially and abaxially (but hairs often denser abaxially); style 2–3.5 mm
            18. Basal lf < 1/2 st; pedicel ± straight in fr; 1100–3700 m
              19. Petiole gen < blade, cottony-hairy; lflets 3–7 per side, ± overlapped, palmately toothed > 1/2 to midvein — gen cottony-hairy throughout . . . . ***P. breweri***
              19′ Petiole gen > blade, glabrous to shaggy-hairy; lflets gen 2–4 per side, occ separated, pinnately toothed ± 1/2 to midvein, occ split to base
                20. Lflets gen gray-hairy, ± overlapped; petiole gen dense-shaggy- or cottony-hairy . . . . [2]***P. bruceae***
                20′ Lflets gen green, glabrous to ± straight-hairy, gen separated; petiole glabrous or strigose . . . . ***P. drummondii***
            18′ Basal lf gen > 1/2 st; pedicel gen ± recurved in fr; < 2000 m
              21. Lflets entire or 2–3-toothed or lobed > 1/2 to midvein, larger 5–10 mm — MP . . . . ***P. basaltica***
              21′ Lflets 2–10-toothed ± 1/2 or more to midvein, larger 5–22 mm
                22. Lflets 5–13 per side, 3–10(–15)-toothed 2/3 to nearly to midvein; pistils +/- 10–30; inland
                  23. St 5–20 cm; lf 2–15 cm, petiole gen < blade; CaRH, ne-most SNH, GB; 900–2000 m . . . . ***P. millefolia***
                  23′ St 25–50 cm; lf 15–32 cm, petiole gen ± = blade; s NCoRO/SnFrB; 30–40 m . . . . ***P. uliginosa***
                22′ Lflets 3–8 per side, 2–6 toothed ± 1/2–2/3 (rarely more) to midvein; pistils (2)5–15; coastal
                  24. Lateral lflets toothed 1/2–2/3 (rarely more) to midvein; st 5–25 cm; c CCo . . . . ***P. hickmanii***
                  24′ Lateral lflets toothed ± 1/2 to midvein; st 20–50(70) cm; c SCo . . . . ***P. multijuga***

15′ Lf (sub)palmate
  25. St ± prostrate or hanging, gen ≤ 25 cm; pedicels gen ± recurved in fr; lflet teeth < 11 — > 1800 m, s CA-FP, n W&I
    26. Lflets white-tomentose abaxially, ± green and strigose adaxially; n W&I . . . . *P. concinna* var. *proxima*
    26′ Lflet surfaces ± equally strigose; s CA-FP
      27. Pl hanging from granite crevices; pistils gen 5–20; fr ± smooth; pedicels gen > 15 mm; SnJt . . . . . . *P. rimicola*
      27′ Pl rosetted to tufted in sandy soil; pistils gen > 15; fr ± veined; pedicels gen < 15 mm; s SNH, SnBr, SnJt . . . . *P. wheeleri*
  25′ St ± ascending to erect, gen 10–100 cm; pedicels ± straight; lflet teeth often > 10
    28. Lflets distally 3–7(9)-toothed, glabrous (exc margins, veins), glaucous (at least when fresh)
. . . . *P. glaucophylla* var. *glaucophylla*
    28′ Lflets evenly 7–35-toothed, gen ± hairy, green to white
      29. Lflets toothed > 3/4 to midvein; st and petiole hairs gen appressed
        30. Longest petiole gen 5–10 cm; central lflet 2–6 cm, teeth often narrowest at base, entire; petals 5–8 mm; SNH, Teh, TR, SnJt, SNE. . . . . *P. gracilis* var. *elmeri*
        30′ Longest petiole gen 10–25 cm; central lflet 5–9 cm, teeth gen widest at base, often secondarily few-toothed; petals 7–10 mm; MP . . . . *P. gracilis* var. *flabelliformis*
      29′ Lflets toothed ± 1/2 or less to midvein; st and petiole hairs spreading to appressed
        31. St (and petioles) shaggy- to cottony-hairy; lvs often subpalmate, lflets often irregularly deeply lobed as well as toothed. . . . . ²*P. bruceae*
        31′ St (and petioles) straight-hairy; lvs palmate, lflets regularly toothed
          32. Lf surfaces strongly contrasting, ± white-tomentose abaxially, green, sparsely hairy adaxially; st hairs gen spreading
            33. Glands gen 0 or hidden; nw KR, n NCoRO, CaRH; 120–1100 m . . . . *P. gracilis* var. *gracilis*
            33′ Glands gen abundant; n W&I, 3000–3100 m . . . . *P. pulcherrima*
          32′ Lf surfaces similarly hairy, gen not white tomentose abaxially; st hairs appressed to spreading
            34. Basal lvs gen prominent in fl; st hairs of 1 length; fr ± smooth; styles gen 1.5–2 mm, < 50
. . . . *P. gracilis* var. *fastigiata*
            34′ Basal lvs gen fallen or withered in fl; st hairs short- and long-spreading; fr strong-veined; styles ± 1 mm, gen > 50 . . . . ²*P. recta*

P. anglica Laichard. ENGLISH CINQUEFOIL Pl tufted from stolons, nonglandular. **LF:** palmate; 2–12 cm, lflets 3–5, central gen 10–25 mm, wedge-shaped to obovate, distally 7–11-toothed ± 1/4 to midvein, sparsely strigose. **INFL:** fls 1 from stolon nodes; pedicels 2–8 cm. **FL:** hypanthium 2–4 mm wide; sepals 4; petals 4, 3–8 mm; filaments 2–3.5 mm, anthers ± 1 mm; pistils 4–20, styles ± 1.5 mm, slender. **FR:** 1.5–2 mm, ± smooth, brown. 2*n*=28,56. Streambanks; 40–160 m. ScV, SnFrB, PR; sporadic in N.Am and elsewhere; native to Eur. Doubtfully naturalized; fl parts gen in 4s (5s in *P. reptans* L., a similar cult sp. not yet confirmed in CA). May–Sep

P. anserina L. (p. 1193) Pl tufted from stolons, nonglandular. **LF:** pinnate, petiole << blade; main lflets 5–10 per side, ± elliptic to oblanceolate, evenly 7–31-toothed 1/2–1/3 to midvein, densely hairy at least abaxially, reduced lflets alternating. **INFL:** fls 1 from stolon nodes. **FL:** hypanthium 4–7 mm wide; filaments 1–3.5 mm, anthers ± 1 mm; styles ± 2 mm, slender. **FR:** ± 2 mm, rough, dark red-brown.

subsp. *anserina* COMMON SILVERWEED **LF:** 3–15 cm, gen densely hairy adaxially; main lflets 10–25 mm. **INFL:** pedicels gen 2–7 cm. **FL:** petals 7–10 mm; pistils 10–50. 2*n*=28,35,42. Shorelines, moist alkaline meadows; 1200–2600 m. SNH, SnBr, GB; circumboreal. May–Sep

subsp. *pacifica* (Howell) Rousi PACIFIC SILVERWEED **LF:** 3–50(75) cm, ± glabrous adaxially; main lflets 10–50 mm. **INFL:** pedicels gen 5–30 cm. **FL:** petals 8–20 mm; pistils gen 20–200. 2*n*=28. Coastal wetlands, often brackish; gen < 150 m. NCo, CCo, SCo; coastal N.Am. Mar–Oct

P. basaltica Tiehm & Ertter BLACK ROCK POTENTILLA Pl rosetted from thick taproot, nonglandular. **ST:** prostrate to decumbent, 10–40 cm, glabrous, glaucous. **LF:** pinnate, petiole << blade; basal gen 5–12 cm, lflets gen 10–15 per side, overlapped, larger 5–10 mm, entire or 2–3-lobed > 1/2 to midvein, ± glabrous. **INFL:** 5–20-fld; pedicels gen ± recurved in fr. **FL:** hypanthium 2–5 mm wide; petals gen 4–6.5 mm; filaments gen 2–3 mm, anthers ± 1 mm; pistils 3–10, styles 2–2.5 mm, slender. **FR:** ± 2 mm, smooth or ± ridged, pale brown. Alkaline meadows; ± 1500 m. MP (Ash Valley); nw NV. May–Aug ★

P. biennis Greene (p. 1193) BIENNIAL OR GREENE'S CINQUEFOIL Ann or bien, taprooted, glandular. **ST:** ascending to erect, 10–70 cm, ± spreading-hairy. **LF:** ternate; basal often withered in fl; cauline gen 2–10 cm, lflets 3, central 5–30 mm, ± obovate, singly and doubly 7–13(19)-toothed ± 1/4 to midvein, ± hairy. **INFL:** few- to many-fld. **FL:** hypanthium 2–4 mm wide; petals 1–2.5 mm, < sepal, oblanceolate-elliptic; stamens 10(15), filaments 0.5–1 mm, anthers ± 0.2 mm; style ± 0.7 mm, tapered from rough-thickened base. **FR:** ± 0.6 mm, smooth, ± white. Moist shores; 1500–3100 m. SNH, TR, GB, DMtns; w N.Am. May–Oct

P. breweri S. Watson (p. 1193) BREWER'S CINQUEFOIL Pl ± tufted from few-branched caudex; glands gen 0. **ST:** decumbent to ascending, gen 10–40 cm, ± cottony-hairy. **LF:** pinnate, petiole gen < blade; basal gen 3–15 cm, lflets 3–7 per side, ± overlapped, larger gen 5–20 mm, obovate, palmately toothed > 1/2 to midvein, gen white-hairy. **INFL:** 3–15-fld. **FL:** hypanthium 4–5 mm wide; petals 5–9 mm; filaments 1.5–4 mm, anthers 0.7–1.2 mm; pistils 15–25, styles 2–3 mm, ± slender. **FR:** 1.5–2 mm, smooth to ± veined, brown. 2*n*=±72–102. Meadows, rocks; 1500–2300 m (KR) or 2700–3700 m. c KR, SNH, Wrn; to WA, NV. [*P. drummondii* subsp. *b.* (S. Watson) Ertter] Jun–Sep

P. bruceae Rydb. (p. 1193) BRUCE'S CINQUEFOIL Pl ± tufted from few-branched caudex; glands 0. **ST:** ascending, gen 15–50 cm, ± shaggy- to cottony-hairy. **LF:** palmate or subpalmate, petiole gen > blade; basal gen 5–20 cm, lflets 5–7, ± overlapped, central gen 15–50 mm, ± obovate to oblanceolate, 7–13-toothed ± 1/2 to midvein, often deeply lobed, gen gray-hairy. **INFL:** 5–50-fld. **FL:** hypanthium ± 5 mm wide; petals 5–8 mm; filaments 1–2.5 mm, anthers 0.7–1.2 mm; style 2–2.5 mm, ± slender. **FR:** ± 1.5 mm, smooth to ± veined, brown. 2*n*=±64–98,129. Seasonally dry meadows; 1200–3700 m. KR, CaRH, SNH, Wrn; adjacent NV, sw OR. [*P. drummondii* subsp. *b.* (Rydb.) D.D. Keck] Lf division, hairs variable, but forming extensive uniform populations esp in n SNH; probably hybridizing with *P. breweri.* Jun–Aug

P. concinna Richardson var. *proxima* (Rydb.) S.L. Welsh & B.C. Johnst. (p. 1197) EARLY CINQUEFOIL Pl tufted from branched

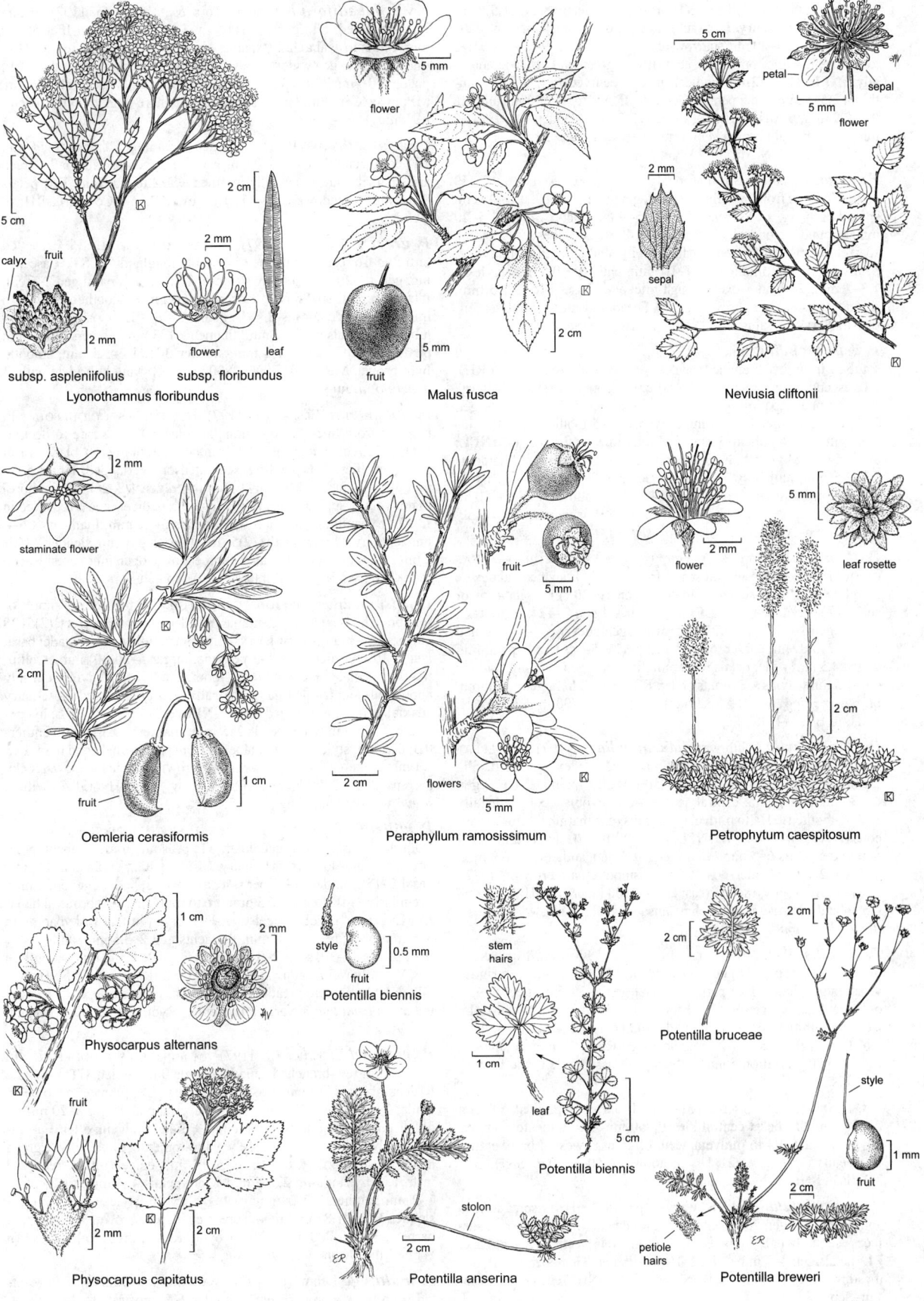

subsp. aspleniifolius   subsp. floribundus

calyx   fruit

flower   leaf

Lyonothamnus floribundus

Malus fusca

petal   sepal

flower

sepal

Neviusia cliftonii

staminate flower

fruit

Oemleria cerasiformis

fruit

flowers

Peraphyllum ramosissimum

flower   leaf rosette

Petrophytum caespitosum

Physocarpus alternans

fruit

Physocarpus capitatus

style   fruit

Potentilla biennis

stem hairs

leaf

Potentilla biennis

stolon

Potentilla anserina

Potentilla bruceae

style   fruit

petiole hairs

Potentilla breweri

caudex; glands 0 or hidden. **ST:** prostrate to decumbent, (1.5)3–15 cm, ± ascending-hairy. **LF:** (sub)palmate; basal gen 1.5–5 cm, lflets gen 5, central 10–20 mm, narrow-obovate, distally 5–9-toothed ± 1/2 to midvein, white-tomentose abaxially, ± green and strigose adaxially. **INFL:** (1)3–12-fld; pedicels gen ± recurved in fr. **FL:** hypanthium 2.5–5 mm wide; petals 2.5–6 mm; filaments 1–2.5 mm, anthers 0.6–0.9 mm; pistils (7)10–30, styles 1.5–2 mm, slender. **FR:** 1.5–2.5 mm, smooth, light brown. Alpine meadows, ridges; ± 3170 m. n W&I (White Mtns); w N.Am. Jul–Aug ★

P. cristae Ferlatte & Strother (p. 1197) CRESTED POTENTILLA Pl loosely matted from branched caudex, glandular. **ST:** erect, 5–20 cm, spreading-hairy. **LF:** ternate; basal 1.5–9 cm, lflets 3, central 5–20 mm, ± round, secondarily 3–5-lobed ± 1/2 to base, lobes 2–5-toothed, sparsely hairy. **INFL:** gen < 7-fld. **FL:** hypanthium 3.5–6 mm wide; petals 3–5.5 mm; filaments 0.7–1.7 mm, anthers ± 0.5 mm; styles ± 1.3–2 mm, ± slender exc ± rough-thickened base. **FR:** 1–1.5 mm, smooth, crested, light brown. 2*n*=42. Seasonally moist, often serpentine-like gravels, talus; 1800–2800 m. KR. Jun–Sep ★

P. drummondii Lehm. (p. 1197) DRUMMOND'S CINQUEFOIL Pl ± tufted from few-branched caudex; glands gen 0 (exc in NCoRH). **ST:** ascending to erect, gen 15–60 cm, glabrous to sparsely hairy. **LF:** ± pinnate, petiole gen > blade; basal gen 5–25 cm, lflets gen 2–4 per side, larger 10–50 mm, unevenly 7–15-toothed ± 1/2 to midvein (often also split to base), glabrous to ± straight-hairy. **INFL:** 3–15-fld. **FL:** hypanthium 3–6 mm wide; petals 7–10 mm; filaments 2.5–3.5 mm, anthers 0.7–1 mm; styles gen 2–3 mm, slender. **FR:** 1.5–2 mm, ± smooth, brown. 2*n*=±64–108. Meadows; 1110–3000 m. KR, NCoRH, SNH; to AK, MT. Pls in s SNH depauperate, possibly distinct. Jun–Aug

P. flabellifolia Torr. & A. Gray (p. 1197) FAN-FOIL, FANLEAF CINQUEFOIL Pl loosely clustered from openly branched caudex, ± nonglandular. **ST:** ascending to erect, gen 10–30 cm, ± glabrous or sparsely spreading-hairy. **LF:** ternate; basal gen 3–12 cm, lflets 3, central gen 10–30 mm, widely obovate, gen unevenly 7–15-toothed ± 1/4 to midvein, ± glabrous. **INFL:** gen 1–5-fld. **FL:** hypanthium gen 3–4.5 mm wide; petals 6–10 mm; filaments gen 2–3 mm, anthers 0.6–1.5 mm; styles ± 2 mm, slender. **FR:** gen 1.2 mm, gen ± smooth, brown or ± red. *n*=14. Moist meadows; 1700–3700 m. KR, CaRH, SNH; to BC, MT. Jul–Sep

P. glaucophylla Lehm. var. ***glaucophylla*** (p. 1197) BLUELEAF CINQUEFOIL Pl tufted from few-branched caudex; glands 0. **ST:** ascending, gen 10–40 cm, base glabrous. **LF:** palmate; basal gen 4–13 cm, lflets gen 5, central gen 15–40 mm, oblanceolate, distally 3–7(9)-toothed ± 1/4 to midvein, glabrous (exc margins, veins), glaucous (at least when fresh). **INFL:** gen < 20-fld. **FL:** hypanthium 3–4 mm wide; petals 6–8 mm; filaments 1–2.5 mm, anthers 0.4–0.9 mm; styles ± 2 mm, slender. **FR:** 1.5 mm, smooth, pale brown. 2*n*=82–±101. Gen ± rocky, moist areas; 2600–3500 m. c&s SNH; w N.Am, Greenland. [*P. diversifolia* Lehm., misappl.] Intergrades with *P. gracilis*, *P. drummondii*. Jul–Sep

P. gracilis Hook. SLENDER CINQUEFOIL Pl tufted from short, thick rhizome; glands gen 0 or hidden. **ST:** ± ascending, strigose to spreading-hairy. **LF:** palmate; basal gen 6–30 cm, lflets 5–9, ± oblanceolate, ± evenly 13–23-toothed. **INFL:** few- to many-fld. **FL:** hypanthium gen 3–5 mm wide; filaments 1.5–2.5 mm, anthers 0.6–1.6 mm; styles 1.5–2 mm, ± smooth, light brown. Variation complex; many pls assignable to the following vars.

var. ***elmeri*** (Rydb.) Jeps. (p. 1197) Hairs gen appressed. **ST:** gen 20–50 cm. **LF:** basal central lflet 20–60 mm, gen tomentose abaxially, toothed > 3/4 to midvein, teeth often narrowest at base, entire. **FL:** petals 5–8 mm. *n*=21. Dry meadows; 1280–3050 m. SNH, Teh, TR, SnJt, SNE; w N.Am. Jun–Aug

var. ***fastigiata*** (Nutt.) S. Watson (p. 1197) Hairs spreading to appressed. **ST:** gen 20–50 cm. **LF:** basal central lflet 20–60 mm, surfaces ± equally hairy, toothed < 1/2 to midvein, teeth widest at base. **FL:** petals gen 4–7 mm. 2*n*=52–109. Common. Gen open forest, dry meadows; 800–3500 m. NW (exc sw), CaR, SNH, Teh, TR, PR, GB. Jun–Sep

var. ***flabelliformis*** (Lehm.) Torr. & A. Gray (p. 1197) Hairs gen appressed. **ST:** gen 50–100 cm. **LF:** basal central lflet 50–90 mm, tomentose abaxially, sparsely hairy adaxially, toothed > 3/4 to midvein, teeth gen widest at base, often secondarily few-toothed. **FL:** petals 7–10 mm. 2*n*=±56–65. Moist or wet meadows; 1050–1600 m. MP; to sw Can. Similar pls sporadic elsewhere (e.g., c SNH, Yosemite Valley). Jul–Aug

var. ***gracilis*** (p. 1197) Hairs gen spreading. **ST:** gen 40–90 cm. **LF:** basal central lflet 30–70 mm, ± white-tomentose abaxially, green and sparsely hairy adaxially, toothed ± 1/2 to midvein. **FL:** petals 7–10 mm. Meadows; 120–1100 m. nw KR, n NCoRO, CaRH; to BC. Jun–Aug

P. grayi S. Watson (p. 1197) GRAY'S CINQUEFOIL Pl ± rosetted or tufted from short rhizome or taproot, nonglandular. **ST:** ± ascending, gen 10–20 cm, sparsely strigose. **LF:** ternate; basal gen 2–6 cm, lflets 3, central 10–25 mm, ± obovate, evenly ± 7-toothed 1/4–1/2 to midvein, ± subglabrous. **INFL:** gen 1–5-fld. **FL:** hypanthium 2.5–4 mm wide; petals gen 4–7 mm; filaments 1–3 mm, anthers ± 0.6 mm; pistils < ± 20, style 1–2.5 mm, slender. **FR:** 1.2–1.5 mm, smooth, light brown. Meadows; 2000–2800 m. SNH. May hybridize with *P. flabellifolia*. Jul–Sep

P. hickmanii Eastw. (p. 1197) HICKMAN'S CINQUEFOIL Pl rosetted from thick taproot, nonglandular. **ST:** prostrate to decumbent, 5–25 cm, ± glabrous. **LF:** pinnate, petiole gen < blade; basal 3–17 cm, lflets 3–6 per side, separated or overlapped, larger 5–20 mm, wedge-shaped, ± evenly 2–5-toothed gen 1/2–2/3 (rarely more) to midvein, ± subglabrous. **INFL:** 2–5-fld; pedicels gen ± recurved in fr. **FL:** hypanthium 3–6 mm wide; petals 6–12 mm; filaments 1.5–4 mm, anthers ± 1 mm; pistils (2)5–15, styles 2–3.5 mm, slender. **FR:** ± 2 mm, smooth, ± tan. Vernally wet meadows, open pine forest; < 100 m. c CCo (San Mateo, Monterey cos.). Apr–Jun ★

P. jepsonii Ertter JEPSON'S CINQUEFOIL Pl tufted from simple or few-branched caudex, ± glandular. **ST:** decumbent to ± erect, (2)5–25 cm, gen ascending-hairy. **LF:** subpinnate, petiole gen ± = blade; basal gen 2–10 cm; lflets gen 2–3 per side, larger 5–20(30) mm, ± elliptic-oblanceolate, ± evenly 7–11-toothed ± 3/4 to midvein, densely appressed- and tangled-hairy abaxially, gen greener and loose-hairy adaxially; margins ± rolled under. **INFL:** ± 3–12(20)-fld. **FL:** hypanthium 2.5–5 mm wide; petals 2–5 mm; filaments 0.5–2 mm, anthers 0.3–0.7 mm; styles ± 1 mm, tapered from rough-thickened base. **FR:** ± 1 mm, ± smooth or faintly veined, ± brown. Alpine meadows, rocky barrens; 2700–3800 m. c SNH, n W&I; w N.Am. Hybridizes with *P. pseudosericea*. Jun–Aug

P. millefolia Rydb. (p. 1197) FEATHER CINQUEFOIL Pl rosetted from thick taproot, occ glandular. **ST:** prostrate to decumbent, 5–20 cm, spreading- to appressed-hairy. **LF:** pinnate, petiole gen < blade; basal 2–15 cm, lflets 5–13 per side, gen overlapped, larger 5–20 mm, irregularly 3–10-toothed 2/3 or more to midvein, ± glabrous to hairy. **INFL:** gen < 10-fld; pedicels gen ± recurved in fr. **FL:** hypanthium 3–6 mm wide; petals 4–8 mm; filaments gen 2–3.5 mm, anthers ± 1 mm; pistils 10–30, style 2–3 mm, slender. **FR:** 1.5–2 mm, smooth, ± tan. Vernally wet meadows; 900–2000 m. CaRH, ne-most SNH, GB; OR, NV. Variable; spreading-hairy pls have been described under the name *P. millefolia* var. *klamathensis* (Rydb.) Jeps.; study needed. Apr–Aug

P. morefieldii Ertter (p. 1197) MOREFIELD'S CINQUEFOIL Pl tufted from few-branched caudex; glands 0 or hidden. **ST:** prostrate to decumbent, 5–15 cm, ± ascending-hairy. **LF:** pinnate, petiole ± = blade; 2–6 cm, lflets 2–4 per side, ± overlapped, larger 5–20 mm, ± oblanceolate, irregularly 7–9-toothed ± 3/4 to midvein, often ± asymmetrically split ± to base, densely white-tomentose abaxially, ± green and strigose adaxially. **INFL:** gen 5–15-fld; pedicels often ± recurved in fr. **FL:** hypanthium 2.5–5 mm wide; petals 4–6 mm; filaments ± 1–2 mm, anthers 0.5–1 mm; pistils ± 15–20, style ± 1.5–2 mm, ± slender. **FR:** ± 1.8 mm, smooth or ± ridged, pale brown. Rocky alpine barrens; 3300–4000 m. c SNH (Coyote Ridge, Inyo Co.), n W&I (n White Mtns, Mono Co.). Jun–Aug ★

P. multijuga Lehm. (p. 1197) BALLONA CINQUEFOIL Pl rosetted from thick taproot, ± nonglandular. **ST:** prostrate to decumbent,

20–50(70) cm. **LF:** pinnate, petiole gen < blade; basal 11–22 cm, lflets 3–8 per side, larger 10–20 mm, ± obovate, ± evenly 3–6-toothed ± 1/2 to midvein. **INFL:** gen < 10-fld; pedicels gen ± recurved in fr. **FL:** hypanthium 4–6 mm wide; petals 4.5–10 mm; filaments 1.5–4.5 mm, anthers ± 1 mm; pistils 5–10, style 2–3 mm, slender. **FR:** 1.8 mm, smooth, ± tan. Brackish marshes; ± 0 m. c SCo (Ballona Marsh, Los Angeles Co.). Apr–Jul ★

P. newberryi A. Gray (p. 1197) NEWBERRY'S CINQUEFOIL Ann to short-lived per, rosetted from taproot, inconspicuously glandular. **ST:** prostrate to decumbent, gen 10–40 cm, spreading- to ascending-hairy. **LF:** pinnate, petiole ± = blade; basal gen 2–10 cm, lflets 4–8 per side, overlapped, larger 3–10 mm, irregularly 2–8- toothed > 1/2 to midvein, ± hairy. **INFL:** many-fld; pedicels gen ± recurved in fr. **FL:** hypanthium 2.5–5 mm wide; petals 3–5.5 mm, cream; filaments 1–1.5 mm, anthers ± 0.5 mm; style 1–1.5 mm, slender or ± tapered from rough-thickened base. **FR:** 1–1.5 mm, veined, brown. Receding shorelines; 1300–2200 m. MP; to WA, NV. May–Aug ★

P. norvegica L. (p. 1197) NORWEGIAN OR ROUGH CINQUEFOIL Ann to short-lived per from taproot, nonglandular. **ST:** ascending to erect, 10–70 cm; hairs spreading, sparse and long abaxially, denser and shorter adaxially. **LF:** gen ternate; basal often withered or fallen in fl; cauline gen 3–12 cm, lflets gen 3, central 15–50 mm, oblanceolate, evenly 11–21-toothed ± 1/3 to midvein, ± hairy. **INFL:** several to many-fld. **FL:** hypanthium 4–10 mm wide; petals 3–4 mm, < sepals; stamens 15–20, filaments 0.5–2 mm, anthers ± 0.3 mm; style ± 0.8 mm, tapered from rough-thickened base. **FR:** ± 1 mm, veined, light brown. 2*n*=56,63,70. Moist, disturbed areas; < 2300 m. c SNH, ScV, PR, SNE; N.Am, native to Eurasia. Distribution of native Am, naturalized Eurasian pls needs study. Jun–Sep

P. pensylvanica L. (p. 1197) PRAIRIE CINQUEFOIL Pl tufted from ± branched caudex, ± glandular. **ST:** ascending to erect, 8–25 cm, densely short-spreading-hairy, more sparsely long-hairy. **LF:** pinnate, petiole gen < blade; basal gen 3–13 cm, lflets gen 3–6 per side, larger 10–25 mm, ± elliptic-oblanceolate, ± evenly 9–13-toothed ± 3/4 to midvein, densely short-hairy abaxially, sparser adaxially; margins ± rolled under. **INFL:** ± 3–10-fld. **FL:** hypanthium 3–5 mm wide, bractlets ± = sepals, margins ± rolled up, petals 3–5 mm; filaments 0.5–2 mm, anthers 0.5–0.9 mm; styles ± 1 mm, tapered from rough-thickened base. **FR:** 1–1.5 mm, ± veined, ± brown. 2*n*=28. Rocky alpine barrens; 2700–3800 m. c&s SNH, W&I; N.Am, Eurasia. Jul–Aug

P. pseudosericea Rydb. (p. 1197) MONO CINQUEFOIL Pl tufted or matted from ± branched caudex; glands 0 or hidden. **ST:** decumbent to ascending, gen 2–15 cm, ± strigose. **LF:** ± palmate; basal gen 2–6 cm, lflets ± 5, central (5)10–15 mm, ± obovate, 5–10-toothed > 3/4 to midvein, tomentose abaxially, densely white-strigose adaxially; margins ± rolled under. **INFL:** 3–10-fld. **FL:** hypanthium 3–4 mm wide; petals 2–4 mm; filaments ± 1 mm, anthers ± 0.4 mm; styles ± 1 mm, tapered from rough-thickened bases. **FR:** ± 1 mm, ± smooth, pale brown. Rocky flats, slopes; 3200–4300 m. c&s SNH, n SNE (Sweetwater Mtns), n W&I. Hybridizes with *P. jepsonii*. Jul–Aug

P. pulcherrima Lehm. BEAUTIFUL CINQUEFOIL Pl tufted from short, thick rhizome; glands gen abundant, minute. **ST:** ± ascending, gen 20–45 cm, spreading-hairy. **LF:** palmate; basal gen 7–15 cm, lflets gen 7, central 20–50 mm, oblanceolate, evenly 15–21-toothed 1/3–1/2 to midvein, white-tomentose abaxially, green and sparsely hairy adaxially. **INFL:** several to many-fld. **FL:** hypanthium 3–5

mm wide; petals 5–6 mm; filaments 1–2 mm, anthers ± 0.7 mm; styles 1.5–2 mm, ± slender. **FR:** ± 1.5 mm, ± veined, light brown. 2*n*=±70,71,108. Dry edges of meadows, streams; 3000–3100 m. n W&I; w N.Am. Jul–Aug ★

P. recta L. (p. 1197) SULPHUR CINQUEFOIL Pl gen ± tufted from taproot or branched caudex, sparsely glandular. **ST:** ± ascending, 20–60 cm, spreading-hairy; base with short and long hairs. **LF:** palmate; basal gen fallen or withered in fl; cauline 8–15 cm, lflets gen 6–7, central 30–80 mm, oblanceolate, evenly 15–35-toothed ± 1/2 to midvein, sparsely hairy. **INFL:** many-fld. **FL:** hypanthium 3–6 mm wide; petals 6–9 mm; stamens ± 25, filaments 1–2 mm, anthers ± 1 mm; styles ± 1 mm, columnar. **FR:** 1–1.5 mm, prominently veined, brown. 2*n*=28,42. Gen disturbed areas; 150–1500 m. CaRH, ScV, SnFrB; N.Am; native to Eurasia. May–Aug ◆

P. rimicola (Munz & I.M. Johnst.) Ertter (p. 1197) CLIFF CINQUEFOIL Pl hanging, taprooted, ± glandular. **ST:** gen 5–20 cm, spreading- to ascending-hairy. **LF:** palmate; basal 2–4 cm, lflets 5, central 10–30 mm, ± obovate, distally 5–9-toothed ± 1/4 to midvein, ± strigose. **INFL:** gen 5–20-fld; pedicels gen > 15 mm, gen ± recurved in fr. **FL:** hypanthium 2–3 mm wide; petals 4–7 mm; filaments 1–2.5 mm, anthers 0.5–1 mm; pistils gen 5–20, style 1.5–2.5 mm, slender. **FR:** ± 1.5 mm, ± smooth, red-tipped. Granite crevices; 2400–2800 m. SnJt; n Baja CA. Jul–Sep ★

P. rivalis Nutt. (p. 1197) BROOK OR RIVER CINQUEFOIL Ann or bien from taproot; glands 0 or hidden. **ST:** gen ascending, gen 10–50 cm, spreading- to ascending-hairy. **LF:** ternate or ± palmate; basal gen withered or fallen in fl; cauline gen 2–12 cm, lflets 3–5, central gen 10–40 mm, oblanceolate to elliptic, irregularly 11–17-toothed ± 1/3 to midvein, ± hairy. **INFL:** many-fld. **FL:** hypanthium 2.5–3.5 mm wide; petals 1.5–2 mm, < sepals, obovate-elliptic; stamens ± 15, filaments 0.5–1 mm, anthers ± 0.3 mm; styles 0.5–1 mm, tapered from rough-thickened bases. **FR:** ± 0.8 mm, smooth, light brown. Moist, ± disturbed areas; 40–2100 m. s NCoRO, CaRH, SN, GV, CW, SnBr, PR, MP, DSon; w&c N.Am. Mar–Nov

P. uliginosa B.C. Johnst. & Ertter CUNNINGHAM MARSH CINQUEFOIL Pl rosetted to tufted from thick taproot, ± nonglandular. **ST:** prostrate to decumbent, 25–50 cm, ± glabrous. **LF:** pinnate, petiole gen ± = blade; basal 15–32 cm, lflets 6–10(12) per side, separated or overlapped, larger 1.2–2.2 mm, wedge-shaped to fan-shaped, unevenly (300)5–10(15)-toothed 3.4 to nearly to midvein, ± subglabrous. **INFL:** 6–10-fld; pedicels gen ± recurved in fr. **FL:** hypanthium 5–6 mm wide; petals 6–10 mm; filaments (1.5)2–3 mm, anthers ± 1 mm; pistils ± 10, styles 2.5–3.5 mm, slender. **FR:** 2–2.6 mm, smooth, ± tan. Presumed extinct. Low-nutrient wetlands; 30–40 m. s NCoRO (Cunningham Marsh, Sonoma Co., at SnFrB boundary). Formerly incl in *P. hickmanii* (Johnston & Ertter 2010 J Bot Res Inst Texas 4:14). May–Aug ★

P. wheeleri S. Watson (p. 1197) WHEELER'S CINQUEFOIL Pl rosetted to tufted from thick taproot or few-branched caudex, inconspicuously glandular. **ST:** prostrate to decumbent, 2–25 cm, ± appressed-hairy abaxially, spreading-hairy adaxially. **LF:** palmate; basal 2–10 cm, lflets 5, central 5–25 mm, wedge-shaped, distally 5–9(11)-toothed ± 1/4 to midvein, ± densely strigose. **INFL:** gen > 5-fld; pedicels gen < 15 mm, gen ± recurved in fr. **FL:** hypanthium 3–4 mm wide; petals 3–6 mm; filaments 1–2 mm, anthers ± 0.5 mm; style gen 1.5–2 mm, ± slender. **FR:** 1–1.5 mm, ± veined, pale. Sandy, ± moist flats; 1800–3500 m. s SNH, SnBr, SnJt. May–Aug

POTERIDIUM

1 sp. (Diminutive of *Poterium*)

P. annuum (Nutt.) Spach (p. 1197) WESTERN BURNET Ann, bien, taprooted, nonglandular. **ST:** gen ascending to erect, gen 10–70 cm. **LF:** alternate, odd-1-pinnately compound; basal withered at fl; largest cauline gen 3–12 cm; lflets 4–7 per side, largest blade 5–20 mm, ± sessile, ± obovate-elliptic, lobes < 15, > 2/3 to midvein, linear. **INFL:** spike, head-like, 5–35 mm, 5–10 mm wide, cylindric-ovoid, ± 10–50-fld; peduncle 3–15 cm; pedicel bractlets 2, subtended by

1 bract, all 3 2–3 mm wide. **FL:** bisexual; hypanthium urn-shaped, bractlets 0; sepals gen 4, 2–3 mm, ovate, green; petals 0; stamens gen 2, filaments thread-like; pistil 1, ovary superior, continuous to style at top, stigma ± bushy, exserted. **FR:** hypanthium enclosing achene, 2–4 mm, 4-angled, ± winged, hard, faces wrinkled. 2*n*=14. Open, esp disturbed areas; 225–1890 m. KR, CaRH, n&c SNH, n CCo, PR, MP; to BC, MT. [*Sanguisorba occidentalis* Torr. & A. Gray, inval.] Apr–Jul

POTERIUM

Per, nonglandular. **LF**: alternate, odd-1-pinnately compound; lflets toothed < 1/3 to midvein. **INFL**: spike, head-like; pedicel bractlets 2, subtended by 1 bract. **FL**: bisexual or pistillate; hypanthium urn-shaped, bractlets 0; sepals gen 4; petals 0; stamens [0]many; pistils (1)2(3), ovaries superior, continuous to style at top, stigma gen ± bushy, exserted. **FR**: hypanthium hard, 4-angled, enclosing achene(s). 2*n*=28,56. 13 spp.: Eur, Asia. (Greek: goblet or beaker)

P. sanguisorba L. (p. 1197)   GARDEN BURNET Tufted, tap-rooted. **ST**: erect, gen 20–70 cm. **LF**: basal present at fl, largest gen 4–20 cm; lflets 4–10 per side, largest blade gen 5–20 mm, round-oblong, stalk gen 1–4 mm, teeth gen < 15. **INFL**: 7–30 mm, 6–20 mm wide, ovoid-spheric, 5–30-fld; peduncle 5–15 cm; bract, pedicel bractlets ± 2 mm wide. **FL**: sepals 3–6 mm, elliptic, green or ± purple; filaments thread-like. **FR**: ± 5 mm; angles short-winged; faces with raised bumpy network. Open, esp disturbed areas; 22–1830 m. CA-FP (exc SNH); to e US; native to Eur. [*Sanguisorba minor* Scop. subsp. *muricata* (Bonnier & Layens) Briq.] Often used in seeding mixtures after fires and in pastures. Mar–Jul

PRUNUS

Joseph R. Rohrer

Shrub, tree. **LF**: simple, alternate or clustered on short-shoots, entire to serrate, gen glabrous, gen glandular on teeth and at blade-petiole junction, veins pinnate; stipules deciduous. **INFL**: raceme, umbel-like or subsessile cluster, or fls 1; pedicel bractlets 0. **FL**: hypanthium cup- to urn-shaped, deciduous in fr, bractlets 0; sepals erect to reflexed; stamens gen 10–30, gen in 2+ whorls; pistil 1, ovary superior, chamber 1, ovules 2, style 1, stigma ± spheric or disk-like. **FR**: drupe, gen ovoid to spheric. 200+ spp.: worldwide, esp n temp. (Greek: plum, prune) Seeds of many spp. ± TOXIC from production of hydrocyanic acid. Many cult for wood, orn, edible fr; some persisting near human habitations, some possibly naturalized (e.g., *P. laurocerasus* L.).

1. Infl an elongate raceme, fls 15–many
  2. Lvs deciduous, finely serrate; infl lfy at base; fr 6–14 mm . . . . ***P. virginiana*** var. ***demissa***
  2′ Lvs evergreen, entire or spiny-serrate; infl lfless at base; fr 12–25 mm . . . . ***P. ilicifolia***
    3. Lf margin spiny-serrate, ± wavy, blade widely ovate to round; petiole 3–10 mm; s NCoR, CW, SW (exc ChI) . . . . subsp. ***ilicifolia***
    3′ Lf margin gen entire, flat, blade gen ovate; petiole 8–25 mm; ChI . . . . subsp. ***lyonii***
1′ Infl a short, ± flat-topped raceme of (3)6–12 fls, or umbel-like to subsessile cluster of 2–5 fls, or fls 1
  4. Ovary, fr gen glabrous
    5. Infl a raceme; sepal entire; fr 7–14 mm; twigs with true terminal bud . . . . ***P. emarginata***
    5′ Infl an umbel-like cluster or fls 1; sepal gland-toothed to nearly entire; fr 15–30 mm; twigs with false terminal bud
      6. Lf blade elliptic to (ob)ovate, base obtuse; fls 1(2), appearing before lvs; tree, roadsides, streambanks, chaparral as waif . . . . ***P. cerasifera***
      6′ Lf blade oblong-ovate to ± round, base rounded to subcordate; fls 2–5, appearing with lvs; shrub, mixed-evergreen or conifer forest . . . . ***P. subcordata***
  4′ Ovary, fr densely puberulent or velvety
    7. Tree 3–10 m, not thorny; lf blade 25–150 mm; fr 25–80 mm; petals 10–25 mm
      8. Petiole (8)10–25 mm; lf blade 25–100 mm; petals pink to nearly white; fr pulp leathery . . . . ***P. dulcis***
      8′ Petiole 5–10(15) mm; lf blade (50)70–150 mm; petals dark pink; fr pulp fleshy . . . . ***P. persica***
    7′ Shrub < 4 m, often thorny; lf blade 5–30 mm; fr 7–18 mm; petals 1.4–12 mm
      9. Lf blade < 2 × longer than wide, base wedge-shaped to rounded or subcordate
        10. Lf blade ovate to obovate, ± hairy, base wedge-shaped to obtuse; fls gen unisexual; hypanthium, sepals densely puberulent abaxially . . . . ***P. eremophila***
        10′ Lf blade ovate to round, glabrous, base obtuse to rounded or subcordate; fls bisexual; hypanthium, sepals glabrous abaxially . . . . ***P. fremontii***
      9′ Lf blade ≥ 2 × longer than wide, base long-tapered
        11. Petals (5)8–11 mm, dark pink to nearly white; pedicel (1)4–12 mm; lf blade 2–6 mm wide, finely serrate . . . . ***P. andersonii***
        11′ Petals 1.4–4 mm, white to ± yellow; pedicel 0–4 mm; lf blade 1–2(4) mm wide, gen entire . . . . ***P. fasciculata***
          12. Lf puberulent; s SNF, Teh, s SCoRI, n TR, e PR, D . . . . var. ***fasciculata***
          12′ Lf glabrous to low-papillate; s CCo, s SCoRO . . . . var. ***punctata***

P. andersonii A. Gray (p. 1203) DESERT PEACH Shrub < 3 m, much-branched, thorny. **LF**: deciduous; petiole 0–7 mm; blade 9–30 mm, 2–6 mm wide, elliptic to oblanceolate, finely serrate, base long-tapered, tip gen acute. **INFL**: umbel-like cluster or not; fls 1–2; pedicels (1)4–12 mm. **FL**: sepals sparsely gland-toothed, ciliate; petals (5)8–11 mm, dark pink to nearly white. **FR**: 10–18 mm, obovoid to ± spheric, densely puberulent, green-yellow to red-orange; pulp dry, thin, often splitting to reveal stone. Rocky slopes, flats, scrub, pinyon/juniper woodland; 900–2600 m. SNH (e slope), GB, DMtns; NV. Mar–May

P. cerasifera Ehrh. CHERRY PLUM Tree 4–8 m, not thorny. **LF**: deciduous; petiole 5–20 mm; blade 30–70 mm, elliptic to (ob)ovate, crenate-serrate, base obtuse, tip acute to obtuse. **INFL**: (umbel-like clusters) or not; fls 1(2); pedicels (4)10–18 mm. **FL**: sepals glabrous, gland-toothed to nearly entire; petals 7–14 mm, white. **FR**: 15–30 mm, glabrous, yellow to red; pulp fleshy. 2*n*=16. Roadsides, streambanks, chaparral as waif; < 1000 m. s NCoR, n SNF, CCo, SnFrB; native to se Eur. Variants with red-pink fls, dark red to ± purple lvs, fr, commonly cult. Feb–Mar ❖

P. dulcis (Mill.) D.A. Webb ALMOND Tree 5–8 m, not thorny. **LF**: deciduous; petiole (8)10–25 mm; blade 25–100 mm, oblong to lanceolate, crenate-serrate, base obtuse, tip acuminate. **INFL**: subsessile cluster or not; fls 1–2; pedicels 1–5 mm. **FL**: sepals tomentose,

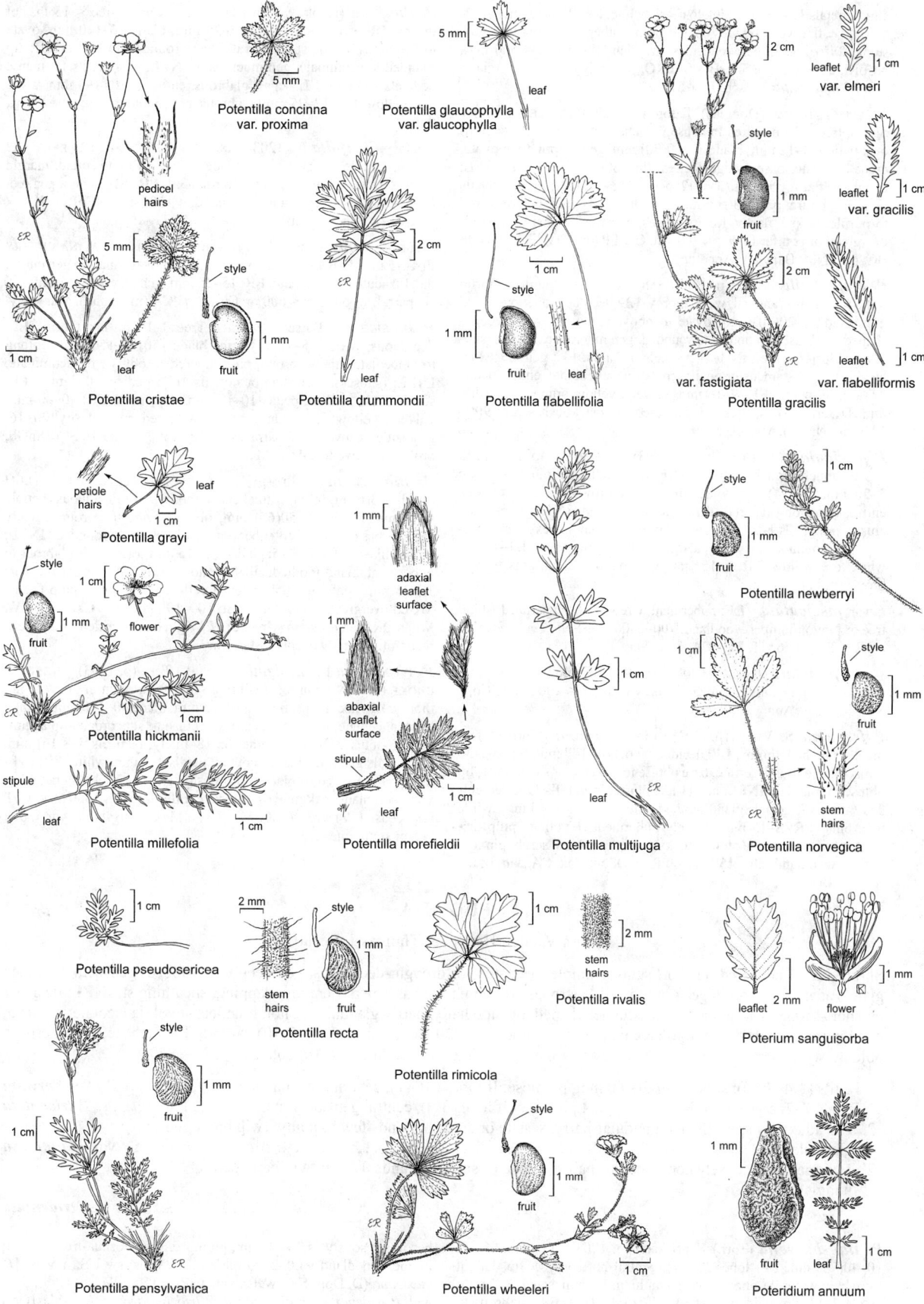

Potentilla concinna var. proxima

Potentilla glaucophylla var. glaucophylla

var. elmeri

var. gracilis

Potentilla cristae

Potentilla drummondii

Potentilla flabellifolia

var. fastigiata

var. flabelliformis

Potentilla gracilis

Potentilla grayi

Potentilla hickmanii

Potentilla newberryi

Potentilla millefolia

Potentilla morefieldii

Potentilla multijuga

Potentilla norvegica

Potentilla pseudosericea

Potentilla recta

Potentilla rivalis

Poterium sanguisorba

Potentilla rimicola

Potentilla pensylvanica

Potentilla wheeleri

Poteridium annuum

entire; petals 12–25 mm, pink to nearly white. **FR**: 25–40 mm, ovoid-oblong, ± flat, velvety, gray-green; pulp leathery, splitting to reveal stone. $2n=16$. Canyons, roadsides, grassland (as waif); < 500 m. s NCoRI, c SNF, GV, e SnFrB, s SCoRO; probably native to w Asia, n Afr. [*P. amygdalus* Batsch] Feb–Mar

P. emarginata (Douglas) Eaton (p. 1203) BITTER CHERRY Shrub, tree < 15 m, often in dense thickets, not thorny. **LF**: deciduous; petiole 3–12 mm; blade 15–60(80) mm, gen elliptic to obovate, crenate-serrate, base wedge-shaped, tip obtuse to rounded. **INFL**: raceme, ± flat-topped; fls (3)6–12; pedicels 3–12(18) mm. **FL**: sepals glabrous, entire; petals 3–8 mm, white. **FR**: 7–14 mm, glabrous, red to purple; pulp ± fleshy. Rocky slopes, canyons, chaparral, mixed-evergreen or conifer forest; < 3000 m. CA-FP (exc GV, ChI), GB; to BC, MT, NM, Baja CA. Apr–Jun

P. eremophila Prigge (p. 1203) MOJAVE DESERT PLUM Shrub < 2.5 m, much-branched, weak-thorny. **LF**: deciduous; petiole 0.5–5 mm; blade 5–20(30) mm, ovate to obovate, serrate, base wedge-shaped to obtuse, tip obtuse to rounded, gen mucronate. **INFL**: subsessile cluster or not; fls 1–2; pedicels 0–3 mm. **FL**: gen unisexual by abortion of stamens or pistil; sepals sparsely ciliate, entire; petals 2.5–6 mm, white. **FR**: 9–16 mm, velvety, yellow-orange; pulp dry, thin. Desert washes, rocky slopes, creosote-blackbush scrub; 900–1200 m. e DMoj. Mar–Apr ★

P. fasciculata (Torr.) A. Gray (p. 1203) DESERT ALMOND Shrub < 3 m, much-branched, thorny. **LF**: deciduous; petiole ± 0; blade 5–20 mm, 1–2(4) mm wide, linear to narrow-oblanceolate, gen entire, base long-tapered, tip acute to rounded. **INFL**: subsessile cluster or not; fls 1–2; pedicels 0–4 mm. **FL**: gen unisexual by abortion of stamens or pistil; sepals glabrous, entire; petals 1.4–4 mm, white to ± yellow. **FR**: 7–15 mm, densely puberulent, gray to red-brown; pulp dry, thin.

var. *fasciculata* **LF**: puberulent. Creosote-bush scrub, Joshua-tree or pinyon/juniper woodland; 700–2200 m. s SNF, Teh, s SCoRI, n TR, e PR, D; to UT, Baja CA. Mar–May

var. *punctata* Jeps. SAND ALMOND **LF**: glabrous to low-papillate. Sandy soils, scrubland, oak woodland; < 200 m. s CCo, s SCoRO. Mar–Apr ★

P. fremontii S. Watson (p. 1203) DESERT APRICOT Shrub < 4 m, much-branched, thorny. **LF**: deciduous; petiole 1–7 mm; blade 6–30 mm, ovate to round, serrate, base obtuse to rounded or subcordate, tip obtuse to rounded. **INFL**: umbel-like cluster or not; fls 1–3; pedicels 2–12 mm. **FL**: sepals gland-toothed, ciliate; petals 3–10 mm, white to ± pink. **FR**: 8–15 mm, densely puberulent, ± yellow; pulp dry, thin, splitting to reveal stone. Rocky slopes, canyons, scrub, pinyon/juniper woodland; 200–1500 m. e PR, w DSon; Baja CA. Jan–Mar

P. ilicifolia (Hook. & Arn.) D. Dietr. Shrub, tree < 15 m, not thorny. **LF**: evergreen; petiole 3–25 mm; blade 16–120 mm, ovate to round, entire or spiny-serrate, base rounded to subcordate, tip rounded to acuminate, gen mucronate. **INFL**: raceme; fls 15–many; pedicels 1–5 mm. **FL**: sepals glabrous, entire; petals 1–3 mm, white to ± yellow. **FR**: 12–25 mm, glabrous, red to blue-black; pulp fleshy, thin.

subsp. *ilicifolia* (p. 1203) ISLAY, HOLLY-LEAFED CHERRY Pl < 9 m. **LF**: petiole 3–10 mm; blade widely ovate to round, margin spiny-serrate, ± wavy, tip acute to rounded. **FR**: 12–18 mm, gen red. Canyons, slopes, scrubland, woodland; < 1600 m. s NCoR, CW, SW (exc ChI); Baja CA. Apr–May

subsp. *lyonii* (Eastw.) P.H. Raven (p. 1203) CATALINA CHERRY Pl 4–15 m. **LF**: petiole 8–25 mm; blade gen ovate, margin gen entire, flat, tip acute to acuminate. **FR**: 15–25 mm, gen blue-black. Canyons, chaparral, woodland; < 600 m. ChI; Baja CA (mainland). Mar–May

P. persica (L.) Batsch PEACH Tree 3–10 m, not thorny. **LF**: deciduous; petiole 5–10(15) mm; blade (50)70–150 mm, oblong to lanceolate, finely serrate, base tapered to obtuse, tip acuminate. **INFL**: (subsessile cluster) or not; fls 1(2); pedicels 0–3 mm. **FL**: sepals entire, ciliate; petals 10–17 mm, dark pink. **FR**: 40–80 mm, velvety, yellow to ± orange tinged with red; pulp fleshy. $2n=16$. Roadsides, canyons, chaparral as waif; < 1300 m. CaRH, SCo, SnGb; probably native to e Asia. Mar

P. subcordata Benth. (p. 1203) PACIFIC PLUM, SIERRA PLUM Shrub < 3(6) m, often in thickets, ± thorny. **LF**: deciduous; petiole 4–18 mm; blade 20–50(65) mm, oblong-ovate to ± round, finely serrate, base rounded to subcordate, tip obtuse to rounded. **INFL**: umbel-like cluster; fls 2–5; pedicels 5–15 mm. **FL**: sepals glabrous to puberulent, gland-toothed, ciliate or not; petals 5–10 mm, white. **FR**: 15–25 mm, glabrous (puberulent), yellow to dark red; pulp fleshy. Mixed-evergreen or conifer forest; 100–1900 m. NW, CaR, SN, CW, MP; s OR. Var. *oregana* (Greene) W. Wight (fr puberulent) may merit recognition; study needed. Mar–May

P. virginiana L. var. *demissa* (Nutt.) Torr. (p. 1203) WESTERN CHOKE CHERRY Shrub, small tree < 6(10) m, often in thickets, not thorny. **LF**: deciduous; petiole 10–25 mm; blade 30–100 mm, elliptic to oblanceolate, finely serrate, base obtuse to subcordate, tip acuminate to obtuse. **INFL**: raceme; fls 18–many; pedicels 5–8(16) mm. **FL**: sepals glabrous, gland-toothed; petals 4–7 mm, white. **FR**: 6–14 mm, glabrous, red to black; pulp fleshy. $2n=16$. Rocky slopes, canyons, scrubland, oak/pine woodland, conifer forest; < 3000 m. CA-FP (exc coast, GV), GB; to BC, MT, TX, n Mex. *P. virginiana* var. *v.* of e N.Am sometimes cult. May–Jun

PURSHIA

Brian Vanden Heuvel & Thomas J. Rosatti

Shrub. **LF**: ± clustered on short-shoots, simple, persistent or drought-deciduous, gen deeply 3–9-lobed, gen with ± sunken glands adaxially, margin gen not toothed, ± strongly rolled under; bases persistent, overlapping, sheathing st. **INFL**: fls gen 1 on short-shoots. **FL**: hypanthium ± funnel-shaped, outside hairy, partly glandular or not, bractlets small, lanceolate; sepals 5, overlapping; petals 5, white to cream [yellow]; stamens (15)20–80(125); pistils 1–7(10), simple. **FR**: achene, ± fusiform to oblong, styles persistent, ± hairy. 6 spp.: sw US, n Mex. (Frederick T. Pursh, N.Am botanist, 1774–1820)

1. Pistils (3)4–7(10); styles in fr 20–60 mm, plumose; lf lobes (3)5(7), central not spiny at tip . . . . *P. stansburyana*
1′ Pistils 1–2; styles in fr 5–7(10) mm, not plumose; lf lobes 3(5), central gen spiny at tip. . . . . *P. tridentata*
  2. Lvs adaxially sparsely nonglandular-hairy, sessile or sunken glands few to many; twig hairs gen glandular. . . . . var. *glandulosa*
  2′ Lvs adaxially densely nonglandular-hairy, sessile or sunken glands 0–few; twig hairs gen nonglandular
  . . . . var. *tridentata*

P. stansburyana (Torr.) Henrickson (p. 1203) CLIFFROSE Pl 10–40(75) dm. **LF**: lobes (3)5(7), central not spiny at tip, lateral from below middle, above. **FL**: hypanthium ± 5 mm; sepals 3–5 mm; petals 7–13 mm, widely ovate; pistils (3)4–7(10). **FR**: glabrous to becoming so; styles 20–60 mm, plumose. $n=9$. Joshua-tree, pinyon/juniper woodland; 900–2600 m. W&I, DMtns; sw US, n Mex. [*P. mexicana* (D. Don) S.L. Welsh var. *s.* (Torr.) S.L. Welsh] Hybridizes with *P. tridentata* var. *glandulosa*, *P. tridentata* var. *t.* Apr–May(Oct)

P. tridentata (Pursh) DC. (p. 1203) BITTERBRUSH Pl 2–25(40) dm. **LF:** lobes 3(5), central gen spiny at tip, lateral from gen above middle. **FL:** hypanthium ± 2.5–5 mm; sepals 2–4 mm; petals 4–8 mm, ± obovate; pistils 1–2. **FR:** canescent; style 5–7(10) mm, not plumose.

var. ***glandulosa*** (Curran) M.E. Jones (p. 1203) **ST:** twig hairs gen glandular. **LF:** adaxially sparsely nonglandular-hairy, sessile or sunken glands few to many. *n*=9. Chaparral at desert margins, Joshua-tree, pinyon/juniper woodland; 500–3505 m. c&s SNH (e slope), Teh, n TR, e edge PR, SNE, DMtns; NV, UT, AZ, Mex. Apr–Jun

var. ***tridentata*** (p. 1203) **ST:** twig hairs gen nonglandular. **LF:** adaxially densely nonglandular-hairy, sessile or sunken glands 0–few. *n*=9. Sagebrush scrub, pinyon/juniper woodland, conifer forest; 900–3400 m. KR, NCoRH, CaR, SNH (e slope), GB; to BC, MT, NM. Mar–Jul

PYRACANTHA

Peter F. Zika

Shrub, ± evergreen; thorns often lfy or branched. **LF:** simple, ± evergreen, margin gen rolled under, unlobed, entire to toothed; stipules early-deciduous; petiole short. **INFL:** raceme or panicle at ends of short-shoots; pedicel bractlets 0. **FL:** hypanthium bractlets 0; sepals persistent; petals white; stamens 20, fused at base; ovary 1/2-inferior, chambers 5. **FR:** pome, drupe-like, ± spheric or depressed-spheric, open at top, stones 5, free, 2-ovuled. 10 spp.: Medit, Asia. (Greek: fire thorn, from fr color, thorns) Fruiting Jun–Apr.

1. Lf gen serrate to crenate over much of margin
  2. Lf tip obtuse or ± notched, oblong-obovate to obovate; pedicels glabrous or sparsely hairy in fl; stones ± black . . . . ***P. fortuneana***
  2′ Lf tip gen acute, gen oblong-elliptic to lanceolate, elliptic, or ovate (rarely oblanceolate in some *Pyracantha coccinea*); pedicels gen hairy in fl; stones brown
    3. Hypanthium hairy in fl; new growth gray-hairy; lvs on vigorous shoots gen elliptic . . . . [***P. coccinea***]
    3′ Hypanthium glabrous in fl; new growth rusty-hairy; lvs on vigorous shoots narrow-oblong-elliptic . . . [***P. crenulata***]
1′ Lf gen entire or with a few well separated low or rounded teeth toward tip
  4. Sepals in fr, young lvs densely gray-hairy abaxially; lvs gen narrowly oblong to ± oblanceolate . . . . ***P. angustifolia***
  4′ Sepals in fr, young lvs ± glabrous or ± rusty- or yellow-brown-hairy abaxially; lvs narrowly oblanceolate to elliptic
    5. Lvs gen widest near middle, tips gen obtuse or abruptly soft-pointed; petals gen ovate, 4–5 mm . . . [***P. atalantioides***]
    5′ Lvs gen widest above middle, tips gen truncate or ± notched; petals gen round or widely elliptic, 3–4 mm . . . . ***P. koidzumii***

P. angustifolia (Franch.) C.K. Schneid. SLENDER OR WOOLLY FIRETHORN Pl < 4 m, gen erect, ± gray-hairy. **LF:** 25–50 mm, gen narrowly oblong to ± oblanceolate, hairs gen thinning in age, teeth 0. **INFL:** ± several-fld, densely gray-hairy. **FR:** 4–6 mm wide, red. Disturbed areas, fencerows, abandoned fields, roadsides; < 200 m. NCoRO, CCo, expected elsewhere; native to China. Feb–Jun ❖

P. fortuneana (Maxim.) H.L. Li CHINESE FIRETHORN Pl < 4 m, erect, gen rusty-hairy, glabrous in age. **LF:** 15–60 mm, oblong-obovate to obovate, glabrous or ± hairy abaxially, glabrous in age, not glaucous, tip obtuse or ± notched, less often abruptly soft-pointed, teeth short, many. **INFL:** ± many-fld, glabrous or sparsely ± brown-hairy. **FR:** 3–6 mm wide, orange-red to dark red. Disturbed ground, canyons, riparian areas; < 1500 m. KR, NCoRO, ScV, CW (exc SCoRI), SCo, SnGb, SnBr, expected elsewhere; to WA; se US; native to China. Feb–Jun

P. koidzumii (Hayata) Rehder TAIWAN FIRETHORN Pl < 4 m, erect, rusty-hairy, glabrous in age. **LF:** 25–45 mm, gen narrow-elliptic to -obovate, abaxially rusty-hairy, glabrous in age, glaucous, tip gen truncate or ± notched, teeth 0–few, gen distal, inconspicuous. **INFL:** ± many-fld, sparsely brown-hairy. **FR:** 4–7 mm wide, orange-red. Disturbed forest, beach bluffs, riparian areas; < 1500 m. ScV, CCo, SnFrB, SCo, SnGb, SnBr, DSon, expected elsewhere; AZ, se US; native to Taiwan. Feb–Jun

PYRUS PEAR

Michael A. Vincent

Tree [(shrub)], thorny or not. **LF:** simple, toothed (entire). **INFL:** few-fld clusters at ends of short-shoots; pedicel bractlets gen 2–3, deciduous. **FL:** hypanthium bractlets 0; stamens 20–30; ovary inferior, chambers 2–5, 2-ovuled, styles 2–5, ± free. **FR:** pome, gen ± obovoid; flesh gritty from stone cells. ± 25 spp.: n temp. (Latin: pear) *P. calleryana* Dcne. (callery pear) possibly naturalized in CA.

P. communis L. COMMON PEAR Pl ± thorny (escaped pls). **LF:** 2–7 cm, ovate to wide-ovate or -elliptic, hairy in youth, glabrous in age, teeth rounded; petiole 22–45 mm. **FL:** petals 10–15 mm, white; odorous. **FR:** 3–15 cm, 2–12 cm diam, calyx persistent. 2*n*=34. Disturbed places; < 1600 m. NCoRO, n SNF, ScV, TR, PR, W&I, DMtns; native to Eurasia. Feb–Apr

ROSA

Barbara Ertter

Shrub to vine, often thicket-forming, gen prickly. **LF:** gen odd-pinnately compound; stipules gen attached to petiole, gen gland-margined. **INFL:** gen ± cyme or fls 1; pedicel bractlets 0. **FL:** hypanthium urn-shaped, bractlets 0; sepals often with long expanded tip; petals gen 5 (exc cult), gen pink in CA (white to red or yellow); stamens gen > 20; pistils gen many, ovaries superior, styles attached at tip, gen hairy. **FR:** bony achenes gen enclosed in fleshy, gen ± red hypanthium (hip). 100+ spp.: gen n temp. (Latin: ancient name) [Ertter & Lewis 2008 Madroño 55:170–177] Spp. hybridize freely; other non-natives established locally. FNANM treatment by Lewis & Ertter uses both subspp. and vars., the latter mostly reserved for localized variants within a subsp.

1. Lflets << 1 cm, toothed ± 1/2 to base; hypanthium densely prickly — PR . . . . **R. minutifolia**
1′ Lflets gen > 1 cm, toothed < 1/4 to base; hypanthium glabrous to stalked-glandular
  2. Sepals with toothed lateral lobes; prickles compressed side-to-side, curved
    3. Hypanthium 1–2 mm wide at fl, 5–7 mm wide at fr; pistils gen < 10 . . . . **R. multiflora**
    3′ Hypanthium 4–5 mm wide at fl, 10–20 mm wide at fr; pistils > 10
      4. Lvs and sepals ± glandless; petals white to pink; terminal lflet ± ovate . . . . **R. canina**
      4′ Lvs and sepals glandular; petals pink; terminal lflet gen elliptic to ± widely obovate . . . . **R. rubiginosa**
  2′ Sepals gen entire (tip sometimes toothed); prickles compressed or slender, straight or curved
    5. Sepals deciduous in fr; hypanthium 1.5–2 mm wide at fl; pistils 5–10; sepal tip gen << body; petals ± 10
      mm — lflets glabrous, fls 1–3; pedicels gen stalked-glandular . . . . **R. gymnocarpa**
      6. Lflets gen (5)7–9; terminal lflet tip gen ± obtuse, ± 10–30 mm; pedicels ± 15–30 mm; pls to ± 5–20 dm,
        gen in shade on non-ultramafic substrates; widespread . . . . var. **gymnocarpa**
      6′ Lflets gen 5(7); terminal lflet tip widely obtuse to rounded, 4–20 mm; pedicels ± 10–15 mm; pls to gen
        3–6 dm, full sun on ultramafic substrates; KR . . . . var. **serpentina**
    5′ Sepals persistent; hypanthium 2.5–7 mm wide at fl; pistils gen > 10; sepal tip ± = or often > body;
      petals 10–25 mm
      7. Dwarf, openly rhizomed shrubs, gen 1–5(10) dm; lflet tip often ± truncate — lf margins gen ±
        glandular, gen double-toothed
        8. Hypanthium stalked-glandular — prickles gen ± slender, often many; NW, CW, MP(1 site),
          150–1550(1950) m . . . . **R. spithamea**
        8′ Hypanthium gen glandless, rarely sparsely glandular
          9. Prickles few, paired, ± thick-based; terminal lflet gen widely obovate; fl 1–2(7); CaR, SN,
            700–2500 m . . . . **R. bridgesii**
          9′ Prickles gen many, gen not paired, both slender and ± thick-based; terminal lflet gen ± elliptic;
            fls gen 1–5; CW, gen < 300 m . . . . **R. pinetorum**
      7′ Open or thicket-forming shrubs, gen > 5 dm; lflet tip not truncate
        10. Terminal lflet ± obovate-elliptic, widest at or above middle; lf axis finely velvety, hairs ± 0.1 mm
          (glabrous), glandless; prickles slender to ± thick-based, gen ± straight; sepals glandless (in CA);
          CaRH, SNH, Teh, TR, GB, DMtns . . . . **R. woodsii**
          11. Prickles gen many, 2–10(13) mm, often ± thick-based, internodal prickles gen present in infl;
            c&s SNH, Teh, TR, SNE, DMtns . . . . subsp. **gratissima**
          11′ Prickles few to many, 2–7 mm, slender, internodal prickles gen 0 in infl; CaRH, n SNH, MP,
            n SNE . . . . subsp. **ultramontana**
        10′ Terminal lflet gen ovate to elliptic, gen widest at or below middle; lf axis variously hairy, hairs to 1
          mm, sometimes glandular; prickles gen ± thick-based, often curved; sepals often glandular; CA-FP,
          adjacent DMoj
          12. Fls gen 1(6); hypanthium gen 5–7 mm wide at fl, (10)13–20 mm wide at fr, gen ± spheric; achenes
            gen 4.5–6 mm — prickles to 20 mm, gen ± compressed, ± straight to ± curved; sepal tips gen >
            body, toothed . . . . **R. nutkana**
            13. Lflets gen ± single-toothed, gen glandless; prickles gen few; NCoRO, KR, CaRH; 750–1500 m
              . . . . subsp. **macdougalii**
            13′ Lflets double-toothed, ± glandular esp beneath; prickles gen many; NCo, NCoRO, CCo;
             < 700 m . . . . subsp. **nutkana**
          12′ Fls 1–many; hypanthium 2.5–5.5 mm wide at fl, 7–15(20) mm wide at fr, spheric to (ob)ovoid;
            achenes 3–5 mm
            14. Prickles gen thick-based, gen curved (straight), few to many; lflets and lf axis gen ± shaggy-hairy,
              hairs to 1 mm; pedicel gen ± hairy; hypanthium gen (ob)ovoid, glabrous to sparsely hairy, neck
              2–4.5 mm wide at fl . . . . **R. californica**
            14′ Prickles slender to ± thick-based, straight (± curved), 0–few; lflets and lf axis ± glabrous to ±
              hairy, hairs 0.1–1 mm; pedicels gen glabrous (exc for glands); hypanthium spheric to ovoid,
              glabrous, neck ± 2 mm wide at fl . . . . **R. pisocarpa**
              15. Fls 1–3(10); sepals gen glandless; prickles 0 to paired at nodes; lflets gen 5(7) . . . . subsp. **ahartii**
              15′ Fls (1)3–10; sepals gen glandular; prickles gen paired at nodes; lflets gen (5)7(9) . . . . subsp. **pisocarpa**

R. bridgesii Rydb. (p. 1203) SIERRAN DWARF ROSE Dwarf shrub, openly rhizomed, gen 1–4(8) dm. **ST:** prickles few, gen paired, 3–10 mm, ± thick-based, straight. **LF:** axis glabrous to finely hairy, glandular; lflets gen 5–7, ± hairy, glandular; terminal lflet gen 10–30(50) mm, gen widely obovate (elliptic), widest above middle, tip gen ± truncate, margins double-toothed, glandular. **INFL:** 1–2(7)-fld; pedicels 4–17 mm, glabrous and glandless to ± glandular. **FL:** hypanthium ± 3–4 mm wide at fl, glabrous, gen glandless, neck 1.5–3 mm wide; sepals glandular, margins entire, tip gen < body, entire; petals 10–20 mm, pink to red; pistils gen 10–30. **FR:** 7–14 mm wide, ± ovoid; sepals erect, persistent; achenes 4–6.5 mm. 2*n*=14,28. Open forest, rocky areas; 700–2500 m. CaR, SN; s OR. May–Aug

R. californica Cham. & Schltdl. (p. 1203) CALIFORNIA ROSE Shrub or thicket-forming, 8–25 dm. **ST:** prickles few to many, paired or not, 3–15 mm, thick-based and compressed, gen curved (straight). **LF:** axis ± shaggy-hairy (± glabrous), hairs to 1 mm, glandless or glandular; lflets 5–7(9), ± hairy, sometimes glandular; terminal lflet gen 15–50 mm, ± ovate-elliptic, gen widest at or below middle, tip rounded to acute, margins single- or double-toothed, glandular or not. **INFL:** (1)3–30(50)-fld; pedicels gen ± 5–20 mm, gen ± hairy, glandless. **FL:** hypanthium 3–5.5 mm wide at fl, glabrous to sparsely hairy, glandless, neck 2–4.5 mm wide; sepals glandular or not, entire, tip gen ± = body, entire; petals gen 15–25 mm, pink; pistils 20–40. **FR:** gen 8–15(20) mm wide, gen (ob)ovoid; sepals gen erect, persistent;

achenes gen 3.5–4.5 mm. *n*=14. Gen ± moist areas, esp streambanks; < 1800 m. CA-FP (exc CaRH, SNH, Teh); s OR, n Baja CA. Variable; needs study. Feb–Nov

R. canina L. DOG ROSE Shrub or thicket-forming, gen 8–40 dm. **ST:** prickles ± few, gen not paired, 3–10 mm, thick-based and compressed, curved. **LF:** axis glabrous or with hairs to 1 mm, glandless; lflets gen 5–7, (±) glabrous; terminal lflet gen 15–40 mm, ± ovate, widest below middle, tip ± acute, margins ± single-toothed, sparsely glandular. **INFL:** gen 1–5-fld; pedicels gen 10–20 mm, gen glabrous, glandless. **FL:** hypanthium gen 4–5 mm wide at fl, glabrous, glandless, neck 2–3 mm wide; sepals ± glandless, margins with toothed lateral lobes, tip gen ± = body, ± toothed; petals 15–30 mm, white to pale pink; pistils 25–40. **FR:** 10–20 mm wide; ± ellipsoid; sepals reflexed, unevenly deciduous; achenes 4.5–6 mm. 2*n*=35. Gen ± dry open areas; 100–1500 m. NW, n SN; to WA, ID, UT, also e US; native to Eurasia. May–Jul

R. gymnocarpa Nutt. WOOD ROSE Loose shrub. **ST:** prickles few to many, gen not paired (exc SnFrB), 2–8 mm, ± slender, straight. **LF:** axis gen glabrous ± glandular; lflets glabrous; terminal lflet margins ± double-toothed, glandular. **FL:** hypanthium 1.5–2 mm wide at fl, glabrous and glandless, neck ± 1.5 mm wide; sepals glandular or not, entire, tip gen << body, entire; petals ± 10 mm, pink to red; pistils 5–10. **FR:** 4–12 mm wide, ellipsoid to ± spheric; sepals erect to reflexed, evenly deciduous; achenes (3)4–7 mm. *n*=7.

var. ***gymnocarpa*** (p. 1203) Pl gen ± 5–20 dm. **LF:** lflets gen (5)7–9; terminal lflet ± 10–30 mm, elliptic to (ob)ovate, tip ± obtuse. **INFL:** pedicels gen ± 15–30 mm, gen stalked-glandular. **FR:** 5–12 mm wide, ellipsoid to ± spheric; achenes (1)4–10. *n*=7. Common. Gen in shade of forest, scrub, gen not ultramafic substrates; 30–2000 m. NW, CaR, n&c SN, CW, PR, MP; to BC, MT. (Feb)Apr–Jul

var. ***serpentina*** Ertter & W.H. Lewis GASQUET ROSE Pl (1)3–6(13) dm. **LF:** lflets gen 5(7); terminal lflet 4–20 mm, widely elliptic to ± round, tip widely obtuse to rounded. **INFL:** pedicels ± 10–15 mm, stalked-glandular or not. **FR:** 4–8 mm wide, irregularly ovoid to ellipsoid; achenes 1–4. Full sun in chaparral, dwarf forest on ultramafic substrates; 400–1500 m. KR; s OR. Apr–Jun

R. minutifolia Engelm. (p. 1203) SMALL-LEAVED ROSE Dense shrub or thicket-forming, ± 3–10 dm. **ST:** prickles many, gen not paired, 2–12 mm, slender, straight. **LF:** axis finely short-hairy, sparsely glandular; lflets 5–7, hairy; terminal lflet ± 3–6 mm, ± round, widest near middle, tip ± obtuse, margins toothed ± 1/2 to midvein, ± glandular. **INFL:** gen 1-fld; pedicels ± 2–10 mm, hairy, glandless. **FL:** hypanthium ± 3 mm wide at fl, densely prickly, neck ± 2 mm wide; sepals glandless, with toothed lateral lobes, tip ± = body, toothed; petals ± 10–20 mm, dark pink; pistils gen ± 10. **FR:** ± 5 mm wide, ± spheric; sepals erect to spreading, persistent; achenes unknown. *n*=7. Chaparral; ± 160 m. s PR (Otay Mesa); n Baja CA. Feb–Apr ★

R. multiflora Thunb. MULTIFLORA ROSA Shrub, thicket-forming, or climbing, 15–30(75) dm. **ST:** prickles gen ± few, gen not paired, ± 4–6 mm, thick-based and compressed, curved. **LF:** axis ± hairy, sparsely glandular; lflets gen 7–9, ± hairy and glandular; terminal lflet 10–45 mm, elliptic to obovate, widest above middle, tip acute to acuminate, margins gen single-toothed, glandless. **INFL:** gen 5–30-fld; pedicels 5–15 mm, sparsely hairy and/or glandular. **FL:** hypanthium 1–2 mm wide at fl, glabrous to ± hairy, glandular or not, neck 1.5–2 mm wide; sepals gen glandular, with toothed lateral lobes, tip < body, entire; petals 7–13 mm, gen white; pistils gen < 10. **FR:** 5–7 mm wide; ovoid to spheric; sepals reflexed, unevenly deciduous; achenes 3.5–5 mm. 2*n*=14. Gen ± disturbed open sites; 20–700 m. NCoRO, CaR, ScV, SnGb; widely naturalized in N.Am; native to e Asia. Apr–Jun

R. nutkana C. Presl Shrub or thicket-forming, gen 5–20 dm. **ST:** prickles paired or not, 10–20 mm, gen ± compressed and thick-based, ± straight to ± curved. **LF:** axis ± hairy, glandular; lflets gen 5–7, sparsely hairy; terminal lflet ± 15–50(60) mm, ± wide-elliptic to -ovate, widest at or below middle, tip ± obtuse. **INFL:** gen 1(6)-fld; pedicels gen ± 10–20 mm, variously glabrous, hairy, and/or glandular. **FL:** hypanthium gen 5–7 mm wide at fl, glabrous, glandular or

not, neck 3–6 mm wide; sepals gen glandular, entire, tip gen > body, toothed; petals 15–25 mm, pink; pistils gen 30–60. **FR:** (10)13–20 mm wide, gen ± spheric; sepals gen erect, ± persistent; achenes gen 4.5–6 mm. *n*=21.

subsp. ***macdougalii*** (Holz.) Piper **ST:** ± openly branched; prickles gen few. **LF:** lflets gen glandless, margins gen ± single-toothed. Gen ± moist flats; 750–1500 m. KR, NCoRO, CaRH; to BC, MT, WY, CO. May–Jul

subsp. ***nutkana*** (p. 1203) NOOTKA ROSE **ST:** densely branched; prickles gen many. **LF:** lflets ± glandular, esp beneath, margins double-toothed. Gen ± moist flats; < 700 m. NCo, NCoRO, CCo; coastal to AK. [*R. n.* var. *n.*] CCo pls may be mostly hybrids with other spp. Apr–Jul

R. pinetorum A. Heller PINE ROSE Dwarf shrub, openly rhizomed, gen < 10 dm. **ST:** prickles gen many, gen not paired, 3–10 mm, both slender and ± thick-based, straight. **LF:** axis glabrous or finely hairy, glandular; lflets 5–7, glabrous to hairy; terminal lflet 10–30 mm, gen ± elliptic, widest near middle, tip ± obtuse, margins ± single- or double-toothed, ± glandular. **INFL:** gen 1–5-fld; pedicels gen 10–30 mm, glabrous, glandular or not. **FL:** hypanthium gen ± 4 mm wide at fl, glabrous, glandless, neck ± 3 mm wide; sepals gen ± glandular, entire, tip gen ± = body, entire or toothed; petals ± 15–20 mm, pink; pistils ± 10–20. **FR:** ± 12 mm wide, spheric; sepals ± erect, persistent; achenes 3–4 mm. 2*n*=14,21. Pine woodland; gen < 300 m. w-c CW. Hybrids of *R. gymnocarpa* × *R. spithamea* also key here. May–Jun ★

R. pisocarpa A. Gray CLUSTER ROSE Shrub or thicket-forming. **ST:** prickles 0–few, ± thick-based to slender, straight (± curved). **LF:** axis ± glabrous to ± hairy, hairs 0.1–1 mm, glandless or sparsely glandular; lflets sparsely hairy (± glabrous); terminal lflet ± ovate-elliptic, widest at or below middle, tip ± obtuse, margins gen single-toothed, glandless. **INFL:** pedicels ± 10–20 mm, gen glabrous. **FL:** hypanthium gen 2.5–4 mm wide at fl, glabrous, glandless, neck ± 2 mm wide; sepal margin entire, tip gen > body, entire; petals 12–18 mm, pink; pistils 20–30. **FR:** 7–13 mm wide, spheric to ovoid; sepals erect, persistent; achenes 3–4.5 mm.

subsp. ***ahartii*** Ertter & W.H. Lewis AHART ROSE Pl gen 4–15 dm. **ST:** prickles 0, single or paired at nodes, often 0, 2–5 mm. **LF:** lflets gen 5(7); terminal lflet gen 20–45(60) mm. **FL:** 1–3(10); sepals gen glandless. **FR:** 8–13 mm wide, ± ovoid, neck 2.5–3.5 mm wide. Gen ± moist areas; 150–1700 m. CaR, n SN; s OR. May–Aug

subsp. ***pisocarpa*** (p. 1203) CLUSTER OR PEA ROSE Pl gen 10–25 dm. **ST:** prickles gen paired at nodes, 2–10 mm. **LF:** lflets gen (5)7(9); terminal lflet 15–35 mm. **FL:** (1)3–10; sepals gen glandular. **FR:** 7–10 mm wide, spheric, neck 1.5–3 mm wide. *n*=7. Gen ± moist areas; 30–2100 m. NW, CaR; to BC. May–Aug

R. rubiginosa L. (p. 1203) SWEET-BRIER Shrub or thicket-forming, 8–30 dm. **ST:** prickles gen few, gen not paired, ± 5–15 mm, thick-based, compressed, curved. **LF:** axis ± hairy, hairs 0.2–1 mm, glandular; lflets gen 5–7, ± hairy, glandular; terminal lflet gen 10–35 mm, elliptic to ± widely obovate, widest near middle, tip obtuse, margins double-toothed, glandular. **INFL:** gen 1–8-fld; pedicels ± 10 mm, glabrous, stalked-glandular. **FL:** hypanthium gen 4–5 mm wide at fl, ± glabrous, glandless to sparsely stalked-glandular, neck ± 3 mm wide; sepals glandular, lateral lobes toothed, tip ± = body, gen toothed; petals gen 10–20 mm, pink; pistils 20–45. **FR:** 10–18 mm wide; ± ellipsoid; sepals spreading to reflexed, unevenly deciduous; achenes 4–5 mm. 2*n*=35. Gen ± dry, often disturbed open sites; 30–1400 m. NW (exc NCoRH), CaR, n&c SN, Teh, CCo, SnFrB; to e N.Am; native to Eur. [*R. eglanteria* L., nom. rej.] May–Aug

R. spithamea S. Watson (p. 1203) COAST GROUND ROSE Dwarf shrub, openly rhizomed, gen < 5 dm. **ST:** prickles few to many, gen not paired, 3–8(12) mm, gen slender (thick-based), ± straight. **LF:** axis gen glabrous (finely hairy), glandular; lflets 5–7(9), 2–4 per side, (±) glabrous; terminal lflet ± 10–30 mm, ± widely elliptic (obovate), widest near middle, tip obtuse to truncate, margins ± double-toothed, glandular. **INFL:** 1–10-fld; pedicels gen 5–15 mm, glabrous, ± stalked-glandular. **FL:** hypanthium gen 4–5 mm wide at fl, stalked-glandular, neck 3–4 mm wide; sepals gen glandular, entire,

tip gen ± = body, entire; petals 10–15 mm, pink to red; pistils 10–20. **FR:** 7–12(15) mm wide, ± spheric; sepals ± erect, persistent; achenes 3.5–5 mm. 2*n*=28. Open forest, chaparral, esp after fire; gen 150–1550(1950) m. NW, CW, MP; OR. Gen blooms after fires. Pls in s CW with larger prickles, described as *R. granulata* Greene, may be hybrids with *R. californica*; study needed. Apr–Aug

R. woodsii Lindl.   Shrub, open or thicket-forming, gen 5–30 dm. **ST:** prickles paired or not, gen ± straight (± curved) (in CA). **LF:** axis finely velvety (glabrous), hairs ± 0.1 mm, glandless; lflets 5–7, (±) glabrous; terminal lflet 10–40 mm, ± obovate-elliptic, widest at or above middle, tip ± obtuse, margins single-toothed, glandless. **INFL:** 1–12-fld; pedicels gen 10–20 mm, ± glabrous, glandless. **FL:** hypanthium gen 3–5 mm wide at fl, glabrous, glandless, neck 2–4 mm wide; sepals glandless (in CA), gen entire (or with simple, linear lobes), tip ± = body, entire; petals gen 15–20 mm, pink; pistils 20–35. **FR:**

gen 9–12 mm wide; sepals gen erect, persistent; achenes 3–4 mm. Yosemite Valley pls ambiguous.

subsp. **gratissima** (Greene) W.H. Lewis & Ertter (p. 1203) MOJAVE ROSE   **ST:** ± densely branched; prickles gen many with internodal prickles gen present in infl, 2–10(13) mm, often ± thick-based, sometimes slender. **FL:** gen 1–3. **FR:** ovoid to spheric. 2*n*=14. Gen ± moist areas; 800–3400 m. c&s SNH, Teh, TR, SNE, DMtns; NV. May–Aug

subsp. **ultramontana** (S. Watson) Roy L. Taylor & MacBryde (p. 1203)   INTERIOR ROSE   **ST:** openly branched; prickles few to many with internodal prickles gen 0 in infl, 2–7 mm, slender. **FL:** 1–12. **FR:** gen ± ovoid. 2*n*=14. Gen ± moist areas; 1000–2500 m. CaRH, n SNH, MP, n SNE; to BC, MT, NV. [*R. w.* var. *u.* (S. Watson) Jeps.] Most CaRH reports referable to *R. pisocarpa* subsp. *ahartii.* May–Jul

RUBUS

Lawrence A. Alice

Gen shrub; (dioecious). **ST:** persisting 1–2 yrs, rooting at tips and/or nodes or not, erect or arched to mounded or prostrate, 5-angled or not, hairy or glabrous, glaucous or not, stalked glands present or not; bristles or prickles 0–many, prickles stout and wide-based or weak and slender, straight or curved. **LF:** simple, palmately lobed, to palmately compound, lflets 3 or 5(11), toothed, abaxially ± glabrous to densely hairy; stipules thread-like to ovate or elliptic. **INFL:** raceme- or panicle-like cyme, axillary or terminal; pedicel bractlets 0. **FL:** gen bisexual; hypanthium flat to saucer-shaped, bractlets 0; sepals persistent, reflexed to ascending, ovate or lance-ovate, hairy or glabrous, stalked or sessile glands present or not, tip pointed, prickly or not; petals widely obovate, spoon-shaped, or elliptic, white to ± pink or magenta; stamens gen >> 20, filaments thread- or strap-like; pistils 5–150, receptacle flat or convex to conical, spongy, gen elongated in fr, ovaries superior, hairy or glabrous, styles long, slender or short, thick, glabrous or hairy; ovules 2, 1 maturing. **FR:** fleshy-coated achenes, aggregate of few to many, yellow, orange, red, or black, gen falling as unit, separating with (blackberry-type) or without (raspberry-type) receptacle attached. 400–750 spp.: worldwide exc Antarctica, esp n temp. (Latin: red; ancient name for bramble, blackberry)

1. Stipules triangular-lanceolate to ovate or elliptic, 1.5–4(10) mm wide
  2. Prickles small, ± wide-based, curved . . . . **R. nivalis**
  2′ Prickles 0
    3. Sts prostrate; lf blade < 4 cm; pistils < 15; styles long, slender . . . . **R. lasiococcus**
    3′ Sts erect; lf blade > 5 cm; pistils > 30; styles short, thick . . . . **R. parviflorus**
1′ Stipules thread-like to linear, ≤ 1 mm wide
  4. Sts not angled; fls gen ≤ 10; fr raspberry-type (blackberry-type in *Rubus ursinus*)
    5. Sts erect, not glaucous; petals magenta. . . . . **R. spectabilis**
    5′ Sts prostrate to decumbent or arched to mounded, glaucous; petals white or ± pink
      6. Lvs gray-hairy abaxially; fls gen unisexual; ovaries glabrous or hairy . . . . **R. ursinus**
      6′ Lvs densely white-tomentose abaxially; fls bisexual; ovaries densely hairy
        7. Prickles weak, straight or ± curved; pistils < 10 . . . . **R. glaucifolius**
        7′ Prickles stout, straight or gen curved; pistils gen > 30. . . . . **R. leucodermis**
  4′ Sts 5-angled; fls gen ≥ 10 (as few as 5 in *Rubus pensilvanicus*); fr blackberry-type
    8. Prickles 0 . . . . **R. ulmifolius** var. **anoplothyrsus**
    8′ Prickles many, stout, wide-based
      9. Lvs compound, lflets deeply dissected . . . . **R. laciniatus**
      9′ Lvs simple or compound, lflets 0 or not dissected
        10. Lvs gray-hairy abaxially; infl a raceme-like cyme . . . . **R. pensilvanicus**
        10′ Lvs densely white-tomentose abaxially; infl a panicle-like cyme
          11. Sts not glaucous; terminal lflets of 1st-yr sts gen wide-elliptic to -obovate, unevenly coarse-toothed
            . . . . **R. armeniacus**
          11′ Sts glaucous; terminal lflets of 1st-yr sts oblong to narrow-obovate, evenly finely toothed
            . . . . **R. ulmifolius** var. **ulmifolius**

R. armeniacus Focke   HIMALAYAN BLACKBERRY   Pl to 3 m, arched to mounded; prickles many, stout, wide-based, straight or curved. **ST:** to 20(25) mm diam, 5-angled, finely hairy or gen glabrous, not glaucous, persisting 2 yrs, rooting at tips. **LF:** 1st-yr st lvs compound, lflets (3)5, terminal gen wide-elliptic to -obovate, unevenly coarse-double-toothed, tip abruptly pointed, abaxially densely white-tomentose; fl st lvs simple or compound, lflets 3(5); stipules ≤ 1 mm wide, thread-like to linear. **INFL:** panicle-like cyme, terminal, fls many. **FL:** sepals hairy, nonglandular; petals 10–15 mm, obovate, white to pink; filaments thread-like; pistils > 30, styles long,

slender, ovaries glabrous. **FR:** blackberry-type, black. 2*n*=28. Common. Disturbed areas, roadsides; < 1600 m. CA-FP; to BC, w N.Am, e N.Am; native to Eurasia. [*R. discolor* Weihe & Nees, misappl.] Mar–Jun ❖

R. glaucifolius Kellogg (p. 1207)   WAXLEAF RASPBERRY   Pl prostrate to decumbent; prickles few, weak, slender, straight or ± curved. **ST:** 2–3 mm diam, not angled, glabrous, ± glaucous, persisting 2 yrs, not rooting at nodes, tips. **LF:** compound, lflets 3, terminal ovate to elliptic, unlobed or shallow-3-lobed, coarse-toothed, tip acute to rounded, abaxially densely white-tomentose; stipules ≤ 1

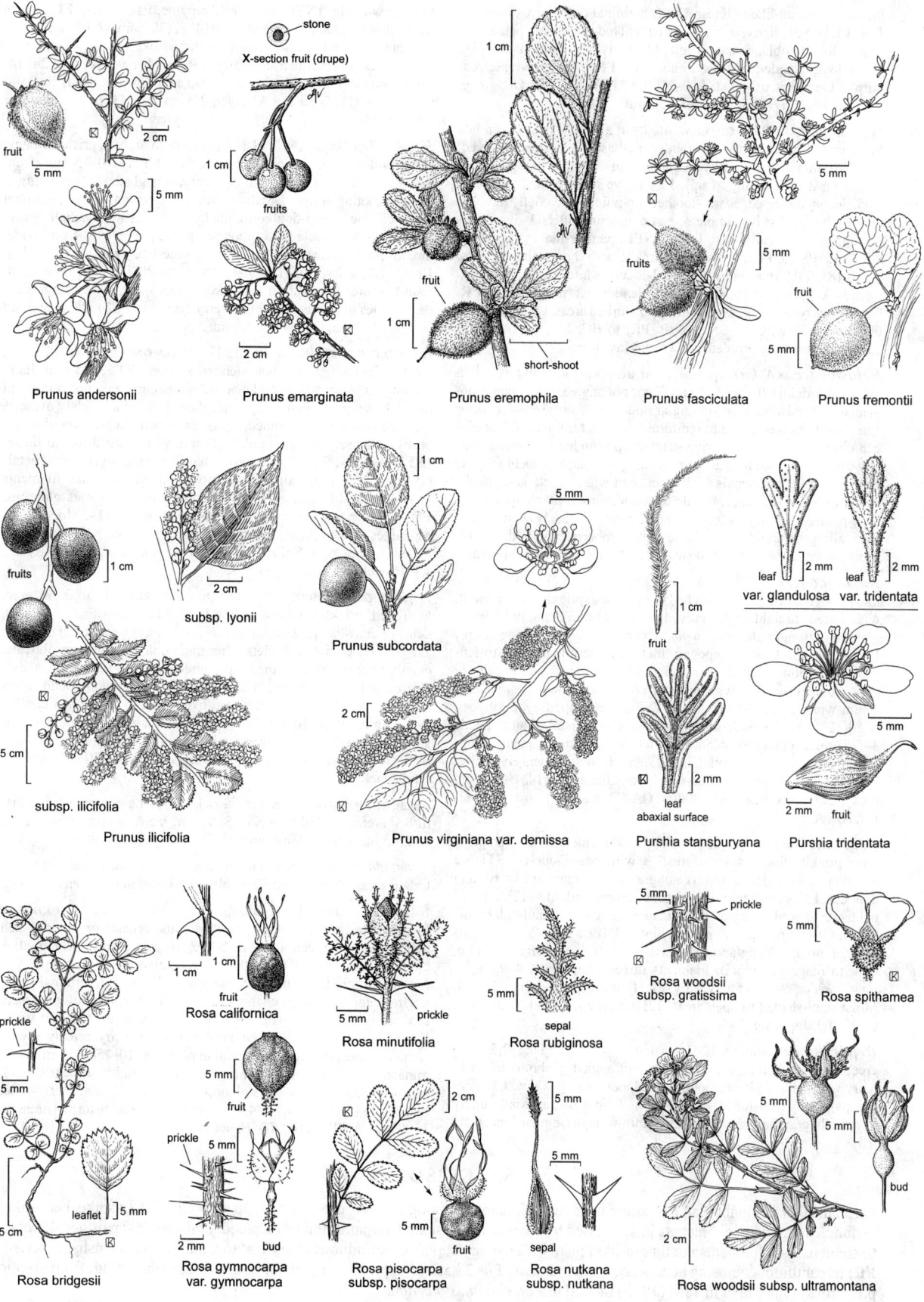

X-section fruit (drupe)
stone
fruits

Prunus andersonii

Prunus emarginata

fruit
short-shoot

Prunus eremophila

fruits

Prunus fasciculata

fruit

Prunus fremontii

fruits

subsp. lyonii

subsp. ilicifolia

Prunus ilicifolia

Prunus subcordata

Prunus virginiana var. demissa

fruit
leaf

var. glandulosa   var. tridentata

leaf
abaxial surface

Purshia stansburyana

fruit

Purshia tridentata

prickle

leaflet

Rosa bridgesii

fruit
Rosa californica

fruit

prickle
bud

**Rosa gymnocarpa
var. gymnocarpa**

prickle

Rosa minutifolia

sepal
Rosa rubiginosa

prickle

**Rosa woodsii
subsp. gratissima**

Rosa spithamea

fruit

**Rosa pisocarpa
subsp. pisocarpa**

sepal

**Rosa nutkana
subsp. nutkana**

bud

Rosa woodsii subsp. ultramontana

mm wide, thread-like to linear. **INFL:** terminal or axillary cymes, fls 1–4. **FL:** sepals hairy, ± prickly, gen stalked-glandular; petals 4–8 mm, oblong to oblanceolate, white; filaments strap-like; pistils < 10, styles long, slender, ovaries white-hairy. **FR:** raspberry-type, red-purple. Openings in forest; 800–2100 m. NW, SN, PR; OR. [*R. g.* var. *ganderi* (L.H. Bailey) Munz] Jun–Jul

R. laciniatus Willd. CUTLEAF BLACKBERRY Pl to 3 m, arched to mounded; prickles many, stout, wide-based, gen strong-curved. **ST:** 3–10 mm diam, 5-angled, ± glabrous to hairy, not or ± glaucous, persisting 2 yrs, rooting at tips. **LF:** 1st-yr st lvs compound, lflets (3)5, deeply dissected, coarse-toothed, tip pointed, abaxially finely to densely hairy; fl st lvs simple or gen compound, lflets 3–5; stipules ≤ 1 mm wide, thread-like to linear. **INFL:** panicle-like cyme, terminal, fls (5)10–25. **FL:** sepals hairy, prickly, with short stalked glands; petals (8)10–15 mm, obovate, gen 3-lobed, white (pink); filaments thread-like; pistils > 30, styles long, slender, ovaries glabrous. **FR:** blackberry-type, black. 2*n*=28. Moist, disturbed areas, garden escape; 60–1500 m. NW, CaR, SN, SnFrB, PR; to BC, MT, e N.Am. Eur cultivar; potentially problematic weed. May–Jul

R. lasiococcus A. Gray (p. 1207) ROUGHFRUIT RASPBERRY Per, prostrate; prickles 0. **ST:** 1–3 mm diam, not angled, finely hairy, not glaucous, persisting 1 yr, rooting at nodes. **LF:** simple, 3-lobed, or compound, lflets 3, round to reniform, toothed, terminal lobe or lflet gen obovate to obovate-wedge-shaped, tip rounded to acute, abaxially finely hairy; stipules 2–3 mm wide, gen ovate to widely elliptic. **INFL:** fls 1–2. **FL:** sepals ± glabrous, gen with few stalked glands; petals (5)8–10(12) mm, obovate to round, white; filaments thread-like; pistils < 15, styles long, slender, ovaries densely white-hairy. **FR:** falling separately or as loose unit, raspberry-type, red. 2*n*=14. Moist, open forest; 1100–2000 m. KR, n NCoRH; to BC. Jun–Aug

R. leucodermis Torr. & A. Gray (p. 1207) WHITEBARK RASPBERRY Pl 1–2(3) m, arched to mounded; prickles many, stout, wide-based, straight or gen curved. **ST:** 4–10 mm diam, not angled, glabrous, strongly glaucous in youth, persisting 2 yrs, rooting at tips. **LF:** 1st-yr st lvs gen compound, lflets (3)5, terminal ovate to lanceolate, shallow-3-lobed, coarse-toothed, tip acute, abaxially densely white-tomentose; fl st lvs simple or compound, lflets 3; stipules ≤ 1 mm wide, thread-like to linear. **INFL:** flat-topped cyme, fls gen 3–10(12). **FL:** sepals hairy, prickly, ± with stalked glands; petals 3–5(8) mm, oblong to oblanceolate-elliptic, white; filaments strap-like; pistils gen > 30, styles long, slender, ovaries densely white-hairy. **FR:** raspberry-type, red-purple to ± black. 2*n*=14. Gen open, rocky, esp moist areas; 40–2400 m. CA-FP (exc GV); to s AK, MT, UT, AZ. Apr–Jul

R. nivalis Douglas (p. 1207) SNOW DWARF BRAMBLE Per, prostrate; prickles few to several, small, ± wide-based, curved. **ST:** 1–2 mm diam, not angled, glabrous, not glaucous, persisting 1 yr, rooting at nodes. **LF:** gen simple, ± 3-lobed, or compound, lflets (2)3, terminal ovate to cordate, toothed, tip acute to obtuse, abaxially glabrous; stipules 2–4 mm wide, ovate to elliptic. **INFL:** fls 1–2. **FL:** sepals hairy, ± prickly, nonglandular; petals (4)6–10 mm, narrow-elliptic, pink to magenta (white); filaments thread-like; pistils 4–9, styles long, slender, ovaries finely hairy. **FR:** falling separately, red. 2*n*=14. Moist semi-shaded to open areas; 1250 m. nw KR (Del Norte Co.); to BC, ID. Jun–Aug ★

R. parviflorus Nutt. (p. 1207) THIMBLEBERRY Pl 0.5–2(2.5) m, erect; prickles 0. **ST:** to 6 mm diam, not angled, glabrous to finely hairy with stalked glands, not glaucous, not rooting at tips. **LF:** simple, palmately (3)5-lobed, coarse-toothed, tip acute, abaxially finely to densely gray-hairy; stipules 1.5–3 mm wide, triangular-lanceolate

to narrow-ovate. **INFL:** panicle-like cyme, fls 3–7(15). **FL:** sepals hairy, stalked-glandular; petals (10)14–22(30) mm, widely elliptic to obovate to round, white; filaments thread-like; pistils > 30, styles short, thick, ovaries densely hairy. **FR:** raspberry-type, red. 2*n*=14. Common; moist semi-shaded areas, esp edges of woodland; 20–2500 m. CA (exc GV, D); ± to s AK, ON, NM, n Mex. [*R. p.* var. *velutinus* Greene] Mar–Aug

R. pensilvanicus Poir. Pl 1–3 m, erect to arched; prickles many, stout, wide-based, straight or gen curved. **ST:** (5)10–15 mm diam, 5-angled, glabrous to gen finely hairy, not glaucous, persisting 2 yrs, not rooting at tips. **LF:** 1st-yr st lvs compound, lflets 5, terminal ovate-elliptic, sharp-double-toothed, tip acuminate, abaxially gray-hairy; fl st lvs simple or compound, lflets 3; stipules ≤ 1 mm wide, thread-like to linear. **INFL:** short, raceme-like cyme, fls 5–10(16). **FL:** sepals hairy, nonglandular; petals 10–20(30) mm, obovate to round, white; filaments thread-like; pistils > 30, styles long, slender, ovaries glabrous. **FR:** blackberry-type, black. 2*n*=28. Disturbed areas; < 1500 m. CA-FP; ne N.Am. Apr–Jul

R. spectabilis Pursh (p. 1207) SALMONBERRY Pl 2–4 m, erect; prickles few, ± stout, slender, straight. **ST:** 3–15 mm diam, not angled, glabrous, not glaucous, persisting 2 yrs, not rooting at tips. **LF:** simple to compound, lflets 3, terminal widely ovate, ± lobed, coarse-double-toothed, tip acute to acuminate, abaxially glabrous to densely hairy; stipules ≤ 1 mm wide, thread-like to linear. **INFL:** axillary, fls 1–2. **FL:** sepals hairy, ± stalked-glandular; petals (10)15–22(30) mm, elliptic to narrow-obovate, magenta; filaments strap-like; pistils gen > 30, styles long, slender, ovaries glabrous. **FR:** raspberry-type, ± yellow to orange to red. 2*n*=14. Moist areas, esp edges of woodland, streambanks; per < 500 m. NCo, nw KR, n NCoRO, n CCo, w SnFrB; to AK. [*R. s.* var. *franciscanus* (Rydb.) J.T. Howell] Mar–Jun

R. ulmifolius Schott ELMLEAF BLACKBERRY Pl to 3 m, erect to arched; prickles present or 0. **ST:** 5–10(15) mm diam, 5-angled, densely short-hairy, glaucous, persisting 2 yrs, gen rooting at tips. **LF:** 1st-yr st lvs compound, lflets 5, terminal oblong to narrow-obovate, evenly finely double-toothed, tip rounded to short-acuminate, abaxially densely white-hairy; fl st lvs simple or gen compound, lflets 3(5); stipules ≤ 1 mm wide, thread-like to linear. **INFL:** panicle-like cyme, terminal, fls 10–60. **FL:** sepals short-hairy, nonglandular; petals (5)10–12(15) mm, obovate, white to pink; filaments thread-like; pistils > 30, styles long, slender, ovaries glabrous. **FR:** blackberry-type, black.

var. **anoplothyrsus** Sudre Prickles 0. Uncommon. Moist, disturbed areas; < 1050 m. s NW, ScV, CW, n SW; native to Eur. [*R. u.* var. *inermis* Focke] May–Jul

var. **ulmifolius** Prickles many, stout, wide-based, gen straight. Disturbed areas. NCoRI; Eur. Probably only a waif in CA. Gen Apr–Aug

R. ursinus Cham. & Schltdl. (p. 1207) CALIFORNIA BLACKBERRY Pl prostrate to decumbent; gen dioecious; bristles or prickles gen many, weak, slender, straight. **ST:** 2–10 mm diam, not angled, ± glabrous to hairy, ± with stalked glands, glaucous, persisting 2 yrs, rooting at tips. **LF:** simple or compound, lflets 3(5), terminal triangular-ovate, irregularly coarse-toothed, tip acute, abaxially sparsely to densely gray-hairy, stipules thread-like to linear, ≤ 1 mm wide. **INFL:** cyme, fls 1–5. **FL:** gen unisexual; sepals hairy, prickly, ± with stalked glands; petals 6–8(11) mm in pistillate, 10–15(18) mm in staminate, elliptic to round, white; filaments thread-like; pistils > 30, styles long, slender, ovaries glabrous or hairy. **FR:** blackberry-type, black. 2*n*=42,49,56,63,77,84,91. Common. Open, disturbed areas; < 1500 m. CA-FP; to BC, ID, Baja CA. Mar–Jul

SANGUISORBA

Per; hairs 0 or not glandular. **LF:** alternate, odd-1-pinnately compound; lflets toothed < 1/3 to midvein. **INFL:** spike, head-like; peduncle long; pedicel bractlets 2, subtended by 1 bract. **FL:** bisexual; hypanthium urn-shaped, bractlets 0; sepals gen 4; petals 0; stamens 2–4, filaments not thread-like; pistils 1(2), ovary superior, continuous to style at top, stigma gen ± bushy, exserted. **FR:** hypanthium enclosing achene(s), 4-angled, hard. 2*n*=28,56. 15 spp.: n temp, arctic. (Latin: blood-absorbing, from styptic properties) Other taxa in TJM (1993) moved to *Poteridium, Poterium*.

S. officinalis L. (p. 1207) GREAT BURNET Rhizome thick, creeping. **ST**: erect, gen 50–140 cm. **LF**: basal present at fl, largest gen 20–40 cm; lflets 3–6 per side, largest blade 25–50 mm, ovate-oblong, stalk 3–25 mm, teeth gen > 15. **INFL**: gen 12–20 mm, 7–10 mm wide, ± elliptic-ovoid, > 20-fld; peduncle 3–15 cm; bract, pedicel bractlets ± 1 mm wide. **FL**: sepals 2–3.5 mm, elliptic-ovate, dark ± purple. **FR**: 2.5–3.5 mm, short-winged on 2 angles, ± winged on others; faces ± smooth. Bogs, streams, often serpentine; 120–1400 m. c NCo, nw KR, n NCoRO; to AK. Study needed to evaluate taxonomic viability of *S. officinalis* subsp. *microcephala* (C. Presl) Calder & Roy L. Taylor. Jun–Sep ★

SIBBALDIA

Barbara Ertter

Per, low, ± matted; caudex branched. **LF**: gen basal, gen 1-ternately compound. **INFL**: cyme; pedicel bractlets 0. **FL**: hypanthium shallow, bractlets gen 5; stamens [4]5[10], pollen sac 1, ± horseshoe-shaped; pistils few to many, ovaries superior, styles slender, attached near middle of fr. **FR**: achene. 6 spp.: n hemisphere. (R. Sibbald, Scottish naturalist, physician, 1641–1722)

S. procumbens L. (p. 1207) Pl hairs appressed, gen ± sparse. **ST**: 2–15 cm, ± spreading. **LF**: petiole 1–7 cm; lflets 5–25 mm, ± wedge-shaped, teeth gen 3, at tip. **INFL**: pedicels gen 3–10 mm, straight. **FL**: hypanthium bractlets linear, < sepals; sepals 2–4 mm, triangular; petals ± 1 mm, widely oblanceolate, yellow; stamens 5. **FR**: ± 1 mm, smooth, brown, often retained in disintegrating fl. *n*=7. Moist rocky areas; 1820–3700 m (lowest in n). KR, CaRH, SNH, SnBr, Wrn, SNE; to ne N.Am, arctic; Mex, Eurasia. Jun–Aug

SORBUS MOUNTAIN ASH

Peter F. Zika

Shrub, tree. **LF**: odd-1-pinnately compound [not], petioled, deciduous; lflets gen toothed, terminal partly fused to uppermost lateral(s) or not. **INFL**: panicle, many-fld; pedicel bractlets 0. **FL**: hypanthium bractlets 0, stamens gen 20; ovary gen inferior, chambers 1–5, styles 1–5, gen free. **FR**: pome, gen spheric, chambers 1–5, 1–2-seeded. ± 130 spp.: n temp, trop Asia. (Latin: ancient name) Spp. intergrade.

1. Hairs of st-buds, lflet axils, lf axils, infl ± white or gray
  2. Lflets dull adaxially, 11–17; st-bud not shiny, not sticky, hairs dense . . . . **[*S. aucuparia*]**
  2′ Lflets shiny adaxially, 7–13; st-bud shiny, sticky, hairs gen sparse . . . . ***S. scopulina***
1′ Hairs of st-buds, lflet axils, lf axils, infl primarily red-brown
  3. Lflets shiny adaxially, 2–4 cm; st-bud shiny, sticky, not glaucous, hairs gen only on scale margins, tips;
    fl, fr pedicel ± glabrous . . . . ***S. californica***
  3′ Lflets dull adaxially, 2–6 cm; st-bud glaucous, not shiny, not sticky, hairs on scale margins, tips, faces;
    fl, fr pedicel glabrous or ± hairy . . . . ***S. sitchensis***
    4. Lateral lflets ± entire or toothed above middle . . . . var. ***grayi***
    4′ Lateral lflets toothed below middle . . . . var. ***sitchensis***

S. californica Greene (p. 1207) CALIFORNIA MOUNTAIN ASH Shrub 1–2(4) m; st-buds 5–12 mm, shiny, sticky, hairs gen sparse, gen only on scale margins, tips, red-brown. **LF**: lflets 7–9(11), 2–4 cm, 1–2 cm wide, oblong-ovate, shiny adaxially, axillary hairs red-brown. **INFL**: pedicels ± glabrous in fl, fr. **FL**: petals 3–4 mm, widely ovate. **FR**: 6–9 mm, bright red, glaucous. **SEED**: 4 mm, lanceolate, barely flat, red-brown. Moist conifer forest; 1200–4300 m. KR, NCoRH, CaR, SNH; s OR, w NV. Intergrades with *S. scopulina*. May–Jun

S. scopulina Greene (p. 1207) ROCKY MOUNTAIN OR CASCADE MOUNTAIN ASH Shrub 1–5 m; st-buds 8–14 mm, shiny, sticky, hairs gen sparse, ± white. **LF**: lflets 7–13, 3–8 cm, 1.5–3 cm wide, lanceolate, oblong, or ovate, shiny adaxially, axillary hairs ± white. **INFL**: pedicels hairy. **FL**: petals 4–6 mm, ovate. **FR**: 8–12 mm, orange to red-orange, glaucous or not. **SEED**: 4–5 mm, ovate, flat, brown. 2*n*=34, 68. Canyons, wooded slopes, moist places, conifer forest; 1000–2800 m. KR, NCoR, CaR, SNH, MP; to AK, SK, NM. [*S. s.* var. *cascadensis* (G.N. Jones) C.L. Hitchc.] Jun–Jul

S. sitchensis M. Roem. Shrub 1–4 m; st-buds 8–13 mm, glaucous, not shiny, not sticky, hairs evenly distributed, red-brown. **LF**: lflets 7–13, 2–6 cm, 0.8–2.5 cm wide, ovate to oblong, dull adaxially, axillary hairs red-brown. **INFL**: pedicels glabrous or ± red-brown-hairy. **FL**: petals 3–5 mm, ovate to diamond-shaped. **FR**: 7–13 mm, red, glaucous. **SEED**: 3.5–5 mm, ovate to elliptic, ± flat, red-brown. 2*n*=34,68.

var. ***grayi*** (Wenz.) C.L. Hitchc. (p. 1207) DWARF OR GRAY'S MOUNTAIN ASH **LF**: lateral lflets ± entire or toothed above middle. **FL**: petals white or ± pink. 2*n*=34. Ridgelines; 1500–1700 m. c KR (w Siskiyou Co.); to AK. Jun–Sep

var. ***sitchensis*** (p. 1207) SITKA MOUNTAIN ASH **LF**: lateral lflets toothed below middle. **FL**: petals white. 2*n*=68. Shores, cirques, talus, conifer forest; 1700–2300 m. KR; to AK, YT, MT. May–Jul

SPIRAEA

Daniel Potter & Thomas J. Rosatti

Shrub, unarmed. **LF**: simple, oblong to (ob)ovate, gen serrate, deciduous; stipules gen 0; petiole 0–short. **INFL**: raceme or panicle, many-fld, bracted; pedicel bractlet gen 1, at top, gen linear. **FL**: hypanthium obconic to bell-shaped, bractlets 0; sepals spreading to erect; petals spreading, pink to rose; stamens 15–many; pistils 5, opposite petals, free or fused at base, surrounded by hypanthium, ovaries superior, styles ± terminal, beak-like in fr, stigmas head-like. **FR**: follicles 5, dehiscent along adaxial, top of abaxial suture. **SEED**: ± fusiform; coats membranous. ± 50 spp.: n temp. (Greek: shrub)

1. Infl longer than wide, not flat-topped; lf tomentose abaxially, darker, ± glabrous adaxially; sepals gen soon
   reflexed . . . . ***S. douglasii***
1′ Infl gen wider than long, ± flat-topped; lf glabrous to sparsely hairy, of ± same color adaxially, abaxially;
   sepals spreading to erect . . . . ***S. splendens***

S. douglasii Hook. (p. 1207)   Pl 10–20 dm, ± tomentose. **LF:** gen oblong to elliptic, gen 3–9 cm; petiole < 10 mm. **FL:** hypanthium ± 1 mm; sepals 0.5–1 mm; petals ± 1.5 mm, pink to rose. *n*=18. Moist areas, conifer forest; < 2060 m. n NCo, KR, CaR, n SNH, w MP; to BC. Jun–Sep

S. splendens K. Koch (p. 1207)   Pl 2–9 dm, glabrous to sparsely fine-hairy. **LF:** gen ovate, gen 1–7 cm; petiole < 3 mm. **FL:** hypanthium 2–2.5 mm; sepals ± 1 mm; petals ± 1.5 mm, rose. Moist, rocky areas incl serpentine, conifer forest; 550–3400 m. KR, CaR, SNH; to BC, NV. [*S. densiflora* Rydb.; *S. d.* subsp. *s.* (K. Koch) Abrams; *S. d.* Torr. & A. Gray, inval.] Jun–Sep

· **RUBIACEAE**   MADDER FAMILY

Robert E. Preston & Lauramay T. Dempster, except as noted

Ann to tree, vine. **LF:** gen opposite (whorled), entire; stipules gen fused to st, adjacent pairs occ fused, or occ lf-like and appearing like whorled lvs. **INFL:** cyme, panicle, spike, cluster, or fl 1, gen terminal and ± axillary. **FL:** gen bisexual; calyx ± 4(5)-lobed, occ 0 (*Galium, Crucianella*) or 6 (*Sherardia*); corolla gen radial, 4(5)-lobed; stamens epipetalous, alternate corolla lobes, gen incl; ovary gen inferior, chambers gen 2 or 4, style 1(2). **FR:** drupe, berry, or 2 or 4 nutlets [capsule]. ± 500 genera, 6000 spp.: worldwide, esp trop; many cult, incl *Coffea*, coffee; *Cinchona*, quinine; many orn. [Robbrecht & Manen 2006 Syst & Geogr Pl 76:85–146] *Diodia teres* Walter doubtfully in CA. Scientific Editors: Douglas H. Goldman, Bruce G. Baldwin.

1. Shrub or small tree
　2. Infl a dense, spheric head, fls >> 50; fr 2–4 dry, brown nutlets; lvs not fleshy . . . . **CEPHALANTHUS**
　2′ Infl a short, compound cyme, fls 1–5; fr drupe, orange-red; lvs fleshy . . . . **[COPROSMA]**
1′ Ann to subshrub
　3. Lvs all opposite
　　4. Fls sessile, axillary, 1–many per node . . . . **[DIODIA]**
　　4′ Fls long-pedicelled; infl terminal cyme, open, few-fld . . . . **KELLOGGIA**
　3′ Lvs gen whorled
　　5. Infl a dense spike, bracts overlapping; calyx 0 . . . . **CRUCIANELLA**
　　5′ Infl not a spike; calyx present or 0
　　　6. Fls pedicelled (exc *Galium murale*), in panicles, clusters of 2–many, occ 1; involucre 0; calyx 0 . . . . **GALIUM**
　　　6′ Fls sessile, in few-fld heads; involucral bracts ± free, ± = lvs; calyx 6-lobed . . . . **SHERARDIA**

CEPHALANTHUS   BUTTON BUSH

Shrub, small tree. **LF:** opposite or in whorls of 3; stipules not lf-like. **INFL:** head, dense-spheric, peduncled. **FL:** calyx 4-lobed; corolla narrowly funnel-shaped; style thin, exserted, stigma ± spheric. **FR:** nutlets 2–4, dry, brown, wider distally. ± 17 spp.: warm-temp Am, Asia, s Afr. (Greek: head fl)

C. occidentalis L. (p. 1207)   Sts round, ± red, glabrous. **LF:** 7–20 cm, elliptic to ovate, petioled, glabrous in age. **INFL:** 3–3.5 cm; peduncle 2.5–5 cm. **FL:** corolla white or ± yellow, tube slender, lobes obtuse; stigma long-exserted. Lake, stream edges; < 1000 m. se KR, NCoRI, SNF, c SNH, GV; reported from SCoRO, MP; AZ, e N.Am, Mex, C.Am. If recognized taxonomically, CA pls assignable to *C. occidentalis* var. *californicus* Benth. May–Sep

CRUCIANELLA   CROSS-WORT

Ann. **LF:** whorled, opposite in infl. **INFL:** spike; bracts overlapping, 3 per fl. **FL:** calyx 0; corolla tubular, lobes 4–5, erect; style divided distally, stigmas spheric. **FR:** nutlets 2, wider distally. 35–40 spp.: Medit to c Asia. (Latin: little cross, presumably from corolla lobes)

C. angustifolia L. (p. 1213)   Pl erect or ± decumbent. **ST:** 12–30 cm, 4-angled. **LF:** in whorls of 4, 8–25 mm, linear. **INFL:** 2.5–7.5 cm, dense, narrow, 4-angled, ± grass-like; bracts scarious, rib green, tip sharp. **FL:** corolla white or yellow. **FR:** glabrous. 2*n*=22. Disturbed areas in grassland, foothill woodland, yellow-pine forest; 30–1100 m. s KR, NCoRI, s CaR, n SN, e ScV; ID; native to Eur. Apr–Jun

GALIUM   BEDSTRAW

Valerie Soza

Ann, per, occ subshrub, glabrous to hairy, gen scabrous; dioecious, bisexual, or fls unisexual and bisexual. **ST:** 4-angled, occ ridged lengthwise. **LF:** gen in whorls of ≥ 4, incl lf-like stipules. **INFL:** panicles, axillary clusters (cymes), or occ 1 in axils. **FL:** bisexual, or unisexual with sterile stamens or pistils; calyx 0; corolla gen rotate, occ ± bell-shaped, gen ± green, yellow to white, occ pink or red, lobes gen 4; ovary 2-lobed, styles 2, bases ± fused. **FR:** 2 nutlets or berry. ± 650 spp.: worldwide, esp temp. (Greek: milk, from use of some spp. for curdling) [Dempster 1978 Univ Calif Publ Bot 73:1–33; Soza & Olmstead 2010 Amer J Bot 97:1630–1646] Ovary and fr gen ± equally hairy on a pl; staminate pls gen identified by vestigial ovaries, pistillate pls gen by vestigial anthers. *G. saxatile* L., *G. schultesii* Vest, and *G. verum* L. are lawn weeds in CA.

1. Ann
　2. Corolla lobes gen 3
　　3. Fr hairs hooked; lvs in whorls of 4, in 2 unequal pairs, or distal-most gen opposite; fls gen 1 in axils . . . . ***G. bifolium***
　　3′ Fr glabrous; lvs in whorls of 4–6; fls several in axils . . . . [2]***G. trifidum***

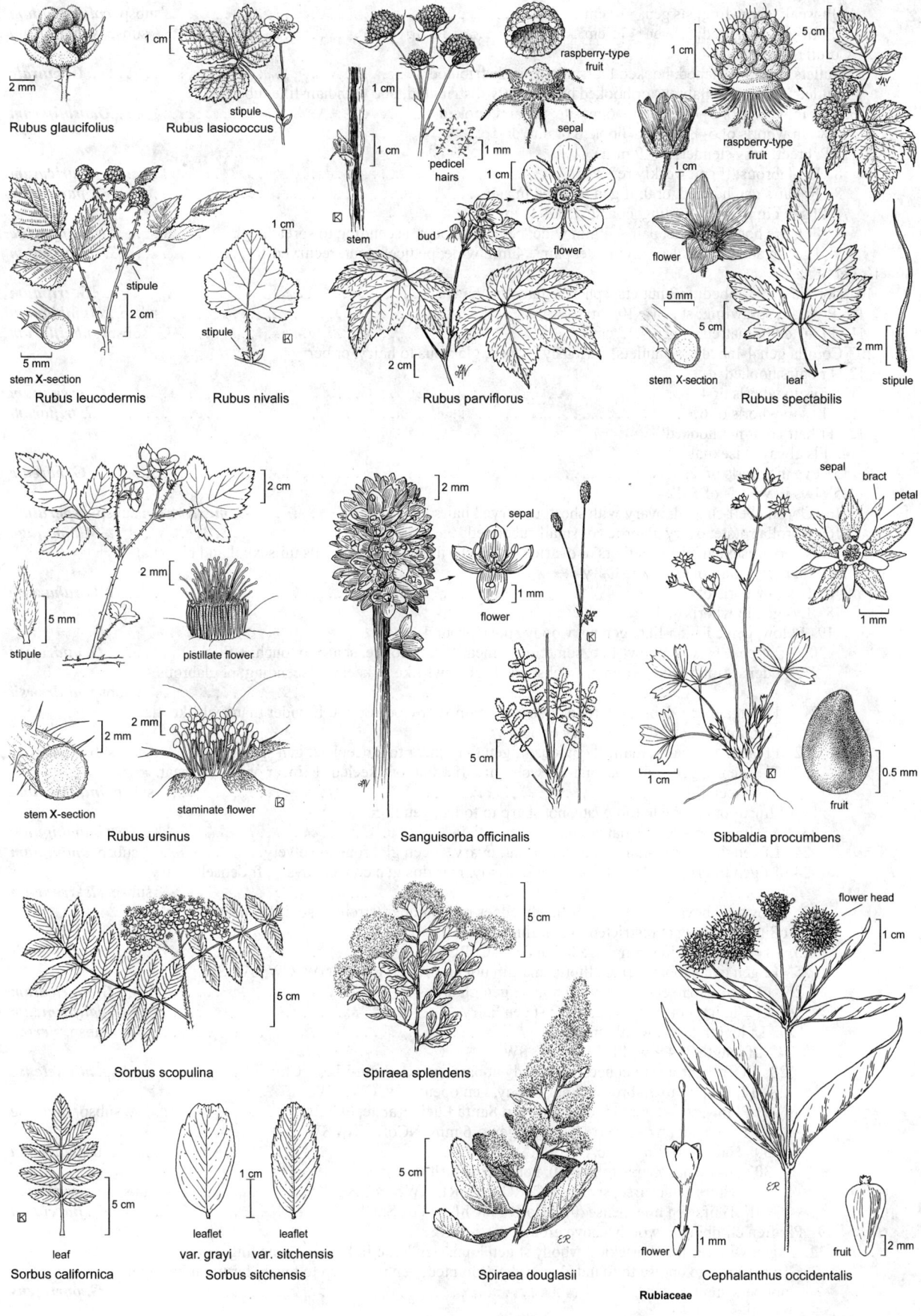

Rubus glaucifolius

2 mm

Rubus lasiococcus
1 cm
stipule

Rubus leucodermis
stipule
2 cm
5 mm
stem X-section

Rubus nivalis
1 cm
stipule

Rubus parviflorus
stem
1 cm
1 cm
pedicel hairs
1 mm
1 cm
bud
flower
raspberry-type fruit
sepal
2 cm

Rubus spectabilis
5 cm
1 cm
raspberry-type fruit
flower
5 mm
5 cm
5 mm
stem X-section
leaf
2 mm
stipule

Rubus ursinus
2 cm
stipule
5 mm
pistillate flower
2 mm
stem X-section
staminate flower
2 mm

Sanguisorba officinalis
2 mm
sepal
flower
1 mm
5 cm

Sibbaldia procumbens
sepal
bract
petal
1 mm
1 cm
fruit
0.5 mm

Sorbus scopulina
5 cm

Spiraea splendens
5 cm

Cephalanthus occidentalis
flower head
1 cm
flower
1 mm
fruit
2 mm

Sorbus californica
5 cm
leaf

Sorbus sitchensis
1 cm
leaflet
var. grayi
leaflet
var. sitchensis

Spiraea douglasii
5 cm

Rubiaceae

    4. Pl weak, sprawling; sts gen ≥ 10 cm . . . . ²subsp. *columbianum*

    4′ Pl tufted or matted; sts gen ≤ 15 cm . . . . ²subsp. *subbiflorum*

2′ Corolla lobes gen 4

    5. Nutlets sausage-shaped, hooked hairs unevenly distributed . . . . **G. murale**

    5′ Nutlets ± spheric, glabrous or hooked hairs evenly distributed, occ with hair-like tubercles

        6. Lvs in whorls of 3–4 and some opposite; fr hairs hooked. . . . . **G. proliferum**

        6′ Lvs in whorls of 5–8; fr hairs hooked or not, or 0

            7. Pl erect, very slender; lf ≤ 9 mm

                8. Fr glabrous; lf gen weakly reflexed . . . . **G. divaricatum**

                8′ Fr hairs gen hooked, or 0; lf gen reflexed in age . . . . **G. parisiense**

            7′ Pl gen climbing to trailing, coarse; lf 12–31 mm

                9. Fr hairs hooked; nutlet pair < 4 mm wide; pedicel slender, ascending to spreading . . . . **G. aparine**

                9′ Fr with acute tubercles only; nutlet pair > 4 mm wide; pedicel stout, recurved. . . . . **G. tricornutum**

1′ Per or shrub

10. Corolla gen 3-lobed; fr 2 nutlets, spheric, glabrous, hard . . . . ²**G. trifidum**

    11. Pl weak, sprawling; st gen ≥ 10 cm . . . . ²subsp. *columbianum*

    11′ Pl tufted or matted; st gen ≤ 15 cm . . . . ²subsp. *subbiflorum*

10′ Corolla gen 4-lobed; fr 2 nutlets, variously shaped, glabrous to hairy, or berry

    12. Fr hairs hooked

        13. Lvs in whorls of 4. . . . . **G. oreganum**

        13′ Lvs in whorls of 6. . . . . **G. triflorum**

    12′ Fr hairs 0 or not hooked

        14. Fls always bisexual

            15. Lvs in whorls of 4. . . . . **G. boreale**

            15′ Lvs in whorls of 5–12

                16. Corolla bell-shaped; ovary with short, upcurved hairs . . . . **G. mexicanum** subsp. *asperulum*

                16′ Corolla rotate; ovary glabrous or small-tubercled. . . . . **G. mollugo**

    14′ Fls gen unisexual, either anthers or ovaries small, sterile, or sometimes fls unisexual and bisexual

        17. Fr a berry, glabrous to short-hairy

            18. Lvs in whorls of 6 . . . . **G. hardhamiae**

            18′ Lvs gen in whorls of 4

                19. Pl low, gen cushion-like, gen not woody above ground

                    20. Lf often ± bristle- or awl-like, sometimes linear to lanceolate, sharp to touch, ± thick . . . . **G. andrewsii**

                    21. Pl dense, internode < lf, main st obscure, lf gen awl-like, ± keeled, ascending; pl glabrous

                        . . . . subsp. *andrewsii*

                    21′ Pl elongate or ± dense, internode gen > lf, main st obvious or not, lf wider than awl-like, flat,

                        ascending or spreading; pl hairy or not

                        22. Pl open, occ hairy, main st obvious; lf gen flat, linear to lanceolate, gen spreading. . . . . subsp. *gatense*

                        22′ Pl ± dense, gen glabrous, main st ± obscure; lf ± flat to ± keeled, ± linear or lance-linear,

                        gen ascending . . . . subsp. *intermedium*

                  20′ Lf linear or elliptic to (ob)ovate, not sharp to touch, gen thin

                    23. Lf ± linear, margins ± flat or convex, gen ciliate . . . . **G. ambiguum**

                      24. Lf gen thin, ± fine-hairy, margins ± flat; ovary, fr gen glabrous to velvety. . . . . subsp. *ambiguum*

                      24′ Lf gen thick, gen glabrous or margin bristly, margins gen convex; ovary, fr densely hairy

                        . . . . subsp. *siskiyouense*

                  23′ Lf ovate (obovate) to elliptic or linear, edges rolled under (strongly so, if lf linear) or not, not

                    ciliate or hairs not restricted to margin

                    25. Lf linear, margin strongly rolled under; SCoRO . . . . **G. clementis**

                    25′ Lf gen ovate (obovate) to elliptic, margin not or ± rolled under; NW, CW, SW, SN

                      26. Lf shiny, sparsely hairy, 6–10 mm; st gen glabrous . . . . **G. muricatum**

                      26′ Lf ± hairy, not shiny, 2–25 mm; st gen hairy . . . . ²**G. californicum**

                        27. Lf length 5–8 × width; n SNF . . . . ²subsp. *sierrae*

                        27′ Lf length 2–4 × width; NW, CW, SW

                          28. Fl, fr glabrous; pl congested, woody at base, gen cushion-like; n ChI . . . . ²subsp. *miguelense*

                          28′ Fl, fr hairy to glabrous; pl not woody, gen open; NW, CW, SW

                            29. Pl low, st ≤ 16 cm; lf gen ≤ 6 mm; n Santa Lucia Range, SCoRO. . . . . subsp. *luciense*

                          29′ Pl gen not low, st gen > 16 cm; lf gen > 6 mm; NCoR, CW, SW

                            30. Hair short, fine; SCo, SnJt . . . . ²subsp. *primum*

                            30′ Hair long, coarse to fine (0); not SCo or SnJt

                              31. Hairs gen coarse, sparse; NCoRO, NCoRI, CW. . . . . ²subsp. *californicum*

                              31′ Hairs gen fine, dense (0); SCoRO, n ChI, WTR, SnGb . . . . ²subsp. *flaccidum*

         19′ Pl often climbing, ± woody above ground

            32. Pl gen with recurved prickles; woody st gen long, slender, climbing or sprawling

                33. Lf tip acute to obtuse to rounded, gen short-pointed, gen not sharp to touch, terminal hair gen

                not persistent . . . . **G. porrigens**

34. Lf widely oblong to ovate . . . . var. ***porrigens***
34′ Lf ± linear . . . . var. ***tenue***
33′ Lf tip gen tapered, gen sharp to touch, terminal hair persistent
  35. Lf elliptic to ovate, acute to acuminate; pl not densely tangled, green; SCoRO, WTR
    36. Pl not shiny; st wiry; lf faces hairy; gen < 200 m, CCo, SCoRO . . . . . . ***G. californicum*** subsp. ***maritimum***
    36′ Pl shiny; st stout; lf faces ± glabrous; ≥ 200 m, SCoRO, WTR . . . . ***G. cliftonsmithii***
  35′ Lf ± linear to narrowly ovate, acute; pl densely tangled, red in age; SCo, ChI, WTR, PR . . . . ***G. nuttallii***
    37. Pl hairs gen 0; ChI . . . . subsp. ***insulare***
    37′ Pl hairs minute, sharp, curved; SCo, WTR, PR . . . . subsp. ***nuttallii***
32′ Pl prickles gen 0; woody st gen short, climbing or not
  38. Lf gen ± glabrous to puberulent or scabrous
    39. Lf narrow to elliptic, ± sessile; infl branchlets ± stout; corolla gen red . . . . ²***G. bolanderi***
    39′ Lf widely elliptic to ovate, ± petioled; infl branchlets thread-like; corolla gen ± yellow . . . . ***G. sparsiflorum***
      40. Lf glabrous, gen ovate, ± leathery. . . . . subsp. ***glabrius***
      40′ Lf puberulent, elliptic to ± ovate, gen thin . . . . subsp. ***sparsiflorum***
  38′ Lf gen hairy
    41. St ± stout, ± stiff; montane
      42. Pl sprawling or climbing, forming dense clumps; fl yellow, hairy externally, throughout;
        fr hairy; lf gen widely elliptic; SnGb . . . . ***G. grande***
      42′ Pl tufted or ± climbing, not forming dense clumps; fl gen red, glabrous to sparsely hairy at
        base; fr glabrous to hairy; lf gen elliptic; n of SnGb. . . . . ²***G. bolanderi***
    41′ St wiry, slender; often coastal, insular, sometimes montane . . . . ²***G. californicum***
      43. Lf length 5–8 × width; n SNF . . . . ²subsp. ***sierrae***
      43′ Lf length 2–4 × width; NW, CW, SW
        44. Fl, fr glabrous; pl congested, woody at base, gen cushion-like; n ChI . . . . ²subsp. ***miguelense***
        44′ Fl, fr hairy to glabrous; pl not woody, gen open; NW, CW, SW
          45. Hair short, fine; SCo, SnJt . . . . ²subsp. ***primum***
          45′ Hair long, coarse to fine (0); not SCo or SnJt
            46. Hairs gen coarse, sparse; NCoRO, NCoRI, CW . . . . ²subsp. ***californicum***
            46′ Hairs gen fine, dense (0); SCoRO, n ChI, WTR, SnGb. . . . . ²subsp. ***flaccidum***
17′ Fr 2 nutlets, hair gen long, straight to strongly curved
 47. Fls unisexual and bisexual
  48. Herb; st gen slender, flexible, base woody; mainland
    49. Lf ovate to round, acute to obtuse; infl narrow, branches short, ± erect . . . . ***G. parishii***
    49′ Lf linear to oblanceolate, acute to acuminate; infl open, branches longer, spreading . . . . ***G. wrightii***
  48′ Erect shrub; st rigid, stout, woody; ChI
    50. Ovary hairs short, curved, ascending; n ChI . . . . ***G. buxifolium***
    50′ Ovary hairs long, straight, spreading (or ± 0); s ChI . . . . ***G. catalinense***
      51. St, lf hairs sparse, gen very short, straight to curved, not curly; San Clemente Island . . . . . . subsp. ***acrispum***
      51′ St, lf hairs ± dense, short, curly; Santa Catalina Island . . . . subsp. ***catalinense***
 47′ Fls unisexual only
  52. Ovary, fr hairs all short, strongly curved, ascending; corolla bell-shaped, divided ≤ 1/2 to base . . . . ***G. jepsonii***
  52′ Ovary, fr hairs gen long, straight, spreading; corolla gen rotate, or bell-shaped and gen
    divided > 1/2 to base
    53. Lf ± linear to strap-shaped to linear-oblong, gen widest at or distal to middle
      54. Distal internodes >> proximal; pedicel gen > fr; fr hairs < fr . . . . ***G. johnstonii***
      54′ Distal internodes ± = proximal; pedicel gen < fr; fr hairs gen ≥ fr . . . . ***G. angustifolium***
        55. Corolla gen glabrous, or not more hairy than st or lf
          56. Woody st internode gen < lf; pl gen glabrous exc lf margin; n ChI . . . . subsp. ***foliosum***
          56′ Woody st internode gen > lf; pl hairy or not; s SN, SCoR, SW (exc n ChI), DMtns
            57. Pl tall to low, ± stout, gen glabrous to white-hairy; lvs persistent; hills, mtns, not in D
              . . . . subsp. ***angustifolium***
            57′ Pl tall, slender, gen glabrous exc lvs scabrous; lvs deciduous; SnBr, e SnJt, DMtns, DSon
              . . . . subsp. ***gracillimum***
        55′ Corolla gen hairy (sometimes glabrous in subsp. jacinticum)
          58. St gen hairy, occ scabrous
            59. St, lf hairs ± equal; st ridges gen narrower than grooves; SnGb. . . . . subsp. ***gabrielense***
            59′ St hairs < lf hairs; st ridges gen wider than grooves; s SNH . . . . subsp. ***onycense***
          58′ St ± glabrous
            60. Pl 35–60 cm, infl pyramid-shaped, many-fld, much-branched; DSon (Borrego Desert)
              . . . . subsp. ***borregoense***
            60′ Pl ≤ 35 cm; infl narrow, few-fld, few-branched; SW
              61. Pl 17–35 cm; lf gen 11–26 mm; internodes ± = lvs; SnJt. . . . . subsp. ***jacinticum***
              61′ Pl 6–20 cm; lf gen 2–15 mm; internodes < 2 × lvs; SnGb, SnBr . . . . subsp. ***nudicaule***
    53′ Lf gen (ob)ovate to lanceolate, elliptic, or ± round (to needle-like in *Galium stellatum*), gen widest
    proximal to middle

62. Pl woody at base and extending distally to ≥ 50 cm
  63. St flexible, few-branched; lf ovate to elliptic, tip obtuse to acute, not sharp at tip; s SNH, TR . . . . . . ***G. hallii***
  63′ St rigid, many-branched; lf lanceolate to needle-like, tip acute to acuminate, sharp at tip; D
    . . . . ***G. stellatum***
62′ Pl woody near base only
  64. St glabrous
    65. Fr ≤ 4 mm wide, incl hairs
      66. Pl flexible; lf 8–23 mm, flat, tip tapered to acute, not sharp to touch. . . . . ***G. magnifolium***
      66′ Pl stiff; lf 2–10 mm, gen arched, tip acuminate, sharp to touch. . . . . ***G. matthewsii***
    65′ Fr gen > 4 mm wide, incl hairs
      67. Infl few-fld, axillary, little exserted from lvs; fr 7–10 mm; lf tip acute to obtuse, not sharp to
        touch; KR, CaRH. . . . . ²***G. glabrescens*** subsp. ***glabrescens***
      67′ Infl many-fld, terminal, much exserted from lvs; fr gen 4–7 mm; lf tip obtuse to acuminate,
        sharp; n SNH, GB, DMtns
        68. Lvs at a node ± equal, lanceolate to ovate, tapered to tip, flat to arched; DMtns. . . . . ***G. argense***
        68′ Lvs at a node ± unequal, gen ± arched, larger 2 ovate to round, abruptly rounded to
          acuminate tip; n SNH, GB, DMtns . . . . ²***G. multiflorum***
  64′ St puberulent to hairy
    69. Lvs at a node unequal, lf tip gen sharp to touch, abruptly rounded to tapered to sharp tip
      70. Corolla rotate . . . . ***G. munzii***
      70′ Corolla ± bell-shaped
        71. Lf flat, larger lvs lanceolate to ovate; DMtns (Panamint Range) . . . . ***G. hilendiae*** subsp. ***carneum***
        71′ Lf ± arched, larger lvs ovate to round; SNH, GB, DMtns. ²***G. multiflorum***
    69′ Lvs at a node equal or lf tip not sharp to touch, acute to obtuse
      72. Lf lanceolate to narrowly elliptic. . . . . ***G. serpenticum***
        73. Corolla gen ± cup-shaped; lf tip gen flat; KR . . . . subsp. ***scotticum***
        73′ Corolla rotate; lf tip gen reflexed; Wrn . . . . subsp. ***warnerense***
      72′ Lf lanceolate to (ob)ovate or round
        74. Fr ≥ 6 mm, incl hairs
          75. Pl gen green, minute-puberulent; lf ovate to obovate . . . . ²***G. glabrescens*** subsp. ***glabrescens***
          75′ Pl ± gray, densely short-gray hairy; lf widely ovate to round
            76. Lf gen fleshy; st ≤ 9 cm; DMtns (Panamint Range). . . . . ²***G. hypotrichium*** subsp. ***tomentellum***
            76′ Lf not fleshy; st ≤ 27 cm; KR, NCoRH, CaRH, n SNH . . . . ***G. grayanum***
              77. Pl 5–27 cm, ± open . . . . var. ***grayanum***
              77′ Pl 2.5–14 cm, ± compact. . . . . var. ***nanum***
        74′ Fr ≤ 6 mm, incl hairs
          78. Pl woody gen for 3–4 cm above base; infl ± terminal, diffuse, branches many, long,
            spreading, many-fld, lvs reduced; corolla bell-shaped . . . . ***G. hilendiae***
            79. Corolla throat wide; lf ovate to ± round; DMoj . . . . subsp. ***hilendiae***
            79′ Corolla throat narrow; lf lanceolate to ovate; DMtns (Kingston Range) . . . . subsp. ***kingstonense***
          78′ Pl not woody above base; infl axillary, narrow, dense, branches few, short, ascending,
            few-fld, lvs often not reduced; corolla rotate to bell-shaped
            80. Lf thin; corolla rotate; Wrn . . . . ***G. glabrescens*** subsp. ***modocense***
            80′ Lf gen fleshy; corolla rotate to bell-shaped; SNH, SNE, DMtns . . . . ***G. hypotrichium***
              81. Pl gen open, tall, 12–30 cm, scabrous, base woody . . . . subsp. ***inyoense***
              81′ Pl gen dense or ± open, short, gen 2–15 cm, scabrous to hairy, base not woody
                82. Pl minutely velvety . . . . subsp. ***hypotrichium***
                82′ Pl scabrous to hairy, not velvety
                  83. Pl green, scabrous to hairy; pl ± open; st ≤ 15 cm; s SNH. . . . . subsp. ***subalpinum***
                  83′ Pl gray, ± densely fine-hairy; pl matted; st ≤ 9 cm; DMtns (Panamint Range) . . . ²subsp. ***tomentellum***

G. ambiguum W. Wight YOLLA BOLLY BEDSTRAW Per, low, matted, not woody above ground, gen hairy; dioecious. **ST:** erect, 5–16 cm. **LF:** in whorls of 4, 6–16 mm, ± linear, tip ± acute, with long bristle, margin gen ciliate. **STAMINATE INFL:** few-fld clusters. **PISTILLATE INFL:** fls gen 1 in axils. **FL:** corolla rotate, ± yellow, ± glabrous to short-fine-ciliate or papillate along margins. **FR:** berry, glabrous to hairy.

  subsp. ***ambiguum*** (p. 1213) Hairy. **LF:** gen thin, ± finely hairy, margins ± flat. **FR:** gen glabrous to velvety. $2n=22,44$. Light shade of conifers, oaks, in serpentine or clay loam; 350–2150 m. s KR, NCoRH, n SNF (El Dorado Co.). Jun–Jul

  subsp. ***siskiyouense*** (Ferris) Dempster & Stebbins (p. 1213) Coarse-hairy to glabrous. **LF:** gen thick, glabrous or margins bristly, faces occ irregularly or minutely hairy, margins gen convex. **FR:** densely hairy. $2n=66$. Dry, sunny banks and hillsides, open forest, in serpentine or clay loam; 160–1750 m. KR (Del Norte Co.), NCoRO; s OR. [*G. a.* var. *s.* Ferris] Jun–Jul

G. andrewsii A. Gray PHLOX-LEAVED BEDSTRAW Per, gen low, gen cushion-like, green to silvery; dioecious. **ST:** 5–22 cm. **LF:** in whorls of 4, 4–11 mm, gen bristle- to awl-like, sharp to touch, ± thick, tip with persistent hair. **STAMINATE INFL:** few-fld clusters. **PISTILLATE INFL:** fls 1 in axils. **FL:** corolla rotate, ± yellow, glabrous. **FR:** berry, black, glabrous. Subspp. difficult to distinguish but distinct based on molecular data, chromosome numbers.

  subsp. ***andrewsii*** (p. 1213) Pl low, dense, cushion-like, glabrous; main st obscure. **ST:** 5–15 cm. **LF:** > internodes, gen awl-like, ± keeled (not flat), gen ascending. $2n=22$. High chaparral, open woodland, gen serpentine or sandy-loam soil; 250–2580 m. NCoRI, s SNF, SCoR, TR, PR; mtns of Baja CA. Apr–Jun

subsp. *gatense* (Dempster) Dempster & Stebbins PHLOX-LEAF SERPENTINE BEDSTRAW Pl open, occ hairy; main st obvious. **ST:** < 22 cm. **LF:** gen < internodes, linear to lanceolate, gen flat, gen spreading. $2n=88$. Dry, rocky places in serpentine soil, chaparral or open oak/pine woodland; 220–1450 m. SnFrB, SCoRI. Apr–Jun ★

subsp. *intermedium* Dempster & Stebbins Pl gen low, ± dense, gen glabrous; main st ± obscure. **ST:** 6–20 cm. **LF:** gen < internodes, ± linear or lance-linear, gen ± flat to ± keeled, gen ascending. $2n=44$. High chaparral, open oak/pine woodland, in various soils, incl serpentine; 300–1615 m. NCoRI, c SNF, SCoRO, WTR, SnGb. Apr–Jun

G. angustifolium A. Gray (p. 1213) NARROWLY LEAVED BEDSTRAW Per, low, tufted or sts elongate, glabrous to hairy, at least base woody. **ST:** 6–100 cm. **LF:** in whorls of 4, ± strap-shaped, 3-veined. **INFL:** panicle of few- to many-fld clusters. **FL:** corolla rotate, gen red to yellow. **FR:** nutlets, gen > pedicel; hairs dense, long, straight, spreading.

subsp. *angustifolium* Pl woody above base, gen glabrous to white-hairy. **ST:** 15–100 cm, ridges narrower than grooves. **LF:** 5–27 mm. **INFL:** many-fld, gen open. **FL:** corolla ≤ hairy than pl. $2n=22,44$. Cliffs, canyons, protected places on hillsides; 15–2650 m. s SN, SCoR, SW (exc n ChI); n Baja CA. Apr–Jun

subsp. *borregoense* Dempster & Stebbins BORREGO BEDSTRAW Pl glabrous, woody above base. **ST:** 35–60 cm, slender, ridges wider than grooves. **LF:** < 8 mm. **INFL:** many-fld, ± dense. **FL:** corolla hairy externally. Among boulders, granitic n slopes; 350–1250 m. DSon (Palm, Hellhole canyons, Pinyon Mtn Valley, San Diego Co.). Apr–Jun ★

subsp. *foliosum* (Hilend & J.T. Howell) Dempster & Stebbins Pl ± woody above base, gen glabrous exc lf margin. **ST:** 30–60 cm, ridges ± as wide as surfaces between; nodes short, obvious after lf-fall. **LF:** gen 3–17 mm. **INFL:** many-fld, ± dense. **FL:** corolla glabrous to puberulent. $2n=22$. Rocky slopes; 30–60 m. n ChI. Mar–Jul

subsp. *gabrielense* (Munz & I.M. Johnst.) Dempster & Stebbins (p. 1213) SAN ANTONIO CANYON BEDSTRAW Pl tufted, woody at base, gen hairy. **ST:** 6–36 cm, ridges gen narrower than grooves. **LF:** gen 2–14 mm. **INFL:** narrow, few-fld, ± dense. **FL:** corolla yellow or ± red, hairy externally. $2n=44$. Slopes, ridges, open forest, high chaparral; 1200–2650 m. SnGb. Jun–Aug ★

subsp. *gracillimum* Dempster & Stebbins SLENDER BEDSTRAW Pl slender, gen glabrous exc lvs scabrous. **ST:** ± 40 cm, woody, ridges ± wide as grooves. **LF:** gen 4–18 mm, short-lived. **INFL:** few-fld, ± open. **FL:** corolla hairs gen 0. **FR:** < 2 mm, > hairs. $2n=22$. Shaded places among granite boulders in canyons, on outcrops; 130–1550 m. SnBr, e SnJt, DMtns, DSon. Apr–Jul ★

subsp. *jacinticum* Dempster & Stebbins SAN JACINTO MOUNTAINS BEDSTRAW Pl ± low, gen glabrous or lvs ± hairy. **ST:** 17–35 cm, ridges ± as wide as grooves. **LF:** gen 11–26 mm. **INFL:** narrow, few-fld, ± open. **FL:** corolla glabrous to sparsely hairy. $2n=66$. Open, mixed forest; 1350–2100 m. w SnJt. May–Jul ★

subsp. *nudicaule* Dempster & Stebbins Pl low, mat-forming, rhizomed, not woody, gen glabrous or ± hairy on lf margins, veins. **ST:** 6–20 cm, ridges ± as wide as grooves. **LF:** gen 2–15 mm. **INFL:** narrow, few-fld, open. **FL:** corolla gen red, occ yellow, sparsely hairy externally. $2n=22$. Steep slopes, open mixed forest and chaparral; 1650–2650 m. SnGb, SnBr. Jun–Aug

subsp. *onycense* (Dempster) Dempster & Stebbins ONYX PEAK BEDSTRAW Pl slender, ± scabrous, ± gray-green, base woody. **ST:** 12–30 cm, ridges gen wider than grooves. **LF:** gen 5–14 mm. **INFL:** ± narrow, few-fld, open. **FL:** corolla gen pink, hairy externally. $2n=22$. Granite outcrops, open oak/pine woodland; 950–2300 m. s SNH (Onyx Peak area, e Kern Co.). [*G. a.* var. *o.* Dempster] Apr–Jul ★

G. aparine L. (p. 1213) GOOSE GRASS Ann, climbing or prostrate, occ short, erect; clings by small, hooked prickles. **ST:** 3–9 dm, weak, brittle. **LF:** in whorls of 6–8, 13–31 mm, proximal-most petioled, ± round, distal sessile, ± narrowly oblanceolate. **INFL:** fls few on branchlets in most axils. **FL:** bisexual; corolla rotate, ± white. **FR:** nutlets;

hairs many, short, hooked. $2n=20,22,42,44,63,64,66,\pm86,88$. Grassy, ± shady places; 30–1500 m. CA (exc DSon); to AK, e N.Am; Eur. Weedy in gardens; if recognized taxonomically, small pls with slender, pointed lvs, ± yellow petals assignable to *G. spurium* L. Mar–Jul

G. argense Dempster & Ehrend. Per, erect, 20–60 cm, stiff, open, glabrous, woody near base; dioecious. **LF:** in whorls of 4, 8–21 mm, lanceolate to ovate, tip sharp, tapered. **INFL:** clusters, terminal on branchlets, much exserted from lvs. **FL:** corolla ± bell-shaped to rotate, glabrous to sparsely hairy, cream. **FR:** nutlets, ± 6 mm incl hairs; hairs long, straight, spreading. $2n=44$. Loose, stony e-facing slopes with *Pinus monophylla*; 1250–2400 m. DMtns (Argus, Nelson ranges, Inyo Co.). Apr–Oct

G. bifolium S. Watson (p. 1213) LOW MOUNTAIN BEDSTRAW Ann, erect, gen 5–15 cm, glabrous; branches few or 0. **LF:** in whorls of 4, in 2 unequal pairs, or distal-most gen opposite, 10–21 mm, lanceolate to narrowly elliptic. **INFL:** fls gen 1 on thin pedicels; pedicels nodding. **FL:** bisexual; corolla white, lobes gen 3, ovate, ascending, glabrous. **FR:** nutlets; hairs short, hooked. Open conifer forest, gravelly slopes, meadows; 1500–3700 m. KR, NCoRH, NCoRI, CaRH, SNH, WTR, MP, n SNE (Bodie Hills), DMtns (Cottonwood Mtns); to BC, MT, CO. Jun–Sep

G. bolanderi A. Gray (p. 1213) BOLANDER'S BEDSTRAW Per, tufted or ± climbing, woody at base and gen distally, glabrous to scabrous or coarsely gray-hairy; dioecious. **ST:** gen 15–36 cm, stout. **LF:** in whorls of 4, in 2 ± unequal pairs, 6–27 mm, narrow to elliptic, tapered at base, acute or abruptly soft-pointed at tip, obscure-3-veined. **STAMINATE INFL:** axillary clusters. **PISTILLATE INFL:** fl 1 in axils. **FL:** corolla rotate, gen glabrous occ sparsely hairy at base, gen red, occ ± yellow. **FR:** berry, hairy to glabrous. $2n=66$. Open mixed or conifer forest, dense chaparral, gen rocky slopes or banks; 150–2600 m. KR, NCoRH, NCoRI, CaR, SNF, n SNH, Teh, n SCoRI, MP. May–Aug

G. boreale L. NORTHERN BEDSTRAW Per, erect, 3–6 dm, ± glabrous to puberulent. **LF:** in whorls of 4, 13–31 mm, linear to widely lanceolate, ± leathery, 3-veined, margins rolled under. **INFL:** panicle, terminal, fls many. **FL:** bisexual; corolla rotate, white, glabrous. **FR:** nutlets; short hairs gen present, upcurved, not hooked. $2n=66$. Wet places or disturbed roadsides; 15–2000 m. KR, CaRH, c SNH, MP; N.Am exc se US, Eurasia. [*G. b.* subsp. *septentrionale* (Roem. & Schult.) H. Hara] Jun–Aug

G. buxifolium Greene (p. 1213) BOX BEDSTRAW Shrub, erect, 3–6 dm, glabrous to scabrous or sparsely hairy, fr hairs not like those of st, lf. **LF:** in whorls of 4, 8–30 mm, elliptic to widely oblanceolate or obovate, rounded, tip occ small-pointed, ± leathery; petiole base swollen. **INFL:** dense, lfy fl clusters from woody st nodes. **FL:** bisexual or unisexual, corolla rotate, white. **FR:** nutlets; hairs curved toward tip. Rocky bluffs, slopes; 10–400 m. n ChI. Similar to *G. catalinense.* Mar–Jul ★

G. californicum Hook. & Arn. CALIFORNIA BEDSTRAW Per, base occ ± woody, low, cushion-like to ± climbing, gen hairy; dioecious. **ST:** 5–90 cm. **LF:** in whorls of 4, 2–25 mm, ovate to elliptic; tip acute to obtuse; ± petioled. **STAMINATE INFL:** clusters, fls few. **PISTILLATE INFL:** fls gen 1 in axils. **FL:** corolla rotate, ± yellow. **FR:** berry, ± hairy, occ glabrous.

subsp. *californicum* (p. 1213) Open, in mats or tufts (forest, chaparral) or tangled masses (sea cliffs), not woody to occ woody at base; hairs ± coarse. **ST:** 8–32 cm. **LF:** 6–18 mm, ovate to elliptic, tip obtuse to abruptly acute, not sharp to touch. **FL:** glabrous to hairy. **FR:** hairy. $2n=132$. Shady to open places, conifer or mixed forest, chaparral, sea cliffs, hillsides; 15–1520 m. NCoRO, NCoRI, CW. Mar–Jul

subsp. *flaccidum* (Greene) Dempster & Stebbins (p. 1213) Pl not woody; hairs gen dense, soft, fine (0). **ST:** (7)16–61 cm. **LF:** 6–25 mm, gen elliptic, tip oblong, short-pointed. **FL:** corolla hairy (glabrous). **FR:** gen hairy, occ glabrous. $2n=88$. Open or dense noncoastal woodland; 30–1500 m. SCoRO, n ChI, WTR, SnGb. Mar–Jul

subsp. *luciense* Dempster & Stebbins CONE PEAK BEDSTRAW Pl low, spreading mat, not woody, ± gray-hairy. **ST:** 5–16 cm. **LF:**

gen 3–6(8) mm, gen elliptic, tip acute, not sharp to touch. **FL:** corolla densely coarse-hairy. **FR:** densely hairy. 2*n*=44. Pine, oak forests; 1100–1370 m. SCoRO (n Santa Lucia Range). Mar–Jul ★

subsp. ***maritimum*** Dempster & Stebbins (p. 1213) Pl spreading to ± climbing, ± woody; st, lf-margin hairs ± stout, straight to curved. **ST:** ≤ 90 cm, wiry. **LF:** 4–9 mm, ovate, tip acuminate, gen sharp to touch, faces hairy. **FL:** corolla hairy to glabrous. **FR:** hairy to glabrous. 2*n*=88. Coastal bluffs, canyons; 15–200(230) m. CCo, SCoRO. Mar–Jul

subsp. ***miguelense*** (Greene) Dempster & Stebbins SAN MIGUEL ISLAND BEDSTRAW Pl gen low, cushion-like (± climbing), woody at base; hairs ± stout, recurved-prickly. **ST:** 15–45 cm, ± glabrous to hairy; internodes short, lvs congested. **LF:** 2–8 mm, widely ovate, leathery, margin ± rolled under, tip acute to obtuse, occ small-pointed, faces glabrous to hairy. **FL:** corolla glabrous, yellow. **FR:** glabrous. Grassy or sandy coastal slopes; 30–60 m. n ChI. Mar–Jul ★

subsp. ***primum*** Dempster & Stebbins ALVIN MEADOW BEDSTRAW Pl low, weak, tufted, to decumbent (± climbing in hybrids), woody or not, hairs ± sparse, fine. **ST:** (9)16–29 cm. **LF:** 4–12 mm, elliptic, tip acute to obtuse. **FR:** ± glabrous to minute-hairy. 2*n*=22. Shade, lower elevations in Jeffrey-, Coulter-pine forests; 1350–1700 m. SCo, SnJt. Most populations affected by hybridization with *G. porrigens*. Mar–Jul ★

subsp. ***sierrae*** Dempster & Stebbins EL DORADO BEDSTRAW Pl weak, slender, cushion-like to weakly tufted, not woody, hairs many, straight, soft. **ST:** 7–14 cm. **LF:** ≤ 20 mm, length 5–8 × width, tip acute, with a terminal hair. **FL:** corolla hairy externally. **FR:** minute-hairy. Open pine, oak forests, chaparral; 100–500 m. n SNF (El Dorado Co.). Mar–Jul ★

G. catalinense A. Gray (p. 1213) Shrub, erect, < 12 dm. **ST:** stout; rigid, nodes enlarged. **LF:** in whorls of 4, 13–25 mm, lanceolate to oblanceolate, gen 1-veined, tip gen obtuse to round; petiole short, base gen swollen, persistent. **INFL:** lfy clusters, axillary, dense. **FL:** bisexual or unisexual, corolla rotate, ± white, glabrous. **FR:** nutlets; hairs long, straight, spreading (glabrous).

subsp. ***acrispum*** Dempster SAN CLEMENTE ISLAND BEDSTRAW Pl gen scabrous, hairs sparse, short or longer and curved but not curly. 2*n*=22. Rocky slopes, cliffs; 3–453 m. s ChI (San Clemente Island). Apr–Jul ★

subsp. ***catalinense*** SANTA CATALINA ISLAND BEDSTRAW Pl not scabrous, hairs ± dense, short, curly. 2*n*=22. Rocky slopes, cliffs; 5–305 m. s ChI (Santa Catalina Island). Apr–Jul ★

G. clementis Eastw. (p. 1213) SANTA LUCIA BEDSTRAW Per, low, cushion-like, gray-hairy; dioecious. **ST:** 8–13 cm. **LF:** gen whorls of 4, occ 6, 2–7 mm, linear-like, margins rolled under, tip tapered to ± obtuse. **STAMINATE INFL:** few-fld clusters in distal axils. **PISTILLATE INFL:** fl 1 in distal axils. **FL:** corolla rotate, hairy externally, ± yellow. **FR:** berry, hairy. 2*n*=22. N-facing slopes, open woodland; 1130–1780 m. SCoRO (n Santa Lucia Range). Jun–Jul ★

G. cliftonsmithii (Dempster) Dempster & Stebbins SANTA BARBARA BEDSTRAW Per, ± climbing; sts stout, woody, shiny, ± scabrous; dioecious. **ST:** 3–18 dm. **LF:** in whorls of 4, 7–15 mm, ovate to elliptic, 1-veined, margin hairs recurved, tip acute to acuminate, with a sharp hair. **INFL:** few-fld clusters on branchlets. **FL:** corolla rotate, ± yellow, glabrous. **FR:** berry, white; hairs 0. 2*n*=±188. Light shade, coastal canyons, dry banks, chaparral; 200–1220 m. SCoRO, WTR. Apr–Jun ★

G. divaricatum Lam. LAMARCK'S BEDSTRAW Ann, erect, spreading, < 30 cm, glabrous to short-hairy. **ST:** slender. **LF:** in whorls of 5–8, gen weak-reflexed, < 7 mm, lanceolate to oblanceolate. **INFL:** panicle, terminal, open. **FL:** bisexual; corolla white, lobes ± acute. **FR:** nutlets; hairs 0. 2*n*=44. Fields, slopes; 10–700 m. NCoRO, n&c SNF, SnFrB; native to Medit. Similar to *G. parisiense*. May–Jul

G. glabrescens (Ehrend.) Dempster & Ehrend. Per, ascending, 5–31 cm, gen green, ± glabrous to puberulent, base ± woody; dioecious. **LF:** in whorls of 4, ovate to obovate. **INFL:** narrow, lfy, fls few on short, axillary branchlets, little-exserted from lvs. **FL:** corolla gen rotate (shallowly bell-shaped), ± yellow to ± red. **FR:** nutlets; hairs long, straight, spreading.

subsp. ***glabrescens*** Pl glabrous to puberulent, tufted. **ST:** 5–24 cm. **LF:** 6–15 mm, ovate to obovate, tip acute to obtuse, occ blunt. **FL:** corolla gen rotate (shallowly bell-shaped), ± glabrous to sparsely hairy. **FR:** 7–10 mm incl hairs. 2*n*=44. Gravelly slopes, talus; 1520–2590 m. KR, CaRH. Jul–Sep

subsp. ***modocense*** Dempster & Ehrend. (p. 1217) MODOC BEDSTRAW Pl puberulent. **ST:** 8–31 cm. **LF:** 7–20 mm, ovate, tip acute to acuminate. **FL:** corolla rotate, cream, glabrous to sparsely hairy. **FR:** 4–5 mm incl hairs, hairs long, spreading. 2*n*=66. Volcanic talus and gravelly slopes; 1680–2590 m. Wrn. Jun–Aug ★

G. grande McClatchie SAN GABRIEL BEDSTRAW Per, sprawling, climbing, forming dense clumps, woody, densely gray-hairy. **ST:** gen 15–152 cm, stout, rooting at nodes. **LF:** in whorls of 4, 5–15 mm, gen widely elliptic, tip acute to abruptly soft-pointed, margins ± rolled under; petioled. **INFL:** pistillate fls 1 in axils, staminate and bisexual fls in terminal, few-fld panicles. **FL:** bisexual or unisexual, corolla rotate, hairy externally, yellow. **FR:** berry, hairy. 2*n*= ± 220. Oak woodland, chaparral; 425–1220 m. SnGb. May–Jul ★

G. grayanum Ehrend. GRAY'S BEDSTRAW Per 2.5–27 cm, tufted or matted, densely hairy; dioecious. **LF:** in whorls of 4, 4–17 mm, widely ovate to round, tip obtuse to acuminate, not sharp to touch. **INFL:** narrow, lfy, fls few on ascending branchlets. **FL:** corolla rotate, ± cupped at base, ± yellow to ± red, puberulent to glabrous. **FR:** nutlets; hairs long, straight, ± brown. 2*n*=22.

var. ***grayanum*** (p. 1217) Pl 5–27 cm, ± open, densely short-hairy. **LF:** 5–17 mm, tip acute to obtuse. **FR:** 7–10 mm incl hairs. Rocky slopes, ridges; 1830–3500 m. KR, NCoRH, CaRH, n SNH. Jul–Aug

var. ***nanum*** Dempster & Ehrend. Pl 2.5–14 cm, ± compact. **LF:** 4–9 mm, tip acute to acuminate. **FR:** 6–8 mm incl hairs. Open fir forest or chaparral, gravelly or rocky slopes and ridges; 1830–2500 m. NCoRH; OR, NV. Jul–Aug

G. hallii Munz & I.M. Johnst. (p. 1217) NODDING BEDSTRAW Subshrub 30–60 cm, climbing to sprawling, ± gray-hairy, base, sts woody; dioecious, occ monoecious. **LF:** in whorls of 4, 6–17 mm, widely ovate to elliptic, 3-veined, tip obtuse to acute, not sharp to touch; petiole base expanded. **INFL:** lfy clusters, axillary, drooping. **FL:** corolla rotate, cream, hairy externally. **FR:** nutlets; hairs long, straight, white, red-brown to ± blue or gray. 2*n*=22. N-facing slopes, canyon bottoms, mixed conifer forest, oaks or sagebrush; 820–2350 m. s SNH, s SCoRI (Caliente Range), s SCoRO (Sierra Madre Range), TR. May–Aug

G. hardhamiae Dempster (p. 1217) HARDHAM'S BEDSTRAW Per, low, matted, hairs stiff, ± dense on sts, sparser on lvs, fls; dioecious. **ST:** < 30 cm, internodes >> lvs. **LF:** in whorls of 6, 1–5 mm, lanceolate to ovate, fleshy, bright green, margin strongly rolled-under; petiole distinct, ± white. **INFL:** long, lfy, fls few in axillary and terminal clusters. **FL:** corolla rotate, sparsely hairy externally, yellow to green or ± pink. **FR:** berry, glabrous to hairy. 2*n*=22. Serpentine soil with Sargent cypress; 400–950 m. SCoRO (Santa Lucia Range). May–Sep ★

G. hilendiae Dempster & Ehrend. HILEND'S BEDSTRAW Per, hairy to occ scabrous; woody for 3–4 cm distal to base; dioecious. **ST:** 13–45 cm. **LF:** in whorls of 4, flat, lanceolate to ovate or ± round. **INFL:** panicle, ± terminal, lvs reduced. **FL:** corolla bell-shaped, gen pink. **FR:** nutlets; hairs long, straight.

subsp. ***carneum*** (Hilend & J.T. Howell) Dempster & Ehrend. PANAMINT MOUNTAINS BEDSTRAW **ST:** ± erect, gen 13–45 cm, stiff, internodes 2–6 × lvs. **LF:** 7–11 mm, unequal, lanceolate to ovate, tip tapered to acute or sharp to touch. **FL:** corolla bell-shaped, throat open, white to pink, ± glabrous to sparsely hairy. **FR:** 4.5–5.5 mm incl hairs. 2*n*=44. Rocky e slopes, open flats, pinyon-pine woodland, sagebrush scrub; 1650–3400 m. DMtns (Panamint Range). May–Jul ★

subsp. ***hilendiae*** Pl long-hairy. **ST:** 13–33 cm, not stiff; internodes 1.5–4.5 × lvs. **LF:** 5–18 mm, ovate to ± round, tip tapered

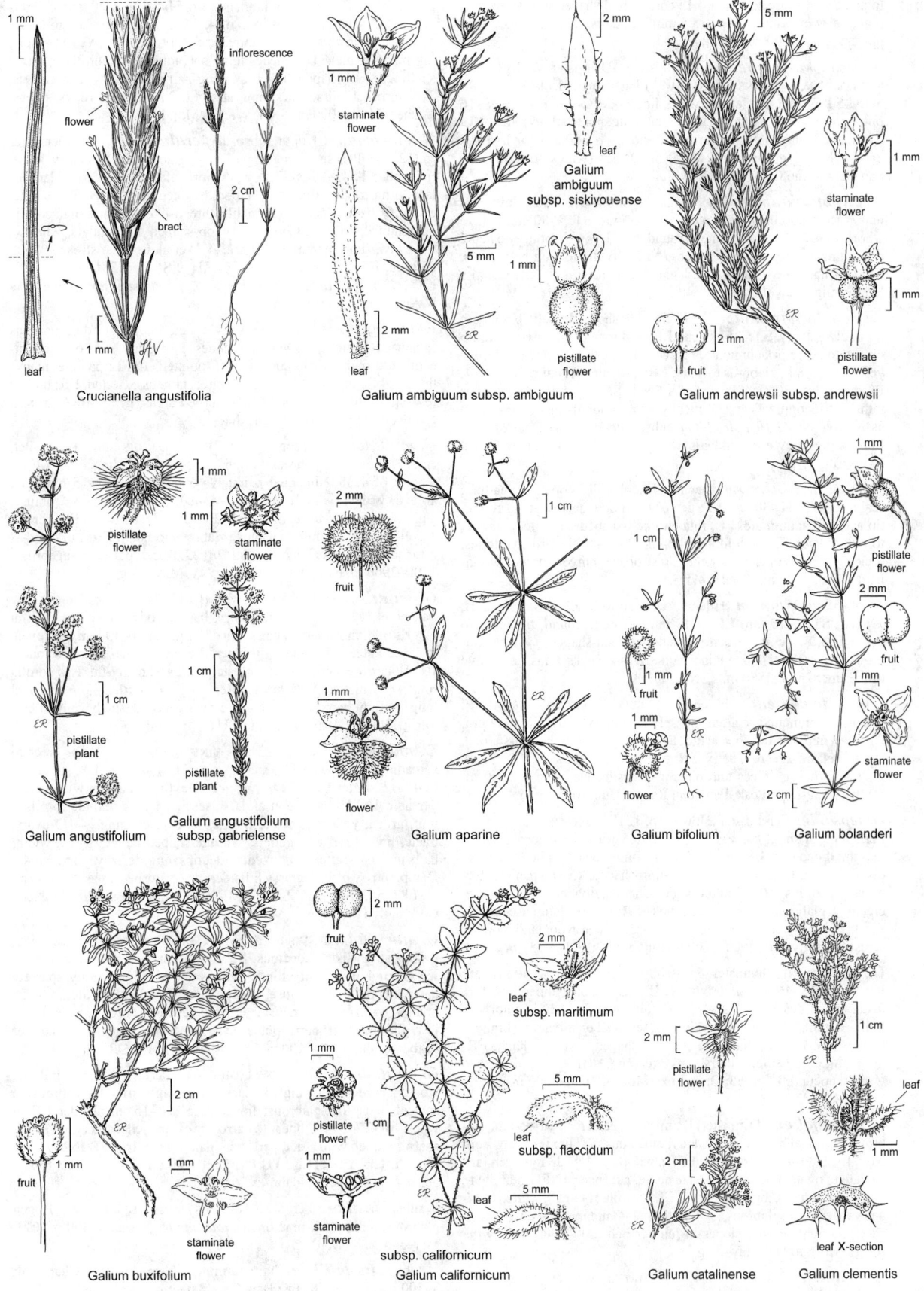

1 mm

inflorescence

flower

2 cm

bract

1 mm

leaf

Crucianella angustifolia

staminate flower

1 mm

2 mm

leaf

Galium ambiguum subsp. siskiyouense

5 mm

leaf

2 mm

pistillate flower

1 mm

Galium ambiguum subsp. ambiguum

5 mm

staminate flower

1 mm

pistillate flower

1 mm

fruit

2 mm

Galium andrewsii subsp. andrewsii

pistillate flower

1 mm

staminate flower

1 mm

pistillate plant

1 cm

Galium angustifolium

1 cm

pistillate plant

Galium angustifolium subsp. gabrielense

2 mm

fruit

1 cm

1 mm

flower

Galium aparine

1 cm

fruit

1 mm

flower

1 mm

Galium bifolium

1 mm

pistillate flower

2 mm

fruit

1 mm

staminate flower

2 cm

Galium bolanderi

2 cm

fruit

1 mm

staminate flower

1 mm

Galium buxifolium

fruit

2 mm

pistillate flower

1 mm

staminate flower

1 mm

1 cm

leaf subsp. maritimum

2 mm

leaf subsp. flaccidum

5 mm

leaf

5 mm

subsp. californicum

Galium californicum

pistillate flower

2 mm

2 cm

Galium catalense

1 cm

leaf

1 mm

leaf X-section

Galium clementis

to acute, not sharp to touch. **FL:** corolla bell-shaped, throat open, ± pink, hairy externally. **FR:** ± 5.5 mm incl hairs. Dry canyons, rocky places; 1300–2700 m. DMoj. May–Jul

subsp. ***kingstonense*** (Dempster) Dempster & Ehrend. KINGSTON MOUNTAINS BEDSTRAW Pl gen matted, densely long-hairy. **ST:** ≤ 35 cm, slender, weak, internodes 2–4 × lvs. **LF:** 8–16 mm, lanceolate to ovate, tip obtuse or acute, not sharp to touch. **FL:** corolla bell-shaped, throat narrow, ± = lobes, pink, hairy to glabrous. 2*n*=44. Rocky places; 1200–2100 m. DMtns (Kingston Range, ne San Bernardino Co.); c NV. May–Jul ★

G. hypotrichium A. Gray ALPINE BEDSTRAW Per, low, gen hairy, occ scabrous, base ± woody; dioecious. **ST:** 2–30 cm. **LF:** in whorls of 4, ± crowded, ovate to round, gen fleshy, tip obtuse to acute or acuminate, not sharp to touch. **INFL:** lfy, fls few on branchlets. **FL:** corolla rotate to bell-shaped, glabrous to hairy. **FR:** nutlets; hairs long, straight, ± yellow.

subsp. ***hypotrichium*** (p. 1217) Pl short, densely branched, minutely velvety. **ST:** 2.5–12 cm. **LF:** 3–8 mm, ovate to ± round, tip obtuse to acute or acuminate. **FL:** corolla rotate to bell-shaped, pale yellow to ± pink, glabrous to hairy. **FR:** 4–6 mm incl hairs. 2*n*=22,44. Ridges, talus; 3000–4200 m. c&s SNH, SNE (White, Sweetwater mtns). If recognized taxonomically, tetraploids (n of Sonora Pass) assignable to *G. hypotrichium* subsp. *ebbettsense* Dempster & Ehrend. but they are ± indistinct from diploids (s of Sonora Pass). May–Aug

subsp. ***inyoense*** Dempster & Ehrend. Pl ± open, spreading, scabrous. **ST:** 12–30 cm. **LF:** gen 6–14 mm, widely ovate to round, tip acute to acuminate. **FL:** rotate to bell-shaped, ± white, glabrous to puberulent. **FR:** 4–5 mm incl hairs. 2*n*=44. Talus, rocky slopes, rocky washes with Jeffrey-pine forest or occ pinyon/juniper woodland; 1950–3200 m. s SNH. May–Aug

subsp. ***subalpinum*** (Hilend & J.T. Howell) Ehrend. Pl short, ± open. **ST:** 2–15 cm. **LF:** 5–10 mm, ovate to round, tip acute to acuminate. **FL:** corolla rotate to shallowly bell-shaped, ± white, glabrous to hairy. **FR:** 4–6 mm incl hairs. 2*n*=44. Rocks, talus, gen above timberline; 2650–3880 m. s SNH. Jun–Aug

subsp. ***tomentellum*** Ehrend. TELESCOPE PEAK BEDSTRAW Pl ± densely branched, matted, densely fine-hairy, gray. **ST:** ≤ 9 cm. **LF:** 5–9 mm, ovate, tip ± acute. **FL:** corolla shallowly bell-shaped, cream-yellow to ± red, hairy. **FR:** 5–8 mm incl hairs. 2*n*=22. Talus, e slopes, rock crevices and outcrops, bristlecone pine; ± 3300 m. DMtns (Telescope Peak, Panamint Range). Jun–Aug ★

G. jepsonii Hilend & J.T. Howell (p. 1217) JEPSON'S BEDSTRAW Per, erect, in small clumps, 8–16 cm, not woody, ± glabrous exc lf margin; dioecious. **LF:** in whorls of 4, congested near pl base, sparse distally on st, 6–15 mm, linear to strap-shaped. **INFL:** panicle, terminal, open, lvs ± 0. **FL:** corolla bell-shaped, divided ≤ 1/2 to base, cream, ± glabrous, lobe tips gen pink. **FR:** nutlets; hairs ascending, short, strongly curved, not hooked. 2*n*=22. Open woodland, conifer forest, gravelly soil; 2000–2500 m. SnGb, SnBr. Jul–Aug ★

G. johnstonii Dempster & Stebbins JOHNSTON'S BEDSTRAW Per, erect, 18–46 cm, glabrous exc lf, base woody; dioecious. **ST:** distal internodes 2–5 × lvs, >> proximal internodes. **LF:** in whorls of 4, 14–30 mm, ± linear. **INFL:** clusters, terminal on ascending branchlets, fls few. **FL:** corolla rotate or ± bell-shaped, ± yellow, hairy. **FR:** nutlets, gen < pedicel; hairs gen spreading, < fr, not hooked, occ weakly upcurved. 2*n*=66. Open mixed forest; 1650–2300 m. SnGb, SnBr. May–Aug ★

G. magnifolium (Dempster) Dempster Per, erect, 10–45 cm, glabrous exc distal bracts, base woody; dioecious. **ST:** flexible, few from base, lfy, internodes 1–4 × lvs. **LF:** in whorls of 4, 8–23 mm, gen lanceolate to ovate, flat, tip tapered to acute, not sharp. **INFL:** staminate clustered in axils, pistillate fls 1. **FL:** corolla rotate, ± white, long-hairy externally (glabrous). **FR:** nutlets, 2–4 mm incl hairs; hairs few, short, straight. Rocky slopes in juniper belt; 800–2000 m. DMtns (Clark Mtn); to UT. May–Aug

G. matthewsii A. Gray (p. 1217) MATTHEWS' BEDSTRAW Per, erect, 13–30 cm, open, glabrous exc distal bracts, base woody; dioe-

cious. **ST:** stiff, gen many-branched from base, tangled or not, internodes 2–7 × lvs. **LF:** in whorls of 4, 2–10 mm, lanceolate to ovate, gen arched, leathery, tip acuminate, sharp to touch. **INFL:** panicle; infl many-branched, branches long, spreading. **FL:** corolla rotate to shallowly bell-shaped, ± yellow, long-hairy externally. **FR:** nutlets, 3–4 mm incl hairs; hairs long, straight. 2*n*=22. Dry, rocky slopes, washes; 1100–3000 m. SNH, W&I, DMoj. May–Aug

G. mexicanum Kunth subsp. ***asperulum*** (A. Gray) Dempster (p. 1217) Per, spreading to reclining, glabrous to hairs few, short, reflexed. **ST:** 45–90 cm. **LF:** gen whorls of 5–8, 13–25 mm, lanceolate to narrowly elliptic, tip tapered to acuminate. **INFL:** panicle, open, lfy, fls few; pedicel, branchlet thread-like. **FL:** bisexual; corolla bell-shaped, > ovary, white; ovary top-shaped. **FR:** nutlets, ± fleshy; hairs few, short, upcurved. 2*n*=22,44. Wet places near streams; 550–2500 m. KR, NCoRO, NCoRH, CaRH, c SNH, MP; to WA, w MT, UT. [*G. m.* var. *a.* (A. Gray) Dempster] 4 subspp. Sp. native to w N.Am, Mex, c Am. Jun–Aug

G. mollugo L. HEDGE BEDSTRAW Per, erect, 3–12 dm, glabrous to hairy. **ST:** stout, ± swollen at nodes. **LF:** gen whorls of 8, oblanceolate to obovate, tip abruptly soft-pointed. **INFL:** panicle, many-fld. **FL:** bisexual; corolla rotate, white; stamens exserted. **FR:** nutlets few, small; hairs 0. 2*n*=22,44. Disturbed areas; gen < 1000 m. KR, NCoRO, SnFrB; native to Eur. Jun–Jul

G. multiflorum Kellogg (p. 1217) KELLOGG'S BEDSTRAW Per, erect, 15–35 cm, glabrous to hairy, base woody; dioecious. **LF:** in whorls of 4, in 2 unequal pairs; lvs of larger pair 8–15 mm, gen arched, widely ovate to round, 3-veined, base round, tip ± obtuse, gen small-pointed to acuminate. **INFL:** panicle, terminal. **FL:** corolla ± bell-shaped, ± white to ± pink, glabrous to hairy. **FR:** nutlets, 4–7 mm incl hairs; hairs long, straight. 2*n*=22. Rocky places in sagebrush; 1300–2900 m. n SNH, GB, DMtns; NV. May–Aug

G. munzii Hilend & J.T. Howell (p. 1217) MUNZ'S BEDSTRAW Per, erect, 10–30 cm, coarse-hairy, base woody; dioecious. **LF:** in whorls of 4, in 2 unequal pairs; lvs of larger pair 6–19 mm, gen ovate to lanceolate, ± 3-veined, tip tapered to acute, gen sharp to touch. **INFL:** narrow panicle, open, clusters axillary, many-fld. **FL:** corolla rotate, ± white to pink, hairy externally. **FR:** nutlets, 3–6 mm incl long, straight hairs. 2*n*=44. N- or e-facing slopes, shady canyon bottoms; 1100–2250 m. DMtns; to s UT. May–Jul ★

G. murale (L.) All. (p. 1217) TINY BEDSTRAW Ann, erect to spreading, glabrous exc lf margin, tip. **ST:** 1–12 cm. **LF:** in whorls of 4–6, 1–4 mm, obovate to oblanceolate, tip acute, gen with a slender hair. **INFL:** fls 1–2 in axils, ± sessile. **FL:** bisexual; corolla 1 mm, green, yellow to white in age, lobes ascending, < 1/2 ovary, ovate, tip obtuse. **FR:** nutlets, sausage-shaped, length 3–4 × width; hairs uneven-scattered, hooked, nutlet tips long-densely hairy. 2*n*=44. Damp, mossy places, grassy hillsides, dry disturbed areas; 20–650 m. NCoRI, n&c SNF, GV, SnFrB, SCo (Riverside Co.), PR (San Diego Co.); native to Eur. Apr–May

G. muricatum W. Wight (p. 1217) HUMBOLDT BEDSTRAW Per, low, cushion-like; dioecious. **ST:** 7–20 cm, gen glabrous. **LF:** in whorls of 4, 6–10 mm, elliptic to obovate, ± leathery, shiny, sparsely hairy adaxially; tip obtuse, abruptly soft-pointed; petioled. **INFL:** open, few-fld clusters in distal axils; fls few. **FL:** corolla rotate, ± yellow, glabrous. **FR:** berry, puberulent. 2*n*=22,44. Shady, damp conifer or mixed forest; 50–1100 m. NCoRO, NCoRH; s OR. May–Oct

G. nuttallii A. Gray SAN DIEGO BEDSTRAW Per, climbing, woody, pl densely tangled, dark red in age, minute-scabrous to sparsely hairy, occ glabrous; dioecious. **ST:** 6–15 dm, slender. **LF:** in whorls of 4, 3–8 mm, linear to narrowly ovate, gen leathery, tip acute, sharp to touch with stout, persistent hair. **STAMINATE INFL:** axillary clusters. **PISTILLATE INFL:** fl 1 in axils. **FL:** corolla rotate, gen ± red. **FR:** berry, glabrous. 2*n*=22.

subsp. ***insulare*** Ferris NUTTALL'S ISLAND BEDSTRAW Pl gen glabrous. Chaparral, pine or *Lyonothamnus* groves; 3–400 m. ChI. Mar–Jun ★

subsp. ***nuttallii*** Pl hairs minute, sharp, curved. Chaparral; 3–500 m. SCo, WTR, PR; Baja CA. Mar–Jun

G. oreganum Britton OREGON BEDSTRAW Per, erect, 20–30 cm, glabrous exc lf margin, veins. **LF:** in whorls of 4, 20–40 mm, widely ovate to elliptic, 3-veined. **INFL:** clusters, terminal on distal branches, fls few. **FL:** bisexual; corolla rotate, ± yellow to ± green, glabrous. **FR:** nutlets, hairs long, hooked. Open conifer forest; 1500 m. KR (Del Norte Co.); to WA. Jun–Aug ★

G. parishii Hilend & J.T. Howell (p. 1217) PARISH'S BEDSTRAW Per, erect or matted, gen puberulent, base woody; not dioecious. **ST:** 5–40 cm. **LF:** in whorls of 4, in 2 unequal pairs, 1–9 mm, ovate to round, gen 3-veined. **INFL:** narrow, open between dense clusters. **FL:** bisexual or unisexual; ± sessile, corolla rotate, gen red to pink or occ yellow, ± hairy externally. **FR:** nutlets, hairs long, straight. 2*n*=22. Steep slopes, rocky washes, talus, among boulders; 1675–3400 m. SnGb, SnBr, e PR, DMtns; s NV. Jun–Aug

G. parisiense L. WALL BEDSTRAW Ann, erect, 15–68 cm, scabrous. **ST:** slender. **LF:** in whorls of 6, 4–9 mm, lanceolate to oblanceolate, gen reflexed in age. **INFL:** panicle, open, few-fld, pedicels thread-like. **FL:** bisexual; corolla basally rotate, ± white to ± purple, lobes erect, glabrous to sparsely hairy. **FR:** nutlets; hairs short, hooked, or 0. 2*n*=44,66. Warm, dry, gen rocky soil to moist areas, coastal-sage scrub, chaparral, grassy hillsides with oaks, roadsides; 110–2200 m. KR, NCoR, SN, SnFrB, SCoR, SCo, WTR, PR (Orange Co.); native to Medit. Apr–Aug

G. porrigens Dempster CLIMBING BEDSTRAW Per, climbing, woody, scabrous, clinging by recurved hairs; dioecious. **ST:** 1–15 dm, slender. **LF:** in whorls of 4, 2–18 mm; tip acute to obtuse or round, terminal hair weak, gen not persistent. **STAMINATE INFL:** axillary clusters. **PISTILLATE INFL:** fls gen 1 in axils. **FL:** corolla rotate, ± yellow to ± red, glabrous. **FR:** berry, glabrous. 2*n*=22.

var. ***porrigens*** (p. 1217) **LF:** 2–18 mm, ± widely oblong to ovate, tip obtuse to round, terminal hair weak, ephemeral, obscurely 3-veined. **FR:** translucent-white, glabrous. Among shrubs in chaparral, forest; 15–2130 m. KR, NCoRO, NCoRI, SN, ScV (Sutter Buttes), CW, SCo, n ChI, TR, PR; s OR, n Baja CA. May–Aug

var. ***tenue*** (Dempster) Dempster (p. 1217) **LF:** 3–8 mm, ± linear, tip acute to obtuse. **FL:** pale yellow to red-green. Chaparral, oak/pine woodland; 150–1050 m. KR, NCoR, CaRF, SN, SnFrB, SCoRI, SCo; s OR. Feb–Jul

G. proliferum A. Gray DESERT BEDSTRAW Ann, erect, 5–30 cm, ± hairy, gen scabrous. **LF:** in whorls of 3–4 and some opposite, sessile, 3–10 mm, gen lanceolate, tip obtuse to abruptly soft-pointed; proximal-most wide, petioled. **INFL:** fls gen 1 in axils, subsessile. **FL:** bisexual; corolla rotate, ± yellow. **FR:** nutlets; hairs short, hooked. Rocky banks, limestone ledges; 1100–1400 m. DMtns; to AZ, Mex. Apr–May ★

G. serpenticum Dempster Per, erect, 5–32 cm, tufted, few-branched, puberulent, base woody; dioecious. **LF:** in whorls of 4, ≤ 15 mm, lanceolate to narrowly elliptic. **INFL:** panicle, terminal, narrow, lfy, on erect to ascending branchlets. **STAMINATE INFL:** axillary clusters. **PISTILLATE INFL:** fls gen 1 in axils. **FL:** corolla ± white, ± glabrous. **FR:** nutlets; hairs long, straight, ± yellow.

subsp. ***scotticum*** Dempster & Ehrend. (p. 1217) SCOTT MOUNTAIN BEDSTRAW **ST:** 12–27 cm. **LF:** 9–15 mm, narrowly elliptic, tapered to acute, gen flat tip, tip not sharp. **FL:** corolla gen ± cup-shaped. 2*n*=22. Steep slopes in open pine forest; 1000–2000 m. KR. Jun–Jul ★

subsp. ***warnerense*** Dempster & Ehrend. (p. 1217) WARNER MOUNTAINS BEDSTRAW **ST:** 5–32 cm. **LF:** 6–15 mm, lanceolate to narrowly elliptic, tip acute, gen reflexed. **FL:** corolla rotate. **FR:** 3–6 mm incl hairs. 2*n*=22. Steep slopes, rocky areas, meadows, juniper woodland; 1450–2750 m. Wrn. Jun–Jul ★

G. sparsiflorum W. Wight SEQUOIA BEDSTRAW Per, erect, 20–50 cm, loose-tufted, base woody; dioecious. **LF:** in whorls of 4, 6–25 mm, widely elliptic to ovate, 3-veined, tip abruptly soft-pointed; ± petioled. **STAMINATE INFL:** panicle, open, lfy, clusters many-fld. **PISTILLATE INFL:** fls gen 1 in axils. **FL:** corolla rotate, gen ± yellow. **FR:** berry, occ hairy.

subsp. ***glabrius*** Dempster & Stebbins Pl occ ± red. **LF:** gen ovate, ± leathery; hairs 0. Shady places in mixed forest; 350–1600 m. KR, NCoRH, NCoRI, CaRF. May–Jul

subsp. ***sparsiflorum*** (p. 1219) Pl green. **LF:** elliptic, occ ± ovate, gen thin, puberulent. 2*n*=22. Shady places in conifer forest; 800–2300 m. SN. Jun–Aug

G. stellatum Kellogg (p. 1219) Shrub, ± erect, 30–90 cm, scabrous; dioecious. **ST:** stout, many-branched, spreading, brittle. **LF:** in whorls of 4, gen 4–8(20) mm, lanceolate to needle-like, gray-green, tip sharp to touch. **INFL:** panicle, axillary, lfy, fls few to many. **FL:** corolla rotate, ± white, hairy externally. **FR:** nutlets, < 5 mm incl hairs; hairs dense, long, straight, white. 2*n*=22,44. Rocky slopes; 130–1600 m. D; to NV, nw Mex. [*G. s.* var. *eremicum* Hilend & J.T. Howell] Mar–Apr

G. tricornutum Dandy ROUGH CORN BEDSTRAW Ann, sprawling to trailing, dense in fl, open in fr. **ST:** 10–35 cm, stout, ± glabrous. **LF:** in whorls of 6–8, 12–19 mm, linear to narrowly oblanceolate, margin short-recurved-prickly, thickened. **INFL:** clusters, axillary, fls few, peduncle < lf, pedicel recurved. **FL:** bisexual; corolla rotate, white. **FR:** nutlets, pairs > 4 mm wide; tubercles acute. 2*n*=44. Grassland, vernal pools, freshwater marsh, roadsides; 100–2000 m. NCoR, n&c SNF, n SNH, GV, SnFrB, SCoRO; native to Eur. Apr–May

G. trifidum L. Per (ann), minutely scabrous. **ST:** 10–50 cm, slender, weak, tangled. **LF:** in whorls of 4–6, 4–19 mm, linear to elliptic or oblong; tip rounded; petioled. **INFL:** several-fld clusters, pedicels slender. **FL:** bisexual; corolla gen 3-lobed, rotate, white to ± pink, glabrous. **FR:** nutlets, spheric, hard, glabrous, smooth, black when dry.

subsp. ***columbianum*** (Rydb.) Hultén (p. 1219) Pl sprawling. **ST:** gen 10–50 cm. 2*n*=24. Wet places, yellow-pine forest; < 3030 m. CA-FP, W&I; to AK, MT, NV, Mex. [*G. t.* var. *pacificum* Wiegand] Jul–Aug

subsp. ***subbiflorum*** (Wiegand) Puff Pl dwarfed, tufted or matted. **ST:** gen ≤ 15 cm. 2*n*=24. Montane meadows, lake margins; 1700–3200 m. KR, CaRH, SNH, SnBr, SnJt, Wrn; to AK, Rocky Mtns. [*G. t.* var. *pusillum* A. Gray] Jun–Aug

G. triflorum Michx. (p. 1219) SWEET-SCENTED BEDSTRAW Per, gen decumbent, radiating from base, glabrous to ± scabrous. **ST:** 20–76 cm. **LF:** in whorls of 6, gen parallel to ground, 6–38 mm, ovate to obovate; tip acute to acuminate, occ small-pointed. **INFL:** clusters, peduncled, axillary, fls 2–3. **FL:** bisexual; corolla rotate, cream to pink, glabrous to sparsely hairy. **FR:** nutlets; hairs hooked, soft, white to brown. 2*n*=22,44,66. Damp, shady forest; 10–3000 m. NW, CaRH, SN, CW, SnBr, MP; to AK, e N.Am, Greenland, Mex, Eur, Asia. May–Jul

G. wrightii A. Gray WRIGHT'S BEDSTRAW Per, erect to sprawling, ± glabrous, puberulent to sparsely hairy; proximal sts woody; not dioecious. **ST:** 15–50 cm. **LF:** in whorls of 4, sessile, ≤ 12 mm, linear to oblanceolate, tip acute to acuminate. **INFL:** panicle, terminal, open. **FL:** bisexual or unisexual; corolla rotate, gen red to pink, glabrous (sparsely hairy). **FR:** nutlets, ± 5 mm incl hairs; hairs few, long, straight. 2*n*=22. Shady, rocky canyons, pinyon-pine woodland; 1600–2000 m. DMtns (Clark Mtn); to w TX, Mex. May–Jun, Aug–Sep ★

KELLOGGIA

Per from thin rhizome. **LF:** opposite; stipules ± scale-like, base fused. **INFL:** cyme, terminal, open, fls few, pedicel long, slender. **FL:** calyx 4–5-lobed; corolla funnel-shaped, 4–5-lobed; styles 2, fused proximally. **FR:** nutlets 2, wider distally. 2 spp.: w N.Am, China. (Albert Kellogg, pioneer CA botanist, 1813–1887)

K. galioides Torr. (p. 1219) **ST:** erect, 15–40 cm, ± 4-angled. **LF:** 19–38 mm, lanceolate to narrowly ovate, occ with smaller axillary lvs. **FL:** calyx inconspicuous; corolla pink or white, tube slender, throat short, lobes lanceolate. **FR:** hairs hooked. ± open places in conifer forest; 700–3110 m. KR, NCoRH, CaRH, SNH, SnBr, PR, MP; to WA, MT, WY, NM. May–Aug

SHERARDIA

1 sp. (Wm. Sherard, Dillenius' patron, John Ray's friend, 1659–1728)

S. arvensis L. (p. 1219) FIELD MADDER Ann, matted. **ST:** decumbent, gen many-branched at base, 7–16 cm, 4-angled. **LF:** whorls of 5–6; 4–13 mm, lanceolate or oblanceolate, margin thick, tip acute or weak-spined, stipules lf-like. **INFL:** head, axillary, involucred; fls sessile, gen 2–3, gen incl in involucre; involucral bracts ± free, ± = lvs, gen < peduncle. **FL:** calyx lobes 6, persistent, tips ± dissected; corolla salverform, gen 4-lobed, pink or lavender; styles 2, thread-like, fused proximally. **FR:** 2 nutlets; hairs soft. 2*n*=22. Pastures, disturbed areas, grassland, dry meadows, oak woodland; 10–1160 m. NCo, NCoRO, NCoRI, CaRF, n SN, ScV, SnFrB, SCoRO, SCo; to ID, AZ, e US; native to Medit. Mar–Jul

RUTACEAE RUE FAMILY

Lindsay P. Woodruff & James R. Shevock, except as noted

Per, shrub, tree, strongly aromatic, occ thorny. **LF:** gen alternate, simple or compound, dotted with minute, translucent glands; stipules 0. **INFL:** cyme, raceme, or fls 1, gen bracted. **FL:** gen bisexual; sepals, petals each 4 or 5, free or fused at base; sepals gen persistent; petals gen ± white or ± green; stamens gen 2–4 × petal number; ovary superior, gen lobed, chambers 1–5, ovules 1–several per chamber. **FR:** berry, drupe, winged achene, or capsule. **SEED:** gen oily. ± 158 genera, ± 1900 spp.: esp trop, warm temp, esp s Afr, Australia; used or cult for food (*Citrus*, 20–25 spp.), perfume, medicine, timber, orn (*Choisya*, *Skimmia*, etc). Some TOXIC: oils may cause sunburn or dermatitis. Scientific Editors: Douglas H. Goldman, Bruce G. Baldwin.

1. Lf simple or 0
  2. Lvs opposite; ovary chamber 1; fr drupe-like . . . . **CNEORIDIUM**
  2′ Lvs alternate; ovary chambers 2; fr a capsule . . . . **THAMNOSMA**
1′ Lf deeply dissected or compound
  3. Shrub or small tree; petals ± green-white, entire; fr a round, ± flat, wing-encircled achene . . . . **PTELEA**
  3′ Per, subshrub; petals yellow, wavy to fringed; fr a capsule . . . . **RUTA**

CNEORIDIUM

1 sp. (Greek: diminutive of *Cneorum*, spurge olive)

C. dumosum (Torr. & A. Gray) Baill. (p. 1219) BUSHRUE Shrub, gen < 1.5 m, rounded, evergreen. **ST:** intricately branched. **LF:** simple, opposite, 1–2.5 cm, linear. **INFL:** cyme or cluster; fls 1–3. **FL:** sepals 4, fused at base, 1–1.5 mm; petals 4(5), 5–6 mm, obovate, white; stamens 8, longest opposite sepals, filaments wider at base; ovary sessile, chamber 1, ovules 2, style short, flat, from near ovary base, stigma head-like. **FR:** drupe-like, 5–6 mm diam, gland-dotted, ± green, red in age. **SEED:** 1–2, 5–6 mm, ± spheric, dark brown. Mesas, coastal bluffs; < 1000 m. s SCo, s ChI (San Clemente Island); Baja CA. Feb–May

PTELEA HOP TREE

Michael A. Vincent & James R. Shevock

Shrub, small tree, fls gen bisexual and pistillate. **LF:** pinnately compound; lflets 3–5, ± sessile, entire to finely serrate. **INFL:** panicle-like. **FL:** sepals 4–5; petals 4–5, entire, ± green-white; stamens 4–5, filaments hairy on inner side; ovary chambers 2, style short, stigmas 2. **FR:** achene, ± flat, round, gland-dotted, winged. **SEED:** 2. 3 spp.: US, Mex. (Greek: elm, from similar fr)

P. crenulata Greene (p. 1219) Pl gen < 5 m. **LF:** deciduous; petiole 2–5 cm; lflets 3, 2–7 cm, lanceolate to obovate, glabrous adaxially, ± hairy abaxially. **INFL:** ± flat-topped. **FL:** sepals minute; petals 4–5 mm, fragrant. **FR:** 1–2 cm, ± straw-colored; wing gen ± notched at tip, base; style persistent. Scrub, woodland; gen < 1050 m. e KR, NCoRI, CaR, SNF, ScV (Sutter Buttes), SnJV, SnFrB, PR. Apr–May

RUTA RUE

Per, subshrub. **LF:** pinnately or ternately compound. **INFL:** panicle or cluster, erect, terminal. **FL:** petals 4–5, wavy to fringed; stamens 8 or 10, in 2 series; ovary chambers 4–5. **FR:** capsule, opening at tip, occ indehiscent, 4–5 lobed, leathery. **SEED:** several. 7 spp.: nw Afr islands, Medit, sw Asia. (Latin: classical name) Cult for orn, flavoring, medicine.

R. chalepensis L. **ST:** 4–8 dm. **LF:** 2–3-pinnately divided, gen 10–20 cm, oblong in outline; segments gen 1–1.5 cm, narrowly elliptic, entire. **FL:** petals 6–8 mm, yellow, margins inrolled, fringed. **FR:** lobes 7–8 mm long, tips pointed. **SEED:** angled, tubercled, ± brown. Uncommon. Disturbed places; gen < 500 m. s CA-FP, esp near coast; native to Medit. *R. graveolens* L. an occ escape from cult. Mar–Aug

THAMNOSMA TURPENTINE-BROOM

Lindsay P. Woodruff

Subshrub or shrub. **LF:** simple or divided into 3 segments, alternate, minute, seasonally deciduous. **INFL:** panicle (raceme-like or fls scattered along sts). **FL:** bisexual; sepals 4, fused at base, persistent; petals 4, erect in fl; stamens 8, in 2 series; ovary

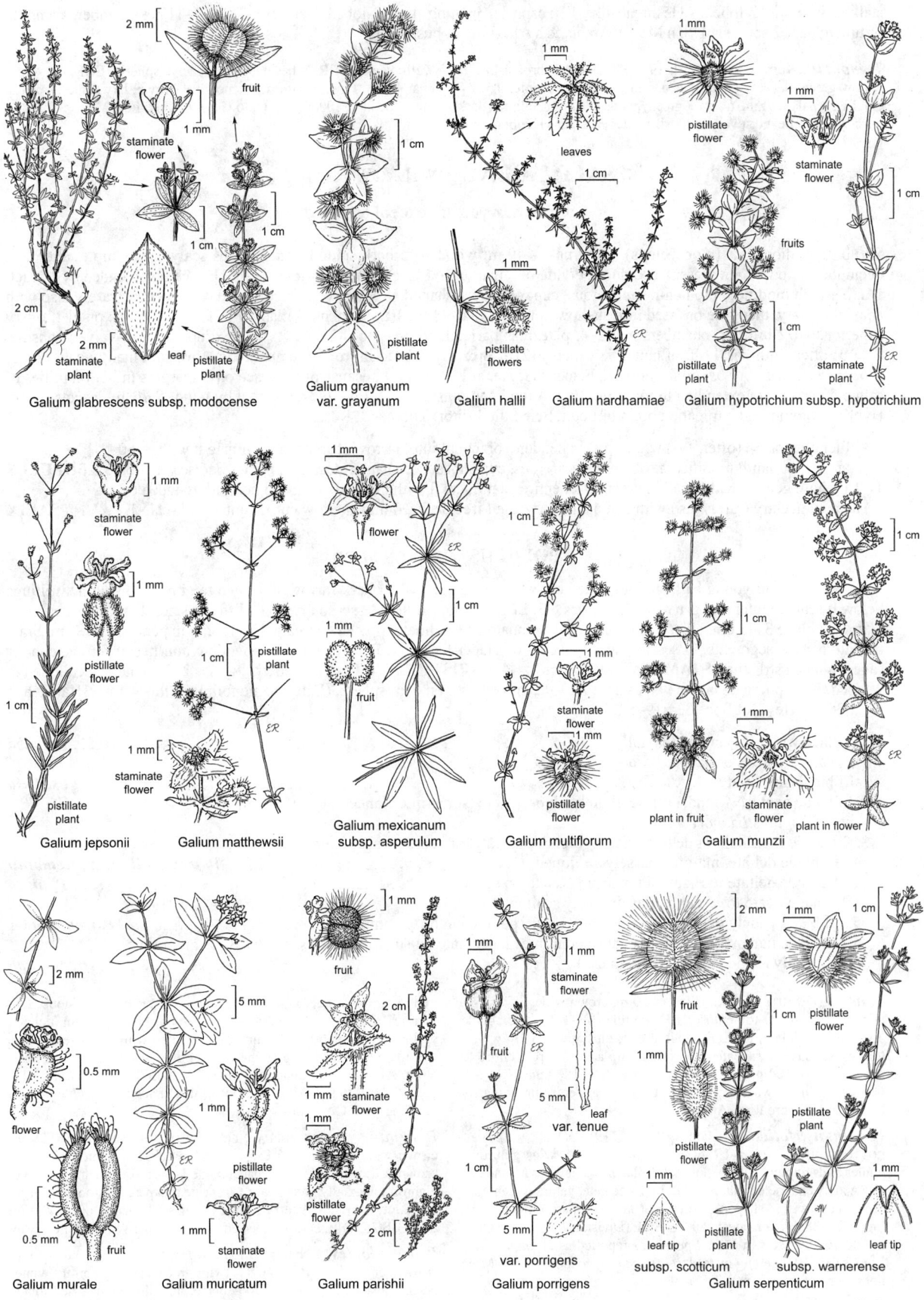

Galium glabrescens subsp. modocense

Galium grayanum var. grayanum

Galium hallii

Galium hardhamiae

Galium hypotrichium subsp. hypotrichium

Galium jepsonii

Galium matthewsii

Galium mexicanum subsp. asperulum

Galium multiflorum

Galium munzii

Galium murale

Galium muricatum

Galium parishii

Galium porrigens

Galium serpenticum

stalked or sessile, 2-lobed, style thread-like. **FR**: capsule, opening at tip, 2-lobed, leathery. **SEED**: 1–11 per chamber, ± smooth or tubercled. 12 spp.: sw US, n Mex, s Arabia to s Afr. (Greek: bush odor)

T. montana Torr. & Frém. (p. 1219)   **ST**: 3–6 dm, broom-like, ± yellow-green, densely gland-dotted, gen lfless. **LF**: simple, 0.5–1.5 cm. **FL**: sepals ± 2 mm, ovate, ± green; petals 8–12 mm, elliptic, dark purple, tips reflexed; ovary stalked, ovules 3–8 per chamber, style well-exserted. **FR**: lobes ± 5 mm wide, ± spheric. **SEED**: 1–4 per chamber, ± 4 mm, reniform, ± smooth, ± brown. Dry slopes, washes, mesa tops; < 2100 m. PR, D; to UT, AZ, Mex. Feb–May

SALICACEAE WILLOW FAMILY

John O. Sawyer, Jr., except as noted

Shrub, tree; dioecious (monoecious). **ST**: trunk < 40 m; wood soft; bark smooth, bitter; buds scaly. **LF**: simple, alternate, deciduous; stipules gen present, deciduous or not, often large. **INFL**: catkin [or various, or fls 1]; each fl subtended by 1 bract. **FL**: perianth modified into non-nectariferous, cup- or saucer-shaped structure or reduced to adaxial nectary (rarely also with abaxial nectary, then free or fused into shallow cup). **STAMINATE FL**: stamens 2–many. **PISTILLATE FL**: pistil 1, ovary superior to 1/2-inferior, chambers gen 2–4, placentas parietal, stigma lobes 2–4. **FR**: berry, drupe, or 2–4-valved capsule. **SEED**: often with basal tuft of hairs. 58 genera, 1210 spp.: widespread in trop, n temp, arctic. Now incl many genera (e.g., *Flacourtia, Idesia, Xylosma*) formerly in Flacourtiaceae, at least in part because of presence on lf margins in both families of salicoid teeth (vein extending to tooth tip). In CA (and gen outside CA), *Populus* pollinated by wind, *Salix* by insects, wind. Hybrids common; identification often difficult. Scientific Editor: Thomas J. Rosatti.

1. Lf length not >> (often ± =) width; catkin pendent, bract cut into narrow segments; staminate fl with 8–60
   stamens; perianth modified into non-nectariferous, cup- or saucer-shaped structure; winter bud scales > 3 . . . **POPULUS**
1′ Lf length gen >> width; catkin erect, bract entire; staminate fl with (1)2(10) stamens; perianth reduced to
   adaxial nectary (rarely also with abaxial nectary, then free or fused into shallow cup); winter bud scale 1 . . . . . . **SALIX**

POPULUS COTTONWOOD

Tree. **ST**: < 40 m; young bark smooth, pale yellow-green to gray; older bark furrowed, brown to gray; twigs with swellings below lf scars; winter bud gen resinous, scales > 3. **LF**: juvenile, adult, late-season lvs may differ in size, shape, hairiness; gen glabrous; blade 3–11 cm, elliptic to deltate, veins pinnate or ± palmate, tip gen elongate. **INFL**: catkin pendent, 3–8 cm; bract cut into narrow segments; fls sessile; nectary a cup- or saucer-like disk. **FL**: perianth modified into non-nectariferous, cup- or saucer-shaped structure. **STAMINATE FL**: stamens 8–60. **PISTILLATE FL**: style short, stigmas 2–3(4), large, scalloped to 2-lobed. **FR**: spheric to conic; valves 2–3(4), 3–12 mm. 40 spp.: n hemisphere. (Latin: name for pls of this genus) [Hamzeh et al. 2006 J Torrey Bot Soc 133:519–527]

1. Lf blade white-tomentose abaxially. . . . . *P. alba*
1′ Lf blade glabrous or hairy abaxially
   2. Lf blade lanceolate . . . . *P. angustifolia*
   2′ Lf blade ovate, ± round, elliptic, rhomboid, or deltate (sometimes lanceolate on suckers or stressed
      pls of *Populus trichocarpa*)
      3. Lf blade rhomboid or deltate, margin crenate, serrate, or coarsely scalloped
         4. Lf blade deltate, margin coarsely scalloped . . . . *P. fremontii* subsp. *fremontii*
         4′ Lf blade deltate to rhomboid, margin crenate to serrate. . . . . *P. nigra*
      3′ Lf blade ovate to ± round, margin finely scalloped
         5. Lf blade widely ovate to ± round, 2–4(7) cm, petiole laterally compressed. . . . . *P. tremuloides*
         5′ Lf blade narrowly to widely ovate (sometimes lanceolate on suckers or stressed pls), 3–7 cm; petiole
            abaxially round, adaxially channeled . . . . *P. trichocarpa*

P. alba L.   WHITE POPLAR   Tree < 20 m; crown wide. **ST**: twigs, winter buds white-tomentose. **LF**: petiole 1/3–1/2 blade; blade 3–9 cm, 3–5-lobed, adaxially blue-green, glossy, abaxially white-tomentose, base ± truncate to ± cordate, tip acute, margin entire to toothed. Disturbed places near settlements; 600–1800 m. KR, CaR, GB, expected elsewhere; native to c Eur, c Asia. Persisting primarily by clonal root-sprouting. Apr–Jun

P. angustifolia E. James (p. 1219) NARROW-LEAVED COTTONWOOD   Tree < 15 m; crown slender. **ST**: twigs glabrous, winter buds gummy. **LF**: petiole ≤ 1/3 blade, abaxially round, adaxially widely channeled; blade 4–11 cm, lanceolate, yellow-green, glabrous, base wedge-shaped, tip acute or tapered, margin finely scalloped. 2n=38. Streamsides; 1500 m. SNE (Division Creek, Inyo Co.); to OR, Rocky Mtns, n Mex. Specimens reported as *P. angustifolia* from SnBr, W&I belong instead to *P. trichocarpa*; those from DMtns not seen. Apr–May   ★

P. fremontii S. Watson subsp. *fremontii* (p. 1219) ALAMO OR FREMONT COTTONWOOD   Tree < 20 m; crown wide. **ST**: twigs yellow, gray in age, glabrous to hairy; winter buds resinous. **LF**: petiole 1/2 to = blade, laterally compressed; blade 3–7 cm, deltate, yellow-green, glabrous to hairy, often stained with milky resin, base ± cordate to truncate, tip ± tapered, margin coarsely scalloped. Scattered. Alluvial bottomland, streamsides; < 2000 m. CA (exc MP); to c Rocky Mtns, n Mex. Hybrids with *p. trichocarpa* (*P.* ×*parryi*) reported from CA. Mar–Apr

P. nigra L.   BLACK POPLAR, LOMBARDY POPLAR   Tree < 20 m; crown wide or slender. **ST**: twigs yellow-brown, winter buds red-brown. **LF**: petiole 1/3–2/3 blade; blade 3–5 cm, deltate to rhomboid, glabrous, adaxially green, base ± truncate to tapered, tip acute, margin crenate to serrate. Disturbed places near settlements; 600–1800 m. SnFrB, SCo, SNE, DMoj, expected elsewhere; native to Eur. Feb–May

P. tremuloides Michx. (p. 1219)   QUAKING ASPEN   Tree < 15 m; crown slender; highly clonal. **ST**: twigs green-white, glabrous; winter buds shiny. **LF**: petiole 2/3 to = blade, laterally compressed; blade 2–4(7) cm, widely ovate to ± round, glabrous, adaxially green, abaxially glaucous, base rounded to cordate, tip tapered, margin finely

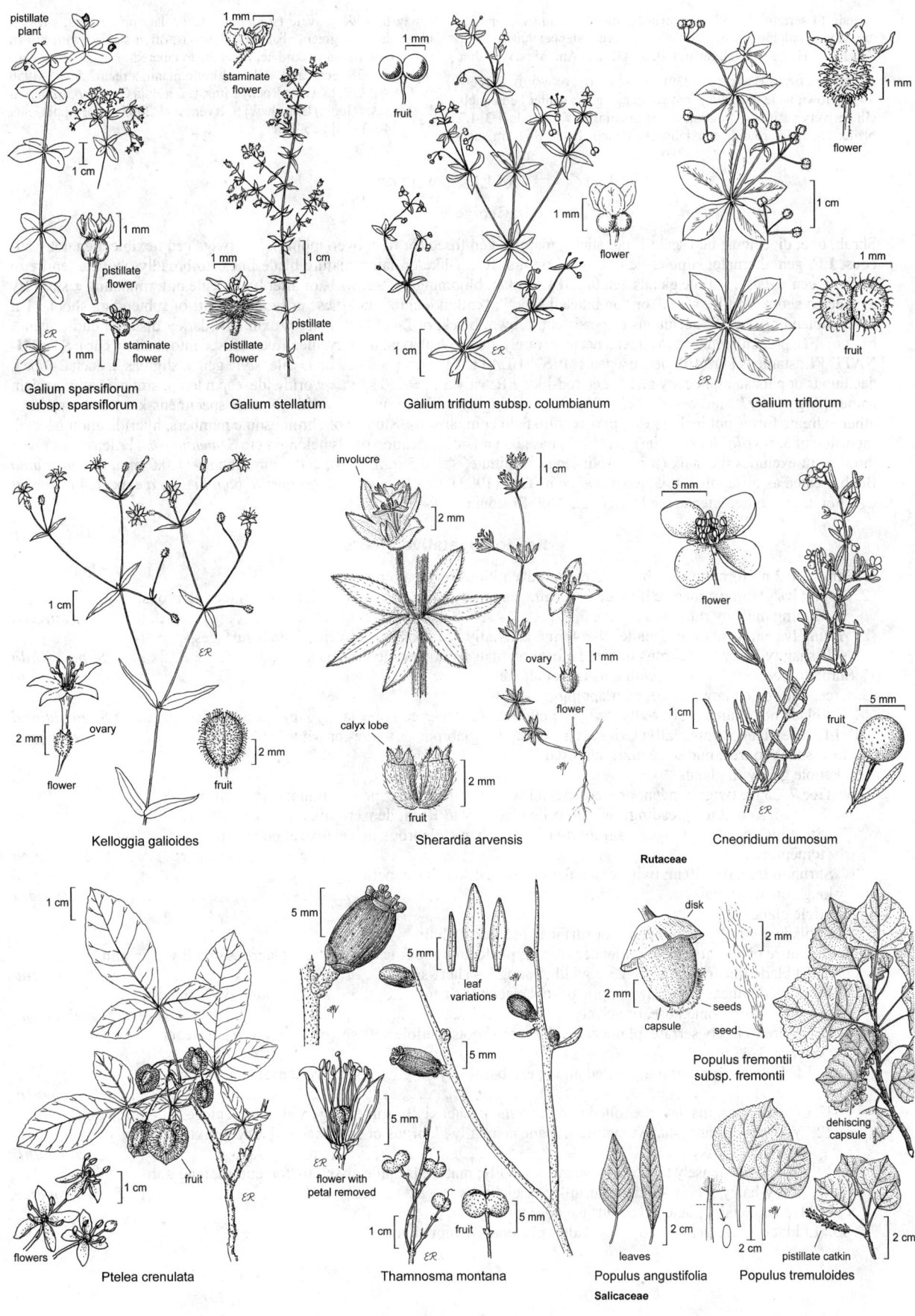

pistillate
plant

1 cm

pistillate
flower

1 mm

staminate
flower

1 mm

ER

**Galium sparsiflorum
subsp. sparsiflorum**

staminate
flower

1 mm

pistillate
plant

1 cm

pistillate
flower

1 mm

Galium stellatum

fruit

1 mm

flower

1 mm

1 cm

ER

Galium trifidum subsp. columbianum

flower

1 mm

fruit

1 mm

1 cm

ER

Galium triflorum

1 cm

ER

ovary

flower

2 mm

fruit

2 mm

Kelloggia galioides

involucre

2 mm

calyx lobe

fruit

2 mm

1 cm

ovary

1 mm

flower

Sherardia arvensis

5 mm

flower

1 cm

fruit

5 mm

ER

Cneoridium dumosum

Rutaceae

1 cm

fruit

1 cm

flowers

ER

Ptelea crenulata

5 mm

5 mm

leaf
variations

5 mm

5 mm

ER

flower with
petal removed

5 mm

1 cm

fruit

5 mm

ER

Thamnosma montana

disk

2 mm

2 mm

capsule

seeds

seed

**Populus fremontii
subsp. fremontii**

dehiscing
capsule

leaves

2 cm

Populus angustifolia

2 cm

pistillate catkin

2 cm

Populus tremuloides

Salicaceae

crenate to serrate. 2*n*=38. Streamsides, moist openings, slopes in montane, subalpine forest, woodland, sagebrush steppe; 900–3150 m. KR, NCoRH, CaR, SNH, SnBr, GB; to AK, e N.Am, Mex. Apr–Jun

P. trichocarpa Hook. (p. 1229)   BLACK COTTONWOOD   Tree < 30 m; crown wide. **ST**: twigs brown, in age gray; winter buds finely ciliate, very resinous, fragrant when opening. **LF**: petiole 1/3–1/2 blade, abaxially round, adaxially channeled; blade 3–7 cm, nar-

rowly to widely ovate (sometimes lanceolate on suckers, stressed pls), adaxially green, abaxially glaucous, often stained with brown resin, base round to cordate, tip acute to tapered, margin finely scalloped. 2*n*=38. Scattered. Alluvial bottomland, streamsides; 5–3050 m. CA-FP, GB; to AK, n Rocky Mtns, UT, n Baja CA. [*P. balsamifera* L. subsp. *t.* (Hook.) Brayshaw] [Cervera et al. 2005 Theor Appl Genet 111:1440–1456] Feb–Apr

SALIX   WILLOW

George W. Argus

Shrub, tree; dioecious; bud scale 1, not sticky, margins gen fused (or free, overlapping). **ST**: twigs gen flexible, gen not glaucous. **LF**: gen alternate; stipules gen vestigial on first lvs, lf-like on later; mature blade linear to broadly obovate, entire to toothed, gen ± hairy; petiole glands gen 0. **INFL**: catkin, blooming before, with, or after lvs, sessile or terminating ± short lfy or bracted shoot ("on lfy shoot" or "on bracted shoot"; "catkin length" incl lfless or bractless part of subtending shoot); 1 fl bract subtending each fl, deciduous or persistent, brown, black, or 2-colored (paler proximally, darker distally; darker gen ± brown). **FL**: perianth reduced to adaxial nectary (rarely also with abaxial nectary, then free or fused into shallow cup). **STAMINATE FL**: stamens (1)2(10); nectary gen 1. **PISTILLATE FL**: ovary stalked or sessile, style gen 1, stigmas 2, each 2-lobed, deciduous or persistent; nectary gen 1, gen rod-like. **FR**: valves 2. ± 450 spp.: ± worldwide, esp n temp, arctic. (Latin: ancient name) [Argus 1997 Madroño 44:115–136] Difficult, highly variable, many hybrids. Not all specimens key easily; sprouts, other extreme forms not incl in keys, may require field comparisons. Studies of chromosome numbers, hybridization needed. Inclusion of *S. sessilifolia* Nutt. in TJM (1993) based on misidentification of pls belonging to *S. melanopsis*. Fr length as given throughout excludes the stalk (stipe). Hair lengths: minute, < ± 0.5 mm; short, ± 0.5 mm; long, > ± 0.5 mm. *S. commutata* Bebb, treated as misapplied to *S. eastwoodiae* in TJM (1993), may occur in n CA mtns; *S. bonplandiana* expected in s-most CA. For alternate treatments, see Dorn (e.g., 2000 Brittonia 52:1–19).

Key to Vegetative Plants

1. Shrub < 0.2 m, forming mats by rhizomes or adventitious roots, alpine
  2. Young lvs glabrous; mature lf blade 6–22 mm, abaxially strongly net-veined, both surfaces glabrous; pl forming mats by rhizomes . . . . ***S. nivalis***
  2′ Young lvs hairy; mature lf blade 19–44 mm, abaxially not strongly net-veined, both surfaces sparsely soft-shaggy-hairy or glabrous in age; pl forming mats by adventitious roots . . . . ***S. petrophila***
1′ Shrub to tree, > 0.2 m, not forming mats, not alpine
  3. Tree; bud scale margins free, overlapping
    4. Lf blade not glaucous abaxially; twigs ± yellow or yellow-green or ± tan; young lvs white-hairy . . . . . . ***S. gooddingii***
    4′ Lf blade glaucous abaxially; twigs ± tan; young lvs glabrous or white- or white-and-rusty-hairy . . . . ***S. laevigata***
  3′ Low shrub to tree; bud scale margins fused
    5. Petiole gen with glands
      6. Tree 7–25 m; twigs pendent or erect, flexible or ± brittle at base; petiole tomentose or silky
        7. Sts, twigs erect to spreading; twigs ± yellow or gray to ± tan, densely silky; petiole silky . . . . ***S. alba***
        7′ Sts, twigs pendent; twigs ± tan, short-silky to velvety, glabrous in age (exc at nodes); petiole tomentose . . . . ²***S. babylonica***
      6′ Shrub to tree, 0.6–10 m; twigs erect, flexible or brittle at base; petiole glabrous or long-soft-wavy-hairy . . . . ***S. lasiandra***
    5′ Petiole glands 0
      8. Lf blade abaxially not glaucous or surface obscured by hair
        9. Lf entire or ± sparsely short-slender-serrate; pl clonal by root-shoots; petiole glabrous or silky, 2–6 mm
          10. Lf blade linear, length 10–23 × width; young lvs short-silky . . . . ²***S. exigua***
          10′ Lf blade linear to narrow-elliptic or -oblanceolate, length 2.8–8.5 × width; young lvs long-soft-wavy-hairy, tomentose, or velvety . . . . ***S. melanopsis***
        9′ Lf entire to finely serrate; pl not clonal by root-shoots; petiole soft-shaggy-hairy, tomentose, or velvety, 3–17 mm
          11. Lf blade margins strongly rolled under, esp basally; young lvs abaxially densely silky or woolly . . . . ²***S. sitchensis***
          11′ Lf blade margins flat or ± rolled under; young lvs abaxially soft-shaggy-hairy, tomentose, silky, or woolly
            12. Young lvs shaggy-hairy; mature lf blade abaxially glabrous or soft-shaggy-hairy, adaxially shiny to highly glossy . . . . ***S. boothii***
            12′ Young lvs densely tomentose, silky, or woolly; mature lf blade abaxially tomentose, long-soft-wavy-hairy, woolly-tomentose, silky, or glabrous in age, adaxially dull . . . . ***S. eastwoodiae***
      8′ Lf blade abaxially glaucous or surface obscured by hair
        13. Lf blade linear, length 6.6–25 × width; pl clonal by root-shoots . . . . ²***S. exigua***

13′ Lf blade narrowly elliptic or wider, length 1.3–8(14.4) × width; pl not clonal by root-shoots (but
    sometimes clonal by layering or st fragmentation)

  14. Lvs (at least some) ± opposite. . . . . *S. purpurea*

  14′ Lvs alternate

    15. Tree 7–25 m; twigs pendent . . . . [2]*S. babylonica*

    15′ Shrub to tree, 0.2–10 m; twigs erect or decumbent

      16. Lf blade margins gen strongly rolled under, at least basally

        17. Lf blade adaxially gen highly glossy. . . . . [9]*S. lasiolepis*

        17′ Lf blade adaxially shiny or dull

          18. Lf blade abaxially tomentose or woolly or gen sparsely silky, often with white and rusty hairs,
            glaucous, adaxially shiny . . . . *S. scouleriana*

          18′ Lf blade abaxially very densely silky or woolly, surface obscured by white hairs, adaxially dull

            . . . . [2]*S. sitchensis*

      16′ Lf blade margins flat or ± rolled under

       19. Lf blade adaxially highly glossy

        20. Lf blade abaxially moderately to densely shaggy, tomentose, or woolly, at least on midrib and
          veins, rarely ± glabrous

          21. Twigs gen thinly glaucous; lf blade abaxially gen woolly or shaggy-hairy, 18–55 wide. . . . . . *S. hookeriana*

          21′ Twigs not glaucous; lf blade abaxially gen tomentose, 6–32 mm wide . . . . [9]*S. lasiolepis*

        20′ Lf blade abaxially glabrous (or soon becoming so) to sparsely silky

          22. Petiole tomentose or velvety; lf blade 35–125 mm; twig hairs spreading; lf blade glands
           adaxially submarginal or nearer blade center; 0–2500 m . . . . [9]*S. lasiolepis*

          22′ Petiole sparsely soft-shaggy-hairy; lf blade 19–75(115) mm; twig hairs appressed or curved
           toward st; lf blade glands gen marginal; 2500–4000 m. . . . . *S. planifolia*

       19′ Lf blade adaxially dull to shiny

        23. Lf blade abaxially glabrous, downy, or soft-shaggy-hairy

          24. Lf blade white- or white-and-rusty-hairy

            25. Stipules gen lf-like; young lvs, mature blades densely silky or gen short-tomentose; 0–2500 m

              26. Stipule tip acute; lf glands submarginal or distinctly on adaxial surface; young lvs thick,
                yellow-green (or ± red). . . . . [9]*S. lasiolepis*

              26′ Stipule tip convex to rounded; lf glands marginal; young lvs thin, ± red . . . . *S. tracyi*

            25′ Stipules 0 or vestigial on first lvs, lf-like on later lvs or not; young lvs, mature blades gen
             densely silky; 1400–3600 m

              27. Lf blade persistently long-(often white-)hairy, 32–89 mm; petiole 3–9 mm; stipules gen 0
                or vestigial . . . . [5]*S. geyeriana*

              27′ Lf blade in age ± glabrous or sparsely short-(gen rusty-)hairy, 44–110 mm; petiole 5–16
                mm; stipules vestigial on first lvs, lf-like on later. . . . . [3]*S. lemmonii*

          24′ Lf blade white-hairy or glabrous

            28. Stipules 0 or vestigial on first lvs, lf-like on later

              29. Young lvs, mature blades gen densely silky; twigs gen glaucous; 1450–3600 m. . . . . [5]*S. geyeriana*

              29′ Young lvs, mature blades densely silky or gen short-tomentose; twigs not glaucous;
                0–2500 m . . . . [9]*S. lasiolepis*

           28′ Stipules lf-like (vestigial or not in *Salix bebbiana*)

             30. Lf blade abaxially gen with raised, net veins, length 1.7–3.9 × width, entire to crenate;
               young lvs densely silky — n MP. . . . . [2]*S. bebbiana*

             30′ Lf blade abaxially smooth, length 2.2–9.6 × width, gen finely serrate (entire or not in *Salix
               lasiolepis*); young lvs gen glabrous or sparsely hairy (densely silky in *Salix lasiolepis*)

              31. Lf base wedge-shaped or convex; blade shiny adaxially; twigs tomentose, glabrous in age

                . . . . [9]*S. lasiolepis*

              31′ Lf base rounded, subcordate, cordate, or convex; blade dull adaxially; twigs glabrous to
                sparsely long-soft-wavy-hairy

                32. Twigs yellow-gray or yellow- or gray-brown, not peeling. . . . . *S. lutea*

                32′ Twigs gen ± tan, peeling

                  33. Lf blade strap-shaped to narrowly elliptic, length 2.9–6.4 × width, base convex to
                    subcordate; stipule tip rounded to acuminate. . . . . *S. ligulifolia*

                33′ Lf blade narrowly oblong or lanceolate to obovate, length 2.4–4.5 × width, base rounded
                    to cordate; stipule tip convex to rounded . . . . *S. prolixa*

        23′ Lf blade abaxially long-soft-wavy-hairy, tomentose, silky, or velvety

          34. Twigs glaucous

            35. Lf blade dull adaxially

              36. Lf blade abaxially very densely white- or white-and-rusty-silky; young lvs short-silky;
               twigs brittle at base . . . . [2]*S. drummondiana*

              36′ Lf blade abaxially sparsely to moderately densely white-silky; young lvs long-silky; twigs
               flexible at base . . . . [3]*S. orestera*

35′ Lf blade shiny adaxially
 37. Lf blade hairier abaxially than adaxially (exc midrib abaxially ± glabrous), gen densely white-silky to -tomentose; young lvs short-silky . . . . [2]*S. drummondiana*
 37′ Lf blade equally hairy abaxially and adaxially (midrib abaxially gen ± as hairy as blade), white-, rusty-, or white-and-rusty-hairy, glabrous in age or not; young lvs short- or long-silky
  38. Lf blade hairs long, gen white or white and rusty, persistent; petiole 2–9 mm; stipules gen 0 or vestigial . . . . [5]*S. geyeriana*
  38 Lf blade hairs short, gen rusty, or in age 0; petiole 5–16 mm; stipules vestigial on first lvs, lf-like on later lvs. . . . . [3]*S. lemmonii*
34′ Twigs not glaucous
 39. Young lvs sparsely or moderately hairy; petiole puberulent; twigs soft-shaggy-hairy — n MP
. . . . [2]*S. bebbiana*
 39′ Young lvs very densely hairy; petiole tomentose, silky, or velvety; twigs soft-shaggy-hairy, tomentose, silky, velvety, or glabrous in age
  40. Lf blade elliptic to obovate, length 1.3–2.8 × width; petiole velvety or tomentose; young lvs velvety or silky; lf tip acute to convex. . . . . *S. delnortensis*
  40′ Lf blade linear to oblanceolate, length 2.8–7.3 × width; petiole tomentose (or silky or velvety in *Salix jepsonii*); young lvs silky; lf tip acute to acuminate (convex or not in *Salix jepsonii*)
   41. Shrub < 1 m, subalpine; lf blade 10–95 mm; 2100–4000 m
    42. Lf blade 10–20 mm, shiny adaxially; petiole silky; young lvs very densely hairy; rare, c SNH (e slope, Mono Co.) . . . . *S. brachycarpa* var. **brachycarpa**
    42′ Lf blade 35–95 mm, dull adaxially; petiole soft-shaggy-hairy; young lvs long-silky; not rare, SN, SNE . . . . [3]*S. orestera*
   41′ Shrub > 1 m, subalpine and not; lf blade 32–144 mm; 0–2800 m (to 3400–3600 m in *Salix geyeriana, Salix jepsonii, Salix lemmonii*)
    43. Lf blade white-and-rusty-hairy
     44. Stipules gen lf-like (or vestigial); young lvs, mature blades densely silky or gen short-tomentose; 0–2800 m . . . . [9]*S. lasiolepis*
     44′ Stipules 0 or vestigial on first lvs, gen lf-like on later; young lvs, mature blades gen densely silky; 1450–3600 m
      45. Lf blade hairs long (gen white or white and rusty), persistent; petiole 2–9 mm; stipules gen 0 or vestigial . . . . [5]*S. geyeriana*
      45′ Lf blade hairs short (gen rusty) or 0 in age; petiole 5–16 mm; stipules vestigial on first lvs, lf-like on later . . . . [3]*S. lemmonii*
    43′ Lf blade white-hairy
     46. Petiole silky, tomentose, or velvety
      47. Lf blade gen densely silky on both surfaces, 32–89 mm; stipules gen 0; petiole velvety
. . . . [5]*S. geyeriana*
      47′ Lf blade gen glabrous in age or abaxially hairier, 35–144 mm; stipules gen lf-like; petiole silky, tomentose, or velvety
       48. Lf blade abaxially densely silky, adaxially dark green; margin of first lvs on shoot gland-dotted . . . . *S. jepsonii*
       48′ Lf blade abaxially sparsely or moderately hairy, adaxially not dark green; margin of first lvs on shoot glandless . . . . [9]*S. lasiolepis*
     46′ Petiole tomentose or soft-shaggy-hairy
      49. Lf blade on both surfaces silky; stipule tip gen acute . . . . [3]*S. orestera*
      49′ Lf blade abaxially tomentose, adaxially long-soft-wavy-hairy or tomentose; stipule tip gen convex
       50. Lf blade abaxially gen densely tomentose; twigs gen velvety; on serpentine . . . . *S. breweri*
       50′ Lf blade abaxially gen sparsely or moderately tomentose; twigs gen sparsely tomentose; on serpentine or not . . . . [9]*S. lasiolepis*

Key to Pistillate Plants

1. Ovary (and gen fr, at least when fr young) glabrous (sometimes hairy on beak or when ovary young)
 2. Fl bracts deciduous, ± tan, sparsely hairy, tips often ragged
  3. Bud scale margins free, overlapping; tree
   4. Lf blade not glaucous abaxially; beak of ovary abruptly tapered to style . . . . [2]*S. gooddingii*
   4′ Lf blade glaucous abaxially; beak of ovary gradually tapered to style . . . . *S. laevigata*
  3′ Bud scale margins fused; shrub to tree
   5. Petiole with glands; pl not clonal; stigmas persistent
    6. Tree 10–25 m; stipules not conspicuously glandular; fr 3.5–5 mm; twigs ± yellow or gray to ± tan . . . . *S. alba*
    6′ Shrub to tree, 1–11 m; stipules conspicuously glandular; fr 6–11 mm; twigs yellow-brown . . . . [2]*S. lasiandra*
   5′ Petiole glands 0; pl clonal by root-sprouting; stigmas deciduous
    7. Petiole hairy; lf margins ± rolled under; style ± green or ± tan . . . . [2]*S. exigua*
    7′ Petiole ± glabrous; lf margins flat; styles ± red or ± brown . . . . *S. melanopsis*

2′ Fl bracts gen persistent, ± tan to black, gen hairy, tips entire
  8. Tree; twigs pendent; nectary > ovary stalk . . . . **S. babylonica**
  8′ Shrub to tree; twigs erect; nectary < ovary stalk
    9. Fl bract dark brown, wider distally, tip rounded, hairs dense, short, gen wavy; lvs white- or
      white-and-rusty-hairy
      10. Stipule tip acute; lf glands submarginal or distinctly on adaxial surface; young lvs thick, yellow-
        green (or ± red) . . . . [2]**S. lasiolepis**
      10′ Stipule tip convex to rounded; lf glands marginal; young lvs thin, ± red. . . . . **S. tracyi**
    9′ Fl bract not as above in all ways; young lvs white-hairy
      11. Lf blade not glaucous abaxially
        12. Catkins blooming before or with lvs, on lfy shoots 1–9 mm; lf blade adaxially glabrous (or
          becoming so) or soft-shaggy-hairy and highly glossy; ovary gen glabrous . . . . **S. boothii**
        12′ Catkins blooming with lvs, on lfy shoots 2–12 mm; lf blade adaxially tomentose or silky and dull to
          shiny; ovary glabrous or gen silky . . . . [3]**S. eastwoodiae**
      11′ Lf blade glaucous abaxially
        13. Stipules 0 or vestigial on early lvs, 0 or prominent on later; fl bract hairs straight or wavy, rarely ± 0
          14. Style 0.6–1.5 mm; fr 5–10 mm; twigs gen thinly glaucous; lf blade abaxially soft-shaggy or woolly,
            length 1.5–4.2 × width; fl bract with long, straight hairs, gen not wider distally, tip pointed or not
            . . . . [3]**S. hookeriana**
          14′ Style 0.1–0.6 mm; fr 2.5–5.5 mm; twigs not glaucous; lf blade abaxially tomentose or silky
            (glabrous in age), length 2.2–9.6 × width; fl bract with short, gen wavy hairs, gen wider distally,
            tip rounded . . . . [2]**S. lasiolepis**
        13′ Stipules prominent on all lvs; fl bract hairs gen curly, rarely straight or wavy
          15. Twigs yellow-gray or yellow- or gray-brown, not peeling, glabrous or sparsely soft-shaggy-hairy . . . . . **S. lutea**
          15′ Twigs red- or yellow-brown, gen peeling, glabrous or long-soft-wavy-hairy
            16. Ovary stalk 0.9–2.5 mm; lf blade strap-shaped to very narrowly elliptic, length 2.9–6.4 × width,
              base convex to subcordate; stipule tip rounded to acuminate . . . . **S. ligulifolia**
            16′ Ovary stalk 1.3–4.2 mm; lf blade narrowly oblong or lanceolate to obovate, length 2.4–4.5 ×
              width, base rounded to cordate; stipule tip convex to rounded . . . . **S. prolixa**
1′ Ovary (and gen fr, at least when fr young) hairy (sometimes hairy only on beak or when ovary young)
  17. Dwarf shrub, forming mats by rhizomes or adventitious roots, alpine
    18. Catkins ± 4–17-fld; fl bract ± glabrous (hairy or not on margin); nectary 0–0.5 mm; style 0.2–0.4 mm;
      lf abaxially strongly net-veined . . . . **S. nivalis**
    18′ Catkins ± 50-fld; fl bract hairy; nectary 0.5–1.2 mm; style 0.4–1.6 mm; lf abaxially not strongly
      net-veined . . . . **S. petrophila**
  17′ Low to tall shrub to tree, not forming mats, not alpine
    19. Fl bracts deciduous, ± tan, sparsely to moderately hairy
      20. Petioles not glandular at distal end; lf margins with 0–4 teeth/cm, abaxial surface hairy . . . . [2]**S. exigua**
      20′ Petioles glandular at distal end; lf margins with 5–14 teeth/cm, abaxial surface sparsely hairy or
        gen glabrous
        21. Bud scale margins free, overlapping; petioles with a pair of glands; lf with white hairs. . . . . [2]**S. gooddingii**
        21′ Bud scale margins fused; petioles gen with a cluster of glands; lf with gen a mixture of white and
          rusty hairs . . . . [2]**S. lasiandra**
    19′ Fl bracts persistent, black, brown, or ± tan, gen moderately to densely hairy
      22. Catkins blooming with lvs, on lfy shoots
        23. Lf blade abaxially not glaucous or surface obscured by hair
          24. Lf blade abaxially densely silky or silky-woolly, adaxially dark green, sparsely silky (glabrous in
            age); twigs gen brittle at base
            25. Lf blade 8–25 mm wide, length 3–7.3 × width, margins flat or ± rolled under; nectary 0.3–0.6 mm;
              twigs very brittle at base; KR, CaRH, SNH; 1000–3400 m . . . . [3]**S. jepsonii**
            25′ Lf blade 17–48 mm wide, length 2–4 × width, margins strongly rolled under, esp basally; nectary
              0.5–0.9 mm; twigs flexible or ± brittle at base; NW, n SNH, CW (exc SCoRI), w WTR; 0–400 m
              (1800–2500 m in Siskiyou, Humboldt cos.) . . . . [4]**S. sitchensis**
          24′ Lf blade on both surfaces sparsely to densely tomentose or silky, gray-green; twigs flexible at base
            26. Lvs on catkin shoots and first lvs on vegetative twigs entire or finely short-slender-serrate . . . . [3]**S. eastwoodiae**
            26′ Lvs on catkin shoots and first lvs on vegetative twigs entire or margins gland-dotted
              27. Lf blade abaxially sparsely to moderately woolly-tomentose or silky, glabrous in age
                28. Lf blade abaxially gen shaggy (sometimes appressed)-hairy, base rounded to cordate; fl bract
                  brown or black; KR, NCoR, CaRH, SNH, SnJV, MP, W&I . . . . [3]**S. eastwoodiae**
                28′ Lf blade abaxially silky, base wedge-shaped to convex; fl bract brown; SN, SNE. . . . . [2]**S. orestera**
              27′ Lf blade abaxially densely silky to woolly
                29. Lf blade 8–25 mm wide, length 3–7.3 × width, margins flat or ± rolled under; nectary 0.3–0.6
                 mm; twigs very brittle at base; KR, CaRH, SNH; 1000–3400 m . . . . [3]**S. jepsonii**

 29′ Lf blade 17–48 mm wide, length 2–4 × width, margins strongly rolled under, esp basally;
   nectary 0.5–0.9 mm; twigs flexible or ± brittle at base; NW, n SNH, CW (exc SCoRI), w WTR;
   0–400 m (1800–2500 m in Siskiyou, Humboldt cos.) . . . . [4]***S. sitchensis***

23′ Lf blade abaxially glaucous
 30. Ovary stalk 0–0.3 mm; shrub, gen < 0.5 m; subalpine . . . . ***S. brachycarpa*** var. ***brachycarpa***
 30′ Ovary stalk 0.4–5 mm; shrub to tree, > 0.5 m; alpine, subalpine, or montane
  31. Young lvs rusty- or white-and-rusty-hairy
   32. Catkin ± spheric, 8–21 mm; style 0.1–0.2 mm; fr 3–6 mm; fl bract pale; lvs gen persistently silky
     on both surfaces; petiole 2–9 mm; stipules 0 or vestigial . . . . [4]***S. geyeriana***
   32′ Catkin cylindric, 19–44 mm; style 0.3–1 mm; fr 5–7 mm; fl bract dark; lvs silky but glabrous in
     age on both surfaces; petiole 5–16 mm; stipules vestigial on first lvs, lf-like on later . . . . [2]***S. lemmonii***
  31′ Young lvs white-hairy
   33. Style 0.1–0.4 mm; ovary stalk 1–5 mm
    34. Catkin cylindric, 16–60 mm; twigs shaggy, not glaucous; petiole puberulent; lf blade dull
      adaxially; stipules lf-like; ovary stalk 2–6 mm; 1000–1400 m . . . . [2]***S. bebbiana***
    34′ Catkin ± spheric, 8–21 mm; twigs tomentose or velvety, gen glaucous; petiole velvety; lf blade
      shiny adaxially; stipules 0 or vestigial; ovary stalk 1–2.8 mm; 1450–3600 m . . . . [4]***S. geyeriana***
   33′ Style 0.4–1 mm; ovary stalk 0.4–2 mm
    35. Style 0.4–0.6 mm; ovary short-silky; twigs very brittle at base; lf blade abaxially densely silky;
      petiole silky or velvety . . . . [3]***S. jepsonii***
    35′ Style 0.6–1 mm; ovary long-silky; twigs flexible at base; lf blade abaxially sparsely to
      moderately silky; petiole tomentose . . . . [2]***S. orestera***
22′ Catkins blooming before lvs (sometimes ± so), sessile or on bracted shoots, rarely on lfy shoots to 5 mm
 36. Fl bract ± tan or light rose
  37. Fl bract sparsely hairy; ovary sparsely or moderately silky, long-beaked; style 0.1–0.4 mm
   38. Catkin cylindric, 16–60 mm; twigs shaggy, not glaucous; petiole puberulent; lf blade dull
     adaxially; stipules lf-like; ovary stalk 2–6 mm; 1000–1400 m . . . . [2]***S. bebbiana***
   38′ Catkin ± spheric, 8–21 mm; twigs tomentose or velvety, gen glaucous; petiole velvety; lf blade
     shiny adaxially; stipules 0 or vestigial; ovary stalk 1–2.8 mm; 1450–3600 m . . . . [4]***S. geyeriana***
  37′ Fl bract moderately to densely hairy; ovary densely hairy, short-beaked; style 0.4–0.8 mm
   39. Ovary tomentose, stalk 0–0.4 mm, < nectary; nectary gen narrowly oblong; lf margins ± rolled
     under, blade abaxially sparsely or moderately tomentose; stipule tip rounded; on serpentine. . . . . ***S. breweri***
   39′ Ovary silky, stalk 0.4–1.4 mm, < to > nectary; nectary square to flask-shaped; lf margins strongly
     rolled under, esp basally, blade abaxially densely silky or woolly; stipule tip gen acute; not on
     serpentine . . . . [4]***S. sitchensis***
 36′ Fl bract brown, black, or 2-colored
  40. Catkins and lvs (at least some) ± opposite . . . . ***S. purpurea***
  40′ Catkins and lvs alternate
   41. Twigs gen strongly glaucous. . . . . ***S. drummondiana***
   41′ Twigs not glaucous or weakly so
    42. Young lvs gen white-hairy (hairs sometimes ± yellow, gray, or rusty)
     43. Catkin stout; ovary stalk 1–2 mm. . . . . [3]***S. hookeriana***
     43′ Catkin slender; ovary stalk 0–1.4 mm
      44. Ovary stalk 0–0.3 mm; catkin 17–53 mm; lf blade abaxially densely tomentose; on serpentine,
       nw KR . . . . ***S. delnortensis***
      44′ Ovary stalk 0.4–1.4 mm; catkin 25–73 mm; lf blade abaxially densely silky or woolly; not on
       serpentine, NW, n SNH, CW (exc SCoRI), w WTR . . . . [4]***S. sitchensis***
    42′ Young lvs white- or white-and-rusty-hairy
     45. Stigma 0.2–0.4 mm; catkin < 45 mm
      46. Catkin ± spheric, 8–21 mm; style 0.1–0.2 mm; fr 3–6 mm; fl bract pale; lvs gen silky
       persistently on both surfaces; petiole 2–9 mm; stipules 0 or vestigial . . . . [4]***S. geyeriana***
      46′ Catkin cylindric, 19–44 mm; style 0.3–1 mm; fr 5–7 mm; fl bract dark; lvs silky but glabrous
       in age on both surfaces; petiole 5–16 mm; stipules vestigial on first lvs, lf-like on later . . . . [2]***S. lemmonii***
     45′ Stigma 0.4–1 mm; catkin 15–117 mm
      47. Young lvs gen sparsely hairy; ovary stalk 0.3–0.8 mm; fr ≤ 6 mm . . . . ***S. planifolia***
      47′ Young lvs gen densely hairy; ovary stalk 0.8–2.3 mm; fr gen > 6 mm
       48. Style 0.6–1.5 mm; stigma 0.3–0.6 mm, gen broad-cylindric to plump . . . . [3]***S. hookeriana***
       48′ Style 0.2–0.6 mm; stigma 0.4–1 mm, slender-cylindric. . . . . ***S. scouleriana***

Key to Staminate Plants

1. Stamens > 2 per fl
 2. Bud scale margins fused; anthers 0.6–1 mm. . . . . ***S. lasiandra***
 2′ Bud scale margins free, overlapping; anthers 0.4–0.6 mm

3. Lf blade not glaucous abaxially; twigs ± yellow or yellow-green or ± tan; margins of lvs on catkin shoots finely serrate; young lvs white-hairy . . . . **S. gooddingii**

3′ Lf blade glaucous abaxially; twigs ± tan; margins of lvs on catkin shoots entire; young lvs glabrous or white- or white-and-rusty-hairy . . . . **S. laevigata**

1′ Stamens 1–2 per fl

  4. Stamen 1 per fl

    5. Catkins and lvs (at least some) ± opposite . . . . **S. purpurea**

    5′ Catkins and lvs alternate

      6. Lf blade 8–25 mm wide, length 3–7.3 × width; fl bract sparsely hairy; filaments gen hairy basally; twigs very brittle at base; KR, CaRH, SNH; 1000–3400 m . . . . [2]**S. jepsonii**

      6′ Lf blade 17–48 mm wide, length 2–4 × width; fl bract moderately hairy; filaments glabrous; twigs flexible or ± brittle at base; NW, n SNH, CW (exc SCoRI), w WTR; 0–400 m (1800–2500 m in Siskiyou, Humboldt cos.) . . . . **S. sitchensis**

  4′ Stamens 2 per fl

    7. Shrub < 0.2 m, forming mats by rhizomes or adventitious roots; alpine

      8. Catkin and lfy shoot terminal; catkin 7–19 mm; fl bract ± glabrous (hairy or not on margin); young lvs glabrous; nectaries adaxial, abaxial . . . . **S. nivalis**

      8′ Catkin and lfy shoot lateral; catkin 18–32 mm; fl bract hairy; young lvs long-soft-wavy-hairy; nectaries gen only adaxial . . . . **S. petrophila**

    7′ Shrub to tree, > 0.2 m, not forming mats; not alpine

      9. Nectaries adaxial and abaxial; catkins blooming with or after lvs, on lfy shoots

        10. Tree; twigs flexible or ± brittle at base

          11. Sts and twigs erect to spreading; twigs ± yellow or gray to ± tan, densely silky; petiole silky . . . . **S. alba**

          11′ Sts and twigs pendent; twigs ± tan, short-silky to velvety, glabrous in age (exc at nodes); petiole tomentose . . . . **S. babylonica**

        10′ Shrub; twigs flexible at base

          12. Shrub gen < 0.5 m, not clonal by root-sprouting, subalpine; catkin 4–5 mm on lfy shoot ± 1 mm; anther 0.3–0.5 mm . . . . **S. brachycarpa** var. **brachycarpa**

          12′ Shrub 0.8–7 m, clonal by root-sprouting, riparian; catkin 13–54 mm on lfy shoot 5–70 mm; anther 0.6–1.1 mm

            13. Lf blade linear to narrow-elliptic or -oblanceolate, length 2.8–8.5 × width; twigs gray-brown; young lvs long-soft-wavy-hairy, tomentose, or velvety; petiole glabrous . . . . **S. melanopsis**

            13′ Lf blade linear, length 10–23 × width; twigs ± tan; young lvs short- and long-silky; petiole hairy . . . . **S. exigua**

      9′ Nectaries adaxial; catkins blooming before or with lvs, sessile or on lfy shoots; native

        14. Catkins blooming before lvs (sometimes ± so), sessile or on bracted shoots or on lfy shoots to 4 mm

          15. Fl bract ± tan or light rose; on serpentine . . . . **S. breweri**

          15′ Fl bract brown to black; not on serpentine (exc *Salix delnortensis*)

            16. Anthers ≥ 0.6 mm

              17. Twigs tomentose or velvety, brittle at base; serpentine . . . . **S. delnortensis**

              17′ Twigs gen hairy, not velvety, brittle or flexible at base; non-serpentine

                18. Anthers yellow; catkins sessile or on lfy shoots to 8 mm, length 1.6–6.4 × width; twigs brittle at base . . . . [2]**S. hookeriana**

                18′ Anthers purple, in age yellow; catkins sessile or on lfy shoots to 4 mm, length 1.3–3.2 × width; twigs flexible at base

                  19. Low shrubs gen restricted to subalpine meadows, SNH, SNE; twigs gen glabrous to sparsely hairy; adaxial nectaries narrowly oblong to oblong . . . . [2]**S. planifolia**

                  19′ Tall shrubs to trees in forest openings, NW, CaR, SN, n CCo, SnFrB, SnGb, SnBr, SnJt, GB; twigs sparsely hairy to densely velvety; adaxial nectaries oblong to square . . . . **S. scouleriana**

            16′ Anthers gen ≤ 0.6 mm

              20. Pls of SNH, SNE (*Salix lasiolepis* also in SNH, W&I, but in staminate pls morphologically indistinguishable from *Salix drummondiana* and *Salix planifolia*)

                21. Twigs brittle at base, gen strongly glaucous; c&s SNH, W&I . . . . **S. drummondiana**

                21′ Twigs flexible at base, gen not glaucous; SNH, SNE . . . . [2]**S. planifolia**

            20′ Pls not of SNH, SNE

              22. Twigs glabrous, long-soft hairy, tomentose, or woolly; filaments glabrous or hairy; anthers yellow; NCo (< 100 m), n NCoRO (500–1000 m), CaRF, n CCo (Sonoma Co.) . . . . [2]**S. hookeriana**

              22′ Twigs glabrous, densely short-soft-spreading-hairy, or tomentose; filaments glabrous; anthers purple, in age yellow; CA

                23. Stipule tip acute; lf glands submarginal or distinctly on adaxial surface; young lvs thick, yellow-green (or ± red) . . . . **S. lasiolepis**

                23′ Stipule tip convex to rounded; lf glands marginal; young lvs thin, ± red . . . . [2]**S. tracyi**

        14′ Catkins blooming with lvs, on (sometimes very short) lfy shoots

          24. Young lvs rusty- or white-and-rusty-hairy

            25. Fl bracts widest above middle; lf abaxial hairs spreading; young lvs glabrous, tomentose, or puberulent . . . . [2]**S. tracyi**

25′ Fl bracts widest at middle; lf abaxial hairs appressed; young lvs silky

  26. Catkin spheric, 11–18 mm; fl bract glabrous or hairy, ± tan to dark brown; anthers 0.4–0.5 mm

    . . . . ³***S. geyeriana***

  26′ Catkin stout-cylindric, 19–44 mm; fl bract hairy, brown to 2-colored; anthers 0.5–0.9 mm. . . . . . ***S. lemmonii***

24′ Young lvs white-hairy or glabrous

  27. Filaments hairy

    28. Lf blade abaxially not glaucous; lvs on catkin shoots gen finely short-slender-serrate (or entire

      or gland-dotted). . . . . ²***S. eastwoodiae***

    28′ Lf blade abaxially glaucous or surface obscured by hair; lvs on catkin shoots entire or gland-dotted

      29. Stipules 0 or vestigial

        30. Catkin spheric, 11–18 mm; anthers 0.4–0.5 mm; lf blade shiny adaxially; petiole velvety . . . . ³***S. geyeriana***

        30′ Catkin stout-cylindric, 16–34 mm; anthers 0.6–1 mm; lf blade dull adaxially; petiole

          tomentose. . . . . ²***S. orestera***

      29′ Stipules lf-like

        31. Lf blade abaxially densely silky; petiole silky or velvety; twigs very brittle at base . . . . ²***S. jepsonii***

        31′ Lf blade abaxially sparsely to moderately tomentose or silky; petiole puberulent or tomentose;

          twigs flexible at base

          32. Fl bract ± tan; twigs moderately to densely long-soft-wavy-hairy; petiole puberulent;

            1000–1400 m . . . . ²***S. bebbiana***

          32′ Fl bract brown; twigs sparsely soft-shaggy-hairy or silky, glabrous in age; petiole tomentose;

            2100–4000 m. . . . . ²***S. orestera***

27′ Filaments glabrous

  33. Later stipules 0 or vestigial, scars 0 or minute. . . . . ³***S. geyeriana***

  33′ Later stipules lf-like, scars not 0, not minute

    34. Lf blade not glaucous abaxially

      35. Young lvs shaggy-hairy; mature lf blade abaxially glabrous or soft-shaggy-hairy, adaxially

        shiny to highly glossy . . . . ***S. boothii***

      35′ Young lvs densely tomentose, silky, or woolly; mature lf blade abaxially tomentose, long-soft-

        wavy-hairy, woolly-tomentose, silky, or glabrous in age, adaxially dull . . . . ²***S. eastwoodiae***

    34′ Lf blade glaucous abaxially

      36. Twigs yellow-gray or yellow- or gray-brown, not peeling; anthers 0.4–0.8 mm. . . . . ***S. lutea***

      36′ Twigs yellow-green or gen ± tan, peeling; anthers 0.5–0.8 mm

        37. Lf blade margin entire to crenate; fl bract 1.2–3.2 mm, hairs wavy; catkin lfy shoots

          0.5–11 mm . . . . ²***S. bebbiana***

        37′ Lf blade margin gen serrate; fl bract 0.8–1.6 mm, hairs curly; catkin lfy shoots 0–3 mm

          38. Lf blade strap-shaped to narrowly elliptic, base convex to subcordate; stipule tip rounded

            to acuminate. . . . . ***S. ligulifolia***

          38′ Lf blade narrowly oblong or lanceolate to obovate, base rounded to cordate; stipule tip

            convex to rounded . . . . ***S. prolixa***

S. alba L. WHITE WILLOW Tree < 25 m. **ST:** erect to spreading; twigs ± yellow or gray to red-brown, flexible or ± brittle at base, silky, glabrous in age. **LF:** later stipules lf-like; petiole 3–13 mm, with glands; young lvs densely silky; mature blade 63–115 mm, lanceolate to narrowly oblong, acuminate (base wedge-shaped or acute), finely serrate, abaxial hairs gen densely long-silky, straight, to ± 0. **INFL:** blooming with lvs, pistillate 31–51 mm, on lfy shoots 3–14 mm; fl bract ± tan, sparsely hairy, tip rounded; pistillate bracts deciduous. **STAMINATE FL:** stamens 2; nectaries adaxial, abaxial. **PISTIL-LATE FL:** ovary glabrous, stalk 0.2–0.8 mm, style 0.16–0.44 mm. 2*n*=76. Disturbed places, gen near settlements; probably < 20 m. CA; to e N.Am; native to Eur. Mostly cult as orn, but many cultivars, hybrids ± naturalized. May–Jun

S. babylonica L. (p. 1229) WEEPING WILLOW Tree < 16 m. **ST:** pendent; twigs yellow- to red-brown, short-silky to velvety, glabrous in age (exc at nodes). **LF:** later stipules lf-like; petiole 7–9 mm, gen with glands; young lvs glabrous to moderately densely silky; mature blade 90–160 mm, lance-linear, acuminate (base wedge-shaped), gen finely sharp-serrate-spiny, abaxial hairs gen sparsely short-silky, straight, to ± 0. **INFL:** blooming with lvs, pistillate 7–22 mm, on lfy shoots (0)2–9 mm; fl bract ± tan. **STAMINATE FL:** stamens 2. **PISTILLATE FL:** ovary glabrous, stalk 0–0.2 mm, style ± 0.2 mm, stigma plump-lobed. 2*n*=76. Disturbed places, around settlements; probably < 50 m. SnFrB, SCo (see note); to se US; native to Asia. Mostly cult as orn; nearly all seen ± naturalized pls in CA are *Salix* ×*sepulcralis* Simonk. (*S. alba* × *S. babylonica*) and *S.* ×*pendulina* Wender. (*S. babylonica* × *S.* ×*fragilis* L.). Feb–May

S. bebbiana Sarg. (p. 1229) BEBB'S WILLOW Shrub, small tree, < 10 m. **ST:** twigs spreading widely, yellow-green or red-brown, soft-shaggy-hairy. **LF:** later stipules lf-like, vestigial or not; petiole 2–13 mm; young lvs hairy; mature blade 20–87 mm, elliptic to narrowly obovate, acuminate (base convex to wedge-shaped), entire to crenate, abaxially gen tomentose or hairs long-silky on veins, wavy, to ± 0. **INFL:** blooming just before or with lvs, pistillate 16–75 mm, on lfy shoots 1–26 mm; fl bract ± tan. **STAMINATE FL:** stamens 2. **PIS-TILLATE FL:** ovary long-beaked, silky, stalk 2–6 mm, style 0.1–0.4 mm. 2*n*=38. Streamsides, lakeshores; 1000–1400 m. n MP (Lower Klamath Lake, Siskiyou Co.; Goose Lake, Modoc Co.); to AK, e N.Am, NM. Apr–Jun ★

S. boothii Dorn (p. 1229) BOOTH'S WILLOW Shrub < 6 m. **ST:** twigs yellow-, gray-, or red-brown, glabrous or shaggy-hairy. **LF:** stipules lf-like; petiole 3–17 mm; young lvs shaggy-hairy; mature blade 26–102 mm, strap-shaped to broadly elliptic, acute to acuminate (base convex to subcordate), entire to finely serrate, abaxial hairs gen long-soft-shaggy or ± densely short-silky, white or white and rusty, wavy, to 0. **INFL:** blooming before or with lvs, pistillate 12–62 mm, on lfy shoots 1–9 mm (margins of shoot-lvs gland-dotted or finely serrate); fl bract dark brown, hairs wavy. **STAMINATE FL:** stamens 2. **PISTILLATE FL:** ovary glabrous, stalk 0.5–2.5 mm, style 0.3–1.4 mm. 2*n*=76. Uncommon. Wet subalpine meadows, shores; 1525–3200 m. KR, CaRH, n&s SNH, Wrn, SNE; to w Can, CO. Apr–Jul

S. brachycarpa Nutt. var. ***brachycarpa*** (p. 1229) SHORT-FRUITED WILLOW Shrub gen < 0.5 m. **ST:** twigs red-brown, silky, glabrous in age. **LF:** stipules gen vestigial; petiole 0.8–4 mm; young

lvs densely silky; mature blade 10–20 mm, ± elliptic, acute (base convex), entire; abaxial hairs gen dense, long-soft-fine or long-silky. **INFL:** blooming with lvs, densely fld, pistillate 6–28 mm, ± spheric, on lfy shoots 0.5–20 mm; fl bract ± tan. **STAMINATE FL:** stamens 2; nectaries adaxial, abaxial. **PISTILLATE FL:** ovary densely woolly or silky, stalk 0–0.3 mm, style 0.5–1.5 mm; nectaries adaxial, abaxial. $2n$=38. Subalpine meadows (esp on limestone); 3200–3500 m. c SNH (e slope, Mono Co.); to n&e Can, CO. [*S. b.* subsp. *b.*] Jun–Aug ★

S. breweri Bebb (p. 1229) BREWER'S WILLOW Shrub < 4 m. **ST:** twigs ± yellow to yellow-brown, velvety or silky-tomentose. **LF:** later stipules lf-like; petiole 3–7 mm; young lvs silky; mature blade 58–144 mm, strap-shaped to oblanceolate, convex or acuminate (base convex to wedge-shaped), entire to irregularly toothed, ± rolled under, abaxially gen sparse- to dense-tomentose or -woolly, hairs wavy. **INFL:** blooming before lvs, pistillate 19–59 mm, on lfy shoots 0–1 mm; fl bract ± tan to light rose. **STAMINATE FL:** stamens 2. **PISTILLATE FL:** ovary tomentose, stalk 0–0.4 mm, style 0.4–0.8 mm; nectary narrow. Serpentine streamsides; 300–1300 m. NCoR, SnFrB, n&c SCoR. Mar–Apr

S. delnortensis C.K. Schneid. (p. 1229) DEL NORTE WILLOW Shrub < 2 m. **ST:** twigs red- or yellow-brown, brittle at base, tomentose or velvety, glabrous in age. **LF:** later stipules lf-like; petiole 6–16 mm; young lvs velvety or silky; mature blade 53–102 mm, elliptic to obovate, acute to convex (base convex to wedge-shaped), entire, ± rolled under, abaxially gen velvety or tomentose or hairs long-soft-fine or short-silky, wavy. **INFL:** blooming before lvs, pistillate 17–53 mm, on lfy shoots 0–3 mm; fl bract brown. **STAMINATE FL:** stamens 2. **PISTILLATE FL:** ovary silky, stalk 0–0.3 mm, style 0.6–1.2 mm. Serpentine banks of rivers, streams; 90–500 m. nw KR (near Gasquet, Del Norte Co.); sw OR. Like *S. breweri* Bebb; may hybridize with *S. lasiolepis* Benth. Mar–May ★

S. drummondiana Hook. (p. 1229) DRUMMOND'S WILLOW Shrub < 5 m. **ST:** twigs red- or mottled yellow-brown, brittle at base, gen strongly glaucous, hairy, glabrous in age. **LF:** later stipules lf-like; petiole 2–12 mm; young lvs white- or white-and-rusty-silky; mature blade 40–85 mm, strap-shaped, elliptic, or oblanceolate, acuminate to convex (base convex to wedge-shaped), entire to shallowly crenate, ± rolled under, abaxial hairs gen dense, short- or long-silky or woolly, white or white and rusty, straight or wavy. **INFL:** blooming before lvs, pistillate 22–87 mm, on lfy shoots 0–3(6) mm; fl bract brown to black. **STAMINATE FL:** stamens 2. **PISTILLATE FL:** ovary silky, stalk 0.3–2 mm, style 0.5–1.5 mm. $2n$=38,57,76. Streamsides, wet meadows, subalpine red-fir forest; 2200–3000 m. c&s SNH, W&I; to w Can, NV, NM. Apr–Jul

S. eastwoodiae A. Heller (p. 1229) SIERRA WILLOW Shrub < 4 m. **ST:** twigs yellow-green or red-brown, glaucous or not, shaggy-hairy, glabrous in age. **LF:** stipules lf-like, early-deciduous or not; petiole 1–14 mm; young lvs hairy; mature blade 23–99 mm, narrowly oblong to elliptic, short-acuminate to convex (base rounded to subcordate), entire to finely short-slender-serrate, abaxially gen densely woolly-tomentose or hairs long-soft-shaggy or short-silky, wavy, to ± 0. **INFL:** blooming with lvs, pistillate 11–51 mm, on lfy shoots 2–12 mm; margin of shoot lvs gen finely short-slender-serrate (or entire or gland-dotted); fl bract brown or black. **STAMINATE FL:** stamens 2. **PISTILLATE FL:** ovary glabrous or gen silky, stalk 0.2–1.6 mm, style 0.5–1.5 mm. $2n$=76. Alpine, subalpine meadows, streams, talus; 1600–3800 m. KR, NCoR, CaRH, SNH, MP, W&I; to w Can, MT, WY, NM. May–Jul

S. exigua Nutt. Shrub or tree < 5 m, clonal by root-shoots. **ST:** twigs yellow- to red-brown. **LF:** later stipules lf-like; petiole 1–7 mm; mature blade 30–147 mm, linear or strap-shaped, acuminate (base wedge-shaped), entire or ± sparsely short-slender-serrate. **INFL:** blooming with or after lvs, branched or not, pistillate 22–70 mm, on lfy shoots 2–70 (400) mm; fl bract ± tan; pistillate bracts deciduous. **STAMINATE FL:** stamens 2; nectaries adaxial, abaxial. **PISTILLATE FL:** ovary stalk 0.2–0.9 mm, stigma deciduous.

1. Twig, lf hairs ± appressed; ovary glabrous (hairy on
   beak or not), stalk 0.2–0.9 mm; style < 0.2 mm or gen 0;
   fl bract tip rounded, staminate 1.2–1.6 mm; young lvs
   short-silky . . . . var. *exigua*

1′ Twig, lf hairs ± spreading, fewer ± appressed; ovary
   hairy (at least when young), stalk 0–0.2 mm; style
   0.1–0.4 mm; fl bract tip acute to convex, staminate
   1.4–2.6 mm; young lvs long-silky . . . . var. *hindsiana*

var. **exigua** (p. 1229) NARROW-LEAVED WILLOW, COYOTE WILLOW **LF:** petiole densely short-soft-spreading-hairy; mature blade abaxial hairs gen densely long-soft-fine-silky, straight or wavy. **PISTILLATE FL:** stigmas flat or wide-cylindric, 0.25–0.5 mm. **FR:** 4–8 mm. $2n$=38. Common. Shores, bars, silt, sand, gravel; < 2800 m. SN, MP, D; to BC, SK, SD, TX, Mex. Hybridizes with *S. exigua* var. *hindsiana*, *S. sessilifolia* (latter exc from CA since TJM (1993)). Mar–Jun

var. **hindsiana** (Benth.) Dorn HINDS' WILLOW **LF:** petiole puberulent, long-soft-wavy-hairy or -silky; mature blade abaxial hairs gen densely long-soft-fine-silky, straight, to 0. **PISTILLATE FL:** stigmas flat or slender-cylindrical, 0.3–1 mm. **FR:** 3–4.5 mm. Common. Floodplains, sandy gravel; < 640 m. NW, GV, CW, SW; sw OR. Hybridizes with *S. exigua* var. *e.* Apr–May

S. geyeriana Andersson (p. 1229) GEYER'S WILLOW Shrub < 5 m. **ST:** twigs ± yellow or yellow- or red-brown, gen glaucous, tomentose or velvety, brittle at base or not. **LF:** stipules 0 or vestigial; petiole 2–9 mm; young lvs silky; mature blade 32–89 mm, strap-shaped to linear, acuminate (base wedge-shaped), entire, flat to ± rolled under, abaxial hairs gen dense, short- or long-silky, white or white and rusty, straight, to 0. **INFL:** blooming just before or with lvs, pistillate 8–21 mm, ± spheric, on lfy shoots 0–8 mm; fl bract ± tan to brown. **STAMINATE FL:** stamens 2. **PISTILLATE FL:** ovary white- or white-and-rusty-silky, stalk 1–2.8 mm, style 0.1–0.2 mm. $2n$=38. Subalpine streams, meadows; 1450–3600 m. s CaRH, n&c SNH, s SNH (esp Kern Plateau), SnBr, GB; to BC, MT, NM. Rare hybrids with *S. lemmonii*, *S. drummondiana* in Sierra, Lassen cos. Apr–Jun

S. gooddingii C.R. Ball (p. 1229) GOODDING'S BLACK WILLOW Shrub or tree < 30 m. **ST:** twigs ± yellow or yellow-green or red-brown, velvety or soft-shaggy-hairy, glabrous in age, brittle at base or not; bud scale margins free, overlapping. **LF:** stipules gen lf-like; petiole 4–10 mm, gen with glands; young lvs white-hairy; mature blade 67–130 mm, linear to narrowly elliptic, acuminate (base wedge-shaped), finely serrate, abaxial hairs gen minute, spreading, wavy, to 0. **INFL:** blooming with lvs, pistillate 23–82 mm, on lfy shoots 2–48 mm (margins of shoot-lvs finely serrate); fl bract ± tan; pistillate bracts deciduous. **STAMINATE FL:** stamens 4–6; nectaries adaxial, abaxial. **PISTILLATE FL:** ovary glabrous or hairy, stalk 1.2–3.2 mm, style 0.1–0.3 mm. $2n$=38. Common. Streamsides, marshes, seepage areas, washes, meadows; 20–2500 (gen < 500) m. NCoRI, CaRF, SNF, GV, SCo, PR, GB, D (esp GV, D); to TX, Mex. Mar–Apr

S. hookeriana Hook. (p. 1229) COASTAL WILLOW Shrub, small tree, < 8 m. **ST:** twigs gray-, red-, or yellow-brown, gen thinly glaucous, glabrous, long-soft hairy, tomentose, or woolly, brittle at base. **LF:** later stipules lf-like; petiole (4)6–29 mm; young lvs white- or white-and-rusty-hairy; mature blade 46–113 mm, elliptic to oblanceolate or broadly obovate, entire, coarsely crenate, or finely serrate, convex to acuminate (base convex to cordate), abaxially gen ± densely tomentose or woolly or hairs long-soft-shaggy or -fine, white or white and rusty, wavy or straight, to ± 0 (exc midrib). **INFL:** blooming before lvs, pistillate 36–117 mm, on lfy shoots 0–10(35) mm; fl bract dark brown. **STAMINATE FL:** stamens 2. **PISTILLATE FL:** ovary glabrous, woolly, or tomentose, stalk 1–2 mm, style 0.6–1.5(2.3) mm. $2n$=57, 114. Coastal dunes, floodplains, meadows; < 100, 500–1000 m. NCo, n NCoRO, CaRF, n CCo (Sonoma Co.); to AK. Glabrous and densely tomentose pls intergrade, may occur together (glabrous pls at 500–1000 m in Humboldt Co. need study). May hybridize with *S. lasiolepis*, *S. scouleriana*. Apr–Jun

S. jepsonii C.K. Schneid. (p. 1229) JEPSON'S WILLOW Shrub 1–3 m. **ST:** twigs gray- or red-brown, velvety or silky, brittle at base. **LF:** stipules lf-like; petiole 4–12 mm; young lvs silky; mature blade 55–103 mm, oblanceolate or narrowly so, acute or convex, (base convex to wedge-shaped), entire, rarely minute-crenate or -serrate, abaxial hairs gen densely short-silky, straight. **INFL:** blooming with

lvs, pistillate 13–55 mm, on lfy shoots 1.5–7 mm; fl bract ± brown. **STAMINATE FL**: stamens 1, rarely 2. **PISTILLATE FL**: ovary short-silky, stalk 0.4–1.2 mm, style 0.4–0.6 mm. Margins of lakes and streams, wet meadows; 1000–3400 m. KR, CaRH, SNH; OR, NV. Like *S. sitchensis* but smaller, lvs narrower (variable), stamens 2 vs 1 (variable), ovary narrower. Jun

S. laevigata Bebb (p. 1229)   RED WILLOW   Tree < ± 20 m. **ST**: twigs yellow- or red-brown, hairy, glabrous in age gen exc at nodes, brittle at base; bud scale margins free, overlapping. **LF**: stipules gen lf-like; petiole 3.5–18 mm, gen with glands; young lvs glabrous or white- or white-and-rusty-hairy; mature blade 53–190 mm, strap-shaped to lanceolate or obovate, acuminate to caudate (base convex to wedge-shaped), ± finely crenate, abaxial hairs gen densely short-soft-spreading, white or white and rusty, to 0. **INFL**: blooming with or after lvs, pistillate 28–79 mm, on lfy shoots 3–14 mm (shoot lvs entire); fl bract ± tan; pistillate bracts deciduous. **STAMINATE FL**: stamens 5; nectaries adaxial, abaxial. **PISTILLATE FL**: ovary glabrous, stalk 1.4–2.8 mm, style 0.12–0.24 mm. 2*n*=38. Common. Riverbanks, seepage areas, lakeshores (subalkaline or brackish), canyons; < 1700 m. CA (exc MP, DSon); s OR, n NV, AZ, Mex, n C.Am. [*S. bonplandiana* Kunth var. *l.* (Bebb) Dorn] Dec–Jun

S. lasiandra Benth.   Shrub to tree, 1–11 m. **ST**: twigs yellow-, gray-, or red-brown, glabrous or soft-shaggy-hairy, brittle at base or not. **LF**: stipules lf-like, glandular; petiole with glands; young lvs glabrous or white- or white-and-rusty-hairy; mature blade 53–170 mm, lanceolate, acuminate to long-acuminate (base convex to rounded), finely serrate. **INFL**: blooming with lvs, pistillate 18.5–103 mm, on lfy shoots 6–56 mm; fl bract ± tan; pistillate bracts deciduous. **STAMINATE FL**: stamens 3–5; nectaries adaxial, abaxial. **PISTILLATE FL**: ovary glabrous, stalk 0.8–4 mm, style 0.2–0.8 mm; nectary adaxial.

1. Lf blades not glaucous abaxially, with stomata adaxially, convex at base; staminate abaxial nectary 2–3-lobed; staminate abaxial, adaxial nectaries free or united at base into cup . . . . var. ***caudata***
1′ Lf blades rarely not glaucous abaxially, lacking stomata adaxially, convex or rounded at base; staminate abaxial nectary unlobed; staminate abaxial, adaxial nectaries free . . . . var. ***lasiandra***

var. ***caudata*** (Nutt.) Sudw.   TAIL-LEAF WILLOW   **LF**: petiole (1)4–15 mm, with pair or cluster of ± lf-like glands; mature blade sparsely long-soft-shaggy, hairs white or white and rusty, straight or curved, to ± 0. 2*n*=76. Wet meadows, lakeshores, riverbanks; 35–3050 m. SNH, SnBr, GB; to AK, SD, NM. [*S. lucida* Muhl. subsp. *c.* (Nutt.) E. Murray] May–Jun

var. ***lasiandra*** (p. 1233)   PACIFIC WILLOW   **LF**: petiole (2)4–30 mm, with cluster of spheric or ± lf-like glands; mature blade hairs white or white and rusty, straight or wavy, to 0. 2*n*=76. Common. Wet meadows, shores, seepage areas; < 2715 m. CA (less common s CA, D); to AK, NM. [*S. lucida* Muhl. subsp. *l.* (Benth.) E. Murray] Mar–Jun

S. lasiolepis Benth. (p. 1233)   ARROYO WILLOW   Shrub, small tree, < 10 m. **ST**: twigs ± yellow, yellow-green, or yellow- or red-brown, glabrous, densely short-soft-spreading-hairy, or tomentose, gen brittle at base. **LF**: later stipules gen lf-like; petiole 3–16 mm, tomentose to velvety; young lvs white- or white-and-rusty-hairy; mature blade 35–125 mm, strap-shaped to elliptic or obovate, acute to convex (base wedge-shaped to convex), entire to irregularly serrate, ± to strongly rolled under, abaxially gen ± dense-tomentose or -woolly-tomentose or hairs sparsely short-soft-spreading or short- or long-silky, white or white and rusty, wavy, to ± 0. **INFL**: blooming before lvs, pistillate 18–72 mm, on lfy shoots 0–6 mm; fl bract dark brown, with gen wavy hairs, tip broadly rounded. **STAMINATE FL**: stamens 2. **PISTILLATE FL**: ovary glabrous, stalk 0.5–2.4 mm, style 0.1–0.6 mm. 2*n*=76. Abundant. Shores, marshes, meadows, springs, bluffs; < 2800 m. CA; to WA, ID, TX, Mex. [*S. l.* var. *bigelovii* (Torr.) Bebb] Highly variable; several weak vars. described. NCo populations suggest intergradation with *S. hookeriana*. Jan–Jun

S. lemmonii Bebb (p. 1233)   LEMMON'S WILLOW   Shrub 1–4 m. **ST**: twigs yellow- to red-brown, strongly glaucous or not, puberu-

lent, brittle at base or not. **LF**: later stipules lf-like; petiole 5–16 mm; young lvs silky; mature blade 44–110 mm, strap-shaped to narrowly elliptic, acuminate (base convex), entire or finely serrate, abaxial hairs gen sparse, short- or long-silky, white and rusty, straight or wavy, to ± 0. **INFL**: blooming just before or with lvs, pistillate 19–44 mm, on lfy shoots 0.5–6 mm; fl bract dark brown to 2-colored. **STAMINATE FL**: stamens 2. **PISTILLATE FL**: ovary silky, stalk 1.1–2.1 mm, style 0.3–1 mm. 2*n*=76. Streams, wet meadows, burns in subalpine pine forest; 1400–3500 m. KR, s NCoRO (Lake Co.), s NCoRI (Sonoma Co.), CaRH, SNH, SnBr, SnJt; to BC, MT, CO (exc UT). Like *S. geyeriana*. May–Jun

S. ligulifolia C.K. Schneid. (p. 1233)   STRAP-LEAFED WILLOW   Shrub < 8 m. **ST**: twigs yellow-green or -brown, glabrous or long-shaggy-hairy. **LF**: stipule lf-like, tip rounded to acuminate; petiole 3–18 mm; young lvs glabrous; mature blade 60–133 mm, strap-shaped to narrowly elliptic, acute to acuminate (base convex to sub-cordate), margin gland-dotted to finely serrate, abaxial hairs sparsely short-silky or densely short-soft-spreading, straight or wavy, to 0. **INFL**: blooming just before or with lvs, pistillate 15.5–49 mm, on lfy shoots 0–6 mm; fl bract brown, hairs curly. **STAMINATE FL**: stamens 2. **PISTILLATE FL**: ovary glabrous, stalk 0.9–2.5 mm, style 0.2–0.6 mm. 2*n*=38. Rivers, streams; 1100–2500 m. KR, CaRH, SNH, Wrn; to ID, WY, NM. [*S. eriocephala* Michx. var. *l.* (C.K. Schneid.) Dorn] Lf margins often incorrectly called entire. Mar–Jun

S. lutea Nutt. (p. 1233)   YELLOW WILLOW   Shrub < 7 m. **ST**: twigs yellow-gray or yellow- or gray-brown, glabrous or hairy. **LF**: stipules gen lf-like; petiole 4–19 mm; young lvs glabrous or silky; mature blade 42–90 mm, strap-shaped to elliptic, lanceolate, or narrowly oblanceolate, acuminate (base convex to rounded), entire to fine-serrate or -crenate, abaxial hairs long-soft-shaggy or sparsely long-silky, straight, to 0. **INFL**: blooming just before or with lvs, pistillate 13–38 mm, on lfy shoots 0.5–7 mm; fl bract brown to ± tan, sparsely curly-hairy. **STAMINATE FL**: stamens 2. **PISTILLATE FL**: ovary glabrous, stalk 0.9–3.8 mm, style 0.13–0.6 mm. 2*n*=38. River, creek margins, wet meadows; 640–3100 m. c&s SNH (esp e slope), SnBr, SnJt, GB, w DMoj; to OR, MT, WY, AZ. [*S. eriocephala* var. *watsonii* (Bebb) Dorn] Mar–May

S. melanopsis Nutt. (p. 1233)   DUSKY WILLOW   Shrub < 4 m, clonal by root-shoots. **ST**: twigs gray-brown, silky, glabrous in age. **LF**: stipules gen lf-like; petiole 2–5 mm; young lvs long-soft-wavy-hairy, tomentose, or velvety; mature blade 30–85 mm, linear to narrow-elliptic or -oblanceolate, acuminate to convex (base wedge-shaped to convex), entire or finely short-slender-serrate, abaxial hairs long-soft-shaggy or -fine or long-silky, appressed to spreading, wavy, to ± 0. **INFL**: blooming with or after lvs, branched or not, pistillate 22–58 mm, on lfy shoots 4–70 mm; fl bract ± tan, pistillate deciduous. **STAMINATE FL**: stamens 2; nectaries adaxial, abaxial. **PISTILLATE FL**: ovary glabrous, stalk 0–0.7 mm, style 0–0.5 mm, stigmas deciduous. Streambanks, often among rocks; 620–2700 m. NCo, KR, CaR, ScV, MP; to w Can, CO, NV. May–Jul

S. nivalis Hook. (p. 1233)   SNOW WILLOW   Shrub < 0.1 m, mat-forming. **ST**: twigs yellow- or red-brown, ± glabrous. **LF**: stipules 0 or vestigial; petiole 1.5–7 mm; young lvs glabrous; mature blade 6–22 mm, ± elliptic to obovate, convex to rounded (base subcordate or not), entire, abaxial hairs long-silky to 0. **INFL**: blooming with lvs, pistillate 7–21 mm, on lfy shoots 1–10 mm; fl bract ± tan to light rose; pistillate catkin 4–17-fld. **STAMINATE FL**: stamens 2; nectaries adaxial, abaxial. **PISTILLATE FL**: ovary silky, stalk 0–0.8 mm, style 0.2–0.4 mm; nectaries adaxial, abaxial. 2*n*=38. Alpine cirques; 3100–3500 m. c SNH (near Mount Dana); to w Can, CO, NM. [*S. reticulata* L. subsp. *n.* (Hook.) Á. Löve et al.] Jun–Aug   ★

S. orestera C.K. Schneid. (p. 1233)   GRAY-LEAFED SIERRA WILLOW   Shrub < 2 m. **ST**: twigs yellow- to red-brown, glaucous or not, silky, glabrous in age. **LF**: stipules gen lf-like; petiole 4–9 mm; young lvs long-silky; mature blade 35–95 mm, strap-shaped to narrowly elliptic or oblanceolate, acuminate to convex (base wedge-shaped to convex), ± entire, abaxial hairs sparsely to ± densely long- to short-silky, white or white and rusty, straight or wavy. **INFL**: blooming with lvs, pistillate 20–55 mm, on lfy shoots 2–15 mm (shoot lvs entire or gland-dotted); fl bract brown. **STAMINATE FL**: stamens 2. **PISTILLATE FL**: ovary long-silky, stalk 0.8–2 mm,

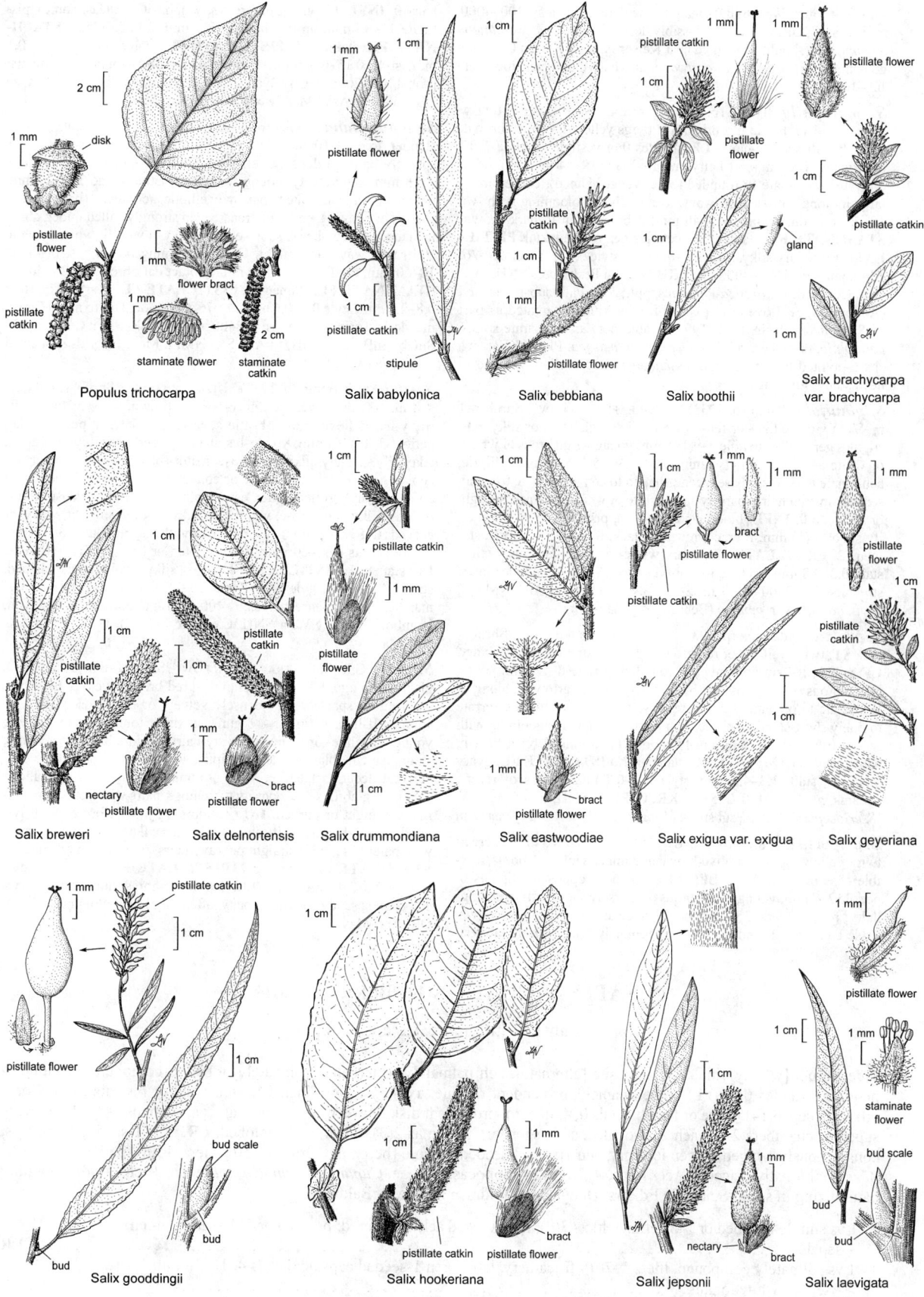

Populus trichocarpa

Salix babylonica

Salix bebbiana

Salix boothii

Salix brachycarpa
var. brachycarpa

Salix breweri

Salix delnortensis

Salix drummondiana

Salix eastwoodiae

Salix exigua var. exigua

Salix geyeriana

Salix gooddingii

Salix hookeriana

Salix jepsonii

Salix laevigata

style 0.6–1 mm. Wet alpine, subalpine meadows, streams; 1100–4000 m. SN, SNE; OR, w NV, UT. Possibly derived from *S. eastwoodiae* × *S. lemmonii* (should be studied near Kaiser Pass, ne Fresno Co.), but does not hybridize with *S. eastwoodiae* where they occur together. Jun–Jul

S. petrophila Rydb. (p. 1233) ROCKY MOUNTAIN WILLOW Shrub < 0.1 m, trailing, rooting. **ST:** twigs yellow-green or -brown, sparsely soft-shaggy-hairy. **LF:** stipules 0 or vestigial; petiole 2–13 mm; young lvs long-soft-hairy; mature blade 19–44 mm, elliptic to obovate, acuminate to rounded (base wedge-shaped), entire, abaxial hairs long-soft-shaggy, wavy, to ± 0. **INFL:** blooming with lvs, 18–60 mm, on lfy shoots 2–40 mm; fl bract ± tan to dark brown. **STAMINATE FL:** stamens 2; nectaries gen only adaxial. **PISTILLATE FL:** ovary silky, stalk 0.2–0.8 mm, style 0.4–1.6 mm. 2*n*=76. Alpine tundra; 1670–4000 m. CaRH (Lassen Peak), c&s SNH; to w Can, CO, NM. [*S. arctica* Pall., misappl.; *S. a.* Pall. subsp. *petraea* (Andersson) Á. Löve et al.] CA, s Rocky Mtns pls treated as part of *S. arctica* Pall. in TJM (1993) treated here as a separate sp., *S. petrophila*, elsewhere as *S. arctica* subsp. *petraea*. Pls with narrow, sharp-pointed lvs approach *S. cascadensis* Cockerell of WA and n, need study. Jul–Aug

S. planifolia Pursh (p. 1233) TEA-LEAFED WILLOW Shrub < 1 m. **ST:** twigs yellow- or red-brown or violet, glabrous or silky. **LF:** stipules gen lf-like; petiole 2–9(14) mm; young lvs glabrous or white- or white-and-rusty-silky; mature blade 19–75(115) mm, ± elliptic, acuminate to convex (base wedge-shaped to convex), entire to finely serrate, abaxial hairs sparsely silky, white or white and rusty, straight or wavy, to 0. **INFL:** blooming before lvs, pistillate 15–67 mm, on lfy shoots 0(5) mm; fl bract brown to black. **STAMINATE FL:** stamens 2. **PISTILLATE FL:** ovary white- or white-and-rusty-silky, stalk 0.3–0.8 mm, style 0.5–2 mm. 2*n*=76, 57. Subalpine meadows, streambanks; 2500–4000 m. SNH, SNE; to n&e N.Am, NM. [*S. phylicifolia* L. var. *monica* (Bebb) Jeps.] May–Jun

S. prolixa Andersson (p. 1233) MACKENZIE'S WILLOW Shrub < 5 m. **ST:** twigs yellow- or red-brown, glabrous or long-shaggy-hairy. **LF:** stipule lf-like, tip convex to rounded; petiole 6–12 mm; young lvs glabrous or hairy; mature blade 50–150 mm, narrowly oblong or lanceolate to obovate, acuminate (base rounded to cordate), serrate to finely short-slender-serrate, abaxial hairs 0. **INFL:** blooming with lvs, pistillate 19–66 mm, on lfy shoots 0.5–6 mm; fl bract brown, hairs wavy. **STAMINATE FL:** stamens 2. **PISTILLATE FL:** ovary glabrous, stalk 1.3–4.2 mm, style 0.3–0.7 mm. Banks of rivers, streams, marshes; 110–2255 m. KR, CaRH, n SNH; to n Can, WY. [*S. eriocephala* Michx. subsp. *mackenzieana* (Hook.) Dorn] Mar–Jun

S. purpurea L. BASKET WILLOW Shrub to tree < 5 m. **ST:** erect; twigs yellow-, gray-, or olive-brown and tinged violet, glabrous, flexible or ± brittle at base. **LF:** at least some ± opposite; stipules 0; petiole 2–7 mm; young lvs glabrous or sparsely short-soft-spreading-hairy; mature blade 35–77 mm, oblanceolate, acuminate or convex (base convex to rounded), entire to remotely finely serrate, abaxial hairs 0. **INFL:** blooming before lvs, ± opposite (at least some), pistillate 13–35 mm, on lfy shoots 0.5–3 mm; fl bract black. **STAMINATE FL:** stamens 1. **PISTILLATE FL:** ovary silky, stalk 0–0.1 mm, style 0.15–0.3 mm, stigma flat. 2*n*=38. Around settlements; probably < 20 m. NCo (Ryan's Slough, Humboldt Bay); to Can, se US; native to Asia. Mostly cult as orn. Mar–May

S. scouleriana Hook. (p. 1233) SCOULER'S WILLOW Shrub, slender tree, < 10 m. **ST:** twigs yellow-green or -brown, twigs sparsely hairy to densely velvety. **LF:** stipules gen lf-like; petiole 2–13 mm, gen velvety; young lvs hairy; mature blade 29–100 mm, obovate or oblanceolate to narrowly elliptic, acuminate (base wedge-shaped or convex), entire or crenate, gen strongly rolled under, abaxial hairs sparse to dense, short- or long-silky or woolly, white or white and rusty, wavy or straight. **INFL:** blooming before lvs, pistillate 18–60 mm, on lfy shoots 0(8) mm; fl bract dark brown or 2-colored. **STAMINATE FL:** stamens 2. **PISTILLATE FL:** ovary silky, stalk 0.8–2.3 mm, style 0.2–0.6 mm. 2*n*=76. Common. Dry to moist forest, meadows, springs, swamps; 1–3400 m. NW, CaR, SN, n CCo, SnFrB, SnGb, SnBr, SnJt, GB; to AK, SK, NM, n Mex. Hybrids suspected with *S. hookeriana*. Feb–Jun

S. sitchensis Bong. (p. 1233) SITKA WILLOW Shrub, small tree, < 8 m. **ST:** twigs yellow-, gray-, or red-brown, silky or long-soft-wavy-hairy, flexible or ± brittle at base. **LF:** later stipules lf-like; petiole 3–13(16) mm; young lvs abaxially densely silky or woolly, adaxially sparsely-silky tomentose; mature blade 31–120 mm, oblanceolate to obovate, acuminate or convex (base wedge-shaped), gen entire, strongly rolled under, esp basally, abaxial hairs dense, short- or long-silky, tomentose, woolly, or silky-woolly, straight, wavy, or curved. **INFL:** blooming just before or with lvs, pistillate 25–73 mm, on lfy shoots (0)1–20 mm; fl bract ± tan or brown. **STAMINATE FL:** stamen 1. **PISTILLATE FL:** ovary silky, ovoid to pear-shaped, stalk 0.4–1.4 mm, style 0.4–0.8 mm. 2*n*=38. Common. Tidal swamps, marshes, springs, streambeds; < 400 m (1800–2500 m in Siskiyou, Humboldt cos.). NW, n SNH, CW (exc SCoRI), w WTR; to AK, AB. Mar

S. tracyi C.R. Ball TRACY'S WILLOW Shrub 1–6 m; lvs, twigs ± glabrous in age. **ST:** twigs yellow- to red-brown, weakly glaucous, glabrous or sparsely to ± densely velvety or tomentose. **LF:** later stipules lf-like; petiole 5–11 mm, adaxially tomentose to velvety; young lvs white- or white-and-rusty-hairy; mature blade 55–96 mm, strap-shaped, oblanceolate, or elliptic, acute to acuminate (base convex or ± decurrent), entire, wavy, or minutely serrate, ± rolled under, abaxially glabrous to ± densely tomentose, hairs white or white-and-rusty, straight or curved. **INFL:** blooming just before or with lvs, pistillate 17–42 mm, on lfy shoots 1.5–3 mm; fl bract brown or brown with paler base, with straight or wavy hairs, tip rounded to truncate. **STAMINATE FL:** stamens 2. **PISTILLATE FL:** ovary glabrous, stalk 1–1.7(2.4) mm, style 0.12–0.6 mm. Shores, floodplains of rivers, creeks; sandy, gravelly, rocky, often serpentine; 90–460 m. NCo; OR. Apr–May

SAPINDACEAE SOAPBERRY FAMILY

Alan T. Whittemore, except as noted

Tree, shrub, [woody vine]. **LF:** opposite [alternate], gen palmately or ternately [pinnately] lobed to compound, deciduous, petioled; stipules 0. **INFL:** umbel, panicle, or pendent raceme, axillary or terminal. **FL:** unisexual or bisexual, radial or ± bilateral; sepals (4)5, free or fused; petals 0, 4, or 5(6); prominent disk between petals and stamens; stamens 5–12, free; ovary superior, chambers 2–3, each 2[1]-ovuled, style short or 0, stigmas 2(3), linear, or 1, unlobed. **FR:** 2(3) 1-seeded mericarps, conspicuously winged, or gen leathery, gen 1[many]-seeded capsule [berry, nut, drupe]. 150 genera, 1500 spp.: ± worldwide. *Acer* traditionally placed in Aceraceae, *Aesculus* in Hippocastanaceae. *Cupaniopsis anacardioides* (A. Rich.) Radlk. possibly naturalizing in s CA. Scientific Editors: Douglas H. Goldman, Bruce G. Baldwin.

1. Lvs simple to lobed or 1–2-ternate, lflets 3(9); fr 1-winged achene, paired; petals 0 or 5, 1–7 mm, green, ≤ sepals . . . . **ACER**
1′ Lvs palmately compound, lflets 5–7(9); fr leathery, large, gen 1-seeded capsule; petals 4, 12–18 mm, ≫ sepals, white to pink . . . . **AESCULUS**

ACER MAPLE

Shrub, tree; occ monoecious. **INFL**: umbel, panicle, or pendent raceme. ± 130 spp.: n hemisphere. (Latin name for *Acer campestre*) Many spp. monoecious or dioecious.

1. Lf compound, lflets stalked; infl axillary, pendent panicle . . . . *A. negundo*
1′ Lf simple, palmately lobed, or compound with lflets sessile; infl not a pendent panicle
  2. Abaxial lf surface white or ± white-green; fls present before lvs, in dense clusters; petals 0 . . . .*A. saccharinum*
  2′ Abaxial lf surface green; fls present with lvs, infl cyme, raceme, or panicle; petals present
    3. Lf lobes 7–9 . . . . *A. circinatum*
    3′ Lf lobes 3–5
      4. Lf 10–25 cm wide; infl pendent, fls 20–90 . . . . *A. macrophyllum*
      4′ Lf 1.4–10 cm wide; infl ascending, fls 2–30
        5. At least lf veins, margins pubescent; each lobe with 0–1(2), large, rounded teeth per side; fr wings
          widely spreading 160–200° . . . . [*A. campestre*]
        5′ Lf glabrous; each lobe with 3–many, small, sharp teeth per side; fr wings overlapping or spreading
          (40)70–120° . . . . *A. glabrum*
          6. Petiole 0.9–2.4 cm, blade 1.4–2.8 cm, 2–3.5 cm wide; lobes rounded or obtuse, lateral lobes with
            3–7 obtuse to acute teeth on outer side . . . . var. *diffusum*
          6′ Petiole 2.8–9.7 cm, blade 2.9–5.6 cm, 3.6–6.3 cm wide; lobes acute, lateral lobes with 5–22 acute
            to apiculate teeth on outer side. . . . . var. *glabrum*

A. circinatum Pursh (p. 1233)   VINE MAPLE   Shrub, small tree, 1–6 m, branches gen sprawling, occ rooting. **LF**: 5.5–9 cm, 7–13 cm wide, 7–9-lobed for 1/4–1/2 of lf length, lobes sharply toothed, abaxially pale green, glabrous to long-hairy at least on lf base, major veins. **INFL**: terminal, ascending, fls 6–22, appearing after lvs. **FL**: petals 1–3 mm, << sepals. **FR**: wings spreading 160–210°. Shaded streambanks; < 1500 m. n&c NW, CaRH, n SN; to AK. May–Jun

A. glabrum Torr.   MOUNTAIN MAPLE   Shrub, small tree, < 6 m; dioecious (or staminate pl with some bisexual fls). **LF**: 3-lobed (or 3 sessile lflets) 1/4–3/4(1) of lf length, at least outer side toothed, teeth 3–22, acute to obtuse; abaxial surface pale green, glabrous. **INFL**: terminal, ascending, fls 3–8, appearing after lvs. **FL**: petals 2–3 mm, ± = sepals. **FR**: wings spreading (0)70–120°, rarely touching, parallel.

  var. *diffusum* (Greene) Smiley   **LF**: petiole 0.9–2.4 cm; blade 1.4–2.8 cm, 2–3.5 cm wide, leathery; lobes rounded to obtuse, outer side toothed, teeth 3–7, acute to obtuse. Dry montane rocky slopes, canyons; 1600–3100 m. CaRH, s SNH, SnBr, W&I, DMtns; to c UT. Apr–May

  var. *glabrum* (p. 1233)   **LF**: petiole 2.8–9.7 cm; blade 2.9–5.6 cm, 3.6–6.3 cm wide, thin; lobes acute, outer side toothed, teeth 5–22, acute to abruptly soft-pointed. Moist to ± dry, montane rocky slopes, canyons; 1400–3000 m. KR, NCoRH, CaRH, n&c SNH; to OR, MT,

CO. [*A. g.* var. *greenei* A.C. Keller; *A. g.* var. *torreyi* (Greene) Smiley] May–Jun

A. macrophyllum Pursh (p. 1233)   BIG-LEAF MAPLE   Tree < 30 m; monoecious. **LF**: 8–15 cm, 10–25 cm wide, 5-lobed to 1/2–3/4 of lf length, lobes with 1–3 rounded-acute 2° lobes on at least one side, abaxially green, pubescent at least near base. **INFL**: pendent raceme or panicle, terminal, fls 20–90, appearing after lvs. **FL**: petals 3–7 mm, ± = sepals. **FR**: wings spreading 50–90°. Common. Streambanks, canyons; < 1500 m. CA-FP (exc GV); to AK. Mar–Jun

A. negundo L. (p. 1237)   BOX ELDER   Tree < 20 m; dioecious. **LF**: 1–2-ternate, lflets 3(9); terminal lflets 4–11 cm, 3–9 cm wide, toothed, gen 1–2-lobed, abaxial surface green, gen felty, occ sparsely pubescent. **INFL**: panicle axillary, pendent, appearing with lvs, fls 10–250. **FL**: petals 0. **FR**: wings spreading 60–90°. Streamsides, bottomland; < 1800 m. CA-FP; US, s Can, s to S.Am. Widely planted, esp GV, as orn or street tree. Mar–Apr

A. saccharinum L.   SILVER MAPLE   Tree, < 20 m; monoecious or dioecious. **LF**: 8–13 cm, 9–13 cm wide, 5-lobed for 1/2–3/4 of lf length, lobes coarsely sharp-toothed or -lobed again, abaxial surface white or white-green, inconspicuously appressed-pubescent, occ sparsely long-hairy on major veins and lf base. **INFL**: axillary, dense, fls ± 15–100, appearing before lvs. **FL**: petals 0. **FR**: wings spreading 60–120°. Riparian woodland; < 50 m. ScV; e US, Can. Feb–Mar

AESCULUS BUCKEYE

William J. Stone

Large shrub or tree. **LF**: palmate, lflets 5–7[9]. **FL**: petals 4[5], >> sepals. **FR**: capsule leathery. **SEED**: 1, large. ± 15 spp.: n hemisphere. (Latin name for a sp. of oak)

A. californica (Spach) Nutt. (p. 1237)   CALIFORNIA BUCKEYE   Pl 4–12 m, broad, rounded. **LF**: lflets 5–7, 6–17 cm, lance-oblong, finely serrate, acute to acuminate; petiole 1–12 cm. **INFL**: panicle-like, erect, 1–2 dm, finely hairy; pedicel 3–10 mm. **FL**: calyx 5–8 mm, 2-lobed; petals 12–18 mm, white to pale rose; stamens 5–7, 18–30 mm, exserted, anthers orange. **FR**: gen 1 at infl tip, occ 2–9,

5–8 cm diam. **SEED**: gen 1, 2–5 cm, glossy brown. 2*n*=40. Dry slopes, canyons, borders of streams; < 1700 m. c&s NW, s CaR, SNF, n&c SNH, Teh, GV (scattered near foothills), n&c CW, WTR, sw DMoj; sw OR. All parts TOXIC. Native Americans used ground seed as fish poison; nectar and pollen TOXIC to honeybees. Gen deciduous Jun–Feb. May–Jun(Aug)

SARCOBATACEAE SARCOBATUS or GREASEWOOD FAMILY

Matthew H. Hils

Shrub, erect; gen monoecious. **ST**: branches many, 90° to main st, thorny, gen interlocking, gen thorn-tipped. **LF**: many, simple, cauline, gen alternate, deciduous, sessile, ± linear, flattened to subcylindric, fleshy; stipules 0. **INFL**: staminate spikes cylindric, dense, erect, terminal; pistillate fls 1–4, axillary; bracts lf-like. **STAMINATE FL**: perianth 0, stamens 1–4, sessile on spike axis, concealed by peltate, angular, persistent, spirally arranged bracts. **PISTILLATE FL**: carpels 2, fused, stigmas

2, ovary 1/2 inferior; perianth cup-like, fused to ovary, persistent, forming wing in fr. **FR**: achene, tapering above encircling, irregularly-edged wing; ± glabrous. **SEED**: coat thin; embryo coiled, green. 1 genus, 2 spp.: to w Can, Great Plains, n Mex. [Cuienoud et al. 2002 Amer J Bot 89:132–144] Formerly treated in Chenopodiaceae. Scientific Editor: Bruce G. Baldwin.

SARCOBATUS GREASEWOOD

(Greek: fleshy bramble)

1. Lf blade 0.5–1.5 cm, dull to ± gray-green; lvs gen clustered in short shoots on cushion-like pads of longer or older twigs; pistillate fls and staminate spikes on short branches with 1–3 minute internodes; staminate spikes at maturity gen < 10 mm; fr 5–11.5 mm, wing 7–16 mm wide . . . . *S. baileyi*
1′ Lf blade 1–4 cm, bright to ± yellow-green; lvs gen 1 on elongate shoots of current season; pistillate fls and staminate spikes on long lateral branches with 3–9 obvious internodes; staminate spikes at maturity gen 10–40 mm; fr 2.5–5.2 mm, wing 4–10 mm wide . . . . *S. vermiculatus*

S. baileyi Coville Pl 5–10 dm, gen ± spheric. **ST**: light gray. *n*=54. Alkaline soils, dry lakes, washes, scrub, roadsides; gen > 1200 m. e SNE (Fish Lake Valley, Mono Co.), DMtns (Coso Range); NV. [*S. vermiculatus* (Hook.) Torr. var. *b.* (Coville) Jeps.] Apr–Jul ★

S. vermiculatus (Hook.) Torr. (p. 1237) Pl 5–21 dm, erect (± spheric). **ST**: ± yellow-cream to ± white or light gray. *n*=18,36. Alkaline soils, dry lakes, washes, scrub, roadsides; 100–2300 m. GB, DMoj; to w Can, Great Plains, n Mex. Apr–Aug

SARRACENIACEAE PITCHER-PLANT FAMILY

Barry A. Rice

Per, gen from slender rhizome, short caudex, or stolon; carnivorous; roots poorly developed. **LF**: in basal rosette, prostrate to erect, each forming a tubular pitcher with fluid that digests captured prey by enzymes, bacteria, or other organisms, with stiff, reflexed hairs within. **INFL**: scapose, fl gen 1. **FL**: bisexual, radial, nodding; sepals 5 [4–6], gen free; petals 5 [0]; stamens many; pistil 1, ovary superior, chambers gen 5, incomplete above or not, placentas gen axile, style 1, 5-lobed, umbrella-like or not, stigma terminal or under tips of style lobes. **FR**: capsule, loculicidal; valves gen 5. **SEED**: many, flattened-ovoid, smooth, or club-like, papillate [winged]. 3 genera, 24 spp.: n CA, OR, BC, e N.Am, n S.Am, esp acidic bogs, streamsides, moist areas; often planted outside native ranges by horticulturists but gen not invasive. [Schnell 2002 Carnivorous Pls of US and Can. Timber Press] Scientific Editor: Thomas J. Rosatti.

1. Pitcher with 2 ± pendent, diverging, fang-like appendages, each 5+ cm (often damaged or imperfect); pitcher top dome-like, with many glassy-transparent windows; scape bracts few to several, scattered . . . . **DARLINGTONIA**
1′ Pitcher without pendent, diverging, fang-like appendages; pitcher top not dome-like, gen with vertical or overhanging lid; scape bracts 3, whorled, subtending fl . . . . **SARRACENIA**

DARLINGTONIA CALIFORNIA PITCHER-PLANT

1 sp. (William Darlington, Philadelphia botanist, 1782–1863)

D. californica Torr. (p. 1237) Rhizomed; stolons ± 1 m. **LF**: nearly erect, 1–6(10) dm, green-yellow to deep red, enlarged upward; pitcher top opening underneath, tube with non-digestive fluids within; appendages yellow or green to purple. **INFL**: < 1 m; scape bracts yellow. **FL**: sepals 4–6 cm, oblong to oblanceolate, yellow-green, purple-tinged or not; petals 2–4 cm, narrowly ovate, yellow-green between wide, dark purple veins (variant lacking purple veins in Nevada Co.); stamens 12–15 in 1 whorl; ovary tip truncate or concave, style 2–3 mm, deeply 5-lobed, stigmas 5. **FR**: 2.5–4.5 cm, obovoid. **SEED**: ± 2 mm, papillate, light red-brown. 2*n*=30. Seeps, boggy places with running water, gen serpentine; 60–2200 m. KR, NCoRO (introduced, Mendocino Co.), n SNH (c Plumas, Sierra, Nevada cos.); w OR; planted elsewhere. Pollinator the bee *Andrena nigrihirta*, possibly also an arachnid; digestion in lvs by bacteria, arthropods, not by pl enzymes. Pitcher resembling cobra with fangs or forked tongue. Apr–Jun ★

SARRACENIA PITCHER-PLANT

LF: pitcher top gen with vertical or overhanging lid [top dome-like, without glassy-transparent windows, only in *Sarracenia psittacina* and its hybrids], opening gen upward, tube with digestive fluids within. **FL**: sepals, petals 5; ovary chambers 5, style tip umbrella-like, peltate. 11 spp.: e US, e Can to BC. (Michel S. Sarrazin, Quebec physician, naturalist, 1659–1735) Many spp., hybrids (traits intermediate) planted in n CA (Mendocino, Del Norte cos.; expected elsewhere), proliferating slowly if at all.

S. purpurea L. (p. 1237) **LF**: ascending to nearly erect, < 20 cm, green-yellow to deep red, enlarged upward. **INFL**: 20–60 cm. **FL**: sepals 2–6 cm, ovate to rhombic, abaxially green to dark purple-red, adaxially pale green, dark purple-red near margins or not; petals 2–6 cm, obovate, tapered to a short claw, purple-red (green); style tip ± 3–5 cm wide. **FR**: 1–2.5 cm, ± spheric to ovoid, 5-lobed. **SEED**: 1–2 mm, flattened, ovate-oblong, brown to ± purple. 2*n*=26. Acidic seeps, marshes, bogs; < 1200 m. NCo (Mendocino Co.), KR (Del Norte Co.), n SNH (Butterfly Valley), possibly elsewhere; native to e N.Am. Despite eradication (easily accomplished by hand), pls continue to appear outside cult, esp in NCo, n SNH, due to replanting. *S. purpurea* subsp. *p.* (pitchers smooth, glabrous abaxially) and *S. purpurea* subsp. *venosa* (Raf.) Wherry (pitchers rough or hairy abaxially), both recognized taxonomically, as well as intermediates between the two, in addition to other sp., often planted outside cult in CA, where they persist and reproduce vegetatively at slow rates. May–Jul

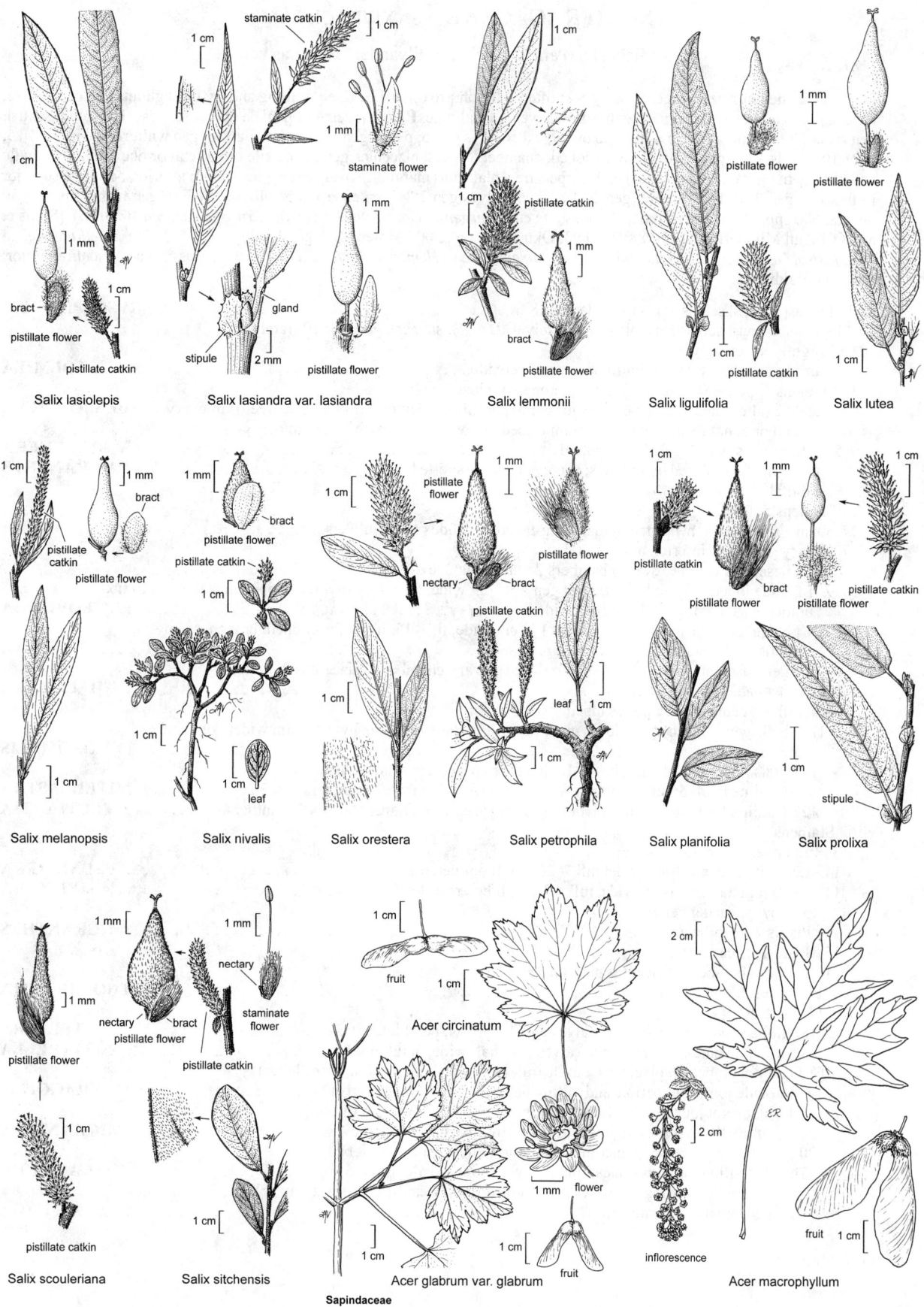

Salix lasiolepis

Salix lasiandra var. lasiandra

Salix lemmonii

Salix ligulifolia

Salix lutea

Salix melanopsis

Salix nivalis

Salix orestera

Salix petrophila

Salix planifolia

Salix prolixa

Salix scouleriana

Salix sitchensis

Acer circinatum

Acer glabrum var. glabrum

Acer macrophyllum

Sapindaceae

SAXIFRAGACEAE SAXIFRAGE FAMILY

Michael S. Park & Patrick E. Elvander, except as noted

Per from caudex or rhizome, gen ± hairy. **ST**: often ± lfy on proximal 1/2, rarely trailing and lfy throughout. **LF**: gen simple, basal and/or cauline, gen alternate, gen petioled; veins ± palmate. **INFL**: raceme or panicle, gen ± scapose. **FL**: gen bisexual, gen radial; hypanthium free to ± fused to ovary; calyx lobes gen 5; petals gen 5, free, gen clawed, gen white; stamens 3, [4], 5, 8, or 10; pistils 1 (carpels ± fused, ovary lobed, chambers 1 or 2, placentas gen 2(3), axile or parietal or occ proximally axile and distally marginal in ovary lobes) or 2 (carpels free, placentas marginal), ovary nearly superior to inferior, occ more superior in fr, styles gen 2(3). **FR**: capsule (gen 2(3)-beaked, valves gen 2(3), gen equal) or 2 follicles. **SEED**: gen many, small. ± 30 genera, 600 spp.: esp n temp, arctic, alpine; some cult (*Bergenia, Darmera, Heuchera, Saxifraga, Tellima, Tolmiea*). [Soltis et al. 2001 Ann Missouri Bot Gard 88:669–693; Okuyama et al. 2008 Molec Phylogen Evol 46:560–575] CA *Mitella* moved to *Mitellastra, Ozomelis, Pectiantia*; *Suksdorfia ranunculifolia* to *Hemieva*. *Parnassia* moved to Parnassiaceae. Scientific Editor: Bruce G. Baldwin.

1. Fl inconspicuous; sepals 4; petals 0; stamens 8 . . . . **CHRYSOSPLENIUM**
1′ Fl ± conspicuous; sepals 5; petals gen 5 (rarely 0, 2, or 4); stamens gen 5 or 10 (rarely 3)
  2. Fl slightly to strongly bilateral
    3. Stamens 3; petals 4; hypanthium open on lower side . . . . **TOLMIEA**
    3′ Stamens 5, petals 0, 2, or 5; hypanthium open or closed on lower side
      4. Petals 2 (alternate upper sepals), recurved; hypanthium open on lower side, ± free from ovary . . . **BENSONIELLA**
      4′ Petals 0 or 5, not recurved; hypanthium closed on lower side, partly fused to ovary
        5. Infl bulblets 0; petals 0 or not spotted . . . . **²HEUCHERA**
        5′ Proximal fls replaced by bulblets; some petals ± 2-spotted . . . . **³MICRANTHES**
  2′ Fl ± radial
    6. Stamens 5
      7. Ovary superior — petals triangular, long-tapered; caudex with bulblets . . . . **BOLANDRA**
      7′ Ovary gen > 1/2-inferior, at least in fl
        8. Lvs basal and cauline; ovary chambers 2, placentas 2, axile
          9. Lf blade lobed < 3/4 to base, largest gen >> 4 cm wide; fls > 15; infl ± flat-topped or ± tapered to tip, branches gen > 3, with fls gen on 1 side . . . . **BOYKINIA**
          9′ Lf blade lobed nearly to base, largest 1–4 cm wide; fls < 15; infl ± flat-topped, branches gen < 3, with fls on all sides . . . . **HEMIEVA**
        8′ Lvs gen basal (at most 3 cauline and reduced); ovary chamber 1, placentas 2, parietal
          10. Infl a panicle; petals ± entire . . . . **²HEUCHERA**
          10′ Infl a raceme (spike); petals lobed
            11. Petals gen 3-lobed at tip, lobes narrowly triangular to elliptic; hypanthium widely obconic to bell-shaped . . . . **OZOMELIS**
           11′ Petals pinnately lobed, lobes 4–10, linear; hypanthium saucer-shaped
              12. Cauline lvs 1–3; infl blooming from tip to base; filaments >> anthers, ± 2/3 sepal . . . . **MITELLASTRA**
              12′ Cauline lvs 0 (occ 1); infl blooming from base to tip; filaments (short) < anthers . . . . **PECTIANTIA**
    6′ Stamens 10
      13. Infl appearing before lvs
        14. Lvs peltate, gen > 1 dm wide; infl 7–20 dm; fl not heterostylous . . . . **DARMERA**
        14′ Lvs not peltate, << 1 dm wide; infl < 3 dm; fl heterostylous . . . . **JEPSONIA**
      13′ Infl not appearing before lvs
        15. Pistils 2; fr follicle . . . . **³MICRANTHES**
        15′ Pistil 1; fr capsule
          16. Ovary chamber 1, placentas parietal
           17. Styles 3 . . . . **LITHOPHRAGMA**
           17′ Styles 2
              18. Infl a spike-like raceme; fr valves equal, narrow; petal lobes ± 5–7, linear . . . . **TELLIMA**
              18′ Infl a raceme-like panicle; fr valves unequal, wide; petal lobes 0 . . . . **TIARELLA**
          16′ Ovary chambers 2, placentas ± axile (rarely distally marginal in ovary lobes)
           19. Lf blade jointed to petiole and falling before it; styles gen fused at base, at least in fl . . . . . . **SAXIFRAGOPSIS**
           19′ Lf blade not jointed to petiole nor falling before it; styles free throughout
              20. Cauline lvs 0 (if present, then all lvs entire); lvs not lobed . . . . **³MICRANTHES**
              20′ Cauline lvs present, sometimes strongly reduced; some lvs lobed
                21. Sts trailing; basal lvs 0; cauline lvs evenly spaced along st; seeds spiny . . . . **CASCADIA**
                21′ Sts erect or trailing; basal lvs present or if 0, then cauline lvs mostly near base; seeds finely papillate . . . . **SAXIFRAGA**

BENSONIELLA

Michael S. Park, Gerald D. Carr & Patrick E. Elvander

1 sp. (G.T. Benson, Stanford botanist, 1896–1928) [Abrams & Bacigalupi 1929 Contr Dudley Herb 1:95]

B. oregona (Abrams & Bacig.) C.V. Morton (p. 1237) BENSONIELLA Pl 20–40 cm; rhizome scaly; bulblets 0. **LF**: basal; petiole 2–15 cm, hairs dense, long, brown; blade 4–20 cm, round-ovate, base cordate, unevenly crenate, ± glabrous adaxially, veins hairy abaxially. **INFL**: spike-like raceme, glandular; bracts 0; pedicel < 1 mm. **FL**: strongly bilateral, hypanthium free of ovary; calyx lobes 1.5–2.5 mm, obovate; petals 2, alternate upper sepals, > calyx lobes, narrowly awl-like, recurved; stamens 5, opposite and > calyx lobes; pistil 1, ovary superior, chamber 1, placentas 2, parietal, styles 1.5–2.5 mm, stigmas ± unlobed, disk-like to head-like. **FR**: capsule, opening before seeds mature, becoming widely dehiscent, forming splash cup. **SEED**: red-brown to black. Wet meadows, bogs; > 750 m. n NCoRO (Humboldt Co.), expected KR; sw OR. May–Jun ★

BOLANDRA

Caudex scaleless, bearing bulblets. **LF**: basal and cauline, basal and proximal petioled, cauline stipuled, distal sessile, reduced; blade ± round, base cordate to reniform, lobes 3–7, deep, teeth sharp. **INFL**: panicle-like, glandular; fls few; bracts sessile, proximal often clasping, lf-like. **FL**: hypanthium free from ovary; petals not clawed; stamens 5, opposite sepals; pistil 1, ovary superior, chambers 2, placentas 2, axile. **FR**: capsule. 2 spp.: WA, OR, CA. (H.N. Bolander, CA botanist, 1831–1897) [Gornall & Bohm 1985 Bot J Linn Soc 90:1–71] Closely related to *Suksdorfia* and *Hemieva*.

B. californica A. Gray (p. 1237) SIERRA BOLANDRA Pl 15–60 cm. **LF**: petiole 3–10 cm; blade 1–5 cm wide. **INFL**: pedicel 0.5–3 mm. **FL**: calyx lobes 3–6 mm, ovate to triangular; petals 4–7 mm, narrowly triangular, long-tapered, 1/3–1/2 as wide as sepals, green, margin purple; stamens << calyx lobes, opposite sepals. **SEED**: brown, minutely papillate. 2*n*=14. Rock crevices, wet cliffs; 1000–3000 m. n&c SNH (El Dorado to Mariposa cos.). May–Aug ★

BOYKINIA

Pl glandular; rhizome scaly; bulblets 0. **LF**: basal and cauline, distal reduced, becoming bract-like; stipules inconspicuous to lf-like; blade round to ovate, base cordate to reniform, occ truncate, obtuse, or tapered, primary lobes 3–many, shallow to deep, teeth sharp-tipped. **INFL**: ± flat-topped or ± tapered to tip, > 15-fld, gen 1-sided; bracts sessile to short-petioled, proximal lf-like. **FL**: hypanthium partly fused to ovary; petals often ephemeral; stamens 5; pistil 1, ovary > 1/2-inferior, chambers 2, placentas 2, axile. **FR**: capsule. 8 spp.: N.Am, Asia. (S. Boykin, GA naturalist, 1786–1848) [Gornall & Bohm 1985 Bot J Linn Soc 90:1–71]

1. Stipules ± 1 cm or more, distal lf-like; petal ≥ 5 mm; lf blade divided > 1/2 to base; infl ± flat-topped . . . . ***B. major***
1′ Stipules ≤ 0.4 cm, distal brown-bristly; petal gen < 4 mm; lf blade divided < 1/8–1/3 to base; infl ± tapered to tip
  2. Petal gen > 3 mm, > calyx lobe; fls not crowded; lf blade divided 1/4–1/3 to base . . . . ***B. occidentalis***
  2′ Petal gen < 3 mm, ± = calyx lobe; fls crowded, ± overlapping; lf blade divided < 1/8 to base . . . . ***B. rotundifolia***

B. major A. Gray Pl (2)5–1.3 dm, stout. **LF**: 9–50 cm; stipules conspicuous, ± 1 cm or more, green, entire or teeth gen sharp-tipped, distal lf-like; petiole 5–35 cm; blade < 20 cm wide, divided > 1/2 to base, lobes and teeth ± straight-sided. **INFL**: ± flat-topped; branches occ coiled at tip; fls crowded. **FL**: calyx lobes 2–5 mm, triangular to elliptic; petals 5–7 mm, elliptic to round. Shaded, moist meadows, streambanks; < 2500 m. KR, n NCoRO (Grouse Mtn), n&c SNH; to WA, MT. Jun–Jul

B. occidentalis Torr. & A. Gray (p. 1237) Pl 1.5–6(10) dm. **LF**: 6–45 cm; stipules ≤ 0.4 cm, gen green, distal brown-bristled; petiole 3–30 cm; lf < 12 cm wide, divided 1/4–1/3 to base, lobes and teeth ± round-sided. **INFL**: ± tapered to tip; branches occ coiled at tip; fls not crowded. **FL**: calyx lobes 1–3 mm, triangular, tapered; petals 3–4 mm, obovate. 2*n*=14. Shady wet banks; < 1500 m. NW (exc NCoRH), CaRF, n SNH, CCo, SnFrB, n SCoRO, WTR (Santa Monica Mtns); to BC. May–Aug

B. rotundifolia A. Gray Pl (3)5–8(13) dm, ± stout. **LF**: 10–30 cm; stipules inconspicuous, distal gen brown-bristled; petiole 5–18 cm; lf blade to 18 cm wide, divided < 1/8 to base, lobes and teeth ± round-sided. **INFL**: ± tapered to tip; branches coiled at tip; fls crowded, ± overlapping. **FL**: calyx lobes 2–3 mm, triangular; petals 2–3 mm, obovate. 2*n*=26. Streambanks; < 2000 m. s SCoRO, TR, PR. May–Jul

CASCADIA

1 sp. (Cascade Ranges) [Johnson 1927 Amer J Bot 14:38–43]

C. nuttallii (Small) A.M. Johnson (p. 1237) Rhizome < 1 mm diam; bulblets 0. **ST**: 5–25 cm, trailing. **LF**: cauline, 3–20 mm, larger along middle of st; petiole 1–5 mm; blade obovate to elliptic, gen 3-lobed to occ entire near tip of st. **INFL**: raceme or panicle; fls few. **FL**: hypanthium ± fused to ovary, sepals erect, << petals, triangular; petals 3–6 mm, elliptic; stamens 10, filaments thread-like; pistil 1, ovary ± inferior, chambers 2, placentas 2, axile, styles free throughout. **FR**: capsule. **SEED**: spiny. 2*n*=16. Wet, shaded cliffs, ledges; < 700 m. KR (Del Norte Co.); sporadically to WA. [*Saxifraga n.* Small] Apr–May ★

CHRYSOSPLENIUM GOLDEN SAXIFRAGE

Michael S. Park

Pl glabrous; caudex without scales or bulblets. **ST**: prostrate to ascending, rooting at nodes. **LF**: cauline, opposite [or alternate] throughout or distal alternate; blade round to ovate, base gen truncate, teeth round. **INFL**: fls 1, axillary, inconspicuous; bracts lf-like. **FL**: hypanthium fused to ovary; calyx lobes 4, petal-like; petals 0; stamens [4]8; pistil 1, ovary inferior to 1/2-inferior, chamber 1, placentas 2, parietal. **FR**: capsule, becoming widely dehiscent, forming splash cup. **SEED**: many, red-brown (to black), shiny. 15 spp.: n temp, s S.Am. (Greek: golden spleen)

C. glechomifolium Nutt. (p. 1237) **LF:** 7–30 mm; petiole < 10 mm; blade 5–20 mm wide. **FL:** calyx lobes 1.5–2 mm, elliptic; stamens ± 1 mm; nectaries 2, surrounding ovary, each with 4 lobes alternate stamens. 2*n*=18. Shady wet areas; < 200 m. NCo; to BC. Feb–May ★

DARMERA UMBRELLA PLANT

Michael S. Park

1 sp. (Karl Darmer, German horticulturist, 19th century) [Schmid & Turner 1977 Madroño 24:68–74]

D. peltata (Benth.) Voss (p. 1237) **Pl** 3–15 dm; rhizome < 5 cm thick, fleshy, scaly; bulblets 0. **LF:** basal, < 1.5 m; blade peltate, gen > 1 dm, ± round, lobes deep, teeth irregular. **INFL:** appearing before lvs, 7–20 dm, ± flat-topped; bracts gen scale-like or 0. **FL:** hypanthium minute, free from ovary; calyx lobes reflexed, 3–4 mm, elliptic; petals not clawed, 5–7 mm, obovate, white to pink; stamens 10, 3–4 mm, filaments tapered; pistils 2, appearing fused early in fl, ovary superior, placentas marginal. **FR:** 2 follicles, 8–12 mm. 2*n*=34. Rocky streambanks; < 2000 m. KR, n NCoRO (Humboldt Co.), CaR, n SNF, SNH; sw OR. Apr–Sep

HEMIEVA

Michael S. Park

1 sp. (Greek: 1/2 of well, from hypanthium) [Soltis et al. 1996 Syst Bot 21:169–185] Close to *Bolandra* and *Suksdorfia*.

H. ranunculifolia (Hook.) Raf. (p. 1243) **Pl** 1–3 dm; caudex scaleless, bearing bulblets. **LF:** basal and cauline, basal and proximal petioled, petioles 5–10 cm, distal sessile, reduced; stipules membranous, entire, ± sheathing; blade 1–4 cm wide, ovate, base cordate to reniform, ± glabrous, 1° lobes 3, deep, teeth coarse, round, tips sharp. **INFL:** cyme, ± flat-topped, < 15-fld, glandular; fls gen crowded, pedicels gen < 3 mm; bracts sessile, linear. **FL:** hypanthium partly fused to ovary; calyx lobes 1–3 mm, ovate to triangular; petals 2–4 mm, ± white, elliptic to obovate, short-clawed; stamens 5, opposite calyx lobes; pistils 2, ovary > 1/2-inferior, placentas marginal. **FR:** 2 follicles. Moist rocky slopes; 1500–2500 m. KR, n SNH/CaRH (Plumas Co.); to BC, AB, MT. [*Suksdorfia r.* (Hook.) Engl.] Jun–Aug ★

HEUCHERA ALUMROOT

Michael S. Park

Rhizome scaly; bulblets 0. **LF:** basal, sometimes a few cauline; blade ovate, base cordate to reniform, lobes and teeth gen shallow, irregular. **INFL:** panicle, often spike- or raceme-like; bracts gen scale-like. **FL:** radial or ± bilateral; hypanthium partly fused to ovary; calyx lobes equal or not; petals 0 or 5, gen equal, clawed, gen white; stamens 5, gen equal; pistil 1, ovary > 1/2-inferior, chamber 1, placentas 2, parietal. **FR:** capsule. **SEED:** red-brown, minutely spiny. 50 spp.: N.Am. (J.H. von Heucher, German professor of medicine, 1677–1747) Highly variable; needs study.

1. Styles and gen all stamens incl in full fl
  2. Petals gen 0; infl narrow, spike-like . . . . ***H. cylindrica***
  2′ Petals present, ± conspicuous; infl more open, panicle or raceme-like
    3. Hypanthium + calyx 2–3(4.2) mm, radial; calyx lobes equal, cream to ± yellow, tips pink to ± green . . . ***H. parvifolia***
    3′ Hypanthium + calyx gen > 3.5 mm, ± bilateral; calyx lobes ± unequal, light pink to red-purple, tips green or not (*Heuchera caespitosa* complex)
      4. Infl glands nearly sessile, stalk ± = gland; lf blade gen < 15 mm, gen broadly ovate, lobed ± 1/2 to base . . . . ***H. abramsii***
      4′ Infl glands nearly sessile and long-stalked, stalk >> gland; lf blade gen > 15 mm, gen round to reniform, lobed 1/4–1/3 to base
        5. Stamens gen < 1/2 calyx lobes; part of hypanthium free from ovary ± 0.7 mm on short side, ± 1.5 mm on long side — PR (San Diego Co.) . . . . ***H. brevistaminea***
        5′ Stamens ± = calyx lobes; part of hypanthium free from ovary > 0.8 mm on short side, > 1.8 mm on long side
          6. Calyx lobes all well developed . . . . ***H. caespitosa***
          6′ Calyx lobes on long side of hypanthium minute — PR . . . . ***H. hirsutissima***
1′ Styles and stamens exserted in full fl
  7. Fl ± bilateral; calyx lobes ± equal or not; part of hypanthium fused to ovary ± = long side of free part
    8. Petals 2–3 mm, > 0.3 mm wide, oblanceolate, unequal; hypanthium strongly oblique . . . . ***H. parishii***
    8′ Petals 3–6 mm, gen << 0.3 mm wide, thread-like to narrowly oblanceolate, equal; hypanthium slightly oblique . . . . ***H. rubescens***
  7′ Fl ± radial; calyx lobes ± equal; part of hypanthium fused to ovary gen > free part
    9. Styles 2–4 mm, exserted; hypanthium broadly obconic, base tapered; infl gen open
      10. Hypanthium + calyx 3.5–4.5 mm; lf blade 6–18 cm wide; n ChI . . . . ***H. maxima***
      10′ Hypanthium + calyx 1.5–3(5) mm; lf blade 2–10 cm wide; mainland . . . . ***H. micrantha***
    9′ Styles gen ≤ 1.8 mm, barely exserted; hypanthium bell-shaped, base truncate; infl ± dense or open
      11. Lf blade gen round, base ± truncate; petiole 2–5 cm, hairs short, stiff; ≥ 1500 m . . . . ***H. merriamii***
      11′ Lf blade gen ovate, base cordate; petiole 7–20 cm, hairs long, soft; < 500 m . . . . ***H. pilosissima***

H. abramsii Rydb. ABRAMS' ALUMROOT **Pl** 5–15 cm. **LF:** blade gen < 15 mm, 5-lobed ± 1/2 to base. **INFL:** dense; glands nearly sessile, stalk ± = gland. **FL:** ± bilateral; hypanthium + calyx 4–7 mm; part of hypanthium fused to ovary 1–2.2 mm, < long side of free part; calyx lobes 1–2 mm, unequal, gen red-purple; petals 4–5 mm, spoon-shaped, claw > blade; stamens < calyx lobes, equal, incl. Dry, rocky areas; 2800–3500 m. SnGb. Jul–Aug ★

H. brevistaminea Wiggins LAGUNA MOUNTAINS ALUMROOT **Pl** 18–25 cm. **INFL:** open; glands nearly sessile and long-stalked, stalk

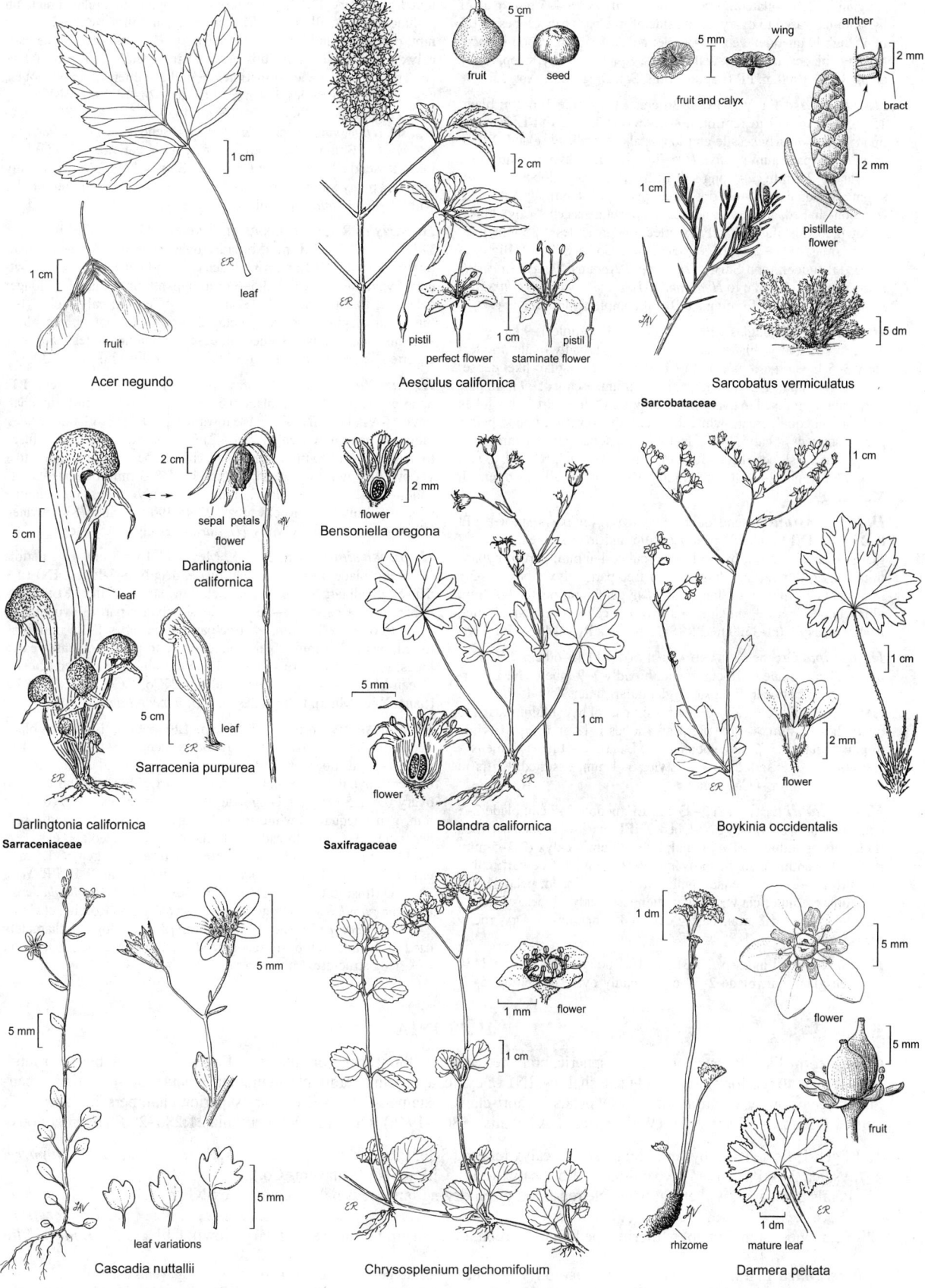

Acer negundo

fruit
seed
pistil
perfect flower
staminate flower
pistil
leaf

Aesculus californica

fruit and calyx
wing
anther
bract
pistillate flower

Sarcobatus vermiculatus
Sarcobataceae

sepal petals
flower
Darlingtonia californica

leaf
leaf

Darlingtonia californica
Sarraceniaceae

Sarracenia purpurea

flower
Bensoniella oregona

flower

Bolandra californica
Saxifragaceae

flower

Boykinia occidentalis

leaf variations
Cascadia nuttallii

flower

Chrysosplenium glechomifolium

flower
fruit
rhizome
mature leaf
Darmera peltata

>> gland. **FL**: ± bilateral; hypanthium + calyx (3.5)4–5 mm; part of hypanthium fused to ovary > long side of free part; calyx lobes 1–1.5 mm, slightly unequal, gen red-purple; petals 4–5 mm, spoon-shaped, claw ± = blade; stamens << calyx lobes, equal, incl. Dry, steep, rocky areas; 1400–1900 m. PR (Laguna Mtns, San Diego Co.). Apr–Jul ★

H. caespitosa Eastw. Pl 10–36 cm. **LF**: petiole 1–6 cm; blade 15–40 mm, round to reniform, shallowly 5-lobed. **INFL**: narrow, open; glands nearly sessile and long-stalked, stalk >> gland. **FL**: ± bilateral; hypanthium + calyx (4)5–7 mm; part of hypanthium fused to ovary 1–2.2 mm, ± = long side of free part; calyx lobes 1–3 mm, slightly unequal, pink-red, tips green; petals 4.5–6 mm, oblanceolate to spoon-shaped, claw ± = blade, unequal; stamens gen < calyx lobes, unequal, longest incl to barely exserted; mature styles > 1.5 mm, incl. Rocky areas; 1900–2300 m. s SNH (Tulare Co.), s SCoRO (Big Pine Mtn), TR (uncommon SnBr). [*H. elegans* Abrams; *Heuchera cespitosa*, orth. var.] Similar to *H. abramsii*, *H. brevistaminea*, *H. hirsutissima*; these together forming a difficult complex. May–Jul ★

H. cylindrica Douglas Pl 8.5–55 cm. **LF**: petiole 1–9 cm, glandular to stiff-hairy; blade gen < 3 cm wide, broadly ovate, moderately 3–5-lobed, gen truncate to ± cordate. **INFL**: spike-like, dense, ± short-glandular. **FL**: ± bilateral; hypanthium + calyx 5–9 mm; part of hypanthium fused to ovary 2–2.5 mm, < to > free part; calyx lobes 2–4 mm, unequal, cream-white to green, tips sometimes pink; petals gen 0; stamens < calyx lobes, incl; mature styles gen < 1 mm, incl. Rocky banks, slopes; 1400–3000 m. MP; to WA, ID, NV. [*H. c.* var. *alpina* S. Watson] Many intergrading vars. historically recognized. May–Aug

H. hirsutissima Rosend. et al. SHAGGY-HAIRED ALUMROOT Pl 8–25 mm. **INFL**: glands nearly sessile and long-stalked, stalk >> gland. **FL**: ± bilateral; hypanthium + calyx 4–6 mm; part of hypanthium fused to ovary << long side of free part; calyx lobes 0.8–2.5 mm, unequal, minute on long side of hypanthium; petals 4–5 mm, oblanceolate, claw ± = blade; stamens ± = calyx lobes, equal, incl. Rocky areas; 2200–3500 m. PR (SnJt, n Santa Rosa Mtns). Jul ★

H. maxima Greene ISLAND ALUMROOT Pl 45–60 cm. **LF**: petiole 8–20 cm; blade 6–18 cm, round, broadly 7–9-lobed. **INFL**: narrow, ± open between fl clusters, glandular; bracts few, lf-like. **FL**: radial; hypanthium + calyx 3.5–4.5 mm; part of hypanthium fused to ovary 2.5–3.5 mm, >> free part; calyx lobes 1–1.5 mm, equal, white to pink; petals 3.5–6 mm, spoon-shaped, claw ± = blade; stamens > calyx lobes, ± exserted; mature styles 3–4 mm, exserted. Cliffs in canyons; < 500 m. n ChI. Apr ★

H. merriamii Eastw. Pl 7–23 cm. **LF**: petiole 2–5 cm; blade 2–5 cm, gen round, shallowly 5–7-lobed. **INFL**: narrow, ± open between fl clusters, glandular. **FL**: ± radial; hypanthium + calyx (2)3–5 mm; part of hypanthium fused to ovary 1.5–2.5 mm, > free part; calyx lobes (0.5)1–2 mm, ± equal, pink, often green-tipped; petals 1.5–3 mm, oblanceolate, claw < blade; stamens > calyx lobes, exserted; mature styles < 1.8 mm, barely exserted. Uncommon. Dry, rocky areas; 1500–2500 m. KR; sw OR. Jul–Aug

H. micrantha Lindl. (p. 1243) Pl 10–100 cm. **LF**: petiole 3–30 cm, gen glandular; blade 2–12 cm, broadly ovate to oblong, 5–7-

lobed, gen hairy. **INFL**: wide to narrow, open, glandular; bracts on peduncle few, small, lf-like. **FL**: radial; hypanthium + calyx 1.5–3(5) mm; part of hypanthium fused to ovary 0.5–2.5 mm, gen > free part; calyx lobes 0.5–1.5 mm, tips green to red; petals 2–3 mm, oblanceolate, claw < blade, sometimes ± pink; stamens > calyx lobes, exserted; mature styles 2–4 mm, exserted. 2*n*=14,28. Moist, rocky banks and cliffs; < 2500 m. NW, CaRH, SNH, CW (exc SCoRI); to BC, ID. [*H. m.* var. *diversifolia* (Rydb.) Rosend. et al.; *H. m.* var. *erubescens* (A. Braun & Bouché) Rosend.; *H. m.* var. *hartwegii* (Wheelock) Rosend.; *H. m.* var. *macropetala* Shipes & E.F. Wells] Highly variable; many intergrading vars. historically recognized. Intergrades with *H. pilosissima*. Apr–Jul

H. parishii Rydb. PARISH'S ALUMROOT Pl 5–27 cm. **LF**: petiole 1–10 cm; blade 5–40 mm, broadly ovate to round-reniform, shallowly 5-lobed. **INFL**: narrow, ± dense, glandular-hairy. **FL**: ± bilateral; hypanthium + calyx 3.5–6 mm; hypanthium inflated on longer side, part fused to ovary ± 1.4 mm, ± = free part; calyx lobes 1–2 mm, ± equal, pink, tips green; petals 2–3 mm, unequal, oblanceolate; stamens > calyx lobes, ± unequal, exserted; mature styles > 1.5 mm, exserted. Rocky places; 1500–3800 m. SnBr. Jun–Jul ★

H. parvifolia Torr. & A. Gray (p. 1243) Pl 14–20(75) cm. **LF**: petiole (1)2.5–5(15) cm; blade 0.5–2(7.5) cm, round-reniform, shallowly 5–9-lobed. **INFL**: dense toward tips, ± short-glandular. **FL**: radial; hypanthium + calyx 2–3(4.2) mm; part of hypanthium fused to ovary 0.8–1(1.5) mm, < free part; calyx lobes ± 1 mm, cream to ± yellow, tips pink to ± green; petals (0.7)1.5–3 mm, oblanceolate to spoon-shaped, claw < blade; stamens << calyx lobes, incl; mature styles << 1 mm, incl. Rocky areas; 2200–3900 m. SNE (Sweetwater, White mtns); to Rocky Mtns. [*H. duranii* Bacig.] May–Sep

H. pilosissima Fisch. & C.A. Mey. Pl 15–55 cm. **LF**: petiole 7–20 cm; blade 45–90 mm, gen ovate, deeply 5–7-lobed. **INFL**: ± dense, glandular; bracts on peduncle few, small, lf-like. **FL**: radial; hypanthium + calyx 2.5–4.5 mm, densely hairy; part of hypanthium fused to ovary ± 1.5 mm, >> free part; calyx lobes 1–1.5 mm, pink to red; petals 2–3 mm, oblanceolate, white to pink; stamens > calyx lobes, exserted; mature styles 1–1.5 mm, slightly exserted. 2*n*=14. Ocean bluffs, shaded slopes; < 500 m. NCo, CCo, SCoRO, WTR (Santa Ynez Mtns). Intergrades with *H. micrantha*. Mar–Jun

H. rubescens Torr. Pl 7–55 cm. **LF**: petiole 1–15 cm; blade 8–60 mm, broadly ovate to ± round, ± deeply 5–9-lobed. **INFL**: often 1-sided, open or dense, cylindric to conic, glandular. **FL**: ± bilateral; hypanthium + calyx 3–6 mm; part of hypanthium fused to ovary 0.9–2.5 mm, ± = long side of free part; calyx lobes (0.5)1–2 mm, gen unequal, ± white to pink-red, tips green, becoming redder; petals 3–6 mm, thread-like to narrowly oblanceolate, claw ± = blade; stamens > calyx lobes, exserted; mature styles gen >> 1.5 mm, exserted. 2*n*=14,28. Dry, rocky areas; 1000–4000 m. SNH, PR, Wrn, SNE, DMtns; to OR, ID, CO, TX, n Mex. [*H. r.* var. *alpicola* Jeps.; *H. r.* var. *glandulosa* Kellogg; *H. r.* var. *rydbergiana* Rosend. et al.; *H. r.* var. *versicolor* (Greene) M.G. Stewart] Morphological characters used to delimit previously recognized vars. vary ± continuously and lack geog structure. May–Sep

JEPSONIA

Caudex corm-like, flattened or ovoid to spheric, branched or not, without scales or bulblets. **LF**: gen 1–3, basal; blade ± round, base cordate to reniform, lobes and teeth shallow. **INFL**: cyme, appearing before lvs, glandular; bracts scale-like. **FL**: hypanthium free of ovary, gen truncate at base; petals 5, short-clawed; stamens 10; pistil 1, ovary superior, chambers 2, placentas 2, axile. **FR**: capsule. 3 spp.: CA. (W.L. Jepson, CA botanist, 1867–1946) [Ornduff 1969 Brittonia 21:286–298] Heterostylous.

1. Lf gen 1; fls gen < 4; hypanthium gen > 2 × calyx lobes; s SCo, PR . . . . ***J. parryi***
1′ Lvs gen 2–3; fls gen > 4; hypanthium < 1.5 × calyx lobes; ChI, n SNF (Calaveras Co.), c SNF
  2. Petals persistent, withering; peduncle red-pink, often drying tan; seeds pale brown; n&c SNF, n SnJV
  . . . . ***J. heterandra***
  2′ Petals persistent but not withering; peduncle ± green, often drying brown; seeds dark brown; ChI . . . . . ***J. malvifolia***

J. heterandra Eastw. (p. 1243) FOOTHILL JEPSONIA Caudex gen flattened, branched. **LF**: gen 2–3. **INFL**: fls gen 4–17; peduncle 5–23 cm, red-pink, often drying tan. **FL**: hypanthium 1.5–3 mm, gen > calyx lobes, truncate to tapered at base; calyx lobes 1.3–2 mm, pink;

petals gen 3.5–6 mm, persistent, withering, veins deep pink. **FR**: gen ± green or rose, ± red-striped. **SEED**: pale brown. 2*n*=14. Crevices, esp in slate-like rock; dry, rocky slopes; < 700 m. n&c SNF (El Dorado to Mariposa cos.), n SnJV (ne Stanislaus Co.). Aug–Jan ★

J. malvifolia (Greene) Small (p. 1243) ISLAND JEPSONIA Caudex flattened and branched, or ovoid to spheric and unbranched. **LF:** gen 2–3. **INFL:** fls gen 4–17; peduncle 5–25 cm, ± green, often drying brown. **FL:** hypanthium 1–2 mm, gen ± = calyx lobes, gen truncate at base; calyx lobes 1–2 mm, yellow-green to pink; petals gen 3–3.5 mm, persistent but not withering, veins red. **FR:** yellow-green, gen tan-striped. **SEED:** dark brown. 2*n*=14. Rocky outcrops, clay slopes; < 700 m. ChI; Baja CA (Guadalupe Island). Sep–Nov ★

J. parryi (Torr.) Small (p. 1243) Caudex gen ovoid (rarely spheric), unbranched. **LF:** gen 1. **INFL:** fls gen < 4; peduncle 3–28 cm, brown. **FL:** hypanthium 2.5–4 mm, >> calyx lobes, gen truncate at base; calyx lobes 0.8–2 mm, ± green; petals gen 3.5–6 mm, veins tan or ± purple. **FR:** green or tan, brown-striped. **SEED:** dark brown. 2*n*=14. Shrubby, rocky to clay slopes; < 1200 m. s SCo (Orange, Riverside, San Diego cos.), PR; Baja CA. Oct–Feb

LITHOPHRAGMA WOODLAND STAR

Rhizome slender, scaleless, bearing bulblets. **LF:** basal and cauline, reduced distally on st, gen alternate, increasingly more deeply lobed from younger basal to distal cauline; blade round, base cordate to reniform, ± lobed, gen toothed. **INFL:** raceme; bracts scale-like or 0. **FL:** hypanthium gen partly fused to ovary; petals gen lobed or toothed; stamens 10; pistil 1, ovary superior to ± inferior, chamber 1, placentas 3, parietal, styles 3. **FR:** capsule, 3-beaked, valves 3. 12 spp.: w N.Am. (Greek: rock hedge, from habitats) [Kuzoff et al. 1999 Syst Bot 24:598–615]

1. Cauline lvs 2, opposite. . . . . *L. cymbalaria*
1′ Cauline lvs ≥ 1, alternate
  2. Hypanthium long-obconic, base acutely tapered to pedicel; ovary > 1/2-inferior
    3. Basal lvs lobed ± 1/2 or less to base; most proximal cauline lf 3-lobed; petals white . . . . *L. affine*
    3′ Basal lvs deeply lobed (> 3/4 to base) to 3-parted; most proximal cauline lf 3-parted; petals white
        or pink . . . . *L. parviflorum*
      4. Hypanthium ± 2 × longer than wide; fl not fragrant; petals white or pink . . . . var. *parviflorum*
      4′ Hypanthium 3–4 × longer than wide; fl fragrant; petals pink . . . . var. *trifoliatum*
  2′ Hypanthium hemispheric to bell-shaped, base truncate, rounded, or obtusely tapered to pedicel; ovary
      superior or < 1/2-inferior
    5. Basal lvs 3-parted or deeply 3-lobed (> 1/2 to base); petals white or pink
      6. Pl 40–60 cm; petals white; s ChI (San Clemente Island) . . . . *L. maximum*
      6′ Pl 8–30 cm; petals pink, occ white; mainland
        7. Petal lobes deep, 3–5; infl often with bulblets in axils of proximal bracts; seed spiny . . . . *L. glabrum*
        7′ Petal lobes shallow, gen 5–7; infl without bulblets; seed smooth . . . . *L. tenellum*
    5′ Basal lvs ± shallowly lobed (< 1/2 to base); petals white
      8. Petals obovate, lobes 3; hypanthium bell-shaped, base truncate; infl often with bulblets in axils of
          distal bracts . . . . *L. heterophyllum*
      8′ Petals ovate-elliptic, entire or lobes > 3; hypanthium hemispheric to bell-shaped, base obtusely tapered
          to round; infl bulblets gen 0
        9. Petals spreading, entire or ± lobed, lobes ± equal (sometimes tooth-like); hypanthium hemispheric;
            ovary < 1/2-inferior; lf teeth ± round . . . . *L. bolanderi*
        9′ Petals ± erect, ± lobed, lobes ± unequal (becoming smaller and tooth-like near base); hypanthium
            broadly bell-shaped; ovary superior; lf teeth ± sharp . . . . *L. campanulatum*

L. affine A. Gray (p. 1243) Pl 10–60 cm. **LF:** basal blade ± 3–5-lobed, teeth ± sharp-tipped. **INFL:** fls 3–15; pedicel 3–10 mm. **FL:** hypanthium obconic, ± inflated above, part fused to ovary ± = free part; petals 5–13 mm, ovate-elliptic, 3-lobed at tip, white; ovary > 1/2-inferior. **SEED:** smooth. 2*n*=14,21,28,35. Open, grassy slopes; < 2000 m. KR, NCoRO, NCoRI, c SN, ScV (Sutter Buttes), n SnJV (Antioch), n CCo, SnFrB, SCoR, SCo (inland), s ChI, TR, PR; sw OR, Baja CA. Variable. Many forms, esp inland and in s CA, approaching and intergrading with *L. parviflorum*. Mar–Apr

L. bolanderi A. Gray (p. 1243) Pl 20–80 cm. **LF:** basal blade ± 3–5-lobed, teeth ± round. **INFL:** fls 5–25; pedicel 1.5–4 mm. **FL:** hypanthium hemispheric, part fused to ovary < free part; petals 4–7 mm, ovate-elliptic, white, entire or lobes > 3 (sometimes tooth-like), ± equal; ovary < 1/2-inferior. **SEED:** spiny. 2*n*=14,28,35,42. Open slopes, riparian, woodland; < 2000(3000) m. KR, NCoRO, NCoRI, CaR, SN, s ScV (Sutter Buttes, rare elsewhere), n SnJV, SnFrB, WTR, SnGb. Feb–Jul

L. campanulatum Howell (p. 1243) Pl 15–55 cm. **LF:** basal blade ± 3–5-lobed, teeth ± sharp. **INFL:** fls 2–11; pedicel 1.5–3 mm. **FL:** hypanthium broadly bell-shaped, part fused to ovary << free part; petals 3–7 mm, ovate-elliptic, white, lobes > 3, ± unequal (becoming smaller and tooth-like near base); ovary superior. **SEED:** spiny. Shady, well-drained slopes; (100)500–2500 m. KR, NCoR, CaRH, n SNH; sw OR. May–Jul

L. cymbalaria Torr. & A. Gray Pl 10–35 cm. **LF:** basal blade shallowly 3-lobed, teeth 0; cauline 2, opposite. **INFL:** fls 2–8; pedicel

4–10 mm. **FL:** hypanthium ± long-obconic, part fused to ovary ± = free part; petals 4–8 mm, ovate-elliptic, entire to shallowly toothed, white; ovary ± 1/2-inferior. **SEED:** spiny. 2*n*=14+. Shady, moist areas; < 1200(2500) m. SnJV (Antioch), SnFrB, SCoR, SCo/WTR (Ventura Co.), n ChI, WTR. Mar–May

L. glabrum Nutt. (p. 1243) Pl 8–25 cm. **LF:** basal blade deeply 3-lobed or ± 3-parted, lobes and lflets lobed, teeth ± sharp-tipped. **INFL:** fls 1–7; pedicel 3–6 mm; proximal bracts often with axillary bulblet. **FL:** hypanthium spheric to bell-shaped, part fused to ovary gen < free part; petals 3–7 mm, ovate, deeply 3–5-lobed, pink (white); ovary < 1/2-inferior. **SEED:** spiny. 2*n*=14,28. Dry, gravelly places; < 3200(3750) m. KR, CaRH, SNH, MP, n SNE (Alpine Co.); to BC, SD, CO. Mar–Jul

L. heterophyllum (Hook. & Arn.) Torr. & A. Gray Pl 15–50 cm. **LF:** basal blade ± 3–5-lobed, teeth gen round. **INFL:** fls 3–12; pedicel 0.5–2 mm; distal bracts often with axillary bulblets. **FL:** hypanthium bell-shaped, base truncate, part fused to ovary << free part; petals 5–12 mm, obovate, 3-lobed, white; ovary superior. **SEED:** spiny. 2*n*=14. Shaded slopes; < 1500 m. NCoRO, s NCoRI (Solano, Napa cos.), n SNH (Calaveras Co.), SnFrB, SCoR. Feb–Jun

L. maximum Bacig. SAN CLEMENTE ISLAND WOODLAND STAR Pl 40–60 cm. **LF:** basal blade 3-parted, lflets lobed, teeth sharp. **INFL:** fls 6–25; pedicel ± 1 mm. **FL:** hypanthium hemispheric to bell-shaped, part fused to ovary gen < free part; petals 3.5–4.5 mm, ovate, ± 5-lobed, white; ovary < 1/2 inferior. **SEED:** spiny. Steep, moist, n-facing canyon slopes; < 400 m. s ChI (San Clemente Island). Mar–May ★

L. parviflorum (Hook.) Torr. & A. Gray (p. 1243)   Pl 10–50 cm. **LF:** basal blade deeply 3-lobed (> 3/4 to base) to ± palmately compound, lobes lobed, teeth ± sharp-tipped. **INFL:** fls 4–14; pedicel 3–7 mm. **FL:** hypanthium long-obconic, part fused to ovary > free part; petals 7–16 mm, gen obovate, 3-lobed, white or pink; ovary > 1/2-inferior. **SEED:** smooth. 2n=14,21,28,35.

var. ***parviflorum***   **FL:** not fragrant; hypanthium ± 2 × longer than wide; petals white or pink. **SEED:** 0.5–0.6 mm. Open areas; < 3000 m. KR, NCoRI, NCoRH, CaRH, SNH, Teh, ScV (Sutter Buttes), SnFrB, SCoRI, WTR, MP. Mar–Jul

var. ***trifoliatum*** (Eastw.) Jeps.   **FL:** fragrant; hypanthium 3–4 × longer than wide; petals pink. **SEED:** 0.6–0.8 mm. 2n=28. Open areas; < 600 m. CaRF, n SNF, SnJV (Oakdale). Seeds gen sterile. [*L. t.* Eastw.] May–Jul

L. tenellum Nutt.   Pl 8–30 cm. **LF:** basal blade deeply 3-lobed, lobes lobed, teeth ± round. **INFL:** fls 3–12; pedicel 3–10 mm. **FL:** hypanthium hemispheric to bell-shaped, part fused to ovary < free part; petals 3–7 mm, ovate, gen 5–7-lobed, pink or sometimes white; ovary < 1/2-inferior. **SEED:** smooth. 2n=14,35. Dry areas; 1300–3000 m. CaRH, n SNH, SnGb, SnBr, MP; to BC, CO, NM. May–Jul

MICRANTHES   SAXIFRAGE

Michael S. Park

Pl gen ± hairy, often glandular; caudex or rhizome gen not woody, gen scaly. **LF:** basal (cauline); blade linear to (ob)ovate or ± round, base tapered to reniform, margin entire or toothed. **INFL:** fls few to many; bracts scale-like. **FL:** gen radial; hypanthium free or ± fused to ovary; petals 5, white, sometimes with yellow spots at base; stamens 10, filaments flat or variously inflated; pistils 1 (chambers 2, placentas 2, axile or occ proximally axile and distally marginal) or 2, ovary superior to ± inferior (sometimes more superior in fr), styles free throughout. **FR:** capsule or 2 follicles. ± 80 spp.: N.Am, Eurasia, S.Am, esp cool temp n hemisphere. (Latin: small fl) [Elvander 1984 Syst Bot Monogr 3:1–44] Intermediates common between *M. integrifolia, M. nidifica, M. fragosa, M. aprica*; some may be vegetatively reproducing, sterile hybrids. Study needed.

1. Lvs cauline on ± trailing st — blade entire; fls in head-like cluster . . . . **M. tolmiei**
1′ Lvs basal (bracts occ lf-like)
  2. Proximal fls replaced by bulblets; fls ± bilateral, 3 petals broader than other 2
    3. Base of all petals ± 2-spotted; lf blade linear-elliptic, margin ± entire or with minute teeth distally. . . . **M. bryophora**
    3′ Base of only 3 broad petals (rarely all) ± 2-spotted; lf blade obovate, teeth gen coarse . . . . **M. ferruginea**
  2′ Infl bulblets 0; fls radial, petals ± equal
    4. Filaments inflated distally; petal spots 2
      5. Lf blade ovate to elliptic, base truncate to tapered; ovaries fused only at base . . . . **M. marshallii**
      5′ Lf blade ± round, base cordate to reniform; ovaries fused throughout — fr tip purple, base yellow-banded . . . . **M. odontoloma**
    4′ Filaments inflated proximally, or ± flat and not inflated; petal spots 0
      6. Filaments inflated proximally; ovary superior in fl and fr . . . . **M. howellii**
      6′ Filaments not inflated proximally, ± flat, narrowed at tip; ovary ± inferior to 1/2-inferior in fl, superior in fr
        7. Pl gen > 4 dm; lf blade linear to oblanceolate, gen > 10 cm, long-tapered to an indistinct petiole; bogs, marshes, lake margins . . . . **M. oregana**
        7′ Pl ≤ 3.5 dm; lf blade elliptic, obovate, ovate, or triangular, gen < 10 cm, short-tapered to a distinct petiole; not in bogs, marshes or lake margins
          8. Lf blade shallowly toothed; infl ± open, often 1-sided; fls not clustered to loosely clustered; anthers purple to red (orange); petals ± 2 × sepals; styles on fr gen > 1.5 mm . . . . **M. californica**
          8′ Lf blade ± entire to minutely toothed; infl not 1-sided; fls loosely to densely clustered; anthers gen yellow to orange; petals < 1.7 × sepals; styles on fr gen < 1 mm
            9. Petals 1–2 mm; sepals ± reflexed . . . . **M. nidifica**
            9′ Petals gen > 2 mm; sepals gen ± erect to spreading (exc *Micranthes fragosa*)
              10. Pedicels ± glabrous to sparsely glandular; gen 1 head-like cluster of fls at tip of infl, sometimes with 1 or 2 smaller clusters on nearby short lateral branches; moist, sandy and gravelly montane to subalpine habitats ≥ 1600 m elevation . . . . **M. aprica**
              10′ Pedicels densely glandular; lateral branches of infl many, each with a terminal cluster of fls; moist meadows, rocky slopes and outcrops
                11. Lateral branches of infl spreading to ascending; infl cluster at tip of main axis open to ± congested, hemispheric to ± flat-topped; sepals ± reflexed to spreading. . . . . **M. fragosa**
                11′ Lateral branches of infl ascending to erect; infl cluster at tip of main axis dense, conical to cylindric; sepals ± erect to ascending . . . . **M. integrifolia**

M. aprica (Greene) Small (p. 1243)   Pl 5–7(13) cm; caudex gen with bulblets. **LF:** 10–45 mm; petiole 5–15 mm; blade obovate to elliptic, base tapered, entire or minutely toothed distally. **INFL:** gen 1 head-like cluster of fls at tip and sometimes 1–2 more nearby; pedicel ± glabrous to sparsely glandular. **FL:** sepals ± spreading to erect, ovate; petals 1.8–3 mm, > sepals, gen ovate; filaments ± flat, narrowed at tip; nectaries disk-like, lobed; pistils 2, ovary > 1/2-inferior in fl. **FR:** 2 follicles. 2n=20. Rocky, wet alpine meadows, vernally moist flats; 1600–3600 m. KR, CaRH, SNH; sw OR, w NV. [*Saxifraga a.* Greene] Pls in KR transitional in size and infl to *M. fragosa, M. californica.* May–Aug

M. bryophora (A. Gray) Brouillet & Gornall (p. 1243)   Pl 3–25 cm, open; rhizome < 1 mm diam; bulblets 0. **LF:** 1–4 cm, ± fleshy, sessile, linear-elliptic, base tapered, margin ± entire or minutely toothed distally. **INFL:** proximal fls replaced by bulblets. **FL:** ± bilateral; sepals reflexed, < petals, ± elliptic; petals 3–5 mm with 2 yellow spots at base, ± of 2 kinds, 3 broad petals spade-shaped to sagittate, 2 narrow petals lanceolate to elliptic; filaments slightly widened toward base; nectaries 0; pistil 1, ovary superior. **FR:** capsule. Sandy meadows, ledges; 1600–3600 m. KR, CaRH, SNH. [*Saxifraga b.* A. Gray] (Jun)Jul–Aug

M. californica (Greene) Small (p. 1243) Pl 15–35 cm; caudex gen with long slender rhizomes and bulblets. **LF:** 4–10 cm; petiole 2–5 cm; blade ovate, base tapered, teeth shallow, ± round or sharp. **INFL:** often 1-sided, ± open. **FL:** sepals gen reflexed, elliptic; petals 2.5–4.5 mm, >> sepals, elliptic to round; filaments ± flat, gen narrowed at tip; nectaries disk-like, lobed; pistils 2, ovary ± 1/2-inferior in fl. **FR:** 2 follicles. $2n=20$. Moist, shady places; < 1200 m. KR, NCoRO, NCoRI, CaRF, SN (expected Teh), GV (Sutter to Stanislaus cos.), CW, SW; sw OR, nw Baja CA. [*Saxifraga c.* Greene] Pls in far n CA difficult to assign; see *M. fragosa*. Feb–May(Jun)

M. ferruginea (Graham) Brouillet & Gornall Pl 5–40 cm; caudex slender, gen producing short rhizomes; bulblets 0. **LF:** 1.5–6 cm, ± fleshy, ± sessile; blade obovate, teeth gen coarse, sharp. **INFL:** open; proximal fls replaced by bulblets. **FL:** ± bilateral; sepals reflexed, < petals, ovate; petals 3.5–5 mm, ± of 2 kinds, 3 broad petals spade-shaped to sagittate with 1–2 yellow spots at base, 2 narrow petals lanceolate to elliptic; filaments thread-like; nectaries inconspicuous; pistils 2, ovary superior. **FR:** 2 follicles. $2n=38$. Wet banks, gravel; 1500–2500 m. KR; to AK, MT. [*Saxifraga f.* Graham] Rare intermediates with *M. bryophora* occur, with petals all spotted, and not strongly of 2 kinds. Jul–Aug

M. fragosa (Small) Small Pl 15–35 cm; caudex gen with rhizomes and bulblets. **LF:** 4–11 cm; petiole 2–5 cm; blade triangular to ovate, base tapered, entire to minutely toothed. **INFL:** open to ± congested at tips, branches spreading to ascending; pedicel densely glandular. **FL:** sepals ± reflexed to spreading, < petals; petals 2–3 mm, obovate; filaments ± flat, narrowed near tip; nectaries disk-like, lobed; pistils 2, ovary >> 1/2-inferior in fl. **FR:** 2 follicles. $2n=20,38$. Moist, rocky slopes and outcrops; 500–2000 m. KR, CaR, MP (exc Wrn). [*Saxifraga nidifica* Greene var. *claytoniifolia* (Small) Elvander] Intermediate between *M. californica* and *M. integrifolia*. Pls with $2n=38$ chromosomes possibly best treated as variant of *M. integrifolia*. Mar–Jun

M. howellii (Greene) Small HOWELL'S SAXIFRAGE Pl 5–20 cm; caudex slender, gen producing short rhizomes; bulblets 0. **LF:** 2–6 cm; petiole 1–4 cm; blade oblong to ovate, base tapered, teeth coarse, round or sharp. **INFL:** open to dense; fls 5–15. **FL:** sepals reflexed, < petals, elliptic to ovate; petals 2.5–4.5 mm, ± obovate; filaments inflated proximally; pistils 2, ovary superior. **FR:** 2 follicles. Moist ledges, crevices; < 900 m. KR; sw OR. [*Saxifraga h.* Greene] Feb–Apr ★

M. integrifolia (Hook.) Small Pl 12–35 cm; caudex with short rhizomes, occ with bulblets. **LF:** 2–7 cm; petiole 0.5–4 cm; blade ± ovate, base tapered, entire or minutely toothed. **INFL:** gen 1 head-like cluster at tip or sometimes 1–3-branched, lateral branches ± open; pedicel densely glandular. **FL:** sepals gen ± erect to ascending, < petals, ovate; petals 2–4 mm, obovate; filaments ± flat, narrowed toward tip; nectaries disk-like, lobed; pistils 2, ovary >> 1/2-inferior in fl. **FR:** 2 follicles. $2n=38$. Uncommon. Vernally moist meadows; 100–1700 m. KR, NCoRO, NCoRI, CaRF, ScV; to BC, ID, NV. [*Saxifraga i.* Hook.] Often intergrades with *M. nidifica* and *M. fragosa* in KR, CaR, MP. Mar–Jun

M. marshallii (Greene) Small Pl 10–40 cm; caudex gen producing rhizomes; bulblets 0. **LF:** 5–15 cm; petiole 3–10 cm; blade ovate to elliptic, base truncate to tapered, teeth coarse, sharp to round. **INFL:** open. **FL:** sepals reflexed, gen < petals, ovate to elliptic; petals 2.5–4.5 mm, elliptic or obovate, 2-spotted, ephemeral; filaments inflated distally; nectaries band-like; pistils 2, ovary superior. **FR:** 2 follicles. Mossy rocks, cliffs; < 2000 m. NCo, KR, NCoRO; sw OR. [*Saxifraga m.* Greene] Pls from KR (Marble Mtns) previously incorrectly identified as *M. rufidula* Small [*Saxifraga rufidula* (Small) Fedde]. Apr–May

M. nidifica (Greene) Small Pl 10–50 cm; caudex with rhizomes and gen bulblets. **LF:** 3–10 cm; petiole 1.5–5 cm; blade ovate, base gen tapered, entire or minutely toothed. **INFL:** gen open between ± head-like clusters. **FL:** sepals ± reflexed, ≤ petals, triangular to ovate; petals 1–2 mm, narrowly elliptic to round; filaments ± flat, narrowed near tip; nectaries disk-like, lobed; pistils 2, ovary > 1/2-inferior in fl. **FR:** 2 follicles. $2n=38$. Open wet meadows, slopes; (1000)1800–3500 m. KR, NCoRO (Buck Mtn, Humboldt Co.), SNH, MP; to BC, MT, NV. [*Saxifraga n.* Greene var. *n.*] Variable. Intergrades with *M. integrifolia* and *M. fragosa* in KR, CaR, MP. Pls in SNH with open infl that are similar to *M. californica* and *M. fragosa* may be distinct (*M. montana* Small). May–Aug

M. odontoloma (Piper) A. Heller (p. 1243) Pls 20–50 cm; caudex producing rhizomes; bulblets 0. **LF:** 4–40 cm; petiole 2–30 cm, base gen ± expanded, sheathing, membranous; blade ± round, base cordate to reniform, teeth coarse, sharp. **INFL:** open. **FL:** sepals reflexed, gen ± = petals, ovate to elliptic; petals 3–4.5 mm, round to elliptic, 2-spotted, ephemeral; filaments club-shaped; nectaries band-like; pistil 1, ovary superior, placentas proximally axile and distally marginal in ovary lobes. **FR:** capsule, follicle-like. $2n=48$. Wet meadows, ledges; > 1500 m. KR, NCoRH, CaRH, SNH, SnBr; to BC, MT, NM. [*Saxifraga o.* Piper] Can be confused with *Saxifraga mertensiana* Bong., which has 2° lf teeth and often reduced cauline lvs present. Jul–Aug

M. oregana (Howell) Small Pls 25–125 cm; caudex > 15 cm, thick, fleshy, sometimes branched; bulblets 0. **LF:** 7–25 cm; petiole ± indistinct; blade linear to oblanceolate, base tapered, entire or sharp-toothed. **INFL:** dense toward tips of main axis and branches, otherwise open. **FL:** sepals reflexed, gen < petals, ovate to triangular; petals 2–4 mm, linear to elliptic; filaments ± flat, narrowed at tip; nectaries disk-like, lobed; pistils 2, ovary > 1/2-inferior in fl. **FR:** 2 follicles. $2n=38,76$. Bogs, marshes, lake margins; 1000–2500 m. KR, CaRH, SN, MP; to WA, ID, NV. [*Saxifraga o.* Howell] Lvs and infl variable, but habitat distinctive. Like *M. pensylvanica* (L.) Haw. of e US. Jun–Aug

M. tolmiei (Torr. & A. Gray) Brouillet & Gornall (p. 1243) Pl ± glabrous; caudex slender, ± woody; bulblets 0. **ST:** ± trailing, ± woody. **LF:** cauline, gen crowded, 8–15 mm, fleshy, sessile, elliptic or obovate, entire. **INFL:** head-like cluster; peduncle 3–12 cm; bracts gen 1–3, lf-like. **FL:** sepals spreading to ascending, < petals, ± ovate; petals 2.5–5 mm, linear to oblanceolate; filaments gen much-widened exc near base and tip; pistil 1, ovary ± superior. **FR:** capsule. **SEED:** winged. $2n=30$. Alpine tundra, fell-fields; > 2000 m. KR, NCoRH, CaRH, SNH; to AK. [*Saxifraga t.* Torr. & A. Gray] Jul–Sep

MITELLASTRA

Michael S. Park & Douglas E. Soltis

1 sp. (Latin: small cap, from fr; star, from fl outline)

M. caulescens (Nutt.) Howell (p. 1247) Pl 1.5–4.5 dm; rhizome scaly; bulblets 0. **LF:** basal and 1–few cauline; petiole glabrous to ± hairy; blade 2–7 cm wide, ± round, lobes 3–7, teeth sharp. **INFL:** blooming from tip to base; pedicel 2–8 mm. **FL:** hypanthium 2.5–4 mm wide, saucer-shaped, ± fused to ovary; petals yellow-green, lobes 4–7, alternate, linear; stamens 5, alternate petals; filaments >> anthers, ± 2/3 calyx lobes; pistil 1, ovary > 1/2-inferior, chamber 1, placentas 2, parietal, styles 2, ± 0.2 mm, stigmas unlobed, head-like. **FR:** capsule, becoming widely dehiscent, forming splash cup. **SEED:** many, red-brown (to black), shiny. $2n=14$. Wet shaded areas; < 1700 m. NW (exc NCoRI); to BC, MT. [*Mitella c.* Nutt.] May–Jul ★

OZOMELIS

Michael S. Park & Douglas E. Soltis

Pl minutely glandular; rhizome scaly; bulblets 0. **LF**: basal, occ 1 cauline; petioles with long, downward-pointing, gland-tipped, white hairs, occ lacking in youngest lvs; blade ovate to round, ± lobed, base ± cordate. **INFL**: spike-like raceme, ± 1-sided, blooming from base to tip; bracts gen scale-like. **FL**: hypanthium widely obconic to bell-shaped, ± fused to ovary; sepals mucronate, gland at tip; petals not clawed, lobes [0]3[5] at tip; stamens 5, alternate petals; pistil 1, ovary gen > 1/2-inferior, chamber 1, placentas 2, parietal, stigmas unlobed, ± sessile. **FR**: capsule, becoming widely dehiscent. **SEED**: many, red-brown to black, minutely papillate. ± 3 spp.: temperate w N.Am. (Greek: odor of honey, from fl)

1. Lf blade ovate, lobes ± deep, gen ± entire; petal lobes narrowly triangular; occ 1 lf cauline . . . . *O. diversifolia*
1′ Lf blade ± round, lobes ± shallow, teeth gen round; petal lobes ± elliptic; lvs all basal. . . . . *O. trifida*

O. diversifolia (Greene) Rydb. (p. 1247) Pl 2–5 dm. **LF**: blade 3–6 cm wide, lobes 5–7. **INFL**: pedicel (0)0.5–2 mm. **FL**: hypanthium 2.5–3.5 mm wide; petals white to light pink, lobes acute at tip. 2*n*=14. Uncommon. Moist woodland, streambanks; 1000–2000 m. KR, NCoRO, NCoRH, CaRH, n SNH; to WA. [*Mitella d.* Greene] Apr–Jul

O. trifida (Graham) Rydb. (p. 1247) Pl 1.5–4.5 dm. **LF**: blade 2–8 cm wide, lobes 5–7. **INFL**: pedicel (0)0.5–2 mm. **FL**: hypanthium 2.5–3.5 mm wide; petals white to pale purple (± purple), lobes rounded at tip. 2*n*=14. Uncommon. Wet shaded slopes; 1400–2300 m. KR, NCoRH, n SNH; to BC, AB. [*Mitella t.* Graham] May–Jul

PECTIANTIA

Michael S. Park & Douglas E. Soltis

Rhizome scaly; bulblets 0. **LF**: basal (occ 1 cauline); blade ± ovate to round, base ± cordate, ± lobed, gen toothed. **INFL**: raceme or spike, ± 1-sided, blooming from base to tip; bracts gen scale-like. **FL**: hypanthium saucer-shaped, ± fused to ovary; petals pinnately lobed, not clawed; stamens 5, filaments < anthers; pistil 1, ovary gen > 1/2-inferior, chamber 1, placentas 2, parietal; styles 2, ≤ 0.3 mm; stigmas 2-lobed. **FR**: capsule, becoming widely dehiscent, forming splash cup. **SEED**: many, red-brown to black, shiny. ± 13 spp.: temp w N.Am, Japan, Taiwan. (Latin: comb, from petals; opposing, for stamens opposite petals in *Pectiantia pentandra*)

1. Stamens opposite petals. . . . . *P. pentandra*
1′ Stamens alternate petals
  2. Lf blade ± round, 3–8 cm wide, lobes indistinct, 7–11, teeth ± round; pedicel 2–7 mm; petiole glabrous to ± hairy; montane, > 1500 m . . . . *P. breweri*
  2′ Lf blade ovate-elliptic, gen < 5 cm wide, lobes distinct, 5–9, teeth ± sharp; pedicel ≤ 1.5 mm; petiole densely hairy; ± coastal n CA, < 1000 m. . . . . *P. ovalis*

P. breweri (A. Gray) Rydb. (p. 1247) Pl 1–3 dm. **LF**: petiole glabrous to ± hairy; blade 3–8 cm wide, ± round, base gen slightly reniform, lobes 7–11, indistinct, teeth ± round. **INFL**: pedicel 2–7 mm. **FL**: hypanthium 3–3.5 mm wide; petals yellow-green, lobes 5–9, gen opposite, linear; stamens alternate petals; styles ± 0.2 mm. 2*n*=14. Moist woodland; 1500–3500 m. s CaRH, SNH; to BC, MT, NV. [*Mitella b.* A. Gray] May–Sep

P. ovalis (Greene) Rydb. Pl 1–3 dm. **LF**: petiole densely hairy; blade 1.5–5 cm wide, ovate-elliptic, lobes 5–9, teeth ± sharp, shallow. **INFL**: pedicel (0)0.5–1.5 mm. **FL**: hypanthium 2.5–3.5 mm wide; petals yellow-green, lobes 4–7, alternate, linear; stamens alternate petals; styles ± 0.3 mm. 2*n*=14. Wet woodland, shaded streambanks; < 1000 m. NCo, NCoRO, n CCo (Marin Co.); to BC. [*Mitella o.* Greene] Apr–Jun

P. pentandra (Hook.) Rydb. Pl 1–4 dm. **LF**: petiole gen glabrous; blade 1.5–8 cm wide, ovate, lobes 5–9, teeth sharp. **INFL**: pedicel 1–4 mm. **FL**: hypanthium 3–5 mm wide; petals green, lobes 5–10, gen opposite, linear; stamens opposite petals; stigmas unlobed or each shallowly 2-lobed, ± sessile. 2*n*=14. Streambanks, wet meadows; 1500–2500 m. KR, NCoRH, CaRH, SNH, Wrn; to AK, CO. [*Mitella p.* Hook.] Jun–Aug

SAXIFRAGA SAXIFRAGE

Michael S. Park

Pl gen ± hairy, often glandular; caudex or rhizome gen not woody, gen scaly. **LF**: blade obovate to round, base tapered to reniform or cordate, margin lobed. **INFL**: raceme or panicle; fls 1–many; bracts scale-like. **FL**: hypanthium free or ± fused to ovary; [petals sometimes spotted]; stamens 10, filaments gen flat; pistil 1, ovary superior to ± inferior (sometimes more superior in fr), chambers 2, placentas 2, axile, styles 2, free throughout. **FR**: capsule. ± 300 spp.: cool n temp, boreal, arctic. (Latin: rock-breaking) *S. stolonifera* W. Curtis cult only. Other taxa in TJM (1993) moved to *Cascadia, Micranthes*.

1. Lvs cauline on ± trailing st, lobed only at blade tip, lobes ± linear. . . . . *S. cespitosa*
1′ Lvs mostly basal, cauline smaller and fewer, lobed from blade base to tip, lobes not linear
  2. Lf blade gen 5–8 mm, divided > 1/2 to base; lf lobes 3–5, elliptic to round, not toothed . . . . *S. hyperborea*
  2′ Lf blade gen > 15 mm, divided < 1/4 to base; lf lobes >> 5, triangular to semicircular, toothed. . . . . . . *S. mertensiana*

S. cespitosa L. TUFTED SAXIFRAGE Pl glandular; caudex slender, ± woody; bulblets 0. **ST**: short, ± trailing, ± woody. **LF**: cauline, gen crowded, 5–10 mm, sessile; blade obovate; lobes 3–5 at tip, ± linear. **INFL**: raceme 2–5 cm; peduncle long, gen with 1 lf-like bract near middle, gen 1–2 fls at tip. **FL**: sepals erect to spreading, << petals, elliptic; petals 3–5 mm, elliptic to obovate, ephemeral; filaments slightly widened toward base; nectaries inconspicuous or 0; ovary inferior in fl. 2*n*=52,56,60,63,65,80. Damp rocky places; > 1900 m. KR (Marble Mtns), Wrn; to arctic, e N.Am, circumboreal. Jul–Aug ★

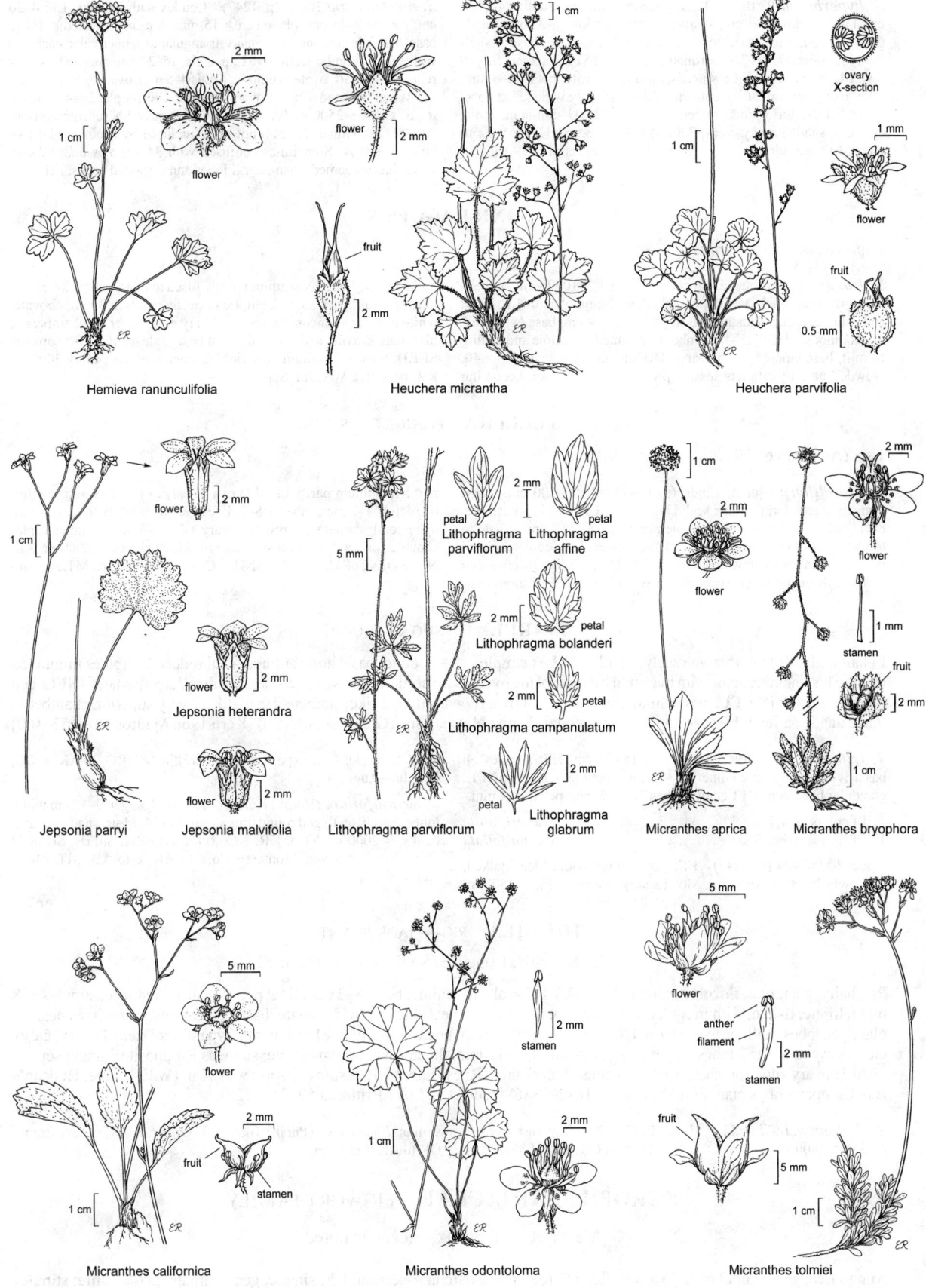

Hemieva ranunculifolia

Heuchera micrantha

Heuchera parvifolia

ovary
X-section

flower

fruit

Jepsonia parryi

Jepsonia heterandra

flower

Jepsonia malvifolia

flower

Lithophragma parviflorum

petal
Lithophragma
parviflorum

petal
Lithophragma
affine

petal
Lithophragma bolanderi

petal
Lithophragma campanulatum

petal
Lithophragma
glabrum

Micranthes aprica

flower

Micranthes bryophora

flower

stamen

fruit

Micranthes californica

flower

fruit
stamen

Micranthes odontoloma

stamen

flower

Micranthes tolmiei

flower

anther
filament
stamen

fruit

S. hyperborea R. Br. (p. 1247) Caudex fragile; bulblets 0. **LF:** basal and cauline, < 3 cm, smaller upward; petiole 5–25 mm; blade gen 5–8 mm, wider than long, divided > 1/2 to base, base ± shallowly reniform; lobes 3–5, ± round to elliptic. **INFL:** raceme; fls gen few. **FL:** sepals erect, gen < petals, elliptic to ovate; petals 2–6 mm, elliptic to ± obovate; filaments thread-like, slightly widened at base; ovary > 1/2-inferior, more superior in fr. $2n=26$. Uncommon. Moist crevices, shaded rocky areas; 3000–4500 m. c&s SNH; sporadic to arctic, CO, AZ, circumboreal. [*S. rivularis* L., misappl.] Jul–Aug

S. mertensiana Bong. (p. 1247) Caudex with bulblets. **LF:** 4–20 cm; petiole 2–15 cm; blade gen > 15 mm, ± round, divided < 1/4 to base, base gen cordate; lobes many, triangular to semicircular, each gen with ± 2–5 irregular teeth. **INFL:** panicle, 15–30 cm, open. **FL:** sepals reflexed, < petals, ovate to elliptic; petals 4–5 mm, ovate to elliptic; filaments gen widened distally; ovary superior. **FR:** purple. $2n=48$. Mossy rocks, cliffs; < 2500 m. NCo, KR, NCoRO, n&c SN (uncommon); to AK, MT. Proximal fls seem not to be replaced by ± pink vegetative bulblets in CA. Sometimes confused with *Micranthes odontoloma*, which has untoothed, triangular lf lobes and 2-spotted petals. Feb–Jul

SAXIFRAGOPSIS

1 sp. (Greek: like *Saxifraga*)

S. fragarioides (Greene) Small (p. 1247) Pl 10–25 cm, ± glandular; rhizome thick, woody, branched, scaly; bulblets 0. **LF:** basal and cauline, reduced distally on st; petiole 1.5–4 cm, base expanded, membranous; blade 1.5–4 cm, obovate, jointed to petiole and falling from it, base tapered, teeth coarse. **INFL:** panicle, ± open; fls > 40, crowded near tips; bracts sessile, proximal often lf-like, becoming linear distally. **FL:** hypanthium partly fused to ovary; calyx lobes 2–3 mm, spreading to reflexed, elliptic to ovate; petals 2–3 mm, obovate, white to pink; stamens 10; pistil 1, ovary ± 1/2-inferior, chambers 2, placentas 2, axile, styles gen fused at base, at least in fl. **FR:** capsule. **SEED:** brown, longitudinally ribbed. Rock crevices; 1500–3000 m. KR; sw OR, c WA. Jul–Sep

TELLIMA FRINGE CUPS

1 sp. (Anagram of *Mitella*)

T. grandiflora (Pursh) Lindl. (p. 1247) Pl 40–100 cm, hairy, sparsely glandular; rhizome scaly; bulblets 0. **LF:** basal and cauline, distal reduced; stipules ± 5 mm, membranous, ± sheathing, ciliate to toothed; petiole 3–30 cm; blade 2–10 cm, ovate, base cordate, clasping above, lobes shallow, teeth sharp. **INFL:** spike-like raceme, gen 1-sided; fls many; bracts scale-like; pedicel 2–5 mm. **FL:** fragrant or not; hypanthium partly fused to ovary; calyx lobes 2–3 mm, ± elliptic; petals 3–7 mm, lobes ± 5–7, linear, ± green-white to rose or red, ephemeral; stamens 10; pistil 1, ovary > 1/2-inferior, chamber 1, placentas 2, parietal. **FR:** capsule. $2n=14$. Moist slopes; < 2000 m. KR, NCoRO, NCoRH, CaRH, n SNH, n CCo, SnFrB; to AK, MT. Apr–Jul

TIARELLA SUGAR-SCOOP

Pl hairy, glands few; rhizome scaly; bulblets 0. **LF:** simple or compound, basal and cauline, distal reduced; stipules minute on cauline lvs, membranous, with marginal bristles; blade ovate, base cordate, lobes 3–5, ± deep, or lflets 3, teeth sharp. **INFL:** gen raceme-like panicle. **FL:** hypanthium minute, free of ovary; petals thread-like; stamens 10; pistil 1, ovary superior, chamber 1, placentas 2, parietal. **FR:** capsule, valves unequal. 2 spp.: N.Am, Asia. (Greek: small tiara) [Kern 1966 Madroño 18:152–160]

T. trifoliata L. LACE FLOWER Pl 15–40 cm. **LF:** stipules on basal lvs < 1 cm, on cauline lvs < 1 mm; petiole > blade. **INFL:** glandular hairs dense. **FL:** calyx lobes 1.5–2.5 mm; petals 3–4 mm.

1. Lf compound; lflets 3 . . . . .var. ***trifoliata***
1′ Lf simple; lobes 3–5 . . . . .var. ***unifoliata***

  var. ***trifoliata*** (p. 1247) **LF:** gen 3–9 cm wide; lflets stalked, shallowly lobed, teeth sharp. Moist shady streambanks; < 1500 m.

KR (Humboldt Co., expected elsewhere), n NCoRO; to AK, e BC, MT. Jun–Aug ★

  var. ***unifoliata*** (Hook.) Kurtz (p. 1247) **LF:** gen 9–12 cm wide; lobes ± deep, teeth gen round, tips sharp. $2n=14$. Moist shady streambanks; < 2000 m. NCo, KR, NCoRO, n NCoRH, SnFrB, SCoRO/CCo (Big Sur River, Monterey Co.); to AK, e to AB, MT. May–Aug

TOLMIEA PIG-A-BACK PLANT

Douglas E. Soltis, Pamela S. Soltis & Walter S. Judd

Pl ± hairy, glandular; rhizome scaly; bulblets 0. **LF:** basal and cauline, basal 8–38 cm (incl petiole), distal reduced; stipules 3–8 mm, elliptic, lf-like, with irregular teeth and short bristles; blade 2–11.5 cm wide, ovate, base gen cordate, sometimes bearing plantlets, lobes shallow, teeth sharp. **INFL:** raceme, 10–80 cm; bracts scale-like. **FL:** bilateral; hypanthium free of ovary; calyx cleft between lower 2 lobes, upper 3 larger; petals 4, 8–12 mm, thread-like, brown-purple; stamens 3, opposite 3 upper sepals; pistil 1, ovary superior, chamber 1, placentas 2, parietal. **FR:** capsule. **SEED:** spiny. 2 spp.: w N.Am. (W.F. Tolmie, Hudson's Bay Company physician, Fort Vancouver, 1812–1886) [Judd et al. 2007 Brittonia 59: 217–225]

T. diplomenziesii Judd et al. (p. 1247) $2n=14$. Moist streambanks; < 1800 m. NW, n CCo, SnFrB; to c OR. Morphologically similar *T. menziesii* (Pursh) Torr. & A. Gray ($2n=28$) occurs from c OR to AK. May–Aug

SCROPHULARIACEAE FIGWORT FAMILY

Margriet Wetherwax, except as noted

Ann to tree, gen glandular, some ± aquatic. **ST:** round to square in ×-section. **LF:** simple, gen alternate, gen ± entire; stipules gen 0 (present in *Limosella*). **INFL:** spike to panicle (head-like), gen bracted, or fls 1–4 in axils. **FL:** gen bisexual; calyx lobes 4–5; corolla bilateral to radial, lobes 4–5; stamens epipetalous, 4–5, 5th a staminode in *Scrophularia*; pistil 1, ovary superior,

chambers gen 2, placentas axile, style 1, stigma lobes gen 2. **FR**: capsule, gen ± ovoid, loculicidal or septicidal, or drupe-like. ± 65 genera, 1700 spp.: ± worldwide; some cult as orn *(Verbascum)*. [Olmstead et al. 2001 Amer J Bot 88:348–361] Other taxa moved to Plantaginaceae *(Antirrhinum, Bacopa, Collinsia, Cymbalaria, Digitalis, Dopatrium, Gambelia, Gratiola, Hebe, Holmgrenanthe, Keckiella, Kickxia, Limnophila, Linaria, Lindernia, Maurandella, Mohavea, Nothochelone, Penstemon, Pseudorontium, Stemodia, Synthyris, Tonella, Veronica),* Orobanchaceae *(Bellardia, Castilleja, Cordylanthus, Orthocarpus, Parentucellia, Pedicularis, Triphysaria),* Phrymaceae *(Mimulus).* Scientific Editors: Douglas H. Goldman, Bruce G. Baldwin.

1. Shrub or tree
  2. Fls white to yellow or purple; parts gen in 4s; lvs gen opposite . . . . **BUDDLEJA**
  2′ Fls white, purple-dotted; parts gen in 5s; lvs alternate . . . . **MYOPORUM**
1′ Ann to per (shrubby)
  3. Ann, pl gen < 30 cm, scapose; gen ± aquatic or of wet habitats . . . . **LIMOSELLA**
  3′ Bien, per (shrubby), 30–200 cm, not scapose, infl a raceme or panicle; pls gen of drier habitats
    4. St square in ×-section; corolla bilateral, pale green to dark red or black; stamens 4, staminode 1;
      infl gen a panicle . . . . **SCROPHULARIA**
    4′ St round in ×-section; corolla ± radial, yellow or white (purple); stamens 5, staminode 0; infl a raceme
      or panicle . . . . **VERBASCUM**

BUDDLEJA   BUTTERFLY BUSH

Margriet Wetherwax & Elizabeth McClintock

Shrub [tree], deciduous to evergreen. **LF**: opposite [alternate or ± whorled], short-petioled to sessile, linear to oblong or lanceolate. **FL**: gen fragrant; calyx ± bell-shaped, lobes gen 4, ± ≤ tube; corolla bell- to funnel-shaped or salverform, lobes gen 4, < tube, abruptly spreading; stamens gen 4, anthers ± sessile. **FR**: 2-parted; calyx persistent. **SEED**: many, gen winged. ± 100 spp.: Am, Afr, Asia. (Rev. Adam Buddle, England, 1660–1715) [Norman 1967 Gentes Herb 10:47–114] Treated in Buddlejaceae in TJM (1993).

1. Pl densely hairy ± throughout; infl spheric, head-like . . . . *B. utahensis*
1′ Pl densely hairy only on abaxial lf surfaces; infl cylindric or wide-spreading, panicle-like
  2. Lf lanceolate, 5–30 cm; corolla lilac to purple; stamens incl within corolla tube . . . . *B. davidii*
  2′ Lf linear to oblong, 1.5–10 cm; corolla white or cream; stamens exserted from corolla tube . . . . *B. saligna*

B. davidii Franch. BUTTERFLY BUSH, SUMMER LILAC Shrub < 5 m, deciduous to semi-evergreen. **LF**: 5–30 cm, lanceolate, serrate to ± entire, adaxially dark green, hairs sparse, gen branched, abaxially white, hairs dense, gen branched. **INFL**: panicle-like, 15–25 cm, slender, cylindric. **FL**: calyx ± 3 mm; corolla salverform, lilac to purple with central orange spot, tube ± 10 mm, lobes 2–3 mm; stamens incl. **FR**: 5–6 mm. **SEED**: winged at both ends. Disturbed, gen ± damp soil; < 200 m. KR, SnFrB; native to China. Commonly cult in CA; potentially problematic weed. May–Sep

B. saligna Willd. FALSE OLIVE Shrub, small tree < 10 m, deciduous. **ST**: twig ± 4-angled. **LF**: 1.5–10 cm, linear to oblong, entire, margins ± rolled under, adaxially glabrous, shiny, olive-green, abaxially white, hairs dense, stellate. **INFL**: panicle-like, ± 12 cm, wide-spreading. **FL**: corolla ± salverform, white or cream, occ with ± red spot in center, tube ± 2.5 mm, lobes ± 1 mm; stamens exserted. **FR**: ± 2 mm, ovoid, exserted from persistent calyx. **SEED**: not winged. Disturbed areas, chaparral; 500–1000 m. WTR (Santa Monica Mtns), SnGb; native to S.Afr. Sep–Oct

B. utahensis Coville (p. 1247) PANAMINT BUTTERFLY BUSH Shrub < 5 dm, densely branched, deciduous; hairs dense, stellate or branched ± throughout; dioecious. **LF**: 1.5–3 cm, linear-oblong, ± thick, margin entire to wavy, rolled under. **INFL**: paired at distal nodes into single, head-like, spheric, dense clusters ± 10–15 mm wide. **FL**: unisexual; calyx 3–4 mm; corolla 4–5 mm, salverform, cream-yellow, ± purple to brown-purple in age, lobes ± 1 mm; stamens incl. **FR**: spheric to oblong. **SEED**: not winged. Uncommon. Slopes, gen on dolomite, limestone, volcanic rocks; < 1900 m. DMoj; to UT. May–Oct

LIMOSELLA   MUDWORT

Ann, stoloned, occ partly submersed. **LF**: basal, erect; stipules gen present; petioles present or not. **INFL**: scapose; fl 1. **FL**: calyx bell-shaped, lobes 5, ovate, acute; corolla radial, bell-shaped, lobes 5, adaxially sparsely papillate; stamens 4; style gen ± terminal, stigmas gen fused, head-like. **FR**: elliptic to spheric, chambers 2 proximally, 1 distally. **SEED**: many, minute. ± 15 spp.: worldwide. (Latin: mud seat, from habitat)

1. Lvs awl-like, cylindric; style > ovary . . . . *L. australis*
1′ Lvs flat, blade linear to ovate; style ≤ ovary
  2. Lf linear to ± spoon-shaped, petiole gen indistinct; corolla lobes rounded . . . . *L. acaulis*
  2′ Lf blade spoon-shaped to ovate, petiole gen distinct; corolla lobes acute . . . . *L. aquatica*

L. acaulis Sessé & Moc. (p. 1251) Cespitose, gen mat-forming. **LF**: 1–6 cm, 0.5–2 mm wide, flat, linear to ± spoon-shaped; stipules gen ear-shaped, transparent; petiole gen 0. **INFL**: pedicels < 2/3 lf length, spreading in fr. **FL**: calyx lobes ± = tube; corolla 2–3 mm, white to lavender, lobes rounded; stamens attached at 1 level; style 0.2–0.7 mm, erect to ± curved. **FR**: 3–5 mm, spheric. Wet, muddy places, gen fresh water; < 3300 m. KR, SNH, GV, CW, SW, GB; to NM, Mex. May–Oct

L. aquatica L. (p. 1251) Tufted. **LF**: petiole 3–10(30) cm; blade 5–30 mm, flat, spoon-shaped to ovate; stipules 0 or sheathing lf base,

transparent. **INFL**: pedicels ≪ petioles, spreading in fr. **FL**: calyx lobes 2 × tube; corolla ± 2 mm, white to ± pink, outside occ blue, lobes acute; stamens attached at different levels; style 0.2–0.4 mm, stout, ± curved, stigma ± 2-lobed. **FR**: 3–5 mm, elliptic to spheric. 2*n*=40. Wet, muddy, periodically flooded places, fresh water; < 3200 m. CaR, SNH, CCo, SnFrB, SnBr, PR, MP; to AK, e N.Am; Eurasia. Jun–Sep

L. australis R. Br. (p. 1251) DELTA MUDWORT Tufted. **LF**: 1–3 cm, awl-like, cylindric, obtuse; stipules 0 or tapered; petiole 0. **INFL**: pedicels < lvs, curved in fr. **FL**: calyx lobes ≤ tube; corolla ± 3 mm,

lobes rounded, white to lavender-blue; stamens attached at 1 level; style > ovary, ± 1 mm, head-like, straight to ± curved, ± terminal. **FR:** spheric. $2n$=20. Muddy or sandy intertidal flats, brackish water; < 10 m. Deltaic GV (near Antioch), CCo (Abbott's Lagoon, Marin Co.); native to e coast N.Am. [*L. subulata* E. Ives] Poorly known in w N.Am, seeds possibly brought in with ship ballast. Apr

MYOPORUM

Robert E. Preston & Elizabeth McClintock

Erect [prostrate] shrub, tree; glabrous. **LF:** alternate [opposite], evergreen, petioled [sessile], entire or toothed, with conspicuous, embedded, gen translucent glands. **INFL:** axillary clusters [or 1]. **FL:** bisexual, radial [to bilateral]; calyx 5-lobed, lobes ± equal; corolla lobes equal [distal < proximal]; stamens 4 [5–8], anthers gen exserted; ovary [2]4-chambered, ovules 1–2 per chamber, style 1, stigmas 1–2. **FR:** indehiscent, dry to drupe-like. ± 30 spp.: esp Australia. (Greek: from glandular pores on lvs) [Chinnock 2007 Myoporaceae. Rosenberg Publishing] *M. acuminatum* R. Br., *M. parvifolium* R. Br. reported in San Diego Co., cult, doubtfully naturalized. Treated in Myoporaceae in TJM (1993).

M. laetum G. Forst. (p. 1251) MYOPORUM, NGAIO TREE Pl 3–10 m, much-branched, broadly spreading; twig tips, young lvs ± bronze-green, sticky. **LF:** < 10 cm, gen lanceolate, finely serrate distally, bright green, ± fleshy. **INFL:** fls 2–4 per axil. **FL:** corolla ± 10 mm diam, ± bell-shaped, white, purple-spotted, tube 3.5–4.5 mm, lobes 4–5.5 mm, glabrous abaxially, long-hairy adaxially. **FR:** 5–10 mm, ovoid, fleshy, pale to dark red-purple. **SEED:** 1 per chamber, 3–3.5 mm, oblong. Open areas in grassland, scrub, riparian habitats, gen coastal; < 460 m. CCo, SCo, ChI; reported from NCo (Sonoma Co.); native to New Zealand. TOXIC: lvs, frs may be fatal to livestock. Commonly cult near coast. Early spring ❖

SCROPHULARIA FIGWORT

Kim R. Kersh

Ann, per, subshrub, erect, gen glandular-puberulent to -hairy. **ST:** square in ×-section; (15)70–180 cm. **LF:** petioled; opposite, pairs at right angles to each other, lanceolate to triangular-ovate, serrate, dentate, or deeply cut, dark to light green, yellow-green, or gray-green (dull green), base cordate to truncate or occ ± wedge-shaped. **INFL:** gen panicle of cymes, occ axillary cymes or fl 1; axes, pedicels gen slender, occ glabrous. **FL:** calyx lobes 5, 2–4 mm, triangular-ovate to lanceolate, acuminate to acute or rounded, green, persistent, margins scarious or not; corolla 6–14 mm, inflated proximally, gen 2-colored, upper lip 2-lobed, gen darker than lower, lower lip < upper, 3-lobed, middle lobe reflexed, lateral lobes erect, mouth constricted; fertile stamens 4, incl, 2 gen longer; staminode 0 or proximally fused to corolla, much-reduced, or elongated with expanded tip; stigma head-like or 2-lobed; nectary disk fleshy, at ovary base. **FR:** septicidal. **SEED:** oblong-ovoid, ridged. 150–200 spp.: N.Am, temp Asia, Medit. (Latin: associated with the disease scrofula) [Shaw 1962 Aliso 5:147–178]

1. Ann, 15–60(90) cm; fls gen 1–5 in lf axils; calyx lobes lanceolate, acuminate, margins not scarious . . . . *S. peregrina*
1′ Per, occ shrubby, 70–180 cm; fls 3–10(24) per panicle branch; calyx lobes triangular-ovate, acute to rounded, margins gen scarious
  2. Infl glandular-long-soft-hairy. . . . . *S. villosa*
  2′ Infl glandular-puberulent to -hairy
    3. Corolla dark red to ± black, appearing much-inflated proximally, mouth much-constricted . . . . *S. atrata*
    3′ Corolla gen less dark, inflated proximally, mouth ± constricted
      4. Lvs yellow- to gray-green (dull-green); base wedge-shaped; pedicels slender, gen curved upward; corolla 7–9 mm, 2-colored, upper lobes maroon, lower cream, edges ± pink . . . . *S. desertorum*
      4′ Lvs dark to light green; base truncate to cordate or occ ± wedge-shaped; pedicels slender to ± stout, straight, not curved upward; corolla 8–14 mm, 1–2-colored, red to maroon or ± green-brown throughout, or upper lobes red to maroon and lower paler or ± yellow-green
        5. Staminode tip club-shaped to obovate; CA-FP . . . . *S. californica*
        5′ Staminode tip fan-shaped; KR, CaR, MP . . . . *S. lanceolata*

S. atrata Pennell (p. 1251) BLACK-FLOWERED FIGWORT Per. **ST:** 100–150 cm. **LF:** larger blades 6–9(11) cm, dark to light green. **INFL:** axes, pedicels glandular-puberulent to -hairy. **FL:** calyx lobes 3 mm, triangular-ovate, green, tip acute to rounded, margins gen scarious; corolla 9–11 mm, appearing much-inflated, dark red to ± black, mouth much-constricted; staminode tip triangular; stigma head-like. Calcium-, diatom-rich soils; < 400 m. CCo, SCoRO (Santa Barbara, s San Luis Obispo cos.). Apr–Jul ★

S. californica Cham. & Schltdl. (p. 1251) CALIFORNIA FIGWORT Per. **ST:** 80–120 cm. **LF:** larger blades 8–17 cm, dark to light green, base cordate to truncate, occ ± wedge-shaped. **INFL:** axes, pedicels glandular-puberulent. **FL:** calyx lobes 3–4 mm, triangular-ovate, green, tip acute to rounded, margins gen scarious; corolla 8–12 mm, inflated, mouth ± constricted, upper lobes red to maroon, lower paler or ± yellow-green; staminode tip club-shaped to obovate; stigma head-like. Common; damp places, chaparral, roadsides; < 2500 m. CA-FP; to BC, AZ. [*S. c.* subsp. *floribunda* (Greene) R.J. Shaw] Variable, study needed. Mar–Jul

S. desertorum (Munz) R.J. Shaw (p. 1251) Per. **ST:** 70–120 cm. **LF:** larger blades 4–8(12) cm, yellow- to gray-green (dull-green), base wedge-shaped, narrowed to petiole. **INFL:** axes, pedicels glandular-puberulent, gen curved upward. **FL:** calyx lobes 2–3 mm, triangular-ovate, green, tip acute to rounded, margins gen scarious; corolla 7–9 mm, inflated, mouth ± constricted, upper lobes maroon, lower lobes cream, edges ± pink; staminode tip club-shaped; stigma head-like. Dry rocky slopes, gen in crevices, among boulders, canyons, gravelly washes; 850–3000 m. SN, GB, DMoj; NV. Apr–Aug

S. lanceolata Pursh (p. 1251) Per. **ST:** 80–150 cm. **LF:** larger blades 9–13 cm, dark to light green, base cordate to truncate, occ ± wedge-shaped. **INFL:** axes, pedicels ± stout, glandular-puberulent. **FL:** calyx lobes 2–4 mm, triangular-ovate, green, tip acute to rounded, margins gen scarious; corolla 8–14 mm, inflated, red to maroon or ± green-brown, mouth ± constricted; staminode broad, tip fan-shaped; stigma head-like. Moist streambanks, meadows, thickets, woodland; 1000–2800 m. KR, CaR, MP; n&c US, Can. May–Jul

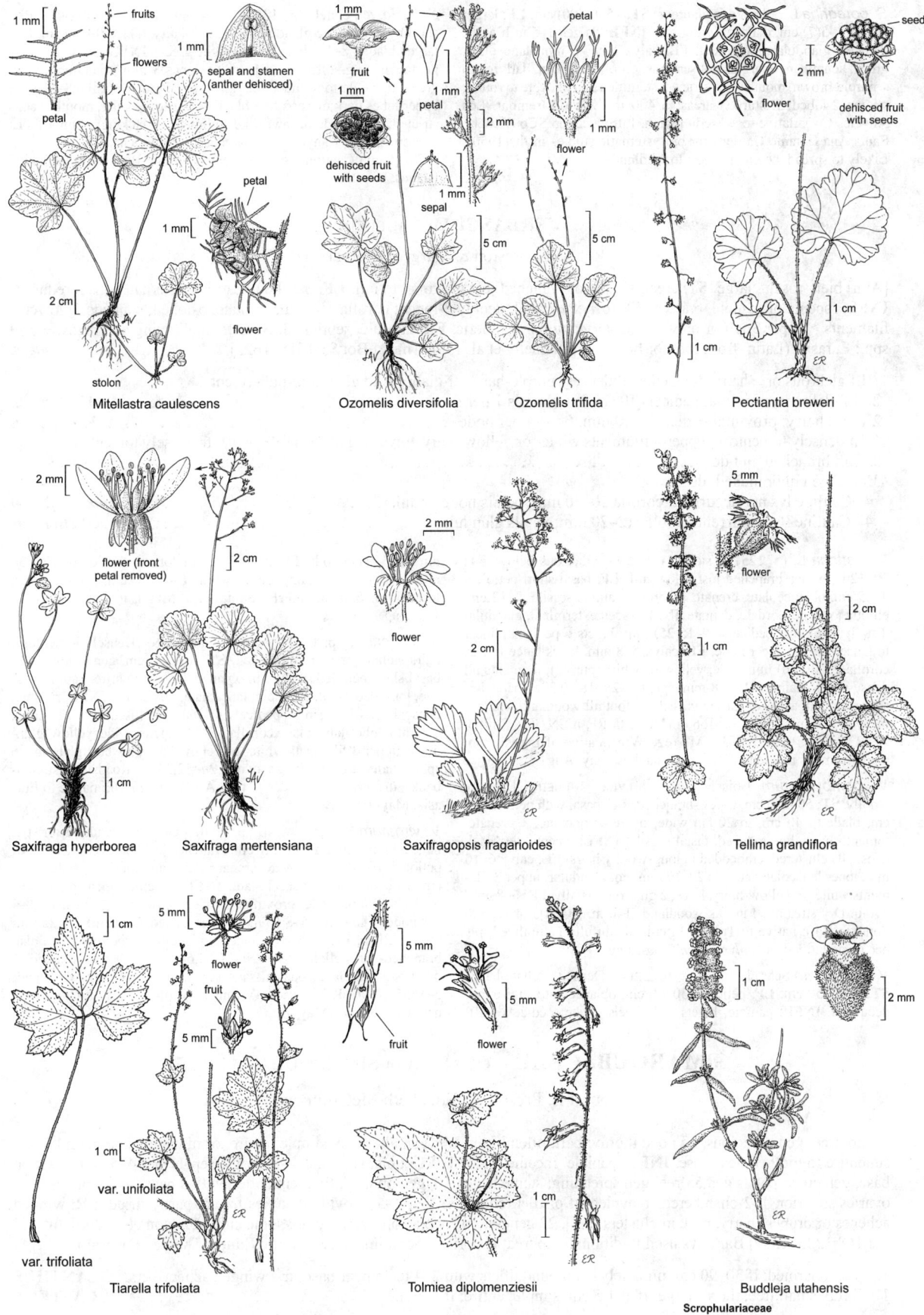

1 mm — fruits

flowers

1 mm — sepal and stamen (anther dehisced)

petal

petal

1 mm

2 cm

flower

stolon

Mitellastra caulescens

1 mm

fruit

1 mm

dehisced fruit with seeds

1 mm — petal

2 mm

1 mm — sepal

5 cm

petal

flower

1 mm

5 cm

Ozomelis diversifolia

Ozomelis trifida

seed

2 mm

flower

dehisced fruit with seeds

1 cm

1 cm

Pectiantia breweri

2 mm

flower (front petal removed)

2 cm

Saxifraga hyperborea

1 cm

Saxifraga mertensiana

2 mm

flower

2 cm

Saxifragopsis fragarioides

5 mm

flower

1 cm

2 cm

Tellima grandiflora

1 cm

5 mm

flower

fruit

5 mm

var. unifoliata

1 cm

var. trifoliata

Tiarella trifoliata

5 mm

fruit

5 mm

flower

1 cm

Tolmiea diplomenziesii

1 cm

2 mm

Buddleja utahensis

Scrophulariaceae

S. peregrina L.   Ann, occ glabrous. **ST:** 15–60(90) cm. **LF:** larger blades 3–5(7) cm, dark to light green. **INFL:** fls gen 1–5 in lf axils; pedicels glandular-puberulent. **FL:** calyx lobes 3 mm, lanceolate, green, acuminate, margins not scarious; corolla 6–9 mm, dark red to ± purple-brown, mouth ± constricted; staminode tip obovate to round; stigma 2-lobed. Disturbed areas; ± 400 m. SCo (Claremont, Los Angeles Co.); native to ± Medit region. Introduced in SCo (Rancho Santa Ana Botanic Garden) for biosystematic studies in the 1950s. Likely to spread, potential threat to wildlands. Mar–May

S. villosa Pennell (p. 1251)   SANTA CATALINA FIGWORT   Per, shrubby in age; occ glandular-long-soft-hairy. **ST:** 120–180 cm. **LF:** larger blades 8–15 cm, dark to light green. **INFL:** axes, pedicels glandular-long-soft-hairy. **FL:** calyx lobes 3 mm, triangular-ovate, green, tip acute to rounded, margins gen scarious; corolla 8–11 mm, upper lobes dark maroon to ± black, lower lobes paler, mouth ± constricted; staminode an awn-like rudiment or lacking; stigma lobes 2, ± spreading. Canyon bottoms, coastal scrub, chaparral; < 400 m. s ChI (Santa Catalina, San Clemente islands); Baja CA (Guadalupe Island). Feb–May   ★

VERBASCUM   MULLEIN

Robert E. Preston & Margriet Wetherwax

[Ann] bien, rosette large. **ST:** erect, simple or branched just proximal to infl. **LF:** basal and cauline, alternate, distal reduced. **INFL:** raceme or panicle, bracted. **FL:** calyx ± radial, deeply 5-lobed; corolla ± radial, ± rotate, 5-lobed; stamens 5, lower 2 filaments > upper 3, all or only upper hairy; stigma ± spheric. **FR:** capsule, septicidal. **SEED:** small, wingless, many. ± 360 spp.: Eurasia. (Latin: from root for bearded) [Donnelly et al. 1998 Amer J Bot 85:1618–1625]

1. Lf glabrous or ± hairy; filaments all densely purple-hairy; fr glabrous to glandular-puberulent
  2. Lf glabrous; proximal pedicels 10–15(25) mm, fls 1 per node . . . . *V. blattaria*
  2′ Lf ± hairy; proximal pedicels < 10 mm, fls 1–4 per node . . . . *V. virgatum*
1′ Lf densely tomentose; upper 3 filaments white- or yellow-hairy, lower 2 glabrous to ± hairy; fr densely tomentose
  3. Infl branched, not dense . . . . *V. speciosum*
  3′ Infl gen unbranched, dense
    4. Cauline lvs not decurrent; corolla 20–30 mm; petals not glandular. . . . . *V. bombyciferum*
    4′ Cauline lvs decurrent; corolla 12–20 mm; petals glandular. . . . . *V. thapsus*

V. blattaria L. (p. 1251)   MOTH MULLEIN   Glabrous below. **ST:** 30–120 cm, occ branched just below infl. **LF:** basal short-petioled, 4–25 cm, oblanceolate, crenate, glabrous; cauline sessile, 2–12 cm, elliptic to ovate, cordate, dentate. **INFL:** raceme, terminal, glandular distally; proximal pedicels 10–15(25) mm; bracts < pedicels, linear to lanceolate; fls 1 per node. **FL:** calyx 5–8 mm, lobes lance-linear; corolla 25–30(40) mm wide, yellow or white (purple); filaments all dense-purple-hairy. **FR:** 6–8 mm, spheric. 2*n*=18,30,32. Roadsides, seeps, streambanks, disturbed grassland, foothill woodland, chaparral, yellow-pine forest; 10–1660 m. NW, CaRF, n SN, GV, SnFrB, WTR, SnGb, PR (exc SnJt), MP (exc Wrn); native of Eurasia. Sp. name from lvs reputed to repel cockroaches. May–Aug

V. bombyciferum Boiss.   GIANT SILVER MULLEIN   Densely woolly. **ST:** 50–150 cm, 0–few-branched. **LF:** basal with petiole 1–5 cm, blade to 40 cm, to 20 cm wide, ovate or obovate, ± crenate, tomentose; cauline reduced, distal sessile. **INFL:** raceme, terminal, dense; fls clustered, embedded in long-woolly hairs. **FL:** calyx 6–10 mm, lobes lanceolate; corolla 20–30 mm, not glandular; upper 3 filaments white- or yellow-hairy, lower 2 glabrous distally. **FR:** 6–8 mm, ovoid. Dry streambed in oak woodland; 140 m. NCoRO (1 record, Sonoma Co.); native to Turkey. In cult, doubtfully naturalized; pls occ misidentified as *V. olympicum* Boiss. Jun

V. speciosum Schrad.   SHOWY MULLEIN   Densely stellate-hairy. **ST:** 100–150 cm. **LF:** petioled, 30–40 cm, oblanceolate, entire, not decurrent. **INFL:** panicle; bracts < pedicels, ovate; pedicels 5–10 mm; fls 5–9 per node. **FL:** calyx 3–6 mm, lobes linear; corolla 20–30 mm wide, yellow; filament hairs white. **FR:** 3–7 mm, ovoid-oblong. Roadsides; 550 m. SnFrB (Santa Cruz Mtns); native to Eurasia. Jun–Aug

V. thapsus L. (p. 1251)   WOOLLY MULLEIN   Densely tomentose, hairs stellate or many-branched. **ST:** 30–200 cm, gen simple. **LF:** basal short-petioled, 8–50 cm, oblanceolate, gen entire; cauline sessile, long-decurrent, 5–30 cm, lanceolate, gen entire. **INFL:** raceme, dense; bracts 12–18 mm; pedicels < 2 mm, gen fused to st. **FL:** calyx 7–9 mm, lobes lanceolate; corolla 15–25(30) mm wide, yellow, glandular; upper 3 filaments white- or yellow-hairy, lower 2 glabrous to sparse-hairy. **FR:** 7–10 mm, ovoid. 2*n*=32,34,36. Roadsides, streambanks, disturbed areas; < 2470 m. CA-FP, n SNE, MP; native to Eurasia. May–Sep   ❖

V. virgatum Stokes   WAND MULLEIN   Gen nonglandular, bristly, hairs ± branched. **ST:** 60–120 cm, branched just below infl. **LF:** basal petioled, 10–30 cm, obovate, crenate to dentate; cauline sessile, 7–15 cm, lanceolate, cordate, crenate. **INFL:** raceme, open, glandular; bracts < 8 mm, ovate; proximal pedicels < 10 mm, ≤ bract; fls 1–4 per node. **FL:** calyx 5–8(9) mm, lobes lanceolate; corolla ± 25 mm wide, yellow. **FR:** 7–8 mm, spheric, glandular-puberulent, occ also branched- or stellate-hairy. 2*n*=32,64,66. Streambanks, roadsides, disturbed grassland, coastal-sage scrub, chaparral, oak woodland; 5–1555 m. NCoRO, SNF, SnJV, SnFrB, SCoRO, SW (exc ChI, SnJt); native to Eurasia. May–Aug

SIMAROUBACEAE   QUASSIA or SIMAROUBA FAMILY

Robert E. Preston & Elizabeth McClintock

Shrub, tree; gen dioecious. **ST:** occ thorny; bark often bitter. **LF:** gen alternate, simple, entire, or pinnately compound, lflets subentire to toothed near base. **INFL:** panicle, raceme, or fls 1. **FL:** unisexual, inconspicuous; sepals gen 4–5, gen fused at base, gen erect; petals gen 5, free, gen spreading; stamens gen 10, gen on disk, filaments often with a basal scale; pistils 1–8, ovaries superior, 1–2-chambered, 1-ovuled [if pistil 1, chambers gen 2–5, 1-ovuled], styles free or partly fused. **FR:** winged achenes or drupes [berry, nut], in clusters [not]. 22 genera, 100 spp.: trop, warm temp; some cult. [Clayton et al. 2007 Int J Pl Sci 168:1325–1339] Bark, lvs used traditionally to treat malaria, other ailments. Scientific Editor: Thomas J. Rosatti.

1. Tree, unarmed; lf 30–90 cm, pinnately compound, lflets with 2–4 teeth near base; fr a winged achene . . . . **AILANTHUS**
1′ Shrub, small tree, thorns large; lf 1–1.5 cm, simple, entire; fr a drupe. . . . . **CASTELA**

AILANTHUS

Pl ± dioecious, with a few bisexual fls. **LF:** gen ± odd-pinnate, ill-smelling when crushed, deciduous. **INFL:** large panicle, terminal. **FL:** calyx lobes 5–6; petals 5–6; stamens 10–12; ovaries very compressed, adherent near middle, styles ± free but twisted together. **FR:** 1–5, ± pendent, seed near middle. 5 spp.: se Asia, Australasia. (Moluccan: sky tree) [Corbett & Manchester 2004 Int J Pl Sci 165:671–690]

A. altissima (Mill.) Swingle (p. 1251)   TREE OF HEAVEN   Pl < 20 m; young parts ± glandular-puberulent. **LF:** 3–9 dm; lflets 13–25, 8–13 cm, lanceolate, base gen ± truncate, with 2–4 teeth, each with a large gland abaxially. **INFL:** 10–20 cm. **FL:** sepals < 1 mm, petals 2–3 mm, spreading. **FR:** < 5 cm, linear or oblong. Disturbed areas, grassland, oak woodland, riparian areas; < 1860 m. KR, NCoRO, NCoRI, CaRF, SN, GV, CW, SW (exc ChI), W&I, DMtns; widely naturalized in temps; native to China. Cult as street tree, fast-growing, spreading by seeds, invasive roots. Common near old habitations. Jun ◆

CASTELA

LF: linear to lanceolate on young sts, scale-like on mature. **FL:** calyx lobes 4–8; petals 4–8; stamens 8–24; ovaries 4–8, adherent near middle, style bases fused, tips spreading. **FR:** dry, 4–8, spreading. ± 15 spp.: sw&s-c US, to S.Am. (René R.L. Castel, French botanist, poet, editor, opera librettist, 1759–1832) *Holacantha* still recognized by some (e.g., Clayton et al. 2007).

C. emoryi (A. Gray) Moran & Felger (p. 1251)   EMORY'S CRUCIFIXION-THORN   Pl often < 1 m, occ to 4+ m, intricately branched; young parts densely puberulent. **LF:** ephemeral, rarely seen. **INFL:** panicle, much-branched, 2.5–5 cm, stiff. **FL:** 6–8 mm diam. **FR:** ± 6 mm, flat-topped; base ± rounded, sometimes persisting several yrs. **SEED:** 1. Dry, gravelly washes, slopes, plains; ± 650 m. s DMoj (exc DMtns), DSon; AZ, nw Mex. Common name used for 2 other desert pls. Jun–Jul ★

SIMMONDSIACEAE   JOJOBA FAMILY

Robert E. Preston & William J. Stone

Shrub, evergreen, much-branched; dioecious. **ST:** bark smooth. **LF:** opposite, simple, leathery; base jointed; stipules 0. **INFL:** staminate fls in axillary clusters, pistillate gen 1. **FL:** small, radial; sepals gen 5, overlapped, becoming larger in pistillate, disk 0; corolla 0; stamens 8–12, free, anthers elongate with longitudinal slits; ovary superior, chambers 3, styles 3, stigmas long, ± not persistent in fr. **FR:** capsule, loculicidal. **SEED:** 1. 1 genus, 1 sp.: sw US, Mex. [Carlquist 2002 Madroño 49:158–164] Recently shown (e.g., Carlquist 2002) to be not closely related to Buxaceae, in which it was sometimes placed; 2° growth unusual, produced by successive cambia. Scientific Editor: Thomas J. Rosatti.

SIMMONDSIA   GOAT-NUT, JOJOBA

(T.W. Simmonds, English botanist, died exploring Trinidad, 1767–1804)

S. chinensis (Link) C.K. Schneid. (p. 1251)   Hairs short, dense, appressed, less dense in age. **ST:** 1–2 m; new growth ± hairy; branches stiff. **LF:** 2–4 cm, oblong-ovate, dull green, subsessile. **INFL:** peduncle 3–10 mm. **FL:** staminate sepals 3–4 mm, ± green, pistillate 10–20 mm in age; anthers yellow. **FR:** ± 1–2.5 cm, nut-like, ovoid, tough, leathery, obtusely 3-angled. **SEED:** large, incl liquid wax. 2*n*=26. Creosote-bush scrub, desert wash scrub, chaparral, coastal scrub; < 1350 m. s SCo, PR, s DMoj, DSon; n Mex. Important as forage pl; seed wax a substitute for sperm whale oil; fr edible. Gen Mar–May

SOLANACEAE   NIGHTSHADE FAMILY

Michael H. Nee

Ann to shrub. **LF:** gen simple, gen alternate, gen petioled; stipules 0; blade entire to deeply lobed. **INFL:** various. **FL:** bisexual; calyx lobes gen 5; corolla ± radial, cylindric to rotate, lobes gen 5; stamens 5, on corolla tube, alternate lobes; ovary superior, gen 2-chambered, style 1. **FR:** berry, loculicidal or septicidal capsule, [(drupe)], 2–5-chambered. 75 genera, 3000 spp.: worldwide, esp ± trop; many alien weeds in CA; many cult for food, drugs, or orn (potato, tomato, peppers, tobacco, petunia); many TOXIC. [Hunziker 2001 Genera Solanacearum. Koeltz Scientific Books] *Nicandra physalodes* (L.) Gaertn. is a waif. Scientific Editor: Thomas J. Rosatti.

1. Fr a capsule
2. Fr prickly. . . . . **DATURA**
2′ Fr not prickly
   3. Calyx lobes > tube, winged toward base. . . . . **[NICANDRA]**
   3′ Calyx lobes gen < tube, not wing-like at base
      4. Corolla narrowly urn-shaped; seeds flat, wing 0.5 mm wide. . . . . **ORYCTES**
      4′ Corolla funnel-shaped to salverform; seeds angled but not flat, wing 0
         5. Fls in racemes or panicles, only lower in axils of lf-like bracts . . . . **NICOTIANA**
         5′ Fls 1 in axils of lvs or lf-like bracts. . . . . **PETUNIA**
1′ Fr a berry (dry or fleshy)

6. Corolla salverform to funnel-, bell-, or urn-shaped, or lobes reflexed
  7. Shrub, gen with thorns. . . . . **LYCIUM**
  7′ Per or shrub, unarmed
    8. Pl erect shrub; corolla narrowly funnel-shaped . . . . **CESTRUM**
    8′ Pl decumbent or climbing per; corolla urn-shaped. . . . . **SALPICHROA**
6′ Corolla ± rotate to shallowly bell-shaped or lobes reflexed
  9. Anthers opening by pores or short slits near tip . . . . ²**SOLANUM**
  9′ Anthers opening by slits from tip (exc any sterile part) to near base
    10. Anthers gen > filaments
      11. Anthers adherent, tube-like around style; lvs ± odd-1–2-pinnate . . . . **LYCOPERSICON**
      11′ Anthers free, not tube-like around style; lvs entire to deeply pinnately lobed. . . . . ²**SOLANUM**
    10′ Anthers gen < filaments
      12. Calyx in fr ± enlarged but not bladder-like; corolla tube tomentose between stamen bases
      . . . . **CHAMAESARACHA**
      12′ Calyx in fr bladder-like; corolla tube hairy but not tomentose between stamen bases . . . . **PHYSALIS**

CESTRUM

Shrub; hairs branched or not. **LF**: entire, glabrous to hairy. **INFL**: panicle. **FL**: corolla narrowly funnel-shaped; stamens incl, filaments equal or not. **FR**: berry, ± dry or not. 250 spp.: trop Am. (Greek: derivation unknown)

1. Corolla bright red; hairs not minute, ± dense, esp on axes . . . . [*C. elegans*]
1′ Corolla yellow- or green-white; hairs minute, sparse
  2. Calyx 2–3 mm; fr white . . . . *C. nocturnum*
  2′ Calyx 5–7 mm; fr dark purple-black . . . . [*C. parqui*]

C. nocturnum L. Pl 2–5 m; hairs minute, sparse. **ST**: glabrous. **LF**: 4–10 cm, ± elliptic, acuminate. **INFL**: 5–10 cm, open, many-fld. **FL**: calyx 2–3 mm; corolla 20 mm, green-white, lobes 2.5–3 mm. **FR**: 6–10 mm, white. **SEED**: 4–5, 4–4.5 mm, black. Disturbed areas; < 1000 m. SCo; native to Mex, C.Am. Feb–Oct

CHAMAESARACHA

Per; hairs ± scale-like. **ST**: decumbent, branched. **LF**: entire to ± deeply pinnately lobed. **INFL**: axillary, cluster, 1–5-fld. **FL**: calyx in fr ± enlarged but not bladder-like, open at top; corolla ± rotate, tomentose between stamen bases; anthers free, gen < filaments, opening by slits; style 1. **FR**: berry, spheric, partly enclosed by calyx. **SEED**: ± flat, reniform. ± 9 spp.: esp sw US, Mex. (Greek: low *Saracha*, a S.Am genus in family)

1. Lf ± linear to lanceolate, margin ± entire to ± deeply lobed, not wavy; corolla 10–15 mm wide; e DMtns
  (New York Mtns). . . . . *C. coronopus*
1′ Lf ± ovate, margin entire, wavy or not; corolla 15–25 mm wide; CaR, SNH, GB. *C. nana*

C. coronopus (Dunal) A. Gray (p. 1257) **ST**: many from base, 10–50 cm. **LF**: 20–65 mm. **INFL**: pedicel ± 1 cm, in fr < 2 cm, reflexed. **FL**: calyx 3–5 mm, in fr 5–10 mm, lobes in fr 2 mm; corolla dirty- or green-white; filaments 3 mm. **FR**: 5–8 mm wide. **SEED**: 3 mm. *n*=12,24,36. Dry, clay soil; ± 1500 m. e DMtns (New York Mtns); to UT, TX, n Mex. May–Jul

C. nana (A. Gray) A. Gray (p. 1257) **ST**: 1–several from base, 5–25 cm. **LF**: 15–50 mm. **INFL**: pedicel 8–18 mm, in fr < 3 cm, recurved. **FL**: calyx ± 5 mm, in fr < 10 mm, lobes in fr 2–3 mm; corolla ± white; filaments 5 mm. **FR**: 1–1.2 cm wide. **SEED**: 1.5–2 mm. *n*=12. Sandy soils, slopes, conifer forest; 1500–2800 m. CaR, SNH, GB; OR, NV. May–Jul

DATURA   JIMSON WEED

Ann to subshrub, hairs ± 0 or simple, ill-smelling. **LF**: entire to deeply lobed. **INFL**: fls 1 in branch forks. **FL**: calyx circumscissile near base, leaving ± rotate collar in fr; corolla funnel-shaped, white or ± purple, lobes 5(10); stamens attached below tube middle; ovary 2- or 4-chambered. **FR**: capsule, leathery or woody, prickly; valves 2–4 or irregular. **SEED**: ± flat, black, brown, gray-brown, or tan. ± 13 spp.: warm regions, esp Mex; several orn, some source of drugs. (Hindu: ancient name) All spp. TOXIC.

1. Corolla 15–20 cm; calyx 8–12 cm, ribbed; seeds tan . . . . ***D. wrightii***
1′ Corolla 4.5–16 cm; calyx 2–9 cm, angled or winged; seeds black or gray-brown
  2. Fr nodding, prickles many, weak; st gray-hairy . . . . ***D. discolor***
  2′ Fr erect, prickles few, stout; st ± glabrous to sparsely hairy
    3. Fr prickles 8–22 mm; calyx 2–2.5 cm. . . . . ***D. ferox***
    3′ Fr prickles 3–10 mm; calyx 3.5–4.5 cm . . . . ***D. stramonium***

D. discolor Bernh. Ann < 5 dm. **ST**: gray-hairy. **LF**: 6–12 cm, 4–10 cm wide, widely ovate, coarsely toothed. **FL**: erect; calyx 5–9 cm, 5-winged toward base, lobes 1–1.5 cm; corolla 10–16 cm, glabrous, white with purple markings in tube, lobes shallow; filaments 4.5 cm, anthers 5–6 mm; style 10–14 cm. **FR**: 4-valved, nodding, 35 mm wide, puberulent; prickles many, < 2 cm, weak. **SEED**: 3–3.5 cm, coarsely wrinkled, black with a white outgrowth near attachment scar. *n*=12. Sandy, gravelly soils, washes; < 500 m. DSon; Mex. Apr–Oct

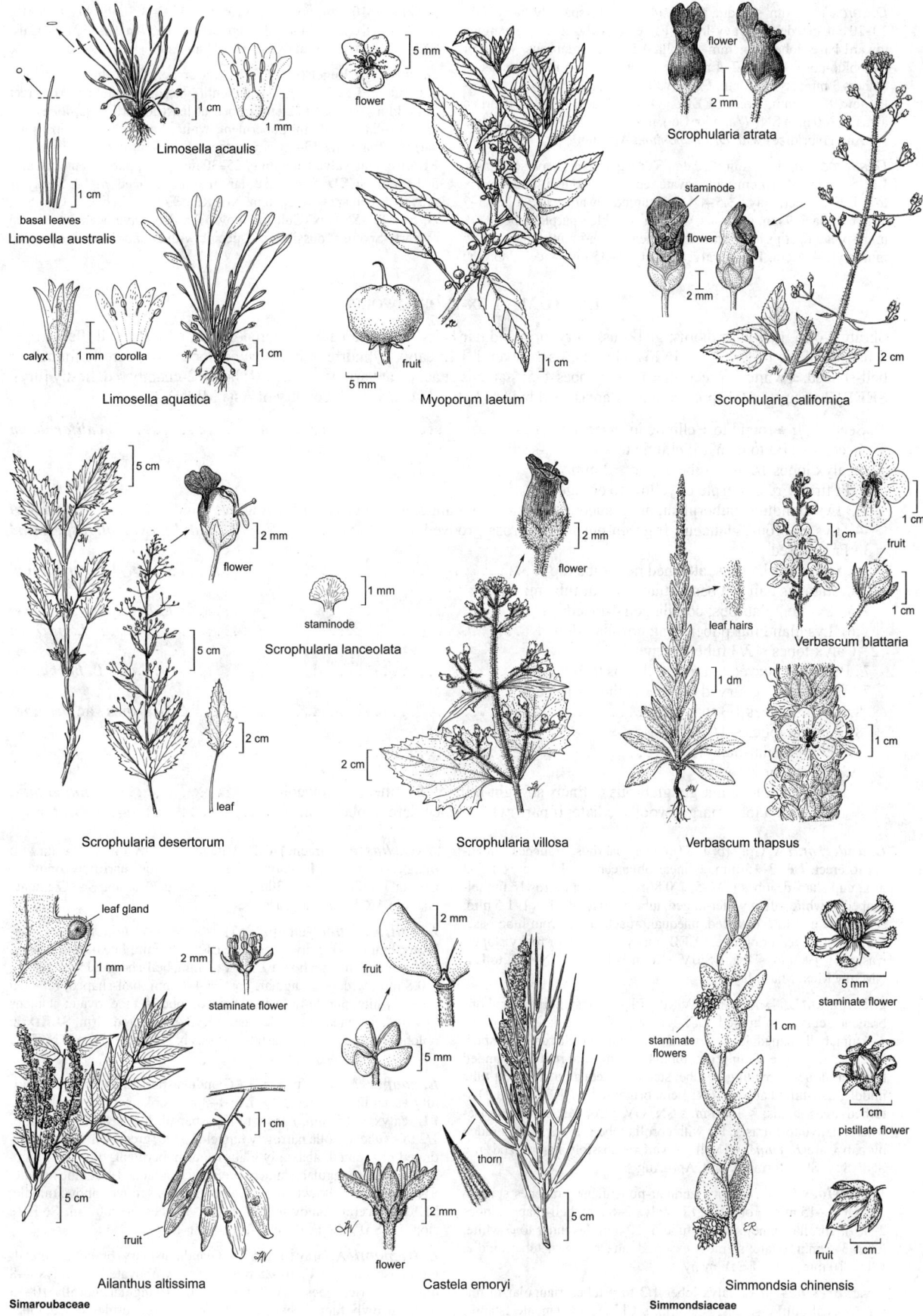

Limosella acaulis

basal leaves
Limosella australis

calyx   1 mm   corolla
Limosella aquatica

flower

fruit

Myoporum laetum

flower
Scrophularia atrata

staminode

flower

Scrophularia californica

5 cm

flower

5 cm

leaf

2 cm

Scrophularia desertorum

staminode
Scrophularia lanceolata

flower

2 cm

Scrophularia villosa

leaf hairs

1 dm

Verbascum thapsus

fruit

Verbascum blattaria

leaf gland

staminate flower

fruit

Ailanthus altissima

Simaroubaceae

fruit

thorn

flower

Castela emoryi

staminate flowers

fruit

Simmondsia chinensis

Simmondsiaceae

staminate flower

pistillate flower

fruit

D. ferox L.   Ann < 6 dm. **ST:** ± glabrous to sparsely hairy. **LF:** 10–20 cm, gen deeply wavy-lobed. **FL:** erect; calyx 2–2.5 cm, angled toward base, lobes 2–4 mm; corolla 4.5–5 cm, glabrous, white or pale blue-purple, lobes 2–4 mm, tips short, narrow; filaments 2 cm, anthers 3 mm; style 2.5 cm. **FR:** 4-valved, erect, 25 mm wide; prickles few, 8–22 mm, stout. **SEED:** ± 4 mm, black. *n*=12. Disturbed areas; < 200 m. n SNF (Amador Co.), n SnJV, SCoRO; native to Mex, s S.Am. Hybridizes with *D. stramonium*. Apr–Jun

D. stramonium L.   Ann < 10 dm. **ST:** ± glabrous to sparsely hairy. **LF:** 5–15 cm, 4–10 cm wide, ovate, coarsely toothed to shallowly lobed. **FL:** erect; calyx 3.5–4.5 cm, angled toward base, lobes 5–7 mm; corolla 6–9 cm, glabrous, white or pale blue-purple, lobes 8–10 mm, spreading, tips long, narrow; filaments 22–25 mm, anthers 3.5–5 mm; style 4–6 cm. **FR:** 4-valved, erect, 25–35 mm wide, glabrous;

prickles 3–10 mm, upper > lower. **SEED:** 3–3.5 mm, black. *n*=12. Sandy soils, open, often disturbed areas; < 1500 m. CA-FP (exc CaR, SNH), MP; to e US; native to Mex. Jun–Aug

D. wrightii Regel (p. 1257)   Ann or per, 5–15 dm. **ST:** white-puberulent. **LF:** 7–20 cm, ovate, entire or coarsely lobed. **FL:** erect to nodding; calyx 8–12 cm, ribbed at least toward base, lobes ± 2 cm; corolla 15–20 cm, puberulent, white, lobes 1–2 cm, tips long, narrow; filaments 13–15 cm, anthers 12–15 mm; style 15–18 cm. **FR:** irregular-valved, nodding, 25–30 mm wide, puberulent; prickles 5–12 mm. **SEED:** 5 mm, flat, tan; margin grooved. *n*=12. Sandy or gravelly open areas; < 2200 m. NCoRI, c&s SNF, Teh, GV, CW, SW, D; to UT, TX, Mex. Cult for showy fls; may have been introduced by early Spanish; possibly conspecific with *D. inoxia* Mill. Apr–Oct

LYCIUM   BOX-THORN, WOLFBERRY

Shrub 1–4 m, gen with thorns, glabrous, hairy, or glandular-hairy. **LF:** alternate, clustered or not, entire, small, fleshy, gen ± flat to elliptic in ×-section. **INFL:** clusters; fls 1–several. **FL:** calyx cylindric to bell-shaped, lobes 2–5; corolla funnel- or bell-shaped, ± white, ± green, or ± purple, lobes 4–5; stamens attached at various levels. **FR:** berry, 2-chambered, fleshy [dry]. **SEED:** 2–many. ± 100 spp.: warm, dry areas worldwide. (Latin: Lycia, ancient country of Asia Minor)

1. Seeds 2; lf ± round to ± elliptic in ×-section . . . . ***L. californicum***
1′ Seeds several to many; lf elliptic to ± flat in ×-section
  2. Calyx lobes 1/2 to > tube or gen > 2 mm
    3. Fr firm, green-purple or yellow to orange
      4. Lvs glandular-puberulent, not glaucous; fr yellow to orange, with 2 cross-grooves above middle . . . . ***L. cooperi***
      4′ Lvs glabrous, glaucous; fr green-purple, not cross-grooved . . . . ***L. pallidum*** var. ***oligospermum***
    3′ Fr soft, red
      5. Stamens glabrous, attached near tube top . . . . ***L. verrucosum***
      5′ Stamens hairy at base, attached near tube middle
        6. Lvs gen glabrous; corolla gen 4-lobed . . . . ***L. brevipes*** var. ***hassei***
        6′ Lvs glandular-puberulent; corolla 5-lobed . . . . ***L. parishii***
  2′ Calyx lobes < 2/3 tube or gen < 2 mm
    7. Branches curved to arched, thorns 0–1(2) cm . . . . ***L. barbarum***
    7′ Branches not curved to arched, thorns 1–5 cm
      8. Corolla lobes 1/3 to = tube . . . . ***L. brevipes*** var. ***brevipes***
      8′ Corolla lobes < 1/3 tube
        9. Pl glandular-hairy . . . . ***L. fremontii***
        9′ Pl ± glabrous
          10. Corolla lobe margin glabrous to finely straight-ciliate; lf ± linear-oblanceolate . . . . ***L. andersonii***
          10′ Corolla lobe margin woolly-ciliate; lf narrowly oblanceolate to obovate . . . . ***L. torreyi***

L. andersonii A. Gray (p. 1257)   Pl ± glabrous; branches spreading to erect. **LF:** 3–15 mm, ± linear-oblanceolate. **FL:** calyx 1.5–3 mm, cup-shaped, lobes (2)4–5, ± 0.8 mm; corolla narrowly funnel-shaped, ± white, often violet-tinged, tube 5–10 mm, lobes 1–1.5 mm; stamens ± incl to ± exserted, unequal, attached 1/3 from tube base. **FR:** 3–8 mm, red or orange. **SEED:** many. Gravelly or rocky slopes, washes; < 1900 m. s Teh, s SnJV, SCo, n WTR, PR, SNE, D; to UT, NM, nw Mex. Mar–May

L. barbarum L.   MATRIMONY VINE   Pl glabrous; thorns 0 or short; branches curved to arched. **LF:** 2–6 cm, (ob)lanceolate. **FL:** calyx 3–4 mm, bell-shaped, lobes gen 2–3, 1–2 mm; corolla funnel-shaped, lavender to purple, fading tan, tube 4–6 mm, ± abruptly expanded to throat, lobes spreading, < tube; stamens ± equal, attached at tube middle, hair-tufted at base. **FR:** 1 cm, bright red. **SEED:** 10–20. Disturbed areas, fields; < 1500 m. s SN, GV, SCo, MP, expected elsewhere; native to Eurasia. Pls with corolla tube < lobes and ± hidden by calyx are *L. chinense* Mill., a waif or possibly naturalized in c SNF, ScV, SCo; formerly cult. Apr–Aug

L. brevipes Benth.   Gen glandular-puberulent; branches spreading. **LF:** 5–15 mm, ± obovate. **FL:** calyx 2–6 mm, bell-shaped, lobes 2–4(6); corolla funnel-shaped, tube 6–10 mm, lavender to ± white, lobes 3–5 mm, ovate; stamens exserted, attached near tube middle. **FR:** ± 10 mm, red. **SEED:** many.

var. *brevipes*   **FL:** calyx lobes 1/3 to = tube, triangular to linear. Coastal bluffs, slopes; < 600 m. s ChI (San Clemente Island), w DSon; nw Mex. Intergrades with *L. torreyi* in w DSon. Mar–Apr

var. *hassei* (Greene) C.L. Hitchc.   SANTA CATALINA ISLAND DESERT-THORN   **FL:** calyx lobes 1–3 × tube, narrowly obovate. Coastal bluffs, slopes; < 300 m. s ChI (Santa Catalina, San Clemente islands). [*L. h.* Greene] Jun   ★

L. californicum Nutt. (p. 1257)   CALIFORNIA BOX-THORN   Glabrous; branches rigidly spreading. **LF:** 3–10 mm, linear to oblanceolate, ± round in ×-section. **FL:** calyx 3 mm, bell-shaped, lobes 2–4(5), ± 0.8 mm, widely triangular; corolla 4–6 mm, bell-shaped, reflexed or not, white, purple-tinged or -veined, lobes = tube, ovate; stamens exserted, attached near tube base. **FR:** 3–6 mm, red, firm. **SEED:** 2, oblong. Coastal bluffs, coastal-sage scrub; < 150 m. s SCo, ChI; Baja CA. (Dec)Mar–Aug   ★

L. cooperi A. Gray (p. 1257)   Glandular-puberulent; branches rigidly ascending to erect, lfy. **LF:** 1–3 cm, oblanceolate to obovate. **FL:** calyx 8–15 mm, narrowly bell-shaped, lobes 4–5, 1.5–3 mm, 1/2 to = tube; corolla narrowly funnel-shaped, green-white, lavender-tinged or -veined, abaxially glabrous or puberulent, tube 9–12 mm, lobes ovate-triangular; stamens ± exserted, attached ± at tube middle. **FR:** 5–9 mm, yellow to orange, with 2 cross-grooves above middle. **SEED:** several. Sandy to rocky flats, washes; < 2000 m. SNH (e slope), s SnJV, SNE, D; to UT, AZ. Mar–May

L. fremontii A. Gray (p. 1257)   Glandular-hairy; branches spreading to ascending. **LF:** 10–25 mm, narrowly obovate. **FL:** calyx 4–8 mm, cylindric, lobes 5, 1–2 mm, < tube, triangular; corolla 10–15 mm, narrowly funnel-shaped, violet or ± white, purple-veined, lobes < 1/3 tube; stamens gen incl, unequal, attached in tube lower 1/3.

FR: 6–8 mm, red. **SEED**: 40–60. Alkaline soils, flats; < 500 m. PR (e slope), s DSon; s AZ, nw Mex. Mar–Apr

L. pallidum Miers var. ***oligospermum*** C.L. Hitchc. (p. 1257) Very thorny, glabrous; branches many, spreading to ascending. **LF**: 10–50 mm, oblong to narrowly obovate, glabrous, glaucous. **FL**: calyx 5–8 mm, bell-shaped, lobes 5, ± = tube; corolla narrowly funnel-shaped, green-white, purple-tinged or -veined, tube 8–12 mm; stamens exserted, attached at tube middle. **FR**: 8–10 mm, green-purple, firm. **SEED**: 5–7. Flats, washes, slopes; < 1200 m. DMoj, n DSon; s NV. Mar–May

L. parishii A. Gray PARISH'S DESERT-THORN Glandular-hairy, intricately spreading-branched. **LF**: 5–30 mm, oblanceolate. **FL**: calyx 2.5–5 mm, bell-shaped, lobes 5, 2–4 mm, ± = tube, oblong-ovate; corolla narrowly funnel-shaped, purple, tube 2.5–6 mm, lobes ± 1 mm, ovate; stamens ± exserted, attached ± at tube middle. **FR**: 4–6 mm, red. **SEED**: 7–12. Sandy to rocky slopes, canyons; < 1000 m. e SCo, w DSon; AZ, nw Mex. Mar–Apr ★

L. torreyi A. Gray Pl ± glabrous (exc lf clusters hair-tufted); twigs slender, ± climbing. **LF**: 1–5 cm, narrowly oblanceolate to obovate. **FL**: calyx 2.5–4.5 mm, lobes 5, 1–1.5 mm; corolla 8–15 mm, narrowly funnel-shaped, green-lavender or ± white, lobes 3–4 mm, lanceolate to ovate; stamens ± exserted, attached at tube middle. **FR**: 6–10 mm, red or orange. **SEED**: 8–30. Washes, streambanks; < 700 m. s DMoj, DSon; to UT, TX, n Mex. Mar–May ★

L. verrucosum Eastw. SAN NICOLAS ISLAND DESERT-THORN Hairy; branches spreading; thorns thick. **LF**: 5–10 mm, narrowly obovate. **FL**: calyx 4 mm, bell-shaped, lobes ± 3–4, 2–3.5 mm, ± = tube, lanceolate; corolla 8–10 mm, lavender, tube cylindric, lobes narrowly obovate; stamens ± exserted, attached near tube top. **FR**: red. **SEED**: many. Habitat unknown; < 100 m. s ChI (San Nicolas Island). Known only from type specimen; perhaps merely a form of *L. brevipes*. Apr ★

LYCOPERSICON

Ann or per, sticky-glandular, aromatic. **LF**: ± odd-1–2-pinnate. **INFL**: raceme; pedicels jointed. **FL**: often nodding; parts gen 5; sepals free ± to base; corolla yellow, ± rotate or lobes reflexed; anthers > filaments, tube-like around style, tapered, opening by slits from tip (exc sterile part); stigma exserted from anthers. **FR**: berry, green to red, fleshy. **SEED**: coat gelatinous. ± 6 spp.: w S.Am, C.Am. (Greek: wolf peach, from supposed toxic properties) [Peralta et al. 2008 Syst Bot Monogr 84:1–186] Recently incl in *Solanum*.

1. Herbage green, hairs to 3.5 mm; lflets 3–10 cm, 10–45 mm wide; fr glabrous. . . . . ***L. esculentum***
1′ Herbage gray-green, hairs to 0.5 mm; lflets gen ± 3.5 cm, ± 10 mm wide; fr puberulent . . . . ***L. peruvianum***

L. esculentum Mill. TOMATO Erect or reclining, often fleshy, from taproot. **LF**: 10–20 cm. **INFL**: few-fld; peduncle 1.5–4 cm; bracts gen 0. **FR**: 3–12 cm wide, spheric, compressed, or pear-shaped, yellow-green to red. Disturbed areas, abandoned fields, roadsides; < 100 m. GV, SnFrB, SCo; native to S.Am. Summer–fall

L. peruvianum (L.) Mill. Spreading to reclining from thick taproot. **LF**: 5–12 cm. **INFL**: few- to many-fld; peduncle 3–15 cm; bracts lf-like, simple. **FR**: 1.5 cm wide, spheric, green. Disturbed sites; < 100 m. s SCo; native to S.Am. [*Solanum p.* L.] Summer–fall

NICOTIANA TOBACCO

Ann to small tree. **LF**: ± basal or not; margin entire, ± wavy or not. **INFL**: raceme or panicle, terminal; lower bracts lf-like. **FL**: calyx 5-lobed, ± enlarging, not fully enclosing fr; corolla gen radial, gen funnel-shaped to salverform; stamens 5, equal or 1 smaller. **FR**: septicidal capsule. **SEED**: many, minute, angled. ± 60 spp.: gen Am; *Nicotiana tabacum* L. widely cult, tobacco of commerce. (J. Nicot, said to have introduced tobacco to Eur, 1530–1600) Seriously TOXIC to livestock.

1. Shrub, small tree; herbage glabrous, glaucous . . . . ***N. glauca***
1′ Ann, per; herbage gen glandular-hairy
  2. Corolla 50–75 mm; cauline lvs gen elliptic; basal lf petioles winged. . . . . [***N. sylvestris***]
  2′ Corolla 15–50(65) mm; cauline lvs gen lanceolate; basal lf petioles winged or not
    3. Upper cauline lvs clasping; per, base often ± woody; corolla open during day. . . . . ***N. obtusifolia***
    3′ Upper cauline lvs not clasping; ann; corolla gen ± closed during day
      4. Cauline lvs gen ± sessile (exc lowest)
        5. Corolla tube 15–20 mm, limb 8–10 mm wide; stamens attached ± at 1 level below tube middle; calyx
            lobes very unequal . . . . ***N. clevelandii***
        5′ Corolla tube 25–50 mm, limb 20–50 mm wide; stamens attached at various levels gen above tube
            middle; calyx lobes ± unequal . . . . ***N. quadrivalvis***
      4′ Cauline lvs petioled
        6. Calyx dark-striped, lobes narrowly lanceolate, unequal, gen ± ≥ tube, not pock-marked; pl densely
            glandular-hairy. . . . . ***N. acuminata*** var. ***multiflora***
        6′ Calyx not striped, lobes triangular, ± equal, < tube, pock-marked; pl glabrous to ± sparsely glandular-
            hairy . . . . ***N. attenuata***

N. acuminata (Graham) Hook. var. *multiflora* (Phil.) Reiche Ann 5–15 dm, densely glandular-hairy. **LF**: 5–25 cm, petioled; basal ovate; cauline reduced. **INFL**: bracts < 50 mm, linear. **FL**: calyx 15–20 mm, dark-striped, lobes ± ≥ tube, unequal, narrowly lanceolate, not pock-marked; corolla ± salverform, green-white, tube 30–40 mm, limb 10–20 mm wide; stamens unequal, attached near tube middle. **FR**: 10–12 mm. Open, sandy or gravelly areas; < 1600 m. CA-FP, MP; to WA, NV; native to S.Am. May–Oct

N. attenuata S. Watson (p. 1257) Ann 5–15 dm, ± glandular-hairy. **LF**: 3–10 cm, petioled; basal elliptic to ovate; cauline reduced.

INFL: bracts < 30 mm, linear. **FL**: calyx 6–10 mm, lobes < tube, ± equal, triangular, acute, pock-marked; corolla ± salverform, tube 20–27 mm, ± green, pink-tinged, limb 4–8 mm wide, white; stamens unequal, attached below tube middle. **FR**: 8–12 mm. Open, well-drained slopes; 200–2800 m. CA (exc coast); to BC, MT, NM, nw Mex. May–Oct

N. clevelandii A. Gray (p. 1257) Ann 2–6 dm, slender, ± glandular-hairy. **LF**: 3–18 cm; lowermost petioled, elliptic to ovate; upper sessile, lanceolate. **INFL**: bracts < 30 mm, linear to lanceolate. **FL**: calyx 8–10 mm, lobes 4, very unequal (1 > tube), linear; corolla ± sal-

verform, green-white, tube 15–20 mm, limb 8–10 mm wide; stamens unequal, attached ± at 1 level below tube middle. **FR:** 4–6 mm. Gen sandy washes, scrub; < 500 m. SCo, ChI (Santa Catalina, Santa Cruz islands), DSon; AZ, nw Mex. Mar–Jun

N. glauca Graham TREE TOBACCO Shrub, small tree, glabrous, glaucous; wood soft. **LF:** 5–21 cm, petioled, gen ± ovate. **INFL:** bracts < 5 mm, linear. **FL:** calyx ± 10 mm, lobes < tube, ± unequal, triangular; corolla 30–35 mm, ± cylindric, yellow; stamens ± equal, attached below tube middle. **FR:** 7–15 mm. Open, disturbed flats or slopes; < 1100 m. NCoRI, c&s SNF, GV, CW, SW, D; to s US, Mex; Afr, Medit; native to S.Am. Apr–Aug ❖

N. obtusifolia M. Martens & Galeotti (p. 1257) Per 2–8 dm, glandular-hairy, base often ± woody. **LF:** 2–10 cm; lower short-petioled, (ob)ovate; upper ± narrowly ovate, clasping. **INFL:** bracts < 20 mm, linear to lanceolate. **FL:** calyx 10–15 mm, lobes ± = tube, ±

equal, narrowly triangular; corolla ± funnel-shaped, green-white or dull white, tube + throat 15–26 mm, limb 8–10 mm wide; stamens unequal, attached near tube base. **FR:** 8–10 mm. Gravelly or rocky washes, slopes; < 1600 m. s SNE, D; to UT, TX, Mex. Mar–Jun

N. quadrivalvis Pursh Ann 3–20 dm, glandular-hairy. **LF:** 4–15 cm; basal, lower cauline short-petioled, elliptic to ovate; upper sessile, reduced. **INFL:** bracts < 35 mm, ± linear. **FL:** calyx 15–20 mm, 10-ridged, lobes narrowly lanceolate, 1 > tube or not; corolla ± salverform, white, green- or violet-tinged, tube 25–50 mm, limb 20–50 mm wide; stamens unequal, attached at various levels above tube middle. **FR:** 15–20 mm. Open, well-drained washes, slopes; < 1500 m. CA-FP, D; to WA, c US. [*N. bigelovii* (Torr.) S. Watson; *N. b.* (Torr.) S. Watson var. *wallacei* A. Gray; *N. q.* var. *b.* (Torr.) DeWolf] Widely cult by w Am native people; cult pls sometimes recognized, with *N. quadrivalvis* var. *bigelovii* (Torr.) DeWolf for wild pls. May–Oct

ORYCTES

1 sp. (Greek: digger)

O. nevadensis S. Watson (p. 1257) NEVADA ORYCTES Ann 5–20 cm, branched from slender taproot, lfy, sticky; some hairs scale-like. **LF:** 1–3 cm, linear to ovate, entire to shallowly lobed, ± wavy; petiole 5–10 mm, narrowly winged. **INFL:** umbels in axils, few-fld. **FL:** calyx 2–3 mm, < 10 mm in fr, lobes < tube; corolla 5–8 mm, narrowly

urn-shaped, ± purple, lobes small; stamens ± unequal, attached at tube base, some ± exserted; style ± ≤ stamens. **FR:** capsule, 2-valved, 6–7 mm wide, spheric. **SEED:** 10–15, round, flat; body 2 mm wide; wing 0.5 mm wide. Sandy soils, dunes; 1200–1500 m. s SNE (Inyo Co.); w NV. Seriously threatened by grazing. May ★

PETUNIA

Ann, per, sticky-glandular. **ST:** main branches from base, with ± long internodes. **LF:** ± opposite near fls, entire. **INFL:** fls 1 in axils of lvs or lf-like bracts. **FL:** calyx divided ± to base, lobes ± lf-like, esp in fr; corolla funnel-shaped, 5-lobed; stamens ± equal or 1 short, 2 medium, 2 long. **FR:** capsule. **SEED:** many, minute, angled. ± 30 spp.: gen S.Am; some cult for orn, sometimes waifs. (Petun, Native American word for tobacco) *P. axillaris* (Lam.) Britton et al., *P. violacea* Lindl. may be naturalized in CA.

P. parviflora Juss. (p. 1257) **ST:** prostrate to decumbent, < 4 dm, rooting at nodes; axillary branches short, lfy. **LF:** 5–14 mm, ± oblanceolate, fleshy, ± sessile. **FL:** calyx lobes 3–6 mm, < 11 mm in fr, elliptic to narrowly obovate; corolla 4–6 mm, ± purple, tube ± white. **FR:** 2–3 mm. **SEED:** pale brown. Open washes, dry streambeds; < 1300 m. CW, SCo, n ChI (Santa Rosa Island), PR, DSon; to

sw UT, se US, Mex; also S.Am; waif elsewhere in US. [*Calibrachoa p.* (Juss.) D'Arcy] Widely treated in *Calibrachoa*, which is the sister group of *Petunia* [Kulcheski et al. 2006 Genetica 126:3–14]; treated in *Calibrachoa* also by Fregonezi et al. [Taxon 61:120–130. 2012], who regarded it as probably native to S.Am, naturalized in N.Am, Eur. Apr–Aug

PHYSALIS GROUND-CHERRY

Ann from taproot or per from rhizome; hairs branched or not, glandular or not. **LF:** ± opposite or not, entire to pinnate-lobed. **INFL:** fls 1–few per axil, pedicelled. **FL:** gen nodding; calyx 5-lobed, enlarged and persistent, bladder-like in fr; corolla ± rotate to shallowly bell-shaped, gen ± yellow, often dark-spotted adaxially; stamens 5, attached to hairy band in tube, anthers free, gen < filaments, opening by slits; style gen straight. **FR:** berry, fleshy [dry]. **SEED:** many, 2–2.5 mm, ± spheric to reniform. ± 85 spp.: Am, Eurasia, Afr, Australia. (Greek: bladder, from calyx in fr) Unripe fr often TOXIC. Needs study in w US. Some spp. cult for edible or orn fr.

1. Fl ± erect; corolla purple; style curved . . . . ***P. lobata***
1′ Fl gen nodding or otherwise not ± erect; corolla ± yellow to pale yellow; style straight
  2. Lf hairs branched; per
    3. Corolla with 5 purple-brown spots at base adaxially or not . . . . ***P. hederifolia*** var. ***fendleri***
    3′ Corolla gen with 0 spots but with darker green veins adaxially . . . . [***P. viscosa***]
  2′ Lf hairs 0 or simple; ann, per, subshrub
    4. Per or subshrub from rhizome
      5. Anthers gen with some blue or purple . . . . [***P. longifolia***]
      5′ Anthers yellow
        6. Pedicel > fl, in fr > calyx . . . . ***P. crassifolia***
        6′ Pedicel < fl, in fr < calyx . . . . ***P. hederifolia*** var. ***palmeri***
    4′ Ann from taproot
      7. Pl ± densely hairy . . . . ***P. pubescens***
        8. Lf 5–9 cm, ± thick, teeth many . . . . var. ***grisea***
        8′ Lf 3–5 cm, thin, teeth 0–few . . . . var. ***integrifolia***
      7′ Pl glabrous to sparsely hairy
        9. Corolla with 5 dark-purple spots adaxially; anthers twisted after opening . . . . ***P. philadelphica***
        9′ Corolla without spots adaxially; anthers not clearly twisted after opening
          10. Corolla pale yellow with dark yellow center, 15–23 mm wide; lf teeth < 7 mm; anthers 3 mm . . . . . ***P. acutifolia***
          10′ Corolla yellow, 7–8 mm wide; lf teeth 0 or < 3 mm; anthers 1–2 mm . . . . ***P. lancifolia***

P. acutifolia (Miers) Sandwith (p. 1257) Ann 2–10 dm, branched; hairs simple, short, appressed. **LF:** 4–12 cm, lanceolate to ± ovate, tapered to base, teeth < 7 mm, prominent, slender. **INFL:** pedicel 15–25 mm, in fr < 40 mm. **FL:** calyx 3–4.5 mm, in fr 20–25 mm, spheric, with 10 ± equal veins; corolla 15–23 mm wide, rotate, pale yellow with dark yellow center; anthers 3 mm, each yellow and blue-green. Disturbed places, roadsides; < 200 m. s SnJV, SCo, DSon; to TX, n Mex. Jul–Oct

P. crassifolia Benth. (p. 1257) Per, subshrub, < 8 dm; hairs simple, dense, short, gen glandular. **ST:** often zigzagged, ridged. **LF:** 1–3 cm, gen ovate, fleshy, entire or ± wavy; petiole ± = blade. **INFL:** pedicel 15–30 mm, in fr > calyx. **FL:** calyx 4–7 mm, in fr 20–25 mm, weak-angled; corolla 15–20 mm diam, widely bell-shaped, yellow; anthers 2–3 mm, yellow. Gravelly to rocky flats, washes, slopes; < 1300 m. PR, s SNE, D; NV, AZ, n Mex. Mar–May

P. hederifolia A. Gray Per 1–8 dm, rhizome fleshy. **LF:** 2–4 cm, ovate, entire to coarsely toothed, gen gray-green; base tapered or ± cordate. **INFL:** pedicel 3–5(10) mm, in fr < 15 mm. **FL:** calyx 6–7 mm, in fr 20–30 mm, ± spheric, with 10 green veins; corolla 10–15 mm, widely bell-shaped, yellow, with 5 purple-brown spots at base adaxially or not; anthers yellow.

var. ***fendleri*** (A. Gray) Cronquist Hairs gen branched, nonglandular. *n*=12. Gravelly to rocky slopes; 900–1800 m. s SNE (incl Inyo Mtns), DMoj; to TX, n Mex. May–Jul

var. ***palmeri*** (A. Gray) C.L. Hitchc. Hairs simple, many glandular. *n*=12. Gravelly to rocky slopes; 700–1600 m. s PR, DMtns; to UT, AZ, Baja CA. May–Jul

P. lancifolia Nees Ann < 8 dm; hairs simple, few, minute, appressed, nonglandular. **LF:** 3–13 cm, lanceolate, tapered to base, teeth 0 or < 3 mm. **INFL:** pedicel 10–25 mm, in fr < 40 mm. **FL:** calyx 2–3 mm, in fr 20–25 mm, 10-veined; corolla 7–8 mm wide, widely bell-shaped, yellow; anthers 1–2 mm, blue-tinged. Wet places,

fields, disturbed places; < 200 m. GV, SnFrB, SCoRI, DSon; to TX, c Mex; native to S.Am. [*P. angulata* L. var. *l.* (Nees) Waterf.] Jun–Sep

P. lobata Torr. (p. 1257) LOBED GROUND-CHERRY Per < 5 dm, decumbent to spreading, few-branched, glabrous to minute-papillate. **LF:** 1–7 cm, lanceolate to ovate, entire to lobed, tapered to base. **INFL:** pedicel 3–4.5 mm, gen not longer in fr. **FL:** ± erect; calyx 3–4.5 mm, in fr 15–20 mm; corolla 15–20 mm wide, rotate, purple, tube white inside; style curved; anthers 1–2.5 mm, ± yellow. *n*=11,22. Granitic soils, dry lake margins; 500–800 m. se DMoj, ne DSon; to KS, TX, n Mex. Sep–Jan ★

P. philadelphica Lam. TOMATILLO Ann < 10 dm; hairs 0 or sparse, simple, gen glandular. **LF:** gen 4–8 cm, ± ovate, entire to toothed. **INFL:** pedicel 3–8 mm, < calyx, in fr not longer. **FL:** calyx 4.5–6 mm, in fr 20–30+ mm, green with 10 purple veins; corolla 8–15 mm wide, ± rotate, yellow with 5 dark purple spots adaxially, tube hairy inside; anthers 2.5–3 mm, twisted after opening, each yellow and blue-green or 1-colored. *n*=12. Disturbed places, cult fields, roadsides; < 700 m. SNF, GV, CW, SCo, PR; to e US; native to Mex. Widely cult for food. May–Sep

P. pubescens L. Ann < 8 dm; hairs simple, ± dense, spreading, most with small glands. **LF:** 3–9 cm, widely ovate to ± cordate, entire to coarsely toothed. **INFL:** pedicel 3–12 mm, in fr ± longer. **FL:** calyx 5 mm, in fr 20–40 mm, sharp-5-angled with ribs between; corolla ± 10 mm wide, widely bell-shaped, yellow, tube with 5 dark spots adaxially; anthers 1.5–2 mm, blue.

var. ***grisea*** Waterf. **LF:** 5–9 cm, ± thick, often drying red-brown; teeth many. *n*=12. Disturbed places, cult fields; < 1500 m. s SNE, D; native to c&e US. Often cult. Aug–Sep

var. ***integrifolia*** (Dunal) Waterf. **LF:** 3–5 cm, thin, drying ± translucent; teeth 0–few. *n*=12. Disturbed places, cult fields; < 1500 m. s NCoRO, SnJV, expected elsewhere; native to e US. Jul–Sep

SALPICHROA

Per from rhizome. **LF:** fleshy, entire. **INFL:** fls 1 in axils. **FL:** calyx deeply lobed; corolla lobes < tube; stamens ± incl, anthers converging toward style, oblong; style 1, thread-like, stigma head-like. **FR:** berry, fleshy. **SEED:** many, compressed. 15 spp.: S.Am, esp Andes. (Greek: trumpet color)

S. origanifolia (Lam.) Baill. Pl < 2 m, decumbent to climbing, hairy. **LF:** 10–35 mm; petiole 5–10 mm; blade widely elliptic to ovate, ± glabrous. **INFL:** pedicel 3–14 mm. **FL:** calyx lobes 2–3 mm; corolla urn-shaped, white or ± green, with a densely woolly ring adaxially, lobes 2–3 mm, triangular; anthers 1.5–2 mm, ± exserted; disk below ovary, brick-red. **FR:** 10–15 mm, ovoid, white or pale yellow, ill-smelling. *n*=12. Disturbed places; < 100 m. NCo, w CW, SCo; to TX, n Mex; Eur; native to S.Am. Jul–Oct

SOLANUM NIGHTSHADE

Ann to shrub, vine or not, prickly or not, often glandular. **LF:** alternate to ± opposite, often unequal, entire to deeply pinnately lobed. **INFL:** panicle or umbel-like, often 1-sided. **FL:** calyx ± bell-shaped; corolla ± rotate, white to purple (yellow), lobes gen of different color toward base of midrib, gen without spots at base; anthers free, > filaments, oblong or tapered, opening by 2 pores or short slits near tip, in age often splitting to base; ovary 2-chambered, style 1, gen ± straight, stigma head-like. **FR:** berry, gen spheric, fleshy (or dry, capsule-like). **SEED:** many, compressed, gen reniform. ± 1500 spp.: worldwide, esp trop Am. (Latin: quieting, from narcotic properties) many TOXIC. Many cult for food (incl potato, *S. tuberosum*), orn; *S. dimidiatum* ◆ in CA an urban weed. See *Lycopersicon* for other taxa recently incl here.

1. Hairs (some or all) stellate; anthers tapered, opening by terminal pores; pl often prickly
  2. Stellate hairs scale-like, rays fused at base . . . . ***S. elaeagnifolium***
  2′ Stellate hairs not scale-like, rays free
   3. Anthers unequal; calyx tightly enclosing fr . . . . ***S. rostratum***
   3′ Anthers gen ± equal; calyx not tightly enclosing fr
    4. Lf abaxially densely white-tomentose or velvety
     5. Calyx not prickly; infl branches 2–4; frs several to many per infl, 7–15 mm diam . . . . ***S. lanceolatum***
     5′ Calyx prickly; infl branches 0; frs 1 per infl, 35–50 mm diam . . . . ***S. marginatum***
    4′ Lf abaxially sparsely tomentose, neither clearly white nor velvety
     6. Lf deeply pinnately lobed, lobes entire to lobed ± to midrib; ann; fr red . . . . ***S. sisymbriifolium***
     6′ Lf ± entire to wavy-lobed, teeth or lobes ± entire; per from rhizome; fr yellow . . . . ***S. carolinense***
1′ Hairs branched or not but not stellate; anthers oblong, opening by terminal pores that become short slits; pl not prickly

7. Lf deeply lobed; ann . . . . *S. triflorum*
7′ Lf entire to shallowly lobed, if deeply 1–2-lobed below middle, shrub
  8. Corolla lobes < tube + throat, gen spreading
    9. Shrub 2–4 m; lf ± entire to deeply 1–2-lobed below middle; fr orange-red . . . . *S. aviculare*
    9′ Per to subshrub gen < 1 m; lf entire to 1–2-lobed at base; fr green to ± purple
      10. Upper st hairs branched . . . . *S. umbelliferum*
      10′ Upper st hairs 0 or gen simple
        11. St ± glabrous, ridged; lf sessile or gradually tapered to acute base . . . . *S. parishii*
        11′ St gen soft- or glandular-hairy, not or weak-ridged; lf petioled, blade base rounded or obtuse to ± cordate
          12. Lf gen 6–10 cm, oblong to ovate; fr gen 15–25 mm; ChI . . . . *S. wallacei*
          12′ Lf 2–7 cm, lanceolate to ovate; fr 10–15 mm; mainland . . . . *S. xanti*
  8′ Corolla lobes ≥ tube + throat, often reflexed
    13. Infl of gen 2 umbel-like clusters, axis gen forked
      14. Corolla purple to violet, lobes each with 2 green spots at base; pl ± woody, ± climbing . . . . *S. dulcamara*
      14′ Corolla white to pale violet, lobes not spotted; pl not woody, gen ± reclining . . . . *S. furcatum*
    13′ Infl umbel- or raceme-like, axis not forked
      15. Calyx enlarged, enclosing fr base; fr ± yellow or green . . . . *S. physalifolium* var. *nitidibaccatum*
      15′ Calyx not enlarged, not enclosing fr base; fr black or ± green
        16. Anthers gen 2.5–4 mm . . . . *S. douglasii*
        16′ Anthers 1.4–2.2 mm
          17. Seeds 1–1.5 mm; hairs ± appressed or curved, nonglandular . . . . *S. americanum*
          17′ Seeds 1.5–2.5 mm; hairs ± spreading or curved, glandular and not . . . . *S. nigrum*

S. americanum Mill. (p. 1257)   Ann to subshrub, 3–8 dm, hairs ± 0 or short, ± appressed or curved, nonglandular. **LF:** 2–15 cm, ovate, entire to coarsely wavy-toothed. **INFL:** umbel- or raceme-like. **FL:** calyx 1–2 mm, lobes in fr recurved; corolla 3–6 mm wide, deeply lobed, white; anthers 1.4–2.2 mm; style 2.5–4 mm. **FR:** 5–8 mm, black or ± green. **SEED:** 1–1.5 mm. *n*=12. Open, often disturbed places; < 1000 m. CA-FP, DMoj (uncommon); to Can, e US, Mex. Much like *S. nigrum*. Apr–Nov

S. aviculare G. Forst.   Shrub 2–4 m, gen glabrous. **LF:** 10–30 cm, ± entire to deeply 1–2-lobed below middle. **INFL:** gen raceme-like. **FL:** calyx 3–4 mm; corolla 30–40 mm wide, blue-violet, lobes < tube + throat; anthers ± 4 mm; style 7–10 mm. **FR:** 10–15 mm, orange-red. **SEED:** ± 1.5 mm. Uncommon. Open, gen disturbed places; < 100 m. NCo, n CCo; native to Australasia. *S. mauritianum* Scop. (herbage densely hairy; corolla 11–15 mm wide) reportedly naturalized in CA. Jun–Oct

S. carolinense L. (p. 1261)   CAROLINA HORSE-NETTLE   Per 2–9 dm, from rhizome, little-branched; prickles slender, spreading; stellate hairs fine, gen sessile. **LF:** 5–15 cm, ovate, ± entire to wavy-lobed, teeth or lobes entire. **INFL:** gen raceme-like. **FL:** calyx 5–6 mm, lobes ± = tube; corolla 20–30 mm wide, gen pale violet (white); anthers 7–9 mm, ± equal. **FR:** 8–20 mm, yellow. **SEED:** ± 2.5 mm. *n*=12. Uncommon. Disturbed places, fields, roadsides; < 200 m. s NCoRO, n SN, GV, SCo; native to c&e US, n Mex. May–Aug ◆

S. douglasii Dunal   Per to subshrub < 20 dm, much-branched; hairs gen simple, < 1 mm, ± curved, white. **LF:** 1–9 cm, ovate, entire to coarsely irregularly toothed. **INFL:** gen umbel-like. **FL:** calyx ± 2 mm; corolla ± 10 mm wide, white, lavender, or lavender-tinged, lobes deep; anthers gen 2.5–4 mm; style 4–5 mm, puberulent below. **FR:** 6–9 mm, black. **SEED:** 1.2–1.5 mm. *n*=12. Dry scrub, woodland; < 1000 m. s NCo, Teh, CW, SW, DMoj; n Mex. ± all yr

S. dulcamara L.   Per to subshrub gen < 10 dm, ± woody, ± climbing, hairs ± 0 or soft. **LF:** 5–12 cm, ± cordate to deeply 1–2-lobed near base. **INFL:** axis forked; clusters gen 2, umbel-like. **FL:** calyx 3–4 mm; corolla 8–12 mm wide, deeply lobed, purple to violet, lobes each with 2 green spots at base. **FR:** 8–12 mm, ± ovoid, red. **SEED:** 2–3 mm. *n*=12. Moist disturbed places, marshes; < 1000 m. CCo, SnFrB, MP; to Can, e US; native to n Eurasia. Jun–Sep

S. elaeagnifolium Cav. (p. 1261)   WHITE HORSE-NETTLE   Per < 10 dm, from rhizome, forming colonies, gen prickly, stellate hairs dense, scale-like, rays fused at base. **LF:** 2–15 cm, oblong, entire to ± lobed, ± yellow. **INFL:** raceme-like. **FL:** calyx 5–8 mm, lobes ± = tube; corolla 20–30 mm wide, purple or blue; anthers 8–10 mm, ± equal. **FR:** 8–15 mm, orange, often persistent. **SEED:** 2.5–3 mm.

n=12. Common. Dry, disturbed places, fields; < 1200 m. CA (exc NCo, KR, GB); to S.Am; native to c US, n Mex. May–Sep ◆

S. furcatum Dunal   Per 5–12 dm, gen ± reclining; hairs simple, sparse, ± curved. **LF:** 3–7 cm, ovate, ± entire or irregularly few-toothed. **INFL:** axis forked; clusters 2, umbel-like. **FL:** calyx 2.5–3 mm; corolla 10–18 mm wide, white to pale violet; anthers 3–3.5 mm. **FR:** 5–6 mm, dark purple. **SEED:** ± 2 mm. *n*=36. Open, often disturbed places; < 200 m. NCo, CCo, SnFrB; OR, Baja CA; native to S.Am. May–Oct

S. lanceolatum Cav.   LANCELEAF NIGHTSHADE   Shrub 8–50 dm, sparsely prickly, hairs stellate, some glandular. **LF:** 3–15 cm, gen lanceolate, entire to lobed. **INFL:** branches 2–4; peduncle 1–3 cm. **FL:** calyx 5–8 mm, lobes ± = tube, tips slender, dark; corolla 25–30 mm wide, blue-purple; anthers 6–8 mm, gen ± equal. **FR:** 7–15 mm, yellow-orange. Disturbed places; < 200 m. GV, SnFrB, SCo; native to Mex, C.Am. Apr–Aug ◆

S. marginatum L. f.   WHITE-MARGINED NIGHTSHADE   Shrub < 2 m, prickly, hairs dense, stellate. **LF:** 10–18 cm, widely ovate, wavy-lobed, abaxially densely white-tomentose. **INFL:** raceme-like; lowermost fl bisexual, reflexed, others staminate. **FL:** calyx ± 10 mm, prickly; corolla 15–30 mm wide, white; anthers 6–7 mm. **FR:** 35–50 mm, glabrous, tough, yellow. **SEED:** ± 3 mm. *n*=12. Disturbed places; gen < 1000 m. CCo, SCo; native to Afr. May–Aug ◆

S. nigrum L.   BLACK NIGHTSHADE   Ann to subshrub, 3–8 dm; hairs ± spreading or curved, glandular and not. **LF:** 4–7 cm, ovate, entire to coarsely wavy-toothed. **INFL:** raceme-like. **FL:** calyx 2–3 mm; corolla 10 mm wide, white, lobes deep; anthers 1.4–2.2 mm; style 3–5 mm. **FR:** 6–8 mm, black. **SEED:** 1.5–2.5 mm. *n*=36. Disturbed places; < 200 m. NCo, n CCo, SnFrB, expected elsewhere; to WA, e US; native to Eurasia. Much like *S. americanum*. Mar–Oct

S. parishii A. Heller (p. 1261)   Per to subshrub, < 10 dm, much-branched, hairs 0 to sparse. **ST:** angled to ribbed. **LF:** 2–7 cm, lanceolate to elliptic, entire, wavy or not, tapered to base. **INFL:** ± umbel-like; pedicels 13–18 mm, > peduncle. **FL:** calyx 4 mm; corolla 17–22 mm wide, gen blue-purple (white), lobes << tube, each with 2 green spots at base; anthers 3.5–5 mm. **FR:** 7–10 mm. **SEED:** ± 2 mm. Dry chaparral, oak/pine woodland, pine forest; < 2000 m. NW, CaR, SN, ScV (Sutter Buttes), CW, SW, MP; s OR, Baja CA. May hybridize with *S. xanti*; needs study. Apr–Jul

S. physalifolium Rusby var. *nitidibaccatum* (Bitter) Edmonds   Ann 1–9 dm, decumbent, sticky; hairs spreading, some glandular. **LF:** 2–6 cm, ovate, entire to irregular-toothed or shallowly lobed. **INFL:** gen raceme-like, few-fld. **FL:** calyx 2–2.5 mm, in fr 4–6 mm, enclosing fr base; corolla 3–5 mm wide, white, lobes > tube; anthers 1.5–2 mm.

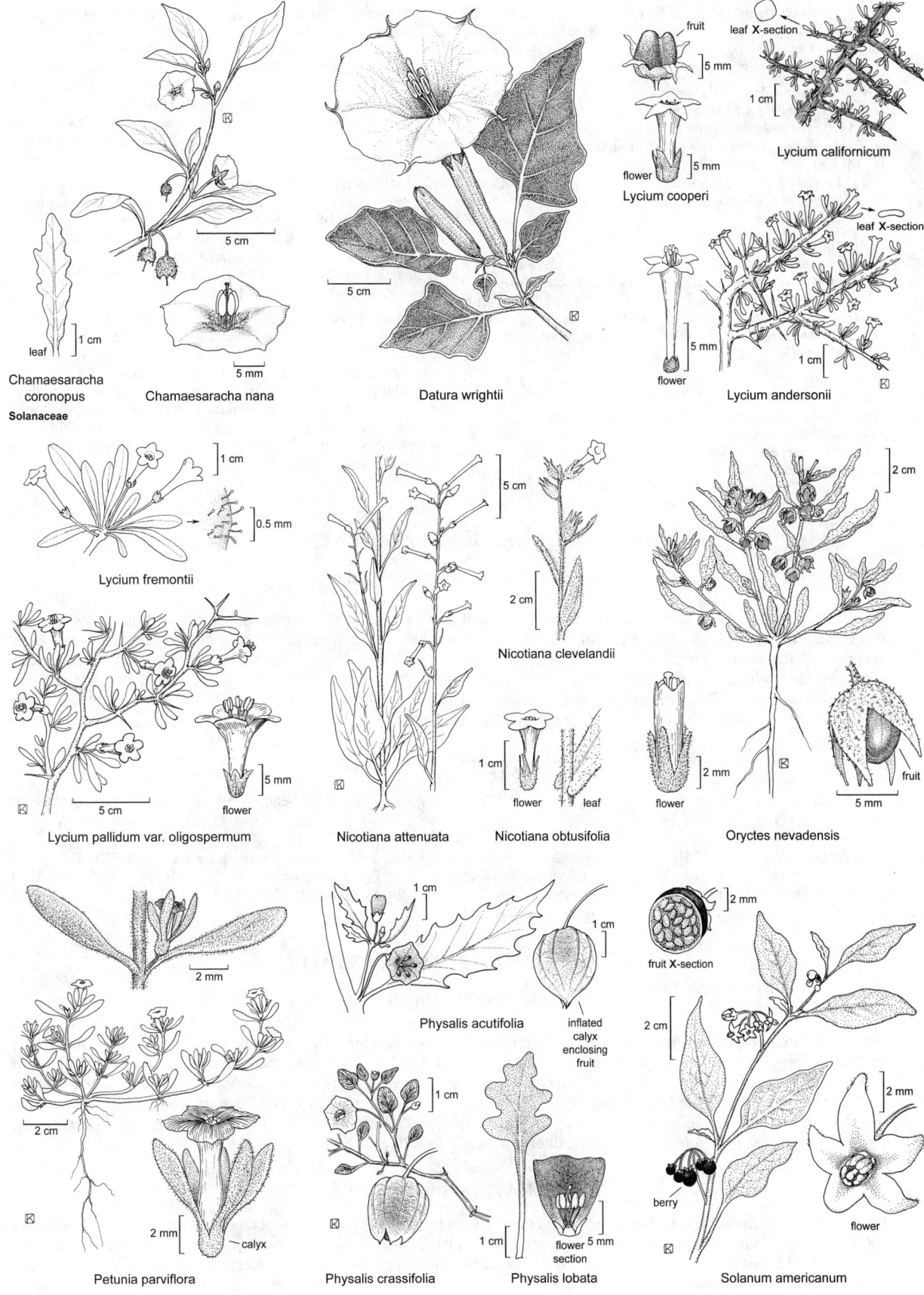

Chamaesaracha
coronopus

Solanaceae

Chamaesaracha nana

Datura wrightii

Lycium cooperi

Lycium californicum

Lycium andersonii

Lycium fremontii

Nicotiana clevelandii

Nicotiana attenuata

Nicotiana obtusifolia

Oryctes nevadensis

Lycium pallidum var. oligospermum

Petunia parviflora

Physalis acutifolia

Physalis crassifolia

Physalis lobata

Solanum americanum

FR: 6–7 mm, ± yellow or green. **SEED:** ± 2 mm, yellow. *n*=12. Disturbed areas; gen < 1000 m. NCo, KR, CaRF, GV, w CW, SCo, n ChI (Santa Cruz Island), SNE; to Can, e US, Mex; native to S.Am. [*S. sarrachoides* Sendtn., misappl.] May–Oct

S. rostratum Dunal (p. 1261) BUFFALOBUR Ann 1–7 dm, densely prickly, hairs stellate. **LF:** 5–15 cm, lobes ± to midrib. **INFL:** raceme-like. **FL:** calyx 6–7 mm, tightly enclosing fr, prickly; corolla 23–35 mm wide, yellow; anthers unequal (4 straight, 6–8 mm; 1 curved, 10–14 mm); style curved. **FR:** 9–12 mm, ± dry. **SEED:** 2–2.5 mm. *n*=12. Disturbed places, roadsides, fields; < 1400 m. s NCoRO, CaRF, GV, CCo, s SCoRO, SCo, MP; to e US, Mex; native to Great Plains. May–Sep ❖

S. sisymbriifolium Lam. Ann 5–15 dm, ± sticky, prickles gen yellow esp distally, hairs stellate. **LF:** 4–15 cm, deeply pinnately lobed, lobes entire to lobed ± to midrib. **INFL:** raceme-like. **FL:** calyx 7–9 mm, ± enlarged in fr; corolla 25–30 mm wide, pale blue to white; anthers 10 mm. **FR:** 10–20 mm, red. **SEED:** ± 2 mm, orange-yellow. Disturbed places, roadsides; < 100 m. GV, expected elsewhere; native to S.Am. Feb–Oct

S. triflorum Nutt. Ann 1–5 dm, decumbent; hairs ± curved or spreading, glandular or not. **LF:** 2–5 cm, oblong to ovate, deeply lobed. **INFL:** umbel-like; fls 2–3. **FL:** calyx 2.5–3 mm, enlarged in fr; corolla 7–9 mm wide, white, lobes ≥ tube; anthers ± 3 mm. **FR:** 8–12 mm, green. **SEED:** many, 2.5–3 mm, yellow. *n*=12. Dry scrub, juniper woodland; 100–2300 m. n SNH, s SNH (e slope), SCo, GB, n DMoj; to c US; native to S.Am. Jun–Sep

S. umbelliferum Eschsch. (p. 1261) Per to subshrub, gen < 10 dm, much-branched; hairs gen dense, branched. **LF:** 1–4 cm, ± elliptic to (ob)ovate, gen entire. **INFL:** ± umbel-like, forked or not; pedicels 12–15 mm, > peduncle. **FL:** calyx 3–3.5 mm; corolla 16–25 mm wide, lavender to blue-purple, lobes < tube + throat, each with 2 ± green spots at base; anthers 3–5 mm. **FR:** 12–14 mm. **SEED:** ± 2 mm. Shrubland, mixed-evergreen forest, woodland; < 1600 m. NW (uncommon), deltaic GV, CW, SW; AZ, Baja CA. May hybridize with *S. parishii*, *S. xanti*; needs study. All yr

S. wallacei (A. Gray) Parish (p. 1261) WALLACE'S NIGHTSHADE Per to subshrub, < 10 dm, much-branched, gen densely soft-hairy, often glandular. **LF:** gen 6–10 cm, oblong to ovate, entire to ± wavy; base rounded to ± cordate. **INFL:** branches gen 2. **FL:** calyx 4.5–6 mm; corolla 30–40 mm wide; anthers 4.5–5.5 mm. **FR:** gen 15–25 mm, yellow. **SEED:** 1.5–2 mm. Canyons, chaparral; < 800 m. ChI; Baja CA (Guadalupe Island). Mar–Aug ★

S. xanti A. Gray (p. 1261) Per to subshrub, 4–9 dm, much-branched, ± hairy, glandular or not. **LF:** 2–7 cm, lanceolate to ovate, ± entire to 1–2-lobed at base; base obtuse to ± cordate. **INFL:** umbel-like, branched or not. **FL:** calyx 4–5 mm; corolla 15–30 mm wide, dark blue or lavender, lobes < tube, with 2 green spots at base; anthers 4–5.5 mm. **FR:** 10–15 mm, ± green. **SEED:** 1.5–2 mm. Shrubland, oak/pine woodland, conifer forest; < 2700 m. CA-FP (exc CaR, GV), n DMtns; NV, Baja CA. Variation complex; may hybridize with *S. parishii*, *S. umbelliferum*. Feb–Jun

STAPHYLEACEAE BLADDERNUT FAMILY

Robert E. Preston & James R. Shevock

[Per], shrub, tree; [some monoecious or dioecious]. **LF:** gen opposite, pinnately compound, gen stipuled; lflets gen toothed. **INFL:** panicle [raceme], drooping. **FL:** radial; sepals 5, free or fused at base, often petal-like; petals 5, free; stamens 5, alternate petals, often attached to disk; ovary superior, chambers 2–4, each with 1–12 ovules in 2 rows. **FR:** gen inflated capsule with open top [follicles, drupe, berry]. **SEED:** gen 1–2 per chamber. 5 genera, 50–60 spp.: n temp, Asia, C.Am, S.Am; some cult for showy fr (*Staphylea*), timber (*Turpinia*). [Matthews & Endress 2005 Bot J Linn Soc 147:1–46] Most closely related to Crossosomataceae, Stachyuraceae. Scientific Editor: Thomas J. Rosatti.

STAPHYLEA BLADDERNUT

LF: deciduous; lflets gen 3. **FL:** petals white or pink. **FR:** bladdery, deeply 3-lobed, pendent. **SEED:** spheric, light brown. 10 spp.: n temp. (Greek: cluster, from infl)

S. bolanderi A. Gray (p. 1261) SIERRA BLADDERNUT Shrub, small tree, 2–6 m, glabrous. **LF:** lflets 2.5–6 cm, widely ovate to round, finely serrate. **INFL:** gen appearing before or with lvs. **FL:** sepals 8–10 mm, white; petals 10–12 mm, white; stamens well exserted. **FR:** 2.5–5 cm, prominently horned. **SEED:** 5–7 mm, ± obovoid, light brown, smooth. Wooded or shrubby slopes; 240–1720 m. e KR, CaR, c&s SN. Mar–May

STYRACACEAE STORAX FAMILY

Peter W. Fritsch

Shrub, tree, stellate-[peltate-]hairy. **LF:** simple, alternate, deciduous [evergreen]; stipules 0. **INFL:** panicle, raceme, cyme, or fls 1. **FL:** gen bisexual, radial, gen showy, fragrant; sepals fused, calyx [(0)]5(9)-toothed, persistent; petals (4)5(10), fused below; stamens gen 2 × petal number, fused to corolla tube, in 1 series, anthers oblong, 2-chambered, longitudinally dehiscent, yellow; ovary ± superior [to inferior], septa [2]3[5], style 1, thread-like, stigma dot-like or minutely lobed. **FR:** capsule [indehiscent (nut-like, winged or not) or drupe]. **SEED:** 1(3)[4–50+]. 11 genera, ± 160 spp.: US to S.Am; also Medit, e Asia; some cult as orn (*Styrax*, *Halesia*). Scientific Editor: Thomas J. Rosatti.

STYRAX SNOWDROP BUSH

Buds superposed, naked. **INFL:** gen raceme, drooping. **FL:** calyx minutely [(0)]5(9)-toothed; corolla lobes 5(10); stamens 10(16), filaments gen fused at base; ovary ± superior. **FR:** 3-valved, partly dehiscent capsule [indehiscent (nut-like) or drupe], ± spheric. ± 130 spp.: range of family; some bark resins medicinal. (Greek: name of Theophrastus for pl providing gum storax) [Fritsch 2001 Molec Phylogen Evol 19:387–408]

S. redivivus (Torr.) L.C. Wheeler (p. 1261) Shrub 1–4 m. **LF**: petiole 3–14 mm; blade 2–8 cm, ovate to ± round or obovate, entire. **INFL**: 2–5 cm, 1–6-fld, terminal on branchlets; pedicel 4–9 mm. **FL**: calyx teeth 6–9, unequal; corolla 12–26 mm, 5–10-lobed, bell-shaped, white; filaments white, anthers 4–6 mm. **FR**: 11–15 mm. **SEED**: 10–12 mm, spheric-ovoid, light brown, smooth. Uncommon.

Dry places in chaparral, woodland; gen < 1500 m. KR, NCoRH, NCoRI, CaR, SN, ScV, SCoRO, SCo, TR, PR, DMoj; sw OR. [*S. officinalis* L. var. *californicus* (Torr.) Rehder; *S. o.* L. var. *fulvescens* (Eastw.) Munz & I.M. Johnst.; *S. o.* L. var. *r.* (Torr.) R.A. Howard] Lf blade, calyx hair density increases n to s CA. Apr–Jun

TAMARICACEAE TAMARISK FAMILY

John F. Gaskin

Shrub, tree, much-branched. **ST**: trunk bark rough. **LF**: alternate, sessile, entire, often scale-like, gen with salt-excreting glands. **INFL**: [spike], raceme, compound raceme, [fls 1]; bracts scale-like. **FL**: sepals 4–5, gen free, overlapping; petals 4–5, free, overlapping, gen attached below nectary; stamens 4–5[many], attached below or to nectary; ovary superior, 1-chambered, placentas basal or parietal, intrusive (simulating chambers) or not, ovules 2–many; styles [0,2]3–4[5]. **FR**: capsule, loculicidal. **SEED**: many, hairy. ± 4 genera, 80 spp.: Eurasia, Afr. [Gaskin 2003 Ann Missouri Bot Gard 90:109–118] Often in saline habitats. Scientific Editor: Thomas J. Rosatti.

TAMARIX TAMARISK, SALTCEDAR

ST: young sts often ± pendent, slender, ± covered by lvs, hairy or glabrous. **LF**: small, awl- or scale-like, sessile, gen ± clasping st, gen encrusted with excreted salt. **INFL**: raceme or compound raceme on current or previous yr's twigs; bract gen ± clasping. **FL**: sepals 4–5, gen ± united at base, persistent; petals 4–5, free, deciduous to persistent, white, pink, red; stamens 4–5[15], free; nectary disk lobes 4–5[15], alternate or confluent with filaments; styles 3–4. **FR**: valves ± lanceolate. **SEED**: hairs in tuft at tip, > seed. ± 60 spp.: Eurasia, Afr. (Latin: Tamaris River, Spain) [Beauchamp et al. 2005 Pl & Soil 275:221–231] Invasive weeds with deep roots, esp along streams, irrigation canals. Most CA spp. originally cult for orn, windbreaks; some hybridize. *T. africana* Poir. excluded.

1. Lf united completely around st (giving st ± jointed appearance), ± mucronate . . . . ***T. aphylla***
1′ Lf ± clasping st, acute to long-acuminate
  2. Sepals, petals, stamens gen 4 . . . . ***T. parviflora***
  2′ Sepals, petals, stamens 5
    3. Nectary disk lobes confluent with stamens, sometimes very shallow or ± 0 . . . . ***T. gallica***
    3′ Nectary disk lobes alternate stamens
      4. Sepals entire; fls with 1–2 stamens attached between lobes of nectary disk, 3–4 below nectary disk . . . ***T. chinensis***
      4′ Sepals minutely dentate; fls with 5 stamens attached below nectary disk . . . . ***T. ramosissima***

T. aphylla (L.) H. Karst. (p. 1261) ATHEL Large shrub or tree, < 25 m. **LF**: united completely around st (giving st ± jointed appearance), ± 2 mm, ± mucronate. **INFL**: 2° raceme 2–6 cm; bract triangular, acuminate. **FL**: sepals 5, 1–1.5 mm, ± round, tip obtuse, entire; petals 5, 2–2.5 mm, oblong to elliptic; stamens 5, alternate nectary disk lobes. Uncommon. Washes, roadsides; < 200 m. e SCo, D; to UT, TX, n Mex; native w India to n&w Afr. Hybridizes with *T. ramosissima, T. chinensis* (rarely). May–Nov ❖

T. chinensis Lour. FIVESTAMEN TAMARISK Shrub < 8 m. **LF**: 1.5–3 mm, lanceolate. **INFL**: 2° raceme 2–6 cm; bract narrowly triangular, acuminate. **FL**: sepals 5, 0.5–1.5 mm, ovate, acute, entire; petals 5, 1.5–2 mm, oblong to elliptic; stamens 5, alternate nectary disk lobes, at least some attached below disk. Common. Canyons, riverbanks, roadsides; < 1300 m. SCo, SnBr, PR, SNE, D; to MT, TX, n Mex; native to e Asia. Similar in morphology to *T. ramosissima*; hybridizes with *T. aphylla* (rarely), *T. ramosissima* (commonly). Mar–Nov ◆

T. gallica L. FRENCH TAMARISK Shrub or tree, < 5 m. **LF**: 1.5–2 mm, lanceolate **INFL**: 2° raceme 1.4–5 cm; bract oblong to narrowly triangular, acute to acuminate. **FL**: sepals 5, 0.5–1.5 mm, ovate, acute, ± entire; petals 5, 1.5–2 mm, elliptic to ovate; stamens 5, confluent with nectary disk lobes. Uncommon. Washes, flats, roadsides; < 300 m. CCo, SnFrB, SCo; to SC, n Mex; native to s Eur. Jun–Aug ◆

T. parviflora DC. (p. 1261) SMALLFLOWER TAMARISK Shrub or tree, < 5 m. **LF**: 2–2.5 mm, lanceolate, long-acuminate. **INFL**: 2° raceme 1.5–4 cm; bract triangular, gen obtuse. **FL**: sepals 4, 1–1.5 mm, elliptic to ovate, entire to finely toothed; petals 4, ± 2 mm, oblong to ovate; stamens gen 4, confluent with nectary disk lobes. Common. Washes, streambanks, slopes, roadsides; < 1300 m. s NCoR, s SNF, Teh, GV, CW (exc SCoRO), SCo, WTR, GB, D; to WA, MS, NC, n Mex; native to se Eur. Mar–Apr ◆

T. ramosissima Ledeb. (p. 1261) SALTCEDAR Shrub or tree, < 8 m. **LF**: 1.5–3.5 mm, lanceolate, acute to acuminate. **INFL**: 2° raceme 1.5–7 cm; bract triangular, acuminate. **FL**: sepals 5, 0.5–1 mm, ± ovate, minutely dentate; petals 5, 1.5–2 mm, obovate to elliptic; stamens 5, alternate nectary disk lobes, attached to edge of disk. Common. Washes, streambanks; < 2000 m. KR, SnJV, CCo, SCoRO, SW, SNE, D; to WA, LA, n Mex; native to Asia. Very similar in morphology to *T. chinensis*; hybridizes with *T. aphylla* (rarely), *T. chinensis* (commonly). Apr–Aug ◆

THEOPHRASTACEAE THEOPHRASTA FAMILY

Anita F. Cholewa

Per [shrub, tree], [gen glandular-hairy]. **LF**: simple, alternate but gen clustered at st tips, petioled, stipules 0, margins entire [spiny-toothed]. **INFL**: gen terminal, [rarely fls 1 in axils]. **FL**: bisexual [unisexual, pls dioecious], radial, parts gen in 5s, gen with glandular dots, lines, or pits; sepals free or united at base; corolla united at base, lobes spreading; stamens epipetalous, opposite corolla lobes, anthers forming cone as fl opens, then spreading; staminodes 5, alternate stamens, epipetalous, scale-,

petal-, or stamen-like; ovary partly inferior, 1-chambered, placenta free-central or basal, style 1, stigma head-like. **FR**: capsule [gen dry berry or drupe]. **SEED**: 1–many, gen yellow to orange-red. 7–9 genera, ± 100 spp.: gen trop Am. [Caris & Smets 2004 Amer J Bot 91:627–643] Incl in Primulaceae in TJM (1993). Scientific Editor: Thomas J. Rosatti.

SAMOLUS

Erect, glabrous. **LF**: cauline or basal and cauline, alternate. **INFL**: raceme, terminal and axillary. **FL**: corolla > calyx, white. **FR**: spheric, valves 5, opposite calyx lobes. ± 10 spp.: ± worldwide. (Celtic: from curative properties)

S. parviflorus Raf. (p. 1261) SEASIDE BROOKWEED **ST**: 1.5–4 dm. **LF**: 2–5 cm; blade lance-oblong to obovate, gen tapered to winged petiole, tip obtuse to rounded. **INFL**: pedicels ascending to spreading, 1–2 cm, with 1 bractlet ± near middle. **FL**: calyx 1–2 mm, lobes triangular; corolla ± 1.5 mm across. **FR**: ± 2.5 mm wide. 2*n*=26. Moist sites; gen < 1300 m. GV, CW, SCo, n ChI, WTR, SnBr, PR; to BC, e N.Am, S.Am. [*S. floribundus* Kunth] *S. ebracteatus* Kunth (lvs crowded near st bases; bractlets, staminodes 0), from Baja CA and Clark Co., NV, expected in extreme se CA. Spring–summer

THYMELAEACEAE DAPHNE FAMILY

Lorin I. Nevling, Jr. & Kerry Barringer

Shrub [herb, tree, vine], often ill-smelling, poisonous. **ST**: erect, flexible, branched, gen with raised, woody lf scars; bark fibrous, tear-resistant, often green and/or ± white, at least in patches, esp when young. **LF**: simple, alternate [opposite, ± whorled], pinnately veined, entire, petioled; stipules ± 0. **INFL**: axillary cluster [umbel, spike, raceme, or fls 1]. **FL**: bisexual [not], [3]4[6]-parted; hypanthium tubular or funnel- to bell-shaped, corolla-like; calyx lobes overlapped or not, or ± 0; petals small, alternate sepals; stamens [2,4]8[many]; disk gen present; ovary superior [1/2-inferior], 1-chambered, 1-ovuled, style 1, stigma dot-like. **FR**: berry [capsule]. **SEED**: 1. ± 50 genera, 750 spp.: worldwide, esp Australia, trop Afr; some cult (*Daphne*, *Edgeworthia*), weedy (*Thymelaea*). [Van der Bank et al. 2002 Taxon 51:329–339] Scientific Editor: Thomas J. Rosatti.

DIRCA LEATHERWOOD

LF: deciduous; broad-ovate to obovate; petioles covering buds. **INFL**: nodding. **FL**: open with or before lvs; calyx lobes short; petals minute, ± scale-like, incl, forming a ring; filaments ± = style, exserted. **FR**: fleshy. 4 spp.: temp N.Am, Mex. (Greek: name from mythology) [Nesom & Mayfield 1995 Sida 56:21–42]

D. occidentalis A. Gray (p. 1261) WESTERN LEATHERWOOD **ST**: 1–3 m, silky-hairy in youth. **LF**: petioles 3–6 mm; blade 2–7 cm, silky-hairy, base, tip rounded. **INFL**: 1–4 fld. **FL**: short-pedicelled; hypanthium yellow. **FR**: 8–10 mm, yellow-green, glabrous. 2*n*=36. Gen n or ne facing slopes, mixed-evergreen forest to chaparral, gen in fog belt; 50–400 m. SnFrB. Nov–Mar ★

TROPAEOLACEAE NASTURTIUM FAMILY

Robert E. Preston & Elizabeth McClintock

Ann, per, ± fleshy, gen ± glabrous; often twining by petioles; roots tuber-like or not. **ST**: prostrate. **LF**: gen simple, gen alternate, peltate [(or not)]; blade << petiole, entire, toothed, lobed, or dissected, veins gen palmate; stipules often 0. **INFL**: fl gen 1 on long, axillary pedicel. **FL**: showy, bisexual, gen bilateral; sepals 5, uppermost (adaxial) gen long-spurred; petals gen 5, clawed, upper 2 unlike lower 3; stamens 8, in 2 whorls, unequal; ovary superior, lobes 3, chambers 3, placentas axile, style 1, >> ovary, 3-parted near tip, stigmas 3. **FR**: separating into 3 gen nut-like, 1-seeded segments (= mericarps). 1 genus, ± 90 spp.: Mex to S.Am; some cult. Scientific Editor: Thomas J. Rosatti.

TROPAEOLUM

(Greek: trophy, from shield-like lvs) [Andersson & Andersson 2000 Taxon 49:721–736]

T. majus L. GARDEN NASTURTIUM **ST**: > 1 m. **LF**: petiole 5–25 cm; blade 3–12 cm wide, round to ± reniform, veins palmate. **INFL**: peduncle < 20 cm. **FL**: 2.5–6 cm diam; sepals > 2 cm, ± ovate, ± tan, nectar spur > 3 cm; petals > 4.5 cm, gen orange or yellow with ± orange marks, claws of lower 3 > 1.5 cm, fringed near top. **FR**: > 1.5 cm, > 2 cm wide; segments broadly ovate, deeply lobed. Disturbed areas, moist or shaded areas in coastal scrub, wooded flats or slopes; < 450 m. NCo, CCo, SnFrB, SCo, s ChI; Abundantly cult, often naturalized. Originated in cult. Mar–Jun

ULMACEAE ELM FAMILY

Alan T. Whittemore

Tree. **LF**: simple, alternate, 2-ranked; veins pinnate; stipules deciduous. **FL**: radial; sepals 4–9, free to fused; corolla 0; stamens 4–9, opposite sepals; ovary superior, chamber 1, ovule 1, style branches 2. **FR**: 2-winged nutlet. 7 genera, ± 60 spp.: temp to trop; some cult for orn (*Ulmus*, *Zelkova*), used for wood (esp *Ulmus*). [Sytsma et al. 2002 Ann Missouri Bot Gard 89:1531–1546] *Celtis* moved to Cannabaceae. Scientific Editor: Bruce G. Baldwin.

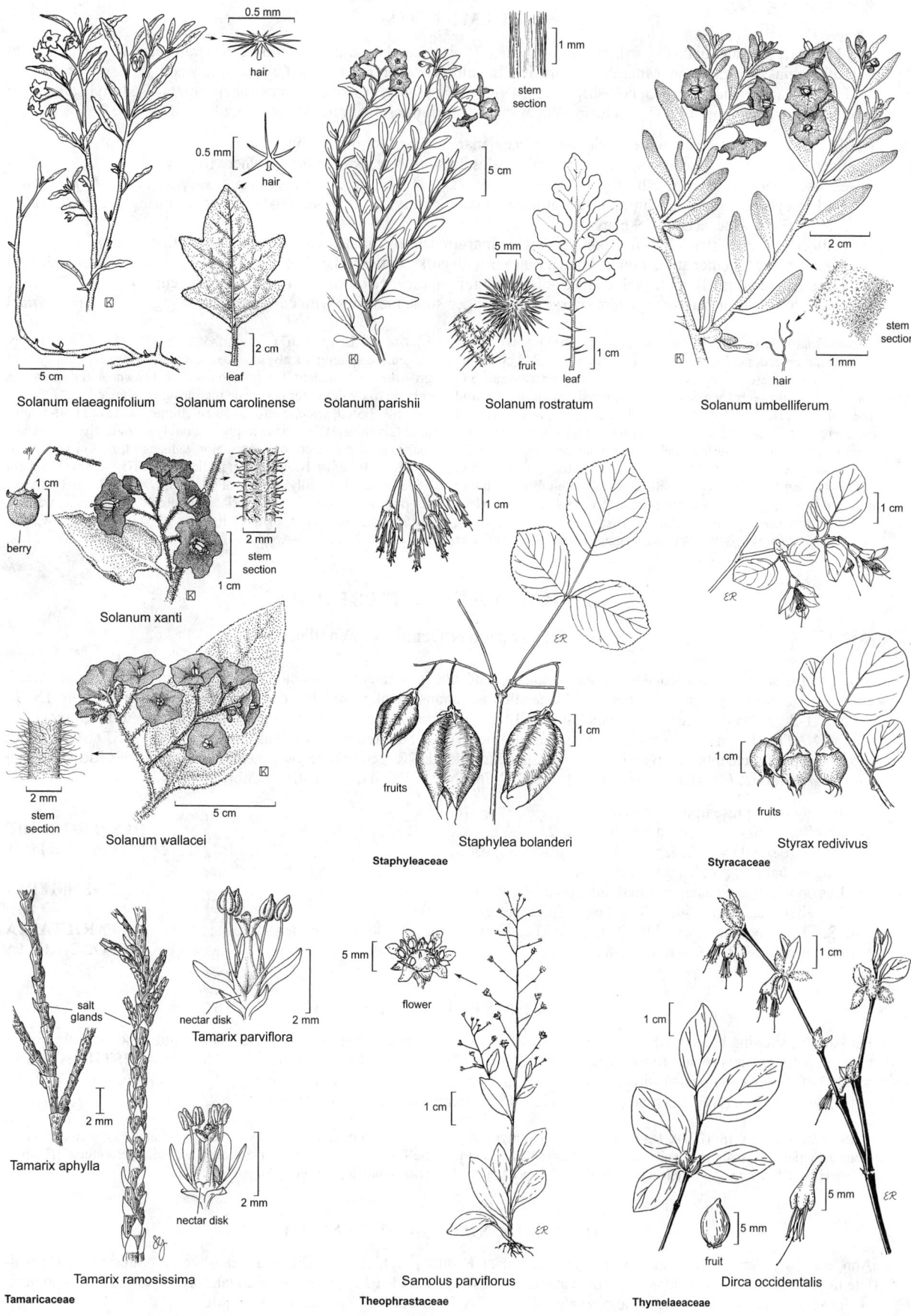

Solanum elaeagnifolium   Solanum carolinense   Solanum parishii   Solanum rostratum   Solanum umbelliferum

Solanum xanti

Solanum wallacei

Staphylea bolanderi
Staphyleaceae

Styrax redivivus
Styracaceae

Tamarix parviflora

Tamarix aphylla

Tamarix ramosissima
Tamaricaceae

Samolus parviflorus
Theophrastaceae

Dirca occidentalis
Thymelaeaceae

ULMUS ELM

Deciduous. **LF**: serrate (or doubly so), base gen oblique, 2° veins straight, parallel, extending to margin, each ending in a tooth; axils of 2° veins gen with prominent tufts of hairs. **INFL**: umbels or short racemes in lf axils on old wood; fls sessile or pedicels 7–17 mm. **FL**: bisexual; calyx gen bell-shaped, lobes 4–9; stamens 4–9, exserted; ovary strongly compressed; style divided to base, branches spreading. ± 40 spp.: n temp. Widely cult as street trees; fls, fr needed for identification.

1. Lf 1–3.5 cm wide, obtuse, acute, or short-acuminate, margin simply or occ doubly serrate
  2. Bark not furrowed, peeling as irregular woody scales 2–8 cm wide, orange-brown where freshly
    exposed, soon weathering ash-gray; fls, fr in autumn. . . . . *U. parvifolia*
  2′ Bark deeply furrowed, remaining firmly attached, medium gray; fls, fr appearing before lvs in spring . . . . . *U. pumila*
1′ Lf 3.5–8.3 cm wide, acuminate, margin doubly serrate
  3. Pedicel 7–17 mm, slender; fl, fr drooping; fr wing margin densely ciliate; bark split into ridges that
    mostly curve together at their ends; branches never with corky outgrowths . . . . [*U. americana*]
  3′ Fl, fr sessile or nearly so; fr incl wings glabrous exc for pubescence on stigmatic surface in notch; bark
    split into ridges or plates that seldom curve together; gen some branches with corky warts or wings . . . . . . [*U. minor*]

U. parvifolia Jacq. (p. 1267) CHINESE ELM, LACEBARK ELM To 25 m; bark not ridged, peeling as irregular woody scales 2–8 cm wide, orange-brown where freshly exposed, soon weathering ash-gray. **ST**: corky outgrowths on branches 0; winter buds red-brown, short-ovoid or ± spheric, glabrous. **LF**: 2.5–5 cm, 1–2 cm wide, lance-ovate to narrowly elliptic, acute to obtuse, margins obtusely and irregularly simply serrate, abaxial surface glabrous or pubescent only on major veins and tufted in vein axils. **INFL**: fl, fr in autumn; pedicel ± 0. **FR**: 1–1.3 cm, 0.6–0.8 cm wide, elliptical to ovate-elliptical, tan to dark red-brown, glabrous exc for pubescence on stigmatic surface in notch. Streams, springs, wetlands, roadsides, disturbed areas; 10–1200 m. CCo, SW, MP, W&I; to e US; native to e Asia. Aug–Oct

U. pumila L. (p. 1267) SIBERIAN ELM To 25 m; bark deeply ridged, remaining firmly attached, medium gray. **ST**: corky outgrowths on branches 0; winter buds dark brown or red-brown, ± spheric to ovoid; inner bud scale margins gen white ciliate. **LF**: 2–8 cm, 1.2–3.5 cm wide, ovate- to lance-elliptic, acute or short-acuminate (acuminate), margins simply or doubly serrate, abaxial surface glabrous or pubescent only on major veins and tufted in vein axils. **INFL**: fl, fr before lvs in spring; pedicel ± 0. **FR**: 1–2 cm, 1–1.5 cm wide, ± round (broadly obovate or elliptic), ± white-tan, glabrous exc for pubescence on stigmatic surface in notch. Streambanks, washes, bottomland, roadsides, disturbed areas; 20–1500 m. SN, SW, GB; native to n Asia. Mar–Apr

URTICACEAE NETTLE FAMILY

Robert E. Preston & Dennis W. Woodland

Ann, per [to shrub, soft-wooded tree], hairs stinging and not [glabrous]; monoecious or dioecious; wind-pollinated. **LF**: alternate or opposite, gen stipuled, petioled, blade often with translucent, raised dots due to crystals in epidermal cells. **INFL**: axillary, 1-fld or head-, raceme-, or panicle-like. **FL**: gen unisexual, small, ± green; sepals gen 4–5, free to fused; petals 0. **STAMINATE FL**: stamens gen 4–5, opposite sepals, incurved in bud, reflexing suddenly when fl opens. **PISTILLATE FL**: ovary 1, superior, chamber 1, style 0–1, stigma 1, gen hair-tufted. **FR**: gen achene. 50 genera, 700 spp.: worldwide; some cult (*Boehmeria*, ramie; *Pilea*, clearweed). [Boufford 1997 FNANM 3:400–413] Scientific Editor: Thomas J. Rosatti.

1. Stinging hairs present; lvs opposite
  2. Pistillate sepals 2–4, fused ± to tip, ± equal . . . . **HESPEROCNIDE**
  2′ Pistillate sepals 4, ± free, outer 2 < inner 2 . . . . **URTICA**
1′ Stinging hairs 0; lvs alternate or opposite
  3. Lvs opposite (alternate), toothed; infl spike-like . . . . **BOEHMERIA**
  3′ Lvs alternate, entire; infl 1-fld or head-, spike-, or panicle-like
    4. Sts decumbent to erect; lf 10–90 mm; infl head-, spike-, or panicle-like, gen few-fld . . . . **PARIETARIA**
    4′ Sts prostrate; lf 3–8 mm; infl 1-fld . . . . **SOLEIROLIA**

BOEHMERIA FALSE NETTLE

Per [shrub], stinging hairs 0. **LF**: opposite (alternate), ± ovate, toothed, 3-veined from base; base round to cordate; tip obtuse to acuminate; crystals round. **INFL**: spike-[panicle-]like. **STAMINATE FL**: sepals 4, ± free; stamens 4. **PISTILLATE FL**: sepals 2–4, fused ± to tip, enclosing ovary. **FR**: flat to lenticular. ± 50 spp.: esp trop, subtrop; *Boehmeria nivea* (L.) Gaudich., ramie, cult for fiber. (G.R. Boehmer, Saxony, 1723–1803) [Kravtsova et al. 2000 Kew Bull 55:43–62]

B. cylindrica (L.) Sw. (p. 1267) Per to subshrub, 1–16 dm; monoecious. **LF**: glabrous, scabrous, or abaxially short-hairy; blades elliptic to lanceolate, 5–18 cm, 2–10 cm wide. **INFL**: interrupted, gen of mixed staminate, pistillate fls. Levees, riparian areas; < 10 m. Deltaic ScV; c&e US, Mex, C.Am, S.Am, West Indies, Bermuda. [*B. nivea* (L.) Gaudich., misappl.] Aug–Nov

HESPEROCNIDE WESTERN NETTLE

Ann, erect, slender, stinging hairs present; monoecious. **LF**: opposite, toothed. **INFL**: gen head-like, of both staminate, pistillate fls. **STAMINATE FL**: sepals 4, ± free; stamens 4. **PISTILLATE FL**: sepals 2–4, ± equal, fused ± to tip. **FR**: lenticular, enclosed by ovate, sac-like, membranous calyx. 2 spp.: CA, Baja CA, HI. (Greek: western nettle)

H. tenella Torr. (p. 1267) **ST:** < 5 dm. **LF:** 4–40 mm, ovate, thin, blunt-serrate; stipules small; petiole ± < blade, slender; crystals gen elongate, rounder in fully exposed pls. **INFL:** < petiole, ± spheric. **FL:** calyx 1–1.5 mm. **FR:** ovate. Moist, shaded areas, often at base of rocks or shrubs, in chaparral, coastal scrub, riparian woodland, mesic oak woodland; < 1200 m. CW, SW (uncommon NCo, NCoRI, SNF); n Baja CA. Feb–Jun

PARIETARIA PELLITORY

Hairs sparse to dense, stinging 0. **ST:** branches from base, decumbent to erect, gen herbaceous. **LF:** alternate, blade 1–9 cm, lanceolate to round, entire; stipules 0; crystals round. **INFL:** head-, spike-, or panicle-like, gen few-fld; fls subtended by involucre of 1–3 lance-linear bracts. **FL:** sepals 4, fused below. **STAMINATE FL:** stamens 4. **FR:** ovoid, shiny. 20–30 spp.: worldwide temp, subtrop. (Latin: wall, from habitat of some)

1. Fr ± black, tip acute; st becoming woody . . . . *P. judaica*
1' Fr tan to ± brown or red-brown, tip obtuse; st herbaceous
  2. Lf blade on older sts lanceolate to linear, base long-tapered, lowest veins from midrib well above
     blade base . . . . *P. pensylvanica*
  2' Lf blade round to ovate or lanceolate on older sts, base truncate to wedge-shaped, lowest veins from
     midrib at blade base . . . . *P. hespera*
    3. Lf blade round to ovate on older sts, width = length; calyx lobes spreading to recurved, acuminate. . . var. *californica*
    3' Lf blade ovate to lanceolate on older sts, width < length; calyx lobes erect, acute or abruptly narrowed
     below mucronate tip. . . . . var. *hespera*

P. hespera Hinton Ann 2–55 cm, decumbent to erect, matted or not. **LF:** blade 5–20 mm, ± round on young sts, round to ovate or lanceolate on older sts, base truncate to wedge-shaped, lowest veins from midrib at blade base. **FL:** calyx lobes gen 2–3 mm, acute to acuminate. **FR:** hidden between calyx lobes, 0.9–1.3 mm, ovate, tan to ± brown, tip obtuse.

  var. *californica* Hinton (p. 1267) **LF:** blade round to ovate on older sts, width = length. **FL:** calyx green to yellow- to light red-brown, lobes spreading to recurved, acuminate. Rocky slopes, canyons, among boulders, in coastal scrub, chaparral, oak woodland; 30–1220 m. c&s CCo, SnFrB, SCo, ChI, PR; n Baja CA. Feb–May

  var. *hespera* (p. 1267) **LF:** blade ovate to lanceolate on older sts, width < length. **FL:** calyx green to dark red-brown, lobes erect, acute or abruptly narrowed below mucronate tip. Rocky slopes, dry washes, in oak woodland, coastal scrub, chaparral, desert scrub; 5–1375 m. s SNF, c&s CCo, SCoRO, SW, s W&I, DMoj exc DMtns, DSon; to UT, NM, nw Mex. Pls with narrowly lanceolate lvs sometimes confused with *P. pensylvanica*. Mar–Jun

P. judaica L. Per 1–8 dm, decumbent to erect. **ST:** becoming woody. **LF:** blade 11–90 mm, narrowly lanceolate to wide-ovate, base tapered to round; tip acuminate to long-tapered. **FL:** calyx lobes 2–3.5 mm. **FR:** 1–1.5 mm, 0.5–0.9 mm wide, ± black, tip acute. Roadsides, disturbed areas on coastal bluffs; < 125 m. s ScV, CCo, SnFrB, SCo; native to Eurasia, n Afr. Invasive in coastal urban settings. Nov–Aug

P. pensylvanica Willd. (p. 1267) Ann < 6 dm, decumbent to erect. **LF:** blade 10–90 mm, on older sts lanceolate to linear, base long-tapered, tip gen acuminate to long-tapered, lowest veins from midrib well above blade base. **FL:** calyx lobes ± erect, 1.5–2 mm, acute, dark red-brown. **FR:** hidden between calyx lobes or not, 0.9–1.5 mm, 0.5–1 mm wide, ovate, ± (red-) brown, tip obtuse. 2*n*=16. Desert scrub, coastal scrub, chaparral, oak forest, riparian woodland; 200–730 m. KR, NCoRI, s SN, SnJt; to s Can, e US, Mex. Possibly alien; CA collections historic, widely scattered. Apr–Jun

SOLEIROLIA BABY'S TEARS

1 sp. (Captain J.F. Soleirol, collector esp in Corsica, 1781–1863)

S. soleirolii (Req.) Dandy (p. 1267) Ann 2–30 cm, sts prostrate, gen mat-forming, stinging hairs 0. **LF:** alternate, 3–8 mm, oblong to ± round, oblique, entire; crystals gen elongate. **INFL:** fls 1, subtended by 3 winged bractlets, lower pistillate, upper staminate. **STAMINATE FL:** sepals 4, ± free; stamens 4. **PISTILLATE FL:** sepals fused. **FR:** achene, 0.8–1 mm, ovoid, shiny, enclosed by calyx, bractlets. 2*n*=20. Damp, shaded places; < 215 m. CCo, SnFrB, SCo, s ChI (San Nicolas Island); Baja CA; native to Corsica, Sardinia. Orn, occ escaping from cult. Mar–Jun

URTICA STINGING NETTLE

Ann, per [to shrub], weak, stinging hairs 0 or few to many; monoecious or dioecious. **ST:** branched or not, erect, spreading, or decumbent. **LF:** opposite, lanceolate to cordate, toothed, prominently 3–5-veined from base; crystals round to elongate. **INFL:** head-, raceme-, or panicle-like. **STAMINATE FL:** sepals 4, ± free, green, sharp-bristly; stamens 4. **PISTILLATE FL:** sepals 4, ± free, outer 2 < inner 2. **FR:** lenticular to deltate, enclosed by 2 inner sepals. ± 45 spp.: esp temp. (Latin: to burn, from stinging hairs)

1. Ann 1–6(8) dm; lf blade 18–40(90) mm; infl gen ± head- or spike-like, often < petiole, of staminate and
  pistillate fls; fr deltate. . . . . *U. urens*
1' Per 5–30 dm; lf blade 60–200 mm; infl raceme- or panicle-like, gen > petiole, of staminate or pistillate fls;
  fr ovate. . . . . *U. dioica*
  2. Non-stinging hairs on st, lf 0 to ± dense, occ > 1 mm, those on abaxial veins appressed; st green. . . . . . subsp. *gracilis*
  2' Non-stinging hairs on st, lf dense, < 1 mm, those on abaxial veins erect; st gray-green. . . . . subsp. *holosericea*

U. dioica L. Per 5–30 dm, from rhizome, ± erect, stinging hairs few to many, non-stinging 0 to dense, gen shorter. **LF:** blade 6–20 cm, narrow-lanceolate to wide-ovate, base tapered to cordate. **INFL:** spike-, raceme-, or panicle-like, 1–7 cm, gen > petiole, of staminate or pistillate fls. **FR:** ovate. *U. dioica* subsp. *d.* dioecious, native to Eurasia; naturalized in N.Am; report from CA in FNANM based on an unconfirmed collection.

subsp. ***gracilis*** (Aiton) Selander (p. 1267) AMERICAN STINGING NETTLE Gen monoecious. **ST:** 10–25 dm. **LF:** gen wide-ovate. 2*n*=26,52. Moist or riparian areas, willow scrub; < 245 m. NCo, KR (Trinity Alps), deltaic SnJV, n CCo; US, Can. Mar–Oct

subsp. ***holosericea*** (Nutt.) Thorne (p. 1267) HOARY NETTLE Gen monoecious. **ST:** 10–30 dm. **LF:** narrow-lanceolate to wide-ovate. 2*n*=26. Meadows, seeps, springs, margins of marshes, streams, lakes, moist areas in chaparral, coastal scrub; < 3370 m. CA-FP (exc but expected NCo, SNF), GB, DMoj (uncommon); w US, n Mex. Pls in shade tend to have wider lvs, fewer hairs, so approach *U. dioica* subsp. *gracilis*. Jun–Sep

U. urens L. (p. 1267) DWARF NETTLE Ann 1–6(8) dm, from slender taproot, monoecious; non-stinging hairs 0 to moderate. **ST:** simple or branched, erect. **LF:** blade 18–40(90) mm, elliptic to broad-elliptic, base wedge-shaped, margins coarse-serrate, tip acute. **INFL:** gen head- or spike-like, 5–25 mm, often < petiole, of staminate and pistillate fls. **FR:** 1.5–2.5 mm, deltate. 2*n*=24,26,52. Disturbed areas, stream banks, shaded areas in grassland, oak woodland, chaparral, coastal-sage scrub, riparian woodland; < 1000 m. NCo, NCoRO, n&s SNF, Teh, GV, CW, SW, w DSon (Coachella Valley), reported from MP; native to Eur. Jan–Jun

VALERIANACEAE VALERIAN FAMILY

Abigail J. Moore & Lauramay T. Dempster, except as noted

Ann, per, occ strongly scented, odor gen unpleasant. **LF:** simple to pinnately lobed or compound; petioles gen sheathing; basal ± whorled; cauline opposite, petioled to sessile. **INFL:** cyme, panicle, or head-like, gen ± dense. **FL:** gen bisexual; calyx fused to ovary tip, limb 0 or lobes gen 5–15, coiled inward, plumose in age, pappus-like, spreading in fr; corolla radial to 2-lipped, lobes gen 5, throat gen > lobes, > tube, base gen spurred or swollen, tube slender, long or short; stamens gen 1–3, fused to petals; ovary inferior, chamber gen 1, or occ 3 but 2 empty or vestigial. **FR:** achene, smooth, ribbed, or winged. ± 17 genera, 300 spp.: gen temp, worldwide exc Australia. Some spp. cult (*Centranthus*), some medicinal (*Valeriana*). [Bell & Donoghue 2005 Organisms Diversity Evol 5:147–159] Scientific Editors: Douglas H. Goldman, Bruce G. Baldwin.

1. Per; calyx lobes rolled inward, plumose in age, pappus-like in fr
  2. Corolla gen ± purple-red, spur 4–6 mm; stamen 1; cauline lvs entire, base occ with small lobes . . **CENTRANTHUS**
  2′ Corolla white to pink, throat swollen at base, spur 0; stamens 3; cauline lvs pinnately lobed to
    compound . . . . **VALERIANA**
1′ Ann; calyx gen 0
  3. St simple or unequally few-branched; corolla gen spurred; ovary 1-chambered . . . . **PLECTRITIS**
  3′ St equally, repeatedly forked; corolla throat ± swollen at base, spur 0; ovary(2)3-chambered . . . . . **VALERIANELLA**

CENTRANTHUS

Ann, per (in CA). **ST:** 1–many; base occ woody. **LF:** gen cauline, simple, entire to lobed [toothed]. **INFL:** cyme, clustered, ± dense, open in age; terminal or axillary. **FL:** calyx lobes 5–15, coiled inward, plumose in age, spreading, persistent in fr; corolla ± funnel-shaped, lobes ± unequal, spreading, tube long, slender, long-spurred; stamen 1. **FR:** ± compressed; adaxial surface 1-veined, abaxial surface 3-veined. 12 spp.: Medit. (Greek: spurred fl)

C. ruber (L.) DC. RED VALERIAN Pl glabrous, glaucous; base gen woody. **ST:** decumbent to erect, simple or branched, 3–9 dm, hollow. **LF:** 5–8 cm, blades widely oblong to lance-elliptic, acute to rounded, entire, occ lobed at base; proximal petioled, distal sessile. **FL:** corolla 14–18 mm, gen ± purple-red, occ lavender or white; spur (3)4–6 mm. **FR:** 3–4 mm, glabrous. 2*n*=14. Disturbed places, rock or wall crevices, roadsides; < 1500 m. s NCo, KR, NCoRO, SNF, GV, CCo, SnFrB, SW, cult elsewhere; native to Medit Eur. Apr–Jul

PLECTRITIS

Abigail J. Moore

Ann, glabrous to subglabrous. **ST:** gen erect, 5–80 cm, ×-section gen angled, branches 0–few. **LF:** simple, basal and cauline, opposite, gen entire; basal short-petioled, spoon-shaped; cauline gen sessile, oblong to ovate or obovate. **INFL:** clustered, head-like or interrupted spike, terminal; bracts palmately divided into 3–5 linear segments. **FL:** calyx 0; corolla 2-lipped to ± radial and funnel-shaped, white to dark pink, tube base gen spurred; stamens 3. **FR:** achene; body ± triangular, 2–4 mm, strongly winged or not, wings lateral, wide, ± glabrous to densely hairy. 5 spp.: w N.Am, sw S.Am. (Greek: spur) [Morey 1962 Ph.D. Dissertation, Stanford Univ] Self-fertile; large-fld taxa cross- and self-pollinated, small fld taxa self-pollinated only. Wing shape, color, hairiness vary in some spp.

1. Fr winged or not, if winged then wing margins ± = thickness as wing body; spur tip gen enlarged; spur gen
    < 1/2 corolla tube length (occ ± 0). . . . . ***P. congesta***
  2. Corolla 1.5–3.5 mm, white to pale pink; fl fragrance weak to 0 . . . . subsp. ***brachystemon***
  2′ Corolla 4–9.5 mm, pale to dark pink; fls fragrant . . . . subsp. ***congesta***
1′ Fr winged, wing margins ≥ 2 × as thick as wing body; spur tip enlarged or not; spur gen ≥ 1/2 corolla tube length
  3. Corolla 2-lipped, pink to dark pink, lower corolla lobes > upper, when fresh lower lip with 2 red spots;
    spur slender, pointed; immature fr pink to brown; mature fr wing margins > 3.5 × as thick as wing body,
    wing margins rolled inwards . . . . ***P. ciliosa***
  3′ Corolla ± radial, white to pale pink, corolla lobes ± equal, red spots 0; spur thick, blunt; immature
    fr white; mature fr wing margins 2–3.5 × as thick as wing body, wing margins spreading, not rolled
    inwards . . . . ***P. macrocera***

P. ciliosa (Greene) Jeps. (p. 1267)   **INFL:** bracts ± red. **FL:** corolla 1.5–8.5 mm, pink to dark pink, 2-lipped, lower lip with 2 red spots; spur slender, pointed. **FR:** convex side of mature fr grooved lengthwise down keel, winged, pink to brown when immature; wing hairs in vertical bands esp near body, wing margin ≥ 2 × as thick as wing body when immature, > 3.5 × as thick as wing body when mature, wing margin rolled inwards. 2*n*=32. Common. Open, partly shaded slopes; < 2100 m. CA-FP; to WA, AZ, n Baja CA. [*P. c.* subsp. *insignis* (Suksd.) Morey] Mar–Jun

P. congesta (Lindl.) DC.   **FR:** convex side keeled lengthwise, grooves few, small; winged or not, wings hairy near tip, near margins, or ± throughout, margin as thick as wing body. 2*n*=32.

subsp. ***brachystemon*** (Fisch. & C.A. Mey.) Morey (p. 1267)   **FL:** corolla 1.5–3.5 mm, uniformly white to pink, ± radial to ± 2-lipped;

spur a minute swelling, or slender, tip gen enlarged; undehisced anthers ≤ 0.7 mm. Common. Coastal bluffs, open, partly shaded slopes; < 1900 m. CA-FP, MP; to BC. [*P. b.* Fisch. & C.A. Mey.] Mar–Jun

subsp. ***congesta***   SEA BLUSH   **FL:** corolla 4–9.5 mm, pale to dark pink, 2-lipped, lower lip uniformly colored; spur slender, tip gen enlarged; undehisced anthers > 0.7 mm. Coastal bluffs, open, partly shaded slopes; < 1700 m. NW, SN, ScV, CW; to BC. Mar–Jun

P. macrocera Torr. & A. Gray (p. 1267)   **FL:** corolla 2–3.5 mm, uniformly white to pale pink, ± radial; spur thick, blunt. **FR:** convex side with lengthwise groove down keel; white when immature; winged, wing glabrous to hairy, margin ≥ 2 × as thick as wing body when immature, 2–3.5 × as thick as wing body when mature, spreading, not rolled inwards. 2*n*=32. Common. Open, partly shaded slopes; < 2000 m. CA-FP, MP; to BC, MT, UT. Mar–Jun

VALERIANA   VALERIAN

[Ann] per from rhizome or short underground caudex, glabrous to soft-hairy. **ST:** gen erect, 1–several. **LF:** basal simple to pinnately lobed or compound, tapered to petiole; cauline ± sessile to ± clasping, pinnately lobed to compound, distal lobe gen > others. **INFL:** cyme, clustered, ± dense to open, terminal or axillary. **FL:** calyx lobes 5–15, rolled inward, plumose in age, spreading, persistent in fr; corolla ± funnel-shaped, white to pink, lobes ± equal, throat >> tube, occ swollen near base, tube slender, occ obscured by swollen throat; stamens 3; ovary ± 1-chambered. **FR:** gen compressed, gen 6-veined vertically. ± 200 spp.: temp worldwide exc Australia. (Latin: strength, from use in folk medicine, or after Valerian, a Roman emperor) [Bell & Donoghue 2005 Organisms Diversity Evol 5:147–159]

1. Corolla 3–4.5 mm, throat ± > lobes
  2. Corolla throat asymmetric, with small swelling on 1 side just distal to tube; lvs gen short-hairy . . . . ***V. californica***
  2′ Corolla throat symmetric, evenly tapered to tube on all sides; lvs glabrous to sparsely hairy — Wrn. . . ***V. occidentalis***
1′ Corolla 4.5–8 mm, throat ± 2 × lobes
  3. All basal lvs gen simple; W&I . . . . ***V. pubicarpa***
  3′ At least some basal lvs lobed to compound; NW . . . . ***V. sitchensis***
    4. Lflets not lobed, fine-crenate to -dentate; < 1200 m . . . . subsp. ***scouleri***
    4′ Lflets irregularly lobed to sparsely toothed; 1500–2200 m . . . . subsp. ***sitchensis***

V. californica A. Heller (p. 1267)   Pls ± glabrous to short-hairy. **ST:** 2.5–5 dm. **LF:** 2–13 cm; basal simple or compound, occ deeply 3–5-lobed, terminal lobe > lateral lobes, blade or terminal lobe ovate-obovate, obtuse to rounded; cauline gen 3–9-lobed, lobe margin entire, finely dentate or few-toothed, terminal lobe acute to rounded. **FL:** bisexual; corolla 3–4.5 mm, cream-white, lobes ± < throat. **FR:** 4–6 mm, ± ovoid. Moist places, conifer forest; 1500–3700 m. KR, NCoRH, CaR, SNH, MP, SNE (exc W&I); OR, NV. Jun–Aug

V. occidentalis A. Heller (p. 1267)   WESTERN VALERIAN   Pls glabrous to sparsely hairy. **ST:** 3–7.5 dm, nodes short-hairy. **LF:** 5–30 cm; petiole base, occ sinuses between lobes short-hairy; basal lvs simple to compound, blade ovate to round, occ deeply 3-lobed, terminal lobe > lateral lobes; cauline deeply lobed to compound, lobes or lflets 3–7, margin entire to fine-crenate or -dentate, terminal lobe obtuse to acute. **FL:** bisexual or pistillate; corolla 3.5–4.5 mm, white, lobes ± < throat. **FR:** 3–5 mm, ovoid. Moist places, conifer forest; 1500–2200 m. MP; to OR, MT, CO. CA pls approach *V. californica.* Jun–Jul ★

V. pubicarpa Rydb. (p. 1267)   Pls gen glabrous. **ST:** 1–7 dm. **LF:** 4–30 cm; basal gen simple, blade elliptic to spoon-shaped; cauline

few, reduced, simple to few-lobed, terminal lobe > 3 × lateral lobes, margin ± entire to coarsely toothed. **FL:** gen bisexual; corolla 4.5–6 mm, white, occ ± pink, throat ± 2 × lobe. **FR:** 3.5–5 mm, lanceolate. Moist, rocky slopes, conifer forest; 2700–3200 m. c SNH, n SNE, expected W&I; to OR, MT, UT. Jul–Aug

V. sitchensis Bong.   Pls glabrous or sparsely hairy. **ST:** 1–7 dm; nodes short-hairy. **LF:** 3–30 cm; at least some basal lvs lobed to compound; cauline deeply lobed to compound, lobes or lflets gen 3–7, lanceolate to round. **FL:** gen bisexual; corolla 5–8 mm, white or ± pink, throat ± 2 × lobe length. **FR:** 4–6 mm, lanceolate to ± oblong.

subsp. ***scouleri*** Piper (p. 1273)   **ST:** 1–6 dm. **LF:** gen basal, < 15 cm; lflets ovate to round, margin ± entire, fine-crenate to -dentate; cauline reduced. **FR:** 4–6 mm, lanceolate to ± oblong. Moist cliffs, streambanks; < 1200 m. NCo, KR, n NCoRO; to BC. [*V. hookeri* Shuttlew.; *V. scouleri* Rydb., illeg.] Apr–Jun

subsp. ***sitchensis*** (p. 1273)   **ST:** 3–7 dm. **LF:** gen cauline, 4–30 cm; lflets irregularly lobed to sparsely toothed. **FR:** 4–5 mm, ovoid or narrowly so. Moist places, meadows; 1500–2200 m. KR; to AK, MT. Jun–Aug

VALERIANELLA   CORN SALAD
Abigail J. Moore

Ann. **ST:** erect, equally, repeatedly forked. **LF:** basal and cauline, gen simple, entire to toothed. **INFL:** cymes, dense, terminal, peduncled, gen paired, subtended by involucre-like ring of bracts. **FL:** calyx gen 0; corolla funnel-shaped, lobes unequal, throat ± swollen at base; stamens 3; ovary (2)3-chambered, 1 chamber fertile, others empty or occ fused into 1. **FR:** ± compressed, grooved lengthwise. ± 80 spp.: Eurasia, n Afr. (Latin: diminutive of *Valeriana*) [Bell 2007 Molec Phylogen Evol 44:929–941]

1. Fr length ± 2 × width, rounded in ×-section; groove between sterile chambers 0.5–0.8 mm wide; n SNF . . . . . ***V. carinata***
1′ Fr length ± = width, compressed perpendicular to wall between sterile and fertile chambers; groove between
  sterile chambers 0.1–0.3 mm wide; KR, NCoRO, n SN, n ScV, SnFrB . . . . ***V. locusta***

V. carinata Loisel.   **ST:** 1–3 dm, sparsely hairy; hairs pointed down. **LF:** 0.4–3 cm; proximal petioled; distal ± sessile; blade obovate to narrowly oblong, entire, distal occ dentate. **FL:** corolla 1.5–2 mm,

white, lobes ± blue. **FR:** 2–3 mm, 1–1.5 mm wide. 2*n*=16. Moist, gen shaded sites; < 700 m. n SNF; to WA, ID; native to Eur. Cult for edible lvs. May

V. locusta (L.) Betcke **ST**: 1–4.5 dm, sparsely hairy; hairs pointed down. **LF**: 0.5–3 cm; proximal petioled; distal ± sessile; blade obovate to narrowly oblong, entire, distal occ dentate. **FL**: corolla 1.5–2 mm, white, lobes ± blue. **FR**: 2–3 mm, 1.5–2 mm wide. 2*n*=16. Moist, gen shaded sites; < 1400 m. KR, NCoRO, CaRH, n SN, n ScV, CCo, SnFrB; to MT, UT; e Can, e US; native to Eur. [*V. olitoria* (L.) Pollich] Cult for edible lvs. Apr–Jun

VERBENACEAE VERVAIN FAMILY

Dieter H. Wilken

Ann to shrub [tree], gen hairy. **LF**: cauline, opposite, gen toothed; stipules 0. **INFL**: raceme, spike, or head, gen elongated in fr; bract gen 1 per fl. **FL**: bisexual; calyx gen 4–5-toothed; corolla 4–5-lobed, radial to bilateral, salverform to 2-lipped; stamens 4–5 (if 4, gen in unequal pairs), epipetalous; ovary superior, 2- or 4-lobed, gen 2- chambered, style 1, often with 2 unequal lobes, only 1 stigmatic, lateral. **FR**: 2 or 4 nutlets, drupe-like, or capsule. ± 31 genera, ± 920 spp.: esp Am trop. Some cult (*Lantana*, *Verbena*, *Vitex*); some weedy worldwide (*Lantana*); some used for wood (*Tectona*, teak). *Avicennia* incl in Acanthaceae. [Marx et al. 2010 Amer J Bot 97:1647–1663] Scientific Editor: Bruce G. Baldwin.

1. Ann to per
2. Calyx 2–4-toothed, ± compressed; corolla ± 2-lipped; nutlets 2 . . . . **PHYLA**
2′ Calyx 5-toothed, cylindric; corolla gen ± radial; nutlets 4 . . . . **VERBENA**
1′ Shrub
3. Infl longer than wide; corolla 2.5–3.5 mm, white . . . . **ALOYSIA**
3′ Infl wider than long; corolla 8–12 mm, yellow to purple . . . . **[LANTANA]**

ALOYSIA

Shrub, strong-smelling. **LF**: blade lanceolate to ovate. **INFL**: raceme or spike, longer than wide. **FL**: calyx 4-toothed; corolla 5-lobed, ± 2-lipped; stamens 4; ovary 2-chambered, ovules 2, style unlobed, stigma ± spheric, terminal. **FR**: nutlets 2. ± 35 spp.: Am. Some used for food flavoring, tea. (Maria Louisa, princess of Asturias and, later, queen of Spain, 18th century)

A. wrightii (Torr.) Abrams (p. 1273) WRIGHT'S BEEBRUSH < 2 m, ± rounded. **ST**: branches many; twigs brown, angles white. **LF**: petiole < 4 mm; blade 4–17 mm, ovate to ± round, crenate, abaxial face densely hairy. **INFL**: spike 1.5–6 cm; axis densely and finely tomentose; bract slightly < calyx, lanceolate. **FL**: calyx 1.5–3 mm, puberulent; corolla 2.5–3.5 mm, white, lobes rounded, upper 2 larger. **FR**: < 2 mm. Rocky, often limestone, slopes, Joshua-tree or pinyon/juniper woodland; 900–1600 m. e&s DMtns, e DSon; to TX, n Mex. Aug–Oct ★

PHYLA

Per, gen mat-like. **ST**: central gen stolon-like; branches decumbent to erect, glabrous or ± strigose. **LF**: opposite or clustered, strigose to appressed-hairy; hairs forked. **INFL**: spike, ± spheric, becoming cylindric in fr, dense; bracts ovate to wedge-shaped. **FL**: calyx ± compressed, 2–4-toothed; corolla ± 2-lipped, tube gen > calyx; stamens 4; ovary 2-chambered, ovules 2, style lobes 2, stigma lateral. **FR**: nutlets 2. ± 15 spp.: warm temp, subtrop Am. (Greek: clan or tribe, from clustered fls)

1. Lf blade widest proximal to middle, 15–25 mm wide, teeth 11–21 . . . . *P. lanceolata*
1′ Lf blade gen widest at or distal to middle, 5–10 mm wide, teeth 5–11 . . . . *P. nodiflora*

P. lanceolata (Michx.) Greene (p. 1273) **ST**: internodes gen 3–10 cm; branches 15–50 cm. **LF**: blade 25–60 mm, lanceolate to ovate, margin serrate from proximal to mid-blade to tip. **INFL**: 7–18 mm; peduncle 4–9 cm. **FL**: corolla white or pale blue to ± purple. 2*n*=32. Wet places, marshes; < 400 m. GV, CCo, SnFrB, SCo, D; to e N.Am, n Mex. May–Nov

P. nodiflora (L.) Greene (p. 1273) **ST**: internodes gen < 4 cm; branches gen < 15 cm. **LF**: blade 5–30 mm, margin ± entire or gen serrate from mid-blade to tip. **INFL**: 6–10 mm; peduncle 1.5–9 cm. **FL**: corolla white to ± red. 2*n*=36. Wet places, pond margins; < 400 m. NW (exc KR, NCoRH), GV, CCo, SnFrB, SCo, ChI (Santa Cruz, Santa Catalina islands), PR, se DMoj, DSon; warm temp, trop ± worldwide. [*P. n.* var. *canescens* (Kunth) Moldenke; *P. n.* var. *incisa* (Small) Moldenke; *P. n.* var. *reptans* (Kunth) Moldenke; *P. n.* var. *rosea* (D. Don) Moldenke] Questionably native; variation in lf margin, lf hairiness may reflect multiple introductions from elsewhere, incl S.Am. May–Nov

VERBENA VERVAIN

Ann to per [shrub]. **ST**: often 4-angled; hairs gen short, stiff. **LF**: reduced distally on st; blade entire to pinnately lobed. **INFL**: spike, often in panicle-like clusters, gen terminal, gen elongated in fr. **FL**: calyx 5-ribbed, 5-toothed, hairs gen strigose or appressed; corolla 4–5-lobed, gen ± radial, sometimes bilateral and 2-lipped; stamens 4; ovary 4-chambered, ovules 4, style 1, lobes 2, 1 tooth-like, 1 with ± spheric stigma. **FR**: nutlets 4, gen oblong. ± 250 spp.: temp, trop Am, Medit Eur. (Latin: ancient name) [Munir 2002 J Adelaide Bot Gard 18:21–103; Yeo 1990 Kew Bull 45:101–120; Yuan & Olmstead 2008 Molec Phylogen Evol 48:23–33] *V. gooddingii*, *V. pulchella* often placed in *Glandularia* (sister to *Verbena* in strictest sense).

1. Corolla 8–14 mm, limb 6–10 mm wide; calyx 5–9.5 mm
2. Lvs 3–5-lobed; calyx hairs spreading . . . . *V. gooddingii*
2′ Lvs 1–2-pinnately dissected; calyx hairs appressed . . . . *V. pulchella*

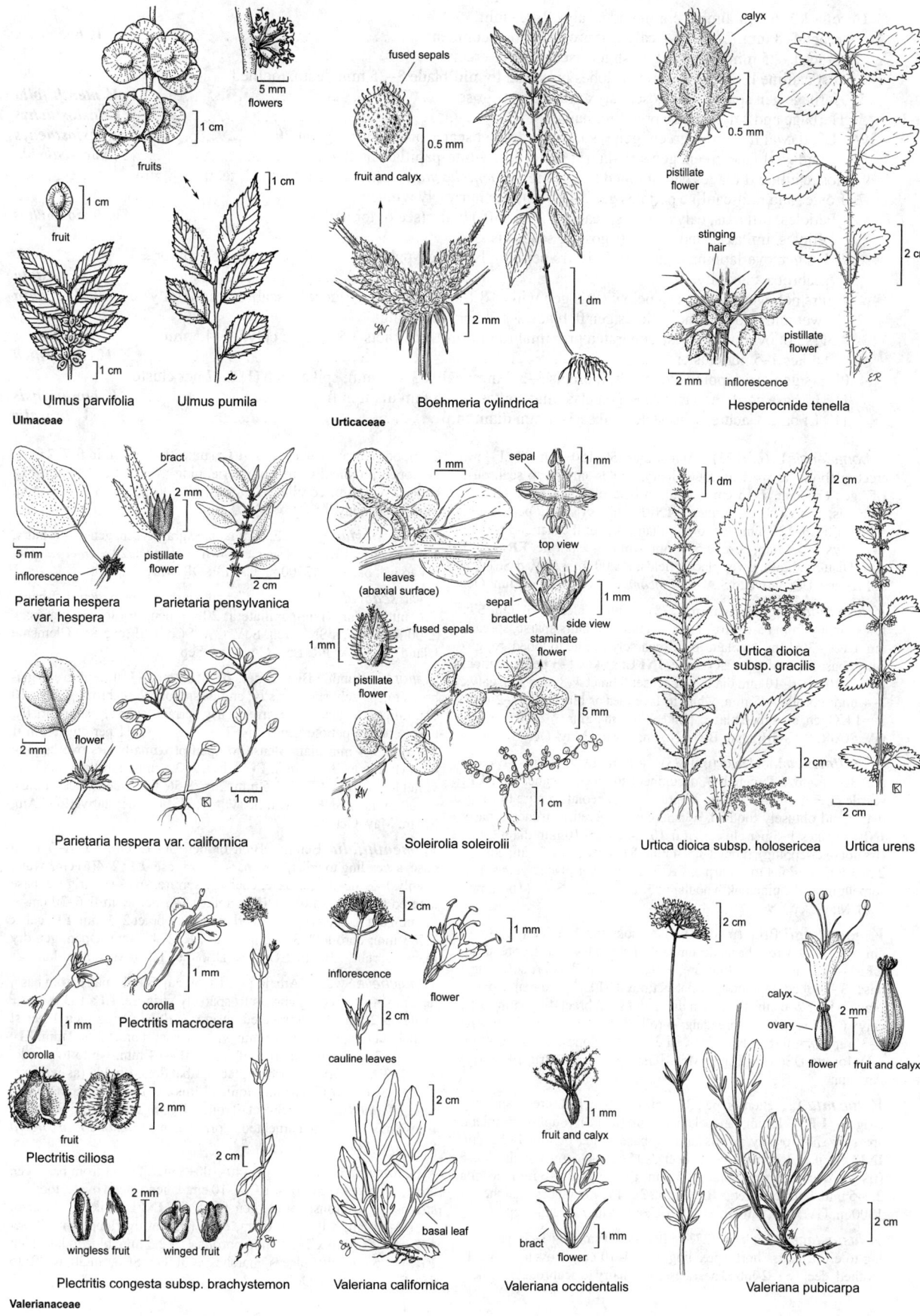

Ulmaceae

Ulmus parvifolia

Ulmus pumila

Urticaceae

Boehmeria cylindrica

Hesperocnide tenella

Parietaria hespera
var. hespera

Parietaria pensylvanica

Parietaria hespera var. californica

Soleirolia soleirolii

Urtica dioica
subsp. gracilis

Urtica dioica subsp. holosericea   Urtica urens

Valerianaceae

Plectritis macrocera

Plectritis ciliosa

Plectritis congesta subsp. brachystemon

Valeriana californica

Valeriana occidentalis

Valeriana pubicarpa

1′ Corolla 1.5–6 mm, limb < 5 mm wide; calyx 2–4.5 mm
  3. Fl bract 4–8 mm, 2–4 mm > calyx; st prostrate to decumbent . . . . ***V. bracteata***
  3′ Fl bract 1–4.5 mm, gen ≤ calyx; st gen ascending to erect
    4. Most cauline lf blades 1–2-lobed (lobes proximal to mid-blade 8–25 mm deep), toothed
      5. Herbage gen sparsely strigose; calyx minutely strigose . . . . ***V. menthifolia***
      5′ Herbage and calyx short-spreading-hairy . . . . ²***V. lasiostachys***
        6. Lf adaxial face gray-green, gen ± soft-hairy; nutlet scar ± brown, ± rough at 20× . . . . var. ***lasiostachys***
        6′ Lf adaxial face green, gen stiff-hairy; nutlet scar white-papillate at 20×. . . . . var. ***scabrida***
    4′ Most cauline lf blades gen unlobed (exc some *Verbena hastata*), entire or gen toothed, teeth 2–8 mm
      7. Lower and mid-cauline petioles gen 1–2.5 cm, occ narrowly winged
        8. Petioles, infl axis, calyx short-spreading-(gen soft-) hairy (see 6. for vars.) . . . . ²***V. lasiostachys***
        8′ Petioles, infl axis, and calyx strigose or scabrous
          9. Frs gen overlapping (axis with 8–10 frs per cm); lvs rough-puberulent, sparsely short-bristly or ±
            scabrous adaxially. . . . . ***V. hastata***
         9′ Frs not overlapping, esp below (axis gen with < 8 frs per cm); lvs strigose to scabrous adaxially . . . . ***V. scabra***
      7′ Lower and mid-cauline petioles gen 0, blade sessile or ± clasping
        10. Frs not overlapping (esp separated proximal to mid-infl); fl bracts 3.5–4 mm; calyx 4–4.5 mm;
          spikes 1–5 per cluster . . . . ***V. californica***
        10′ Frs gen overlapping along infl; fl bracts 3–3.5 mm; calyx 3–3.5 mm; spikes gen (1)3–17 per cluster
          11. Lf base cordate to truncate, gen clasping; spikes ± 5–6 mm diam in fr . . . . ***V. bonariensis***
          11′ Lf base ± acute, subsessile; spikes 3–4 mm diam in fr . . . . ***V. litoralis***

V. bonariensis L. (p. 1273) Ann or bien, 50–150+ cm. **ST:** 1–few, erect, glabrous to sparsely rough-hairy, angles smooth to scabrous. **LF:** gen clasping, 7–15 cm, elliptic to lanceolate, coarsely serrate, scabrous, base cordate to truncate. **INFL:** spikes (3)8–17 per cluster, spike in fr 15–125 mm, ± 5–6 mm diam, dense; fl bract 3–3.5 mm. **FL:** calyx 3–3.5 mm; corolla 5–6 mm, white or ± purple. **FR:** 1–1.5 mm. Disturbed, often wet places, fields; < 200 m. s ScV, n SnJV, n SnFrB; to s US; native to S.Am. [*V. incompta* P.W. Michael] Jun–Oct

V. bracteata Lag. & Rodr. (p. 1273) Ann or bien, 8–30 cm. **ST:** few to many from base, prostrate to decumbent; hairs sparse, spreading. **LF:** 1–3(6) cm, ± oblanceolate, coarsely serrate to lobed, rough-hairy; base tapered to ± flat petiole. **INFL:** spikes 1–3 per st branch, in fr 2–10 cm, 6–10 mm diam, gen dense; fl bract 4–8 mm. **FL:** calyx 2–4 mm; corolla 4–5 mm, white to lavender or blue. **FR:** 1–2 mm. 2*n*=14. Open, disturbed places, pond or lake margins; < 2200 m. CaR, GV, SCoR, SW, GB, D; to BC, e N.Am, n Mex. May–Oct

V. californica Moldenke (p. 1273) RED HILLS VERVAIN Bien or per, 30–75 cm. **ST:** gen 1–3, decumbent to erect, ± canescent. **LF:** sessile or ± clasping, 2–9 cm, elliptic to oblanceolate, entire to irregularly and obtusely toothed, ± canescent, tapered to truncate base. **INFL:** spikes 1–5 per cluster, in fr 10–24 cm, 5–10 mm diam, open (frs not overlapping); fl bract 3.5–4 mm. **FL:** calyx 4–4.5 mm; corolla 2.5–3.5 mm, violet to ± purple. **FR:** ± 2 mm. Wet places, seeps, gen serpentine soils, pine/oak woodland; 300–400 m. c SNF (Tuolumne Co.). May–Sep ★

V. gooddingii Briq. (p. 1273) Per (sometimes fl 1st yr), 10–45 cm. **ST:** 3–10+ from base, decumbent to erect; hairs soft, spreading. **LF:** 1–4 cm, lanceolate to ovate, obtusely toothed, short-soft-hairy; basal 3–5-lobed, tapered to ± flat petiole. **INFL:** spike gen 1 per st branch, in fr 2–6 cm, 10–15 cm diam, dense; fl bract 4.5–6 mm. **FL:** calyx 6–9.5 mm, hairs spreading; corolla 8–14 mm, purple-blue. **FR:** 2–3 mm. 2*n*=30. Sandy soils, washes, rocky slopes; 1200–2000 m. e DMoj, ne DSon; to UT, n Mex. [*Glandularia g.* (Briq.) Solbrig] Apr–Jun

V. hastata L. Bien or per, 35–150 cm. **ST:** 1–2, erect, sparsely strigose. **LF:** 9–15 cm, lanceolate, serrate (lobed), rough-puberulent, sparsely short-bristly, or ± scabrous, base acute; petiole 1–2.5 cm. **INFL:** spikes 1–8 per cluster, in fr 3–15 cm, 4–5 mm diam, dense (frs gen overlapping); fl bract 2–3 mm. **FL:** calyx 2.5–3 mm; corolla 2.5–5 mm, blue to violet. **FR:** ± 2 mm. 2*n*=14. Wet places, marshes; < 1300 m. GV, CCo, SnFrB, MP; to BC, e N.Am, AZ. Jun–Sep

V. lasiostachys Link (p. 1273) Per 35–80 cm. **ST:** 1–5+, ascending to erect; hairs short, spreading. **LF:** 4–10 cm, ± ovate, coarsely toothed, deeply 1–2-lobed near base, soft-hairy to scabrous; petiole < 2 cm, occ narrowly winged. **INFL:** spikes 1–3 per st, in fr 7–25 cm, gen open proximally (frs not overlapping); fl bract 3–4.5 mm. **FL:** calyx 2.5–4 mm; corolla 2.5–5 mm, blue to purple. **FR:** 1–2 mm. Vars. need study.

var. ***lasiostachys*** **LF:** adaxial face gray-green, gen ± soft-hairy, esp near base. **FR:** nutlet scar ± brown, ± rough at 20×. 2*n*=14. Open, dry to wet places; < 2500 m. CA-FP; OR. May–Sep

var. ***scabrida*** Moldenke **LF:** adaxial face green, gen stiff-hairy. **FR:** nutlet scar white-papillate at 20×. Open, dry to wet places; < 2300 m. s NW, s SNF, Teh, ScV, CW, SCo, ChI (exc San Clemente Island), WTR, w PR; Baja CA. May–Sep

V. litoralis Kunth Bien or per, 40–150+ cm. **ST:** 1–few, erect, glabrous to sparsely short-bristly; angles smooth to scabrous. **LF:** 3–10 cm, elliptic to lanceolate, irregularly serrate, sparsely short-bristly, base ± acute; petiole gen 0. **INFL:** spikes (1)3–11 per cluster, in fr 3–7 cm, 3–4 mm diam, dense to open proximally (frs overlapping or not); fl bract 3–3.5 mm. **FL:** calyx 3–3.5 mm; corolla 4–5.5 mm, violet to ± purple. **FR:** 1–1.5 mm. 2*n*=28,56. Disturbed places, fields; < 200 m. n&c SNF, ScV, n SnJV; to s US, Pacific; native to C.Am, S.Am. May–Oct

V. menthifolia Benth. Bien or per, 30–75 cm. **ST:** 1–3 from base, ascending to erect, gen sparsely strigose. **LF:** 2–4(6) cm, ovate, deeply 1–2-lobed near base, coarsely serrate, sparsely strigose, base tapered to ± flat petiole. **INFL:** spikes 1–3 per st, in fr 6–30 cm, < 0.5 mm diam, open (frs not overlapping); fl bract 2–3 mm. **FL:** calyx 2.5–3 mm; corolla 2–3 mm, purple. **FR:** 1–1.5 mm. Open, gen dry places, scrub; < 300 m. SCo, PR, DSon; to TX, n Mex. Apr–Jun

V. pulchella Sweet Ann to per, 15–60 cm. **ST:** 1–many from base, decumbent to erect, glabrous to sparsely strigose. **LF:** 1–3.5 cm, ± ovate, 1–2-pinnately dissected, strigose. **INFL:** spike gen 1 per st branch, in fr 3–8 cm, 10–15 mm wide, dense; fl bract 3.5–9 mm. **FL:** calyx 5–7 mm, hairs appressed; corolla 9–14 mm, white to purple. **FR:** 2–3 mm. Dry, disturbed places, abandoned fields (as waif); < 300 m. SCo, s WTR (Santa Monica Mtns), w PR; to s US; native to S.Am. [*Glandularia p.* (Sweet) Tronc.; *V. tenuisecta* Briq.] Reported from SnJV but not documented. Correct name might be *V. aristigera* S. Moore. May–Aug

V. scabra Vahl Bien or per, 40–100+ cm. **ST:** 1–3 from base, gen erect, strigose to scabrous. **LF:** 4–10 cm, lanceolate to ovate, toothed, strigose to scabrous; petiole gen 1–2.5 cm. **INFL:** spikes 1–5 per st, in fr 4–12 cm, < 0.5 mm diam, open (frs not overlapping); fl bract 1–2 mm. **FL:** calyx 2–2.5 mm; corolla 1.5–2.5 mm, blue to lavender. **FR:** 1–1.5 mm. Wet places, marshes; < 300 m. SCo, SnGb, w PR; to se US, n Mex. Sep–Oct

VIOLACEAE VIOLET FAMILY

R. John Little

Ann, per, [small shrub, tree, vine], from caudices, taproots, rhizomes, or stolons; hairs 0 or simple. **ST**: 0 or prostrate to erect. **LF**: basal, cauline, or both, alternate, [opposite], simple to compound, petioled; stipules gen small; blade linear to round, entire to toothed or lobed. **INFL**: fls 1 [raceme], axillary or scapose; peduncle bractlets 2, gen alternate. **FL**: bisexual, bilateral [radial]; sepals 5, free, basal lobes present [0], gen not prominent; petals 5, free, lowest often largest, base ± elongated into a spur; stamens 5, alternate petals, filaments short, wide, with large in-pointing hairs, lowest 2 anthers with basal nectaries extending into petal spur; ovary superior, chamber 1, placentas parietal, 3, ovules [1] gen many, style 1, often enlarged distally, stigma often oblique or hooked, hairy or not. **FR**: capsule [berry], 3-valved, loculicidal, explosively dehiscent or not. **SEED**: gen with outgrowth, attractive to ants. 23 genera, 830 spp.: worldwide, gen temp, trop (esp higher elevations). [Munzinger & Ballard 2003 Syst Bot 28:345–351] Lengths of lowest petal incl spur. Scientific Editor: Thomas J. Rosatti.

VIOLA VIOLET

LF: gen deciduous. **FL**: sepals ± equal, entire; petals unequal, lowest gen largest, with spur gen < 3 [20] mm, lateral 2 equal, gen spreading, upper 2 equal, erect or reflexed, overlapped or not, lateral 2 gen, others sometimes with beard of variously shaped hairs basally; cleistogamous fls gen present, petals 0. **FR**: ovoid to oblong, hairy or not. **SEED**: 8–75. ± 500 spp.: temp, worldwide, HI, Andes. (Latin: classical name) Important orns incl *V. odorata*, *V. tricolor* L. (Johnny-jump-up, wild pansy), *V.* ×*wittrockiana* Gams (garden pansy).

1. St 0, internodes 0
  2. Rhizomes thick, fleshy; pls without stolons. . . . . *V. nephrophylla*
  2′ Rhizomes thin, not fleshy; pls with stolons, at least in summer
    3. Fr puberulent; lf blade crenate; stolons green, lfy . . . . *V. odorata*
    3′ Fr glabrous; lf blade crenate to serrate; stolons pale, not lfy
      4. Lf blade lanceolate, narrowly ovate, or elliptic; petals white . . . . *V. primulifolia* subsp. *occidentalis*
      4′ Lf blade reniform, ovate, or round; petals white or lilac to pale blue or ± white, with purple lines
        5. Lf blade glabrous to coarsely hairy; petals white; pls in dense patches . . . . *V. macloskeyi*
        5′ Lf blade gen glabrous; petals lilac, pale blue, or ± white; pls not in dense patches . . . . *V. palustris*
1′ St present, internodes short or long
  6. Stipules ± = lf; ann; lateral petals = or gen < sepals . . . . *V. arvensis*
  6′ Stipules << lf; per; lateral petals >>sepals
    7. Lvs compound
      8. Petals of 1 color, light gold- or deep lemon-yellow; cleistogamous fls present or 0
        9. Lf blades longer than wide; lf segments 3–5, 1–2.5(5) mm wide; cleistogamous fls 0; lateral 2 petals
          bearded with cylindric hairs. . . . . *V. douglasii*
        9′ Lf blades wider than long; lf segments 3, 2–10 mm wide; cleistogamous fls present, exclusively or not;
          lateral 2 petals ± bearded or not with club-shaped hairs. . . . . *V. sheltonii*
      8′ Petals of 2 colors, upper dark red-violet, lower lilac, pale yellow, or cream to ± white; cleistogamous fls 0
        10. Lower 3 petals (apart from veins, spots) lilac to ± white; lvs gen puberulent (glabrous). . . . . *V. beckwithii*
        10′ Lower 3 petals (apart from veins, spots) pale yellow, cream, or ± white, deep yellow to orange
          basally; lvs glabrous exc veins, margins gen ciliate, stipules occ ciliate . . . . *V. hallii*
    7′ Lvs simple
      11. Lower st erect, without lvs, peduncles
        12. Upper 2 petals deep lemon-yellow abaxially; st green; cauline lf ovate to deltate, entire; moist to wet
          gen shady places. . . . . *V. glabella*
        12′ Upper 2 petals dark red-brown abaxially; st gen purple; cauline lf reniform, deltate, ± ovate, or
          diamond-shaped, deeply divided or not; gen dry shady or open places . . . . *V. lobata*
          13. Cauline lf diamond-shaped, reniform-cordate, or gen deltate, not lobed or dissected . . . . subsp. *integrifolia*
          13′ Cauline lf ± ovate, deltate, or gen reniform, palmately lobed or deeply dissected. . . . . subsp. *lobata*
      11′ Lower st prostrate to erect, with lvs, peduncles
        14. Pl with stolon-like st; lvs evergreen . . . . *V. sempervirens*
        14′ Pl without stolon-like st; lvs deciduous
          15. Petals white at least adaxially, lateral with large purple spot basally above smaller yellow area
            16. Lf gen shiny, cauline lf base wedge-shaped; pl glabrous. . . . . *V. cuneata*
            16′ Lf not shiny, cauline lf base ± cordate to truncate; pl ± glabrous to gen puberulent . . . . *V. ocellata*
          15′ Petals not white
            17. Petals light to deep violet, lavender, or blue-violet
              18. Rhizome fleshy; style head glabrous; sepals not ciliate. . . . . *V. langsdorffii*
              18′ Rhizome woody; style head bearded; sepals ciliate or not
                19. Lf ovate to ovate-triangular, entire to crenate; sepals not ciliate; spur gen longer than wide,
                  straight or tip hooked; CA-FP, Wrn. . . . . *V. adunca* subsp. *adunca*
                19′ Lf ovate to reniform, crenate; sepals ciliate or not; spur ± as long as wide, straight, tip not
                  hooked; KR (near Joe Bar, Siskiyou Co.) . . . . *V. howellii*

17′ Petals yellow to orange

  20. Petals gold-yellow, lowest 10–20 mm; basal lf 0 per caudex; cleistogamous fl 0 . . . . *V. pedunculata*

  20′ Petals deep lemon-yellow, lowest 6–16 mm; basal lf 1–6 per caudex; cleistogamous fl present
     (0 in *Viola tomentosa*)

    21. Fr glabrous to minutely puberulent

      22. Basal lf blade base gen ± truncate . . . . *V. praemorsa* subsp. *praemorsa*

      22′ Basal lf blade base gen tapered, obliquely so or not

        23. Lf gen entire . . . . *V. bakeri*

        23′ Lf ± entire, ± crenate, serrate, wavy, or gen irregularly toothed . . . . *V. praemorsa* subsp. *linguifolia*

    21′ Fr puberulent to tomentose

      24. Basal lvs entire; cleistogamous fls 0; fr tomentose; n-c SNH . . . . *V. tomentosa*

      24′ Basal lvs not entire (*Viola purpurea* subsp. *integrifolia*, *Viola pinetorum* subsp. *pinetorum* occ
        entire); cleistogamous fls present; fr puberulent

        25. Lvs canescent to tomentose

          26. Basal lvs oblanceolate to obovate or gen linear to narrowly lanceolate, canescent to
            gray-tomentose; s SNH, Teh, WTR, SnBr . . . . *V. pinetorum* subsp. *grisea*

          26′ Basal lvs oblong, ovate, or round, green-tomentose; SNE, DMoj . . . . *V. purpurea* subsp. *aurea*

        25′ Lvs ± glabrous to puberulent

          27. Sts mostly buried, not much elongated by end of season; pl 3–8.5(12) cm

            28. Cauline lvs gen entire, puberulent; basal lvs ± crenate to irregularly shallowly dentate or
              entire, unlobed; NW, CaR, n&c SNH . . . . *V. purpurea* subsp. *integrifolia*

            28′ Cauline lvs coarsely crenate or dentate, ± glabrous; basal lvs coarsely serrate or gen
              irregularly dentate or crenate with 2–4 rounded lobes per side; NCoRH, NCoRI, SnGb,
              GB, DMtns (Panamint Range) . . . . *V. purpurea* subsp. *venosa*

        27′ Sts gen not buried, gen elongated by end of season; pl 3–25(34.5) cm

          29. Basal lf bases cordate to truncate

            30. Basal lf gen gray-green, occ purple-tinted abaxially

              31. Basal lf with 4–5(6) prominent lobes per side; TR, PR, SNE, DMoj
                . . . . [3]*V. purpurea* subsp. *mohavensis*

              31′ Basal lf without prominent lobes; CaR, SN, CW, SW, MP . . . . [2]*V. purpurea* subsp. *quercetorum*

            30′ Basal lf gen purple-tinted, occ gray-green abaxially

              32. Cauline lf not prominently lobed; basal petiole 1.8–8.5 cm; peduncle 4.6–6 cm;
               n SNH, GB . . . . *V. purpurea* subsp. *dimorpha*

              32′ Cauline lf with 3–4(5) prominent lobes per side; basal petiole 4.5–14.5 cm;
               peduncle 1.7–14 cm; TR, PR, SNE, DMoj . . . . [3]*V. purpurea* subsp. *mohavensis*

          29′ Basal lf bases tapered

            33. Basal lf gen gray-green, occ purple-tinted abaxially . . . . [2]*V. purpurea* subsp. *quercetorum*

            33′ Basal lf gen purple-tinted abaxially

              34. Cauline lf 2–10 × longer than wide

                35. Cauline lf blade 3.5–9.6 cm, serrate or dentate with 2–4 short projections per side to ±
                  entire; CaR, SN, WTR, SnBr, SnJt . . . . *V. pinetorum* subsp. *pinetorum*

                35′ Cauline lf blade 1.5–4.8 cm, wavy, shallowly dentate, sharp-angled, or entire;
                  CaR, SNH, WTR, SnBr, SnJt . . . . *V. purpurea* subsp. *mesophyta*

             34′ Cauline lf gen < 2 × longer than wide

              36. Basal lf with 4–5(6) prominent lobes per side, not shiny adaxially;
               TR, PR, SNE, DMoj . . . . [3]*V. purpurea* subsp. *mohavensis*

             36′ Basal lf without prominent lobes, occ shiny adaxially; NW, CaR, SN,
               CW, SW, MP . . . . *V. purpurea* subsp. *purpurea*

V. adunca Sm. subsp. *adunca* (p. 1273) WESTERN DOG VIOLET, EARLY BLUE VIOLET Per 3.5–30(35) cm, glabrous to puberulent. **ST:** prostrate to erect, gen many, often woody at base in age, gen some much elongated by end of season, clustered on 1–several caudices gen at ground level, from woody rhizome. **LF:** simple; basal 1–4 per caudex, glabrous to hairy, petiole 0.5–13.5 cm, blade 0.5–6.6 cm, 3.6–4.8 cm wide, ovate to ovate-triangular, entire to crenate, base cordate, truncate, or long-tapered, tip acute to obtuse; cauline petiole 0.5–12.1 cm, blade 0.6–4.2 cm, 0.4–4.4 cm wide, ± like basal. **INFL:** axillary; peduncle 1–10.3 cm. **FL:** sepals lanceolate, not ciliate; petals light to deep violet, lower 3 white basally, veined dark violet, lateral 2 bearded with cylindric hairs, lowest 7–21 mm, spur 3–7 mm, gen elongate, conspicuous, straight or tip hooked. **FR:** 6–11 mm, short-ovoid, glabrous. **SEED:** 1.5–2 mm, dark brown to olive-black. $2n=20,30,40$. Vernally moist meadows, damp streambanks, meadow edges in conifer forest, gen shade; < 3570 m. CA-FP (gen mtns; to 0 m NCo), Wrn; to AK, YT, e N.Am, NM. Polymorphic, with many named variants. Often confused with *V. nephrophylla* (st 0). Larval food pl for 3 federally endangered or threatened butterflies. Apr–Aug

V. arvensis Murray FIELD PANSY, EUROPEAN FIELD-PANSY, WILD PANSY Ann 5–35 cm, ± glabrous to puberulent. **ST:** prostrate to erect, gen several, branched at ground level, from taproot. **LF:** simple; basal 0; cauline stipules ± = lf, palmately lobed, petiole 0.5–2.3 cm, upper lf 0.8–3.4 cm, 0.3–1.9 cm wide, blade lanceolate or ± oblong to ovate, gen glabrous adaxially, hairy abaxially at least on major veins, coarsely crenate-serrate, ciliate or not, base rounded to wedge-shaped or lowest truncate, tip blunt to acute. **INFL:** axillary; peduncle 2–8 cm. **FL:** sepals lanceolate, = or gen > lateral petals, basal lobes prominent, to 4 mm, truncate; petals white to pale yellow, upper 4 ± violet, lower 3 with yellow basal area, often veined violet, lateral 2 bearded with club-shaped hairs, lowest 7–15 mm with dark yellow area basally, spur gen > 3 mm; cleistogamous fls 0. **FR:** 5–9 mm, ± round, glabrous. **SEED:** 1.5–1.9 mm, brown. $2n=34$. Aban-

doned fields; < 1333 m. KR, CaR, SNF, GV, waif from cult elsewhere; widespread N.Am; native to Eur; Siberia, n Afr. Vegetatively similar to *V. tricolor* subsp. *t.* May–Jul

V. bakeri Greene (p. 1273) BAKER'S VIOLET Per 3–30 cm, glabrous or puberulent. **ST:** prostrate to erect, gen several, clustered on 1–several subterranean caudices from woody rhizome. **LF:** simple; basal 1–4 per caudex, petiole 1–14 cm, blade 1.8–8.8 cm, 0.7–3.9 cm wide, lanceolate, oblanceolate, elliptic (ovate), thin, gen entire, base obliquely tapered, tip obtuse or acute; cauline petiole 1.5–7.5 cm, blade 1.9–6.7 cm, 0.5–1.6 cm wide, ± like basal. **INFL:** axillary; peduncle 1.5–11.6 cm. **FL:** sepals lanceolate, not ciliate; petals deep lemon-yellow, upper 2 often maroon or ± brown abaxially, lower 3 veined brown-purple, lateral 2 bearded with cylindric hairs, lowest 6–14 mm. **FR:** 5–10 mm, round to ovoid, glabrous to minutely puberulent. **SEED:** 2.8–3.1 mm, tan to dark red-brown, outgrowth globular. $2n=48$. Vernally moist openings in conifer forest; 900–3800 m. KR, NCoRH, CaRH, SNH, MP; to WA, NV. May–Jul

V. beckwithii Torr. & A. Gray (p. 1273) BECKWITH'S VIOLET, GREAT BASIN VIOLET Per 2–22 cm, glabrous to puberulent. **ST:** decumbent to erect, gen several, up to 1/2 subterranean, clustered on deep caudex. **LF:** ternate-compound; basal 2–6 per caudex, gen puberulent (glabrous), petiole 2–10.5 cm, blade 2.4–5 cm, 3.5–4.5 cm wide, ovate to deltate, ± fleshy, lflets dissected into several 1–7 mm wide, narrow oblong, elliptic, lanceolate, or oblanceolate segments, tip obtuse to acute; cauline petiole 2–5.7 cm, blade 1–2.7 cm, 1.5–3 cm wide, ± like basal. **INFL:** axillary; peduncle 2–15.7 cm. **FL:** sepals lanceolate, not ciliate; upper 2 petals dark red-violet, often overlapped, lower 3 lilac to ± white, veined dark violet, yellow to orange basally, lateral 2 bearded with cylindric hairs, lowest 10–22 mm; cleistogamous fls 0. **FR:** 7–12 mm, oblong-ovoid, glabrous. **SEED:** 3–4 mm, brown. Vernally moist places among shrubs or beneath pines; 900–2700 m. KR, CaRH, n SNH, MP; to OR, ID, UT. 3 lower petals white in some populations (V.B. Baird 1942). Mar–May

V. cuneata S. Watson (p. 1273) NORTHERN TWO-EYED VIOLET, WEDGE-LEAVED VIOLET Per 2–25 cm, glabrous. **ST:** prostrate to erect, gen several from thin, shallow rhizome or a deep caudex with fleshy roots. **LF:** simple; basal 2–6 per caudex, petiole 4.5–9.8 cm, blade 1–4 cm, 2.7–3.8 cm wide, round-ovate to deltate, gen shiny, ± leathery, veined purple, ± serrate, base wedge-shaped, tip acute; cauline petiole 0.5–20 cm, blade 0.9–2.6 cm, 0.7–1.8 cm wide, round to gen diamond-shaped, crenate to ± serrate, base wedge-shaped, tip acute to obtuse. **INFL:** axillary; peduncle 1–10.5 cm. **FL:** sepals lanceolate with purple midvein, not ciliate; petals white adaxially, deep red-violet abaxially, upper 2 often overlapped, with yellow area basally, lateral 2 with large purple spot basally above smaller yellow area, bearded with club-shaped hairs, lowest 8–14 mm, veined purple, with large yellow spot basally. **FR:** 5–9 mm, ± round, glabrous to ± minutely scabrous. **SEED:** 2.5–3 mm, deep brown-purple. Uncommon. Vernally moist areas in open pine or oak forests, often on serpentine; 120–2200 m. NW, n SNH; sw OR. Mar–Sep

V. douglasii Steud. (p. 1273) DOUGLAS' VIOLET, GOLDEN VIOLET Per 3–20 cm, glabrous to puberulent. **ST:** decumbent to erect, gen several, up to 1/2 subterranean, branched, clustered on deep caudex. **LF:** 2-pinnate-compound; basal 1–6 per caudex, petiole 5–6.8 cm, blade 3.5–5 cm, 2.4–3.5 cm wide, ovate, lflets 3–5, divided into 3–5 segments, segments 1–2.5(5) mm wide, linear, narrow-elliptic or oblong, gen densely ciliate, tip acute to obtuse; cauline petiole 0.9–4 cm, blade 1.1–4.1 cm, 1–3.6 cm wide, ± like basal. **INFL:** axillary; peduncle 2–12.5 cm. **FL:** sepals lanceolate, ciliate; petals light gold-yellow, upper 2 dark red-brown to ± black abaxially, lower 3 veined dark brown, lateral 2 bearded with cylindric hairs, lowest 8–21 mm; cleistogamous fls 0. **FR:** 5–12 mm, round to oblong, glabrous. **SEED:** ± 2.5 mm, pale brown. $2n=24,48$. Vernally moist flats, grassy slopes, often on serpentine; 20–2300 m. NW, CaRH, SNF, n SNH, Teh, GV, CW, SnGb, SnBr, PR; OR, n Baja CA. Feb–Jul

V. glabella Nutt. (p. 1273) STREAM VIOLET, SMOOTH YELLOW VIOLET Per 3–38 cm, glabrous or finely puberulent. **ST:** erect, 1–3, green, from a thick, shallow, vertical or gen horizontal rhizome. **LF:** simple; basal 0–4 per caudex, petiole 7–27.5 cm, blade 3.3–8.5 cm, 2–9.3 cm wide, ovate to reniform or round, thin, crenate to serrate,

ciliate or not, base cordate, tip acute to rounded; cauline only near st tip, petiole 0.2–2.9 cm, blade 1.4–5.7 cm, 0.8–4.7 cm wide, ovate to deltate, thin, crenate to ± serrate, base cordate to truncate, tip gen acute. **INFL:** from upper axils; peduncle 2–8 cm. **FL:** sepals lance-linear, not ciliate; petals deep lemon-yellow on both surfaces, lower 3, sometimes upper 2, veined deep purple, lateral 2 bearded with cylindric hairs, lowest 6–18 mm. **FR:** 7–13 mm, ovate to elliptic, glabrous. **SEED:** ± 2 mm, shiny, pale brown. $2n=24$. Moist to wet gen shady places in forest, streambanks, etc; < 2600 m. NW, CaR, SN, CW, Wrn; to AK, MT; Asia. Mar–Aug

V. hallii A. Gray (p. 1273) HALL'S VIOLET, WILD PANSY Per 5–22 cm, glabrous or lf veins, lf margins, sepals gen ciliate, stipules occ ciliate. **ST:** decumbent to erect, gen several, up to 1/2 subterranean, clustered on deep caudex. **LF:** ternate-compound; basal 2–4 per caudex, petiole 5–8 cm, blade 2.8–6 cm, to 2.6–6.5 cm wide, ovate to deltate, fleshy, lflets dissected into several narrowly elliptic, lanceolate, or oblanceolate segments 1–7 mm wide, tip acute; cauline petiole 1.3–6 cm, blade 2–4.8 cm, 1.2–5.5 cm wide, ± like basal. **INFL:** axillary; peduncle 2.5–11 cm. **FL:** sepals ovate to lanceolate; upper 2 petals dark red-violet, ± black abaxially, lower 3 pale yellow, cream, or ± white, deep yellow to orange basally, veined dark violet, lateral 2 bearded with club-shaped hairs, lowest 5–18 mm; cleistogamous fls 0. **FR:** 4–12 mm, elliptic, glabrous. **SEED:** ± 3.5 mm, shiny, pale brown. $2n=60,72$. Vernally moist areas, open forest, grassy hills, flats, chaparral, often serpentine or gravelly soil; 150–2100 m. NW; sw OR. Apr–Jul

V. howellii A. Gray HOWELL'S VIOLET Per 2–44 cm, glabrous to sparsely hairy. **ST:** prostrate to erect, 1–4, not woody at base, often much elongated by season end, clustered on 1–3 caudices gen at ground level, from woody rhizome. **LF:** simple; basal 1–6 per caudex, petiole 4–15 cm, blade 1.9–6.8 cm, 2.1–6.4 cm wide, ovate to reniform, thin, crenate, base cordate, tip acute to gen obtuse; cauline petiole 1–4.5 cm, blade 2–5.1 cm, 1.2–5.8 cm wide, ± like basal. **INFL:** axillary; peduncle 2.8–14 cm. **FL:** sepals lanceolate, ciliate or not; petals violet to soft blue-violet, ± white at base to ± white, lower 3 veined dark violet, lateral 2 densely bearded with cylindric hairs, lowest 14–23 mm, spur 2.4–5 mm, ± as long as wide. **FR:** 7–11 mm, elliptic, glabrous. **SEED:** light brown, 1.5 mm. $2n=40,80$. Moist, shady areas, conifer forest; 50–1295 m. KR (near Joe Bar, Siskiyou Co.); BC, OR, WA. Apr–Jul ★

V. langsdorffii Ging. ALASKA VIOLET, LANGSDORFF'S VIOLET Per 2–30 cm, gen glabrous. **ST:** ascending to erect, gen < 3, from a shallow, thin, fleshy rhizome. **LF:** simple; basal 2–3 per caudex, glabrous or hairs scattered along abaxial veins, petiole 0.8–21 cm, blade 0.9–5.8 cm, 1–6 cm wide, ovate to reniform, crenate to crenate-serrate, base cordate to truncate, tip rounded or acute to gen obtuse; cauline petiole 2.2–12.1 cm, blade 1.9–4.2 cm, 1.8–5.9 cm wide, ± like basal. **INFL:** axillary; peduncle 2.2–20.7 cm. **FL:** sepals lanceolate to ovate, not ciliate; petals light to deep violet, lower 3 ± white basally, lateral 2 bearded with cylindric hairs, lowest 12–24 mm, spur 2–5 mm, conspicuous, as long as wide; style head glabrous [occ bearded]. **FR:** 7.5–13 mm, ovate to oblong, glabrous. **SEED:** 2.5–2.8 mm, dark olive to ± black. $2n=12$. In bogs of coastal sand dunes and sandy, stabilized areas dominated by per spp. of *Carex, Juncus*; < 10 m. n NCo (Lake Earl, Del Norte Co.); to AK, YT; Asia. Apr–Aug ★

V. lobata Benth. PINE VIOLET, YELLOW WOOD VIOLET Per 5–46 cm, glabrous or puberulent, glaucous or not. **ST:** erect, 1–3, gen purple, lfless proximally, from thick, shallow or deep, horizontal or vertical rhizome. **LF:** simple, deeply divided or not; basal 0–2 per caudex, petiole 5–24 cm, blade 3.5–8.5 cm, 4.5–13.5 cm wide, deltate to reniform, entire to irregularly serrate, ciliate, base cordate, wedge-shaped, or truncate, tip blunt, obtuse, or acute; cauline only near st tip, petiole 0.2–7.4 cm, blade 1.5–5.5 cm, 1.4–10 cm wide, reniform, deltate, ± ovate, or diamond-shaped, base cordate, wedge-shaped, or truncate, entire to coarsely toothed, tip acute to obtuse. **INFL:** axillary from upper lf axils; peduncle 2–13 cm. **FL:** sepals lanceolate, gen ciliate; petals deep lemon-yellow, gen upper 2, sometimes lateral 2 dark red-brown abaxially, lower 3, sometimes upper 2 veined brown-purple basally, lateral 2 bearded with cylindric hairs, lowest 8–19 mm. **FR:** 6–16 mm, elliptic-ovate, glabrous. **SEED:** 2.1–2.7 mm, shiny, pale brown, blotched or streaked with brown. $2n=12$.

subsp. *integrifolia* (S. Watson) R.J. Little (p. 1277) **LF:** cauline blade gen longer than wide, diamond-shaped, reniform-cordate, or gen deltate, proximally serrate to dentate, distally often entire, tip acute. Uncommon. Dry shaded or open forest, chaparral, often on serpentine; 450–2070 m. NW, CaR, SN, PR; sw OR. Mar–Jul

subsp. *lobata* (p. 1277) **LF:** cauline blade gen wider than long, ± ovate, deltate, or gen reniform, palmate-lobed or -dissected, lobes or segments 3–12, gen entire to coarsely serrate. Dry shady or open places in chaparral, oak woodland, yellow-pine, mixed-conifer, or redwood forest, occ moist, often serpentine; 45–2300 m. NW, CaR, SN, PR; sw OR, n Baja CA. Apr–Aug

V. macloskeyi F.E. Lloyd (p. 1277) MACLOSKEY'S VIOLET, SMALL WHITE VIOLET, SMOOTH WHITE VIOLET   Per to 10 cm, from short, fleshy rhizomes, forming dense patches by late season stolons, glabrous to coarsely hairy. **ST:** 0. **LF:** simple, basal, ascending to erect, petiole 1–10 cm, blade 1–6.5 cm, 1–5.5 cm wide, reniform to ovate, ± entire to shallowly crenate, not ciliate, base cordate, tip rounded to acute. **INFL:** scapose; peduncle 2.5–21 cm. **FL:** sepals ovate to lanceolate, not ciliate; petals white, lower 3 veined purple, lateral 2 bearded with cylindric hairs or not, lowest 6–12 mm. **FR:** 5–7 mm, elliptic, glabrous. **SEED:** 1–1.5 mm, beige to bronze. 2*n*=24. Bogs, wet meadows, seeps, lake margins, streamsides, mesic roadside depressions, often with mosses; 600–3370[3600] m. n&c CA-FP (mtns), SnBr, SnJt, Wrn, SNE; to YT, e Can, NC, NV. Mar–Sep

V. nephrophylla Greene (p. 1277) LECONTE VIOLET   Per to 50 cm, from thick, fleshy rhizomes, gen glabrous. **ST:** 0. **LF:** simple, basal, ascending to erect, petiole 5–25 cm, blade in midseason 1–7 cm, 1–7 cm wide, ovate to widely ovate or reniform, crenate or serrate, base cordate to widely ovate, tip acute or obtuse. **INFL:** scapose; peduncle 3–25 cm. **FL:** sepals ovate to lanceolate, ciliate or not; petals deep blue-violet to white, lower 3 white basally, veined dark violet, lateral 2, sometimes lowest bearded with cylindric or club-shaped hairs, lowest 10–22 mm; cleistogamous fls 0. **FR:** 5–12 mm, ellipsoid, glabrous. **SEED:** 1.5–2.5 mm, beige, mottled to bronze. 2*n*=54. Shady areas in moist or swampy ground, lake margins, yellow-pine forest; 335–2302 m. KR, CaRH, n SNH, SCo, SnGb, SnBr, GB; to BC e to NS; to e US. [*V. sororia* subsp. *affinis* (Leconte) R.J. Little] Jan–Sep

V. ocellata Torr. & A. Gray (p. 1277) WESTERN HEART'S EASE, TWO-EYED VIOLET   Per 2–37 cm, ± glabrous to gen puberulent. **ST:** ascending to erect, gen several, from shallow rhizome or a deep rhizomatous caudex with fleshy roots. **LF:** simple; basal 1–6 per caudex, ± glabrous to gen puberulent, petiole 3.7–10 cm, blade 1–6 cm, 1.2–4 cm wide, ovate, deltate, ± reniform, crenate, base gen ± cordate, tip acute; cauline petiole 0.4–9 cm, blade 1.6–4.4 cm, 1.1–3.6 cm wide, ovate, deltate, triangular, crenate to ± serrate, base ± cordate to truncate, tip acute to obtuse. **INFL:** axillary; peduncle 1–10 cm. **FL:** sepals lanceolate, ± ciliate; petals white adaxially, upper 2 deep red-violet abaxially with yellow basally, lateral 2 with large purple spot basally above smaller yellow area, bearded with club-shaped hairs, lowest petal 5–15 mm with large yellow spot basally, veined purple. **FR:** 5–8 mm, round-ovoid, minutely scabrous. **SEED:** ± 2 mm, brown-purple. 2*n*=12. Moist or vernally moist areas, rocky or grassy banks, thickets, yellow-pine, redwood forest, often on serpentine; 75–1280 m. NW, CaRF, n&s CW; sw OR. Mar–Jul

V. odorata L. ENGLISH VIOLET, SWEET VIOLET   Per to 12 cm, from thin rhizomes, spreading by lfy, green stolons; finely hairy. **ST:** 0. **LF:** simple, basal, ascending to erect, petiole 2–17 cm, blade 1.5–7 cm, 1.5–5 cm wide, elliptic or ovate, crenate, base cordate, tip obtuse to round. **INFL:** scapose; peduncle 4–15 cm. **FL:** sepals lanceolate, ciliate; petals deep to pale blue-violet, pale blue, or white, gen white basally, lateral 2 gen bearded with cylindric hairs, lowest 12–22 mm, gen veined purple, spur ± 3 mm. **FR:** 5–8 mm, ovoid, flecked purple, puberulent. **SEED:** 3–4 mm, brown. 2*n*=20. Garden escape reported in shady, moist areas gen in riparian habitats; 30–1255 m. NCo, ScV, CCo, SnFrB, SnGb; to BC, e N.Am; Australia, native to Eurasia. Jan–May

V. palustris L. (p. 1277) ALPINE MARSH VIOLET   Per to 19 cm, from thin rhizomes, stoloned in summer, gen glabrous. **ST:** 0. **LF:** simple, basal, ascending to erect, gen glabrous, petiole 1–17 cm, blade 0.5–6.4 cm, 0.5–5.5 cm wide, reniform, ovate, or round, finely crenate, not ciliate, base cordate, tip obtuse to acuminate. **INFL:** scapose; peduncle 2–20.7 cm. **FL:** sepals ovate to lance-ovate, not ciliate; petals lilac, pale blue, or ± white, lower 3 veined purple, lateral 2 bearded with cylindric hairs, lowest 8–16 mm. **FR:** 6–10 mm, elliptic, glabrous. **SEED:** 1–2 mm, brown. 2*n*=48. Marshes, swamps, streambanks, often beneath shrubs; < 75 m. NCo; to AK, AB, ne N.Am, SD, CO, AZ; also in Eur. Apr–Jul ★

V. pedunculata Torr. & A. Gray (p. 1277) JOHNNY-JUMP-UP   Per 5–39 cm, glabrous to puberulent. **ST:** decumbent to erect, gen many, branched, clustered on a shallow to deep caudex with fleshy roots. **LF:** simple; basal 0; cauline petiole 2.7–7.2 cm, blade 1–5.5 cm, 1–5.5 cm wide, deltate to ovate, crenate to serrate, base truncate, ± cordate, or tapered, tip acute to obtuse. **INFL:** axillary; peduncle 2.9–20 cm. **FL:** sepals lanceolate, ciliate or not; petals gold-yellow, upper 2 red-brown abaxially, lower 3 veined dark brown, lateral 2 bearded with club-shaped hairs, lowest 10–20 mm; cleistogamous fls 0. **FR:** 5–11 mm, elliptic, glabrous. **SEED:** 2.7 mm, shiny, dark brown. 2*n*=12. Open, grassy slopes, hillsides, chaparral, oak woodland, gen full sun; < 1540 m. NCoRO, NCoRI, CW, SW; n Baja CA. [*V. p.* subsp. *tenuifolia* M.S. Baker & J.C. Clausen] Often confused with *V. purpurea* subsp. *p.*, *V. purpurea* subsp. *quercetorum.* Only larval food pl for federally endangered butterfly. Feb–Apr

V. pinetorum Greene MOUNTAIN YELLOW VIOLET   Per 4.5–22 cm, glabrous to puberulent, canescent, or gray-tomentose. **ST:** prostrate to erect, gen several, clustered on 1–several subterranean caudices from woody rhizome. **LF:** simple; basal 1–4 per caudex, petiole 2.3–9.5 cm, blade 1.3–5 cm, 0.3–2.5 cm wide, linear to elliptic or ovate to lance-elliptic, or oblanceolate to obovate, thin or not, jagged to irregularly serrate or dentate, occ entire, sometimes wavy, base long-tapered or wedge-shaped, tip acute; cauline petiole 0.9–8.3 cm, blade 0.3–1.4 cm wide, ± like basal in shape. **INFL:** axillary; peduncle 2.9–11.5 cm. **FL:** sepals lanceolate, not ciliate; petals deep lemon-yellow, upper 2 red- to purple-brown abaxially, lower 3 veined dark brown, lateral 2 bearded with cylindric hairs, lowest 5–12 mm. **FR:** 3.5–7 mm, ovoid, puberulent. **SEED:** 2–3.5 mm, medium to dark brown. 2*n*=12.

subsp. *grisea* (Jeps.) R.J. Little (p. 1277) GRAY-LEAVED VIOLET   Pl 4.5–9.5 cm, canescent to gray-tomentose. **ST:** gen buried, ± not elongated. **LF:** basal petiole 2.3–6.5 cm, blade 1.7–4 cm, 0.3–1 cm wide, oblanceolate to obovate or gen linear to narrowly lanceolate, irregularly serrate to jagged; cauline petiole 0.9–3.9 cm, blade 1.5–4 cm, 0.3–0.5(0.9) cm wide, ± entire to irregularly serrate or jagged. **INFL:** peduncle 2.9–6(7) cm. **FL:** lowest petal 5–9 mm. Mtn peaks, alpine zones; 1980–3700 m. s SNH, Teh, WTR, SnBr. Jun–Jul ★

subsp. *pinetorum* Pl 6.5–22 cm, glabrous to gen puberulent, canescent. **ST:** gen not buried, spreading to erect. **LF:** basal petiole 3.2–9.5 cm, blade 1.3–5 cm, 0.7–2.5 cm wide, linear to ovate or lance-elliptic, jagged or irregularly serrate or dentate, occ entire, puberulent to canescent, gen purple-tinted abaxially; cauline petiole to 8.3 cm, blade 3.5–9.6 cm, 0.3–1.4 cm wide, serrate or dentate with 2–4 short projections per side to ± entire. **INFL:** peduncle 3.4–11.5 cm. **FL:** lowest petal 6–12 mm. Vernally moist soil, often under conifers; 1400–3100 m. CaR, SN, WTR, SnBr, SnJt. Late May–late Jul

V. praemorsa Lindl.   Per 7.5–30 cm, glabrous to densely puberulent. **ST:** prostrate to erect, gen several, clustered on 1–several subterranean caudices from woody rhizome. **LF:** simple; basal 1–5 per caudex, petiole 4.3–19.2 cm, blade 2.3–8.5 cm, 1.4–3.7 cm wide, elliptic or ovate, entire, wavy, or gen irregularly crenate to serrate, base tapered, often oblique to ± truncate, tip acute or obtuse; cauline blade 2.6–5.8 cm, 1.3–3.5 cm wide, like basal. **INFL:** axillary; peduncle 2.7–26 cm. **FL:** sepals lanceolate, ciliate or not; petals deep lemon-yellow, upper 2, sometimes lateral 2 maroon or ± brown abaxially, lower 3 veined brown-purple, lateral 2 bearded with cylindric hairs, lowest 12–20 mm. **FR:** 6–12 mm, elliptic to oblong, glabrous to minutely puberulent. **SEED:** 2–3 mm, medium to dark brown, ± 1/3 length covered by outgrowth. 2*n*=36,48. Study needed; 1 other subsp., ± nw US.

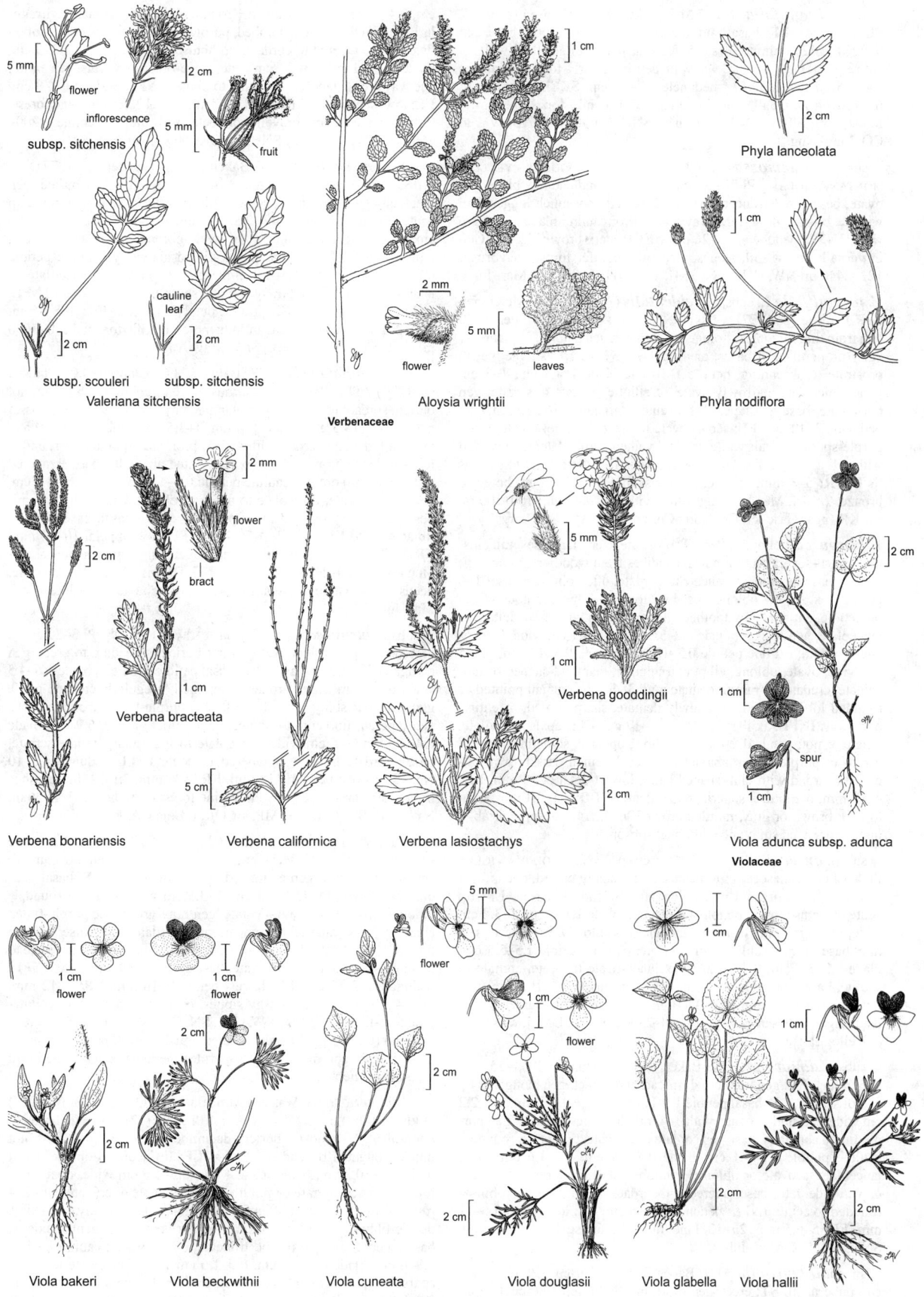

5 mm
flower

inflorescence
5 mm

subsp. sitchensis

fruit

cauline
leaf

2 cm

2 cm

subsp. scouleri   subsp. sitchensis

Valeriana sitchensis

1 cm

2 mm

flower

5 mm

leaves

Aloysia wrightii

Verbenaceae

2 cm

Phyla lanceolata

1 cm

Phyla nodiflora

2 cm

2 mm

flower

bract

Verbena bracteata

1 cm

5 cm

Verbena bonariensis   Verbena californica

Verbena lasiostachys

5 mm

1 cm

2 cm

Verbena gooddingii

2 cm

1 cm

spur

1 cm

Viola adunca subsp. adunca

Violaceae

1 cm

flower

1 cm

flower

2 cm

2 cm

Viola bakeri   Viola beckwithii

5 mm

flower

1 cm

flower

2 cm

2 cm

Viola cuneata   Viola douglasii

5 mm

1 cm

1 cm

2 cm

2 cm

Viola glabella   Viola hallii

subsp. *linguifolia* (Nutt.) M.S. Baker & J.C. Clausen (p. 1277) Pl 11–30 cm. **LF:** basal and cauline, blade gen elliptic, base gen tapered, often oblique; basal 5–8.5 cm, much longer than wide, ± entire, ± crenate, serrate, wavy, or gen irregularly toothed; cauline 1.4–3.5 cm wide. **INFL:** peduncle 2.7–21 cm. **SEED:** medium to red-brown. 2*n*=36,48. Gen in vernally moist soil, slopes, meadows, forest; 750–2400 m. NW, CaR, n&c SNH, MP; to WA, AB, WY, to CO. May–Aug

subsp. *praemorsa* (p. 1277) ASTORIA VIOLET, YELLOW MONTANE VIOLET Pl 7.5–21 cm. **LF:** basal and cauline, blade gen ovate, base gen ± truncate; basal 2.3–6.7 cm, not much longer than wide, ± crenate or serrate, wavy, or entire; cauline blade 1.3–3 cm wide. **INFL:** peduncle 5.5–26 cm. **SEED:** dark brown. 2*n*=36. Gen in vernally moist soil, slopes, meadows, conifer forest, sagebrush; 122–2440 m. NW, CaR, n&c SNH, Wrn; to BC, AB, UT. Mar–Jul

V. primulifolia L. subsp. ***occidentalis*** (A. Gray) L.E. McKinney & R.J. Little (p. 1277) WESTERN WHITE BOG VIOLET Per to 20 cm, from thin rhizomes, stoloned in summer, gen glabrous or sparsely hairy on petioles, lf bases, or peduncles. **ST:** 0. **LF:** simple, basal, prostrate to ascending, petiole 1–10 cm, blade 1.5–7 cm, 1–3 cm wide, lanceolate, narrowly ovate, or elliptic, crenate to serrate, gen not ciliate, base gen tapered, tip acute or rounded. **INFL:** scapose; peduncle 3–12 cm. **FL:** sepals ovate to lanceolate, not ciliate, often purple-spotted; petals white, lower 3 veined purple, lateral 2 bearded with cylindric or club-shaped hairs, lowest 10–14 mm; cleistogamous fls 0. **FR:** 5–8 mm, elliptic, glabrous. **SEED:** 1.5–2 mm, beige to bronze. 2*n*=24. Marshes, bogs often with *Darlingtonia*; 100–500 m. nw KR (near Gasquet, Del Norte Co.); sw OR. Apr–Sep ★

V. purpurea Kellogg Per. **ST:** prostrate to erect, gen several, clustered on 1–several subterranean caudices from woody rhizome. **LF:** simple, base tapered or truncate to cordate, often oblique; basal 1–6 per caudex, stipule oblong, fused to, forming 2 membranous wings on petiole, entire or few-toothed, the tip of each wing free, deltate or lanceolate, gen fringed, blade 0.8–5.3 cm, 0.4–4.1 cm wide, lanceolate to round; cauline petiole 0.3–12.3 cm, blade 0.3–4.8 cm, 0.3–5 cm wide, ovate, oblong, elliptic, triangular, diamond-shaped or lanceolate, crenate, serrate or dentate with 3–4(5) prominent pointed or rounded lobes per side, shallowly dentate, sharp angled, or entire, occ wavy. **INFL:** axillary; peduncle 1–17 cm. **FL:** sepals lanceolate, ciliate or not; petals deep lemon-yellow, upper 2, sometimes lateral 2 red- to purple-brown abaxially, lower 3 veined dark brown, lateral 2 bearded with club-shaped hairs, lowest petal 6–16 mm. **FR:** 4–12 mm, ovoid to ± spheric, puberulent. **SEED:** 2.1–3.1 mm, light to dark brown or gray, mottled brown. 2*n*=12,24. Subspp. variable, intergrade, need study; no additional subspp.

subsp. *aurea* (Kellogg) J.C. Clausen (p. 1277) GOLDEN VIOLET Pl 4–12 cm, canescent to tomentose. **ST:** mostly buried, decumbent or erect, gen short, not much elongated by end of season. **LF:** tip acute or obtuse; basal petiole 4.6–9.5 cm, blade 1.1–5 cm, 1–3.4 cm wide, oblong, ovate, or round, crenate to shallowly, irregularly serrate, base tapered, oblique, or truncate; cauline petiole 1.5–5.5 cm, blade 1.3–3.7 cm, 0.7–2 cm wide, lance-ovate to ovate, crenate to serrate, base tapered to truncate. **INFL:** peduncle 2.4–10.5 cm. **FL:** lowest petal 8–13 mm. **FR:** 4–7 mm. 2*n*=12. Pinyon/juniper woodland, sagebrush, sandy slopes; 1000–2300 m. SNE, DMoj; w NV. [*V. a.* Kellogg] Apr–Jun ★

subsp. *dimorpha* M.S. Baker & J.C. Clausen Pl 3–25 cm, puberulent. **ST:** gen not buried, spreading to erect, gen elongated by end of season. **LF:** basal petiole 1.8–8.5 cm, blade 0.8–3 cm, 0.5–2.2 cm wide, ± round to ovate, with 3–4 rounded lobes per side, gen purple-tinted abaxially, base gen cordate to truncate, tip obtuse-rounded; cauline petiole 0.5–4.1 cm, blade 0.7–1.9 cm, 0.4–1.6 cm wide, lanceolate to ovate or oblong, ± unlobed, ± entire, crenate, serrate, or wavy-dentate, base tapered, subcordate or truncate, tip obtuse-rounded to acute. **INFL:** peduncle 4.6–6 cm. **FL:** lowest petal 6–11 mm. **FR:** 5–6.5 mm. 2*n*=12. Pine, fir, cedar forest; 1220–2440 m. n SNH, GB; OR. May–Jul

subsp. *integrifolia* M.S. Baker & J.C. Clausen Pl 1.5–9(12) cm, puberulent. **ST:** erect, gen short, mostly buried, not much elongated by end of season. **LF:** basal petiole 2.9–7.5 cm, blade 0.8–3.1

cm, 0.4–2.3 cm wide, ovate to ± round, entire, or ± crenate to irregularly shallowly dentate, unlobed, purple-tinted, green adaxially, often fleshy, base tapered to cordate, tip obtuse; cauline petiole 1.3–3 cm, blade 1.2–3.1 cm, 0.3–2.5 cm wide, lance-oblong to ovate, gen entire, occ wavy, base tapered, tip acute to obtuse **INFL:** peduncle 1–9 cm. **FL:** lowest petal 8–11 mm. **FR:** 5–7 mm. 2*n*=12. Red fir, pine forests to timberline, sandy, gravelly, or rocky soil, incl serpentine; 1200–2600 m. NW, CaR, n&c SNH; sw OR. May–Aug

subsp. *mesophyta* M.S. Baker & J.C. Clausen (p. 1277) Pl 9–18.5 cm, ± glabrous to gen puberulent. **ST:** gen not buried, gen erect, elongated by end of season. **LF:** base tapered, gen oblique, tip acute or obtuse; basal petiole 2.8–13.8 cm, blade 1.7–4.2 cm, 0.5–2.3 cm wide, lanceolate, oblong, or ovate-oblong, irregularly dentate, with pointed lobes or not, purple-tinted abaxially; cauline petiole 0.3–12.3 cm, blade 1.5–4.8 cm, 0.5–1.7 cm wide, lanceolate to lance-ovate, wavy, shallowly dentate, sharp-angled, or entire. **INFL:** peduncle 2.6–6.7 cm. **FL:** lowest petal 7–10 mm. **FR:** 4–5.5 mm. 2*n*=12. Damp, shady areas in lodgepole pine, fir forest; 1400–3600 m. CaR, SNH, WTR, SnBr, SnJt. May–Aug

subsp. *mohavensis* (M.S. Baker & J.C. Clausen) J.C. Clausen (p. 1277) Pl 5–24 cm, ± glabrous or gen puberulent. **ST:** gen not buried, spreading to erect, gen elongated by end of season. **LF:** basal petiole 4.5–14.5 cm, blade 1–4 cm, 1–3.5 cm wide, ovate, deltate, or round, dentate-serrate with 4–5(6) prominent pointed or rounded lobes per side, gray-green to purple-tinted abaxially, base ± tapered or truncate, tip obtuse; cauline petiole 0.8–11 cm, blade 1.5–3.7 cm, 0.5–2.5 cm wide, lanceolate to ovate or elliptic with 3–4(5) prominent pointed or rounded lobes per side, ± like basal, base tapered, tip acute. **INFL:** peduncle 1.7–14 cm. **FL:** lowest petal 10–14 mm. **FR:** 5–7 mm. 2*n*=12. Desert scrub, sagebrush, dry areas in yellow-pine forest; 900–2600 m. TR, PR, SNE, DMoj; NV, sw AZ. Variable, needs study; cauline lf margins in many populations ± like basal. Mar–Jul

subsp. *purpurea* (p. 1277) MOUNTAIN VIOLET Pl 3–25 cm, ± glabrous or puberulent. **ST:** gen not buried, spreading to erect, gen elongated by end of season. **LF:** basal petiole 4–11 cm, blade 1.6–4.5 cm, 1.6–4.1 cm wide, ± round, ± unlobed, irregularly crenate, gen ± glabrous, occ shiny adaxially, ± fleshy, purple-tinted abaxially, base gen tapered, tip acute to obtuse; cauline petiole 1.5–6.9 cm, blade 0.3–2.9 cm, 1–5 cm wide, lanceolate to triangular, crenate-serrate, base ± cordate, truncate, or tapered, tip acute. **INFL:** peduncle 4.5–10 cm. **FL:** lowest petal 10–12 mm. **FR:** 5–6 mm. 2*n*=12. In openings or beneath shrubs, gen in yellow-pine forest or higher; 213–2896 m. NW, CaR, SN, CW, SW, MP; sw OR, n Baja CA. Mar–Jul

subsp. *quercetorum* (M.S. Baker & J.C. Clausen) R.J. Little (p. 1277) Pl 4–25(34.5) cm, puberulent. **ST:** gen not buried, spreading to erect, gen elongated by end of season. **LF:** basal petiole 1.9–9.5 cm, blade 1–5.3 cm, 1–3.5 cm wide, ovate to round, ± unlobed, irregularly wavy-dentate, gen gray-green, occ purple-tinted abaxially, base tapered to truncate or ± cordate, tip obtuse; cauline petiole 1.3–5.3 cm, blade 1.4–2.2 cm, 0.8–1.3 cm wide, lanceolate to diamond-shaped, crenate-serrate, base gen tapered, tip acute. **INFL:** peduncle 2.7–17 cm. **FL:** lowest petal 10–16 mm. **FR:** 8–12 mm. 2*n*=24. Dry, grassy or brushy slopes, chaparral, gen below yellow-pine forest; 304–1981 m. NW, CaR, SN, CW, SW, MP; sw OR. Fresh material (light gray) ± needed to distinguish from *V. purpurea* subsp. *p.* (dull to bright green tinted ± purple); perhaps better treated as a separate sp. Feb–Jul

subsp. *venosa* (S. Watson) M.S. Baker & J.C. Clausen (p. 1277) PURPLE-MARKED YELLOW VIOLET Pl 3–8.5(12) cm, ± glabrous to puberulent. **ST:** mostly buried, decumbent or erect, gen short, not much elongated by end of season. **LF:** tip acute to obtuse; basal petiole 5–10.3 cm, blade 0.8–2.1 cm, 0.5–3.6 cm wide, ovate to ± round, coarsely serrate or gen irregularly dentate or crenate with 2–4 rounded lobes per side, often fleshy, ± glabrous adaxially, occ shiny, purple-tinted, puberulent abaxially, major veins prominent adaxially, base tapered, oblique or not, truncate, or ± cordate; cauline petiole 0.8–6 cm, blade 0.9–2.3 cm, 0.6–1.6 cm wide, lanceolate to ovate, coarsely crenate or dentate, base tapered or oblique. **INFL:** peduncle 2.9–7 cm. **FL:** lowest petal 6–14 mm. **FR:** 4–5.5 mm. 2*n*=12. Many

habitats, substrates, incl pine forest, desert, gravelly plains, edges of wet meadows, grassy or rocky slopes, shaded or exposed areas, dry to moist soil, near snowdrifts; 1300–3350 m. NCoRH, NCoRI, SnGb, GB, DMtns (Panamint Range); to BC, WY, CO, AZ. Most widely ranging member of sp. Apr–Sep

V. sempervirens Greene (p. 1277) EVERGREEN VIOLET, REDWOOD VIOLET Per, with stolon-like sts, glabrous or with scattered bristles on 1 or both lf surfaces. **ST:** prostrate, 1–several from current and/or previous yr's growth, lfy, ± woody in age, often forming rosettes near tip; rooted rosettes gen form vertical, rhizomatous caudex that produces new sts. **LF:** simple, evergreen; basal 1–5 per caudex, petiole 2–16 cm, blade 1–4.5 cm, 2–3.9 cm wide, ovate to round, often purple-spotted on 1 or both surfaces, crenate, base cordate to truncate, tip blunt to obtuse; cauline petiole 0.3–3 cm, blade 1.2–2.2 cm, 1.2–2 cm, ± like basal. **INFL:** axillary; peduncle 5–10 cm. **FL:** sepals lanceolate, often purple-streaked or spotted, not ciliate; petals lemon-yellow, lower 3 veined brown-purple, lateral 2 bearded with cylindric hairs, lowest 8–17 mm. **FR:** 6–7 mm, ovoid, tan, glabrous. **SEED:** ± 2 mm, brown, tinged purple. 2*n*=24,48. Shady areas in coastal forest; 5–1400 m. NW, CW; to AK, ID. Jan–Jul

V. sheltonii Torr. (p. 1281) SHELTON'S VIOLET Per 2–10 cm, glabrous to sparsely puberulent. **ST:** prostrate to erect, gen several, clustered on a deep or gen shallow, often vertical woody rhizome. **LF:** ternate-compound, wider than long; basal 1–2 per caudex, petiole 8.6–21 cm, blade 2–7 cm, 2–11 cm wide, reniform to semicircular, ± fleshy, lflets ± obovate, cleft, lobed, or 2–3-divided, segments 2–10 mm wide, oblanceolate, fiddle-shaped, lanceolate to spoon-shaped, gen not ciliate, tips acute to obtuse; cauline petiole 5.5–12 cm, blade 1.2–6.3 cm, 1.2–10.5 cm wide, ± like basal. **INFL:** axillary; peduncle 5–19 cm. **FL:** sepals lanceolate, gen not ciliate; petals deep lemon-yellow, upper 2 dark brown to purple-brown abaxially, lower 3 veined purple-brown, lateral 2 bearded with club-shaped hairs or not, lowest 7–18 mm. **FR:** 6–8 mm, oblong to ovoid, glabrous to puberulent. **SEED:** 2.5 mm, ± brown, shiny. 2*n*=12. Fir, pine, or oak woodland, rich or gravelly soil; 484–2500 m. NW, N&s SNH, SnFrB, SCoR, TR, nw PR, Wrn; to BC, ID, CO. Some populations with cleistogamous fls only. Mar–Jul

V. tomentosa M.S. Baker & J.C. Clausen (p. 1281) FELT-LEAVED VIOLET Per 7–10 cm, densely white-tomentose. **ST:** prostrate to erect, gen several, clustered on 1–several subterranean caudices from woody rhizome. **LF:** simple; basal 1–6 per caudex, petiole 2–6 cm, blade 1.5–5 cm, 1.4–2.1 cm wide, elliptic to narrowly ovate, entire, base obliquely tapered, tip acute to gen obtuse; cauline petiole 1.5–3.5 cm, blade 1.8–4 cm, 0.6–1.1 cm wide, ± like basal. **INFL:** axillary; peduncle 1–4 cm. **FL:** sepals lanceolate, not ciliate; petals deep lemon-yellow, upper 2 often maroon or ± brown abaxially, lower 3 veined brown-purple, lateral 2 bearded with cylindric hairs, lowest 6–11 mm; cleistogamous fls 0. **FR:** to 5 mm, ± round, white-tomentose. **SEED:** 2.5–2.8 mm, brown, mottled lighter. 2*n*=12. Dry, gravelly places in open pine forest (Jeffrey, lodgepole, ponderosa); 1350–2030 m. n-c SNH. Hybrids with *V. purpurea* appear sterile. May–Aug ★

VISCACEAE MISTLETOE FAMILY

Job Kuijt

Per, shrub, gen ± green, parasitic on aboveground parts of woody pls; dioecious [monoecious]. **ST:** brittle; 2° branches gen many. **LF:** simple, entire, opposite, 4-ranked, with blade or scale-like (then each pair gen fused). **INFL:** spikes or cymes, axillary or terminal; bracts opposite, 4-ranked, scale-like, each pair fused. **FL:** unisexual, radial, 2–4 mm; perianth parts in 1 series. **STAMINATE FL:** perianth parts 3–4(7); anthers gen sessile, opposite and gen on perianth parts. **PISTILLATE FL:** perianth parts gen 2–4; ovary inferior, 1-chambered, style unbranched, stigma ± obscure. **FR:** berry, shiny. **SEED:** 1(2), without thickened coat, gelatinous. 7 genera, ± 450 spp.: trop, gen n temp. All parts of most members may be TOXIC. [Kuijt 2003 Syst Bot Monogr 66:1–643] Sometimes incl in Loranthaceae; parasitic on pls in many other families. Scientific Editor: Thomas J. Rosatti.

1. St gen < 20 cm, ± angled at least when young, straw-colored, yellow, yellow-green, olive-green, green, brown, purple; lf scale-like, < 1 mm; fr ± broadly fusiform-spheric, 2-colored (1 color below, 1 above), not white, not pink, not ± red, explosive, pedicel recurved; pistillate perianth parts 2; anthers ± 1-chambered; on *Abies, Pinus, Pseudotsuga, Tsuga, Picea* . . . . **ARCEUTHOBIUM**
1′ St gen > 20 cm, not angled, green, yellow-green, gray-green, less often ± red; lf blade 5–70 mm, or lf scale-like, < 1 mm; fr ± spheric, 1-colored, white, pink, or ± red, not explosive, pedicel ± straight or 0; pistillate perianth parts gen 3–4; anthers 2–several-chambered; on woody angiosperms or *Abies, Calocedrus, Juniperus, Hesperocyparis.*
   2. Perianth parts gen 3, pistillate persistent; infl few- to many-fld spikes; fls ± sunken into axis (by proliferation of axis tissue); fr ± 3–6 mm; lf blade 5–47 mm, or lf scale-like, < 1 mm, glabrous or minutely tomentose; anthers 2-chambered; on esp *Abies, Acacia, Adenostoma, Alnus, Arctostaphylos, Calocedrus, Cercidium, Fraxinus, Hesperocyparis, Juglans, Juniperus,* (*Larrea*)*, Olneya, Platanus, Populus, Prosopis, Quercus, Rhus, Robinia, Salix, Umbellularia,* reportedly on *Pinus monophylla*
   . . . . **PHORADENDRON**
   2′ Perianth parts gen 4, pistillate gen deciduous; infl few-fld cymes; fls not sunken into axis; fr 6–10 mm; lf blade gen 50–80 mm, fleshy, glabrous; anthers several-chambered; on esp *Acer, Alnus, Betula, Crataegus, Malus, Populus, Robinia, Salix, Ulmus* . . . . **VISCUM**

ARCEUTHOBIUM DWARF MISTLETOE

Per, shrub, glabrous. **ST:** gen < 20 cm, ± angled at least when young, ± yellow, ± green, brown, purple; branches whorled or in 1 plane. **LF:** < 1 mm, scale-like. **INFL:** gen spikes, peduncle short; fls gen opposite, 4-ranked, less often whorled or 1, terminal. **STAMINATE FL:** perianth parts gen 3–4; anthers ± 1-chambered. **PISTILLATE FL:** perianth parts 2, persistent, minute. **FR:** gen 2–5 mm, broadly fusiform-spheric, 2-colored (1 color below, 1 above); pedicel short, recurved; seeds projected to 15 m by fr explosion. 26 spp.: temp, trop n hemisphere. (Greek: juniper, life) [Nickrent et al. 2004 Amer J Bot 91:125–138] Most important of timber pathogens; most spp. cause abnormal branching (witches' brooms) in hosts. Recent molecular studies support reunification of many w N.Am spp. under *A. campylopodum.*

1. St gen << 2 cm, 2° branches rare; on *Pseudotsuga menziesii*, rarely on associated *Abies* . . . . ***A. douglasii***
1′ St > 2 cm, 2° branches many; on *Abies, Pinus, Tsuga,* rarely *Picea breweriana*
  2. Larger branches whorled; on *Pinus contorta* subsp. *murrayana,* rarely on associated *Pinus* . . . . ***A. americanum***
  2′ Larger branches not whorled, in 1 plane; on *Abies, Pinus* (exc *Pinus contorta* subsp. *murrayana*),
    Tsuga, rarely *Picea.* . . . . ***A. campylopodum***

A. americanum Engelm. LODGEPOLE-PINE DWARF MISTLETOE **ST**: 6–18(25) cm, 1–2 mm wide at base; staminate yellow, pistillate yellow-green; larger branches whorled. **SEED**: mature Aug–Sep. *n*=14. Lodgepole-pine forest, on *Pinus contorta* subsp. *murrayana,* rarely on associated *Pinus*; 1300–2500 m. CaRH, SNH; to BC, CO. Apr–May

A. campylopodum Engelm. (p. 1281) WESTERN DWARF MISTLETOE **ST**: 3–14 cm, 1–6 mm wide at base, yellow, olive-green, or brown. **SEED**: mature Aug–Dec. *n*=14. Common. Conifer forest, on *Abies, Pinus* (exc *Pinus contorta* subsp. *murrayana*), *Tsuga,* rarely *Picea*; < 2800 m. NW, CaR, SN, CW, TR, PR, GB, DMtns; to se AK, MT, CO, NM, n Baja CA. [*A. abietinum* (Engelm.)

Hawksw. & Wiens; *A. californicum* Hawksw. & Wiens; *A. cyanocarpum* J.M. Coult. & A. Nelson; *A. divaricatum* Engelm.; *A. littorum* Hawksw. et al.; *A. monticola* Hawksw. et al.; *A. occidentale* Engelm.; *A. siskiyouense* Hawksw. et al.; *A. tsugense* (Rosend.) G.N. Jones; *A. t.* subsp. *mertensianae* Hawksw. & Nickrent; *A. t.* subsp. *t.*] Gen Jul–Nov

A. douglasii Engelm. DOUGLAS-FIR DWARF MISTLETOE **ST**: gen < 2 cm, 1 mm wide at base, green; 2° branches rare. **SEED**: mature Sep–Oct. *n*=14. Uncommon. Mixed-conifer woodland, on *Pseudotsuga menziesii,* rarely on associated *Abies*; 1500–2000 m. KR, CaRH; to BC, CO. May–Jun

PHORADENDRON MISTLETOE

Per, shrub, woody at least at base, glabrous or short-hairy. **ST**: gen > 20 cm, not angled, green, less often ± red. **LF**: with blade or < 1 mm, scale-like. **INFL**: spikes, few- to many-fld, peduncled; fls ± sunken into axis. **FL**: perianth parts gen 3. **STAMINATE FL**: anthers 2-chambered. **PISTILLATE FL**: perianth parts persistent. **FR**: ± 3–6 mm, ± spheric, 1-colored, white, pink, or ± red, bird-dispersed; pedicel 0. *n*=14. ± 240 spp.: temp, trop Am. (Greek: tree thief) [Kuijt 2003 Syst Bot Monogr 66:1–643]

1. Lf scale-like, < 1 mm
  2. St canescent, esp at tip, ± red to green, rigid; spike with 1–3(7) fertile internodes; on (*Larrea*), *Olnyea,*
    Parkinsonia, Prosopis, Senegalia, Simmondsia; < 1200 m. . . . . ***P. californicum***
  2′ St glabrous, ± green or yellow-green, not rigid; spike with 1(2) fertile internodes; on *Calocedrus,*
    Juniperus; 1300–2600 m . . . . ***P. juniperinum***
1′ Lf with blade, gen 5–60 mm
  3. Lf length gen > 3 × width, (5)10–25 mm, gen < 20 mm; st glabrous; spike with gen 1 fertile internode,
    pistillate 2-fld; on conifers . . . . ***P. bolleanum***
  3′ Lf length gen < 1.5 × width, 15–60 mm, gen > 20 mm; st gen short-hairy, at least at tip; spike with 2–5(7)
    fertile internodes, pistillate 6–15(20)-fld; on woody pls exc conifers. . . . . ***P. serotinum***
    4. Lf ≤ 6 cm, width ≤ 5 cm; lf yellow-green, gen shiny . . . . subsp. ***macrophyllum***
    4′ Lf ≤ 3 cm, width ≤ 2 cm; lf gray-green, dull. . . . . subsp. ***tomentosum***

P. bolleanum (Seem.) Eichler **ST**: 3–6 dm, erect-spreading, green to olive-green, glabrous; internodes 6–22 mm. **LF**: (5)10–25 mm, 2–8 mm wide, oblanceolate-oblong. **STAMINATE INFL**: fertile internodes 1(2), 6–20-fld. **PISTILLATE INFL**: fertile internodes gen 1, 2-fld. **FR**: ± 4 mm, white to straw-colored or ± pink, glabrous. Pinyon/juniper woodland, on *Hesperocyparis, Juniperus,* locally on *Abies concolor*; 200–2500 m. KR, NCoR, CaR, n SNH, c&s SN, CW, TR, PR, GB, D; to s OR, c&s AZ, nw Mex. [*P. densum* Trel.; *P. pauciflorum* Torr.] Jun–Aug

P. californicum Nutt. (p. 1281) DESERT MISTLETOE **ST**: 4–10 dm, erect to spreading, ± red to green, canescent, esp at tip, ± glabrous in age; internodes 13–28 mm. **LF**: < 1 mm, scale-like. **STAMINATE INFL**: fertile internodes 1–3(6), 6–14-fld. **PISTILLATE INFL**: fertile internodes 1–3(7), 2(3)-fld. **FR**: ± 3 mm, white to red-pink, glabrous. Desert, on (*Larrea*), *Olneya, Parkinsonia, Prosopis, Senegalia, Simmondsia*; < 1200 m. WTR, D; NV, AZ, Baja CA. Jan–Mar

P. juniperinum A. Gray JUNIPER OR INCENSE CEDAR MISTLETOE **ST**: 2–8 dm, erect or pendent, gen woody only at base, ± green or yellow-green, glabrous; internodes 5–20 mm. **LF**: < 1 mm, scale-like. **STAMINATE INFL**: fertile internodes 1(2), 6(8)-fld. **PISTILLATE INFL**: fertile internode 1, 2-fld. **FR**: ± 4 mm, pink-white, glabrous. Pinyon/juniper woodland, ponderosa-pine forest, on *Calocedrus,*

Juniperus; 1700–2600 m. KR, NCoR, CaRH, n SNF, SNH, SCoRI, TR, PR, GB, DMoj; to OR, CO, TX, Mex. [*P. libocedri* (Engelm.) Howell] Jul–Sep

P. serotinum (Raf.) M.C. Johnst. AMERICAN MISTLETOE **ST**: erect to spreading, green, gen short-hairy, at least at tip, ± glabrous in age; internodes 15–59 mm. **LF**: 15–60 mm, 10–25 mm wide, obovate to elliptic-round, ± petioled or not, ± glabrous to densely short-hairy. **STAMINATE INFL**: fertile internodes 2–5(7), gen 25–35-fld. **PISTILLATE INFL**: fertile internodes 2–4(5), 6–15(20)-fld. **FR**: 4–5 mm, white, pink-tinged or not, glabrous to short-hairy near tip. 2 other subspp., in se US, Mex.

    subsp. ***macrophyllum*** (Engelm.) Kuijt **ST**: ≤ 1 m. **LF**: ≤ 6 cm, width ≤ 5 cm, yellow-green, gen shiny. On trees other than *Quercus* (esp *Alnus, Fraxinus, Juglans, Platanus, Populus, Robinia, Salix*); < 1200 m. NCoRO, NCoRI, SNF, GV, CW, SCo, TR, PR, D; to CO, w TX, Baja CA. [*P. m.* (Engelm.) Cockerell] Dec–Mar

    subsp. ***tomentosum*** (DC.) Kuijt (p. 1281) **ST**: gen ≤ 80 cm. **LF**: ≤ 3 cm, width ≤ 2 cm, gray-green, dull. Gen on *Quercus,* rarely on *Adenostoma, Arctostaphylos, Cercocarpus ledifolius, Rhus, Umbellularia*; 60–2100 m. KR, NCoR, CaR, SN, GV, SnFrB, SCoR, SCo, TR, PR, DMoj; to n OR, TX, Mex. [*P. villosum* (Nutt.) Engelm.] Jul–Sep

VISCUM MISTLETOE

Shrub, glabrous, evergreen. **ST**: gen < 20 cm, rounded, green, less often ± red; 2° branches opposite or whorled. **LF**: with blade. **INFL**: few-fld cyme, dense, subtended by pair of fused bracts; peduncle 0 or short. **FL**: perianth parts gen 4. **STAMINATE FL**: anthers several-chambered, cushion-like. **PISTILLATE FL**: perianth parts gen deciduous. **FR**: 6–10 mm, spheric, white in CA, bird-dispersed; pedicel short, ± straight, or 0. ± 125 spp.: temp, trop, Old World. (Latin: from viscid seed cover) [Hawksworth & Scharpf 1987 Eur J Forest Pathol 16:1–5]

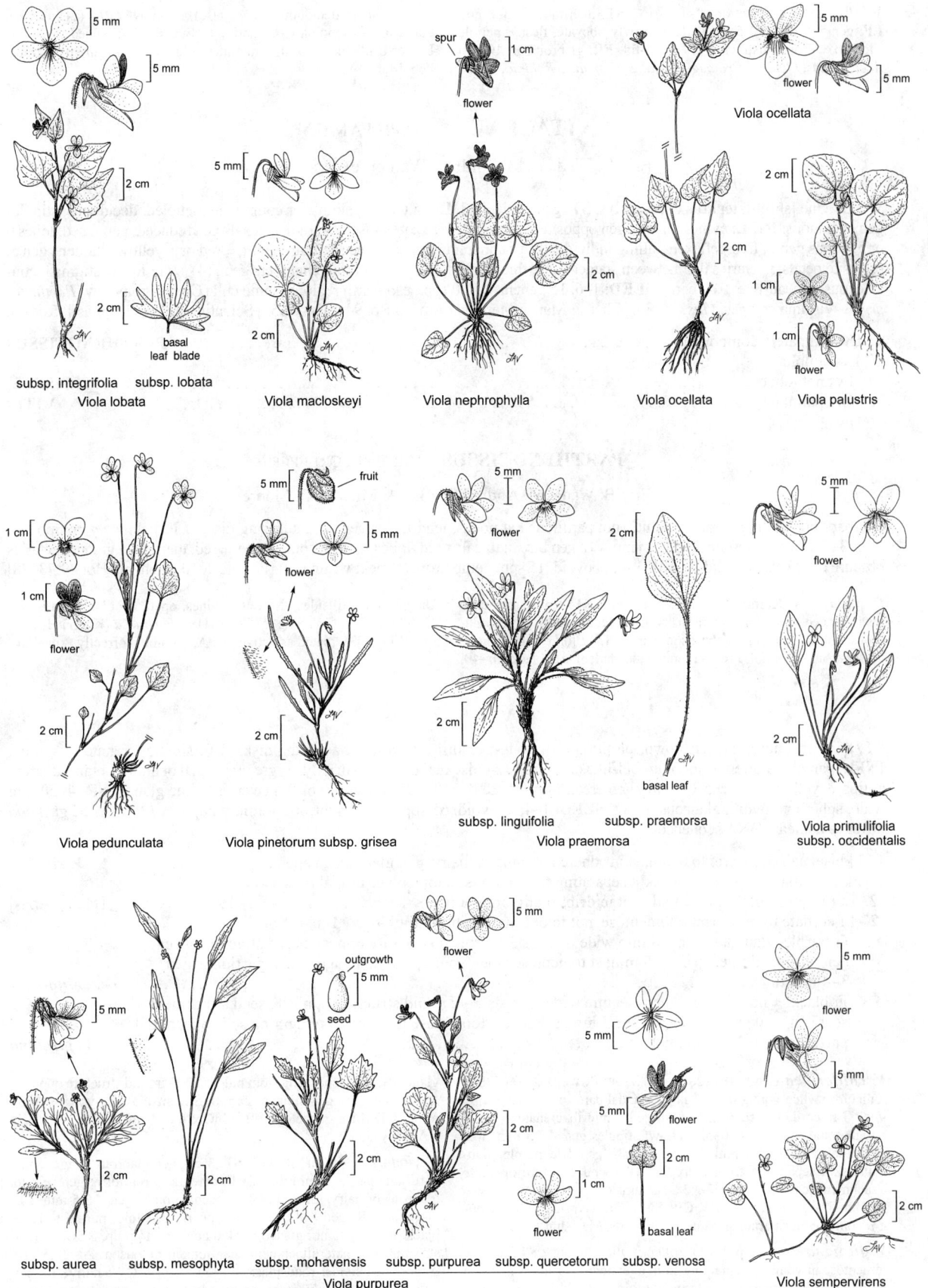

5 mm

5 mm

2 cm

2 cm

basal
leaf blade

subsp. integrifolia · subsp. lobata
Viola lobata

5 mm

2 cm

Viola macloskeyi

spur

flower

1 cm

2 cm

Viola nephrophylla

5 mm

flower

2 cm

Viola ocellata

5 mm
flower
5 mm

Viola ocellata

2 cm

1 cm

1 cm
flower

Viola palustris

1 cm

1 cm
flower

2 cm

Viola pedunculata

5 mm

fruit
5 mm
flower

2 cm

Viola pinetorum subsp. grisea

5 mm
flower

2 cm

2 cm

basal leaf

subsp. linguifolia   subsp. praemorsa
Viola praemorsa

5 mm
flower

2 cm

Viola primulifolia
subsp. occidentalis

5 mm

2 cm   2 cm

subsp. aurea   subsp. mesophyta

outgrowth
5 mm
seed

subsp. mohavensis

flower
5 mm

2 cm

subsp. purpurea

5 mm

5 mm
flower

2 cm

1 cm
flower

subsp. quercetorum

2 cm

basal leaf

subsp. venosa

Viola purpurea

5 mm
flower
5 mm

2 cm

Viola sempervirens

V. album L.   EUROPEAN MISTLETOE   **ST**: internodes ± 3–8 cm. **LF**: gen 5–8 cm, ± 1.5 cm wide, narrowly obovate, fleshy; petiole short to ± 0 or indistinct. **INFL**: 3–5-fld. **FR**: glabrous. *n*=10. On *Acer, Alnus, Betula, Crataegus, Malus, Populus, Robinia, Salix,* *Ulmus*, other deciduous trees; 60–100 m. NCoRO (Sebastopol, Santa Rosa, Sonoma Co.); native to Eurasia. Introduced to CA by Luther Burbank, ± 1900, sometimes sold locally in Christmas trade. Feb–Mar ◆

VITACEAE   GRAPE FAMILY

Eric B. Wada & M. Andrew Walker, except as noted

Woody vine [shrub]; tendrils opposite lvs. **ST**: gen lenticelled. **LF**: alternate, simple or compound, petioled, deciduous; stipules gen deciduous. **INFL**: cyme, panicle, gen opposite lf, peduncled. **FL**: gen bisexual, radial; sepals gen reduced, gen fused, lobes 0 or 5; petals gen 5, free, reflexed, falling individually, or adherent at tips, ± erect, falling as unit, ± red or ± yellow; stamens gen 5, opposite petals; nectaries 0 or between stamens as ± free glands; ovary 1, superior, chambers gen 2(4), style 1 or 0, stigma inconspicuous or head-like. **FR**: berry. **SEED**: 1–6. 15 genera, ± 800 spp.: esp warm regions; some cult (*Cissus*, grape ivy; *Parthenocissus*, Virginia creeper; *Vitis*, grape). [Chen & Manchester 2007 Amer J Bot 94:1534–1553] Scientific Editor: Thomas J. Rosatti.

1. Lvs palmately compound.. **PARTHENOCISSUS**
1′ Lvs simple
  2. Lvs not lobed.. **[CISSUS]**
  2′ Lvs palmately lobed... **VITIS**

PARTHENOCISSUS   VIRGINIA CREEPER

Eric B. Wada, M. Andrew Walker & Michael O. Moore

ST: bark not peeling; st center white, not partitioned at nodes; tendril tips gen with adhering disks. **LF**: palmately compound; lflets 3–7, coarsely serrate. **INFL**: cyme. **FL**: gen bisexual; calyx red, lobes shallow; petals free, ± red, margins ± green; nectaries obscure or 0. **FR**: obovoid. **SEED**: 1–4, obovoid. 15 spp.: temp, trop. (Greek: virgin ivy) [Pringle 2010 Michigan Bot 49:73–78]

P. inserta (A. Kern.) Fritsch   WOODBINE   **ST**: tendril branches few, tips gen without adherent disks. **LF**: lflet adaxially glossy, glabrous, abaxially ± dull, glabrous to hairy. **INFL**: forked at peduncle tip, again above or not. **FR**: 9–12 mm wide, dark blue to black. 2*n*=40. Uncommon. Hillsides, thickets, ravines, open woodland, roadsides; < 1000 m. GV, CW, SW; to TX, ne US. [*P. vitacea* (Knerr) Hitchc.] In TJM (1993) treated as native to CA, possibly correctly. Apr–Jun

VITIS   GRAPE

ST: bark peeling; st center brown, partitioned at nodes; tendril tips without adhering disks. **LF**: simple, crenate to serrate. **INFL**: panicle of often head- or umbel-like clusters. **FL**: unisexual or bisexual; calyx ± green, lobes 0 or short; petals adherent at tips, ± yellow; stamens 3–9 mm, gen erect, in pistillate fls reflexed and sterile or 0; nectaries ± free glands. **FR**: 4–20 mm wide, spheric to ovoid, glaucous or not. **SEED**: 1–4, obovoid. 65 spp.: temp, subtrop. (Latin: vine) *V. californica, V. girdiana* differ in nuclear rDNA sequences.

1. Fl bisexual; fr spheric to ovoid, skin adherent to pulp; st hairy, gen glabrous in age .... *V. vinifera*
1′ Fl unisexual; fr spheric or 0, skin separating from pulp; st ± tomentose or glabrous
  2. Lf reniform, glabrous, ± folded at midrib; st climbing or not, < 1 m .... *[V. rupestris]*
  2′ Lf cordate to reniform, ± tomentose, not folded at midrib; st climbing, > 1 m
    3. Fr purple, glaucous, gen > 8 mm wide if seeds 3–4; round structure opposite seed attachment scar gen raised; stipules gen < 3.5 mm; st tomentose when young, less so in age, nodal partitions gen 3–4 mm thick... *V. californica*
    3′ Fr black, ± not glaucous, gen < 8 mm wide if seeds 3–4; round structure opposite seed attachment scar gen sunken; stipules gen > 3.5 mm; st densely tomentose in youth, ± remaining so in age, nodal partitions gen 2–3 mm thick .... *V. girdiana*

V. californica Benth. (p. 1281)   CALIFORNIA WILD GRAPE   **ST**: tomentose when young, less so in age; nodal partitions gen 3–4 mm thick. **LF**: cordate to reniform, not folded at midrib, crenate to ± serrate, ± tomentose, lobes 0–5, shallow; stipules gen < 3.5 mm. **FL**: unisexual. **FR**: gen > 8 mm wide if seeds 3–4, spheric, purple, glaucous; skin separating from pulp. **SEED**: round structure opposite attachment scar gen raised. 2*n*=38. Streamsides, springs, canyons; < 1250 m. NW, CaRF, SNF, GV, CW, SNE; OR. Hybrids with *V. vinifera* common in riparian areas near vineyards. May–Jun

V. girdiana Munson (p. 1281)   DESERT WILD GRAPE   **ST**: densely tomentose in youth, ± remaining so in age; nodal partitions gen 2–3 mm thick. **LF**: cordate to reniform, not folded at midrib, gen serrate, ± tomentose, lobes 0 or 3–5, shallow; stipules gen > 3.5 mm. **FL**: unisexual. **FR**: gen < 8 mm wide if seeds 3–4, spheric, black, ± not glaucous; skin separating from pulp. **SEED**: round structure opposite attachment scar gen sunken. 2*n*=38. Streamsides, canyons; < 1750 m. SW, DMtns, expected SnBr, SnGb; s NV, nw AZ, sw UT, Baja CA. May–Jun

V. vinifera L.   WINE GRAPE   **ST**: hairy, gen glabrous in age; nodal partitions gen 3–5 mm thick. **LF**: cordate to reniform, gen serrate, glabrous or 3–5, deep; stipules gen < 3.5 mm. **FL**: bisexual. **FR**: gen > 8 mm wide, spheric to ovoid, purple to blue-black, densely to not glaucous; skin adherent to pulp. **SEED**: round structure opposite attachment scar sunken or raised. 2*n*=38,57,76. Abandoned fields, roadsides; < 1000 m. GV, CW; native to Eur. Hybrids with *V. californica* commonly found in riparian areas near vineyards. May–Jun

ZYGOPHYLLACEAE   CALTROP FAMILY

Duncan M. Porter

Ann, per, shrub, often armed; caudex present or not. **ST:** branched; nodes often angled, swollen. **LF:** 1-compound, opposite, petioled; stipules persistent or not; lflets entire. **INFL:** fls 1–2 in axils. **FL:** bisexual; sepals 5, free, persistent or not; petals 5, free, gen spreading, twisted (corolla propeller-like) or not; stamens 10, appendaged on inside base or not; ovary superior, chambers (and lobes) 5–10, each with 1–several ovules, placentas axile. **FR:** capsule or splitting into 5–10 nutlets (= mericarps). 27 genera, ± 250 spp.: widespread esp in warm, dry regions; some cult (*Guaiacum*, lignum vitae; *Tribulus*, caltrop). [Sheahan & Chase 2000 Syst Bot 25:371–384] Scientific Editor: Thomas J. Rosatti.

1. Lflets 2
  2. Lflets fused at base . . . . **LARREA**
  2′ Lflets free at base . . . . **ZYGOPHYLLUM**
1′ Lflets 3 or more
  3. Lflets 3, palmate, spine-tipped; stipules spine-tipped . . . . **FAGONIA**
  3′ Lflets 6–18, pinnate, not spine-tipped; stipules not spine-tipped
    4. Fr with many tubercles, 0 spines, 10 nutlets . . . . **KALLSTROEMIA**
    4′ Fr with many tubercles, 2–4 stout spines, 5 nutlets . . . . **TRIBULUS**

FAGONIA

Per, shrub. **ST:** < 1 m, spreading, angled or ridged. **LF:** palmately compound; stipules stiff, spine-tipped; lflets 3, spine-tipped, terminal largest. **INFL:** fls 1 in axils. **FL:** sepals deciduous; petals clawed, twisted, purple to pink, deciduous. **FR:** capsule, deeply 5-lobed, obovoid, loculicidal; style persistent; pedicel reflexed. **SEED:** 1 per chamber. 35 spp.: sw N.Am, sw S.Am, Canary I. to India, sw Afr. (Guy-Crescent Fagon, French botanist, chemist, physician to Louis XIV, 1638–1718) [Beier 2005 Syst Biodivers 3:221–263]

1. St ascending to erect, scabrous; glands only on youngest herbage, << 0.1 mm wide; stipules above base
    curved; lflets lanceolate . . . . *F. laevis*
1′ St prostrate, not scabrous; glands on youngest herbage and elsewhere, ± 0.15 mm wide; stipules above
    base straight; lflets elliptic to ovate . . . . *F. pachyacantha*

F. laevis Standl. (p. 1281)   Shrub < 1 m, intricately branched. **LF:** lflets 3–9 mm, gen < petiole, 1–4 mm wide. **FL:** ± 1 cm wide. **FR:** 4–5 mm wide, minutely strigose or hairy, rarely glandular or glabrous; style 1–2 mm, wider at base. Rocky hillsides, sandy washes; < 1200 m. D; to sw UT, nw Mex. Further study needed to determine if *F. longipes* Standl. [*F. californica* Benth. subsp. *longipes* (Standl.) Felger & C.H. Lowe] (minute, glandular hairs on frs) represents a distinct evolutionary lineage. Mar–May, Nov–Jan

F. pachyacantha Rydb. (p. 1281)   Per; caudex woody. **LF:** lflets < 25 mm, ± ≥ petiole, < 9 mm wide. **FL:** ± 1.5 cm wide. **FR:** 5 mm wide, hairy, gen with some glands; style 2–3 mm, not or barely wider at base. Flat, sandy or rocky habitats; < 500 m. DSon; to sw AZ, Mex. [*F. californica* subsp. *p.* (Rydb.) Wiggins] Mar–May, Nov–Jan

KALLSTROEMIA

Ann, unarmed. **ST:** spreading radially, prostrate to ascending, < 1 m. **LF:** even-1-pinnate; stipules narrow, green. **INFL:** fls 1 in axils. **FL:** sepals persistent or not; petals yellow to orange, persistent, withering. **FR:** 10-lobed, splitting into 10 tubercled nutlets; style persistent; pedicel reflexed. **SEED:** 1 per chamber. 17 spp.: warm and trop Am. (Anders Kallstroem, obscure contemporary of Scopoli, author of genus, 1733–1812) [Porter 1969 Contr Gray Herb 198:41–153]

1. Style < fr body; peduncle gen < to ± = subtending lf; petals 3–5 mm; sepals deciduous . . . . *K. californica*
1′ Style > fr body; peduncle > to >> subtending lf; petals 6–30 mm; sepals persistent
  2. Sepals >> fr body, spreading; petals 15–30 mm, yellow to orange, darker at base . . . . [*K. grandiflora*]
  2′ Sepals ± = to slightly > fr body, appressed; petals 6–12 mm, orange . . . . *K. parviflora*

K. californica (S. Watson) Vail (p. 1281)   **ST:** prostrate to decumbent, < 0.7 m, strigose to glabrous. **LF:** stipules 1.5–5 mm; lflets 6–12. **FL:** pedicel gen < to ± = subtending lf; sepals spreading, deciduous; petals 3–5 mm, yellow. **FR:** 3–5 mm wide, ovoid; style < body. Flat, sandy or disturbed areas; < 600 m. D; to TX, Mex. Aug–Oct

K. parviflora Norton (p. 1281)   **ST:** prostrate to decumbent, < 1 m, strigose to glabrous. **LF:** stipules 5–7 mm; lflets 6–10. **FL:** pedicel > subtending lf; sepals persistent, ± = to slightly > fr body, appressed; petals 6–12 mm, orange. **FR:** 4–6 mm wide, ovoid; style to 3 × fr body. Uncommon. Sandy roadsides, slopes; 1500–2000 m. c PR (near Warner Hot Springs, San Diego Co.), e DMoj, expected elsewhere; to c US, c Mex. Recently collected in San Bernardino Co., near site of 1950 collections. Aug–Oct

LARREA CREOSOTE BUSH

Shrub, unarmed. **ST**: branched, erect to prostrate, < 4 m, ± red becoming gray; nodes swollen, darker; hairs 0 or appressed. **LF**: stipules persistent; lflets 2, fused at base. **INFL**: fls 1 in axils. **FL**: sepals unequal, overlapping, deciduous; petals clawed, twisted, yellow, deciduous; stamen appendages bract-like, coarsely toothed. **FR**: 5-lobed, spheric, short-stalked, hairy, splitting into 5 hairy, 1-seeded nutlets. 5 spp.: warm, dry Am. (J.A. Hernández de Larrea, Spanish bishop, 1730–1803) [Lia et al. 2001 Molec Phylogen Evol 21:309–320]

L. tridentata (DC.) Coville (p. 1281) **LF**: lflets < 18 mm, < 8.5 mm wide, obliquely lanceolate to curved; awn between lflets < 2 mm, ± deciduous. **FL**: < 2.5 cm wide; sepals ovoid, appressed-hairy; petal claw ± brown; stamens > appendages; ovary hairs dense, straight, stiff, silvery (red-brown in fr); style 4–6 mm, persistent on young fr. **FR**: 4.5 mm wide (exc hairs), hairs ± 2–4 mm, dense, spread-ing. Common. Desert scrub; < 1000 m. SNE, D, (uncommon Teh, SnJV, SCo, SnJt); to sw UT, TX, c Mex. Closely related to s S.Am *L. divaricata*. Clones may live > 11000 yrs, longest among extant pls; resinous odor characteristic; dominant shrub over vast areas of desert. Apr–May

TRIBULUS PUNCTURE VINE, CALTROP

Ann. **ST**: prostrate, spreading radially, gen < 1 m. **LF**: even-1-pinnate; stipules ± lf-like. **INFL**: fls 1 in axils. **FL**: sepals decidu-ous; petals yellow, deciduous. **FR**: 5-lobed, splitting into 5 nutlets, each with many tubercles, 2–4 stout spines; style decidu-ous; pedicel reflexed. **SEED**: 3–5 per chamber. ± 12 spp.: esp dry Afr. (Latin: weapon used to impede cavalry, from armed fr)

T. terrestris L. (p. 1281) **ST**: ± silky or appressed-hairy, also sharply bristly. **LF**: stipules 1–5 mm; lflets 6–12. **FL**: < 5 mm wide; pedicel gen < subtending lf. **FR**: 5 mm, < 1 cm wide, ± flat, hairy, gray or ± yellow; spines 4–7 mm, spreading, hairy to glabrous. Dry, disturbed areas incl roadsides, railways, vacant lots; gen < 1000 m. CA; to WY, e US, c Mex; native to Medit. TOXIC to livestock in vegetative condition, frs cause mechanical injury. First collected in CA in 1902; long a pernicious weed, now controlled by introduced weevils. Apr–Oct ◆

ZYGOPHYLLUM BEAN-CAPER

Per, unarmed; caudex woody. **ST**: branched, fleshy. **LF**: stipules ± lf-like or membranous; lflets 2 [> 2], free, fleshy. **INFL**: fls 1–2 in axils. **FL**: sepals often unequal, overlapping, deciduous or not; petals ± erect, white, yellow, or orange, base sometimes orange or red. **FR**: capsule, cylindric to spheric, 5-angled, septicidal. **SEED**: 1–few per chamber. ± 70 spp.: Eurasia, s Afr, Australia. (Greek: yoke lf, from sometimes oblique lflets)

Z. fabago L. (p. 1281) **ST**: erect, glabrous, < 1 m. **LF**: lflets < 4.5 cm, 3 cm wide, obliquely ovate; awn between lflets 1 mm, ± persis-tent. **FL**: 6–7 mm; sepals ± = petals; stamens exserted, appendages < stamens, ± linear, divided at tip. **FR**: 25–35 mm, oblong-cylindric; style persistent, thread-like, < 7 mm; pedicel reflexed. Uncommon. Disturbed areas; < 1000 m. s NCoRO, SnJV, D; to KS, TX, also in NY, PA, native to Medit, c Asia. Forms large colonies; fl buds have been used in place of capers. Summer ◆

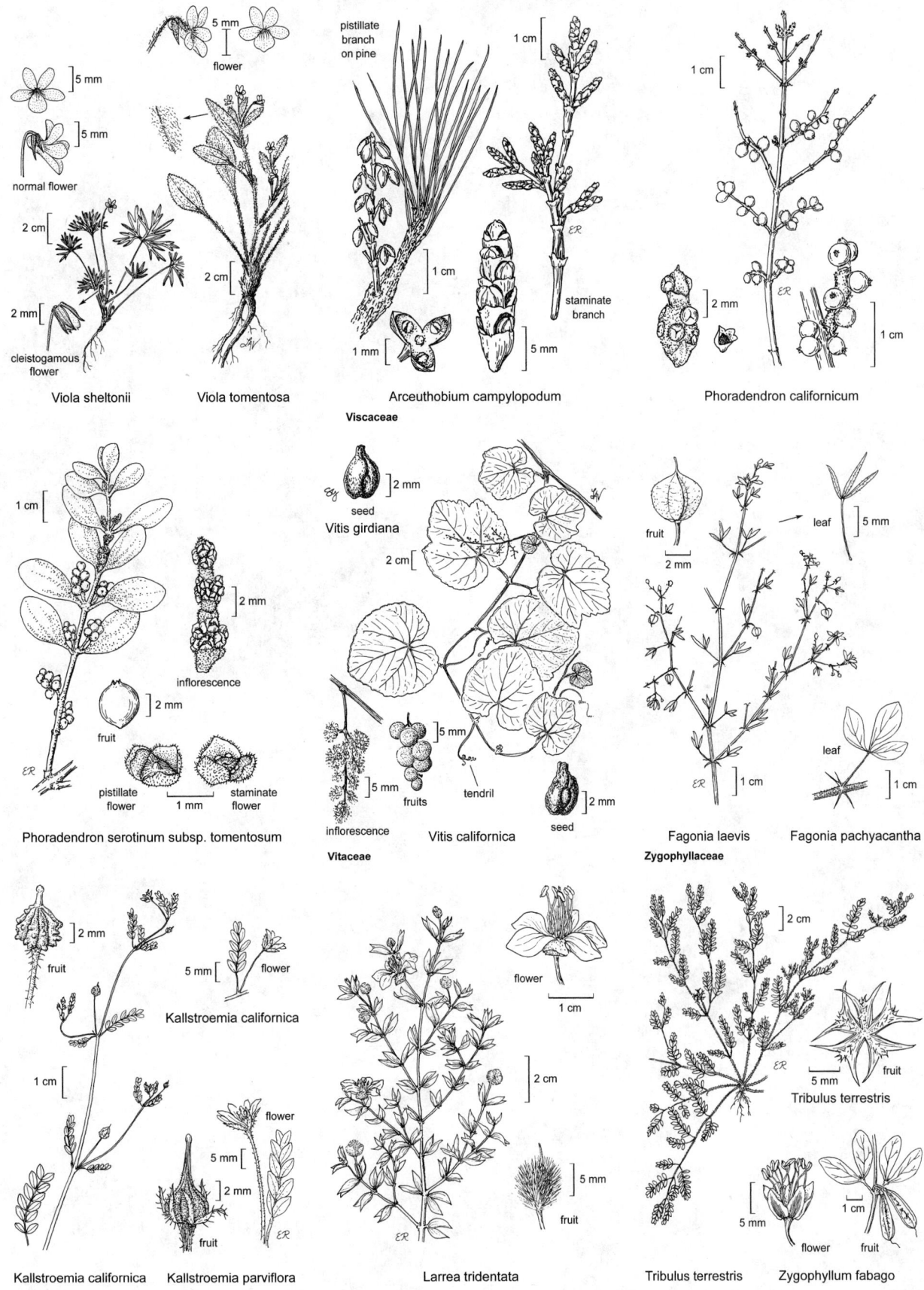

5 mm
flower

5 mm
normal flower

5 mm

2 cm

2 mm
cleistogamous flower

Viola sheltonii

2 cm

Viola tomentosa

pistillate branch on pine

1 cm

1 cm

staminate branch

1 mm

5 mm

Arceuthobium campylopodum

Viscaceae

1 cm

2 mm

1 cm

Phoradendron californicum

1 cm

2 mm
inflorescence

2 mm
fruit

1 mm
pistillate flower   staminate flower

Phoradendron serotinum subsp. tomentosum

2 mm
seed
Vitis girdiana

2 cm

5 mm
fruits

5 mm
inflorescence

tendril

2 mm
seed

Vitis californica

Vitaceae

fruit
2 mm

5 mm
leaf

leaf

1 cm

Fagonia laevis

1 cm

Fagonia pachyacantha

Zygophyllaceae

2 mm
fruit

5 mm
flower

Kallstroemia californica

1 cm

flower

5 mm

2 mm
fruit

Kallstroemia californica   Kallstroemia parviflora

flower

1 cm

2 cm

5 mm
fruit

Larrea tridentata

2 cm

5 mm
fruit
Tribulus terrestris

5 mm
flower

1 cm
fruit

Tribulus terrestris   Zygophyllum fabago

MONOCOTS

Fl pls; cotyledon 1; lf veins gen parallel from base or midrib; fl parts gen in 3s; pollen aperture 1 [*Halodule wrightii* Asch. (Cymodoceaceae), once naturalized in Salton Sea, presumed extirpated in CA.]

AGAVACEAE CENTURY PLANT FAMILY

Dale W. McNeal, except as noted

Per, shrub, tree, fibrous succulent or not, from bulbs or rhizomes. **ST**: above ground or not, branched or not. **LF**: simple, deciduous or not, basal or in terminal rosettes, gen sessile, linear, lanceolate, oblanceolate or ovate, fibrous or not, thin and flexible or thick and rigid or succulent; margin entire, fine-serrate, dentate, or with filaments, tips rigid or flexible, with a spine or not. **FL**: bisexual; perianth parts 6, in 2 petal-like whorls, free or ± fused; stamens 6, ± fused to perianth, filaments often wide, succulent; ovary superior or inferior, chambers 3, style 1 (thick, poorly defined), stigma head-like or 3-lobed. **FR**: capsule, indehiscent, loculicidal, or septicidal. **SEED**: few to many, ± flat or ovoid, gen black. 23 genera, 637 spp.: worldwide. Scientific Editors: Dale W. McNeal, Thomas J. Rosatti.

1. Gen shrub- or tree-like; lvs thick, gen rigid
   2. Ovary inferior; perianth yellow to green-yellow . . . . **AGAVE**
   2′ Ovary superior; perianth white or cream to ± green, purple-tinged or -tipped or not, or red-brown abaxially, ± white adaxially
      3. Fr erect, loculicidal; peduncle gen > 2.5 cm wide, bracts gen reflexed; stigma head-like. . . . . **HESPEROYUCCA**
      3′ Fr erect or pendent, tardily septicidal or gen indehiscent; peduncle gen < 2 cm wide, bracts gen ascending; stigma 3-lobed. . . . . **YUCCA**
1′ Per, gen not shrub- or tree-like; lvs thin, flexible
   4. St not evident above ground — lvs, fls tufted . . . . **LEUCOCRINUM**
   4′ St evident above ground
      5. Infl a panicle
         6. Style thread-like, deciduous; perianth twisted together above ovary in fr. . . . . **CHLOROGALUM**
         6′ Style ± stout, persistent; perianth not twisted together above fr. . . . . ²**HASTINGSIA**
      5′ Infl a raceme
         7. Perianth parts < 1 cm, white (green-white) to ± yellow or ± purple. . . . . ²**HASTINGSIA**
         7′ Perianth parts 1.2–6 cm, white or blue to ± purple
            8. Perianth parts blue to ± purple (white), fused < 5% of length; scapose; lf margins not wavy. . . . . **CAMASSIA**
            8′ Perianth parts white with silver-green midstripe esp abaxially, fused ± 25% of length; not scapose; lf margins wavy . . . . **HESPEROCALLIS**

AGAVE CENTURY PLANT

James L. Reveal

Shrub-like, st often short, forming basal rosettes, scapose, often rhizomed, gen dying after fl. **LF**: long-lived, sessile, lance-linear to ovate, thick, gen rigid, often fleshy, glabrous, armed with marginal teeth, spine at tip. **INFL**: panicle-, raceme-, or spike-like, bracted, often with bulblets. **FL**: [1] paired, or in clusters of 3–40; perianth gen funnel- or bell-shaped, yellow to green-yellow, parts petal-like, 6 in 2 whorls, fused basally, lobes erect to ascending or ± spreading distally; stamens 6; ovary inferior, 3-chambered. **FR**: oblong to ovoid, often beaked, loculicidal. **SEED**: many, ± flat, black. ± 200 spp.: warm, trop Am. (Greek: noble, for imposing stature) [Reveal & Hodgson 2002 FNANM 26:442–461]

1. Infl raceme-like; filaments white, ± from base of perianth tube; lf tip-spine 4–20 cm. . . . . *A. utahensis*
   2. Lf tip-spine ivory-white, 10–20 cm . . . . var. *eborispina*
   2′ Lf tip-spine brown to ± white, 4–8.5 cm . . . . var. *nevadensis*

1′ Infl panicle-like; filaments yellow, from middle of perianth tube or above; lf tip-spine 2–4 cm
  3. Lf narrow-ovate; perianth 6–10 cm, lobes 17–40 mm; fr 5.5–7 cm; s SCo (sw San Diego Co.) . . . *A. shawii* var. *shawii*
  3′ Lf lance-linear to lanceolate; perianth 3–6 cm, lobes 13–20 mm; fr 3–6 cm; DMtns (San
    Bernardino Co.), DSon . . . . *A. deserti*
    4. Perianth tube 3–5 mm, bell-shaped; filaments ± from top of perianth tube; rosettes many; DSon . . . . var. *deserti*
    4′ Perianth tube (3)5–10 mm, funnel-shaped; filaments from above middle of perianth tube; rosette 1;
      DMtns (San Bernardino Co.) . . . . var. *simplex*

A. deserti Engelm. St 0–0.4 m; rosettes 1 or many, 3–7 dm. **LF:** lance-linear to lanceolate, marginal teeth 2–3 mm or 5–8 mm, gen 1.5–3 cm apart, tip-spine 2–4 cm. **INFL:** panicle-like, 2–6 m incl peduncle; bracts 8–15 cm, triangular, persistent; branches 6–15, gen 10–20 cm; fls 12–48 per cluster. **FL:** perianth 3–6 cm, lobes 13–20 mm, equal; stamens long-exserted, filaments 2.5–3.5(4.2) cm, yellow, anthers 13–21 mm; ovary 1.6–4 cm. **FR:** 3–6 cm. **SEED:** 5–6 mm.

  var. **deserti** (p. 1287) DESERT AGAVE Rosettes many. **LF:** 25–40 cm, ± gray. **FL:** perianth tube 3–5 mm, bell-shaped; filaments ± from top of tube. Sandy to rocky slopes, washes in desert scrub; 300–1500 m. DSon; Baja CA. May–Jul

  var. **simplex** (Gentry) W.C. Hodgs. & Reveal SIMPLE DESERT AGAVE Rosette 1. **LF:** (20)25–60 cm, ± green. **FL:** perianth tube (3)5–10 mm, funnel-shaped; filaments from above middle of tube. Sandy to rocky slopes, washes in desert scrub; 300–1500 m. DMtns (San Bernardino Co.); AZ, SON. [*A. d.* subsp. *s.* Gentry] May–Jul

A. shawii Engelm. var. **shawii** (p. 1287) SHAW AGAVE St 0.5–2 m; rosette 1, 0.8–20 dm. **LF:** 20–50 cm, narrow-ovate, marginal teeth 5–6 mm, 1–2(5) cm apart, tip-spine 2–4 cm. **INFL:** panicle-like, 2–4 m incl peduncle; bracts 10–25 cm, lanceolate to triangular, persistent; branches gen 8–14, gen 10–20 cm; fls 35–75[100] per cluster. **FL:** yellow to ± red; perianth 6–10 cm, tube 12–19 mm, wide-funnel-shaped, lobes 17–40 mm, unequal; stamens long-exserted, filaments 4.3–7 cm, from middle of tube, yellow, anthers 20–35 mm; ovary 3–5 cm. **FR:** 5.5–7 cm. **SEED:** 4–7 mm. Coastal bluffs, historically nearby mesas, foothills; < 100[< 300] m. s SCo (sw San Diego Co.); Baja CA. Sep–May ★

A. utahensis Engelm. UTAH AGAVE St 0; rosette gen 1. **LF:** [12]15–30[50] cm, lance-linear, marginal teeth 4–12(15) mm, 1–4 cm apart, tip-spine [2]4–20 cm. **INFL:** raceme-like, 2–4 m incl peduncle; bracts 4–8 cm, narrow-triangular, not persistent; branches 50–75[+], gen 0.5–5 cm; fls 2–8 per cluster. **FL:** perianth 2.3–4.3 cm, tube 6.5–10[11.5] mm, bell-shaped, lobes ± equal, 7–12 mm; stamens ± exserted, filaments 1.3–2.4 cm, ± from base of tube, white, anthers 5–12 mm; ovary 1.2–2.5 cm. **FR:** 1–2.5 cm. **SEED:** 2–4 mm. Other subspp., vars. in NV, UT, AZ.

  var. **eborispina** (Hester) Breitung IVORY-SPINED AGAVE Rosettes 1.5–3 dm. **LF:** 15–30 cm, olive-green, marginal teeth 6–12(15) mm, tip-spine 10–20 cm, ivory-white. Desert scrub [to conifer woodland] on calcareous outcrops; 1100–1200 m. ne DMoj (Nopah Range); NV. May–Jul ★

  var. **nevadensis** Greenm. & Roush CLARK MOUNTAIN AGAVE Rosettes 1.5–2.5 dm. **LF:** 15–25 cm, glaucous-green, marginal teeth 4–6 mm, tip-spine 4–8.5 cm, brown to ± white. Desert scrub [to conifer woodland] on calcareous outcrops; 1200–1500 m. n&e DMtns (Clark, Ivanpah mtns, Kingston Range); NV. May–Jul ★

CAMASSIA CAMAS, QUAMASH

Per; bulb 1 [or clustered (each bulb has 1 lf-cluster, 1 scape, or both, so no need to dig up bulbs to make this determination)], coat black or brown. **LF:** basal, linear, keeled, glabrous. **INFL:** scapose raceme; fls 3–many; bracts 1–6 cm, narrow-lanceolate, scarious in age; pedicels 1–5 cm, spreading or incurved in fr. **FL:** ± radial (or bilateral); perianth parts 6, in 2 petal-like whorls, fused < 5% of length, 12–40 mm, lanceolate, blue to ± purple (white), 3–9 veined, twisted together above ovary in fr [or not]; stamens 6, anthers gen 4–7 mm, attached at middle; ovary chambers 3. **FR:** loculicidal. **SEED:** 6–36, black. *n*=15. ± 6 spp., 4 in nw N.Am. (chamass, qám′es, or quamash, Native American word) [Uyeda & Kephart 2006 Syst Bot 31:643–655] CA spp. highly variable, may hybridize, in need of study; bulbs traded among, eaten by Native Americans, perhaps creating local forms.

1. Perianth parts gen 20–40 mm, in fr early-deciduous; pedicels in fr gen ± spreading . . . . . . *C. leichtlinii* subsp. *suksdorfii*
1′ Perianth parts gen 10–20 mm, in fr persistent; pedicels in fr erect to incurved . . . . *C. quamash* subsp. *breviflora*

C. leichtlinii (Baker) S. Watson subsp. **suksdorfii** (Greenm.) Gould (p. 1287) Bulb 1, ovoid, 1.5–3 cm. **ST:** 2–10 dm. **LF:** 3–9, 2–6 dm, 5–25 mm wide. **INFL:** ± open; fls 15–40, gen < 5 open at once. **FL:** radial or ± so, blue violet to bright blue (white); perianth parts gen 5-veined, early-deciduous. **FR:** 1–3 cm, oblong to ovoid, often early-deciduous. **SEED:** 6–12 per chamber. Common; wet meadows; 1000–2600 m. NW, CaR, n&c SNH; to BC. [*C. quamash* subsp. *q.*, misappl.] Hybrids with *C. quamash* complicate identification. May–Aug

C. quamash (Pursh) Greene subsp. **breviflora** Gould (p. 1287) Bulb gen 1, 1–5 cm. **ST:** 2–5 dm. **LF:** gen < 10, 1.5–3 dm, 6–17 mm wide. **INFL:** often dense; fls several to many, (4)10–20+ open at once. **FL:** bilateral to ± radial, blue violet to bright blue; perianth parts gen 3-veined, persistent, forced apart by expanding fr. **FR:** ovoid, 8–20 mm. **SEED:** 5–10 per chamber. Wet meadows; < 2500 m. NW, CaR, n SNH; to WA, NV. [*C. q.* (Pursh) Greene subsp. *linearis* Gould] Several proposed subspp. (only 1 from CA) evidently are ± minor geog variants. May–Jul

CHLOROGALUM SOAP PLANT, AMOLE

Judith A. Jernstedt

Per; bulb ovoid to ± elongate, outer coat white to brown, often fibrous; gen not heterostylous. **LF:** basal, linear. **INFL:** panicle, peduncled; fls, buds 1–several per node; bracts scarious. **FL:** perianth parts 6 in 2 petal-like whorls, free, white, purple, or ± pink, twisted together above ovary in fr; stamens 6, attached to base of perianth, anthers attached at middle; ovary superior, chambers 3, style thread-like, deciduous, stigma ± 3-lobed. **FR:** stalked, loculicidal. **SEED:** 1–2 per chamber, ovoid, black. 5 spp.: w N.Am, esp CA. (Greek: green milk or juice) [Jernstedt 2006 FNANM 26:307–310]

1. Fls open during day
  2. Perianth white or ± pink; fls, buds 2–several per node . . . . *C. parviflorum*
  2′ Perianth deep blue to purple; fls, buds 1 per node . . . . *C. purpureum*
    3. Infl 25–40 cm . . . . var. *purpureum*
    3′ Infl 10–20 cm . . . . var. *reductum*
1′ Fls open in evening
  4. Lf 2–5 mm wide, margin ± flat . . . . *C. angustifolium*
  4′ Lf 4–25 mm wide, margin gen wavy
    5. Pedicels 2–5 mm, << perianth . . . . *C. grandiflorum*
    5′ Pedicels 5–35 mm, ± ≥ perianth . . . . *C. pomeridianum*
      6. Bulb coat membranous or with few coarse fibers . . . . var. *minus*
      6′ Bulb coat not membranous, with many coarse fibers
        7. Infl < 40 cm, ± prostrate or branches spreading from base . . . . var. *divaricatum*
        7′ Infl 50–250 cm, erect . . . . var. *pomeridianum*

C. angustifolium Kellogg (p. 1287)   Bulb 3–5 cm; coat red-brown, membranous, fibers delicate; heterostylous. **LF**: 2–5 mm wide, margin ± flat. **INFL**: (15–20)30–70 cm; branches ascending; pedicels 2–3 mm, slender. **FL**: open in evening; perianth parts spreading, 8–12 mm, oblong, white, midvein green-yellow; stamens 8–12 mm, ≤ perianth, anthers 1.5–3 mm, yellow; style 2–4 mm or 6–8 mm. **FR**: 1.5–3 mm. *n*=17,17+. Heavy soils in grassland or woodland; < 500 m. NCoRO, NCoRI, CaRF, SNF (esp n s OR. Apr–Jul

C. grandiflorum Hoover   RED HILLS SOAPROOT   Bulb 5–7 cm; coat ± red to brown, membranous, outer bulb scales with few delicate fibers. **LF**: basal, 4–12 mm wide, margin wavy. **INFL**: 30–100 cm; branches ascending; pedicels 2–5 mm, << perianth, stout. **FL**: open in evening; perianth parts recurved, 15–30 mm, linear, white, midvein ± purple; stamens gen ± < perianth, anthers ± 3 mm, yellow; style 12–28 mm, ≥ perianth. **FR**: 5–8 mm. Serpentine outcrops, open shrubby or wooded hills; 300–500 m. n&c SNF (Placer, El Dorado, Tuolumne cos.). May–Jun   ★

C. parviflorum S. Watson   Bulb 4–7 cm; coat dark brown, membranous. **LF**: 3–9 mm wide, margin wavy. **INFL**: 30–90 (200) cm, branches ascending to erect; fls, buds 2–several per node; pedicels 2–8 mm, gen < perianth, slender. **FL**: open during 1 day; perianth parts spreading from above base, 7–8 mm, white or ± pink, midvein darker; stamens 3–4 mm, < perianth, anthers 1.5–2 mm, yellow; style 7–9 mm, > perianth. **FR**: 3–4 mm. *n*=30. Dry, open coastal-sage scrub; < 750 m. c&s SCo; n Baja CA. May–Aug

C. pomeridianum (DC.) Kunth   Bulb 7–15 cm, coat membranous or with coarse fibers. **LF**: 20–70 cm, 6–25 mm wide; margin gen wavy. **INFL**: 30–250 cm, much-branched; pedicels 5–35 mm, slender, ± ≥ perianth. **FL**: open in evening; perianth parts spreading or recurved, 15–25 mm, linear, white, midvein green or purple; stamens < perianth, anthers ± 2 mm, purple or yellow, pollen yellow or cream; style 10–15 mm, < to > perianth. **FR**: 5–7 mm, spheric to ± lobed.

var. ***divaricatum*** (Lindl.) Hoover   Bulb coat with many coarse fibers. **INFL**: < 40 cm, ± prostrate or branches spreading from base. *n*=18. Coastal bluffs, hills; gen < 100 m. s NCo, CCo. May–Jul

var. ***minus*** Hoover (p. 1287)   DWARF SOAPROOT   Bulb coat membranous or with few coarse fibers. **INFL**: 30–170 cm, erect or branches ascending. *n*=18. Serpentine outcrops in chaparral; gen < 750 m. NCoRI, SnFrB, SCoRO. May–Jun   ★

var. ***pomeridianum*** (p. 1287)   Bulb coat with many coarse fibers. **INFL**: 50–250 cm, erect, branches many. *n*=15,18,18+. Common. Open grassland, chaparral, woodland; < 1500 m. NW, SNF, n SNH, w GV, CW, SW; sw OR. Bulb juices lather in water; roasted bulbs a Native American food. May–Aug

C. purpureum Brandegee   Bulb 2.5–3 cm; scales white to brown, membranous. **LF**: basal, 2–5 mm wide, linear, margin wavy. **INFL**: 10–40 cm; branches few; pedicels 4–10 mm, gen > fls, slender. **FL**: open during day; perianth parts recurved, 5–7 mm, deep blue to purple; stamens ± = perianth, anthers ± 1 mm, yellow; style exserted, 5–6 mm. **FR**: ± 3 mm. Vars. intergrade but apparently retain distinctions in common garden experiments, produce fertile seed when artificially crossed.

var. ***purpureum*** SANTA LUCIA PURPLE AMOLE   **INFL**: 25–40 cm. *n*=30. Open woodland; ± 300 m. ne SCoRO (e side Santa Lucia Range, Monterey Co.). Threatened by trampling, grazing, vehicles. May–Jun   ★

var. ***reductum*** Hoover (p. 1287)   CAMATTA CANYON AMOLE   **INFL**: 10–20 cm. Serpentine woodland; ± 600 m. se SCoRO (ne La Panza Range, San Luis Obispo Co.). May–Jun   ★

HASTINGSIA

Per; bulb ovoid to ± elongate, outer coat black. **ST**: 25–90 cm, slender. **LF**: basal, grass-like, keeled, ± glaucous. **INFL**: ± scapose raceme or panicle, bracted; fls 20–70+. **FL**: perianth parts 6, in 2 petal-like whorls, fused at base; stamens 6, fused to perianth, outer 3 > perianth, opening before inner, all equal after fls open; ovary superior, chambers 3, style 1, stigma 3-lobed. **FR**: short-stalked, loculicidal. **SEED**: ovoid, flat, black. 4 spp.: CA, OR. (S.C. Hastings, first Chief Justice of CA Supreme Court, 1814–1893) [Becking 2002 FNANM 26:309–312]

1. St 40–89 cm; lvs (28)35–41(53) cm; perianth parts spreading to ascending in distal 3/4; pls in areas with water all yr . . . . *H. alba*
1′ St 28–52 cm; lvs (19)21–27(35) cm; perianth parts spreading to reflexed in distal 3/4; pls in areas moist in spring, dry in summer . . . . *H. serpentinicola*

H. alba (Durand) S. Watson (p. 1287)   Bulb 26–56 mm, 17–31 mm wide. **INFL**: dense; branches gen 2–3. **FL**: 6–8 mm; perianth parts elongating as anthers mature, equal, white to ± yellow, outer ± 1 mm wide, linear, blunt, inner ± 2 mm wide, ovate, acute. **FR**: 6–9 mm, oblong. *n*=26. Wet meadows, bogs, rocky seeps; 500–2300 m. NW, CaR, n SNH; sw OR. Jun–Jul

H. serpentinicola Becking (p. 1287)   Bulb 23–40 mm, 14–21 mm wide. **INFL**: open; branches gen 0(3). **FL**: 5–6 mm; perianth parts 1–2 mm, linear, (green-white to) ± yellow or ± purple, midvein darker. **FR**: 5–8 mm, oblong. *n*=26,27?. Well-drained, exposed serpentine; 1800–2200 m. n&c NW, CaRH; sw OR. May–Jun

HESPEROCALLIS DESERT LILY

1 sp. (Greek: western beauty) [Utech 2002 FNANM 26:221; Bogler et al. 2006 Aliso 22:313–328]

H. undulata A. Gray (p. 1287)   Per; bulb 4–6 cm, ovoid, deep. **ST**: 30–180 cm, gen simple. **LF**: mostly basal (cauline reduced), 20–50 cm, 8–15 mm wide, blue-green; margins wavy, white. **INFL**: raceme; bracts conspicuous, ± ovate, papery; pedicels ± 1 cm. **FL**: jointed to pedicel, 4.5–6 cm, funnel-shaped; perianth parts 6, fused below, petal-like, white with silver-green midstripe esp abaxially, lobes 3–4 cm, ± oblanceolate, spreading; stamens 6, attached to perianth; ovary superior, chambers 3, style slender, stigma ± 3-lobed. **FR**: 12–16 mm, 3-lobed, loculicidal. **SEED**: ± 5 mm, flat, black. $n=24$. Sandy flats; < 800 m. D; w AZ. Feb–May

HESPEROYUCCA QUIXOTE PLANT, CHAPARRAL YUCCA

William J. Hess

Shrub-like, dying after fr or not; rosettes ≥1, clumped or not. **LF**: linear to narrow-lanceolate, thick, ± rigid, glaucous, spine-tipped, margins ± yellow, fine-toothed. **INFL**: scapose, panicle, cylindric, glabrous, exceeding rosettes; peduncle gen > 2.5 cm wide; bracts gen reflexed. **FL**: bisexual; perianth bell-shaped or spheric, parts free, wide-lanceolate, white to ± green or purple-tinged; stamens thick, gen > pistil; ovary superior, stigma head-like. **FR**: erect, obovoid, loculicidal. **SEED**: many per chamber, 6–9 mm diam, thin, flat, dull black. $x=30$. 3 spp.: AZ, CA, n Mex. (Greek: western yucca)

H. whipplei (Torr.) Trel. (p. 1287)   **ST**: ± 0. **LF**: 40–100 cm, 0.7–2 cm wide, ± gray-green; expanded base 4–7 cm, 4–7 cm wide, ± white to ± green. **INFL**: 1 per rosette; 2–40 dm, dense; peduncle 15–35 dm; branches, fls many. **FL**: perianth ± 4 cm, white or cream to ± green, purple-tinged or -tipped or not; filaments linear below, tip club-like; pistil 1–1.5 cm, stigma domed, green, clear-papillate. **FR**: 3–5 cm, 1.5–4 cm wide. Chaparral, coastal, desert scrub; < 2500 m. s SN (esp e slope), SCoR, SW, SNE (Deep Springs Valley), sw edge DMoj; n Baja CA. [*Yucca w.* Torr.] Apr–May

LEUCOCRINUM SAND LILY

1 sp. (Greek: white lily) [Pires et al. 2006 Aliso 22:287–304]

L. montanum A. Gray (p. 1287)   Per from short, often deep caudex; roots fleshy. **ST**: ± 0. **LF**: basal, spreading, 10–20 cm, linear, base sheathed by membranous bracts. **INFL**: pedicels underground. **FL**: 5–10 cm; perianth ± salverform, parts 6, fused into slender tube, petal-like, white, persistent, lobes 2–2.5 cm, ± oblong, spreading; stamens 6, attached near top of perianth tube, anthers 4–6 mm, linear; ovary underground, superior, chambers 3, style 1, persistent, stigma ± 3-lobed. **FR**: 5–8 mm, 3-angled, ± wrinkled, loculicidal. **SEED**: 3–4 mm, angled, black. $2n=28$. Sandy flats, sagebrush scrub, juniper woodland, montane forest; 1000–1500 m. KR, CaRH, MP; to OR, NE, UT. May–Jun

YUCCA SPANISH BAYONET, YUCCA

William J. Hess

Shrub- or tree-like, gen branched from woody caudex, occ dying after fr. **LF**: rosette basal or at branch tips, 2–15 dm, linear, thick, ± rigid, stout-spine-tipped, bases ± expanded, margins gen curved up, entire or dentate, often fibrous-shredding. **INFL**: panicle [raceme], erect (pendent), dense; peduncle gen < 2 cm wide; bracts gen ascending. **FL**: gen pendent, 3–13 cm; perianth parts 6 in 2 whorls, gen ± fused, ± white, fleshy, waxy; stamens 6, filaments ± thick, fleshy; ovary superior, 3(6)-chambered, gen green, style short, often thick, poorly defined, stigmas 3-lobed, white to pale green. **FR**: berry-like or gen capsule, erect or pendent, tardily septicidal or gen indehiscent. **SEED**: ± many in 2 rows per chamber, often flat, black. ± 40 spp.: e coastal plain, se, sc, & esp dry sw N.Am; n, c-w Mex, n C.Am. (Haitian: yuca or manihot, because young infls occ roasted for food) Pollinated at night by small moths while laying eggs in ovary. *Y. whipplei* moved to *Hesperoyucca.*

1. Lf margins minute-serrate, not fibrous-shredding; lf 15–35 cm; fr dry, spongy, or leathery in youth, spreading to erect in age; s SNH (e slope), Teh, e SNE, DMoj . . . . ***Y. brevifolia***
1′ Lf margins fibrous-shredding, not serrate; lf 30–150 cm; fr fleshy in youth, pendent in age; s SW, s DMoj, e DMtns, nw DSon
  2. St ± 0 or short-decumbent, rosettes at tips, ± at ground; perianth parts 5–13 cm, fused > 1 mm; e DMtns . . . . ***Y. baccata*** var. ***baccata***
  2′ St erect, rosettes at tips, well above ground; perianth parts 3–5 cm, free or fused < 1 mm; s SW (San Diego Co.), s DMoj, nw DSon . . . . ***Y. schidigera***

Y. baccata Torr. var. ***baccata*** (p. 1287)   BANANA YUCCA, BLUE YUCCA   Pl < 2.5 m. **ST**: ± 0 or short-decumbent, branched or not, rosettes at tips, ± at ground. **LF**: 50–100 cm, 2–6 cm wide, glaucous, ± dark green, expanded base ± 10 cm, 5 cm wide, ± red, margins fibrous-shredding. **INFL**: 6–8 dm, purple-tinged, within to distal ± 1/4 exserted from rosettes, gen glabrous. **FL**: pendent; perianth 5–13 cm, bell-shaped, parts lanceolate, fused > 1 mm, red-brown abaxially, ± white adaxially; pistil 5–8 cm. **FR**: berry-like, pendent in age, 15–17 cm, fleshy in youth. Uncommon. Dry Joshua-tree woodland; 800–1300 m. e DMtns; to CO, TX, Mex. Reported to hybridize with *Y. schidigera*. A 2nd var. in AZ, NM, Mex. May–Jun

Y. brevifolia Engelm. (p. 1295)   JOSHUA TREE   Pl 1–15 m. **ST**: erect, above ground, gen branched above, rosettes at tips, well above ground. **LF**: 15–35 cm, 0.7–1.5 cm wide, dark green, expanded base 2–4 cm, 4–5 cm wide, ± white, margins minute-serrate, yellow. **INFL**: 3–5 dm, distal gen ± 1/2 exserted from rosettes. **FL**: erect; perianth 4–7 cm, ± bell-shaped, parts lanceolate to oblong, ± fused at base, cream to ± green; filaments thick; pistil ± 3.5 cm. **FR**: capsule, spreading to erect in age, 6–8.5 cm, ellipsoid, dry, spongy, or leathery in youth. Desert flats, slopes; 400–2000 m. s SNH (e slope), Teh, e SNE, DMoj; to sw UT, w AZ, s NV. Growth form variable. Apr–May

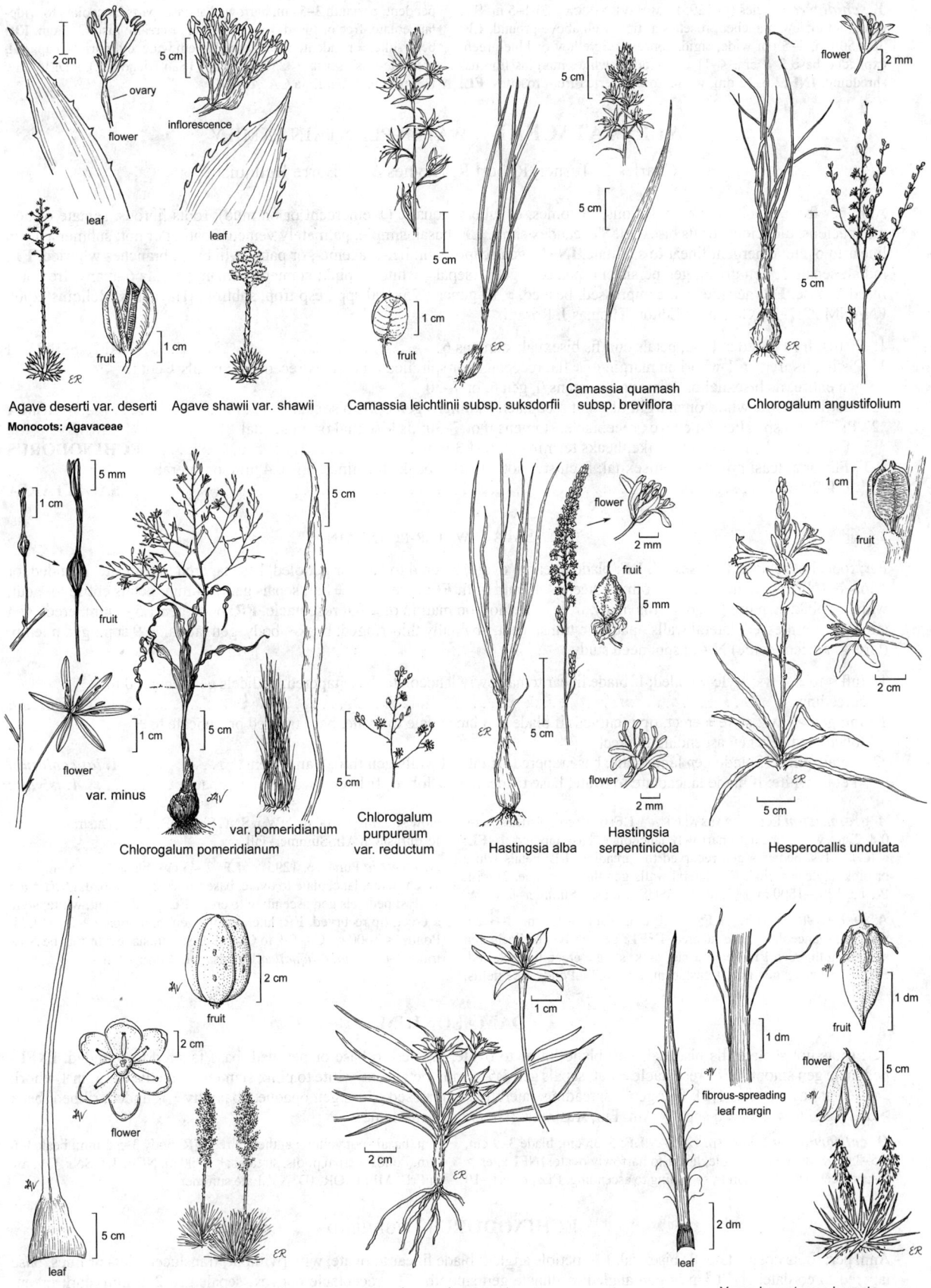

Agave deserti var. deserti

Agave shawii var. shawii

Monocots: Agavaceae

Camassia leichtlinii subsp. suksdorfii

Camassia quamash subsp. breviflora

Chlorogalum angustifolium

var. minus

Chlorogalum pomeridianum

var. pomeridianum

Chlorogalum purpureum var. reductum

Hastingsia alba

Hastingsia serpentinicola

Hesperocallis undulata

Hesperoyucca whipplei

Leucocrinum montanum

Yucca baccata var. baccata

Y. schidigera Ortgies (p. 1295) MOJAVE YUCCA Pl 1–5 m. **ST**: erect, not or few-branched, rosettes at tips, well above ground. **LF**: 30–150 cm, 3–5 cm wide, rigid, concave, ± yellow or blue-green, expanded base 2–8 cm, 4–11 cm wide, ± white, margins fibrous-shredding. **INFL**: 6–12 dm, within to ± exserted from rosettes. **FL**: pendent; perianth 3–5 cm, narrowed at base, parts lanceolate to wide-lanceolate, free or fused < 1 mm, white or cream; pistil 2–3 cm. **FR**: berry-like, ± pendent, 5–11.5 cm, 3–4 cm wide, cylindric. Chaparral, creosote-bush scrub; < 2500 m. s SW (San Diego Co.), s DMoj, nw DSon; NV, AZ, n Baja CA. Apr–May

ALISMATACEAE WATER-PLANTAIN FAMILY

Charles E. Turner, Robert R. Haynes & C. Barre Hellquist

Ann, per from caudices, corms, stolons, rhizomes, or tubers, aquatic (± emergent or on mud); roots fibrous, septate or not; monoecious, dioecious, or fls bisexual. **ST**: caudex short. **LF**: basal, simple, palmately veined, floating or not; submersed gen linear to ovate; emergent linear to sagittate. **INFL**: gen scapose, umbel-, raceme-, or panicle-like; fls, branches whorled. **FL**: radial; sepals 3, gen green, gen persistent; petals 3, gen > sepals, white or pink; stamens 6–many; pistils 6–many, free or ± fused at base. **FR**: achene, gen compressed, beaked. ± 12 genera, 75–100 spp.: esp trop, subtrop. [Haynes & Hellquist 2000 FNANM 22:7–25] Scientific Editor: Thomas J. Rosatti.

1. Pistils, frs ± fused at base; petals cut; fls bisexual; stamens 6 . . . . **DAMASONIUM**
1′ Pistils, frs free, in 1 whorl on margin of ± flat receptacle or spiralled on convex receptacle; petals ± cut to gen entire; fls bisexual or unisexual; stamens 0, gen 6, or 7–30
  2. Pistils, frs in 1 whorl on margin of ± flat receptacle; stamens gen 6; fls bisexual . . . . **ALISMA**
  2′ Pistils, frs spiralled on convex receptacle; stamens 0 or 7–30; fls bisexual or unisexual
    3. Fls bisexual; fr cluster bur-like, beaks terminal, 0.6–1.3 mm . . . . **ECHINODORUS**
    3′ Fls, or at least proximal, unisexual; fr cluster not bur-like, beaks terminal, 0.1–0.4 mm, or lateral,
      0.1–2 mm . . . . **SAGITTARIA**

ALISMA WATER-PLANTAIN

Per; roots not septate; fls bisexual. **LF**: blade linear to ovate, tapered to base or petioled, base tapered to truncate, rounded, or ± lobed. **INFL**: peduncle gen smooth; pedicels < 45 mm in fr. **FL**: receptacle ± flat; sepals gen 1–4 mm; petals entire to ± cut, white or pink; stamens gen 6; pistils many, free, in 1 whorl on margin of ± flat receptacle. **FR**: body gen 1.5–3 mm, erect, gen strongly compressed, lateral walls opaque to translucent, abaxially thin-ridged; beak < body, gen lateral. ± 9 spp.: gen n temp. (Greek: ancient name) N.Am spp. need study.

1. Infl < to ± > lvs; style ± coiled; lf blade linear to narrowly lanceolate, base tapered; pedicels gen recurved to
  spreading . . . . *A. gramineum*
1′ Infl gen >> lvs; style ± erect, tip recurved; lf blade gen lanceolate to ovate, base tapered or truncate to ±
  lobed; pedicels gen ascending to erect
  2. Petals pink; lf blade gen lanceolate, base tapered; fr lateral walls gen thin, translucent . . . . *A. lanceolatum*
  2′ Petals white; lf blade lanceolate to ovate, base truncate to ± lobed; fr lateral walls thick, opaque . . . . *A. triviale*

A. gramineum Lej. GRASS ALISMA **LF**: 6–30 cm; blade 3–7 cm, 0.8–2 cm wide, linear to narrowly lanceolate, base tapered. **INFL**: < to ± > lvs; pedicels gen recurved to spreading. **FL**: petals white or pink; style ± coiled. **FR**: lateral walls gen thick, opaque. 2*n*=14. Ponds; 1200–1800 m. MP; to Can, e US; Eurasia. Summer–fall ★

A. lanceolatum With. **LF**: 12–40 cm; blade 6–12 cm, 1–3 cm wide, gen lanceolate, base tapered. **INFL**: gen >> lvs; pedicels gen ascending to erect. **FL**: petals ± cut, pink; style ± erect, tip recurved. **FR**: lateral walls gen thin, translucent. 2*n*=26,28. Ponds, rice fields, slow streams; < 500 m. NW, n SNF, ScV; OR; Chile, Australia; native to Eurasia, n Afr. Summer–fall

A. triviale Pursh (p. 1295) **LF**: 7–45 cm; blade 5.5–15 cm, 1.5–10 cm wide, lanceolate to ovate, base truncate to ± lobed. **INFL**: gen >> lvs; pedicels gen ascending to erect. **FL**: petals ± cut, white; style ± erect, tip recurved. **FR**: lateral walls gen thick, opaque. 2*n*=14,28. Ponds; < 1600 m. CA-FP; to Can, se US; Eurasia, e Afr, maybe Australia. [*A. plantago-aquatica* L., misappl.] Spring–fall

DAMASONIUM

Per; roots not septate; fls bisexual. **LF**: blade linear to ovate, tapered to base or petioled, base tapered to rounded. **INFL**: peduncle gen smooth. **FL**: receptacle ± flat; sepals gen 3.5–6 mm; petals cut, white to pink; stamens 6; pistils 6–15, in 1 whorl, ± fused at base, ovules 1[2]. **FR**: gen ± spreading, laterally compressed, sides gen opaque, abaxially ± rounded, ribbed; beak ≥ body, gen terminal. ± 5 spp.: N.Am, Eur, Australia. (Greek: ancient name)

D. californicum Benth. (p. 1295) **LF**: 5–35 cm; blade 3–9 cm, 0.5–3 cm wide, gen < petiole, linear to narrowly ovate. **INFL**: gen > lvs; pedicels 15–65 mm in fr, spreading to ascending. **FL**: petals 6–10 mm, basal spot yellow; anthers ± red. **FR**: body 3–5.5 mm; beak 3–6 mm. Ponds, vernal pools, streams; < 1700 m. NCoRI, n SNF, GV, nw SnFrB, MP; to OR, ID, NV. Late summer

ECHINODORUS BURHEAD

Ann, per; roots not septate; fls bisexual. **LF**: petiole angled; blade linear to ovate, with [without] translucent dots or lines, base tapered to cordate. **INFL**: axes gen angled; peduncle gen smooth. **FL**: receptacle convex; sepals gen 2–6 mm, dark green;

petals gen entire, white; stamens 9–15; pistils many, free, spiralled on convex receptacle. **FR**: body ± compressed, gen ribbed; beak terminal [lateral]. ± 26 spp.: Am, esp trop. (Greek: spiny, leathery container, from fr)

E. berteroi (Spreng.) Fassett (p. 1295)   Ann (short-lived per). **LF**: 8–30 cm; blade coarsely veined; submersed blades linear, wavy, or gen 0; floating, emergent blades 6–14 cm, 3–15 cm wide, elliptic to cordate. **INFL**: gen > lvs; fls 1–3(4) per node; peduncle angled; pedicels 6–28 mm, gen ascending. **FL**: petals 6–9 mm. **FR**: cluster bur-like; body 1.5–3 mm, ribs gen 5. $2n=22$. Ponds, ditches; < 300 m. NCoRI, GV, CW, SW; to se US, S.Am. Mid-summer–fall

SAGITTARIA ARROWHEAD

Ann, per; roots septate; gen monoecious; scape gen straight at infl. **LF**: petiole cylindric to 3-angled; submersed blades tapered to base; floating or emergent blades gen sagittate (linear to ovate). **INFL**: lowest node gen with 3 pistillate fls, those above gen staminate. **FL**: sepals 3–10 mm, reflexed to appressed in fr; petals gen entire. **STAMINATE FL**: stamens 7–30. **PISTIL-LATE FL**: receptacle convex; pistils many, spiralled on convex receptacle. **FR**: body gen 2–3.5 mm, strongly compressed, abaxially winged or ridged; beak gen lateral, spreading to erect. ± 30 spp.: worldwide, esp Am. (Latin: arrow, from lf shape) Some spp. weedy; tubers of some eaten by humans, wildlife; *S. brevirostra* Mack. & Bush reportedly persisting at Stafford Lake and Chileno Laguna, Marin Co.

1. Pistillate fls ± sessile; scape bent at infl. . . . . ***S. rigida***
1′ Pistillate fls pedicelled; scape straight at infl
  2. Stolons, tubers 0; lowest infl node with 2 bisexual fls . . . . ***S. montevidensis*** subsp. ***calycina***
  2′ Stolons, tubers present; lowest infl node with 3 pistillate fls
    3. Emergent lf blades linear to lanceolate; pedicels recurved in fr, thickened . . . . ***S. sanfordii***
    3′ Emergent lf blades gen sagittate; pedicels ascending in fr, not thickened
      4. Basal lobes of emergent lf blades gen > terminal; tubers spheric, tan . . . . ***S. longiloba***
      4′ Basal lobes of emergent lf blades ≤ terminal; tubers oblong, ± white or ± blue
        5. Fr beak terminal, erect, 0.1–0.4 mm; emergent lvs with petioles recurved, basal lobes of blades
          < terminal . . . . ***S. cuneata***
        5′ Fr beak lateral, spreading, 1–2 mm; emergent lvs with petioles ascending to erect, basal lobes of
          blades ± = terminal . . . . ***S. latifolia***

S. cuneata E. Sheld. (p. 1295)   Per; tuber oblong, ± white or ± blue. **LF**: petioles of emergent lvs recurved, blades 2.5–17 cm, sagittate, basal lobes < terminal. **STAMINATE FL**: filaments glabrous. **PISTILLATE FL**: pedicel ascending in fr; sepals reflexed in fr. **FR**: beak terminal, erect, 0.1–0.4 mm. $2n=22$. Ponds, slow streams; < 2500 m. NW, CaR, SN, SnBr, GB; to s Can, n US, TX. Jun–Aug

S. latifolia Willd. (p. 1295)   Per; tubers oblong, ± white or ± blue. **LF**: petioles of emergent lvs ascending to erect, blades 1.5–30 cm, sagittate, basal lobes ± = terminal. **STAMINATE FL**: filaments glabrous. **PISTILLATE FL**: pedicel ascending in fr; sepals reflexed in fr. **FR**: beak spreading, 1–2 mm. $2n=22$. Ponds, slow streams; < 1500 m. CA-FP, GB; to s Can, e US, n S.Am. Jul–Aug

S. longiloba J.G. Sm. (p. 1295)   Per; tubers spheric, tan. **LF**: petioles of emergent lvs ascending to erect, blades 13–27 cm, sagittate, basal lobes gen > terminal. **STAMINATE FL**: filaments glabrous. **PISTILLATE FL**: pedicel ascending in fr; sepals reflexed in fr. **FR**: beak erect, 0.1–0.6 mm. Ponds, rice fields; < 300 m. GV; to c US, n Mex, C.Am. May–Jun

S. montevidensis Cham. & Schltdl. subsp. ***calycina*** (Engelm.) Bogin (p. 1295)   Ann. **LF**: petioles of emergent lvs erect to ascending, blades 5–15 cm, sagittate, basal lobes ± = terminal. **INFL**: low-est node with 2 bisexual fls. **BISEXUAL FL**: pedicel recurved in fr, thickened; sepals appressed in fr; petals white, gen with green-yellow spot at base. **STAMINATE FL**: filaments papillate. **FR**: side oil-streaked when fresh; beak spreading, 0.4–0.8 mm. Ponds, rice fields; < 300 m. GV, SnFrB (Sonoma Co.), c SCo (Los Angeles Co.); to e US, n Mex. Jul–Aug

S. rigida Pursh SESSILE-FRUITED ARROWHEAD   Per, tubered; scape bent at infl. **LF**: submersed lvs with petioles ± blade-like, linear, blades gen 0; petioles of emergent lvs blade-like or 3-angled, blade linear to elliptic (hastate to sagittate). **STAMINATE FL**: filaments dilated, pubescent. **PISTILLATE FL**: ± sessile; sepals reflexed in fr. **FR**: beak recurved, 0.8–1.4 mm. $2n=22$. Ponds, streams, marshes; 6–1385 m. CaRH (Plumas Co.), ScV (Tehama Co.), CCo (Marin Co.); WA, ID, SK, e N.Am. May–Aug

S. sanfordii Greene (p. 1295) SANFORD'S ARROWHEAD   Per; tubers spheric. **LF**: emergent lvs with petioles ± blade-like, ± flat, or 3-angled, blades 14–25 cm, linear to lanceolate. **STAMINATE FL**: filaments papillate. **PISTILLATE FL**: pedicel recurved in fr, thickened; sepals appressed in fr. **FR**: side oil-streaked when fresh; beak erect, 0.2–0.6 mm. Ponds, ditches; < 300 m. n NCo (Del Norte Co.), KR, CaRF, GV, n SCo (Ventura Co.). Threatened by develop-ment. May–Oct ★

ALLIACEAE ONION or GARLIC FAMILY

Dale W. McNeal

Per; bulb 1 or on rhizomes, reforming each yr; bulblets at bulb bases or on rhizomes; outer bulb coat brown, red-brown, yellow-brown, or gray; inner coats gen white (pink, red, or yellow); onion odor, taste present (exc *Nothoscordum*). **ST**: scapose, cylindric, sometimes flat or triangular. **LF**: basal, sheathing st, linear [or not], cylindric, channeled or flat, gen ± withering from tip by fl. **INFL**: umbel (1-fld in *Ipheion*), bracts gen 2, splitting and appearing 2+ or not, ± fused, enclosing fl buds, scarious. **FL**: perianth parts 6 in 2 petal-like whorls, ± free to fused in lower 1/3–1/2; stamens 6, fused to perianth, filaments widened at base, anthers attached at middle; ovary superior, 3-lobed, chambers 3, ovules 2+ per chamber, style 1, stigma entire or ± 3-lobed. **FR**: capsule, loculicidal. **SEED**: black, sculpture net-like, smooth, or granular. 13 genera, 750–800 spp.: worldwide. Many cult for food, orn. Scientific Editor: Thomas J. Rosatti.

1. Perianth parts ± free; onion odor present; infl 3–many-fld, fls rarely partly or completely replaced by bulblets
. . . . **ALLIUM**
1′ Perianth parts fused in lower 1/3 or 1/2; onion odor present or 0; infl 1–several-fld, fls not replaced by bulblets
  2. Onion odor present; infl 1-fld; perianth parts fused in lower 1/2 . . . . **IPHEION**
  2′ Onion odor 0; infl several-fld; perianth parts fused in lower 1/3 . . . . **NOTHOSCORDUM**

ALLIUM ONION, GARLIC

Outer bulb coat gen brown to gray, inner gen white. **ST**: scapose, cylindric, triangular in ×-section, or flat. **LF**: basal, 1–5[12] per st, linear, cylindric, channeled, or flat, gen withering from tip before fl. **INFL**: umbel, fls 3–many, rarely all or in part replaced by bulblets; bracts gen 2–4, obvious, ± fused, scarious. **FL**: perianth parts ± free, gen with darker or contrasting midvein, outer gen wider; filaments fused into a ring; ovary with 0, 3, or 6 crests, ovules gen 2 per chamber. **SEED**: obovoid, gen unappendaged. 700 spp.: gen n temp, esp CA. (Latin: garlic) [McNeal & Jacobsen 2002 FNANM 26:224–276] Replanting bulbs after study essential for survival of pl; shape, arrangement of cells of outer bulb coat (outer bulb coat sculpture) gen important in identification, gen determined only with magnification; color of outer bulb coat may be masked by substrate; st lengths from top of bulb to base of infl, not from substrate surface.

1. Bulblets present in infl, present at bulb bases; disturbed places
  2. Bulb 1–2 cm diam; lvs 20–60 cm, 2–4 mm wide, hollow below middle . . . . ***A. vineale***
  2′ Bulb (1.5)3–8 cm diam; lvs 20–100 cm, 5–20 mm diam, solid. . . . . [***A. sativum***]
1′ Bulblets not present in infl, present at bulb bases or on rhizomes or not; disturbed places or not
  3. Lvs 1 per st, cylindric; ovary crests 6, obvious, ± triangular
    4. Stigma lobes ± 0
      5. Stamens exserted; perianth parts 5–9 mm, inner > outer, margins ± irregular to jagged; st 20–50 cm
. . . . ***A. sanbornii*** var. ***sanbornii***
      5′ Stamens incl; perianth parts 8–20 mm, ± equal, entire; st gen < 20 cm
        6. Pls of DMtns, fls white; outer bulb coat sculpture ± obvious, cells transversely elongate, intricately contorted; DMtns . . . . ***A. nevadense***
        6′ Pls not of DMtns or fls not white; outer bulb coat sculpture 0 or cells ± square, in 2–3 rows basally, or obscurely polygonal
          7. Perianth parts 8–12 mm; pedicels slender, > fls. . . . . ***A. atrorubens***
            8. Perianth parts gen deep red-purple (white), lanceolate, appearing long-acuminate by inrolled margins . . . . var. ***atrorubens***
            8′ Perianth parts pale pink, dark-veined, ± ovate, acute to acuminate . . . . var. ***cristatum***
          7′ Perianth parts 12–20 mm; pedicels stout, gen < fls
            9. Perianth parts lance-linear, long-tapered; bulblets 1–2, stalked, at bulb base; TR, w PR . . . . ***A. monticola***
            9′ Perianth parts lanceolate, acute; bulblets 0; n SnBr, DMoj. . . . . ***A. parishii***
    4′ Stigma lobes 3, often slender, recurved
      10. Stamens exserted or ± at same level as perianth
        11. Inner perianth parts 1.5 × outer; ovary crests entire or notched . . . . ***A. sanbornii*** var. ***congdonii***
        11′ Inner perianth parts ± = outer; ovary crests irregularly dentate to deeply cut . . . . ***A. howellii***
          12. St (10)± 20(35) cm, slender; fls gen 10–30; perianth gen pink to lavender (white); ovary crests purple; fl Mar–Apr; s SNF, Teh, SnJV, SnFrB, SCoR, WTR . . . . var. ***howellii***
          12′ St 20–60 cm, stout; fls gen > 50; perianth gen white; ovary crests white or green; fl Apr–Jun; SCoRI, n WTR
            13. Stamens exserted 0–2 mm; n WTR . . . . var. ***clokeyi***
            13′ Stamens exserted 2–4 mm; SCoRI . . . . var. ***sanbenitense***
      10′ Stamens incl
        14. Perianth parts (at least inner) dentate or jagged
          15. St 25–40 cm . . . . ***A. jepsonii***
          15′ St 5–20 cm
            16. Outer bulb coats brown to gray; perianth ± spreading, tips reflexed, inner parts jagged. . . . . ***A. abramsii***
            16′ Outer bulb coats red-brown; perianth erect, ± straight, inner parts dentate near tip . . . . ***A. denticulatum***
        14′ Perianth parts ± entire
          17. Ovary crests entire, notched, or jagged, not dentate to deeply cut
            18. Perianth 10–18 mm, maroon or deep red-purple
              19. Perianth deep red-purple, tips ± spreading to ± erect; pls of serpentine, se SnFrB (Mount Hamilton Range) . . . . ***A. sharsmithiae***
              19′ Perianth maroon near tips, white or ± green basally, tips reflexed; pls not of serpentine, s SNH, Teh . . . . ***A. shevockii***
            18′ Perianth 6–9 mm, white to pink, in age ± red
              20. Infl dense, pedicels straight in fr; perianth parts elliptic to ovate, obtuse to acute. . . . . ***A. munzii***
              20′ Infl open, pedicels curved in fr; perianth parts lanceolate, acuminate. . . . . ***A. parryi***
          17′ Ovary crests dentate to deeply cut
            21. Perianth dark red-purple, tips gen recurved to spreading . . . . ***A. fimbriatum*** var. ***fimbriatum***

[MONOCOTS] **Alliaceae** (Allium)   1291

21′ Perianth white, pink, or lavender, midveins darker, tips erect
   22. Fls (6)8–12 mm . . . . *A. fimbriatum*
     23. St slender, 10–25 cm; pls not of serpentine; SNE, DMoj . . . . var. *mohavense*
     23′ St stout, 10–37 cm; pls of serpentine; c NCoRI . . . . var. *purdyi*
   22′ Fls gen 6–9 mm
     24. St slender, 7–20 cm; perianth erect; SnFrB (Mount Hamilton Range), SCoR, WTR . . . . *A. diabolense*
     24′ St stout, 25–50 cm; perianth spreading; c SNF (Tuolumne Co.) . . . . *A. tuolumnense*
3′ Lvs ≥ 2 per st (if 1, lf flat or channeled); ovary crests 6, obvious, ± triangular, or ± 0
  25. St 3–30 mm diam, hollow, inflated below middle; bulb gen 5–10 cm diam; fls 100–500 . . . . [*A. cepa*]
  25′ St < 3 mm diam unless flattened and/or winged, solid or hollow, not inflated below middle; bulb < 3 cm
     diam; fls 5–150
    26. Bulbs oblong or oblong-ovoid, clustered on stout rhizome
     27. St 50–100 cm; perianth parts narrowly lanceolate, acuminate; stamens exserted . . . . *A. validum*
     27′ St 10–40 cm; perianth parts narrowly ovate, acute; stamens incl
       28. Ovary topped with a low, papillate cap covering lobes . . . . *A. haematochiton*
       28′ Ovary crests 6, obvious, triangular, margins papillate or not . . . . *A. marvinii*
    26′ Bulbs ovoid to ± spheric, not clustered on short rhizome
     29. Ovary crests 6, obvious, triangular; bulblets > 3, gen << bulb, gen clustered at bulb base or on rhizome
       30. Lvs gen withered in fl; perianth red-purple, rigid-keeled, ± shiny in fr, tip margins inrolled, parts
          with a purple crescent adaxially just above base . . . . *A. campanulatum*
       30′ Lvs gen not withered in fl; perianth white or pink to rose-purple, not keeled, papery in fr, tip
          margins ± flat, parts without darker crescent adaxially
         31. Perianth parts lanceolate, acuminate; CaRH, SNH, GB . . . . *A. bisceptrum*
         31′ Perianth parts ovate to elliptic, acute; KR, CaRF, n&c SNF, s SNH . . . . *A. membranaceum*
     29′ Ovary crests 0 or obscure, not triangular; bulblets 1–3, ± = bulb, within bulb coats or on rhizome
       32. Lvs flat or widely channeled, ± sickle-shaped; pl in fr dry, breaking at ground
         33. Outer bulb coat sculpture ± net-like throughout
           34. Outer bulb coat cells rectangular, transversely elongate, in ± regular rows
             35. Perianth parts lance-linear . . . . *A. anceps*
             35′ Perianth parts ovate to oblanceolate
               36. St 15–20 cm, infl held well above ground; pedicel > fl . . . . *A. lemmonii*
               36′ St 3–10 cm, infl held near ground; pedicel ± = fl . . . . *A. punctum*
          34′ Outer bulb coat cells ± square or polygonal, if transversely elongate then curved and not in regular rows
           37. Outer bulb coat cells curved, ± transversely elongate . . . . *A. tribracteatum*
           37′ Outer bulb coat cells ± rectangular, ± transversely elongate, square, or polygonal . . . . *A. obtusum*
             38. Lvs 1 per st; perianth parts pink, lanceolate, acute . . . . var. *conspicuum*
             38′ Lvs 1–2 per st; perianth parts white, oblong-elliptic, obtuse . . . . var. *obtusum*
        33′ Outer bulb coat sculpture 0 or cells ± square, in 2–3 rows basally
         39. St cylindric or ± flat, not winged
           40. Stamens well incl
             41. Lvs straight or ± sickle-shaped; fls 20–30; KR, NCoR, n&c SNF, s SNH, Teh, SCoRI,
                WTR, SnJt . . . . *A. cratericola*
             41′ Lvs sickle-shaped; fls gen 5–10; KR, CaRH, SNH, GB . . . . [2]*A. parvum*
           40′ Stamens exserted or ± at same level as perianth
             42. Lvs 2 per st . . . . *A. yosemitense*
             42′ Lvs 1 per st
               43. Filaments smooth . . . . *A. burlewii*
               43′ Filaments bumpy or warty near base . . . . *A. hoffmanii*
        39′ St ± flat, gen winged above
          44. Perianth parts lance-linear, long-acuminate; stamens exserted . . . . *A. platycaule*
          44′ Perianth parts lanceolate to ovate, obtuse to long-acuminate; stamens incl
           45. Lvs gen 6–8 mm wide; perianth parts 9–15 mm, long-acuminate . . . . *A. falcifolium*
           45′ Lvs < 6 mm wide; perianth parts 6–10 mm, obtuse to acute
             46. Perianth parts lanceolate, acute, appearing acuminate by inrolled margins . . . . *A. tolmiei* var. *tolmiei*
             46′ Perianth parts widely lanceolate to ovate, obtuse (tip margins ± flat)
               47. Inner bulb coats white; perianth parts not keeled in fr . . . . [2]*A. parvum*
               47′ Inner bulb coats gen pink to red; perianth parts keeled in fr . . . . *A. siskiyouense*
       32′ Lvs narrowly channeled to ± cylindric, if flat or widely channeled, not sickle-shaped; pl in fr dry
         or not, not breaking at ground
        48. Outer bulb coat sculpture 0 or obscure
         49. Bulb symmetric, present at fl; rhizomes 0
          50. Longest infl bracts >> pedicels; fl 4.5–7 mm . . . . *A. paniculatum* var. *paniculatum*
          50′ Longest infl bracts ≤ pedicels; fl 7–18 mm
           51. Infl not ± 1-sided; fls erect; perianth parts elliptic, obtuse . . . . *A. neapolitanum*
           51′ Infl gen ± 1-sided; fls pendent; perianth parts lanceolate, acute . . . . *A. triquetrum*

49′ Bulb symmetric or not, ± 0 at fl; rhizomes obscure or obvious
  52. Rhizomes obvious, gen 3–5 cm; lvs widely channeled or ± flat, keeled; perianth parts ovate to obovate, ascending, entire . . . . *A. unifolium*
  52′ Rhizomes obscure, < 2 cm; lvs ± cylindric; perianth parts ± lanceolate, erect, at least inner serrate . . . . *A. bolanderi*
    53. Bulb oblique-ovoid, not tuber-like, outer coat breaking with serrate edges; perianth parts 8–12 mm, narrowly ovate . . . . var. *bolanderi*
    53′ Bulb oblique-oblong, tuber-like, outer coat breaking with irregular edges; perianth parts 9–14 mm, narrowly lanceolate . . . . var. *mirabile*
48′ Outer bulb coat sculpture obvious, ± net-like
  54. Outer bulb coat cells square, polygonal, in regular rows, not in ± herringbone pattern, not ± contorted
    55. Perianth parts 8–15 mm, spreading (or tips reflexed), inner dentate . . . . *A. acuminatum*
    55′ Perianth parts 4–9 mm, erect or spreading, entire . . . . *A. lacunosum*
      56. St 10–25 cm; infl dense, pedicel 0.7–1.5 × fl
        57. Lvs gen < st; st 15–25 cm; fls 5–7 mm; s SNF, Teh, w DMoj . . . . var. *kernense*
        57′ Lvs > st; st 10–20 cm; fls 6–9 mm; n CCo, s SnJV, SCoR, SCo, n ChI . . . . var. *lacunosum*
      56′ St 15–35 cm; infl open, pedicel 1.5–3.5 × fl
        58. Bracts 2; fl 6–8 mm; WTR, SnBr, SNE, DMoj . . . . var. *davisiae*
        58′ Bracts 3; fl 4–6 mm; SCoR, SnBr, PR . . . . var. *micranthum*
  54′ Outer bulb coat cells transversely elongate, in ± herringbone pattern, or ± contorted
    59. Outer bulb coat cells in obscure, wavy, herringbone pattern; perianth parts spreading, outer ± = inner
      60. St 5–17 cm; perianth parts erect in fr; fls persistent — c CCo . . . . *A. hickmanii*
      60′ St 15–50 cm; perianth parts folded over fr; fls, pedicels deciduous
        61. Ovary crests 6, lateral, ± rectangular; infl dense, pedicels 0.7–2 × fl . . . . *A. amplectens*
        61′ Ovary crests 0 or 3, central, minute; infl open, pedicels 1.5–4 × fl
          62. Lvs 1–3(4) mm wide, widely channeled to cylindric, not keeled; inner bulb coats light yellow; perianth translucent in fr . . . . *A. hyalinum*
          62′ Lvs (3)4–6 mm wide, widely channeled or flat, keeled; inner bulb coats white; perianth ± opaque, papery in fr . . . . *A. praecox*
    59′ Outer bulb coat cells in obvious, not wavy, herringbone pattern; perianth parts ± erect, outer > inner
      63. Perianth parts folded over ovary in fr; fls, pedicels deciduous . . . . *A. serra*
      63′ Perianth parts erect, rigid in fr; fls, pedicels persistent
        64. Lvs 3–6 per st, curved to curled; infl dense, pedicels erect, 0.7–2 × fl; on or near sea cliffs . . . . *A. dichlamydeum*
        64′ Lvs 2–3 per st, straight to ± curved; infl open, pedicels spreading, 1–3 × fl; not on or near sea cliffs
          65. Inner perianth part margins dentate, curled . . . . *A. crispum*
          65′ Inner perianth part margins entire to dentate, not curled . . . . *A. peninsulare*
            66. Lvs curved; fls 8–12 mm; stigma minute, entire; CCo, SnFrB . . . . var. *franciscanum*
            66′ Lvs straight; fls 10–15 mm; stigma 3-lobed or head-like; widespread . . . . var. *peninsulare*

A. abramsii (Traub) McNeal Bulb 10–15 mm, ovoid to ± spheric; outer coat sculpture 0 or cells ± square, in 2–3 rows basally. **ST:** 5–15 cm. **LF:** 1, < 3 × st, cylindric. **INFL:** fls 6–40; pedicels 6–15 mm. **FL:** 8–15 mm; perianth parts ± spreading with reflexed tips, linear to narrowly lanceolate, rose-purple, inner < outer, jagged; ovary tip entire or notched, crests 6, obvious, triangular. *n*=7. Uncommon. Granitic sand; 1400–2000 m. c&s SNH. May–Jul ★

A. acuminatum Hook. (p. 1295) Bulb 8–16 mm, ovoid to ± spheric; outer coat yellow-brown, sculpture obvious, cells square or polygonal, walls thick. **ST:** 10–35 cm. **LF:** 2–3, ± = st, ± cylindric. **INFL:** fls 10–40; pedicels 6–25 mm. **FL:** 8–15 mm; perianth parts ascending to erect, with reflexed to spreading tips, lanceolate to narrowly ovate, white to rose-purple, inner dentate; ovary crests 3, minute, 2-lobed, central. *n*=7. Hills, plains; < 1900 m. NW, CaR, ne SnFrB (Mount Diablo), MP; to BC, Rocky Mtns, NM. Apr–Jul

A. amplectens Torr. (p. 1295) Bulb 6–15 mm, ovoid to ± spheric; outer coat cells transversely elongate, in obscure, wavy herringbone pattern; inner coats gen red. **ST:** 15–50 cm. **LF:** 2–4, < st, ± cylindric. **INFL:** fls 10–50; pedicels 4–16 mm. **FL:** 5–9 mm; perianth parts ascending to ± erect, lanceolate, white to pink, in fr folded over ovary; ovary crests 6, ± obvious, lateral. *n*=7,14; 2*n*=21. Clays incl serpentine, open or wooded places; < 1800 m. CA-FP, Wrn; to BC. Apr–Jul

A. anceps Kellogg Bulb 15–20 mm, ovoid; outer coat brown to yellow-brown, sculpture delicate, cells rectangular, transversely elongate, walls thin. **ST:** 10–15 cm. **LF:** 2, 2 × st, flat, sickle-shaped.

INFL: fls 15–35; pedicels 15–30 mm. **FL:** 6–12 mm; perianth parts lance-linear, entire, ± pink; ovary crests 6, minute, central. *n*=7. Barren clay, rocky slopes; 1200–1550 m. CaRH, n&c SNH, MP, SNE (Masonic Hills); s OR, sw ID, w NV. Apr–May

A. atrorubens S. Watson Bulb 10–16 mm, ovoid to ± spheric; outer coat brown, sculpture 0 or cells ± square, in 2–3 rows basally; inner coats pink to white. **ST:** 5–17 cm. **LF:** 1, < 2 × st, cylindric, tip tightly coiled (often breaking off in age or when pressed). **INFL:** fls 5–50; pedicels 6–20 mm, > fls. **FL:** 8–12 mm; perianth parts lanceolate to ovate, entire, red-purple or pink (white); ovary crests 6, obvious, triangular, tip entire or notched.

  var. *atrorubens* GREAT BASIN ONION **FL:** perianth parts lanceolate, gen deep red-purple (white), appearing long-acuminate by inrolled margins. *n*=7. Rocky or sandy soil; 1200–2100 m. GB, DMtns; to UT, AZ. May–Jun ★

  var. *cristatum* (S. Watson) McNeal INYO ONION **FL:** perianth parts ± ovate, acute, pale pink, acute to acuminate, tip margins flat. *n*=7. Sands; 1200–2100 m. SNE, DMtns; w NV, sw UT. May–Jun ★

A. bisceptrum S. Watson (p. 1295) Bulb 10–15 mm, ovoid to ± spheric, bulblets gen clustered at base; outer coat cells ± square, walls wavy. **ST:** 1–3, 10–35 cm. **LF:** 2–5, ± = st, widely channeled. **INFL:** fls 15–40; pedicels 10–25 mm. **FL:** 6–10 mm; perianth parts lanceolate, acuminate, entire, white or pink to rose-purple; ovary crests 6, low, ± triangular, dentate. *n*=7. Meadows, aspen groves; 2000–2900 m. CaRH, SNH, GB; to ID, UT, AZ. May–Jul

A. bolanderi S. Watson (p. 1295) Bulb 7–20 mm oblique-ovoid to -oblong, ± 0 at fl; outer coat sculpture obscure, cells transversely elongate, in herringbone pattern; rhizomes 1–2, obscure, < 2 cm incl new bulbs at tips. **ST:** 10–35 cm. **LF:** 2–3, ± = st, ± cylindric. **INFL:** fls 10–20; pedicels 10–20 mm. **FL:** perianth parts erect with spreading tips, ± lanceolate, serrate distally, tip obtuse to acute, appearing narrower by inrolled margins, ± entire, red-purple (white); ovary crests 3, minute, 2-lobed, central.

var. ***bolanderi*** Bulb 7–12 mm, oblique-ovoid; outer coat breaking with serrate edges. **FL:** perianth parts 8–12 mm, narrowly ovate. *n*=7. Uncommon. Rocky clays incl serpentine; < 1000 m. NW, CaR, se SnFrB; sw OR. May–Aug

var. ***mirabile*** (L.F. Hend.) McNeal Bulb 10–20 mm, oblique-oblong, tuber-like; outer coat breaking with irregular edges. **FL:** perianth parts 9–14 mm, narrowly lanceolate. *n*=7. Uncommon. Rocky clays incl serpentine; < 1000 m. KR, n NCoR; sw OR. May–Jul

A. burlewii Davidson Bulb 15–25 mm, ovoid; outer coat sculpture 0 or cells ± square, in 2–3 rows basally. **ST:** 2–8 cm. **LF:** 1, 1–2 × st, widely channeled. **INFL:** fls 8–20; pedicels 6–10 mm. **FL:** 7–10 mm; perianth parts ovate, erect, entire, white or tinged pink basally, in age dull purple; filaments smooth; ovary crests 0 or 6, minute, central. *n*=7. Sandy, gravelly soils, granite, serpentine; 1800–2800 m. s SNH, Teh, SCoR, SW. Apr–Jul

A. campanulatum S. Watson (p. 1295) Bulb 1–2 cm, ovoid; bulblets gen clustered at base or on slender rhizomes; outer coat cells ± square, walls wavy; inner pink to white. **ST:** 10–30 cm. **LF:** 2, ± = st, widely channeled, gen withered in fl. **INFL:** fls 10–50; pedicels 1–2 cm. **FL:** 5–8 mm; perianth parts ± spreading, lanceolate to ovate, rigid-keeled in fr, acuminate, entire, rose to purple (white), with a purple crescent adaxially just above base; ovary crests 6, low, triangular, dentate. *n*=7,14. Dry mtn slopes; 600–2600 m. KR, NCoRO, CaR, SNH, SCoRO, SW, MP; OR, NV. May–Aug

A. cratericola Eastw. Bulb 15–25 mm, ovoid; outer coat sculpture 0 or cells ± square, in 2–3 rows basally. **ST:** 3–10 cm. **LF:** 1–2, 1.5–4 × st, ± flat to widely channeled. **INFL:** fls 20–30; pedicels 5–18 mm. **FL:** 7–14 mm; perianth parts ± oblong to elliptic, entire, ± pink; ovary crests 3, minute, central. *n*=7,14. Open, serpentine, volcanic, or granitic places; 300–1800 m. KR, NCoR, n&c SNF, s SNH, Teh, se SnFrB, SCoRI, WTR, SnJt. Most n CA pls have 1 lf; s CA pls and 2 populations in Mariposa Co. have 2 lvs; some NCoR populations mixed. Mar–Jun

A. crispum Greene (p. 1295) Bulb 9–15 mm, ovoid to ± spheric; outer coat cells transversely elongate, in herringbone pattern. **ST:** 15–35 cm. **LF:** 2–3, < st, ± cylindric. **INFL:** fls gen 10–40; pedicels 10–35 mm. **FL:** 8–13 mm; perianth parts rose-purple, ± erect, tips spreading, outer ovate, ± entire, inner with dentate, curled margins; ovary crests 3, minute, 2-lobed, central. *n*=7. Clay slopes incl serpentine; < 800 m. s SnJV, CW. Mar–Jun

A. denticulatum (Traub) McNeal Bulb 10–14 mm, ovoid to ± spheric; outer coat red-brown, sculpture 0 or cells ± square, in 2–3 rows basally; inner coats pale brown to white. **ST:** 5–18 cm. **LF:** 1, 1.5–2 × st, cylindric. **INFL:** fls 5–30; pedicels 5–20 mm. **FL:** 9–17 mm; perianth parts rose-purple, inner dentate near tip; ovary crests 6, obvious, entire to finely, irregularly dentate. *n*=7. Dry slopes; 900–1600 m. s SN, Teh, WTR, w DMoj. Apr–Jul

A. diabolense (Traub) McNeal Bulb 10–16 mm, ovoid to ± spheric; outer coat red-brown, sculpture 0 or cells ± square, in 2–3 rows basally; inner coats pale brown to white. **ST:** 7–20 cm. **LF:** 1, 1.5–3 × st, cylindric. **INFL:** fls 10–50; pedicels 7–20 mm. **FL:** 6–10 mm; perianth parts entire, white, midveins or tips pink; ovary crests 6, obvious, jagged to deeply cut. *n*=7. Dry serpentine; 500–1500 m. SnFrB (Mount Hamilton Range), SCoR, WTR. [*Allium diabloense*, orth. var.] Apr–Jun

A. dichlamydeum Greene Bulb 10–15 mm, ovoid to ± spheric; outer coat cells transversely elongate, in herringbone pattern. **ST:** 10–30 cm, stout. **LF:** 3–6, ≥ st, channeled to ± cylindric. **INFL:** fls gen 5–30; pedicels 5–20 mm. **FL:** 9–12 mm; perianth parts ovate, deep red-purple, entire or inner dentate; ovary crests 3, minute,

2-lobed, central. *n*=7. Dry clays on or near sea cliffs; 50–150 m. NCo, n CCo. May–Jul

A. falcifolium Hook. & Arn. (p. 1299) Bulb 15–25 mm, ovoid; outer coat brown to red-brown, sculpture 0 or cells ± square, in 2–3 rows basally; inner coats white or pink. **ST:** 5–20 cm, flat, winged. **LF:** 2, 1.5–3 × st, flat, sickle-shaped. **INFL:** fls 10–30; pedicels 8–15 mm. **FL:** 9–15 mm; perianth parts lanceolate, rose-purple or dingy white, long-acuminate, at least inner gen dentate with minute glands; ovary crests 3, low, wide. *n*=7. Common. Heavy clays incl serpentine; 100–2100 m. NW, SnFrB; sw OR. Apr–Jun

A. fimbriatum S. Watson Bulb 10–17 mm, ovoid to ± spheric; outer coat red-brown, sculpture 0 or cells ± square, in 2–3 rows basally; inner coats pale brown to white. **ST:** 10–37 cm. **LF:** 1, 1.5–2 × st, cylindric. **INFL:** fls 6–75; pedicels 6–20 mm. **FL:** (6)8–12 mm; perianth parts lanceolate to ovate, entire, dark red-purple to white; ovary crests 6, obvious, dentate to deeply cut.

var. ***fimbriatum*** (p. 1299) **ST:** 10–20 cm. **INFL:** fls 6–35. **FL:** perianth dark red-purple; ovary crests finely dentate to deeply cut. *n*=7. Common. Dry slopes, flats; 300–2700 m. s NCoR, s SNF, Teh, CW, SW, D; n Baja CA. Apr–Jun

var. ***mohavense*** Jeps. **ST:** 10–25 cm, slender. **INFL:** fls 12–60. **FL:** perianth white, pink, or light lavender; ovary crests deeply cut, additional outgrowths on ovary present or not. *n*=7. Common. Dry slopes, flats; 700–1400 m. SNE, w DMoj. Pls from n base SnBr intermediate to *A. fimbriatum* var. *f.* placed here provisionally. Apr–May

var. ***purdyi*** (Eastw.) McNeal PURDY'S ONION **ST:** 10–37 cm, stout. **INFL:** fls 20–75. **FL:** perianth white to lavender; ovary crests finely, irregularly dentate to deeply cut, rarely 0. *n*=7. Serpentine outcrops; 300–600 m. c NCoRI. Apr–Jun ★

A. haematochiton S. Watson (p. 1299) Bulbs clustered on stout rhizome, 2–3 cm, oblong; outer coat ± red-brown, sculpture finely striate, cells vertically elongate in narrow rows; inner coats deep red to white. **ST:** 10–40 cm. **LF:** 4–6, < or ± = st, flat. **INFL:** fls 10–30; pedicels 7–15 mm. **FL:** 6–8 mm; perianth parts narrowly ovate, entire, white to rose; ovary topped with a low, papillate cap covering lobes. *n*=7. Common. Dry slopes, ridges; < 800 m. SCoRO, SCo, WTR, PR; n Baja CA. Mar–May

A. hickmanii Eastw. HICKMAN'S ONION Bulb 8–12 mm, ovoid to ± spheric; outer coat cells transversely elongate, in obscure, wavy herringbone pattern. **ST:** 5–17 cm. **LF:** 2, 1.5 × st, ± cylindric. **INFL:** fls 4–15; pedicels 4–12 mm. **FL:** 5–7 mm, persistent; perianth parts spreading, erect in fr, lanceolate to narrowly ovate, entire, white to pale pink; ovary crests 0 or 3, minute, central. *n*=7. Grassy, wooded slopes; ± 50 m. c CCo (Monterey Peninsula; Arroyo de la Cruz, San Luis Obispo Co.). Mar–May ★

A. hoffmanii Traub BEEGUM ONION Bulb 15–25 mm, ovoid; outer coat sculpture 0 or cells ± square, in 2–3 rows basally. **ST:** 5–10 cm. **LF:** 1, 1–2 × st, widely channeled. **INFL:** fls gen 10–40; pedicels 8–15 mm. **FL:** 8–10 mm, ± narrowed above ovary; perianth parts erect, lance-linear, long-tapered, pink to purple, midveins obvious, ± green; stamens exserted, bumpy or warty near base; ovary crests 6, obscure, central. *n*=7. Locally common. Serpentine outcrops; 1100–1800 m. s KR, n NCoRH. Jun–Jul ★

A. howellii Eastw. Bulb 9–17 mm, ovoid to ± spheric; outer coat red-brown, sculpture 0 or cells ± square, in 2–3 rows basally; inner coats pale brown to white. **ST:** (10)15–60 cm. **LF:** 1, 0.7–1.5 × st, cylindric. **INFL:** fls gen 10–100; pedicels 7–25 mm. **FL:** 5–8 mm; perianth parts ovate, entire, white to pale lavender, inner ± = outer; ovary crests 6, obvious, irregularly dentate to deeply cut.

var. ***clokeyi*** Traub MOUNT PINOS ONION **ST:** 20–40 cm, stout. **INFL:** fls gen 50–100; pedicels 15–25 mm. **FL:** perianth parts white; stamens exserted 0–2 mm; ovary crests white or green. *n*=7. Locally common. Open slopes, sagebrush scrub; 1300–1850 m. n WTR. May–Jun ★

var. ***howellii*** **ST:** (10)± 20(35) cm, slender. **INFL:** fls gen 10–30; pedicels 7–15 mm. **FL:** perianth parts gen pink to lavender (white); stamens exserted 0–2 mm; ovary crests purple. *n*=7. Com-

mon. Grassy slopes, incl serpentine; 200–900 m. s SNF, Teh, SnJV, SnFrB, SCoR, WTR. Mar–Apr

var. ***sanbenitense*** (Traub) McNeal & T.D. Jacobsen **ST:** 25–60 cm, stout. **INFL:** fls gen 50–90; pedicels 10–20 mm. **FL:** perianth parts white or pale pink; stamens exserted 2–4 mm; ovary crests white or green. *n*=7. Uncommon. Grassy openings in chaparral, vertic clay; 300–1000 m. SCoRI (se San Benito, w Fresno cos.). Apr–May

A. hyalinum Curran (p. 1299) Bulb 5–12 mm, ovoid to ± spheric, gen clustered; outer coat cells transversely elongate, in obscure, wavy, herringbone pattern; inner coats light yellow. **ST:** 15–45 cm. **LF:** 2–3, 0.7–1.5 × st, widely channeled to cylindric, not keeled. **INFL:** fls gen 5–25; pedicels 10–35 mm. **FL:** 6–10 mm; perianth parts spreading, white or pale pink, in fr translucent, folded over ovary; ovary crest 0. *n*=7. Common. Grassy slopes, outcrops; 50–1500 m. SNF, GV, SnFrB. Mar–May

A. jepsonii (Traub) S.S. Denison & McNeal JEPSON'S ONION Bulb 15–25 mm, ovoid; outer coat sculpture 0 or cells ± square, in 2–3 rows basally. **ST:** 25–40 cm. **LF:** 1, ± = st, cylindric. **INFL:** fls gen 20–60; pedicels 7–20 mm. **FL:** 7–8.5 mm; perianth parts erect, ovate-elliptic, jagged, white, midveins deep pink, tip reflexed; stamens incl; ovary crests 6, obvious, ± jagged. *n*=7. Open, serpentine or volcanic slopes, flats; 300–600 m. n&c SNF (Butte, El Dorado, Placer, Tuolumne cos.). May–Jul ★

A. lacunosum S. Watson (p. 1299) Bulb 1–2 cm, ovoid; outer coat often many, thickly surrounding bulb, cells ± square, polygonal, or transversely rectangular, but not forming a herringbone pattern, walls thick, wavy. **ST:** 10–35 cm. **LF:** 2, 0.7–2 × st, ± cylindric or flat. **INFL:** fls 5–45; pedicels 5–25 mm. **FL:** 4–9 mm; perianth parts erect or spreading, oblanceolate to narrowly ovate, entire, white or pale pink; ovary crests 3, minute, 2-lobed, central, crests and upper ovary densely papillate.

var. ***davisiae*** (M.E. Jones) McNeal & Ownbey **INFL:** open; bracts 2; fls 10–35; pedicels 10–25 mm. **FL:** 6–8 mm. *n*=7. Uncommon. Open, sandy slopes, ridges; 600–2100 m. WTR, SnBr, SNE, DMoj. Apr–May

var. ***kernense*** McNeal & Ownbey **INFL:** dense; bracts 2; fls 10–45; pedicels 10–15 mm. **FL:** 5–7 mm. *n*=7. Uncommon. Open sandy slopes; 700–1300 m. s SNF, Teh, w DMoj. [*Allium lacunosum* var. *kernensis*, orth. var.] Apr–May

var. ***lacunosum*** **INFL:** dense; bracts 2; fls 5–25; pedicels 5–12 mm. **FL:** 6–9 mm. *n*=7. Serpentine outcrops, clay; 50–1000 m. n CCo, s SnJV, SCoR, SCo, n ChI. Apr–May

var. ***micranthum*** Eastw. **INFL:** open; bracts 3; fls gen 10–30; pedicels 6–13 mm. **FL:** 4–6 mm. *n*=7. Uncommon. Dry rocky hillsides, incl serpentine; 300–600 m. SCoR, SnBr, PR. Apr–May

A. lemmonii S. Watson (p. 1299) Bulb 15–22 mm, ovoid; outer coat sculpture ± obvious, cells narrowly rectangular, transversely elongate, ± in vertical rows. **ST:** 15–20 cm, flat, narrowly winged. **LF:** 2, ± = st, flat, sickle-shaped. **INFL:** fls 10–40; pedicels 8–16 mm. **FL:** 6–9 mm; perianth parts narrowly ovate, entire, white to pink; ovary crests 6, obscure, ridge-like. *n*=7. Common. Drying clay soils; 1200–1900 m. n&c SNH, MP; to OR, ID, NV. May–Jun

A. marvinii Davidson (p. 1299) YUCAIPA ONION Bulbs clustered, 2–3 cm, oblong, on short rhizome; outer coat ± red-brown, sculpture finely striate, cells vertically elongate in narrow rows; inner coats white, pink, or deep red. **ST:** 10–40 cm. **LF:** 4–6, < st, flat. **INFL:** fls 10–30; pedicels 7–15 mm. **FL:** 6–8 mm; perianth parts > stamens, narrowly ovate, entire, white, midveins rose; ovary crests 6, obvious, triangular, margins papillate or not. Dry slopes, ridges; 300–1250 m. SCo, SnBr, PR. Previously incl in *A. haematochiton.* Mar–Apr ★

A. membranaceum Traub Bulb 10–16 mm, ovoid, bulblets in tight cluster at base or not; outer coat sculpture obscure, cells ± square, walls very wavy. **ST:** 15–40 cm. **LF:** 2–3, ± = st, flat. **INFL:** fls 15–35; pedicels 10–20 mm, slender. **FL:** 7–12 mm; perianth parts spreading, ovate to elliptic, acute, entire, white to pale pink, in fr papery; ovary crests 6, short, triangular. *n*=7. Uncommon. Wooded slopes; 150–1400 m. KR, CaRF, n&c SNF, s SNH. May–Jun

A. monticola Davidson Bulb 10–22 mm, ovoid, bulblets 1–2, large, stalked, at base; outer coat sculpture 0 or cells ± square, in 2–3 rows basally; inner coats white or pink. **ST:** 6–25 cm, glaucous. **LF:** 1, < 2 × st, cylindric, glaucous. **INFL:** fls 8–25; pedicels 5–12 mm. **FL:** 12–19 mm; perianth parts erect, lance-linear, long-tapered, entire, pink to rose-purple near tip, often white basally; ovary crests 6, obvious, ± linear to narrowly triangular, entire. *n*=7. Uncommon. Rocky ridges, talus slopes; 1400–3200 m. TR, nw PR (Orange Co.). May–Jul

A. munzii (Traub) McNeal MUNZ'S ONION Bulb 10–15 mm, ovoid; outer coat red-brown, sculpture 0 or cells ± square, in 2–3 rows basally; inner coats pale brown, white, or pink. **ST:** 15–35 cm. **LF:** 1, ± 1.5 × st, cylindric. **INFL:** fls 10–35; pedicels 7–12 mm. **FL:** 6–8 mm; perianth parts ascending to ± erect, elliptic to ovate, obtuse to acute, entire, white, red in fr; ovary crests 6, obvious, jagged. *n*=7. Grassy openings in coastal-sage scrub; 300–900 m. e SCo, nw PR (w Riverside Co.). Threatened by urbanization, citrus culture. Apr–May ★

A. neapolitanum Cirillo Bulb 1–2 cm, ± spheric; outer coat ± crusted, sculpture obscure, cells ± square, walls thick. **ST:** 20–60 cm, 3-angled, 2 angles slightly winged. **LF:** 2–3, ≤ st, flat, linear to narrowly lanceolate. **INFL:** fls 10–25; pedicels 15–35 mm. **FL:** 7–12 mm; perianth parts ± spreading, elliptic, obtuse, entire, white; ovary crest 0. Disturbed ± urban places; < 100 m. NCo, NCoR, ScV, CCo, SnFrB, SCo; native to Medit. Probably escaped from cult as orn. Mar–Apr

A. nevadense S. Watson (p. 1299) NEVADA ONION Bulb 9–15 mm, ovoid, bulblets gen 1–2, stalked, at base; outer coat sculpture gen obvious, cells transversely elongate, intricately contorted; inner coats white or pink. **ST:** 5–15 cm. **LF:** 1, 1.5–2 × st, cylindric, tip tightly coiled before withering. **INFL:** fls gen 5–25; pedicels 6–17 mm. **FL:** 7–12 mm; perianth parts ascending to ± erect, lanceolate to ovate, entire, white or fading pink; stamens incl; ovary crests 6, obvious, triangular, entire or tip notched. *n*=7,14. Sandy or gravelly slopes; 1300–1700 m. DMtns; to OR, ID, CO, AZ. Apr–Jun ★

A. obtusum Lemmon (p. 1299) Bulb 1–2 cm, ovoid; outer coat cells ± rectangular, ± transversely elongate, or square or polygonal. **ST:** 1.5–17 cm. **LF:** 1–2, 1–4 × st. **INFL:** fls 6–60; pedicels 2–15 mm. **FL:** 4–12 mm; perianth parts ± spreading to ± erect, lanceolate to oblong-elliptic, entire, white or pink; ovary crests 3, obscure to ± obvious, 2-lobed, central. NCoRI, CaRH, n&c SNF, SNH; sw OR, NV.

var. ***conspicuum*** Mortola & McNeal **LF:** 1, 1–2.5 × st, widely channeled. **INFL:** fls 10–60. **FL:** perianth parts lanceolate, acute, pink. *n*=7. Uncommon. Granitic sands; 800–3000 m. n&c SN. May–Jun

var. ***obtusum*** **LF:** 1–2, 1.5–4 × st, channeled. **INFL:** fls 6–30. **FL:** perianth parts oblong-elliptic, obtuse, white. *n*=7. Common. Granitic sands; 1500–3500 m. NCoRI, CaRH, n&c SNF, SNH; sw OR, NV. Small, easily overlooked. May–Jun

A. paniculatum L. var. ***paniculatum*** Bulb 10–15 mm, ovoid; outer coat sculpture obscure, cells rectangular, vertically elongate; inner coats light brown. **ST:** 30–70 cm. **LF:** 3–5, ≤ st, channeled, sheathing st in lower 30–50%. **INFL:** fls 25–100; bracts 2, long-tapered, unequal, longer 5–14 cm, >> pedicels; pedicels unequal, 10–45 mm. **FL:** 4.5–7 mm, bell-shaped; perianth parts ovate, entire, white to lilac-pink; ovary crest 0. *n*=8,16. Disturbed areas; ± 50 m. s NW, n CW; native to s Eur. Jun–Jul ◆

A. parishii S. Watson PARISH'S ONION Bulb 10–15 mm, ovoid; outer coat brown to red-brown, sculpture 0 or cells ± square, in 2–3 rows basally; inner coats pink. **ST:** 5–25 cm. **LF:** 1, 1–2 × st, cylindric. **INFL:** fls 6–25; pedicels 5–15 mm. **FL:** 12–18 mm; perianth parts spreading, lanceolate, acute, entire, pale pink; ovary crests 6, entire or finely, irregularly dentate. *n*=7. Open rocky slopes; 900–1400 m. n SnBr, DMoj; w AZ. Populations scattered. Apr–May ★

A. parryi S. Watson Bulb 8–14 mm, ovoid to ± spheric; outer coat ± red-brown, sculpture 0 or cells ± square, in 2–3 rows basally; inner coats lighter. **ST:** 5–20 cm. **LF:** 1, < 1.5 × st, cylindric. **INFL:** fls 8–50; pedicels 6–20 mm, curved in fr. **FL:** 6–9 mm; perianth parts

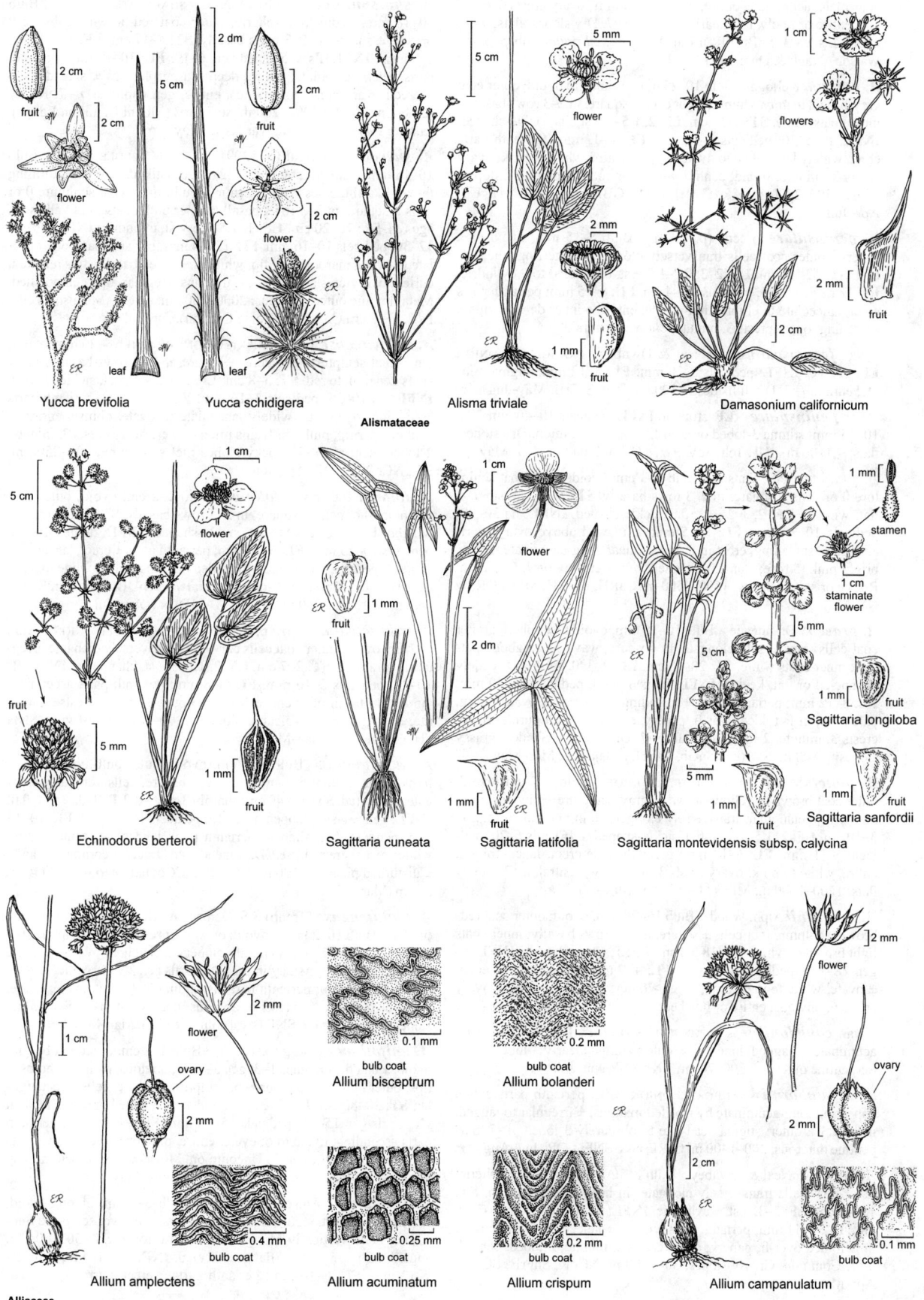

Yucca brevifolia

Yucca schidigera

Alisma triviale

Alismataceae

Damasonium californicum

Echinodorus berteroi

Sagittaria cuneata

Sagittaria latifolia

Sagittaria montevidensis subsp. calycina

Sagittaria longiloba

Sagittaria sanfordii

bulb coat
Allium bisceptrum

bulb coat
Allium bolanderi

Allium amplectens

bulb coat
Allium acuminatum

bulb coat
Allium crispum

Allium campanulatum

Alliaceae

lanceolate, acuminate, entire, white, ± red in fr; ovary crests 6, obvious, entire or finely, irregularly dentate. *n*=7. Dry slopes, flats; 900–2200 m. s SNH, SnBr, PR; n Baja CA. [*A. fimbriatum* subsp. *p*. (S. Watson) Traub & Ownbey] May–Jul

A. parvum Kellogg   Bulb 10–25 mm, oblique to ovoid; outer coat gray-brown to gray, sculpture 0 or cells ± square, in 2–3 rows basally; inner coats white. **ST:** 3–12 cm. **LF:** 2, 1.5–3 × st, flat, sickle-shaped. **INFL:** fls 5–30; pedicels 3–12 mm. **FL:** 6–12 mm; perianth parts erect, widely lanceolate to ovate, obtuse, entire, white or pink; ovary crests 3, minute, round, central. *n*=7. Common. Stony clay slopes, talus; 1200–2800 m. KR, CaRH, SNH, GB; to s&e OR, MT, UT. Apr–Jun

A. peninsulare Greene (p. 1299)   Bulb 8–15 mm, ovoid to ± spheric; outer coat cells transversely elongate, in herringbone pattern. **ST:** 12–45 cm. **LF:** 2–3, 0.7–1.5 × st, channeled to ± cylindric. **INFL:** fls gen 5–35; pedicels 0.8–4 cm. **FL:** 8–15 mm; perianth parts erect, lanceolate to elliptic, red-purple, entire to inner dentate, tips ± spreading; ovary crests 3, minute, 2-lobed, central.

  var. ***franciscanum*** McNeal & Ownbey   FRANCISCAN ONION **LF:** curved. **INFL:** pedicels 8–20 mm. **FL:** 8–12 mm; stigma minute, entire. *n*=7. Dry hillsides; < 300 m. CCo, SnFrB. May–Jun ★

  var. ***peninsulare***   **LF:** straight. **INFL:** pedicels 10–40 mm. **FL:** 10–15 mm; stigma 3-lobed or head-like. *n*=7. Common. Dry slopes, flats; < 1100 m. SNF, Teh, ScV, SW; s OR, n Baja CA. Apr–May

A. platycaule S. Watson   Bulb 2–3 cm, ovoid; outer coat sculpture 0 or cells ± square, in 2–3 rows basally. **ST:** 7–25 cm, stout, ± flat, winged. **LF:** 2, ± 2 × st, flat, sickle-shaped. **INFL:** fls 30–90; pedicels 10–25 mm. **FL:** 8–15 mm, narrowed above ovary; perianth parts spreading, erect in fr, lance-linear, long-acuminate, entire, bright pink to rose; stamens exserted; ovary crest 0. *n*=7. Common. Rocky or sandy slopes; 1500–2500 m. CaRH, n SNH, MP; s OR, w NV. May–Jun

A. praecox Brandegee   Bulb 10–18 mm, ovoid to ± spheric; outer coat cells transversely elongate, in obscure, wavy herringbone pattern; inner coats white. **ST:** 20–60 cm. **LF:** 2–3, 0.7–1.5 × st, widely channeled or flat, keeled. **INFL:** fls gen 5–40; pedicels 15–40 mm. **FL:** 8–12 mm; perianth parts ± spreading to ± erect, ± ovate, entire, pale pink, in fr folded over fr, papery, ± opaque, dull purple; ovary crests 3, minute, 2-lobed, central. *n*=7. Uncommon. Shaded, grassy slopes; < 800 m. SW (exc SnGb, SnBr); n Baja CA. Mar–May

A. punctum L.F. Hend.   DOTTED ONION   Bulb 1–2 cm, ovoid; outer coat brown to yellow-brown or gray, sculpture ± obvious, cells narrowly rectangular, transversely elongate, ± in vertical rows. **ST:** 3–10 cm. **LF:** 2, 1.5–2 × st, flat, sickle-shaped. **INFL:** fls 6–20; pedicels 5–11 mm. **FL:** 6–13 mm; perianth parts erect, lance-oblong, entire, white to pink; ovary crests 3, wide, low, central. *n*=7. Rocky flats; 1200–1600 m. MP; s OR, nw NV. May–Jun ★

A. sanbornii Alph. Wood   Bulb 15–25 mm, ovoid; outer coat red-brown, sculpture 0 or cells ± square, in 2–3 rows basally; inner coats light brown or white. **ST:** 18–60 cm. **LF:** 1, = st, cylindric. **INFL:** fls gen 18–150; pedicels 5–22 mm. **FL:** 4–9 mm; perianth parts erect, ± ovate, white to deep pink, inner > outer; stamens exserted; ovary crests 6, obvious, entire or notched.

  var. ***congdonii*** Jeps.   CONGDON'S ONION   **FL:** perianth parts acuminate, ± jagged, inner 1.5 × outer; stigma clearly 3-lobed. *n*=7. Serpentine outcrops; 300–700 m. n&c SNF. Jun–Jul ★

  var. ***sanbornii***   SANBORN'S ONION   **FL:** perianth parts acute, appearing long-acuminate by inrolled margins, ± irregular to jagged, inner 1.3 × outer; stigma head-like to obscurely 3-lobed. *n*=7. Serpentine outcrops; 300–1400 m. CaRF, n&c SNF; s OR. Jun–Aug ★

A. serra McNeal & Ownbey   Bulb 8–12 mm, ovoid to ± spheric; outer coat cells transversely elongate, in herringbone pattern. **ST:** 15–40 cm. **LF:** 2–3, ≤ st, ± cylindric. **INFL:** fls 10–40; pedicels 7–15 mm. **FL:** 8–11 mm; perianth parts ± erect, ± lanceolate, pink to rose, in fr folded over fr, papery; ovary crests 3, minute, 2-lobed, central. *n*=7. Common. Grassy slopes; 30–1200 m. NCoR, SnFrB, SCoRI. Apr–May

A. sharsmithiae (Traub) McNeal   SHARSMITH'S ONION   Bulb 10–18 mm, ovoid to ± spheric; outer coat red-brown, sculpture 0 or cells ± square, in 2–3 rows basally. **ST:** 4–17 cm. **LF:** 1, ± 2 × st, cylindric. **INFL:** fls 5–50; pedicels 6–19. **FL:** 10–18 mm; perianth parts erect, lanceolate, entire, deep red-purple, tips ± spreading to ± erect; ovary crests 6, obvious, entire, gen papillate. *n*=7. Rocky serpentine slopes; 400–1200 m. se SnFrB (Mount Hamilton Range). [*Allium sharsmithiae*, orth. var.] Apr–May ★

A. shevockii McNeal (p. 1299)   SPANISH NEEDLE ONION   Bulb 10–15 mm, ± spheric, bulblets 1–2, large, stalked, near base, forming thread-like rhizomes with 1 terminal bulb; outer coat sculpture 0 or cells ± square, in 2–3 rows basally; inner bulb coats light yellow, in age red. **ST:** 10–20 cm. **LF:** 1, 1.5–2.5 × st, cylindric. **INFL:** fls gen 12–30; pedicels 10–16 mm. **FL:** 12–14 mm; perianth parts oblanceolate to ovate, maroon near tip, white or ± green basally, tips reflexed, curled; ovary crests 6, obvious, margins irregular, tips gen notched. Metamorphic outcrops, talus; 2000–2500 m. s SNH (Spanish Needle Peak, ne Kern Co.), Teh (Horse Canyon). Jun–Jul ★

A. siskiyouense Traub   SISKIYOU ONION   Bulb 8–20 mm, ovoid; outer coat sculpture 0 or cells ± square, in 2–3 rows basally; inner coats gen pink to red. **ST:** 3–8 cm. **LF:** 2, ± 2 × st, flat, sickle-shaped. **INFL:** fls 10–35; pedicels 5–16 mm. **FL:** 8–11 mm; perianth parts ascending to ± erect, widely lanceolate to ovate, obtuse, entire or dentate near tip, pink, midveins often darker; ovary crests 3, minute, 2-lobed, central. *n*=7. Rocky slopes incl serpentine; 900–2500 m. KR, NCoR; sw OR. Apr–Jun ★

A. tolmiei Baker var. ***tolmiei***   Bulb 1–2 cm, ovoid; outer coat sculpture 0 or cells ± square, in 2–3 rows basally. **ST:** 5–15 cm, flat, winged. **LF:** 2, < 2 × st, flat, sickle-shaped. **INFL:** fls gen 10–40; pedicels 10–25 mm. **FL:** 6–10 mm; perianth parts ± erect, lanceolate, acute, appearing acuminate by inrolled margins, entire, white to pink; ovary crests 3, obscure, 2-lobed, central. *n*=7,14. Uncommon. Rocky clay flats; 1500–2200 m. MP; to WA, ID, NV. Apr–Jul

A. tribracteatum Torr. (p. 1299)   THREE-BRACTED ONION   Bulb 1–2 cm, ovoid; outer coat cells curved, ± transversely elongate, irregularly arranged. **ST:** 2–7 cm. **LF:** 2, 1.5–3 × st, channeled. **INFL:** fls 10–30; pedicels 6–10 mm. **FL:** 6–8 mm; perianth parts ascending, lanceolate to elliptic, entire, white to purple; ovary crests 3, minute, ± lateral. *n*=7. Volcanic slopes; 1300–1900(3000) m. c SNH (Tuolumne Co.). Mar–May ★

A. triquetrum L.   Bulb 1–2 cm, ovoid; outer bulb coats thin, ± translucent, yellow-brown, sculpture obscure, cells vertically elongate, contorted. **ST:** 10–40 cm; sharply 3-angled. **LF:** 2–3, ± = st, flat. **INFL:** gen ± 1-sided, open; fls 3–15; pedicels 15–25 mm. **FL:** 10–18 mm, pendent, bell-shaped; perianth parts lanceolate, acute, entire, white; ovary crest 0. **SEED:** appendaged. Locally common. Shady ± disturbed places; < 100 m. NCo, KR, CCo; native to w Medit. Cult as orn. Mar–Apr

A. tuolumnense (Traub) S.S. Denison & McNeal   RAWHIDE HILL ONION   Bulb 10–25 mm, ovoid; outer coat red-brown, sculpture 0 or cells ± square, in 2–3 rows basally; inner coats light brown. **ST:** 25–50 cm, stout. **LF:** 1, ± = st, cylindric. **INFL:** fls gen 20–60; pedicels 7–20 mm. **FL:** 6–8 mm; perianth parts ascending to ± erect, ovate, entire, white or pink; ovary crests 6, obvious, deeply cut. *n*=7. Serpentine slopes; 300–600 m. c SNF (sw Tuolumne Co.). Mar–May ★

A. unifolium Kellogg (p. 1299)   Bulb 1–2 cm, ovoid to oblique-ovoid, ± 0 at fl; rhizomes 1–3, gen 3–5 cm, obvious, with new bulbs at tips; outer bulb coat pale brown, sculpture obscure, cells ± rectangular. **ST:** 30–80 cm. **LF:** 2–3, < st, widely channeled or ± flat, keeled. **INFL:** fls gen 15–35; pedicels 15–40 mm. **FL:** 11–15 mm; perianth parts ascending, ovate to obovate, entire, pink or white; ovary crests 6 longitudinal ridges. *n*=7. Uncommon. Moist clay or serpentine, esp grassy streambanks; < 1100 m. NW, CW; OR. May–Jun

A. validum S. Watson (p. 1299)   Bulb 3–5 cm, oblong-ovoid, clustered on short, stout rhizome; outer coat ± brown, sculpture fine-striate, cells vertically elongate in narrow rows. **ST:** 50–100 cm, angled. **LF:** 3–6, ± = st, flat or ± keeled. **INFL:** fls 15–40; pedicels 7–12 mm. **FL:** 6–10 mm; perianth parts ± erect, narrowly lanceo-

late, acuminate, entire, rose to white; stamens exserted; ovary crest 0. n=14,28. Common. Wet meadows, often with *Salix*; 1200–3400 m. NW, CaRH, SNH, Wrn; to BC, ID, NV. Jun–Aug

A. vineale L. Bulb 1–2 cm, ovoid, bulblets several, stalked, hard-shelled, exposed as bulb coats disintegrate; outer coat ± brown to ± yellow, striate, splitting into strips, cells vertically elongate in narrow, wavy rows. **ST:** 30–100 cm. **LF:** 2–4, << st, 20–60 cm, 2–4 mm diam, sheathing lower st. **INFL:** fls gen few, sterile, all or in part replaced by bulblets; pedicels 12–15 mm. **FL:** 3–4 mm; perianth parts ovate, entire, green-white to purple; inner filaments with 2 obvious, lateral appendages; ovary crest 0. Uncommon. Disturbed places; < 100 m. s NW, n SNF, ScV, n CW, SCo; native to Eur. Jun–Aug ◆

A. yosemitense Eastw. (p. 1299) YOSEMITE ONION Bulbs 2–3 cm, ovoid, in clusters of 2–12 or more; outer coat sculpture 0 or cells ± square, in 2–3 rows basally. **ST:** 6–23 cm. **LF:** 2, 1–3 × st, widely channeled to ± flat. **INFL:** fls 10–50; pedicels 7–34 mm. **FL:** 7–15 mm; perianth parts ascending, linear-oblong, entire, white to pink; ovary crests 3, minute, 2-lobed, central. n=7. Open, rocky slopes; 800–2200 m. c SN (Tuolumne, Mariposa cos.). May–Jun ★

IPHEION STAR FLOWER

Per from deep, membrane-coated bulb, onion odor present. **LF:** several, basal, unkeeled. **INFL:** scapose; fls 1; bracts 2, fused at base. **FL:** showy; perianth parts fused in lower 1/2, lobes spreading, ± equal; anthers attached at middle; style thread-like, stigma ± 3-lobed. **FR:** capsule. ± 6 spp.: s S.Am. (Greek: origin obscure)

I. uniflorum (Graham) Raf **LF:** < 30 cm, ± flat, glaucous. **FL:** perianth white or ± blue, lobes 15–20 mm; filaments attached at 2 levels. Uncommon. Disturbed places; gen < 300 m. GV, SnFrB, SW; native to Argentina. Feb–Mar

NOTHOSCORDUM FALSE GARLIC

Per; bulb with basal bulblets or not; outer coat membranous, brown; onion odor absent. **ST:** scapose, cylindric. **LF:** basal, linear, flat or channeled, withering from tip before fl; bases sheathing. **INFL:** umbel, fls several; bracts obvious, gen 2, ± fused, scarious. **FL:** perianth parts fused in lower 1/3, petal-like, lobes ± spreading; stigma head-like. **FR:** capsule, loculicidal. **SEED:** several(to 12), angled or flat, black. ± 20–35 spp. (Greek: false garlic) [Jacobsen & McNeal 2002 FNANM 26:276–278]

N. gracile (Aiton) Stearn Bulb ± 1.5 cm. **ST:** 20–40 cm. **LF:** 2–several, 25–30 cm, 4–10 mm wide. **INFL:** < 4 cm diam; bracts < umbel, persistent; pedicels erect to spreading, unequal. **FL:** 8–14 mm; perianth white with ± green bases, ± red midveins, lobes oblanceolate, obtuse; stamens incl; style persistent in fr. **FR:** 6–7 mm, obovoid. Disturbed areas; < 100 m. GV, CW, SCo; native to S.Am. [*N. inodorum* (Aiton) G. Nicholson, misappl.] Apr–Jun ◆

AMARYLLIDACEAE AMARYLLIS FAMILY

Dale W. McNeal

Per, gen from bulb, coat membranous. **ST:** erect, gen cylindric, solid. **LF:** basal, sessile. **INFL:** scapose, umbel-like or 1-fld; bracts gen 2(8), conspicuous, ± fused. **FL:** perianth often with a conspicuous, ± tubular crown, parts 6, in 2 whorls, petal-like, [free to] ± fused, radial or not; stamens 6, ± fused to perianth, ± united, anthers attached at base; ovary inferior, chambers 3, each many-ovuled, style 1, stigmas 1 or 3. **FR:** gen capsule, dry, loculicidal, or fleshy berry. ± 60 genera, 800 spp. [Meerow & Snijman 2006 Aliso 22:355–366] Scientific Editor: Thomas J. Rosatti.

1. Perianth crown a conspicuous tube
2. Stamens free from, gen incl in crown . . . . **NARCISSUS**
2′ Stamens fused to, exserted from crown. . . . . **PANCRATIUM**
1′ Perianth crown 0
3. Fls appearing after lvs withered; perianth parts ± pink, sometimes white or yellow basally; filaments
>> anthers . . . . **AMARYLLIS**
3′ Fls appearing before lvs withered; perianth parts white, each with a green spot below thickened tip; filaments < anthers . . . . **[LEUCOJUM]**

AMARYLLIS

1 sp. (Greek: a shepherdess name in classical poetry)

A. belladonna L. NAKED LADIES Bulb 5–10+ cm. **ST:** 30–60 cm, stout. **LF:** linear, 30–45 cm, 1.5–3 cm wide. **INFL:** umbel-like, appearing after lvs withered; bracts scarious; pedicels 2–4 cm. **FL:** perianth spreading or ± reflexed, funnel-shaped, parts 5–8 cm, fused ± 1 cm, ± pink, sometimes white or yellow basally, crown 0; filaments >> anthers; stigma ± spheric or minutely 3-lobed. **FR:** capsule, ± spheric, dehiscing irregularly. **SEED:** 0 [few, fleshy, white or ± pink]. Disturbed sites, often around abandoned home sites; < 750 m. CCo, SnFrB, SCo, SnBr; LA, temp trop, S.Am, Afr, Medit. Jul–Sep

NARCISSUS

Bulb ovoid. **LF:** (1)several, ± linear, flat. **INFL:** umbel-like, 1–20-fld; bract 1, membranous or ± papery. **FL:** pedicelled or not, gen fragrant; perianth parts fused below, reflexed to erect above, crown a conspicuous tube; stamens free from, gen incl in crown; style 1, stigma minutely 3-lobed. **FR:** capsule, loculicidal, papery to leathery. **SEED:** many, black. ± 26 spp.: Eur,

n Afr, Asia. (Greek: mythological youth) Extreme variation in perianth color from long history of cult; only most common colors indicated here. Pls with perianth parts, crown gen white belong to *N. papyraceus* Ker Gawl., reported (but evidently not documented) for CA in FNANM, similar to, possibly same sp. as, *N. tazetta*.

1. Infl 1-fld; perianth parts 2.5–3.5 cm, crown tubular, ≥ perianth parts . . . . ***N. pseudonarcissus***
1′ Infl (2)3–15-fld; perianth parts 1–2 cm, crown cup-shaped, << perianth parts . . . . ***N. tazetta***

N. pseudonarcissus L.   DAFFODIL   Bulb 2–5 cm, 3–4 cm wide, coats gen brown. **ST**: 25–50 cm. **LF**: 4, 20–45 cm, flat, glaucous. **INFL**: bract 2–3 cm, pale brown, papery. **FL**: fragrant; perianth 5–7 cm wide, tube 1.5–2 cm, abruptly tapered to base, parts erect to spreading, often twisted, oblanceolate, yellow, acute, crown 1–1.5 cm wide, yellow, ruffled. Disturbed places; < 350 m. NCoRI, ScV, CCo; to WA, e US; Eur, w Asia, n Afr. Highly variable, long cult. Spring

N. tazetta L.   PAPER WHITE   Bulb 4–6 cm, 3–5 cm wide, coats pale to dark brown. **ST**: 25–45 cm. **LF**: 4, 20–45 cm, flat, glaucous. **INFL**: bract 4–6 cm, pale brown, papery. **FL**: very fragrant; perianth 2–4 cm wide, tube 1.5–2 cm, gradually tapered to base, parts spreading to reflexed, linear-ovate to oblanceolate, acute, gen white to cream, crown 5–10 mm wide, crenate to ruffled, gen yellow. Disturbed places, former home sites; < 600 m. SnFrB, SCo, n ChI, WTR, PR; Iran, widely naturalized elsewhere. Highly polymorphic, cult for centuries. Spring

PANCRATIUM

Bulb very deep, 4–8 cm, tapered to a long neck. **ST**: 40–50+ cm, compressed. **LF**: ± = st, 20–30 mm wide. **INFL**: umbel-like, 3–14-fld; bracts 2, 4–7 cm; pedicels 5–10 mm. **FL**: very fragrant, perianth parts 3–5 cm, lance-linear, white, tube 1.5–10 cm, slender; crown a conspicuous tube with 12 triangular teeth; stamens fused to, exserted from crown, alternate crown teeth; stigma obscurely 3-lobed. **FR**: loculicidal capsule. **SEED**: 12–40, black. 15 spp.: s Eur to trop Afr. (Greek: athletic contest)

P. maritimum L.   St, lf glaucous. **FL**: crown ≥ perianth parts; anthers ≤ filaments. **FR**: ± spheric, obscurely triangular in ×-section.   Coastal dunes; < 10 m. SCo (El Segundo Dunes, just w of Los Angeles International Airport, site of former subdivision); s Eur. Jun–Jul

APONOGETONACEAE   CAPE-PONDWEED FAMILY

Robert F. Thorne, C. Barre Hellquist & Robert R. Haynes

Per from corm, tuber, or short, thick rhizome, aquatic, glabrous; sap milky. **LF**: alternate, simple, [gen submersed] or floating, petioled, basal in opposite pairs; blade expanded or not, elliptic to lanceolate if floating. **INFL**: terminal, scape above water, gen spike or panicle with 2–3(10) spike-like branches, subtended by conspicuous, deciduous bract. **FL**: bisexual [unisexual], bilateral; perianth parts [0]1[6], gen petal-like, white or yellow; stamens 6–18 in 2 series or 6–50 in 3–4 series; pistils 2–6(9), fused ± 1/2–2/3, separating in fr, ovary superior, chamber 1, ovules 1–12, style short, stigmatic surface grooved. **FR**: follicle, leathery. **SEED**: 4; embryo straight. 1 genus, ± 54 spp.: trop Asia, Australia, s Afr. [Les et al. 2005 Syst Bot 30:503–519] Scientific Editor: Thomas J. Rosatti.

APONOGETON

(Greek: from aquatic habitat)

A. distachyos L. f. (p. 1299)   CAPE-PONDWEED   **LF**: petiole to 100 cm; blade 6–23 cm, ± narrow-lanceolate to -elliptic. **INFL**: panicle of gen 2 spike-like branches, emergent; bract ± 3 cm; fls in 2 rows, ± on 1-side of axis. **FL**: perianth part 1, white, enlarging, turn-ing green in fr, base wide; stamens many. **FR**: to 22 mm, curved or straight; beak terminal, ± 5 mm. Ponds; < 150 m. c-w SnFrB, s SCo, expected elsewhere; native to s Afr. [*Aponogeton distachyon*, orth. var.] Widely cult for aquaria, often escaping but gen a waif. Feb–Apr

ARACEAE   ARUM FAMILY

Thomas J. Rosatti, except as noted

Per, [shrub, vine], terrestrial [growing on other pls or not], or aquatic, sometimes free-floating, then sometimes much reduced, in dense, clonal populations, 0.4–10 mm, flat and tongue-shaped to spheric, not differentiated into sts and lvs, new pls produced in budding pouch at base or along margins, sometimes overwintering on bottom as dense, rootless, starch-filled daughter pl (winter bud); often from short, gen erect caudex; roots 0–many; often monoecious. **ST**: sometimes above ground in addition to caudex, or not differentiated from pl body. **LF**: simple or compound, basal (or cauline, 2-ranked), or not differentiated from pl body. **INFL**: gen spike, fleshy, gen ill-smelling, or fl 1, rarely seen, minute, appearing like 2–3 unisexual fls, often sheathed by minute membrane; fls bisexual or pistillate below, staminate above; bract subtending spike 1, gen showy (petal-like), gen > spike, sheathing or not. **FL**: perianth parts 0, 4, 6, free or fused; stamens 0–4, 6, free or fused; ovary superior to 1/2-inferior and sunken in infl axis, chambers 1–3, stigma ± sessile. **FR**: berry or achene-like, winged or not. **SEED**: 1–many, often ribbed. ± 114 genera, 1850 spp.: gen trop, subtrop; some cult for food, orn in ponds, aquaria (*Colocasia*, taro) or orn (*Philodendron*, *Anthurium*). [Les et al. 2002 Syst Bot 27:221–240; Thompson 2000 FNANM 22:128–142] Since TJM (1993), incl Lemnaceae, and exc *Acorus*, now in Acoraceae (the sole member in CA, *Acorus calamus* L., is an historical waif). *Pistia stratiotes* L. ◆ is a waif. *Pinellia ternata* (Thunb.) Breitenbach possibly naturalized in CA. In taxa once incl in Lemnaceae, vein number per pl body best determined using backlight. Scientific Editors: Bruce G. Baldwin, Thomas J. Rosatti.

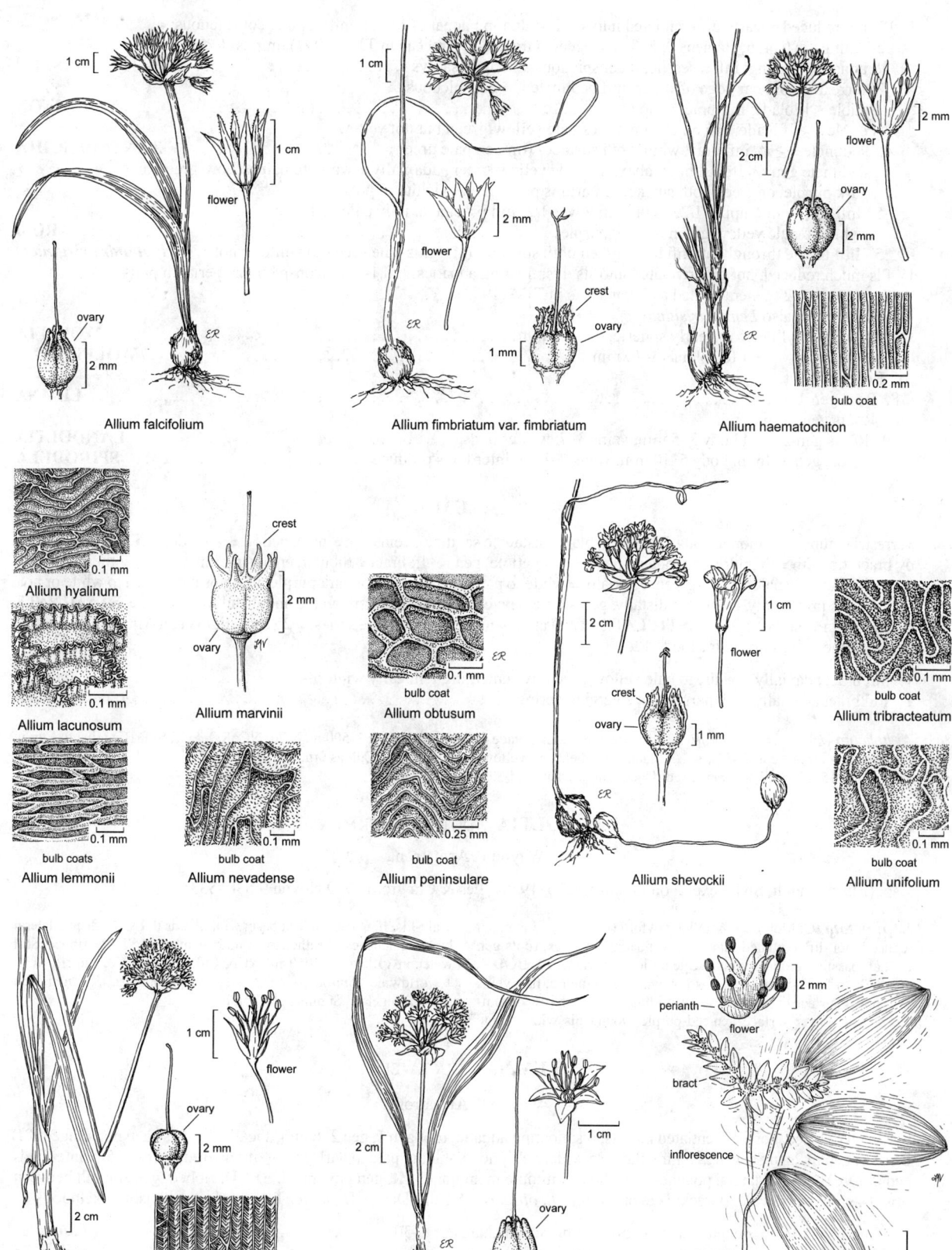

Allium falcifolium

Allium fimbriatum var. fimbriatum

Allium haematochiton

Allium hyalinum

Allium lacunosum

Allium lemmonii

Allium marvinii

Allium nevadense

Allium obtusum

Allium peninsulare

Allium shevockii

Allium tribracteatum

Allium unifolium

Allium validum

Allium yosemitense

Aponogeton distachyos

Aponogetonaceae

1. Pls not reduced, clearly differentiated into sts, lvs; floating aquatics or not; infl a spike, conspicuous;
   perianth parts 0, 4, 6; stamens 0, 2, 3, 4, 6 (genera treated as Araceae in TJM (1993), minus *Acorus*)
   2. Pl aquatic, floating; infl ± sessile, inconspicuous among basal lvs . . . . **[PISTIA]**
   2′ Pl terrestrial or near edge of water; infl peduncled, ± conspicuous
      3. Petiole < lf blade; infl bract yellow . . . . **LYSICHITON**
      3′ Petiole ± ≥ lf blade; infl bract sometimes pale yellow-green but not yellow
         4. Peduncle ± ≥ lf; infl bract white both surfaces (tip gen pale green) . . . . **ZANTEDESCHIA**
         4′ Peduncle gen << lf; infl bract abaxially pale yellow-green, adaxially ± white to pale yellow-green or
            dark purple, or green both surfaces (margins pale green to white or not)
            5. Infl sterile in ± upper 1/2, fls only in lower 1/2; infl bract abaxially pale yellow-green, adaxially ±
               white to pale yellow-green or dark purple . . . . **ARUM**
            5′ Infl fertile throughout; infl bract green both surfaces (margins pale green to white or not) . . . [*Peltandra virginica*]
1′ Pls much reduced, not differentiated into sts, lvs; floating aquatics; infl 1-fld, inconspicuous; perianth parts
   0; stamens 1, 2 (genera treated as Lemnaceae in TJM (1993))
   6. Roots 0 (see also *Lemna trisulca*)
      7. Pl body ± cylindric to nearly spheric, 0.4–1.3 mm . . . . **WOLFFIA**
      7′ Pl body flat, linear to oblong, 3–10 mm . . . . **WOLFFIELLA**
   6′ Roots present
      8. Root gen 1 . . . . **LEMNA**
      8′ Roots gen 2–16
         9. Roots gen 2–7; pl body 3–5 mm, veins 3–7; winter buds 0 . . . . **LANDOLTIA**
         9′ Roots gen 7–16; pl body 5–10 mm, veins 7–12; winter buds produced . . . . **SPIRODELA**

ARUM

Terrestrial, tuberous; monoecious. **LF:** basal; blade hastate to sagittate, veins gen ± pale; petiole ± ≥ blade. **INFL:** gen exceeded by bract, tip with ± cylindric appendage; peduncle < petiole, gen << lf; bract ± tubular, enclosing infl at base, withering before fr, blade abaxially pale yellow-green, adaxially ± white to pale yellow-green or dark purple, margins pale green to white or not; fls pistillate proximally, staminate distally, gen sterile between pistillate and staminate and distal to staminate. **STAMINATE FL:** perianth 0; stamens 3–4. **PISTILLATE FL:** ovary chamber 1, ovules in 2 series. ± 25 spp.: Eurasia, n Afr. (Greek: ancient name) Some cult as orn or for food.

1. Infl bract adaxially ± white to pale yellow-green; lvs emerging fall, early winter . . . . *A. italicum*
1′ Infl bract adaxially dark purple; lvs emerging spring . . . . [*A. palaestinum*]

A. italicum Mill. ITALIAN ARUM Tuber horizontal. **LF:** blade < 35 cm; petiole < 42 cm. **INFL:** < 1/2 bract, tip appendage yellow; bract blade 15–32 cm. **FR:** orange-red. Uncommon. Disturbed, gen shaded areas; < 500 m. NCo, NCoRO, ScV, SnFrB; native to Eur, w Asia, n Afr. Cult as orn. Mar–Jun

LANDOLTIA DOTTED DUCKMEAT

Wayne P. Armstrong

1 sp. (Elias Landolt, Swiss expert on Lemnaceae, b. 1926) [Les & Crawford 1999 Novon 9:530–533]

L. punctata (G. Mey.) Les & D.J. Crawford (p. 1307) Pls much reduced, not differentiated into sts, lvs; floating aquatics; roots gen 2–7, all passing through minute scale on lower surface. **PL BODY:** gen in 2s to 5s; 3–5 mm, oblong to obovate, asymmetric, flat; veins 3–7; upper surface dark green, shiny, midline often with row of minute bumps, lower surface gen red-purple; young pls with minute scale-like lf on each side at base; winter buds 0. **INFL:** fls in 2 lateral budding pouches, sheathed by minute membrane. **SEED:** ribbed. Still water in valleys; < 2050 m. NCo, CaRH, GV, CCo, SnFrB, SCo; ± worldwide. [*Spirodela p.* (G. Mey.) C.H. Thomps.; *S. oligorrhiza* (Kurz) Hegelm.] Summer–early fall

LEMNA DUCKWEED

Wayne P. Armstrong

Pls much reduced, not differentiated into sts, lvs; floating aquatic; root gen 1, gen 2–6 mm, sheath near base gen not winged. **PL BODY:** gen in 2s to 8s; gen 2–5 mm, flat, gen widely elliptic to oblong, pale to dark green, often ± red; veins 1–5; winter buds gen 0. **INFL:** fls in 2 lateral pouches, sheathed by minute membrane. **FR:** gen unwinged. **SEED:** ribbed, gen smooth between ribs. 14 spp.: worldwide. (Greek: lake or swamp) *L. obscura* (Austin) Daubs. [*L. minor* L. var. *obscura* Austin] excluded.

1. Pl body 6–10 mm, on tapered stalk, often in branched chains of 8–30 . . . . *L. trisulca*
1′ Pl body 1–6 mm, appearing sessile, single or gen in 2s to 8s
   2. Veins per pl body 1 (sometimes obscure)
      3. Pl body 1–2.5 mm, gen thinner at margin, gen in 2s, not attached; vein < 2/3 distance from root
         attachment to tip of pl body . . . . *L. minuta*
      3′ Pl body 2–4 mm, thin ± throughout, often in 4s to 8s, attached; vein up to 3/4 distance from root
         attachment to tip of pl body . . . . *L. valdiviana*

2′ Veins per pl body 3–5
    4. Root sheath with 2 obvious, wing-like appendages at base . . . . *L. aequinoctialis*
    4′ Root sheath not winged
      5. Pl body widely elliptic to round, tip asymmetric, lower surface gen convex. . . . . *L. gibba*
      5′ Pl body widely elliptic to obovate, tip gen symmetric, lower surface flat
        6. Upper surface gen smooth; lower surface green; winter buds 0. . . . . *L. minor*
        6′ Upper surface midline gen with row of minute bumps; lower surface often ± red in age; winter buds
          dark green or brown in fall . . . . *L. turionifera*

L. aequinoctialis Welw.   Root-sheath winged. **PL BODY**: in 2s; 2–3.5 mm, obovate-elliptic, light green, base asymmetric, tip symmetric, upper surface tip and root attachment with minute bumps, lower surface light green, smooth. **SEED**: cross-lined between ribs. Freshwater in hot regions; < 200 m. n SNF, GV, CCo, SCo, DSon; worldwide. Sep–Dec

L. gibba L. (p. 1307)   **PL BODY**: gen in 2s or 3s; 3–6 mm, widely elliptic to round, glossy green or yellow-green mottled red, base symmetric, tip asymmetric, upper surface barely convex, midline bumps gen 0, lower surface gen convex, with enlarged air spaces often bordered in red. **FR**: strongly winged. Common. Fresh or brackish water; gen < 1500 m. CA (exc KR, SNE); worldwide. Variable; vegetative pl bodies with lower surface flat appear much like *L. minor*. Gen replaced > 1500 m by *L. turionifera*. Summer

L. minor L.   **PL BODY**: in 2s or 3s; 2–5 mm, elliptic-obovate, glossy green, base, tip gen symmetric, upper surface gen smooth, midline bumps gen 0, lower surface flat, air spaces ± obscure. Common. Freshwater; < 2000 m. NCo, NCoR, n SNF, GV, CCo, SnFrB, SCo, PR, MP, D; worldwide. Aug

L. minuta Kunth   **PL BODY**: gen in 2s (esp when crowded, in full sun); 1–2.5 mm, widely elliptic to oblong, pale green, gen thinner at margin, base, tip gen symmetric; vein 1, obscure, < 2/3 distance from root attachment to tip of pl body, gen not exceeding region of visible air spaces between cells. **SEED**: cross-lined between ribs. Common.

Freshwater; < 2200 m. NCo, NCoR, SN, SnJV, CW (exc SCoRO), SCo, SnGb, PR, D; w US, S.Am, Eur, n Asia. Transparent green pls, 2–4 mm, in 4s, much like *L. valdiviana*, possibly in response to growth conditions, need study. Aug

L. trisulca L.   Root often 0. **PL BODY**: often in branched chains of 8–30, on tapered stalk; 6–10 mm, lanceolate to oblong, transparent green, base, tip symmetric, surfaces smooth. Meadows, mtn streams; < 3000 m. CaR, SN, ScV, SnBr, GB, MP, DMoj; ± worldwide. Gen forming dense, tangled masses below water surface. Aug

L. turionifera Landolt   **PL BODY**: gen in 2s or 3s; 2–5 mm, widely elliptic to obovate, glossy green, base, tip gen symmetric, upper surface shiny green, midline gen with row of minute bumps, lower surface flat, often ± red; winter buds 1–2 mm, dark green or brown in fall. Freshwater; < 3000 m. NW, CaR, SN, SnJV, CW, SCo, PR, GB, D; N.Am, Asia. Like *L. minor*, exc for winter buds. Aug

L. valdiviana Phil.   **PL BODY**: often in 4s to 8s, attached; 2–4 mm, thin ± throughout, elliptic to narrowly oblong, uniformly transparent green; base often asymmetric; tip symmetric; surfaces smooth; vein up to 3/4 distance from root attachment to tip of pl body, gen exceeding region of visible air spaces between cells. **SEED**: cross-lined between ribs. Freshwater; < 1500 m. NCo, NCoR, SN, SCo, SnBr, PR; to e US, S.Am. Variable. Apparently uncommon in CA, esp in mtns. Forms tangled masses under other pls. Spring–fall

LYSICHITON

Thomas J. Rosatti & Elizabeth McClintock

Terrestrial. **LF**: basal; petiole < blade, stout; blade ovate to oblanceolate. **INFL**: at first exceeded by bract, then exceeding bract by peduncle elongation; peduncle < lvs, stout; bract boat-shaped, base sheathing peduncle, partly enclosing infl, deciduous. **FL**: bisexual; perianth parts 4, fused at base; stamens 4; ovary partly embedded in infl axis, chambers (1)2, 1–2-ovuled. 2 spp.: ne Asia, nw N.Am. (Greek: loosening tunic, from deciduous bract) [Nie et al. 2006 Molec Phyl Evol 40:155–165]

L. americanus Hultén & H. St. John (p. 1307)   YELLOW SKUNK-CABBAGE   **LF**: present gen in spring after fl; blade 30–150 cm, ± fleshy, midvein thick, grading into petiole. **INFL**: < 12 cm, ± 2.5 cm diam; peduncle 3–5 dm; bract < 20 cm, yellow. **FL**: yellow-green, ill-smelling. **FR**: green-white. $2n=28$. Uncommon. Marshy areas, stream edges, gen in conifer forest; < 1300 m. NCo, CCo; to AK, MT; naturalized in Eur. [*Lysichiton americanum*, orth. var.] Mar–Jun

SPIRODELA DUCKMEAT

Wayne P. Armstrong

Pls much reduced, not differentiated into sts, lvs; floating aquatics; roots gen 7–16, some through minute scale on lower surface. **PL BODY**: gen in 2s to 5s; 5–10 mm, oblong to round, flat, upper surface shiny green, lower gen red-purple; veins 7–12; young pls with minute scale-like lf on each side at base; winter buds produced. **INFL**: fls in 2 lateral budding pouches, sheathed by minute membrane. **SEED**: ribbed. 2 spp.: worldwide. (Greek: visible thread, from roots)

S. polyrhiza (L.) Schleid. (p. 1307)   Roots 5–16, 1–2 through minute scale. **PL BODY**: 5–10 mm, round-ovate, symmetric; veins 7–12; upper surface smooth; winter buds produced in autumn. Freshwater; < 2500 m. NCoR, SN, GV, CCo, SnFrB, GB, DMoj; ± worldwide. Aug

WOLFFIA WATERMEAL

Wayne P. Armstrong

Pls much reduced, not differentiated into sts, lvs; floating aquatics; roots 0. **PL BODY**: gen in unequal 2s; 0.4–1.3 mm, ± cylindric to nearly spheric, floating on or partially below water surface; veins 0; budding pouch funnel-shaped; winter buds often produced. **FL**: produced in cavity on upper surface; sheathing membrane 0; stamen 1; pistil 1. **SEED**: smooth. 11 spp.: worldwide, esp temp, trop. (J.F. Wolff, German botanist, physician, 1778–1806) [Armstrong & Thorne 1984 Madroño 31:172–179] Key for use at > 10×.

1. Tip of pl body acute, upper surface ± concave or flat — dead pl with brown cells . . . . *W. borealis*
1′ Tip of pl body rounded, upper surface ± convex (exc *Wolffia arrhiza*)
  2. Upper surface of vegetative pl body with a conical bump near center; dead pl with brown cells . . . . *W. brasiliensis*
  2′ Upper surface of pl body smooth; dead pls without brown cells
    3. Upper surface dark green, flat; stomates gen 15–100 . . . . *W. arrhiza*
    3′ Upper surface transparent green, ± convex; stomates 1–10
      4. Pl body 0.8–1.3 mm, ± spheric, upper surface mostly convex, center flat . . . . *W. columbiana*
      4′ Pl body 0.4–0.8 mm, longer than wide, upper surface barely convex throughout . . . . *W. globosa*

W. arrhiza (L.) Wimm. **PL BODY:** 0.8–1.3 mm, nearly spheric, basally transparent, upper surface flat, dark green, only center floating above water, stomates gen 15–100. Uncommon. Freshwater; < 200 m. SCo (San Diego Co.); Eur, sw Asia, Afr. In backlight, pls appear slightly darker than other transparent green spp. Easily confused with *W. columbiana*. Spring–summer

W. borealis (Hegelm.) Landolt & Wildi (p. 1307) **PL BODY:** 0.7–1.2 mm, longer than wide, upper surface ± concave or flat, dark green (transparent below), floating above water, stomates 50–100, tip acute, ± upturned in side view; dead pl with brown cells. Uncommon. Freshwater; < 1000 m. NCoR, CaRH, SCo; to n US, Can. Widely scattered localities. Possibly introduced in s CA. In backlight, appears darker than other transparent green spp. Summer–early fall

W. brasiliensis Wedd. BRAZILIAN WATERMEAL **PL BODY:** 0.7–1.2 mm, longer than wide, upper surface flat near margin, elevated to conical bump near center, dark green (transparent below), floating above water, stomates 50–100, tip round; dead pl with brown cells. Ponds; < 100 m. ScV (near Sacramento River), SnFrB; to e US, S.Am. In backlight, appears darker than other transparent green spp.; in CA may be a recent introduction. Late spring–early fall ★

W. columbiana H. Karst. (p. 1307) **PL BODY:** 0.8–1.3 mm, ± spheric, transparent green, upper surface mostly convex, center flat, floating above water, stomates 1–10(30), tip round. Freshwater; < 200 m. SnJV, CCo, SnFrB, SCoRO, SCo; to Can, S.Am. Small pls much like *W. globosa*. Summer–fall

W. globosa (Roxb.) Hartog & Plas (p. 1307) **PL BODY:** 0.4–0.8 mm, longer than wide, ± cylindric or not, transparent green, upper surface barely convex throughout, darker green or not, only center floating above water, stomates 1–10, tip round; budding pouch gen with collar of long cells at junction with daughter pl. Ponds of interior valleys; < 200 m. SNF, SnJV, SCo; worldwide, esp trop. Smallest of all known angiosperms; possibly introduced in CA. Late spring–fall

WOLFFIELLA   MUD-MIDGET

Wayne P. Armstrong

Pls much reduced, not differentiated into sts, lvs; floating aquatics just below water surface; roots 0. **PL BODY:** gen in unequal 2s; 3–10 mm, linear to oblong, thin-membranous, flat, free ends gen recurved; budding pouch triangular, with track of long cells on lower surface between midline, margin; winter buds 0. **FL:** produced in cavity on upper surface; sheathing membrane 0; stamen 1; pistil 1. **SEED:** smooth. 10 spp.: worldwide, esp warm temp, trop. (Diminutive of *Wolffia*) [Landolt 1984 Veröff Geobot Inst ETH Stiftung Rübel Zürich 51:164–172]

1. Pl body 4–10 mm, widely oblong, ends strongly recurved; budding pouch angle 80–120° . . . . *W. lingulata*
1′ Pl body 3–5 mm, linear to narrowly oblong, ends gen not to ± recurved; budding pouch angle gen 40–70° . . . *W. oblonga*

W. lingulata (Hegelm.) Hegelm. (p. 1307) **PL BODY:** a semicircle in side view; upper surface gen concave; budding pouch with track of long cells between middle and edge of lower wall. Coastal, interior valleys; < 200 m. NCo, SNF, GV, CCo, SnFrB, SCo; to se US, S.Am. Variable; small pls much like *W. oblonga*. All yr

W. oblonga (Phil.) Hegelm. (p. 1307) **PL BODY:** not a semicircle in side view; upper surface gen flat; budding pouch with track of long cells along edge of lower wall. Uncommon. Coastal and interior valleys; < 200 m. NCo, SNF, GV, CCo, SnFrB, SCo; to se US, S.Am. Daughter pl often angled from axis of parent, so pair appears boomerang-like. All yr

ZANTEDESCHIA

Pl from rhizomes, terrestrial, gen fragrant; monoecious. **LF:** basal, clumped; petiole ± ≥ blade, spongy; blade hastate or sagittate to ovate, base often deeply lobed. **INFL:** gen exceeded by bract; peduncle ± ≥ lf, spongy; bract white [brightly colored]. **FL:** perianth 0. **STAMINATE FL:** stamens 2–3, free. **PISTILLATE FL:** ovary chambers 3. ± 6 spp.: s Afr. (F. Zantedeschi, Italian botanist, 1773–1846)

Z. aethiopica (L.) Spreng. CALLA-LILY **LF:** present with infl; petiole < 9 dm; blade 15–45 cm, 10–25 cm wide. **INFL:** yellow; bract < 25 cm, funnel-shaped, tip linear, ± recurved, gen pale green. 2*n*=26. Disturbed areas, near former habitations; < 300 m. NCo, NCoRO, CCo, SnFrB, SCoRO, SCo; native to s Afr. Commonly cult for showy infl, bract. Mar–Jun ❖

ARECACEAE (Palmae)   PALM FAMILY

Scott Simono

[Subshrub, shrub], tree, evergreen; dioecious (monoecious), or fls bisexual. **ST:** erect [subterranean, creeping, climbing], slender to massive, smooth or covered with fibrous or prickly remains of lf bases. **LF:** palmately or pinnately dissected or compound, alternate, forming a terminal crown; petiole gen long; blade sheathing; blade incl lflets folded lengthwise. **INFL:** panicle (spike), axillary; peduncle sheathed by 1+ large bracts; fls many, gen ± sessile. **FL:** gen small, ± radial; sepals, petals each gen 3, similar or not, fused at base or free; stamens gen 6; pistils 1, compound, or 3, simple, ovaries superior, if 1, gen 3-chambered, styles free or fused. **FR:** gen a drupe, fleshy or dry. **SEED:** gen 1. 1914 genera, 2500 spp.: trop, subtrop. [Zona 2000 FNANM 22:95–123] Cult worldwide for food, orn, building material. Scientific Editor: Bruce G. Baldwin.

1. Lf segments loosely arching to drooping, tent-like, margins folded downward; petioles gen unarmed, bases not persisting . . . . **[SYAGRUS]**
1′ Lf segments stiffly spreading, V-shaped, margins folded upward; petioles gen armed, bases persisting on trunk
    2. Lf blade pinnate, elongate; petioles not split at base; dioecious . . . . **PHOENIX**
    2′ Lf blade ± palmate, rounded; petioles split at base; fl bisexual . . . . **WASHINGTONIA**

PHOENIX

Dioecious. **LF**: pinnately compound; petiole gen armed, bases persistent on trunk; lflet margins folded upward, proximal lflets occ reduced, spine-like. **INFL**: within crown, < lvs, fls borne singly. **FL**: perianth ± yellow; sepals fused proximally; petals gen free; pistils 3. **FR**: berry-like drupe, gen fleshy, sweet. 137 spp.: s Eur, Afr, s Asia, Philippines. (Greek: name for date palm, meaning uncertain)

1. Trunk gen 1, massive, < 20 m, basal shoots 0; lvs ± 50–200, bright green, in broad, dense crown . . . . *P. canariensis*
1′ Trunks often several, slender, to 35 m, basal shoots present when young; lvs 20–40, ± gray-green, in ± open, sparse crown . . . . *[P. dactylifera]*

P. canariensis Chabaud CANARY ISLAND PALM **LF**: gen 5–7 m. **FR**: ± 2 cm, rounded to ovate, brown, pulp thin. Uncommon. Near development, disturbed areas; < 1000 m. SnFrB, SCo; native to Canary Islands. Oct–Apr ❖

WASHINGTONIA FAN PALM

LF: petiole 1–2 m, gen armed, bases persistent on trunk; blade 1–2 m, gen persistent as brown "skirt", palmately divided nearly to middle, segments 40–60, margins folded upward, with thread-like fibers, tips ± reflexed. **INFL**: within crown, > lvs, fls borne singly. **FL**: bisexual; calyx lobes ± erect; corolla lobes reflexed, white; pistil 1, ovary 3-lobed. **FR**: drupe, oblong or ovate, black, ± fleshy. 2 spp.: deserts of s CA, AZ, n Mex. (George Washington, 1st president of USA, 1732–1799)

1. Mature trunk thick, ± gray, base gen not swollen; lf blade gray-green, forming a loose, open crown . . . . *W. filifera*
1′ Mature trunk slender, ± brown, gen tapering from swollen base; lf blade bright green, forming a compact crown . . . . *W. robusta*

W. filifera (André) de Bary (p. 1307) CALIFORNIA FAN PALM **ST**: trunk < 20 m, ± 100 cm diam. **LF**: 1.5–3 m; petiole green, sharply toothed at base; thread-like fibers of lf segment margins many. **INFL**: to 5 m. Groves, moist places, seeps, springs, streamsides; < 1200 m. DSon, introduced s SNF (Kern River), SCo (Santa Ana River), DMoj (Death Valley National Park), expected elsewhere; se AZ, n Baja CA. Feb–Jun

W. robusta H. Wendl. MEXICAN FAN PALM **ST**: trunk < 30 m, < 80 cm diam. **LF**: to 1 m; petiole red-brown, sharply toothed throughout; thread-like fibers of lf segment margins inconspicuous or 0. **INFL**: to 3 m. Desert washes, disturbed areas, riparian corridors; < 500 m. SCo; native to Baja CA, SON. Apr–Jun ❖

ASPARAGACEAE ASPARAGUS FAMILY

Dale W. McNeal

Per [to shrub] or vine; rhizome short, gen with fleshy tubers. **ST**: branchlets lf-like, many, thread-like or flat, functioning as lvs. **LF**: scale-like, papery, with a spiny spur at base or not. **INFL**: raceme, umbel, or fls 1. **FL**: bisexual or not, ± white to green-yellow; perianth segments 6, in 2 petal-like whorls, free or ± fused; stamens 6, epipetalous, anther attached near middle; ovary superior, chambers 3. **FR**: berry, spheric, often red or blue. **SEED**: 1–6, black. 3+ genera depending on interpretation, 320 spp.: esp n temp. Scientific Editor: Thomas J. Rosatti.

ASPARAGUS

Per. **INFL**: pedicels slender, jointed near middle. **FL**: stamens ± ≥ perianth, exserted; style 1, short, slender, stigmas 3. ± 300 spp.: esp n temp. (Greek: ancient name) *A. densiflorus* (Kunth) Jessop misappl. to *A. aethiopicus* L., a waif.

1. Lf-like branchlets 6–20 mm wide . . . . *A. asparagoides*
1′ Lf-like branchlets < 1 mm wide [≤ 3 mm wide in *A. aethiopicus*]
    2. Pl erect; fls unisexual and bisexual . . . . *A. officinalis* subsp. *officinalis*
    2′ Pl climbing or sprawling; fls bisexual
        3. Fls 5–9(15); fr red . . . . *[A. aethiopicus]*
        3′ Fls 1–4; fr purple-black . . . . *A. setaceus*

A. asparagoides (L.) Druce (p. 1307) Climbing or sprawling; roots tuberous. **ST**: 1–5 m; branchlets 6–20 mm wide. **LF**: not spurred. **INFL**: axillary, umbels or fls 1. **FL**: bisexual; perianth 5–7 mm, bell-shaped, green-white. **FR**: 6–8 mm, red or blue or purple. Disturbed places, fields; < 1700 m. CCo, s SCo, PR; native to s Afr. Dec–Apr ❖

A. officinalis L. subsp. *officinalis* (p. 1307) Erect; rhizome thick, matted; tubers 0. **ST**: 1–3 m; branchlets < 1 mm wide. **LF**: spurred. **INFL**: axillary racemes. **FL**: unisexual and bisexual, nodding; perianth 3–7 mm, bell-shaped, green-white. **FR**: 6–8 mm, red. *n*=10. Disturbed places, roadsides, fields; gen < 200 m. CA-FP, W&I; native to Eur. Cult for food; naturalized populations scattered. Mar–Sep

A. setaceus (Kunth) Jessop   Climbing or sprawling, woody; roots fibrous. **ST:** ≤ 4 m, wiry; branchlets < 1 mm wide. **LF:** spurred, 4–10 mm. **INFL:** gen terminal umbels; fls 1–4. **FL:** bisexual, nodding. **FR:** 4–5 mm, purple-black. *n*=10. Disturbed places (sites of former gardens); < 100 m. SCo (urban canyons, San Diego); FL; native to s Afr. Apr–Sep

ASPHODELACEAE   ASPHODEL FAMILY

Dale W. McNeal

Ann, per, shrub [tree], fleshy or not, rhizomed. **ST:** branched or not. **LF:** basal or ± so or in terminal clusters, alternate, linear to widely lanceolate [ovate], small (1.5–27 cm, 0.5–3 mm wide) [to large (1.9–4 m, 3–6 cm wide)], leathery or not, fleshy or not, entire or with minute sharp teeth or coarse prickles. **INFL:** panicle [spike, raceme], bracts scattered. **FL:** perianth parts 6, in 2 petal-like whorls, radial or ± bilateral, free or ± fused; stamens 6, equal or not, anthers attached at base or middle; ovary superior, chambers 3. **FR:** capsule, loculicidal. **SEED:** 6–many, flat, angled, or winged, gray, brown, ± black, black-brown. ± 17 genera, 800 spp.: Afr, Medit to c Asia. Scientific Editor: Thomas J. Rosatti.

1. Lvs fleshy, margins with coarse prickles . . . . **ALOE**
1′ Lvs not or ± fleshy, margins entire or with minute sharp teeth
  2. Fls 25–50 mm; perianth parts fused >> 1/2, red and yellow . . . . **KNIPHOFIA**
  2′ Fls 5–12 mm; perianth parts ± free or fused < 1/2, white, ± pink, or yellow
    3. Fls white or ± pink . . . . **ASPHODELUS**
    3′ Fls yellow . . . . **[BULBINE]**

ALOE

Per, shrub [tree], fleshy, sap bitter. **ST:** branched or not. **LF:** basal and cauline, reduced upward, simple, base clasping, margin with coarse prickles [entire]. **INFL:** gen axillary, gen ± scapose, [spike] raceme or panicle with many raceme-like branches, bracted. **FL:** gen nodding, odor 0; perianth parts 6 in 2 petal-like whorls, fused exc near tip [± to base]; stamens [3 or] 6, unequal; ovary superior, chambers 3, style slender. **SEED:** many, flat, gen winged. ± 300 spp.: Medit Eur, s Asia, esp Afr. (Arabic name for these pls) *A. striatula* Haw. may be naturalizing locally in SnFrB (Berkeley Hills).

1. Perianth lobes 5–10 mm; stamens exserted . . . . *A. maculata*
1′ Perianth lobes < 2 mm; stamens incl . . . . *A. ×schoenlandii*

A. maculata All.   Erect, densely clumped. **ST:** 0–0.2 m, simple or branched. **LF:** 1–3 dm, 5–12 cm wide, lanceolate to ovate, adaxially pale to dark green with white spots, abaxially greenish. **INFL:** branches 0 or 2–5, raceme-like, 20–65-fld, 10–30 cm, dense. **FL:** perianth 25–40 mm, 2–10 mm wide, constricted above ovary, orange-red to pink; stamens unequal, outer 25–30 mm, inner 30–35 mm, anthers 4–5 mm. **FR:** 20–35 mm, 8–16 mm wide, oblong, red-brown. **SEED:** ± 2 mm. Coastal bluffs; < 100 m. CCo (Point Conception); widely naturalized; native to S.Afr. [*A. saponaria* (Aiton) Haw., poss. illeg.] Apr

A. ×schoenlandii Baker   Erect to decumbent in age. **ST:** 0.5–9 dm, loosely clustered, branched. **LF:** 2–2.5 dm, 9–12 cm wide, triangular-ovate to lanceolate, pale green, white-spotted, glaucous. **INFL:** branches 5–10, raceme-like, 25–90-fld, 4–9 dm, dense. **FL:** perianth 25–32 mm, 3–10 mm wide, ± constricted above ovary, red; stamens ± equal, 25–35 mm, anthers 3–5 mm. **FR:** 20–35 mm, 6–15 mm wide, oblong, brown. **SEED:** ± 5 mm, sterile. Coastal-sage scrub; < 100 m. SCo (La Jolla, San Diego Co.); widely naturalized; native to S.Afr. [*A. saponaria* (Aiton) Haw. × *A. striata* Haw.] Sterile natural hybrid between *A. maculata* All. and *A. striata* Haw. of S. Afr, apparently planted in La Jolla in early 1900s, now well established, reproducing vegetatively. Apr

ASPHODELUS   ASPHODEL

Ann, per. **LF:** basal, alternate, linear, entire. **INFL:** bracts persistent, narrowly lanceolate, scarious; pedicels jointed. **FL:** perianth parts ± free, white or ± pink; filaments widened, esp toward base; stigma head-like, ± 3-lobed. **SEED:** 6, black. 12 spp.: s Eur. (Greek: ancient name, fl of Hades and the dead)

A. fistulosus L. (p. 1307)   Ann or short-lived per; roots many from ± tuber-like st bases. **ST:** 15–70 cm, branched, hollow. **LF:** many, 10–30 cm, ± 4 mm diam, ± cylindric, hollow, bases wide, membranous. **INFL:** open. **FL:** 5–12 mm; perianth parts oblong, obtuse, white, midrib white or brown-orange; stamens exserted, of 2 lengths, longest = style. **FR:** 5–7 mm, ± spheric. **SEED:** 3–4 mm. 2*n*=28,56. Disturbed areas, fields; < 50 m. s SnJV, CCo, SCo, s ChI (San Clemente Island), expected elsewhere; native to s Eur. Mar–Jul   ◆

KNIPHOFIA   RED-HOT POKER

Per, rhizome short, stout, erect; roots long, cord-like. **ST:** scapose, erect, glaucous. **LF:** basal, linear, grass-like, 30–100 cm, 1–2 cm wide ± folded, entire or with minute sharp teeth, glaucous. **INFL:** spike-like raceme, cylindric, fls reflexed. **FL:** perianth parts petal-like, tube 25–45 mm, lobes 2–5 mm, rounded to acute; stamens (4 or) 6, incl to ± exserted, outer 3 shorter, anthers attached at middle; style 1, exserted, stigma small. **SEED:** many. 60 spp.: trop and s Afr. (J.H. Kniphof, German botanist, professor, 1704–1763)

K. uvaria (L.) Oken   **ST:** 0.5–1.3 m. **LF:** 60–100 cm, long-pointed. **INFL:** 15–25 cm; bracts lanceolate, acute, membranous. **FL:** perianth 25–50 mm, red, lower (on pl) yellow in age, lobes rounded; stamens exserted; style exserted. 2*n*=12. Marsh, creek margins, chaparral, canyons, disturbed places, often sandy; < 100 m. NCo, CCo, expected elsewhere; s Afr. Apr–Jul

COMMELINACEAE SPIDERWORT FAMILY

Robert E. Preston & Elizabeth McClintock

Ann, per, gen glabrous. **ST**: prostrate to erect or climbing; nodes often rooting. **LF**: alternate, entire, simple, linear to ovate, closed basal sheath or lower lf clasping st. **INFL**: cyme, umbel-like or not, terminal or terminal and axillary, subtended by 1–2 bracts [not]. **FL**: gen bisexual, bilateral or radial, gen insect-pollinated; sepals 3, gen green; petals 3, blue, white, rose, purple, or pale violet, gen ephemeral; stamens 6 (3 sterile or not), filaments gen slender, often hairy; ovary superior, chambers 3, style 1. **FR**: gen capsule. **SEED**: 1–few per chamber. 40 genera, ± 630 spp.: esp trop, subtrop; some cult as orn. [Faden 2000 FNANM 22:170–197] Scientific Editor: Thomas J. Rosatti.

1. Petals unequal; fertile stamens 3, sterile stamens 3 . . . . **COMMELINA**
1′ Petals equal; fertile stamens 6 . . . . **TRADESCANTIA**

COMMELINA

Ann [per]. **INFL**: fls 1 or in few-fld clusters, subtended by 1 ± lf-like bract. **FL**: bilateral; 2 petals larger, [gen blue or] pale violet, 1 smaller, paler; filaments glabrous. ± 170 spp.: trop, warm temp. (Jan, 1629–1692, and nephew Kaspar, 1667–1731, Commelijn, Holland)

C. benghalensis L. TROPICAL SPIDERWORT Ann. **ST**: ascending to decumbent or occ straggling, nodes rooting. **LF**: 2–9 cm, ovate to lance-elliptic; sheath with ± red hairs at tip; margins gen wavy, hairy. **FL**: of 2 kinds, on upper st open, ± 1 cm diam, on underground sts cleistogamous. **FR**: 4–6 mm. Disturbed areas; < 610 m. SCo; trop weed; native to trop Asia, Afr. Orn; persisting near former residences, establishing from yard waste, escaping from cult. May–Sep

TRADESCANTIA

Per. **INFL**: umbel-like, subtended by 2 lf-like bracts. **FL**: radial; petals equal, blue, rose, purple, or white; filament hairs gen from base, each a row of unusually large cells. ± 70 spp.: N.Am, S.Am. (John Tradescant, British naturalist, 1570s–1638)

1. St prostrate or decumbent, nodes rooting; lvs oblong to ovate; petals white, 8–9 mm . . . . *T. fluminensis*
1′ St erect or ascending, nodes rarely rooting; lvs linear to lance-linear; petals blue to rose (white), 8–20 mm
. . . . [*T. ohiensis*]

T. fluminensis Vell. (p. 1307) **LF**: 25–60 mm, glabrous, not glaucous. **INFL**: terminal, occ axillary, fls few to many. **FL**: sepals 5–7 mm. Shaded woodland, streambanks; 70–255 m. SnJV, SnFrB, SCoRO, SCo; uncommonly naturalized; native to S.Am. Apr–Jul

CYPERACEAE SEDGE FAMILY

S. Galen Smith, except as noted

Ann, per, often rhizomed or stoloned, often of wet open places; roots fibrous; gen bisexual. **ST**: gen 3-sided, gen solid. **LF**: gen 3-ranked; base sheathing, sheath gen closed, ligule gen 0; blade (0 or) linear, parallel-veined. **INFL**: spikelets gen arranged in head-, spike-, raceme-, or panicle-like infls; fl gen sessile in axil of fl bract, enclosed in a sac-like structure (perigynium) or gen not. **FL**: small, gen wind-pollinated; perianth 0 or gen bristle like; stamens gen 3, anthers attached at base, 4 chambered; ovary superior, chamber 1, ovule 1, style 2–3-branched. **FR**: achene, 2–3 sided. ± 100 genera, 5000 spp.: esp temp. [Ball et al. 2002 FNANM 23:1–608] Difficult; taxa differ in technical characters of infl, fr. In *Carex* and *Kobresia*, what appear to be pistillate fls in fact are highly reduced infls (whether or not the same applies to staminate fls is still under debate). In some other works (e.g., FNANM) these are called spikelets, and they are treated as being arranged in spikes. Here and in TJM (1993), what appear to be pistillate fls are called pistillate fls in *Carex* (and they are treated as being arranged in spikelets), but spikelets in *Kobresia* (and they are treated as being arranged into spikes). Though internally inconsistent, the approach here is consistent with traditional usage, and reflects a preference for character states that may be determined in the field. Scientific Editors: S. Galen Smith, Thomas J. Rosatti, Bruce G. Baldwin.

1. Fl, fr enclosed in sac-like structure (perigynium); fls unisexual
 2. Perigynium open at tip . . . . **CAREX**
 2′ Perigynium open on 1 side . . . . **KOBRESIA**
1′ Fl, fr not enclosed in sac-like structure; fls bisexual or some staminate
 3. Fl bracts 2-ranked; spikelets gen flat
  4. Infl in lf axils; lvs cauline; st internodes hollow . . . . **DULICHIUM**
  4′ Infl terminal; lvs basal or basal and cauline; st internodes solid or spongy with air cavities
   5. Spikelets with ≥ 2 sterile proximal fl bracts; lf sheaths ± black; infl head-like . . . . **SCHOENUS**
   5′ Spikelets with 0–1 sterile proximal fl bracts; lf sheaths not ± black; infl head-like or not
    6. Spikelets with 2–36 fl bracts . . . . **CYPERUS**
    6′ Spikelets with 2(3) fl bracts . . . . **KYLLINGA**
 3′ Fl bracts spiraled (exc *Isolepis levynsiana*); spikelets not flat (± flat in *Isolepis levynsiana*)

7. Fl sts gen with cauline lvs 0 (1–2 cauline in *Schoenoplectus saximontanus*)
   8. Infl bracts 0; spikelet 1; lvs 2, blade 0 or tooth-like, ≤ 1 mm; fr tubercle gen present . . . . **ELEOCHARIS**
   8′ Infl bracts ≥ 1; spikelets ≥ 1; lvs 1–3, gen some clearly bladed; fr tubercle 0
     9. Fl bract (outer if 2) with ≥ 3 veins; st ≤ 40 cm
      10. Fl bract 1 per fl . . . . **ISOLEPIS**
      10′ Fl bracts (1)2 per fl (a 2nd, inner bract between fl, spikelet axis gen present) . . . . **LIPOCARPHA**
     9′ Fl bract with 1 vein, at least in distal-most part of spikelet; st ≤ 400 cm
      11. Sts ≤ 15 cm, < 1 mm diam; spikelets 1, 3–4.6 mm, 1.5–2.8 mm wide . . . . **TRICHOPHORUM**
      11′ Sts (1)10–400 cm, ≤ 10 mm diam; spikelets 1–200, 3–23 mm, 2–7 mm wide
       12. Ligule ciliate; st, lf air cavities 0; st wiry; fl bracts shiny, in proximal part of spikelet 3–9-veined,
         at least in distal-most 1-veined, tip not notched . . . . **AMPHISCIRPUS**
       12′ Ligule glabrous; st, lf gen with air cavities; st rarely wiry; fl bracts dull, 1-veined, tip gen notched
        . . . . **SCHOENOPLECTUS**
7′ Fl sts with cauline lvs ≥ 1 (see also *Schoenoplectus saximontanus*)
   13. Infls gen > 1, gen ≥ 1 in lf axil
     14. Fls 10–50 per spikelet . . . . **SCIRPUS**
     14′ Fls < 6 per spikelet
      15. St 5–10 mm diam; lf blades 5–10 mm wide, margins saw-toothed . . . . **CLADIUM**
      15′ St ≤ 2 mm diam; lf blades 0.5–5 mm wide, margins gen scabrous . . . . **RHYNCHOSPORA**
   13′ Infl 1, terminal
     16. Lf sheath tip margins scabrous or ciliate; st, lf blades glabrous or gen ± scabrous or puberulent
      17. Lf sheath tip ciliate, hairs >> 1 mm, soft . . . . **BULBOSTYLIS**
      17′ Lf sheath tip scabrous, hairs << 1 mm, stiff . . . . **FIMBRISTYLIS**
     16′ Lf sheath tip margins glabrous; st, lf blades glabrous or on keels or angles scabrous
      18. Fl bracts puberulent (glabrous in age), tip notched, gen with curved awn often broken off; st sharply
        3-angled; tubers durable . . . . **BOLBOSCHOENUS**
      18′ Fl bracts glabrous, tip entire, awn 0; st 3-angled or cylindric; tubers 0 . . . . **ERIOPHORUM**

AMPHISCIRPUS NEVADA BULRUSH

1 sp. (Greek, Latin: doubtful bulrush) [Smith 2002 FNANM 23:27–28]

A. nevadensis (S. Watson) Oteng-Yeb. (p. 1323)   Per, 10–70 cm, smooth, tough, wiry; rhizomes long, 1–4 mm diam, tough, hard; st, lf air cavities 0. **ST**: simple, 0.5–2 mm diam, ± cylindric, ridged. **LF**: basal, spiraled; sheath often disintegrating to fibers; ligule ciliate; blades 5–10, C-shaped in ×-section, distal > sheath, 0.5–2 mm wide, tip sharp, margin sparse-scabrous. **INFL**: terminal, head-like, bracts lf-like, main bract spreading or erect, 1–15 mm; spikelets 1–6(10), 5–20 mm, 3–5 mm wide, ovate to lanceolate, not ± flat, fls many, 1 per fl bract; fl bracts spiraled, pale to dark red-brown, shiny, smooth, glabrous, papery to tough, hard, margins ciliate, basal 1–2 often like involucre, to 15 mm, with awn-like blade, others ± 4 mm, 3 mm wide, ovate, in proximal part of spikelet 3–9-veined, at least in distal-most 1-veined, tip not notched. **FL**: bisexual; perianth of 1–6 barbed bristles; stamens 3, anthers ± 2 mm; style 1, thread-like, base not enlarged; stigmas 2. **FR**: 2-sided, 2–2.3 mm, 1.5–1.7 mm wide, obovate, brown, smooth, beak 0; tubercle 0. Saline, often alkaline seasonal wetlands; 400–2400 m. CaR, GB, D; s S.Am. [*Scirpus n.* S. Watson] Superficially like *Schoenoplectus pungens*. Summer

BOLBOSCHOENUS TUBEROUS BULRUSHES

Per, erect, 50–200 cm, rhizomed, tubers durable. **ST**: simple, sharply 3-angled, glabrous or angles scabrous, evident internal air cavities 0, not hollow. **LF**: basal and cauline, 3-ranked; sheath closed, long; ligule 0; blade gen present, long, thin, flat, V-shaped near base, keeled abaxially, margin, keel ± scabrous. **INFL**: 1, terminal, panicle- (or head-) like, appearing with lvs; branches often scabrous; infl bracts like lf blades, main 1 > infl; spikelets ± ovate, not ± flat, fl bracts spiraled, ≥ 25, each with 1 fl in axil, ± ovate, membranous to papery, puberulent (glabrous in age), brown to ± colorless, tip notched 0.5–1 mm, gen with curved awn often broken off. **FL**: bisexual; perianth of 3–6 bristles, ≤ fr, ± straight, stout, barbed; stamens 3, anthers ≥ 1.5 mm; style 1, thread-like, base not enlarged; stigmas 2–3. **FR**: gen obovate, smooth, brown, mucronate; wall cells small, solid or large, hollow (under dissecting microscope). Wetlands, often emergent. 7–15 spp.: temp, subtrop. (Greek: bulb rush, for tubers) [Browning et al. 1995 Brittonia 47:433–445; Smith 2002 FNANM 23:37–44] Intermediates (putative hybrids) between spp. cause major problems in classification, identification.

1. Stigmas 2; fr 2-sided, tightly attached bristles 0 . . . . ***B. maritimus*** subsp. ***paludosus***
1′ Stigmas gen 3; fr gen 3-sided, tightly attached bristles 0–6
   2. Fl bract awn base ± 0.5 mm wide; anthers orange; fr floating on water, tightly attached bristles 0 . . . . ***B. robustus***
   2′ Fl bract awn base 0.2–0.3 mm wide; anthers yellow; fr sinking in water, tightly attached bristles 3–6
     3. Widest lf blade ≥ 7 mm wide; spikelet 6–10 mm wide; lf sheath tip papery, veined; fr 3.8–5.5 mm, as
       deep as wide, mucro 0.2–0.8 mm . . . . ***B. fluviatilis***
     3′ Widest lf blade < 7 mm wide; spikelet 3–5 mm wide; lf sheath tip with triangular, membranous, veinless
       area; fr 2.5–3.3 mm, shallower than wide, mucro ≤ 0.1 mm . . . . ***B. glaucus***

B. fluviatilis (Torr.) Soják (p. 1323) RIVER BULRUSH Pl 1–2 m. **ST**: 5–15 mm diam. **LF**: sheath tip papery, veined; widest blade 7–22 mm wide. **INFL**: proximal infl bract 4–15 mm wide; spike-lets 10–40, > 1/2 on branches, 10–25 mm, 6–10 mm wide; fl bracts 7–10 mm, awn 2–3 mm, base ± 0.5 mm wide. **FL**: perianth bristles tightly attached to, = fr; anthers yellow; stigmas 3. **FR**: 3.8–5.5 mm,

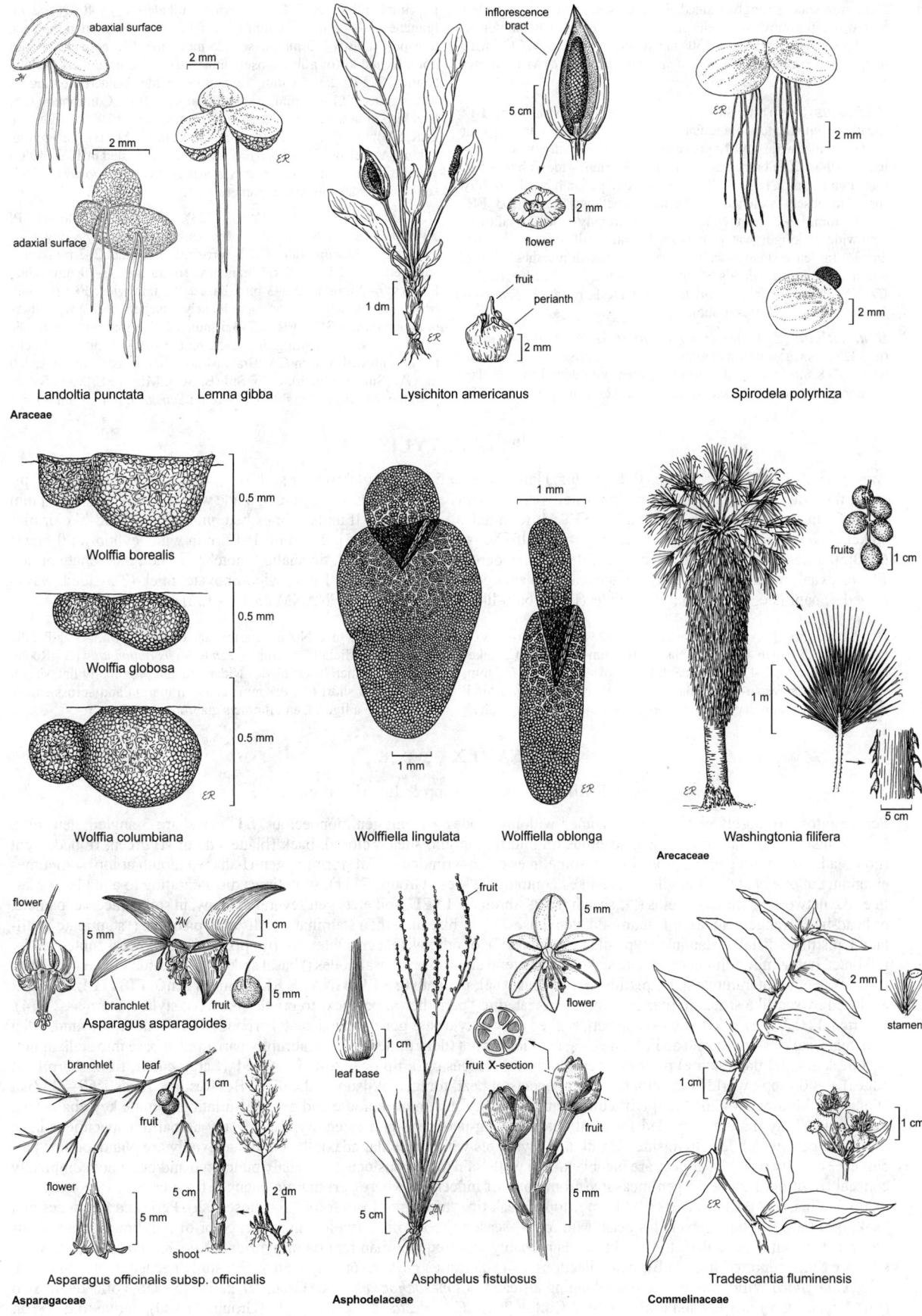

abaxial surface

2 mm

2 mm

adaxial surface

Landoltia punctata   **Lemna gibba**
Araceae

inflorescence bract

5 cm

2 mm

flower

1 dm

fruit

perianth

2 mm

Lysichiton americanus

2 mm

ER

2 mm

Spirodela polyrhiza

0.5 mm

Wolffia borealis

0.5 mm

Wolffia globosa

0.5 mm

Wolffia columbiana

1 mm

1 mm

Wolffiella lingulata   **Wolffiella oblonga**

fruits

1 cm

1 m

5 cm

Washingtonia filifera
Arecaceae

flower

5 mm

1 cm

branchlet   fruit   5 mm

Asparagus asparagoides

branchlet   leaf

1 cm

fruit

flower

5 cm

5 mm

shoot

2 dm

Asparagus officinalis subsp. officinalis
Asparagaceae

fruit

5 mm

flower

leaf base

1 cm

fruit X-section

fruit

5 mm

5 cm

5 mm

Asphodelus fistulosus
Asphodelaceae

2 mm

stamen

1 cm

1 cm

Tradescantia fluminensis
Commelinaceae

2–2.9 mm wide, strongly 3-sided, as deep as wide, sinking in water; mucro 0.2–0.8 mm; wall cells small, solid. 2*n*=94. Uncommon in CA. Fresh water marshes; < 1300 m. NCoRI, n SNH, ScV, CCo, MP; temp N.Am exc most of s US. [*Schoenoplectus f.* (Torr.) M.T. Strong; *Scirpus f.* (Torr.) A. Gray] Summer

B. glaucus (Lam.) S.G. Sm.   Pl ≤ 150 cm. **ST:** 1.5–3 mm diam. **LF:** sheath tip with acutely triangular, membranous, veinless area; widest blade 2–6 mm wide. **INFL:** proximal infl bract 2–3 mm wide; spikelets 3–30, > 1/2 on branches, 10–40 mm, 3–5 mm wide; fl bracts 5–6 mm, awn 1–2 mm, base ± 0.25 mm wide. **FL:** perianth bristles tightly (or some loosely) attached to, ≤ fr; anthers yellow; stigmas (2)3. **FR:** 2.5–3.3 mm, 1.3–2.3 mm wide, weakly or strongly 3-sided, shallower than wide, sinking in water; mucro ≤ 0.1 mm; wall cells small, solid. 2*n*=94. In dense stands locally. Fresh to brackish marshes, shores, wildlife refuges, rice-fields; < 800 m. GV, CCo, SnFrB, SCo; to OR, ID; NY; Eurasia, Afr. [*Scirpus tuberosus* Desf., misappl.; *Schoenoplectus g.* (Lam.) Kartesz] Summer

B. maritimus (L.) Palla subsp. ***paludosus*** (A. Nelson) T. Koyama (p. 1323)   SALTMARSH BULRUSH, ALKALI BULRUSH   Pl ≤ 150 cm. **ST:** (1)3–8 mm diam. **LF:** sheath tip gen with acutely triangular, membranous, veinless area; widest blade 2–12 mm wide. **INFL:** proximal infl bract 2–12 mm wide; spikelets (1)2–40, ≤ 1/2 on branches, 7–40 mm, 4–10 mm wide; fl bracts 5–8 mm, membranous, transparent, awn 1–3 mm, base 0.25 mm wide. **FL:** perianth bristles not attached to (or a few loosely attached), ≤ 1/2 fr; anthers yellow; stigmas 2. **FR:** 2.3–4.1 mm, 1.9–2.8 mm wide, 2-sided, floating on water; mucro 0.1–0.4 mm; wall cells large, hollow. Common in CA. Brackish to saline coastal, inland marshes, shores; < 2900 m. CA (exc CaR, SN); to AK, NS, TX, temp S.Am; Eurasia, Afr, HI. [*Scirpus var. p.* (A. Nelson) Kük.] *Scirpus maritimus* L. treated in TJM (1993) without infraspecific taxa, with *S. maritimus* var. *paludosus* (A. Nelson) Kük. as a synonym. Summer

B. robustus (Pursh) Soják (p. 1323)   SEACOAST BULRUSH   Pl 50–150 cm. **ST:** 4–8 mm diam. **LF:** sheath tip gen papery, veiny; widest blade 4–12 mm wide. **INFL:** proximal infl bract 2–7 mm wide; spikelets (1)5–25, ≤ 1/2 on branches, 10–30 mm, 6–10 mm wide; fl bracts 6–9 mm, awn 2–3 mm, base ± 0.5 mm wide. **FL:** perianth bristles not attached to (or some loosely attached), ± 1/2 fr; anthers orange, stigmas 3(2). **FR:** 2.7–3.5 mm, 1.8–2.8 mm wide, ± flat-3-sided or 2-sided, floating on water; mucro 0.1–0.3 mm; wall cells large, hollow. Local in CA. Brackish to saline coastal marshes; ± 0 m. GV (Suisun Marsh), CCo, SnFrB, SCo; ME to FL; Mex, S.Am. [*Schoenoplectus r.* (Pursh) M.T. Strong; *Scirpus r.* Pursh] Summer

BULBOSTYLIS

Ann [per], 2–30[50] cm, rhizomes 0. **ST:** < 0.5[1] mm wide, ± 5-angled, glabrous or scabrous, solid. **LF:** several, basal, spiraled, fl st with 1+ cauline; sheath margin distally ciliate, with hairs ± 1 mm, soft; ligule 0; blade well developed, < 0.5[1] mm wide, C-shaped in ×-section proximally. **INFL:** 1, terminal [(> 1, some in lf axils)], branched, umbel-like [head-like or of 1 spikelet]; infl bracts (1)several, longest < to > infl, lf-like; spikelets 1–5 [> 5], 2–5 mm, 1–1.5 mm wide, cylindric; fl bracts 6–30[50], 1.5–2 mm, spiraled, each with 1 fl in axil, puberulent, midrib keeled, tip shallow-notched or not, mucronate or not. **FL:** bisexual; perianth 0; style [2]3-branched, base enlarged, persistent. **FR:** > 1 mm, wide-obovate, thick-(2)3-sided, wavy-ridged or papillate. ± 100 spp.: worldwide. (Latin: bulb-like style) [Kral 2002 FNANM 23:131–136]

B. capillaris (L.) C.B. Clarke (p. 1323)   THREAD-LEAVED BEAKSEED   **ST:** gen > lvs. **LF:** blade < 0.5 mm wide. **INFL:** spikelets 1–7, 2–5 mm, 1–1.5 mm wide; fl bracts ovate. **FR:** 0.5–0.7 mm, pale brown, transverse-wavy-ridged, angles sharp, tubercle ± round. 2*n*=84. Open damp/dry sandy-gravelly soil; 300–2200 m. CaRH, n SNF, SNH; to e N.Am, Caribbean, C.Am, S.Am, Asia, Pacific Islands. Superficially resembles *Fimbristylis autumnalis* (L.) Roem. & Schult., which has lf blades wider and gen proximally flat, sheath summit with short (<< 0.5 mm) hairs on margin and across sheath (resembling a ligule), and fl bracts glabrous. Jun–Jul ★

CAREX   SEDGE

Peter F. Zika, Andrew L. Hipp & Joy Mastrogiuseppe

Per, cespitose to loosely cespitose to rhizomed with internodes > 1 cm; gen monoecious. **ST:** gen sharp-3-angled, gen solid. **LF:** 3-ranked, gen glabrous exc gen scabrous on midrib, margin; sheath closed, back (blade side of st) green, ribbed, front (non-blade side of st) gen thin, translucent, sometimes cross-wrinkled or flat, forming gen U-shaped mouth at top, sometimes extending above blade as a fragile sleeve-like "contraligule" (esp Groups 7, 11), sometimes disintegrating to a ladder- or lattice-like network or fringe of veins ("lf sheath fronts fibrous"). **INFL:** spikelets gen several to many, in spike, raceme, panicle, or head-like arrangement, each 1–many-fld, gen unisexual, or bisexual, then staminate fls distal to pistillate ("staminate/pistillate"), pistillate distal to staminate ("pistillate/staminate"), or otherwise, gen subtended by spikelet bract, lowest subtended by infl bract, occ some additional pistillate spikelets on lateral shoots from basal nodes ("basal spikelets"); fls subtended by fl bract ("scale" in other literature, esp for pistillate). **FL:** unisexual; perianth 0. **STAMINATE FL:** stamens gen 3. **PISTILLATE FL:** enclosed by sac-like structure (perigynium, abbreviated to "peri" here), occ next to bristle-like axis; style 1, stigmas 2–3(4), exserted. **FR:** 2–3(4)-sided, enclosed in peri, stalked or not, style base gen not persistent; peri body 2–3(4)-sided or round, often with marginal ribs, some with additional veins, papillate or not (determined at 20×), abruptly narrowed at base into stalk or not; peri beak abaxial flap (suture) prominent or gen inconspicuous or 0, tip open, often notched. (Latin: cutter, from sharp lf, st edges) ± 2000 spp.: worldwide; important components of peat, forage. [Wilson et al. 2007 J Bot Res Inst Texas 1:69–77; Zika et al. 1998 Madroño 45:261–270] Difficult because of many spp., morphologic and genetic variation, minute key characters. Peri around fully mature fr needed for identification (long-persistent peri often atypical). Many herbarium specimens have immature peri, which lead to misidentification. 2-styled pls with peri ± flat adaxially, curved abaxially are planoconvex; peri curved ± equally on both surfaces are biconvex. Peri walls said to be translucent are easily punctured and/or do not completely conceal fr within. Peri beaks gen measured from point of inflection, where peri margin changes from convex to concave, to its tip, but in a few taxa it is measured from fr top to beak tip ("measured from fr top" for those taxa). Peri (and fr) shapes incl beak; peri (and fr) "body" excludes beak. Mid to late season shoots often atypical in shape, color of infl, bracts, peri. Number of peri given is per spikelet. Actual hybrids probably less frequent than reports of hybrids. *C. pityophila* Mack, native to s Rocky Mtns, reported from SnBr, but collections also suggest *C. globosa* or may be distinct; study needed. In TJM (1993), *C. cephalophora* Willd. misappl. to pls belonging instead to *C. mesochorea* Mack. (Group 9), native to e US, collected in SCo (Los Angeles Co.) in 1929 and in ScV (Butte Co.) in 2010. *C. molesta* Mack. ex Bright (Groups 11A,G), native to e US, an historical urban weed, *C. leavenworthii* Dewey an urban weed.

Key to Groups

1. Pl hairy at least in part (check sheaths, blades, peri)
    2. Lvs hairy; peri hairy or glabrous . . . . **Group 1**
    2′ Lvs glabrous; peri hairy . . . . **Group 2**
1′ Pl completely glabrous throughout
    3. Infl 1 terminal spikelet . . . . **Group 3**
    3′ Infl gen ≥ 2 spikelets
        4. Stigmas 3; fr 3-angled
            5. Infl bract sheath ≥ 6 mm . . . . **Group 4**
            5′ Infl bract sheath < 5 mm
                6. Peri inflated at least near beak . . . . **Group 5**
                6′ Peri not inflated . . . . **Group 6**
        4′ Stigmas 2; fr lenticular
            7. Pistillate and staminate fls on different sts (unisexual sts) . . . . **Group 7**
            7′ Pistillate and staminate fls on same st (bisexual sts)
                8. Terminal spikelet(s) gen staminate; lateral spikelets pistillate (rarely tips staminate) . . . . **Group 8**
                8′ Terminal and lateral spikelets all a mix of staminate and pistillate fls
                    9. Spikelets each with staminate fls above pistillate fls (often easiest to see in terminal spikelet; look for
                       remnants of stamens) . . . . **Group 9**
                    9′ Spikelets each with pistillate fls above staminate fls
                        10. Peri not winged (occ thin-edged above); st gen solid; cespitose or rhizomed . . . . **Group 10**
                        10′ Peri winged; st hollow; cespitose . . . . **Group 11**

Group 1

Sheaths and/or blades hairy

1. Pls dioecious
    2. Spikelets 1–4; pl loosely cespitose; fr << peri body . . . . ***C. scabriuscula***
    2′ Spikelet gen 1; pl rhizomed; fr ± = peri body . . . . ***C. scirpoidea*** subsp. ***pseudoscirpoidea***
1′ Pls monoecious
    3. Lf blades glabrous, sheaths hairy at least at mouth
        4. Peri glabrous; lvs 1.5–23 mm wide
            5. Longer teeth of peri beak 1.5–3.3 mm, curved . . . . [2]***C. atherodes***
            5′ Longer teeth of peri beak < 0.5 mm, straight
                6. Pl rhizomed; lvs 8–23 mm wide; peri abruptly narrowed to beak . . . . ***C. amplifolia***
                6′ Pl cespitose; lvs 1.5–5.5 mm wide; peri gradually narrowed to beak . . . . [3]***C. mendocinensis***
        4′ Peri hairy; lvs 1.5–8 mm wide
            7. Pl rhizomed; often in standing water . . . . couplet 5
            7′ Pl cespitose; upland or damp ground, rarely in standing water
                8. Pistillate fl bract face hairy; lowest infl bract sheath < 5 mm; spikelets, upper 1/2 of peri dark purple;
                   talus, 2600–3900 m . . . . [2]***C. congdonii***
                8′ Pistillate fl bract face glabrous; lowest infl bract sheath 20–30 mm; spikelets, peri green to gold or
                   pale brown; moist ground, often on serpentine, 150–1600 m. . . . . [3]***C. mendocinensis***
    3′ Lf blades hairy, sheaths hairy or glabrous
        9. Pl long-rhizomed; sheath fronts occ fibrous
            10. Peri sparse-hairy on main veins or glabrous; longer teeth of peri beak 1.5–3.3 mm . . . . [2]***C. atherodes***
            10′ Peri dense-hairy; longer teeth of peri beak 0.2–1.4 mm
                11. Peri gradually narrowed to long beak; fr << peri body; lf sheaths white, not red-dotted, not fibrous;
                    mature spikelets 8–10 mm wide . . . . ***C. sheldonii***
                11′ Peri ± abruptly narrowed to beak; fr ± = body; lf sheaths often red-dotted, fibrous; mature spikelets
                    4–7 mm wide
                    12. Lf blades inrolled to triangular-channeled, 0.7–2(2.2) mm wide, those of vegetative shoots esp
                        prolonged into curled, thread-like tips; midvein of lf, infl bract a low, rounded, inconspicuous keel
                        at least toward base. . . . . ***C. lasiocarpa***
                    12′ Lf blades flat or folded into an M-shape exc at base and near tip, (2)2.2–4.5(6) mm wide, not
                        prolonged into thread-like tips; midvein of lf, infl bract a prominent, sharp keel . . . . ***C. pellita***
        9′ Pl cespitose to loosely so, forming clumps; sheath fronts not fibrous
            13. Lowest infl bract sheath < 5 mm; terminal spikelet staminate; not on serpentine
                14. Peri glabrous, face strongly 5-veined; pistillate fl bracts glabrous; st brown (red-brown) at base. . . . . . ***C. whitneyi***
                14′ Peri hairy, face faintly 0–10 veined; pistillate fl bracts hairy near tips; st red to purple at base
                    15. Pistillate fl bracts dark purple with 1–5 pale brown, strong veins; spikelets dark; peri > pistillate fl
                        bracts, purple on upper 1/2 of body, gradually narrowed above; talus, 2600–3900 m . . . . [2]***C. congdonii***

15′ Pistillate fl bracts with 1 weak vein in wide central green zone; spikelets light; peri < or > pistillate fl bracts, green to dark brown (dark purple) on upper 1/2 of body, gen abruptly narrowed above; moist or wet sites, 1200–2600 m . . . . ***C. sartwelliana***
  13′ Lowest infl bract sheath > 5 mm; terminal spikelet staminate or staminate and pistillate; gen on serpentine
    16. Peri gen glabrous (sparse-hairy toward tip); lf blade glabrous or sparsely hairy at base; basal lvs 1.5–5.5 mm wide. . . . . ³***C. mendocinensis***
    16′ Peri hairy; lf blade hairy; basal lvs occ > 7 mm wide
      17. Pistillate fl bracts obovate to wide-elliptic, red-brown or ± purple, midrib green; peri purple-splotched; on serpentine, < 600 m, NCo, KR, NCoRO, n CCo (Marin, Santa Cruz cos.), SnFrB, SCoRO. . . . . ***C. gynodynama***
      17′ Pistillate fl bracts elliptic to elliptic-obovate, green or pale red-brown, midrib green; peri green or gold-green to pale brown, occ red-dotted but not purple-splotched; often on serpentine, 60–1200 m, NCoR (Mendocino, Lake cos.), CaR, n&c SN. . . . . ***C. hirtissima***

Group 2

Peri hairy

1. Fl sts all unisexual, either staminate or pistillate
  2. Fl sts staminate
    3. Fl st base scaly, not fibrous; spikelets 1–4; anthers 2–2.8 mm . . . . ²***C. scabriuscula***
    3′ Fl st base ± fibrous, scaly or not; spikelet 1; anthers 3.1–4.5 mm
      4. Lf sheath front occ hairy, not coarse-veined, fine-red-dotted; ligule occ longer than wide . . . . ²***C. scirpoidea*** subsp. ***pseudoscirpoidea***
      4′ Lf sheath front glabrous, coarse-veined, gen not red-dotted; ligule wider than long . . . . ³***C. serpenticola***
  2′ Fl sts pistillate
    5. Peri flat exc over relatively small fr, peri veins strong on lower 1/2 (obscure if dark purple) . . . . ²***C. scabriuscula***
    5′ Peri plump, filled by fr, peri veins 0 or short, weak on lower 1/2
      6. Spikelets gen 1 (2, overlapped), oblong-cylindric; mature sts stiffly erect . . . ²***C. scirpoidea*** subsp. ***pseudoscirpoidea***
      6′ Spikelets 2–4, lowest gen well-separated or obvious, gen elliptic-ovate to ± spheric; mature sts weak, drooping. . . . . ³***C. serpenticola***
1′ Some or all fl sts bisexual, with both staminate and pistillate fls
  7. Stigmas 4, warty, not plumose, at 15×; fr base 4-angled; lf sheath mouth V-shaped; lvs ± sickle-shaped — KR, NCoR. . . . . ***C. concinnoides***
  7′ Stigmas 3, fine-plumose at 15×; fr base 3-angled; lf sheath mouth U-shaped; lvs straight (exc *Carex brevicaulis* sickle-shaped)
    8. Basal spikelets present
      9. Peri with 12–20 strong veins at least to midbody; pistillate fl bracts gen prominently 3–5-veined
        10. Lvs pale blue-green or glaucous; peri stalks ± = beaks; peri bodies barrel-shaped; lvs gen papillate abaxially at 40×; basal spikelet on erect, gen short, stout stalk. . . . . ***C. brainerdii***
        10′ Lvs green, not glaucous; peri stalks 1.5–2 × beaks; peri bodies obovoid; lvs not papillate or papillate only on veins abaxially at 40×; basal spikelet on gen arching, long, slender stalk. . . . . ***C. globosa***
      9′ Peri with 2 strong veins, veins otherwise 0 or faint; pistillate fl bracts prominently 1(3)-veined
        11. Infl bracts on taller (non-basal) sts inconspicuous, scale-like, < infl; old lvs occ persisting as shredded fibrous tufts; peri 1.5–2.1 mm wide; coastal dunes, headlands, NCo, CCo . . . . ***C. brevicaulis***
        11′ Infl bracts on taller (non-basal) sts prominent, lf-like, gen ≥ infl; old lvs not persisting as shredded fibrous tufts; peri 1–1.7 mm wide; NCo, CCo, and elsewhere
          12. Peri 2.3–3.1 mm, beak 0.4–0.8 mm, teeth 0.1–0.2 mm; sts gen spreading or arching, ± smooth exc near infl; pl loosely cespitose; rhizomes slender . . . . ²***C. deflexa*** var. ***boottii***
          12′ Peri 3.1–4.5 mm, beak 0.7–1.7 mm, teeth 0.2–0.4 mm; sts gen ascending, scabrous; pl loosely to densely cespitose; rhizomes stout . . . . ***C. rossii***
  8′ Basal spikelets 0
    13. Staminate spikelets 2–3
      14. Sts 10–50 cm; dry uplands, Siskiyou Co.. . . . . ²***C. halliana***
      14′ Sts 60–180 cm; wet soil, Siskiyou Co. and elsewhere
        15. Pls cespitose (*Carex spissa* cespitose to rhizomed); peri sparse-hairy, sharp-3-angled or 2-edged, gen fine-red-dotted; springs, seeps, streambanks, gen not in standing water, often on serpentine; CCo, SCoRO, SCo
          16. Peri sharp-3-angled, lanceolate, gradually narrowed to long beak ± 2 mm; lateral spikelets with 15–45 peri; gen on serpentine; SCoRO (Monterey, San Luis Obispo cos.), PR . . . . ²***C. obispoensis***
          16′ Peri often ± flat, 2-edged, or strongly inflated, but not 3-angled, obovoid, abruptly narrowed to beak ± 0.5 mm; lateral spikelets with 150–300 peri; on serpentine or not; CCo, SCo. . . . . ***C. spissa***
        15′ Pls rhizomed; peri dense-hairy, plump, not sharp-3-angled or 2-edged, green or dark-purple-splotched; often emergent in shallow water, not on serpentine; widespread
          17. Lf blades inrolled to triangular-channeled, 0.7–2(2.2) mm wide, those of vegetative shoots esp prolonged into curled, thread-like tips; midvein of lf, infl bract a low, rounded, inconspicuous keel at least toward base . . . . ***C. lasiocarpa***

17′ Lf blades flat or folded into an M-shape exc at base and near tip, (2)2.2–4.5(6) mm wide, not
   prolonged into thread-like tips; midvein of lf, infl bract a prominent, sharp keel . . . . *C. pellita*
13′ Staminate spikelet 1 (exc *Carex triquetra, Carex obispoensis* with 1–2 upper lateral spikelets occ
   staminate/pistillate, but not entirely staminate)
   18. Spikelets 1, terminal . . . . *C. filifolia*
      19. Longest pistillate fl bracts (exc awns) gen < 2.5 mm, gen << peri; peri body gen abruptly narrowed
         to beak 0–0.4 mm; fr 1.6–2.4 mm . . . . var. *erostrata*
      19′ Longest pistillate fl bracts (exc awns) gen > 3 mm, gen ≥ peri; peri body ± gradually narrowed to
         beak 0.1–0.8 mm; fr 2.2–3 mm . . . . var. *filifolia*
   18′ Spikelets > 1, terminal and lateral
      20. Lateral spikelets gen < 1 cm; mouth of lf sheath fine-toothed
         21. Rhizomes short, pls loosely cespitose; peri 1–1.4 mm wide . . . . ²*C. deflexa* var. *boottii*
         21′ Rhizomes elongate, pls gen scattered; peri 1.5–2.2 mm wide
            22. Pistillate fl bracts, lower staminate fl bracts green, red, or purple with white margin 0.4–0.8 mm
               wide; terminal spikelet staminate on taller infls; sts bisexual; CaRH . . . . *C. inops* subsp. *inops*
            22′ Pistillate fl bracts, lower staminate fl bracts dark purple with white margin 0.1–0.2 mm wide;
               terminal spikelet on taller infls staminate, pistillate, or pistillate/staminate; sts unisexual or
               bisexual; KR . . . . ³*C. serpenticola*
      20′ Lateral spikelets gen > 1.5 cm; mouth of lf sheath entire or fine-ciliate
         23. Pl rhizomed; lf blade smooth; often in pumice — KR, CaRH . . . . ²*C. halliana*
         23′ Pl cespitose; lf blade smooth, scabrous, or minute-papillate; not in pumice
            24. Peri hairs obvious, not sparse
               25. Peri lanceolate, ± gradually narrowed to beak ± 2 mm; gen on serpentine; SCoRO (Monterey,
                  San Luis Obispo cos.), PR . . . . ²*C. obispoensis*
               25′ Peri elliptic, abruptly narrowed to beak < 1 mm; not on serpentine; CCo, SCo, s ChI, TR,
                  PR . . . . *C. triquetra*
            24′ Peri hairs 0 or occ inconspicuous, sparse
               26. Spikelets ≤ 60 mm; peri, pistillate fl bract green to brown, occ red-marked; gen on serpentine
                  . . . . *C. mendocinensis*
               26′ Spikelets ≤ 32 mm; peri, pistillate fl bract red-brown to dark purple, or green, marked with
                  red-brown to dark purple; gen not on serpentine
                  27. Peri, at least distally on faces, with sparse, spreading-ascending, short, stiff, stout hairs or
                     bristles; fl bract midvein distal margin scabrous to ciliate . . . . *C. fissuricola*
                  27′ Peri faces gen glabrous or with sparse, spreading to appressed, gen long, soft, thin hairs;
                     fl bract midvein, margin glabrous . . . . *C. luzulina*

Group 3

Spikelet 1

1. Lowest pistillate fl bracts green, lf-like, 1–15 cm, base clasping peri; peri with bristle-like axis inside
   2. Upper st round or blunt-3–angled, gen smooth toward tip; peri with 2 strong marginal veins, several faint
      veins; longest lvs much exceeded by infl, blades inrolled or flat, 0.8–1.5 mm wide; pl cespitose . . . . *C. multicaulis*
   2′ Upper st sharp-3–angled, scabrous toward tip; peri with 2 strong marginal veins, gen 0 faint veins;
      longest lvs ± equaling infl, blades flat, 1.5–3 mm wide; pl loosely cespitose . . . . *C. tompkinsii*
1′ Lowest pistillate fl bracts not green, not lf-like, ≤ 1.5 cm, base clasping peri or not; peri with bristle-like
   axis inside or not
   3. Peri body wide-elliptic to ± round
      4. Peri 2–4.8 mm wide; gen some lower pistillate fl bracts 3- or 5-veined; dry soil . . . . *C. breweri*
      4′ Peri 1.5–1.8 mm wide; lower pistillate fl bracts 1-veined; wet or seasonally wet soil
         5. Stigmas 2; lvs green, ≤ 1 mm wide; peatland, moist alpine meadows . . . . *C. capitata*
         5′ Stigmas 3; lvs ± glaucous, 2–5 mm wide; meadows . . . . ²*C. idahoa*
   3′ Peri body narrowly elliptic to oblanceolate
      6. Peri green, tip blunt, beak 0
         7. Peri obovoid, marginal veins 2, fine veins 0 . . . . ²*C. geyeri*
         7′ Peri elliptic to narrow-oblong-elliptic, with many fine veins . . . . *C. leptalea*
      6′ Peri green or straw to dark purple, tip pointed or distinctly beaked
         8. Peri 1–3 per st, 4.9–7 mm; lvs evergreen . . . . ²*C. geyeri*
         8′ Peri 8–50 per st, 2–5 mm; lvs deciduous
            9. Mature peri gradually narrowed to ± long stalk and beak; peri narrow-elliptic, often ± ascending to
               spreading or reflexed; pistillate fl bracts shed before peri . . . . *C. nigricans*
            9′ Mature peri abruptly narrowed to short stalk or beak or both; peri elliptic to obovate, ascending or
               erect; pistillate fl bracts persisting with peri
               10. Lvs ± glaucous, flat, 2–5 mm wide; peri without bristle-like axis inside . . . . ²*C. idahoa*
               10′ Lvs green, not glaucous, inrolled or channeled, 0.4–1 mm wide; peri with bristle-like axis inside
                  . . . . *C. subnigricans*

Group 4

Stigmas 3; infl bract sheath ≥ 6 mm

1. Pl rhizomed
   2. Peri inflated; style persisting on fr; st base spongy-thickened; lf sheaths, lower lf blades with
      well-developed brickwork pattern of crosswalls . . . . *C. utriculata*
   2′ Peri not inflated; style breaking off near fr; st base not spongy-thickened; lf sheaths, lower lf blades
      without well-developed brickwork pattern of crosswalls
      3. Stigmas 2–3 (gen an even mix on a st), ripe fr lenticular or 3-angled; serpentine wetlands, KR . . . . *C. hassei*
      3′ Stigmas 3, ripe achenes 3-angled; various (incl serpentine) wetlands or meadows, KR and elsewhere
         4. Peri beak distinct or indistinct, 0.5–1.5 mm; lvs green, not glaucous
            5. St base red or purple; peri abruptly narrowed to distinct beak; lower lvs on fl sts reduced to bladeless
               sheaths. . . . . *C. californica*
            5′ St base brown; peri gradually narrowed to indistinct beak; lower lvs on fl sts well developed. . . . . [4]*C. luzulina*
         4′ Peri beak indistinct, < 0.5 mm; lvs glaucous
            6. Peri beak bent to 1 side; peri obovate, veins gen strong; lvs flat or folded; serpentine fens;
               KR, NCoRI . . . . *C. klamathensis*
            6′ Peri beak straight; peri elliptic, veins weak or 0; lvs channeled; coastal peatland, not on serpentine;
               NCo (Mendocino Co.) . . . . *C. livida*
1′ Pl cespitose to loosely so (*Carex spissa* cespitose to rhizomed)
   7. Peri beak teeth spreading, gen curved, 1.3–2.8 mm . . . . *C. comosa*
   7′ Peri beak teeth 0 or erect, straight, < 1 mm
      8. Terminal spikelet gen pistillate/staminate
         9. Pistillate fl bracts white; peri 1.3–2.3 mm. . . . . [2]*C. tiogana*
         9′ Pistillate fl bracts red-brown or purple; peri 3–5.5 mm
            10. Peri plump, 0.9–2 mm wide
               11. Peri not papillate, ovate to lanceolate. . . . . [4]*C. luzulina*
               11′ Peri papillate, wide-elliptic to ovate. . . . . *C. serratodens*
            10′ Peri flat, 1.7–3.5 mm wide
               12. Lateral spikelets pistillate, larger with < 50 peri . . . . *C. heteroneura*
               12′ Lateral spikelets gen pistillate/staminate, larger with > 60 peri . . . . *C. mertensii*
      8′ Terminal spikelet gen staminate or staminate/pistillate (pistillate/staminate)
         13. Wider lvs 9–20 mm wide; sts often weak, leaning
            14. Peri gradually narrowed to indistinct beak; spikelets erect or spreading
               15. Peri 4.3–6.5 mm, 20–32-veined, not red-dotted; infl bract sheaths inflated; fr sts often prostrate;
                  coastal forest, not on serpentine, < 900 m . . . . *C. hendersonii*
               15′ Peri 2.5–4.3 mm, finely 8–10-veined, occ red-dotted; infl bract sheaths not inflated; fr sts erect or
                  ascending; openings, mixed-evergreen forest, disturbed wet places, often on serpentine,
                  60–1200 m . . . . *C. hirtissima*
            14′ Peri abruptly narrowed to distinct beak; spikelets erect to spreading or drooping
               16. Spikelets 1–3 cm, dark purple; peri < 30, margins flat; lvs green; 1900–3100 m . . . . [2]*C. luzulifolia*
               16′ Spikelets 3–20 cm, green; peri gen > 75, margins not flat; lvs blue-green, at least when young; < 1200 m
                  17. Peri body elliptic, 1.1–1.5 mm wide; st base red to purple; pistillate spikelets 5–8 mm wide . . . . *C. pendula*
                  17′ Peri body obovate, 1.5–2.5 mm wide; st base brown; pistillate spikelets 7–12 mm wide . . . . *C. spissa*
         13′ Wider lvs gen 1–8 mm wide; sts erect or leaning
            18. Spikelets ± sessile, erect; infl bracts ± stiff, ascending or spreading; peri green to ± yellow or brown
               19. Peri body lanceolate, gradually narrowed to indistinct beak . . . . [2]*C. lemmonii*
               19′ Peri body elliptic to ovate or obovate, abruptly narrowed to distinct beak . . . . *C. viridula* subsp. *viridula*
            18′ Spikelets stalked, erect or arching to drooping; infl bracts ± weak, spreading; peri white, green,
               brown, red-brown, or purple
               20. Peri beak distinct, ≤ 0.5 mm, or indistinct
                  21. Peri 1.3–2.3 mm, 0.4–0.9 mm wide; pistillate fl bract white. . . . . [2]*C. tiogana*
                  21′ Peri 3–5.5 mm, 0.9–2 mm wide; pistillate fl bract red-brown or dark purple
                     22. Peri body ovate to lanceolate, gradually narrowed to indistinct beak. [4]*C. luzulina*
                     22′ Peri body obovate to wide-elliptic, abruptly narrowed to distinct beak . . . . *C. raynoldsii*
               20′ Peri beak distinct, gen > 0.5 mm
                  23. Pistillate spikelets (15)20–60 mm, linear; peri beak ciliate on inner side of teeth . . . . *C. mendocinensis*
                  23′ Pistillate spikelets gen ≤ 20 mm, oblong; peri beak gen not ciliate on inner side of teeth, but beak
                     margins and outer side of teeth gen with spreading, stiff bristles
                     24. Peri faces, at least near beak, with sparse, spreading-ascending, short, stiff, bristles; pistillate fl
                        bract margins ciliate distally . . . . *C. fissuricola*
                     24′ Peri faces gen glabrous (or with sparse, weak hairs); pistillate fl bract margins glabrous
                        25. Peri flat margin > 1/2 fr width; peri 1.7–2.5 mm wide; infl bract sheath expanded toward
                           shallow-U-shaped mouth (1.8)2–3 mm wide . . . . [2]*C. luzulifolia*

25′ Peri flat margin 0 or < 1/2 fr width; peri 0.9–2 mm wide; infl bract sheath not (or ±) expanded
toward shallow Y- or V-shaped mouth 0.6–2 mm wide
26. Peri beak tip to fr tip < 1.5 mm; lowest spikelet stalk gen exserted < 10 mm; pistillate fl bracts
white, green, gold, or red-brown to dark purple, margin white . . . . **²*C. lemmonii***
26′ Peri beak tip to fr tip ≥ 1.5 mm; lowest spikelet stalk exserted > 10 (> 5) mm; pistillate fl bracts
gen red-brown to dark purple . . . . **⁴*C. luzulina***

Group 5

Stigmas 3; infl bract sheath < 6 mm; peri inflated at least near beak

1. Lower lateral spikelets nodding on drooping stalks
2. Peri strongly 7–11-veined, beak teeth without eroded tips, 0.3–2.8 mm
3. Peri beak teeth spreading or strongly curved, 1.3–2.8 mm, not prickle-like; st base brown . . . . **C. comosa**
3′ Peri beak teeth erect, straight, 0.3–0.9 mm, prickle-like; st base red-purple . . . . **C. hystericina**
2′ Peri weakly 0–7-veined, beak teeth with eroded tips or < 0.5 mm
4. Peri obovate; lf margins very sharply scabrous . . . . **²*C. spissa***
4′ Peri elliptic to ovate; lf margins smooth to ± serrate
5. Pl rhizomed; peri beak 0.7–1.2 mm . . . . **C. amplifolia**
5′ Pl cespitose; peri beak 0.5 mm . . . . **C. pendula**
1′ Lower lateral spikelets gen erect, stalks sometimes 0
6. Lateral spikelets 4–20 mm; peri 1.8–4.5 mm
7. Pistillate fl bracts dark purple or black; peri 1.6–2.1 mm wide . . . . **C. raynoldsii**
7′ Pistillate fl bracts green to rusty; peri 0.8–1.6 mm wide . . . . **C. viridula** subsp. **viridula**
6′ Lateral spikelets 20–100 mm; peri 3.6–12.7 mm
8. Peri widest above middle, weakly veined; lf margins very sharply scabrous; young lvs blue-green . . . . **²*C. spissa***
8′ Peri widest below middle, strongly veined; lf margins entire or weakly serrate; young lvs green
9. Pl long-rhizomed; st base spongy-thickened, brown or red-tinged, crosswalls between veins many, well
developed, yielding brick-like pattern, sheaths gen not fibrous; ligule length gen < width on lower lvs;
widest lvs to 15 mm wide . . . . **C. utriculata**
9′ Pl ± cespitose; st base not spongy-thickened, gen red- or purple-tinged, crosswalls between veins gen
few, weakly developed, not yielding brick-like pattern; sheaths often fibrous; ligule length > width on
lower lvs; widest lvs to 6.5 mm wide
10. Larger peri 7.5–12.7 mm, length 3.4–7.4 × width, tip gradually narrowed to an indistinct beak;
< 1800 m . . . . **C. exsiccata**
10′ Larger peri 3.6–8.2 mm, length gen 2–3.5 × width, tip ± abruptly narrowed to ± distinct beak;
< 3300 m . . . . **C. vesicaria**

Group 6

Stigmas 3; infl bract sheath < 6 mm; peri not inflated

1. Peri beak teeth curved, 1.3–2.8 mm . . . . **C. comosa**
1′ Peri beak teeth 0 or inconspicuous, straight, 0–1 mm
2. Lvs 8–23 mm wide; lateral spikelets often with > 100 peri
3. St brown at base; upper 2–4 spikelets staminate; sheaths, peri red-dotted . . . . **C. spissa**
3′ St ± red or ± purple at base; uppermost spikelet staminate; sheaths, peri not red-dotted
4. Pl rhizomed; peri beak 0.7–1.1 mm . . . . **C. amplifolia**
4′ Pl cespitose; peri beak 0.5 mm . . . . **C. pendula**
2′ Lvs < 8.5 mm wide; lateral spikelets with < 70 peri
5. Pl long-rhizomed; peri beak 0–0.5 mm; lvs glaucous; peri papillate
6. Peri abruptly narrowed to distinct beak
7. Peri beak straight, beak, upper body often minute-serrate or bristly; moist mtn meadows over
dolomite, 2800–3400 m, W&I (White Mtns) . . . . **³*C. idahoa***
7′ Peri beak bent to side, beak, upper body not minute-serrate, not bristly; serpentine fens, 450–900 m,
KR, NCoRI . . . . **C. klamathensis**
6′ Peri tapered to indistinct beak
8. Spikelets nodding; roots with dense felt-like hairs . . . . **C. limosa**
8′ Spikelets erect or ascending; roots without dense felt-like hairs
9. Pistillate fl bracts prominently awned; sheaths fibrous; terminal spikelet pistillate/staminate . . . . . . **C. buxbaumii**
9′ Pistillate fl bracts awnless; sheaths not fibrous; terminal spikelet staminate . . . . **C. livida**
5′ Pl cespitose to loosely so; peri beak < to > 0.5 mm; lvs green or glaucous; peri papillate or not
10. Spikelets green to gold, pistillate fl bracts without dark purple or black
11. Peri 1.3–2 mm, beak 0.2–0.4 mm; pistillate fl bracts blunt or mucronate; 3100–3350 m, c SNH
(Mono Co.) . . . . **C. tiogana**

11′ Peri 4.5–7.3 mm, beak 0 or 1.9–2.8 mm; pistillate fl bracts long-awned; < 1800 m, KR, SN

  12. Terminal spikelet staminate; lateral spikelets 2–4, gen nodding, dense; peri rounded, > 25, ribs 3, veins 13–21, beak 1.9–2.8 mm, teeth prominent, 0.3–0.9 mm . . . . ***C. hystericina***

  12′ Terminal spikelet staminate fls above pistillate; lateral spikelet 1, erect, not dense; peri 3-angled, 1–5, ribs 3, veins 0, beak 0, teeth 0 . . . . ***C. tompkinsii***

10′ Spikelets darker than green to gold, pistillate fl bracts at least partly dark brown to dark purple or black

  13. Terminal spikelet staminate (some spp. variable, keying here or at 13′)

    14. Peri ascending, ≤ 1/2 filled by fr; lower spikelets often nodding on drooping stalks . . . . ***C. spectabilis***

    14′ Peri spreading, > 1/2 filled by fr; lower spikelets gen erect

      15. Peri body ovate to widely elliptic, beak 0.3–1 mm, teeth gen bristly, 0.2–0.5 mm; < 1800 m, often on serpentine . . . . **²*C. serratodens***

      15′ Peri body elliptic to obovate, beak 0.1–0.5 mm, teeth 0 or smooth to bristly, < 0.2 mm; 1800–3200 m, not on serpentine

        16. Lateral spikelets < terminal; peri 2–3 mm . . . . ³***C. idahoa***

        16′ Lateral spikelets ≥ terminal; peri 3.5–4.5 mm . . . . ***C. raynoldsii***

  13′ Terminal spikelet pistillate or pistillate/staminate (or staminate/pistillate/staminate in some *Carex helleri*)

    17. Peri flat, < 1/2 filled by fr, gen not papillate (± papillate near beak)

      18. Lateral spikelets spreading or drooping, often > 2 cm, gen pistillate/staminate, basal fls with remnant filaments . . . . ***C. mertensii***

      18′ Lateral spikelets gen erect or ascending, gen < 2 cm, gen pistillate

        19. Lower pistillate fl bracts acuminate, awns gen 0.5–2 mm; peri elliptic to obovate; st 5–40 cm . . . . . . ***C. helleri***

        19′ Lower pistillate fl bracts acute to acuminate, awns < 0.5 mm or gen 0; peri ovate to ± round or obovate; st 25–100 cm . . . . ***C. heteroneura***

    17′ Peri plump, > 1/2 filled by fr, papillate or not

      20. Peri ovate to widely elliptic, gen 3–5 mm . . . . ²***C. serratodens***

      20′ Peri elliptic to obovate, gen 2–3 mm

        21. Peri green to olive, obvious, strongly contrasting with, not hidden by shorter, darker pistillate fl bracts . . . . ***C. stevenii***

        21′ Peri yellow to dark brown to dark purple, hidden by or not contrasting with pistillate fl bracts

          22. Peri upper body, beak not minute-serrate or bristly; pistillate fl bracts oblong or lanceolate to widely elliptic, tips not bristly . . . . ***C. albonigra***

          22′ Peri upper body, beak minute-serrate or bristly; pistillate fl bracts obovate to widely elliptic, tips or awns bristly or not . . . . ³***C. idahoa***

Group 7

Spikelets 2+; stigmas 2; sts gen unisexual

1. Sts staminate or mostly so

  2. Pl cespitose to loosely so (*Carex alma* cespitose to rhizomed); spikelets ≥ 1 per lower node or branch; lf sheath front red-dotted or not

    3. Lf sheath front not red-dotted; anther tips acute, awn 0 . . . . ²***C. infirminervia***

    3′ Lf sheath front red-dotted; anther tips with bristle-like awn ≤ 0.3 mm

      4. Lf sheath front not cross-wrinkled, colorless at mouth, rim not thick or dark, gen with contraligule . . . . . . ²***C. alma***

      4′ Lf sheath front occ cross-wrinkled, copper or ± purple at mouth, rim often thick or dark, without contraligule . . . . ²***C. cusickii***

  2′ Pl rhizomed; spikelets gen 1 per node or branch; lf sheath front gen not red-dotted

    5. St angles smooth near tip; rhizomes 0.6–1.9 mm thick

      6. Anther awns 0.2–1 mm, bristly at 30×; lf blades 1–3.5 mm wide; infl 12–45 mm . . . . ²***C. douglasii***

      6′ Anther awns < 0.1 mm, smooth to warty at 30×; lf blades < 1.5 mm wide; infl ≤ 20 mm . . . . ²***C. duriuscula***

    5′ St angles ± scabrous near tip; rhizomes 1.5–5 mm thick

      7. Longest anther awns 0.2–0.4 mm, filaments gen incl; st gen < 40 cm; coastal, islands . . . . ²***C. pansa***

      7′ Longest anther awns 0.1–0.2 mm, filaments exserted; st 20–75 cm; coastal or inland

        8. Anther awns gen hairy at 20×; lf sheath mouth occ with dark or ± thick rim; lf blades 1.5–3 mm wide . . . . ²***C. praegracilis***

        8′ Anther awns glabrous (warty) at 20×; lf sheath mouth with or without dark stripe, gen without thick rim; lf blades 2–5 mm wide . . . . ²***C. simulata***

1′ Sts pistillate or mostly so

  9. Pl cespitose (*Carex alma* cespitose to rhizomed)

    10. Lf sheath front copper or ± purple at mouth, often with thick or dark rim, mouth gen concave or truncate . . . . ²***C. cusickii***

    10′ Lf sheath front colorless at mouth, without thick or dark rim, mouth concave to convex

      11. Lf sheath front red-dotted; peri ovate to triangular-ovate; infl a dense panicle of spikelets, spikelets gen several per lower node, or lower node long-branched with 3+ spikelets . . . . ²***C. alma***

      11′ Lf sheath front not red-dotted; peri narrow-lanceolate; infl a raceme of spikelets, spikelets 1 per node . . . . ²***C. infirminervia***

9′ Pl rhizomed
  12. Upper st smooth on angles; rhizomes 0.6–1.8 mm thick; shoots often 2–several in small clusters, 0 at many nodes
    13. Pistillate fl bracts 4.3–7.5 mm, pale brown to ± white; peri beak 0.9–1.9 mm; spikelets gen unisexual; anthers 2.5–3.9 mm, tip 0.2–1 mm, bristly at 30× ....²*C. douglasii*
    13′ Pistillate fl bracts 2.4–4.1 mm, red-brown; peri beak 0.3–0.9 mm; spikelets occ with staminate fls above pistillate (as in Group 9); anthers 1.4–3 mm, awn < 0.1 mm, smooth to warty at 30× .... ²*C. duriuscula*
  12′ Upper st ± scabrous on angles; rhizomes 1.5–5 mm thick; shoots gen 1 every few nodes
    14. Peri beak 0.25–0.5 mm; contraligule 0.3–1.6 mm; anthers occ present, tip smooth to warty at 30× ... ²*C. simulata*
    14′ Peri beak 0.5–1.5 mm; contraligule 0; anthers occ present, tip bristly-hairy at 30×
      15. Pistillate fl bracts, base of spikelet bracts dark brown or purple-brown, gen shiny, darker than peri; pistillate infl not elongate, ovate to elliptic; coastal, islands .... ²*C. pansa*
      15′ Pistillate fl bracts, base of spikelet bracts pale brown to pale red-brown, ± dull, lighter than peri; pistillate infl elongate, gen oblong-elliptic; gen inland .... ²*C. praegracilis*

Group 8

Stigmas 2, terminal spikelets gen staminate, lateral gen pistillate, rarely staminate-tipped

1. Peri fleshy, plump, often dense-papillate, tip blunt or with beak ≤ 0.3 mm; sheath of infl bract gen > 4 mm; pls long-rhizomed
  2. Mature peri, just before dropping, bright orange (green to white when immature, as in many herbarium specimens), ± spheric, ± spreading; pistillate fl bracts spreading, often falling before peri; ripe peri smooth (often papillate when immature) .... *C. aurea*
  2′ Mature peri, just before dropping, green to white, occ purple- or red-brown-marked, elliptic to elliptic-obovate, ascending; pistillate fl bracts ascending, gen ± persistent with peri; ripe peri dense-papillate at 10×
    3. Upper 1/2 of peri gen green or white, ± abruptly rounded, blunt, or only slightly narrowed, red-brown gen only on small, occ bent beak, if present; lower staminate fl bracts blunt or short-awned, gen < 4 mm, not like infl bract; spikelets 1 per lower node .... *C. hassei*
    3′ Upper 1/2 of peri red-brown- or purple-splotched, gradually narrowed to tip or short straight beak; lower staminate fl bracts gen long-awned, ≥ 4.5 mm, like infl bract; spikelets occ > 1 per lower node ...... *C. saliniformis*
1′ Peri not fleshy, ± flat, papillate or not, tip distinctly beaked < or > 0.3 mm; sheath of infl bract gen < 2 mm; pls long-rhizomed or cespitose
  4. Pl cespitose; peri veined
    5. Lf sheath fronts fibrous; lower sheaths scabrous, red-brown to shiny dark brown
      6. St 30–70 cm, < 1 cm wide at base; staminate spikelets 1–2, terminal < 5 cm; peri purple-splotched on upper 2/3; below high water mark on rocky streams .... *C. nudata*
      6′ St 75–150 cm, ± 1 cm wide at base; staminate spikelets 2–7, terminal gen > 5 cm; peri red-spotted, gen more densely so toward base; creekbanks, wet meadows, swamps .... *C. schottii*
    5′ Lf sheath fronts not fibrous; lower sheaths glabrous, dull light brown .... *C. lenticularis*
      7. Lower spikelets 4–6 mm wide; peri stalks 0.4–0.7 mm; lower spikelets crowded, overlapping; peri = pistillate fl bracts; NCo, < 20 m .... var. *limnophila*
      7′ Lower spikelets gen 3–4 mm wide; peri stalks 0.1–0.5 mm; lower spikelets gen more separated; peri > pistillate fl bracts; CCo, SnFrB or montane, < 3000 m
        8. Upper 1/2 of peri purple-dotted, abaxially 1–3-veined, beak gen purple; peri stalks 0.1–0.2 mm .... var. *impressa*
        8′ Upper 1/2 of peri gen green, abaxially gen 5–7-veined, beak gen green, tip purple; peri stalks 0.2–0.5 mm .... var. *lipocarpa*
  4′ Pl rhizomed; peri veined or not
    9. Lf sheath fronts gen fibrous; lower sheaths gen scabrous
      10. Peri ± soft, thin
        11. Bladeless sheaths present on lower st; peri 2.2–3.1 mm, 1.2–2 mm wide, weakly 1–3-veined adaxially on lower 1/2 .... *C. angustata*
        11′ Bladeless sheaths 0 on lower st; peri 3–4 mm, 2–2.2 mm wide, strongly 3–7-veined abaxially, adaxially .... *C. senta*
      10′ Peri ± tough, thick
        12. Peri gen red-spotted, faintly veined, dull, beak teeth bristly; pistillate fl bract awns bristly; fr 0 or lenticular, not indented .... ²*C. barbarae*
        12′ Peri gen not red-spotted, not veined, shiny, beak teeth 0 or minute, not bristly; pistillate fl bract awns gen entire; fr gen indented on 1–2 sides, like a bent beer can .... *C. obnupta*
    9′ Lf sheath fronts not fibrous (gen fibrous in *Carex barbarae*); lower sheaths gen not scabrous
      13. Spikelets drooping or nodding (gen erect when immature)
        14. Longest spikelets 1.8–5 cm; fr often indented on 1–2 sides, like a bent beer can; coastal salt marshes, freshwater intertidal zones .... *C. lyngbyei*
        14′ Longest spikelets 4.55–11.5 cm; fr not indented; coastal or inland, freshwater .... *C. aquatilis* var. *dives*
      13′ Spikelets gen erect or ascending

15. Mature peri faces veined, walls tough or thick
    16. Lvs green, gen not glaucous, sheath front red-dotted, gen dark copper or purple near tip, lower gen fibrous . . . . ²*C. barbarae*
    16′ Lvs gen blue-green, ± glaucous, sheath fronts gen white, occ sparse-red-dotted or ± coppery, lower not fibrous . . . . *C. nebrascensis*
15′ Mature peri faces not veined, walls delicate or thin
    17. Lf sheaths red-dotted or white; infl bract gen > infl; fl st base spongy-thickened . . . . . *C. aquatilis* var. *aquatilis*
    17′ Lf sheaths brown- or purple-dotted; infl bract < infl; fl st base thin, not spongy-thickened
        . . . . *C. scopulorum* var. *bracteosa*

Group 9

Stigmas 2; spikelets with staminate fls above pistillate fls

1. Pls rhizomed (loosely cespitose)
  2. Infl ovoid to oblong, open to dense; lowest spikelets gen obvious at arm's length
    3. Infl 3–4 mm wide; lateral spikelets with 1–3 peri . . . . *C. disperma*
    3′ Infl > 5 mm wide; lateral spikelets with > 3 peri
      4. Peri bulging, marginal veins occ not visible abaxially; pistillate fl bracts often with awns > 1 mm, the bracts covering peri . . . . ²*C. tumulicola*
      4′ Peri not bulging, marginal veins visible abaxially; pistillate fl bracts with awns 0 or < 1 mm, the bracts < or > peri, covering peri or not . . . . **Group 7, couplet 12**
  2′ Infl spheric to ovoid, dense; lowest spikelets not obvious at arm's length
    5. Spikelets gen pistillate . . . . **Group 7, couplet 12**
    5′ Spikelets staminate/pistillate
      6. Lvs inrolled, ± = st, 0.5–1.5 mm wide; st curved; peri ± not inflated, > pistillate fl bracts, beak indistinct, 0.4–0.9 mm; anthers 0.9–1.4 mm . . . . *C. incurviformis*
      6′ Lvs flat, < st, (1.5)2–3.5(4) mm wide; st straight; peri not inflated, ± = pistillate fl bracts, beak distinct, 0.9–1.5 mm; anthers 1.5–2.8 mm . . . . *C. vernacula*
1′ Pls cespitose to loosely so
  7. Infl branched, lower nodes or branches gen with > 1 spikelet; spikelets often > 10
    8. Lf sheath faces red-dotted
      9. Peri pale to medium brown
        10. Peri, beak, pistillate fl bract dull yellow-brown to medium- or red-brown; pistillate fl bract awns 0.5–2 mm; peri adaxial veins 0–5, abaxial 3–7; infl often continuous, obvious bracts gen 0–few; fl sts gen longer than lvs . . . . *C. densa*
        10′ Peri, beak, pistillate fl bract dull yellow-green to pale gray-brown; pistillate fl bract awns to 3 mm; peri adaxial veins 0, abaxial 0–3; infl often interrupted, obvious bracts often > few; fl sts gen < lvs . . . *C. vulpinoidea*
      9′ Peri dark brown to black
        11. Lf sheath faces widely dark or coppery at top . . . . *C. cusickii*
        11′ Lf sheaths faces pale, not dark or coppery at top (narrowly copper-margined)
          12. Peri 3–4.5 mm, 1.4–2.1 mm wide, beak 1–1.5 mm fr 1–1.3 mm wide . . . . *C. alma*
          12′ Peri 2–2.9 mm, 0.9–1.4 mm wide, beak 0.9–1.1 mm; fr 0.7–1 mm wide . . . . *C. diandra*
    8′ Lf sheath faces white, not red-dotted
      13. Peri widest near base, gradually narrowed to indistinct beak, length > 2 × width; sheath faces gen cross-wrinkled
        14. St not winged, mid-st gen ≤ 1.5 mm wide, not easily compressed between fingers; lower infl branches < 1 cm . . . . ²*C. neurophora*
        14′ St wide-3-winged, mid-st gen 2.5–7 mm wide, easily compressed between fingers; lower infl branches gen 1–3+ cm . . . . *C. stipata* var. *stipata*
      13′ Peri gen widest near middle, ± abruptly narrowed to distinct beak, length gen < 2 × width; sheath faces cross-wrinkled or not
        15. Infl open, lowest internode gen 2+ × lowest spikelet . . . . ²*C. divulsa* subsp. *divulsa*
        15′ Infl dense, lowest internode < lowest spikelet (see couplet 12)
  7′ Infl not branched, lower nodes with 1 spikelet; spikelets gen < 10
    16. Peri widest near base, gradually narrowed to indistinct beak
      17. Peri beak pale or green; introduced weeds at low elevations . . . . *C. texensis*
      17′ Peri beak dark brown; native, often montane
        18. Lf sheath front gen cross-wrinkled; peri beak serrate at least on 1 side . . . . ²*C. neurophora*
        18′ Lf sheath front not cross-wrinkled; peri beak entire
          19. Peri adaxial veins 5–7; lower lvs gen overlapped and hiding sheaths; sheath mouth without thick rim . . . *C. jonesii*
          19′ Peri adaxial veins 7–12; lower lvs gen not overlapped and hiding sheaths; sheath mouth gen with thick rim . . . . *C. nervina*
    16′ Peri widest near middle, often abruptly narrowed to distinct beak
      20. Peri pale green or light brown; peri > pistillate fl bracts, easily visible
        21. Infl lowest internode gen 2+ × lowest spikelet . . . . ²*C. divulsa* subsp. *divulsa*
        21′ Infl lowest internode ≤ lowest spikelet . . . . *C. vallicola*

20′ Peri center brown or copper, margins gen contrasting pale or green; peri ≥ pistillate fl bracts, often obscured

  22. Infl ovoid or spheric, 0.8–2 cm, 6–15 mm wide; lower spikelets gen not distinct . . . . *C. hoodii*

  22′ Infl oblong, 1.5–5 cm, 6–10 mm wide; lower spikelets ± distinct

    23. Peri ascending to spreading, beak 0.6–1.3 mm; pistillate fl bracts 3.4–4 mm, 1.6–2 mm wide; lower bracts gen < spikelets, infl . . . . *C. occidentalis*

    23′ Peri erect, beak 1.2–3 mm; pistillate fl bracts 4–5.2 mm, 2–2.8 mm wide; lower bracts > spikelets, often > infl . . . . ²*C. tumulicola*

Group 10

Stigmas 2; spikelets with pistillate fls above staminate fls; peri not winged; sts gen solid

1. Infl bracts >> infl; pl rhizomed; < 100 m. . . . . **C. inversa**

1′ Infl bracts gen < infl; pl cespitose to loosely so; gen < 3500 m

  2. Peri spreading or reflexed, beak obvious, spikelet outline jagged at arm's length

    3. Staminate fl bracts inconspicuous in wide terminal spikelet

      4. Pistillate fl bracts green, white, or light brown, spikelets green or pale brown; infl oblong, elliptic, or lanceolate, ≥ 1.5 cm . . . . ²*C. arcta*

      4′ Pistillate fl bracts ± black, spikelets dark; infl ovoid, ≤ 1.5 cm . . . . *C. illota*

    3′ Staminate fl bracts obvious at narrowed base of terminal spikelet

      5. Peri ascending to spreading, beak 15–40% peri length, peri attached at base. . . . . ²*C. laeviculmis*

      5′ Peri widely spreading or reflexed, beak gen 20–85% peri length, peri attached near base abaxially

        6. Lower peri in spikelet ± abruptly narrowed to short beak, margin with slight bulge or shoulder; beak 0.4–0.95, 18–44% length of body, peri gen 1.9–3.3 mm, length 1–2(2.2) × width. . . . . *C. interior*

        6′ Lower peri in spikelet more gradually narrowed to longer beak, margin without bulge or shoulder; beak 0.95–2 mm, 35–86% length of body, peri gen 2.9–4.75 mm, length 1.7–3.6 × width . . . . *C. echinata*

          7. Peri gen 2.9–4 mm, veins gen 0 over fr adaxially; widest lvs gen 0.7–2.7 mm wide; infl open to dense; KR, NCoRO, CaRH, SNH, SnBr, < 3200 m. . . . . subsp. *echinata*

          7′ Peri gen 3.5–4.75 mm, veins 2–12 over fr adaxially; widest lvs gen 1.7–3.8 mm wide; infl dense; NCo, CCo, < 200 m . . . . subsp. *phyllomanica*

  2′ Peri ascending or beak inconspicuous, spikelet outline gen smooth at arm's length

    8. Peri beak 0–0.5 mm; peri widest near middle

      9. Peri fleshy; terminal spikelet gen obviously staminate at base, some terminal spikelets entirely staminate (see Group 8, couplet 2)

      9′ Peri not fleshy; terminal spikelet inconspicuously staminate at base, not entirely staminate

        10. Peri beak abaxial flap pale, non-contrasting, not reaching to just below base of beak; st gen erect, (10)30–90 cm . . . . *C. canescens* subsp. *canescens*

        10′ Peri beak abaxial flap dark, contrasting, reaching to just below base of beak; sts gen leaning at 45° angle, 10–30 cm . . . . *C. praeceptorum*

    8′ Peri beak gen 0.5–2.5 mm; peri widest below middle or near base

      11. Infl straight; spikelets clustered, lowest > lowest internode . . . . ²*C. arcta*

      11′ Infl gen bent at lowest spikelet; spikelets separated, lowest gen < lowest internode

        12. Peri beak 15–38% peri length (beak measured from fr top, on peri from middle of spikelet)

          13. Peri ascending or spreading, 2.3–3.7 mm, beak occ recurved; lvs 0.9–2.3 mm wide . . . . ²*C. laeviculmis*

          13′ Peri erect or ascending, 2.8–4.5 mm, beak not recurved; lvs 2.4–5.9 mm wide . . . . *C. leptopoda*

        12′ Peri beak 39–53% peri length (beak measured from fr top, on peri from middle of spikelet)

          14. Peri beak teeth prominent, 0.2–1 mm; mid-st glabrous or ± scabrous at 20×. . . . . *C. bolanderi*

          14′ Peri beak teeth not prominent, 0–0.2(0.4) mm; mid-st papillate at 20× . . . . *C. infirminervia*

Group 11

(sect. *Ovales*) Stigmas 2; spikelets with pistillate fls above staminate; peri winged; st hollow. Infl bracts gen elongate on first flush of growth of some spp., but other spp. sporadically produce elongate infl bracts on late season shoots (e.g., *C. fracta, C. specifica, C. subfusca*). Late season shoots often with atypically elongate or compact infl, and/or atypically shaped peri. Peri beak shape variable in some spp. (e.g., *C. fracta, C. preslii, C. subfusca*), ± uniform, important in others, e.g., *C. specifica* (see illustration), with beak margin either ± flat and serrate or ± so, exc tip entire 0–0.3 mm, or e.g., *C. abrupta* (see illustration), beak tips ± cylindric, tip entire for ≥ 0.4 mm. [Note: For keying, use subbasal peri in mature spikelets from first flush of growth; avoid peri from upper 1/2 of spikelet, they are narrow and not diagnostic; similarly basal-most peri are occ malformed; use ripe peri, immature peri gen do not show critical ventral veins]

1. Lf sheath front green ± to top, white-translucent triangular area extending < 6 mm below top . . . . **Group 11A**

1′ Lf sheath front with extensive white-translucent zone extending > 10 mm below top

  2. Contraligule 3–13 mm . . . . **Group 11B**

  2′ Contraligule 0–2 mm

    3. Peri margin entire and marginal wings 0 or < 0.2 mm wide . . . . **Group 11C**

    3′ Peri margin minute-serrate, esp upper body or beak base or marginal wings gen > 0.2 mm wide

      4. Pistillate fl bracts uniformly covering and ± obscuring peri; peri gen with strong abaxial veins. . . . . **Group 11D**

4′ Pistillate fl bracts < or narrower than peri, peri easily visible; peri with strong abaxial veins or not
  5. Larger peri 6–9.3 mm. . . . . **Group 11E**
  5′ Larger peri < 6 mm
    6. Infl bract obvious, > infl. . . . . **Group 11F**
    6′ Infl bract inconspicuous, < infl
      7. Peri beak tip broadly winged and flattened, minute-serrate throughout or ± so (cylindric entire tip ≤
        0.3 mm), gen pale or only slightly darkened . . . . **Group 11G**
      7′ Peri beak tip cylindric, entire ≥ 0.4 mm, often darkened or ± black. . . . . **Group 11H**

Group 11A

Lf sheath front green ± to top

1. Peri gradually narrowed to indistinct beak, gen lanceolate; pistillate fl bract tip acuminate to short-awned
. . . . *C. scoparia* var. *scoparia*
1′ Peri abruptly narrowed to distinct beak, ovate to ± round to obovate or widely elliptic; pistillate fl bract tip
blunt to acute
  2. Peri body widest at or below middle; vegetative sts not leaning or branching or rooting at nodes . . . . *C. feta*
  2′ Peri body widest above middle; vegetative sts occ leaning or branching or rooting at nodes . . . . *C. longii*

Group 11B

Ligules long, sleeve-like

1. Infl gen short, gen ± ovate, dense, ± head-like . . . . *C. pachycarpa*
1′ Infl elliptic to oblong, dense or not, not head-like
  2. Peri wings gen 0.1–0.25 mm wide at mid-body; contraligule gen 5–13 mm; infl often ±white or green
    (pale brown); peri abruptly narrowed to distinct beak . . . . *C. fracta*
  2′ Peri wings gen 0.3–0.5 mm wide at mid-body; contraligule gen 1–5 mm; infl green or light- to
    rusty-brown; peri abruptly or gradually narrowed to beak
    3. Peri adaxial veins gen 1–6, 0 exceeding fr; peri ovate to narrow-ovate, 3.7–5.5 mm, walls translucent
      to opaque . . . . *C. multicostata*
    3′ Peri adaxial veins gen 6–10, ≥ 3 exceeding fr; peri gen lanceolate, 4.9–6.6 mm, walls gen translucent . . . *C. specifica*

Group 11C

Peri marginal wings entire (gen 0 in *Carex illota*)

1. Peri spreading, margins sharp-edged but not winged at mid-body; pistillate fl bracts and spikelets dark
brown to ± black; infl dense, length ≤ 2 × width, often wider than long; peri gradually narrowed to indistinct
beak (beak base occ winged ± 0.1 mm); peri base with spongy or pithy interior around fr . . . . *C. illota*
1′ Peri ascending, margins winged ≤ 0.2 mm at mid-body; pistillate fl bracts and spikelets pale to dark brown;
infl often ± open, length 3+ × width; peri abruptly or gradually narrowed to ± distinct beak; peri base
without spongy or pithy interior around fr
  2. Peri wings often inrolled, beak gen with inconspicuous flap; fr gen 1.7–2.3 mm, gen 1.1–1.5 mm wide;
    < 850 m . . . . *C. gracilior*
  2′ Peri wings flat, beak with obvious white-margined flap; fr 1.1–1.4 mm, 0.7–1 mm wide; 1000–3400 m . . . . . *C. integra*

Group 11D

Peri covered by pistillate fl bracts; peri abaxially strongly 3–14-veined

1. Infl ± bent, flexible, or nodding
  2. Peri wing gen 0.1–0.2 mm wide, inrolled, not visible in abaxial view; peri gen without adaxial veins,
    margins often entire above widest point . . . . [2]*C. gracilior*
  2′ Peri wing gen 0.2–0.6 mm wide, flat, visible in abaxial view; peri often with strong adaxial veins,
    margins serrate above widest point
    3. Peri beak tip brown, cylindric and entire > 0.4 mm, or flat and entire < 0.4 mm; peri gen ± flat exc over
      fr, body gen wide-elliptic. . . . . [2]*C. leporina*
    3′ Peri beak tip white, gen cylindric, entire > 0.4 mm; peri gen ± planoconvex, body gen narrow-elliptic . . . *C. praticola*
1′ Infl straight, stiff, erect
  4. Longer peri 6–8.1 mm
    5. Infl open; peri beak with obvious white abaxial flap; peri walls thick, opaque. . . . . *C. petasata*
    5′ Infl dense; peri beak with inconspicuous abaxial flap; peri walls thin, ± translucent. . . . . *C. specifica*
  4′ Longer peri < 6 mm
    6. Peri 0.8–1.2 mm wide, beak indistinct, wings gen incurved . . . . *C. leporinella*

6′ Peri 1.3–2.6 mm wide, beak distinct or not, wings not incurved
  7. Peri gen widest near or above middle; infl gen dense, spikelets overlapped; lf sheath mouth white; subalpine to alpine, 2500–4000 m
    8. Strong adaxial peri veins gen 0(4); peri walls thin, ± translucent . . . . ***C. phaeocephala***
    8′ Strong adaxial peri veins 3–8; peri walls thick, opaque. . . . . ***C. tahoensis***
  7′ Peri gen widest below middle; infl dense or open, spikelets overlapped or not; lf sheath mouth occ brown-tinged; coastal, montane, 0–3500 m
    9. Peri adaxial veins gen reaching beak base; peri walls sometimes thin, ± translucent
      10. Peri not shiny at 10×, beak tip ± flat, serrate; infl elongate, gen open, green or gold; montane, 800–3500 m. . . . . **Group 11B, couplet 2**
      10′ Peri ± shiny at 10×, beak tip cylindric, entire; infl ovoid, dense, gen brown or coppery; coastal, insular, or montane, < 900 m . . . . ***C. harfordii***
    9′ Peri adaxial veins 0 or gen reaching at most to fr tip; peri walls gen ± thick, opaque
      11. Well-developed peri with wing at mid-body gen 0.4–0.6 mm wide . . . . ²***C. leporina***
      11′ Well-developed peri with wing at mid-body ≤ 0.4 mm wide
        12. Infl gen open, lower spikelets distinct; peri wings gen 0.1–0.2 mm wide at mid-body, gen inrolled, not visible abaxially, margins at 16× entire from base to above widest point; peri upper body entire or not, beak minute-serrate; lf blades 1.2–3 mm wide. . . . . ²***C. gracilior***
        12′ Infl ± dense (exc on late season sts), lower spikelets often clustered or overlapped, indistinct; peri wings gen 0.2–0.3(0.4) mm wide at mid-body, flat, visible abaxially, margins at 16× entire from base to ± widest point; peri upper body, most of beak minute-serrate; lf blades 1.3–4.6 mm wide
        . . . . ***C. subbracteata***

Group 11E

Larger peri 6–9.3 mm

1. Infl ± ovate (elliptic, triangular), dense, head-like
  2. Pistillate fl bracts gen dark brown to ± black; peri flat exc over fr . . . . ***C. haydeniana***
  2′ Pistillate fl bracts light brown or red-brown; peri planoconvex . . . . ***C. pachycarpa***
1′ Infl elliptic to oblong, dense or not, not head-like
  3. Infl bent or flexed at lower node, curved . . . . ***C. praticola***
  3′ Infl straight
    4. Peri base abruptly narrowed to stalk 0.5–1 mm (measured below fr base); infl gen dense, elliptic, spikelets (4)6–14; peri margin crinkled or flat; pistillate fl bracts green to light brown or red-brown . . . . . ***C. specifica***
    4′ Peri base abruptly narrowed to stalk 1–2 mm (measured below fr base); infl open, oblong, spikelets 1–7; peri margin flat; pistillate fl bracts red-brown
      5. Translucent pistillate fl bract margins ≤ 0.4 mm; peri beak abaxial flap inconspicuous . . . . ***C. davyi***
      5′ Translucent pistillate fl bract margins gen > 0.4 mm wide; peri beak abaxial flap obvious . . . . ***C. petasata***

Group 11F

Infl bract gen 2–3 × infl

1. Infl ± lateral, lowest bract wide, lf-like, erect, appearing to be continuation of st; most peri beaks flat, winged, minute-serrate (cylindric, unwinged, entire portion < 0.4 mm). . . . . ***C. unilateralis***
1′ Infl ± terminal, lowest bract often narrower, less lf-like, ascending or spreading, not appearing to be continuation of st; peri beaks ± flat or cylindric (unwinged, entire portion > 0.4 mm)
  2. Peri adaxial veins exceeding fr top
    3. Contraligule gen 5–13 mm; infl 2.5–8 cm, gen ± white or green (to pale brown). . . . . ***C. fracta***
    3′ Contraligule 0 or < 3 mm; infl 1.5–4 cm, gen dark brown or ± black . . . . ***C. harfordii***
  2′ Peri adaxial veins 0 or not exceeding fr top
    4. Peri base abruptly narrowed to distinct stalk (on dried material often bent or crumpled); peri strongly flattened exc over fr, or occ biconvex . . . . ***C. athrostachya***
    4′ Peri stalk indistinct or 0; peri planoconvex
      5. Peri gen 3.5–4.7(5.7) mm; infl coarse, parts large at arm's length; < 900 m . . . . ***C. subbracteata***
      5′ Peri gen 2.3–3.5 mm; infl fine, parts small at arm's length; (75)700–3800 m . . . . ***C. subfusca***

Group 11G

Peri beak wide-winged, minute-serrate; tip cylindric, entire < 0.4 mm

1. Infl bent or flexed at lowest node. . . . . ²***C. scoparia*** var. ***scoparia***
1′ Infl erect, not bent or flexed at lowest node
  2. Widest peri 2.2–3.4 mm wide
    3. Peri wide-ovate or -elliptic or ± round, winged margin gen 0.5–1 mm wide . . . . ***C. straminiformis***

3′ Peri lanceolate to ovate, winged margin gen 0.3–0.5 mm wide
  4. Infl dense, ± spheric, 10–22 mm; peri 4.9–5.9(6) mm . . . . ²*C. pachycarpa*
  4′ Infl open, linear-oblong, 20–60 mm; peri (5.7)6–8.1 mm . . . . ²*C. petasata*
2′ Widest peri ≤ 2.1 mm wide
  5. Wider peri wings gen 0.1–0.3 mm at midbody
    6. Infl 2.5–8 cm; contraligule gen 5–13 mm; peri adaxial veins exceeding fr gen ≥3 . . . . *C. fracta*
    6′ Infl 1.1–3 cm; contraligule gen < 3 mm; peri adaxial veins gen 0 or not exceeding fr . . . . *C. subfusca*
  5′ Wider peri wings gen 0.3–0.5 mm at midbody
    7. Infl ± ovate (elliptic, triangular), ± head-like
      8. Peri thick, planoconvex, abaxial veins 8–17, wide, tissue often sunken between veins; adaxial veins
        0–12; individual spikelets gen indistinct at arm's length . . . . ²*C. pachycarpa*
      8′ Peri thin, flat, abaxial veins 0–8, thin, tissue not sunken between veins; adaxial veins 0–5; individual
        spikelets distinct at arm's length . . . . ²*C. scoparia* var. *scoparia*
    7′ Infl oblong, lanceolate, or elliptic, gen not head-like
      9. Infl gen open, linear-oblong, spikelets distinct at arm's length . . . . ²*C. petasata*
      9′ Infl often dense, elliptic to lanceolate or oblanceolate, at least upper spikelets indistinct at arm's length
        10. Peri adaxial veins exceeding fr; peri gen 4.9–6.6 mm, gen lanceolate . . . . *C. specifica*
        10′ Peri adaxial veins gen shorter, not exceeding fr or 0; peri 3.3–5.5 mm, gen ovate
          11. Peri gen gradually narrowed to indistinct beak, adaxial veins obvious, (1)3–13; peri walls
            translucent to opaque . . . . *C. multicostata*
          11′ Peri abruptly narrowed to distinct beak, adaxial veins 0 (3, inconspicuous); peri walls opaque . . . . *C. preslii*

Group 11H

Peri beak tip gen cylindric, gen entire ≥ 0.4 mm

1. Peri flat, thin exc over fr (± inflated, convex both faces)
  2. Peri wide-elliptic to ± round, margins occ crinkled or ruffled, beak distinct, 20–25% of total peri length;
    anthers early-deciduous to persistent; pistillate fl bracts 1/2 peri width
    3. Peri beak ≤ 1.2 mm; lvs gen 0.5–2.7 mm wide when flat, in life gen folded or channeled toward base; infl
      dense to more open, ovate, elliptic, or oblong . . . . *C. proposita*
    3′ Peri beak gen 1.2–1.5 mm; lvs (1.7)3–4 mm wide when flat, in life flat or folded; infl gen dense, ovate
    . . . . *C. straminiformis*
  2′ Peri ovate to lanceolate, margins gen flat, beak indistinct or > 25% of total peri length; anthers gen
    deciduous; pistillate fl bracts > 1/2 peri width
    4. Most peri distinctly short-stalked, dried stalks often bent or crumpled; infl bract on some heads
      long-awned, wide at base . . . . *C. athrostachya*
    4′ Most or all peri not distinctly stalked; infl bract rarely long-awned or wide at base
      5. Peri 4–6.5 mm, (2.3)2.6–4.2 mm beak tip to fr; peri green, dark brown, or coppery, margins often
        darker than wings; st 9–20(40) cm; subalpine to alpine . . . . *C. haydeniana*
      5′ Peri (2.8)3.4–4.5(5.2) mm, 1.5–2.5(2.8) mm beak tip to fr; peri green to light brown, center often
        brown, margin color gen same as wings; st 20–110 cm; montane to alpine . . . . *C. microptera*
1′ Peri gen planoconvex, thick
  6. Peri adaxial veins gen 3–10, gen exceeding fr
    7. Infl elongated, oblong to elliptic, not head-like, lower spikelets gen distinct, bases gen tapered; montane
      8. Peri 2.5–4.8(5.3) mm . . . . *C. mariposana*
      8′ Peri (5.7)6–8.5 mm . . . . *C. petasata*
    7′ Infl compact, often ovoid, head-like, lower spikelets indistinct, bases gen blunt; coastal or montane
      9. Peri margins entire from base to ± widest point, minute-serrate on beak and upper body; sts erect, not
        branched at nodes; sheath face tops gen pale; gen without lf-like infl bract; 1200–3450 m . . . . *C. abrupta*
      9′ Peri margins entire from base often to above widest point, minute-serrate on beak and occ uppermost
        body; sts often leaning, occ branched at nodes, branches lfy or fertile; sheath face tops often
        brown-tinged; infl bract occ lf-like; < 900 m . . . . *C. harfordii*
  6′ Peri adaxial veins 0 or short, 0 (1) exceeding fr
    10. Peri margins at 16× entire from base to above widest point; marginal wings at mid-body often inrolled
      so not visible from abaxial side . . . . *C. gracilior*
    10′ Peri margins at 16× entire from base to ± widest point; marginal wings at mid-body gen not inrolled,
      visible from abaxial side
      11. Pistillate fl bracts gen 4.2–5.8 mm, ± hiding peri
        12. Infl gen open, often bent or flexed at lowest node, most spikelets distinct at arm's length . . . . *C. praticola*
        12′ Infl gen dense, erect, most spikelets indistinct at arm's length . . . . ²*C. subbracteata*
      11′ Pistillate fl bracts gen < 4.2 mm, not hiding peri
        13. Peri gen 2.3–3.5 mm, wing margin at midbody gen 0.1–0.25 mm wide; infl at arm's length
          fine-textured, parts small . . . . *C. subfusca*

13′ Peri gen ≥ 3.5 mm, wing margin at midbody gen 0.3–0.4 mm wide; infl at arm's length coarse-
    textured, parts large

14. Peri spreading; spikelet ± star-shaped, outline jagged at arm's length; pistillate fl bract dark
    throughout, or white margins narrow, < 0.1 mm wide. . . . . ***C. pachystachya***

14′ Peri ± ascending; spikelet not star-shaped, outline ± smooth at arm's length; pistillate fl bract gen
    with white margins 0.2–0.5 mm wide

15. Peri beaks of 2 kinds in an infl, some ± flat, with smooth entire part < 0.4 mm, some cylindric,
    with smooth entire part > 0.4 mm; gen dry soil; 1800–3400 m, KR, NCoR, CaRH, SNH, SnJt. . . . . . . ***C. preslii***

15′ Peri beaks of 1 kind, gen cylindric, smooth entire part > 0.4 mm; seasonally wet soil; < 900 m,
    NCo, NCoRO, CW, SCo, n ChI, WTR. . . . . ²***C. subbracteata***

C. abrupta Mack. (p. 1323) ABRUPT-BEAKED SEDGE (Group
11H) **ST:** erect, 18–66 cm, without nodal branching. **LF:** blade 1.5–
3.7(4.9) mm wide; sheath face top gen pale; contraligule gen < 3
mm, shorter than wide. **INFL:** dense, 10–22 mm, spheric to ovoid,
green to brown; lowest internode < 3 mm; lowest spikelet indistinct,
base obtuse, gen without lf-like bract; pistillate fl bract < peri, brown
or coppery, obtuse. **FR:** 1.2–1.8 mm, 0.7–1.2 mm wide; peri 2.8–
4.5(5.2) mm, 1–1.5(2.1) mm wide, elliptic to lance-ovate, flat margin
incl wing < 0.3 mm wide, entire from base to ± widest point, body
planoconvex, occ biconvex, pale to light brown (dark brown), upper
margin green to gold, abaxial veins (5)7–10, adaxial 3–8, reaching
beak base, beak (0.7)1–1.5(1.7) mm, gen tapered from body, tip
cylindric, entire, unwinged for gen > 0.5 mm, gen dark brown. Meadows, open forest, rocky slopes; 1200–3450 m. KR, NCoRH, CaRH,
SNH, TR, PR, Wrn, W&I, DMtns; to OR, ID, NV. Intergrades with
C. mariposana, a more delicate sp. that fl 2 weeks later. Occ confused
with *C. microptera*, which gen has peri flat or biconvex, adaxial veins
0. Jul–Aug

C. albonigra Mack. (p. 1323) BLACK-AND-WHITE SEDGE
(Group 6) Cespitose. **ST:** 10–30 cm. **LF:** blade 2.5–5 mm wide, gray-
green. **INFL:** gen head-like; terminal spikelet staminate at base, pistillate above; pistillate fl bract ± = peri, oblong or lanceolate to widely
elliptic, dark purple, margin, tip ± white, obtuse to mucronate. **FR:**
1.3–2 mm, 0.7–1.3 mm wide, < peri body; peri not obvious in infl,
2.5–3.4 mm, 1.3–2 mm wide, plump, papillate above, dark purple,
beak 0.1–0.5 mm. 2*n*=52. Wet or ± dry rocky slopes, bowls, snow-
melt channels, summits; 3000–4200 m. c&s SNH, W&I; to w Can,
CO. Pls in CA, Rocky Mtns differ; study needed. Jul–Aug

C. alma L.H. Bailey (p. 1323) STURDY SEDGE (Groups 7,9)
Cespitose to rhizomed; (dioecious). **LF:** blade 3–6 mm wide; sheath
front red-dotted, occ cross-wrinkled. **INFL:** open or dense, 2.5–15 cm,
1–2 cm wide; spikelets >> 10, > 1 per lower node or branch; spikelet
bracts obvious; pistillate fl bract ± ≥ peri, awned, ± white or margin
wide, white. **STAMINATE FL:** anther > 1.9 mm, awn < 0.1 mm. **FR:**
1.5–2.5 mm, 1–1.3 mm wide, ovate; peri 3–4.5 mm, 1.4–2.1 mm wide,
body ovate to deltate, tapered or abruptly narrowed above, gold to
dark brown, veins abaxially few, adaxially 0–few, wall rounded over
fr, filled with pithy tissue below, beak 0.6–1.4 mm, serrate or ciliate.
Springs, streambanks; 120–2600 m. c&s SNH, Teh, SnJV, SCoRO,
SCo, TR, PR, DMoj; NV, AZ, Baja CA. May–Jul

C. amplifolia Boott (p. 1323) BIG-LEAF SEDGE (Groups 1,5,6)
Rhizomed. **ST:** 50–100 cm, sharp-3-winged. **LF:** blade 8–23 mm
wide, with small, raised, irregularly spaced cross-walls; sheath back
± hairy. **INFL:** lateral spikelets 3.5–14 cm; lower pistillate fl bracts ±
purple, tip white, glabrous-awned. **FR:** 1.3–2 mm, 1–1.5 mm wide;
peri > 60, spreading, 2.5–3.5 mm, 1.2–2 mm wide, glabrous, green
and brown, abruptly narrowed to beak, beak 0.7–1.2 mm, curved,
tip ± unnotched. Seasonally wet areas; < 2200 m. KR, NCoR, CaR,
SNH, CCo, SnFrB, MP; to s BC, nw MT. May–Sep

C. angustata Boott (p. 1323) NARROW-LEAVED SEDGE (Group
8) Rhizomed, forming dense stands but not large, raised clumps. **ST:**
< ± 1 m. **LF:** blade 2–7 mm wide; bladeless sheaths prominent on
lower st; sheath backs scabrous, fronts with ± red prickles, fibrous
on lower lvs. **INFL:** staminate spikelets 1–2, terminal < 5 cm; lateral spikelets 2–7 cm, 3–5 mm wide, pistillate exc tip of upper occ
staminate; infl bract ± ≥ infl; pistillate fl bract often > peri body, gen
< 0.5 mm wide, acute to gen pointed at tip, occ hairy-awned. **FR:**
1–1.5 mm, 0.7–1.3 mm wide; peri 2.2–3.1 mm, 1.2–2 mm wide, front

weakly 1–3-veined on lower 1/2, ± papillate, green, purple-dotted
above or not, upper margin occ with stiff, curved hairs, beak 0.2–
0.5 mm, tip notched < 0.1 mm, glabrous. 2*n*=66,68. Wet meadows,
streambanks; 300–2300 m. KR, NCoRO, NCoRH, CaRH, SNH, MP;
to e WA; also c ID. Jun–Sep

C. aquatilis Wahlenb. (Group 8) **INFL:** gen open; infl bract lf-
like, > 1/2 infl; pistillate fl bract tip often white. **PISTILLATE FL:**
style often exserted. **FR:** 1.1–1.8 mm, 0.7–1.6 mm wide, shiny;
peri faces unveined, ± papillate, beak notched < 0.1 mm, glabrous.
2*n*=72,74,76,80.

var. **aquatilis** WATER SEDGE **LF:** blade 3–8 mm wide; sheath
front red-dotted or white, mouth white or pale brown. **INFL:** lowest
internode 2–18 cm; lateral spikelets 1–10 cm, 3–7 mm wide, stalks
0 or gen erect, < 4.2 cm, tip gen pistillate; infl bract gen > infl. **FR:**
peri 2–3.6 mm, 1.2–2.3 mm wide, ± red-dotted or -blotched, beak
0.1–0.2 mm, thick. 2*n*=72,76,78,80. Wet places; < 3200 m. KR,
NCoR, CaRH, SNH, SnBr, Wrn, W&I; to AK, e Can; n Eur. Jul–Sep

var. **dives** (Holm) Kük. (p. 1323) SITKA SEDGE **LF:** blade 5–18
mm wide; sheath front purple-dotted, mouth purple-brown. **INFL:**
lateral spikelets 4.5–11.5 cm, 4–7 mm wide, lower stalks gen nodding,
< 11 cm, tip gen staminate; infl bract gen > infl. **FR:** peri 1.9–3.5 mm,
1–1.2 mm wide, ± yellow- or purple-dotted, beak 0.2–0.4 mm, gen not
thick. 2*n*=72,76,78,80. Wet meadows, shores, fresh water; < 1100 m.
NCo, KR, NCoRO, CaRH (Butte Co.), CCo; to AK. Hybridizes with
C. lyngbyei. Jun–Sep

C. arcta Boott (p. 1323) NORTHERN CLUSTERED SEDGE (Group
10) Cespitose. **LF:** blade 2–4 mm wide; sheath fronts of lower lvs at
least sparsely minute-red-dotted. **INFL:** ± dense, 1.5–3 cm, green to
brown; spikelets 7–15, distinct, 5–10 mm, upper occ narrow-tapered
at base, lower ± separate; infl bract < or > infl; pistillate fl bract green,
white, or light brown, obtuse to mucronate. **FR:** 1.2–1.6 mm, 0.8–1.1
mm wide; peri spreading to ascending, 2–3.4 mm, 1.1–1.5 mm wide,
ovate, green, upper body ± papillate, wall filled with pithy tissue at
base, gen upper body, beak 0.6–1.2 mm, > 1/4 body, conic, minute-
serrate at 10×, not recurved, tip ± red. 2*n*=60. Wet places, esp sphagnum bogs; < 1400 m. NCo, NCoRO; to w Can; also e Can, e US.
Jun–Aug ★

C. atherodes Spreng. (p. 1323) WHEAT SEDGE (Group 1) **ST:**
30–150 cm. **LF:** blade 3–10 mm wide, hairy abaxially; sheath hairy
at least near mouth. **INFL:** pistillate spikelets stalked or not; infl bract
long-sheathing; pistillate fl bract red-brown, margin white. **PISTIL-
LATE FL:** style straight. **FR:** 2.3–3.2 mm, 1.2–1.5 mm wide; peri
ascending, 7–10 mm, 1.7–2.5 mm wide, stalked, green, glabrous
or sparse-hairy on main veins, beak > 1 mm, straight, longer teeth
1.5–3.3 mm, curved. Marshes, seasonally wet meadows; 1300–1600
m. CaRH, MP (exc Wrn); to n Can, ne US; Eurasia. Jul–Aug ★

C. athrostachya Olney (p. 1323) LONG-BRACTED SEDGE
(Groups 11F,H) **LF:** blade 1.5–4 mm wide. **INFL:** dense, (8)15–22
mm, green to light brown; axis ± erect; spikelets ± indistinct; lowest
2–3 spikelet bracts ascending, >> (<) infl, gen < 1.8 mm wide, lf-like,
base expanded ± around st; pistillate fl bract < peri, often short-awned.
FR: 1.1–1.7 mm, 0.8–1(1.2) mm wide; peri appressed to ascending,
(2.8)3.5–4(4.8) mm, (0.8)1–1.5(1.8) mm wide, 0.35–0.45 mm deep,
lanceolate to narrow-ovate, flat around fr, green to light brown, flat
margin incl wing 0.1–0.2(0.4) mm wide, veins weak, < 8 abaxially,
0–4 adaxially, beak gen cylindric, entire, unwinged for > 0.4 mm, gen
green above, tip white, stalk gen distinct, short, narrow or spongy-

thickened, gen bent when dry. 2*n*=68. Common. Seasonally moist meadows, marshes; 400–3200 m. KR, NCoR, CaRH, SNH, SnBr, SnGb, PR, GB; w N.Am. Intergrades with *C. unilateralis*. Can be distinguished from pls of *C. harfordii* with elongate bracts by peri adaxial veins and peri length/width. May–Sep

C. aurea Nutt. (p. 1323) GOLDEN SEDGE (Group 8) **LF**: blade 2–4 mm wide. **INFL**: terminal spikelet occ pistillate above, 0.9–2 mm wide in staminate part, at least lowest non-basal spikelet separate, lateral ones erect to nodding, 4–20 mm, 3–5 mm wide; infl bract sheath > 4.5 mm, mouth U-shaped; lower staminate fl bracts 2–4 mm; pistillate fl bract <, not appressed against, often falling before fully expanded peri, white to red-brown, awn acute to narrow. **FR**: 1.3–2 mm, 1–1.6 mm wide; peri gen 4–10, ± ascending to spreading, 1.8–3 mm, 1–2 mm wide, spheric, gen rounded or wide-tapered at base, gen sessile, papillate or not, green to white when immature, maturing bright orange (unique in CA *Carex*), smooth, walls translucent, fleshy just before falling, when dry orange or gold to ± purple, body tip wide, blunt, beak ± 0, tip unnotched, often red-brown. 2*n*=52. Wet meadows, streambanks; 1100–3300 m. KR, NCoRH, NCoRI, CaR, SNH, TR, MP, n SNE (Bodie), W&I, DMtns; to BC, ne N.Am, UT, NM. Dried, immature peri ± indistinguishable from those of *C. hassei*. Jun–Aug

C. barbarae Dewey (p. 1323) WHITEROOT OR SANTA BARBARA SEDGE (Group 8) St base brown to purple. **LF**: blade 3.5–9 mm wide; sheath fronts red-dotted, gen dark copper or purple near tip, gen fibrous on lower lvs. **INFL**: lateral spikelets 2.5–8 cm, 5–8 mm wide, brown, not pressing flat when fully mature, tips staminate or not; infl bract < to > infl; pistillate fl bract awn or tip hairy. **FR**: 1.7–2 mm, 1.2–1.7 mm wide; peri 50–200, spreading, 3–4.5 mm, 1.9–2.5 mm wide, brown, red-dotted, wall thick, tough, beak 0.2–0.5 mm, tip notched > 0.1 mm in front, teeth bristly. Seasonally wet places; < 1000 m. KR, NCoR, CaRF, SNF, c SNH, GV, CW, SCo, n ChI (Santa Cruz Island), WTR, PR; s OR. Important traditional basket fiber pl. May–Aug

C. bolanderi Olney (p. 1323) BOLANDER'S SEDGE (Group 10) Loosely cespitose. **ST**: 15–90 cm. **LF**: blade 2–5 mm wide. **INFL**: 3–8 cm; spikelet gen > 13 mm, linear to oblong, gen not jagged-sided, base of lower staminate, narrow-tapered; infl bract lf-like or not, gen < infl; pistillate fl bract gen covering peri, gen gold-brown at least near center, awn often > 1.2 mm. **FR**: 1.5–1.8 mm, 1–1.3 mm wide; peri ascending, 3.3–5 mm, 1–1.4 mm wide, length gen > 3 × width, lanceolate, green, lower wall filled with pithy tissue, body narrow-elliptic, beak 1.5–2.5 mm (measured from fr top), 40–53% peri length, long-tapered, teeth 0.2–1 mm. Moist meadows, springs, shores, forest; < 2500 m. NW, CaR, n SNF, SNH, CCo, SnFrB, SCoRO, SW, MP; to BC, MT, NM. May–Aug

C. brainerdii Mack. BRAINERD'S SEDGE (Group 2) In clumps connected by rhizomes. **ST**: gen 10–30 cm. **LF**: blade gen > bisexual infl, 1.5–3 mm wide, pale blue-green to gray-green or glaucous, papillate abaxially at 40×; basal sheaths dark purple-red, scabrous. **INFL**: staminate spikelet 10–17 mm, 1.5–3.5 mm wide; pistillate spikelets 4–6, basal 1, on erect, gen short, stout stalk; infl bract often ≥ infl; pistillate fl bract gen 3–5 veined, green or red-banded, at least some awned. **FR**: 2.1–2.5 mm, 1.5–2.1 mm wide; peri gen tan or brown; peri 1–6, 4–5.2 mm, 1.5–2.2 mm wide, hairy, faces veined or several-ribbed at least below, body barrel-shaped, stalk ± = beak. Dry rocky areas, open forest; 600–2800 m. KR, NCoRH, NCoRO, CaRH, n&c SNH, SCoRO (San Luis Obispo Co.); s OR. Pls with green lvs from n SN may be distinct; study needed. May–Jul

C. brevicaulis Mack. (p. 1327) SHORT-STEMMED SEDGE (Group 2) Forming dense turf, short-rhizomed; st base occ fibrous. **ST**: 3–15 cm, scabrous at least on 1 angle exc near base. **LF**: exceeding infl; blade 1–3.5 mm wide, gen sickle-shaped, stiff, pen folded, bright green. **INFL**: infl bract gen < infl; staminate spikelet of taller infls 5–17 mm, 1.5–2 mm wide; pistillate spikelets 2–4; pistillate fl bract tinged red-brown, obtuse to short-awned. **FR**: 1.5–2.6 mm, 1.5–2 mm wide; peri 1–6, 3.8–4.5 mm, 1.5–2.1 mm wide, body with 2 marginal veins only, beak notched in front < 0.1 mm. 2*n*=28. Coastal dunes, headlands, rocky or sandy soil; < 400 m. NCo, CCo; to BC. Apr–May

C. breweri Boott (p. 1327) BREWER'S SEDGE (Group 3) Rhizomed. **ST**: 10–25 cm. **LF**: exceeded by infl; blade 1 mm wide, rolled, quill-like. **INFL**: spikelet 6–10 mm wide, wide-conic; lower pistillate fl bract wide, gold-brown to black, gen 3–5-veined, margin wide, ± white. **FR**: 1.7–2.3 mm, 0.8–1 mm wide (width < lateral distance from fr to peri), << peri body, 3-sided, next to bristle-like axis; peri 10–40, ascending to spreading, 4–7 mm, 2–4.8 mm wide, flat, thin, gold-brown. Dry gravelly to sandy open areas; 2000–3900 m. CaRH, SNH, W&I; to WA. Jul–Sep

C. buxbaumii Wahlenb. (p. 1327) BUXBAUM'S SEDGE (Group 6) Rhizomed. **ST**: 25–100 cm. **LF**: blade 1.5–4 mm wide, basal minute; sheath front fibrous on lower lvs. **INFL**: terminal spikelet pistillate above; lowest spikelets erect, often sessile; pistillate fl bract purple or brown, awn glabrous. **FR**: 1.4–2.1 mm, 1.1–1.5 mm wide; peri ascending to ± spreading, 2.5–4.3 mm, 1.4–2.1 mm wide, papillate, gray-green, beak < 0.3 mm. 2*n*=±74,±100,±105. Peatland, wet meadows; < 3300 m. NW, SNH, CCo; to AK, e N.Am, Eurasia. Jun–Aug ★

C. californica L.H. Bailey (p. 1327) CALIFORNIA SEDGE (Group 4) Rhizomed. **ST**: 20–70 cm, base red or purple. **LF**: basal blades minute, on lower fl sts 0, others 2–5 mm wide, green, abaxially papillate; sheath front red or red-dotted. **INFL**: infl bract long-sheathing; pistillate fl bract purple-brown, glandular-papillate on midrib, awned or not. **FR**: 1.7–3.1 mm, 1.4–2.1 mm wide; peri appressed to ascending, 2.7–5.1 mm, 1.7–2.3 mm wide, papillate, green, abruptly narrowed to distinct beak 0.5–1 mm. Pygmy forest, meadows, swamps, damp roadbanks; < 350 m. NCo, NCoRO; to WA; also ID. May–Jul ★

C. canescens L. subsp. **canescens** (p. 1327) SILVERY SEDGE (Group 10) Cespitose. **ST**: (10)30–90 cm. **LF**: blade 1.5–4 mm wide, glaucous. **INFL**: ± open, 2–15 cm; lowest internode > 2 mm; spikelets < 10, 3–12 mm, green to light brown, upper narrow-tapered at base or not, lowest gen < 2 × lowest internode; infl bract < infl; pistillate fl bract white, occ gold or pale brown, obtuse to acute. **FR**: 1.2–1.5 mm, 0.7–1 mm wide; peri ascending (spreading-ascending), 1.7–2.8 mm, 0.9–1.8 mm wide, green to gold or silvery, wall filled with pithy tissue to upper body, upper body papillate, beak inconspicuous, 0.2–0.5 mm, < 1/4 body, wide-conic, ± entire, tip at most minute-notched, ± red. 2*n*=54,56,62. Wet, open places; 1000–3200 m. NCoRO, CaRH, n&c SNH; to AK, e N.Am; also S.Am, Eurasia, Australia. Jun–Aug

C. capitata L. (p. 1327) CAPITATE SEDGE (Group 3) Loosely cespitose. **ST**: 10–35 cm. **LF**: exceeded by infl; blade ≤ 1 mm wide, green, rolled, quill-like. **INFL**: spikelet bisexual; spikelet bract 0; pistillate fl bract brown, margin wide, white. **PISTILLATE FL**: stigmas 2. **FR**: 1–1.8 mm, 0.5–1.2 mm wide, < peri body, 2-sided, next to bristle-like axis; peri 6–25, ascending to spreading, 2–3.5 mm, 1.5–1.8 mm wide, flat around fr, fine-veined abaxially, pale green, edge sharp to base, entire, faces glabrous, beak dark. 2*n*=50. Peatlands, moist alpine meadows; 1200–3900 m. CaRH, SNH, W&I; to AK, ne N.Am, S.Am; Eurasia. Jul–Sep

C. comosa Boott (p. 1327) BRISTLY SEDGE (Groups 4,5,6) Loosely cespitose. **ST**: 50–100 cm. **LF**: ligule length ± ≤ width. **INFL**: lower spikelets on long, nodding stalks; pistillate fl bract gen white or cream with pale ± red center, awn linear, > body. **FR**: 1.5–2 mm, 0.7–1.1 mm wide, ± = peri body; peri spreading, 5–7.5 mm, 1.1–1.6 mm wide, ± angled, stalked, shiny, green to gold, beak teeth 1.3–2.8 mm, spreading or curved. Wet places; < 400 m. NCoRI, CaRH, GV, n CCo (Bodega Bay), SnFrB, SnBr, MP (Shasta Co.); to BC, e N.Am. Jul–Sep ★

C. concinnoides Mack. (p. 1327) NORTHWESTERN SEDGE (Group 2) Rhizomed. **ST**: 15–35 cm, base dark purple-red. **LF**: exceeded by to exceeding infl; blade 2–5 mm wide, ± sickle-shaped, uppermost short; sheath mouth V-shaped. **INFL**: staminate spikelet 2–4 mm wide; pistillate spikelets 1–2, linear to oblong; spikelet bracts purple, ± sheathing; pistillate fl bract purple, margin ciliate. **PISTILLATE FL**: stigmas 4, warty, not plumose, at 15×. **FR**: 1.9–2.8 mm, 0.8–1.7 mm wide, base 4-angled; peri 5–10, ascending, 2.5–3.5 mm, 1.1–1.6 mm wide, 2-ribbed, white to light brown, beak 0.1–0.5 mm, often purple distally, tip white. Dry to moist open forest, meadows, often serpentine; 15–900 m. KR, NCoR; to w Can, MT. May–Jul

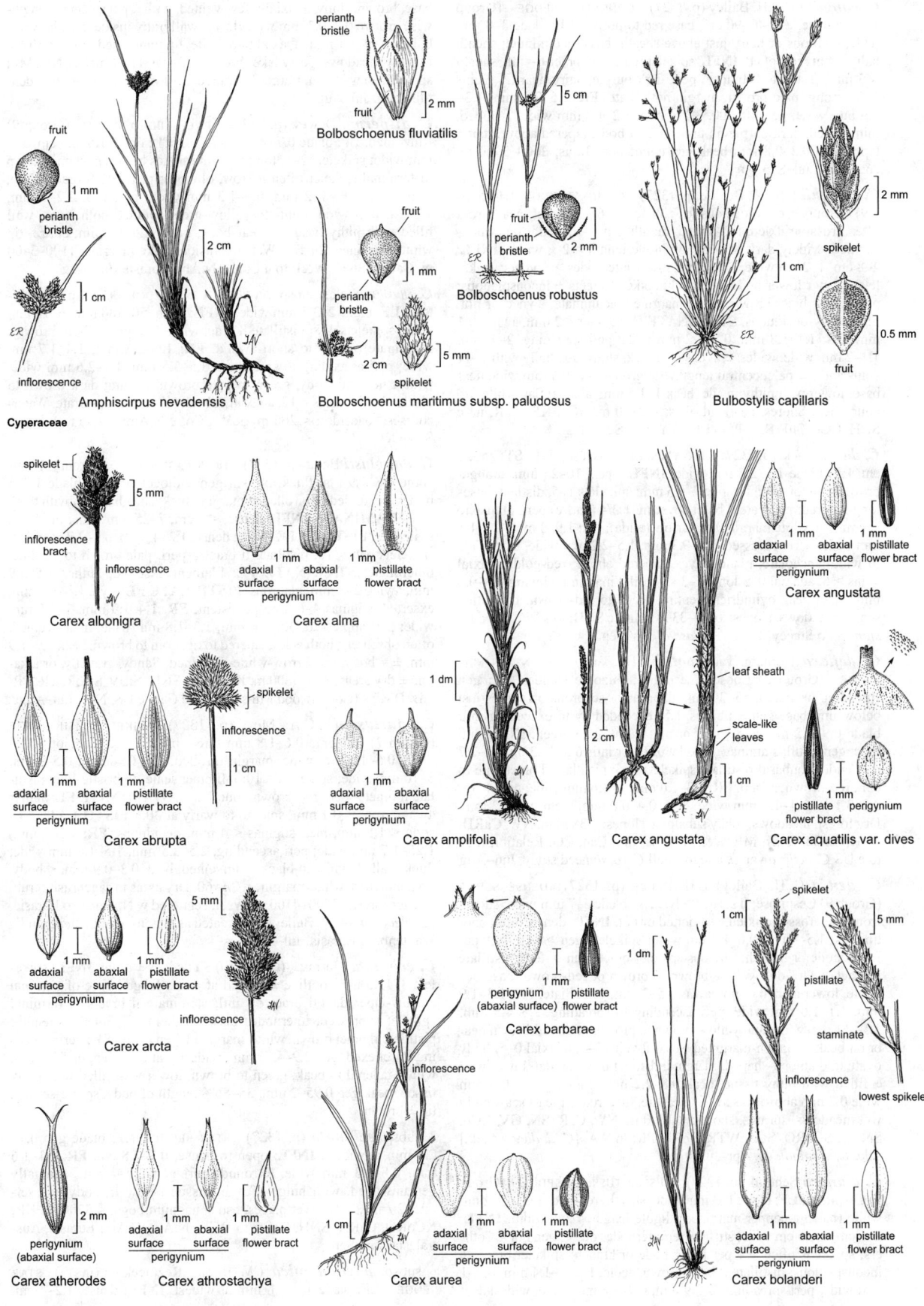

Cyperaceae

Amphiscirpus nevadensis

fruit
perianth bristle
1 mm
2 cm
1 cm
ER
inflorescence

Bolboschoenus fluviatilis
perianth bristle
fruit
2 mm

Bolboschoenus maritimus subsp. paludosus
fruit
perianth bristle
1 mm
2 cm
5 mm
spikelet

Bolboschoenus robustus
fruit
perianth bristle
5 cm
2 mm
ER

Bulbostylis capillaris
spikelet
2 mm
1 cm
ER
fruit
0.5 mm

Carex albonigra
spikelet
5 mm
inflorescence bract
inflorescence

Carex alma
adaxial surface
abaxial surface
1 mm
perigynium
pistillate flower bract
1 mm

Carex angustata
adaxial surface
abaxial surface
1 mm
perigynium
pistillate flower bract
1 mm

Carex abrupta
adaxial surface
abaxial surface
1 mm
pistillate flower bract
1 mm
perigynium
spikelet
inflorescence
1 cm

Carex amplifolia
1 dm
adaxial surface
abaxial surface
1 mm
perigynium

Carex angustata
2 cm
leaf sheath
scale-like leaves
1 cm

Carex aquatilis var. dives
pistillate flower bract
perigynium
1 mm

Carex arcta
adaxial surface
abaxial surface
1 mm
perigynium
pistillate flower bract
1 mm
5 mm
inflorescence

Carex barbarae
perigynium (abaxial surface)
pistillate flower bract
1 mm

Carex bolanderi
spikelet
1 cm
5 mm
pistillate
staminate
1 dm
inflorescence
lowest spikelet
adaxial surface
abaxial surface
1 mm
perigynium
pistillate flower bract
1 mm

Carex atherodes
perigynium (abaxial surface)
2 mm

Carex athrostachya
adaxial surface
abaxial surface
1 mm
perigynium
pistillate flower bract
1 mm

Carex aurea
inflorescence
1 cm
adaxial surface
abaxial surface
1 mm
perigynium
pistillate flower bract
1 mm

C. congdonii L.H. Bailey (p. 1327) CONGDON'S SEDGE (Group 1) Cespitose. **ST:** 40–90 cm, base red to purple. **LF:** blade 3–8 mm wide, glabrous or hairy just above sheath, basal occ minute; sheath hairy, front red-dotted. **INFL:** spikelets dark; infl bract ± = infl, sheath < 5 mm; pistillate fl bract < peri, dark purple, hairy near tip, veins 1–5, strong, pale brown, margin not ciliate. **FR:** 1.5–2.5 mm, 1.3–1.9 mm wide; peri 40–200, 3.1–4 mm, 1.2–1.8 mm wide, 2-ribbed, thin-walled, dark purple on upper 1/2 of body, tapered above, hairy, tapered to 0.3–0.8 mm beak gen unnotched. Talus; 2600–3900 m. c&s SNH. Jul–Sep ★

C. cusickii Piper & Beattie (p. 1327) CUSICK'S SEDGE (Groups 7,9) Cespitose; occ dioecious. **LF:** blade 2.5–6 mm wide; sheath front often cross-wrinkled, red-dotted, mouth gen U-shaped, often thick-rimmed, widely dark or coppery; ligule length > 2 × width. **INFL:** 4–8 cm, 1–2 cm wide, ± flexible; lower internodes >> others; spikelets > 1 per lower node or branch; spikelet bracts ± inconspicuous; pistillate fl bract brown, white-margined, acuminate, awned or not, often early-deciduous. **STAMINATE FL:** anther < 2 mm, awn ≤ 0.1 mm. **FR:** 1.1–1.8 mm, 0.8–1.2 mm wide; peri spreading, 2–4 mm, 1.1–2 mm wide, widest ± at base, ± black, shiny, abaxially with 7–11 veins and a ± pale, central lengthwise groove 0.1–0.3 mm wide near base, lower wall pithy inside, beak 1–1.5 mm, abaxial flap white, tip unnotched. Shores, peatland, fens; < 2100 m. NCo, KR, CaR, n&c SNH, CCo, SnFrB, MP; to BC, WY. Jun–Sep

C. davyi Mack. (p. 1327) DAVY'S SEDGE (Group 11E) **ST:** 25–35 cm. **LF:** blade 1.5–2.5 mm wide. **INFL:** open, 10–25 mm, orange-brown; 2nd lowest internode 2–6.5 mm; spikelets 1–5, distinct, bases long-tapered; pistillate fl bract covering ± 3/4 (entire) peri, obtuse to acuminate, white margin 0 or narrow (wide). **FR:** 1.9–3 mm, 1.1–1.5 mm wide; peri appressed, 5.9–9.3 mm, 1.5–2.2 mm wide, lanceolate to wide-lanceolate, planoconvex, tapered above, red-gold, abaxial veins > 6, adaxial 0–3 long, 1–3 short, flat margin incl wing 0.3–0.6 mm wide, beak cylindric, ± entire, green to red-brown. Dry often sparse meadows, slopes; 1400–3300 m. n&c SNH; to s WA. [*C. constanceana* Stacey] Like *C. petasata.* Jun–Sep ★

C. deflexa Hornem. var. ***boottii*** L.H. Bailey MOUNTAIN MAT SEDGE (Group 2) Loosely cespitose; rhizome slender. **ST:** gen spreading or arching, 3–30 cm, glabrous or scabrous on angles just below infl, base gen ± fibrous. **LF:** exceeded by to exceeding infl; blade 0.9–3.2 mm wide, stiff or not, ± flat, pale green. **INFL:** infl bract gen ≥ infl; staminate spikelet of taller infls 6.3–11.3 mm, 1–2.7 mm wide; nonbasal pistillate spikelets 1–4; pistillate fl bract pale to dark red-brown, awn 0. **FR:** 1.3–1.6 mm, 1–1.6 mm wide; peri 4–15, 2.2–3.1 mm, 1–1.4 mm wide, beak 0.4–0.8 mm, teeth 0.1–0.2 mm. Dry forest, meadows, rocky subalpine slopes; < 3800 m. NW, CaRH, SNH, SnGb, SnBr, MP, W&I, DMtns; to n Can, CO. Relationships to e US *C. deflexa* var. *d.* and to small *C. rossii* need study. Jun–Aug

C. densa (L.H. Bailey) L.H. Bailey (p. 1327) DENSE SEDGE (Group 9) Cespitose; fl sts gen > lvs. **LF:** blade 3–7 mm wide; sheath front gen cross-wrinkled, red-dotted or not. **INFL:** dense, often continuous, 1.5–8 cm, 0.8–1.5 cm wide; spikelets gen >> 10, > 1 per lower node or branch; obvious spikelet bracts gen 0–few; pistillate fl bract < peri, dull yellow-brown to brown or red-brown, base occ white, lower awned or not, awns 0.5–2 mm, gen ciliate. **FR:** 1.3–1.9 mm, 1.1–1.6 mm wide; peri ascending to spreading, 2.5–4.5 mm, 1.3–2.1 mm wide, dull yellow-brown to brown, often red-brown near or on beak, or green-margined, abaxial veins 3–7, adaxial 0–5, body ovate to diamond-shaped, margin indistinct or wide, flat, lower wall ± filled with pithy tissue, beak 0.8–2.2 mm, conic, gen > 0.6 mm wide 0.2 mm above fr, serrate or ciliate, tip ± red. At least seasonally wet meadows, springs, shores; < 1500 m. NW, CaR, SN, GV, CCo, SnFrB, SCoRO, SCo, WTR, SnGb, PR; to WA. [*C. dudleyi* Mack.] Like *C. vulpinoidea.* Apr–Jul

C. diandra Schrank (p. 1327) LESSER TUSSOCK SEDGE (Group 9) Cespitose. **LF:** blade 1–3 mm wide; sheath front red-dotted, mouth white (narrow-copper-margined); ligule length < 2 × width. **INFL:** 2–5 cm, < 1.5 cm wide, stiff; lower internodes occ >, gen not >> others; spikelets often > 1 per lower node or branch; spikelet bracts ± inconspicuous; pistillate fl bract brown, acute. **FR:** 1–1.4 mm, 0.7–1 mm wide; peri spreading, 2–2.9 mm, 0.9–1.4 mm wide, widest ± at

base, brown, shiny, abaxially few-veined, with ± pale, central lengthwise groove 0.2–0.3 mm wide, lower wall pithy inside, adaxial veins 0, beak 0.9–1.1 mm, abaxial flap white, tip unnotched. 2*n*=48,50,60. Marshy meadows, peaty lake shores; 150–2400 m. CaRH, SNH, MP; sporadic in w N.Am. Larger CA range in TJM (1993) due to misidentification. Jul–Aug

C. disperma Dewey (p. 1327) TWO-SEEDED SEDGE (Group 9) Rhizomed. **LF:** blade 0.7–2 mm wide. **INFL:** open, 1.5–2.5 cm, 3–4 mm wide; spikelets 2–4, lower << internodes between; staminate tip of terminal spikelet often narrow, obvious; pistillate fl bract white, acuminate. **FR:** 1–2 mm, 0.9–1.3 mm wide; peri 1–3, 2.2–3 mm, 1.1–1.6 mm wide, light- to yellow-green, veined both sides, wall filled with pithy tissue to near beak, beak 0.2–0.3 mm, entire, tip white-margined. 2*n*=70. Wet streamsides, lake margins; 1100–3400 m. CaRH, SNH, W&I; to n Can, e N.Am; Eurasia. Jul–Aug

C. divulsa Stokes subsp. ***divulsa*** GRAY SEDGE (Group 9) Cespitose. **LF:** blade 2–3.5 mm wide. **INFL:** open, 50–180 mm, 6–10 mm wide; spikelets 4–8; pistillate fl bract white to ± green (light brown), acuminate (acute) to short-awned. **FR:** 1.9–2.5 mm, 1.4–1.7 mm wide; peri ascending to spreading, 3.5–5.5 mm, 1.7–2.6 mm wide, widest near midbody, green to pale brown, turning dark brown to black, abaxial veins < 12, adaxial 0, beak 0.7–1.5 mm, serrate. Watercourses, roadsides; < 200 m. ScV, CCo; e N.Am; native to Eurasia. Sep–Oct

C. douglasii Boott (p. 1327) DOUGLAS' SEDGE (Group 7) Rhizome 0.8–1.9 mm thick, brown; gen ± dioecious. **LF:** blade 1–3.5 mm wide, folded or inrolled at margin, thick; sheath mouth with thick rim. **STAMINATE INFL:** dense, < 3 cm, 7–25 mm wide; infl bract scale-like. **PISTILLATE INFL:** dense, 1.2–4.5 cm, 0.8–2.7 cm wide; spikelets gen < 10; pistillate fl bract > peri, pale brown to ± white, mucronate. **STAMINATE FL:** filament exserted, anther 2.5–3.9 mm, awn 0.2–1 mm, glabrous. **PISTILLATE FL:** style 1.8–3.5 mm, exserted, stigmas 4–6 mm, persistent. **FR:** 1.4–1.9 mm, 1–1.5 mm wide; peri appressed, 3.5–4.6 mm, 1.3–1.8 mm wide, many-veined, often obscurely, both sides, tapered to tip, gold to brown, beak 0.9–1.9 mm, ± = body, tip narrow-white-margined. Sandy, gravelly, or alkaline ± dry areas; 300–3800 m. KR, CaR, SNH, SnJV, SCoR, TR, PR, GB, DMtns (Cottonwood Mtns); to w&c Can, c US, NM. Jun–Aug

C. duriuscula C.A. Mey. (p. 1327) SPIKELETRUSH SEDGE (Group 7) Rhizome 0.6–1.8 mm thick, brown; occ dioecious. **LF:** blade 0.5–1.5 mm wide, margin inrolled. **INFL:** dense, 0.5–2 cm, 5–10 mm wide; spikelets < 10; infl bract scale- to bristle-like; pistillate fl bract > peri, red-brown, mucronate. **STAMINATE FL:** anther awn narrow, < 0.1 mm, smooth to warty at 30×. **PISTILLATE FL:** style < 1.5 mm, incl, stigmas < 4 mm, deciduous. **FR:** 1.5–2 mm, 1.25–1.7 mm wide; peri ascending, 2.5–3.5 mm, 1.5–1.8 mm wide, thick-walled, white to black, ± unveined, beak 0.3–0.9 mm, < body, tip unnotched, white-margined. 2*n*=60. Dry areas in sagebrush scrub, conifer forests; 3500–4100 m. W&I; scattered w N.Am; also Eurasia. [*C. eleocharis* L.H. Bailey] Occ treated as synonym of *C. stenophylla* Wahlenb. of Eurasia. Jul–Aug ★

C. echinata Murray (Group 10) **ST:** edges ± blunt. **INFL:** spikelets 3–15.5 mm, outline ± jagged at arm's length, base of terminal narrow-tapered; infl bract << infl; staminate fl bracts of terminal spikelet appressed, internodes between them 2/3–3/4 their length; pistillate fl bract brown, white-margined, gen acute. **FR:** peri spreading to reflexed, gen 2.9–4.75 mm, widest ± at base, length 1.7–3.6 × width, tapered to beak, green to brown, lower wall filled with pithy tissue, beak gen 0.95–2 mm, 35–86% length of body, sparse-serrate, tip ± brown.

subsp. ***echinata*** (p. 1327) STAR SEDGE **LF:** blade gen 0.7–2.7 mm at widest. **INFL:** open to dense, 0.7–7.8 cm. **FR:** 1.1–1.6 mm, 0.8–1.3 mm wide, ± round; peri gen 2.9–4 mm, adaxially gen unveined, with bulge of pithy tissue below fr, body gen serrate. 2*n*=50,52,58. Wet places, esp sphagnum bogs; < 3200 m. KR, NCoRO, CaRH, SNH, SnBr; to BC, e N.Am, C.Am, Eurasia, Australasia. Jun–Sep

subsp. ***phyllomanica*** (W. Boott) Reznicek COASTAL STAR SEDGE **LF:** blade 1.7–3.8 mm at widest. **INFL:** dense, 1.2–4 cm.

FR: 1.5–2.2 mm, 1–1.5 mm wide, ovate; peri gen 3.5–4.75 mm, adaxially 2–12-veined, without bulge of pithy tissue, body gen entire. 2*n*=54,70. Peatland, shores, swamps, occ brackish places; < 200 m. NCo, CCo; to AK. May–Aug

C. exsiccata L.H. Bailey (p. 1327) WESTERN INFLATED SEDGE (Group 5) Rhizome short or 0. **ST**: 30–100 cm, angles sharp, scabrous near infl. **LF**: gen much exceeded by infl; sheath fronts gen fibrous on lower lvs; blade 2.5–6.2 mm wide, ligule length > width. **INFL**: infl bract > infl, < 2.5 × infl; lateral spikelets short-stalked, erect; pistillate fl bract gold to ± purple, not awned. **FR**: 1.7–3 mm, 1.1–1.8 mm wide; peri 20–150, in few ± open rows, ascending, gen 7.5–12.7 mm, 1.5–3 mm wide, green to brown, tapered above, beak 1.5–3 mm, teeth erect, 0.2–1.5 mm. Wet places; < 1800 m. NCo, s KR, NCoRO, CCo, SnFrB; to AK, MT. [*C. vesicaria* var. *major* Boott] Intergrades with *C. vesicaria*; study needed. Jul–Aug

C. feta L.H. Bailey (p. 1329) GREEN-SHEATHED SEDGE (Group 11A) Erect, not stoloned. **LF**: blade 2.5–5 mm wide; sheath front firm, green-ribbed ± to top, with a thin, ± triangular white area extending < 6 mm below top; contraligule (3)4–8 mm. **INFL**: open, 30–80 mm, white-green to pale gold; spikelets distinct (or > 1 per lower node or branch); infl bract bristle- or lf-like; pistillate fl bract gen white or pale gold with green or light brown center, ± covering peri. **FR**: 1.3–2 mm, 0.9–1.2 mm wide; peri ascending to ascending-spreading, 3–4.2 mm, 1.7–2.1 mm wide, 0.4–0.5 mm deep, ovate to wide-ovate, flat or planoconvex, green to pale gold, flat margin incl wing 0.3–0.6 mm wide, veined abaxially, ± unveined adaxially, beak cylindric, ± entire for < 0.4 mm, tip ± red. Meadows, streambanks, open forest, seeps, damp or wet soil; 30–3100 m. KR, NCoR, CaR, n SNF, SNH, SnFrB, SnBr, PR, MP; to BC. Like *C. fracta* but spikelets gen smaller, more distinct. May–Aug

C. filifolia Nutt. (Group 2) Cespitose. **LF**: blade 0.2–0.7 mm wide, rolled, quill-like. **INFL**: 1 terminal spikelet; lower pistillate fl bracts pale red-brown, white- or yellow-brown-margined, obtuse to awned, clasping peri base. **PISTILLATE FL**: style exserted, black, persistent. **FR**: 1.6–3 mm, 1.2–1.8 mm wide, ± = peri body, next to bristle-like axis; peri appressed to ascending, 1.9–3.7 mm, 1.3–2.1 mm wide, 3-sided to round, hairy at least just below beak, unveined, ± white to gold. Vars. possibly indistinct; study needed.

var. ***erostrata*** Kük. (p. 1329) SAGEBRUSH SEDGE **ST**: 5–25 cm. **LF**: exceeded by to exceeding infl. **INFL**: longest pistillate fl bracts (exc awns) gen < 2.5 mm, gen << peri. **FR**: 1.6–2.4 mm; peri 1.9–3 mm, occ ruptured by fr, body gen abruptly narrowed to beak 0–0.4 mm. Meadows; 1500–3700 m. SNH, SnBr, Wrn, W&I; s OR, w NV. May–Aug

var. ***filifolia*** THREAD-LEAVED SEDGE **ST**: 8–30 cm. **LF**: exceeded by to equaling infl. **INFL**: longest pistillate fl bracts (exc awns) gen > 3 mm, gen ≥ peri. **FR**: 2.2–3 mm; peri 2.9–3.7 mm, body ± tapered to beak 0.1–0.8 mm. ± dry areas with subsurface moisture; ± 3200 m. c SNH (Inyo Co.); scattered w N.Am. Seldom collected, most CA pls identified as this are instead vigorous pls of *C. filifolia* var. *erostrata*. Jul–Aug

C. fissuricola Mack. (p. 1329) CLEFT SEDGE (Groups 2,4) Cespitose to loosely so. **ST**: 50–80 cm. **LF**: blade 3–8 mm wide. **INFL**: terminal spikelet staminate at least at tip; lateral spikelets gen ± purple to brown, tips tapered, stalks long-exserted, ± nodding; infl bract blade << infl, sheath linear or ± expanded to mouth, mouth 0.7–1.8 mm wide, gen Y- or V-shaped, purple bordered; pistillate fl bract brown or dark purple, midstripe pale, hairy or scabrous, margin ciliate distally, tip acute or mucronate. **FR**: 1.5–2 mm, 0.8–1 mm wide; peri ascending, 3.2–5.5 mm, 0.9–2 mm wide, gen ± red-brown or purple, 2-ribbed, flat margin > 1/2 fr width, at least faces distally with sparse, spreading-ascending, short, stiff, bristles, occ shed in age, bristly-ciliate, beak 0.8–2 mm, margin bristly-ciliate. Meadows, rocky streamsides; 1500–3300 m. SNH; ID, ne NV, UT. Intermediate between *C. luzulifolia*, *C. luzulina*; study needed. Jul–Sep

C. fracta Mack. (p. 1329) FRAGILE-SHEATHED SEDGE (Groups 11B,F,G) **LF**: blade 2.5–6 mm wide; sheath front ribbed at least in lower 1/2, translucent > 10 mm below mouth; contraligule gen 5–13 mm. **INFL**: dense above or open, 2.5–8 cm, gen ± white or green

(to pale brown); spikelets often > 1 per lower node or branch; infl bract bristle- or lf-like, base clasping st or not; pistillate fl bract ± covering peri, gen white, occ gold or ± brown, with green to gold center, awned. **FR**: 1.2–1.6(2) mm, 0.8–1.2 mm wide; peri ascending, 2.9–4(4.8) mm, 1–1.7(1.9) mm wide, ovate, planoconvex (biconvex), green to pale gold, ± abruptly narrowed above, flat margin incl wing gen 0.1–0.25 mm wide at midbody, veins gen strong both sides, beak gen cylindric, ± entire gen for < 0.4 mm, pale. Common. Montane meadows, open forests, edges, roadsides, moist or dry soil; 250–3300 m. KR, NCoRH, CaRH, SN, SCoRO, TR, PR; to WA. [*C. amplectens* Mack.] Like *C. feta* but spikelets gen more clustered. May–Sep

C. geyeri Boott (p. 1329) GEYER'S SEDGE (Group 3) In clumps connected by rhizomes. **ST**: 10–40 cm. **LF**: exceeded by to exceeding infl; blade 2–3.5 mm wide, flat or folded, evergreen. **INFL**: lower pistillate fl bracts > peri, green, short-awned. **FR**: 4.5–6.2 mm, 1.7–2.5 mm wide, ± = peri body, next to bristle-like axis; peri 1–3, appressed to ascending, 4.9–7 mm, 1.7–2.5 mm wide, obovoid, planoconvex, green, marginal veins 2, fine veins 0, beak minute. Open forest, slopes; 900–2100 m. KR, CaR, n SN; to w Can, CO; possibly alien in PA. May–Aug ★

C. globosa Boott (p. 1329) ROUND-FRUITED SEDGE (Group 2) Loosely cespitose. **ST**: 15–47 cm. **LF**: blade 1.5–2.5 mm wide, green, not glaucous, not papillate or papillate only on veins abaxially at 40×; basal sheaths brown to red. **INFL**: staminate spikelet of taller infls 10–20 mm, 1.5–4 mm wide; pistillate spikelets 2–3, basal gen arching on long, slender stalk; infl bract sheath of bisexual infl gen < 2 mm; pistillate fl bract tinged red-brown or purple, obtuse to awned. **FR**: 1.9–2.5 mm, 1.6–1.9 mm wide, gen white; peri 1–10, 3.9–5.1 mm, 1.8–2.3 mm wide, strongly veined at least below, body obovoid, stalk 1.5–2 × beak. Well-drained soil of wooded areas, edges; < 1800 m. NCo, NCoRO, CW, SW. Pls from SnBr similar to *C. pityophila* Mack. of s Rocky Mtns may be distinct; study needed. Apr–Jun

C. gracilior Mack. SLENDER SEDGE (Groups 11C,D,H) **ST**: self-supporting. **LF**: blade 1.2–3 mm wide; ligule 1.5–2(3) mm. **INFL**: gen open, (12)15–30 mm; lowest internode often > 4 mm, often > lowest spikelet, lowest 2 together < 14 mm, gen ± 1/3 infl; lower spikelets gen distinct, wedge-shaped or rounded, upper ± indistinct; infl bract often bristle-like, < to > lowest spikelets, < to > infl, base occ clasping st; pistillate fl bract gen > 3.4 mm, < to > peri, gold to red-brown, center often green, gold, or ± white. **FR**: gen 1.7–2.3(2.8) mm, gen 1.1–1.5 mm wide; peri appressed-ascending, (2.9)3.5–4.5(5.4) mm, gen 1.5–1.9(2.2) mm wide, narrow- to occ wide-ovate, planoconvex or biconvex, gen gold with brown margin, shiny, flat margin incl wing gen 0.1–0.2 mm wide at midbody, often inrolled, entire exc on or just below beak, occ filled with pith-like tissue toward base, abaxial veins 5–9, adaxial 0 (3 but not exceeding fr top), beak gen ± 2/5 peri, brown to red-brown, cylindric, ± entire for 0.5–1 mm. Grassland, creekbanks, open forest, at least seasonally moist soil; < 850 m. NCoRO, CaR, in SN, ScV, CW (exc SCoRI), nw SCo, w WTR. Intergrades with *C. subbracteata*. Apr–Jun

C. gynodynama Olney (p. 1329) WONDER-WOMAN SEDGE (Group 1) Loosely cespitose. **ST**: 20–90 cm. **LF**: hairy; blade 3–9 mm wide. **INFL**: terminal spikelet staminate at least at tip; infl bract < infl, sheath > 5 mm; pistillate fl bract obovate to wide-elliptic, red-brown or ± purple, obtuse to mucronate, hairy, margin not ciliate, midrib green. **FR**: 2.3–2.5 mm, 1.3–1.8 mm wide; peri 20–40, 3.8–5.5 mm, 1.4–2 mm wide, hairy, many-veined, purple-splotched, red-dotted, base long-tapered, beak 0.5–1.1 mm, conic, white at tip, teeth erect, < 0.5 mm. Moist meadows in open forest, serpentine; < 600 m. NCo, KR, NCoRO, n CCo (Marin, Santa Cruz cos.), SnFrB, SCoRO; to s OR. Hybridizes with *C. hendersonii*, *C. mendocinensis*. Apr–Jun

C. halliana L.H. Bailey (p. 1329) OREGON SEDGE (Group 2) Rhizomed. **ST**: 10–50 cm. **LF**: blade ascending, 3–5 mm wide, strongly V-folded at least near base, glabrous. **INFL**: spikelets erect; infl bract > infl, sheath < or > 0.5 mm; pistillate fl bract ± ≤ peri, gen ± purple, acute, margin white, not ciliate. **FR**: 1.9–2.5 mm, 1.3–1.8 mm wide; peri 20–40, 3.6–5 mm, 1.7–2.3 mm wide, thick-walled, strongly many-ribbed, green to gold, beak 1–1.7 mm, teeth erect, 0.2–0.5 mm. Dry upland, forest edges, often on pumice; 1300–2200 m. KR, CaRH; to WA. Jul–Aug ★

C. harfordii Mack. (p. 1329) HARFORD'S SEDGE, MONTEREY SEDGE (Groups 11D,F,H) **ST:** often leaning, occ branched late season, branches fertile or not. **LF:** blade gen 2–5(9.9) mm wide; contraligule 1–4 mm; ligule gen < 2 mm. **INFL:** dense, 1.5–4 cm, gen dark brown or ± black; spikelet stalks 12–36 cm; spikelets ± indistinct; lowest infl internode < 1.5–2.5 mm; infl bract lf- or bristle-like, < or > infl; pistillate fl bract (2.5)3.5–4.7 mm, < to ± covering peri, green to dark brown, center often green to gold, margins dark or white, acute to short-awned. **FR:** (1.3)1.5–2.1 mm, 0.9–1.4 mm wide; peri ascending to spreading-ascending, (2.6)3.3–4(4.6) mm, (1.3)1.4–1.6(2) mm wide, ovate to wide-ovate, planoconvex; green to brown, shiny, margin gen green, veins both sides, gen exceeding fr, wall ± thin, flat margin incl wing 0.2–0.3(0.4) mm wide, beak cylindric, ± entire to gen > 0.4 mm, ± dark. Wet or seasonally wet shores, meadows, open forest; < 900 m. NCo, NCoRO, CW, ChI; s OR. Differs from *C. subbracteata* in peri texture, adaxial veins; putative hybrids ± sterile. Feb–Sep

C. hassei L.H. Bailey (p. 1329) HASSE'S SEDGE (Groups 4,8) Rhizomed (loosely cespitose). **LF:** blade 2–4 mm wide. **INFL:** terminal spikelet staminate (pistillate) at tip, staminate part 1.2–3.5 mm wide, lowest non-basal spikelet gen separate, lateral ones erect to nodding, dense-clustered or not, 3–5, 7–25 mm, 3.5–4.5 mm wide, long-stalked or not, lower gen from near pl base; infl bract >> infl, sheath > 4.5 mm, mouth V- or U-shaped; lower staminate fl bracts 3–6 mm; pistillate fl bract < to >, appressed against, falling after fully expanded peri, obtuse to awned, red-brown to dark purple, margin, tip white. **PISTILLATE FL:** stigmas 2(3). **FR:** 1.4–2 mm, 1–1.5 mm wide, lenticular (3-angled); peri gen 10–20, ascending to spreading, 1.9–3.1 mm, 1–1.6 mm wide, elliptic to ± obovate, stalked, gen veined, when fresh white to green (purple splotched, or base gold), walls ± translucent, gen fleshy at least at base, when dry green-white or pale gold, base tapered, tip wide, blunt, papillate at 10×, beak < 0.2 mm, tip often red-brown, notch 0. Springs, peatland, fens, moist meadows, serpentine or not; < 2900 m. NCo, KR, NCoRH, CaRH, SNH, CCo, SnFrB, TR, PR, SNE, DMtns; to BC, UT, AZ. [*C. garberi* Fernald misappl.] Like *C. aurea*, variable in sex of terminal spikelet, but with peri greener, more elliptic, veined, papillate. Pls with 2–3 stigmas gen in KR serpentine fens. May–Aug

C. haydeniana Olney (p. 1329) CLOUD OR HAYDEN'S SEDGE (Groups 11E,H) **ST:** gen decumbent, 9–20(40) cm. **LF:** blade 1.5–4 mm wide, flat; sheath front occ cross-wrinkled above; ligule < 2 mm. **INFL:** dense, 9–18 mm, ± spheric, gen dark brown, less often ± green or gold; lowest infl internode gen < 2 mm; spikelets gen indistinct; pistillate fl bract 3–4.8 mm, < peri, gen dark brown to ± black with occ green or coppery midveins, gen contrasting with lighter peri. **STAMINATE FL:** anthers early-deciduous. **FR:** (1.2)1.4–1.8(2) mm, 0.8–1.1(1.3) mm wide, stalk 0.4–0.7 mm; peri 4–6.5 mm, 1.5–2.6 mm wide, lanceolate- to wide-ovate, flat exc over fr, green, dark brown or coppery, flat margin incl wing 0.3–0.7 mm wide, margin often dark, gen not crinkled, abaxial veins < 8, adaxial gen 0, beak cylindric, entire 0.4–0.6 mm, green to black. 2n=82. Uncommon. Rocky slopes, flats, moist soil; 2400–4200 m. SNH, Wrn, W&I, DMtns; to BC, Rocky Mtns. CA pls much like *C. microptera*. Intergrades with *C. macloviana* d'Urv. (± n N.Am, S.Am, n Eur), which has pistillate fl bracts, peri coppery; study needed. Jul–Sep

C. helleri Mack. (p. 1329) HELLER'S SEDGE (Group 6) Cespitose. **ST:** 5–40 cm. **LF:** blade 2–3.5 mm wide. **INFL:** gen head-like; lowest spikelet occ separate; terminal spikelet pistillate above (or staminate/pistillate/staminate); lower spikelets sessile; pistillate fl bract gen < 1 mm wide, > but width << peri, red-brown to purple, tips acuminate, lower pistillate fl bract awns gen 0.5–2 mm. **FR:** 1.4–1.8 mm, 0.7–1.1 mm wide; peri 2.5–3.8 mm, 1.5–2.8 mm wide, gen red-brown to purple at least above, gen not papillate (± papillate near beak at 20×), elliptic to obovate, tapered or abruptly narrowed to beak 0.2–0.5 mm. Dry, rocky or gravelly slopes; 2400–4100 m. CaRH, SNH, SNE (Sweetwater Mtns), W&I; NV (Elko Co.). Much like *C. heteroneura*, but with denser infls, gen higher elevations. Jul–Sep

C. hendersonii L.H. Bailey (p. 1329) TIMBER OR HENDERSON'S SEDGE (Group 4) Cespitose. **ST:** 40–100 cm, erect or decumbent, in fr often prostrate. **LF:** blade 6–16 mm wide. **INFL:** infl bract sheath

> 6 mm, inflated; pistillate fl bract white, ± awned. **FR:** 2.7–3.3 mm, 1.6–2 mm wide; peri 4.3–6.5 mm, 1.3–2 mm wide, ± stalked, green, veins 20–32, body tip wide-conic, beak gen < 0.2 mm or indistinct, tip unnotched. Coastal forests; < 900 m. NCo, NCoRO, CCo, SnFrB; to BC; also ID. Hybridizes with *C. gynodynama.* May–Jun

C. heteroneura W. Boott (p. 1329) SMOOTH-FRUITED SEDGE (Groups 4,6) Cespitose. **ST:** 25–100 cm. **LF:** blade 2–7 mm wide. **INFL:** ± dense, gen not head-like; terminal spikelet gen pistillate above; lowest spikelets gen erect (nodding), often long-stalked, ± distant from others, obviously 2-colored or not; pistillate fl bract gen > 1.2 mm wide, brown to dark purple, midrib pale or dark, acute to acuminate, awns < 0.5 mm or gen 0. **FR:** 1–2 mm, 0.7–1.4 mm wide, stalk < 1.2 mm; peri ascending, 2.5–4.5 mm, 1.4–3.2 mm wide, green, brown, or dark purple, margin gen pale, gen not papillate (± papillate near beak at 20×), ± flat exc over fr, flat margin < 0.8 mm, body ovate, round, or obovate, ± rounded at tip, beak 0.2–0.5 mm. Meadows, forest openings, rocky slopes; 1300–4000 m. KR, CaRH, SNH, SnGb, SnBr, SnJt, Wrn, SNE (Sweetwater Mtns), W&I; to BC, Rocky Mtns. [*C. epapillosa* Mack.; *C. h.* var. *e.* (Mack.) F.J. Herm.; *C. bella* L.H. Bailey, misappl.] Jun–Sep

C. hirtissima W. Boott (p. 1329) FUZZY SEDGE (Groups 1,4) Loosely cespitose. **ST:** 30–60 cm, sparse-hairy (glabrous), in fr erect or ascending. **LF:** dense-hairy (glabrous); blades 3–12 mm wide, basal minute. **INFL:** terminal spikelet gen pistillate at least at tip; infl bract < infl, sheath > 5 mm, not inflated; pistillate fl bract ± hairy (glabrous), elliptic to elliptic-obovate, green or pale red-brown, midrib green, obtuse to mucronate, margin white, not ciliate. **FR:** 2–2.4 mm, 1.3–1.6 mm wide, ± = peri body; peri 20–30, 2.5–4.3 mm, 1.4–1.8 mm wide, green or gold-green to pale brown, occ red-dotted but not purple-splotched, 2-ribbed, finely 8–10-veined, hairy (glabrous), base short-tapered, beak > 1 mm, conic, white at tip, teeth erect, 0.3–0.8 mm. Openings, mixed-evergreen forest, disturbed wet places, often on serpentine; 60–1200 m. NCoR (Mendocino, Lake cos.), CaR, n&c SN. Much like *C. gynodynama.* Apr–Jul

C. hoodii Boott (p. 1329) HOOD'S SEDGE (Group 9) Cespitose. **LF:** blade 1.5–3.5 mm wide. **INFL:** dense, 0.8–2 cm, 6–15 mm wide; spikelets 4–8, indistinct exc occ lowest; infl bracts gen < spikelets; pistillate fl bract < peri, acute to long-acuminate, red-brown, midrib green, margin white. **FR:** 1.7–2.1 mm, 1.3–1.7 mm wide; peri spreading, 3.4–5 mm, 1.4–2 mm wide, sessile, copper-brown, margin green, veins gen 0, beak 0.8–1.8 mm, tip notched. 2n=60. Rocky or gravelly slopes, meadow edges; 650–3600 m. KR, NCoR, CaRH, SNH, SnBr, MP, W&I; to w Can, SD, CO. Jun–Sep

C. hystericina Willd. (p. 1329) PORCUPINE SEDGE (Groups 5,6) Rhizome short. **ST:** 15–100 cm. **LF:** ligule length >> width. **INFL:** lower spikelets on long, nodding stalks; pistillate fl bract gen white or cream, center pale ± red, lanceolate, awn > body. **FR:** 1.2–2 mm, 0.9–1.3 mm wide, << peri body; peri ascending-spreading to spreading, 5–7 mm, 1.5–2.1 mm wide, round in ×-section, ± sessile, green to gold, beak 1.9–2.8 mm, teeth erect, 0.3–0.9 mm. 2n=58. Wet places; < 500 m. KR; to w Can, e N.Am. [*Carex hystricina,* orth. var.] May–Jun ★

C. idahoa L.H. Bailey (p. 1329) IDAHO SEDGE (Groups 3,6) Loosely cespitose. **ST:** 25–40 cm. **LF:** blade 2–5 mm wide, flat, ± glaucous. **INFL:** ± open; terminal spikelet pistillate, occ staminate at base or tip (throughout), 10–25 mm, lateral 0–3, 5–13 mm, stalk 0 or short; pistillate fl bract gen > peri, obovate to widely elliptic, gen dark purple, brown, or black, margin white, tips or awns bristly or not. **PISTILLATE FL:** stigmas 3. **FR:** 1.3–2 mm, 0.9–1.5 mm wide, ± = peri body; peri 2–3 mm, 1.5–1.8 mm wide, body elliptic to obovate, pale yellow or brown, upper body papillate, hairs or bristles on and just below beak few, stiff, beak 0.2–0.3 mm. Meadows; 2800–3400 m. W&I (White Mtns); to OR, MT, UT. [*C. parryana* var. *hallii* (Olney) Rydb., misappl.] Relationships to *C. hallii, C. parryana,* and n US pls need study. Jul ★

C. illota L.H. Bailey (p. 1329) SHEEP SEDGE (Groups 10,11C) **LF:** blade 1–3 mm wide. **INFL:** dense, 6–15 mm, dark brown; lowest internode < 1 mm; spikelets indistinct exc occ lower; pistillate fl bract < peri, ± black, shiny, ± obtuse. **FR:** 1.2–1.5 mm, 0.7–1 mm

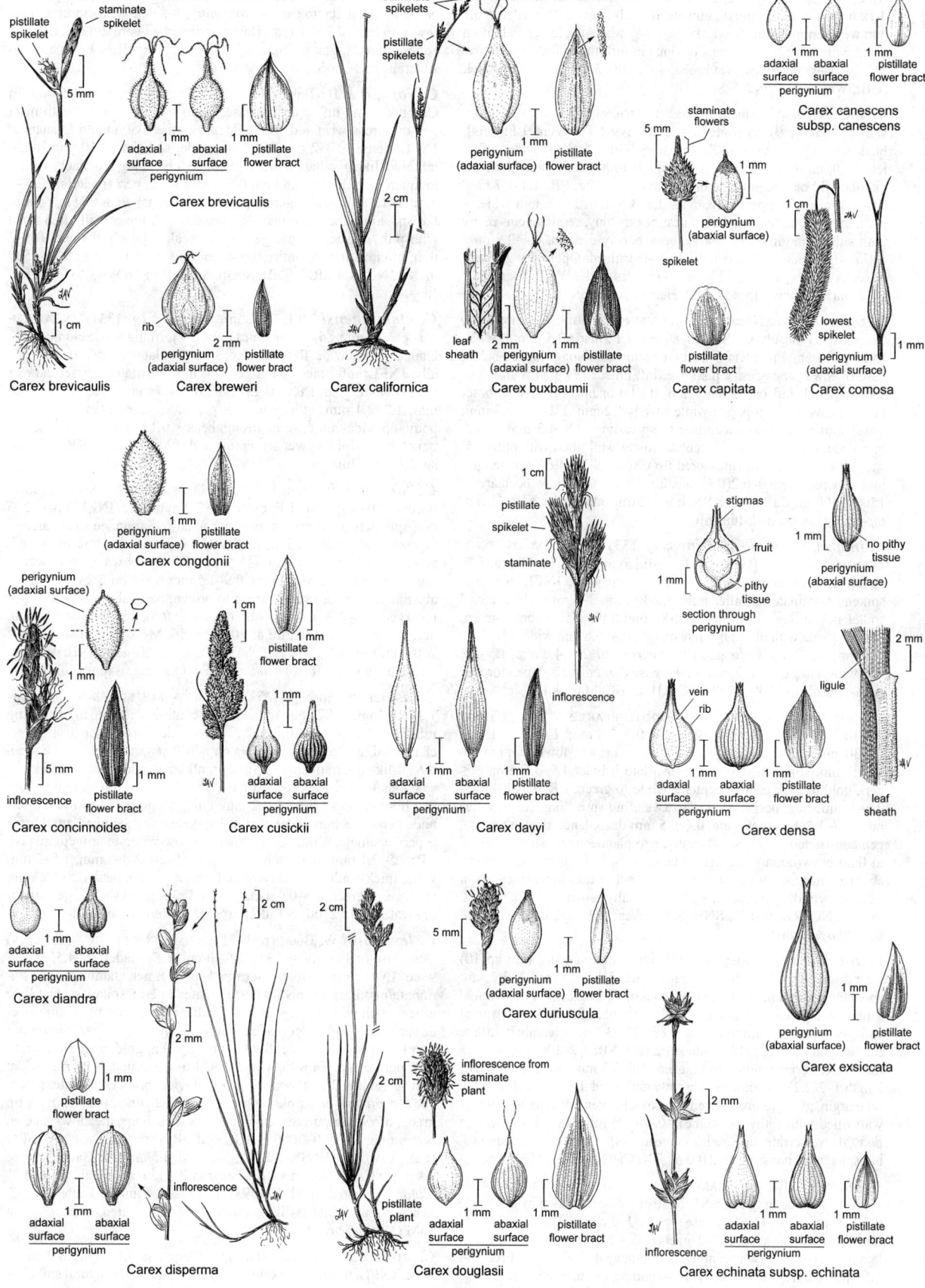

pistillate spikelet / staminate spikelet
5 mm
adaxial surface / abaxial surface — perigynium
1 mm pistillate flower bract
Carex brevicaulis

1 cm
Carex brevicaulis

rib
perigynium (adaxial surface)
2 mm
pistillate flower bract
Carex breweri

staminate spikelets / pistillate spikelets
2 cm
Carex californica

perigynium (adaxial surface)
1 mm
pistillate flower bract
Carex buxbaumii

leaf sheath
2 mm perigynium (adaxial surface)
1 mm pistillate flower bract

adaxial surface / abaxial surface — perigynium
1 mm
1 mm pistillate flower bract
Carex canescens subsp. canescens

staminate flowers
5 mm
perigynium (abaxial surface)
1 mm
spikelet
pistillate flower bract
Carex capitata

1 cm
lowest spikelet
perigynium (adaxial surface)
1 mm
Carex comosa

perigynium (adaxial surface) / pistillate flower bract
1 mm
Carex congdonii

perigynium (adaxial surface)
1 mm
inflorescence
5 mm
pistillate flower bract
1 mm
Carex concinnoides

1 cm
pistillate flower bract
1 mm

adaxial surface / abaxial surface — perigynium
1 mm
Carex cusickii

adaxial surface / abaxial surface — perigynium
1 mm
pistillate flower bract
1 mm
Carex davyi

1 cm
pistillate spikelet
staminate
inflorescence

stigmas
fruit
1 mm
pithy tissue
1 mm
section through perigynium

no pithy tissue
perigynium (abaxial surface)
1 mm

vein / rib
adaxial surface / abaxial surface — perigynium
1 mm
pistillate flower bract
1 mm
ligule
2 mm
leaf sheath
Carex densa

adaxial surface / abaxial surface — perigynium
1 mm
Carex diandra

pistillate flower bract
1 mm

adaxial surface / abaxial surface — perigynium
1 mm
2 cm
2 mm
inflorescence
Carex disperma

2 cm
2 cm
inflorescence from staminate plant
pistillate plant
adaxial surface / abaxial surface — perigynium
1 mm
pistillate flower bract
1 mm
Carex douglasii

5 mm
perigynium (adaxial surface)
1 mm
pistillate flower bract
Carex duriuscula

perigynium (abaxial surface)
1 mm
pistillate flower bract
Carex exsiccata

2 mm
inflorescence
adaxial surface / abaxial surface — perigynium
1 mm
pistillate flower bract
1 mm
Carex echinata subsp. echinata

wide; peri spreading, 2.5–3.5 mm, 1–1.5 mm wide, planoconvex, dark brown, ± black above, unwinged but body with thin edge < 0.1 mm wide, entire, veins 0–6 abaxially, 0(6) adaxially, lower wall often filled with pithy tissue, beaks obvious in infl, tip cylindric, ± entire. 2*n*=64. Marshes, bogs, wet meadows; 2100–3400 m. CaRH, SNH, SnBr, Wrn; w N.Am. Jul–Sep

C. incurviformis Mack. (p. 1329) MOUNT DANA OR INCURVED SEDGE (Group 9) Rhizome long, gen wavy. **ST:** < 6 cm. **LF:** > infl; blade 0.5–1.5 mm wide. **INFL:** dense, 6–9 mm, 5–8 mm wide; spikelets < 10, indistinct; infl bract like pistillate fl bract in length, color; pistillate fl bract acuminate, brown, margin white. **FR:** 1.4–1.7 mm, 0.8–1.3 mm wide; peri spreading, 2.8–3.5 mm, 1–1.6 mm wide, > pistillate fl bract, brown, darker near beak, shiny, sessile, veins many both sides, margin entire above, beak poorly defined, 0.4–0.9 mm, entire, tip unnotched, narrowly white-margined. Open, dry gravelly or rocky slopes, alpine; 3700–4000 m. c&s SNH, W&I; AK to CO. [*C. i.* var. *danaensis* (Stacey) F.J. Herm.] Jul–Aug

C. infirminervia Naczi WEAKLY VEINED SEDGE (Groups 7,10) Cespitose to loosely so. **ST:** 20–80 cm. **LF:** blade 2.5–5 mm wide. **INFL:** 2–4 cm; spikelets 5–15 mm, round to oblong, jagged-sided, base of lower ones gen ± pistillate (infl unisexual), gen rounded or short-tapered; infl bract gen < infl, lf-like or not; pistillate fl bract gen not covering fr top, gen white, awn < 1.2 mm. **FR:** 1.5–1.9 mm, 1–1.3 mm wide; peri ascending to spreading, 2.8–4.5 mm, 1–1.5 mm wide, green, narrow-lanceolate, lower wall filled with pithy tissue, beak 1.5–2.2 mm (measured from fr top), 39–49% peri length, long-tapered, teeth 0–0.2(0.4) mm. 2*n*=54. Moist soil, wooded areas; 1500–2200 m. CaRH, n&s SNH; w N.Am. Intergrades with *C. leptopoda*; study needed. Jun–Aug

C. inops L.H. Bailey subsp. **inops** (p. 1331) VOLCANO OR LONG-RHIZOMED SEDGE (Group 2) Long-rhizomed. **ST:** 10–50 cm. **LF:** equaling to exceeding infl; blade 1–3 mm wide. **INFL:** terminal spikelet staminate on taller infls, 12–30 mm, 2–4 mm wide; lateral spikelets pistillate, 1–4, oblong to ± round; pistillate fl bract green, red, or purple with obvious white margin 0.4–0.8 mm wide. **FR:** 1.5–2.5 mm, 1.1–2 mm wide, gen white; peri 4–20, 2.6–4.8 mm, 1.5–2.2 mm wide, strong veins 2, veins otherwise 0 or faint. Dry open forests, edges, meadows; 700–2000 m. CaRH; to BC. May–Jul

C. integra Mack. (p. 1331) SMOOTH-BEAKED SEDGE (Group 11C) **LF:** blade 1–2 mm wide; ligule 0.5–2.5 mm. **INFL:** ± dense, 15–30 mm, gold to brown or occ dark brown; lowest internode 2–9.5 mm; spikelets ± distinct; pistillate fl bract 1.9–3.5 mm, ± ≤ peri, gold to dark brown, center ± white to green. **FR:** 1.1–1.4 mm, 0.6–1 mm wide; peri appressed to ascending-spreading, 2.2–3.3(3.6) mm, 0.7–1.2(1.4) mm wide, 0.4–0.5 mm deep, lance-ovate to ovate, gen constricted near base when dry, gen planoconvex, straw-colored to light brown, margin green distally, wing < 0.1 mm wide, entire, abaxial veins 0–4, weak, adaxial 0–1, ± = fr, beaks inconspicuous in infl, tip cylindric, entire. 2*n*=82. Seasonally moist soil; 1000–3400 m. KR, NCoRH, CaRH, SNH, SnBr, Wrn; OR, NV. Like *C. illota*, *C. subfusca*. Jun–Sep

C. interior L.H. Bailey (p. 1331) INLAND SEDGE (Group 10) Cespitose. **LF:** blade 0.6–2.7 mm wide. **INFL:** open, 1–3.7 cm; spikelets 3–20 mm, outline ± jagged at arm's length, base of terminal 1 narrow-tapered; infl bract << infl; staminate fl bracts of terminal spikelet appressed, internodes between 2/3–3/4 their length; pistillate fl bract gen obtuse, gold, white-margined. **FR:** 1.2–1.8 mm, 0.9–1.5 mm wide; peri spreading to reflexed, 1.9–3.3 mm, widest ± at base, length 1–2(2.2) × width, gen abruptly narrowed to beak, with ± bulge on margin, green to brown, veined abaxially, gen not adaxially, lower wall filled with pithy tissue, beak 0.4–0.95 mm, 18–44% length of body, dense-serrate, tip ± red. Wet meadows, shores, often calcareous, not sphagnum bogs; 1100–2100 m. KR, CaRH; to AK, ME, Mex. Jul

C. inversa R. Br. KANGAROO OR KNOB SEDGE (Group 10) Rhizomed, often forming tufts. **ST:** 5–50 cm. **LF:** exceeded by sts; blade 0.7–3.8 mm wide. **INFL:** dense (open), 1–2 cm; spikelets 5–10 mm, oblong, staminate at base; 1–2 bracts (exc fl bracts) gen >> infl; pistillate fl bracts < peri, white, midvein green, awns 0–0.4 mm. **FR:** 1.3–1.5 mm, 1.2–1.3 mm wide; peri ascending or spreading-ascending,

2.4–4 mm, 1–2 mm wide, planoconvex, body elliptic, sharp-edged or ± winged, green to pale brown, shiny, 3–8 veins > fr, abruptly narrowed to beak 0.7–1.1 mm. Damp or disturbed bare ground, riparian zones, thickets, grassy areas; < 750 m. ScV, PR; HI, w Eur; native to Australasia. Mar–Sep

C. jonesii L.H. Bailey (p. 1331) JONES' SEDGE (Group 9) Cespitose. **LF:** tufted near st base; blade 1–3 mm wide; sheath front gen not cross-wrinkled, gen hidden by other lvs, mouth U-shaped. **INFL:** dense, 0.8–2 cm, 6–12 mm wide; spikelets < 10, indistinct; infl bract like pistillate fl bract; pistillate fl bract gold to black, obtuse to mucronate. **FR:** 1–1.8 mm, 0.5–1 mm wide; peri spreading, 2.5–4 mm, 1–1.5 mm wide, gen widest ± at base, tapered above, gold to brown, shiny, abaxial veins 7–11, adaxial 5–7, lower wall often filled with pithy tissue, margin gen entire, beak ± poorly defined, 0.5–2 mm, obvious in infl, entire, tip unnotched. Moist places; 900–3500 m. KR, NCoR, CaRH, SNH, SnGb, SnBr, Wrn; to WA, Rocky Mtns. Jun–Sep

C. klamathensis B.L. Wilson & Janeway (p. 1331) KLAMATH SEDGE (Groups 4,6) Rhizomes long, pls forming dense clumps. **LF:** blade 2–6 mm wide, flat or folded, strongly glaucous, papillate abaxially. **INFL:** infl bract sheath gen > 4 mm; pistillate fl bract dark or red-brown to gold. **FR:** 1.6–2.7 mm, 0.7–1.7 mm wide; peri 1.7–3.6 mm, 1.2–2.4 mm wide, obovoid, papillate, green-glaucous or tan, body top wide-conic, veins strong, beak < 0.15 mm, bent to side, tip unnotched. Moist to wet serpentine soils; 900–1600 m. KR, NCoRI; sw OR. Jun–Jul ★

C. laeviculmis Meinsh. (p. 1331) SMOOTH-STEMMED SEDGE (Group 10) Cespitose. **LF:** blade 0.9–2.3 mm wide. **INFL:** open, 2–6 cm; spikelets 3–10 mm, base of terminal one rounded to ± narrow-tapered; infl bract << infl; pistillate fl bract gold, white-margined, acute, midrib raised. **FR:** 1.3–1.5 mm, 0.7–1 mm wide; peri ascending or spreading, 2.3–3.7 mm, 0.9–1.5 mm wide, body ovate, veined abaxially, often adaxially, green to brown, beak obvious in infl, occ recurved, 0.5–1.3 mm (measured from fr top), 15–40% peri length, narrow-conic, often entire at 10×. 2*n*=56. Moist soil in woodland; 700–1800 m. KR, NCoRH, NCoRO, s CaRH, n&c SNH, Wrn; to AK, Rocky Mtns. [*C. brunnescens* (Pers.) Poir. misappl.] Jun–Sep

C. lasiocarpa Ehrh. (p. 1331) WOOLLY-FRUITED SEDGE (Groups 1,2) Rhizomed. **ST:** 60–120 cm. **LF:** cauline blade much exceeding infl or basal minute, 0.7–2(2.2) mm wide, inrolled to triangular-channeled, glabrous (hairy), esp on non-fl sts prolonged into curled, thread-like tip; midvein of lf (and of infl bract) a low, rounded, inconspicuous keel at least toward base; sheath mouth often ciliate, front gen fibrous; upper ligules < 2 mm, thick, tough. **INFL:** lowest internode gen < 0.8 mm wide; pistillate spikelets sessile; pistillate fl bract > peri, ± ciliate, ± purple or brown, narrow-white-margined, awned. **FR:** 1.5–2.1 mm, 1–1.5 mm wide; peri 15–50, 2.8–5 mm, 1.5–2 mm wide, thick-walled, ± many-ribbed, gen ± purple, beak 0.5–1.2 mm, teeth gen ± erect, 0.2–0.9 mm. 2*n*=54. Lake, pond shores, gen standing water; 600–2100 m. CaRH, n SNH (s Plumas Co.). Jul–Aug ★

C. lemmonii W. Boott (p. 1331) LEMMON'S SEDGE (Group 4) Cespitose or loosely so. **ST:** 20–80 cm. **LF:** blade 1.5–4.5(7) mm wide. **INFL:** gen < 40 cm; terminal spikelet gen staminate, occ staminate/pistillate or pistillate/staminate; lateral spikelets pistillate at least below, longest < 20 mm, stalks gen exserted < 1 cm, erect or spreading; infl bract blade << infl, sheath long, linear, mouth not purple-banded; pistillate fl bract white, green, gold, or red-brown to dark purple, margin wide, white. **FR:** 1.5–1.9 mm, 0.9–1.4 mm wide; peri spreading, 2.7–4.5 mm, 1–1.7 mm wide, body lanceolate, green, occ red-brown- or purple-blotched, beak indistinct, 0.6–1.2 mm, tip green, brown, or purple, gen ciliate. Marshes, bogs, meadows, occ on serpentine; (30)700–3000 m. KR, s NCoRI (Sonoma Co.), s NCoRO (Lake Co.), CaRH, SNH, SnBr, SnJt. [*C. albida* L.H. Bailey] Some pls in s NCoRI (Sonoma Co.), SnBr in lf width approach, exceed *C. albida* as treated in TJM (1993); in Lassen, Plumas, Butte cos., *C. lemmonii* has white pistillate fl bracts, peri attributed to *C. albida* in TJM (1993). May–Aug

C. lenticularis Michx. (Group 8) Cespitose. **LF:** blade 1–3.5 mm wide. **INFL:** terminal spikelet rarely pistillate above; lateral spikelets

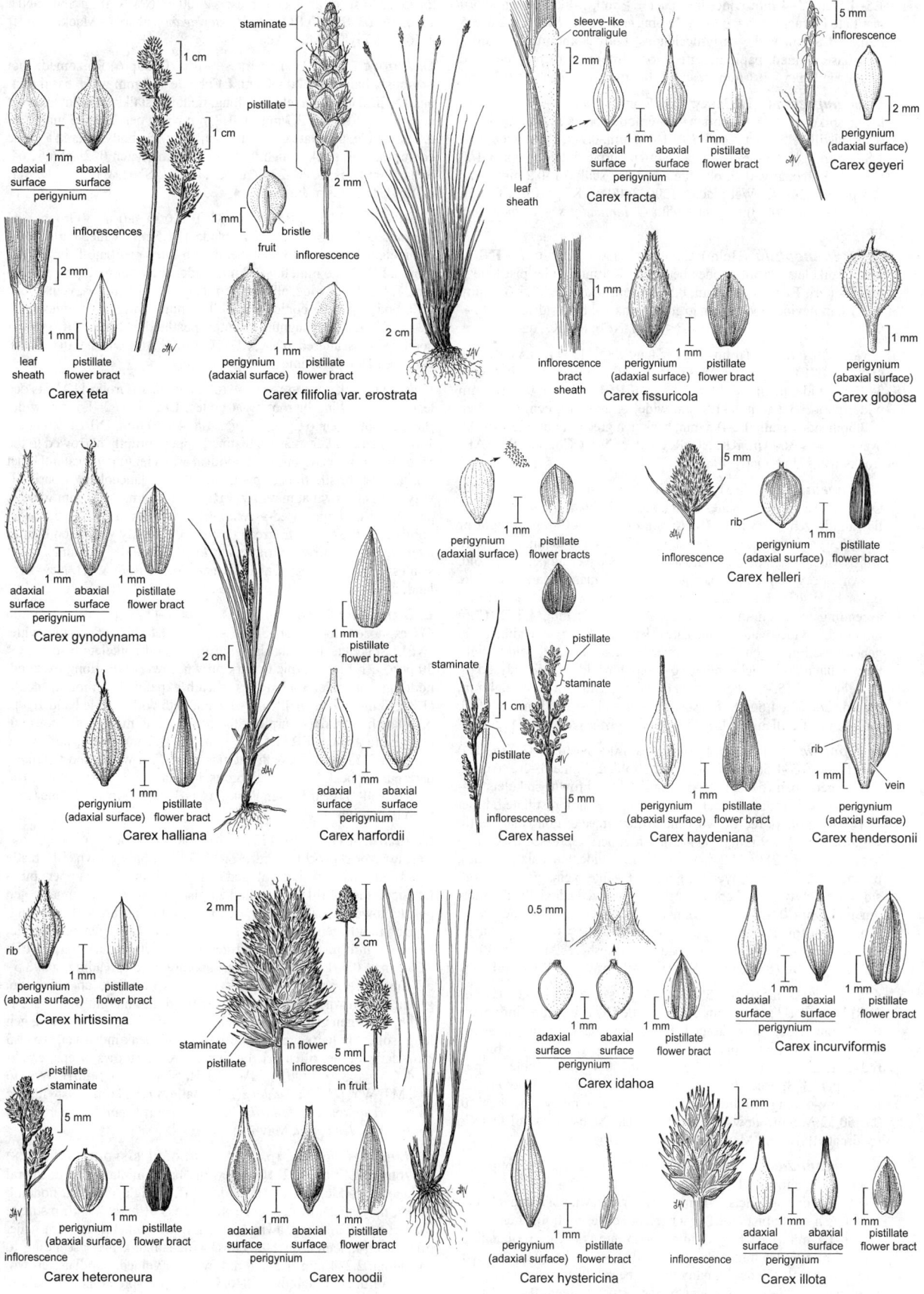

Carex feta

Carex filifolia var. erostrata

Carex fracta

Carex geyeri

Carex fissuricola

Carex globosa

Carex gynodynama

Carex halliana

Carex harfordii

Carex hassei

Carex helleri

Carex haydeniana

Carex hendersonii

Carex hirtissima

Carex heteroneura

Carex hoodii

Carex idahoa

Carex incurviformis

Carex hystericina

Carex illota

1.5–4.5 cm, 3–4 mm wide; infl bract ± ≥ infl, lf-like, sheath short; pistillate fl bract << to = peri, wide-obtuse or acute. **FR:** peri 1.8–3.5 mm, 1–1.8 mm wide, early-deciduous, body ± abruptly narrowed near base, veined, papillate, green, or white over fr, purple-dotted or not, stalk gen > 0.1 mm, beak 0.1–0.5 mm, tip notched < 0.1 mm.

var. ***impressa*** (L.H. Bailey) L.A. Standl. TARN SEDGE **INFL:** lower spikelets < or less often > internodes between, gen 3–4 mm wide; pistillate fl bract < peri. **FR:** 1–1.4 mm, 0.9–1.3 mm wide; peri 1.8–3 mm, 1.1–1.5 mm wide, body often purple-dotted in upper 1/2, veins 1–3 abaxially, 0–2, obscure adaxially, stalk 0.1–0.2 mm, beak gen purple. $2n=92$. Wet places; 1200–3000 m. KR, NCoR, CaRH, SNH, Wrn; to WA. Intergrades with *C. lenticularis* var. *lipocarpa*. Jun–Sep

var. ***limnophila*** (Holm) Cronquist LAGOON SEDGE **INFL:** lower spikelets >> internodes between, 4–6 mm wide; pistillate fl bract = peri. **FR:** 1.1–1.5 mm, 0.8–1.6 mm wide; peri 2.5–3.5 mm, 1.3–1.6 mm wide, body green to gold, veins 5–7 both sides, stalk 0.4–0.7 mm, beak purple. Wet places; < 20 m. NCo; to AK. Jun–Aug ★

var. ***lipocarpa*** (Holm) L.A. Standl. (p. 1331) LAKESHORE SEDGE **INFL:** lower spikelets < or less often > internodes between, 3–4 mm wide; pistillate fl bract < peri. **FR:** 1–1.7 mm, 0.8–1.5 mm wide; peri 2–3.2 mm, 1–1.8 mm wide, body gen green, veins gen 5–7 both sides, stalk 0.2–0.5 mm, beak gen green, tip purple. $2n=92$. Wet places; < 3600 m. KR, NCoR, CaRH, SNH, CCo, SnFrB; to AK, Rocky Mtns. Jun–Oct

C. leporina L. (p. 1331) HARE OR OVAL SEDGE (Group 11D) Occ late season lfy branches from nodes. **LF:** blade 1.5–4 mm wide, flat; ligule gen < 2.5 mm. **INFL:** gen open, 15–40 mm, straight or bent; lowest internode gen > 4 mm; spikelets gen distinct, ovate to wide-ovate, base tapered; pistillate fl bract ± covering peri, red-gold, brown, or ± green, center pale gold or green, margin dark exc white tip or base. **FR:** 1.1–1.8(2) mm, 0.9–1.2 mm wide, ± sessile; peri ascending to ascending-spreading, 3.4–4.7(5.2) mm, 1.3–2.1(2.5) mm wide, wide-ovate (lanceolate), length 1.8–2.6 × width, body gen wide-elliptic, gen ± flat exc over fr (planoconvex), gold to light brown, flat margin incl wing gen 0.4–0.6 mm wide at midbody, adaxial veins gen 2–5, = fr, beak flat or cylindric, entire for < or > 0.4 mm, brown. $2n=62,64,66,68$. Seasonally wet soil; < 1200 m. NCoRO, SnFrB; to BC; alien in e US, Eurasia. [*C. ovalis* Gooden] May–Aug

C. leporinella Mack. (p. 1331) BOG HARE SEDGE (Group 11D) **LF:** blade 0.5–1.5(2) mm wide, gen folded; ligule 1–2(2.5) mm. **INFL:** gen open (dense), 15–35 mm, gold to brown; spikelets distinct, gen fusiform; infl bract gen bristle-like, < infl; pistillate fl bract gen > peri, gold or red-brown, center often green or gold, margin, tip white. **FR:** 1.4–1.9 mm, 0.7–1 mm wide; peri appressed-ascending to ascending, (3.2)3.5–4.2 mm, 0.8–1.2 mm wide, lanceolate (ovate), planoconvex (flat exc over fr), gold to ± white, veins gen 3–7 both sides, shallowly boat-shaped, wings gen incurved adaxially, flat margin incl wing 0.05–0.1(0.2) mm wide, tapered to indistinct beak, beaks inconspicuous in infl, cylindric, entire for > 0.4 mm, gold, ± red, or brown. Moist or wet meadows, lakeshores; 1900–4000 m. KR, SNH, SNE (Sweetwater Mtns), W&I; to WA, MT, UT. Jul–Sep

C. leptalea Wahlenb. (p. 1331) BRISTLE-STALKED SEDGE (Group 3) Rhizomed. **ST:** 10–40 cm. **LF:** exceeded by to equaling infl; blade 0.5–1 mm wide, flat or folded. **INFL:** spikelet bisexual, 2–3 mm wide; spikelet bract 0; pistillate fl bract < 1/2 peri, white to ± brown, red-dotted, persistent. **FR:** 1.3–1.8 mm, 0.8–1 mm wide, < peri body; peri 3–8, ascending, 2.5–4.5 mm, 0.9–1.5 mm wide, elliptic to narrow-oblong-elliptic, green to straw, veins many, fine, beak 0. $2n=50,52$. Wet meadows, swamps; < 700 m. NCo, KR, NCoRO, CCo (extirpated); N.Am, Mex, Caribbean. Jun–Aug ★

C. leptopoda Mack. SLENDER-FOOTED SEDGE (Group 10) Loosely cespitose. **ST:** 20–80 cm. **LF:** blade 2.4–5.9 mm wide. **INFL:** 2–4 cm; spikelets 5–15 mm, longer than wide; infl bract lf-like or not, gen < infl; pistillate fl bract gen not covering fr top, gen white, awn < 1.2 mm. **FR:** 1.5–1.9 mm, 1–1.3 mm wide; peri ascending, 2.8–4.5 mm, 1–1.5 mm wide, gen length < 3 × width, green, lanceolate, lower wall filled with pithy tissue, beak 0.9–1.7 mm (measured from fr top), 28–38% peri length, long-tapered, teeth 0.1–0.3 mm.

$2n=54$. Moist soil, wooded areas; < 2400 m. NW, CaR, n SNF, SNH, CCo, SnFrB, SW; to BC, MT. [*C. deweyana* subsp. *l.* (Mack.) Calder & Roy L. Taylor] May–Aug

C. limosa L. (p. 1331) MUD SEDGE (Group 6) Rhizomed; root obviously hairy. **ST:** 20–60 cm. **LF:** blade 1–3 mm wide, basal minute. **INFL:** lateral spikelets on long, nodding stalks; pistillate fl bract gen brown. **FR:** 1.5–2.7 mm, 1–1.8 mm wide; peri 2.3–4.5 mm, 1.6–2.6 mm wide, papillate above, brown-glaucous, body tapered to wide conic tip, wall thick, tough, beak < 0.2 mm or gen 0. $2n=56,62,64$. Sphagnum bogs; 1200–2700 m. KR, CaRH, SNH, Wrn; to AK, MT, e Can, UT; Eurasia. Jul–Sep ★

C. livida (Wahlenb.) Willd. (p. 1331) LIVID SEDGE (Groups 4,6) Rhizomed. **ST:** 15–60 cm. **LF:** blade 1–3.5 mm wide, ± glaucous, channeled. **INFL:** infl bract sheath < 6 mm; pistillate fl bract red-brown. **FR:** 2–3.5 mm, 0.9–1.8 mm wide, occ 2 per peri; peri 3.5–4.6 mm, 1.2–2.4 mm wide, elliptic, papillate, green-glaucous, veins weak or 0, body tip wide-conic, beak < 0.15 mm, straight, tip unnotched. $2n=32,50,52,76$. Sphagnum swamps, peatland; < 100 m. NCo (Mendocino Co.); scattered to AK, e US, C.Am, S.Am, Eurasia. In CA, last collected 1866, presumed extirpated. May ★

C. longii Mack. GREEN-AND-WHITE SEDGE (Group 11A) Sts occ leaning, branching or rooting at nodes. **LF:** blade 2–4.5 mm wide; sheath front green to tip; ligule gen 1.8–4.4(6) mm. **INFL:** gen open, 1–4.5(6) cm; spikelets 3–10, distinct, upper abruptly narrowed to the base (± club-shaped), outline smooth at arm's length; lowest infl bract < infl, often bristle-tipped; pistillate fl bracts lanceolate, < peri, silvery to light brown at maturity. **FR:** 1.3–1.7 mm, 0.7–1 mm wide, < peri body; peri appressed-ascending, 3–4.6 mm, 1.6–2.8 mm wide, obovate, flat exc over fr, green to brown, strongly veined both sides over fr, beak flat, wide-triangular. $2n = 58,60,62$. Seasonally wet swales; 200 m. NCoRI, CaRF (Shasta Co.); OR, WA, HI, New Zealand; native to e US, s Great Plains, Mex, C.Am. May–Jun

C. luzulifolia W. Boott (p. 1331) LITTLE-LEAVED SEDGE (Group 4) Cespitose to loosely so. **ST:** 40–100 cm. **LF:** blade 5–20 mm wide. **INFL:** spikelets 1–3 cm, dark purple; lateral spikelets pistillate, < 30 peri, > 1.2 cm, purple to dark brown, lower stalks long-exserted, nodding; infl bract ± bladeless, sheath expanded to mouth, mouth (1.8)2–3 mm wide, shallow-U-shaped, with wide purple band at top; pistillate fl bract dark purple with narrow white margin, glabrous but occ ciliate-awned. **FR:** 1.5–2.2 mm, 0.8–1.2 mm wide; peri 4–5.3 mm, 1.7–2.5 mm wide, with flat margin > 1/2 fr width, short-stalked, dark purple, beak 1–2 mm, glabrous. Montane meadows; 1900–3100 m. KR, CaRH (near Lassen Peak), n SNH, c SNH (n Tuolumne Co.); to MT. Jul–Sep

C. luzulina Olney (p. 1331) WOODRUSH SEDGE (Groups 2,4) Cespitose or short-rhizomed. **ST:** 15–90 cm, base brown. **LF:** blade 1.5–10 mm wide, well developed on lower fl sts. **INFL:** upper spikelets occ clustered, lateral spikelets pistillate or tips staminate, gen purple, lowest stalk exserted > 10 (> 5) mm; infl bract sheath long, linear, mouth gen shallow-U-shaped or truncate, gen not purple, blade << infl; pistillate fl bract gen red-brown to dark purple. **FR:** 1.3–2 mm, 0.7–1.1 mm wide; peri ascending to spreading, gen 3.5–5.5 mm, 0.9–1.6 mm wide, body ovate to lanceolate, tapered to top, green, red-brown, or dark purple, 2-ribbed, flat margin 0 or < 1/2 fr width, faces gen glabrous or with sparse, spreading to appressed, gen long, soft, thin hairs, gen with raised veins, beak indistinct, 0.5–1.5 mm, dark purple, often bristly-ciliate. Wet meadows, seeps, creekbanks, bogs; 800–3200 m. NW, CaRH, SNH, CCo, SnFrB, Wrn; to BC, MT, WY. [*C. l.* var. *ablata* (L.H. Bailey) F.J. Herm.] Pls with occ hairy peri approach *C. fissuricola*; pls with purple peri but short beaks approach *C. lemmonii*. May–Sep

C. lyngbyei Hornem. (p. 1331) LYNGBYE'S SEDGE (Group 8) Rhizomed. **LF:** blade 2–10 mm wide, basal minute. **INFL:** lateral spikelets 1.8–5 cm, 5–8 mm wide, stiff, straight on long, nodding stalks, tip gen staminate, base ± truncate; infl bract > infl; pistillate fl bract tip often white, awned. **FR:** 1.7–2.5 mm, 1.2–1.8 mm wide, often deep-indented on 1–2 sides, wide-stalked; peri ascending to spreading, 2.2–4 mm, 1.5–2.8 mm wide, ± veined, papillate, brown, dull, wall thick, tough, beak 0–0.3 mm, tip glabrous, notch shallow

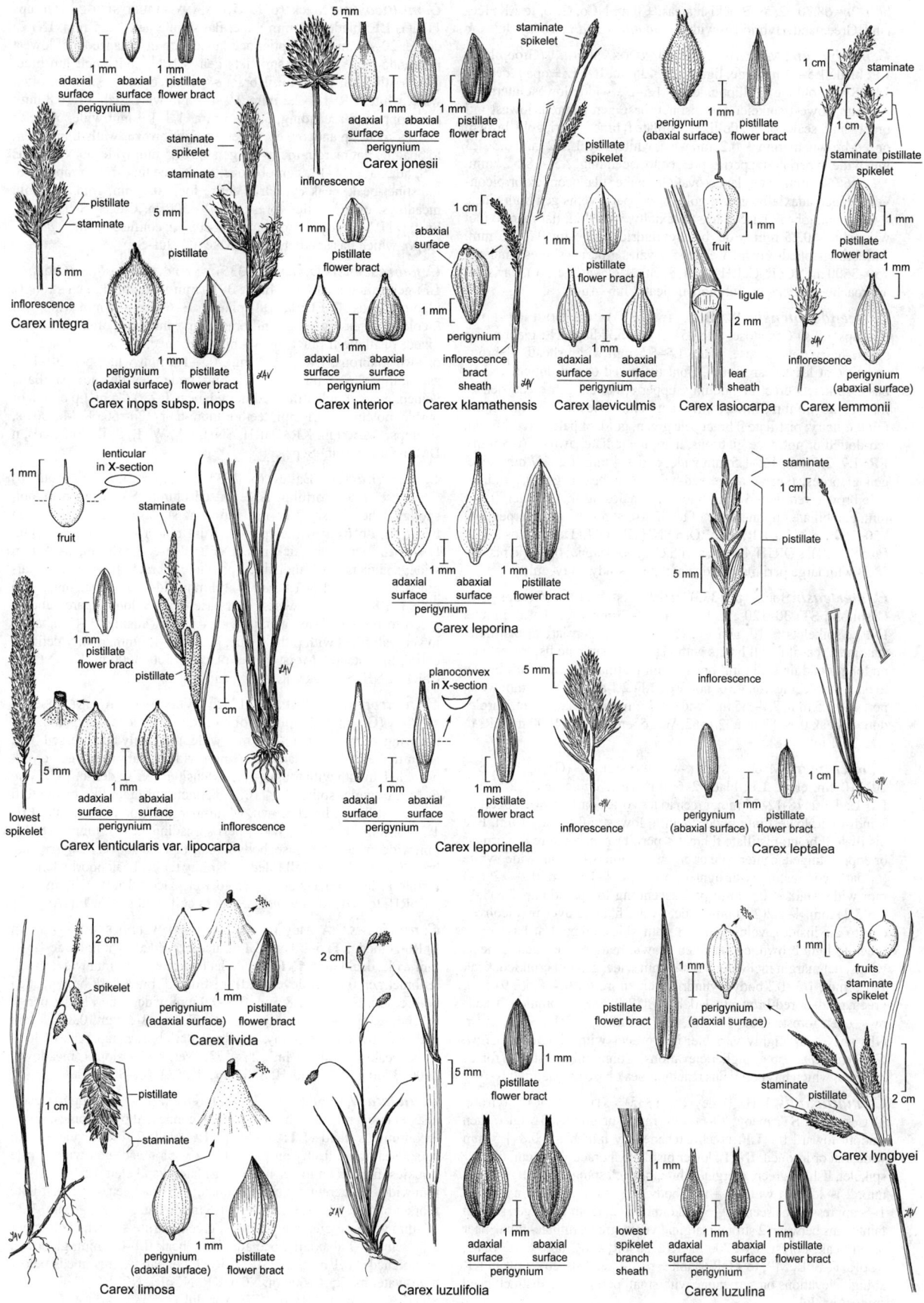

adaxial surface · abaxial surface · 1 mm · pistillate flower bract · perigynium

inflorescence

pistillate · staminate

5 mm

inflorescence

Carex integra

staminate spikelet

staminate

5 mm

pistillate

perigynium (adaxial surface) · 1 mm · pistillate flower bract

Carex inops subsp. inops

5 mm

inflorescence

adaxial surface · abaxial surface · 1 mm · pistillate flower bract · perigynium

Carex jonesii

1 mm · pistillate flower bract

adaxial surface · abaxial surface · 1 mm · perigynium

Carex interior

staminate spikelet

pistillate spikelet

abaxial surface · 1 cm

1 mm · perigynium · inflorescence bract sheath

Carex klamathensis

1 mm · pistillate flower bract

adaxial surface · abaxial surface · 1 mm · perigynium

Carex laeviculmis

fruit · 1 mm · ligule · 2 mm · leaf sheath

perigynium (abaxial surface) · 1 mm · pistillate flower bract

Carex lasiocarpa

1 cm · staminate

1 cm · staminate pistillate spikelet

1 mm · pistillate flower bract

inflorescence

perigynium (abaxial surface) · 1 mm

Carex lemmonii

1 mm · fruit · lenticular in X-section

staminate

1 mm · pistillate flower bract · pistillate

1 cm

lowest spikelet · 5 mm · adaxial surface · abaxial surface · 1 mm · perigynium · inflorescence

Carex lenticularis var. lipocarpa

adaxial surface · abaxial surface · 1 mm · pistillate flower bract · perigynium

Carex leporina

planoconvex in X-section

adaxial surface · abaxial surface · 1 mm · pistillate flower bract · perigynium

5 mm · inflorescence

Carex leporinella

staminate · 1 cm · pistillate · 5 mm

inflorescence

perigynium (abaxial surface) · 1 mm · pistillate flower bract · 1 cm

Carex leptalea

2 cm · spikelet · perigynium (adaxial surface) · 1 mm · pistillate flower bract

Carex livida

2 cm · pistillate · 1 cm · staminate · perigynium (adaxial surface) · 1 mm · pistillate flower bract

Carex limosa

2 cm · 5 mm · 1 mm · pistillate flower bract · adaxial surface · abaxial surface · 1 mm · perigynium

Carex luzulifolia

pistillate flower bract · 1 mm · perigynium (adaxial surface) · 1 mm · fruits staminate spikelet

staminate · pistillate · 2 cm

Carex lyngbyei

lowest spikelet branch sheath · 1 mm · adaxial surface · abaxial surface · 1 mm · perigynium · 1 mm · pistillate flower bract

Carex luzulina

or 0. $2n=68,70,72,76$. Brackish areas; ± 0 m. NCo, CCo; to AK; Iceland, Greenland. Hybridizes with *C. aquatilis* var. *dives*. May–Jul ★

C. mariposana Mack. (p. 1335) MARIPOSA SEDGE (Group 11H) **LF**: blade 1.5–4 mm wide; ligule gen < 3 mm. **INFL**: ± open (dense), 14–48 mm, oblong to elliptic, length 1.5–2 × width; lowest internode ≥ 2 mm; lowest spikelet gen distinct, base gen tapered; lowest infl bract < infl, scale- or bristle-like; pistillate fl bract < peri, gold, brown, or red-brown, margin 0–0.2 mm wide, white. **FR**: 1.5–2 mm, 0.8–1.3 mm wide, = peri body; peri appressed to ascending, 2.5–4.8(5.3) mm, 1.1–1.6(2.2) mm wide, lance-ovate to ovate, planoconvex or biconvex, bulged adaxially, green, gold, or coppery, veins gen both sides, > fr, reaching beak base at least adaxially, wall thin, flat margin incl wing gen < 0.25 mm wide, beak cylindric, entire for 0.5–0.6 mm, dark brown to red-brown. Meadows, swales, riparian shores, thickets; 750–3600 m. NCoR, CaRH, SNH, SnBr; NV (Washoe Co.). Variable, approaches *C. abrupta* but more delicate. Jun–Aug

C. mendocinensis W. Boott (p. 1335) MENDOCINO SEDGE (Groups 1,2,4) Cespitose. **ST**: 25–80 cm, nodding. **LF**: glabrous or sparse-hairy toward base; blade 1.5–5.5 mm wide; sheath glabrous or hairy, at least near mouth, front often red or red-dotted. **INFL**: basal spikelets erect or nodding, upper spikelets gen erect, dense-fld, (15)20–60 mm, linear; infl bract < infl, sheath 2–3 cm, mouth minute-hairy; pistillate fl bract pale green, gold, or pale to red-brown, red-dotted or not, face glabrous, margin ± ciliate, awns 0 (0.3 mm). **FR**: 1.7–2.5 mm, 1.2–1.5 mm wide; peri < 4 mm, 1.2–1.7 mm wide, gen glabrous, (sparse-appressed-hairy near beak), green, gold, or pale brown, red-dotted or not, loosely enclosing fr, beak gen 0.5–1 mm, gen ciliate on inner side of teeth. Moist areas, often serpentine; 150–1600 m. NCo, KR, NCoRO, n SNH (Butte, El Dorado cos.), CW (exc SCoRI); s OR. Hybridizes with *C. gynodynama*; open-fld pls in n SNH with large peri may be distinct, need study. May–Jul

C. mertensii Bong. (p. 1335) MERTENS' SEDGE (Groups 4, 6) Cespitose. **ST**: 30–120 cm. **LF**: blade 4–7 mm wide, basal minute. **INFL**: spikelets 6–10, gen > 2 cm, terminal, gen lateral pistillate/staminate, basal few fl bracts with vestigial staminate fls; lowest gen on long, nodding stalk; infl bract > infl; pistillate fl bract dark brown to dark purple, mucronate or not. **FR**: 1.6–2.1 mm, 0.7–1.1 mm wide; peri 3.9–5.5 mm, 2.1–3.5 mm wide, > 3 × fr width, flat, green, purple-dotted, beak 0.1–0.7 mm. $2n=62$. Wet open areas; ± 1500 m. KR; to AK, MT. Jul–Aug

C. microptera Mack. SMALL-WINGED SEDGE (Group 11H) **ST**: 20–110 cm, erect. **LF**: blade 2–6 mm wide; ligule gen < 2.5 mm. **INFL**: dense, (8)12–25 mm, green to less often dark brown; spikelets ± indistinct, lowest internode < 2 mm; lowest infl bract << infl, like pistillate fl bract; pistillate fl bract < peri, brown, often red-, purple-, or copper-tinged, center pale or green, margin 0–0.1 mm wide, white (or dark, contrasting with lighter peri). **FR**: 1–1.6 mm, 0.7–1.2(1.4) mm wide, stalk < 0.5 mm; peri ascending to spreading, (2.8)3.4–4.5(5.2) mm, 1–2.2(2.4) mm wide, ovate, flat exc over fr (biconvex due to air inside), veins gen < 8 both sides, adaxial at base or 0, green to light brown, center often brown, margin color gen same as wings, flat margin incl wing 0.2–0.5 mm wide, gen not crinkled, beak 1–1.5 mm, 1/3–1/2 body, cylindric, gen entire for 0.4–0.5(0.9) mm, green, gold, or red-brown to dark brown. $2n=80,82$. Common. Meadows, open forest; 1500–3400 m. KR, NCoRH, CaRH, SNH, SnBr, GB; w N.Am. Highly variable; intergrades with *C. pachystachya*, *C. haydeniana* esp at higher elevations. Commonly mistaken for *C. abrupta*, which has peri veins reaching beak base adaxially. Jun–Aug

C. multicaulis L.H. Bailey (p. 1335) STICK SEDGE (Group 3) Cespitose. **ST**: many, 20–60 cm, round or bluntly 3-angled, gen smooth toward tip. **LF**: much exceeded by infl; blade 0.8–1.5 mm wide, flat or inrolled. **INFL**: lower pistillate fl bracts 1–15 cm, often > spikelet, lf-like, green, margins white, base clasping peri. **FR**: 4.5–5.4 mm, 2.1–2.3 mm wide, ± = peri body, next to bristle-like axis; peri 1–6, appressed to ascending, 5–7.2 mm, 2.2–2.5 mm wide, gen several faint veins between 2 strong marginal veins, green, minute-ciliate near or on beak. Forest; 15–2200 m. KR, NCoR, CaR, n&c SNF, SNH, SCoR, TR, PR, MP; to s OR. Habit ± rush-like, distinctive; pls in KR at high elevations on serpentine with small peri may be distinct, need study. Apr–Jul

C. multicostata Mack. (p. 1335) MANY-RIBBED SEDGE (Groups 11B,G) **LF**: blade 2–6 mm wide, flat; ligule gen < 2.5 mm. **INFL**: dense, 20–40 mm, ± elliptic, occ lanceolate or oblanceolate; lowest internode gen < 5 mm; spikelets gen < 10, ± indistinct; infl bract gen bristle-like; pistillate fl bract < peri, red-brown, margin often wide, white. **FR**: 1.7–2.2 mm, 0.9–1.4 mm wide, stalk 0.4–0.8 mm; peri appressed-ascending, 3.7–5.5 mm, 1.4–1.9 mm wide, ovate to narrow-ovate, planoconvex, green to light brown, walls translucent to opaque, flat margin incl wing gen 0.3–0.5 mm wide, abaxial veins > 8, adaxial (1)4–13, gen obvious, not exceeding fr, gen tapered to indistinct beak, beak cylindric, entire for < 0.4 mm, gold. Dry soil, meadows, open conifer forest; 1900–3500 m. CaRH, SNH, SnBr, SnJt; w NV. Large pls approach *C. specifica*; confused with *C. pachycarpa*, which has smaller, more ovate infl. Jul–Sep

C. nebrascensis Dewey (p. 1335) NEBRASKA SEDGE (Group 8) **LF**: gen tufted at st base; blade 3–12 mm wide, thick, gen strongly blue-glaucous. **INFL**: lateral spikelets 3–6 cm, 5–9 mm wide, gen 2-colored, pressing ± flat; infl bract gen > infl, base often dark-margined; pistillate fl bract often white-margined, obtuse to awned, tip or awn glabrous. **FR**: 1.3–2 mm, 0.9–1.8 mm wide; peri 30–150, ascending, 2.6–4 mm, 1.6–2.5 mm wide, sessile, not papillate, veined, gen ± tough, often red-dotted, beak 0.2–0.6 mm, tip brown or purple, notched > 0.1 mm, teeth minute-hairy. $2n=66,68$. Meadows, swamps; < 2500 m. KR, CaRH, SNH, GV, WTR, SnBr, SnJt, GB, n DMtns; w US. Jun–Sep

C. nervina L.H. Bailey (p. 1335) SIERRA SEDGE (Group 9) Cespitose, often forming large raised clump. **ST**: winged or not, spongy when fresh. **LF**: not tufted at st base; blade 3.5–5 mm wide; sheath back white-spotted, front not cross-wrinkled, mouth U-shaped, gen with thick rim. **INFL**: dense, 1.5–3 cm, 6–18 mm wide; spikelets < 10, indistinct; pistillate fl bract brown, white-margined, acute. **FR**: 1.5–1.8 mm, 1–1.3 mm wide; peri spreading, 3–4.5 mm, 1–1.8 mm wide, widest ± at base, body ± long-tapered above, green to brown, shiny, abaxial veins 12–15, adaxial 7–12, abaxial lower wall filled with pithy tissue, beak 0.8–1.7 mm, poorly defined, entire, tip notched. Moist to wet places; 1200–3000 m. KR, NCoR, CaRH, SNH, SnJV; s OR, w NV. Jun–Sep

C. neurophora Mack. (p. 1335) VEINED OR ALPINE NERVED SEDGE (Group 9) Cespitose, often forming large raised clump. **ST**: not winged, mid-st gen ≤ 1.5 mm wide, not easily compressed. **LF**: not tufted at st base; blade 1.5–3 mm wide; sheath fronts gen cross-wrinkled, mouth with tongue-like extension. **INFL**: dense, 1–2.5 cm, 6–15 mm wide; spikelets < 10, indistinct; infl bract like pistillate fl bract; pistillate fl bract brown, white-margined, obtuse to acute. **FR**: 1.1–1.6 mm, 0.8–1 mm wide; peri spreading, 2.9–3.9 mm, 0.8–1.5 mm wide, widest ± at base, body ± long-tapered above, many-ribbed, brown, shiny, lower wall filled with pithy tissue, beak poorly defined, serrate at least on 1 side. Moist to wet places; 1500–2100 m. KR, NCoRH; to WA, Rocky Mtns. Seldom collected in CA. Jul–Aug

C. nigricans C.A. Mey. (p. 1335) BLACK ALPINE SEDGE (Group 3) Rhizomed. **ST**: 5–30 cm. **LF**: much exceeded by infl; blade 1.5–3 mm wide, deciduous. **INFL**: spikelet bisexual, 6–10 mm wide; spikelet bract gen 0 (well-developed); pistillate fl bract dark brown, shed before peri. **FR**: 1.1–1.8 mm, 0.6–0.9 mm wide, < peri body; peri ± ascending, spreading or esp lowest reflexed, 3–5 mm, 0.8–1.5 mm wide, narrow-elliptic, shiny, unveined, dark brown, tapered to ± long stalk, beak obvious in infl. $2n=±72$. Wet, rocky slopes, meadows; 1900–3700 m. KR, CaRH, SNH; to AK, CO. Jul–Sep

C. nudata W. Boott (p. 1335) TORRENT SEDGE (Group 8) Cespitose, in large, raised, dense clumps not connected by rhizomes, not in ± continuous stands. **ST**: 30–70 cm. **LF**: blade 2–4 mm wide; sheath backs scabrous, fronts purple-dotted or -blotched, gen with purple prickles, fibrous on lower lvs. **INFL**: lateral spikelets 1.5–6 cm, 4–6 mm wide, stalk erect, tip occ staminate; infl bract gen < infl; pistillate fl bract obtuse. **FR**: 1.3–2 mm, 1–1.5 mm wide; peri 2.2–4 mm, 1.2–1.8 mm wide, sessile, not papillate, green, purple-splotched on upper 2/3, veins 3–9 abaxially, occ adaxially, beak 0.1–0.3 mm, glabrous, tip notched < 0.1 mm. $2n=70,72$. Rocky or sandy streambeds below high-water mark; < 1600 m. NCo, KR, NCoRI, NCoRO, CaRF, SNF, SNH, ScV, CW, WTR; to s WA. Apr–Jul

C. obispoensis Stacey (p. 1335) SAN LUIS OBISPO SEDGE (Group 2) Cespitose. **ST:** 60–180 cm. **LF:** blade 2.5–8 mm wide, V-folded. **INFL:** 1–2 upper lateral spikelets occ staminate over pistillate; pistillate spikelets nodding, lowest spikelet stalk long-exserted; infl bract < infl, sheath > 9 mm; pistillate fl bract brown, margin wide, white. **FR:** 2.7–3.3 mm, 1.4–1.6 mm wide; peri 15–45 per lateral spikelet, 5–8 mm, 1.4–1.5 mm wide, lanceolate, sparse-hairy, sharply 3-angled, gen fine-red-dotted, tapered to stout beak ±2 mm, teeth erect, 0.1–0.5 mm. Springs, streamsides in chaparral, gen on serpentine; < 800 m. SCoRO (Monterey, San Luis Obispo cos.), PR. Mar–Jun ★

C. obnupta L.H. Bailey (p. 1335) SLOUGH SEDGE (Group 8) Rhizomed, forming beds or large, dense, raised clumps. **LF:** blade 3–7 mm wide; sheath front with ± purple prickles, thickened mouth, fibrous on lower lvs. **INFL:** lateral spikelets gen staminate at tip, 2.5–25 cm, 4.5–10 mm wide, tapered at base, flexible, nodding, stalk 0 or gen erect, ± short; lowest spikelet bract > infl; pistillate fl bract margin white, tip long-tapered. **FR:** 1.4–2.5 mm, 0.7–1.7 mm wide, often deep-indented on 1–2 sides, sessile; peri ascending to spreading, 2–3.8 mm, 1.4–2.2 mm wide, not papillate, unveined, dark brown, shiny, wall thick, tough, beak 0.1–0.3 mm, tip notch 0 or < 0.1 mm. $2n=70,72,74,\pm76$. Moist openings, shores, redwood forest; < 1100 m. NCo, nw KR, NCoRO, CCo, SnFrB; to BC. Jun–Sep

C. occidentalis L.H. Bailey (p. 1335) WESTERN SEDGE (Group 9) Cespitose to loosely so. **LF:** blade 1.5–2.5 mm wide. **INFL:** ± open, 1.5–3 cm, 6–8 mm wide; spikelets gen < 10, ± distinct, lower ± separate; lowest spikelet bract < infl, like pistillate fl bract; pistillate fl bract brown, short-awned. **FR:** 1.3–2 mm, 1.1–1.7 mm wide; peri ascending to spreading, 2.5–3.5 mm, 1.5–1.9 mm wide, ± = pistillate fl bract, green to brown, veins ± 0, beak 0.6–1.3 mm, tip notched. Dry woodland, meadows; 1600–3200 m. SnGb, SnBr, SnJt, W&I; to SD, TX. Jul–Aug ★

C. pachycarpa Mack. FURROWED BROOMSEDGE (Groups 11B,E,G) **LF:** blade 2–5 mm wide. **INFL:** dense, ± head-like, 10–22 mm, ± ovate (elliptic, triangular); lowest internode gen < 5 mm; spikelets gen < 6, ± indistinct at arm's length; infl bract bristle-like; pistillate fl bract < peri, light brown or red-brown, margin often wide, white. **FR:** 2–2.3 mm, 1.2–1.5 mm wide, stalk 0.3–0.4 mm; peri appressed-ascending, 4.9–5.9(6) mm, 1.8–2.4 mm wide, body ovate, planoconvex, green to light brown, flat margin incl wing gen 0.3–0.5 mm wide, abaxial veins 8–17, wide, tissue often sunken between, adaxial veins (0)4–12, beak cylindric, entire for < 0.4 mm, gold. Meadows, dry soil; 1300–2300 m. KR, NCoRH, CaRH, SNH; to WA, ID, NV. Jul–Aug

C. pachystachya Steud. (p. 1335) STARRY BROOMSEDGE (Group 11H) **LF:** blade 1.2–6.5 mm wide; ligule gen < 2.5 mm. **INFL:** ± dense (open), 9–28 mm; spikelets gen > 5, at most only lower distinct, outline gen jagged at arm's length, spikelet base gen rounded or wedge-shaped; lowest 2 infl internodes together < 1/2, gen ± 1/3, total infl; lowest bract gen scale-like (bristle-like); pistillate fl bract gen < 3.5 mm (> 3.5 mm), gen < peri, dark gold or ± red to brown ± throughout or center pale or green, margin dark or white margin < 0.1 mm. **FR:** gen (1.2)1.4–1.9(2.2) mm, (0.7)1–1.5 mm wide, ovate, 2/3–4/5 peri body; peri gen spreading, 2.8–4.7(5.1) mm, 1.1–2.3 mm wide, ovate to elliptic-ovate, planoconvex, gold to brown or coppery to ± black, shiny, upper margin gen dark brown or coppery (narrowly green or gold), ± equal in color to pistillate fl bract, often filled with pith-like tissue near base, flat margin incl wing 0.2–0.4(0.5) mm wide, minute-serrate, not incurved to front, abaxial veins (0)2–9, adaxial 0, ± 4 gen to fr middle (6 reaching beak base), beak < 2/5 peri, cylindric, entire for 0.4–0.7 mm, gold or red-brown to black, obvious in infl. $2n=74,76,78,80,82$. Dry to wet meadows, shores, open forest; (75)900–3500 m. KR, NCoRH, CaR, n SNH, ScV, CCo, Wrn; to AK, SK, CO; also Siberia. Variable; intergrades with *C. microptera*; pls with adaxial veins 6, reaching beak base may belong instead to *C. abrupta*. Jun–Sep

C. pansa L.H. Bailey (p. 1335) SAND DUNE SEDGE (Group 7) Rhizome 2–5 mm thick, dark brown; occ dioecious. **LF:** blade 1–3 mm wide, flat or V-shaped; sheath mouth with thick rim or not. **INFL:** 1.5–2.5 cm, 1–2.5 cm wide, ± head-like, ovate to elliptic; spikelets gen < 10, occ > 1 per lower node or branch; infl bract gen like pistil-late fl bract; pistillate fl bract dark brown or purple-brown, white-margined or not, gen shiny, mucronate to awned. **STAMINATE FL:** filament gen incl, longest anther awns 0.2–0.4 mm, minute-hairy. **PISTILLATE FL:** style incl or not. **FR:** 1.4–2.1 mm, 1.1–1.6 mm wide; peri appressed to ascending, 3–4.5 mm, 1.3–2 mm wide, body tapered above, brown, gen shiny, veins 0 adaxially, many abaxially, lower wall occ filled with pithy tissue, base narrowed for > 0.3 mm below fr, beak 0.5–1.5 mm, serrate, tip unnotched. Coastal sand; < 10 m. NCo, CCo, SCo, n ChI; to BC. Apr–Jul

C. pellita Willd. (p. 1335) WOOLLY SEDGE (Groups 1,2) Rhizomed. **ST:** 30–100 cm. **LF:** blade (2)2.2–4.5(6) mm wide, flat or folded in M-shape exc at base, near tip, glabrous, basal minute; midvein of lf, infl bract a prominent, sharp keel; upper ligules > 2 mm, thin, membranous; sheath mouth glabrous, not ciliate, front fibrous on lower lvs. **INFL:** lowest internode 0.5–0.9 mm wide; pistillate spikelets gen sessile; pistillate fl bract > peri, ± purple or brown, with narrow white margin, ± ciliate, awned. **FR:** 1.5–2.1 mm, 1–1.5 mm wide; peri 25–75, 2.8–5 mm, 1.5–2 mm wide, thick-walled, ± many-ribbed, gen ± purple, beak 0.5–1.2 mm, teeth ± outcurved, 0.2–0.9 mm. Gen marshy places, creekbanks; 60–3300 m. NCo, KR, CaRH, SNH, CCo, SnFrB, SnGb, SnBr, GB, DMoj; to AK, e N.Am. [*C. lanuginosa* Michx., misappl.] May–Sep

C. pendula Huds. PENDULOUS SEDGE (Groups 4,5,6) Cespitose. **ST:** 1–2(3) m, base red to purple. **LF:** 7–20 mm wide. **INFL:** bract sheath 4–65 mm; 1(2) upper spikelets staminate; lateral spikelets pistillate or staminate over pistillate, spikelets nodding, 3–20 cm, green, stalks gen < 10 cm; pistillate fl bract red-brown to ± purple, midvein green. **FR:** 1.3–1.6 mm, 1–1.3 mm wide; peri gen > 75, spreading, 2.6–4 mm, 1.1–1.5 mm wide, body elliptic, plump, ± inflated, glabrous, green, brown-streaked, margin not flat, beak smooth, 0.5 mm. Moist forest, waterways; < 500 m. ScV, SnFrB; OR, WA, e N.Am; native to Eur. Potentially problematic weed. Apr–Jun

C. petasata Dewey (p. 1335) LIDDON'S SEDGE (Groups 11D,E, G,H) **LF:** blade 2–5 mm wide; contraligule gen < 2 mm; ligule 2–5 mm. **INFL:** open, erect, (2)2.5–6 cm, linear-oblong, green, gold, pale brown, or often ± white; lowest internode 4.5–9 mm; spikelets 3–7, distinct at arm's length, base long-tapered; infl bract scale-like (bristle-like); staminate fl bract with white margin (0.2)0.3–0.7 mm wide; pistillate fl bracts 5.8–7.6 mm, ± covering peri, gen hiding beak, margin gen translucent, > 0.4 mm wide. **FR:** 2.2–3 mm, (1.1)1.3–1.8 mm wide; peri appressed to ascending, (5.7)6–8.1 mm, (1.5)1.7–2.5 mm wide, lanceolate to wide-lanceolate, planoconvex, tapered to beak, cream-white to brown, flat margin incl wing (0.2)0.3–0.5 mm wide, veins 5–8 both sides, walls opaque, beak green, gold, red-brown, or brown, tip white, cylindric, ± entire < 1 mm or flat, ± serrate, abaxial flap white-margined. Dry to wet meadows, open forest; 600–3400 m. CaRH (Lassen Co.), n SNH (Alpine Co.), MP; to AK, SK, Rocky Mtns. Jun–Jul ★

C. phaeocephala Piper (p. 1335) DUNHEAD OR ALPINE HARE SEDGE (Group 11D) St 5–40 cm. **LF:** 1.2–2.5(3) mm wide, folded or margin downcurved; contraligule < 3 mm; ligule gen < 2.5 mm. **INFL:** dense or open, 10–35(40) mm, 10–15 mm wide, gold to brown or brown and green; lowest spikelet internode 4–10 mm, infl bract scale- or bristle-like, < infl; spikelets 3–7, distinct or not, ascending, club-shaped or fusiform; pistillate fl bract 3.7–5.1 mm, ± equal, hiding peri, dark brown to ± orange, white margin 0.1–0.3 mm. **FR:** 1.1–1.9(2) mm, 0.7–1.1(1.2) mm wide, stalked; peri ascending, (3.5)3.8–5.2 mm, 1.5–2.5 mm wide, length 2.2–4 × width, narrow- to wide-obovate, ± planoconvex (flat), ± green or pale gold to brown (cream-white), green-margined, walls gen thin, ± translucent, flat margin incl wing 0.3–0.6 mm wide, abaxially veined, strong adaxial veins gen 0(4), beak 0.5–1.2 mm, brown to red-brown or gold, cylindric, ± entire for 0.4–1 mm, tip white, abaxial flap white-margined. $2n=84$. Dry often rocky soils, slopes; 2500–4000 m. CaRH, SNH, SnGb, SnBr, SnJt, Wrn, SNE; to AK, w Can, CO. Jul–Sep

C. praeceptorum Mack. (p. 1335) TEACHER SEDGE (Group 10) Cespitose. **ST:** 10–30 cm, gen leaning at 45° angle. **LF:** blade 1.2–2.5 mm wide. **INFL:** often dense, 1–2.5 cm; spikelets < 10, ± distinct, 4–7 mm, lowest 1 gen > 2 × lowest internode, terminal 1 narrow-tapered at base; infl bract << infl; pistillate fl bract light brown with

white margin, tip, obtuse to acute. **FR**: 1.2–1.5 mm, 0.8–1 mm wide; peri ascending, 1.5–2.5 mm, 1–1.2 mm wide, gold, upper body papillate, wall filled with pithy tissue to upper body, beaks interrupting outline of spikelets, 0.2–0.4 mm, wide-conic, abaxial flat reaching to just below base, dark brown. Wet places; 2200–3500 m. CaRH, c&s SNH; to BC, CO. Jul–Aug

C. praegracilis W. Boott (p. 1339) BLACK CREEPER OR FREEWAY SEDGE (Group 7) Rhizome 2–5 mm thick; occ dioecious. **LF**: blade 1.5–3 mm wide, flat or V-shaped; sheath mouth occ with dark or ± thick rim. **INFL**: gen dense, 1–5 cm, 6–10 mm wide; spikelets gen < 10, occ > 1 per lower node or branch; pistillate fl bract ± dull, gold to brown, white-margined or not, mucronate to awned. **STAMINATE FL**: filament exserted, anther awns gen minute-hairy at 20×, longest 0.1–0.2 mm, slender. **PISTILLATE FL**: style gen incl. **FR**: 1.2–1.9 mm, 1–1.4 mm wide; peri ascending, 2.8–4 mm, 1.3–1.6 mm wide, widest above base, tapered above, dark brown, gen dull, veins several abaxially, 0 adaxially, lower wall occ filled with pithy tissue, stalk < 0.3 mm, beak 0.6–1.5 mm, serrate, tip unnotched, white. Common. Often alkaline, ± moist places; < 2700 m. CA exc DSon; w&c N.Am, S.Am; alien in e N.Am. Apr–Aug

C. praticola Rydb. (p. 1339) NORTHERN MEADOW SEDGE (Groups 11D,E,F) St 25–80 cm. **LF**: blade 1.5–4 mm wide; ligule gen < 2.5 mm. **INFL**: gen open, (17)25–50 mm, often bent or flexed at lowest node, ± white, green, gold, or dark brown; lowest internode gen (4)5–14 mm; spikelets 4–10, spaced, gen distinct at arm's length, oblanceolate to wide-ovate, base often long-tapered, lowest < lowest internodes; infl bract gen bristle-like, < infl; pistillate fl bract (3.4)4.2–5.8 mm, gen covering peri, white, gold, ± red, coppery, or brown, white margin 0.1–0.3 mm wide. **FR**: 1.4–2.1(2.7) mm, (0.8)1–1.5(1.7) mm wide; peri appressed to ascending-spreading, (3.7)4.5–6.5 mm, 1.2–2(2.5) mm wide, lanceolate to ovate, body planoconvex (± flat exc over fr), gen narrow-elliptic, gen cream-white, green, or gold, walls gen ± translucent, brown or coppery over fr, gen shiny, flat margin incl wing gen 0.2–0.5 mm wide, abaxial veins 0–7, weak, adaxial veins gen prominent on some peri, < 1/2 fr, body ± abruptly narrowed to beak, beak 1–1.6 mm, brown, tip white, cylindric, ± entire, abaxial flap white-margined. 2*n*=76,78. Moist to wet meadows, riparian edges, open forest; (20)500–3200 m. NCo (Humboldt Co.), NCoRO, NCoRH, c SNH; Rocky Mtns, n N.Am. May–Jul ★

C. preslii Steud. (p. 1339) PRESL'S SEDGE (Groups 11G,H) **LF**: blade 2–4 mm wide, flat or folded; ligule gen < 2.5 mm. **INFL**: dense or open, 17–25 mm, green and brown, gold, or brown; spikelets 3–7, lower gen distinct; infl bract bristle-like or not, < lowest spikelet; pistillate fl bract ± ≤ peri, white, gold, or red-brown, center gen pale to green, tip brown to gold. **FR**: (1.5)1.7–2.1(2.3) mm, 0.9–1.7 mm wide, ± squared, = peri body, ± sessile; peri appressed- to spreading-ascending, 3.3–4.3(5) mm, 1.4–2 mm wide, ovate, gen planoconvex (biconvex), green to gold, upper body margin green to gold, flat margin incl wing 0.1–0.5 mm wide, walls thin but opaque, abaxially veined, adaxial veins 0 (3, inconspicuous), not exceeding fr, abruptly narrowed to distinct beak 1–1.6 mm, beaks mixed in infl, some ± flat, with smooth entire part < 0.4 mm, others cylindric, smooth entire part > 0.4 mm, gen green to red-brown to tip. 2*n*=±78,80. Meadows, open forest, dry rocky slopes; 1800–3400 m. KR, NCoR, CaRH, SNH; w N.Am. Like *C. pachystachya* but in gen drier sites, with more open infl, some beaks flat-tipped. At high elevations sometimes like *C. phaeocephala* but with peri ovate. Jun–Sep

C. proposita Mack. POTATO CHIP SEDGE (Group 11H) **LF**: blade gen 0.5–2.7 mm wide, gen folded or channeled toward base; ligule 0.9–2 mm. **INFL**: dense, ovate, to more open, elliptic to oblong, (15)17–25(30) mm; lowest internode 4–8.6 mm; spikelets distinct or not, wide-ovate to spheric; infl bract scale- or bristle-, occ lf-like; pistillate fl bract < peri but << narrower. **STAMINATE FL**: anthers ± persistent. **FR**: (1.2)1.5–2.1 mm, 0.7–1.1(1.3) mm wide, often stalked; peri ascending to spreading, 4.3–5.3 mm, gen 2–3 mm wide, length 1.7–2.6 × width, round to ovate, 2–3 × pistillate fl bract width, flat exc over fr, gold to coppery or brown, green-margined, flat margin incl wing 0.4–0.9 mm wide, crinkled or not, abaxial veins 3–11, obvious, adaxial 0–13, abruptly narrowed to short beak ≤ 1.2 mm, beak cylindric, entire for 0.4–0.9 mm, green to red-brown, tip white,

abaxial flap white-margined. Uncommon. Rocky places; 3000–4100 m. c&s SNH; WA, c ID. Much like *C. straminiformis*, exc in lf width, peri beak length, taper. Jul–Aug

C. raynoldsii Dewey (p. 1339) RAYNOLD'S SEDGE (Groups 4,5,6) Cespitose to loosely so. **ST**: 20–75 cm. **LF**: blade 3–8 mm wide. **INFL**: pistillate fl bract purple or with pale midrib not extending to tip, awn 0. **FR**: 1.8–2.4 mm, 1.1–1.8 mm wide, ± = peri in width; peri spreading, 3.5–4.5 mm, 1.6–2.1 mm wide, body obovate to wide-elliptic, with flat margin above, green, faces veined, beak distinct, 0.1–0.5 mm, ± purple. 2*n*=58. Meadows; 1800–3100 m. KR, CaRH, SNH, Wrn; to BC, Rocky Mtns. Jun–Aug

C. rossii Boott (p. 1339) ROSS' SEDGE (Group 2) Densely to loosely cespitose; rhizomes stout. **ST**: 7–40 cm, gen ascending, scabrous on angles, base gen not very fibrous. **LF**: exceeded by to exceeding infl; blade 0.8–4 mm wide, stiff or not, ± flat, bright or gray-green. **INFL**: infl bract gen > infl; staminate spikelet of taller infls 4–10(12.8) mm, 0.7–2.5 mm wide; nonbasal pistillate spikelets 2–4; pistillate fl bract green or tinged ± red, awned or not. **FR**: 1.9–2.4 mm, 1.3–1.7 mm wide; peri 3–15, 3.1–4.5 mm, 1.1–1.7 mm wide, beak 0.7–1.7 mm, teeth 0.2–0.4 mm. Dry forest, meadows; < 3800 m. NW, CaRH, SNH, SnGb, SnBr, MP, W&I, DMtns; to n Can, CO, AZ, scattered to n-c US. [*C. geophila* Mack., misappl.] Report of *C. geophila* in SN evidently based on misidentification. May–Aug

C. saliniformis Mack. (p. 1339) DECEIVING SEDGE (Group 8) Rhizomed. **ST**: upper angles blunt, glabrous. **LF**: blade 2–5 mm wide. **INFL**: lateral spikelets 2–4, 6–15 mm, 3–5 mm wide, occ > 1 per lower branch, stalk long; infl bract gen > infl, sheath > 4.5 mm; lowest staminate fl bract often > 1/3 spikelet; lower pistillate fl bracts occ lf-like, green, margin white or red-brown, awn stout, ciliate, < 2.4 mm. **FR**: 1.4–1.9 mm, 0.8–1.5 mm wide; peri 8–20, 2.5–4 mm, 1.3–2.1 mm wide, sessile, papillate, body gen tapered to dark red-brown or purple tip, beak indistinct or 0–0.3 mm, tip unnotched. Marshes, pond shores, wet openings; < 250 m. NCo, SnFrB (extirpated). Weakly separated from *C. hassei*; study needed. May–Jul ★

C. sartwelliana Olney (p. 1339) SARTWELL'S OR YOSEMITE SEDGE (Group 1) Cespitose. **ST**: 30–145 cm, base red-purple. **LF**: hairy; blade 3–7 mm wide. **INFL**: pistillate spikelets ± sessile; infl bract ± = infl; pistillate fl bract < to > peri, with 1 weak vein in wide green central zone, margin white, ciliate. **FR**: 1.5–2.1 mm, 1.2–1.5 mm wide; peri 40–200, ascending, 2.3–3.9 mm, 1.2–1.8 mm wide, 2-ribbed, ± veined, green to dark brown (dark purple) on upper 1/2 of body, gen abruptly narrowed above, thin-walled, beak 0.4–1 mm, tip notch 0 or abaxial, < 0.2 mm. Moist or wet meadows, roadside seeps, damp granite ledges, creekbanks, open forest; 1200–2600 m. c&s SNH, PR. May–Aug

C. scabriuscula Mack. (p. 1339) SISKIYOU SEDGE (Groups 1,2) Loosely cespitose; dioecious. **ST**: 30–45 cm, fl st base scaly, not fibrous. **LF**: much exceeded by infl; blade 2–3 mm wide. **INFL**: spikelets 1–4; infl bract < or > infl, lf-like, spikelet bract bristle- or scale-like; pistillate fl bract in width < peri, glabrous, brown, margin narrow-white or not, ciliate at least at tip, acute to awned. **STAMINATE FL**: anthers 2–2.8 mm. **PISTILLATE FL**: stigmas occ 4. **FR**: 1.5–1.8 mm, 0.7–1.2 mm wide, << peri body, occ 4-sided; peri 20–40, ascending, 2.8–3.5 mm, 1.4–2 mm wide, flat exc over fr, often strongly veined on lower 1/2, brown to ± black, faces minute-hairy at least near beak, beak 0.3–0.5 mm. Wet serpentine meadows; 850–2300 m. KR, n SNH (Plumas Co.); to sw OR. [*C. gigas* (Holm) Mack.] Jun–Jul ★

C. schottii Dewey (p. 1339) SCHOTT'S SEDGE (Group 8) Cespitose. **ST**: 75–150 cm. **LF**: blade 4–12 mm wide, basal minute; sheath back scabrous, front red-dotted, fibrous; bladeless sheaths > 7 mm wide at midlength, keeled, scabrous. **INFL**: staminate spikelets 2–7, terminal 1 gen > 5 cm; lateral spikelets staminate at least at tip, gen 5–20 cm, 5–7 mm wide, ± nodding, stalks 0 or long; infl bract gen > infl, lf-like; pistillate fl bract obtuse to mucronate. **FR**: 1.6–2.3 mm, 1–1.6 mm wide; peri 2.7–4.5 mm, 1.5–2.5 mm wide, sessile, veined, papillate or not, gold to brown, beak 0.1–0.5 mm, glabrous, tip notched < 0.5 mm. Streambanks, wet meadows, swamps; < 2500 m. NCoRO, SnFrB, SCo, TR, PR. Apr–Jul

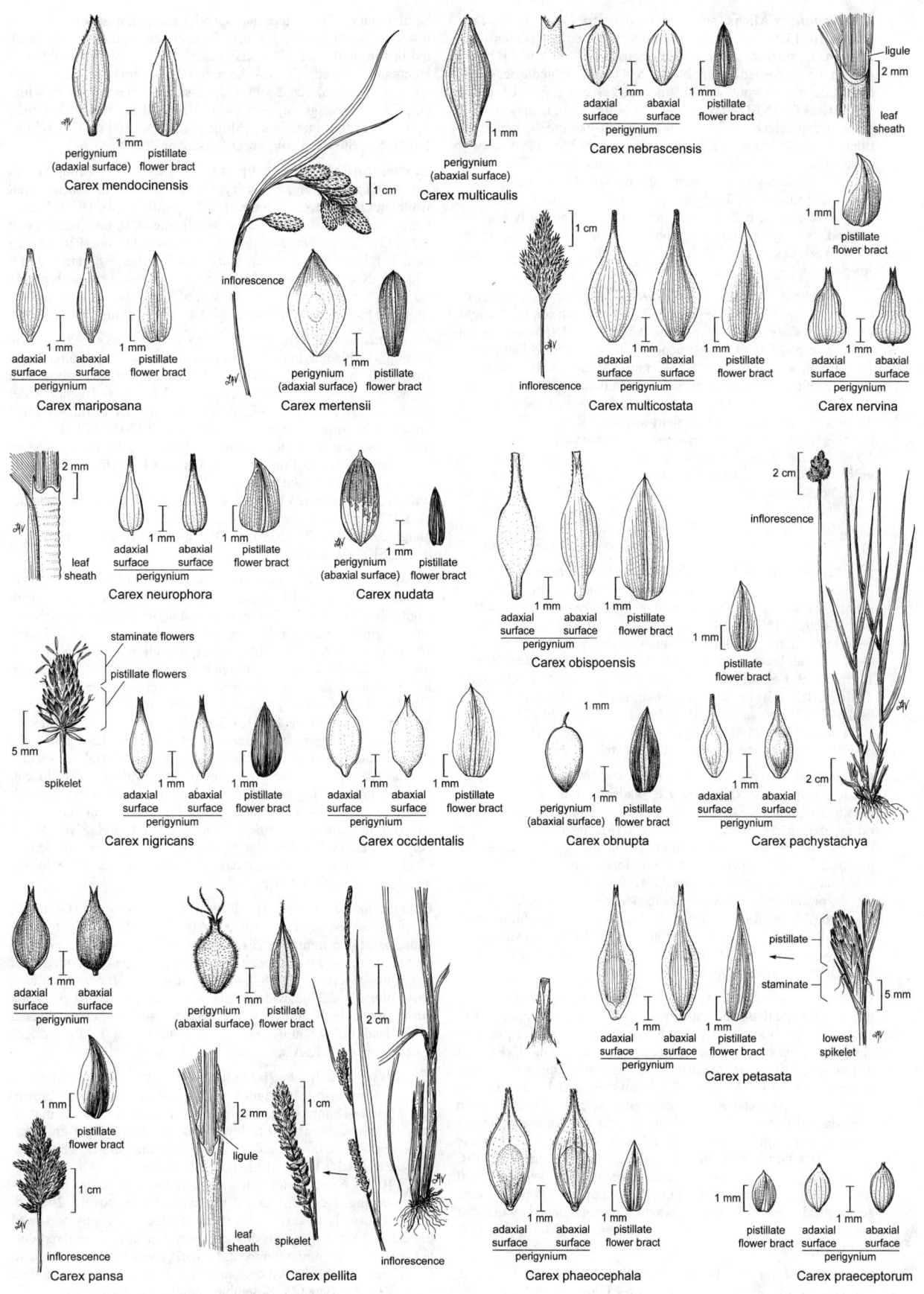

perigynium
(adaxial surface)   pistillate
flower bract
Carex mendocinensis

perigynium
(abaxial surface)
Carex multicaulis

adaxial   abaxial
surface   surface   pistillate
flower bract
Carex nebrascensis

ligule
leaf
sheath

pistillate
flower bract

adaxial   abaxial
surface   surface
perigynium
Carex mariposana

inflorescence

perigynium
(adaxial surface)   pistillate
flower bract
Carex mertensii

inflorescence

adaxial   abaxial   pistillate
surface   surface   flower bract
perigynium
Carex multicostata

adaxial   abaxial
surface   surface
perigynium
Carex nervina

leaf
sheath
Carex neurophora

adaxial   abaxial   pistillate
surface   surface   flower bract
perigynium

perigynium
(abaxial surface)   pistillate
flower bract
Carex nudata

adaxial   abaxial   pistillate
surface   surface   flower bract
perigynium
Carex obispoensis

inflorescence

pistillate
flower bract

staminate flowers
pistillate flowers

spikelet

adaxial   abaxial   pistillate
surface   surface   flower bract
perigynium
Carex nigricans

adaxial   abaxial   pistillate
surface   surface   flower bract
perigynium
Carex occidentalis

perigynium
(abaxial surface)   pistillate
flower bract
Carex obnupta

adaxial   abaxial
surface   surface
perigynium
Carex pachystachya

adaxial   abaxial
surface   surface
perigynium

pistillate
flower bract
Carex pansa

inflorescence

perigynium
(abaxial surface)   pistillate
flower bract

ligule

leaf
sheath   spikelet

inflorescence
Carex pellita

adaxial   abaxial   pistillate
surface   surface   flower bract
perigynium
Carex phaeocephala

adaxial   abaxial   pistillate
surface   surface   flower bract
perigynium
Carex petasata

pistillate

staminate

lowest
spikelet

pistillate
flower bract   adaxial   abaxial
surface   surface
perigynium
Carex praeceptorum

C. scirpoidea Michx. subsp. ***pseudoscirpoidea*** (Rydb.) D.A. Dunlop (p. 1339) WESTERN SINGLE-SPIKED SEDGE (Groups 1,2) Rhizomed; ripe fr sts stiff-erect; dioecious. **ST**: 10–40 cm, fl st base ± fibrous. **LF**: exceeded by infl; blade 1.5–4 mm wide; ligule occ longer than wide; sheath front occ hairy, not coarse-veined, fine-red-dotted. **PISTILLATE INFL**: spikelets gen 1 (2, upper gen larger), oblong-cylindric; spikelet bract 0 or < spikelet, occ lf-like and sheathing; pistillate fl bract ≥ peri in width, often minute-hairy, brown to black with wide white margin, ciliate or not, obtuse or mucronate. **FR**: 1.5–2.1 mm, 0.7–1.1 mm wide, ± = peri body; peri 20–40, ascending, 2–3.5 mm, 1–1.7 mm wide, 3-sided, faint-veined, ± purple, minute-hairy, at least near beak, beak 0.4–0.7 mm, entire or irregularly shallow-toothed. Rocky, occ limey seasonally wet places; 2100–3700 m. SNH, W&I; OR to BC, CO. [*C. s.* Michx. var. *p.* (Rydb.) Cronquist] Jul–Sep ★

C. scoparia Willd. var. ***scoparia*** POINTED BROOMSEDGE (Groups 11A,G) **LF**: blade gen 1.5–3 mm wide; sheath front white-membranous or green ± to top; ligule 2.3–4.8 mm. **INFL**: open, often flexible or nodding at lowest node, or dense, erect on late shoots, 15–70 mm, lowest internode 2–12 mm, spikelets 5–10, distinct at arm's length, fusiform to ellipsoid, sides smooth; infl bract gen scale-like (lf-like); pistillate fl bract < peri, gold or brown, some with green or gold center, acuminate to short-awned. **FR**: 1.3–1.7(1.9) mm, 0.7–0.9(1.2) mm wide; peri ± appressed or ascending, 4.2–5.8 mm, 1.2–2(2.5) mm wide, lanceolate (elliptic), tapered to indistinct beak, green to gold-brown, thin, flat exc over fr, abaxial veins thin, 0–8, tissue not sunken between, adaxial veins weak, 0–5, flat margin incl wing 0.2–0.5(0.6) mm wide, beak pale- to gold-brown, cylindric, entire for < 0.4 mm. 2*n*=56–70. Wet, open places, lakeshores; ± 1200 m. n SNH (Plumas Co.); N.Am. Evidently only 1 CA record. Jul ★

C. scopulorum Holm var. ***bracteosa*** (L.H. Bailey) F.J. Herm. (p. 1339) MOUNTAIN SEDGE (Group 8) Rhizomed; fl sts thin, not spongy-thickened. **LF**: blade 3–6 mm wide; sheath front brown- or purple-dotted. **INFL**: ± dense, gen dark purple; lowest internode 5–45 mm; lateral spikelet 8–30 mm, 4–6 mm wide, gen erect, stalk 0–15 mm; infl bract < infl; pistillate fl bract obtuse to acute. **FR**: 1.2–1.8 mm, 0.9–1.5 mm wide, dull; peri 8–50, ascending to spreading, 2–4 mm, 1.2–2.3 mm wide, unveined, papillate, gen purple above, beak 0.1–0.3 mm, glabrous, tip notched < 0.1 mm. 2*n*=72,76,78,80. Wet places; 1200–3400 m. KR, NCoRO, NCoRH, CaRH, SNH, Wrn, SNE (Sweetwater Mtns); to BC, Rocky Mtns. Jul–Sep

C. senta Boott (p. 1339) WESTERN ROUGH SEDGE (Group 8) Loosely cespitose or rhizomed. **LF**: blade 3–5 mm wide; bladeless sheaths 0 on lower st; sheath back scabrous, on lower lvs with small, raised, irregularly spaced cross-walls. **INFL**: lateral spikelet 2.5–5 cm, ± 5 mm wide, stalk 0 or short; infl bract base dark-margined; pistillate fl bract ± ≤ peri, gen dark purple, obtuse to acuminate. **FR**: 1.5–2 mm, 1–1.5 mm wide; peri 25–100, 3–4 mm, 2–2.2 mm wide, sessile, prominently 3–7-veined both sides, gen not papillate, green, often purple-dotted or -blotched, beak 0.2–0.4 mm, glabrous, dark purple or not, tip notched < 0.1 mm. Streambanks, marshy areas, meadows; < 2900 m. NCoRI, SN, s ScV, CW, SW; AZ, Baja CA. Much like *C. nudata.* Apr–Aug

C. serpenticola Zika SERPENTINE SEDGE (Group 2) Rhizomed, occ forming small tufts or dense mats; ripe fr sts weak, drooping. **LF**: blade 1.5–3.5(5) mm wide; sheath front fibrous, glabrous, coarse-veined, gen not red-spotted; ligule gen wider than long. **INFL**: infl gen either pistillate or staminate, occ bisexual; staminate infl with terminal spikelet 13–24 mm, 1.4–4.1 mm wide, and 1–3 aborted lateral spikelets; pistillate infl with terminal spikelet 8–47 mm, 6–9 mm wide, lateral spikelets 1–3, 5–8 mm, lowest gen well-separated or obvious, gen elliptic-ovate to ± round, basal spikelets 0–1; pistillate fl bract dark purple with thin white margin 0.1–0.2 mm wide. **FR**: 1.9–2.2 mm, 1.4–1.8 mm wide, pale to brown; peri 4–9 per lateral spikelet, 3.1–3.6 mm, 1.5–1.8 mm wide, plump, filled by fr, veins gen 0 or short, inconspicuous, beak with 2 obvious teeth. Dry to moist savanna, riparian, spring margins, on serpentine; 100–1200 m. KR; sw OR. Apr–Jun ★

C. serratodens W. Boott (p. 1339) SAW-TOOTHED SEDGE (Groups 4,6) Cespitose. **ST**: 30–120 cm. **LF**: blade 1.7–4 mm wide,

basal minute. **INFL**: terminal spikelet staminate at least at base; lower spikelets sessile; pistillate fl bract < peri, dark purple (dark red-brown) with pale midrib, hairy-awned (occ only a few pistillate fl bracts short-awned). **FR**: 1.8–2.5 mm, 1.1–1.8 mm wide; peri spreading, gen 3–5 mm, 1.2–2 mm wide, 3-sided, ± flat around fr, wide-elliptic to ovate, green, purple-dotted, papillate, beak 0.3–1 mm, ± purple, teeth minute-hairy. Moist places; < 1800 m. KR, NCoR, CaRH, SN, SnFrB, SCoR; sw OR, AZ. Apr–Jul

C. sheldonii Mack. (p. 1339) SHELDON'S SEDGE (Group 1) Rhizomed. **ST**: 40–80 cm. **LF**: hairy; blade 3–6 mm wide; sheath front white, not red-dotted, ± fibrous. **INFL**: pistillate spikelets 8–10 mm wide, stalks 0 or short; infl bract > infl, sheath 0; pistillate fl bract green to ± purple, white-margined, short-awned or not. **FR**: 1.8–2.5 mm, 1–1.3 mm wide, << peri body; peri 25–100, 5–8 mm ± 2 mm wide, body ovate, many-ribbed, tapered to 2 mm beak with purple tip, teeth outcurved, 0.7–1.4 mm. Wet places; 1200–1500 m. n SNH (Plumas, Placer cos.), MP; to OR, ID, UT. May–Aug ★

C. simulata Mack. (p. 1339) SHORT-BEAKED OR ANALOG SEDGE (Group 7) Rhizome long, 1.5–3 mm thick, dark brown; often dioecious. **LF**: blade 2–5 mm wide; sheath mouth fragile, often with brown stripe across. **INFL**: dense, 1.2–3.5 cm, 6–10 mm wide; spikelets < 10; infl bract < infl, like pistillate fl bract; pistillate fl bract brown, white-margined, lower ± awned. **STAMINATE FL**: filaments exserted, anther awns stout, glabrous (warty) at 20×, obtuse at tip, longest 0.1–0.2 mm. **PISTILLATE FL**: style exserted. **FR**: 1.1–2.3 mm, 0.7–1 mm wide; peri 1.7–2.6 mm, 1.3–1.6 mm wide, brown, shiny, veins few both sides, lower wall filled with pithy tissue, beak 0.25–0.5 mm, gen < 1/4 body, serrate, tip unnotched, white-margined. Moist soil; < 3300 m. KR, CaRH, SNH, CCo, SnFrB, GB; to BC, SK, NM. Jun–Sep

C. specifica L.H. Bailey (p. 1339) NARROW-FRUITED SEDGE (Groups 11B,D,E,G) **LF**: blade 2–5 mm wide, flat or folded; sheath front often ± firm, ribbed, with at least a narrow, thin, membranous, white strip extending > 10 mm below top; contraligule 1–4 mm. **INFL**: gen dense, (15)19–40(50) mm, gen elliptic (oblong); lowest internode 2.5–5(7) mm, green to light brown; spikelets (4)6–14, gen indistinct at arm's length, fusiform; infl bract scale- or bristle-like; pistillate fl bract < peri, green, light brown, or red-brown with white margin 0–0.5 mm wide. **FR**: 1.6–2.1(2.3) mm, 1–1.5 mm wide, stalk 0.5–1 mm; peri appressed to ascending, 4.9–6.6 mm, 1.5–2 mm wide, gen lanceolate (fusiform or ovate), planoconvex, abaxial veins 9–18, adaxial gen 6–10, gen 3+ > fr, green to gold, walls gen translucent, flat margin incl wing gen 0.3–0.6 mm wide, crinkled or flat, beak cylindric, entire for < 0.4 mm, green, light brown, or red-brown. Dry soil, meadows, open forest; 1200–3500 m. CaRH, SNH; w NV. Peri occ narrow like those of *C. petasata* but smaller, infl denser. Confused with *C. fracta* (both have contraligule), but differs in peri length, wing width. Jul–Sep

C. spectabilis Dewey (p. 1339) SHOWY SEDGE (Group 6) Cespitose or loosely cespitose. **ST**: 20–90 cm. **LF**: blade 2–5 mm wide, basal gen minute. **INFL**: lowest spikelets often on nodding stalks, purple; pistillate fl bract dark purple with pale midrib, awned or tip long-tapered. **FR**: 1.6–2 mm, 0.7–1 mm wide; peri ascending, 2.6–5 mm, 1.6–2.3 mm wide, ± flat exc over fr or ± inflated, papillate, purple above, body elliptic, ovate, oblanceolate, or obovate, tapered to tip, beak 0.2–0.5 mm. 2*n*=±42. Wet places; 1800–3700 m. KR, CaRH, SNH; to AK, AB, UT. Jul–Sep

C. spissa L.H. Bailey (p. 1339) SAN DIEGO SEDGE (Groups 2,4,5,6) Cespitose to rhizomed. **ST**: 1–1.8 m, stout, base brown. **LF**: blade 6–18 mm wide, blue-green at least when young, margin very sharply scabrous. **INFL**: lower spikelets staminate at tip, erect, stalked; pistillate fl bract gold to red-brown, margin gen white, awn hairy. **PISTILLATE FL**: style bent, tough, not jointed to fr, persistent. **FR**: 2–2.6 mm, 1–1.3 mm wide; peri 150–300 per lateral spikelet, spreading, 3–5 mm, 1.5–2.5 mm wide, body obovate, 2-edged, ± flat or strongly inflated, veins 0–few on faces, abruptly narrowed to beak ± 0.5 mm, glabrous or short-hairy, glaucous, green, red-dotted, beak 0.1–0.8 mm, often bent, tip dark, teeth < 0.5 mm, straight. Creekbanks, seeps, canyon bottoms, on serpentine or not; < 1200 m. CCo, SCo, PR; Baja CA. Resembles small cattails. Apr–Sep

C. stevenii (Holm) Kalela (p. 1339) STEVEN'S SEDGE (Group 6) Cespitose. **ST**: 15–60 cm. **LF**: blade 3–4 mm wide. **INFL**: open, spikelets gen overlapped, 5–15 mm, 3–4.5 mm wide, ± erect on short stalks, pistillate or pistillate above; pistillate fl bract < peri, ovate, dark purple (dark brown), blunt to acute (short-awned). **FR**: 1.8–2 mm, 0.9–1 mm wide; peri ascending, 2–2.5 mm, 1.2–1.5 mm wide, green to olive, papillate, body obovate, abruptly narrowed to beak 0.2–0.3 mm. Uncommon. Alpine creekbanks; 2800–2900 m. W&I (White Mtns); s Rocky Mtns. [*C. norvegica* Retz., misappl.] Jul–Aug ★

C. stipata Willd. var. ***stipata*** (p. 1339) AWL-FRUITED SEDGE (Group 9) Cespitose. **ST**: deep-concave between sharp or winged angles, spongy when fresh. **LF**: blade 5–11 mm wide; sheath front cross-wrinkled. **INFL**: 2–10 cm, 1–3 cm wide; spikelets >> 10, > 1 per lower node or branch, indistinct; pistillate fl bract pale, often awned. **FR**: 1.3–2.3 mm, 1.5–1.8 mm wide; peri spreading, 3.6–6 mm, 1.5–1.8 mm wide, body widest ± at base, many-ribbed both sides, brown, long-tapered to poorly defined beak, lower wall filled with pithy tissue. Wet places; < 1700 m. NW, n&c SNH, SnFrB; N.Am, Japan. May–Aug

C. straminiformis L.H. Bailey (p. 1341) MOUNT SHASTA SEDGE (Groups 11G,H) **LF**: blade (1.7)3–4 mm wide, flat or folded; sheath front often cross-wrinkled; ligule 1.5–2.5(3.5) mm. **INFL**: gen dense, 15–25(30) mm, ovate, lowest internode 0.5–3.9(5) mm; spikelets distinct, ± spheric; infl bract scale- or bristle-like; pistillate fl bract < peri, gen red-brown (white or gold), center pale to green, white margin 0–0.15(0.3). **STAMINATE FL**: anthers early-deciduous to ± persistent. **FR**: (1.4)1.7–2.4 mm, 1–1.6 mm wide; peri ascending to spreading, 4–5.8 mm, 1.8–3.4 mm wide, wide-ovate or -elliptic or ± round, green or gold, flat exc over fr, flat margin incl wing gen 0.5–1 mm wide, edge often crinkled or ruffled, at least above, abaxial veins 10–20, weak, beak gen 1.2–1.5 mm, cylindric, entire for < 0.3(0.7) mm, gold or red-brown to dark brown, tip occ white. Common. Dry rocky or gravelly slopes; 1700–4100 m. KR, NCoRH, CaRH, SNH, Wrn, SNE; to WA; also MT, UT. Jul–Sep

C. subbracteata Mack. (p. 1341) SMALL-BRACTED SEDGE (Groups 11D,F,H) **ST**: often leaning. **LF**: blade 1.3–4.6 mm wide; sheath front membranous, white strip extending > 10 mm below top; contraligule 1.5–3(5.5) mm. **INFL**: ± dense, erect, 13–35 mm; lowest internode 2–3(4.5) mm, gen < lowest spikelet, lower spikelets often clustered or overlapped, indistinct, wedge-shaped or rounded; infl bract bristle- or lf-like, < to > lowest spikelets, < to > infl, base occ clasping st; pistillate fl bract gen 3.4–4.5 mm, occ > peri, brown, ± red, or coppery with paler or green center. **FR**: (1.3)1.5–2.1 mm, 0.9–1.7 mm wide; peri ascending, gen 3.5–4.7(5.7) mm, gen 1.3–1.7(2.2) mm wide, ± ovate, planoconvex, leathery, gen gold with brown margin, shiny, margin incl wing 0.2–0.3(0.4) mm wide at midbody, entire from base to ± widest point at 16×, most of beak and upper body minute-serrate, wall ± tough, veined abaxially, adaxial veins 0–7, = or gen << fr (1 > fr), beak gen ± 2/5 peri, gen cylindric, ± entire for > 0.4 mm, dark brown to red-brown, tip occ white. At least seasonally moist soil, grassland to open forest; < 900 m. NCo, NCoRO, CW, SCo, n ChI, WTR; s OR, alien in WA, BC. Much like *C. gracilior*, which has more open infl, more entire peri margin; putative hybrids with similar *C. harfordii* in CCo gen sterile. Apr–Jun

C. subfusca W. Boott (p. 1341) RUSTY OR PALE BROOMSEDGE (Groups 11F,G,H) **LF**: blade 1.2–3.7 mm wide; contraligule gen < 3 mm. **INFL**: open or dense, 11–30 mm, green to brown, fine-textured at arm's length, lowest internode (1.5)2.5–9.5 mm; spikelets gen distinct, base round or wedge-shaped; lowest bract bristle-like or not, < or > infl; pistillate fl bract gen 2.1–3.5 mm, gen < peri, gen white or pale brown to ± red, with pale or green center. **FR**: 1–1.6 mm, 0.7–1.25 mm wide, ± = peri body; peri appressed to spreading, 2.3–3.5 mm, 0.9–1.9 mm wide, ± ovate, planoconvex, pale green, pale yellow, or light brown, upper margin green or gold, flat margin incl wing gen 0.15–0.25 mm wide, abaxial veins thin, adaxial gen 0 or not exceeding fr, beak flat, ± serrate, or cylindric, entire for 0.4(0.7) mm, green or pale yellow to brown. 2*n*=84. Common. Seasonally wet meadows, creeks, seeps; (75)700–3800 m. KR, NCoR, CaRH, n SNF, SNH, SnFrB, SCoRO, TR, PR, MP, W&I, DMtns; OR,

AZ. Small peri, pale infl gen diagnose this sp.; occ pls with distinct adaxial peri veins but these rarely to beak base. May–Sep

C. subnigricans Stacey (p. 1341) DARK MOUNTAIN SEDGE (Group 3) Rhizomed. **ST**: 5–20 cm. **LF**: exceeded by infl; blade 0.4–1 mm wide, green, inrolled or channeled. **INFL**: spikelet bisexual, 3–6 mm wide; pistillate bract 0; pistillate fl bract wide, gen 1-veined, light brown. **FR**: 1.3–1.8 mm, 0.7–1 mm wide (width > lateral distance from fr to peri), < peri body, next to bristle-like axis; peri 10–40, ascending, 2.5–4.1 mm, 0.9–2 mm wide, flat, gold-brown, glabrous. Meadows, gen dry, rocky slopes; 2600–3800 m. SNH, SNE (Sweetwater Mtns), W&I; to ID, UT. [*C. micropoda* C.A. Mey., misappl.; *C. pyrenaica* Wahlenb., misappl.] Jul–Sep

C. tahoensis Smiley (p. 1341) TAHOE SEDGE (Group 11D) **LF**: blade (0.7)1.5–2(3) mm wide, folded or margin downcurved; ligule 0.8–1.5 mm. **INFL**: open or dense, (15)20–30(37) mm, 5–12 mm wide, ± white, gold, or brown; spikelets 3–7, distinct, appressed to ascending, 4–6 mm wide, lowest bract like pistillate fl bract or bristle-like, < infl; pistillate fl bract ± covering peri (at least body), red-brown with straw or tan center, white margin (0.1)0.3–0.6 mm wide. **FR**: 1.9–2.4 mm, 1.2–1.6 mm wide; peri appressed or appressed-ascending, 3.7–6 mm, 1.3–2.2(2.6) mm wide, gen widest near or above middle, strongly planoconvex, green, gold, coppery, or brown, walls leathery, opaque, flat margin incl wing 0.2–0.4(0.5) mm wide, abaxial veins 7–14, adaxial gen strong, (3)5–8, = fr, beak cylindric, ± entire for 0.2–0.7 mm, gold, ± red, or brown, tip gen white, abaxial flap white-margined or inconspicuous. Open, rocky slopes; 3200–3700 m. c&s SNH; scattered in montane w N.Am. [*C. eastwoodiana* Stacey; *C. phaeocephala* f. *e.* (Stacey) F.J. Herm.] Once thought endemic to CA. Jul–Aug ★

C. texensis (L.H. Bailey) L.H. Bailey TEXAS SEDGE (Group 9) **LF**: blade 0.7–1.5 mm wide. **INFL**: open, 1–3 cm, < 5 mm wide; spikelets gen < 10; pistillate fl bract white to ± green, acuminate to awned, deciduous before peri. **FR**: 1.3–1.6 mm, 0.8–1 mm wide; peri spreading to reflexed, 2.4–3 mm, 0.8–1.2 mm wide, sessile, tapered above, unveined, green to pale gold, lower wall filled with pithy tissue, margin entire above, beak 0.5–1 mm, conic, entire, tip narrow-white-margined. Creekbanks, disturbed ground; < 100 m. ScV (Butte Co.), SCo (Santa Barbara, Los Angeles cos.); native to c&se US. [*C. retroflexa* var. *t.* (L.H. Bailey) Fernald] Apr–Jun

C. tiogana D.W. Taylor & J.D. Mastrog. (p. 1341) TIOGA PASS SEDGE (Groups 4,6) Cespitose. **ST**: 1.8–14 cm. **LF**: exceeding infl; blade 1.5–3 mm wide, ± sickle-shaped, margin, midrib dense-coarse-serrate at 10×. **INFL**: lateral spikelets erect to nodding; infl bract sheath 3–5 mm; pistillate fl bract white, midrib of at least some raised, spiny or serrate, early-deciduous. **FR**: 0.8–1.4 mm, 0.4–0.8 mm wide; peri 3–8, ascending to spreading, 1.3–2.3 mm, 0.4–0.9 mm wide, with a few curved spines on upper body margin, shiny brown, beak 0.2–0.5 mm, ciliate. Coarse, wet, limey soil, subalpine to alpine; 3000–3350 m. c SNH (Mono, Tuolumne cos.); se OR (Steens Mtn). Here treated as distinct from *C. capillaris* L., which occurs no nearer CA than ne OR, ne NV; study needed. Jul–Sep ★

C. tompkinsii J.T. Howell (p. 1341) TOMPKINS' SEDGE (Groups 3, 6) Loosely cespitose. **ST**: many, 10–60 cm, upper angles sharp, scabrous toward tip. **LF**: longest ± equaling infl; blade 1.5–3 mm wide, flat. **INFL**: spikelets 1, occ 2; lower pistillate fl bract 1–15 cm, > infl, lf-like, green, margin white, base clasping peri. **FR**: 3.1–4.9 mm, 1.2–2.5 mm wide, enlarging peri body, next to bristle-like axis; peri 1–5, appressed to ascending, 4.5–6 mm, 1.6–2.6 mm wide, green, with 2 strong marginal veins, gen unveined between, minute-ciliate near or on beak. Open forest, slopes; 400–1900 m. c&s SN. Apr–Jun ★

C. triquetra Boott (p. 1341) TRIGONOUS SEDGE (Group 2) Cespitose. **ST**: 30–60 cm. **LF**: blade 2.5–6 mm wide, adaxially minute-papillate; sheath front red or red-dotted. **INFL**: 1–2 upper lateral spikelets occ staminate over pistillate; lateral spikelets often staminate at base, nodding; infl bract gen < infl, sheath > 5 mm; pistillate fl bract copper, acute to mucronate, midrib occ hairy. **FR**: 3.2–3.5 mm, 1.8–2.5 mm wide, ± = peri body; peri 5–30, 3.3–5 mm, 1.6–2.8 mm wide, elliptic, ± veined, hairs spreading, abruptly narrowed to beak < 1 mm, teeth erect, < 0.3 mm. Clay, often rocky non-serpentine soil; < 1800 m. CCo, SCo, s ChI, TR, PR; n Baja CA. Mar–Jun

C. tumulicola Mack. (p. 1341) FOOTHILL SEDGE (Group 9) Loosely cespitose. **LF:** blade 1–2.5 mm wide. **INFL:** open, often flexible, 2–5 cm, 6–8 mm wide; infl bract bristle-like, often > infl; pistillate fl bract ± = peri, red, white-margined, awned. **FR:** 1.8–2.3 mm, 1.3–2 mm wide; peri erect, 3.5–5 mm, 1.5–2 mm wide, tapered above or not, light green to brown, veins several abaxially, 0 adaxially, beak 1.2–3 mm, teeth ± red, tip notched. Meadows, open woodland; < 1800 m. NCo, NCoR, SN, ScV, CCo, SnFrB, ChI; to BC. Apr–Jul

C. unilateralis Mack. (p. 1341) ONE-SIDED SEDGE (Group 11F) **LF:** blade 2–5 mm wide; ligule 2.5–8 mm. **INFL:** dense, 10–25 mm, axis gen ascending, infl appearing ± lateral, infl bract like a continuation of st, lf-like, base ± clasping st; spikelets ± indistinct; lowest spikelet internode 1.5–4 mm, lowest 2–3 spikelet bracts ± erect to ascending, at least the lowest >> infl, often > 2 mm wide; pistillate fl bract < peri, gold to ± red, center pale or green, occ white-margined, often awned. **FR:** (1.2)1.5–1.9 mm, 0.8–1(1.2) mm wide, near middle of peri body; peri ascending to ascending-spreading, 3.5–5 mm, 1.3–1.75 mm wide, ovate to lanceolate, gen flat around fr, gen sessile, body green, gold, or occ coppery, flat margin incl wing 0.2–0.3 mm wide, veined at least abaxially, beak cylindric, unwinged, entire for < 0.4 mm or gen flat, winged, serrate, gen gold or red-brown. Seasonally wet places; < 1000 m. NCo, NCoRO; to BC. Much like *C. athrostachya.* Jun–Aug

C. utriculata Boott (p. 1341) SOUTHERN BEAKED SEDGE (Groups 4,5) Rhizomed. **ST:** 30–120 cm, angles blunt, base spongy-thickened. **LF:** ± equaling infl; blades of lower lvs, sheaths with well-developed brickwork pattern of crosswalls; ligule length gen < width. **INFL:** lateral spikelets erect, sessile or short-stalked; pistillate fl bract green with ± red margin, awn 0 or < body. **FR:** 1.1–2 mm, 0.9–1.3 mm wide; peri 50–200, in many dense rows, ascending-spreading to reflexed, 3.9–7 mm, 1.3–3 mm wide, inflated, abruptly narrowed to 1–2 mm beak, shiny, gold to brown, teeth erect, 0.1–0.8 mm. 2*n*=72,76,82. Common. Wet places, shallow water; < 3400 m. KR, NCoR, CaRH, SNH, CCo, SnFrB, SnBr, SNE; N.Am, Eurasia. [*C. rostrata* Stokes, misappl.] Jun–Sep

C. vallicola Dewey (p. 1341) WESTERN VALLEY SEDGE (Group 9) Cespitose (loosely so). **ST:** 15–60 cm. **LF:** blade 0.5–2 mm wide. **INFL:** 1–3 cm, 4–8 mm wide; spikelets < 10; infl bract like pistillate fl bract or bristle-like; pistillate fl bract < peri, white, mucronate. **FR:** 1.6–2.5 mm, 1.4–2.1 mm wide; peri spreading, 2.5–4 mm, 1.5–2.2 mm wide, sessile, green to light brown, shiny, abaxially bulged so that marginal ribs instead adaxial, ± veined, adaxial veins 0, beak 0.6–1 mm, teeth ± red. Moist to ± dry slopes, montane; 1800–3100 m. n&c SNH, SNE (Sweetwater Mtns), W&I, MP (exc Wrn); to BC, SD, NM, Mex. Jul–Aug ★

C. vernacula L.H. Bailey (p. 1341) FOETID SEDGE (Group 9) Rhizomed (loosely cespitose). **ST:** 5–30 cm. **LF:** < infl; blade (1.5)2–3.5(4) mm wide. **INFL:** dense, 8–16 mm, 8–16 mm wide; spikelets < 10, indistinct; infl bract like pistillate fl bract; pistillate fl bract dark brown with narrow white tip, shiny, long-tapered or awned. **FR:** 1.2–1.6 mm, 0.7–1.1 mm wide, ovate; peri ascending to spreading, 3.3–4.8 mm, < pistillate fl bract, 1.2–1.8 mm wide, brown, often green-margined, ± shiny, abaxially veined, adaxially veined or not,

margin entire above, beak well-defined, 0.9–1.5 mm, entire. Open, often rocky areas wet from snow-melt; 1800–4000 m. CaRH, SNH, Wrn, SNE (Sweetwater Mtns), W&I; scattered in montane w US. Jul–Sep

C. vesicaria L. (p. 1341) INFLATED SEDGE (Group 5) Cespitose, rhizome short. **ST:** 15–105 cm, angles sharp, scabrous near infl. **LF:** gen much exceeded by infl; blade 1.8–6.5 mm wide; sheath front gen not fibrous on lower lvs; ligule length > width. **INFL:** lateral spikelets short-stalked, erect or lower ones ascending; pistillate fl bract gold to ± purple, awned or not. **FR:** 1.7–2.3 mm, 1.1–1.5 mm wide; peri 20–150, in few ± open rows, ascending, (3.6)4–7(8.2) mm, 1.7–3.5(4.5) mm wide, green to brown,± abruptly narrowed above, beak 1–2.6 mm, teeth erect to outcurved, 0.3–1 mm. 2*n*=74,82,86. Wet places, shallow water; < 3300 m. NW, CaRH, SNH, CCo, SnFrB, MP; to BC, e N.Am; Eurasia. [*C. inflata* Huds., misappl.] Jun–Sep

C. viridula Michx. subsp. ***viridula*** (p. 1341) GREEN YELLOW SEDGE (Groups 4,5) Cespitose. **ST:** 5–40 cm. **LF:** blade 1.5–3 mm wide. **INFL:** dense; terminal spikelet staminate (staminate/pistillate), 1–2 mm wide, linear; lateral spikelets sessile or lowest 1 stalked, erect; infl bracts >> infl, gen long-sheathing, blade ascending-spreading; pistillate fl bract ± red, center green. **FR:** 1.1–1.8 mm, 0.8–1.1 mm wide; peri crowded, spreading to reflexed, 2–3.5 mm, 0.9–1.5 mm wide, body elliptic to ovate or obovate, many-ribbed, yellow-green to brown, beak distinct, 0.6–1.2 mm, narrow-conic. 2*n*=70,72. Sphagnum bogs, wet meadows, dune swales, lakeshores, serpentine fens; < 1800 m. NCo, KR, c SNH (1 site); to AK, e N.Am; Eur, e Asia. [*C. v.* var. *v.*] Jul–Sep ★

C. vulpinoidea Michx. (p. 1341) BROWN FOX SEDGE (Group 9) Cespitose; fl sts gen < lvs. **LF:** blade 2–5 mm wide; sheath front gen cross-wrinkled, minute-red-dotted or not. **INFL:** dense, often ± interrupted, 5–10 cm, 1–1.5 cm wide; spikelets >> 10, > 1 per lower node or branch, often with several to many obvious thin spikelet bracts; pistillate fl bract dull yellow-green to pale gray-brown, base, margin white or not, awns often glabrous, lower gen 1–3 mm. **FR:** 1.2–1.6 mm, 0.8–1 mm wide; peri spreading, gen 2–3 mm, 1–2 mm wide, length ≤ 2 × width, often entire above, dull yellow-green to pale gray-brown, walls ± translucent, abaxial veins 0–3, adaxial 0, adaxial wall ± filled with pithy tissue in ± white U-shaped area around fr, ± flat over fr, beak 0.7–1.8 mm, teeth gold. 2*n*=52,54. Wet areas, marshes, shores, floodplains; < 1200 m. se KR, CaR, n SNF, GV, CCo; e OR to BC, NL, FL, Mex; alien in New Zealand, Eur. First collected outside cult in CA in redwood grove in Golden Gate Park in San Francisco in 1958; first documented in n GV in 1982, now spreading. Much like *C. densa,* but peri with fewer veins; scales and peri more gray-brown than yellow- or red-brown of *C. densa.* Jun–Jul

C. whitneyi Olney (p. 1341) WHITNEY'S SEDGE (Group 1) Cespitose. **ST:** 25–100 cm, base brown (red-brown). **LF:** blade 2–6 mm wide, with long, soft hairs, basal minute; sheath front ± brown. **INFL:** lateral spikelets sessile; infl bract sheath < 5 mm; pistillate fl bract spreading, white. **FR:** 2.8–3.9 mm, 1.7–2.4 mm wide, ± = peri body; peri 3.7–5.5 mm, 1.5–2.5 mm wide, strongly 5-veined, green, glabrous, beak 0.2–1 mm, tip with narrow white margin. Dry, sandy or gravelly meadows, open forest; 1200–3400 m. KR, CaRH, n&c SNH, Wrn; OR, w NV. Jun–Aug

CLADIUM TWIG RUSH

Gordon C. Tucker

Per, glabrous. **ST:** cylindric, stiff, hollow. **LF:** basal and cauline; blade [1]5–10 mm wide, flat [or not], edges, midrib sharp-serrate [or smooth]; sheath permanently sheathing; ligule 0. **INFL:** 1 terminal and gen ≥ 1 axillary, panicle-like, much-branched; infl bracts lf-like; spikelets many, ± 3 mm, not flat; fl bracts spiraled, 5–6, ovate, proximal 1–3 sterile, distal 3–4 each with 1 fl in axil. **FL:** distal-most bisexual, others staminate; perianth 0; stamens 2[3]; style 3-branched, base ± thick, persistent as beak. **FR:** ovoid, not angled, base truncate, tip acute, beak ± 0.2 mm, tubercle 0. 4 spp.: worldwide. (Greek: branch, for infl) [Tucker 2002 FNANM 23:240–242]

C. californicum (S. Watson) O'Neill (p. 1347) CALIFORNIA SAWGRASS Pl 1–2 m, stout, rhizomes < 20 cm. **ST:** 5–10 mm diam. **LF:** blade 1–2 m, 5–10 mm wide, margin saw-toothed. **FR:** 1.5–2 mm, brown. 2*n*=36. Gen alkaline marshes, swamps; 2150 m. s CCo, SCoRO, SCo, WTR, D; to UT, AZ, TX, n Mex. Jun–Sep ★

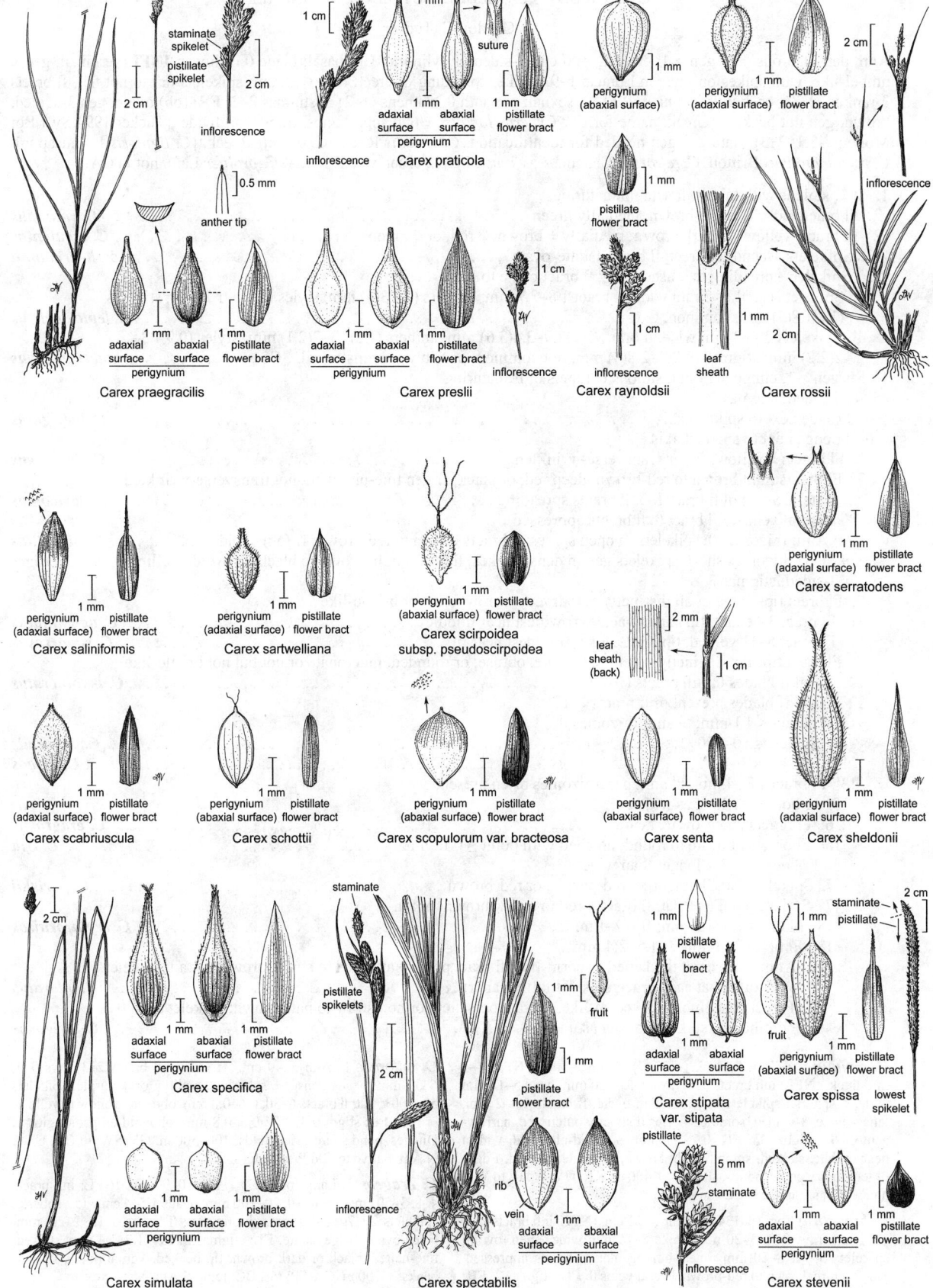

CYPERUS FLATSEDGE, NUTSEDGE, GALINGALE

Gordon C. Tucker

Ann, per, glabrous. **ST:** gen > 1, erect, 2–100 cm, 3-sided or cylindric. **LF:** basal; blade 0 or linear. **INFL:** terminal, gen ± umbel-like, with spikes on rays; infl bracts 1–9, lf-like, spreading or erect; rays ≤ 20 cm; spikelets flat to not flat; fl bracts 2-ranked, 2–36, each with 1 fl in axil. **FL:** bisexual; perianth 0; stamens (1–2) 3; stigmas 2–3. **FR:** (ob) ovoid, gen 3-angled, brown, gen not beaked. (Greek: name for Eur *Cyperus longus*) ± 600 spp.: temps, trops worldwide. [Tucker 1994 Syst Bot Monogr 43:1–213] Mature fr gen needed for identification. *C. gracilis* R. Br., *C. owanii* Boeck. [*C. ligularis* L., misappl.], *C. regiomontanus* Britton, *C. retrorsus* Chapm., *C. virens* Michx., urban weeds in CA. *C. prolifer* Lam. not in CA.

1. St 1; stolons with tubers; fr seldom maturing
  2. Fl bracts purple to red-brown, medially green . . . . *C. rotundus*
  2′ Fl bracts yellow- to dark brown, medially ± brown, ± red, or ± green . . . . *C. esculentus*
    3. Spikelets ascending-erect; fl bract deltate-ovate . . . . var. *heermannii*
    3′ Spikelets spreading to ± ascending; fl bract ovate to lance-ovate
      4. Spikelets (1.2)1.5–2 mm wide; fl bract 1.8–2.7 mm; anthers (1)1.3(6) mm; styles (0.7)1–1.2 mm; stigmas
        (1.2)1.8(5) mm; common, CA. . . . . var. *leptostachyus*
      4′ Spikelets 2.4–3 mm wide; fl bract (2.7)2.9–3.4(3.6) mm; anthers (1.2)1.3–2(2.1) mm; styles (0.9)1.3–
        2(2.2) mm; stigmas (2)2.3–2.8(4) mm; not common, possibly only historical, SCo . . . . var. *macrostachyus*
1′ Sts gen > 1, clumped or not; stolons 0, tubers 0; fr maturing
  5. Fr 2-sided; stigmas 2
    6. Fr face next to spikelet axis . . . . *C. laevigatus*
    6′ Fr edge next to spikelet axis
      7. Fl bracts ± yellow; fr gen transverse-wrinkled . . . . *C. flavescens*
      7′ Fl bracts light brown to red-brown, deep red, or black; fr gen fine-pitted but not transverse-wrinkled
        8. Lateral veins of fl bract 1–2; fl bracts spreading . . . . *C. flavicomus*
        8′ Lateral veins of fl bract 0; fl bracts appressed
          9. Ann; rhizomes 0; spikelets in open spikes; fl bracts brown to red-brown; fr (ob) ovoid . . . . *C. bipartitus*
          9′ Per; rhizomes short; spikelets gen in dense spikes; fl bracts light brown to black; fr ovoid to elliptic . . . . . *C. niger*
  5′ Fr 3-sided; stigmas 3
    10. Fl bract tips some or all distinctly outcurved, narrow-acute or bristle-like
      11. Fl bract 3-veined, tip narrow-acute; fr widest near middle . . . . *C. acuminatus*
      11′ Fl bract 5–11-veined, tip bristle-like; fr widest near tip . . . . *C. squarrosus*
    10′ Fl bract tips not distinctly outcurved, acute, obtuse, or rounded, mucronate or not but not bristle-like
      12. Basal lf blades 0; infl bracts (10)14–22 . . . . *C. involucratus*
      12′ Basal lf blades present; infl bracts 1–12
        13. Fl bract < 1.1 mm; pl ann; rhizomes 0
          14. Spikelets 50–100+. . . . . *C. difformis*
          14′ Spikelets 3–15. . . . . *C. fuscus*
        13′ Fl bract 1.3–4 mm; pl ann, per; rhizomes often present
          15. Fr body ± as long as wide
            16. Fl bracts lance-ovate, acute . . . . *C. eragrostis*
            16′ Fl bracts obovate to round, notched with point ≤ 0.12 mm . . . . *C. iria*
          15′ Fr body 1.5–3 × longer than wide
            17. Spikelets flat; fl bract red, red-purple, or red-brown . . . . *C. parishii*
            17′ Spikelets not to ± flat; fl bract ± red to straw, brown, or tan
              18. Fl bract 1.3–1.5 mm; fr 0.7–1 mm. . . . . *C. erythrorhizos*
              18′ Fl bract 2–4 mm; fr 1.5–2.4 mm
                19. Ann; st base not thickened or corm-like; fl bract persistent, ± red to straw, brown, or tan; spikelet
                  axis jointed at each bract, spikelet falling apart . . . . *C. odoratus*
                19′ Per; st base thickened, corm-like; fl bract persistent or not, straw to pale brown; spikelet axis
                  continuous, spikelet falling as unit. . . . . *C. strigosus*

C. acuminatus Torr. & Hook. (p. 1347) Ann 5–40 cm. **ST:** 0.4–1 mm thick. **INFL:** infl bracts 3–6; rays 0–5, < 20 mm; head 5–18 mm wide, ± spheric; spikelets 4–7 mm, flat, ovate; fl bracts 1.3–2 mm, lance-ovate, 3-veined, some or all tips distinctly outcurved, narrow-acute. **FL:** stigmas 3. **FR:** 0.7–1.1 mm, obovoid-ellipsoid, widest near middle, stalked, smooth, light brown, tip acute, beaked. Edges of temporary pools, ponds, streams; < 400 m. KR, GV, WTR; to WA, n c&e US, s Can, ne Mex. Jun–Oct

C. bipartitus Torr. (p. 1347) Ann 3–25 cm. **INFL:** infl bracts 2–3, 1–12 cm; rays 0–4, 3–20 mm; spike 8–15 mm wide, open, ovoid; spikelets 3–10, 8–20 mm, linear-oblong, flat; fl bracts appressed, 1.8–2.3 mm, brown to red-brown, lateral veins 0. **FL:** stigmas 2. **FR:** 1–1.5 mm, (ob) ovoid, 2-sided, fine-pitted. Sandbars, pond shores; 100–1500 m. NW, c SNF, n SNH; to Can, S.Am. Jul–Nov

C. difformis L. Ann 3–40 cm. **INFL:** infl bracts 2–3; rays 0–4, 6–30 mm; heads dense, ± spheric, ± yellow or ± purple; spikelets 50–100+, flat; fl bracts 6–30, 0.6–0.8 mm, obovate, yellow to yellow-brown. **FL:** stigmas 3. **FR:** 0.6–0.8 mm, obovoid-ellipsoid, glossy. Ditches, pond shores, rice fields; 100–500 m. GV, SW; to OR, UT, e N.Am; native to Old World. Jul–Nov

C. eragrostis Lam. (p. 1347) Per 10–90 cm. **INFL:** infl bracts 4–8, 3–50 cm; rays 0–10, 20–100 mm; heads 1.5–4 cm wide, spheric; spikelets 20–70, 5–20 mm, oblong, flat; fl bracts 6–12, 2–2.3 mm, lance-ovate, beige, acute. **FL:** stigmas 3. **FR:** 1.2–1.4 mm, stalked, fine-netted, black or dark brown, tip beaked. Vernal pools, streambanks; < 700 m. CA-FP; to s BC; temp S.Am. May–Nov

C. erythrorhizos Muhl. (p. 1347) Ann 5–100 cm; roots ± red. **INFL:** infl bracts 3–11, 5–70 cm; rays 2–6(12), 1–8(28) cm; spikes

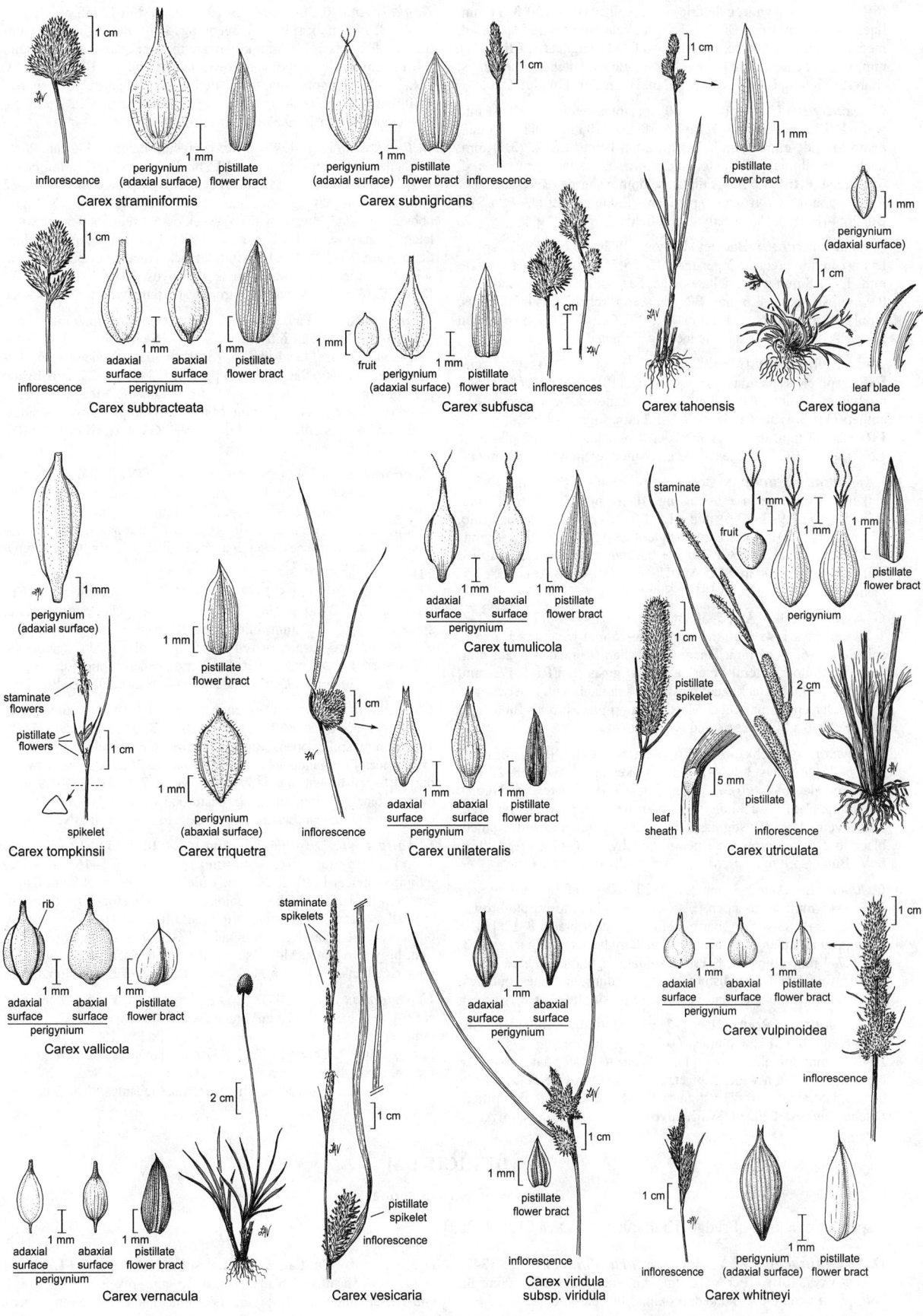

Carex straminiformis — inflorescence; perigynium (adaxial surface); pistillate flower bract

Carex subnigricans — perigynium (adaxial surface); pistillate flower bract; inflorescence

Carex tahoensis — pistillate flower bract; perigynium (adaxial surface)

Carex tiogana — leaf blade

Carex subbracteata — inflorescence; adaxial surface / abaxial surface (perigynium); pistillate flower bract

Carex subfusca — fruit; perigynium (adaxial surface); pistillate flower bract; inflorescences

Carex tompkinsii — perigynium (adaxial surface); staminate flowers; pistillate flowers; spikelet

Carex triquetra — pistillate flower bract; perigynium (abaxial surface)

Carex unilateralis — inflorescence; adaxial surface / abaxial surface (perigynium); pistillate flower bract

Carex tumulicola — adaxial surface / abaxial surface (perigynium); pistillate flower bract

Carex utriculata — staminate; fruit; perigynium; pistillate flower bract; pistillate spikelet; leaf sheath; pistillate; inflorescence

Carex vallicola — rib; adaxial surface / abaxial surface (perigynium); pistillate flower bract

Carex vernacula — adaxial surface / abaxial surface (perigynium); pistillate flower bract

Carex vesicaria — staminate spikelets; pistillate spikelet; inflorescence

Carex viridula subsp. viridula — adaxial surface / abaxial surface (perigynium); pistillate flower bract; inflorescence

Carex vulpinoidea — adaxial surface / abaxial surface (perigynium); pistillate flower bract; inflorescence

Carex whitneyi — inflorescence; perigynium (adaxial surface); pistillate flower bract

(6)10–16(23) mm wide, cylindric-ovoid; spikelets 20–150, 3–11 mm, linear, ± flat; fl bracts 6–30, 1.3–1.5 mm, light brown, ± red-speckled, medially ± green, obtuse, lateral veins 0. **FL:** stigmas 3. **FR:** 0.7–1 mm, sides unequal, light gray to brown, glossy. Ditches, riverbanks, shores; < 500 m. CA-FP, DSon; s Can, US, n Baja CA. Jul–Oct

C. esculentus L.   Per (10)20–80 cm; stolons with tubers < 11 mm wide. **INFL:** infl bracts 3–7; rays 4–10, 20–110 mm; spikes ± open, ovate to wide-elliptic; spikelets linear; fl bracts 1.8–2.7(3.4) mm, ovate to ± elliptic, yellow- to dark brown, medially ± brown, ± red, or ± green. **FR:** 1–1.6 mm, elliptic, seldom maturing. Recognition of vars. supported by growth experiments (Schippers et al. 1995 Syst Bot 20:461–481). Occ weedy in cult fields.

var. *heermannii* (Buckley) Britton   Pl 20–25 cm. **INFL:** spikelets ascending-erect, ≥ 2 per mm, 15–25(55) mm; fl bract 2.4–2.6 mm, 1.7–1.8 mm wide, deltate-ovate. **FL:** anthers 1.2 mm; styles ± 0.9 mm, stigmas ± 1.6 mm. **FR:** not seen. Probably disturbed soils, cropland, riverbanks; < 1200 m. PR; UT, FL, Mex. Type specimen from CA, but without specific locality. Summer

var. *leptostachyus* Boeck. (p. 1347)   Pl (10)30–40(65) cm. **INFL:** spikelets spreading to ± ascending, (10)15–20 mm, (1.2)1.5–2 mm wide; fl bract 1.8–2.7 mm, (1)1.5–2 mm wide, lance-ovate. **FL:** anthers (1)1.3(6) mm; styles (0.7)1–1.2 mm, stigmas (1.2)1.8(5) mm. **FR:** 1.2–1.3 mm, 0.6–0.7 mm wide. Cropland, disturbed places; < 1200 m. CA; BC, US, Mex, C.Am, S.Am; alien in w Eur. Summer

var. *macrostachyus* Boeck.   Pl (20)40–70(80) cm. **INFL:** spikelets spreading to ± ascending, 10–40 mm, 2.4–3 mm wide; fl bract (2.7) 2.9–3.4(3.6) mm, (1.8)1.9–2.3(2.4) mm wide, ovate. **FL:** anthers (1.2)1.3–2(2.1) mm; styles (0.9)1.3–2(2.2) mm, stigmas (2)2.3–2.8(4) mm. **FR:** ± 1.7 mm, ± 0.6 mm wide. Dunes, disturbed soil, cropland; < 200 m. SCo; e&s US, Mex, C.Am, S.Am. 1 CA collection (San Diego, Orcutt 1314). Summer

C. flavescens L.   Ann 5–30 cm. **ST:** 3–6. **INFL:** infl bracts 2–3, 1–12 cm; rays 1–4, 3–30 mm; spikes 8–25 mm wide, open, ovoid; spikelets 1–6, 5–15 mm, lance-oblong, flat; fl bracts 1.8–2.2 mm, ovate, ± yellow, lateral veins 0. **FL:** stigmas 2. **FR:** 1–1.2 mm, obovoid, 2-sided, black, gen transverse-wrinkled, with a network of rectangular epidermal cells. Marshy, damp places; 500–1000 m. c SNF, ScV; to s Can; S.Am, Old World. Jun–Oct

C. flavicomus Michx.   Ann 30–75 cm. **ST:** 1–3. **INFL:** infl bracts 3–7, 5–35 cm; rays 3–6, 4–20 mm; spikes open, ovoid, 8–22 mm wide; spikelets 6–20(60), 7–30 mm, linear to lanceolate, flat; fl bracts spreading, (1.4)1.7–2.3 mm, ovate-obovate, light brown to deep red, lateral veins 1–2. **FL:** stigmas 2. **FR:** 1.2–1.6 mm, obovoid, 2-sided, black to dark red-brown, ± smooth. Marshy, damp places; ± 100 m. ScV (Butte Co.); to e US, Mex, S.Am, Old World. Aug–Oct

C. fuscus L.   Ann 2–30 cm. **ST:** 3–20. **INFL:** infl bracts 2–3; rays 1–3, 2–15 mm; heads open, ± spheric, ± yellow to purple-brown; spikelets 3–15, 3–7 mm, narrow-elliptic, flat; fl bracts 8–12(16), ± 1 mm, round, mucronate, lateral veins 0, midveins 3. **FL:** stigmas 3. **FR:** 0.7–1 mm, ellipsoid, ± sessile, smooth, light brown, base wedge shaped, tip acute. Damp disturbed soil, receding shorelines, puddles; < 50 m. GV; to e N.Am; native to temp Eurasia. Jul–Oct

C. involucratus Rottb.   Per 1–2 m. **ST:** 1–20, ± 3-angled. **LF:** basal blades 0. **INFL:** infl bracts (10)14–22, blades long; rays 14–22, 20–200 mm; spikelets 5–15 mm, ± linear to ovate, flat, in stellate clusters 15–30 mm wide; fl bracts 1.6–2.4 mm, ovate-deltate, light brown, lateral veins 0. **FL:** stigmas 3. **FR:** 0.6–0.8 mm, fine-pitted. Ditches, shores; < 200 m. SCo; native to e Afr, sometimes cult. All yr

C. iria L.   Ann 20–30 cm. **ST:** 3–10, 3-sided. **INFL:** infl bracts 4–7; rays 4–8, 1–50 mm; spikelets ascending, 4–20 mm, ± linear to elliptic, flat; fl bracts 1.3–1.8 mm, obovate to round, notched with point ≤ 0.12 mm, brown to golden brown, lateral veins 2. **FL:** stigmas 3. **FR:** 1.2–1.4 mm, obovoid, fine-pitted, glossy. Ditches, pond shores; < 100 m. ScV (Yuba Co.); e US; Mex, C.Am, S.Am, Old World. 1st collected in CA in 1999. Aug–Oct

C. laevigatus L. (p. 1347)   Per 1–60 cm; rhizomes 1–2 mm thick, horizontal. **ST:** clumped or not. **LF:** blade 0–7 cm. **INFL:** infl bracts 1–3, 1–10 cm; rays 0–1, to 2 mm; heads ± ovoid; spikelets 1–14, 4–12 mm, elliptic to oblong lanceolate, ± flat; fl bracts 8–24, 1.5–2 mm, obovate to round, ± white with red speckles or 1 red spot 0.5 mm wide, lateral veins 0–several. **FL:** stigmas 2. **FR:** 1.2–1.8 mm, oblong-elliptic to ovate, 2-sided, dark brown to ± black. Alkaline or brackish, wet soils, hot springs, permanent pools in arroyos; 30–1000 m. SW, GB, D; to TX, NC; Mex, scattered in trop, warm temp worldwide. Jul–Dec

C. niger Ruiz & Pav. (p. 1347)   Per 1–50 cm; rhizomes short, ± 1 mm thick. **INFL:** infl bracts 2–3, 1–15 cm; rays 0–2, 0–5(40) mm; spikes dense, head-like, 5–15 mm wide; spikelets 3–25, 3–9 mm, linear-oblong, flat; fl bracts appressed, 3–9, 1.5–2.2 mm, ovate, light brown to black, lateral veins 0. **FL:** stigmas 2. **FR:** 1.2–1.4 mm, ovoid to elliptic, 2-sided, brown. Marshes, swamps, moist roadsides; < 1500 m. KR, NCoR, CaR, SNF, n SNH, GV, CW, SCo, PR, SNE; to OK; S.Am, S.Afr. Jul–Nov

C. odoratus L. (p. 1347)   Ann 10–50 cm. **INFL:** infl bracts 5–9, 5–24 cm; rays 6–12, 10–100 mm; spikes ± open, ovoid; spikelets linear, not to ± flat; fl bracts 6–24, 2.1–3.5 mm, elliptic to ovate, ± red to straw, brown, or tan, medially green. **FL:** stigmas 3. **FR:** 1.5–1.9 mm, ± flat front-to-back. Wet disturbed soils; < 500 m. GV, SCo, D; trop, warm temp. Jul–Oct

C. parishii Britton (p. 1347)   Ann 5–25 cm, cespitose. **INFL:** infl bracts 2–5, 3–20 cm; rays 1–6, 20–70 mm; spikes open, spheric; spikelets 5–30, 6–22 mm, linear, flat; fl bracts 8–10, gen ± 3 mm, elliptic, red, red-purple, or red-brown, medially green, deciduous, 7–9-veined. **FL:** stigmas 3. **FR:** 1–1.3 mm, wide-elliptic, (dark ± purple) brown. Streambanks, roadsides; < 800 m. SW, D; AZ. Jul–Oct

C. rotundus L.   PURPLE NUTSEDGE   Per 10–40 cm; stolons with tubers < 1 cm thick. **INFL:** infl bracts 2–5, 5–25 cm; rays 5–10, 10–150 mm; spikes open, elliptic, 1–5 cm wide; spikelets 3–12, 4–40 mm, linear to ± lanceolate, flat; fl bracts 6–36, 2–3.4 mm, ovate, purple to red-brown, medially green, veins 7–9. **FR:** 1.4–1.9 mm, elliptic, black, seldom maturing. Disturbed soils, cropland; < 250 m. GV, SCo, DSon; trop, warm temp; native to Eurasia. Jul–Nov   ◆

C. squarrosus L. (p. 1347)   Ann 1–10(20) cm. **INFL:** infl bracts 1–4, 1–15 cm; rays 0–6, 4–40 mm; spikes open, 5–15 mm wide, ovoid; spikelets 1–30, 2–20 mm, wide-lanceolate to oblong, flat; fl bracts 2–12, 1.2–2 mm, lance-oblong, ± green to straw or brown-red, medially ± green, 5–11 veined, tip bristle-like, 0.5–1.3 mm, strongly outcurved. **FL:** stamen 1; stigmas 3. **FR:** 0.7–1.1 mm, ± obovoid, light brown to black. Moist, sunny, disturbed places, esp pond margins, riverbanks; < 1500 m. CA; temp, trop ± worldwide. Jun, Nov

C. strigosus L. (p. 1347)   FALSE NUTSEDGE   Per 5–70 cm. **ST:** corm-like at base. **INFL:** infl bracts 3–6, 5–30 cm; rays 0–7, 20–150 mm; spikes open, 1–5 cm wide, ovoid; spikelets 15–50, 6–30 mm, linear, ± flat; fl bracts 3–11, 2.8–4 mm, lanceolate, straw to pale brown, medially green, veins 7–11. **FL:** stigmas 3. **FR:** 1.8–2.4 mm, narrow-elliptic. Moist soils, pond margins, roadsides; < 1000 m. CA; US, s Can. Jul–Oct

DULICHIUM

Joy Mastrogiuseppe

1 sp. (Latin: a kind of sedge) [Ball 2002 FNANM 23:252–253]

D. arundinaceum (L.) Britton var. *arundinaceum* (p. 1347)   THREE-WAY SEDGE   Per < 10 dm, rhizomes long. **ST:** cylindric, hollow. **LF:** cauline, 3-ranked, proximal bladeless; blade 0–15 cm, (1.5)3–8 mm wide, flat; sheath not splitting, green. **INFL:** spikes 1–17, from sheaths of bladed lvs, 6–30 cm; spikelets 3–10, 1.5–3 cm, 1.2–2.5 mm wide, flat; fl bracts 2-ranked, 4–8, 3–9.5 mm, lance-

linear, veined, each with 1 fl in axil. **FL:** bisexual; perianth bristles 6–9, ± 4–7.5 mm, ± > fr, barbs reflexed; stamens 3; style 2-branched, base persistent. **FR:** 2–4 mm (exc style base), linear-ellipsoid, flat, yellow. 2*n*=32. Uncommon. Lake, pond stream margins, bogs, wet meadows, often emergent; 700–2400 m. KR, NCoRO, CaR, SN; to BC, e N.Am. Known as fossil from Eur, e Asia. Jun–Oct

ELEOCHARIS SPIKERUSH

Ann, per, gen forming mats, glabrous, internal air cavities evident; caudex gen 0; rhizomes gen evident, long, scaly, bulb or tuber at tip gen 0. **ST:** simple, gen erect, smooth, gen not hollow; tip gen not rooting. **LF:** 2, basal, blades 0 or tooth-like, ≤ 1 mm. **INFL:** infl bracts 0; spikelet terminal, 1, gen ovate, not ± flat [(± flat)], gen not forming plantlets, fls 3–100+; fl bracts spiraled [(2-ranked)], each with 1 fl in axil, gen ovate, gen brown, gen membranous, smooth, tip gen acute to obtuse, notch 0; basal fl bract gen encircling st, gen < 1/2 spikelet, fl gen 0. **FL:** bisexual; perianth parts reduced to bristles, 0–8, gen ± ≤ fr, barbs gen recurved; stamens gen 3; style 1, thread-like, base enlarged, gen persistent on fr as tubercle. **FR:** gen obovate, gen brown; tubercle (0 or) gen distinct, gen pyramidal. Wetland obligates. ± 200 spp.: trop to boreal. (Greek: marsh-dwelling grace) [Smith et al. 2002 FNANM 23:60–120] *E. lanceolata* Fernald, *E. equisetoides* Torr. not in CA.

1. St 4-angled, 1–5.4 mm diam . . . . *E. quadrangulata*
1′ St rarely 4-angled, if so then < 1 mm diam
  2. Rhizomes, caudices 0; ann (per)
    3. Distal lf sheath delicate, often disintegrating; basal fl bract with a fl; stigmas 3; fr with ± 6–10
      longitudinal ridges, many fine transverse bars; pls 1–7 cm . . . . ²*E. bella*
    3′ Distal lf sheath firm, persistent; basal fl bract with a fl or not; stigmas 2–3; fr smooth; pls 1–50 cm
      4. Fr black (brown when immature); tubercle not completely adherent to fr, not flat
        5. Fr 0.3–0.5 mm; perianth bristles gen ± white; fl bracts membranous. . . . . *E. atropurpurea*
        5′ Fr 0.5–1.1 mm; perianth bristles gen red-brown; fl bracts membranous to papery . . . . *E. geniculata*
      4′ Fr brown; tubercle completely adherent to fr, flat
        6. Tubercle ≤ 1/4 fr; perianth bristles 0 or 5–8, vestigial, << fr, or ± = fr. *E. engelmannii*
          7. Perianth bristles 0 or vestigial, << fr . . . . var. *detonsa*
          7′ Perianth bristles ± = fr . . . . var. *engelmannii*
        6′ Tubercle ≥ 1/3 fr; perianth bristles (0 or) 5–7, to gen much exceeding tubercle
          8. Tubercle 0.5–0.8 mm wide, 2/3–9/10 as wide as fr . . . . *E. obtusa*
          8′ Tubercle 0.3–0.5 mm wide, 1/2–3/4 as wide as fr . . . . *E. ovata*
  2′ Rhizomes or caudices present; per (ann)
    9. Distal lf sheath delicately membranous, often disintegrating; st 0.2–1 mm diam
      10. Basal fl bract without a fl; fr with 2–3 longitudinal angles, 0 transverse bars
        11. Fr 2-sided; distal lf sheath tip much inflated, gen wrinkled; tubercle distinct from fr; rhizome tip
          without tuber. . . . . *E. flavescens* var. *flavescens*
        11′ Fr (2)3-sided; distal lf sheath not inflated or wrinkled; tubercle often merging with fr or vestigial;
          rhizome tip often with tuber
          12. Perianth bristles 0 or vestigial or ± 1/2 fr; fr finely roughened; rhizome tip tubers 0 or wide-oblong
            to spheric, not curved; gen inland . . . . *E. coloradoensis*
          12′ Perianth bristles (vestigial or) gen = fr to ± exceeding tubercle; fr smooth; rhizome tip tubers 0 or
            oblong, often curved; gen coastal . . . . *E. parvula*
    10′ Basal fl bract with a fl; fr with 6–12 longitudinal ridges, many fine transverse bars
      13. Fl bract 1–1.5 mm, tip narrow-acute to acuminate, ± recurved; anthers 0.3–0.5 mm . . . . ²*E. bella*
      13′ Fl bract 1.5–2.5(3.5) mm, tip blunt to acute, not recurved; anthers 0.3–1.5 mm
        14. Anthers ≤ 0.5 mm; st very spongy, without angles or ridges, 0.4–1 mm diam. . . . . *E. radicans*
        14′ Anthers 0.7–1.5 mm; st not very spongy, often with 3–12 angles or ridges, 0.2–0.5(0.7) mm diam
          . . . . *E. acicularis*
          15. Fr 2 × longer than wide; st often 3–4-angled, base not corm-like. . . . . var. *acicularis*
          15′ Fr < 2 × longer than wide; st often 5–12-ridged, base corm-like or not
            16. St to 60 cm, base not corm-like . . . . var. *gracilescens*
            16′ St to 6 cm, base corm-like. . . . . var. *occidentalis*
    9′ Distal lf sheath firmly membranous to papery, persistent; st 0.2–4 mm diam
      17. Distal lf sheath tips some or all with tooth-like projection to 1 mm
        18. Stigmas 2; fr 2-sided; basal fl bracts not completely clasping st and/or at least some spikelets with
          next-to-basal bract without fl . . . . ²*E. macrostachya*
        18′ Stigmas 3; fr 3-sided; basal fl bracts completely clasping st, next-to-basal bract with fl
          19. Fl bracts 5–10 per 1 mm of spikelet (spikelet 4–12 mm, fl bracts 30–100), tips wide-rounded,
            often transversely wrinkled and recurved . . . . ²*E. montevidensis*
          19′ Fl bracts 3–4 per 1 mm of spikelet (spikelet 3–20 mm, fl bracts 15–40), tips rounded to acute,
            not transversely wrinkled, not recurved . . . . ²*E. parishii*
      17′ Distal lf sheath tips without tooth-like projection
        20. Stigmas 2; fr 2-sided; basal fl bract without fl
          21. Basal fl bracts, or at least some, clasping ≥ 3/4 of st, next-to-basal with a fl or not . . . . ²*E. macrostachya*
          21′ Basal fl bracts clasping 2/3 to 3/4 of st, next-to-basal without a fl . . . . *E. palustris*

20′ Stigmas (2)3; fr (2)3-sided; basal fl bract with or without fl
  22. Rhizome mostly hidden by sts and/or ≥ 3 mm diam, often caudex-like, tip without enlarged bud or bulb
    23. St sharp-4-angled; spikelets often forming plantlets . . . . ***E. pachycarpa***
    23′ St not sharp-4-angled; spikelets not forming plantlets (st tips often rooting in *Eleocharis rostellata*)
      24. Sts ± flat, some elongating and arching, tips often rooting; fl bracts 3.5–6 mm; fr 1.5–2.5 mm . . . ***E. rostellata***
      24′ St subcylindric, not elongating and arching, tips not rooting; fl bracts 2–3.5 mm; fr 0.8–1.2 mm
        25. Rhizome 1.5–3 mm diam, hidden by sts; fl bracts dark brown; st 0.3–0.5 mm diam
        . . . . ***E. bolanderi***
        25′ Rhizome 3–5 mm diam, mostly not hidden by sts; fl bracts medium brown; st 0.3–2 mm diam. .
        E. decumbens
  22′ Rhizome not hidden by sts, ≤ 2 mm diam, tip with enlarged bud or bulb
    26. Fl bracts 1.5–2.5 mm; rhizome tough, tip without enlarged bud or bulb; tubercle distinct from fr;
      basal fl bract < 1/2 spikelet
      27. Fl bract tips in proximal 1/2 of spikelet wide-rounded, gen transversely wrinkled and recurved;
        fr fine-wrinkled at 10–30× . . . . ²***E. montevidensis***
      27′ Fl bract tips in proximal 1/2 of spikelet rounded to acute, gen not transversely wrinkled and
        recurved; fr smooth, wrinkled, or lattice-like at 10–30× . . . . ²***E. parishii***
    26′ Fl bracts 2.5–6 mm; rhizome weak, tip often with enlarged bud or bulb; tubercle often ± not
      distinct from fr; basal fl bract gen ≥ 1/2 spikelet
      28. St-tuft base gen swollen, often with bulbs; rhizome tip often with bulb; caudex gen 0 . . . . ***E. quinqueflora***
      28′ St-tuft base not swollen, without bulbs; rhizome tip without bulb; caudex present
        29. St 1.5–2.5 mm wide, flat, spirally twisted; n SNH . . . . ***E. torticulmis***
        29′ St 0.5–1.2 mm wide, cylindric to ± flat, not spirally twisted; NW, CaRH, SN, SnFrB, TR, DMtns
          30. St strongly arched; perianth bristles 4–7, some < 1/2 fr, others equaling or exceeding tubercle;
          TR . . . . ***E. bernardina***
          30′ St gen straight; perianth bristles 6, longest = fr to exceeding tubercle; NW, CaRH, SN, SnFrB,
          TR, DMtns . . . . ***E. suksdorfiana***

E. acicularis (L.) Roem. & Schult. NEEDLE SPIKERUSH Per 1–60 cm, often forming mats; rhizome weak, to 0.5 mm diam. **ST**: 0.2–0.5(7) mm diam, subcylindric, often 3–12- angled or -ridged. **LF**: distal sheath delicate, often disintegrating, tip inflated or not. **INFL**: spikelet 2–8 mm, 1–2 mm wide, often 0 on submersed pls; fl bracts to 25, 1.5–2.5(3.5) mm, tip blunt to acute, not recurved, basal with fl. **FL**: anthers 0.7–1.5 mm; stigmas 3. **FR**: 0.7–1 mm, 0.4–0.6 mm wide, 3-sided to ± round in ×-section, often ± white, longitudinal ridges ± 8–12, crossbars 30–60 per ridge; perianth bristles 0 (2–4, ≤ fr). Deeply submersed variants (*E. acicularis* var. *submersa* (Nilsson) Svenson), which often form large vegetative mats and often long thread-like sts without spikelets, may closely resemble *Schoenoplectus subterminalis*. Vars. often not identifiable.

  var. ***acicularis*** (p. 1347) **ST**: to 60 cm, often 3–4-angled, base not corm-like. **FR**: 2 × longer than wide. Common. Fresh wet soil to deeply submersed; < 3300 m. CA-FP, MP; to AK, e N.Am, S.Am (Ecuador), Eurasia. Late spring–summer

  var. ***gracilescens*** Svenson **ST**: to 60 cm, often 5–12-ridged, base not corm-like. **INFL**: spikelet to 8 mm; fl bracts to 3.5 mm. **FR**: < 2 × longer than wide. Fresh wet soil to deeply submersed; < 3300 m. NCo, NCoR, SN, ScV; to OR, TN. Late spring–summer

  var. ***occidentalis*** Svenson **ST**: to 6 cm, often 5–8-ridged, base corm-like. **FR**: < 2 × longer than wide. Common. Fresh wet soil to deeply submersed; < 3300 m. NCoRO, CaR, SN, ScV; to WA, WY, NM. Late spring–summer

E. atropurpurea (Retz.) J. Presl & C. Presl (p. 1347) Ann 2–12(19) cm, tufted; rhizome 0. **ST**: to 0.3 mm diam, subcylindric. **LF**: distal sheath firm, persistent, tip acute. **INFL**: spikelet 2–6 mm, 1–2.5 mm wide; fl bracts to 100, 0.6–1.3 mm, basal with fl or not. **FL**: anthers to 0.5 mm; stigmas 2. **FR**: 0.3–0.5 mm, 0.3–0.4 mm wide, 2-sided, smooth, black (brown when immature); perianth bristles (0)4(6), vestigial or ≤ fr, gen ± white. 2*n*=20. Very local. Fresh wet bare soil; < 800 m. GV, SCo, SnBr; to BC; also NM to MI, FL. Possibly introduced from e N.Am; several CA collections are weeds in rice fields; earliest CA collection in 1932. Summer–fall

E. bella (Piper) Svenson (p. 1347) Ann (per?) 1–7 cm, tufted or forming small mats; rhizome to 0.3 mm diam, weak, or gen 0. **ST**: 0.2–0.3 mm diam, 4-angled or cylindric. **LF**: distal sheath delicate, often disintegrating, tip acute. **INFL**: spikelet 1.5–4 mm, 1–2 mm wide; fl bracts 4–15, 1–1.5 mm, narrow-acute to acuminate, ±

recurved, basal with fl. **FL**: anthers 0.3–0.5 mm; stigmas 3. **FR**: 0.6–0.8 mm, 0.3–0.4 mm wide, obscure-3-sided or ± round in ×-section, ± white or pale, longitudinal ridges ± 6–10, crossbars 20–30 per ridge; perianth bristles 0. Common. Fresh wet bare soil; < 2300 m. NW, CaR, SN, ScV, SCoRO, SW, GB; to WA, MT, NM; Mex (Chihuahua). [*E. acicularis* var. *b.* Piper] Spring–summer

E. bernardina (Munz & I.M. Johnst.) Munz & I.M. Johnst. Per 5–15 cm; rhizome 0.5–1 mm diam, weak, tip with elliptic bud 10 mm, 2–5 mm wide; caudex hard. **ST**: strongly arched, 0.5–1 mm diam, subcylindric. **LF**: distal sheath firm, persistent, tip subtruncate or wide-obtuse. **INFL**: spikelet 3–6 mm, 2–4 mm wide; fl bracts 4–9, 3.5–5 mm, basal ≤ 1/2 spikelet. **FL**: anthers 1–1.5 mm; stigmas 3. **FR**: 1.5–2.1 mm, 0.9–1.2 mm wide, 2–3-sided, smooth or finely net-veined, tip often beak-like; tubercle often merging with fr; perianth bristles 4–7, unequal, some < 1/2 fr, others equaling or exceeding tubercle. Very local. Fresh wet meadows, fens in conifer forest; 2100–2700 m. TR; Baja CA. Summer

E. bolanderi A. Gray (p. 1347) Per 10–30 cm, in ± round colonies; rhizome caudex-like, hidden by sts. **ST**: 0.3–0.5 mm diam, subcylindric. **LF**: distal sheath firm, persistent, tip obtuse. **INFL**: spikelet 3–8 mm, 2–3 mm wide; fl bracts 8–30, 2.5–4 mm. **FL**: anthers 0.9–1.4 mm; styles 3. **FR**: 0.9–0.7 mm, 0.7–0.8 mm wide, (2)3-sided, fine-wrinkled or -net-like, straw; tubercle often merging with fr; perianth bristles 3–6, vestigial or ≤ 1/2 fr. Uncommon, local. Fresh wet meadows, fens, springs, stream margins; 410–1580 m. SN, MP; to OR, CO. Because specimens without rhizomes confused with *E. decumbens*, CA distributions of both spp. need study. Late spring–summer

E. coloradoensis (Britton) Gilly Per 2–9 cm; rhizome weak, to 1.5 mm diam, tip often with tuber 2.5–4 mm, 0.7–1.5 mm wide, wide-oblong to spheric. **ST**: 0.2–0.5 mm diam, cylindric. **LF**: distal sheath delicate, often disintegrating, tip rounded. **INFL**: spikelet 3–6 mm, 1–1.5 mm wide; fl bracts 6–25, 1.7–2.5 mm. **FL**: anthers 0.6–0.9 mm; stigmas (2)3. **FR**: 0.8–1 mm, 0.6–0.7 mm wide, (2)3-sided, fine-roughened; tubercle often merging with fr or ± 0; perianth bristles 0 or vestigial (5, unequal, ± 1/2 fr). *n*=5. Very local. Fresh to brackish bare wet soil, inland; < 1500 m. CaRH, GV, SCoRO, s SCo, TR, PR, MP, DSon; to OR, SK, MS, Mex. May be used to control weeds in irrigation canals (Yeo 1980 Weed Sci 28:263–272). Spring–fall

E. decumbens C.B. Clarke Per 10–50 cm; rhizome 3–5 mm diam, mostly not hidden by sts, tough. **ST**: 0.3–2 mm diam, sub-

cylindric. **LF:** distal sheath firm, persistent, tip subtruncate to obtuse. **INFL:** spikelet 3–8 mm, 2–2.5 mm wide; fl bracts 10–20, 3–3.5 mm. **FL:** anthers 1.2–1.5 mm; stigmas 3. **FR:** 1–1.3 mm, 0.8–0.9 mm wide, 3-sided, fine-roughened; perianth bristles 6, gen equaling or exceeding tubercle. Fresh wet meadows, fens, seeps, lake shores; 500–2500 m. KR, NCoRI, CaRH, SN; s OR. See note under *E. bolanderi.* Summer

E. engelmannii Steud. Ann 2–40 cm, tufted; rhizome 0. **ST:** 0.5–2 mm diam, cylindric. **LF:** distal sheath firm, persistent, tip obtuse to acute, with tooth-like projection to 0.3 mm. **INFL:** spikelet 5–10(20) mm, 2–3(4) mm wide; fl bracts 25–100+, 2–2.5 mm. **FL:** anthers 0.3–1 mm; stigmas 2–3. **FR:** 0.9–1.1 mm, 0.7–1 mm wide, 2–3-sided, smooth, tip wide-truncate; tubercle adherent to fr, flat, < 1/4 fr, 9/10 as wide as fr; perianth bristles 0 or 5–8, vestigial to ± = fr [± exceeding tubercle]. $2n=10$. [*E. obtusa* var. *e.* (Steud.) Gilly] Because of confusion with *E. obtusa, E. ovata,* CA distributions of all 3 spp. need study.

   var. ***detonsa*** A. Gray **FR:** perianth bristles 0 or vestigial, << fr. Fresh wet bare soil; 30–2400 m. NW, CaRH, SN, MP (Grasshopper Valley, Lassen Co.). Spring–fall

   var. ***engelmannii*** (p. 1347) **FR:** perianth bristles ± = fr. Fresh wet bare soil; 30–2400 m. NCoR, CaRF (3 mi n of Redding, Shasta Co.), SN, n CCo (Marin Co.), SCo (Riverside Co.), n MP; to s Can, e US. Spring–fall

E. flavescens (Poir.) Urb. var. ***flavescens*** (p. 1347) Per 5–42 cm; rhizome 1 mm diam, firm. **ST:** 0.3–0.6 mm diam, cylindric. **LF:** distal sheath delicate, often disintegrating, tip inflated, obtuse, often wrinkled. **INFL:** spikelet 1.5–9 mm, 1–3.5 mm wide; fl bracts to 65, 1–3 mm. **FL:** anthers ± 0.8 mm; stigmas 2(3). **FR:** 0.4–0.8(1.1) mm, 0.3–0.6 mm wide, 2-sided, smooth; perianth bristles (0)5–8, vestigial to ± > fr (> fr). $2n=30$. Rare in CA. Wet sand, gravel, marsh, brackish canals, hot springs, maritime mud flats, salt marshes; < 1600 m. n SN, GV; to MT, VA, FL; W. Indies, S.Am. [*E. thermalis* Rydb.] Summer–fall

E. geniculata (L.) Roem. & Schult. (p. 1347) Ann 5–45 cm, tufted; rhizome 0. **ST:** 0.2–1 mm diam, cylindric. **LF:** distal sheath firm, persistent, tip acute. **INFL:** spikelet 1–9 mm, 1–4 mm wide; fl bracts to 125, 0.8–3 mm, membranous to papery, rounded to acute. **FL:** anthers ± 0.5 mm; stigmas 2. **FR:** 0.5–1.1 mm, 0.3–0.7 mm wide, 2-sided, smooth, black (brown when immature); perianth bristles (0)4–8, vestigial to gen = fr (much exceeding tubercle), gen red-brown. $2n=10$. Uncommon, very local in CA. Wet, often brackish places; < 1500 m. SCo/PR, SnBr, DSon; to s ON, FL; Mex, West Indies, S.Am, Asia, Afr, Pacific Islands. Spring–winter

E. macrostachya Britton (p. 1347) Per 20–100 cm; rhizome 1–2 mm diam, tough. **ST:** 0.5–2.5(3.5) mm diam, cylindric to ± flat. **LF:** distal sheath firm, persistent, tip truncate to obtuse, sometimes with tooth-like projection < 1 mm. **INFL:** spikelet 5–40 mm, 2–5 mm wide; fl bracts 30–80, 2.5–5.5 mm, basal without fl, at least some clasping ≥ 3/4 of st, next-to-basal with fl or not. **FL:** anthers 1.3–2.7 mm; stigmas 2. **FR:** 1.1–1.9 mm, 0.8–1.5 mm wide, 2-sided, smooth or fine-roughened; perianth bristles 0 or 4(5), vestigial to equaling tubercle. $2n=18,19,38$. Common. Fresh to brackish wetland; 10–2300 m. CA; to AK, ne Can, c US; Mex, S.Am. Very variable, esp outside CA; CA pls intermediate between *E. palustris* and *E. erythropoda* Steud. or *E. palustris* and *E. uniglumis* (Link) Schult.; more study needed. Spring–summer

E. montevidensis Kunth (p. 1347) Per 25–50 cm; rhizome 0.7–2 mm diam, tough. **ST:** 0.5–1.2 mm diam, subcylindric or ± flat. **LF:** distal sheath firm, persistent, tip subtruncate to obtuse, gen some or all with tooth-like projection to 1 mm. **INFL:** spikelet 4–12 mm, 2–3 mm wide; fl bracts 30–100, 1.5–2.5 mm, tip wide-rounded, often transversely wrinkled, recurved. **FL:** anthers 0.8–1.5 mm; stigmas 3. **FR:** 0.7–1 mm, 0.7–0.8 mm wide, 3-sided, fine-wrinkled or -net-like; perianth bristles 5–6(7), (vestigial to) << to = fr. $2n=20$. Locally common. Fresh wet places in riparian and foothill woodland, pine forest, streams; < 2000 m. NCoRI, CaR, n SN, ScV, CCo, SCoR, SW, DSon; to FL, Mex, S.Am. Variable, esp outside CA; often resembles *E. parishii.* Spring–summer

E. obtusa (Willd.) Schult. (p. 1347) Ann 3–50 cm, tufted; rhizome 0. **ST:** 0.2–2 mm diam, cylindric. **LF:** distal sheath firm, persistent, tip obtuse to acute, with tooth-like projection to 0.3 mm. **INFL:** spikelet (2)5–13 mm, (2)3–4 mm wide; fl bracts 15–150+, (2)3–4 mm. **FL:** anthers 0.3–0.6 mm; stigmas 2–3. **FR:** 0.9–1.2 mm, 0.7–0.9 mm wide, 2–3-sided, smooth; tubercle 1/3–1/2 as high as wide, 2/3–9/10 as wide as fr, flat, adherent to fr; perianth bristles (0 or) 5–7, ± to gen much exceeding tubercle. $2n=10$. Fresh shores, marshes, seepages, stream beds, disturbed places; < 1600? m. NW, SN, ScV/n SNF, CCo; to s BC, QC, FL; HI. See note under *E. engelmannii.* Late spring–fall

E. ovata (Roth) Roem. & Schult. Ann 2–35 cm, tufted; rhizome 0. **ST:** 0.3–1 mm diam, cylindric. **LF:** distal sheath firm, persistent, tip obtuse to acute, with tooth-like projection to 0.2 mm. **INFL:** spikelet 2–8 mm, 2–4 mm wide; fl bracts 25–100+, 1.5–2.5 mm. **FL:** anthers 0.3–0.6 mm; stigmas 2–3. **FR:** 0.7–1 mm, 0.6–0.9 mm wide, 2–3-sided, smooth, tip rounded; tubercle 3/5 to as high as wide, 1/2–3/4 as wide as fr, flat, adherent to fr; perianth bristles (0 or) 5–7, exceeding tubercle. $2n=10$. Rare in CA. Fresh drying soil (sand) of lake shores; < 2000 m. CCo, SNE; to BC, AB; MN to NL, NC, OK; Eur. See note under *E. engelmannii.* Summer–fall

E. pachycarpa E. Desv. (p. 1351) Per 7–50 cm, in dense clumps; rhizome caudex-like, 2 mm diam, tough, mostly hidden by sts. **ST:** 0.3–0.5 mm diam, sharp-4-angled. **LF:** distal sheath firm, persistent, tip acute to subacute. **INFL:** spikelets 3–10 mm, 2–3 mm wide, often forming plantlets; fl bracts 8–15, 2–3 mm. **FL:** anthers 1.2–1.5 mm; stigmas 3. **FR:** 0.9–1.1 mm, 0.7–0.9 mm wide, 3-sided, smooth or ± rough, tip ± truncate; perianth bristles ≤ 6, some or most = fr. Local, uncommon. Fresh shores, ponds, streambeds, seeps, springs, fens; 100–1400 m. NCo, NCoRO, n SN, c SNF, GV, SCoRO, MP; NV; native to Argentina, Chile. Late spring–summer

E. palustris (L.) Roem. & Schult. Per 30–100 cm; rhizome 1.5–4 mm diam, tough. **ST:** 0.5–2.5 mm diam, ± cylindric. **LF:** distal sheath firm, persistent, tip obtuse to acute. **INFL:** spikelet 5–25 mm, 3–7 mm wide; fl bracts 30–100, 3–5 mm, basal clasping 0.6–0.75 st, next-to-basal without fl. **FL:** anthers 1.5–2.2 mm; stigmas 2. **FR:** 1.1–2 mm, 1–1.5 mm wide, 2-sided, smooth or fine-roughened; perianth bristles 0 or 4(5), << fr to equaling tubercle. $2n=16,17,36$. Locally common. Fresh ponds, shores, marshes; < 3000 m. NW, SNH, SCo, ChI, SnBr, MP; N.Am; Mex, Eurasia, New Zealand. Variable outside CA. Resembles *E. macrostachya.* Summer

E. parishii Britton (p. 1351) Per 10–50 cm; rhizome 0.7–2 mm diam, tough. **ST:** 0.2–1 mm diam, ± cylindric or ± flat. **LF:** distal sheath firm, persistent, tip subtruncate to subacute, some or all with tooth to 1 mm. **INFL:** spikelet 3–20 mm, 1.5–2.5 mm wide, narrow-lanceolate or cylindric; fl bracts 15–40, 2–3 mm, tips rounded to acute, not transversely wrinkled, not recurved. **FL:** anthers 1–2 mm; stigmas 3. **FR:** 0.8–1.4 mm, 0.5–0.7 mm wide, 3-sided, smooth or minute-net-like; perianth bristles 3–7, vestigial to ± exceeding tubercle. $2n=10$. Locally common. Fresh streams, vernal pools, foothill and riparian woodland, pine forest; 500–2300 m. KR, NCoRI, CaRF, n SN, c SNH, s SNF, GV, SnFrB, SCoRO, SCo, WTR, SnGb, PR, GB, DMoj, w DSon; to OR, KS, TX; Mex. Some pls resemble *E. montevidensis.* Late spring–fall

E. parvula (Roem. & Schult.) Bluff et al. (p. 1351) SMALL SPIKERUSH Per 2–9 cm; rhizome to 1.5 mm diam, weak, tip often with tuber 2–2.5 mm, 0.5–1 mm wide, oblong, often markedly curved. **ST:** 0.2–0.5 mm diam, cylindric. **LF:** distal sheath delicate, often disintegrating, tip rounded. **INFL:** spikelet 2–4 mm, 1–2 mm wide; fl bracts 6–10, 1.4–2.7 mm. **FL:** anthers 0.7–1.2 mm; stigmas (2)3. **FR:** 0.9–1.2 mm, 0.6–0.8 mm wide, 3-sided, smooth; tubercle often merging with fr or vestigial; perianth bristles 6, (vestigial or) gen = fr to ± exceeding tubercle. $n=10$ (Europe). Brackish wet soil, coastal; < 50 m. NCo, SnFrB, SCo; to BC; KS to NL, FL, LA; Mex, C.Am, Eurasia. Late winter–fall ★

E. quadrangulata (Michx.) Roem. & Schult. (p. 1351) SQUARESTEM SPIKERUSH Per 40–100 cm; rhizome 1.5–4 mm diam. **ST:** 1–5.4 mm diam, 4-angled. **LF:** distal sheath firm, persistent, tip acute to acuminate, blade-like or not. **INFL:** spikelet 15–76 mm, 3–6 mm wide,

cylindric; fl bracts 28–100+, (4.8)5.2–7 mm. **FL:** anthers 2.3–2.9 mm; stigmas (2)3. **FR:** 1.8–3 mm, 1.3–2 mm wide, 2-sided, each face often with 19–38 rows of transverse-elongate cells; perianth bristles 6–7, < fr to equaling tubercle. Local. Freshwater ponds; 10–600 m. GV; OR; WI to MA, TN, FL. Native to e US. Summer

E. quinqueflora (Hartmann) O. Schwarz   Per 5–30 cm, forming small mats, st-tuft base gen swollen, often with bulbs, 2-veined scales present or not; rhizome 0.5–1 mm diam, weak, tip often with bulb 3–11 mm, 1–5 mm wide, ovate; caudex 0 or present. **ST:** 0.2–0.8 mm diam, cylindric. **LF:** distal sheath firm, persistent, tip ± truncate to acute. **INFL:** spikelet 4–8 mm, 1.5–4 mm wide, or 0; fl bracts 3–8, 2.5–6 mm, basal ≥ 1/2 spikelet, with fl or not. **FL:** anthers 1–2.5 mm; stigmas 3. **FR:** 1.3–2.3 mm, 0.8–1.4 mm wide, 3-sided, smooth or fine-longitudinal-ridged, distally tapered, tip often beak-like; tubercle often merging with fr; perianth bristles (0)3–6(7), vestigial to equaling tubercle. Uncommon, local. Wet meadows, fens, seeps, shores; 40–3600 m. KR, NCoRI, CaR, SN, ScV, SCoRO, SnBr, SnJt, GB; to AK, e Can, Greenland, NJ; Eurasia. [*E. pauciflora* (Lightf.) Link] 2 additional taxa possibly to be named in CA; study needed. Spring–summer

E. radicans (Poir.) Kunth (p. 1351)   Per 1–12 cm; rhizome to 0.5 mm diam, weak. **ST:** 0.4–1 mm diam, cylindric, without angles or ridges, very spongy. **LF:** distal sheath delicate, often disintegrating, tip blunt. **INFL:** spikelet 2–3 mm, 1–1.5 mm wide; fl bracts 5–15, 1.5–2(2.5) mm, colorless to straw, tip blunt, not recurved, basal with fl. **FL:** anthers 0.3–0.5 mm; stigma 3. **FR:** 0.8–0.9 mm, 0.4 mm wide, 3-sided to round in ×-section, often ± white, longitudinal ridges ± 7, 50 crossbars per ridge; perianth bristles ± 4, = fr. Uncommon. Exposed soil or shallow water of shores, streams, marshes, seeps; 100–1400 m. NCo, NCoR, GV, CCo, SW; to MI, FL, W. Indies, S.Am, HI. Spring–winter

E. rostellata (Torr.) Torr. (p. 1351)   Per 20–50 cm, forming dense mats; rhizome 3 mm diam, caudex-like, tough, hidden by sts. **ST:** 0.3–2 mm wide, ± flat, ridges gen ≤ 8, sharp, firm to hard, tips often rooting. **LF:** distal sheath firm, persistent, tip obtuse to subacute. **INFL:** spikelet 5–17 mm, 2.5–5 mm wide; fl bracts 20–40, 3.5–6 mm. **FL:** anthers 2–2.4 mm; stigmas (2)3. **FR:** 1.5–2.5 mm, 1–1.2 mm wide, (2)3-sided, smooth or fine-roughened, tip gen beak-like; tubercle often merging with fr or 0; perianth bristles to (0)4–6, ± = fr or equaling tubercle. Uncommon. Mineral-rich fens, springs, hot springs, coastal marshes; -50–2400 m. NW (exc NCoRH), CaR, CCo, SnFrB, SCo, WTR, SnGb, GB, DMoj; to BC, NS, FL, W. Indies. Late spring–summer

E. suksdorfiana Beauverd   Per 4–40 cm, st-tuft base not swollen, without bulbs or 2-veined scales; rhizome 0.5–1.5 mm diam, weak, tip with bud 10 mm, 2–5 mm wide, elliptic; caudex present, hard. **ST:** erect, 0.5–1.2 mm wide, cylindric to ± flat. **LF:** distal sheath firm, persistent, tip subtruncate to obtuse. **INFL:** spikelets 5–10 mm, 2–4 mm wide, often 0; fl bracts 8–12, 3.5–5 mm, basal gen ≥ 1/2 spikelet. **FL:** anthers 1.6–3.5 mm; stigmas 3. **FR:** 2–2.7 mm, 0.7–1.3 mm wide, 3-sided, smooth or fine-longitudinal-ridged or -net-like, tip narrowed, beak-like; tubercle often merging with fr; perianth bristles 6, longest = fr to exceeding tubercle. Often subdominant but local. Wet meadows, fens, springs; < 3400 m. NW, CaRH, SN, SnFrB, TR, DMtns; to BC, nw MT, CO. Herbarium specimens often confused with *E. rostellata*. Summer

E. torticulmis S.G. Sm.   CALIFORNIA TWISTED SPIKERUSH   Per 20–40 mm, st-tuft with hard caudex; rhizome 1.5–2 mm diam, weak, tip with bud ± 10 mm, ± 5 mm wide, ovate. **ST:** 1.5–2.5 mm wide, spirally twisted, flat, obliquely contracted 5–40 mm below tip. **LF:** distal sheath firm, persistent, tip obtuse. **INFL:** spikelet 6–8 mm, 2–3 mm wide, often 0; fl bracts 8–10, 3.5–5 mm. **FL:** anthers 1.8–3 mm; stigmas 3. **FR:** 1.8–2.8 mm, 1–1.3 mm wide, (2)3-sided, tip beak-like; tubercle often merging with fr; perianth bristles 0–5, unequal, vestigial to = fr. Fen (wet meadow) in mixed-conifer forest; < 1200 m. n SNH. Summer   ★

ERIOPHORUM   COTTON GRASS

Peter W. Ball

Per, erect. **ST:** cylindric or 3-angled, solid. **LF:** basal and cauline; ligule present; blade ± scabrous on keel or angles. **INFL:** 1, terminal, subumbel- or head-like [or spikelet 1]; infl bracts 1–several, lf- or scale-like; spikelets several [1]; fl bracts spiraled, > 10, ovate, membranous, glabrous, tip entire. **FL:** bisexual; perianth bristles 6–7 or 10–25, >> fr, gen > fl bracts, >> fl bracts in fr; stamens 3; style 3-branched. **FR:** 3-sided, ± flat. ± 25 spp.: n temp. (Greek: wool-bearing)

1. St 3-angled; perianth bristles 6–7, barbed, yellow-brown; infl gen head-like (1–2-branched) . . . . ***E. crinigerum***
1′ St cylindric; perianth bristles 10–25, smooth, ± white; infl subumbel-like. . . . . ***E. gracile***

E. crinigerum (A. Gray) Beetle (p. 1351)   Pl 2–10 dm, cespitose. **ST:** ± scabrous. **LF:** 4–40 cm, 1–6 mm wide, blade flat, scabrous. **INFL:** spikelets 5–30 +, 3–10 mm; fl bracts 2.5–4 mm. **FL:** perianth bristles 3–9 mm, irregularly bent to ± straight, barbs gen distal, dense, upward-pointing. Wet meadows, streambanks, seepage slopes; gen on serpentine; gen > 2000 m. NW, SNH; sw OR. [*Eriophorum criniger*, orth. var.] Jun–Aug

E. gracile Roth (p. 1351)   SLENDER COTTONGRASS   Pl 30–60 cm; rhizomes long. **ST:** smooth. **LF:** 2–30 cm, 1–2 mm wide, blade distally 3-sided, smooth. **INFL:** spikelets (1)2–5, 7–11 mm; fl bracts 3–4 mm. **FL:** perianth bristles 10–20 mm, ± straight, barbs 0. 2*n*=30,38. Wet meadows, bogs; gen 600–2900 m. KR, CaR, n&c SNH, SnFrB (extirpated); circumboreal. May–Jul   ★

FIMBRISTYLIS

Ann, per, scapose, rhizomed or not. **ST:** ± cylindric or flat, often prominent-ridged, ≤ 1 mm diam. **LF:** several, basal, spiraled, fl st with 1+ cauline; blade flat proximally, longest > sheath, ≤ 4 mm wide; sheath distally open, margin scabrous, tip hairs << 1 mm, stiff; ligule 0 or of short hairs. **INFL:** 1, terminal, branched or head-like, infl bracts (1)2–8; spikelets (1)80+, cylindric [(flat)]; fl bracts 8–100, spiraled, each with 1 fl in axil or occ 0 in proximal 1–2, membranous, glabrous or puberulent, tip entire, mucronate, or short-awned. **FL:** bisexual; perianth 0; style often flat, 2[3]-branched, base enlarged, not persistent. **FR:** 2–3-sided, wide-obovate, minute-netted-honeycombed, occ warty, brown or ± white, mucronate. Gen moist to wet places. 100+ spp.: warm temp to trop. (Latin: fringe style) [Kral 2002 FNANM 23:121–131] *F. miliacea* record (1866) undocumented for CA.

1. Spikelets 3–5 mm wide; per, rhizomed . . . . ***F. thermalis***
1′ Spikelets 1–1.5 mm wide; ann, rhizomes 0
   2. Infl branched. . . . . ***F. autumnalis***
   2′ Infl head-like. . . . . ***F. vahlii***

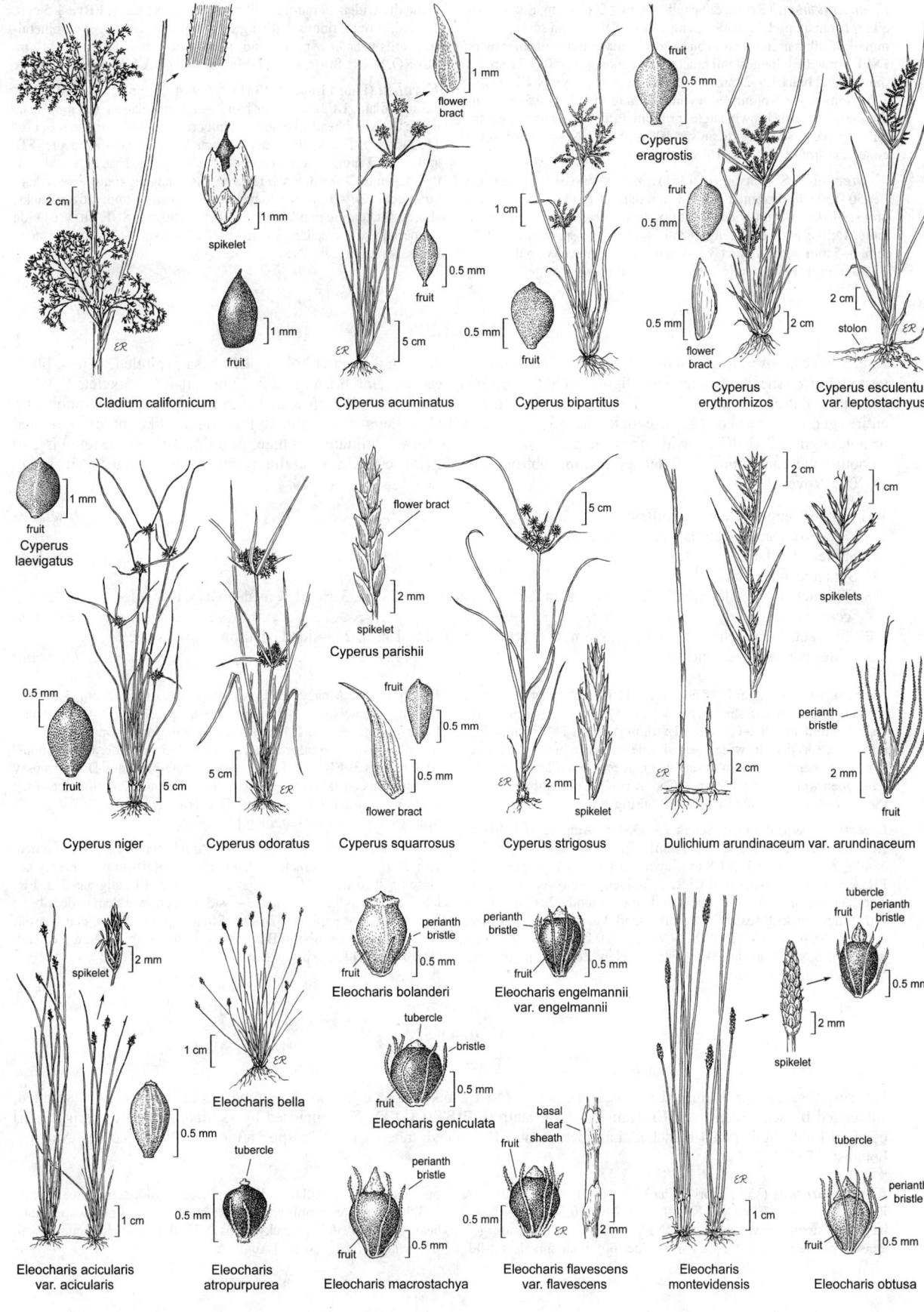

Cladium californicum

spikelet

fruit

Cyperus acuminatus

flower bract

fruit

Cyperus bipartitus

fruit

Cyperus eragrostis

fruit

flower bract

Cyperus erythrorhizos

stolon

Cyperus esculentus var. leptostachyus

fruit

Cyperus laevigatus

Cyperus niger

fruit

Cyperus odoratus

flower bract

spikelet

Cyperus parishii

fruit

flower bract

Cyperus squarrosus

spikelet

Cyperus strigosus

spikelets

perianth bristle

fruit

Dulichium arundinaceum var. arundinaceum

spikelet

Eleocharis acicularis var. acicularis

Eleocharis bella

tubercle

Eleocharis atropurpurea

perianth bristle

fruit

Eleocharis bolanderi

perianth bristle

fruit

Eleocharis engelmannii var. engelmannii

tubercle

bristle

fruit

Eleocharis geniculata

perianth bristle

fruit

Eleocharis macrostachya

basal leaf sheath

fruit

Eleocharis flavescens var. flavescens

spikelet

Eleocharis montevidensis

tubercle

fruit

perianth bristle

tubercle

perianth bristle

fruit

Eleocharis obtusa

F. autumnalis (L.) Roem. & Schult.   Ann ≤ 20(30) cm, rhizomes 0. **ST:** ≤ 1 mm diam. **LF:** blade ≤ 2 mm wide, flat, margin scabrous; > 1 mm of sheath margins next to blade short-ciliate; ligule of short hairs. **INFL:** branched; longest infl bract ≤ infl; spikelets (1)50, ≤ 7 mm, ± 1 mm wide; fl bract 1.5–2 mm, lanceolate, keeled, glabrous. **FL:** anther ± 0.2 mm; style 3-branched, cylindric, glabrous. **FR:** 0.5–0.7 mm, thick-3-sided, pale brown, angles prominent. 2*n*=10. Damp open disturbed places, esp along streams; < 100 m. ScV; e N.Am. Earliest CA collection from Butte Co., 1979. Jul–Sep

F. thermalis S. Watson (p. 1351)   HOT SPRINGS FIMBRISTYLIS Per 50–150 cm, rhizomed. **ST:** ± 1 mm diam. **LF:** puberulent or scabrous; blade ≤ 4 mm wide; sheath often puberulent; ligule of short hairs. **INFL:** branched; longest infl bract < infl; spikelet (5)10–12 mm, 3–5 mm wide; fl bract 3.5–4 mm, ovate, not keeled, puberulent. **FL:** anther ± 1 mm; style 2-branched, flat, base enlarged, distally

markedly widened, margin incl branches dense-hairy. **FR:** ± 1.5 mm, 2-sided, brown, fine-pitted, angles prominent. 2*n*=20. Wet mineralized soils near hot springs and in seepage meadows; 110–1340 m. SCoRO, SnGb, SnBr, W&I, DMoj; s NV, UT, AZ, Mex. Jul–Sep ★

F. vahlii (Lam.) Link (p. 1351)   Ann 4–15 cm, rhizome 0. **ST:** < 0.5 mm diam. **LF:** smooth or hairy-scabrous; blade < 0.5 mm wide, ligule 0. **INFL:** head-like; longest infl bracts >> infl; spikelets ± (1)2–15, 2–5 mm, 1–1.5 mm wide; fl bracts keeled, glabrous, acute. **FL:** anther ± 0.2 mm; style 2-branched, flat, glabrous, base enlarged. **FR:** 0.5–0.7 mm, 2-sided, ± white, with sharp angles, transverse-rectangular cells. 2*n*=20. Moist to wet soil, alluvial or mineralized banks, shores, fluctuating pond, lake margins; < 200 m. s SNF, SnJV, e DSon (Imperial Co.); to se US, S.Am. Often a "drawdown" pl around stock tanks, reservoirs. Jul–Nov

ISOLEPIS

Ann, per, cespitose, rhizomed or not. **ST:** simple, cylindric, ≤ 0.5 mm diam, not hollow. **LF:** basal, spiraled, 1–few; blade vestigial to >> sheath, ≤ 1 mm wide; ligule 0. **INFL:** terminal, head-like, infl bracts 1–2, 2–33 mm, lf-like; spikelets 1–3(15), ± ovate, ± flat or not, many-fld, 1–10 mm, 1–2 mm wide; fl bracts spiraled, each with 1 fl in axil, ovate, membranous, tip entire, gen short-awned. **FL:** bisexual; perianth 0; stamens 1–3, anthers ≥ 0.6 mm; style 1, thread-like, base ± enlarged or not, stigmas 2–3. **FR:** ± wide-obovate, gen ± white to dark brown, minute-papillate, prominent-ridged lengthwise, or smooth, mucronate; tubercle 0. 69 spp.: temp, subtrop, ± worldwide, esp Afr, Australia. (Greek: equal fl bract) [Muasya et al. 2007 Novon 17:59]

1. Fr ribbed lengthwise, not papillate . . . . ***I. setacea***
1′ Fr fine- to obscure-papillate or smooth, not ribbed
  2. Spikelets ± flat . . . . ***I. levynsiana***
  2′ Spikelets not ± flat
    3. Fl bracts in spikelet middle 1.8–2 mm, awn 0.2–0.5 mm; fr 1–1.5 mm, 3-sided, ± as deep as wide, sides
      concave . . . . ***I. carinata***
    3′ Fl bracts in spikelet middle 1.2–1.8 mm, awn < 0.1 mm; fr 0.8–1 mm, 2–3-sided, shallower than wide,
      sides convex or ± concave . . . . ***I. cernua***

I. carinata Torr.   Ann 1–25 cm. **LF:** 1. **INFL:** infl bract 5–33 mm; fl bracts often clasping shed fr, basal ≤ 2.5(5) mm, awn ≤ 2 mm, others 1.8–2 mm, awn 0.2–0.5 mm. **FL:** stigmas 3. **FR:** 1–1.5 mm, 0.7–1 mm wide, ± as deep as wide, 3-sided, sides concave, not ribbed, fine- to obscure-papillate. Often drying wet places in grassland, rock barrens, open woodland; < 800 m. NCo, NCoRO, CCo, SnFrB; se US. [*Scirpus koilolepis* (Steud.) Gleason] Spring

I. cernua (Vahl) Roem. & Schult. (p. 1351)   Ann (per?) 4–40 cm. **LF:** 1. **INFL:** infl bracts 2–6(23) mm; fl bracts not clasping shed fr, basal ≤ 2 mm, others 1.2–1.8 mm, awn < 0.1 mm. **FL:** stigmas 2–3. **FR:** 0.8–1 mm, 0.5–0.7 mm wide, 2–3-sided, shallower than wide, sides convex or ± concave, fine-papillate or smooth. 2*n*=30. Sandy, sometimes brackish sea shores, bluffs, sand dunes, creeks, marshes; < 800 m. NCo, GV, CCo, SnFrB, SCo, SnBr/DMtns; to BC; TX (naturalized); S.Am, Eurasia, Afr, Australia, New Zealand. [*Scirpus c.* Vahl] Late spring–winter

I. levynsiana Muasya & D.A. Simpson   Ann 1–6 cm. **LF:** few. **INFL:** infl bracts 1–2, 3–16 mm, ± erect; spikelets 1–3(4), 4–10 mm, linear-oblong, ± flat; fl bracts 1.3–1.6 mm, lance-ovate, dull white tinted ± red or ± yellow, lateral veins 4–5 per side, conspicuous. **FL:** stigmas 3. **FR:** 0.9–1 mm, 3-sided, fine-papillate. Damp, mossy openings in coastal forest; ± 100 m. NCo (Mendocino, Sonoma cos.); s Afr, s Australia, New Zealand. [*Cyperus tenellus* L. f.; *I. t.* (L. f.) Muasya & D.A. Simpson, illeg.] Early spring

I. setacea (L.) R. Br. (p. 1351)   (Ann?) per 3–25 cm, rhizomed. **LF:** 2. **INFL:** infl bracts 1–2, lowest 3–10(20) mm; fl bracts not clasping shed fr, 1.2–1.6 mm, awn ≤ 0.1 mm. **FL:** stigmas 2–3. **FR:** 0.8–1 mm, 0.5–0.6 mm wide, 2–3-sided, shallower than wide, ribbed lengthwise, not papillate. 2*n*=30. Stream, pond shores, gen coastal; < 1500 m. NCo, n SN; to BC; Eurasia, Afr, Australia, New Zealand. [*Scirpus s.* L.] Late spring–fall

KOBRESIA

Peter W. Ball

Per, cespitose, short-rhizomed. **LF:** basal, sheathing. **INFL:** spikelets in 1 [or more] terminal spikes [or panicles], spiraled, subtended by scale-like bract. **FL:** unisexual, perianth 0. **PISTILLATE FL:** enclosed by sac-like structure (perigynium) open on 1 side; style gen 3-branched, base persistent. **FR:** 3-sided; tubercles 0. ± 35 spp.: n temp. (J.P. von Kobres, Austrian botanist, 1747–1823)

K. myosuroides (Vill.) Fiori & Paol. (p. 1351)   SEEP KOBRESIA Pl 1–3.5 dm, slender, wiry. **LF:** ≤ st, thread-like, 0.2–0.5 mm wide, base persistent, brown, ± glossy. **INFL:** spike 1, 10–30 mm, gen dense; spikelets 2–3.5 mm, ± 1 mm wide, proximal spikelets 2-fld,

bisexual (1-fld, pistillate), distal spikelets 1-fld, staminate; fl bract 2–3.5 mm, ovate, membranous. **FR:** 2–2.7 mm, ± 1 mm wide, short-beaked. 2*n*=52,56–58. Rocky seeps; > 2700 m. c SNH; circumboreal. [*K. bellardii* (All.) Loisel.] Aug ★

KYLLINGA

Gordon C. Tucker

Per [ann], glabrous; rhizome long [0]. **ST:** 3-angled. **LF:** basal; ligule 0; blade flat or V-shaped in ×-section. **INFL:** terminal, head-like, of 1–3[4] dense spikes; infl bracts 2–3[4], lf-like; spikelets 50–100+, 1–2-fld; fl bracts 2(3), 2-ranked, proximal sterile or with bisexual fl, distal sterile or with staminate fl. **FL:** perianth 0; stamens 1–3; stigmas 2. **FR:** 1 per spikelet, dispersed in entire spikelet, 2-sided, lenticular, ± smooth; tubercle 0. 40–45 spp.: trop to warm temp worldwide. (P. Kylling, Danish botanist, 1640–1696) [Roalson 2008 Bot Rev 74:345; Tucker 2002 FNANM 23:193–194]

K. brevifolia Rottb. (p. 1351) **ST:** erect, 2–55 cm. **LF:** blade 1.5–3.5 mm wide. **INFL:** longest infl bract gen erect; spikes 1–3, 4–7 mm; spikelets 2–3 mm, lance-oblong; fl bracts 2–3 mm, midrib prominent, sides transparent, veined. **FR:** 1–1.3 mm, ± elliptic-oblong, brown. 2*n*=120. Pond edges, moist disturbed areas; < 300 m. GV, SnFrB, SW; trop, warm temp; native to trop Am. May–Sep

LIPOCARPHA

Gordon C. Tucker

Ann, glabrous. **ST:** ± erect, 1–20 cm. **LF:** basal, 1–3. **INFL:** infl bracts 1–3, lf-like; spikelets in dense, spheric to cylindric spikes, 50–150, dense, spiraled, sessile; fl bracts spiraled, 100–400, (1)2 per fl (a 2nd, inner bract between fl, spikelet axis gen present), outer > inner, mucronate to awned, brown, with 1 central green, 2–10 lateral ± white veins, inner gen colorless, gen veinless. **FL:** bisexual; perianth 0; stamens 1–3, anthers 0.2–0.3 mm; styles 2-branched. **FR:** 3-angled to ± flat, abruptly soft-pointed, papillate, brown. ± 35 spp.: N.Am, trop, warm temp. (Greek: falling chaff, from translucent inner fl bract) [Tucker 2002 FNANM 23:195–197]

1. Outer fl bract awn (0.6)1–1.5 mm, outcurved; fr light brown . . . . *L. occidentalis*
1′ Outer fl bract awn ≤ 1 mm, ± straight to outcurved; fr light red-brown to black or light to dark brown
  2. Inner fl bract ± = fr; outer fl bract awn 0.5–1 mm; fr light red-brown to black . . . . *L. aristulata*
  2′ Inner fl bract 0 or < 1/2 fr; outer fl bract awn < 0.2 mm; fr light to dark brown . . . . *L. micrantha*

L. aristulata (Coville) G.C. Tucker (p. 1351) **ST:** 2–15 cm. **INFL:** spikes 1–2, 3–7 mm, gen ± open, ovoid to gen cylindric; outer fl bract body 0.8–1 mm, awn 0.5–1 mm, inner fl bract ± = fr. **FR:** 0.7–0.9 mm, widest just below tip, papillate, light red-brown to black; faces gen flat to ± concave. Wet soil; 100–400 m. NCoRO, ScV; to WA, se US. Aug–Sep

L. micrantha (Vahl) G.C. Tucker (p. 1351) **ST:** 1–10 cm. **INFL:** spikes 1–3, 3–6 mm, dense, cylindric; outer fl bract body 0.8–1 mm, awn < 0.2 mm, inner fl bract 0 or < 1/2 fr. **FR:** 0.7–0.8 mm, widest near middle, fine-pitted, light to dark brown; faces convex. Wet soil; < 1500 m. NCoRO, SNF, GV; s Can, US, trop Am. Aug–Oct

L. occidentalis (A. Gray) G.C. Tucker (p. 1351) **ST:** 1–7 cm. **INFL:** spikes 1–2, 2–3 mm, ± dense, gen ± spheric; outer fl bract body 0.5–0.8 mm, awn (0.6)1–1.5 mm, inner fl bract ± = fr. **FR:** 0.5–0.6 mm, widest at or just below tip, fine-pitted, light brown; faces convex. Wet soil, esp emergent shorelines; 1200–1900 m. NCoRO, SN, PR; to WA. Jun–Aug

RHYNCHOSPORA BEAKED-RUSH

Per [ann] 5–100(150) cm; rhizomes gen 0. **ST:** gen 3-angled, ≤ 2 mm diam, glabrous, margin often minute-scabrous. **LF:** few to many, basal and cauline, spiraled; blade > sheath, 0.5–3 wide, proximally flat, midrib abaxially keeled or not, margin gen, midrib occ scabrous; sheath permanently tubular (not splitting), glabrous; ligule present or 0. **INFL:** gen several, terminal and axillary, branched [head]; spikelets several to many, cylindric, 3.5–5.5 mm, on short stalks with spikelet bracts; fl bracts ± 5, spiraled [2-ranked], each with 1 fl in axil or proximal 1 or more smaller, with 0 fl, and intergrading with spikelet bracts, 1.7–3.5(4) mm, ovate, membranous, glabrous, gen mucronate or short-awned. **FL:** bisexual or occ distal-most staminate; perianth bristles 2–12[20], to ± exceeding tubercle, barbed, persistent on fr [not]; style 2[3]-branched, base much enlarged, persistent on fr as a prominent tubercle. **FR:** 2-sided [± cylindric], body (exc tubercle) obovate to pear-shaped, smooth or winkled [variously sculptured otherwise], brown. > 250 spp.: esp trop. (Greek: snout seed) [Kral 2002 FNANM 23:200–239] Kral's (2002) statement that *R. kunthii* Kunth, *R. recognita* (Gale) Kral (treated here as a synonym of *R. globularis*) are both in CA is evidently undocumented.

1. Perianth barbs reflexed; fr ± smooth, base narrow-stalk-like
  2. Perianth bristles 10–12; spikelets pale-brown to ± white . . . . *R. alba*
  2′ Perianth bristles 5–7; spikelets brown to dark brown . . . . *R. capitellata*
1′ Perianth barbs ascending; fr wrinkled, base wide-stalk-like or not stalk-like
  3. Perianth bristles gen exceeding tubercle tip; fr base wide-stalk-like . . . . *R. californica*
  3′ Perianth bristles gen exceeded by tubercle base; fr base not stalk-like . . . . *R. globularis*

R. alba (L.) Vahl (p. 1355) WHITE BEAKED-RUSH **ST:** ≤ 1 mm diam, angles 3, sharp, minute-scabrous below infl. **LF:** blade 0.5–1.5 mm wide, margin minute-scabrous. **INFL:** spikelets 3.5–5.5 mm, elliptic, pale brown to ± white, tip acute; fl bracts (3)3.5–4 mm. **FL:** perianth bristles 10–12, equal to or exceeding tubercle tip, barbs reflexed. **FR:** (2.3)2.5–3 mm incl tubercle, ± smooth, base narrow-stalk-like, tubercle ± 1 mm. 2*n*=26,42. Boggy open sites; < 2000 m. NCo (Mendocino Co.), KR (Trinity Co.), CaRH, n SNH (Plumas Co.), c SNH (Tuolumne Co.); to AK, e Can; e US, PR, Eurasia. Jul–Aug ★

R. californica Gale (p. 1355) CALIFORNIA BEAKED-RUSH **LF:** blade 2–3 mm wide. **INFL:** spikelets ± 4–5 mm, wide-ovate, brown; fl bracts ± 3 mm. **FL:** perianth bristles 6, gen exceeding tubercle tip, barbs ascending. **FR:** ± 3 mm incl tubercle, wrinkled, base wide-stalk-like, tubercle ± 1 mm. Marshes, seeps; < 200 m. NCoRO (Pitkin Marsh, Sonoma Co.), n SNF (Butte Co.), CCo (Point Reyes, Marin Co.). May–Jul ★

R. capitellata (Michx.) Vahl BROWNISH BEAKED-RUSH **ST:** angles 3, blunt, minute-scabrous near infl. **LF:** blade ≤ 3 mm wide,

margin distally minute-scabrous. **INFL:** spikelets 3.5–4(5) mm, brown to dark-brown, lance-elliptic; fl bracts 2.7–3 mm. **FL:** perianth bristles 5–7, ± reaching tubercle tip, barbs reflexed. **FR:** (2)2.5–3 mm incl tubercle, smooth, base narrow-stalk-like, tubercle 1–1.5 mm. Wet meadows, fens, seeps, marshes; < 2000 m. KR (Trinity Co.), NCoRO (Sonoma Co.), n&c SN; e N.Am. Jul–Aug ★

R. globularis (Chapm.) Small (p. 1355) **ST:** ≤ 3 mm diam, angles 3, sharp, minute-scabrous. **LF:** blade 1–3 mm wide, margin, midrib

minute-scabrous. **INFL:** spikelets gen 2 mm, ± spheric to ovate, dark brown; fl bracts 1.7–2.3 mm. **FL:** perianth bristles ≤ 6, gen exceeded by tubercle base, barbs ascending. **FR:** 1.5–1.8 mm incl tubercle, wrinkled, base not stalk-like, tubercle ± 0.3 mm. Marshes; ± 50 m. s NCoRO (Pitkin, Perry, Cunningham marshes, Sonoma Co.); e N.Am, C.Am, W. Indies. [*R. recognita* (Gale) Kral] Disjunct from e N.Am range. Jul–Aug ★

SCHOENOPLECTUS NAKED-STEMMED BULRUSHES

Gen per, gen erect, gen with long, scaly rhizomes; st, lf gen with air cavities. **ST:** simple, smooth, (wiry). **LF:** gen all basal, whorled or 3-ranked; blade gen present, at least on distal sheath, smooth, or margin minute-scabrous; sheath closed, long; ligule glabrous. **INFL:** terminal, branch sts often scabrous, main infl bract like lf blade; spikelets ovate, not ± flat, many-fld; fl bracts spiraled, each with 1 fl in axil, ovate, 1-veined, brown to straw, dull, often fine-lined-spotted, membranous, gen ± scabrous, tip gen notched, gen with short awn. **FL:** bisexual; perianth bristles ± straight, ± ≤ fr, gen brown, reflexed-barbed (or with soft hairs) [(smooth)]; stamens gen 3; style 1, thread-like, base not enlarged, stigmas 2–3. **FR:** gen obovate, brown, gen smooth, mucronate; tubercle 0. Wetlands, often emergent (or submersed). ± 77 spp.: temp, subtrop, worldwide. (Greek: rush woven, from use of sts in baskets, etc.) [Smith 2002 FNANM 23:44–60]

1. St 3-sided, at least near infl, where sometimes obscure
  2. Infl panicle-like . . . . ***S. californicus***
  2′ Infl head-like or spikelet 1
    3. Lf blades 0; fl bract tips not notched; fr with many wavy transverse ridges; pl tufted, rhizomes short, inconspicuous . . . . ***S. mucronatus***
    3′ Lf blades 1–6; fl bract tips notched; fr smooth; pl mat-forming, rhizomes long, conspicuous
      4. Distal lf blade < 1.5 × sheath; fl bract tip notch 0.1–0.4 mm, awn 0.2–0.6 mm . . . . ***S. americanus***
      4′ Distal lf blade (1)2–5 × sheath; fl bract tip notch (0.3)0.5–1 mm, awn 0.5–2.5 mm . . . ***S. pungens*** var. ***longispicatus***
1′ St cylindric
  5. St ≤ 1.5 mm diam; pl to 1.5 m; fl bract tip notch 0; spikelets 1–20
    6. Spikelets 1–20, spreading; fr with sharp wavy transverse ridges; pl tufted, rhizomes short, gen not evident . . . . ***S. saximontanus***
    6′ Spikelets 1, erect; fr smooth; pl mat-forming, rhizomes long, evident . . . . ***S. subterminalis***
  5′ St 2–10 mm diam; pl to 4 m; fl bract tip notch present; spikelets 3–100+
    7. Fl bract awns some or all markedly contorted, 0.5–2 mm (often broken off); spikelets 3–100+, in clusters of 1–8, not all 1 on 1 pl . . . . ***S. acutus*** var. ***occidentalis***
    7′ Fl bract awns not or ± contorted, 0.2–1 mm; spikelets 5–200, in clusters of 1(2) or 1–4(7), all 1 on 1 pl or not
      8. Fr 3-sided, 1.8–2.2 mm; stigmas 3; spikelets in clusters of 1(2) . . . . ***S. heterochaetus***
      8′ Fr 2-sided, 1.5–1.8 mm; stigmas 2(3); spikelets in clusters of 1–4(7) . . . . ***S. tabernaemontani***

S. acutus (Bigelow) Á. Löve & D. Löve var. ***occidentalis*** (S. Watson) S.G. Sm. (p. 1355) COMMON TULE Per 1–4 m; rhizome long, 5–15 mm diam. **ST:** 2–10 mm diam, cylindric; wider air cavities in distal 1/4 ± 1–2.5 mm wide. **LF:** blades 1–2, << sheath, 3–7 mm wide; sheath splitting, leaving fibers. **INFL:** panicle-like; infl bract erect or spreading, 1–9 cm; spikelets 3–100+, in clusters of 1–8, not all 1 on 1 pl, 6–18 mm, 3–4 mm wide; fl bract 3–4 mm, 2–3 mm wide, sparse- (to dense-) scabrous, woolly-ciliate, often deciduously, tip notch 0.3–0.5 mm, awn 0.5–2 mm, often broken off, some or all markedly contorted. **FL:** perianth bristles (4)6(8), (vestigial or) ≤ fr; stigmas 2–3. **FR:** 2–3 mm, 1.2–1.7 mm wide, smooth, 2–3-sided. Common. Marshes, shores, fens, shallow lakes, often emergent; < 2500 m. CA (exc e D); to BC, MT, TX. [*Scirpus a.* var. *o.* (S. Watson) Beetle] Pls with stigmas 2, fr 2-sided (*S. acutus* var. *a.*) in e N.Am w to AK, WA, ID, TX, possible waif in CA. Pls in s CA with stigmas gen 3, fr gen 3-sided may deserve recognition as var. Intermediates to, putative hybrids with *S. californicus* collected in CCo, SnFrB, deltaic GV; putative hybrids with *S. heterochaetus* locally common, with *S. tabernaemontani* locally common in e N.Am. Summer

S. americanus (Pers.) Schinz & R. Keller OLNEY'S THREE-SQUARE BULRUSH Per 0.4–2.5 m; rhizome long, 2–5 mm diam. **ST:** 3–10 mm diam, sides 3, deep-concave (flat), edges sharp. **LF:** blades 1–3, < 1.5 × sheath, 2–8 mm wide, 3-sided toward tip, ± flat; sheath not splitting. **INFL:** head-like; infl bract gen erect, 1–6 cm; spikelets 2–20, 5–15 mm, 3–5 mm wide; fl bract 3–4 mm, tip notch 0.1–0.4 mm, awn 0.2–0.6 mm, often sparse-scabrous. **FL:** perianth bristles (2)5–6(7), ≤ fr; stigmas 2–3. **FR:** 1.8–2.8 mm, 1.3–2 mm wide, 2- or obscure-3-sided, smooth. 2*n*=78. Mineral-rich or brackish marshes,

shores, fens, springs; < 2200 m. KR, NCoRO, CaRH, GV, SnFrB, SW, GB, D; to BC, NS, FL; Mex to S.Am. [*Scirpus a.* Pers.] Hybrids with *S. pungens* intermediate between parental spp., forming persistent clones, rare in CA. Summer

S. californicus (C.A. Mey.) Soják (p. 1355) SOUTHERN BULRUSH Per 1–4 m; rhizome long, 10–15 mm diam. **ST:** 4–10 mm diam, blunt-3-sided throughout to cylindric proximally, obscure-3-sided near infl. **LF:** blades 0–1, ≤ 2 cm, 2 mm wide, flat; sheath splitting, leaving coarse fibers. **INFL:** panicle-like; infl bract gen erect, 1–8 cm; spikelets 25–150+, in clusters of 1–2+, 5–11 mm, ± 3 mm wide; fl bract 2.5 mm, ciliate, midrib, awn sparse-scabrous, tip notch 0.1–0.2 mm, awn to 0.5 mm, not contorted. **FL:** perianth bristles 2–4, dark red-brown, = fr, soft hairs on margins; stigmas 2. **FR:** 1.8–2.2 mm, 1.3 mm wide, 2-sided, smooth. 2*n*=68. Common. Brackish to fresh marshes, shores; < 1500 m. NCo, GV, CCo, SnFrB, SCo, n ChI, PR, w DMoj, e D (Colorado River); to NC, FL; S.Am, W. Indies, Pacific Islands. [*Scirpus c.* (C.A. Mey.) Steud.] Spring–summer

S. heterochaetus (Chase) Soják (p. 1355) SLENDER BULRUSH Per 1.5–2.5 m; rhizome long, 5–8 mm diam. **ST:** 4–8 mm diam, cylindric. **LF:** blades 1–2, << to > sheath, ≤ 200 mm, ± 5 mm wide, thin-C-shaped in ×-section to flat; sheath splitting, leaving delicate fibers. **INFL:** panicle-like; infl bract gen erect, 1–15 cm; spikelets 5–30, in clusters of 1(2), gen all 1 on 1 pl, 5–15 mm, 3–4 mm wide; fl bract 3–4 mm, ciliate, midrib, awn sparse-scabrous, tip notch 0.5 mm, awn 0.3–1 mm, often ± contorted. **FL:** perianth bristles 4(5), << to = fr; stigmas 3. **FR:** 1.8–2.2 mm, 1.3 mm wide, 3-sided, smooth. 2n = 38. Fresh marshes, lakes, often emergent; < 1600 m. CaRH; to AB,

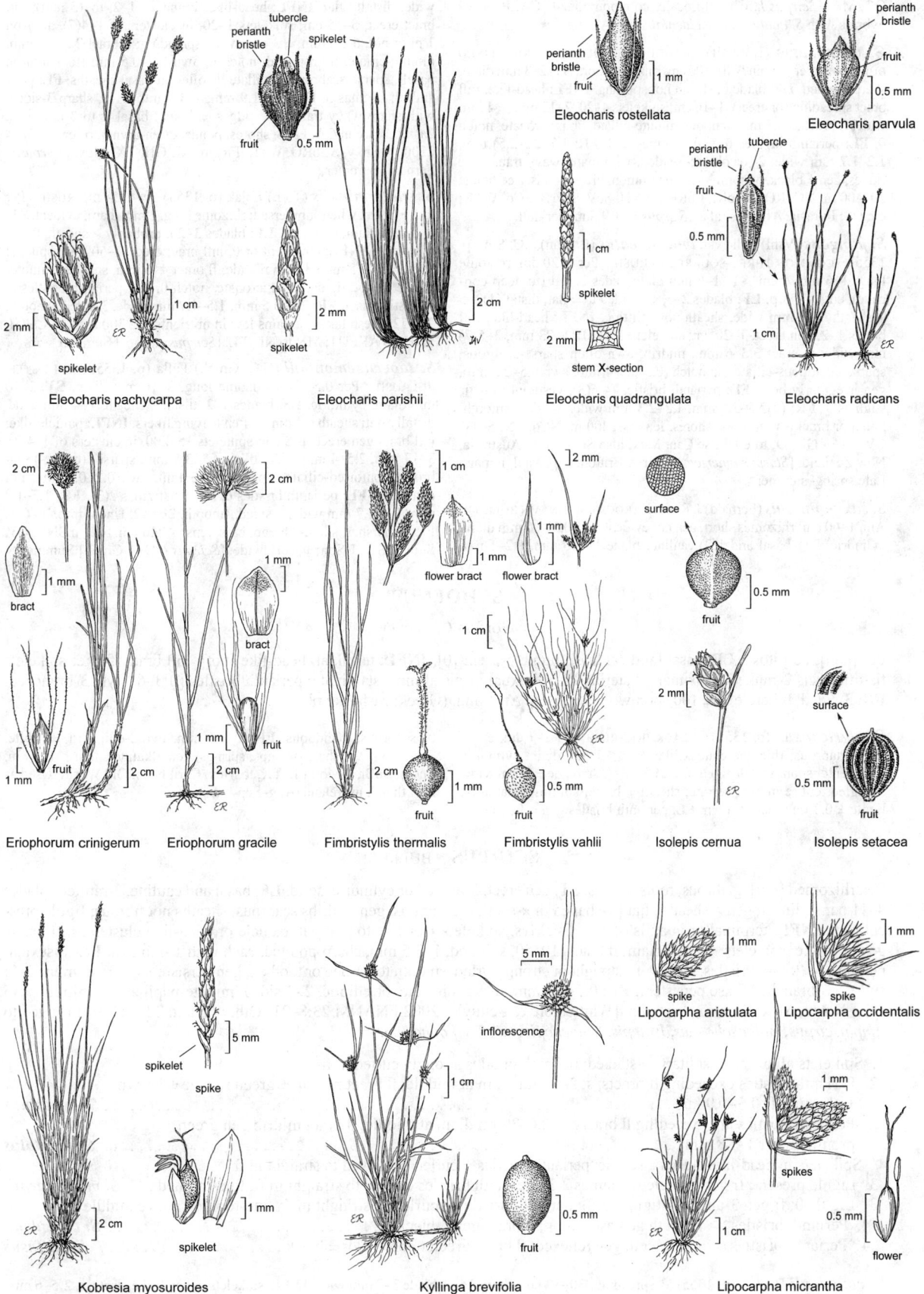

Eleocharis rostellata

perianth bristle
fruit
1 mm

Eleocharis parvula

perianth bristle
fruit
0.5 mm

perianth bristle
spikelet
tubercle
fruit
0.5 mm

Eleocharis pachycarpa

perianth bristle
tubercle
spikelet
fruit
0.5 mm
2 mm
1 cm
spikelet
ER

Eleocharis parishii

spikelet
2 mm
2 cm
JAV

Eleocharis quadrangulata

spikelet
1 cm
stem X-section
2 mm

Eleocharis radicans

1 cm
ER

Eriophorum crinigerum

2 cm
bract
1 mm
fruit
1 mm
2 cm

Eriophorum gracile

2 cm
bract
1 mm
fruit
1 mm
2 cm
ER

Fimbristylis thermalis

1 cm
flower bract
1 mm
2 cm
fruit
1 mm

Fimbristylis vahlii

flower bract
1 mm
2 mm
1 cm
fruit
0.5 mm
ER

Isolepis cernua

surface
fruit
0.5 mm
2 mm
ER

Isolepis setacea

surface
fruit
0.5 mm

Lipocarpha aristulata

spike
1 mm

Lipocarpha occidentalis

spike
1 mm

Kobresia myosuroides

spikelet
spike
5 mm
spikelet
1 mm
2 cm
ER

Kyllinga brevifolia

inflorescence
5 mm
1 cm
fruit
0.5 mm
ER

Lipocarpha micrantha

spikes
1 mm
flower
0.5 mm
1 cm
ER

QC, MA. [*Scirpus h.* Chase] Locally common outside CA. Putative hybrids with *S. acutus* var. *occidentalis* in CA. Aug ★

S. mucronatus (L.) Palla ROUGH-SEED BULRUSH, RICE-FIELD BULRUSH Per (or ann?) 40–100 cm, rhizome short. **ST:** 2–3 mm diam, sharp-3-sided. **LF:** blade 0; sheath not splitting. **INFL:** head-like; infl bract spreading (or erect), 1–10 cm; spikelets 4–20, 7–12 mm, ± 4 mm wide; fl bract 3–3.5 mm, smooth, minute-ciliate, tip mucronate, notch 0. **FL:** perianth bristles 6, = fr; stigmas 2–3. **FR:** 1.7–2.2(2.5) mm, 1.2–1.7 mm wide, 2- or obtuse-3-sided, with many wavy transverse ridges, gen ± black, awn minute. Uncommon. Fresh ponds, rice fields; 20–400 m. NCoRO, NCoRI, CaRF, n SNH, ScV, SnFrB, SCo, WTR; e-c US; Eurasia, Afr, Australia. [*Scirpus m.* L.] Summer–fall

S. pungens (Vahl) Palla var. ***longispicatus*** (Britton) S.G. Sm. (p. 1355) COMMON THREE-SQUARE BULRUSH Per 1–20 dm; rhizome long, 1–6 mm diam. **ST:** 1–6 mm diam, sides 3, flat (to deep-concave), edges sharp. **LF:** blades 2–6, ± 3-sided to ± flat, distal (1)2–5 × sheath, 2–9 mm wide, sheath not splitting. **INFL:** head-like; infl bracts 1–2, gen erect, 1–20 cm; spikelets 1–5(10), 5–23 mm, 3–5(7) mm wide; fl bract 3.5–6 mm, midrib, awn often sparse-scabrous, sparse-deciduous-ciliate, tip notch (0.3)0.5–1 mm, awn 0.5–2.5 mm, gen irregularly bent. **FL:** perianth bristles (4)6(8), vestigial to = fr; stigmas 3. **FR:** (2)2.5–3.5 mm, 1.3–2.3 mm wide, 3-sided, smooth. Fresh or brackish marshes, shores, fens; < 2400 m. NCo, SN, SnJV, CW, SCo, GB, D; to e US, se Can, Mex, also S.Am, Eur, Australia, New Zealand. [*Scirpus americanus* var. *l.* Britton; *S. p.* Vahl, in part] Late spring–summer

S. saximontanus (Fernald) J. Raynal ROCKY MOUNTAIN BULRUSH Ann 1–60 cm; rhizomes short, gen not evident. **ST:** 0.5–1.5 mm diam, cylindric. **LF:** basal and 1(2) cauline; blade 1–200 mm, 0.2–1 mm

wide, distally flat. **INFL:** head-like (branches 1–2, to 15 mm); infl bract erect, 5–15 cm; spikelets 1–20, in clusters of 1–4(7), all 1 on 1 pl or not, 6–20 mm, ± 3 mm wide, spreading; fl bract 2.2–3.5 mm, minute-ciliate, tip rounded, notch 0, awn 0.2–0.5 mm (to 1 mm on basal), sparse-scabrous; pistillate fls often in basal sheaths. **FL:** perianth 0; stigmas 3. **FR:** 1.3–1.9 mm, 1–1.4 mm wide, sharp-3-sided, with sharp wavy transverse ridges, ± black; basal fr to 3 mm or 0. 2*n*=50. Rare in CA. Fresh shores, ponds, often drying, often sandy; < 2200 m. SnJV, SCoRO, WTR, PR; to BC, OH, TX; Mex. [*Scirpus s.* Fernald] Summer

S. subterminalis (Torr.) Soják (p. 1355) WATER BULRUSH Per to 1.5 m, lax when submersed; rhizome long, 1 mm diam, evident. **ST:** to 1 mm diam, cylindric. **LF:** blades 3–20, probably > sheath, 0.2–1 mm wide. **INFL:** emergent or 0, infl bract erect, 7–60 mm; spikelet 1, erect, 5–15 mm, 3–7 mm wide; fl bract 4–6 mm, smooth, minute-ciliate or not, tip acute, mucronate, notch 0. **FL:** perianth bristles 6, ≤ fr; stigmas 3. **FR:** 2.5–3.5 mm, 1.5–1.7 mm wide, 3-sided, smooth. 2*n*=72. Fresh lakes, streams low in nutrients; < 2300 m. KR, CaRH, SNH; to AK, UT; MN to NL, FL. [*Scirpus s.* Torr.] Summer ★

S. tabernaemontani (C.C. Gmel.) Palla (p. 1355) SOFT-STEM BULRUSH Per 0.5–2 m; rhizome long, 3–10 mm diam. **ST:** 2–10 mm diam, cylindric. **LF:** blades 1–2, distal ≤ sheath, 1–4 mm wide, distally flat; sheath splitting, often leaving fibers. **INFL:** panicle-like; infl bract gen erect, 1–8 cm; spikelets 15–200, in clusters of 1–4(7), 3–17 mm, 2.5–4 mm wide; fl bract 2–3.5 mm, sparse- (dense-) scabrous, contorted-ciliate, tip notch 0.3 mm, awn 0.2–0.8 mm, not contorted. **FL:** perianth bristles 6, ± = fr; stigmas 2(3). **FR:** 1.5–1.8 mm, 1.2–1.7 mm wide, 2-sided, smooth. 2*n*=42. Uncommon in CA. Fresh marshes, shores, stream bars, fens; < 2400 m. NW, n SNH, GV, SnFrB, SW, DSon; ± worldwide. [*Scirpus t.* C.C. Gmel.] Summer

SCHOENUS

Gordon C. Tucker

Per [ann], cespitose. **LF:** basal [and cauline]; ligule present [0]. **INFL:** terminal, head-like [not]; infl bract lf-like; spikelets flat; fl bracts 2-ranked, proximal ≥2 sterile. **FL:** bisexual or distal-most staminate; perianth bristles [0]3–6; style 3-branched. **FR:** 3-sided; tubercle 0. ± 100 spp.: worldwide, esp Australia. (Greek: rush-like pl)

S. nigricans L. (p. 1355) BLACK BOG-RUSH Pl 2–7 dm, erect, wiry, glabrous; rhizome caudex-like. **ST:** ≤ 1 mm diam, cylindric, stiff, center spongy with air cavities. **LF:** > 1/2 st; blade ± 1 mm wide, 3-angled, entire minute-serrate; sheath ± black, splitting adaxially; ligule ± 0.3 mm, dark brown. **FL:** perianth bristles gen < fr, smooth

or scabrous, deciduous. **FR:** 1–1.5 mm, ovoid-ellipsoid, ± white. 2*n*=44,54. Marshes, swamps, springs, gen alkaline soils; < 1500 m. SnGb, SnBr, DMoj; FL, TX, to Mex, Caribbean, Old World. Used as roof thatch in Ireland. Aug–Sep ★

SCIRPUS BULRUSH

Per, rhizomed [not], glabrous; roots fibrous. **ST:** gen erect, 3-angled or cylindric, solid. **LF:** basal and cauline, 3-ranked; blades 4–11 per st, linear, gen > sheaths, flat [V-shaped in ×-section], margins, gen midribs scabrous; sheaths not fibrous; ligule present or 0. **INFL:** terminal or in axils of 1–3 distal lvs, spikelets 1 or few to many in panicle or head-like clusters; infl bracts gen 3, lf-like; spikelets < 5 mm diam; fl bracts 10–50, spiraled, 1–3.5 mm, sharp-pointed, each with 1 fl in axil. **FL:** bisexual; perianth of (0)3–6 bristles, < or >> fr, straight or strong-curled, smooth to barbed or toothed, gen persistent on fr; stamens 1–3; style 2–3-branched, base persistent. **FR:** 0.5–1.8 mm, ovate, obovate, or elliptic, 2–3-sided, minute-papillate or -pitted. ± 35 spp.: N.Am, Mex, Eurasia, Australia. [Whittemore & Schuyler 2002 FNANM 23:8–21] Other taxa in TJM (1993) moved to *Amphiscirpus, Bolboschoenus, Isolepis, Schoenoplectus, Trichophorum*.

1. Spikelets all or many solitary, ± stalked; perianth bristles strongly curled in fr
  2. Perianth bristles exceeding fl bracts; st 3–5 mm diam at middle; fl bract midrib ± green to pale ± brown, ± obscure; fr 0.6–0.9 mm . . . . ***S. cyperinus***
  2′ Perianth bristles not exceeding fl bracts; st 0.6–3 mm diam at middle; fl bract midrib gen green, prominent; fr 1–1.8 mm . . . . ***S. pendulus***
1′ Spikelets in head-like clusters, sessile; perianth bristles ± curled or curved to straight in fr
  3. Ligule present; fr gen 2-angled; stigmas 2(3); perianth bristles curved to straight in fr, fine-toothed . . . ***S. microcarpus***
  3′ Ligule 0; fr gen 3-angled; stigmas 3(2); perianth bristles ± curled to straight in fr, fine-toothed above middle
    4. Perianth bristles >> fr, teeth gen ascending; fl bract gen ± black . . . . ***S. congdonii***
    4′ Perianth bristles < to > fr, teeth gen reflexed; fl bract green to brown or ± black . . . . ***S. diffusus***

S. congdonii Britton (p. 1355) Spreading, 30–50 cm; rhizome long. **ST:** 2–4 mm diam at middle, distally 3-angled. **LF:** ligule 0;

blade 3–7 mm wide. **INFL:** spikelets in head-like clusters, 2.5–6 mm, 1–3 mm wide; fl bract 1.5–3 mm, gen ± black, midrib green to straw.

FL: perianth bristles >> fr, not or ± exceeding fl bracts, ± curled; teeth above middle, fine, gen ascending. **FR:** 0.9–1.3 mm, 3(2)-angled. Meadows, marshes, lake shores, streambanks; 700–2600 m. KR, NCoR, CaR, MP; s OR, NV. Jul–Aug

S. cyperinus (L.) Kunth Cespitose, ± 1–2 m; rhizome short. **ST:** 3–5 mm diam at middle, cylindric to 3-angled. **LF:** ligule < 1 mm wide; blade 3–10 mm wide. **INFL:** spikelets 3.5–8 mm, 2.5–3.5 mm wide; fl bracts 1–2.2 mm, red-brown to ± black, midrib ± green to pale ± brown, ± obscure. **FL:** perianth bristles >> fr, exceeding fl bracts, curled, smooth. **FR:** 0.6–0.9 mm, (2)3-angled. 2*n*=66. Fresh moist streambanks, drying alluvial flats; ± 1200 m. c SNH (Yosemite Valley, Mariposa Co.); to WA, e U.S., s Can. Evidently first collected in CA in 1976; study of nativity needed. Aug–Sep

S. diffusus Schuyler (p. 1355) Spreading, ± 70–80 cm; rhizome long. **ST:** 2–5 mm diam at middle, 3-angled. **LF:** ligule 0; blade 5–8 mm wide. **INFL:** spikelets in head-like clusters, 3–6 mm, 1.5–2.5 mm wide; fl bract 1.4–2.6 mm, green to brown or ± black. **FL:** perianth bristles < to > fr, not exceeding fl bracts, ± curled to straight, teeth above middle, fine, gen reflexed. **FR:** 0.9–1.3 mm, 3-angled. Meadows, marshes, streambanks; 200–2400 m. KR, s CaR, SN. Jul

S. microcarpus J. Presl & C. Presl (p. 1355) Spreading, 60–200 cm; rhizome long. **ST:** 2–5 mm diam at middle, 3-angled. **LF:** blade 5–15(20) mm wide; ligule present. **INFL:** spikelets in head-like clusters, 2–8 mm, 1–3.5 mm wide; fl bract 1–3.5 mm, green or ± black, midrib gen prominent-green. **FL:** perianth bristles < to ± > fr, straight to curved, stout, teeth fine, reflexed. **FR:** ± 0.5–1.5 mm, gen 2-angled. 2*n*=64,66. Marshes, wet meadows, streambanks, pond margins, sometimes weedy; < 3000 m. CA; to AK, e N.Am, e Asia. Jun–Jul

S. pendulus Muhl. PENDULOUS BULRUSH Cespitose, 50–100 cm; rhizome short. **ST:** 0.6–3 mm diam at middle. **LF:** blade 4–8(12) mm wide; ligule 0. **INFL:** spikelets all or many solitary, ± stalked, 5–10(12) mm, 2–3 mm wide, fl bract ± 1.5–2 mm, brown to red-brown, midrib green, prominent. **FL:** perianth bristles >> fr, not exceeding fl bracts, curled. **FR:** 1–1.8 mm, 3-angled. 2*n*=40. Marshes, wet meadows; < 900 m. n CaRH (near Yreka); s OR, se Can, e US, Mex. Jun–Aug ★

TRICHOPHORUM

Per, cespitose, < 15 cm, smooth. **ST:** simple, < 1 mm diam, cylindric, grooved, not hollow. **LF:** basal or subbasal, spiraled; blade ≤ 1 mm wide; sheath brown; ligule present. **INFL:** terminal, spikelet 1; infl bract 1, like fl bracts, with lf-like blade or not; spikelets 1, 3–4.6 mm, 1.5–2.8 mm wide, ovate, not ± flat, 2–6-fld; fl bracts spiraled, each with 1 fl in axil, ovate, brown, membranous, tip entire. **FL:** bisexual; perianth of 0–6 bristles, < fr; stamens 3, anthers 0.8–1.5 mm. **FR:** obovate to elliptic, smooth, minute-mucronate or not; tubercle 0. 9 spp.: circumpolar or circumboreal. (Greek: hair stalk) [Crins 2002 FNANM 23:28–31] May be mistaken for *Eleocharis*.

1. Infl bract > spikelet; rhizomes 0 .... ***T. clementis***
1′ Infl bract < spikelet; rhizomes long.. ***T. pumilum***

T. clementis (M.E. Jones) S.G. Sm. (p. 1355) Clumped. **LF:** blade 4–17 mm. **FL:** perianth bristles 0–6, brown, ± flat, smooth or scabrous. **FR:** 3-sided. Dry to wet meadows, streambanks; 2400–3600 m. c&s SNH. [*Scirpus c.* M.E. Jones] Summer

T. pumilum (Vahl) Schinz & Thell. LITTLE BULRUSH Forming mats. **LF:** blade 2–13 mm. **FL:** perianth 0. **FR:** 2–3-sided. Wet sites, limestone soils; 3100–3250 m. c SNH, W&I; AK to MT, also QC, Eur, c Asia. [*Scirpus p.* Vahl] Summer ★

HYDROCHARITACEAE WATERWEED FAMILY ·

Robert F. Thorne, C. Barre Hellquist & Robert R. Haynes

Ann, per, aquatic, freshwater or marine, glabrous or hairy; monoecious, dioecious, or fls bisexual. **LF:** basal, alternate, opposite, or whorled, gen sessile, gen ± sheathing at base. **INFL:** axillary, terminal, or scapose, cyme or fls 1, subtended by ± sheathing, entire or lobed bract; staminate fls sometimes deciduous, free-floating. **FL:** gen radial; perianth 0 or tube 0 or elongate, peduncle-like in fl; sepals (0)3(4), green; petals (0)3(4), white or not; stamens (0)2–many, gen in 1+ series; ovary inferior, chamber 1 or falsely 6–9, placentas parietal, ovules 1–many, style lobes gen 3, linear, lobed or notched. **FR:** achene or berry-like and dehiscing irregularly, linear to spheric, submersed. ± 17 genera, ± 130 spp.: worldwide; some cult for aquaria, others noxious weeds. [Haynes 2000 FNANM 22:26–38] *Ottelia alismoides* (L.) Pers. is a possibly extirpated alien. Scientific Editor: Thomas J. Rosatti.

1. Pls floating (rooted in mud when stranded); lvs (when floating) with swollen spongy cells abaxially..... **LIMNOBIUM**
1′ Pls submersed to emergent; lvs without swollen spongy cells abaxially
  2. Lvs basal, petioled, alternate, lower, submersed blades elliptic to ovate, upper, floating or emergent
    blades ovate to cordate or reniform — fl bisexual, emergent .... **[OTTELIA]**
  2′ Lvs cauline, sessile, opposite to whorled, blade oblong or linear
    3. Perianth 0; lvs ± opposite to ± whorled; seeds smooth or pitted.................................... **NAJAS**
    3′ Perianth present; lvs opposite or gen whorled; seeds smooth
      4. Lvs 1.2–4 cm, 2–5 mm wide; staminate fls 2–3, petals 8–10 mm .... **EGERIA**
      4′ Lvs < 1.5(2) cm, < 3 mm wide; staminate fls 1, petals gen ≤ 5 mm
        5. Lvs often 2 per node proximally, 3–7 distally, minutely serrate, midrib abaxially ± smooth .... **ELODEA**
        5′ Lvs gen 4–8 per node, coarsely serrate, midrib abaxially prickled or not .... **HYDRILLA**

EGERIA BRAZILIAN WATERWEED

Per, submersed, rooted in mud; dioecious. **ST:** slender, branched or not. **LF:** opposite below, crowded and in whorls of 3–6 above, sessile, finely to coarsely serrate. **STAMINATE INFL:** axillary, sessile; fls 2–4(5), 1 open at a time; bract slender. **PISTILLATE INFL:** [axillary, sessile; fl 1; bract slender, ± tubular]. **STAMINATE FL:** floating; pedicel elongate, peduncle-like, persistent on pl; stamens 9. **PISTILLATE FL:** [floating; perianth tube elongate, peduncle-like; stigmas 3, 3–4-lobed].

2 spp.: S.Am. (Latin: mythical water nymph) Only staminate pls observed outside native range. *E. najas* Planch. ◆ expected but not yet found.

E. densa Planch. (p. 1355)   **ST:** 2–3 mm thick. **LF:** 1.2–4 cm, 2–5 mm wide, recurved, narrowly oblong, finely serrate. **STAMINATE FL:** pedicel ≤ 8 cm; petals 8–10 mm, white, ± round. **PISTILLATE FL:** [as staminate; petals 6–7 mm]. $2n$=46. Streams, ponds, sloughs; < 2200 m. SNF, GV, SnFrB, SnJt, GB; to e US; Eur; native to S.Am. Staminate pls widely cult for aquaria. Jul–Aug ◆

ELODEA

Per, rooting at nodes, submersed; dioecious or some fls bisexual. **ST:** slender, gen branched. **LF:** often 2 per node proximally, 3–7 distally, sessile, minutely serrate, midrib abaxially ± smooth. **INFL:** axillary, sessile or not; fl 1; bract gen notched. **STAMINATE FL:** floating, deciduous or not; perianth tube elongate, slender, peduncle-like; stamens 3–9. **PISTILLATE FL:** floating; perianth tube elongate, slender, peduncle-like; style slender, stigmas 3, lobes 0 or 2. **FR:** cylindric to ovoid. **SEED:** several, smooth. ± 12 spp.: temp, trop Am. (Greek: of marshes)

1. Mid-st lvs 2–3 per node, tip rounded or obtuse; staminate infl bract 10–42 mm . . . . *E. bifoliata*
1′ Mid-st lvs 3(4) per node, tip obtuse, abruptly pointed, or acute; staminate infl bract 2.2–13.5 mm
  2. Lvs 2–5 mm wide, tip obtuse to abruptly pointed; staminate fls deciduous in late bud or when fls open,
    petals ± 5 mm; pistillate sepals ± 2–3 mm. . . . . *E. canadensis*
  2′ Lvs 0.3–2 mm wide, tip acute; staminate fls deciduous in bud, petals gen 0 or ± 0.5 mm; pistillate sepals
    ± 1 mm . . . . *E. nuttallii*

E. bifoliata H. St. John TWO-LEAF WATERWEED **LF:** gen not crowded at st tips, 2 per node on lower sts, 2–3 on mid sts; blade gen 4.7–24.8 mm, 0.8–2.5 mm wide, linear to narrowly elliptic, rounded to obtuse at tip, flaccid. **STAMINATE INFL:** bract 10–42 mm, proximally slender, distally 6–10 mm, 2.5–5 mm wide, elliptic, inflated, deciduous; peduncles, fls deciduous soon after fls open. **STAMINATE FL:** sepals 3.5–5 mm, 2–2.5 mm wide, petals 4.5–5 mm. ± 0.5 mm wide. **PISTILLATE FL:** sepals 1.4–2.8 mm, 1–1.3 mm wide; petals 1.8–4 mm, 1.1–1.3 mm wide. **SEED:** 2.8–3 mm, ellipsoid, dense-long-hairy. Shallow waters, streams; 1300–1800 m. MP; to BC; w US to MN. [*E. nevadensis* H. St. John] Jun–Aug

E. canadensis Michx. (p. 1355) COMMON WATERWEED **LF:** gen crowded at st tips, 3 per node on mid, upper sts; blade gen 5–15 mm, 2–5 mm wide, linear, obtuse or abruptly pointed at tip, flaccid or not. **STAMINATE INFL:** bract 8.2–13.5 mm; peduncles, fls deciduous in late bud or when fls open. **STAMINATE FL:** sepals 3.5–5 mm, ± 4 mm wide; petals ± 5 mm. **PISTILLATE FL:** sepals ± 2–3 mm. **SEED:** glabrous. $2n$=24,48. Shallow water, sloughs, ponds, lakes; 300–2600 m. NCoRO, CaRH, SNH, GV, SnFrB, SnGb, SnBr, GB; to BC, e US; naturalized in Eur, Australia. Jul–Aug

E. nuttallii (Planch.) H. St. John (p. 1355) NUTTALL'S WATERWEED **LF:** gen not crowded at st tips, 3(4) per node on mid, upper sts; blade gen 6–13 mm, 0.3–2 mm wide, linear to narrowly lanceolate, acute at tip, flaccid or not. **STAMINATE INFL:** bract 2.2–4 mm; peduncles, fls deciduous in bud. **STAMINATE FL:** sepals ± 2 mm, ± 1.5 mm wide; petals gen 0 or ± 0.5 mm. **PISTILLATE FL:** sepals ± 1 mm. **SEED:** short-soft-hairy. $2n$=48. Shallow water, streams, lakes, ponds; 500–2800 m. KR, SNH, ScV, SnBr, GB; to WA, c US; naturalized in Eur. Jul–Aug

HYDRILLA

1 sp. (Greek: in water)

H. verticillata (L. f.) Royle (p. 1355) Per, rooting at nodes, submersed; dioecious. **ST:** tuber-like; erect sts elongate, much-branched. **LF:** gen 4–8 per node, sessile, 1–2 cm, 1.5–2 mm wide, oblong, coarsely serrate, midrib prickled abaxially or not, tip gen sharp-pointed. **STAMINATE INFL:** fl 1, bract at base of elongate perianth tube, spheric. **PISTILLATE INFL:** fl 1, bract at base of elongate perianth tube, ± cylindric. **STAMINATE FL:** deciduous, free-floating; perianth parts 6 in 2 whorls, 3–5 mm; stamens 3. **PISTILLATE FL:** persistent, floating; perianth parts 6 in 2 whorls, 3–5 mm; stigmas 3. **FR:** 5–6 mm, ± fusiform. $2n$=32. Canals, ponds, lakes; < 200 m. NCoRI, n&c SNF, ScV, SCo, D; se&e US, C.Am; native to Eurasia, Australia. Jun–Aug ◆

LIMNOBIUM FROGBIT

Per, floating in dense mats, rooted in mud when stranded; monoecious. **ST:** stolons floating or suspended in water. **LF:** basal, emergent or floating, petioled; floating with swollen spongy cells abaxially. **INFL:** cyme, peduncle 0 or short. **FL:** unisexual, emergent, pedicelled; petals green-white to yellow. **STAMINATE FL:** filaments fused at least 1/2, anthers elongate. **PISTILLATE FL:** ovary chambers 1, styles 3–9, 2-lobed ± to base. **FR:** fleshy berry, ellipsoid to spheric, smooth to ridged, dehiscing irregularly. **SEED:** ellipsoid, spiny. 2 spp.: Am. (Greek: living in pools)

L. spongia (Bosc) Steud. (p. 1363) AMERICAN FROGBIT **LF:** 1–10 cm, 0.9–7.8 cm wide; petioles to 15 cm; blades ovate to ± round, bases reniform to cordate. **FL:** sepals 3; petals 3. **STAMINATE FL:** stamens 9–12(18), filaments fused into column, anthers 6–12. **PISTILLATE FL:** stigmas 6–18. **FR:** 4–12 mm diam. Streams, ponds, lagoons; < 200 m. NCo, SnFrB; se US, n to NY. [*L. laevigatum* (Willd.) Heine, misappl.] Jun–Sep ◆

NAJAS

Ann, aquatic, submersed, mat-like or not; monoecious or dioecious. **ST:** several, often much-branched, slender. **LF:** simple, cauline, ± opposite to ± whorled, sessile; sheath gen wider than, expanded abruptly at junction with blade; blade gen linear, margin entire to coarsely spine-toothed. **INFL:** axillary; fls 1–few, clustered, inconspicuous. **STAMINATE FL:** subtended by (0)2 minute involucres, inner membranous, flask-shaped, outer cup-like, tubular, or with free scales; perianth 0; stamen 1,

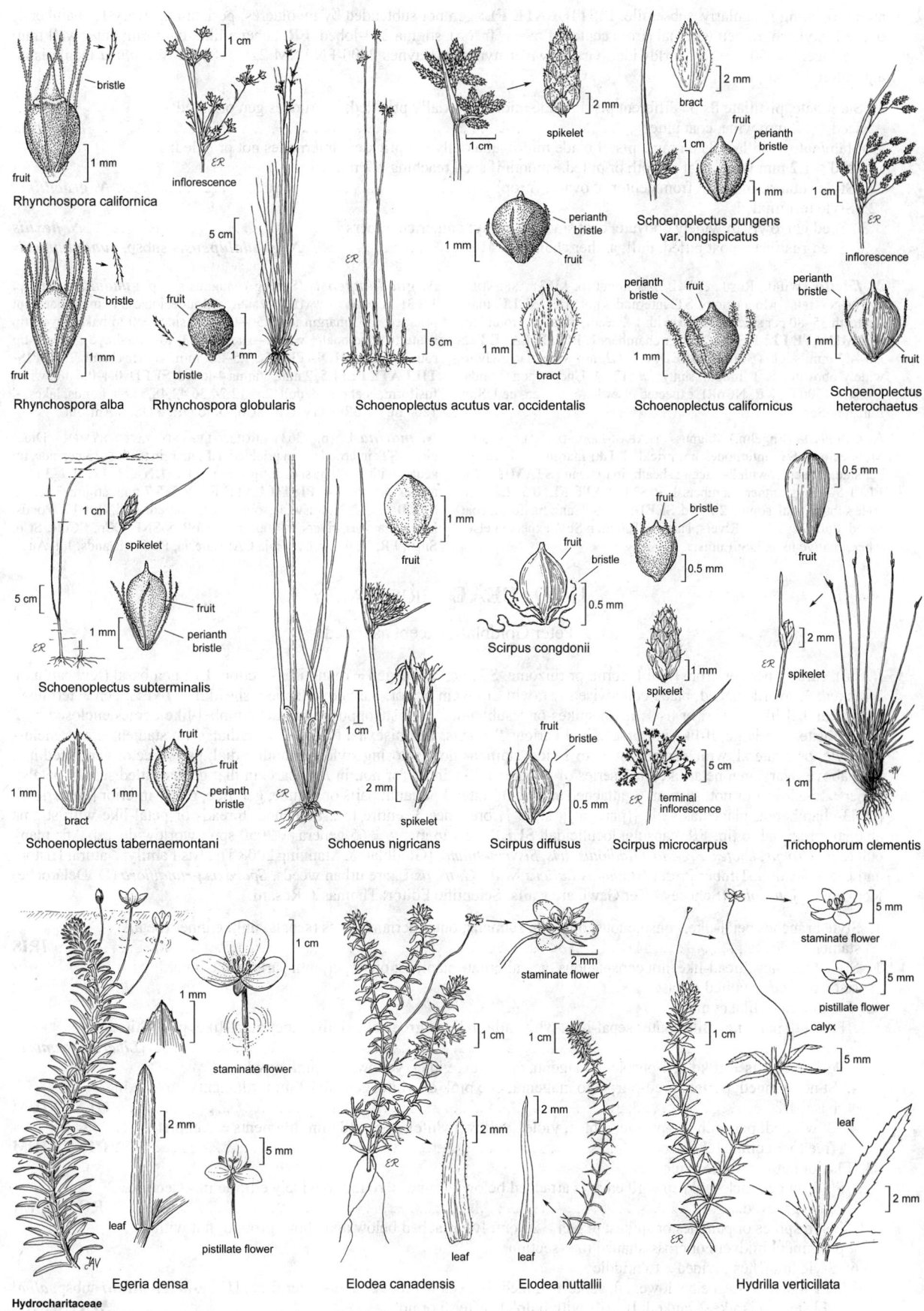

Rhynchospora californica

Rhynchospora alba

Rhynchospora globularis

Schoenoplectus acutus var. occidentalis

Schoenoplectus pungens var. longispicatus

Schoenoplectus californicus

Schoenoplectus heterochaetus

Schoenoplectus subterminalis

Schoenoplectus tabernaemontani

Schoenus nigricans

Scirpus congdonii

Scirpus diffusus

Scirpus microcarpus

Trichophorum clementis

Egeria densa

Elodea canadensis

Elodea nuttallii

Hydrilla verticillata

Hydrocharitaceae

anther opening irregularly, subsessile. **PISTILLATE FL**: gen not subtended by involucres; perianth 0; ovary 1, chamber 1, ovule 1, style short, gen terminal (from center of ovary, fr top), stigma 2–4-lobed. **FR**: achene-like, fusiform; outer wall thin, ± translucent. ± 50 spp.: ± worldwide. (Greek: water nymph) [Haynes 2000 FNANM 22:77–83] *N. graminea* Delile is an agricultural weed.

1. Staminate, pistillate fls on different pls; lf blade midrib abaxially prickled; internodes gen prickled;
   seed ≥ 1.2 mm wide, coat pitted . . . . ***N. marina***
1′ Staminate, pistillate fls on same pls; lf blade midrib abaxially not prickled; internodes not prickled;
   seed ≤ 1.2 mm wide, coat smooth or pitted, smooth if seed reaching 1.2 mm wide
 2. Style subterminal (not from center of ovary, fr top) . . . . ***N. gracillima***
 2′ Style terminal
  3. Seed narrowly to widely obovate, coat smooth, shiny; anther chambers 1 . . . . ***N. flexilis***
  3′ Seed fusiform, coat pitted, dull; anther chambers 4 . . . . ***N. guadalupensis*** subsp. ***guadalupensis***

N. flexilis (Willd.) Rostk. & W.L.E. Schmidt (p. 1363) SLENDER WATER-NYMPH Monoecious. **ST**: internodes not prickled. **LF**: margin teeth 35–80 per side, tip acute, with 1–2 teeth; sheath tip rounded. **STAMINATE FL**: 1–3 mm; anther chambers 1. **PISTILLATE FL**: 2.5–4.7 mm; stigma 3-lobed. **SEED**: 0.2–1.2 mm wide, narrowly to widely obovate, coat smooth, shiny. 2*n*=12,24. Uncommon. Ponds, lakes; < 1500 m. KR, NCoRI, expected elsewhere; to Can, ne US; n Eur. Jun–Sep

N. gracillima (Engelm.) Magnus THREAD-LEAVED WATER-NYMPH Monoecious. **ST**: internodes not prickled. **LF**: margin teeth 13–17 per side, tip acute, with 2–3 teeth; sheath tip truncate. **STAMINATE FL**: 1.5–2 mm; anther chambers 1. **PISTILLATE FL**: 0.5–2.7 mm; style subterminal, stigma 2-lobed. **SEED**: 0.4–0.7 mm, fusiform, coat pitted, dull. 2*n*=24,36. Rivers, lakes; ± 100 m. n ScV, expected elsewhere; native to ne US; Eurasia. Jun–Aug

N. guadalupensis (Spreng.) Magnus subsp. ***guadalupensis*** (p. 1363) COMMON WATER-NYMPH Monoecious. **ST**: internodes not prickled. **LF**: margin teeth 50–100 per side (± 0 to naked eye), tip acute to mucronate, with 1–3 teeth (± 0 to naked eye); sheath tip rounded. **STAMINATE FL**: 1.5–2.5 mm; anther chambers 4. **PISTILLATE FL**: 1.5–2 mm; stigma 4-lobed. **SEED**: 0.4–0.6 mm wide, fusiform, coat pitted, dull. 2*n*=12,24,36,42,48,54,60. Ponds, lakes; < 1200 m. NCoRO, GV, CW, SCo, D; to OR, e US, S.Am. Jul

N. marina L. (p. 1363) HOLLY-LEAVED WATER-NYMPH Dioecious. **ST**: internodes gen prickled. **LF**: margin teeth 8–13 per side, tip acute, with 1 tooth; sheath tip acute. **STAMINATE FL**: 1.7–3 mm; anther chambers 4. **PISTILLATE FL**: 2.5–5.7 mm; stigma 3-lobed. **SEED**: 1.2–2.2 mm wide, ovate, coat pitted, dull. 2*n*=12. Ponds, lakes, marshes, rivers; < 1000 m. NCoR, s SNF, CCo, SCoR, SCo, SnBr, PR, D; to n-c US, Baja CA; Eurasia, Pacific islands. Jul–Aug

IRIDACEAE IRIS FAMILY

Peter Goldblatt, except as noted

[(Ann), shrub], per gen from [bulb], corm, or rhizome. **ST**: gen erect, gen ± round in ×-section. **LF**: gen basal (few cauline), 2-ranked, ± sword-shaped, blade edge-wise to st, with midvein or not; bases overlapped, sheathing. **INFL**: gen ± terminal; spikes, umbel-like cymes, or fls 1; fls in spikes or 1 subtended by 2 subopposite fl bracts; umbel-like cymes enclosed by 2 subopposite, gen large, lf-like infl bracts, incl various fl bracts. **FL**: bisexual (unisexual), radial, with stamens erect, enclosing style, or bilateral, with stamens, style to 1 side, stamens not enclosing style; perianth radial, parts free or gen fused into tube above ovary, gen petal-like, in 2 series of 3, outer ± like inner (or not, in *Iris*, parts in that genus called sepals, petals), upper ± like lower or not; stamens 3, attached at base of outer 3 perianth parts or in tube, gen free; ovary inferior [(superior)], [(1)]3-chambered, placentas axile [(parietal)], style 1, branches 3, entire to 2-branched, thread- or petal-like with stigma abaxial, proximal to tip. **FR**: capsule, loculicidal. **SEED**: few to many. ± 65 genera, ± 2050 spp.: worldwide, esp Afr; many cult (e.g., *Crocus*, *Dietes*, *Freesia*, *Gladiolus*, *Iris*, *Sisyrinchium*). [Goldblatt & Manning 2008 The Iris Family: Natural History and Classification. Timber Press] *Gladiolus italicus* Mill., *G. tristis* L. are urban weeds. *Sparaxis grandiflora* (D. Delaroche) Ker Gawl., *S. tricolor* (Schneev.) Ker Gawl. are waifs. Scientific Editor: Thomas J. Rosatti.

1. Style branches petal-like, conspicuous, opposite stamens; outer perianth parts (sepals) unlike inner (petals);
   stamens free . . . . **IRIS**
1′ Style branches thread-like, not conspicuous, gen alternate stamens; outer perianth parts ± like inner;
   stamens free or united at base
 2. Fls in umbel-like cymes
  3. Perianth parts not alike, outer sepal-like, white adaxially, ± green abaxially, inner petal-like, gen white .
     . . . . **[*Libertia formosa*]**
  3′ Perianth parts ± alike, red-purple to magenta or ± blue, violet, yellow, occ pink or white
   4. St not winged; perianth red-purple to magenta, occ pink or white, gen 18–33 mm; filaments fused ± in
      basal 1/2 . . . . **OLSYNIUM**
   4′ St winged; perianth red-purple, ± blue, violet, yellow (white), gen < 18 mm; filaments ± completely
      free to ± completely fused . . . . **SISYRINCHIUM**
 2′ Fls not in umbel-like cymes
  5. Fls 1 on peduncles; pl gen < 10 cm; lvs attached below ground, ± round to widely elliptic in ×-section,
     gen 4-grooved. . . . . **ROMULEA**
  5′ Fls in spikes or panicles of spikes; pl gen > 20 cm; lvs attached below and above ground, flat with
     prominent midvein or cross-shaped in ×-section
   6. Style branches divided ± to middle
    7. Fl st bent ± 90° below lowest fl, spike 1-ranked; lf axils without bulblets after fl . . . **[*Freesia leichtlinii* subsp. *alba*]**
    7′ Fl st erect, spike 2-ranked; lf axils with bulblets after fl or not . . . . **WATSONIA**

6′ Style branches entire or notched at tip

  8. Fl bracts membranous or papery and crinkled, outer 3-lobed to -toothed or often cut into lance-linear segments in upper 1/2 or at tip, inner 2-lobed or -toothed

    9. Fl bracts membranous, outer 3-lobed to -toothed; stamens attached at mouth. . . . . *[Ixia maculata]*

    9′ Fl bracts papery, crinkled, outer often cut into lance-linear segments in upper 1/2 or at tip; stamens attached in tube . . . . **[SPARAXIS]**

  8′ Fl bracts green, soft or leathery, outer entire, inner notched at tip

    10. Perianth tube abruptly enlarged above base, lobes unequal, upper 1 > lower 5; infl gen an erect spike (panicle). . . . . **CHASMANTHE**

    10′ Perianth tube not abruptly enlarged above base, lobes subequal; infl gen an arched spike or panicle of spikes . . . . **CROCOSMIA**

CHASMANTHE

Corm depressed, cover fibrous. **ST:** often branched. **INFL:** gen spike (panicle); fls each subtended by 2 ± equal bracts, inner forked at tip. **FL:** bilateral; perianth trumpet-shaped, tube > bracts, curved, ± cylindric, abruptly enlarged above base, upper lobe > lower 5, ± straight, lower 5 reflexed; stamens free, exserted, arched under upper perianth lobe; style exserted, branches notched at tip. **SEED:** many, spheric, orange. 3 spp.: s Afr, all orn. (Greek: gaping fl) [Goldblatt et al. 2004 *Crocosmia* and *Chasmanthe*. Timber Press]

C. floribunda (Salisb.) N.E. Br. **ST:** erect, gen < 1 m. **LF:** 20–65 cm, 2.5–5 cm wide, flat, midvein prominent. **INFL:** fls 20–30, 2-ranked. **FL:** < 4 cm, scarlet or yellow. Uncommon. Disturbed areas; < 50 m. NCo, CCo, SnFrB, SCo; native to s Afr. Often mistaken for *C. aethiopica* (L.) N.E. Br.; pls sometimes persist from garden waste. Feb–May

CROCOSMIA

Corm round to ovoid, often making stolons, cover fibrous. **ST:** gen branched. **INFL:** spike or panicle of spikes, > lvs. **FL:** radial or bilateral; perianth funnel- to trumpet-shaped, orange to brick-red, tube gen curved, lobes ± spreading, ± equal [unequal], spreading; stamens free, exserted from tube; style ± > stamens, branches notched at tip [entire]. **SEED:** gen 12–many, some occ aborted. 8 spp.: ± sub-Saharan Afr, Madagascar. (Greek: saffron odor, from dried fls in warm water) [Goldblatt et al. 2004 *Crocosmia* and *Chasmanthe*. Timber Press]

C. ×*crocosmiiflora* (Lemoine) N.E. Br. **ST:** < 1 m. **LF:** 30–60 cm, 1–2.5 cm wide; midvein prominent. **FL:** perianth 2–5 cm, lobes < 5 cm wide. Common. Gen in disturbed coastal areas, roadsides, often from garden waste; < 50 m. NCo, SnFrB, SCo, expected elsewhere; alien in nw&se N.Am, Australia, New Zealand, HI, Madagascar, S.Am. Made in France from *C. pottsii* (Baker) N.E. Br. × *C. aurea* (Hook.) Planch., both s Afr; spreading by stolons. Jul–Aug ❖

IRIS IRIS

Carol A. Wilson

Rhizome [bulbs, fleshy roots]. **LF:** 2-ranked in basal fan; cauline 0–few, reduced, often bract-like, without development of distal portion. **INFL:** ± flat cyme, fls 1–many. **FL:** perianth parts ± clawed; sepals gen wider than petals, spreading or reflexed, occ with white area in basal 3/4, this gen with smaller yellow area; petals erect; stamens free [(not)]; ovary inferior, style branches petal-like [(not)], arched over stamens, each with scale-like flap (with stigmas on inner surface) opposite stamen and just below 2-lobed tip (crest), with sepals forming 3, 2-lipped units [(not)]. **FR:** loculicidal capsule, rounded or triangular, chambers 3. **SEED:** gen compressed, pitted, light to dark brown (red). ± 160 spp.: gen n temp. (Greek: rainbow, from fl colors) [Wilson 2003 Syst Bot 28:39–46] Hybrids between some sympatric spp.; *I. germanica* only sp. in CA with bearded sepals.

1. Sepal obviously bearded . . . . ***I. germanica***

1′ Sepal not obviously bearded

  2. Perianth tube > 30 mm

    3. Cauline lvs bract-like

      4. Perianth pale cream-yellow with prominent brown-violet veins, rose-violet with darker veins, or ± white with rose veins; stigmas truncate, ± rounded, or 2-lobed . . . . ***I. purdyi***

      4′ Fls pale yellow or cream with same color (or ± red) veins; stigmas triangular or tongue-shaped
         . . . . ***I. tenuissima*** subsp. ***purdyiformis***

    3′ Cauline lvs similar to basal

      5. Stigmas tongue-shaped, tip rounded; rhizome 2–5 mm diam

        6. Stigmas entire; perianth parts elliptic or narrowly elliptic; st gen < 1 dm . . . . ***I. chrysophylla***

        6′ Stigmas finely toothed; perianth parts widely oblanceolate; st ≥ 1 dm . . . . ***I. tenuissima*** subsp. ***tenuissima***

      5′ Stigmas triangular, tip acute; rhizome 4–9 mm diam

        7. Perianth tube funnel-shaped; basal lvs ≥ 6 mm wide . . . . ***I. fernaldii***

        7′ Perianth tube abruptly inflated at top (bowl-like); basal lvs ≤ 6 mm wide. . . . . ***I. macrosiphon***

  2′ Perianth tube < 30 mm

    8. Stigmas 2-lobed; rhizomes 9–30 mm diam; cauline lvs bract-like for at least 2/3 length

9. Lowest bracts alternate; moist, coastal prairie or open coastal forest; fls 3–6, lilac-purple . . . . *I. longipetala*
9′ Lowest bracts opposite; seasonally wet sites; fls 1–2, pale lilac to ± white, or 2–9, brown, red-brown, or green
   10. Basal lvs 25–35 mm wide; bracts not scarious; fls 2–9, pale yellow to brown, red-brown, or green,
      veined green or red-brown . . . . *I. foetidissima*
   10′ Basal lvs 3–9 mm wide; bracts gen scarious; fls 1–2, pale lilac to ± white, veined lilac-purple
      . . . . *I. missouriensis*
8′ Stigmas triangular, tongue-shaped, or rounded; rhizomes 1–12 mm or 30–40 mm diam; cauline lvs
  bract-like or similar to basal
   11. Stigmas rounded; basal lvs 20–35 mm wide; rhizomes 30–40 mm diam; wetlands . . . . *I. pseudacorus*
   11′ Stigmas triangular or tongue-shaped; basal lvs 2–20 mm wide; rhizomes 1–12 mm diam; dry slopes
     (moist meadows)
     12. Cauline lvs bract-like for at least 2/3 of length
       13. Perianth tube ≤ 10 mm; cauline lvs overlapped; fls gen 2 . . . . *I. bracteata*
       13′ Perianth tube > 10 mm; cauline lvs gen not overlapped; fls 1–2
        14. Lowest bracts spreading; ovary triangular, clearly winged
         15. Ovary clearly winged; perianth with deep maroon or ± brown veins . . . . *I. tenax* subsp. *klamathensis*
         15′ Ovary triangular but not winged; perianth lacking dark veins . . . . *I. hartwegii* subsp. *pinetorum*
        14′ Lowest bracts enclosing perianth tube; ovary rounded or triangular, not winged
         16. Fls gen golden yellow; ovary rounded; petals elliptic . . . . *I. innominata*
         16′ Fls gen lavender to blue, occ ± white; ovary triangular; petals obovate . . . . *I. thompsonii*
     12′ Cauline lvs similar to basal
       17. Fls pale cream to yellow or gold-yellow, veined gold or bright yellow or not
        18. Fls 3; outer bract 9–15 cm; perianth tube 11–15 mm . . . . *I. hartwegii* subsp. *columbiana*
        18′ Fls 2; outer bract 6–11 cm; perianth tube 5–10 mm . . . . *I. hartwegii* subsp. *hartwegii*
       17′ Fls blue to lavender or purple (white or pale cream), with cream or ± white throat or not
        19. Basal lvs 7–10 mm wide; fls 2; dry slopes . . . . *I. hartwegii* subsp. *australis*
        19′ Basal lvs 10–22 mm wide; fls gen > 2; grassy sites
         20. St gen branched, fls 2/branch; coastal plains . . . . *I. douglasiana*
         20′ St unbranched, fls 3; wet grassy sites in s SN . . . . *I. munzii*

I. bracteata S. Watson (p. 1363) SISKIYOU IRIS Rhizome 6–9 mm diam. **ST:** unbranched, 5–40 cm. **LF:** basal 4–18 mm wide, glossy; cauline 1–6, bract-like, overlapped. **INFL:** fls gen 2; lowest 2 bracts opposite, enclosing perianth tube, outer 5–9 cm, 6–15 mm wide. **FL:** perianth yellow, veined deep maroon or brown, tube 4–10 mm, funnel-shaped; sepals 4–8 cm, 15–25 mm wide, obovate; petals 4–5 cm, 7–15 mm wide, obovate; ovary rounded, style branches 29–45 mm, crests 9–17 mm, stigmas triangular; nectar abundant. 2*n*=40. Partly shady places, gen in yellow-pine forest; 350–1100 m. KR (n Del Norte, nw Siskiyou cos.); sw OR. May ★

I. chrysophylla Howell (p. 1363) YELLOW-FLOWERED IRIS Rhizome 2–5 mm diam. **ST:** unbranched, 1–12 cm. **LF:** basal 4–7 mm wide; cauline 0–4, similar to basal. **INFL:** fls 2; lowest 2 bracts opposite, enclosing perianth tube, 6–14 cm, gen < inner, 10–20 mm wide. **FL:** perianth pale cream-yellow to ± white, with faint ± blue tinge or not, veined ± burgundy, tube 45–90 mm, narrow, abruptly funnel-shaped distally; sepals 4–8 cm, 14–20 mm wide, elliptic; petals 4–6 cm, 7–10 mm wide, narrowly elliptic; ovary triangular, style branches 35–60 mm, crests 15–25 mm, stigmas tongue-shaped, entire. 2*n*=40. Uncommon. Open conifer forest, roadbanks; 600–1000 m. KR (n Del Norte, nw Siskiyou cos.); w OR. May–Jun

I. douglasiana Herb. (p. 1363) Rhizome 8–9 mm diam. **ST:** gen branched, 15–50 cm. **LF:** basal 10–22 mm wide; cauline 1–3, similar to basal. **INFL:** fls 2–6; lowest 2 bracts ± opposite, enclosing perianth tube, outer 6–16 cm, 4–12 mm wide. **FL:** perianth light to dark lavender, deep red-purple, or pale cream, veined purple, tube 10–24 mm, funnel-shaped; sepals 5–8 cm, 14–30 mm wide, obovate; petals 5–7 cm, 10–18 mm wide, obovate; ovary triangular, with nipple-like projection at tip, style branches 30–50 mm, crests 9–20 mm, stigmas triangular. 2*n*=40. Common. Grassy places, esp near coast; gen < 200 m. NW, CW; sw OR. Highly variable. Thrives in pastures, lvs unpalatably bitter to livestock. May–Jul

I. fernaldii R.C. Foster Pl drying gray-green; rhizome 4–7 mm diam. **ST:** unbranched, 20–40 cm. **LF:** basal 6–8 mm wide; cauline 1–3, similar to basal. **INFL:** fls 2–3; lowest 2 bracts opposite, enclosing perianth tube, outer 6–14 cm, 12–20 mm wide. **FL:** perianth yellow, veined darker, tube 35–60 mm, narrow, gradually funnel-shaped

distally; sepals 5–7 cm, 7–21 mm wide, elliptic, clawed; petals 4–6 cm, 6–12 mm wide, elliptic; ovary triangular, style branches 30–50 mm, crests 13–17 mm, stigmas triangular. 2*n*=40. Common. Shady places; 50–2000 m. s NCoRO, s NCoRI, n CCo, SnFrB. Some confusion with *I. macrosiphon* with cream to gold-yellow perianths. Apr

I. foetidissima L. CORAL IRIS Rhizome 9–14 mm diam. **ST:** gen branched, 25–40 cm. 2–4. **LF:** odor fetid when crushed; basal 25–35 mm wide; cauline 2–4, similar to basal esp near tip. **INFL:** fls 2–9; lowest 2 bracts opposite, enclosing perianth tube, outer 6–8 cm, 16–22 mm wide. **FL:** perianth pale yellow to brown, red-brown, or green, veined green or red-brown, tube 11–14 mm, constricted near ovary; sepals 3–5 cm, 18–20 mm wide, obovate; petals 3–4 cm, 11–18 mm wide, narrowly obovate; ovary triangular, style branches 22–30 mm, crests 5–8 mm, stigmas 2-lobed. **SEED:** globular, gen red. Damp areas in light shade; 300–1400 m. s ScV (ranches only), SnFrB; Eur, n Afr, c Asia. Jun–Jul

I. germanica L. Rhizome 10–20 mm diam. **ST:** branched, 30–60 cm. **LF:** basal 20–35 mm wide; cauline 0–3, bract-like. **INFL:** fls 2–6; lowest 2 bracts alternate, spreading, inner scarious near tip. **FL:** perianth gen white to blue-purple, tube 20–26 mm; sepals 7–10 cm, 40–60 mm wide, obovate, with white or light blue, gen yellow-tipped beards; petals 5–8 cm, 35–65 mm wide, obovate; ovary triangular, style branches 20–35 mm, crests 12–16 mm, stigmas rounded-triangular. 2*n*=44. Roadsides, old homesteads in open or light shade; 100–1000 m. n SNF, SCoRI, SW, DMoj; e US, Eur, Asia. Variable esp in size, color in cult. Apr–May

I. hartwegii Baker Rhizome 3–10 mm diam. **ST:** unbranched, 0–30 cm. **LF:** basal 3–14 mm wide; cauline 1–4, bract-like or gen similar to basal. **INFL:** fls 2–3; lowest 2 bracts gen alternate, ≤ 9 cm apart, spreading at fl, outer 5–15 cm, 7–23 mm wide. **FL:** perianth tube 5–20 mm, barrel- or funnel-shaped; sepals 4–7 cm, 11–27 mm wide, elliptic; petals 5–7 cm, 5–14 mm wide, narrowly elliptic to elliptic; ovary triangular, style branches 31–45 mm, crests 8–15 mm, stigmas triangular.

  subsp. ***australis*** (Parish) L.W. Lenz Rhizome 6–8 mm diam. **ST:** 8–30 cm. **LF:** basal 7–10 mm wide, base gen ± pink. **INFL:** fls 2; outer bract 7–12 cm, 11–23 mm wide. **FL:** perianth blue-violet to

purple with cream or ± white throat, tube 7–10 mm, barrel-shaped. $2n=40$. Common. Dry slopes in pine forest; 1200–2300 m. e SnGb, SnBr, SnJt. May–Jun

subsp. ***columbiana*** L.W. Lenz TUOLUMNE IRIS Rhizome 7–10 mm diam. **ST:** 7–25 cm. **LF:** basal 7–14 mm wide, base ± red. **INFL:** fls 3; outer bract 9–15 cm, 13–17 mm wide. **FL:** perianth pale cream-yellow, veined gold, tube 11–15 mm, barrel-shaped. Dry slopes in oak woodland; 540–1400 m. c SNH (Tuolumne Co.). May ★

subsp. ***hartwegii*** (p. 1363) Rhizome 7–10 mm diam. **ST:** 7–30 cm. **LF:** basal 5–10 mm wide, base gen green. **INFL:** fls 2; outer bract 6–11 cm, 9–14 mm wide. **FL:** perianth pale to gold-yellow, veined gen darker, tube 5–10 mm, barrel-shaped. $2n=40$. Common. Slopes in open pine forest; 400–2300 m. s CaR, SN. May–Jun

subsp. ***pinetorum*** (Eastw.) L.W. Lenz Rhizome 3–4 mm diam. **ST:** 0–7 cm. **LF:** basal 3–5 mm wide, base faintly pink; cauline bract-like. **INFL:** fls 2; lowest bract 5–7 cm, 7–10 mm wide. **FL:** perianth pale cream-yellow, tube 12–20 mm, funnel-shaped, sepals veined bright yellow. Uncommon. Flats in open pine forest; 400–1500 m. s CaR, n&c SN. May

I. innominata L.F. Hend. DEL NORTE COUNTY IRIS Rhizome 3–4 mm diam. **ST:** unbranched, 5–20 cm. **LF:** basal 3–6 mm wide; cauline 1–3, bract-like basally, lf-like near st tip. **INFL:** fls 1(2); lowest 2 bracts opposite, enclosing perianth tube, outer 3–6 cm, 9–12 mm wide. **FL:** perianth deep gold-yellow, veined darker, tube 15–26 mm, funnel-shaped; sepals 4–7 cm, 16–26 mm wide, obovate; petals 4–6 cm, 8–14 mm wide, elliptic; ovary rounded, style branches 29–40 mm, crests 7–12 mm, stigmas triangular. $2n=40$. Open or partly shaded slopes with well-drained soil; 400–2000 m. n KR; sw OR. May not occur in CA, sometimes confused with *I. thompsonii*, narrow lf forms of *I. bracteata*. May–Jun ★

I. longipetala Herb. COAST IRIS Rhizome 10–25 mm diam. **ST:** rarely branched, 30–60 cm. **LF:** basal 5–11 mm wide; cauline 1–2, bract-like for at least 2/3 st length. **INFL:** fls 3–6; lowest 2 bracts alternate (opposite), enclosing perianth tube, 0.5–10 cm apart, outer 7–15 cm. **FL:** perianth lilac-purple, veined darker, tube 5–13 mm, funnel-shaped; sepals 6–10 cm, 30–50 mm wide, obovate; petals 5–9 cm, 15–21 mm wide, elliptic; ovary rounded, style branches 35–43 mm, crests 12–15 mm, stigmas 2-lobed. $2n=86–88$. Moist, coastal prairie or open coastal forest; < 600 m. c&s NCo, s NCoRO, n&c CCo, SnFrB. May be a coastal form of *I. missouriensis*. Mar–Jun ★

I. macrosiphon Torr. Rhizome 7–9 mm diam. **ST:** unbranched, 0–15 cm. **LF:** basal 3–6 mm wide, base gen colorless; cauline 0–3, similar to basal. **INFL:** fls 1–2; lowest 2 bracts opposite, enclosing perianth tube, outer 5–9 cm, 9–17 mm wide. **FL:** perianth cream to gold-yellow or lavender to deep blue-purple, gen veined darker, tube 34–65 mm, abruptly inflated, bowl-like at top; sepals 4–7 cm, 14–22 mm wide, widely elliptic, clawed, ± flared at base; petals 4–7 cm, 6–11 mm wide, elliptic; ovary triangular, style branches 19–33 mm, crests 8–18 mm, stigmas triangular. $2n=40$. Common. Open to partly shaded slopes in oak or pine woodland; gen < 1000 m. s KR, NCoR, n&c SNH, SnFrB. Highly variable in size, perianth color. Mar–May

I. missouriensis Nutt. (p. 1363) WESTERN BLUE FLAG Rhizome 20–30 mm diam. **ST:** rarely branched, 20–50 cm. **LF:** basal 3–9 mm wide, base ± purple or not; cauline 1–2, bract-like for at least 2/3 st length. **INFL:** fls 1–2; lowest 2 bracts opposite, gen scarious, enclosing perianth tube, outer 4–8 cm, 5–7 mm wide. **FL:** perianth pale lilac to ± white, veined lilac-purple, tube 4–12 mm, funnel-shaped; sepals 4–7 cm, 18–22 mm wide, obovate; petals 4–6 cm, 9–12 mm wide, widely oblanceolate; ovary triangular, style branches 25–35 mm, crests 7–10 mm, stigmas 2-lobed. $2n=38$. Vernally moist grassy or rocky areas; 900–3400 m. KR, NCoR, SN, SCoRI, TR, PR, GB; w N.Am, n Mex. Increases in grazed pastures because lvs unpalatably bitter. May–Jul

I. munzii R.C. Foster (p. 1363) MUNZ'S IRIS Rhizome 10–11 mm diam. **ST:** unbranched, 38–65 cm. **LF:** basal 12–20 mm wide, base gen green; cauline 1–4, similar to basal. **INFL:** fls (2)3(4); lowest 2 bracts alternate, spreading, 1–19 cm apart, outer 7–16 cm, 16–26 mm

wide. **FL:** perianth ± blue to pale lavender, veined blue, tube 7–11 mm, barrel-shaped; sepals 6–9 cm, 18–35 mm wide, elliptic; petals 6–8 cm, 12–17 mm wide, elliptic; ovary triangular, style branches 35–55 mm, crests 13–20 mm, stigmas long-triangular. $2n=40$. Wet, grassy sites, open to part shade; 540–800 m. s SNF, s SNH (Greenhorn Mtns, Kern Co.). Apr ★

I. pseudacorus L. Rhizome 30–40 mm diam. **ST:** branched, 50–150 cm. **LF:** basal 20–35 mm wide, stiff, midvein conspicuous; cauline 3–6, similar to basal. **INFL:** branched, fls 3–6 per branch; lowest 2 bracts gen opposite, outer 5–7 cm, 10–20 mm wide. **FL:** perianth gen bright yellow or cream with prominent brown veins, marks on sepals, tube 12–13 mm, barrel-shaped; sepals widely oblanceolate, 5–8 cm, 35–50 mm wide; petals obovate, clawed, 2–3 cm, 4–8 mm wide; style branches < 25 mm, crests 7–10 mm, stigmas rounded. **FR:** 5–8 cm. $2n=24,30,32,34$. Common. Pond margins, estuaries; 100–1400 m. NCoRI, n&c SNF, w edge c SNH, GV, CW, SCo, TR; most of US; native to Eur, Asia. Invasive in ponds, streams, estuaries. Apr–Jun ◆

I. purdyi Eastw. (p. 1363) Rhizome 4–6 mm diam. **ST:** unbranched, 6–25 cm. **LF:** basal 5–11 mm wide; cauline 0–5, bract-like, gen overlapped. **INFL:** fls 2; lowest 2 bracts opposite, enclosing perianth tube, outer 5–8 cm, 14–22 mm wide. **FL:** perianth pale cream-yellow with prominent brown-violet veins, rose-violet with darker veins, or ± white with rose veins, tube 35–60 mm, funnel-shaped; sepals 5–7 cm, 15–23 mm wide, obovate; petals 4–6 cm, 10–15 mm wide, narrowly obovate; ovary triangular, style branches 35–45 mm, crests 12–16 mm, truncate, ± rounded, or 2-lobed, margins gen minutely, irregularly toothed. $2n=40$. Common. Grassy or rocky slopes on edge of Douglas-fir or redwood forests; gen < 1200 m. NCoRO. Apr–May

I. tenax Lindl. subsp. ***klamathensis*** L.W. Lenz (p. 1363) ORLEANS IRIS Rhizome 1–3 mm diam. **ST:** unbranched, 10–30 cm. **LF:** basal 2–4 mm wide; cauline 1–3, bract-like basally, lf-like near st tip. **INFL:** fls 1–2; lowest 2 bracts opposite to alternate, < 18 mm apart, spreading, outer 4–7 cm, 8–12 mm wide. **FL:** perianth pale buff yellow, with deep maroon or ± brown veins, tube 11–20 mm, funnel-shaped; sepals 5–7 cm, 13–24 mm wide, elliptic; petals 4–7 cm, 7–11 mm wide, narrowly elliptic; ovary triangular-winged, style branches 35–41 mm, crests 9–15 mm, stigmas triangular. Shaded mixed-evergreen forests; 80–800 m. w KR (near Orleans, Humboldt Co.). May ★

I. tenuissima Dykes Rhizome 2–5 mm diam. **ST:** unbranched, 10–25 cm. **LF:** basal 3–8 mm wide; cauline 1–4, bract-like or similar to basal. **INFL:** fls 2; lowest 2 bracts opposite, enclosing perianth tube, outer 4–9 cm, 14–20 mm wide. **FL:** perianth tube 30–58 mm, narrow at base, abruptly wider at mid-throat; sepals 5–7 cm, 11–18 mm wide, widely oblanceolate; petals 4–7 cm, 8–12 mm wide, widely oblanceolate; ovary triangular, style branches 35–50 mm, crests 12–25 mm, stigmas triangular or tongue-shaped, finely toothed.

subsp. ***purdyiformis*** (R.C. Foster) L.W. Lenz **LF:** basal 3–5 mm wide; cauline 3–4, bract-like, ± inflated, gen flushed with pink. **INFL:** outer bract 4–5.6 cm. **FL:** perianth pale yellow or cream with same color (or ± red) veins. Uncommon. Shaded rocky slopes, yellow-pine forest; 500–2000 m. CaRH. Apr–May

subsp. ***tenuissima*** (p. 1363) **LF:** basal 5–8 mm wide; cauline 1–3, similar to basal. **INFL:** outer bract 6–9 cm. **FL:** perianth gen pale cream with purple or ± red veins. $2n=40$. Dry, open woodland, roadcuts; 100–1700 m. NW, n SNF. May

I. thompsonii R.C. Foster (p. 1363) Rhizome 4–6 mm diam. **ST:** unbranched, 10–25 cm. **LF:** basal 3–4 mm wide, base gen green; cauline 1–3, bract-like basally, lf-like near st tip. **INFL:** fls 1(2); lowest 2 bracts opposite, enclosing perianth tube, outer 3–8 cm, 9–18 mm wide. **FL:** perianth pale lavender to violet purple or occ white to gray, tube 17–26 mm, funnel-shaped; sepals 4–7 cm, 15–30 mm wide, obovate; petals 3–6 cm, 8–18 mm wide, obovate; ovary triangular, style branches 30–48 mm, crests 9–13 mm, stigmas triangular. Common. Grassy slopes in part shade on forest edge; 90–600 m. n NCo, KR (Del Norte Co.); sw OR. Apr–May

OLSYNIUM

Anita F. Cholewa

Rhizome indistinct. **ST**: gen tufted, rounded. **LF**: bases overlapped, sheathing. **INFL**: fls in umbel-like cymes; bracts 2, unequal. **FL**: perianth red-purple to magenta, occ pink or white, parts ± alike; filaments fused in basal 1/2. **SEED**: brown, angled, pitted. 12 spp.: w hemisphere. (Greek: little, united, from stamen bases) [Goldblatt et al. 1990 Syst Bot 15:497–510]

O. douglasii (A. Dietr.) E.P. Bicknell var. ***douglasii*** (p. 1363)
PURPLE-EYED-GRASS, GRASS WIDOWS **ST**: strongly tufted, < 31 cm.
FL: perianth gen 18–33 mm; filament tube ± enlarged above base.
n=32. Open, vernally moist, often rocky places; < 2200 m. KR, n NCoRO, CaR, MP; to BC. [*Sisyrinchium d.* A. Dietr var. *d.*] Spring

ROMULEA

Per; corm rounded, gen asymmetric, cover ± woody, split at top into several teeth. **LF**: slender, ± round or widely elliptic in ×-section, gen 4-grooved. **INFL**: fls 1 on erect, basal peduncle, fl bracts green, gen with membranous margins, inner forked at tip. **FL**: radial; perianth gen funnel-shaped, tube gen << lobes; style gen ± = stamens, branches deep-divided. **SEED**: many. ± 90 spp.: esp Afr, Medit. (Romulus, legendary co-founder of Rome) [Manning & Goldblatt 2001 Adansonia ser. 3 23:59–108]

R. rosea (L.) Eckl. var. *australis* (Ewart) M.P. de Vos Pl gen < 10 cm. **LF**: 5–35 cm, 1–2.5 mm diam. **INFL**: peduncle 5–15 cm; outer bract ± 15 mm, inner ± 10 mm. **FL**: 1.5–2 cm; perianth tube yellow, lobes pink or lilac; style 7–10 mm, not exceeding stamens. **FR**: 10–15 mm, thin-walled; bracts persistent. Uncommon. Disturbed areas, dry, sandy or often hard-packed soil; < 50 m. NCo, ScV, CCo, SnFrB; alien in Australia, elsewhere; native to s Afr. Mar–Apr

SISYRINCHIUM

Anita F. Cholewa

[Ann] per; rhizomes compact. **ST**: 1 or tufted, ± flat, winged, nodes well above basal lvs with lvs or not, each with ≥ 1 fl-branch. **LF**: bases overlapped, sheathing. **INFL**: fls in umbel-like cymes; bracts 2, equal in length or not, margins translucent. **FL**: perianth red-purple, ± blue, violet, yellow (white), parts mucronate, ± alike, outer gen wider; filaments ± free to ± fused. **SEED**: ovoid, smooth or pitted. 100+ spp.: w hemisphere. (Latin, Greek: pig, snout) Use of treatments prior to ± 2003 often results in misidentification. *S. douglasii* moved to *Olsynium*.

1. Perianth ± yellow; filaments ± completely free (sect. *Echthronema*)
  2. St gen 2–7 mm wide; perianth gen 11–18 mm . . . . ***S. californicum***
  2′ St gen < 2 mm wide; perianth gen 7–11 mm
    3. Pedicels spreading to drooping in fr; fr gen widest near or above middle . . . . ***S. elmeri***
    3′ Pedicels ascending to erect in fr; fr gen widest at middle . . . . ***S. longipes***
1′ Perianth ± blue to violet (white), parts gen with a basal yellow blotch; filaments ± completely fused (sect. *Sisyrinchium*)
  4. St branched; lvs cauline at ≥ 1 node
    5. Translucent margins of inner bract not extended above tip; perianth deep blue-purple to -violet or pale
      blue, occ white, 10.5–17 mm, outer parts gen widely wedge-shaped; w CA . . . . ***S. bellum***
    5′ Translucent margins of inner bract extended above tip as 2 rounded teeth; perianth gen pale blue, 10–15
      mm, outer parts gen narrowly wedge-shaped; ne DMoj. . . . . ***S. funereum***
  4′ St unbranched; lvs all ± basal
    6. Rhizome elongate; perianth ± purple, base darker . . . . ***S. hitchcockii***
    6′ Rhizome compact; perianth light to dark violet-blue (white), base not darker, with yellow blotch
      7. Inner bract ± = outer, with widely translucent margins extended above tip as 2 rounded teeth; outer
        perianth parts gen wedge-shaped. . . . . ***S. halophilum***
      7′ Inner bract << outer, with ± narrowly translucent margins ended at or below tip; outer perianth parts
        gen elliptic to narrowly oblanceolate . . . . ***S. idahoense***
        8. St 2–2.5 mm wide, distal margins gen with a few small teeth; outer perianth parts 13–17(20) mm . . . var. ***idahoense***
        8′ St 1.5–2 mm wide, distal margins gen entire; outer perianth parts 8–13 mm . . . . var. ***occidentale***

S. bellum S. Watson (p. 1363) WESTERN BLUE-EYED-GRASS **ST**: gen tufted, < 64 cm, 1.5–5.3 mm wide, lf-bearing nodes gen 1. **INFL**: translucent margins of inner bract wider just below tip, not extended above tip. **FL**: perianth 10.5–17 mm, blue-purple to -violet or pale blue, occ white, tips truncate to notched. n=16. Common. Open, gen moist, grassy areas, woodland; < 2400 m. w CA; to OR, Mex. Highly variable; gen self-sterile. Mar–May

S. californicum (Ker Gawl.) W.T. Aiton (p. 1363) GOLDEN-EYED-GRASS **ST**: tufted, < 62 cm, gen 2–7 mm wide, dull green, drying ± black, lf-bearing nodes 0. **INFL**: translucent margins of inner bract widened toward tip. **FL**: perianth gen 11–18 mm, medium to bright yellow, veins gen ± brown. n=17. Gen moist places near coast; gen < 200 m. NCo, KR, NCoRI, n SNH, n&c CCo, SnFrB; to BC. Spring–late summer

S. elmeri Greene ELMER'S BLUE-EYED-GRASS **ST**: tufted, gen < 25 cm, < 2 mm wide, medium green, drying dark to olive-green but not ± black, lf-bearing nodes 0. **INFL**: translucent margins of inner bract extended beyond tip as 2 rounded or dissected teeth. **FL**: perianth 7–8.5 mm, deep- to orange-yellow, veins dark ± brown. n=17. Wet meadows; 350–2700 m. s KR, s CaRH, SN, SnBr. May–Aug

S. funereum E.P. Bicknell DEATH VALLEY BLUE-EYED-GRASS **ST**: tufted, < 70 cm, pale green, glaucous, lf-bearing nodes gen 1. **INFL**: translucent margins of inner bract widest at tip, extended above tip as 2 rounded teeth. **FL**: perianth 10–15 mm, gen pale blue, base yellow, tips truncate to notched. n=16. Gen strongly alkaline margins of wet areas; < 800 m. ne DMoj (Death Valley region); adjacent NV. Self-sterile. Feb–Apr ★

S. halophilum Greene (p. 1363) NEVADA BLUE-EYED-GRASS
ST: gen tufted, < 40 cm, glaucous, lf-bearing nodes 0. **INFL**: translucent margins of inner bract widest at tip, extended above tip as 2 rounded teeth. **FL**: perianth gen 9–12 mm, gen medium violet-blue, base yellow, outer parts gen wedge-shaped, tips truncate (to notched). $n=16$. Gen moist, alkaline meadows; < 2600 m. n&s SNH, GB, adjacent DMoj; to UT. Self-sterile. May–Jun

S. hitchcockii Douglass M. Hend. HITCHCOCK'S BLUE-EYED-GRASS Rhizomes elongate. **ST**: > 50 cm, lf-bearing nodes gen 0. **INFL**: translucent margins of inner bract obtuse or truncate at tip. **FL**: perianth 15–19 mm, ± purple, base darker. $n=32$. Grassy, vernally moist areas; 200–300 m. NCo (Cape Ridge area); to sw OR. May–Jun ★

S. idahoense E.P. Bicknell (p. 1363) IDAHO BLUE-EYED-GRASS
ST: tufted, < 45 cm, green to glaucous, lf-bearing nodes 0(1). **INFL**: translucent margins of inner bract ± uniformly narrow. **FL**: perianth 8–17(20) mm, gen blue to blue-violet, base yellow, outer parts gen elliptic to narrowly oblanceolate, tips rounded to deep-notched.

Highly variable, self- or cross-pollinated. Vars. intergrade; 2 others, in WA, BC.

var. ***idahoense*** **ST**: distal margin gen with a few small teeth. **INFL**: margins of outer bract fused in basal 4–7 mm. **FL**: outer perianth parts 13–17(20) mm. $n=32,48$. Common. Open, moist, grassy places; 100–2800 m. KR, CaR, MP; to BC, MT, nw WY. Jun–Aug

var. ***occidentale*** (E.P. Bicknell) Douglass M. Hend. **ST**: distal margin gen entire. **INFL**: margins of outer bract fused gen in basal 3–6 mm. **FL**: outer perianth parts 8–13 mm. $n=32$. Open, moist, grassy places; 100–2850 m. SN, SNE; to e WA, sw MT, CO, NM. Jun–Aug

S. longipes (E.P. Bicknell) Kearney & Peebles TIMBERLAND BLUE-EYED-GRASS **ST**: ≤ 45 cm, 0.6–2.3 mm wide, drying olive-green, lf-bearing nodes 0. **INFL**: translucent margins of inner bract uniformly narrow, extended above tip as 2 rounded, sometimes dissected teeth. **FL**: perianth 7–11 mm, yellow to orange, veins dark ± brown. $n=17$. Wet to moist meadows, streambanks, similar places; 700–1060 m. SnBr; AZ, NM. May–Aug ★

WATSONIA

Corm depressed, cover fibrous. **INFL**: spike, branched below or not; fls gen ± 2-ranked, fl bracts green, leathery. **FL**: radial or gen bilateral; perianth funnel- or trumpet-shaped, tube curved, lobes ± equal or unequal, oblong or lanceolate; style branches divided ± to middle, thread-like. **FR**: gen oblong. **SEED**: gen winged at 1 or both ends. ± 50 spp.: s Afr. (Sir William Watson, English botanist-physician, 1715–1787) [Goldblatt 1989 Ann Kirstenbosch Bot Gard 17:1–148] Many spp. cult as orn.

1. Perianth rose-pink to white, tube < 2 cm, ± curved; fl radial; infl, lf axils without bulblets . . . . [*W. marginata*]
1′ Perianth brick-red, tube 4–5 cm, sharp-curved; fl bilateral; infl, occ lf axils with bulblets after fl or not . . . . . ***W. meriana***

W. meriana (L.) Mill. **ST**: 1–1.5 m. **LF**: ± 5–6, < 60 cm, < 6 cm wide. **INFL**: 10–15-fld, ± open. **FL**: perianth 6–7 cm, lobes ± 2.5 cm, oblong to obovate. Uncommon, but sometimes locally abundant.

Disturbed roadsides, fields; < 50 m. NCo, CCo; native to s Afr. [*W. bulbillifera* J.W. Mathews & L. Bolus] Reproduces by bulblets, can be invasive. May–Jul ❖

JUNCACEAE RUSH FAMILY

Peter F. Zika, except as noted

Ann, per gen from rhizomes. **ST**: round or flat. **LF**: gen basal; sheath margins fused, or overlapping and gen with 2 ear-like extensions at blade junction; blade round, flat, or vestigial, glabrous or margin hairy. **INFL**: head-like clusters or fls 1, variously arranged; bracts subtending infl 2, gen lf-like; bracts subtending infl branches 1–2, reduced; bractlets subtending fls gen 1–2, gen translucent. **FL**: gen bisexual, radial; sepals and petals similar, persistent, scale-like, green to brown or ± purple-black; stamens gen 3 or 6, anthers linear, persistent; pistil 1, ovary superior, chambers gen 1 or 3, placentas 1 and basal or 3 and axile or parietal, stigmas gen > style. **FR**: capsule, loculicidal. **SEED**: 3–many, gen with white appendages on 1 or both ends. 7 genera, 440 spp.: temp, arctic, and trop mtns. [Kirschner 2002 Species Plantarum: Fl World, vols. 6–8 (Juncaceae). ABRS] Fls late spring to early fall. Scientific Editors: Douglas H. Goldman, Bruce G. Baldwin.

1. Lvs glabrous, sheath open; seeds many per fr . . . . **JUNCUS**
1′ Lvs gen hairy on margins, sheath closed; seeds 3 per fr . . . . **LUZULA**

JUNCUS RUSH

Rhizome 0 or gen with scale-like lvs. **ST**: gen cylindric or flat. **LF**: blade well developed and cylindric or flat, occ closely resembling st, or reduced to small point; crosswalls gen present; appendages gen present at blade-sheath junction. **INFL**: gen terminal, appearing lateral when pushed aside by infl bract; bractlets 0–2. **FL**: sepals, petals similar; stamens gen 3 or 6(2); pistil 1, ovary chambers 1–3, placentas axile or parietal, stigmas gen 3(2). **SEED**: many. 315 spp.: worldwide, esp n hemisphere. (Latin: to join or bind, from use of sts) [Ertter 1986 Mem New York Bot Gard 39:1–90] All spp. with lf crosswalls may have lvs, sts swollen, deformed by sucking insects. Fruiting time given instead of flowering time. *J. bulbosus* L., *J. dichotomus* Elliott, and *J. elliotti* Chapm. reportedly naturalized in CA.

Key to Groups

1. Ann gen < 10 cm; roots fine, fibrous; lvs narrow, gen inrolled, gen < 1 mm wide . . . . **Group 1**
1′ Per; pl gen larger in all respects; roots coarse or rhizomes present; lvs narrow or wide, inrolled to flat or cylindric
2. Infl bract cylindric, resembling continuation of st; infl appearing lateral . . . . **Group 2**
2′ Infl bract not cylindric, not resembling continuation of st, or if so then channeled along inner side; infl appearing terminal
3. Lf blade gen cylindric or nearly so; crosswalls gen complete . . . . **Group 3**

3′ Lf blade flat or wiry; crosswalls incomplete or 0

  4. Lf blade iris-like, flattened, oriented with edge toward st (at least from middle to tip), crosswalls gen incomplete . . . . **Group 4**

  4′ Lf blade wiry or grass-like, if flattened then oriented with flat side toward st, crosswalls 0

    5. 2 bractlets subtending fl; lvs slender, gen < 1.3 mm wide, gen wiry or channeled; fls 1 or in clusters . . . . . **Group 5**

    5′ 1 bractlet subtending fl; lvs broader, gen 2–7 mm wide, flat; fls in clusters. . . . . **Group 6**

Group 1

Ann

1. St gen branched, lfy; infl or fls scattered along st or congested in terminal cluster; stamens 3 or 6

  2. Sepals 2–4 mm

    3. Fls scattered along st, 1 per node. . . . . *J. bufonius* var. *occidentalis*

    3′ Fls terminal, 3–10(25) per cluster

      4. Perianth green to red or pale brown; lowest bract > infl . . . .[2]*J. capitatus*

      4′ Perianth dark brown to ± black; lowest bract < infl . . . . *J. planifolius*

  2′ Sepals gen 4–10 mm

    5. Fr tip gen truncate, ± = petals; petals ± blunt . . . . **J. ambiguus**

    5′ Fr tip gen rounded or acute, < petals; petals acute to acuminate

      6. Infl open; gen non-saline soil. . . . . *J. bufonius* var. *bufonius*

      6′ Infl congested, dense; gen saline soil. . . . . *J. bufonius* var. *congestus*

1′ St not branched, lfless, lvs basal; infl or fls terminal; stamens 2–3(–5)

  7. Infl bracts lf-like, > infl; sepals >>, more acuminate than petals . . . .[2]*J. capitatus*

  7′ Infl bracts inconspicuous, membranous, < infl; sepals, petals similar in size, shape, occ sepals < petals

    8. Anthers ≥ 0.9 mm, >(=) filaments; style ≥ 0.9 mm, fls outcrossing

      9. Fr gen 11–17 mm, ± 3–4 × > perianth . . . . *J. digitatus*

      9′ Fr 1–5 mm, < or ± = perianth

        10. Perianth parts acuminate; petals colored to tip, margins narrowly scarious; seed striate at 15×. . . . . . . *J. triformis*

        10′ Perianth parts acute; petals gen not colored to tip, sepal and petal margins broadly scarious; seed not striate at 15× . . . . *J. leiospermus*

        11. Fls gen 1(2) per st . . . . var. *ahartii*

        11′ Fls gen 2–4(7) per st . . . . var. *leiospermus*

    8′ Anthers < 0.8 mm, < filaments; style < 0.6 mm, fls gen self-pollinating

      12. Perianth dark brown to ± black; gen 3–6+ fls per st . . . .[2]*J. planifolius*

      12′ Perianth gen green to pink or red; gen 1–2 fls per st

        13. Seed not striate; fls 1 per st; bracts 0 or 1–2, truncate to acute

          14. Infl bractlets 0; st thickened below fl . . . . *J. hemiendytus* var. *abjectus*

          14′ Infl bractlets 1–2; st not thickened below fl

            15. Infl bractlet 1, tip truncate, sheathing st; stamens 2–3. . . . . *J. uncialis*

            15′ Infl bractlets 1–2, tips acute to blunt, ovate to rounded, not sheathing st; stamens 2–4

              16. Perianth parts gen 6, shiny, incurved, gen > fr. . . . . *J. bryoides*

              16′ Perianth parts gen 4, dull, erect or outcurved, gen < fr . . . . *J. hemiendytus* var. *hemiendytus*

        13′ Seed striate; fls 1–2(7) per st; bracts 2, acute to acuminate

          17. Perianth parts gen 4

            18. Fr much paler than perianth; seed 0.5–0.8 mm, < 4 per row, < 10 per capsule; fr gen ± spheric to obovoid, < perianth. . . . .[2]*J. capillaris*

            18′ Fr, perianth both pale green or pink; seed 0.3–0.5 mm, > 4 per row, > 10 per capsule; some fr ± ellipsoid or oblong, ≥ perianth. . . . . *J. tiehmii*

          17′ Perianth parts gen 6

           19. Perianth gen > and darker than capsule; < 10 seeds per capsule; seed 0.5–0.8 mm, striations gen faint at 30×. . . . .[2]*J. capillaris*

          19′ Perianth ± = and same color as capsule; > 40 seeds per capsule; seed 0.3–0.6 mm, striations obvious at 30×

           20. Fls gen (1)2–3(4) per st; fr, perianth gen ± dark red in age; seed 0.4–0.6 mm . . . . *J. kelloggii*

           20′ Fls 1(2) per st; fr, perianth gen pale green to pale red in age, perianth midvein tips dark red; seed 0.3–0.4 mm . . . . *J. luciensis*

Group 2

Per; infl bract erect and st-like

1. Rhizomes long, creeping; st scattered or in lines, in loose colonies

  2. Blades well developed on some upper sheaths, > 5 cm, st-like. . . . . *J. mexicanus*

  2′ Blades 0 or < 1 cm, not st-like

    3. Perianth < 5.5(6) mm; infl open

      4. St smooth, not stiff, gen dark green; fr acute below beak; st gen < 1.1 m . . . . *J. balticus* subsp. *ater*

      4′ St ridged, stiff, gray-green; fr ± obtuse below beak; st to 2 m. . . . . *J. textilis*

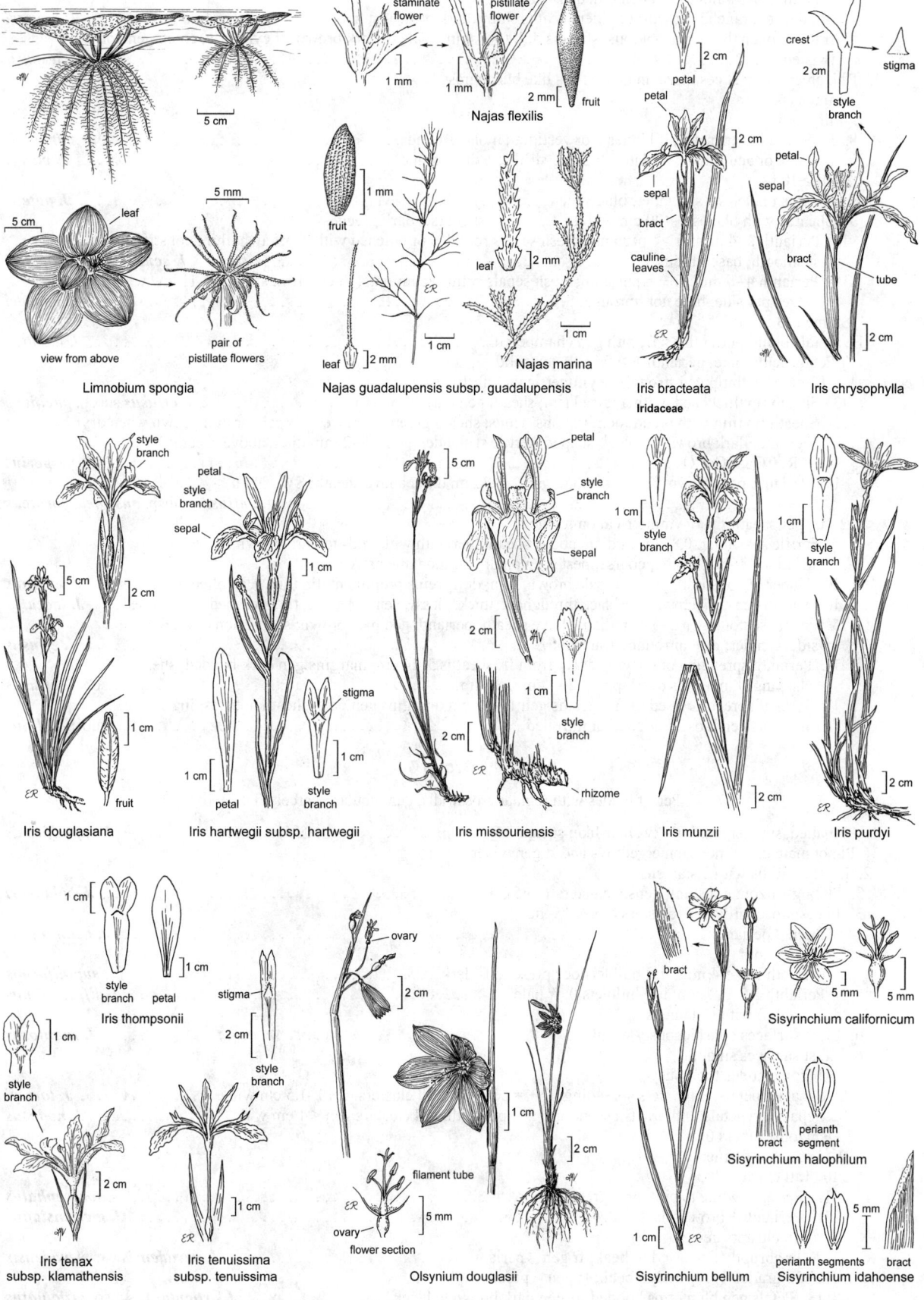

staminate flower
pistillate flower
1 mm
1 mm
2 mm [fruit
Najas flexilis

fruit
1 mm

leaf
2 mm

leaf
2 mm

1 cm

Najas guadalupensis subsp. guadalupensis

leaf
2 mm

1 cm

Najas marina

Limnobium spongia

leaf
view from above
pair of
pistillate flowers
5 cm
5 cm
5 mm

petal
2 cm
petal
sepal
bract
cauline
leaves
2 cm
Iris bracteata

Iridaceae

crest
2 cm
stigma
style
branch
petal
sepal
bract
tube
2 cm
Iris chrysophylla

style
branch
5 cm
2 cm
1 cm
fruit
Iris douglasiana

petal
style
branch
sepal
1 cm
stigma
1 cm
petal
style
branch
1 cm
Iris hartwegii subsp. hartwegii

petal
5 cm
style
branch
sepal
2 cm
1 cm
style
branch
2 cm
rhizome
Iris missouriensis

1 cm
style
branch
2 cm
Iris munzii

1 cm
style
branch
2 cm
Iris purdyi

1 cm
style
branch
petal
1 cm
style
branch
petal
Iris thompsonii
1 cm
style
branch
stigma
2 cm
style
branch
2 cm
1 cm
Iris tenax
subsp. klamathensis
Iris tenuissima
subsp. tenuissima

ovary
2 cm
1 cm
2 cm
filament tube
ovary
flower section
5 mm
Olsynium douglasii

bract
5 mm
5 mm
Sisyrinchium californicum

bract
perianth
segment
Sisyrinchium halophilum

perianth segments
bract
5 mm
1 cm
Sisyrinchium bellum
Sisyrinchium idahoense

3′ Perianth gen > 6 mm; infl open or a dense head
  5. Infl dense, branches obscured; older sheaths shiny, dark brown; st gen flattened, gen arching, twisted . . . ***J. breweri***
  5′ Infl gen open, branches obvious; sheaths dull to ± shiny, pale to dark brown; st cylindric, erect, gen not
    twisted . . . . ***J. lescurii***
1′ Rhizomes short; st cespitose, in dense tufts like bunchgrass
 6. Stamens 6
  7. Infl < 6 fls
   8. Fr notched or truncate; lf blades 0 or vestigial; moist ground . . . . ***J. drummondii***
   8′ Fr acute or acuminate; lf blades > 2 cm, st-like; ± dry ground. . . . . ***J. parryi***
  7′ Infl > 10 fls
   9. Sheaths bladeless; st slender, blue-green. . . . . ***J. patens***
   9′ Sheaths with blades 5–100 cm, st-like, tip sharp; st coarse, dark green
    10. Perianth 2–4 mm; fr >> perianth; fresh sepals rounded or notched with short awn, thin, not stiff;
     st smooth, base fibrous . . . . ***J. acutus*** subsp. ***leopoldii***
    10′ Perianth 4–6 mm; fr ± = perianth; fresh sepals spiny, acuminate, thick, firm; st with 6–11 low wide
     ridges per side, base not fibrous. . . . . ***J. cooperi***
 6′ Stamens 3
  11. Petals blunt, perianth << fr; pith gen chambered . . . . ***J. usitatus***
  11′ Petals acuminate, perianth ± ≥ fr; pith gen solid
   12. Upper sheath tip gen strongly asymmetrical on fertile st
    13. Sheath tip thickened, with a raised rim; sheath gen dark brown to black. . . . . ***J. effusus*** subsp. ***pacificus***
    13′ Sheath tip thin with broad membranous wings; sheath green when fresh to pale or mid-brown when dry
     14. Perianth dark brown- to black-striped; fertile st slender, gen 0.8–2 mm diam above sheath; NCo,
      KR, CCo, SCoRO . . . . ***J. hesperius***
     14′ Perianth pale brown; fertile st stout, gen 2–3.7 mm diam above sheath; SW, DSon
      . . . . ***J. effusus*** subsp. ***austrocalifornicus***
   12′ Upper sheath tip ± symmetrical on fertile st
    15. Fertile st slender, 0.8–2 mm diam above sheath; perianth with mid- to dark-brown stripes; st
     striations 8–12 per side, coarse, best seen on dry st; montane natives
     16. Upper 1/2 of sheath green to pale brown, thin, dull, veins prominent, tip thin, ± inrolled towards st . . . ***J. exiguus***
     16′ Sheath gen dark brown to ± black throughout, thick, glossy, veins obscure, tip thickened, not inrolled . . . . ***J. laccatus***
    15′ Fertile st stout, gen 2–5 mm diam above sheath; perianth gen pale brown; st striations 14–26 per
     side, slender; rare introduced subspp. . . . . ***J. effusus***
    17. Perianth spreading or curving away from fr; sheaths 5–14 cm, margins gen dark-banded; sheath
     clasping st, margins overlapping 2–4 cm from tip. . . . . subsp. ***effusus***
    17′ Perianth erect, pressed to fr; sheaths gen 15–27 cm, margins gen pale; sheath not clasping st,
     margins gen loose, flattened or unrolled . . . . subsp. ***solutus***

Group 3

Per; lf blades with regular crosswalls, gen tubular, scarcely flattened

1. Pl matted, submerged, early lvs hair-like; st capillary. . . . . [3]***J. supiniformis***
1′ Pl not matted, gen not submerged; lvs and st gen wider
 2. Most or all fls with 3 stamens
  3. Pl long-rhizomatous; infl dense, clusters 1–3(5), > 20–fld. . . . . ***J. bolanderi***
  3′ Pl cespitose; infl open, clusters > 5, < 15–fld
   4. Fr gen ≤ perianth . . . . ***J. acuminatus***
   4′ Fr > perianth
    5. Perianth 3.4–5.4 mm; infl bulblets occ present; fr dark . . . . [3]***J. supiniformis***
    5′ Perianth 2.3–3.2 mm; infl bulblets 0; fr pale. . . . . ***J. diffusissimus***
 2′ Most or all fl with 6 stamens
  6. Lf, st surfaces conspicuously wrinkled. . . . . ***J. rugulosus***
  6′ Lf, st surfaces smooth
   7. Rhizome nodes tuberous
    8. Sepals > petals; infl bracts spreading; fr ± = perianth; infl clusters gen 1–1.5 cm wide . . . . ***J. torreyi***
    8′ Sepals ± = petals; infl bracts ascending; fr > perianth; infl clusters gen < 1 cm wide. . . . . ***J. nodosus***
   7′ Rhizome tubers 0
    9. Anthers > filaments
     10. Infl clusters 1–3
      11. Perianth white to pink or ± green . . . . ***J. chlorocephalus***
      11′ Perianth ± brown-black. . . . . [2]***J. mertensianus***
     10′ Infl clusters gen > 5
      12. Fr abruptly narrowed to beak; fr gen ≤ perianth . . . . ***J. nevadensis*** var. ***nevadensis***
      12′ Fr gradually tapered to beak; fr gen > perianth
       13. Petals gen blunt, tips hooded; fr gen dark brown to black. . . . . [2]***J. articulatus*** subsp. ***articulatus***
       13′ Petals acuminate, bristle-tipped; fr light brown to red or dark red . . . . ***J. dubius***

9′ Anthers ≤ filaments
  14. Infl clusters > 5; fr > perianth; infl bulblets occ present; st occ rooting at nodes
    15. Infl clusters > 10; perianth 1.9–3.1 mm; pale petal margins wide, flat, tips gen blunt, hooded
    . . . . [2]***J. articulatus*** subsp. ***articulatus***
    15′ Infl clusters < 10; perianth 3.4–5.4 mm; pale petal margins narrow, gen inrolled, tips acuminate
    . . . . [3]***J. supiniformis***
  14′ Infl clusters gen 1–3; fr gen ≤ perianth; infl bulblets 0, st not rooting at nodes
    16. Fr abruptly tapered to distinct beak; lvs ± to strongly flattened, hollow, blue-green; st with 1–3(4)
      heads; n NCo. . . . . ***J. nevadensis*** var. ***inventus***
    16′ Fr gen notched or truncate, ± beakless; lvs tubular, hollow, green; st with 1(2) heads; montane
      17. Perianth gen < 2.8 mm; perianth brown, stiff; infl bract narrow, gen not sheathing; SW. . . . . ***J. duranii***
      17′ Perianth gen > 3.1 mm; perianth ± purple-black, not stiff; infl bract wide, gen sheathing; not SW
      . . . . [2]***J. mertensianus***

Group 4

Per; lf blades flattened, iris-like, crosswalls gen incomplete

1. Fr gradually tapered to long beak
  2. Perianth green, tips red or purple; infl gen > 15 cm, clusters gen ≤ 8-fld . . . . ***J. oxymeris***
  2′ Perianth gen dark brown to dark purple-brown or ± black; infl gen < 10 cm, clusters > 8-fld . . . . . . ***J. phaeocephalus***
    3. Infl 10+ few-fld clusters; clusters gen 5–10 mm wide. . . . . var. ***paniculatus***
    3′ Infl 1–10 many-fld clusters; clusters gen > 10 mm wide. . . . . var. ***phaeocephalus***
1′ Fr abruptly tapered to beak
  4. Anthers gen 1–1.5 mm, >> filaments; styles ≥ 1 mm; fls outcrossing. . . . . ***J. macrandrus***
  4′ Anthers 0.5–1 mm, gen ≤ filaments; styles ≤ 1 mm; fls gen self-pollinating
    5. Stamens gen 3; perianth gen dark brown to black . . . . ***J. ensifolius***
    5′ Stamens gen 6; perianth gen green to red-tipped or brown
      6. Lvs gen 2–5 mm wide; perianth brown. . . . . ***J. saximontanus***
      6′ Lvs gen 5–14 mm wide; perianth green, red-tipped . . . . ***J. xiphioides***

Group 5

Per; lf blades wiry or slightly flattened and channeled, < 1.3 mm wide

1. Perianth black-striped, tips blunt, hooded, incurved; anthers >> filaments; rhizomatous; coastal salt marshes
  . . . . ***J. gerardi*** subsp. ***gerardi***
1′ Perianth green, brown, or red, tips acute or acuminate, not incurved or hooded; anthers < to > filaments;
  cespitose; inland wetlands
  2. Lf sheath appendages stiff, thick, plastic-like, rounded, shiny, ± yellow . . . . ***J. dudleyi***
  2′ Lf sheath appendages not stiff, thin, rounded to acuminate, dull, white or translucent
    3. Fr chambers 3, partitions straight, united exc at tip
      4. Sepals > petals; style (0.3)0.4–0.8 mm; anthers > filaments; perianth gen green or light brown, pale
        margins narrow; infl open. . . . . ***J. brachyphyllus***
      4′ Sepals ± = petals; style 0.1–0.2 mm; anthers < filaments; perianth with thick, darker brown stripes,
        pale margins broad; infl gen compact . . . . ***J. confusus***
    3′ Fr chamber 1, partitions concave, separated exc at base
      5. Fr ridged, truncate or notched; perianth green or with thick dark brown stripes . . . . ***J. occidentalis***
      5′ Fr ridges 0 or obscure, fr rounded or acute; perianth green or ± red, pale brown in age
        6. Lf sheath appendages gen rounded, ± opaque below, 0.2–0.6 mm; bractlets acuminate, gen bristle-
          tipped; st with 2–6 strong ridges per side; DMtns. . . . . ***J. interior***
        6′ Lf sheath appendages gen acute to acuminate, ± translucent, gen 1–8 mm until late in season when
          gen smaller or absent; bractlets gen acute to blunt; st with or without strong ridges per side; widespread
          7. St with 2–6 strong ridges per side; fr < 2.5 mm, < 75% perianth; infl gen orange-red;
            fl < internodes. . . . . [*J. anthelatus*]
          7′ St with 0–1 strong ridges per side; fr gen > 2.5 mm, > 75% perianth; infl gen green; fl gen > internodes . . . ***J. tenuis***

Group 6

Per; lf blades flattened, grass-like, gen 2–7 mm wide

1. Stamens 3
  2. Pl base with knobby tubers; anthers ± purple-red; cespitose . . . . ***J. marginatus***
  2′ Tubers 0; anthers yellow; rhizomatous or cespitose . . . . ***J. planifolius***
1′ Stamens 6
  3. Seeds slender, 1–2 appendages ≥ 1/2 seed body
    4. Seed body > both appendages; petals gen > sepals; infl clusters gen 5–20, clusters gen 2–5-fld; lvs ≤
      4.5 mm wide. . . . . ***J. howellii***

4′ Seed body < 1–2 appendages; petals gen = sepals; infl clusters 1–5, clusters gen 10–30-fld; lvs < 3 mm
wide . . . . ***J. regelii***
3′ Seeds plump, appendages 0 or short points, << seed body
  5. Perianth gen 2–3.6 mm; fr ≥ perianth
    6. Fr blunt or notched, petals blunt or acute; heads 1–6, branches erect, infl narrow; lvs blue-green . . . . ***J. covillei***
    6′ Fr, petals acuminate; heads 10–50, branches spreading, infl diffuse; lvs green . . . . ***J. cyperoides***
  5′ Perianth gen 4–6 mm; fr gen ≤ perianth
    7. Lf sheath appendages well developed, gen rounded, > 1 mm
      8. Rhizomatous; sepals gen = petals; infl clusters 1–4(5); perianth midveins smooth to occ papillate,
green . . . . ***J. longistylis***
      8′ Loosely cespitose; sepals < petals; infl clusters 8–30; perianth midveins smooth, gen red-streaked
. . . . ***J. macrophyllus***
    7′ Lf sheath appendages 0 or obscure, gen pointed, < 1 mm
      9. Perianth parts < 1.5 mm wide; fr brown; heads 3–12; gen montane . . . . ***J. orthophyllus***
      9′ Perianth parts > 1.5 mm wide; fr ± black; heads 1–3(5); coastal . . . . ***J. falcatus***
        10. Petals acuminate, gen > 5 mm; anthers gen > 1.7; fr tip acute to truncate, inconspicuous,
<< perianth, gen ellipsoid to oblong; CCo . . . . subsp. ***falcatus***
        10′ Petals acute to blunt, gen < 4.5 mm; anthers < 1.5 mm; fr tip notched, conspicuous, ± = perianth,
gen spheric to broadly ellipsoid; NCo . . . . subsp. ***sitchensis***

J. acuminatus Michx. (p. 1369) TAPERED RUSH (Group 3) Per, cespitose, 20–80 cm; rhizome short. **LF:** gen basal; sheath appendages 1.5–5 mm, rounded; blade cylindric, crosswalls complete, obvious. **INFL:** lowest bract < infl; branches spreading; clusters gen 6–50, 5–20-fld. **FL:** perianth parts 2.5–3.5 mm, ± equal, narrowly acuminate, light brown to ± green; stamens gen 3, filaments ± > anthers. **FR:** gen ≤ perianth, light brown to red; chamber 1. **SEED:** 0.3–0.5 mm; appendages minute. 2*n*=40. Shores, swales; gen < 1300 m. NCo, KR, NCoRI, CaRF, n SN, GV, SnFrB; to BC, e N.Am, Mex. Jul–Aug

J. acutus L. subsp. ***leopoldii*** (Parl.) Snogerup (p. 1369) SOUTHWESTERN SPINY RUSH Per, cespitose, 50–140 cm; rhizome many-branched. **ST:** rigid, hardened, cylindric. **LF:** basal, 40–120 cm, rigid; tip stiff, sharp; sheath appendages firm. **INFL:** appearing lateral, gen ± open; lowest bract cylindric, resembling st, ≤ infl; branches uneven; clusters 2–4-fld, each subtended by 1 obvious, clasping bract; bractlets 0. **FL:** perianth 2–4 mm, margins membranous, sepals thin, obtuse, petals ± rounded; anthers 6, >> filaments, ± red-brown. **FR:** gen >> perianth, ± spheric to elliptic, shiny brown. **SEED:** irregular, occ narrowly winged; appendages ≤ 0.3 mm. 2*n*=48. Moist saline places, salt marshes, alkaline seeps; gen < 300 m. CCo, SCo, s ChI, WTR, DSon; to AZ, Baja CA, S.Am, Afr. Typical subsp. in Medit, n Afr, introduced in Australia, New Zealand. Jun–Aug ★

J. ambiguus Guss. (p. 1369) FROG RUSH (Group 1) Ann, densely cespitose or st 1, 3–17 cm. **ST:** 0.5–1 mm wide. **LF:** ± cauline, gen 1–2 per st. **INFL:** lowest bract > infl; upper bracts 1/3 to ± = infl; bractlets 1.5 mm. **FL:** sepals 4–6.8 mm, acute; petals 3.3–5.3 mm, ± blunt or abruptly short-pointed; stamens 6, filaments > anthers. **FR:** < sepals, ± = petals; tip ± truncate to ± acute. **SEED:** 0.3–0.4 mm, ovoid or barrel-shaped; appendages 0. 2*n*=34. Uncommon. Moist san saline places; gen < 500 m. NCo, CaRF, SnJV, CCo, SnFrB, SCo, ChI, SnGb, MP (exc Wrn); scattered in N.Am, native to Eurasia, n Afr. Presumed introduced. Taxonomy, relationship to Eurasian *J. bufonius* complex needs study. Apr–Sep

J. articulatus L. subsp. ***articulatus*** JOINTED RUSH (Group 3) Per 10–60 cm; rhizome short, branching. **ST:** erect or ascending; nodes gen rooting. **LF:** basal lf blades reduced; cauline lvs 1–3, sheaths loose, blades 5–10 cm, cylindric or occ flattened, crosswalls complete, sheath appendages rounded, 1–1.5 mm, firm. **INFL:** 2–15 cm; lowest bract inconspicuous; branches spreading; clusters gen < 25, 3–12-fld. **FL:** perianth parts 1.9–3.1 mm, ± equal, sepals acute to acuminate, petal tips gen blunt, hooded, green to dark brown, stamens 6, filaments ± = anthers. **FR:** > perianth, gen dark brown to black, 3-angled, tapering from near middle. **SEED:** 0.5 mm, oblong; appendages minute. 2*n*=80. Moist ground, seeps, shores, marshes; < 2000 m. NCo, KR, NCoRO, CaR, n SN, s SNF, GV, SnFrB, SCoRO, SnBr, DMoj; to AK, e N.Am, Eurasia, Afr. Another subsp. in e Asia. Jul–Aug

J. balticus Willd. subsp. ***ater*** (Rydb.) Snogerup (p. 1369) BALTIC RUSH (Group 2) Per gen 35–110 cm; rhizome scaly, long, creeping, gen unbranched, slender to stout. **ST:** 1–6 mm wide, gen cylindric.

LF: basal; blades 0; sheaths variable, 2–15 cm. **INFL:** appearing lateral, ± open; lowest bract cylindric, resembling st, gen >> infl; fls 5–50+; bractlets 2, membranous. **FL:** perianth parts 3–6 mm, sepals ± ≥ petals, scarious margins of petals wider than those of sepals; stamens 6, filaments << anthers. **FR:** < to > perianth; acute below obvious beak. **SEED:** 0.4–0.8 mm; appendages 0. 2*n*=40,80. Moist to ± dry sites; gen < 2200 m. CA (exc Teh); to AK, ND, C.Am. [*J. arcticus* Willd. var. *b.* (Willd.) Trautv.] Variable, intergrading complex needing study. Hybridizes with *J. breweri, J. mexicanus, J. lescurii.* Jul–Nov

J. bolanderi Engelm. (p. 1369) BOLANDER'S RUSH (Group 3) Per 30–80 cm; rhizome stout, creeping. **LF:** basal scale-like, cauline 3–4, sheath appendages 3–5 mm, scarious, blade 10–20 cm, 1–1.5 mm wide, cylindric, blade crosswalls complete. **INFL:** lowest bract gen > infl; clusters 1–3(5), > 20-fld. **FL:** perianth parts 3–3.5 mm, ± equal, narrowly acuminate, dark to light brown; stamens 3, filaments > anthers. **FR:** ± = perianth, oblong, short-beaked. **SEED:** ± 0.5 mm, obovoid; appendages minute. Swampy or sandy ground; < 1600 m. NW, CaRH, CCo, SnFrB; to BC, ID. Jul–Sep

J. brachyphyllus Wiegand SHORTLEAVED RUSH (Group 5) Per, cespitose, 30–50 cm. **ST:** stiff, stout, conspicuously grooved. **LF:** ± basal; sheath appendages conspicuous, membranous; blade with flat side toward st, much exceeded by st, stiff, spreading, crosswalls 0. **INFL:** open; lowest bract = to gen > infl; branches ascending; fls 1, many. **FL:** perianth parts 4–5.5 mm, long-acuminate, sepals > petals; stamens 6, anthers > filaments. **FR:** ± = petals, widely oblong, ± 3-angled; chambers 3. **SEED:** ± twisted; appendages minute. 2*n*=80. Meadows moist in spring, washes; 300–1600 m. CaRH, SNF, n SNH, MP; to WA, MT, TX, c US. Typical of sandy prairies of c US; long-anthered montane pls of nw US need further study. Jun–Jul

J. breweri Engelm. (p. 1369) SALT OR BREWER'S RUSH (Group 2) Per 10–130 cm; rhizome thick, creeping, occ vertical. **ST:** slender, twisted or arching, firm. **LF:** basal; sheaths dark brown, shiny, loose; blades 0. **INFL:** appearing lateral, compact; lowest bract ≥ st, cylindric, resembling st; bractlets 2 per fl; fls gen few. **FL:** perianth parts 5–8 mm, sepals gen > petals, dark, middle ± green, sides brown to ± purple, margins membranous. **FR:** < perianth, ovoid, dark brown, shiny; tip acute. **SEED:** 0.7–1.1 mm, ovoid, appendages 0. Coastal dunes and marshes; < 100 m. NCo, CCo, SCo; to BC. Aug–Apr

J. bryoides F.J. Herm. (p. 1369) MOSS RUSH (Group 1) Ann, densely cespitose, 0.1–2.2 cm, ± red-brown in fr; gen self-pollinating. **ST:** hair-like, 0.1–0.2 mm wide. **LF:** basal, 1/4 to = st; sheath appendages 0. **INFL:** fl 1 per st; bracts (1)2, < 1 mm. **FL:** perianth parts 4–8, 1–2.8 mm, sepals ≤ petals, shiny, incurved over fr; stamens 2–4, filaments > anthers. **FR:** < perianth, ovoid to elliptic; chambers 2–3. **SEED:** 0.3–0.5 mm, not striate; appendages small. Common. Wet places, washes, meadows, granitic seeps; 600–3600 m. CaRH, SNH, SCoR, TR, PR, SNE; to OR, ID, CO, Baja CA. Apr–Aug

J. bufonius L. TOAD RUSH (Group 1) Ann gen 2–10(41) cm, gen branched from base. **ST:** gen ± 1 mm wide. **LF:** ± cauline, 1–3 per st,

0.5–1.5 mm wide. **INFL:** open to clustered or congested, fls 1–few in small clusters, ± throughout pl; lowest bracts lf-like; bractlets 1–2.5 mm. **FL:** perianth parts 2–7 mm, sepals gen > petals, acuminate, or petals blunt; stamens 6, filaments < to > anthers. **FR:** < perianth, oblong to obovoid; tip acute. **SEED:** 0.3–0.6 mm, ovoid to elliptic; appendages 0 or small. Vars. weakly delineated. Relationship of N.Am pls to Eurasian *J. ranarius* Songeon & E.P. Perrier, *J. hybridus* Brot. needs study. Native and naturalized.

var. **bufonius** (p. 1369) Pl relatively large in all features. **INFL:** fls 1 at nodes, occ crowded near branch tips, sprawling to erect. **FL:** perianth gen 4–7.3 mm, petals gen acuminate. **FR:** < petals. 2*n*=108. Damp sunny ground, gen disturbed; < 3200 m. CA; ± worldwide. May–Sep

var. **congestus** Wahlb. CLUSTERED TOAD RUSH Pl gen large. **INFL:** fls gen congested on branch tips in dense, ± coiled cymes. **FL:** perianth gen 5–8(10) mm, petals acuminate. **FR:** < petals. Drying pools, shores, disturbed ground, gen saline; < 2500 m. NCoR, CaRF, s SNH, GV, CW, SW (exc s Chl); to WY, n Baja CA; native to n Eur. Taxonomy unclear, relationship to typical *J. bufonius* needs study. Apr–Aug

var. **occidentalis** F.J. Herm. WESTERN TOAD RUSH Pl relatively small in all features. **INFL:** fls 1 at nodes. **FL:** perianth 2–4 mm, petals acuminate. **FR:** < petals. Drying pools, creek banks; < 2000 m. CA; to BC, ID, CO, AZ, e Asia. Relationship to *J. sphaerocarpus* Nees, *J. amuricus* (Maxim.) V.I. Krecz. & Gontsch. of Eur needs study. Apr–Jun

J. **capillaris** F.J. Herm. (p. 1369) HAIRSTEMMED RUSH (Group 1) Ann, densely cespitose, 0.7–5.7 cm, ± red-brown in fr; gen self-pollinating. **ST:** 0.1–0.3 mm wide. **LF:** basal, < 2.2 cm; sheath appendages 0. **INFL:** fls 1–3 per st; bracts 0.4–1.5 mm, acute to acuminate, membranous. **FL:** perianth parts 4–6, 1.8–2.8 mm, petals > sepals; stamens 2–3, filaments > anthers, anthers < 0.7 mm. **FR:** < perianth and much lighter in color, gen ± spheric to obovoid. **SEED:** < 10 per fr, < 4 per row, 0.5–0.8 mm, at least base ± obscurely striate; appendages small. 2*n*=36. Moist, bare to grassy places, meadows, streambanks, granitic seeps; 1200–3200 m. SNH; s OR. Jul–Sep

J. **capitatus** Weigel DWARF RUSH (Group 1) Ann < 10(14) cm. **ST:** thread-like, channeled, angled, or flat. **LF:** basal; blade 1.5–3.5 cm, margins ± inrolled; sheath loose, appendages 0. **INFL:** clusters 1(5) per st; fls 2–6(25); lower bracts 5–15 mm, gen > infl, lf-like, ± keeled. **FL:** sepals 3–4(5) mm, acuminate, petals 2–2.5(3.5) mm, ovate, thin; stamens 3. **FR:** < petals, ovoid. **SEED:** 0.2–0.3 mm, asymmetric. 2*n*=18. Uncommon. Vernal pools, swales, damp meadows, creeks, disturbed dunes; < 1000 m. NCo, NCoR, CaRF, n&c SNF, GV, CCo, SnFrB; scattered in s-c US; native to Eurasia, n Afr. Perianth parts as few as 2 when conditions poor. Apr–Jun

J. **chlorocephalus** Engelm. (p. 1369) GREENHEADED RUSH (Group 3) Per 12–50 cm; rhizome ± stout, mat-forming. **LF:** gen basal, purple-red, blade 0; cauline lvs 3–4, sheath appendages 4–6 mm, scarious, blade ± 1/3 st, < 1 mm wide, cylindric, crosswalls complete. **INFL:** lowest bract inconspicuous, gen reaching lowest cluster; clusters 1–3, ± as wide as long; bractlets scarious, ± 1/3 perianth. **FL:** perianth parts ± 4 mm, ± equal, white to pink or ± green, midvein pale green to red, margins wide, scarious, sepals pointed, petals rounded; stamens 6, filaments < 1/2 anthers; style > perianth, stigmas prominent, 2.7 mm. **FR:** << perianth, 3-angled, brown. **SEED:** 0.6 mm, asymmetric, light brown. Wet areas in montane conifer forest; 1200–3100 m. n&c SN; NV. Jul–Sep

J. **confusus** Coville (p. 1369) COLORADO RUSH (Group 5) Per, cespitose, 30–50 cm. **ST:** slender, light green. **LF:** basal; sheath appendages 0.2–1 mm, thin, rounded, ± white; blades with flat side toward st, ± thread-like, < 2/3 st, thick, crosswalls 0, margins gen inrolled. **INFL:** gen compact; lowest bract 2–8 cm; fls gen clustered, few; bractlets 2 per fl, large, ovate. **FL:** perianth parts 2.6–4 mm, ± equal, midvein area straw-colored, flanking lateral stripes dark, margins wide, scarious; stamens 6, filaments > anthers. **FR:** oblong, ± < perianth, ridged, deeply notched; chambers 3. **SEED:** 0.4–0.5 mm; ends oblique; appendages 0. 2*n*=80. Meadows in conifer forest; 1200–2000 m. KR, NCoRI, CaRH, n&c SNH, Wrn; to w Can, SD, CO, NM. Jul–Sep

J. **cooperi** Engelm. (p. 1369) COOPER'S RUSH (Group 2) Per, ± cespitose, 40–80 cm; rhizome short, thick, many-branched; roots

large, spongy. **LF:** basal; blades short, stiff, cylindric, tips sharp. **INFL:** appearing lateral; lowest bract cylindric, resembling st, > infl, tip sharp; branches unequal; clusters 2–10-fld; bracts within infl obvious, > cluster; bractlets white. **FL:** perianth parts 4–6 mm, sepals > petals, pale ± green-straw-colored, sepal tips gen acuminate, spiny, firm; stamens 6, large, filaments < anthers. **FR:** ± = perianth, narrowly oblong, 3-angled. **SEED:** with a conspicuous white ridge; appendages unequal, minute. Alkaline places; < 700 m. D; NV, Mex. May–Oct ★

J. **covillei** Piper (p. 1369) COVILLE'S RUSH (Group 6) Per, cespitose, 5–25 cm; rhizome long, creeping, heavy. **LF:** gen basal, sheath appendages 0; blade with flat side toward st, gen reaching top of infl, 2–4 mm wide; cauline lvs 0–2. **INFL:** lowest bract < to > infl; branches 1–5, erect; narrow; clusters 1–6, 3–7-fld. **FL:** perianth parts 2–3.5 mm, sepals > petals, midvein area papillate, brown to green; stamens 6, filaments ± < anthers. **FR:** > perianth, widely cylindric, dark; tip blunt or notched, short-beaked. **SEED:** 0.4 mm; appendages small. 2*n*=38. Moist sandy places, montane forest, creekbanks; 300–3000 m. NW, CaR, SN, n CCo, SnFrB, TR; to BC, MT. [*J. c.* var. *obtusatus* (Engelm.) C.L. Hitchc.] Jul–Sep

J. **cyperoides** Laharpe (p. 1369) BOLIVIAN RUSH (Group 6) Per 5–50 cm; rhizome long, ± erect, grading into st. **ST:** lfy; nodes conspicuous. **LF:** evenly spaced on st, occ overlapping; 4–20 cm, 2–8.5 mm wide; sheath appendages 0; blade with flat side toward st, crosswalls 0. **INFL:** lowest bract < infl, resembling cauline lvs; diffuse, branches many, spreading; clusters 10–70, 2–10-fld; bracts membranous. **FL:** perianth parts 2.2–3.9 mm, acuminate, green, sepals gen < petals; stamens 6, filaments ≥ anthers. **FR:** > perianth, gradually tapered to elongate beak 0.5–1 mm. **SEED:** 0.4–0.5 mm; rectangular-veined, appendages minute. Creekbanks, damp meadows, yellow-pine forest; 600–1100 m. n SNH (Butte, Yuba cos.); native to S.Am. Jul–Aug

J. **diffusissimus** Buckley SLIM-POD RUSH (Group 3) Per, cespitose, 20–75 cm. **ST:** thin. **LF:** few; sheath appendages ± purple at base; blade cylindric, crosswalls complete. **INFL:** 10–30 cm; lowest bract inconspicuous; branches spreading; clusters many, 2–5-fld. **FL:** gen not all developing; perianth parts 2.3–3.2 mm, ± equal; stamens 3, filaments gen > anthers. **FR:** gen 2 × perianth, slender, 3-angled. **SEED:** 0.3–0.4 mm; appendages minute. Pond shores, riverbanks; < 100 m. e ScV; OR, WA; native to c, se US. Jul–Oct

J. **digitatus** C.W. Witham & Zika FINGER RUSH (Group 1) Ann, densely cespitose, 4–10.5 cm, ± red in fr. **ST:** < 0.4 mm wide, hairlike. **LF:** basal; sheath appendages 0. **INFL:** fls (1)2–8 per st; bracts 1–1.8 mm, ovate. **FL:** perianth parts 6, 3.5–4.4 mm, sepals gen < petals, acuminate, midvein area green to red; stamens 3, filaments < anthers, anthers ≥ 1.1 mm. **FR:** >> perianth, linear-oblong, gen ± curved, ± red; chambers 2–3. **SEED:** 0.5–0.65 mm, striate at 10×. Vernal pools, swales, volcanic seeps; 650–800 m. CaRF (Shasta Co.), n SNF. May–Jun ★

J. **drummondii** E. Mey. (p. 1369) DRUMMOND'S RUSH (Group 2) Per, densely cespitose, 5–40 cm. **LF:** basal; sheaths 2–7 cm, pale; blades 0 or small, inner bristle-like. **INFL:** appearing lateral; lowest bract cylindric, resembling st, > infl; fls gen 2–3. **FL:** perianth 5–7 mm, sepals wider, gen > petals; stamens 6, filaments << anthers. **FR:** ≥ perianth; tip notched or truncate. **SEED:** body 0.5 mm, narrow; appendages > body. Moist, rocky places in conifer forest; 2000–3500 m. KR, CaR, SN; to AK, NT, CO, NM. Jul–Sep

J. **dubius** Engelm. (p. 1369) MARIPOSA RUSH (Group 3) Per, gen densely matted, 15–70 cm; rhizome stout. **LF:** basal blades 0; cauline lvs few, sheath appendages 4–6 mm, prominent, scarious, blade cylindric, crosswalls complete. **INFL:** lowest bract gen 1–2 cm, inconspicuous; branches spreading; clusters gen many, 4–10-fld; bracts, bractlets scarious. **FL:** perianth parts 2.5–3 mm, bristle-tipped; stamens 6, filaments << anthers. **FR:** barely > perianth, 3-sided. **SEED:** obovate; appendages minute. Wet places; < 2000 m. NCoRO, CaRF, n&c SNH, ScV, SnFrB, SCoRO, SnGb, PR, DMoj; n Mex. Relationship to *J. rugulosus* needs study. Jul–Sep

J. **dudleyi** Wiegand (p. 1369) DUDLEY'S RUSH (Group 5) Per, cespitose, 20–84 cm. **LF:** ± basal; sheath margins scarious to thickened; sheath appendages 0.2–0.5 mm, stiff, thick, plastic-like, ± yellow, glossy, rounded; blade with flat side toward st, gen < 1/12 st,

1–1.3 mm wide, not stiff, margins gen inrolled. **INFL:** lowest bracts lf-like, 1 gen >> infl; branches unequal; fls 1, gen denser near st tips; bractlets 2 per fl, blunt to acuminate. **FL:** perianth 3.5–5 mm, lanceolate, acuminate, conspicuously spreading in fr; stamens 6, filaments > anthers; style 0.2–0.3 mm. **FR:** < perianth, widely ovoid, acute or blunt, without strong ridges; chamber 1, partitions incomplete. **SEED:** 0.4–0.5 mm, oblong; appendages small. $2n=80$. Wet areas in montane conifer forest; < 2000 m. KR, n SNH, MP (exc Wrn); to BC, e N.Am. Jul–Aug ★

J. duranii Ewan (p. 1369) DURAN'S RUSH (Group 3) Per, cespitose, 3–20 cm; rhizome vertical. **ST:** slender, ± flat. **LF:** gen cauline; sheath appendages prominent, rounded; blade cylindric, gen exceeded by st, tip long, crosswalls complete but obscure. **INFL:** lowest bract > infl, narrow, not sheathing, rolled tip > rest of bract blade, gen not present in fr; cluster gen 1; bractlets awned. **FL:** perianth parts gen < 2.8 mm, ± equal, lance-linear, stiff, brown; stamens 6, filaments ≤ anthers. **FR:** ± = perianth, obovoid. **SEED:** 0.5 mm, narrowly lanceolate. Creek banks, wet places, in montane conifer forest; 1800–2750 m. SnGb, SnBr, SnJt. Relationship with *J. mertensianus* needs study. Aug–Sep ★

J. effusus L. SOFT OR LAMP RUSH (Group 2) Per, cespitose, 60–155 cm; rhizome stout. **ST:** 2–5 mm wide at base, above sheath; when fresh, upper st smooth, shiny. **LF:** basal; blades 0. **INFL:** appearing lateral, open; lowest bract cylindric, resembling st, >> infl; fls gen many, single; bractlets 2 per fl. **FL:** perianth parts 1.8–4.2 mm, ± equal, gen pale brown; stamens 3, filaments ≤ to > anthers. **FR:** ± = perianth, obovoid, ± truncate. **SEED:** 0.5 mm; appendage 1, minute. $2n=40$. Native and naturalized.

subsp. ***austrocalifornicus*** Zika (p. 1369) SONORAN OR BAJA RUSH **ST:** 64–155 cm, fertile st gen 2–3.5 mm wide above sheath, 14–22 ridges per side, fresh upper st shiny, smooth, pith solid. **LF:** sheath 5.5–15 cm, base dark brown, papillate, upper 1/2 green to pale brown, veins ± abruptly tapered to broadly asymmetrical winged tip, margins thin, pale or darkened; **INFL:** 2–8 cm, open to dense; bract 5–23 cm, > infl. **FL:** perianth 2.2–2.8 mm, acuminate, pale, spreading or pressed to fr; filaments ≤ anthers. **FR:** 1.7–1.9 mm, shiny, elliptic-oblong, truncate, partitions straight, chambers 3. **SEED:** 0.4–0.6 mm, 0.2–0.25 mm wide, asymmetrical, netted, vertical lines more obvious than horizontal. Riparian, springs, salt marsh edge; < 2400 m. SW (exc ChI), DSon; to AZ, n Baja CA. Aug–Oct

subsp. ***effusus*** (p. 1369) COMMON RUSH **ST:** to 105 cm, 2–5 mm wide above sheath, 14–26 fine ridges per side, pith solid. **LF:** sheath 5–14 cm, deep brown below, papillate, dull, upper 1/2 green to pale brown, veins ± strongly converging at broad ± symmetrical summit, margins thin, gen dark-banded, overlapping near tip. **FL:** perianth 2.2–2.5 mm, pale brown, spreading or curving away from fr; filaments > anthers. **FR:** 1.7–2 mm, < perianth, oblong-truncate, brown, partitions concave, united only at base. **SEED:** ± asymmetrical, appendages minute. Lakeshores, wet pastures, disturbed damp ground; < 1700 m. CaRH, n SN, ScV; to AK, e N.Am, native to Eurasia. Invasive, easily mistaken for native subspp. Jun–Oct

subsp. ***pacificus*** (Fernald & Wiegand) Piper & Beattie (p. 1369) PACIFIC RUSH **ST:** to 112 cm, 2.2–3.3 mm wide above sheath, 14 fine or broad ridges per side, pith solid. **LF:** sheath 6–11 cm, gen dark brown or ± black, papillate, uniformly dark, veins strongly converging at broadly asymmetrical summit, margins thickened, darker, edges overlapping in distal 1–1.5 cm. **INFL:** gen open, 3–5 cm, branches ascending; bracts 7–24 cm, > infl. **FL:** perianth 2.3–2.8 mm, acuminate, gen pale, pressed to fr, ± ≥ fr, margins scarious; filaments = anthers. **FR:** 2.1–2.4 mm, oblong, truncate, inner partitions straight, chambers 3. **SEED:** 0.35–0.6 mm, 0.2–0.25 mm wide, brown, netted, ± asymmetrical. Seeps, shores, marshes, gen damp sunny ground; < 2500 m. CA-FP (exc SW); to BC. [*J. e.* var. *p.* Fernald & Wiegand] May–Oct

subsp. *solutus* (Fernald & Wiegand) Hämet-Ahti EASTERN SOFT RUSH **ST:** to 146 cm, 2–5 mm wide above sheath, 16–26 fine ridges per side, pith gen irregularly chambered if aquatic. **LF:** sheath 15–27 cm, dark brown below, occ papillate, dull, green to pale brown distally, outer veins ± strongly converging at broadly ± symmetrical summit, margins thin, gen pale, overlapping halfway or split to base

and unrolling. **FL:** perianth 2.2–2.9 mm, pale brown, pressed to fr; filaments ≤ anthers. **FR:** 2.2–2.7 mm, elliptic-oblong, truncate, gen ± ≤ perianth; seed ± asymmetrical, appendages minute. Riverbanks, pond shores, wet soil or shallow water; < 250 m. GV; to BC; native to e N.Am. Jul–Sep

J. ensifolius Wikstr. (p. 1369) DAGGER RUSH, SWORDLEAVED RUSH (Group 4) Per 20–60 cm; rhizome slender, creeping. **LF:** gen basal; bases overlapping; sheath appendages 0 or obscure; blade flat, with edge toward st, 2–5 mm wide, gen curved, crosswalls incomplete, tip long. **INFL:** variable; lowest bract ≤ 1/2 infl; clusters gen 2–10, gen many-fld, gen hemispheric. **FL:** variable; perianth parts 2.2–3.3 mm, ± equal, gen dark brown to black; stamens gen 3, filaments > anthers; style 0.1–0.6 mm. **FR:** gen ≥ perianth, oblong, abruptly short-beaked. **SEED:** ± 0.5 mm, widely fusiform; appendages 0 or short. $2n=40$. Common. Wet places; < 2800 m. NW, CaR, SN, CCo, TR, GB; to AK, w Can, UT, ne Asia, Mex. Relationship to *J. saximontanus* needs study; introduced in e N.Am, Europe, HI, New Zealand. Jul–Sep

J. exiguus (Fernald & Wiegand) Snogerup & Zika (p. 1369) KLAMATH OR WEAK RUSH (Group 2) **ST:** to 71 cm, 0.9–1.6 mm wide above sheath, smooth when fresh, ≤ 11 broad ridges per side when dry, pith solid. **LF:** sheath 5–12 cm, dark brown to red-brown, ± shiny below, dull, green to pale brown distally, veins gradually converging at narrow symmetrical truncate summit, margins thin, gen minutely blackened, ± inrolled at summit, overlapping halfway from or near base. **FL:** perianth 1.7–2.4 mm, ± spreading from fr, dark brown stripes flank midvein early in season; 3 stamens, filaments > anthers. **FR:** 1.6–1.9 mm, < perianth, oblong, truncate, internal partitions straight, fused, chambers 3. **SEED:** 0.35–0.6 mm, 0.2–0.25 mm wide, netted. Montane, wet meadows, springs, shores; 700–2300 m. KR, NCoR, CaR, SNH; OR. [*J. effusus* var. *e.* Fernald & Wiegand] Jul–Sep

J. falcatus E. Mey. (Group 6) Per 3–30 cm; rhizome long, scaly. **LF:** gen basal; sheath appendages 0 or small; flat side of blade toward st, gen reaching top of infl, stiff, ± curved; cauline lvs 1–2, < basal, stiff. **INFL:** lowest bract gen > infl; clusters 1–3(5), 5–25-fld, hemispheric. **FL:** perianth parts 3.3–6.4 mm, ± equal, papillate, dark, margins wide, scarious; stamens 6, filaments << anthers. **FR:** < or ± = perianth, ± spheric to longer than wide; tip notched; beak short, obvious. **SEED:** asymmetric; outer covering loose, white; appendages 0. $2n=38$. Circumscription of subspp. needs study.

subsp. ***falcatus*** (p. 1369) SICKLELEAVED RUSH **FL:** petals acuminate, gen 5–6.4 mm, anthers gen 1.7–2.5 mm. **FR:** gen ellipsoid to oblong. Peatland, moist sandy coastal areas; gen < 100 m. CCo; also native to se Australia. [*J. f.* var. *f.*] Jun–Sep

subsp. ***sitchensis*** (Buchenau) Hultén ALASKAN SICKLELEAVED RUSH **FL:** petals acute to blunt, gen 3.3–4.5 mm, anthers gen 0.8–1.4 mm. **FR:** gen spheric to broadly ellipsoid. Peatland, moist sandy coastal areas; gen < 100 m. NCo; to w AK. [*J. f.* var. *s.* Buchenau] Jul–Oct

J. gerardi Loisel. subsp. ***gerardi*** MUD RUSH (Group 5) Per 15–45 cm; rhizome creeping. **LF:** gen basal, flat; sheath appendages blunt, 0.4–0.7 mm. **INFL:** gen loose; lowest bract lf-like, 2–10 cm, gen < infl; fls 1, bractlets 2 per fl. **FL:** perianth parts 3–4 mm, ± equal, black-striped, tips hooded, incurved, blunt; stamens 6, filaments << anthers. **FR:** ± = perianth, ± ellipsoid, ± blunt. **SEED:** 0.5–0.7 mm; ± ellipsoid, netted, appendages 0. $2n=84$. Coastal salt marsh; < 10 m. CCo (Contra Costa, Solano cos.); OR to s BC, e N.Am, Eurasia, n Afr. Possibly native, needs study; introduced in Australia, New Zealand, Greenland. Jun–Aug

J. hemiendytus F.J. Herm (Group 1) Ann, cespitose, 0.1–3.2 cm; gen self-pollinating. **ST:** 0.1–0.5 mm wide. **LF:** basal, < 1.8 cm; sheath appendages 0. **INFL:** fl gen 1 per st; bracts 0–2, < 2 mm, ovate to rounded, not sheathing. **FL:** perianth parts 4–6, 1.9–3.5 mm, sepals ≥ petals, dull; stamens 2–3, filaments > anthers, anthers < 0.7 mm. **FR:** < to > perianth, obovoid to oblong; chambers 2–3. **SEED:** 0.3–0.55 mm, not striate; appendages small.

var. ***abjectus*** (F.J. Herm.) Ertter CENTER BASIN RUSH **ST:** 0.2–0.5 mm wide, wider below fl. **INFL:** bracts 0. **FR:** gen < peri-

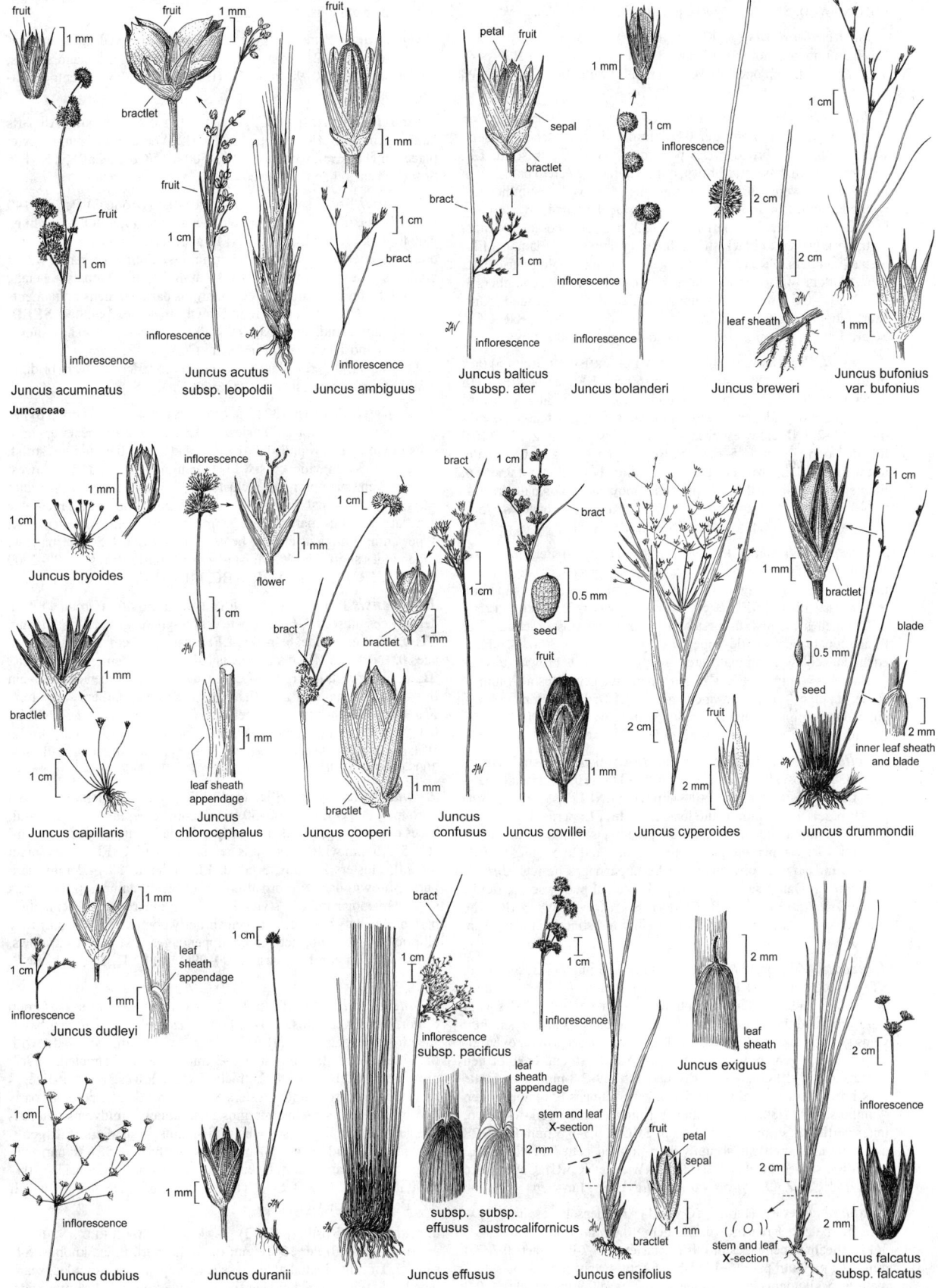

fruit
1 mm
Juncus acuminatus

Juncaceae

fruit
1 mm
fruit
bractlet
fruit
1 cm
inflorescence
fruit
1 cm
Juncus acutus
subsp. leopoldii

fruit
1 mm
1 cm
bract
inflorescence
Juncus ambiguus

petal fruit
sepal
bractlet
bract
1 cm
inflorescence
Juncus balticus
subsp. ater

1 mm
1 cm
inflorescence
inflorescence
Juncus bolanderi

inflorescence
2 cm
2 cm
leaf sheath
Juncus breweri

1 cm
1 mm
Juncus bufonius
var. bufonius

1 mm
1 cm
Juncus bryoides

inflorescence
flower
1 mm
1 cm
bract
1 mm
leaf sheath
appendage
Juncus
chlorocephalus

1 cm
bract
bractlet
1 mm
bractlet
1 mm
Juncus cooperi

bract
1 cm
bractlet
1 mm
1 cm
1 mm
Juncus
confusus

bract
1 cm
bract
seed
0.5 mm
fruit
1 mm
Juncus covillei

2 cm
fruit
2 mm
Juncus cyperoides

1 cm
1 mm
bractlet
0.5 mm
seed
blade
2 mm
inner leaf sheath
and blade
Juncus drummondii

bractlet
1 mm
1 cm
Juncus capillaris

1 mm
1 cm
leaf
sheath
appendage
1 mm
inflorescence
Juncus dudleyi

1 cm
inflorescence
Juncus dubius

1 cm
1 mm
Juncus duranii

bract
1 cm
inflorescence
subsp. pacificus
leaf
sheath
appendage
2 mm
subsp.
effusus
subsp.
austrocalifornicus
Juncus effusus

1 cm
inflorescence
stem and leaf
X-section
2 mm
fruit
petal
sepal
bractlet
1 mm
Juncus ensifolius

2 mm
leaf
sheath
Juncus exiguus

2 cm
inflorescence
2 cm
stem and leaf
X-section
2 mm
Juncus falcatus
subsp. falcatus

anth. Damp open areas, esp vernally wet; 1400–3400 m. CaRH, SNH, MP (exc Wrn), SNE (exc W&I); to OR, ID, NV. Jun–Aug ★

var. *hemiendytus* (p. 1373) HERMANN'S DWARF RUSH **ST:** 0.1–0.3 mm wide, not wider below fl. **INFL:** bracts 1–2. **FR:** > perianth. 2*n*=32. Damp open areas, esp vernally wet; 400–3200 m. NW, CaR, SN, SnBr, MP; to WA, ID, UT. Apr–Aug

J. hesperius (Piper) Lint COAST OR BOG RUSH (Group 2) Per, cespitose. **ST:** to 115 cm; gen 0.8–2 mm wide above sheath, rough when fresh, < 12 broad ridges per side when dry, pith solid. **LF:** sheath 5–15 cm, base mid-brown, dull, papillate, green to pale brown in upper 1/2, veins gradually converging at broadly asymmetrical winged summit, margins thin, flat, pale or darkened, overlapping halfway from base. **FL:** perianth 2.3–2.8 mm, pressed against most of fr, dark brown to black stripes flanking midvein; 3 stamens, filaments ≥ anthers. **FR:** 1.9–2.1 mm, gen < perianth, oblong, truncate, shiny, dark brown to ± black; internal partitions straight, chambers 3. **SEED:** 0.4–0.6 mm, 0.2–0.3 mm wide, netted. Damp ground, bogs, shores, upper edges of salt marshes; gen < 100 m. NCo, KR, CCo, SCoRO; to BC. [*J. effusus* var. *brunneus* Engelm.] Jun–Nov

J. howellii F.J. Herm. (p. 1373) HOWELL'S RUSH (Group 6) Per; rhizome long, branched, stout. **ST:** 15–60 cm. **LF:** gen basal; sheath appendages 0.5–3 mm, acute or rounded; blade with flat side toward st, 1.4–4.5 mm wide, crosswalls 0, margins minutely scabrous; cauline lvs 1–2. **INFL:** lowest bract < 1.5 cm; clusters (3)5–20, 2–5(10)-fld. **FL:** perianth parts 5–6.5 mm, sepals gen < petals, acuminate, midvein area dark brown or green, occ minutely papillate; stamens 6, filaments << anthers. **FR:** << perianth, obovoid; beak small. **SEED:** (0.6)0.7–1 mm; 1–2 appendages ≥ 1/2 seed body. Meadows; 750–2380 m. KR, NCoRH, CaRH, n SNH; OR. Jul–Sep

J. interior Wiegand INLAND RUSH (Group 5) Per, cespitose, 11–60 cm. **LF:** sheath margins membranous, appendages 0.2–0.4(0.6) mm; blade with flat side toward st, 1/4–1/2 st, 0.5–1.1 mm wide, not stiff, crosswalls 0. **INFL:** ± open; lower bracts 1–3, lf-like, < or >> infl; branches unequal; fls 1, ± denser near branch tips; bractlets 2 per fl, acuminate, gen bristle-tipped. **FL:** perianth parts 3.5–4.4 mm, ± equal, lanceolate, acuminate, erect in fr, green to pale brown; stamens 6, filaments > anthers. **FR:** < to gen > perianth, gen ellipsoid, blunt or acute; chamber 1, partitions incomplete. **SEED:** 0.3–0.4 mm, oblong; appendages minute, blunt. 2*n*=80. Granitic washes; 1800–1900 m. DMtns (San Bernardino Co.); AZ to BC, ON, TN. Jul–Aug ★

J. kelloggii Engelm. (p. 1373) KELLOGG'S DWARF RUSH (Group 1) Ann, cespitose, 0.1–6 cm; ± red in fr. **ST:** 0.1–0.3 mm wide, hairlike. **LF:** basal, < 2.5 cm, 0.1–0.4 mm wide. **INFL:** fls gen (1)2–3(4) per st; bracts 1–2.5 mm, acute to acuminate. **FL:** perianth parts gen 6, 1.5–3.2 mm, unequal; stamens 2–3, filaments > anthers, anthers < 0.7 mm. **FR:** ± = perianth, color similar, obovoid to elliptic. **SEED:** 0.4–0.6 mm, ovoid, obviously striate; appendages minute. 2*n*=34. Uncommon. Damp sandy or clay soils, vernal pools, seeps, fields, meadows; < 800 m. NCoR, CaRF, n SNF, ScV, CW; to BC, ID. Gen self-pollinating. Type (CCo) possibly a distinct, extinct subsp. Apr–Jun

J. laccatus Zika (p. 1373) SHINY RUSH (Group 2) Per, cespitose. **ST:** 45–105 cm, 1.3–2.1 mm wide above sheath, 7–11 broad ridges per side, upper st shiny and smooth when fresh, pith solid. **LF:** sheath 3–15 cm, medium to gen dark brown or ± black, shiny, ± smooth, veins gen obscure, tip truncate, thick, gen flat, margins dark, overlapping halfway, gen split to base in age. **INFL:** 1–6 cm, open or gen dense; bract 5–12 cm, > infl. **FL:** perianth 1.9–2.8 mm, acuminate, dark brown stripes flanking midvein, ≥ fr; stamens 3, filaments gen ≤ anthers. **FR:** 1.8–2.4 mm, oblong, truncate, shiny yellow-brown, inner partitions straight, chambers 3. **SEED:** 0.3–0.5 mm, 0.2–0.28 mm wide, asymmetrical, abruptly short-pointed, netted. Peatland, wet meadows, shores, swales; < 2100 m. NW (exc NCoRI), CaRH, n&c SNH; to BC, AZ. [*J. effusus* var. *gracilis* Hook.] Jun–Sep

J. leiospermus F.J. Herm. (Group 1) Ann, densely cespitose, 1.9–11.6 cm, pale to ± red-brown. **ST:** gen > 0.4 mm wide. **LF:** basal, < 3/4 st; sheath appendages 0. **INFL:** cluster 1, 1–7-fld; bracts 0.7–2.4 mm. **FL:** perianth parts 6–10, 1.5–4.6 mm, petals occ > sepals; stamens 3–5, filaments <<(=) anthers, anthers ≥ 0.9 mm, styles ≥ 0.9 mm. **FR:** spheric to oblong, gen > perianth, same color as perianth;

chambers 3–5. **SEED:** 0.3–0.4 mm, ovoid, not striate; appendages minute or 1 obvious. 2*n*=32.

var. *ahartii* Ertter AHART'S DWARF RUSH All parts smaller, gen fewer. **FL:** 1(2) per st. Vernal pool margins; grassland swales, gopher mounds; 30–90 m. e GV (Butte, Placer, Sacramento, Calaveras cos.). Mar–May ★

var. *leiospermus* (p. 1373) RED BLUFF DWARF RUSH All parts larger, gen more. **FL:** 2–7 per st. 2*n*=32. Vernal pool margins, wet places in chaparral, woodland; 280–500 m. NCoRI, CaRF, n SNF, n ScV (Shasta, Tehama, Butte cos.), MP. Apr–Jun ★

J. lescurii Bol. SAN FRANCISCO RUSH (Group 2) Per 30–140 cm; rhizome creeping. **LF:** basal; sheaths variable, dull to ± shiny, pale to dark brown; blades 0. **INFL:** appearing lateral; variable, gen branches open; lowest bract cylindric, resembling st, gen < 1/2 st; bractlets 2 per fl; fls many. **FL:** perianth 5–8 mm, sepals > petals, acuminate, midvein area ± green, margins dark; stamens 6, filaments << anthers. **FR:** < perianth, light brown; tip acute, 3-angled. **SEED:** 0.4–0.7 mm, ovoid; appendages 0. Salt or freshwater marshes, shores of creeks and lakes; < 100 m. NCo, CCo; s OR. [*Juncus lesueurii* Bol., orth. var.] Intermediate between, possibly derived by hybridization of *J. balticus* subsp. *ater, J. breweri.* Aug–Sep

J. longistylis Torr. (p. 1373) LONG-STYLED RUSH (Group 6) Per 20–60 cm, rhizomatous. **ST:** slender. **LF:** gen basal; sheath appendages obvious, gen rounded, > 1 mm; blade with flat side toward st, < 1/2 st, 1–3 mm wide, crosswalls 0; cauline lvs 1–3. **INFL:** lowest bract 1–2 cm, membranous; fl clusters 1–4(5), 3–12-fld. **FL:** perianth parts 5–6 mm, sepals gen = petals, midvein broad, dark green, occ papillate, scarious margin wide; stamens 6, filaments < anthers. **FR:** < perianth, 3-angled, brown; beak short. **SEED:** 0.5 mm, narrow; appendages short. 2*n*=40. Moist places in conifer forest; 1800–2900 m. SNH, TR, PR, Wrn, W&I; to BC, MI, e Can, NM. Jul–Sep

J. luciensis Ertter SANTA LUCIA DWARF RUSH (Group 1) Ann, densely cespitose, 0.4–6.2 cm, pale yellow-green; gen self-pollinating. **ST:** 0.1–0.3 mm wide, hair-like. **LF:** basal, < 1.5 cm; sheath appendages 0. **INFL:** fls 1(2) per st; bracts 0.4–1.6 mm, acute to acuminate. **FL:** perianth parts gen 6, 1.6–4.2 mm, gen ± equal, green exc midvein tips dark red; stamens 2–3, filaments > anthers, anthers < 0.7 mm. **FR:** ± = perianth, gen pale green to pale red; chambers 2–3. **SEED:** 0.3–0.4 mm, obviously striate, elliptic; appendages 0 or small. 2*n*=32. Wet, sandy soils of seeps, meadows, vernal pools, streams, roadsides; 300–1900 m. CaRH, n SNH, SCoRO, TR, PR, MP. Apr–Aug ★

J. macrandrus Coville (p. 1373) LONG-ANTHERED RUSH (Group 4) Per, cespitose, 30–50 cm; rhizome creeping. **LF:** gen basal; bases overlapping; sheath appendages 0; blade flat, with edge toward st, 1.5–3.5 mm wide, crosswalls gen incomplete. **INFL:** lowest bract << infl; clusters 3–many, 3–5-fld. **FL:** perianth 2.8–4.2 mm, dark purple-brown, occ with broad green midvein; stamens 6, filaments << conspicuous anthers; styles 1–1.5 mm; outcrossing. **FR:** body < perianth, widely oblong; beak gen abruptly pointed. **SEED:** narrowly elliptic, 0.4–0.5 mm, net-veined, appendages small. Wet meadows, creekbanks, in conifer forest; 1200–2900 m. SNH, TR, SnJt, SNE. Jul–Oct

J. macrophyllus Coville (p. 1373) LONG-LEAVED RUSH (Group 6) Per, loosely cespitose, 20–100 cm; rhizome short. **LF:** gen basal; sheath appendages (0.7)1.5–3.4 mm, membranous; blade with flat side toward st, gen = st, 1.5–3 mm wide, ± channeled, midrib prominent; cauline lvs 1–2, thick. **INFL:** lowest bract lf-like, < infl; branches ascending; clusters 8–30, 3–5-fld. **FL:** perianth parts 4–6 mm, sepals < petals, margins membranous, midveins gen red-streaked, not papillate; stamens 6, filaments << anthers. **FR:** gen ≤ perianth, obovoid, ± abruptly beaked, shiny brown. **SEED:** narrowly elliptic; appendages short. Uncommon. Wet slopes, creekbanks; 700–2600 m. s SN, SCoRO, SCo, TR, PR (exc SnJt), DMtns; to sw UT, AZ, Baja CA. Jul–Oct

J. marginatus Rostk. (p. 1373) GRASSLEAVED OR RED-ANTHERED RUSH (Group 6) Per 15–70 cm; rhizome short, thick, knobby. **ST:** slender. **LF:** distributed along st; sheath appendages rounded, scarious; blade with flat side toward st, exceeded by infl, 1–6 mm wide, not stiff, crosswalls 0. **INFL:** open to compact; lowest bract << infl;

clusters 2–40, 5–12-fld. **FL:** perianth parts 2.5–3.5 mm, sepals < petals, sharply acute, petals blunter, ± red-brown; stamens 3, filaments >> ± purple-red anthers. **FR:** = perianth, spheric; beak 0. **SEED:** 0.5 mm, ovoid, brown; appendages minute. 2n=38,40. Locally common in sunny seeps, shallow water, abandoned placer mines; 300–950 m. CaRF (Tehama Co.), n SNF (Nevada Co.); OR, native from AZ to NM, e N.Am, C.Am, S.Am. Jul–Aug

J. mertensianus Bong. (p. 1373) MERTENS' RUSH (Group 3) Per 15–45 cm; rhizome vertical, stout. **ST:** flat. **LF:** cauline lvs 2–3; sheath appendages prominent, rounded to acute, opaque; blade cylindric, crosswalls complete. **INFL:** lowest bract short-sheathing, wide, narrow tip > infl; cluster gen 1, gen > 12-fld; bractlets short-awned, dark brown, opaque. **FL:** perianth parts 3–4 mm, not stiff, shiny ± brown-black, sepals narrowly acuminate, petals acute; stamens 6, filaments < to > anthers. **FR:** = or gen < perianth, oblong; tip notched or truncate. **SEED:** ± 0.5 mm, lance-ovate; appendages minute. 2n=40. Common. Alpine, subalpine meadows, streambanks, lake margins; 1200–3500 m. KR, NCoRH, CaRH, SNH, Wrn, SNE; to AK, SK, NM. Important forage for sheep. Jul–Oct

J. mexicanus Willd. (p. 1373) MEXICAN RUSH (Group 2) Per 10–60 cm; rhizome heavy. **ST:** erect or gen spirally twisted, flattened, slender. **LF:** basal; sheaths loose, appendages short, firm; some upper sheaths gen bearing 5–20 cm blades resembling st. **INFL:** appearing lateral, ± compact; lowest bract 1–25 cm, cylindric, resembling st; fls 2–many. **FL:** perianth parts 3–5.5 mm, sepals > petals, gen acuminate, color variable, midstripe intensity and width variable, margins gen scarious; stamens 6, filaments << anthers. **FR:** gen ≥ perianth, ovoid, 3-angled; beak 0.3 mm. **SEED:** 0.5–0.7 mm; appendages 0. Common. Coast to montane meadows; < 3800 m. CA; to s OR, NM, C.Am, S.Am. [*J. arcticus* Willd. var. *m.* (Willd.) Balslev] Sheath blades less common farther n. Aug–Nov

J. nevadensis S. Watson (Group 3) Per gen 10–50 cm; rhizome elongate, creeping. **ST:** slender. **LF:** sheath appendages 2–3.5 mm, membranous; blade gen < 2 mm wide, cylindric or ± flattened, crosswalls complete, green or glaucous. **INFL:** variable in dimensions, density, darkness; lowest bract inconspicuous; clusters 1–many, < 50-fld. **FL:** perianth parts 3–4.5 mm, ± equal, mid- to dark-brown; stamens 6, filaments < to > anthers. **FR:** gen ≤ perianth, oblong, shiny brown; tip abruptly beaked. **SEED:** 0.4–0.6 mm, narrow elliptic, ovate, or ± spherical, brown; appendages ± 0. Important forage for cattle and horses. Narrow-lvd mtn pls need study.

var. ***inventus*** (L.F. Hend.) C.L. Hitchc. DUNE RUSH Per gen 10–30 cm. **LF:** sheath appendages 2–3 mm; blade ± flattened, crosswalls gen obvious, green or glaucous. **INFL:** variable; lowest bract inconspicuous; clusters gen 1–3(4), gen 25–45-fld. **FL:** perianth parts 3–4.5 mm, ± equal, dark brown; filaments ≥ anthers. **FR:** ± = perianth. Coastal peatlands, dune swales; < 20 m. n NCo; to s WA. Aug–Oct ★

var. ***nevadensis*** (p. 1373) NEVADA OR SIERRAN RUSH Per gen 10–50 cm. **LF:** sheath appendages 2–3.5 mm; blade crosswalls gen obvious. **INFL:** variable; lowest bract inconspicuous; clusters gen > 5, < 15-fld. **FL:** perianth parts 3–3.8 mm, ± equal, gen dark brown; filaments << anthers. **SEED:** 0.6 mm, narrow elliptic. Common. Mtn meadows, streambanks; 1200–3300 m. KR, NCoRH, NCoRI, CaRH, n SNF, SNH, SnGb, SnBr, GB; to BC, SK, NM. Jul–Oct

J. nodosus L. (p. 1373) KNOTTED RUSH (Group 3) Per 15–60 cm; rhizome creeping, slender, tuber-bearing. **ST:** slender. **LF:** sheath appendages 0.5–1 mm, rounded, firm; blade cylindric, upper exceeding infl, crosswalls complete, prominent. **INFL:** lowest bract ≥ infl; clusters 2–20, 10–25-fld, spreading, spheric. **FL:** perianth parts 2.5–4.1 mm, acuminate, ± equal; stamens 6, filaments > anthers. **FR:** > perianth, slender, sharply 3-angled, long-tapered. **SEED:** 0.5 mm; appendages small. 2n=40. Streambanks, lakeshores, wet meadows; < 1700 m. se SNH, SnGb, W&I, n DMtns; to AK, TX, VA, ME, e Can, Mex. Jul–Sep ★

J. occidentalis (Coville) Wiegand (p. 1373) WESTERN RUSH (Group 5) Per, cespitose, 30–60 cm, stiff. **LF:** many, basal; sheath appendages < 1.5 mm, membranous; blade with flat side toward st, gen < 1/2 st, 1–1.5 mm wide, crosswalls 0. **INFL:** gen clustered on coast, more open inland; fls 1; lowest bract gen > infl; bractlets 2 per fl, blunt to bristle-tipped. **FL:** perianth parts gen 2.8–5 mm, ± equal, erect or spreading, green inland or gen brown-striped on coast, scarious margins wide; stamens 6, filaments ≥ anthers. **FR:** ≤ 3/4 perianth, oblong to ovate, 3-angled, dark brown; tip gen truncate to notched; chamber 1, partitions concave. **SEED:** oblong; appendages short. Moist gen sunny areas; < 2400 m. NW, CaR, n&c SN, ScV, CW (exc SCoRI), SCo, WTR, PR (exc SnJt); to BC, n Baja CA. Inland, montane pls with pale spreading perianth parts, fls smaller than coastal populations, need study. Hybridizes with *J. tenuis*. May–Sep

J. orthophyllus Coville (p. 1373) STRAIGHTLEAVED RUSH (Group 6) Per 20–50 cm; rhizome scaly, hard, dark, creeping. **LF:** gen basal; sheath appendages 0 or minute; blade with flat side toward st, << to ± = st, 2–6 mm wide, crosswalls 0; cauline lvs 1–3. **INFL:** open; lowest bract inconspicuous; branches papillate; clusters 3–12, 6–10-fld, hemispheric; bractlets prominent, ± 1/2 perianth. **FL:** perianth parts 5–6 mm, minutely papillate, bristle-tipped, midvein area green, margins brown, sepals < petals; stamens 6, filaments < anthers. **FR:** ± < perianth; tip flat; beak small. **SEED:** 0.4–0.6 mm, brown; appendages minute, 0–0.1 mm. Inland wet places, esp meadows, streambanks in forest; 1200–3500 m. NCoRO, NCoRH, CaR, n SNF, SNH, TR, PR (Santa Ana Mtns), MP (exc Wrn), SNE (exc W&I), DMtns; to WA, ID, NV. Aug–Oct

J. oxymeris Engelm. (p. 1373) POINTED RUSH (Group 4) Per 30–150 cm; rhizome creeping. **LF:** bases overlapping; sheath appendages 0 or small; blade flat, edge toward st, 3–7 mm wide, crosswalls gen incomplete. **INFL:** 15–30 cm, open; lowest bract inconspicuous; branches ascending; clusters 10–70, few-fld. **FL:** perianth parts 2.5–4 mm, ± equal, sharp-pointed, green, light brown in age, tips red to purple, margins scarious; stamens 6, filaments << conspicuous anthers. **FR:** gen > perianth, 3-angled, gradually tapered to slender beak, brown. **SEED:** gen 0.5 mm; appendages small, blunt. Uncommon. Swales, montane meadows, damp, sunny ground; < 2100 m. NCoRO, NCoRI, CaRF, SN, GV, TR, PR; to sw BC. Highly variable, possibly from hybridization with *J. phaeocephalus*. Jul–Aug

J. parryi Engelm. (p. 1373) PARRY'S RUSH (Group 2) Per, densely cespitose, 6–30 cm. **LF:** basal; sheaths many, appendages 0, or obscure, membranous; blades on distal lvs only, 3–8 cm, st-like. **INFL:** appearing lateral; lowest bract gen > infl, cylindric; fls 1–3, occ sessile; bractlets 2 per fl, dissimilar. **FL:** perianth parts 5–7 mm, sepals > petals, pale brown, tips acuminate; stamens 6, filaments << anthers. **FR:** gen > perianth, 3-angled; tip acute or acuminate. **SEED:** 1.1–2 mm, narrowly ellipsoid; appendages 0.4–0.7 mm; seed body 0.6–0.8 mm. Dry sunny slopes; 2000–3800 m. KR, CaRH, SN, SnBr, W&I; to sw Can, CO. Jun–Oct

J. patens E. Mey. (p. 1373) SPREADING RUSH (Group 2) Per, densely cespitose, 30–105 cm. **ST:** ± blue-gray-green and distinctly grooved when fresh, < 14 ridges per side, 1.2–1.6 mm wide above sheath, pith solid. **LF:** basal; blades 0 or rudimentary; sheaths 4–16 cm, base dark brown, purple or black, ± shiny, upper 1/2 green to pale brown, dull, veins gradually converging at narrow ± winged tip, margins thin, pale, overlapping halfway or split to base. **INFL:** appearing lateral, open to compact; lowest bract >> infl, 6–19 cm; cylindric, resembling st; fls gen many. **FL:** perianth parts 2.2–3.4 mm, sepals ≥ petals, narrow, acuminate, spreading in fr; stamens 6, filaments ≤ anthers. **FR:** gen < perianth, 1.7–2.5 mm, ± spheric, blunt, internal partitions concave, chambers 1, style 0.2–0.3 mm. **SEED:** 0.4–0.6 mm, 0.2–0.35 mm wide, asymmetric; not or weakly net-veined, appendages minute. Marshy places, creeks, seeps; < 1600 m. KR, NCoR, CaRF, n SNF, CW, SCo, ChI, WTR; to s WA, Mex. Jun–Oct

J. phaeocephalus Engelm. (Group 4) Per 10–50 cm, forming dense stands; rhizome stout, creeping. **ST:** flat. **LF:** bases overlapping; sheath appendages indistinct; blade flat, edge toward st, gen > st, 1.5–4 mm wide, crosswalls gen incomplete, tip fine-pointed. **INFL:** lowest bract ± > lowest cluster; clusters 1–many, > 8-fld. **FL:** perianth parts 3–5 mm, ± equal, widely lanceolate, dark brown to dark purple or ± black, shiny; stamens 6, filaments << anthers, anthers conspicuous; stigmas long-exserted. **FR:** body ≤ perianth; beak long-tapered, occ exceeding perianth. **SEED:** 0.5–0.7 mm, narrowly ovoid. Vars. weakly defined, need study.

var. ***paniculatus*** Engelm. PANICLED RUSH **INFL:** clusters gen > 10, gen 5–10 mm wide, few-fld. Wet places, coastal, inland; < 2200 m. NCo, s ScV, CCo, SnFrB, SCo, SnBr, PR. Relationship to *J. macrandrus* needs study. Jun–Sep

var. ***phaeocephalus*** (p. 1373) BROWNHEADED RUSH **INFL:** clusters 1–10, gen 10–15 mm wide, many-fld. Coastal meadows, dune hollows, marsh edges; gen < 200 m. NCo, NCoRO, CCo, SCo, n ChI; Baja CA. Jun–Aug

J. planifolius R. Br. NEW ZEALAND OR FLAT-LEAVED RUSH (Groups 1, 6) Ann or per, 1–50 cm; rhizome 0, short, or elongate. **ST:** slender. **LF:** basal; blade with flat side toward st, exceeded by infl, 1–6 mm wide, not stiff, crosswalls 0. **INFL:** open or compact; lowest bract < infl; clusters 1–30, 3–10-fld. **FL:** perianth parts 2.3–3.2 mm, acute, dark brown to ± black, petals blunter than sepals; stamens 3, filaments > yellow anthers. **FR:** < perianth, ellipsoid to ovoid; beak small. **SEED:** 0.3–0.4 mm, ovoid, brown, not or faintly striate; appendages minute. Locally common. Sunny damp ditches, freshwater shorelines; < 150 m. NCo; s OR, native to Australia, New Zealand, Chile, introduced in HI, Ireland. Sep–Nov

J. regelii Buchenau (p. 1373) REGEL'S RUSH (Group 6) Per, occ cespitose, 30–60 cm; rhizome stout. **ST:** slender, ± flat. **LF:** gen basal; sheath appendages 0 or minute; blade with flat side toward st, < 3 mm wide, crosswalls 0; cauline lvs 1–3, exceeded by infl. **INFL:** lowest bract 1–4 cm, papillate; branches few, straight; clusters 1–5, gen 10–30-fld. **FL:** perianth parts 4–6 mm, ± equal, papillate, bristle arising below tip, sepals pointed, petals more rounded; stamens 6, filaments ≤ anthers. **FR:** ± = perianth, spheric, dark ± purple-brown; beak ± 0.5 mm. **SEED:** 1.2–1.8 mm; appendages > seed body. Montane meadows; 800–1900 m. KR; to BC, WY, UT. Aug–Sep ★

J. rugulosus Engelm. (p. 1377) WRINKLED RUSH (Group 3) Per, gen densely matted, 15–70 cm; rhizome stout, horizontal; surfaces prominently wrinkled. **LF:** basal blades 0; sheath appendages 4–6 mm, prominent, scarious; cauline blades cylindric, crosswalls complete, obscure. **INFL:** lowest bract gen inconspicuous; branches spreading; clusters gen many (< 150), 4–8-fld; bracts, bractlets scarious. **FL:** perianth parts brown to ± brown-red, scarious margins wide, sepals 2.5 mm, ± bristle-tipped, petals 2.5–3 mm, petal veins prominent at base; stamens 6, filaments ± = anthers. **FR:** > perianth, 3-angled, long-beaked, bright ± brown-red. **SEED:** 0.4 mm, plump, brown; appendages minute. 2*n*=40. Common. Wet places; < 2100 m. s SNF, CCo, s SCoRO, SCo, TR, PR, DMtns, DSon; Baja CA. Relationship to *J. dubius* needs study. Jun–Sep

J. saximontanus A. Nelson (p. 1377) ROCKY MOUNTAIN RUSH (Group 4) Per 30–60 cm; rhizome stout, creeping. **LF:** bases overlapping; sheath margins membranous, appendages 0 or small; blade flat, edge toward st, gen 2–5 mm wide, crosswalls gen incomplete. **INFL:** variable; lowest bract short; clusters gen 8–30, 4–25-fld. **FL:** variable; perianth parts 2.5–3.5 mm, sepals > petals, pale to dark brown; stamens gen 6, filaments gen > anthers; style 0.4–0.9 mm. **FR:** < to > perianth, oblong, abruptly short-beaked. **SEED:** 0.4–0.5(1) mm; appendages minute to 0.3 mm. Wet places, montane conifer forest; 1500–2900 m. KR, CaRH, SNH, SnGb, SnBr, SnJt, GB, DMtns; to AK, SK, TX, Mex. [*J. s.* f. *brunnescens* (Rydb.) F.J. Herm.] Relationship to *J. ensifolius* needs study. Jul–Sep

J. supiniformis Engelm. (p. 1377) HAIR-LEAVED RUSH (Group 3) Per, matted if submerged when young, cespitose, 8–40 cm; rhizome slender, spreading. **ST:** nodes gen rooting, forming new plantlets; erect fl-sts appear as water recedes. **LF:** early submerged lvs < 30 cm, hair-like; sheath appendages 1–2 mm, membranous; cauline blades exceeding st, cylindric, crosswalls complete. **INFL:** lowest bract > infl; clusters gen 5–9, 3–9-fld. **FL:** perianth parts 3.4–5.4 mm, sepals ≤ petals, narrowly lanceolate, 3–4-veined; stamens gen 3(6), filaments > anthers. **FR:** > perianth, oblong, tip acute to acuminate. **SEED:** 0.6–0.7 mm, obovoid, ends pointed; appendages minute. Marshes, ponds; gen < 100 m. NCo (Mendocino Co.); to AK. Jun–Aug ★

J. tenuis Willd. (p. 1377) POVERTY OR SLENDER RUSH (Group 5) Per, cespitose, st 20–60 cm, gen 0–1 strong ridges per side. **LF:** sheath margins membranous, sheath appendages gen acute to acuminate, thin, ± translucent, gen 1–8 mm until late in season when

gen smaller or 0; blade with flat side toward st, 1/2 to = st, 1–1.3 mm wide, not stiff, crosswalls 0. **INFL:** ± open; lower bracts lf-like, gen > infl; branches unequal; fls 1, gen > internodes, ± denser near branch tips; bractlets 2 per fl, acute to blunt. **FL:** perianth parts 3–4.5 mm, ± equal, lanceolate, acuminate, spreading in fr, green to ± red or pale brown; stamens 6, filaments >> anthers. **FR:** < perianth, widely ovoid, blunt to acute; chamber 1, partitions incomplete. **SEED:** 0.4–0.5 mm, oblong; appendages minute, blunt. 2*n*=80. Uncommon. Damp places; gen < 1500 m. NW, CaR, SN (exc Teh), GV, CCo, SnGb, MP; N.Am, introduced worldwide exc Antarctica. Hybridizes with *J. occidentalis*. Jun–Sep

J. textilis Buchenau (p. 1377) MAT OR BASKET RUSH (Group 2) Per 100–200 cm; rhizome heavy, creeping. **ST:** cylindric, striate; base 3–5 mm wide. **LF:** basal; sheaths < 18 cm, brown; blades 0. **INFL:** lateral, open; lowest bract ≥ infl, cylindric, resembling st; branches 5–10 cm, extremely variable; fls many, 1; bractlets 2 per fl, tips rounded. **FL:** perianth parts 3.5–5 mm, sepals ± > petals, acute, green to light brown; stamens 6, filaments << anthers. **FR:** < perianth, obovate-obtuse, dark brown, shiny, gen aborted. **SEED:** 0.5–0.8 mm, plump; appendages 0. Dry or moist soils; < 1800 m. s SCoRO, SW (exc ChI). Jul–Nov

J. tiehmii Ertter (p. 1377) TIEHM'S RUSH (Group 1) Ann densely cespitose, 0.5–6 cm. **ST:** 0.1–0.2 mm wide. **LF:** basal, < 2.5 cm; sheath appendages 0. **INFL:** fls 1–7 per st; bracts 0.6–1.5 mm, acute to acuminate. **FL:** perianth parts gen 4, 1–2.9 mm, ± equal, pale green or pink; stamens 2–3, filaments > anthers, anthers < 0.7 mm. **FR:** gen > perianth, ± ellipsoid to oblong, same color as perianth. **SEED:** 0.3–0.5 mm, obviously striate; appendages small. 2*n*=34. Bare, moist granitic sand of seeps, streambanks, meadows; 300–3100 m. CaRH, SNH, SCoRO, SnGb, SnBr, PR, MP; to WA, ID, NV, Baja CA. May–Aug

J. torreyi Coville (p. 1377) TORREY'S RUSH (Group 3) Per 30–100 cm; rhizome thin, creeping, with narrow tubers. **ST:** stout. **LF:** sheath appendages prominent, thin; blade cylindric, crosswalls complete. **INFL:** lowest bract > infl; clusters 1–20, 25–80-fld, crowded. **FL:** perianth parts slender, tapered to rigid points, sepals 4–5.6 mm, petals 3.5–5 mm; stamens 6, filaments > anthers. **FR:** ± = perianth, thin, 3-angled, tapered to point, seeds below middle only. **SEED:** 0.4–0.5 mm, oblong; appendages 0. 2*n*=40. Meadows, moist woodland; < 1800 m. Teh, SnJV, CCo, SCo, WTR, SnGb, PR (exc SnJt), GB (exc Wrn), D; to BC, e Can, GA. Jun–Sep

J. triformis Engelm. (p. 1377) YOSEMITE DWARF RUSH (Group 1) Ann, densely cespitose, 1.8–16.5 cm, gen ± red in fr. **ST:** < 0.4 mm wide, hair-like. **LF:** basal, < 1/3 st length; sheath appendages 0. **INFL:** fls 1–8 per st; bracts 0.7–2.4 mm. **FL:** perianth parts 4–6, 1.3–4.5 mm, sepals gen < petals, acuminate, midvein area green to red; stamens 2–3, filaments < anthers, anthers 0.9–2 mm. **FR:** gen < perianth, spheric to oblong, same color as perianth; chambers 2–3. **SEED:** 0.3–0.7 mm, striate at 15×. 2*n*=36. Vernal pools, swales, volcanic and granitic seeps; 50–2500 m. CaR, SN, SCo, SnBr, PR. May–Sep

J. uncialis Greene (p. 1377) INCH-HIGH RUSH (Group 1) Ann, densely cespitose, 0.3–3.5 cm, pale, not ± red in fr; gen self-pollinating. **ST:** 0.1–0.5 mm wide. **LF:** basal, < 3/4 st; sheath appendages 0. **INFL:** fl 1 per st; bract 1, sheathing st, 0.2–0.9 mm, truncate. **FL:** perianth parts (4)6(8), 1.7–5.3 mm, sepals gen > petals, dull; stamens 2–3, filaments > anthers, anthers < 0.7 mm. **FR:** ± ≥ perianth, ovoid or elliptic; chambers 2–4. **SEED:** 0.3–0.4 mm, ovoid, not striate; appendages 0 or small. 2*n*=32. Uncommon. Vernal pool margins, other drying places; 45–1700 m. NCoRI, CaRF, c SNF, GV, SCoRO, SCo; to WA, NV. Apr–Jul

J. usitatus L.A.S. Johnson (p. 1377) AUSTRALIAN RUSH (Group 2) Per, densely cespitose. **ST:** < 115 cm, 1–4.7 mm wide above sheath, rough when fresh, < 25 fine ridges per side when dry, pith chambered (unchambered). **LF:** upper sheaths 8–22 cm, base shiny, mid- to dark-brown, distally green to pale brown, dull, gen loose, veins gradually converging at narrowly symmetrical summit, margins thin, flat, pale, overlapping halfway; **FL:** perianth 1–1.9 mm, << and pressed to fruit, green, pale brown, or ± red, scarious margins wide, sepals acute, petals blunt; stamens 3, filaments ≥ anthers. **FR:**

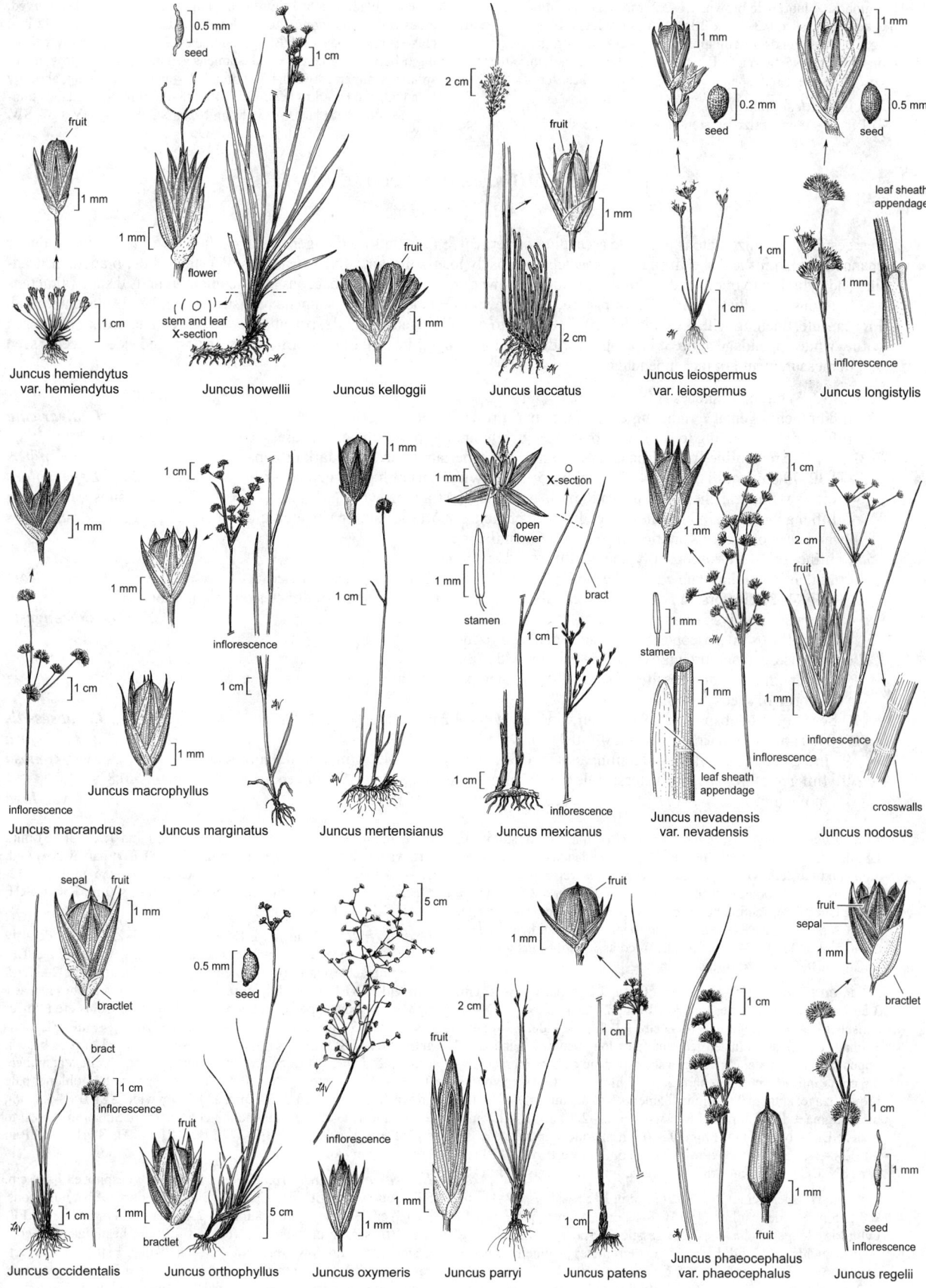

seed
0.5 mm
fruit
1 mm
1 cm
Juncus hemiendytus
var. *hemiendytus*

1 mm
flower
stem and leaf
X-section
Juncus howellii

1 cm

fruit
1 mm
Juncus kelloggii

2 cm
fruit
1 mm
2 cm
Juncus laccatus

1 mm
seed
0.2 mm
1 cm
1 cm
Juncus leiospermus
var. *leiospermus*

1 mm
seed
0.5 mm
leaf sheath
appendage
1 cm
1 mm
inflorescence
Juncus longistylis

1 mm
1 cm
inflorescence
inflorescence
Juncus macrandrus

1 cm
1 mm
1 mm
Juncus macrophyllus

1 cm
1 cm
Juncus marginatus

1 mm
1 cm
Juncus mertensianus

1 mm
open
flower
X-section
1 mm
stamen
bract
1 cm
1 cm
inflorescence
Juncus mexicanus

1 cm
1 mm
stamen
1 mm
1 mm
leaf sheath
appendage
inflorescence
Juncus nevadensis
var. *nevadensis*

1 cm
2 cm
fruit
1 mm
inflorescence
crosswalls
Juncus nodosus

sepal
fruit
1 mm
bractlet
bract
1 cm
inflorescence
1 cm
Juncus occidentalis

0.5 mm
seed
fruit
1 mm
bractlet
5 cm
Juncus orthophyllus

5 cm
inflorescence
1 mm
1 mm
Juncus oxymeris

fruit
2 cm
1 mm
1 cm
Juncus parryi

fruit
1 mm
1 cm
1 cm
Juncus patens

1 cm
1 mm
fruit
Juncus phaeocephalus
var. *phaeocephalus*

fruit
sepal
1 mm
bractlet
1 cm
1 mm
seed
inflorescence
Juncus regelii

± globose, blunt, pale brown, 1.1–2.2 mm, inner partitions concave, chambers 1; seed 0.3–0.5 × 0.15–0.25 mm wide, ± oblong; appendages minute. Ponds, floodplains, seeps, wet disturbed ground; < 350 m. CaRF, n SNF, ScV, n SnJV; Australia, New Zealand, introduced in UK. Invasive, gen mistaken for *J. effusus* and allies. Jun–Oct

J. xiphioides E. Mey. (p. 1377) IRIS-LEAVED RUSH (Group 4) Per 40–80 cm; rhizome stout, creeping. **LF:** sheath appendages obscure; blade gen 5–14 mm wide, flat, with edge toward st, curved, crosswalls gen incomplete, tip long. **INFL:** lowest bract < 1/2 infl; clusters many, 3–10-fld. **FL:** perianth parts 2.5–4 mm, narrow, revealing fr between, green to ± red; stamens gen 6, filaments < to > inconspicuous anthers; style 0.4–0.9 mm. **FR:** ≥ perianth, oblong, abruptly (gradually) tapered to beak. **SEED:** 0.45–0.6 mm; appendages minute. 2*n*=40. Wet places; < 2100 m. NCo, NCoR, SN, GV, CW, SW, W&I, D; to NV, NM, Baja CA. Jul–Oct

LUZULA HAIRY WOOD RUSH

Jan Kirschner

Per, cespitose or rhizomatous, rhizome ascending or vertical. **ST:** cylindric. **LF:** gen basal, cauline few; blades linear, flat or channeled, margins and sheath opening, sparsely to densely long-soft-hairy. **INFL:** panicles of 1–few fls per branch, or head-like and cylindric to ovoid in dense to loose clusters; lower bract lf-like or herbaceous at base, membranous distally, bractlets 1–3, margins gen ciliate. **FL:** perianth parts 6, pale brown to ± black-brown; stamens 6; pistil 1, chamber 1, placenta basal. **FR:** capsule, opening with 3 valves. **SEED:** 3, elliptic to ovoid, ridged on 1 side, occ attached to placenta by tuft of hairs, gen with ± white appendage at tip. ± 115 spp.: worldwide. (Latin: light; Italian: glow worm) Measure unridged side of seeds; seed length measurement not incl appendage.

1. Fls 1(2–3) on infl branches
  2. Infl branches gen at ± right angles; perianth part tip ± reflexed . . . . ***L. divaricata***
  2′ Infl branches nodding to spreading but not at ± right angles; perianth parts appressed to erect, not reflexed
    3. Pl < 30 cm; cauline lvs 2–3, basal lvs 3–4 mm wide; perianth parts gen dark brown . . . . ***L. piperi***
    3′ Pl 30–70 cm; cauline lvs (3)4–5, basal lvs 5–8 mm wide; perianth parts gen straw-brown . . . . ***L. parviflora***
      4. Infl few-fld, branches long; perianth parts gen pale- to straw-brown. . . . . subsp. ***fastigiata***
      4′ Infl many-fld, main branches long, distal 2° branches short; perianth parts medium brown . . . . subsp. ***parviflora***
1′ Fls in dense, ovoid to cylindric clusters, or entire infl congested
  5. Cauline lf tips microscopically gen not thickened or rounded
    6. Infl dense, gen interrupted at base, nodding; basal lvs 1.5–3 mm wide, channeled . . . . ***L. spicata***
    6′ Infl of 4–20 clusters on peduncles of variable length, not nodding, some peduncles occ arching or ±
       flexuous; basal lvs 3–6 mm wide, flat. . . . . ***L. subcongesta***
  5′ Cauline lf tips microscopically gen thickened and rounded
    7. Infl dense, ± black-brown; seeds 0.5–0.6 mm wide . . . . ***L. orestera***
    7′ Infl of peduncled or ± sessile clusters, if dense then pale brown to straw-brown, otherwise not darker
       than brown; seeds > 0.7 mm wide
      8. Seeds ovoid-subspheric, 0.9–1.1 mm wide; style 0.6–1.2 mm . . . . ***L. subsessilis***
      8′ Seeds ovoid or narrowly ovoid; 0.7–0.9 mm wide, style 0.4–0.7 mm . . . . ***L. comosa***
        9. Infl gen dense; style 0.4–0.5 mm; anthers 0.6–1 mm; seeds 1–1.2 mm, 0.8–0.9 mm wide. . . . . var. ***comosa***
        9′ Infl gen of peduncled clusters; style 0.5–0.7 mm; anthers 0.8–1.2(1.5) mm; seeds 1.1–1.3 mm, 0.7–0.8
           mm wide. . . . . var. ***laxa***

L. comosa E. Mey. Cespitose, < 60 cm, rhizome short; stolons 0. **LF:** flat, ciliate, tip obtuse, thickened. **INFL:** peduncled to congested, elongated clusters with basal fls gen remote, lower part of clusters gen loose; lower bract herbaceous. **FL:** perianth parts 2.5–4 mm, ± equal, lanceolate, acuminate, pale straw-brown to pale brown in age. **FR:** ± < perianth. **SEED:** seeds ovoid to narrowly ovoid; appendage 0.4–0.5 mm. 2*n*=12,24. Occ misidentified as *L. subsessilis*. Pls from SnGb, SnBr, SnJt need further study.

var. ***comosa*** (p. 1377) Pl 25–60 cm. **LF:** basal lvs 8–13 cm, 3.5–6 mm wide; cauline 3–4, 6–10 cm, 2–4.5 mm wide. **INFL:** densely to loosely congested, 1.5–3 cm, ± 1–1.5 cm wide, ± 8–15-fld, ± small-lobed, basal cluster gen remote or long-peduncled in axil of upper cauline lf, distal clusters gen short-peduncled; lower bract 1.5–4.5 cm, > infl. **FL:** perianth parts 2.5–3.7 mm, pale straw-brown to ± brown in age; anthers 0.6–1 mm, filaments 0.5–0.6 mm, style 0.4–0.5 mm, stigma ± 1.3–1.8 mm. **FR:** valves ± 2.3–2.8 mm, 1.4–1.8 mm wide. **SEED:** ovoid, 1–1.2 mm, 0.8–0.9 mm wide; appendage ± 0.5 mm. 2*n*=24. Meadows, open woodland, conifer forest; gen < 1500 m. NW, SN, CCo, SnFrB; to AK. [*L. c.* var. *congesta* S. Watson] Jun–Jul

var. ***laxa*** Buchenau Pl 13–35 (50) cm. **LF:** basal lvs ± 7–11 cm, ± 3.5–5 mm wide; cauline (1)2(3), 7–12 cm, 2–6 mm wide. **INFL:** cylindric, elongated clusters, 1(4) ± sessile, 2–6 peduncled, 8–14 mm, 5–7 mm wide, 5–12-fld, basal fls gen ± remote; peduncles erect to erect-spreading; lower bract < 5 cm, gen < infl. **FL:** perianth parts (3.3)3.5–4 mm, straw-brown to pale brown; anthers 0.8–1.2(1.5) mm, filaments 0.4–0.6 mm, style 0.5–0.7 mm, stigma 1.5–2.3(2.8) mm. **FR:** valves 2.5–3 mm, 1.6–1.8 mm wide. **SEED:** narrowly ovoid, 1.1–1.3 mm, 0.7–0.8 mm wide; appendage 0.4–0.5 mm. 2*n*=12. Meadows, open woodland, conifer forest; 50–2800 m. KR, CaRH, SNH, GB; to BC, MT. Jun–Jul

L. divaricata S. Watson (p. 1377) Pl (10)15–25(30) cm, densely cespitose, pale green, ± glabrous; rhizome short, ± ascending. **LF:** blade flat, acute; basal lvs < 20 cm, 4–6 mm wide; cauline 2–3, 3–5 mm wide. **INFL:** many-fld, loose, ± 7–10 cm diam, branches gen at ± right angles, distal peduncles gen > 1–1.5 cm; fls not clustered; lower bract ± 1–2.5 cm, ± purple-straw-brown to ± green, << infl. **FL:** perianth parts (1.8)2–2.2(2.4) mm, equal, straw-colored to pale brown, gen ± red-tinged, ± lanceolate, long-acuminate; tip ± reflexed; anthers 0.4–0.6 mm, filaments ± 0.4–0.5 mm. **FR:** > perianth, ± oblong, pale brown, acute; valves 1.9–2.2 mm, 1–1.2 mm wide. **SEED:** 1–1.2 mm, ± 0.6 mm wide; appendage indistinct, < 0.1 mm. Subalpine forest to alpine granitic slopes; 2100–3700 m. KR, CaRH, SNH; CA-FP in OR, NV. Jun–Aug

L. orestera Sharsm. (p. 1377) Pl densely cespitose, 7–14 cm. **ST:** stout, erect. **LF:** flat, tip obtuse, thickened; basal lvs numerous, 3–5(8) cm, 3–4 mm wide; cauline 1–2, 3–4 cm, 3–4 mm wide. **INFL:** ± 1–4(6) sessile clusters, gen (0.5)0.8–1.2 cm, heads densely congested, many-fld; lower bracts 1–3, > infl, lf-like, ± stiff, ± red, erect-spreading. **FL:** perianth parts ± 2.4–2.9 mm, ± equal, lanceolate, acuminate, dark brown, margins pale, distinct; anthers 0.4–0.6 mm,

filaments 0.5–0.6 mm. **FR**: < perianth, ± obtuse, dark brown; valves 1.8–2 mm, ± 1.1 mm wide. **SEED**: 0.8–0.9(1) mm, 0.5–0.6 mm wide; appendage 0.15–0.2 mm. $2n=20,22$. Alpine and subalpine meadows, fell-fields; 2700–3600 m. SNH. Possibly in NV, OR. Jul–Aug

L. parviflora (Ehrh.) Desv. Pl loosely cespitose, 30–70 cm; rhizome ascending; stolons short (0). **LF**: flat, acute; basal lvs 4–8 mm wide; cauline (3)4–5, to 7–10 cm, 3–8 mm wide, lanceolate, acuminate. **INFL**: loose, occ nodding, main branches occ to 13 cm, fls 1; lower bract 1.5–4 cm, < infl. **FL**: perianth parts 1.8–2.3 (2.4) mm, acute. **FR**: ≥ perianth, oblong-elliptic, acute to acuminate, awn 0.1–0.2 mm; valves 1.9–2.4 mm, 1–1.3 mm wide. **SEED**: 1.1–1.3 mm, 0.6–0.7 mm wide; appendage 0–0.1 mm. $2n=24$. Intermediates known between subspp.

subsp. ***fastigiata*** (E. Mey.) Hämet-Ahti **LF**: basal lvs 4–8 mm wide; cauline to 10 cm, 4–7 mm wide. **INFL**: few-fld; main axis, distal 2° branches long. **FL**: perianth parts 1.8–2.2 mm, gen pale brown to straw-colored; anthers 0.3–0.5 mm, filaments 0.4–0.5 mm. **FR**: valves 2.1–2.4 mm, 1.1–1.3 mm wide. Moist places in conifer woodland, streambanks; 1000–3100 m. KR, NCoR, c SNH; to AK, WY, UT. [*L. divaricata*, misappl., in part] Further study needed. Jun–Aug

subsp. ***parviflora*** (p. 1377) **LF**: basal lvs to 7–9 cm, 5–8 mm wide. **INFL**: many-fld; main branches long, distal 2° branches short. **FL**: perianth parts 1.9–2.3(2.4) mm, medium-brown; anthers 0.4–0.5 mm, filaments ± 0.5 mm. **FR**: valves 1.9–2.3 mm, 1–1.3 mm wide, dark brown. Moist places in conifer woodland, streambanks, subalpine grassland; 1000–3300 m. NW (exc NCoRI), CaRH, SN, Wrn; to AK, Can, ne US; circumboreal. Jun–Aug

L. piperi (Coville) M.E. Jones Pl 20–30(35) cm, cespitose, ± blue-green, ± glabrous. **LF**: tip acute; basal lvs 5–10 cm, 3–4 mm wide; cauline 2–3, to 3–7 cm, 3–5 mm wide. **INFL**: conspicuously nodding, gen ± 5 cm, 3 cm wide, gen 30–60-fld, fls not clustered; lower bract to ± 1 cm, ± brown, ciliate. **FL**: perianth parts (1.8)2–2.3(2.5) mm, equal, ± lanceolate, acute, gen dark brown; anthers (0.3)0.4–0.6(0.8) mm; filaments ± 0.5 mm. **FR**: capsule > perianth, oblong-ellipsoid, ± acuminate to acute, ± dark-brown; valves 1.8–2.2(2.4) mm, 1–1.2 mm wide. **SEED**: seeds pale brown, 1–1.2 mm, 0.5–0.6 mm wide;

appendage indistinct. $2n=24$. Areas of late-melting snow; 2500–2600 m. KR; to AK, w MT; e Asia. Further study needed. Jul

L. spicata (L.) DC. (p. 1377) Pl densely cespitose, 5–30 cm. **LF**: channeled; basal lvs numerous, gen to 1.5–3 mm wide; cauline acute. **INFL**: nodding, gen ± interrupted at base, clusters dense; middle bract lance-ovate, acuminate, membranous; lower bract ≤ infl, ± green- to brown-membranous. **FL**: perianth parts gen 2.2–2.8(3.3) mm, lance-oblong, acuminate, brown; anthers (0.4)0.5–0.6(0.7) mm, ± = filaments. **FR**: valves gen 1.6–2.3 mm, < perianth. **SEED**: 0.9–1.1 mm, ± 0.6(0.7) mm wide; appendage 0.10–0.15 mm. $2n=24$. Alpine slopes, on wind-eroded acidic soils; 2900–3700 m. SNH, Wrn, SNE (Sweetwater Mtns); to AK, MT, CO; NY to ME; circumpolar. Variable, further study needed in CA-FP, GB. Jul–Aug

L. subcongesta (S. Watson) Jeps. (p. 1377) Pl loosely tufted, 20–40 cm, ± glabrous, ± blue-green; rhizome horizontal. **LF**: flat, tip acute; basal lvs to 10(15) cm, 3–6 mm wide; cauline 3–4, 3–5 cm, 3–6 mm wide. **INFL**: to 4–6 cm; not nodding; peduncles branched distally, thin, occ ± flexuous to arching; clusters 4–20, 3–7-fld; lower bracts to 1.5 cm, ciliate, herbaceous. **FL**: perianth parts 1.7–2.1 mm, equal, lance-ovate, dark brown; anthers 0.3–0.6 mm, filaments 0.5–0.7 mm. **FR**: < perianth, ovoid, tip ± conic; valves 1.6–1.8 mm, dark brown, lustrous. **SEED**: 1.1–1.2 mm, 0.6 mm wide, narrowly ellipsoid, medium brown; appendage 0. $2n=24$. Alpine to subalpine moist or wet places; 2000–3500 m. KR, CaRH, SNH. Jul–Aug

L. subsessilis (S. Watson) Buchenau (p. 1377) Pl cespitose, 20–30(40) cm; rhizome short; stolons ± 0. **LF**: flat, tip obtuse, thickened; basal lvs ± 7–14 cm, 3–4.5 mm wide, cauline 1–3, (3.5)5–9(11) cm, to ± 4.5 mm wide. **INFL**: 1–2 clusters ± sessile, (0)1–3(6) peduncled, 0.6–1.3 mm, 0.7–1 mm wide, broadly ovoid to ± cylindric, (3)5–9(13)-fld; lower bract 1.5–7 cm, gen < (>) infl. **FL**: perianth parts (3.1)3.4–4(5) mm, ± equal, lanceolate, acuminate to acute, ± dark chestnut-brown; anthers (1.1)1.5–2 mm, filaments 0.4–0.6 mm; style 0.6–1.2 mm, stigmas 2–3.5 mm. **FR**: < perianth, obovoid, pale brown, ± obtuse; valves ± 2.6–2.9 mm, 1.8–2 mm wide. **SEED**: 1.2–1.3 mm, 0.9–1.1 mm wide, ovoid-subspheric; appendage 0.2–0.3 mm. $2n=12$. Open, drier woodland; < 2600 m. NCoR, c&s SNF, SNH, CCo; to AK. Further study needed. Apr–Jul

JUNCAGINACEAE ARROW-GRASS FAMILY

David J. Keil

Ann, per, rhizomed or dense-tufted, terrestrial or aquatic. **LF**: gen ± basal, alternate, ± flat to narrow-cylindric; sheath open, gen liguled. **INFL**: spike or gen ± scapose raceme; bracts 0. **FL**: gen bisexual; perianth parts gen 6 in 2 whorls (0, 1) [3, 4], free, scale-like, ± green or tinged ± red-purple; stamens 0, 1, 3, or 6, filaments short, ± fused to perianth; pistil 1, carpels 1, 6 [3, 4], ± fused, separating in fr [or not], each with 1 chamber and 1 ovule or 3 carpels fertile, 3 sterile, placentas basal, stigmas gen ± sessile. **FR**: achene or gen mericarps. 3 genera, 30 spp.: temp, circumboreal, Australia, s Afr, S.Am. [Haynes & Hellquist 2000 FNANM 22:43–46; von Mering & Kadereit 2010 *in* Seberg (ed.), Diversity, Phylogeny, and Evolution in the Monocotyledons, Aarhus Univ Press] *Lilaea* incl in *Triglochin*. Scientific Editor: Thomas J. Rosatti.

TRIGLOCHIN ARROW-GRASS

Pl ± glabrous. **LF**: basal, ± tufted; sheath membranous; ligule tip entire to 2-lobed [0]. **FL**: perianth parts gen adaxially concave; anthers ± sessile; stigma papillate or ± plumose. **FR**: achene or gen 3, 6 mericarps. **SEED**: 1, linear, ± flat or angled. ± 20 spp.: temp, circumboreal, Australia, s Afr, S.Am. (Greek: 3 points, from frs of some) TOXIC when fresh, from cyanogenic compounds.

1. Ann; infl a gen emergent spike of bisexual and sometimes staminate fls + 2 submerged, long-styled sessile
    pistillate fls enclosed in lf sheath at base of scape; perianth parts 0 or 1; stamens 0 or 1; carpels 0 or 1, fertile ***T. scilloides***
1′ Per; infl an aerial raceme of bisexual fls; perianth parts gen 6; stamens 1, 3, 6, occ fewer; carpels 6, all or 3 fertile
   2. Fertile carpels, mericarps 3
      3. Fr 5–7(9) mm; mericarps weak-ridged abaxially; rhizome ± stout; pl tufted; c&s SNH, 2100–3450 m . . . . ***T. palustris***
      3′ Fr 1–1.5 mm; mericarps strong-3-keeled abaxially; rhizome slender; pl mat-forming; NCo, CCo, n SCo,
          < 30 m . . . . ***T. striata***
   2′ Fertile carpels, mericarps 6
      4. Pl dense-tufted, (1)4–11 dm; rhizome stout; ligule tip entire to ± notched, (1)1.5–5 mm; lf ± elliptic in
          ×-section, gen 2–5 mm wide . . . . ***T. maritima***

4′ Pl loose-tufted to mat-forming, 1–6 dm; rhizome ± slender, often creeping; ligule tip deep-2-lobed,
　　0.5–1.5 mm; lf ± round or semicircular in ×-section, gen 1–2 mm wide . . . . ***T. concinna***
　5. Pl gen 1–3(4.5) dm; infl gen ≥ lvs; salt marshes; NCo, CCo, SCo . . . . var. ***concinna***
　5′ Pl (1)3–5(6) dm; infl gen ≫ lvs; alkaline meadows, seeps, mudflats, stream and lake margins; ne KR,
　　n CaRH, s SNH, GB, DMoj . . . . var. ***debilis***

T. concinna Burtt Davy　Per 1–6 dm, loose-tufted to mat-forming; rhizomes often creeping, ± slender. **LF:** 5–30 cm, gen 1–2 mm wide, ± round or semicircular in ×-section; ligule 0.5–1.5 mm, tip deep-2-lobed. **INFL:** aerial raceme, ≥ lvs; pedicels 3–7 mm in fr, ascending. **FL:** perianth parts gen 6, 1–2 mm; stamens gen 6; fertile carpels 6. **FR:** mericarps 6, 3–6(7) mm, fully separating, abaxially smooth or weak-ridged. Some authors merge *T. concinna, T. maritima*, but pls of n CCo very different in habit, stature, and grow together without intergrading. Some pls of interior may be more difficult to distinguish from *T. maritima*, need study.

　var. ***concinna***　SEASIDE ARROW-GRASS　Pl gen 1–3(4.5) dm. **INFL:** gen ≥ lvs. 2*n*=48. Coastal salt marshes; < 10 m. NCo, CCo, SCo; to BC, Baja CA. Mar–Aug

　var. ***debilis*** (M.E. Jones) J.T. Howell　Pl (1)3–5(6) dm. **INFL:** gen ≫ lvs. 2*n*=96. Alkaline meadows, seeps, mudflats, stream and lake margins; 400–2500 m. ne KR, n CaRH, s SNH, GB, DMoj; circumboreal, S.Am. Mar–Aug

T. maritima L. (p. 1377)　COMMON ARROW-GRASS　Per (1)4–11 dm, dense-tufted; rhizomes stout. **LF:** 1–8 dm, gen 2–5 mm wide, ± elliptic in ×-section; ligule (1)1.5–5 mm, tip entire to ± notched. **INFL:** aerial raceme, gen > lvs; pedicels 3–6 mm in fr, ascending. **FL:** perianth parts gen 6, ± 2 mm; stamens gen 6; fertile carpels 6. **FR:** mericarps 6, 3–8 mm, fully separating, abaxially smooth or weak-ridged. 2*n*=12,24,30,36,48,60,120,144. Coastal salt marshes, interior saline, brackish, alkaline marshes; < 2800 m. NCo, KR, CaRH, SN, sw ScV, CCo, SnFrB, SCo, SnBr, GB; circumboreal, S.Am. Apr–Aug

T. palustris L. (p. 1377)　MARSH ARROW-GRASS　Per 1–3(5.5) dm, tufted; rhizomes ascending, ± stout. **LF:** 5–30 cm, 0.5–2 mm wide, ± elliptic in ×-section; ligule 0.5–1.5 mm, tip gen deep-2-lobed. **INFL:** aerial raceme, gen 1–2 per pl, gen > lvs; pedicels gen 4–6 mm in fr, erect. **FL:** perianth parts gen 6, 1.5–2 mm; stamens 6; fertile carpels 3. **FR:** mericarps 3, 5–7(9) mm, distally clinging to fr axis, abaxially weak-ridged. 2*n*=24. Wet meadows, wet flats, stream and lake margins; 2100–3450 m. c&s SNH; circumboreal, S.Am, New Zealand. Jul–Aug ★

T. scilloides (Poir.) Mering & Kadereit (p. 1377)　FLOWERING-QUILLWORT　Ann, tufted, in youth submersed, in age ± emergent; roots fibrous. **LF:** 5–20(45) cm, 1–5 mm wide, ± elliptic to ± round in ×-section; ligule ± 1 mm, tip ± obtuse or ± notched. **INFL:** ± scapose, gen emergent, ± dense spike of bisexual and sometimes staminate fls, (2)6–20 cm, + 2 submersed, sessile pistillate fls enclosed in lf sheath at base of scape. **BISEXUAL FL:** perianth part 1, 2–3 mm; stamen 1; carpel 1, style gen short. **STAMINATE FL:** perianth part 1, 2–3 mm; stamen 1. **PISTILLATE FL:** perianth 0; carpel 1, style 6–20 cm, thread-like, stigma floating, head-like. **FR:** achene, 2–10 mm, ribbed, ± flat, ± winged in bisexual fls, angled in pistillate fls, tip truncate with off-center beak. 2*n*=12. Vernal pools, streams, ponds, lake margins; < 1700 m. NCo, NCoRI, SN, GV, CW, SCo, PR, GB; to w Can, MT, Mex, Chile; naturalized in Australia, Iberian Peninsula. [*Lilaea s.* (Poir.) Hauman] Previously in *Lilaea*, yet highly nested in *Triglochin*, a paraphyletic genus made monophyletic by inclusion of this sp. (von Mering & Kadereit 2010). Mar–Oct

T. striata Ruiz & Pav. (p. 1377)　THREE-RIBBED ARROW-GRASS　Per 1–2(4.5) dm, mat-forming; rhizomes spreading to ascending, slender. **LF:** 5–20(45) cm, 1(2) mm wide, ± elliptic in ×-section; ligule 1–2.5 mm, tip rounded to acuminate. **INFL:** aerial raceme, 5–25(32) cm, < to > lvs; pedicels 0.4–2.1(3.5) mm in fr, spreading to ascending. **FL:** perianth parts (< 6)6, (< 0.6)0.6–1 mm; stamens (< 6)6, often unequal; fertile carpels 3. **FR:** mericarps 3, 1–1.5 mm, fully separating, abaxially strong-3-keeled. 2*n*=18,24. Uncommon. Brackish to freshwater coastal marshes, springs; < 30 m. NCo, CCo, n SCo; to WA, se US; Chile, Australasia, Afr, Madagascar, naturalized w Eur. May–Sep

LAXMANNIACEAE　WIRE-LILY FAMILY

Dale W. McNeal

Per, ± palm- or tree-like, trunks [0]1–several. **ST:** [0] tall, slender, woody [or not], with obvious lf scars. **LF:** [basal] cauline, alternate, gen crowded near st tip, sessile or ± narrowed basally, ± clasping. **INFL:** terminal [axillary], panicles [(racemes), umbels, large cymes, or fls 1], sessile or ± so. **FL:** perianth parts 6, in 2 petal-like whorls, fused at base; stamens 6, equal [unequal], anthers attached at middle [base]; ovary superior, 3-chambered, style 1, stigma minutely 3-lobed. **FR:** berry [loculicidal or indehiscent capsule]. 14–15 genera, 178 spp.: Australia, Asia, S.Am. Scientific Editor: Thomas J. Rosatti.

CORDYLINE　CABBAGE TREE

ST: trunks increasing in diam in age. **LF:** long-lived, crowded in tufts of 200+ distally, 30–100 cm, long-linear to narrow-elliptic, narrowed just above base but not petioled. **INFL:** terminal (but soon made lateral by dominance of axillary shoot), erect or drooping; bracts 3, > panicle, fls ± sessile, subtended by 3 scale-like bractlets. **FL:** small, radial, ± white, perianth tube short; stamens 6, fused to perianth base; ovules several to many. **FR:** dry in age. **SEED:** black. ± 15 spp.: Australasia to New Zealand. (Greek: club, from thickened fleshy roots)

C. australis (G. Forst.) Endl.　NEW ZEALAND CABBAGE TREE　**ST:** 1–2[5–15] m. **LF:** 80–100 cm, 2.5–5 cm wide, acute to acuminate, ± flat, stiff. **FL:** 6–10 mm wide, fragrant. **FR:** white. *n*=19. Disturbed and/or formerly cult areas; < 50 m. NCo, CCo, PR; New Zealand. Apr–Jun ❖

LILIACEAE　LILY FAMILY

Dale W. McNeal, except as noted

Per from membranous bulb or scaly rhizome. **ST:** underground or erect, branched or not. **LF:** basal or cauline, alternate, sub-opposite, or whorled. **INFL:** raceme, panicle, ± umbel-like or not. **FL:** perianth parts 6 in 2 gen petal-like whorls, often showy; stamens 3 or 6, filaments free or ± fused to perianth, anthers attached at base or near middle; ovary superior or ± so, style 1,

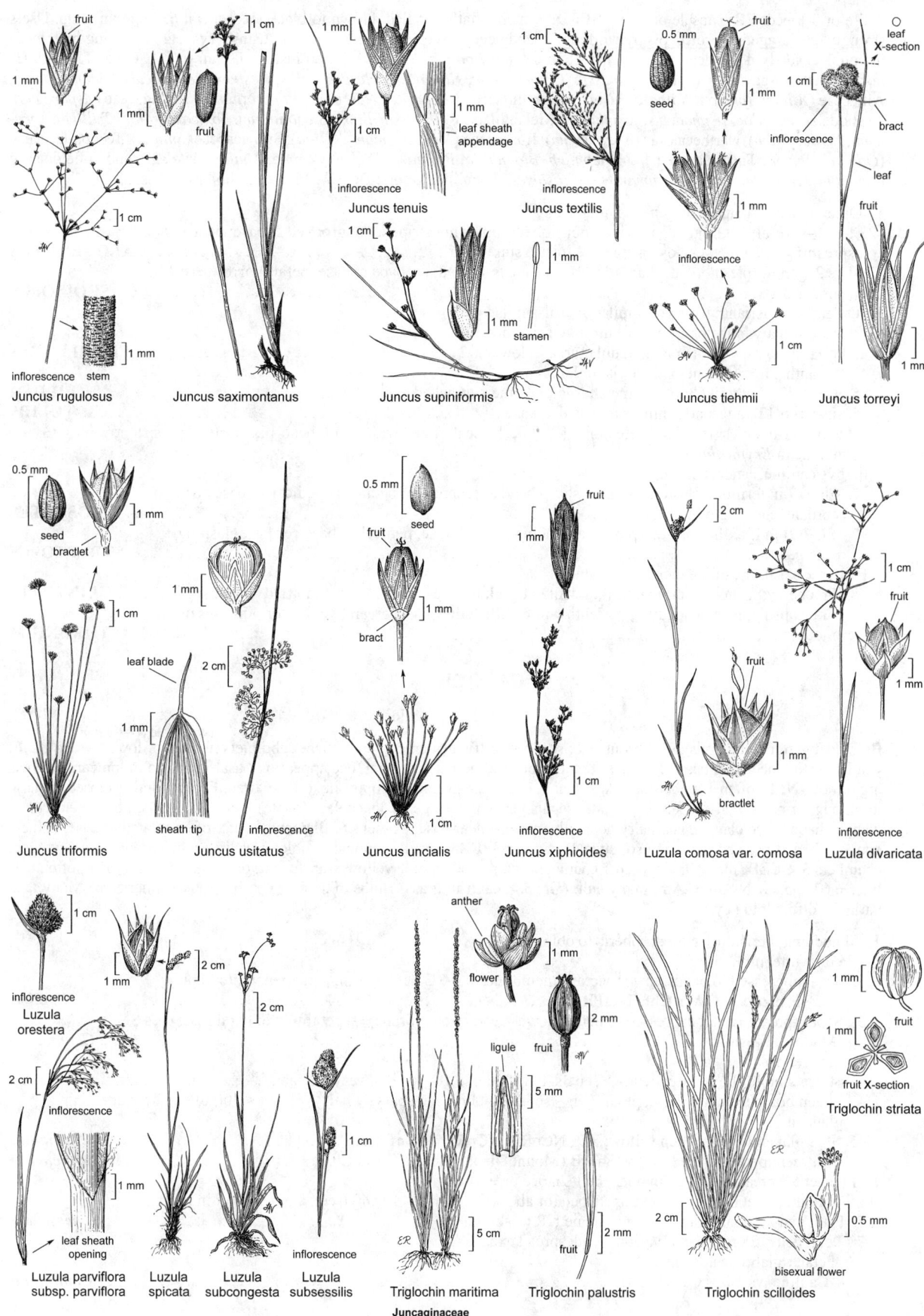

fruit

1 mm

1 cm

inflorescence   stem

Juncus rugulosus

1 mm

fruit

Juncus saximontanus

1 mm

1 cm

inflorescence

Juncus tenuis

1 mm

leaf sheath appendage

1 cm

1 mm

stamen

Juncus supiniformis

1 cm

inflorescence

Juncus textilis

0.5 mm

seed

fruit

1 mm

1 mm

inflorescence

1 cm

Juncus tiehmii

leaf X-section

1 cm

inflorescence

bract

leaf

fruit

1 mm

Juncus torreyi

0.5 mm

seed

bractlet

1 cm

Juncus triformis

1 mm

leaf blade

1 mm

sheath tip

2 cm

inflorescence

Juncus usitatus

0.5 mm

seed

fruit

1 mm

bract

1 cm

Juncus uncialis

fruit

1 mm

inflorescence

1 cm

Juncus xiphioides

2 cm

fruit

1 mm

bractlet

Luzula comosa var. comosa

1 cm

fruit

1 mm

inflorescence

Luzula divaricata

1 cm

inflorescence

Luzula orestera

1 mm

2 cm

2 cm

2 cm

inflorescence

leaf sheath opening

1 mm

Luzula parviflora subsp. parviflora

Luzula spicata

Luzula subcongesta

1 cm

inflorescence

Luzula subsessilis

anther

flower

1 mm

ligule

fruit

2 mm

5 mm

fruit

2 mm

5 cm

Triglochin maritima

Triglochin palustris

1 mm

fruit

1 mm

fruit X-section

Triglochin striata

2 cm

bisexual flower

0.5 mm

Triglochin scilloides

Juncaginaceae

entire or 3-lobed. **FR**: capsule or berry. **SEED**: 3–many, flat or angled, brown to black. 16 genera, 635 spp.: n temp. Users strongly encouraged to protect pls by working around need to see underground parts in using keys, e.g., by trying both leads in couplets solely dependent on such characters. *Muscari botryoides* (L.) Mill. an historical waif in CA. Other TJM (1993) taxa moved to Agavaceae (*Agave, Camassia, Chlorogalum, Hastingsia, Hesperocallis, Hesperoyucca, Leucocrinum, Yucca*), Alliaceae (*Allium, Ipheion, Nothoscordum*), Amaryllidaceae (*Amaryllis, Narcissus, Pancratium*), Asparagaceae (*Asparagus*), Asphodelaceae (*Aloe, Asphodelus, Kniphofia*), Melanthiaceae (*Pseudotrillium, Stenanthium, Toxicoscordion, Trillium, Veratrum, Xerophyllum*), Nartheciaceae (*Narthecium*), Ruscaceae (*Maianthemum, Nolina*), Smilacaceae (*Smilax*), Tecophilaeaceae (*Odontostomum*), Themidaceae (*Androstephium, Bloomeria, Brodiaea, Dichelostemma, Muilla, Triteleia*), and Tofieldiaceae (*Triantha*). N.Am spp. of *Disporum* now in *Prosartes*. Scientific Editors: Dale W. McNeal, Thomas J. Rosatti.

1. Outer, inner perianth parts gen dissimilar
  2. Lvs 1–several, not mottled, abaxially not ribbed; sepals not striped, not grooved at base; petals wide, spreading to ± erect, enclosing stamens, pistil; stamens 6 . . . . **CALOCHORTUS**
  2′ Lvs 2, gen purple-mottled, abaxially ribbed; sepals striped, grooved at base; petals narrow, ± erect; stamens 3 . . . . **SCOLIOPUS**
1′ Outer, inner perianth parts gen similar or ± dissimilar
  3. Pls from scaly bulbs or corms or bulb-like rhizomes
    4. Perianth parts with distinct glandular area in lower 1/2 . . . . **FRITILLARIA**
    4′ Perianth parts without distinct glandular area
      5. Lvs > 12, ± whorled (often some alternate); anthers attached at middle. . . . . **LILIUM**
      5′ Lvs 2–6(12), alternate; anthers attached at base . . . . **TULIPA**
  3′ Pls from rhizomes (from elongate bulb of 1–2 fleshy scales, gen with small, bead-like parts of persistent rhizome in *Erythronium*)
    6. Lvs cauline, ± not reduced
      7. Fls 1–7 in terminal clusters, without sharp bend at juncture of peduncle, pedicel; lf base acute to cordate, clasping or not . . . . **PROSARTES**
      7′ Fls 1(2) in lf axils, with sharp bend at juncture of peduncle, pedicel; lf base cordate, clasping . . . . **STREPTOPUS**
    6′ Lvs ± basal, if cauline then much reduced
      8. Fr a berry; fls not nodding; perianth white or pink to rose-purple; lvs not mottled. . . . . **CLINTONIA**
      8′ Fr a capsule; fls gen nodding; perianth white with yellow base, cream, yellow, or pink; lvs often mottled . . . . **ERYTHRONIUM**

CALOCHORTUS

Peggy L. Fiedler

Bulb coat gen membranous, occ fibrous. **ST**: scapose or lfy, gen erect, gen branched, bulblets in axils of lower lvs or 0. **LF**: gen linear to lanceolate; basal lf 1, persistent or not; cauline lvs 0–several, occ appearing basal, gen smaller upward, withering or not. **INFL**: often ± umbel-like; fls 2–many; bracts 0–several, gen opposite, often paired. **FL**: perianth ± closed, spheric to oblong, or open, bell-shaped or ± rotate; sepals gen < petals, gen ± lanceolate (ovate), gen ± glabrous; petals gen widely wedge-shaped, occ clawed, gen hairy adaxially, nectary near base; stamens 6, filaments ± flat, often dilated at base, anthers gen attached at base or appearing so; style 1, stigmas 3. **FR**: capsule, septicidal; oblong or linear, gen 3-angled or -winged, chambers 3. **SEED**: many in 2 rows per chamber, flat, gen ± tan or ± yellow, translucent, or irregular dark brown, often net-like. ± 67 spp.: w N.Am, C.Am; many cult. (Greek: beautiful grass) Bulbs of some eaten by Native Americans. Many taxa variable, difficult to key.

1. Fl nodding; perianth ± closed, spheric to oblong
  2. Petal white, pink, or rose
    3. Sepal appressed to petals; lower nectary membrane 1/3–2/3 petal width; petal gen white to pink; s CaRF, n&c SN, CW, n ChI, TR, PR . . . . *C. albus*
    3′ Sepal spreading, not appressed to petals; lower nectary membrane ± = petal width; petal rose; c&s SN (Madera, Fresno cos.). . . . . *C. amoenus*
  2′ Petal yellow
    4. St gen simple above base; petals > sepals, hairy; pl gen 10–100 cm, very glaucous. . . . . *C. raichei*
    4′ St gen branched above base; petals gen ≤ sepals, glabrous to sparsely hairy; pl gen < 30(50) cm, not very glaucous
      5. Pl ± glaucous; petal deep yellow; KR, NCoRO, NCoRI, nw SnFrB. . . . . *C. amabilis*
      5′ Pl green; petal light yellow; ne SnFrB (Mount Diablo, Contra Costa Co.) . . . . *C. pulchellus*
1′ Fl erect to spreading; perianth open, bell-shaped or ± rotate
  6. Fr nodding; petal ± not spotted or striped (or above-ground st < 10 cm) (sect. *Calochortus* in part)
    7. Petal gen > 30 mm, pink to ± purple; ne KR (Siskiyou Co.). . . . . *C. persistens*
    7′ Petal gen < 28 mm, color various; widespread exc s CA
      8. Petal ± glabrous adaxially
        9. Cauline lvs ± 0; s CaRH, SNH. . . . . *C. minimus*

9′ Cauline lvs present; NW, n&c CCo, SnFrB
  10. St 8–25 cm, branched, bulblets 0; s NCoRO, SnFrB . . . . ***C. umbellatus***
  10′ St gen < 5 cm, gen simple, bulblets present; NW, CaRH, n&c CCo, SnFrB . . . . ***C. uniflorus***
8′ Petal ± conspicuously hairy adaxially
  11. Petal deep yellow — CaRF, sw CaRH, n&c SN . . . . ***C. monophyllus***
  11′ Petal white, pink, purple, or blue
    12. Petal lanceolate to obovate, ± ciliate on sides but not tip
      13. Petal obovate to wedge-shaped; st simple or branched; NW, CaR, n SN, n ScV, n CCo, SnFrB . . . . . . ***C. tolmiei***
      13′ Petal lanceolate; st simple; s SNH (Greenhorn Mtns) . . . . ***C. westonii***
    12′ Petal ± oblanceolate to obovate, ciliate throughout
      14. Petal ± blue (purple crescent 0), ± smooth (not bumpy) adaxially; NW, CaR, SNH . . . . ***C. coeruleus***
      14′ Petal green-white, base dark ± purple, minute-bumpy adaxially; e KR, NCoRI . . . . ***C. elegans***
6′ Fr erect; petal spotted or striped (exc *Calochortus nudus*, occ *Calochortus palmeri*)
  15. Basal lf ± persistent at fl
    16. Petal light yellow-green, flecked purple-brown; n CCo (Ring Mtn, Marin Co.) . . . . ***C. tiburonensis***
    16′ Petal white to purple; KR, CaR, MP
      17. Petal with red-purple crescent above nectary adaxially . . . . ***C. longebarbatus*** var. ***longebarbatus***
      17′ Petal without crescent adaxially
        18. Petal 30–40 mm with dark purple crescent abaxially; nectary deeply depressed; dry places; < 1100 m . . . ***C. greenei***
        18′ Petal 14–16 mm, without crescent abaxially; nectary ± shallow; moist places; > 1200 m . . . . ***C. nudus***
  15′ Basal lf ± withered by fl (exc *Calochortus excavatus*, *Calochortus syntrophus*)
    19. Nectary surface ± glabrous but ± hidden by bordering hairs; bulb coat fibrous — s CW, SW
      20. Petal ± pink, gen toothed, not ciliate . . . . ***C. plummerae***
      20′ Petal tan, yellow, ± purple to red-brown, gen ± ciliate
        21. Petal tip dark-hair-tufted. . . . . ***C. obispoensis***
        21′ Petal tip fringed, not hair-tufted
          22. Petal margin fringed with 2 rows of hairs; anthers abruptly pointed; SCoRO, WTR . . . . ***C. fimbriatus***
          22′ Petal margin fringed with 1 row of hairs; anthers rounded; SCoRO, SCo, WTR, PR . . . . ***C. weedii***
            23. Petal deep yellow. . . . . var. ***weedii***
            23′ Petal light yellow or tan or dark ± purple to red-brown. . . . . var. ***intermedius***
    19′ Nectary surface hairy; bulb coat membranous
      24. Nectary ± round, clearly to ± depressed, encircled by fringed membrane
        25. Petal gen orange to red . . . . ***C. kennedyi*** var. ***kennedyi***
        25′ Petal yellow, gold, or white gen tinged lilac or lavender
          26. Petal white, tinged lilac, or lavender, green-striped abaxially; sepal occ dark-spotted
            27. Petal base marked dark purple; bracts gen 3–8 cm; anthers red-brown — sw SNE. . . . . ***C. excavatus***
            27′ Petal base white or tinged lilac; bracts 2–7 cm; anthers yellow, maroon, blue, purple, red or red-brown
              28. Sepal base not spotted; petal obovate to wedge-shaped. . . . . ***C. invenustus***
              28′ Sepal base dark-spotted; petal narrowly obovate
                29. Petal with red or purple arch above nectary; nectary in yellow spot; n SNH, e edges c&s SNH,
                  MP, s SNE . . . . ***C. bruneaunis***
                29′ Petal without dark arch above nectary; nectary in red or purple spot; n DMtns (Panamint
                  Range) . . . . ***C. panamintensis***
          26′ Petal yellow or gold, not green-striped abaxially; sepal base gen dark-spotted
            30. Perianth bell-shaped; nectary encircled by slender or club-shaped hairs; st straight or occ twisted
              — SnBr, PR, n&e DMtns
              31. Lvs flat or inrolled; nectary encircled by dense, slender hairs; st straight, gen stout; SnBr
                (s slope), PR . . . . ***C. concolor***
              31′ Lvs channeled; nectary encircled by sparse, club-like hairs; st occ twisted; n&e DMtns
                (Panamint, Clark Mtn ranges, Providence Mtns). . . . . ***C. kennedyi*** var. ***munzii***
           30′ Perianth cup-shaped; nectary encircled by club-shaped hairs; st often zigzag. . . . . ***C. clavatus***
            32. Petal gen < sepal; nectary deep — n SNF (El Dorado, Amador cos.). . . . . var. ***avius***
            32′ Petal ≥ sepal; nectary shallow to ± deep
              33. St gen < 30 cm, slender, straight
                34. Petal 30–40 mm, sparsely hairy; lvs not recurved; WTR, SnGb. . . . . var. ***gracilis***
                34′ Petal 40–50 mm, hairy; lvs strongly recurved; s-c CCo (nw San Luis Obispo Co.). . . . . . var. ***recurvifolius***
              33′ St gen 50–100 cm, coarse, zigzag
                35. Petal deep yellow, hairs very knobby; anther deep purple; s SCoRO, n SCoRI, WTR, SnGb . . . var. ***clavatus***
                35′ Petal light yellow, hairs not very knobby; anther yellow to purple; w SnJV (and adjacent
                e SnFrB, SCoRI), n WTR . . . . var. ***pallidus***
      24′ Nectary not simultaneously ± round, clearly to ± depressed, and encircled by fringed membrane (but
        occ 1 of these)
        36. St bulblets gen 0 (exc some *Calochortus splendens*)
          37. Petal purple-veined, not spotted . . . . ***C. striatus***
          37′ Petal spotted, not obviously veined or striped
            38. St twining or sprawling; perianth yellow-banded . . . . ***C. flexuosus***

38′ St ± erect; perianth not yellow-banded
  39. Petal white or flushed pink, red-brown-spotted above nectary, nectary yellow-hairy; s PR (s San Diego Co.) . . . . **C. dunnii**
  39′ Petal lavender to deep purple, base gen purple-spotted, not yellow-hairy; c&s NCoRI, w edge SnJV, CW, w SW, SnJt, w edge DSon . . . . **²C. splendens**
 36′ St bulblets gen present (exc *Calochortus palmeri* var. *munzii*)
  40. Petal bright or deep yellow
   41. Nectary gen ± square . . . . **²C. venustus**
   41′ Nectary chevron- or crescent-shaped to oblong
    42. Petal pale yellow . . . . **²C. argillosus**
    42′ Petal deep yellow. . . . . **C. luteus**
  40′ Petal white to pink, purple, lavender, or red
   43. Petal light to deep lilac, lavender, or variably colored — gen with sparse white hairs near base; not striped, highly patterned. . . . . **²C. splendens**
   43′ Petal all or variably white, pink, purple, or red
    44. Nectary ± ovate or oblong (occ ± round)
     45. Anther sagittate — CaRH, SNH, MP, SNE . . . . **C. leichtlinii**
     45′ Anther not sagittate
      46. Fr narrowly oblong, not angled . . . . **C. catalinae**
      46′ Fr linear, angled
       47. Petals gen 40–50 mm; nectary hairs slender; ne KR (near Yreka, Siskiyou Co.). . . . . **C. monanthus**
       47′ Petals gen 20–30 mm; nectary hairs thick; Teh, s CW, TR, SnJt. . . . . **C. palmeri**
        48. Nectary glabrous or purple-hairy; st bulblets 0; bracts opposite; SnJt. . . . . var. **munzii**
        48′ Nectary yellow-hairy; st bulblets present; bracts alternate; Teh, s CW, TR, SnJt . . . . var. **palmeri**
    44′ Nectary 1–2 crescents or chevrons, ± triangular, ± square, or ± elliptic
     49. Nectary ± triangular-sagittate; sepals gen > petals . . . . **C. macrocarpus**
     49′ Nectary 1–2 crescents or chevrons or ± square or ± elliptic; sepals ≤ petals
      50. Nectary ± square
       51. Nectary in red spot; 2nd, distal, paler petal spot 0 (present); seed 5–6 mm, 3 mm wide . . . . . **C. simulans**
       51′ Nectary not in red spot; 2nd, distal, paler petal spot gen present; seed 4–5 mm, 1–2 mm wide . . . **²C. venustus**
      50′ Nectary 1–2 crescents, chevrons, or ± elliptic
       52. Nectary 2 crescents or chevrons . . . . **C. vestae**
       52′ Nectary 1 crescent, chevron, or ± elliptic
        53. Nectary hairs orange; petal light yellow basally; basal lvs gen persistent; KR, CaR (The Cove, Shasta Co.). . . . . **C. syntrophus**
        53′ Nectary hairs red-brown or brown; petal highly variable in color basally, often purple-lined near base; basal lvs gen withering; NW, CaRF, SNF, s SNH, CW, SW
         54. Fr lanceolate; w CW. . . . . **²C. argillosus**
         54′ Fr linear; NW, CaRF, SNF, s SNH, CW, SW . . . . **C. superbus**

C. albus (Benth.) Benth. (p. 1385) WHITE GLOBE LILY, FAIRY-LANTERN Pl lfy, ± glaucous. **ST:** 20–80 cm, base bulb-bearing. **LF:** basal 30–70 cm, persistent; cauline lvs 2–6, 5–25 cm, lanceolate to linear. **INFL:** fls 2–many, nodding; bracts gen paired, 1–5 cm, lanceolate. **FL:** perianth ± closed at tip, ± oblong; sepals 10–15 mm, ovate to lanceolate, appressed to petals; petals 20–25 mm, gen white to pink, ± elliptic, sparsely ciliate, hairs above nectary ± yellow, slender; nectary crescent-shaped, depressed, with several fringed membranes with yellow or white glandular hairs 1/3–2/3 petal width; filaments 4–5 mm, dilated at base, anthers 4 mm, oblong, short, abruptly-tipped, white to light pink. **FR:** nodding, 20–40 mm, elliptic-oblong, prominently winged. **SEED:** irregular, dark brown. 2*n*=20. Common. Shady to open woodland, scrub; < 2000 m. s CaRF, n&c SN, CW, n ChI, TR, PR. If recognized taxonomically, pls with deep rose fls from sw SnFrB and n&c SCoRO assignable to *C. albus* var. *rubellus* Greene. Hybridizes with *C. monophyllus*. Apr–Jun

C. amabilis Purdy DIOGENES' LANTERN Pl lfy, ± glaucous. **ST:** 10–50 cm, gen branched. **LF:** basal 20–50 cm, persistent; cauline 2–4, 2–20 cm, lanceolate to linear. **INFL:** fls 2–many, nodding; bracts 2–10 cm, lanceolate. **FL:** perianth spheric, closed or parts crossing each other at tip; sepals 15–20 mm, ovate to lanceolate, spreading; petals 16–20 mm, widely lanceolate, deep yellow, brown-spotted abaxially, glabrous exc margin densely ciliate, nectary deep, bordered above by band of ± yellow hairs; filaments 5 mm, dilated at base, anthers 3–4 mm, white to pale yellow. **FR:** nodding, 20–30 mm, oblong, winged. **SEED:** irregular, dark brown. 2*n*=20. Common. Shady or open woodland, scrub; 100–1000 m. KR, NCoRO, NCoRI, nw SnFrB. Apr–Jun

C. amoenus Greene Pl lfy, ± glaucous. **ST:** 20–50 cm. **LF:** basal 20–50 cm, persistent; cauline 2–5, acuminate. **INFL:** fls 2–many, nodding; bracts gen paired, 1–5 cm, lanceolate. **FL:** perianth spheric, closed at tip; sepals 10–15 mm, ovate to lanceolate, spreading; petals 16–25 mm, rose, ± widely elliptic, sparsely ciliate, sparsely hairy, nectary ± basal, widely elliptic, not deep, with 4–5 ciliate membranes, lowest ± = petal width; filaments not dilated, tapered at anther base, anthers white to pink. **FR:** nodding, 20–30 mm, widely elliptic, ± winged. **SEED:** irregular, dark brown. 2*n*=20. Grassy slopes in partial shade; 500–1500 m. c&s SN (Madera, Fresno cos.). Apr–Jun

C. argillosus (Hoover) Zebell & P.L. Fiedl. (p. 1385) CLAY MARIPOSA LILY **ST:** 40–60 cm, simple, bulblets present. **LF:** basal 20–30 cm, withering; cauline reduced upward. **INFL:** ± umbel-like; fls 1–4, erect; bracts 2–8 cm. **FL:** perianth bell-shaped; sepals 20–40 mm; petals 20–40 mm, ± rounded, white to purple or pale yellow, central red spot within pale yellow, sparsely hairy; nectary 1 crescent or chevron, not depressed, densely short-hairy; filaments not dilated at base, anthers purple, pink to yellow-white. **FR:** erect, 4–6 cm, lanceolate. Hard clay from volcanic or metamorphic rocks; < 800 m. w CW. Fls highly variable, gen showy. Apr–Jun

C. bruneaunis A. Nelson & J.F. Macbr. (p. 1385) **ST:** 10–40 cm, gen simple, bulblets present. **LF:** basal 10–20 cm, withering; upper cauline reduced upward, inrolled. **INFL:** fls 1–4, erect; bracts 2–several, 2–7 cm, unequal in length, linear. **FL:** perianth bell-shaped, parts white, tinged lilac, with median green stripe abaxially and dark spot near base; sepals 10–40 mm, lanceolate, acuminate; petals 20–40 mm, narrowly obovate, ± glabrous or with few short golden

hairs near nectary, dark red or purple arch above nectary, nectary surrounded by yellow spot, round, depressed, densely covered with short, simple or distally-branched hairs, encircled by fringed membrane; filaments 5–6 mm, dilated at base, anthers 5–9 mm, oblong, obtuse, yellow, ± blue, or red-brown. **FR:** 3–7 cm, erect, lance-linear, angled. **SEED:** ± flat, light tan. *n*=7. Dry shrub- or grassland in pinyon/juniper woodland; 1700–3000 m. n SNH, e edges c&s SNH, MP, s SNE; to OR, MT, NV, UT. May–Aug

C. catalinae S. Watson (p. 1385)   CATALINA MARIPOSA LILY   **ST:** 20–70 cm, gen branched above, bulblets present. **LF:** basal 10–30 cm, flat, withering; cauline reduced upward. **INFL:** fls 1–4, erect; bracts 2–10 cm. **FL:** perianth cup-shaped; sepals 20–30 mm, lanceolate, green, purple-spotted near base; petals 20–50 mm, wedge-shaped, white, tinged lilac, purple-spotted near base, ± glabrous, nectary not depressed, oblong, densely branched-hairy; filaments 8–10 mm, deep purple, anthers 4–5 mm, narrowly oblong, obtuse, ± flat, lilac, red. **FR:** 2–5 cm, erect, narrowly oblong, not angled, tip obtuse. **SEED:** ± flat, light tan, net-like. *n*=7. Heavy soil, open grassland or scrub; < 700 m. s CCo, s SCoRO, w SCo, esp ChI, w edge WTR, SnGb, n PR. Threatened by development. Mar–May ★

C. clavatus S. Watson   **ST:** 20–100 cm, coarse, often zigzag, bulblets present. **LF:** basal 10–20 cm, linear, deeply channeled, withering; cauline reduced upward. **INFL:** fls 1–6, erect; bracts ± opposite, 4–8 cm, linear to narrowly lanceolate, dilated at base. **FL:** perianth cup-shaped; sepals 20–40 mm, lance-ovate, gen marked red-brown near base; petals 30–50 mm, widely wedge-shaped to obovate, clawed, yellow, gen banded darker above nectary, hairs near nectary club-shaped, nectary round, ± depressed, surrounded by fringed membrane, densely short-knobby-hairy; filaments ± 10 mm, not dilated, tapered at anther base, anthers 8–10 mm, oblong, obtuse, red-brown to dark purple. **FR:** 6–9 cm, erect, narrowly lanceolate, angled. **SEED:** ± flat, tan. *n*=8.

var. ***avius*** Jeps. (p. 1385)   PLEASANT VALLEY MARIPOSA LILY   **ST:** 50–100 cm. **FL:** petals gen < sepals, nectary deep. Open oak/pine forest; 900–1800 m. n SNF (El Dorado, Amador cos.). May–Jul ★

var. ***clavatus*** (p. 1385)   CLUB-HAIRED MARIPOSA LILY   **ST:** 50–100 cm. **FL:** petals 40–50 mm, deep yellow; anthers 8–10 mm, deep purple. Gen serpentine; < 1300 m. s SCoRO, n SCoRI, WTR, SnGb. [*C. c.* subsp. *c.*] Apr–Jun ★

var. ***gracilis*** Ownbey   SLENDER MARIPOSA LILY   **ST:** 20–30 cm, slender. **FL:** petals 30–40 mm, sparsely hairy, with red-brown line above small, shallow nectary; anthers 4–7 mm. Shaded foothill canyons; < 1000 m. WTR, SnGb. May–Jun ★

var. ***pallidus*** (Hoover) P.L. Fiedl. & Zebell   **ST:** 50–100 cm. **FL:** petals light yellow, hairs not very knobby; anthers yellow to purple. Rocky slopes, chaparral, open forest, often serpentine; < 1300 m. w SnJV (and adjacent e SnFrB, SCoRI), n WTR. [*C. c.* subsp. *p.* (Hoover) Munz] Apr–Jul

var. ***recurvifolius*** (Hoover) P.L. Fiedl. & Zebell (p. 1385)   ARROYO DE LA CRUZ MARIPOSA LILY   **ST:** 8–20 cm, slender. **LF:** strongly recurved. **FL:** petals 40–50 mm, deep yellow; anthers 8–10 mm, deep purple. Rocky slopes; ± 50 m. s-c CCo (nw San Luis Obispo Co.). [*C. c.* S. Watson subsp. *r.* (Hoover) Munz] May–Jul ★

C. coeruleus (Kellogg) S. Watson   BEAVERTAIL-GRASS   **ST:** 3–15 cm, simple, gen slender, wavy or not. **LF:** basal 10–20 cm, persistent; cauline gen 0. **INFL:** fls 1–10, spreading to erect; bracts 2–several, 1–7 cm, lance-linear, acuminate. **FL:** perianth bell-shaped; sepals ± 10 mm, lance-oblong, glabrous; petals 8–12 mm, obovate, clawed, ± blue, ciliate, hairy adaxially, nectary transverse, arched upward, ± depressed, bordered below by ciliate membranes, above by short hairs; filaments ± 4 mm, dilated at base, anthers 4 mm, oblong, acute to short-pointed. **FR:** nodding, 10–15 mm, widely elliptic, not angled. **SEED:** irregular, net-like. Common. Gravelly openings in woodland; 600–2500 m. NW, CaR, SNH. Intermediates to *C. tolmiei* scattered but common in NW, CaR. May–Jun

C. concolor (Baker) Purdy   **ST:** 30–60 cm, straight, gen stout, bulblets gen 0. **LF:** basal 10–20 cm, glaucous, withering; cauline 2–4, 5–15 cm, inrolled. **INFL:** ± umbel-like; fls 1–7, erect; bracts

4–8 cm. **FL:** perianth bell-shaped; sepals 20–30 mm, gen dark-red-blotched near base; petals 30–50 mm, yellow (often drying ± purple), sparsely long-yellow-hairy near nectary, nectary gen small, ± round, depressed, encircled by fringed membrane, densely slender-hairy. **FR:** erect, 5–8 cm, narrowly lanceolate, angled. 2*n*=±14. Dry, often granitic slopes in chaparral, yellow-pine forest; 600–2500 m. SnBr (s slope), PR; n Baja CA. May–Jul

C. dunnii Purdy (p. 1385)   DUNN'S MARIPOSA LILY   **ST:** 20–60 cm, slender, gen branched, bulblets 0. **LF:** basal 10–20 cm, channeled, withering; cauline reduced upward. **INFL:** fls 2–6, erect; bracts 1–2 cm. **FL:** perianth widely bell-shaped; sepals 10–20 mm, ovate, acute, green-white adaxially; petals 20–30 mm, gen rounded to wedge-shaped, white or flushed pink, red-brown-spotted above nectary, yellow-hairy near and in nectary, nectary round, not depressed; filaments 5–6 mm, dilated at base, anthers 4–5 mm, oblong, white. **FR:** erect, 2–3 cm, linear, angled. *n*=7. Dry, stony ridges in chaparral, yellow-pine forest; 1500–1700 m. s PR (s San Diego Co.); n Baja CA. Jun ★

C. elegans Pursh   CAT'S EAR   **ST:** 5–15 cm, gen simple, ± slender, wavy. **LF:** basal 10–20 cm, persistent; cauline gen 0. **INFL:** fls 1–7, spreading to erect; bracts 2–several, 1–3 mm, lanceolate to linear. **FL:** perianth bell-shaped; sepals 12–16 mm, lanceolate to elliptic, ± papillate adaxially; petals 12–20 mm, ± oblanceolate, green-white, base dark ± purple, ciliate, hairy, nectary shallow, bordered below with ± ciliate membranes, above with short hairs; filaments 3–4 mm, dilated at base, anthers 3–5 mm, long-pointed, white to lilac. **FR:** nodding, 1–2 cm, widely elliptic, winged. **SEED:** irregular, light brown. 2*n*=20. Open woodland; 1500–2500 m. e KR, NCoRI; to WA, MT. [*C. e.* var. *nanus* Alph. Wood] May–Jul

C. excavatus Greene (p. 1385)   INYO COUNTY STAR-TULIP   **ST:** 10–30 cm, slender, simple, bulblets gen present. **LF:** basal 10–20 cm, gen persistent; cauline reduced upward. **INFL:** ± umbel-like; fls 1–6, erect; bracts paired, gen 3–8 cm. **FL:** perianth widely bell-shaped; sepals 20–30 mm, lanceolate, gen not spotted; petals 30–40 mm, widely wedge-shaped, white, base marked dark purple, green-striped abaxially, sparsely short-hairy near nectary, nectary round, depressed, encircled by fringed membrane, densely short-branched-hairy; filaments 6–8 mm, anthers 7–10 mm, oblong, obtuse, red-brown. **FR:** erect, 5–8 cm, narrowly lanceolate, angled. Grassy meadows in shadscale scrub; 1300–2000 m. sw SNE. Threatened by groundwater development. Apr–May ★

C. fimbriatus H.P. McDonald (p. 1385)   Bulb coat fibrous. **ST:** erect, 30–90 cm, slender, gen branched, bulblets 0. **LF:** basal 20–40 cm, withering; cauline reduced upward, inrolled. **INFL:** fls 1–8, erect. **FL:** perianth widely bell-shaped; sepals 20–30 mm, narrowly lanceolate, long-tapered; petals < to << sepals, ± square, pale cream, ± yellow, ± purple, or dark red to red-brown, dark-hairy; nectary ± round, depressed, ± glabrous, ± hidden by dense, long, cream-white to yellow bordering hairs; filaments 12–15 mm, dilated at base, anthers > filaments, oblong, abruptly pointed, white, yellow to red-brown. **FR:** erect, 4–5 cm, linear, angled, tip acuminate. *n*=9. Dry, open coastal woodland, chaparral; < 900 m. SCoRO, WTR. [*Calochortus weedii* var. *vestus* Purdy] Taxonomic limits uncertain; study needed. Jul–Aug ★

C. flexuosus S. Watson (p. 1385)   **ST:** 10–40 cm, gen wavy, ± sprawling, often intertwining with other pls, bulblets rarely present. **LF:** basal 10–20 cm, withering; cauline linear, long-tapered, reduced upward. **INFL:** fls 1–6, erect; bracts 1–3 cm. **FL:** perianth bell-shaped, parts purple-spotted, yellow-banded; sepals 20–30 mm, lanceolate to narrowly ovate; petals 30–40 mm, ovate, wedge-shaped, white, lilac-tinged, sparsely short-hairy near nectary, nectary 1 crescent, not depressed, densely short-hairy; filaments 6–10 mm, anthers 5–7 mm, oblong, white to lilac. **FR:** erect, 3–4 cm, stout, angled. **SEED:** flat, light tan, net-like. *n*=7. Dry, rocky sites; creosote-bush or sagebrush scrub; 600–1700 m. e D; to CO, NM. Apr–May

C. greenei S. Watson (p. 1385)   GREENE'S MARIPOSA LILY   **ST:** 10–40 cm, gen branched. **LF:** basal ± 20 cm, persistent; cauline 1–2, reduced upward. **INFL:** fls 1–5, erect; bracts 2–4, 2–4 cm, lance-linear. **FL:** perianth bell-shaped; sepals 25–30 mm, ovate, acuminate, ±

green tinged with lilac, yellow hairy, purple spot at base; petals 30–40 mm, widely obovate, ± purple with darker crescent abaxially above nectary, yellow below, not ciliate, ± hairy, nectary deeply depressed, bordered below by wide ciliate membranes, above by short hairs; filaments 10–14 mm, anthers 8–12 mm, narrowly oblong, white to lilac. **FR:** erect, 20–25 cm, elliptic, winged. **SEED:** irregular. Shrubby hillsides, open woodland; ± 700–1100 m. e KR, nw CaR, w MP; s OR. Jun–Jul ★

C. invenustus Greene (p. 1385)   Pl glaucous. **ST:** 20–50 cm, slender, gen simple, bulblets present. **LF:** basal 10–20 cm, channeled, withering; upper cauline 1–2, reduced upward, inrolled. **INFL:** ± umbel-like; fls 1–6, erect; bracts 2–5 cm. **FL:** perianth bell-shaped; sepals 20–30 mm, lance-ovate, yellow-green adaxially; petals 20–40 mm, narrowly obovate to wedge-shaped, short claw at base, white or tinged lilac, occ purple-spotted below nectary, green-striped abaxially, sparsely short-hairy near nectary, nectary small, round, ± depressed, encircled by fringed membrane, densely short-branched-hairy; filaments 6–7 mm, anthers 7–8 mm, oblong-linear, obtuse, purple or yellow. **FR:** erect, 5–7 cm, linear, angled, tip long-tapered. **SEED:** ± flat, net-like. *n*=7. Dry soil, gen granitic, gen montane conifer forest; 1500–3000 m. s SNF, c&s SNH, Teh, se SnFrB, SCoR, SW, DMoj. May–Aug

C. kennedyi Porter (p. 1385)   Pl glaucous. **ST:** 10–20(50) cm, gen simple, occ twisted, bulblets gen 0. **LF:** basal 10–20 cm, glaucous, channeled, withering; cauline 0–2, reduced upward. **INFL:** ± umbel-like; fls 1–6, erect; bracts 2–4 cm, dilated at base. **FL:** perianth bell-shaped, parts often dark-spotted near base; sepals 20–30 mm, lanceolate; petals 30–50 mm, wedge-shaped to obovate, pointed at tip, yellow to orange-red, sparsely club-like-hairy near nectary, nectary round, depressed, encircled by fringed membrane, densely simple- or forked-hairy; filaments 4–5 mm, anthers oblong-ovate, obtuse, gen purple. **FR:** erect, 4–6 cm, ± lanceolate, angled, striped. **SEED:** flat, net-like. *n*=8.

var. ***kennedyi***   **FL:** petal orange (esp e DMoj) to red (W&I, w DMoj). Heavy or rocky soil in creosote-bush scrub, pinyon/juniper woodland; 600–2200 m. n TR, W&I, DMoj; NV, AZ, Mex. Apr–Jun

var. ***munzii*** Jeps.   **FL:** petal yellow; anthers dark purple-brown. Heavy or rocky soil in creosote-bush scrub, pinyon/juniper woodland; 600–2200 m. n&e DMtns (Panamint, Clark Mtn ranges, Providence Mtns); NV, AZ, Mex. Apr–Jun

C. leichtlinii Hook. f. (p. 1385)   **ST:** 20–60 cm, simple, bulblet gen 1 in sheath at base. **LF:** basal 10–15 cm, flat, withering; cauline 1–2, 2–9 cm, reduced upward. **INFL:** ± umbel-like; fls 1–5, erect; bracts paired, 1–5 cm. **FL:** perianth bell-shaped; sepals 10–20 mm; petals 10–40 mm, white to smoky-blue, often tinged pink, red- to black-spotted above nectary, sparsely short-hairy near nectary, nectary ± depressed, ± ovate, densely short-hairy; filaments 5 mm, anthers 5–7 mm, sagittate, white to light yellow. **FR:** erect, 3–6 cm, narrowly lanceolate, angled, tip acute. **SEED:** flat, inflated, dark brown. 2*n*=20. Common. Open, gravelly places in chaparral or montane conifer forest; 1300–4000 m. CaRH, SNH, MP, SNE; s OR, w NV. Jun–Aug

C. longebarbatus S. Watson var. ***longebarbatus*** (p. 1385) LONG-HAIRED STAR-TULIP   **ST:** 10–30 cm, gen simple, bulblets present. **LF:** basal 20–30 cm, persistent; cauline 1–2. **INFL:** fls 1–4, erect; bracts 2–6 cm. **FL:** perianth bell-shaped; sepals 15–20 mm, lance-ovate; petals 20–30 mm, widely obovate, pink to lavender, with central red-purple crescent, not ciliate, sparsely long-hairy, nectary bordered below by ciliate membrane, above by short hairs; filaments 7–10 mm, dilated at base, anthers 3–4 mm, obtuse to short-pointed, lilac. **FR:** erect, 20–25 mm, widely elliptic to ± spheric, winged. **SEED:** irregular, light brown, net-like. 2*n*=20. Vernal meadows, heavy clay soil; 1200–1900 m. CaRH, MP; to s-c WA. Threatened by grazing. Jun–Aug ★

C. luteus Lindl. (p. 1385)   **ST:** 20–50 cm, slender, bulblets present. **LF:** basal 10–20 cm, narrowly linear, withering. **INFL:** ± umbel-like; fls 1–7, erect; bracts 1–8 cm. **FL:** perianth bell-shaped; sepals 20–30 mm, lanceolate, long acuminate; petals 20–40 mm, deep yellow, gen lined red-brown adaxially, often with central red-brown

blotch adaxially, sparsely slender-hairy near nectary, nectary not depressed, ± crescent-shaped to oblong, densely matted-short-hairy; filaments 7–9 mm, anthers 4–6 mm, linear-oblong, white to light yellow. **FR:** erect, 3–6 cm, narrowly lanceolate, angled, tip long-tapered. **SEED:** flat, tan or light yellow, net-like. 2*n*=14,20,21. Heavy soils in grassland, woodland, mixed-evergreen forest; < 700 m. c&s NW, CaR, SNF, GV, CW, n ChI; sw OR. Hybridizes with *C. superbus*. Fls highly variable. Apr–Jun

C. macrocarpus Douglas (p. 1385)   Pl glaucous. **ST:** 20–50 cm, stout, gen simple, bulblets gen present. **LF:** basal 5–10 cm, withering; cauline 2–5, 5–20 cm, inrolled, tips curled. **INFL:** ± umbel-like; fls 1–6, erect; bracts 3–5 cm. **FL:** perianth bell-shaped; sepals 40–50 mm, lanceolate, long-acuminate, tinged purple, gen > petals; petals 35–60 mm, narrowly obovate, purple, green-striped abaxially, gen purple-banded and bearded above nectary, nectary ± depressed, ± triangular-sagittate, densely slender-hairy; filaments 8–9 mm, anthers ± 10 mm, lance-linear, white, yellow-pink, light purple. **FR:** erect, 4–5 cm, ± lanceolate, angled, tip acuminate. **SEED:** flat, inflated, light yellow or tan, net-like. Common. Sagebrush scrub, yellow-pine forest, gen volcanic soil; 1300–2000 m. n CaR, GB; to BC. Jun–Aug

C. minimus Ownbey   **ST:** < 10 cm, simple. **LF:** basal 10–20 cm, persistent; cauline ± 0. **INFL:** fls 1–5(10), ± erect; bracts gen 2, opposite, 1–2(6) cm, lanceolate. **FL:** perianth bell-shaped; sepals 8–10 mm, lanceolate; petal 10–14 mm, ± erect, obovate to wedge-shaped, white, not ciliate, ± glabrous exc near nectary, nectary bordered below by ± ciliate membrane; filaments 4–5 mm, dilated at base, anthers 3–4 mm, linear-oblong, white, ± blue, or purple. **FR:** nodding, 15–20 mm, elliptic, winged. **SEED:** irregular, yellow to light brown, net-like. 2*n*=20. Common. Moist, open woodland, lake margins; 1200–3000 m. s CaRH, SNH. Derivatives of hybrids with *C. nudus* in n SNH. May–Aug

C. monanthus Ownbey (p. 1385)   SINGLE-FLOWERED MARIPOSA LILY   **ST:** simple, straight, bulblets present. **LF:** basal withering. **INFL:** fl 1, erect; peduncle long; bracts linear, long-tapered. **FL:** perianth ± narrowly bell-shaped; sepals ± 40 mm, lanceolate, long-tapered; petals 40–50 mm, obovate to wedge-shaped, irregularly toothed distally, ± pink with a chevron-shaped, dark red spot above each nectary, irregularly toothed above, sparsely slender-hairy near nectary, nectary not depressed, oblong, densely slender-hairy; anthers > filaments, lance-linear, short-pointed. **FR:** erect, linear, angled. Presumed extinct; vernal meadow; ± 800 m. ne KR (near Yreka, Siskiyou Co.). Evidently collected only once, in 1876. Jun ★

C. monophyllus (Lindl.) Lem.   YELLOW STAR-TULIP   **ST:** 8–20 cm, simple or branched, slender, wavy. **LF:** basal 10–30 cm, persistent; cauline 0–3, 1–7 cm, lanceolate to linear, reduced upward. **INFL:** fls 1–6, ± erect; bracts ≥ 2, paired, 2–6 cm, lanceolate to linear. **FL:** perianth bell-shaped; sepals 16–20 mm, ovate to obovate, green-yellow; petals ± = sepals, narrowly obovate, deep yellow, often red-brown-spotted above base, brown-clawed at base, ciliate, densely branched-hairy, nectary crescent-shaped, bordered below by ciliate membrane, above by short, yellow, club-shaped hairs; filaments 4–5 mm, dilated at base, anthers 3–4 mm, pointed, yellow. **FR:** nodding, 12–20 mm, widely elliptic, winged. **SEED:** irregular, dark brown, net-like. 2*n*=20. Wooded slopes, clay-loam soils; 400–1200 m. CaRF, sw CaRH, n&c SN; to s OR. Hybridizes with *C. albus*. Apr–May

C. nudus S. Watson (p. 1385)   **ST:** 10–25 cm, gen simple. **LF:** basal 5–15 cm, persistent; cauline 0. **INFL:** fls 1–several, erect; bracts gen 2, 1–2(5) cm. **FL:** perianth bell-shaped, white to pale lavender; sepals 10–12 mm, lance-ovate; petals 14–16 mm, not ciliate, ± glabrous, nectary transverse, shallow, bordered below by wide, ciliate membrane; filaments 4–5 mm, dilated at base, anthers 3–4 mm, obtuse to short pointed, purple. **FR:** erect, 15–20 mm, widely elliptic, winged. **SEED:** irregular, light brown, net-like. 2*n*=20. Moist, grassy areas, lake, bog margins; 1200–2500 m. KR, CaRH, n SNH; sw OR. Hybridizes with *C. minimus*. May–Jul

C. obispoensis Lemmon (p. 1385)   SAN LUIS MARIPOSA LILY   Bulb coat fibrous. **ST:** 30–60 cm, slender, gen branched. **LF:** basal 20–30 cm, linear, long-tapered, channeled, withering; cauline 2, 4–12 cm, reduced upward, upper inrolled. **INFL:** fls 2–6, erect. **FL:** perianth ± rotate; sepals 10–30 mm, lanceolate to acuminate, reflexed;

petals 10–20 mm, yellow to deep orange, ± ovate, coarse-hairy adaxially, fringed, tip dark-hair-tufted, nectary round, glabrous, ± depressed, ± hidden by dense, slender, bordering hairs with fused bases; filaments slender, anthers dark red-brown. **FR:** erect, 3–4 cm, linear, angled, tip long-tapered. **SEED:** ± flat, light yellow to tan, finely net-like. *n*=9. Dry serpentine, gen open chaparral; 100–500 m. c CCo (e of Morro Bay), s SCoRO (San Luis Obispo Co.). May–Jun ★

C. palmeri S. Watson Pl glaucous. **ST:** 30–60 cm, straight, gen branched, bulblets gen present. **LF:** basal 10–20 cm, narrowly linear, withering; cauline 1–3, 5–15 cm, reduced upward. **INFL:** fls 1–6, erect; bracts 1–2 cm. **FL:** perianth widely bell-shaped; sepals ± 30 mm, oblong, gen brown-spotted near base; petals 20–30 mm, obovate to wedge-shaped, white to lavender, clawed, occ brown-spotted above nectary, gen yellow-hairy near nectary, nectary not depressed, ± round, gen densely thick-knobby-hairy; filaments 7–8 mm, slender, anthers 5–7 mm, oblong, obtuse, ± white to ± blue. **FR:** erect, 2–5 cm, linear, angled. *n*=7.

var. ***munzii*** Ownbey (p. 1385) SAN JACINTO MARIPOSA LILY **ST:** bulblets 0. **INFL:** pedicels paired; bracts opposite. **FL:** nectary glabrous or purple-hairy. Yellow-pine forest; 1200–2200 m. SnJt. Jun ★

var. ***palmeri*** PALMER'S MARIPOSA LILY **ST:** bulblets present. **INFL:** bracts alternate. **FL:** nectary yellow-hairy. Meadows, vernally moist places in yellow-pine forest, chaparral; 1200–2200 m. Teh, s CW, TR, SnJt. May–Jul ★

C. panamintensis (Ownbey) Reveal PANAMINT MARIPOSA LILY **ST:** 40–60 cm, gen simple, bulblets present. **LF:** basal 10–20 cm, withering; upper cauline 1–2, 2–6 cm, inrolled. **INFL:** fls 1–4, erect; bracts 2–4 cm. **FL:** perianth bell-shaped; sepals 10–40 mm, dark-spotted near base; petals 20–40 mm, narrowly obovate, white tinged lilac, green-striped abaxially, not spotted, ± glabrous, nectary in red or purple spot, round, depressed, encircled by fringed membrane, densely short-hairy; filaments ± 6 mm, ± dilated toward base, anthers 5–7 mm, oblong, ± blue. **FR:** erect, ± 7 cm, ± linear, angled, tip acuminate. *n*=7. Dry pinyon/juniper woodland; 2500–3200 m. n DMtns (Panamint Range); w NV. Jun–Jul ★

C. persistens Ownbey (p. 1385) SISKIYOU MARIPOSA LILY **ST:** ± 10 cm, simple. **LF:** basal ± 20 cm, persistent; cauline 1, 4–8 cm. **INFL:** ± umbel-like; fls 2, erect; bracts 2, 2–3 cm, lanceolate. **FL:** perianth bell-shaped, persistent; sepals 35–40 mm, lance-elliptic, acuminate, ± purple; petals ± = sepals, pink to ± purple, obovate, yellow-ciliate, hairy only above nectary exc nectary bordered below by wide, ciliate membrane, above by short hairs; filaments ± 10 mm, narrow to ± dilated below, anthers ± = filaments, lanceolate, pointed, purple. **FR:** nodding, ± 1 cm, winged. 2*n*=20. Open, rocky areas; 1000–1500 m. ne KR (Siskiyou Co.). Jun–Jul ★

C. plummerae Greene (p. 1385) PLUMMER'S MARIPOSA LILY Bulb coat fibrous. **ST:** 30–60 cm, slender, gen branched, bulblets 0. **LF:** basal 20–40 cm, linear, withering; cauline 4–17 cm, narrowly long-tapered, reduced upward, upper inrolled. **INFL:** fls 2–6, erect; bracts lf-like. **FL:** perianth widely bell-shaped, pale pink to rose drying ± purple; sepals 30–50 mm, lanceolate, long-tapered; petals 30–40 mm, widely wedge-shaped to obovate, margin entire or occ fringed to near tip, gen toothed, long-yellow-hairy in wide central band, nectary round, ± depressed, ± glabrous, ± hidden by dense, orange bordering hairs; filaments 9–11 mm, dilated at base, anthers 10–14 mm, lance-linear, acute to ± long- pointed, yellow to light tan. **FR:** erect, 4–8 cm, linear, angled. **SEED:** flat, finely net-like. *n*=9. Dry, rocky chaparral, yellow-pine forest; < 1700 m. SCo, TR, PR. May–Jul ★

C. pulchellus (Benth.) Alph. Wood (p. 1385) MOUNT DIABLO FAIRY-LANTERN Herbage green. **ST:** erect, 10–30 cm, stout. **LF:** basal 10–40 cm, persistent; cauline 2–3, gen reduced upward. **INFL:** fls 1–many, nodding. **FL:** perianth spheric, closed at tip; sepals 20–30 mm, lanceolate to lance-ovate; petals 25–33 mm, narrowly ovate, light yellow, ciliate with thick hairs, sparsely hairy, nectary deep, bordered above by slender hairs; filaments 6–8 mm, dilated at base, anthers 3–5 mm, oblong, mucronate, yellow. **FR:** nodding, 20–30 mm, winged. **SEED:** irregular, dark brown. 2*n*=20. Wooded slopes,

rarely chaparral, gen n aspect; 200–800 m. ne SnFrB (Mount Diablo, Contra Costa Co.). Hybrids with *C. umbellatus* rare. Apr–Jun ★

C. raichei Farwig & V. Girard (p. 1385) THE CEDARS FAIRY-LANTERN Pl very glaucous, mealy. **ST:** 10–100 cm, stout, gen simple. **LF:** basal 10–40 cm, channeled, persistent; cauline 1–4, 10–20 cm, reduced upward. **INFL:** fls 1–2, nodding; bracts paired, narrowly lanceolate, long-tapered. **FL:** perianth ± spheric, closed at tip; sepals 15–25 mm, pale green tinged pale purple, widely lanceolate; petals 35–45 mm, pale yellow, obovate, ciliate, hairy, nectary ± depressed, bordered above by long, slender hairs; filaments 7–12 mm, dilated at base, anthers ± 6 mm, oblong, obtuse. **FR:** nodding, 25–35 mm, oblong, angled. **SEED:** irregular, dark brown. Open serpentine in woodland; 200–300 m. s NCoRO (The Cedars, Big Austin Creek drainage, Sonoma Co.). Potentially threatened by mining. May–Aug ★

C. simulans (Hoover) Munz (p. 1385) LA PANZA MARIPOSA LILY **ST:** 10–60 cm, gen branched, bulblets present. **LF:** basal 10–20 cm, withering; cauline 1–3, 4–10 cm. **INFL:** ± umbel-like; fls 1–3, erect; bracts 2–8 cm. **FL:** perianth bell-shaped; sepals 20–30 mm, tips recurved; petals 30–50 mm, obovate to wedge-shaped, white to yellow, sparsely hairy near base, nectary in dark red spot (2nd spot gen 0), not depressed, ± square, short-hairy; filaments 9–11 mm, anthers 7–9 mm, lance-linear to oblong, yellow-red. *n*=7. Sand (often granitic), grassland to yellow-pine forest; < 1100 m. se SCoRO (c San Luis Obispo Co.). May–Jul ★

C. splendens Benth. (p. 1385) **ST:** 20–60 cm, bulblets gen 0. **LF:** basal 10–20 cm, withering; cauline 2, 5–17 cm. **INFL:** central axis distinct; fls 1–4, erect; bracts 2–5 cm. **FL:** perianth bell-shaped, base narrowed; sepals 20–30 mm, lanceolate, long-acuminate, gen recurved, deep lilac, often purple-spotted; petals 30–50 mm, widely fan-shaped, tapered to lilac-purple claw, toothed above, gen purple-spotted and sparsely white-hairy near base, nectary not depressed, ± square, gen densely branched-hairy; filaments 7–8 mm, dilated at base, anthers 5–7 mm, obtuse to short-pointed, lilac, blue, purple, or white. **FR:** erect, 5–7 cm, linear, angled, tip long-tapered. **SEED:** flat, light ± yellow to tan. *n*=7, 14. Dry, often granitic soils, grassland, chaparral, yellow-pine forest; < 2800 m. c&s NCoRI, w edge SnJV, CW, w SW, SnJt, w edge DSon; n Baja CA. May–Jul

C. striatus Parish (p. 1385) ALKALI MARIPOSA LILY **ST:** 1–5 cm. **LF:** basal 10–20 cm, gen withering; cauline 1–2, 6–8 cm. **INFL:** ± umbel-like; fls 1–5, erect; bracts 1–3 cm, linear. **FL:** perianth bell-shaped, base narrowed; sepals 10–20 mm, lanceolate; petals 20–30 mm, obovate to wedge-shaped, irregularly toothed above, white to lavender, purple-veined, sparsely hairy near nectary, nectary not depressed, oblong, densely simple-hairy; filaments 5–7 mm, ± dilated below, anthers 4–6 mm, oblong, lilac, purple. **FR:** erect, 4–5 cm, linear, angled. **SEED:** flat, light ± yellow or tan, net-like. Alkaline meadows, moist creosote-bush scrub; 800–1400 m. s SNF, w DMoj; w NV. Threatened by grazing, urbanization. Apr–Jun ★

C. superbus J.T. Howell (p. 1385) **ST:** 40–60 cm, bulblets present. **LF:** basal 20–30 cm, gen withering; cauline reduced upward. **INFL:** ± umbel-like; fls 1–3, erect; bracts 2–8 cm. **FL:** perianth bell-shaped, base narrowed, parts centrally dark-blotched in bright yellow zone; sepals 20–40 mm; petals 20–40 mm, white to ± yellow or lavender, gen purple-lined near base, sparsely short-hairy near nectary, nectary not depressed, 1 crescent or chevron densely short-brown-hairy; filaments 7–9 mm, slender to dilated below, anthers 8–10 mm, lance-linear to -oblong, acute to obtuse, white to light yellow. **FR:** erect, 5–6 cm, linear, angled. **SEED:** flat, light yellow to tan. *n*=6,7. Common. Open grassland, woodland, dry meadows, yellow-pine forest; < 1700 m. NW, CaRF, SNF, s SNH, CW, SW. Hybridizes with *C. luteus*. Fls highly variable. May–Jul

C. syntrophus Callahan (p. 1389) CALLAHAN'S MARIPOSA LILY Pl ± glaucous. **ST:** 40–60 cm, stout, bulblets present. **LF:** basal 20–30+ cm, channeled, gen persistent; cauline occ appearing basal, 15 cm, reduced upward. **INFL:** ± umbel-like; fls 1–5, erect; bracts 5–20 cm. **FL:** perianth bell-shaped, parts centrally dark-blotched; sepals 20–40 mm, lanceolate, long-tapered; petals 30–50 mm, obovate to wedge-shaped, white, margins fringed, basal part yellow;

sparsely short-orange-hairy near nectary, nectary not depressed, ± elliptic to crescent-shaped, densely orange-short-hairy; filaments gen ± 8 mm, dilated at base, anthers ± = filaments, lance-linear, white. **FR**: erect, 6–7 cm, linear, angled. Abundant where found. Stony sandstone (Kilarc series) in blue-oak woodland; 500–1700 m. KR, CaR (The Cove, Shasta Co.). May–Jun ★

C. tiburonensis A.J. Hill (p. 1389) TIBURON MARIPOSA LILY **ST**: 10–60 cm, gen branched. **LF**: basal 10–70 cm, flat, persistent; cauline 1–several, reduced upward. **INFL**: fls 2–7, erect. **FL**: perianth bell-shaped; sepals 20–35 mm; petals ± = sepals, light yellow-green, flecked purple-brown, oblanceolate, ciliate to near tip, hairy, nectary depressed, bordered below by ciliate membrane, above by ≥ 2 rows of short hairs; filaments 10–12 mm, anthers 6–7 mm, lanceolate, dilated at base, short-tipped, red-brown. **FR**: erect, 20–30 mm, angled, tip acute. **SEED**: irregular, dark brown to black. 2*n*=20. Serpentine grassland; 50–150 m. n CCo (Ring Mtn, Marin Co.). May–Jun ★

C. tolmiei Hook. & Arn. (p. 1389) PUSSY EARS **ST**: 10–40 cm, simple or branched, ± slender. **LF**: basal 10–40 cm, persistent; cauline gen 1. **INFL**: ± umbel-like; fls 1–several, spreading or erect; bracts ≥ 2, 1–7 cm, lanceolate to linear. **FL**: perianth bell-shaped; sepals 10–15 mm, lance-oblong, acute or acuminate, glabrous; petals 12–25 mm, obovate to wedge-shaped, white to ± pink or ± purple, ± ciliate, densely hairy, nectary bordered below by ciliate membrane, above by short hairs; filaments 6–8 mm, dilated at base, anthers 3–5 mm, lanceolate, white to purple. **FR**: nodding, 20–30 mm, elliptic-oblong, winged. **SEED**: irregular, dark brown, net-like. 2*n*=20. Common. Dry, grassy slopes, woodland, often in poor soil; 50–2000 m. NW, CaR, n SN, n ScV, n CCo, SnFrB; to WA, ID. Apr–Jul

C. umbellatus Alph. Wood OAKLAND STAR-TULIP **ST**: 8–25 cm, gen 2-branched, bulblets 0. **LF**: basal 20–40 cm, persistent; cauline gen 1, linear. **INFL**: ± umbel-like; fls 3–12, ± erect; bracts ≥ 2, 1–6 cm, linear. **FL**: perianth bell-shaped, white or pale pink-lilac, gen purple-spotted at base; sepals 10–14 mm, lance-elliptic; petals 12–18 mm, widely wedge-shaped to obovate, ± toothed, not ciliate, ± glabrous, nectary rounded below, truncate above, covered by ciliate membrane, bordered above by short hairs; filaments 5 mm, dilated at base, anthers 2–3 mm, oblong, white to light blue or pink. **FR**: nodding, 10–14 mm, widely elliptic, winged. **SEED**: irregular, dark brown, net-like. 2*n*=20. Open chaparral or woodland, gen on serpentine; 100–700 m. s NCoRO, SnFrB. Mar–May ★

C. uniflorus Hook. & Arn. (p. 1389) PINK STAR-TULIP **ST**: gen < 5 cm, gen simple, bulblets present. **LF**: basal 10–40 cm, persistent; cauline 1–3, linear, reduced upward. **INFL**: ± umbel-like; fls 1–5, erect; bracts ≥ 2, 1–4 cm, linear. **FL**: perianth bell-shaped, lilac; sepals 12–16 mm, lance-elliptic, acuminate; petals 15–28 mm, widely wedge-shaped to obovate, ± white to pink, often purple-spotted above nectary, not ciliate, sparsely hairy near nectary, nectary covered by ciliate membrane, bordered above by short hairs; filaments 7–8 mm, dilated at base, anthers 3–4 mm oblong, light blue to lilac, cream, or pink. **FR**: nodding, 15–25 mm, elliptic, narrowly winged. **SEED**: irregular, light brown, net-like. 2*n*=20,40. Moist meadows; < 500 m. NW, CaRH, n&c CCo, SnFrB; OR. Apr–Jun ★

C. venustus Benth. (p. 1389) Pl ± glaucous. **ST**: 10–60 cm, gen branched, bulblets present. **LF**: basal 10–20 cm, withering; cauline 0–2, 3–8 cm. **INFL**: ± umbel-like; fls 1–6, erect; bracts 2–8 cm. **FL**: perianth bell-shaped; sepals 20–30 mm, lance-oblong, long-acuminate, tips recurved; petals 30–50 mm, ± obovate, ± clawed, white, yellow, purple, pink, or dark red, centrally dark-blotched, gen with 2nd paler blotch above, distally dark-margined, sparsely hairy toward base, nectary not in red spot, not depressed, ± square, short-yellow-hairy; filaments 7–10 mm, gen dilated at base, anthers 7–10 mm, oblong, ± yellow-white or pink to lilac. **FR**: erect, 5–6 cm, linear, angled. **SEED**: flat, ± yellow to light tan. *n*=7. Sandy (often granitic) soil in grassland, woodland, yellow-pine forest; 300–2700 m. SNF, c&s SNH, Teh, CW, WTR, SnGb. Fls highly variable, esp in colors and patterns, within and between populations. May–Jul

C. vestae Purdy (p. 1389) **ST**: 30–50 cm, gen branched, bulblets present. **LF**: basal 10–20 cm, withering; cauline 1–3, 7–20 cm. **INFL**: ± umbel-like; fls 1–6, erect; bracts 2–5 cm. **FL**: perianth bell-shaped; sepals 20–30 mm, lance-linear, long-tapered; petals 30–40 mm, wedge-shaped to obovate, white to ± purple, ± red-lined below nectary, centrally red-brown-blotched in pale yellow zone, sparsely short-hairy near nectary, nectary not depressed, 2 side-by-side crescents or chevrons, densely short-hairy; filaments 7–10 mm, anthers 7–10 mm, oblong-linear, white to light yellow or pink. **FR**: erect, linear, angled. *n*=14. Clay soil in mixed-evergreen or yellow-pine forests; 500–900 m. s KR, NCoR. May–Jul

C. weedii Alph. Wood Bulb coat fibrous. **ST**: 30–90 cm, slender, gen branched, bulblets 0. **LF**: basal 20–40 cm, withering; upper cauline reduced upward, inrolled. **INFL**: fls 2–6, erect. **FL**: perianth widely bell-shaped; sepals 20–30 mm, ovate, long-tapered; petals ≤ sepals, ovate to wedge-shaped, cream to light or deep yellow, ± purple, or red-brown, flecked and often margined red-brown, ± fringed, long-hairy, nectary round, ± depressed, ± glabrous, ± hidden by dense, long, yellow bordering hairs; filaments 12–15 mm, dilated at base, anthers 8–10 mm, oblong, yellow, red-brown. **FR**: erect, 4–5 cm, linear, angled, tip acuminate. **SEED**: flat, finely net-like. *n*=9.

var. *intermedius* Ownbey INTERMEDIATE MARIPOSA LILY **FL**: petals rounded, light yellow or tan or dark ± purple to red-brown, dark- or yellow-hairy; anthers rounded. Dry, rocky, open slopes; < 680 m. SCo, n PR. Jun–Jul ★

var. *weedii* (p. 1389) **FL**: petals ± truncate, deep yellow, dark-or yellow-hairy; anthers acute. *n*=9. Dry chaparral, often in heavy or rocky soil; < 1900 m. SCoRO, SCo, PR. May–Aug

C. westonii Eastw. (p. 1389) SHIRLEY MEADOWS STAR-TULIP **ST**: 3–15 cm, simple, slender. **LF**: basal 10–20 cm, persistent; cauline 0. **INFL**: fls 1–10, spreading to erect. **FL**: perianth bell-shaped; sepals ± 10 mm; petals 8–12 mm, ± blue, lanceolate, not ciliate near tip, sparsely hairy, nectary ± depressed, bordered below by ciliate membranes, above by short hairs; filaments ± 4 mm, ± dilated below, anthers sharp-tipped, white to lilac. **FR**: nodding, 10–15 mm, angled. **SEED**: irregular. Meadows, open woodland; ± 1500–2000 m. s SNH (Greenhorn Mtns). [*C. coeruleus* var. *w.* (Eastw.) Ownbey] May–Jun ★

CLINTONIA

Rhizome slender, spreading. **LF**: ± basal, wide, cauline 0 or much reduced; petioles sheathing. **INFL**: scapose, umbel or fls 1. **FL**: perianth cylindric to bell-shaped, parts in 2 similar, petal-like whorls, free, ascending to ± erect; stamens 6, fused to perianth, hairy near base [or not]; ovary superior, chambers 2–3, stigma obscurely 2–3-lobed. **FR**: berry, ovoid, blue [or not]. **SEED**: 2–several, black. 6 spp.: w N.Am, e N.Am, Asia. (De Witt Clinton, naturalist, governor of NY, 1769–1828) [Hayashi et al. 2001 Pl Spec Biol 16:119–137]

1. Fls in umbel; perianth pink to rose-purple; peduncle >> lvs . . . . *C. andrewsiana*
1′ Fls 1(2); perianth white; peduncle gen < lvs . . . . *C. uniflora*

C. andrewsiana Torr. (p. 1389) Rhizome 4–10 mm diam. **LF**: 5(6), 15–30 cm, 5–12 cm wide, elliptic, ± hairy. **INFL**: umbels gen 1 terminal, 0–3 lateral; peduncle 25–50 cm, hairy; bracts 1–3; pedicels 1–3 cm, unequal, hairy. **FL**: 10–18 mm; perianth parts pouch-like at base. **FR**: 8–12 mm. *n*=14. Shaded, damp redwood forest; < 400 m. NW, CCo, w SnFrB, n SCoRO; sw OR. May–Jul

C. uniflora (Schult. & Schult. f.) Kunth (p. 1389) Rhizome 1–3 mm diam. **LF**: 2–3, 7–15 cm, 2–6 cm wide, oblanceolate to oblong or elliptic, ± hairy. **INFL**: peduncle 4–8 cm, hairy; bracts 1–2. **FL**: 18–22 mm; perianth parts not pouched at base. **FR**: 6–10 mm. *n*=14. Shaded conifer forest; 1000–1900 m. KR, NCoRO, CaR, SN; to sw Can. May–Jul

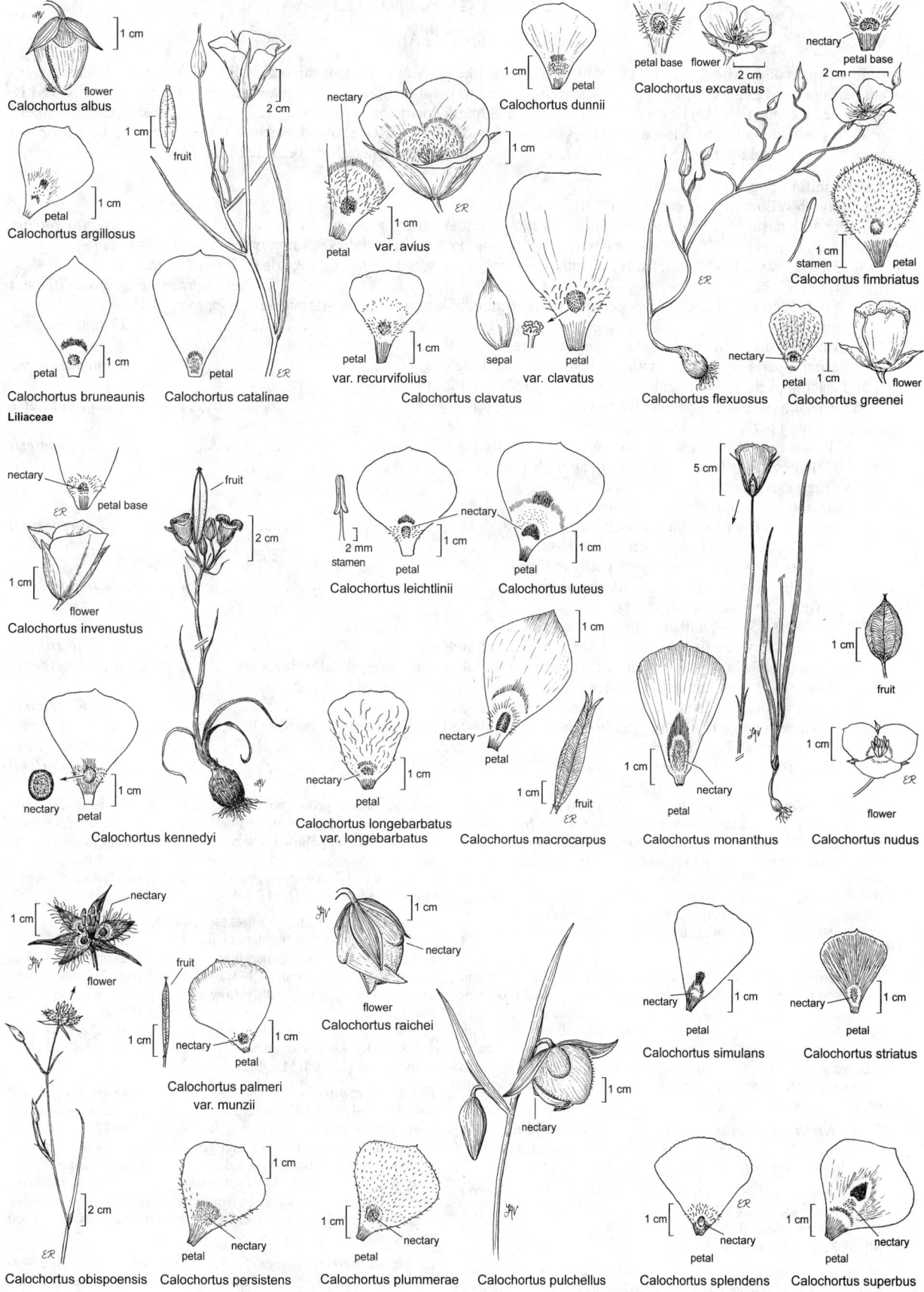

Calochortus albus
flower

Calochortus argillosus
petal

Calochortus bruneaunis
petal

Calochortus catalinae
petal

fruit

Liliaceae

Calochortus dunnii
petal

nectary
petal
var. avius

petal
var. recurvifolius

Calochortus clavatus
sepal
petal
var. clavatus

Calochortus excavatus
petal base   flower   nectary   petal base

Calochortus fimbriatus
stamen   petal

Calochortus flexuosus   Calochortus greenei
nectary   petal   flower

Calochortus invenustus
nectary   petal base   flower

Calochortus kennedyi
nectary   petal

fruit

Calochortus leichtlinii
stamen   petal

Calochortus luteus
nectary   petal

Calochortus longebarbatus
var. longebarbatus
nectary   petal

Calochortus macrocarpus
nectary   petal   fruit

Calochortus monanthus
petal   nectary

Calochortus nudus
fruit
flower

Calochortus obispoensis

Calochortus palmeri
var. munzii
nectary   petal
fruit
flower
nectary

Calochortus persistens
petal   nectary

Calochortus raichei
nectary
flower

Calochortus plummerae
petal   nectary

Calochortus pulchellus
nectary

Calochortus simulans
nectary   petal

Calochortus striatus
nectary   petal

Calochortus splendens
nectary   petal

Calochortus superbus
petal   nectary

ERYTHRONIUM FAWN LILY

Geraldine A. Allen

Bulb elongate of 1–2 fleshy scales, gen with small, bead-like parts of persistent rhizome. **LF:** 2 (1 in non-fl pl), basal, 6–35 cm, lanceolate to ovate (solitary basal lf wider), narrowed to petiole, often mottled, glabrous; margin entire to wavy. **INFL:** peduncled raceme; fls 1–10; bracts 0. **FL:** showy, gen nodding; perianth parts 6, similar, free, ± lanceolate, ± recurved; stamens 6; style 1, stigma entire to 3-lobed. **FR:** capsule, ovoid to oblong. **SEED:** ± ovoid, ± angular, brown. *n*=12. ± 27 spp.: esp temp N.Am. (Greek: red, from fls of some) Lf, fl markings to be noted when fresh, because of fading in pressed specimens.

1. Lvs uniformly green, not mottled
  2. Perianth yellow; stigma entire or lobed
    3. Perianth parts 15–28 mm; style, filaments yellow; stigma entire or lobes < 1 mm; fls 1–10 . . . . ***E. pluriflorum***
    3′ Perianth parts 20–35 mm; style, filaments white; stigma ± entire or lobes < 1 mm or 2–4 mm; fls 1–5
      4. Fls 1(3); style 10–15 mm; fresh perianth parts with narrow, pale zone at base; stigma lobes 2–4 mm; KR, NCoR . . . . ***E. grandiflorum*** subsp. ***grandiflorum***
      4′ Fls 1–5; style 8–10 mm; fresh perianth parts without pale zone at base; stigma ± entire or lobes < 1 mm; c SNF . . . . ***E. tuolumnense***
  2′ Perianth white with yellow base; stigma ± entire
    5. Perianth parts 10–20 mm, without sac-like folds at base . . . . ***E. purpurascens***
    5′ Perianth parts 18–45 mm, inner with small sac-like folds at base
      6. Perianth parts < 1/3 yellow; lvs 6–17 cm; fls 1–3; KR, CaRH . . . . ***E. klamathense***
      6′ Perianth parts 1/3–2/3 yellow; lvs 10–35 cm; fls 1–8; c&s SNH
        7. Filaments ± white, anthers yellow; s SNH (Tulare Co.) . . . . ***E. pusaterii***
        7′ Filaments ± yellow, anthers cream; c SNH (Tuolumne Co.) . . . . ***E. taylorii***
1′ Lvs irregularly mottled with brown or white
  8. Filaments ≥ 1 mm wide; stigma lobed
    9. Perianth parts white with yellow base . . . . ***E. oregonum***
    9′ Perianth parts violet-pink with yellow base . . . . ***E. revolutum***
  8′ Filaments < 0.8 mm wide; stigma entire to lobed
    10. Perianth violet to pink with dark purple base . . . . ***E. hendersonii***
    10′ Perianth ± white with yellow base
      11. Style 10–15 mm; anthers white to cream
        12. Peduncle when fls > 1 branched well above lvs; bulblets 0 . . . . ***E. californicum***
        12′ Peduncle when fls > 1 branched near ground; bulblets from long, slender rhizomes . . . . ***E. multiscapideum***
      11′ Style 4–10 mm; anthers white, cream, yellow, or pink to red tinged
        13. Anthers yellow; style ± bent; bulblets 0 or sessile . . . . ***E. helenae***
        13′ Anthers white to cream, pink to brown- or purple-red; style ± straight; bulblets 0 . . . . ***E. citrinum***
          14. Anthers white to cream . . . . var. ***citrinum***
          14′ Anthers pink to brown- or purple-red . . . . var. ***roderickii***

E. californicum Purdy (p. 1389) Bulb 35–60 mm, ovoid. **LF:** 7–15 cm, oblong to narrowly ovate, mottled brown or white. **INFL:** peduncle 10–30 cm; fls 1–3. **FL:** perianth parts 25–40 mm, ± narrowly ovate, white to cream with yellow base, ± banded brown to red, inner with small sac-like folds at base; stamens 12–25 mm, filaments slender, ± white, anthers ± white to cream; style 10–15 mm, ± white, stigma lobes < 2 mm, spreading. Openings in dry woodland; < 1900 m. KR, NCoR. May intergrade with *E. citrinum*, *E. multiscapideum*. Mar–Apr

E. citrinum S. Watson Bulb 40–50 mm, slender. **LF:** 9–15 cm, lanceolate to narrowly ovate, ± wavy-margined, mottled brown or white. **INFL:** peduncle 12–35 cm, green or tinged red; fls 1–3. **FL:** perianth parts 25–45 mm, lanceolate to narrowly elliptic, white with yellow base, ± pink in age, inner ± with small sac-like folds at base; stamens gen 11–17 mm, filaments slender, white, anthers white to cream or pink to brown- or purple-red; style 6–10 mm, white, stigma entire or lobes < 1 mm.

var. ***citrinum*** (p. 1389) LEMON-COLORED FAWN LILY **FL:** anthers white to cream. Dry woodland, shrubby slopes (± on serpentine); 100–1100 m. KR; sw OR. [*E. howellii* S. Watson] Mar–May ★

var. ***roderickii*** Shevock & G.A. Allen (p. 1389) SCOTT MOUNTAINS FAWN LILY **FL:** anthers pink to brown- or purple-red. Dry conifer woodland (± on serpentine); 850–1300 m. KR (Scott Mtns). Mar–Jun ★

E. grandiflorum Pursh subsp. ***grandiflorum*** GLACIER LILY Bulb 30–50 mm, slender. **LF:** 5–20 cm, lanceolate, ± wavy-margined, green. **INFL:** peduncle 5–30 cm; fls 1(3). **FL:** perianth parts 20–35

mm, lanceolate, yellow with narrow, pale zone at base, inner ± with sac-like folds at base; stamens 11–18 mm, filaments ± slender, white, anthers cream to yellow [dark red]; style 10–15 mm, white, stigma lobes 2–4 mm. Subalpine meadows; 500–2300 m. KR, NCoR; to sw Can, CO. [*Erythronium grandiflorum* var. *pallidum* H. St. John] Other subsp. in WA, ID, MT. Apr–Jul

E. helenae Applegate ST. HELENA FAWN LILY Bulb 30–55 mm, ovoid, making sessile bulblets or not. **LF:** 10–20 cm, widely lanceolate to ovate, ± wavy-margined, mottled brown or white. **INFL:** peduncle 5–30 cm, ± red; fls 1–3. **FL:** perianth parts 25–40 mm, lanceolate to ovate, white with yellow base, ± pink in age, inner with small sac-like folds at base; stamens 8–13 mm, filaments ± slender, ± yellow, anthers yellow; style 5–8 mm, ± bent, ± white, stigma lobes < 1 mm. Dry woodland, on serpentine; 300–1200 m. s NCoRI (near Mount Saint Helena). Mar–May ★

E. hendersonii S. Watson HENDERSON'S FAWN LILY Bulb 40–55 mm, slender. **LF:** 10–25 cm, oblong to ovate, entire to ± wavy-margined, mottled brown or white. **INFL:** peduncle 12–30 cm, ± red to ± purple; fls 1–4. **FL:** perianth parts 18–35 mm, widely lanceolate, violet to pink (darker toward tip) with dark purple zone at base, inner with small sac-like folds at base; stamens 10–14 mm, filaments slender, purple, anthers pale brown to purple; style 6–8 mm, violet, stigma entire or lobes < 1 mm. Dry woodland, openings; 300–1600 m. KR; sw OR. Apr–Jul ★

E. klamathense Applegate KLAMATH FAWN LILY Bulb 25–40 mm, slender. **LF:** 6–17 cm, lanceolate to narrowly elliptic, ± folded along midvein, entire to wavy-margined, green. **INFL:** peduncle

8–20 cm; fls 1–3. **FL:** perianth parts 20–35 mm, widely lanceolate, white with yellow base, ± pink in age, inner with small sac-like folds at base; stamens 8–14 mm, filaments slender, white, anthers ± yellow; style 4–9 mm, white, stigma ± entire. Montane meadows, forest openings; 1200–1850 m. KR, CaRH; s OR. Apr–Jul ★

E. multiscapideum (Kellogg) A. Nelson & P.B. Kenn. Bulb 20–50 mm, ovoid, making bulblets from long, slender rhizomes. **LF:** 4–15 cm, ± elliptic, entire to wavy-margined, mottled brown or white. **INFL:** peduncle 8–20 cm, ± branched just above lvs; fls 1–4. **FL:** perianth parts 16–35 mm, widely lanceolate to elliptic, ± white with yellow base; stamens 10–15 mm, filaments slender, white, anthers white to cream; style 10–13 mm, white, stigma lobes 1–4 mm, recurved. Open woodland, shrubby slopes; 50–1200 m. KR, CaR, n&c SNF, n SNH. [*Erythronium multiscapoideum*, orth. var.] Occ intergrades with *E. californicum*. Mar–May

E. oregonum Applegate   GIANT FAWN LILY   Bulb 35–50 mm, narrowly ovoid. **LF:** 10–22 cm, widely lanceolate to ovate, entire to ± wavy-margined, mottled brown or white. **INFL:** peduncle 15–45 cm, green to ± red; fls 1–3. **FL:** perianth parts 30–45 mm, lanceolate to narrowly elliptic, white [to cream] with yellow base, red-banded or not, inner with sac-like folds at base; stamens 12–22 mm, filaments 1–3 mm wide, white, anthers cream to yellow; style 12–18 mm, white, stigma lobes 3–6 mm, slender, recurved. Openings in woodland; 100–750 m. NCoRO; s OR to sw BC. CA populations are geog separate, may be a white form of *E. revolutum*. Mar–May ★

E. pluriflorum Shevock et al.   SHUTEYE PEAK FAWN LILY   Bulb 40–75 mm, ± ovoid. **LF:** 7–30 cm, oblanceolate to elliptic, ± wavy-margined, green. **INFL:** peduncle 8–35 cm, green to ± red; fls 1–10. **FL:** perianth parts 15–28 mm, lanceolate, yellow, bronze in age, without sac-like folds at base; stamens 8–12 mm, filaments slender, yellow, anthers yellow; style 6–8 mm, yellow, stigma entire or lobes < 1 mm. Open, rocky places; 2300–2550 m. c SNH (Madera Co.). May–Jul ★

E. purpurascens S. Watson (p. 1389)   Bulb 25–40 mm, slender. **LF:** 6–15 cm, lanceolate to narrowly ovate, ± wavy-margined, green. **INFL:** peduncle 7–20 cm, green or faintly ± red; fls 1–6. **FL:** perianth parts 10–20 mm, lanceolate, white with yellow base, pink-purple in age, without sac-like folds at base; stamens 8–12 mm, filaments slen-der, yellow, anthers ± yellow; style 4–5 mm, yellow, stigma ± entire. Open forest, meadows, rocky places; 1100–2700 m. KR, NCoRH, CaRH, n&c SNH. May–Aug

E. pusaterii (Munz & J.T. Howell) Shevock et al.   KAWEAH FAWN LILY   Bulb 40–60 mm, narrowly ovoid. **LF:** 10–35 cm, lanceolate, ± wavy-margined, green. **INFL:** peduncle 12–40 cm, ± red; fls 1–8. **FL:** perianth parts 25–45 mm, lanceolate, white, lower 1/3–2/3 bright yellow, ± pink in age, inner with sac-like folds at base; stamens 8–15 mm, filaments slender, ± white, anthers yellow; style 7–10 mm, ± white, stigma lobes gen < 1 mm. Meadows, rocky ledges; 2100–2775 m. s SNH (Tulare Co.). May–Jul ★

E. revolutum Sm. (p. 1389)   COAST FAWN LILY   Bulb 35–50 mm, narrowly ovoid. **LF:** 10–25 cm, widely lanceolate to ovate, entire to ± wavy-margined, mottled brown or white. **INFL:** peduncle 15–40 cm, green to ± red; fls 1–3. **FL:** perianth parts 25–40 mm, lanceolate to narrowly elliptic, violet-pink with yellow bands at base, inner with small sac-like folds at base; stamens 12–22 mm, filaments 2–3 mm wide, ± lanceolate, white to pink, anthers yellow; style 12–18 mm, white to pink, stigma lobes 4–6 mm, slender, recurved. Streambanks, wet places in woodland; < 1350 m. NCo, KR, NCoRO; to s BC. Mar–Jul ★

E. taylorii Shevock & G.A. Allen   PILOT RIDGE FAWN LILY   Bulb 40–70 mm, narrowly ovoid. **LF:** 18–35 cm, elliptic to lanceolate, ± wavy-margined, green. **INFL:** peduncle 20–40 cm, green; fls 1–8. **FL:** perianth parts 25–45 mm, lanceolate, white, lower 1/3–2/3 bright yellow, ± pink in age, inner with sac-like folds at base; stamens 10–16 mm, filaments slender, ± yellow, anthers cream; style 9–11 mm, white to cream, stigma lobes gen < 1 mm. Forest openings, rocky ledges; 1300–1400 m. c SNH (Tuolumne Co.). Mar–May ★

E. tuolumnense Applegate (p. 1389)   TUOLUMNE FAWN LILY   Bulb 50–100 mm, ovoid, readily making sessile bulblets. **LF:** 15–35 cm, elliptic to ovate, entire to ± wavy-margined, green. **INFL:** peduncle 15–35 cm, green to ± red; fls 1–5. **FL:** perianth parts 20–35 mm, ± narrowly ovate, yellow, inner with small sac-like folds at base; stamens 12–16 mm, filaments 0.4–0.6 mm wide at base, ± white, anthers yellow; style 8–10 mm, ± white, stigma entire or lobes < 1 mm. Open woodland, shady canyons; 600–1000 m. c SNF (Tuolumne Co.). Mar–Jun ★

FRITILLARIA   FRITILLARY

Dale W. McNeal & Bryan D. Ness

Bulb with 1–several large fleshy scales, 0–many small scales. **ST:** erect, simple (0 in non-fl pls). **LF:** cauline, alternate, subopposite, or whorled below, sessile, linear to ± ovate (1 bulb-lf in non-fl pls). **INFL:** raceme; bracts lf-like. **FL:** gen nodding, bell- or cup-shaped; perianth parts 6 in 2 whorls, each part with distinct glandular area in lower 1/2; stamens 6, incl, attached at perianth base, anthers attached ± near middle; ovary ± sessile, style 1, ± entire or 3-branched. **FR:** capsule, loculicidal, thin-walled, ± rounded, 6-angled, or winged, chambers 3. **SEED:** many, 2 rows per chamber, flat, ± brown. ± 100 spp.: n temp. (Latin: dicebox, from fr shape) Bulbs of some eaten by Native Americans.

1. Style entire or with lobes < 1.5 mm
  2. Lvs in 1–3 whorls of 4–8 below, those subtending fls alternate . . . . ***F. brandegeei***
  2′ Lvs alternate to subopposite
    3. Perianth 0.8–2.2 cm, yellow to orange; ne CA . . . . ***F. pudica***
    3′ Perianth 2–3.5 cm, white to pink or purple; not ne CA
      4. Fls nodding, fragrant, bell-shaped, perianth parts white to pink, oblanceolate, tips acute to
        short-tapered, spreading to recurved; s SNF. . . . . ***F. striata***
      4′ Fls spreading to ± erect, nodding in age, not notably fragrant, not bell-shaped, perianth parts
        pink-purple, obovate, tips rounded to acute, not recurved; NCoRI, n SNF, edges of ScV . . . . ***F. pluriflora***
1′ Style with branches > 1.5 mm
  5. Perianth scarlet to ± purple, gen mottled yellow and/or purple; tips strongly recurved or spreading
    6. Perianth bell-shaped, tips spreading, nectary 1/2 perianth. . . . . ***F. gentneri***
    6′ Perianth tubular, tips gen strongly recurved, nectary 1/5–1/4 perianth . . . . ***F. recurva***
  5′ Perianth not both scarlet to ± purple and mottled, tips gen not recurved (gen recurved or flared in
    Fritillaria eastwoodiae)
    7. Lvs > 10 or in whorls of 2–8 on lower st, not sickle-shaped
      8. Fls spreading to ± erect; upper lvs ± 1/3–1/2 lowest lf; lvs often ≥ infl . . . . [2]***F. pinetorum***
      8′ Fls gen nodding (± spreading); upper lvs gen ± = lowest lf; lvs gen < infl
        9. Perianth purple-brown, mottled yellow or white; lvs 2–3 per node; 1000–3200 m, esp inland mtns
    . . . . [2]***F. atropurpurea***

9′ Perianth dark purple, green-white, green-yellow, or red to ± black; lvs gen ≥ 3 per node; < 1800 m, esp coastal mtns (notable exceptions incl *Fritillaria micrantha, Fritillaria eastwoodiae*)

  10. Perianth dull green-yellow, dark-spotted; nectary ± diamond-shaped to ovate . . . . ***F. ojaiensis***

  10′ Perianth green-white, pale green or ± green or yellow to red or dark purple to ± black, mottled (esp *Fritillaria affinis*); nectary lanceolate or narrower

    11. Perianth pale green to ± black (dark purple), not mottled; nectary ± 1/2 perianth; small bulb scales 0–4; SCoRO(?), SCoRI (Monterey, San Benito, San Luis Obispo cos.) . . . . ***F. viridea***

    11′ Perianth green-white or -yellow to red or ± purple, mottled or not; nectary < 1/2 perianth (1/2–2/3 in *Fritillaria affinis*); small bulb scales gen ≥ 10 (< 10 in *Fritillaria affinis*); NW, CaR, SN, CW

      12. Perianth gen > 2 cm, gen clearly yellow- or purple-mottled, margins wavy — NW, CaRF, n SNF, CW (exc SCoRI) . . . . ***F. affinis***

      12′ Perianth gen < 2 cm, not (or faintly) mottled, margins not wavy

        13. Style branches not or ± recurved; perianth ± green to yellow to red, tips gen recurved or flared; nectary green, gold, or yellow; KR, CaR (Shasta, Tehama, Butte cos.) . . . . ***F. eastwoodiae***

        13′ Style branches recurved; perianth ± purple to green-white, tips not recurved or flared; nectary green-white, dotted purple; SN . . . . ***F. micrantha***

7′ Lvs ≤ 10 or alternate (crowded), occ sickle-shaped

  14. Perianth clearly mottled (white with purple spots or lines and pink shading in *Fritillaria purdyi*); small bulb scales gen > 10 (0–3 in *Fritillaria purdyi*)

    15. Lvs sickle-shaped; fls ± erect; SnFrB, SCoRI (San Benito Co.) . . . . ***F. falcata***

    15′ Lvs gen not sickle-shaped; fls nodding or spreading (spreading to ± erect in *Fritillaria pinetorum*); not SnFrB, SCoRI (San Benito Co.)

      16. Lvs 2–10, ovate, sickle-shaped or not; perianth white, spotted or lined purple; small bulb scales 0–3 . . . . ***F. purdyi***

      16′ Lvs gen > 4, not sickle-shaped, perianth purple-brown, mottled yellow or white; small bulb scales 45–50

        17. Fls nodding; upper lvs ± ≤ lowest cauline lf; cauline lvs gen < infl; KR, CaR, SN, GB, DMtns . . . . ²***F. atropurpurea***

        17′ Fls spreading to ± erect; upper lvs ± 1/3–1/2 lowest cauline lf; cauline lvs often ≥ infl; c SNH (Mono, Inyo cos.), Teh, TR . . . . ²***F. pinetorum***

14′ Perianth not clearly mottled; small bulb scales < 10

  18. Lvs 2–4, sickle-shaped . . . . ***F. glauca***

  18′ Lvs gen > 4, not sickle-shaped

    19. Nectary an obscure narow band 1/2–2/3 perianth; perianth white, striped green . . . . ***F. liliacea***

    19′ Nectary a prominent narrow band 2/3 to = perianth; perianth brown, purple-brown, or green-purple at least adaxially

      20. Perianth green-white or yellow abaxially, purple-brown adaxially; fl ill-scented . . . . ***F. agrestis***

      20′ Perianth dark brown, green-purple or yellow-green; fl ill-scented or not . . . . ***F. biflora***

        21. Lvs narrowly ovate to oblong; fl not ill-scented . . . . var. ***biflora***

        21′ Lvs linear to narrowly lanceolate; fl ill-scented . . . . var. ***ineziana***

F. affinis (Schult. & Schult. f.) Sealy (p. 1389) CHECKER LILY Large bulb scales 2–5, small 2–20. **ST:** 1–12 dm. **LF:** in 1–4 whorls of 2–8 below, alternate above, 4–16 cm, lance-linear to ovate. **FL:** nodding; perianth parts 1–4 cm, oblong to ovate, brown-purple mottled yellow to pale yellow-green mottled purple, nectary 1/2–2/3 perianth, lanceolate, yellow with purple dots; style divided 1/2. **FR:** widely winged. 2*n*=24,36,48. Common. Oak or pine scrub, grassland; < 1800 m. NW, CaRF, n SNF, CW (exc SCoRI); to BC, ID. [*F. a.* var. *tristulis* (A.L. Grant) Ness, ined.] Highly variable; needs study; hybridizes with *F. recurva.* Mar–Jun

F. agrestis Greene (p. 1389) STINKBELLS Large bulb scales 2–9, small 0–2. **ST:** 3–6 dm. **LF:** 5–12, alternate, crowded below mid st, 5–15 cm, linear to lance-oblong. **FL:** nodding, ill-scented; perianth parts 1.8–3.5 cm, ovate, green-white or yellow abaxially, purple-brown adaxially, nectary prominent, 2/3 perianth, narrowly linear, green; style divided 1/2. **FR:** ± angled. 2*n*=24. Clay, often vertic, occ serpentine; < 500 m. NCoRO (Mendocino Co.), SNF, GV, CW. Mar–Jun ★

F. atropurpurea Nutt. (p. 1389) Large bulb scales 2–5, small 45–50. **ST:** 1–6 dm. **LF:** 2–3 per node, 4–12 cm, linear to lanceolate. **FL:** nodding; perianth widely open, parts 1–2.5 cm, oblong to ± diamond-shaped, purple-brown mottled yellow or white, nectary indistinct, ± = perianth, elliptic, yellow with dark ± red dots; style divided > 1/2. **FR:** acute-angled. Common. Often in lf mold under trees; 1000–3200 m. KR, CaR, SN, GB, DMtns; to OR, MT, NM. Confused with *F. pinetorum*, which has upper cauline lvs ± 1/3–1/2 (vs ± =) lowest, and gen a different distribution (the 2 spp. ± overlap only in Teh and Mono, Inyo cos.). Apr–Jul

F. biflora Lindl. CHOCOLATE LILY, MISSION BELLS Large bulb scales 2–8, small 0–4. **ST:** 1–4.5 dm. **LF:** 3–7, alternate, often ± crowded just above ground, 5–19 cm, linear or oblong to narrowly ovate. **FL:** nodding; perianth parts 1.8–4 cm, narrowly ovoid, dark brown, green-purple, or yellow-green, nectary prominent, 2/3 perianth, narrowly linear, ± purple to ± green; style divided 1/2–2/3. **FR:** angled.

  var. ***biflora*** (p. 1389) **LF:** 8–40 mm wide, oblong to narrowly ovate. **FL:** 1.8–4 cm, not ill-scented. 2*n*=24. Grassy slopes, mesas, serpentine barrens; < 1300 m. NCo (Mendocino Co.), NCoR (Mendocino, Napa cos.), CW, SW. Mar–May

  var. ***ineziana*** Jeps. HILLSBOROUGH CHOCOLATE LILY **LF:** 3–6 mm wide, linear to narrowly lanceolate. **FL:** 1.5–2 cm, ill-scented. Serpentine soil; ± 150 m. SnFrB (San Mateo Co.). Occ confused with *F. agrestis*, which gen occurs in heavier soils. Mar–Apr ★

F. brandegeei Eastw. (p. 1389) GREENHORN FRITILLARY Large bulb scales 8–12, small 60–200+. **ST:** 4–10 dm. **LF:** in 1–3 whorls of 4–8 below, alternate above, 4–11 cm, lanceolate. **FL:** nodding; perianth parts 1.2–2 cm, lance-oblong, pink to ± purple, nectary 1/3 perianth, lanceolate, green with ± red edges; style entire. **FR:** winged. 2*n*=24. Granitic soils, open forest; 1500–2100 m. s SN (esp Greenhorn Mtns), Teh. [*Fritillaria brandegei*, orth. var.] Apr–Jun ★

F. eastwoodiae R.M. MacFarl. (p. 1389) BUTTE COUNTY FRITILLARY Large bulb scales 2–5, small 10–60. **ST:** 2–8 dm. **LF:** in 1–2 whorls of 3–5 below, alternate above, 5–10 cm, linear to narrowly lanceolate, ± glaucous. **FL:** nodding; perianth parts 1–1.7 cm, flared to gen recurved at tips, narrowly elliptic, ± green to yellow to red, nectary

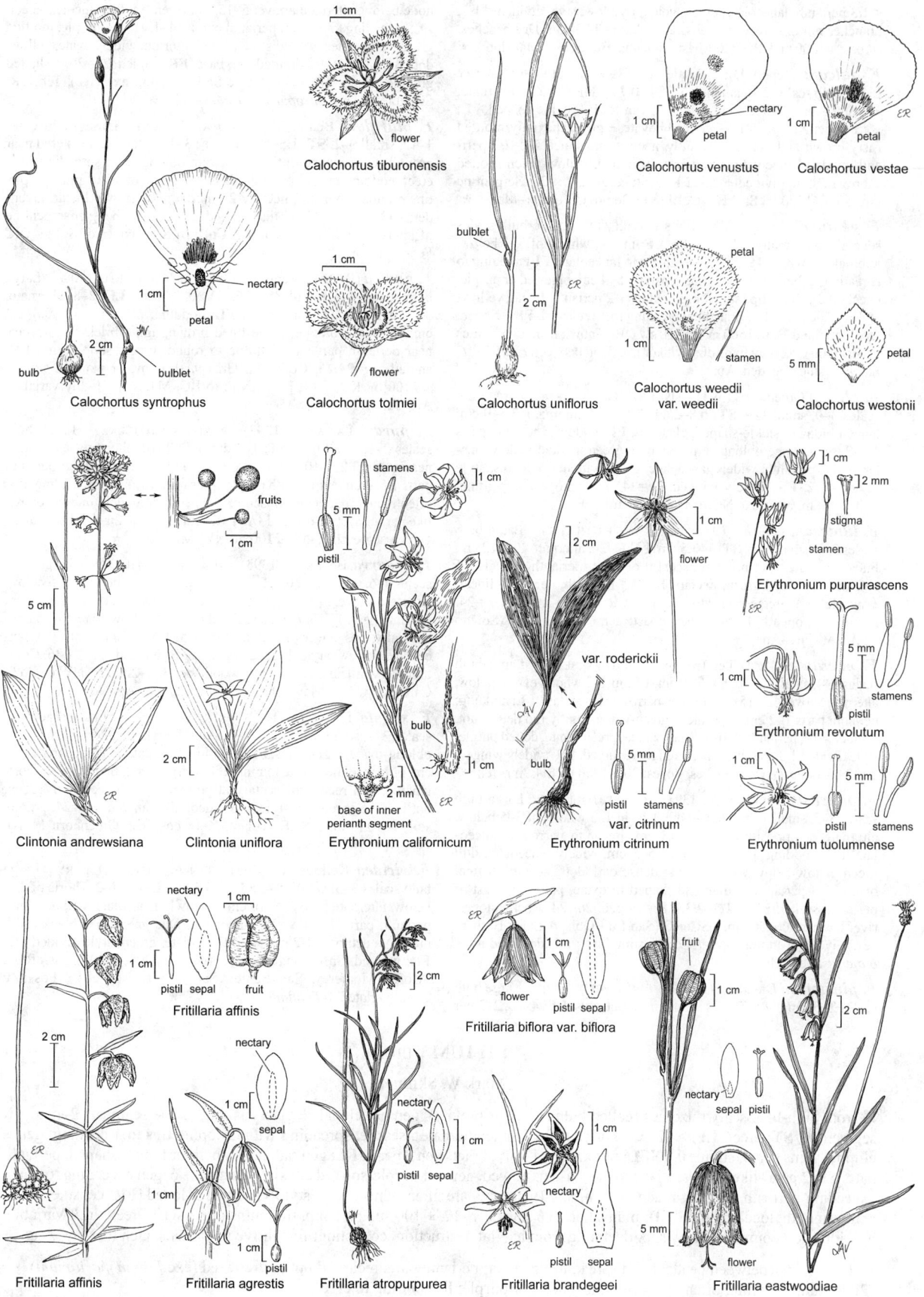

Calochortus tiburonensis

Calochortus venustus

Calochortus vestae

Calochortus syntrophus

Calochortus tolmiei

Calochortus uniflorus

Calochortus weedii var. weedii

Calochortus westonii

Clintonia andrewsiana

Clintonia uniflora

Erythronium californicum

Erythronium purpurascens

var. roderickii

Erythronium citrinum var. citrinum

Erythronium revolutum

Erythronium tuolumnense

Fritillaria affinis

Fritillaria affinis

Fritillaria agrestis

Fritillaria atropurpurea

Fritillaria biflora var. biflora

Fritillaria brandegeei

Fritillaria eastwoodiae

< 1/3 perianth, lanceolate, green, gold, or yellow; style divided < 1/2, branches not or ± recurved. **FR:** angled. 2*n*=24,34,36. Dry benches, slopes; < 1500 m. KR, CaR (Shasta, Tehama, Butte cos.). Mar–Jun ★

F. falcata (Jeps.) D.E. Beetle (p. 1393) TALUS FRITILLARY Large bulb scales 2–4, small 8–32. **ST:** 0.7–2 dm. **LF:** 2–6, alternate, ± fleshy near st, 3.5–8.5 cm, widely linear, folded, sickle-shaped. **FL:** ± erect; perianth parts 1.5–2.2 cm, obovate, ± green abaxially, mottled rusty-brown and yellow adaxially, nectary indistinct, 1/2–2/3 perianth, widely lanceolate to ± diamond-shaped, yellow-green spotted ± brown; style divided ± 2/3. **FR:** acute-angled. 2*n*=24. Serpentine talus; 300–1200 m. SnFrB, SCoRI (San Benito Co.). Mar–May ★

F. gentneri Gilkey GENTNER'S FRITILLARY Large bulb scales several, small many. **ST:** 7–5 dm. **LF:** in 1–3 whorls of 3–5 below, alternate above, 7–15 cm, widely linear to lanceolate. **FL:** nodding to spreading, ± bell-shaped; perianth parts 3.5–4 cm, blue-red or purple, checkered yellow, tips spreading, nectary 1/2 perianth, linear, yellow; style divided 1/4–1/2, branches spreading to ± recurved. **FR:** winged. Dry woodland; 300–1500 m. CaRH; sw OR. Known from only 2 sites in Siskiyou Co.; may hybridize with or is hybrid between *F. affinis*, *F. recurva*; study needed. Apr ★

F. glauca Greene (p. 1393) SISKIYOU FRITILLARIA Large bulb scales 3–9, small 1–9. **ST:** 0.8–2 dm. **LF:** 2–4, alternate, 3.5–9 cm, lance-oblong or sickle-shaped, glaucous. **FL:** nodding; perianth parts 1.5–2.5 cm, lance-oblong, ± purple or ± green marked yellow, nectary < 1/2 perianth, widely lanceolate, green with maroon dots; style divided 1/2. **FR:** widely winged. 2*n*=24. Talus slopes, serpentine; 600–2100 m. NW (exc NCo); s OR. Jun–Jul ★

F. liliacea Lindl. (p. 1393) FRAGRANT FRITILLARY Large bulb scales 2–7, small 1–2. **ST:** 1–3.5 dm. **LF:** 2–20, alternate, 3.5–12 cm, linear to ovate. **FL:** nodding, scent 0 or faint; perianth parts 1–1.6 cm, white, striped green, nectary 1/2–2/3 perianth, narrowly linear, ± purple to ± green; style divided 1/2. **FR:** obtuse-angled. 2*n*=24. Heavy soil, open hills, fields near coast; gen < 200 m. ScV (Solano Co.), CW. Feb–Apr ★

F. micrantha A. Heller (p. 1393) BROWN BELLS Large bulb scales 1–4, small 12–60. **ST:** 4–9 dm. **LF:** in 1–3 whorls of 4–6 below, alternate above, 4.5–15 cm, linear to narrowly lanceolate. **FL:** nodding; perianth parts 1–2 cm, ± purple to green-white, faintly mottled or not, nectary > 1/3 perianth, narrowly lanceolate, green-white, dotted purple; style divided 1/3–2/3, branches strongly recurved. **FR:** widely winged. 2*n*=24. Common. Dry benches, slopes; 300–1800 m. SN. Apr–Jun

F. ojaiensis Davidson (p. 1393) OJAI FRITILLARY Large bulb scales 3–5, small 1–3. **ST:** 4–7 dm. **LF:** in 1–3 whorls of 3–5 below, alternate or opposite above, 4–13 cm, linear to narrowly lanceolate. **FL:** nodding; perianth parts 1.5–3 cm, widely lanceolate, dull green-yellow below with sparse to dense dark dots, nectary distinct or not, 1/3 perianth, ± diamond-shaped to ovate, paler than rest of perianth; style divided 1/2–2/3. **FR:** winged. 2*n*=24. Rocky slopes, river basins; 300–500 m. s SCoRO (San Luis Obispo, Santa Barbara cos.), WTR (Ventura Co.). *F. affinis* sometimes misidentified as *F. ojaiensis*. Feb–May ★

F. pinetorum Davidson (p. 1393) PINE FRITILLARY Large bulb scales 2–5, small 45–50. **ST:** 1–4 dm, ± glaucous. **LF:** 4–20, 2–3 per node below, alternate above, 5–15 cm, often ≥ infl, glaucous, linear. **FL:** spreading to ± erect; perianth parts 1.4–1.9 cm, ± purple, mottled green-yellow, nectary indistinct, 2/3 pinetorum, widely ovate, yellow, dotted brown; styles divided to ± base. **FR:** angled. 2*n*=26. ± shaded granitic slopes; 1800–3200 m. c SNH (Mono, Inyo cos.), Teh, TR. See note under *F. atropurpurea*. May–Jul ★

F. pluriflora Benth. (p. 1393) ADOBE-LILY Large bulb scales 1–12, small 0–2. **ST:** 1.5–4.5 dm. **LF:** 3–10, alternate, clustered near ground, 6–15 cm, elliptic to obovate-oblong. **FL:** spreading to ± erect, nodding in age; perianth parts 2–3.5 cm, obovate, pink-purple, tips rounded to acute, nectary 2/3 perianth, narrowly linear, lavender; style entire. **FR:** obtuse-angled. 2*n*=24. Adobe, gen serpentine of interior foothills; < 900 m. NCoRI, n SNF, edges of ScV; s OR. Feb–Apr ★

F. pudica (Pursh) Spreng. (p. 1393) YELLOW FRITILLARY Large bulb scales 4–5, small 85–125. **ST:** 0.7–3 dm. **LF:** 2–8, alternate, 3–20 cm, linear to lanceolate. **FL:** nodding; perianth parts 0.8–2.2 cm, yellow to orange, some lined brown, aging brick-red, nectary near perianth part base, elliptic to round, green; style entire. **FR:** angled. 2*n*=24,26. Common. Grassy, shrubby, or wooded slopes; < 2100 m. KR, CaR, n SN, MP; to BC, MT, WY. Highly variable. Apr–Jun

F. purdyi Eastw. (p. 1393) PURDY'S FRITILLARY Large bulb scales 2–8, small 0–3. **ST:** 1–4 dm. **LF:** 2–10, alternate, ± crowded near ground, 2.5–10 cm, ovate. **FL:** nodding to spreading; perianth parts 1.5–3 cm, white with purple spots or lines and pink shading, tips often recurved, nectary obscure, ± = perianth, widely linear, colored like perianth; style divided 1/2. **FR:** acute-angled. 2*n*=24. Dry ridges, gen on serpentine; 400–2100 m. NW. Mar–Jun ★

F. recurva Benth. (p. 1393) SCARLET FRITILLARY Large bulb scales 4–6, small 20–30. **ST:** 3–9 dm. **LF:** in 1–3 whorls of 2–5 below, alternate above, 3–15 cm, linear to narrowly lanceolate. **FL:** nodding; perianth parts 1.5–3.7 cm, scarlet, checkered yellow adaxially, purple abaxially, tips recurved, nectary 1/5–1/4 perianth, narrowly lanceolate, yellow; style divided 1/4–1/2. **FR:** winged. 2*n*=24,36. Common. Dry hillsides in scrub or woodland; 300–2200 m. NW (NCo?), CaR, SN; w NV. May hybridize with *F. affinis*. Apr–Jun

F. striata Eastw. (p. 1393) STRIPED ADOBE-LILY Large bulb scales 2–7, small 0–1. **ST:** 2.5–3.8 dm. **LF:** 3–10, alternate, 6–7 cm, oblong-ovate, ± glaucous. **FL:** nodding, fragrant; perianth parts 2–3.5 cm, oblanceolate, white to pink, often striped red, tips acute to short-tapered, gen recurved, nectary at perianth part base, linear, green, outlined in lavender; style ± divided. **FR:** angled. 2*n*=24. Adobe soil; < 1000 m. s SNF (Tulare, Kern cos., esp Greenhorn Mtns). Feb–Apr ★

F. viridea Kellogg (p. 1393) SAN BENITO FRITILLARY Large bulb scales 3–5, small 0–4. **ST:** 3–6.5 dm. **LF:** in 1–2 whorls of 3–4 below, alternate above, 4–10 cm, narrowly lanceolate. **FL:** nodding; perianth parts 0.9–1.8 cm, lanceolate, pale green to ± black (dark purple), nectary ± 1/2 perianth, lanceolate, green; style divided 1/2. **FR:** winged. Shrub understory, serpentine; 200–1500 m. SCoRO?, SCoRI (Monterey, San Benito, San Luis Obispo cos.). Possibly closely related to *F. affinis*. Mar–May ★

LILIUM LILY

Mark W. Skinner

Pl from bulb-like, scaly rhizomes (called bulbs here for brevity), gen not clonal, ± glabrous; bulb scale segments 2–many, if segmented. **ST:** erect. **LF:** > 12, ± whorled (often some alternate), sessile, spreading with drooping tips to ascending, gen ± elliptic; veins gen 3; stipule 0. **INFL:** fls axillary, 1–40+; bracts gen 2 per fl. **FL:** gen radial, gen bell- or funnel-shaped; perianth parts 6 in 2 petal-like whorls, ± lanceolate, base narrowed, gen red-purple-spotted adaxially; stamens 6, gen exceeding to much exceeding perianth, anthers attached at middle (measures are after dehiscence); style 1, stigma 3-lobed. **FR:** capsule, erect, gen ± smooth, loculicidal. **SEED:** many, flat, in 6 stacks. *n*=12. ± 100 spp.: n temp, trop mtns of e Asia. (Greek: lily) Variable, hybridization common. Many spp. declining from habitat destruction, collecting; few thrive in gardens. Gen fls May–Aug.

1. Fl nodding (between pendent and spreading) to erect; perianth parts recurved but not reflexed (exc *Lilium maritimum*)
  2. Perianth ± white, often aging pink, lavender, or purple; fls nodding to erect
    3. Perianth parts 4.2–6.6 cm; fl ascending to erect; NW, SnFrB . . . . **L. rubescens**

3′ Perianth parts gen 6.2–11.3 cm; fl nodding to ascending; KR, CaR, SN . . . . ***L. washingtonianum***

  4. Perianth aging deep pink or lavender; perianth gen 6.2–9.5 cm; KR. . . . . subsp. ***purpurascens***

  4′ Perianth aging ± white (light pink); perianth 7.9–11.3 cm; KR, CaR, SN . . . . subsp. ***washingtonianum***

2′ Perianth yellow to red or pink; fls nodding to ascending

  5. Perianth bright yellow, parts 7.7–10.7 cm; TR, PR . . . . ***L. parryi***

  5′ Perianth ± red, orange, or pink (pale yellow), parts ≤ 5 cm; s NCo, KR, n&c SNH, n CCo

    6. Lf 1.8–7 cm, gen glaucous, margins gen wavy; dry places; KR. . . . . ***L. bolanderi***

    6′ Lf 3–18 cm, not obviously glaucous, margins gen not wavy; ± wet or seasonally inundated places; s NCo, n&c SNH, n CCo

      7. Perianth red or red-orange abaxially, adaxially (exc base green), tips reflexed or rolled; fl nodding (spreading); fr 2.4–4.1 cm; s NCo, n CCo . . . . ***L. maritimum***

      7′ Perianth light orange to red (± pink) on distal 40%, lighter near base, sometimes uniformly light orange (yellow), abaxially paler, tips ± recurved; fl nodding to ascending; fr 1.5–2.7 cm; n&c SNH . . . . ***L. parvum***

1′ Fl pendent (sometimes some nodding); perianth parts reflexed

  8. Perianth pink to white, gen aging pink, midribs yellow — KR, n NCoRO . . . . ***L. kelloggii***

  8′ Perianth yellow to red

    9. Pl of ± dry places; perianth yellow or orange; bulb erect to ascending

      10. Pistil 2.4–3.7 cm; perianth parts 3.4–7.1 cm; anthers 5–13 mm; nw NW. . . . . ***L. columbianum***

      10′ Pistil 4.5–7.1 cm; perianth parts gen 5.6–9.5 cm; anthers 11–19 mm; s CaRH, n SNH, s CW, SW . . . ***L. humboldtii***

        11. Perianth orange, red spots unmargined; s CaRH, SNH. . . . . subsp. ***humboldtii***

        11′ Perianth yellow or light orange, red spots near tip margined with lighter red; s CW, SW. . . . . subsp. ***ocellatum***

    9′ Pl of ± moist places; perianth yellow to ± red; bulb spreading

      12. Perianth ± uniformly yellow, orange, or yellow-orange (green at base)

        13. Fl ± fragrant; anthers 3–6 mm, magenta or dull red; pistil 2.5–3.5 cm; fr 1.5–3 cm; not clonal; c&s SNH . . . . ***L. kelleyanum***

        13′ Fl not fragrant; anthers 5–13 mm, pale yellow; pistil 3–4.3 cm; fr 2.3–4.2 cm; ± clonal; e KR . . . . ***L. pardalinum*** subsp. ***wigginsii***

      12′ Perianth ± 2-toned, with green, yellow, or orange centers, darker orange or red tips

        14. Perianth red to maroon (orange), center green to yellow, abaxially green on basal 40–50%, parts 4.3–8.1 cm; filaments ± parallel; rhizome unbranched; n NCo. . . . . ***L. occidentale***

        14′ Perianth yellow to orange, tips redder, abaxially green only on basal ± 20%, parts 3.4–10.4 cm; filaments not parallel; rhizome often branched; CA-FP . . . . ***L. pardalinum*** (most)

          15. Perianth parts 5.9–10.4 cm; gen in large clones; anthers 11–22 mm, lvs gen ± elliptic; CA-FP . . . . subsp. ***pardalinum***

          15′ Perianth parts 3.4–8.3 cm; gen in small clones; anthers 5–18 mm, lvs gen ± elliptic to ± linear; KR, s NCoRO, CaR, n SNH

            16. Lf ± linear, length 7–34 × width; w KR . . . . subsp. ***vollmeri***

            16′ Lf ± elliptic, length 3–17 × width; e KR and elsewhere

              17. Pollen red- or brown-orange; anthers magenta; s NCoRO . . . . subsp. ***pitkinense***

              17′ Pollen gen yellow to bright orange (red-orange); anthers orange to magenta; KR, CaR, n SNH . . . . subsp. ***shastense***

L. bolanderi S. Watson (p. 1393) BOLANDER'S LILY Pl < 1.1 m, glaucous; bulb erect, ± ovoid, scales unsegmented, longest 3–6 cm. **LF:** in 2–6 whorls, ± ascending, often cupping st, 1.8–7 cm, gen ± obovate or ± oblanceolate, gen distinctly glaucous; margin gen wavy. **INFL:** fls 1–9, nodding to spreading. **FL:** narrowly bell-shaped, not fragrant; perianth parts 3–4.7 cm, ± recurved in distal 20–40%, red or magenta (salmon, pale yellow), adaxially basal 30–50% often ± yellow; stamens exceeded by to held at ± same level as perianth, filaments ± parallel, anthers 3–8 mm, ± red or magenta, pollen red-brown, orange, or yellow; pistil 2–3.5 cm. **FR:** 2–4 cm. Serpentine soil in chaparral, conifer forest, gen with *Xerophyllum*; 150–1500 m. KR; sw OR. Hybridizes with *L. pardalinum*, *L. washingtonianum* subsp. *purpurascens*, *L. rubescens*. Jun–Aug ★

L. columbianum Leichtlin COLUMBIA LILY Pl < 1.7 m, glaucous or not; bulb erect-ovoid to oblique-elongate, scales (1)2–3(4–5)-segmented, longest 3–7.3 cm. **LF:** in 1–9 whorls, gen ascending, 1.5–16 cm, ± oblanceolate to ± obovate or not; margin gen wavy. **INFL:** fls 1–25(45), pendent to nodding. **FL:** widely bell-shaped, not fragrant; perianth parts 3.4–7.1 cm, reflexed in distal 50–60%, orange (often ± red abaxially); stamens ± exceeding perianth, filaments spreading, anthers 5–13 mm, ± yellow, pollen orange or yellow; pistil 2.4–3.7 cm. **FR:** 2–5.4 cm. Dry scrub, coastal prairie, gaps and roadsides in conifer forest esp along coast; < 1300 m. nw NW; to BC, ID. Variable. Hybridizes with *L. pardalinum* subsp. *vollmeri*, *L. pardalinum* subsp. *wigginsii*; possibly with *L. kelloggii* in Del Norte Co. along Highway 199 near OR border. Jun–Jul

L. humboldtii Duch. Pl < 3.1 m; bulb ± erect-ovoid or oblique-elongate, scales segmented or not, longest 3–12 cm. **ST:** brown-purple or not. **LF:** in 2–8 whorls, gen ascending, 4.5–14.5 cm, gen ± oblanceolate; margin gen wavy. **INFL:** fls 1–33(40), pendent. **FL:** ± widely bell-shaped, not fragrant; perianth parts gen 5.6–9.5 cm, reflexed in distal 80%, orange or yellow, adaxially ridged basally; filaments diverging widely, anthers 11–19 mm, purple, pollen red-brown to tan-yellow; pistil 4.5–7.1 cm. **FR:** 2.5–5.4 cm, ribbed.

subsp. ***humboldtii*** HUMBOLDT LILY Bulb scales off-white, purple-speckled or not, unsegmented. **FL:** perianth orange, magenta spots unmargined; pollen red- or rusty-brown or rusty-orange. Yellow-pine forest, chaparral openings; (200)600–1100 m. s CaRH, n SNH. Jun–Aug ★

subsp. ***ocellatum*** (Kellogg) Thorne (p. 1393) OCELLATED HUMBOLDT LILY Bulb scales often purple at tip, obscurely (1)2–5-segmented. **FL:** perianth yellow or light orange, spots margined lighter red (toward tip larger, with wider margins); pollen ± tan or tan-yellow. Oak canyons, chaparral, yellow-pine forest; < 1800 m. s CW, SW. May–Aug ★

L. kelleyanum Lemmon KELLEY'S LILY Pl < 2.2 m; bulb spreading-elongate, scales (1)2–3-segmented, longest 1–3 cm. **LF:** in 1–4 whorls, spreading, drooping at tips, 7.5–17.5 cm; margin not wavy. **INFL:** fls 1–15(25), pendent. **FL:** ± widely bell-shaped, mild-fragrant; perianth parts 4.2–6 cm, reflexed in distal 60%, yellow or

yellow-orange; filaments diverging, anthers 3–6 mm, magenta or dull red, pollen orange, tan-orange, or red-brown; pistil 2.5–3.5 cm. **FR:** 1.5–3 cm. Hillside seeps, wet thickets, streamsides in subalpine forest; 2200–3300 m. c&s SNH. Intergrades with *L. parvum* in s Mono Co. Jul–Aug

L. kelloggii Purdy KELLOGG'S LILY Pl < 2.1 m; bulb ± erect-ovoid, scales unsegmented, longest 3–6.5 cm. **LF:** in 2–7 whorls, 6–16.5 cm, gen ± narrowly elliptic; margin gen not wavy. **INFL:** fls 1–27, pendent. **FL:** ± widely bell-shaped, fragrant; perianth parts 3.2–7.2 cm, reflexed in distal 67%, pink to white, gen aging pink, midrib yellow; filaments diverging, anthers 5–14 mm, pale red-orange or magenta, pollen orange; pistil 2.9–4 .2 cm. **FR:** 3–6 cm. Gaps, roadsides in conifer forest or chaparral; 200–1300 m. KR, n NCoRO; sw OR. Jun–Aug ★

L. maritimum Kellogg (p. 1393) COAST LILY Pl < 2.5 m (< 0.3 m on coastal bluffs); bulb ± spreading-elongate, lumpy, scales 1(2)-segmented, longest 1.5–4 cm. **LF:** basal, alternate, or in 1–3 whorls, 3–18 cm, narrow or not; margin not wavy. **INFL:** fls 1–13, nodding (spreading). **FL:** bell-shaped, not fragrant; perianth parts 3.4–5 cm, elliptic, reflexed or rolled in distal 20–50%, red or red-orange, darker spots concentrated mid-basally, surrounded by light orange (or yellow-green); stamens held at same level as perianth, filaments ± parallel, anthers 4–12 mm, light magenta, pollen orange; pistil 2.2–3.2 cm. **FR:** 2.4–4.1 cm. Coastal prairie or scrub, peatland, gaps in closed-cone-pine forest; < 150 m. s NCo, n CCo. Hybridizes with *L. pardalinum* subsp. *p.* May–Jul ★

L. occidentale Purdy (p. 1393) WESTERN LILY Pl < 2.2 m; bulb spreading-elongate, unbranched, scales 1–2(3)-segmented, longest 0.9–2.5 cm. **LF:** in 1–9 whorls (alternate), 4–27 cm, ± linear or not; margin not wavy. **INFL:** fls 1–13(35), pendent. **FL:** ± widely bell-shaped, not fragrant; perianth parts 4.3–8.1 cm, reflexed in distal 60%, ± 2-toned, adaxially red to maroon (orange) on distal 50–60%, base yellow to green, with a band of orange or yellow between or not, abaxially strongly green on basal 40–50%; filaments ± parallel, anthers 5–14 mm, dull red or magenta, pollen gen red-brown (± orange); pistil 3–5.5 cm. **FR:** 2.1–5.4 cm. Coastal scrub or prairie, gaps in conifer forest; < 100 m. n NCo (near Crescent City, Del Norte Co.; Humboldt Bay, Humboldt Co.); sw OR. In disjunct colonies. Central green star of ± all perianths ages yellow or yellow-orange. Threatened by habitat loss, grazing, competition, collecting. Jun–Aug ★

L. pardalinum Kellogg Pl < 2.8 m, ± clonal; bulb spreading-elongate, often branched, scales (1)2–4-segmented, longest 1–3.3 cm. **LF:** alternate or in 1–8 whorls, 4–27 cm, gen ± elliptic; margin gen not wavy. **INFL:** fls 1–28(35), pendent. **FL:** ± widely bell-shaped, gen not fragrant; perianth parts 3.4–10.4 cm, reflexed in distal 67–75%, gen ± 2-toned, adaxially gen pale orange to red on distal 25–60%, lighter near base, with maroon spots near tip margined yellow or orange, abaxially paler and green on basal ± 20%; filaments ± widely diverging, anthers 5–22 mm, ± magenta to orange or yellow, turning darker, pollen red-brown to yellow, turning lighter; pistil 3–8 cm. **FR:** 2.3–6 cm.

subsp. ***pardalinum*** LEOPARD LILY Pl < 2.8 m, strongly clonal; bulb scales (1)2(3)-segmented. **LF:** whorled. **FL:** rarely fragrant; perianth parts 5.9–10.4 cm, 2-toned, tips darker; anthers 11–22 mm, pale magenta or magenta, turning purple (or ± yellow), pollen red- to brown-orange; pistil 5–8 cm. **FR:** 3–6 cm. Moist places (drier along coast); < 1700 m. CA-FP. Hybridizes with *L. parvum*; ± replaced by *L. parryi* in SW mtns. Variable. Intergrades with *L. pardalinum* subsp. *shastense* in Siskiyou Co. (Salmon Mtns), confusingly with *L. pardalinum* subsp. *wigginsii* in n Humboldt Co. (near Onion Mtn). May–Aug

subsp. ***pitkinense*** (Beane & Vollmer) M.W. Skinner PITKIN MARSH LILY Pl < 2 m, moderately clonal; bulb scales (1)2-segmented. **LF:** whorled. **FL:** perianth parts 4.9–7.1 cm, 2-toned, tips darker; anthers 6–11 mm, magenta, pollen red- or brown-orange; pistil 3.4–4.6 cm. Marshes, valley-oak scrub; 35–60 m. s NCoRO (Pitkin Marsh, Sonoma Co.). Barely distinct from *L. pardalinum* subsp. *p.* Threatened by habitat loss, competition, collecting. Jun–Jul ★

subsp. ***shastense*** (Eastw.) M.W. Skinner SHASTA LILY Pl < 2.1 m, weakly clonal; bulb scales (1)2–4-segmented. **LF:** whorled (or alternate in young pls). **FL:** perianth parts 3.7–7.6 cm, 2-toned, tips darker; anthers 5–14 mm, orange to magenta, pollen yellow to bright orange, occ red-orange; pistil 3.3–4.4 cm. **FR:** 2.2–4.3 cm. Wet meadows, streamsides in conifer forest; 1100–1800 m. KR, CaR, n SNH; sw OR. Intergrades with *L. pardalinum* subsp. *wigginsii* in KR (Marble Mtns, Siskiyou Co.). Hybridizes with *L. parvum* (Sierra Co. near Stampede Reservoir and Yuba Pass). Jul–Aug

subsp. ***vollmeri*** (Eastw.) M.W. Skinner (p. 1393) VOLLMER'S LILY Pl < 1.7 m, weakly clonal; bulb scales 1–2-segmented. **LF:** whorled (or alternate esp in small pls), ± linear. **FL:** perianth parts 4.8–8.3 cm, 2-toned, tips darker; anthers 5–18 mm, magenta or purple, pollen red-orange or orange; pistil 3.5–5.3 cm. **FR:** 2.5–5 cm. Peatland, streams, springs; 100–1200 m. w KR; sw OR. Hybrid swarms with *L. pardalinum* subsp. *wigginsii* in w Siskiyou Co. Pls in deep shade gen have wider lvs, are much like *L. pardalinum* subsp. *p.* Jul–Aug ★

subsp. ***wigginsii*** (Beane & Vollmer) M.W. Skinner (p. 1393) WIGGINS' LILY Pl < 1.7 m, weakly clonal; bulb scales 2–4-segmented. **LF:** whorled (or alternate esp in small pls). **FL:** perianth parts 3.4–7.1 cm, not 2-toned (gen uniformly orange or yellow-orange); stamens often malformed or shrunken, anthers 5–13 mm, pale yellow, pollen yellow or orange; pistil 3–4.3 cm. **FR:** 2.3–4.2 cm. Wet thickets, meadows, streams among conifers; 800–2000 m. e KR; sw OR. Jul–Aug ★

L. parryi S. Watson (p. 1393) LEMON LILY Pl < 1.9 m; bulb spreading-elongate, scales (1)2(4)-segmented, longest 0.8–2.7 cm. **LF:** in 1–5 whorls (or alternate in young pls), 7–29 cm, narrowly linear or not; margin not wavy. **INFL:** fls 1–31, spreading or ± nodding. **FL:** gen ± bilateral, funnel-shaped, strongly fragrant; perianth parts 7.7–10.7 cm (inner wider), ± oblanceolate, recurved in distal 40%, bright yellow, maroon spots gen sparse, minute; stamens ± exceeding perianth, filaments ± parallel, anthers 8–14 mm, pale magenta-brown, pollen rusty- or brown-orange; pistil 5.3–9.3 cm. **FR:** 4–6 cm. Meadows, streams in montane conifer forest; 1300–2600 m. TR, PR; AZ. [*L. p.* var. *kessleri* Davidson] Jun–Sep ★

L. parvum Kellogg ALPINE LILY, SIERRA TIGER LILY Pl < 1.7 m; bulb spreading-elongate, scales (1)2–3(4)-segmented, longest 1–3.5 cm. **LF:** in 2–5 whorls, 4–15 cm; margin gen not wavy. **INFL:** fls 1–26(41), nodding to ascending. **FL:** ± bilateral or not, ± funnel-shaped, not fragrant; perianth parts 3.2–4.2 cm (outer wider), ± recurved in distal 33–40%, adaxially light orange to red (± pink) on distal 40%, lighter near base, uniformly light orange or not (yellow), abaxially paler; stamens ± exceeding perianth, filaments ± diverging, anthers 3–8 mm, pale yellow, ± orange, or magenta, pollen ± yellow to red-orange; pistil 2.3–3.7 cm. **FR:** 1.5–2.7 cm. Wet meadows, willow thickets, streams in conifer forest; 1400–2900 m. n&c SNH. Jun–Aug

L. rubescens S. Watson (p. 1399) REDWOOD LILY Pl < 2 m, often glaucous; bulb ± erect-ovoid, scales unsegmented, longest 4–9 cm. **LF:** in 3–9 whorls, gen ± ascending, 3–13 cm, gen oblanceolate; margin gen wavy. **INFL:** fls 1–40, ascending to erect. **FL:** funnel-shaped, fragrant; perianth parts 4.2–6.6 cm (inner wider, strongly oblanceolate), recurved in distal 33–50%, adaxially white, turning pink-purple, magenta spots minute, abaxially often ± red or ± purple; stamens held at same level as perianth, filaments ± parallel exc distally, anthers 4–8 mm, pale yellow, pollen yellow; pistil 2.7–3.8 cm. **FR:** 2–3.7 cm, gen ribbed. Dry soils in chaparral, gaps in conifer forest; 30–1800 m. NW, SnFrB. May–Aug ★

L. washingtonianum Kellogg WASHINGTON LILY Pl < 2.6 m, often glaucous; bulb oblique-elongate to ± erect-ovoid, scales unsegmented, 2-segmented, or indistinctly 2(3)-segmented, longest 3.3–12 cm. **LF:** in 1–9(14) whorls, spreading to ascending, ± clasping st or not, 3–13 cm, gen oblanceolate; margin wavy or not. **INFL:** fls 1–33, nodding to ascending. **FL:** gen ± bilateral, ± funnel-shaped, strongly fragrant; perianth parts gen 6.2–11.3 cm (inner wider, strongly oblanceolate), recurved in distal 25–33%, white, turning deep pink or not, magenta spots minute; stamens ± exceeding perianth, filaments ± parallel, anthers 8–15 mm, off-white or cream, pollen yellow or cream; pistil 7.5–10.4 cm. **FR:** 2.7–5.8 cm.

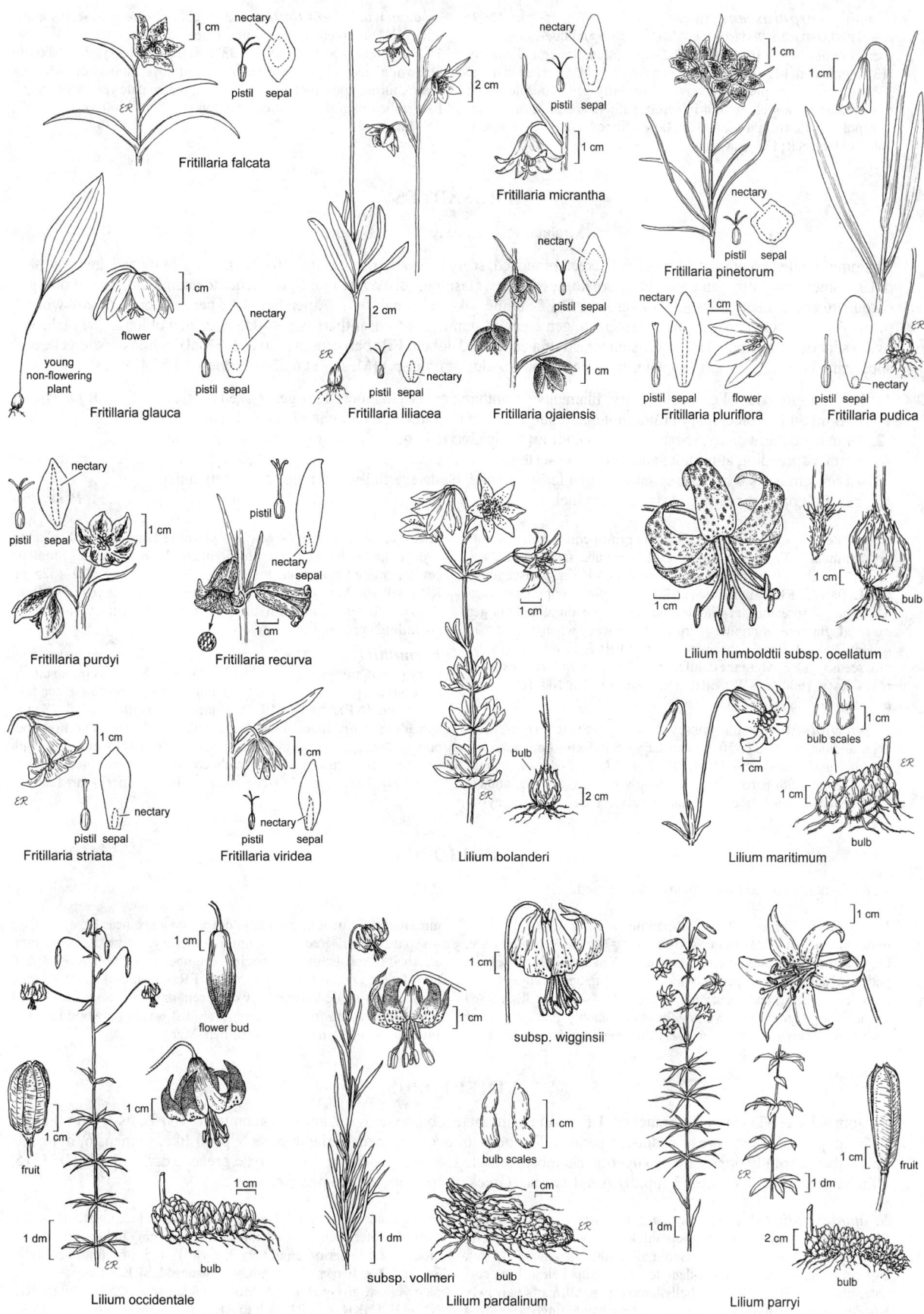

Fritillaria falcata

Fritillaria glauca

Fritillaria micrantha

Fritillaria pinetorum

Fritillaria liliacea

Fritillaria ojaiensis

Fritillaria pluriflora

Fritillaria pudica

Fritillaria purdyi

Fritillaria recurva

Lilium humboldtii subsp. ocellatum

Fritillaria striata

Fritillaria viridea

Lilium bolanderi

Lilium maritimum

Lilium occidentale

Lilium pardalinum

subsp. wigginsii

subsp. vollmeri

Lilium parryi

young non-flowering plant

flower

nectary

pistil sepal

nectary

pistil sepal

nectary

pistil sepal

nectary

pistil sepal

nectary

pistil sepal

nectary

pistil sepal

flower

nectary

pistil sepal

nectary

pistil sepal

nectary

pistil sepal

pistil

nectary sepal

nectary

pistil sepal

nectary

pistil sepal

bulb

bulb

bulb scales

bulb

fruit

bulb

flower bud

bulb scales

bulb

fruit

bulb

subsp. ***purpurascens*** (Stearn) M.W. Skinner (p. 1399) PURPLE-FLOWERED WASHINGTON LILY **Bulb** oblique-elongate to ± erect-ovoid, scales unsegmented, 2-segmented, or gen indistinctly 2(3)-segmented. **FL:** perianth parts gen 6.2–9.5 cm, recurved in distal 33%, turning deep pink or lavender, abaxially gen ± purple, often faintly so, adaxial yellow midrib gen 0; anthers cream, turning yellow, pollen pale (bright) yellow. **FR:** gen ribbed. Douglas-fir forest; 300–2000 m. KR; OR. Jun–Aug ★

subsp. ***washingtonianum*** **Bulb** oblique-elongate, scales indistinctly 2(3)-segmented or gen unsegmented. **FL:** perianth parts 7.9–11.3 cm, recurved in distal 25–33% rarely turning pink, abaxially gen white, adaxial yellow midrib in basal 50%; anthers off-white or cream, turning pale pink or dirty yellow, pollen pale yellow or cream. **FR:** ribbed or not. Mixed conifer forest; 1050–1900 m. KR, CaR, SN. Jun–Aug

PROSARTES

Michael R. Mesler & Robin Bencie

Rhizome slender, creeping to ± erect. **ST:** erect, branched, scaly below, lfy above, hairs 0 or gen sharp-branched, gen not glandular, some or all falling in age. **LF:** alternate, sessile to ± clasping, (ob) ovate to elliptic, acute to acuminate, base ± oblique, cordate to acute, main veins ≥3, converging. **INFL:** umbel-like, terminal; fls 1–7, pendent. **FL:** perianth parts 6, in 2 whorls, petal-like, free, white to ± green, bases green, gen convex; stamens 6, free, anthers gen < filaments, gen oblong, gen glabrous; ovary superior, chambers (1)3, style slender, stigma entire or 3-lobed. **FR:** berry, orange to red. **SEED:** white to pale yellow. 6 spp.: temp N.Am. (Greek: to append, from pendulous ovules of type sp.) [Mesler et al. 2010 Madroño 52:129–135]

1. St, lf margin, pedicel glandular-hairy; filaments << anthers; ovary chambers 1; fr gen 2-seeded . . . . ***P. parvifolia***
1′ St, lf margin, pedicel hairy or not, nonglandular; filaments > anthers; ovary chambers 3; fr gen > 3-seeded
  2. Lf margin hairs many, short, sharp, ascending, blade abaxially minute-scabrous, esp on veins; perianth
    parts ± spreading at middle; anthers gen exserted. . . . . ***P. hookeri***
  2′ Lf margin hairs 0 or long, spreading, gen falling in age, blade abaxially glabrous or sparsely hairy;
    perianth parts spreading at tip; anthers incl. . . . . ***P. smithii***

P. hookeri Torr. (p. 1399) St, lf margin, pedicel hairy or not, nonglandular. **ST:** 30–80 cm. **LF:** 3–15 cm; abaxially minute-scabrous, esp on veins, margin hairs many, short, sharp, ascending. **INFL:** fls 1–3. **FL:** ± funnel-shaped, tube ± cylindric; perianth parts 8–18 mm, ± spreading at middle; filaments spreading, anthers gen exserted, glabrous or minute-sharp-bristly, ± wavy when open; ovary weakly 3-sided, style glabrous or ± hairy. **FR:** 7–12 mm, ± spheric, red; seeds gen > 3. Montane conifer, mixed-evergreen forest, exposed roadsides; < 1600 m. NW, CaRH, SN, CW; to w Can, MT. [*Disporum h.* (Torr.) G. Nicholson] Mar–Jun

P. parvifolia S. Watson SISKIYOU BELLS St, lf margin, pedicel glandular-hairy. **ST:** 30–80 cm. **LF:** 1.5–5.5 cm; both surfaces, margins glandular-hairy. **INFL:** fls 1–4. **FL:** bell-shaped to narrowly so; perianth parts 8–10 mm, widely rounded at base; anthers >> filaments, lanceolate, surrounding style base; ovary ± cylin-

dric, chambers 1, ovules gen 2, style base hairy. **FR:** 10–13 mm, ± depressed-spheric, orange to orange-red; seeds gen 2. Montane conifer, mixed-evergreen forest, exposed roadsides; 600–1525 m. KR (Siskiyou Mtns); sw OR. [*Disporum hookeri* var. *p.* (S. Watson) Britton] Pls fertile, not of hybrid origin, clearly distinct, contrary to prevailing opinion. May–Jun ★

P. smithii (Hook.) Utech et al. (p. 1399) St, lf margin, pedicel hairy or not, nonglandular. **ST:** 30–100 cm. **LF:** 5–12 cm; abaxially glabrous or sparsely hairy, margin hairs 0 or long, spreading, gen falling in age. **INFL:** fls 1–7. **FL:** ± cylindric; perianth parts 15–28 mm, spreading at tip; filaments ± erect, anthers incl, surrounding upper part of style; ovary 3-sided, style hairy. **FR:** 10–15 mm, obovoid, orange; seeds gen > 3. Moist, shady coastal, montane forest; < 1575 m. w NW, SnFrB; to BC. [*Disporum s.* (Hook.) Piper] Mar–Jun

SCOLIOPUS

1 sp. (Greek: crooked foot, from curving pedicel)

S. bigelovii Torr. (p. 1399) **Rhizome** short, slender. **ST:** short, underground. **LF:** 2(3) basal, sheathing, 5–20 cm, 5–10 cm wide, elliptic to oblong, many-veined, mottled. **INFL:** umbel, subsessile; peduncle underground; pedicels 1–12, 8–20 cm, appearing scapose, 3-angled, twisting and recurving so fr touches soil. **FL:** ill-scented when fresh, ± green or ± yellow, mottled, heavily lined purple or dark brown; perianth parts 6 in 2 very different whorls; sepals 14–17

mm, narrowly ovate, spreading, deeply concave near base; petals ± = sepals, linear, erect or arching over ovary; stamens 3, 5–6 mm, attached to sepals; ovary 3-angled, chamber 1, style branches 3, 5–6 mm, linear, ± recurved or spreading. **FR:** capsule-like, 15–18 mm, elliptic, thin-walled, irregularly dehiscent or indehiscent. **SEED:** ± 3 mm, oblong, with food body. *n*=7,8. Moist, shady redwood forest; < 1100 m. NCoRO, n CCo, SnFrB. Feb–Mar

STREPTOPUS

Rhizomes long. **ST:** simple or branched. **LF:** cauline, alternate, oblong to ovate, acute to acuminate. **INFL:** fls 1(2) in lf axils, with sharp bend at juncture of peduncle, pedicel. **FL:** bell- to saucer-shaped; perianth parts 6, petal-like; stamens 6, filaments flat, anthers abruptly tipped; ovary superior, chambers 3, style 1, entire or 3-lobed. **FR:** berry, ± green to dark red. **SEED:** 3–15, elliptic, grooved or wrinkled. 7 spp.: N.Am, Eurasia. (Greek: twisted foot, from peduncles)

S. amplexifolius (L.) DC. var. ***americanus*** Schult. & Schult. f. (p. 1399) **Rhizome** thick; roots thick, fibrous. **ST:** 30–100 cm, much-branched, glabrous to densely hairy. **LF:** ± sessile, 5–15 cm, 2–5 cm wide, entire or minute-dentate, ± glaucous below, base cordate, clasping. **FL:** 9–15 mm, bell-shaped; perianth parts narrowly lance-oblong, spreading to ± recurved at tips, white, tinged (yellow-)

green, outer flat, inner ± keeled, ± narrower, clasping stamens; stamens 6, outer filaments ± 1 mm, inner 2–3 mm, anthers attached at base; ovary superior, chambers 3, style 1, 4–5 mm, thick, entire or 3-lobed. **FR:** berry, 10–15 mm, yellow or red. **SEED:** many, ± 3 mm, pale yellow, grooved. *n*=16. Moist, shaded areas; 250–1700 m. KR, NCoRH, CaRH, n SNH, MP; to AK, c N.Am. May–Jun

TULIPA

Bulb coat gen hairy adaxially. **LF**: cauline, alternate, ± fleshy, ± smaller upward. **INFL**: fls 1[4]. **FL**: perianth parts 6 in 2 petal-like whorls, free; stamens 6, anthers attached at base; ovary superior, chambers 3, ovules many, style 1, short or 0, stigma 3-lobed. **FR**: capsule, loculicidal, spheric or elliptic. **SEED**: many, flat, black. ± 150 spp.: n temp, Eurasia. (Persian or Turkish: turban)

T. clusiana DC.   TULIP   Bulb 2–3.5 cm, 1–2.5 cm wide, coat long-wavy-hairy adaxially, forming a felt that protrudes from top. **LF**: 2–6(12), linear to lance-linear. **FL**: perianth parts gen with a basal, small ± purple blotch adaxially, outer 30–60 mm, lance-elliptic to elliptic, gen white with pink-red tinge or wide central band abaxially or not, inner 25–50 mm, oblong-oblanceolate to obovate, gen white; anthers 4.5–9 mm; ovary ± 10 mm, elliptic. Disturbed places; < 500 m. PR (spontaneous at site of former Desert Nursery near Riverside); e Eur; native Iran to n Pakistan. Spring–summer

MELANTHIACEAE   FALSE-HELLEBORE FAMILY

Dale W. McNeal, except as noted

Per, from rhizome or bulb, or rhizomes ending in weakly developed bulbs, scapose or not. **LF**: alternate, whorled, or mostly basal and spirally arranged, deciduous after 1 yr or not. **INFL**: raceme, panicle, or fls 1. **FL**: perianth parts 6, in 2 petal-like whorls or of sepals and petals, free or fused below, ± spreading; stamens 6, from perianth, anthers attached at base or near middle; ovary superior or partly inferior, chambers 3, styles 3, persistent. **FR**: capsule, loculicidal or septicidal. 10 genera, 130 spp.: n hemisphere. W N.Am *Zigadenus* moved to *Toxicoscordion*. Scientific Editors: Dale W. McNeal, Thomas J. Rosatti.

1. Lvs linear
  2. Lvs persistent after 1 yr, wiry, tough; cauline lvs many, reduced upward . . . . **XEROPHYLLUM**
  2′ Lvs deciduous after 1 yr, not wiry, not tough; cauline lvs 0 or 1–2, much reduced
    3. Perianth parts pale green-yellow to ± purple, without glands; ovary partly inferior (at base). . . . . . **STENANTHIUM**
    3′ Perianth parts white to ± yellow, with 1–2 distinct glands near base; ovary superior. . . . . **TOXICOSCORDION**
1′ Lvs lanceolate to widely ovate
  4. Lvs > 3, alternate; pl 1–2 m . . . . **VERATRUM**
  4′ Lvs 3, in 1 whorl subtending fl; pl < 0.7 m
    5. Lvs petioled (subsessile), leathery; petals white (pink), gen purple-spotted . . . . **PSEUDOTRILLIUM**
    5′ Lvs sessile or subsessile, not leathery; petals white, ± pink, ± yellow, or purple, not spotted . . . . **TRILLIUM**

PSEUDOTRILLIUM

1 sp.: CA, OR. (Like *Trillium*) [Farmer 2006 Aliso 22:579–592]

P. rivale (S. Watson) S.B. Farmer (p. 1399)   BROOK WAKEROBIN   **ST**: 0.5–2 dm. **LF**: 3, in 1 whorl subtending fl, petioled (subsessile), blade 1.7–11 cm, lanceolate to cordate, acute to acuminate, leathery. **INFL**: fls 1 per st; peduncle gen erect as fl opens, becoming recurved and continuing to elongate below bracts after pollination until ovary contacts soil. **FL**: bisexual; sepals 1–2 cm, 4–11 mm wide, lanceolate or ovate to oblong; petals 1.5–3 cm, 0.8–1.9 cm wide, ovate-cordate, white (pink), gen purple-spotted; stamens 6, 6–15 mm; ovary chambers 3 stigmas 3, recurved. **FR**: capsule, berry-like, ± pulpy, rarely dehiscent, gen shed as a unit. *n*=5. Yellow-pine forest along rocky streambanks and in *Darlingtonia* bogs, gen serpentine; 40–1500 m. KR; sw OR. [*Trillium r.* S. Watson] Apr–Jun

STENANTHIUM

ST: ± scapose, simple or branched. **LF**: mostly basal, grass-like; cauline 1–2, much reduced; deciduous after 1 yr. **FL**: nodding, unisexual or not; perianth parts free or ± fused at base, glands 0; filaments wider at base, anthers attached near middle, 1-chambered; ovary partly inferior (at base), ovoid. **FR**: septicidal. **SEED**: oblong, coat loose, winged at both ends. 5 spp.: N.Am, Asia. (Greek: narrow fl) [Zomlefer et al. 2006 Aliso 22:566–578]

S. occidentale A. Gray (p. 1399)   Bulb 2–4 cm. **ST**: 20–50 cm. **LF**: 10–30 cm, 3–15 mm wide. **INFL**: raceme or panicle, 10–20 cm; bracts lance-linear, scarious; pedicels 5–30 mm, ascending. **FL**: 1–2 cm, bell-shaped; perianth parts lance-oblong, pale green-yellow to ± purple, esp near margins or not, tips recurved. **FR**: 15–20 mm (incl persistent styles). **SEED**: ± 3 mm. *n*=8. Moist banks, thickets, meadows; 1500–1900 m. KR; to w Can, MT. Late May–Jul

TOXICOSCORDION   DEATH CAMAS

Dale W. McNeal & Wendy B. Zomlefer

ST: ± scapose. **LF**: ± basal, reduced upward, linear, gen folded, ± curved, entire, deciduous after 1 yr. **INFL**: raceme or panicle; fl bracts 1 (or reportedly 2). **FL**: staminate, sterile, or gen bisexual; perianth parts 6, petal-like, free or ± fused to ovary base, white to ± yellow [or not], adaxially with [0]1[2] glands near base; stamens 6, free to ± attached to perianth; ovary superior, chambers 3, styles 3. **FR**: capsule, septicidal. **SEED**: many. ± 8 spp.: c US, w N.Am. (Greek: poison garlic, for poisonous bulb) All taxa highly TOXIC (gen unpalatable) to livestock, humans from alkaloids (esp in bulbs). As treated in TJM (1993), *Zigadenus* polyphyletic, so CA members transferred to *Toxicoscordion* (Zomlefer & Judd 2002 Novon 12:299–308).

1. Perianth parts 2–8 mm; stamens ≥ perianth
  2. Infl a panicle; fls bisexual or on branches sterile or staminate; outer perianth parts gen ± not clawed;
    bracts 5–15 mm, green; NW, CaR, n SNH, GB. . . . . *T. paniculatum*
  2′ Infl a panicle or raceme exc occ with 1 branch below; fls bisexual; outer perianth parts often clawed;
    bracts 5–40 mm, green to white-membranous; NW, CaR, SN, CW, WTR, PR, GB, DMtns
    3. Pl scabrous at least below; pedicels in fr gen spreading, with tips upturned or not, 12–40 mm; proximal
      infl branches in fr 60–90° from main axis; NW . . . . *T. micranthum*
    3′ Pl glabrous; pedicels in fr spreading to gen ascending, 3–25 mm; proximal infl branches in fr 10–60°
      from main axis; NW, CaR, SN, CW, WTR, PR, GB, DMtns . . . . *T. venenosum* var. *venenosum*
1′ Perianth parts 4–15 mm; stamens gen < perianth (± = perianth in *Toxicoscordion fontanum*)
  4. Bract 2–8 mm, < 1/2 pedicel; filaments curved, thin; fr width 1/2 length . . . . *T. brevibracteatum*
  4′ Bract 3–50 mm, 1/3–3 × pedicel; filaments straight, proximally thick; fr elongate, width gen < 1/2 length
    5. Fls bisexual exc on lower infl branches staminate or with stunted ovary; perianth parts 5–12 mm, outer
      often persistently reflexed below fr; bulbs often clumped; SNF . . . . *T. exaltatum*
    5′ Fls bisexual; perianth parts 4–15 mm, outer not persistently reflexed below fr; bulbs not clumped; not SNF
      6. Perianth ± bell-shaped, parts 4–12 mm, ovate to elliptic; bracts 5–35 mm; infl narrow, often pyramid-
        shaped in youth; NCoR, SnFrB, SCoRI . . . . *T. fontanum*
      6′ Perianth rotate, parts 5–15 mm, widely ovate; bracts 5–50 mm; infl wide; NW, c SNF (just w of
        Chinese Camp, Tuolumne Co.), ScV, CW, SW . . . . *T. fremontii*

T. brevibracteatum (M.E. Jones) R.R. Gates (p. 1399) Bulb 10–40 mm diam, ovoid; outer coat dark brown to black. **ST:** 30–60 cm, glabrous. **LF:** 10–30 cm, 3–10 mm wide, scabrous-ciliate. **INFL:** gen panicle, 10–35 cm, open; bracts 2–8 mm, green; pedicels in fr gen spreading, 10–40 mm. **FL:** on branches gen bisexual or staminate; perianth parts 5–8 mm, ovate to elliptic, ± obtuse, outer ± not clawed, inner clawed, gland distal margins evident; stamens ± 2/3 perianth; styles spreading or recurved. **FR:** 8–20 mm, oblong. *n*=11. Sandy desert; 600–1800 m. Teh, WTR, DMoj. [*Zigadenus b.* (M.E. Jones) H.M. Hall] Apr–May

T. exaltatum (Eastw.) A. Heller Bulb 20–60 mm, 20–50 mm diam, ovoid; outer coat dark brown. **ST:** 30–100 cm, glabrous. **LF:** 20–60 cm, 5–30 mm wide, scabrous-ciliate. **INFL:** panicle, 20–40 cm, open; bracts 3–25 mm, green to white-membranous; pedicels in fr spreading, 10–40 mm. **FL:** bisexual exc on lower branches staminate or with stunted ovary, 3 long styles; perianth parts 5–12 mm, ovate to elliptic, obtuse, outer ± not clawed, inner clawed 1–2 mm, gland distal margin evident, dentate; stamens 2/3–3/4 perianth; styles erect to ± spreading. **FR:** 10–25 mm, oblong. *n*=11. Meadows, wooded slopes; 600–1500 m. SNF. [*Zigadenus e.* Eastw.] Apr–Jun

T. fontanum (Eastw.) Zomlefer & Judd (p. 1399) Bulb 25–45 mm diam; outer coat brown. **ST:** 60–80 cm, glabrous. **LF:** erect, often > st, 10–25 mm wide. **INFL:** gen panicle; bracts 5–35 mm, green; pedicels in fr gen spreading, with tips upturned, 15–45 mm. **FL:** bisexual; perianth parts 4–12 mm, ovate to elliptic, cordate at base, outer gen ± not clawed, inner clawed, gland distal margin evident, serrate to dentate; stamens ± = perianth; styles erect. Vernally moist or marshy areas, often serpentine; < 500 m. NCoR, SnFrB, SCoRI. [*Zigadenus f.* Eastw.; *Z. micranthus* Eastw. var. *f.* (Eastw.) McNeal] Mar–Aug ★

T. fremontii (Torr.) Rydb. (p. 1399) Bulb 20–35 mm diam, ± spheric; outer coat black. **ST:** 40–90 cm, glabrous. **LF:** 20–50 cm, 8–30 mm wide, curved, scabrous-ciliate. **INFL:** panicle or raceme, 5–40 cm, open; bracts 5–50 mm, green; pedicels spreading in fr, 10–50 mm. **FL:** bisexual; perianth parts 5–15 mm, widely ovate, obtuse, outer very short-clawed, inner clawed 2–3 mm, gland dis-

tal margin evident, dentate; stamens ± 1/2 perianth; styles erect to ± spreading. **FR:** 10–35 mm, cylindric. *n*=11. Grassy or wooded slopes, outcrops; < 1000 m. NW, c SNF (just w of Chinese Camp, Tuolumne Co.), ScV, CW, SW; OR, n Baja CA. [*Zigadenus f.* (Torr.) S. Watson] Feb–Jun

T. micranthum (Eastw.) A. Heller Bulb 10–25 mm diam; outer coat dark brown to black. **ST:** 15–50 cm, scabrous at least below. **LF:** < st, 4–10 mm wide. **INFL:** gen panicle; bracts 5–40 mm, green to white-membranous; pedicels gen spreading in fr, with tips upturned or not, 12–40 mm. **FL:** bisexual; perianth parts 3–8 mm, ovate to elliptic, not cordate, outer clawed < 5 mm or ± not clawed, inner clawed, gland distal margin evident, dentate; stamens ± = (>) perianth. Dry slopes, flats; < 1000 m. NW; sw OR. [*Zigadenus m.* Eastw. var. *m.*] Apr–Jul

T. paniculatum (Nutt.) Rydb. (p. 1399) Bulb 30–50 mm, 8–30 mm wide, ovoid; outer coat dark brown to black. **ST:** 20–70 cm, glabrous. **LF:** 20–50 cm, 6–16 mm wide, scabrous-ciliate. **INFL:** panicle, 10–30 cm; bracts 5–15 mm, green; pedicels in fr spreading to ascending, 3–25 mm. **FL:** bisexual or on branches sterile or staminate; perianth parts 2–6 mm, unequal, outer gen ± not clawed, inner clawed, gland distal margin evident or not; stamens ±, occ > perianth; styles ascending to erect. **FR:** 8–20 mm, cylindric. *n*=11. Dry sagebrush scrub to conifer forest; 1200–2300 m. NW, CaR, n SNH, GB; to WA, MT, CO, NM. [*Zigadenus p.* (Nutt.) S. Watson] May–Jun

T. venenosum (S. Watson) Rydb. var. *venenosum* (p. 1399) Bulb 12–25 mm diam, widely ovate; outer coat ± brown. **ST:** 15–70 cm, glabrous. **LF:** 10–40 cm, 4–10 mm wide, scabrous-ciliate. **INFL:** raceme exc occ with 1 branch below, 5–25 cm; bracts 5–25 mm, white-membranous to gen green; pedicels in fr spreading to gen ascending, 3–25 mm. **FL:** bisexual; perianth parts 4–6 mm, ± ovate, obtuse, outer gen with claw < 5 mm, inner clawed, gland distal margin evident or not; stamens =, occ > perianth; styles erect. **FR:** 8–14 mm, cylindric. *n*=11. Moist meadows to dry rocky hillsides; < 2600 m. NW, CaR, SN, CW, WTR, PR, GB, DMtns; to BC, c US, AZ, n Baja CA. [*Zigadenus v.* S. Watson var. *v.*] May–Jul

TRILLIUM WAKEROBIN, TRILLIUM

Dale W. McNeal & Bryan D. Ness

Rhizome short, thick, spreading to erect. **ST:** erect, ≥1 per rhizome. **LF:** 3, in 1 whorl subtending fl, sessile or subsessile, ± ovate. **INFL:** fls 1 per st, erect to nodding, stalked or sessile. **FL:** bisexual; sepals 3, free, persistent, ± green; petals 3, free, withering, white, ± pink, ± yellow, or purple; stamens 6; ovary chambers 3, styles 3. **FR:** capsule, ± berry-like. **SEED:** many, ovoid. ± 30–40 spp.: N.Am, Asia. (Latin: 3, from lvs) [Farmer 2006 Aliso 22:579–592; Freeman 1975 Brittonia 27:1–62] *T. rivale* moved to *Pseudotrillium*.

1. Fl stalked . . . . *T. ovatum*
  2. Petals 5–23 mm (often < 15 mm), 1.8–5.5 mm wide; fl ± nodding; 1200–2000 m; c KR (esp Marble Mtns,
    Siskiyou Co.). . . . . subsp. *oettingeri*
  2′ Petals 1.5–7 cm, 5–45 mm wide; fl gen ± erect; 10–1600 m; NW, CW . . . . subsp. *ovatum*

1′ Fl sessile
  3. Stamens < 1.25 × pistil; fl odor gen musty or fetid; w KR, SN (exc Teh), s CCo, SCoRO . . . . ***T. angustipetalum***
  3′ Stamens ± 2 × pistil; fl odor gen sweet rose-like or ± spicy; NW, CaR, n SN, SnFrB
    4. Ovary, tissue between anther sacs ± green (ovary ± purple); petals white to ± pink, base occ ± purple;
      NW, CaR, n SN, SnFrB . . . . ***T. albidum***
    4′ Ovary, tissue between anther sacs purple; petals yellow to pink to dark purple, occ white; KR, NCoRI,
      SnFrB . . . . ***T. chloropetalum***

T. albidum J.D. Freeman (p. 1399) **ST**: 2–6 dm. **LF**: sessile, 7–20 cm, 12–15 cm wide, rounded to obtuse at tip, gen weakly ± brown- or green-spotted. **FL**: sessile; odor gen sweet rose-like or ± spicy; sepals spreading, 3–6.5 cm, lanceolate; petals erect to ascending, 4–8 cm, oblanceolate to obovate, white to ± pink, base occ ± purple; stamens 15–25 mm, tissue between anther sacs ± green; ovary ± green (± purple). **FR**: green or ± purple, pulpy, juicy. *n*=5. Common. Edges of redwood or mixed-evergreen forest, coastal scrub, chaparral, moist canyon slopes, ravine banks; < 2000 m. NW, CaR, n SN, SnFrB; to WA. Feb–Jun

T. angustipetalum (Torr.) J.D. Freeman (p. 1399) **ST**: 2–7 dm. **LF**: subsessile, 9–25 cm, 8–15 cm wide, rounded to obtuse at tip, weakly light-brown- or dark-green-spotted. **FL**: erect, sessile; odor gen musty or fetid; sepals 3.5–5.5 cm, oblong- to lance-linear; petals erect or leaning inward and ± concealing stamens, 5–11 cm, linear, dark purple; stamens 11–23 mm, tissue between anther sacs purple; ovary dark purple. **FR**: spheric, 6-angled, ± winged, fleshy. *n*=5. Montane conifer forest, foothill woodland, chaparral, riparian woodland; 30–2000 m. w KR, SN (exc Teh), s CCo, SCoRO; s OR. [*T. kurabayashii* J.D. Freeman] Mar–Apr

T. chloropetalum (Torr.) Howell (p. 1399) GIANT TRILLIUM **ST**: 2–7 dm. **LF**: sessile, 7–21 cm, 7–18 cm wide, rounded to obtuse at tip, gen weakly ± brown-green-spotted. **FL**: erect, sessile; odor gen sweet rose-like or ± spicy; sepals 3.5–6.5 cm, lanceolate; petals erect or leaning inward and ± concealing stamens, 6.5–10 cm, linear-oblanceolate to obovate, yellow to pink to dark purple, occ white; stamens 15–30 mm, tissue between anther sacs purple; ovary purple. **FR**: ovoid, obscurely 6-angled, red-purple, pulpy. *n*=5. Edges of red-wood forest, chaparral, gen moist slopes, canyon banks in alluvial soils; 100–2000 m. KR, NCoRI, SnFrB. [*T. c.* var. *giganteum* (Hook. & Arn.) Munz] Highly variable; a population may contain many color variants. Apr–May

T. ovatum Pursh WHITE OR WESTERN TRILLIUM **ST**: 1–7 dm. **LF**: 4–20 cm, 5–20 cm wide, acute to acuminate at tip, gen green, gen not spotted. **FL**: erect to nodding, stalked; petals ascending, 0.5–7 cm, linear to widely obovate, white, aging pink; stamens 1.7–15 mm, tissue between anther sacs yellow or ± white. **FR**: obscurely winged, green or white. *n*=5.

  subsp. ***oettingeri*** Munz & Thorne (p. 1399) SALMON MOUNTAINS WAKEROBIN **ST**: 1–3 dm. **LF**: ± sessile. **FL**: ± nodding; sepals 5–25 mm, 1.5–5.3 mm wide, linear to lance-linear; petals 5–23 mm (often < 15 mm), 1.8–5.5 mm wide, linear to lance-linear. Mixed montane or conifer forest on moist slopes; 1200–2000 m. c KR (esp Marble Mtns, Siskiyou Co.). Feb–Apr ★

  subsp. ***ovatum*** (p. 1399) **ST**: 1.5–7 dm. **LF**: gen sessile. **FL**: gen ± erect; sepals 1–6 cm, 2–24 mm wide, lanceolate; petals 1.5–7 cm, 5–45 mm wide, lance-oblong to widely obovate. Redwood, mixed-evergreen forest on moist wooded slopes; 10–1600 m. NW, CW; to w Can, CO. Feb–Apr

VERATRUM   CORN LILY, FALSE-HELLEBORE

Pl coarse, lfy; rhizome thick. **ST**: erect, 1–2 m, simple, hollow. **LF**: many, alternate, lanceolate to widely ovate, clasping, gen acute at tip, coarse-veined, reduced upward. **INFL**: panicle; fls many. **FL**: bisexual or staminate; perianth parts 6, free, spreading, petal-like, white or ± green to red-brown, nectary glands 1–2 near base; stamens 6, attached to perianth; ovary ± inferior, chambers 3, styles 3, short, stigmas long. **FR**: capsule, septicidal. ± 25 spp.: n temp. (Latin: dark roots) Alkaloids used medicinally and TOXIC to both livestock and humans.

1. Ovary woolly, fr ± woolly; outer perianth parts irregularly shallowly fringed . . . . ***V. insolitum***
1′ Ovary, fr glabrous or with only a few straight hairs, not woolly; perianth parts entire to deeply fringed
  2. Perianth deeply fringed . . . . ***V. fimbriatum***
  2′ Perianth entire to finely dentate
    3. Infl dense, branches crowded, spreading to ascending . . . . ***V. californicum*** var. ***californicum***
    3′ Infl open, branches ± spaced, lower drooping . . . . ***V. viride***

V. californicum Durand var. ***californicum*** (p. 1403) **LF**: ovate; lower 20–40 cm, tomentose-ciliate, abaxially curly-hairy, adaxially glabrous or veins sparsely short-hairy. **INFL**: gen 30–60 cm, dense, tomentose; branches crowded, spreading to ascending; pedicels 1–6 mm. **FL**: 10–15 mm; perianth parts elliptic to obovate, white or ± green, glabrous to sparsely woolly below, entire to finely dentate, glands 1–2, Y-shaped, green; stamens 1/2–2/3 perianth; ovary glabrous. **FR**: 2–3 cm, narrowly ovoid. **SEED**: 10–12 mm, ± winged. *n*=16. Streambanks, moist meadows, forest edges; < 3500 m. NW, CaRH, SNH, TR, PR, MP, SNE (Sweetwater Mtns); to WA, MT, CO, Mex. [*V. c.* var. *caudatum* (A. Heller) C.L. Hitchc.] Jul–Aug

V. fimbriatum A. Gray (p. 1403) FRINGED FALSE-HELLEBORE **LF**: lanceolate; lower 20–50 cm, glabrous or sparsely hairy. **INFL**: gen 15–50 cm, tomentose; branches spreading; pedicels 6–12 mm. **FL**: 6–10 mm; perianth parts diamond-shaped to ovate, white, glabrous, deeply fringed, glands 2, elliptic, yellow; stamens ± 1/2 perianth; ovary glabrous. **FR**: ± 8 mm, obovoid. **SEED**: ± 6 mm, ± margined. *n*=16. Wet meadows in coastal scrub; < ± 100 m. c NCo, w NCoRO (Mendocino, Sonoma cos.). Jul–Sep ★

V. insolitum Jeps. (p. 1403) SISKIYOU FALSE-HELLEBORE **LF**: elliptic; lower 10–23 cm, hairy below and on sheaths. **INFL**: gen 20–50 cm, ± gray woolly; branches ascending; pedicels 6–15 mm. **FL**: 6–9 mm; perianth parts widely ovate, white to ± yellow, hairy, outer irregularly shallowly fringed, glands 2, elliptic, ± green; stamens ± = perianth; ovary woolly. **FR**: 2–3 cm, oblong-ovoid, woolly. **SEED**: 10–15 mm, widely winged. *n*=16. Openings in thickets, mixed-evergreen forest on red clay; > 900 m. KR; to WA. Jun–Aug ★

V. viride Aiton (p. 1403) **LF**: elliptic to widely ovate; lower 15–35 cm, sparsely curly-ciliate, abaxially hairy, adaxially veins sparsely short-hairy. **INFL**: gen 30–70 cm, open, woolly; branches ± spaced, lower drooping; pedicels 2–5 mm. **FL**: 6–10 mm; perianth parts oblanceolate to oblong, green to ± yellow, ± irregularly toothed, adaxially hairy, glands 2, elliptic, dark green or ± yellow; stamens ± 1/2 perianth; ovary glabrous. **FR**: 2–3 cm, oblong-ovoid. **SEED**: 7–10 mm, widely winged. *n*=16. Wet subalpine meadows; 1500–2000 m. KR; to AK, e N.Am. [*V. v.* var. *eschscholzianum* (Roem. & Schult.) Breitung] Aug–Sep

XEROPHYLLUM BEAR-GRASS, BASKET-GRASS

Pl non-fl for several yrs, stalk dying after fr; rhizome woody, tuber-like. **ST:** stout, simple, lfy. **LF:** persistent, wiry, tough, grass-like, strongly scabrous. **INFL:** raceme, dense, ± club-shaped, longer (to 50 cm) in fr. **FL:** perianth parts free, oblong to ovate, persistent, ± white to cream, 5–7-veined; filaments wider at base, anthers attached near middle; style short, stigmas thread-like. **FR:** 3-angled, loculicidal. **SEED:** 9–18, oblong, 3-angled, black. 2 spp.: N.Am. (Greek: dry lf, from tough, persistent lvs) [Fay et al. 2006 Aliso 22:559–565]

X. tenax (Pursh) Nutt. (p. 1403)   **ST:** 15–150 cm. **LF:** basal in dense clump, 30–100 cm, 2–6 mm wide; cauline many, reduced upward. **INFL:** pedicels 2–5 cm, spreading to ± erect. **FL:** 5–10 mm; perianth parts ± oblong; style ± 1 mm, stigma branches 2–4 mm. **FR:** 5–7 mm. n=15. Dry open slopes, ridges, montane conifer forest; < 2300 m. KR, NCoRO, CaRH, n SNH, CCo, SnFrB; to BC, MT, WY. May–Aug

NARTHECIACEAE BOG ASPHODEL FAMILY

Dale W. McNeal

Per; rhizomes slender, creeping. **ST:** simple, glabrous. **LF:** mostly basal, along rhizome, linear, grass-like, densely overlapped, cauline reduced. **INFL:** raceme. **FL:** perianth parts 6 in 2 petal-like whorls, erect in age, lance-linear; stamens 6, epipetalous, filaments densely hairy, anthers attached near middle; ovary superior, shallowly 3-lobed, chambers 3, style minutely 3-lobed. **FR:** capsule, loculicidal, weakly 3-lobed. **SEED:** many, bristle-tipped at both ends, pale yellow to brown. 5 or 6 genera, 17–29 spp.: gen n temp. Scientific Editor: Thomas J. Rosatti.

NARTHECIUM BOG ASPHODEL

ST: erect, ± scapose, 10–70 cm. **LF:** upper 3–6 reduced. **FL:** perianth parts free, yellow to green-yellow; stamens 6, filaments woolly; ovary lobes fused into a stylar beak. **SEED:** brown. 4–8 spp.: n temp. (Greek: ancient name)

N. californicum Baker (p. 1403)   **ST:** 20–60 cm. **LF:** basal 10–30 cm. **INFL:** 8–15 cm; fl bract 1; pedicels overlapped or not, 5–12 mm, bractlets (0)1, gen above middle. **FL:** 5–10 mm; anthers 3–4 mm. **FR:** ± 2 × perianth, lanceolate. **SEED:** 6–9 mm (incl bristle-tips). $2n$=26. Wet meadows, streambanks; 700–2600 m. KR, NCoRO, CaRH, n&c SNH. Jul–Aug

ORCHIDACEAE ORCHID FAMILY

Ronald A. Coleman, Dieter H. Wilken & William F. Jennings, except as noted

Per, terrestrial [growing on other pls], non-green (nutrition from association of roots with fungi) or green, gen from rhizomes or tubers with few to many fleshy to slender roots; cauline lvs ± reduced to sheathing st bracts or not. **LF:** 1–many, basal to cauline, linear to ± round, alternate to opposite (if only 1 pair), gen sessile. **INFL:** fls 1–many, spike or raceme, bracted. **FL:** bisexual, bilateral, in bud gen rotating 180° by twisting ovary (position of parts indicated after twisting); sepals gen 3, gen free, gen petal-like, uppermost gen erect, lateral with chin- or spur-like projection (mentum) or not; petals 3, 1 (lip) different, spurred or not; stamens gen 1 (3 in *Cypripedium*, 2 functional, 1 a staminode), fused with style, stigma into column, pollen gen lumped, gen removed as unit by insect; ovary inferior, 1-chambered, placentas 3, parietal, stigma 3 lobed, gen under column tip. **FR:** capsule. **SEED:** many, minute. ± 800 genera, ± 25000 spp.: esp trop (worldwide exc Antarctica). Many cult for orn, esp *Cattleya, Cymbidium, Epidendrum, Oncidium, Paphiopedilum; Vanilla planifolia* Andrews frs used to flavor food. *Platanthera* may be paraphyletic without inclusion of *Piperia* (Bateman et al. 2009 Ann Bot 104:431–445); study needed. [Romero-Gonzalez et al. 2002 FNANM 26:490–651] Scientific Editors: Ronald A. Coleman, Thomas J. Rosatti.

1. Functional stamens 2, staminodes 1 . . . . **CYPRIPEDIUM**
1′ Functional stamens 1, staminodes 0
  2. Lvs 0 at fl
   3. Fl spurred, without mentum . . . . ²**PIPERIA**
   3′ Fl not spurred, with mentum or not
    4. Fls in spiral, lip pouch-like . . . . ²**SPIRANTHES**
    4′ Fls not in spiral, lip not pouch-like
     5. Pl white, becoming ± yellow or brown . . . . **CEPHALANTHERA**
     5′ Pl brown, tan, ± pink, ± red, ± purple (± green) . . . . **CORALLORHIZA**
  2′ Lvs present at fl
   6. Lvs cauline
    7. Lvs 2, ± opposite, ± at mid st . . . . **LISTERA**
    7′ Lvs > 2, alternate, along st
     8. Lvs with > 1 lengthwise fold, green, white, or pink; fl not spurred . . . . **EPIPACTIS**
     8′ Lvs with 1 lengthwise fold, green; fl spurred . . . . **PLATANTHERA**
   6′ Lvs ± basal
    9. Lvs 1
     10. Fls 1, lip pouch-like . . . . **CALYPSO**
     10′ Fls > 1, lip not pouch-like . . . . **MALAXIS**

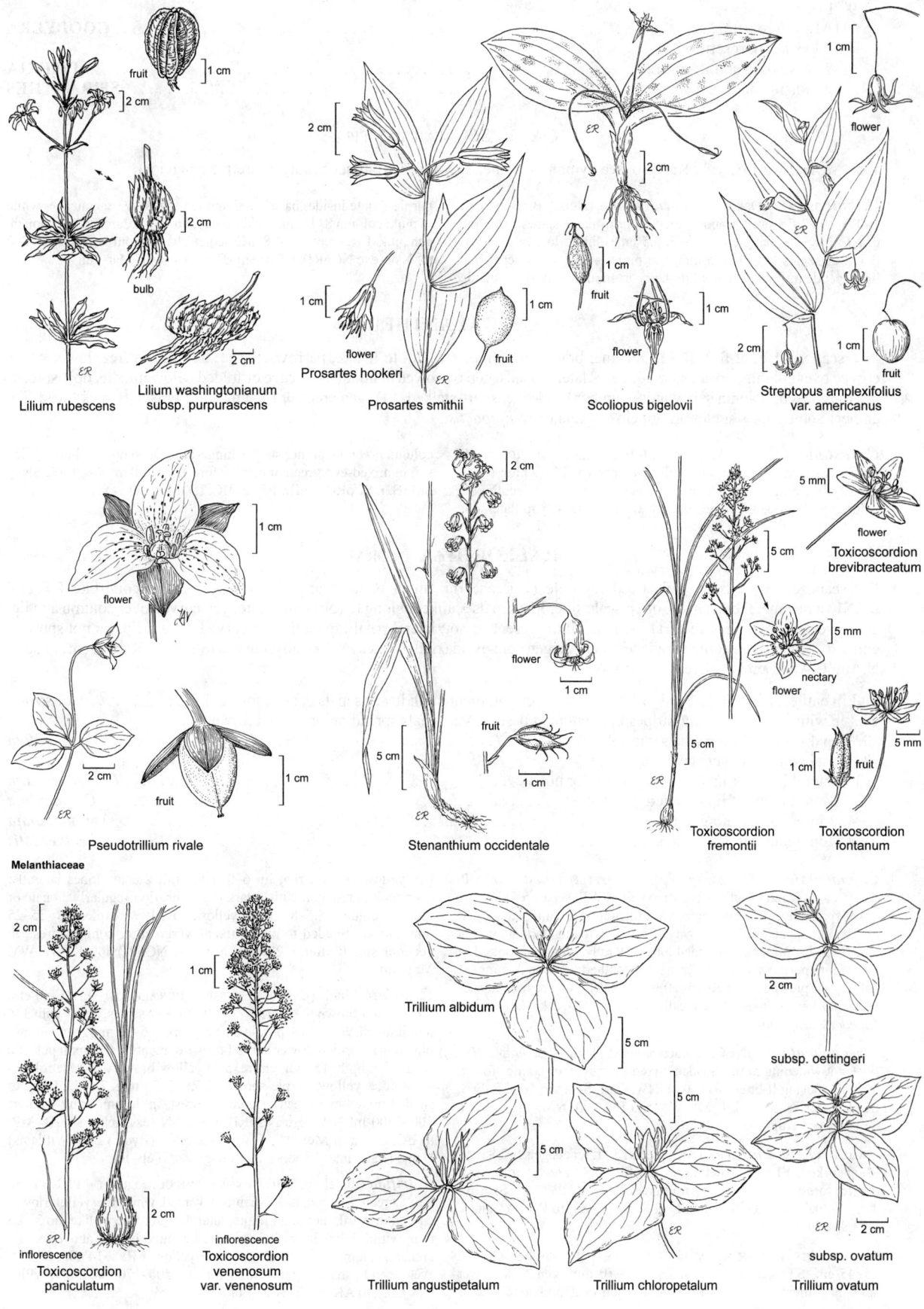

Lilium rubescens

Lilium washingtonianum
subsp. purpurascens

fruit

bulb

Prosartes hookeri

Prosartes smithii

flower

fruit

Scoliopus bigelovii

flower

fruit

Streptopus amplexifolius
var. americanus

flower

fruit

Pseudotrillium rivale

flower

fruit

Stenanthium occidentale

flower

fruit

Toxicoscordion
fremontii

Toxicoscordion
brevibracteatum

flower

nectary

flower

fruit

Toxicoscordion
fontanum

Melanthiaceae

Toxicoscordion
paniculatum

inflorescence

Toxicoscordion
venenosum
var. venenosum

inflorescence

Trillium albidum

Trillium angustipetalum

Trillium chloropetalum

subsp. oettingeri

subsp. ovatum

Trillium ovatum

9′ Lvs > 1
  11. Lvs in rosette . . . . **GOODYERA**
  11′ Lvs not in rosette
    12. Fls not in spiral, spurred . . . . ²**PIPERIA**
    12′ Fls in spiral, not spurred . . . . ²**SPIRANTHES**

CALYPSO FAIRY SLIPPER

1 sp., several vars. (Greek: Kalypso, sea nymph in Homer's *Odyssey*, from her beauty, secretive behavior)

C. bulbosa (L.) Oakes var. ***occidentalis*** (Holz.) B. Boivin (p. 1403) Pl 7–18 cm, ± scapose; caudex corm-like, ± spheric; rhizome 0. **LF**: basal, 1, petioled, blade 3–6.5 cm, elliptic to ovate. **INFL**: fl 1. **FL**: sepals free, 15–25 mm, gen pink; lateral petals sepal-like, lip ± pendent, pouch-like, white-hairy at mouth, ± purple outside, purple-striate inside, base ± red-spotted, distal 1/3 ± concave, white to pink; column 8–11 mm, wide-ovate, hood-like, arched over pouch, gen pink. **FR**: erect. 2*n*=28. Mesic gen shaded conifer forest; < 1800 m. NW (exc NCoRI), CCo, SnFrB; to AK, MT. Mar–Jun

CEPHALANTHERA

Pl ± scapose. **LF**: at fl 0. **INFL**: raceme; bracts sheathing, ± scale- to lf-like, narrow-elliptic. **FL**: sepals free, lower gen ± curved over column; lip not spurred, gen < lateral petals, gen narrowed at middle, concave or folded below middle, tip ± spreading to reflexed; column subcylindric, anther head-like, short-stalked. **FR**: gen erect. ± 15 spp.: esp Eurasia. (Greek: head-like anther) Some spp. ± subterranean; closely related to *Epipactis*.

C. austiniae (A. Gray) A. Heller (p. 1403) PHANTOM ORCHID Pl 20–55 cm, white, becoming ± yellow or brown. **FL**: white; sepals 12–20 mm, elliptic to oblanceolate, acute; lip folded lengthwise, below middle lobed, above reflexed, with a yellow, papillate spot; column 6–9 mm, anther stalk ± hinge-like. Decomposed litter of rich soil in mixed-evergreen or conifer forest; < 2200 m. NW, CaR, SNH, SnFrB, n SCoRO, SnBr, PR; to BC, ID. Mar–Jul

CORALLORHIZA CORALROOT

Pl ± scapose, brown, tan, ± pink, ± red, ± purple, (± green); rhizome branches many, short, scaly, together coral-like. **LF**: at fl 0. **INFL**: raceme; fl bract << fl, often scale-like. **FL**: sepals ± alike, oblong to (ob)lanceolate, gen curved over column and lip, gen 3-veined, lower gen fused at base, mentum present or not; lateral petals spreading or curved toward lip, lip not spurred, entire to 3-lobed, spreading to reflexed; column gen convex adaxially, concave abaxially, curved over lip. **FR**: pendent. 11 spp.: N.Am, C.Am, Eurasia. (Greek: coral root)

1. Lip entire, stripes 3–5, ± red to ± purple, faint or not, mentum 0; lower sepals curved forward . . . . ***C. striata***
1′ Lip with 2 lobes laterally, stripes 0, mentum at base; lower sepals spreading or curved forward
  2. Sepal veins 1; lower sepals curved forward . . . . ***C. trifida***
  2′ Sepal veins 3; lower sepals spreading
    3. Lateral lobes of lip acute, prominent or not . . . . ***C. mertensiana***
    3′ Lateral lobes of lip rounded, prominent . . . . ***C. maculata***
      4. Lip ± not widening to tip . . . . var. ***maculata***
      4′ Lip widening to tip . . . . var. ***occidentalis***

C. maculata (Raf.) Raf. (p. 1403) SPOTTED CORALROOT Pl 17–55 cm. **ST**: red to yellow-brown to yellow. **FL**: sepals 5.5–10 mm, lower spreading, color gen same as sts, mentum < 2.5 mm; lateral petals gen like sepals, yellow-brown or deep pink to red, dark-spotted or not, lip 5–7 mm, with 2 rounded lobes laterally, white, unspotted or gen red- to purple-spotted, tip crenate or toothed; column 3–5 mm, ± yellow, purple-spotted. **FR**: 15–20 mm. 2*n*=42. Where together, *C. maculata* var. *occidentalis* typically flowers 2–4 weeks earlier than *C. maculata* var. *m.*

  var. ***maculata*** **INFL**: fl bracts gen 0.5–1 mm, gen entire. **FL**: lip ± not widening to tip. Shaded mixed-evergreen or conifer forest, in decomposing lf litter; < 2800 m. NW, CaR, SN (exc Teh), SnFrB, SCoRO, SnGb, SnBr, PR, MP, W&I; to BC, CO, NM, also se Can, ne US. May–Aug

  var. ***occidentalis*** (Lindl.) Ames **INFL**: fl bracts gen 1–2.8 mm, often forked. **FL**: lip widening to tip. Shaded mixed-evergreen or conifer forest, in decomposing lf litter; < 2800 m. NW, CaR, SN (exc Teh), SnFrB, SCoRO, SnGb, SnBr, PR, MP, W&I; to BC, se Can, se US, NM. Feb–Aug

C. mertensiana Bong. (p. 1403) WESTERN CORALROOT Pl 15–45 cm. **ST**: gen ± red. **FL**: sepals 7–10 mm, gen pink, lower spreading, mentum 0.5–2.5 mm; lateral petals deep pink to red, veins gen yellow or dark red, lip 6–9 mm, with 2 acute lobes laterally, deep pink to red, gen with 3 dark red veins, tip irregularly crenate or toothed; column 5.5–8 mm, ± yellow, basally ± purple. **FR**: 15–25 mm. 2*n*=40. Shaded to open mixed-evergreen or conifer forest, in decomposing lf litter; < 2200 m. NW (exc NCoRI); to AK, MT, WY. May–Jul

C. striata Lindl. (p. 1403) STRIPED CORALROOT Pl 15–50 cm. **ST**: gen red-brown to ± purple (± yellow). **FL**: sepals, petals with 3–5 longitudinal, ± red to ± purple stripes; sepals 6–17 mm, gen yellow-pink to pale brown, lower curved forward, mentum 0; lateral petals ± sepal-like, lip 8–15 mm, entire, pale yellow-brown to ± red; column 4–7 mm, ± yellow, purple-spotted. **FR**: 12–25 mm. 2*n*=42. Open to shaded mixed-evergreen or conifer forest, in decomposing lf litter; 100–2200 m. NW (exc NCoRI), CaR, SN (exc Teh), SnFrB, MP; to BC, e Can, n Mex. [*C. s.* var. *vreelandii* (Rydb.) L.O. Williams] Named vars. intergrade, so not recognized. Feb–Jul

C. trifida Châtel. (p. 1403) NORTHERN CORALROOT Pl 8–30 cm. **ST**: ± green. **FL**: sepals 4.5–6 mm, 1-veined, white or ± yellow, lower curved forward, mentum minute; lateral petals ± sepal-like, lip 3–4.5 mm, with 2 lobes laterally, white, red-spotted or not, above middle crenate; column 2.5–6 mm, white or ± yellow. **FR**: 8–14 mm. 2*n*=42. Wet, open to shaded, gen conifer forest; 1400–1700 m. n SNH (Plumas Co.); to AK, e US, NM; Eurasia. Jul ★

CYPRIPEDIUM LADY'S-SLIPPER

ST: often puberulent above. **LF**: cauline, lanceolate to ovate. **INFL**: fls 1–20; bracts ± lf-like. **FL**: lower sepals ± united [free], pendent behind lip; lateral petals downcurved to spreading, ± like upper sepal, twisted or not, lip pouch-like, gen obovoid, mouth ± puckered, margin inrolled; column curved over lip mouth, functional stamens 2, staminodes 1. **FR**: pendent to spreading, ribbed. ± 35 spp.: temp N.Am, Eurasia; some cult for orn. (Greek: Venus foot, from lip shape) [Sheviak 1993 Amer Orchid Soc Bull 62:403]

1. Lip yellow . . . . ***C. parviflorum*** var. ***makasin***
1′ Lip yellow-green below, purple above, or white
  2. Lvs 2, opposite . . . . ***C. fasciculatum***
  2′ Lvs > 2, alternate
    3. Upper sepal, lateral petals green to yellow-green . . . . ***C. californicum***
    3′ Upper sepal, lateral petals red- to dark-brown (green) . . . . ***C. montanum***

C. californicum A. Gray (p. 1403) CALIFORNIA LADY'S-SLIPPER Pl 8–130 cm. **LF**: 5–12, alternate, 5–15 cm, lower elliptic, upper ± lanceolate. **INFL**: open; fls (1)4–20. **FL**: upper sepal 15–20 mm, ± elliptic to ovate, green to yellow-green; lateral petals 12–18 mm, lanceolate to oblong, spreading, green to yellow-green, lip 15–20 mm, white; staminode ± round, reflexed, white, gen with green stripe. Streambanks, moist slopes, fens, partial shade to full sun, mixed-evergreen or conifer forest; 50–2200 m. KR, n NCoRO, w CaR, n SN, nw SnFrB; sw OR. Apr–Jul ★

C. fasciculatum Kellogg (p. 1403) CLUSTERED LADY'S-SLIPPER Pl 12–20 cm. **LF**: 2, opposite, 5–12 cm, wide-elliptic. **INFL**: fls 1–14, ± clustered, ± nodding. **FL**: upper sepal 1.5–2.5 cm, lanceolate, ± green to brown, veins dark brown; lateral petals 15–25 mm, lanceolate, downcurved, lip 8–15 mm, yellow-green below, purple above; staminode < 5 mm, pale (± green) yellow. 2*n*=20. Mesic to moist, shady conifer forest; 100–2000 m. NW, n SN, sw SnFrB; to WA, MT, WY, CO. Mar–Jul ★

C. montanum Lindl. (p. 1403) MOUNTAIN LADY'S-SLIPPER Pl 25–70 cm. **LF**: 4–6, alternate, 5–18 cm, elliptic to ovate. **INFL**: open; fls gen 1–3. **FL**: upper sepal 3–6 cm, lanceolate, twisted or wavy, red- to dark-brown (green); lateral petals 25–60 mm, narrow-lanceolate, downcurved, twisted, lip 20–30 mm, white, ± red-striped below or not; staminode 8–12 mm, yellow to yellow-green, red- to purple-spotted. Moist areas, dry slopes, mixed-evergreen or conifer forest; 200–2200 m. NW, CaR, n&c SN, sw SnFrB, MP; to AK, MT, WY. Mar–Jun ★

C. parviflorum Salisb. var. ***makasin*** (Farw.) Sheviak NORTHERN SMALL YELLOW LADY'S-SLIPPER Pl 15–35 cm. **LF**: 2–5, alternate, 5–18 cm, ovate to lance-elliptic. **INFL**: open; fls gen 1–2. **FL**: upper sepal 18–80 mm, ± elliptic to ovate, green to yellow-green, marked dark red-brown or not; lateral petals 24–95 mm, lanceolate to oblong, spreading, gen ± twisted, colored as upper sepal, lip 15–30 mm, yellow; staminode ± cordate or ovate, yellow. 2*n*=20. Mesic to moist, shady conifer forest; 1000–1900 m (estimated). SN; WA to AK, ne US, NL. Only 1 documented occurrence in CA; probably extirpated. May–Jun(est) ★

EPIPACTIS

LF: cauline, alternate, gradually reduced upward, lanceolate to wide-ovate, with > 1 lengthwise fold, green, white, or pink. **INFL**: ± 1-sided, open; fls 4+; fl bract ± lf-like. **FL**: sepals ± alike, lanceolate to ovate, lower spreading to downcurved; lateral petals ascending or curved forward, ± = sepals in shape, color, lip not spurred, abruptly narrowed at ± middle, of 2 very different parts, proximally concave to ± pouch-like, distally grooved to ± not; column curved over lip. **FR**: spreading to pendent. ± 25 spp.: N.Am, Eurasia, n Afr. (Greek: ancient name)

1. Lip 14–20 mm, proximally concave, distally grooved; lateral sepals 16–24 mm; ± wet places . . . . ***E. gigantea***
1′ Lip 9–12 mm, proximally ± pouch-like, distally ± not grooved; lateral sepals 10–13 mm; ± dry places . . . . ***E. helleborine***

E. gigantea Hook. (p. 1407) STREAM ORCHID Pl 30–70(100) cm. **LF**: 5–15 cm, lanceolate to wide-elliptic, green. **INFL**: fls few–20+; fl bract lanceolate to oblong. **FL**: sepals ± green to ± red, ± purple-veined; lateral petals 13–15 mm, lip proximally ± green to ± yellow, veined red-purple, distally yellow, red-tinged or -veined; column 5–9 mm. **FR**: 20–28 mm. 2*n*=40. Seeps, wet meadows, streambanks; < 2600 m. CA-FP (exc GV, s ChI), GB, D; to BC, SD, TX, Mex. [*E. g.* Douglas f. *rubrifolia* P.M. Br.] Mar–Oct

E. helleborine (L.) Crantz BROAD-LEAVED HELLEBORINE Pl 40–100 cm. **LF**: 6–10 cm, lanceolate to ovate, green, white, or pink. **INFL**: fls few–20; fl bract linear to narrow-lanceolate. **FL**: sepals ± green, often purple-tinged or -striped; lateral petals 8–11 mm, lip proximally white to ± pink outside, brown to ± purple inside, distally white to ± pink; column 3–5 mm. **FR**: 1–1.5 cm. 2*n*=36,38,40,44. Gen dry slopes, roadcuts, mixed-conifer forest; < 1300 m. s NCoRO, s NCoRI, c SNH, CCo, SnFrB, expected elsewhere; to e Can, c&ne US; native to Eur. Apr–Dec

GOODYERA

Rhizomes slender. **ST**: ± scapose. **LF**: basal, in rosettes, evergreen, blades tapered to base, white-veined to -mottled or not. **INFL**: ± 1-sided or fls spiralled; bracts lf-like, fl bract ± = fl. **FL**: sepals ± equal, upper adherent to lateral petals, forming hood ± enclosing column and lip, lower spreading to reflexed; lip proximally ± pouch-like [concave], distally deeply grooved. **FR**: ascending to erect. ± 25 spp.: esp temp n hemisphere, also trop. (John Goodyer, English botanist, 1592–1664) [Ackerman 1975 Madroño 23:191–198]

G. oblongifolia Raf. (p. 1407) RATTLESNAKE-PLANTAIN Pl 18–35 cm. **LF**: 4–9 cm, lanceolate to wide-elliptic. **INFL**: dense; bracts 7–11 mm, gen < fls. **FL**: sepals green-brown, upper 6–11 mm, lower 5–9 mm; corolla white, lip 6–10 mm; column 3–5 mm. 2*n*=22,30. Dry to mesic conifer forest, in decomposing lf litter; < 2200 m. NW, CaR, SN (exc Teh), CW, MP; to AK, e N.Am, Mex. May–Sep

LISTERA TWAYBLADE

Ronald A. Coleman, L.K. Magrath, Dieter H. Wilken & William F. Jennings

Rhizomes slender. **LF**: cauline, 2, ± opposite, gen ovate to ± round. **INFL**: raceme, ± open; fl bract < fl. **FL**: sepals ± equal, green to ± purple, lower spreading; lateral petals ± like sepals, ascending to erect; lip gen > sepals, petals, spreading to descending, gen wedge-shaped, flat, tip entire to deeply lobed; column subcylindric, straight to curved, anther at tip. **FR**: gen spreading. ± 25 spp.: temp, arctic N.Am, Eurasia. (Martin Lister, English naturalist, 1638–1711) [Coleman 1995 Wild Orchids Calif. Cornell Univ] Names available in *Neottia* (earlier name), with same epithets, for all CA taxa if *Listera* treated as congeneric with that genus.

1. Lip deeply 2-lobed; lf base cordate . . . . *L. cordata*
1′ Lip not lobed, tip ± entire or notched; lf base abruptly tapered or rounded
  2. Lip 4.5–7 mm, tapered to base, base wider than column base, gen with 2 short, thread-like teeth, tip ± entire . . . . *L. banksiana*
  2′ Lip 8–13 mm, abruptly narrowed to claw, claw ± as wide as column base, teeth indistinct, tip notched . . . . *L. convallarioides*

L. banksiana Lindl. (p. 1407) NORTHWEST TWAYBLADE Pl 10–30 cm. **LF**: blade 2.5–7 cm, base abruptly tapered or rounded. **INFL**: 2–5 cm. **FL**: sepals 3–4 mm; lip 4.5–7 mm, ± oblanceolate, tapered to base, base gen with 2 thread-like teeth, tip rounded, ± entire; column 1.5–2.5 mm. Moist, shady conifer forest; 100–1900 m. NCo, KR; to AK, MT, WY. [*L. caurina* Piper] Apr–Jun

L. convallarioides (Sw.) Elliott (p. 1407) BROAD-LEAVED TWAYBLADE Pl 10–35 cm. **LF**: blade 2.5–7 cm, base abruptly tapered or rounded. **INFL**: 3–4 cm. **FL**: sepals 4–5.5 mm; lip 8–13 mm, ± oblanceolate, abruptly narrowed to claw, teeth indistinct, tip notched; column 2.5–3.5 mm. 2*n*=36. Moist, shady conifer forest; 800–2900 m. NW, CaR, SNH, SnBr, SnJt, MP; to AK, e N.Am, AZ. May–Aug

L. cordata (L.) R. Br. (p. 1407) HEART-LEAVED TWAYBLADE Pl 6–25 cm. **LF**: blade 1–4 cm, base cordate. **INFL**: 2.5–4 cm. **FL**: sepals 2–4 mm, oblong to narrowly elliptic; lip 8–10 mm, oblong below middle, with 2 short teeth at base, deeply 2-lobed; column 0.5–1.5 mm. 2*n*=36,38,40,42. Moist, shady conifer forests; 100–1300 m. NCo, KR; to AK, e N.Am, Eurasia. [*L. c.* var. *nephrophylla* (Rydb.) Hultén] Mar–Jun ★

MALAXIS

Caudex short, slender to bulb-like; rhizome 0. **LF**: ± basal, 1(2[5]), sheathing. **INFL**: raceme [spike]; bracts << fls. **FL**: lower sepals free [basally united], spreading to ascending, gen green; lateral petals gen narrower than sepals, spreading to erect, lip spreading to ± pendent; column gen > petals, oblong to triangular, gen concave. **FR**: spreading to erect. ± 200 spp.: worldwide (exc Afr). (Greek: soft, from lvs)

M. monophyllos (L.) Sw. var. ***brachypoda*** (A. Gray) F. Morris & E.A. Eames (p. 1407) WHITE BOG ADDER'S-MOUTH Pl 10–15 cm; caudex bulb-like. **LF**: 4–5.5 cm, wide-lanceolate to ovate, sheath 2–3 cm. **INFL**: 2–6 cm, open; pedicels < 1.5 mm. **FL**: green to yellow; sepals 1–2 mm, narrow-lanceolate; lateral petals ± = lip, linear, ± curved behind fl, lip 1.5–3 mm, triangular, base cordate, tip beak-like; column < 1 mm. **FR**: 3.5–5.5 mm. 2*n*=28. Wet meadows, shaded places, conifer forest; 2200–2800 m. SnBr (presumed extinct SnJt); to AK, ne N.Am; also CO. [*M. m.* (L.) Sw. subsp. *b.* (A. Gray) Á. Löve & D. Löve] Jul–Aug ★

PIPERIA PIPERIA

James D. Ackerman & Robert Lauri

Pl 10–130 cm; tubers, 1–4 cm, gen ± round; st bracts lance-linear to ovate. **LF**: at fl 0 or ± basal, not in rosette, 2–5, linear to widely oblanceolate. **INFL**: spike or raceme, gen cylindric, fls not in spiral; fl bract gen < fl. **FL**: fragrance, when present, gen at night; perianth white to green; sepals gen 2–5 mm, 1–2 mm wide, 1-veined, upper pointed forward to erect, lower free, spreading to reflexed; lateral petals ± = sepals, spreading to erect, lip spurred, pointed forward, down (or upcurved); column < lip; ovary inferior, gen twisted 180°. **FR**: ascending to erect. 10 spp.: N.Am. (Charles V. Piper, Am botanist, 1867–1926) [Ackerman & Morgan 2002 FNANM 26:571–577] Some spp. difficult to separate.

1. Spur 6–18 mm (2.5–9 mm in *Piperia cooperi*, keyed only under 1′)
  2. Spur ± straight, perpendicular to infl axis; upper sepal pointed forward; st gen < 3 mm diam . . . . *P. transversa*
  2′ Spur ± curved, parallel to infl axis; upper sepal ascending to erect to curved back; st gen > 3 mm diam
    3. Sepals white with dark green midvein . . . . *P. elegans* subsp. *elegans*
    3′ Perianth green to yellow-green
      4. Lip deltate-ovate; lower sepals spreading . . . . *P. michaelii*
      4′ Lip narrowly lanceolate or lance-deltate; lower sepals ± reflexed
        5. Lateral petals ± sickle-shaped, 2 mm wide at base, 2–3 × as long . . . . *P. elongata*
        5′ Lateral petals linear, 1 mm wide at base, 4–5 × as long . . . . ²*P. leptopetala*
1′ Spur 1–6(9) mm
  6. Perianth mostly white or white-margined on upper sepal, lateral petals, mostly white on lower sepals, lip; lip pointed down toward spur
    7. Infl gen ± 1-sided; st bracts < 6; lip ± ovate, ± obtuse . . . . *P. candida*
    7′ Infl ± not 1-sided; st bracts > 6; lip ± ovate or not, ± acute
      8. Lateral petals ovate to lance-oblong, white margins ± equal; st base swollen toward tuber; fragrance at night . . . . *P. elegans* subsp. *decurtata*
      8′ Lateral petals sickle-shaped, white margins unequal, outer much wider than inner; st base narrowed toward tuber; fragrance in daytime . . . . *P. yadonii*

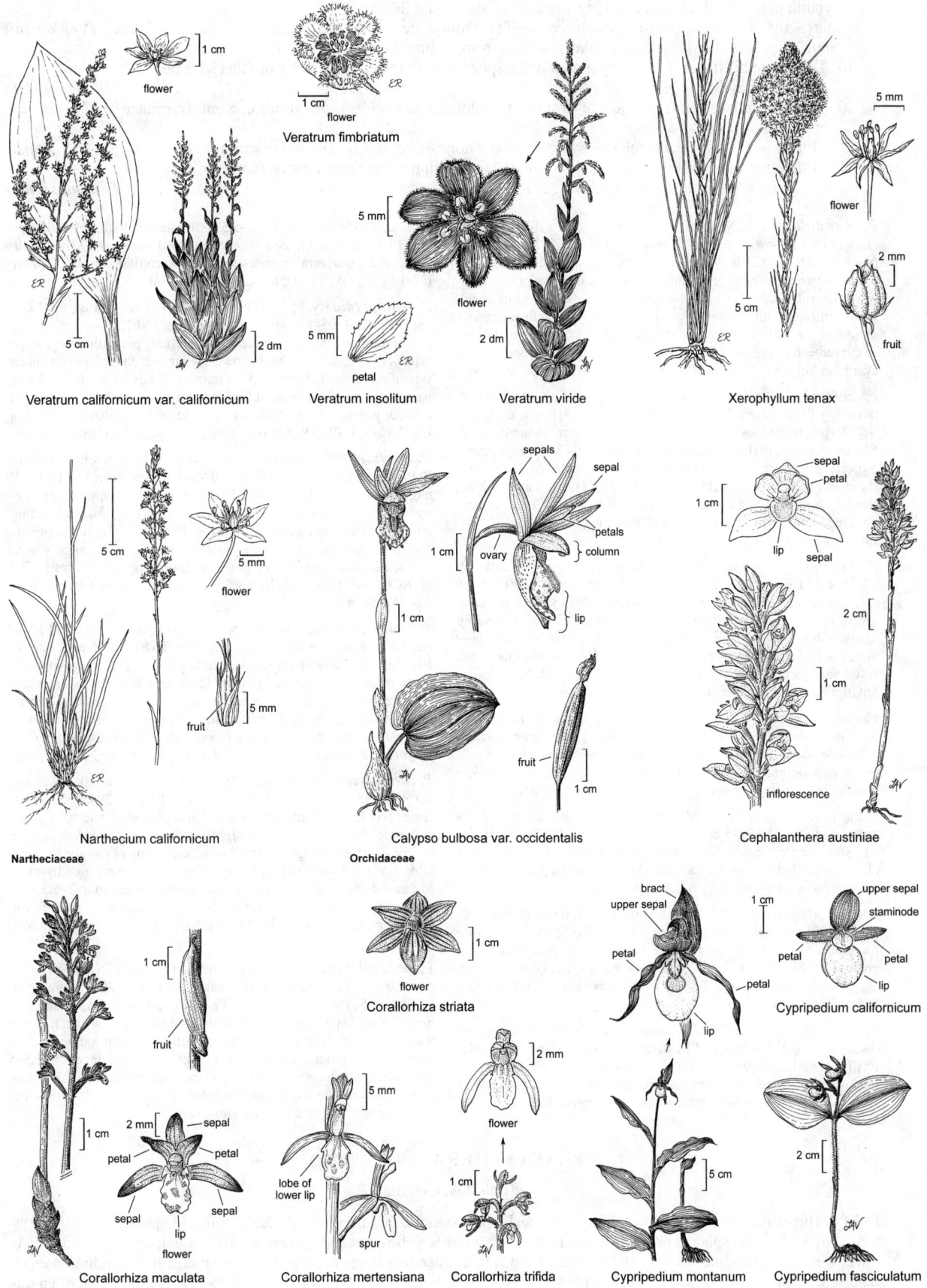

Veratrum fimbriatum

Veratrum californicum var. californicum

Veratrum insolitum

Veratrum viride

Xerophyllum tenax

flower

petal

flower

fruit

Narthecium californicum

Nartheciaceae

sepals
sepal
petals
column
ovary
lip

flower

fruit

Calypso bulbosa var. occidentalis

Orchidaceae

sepal
petal
lip
sepal

inflorescence

Cephalanthera austiniae

Corallorhiza maculata

fruit

sepal
petal
petal
sepal
sepal
lip
flower

Corallorhiza striata

flower

lobe of
lower lip

spur

Corallorhiza mertensiana

flower

Corallorhiza trifida

bract
upper sepal

petal

petal

lip

Cypripedium montanum

upper sepal
staminode
petal
petal
lip

Cypripedium californicum

Cypripedium fasciculatum

6′ Perianth green; lip ± straight to strongly upcurved above middle
  9. Spur < lip, 1–2.5 mm; lvs grass-like, erect, 3–10(19) mm wide . . . . ***P. colemanii***
  9′ Spur ≥ lip, 2–9 mm; lvs not grass-like, prostrate to spreading, (5)10–40 mm wide
    10. Lateral petals linear, 4–5 × longer than wide; spur tapered; fragrance lemony or otherwise pleasant
      but not honey-like . . . . ²***P. leptopetala***
    10′ Lateral petals lanceolate to deltate-ovate, < 3.5 × longer than wide; spur cylindric, blunt; fragrance
      musky, soapy, or honey-like
      11. Infl axis ≤ peduncle; lip deltate-ovate; fragrance honey-like; st long-tapered toward tuber . . . . ***P. cooperi***
      11′ Infl axis gen > peduncle; lip broadly ovate to lance-elliptic; fragrance musky, soapy, or honey-like;
      st uniform or swollen toward tuber . . . . ***P. unalascensis***

P. candida Rand. Morgan & Ackerman (p. 1407) WHITE-FLOWERED REIN-ORCHID Pl 10–60 cm. **LF:** basal 5–18 cm, 11–35 mm wide. **INFL:** (2)10–30 cm, ±.1-sided, ± congested to open. **FL:** sepals white, with faint green midvein or not, upper ± pointed forward; lateral petals ± pointed forward to ± ascending, translucent, white, midvein white or faintly green, lip (1.5)2–3(4) mm, ± ovate, ± obtuse, recurved toward spur, white, spur 1.5–3.5 mm, pointed down or curved ± forward. *n*=21. Open to shady sites, conifer and mixed-evergreen forest; < 1500 m. NW, sw SnFrB; to AK. May–Sep ★

P. colemanii Rand. Morgan & Glicenstein COLEMAN'S REIN-ORCHID Pl 16–53 cm. **LF:** basal 6–16 cm, 3–10(19) mm wide, grass-like. **INFL:** (6)9–30(40) cm, sparsely fld, axis longer than peduncle. **FL:** fragrance 0; perianth translucent green; sepals 2–4 mm, upper ± projecting between lateral petals, lower strongly recurved; lateral petals erect to recurved, ± sickle-shaped, lip 2–3(5) mm, lance-deltate, strongly upcurved above middle, spur ± club-shaped, 1–2.5 mm, < lip. Open conifer forest, scrub; 1300–2000 m. NCoRH, CaRH, SN. Jun–Aug ★

P. cooperi (S. Watson) Rydb. CHAPARRAL REIN-ORCHID Pl 14–90 cm. **LF:** basal 8–20(27) cm, 8–31 mm wide. **INFL:** 3–56 cm, open, axis ≤ peduncle. **FL:** fragrance honey-like; perianth green; upper sepal ascending to pointed forward, lateral spreading to ascending; lip 1.6–4 mm, deltate-ovate, flat, ± straight, spur 2.5–9 mm, pointed back or down. *n*=21. Gen dry sites, scrub, chaparral, woodland, forest; < 1500 m. SCo, s ChI (Santa Catalina Island), SnGb, PR; Baja CA. Mar–Jun ★

P. elegans (Lindl.) Rydb. **INFL:** 3–40 cm, gen dense in full sun, open in shade. **FL:** sepals white, midvein dark green, upper ± erect; lateral petals spreading to erect, ovate to lance-oblong, white to pale green, base or midvein green, white margins ± equal, lip 2.5–7 mm, ± lanceolate, ± curved downward, spur pointed down. *n*=21.

  subsp. ***decurtata*** Rand. Morgan & Glicenstein POINT REYES REIN-ORCHID Pl 15–35 cm. **LF:** basal 6–30 cm, 10–75 mm wide. **FL:** fragrance cinnamon-like; lateral petals ± green, lip white, spur (3)5–6 mm. Gen dry, open sites, coastal scrub, coastal prairie; 100 m. n CCo (Point Reyes, Marin Co.). Aug–Sep ★

  subsp. ***elegans*** (p. 1407) COAST PIPERIA, ELEGANT PIPERIA Pl 12–73(100) cm. **LF:** basal 6–30(38) cm, 10–75(90) mm wide. **FL:** fragrance musky; lateral petals white to pale green, base or midvein green, lip white to pale green, spur 7–14 mm. *n*=21. Gen dry, open sites, scrub, conifer forest; < 500 m. NCo, w KR, NCoRO, CCo, SnFrB; to BC. May–Sep

P. elongata Rydb. (p. 1407) CHAPARRAL ORCHID, WOOD REIN-ORCHID Pl (9)14–130 cm. **LF:** basal 8–30 cm, 10–65 mm wide. **INFL:** 15–30 cm, 10–30 mm wide, open to dense. **FL:** fragrance faint, harsh to honey-like; perianth green; upper sepal ± erect, lower ± reflexed; lateral petals ± sickle-shaped, gen erect, flat, 2 mm wide

at base, 2–3 × as long, lip 2–5.5 mm, lance-deltate, reflexed, spur 6.5–18 mm, slender, gen curved, pointed down. *n*=21. Gen dry sites, scrub, chaparral, mixed-evergreen or conifer forest; < 2200 m. CA-FP (exc GV, s ChI); to BC, MT. May–Jul

P. leptopetala Rydb. NARROW-PETALED REIN-ORCHID Pl 13–70 cm. **LF:** basal 6.5–15 cm, 15–30 mm wide. **INFL:** 4–40 cm, ± open. **FL:** fragrance lemony or otherwise pleasant; perianth green; upper sepal erect to curved back, lower recurved; lateral petals linear, spreading or erect, 1 mm wide at base, 4–5 × as long, lip 2.5–5 mm, narrowly lanceolate, pointed forward or down, spur 4–9 mm, tapered, curved, pointed down. Gen dry sites, scrub, woodland; < 2200 m. KR, NCoR, CaR, SN, SnFrB, SCoR, TR, PR; to WA. May–Jul ★

P. michaelii (Greene) Rydb. (p. 1407) MICHAEL'S REIN-ORCHID Pl 9–70 cm. **LF:** basal 7–30 cm, 10–30 mm wide. **INFL:** 15–30 cm, 10–30 mm wide, ± dense. **FL:** fragrance strong, pleasant; perianth green to yellow-green; upper sepal ascending, lower spreading; lateral petals ± ascending, ± concave, lip 1.7–6 mm, deltate-ovate, spur 8–12 mm, ± curved, gen pointed down. Gen dry sites, coastal scrub, woodland, mixed-evergreen or closed-cone-pine forest; < 700 m. NCo, SNF, CCo, SnFrB, n SCo, n ChI (Santa Cruz Island), WTR. Apr–Aug ★

P. transversa Suksd. (p. 1407) FLAT SPURRED PIPERIA Pl 12–57 cm. **LF:** basal 6–19 cm, 10–45 mm wide. **INFL:** 7–26 cm, ± dense. **FL:** fragrance clove-like; sepals, lateral petals white to ± yellow with green midvein; upper sepal pointed forward; lateral petals spreading, ± curved back, lip 2.2–5.3 mm, oblong to ± ovate, pointed forward or down, white, spur 6–12 mm, ± straight, perpendicular to infl axis. 2*n*=42. Gen dry sites, scrub, oak woodland, mixed-evergreen or conifer forest; < 2600 m. NW, CaR, SN, CW, TR, PR; to BC. May–Aug

P. unalascensis (Spreng.) Rydb. (p. 1407) ALASKA PIPERIA, SLENDER-SPIRE ORCHID Pl 9–70 cm. **LF:** basal 5–20 cm, 5–40 mm wide. **INFL:** 3–44 cm, gen open, dense above or not, axis gen > peduncle. **FL:** fragrance ± at night but lingering, musky, soapy, or honey-like; perianth green; upper sepal ascending or pointed forward; lateral petals ± erect to pointed forward, lip 2–5 mm, broadly ovate to lance-elliptic, gen pointed down, tip upcurved, spur 2–5.5 mm, pointed back or down. *n*=21. Gen dry sites, scrub, woodland, forest; < 3000 m. CA-FP (exc GV, SCoR), MP; to AK, ne Can, SD, NM. May–Aug

P. yadonii Rand. Morgan & Ackerman (p. 1407) YADON'S REIN-ORCHID Pl 10–50(80) cm. **LF:** basal 10–17 cm, 20–39 mm wide. **INFL:** (2)5–15(30) cm, dense. **FL:** fragrant in daytime, harsh to honey-like; upper sepal green with white margin, ± erect, lower white, midvein faintly green; lateral petals ± erect, sickle-shaped, green, white margins unequal, outer much wider than inner, lip 2.5–5 mm, lance-deltate, pointed down toward spur, white, spur 2.5–5 mm, pointed down. Gen sandy soil or sandstone, coastal scrub, Monterey-pine forest; < 150 m. c CCo (n Monterey Co.). Jun–Jul ★

PLATANTHERA BOG-ORCHID

Ronald A. Coleman

Rhizome tuber-like, elongate. **LF:** cauline, alternate, linear to elliptic or lanceolate, gradually reduced upward, with 1 lengthwise fold. **INFL:** gen spike; fl bracts lf-like. **FL:** perianth white to yellow-green or green; sepals ± equal, upper gen hood-like, lower free, gen spreading; lateral petals gen erect, lip spurred, pendent to upcurved; column ± erect. **FR:** ascending to erect. ± 85 spp.: temp N.Am, Eurasia. (Greek: wide anther) [Sheviak & Jennings 2006 Rhodora 108:19–33] Identification often difficult due to intermediates, hybrids; additional spp. expected in CA. *P. hyperborea* (L.) Lindl. not in CA.

1. Perianth white to cream; spur gen ≥ 1.5 × lip . . . . ***P. dilatata*** var. ***leucostachys***
1′ Perianth green to yellow-green; spur << to ± > lip
  2. Column gen 2.5–4 mm, ± 2/3 hood formed by upper sepal, lateral petals; lip linear to lance-linear . . . . . ***P. sparsiflora***
  2′ Column gen 1.5–2 mm, ≤ ± 1/2 hood formed by upper sepal, lateral petals; lip wider than lance-linear
    3. Spur gen cylindric, ± curved — lip lanceolate to oblong or linear-oblong, green-yellow to yellow; lvs
      clustered near base . . . . ***P. tescamnis***
    3′ Spur club- to sac-like
      4. Lip linear to lanceolate, green, red-marked or not; lvs not clustered . . . . ***P. stricta***
      4′ Lip gen lance-rhombic, yellow, not red-marked; lvs clustered below . . . . ***P. yosemitensis***

P. dilatata (Pursh) L.C. Beck var. ***leucostachys*** (Lindl.) Luer (p. 1407) WHITE-FLOWERED BOG-ORCHID Pl 15–150 cm. **LF:** 5–35 cm, 9–30 mm wide. **INFL:** 5–35 cm, gen dense; lower bracts 9–25 mm. **FL:** perianth white to cream; sepals 4–8 mm; lip 5–10 mm, ± lanceolate, spur 5–15 mm, gen ≥ 1.5 × lip, gen cylindric, ± curved; column gen < 1/2 hood formed by upper sepal, petals. 2*n*=42. Wet, gen open places, meadows, seeps, streambanks; < 3400 m. CA-FP (exc GV), GB, n DMtns (Panamint Range); to AK, MT, UT, NV. [*P. l.* Lindl.] Hybridizes with *P. sparsiflora* (*P.* ×*lassenii* W.J. Schrenk), *P. stricta* (*P.* ×*estesii* W.J. Schrenk), *P. tescamnis* (unnamed). May–Sep

P. sparsiflora (S. Watson) Schltr. (p. 1407) SPARSE-FLOWERED BOG-ORCHID Pl 25–55 cm. **LF:** 4–15 cm, 5–30 mm wide. **INFL:** gen 15–40 cm, dense to ± open, lowest fls gen not overlapping. **FL:** perianth yellow-green to green; sepals 5–9 mm; lip 6–10 mm, linear to lance-linear, spur ± = lip, ± cylindric, ± curved, tip acute; column gen 2.5–4 mm. 2*n*=42. Full sun to partial shade, wet meadows, streambanks, seeps, conifer forest; 100–3400 m. NW (exc NCoRI), CaR, SN, SCo, TR, MP, D; to OR, UT, NM, Baja CA. May–Sep

P. stricta Lindl. (p. 1407) SLENDER BOG-ORCHID Pl 20–90 cm. **LF:** 4–11 cm, 5–25 mm wide, oblong to ovate, evenly spaced. **INFL:** 5–22 cm, gen open below; lower bracts 5–35 mm. **FL:** perianth gen green to yellow-green; sepals 3–6 mm, lance-elliptic, twisted, ± reflexed; lateral petals lance-elliptic, lip ± pendent, 5–7 mm, linear to lanceolate, green, red-marked or not, spur 0.5–1 × lip, pendent, club-to sac-like, tip blunt, tinged red-± purple or not; column ± 2 mm, ≤ ± 1/2 hood formed by upper sepal, petals. 2*n*=42. Full sun to part shade, wet meadows, seeps, conifer forest; 1000–2300 m. KR, CaR, MP; to AK, MT, WY. May–Sep ★

P. tescamnis Sheviak & W.F. Jenn. Pl 29–126 cm. **LF:** 4–9, clustered near base, 8–29 cm, 8–50 mm wide, to lance-linear to -elliptic to ovate. **INFL:** 15–61 cm, open to dense. **FL:** sepals 2.7–5 mm, lanceolate, green; petals 4.5–8 mm, sickle-shaped, forming hood with upper sepal, lip ± pendent, gen lanceolate to oblong or linear-oblong, green-yellow to yellow, spur 0.8–1.4 × lip, gen cylindric, ± curved, tip subacute; column ≤ ± 1/2 hood formed by upper sepal, petals. 2*n*=42. Mesic canyons, woodland edge, dry slopes, conifer forest; 1825–2950 m. SN, GB; to CO, AZ. [*P. hyperborea* (L.) Lindl., misappl.] Jun–Aug

P. yosemitensis Colwell et al. YOSEMITE BOG-ORCHID Pl 20–80 cm. **LF:** 5–7, clustered below, 9–25 cm, 1.5–3 cm wide, lanceolate, tip acute. **FL:** strongly pungent; sepals green, upper ovate, lateral oblong; petals 3–4 mm, ± yellow, ovate-deltoid, forming hood with upper sepal, lip 4–6 mm, gen lance-rhombic, often upturned, yellow, spur sac-like, 2–28 mm; column ≤ ± 1/2 hood formed by upper sepal, petals, rounded, anther sacs gen parallel. **FR:** capsule, 0.3–1 cm. 2*n*=42. Wet meadows; 2100–2285 m. c SNH (Yosemite National Park), s SNH. Jul–Aug ★

SPIRANTHES LADIES TRESSES

LF: at fl 0 or ± basal. **INFL:** spike, gen dense, fls in spiral; bracts lf-like, < to > fls, gradually reduced upward, linear to oblong. **FL:** sepals, lateral petals narrow-lanceolate; upper sepal ± fused to lateral petals, together hood-like, enclosing column, lower ± free, ± = lip, adherent to hood; lip not spurred, pouch-like, deeply grooved below middle, concave above; column < lip, tip with anther on back. **FR:** spreading to ascending. ± 40 spp.: esp Am, also Japan, Australia, New Zealand. (Greek: coiled fls) Pls in SN may be hybrids between spp. below, for which *S. stellata* P.M. Brown et al. might be correct.

1. Lip lanceolate to ± ovate, tip puberulent above; perianth ± yellow (cream); upper sepal, lateral petals gen
  spreading, not forming hood . . . . ***S. porrifolia***
1′ Lip ± violin-shaped, tip glabrous above; perianth ± white (cream); upper sepal, lateral petals forming hood
  . . . . ***S. romanzoffiana***

S. porrifolia Lindl. (p. 1407) Pl 18–56 cm. **LF:** basal 8–14 cm. **INFL:** 5–14 cm; fl bracts 7–20 mm. **FL:** sepals, petals 7–12 mm; column 2–4 mm. 2*n*=44[66]. Wet meadows, freshwater marshes, seeps, grassland, oak woodland; < 2500 m. NW (exc NCoRI), CaR, SN, n CCo, SnFrB, SCoRO, SnGb, PR; to WA. Jun–Sep

S. romanzoffiana Cham. (p. 1407) Pl 7–30 cm. **LF:** basal 3–13 cm. **INFL:** 2–14 cm; fl bracts 6–14 mm. **FL:** sepals, petals 8–12 mm; column 1.5–4 mm. 2*n*=44,66,88. Wet meadows, freshwater marshes, seeps; < 3300 m. NW, CaR, SN, CCo, SnFrB, SnBr, SnJt, MP; to AK, ne N.Am, NM; Eur. May–Sep

POACEAE (Gramineae) GRASS FAMILY

James P. Smith, Jr., except as noted

Ann to woody per; roots gen fibrous. **ST:** gen round, hollow; nodes swollen, solid. **LF:** alternate, 2-ranked, gen linear, parallel-veined; sheath gen open; ligule membranous or hairy, at blade base. **INFL:** various (of gen many spikelets). **SPIKELET:** glumes gen 2; florets (lemma, palea, fl) 1–many; lemma gen membranous, sometimes glume-like; palea gen ± transparent, ± enclosed by lemma. **FL:** gen bisexual, minute; perianth vestigial; stamens gen 3; stigmas gen 2, gen plumose. **FR:** grain, sometimes achene- or utricle-like. 650–900 genera; ± 10550 spp.: worldwide; greatest economic importance of any family (wheat, rice, maize, millet, sorghum, sugar cane, forage crops, orn, weeds; thatching, weaving, building materials). [Barkworth et al. 2003 FNANM:25; Barkworth et al. 2007 FNANM:24] Gen wind-pollinated. *Achnatherum, Ampelodesmos, Hesperostipa, Nassella, Piptatherum, Piptochaetium, Ptilagrostis* moved to *Stipa; Elytrigia, Leymus, Pascopyrum, Pseudoroegneria, Taeniatherum* to *Elymus; Hierochloe* to *Anthoxanthum; Lolium, Vulpia* to *Festuca; Lycurus* to *Muhlenbergia; Monanthochloe* to *Distichlis; Pleuraphis* to *Hilaria; Rhynchelytrum* to *Melinis*. The following taxa (in genera not incl here), recorded in CA from historical collections or reported in literature, are extirpated, lacking vouchers, or not considered naturalized: *Acrachne racemosa* (Roth) Ohwi, *Allolepis texana*

(Vasey) Soderstr. & H.F. Decker, *Amphibromus nervosus* (Hook. f.) Baill., *Axonopus affinis* Chase, *A. fissifolius* (Raddi) Kuhlm., *Coix lacryma-jobi* L., *Cutandia memphitica* (Spreng.) K. Richt., *Dinebra retroflexa* (Vahl) Panz., *Eremochloa ciliaris* (L.) Merr., *Eustachys distichophylla* (Lag.) Nees, *Gaudinia fragilis* (L.) P. Beauv., *Miscanthus sinensis* Andersson, *Neyraudia arundinacea* (L.) Henrard, *Phyllostachys aurea* Rivière & C. Rivière, *P. bambusoides* Siebold & Zuccarini, *Rottboellia cochinchinensis* (Lour.) Clayton, *Schedonnardus paniculatus* (Nutt.) Branner & Coville, *Schizachyrium cirratum* (Hack.) Wooton & Standl., *S. scoparium* (Michx.) Nash, *Themeda quadrivalvis* (L.) Kuntze, *Thysanolaena latifolia* (Hornem.) Honda, *Tribolium obliterum* (Hemsl.) Renvoize, *Zea mays* L., *Zizania palustris* L. var. *interior* (Fassett) Dore, *Zoysia japonica* Steud. *Paspalum pubiflorum* E. Fourn., *P. quadrifarium* Lam., are now reported for s CA (J Bot Res Inst Texas 4:761–770). See Glossary p. 30 for illustrations of general family characteristics. Scientific Editors: James P. Smith, Jr., J. Travis Columbus, Dieter H. Wilken.

1. Lf blade base constricted above sheath, forming false petiole; st woody . . . . **[PHYLLOSTACHYS]**
1′ Lf blade base not constricted, not forming false petiole, forming open or closed tubular sheath; st gen herbaceous
  2. Spikelets enclosed in bristly to spiny, bur-like involucre . . . . **CENCHRUS**
  2′ Spikelets not enclosed in bristly to spiny, bur-like involucre
    3. Some or all florets modified into bulblets with conspicuous, awn-like tails . . . . *Poa bulbosa* subsp. *vivipara*
    3′ Florets not modified into bulblets
      4. Lf blades, at least upper, gen stiff at maturity, sharp-pointed and conspicuously arranged in 2 vertical rows
        5. Spikelets unisexual, pls gen dioecious; pls of salt marshes and moist alkaline sites . . . . ²**DISTICHLIS**
        5′ Spikelets bisexual; pls known only from sand dunes in Inyo Co.. . . . . **SWALLENIA**
      4′ Lf blades gen soft, rounded to acute, but not sharp-pointed (exc *Blepharidachne*, *Munroa*), not
        conspicuously arranged in 2 vertical rows (exc *Muhlenbergia asperifolia*)
        6. Lf blade and sheath undifferentiated; st internodes solid; ligules 0
          7. Spikelets 2-ranked on infl axis. . . . . **ORCUTTIA**
          7′ Spikelets spirally inserted on infl axis
            8. Pl glandular-sticky at maturity; infl a dense, cylindric spike, gen fully exserted at maturity;
              spikelets dorsally compressed . . . . **NEOSTAPFIA**
            8′ Pl not glandular-sticky at maturity; infl club-shaped, often ± enclosed by lvs at maturity; spikelets
              laterally compressed. . . . . **TUCTORIA**
        6′ Lf blade and sheath clearly differentiated; st internodes gen hollow; ligules gen present
          9. Basal or subterranean internodes swollen, bulb- or corm-like
            10. Infl spike-like, cylindric or ovoid, branches not evident; glumes winged. . . . . *Phalaris aquatica*
            10′ Infl panicle-like, open with evident branches; glumes not winged
              11. Sheath open; glumes ± enclosing florets; lower lemma awns bent, twisted . . . . ²**ARRHENATHERUM**
              11′ Sheath closed to near tip; glumes = lowest floret; lower lemma awns gen 0 or awns straight,
                untwisted. . . . . **MELICA**
          9′ Basal or subterranean internodes not swollen, not bulb- or corm-like
            12. Robust per, gen 1.5–7 m; st 0.5–3 cm diam, tough, hardened to woody, gen persisting; infl
              terminal, often conspicuously plume- or fan-like . . . . **Group 1**
            12′ Low ann to mid-sized per, gen < 1.5 m; st gen < 5 mm wide, strictly herbaceous, gen dying back
              annually; infl axillary, terminal, or both, but not conspicuously plume- nor fan-like
              13. Pls gen < 15 cm; mature infl not clearly exceeding lvs, often ± enclosed in upper sheath
                14. Floret 1; lemma 1-veined. . . . . **CRYPSIS**
                14′ Florets 2+; lemma ≥ 3-veined
                  15. Lemma awn 0
                    16. Infl subtended by 5–10 bristles; spikelets 1–2 mm; florets 2 . . . . *Pennisetum clandestinum*
                  16′ Infl not subtended by bristles; spikelets 6–10 mm; florets 3–5. . . . . **SCLEROCHLOA**
                15′ Lemma awns 1 or 3
                  17. Lemma 3-lobed, conspicuously ciliate; upper florets 3-awned. . . . . **BLEPHARIDACHNE**
                  17′ Lemma 2-lobed or tapering to awn
                    18. Glumes 6–9 mm, > lemma; lemma 2-lobed . . . . **DASYOCHLOA**
                    18′ Glumes 2–4 mm, < lemma; lemma tapering to point. . . . . **MUNROA**
             13′ Pls gen >> 15 cm; mature infl clearly exceeding upper lvs
              19. Spikelets 1–2 per infl
                20. Glumes < lowest floret; awns (if present) not twisted, nor bent. . . . . *Brachypodium distachyon*
                20′ Glumes > lowest floret; awns twisted and bent . . . . *Danthonia unispicata*
              19′ Spikelets few to many in a well-developed infl
                21. Spikelets subtended by 1+ bristles, sterile branches, or long, silky hairs > spikelet
                  22. Spikelets subtended by long, silky hairs from their bases, rachis joints, or pedicels
                    23. Spikelets dissimilar, sessile bisexual, stalked sterile . . . .′. . . . . **BOTHRIOCHLOA**
                    23′ Spikelets similar, bisexual. . . . . **IMPERATA**
                  22′ Spikelets subtended by 1 or more gen stiff bristles or sterile branches
                  24. Bristles short-stiff-hairy . . . . **SETARIA**
                  24′ Bristles scabrous to long-ciliate
                    25. Infl panicle-like, branches 2–7. . . . . **ANDROPOGON**
                    25′ Infl spike-like, dense, cylindrical . . . . **PENNISETUM**
              21′ Spikelets not subtended by bristles or by long, silky hairs

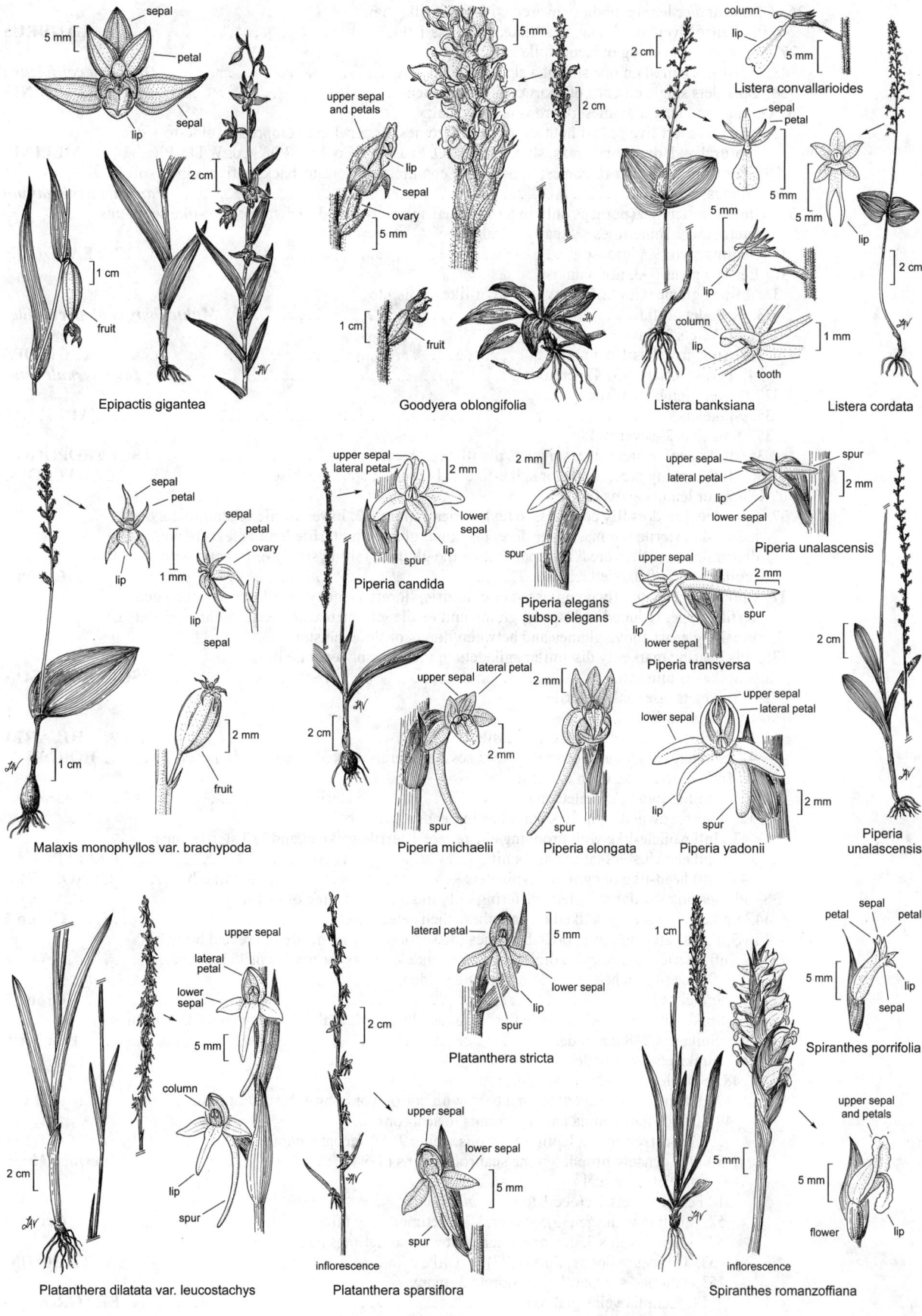

Epipactis gigantea

Goodyera oblongifolia

Listera convallarioides

Listera banksiana

Listera cordata

Malaxis monophyllos var. brachypoda

Piperia candida

Piperia elegans subsp. elegans

Piperia unalascensis

Piperia transversa

Piperia michaelii

Piperia elongata

Piperia yadonii

Piperia unalascensis

Platanthera stricta

Spiranthes porrifolia

Platanthera dilatata var. leucostachys

Platanthera sparsiflora

Spiranthes romanzoffiana

26. Glume and/or lemma bodies covered with long, silky hairs
  27. Infl dense, ovoid or oblong, head-like; spikelets 1-fld . . . . **LAGURUS**
  27′ Infl open, branching evident; spikelets 2-fld
    28. Spikelets paired on one side of a slender rachis . . . . *Digitaria californica*
    28′ Spikelets single on ends of short capillary branches. . . . . **MELINIS**
26′ Glumes and lemma bodies glabrous to short-hairy
  29. Ann or short-live per without stolons or rhizomes; upper floret compressed side-to-side;
      naturalized, disturbed areas, slopes, NCoRO, SnJV, s CCo, SCoRO, SCo, WTR, PR . . . . **MELINIS**
  29′ Per from stolons or rhizomes; upper floret compressed front-to-back; native, sandy soils,
      dunes, e SCo, D . . . .*Panicum urvilleanum*
30. Glumes or lemma (perhaps only in a unisexual spikelet) with 3–9 awns or awn-like segments
    (lateral awns sometimes << than central 1)
  31. Lemma awns 9, plumose . . . . **ENNEAPOGON**
  31′ Lemma awns 0–3, not plumose
    32. Lower glume with (2)3–9 awns or awn-like segments
      33. Spikelets 1–fld. . . . . *Muhlenbergia alopecuroides*
      33′ Spikelets 2–5 fld
        34. Ann; spikelets 1 per node . . . . **AEGILOPS**
        34′ Per; spikelets 2(3–4) per node . . . . *Elymus multisetus*
    32′ Lower glume awn 0 or 1
      35. Spikelets 1-fld . . . . **ARISTIDA**
      35′ Spikelets 2–several-fld
        36. Infl open, panicle-like; spikelets pistillate . . . . [2]**SCLEROPOGON**
        36′ Infl 1–many raceme-like or spike-like 1° branches; spikelets bisexual . . . . **BOUTELOUA**
30′ Glume or lemma awns 0–2
  37. Spikelets gen dorsally compressed (cylindrical); florets 2, lower sterile or staminate, often
    reduced to sterile lemma, upper floret bisexual; glume and fertile lemma texture noticeably
    dissimilar; spikelets breaking apart below the glumes, falling singly or in pairs with
    segments of infl axis attached . . . . **Group 2**
  37′ Spikelets gen laterally compressed or cylindric; florets 1–many, if > 1, lower one(s) gen
    fertile, upper reduced and sterile; glume and fertile lemma texture gen similar; spikelets gen
    breaking apart above glumes and between florets or floret clusters
    38. Pls bearing markedly dissimilar spikelets in the same or separate infls
      39. Spikelets unisexual . . . . [2]**SCLEROPOGON**
      39′ Spikelets bisexual or sterile
        40. Infl spike-like
          41. St internodes solid; spikelets sessile . . . . **HILARIA**
          41′ St internodes hollow; central spikelet sessile, laterals stalked (exc in *Hordeum vulgare*) . . . **HORDEUM**
        40′ Infl panicle-like, open to dense
          42. Lemma awn 0; spikelets in groups of 7 (1 fertile + 6 sterile) . . . . *Phalaris paradoxa*
          42′ Lemma awned; spikelets paired or in pendant clusters
            43. Infl panicle-like with drooping clusters of 1 fertile spikelet and 1–3 sterile ones;
              spikelet clusters falling as 1 unit. . . . . **LAMARCKIA**
            43′ Infl head-like or cylindric, spikelets gen paired; spikelets falling separately . . . . **CYNOSURUS**
    38′ Pls bearing similar spikelets, differing only in size and degree of maturity
      44. Spikelets sessile or stalked on an unbranched central axis . . . . **Group 3**
      44′ Spikelets attached to 1° or 2° branches (dissection may be needed to reveal branches)
        45. Infl a series of 2 or more digitate, raceme-like, clusters, or whorls of 1° branches. . . . . **Group 4**
        45′ Infl much-branched, 2° or 3° branching evident
          46. Spikelets 1-fld . . . . **Group 5**
          46′ Spikelets 2–many-fld (sterile florets reduced to small scales or hairy, awl-like rudiments)
            47. Spikelets 2–8 per node. . . . . **ELYMUS**
            47′ Spikelets 1 per node
              48. Spikelets unisexual; pls dioecious
                49. Lf tips prow-shaped; lemma base with cottony or cobwebby tuft of hairs . . . . **POA**
                49′ Lf tips flat; lemma bases glabrous to scabrous
                  50. Pls rhizomatous; lemma glabrous, veins 9–11, inconspicuous. . . . . [2]**DISTICHLIS**
                  50′ Pls densely tufted; lemma scabrous, veins (3)5(7) . . . . *Festuca kingii*
              48′ Spikelets bisexual
                51. Fertile florets 2+ (occ 1 in *Dissanthelium*)
                  52. Lemma veins gen (3)5–several, sometimes very faint . . . . **Group 6**
                  52′ Lemma veins 3, gen prominent (laterals sometimes near margins)
                    53. Glumes > florets; florets 2(3) — ChI . . . . **DISSANTHELIUM**
                    53′ Glumes < upper florets; florets 3–many
                      54. Lemma veins glabrous. . . . . **ERAGROSTIS**
                      54′ Lemma veins hairy, at least below
                        55. Lf blade margin white; lemma awn 0.5–2.5 mm; stigmas white. . . . . **ERIONEURON**
                        55′ Lf blade margin green; lemma awn 0; stigmas purple . . . . **TRIDENS**

51′ Fertile floret 1
- 56. Spikelet breaking below glumes; lower floret bisexual, awn 0; upper floret staminate, awned . . . . **HOLCUS**
- 56′ Spikelet breaking above glumes; staminate or sterile floret(s) below fertile one (sometimes << fertile and easily overlooked)
  - 57. Fertile floret gen awned
    - 58. Per; nodes green or straw-colored; spikelets breaking apart below lowest floret . . . . [2]**ARRHENATHERUM**
    - 58′ Ann; nodes purple-black; spikelets breaking apart above lowest floret . . . . **VENTENATA**
  - 57′ Fertile floret awn 0
    - 59. Lower florets reduced to sterile lemma < 1/2 fertile floret . . . . **PHALARIS**
    - 59′ Lower staminate or sterile florets ≥ fertile one
      - 60. Spikelets straw-colored or brown at maturity; stamens 2–3; auricles 0 (exc in *Anthoxanthum aristatum, A. odoratum*). . . . . **ANTHOXANTHUM**
      - 60′ Spikelets green or purplish at maturity; stamens 3 or 6; auricles ciliate . . . . . . **EHRHARTA**

Group 1

Robust per; st gen 1.5+ m, 0.5+ cm diam

1. St much-branched above base . . . . [PHYLLOSTACHYS]
1′ St not branched above base (exc infl)
- 2. Lvs predominantly basal
  - 3. St internodes hollow; spikelets unisexual (pls gen pistillate). . . . . **CORTADERIA**
  - 3′ St internodes solid; spikelets bisexual. . . . . *Stipa mauritanica*
- 2′ Lvs well distributed along sts
  - 4. Infl branches and spikelets glabrous
    - 5. Infl spike-like (branched below) . . . . *Elymus condensatus*
    - 5′ Infl panicle-like . . . . **SORGHUM**
  - 4′ Infl branches and/or spikelets hairy
    - 6. St internodes solid; spikelets paired, one or both stalked
      - 7. Spikelets unequally stalked; infl axis remaining intact at maturity . . . . **IMPERATA**
      - 7′ 1 spikelet sessile, 2nd stalked; infl axis breaking apart at maturity, segments with spikelets attached. . . **SACCHARUM**
    - 6′ St internodes hollow; spikelets 1 per node
      - 8. Spikelet axis glabrous; lemma densely hairy . . . . **ARUNDO**
      - 8′ Spikelet axis silky-hairy; lemma glabrous . . . . **PHRAGMITES**

Group 2

Spikelets dorsally compressed, breaking apart below glumes; glumes and fertile lemma of dissimilar texture

1. Glumes leathery to hardened (at least in sessile spikelets), ± equal, 1 or both gen > upper floret (exc awns); fertile and sterile lemmas gen translucent
- 2. Spikelets 1 per node, subtended by a hairy bristle. . . . . **ANDROPOGON**
- 2′ Spikelets 2 or 3 per node, 1 sessile and 1 (2) stalked
  - 3. Terminal spikelets per branch in 3s, 1 sessile and 2 stalked. . . . . **SORGHUM**
  - 3′ Terminal spikelets per branch in 2s, 1 sessile and 1 stalked
    - 4. Infl unbranched, raceme-like. . . . . **HETEROPOGON**
    - 4′ Infl branched, panicle-like
      - 5. Infl internodes and spikelet stalks with a translucent longitudinal groove; lowest pair of spikelets per raceme fertile . . . . **BOTHRIOCHLOA**
      - 5′ Infl internodes and spikelet stalks without a translucent longitudinal groove; lowest pair of spikelets per raceme staminate . . . . **HYPARRHENIA**
1′ Glumes membranous, flexible, lower glume gen < upper (0), upper glume ≤ upper floret; sterile lemma membranous and resembling upper glume, fertile lemma leathery to hardened (membranous in *Digitaria*)
- 6. Infl spike-like; spikelet bases +/- embedded in 1 side of flattened, thickened axis. . . . . **STENOTAPHRUM**
- 6′ Infl panicle-like or a series of paired, digitate, or racemose branches; spikelets not embedded, on 1 or more sides of rounded, unthickened axes
  - 7. Spikelet subtended by a cup- or ring-like structure . . . . **ERIOCHLOA**
  - 7′ Spikelet not subtended by cup- or ring-like structure
    - 8. Ligule 0, at least on upper lvs . . . . **ECHINOCHLOA**
    - 8′ Ligule present
      - 9. Infl a series of digitate or raceme-like branches
        - 10. Fertile floret membranous, flexible at maturity . . . . **DIGITARIA**
        - 10′ Fertile floret leathery to hardened, rigid at maturity. . . . . **PASPALUM**
      - 9′ Infl panicle-like
        - 11. Fertile lemma thin, flexible; lower glume minute or 0. . . . . *Digitaria californica* var. *californica*
        - 11′ Fertile lemma ± thick, rigid; lower glume well-developed . . . . **PANICUM**

Group 3

Infl solitary, spike-like or raceme-like; spikelets sessile or stalked

1. Some or all spikelets stalked
  2. Spikelets 3 per node, sessile and stalked (or only stalk remaining after upper spikelet has fallen); infl spike-like
    3. St internodes solid; auricles 0; central spikelet short-stalked, laterals sessile; glumes awned . . . . **HILARIA**
    3′ St internodes hollow; auricles present or 0; central spikelet sessile, laterals stalked; glumes awnless . . . **HORDEUM**
  2′ Spikelets 1 per node, stalked; infl raceme-like
    4. Lemma veins prominent, parallel; palea keeled, winged on lower 1/2 . . . . **PLEUROPOGON**
    4′ Lemma veins converging at tip; palea not winged
      5. Glumes ≥ florets, ± enclosing them . . . . ²**DANTHONIA**
      5′ Glumes ± = lower florets only
        6. Spikelet axis thickened, falling with florets; lemma ± circular in ×-section . . . . **DESMAZERIA**
        6′ Spikelet axis not thickened, not falling with florets; lemma flattened
          7. Upper lf sheaths closed; lemma tips gen 2-forked or 2-lobed . . . . **BROMUS**
          7′ Upper lf sheaths open; lemma tips obtuse or acute
            8. Upper glume veins (3)5–9; palea keeled, ciliate . . . . ²**BRACHYPODIUM**
            8′ Upper glume veins 3(5); palea glabrous or hairy, but not keeled, nor ciliate . . . . ²**FESTUCA**
1′ All spikelets sessile or nearly so (upper spikelets sometimes stalked in *Scribneria*)
  9. Glume 1 (exc in uppermost spikelets; 0 in *Elymus californicus*)
    10. St internodes hollow; infl axis remaining intact at maturity; spikelets breaking apart above glumes;
        lemma veins 5 . . . . ²**FESTUCA**
    10′ St internodes solid; infl axis breaking apart at maturity, spikelets falling with axis segments; lemma
        veins 3 . . . . **HAINARDIA**
  9′ Glumes 2
    11. Spikelet bases ± embedded in cavities or pits in thickened infl axis
      12. Per from stolons or rhizomes; spikelets attached to one side of flattened infl axis . . . . **STENOTAPHRUM**
      12′ Ann; spikelets attached to both sides of rounded infl axis
        13. Spikelets awnless . . . . **PARAPHOLIS**
        13′ Spikelets awned
          14. Glumes awned; lemmas awnless; florets 2–5 . . . . **AEGILOPS**
          14′ Glumes awnless; lemma awned; floret 1 . . . . **SCRIBNERIA**
    11′ Spikelet bases not embedded in cavities or pits of thickened infl axis
      15. Spikelets borne on one side of infl axis; lemma veins 3 . . . . **BOUTELOUA**
      15′ Spikelets 2-ranked; lemma veins 5–9
        16. Spikelets 2–8 per node . . . . ²**ELYMUS**
        16′ Spikelets 1 per node
          17. Lemma keels ciliate . . . . **SECALE**
          17′ Lemma keels (if present) not ciliate
            18. Glumes > florets and gen enclosing them (exc awns and lemma teeth). . . . . ²**DANTHONIA**
            18′ Glumes < florets
              19. Fertile floret 1, subtended by 2 sterile florets . . . . *Ehrharta longiflora*
              19′ Fertile florets 2+, sterile florets above the fertile ones
                20. Auricles 0. . . . . ²**BRACHYPODIUM**
                20′ Auricles present
                  21. Glumes and lemma backs rounded . . . . ²**ELYMUS**
                  21′ Glumes and lemmas keeled
                    22. Per; lemma awns 1–6 mm. . . . . **AGROPYRON**
                  22′ Ann; lemma awns to (0)4 cm . . . . **TRITICUM**

Group 4

Infl branches digitate, raceme-like, clustered, or in whorls; no further branching evident

1. Infl branches digitate or clustered at tip of infl axis
  2. Glume and lemma awns 0
    3. Per; floret 1; internal spikelet axis extended behind palea as a slender bristle, rarely bearing a sterile
      floret . . . . **CYNODON**
    3′ Ann or short-lived per; florets 2+; internal spikelet axis not extended behind palea. . . . . **ELEUSINE**
  2′ Glumes and/or lemmas awned
    4. Spikelets laterally compressed or cylindric, but not keeled; upper glume awn 0; tip of infl branch not
      extending beyond last spikelet. . . . . ²**CHLORIS**
    4′ Spikelets strongly keeled; upper glume awned; tip of infl branch extending beyond last spikelet
    . . . . **DACTYLOCTENIUM**
1′ Infl branches raceme-like, 1+ per node along a central, unbranched axis, or in whorls
  5. Floret 1 per spikelet; pls of coastal marshes and wet places . . . . **SPARTINA**

5′ Florets 2+ per spikelet (exc basal spikelet per branch in *Bouteloua aristidoides*); pls of various habitats
  6. Fertile floret 1 per spikelet
    7. Infl branches 2+ per node.. ²**CHLORIS**
    7′ Infl branches 1 per node
      8. Spikelet laterally compressed; lowest floret bisexual .... **BOUTELOUA**
      8′ Spikelet dorsally compressed; upper floret bisexual.................................... **ECHINOCHLOA**
  6′ Fertile florets 2+ per spikelet
    9. Lemma 3-veined, sometimes appearing 1-veined.................................... **LEPTOCHLOA**
    9′ Lemma 5–several-veined
      10. Upper glume 3-veined .... **FESTUCA**
      10′ Upper glume 7-veined .... **SCLEROCHLOA**

Group 5

Infl panicle-like, open and spreading to compact, cylindrical or head-like; spikelets 1-fld

1. St nodes purple-black; floret staminate .... **VENTENATA**
1′ St nodes green or straw-colored; floret bisexual
  2. Glumes 0; palea 3-veined; pls of wet sites.. **LEERSIA**
  2′ Glumes 1 or 2; palea, when present, 2-veined; pls gen of drier sites (exc *Oryza* and some *Phalaris*)
    3. Lower glume awns 2 .... *Muhlenbergia alopecuroides*
    3′ Lower glume awn 0 or 1
      4. Spikelets ± round, overlapping in 2 rows on 1 side of infl branch; glumes winged, body transversely
        wrinkled.. **BECKMANNIA**
      4′ Spikelets not round, nor overlapping in 2 rows; glumes not winged, body smooth or veined, but not
        transversely wrinkled
        5. Glumes ciliate-keeled
          6. Glumes united at base; lemma awned; palea 0 .... **ALOPECURUS**
          6′ Glumes separate at base; lemma awnless; palea present.......................... **PHLEUM**
        5′ Glumes glabrous to sparsely hairy, but not ciliate-keeled
          7. Glumes unequal, tapering to a long point, ± swollen at base .... **GASTRIDIUM**
          7′ Glumes equal or unequal, but not tapering to long points, not swollen at base
            8. Glumes keeled, often winged; floret subtended by a membranous flap or tuft of hairs .... **PHALARIS**
            8′ Glumes gen rounded, not winged; floret not subtended by a flap or tuft of hairs
              9. Lemma gen hardened at maturity, margins gen overlapping and enclosing palea and grain; grain
                hardened, round in ×-section; callus well-developed, blunt or sharp-pointed
                10. St internodes ± solid; lemma veins 3, awns gen straight; ligule of hairs or basal membrane
                  long-ciliate.. **ARISTIDA**
                10′ St internodes hollow; lemma veins 3–7, awns gen bent 1–2 ×; ligule gen membranous............. **STIPA**
             9′ Lemma and grain membranous or firm, but not hardened, gen flattened; callus not esp well-developed
              11. Spikelet axis extended beside or above floret as a glabrous or hairy stub or bristle
                12. Spikelets breaking apart below glumes at maturity; stamens 1–2.......................... **CINNA**
                12′ Spikelets breaking apart above glumes at maturity; stamens 3
                  13. Ann; callus glabrous to sparsely hairy .... **APERA**
                  13′ Ann or per from stolons or rhizomes; callus hairy
                    14. Infl dense, cylindrical; spikelets 12–14 mm; lemma awn 0 .... **AMMOPHILA**
                    14′ Infl open, panicle-like; spikelets 2–4(11) mm; lemma awn 0.5–17 mm........... **CALAMAGROSTIS**
              11′ Spikelet axis not extended beside or above floret
              15. Pls of flooded fields and in adjacent ditches; stamens 6..................................... **ORYZA**
              15′ Pls terrestrial; stamens 1 or 3
                16. Glumes with well-developed awns (exc *Polypogon viridis*); spikelets breaking apart below
                  glumes... **POLYPOGON**
                16′ Glumes acute, acuminate, or awn-tipped; spikelets breaking apart above glumes
                  17. Lemma vein 1, awn 0; lf sheaths gen ciliate at tip; seed ejected from grain at maturity ... **SPOROBOLUS**
                  17′ Lemma vein 3 or 5, awn present or 0; lf sheaths not ciliate at tip; seed remaining within
                    grain at maturity
                    18. Lemma veins 5(3), faint; palea gen 0 or << lemma................................... **AGROSTIS**
                    18′ Lemma veins 3, gen obvious; palea gen well-developed .... **MUHLENBERGIA**

Group 6

Infl panicle-like, open and spreading to compact, cylindrical or head-like; fertile florets 2+

1. One or both glumes ≥ lowest floret, sometimes enclosing all of them
  2. Spikelet length ± = width; glumes, lemma papery to translucent, inflated, at right angles to spikelet axis .... **BRIZA**
  2′ Spikelet longer than wide; glumes, lemma not papery to translucent, nor at right angles to axis
    3. Glumes of 2 kinds, lower linear to lanceolate, upper 3–4× wider than lower when spread flat; spikelet
      axis breaking below glumes .... **SPHENOPHOLIS**

3′ Glumes similar in shape, equal in length or not; spikelets breaking above glumes and between florets (exc *Trisetum*)

  4. Lemma awnless

    5. Spikelet axis extended beyond bisexual florets as a slender, hairy bristle, sometimes with a reduced floret at its tip

      6. Spikelets shiny; glume keels minutely ciliate; axis internodes < 1 mm . . . . ²**KOELERIA**

      6′ Spikelets not shiny; glume keels not ciliate; axis internodes 1.5–2 mm . . . . ³**TRISETUM**

    5′ Spikelet axis not extended beyond bisexual florets

      7. Per; palea veins swollen at base; spikelets 7–15 mm . . . . [*Danthonia decumbens*]

      7′ Ann; palea veins not swollen at base; spikelets 4.5–7 mm . . . . SCHISMUS

  4′ Lemma awned (sometimes awnless in *Avena sativa*)

    8. Spikelets 15–50 mm . . . . **AVENA**

    8′ Spikelets < 15(20) mm

      9. Lemma awned at or below middle

        10. Ann; lemma tip 2-lobed; spikelet axis not extended beyond upper floret . . . . **AIRA**

        10′ Per (exc *Deschampsia danthonioides*); lemma tip 2–4 toothed at truncate tip; spikelet axis gen extended beyond upper floret

          11. Lvs gen basal, blades 1–3 mm wide; spikelet axis ≥ 0.75 mm beyond upper floret, gen densely hairy . . . . **DESCHAMPSIA**

          11′ Lvs gen cauline, blades 3–7 mm wide; spikelet axis inconspicuous beyond upper floret, ± 0.5 mm, gen glabrous . . . . **VAHLODEA**

      9′ Lemma awned at or near tip or between 2 teeth

        12. Spikelet axis clearly hairy . . . . ³**TRISETUM**

        12′ Spikelet axis ± glabrous

          13. Lemma hairy, hairs in 1+ transverse rows . . . . **RYTIDOSPERMA**

          13′ Lemma glabrous or if hairy, hairs on margins or evenly distributed . . . . **DANTHONIA**

1′ Lower glume < lowest floret, never enclosing 2 or more florets

  14. Lemma awnless

    15. Lemma veins ± prominent, equally-spaced, parallel

      16. Upper glume 1-veined; lf sheath closed to near top . . . . GLYCERIA

      16′ Upper glume 3-veined; lf sheath open

        17. Rhizomes 0; lemma faintly 5-veined; pls of saline or alkaline soils . . . . PUCCINELLIA

        17′ Rhizomes present; lemma distinctly 7–9-veined; pls of freshwater sites . . . . TORREYOCHLOA

    15′ Lemma veins not esp prominent, converging at tip

      18. Lf tips prow-shaped; callus and/or lemma base gen cobwebby-hairy . . . . POA

      18′ Lf tips flat; callus or lemma base glabrous to scabrous

        19. Spikelet axis segments thickened, falling with florets; lemma ± round in ×-section . . . . **DESMAZERIA**

        19′ Spikelet axis segments not thickened, not falling with florets; lemma flattened

          20. Lower glume broadly translucent distally and along margins; uppermost florets vestigial, densely clustered, sometimes appearing as a club-shaped rudiment; lf sheaths closed . . . . ²**MELICA**

          20′ Lower glume gen membranous or narrowly translucent near tip or margins only; uppermost florets, if vestigial, not densely clustered, club-shaped rudiment 0; lf sheaths open or closed

            21. Lf sheath closed to near top; lemma tip 2-toothed; spikelets gen 15–70 mm . . . . ²**BROMUS**

            21′ Lf sheath open for at least 1/2 its length; lemma tip tapering to a point; spikelets gen 3.5–12(18) mm . . . . ²**FESTUCA**

  14′ Lemma awned

    22. Infl often dense, cylindric, spike-like, or more open in full fl; spikelets shiny, palea colorless . . . . ²**KOELERIA**

    22′ Infl branching evident; spikelets not esp shiny; palea brown or green, at least along veins

      23. Spikelets in dense, 1-sided clumps on distal portions of panicle branches; lemma keels stiff-ciliate . . . **DACTYLIS**

      23′ Spikelets not densely clumped at ends of panicle branches; lemma keels not stiff-ciliate

        24. Lower glume broadly translucent distally and along margins; uppermost florets vestigial, densely clustered, sometimes appearing as a club-shaped rudiment; lf sheaths closed . . . . ²**MELICA**

        24′ Lower glume gen membranous or narrowly translucent near tip or margins only; uppermost florets, if vestigial, not densely clustered, club-shaped rudiment 0; lf sheaths open or closed

          25. Spikelet axis hairy, extended beyond upper florets; upper glume ≥ lowest floret . . . . ³**TRISETUM**

          25′ Spikelet axis gen glabrous, not extended beyond upper florets; upper glume < lowest floret

            26. Lf sheath closed; spikelet gen 1+ cm; lemma tip 2-toothed; ovary tip hairy . . . . ²**BROMUS**

            26′ Lf sheath open for at least 1/2 its length (exc in *Festuca rubra*); spikelet gen < 1 cm; lemma tip entire, occ minutely 2-toothed; ovary tip glabrous or hairy . . . . ²**FESTUCA**

AEGILOPS GOAT GRASS

Ann. ST: gen erect to abruptly bent at base, gen glabrous. **LF**: sheath margins translucent, auricles ciliate; ligule membranous or 0; blade 1.5–15 cm, 1.5–5 mm wide, flat, spreading. **INFL**: 1.5–11(15) cm, spike-like, cylindric or wider at base, ± open

to dense; spikelets 2-ranked, 1 per node, basal gen vestigial, distal spikelets also gen reduced; breaking away as a single unit or in sections with axis segments attached. **FERTILE SPIKELET**: gen not compressed, 5–15 mm; glumes thick, hard, 3 ± veined, tips gen toothed or 1–5-awned; florets 2–5[8]; lemma similar to glumes or firmer, toothed or 1–3-awned at tip; palea papery, 2-keeled. **DISTAL SPIKELET**: 2–5 mm, similar or not to fertile spikelet, gen sterile. 21–23 spp.: Medit, sw&c Asia. (Greek, preferred by goats, or Latin, a sweet-fruited oak) [Saufferer 2007 FNANM 24:261–267] Interfertile with *Triticum* and perhaps not distinct from it. *A. tauschii* Coss. is reported from a single occurrence in Riverside Co., doubtfully naturalized.

1. Spikelets narrowly cylindrical; glumes of fertile spikelets 1-awned or long-toothed . . . . ***A. cylindrica***
1′ Spikelets lance-ovate to urn-shaped; glumes of fertile spikelets 2–5-awned
  2. Distal spikelets > 7 mm; fertile lemmas 2–3-toothed, 1 occ extended as awn to 10 mm . . . . ***A. triuncialis***
  2′ Distal spikelets 2–5 mm; fertile lemmas with 2–3 awns to 40 mm
    3. Basal vestigial spikelets 1(2); spikelets gradually narrowing distally . . . . [***A. geniculata***, incl *A. ovata*]
    3′ Basal vestigial spikelets 3(2); lower spikelets ± ovate, upper abruptly ± oblong . . . . ***A. neglecta***

A. cylindrica Host (p. 1417) JOINTED GOAT GRASS **ST**: 14–50 cm. **LF**: blade 3–15 cm, 2–5 mm wide; sheath margins, occ ciliate. **INFL**: 2–12 cm; spikelets narrowly cylindrical, ± 0.3 cm wide; vestigial basal spikelets 0–2; axis breaking apart in fr; spikelets partly sunken in axis. **FERTILE SPIKELET**: 9–12 mm, narrowly cylindrical, glumes acute, tapered, or short-awned; florets 2–5; lemma abruptly pointed or with awn 1–5 mm. **DISTAL SPIKELET**: 7–10 mm; glume awns 3–6 cm, lemma awn 4–8 cm. 2*n*=28. Disturbed, dry sites, cult fields; CaR, ScV, SW, MP; to WA, Great Plains, Mex; native to Medit Eur, w Asia. Crosses and backcrosses with *Triticum aestivum* L. to yield fertile hybrids. May–Jul ◆

A. neglecta Bertol. THREE-AWNED GOAT GRASS **ST**: 25–35 cm. **LF**: blade 2–8 cm, to 3–4 mm wide, long-hairy, margins gen ciliate. **INFL**: 3–6 cm, vestigial basal spikelets gen 3; distal spikelets abruptly ± oblong; falling as a whole at maturity. **FERTILE SPIKE-LET**: 10–11 mm, laterally compressed, ± inflated or urn-shaped; glumes ± equal, 9–10 mm, 7–9-veined, 2–3-awned to 5 cm; lemma 10–11 mm, 5-veined, 2–4-awned to 2.5 cm. **DISTAL SPIKELET**: 2–5 mm, narrowly cylindrical. 2*n*=28,48. Disturbed fields, roadsides; 30–800 m. KR, NCo, SN, ScV; to WA; native to Medit, w Asia. [*A. triaristata* Willd.] Because this sp. has not appeared in earlier CA floras, specimens have often been misidentified as *A. triuncialis* L. Potentially an aggressive weed. May–Jul

A. triuncialis L. (p. 1417) BARBED GOAT GRASS **ST**: 17–45 cm. **LF**: blade 1.5–7 cm, 2–3 mm wide. **INFL**: 2–5.5 cm; vestigial basal spikelets 2–3; ± cylindrical distally; axis breaking at base of spikelets at maturity; spikelets gen not sunken in axis. **FERTILE SPIKELET**: 7–13 mm, lance-ovate, glumes 2–3 awned; florets gen 3–5, lower 2 gen fertile; lemma 2–3-toothed, central tooth occ extended as an awn to 10 mm. **DISTAL SPIKELET**: 7–9 mm, glumes 3-awned or 1-awned with 2 lateral teeth, awns gen 4–8 cm. 2*n*=28. Disturbed sites, cult fields, roadsides; < 1000 m. s NCoR, CaRF, n&c SNF, ScV, n CW; native to Medit Eur, w Asia. Many collections of *A. neglecta* are misidentified as this. May–Jul ◆

AGROPYRON   CRESTED WHEAT GRASS

Per, gen cespitose. **ST**: erect or bent, 2.5–10 dm. **LF**: sheath open, gen appendaged; ligule membranous; blade flat or rolled. **INFL**: spike-like, axis not breaking apart at maturity; spikelets 1, 2-ranked, strongly overlapping, divergent or spreading. **SPIKELET**: laterally compressed, glumes ± equal, < floret, lanceolate, 1–5-veined, keeled, acute to short-awned; florets 3–8[16]; axis breaking above glumes and between florets; lemma 5–7-veined, keeled, acute to awned; palea ± = lemma; anthers 3, 3–5 mm. 12–15 spp.: Medit, e Eur, c Asia. (Greek: field wheat, perhaps referring to a weed resembling wheat) [Barkworth 2007 FNANM 24:277–279] Siberian wheat grass, *A. fragile* (Roth) P. Candargy, has also been reported for CA, but is doubtfully naturalized; often used for soil stabilization on range and cropland.

A. cristatum (L.) Gaertn. subsp. ***pectinatum*** (M. Bieb.) Tzvelev (p. 1417) Occ rhizomatous. **ST**: gen erect. **LF**: blade 3–12(20) cm, 1.5–6 mm wide, glabrous or pubescent. **INFL**: 1.3–10 cm; internodes gen 1–5 mm, equal or not, glabrous or long-hairy; spikelets diverging at 30–95° angles. **SPIKELET**: 7–16 mm; glumes 3–6 mm, gen 3-veined, gen awned, awns 1.5–3 mm; florets 3–8; lemma 5–9 mm, gen 5-veined, tip acute, gen awned, awn 1–6 mm. 2*n*=14,28,42. Disturbed areas, degraded agricultural sites; 600–1500 m. KR, CaRF, n SNH, s SCoRO, SW, GB, DMoj; most of N.Am; native to Eur, Medit, Asia. [*A. c.* subsp. *desertorum* (Link) Á. Löve; *A. d.* (Link) Schult.] Used to rejuvenate burned or overgrazed areas. Jun–Aug

AGROSTIS   BENT GRASS

Paul M. Peterson & Michael J. Harvey

Ann or per, gen tufted, occ from rhizomes or stolons. **ST**: gen erect. **LF**: sheath gen smooth, glabrous; ligule membranous; blade flat to rolled. **INFL**: panicle-like, densely cylindric to openly ovate. **SPIKELET**: glumes gen subequal, back gen glabrous, vein gen finely scabrous, 1-veined, gen acute; floret 1, < glumes, gen breaking above glumes; callus glabrous to densely hairy; lemma gen 5-veined, veins not converging, occ extended as short teeth, awned from back or not; palea gen 0 or << lemma, translucent; anthers gen 3. ± 220 spp.: esp temp Am, Eurasia. (Greek: pasture) [Harvey 2007 FNANM 24:633–662; 693–697] Some cult in pastures, lawns. *Agropogon lutosus* (Poir.) P. Fourn. is a sterile hybrid between *A. stolonifera* and *Polypogon monspeliensis*. *A. viridis* is treated as *P. viridis*. *A. nebulosa* Boiss. & Reut. is reported for CA (FNANM 24: 661), but no specimens have been located. Generic delimitation adopted here reflects editorial preference.

1. Rhizomes or stolons well developed, clearly present; per
  2. Pl from stolons
    3. Ligule 0.5–2 mm, gen wider than long; stolons < 5 cm; infl widely ovate in outline, 1° branches mostly
      spreading, spikelets not crowded . . . . [3]***A. capillaris***
    3′ Ligule 2–5 mm, longer than wide; stolons 5–100 cm; infl elliptic to lanceolate in outline, 1° branches
      gen all ascending, spikelets overlapping, crowded . . . . ***A. stolonifera***

2′ Pl from rhizomes
  4. Floret callus hairs 1.5–2 mm, gen > 1/2 lemma; ligule 4–7 mm . . . . ³*A. hallii*
  4′ Floret callus hairs gen minute, sparse, or 0; ligule gen < 3 mm (exc *Agrostis gigantea*)
    5. Palea 0 or minute, << lemma
      6. Lower lf blade 1–6 mm wide; lemma tip minutely toothed; rhizome < 10 cm; anthers 0.7–1.8 mm . . . . . ³*A. pallens*
      6′ Lower lf blade < 1 mm wide; lemma tip ± acute; rhizome < 5 cm; anthers 0.5–0.7 mm . . . . ³*A. variabilis*
    5′ Palea well developed, ± 1/2 to slightly < lemma
      7. Infl ± oblong in outline, ± open, branches gen ascending to erect; montane to alpine, 1300–3500 m . . . . ²*A. humilis*
      7′ Infl gen ovate in outline, open, most branches spreading; open, gen disturbed places, < 2000 m
        8. Rhizomes < 5 cm, slender, not clearly scaly; ligule gen wider than long; infl branches with spikelets
          on distal 1/2. . . . . ³*A. capillaris*
        8′ Rhizomes < 25 cm, ± thick, ± scaly; ligule longer than wide; infl branches with spikelets ±
          throughout . . . . ²*A. gigantea*
1′ Rhizomes or stolons 0; ann or per
  9. Lemma awned from back or near tip
    10. Infl open, gen oblong to ovate in outline; spikelets not crowded, infl axes clearly visible
      11. Infl gen oblong to lanceolate in outline, 1° branches gen ascending
        12. Lemma awned below middle; anthers 1–1.5 mm; lower lf sheaths finely tomentose. . . . . *A. hooveri*
        12′ Lemma awned at or above middle; anthers ≤ 0.8 mm; lower lf sheaths gen glabrous
          13. Lower infl branches 1–2 cm; awn < 3.5 mm, straight to bent; palea < 1/3 lemma; anthers 0.3–0.6 mm . . . ³*A. exarata*
          13′ Lower infl branches 2–6 cm; lemma awn < 2 mm, straight; palea 1/5–1/4 lemma; anthers
            0.6–0.8 mm. . . . . ²*A. oregonensis*
      11′ Infl gen ovate in outline, lowest 1° branches spreading, upper branches gen ascending
        14. Lemma back puberulent below middle; floret axis prolonged beyond floret ± 1 mm, short-hairy-
          tufted; palea > 1 mm. . . . . *A. avenacea*
        14′ Lemma back glabrous or fine-scabrous; floret axis not prolonged beyond lemma; palea 0 or minute
          15. Lemma gen awned near tip, awn 3–10 mm, wavy; callus hairs < 0.6 mm, dense; anther 1, persistent
            in fr; ann . . . . *A. elliottiana*
          15′ Lemma awned below middle, awn < 2 mm, ± straight; callus hairs << lemma, sparse; anthers 3,
            deciduous; per. . . . . ²*A. scabra*
    10′ Infl dense, gen cylindric; spikelets crowded, overlapping, infl axes not clearly visible
      16. 1° infl branches gen > 0.5 cm, often evident at base (exc *Agrostis blasdalei*)
        17. Lf blade gen < 1 mm wide, ± inrolled; floret callus glabrous; infl base often partly enclosed by upper
          lf; anthers 1–2 mm . . . . ²*A. blasdalei*
        17′ Lf blade gen 2–10 mm wide, flat; floret callus minutely hairy; infl clearly stalked; anthers ± 0.5 mm
          18. Back of glume fine-scabrous throughout; palea 0.5–0.7 mm; ligule 1.5–2 mm . . . . ³*A. densiflora*
          18′ Back of glume ± glabrous (keel fine-scabrous); palea ± 0.3 mm; ligule 2.5–4 mm . . . . ³*A. exarata*
      16′ 1° infl branches gen < 0.5 cm (< 1.5 cm in *Agrostis microphylla*), obscured by densely clustered spikelets
        19. Lemma teeth 4, two < 1 mm, other two 1–1.5 mm; lemma awned below middle; callus densely
          short-hairy. . . . . *A. tandilensis*
        19′ Lemma teeth 0 or 2, equal; lemma awned at or above middle; callus gen sparsely hairy, hairs minute
          20. Lemma awned above middle, awn < 3.5 mm, straight; glume tips acute; per . . . . ³*A. densiflora*
          20′ Lemma awned ± at middle, awn 3.5–10 mm, gen bent; glume tips narrowly acuminate to awn-like; ann
            21. Lemma 2–4 mm, awn 8–10 mm. . . . . *A. hendersonii*
            21′ Lemma 1.5–2 mm, awn 3.5–8 mm . . . . *A. microphylla*
  9′ Lemma awnless (occ short-awned near tip in *Agrostis capillaris*, from near middle in *Agrostis pallens*)
    22. Spikelets crowded and often overlapping on same branch, spikelet stalks and 2° axes not clearly visible
      23. Palea ± 1/3 lemma . . . . ³*A. densiflora*
      23′ Palea 0 or < 1/3 lemma
        24. Lf blade gen < 1 mm wide, gen inrolled or folded
          25. Anthers 1–2 mm; infl branches gen < 0.5 cm; coastal habitats < 100 m. . . . . ²*A. blasdalei*
          25′ Anthers 0.5–0.7 mm; infl branches 0.5–1.5 cm; inland mtns, 1600–4000 m. . . . . ³*A. variabilis*
        24′ Lf blade gen 2–7 mm wide, gen flat
          26. Floret callus hairs 1.5–2 mm, gen slightly > 1/2 lemma . . . . ³*A. hallii*
          26′ Floret callus hairs 0 or < 0.5 mm, < 1/2 lemma
            27. Lower infl branches 1–2 cm; anthers 0.3–0.6 mm. . . . . ³*A. exarata*
            27′ Lower infl branches 2–5 cm; anthers 0.7–1.8 mm. . . . . ³*A. pallens*
    22′ Spikelets not crowded, gen well spaced on same branch, axes gen clearly visible
      28. 1° infl axes branched 1–2 × above middle, spikelets 0 on lower 1/2
        29. Lower lf blades 2–4 mm wide; glumes 2–3 mm; palea 1/5–1/4 lemma; anthers 0.6–0.8 mm. . . . . ²*A. oregonensis*
        29′ Lower lf blades 0.5–3 mm wide; glumes 1.5–3 mm; palea 0 or minute, << lemma; anthers ≤ 0.7 mm
          30. Lvs basal and cauline; infl ± 2 × longer than wide; 1° axes ± stiff . . . . *A. idahoensis*
          30′ Lvs mostly basal; infl ± long as wide; 1° axes flexible, lower ± arched . . . . ²*A. scabra*
      28′ 1° infl axes branched 1–2 × from base upwards, spikelets distributed throughout
        31. Floret callus hairs 1.5–2 mm, gen slightly > 1/2 lemma; anthers ≥ 1.5 mm . . . . ³*A. hallii*

31′ Floret callus glabrous or hairs minute, << lemma; anthers gen < 1.5 mm (exc some *Agrostis pallens*)
   32. Palea 0 or minute, << lemma
      33. Lf blade 1–6 mm wide; pls gen 10–70 cm; anthers 0.7–1.8 mm . . . . ³*A. pallens*
      33′ Lf blade gen < 1 mm wide; pls 4–30 cm; anthers 0.5–0.7 mm . . . . ³*A. variabilis*
   32′ Palea ≥ 1/2 lemma
      34. Infl branches ascending to erect; anthers < 1 mm; moist to ± dry areas, gen > 1500 m. . . . . ²*A. humilis*
      34′ Infl branches spreading to ascending; anthers 0.8–1.5 mm; disturbed areas, gen < 1500 m
         35. Ligule 0.5–2 mm, gen wider than long; infl branches with spikelets on distal 1/2 . . . . ³*A. capillaris*
         35′ Ligule 2–6 mm, longer than wide; infl branches with spikelets ± throughout . . . . ²*A. gigantea*

A. avenacea J.F. Gmel. PACIFIC BENT GRASS Per 15–65 cm. **LF:** ligule 3–5 mm; lower blades 8–20 cm, 1–3 mm wide, gen flat, finely scabrous. **INFL:** 7–30 cm, widely ovate, open; upper 1° branches gen ascending, lower 1° branches spreading, 5–15 cm, axes branched above middle, thread-like. **SPIKELET:** glumes 2.5–3.6 mm, back puberulent below middle; floret axis prolonged beyond floret ± 1 mm, tip hairy-tufted, hairs ± 0.6 mm; callus hairs < 0.7 mm; lemma 1.3–2(2.3) mm, back puberulent below middle, tip 2-toothed, awned from middle, awn 4–7.5 mm, bent; palea > 1 mm, ± 1/2 lemma; anther ± 0.5 mm. 2*n*=56. Open, often disturbed places; < 300 m. s NCo, s NCoR, SNF, GV, CW, n SCo; to TX, OH, SC; native to s Pacific islands. [*A. filiformis* G. Forst.; *Lachnagrostis f.* (G. Forst.) Trin.] Jun–Jul ❖

A. blasdalei Hitchc. (p. 1417) BLASDALE'S BENT GRASS Per 6–30 cm, decumbent to erect. **LF:** ligule gen 1–1.5 mm; lower blades 2–5 cm, gen < 1 mm wide, ± inrolled. **INFL:** 2–8 cm, cylindric, dense; base often partly enclosed by upper lf; 1° branches ascending to appressed, lower gen < 0.5 cm. **SPIKELET:** glumes 1.8–4 mm; callus glabrous; lemma 1.5–3 mm, occ awned above middle, awn < 0.7 mm, straight; palea ± 0.3 mm, < 1/3 lemma; anthers 1–2 mm. 2*n*=42. Dunes, gravelly soils, coastal bluffs, scrub; < 100 m. s NCo, n CCo, n SnFrB. May intergrade locally with *A. densiflora*; needs study. May–Jul ★

A. capillaris L. (p. 1417) COLONIAL BENT Per 10–75 cm; stolons or rhizomes < 5 cm, slender. **LF:** ligule 0.5–2 mm, gen wider than long; lower blades 3–10 cm, 1–5 mm wide, gen flat. **INFL:** 3–20 cm, widely ovate in outline, open; 1° most branches spreading, lower 1.5–4 cm, axes thread-like. **SPIKELET:** glumes 2–3 mm; callus glabrous or minutely hairy; lemma 1.5–2.5 mm, occ short-awned near tip; palea 1/2–2/3 lemma; anthers 0.8–1.3 mm. 2*n*=28. Roadsides, open, disturbed places; < 1900 m. KR, CaR, n&c SN, CCo, SnFrB, SCo; to AK, w US, Can; native to Eur. Jul–Sep

A. densiflora Vasey (p. 1417) CALIFORNIA BENT GRASS Per 9–85 cm. **LF:** ligule 1.5–2 mm; lower blades 2–12 cm, 2–10 mm wide, flat. **INFL:** 2–10 cm, ± cylindric, dense; 1° branches ± appressed, < 1.5 cm. **SPIKELET:** glumes 2–3 mm, back finely scabrous, tip acute; callus minutely hairy; lemma 1.5–2 mm, occ awned above middle, awn < 3.5 mm, straight; palea 0.5–0.7 mm, ± 1/3 lemma; anthers ± 0.5 mm. 2*n*=42. Coastal bluffs, sandy soils; < 200 m. NCo, CCo, w SnFrB; to OR. May–Aug

A. elliottiana Schult. (p. 1417) SIERRA BENT GRASS Ann 5–45 cm. **LF:** ligule 2–3 mm; lower blades 0.5–4 cm, < 1 mm wide, flat to inrolled. **INFL:** 5–20 cm, gen widely ovate in outline, open; lower 1° branches 1–4.5 cm, spreading, upper ascending, axes thread-like with spikelets clustered near tip. **SPIKELET:** floret axis not prolonged beyond floret; glumes 1.5–2 mm; callus hairs < 0.6 mm, dense; lemma 1–2 mm, back glabrous or fine-scabrous, gen awned from near tip, awn 3–10 mm, wavy; palea 0; anther 1, persistent in fr. 2*n*=14. Vernal pool margins; < 500 m. KR, NCoRI, CaRF, n SNF, n ScV; to NM, KS, TX, e US, Mex. Apr–May

A. exarata Trin. (p. 1417) SPIKE BENT GRASS Per 8–100 cm. **LF:** lower sheaths gen glabrous, ligule 2.5–4 mm; lower blades 4–15 cm, 2–7 mm wide, flat. **INFL:** 5–30 cm, oblong to ± ovate in outline, ± open to dense, occ interrupted near base; 1° branches 1–2 cm, ascending to ± appressed. **SPIKELET:** glumes 1.5–3.5 mm, acute to narrowly acuminate, back ± glabrous, keel fine-scabrous; callus hairs < 0.5 mm; lemma 1–2 mm, awned at or above middle, awn < 3.5 mm, straight to bent; palea ± 0.3 mm, < 1/3 lemma; anthers 0.3–0.6 mm. 2*n*=28,42,56. Common. Moist or disturbed areas, open woodland, conifer forest; < 2000 m. CA-FP, GB, DMtns (Panamint Range); to AK, w US, Mex. Jun–Aug

A. gigantea Roth REDTOP Per 20–100 cm; rhizomes < 25 cm, ± scaly. **LF:** ligule 2–6 mm, longer than wide; lower blades 4–10 cm, 3–8 mm wide, flat. **INFL:** 8–25 cm, widely ovate in outline, open; 1° branches gen spreading, lower 4–7 cm, axes thread-like. **SPIKELET:** glumes 2–3 mm; callus hairs 0 or minute; lemma 1.5–2 mm, awn 0 (short-awned); palea 0.7–1.4 mm; anthers 1–1.4 mm. 2*n*=42. Roadsides, disturbed areas; < 2000 m. CA-FP; to e US; native to Eur. Difficult to separate from *A. stolonifera*. Jun–Sep

A. hallii Vasey (p. 1417) HALL'S BENT GRASS Per 17–100 cm; rhizomes < 50 cm. **LF:** ligule 4–7 mm; lower blades 7–20 cm, 2–5 mm wide, flat. **INFL:** 7–22 cm, lanceolate to narrowly ovate in outline, ± open to dense; 1° branches ascending to ± appressed, lower 1–5 cm. **SPIKELET:** glumes 2.5–4 mm; callus hairs 1.5–2 mm; lemma 2–3 mm, awn 0; palea minute, << lemma; anthers 1.5–2.3 mm. 2*n*=42. Open oak woodland, conifer forest; < 1800 m. w NW, CCo, SnFrB, n SCo, WTR; OR. May–Jul

A. hendersonii Hitchc. (p. 1417) HENDERSON'S BENT GRASS Ann 6–70 cm. **LF:** ligule 1–4 mm; lower blades 1–4 cm, ± 1 mm wide, flat to weakly inrolled. **INFL:** 1–5 cm, cylindric, dense; 1° branches < 0.5 cm, ascending to ± appressed. **SPIKELET:** glumes 5–7 mm, tip narrowly acuminate to awn-like; callus hairs ± 0.7 mm; lemma 2–4 mm, awned ± at middle, awn 8–10 mm, ± bent; palea 0; anthers ± 0.5 mm. 2*n*=42. Vernal pools; < 300 m. CaRF, n SNF, ScV, n SnJV; OR. May–Jul ★

A. hooveri Swallen (p. 1417) HOOVER'S BENT GRASS Per 30–80 cm. **LF:** lower lf sheaths finely tomentose; ligule 4–6 mm; lower blades 10–16 cm, 1–2 mm wide, flat, becoming inrolled. **INFL:** (4)10–17 cm, gen lanceolate in outline, open; 1° branches ± ascending, 15–40 cm, axes thread-like. **SPIKELET:** glumes 2–3 mm; callus hairs < 0.3 mm, dense; lemma 1.5–2 mm, awned below middle, awn < 2.5 mm, bent; palea 0; anthers 1–1.5 mm. Dry sandy soils, open chaparral, oak woodland; < 600 m. s CCo, s SCoRO (San Luis Obispo, Santa Barbara cos.). Apr–Aug ★

A. humilis Vasey (p. 1417) MOUNTAIN BENT GRASS Per, cespitose, occ rhizomatous; 5–50 cm, erect to ascending. **LF:** mostly basal; ligule 0.5–2 mm; lower blades 2–15 cm, 1–4 mm wide, flat to ± folded. **INFL:** 1.5–14 cm, narrowly oblong to ovate, ± open; 1° branches gen ascending to erect, the lower 0.5–7 cm. **SPIKELET:** green to purple, glumes 1.5–2.3 mm; callus hairs minute, sparse; lemma = glumes, awn 0, 5-veined; palea 1–1.5 mm; anthers 0.5–0.7 mm. Moist to dry, subalpine or alpine meadows, slopes; 1500–3350 m. KR, NCoRH, CaRH, c&s SNH; to AK, MT, CO, NM. [*A. thurberiana* Hitchc.; *Podagrostis h.* (Vasey) Björkman; *P. thurberiana* (Hitchc.) Hultén] Jul–Aug ★

A. idahoensis Nash (p. 1417) IDAHO REDTOP Per 8–30 cm. **LF:** basal and cauline; ligule 1–3 mm; lower blades 1–5 cm, 0.5–2 mm wide, flat, often inrolled with age. **INFL:** 3–13 cm, lanceolate to ovate in outline, ± open; 1° branches gen ascending, axes ± stiff, lower 1–4 cm, axes thread-like. **SPIKELET:** glumes 1.5–2.5 mm; callus glabrous or hairs < 0.3 mm; lemma 1–2 mm, awn 0; palea minute, << lemma; anthers 0.3–0.5 mm. Open, wet meadows, conifer forest; < 3500 m. NW, CaR, SN, n SnFrB, SnBr, SnJt, W&I; to BC, w US. Jul–Aug

A. microphylla Steud. (p. 1417) SMALL-LEAF BENT GRASS Ann 8–45 cm. **LF:** ligule 1.5–4 mm; lower blades 3–15 cm, 0.7–2.5 mm wide, fine-scabrous, flat, becoming inrolled. **INFL:** 2–12 cm, ± cylindric, dense; 1° branches ascending to ± appressed, lower 0.3–1.5 cm. **SPIKELET:** glumes 2.5–5 mm, tips narrowly acuminate to awn-like; callus hairs < 0.5 mm; lemma 1.5–2 mm, awned from middle,

awn 3.5–8 mm, slightly bent; palea 0; anther ± 0.5 mm. Thin, rocky soils, cliffs, vernal pools, occ on serpentine; < 200 m. NCo, s NCoR, GV, CCo, SCo; to BC, Baja CA. May–Jul

A. oregonensis Vasey OREGON REDTOP Per 12–75 cm. **LF:** ligule 2–4.5 mm; lower blades 10–30 cm, 2–4 mm wide, gen flat. **INFL:** 8–35 cm, lanceolate to ovate in outline, open; 1° branches ascending, lower 2–6 cm, axes thread-like. **SPIKELET:** glumes 2–3 mm; callus hairs 0 or minute, sparse; lemma 1.5–2.5 mm, occ awned above middle, awn < 2 mm, straight; palea 1/5–1/4 lemma; anthers 0.6–0.8 mm. Moist areas, meadows, streambanks; < 2400 m. KR, NCoR, CaR, SN, SnBr, SnJt; to BC, MT, WY. Jun–Jul

A. pallens Trin. DUNE BENT GRASS Per gen 10–70 cm, occ from rhizomes < 10 cm. **LF:** ligule 1.5–3 mm; lower blades 1.5–5 cm, 1–6 mm wide, flat to inrolled. **INFL:** 5–20 cm, lanceolate to narrowly ovate in outline, ± open; 1° branches gen ascending, lower 2–5 cm. **SPIKELET:** glumes 2–3 mm; callus hairs minute; lemma 1.5–2.5 mm, occ awned from near middle, awn 0.5–2.5 mm, ± straight; palea 0 or minute, << lemma; anthers 0.7–1.8 mm. 2*n*=42,56. Common. Open meadows, woodland, forest, subalpine; 200–3500 m. CA-FP, GB; to BC, MT, Mex. [*A. diegoensis* Vasey; *A. lepida* Hitchc.] Geog and ecological variation need study. Jun–Aug

A. scabra Willd. (p. 1417) ROUGH BENT GRASS Per 20–75 cm. **ST:** ascending to erect. **LF:** mostly basal; ligule 2–5 mm; lower blades 4–14 cm, 1–3 mm wide, flat, finely scabrous. **INFL:** 8–25 cm, ovate in outline, open; 1° upper branches ascending, lower branches spreading, 4–11 cm, ± arched, axes thread-like, branched 1–2 × above middle, often breaking at base in fr. **SPIKELET:** floret axis not prolonged beyond lemma; glumes 1.5–3 mm; callus hairs minute, sparse; lemma 1.5–2 mm, back glabrous or fine-scabrous; occ awned from below middle, awn < 2 mm, ± straight; palea 0 or minute, << lemma; anthers 3, 0.4–0.7 mm, deciduous. 2*n*=42. Open roadsides, meadows,

conifer forest; 100–3500 m. KR, NCoR, SN, TR, SnJt, SNE; to AK, Can, e US. [*A. s.* var. *geminata* (Trin.) Swallen] Jul–Sep

A. stolonifera L. (p. 1417) CREEPING BENT Per 8–60 cm, decumbent to erect, often mat-like; stolons 5–100 cm. **LF:** ligule 2–5 mm, longer than wide; lower blades 2–10 cm, 2–5 mm wide, flat. **INFL:** 3–15 cm, elliptic to lanceolate in outline, ± dense; 1° branches ascending to ± erect, lower gen 2–6 cm. **SPIKELET:** glumes 1.5–3 mm; callus hairs minute, sparse; lemma 1.5–2 mm, awn 0; palea slightly < lemma; anthers 1–1.5 mm. 2*n*=28. Ditches, lake margins, marshes; < 1000 m. NW, CaR, n SNF, SNH, CW, SW (exc ChI), SNE, DMtns; to s Can, e US; native to Eur. [*A. alba* var. *a.*, in part, misappl.; *A. a.* L. var. *palustris* (Huds.) Pers.] Difficult to separate from *A. gigantea.* Jun–Sep ❖

A. tandilensis (Kuntze) Parodi KENNEDY'S BENT GRASS Ann 9–21 cm. **LF:** ligule 2–2.5 mm; lower blades 2–5 cm, < 1 mm wide, flat, inrolled with age. **INFL:** 2–5 cm, cylindric, dense; 1° branches erect to appressed, gen < 0.5 cm. **SPIKELET:** glumes 3–3.5 mm; callus densely short-hairy; lemma ± 1.5 mm, back densely puberulent below middle, awned below middle, awns < 6 mm, bent, lemma tip 4-toothed, 2 teeth < 1 mm, other teeth 1–1.5 mm; palea 0; anther 1, < 0.2 mm. Vernal pools; < 100 m. Deltaic GV (Solano Co.), s SCo (San Diego Co.), expected elsewhere; native to Argentina. [*A. kennedyana* Beetle; *Bromidium t.* (Kuntze) Rúgolo] Apr–May

A. variabilis Rydb. (p. 1417) MOUNTAIN BENT GRASS Per 4–30 cm, occ from rhizomes < 5 cm. **LF:** mostly basal; ligule 1–2.5 mm; lower blades 3–7 cm, < 1 mm wide, flat, becoming folded. **INFL:** 2.5–6 cm, ± cylindric, gen ± dense; 1° branches ascending to erect, lower 0.5–1.5 cm. **SPIKELET:** glumes 2–2.5 mm; callus hairs minute; lemma 1.5–2 mm, awn gen 0; palea 0; anthers 0.5–0.7 mm. Meadows, subalpine forest, talus, alpine; 1600–4000 m. KR, NCoRH, CaRH, SNH, Wrn; to BC, w Can, CO, NM. Jul–Aug

AIRA

James P. Smith, Jr. & Dieter H. Wilken

Ann. **ST:** tufted, glabrous to puberulent. **LF:** ± basal; collar glabrous to puberulent; ligule < 1 mm, membranous; blade 0.3–14 cm, 0.3–2.5 mm wide, flat to rolled, upper gen much reduced. **INFL:** panicle-like, open to dense. **SPIKELET:** bisexual, laterally compressed, 1.5–4 mm; glumes > lower floret, translucent, keel scabrous; callus short-bristly; axis breaking above glumes and between florets; florets 2, fertile; lemma faintly 5-veined, gen glabrous, tip with 2 slender teeth, slightly scabrous, awned at or below middle, awn bent once or straight, exserted (occ reduced or 0 in lower floret) or straight; palea slightly < lemma; anthers 3. 8 spp.: s Eur, Medit, Afr, w Asia. (Greek: a weedy grass, perhaps a *Lolium*) [Wipff 2007 FNANM 24:615–617] *A. cupaniana* Guss. [*A. caryophyllea* var. *cupaniana* (Guss.) Fiori] was collected in Contra Costa Co. in 1995, but does not appear to have persisted.

1. Infl compact, spike-like, 0.3–0.7 cm wide; glumes 3–3.5 mm . . . . ***A. praecox***
1′ Infl open, 1.5–10 cm wide; glumes 1.5–3 mm
  2. Infl branches equal; spikelet stalk gen 1–2 × spikelet; spikelet 2.4–3.5 mm . . . . ***A. caryophyllea***
  2′ Infl branches unequal; spikelet stalk 2–8 × spikelet; spikelet 1.5–2.5 mm. . . . . ***A. elegans***

A. caryophyllea L. (p. 1417) SILVER HAIR GRASS **ST:** 1–10, 6–45 cm, gen glabrous (occ puberulent just below nodes). **LF:** sheath slightly scabrous; ligule minutely scabrous. **INFL:** > 1.5 cm wide, open, branches equal; spikelet stalk gen 1–2 × spikelet. **SPIKELET:** 2.4–3.5 mm; glumes 2–3 mm; lemma 1.5–2 mm, awn ± 3 mm. 2*n*=24. Sandy soils, open or disturbed sites; < 1900 m. NW, w CaR, SN, GV, CW, e SW; to BC, e US, Baja CA; native to Eur. Apr–Jun

A. elegans Roem. & Schult. (p. 1417) ELEGANT HAIR GRASS **ST:** 9–35 cm, gen glabrous. **LF:** sheath and ligule glabrous to slightly scabrous. **INFL:** > 1.5 cm wide, open, branches not equal; spikelet stalk 2–8 × spikelet. **SPIKELET:** 1.5–2.5 mm; glumes 1.5–2.5 mm; lemma ± 2 mm, awn ± 3 mm. 2*n*=14. Sandy to clay soils, open sites;

gen < 400 m. NCoRO, n SNF, ScV, SnFrB, w WTR, PR; to WA, e&s US; native to s Eur. [*A. elegantissima* Schur] Also treated as *A. caryophyllea* var. *capillaris* Bluff, Nees & Schauer, *A. elegantissima* Schur, *A. pulchella* Willd. Until the taxonomy of these European spp. is more clearly understood, it seems best to use *A. elegans* as in Wipff 2007 FNANM 24:615–617. Apr–May

A. praecox L. (p. 1417) EARLY HAIR GRASS **ST:** (1)5–25(35) cm, gen glabrous. **LF:** sheath gen slightly scabrous; ligule 1.4–5.3 mm, slightly scabrous, acute; blade 0.2–5 cm, 0.3–2 mm, prow-tipped **INFL:** gen 1–4 cm, 0.3–0.7 cm wide, narrow. **SPIKELET:** gen 3–4 mm; glumes 3–3.5 mm; lemma 1.5–3 mm, awn 3–4.5 mm. 2*n*=10,14. Sandy soils, open sites; < 100 m. NCo, s NCoRO; to BC, e US; native to s Eur. May–Jul

ALOPECURUS FOXTAIL

William J. Crins

Ann, per, cespitose or from stolons. **ST:** decumbent to erect, 1–8 dm; nodes visible, brown. **LF:** ligule 1–6 mm, membranous, truncate to acute, gen scabrous; blade flat, glabrous or scabrous. **INFL:** panicle-like, gen cylindric, dense; branches short.

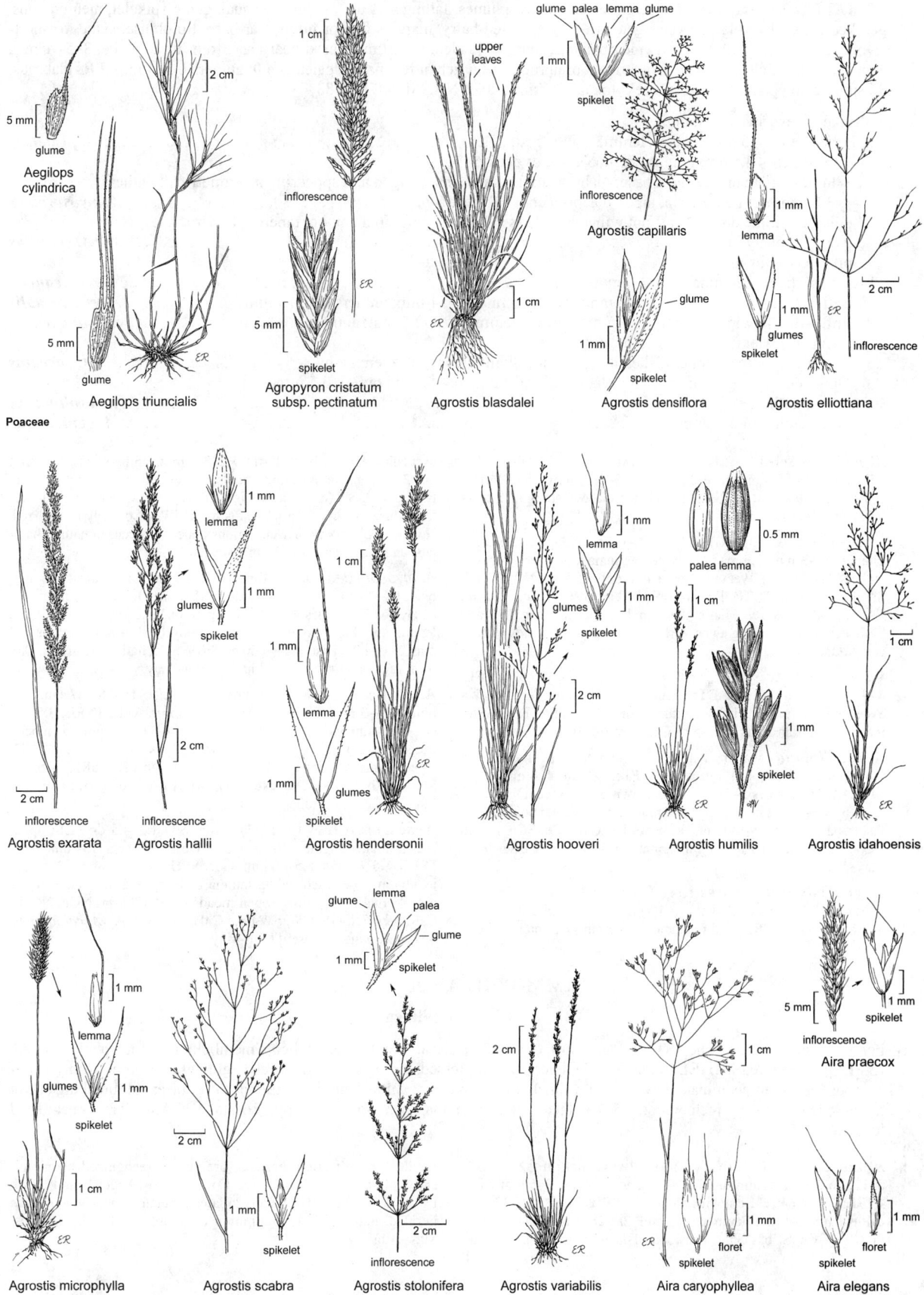

Poaceae

Aegilops cylindrica

Aegilops triuncialis

Agropyron cristatum subsp. pectinatum

Agrostis blasdalei

Agrostis capillaris

Agrostis densiflora

Agrostis elliottiana

Agrostis exarata

Agrostis hallii

Agrostis hendersonii

Agrostis hooveri

Agrostis humilis

Agrostis idahoensis

Agrostis microphylla

Agrostis scabra

Agrostis stolonifera

Agrostis variabilis

Aira caryophyllea

Aira elegans

Aira praecox

SPIKELET: ± laterally compressed, breaking below glumes, falling as 1 unit; glumes ± equal, gen = spikelet, membranous, gen keeled, keel and lateral veins gen stiff- or appressed-hairy, margins free or fused near base, tip obtuse, acute, or short-awned, 3-veined; floret 1; lemma membranous, margins keeled, sometimes fused near base, truncate to acute, 3–5-veined, awned on back below middle, awn straight or abruptly bent gen at lemma tip; palea gen 0; anthers 0.5–4 mm. **FR**: glabrous. 36 spp.: temp N.Am, Eurasia. (Greek: fox tail) [Crins 2007 FNANM 24:780–788]

1. Spikelets exc awns > 4 mm
  2. Per; st gen > 4.5 dm; glumes, lemma acute . . . . ***A. pratensis***
  2′ Ann; st gen < 4.5 dm; glumes, lemma acute or obtuse
    3. Glumes ciliate on keel near base, glabrous on lateral veins, winged and appearing asymmetrical; anthers > 3.5 mm; infl gen > 7 cm, tapered at both ends; lf sheaths not inflated . . . .***A. myosuroides***
    3′ Glumes ciliate on all veins, not winged; anthers < 1.5 mm; infl < 6.5 cm, not tapered; upper lf sheaths conspicuously inflated . . . . ²***A. saccatus***
1′ Spikelets exc awns < 4 mm
  4. Awn straight, > lemma by 0–2 mm; per . . . . ***A. aequalis***
    5. Infl 3–6 mm wide; glumes 1.8–3 mm; awn > lemma by < 1 mm; anthers 0.5–0.9 mm . . . . var. ***aequalis***
    5′ Infl 4–9 mm wide; glumes 2.7–3.7 mm; awn > lemma by 1–2.5 mm; anthers 0.7–1.2 mm. . . . . var. ***sonomensis***
  4′ Awn bent, > lemma by 1.5–5 mm; per or ann
    6. Spikelets > 3 mm; upper lf sheath conspicuously inflated; awn > lemma by 3–5 mm . . . . ²***A. saccatus***
    6′ Spikelets gen < 3 mm; upper lf sheath not inflated; awn > lemma by 1.5–3.5 mm
      7. Ann; anthers < 1 mm . . . . ***A. carolinianus***
      7′ Per; anthers 1.2–2 mm . . . . ***A. geniculatus***

A. aequalis Sobol. SHORT-AWN FOXTAIL Per. **ST**: 0.9–4.7 dm. **LF**: ligule 2–5.5 mm; blade 2.5–10 cm, 1–5(8) mm wide. **INFL**: 1–7.5 cm, 3–9 mm wide. **SPIKELET**: lemma awn straight; anthers 0.5–1 mm. 2*n*=14.

var. ***aequalis*** (p. 1423) **INFL**: 3–6 mm wide. **SPIKELET**: glumes 1.8–3 mm; lemma awn exceeding lemma body by < 1 mm; anthers 0.5–0.9 mm. Wet meadows, shores; 50–3500 m. NW, CaRH, SN (exc Teh), SnJV, TR, PR, GB (exc W&I); to subarctic, e N.Am, Eurasia. [*A. aristulatus* Michx.] Pls from high elevations in SNH and SNE also may have long awns, perhaps representing hybrids with *A. geniculatus*. May–Jul

var. ***sonomensis*** P. Rubtzov SONOMA ALOPECURUS **INFL**: 4–9 mm wide. **SPIKELET**: glumes 2.7–3.7 mm; lemma awn exceeding lemma body by 1–2.5 mm; anthers 0.7–1.2 m. Freshwater marshes, riparian scrub; 5–365 m. CCo, nw SnFrB. May–Jul ★

A. carolinianus Walter Ann. **ST**: 0.6–4.5 dm. **LF**: ligule 3–4.5 mm; blade 1–8 cm, 1–3 mm wide. **INFL**: 1–7 cm, 4–6 mm wide. **SPIKELET**: glumes 2–3 mm; lemma awn bent, exceeding lemma body by 1.5–3.5 mm; anthers 0.5–1 mm. 2*n*=14. Vernal pools, open, disturbed ground; 50–1400 m. NCoR, CaRF, GV, CCo, SCo, PR; to BC, e N.Am. Occurs in natural habitats in SCo, weedy elsewhere. May–Jun

A. geniculatus L. WATER FOXTAIL Per. **ST**: 1.4–5.5 dm. **LF**: ligule 2.5–5 mm; blade 2–8 cm, 1–4(7) mm wide. **INFL**: 1.5–6 cm, 4–7.5 mm wide. **SPIKELET**: glumes 2.5–3 mm; lemma awn bent, exceeding lemma body by (1.5)2–3.5 mm; anthers (1)1.5–2 mm. 2*n*=28. Open, wet meadows, pools, shores, streambanks; < 3200 m. NW, CaRH, SN (exc Teh), CCo, SCo, GB; to BC, e N.Am, Eurasia. [*A. pallescens* Piper] Variable. Pls from NW, CCo likely naturalized. However, pls from other bioregions appear to occur in natural habitats, and may be a native form. May–Jun

A. myosuroides Huds. Ann. **ST**: 4–4.5 dm. **LF**: ligule 4.5–6 mm; blade 8–11 cm, 3.5–6 mm wide. **INFL**: 7–10 cm, 4–7 mm wide. **SPIKELET**: glumes 4.5–5.5 mm, glabrous exc keel ciliate at base; lemma awn bent, exceeding lemma body by 3–4 mm; anthers ± 4 mm. 2*n*=14,28. Pastures; < 100 m. SnJV, SCo; native to Eurasia. Cult for forage but not successful in CA. Mar–May

A. pratensis L. MEADOW FOXTAIL Per. **ST**: 4.8–7.7 dm. **LF**: ligule 1.5–3 mm; blade 2.5–16 cm, 2–6 mm wide. **INFL**: 3.5–7.5 cm, 6–10 mm wide. **SPIKELET**: glumes 4–5 mm; lemma awn bent, exceeding lemma body by 2–5.5 mm; anthers 2–3.5 mm. 2*n*=28,42. Open, damp meadows; 20–2400 m. NCo, NCoR, CaRH, n SN, c SNH, CCo, SCoR, SW (exc ChI), MP; to AK, e N.Am; native to Eurasia. Cult for forage. May–Jun

A. saccatus Vasey (p. 1423) Ann. **ST**: 1.2–4.5 dm. **LF**: upper sheath inflated; ligule 1.5–5.5 mm; blade 1–8(13) cm, 1–4 mm wide. **INFL**: 1.5–6.5 cm, 5.5–10 mm wide. **SPIKELET**: glumes 3–5 mm; lemma awn bent, exceeding lemma body by 3–5 mm; anthers ± 1 mm. Vernal pools, moist, open meadows; < 1700 m. NCo, NCoR, CaR, n&c SN, GV, CW, SW (exc ChI), MP; to WA. [*A. californicus* Vasey; *A. howellii* Vasey] Mar–May

AMMOPHILA BEACHGRASS

Dieter H. Wilken

Per with long, thick rhizomes. **ST**: clumped, stiff, erect. **LF**: basal and cauline, erect; ligule membranous, acute; blade inrolled. **INFL**: panicle-like. **SPIKELET**: subsessile, laterally compressed, strongly keeled; floret 1; glumes ± > floret, firmly membranous, obtuse to acuminate, lower gen 1-veined, upper 3-veined; callus hairs long, tufted; floret bisexual, breaking above glumes; lemma firmly membranous, 5–7-veined; palea = lemma, membranous. 4 spp.: coastal e N.Am, Eur. (Greek: sand loving)

A. arenaria (L.) Link (p. 1423) EUROPEAN BEACHGRASS **ST**: 5–12 dm. **LF**: 4–11 dm; ligule 1–3 cm, acute; blade 2–5 mm wide. **INFL**: 15–30 cm, ± 2 cm wide, cylindric. **SPIKELET**: 10–13 mm; callus hairs 2–4 mm; lemma 8–10 mm. 2*n*=28. Sand dunes; < 240 m. NCo, CCo, SCo, s ChI (San Nicolas Island); native to n Eur. Cult for dune stabilization, baskets, brooms. If recognized taxonomically, pls from c SCo (Orange Co.), n CCo (Angel Island), with ligules 1–5 mm, firm, minutely ciliate, assignable to *A. breviligulata* Fernald, native to e US, common on shores of the Great Lakes. May–Aug ❖

ANDROPOGON BLUESTEM

Christopher S. Campbell & Kelly W. Allred

Per, cespitose. **ST**: erect, branched; nodes gen hairy. **LF**: cauline; ligule membranous, minutely ciliate; blade flat or folded. **INFL**: panicle-like with 2 or more spike-like branches, 1 or compactly clustered, partly enclosed in lf sheaths; axes breaking apart with age; spikelet sessile, subtended by hairy, naked stalk and axis segment, falling with stalk and axis segment as 1 unit. **SPIKELET**: glumes ± = florets, lanceolate; florets 2, lower vestigial, upper fertile; lemma translucent, awned; palea << lemma or 0; stamens 1–3. **FR**: oblong, ± brown or ± purple. ± 100 spp.: warm temp, trop. (Greek: man beard, from hairy staminate spikelets) [Campbell 2003 FNANM 25:649–664] Some spp. cult for forage, revegetation.

1. Lf blade 13–109 cm; ligule > 1 mm; lf sheath minutely scabrous; lower glume keel scabrous in lower 1/2
 . . . . ***A. glomeratus*** var. ***scabriglumis***
1′ Lf blade 11–52 cm; ligule < 1 mm; lf sheath glabrous; lower glume keel glabrous in lower 1/2
 . . . . ***A. virginicus*** var. ***virginicus***

A. glomeratus (Walter) Britton et al. var. ***scabriglumis*** C.S. Campb. (p. 1423) SOUTHWESTERN BUSHY BLUESTEM **ST**: 0.8– 1.5 m. **LF**: lower blades 3–6 dm, 3.5–6 mm wide. **INFL**: many, compactly clustered, plume-like. **SPIKELET**: 4–4.5 mm; callus hairs 1–2 mm; awn 0.5–2 mm. Moist, open, disturbed areas, seeps; < 600 m. ChI, TR, DMoj; naturalized NCo, NCoRO, n SNF, SNH, ScV, SCo; to NM, Baja CA. Sep–Mar

A. virginicus L. var. ***virginicus*** (p. 1423) BROOMSEDGE BLUESTEM **ST**: 0.5–2 m. **LF**: lower blades 1–5 dm, 1.5–5 mm wide. **INFL**: many, compactly clustered. **SPIKELET**: 3.5–4 mm; callus hairs 1–2.5 mm; awn 0.5–2 mm. 2*n*=20. Moist, open, disturbed areas, seeps; < 300 m. NCoRO, CaRF, n&c SNF, n SNH, ScV, SnFrB; native to c&e US, Mex, S.Am, widely naturalized; other vars. in e&se US. Sep–Jan

ANTHOXANTHUM VERNAL GRASS, VANILLA GRASS

Ann, per, cespitose, sometimes rhizomatous. **ST**: ascending to erect, 1–10 dm. **LF**: cauline or mostly basal, fragrant; auricles present or not; ligule membranous; blade flat or rolled, glabrous or hairy. **INFL**: panicle- or spike-like. **SPIKELET**: subsessile, laterally compressed; glumes > florets, = or not, tip acute, 1- or 3-veined; florets 3, lower 2 sterile or staminate, upper bisexual, breaking apart above glumes, florets falling as 1 unit; lemma of lower florets > upper floret, tip 2-forked or -lobed, hairy, 3-veined, awned at or below middle or awn 0; fertile lemma 3–7-veined, glabrous or hairy, awn 0; palea 0 in lower florets, present and < lemma in fertile floret, 1-veined. about 50 spp.: temp Eurasia, Am, Afr, Oceania, subantarctic. (Greek: fl + yellow, referring to golden color of mature infl) [Allred & Barkworth 2003 FNANM 25:758–764] As treated here, the genus incl *Hierochloe*, which is readily distinguishable in N.Am, but not in Asia and s hemisphere. Fresh lvs of some spp. used for fragrance in churches on saints' days and as incense by Native Americans. *A. hirtum* (Schrank) Y. Schouten & Veldcamp reported from n CA; records lacking.

1. Glumes unequal, lower glumes 1-veined, ± 1/2 upper glumes; 2 lower floret awns 3–10 mm
 2. Ann; lower sterile lemma awn bent . . . . ***A. aristatum*** subsp. ***aristatum***
 2′ Per; lower sterile lemma awn straight . . . . ***A. odoratum***
1′ Glumes ± equal, gen 3-veined; 2 lower floret awn 0
 3. Upper lf blades ± spreading; lower lemma hairy, often papillate . . . . ***A. nitens*** subsp. ***nitens***
 3′ Upper lf blades gen stiffly erect; lemma mostly glabrous. . . . . ***A. occidentale***

A. aristatum Boiss. subsp. ***aristatum*** (p. 1423) ANNUAL VERNAL GRASS Ann. **ST**: 1–6 dm, often bent at base. **LF**: upper sheaths 1–6 cm; ligule < 2 mm; auricles gen present; blade 1–6 cm, 1–5 mm wide, ciliate, soft-hairy. **INFL**: 1–4 cm, 3–9 mm wide. **SPIKELET**: 4–8 mm; lower glumes 3–5 mm, upper glumes 5–7 mm; lemma hairs ± stiff, hairs at lemma base brown; sterile floret awns 3–10 mm. 2*n*=10,20. Moist to dry disturbed sites; < 200 m. NW, CCo, SnFrB; to BC, c&e US; native to Eur. May–Jun

A. nitens (Weber) Y. Schouten & Veldkamp subsp. ***nitens*** (p. 1423) VANILLA GRASS Per, cespitose or with elongate rhizomes. **ST**: 2–5 dm. **LF**: sheath glabrous to puberulent; ligule 3–5 mm; blade 10–30 cm, 2–8 mm wide, flat or rolled. **INFL**: gen open, 5–8 cm; lower branches drooping to spreading. **SPIKELET**: 3–7.5 mm, lower florets staminate; glumes ± equal; lower lemmas short-hairy throughout, hairs tan to golden-brown, often papillate, upper lemma hairy, 2.5–4 mm. 2*n*=28. Wet sites, meadows; 1500–1800 m. n CaRH (Shasta, Siskiyou cos.); to AK, Can, e US, Eurasia. [*Hierochloe odorata* (L.) P. Beauv.] Apr–Jul ★

A. occidentale (Buckley) Veldkamp (p. 1423) CALIFORNIA SWEET GRASS Per, cespitose or with elongate rhizomes. **ST**: 6–10 dm. **LF**: sheath minutely scabrous; ligule 2–4 mm; blade 7–30 cm, 5–15 mm wide, upper lvs stiffly erect, narrowed at base. **INFL**: ± open, 7–13 cm; lower branches slender, often drooping. **SPIKELET**: 5–6 mm; glumes ± equal, 3.5–5 mm; lemma rounded at slightly lobed tip, lemmas of lower florets glabrous to short hairy at base, hairs clear, upper lemma 3–5 mm, margins long-hairy. 2*n*=42. Moist to dry, conifer forest; < 750 m. NCo, NCoRO, CCo, SnFrB, SCoRO; to WA. [*Hierochloe o.* Buckley] Jan–Jul

A. odoratum L. (p. 1423) SWEET VERNAL GRASS Per. **ST**: 3–6 dm, erect. **LF**: upper sheaths 4.5–9 cm; ligule 1–3 mm; auricles gen present; upper blades 3–6 cm, 3–10 mm wide, slightly ciliate and soft-hairy at base. **INFL**: spike-like, congested, 2–14 cm, 5–15 mm wide. **SPIKELET**: 7–10 mm; lower glumes 3–4 mm, upper glumes 8–10 mm; lemma hairs ± soft, hairs at lemma base colorless; sterile floret awns 2–9 mm. 2*n*=10,20. Meadows, pastures, openings in conifer forest, disturbed sites; gen < 1600 m. NCo, KR, NCoRO, NCoRI, n&c SN, CCo, SnFrB; widespread US, temp; native to Eur. Vanilla-like odor is from presence of coumarin, which has caused hemorrhaging in cattle when consumed in sufficient quantity. It is also the basis of a drug used to prevent blood clots. May–Jun ❖

APERA

Dieter H. Wilken

Ann. **ST**: erect, 1 to tufted, gen glabrous. **LF**: basal and cauline; ligule membranous, acute to toothed at tip; blade flat to rolled, gen ridged. **INFL**: panicle-like; branches spike-like, minutely scabrous. **SPIKELET**: glumes subequal, > floret, membranous, lower 1-veined, upper 3-veined; floret 1, vestigial florets occ present, bisexual, breaking above glumes; lemma rounded on back, membranous, short bristly at base, minutely scabrous and awned near tip, obscurely 3–5-veined, awn straight, flexible; palea ± = lemma. 3 spp.: temp Eur, Asia. Cult for revegetation. (Greek: not maimed, alluding to occ presence of vestigial florets) [Allred 2007 FNANM 25:788–789]

1. Infl branches ascending, spikelet-bearing to near base; anthers 0.3–0.5 mm . . . . ***A. interrupta***
1′ Infl branches spreading, naked near base; anthers 1–2 mm . . . . ***A. spica-venti***

A. interrupta (L.) P. Beauv.   **ST**: gen < 3, 1–6 dm, sometimes tufted. **LF**: ligule 1–4 mm; blade 1–3 mm wide, minutely scabrous above. **INFL**: 5–15 cm; glumes 1.5–3 mm; lemma 1.5–2 mm, minutely scabrous at tip, awn 5–8 mm. 2*n*=14,28. Disturbed sites; 1220–1700 m. KR, CaR, MP (w of Alturas, Modoc Co.), W&I (Furnace Creek, e Mono Co.); reported scattered elsewhere in w US; native to Eur, Asia. Jun–Jul

A. spica-venti (L.) P. Beauv.   **ST**: gen < 4, 2–12 dm, sometimes tufted. **LF**: ligule 3–9 mm; blade 3–5 mm wide, ± glabrous to minutely scabrous above. **INFL**: 8–25 cm; glumes 1.5–3 mm; lemma 1.5–2.5 mm, minutely scabrous at tip, awn 4–9 mm. 2*n*=14. Disturbed and open sites; 90–430 m. n SnJV (near La Grange, e Stanislaus Co.), s SCoRO; reported elsewhere in N.Am; native to temp Eur. May–Jun

ARISTIDA   THREE-AWN

Rosa Cerros-Tlatilpa & Kelly W. Allred

Ann, per, cespitose. **ST**: ascending to erect, gen glabrous (exc *Aristida californica*). **LF**: basal and cauline; basal often tufted; ligule hairy; blade flat or inrolled. **INFL**: raceme-like or panicle-like; branches spike-like. **SPIKELET**: glumes narrowly lanceolate, thin, 1-veined, awn gen 0; floret 1, breaking above glumes; lemma ± fusiform, hard when mature, 3-veined, tip beak-like or not, awned at tip, awns 3, equal or unequal; palea < lemma, enclosed by lemma, transparent. **FR**: narrowly fusiform. ± 300 spp.: worldwide, arid warm temp. (Latin: awn) [Allred 2007 FNANM 25:315–342] Some spp. noxious weeds.

1. Lower sts densely hairy . . . . ***A. californica***
1′ Lower sts glabrous
 2. Ann
  3. Glumes 10–22 mm; lemma awns gen 10–70 mm . . . . ***A. oligantha***
  3′ Glumes 3–13 mm; lemma awns 3–20 mm
  4. Central awn straight to curved at base . . . . ***A. adscensionis***
  4′ Central awn spirally coiled at base . . . . ***A. dichotoma***
 2′ Per
  5. 1° infl branches spreading at base (upper appressed to ascending in *Aristida purpurea* var. *wrightii*)
   6. Lower infl branches spreading, upper ascending to appressed. . . . . ²***A. purpurea*** var. ***parishii***
   6′ All infl branches spreading
    7. Lf blade base glabrous; lemma tip much twisted; anthers 0.8–1 mm . . . . ***A. divaricata***
    7′ Lf blade base sparsely long-hairy; lemma tip slightly twisted or straight; anthers 1–2 mm
     . . . . ***A. ternipes*** var. ***gentilis***
  5′ 1° infl branches appressed to ascending at base (curving or drooping in *Aristida purpurea* var. *purpurea*) . . . . ***A. purpurea***
   8. Awns gen > 40 mm
    9. Upper glume gen 16–25 mm; lemma 0.3–0.8 mm wide just below awns, awns 0.2–0.5 mm wide at base . . . . var. ***longiseta***
    9′ Upper glume gen 7–16 mm; lemma 0.1–0.3 mm wide just below awns, awns 0.1–0.3 mm wide at base . . . . ²var. ***purpurea***
   8′ Awns gen < 40 mm
    10. 1° infl branches and spikelet stalks thread-like, curving to drooping, appearing S- or U-shaped. . . . ²var. ***purpurea***
    10′ 1° infl branches and spikelet stalks gen stiff, straight, gen appressed to ascending (lower spreading in *Aristida purpurea* var. *parishii*)
     11. Lemma tip, awn base ± 0.1 mm wide at base, awns delicate. . . . . var. ***nealleyi***
     11′ Lemma tip, awn base 0.2–0.3 mm wide, awns ± stiff
      12. Infl gen 3–15 cm; lf blade gen 4–10 cm . . . . var. ***fendleriana***
      12′ Infl gen 12–30 cm; lf blade gen 10–25 cm
       13. Infl dense, ± red when young; lower branches with 10–20 spikelets. . . . . ²var. ***parishii***
       13′ Infl ± open, tan to brown when young; lower branches with 2–10 spikelets. . . . . var. ***wrightii***

A. adscensionis L. (p. 1423) SIXWEEKS THREE-AWN Ann. **ST**: branched below, 0.5–8 dm. **LF**: blade < 15 cm, inrolled. **INFL**: 2–22 cm, narrow. **SPIKELET**: lower glume 4–8 mm, upper 6–12 mm; lemma 6–15 mm, awns ± equal, 7–20 mm, central straight to curved at base. **FR**: ± 15 mm. 2*n*=22,33,44. Disturbed areas, dry, open places, rocky sites, shrubland; < 1400 m. CCo, SCo, s ChI,

WTR, PR, D; to MO, TX, S.Am. Some pls have very short lateral awns. Jan–Nov

A. californica Thurb.   CALIFORNIA THREE-AWN, MOJAVE THREE-AWN Per, ± bushy. **ST**: much-branched, gen 1–4 dm, densely hairy. **LF**: sheath << internodes; blade < 6 cm, gen inrolled. **INFL**: 5–10

cm. **SPIKELET**: lower glume 4–10 mm, upper 7–15 mm; lemma 5–7 mm, narrow beak at tip 4–26 mm, awns 20–50 mm, beak and awns breaking from lemma. **FR**: ± 10 mm. 2*n*=22. Dry sandy sites, dunes, shrubland; < 700 m. D; AZ, TX, nw Mex. Feb–Nov

A. dichotoma Michx. CHURCHMOUSE THREE-AWN Ann. **ST**: gen erect, 1.5–6 dm. **LF**: sheath << internodes; blade 3–10 cm, gen flat. **INFL**: 2–11 cm. **SPIKELET**: lower glume 3–8 mm, upper 4–13 mm; lemma 3–11 mm, central awn 3–8 mm, coiled at base, lateral awns 1–4 mm, straight; anthers 2–3 mm. Uncommon. Granitic outcrops, sandy fields; n SNF; e-c, s US. Jun–Aug

A. divaricata Willd. POVERTY THREE-AWN Per, cespitose. **ST**: erect, 3–12 dm. **LF**: blade 5–20 cm, loosely inrolled, base glabrous. **INFL**: 10–30 cm, 6–25 cm wide, open; stalk flattened, easily broken; 1° infl branches stiffly spreading. **SPIKELET**: glumes ± equal, 8–12 mm; lemma 8–13 mm, tip twisted 4+ times, awns gen 10–20 mm, lateral ≤ central; anthers 0.8–1 mm. **FR**: 8–10 mm. 2*n*=22. Uncommon. Dry slopes, shrubland, grassland; < 1500 m. s SnJV, SCo, s PR; to Great Plains, C.Am. Pl(s) with lateral awns 0 or << central are *A. orcuttiana* Vasey [*A. schiedeana* Trin. & Rupr. var. *orcuttiana* (Vasey) Allred & Valdés-Reyna], once collected near San Diego, believed extirpated. Mar–Oct

A. oligantha Michx. (p. 1423) OLDFIELD THREE-AWN Ann. **ST**: highly branched, 3–6 dm. **LF**: blade 3–15 cm, 0.5–2 mm wide, gen inrolled. **INFL**: 5–20 cm, open, raceme-like. **SPIKELET**: divergent, spreading; glumes ± equal, 10–22 mm, short-awned, lower 3–7-veined; lemma 10–25 mm, awns gen 10–70 mm, straight, spreading. **FR**: 10–15 mm. 2*n*=22. Disturbed places, dry slopes, fields, grassland, shrubland, woodland; < 1000 m. NW, CaRF, SNF, GV, SCoRO, MP; to OR, e US. Jul–Nov

A. purpurea Nutt. Per. **ST**: gen erect, unbranched, 1–10 dm. **LF**: blade 5–25 cm, 1–2 mm wide, gen inrolled. **INFL**: 1° infl branches spreading to appressed, tips sometimes spreading or ascending. **SPIKELET**: glumes thin, lower 4–12 mm, upper 7–25 mm; awns equal or central slightly longer. 2*n*=22,44,66,88.

var. ***fendleriana*** (Steud.) Vasey FENDLER THREE-AWN **LF**: gen basal; blade gen 4–10 cm. **INFL**: 3–15 cm; 1° infl branches stiff, straight, gen appressed. **SPIKELET**: lower glume 5–8 mm, upper 10–15 mm; lemma 8–14 mm, awns gen 18–40 mm, 0.2–0.3 mm wide at base, ± stiff. 2*n*=22,44. Dry, rocky slopes, shrubland; 1000–1800 m. SnBr, PR, SNE, DMoj; to MT, Great Plains, n Mex. [*A. f.* Steud.] Apr–Oct

var. ***longiseta*** (Steud.) Vasey (p. 1423) RED THREE-AWN **LF**: basal or cauline; blade 4–16 cm. **INFL**: 5–15 cm; 1° branches stiff, straight, appressed to ascending at base, delicate, drooping distally.

SPIKELET: lower glumes 8–12 mm, upper gen 16–25 mm; lemma 10–15 mm, 0.3–0.8 mm wide just below awns, awns 40–140 mm, 0.2–0.5 mm wide at base. 2*n*=22,44,66. Disturbed ground, dry slopes, plains, shrubland; < 1800 m. SnBr, D; to sw Can, n Mex. [*A. l.* Steud.; *A. p.* var. *robusta* (Merr.) A.H. Holmgren & N.H. Holmgren] May–Nov

var. ***nealleyi*** (Vasey) Allred (p. 1423) NEALLEY THREE-AWN **LF**: gen basal; blade 5–20 cm. **INFL**: 8–18 cm, light brown; 1° branches stiff, straight, gen appressed to ascending. **SPIKELET**: lower glume 4–7 mm, upper 8–14 mm; lemma 7–13 mm, awns delicate, 15–30 mm, ± 0.1 mm wide at base. 2*n*=22,44. Dry slopes, plains, shrubland; < 2000 m. SCo, SnBr, PR, D; to s UT, OK, n Mex. [*A. glauca* (Nees) Walp.] Mar–Nov

var. ***parishii*** (Hitchc.) Allred (p. 1423) PARISH THREE-AWN **LF**: cauline; blade gen 10–20 cm, gen flat. **INFL**: 15–24 cm, dense; 1° branches straight, stiff, lower spreading, upper ascending to appressed, ± red when young, fading to straw-colored, lower with 10–20 spikelets. **SPIKELET**: lower glume 7–11 mm, upper 10–15 mm; lemma 10–15 mm, awns 20–30 mm, 0.2–0.3 mm wide at base, ± stiff. Dry slopes, plains, chaparral, shrubland; 200–1300 m. SCo, SnBr, D; s NV, Baja CA. [*A. parishii* Hitchc.] Feb–Nov

var. ***purpurea*** PURPLE THREE-AWN **LF**: gen cauline; blade 3–17 cm. **INFL**: 5–25 cm, ± purple; 1° infl branches and spikelet stalks thread-like, curving to drooping, appearing S- or U-shaped. **SPIKELET**: lower glume 4–9 mm, upper 7–16 mm; lemma 6–12 mm, awns gen 20–60 mm, 0.1–0.3 mm wide at base. 2*n*=22,44,66. Dry slopes, shrubland; < 1500 m. SCo, SnBr, PR, DMtns; to AR, n Mex. Feb–Nov

var. ***wrightii*** (Nash) Allred (p. 1423) WRIGHT THREE-AWN **LF**: cauline, 10–25 cm, flat to inrolled. **INFL**: gen 12–30 cm, ± open; 1° branches straight, stiff, erect to ascending, tan to brown when young, fading straw-colored, lower with 2–10 spikelets. **SPIKELET**: lower glume 5–10 mm, upper 10–16 mm; lemma 8–15 mm, awns gen 20–35 mm, 0.2–0.3 wide at base, ± stiff. 2*n*=22,44,66. Sandy to rocky slopes, plains, shrubland; 200–1300 m. PR, D; to s UT, OK, n Mex. [*A. w.* Nash] Feb–Nov

A. ternipes Cav. var. ***gentilis*** (Henrard) Allred (p. 1423) HOOK THREE-AWN Per, sometimes bushy. **ST**: few, prostrate to erect, 25–80 cm. **LF**: blade 5–40 cm, flat to inrolled, base sparsely long-hairy. **INFL**: 15–40 cm, open; 1° branches spreading. **SPIKELET**: glumes ± equal, 9–15 mm; lemma 10–15 mm, tip slightly twisted or straight, awns equal to unequal, central 10–25 mm, lateral 6–23 mm; anthers 1–2 mm. 2*n*=44. Dry hills, slopes; 100–1350 m. NCoRI, s SNF, GV, SCo, n ChI (Santa Cruz Island), TR, PR, DMoj; to TX, C.Am. [*A. hamulosa* Henrard; *A. t.* var. *h.* (Henrard) Trent, illeg.] May–Nov

ARRHENATHERUM OAT GRASS

James P. Smith, Jr. & Dieter H. Wilken

Per, gen cespitose. **ST**: erect, basal internodes sometimes bulbous. **LF**: basal and cauline; ligule membranous, obtuse, minutely soft-hairy; blade flat, soft-hairy. **INFL**: panicle-like, narrow, branches spreading until after fl. **SPIKELET**: laterally compressed; glumes unequal, keeled, acute, lower 1- or 3-veined, upper 3-veined; florets 2, lower < upper, lower gen staminate, upper pistillate or bisexual, breaking above glumes, falling as 1 unit; callus hairy; lemma keeled, 5–7-veined, lemma of lower floret awned below middle, awn bent, twisted, upper lemma awnless or awned at tip, awn straight; palea < lemma. 6 spp.: temp Eur, Asia. (Greek: masculine awn, with reference to awned staminate floret) [Hatch 2007 FNANM 24:740–742]

A. elatius (L.) J. Presl & C. Presl (p. 1423) TALL OAT GRASS Per, cespitose, sometimes rhizomatous. **ST**: ascending to erect, 5–15(18) dm, basal internodes bulbous or not. **LF**: ligule 1–3 mm, obtuse, minutely ciliate; blade 5–30 cm, 3–8 mm wide, flat, glabrous to minutely scabrous. **INFL**: 7–30 cm, 1–6 cm wide. **SPIKELET**: subsessile to stalked; lower glume 4–7 mm, upper 7–10 mm, glume margins translucent, keel soft-hairy; lemma 6–9 mm, awn of lower lemma 10–17 mm, awn of upper < 4 mm (sometimes 0); palea translucent. 2*n*=14,28,42. Disturbed, open sites; 30–1800 m. NW, CaR, n&c SNF, c SNH, CCo, SnFrB, WTR, MP; to AK, s Can, most of US; native to temp Eur. [*A. e.* var. *biaristatum* (Peterm.) Peterm.; *A. e.* subsp. *bulbosum* (Willd.) Schöbl. & G. Martens] Grown for forage; variegated forms cult as orn; potentially problematic weed. May–Jul

ARUNDO

Kelly W. Allred

Per; rhizomes short, thick. **ST**: erect, bamboo-like. **LF**: cauline; sheaths > internodes, glabrous; ligule thinly membranous, fringed; blade flat or folded, glabrous, margin scabrous. **INFL**: panicle-like, plume-like, silvery to ± purple. **SPIKELET**: lat-

erally compressed; glumes > florets, 3–5-veined; axis glabrous; florets 1–several, breaking above glumes and between florets; lemma 3–7-veined, hairy, awn ± 0; palea < lemma; 3 spp.: warm temp, trop. (Latin: a reed grass) Used in construction, for mats and baskets, fishing rods, to make reeds for woodwind instruments, organ-pipes, windbreaks.

A. donax L. (p. 1427) GIANT REED **ST**: 2–10 m; nodes glabrous; internodes 1–3.5 cm thick, ± woody at maturity. **LF**: blade < 1 m, 2–6 cm wide. **INFL**: 2–7(9) dm; branches ascending. **SPIKELET**: 10–14 mm; glumes 10–13 mm, thin, silvery, ± brown, or ± purple; lemma 8–12 mm, tip 2-toothed, hairs < 8 mm, silky; palea 3–5 mm, hairy at base; anthers 2.5–3 mm. 2*n*=110. Moist places, seeps, ditchbanks; < 1500 m. NCo, KR, c SNF, GV, CCo, SCoR, SCo, SnGb, D; s US; native to Eur. Mar–Sep ◆

AVENA OATS

Bernard R. Baum, James P. Smith, Jr. & Dieter H. Wilken

Ann. **ST**: erect, 1–6, ± glabrous. **LF**: basal and cauline; ligule 2–5 mm, membranous, rounded at tip; blade flat. **INFL**: panicle-like, open. **SPIKELET**: 15–50 mm, laterally compressed, gen stalked, ± pendent; glumes unequal or ± equal, gen > florets, membranous, 3–11-veined, gen glabrous; axis occ prolonged behind upper floret, vestigial floret at tip; florets (1)2–6(8), 2+ bisexual, reduced florets distal to proximal ones, breaking above glumes and between florets or not; lemma hard, glabrous to hairy below awn, awned at or slightly below middle, 5–9-veined, tip 2-forked, forks ± tooth-like, awn stiff, gen bent, slightly to often strongly coiled below bend; palea ± < lemma; anthers 3. **FR**: cylindric, longitudinally grooved, pubescent. 29 spp.: temp Eur, n Afr, c Asia. (Latin: oats) [Baum 2007 FNANM 24:734–739] Cult for grain, hay. CA records of *A. strigosa* Schreb. are based on misidentifications of *A. barbata*. CA records of *A. occidentalis* Durieu, are based on a misidentification of *A. fatua*.

1. Callus glabrous; florets remaining attached at maturity; lemma glabrous or nearly so, awn of lower floret gen 0(15) mm. . . . . **A. sativa**
1′ Callus bearded; florets falling singly or as a unit at maturity; lemma densely strigose or soft hairy, esp below, awned
  2. Lemma bristle-tipped, teeth 2–6 mm . . . . **A. barbata**
  2′ Lemma tip ragged or 2-forked, teeth ≤ 1.5 mm
    3. Florets falling separately from glumes; awn of lowest floret 25–40 mm; lf margin glabrous. . . . . *A. fatua*
    3′ Florets falling from glumes as a unit; awn of lowest floret 35–90 mm; lf margin sparsely hairy. . . . . *A. sterilis*

A. barbata Link (p. 1427) SLENDER WILD OAT **ST**: (3)6–8(15) dm, becoming erect. **LF**: blade 6–30 cm, 2–6 mm wide, glabrous to minutely scabrous (sometimes ciliate). **SPIKELET**: 21–30 mm; breaking apart above glumes and between florets; glumes 15–30 mm, 5–9-veined; callus bearded; florets 2–3; lemma 12–26 mm, gen densely soft-hairy below awn, teeth 2–6 mm, bristly, awns 20–45 mm, bent, twisted below bend. 2*n*=28. Disturbed sites; 40–1200 m. CA-FP, MP, DMoj; to WA, MT, AZ; native to s Eur, n Afr to India. Mar–Jun ❖

A. fatua L. (p. 1427) WILD OAT **ST**: 3–16 dm, erect at maturity. **LF**: blade 10–45 cm, 4–15 mm wide, margins glabrous. **SPIKELET**: 18–32 mm; breaking apart above glumes and between florets; glumes 18–32 mm, 9–11-veined; callus bearded; florets 2–3, lemma 14–20 mm, gen glabrous on back to soft-hairy in lower 1/3, tip 2-forked, forks < 1 mm, awn of lowest floret 25–40 mm, bent, twisted below bend. 2*n*=42. Disturbed sites; < 2400 m. CA-FP, MP, DMoj; to AK, e Can, most of US; native to Eurasia. Hybridizes with *A. sativa*. Apr–Jun ❖

A. sativa L. (p. 1427) CULTIVATED OAT **ST**: 4–18 dm. **LF**: blade 8–45 cm, 3–16 mm wide, minutely scabrous. **SPIKELET**: (18)25– 32(50) mm; glumes 15–50 mm equal, 7–9-veined; callus glabrous; florets 1–2(7), remaining attached at maturity; lemma 12–25 mm, ± glabrous on back (sometimes stiff-hairy at base), tip acute to minutely 2-forked, forks < 1 mm, awn present or 0, awn of upper floret when present 15–30 mm, straight, slightly twisted, awn of lower floret < 15 mm or 0. 2*n*=42. Disturbed sites; < 2200 m. CA-FP; to AK, e Can, most of US; origin in Near East. [*A. fatua* L. var. *s.* (L.) Hausskn.] Hybridizes with *A. fatua*, *A. sterilis*. Cult for grain in temperate and cold climates; believed to be derived from wild *A. fatua* by early humans. Mar–May

A. sterilis L. ANIMATED OAT **ST**: 3–12 dm, becoming erect at maturity. **LF**: blade 8–60 cm, 4–8 mm wide, margins sparsely hairy. **SPIKELET**: 24–50 mm; glumes 20–50 mm, 7–11-veined; callus bearded; florets 2–5, falling as a unit at maturity; lemma 18–40 mm, strigose below middle, forks at tip 1–1.5 mm, awn of lowest floret 35–90 mm, bent, twisted and short-hairy below bend. 2*n*=42. Disturbed sites; 30–120 m. CCo (Alameda, Contra Costa cos.), SCo (San Diego Co.); Can, PA, NJ; native Medit to Afghanistan. The awn twists with changes in humidity and moisture. Hybridizes with *A. sativa*. Mar–Jun

BECKMANNIA

Ann [rhizomatous per]. **ST**: gen erect, 2–12(15) dm. **LF**: cauline; sheath glabrous, ribbed; ligule membranous, acute, entire or irregularly cut; blade flat, glabrous. **INFL**: panicle-like; branches short, spike-like, ascending to appressed; spikelets 2-ranked, overlapping. **SPIKELET**: ± round in outline; glumes ± equal, strongly laterally compressed, > adjacent lemmas, round to obovate in side view, keeled, wrinkled, 3-veined; axis breaking below glumes, falling as 1 unit; florets 1–2, bisexual or upper reduced; lemma 5-veined, acuminate, awn 0; palea narrow, ± = lemma. 2 spp.: N.Am, Eurasia. (J. Beckmann, German botanist, 1739–1811) [Hatch 2007 FNANM 24:484–486]

B. syzigachne (Steud.) Fernald (p. 1427) AMERICAN SLOUGH GRASS Ann, tufted. **ST**: 2–12 dm. **LF**: ligule 4–11 mm, pubescent; blade 6–21 cm, 4–10(20) mm wide, flat, scabrous. **INFL**: 7–30 cm, narrow, ± dense; branches 1-sided, gen 1–2 cm, ascending to appressed. **SPIKELET**: 2–3 mm; glumes prominently keeled, gla- brous to finely scabrous, tip short-pointed; floret 1; lemma 2–3 mm, lanceolate, acuminate, awn 0; palea ± equal lemma, acute. 2*n*=14. Margins of ponds, streams, marshes; < 2000 m. NW, CaR, n SN, ScV, n CCo, SnFrB, MP, n SNE; most of N.Am, exc se US. [*B. s.* subsp. *baicalensis* (Kusn.) Hultén] Used for forage. May–Jul

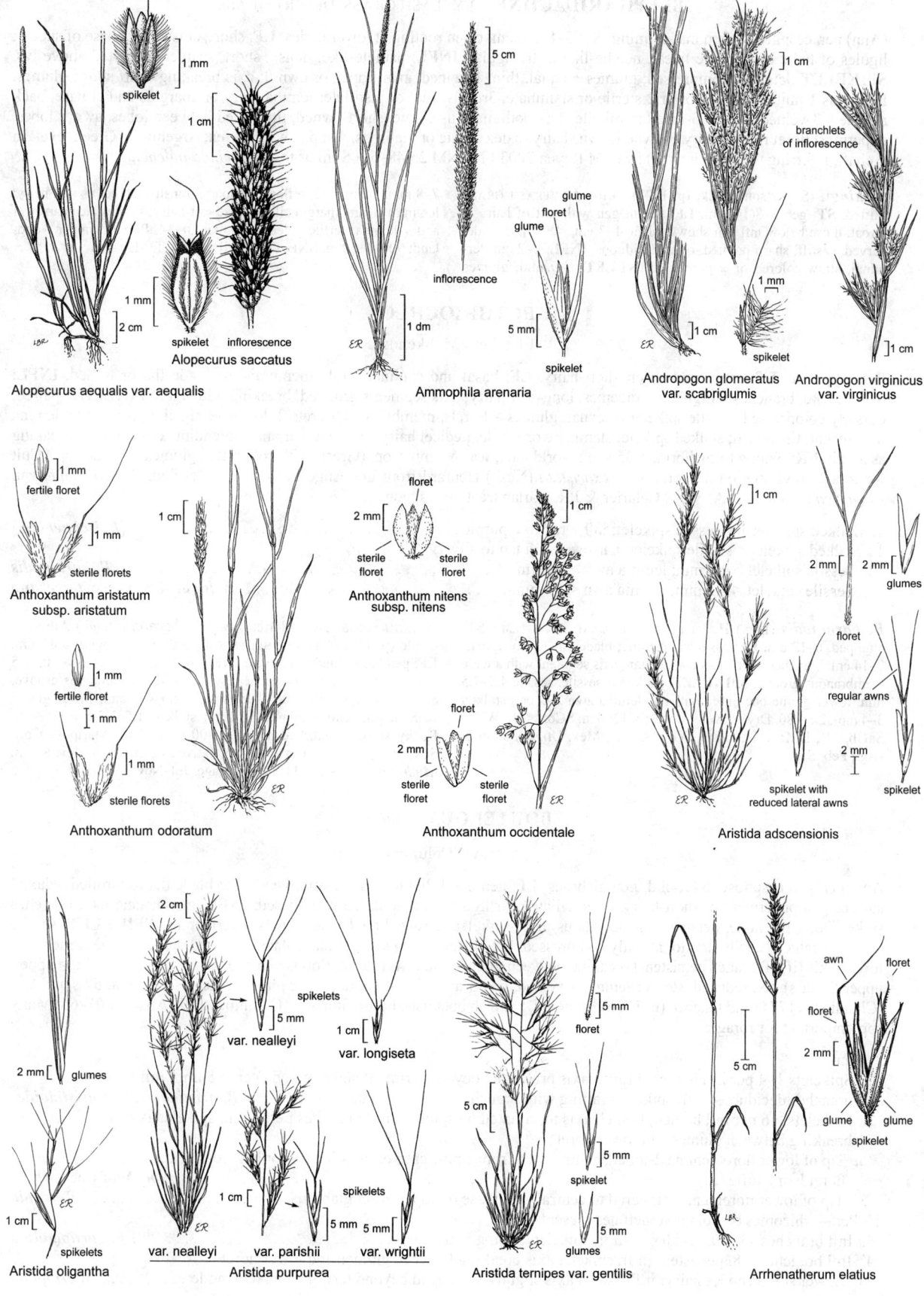

Alopecurus aequalis var. aequalis

Alopecurus saccatus

Ammophila arenaria

Andropogon glomeratus var. scabriglumis

Andropogon virginicus var. virginicus

Anthoxanthum aristatum subsp. aristatum

Anthoxanthum odoratum

Anthoxanthum nitens subsp. nitens

Anthoxanthum occidentale

Aristida adscensionis

Aristida oligantha

Aristida purpurea
var. nealleyi   var. parishii   var. wrightii
var. nealleyi
var. longiseta

Aristida ternipes var. gentilis

Arrhenatherum elatius

BLEPHARIDACHNE EYELASH GRASS, DESERT GRASS

(Ann) per, cespitose, often mat-forming. **ST:** 3–8(20) cm, often rooting at lower nodes. **LF:** short, clustered at base of shoots; ligules of hairs or 0; blade linear, needle-like to triangular. **INFL:** panicle-like, dense, short, scarcely elevated above lvs. **SPIKELET:** laterally compressed; glumes ± equal, thin, 1-veined, awn-tipped or awn 0; axis breaking apart above glumes, falling as 1 unit; florets 4, lower 2 sterile or staminate, 3rd bisexual or pistillate; lemma hairy on margins and at base, back rounded, 3-veined, deeply 3-lobed to middle, 1 or both margins ciliate, short-awned, tip awned between lobes, awn = lobes, uppermost floret rudimentary, 3-awned, awns hairy; palea ciliate or glabrous. 4 spp.: sw US, Mex, Argentina. (Greek: eyelash + chaff, referring to ciliate lemma) [Valdés-Reyna 2003 FNANM 25:48–50] Similar to *Dasyochloa pulchella*.

B. kingii (S. Watson) Hack. (p. 1427) KING'S EYELASH GRASS
Tufted. **ST:** gen 3–8(14) cm. **LF:** sheath gen with tuft of hairs at throat, those below infl gen showy; blade 1–3 cm, < ± 1 mm wide, curved, ± stiff, sharp-pointed, often deciduous. **INFL:** 1–2 cm, terminal, straw-colored or ± purple. **SPIKELET:** 6–9 mm; glumes 7–8 mm, acuminate, > florets, papery, translucent; callus soft-hairy; lemma ± 6 mm, margin ciliate, base soft-hairy; sterile palea narrower and < lemma; fertile palea = lemma. $2n=14$. Pinyon/juniper woodland; 900–2150 m. SNE, DMtns; ID, NV, UT. May–Jun ★

BOTHRIOCHLOA

Elizabeth M. Skendzic

Per, cespitose. **ST:** erect; nodes gen short-hairy. **LF:** basal and cauline; ligule membranous; blade flat or folded. **INFL:** panicle-like; branches ± digitate or racemes, long-soft-hairy; axis segments grooved, breaking with age. **SPIKELET:** paired, dorsally compressed; sessile spikelet bisexual, glumes > florets, membranous, florets 2, lower vestigial, upper fertile, lemma translucent, tip awned; stalked spikelet staminate or sterile, pedicel hairy; spikelet pair and subtending axis segment breaking as a unit. **FR:** oblong to fusiform. ± 35 spp.: worldwide, warm temp, trop. (Greek: pit, from pitted glumes of some spp.) Cult for forage, revegetation. Reports of *B. exaristata* (Nash) Henrard from Los Angeles Co. are unverified. *B. ischaemum* var. *songarica* (Fisch. & C.A. Mey.) Celarier & J.R. Harlan treated as synonym of *B. ischaemum*.

1. Stalked spikelet ± = sessile spikelet; infl ± red to ± purple . . . . [***B. ischaemum***]
1′ Stalked spikelet < sessile spikelet, narrower; infl tan to silvery
　2. Sessile spikelet > 4.5 mm; lemma awn > 18 mm. . . . . ***B. barbinodis***
　2′ Sessile spikelet < 4.5 mm; lemma awn < 18 mm. . . . . ***B. laguroides*** subsp. ***torreyana***

B. barbinodis (Lag.) Herter (p. 1427) CANE BLUESTEM **ST:** clumped, 6–12 dm. **LF:** basal and cauline; blade 20–30 cm. **INFL:** 7–14 cm; branches many, 4–9 cm, gen tan; axis segment with a wide membranous groove. **SPIKELET:** stalked > sessile, sessile 4.5–7.5 mm, lower glume occ pitted on back, lemma awn 2–3 cm; stalked 3–4 mm. $2n=180$. Dry, gravelly slopes; < 1200 m. SCo, s ChI, WTR, SnGb, PR, DMtns, DSon; to CO, OK, TX, Mex. [*Andropogon b.* Lag.] Feb–Sep

B. laguroides (DC.) Herter subsp. *torreyana* (Steud.) Allred & Gould (p. 1427) SILVER BLUESTEM **ST:** gen clumped, 4–13 dm. **LF:** gen basal; blade 5–21 cm. **INFL:** 4–12 cm; branches > 10, 1–5 cm, silvery or tan; axis segment with a wide membranous groove. **SPIKELET:** sessile > stalked, sessile 2.5–4.5 mm, lower glume without pit, lemma awn 7–16 mm; stalked 1.5–3.5 mm. $2n=60$. Rocky slopes, disturbed areas; ± 900 m. c SNF (Mariposa Co.), ScV, SCoRO (Santa Barbara Co.); native to c&s US, n Mex, S.Am. [*Andropogon t.* Steud.] Cult for forage. Jul–Nov

BOUTELOUA GRAMA

J. Travis Columbus

Ann, per, gen cespitose. **ST:** solid, gen glabrous. **LF:** gen basal; ligule gen < 1 mm, gen hairy; blade flat to inrolled, adaxial surface gen puberulent or short-hairy, often ciliate near ligule, hairs long, bulbous-based. **INFL:** gen panicle-like; branches spike-like, 1 per node, persistent or deciduous in fr; spikelets 2-rowed on 1 side of axis, overlapping. **SPIKELET:** sessile or short-stalked, ± cylindric to laterally compressed; glumes gen unequal, gen lanceolate, 1-veined, upper glume firmer than lower; axis (if infl branch persistent) breaking between glumes and lower floret; florets gen 2–3, lower floret bisexual, > upper, upper floret(s) gen reduced, sterile; lemma 3-veined, gen 3-awned, awns straight, scabrous; palea ± = lemma. 57 spp.: Am. (Claudio (b. 1774) and Esteban (b. 1776) Boutelou, Spanish botanists, horticulturists) [Columbus 1999 Aliso 18:61–65] Many spp. important for forage.

1. Ann
　2. Spikelets 1–4 per infl branch; branch axis prolonged beyond terminal spikelet node gen > 5 mm; infl branches deciduous in fr, spikelets falling with branch. . . . . ***B. aristidoides*** var. ***aristidoides***
　2′ Spikelets > 6 per infl branch; branch axis terminated by spikelet; infl branches persistent, spikelet axis breaking between glumes and lower floret
　　3. Tip of lower floret lemma 2-lobed, central awn from sinus; base of middle or, if only 2 florets, upper floret hairy-tufted. . . . . ***B. barbata*** var. ***barbata***
　　3′ Tip of lower floret lemma tapered to central awn; base of upper floret glabrous. . . . . ²***B. trifida***
1′ Per — rhizomes or stolons sometimes present
　4. Infl branches 13–60, deciduous in fr, spikelets falling with branch . . . . ***B. curtipendula***
　4′ Infl branches < 8, persistent in fr, spikelet axis breaking between glumes and lower floret
　　5. Lower st internodes hairy; infl branch axis slightly prolonged beyond terminal spikelet node . . . . ***B. eriopoda***

5′ Lower st internodes glabrous or minutely scabrous; infl branch axis terminated by fertile or vestigial spikelet

  6. Base of middle or, if only 2 florets, upper floret hairy-tufted; tip of lower floret lemma 2-lobed, central
awn from sinus . . . . ***B. gracilis***

  6′ Base of upper floret glabrous; tip of lower floret lemma tapered to central awn . . . . ²***B. trifida***

B. aristidoides (Kunth) Griseb. var. ***aristidoides*** (p. 1427) NEEDLE GRAMA Ann. **ST**: decumbent to erect, 0.5–3.5 dm. **LF**: blade < 7 cm, < 2 mm wide. **INFL**: branches 4–16, 8–25 mm, pendent to appressed, deciduous in fr; branch axis prolonged beyond terminal spikelet node gen > 5 mm, base densely short-hairy; spikelets 1–4 per branch, gen appressed, falling with branch. **SPIKELET**: upper glume 5–7 mm, glabrous or hairy, acute; florets 1–2; lower floret lemma ± = upper glume, glabrous or hairy, lobes 0, awns 0 or 2, lateral, < 1 mm; base of upper floret (if present) hairy-tufted, lobes gen 0 between awn bases, awns 2–7 mm, gen unequal. 2*n*=40. Dry, open, sandy to rocky slopes, flats, washes, disturbed sites, scrub, woodland; < 1800 m. e PR, e&s DMoj, DSon; to UT, TX, s Mex, S.Am. Other var. in AZ, NM, nw Mex. All yr

B. barbata Lag. var. ***barbata*** (p. 1427) SIXWEEKS GRAMA Ann. **ST**: prostrate to erect, 0.3–3 dm. **LF**: blade < 6 cm, < 2 mm wide. **INFL**: branches 2–8, 6–25 mm, spreading to appressed, persistent in fr; branch axis terminated by spikelet, base glabrous or puberulent; spikelets 7–40 per branch, spreading to ascending, breaking apart between glumes and lower floret. **SPIKELET**: upper glume 1.5–3 mm, glabrous or puberulent, tip notched, awned from sinus < 1 mm; florets 2–3, lower floret lemma ± = upper glume, hairy below middle, tip 2-lobed, awns 0.5–3 mm, ± equal, central awn from sinus; base of middle or, if only 2 florets, upper floret hairy-tufted, lobed between awn bases, awns 1–3.5 mm, ± equal; uppermost floret (if present) < 1 mm, awn 0. 2*n*=20,40. Gen open, sandy to rocky slopes, flats, washes, roadsides, disturbed sites, scrub, woodland, pine forest; < 1800 m. SnJV, e SCo, e PR, D; to MT, KS, s Mex, Argentina. Other vars. in AZ, NM, n Mex. All yr

B. curtipendula (Michx.) Torr. (p. 1427) SIDE-OATS GRAMA Per, sometimes rhizomed. **ST**: gen erect, 2–9 dm. **LF**: blade < 25 cm, < 4 mm wide. **INFL**: branches 13–60, 5–20 mm, gen pendent, deciduous in fr; branch axis slightly prolonged beyond terminal spikelet node, base puberulent; spikelets 1–8 per branch, ascending to appressed, falling with branch. **SPIKELET**: upper glume 4–8 mm, glabrous or scabrous, acute or awned < 0.5 mm; florets 1–2; lower floret lemma < to ± = upper glume, glabrous or sparsely hairy, lobes 0, awns 2, lateral, < 2 mm; base of upper floret (if present) glabrous, tip 2-lobed, central awn from sinus, < 6 mm, lateral awns < 4 mm. 2*n*=20,28,35,40–103. Dry, rocky slopes, crevices, sandy to rocky drainages, scrub, woodland; < 1900 m. s ScV (Yolo Co. as roadside waif), e PR, e&s DMtns, sw DSon; to s Can, e US, s Mex, S.Am. If recognized taxonomically, pls with erect sts, rhizomes 0 assignable to *B. curtipendula* var. *caespitosa* Gould & Kapadia. Apr–Nov

B. eriopoda (Torr.) Torr. (p. 1427) BLACK GRAMA Per, gen stoloned. **ST**: decumbent to erect, 1.5–6 dm; internodes, esp lower, hairy. **LF**: blade < 10 cm, < 2 mm wide. **INFL**: branches 2–7, 10–40 mm, spreading to appressed, persistent in fr; branch axis slightly prolonged beyond terminal spikelet node, base densely hairy; spikelets 6–18 per branch, ascending to appressed, breaking apart between glumes and lower floret. **SPIKELET**: upper glume 4–8 mm, glabrous or puberulent, sharply acute; florets 2; lower floret lemma ± ≥ upper glume, base hairy-tufted, glabrous or sparsely hairy above, lobes gen 0, central awn 1.5–4 mm, lateral awns 0–1.5 mm; upper floret base hairy-tufted, lobes gen 0 between awn bases, awns 4–8 mm, ± equal. 2*n*=20,21,28. Dry, open, sandy to rocky slopes, flats, drainages, scrub, woodland; 900–1900 m. e DMtns; to WY, OK, n Mex. May–Oct ★

B. gracilis (Kunth) Griffiths (p. 1427) BLUE GRAMA Per, gen short-rhizomed. **ST**: ascending to erect, 1–6 dm. **LF**: blade < 15 cm, < 2 mm wide. **INFL**: branches 1–3, 10–50 mm, spreading to appressed, persistent in fr; branch axis terminated by vestigial spikelet, base hairy; spikelets 20–80 per branch, spreading to ascending, breaking apart between glumes and lower floret. **SPIKELET**: upper glume 3.5–6 mm, gen with long, bulbous-based hairs on vein, acute or awned < 0.5 mm; florets 2–3; lower floret lemma ± = upper glume, base hairy-tufted, back hairy, tip 2-lobed, awns 1–3 mm, unequal, central awn from sinus; base of middle or, if only 2 florets, upper floret hairy-tufted, lobed between awn bases, awns 2.5–6 mm, ± equal; uppermost floret (if present) < 2 mm, awn 0. 2*n*=20,21,28,35,40,42,6 0,61,77,84. Sandy to rocky slopes, flats, drainages, scrub, woodland, pine forest; < 2700 m. SnBr, e DMtns (Mid Hills, Clark Mtn Range, Ivanpah, New York mtns), waif elsewhere; to s Can, e US, s Mex, Argentina. May–Oct

B. trifida S. Watson (p. 1427) THREE-AWNED GRAMA Per, sometimes fl 1st yr, sometimes short-rhizomed. **ST**: ascending to erect, 1–3 dm. **LF**: blade < 5 cm, < 1.5 mm wide. **INFL**: branches 1–7, 10–35 mm, ascending to appressed, persistent in fr; branch axis terminated by spikelet, base puberulent to hairy; spikelets 8–24 per branch, ascending, breaking between glumes and lower floret. **SPIKELET**: upper glume 2.5–5 mm, glabrous, tip gen notched, awned from sinus < 1 mm; florets 2(3), lower floret lemma < upper glume, glabrous or hairy, tip tapered to central awn, awns 2–8 mm, ± equal; upper floret base glabrous, lobes 0 between awn bases, awns 2–8 mm, ± equal. 2*n*=20,28. Dry, rocky, gen calcareous slopes, crevices, washes, scrub; 200–1600 m. n&e DMoj (mostly DMtns); to UT, TX, c Mex. Mar–Sep ★

BRACHYPODIUM FALSE BROME

Michael B. Piep

Ann, per from rhizomes or cespitose, 5–200 cm. **ST**: decumbent to erect; nodes often hairy. **LF**: gen cauline; blade flat to inrolled, glabrous to short-hairy; ligule membranous. **INFL**: spike- to raceme-like; spikelets gen 1 per node, ± cylindric, ascending to appressed, sessile to short-stalked. **SPIKELET**: glumes unequal, ≤ lowest floret, 3–9-veined, acute to awned; florets 3–24, bisexual; axis breaking above glumes and between florets; lemma back rounded, 5–9-veined, acute to awned from tip; palea slightly < lemma, clearly ciliate or toothed. ± 18 spp.: temp, subtrop worldwide. (Greek: short foot, from short, thick spikelet stalk in some spp.) [Piep 2007 FNANM 24:187–192] Reports of *B. phoenicoides* (L.) Roem. & Schult. from Sonoma Co. have not been verified.

1. Ann; spikelets laterally compressed; anthers 0.5–1.1 mm . . . . ***B. distachyon***

1′ Per; spikelets ± cylindrical; anthers 3–5.5 mm

  2. Lemma awn 0 or 1–7 mm, < lemma . . . . [***B. pinnatum***]

  2′ Lemma awn 7–15 mm, ≥ lemma . . . . ***B. sylvaticum***

B. distachyon (L.) P. Beauv. (p. 1427) Ann, gen loosely tufted, 15–40 cm. **ST**: decumbent to erect, nodes conspicuously pubescent. **LF**: blade 1.5–8 cm, 3–5 mm wide, flat. **INFL**: 2–7 cm; spikelets 1–7 per st, laterally compressed. **SPIKELET**: glumes 5–8 mm; florets 7–15; lemma 7–10 mm, awn 4–17 mm; palea stiff-ciliate to minutely toothed above middle. Disturbed areas, dry slopes; < 900 m. s NCoR, s CaRF, SN, GV, CW, SCo, s ChI (Santa Catalina Island), PR, DSon; OR, AZ, CO, TX, Australia; native to s Eur. Apr–Jul ❖

B. sylvaticum (Huds.) P. Beauv. Per, gen cespitose, 30–200+ cm. **ST**: erect. **LF**: sheaths smooth, blade 8–35 cm, 4–15 mm wide, flat, loose, veins not prominent. **INFL**: 2–20 cm; spikelets 3–12 per st, gen distant. **SPIKELET**: glumes 6–11 mm; florets 6–16(22); lemma 6–12 mm, awn 7–15 mm, ≥ lemma, straight or weakly flexuous; palea ciliate. Forest, woodland, upland prairies; < 600 m. SnFrB; OR, WA, BC, VA; native to Eur and n Afr. Nov–Dec ◆

BRIZA QUAKING GRASS

Dieter H. Wilken

Ann, per. **ST**: ascending to erect, 5–100 cm. **LF**: basal to cauline; ligule membranous to translucent; blade flat. **INFL**: erect to pendent, panicle-like, open. **SPIKELET**: ± pendent, ± laterally compressed, subconic to ovoid; glumes subequal, papery, rounded at tip, 3–9-veined; florets 3–19; axis breaking above glumes and between florets; lemma width > length, papery to translucent, rounded at tip, 7–9-veined; palea ± = lemma. 10–12 spp.: Eur, n Afr. (Greek: a kind of grain) [Snow 2007 FNANM 24:612–614] *B. media* L., a cult per with ligule < 0.5 mm and spikelets 4–6 mm, best treated as waif in CA.

1. Spikelets 10–19 mm. . . . .*B. maxima*
1′ Spikelets 2–5 mm . . . . *B. minor*

B. maxima L. (p. 1435) RATTLESNAKE GRASS, LARGE QUAKING GRASS Ann. **ST**: 20–80 cm. **LF**: ligule 1–4 mm; blade 1–7 mm wide. **INFL**: spreading to pendent, 2–10 cm. **SPIKELET**: 1–14 per infl, ovoid, obtuse at base; glumes 4–7 mm, 5–9-veined; florets 12–19; lemma 6–8 mm. 2*n*=10,14. Shaded sites, roadsides, pastures, weedy on coastal dunes; < 970 m. NW, w CaR, n SN, ScV, CW (exc SCoRI), SCo, WTR, PR (upper San Diego River), ne MP; to BC, e US; native to s Eur. Cult for orn. Apr–Jul ❖

B. minor L. (p. 1435) ANNUAL QUAKING GRASS, SMALL QUAKING GRASS Ann. **ST**: 8–50 cm. **LF**: ligule 3–13 mm, blade 3–10 mm wide. **INFL**: erect, 3–20 cm. **SPIKELET**: gen > 15 per infl, triangular to oval, truncate at base; glumes 1.5–4 mm, 3–5-veined; florets 4–6(13); lemmas 1–2 mm, veins indistinct. 2*n*=10,14. Shaded or moist, open sites; 20–600 m. NW, CaRF, SN, GV, CW (exc SCoRI), SCo, WTR, PR, DSon (Rancho Mirage); to BC, e US; native to s&w Eur. Apr–Jul

BROMUS BROME, CHESS

Jeffery M. Saarela & Paul M. Peterson

Ann to per. **LF**: basal and cauline; sheath closed to near top, hairy or glabrous; ligule ≤ 7 mm, membranous, entire to fringed; blade flat to inrolled. **INFL**: gen raceme- or panicle-like, open to dense; pedicels gen stiff, rigid. **SPIKELET**: strongly laterally compressed to cylindric; florets 3–30; axis breaking above glumes and between florets; glumes unequal, gen < lower floret, lower 1–3-veined, upper 3–7-veined, back rounded to strongly keeled, tip acute; lemma 5–9-veined, tip 2-toothed or entire, acute to obtuse, awned from between teeth or awns 0; palea gen < lemma. ± 160 spp.: temp worldwide. (Greek: ancient name) [Pavlick and Anderton 2007 FNANM 24:193–237; Saarela et al. 2007 Aliso 23:450–467] *B. scoparius* L., *B. erectus* Huds. not known to be naturalized in CA. *B. pacificus* Shear not in CA.

1. Spikelet strongly flattened; lemma strongly keeled (sect. *Ceratochloa*)
   2. Lemma awn 0–3.5 mm; lemma veins prominent . . . . *B. catharticus* var. *catharticus*
   2′ Lemma awn 4–15 mm; lemma veins obscure or prominent
      3. Some lower infl branches > 10 cm, spreading to nodding . . . . *B. sitchensis*
      3′ Lower infl branches < 10 cm, or if longer, erect to ascending
         4. Upper glume ± = lowermost lemma
            5. Ann; lower glume 3–5-veined, upper 9–15 mm, 5–9-veined; lemma 7–veined, margin gen hairy (occ glabrous), backs hairy or glabrous, awn 7–15 mm; anthers ≤ 0.5 mm . . . . *B. arizonicus*
            5′ Per; lower glume 5–7(9)-veined, upper 11–20 mm, 7–9-veined; lemma 9–11-veined, gen hairy at least distally, awn 6–12 mm; anthers 0.6–1 mm . . . . ²*B. catharticus* var. *elatus*
         4′ Upper glume gen < lowermost lemma — lemma scabrous or variously hairy, marginal hairs if present similar in length to those on back
            6. Infl dense, pedicel gen < spikelets; spikelets crowded, overlapping, stalks not visible; st 20–80 cm, occ bent at base; st, lvs glabrous or occ scabrous; ligule ≤ 6 mm . . . . *B. maritimus*
            6′ Infl loose to compact, at least some pedicels > spikelets; spikelets not crowded or overlapping, stalk often visible; st 30–120 cm, erect; st, lvs often hairy; ligule gen < 4 mm
               7. Most awns ≥ (6)7 mm
                  8. Lemma gen uniformly hairy, occ scabrous, veins 7, obscure . . . . *B. carinatus* var. *carinatus*
                  8′ Lemma gen hairy at least distally, veins 9–11 on distal 1/2, prominent . . . . ²*B. catharticus* var. *elatus*
               7′ Most awns < 7 mm
                  9. Sheath throat and/or lemma hairy . . . . *B. carinatus* var. *marginatus*
                  9′ Sheath throat, lemma glabrous. . . . . *B. polyanthus*
1′ Spikelet not strongly flattened; lemma rounded over midrib and not strongly keeled
   10. Lemma tip conspicuously 2-toothed, teeth translucent, awn-like to acuminate, 1–7 mm; largest lemmas gen < 2 mm wide
      11. Lemma awn bent and/or twisted (sect. *Neobromus*) . . . . *B. berteroanus*
      11′ Lemma awn straight, not twisted (sect. *Genea*)
         12. Lemma mostly > 20 mm; awn 30–65 mm . . . . *B. diandrus*

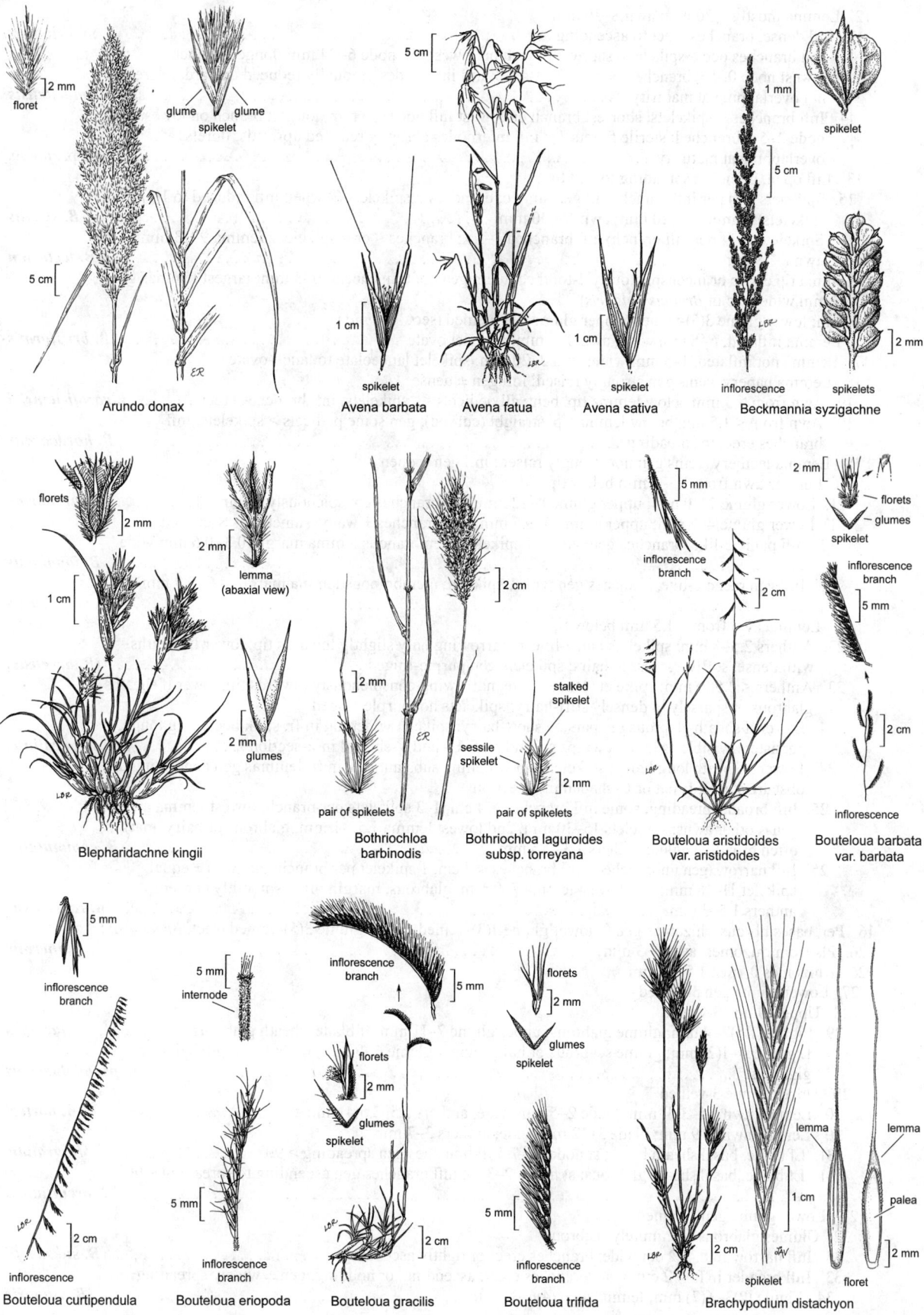

Arundo donax

Avena barbata

Avena fatua

Avena sativa

Beckmannia syzigachne

Blepharidachne kingii

Bothriochloa
barbinodis

Bothriochloa laguroides
subsp. torreyana

Bouteloua aristidoides
var. aristidoides

Bouteloua barbata
var. barbata

Bouteloua curtipendula

Bouteloua eriopoda

Bouteloua gracilis

Bouteloua trifida

Brachypodium distachyon

12′ Lemma mostly < 20 mm; awn 8–30 mm
  13. Infl dense, branches erect to ascending . . . . *B. madritensis*
    14. Infl branches occ > spikelets, shortest branch on lowest infl node 6–24 mm, longest branch on
      lowest node 0–1 × branched; sterile florets ≤ 3; infl internodes gradually reduced upwards; florets
      not overlapping at maturity . . . . subsp. *madritensis*
    14′ Infl branches < spikelets, shortest branch on lowest infl node ≤ 6 mm, longest branch on lowest
      node 2–5 × branched; sterile florets 3+; infl internodes abruptly reduced upwards; florets
      overlapping at maturity . . . . subsp. *rubens*
  13′ Infl open, branches spreading to nodding
    15. Spikelets 1(3) per infl branch; infl gen simple; branches > spikelets (< when infl reduced to 1(3)
      spikelets); lemma 13–20 mm; awn 15–30 mm. . . . . *B. sterilis*
    15′ Spikelets 1–14 per infl branch; infl branched 1–5 ×; branches < or > spikelets; lemma 9–13 mm;
      awn 8–18 mm . . . . *B. tectorum*
10′ Lemma tip entire or inconspicuously 2-toothed, teeth gen not translucent, 0–3 mm; largest lemmas gen
  > 2 mm wide (exc in *Bromus vulgaris*)
  16. Ann; lower glume 3(5)-veined; upper glume 5–9-veined (sect. *Bromus*)
    17. Lemma inflated, 6–8 mm wide; awn 0–1 mm; spikelet ovate . . . . *B. briziformis*
    17′ Lemma not inflated, 1–5 mm wide; awn 2–20 mm; spikelet lanceolate to lance-ovate
      18. Lemma papery, veins gen strongly raised; infl gen ± dense
        19. Awn from > 3 mm below lemma tip, bent; all pedicels << spikelets; infl branches erect . . . . . . *B. caroli-henrici*
        19′ Awn from < 1.5 mm below lemma tip, straight (curved); gen some pedicels > spikelets; infl
          branches erect to spreading . . . . *B. hordeaceus*
      18′ Lemma leathery, veins gen not strongly raised; infl gen ± open
        20. Lemma awn from 1.5–5 mm below tip
          21. Lower glume 7–10 mm; upper glume 8–12 mm; infl branches conspicuously S-curved . . . . *B. arenarius*
          21′ Lower glume 4–7 mm; upper glume 5–9.5 mm; infl branches ± wavy, sometimes S-curved
            22. Infl panicle-like, branches gen with > 1 spikelet; membranous lemma margin 0.3–0.6 mm wide
            . . . . *B. japonicus*
            22′ Infl gen raceme-like, branches gen with 1 spikelet; membranous lemma margin 0.6–0.9 mm wide
            . . . . *B. squarrosus*
        20′ Lemma awn from < 1.5 mm below tip
          23. Anthers 2.5–5 mm; spikelets lance-linear, narrowing only slightly towards tip; lower lf sheaths
            with dense, softly appressed hairs; spikelets gen purple-tinged . . . . [*B. arvensis*]
          23′ Anthers < 2.5(3) mm; spikelets lance-ovate, narrowing conspicuously towards tip; lower lf sheaths
            glabrous or sparsely to densely stiff-hairy; spikelets not purple-tinged.
            24. Lower lf sheath glabrous or sparsely short-hairy; spikelet widening in fr, stalk becoming tough,
              persistent, visible as lemma wraps around fr; fr U- and V-shaped in ×-section . . . . *B. secalinus*
            24′ Lower lf sheath long-hairy; spikelets not widening substantially in fr, lemmas gen continuing to
              obscure stalks; fr flat or C-shaped in ×-section
              25. Infl broad, spreading, some infl branches > 4 cm, 1–3 spikelets per branch; lowest lemma awn
                gen < other awns; spikelets 15–30 mm; 2nd lowest lemma 7.5–11 mm, glabrous or hairy, margin
                often broadly angled; anthers 1.3–2.5 mm . . . . *B. commutatus*
              25′ Infl narrow, gen unbranched, infl branches < 4 cm, 1 spikelet per branch; all awns ± equal;
                spikelet 11–18 mm; 2nd lowest lemma 7–9 mm, glabrous, margin often smoothly curved;
                anthers 1.5–3 mm . . . . *B. racemosus*
16′ Per, bases fibrous, rhizomes gen 0; lower glume 1(3)-veined; upper glume 3(5)-veined (sect. *Bromopsis*)
  26. Pls from rhizomes; awn 0–3 mm. . . . . *B. inermis*
  26′ Rhizomes 0; awn 1.5–11 mm
    27. Lower glume gen 3-veined
      28. Upper glume 5-veined
        29. Ligule (1.5)2–4 mm; glume glabrous; upper glume 7–11 mm; lf blade, sheath glabrous. . . . . *B. laevipes*
        29′ Ligule 0.4–1(2) mm; glume scabrous or hairy; upper glume 6–9 mm; lf blade, sheath hairy or
          glabrous. . . . . *B. pseudolaevipes*
      28′ Upper glume 3-veined
        30. Lemma awn 1.5–3(4) mm; blade 2–5 mm wide; anthers 1.5–3.5(4) mm . . . . *B. porteri*
        30′ Lemma awn 3–9 mm; blade 3–12 mm wide; anthers 3–7 mm
          31. Lf blade, basal sheath hairy; st nodes 3–7; infl branches gen spreading > 90° . . . . [2]*B. grandis*
          31′ Lf blade, basal sheath glabrous; st nodes 2–3(4); infl branches gen ascending to spreading ≤ 90°
          . . . . [3]*B. orcuttianus*
    27′ Lower glume gen 1-veined
      32. Glumes glabrous or minutely scabrous
        33. Infl narrow in fl, ≤ 2 cm wide; branches erect or tightly ascending . . . . [2]*B. suksdorfii*
        33′ Infl broader in fl, > 2 cm wide; branches erect, ascending, or nodding, gen ± widely spreading
          34. Ligule (2)3–6(7) mm; lemma awn (4)6–11 mm. . . . . [2]*B. vulgaris*

34′ Ligule 0.5–3 mm; lemma awn (2)3–7(9) mm

  35. Lemma margin not densely hairy; lemma awn (4)5–7(9) mm; infl branches gen ascending to spreading ≤ 90° . . . . ³***B. orcuttianus***

  35′ Lemma margin densely hairy; lemma awn (2)3–5(6.5) mm; infl branches spreading to nodding > 90°

    36. Lemma back glabrous or sparsely hairy with hairs to 0.1 mm; anthers(0.9)1–1.4(1.6) mm; upper glumes (6.2)7.1–8.5(9.5) mm; basal sheath glabrous or long-hairy; upper st blade hairy on upper surface; upper st sheath, nodes gen hairy; fr (5.4)6.2–7.2(7.5) mm . . . . ***B. ciliatus***

    36′ Lemma back sparsely to densely hairy with hairs > 0.1 mm; anthers (1.2)1.6–2.7(3.4) mm; upper glumes (7.8)8.9–11.3(13.2) mm; basal sheath densely, ± short-hairy, upper st sheath glabrous; upper st blade upper surface glabrous; upper st nodes gen glabrous; fr (6.9)7.7–9.7(10.5) mm . . . . ***B. richardsonii***

32′ Glumes hairy

  37. Infl narrow at fl, ≤ 2 cm wide, branches erect or tightly ascending; lf blade, sheath glabrous . . . . . ²***B. suksdorfii***

  37′ Infl broader at fl, > 2 cm wide, branches erect, ascending, or nodding; lf blade, sheath hairy or glabrous

    38. Ligule (2)3–6(7) mm; lemma awn (4)6–11 mm; infl branches glabrous or scabrous . . . . ²***B. vulgaris***

    38′ Ligule 1–3 mm; lemma awn 3–7(9) mm; infl branches hairy

      39. Lower sheath long-soft-hairy, hairs 2–4 mm, or glabrous; lf blade glabrous . . . . ³***B. orcuttianus***

      39′ Lower sheath densely short-stiff-hairy, hairs ≤1 mm; lf blade hairy

        40. Longest lf blade (13)18–38 cm; nodes 3–7 per st . . . . ²***B. grandis***

        40′ Longest lf blade 7.5–16.5 cm; nodes 1–2(3) per st . . . . ***B. hallii***

B. arenarius Labill. (p. 1435) AUSTRALIAN CHESS Ann 15–60 cm. **LF:** hairy; ligule 1.5–2.5 mm; blade 2–5 mm wide. **INFL:** 4–19 cm, open; branches nodding to spreading, conspicuously S-curved, gen > spikelets. **SPIKELET:** 10–20 mm, ± compressed dorsally; glumes hairy, lower 7–10 mm, 3-veined, upper 8–12 mm, 5–7-veined; lemma 7–11 mm, back rounded, hairy, veins not strongly raised, teeth 0.5–2 mm, awn (6)9–14 mm, from 1.5+ mm below tip; anthers 0.7–1 mm. Open, disturbed places; < 2000 m. CA-FP, D; to OR, AZ; native to Australia. Apr–Jul

B. arizonicus (Shear) Stebbins (p. 1435) ARIZONA BROME Ann 40–90 cm. **LF:** glabrous or hairy; ligule 1–2 mm; blade 1.5–10 mm wide. **INFL:** 5–30 cm; branches ascending to erect, lower ± spreading. **SPIKELET:** 18–25 mm, strongly flattened; glumes glabrous or scabrous, lower 8–13 mm, 3–5-veined, upper 9–15 mm, 5–9-veined, ± = lowest lemma; lemma 9.5–14 mm, 7-veined, strongly keeled, margin gen hairy (occ glabrous), back hairy or glabrous, awn 7–15 mm; anthers ≤ 0.5 mm. 2*n*=84. Open, disturbed places, fields; < 2200 m. s NCoRO, GV, SnFrB, SCoR, SCo, ChI, D; AZ, ne Mex, Baja CA. [*B. trinii* var. *excelsus* Shear] Mar–Jun

B. berteroanus Colla (p. 1435) CHILEAN CHESS Ann 30–60 cm, often tufted. **LF:** sparsely to densely hairy; ligule 1–3 mm; blade 2–9 mm wide. **INFL:** 8–30 cm, ± open; branches gen ascending. **SPIKELET:** 15–20 mm, not strongly flattened; glumes glabrous, lower 8–16 mm, 1(3)-veined, upper 10–18 mm, 3(5)-veined; florets 3–9; lemma 6–15 mm, back rounded, hairy, teeth 1–3 mm, awn-like to acuminate, awn (7)14–22 mm, bent, twisted below middle. 2*n*=42. Open, sandy or gravelly soils; < 1700 m. NCoRI, CaRF, SN, SnJV, CW, SW, SNE, D; to OR, NE, AZ; also in n Mex, S.Am. [*B. trinii* Desv. var. *t.*] Mar–Jun

B. briziformis Fisch. & C.A. Mey. RATTLESNAKE CHESS Ann 17–70 cm. **LF:** sheath hairy; ligule 0.5–2 mm; blade 1.5–5 mm wide, hairy. **INFL:** 2.5–15 cm, open, nodding. **SPIKELET:** 15–27 mm, not strongly flattened; glumes glabrous, lower 4.6–6 mm, 3–5-veined, upper 5–8 mm, 5–9-veined; lemma 6–9 mm, 6–8 mm wide, inflated, back rounded, glabrous or hairy, veins strongly raised, teeth < 1 mm, awn 0–1 mm; anthers 0.7–1 mm. 2*n*=14. Open places; 1250–1830 m. KR, CaR, n SN, MP; to AK, BC, ne US; native to Eurasia. Often confused with *Briza maxima*. May–Jul

B. carinatus Hook. & Arn. (p. 1435) CALIFORNIA BROME Per, occ fl in first yr. **LF:** glabrous or hairy; ligule 2–3(4) mm; blade 3–12 mm wide. **INFL:** 9–40 cm; upper branches erect to ascending, lower branches ascending to spreading; sometimes with cleistogamous florets. **SPIKELET:** 20–40 mm, strongly flattened. 2*n*=56.

  var. ***carinatus*** Pl 50–100 cm. **INFL:** 15–40 cm. **SPIKELET:** 20–40 mm; glumes glabrous or hairy, lower 6.5–12 mm, 3–7-veined, upper 6–12 mm, 5–9-veined; lemma 12–20 mm, strongly keeled, gen

uniformly hairy, 7-veined, veins obscure, awn (6)8–15 mm; anthers 0.5–4.5 mm. Coastal prairies, openings in chaparral, plains, open oak and pine woodland; < 3500 m. CA (exc SnJV, DSon); to BC. Apr–Aug

  var. ***marginatus*** (Steud.) Barkworth & Anderton MOUNTAIN BROME Pl 45–120 cm. **LF:** sheath, blade gen hairy, or just hairy at throat; ligule 2–3.5 mm; blade 6–12 mm wide, sometimes inrolled. **INFL:** 9–30 cm. **SPIKELET:** 25–40 mm; glumes hairy, lower 7–9 mm, 3–5-veined, upper 9–11 mm, 9–11-veined; lemma 11–14 mm, strongly keeled, obscurely 7–9-veined, margin hairy, back glabrous or hairy, awn 4–7 mm; anthers 1–4 mm. Open slopes, meadows, forest; < 3500 m. CA; to BC, WY, w TX, Mex. [*B. breviaristatus* Buckley; *B. luzonensis* J. Presl; *B. m.* Steud.; *B. subvelutinus* Shear] May–Jul

B. caroli-henrici Greuter Ann 15–45 cm. **LF:** hairy; ligule 0.5–1 mm; blade 1–4 mm wide. **INFL:** 4–10 cm, dense; branches and spikelets erect. **SPIKELET:** 25–45 mm, not strongly flattened; glumes glabrous or hairy, lower 8–10 mm, 3-veined, upper 9.5–12 mm, 3–5-veined, lemma 11–18 mm, back rounded, glabrous or hairy, veins strongly raised, teeth < 1 mm, awn 10–18 mm, bent, from > 3 mm below lemma tip; anthers 0.5–1.2 mm. 2*n*=14,28. Open, disturbed places; < 100 m. ScV (Butte, Yolo cos.); native to Medit Eur. [*B. alopecuros* Poir., misappl.] Reported from ne US. Apr–Nov

B. catharticus Vahl Ann or short-lived per, 20–120 cm. **LF:** glabrous or hairy. **INFL:** 8–30 cm, ± open. **SPIKELET:** 15–30 mm, strongly flattened. 2*n*=42.

  var. ***catharticus*** RESCUE GRASS **LF:** glabrous or sparingly hairy; ligule 2–5 mm; blade 2–9 mm wide. **INFL:** 10–30 cm, upper branches erect to ascending, lower branches ascending to spreading. **SPIKELET:** 15–30 mm; glumes glabrous, scabrous, or occ hairy, lower 6–12 mm, upper 8–14 mm, 5–7(9)-veined; lemma 10–17 mm, glabrous or slightly hairy, veins (9)11–13-veined, prominent for most of their length; awn 0–3.5 mm; anthers 0.5–1.3 mm. Open, disturbed places; < 1500 m. CA; to e US, Eur, Australia; native to S.Am. Apr–Nov

  var. ***elatus*** (E. Desv.) Planchuelo CHILEAN BROME Per 30–110 cm. **LF:** glabrous or hairy; ligule 1–4 mm; blade 3–10 mm wide. **INFL:** 8–20 cm; branches ascending to erect. **SPIKELET:** 15–25 mm; glumes glabrous or scabrous, lower 8–13 mm, upper 11–20 mm, 7–9-veined, lemma 10–16 mm, 9–11-veined, veins prominent on distal portion, gen hairy at least distally, awn 6–12 mm; anthers 0.6–1. Disturbed areas; < 200 m. s NCo, s NCoR, n&c SNF, deltaic GV, CCo, SnFrB, n ChI; ballast in OR, presumed extirpated; New Zealand; native to S.Am. [*B. stamineus* E. Desv.] Apr–Aug

B. ciliatus L. (p. 1435) FRINGED BROME Per 55–149 cm. **LF:** upper st node and sheath gen hairy; adaxial surface of upper st blade hairy, basal sheath glabrous or long-hairy; ligule 0.5–1.5 mm; blade

4–12 mm wide. **INFL:** 8–21 cm, open; branches spreading to nodding. **SPIKELET:** 15–25 mm, not strongly flattened; glumes glabrous or minutely scabrous, lower 6–7.5 mm, 1(3)-veined, upper (6.2)7.1–8.5(9.5) mm, 3-veined; lemma 8–14 mm, margin hairy along lower 1/2–3/4, back rounded, glabrous or sparsely hairy, hairs to 0.1 mm; anthers (0.9)1–1.4(1.6) mm; fr (5.4)6.2–7.2(7.5) mm 2*n*=14. Damp meadows, woodland, thickets, streambanks, roadsides; 1100–3230 m. CaRH, SNH, SnBr, n W&I; to AK, ne N.Am, n Mex. Apr–Aug

B. commutatus Schrad. HAIRY CHESS, MEADOW BROME Ann 40–120 cm. **LF:** lower sheath long-hairy; ligule 1–4 mm; blade 3–9 mm wide. **INFL:** 6–18 cm; broad, spreading, some branches gen > 4 cm, 1–3 spikelets per branch. **SPIKELET:** 15–30 mm, not strongly flattened, not widening in fr, lemma obscuring most stalks in fr; glumes glabrous, lower 5–7 mm, 3–5-veined, upper 6–9 mm, 5–9-veined; lemma 7.5–11 mm, margin gen broadly angled, back rounded, glabrous or hairy, veins not strongly raised, awn 4–10 mm, from < 1.5 mm below lemma tip, teeth < 1 mm, lowest lemma awn gen < others; anthers 1.3–2.5 mm; fr flat or C-shaped in ×-section. 2*n*=14,28,56. Disturbed areas; < 2200 m. NW, CaR, n SN, c&s SNH, SNE, ScV, CCo, SnFrB, GB; to BC, e N.Am; native to Eur. May–Jul

B. diandrus Roth RIPGUT GRASS Ann 15–120 cm. **LF:** sheath glabrous or hairy; ligule 1–3 mm; blade 2–7 mm wide, hairy. **INFL:** 6–25 cm, ± open; lower branches gen nodding, upper branches spreading to ascending; infl branches gen 1, < or > spikelets, lower (longest) 1-branched. **SPIKELET:** 25–70 mm, not strongly flattened; glumes glabrous to scabrous, lower 12–25 mm, 1(3)-veined, upper 18–35 mm, 3(5)-veined; florets 5–8; lemma 18–35 mm, 1–2 mm wide, back rounded, teeth 3–7 mm, awn 30–65 mm, straight; anthers 0.5–1.3 mm. 2*n*=28,42,56. Open, gen disturbed areas; < 2170 m. CA; to BC, CO, TX, S.Am; native to Eur. [*B. rigidus* Roth; *B. d. var. r.* (Roth) Sales] Feb–Jul ❖

B. grandis (Shear) Hitchc. TALL BROME Per 70–180 cm; st nodes 3–7. **LF:** blade, st sheath, basal sheath hairy; ligule 1–3 mm; longest blade (13)18–38 cm, 3–12 mm wide. **INFL:** 15–26 cm, open; branches spreading to nodding, hairy. **SPIKELET:** 25–35(45) mm, not strongly flattened; glume hairy, back occ glabrous, lower 5–8.5 mm, 1(3)-veined, upper 7–10(12) mm, 3-veined; lemma 11–14 mm, margin hairy, back rounded, hairy, awn 3–6 mm; anthers 3–5 mm. 2*n*=14. Dry, open places, shrubland, oak woodland, conifer forest; 365–2400 m. n SNF, c&s SN, CW, TR, PR; Baja CA. [*B. porteri* (J.M. Coult.) Nash var. *assimilis* Burtt Davy] May–Jul

B. hallii (Hitchc.) Saarela & P.M. Peterson HALL'S BROME Per 90–150 cm, st nodes 1–2(3). **LF:** st sheath and basal sheath hairy; ligule 1–2 mm; longest blade 7.5–16.5 cm, 3–12 mm wide, hairy. **INFL:** 5–16 cm, open; branches erect, ascending or spreading. **SPIKELET:** 25–35(45) mm, not strongly flattened; glumes scabrous or hairy, lower 5–8(9) mm, 1(3)-veined, upper (7)8–9 mm, 3-veined; lemma 10–14 mm, back rounded, scabrous or puberulent, awn 3.5–7 mm; anthers 3–6 mm. Montane to subalpine forest; 1580–2680 m. s SN, SCoR, TR. [*B. orcuttianus* Vasey var. *h.* Hitchc.] Jun–Jul

B. hordeaceus L. (p. 1435) SOFT CHESS Ann 11–65 cm. **LF:** hairy; ligule 1–1.5 mm; blade 1.5–5 mm wide. **INFL:** 2.5–13 cm, dense, branches erect to spreading, some pedicels > spikelet. **SPIKELET:** 12–22 mm, not strongly flattened; glumes glabrous or hairy, lower 5–8 mm, 3–5-veined, upper 6–9 mm, 5–9-veined; lemma 6.5–10 mm, 1.9–2.5 mm wide, back rounded, glabrous or hairy, veins strongly raised, teeth < 1 mm, awn 4–10 mm, from < 1.5 mm below lemma tip, straight, occ curved; anthers 0.2–2 mm. 2*n*=28. Fields, disturbed areas; < 1000(2560) m. CA; to BC, e US. [*B. h.* subsp. *thominei* (Nyman) Braun-Blanquet] If recognized taxonomically, pls with awn outcurved in fr, ± flat near base, assignable to *B. hordeaceus* subsp. *divaricatus* (Bonnier & Layens) Kerguélen, *B. molliformis* Billot, *B. hordeaceus* subsp. *molliformis* (J. Lloyd) Maire & Weiller, inval. Apr–Jul ❖

B. inermis Leyss. SMOOTH BROME, HUNGARIAN BROME Per 45–130 cm, rhizomatous. **LF:** sheath, blade gen glabrous; ligule 0.8–3 mm; blade 5–15 mm wide. **INFL:** 10–20 cm; ± open, branches erect. **SPIKELET:** 20–33 mm, not strongly flattened; glumes gla-

brous, lower 6–8 mm, 1(3)-veined, upper 7–10 mm, 3-veined; lemma 9–13 mm, back rounded, glabrous, scabrous, or margin hairy, awn 0–3 mm; anthers 3.5–6 mm. 2*n*=28,42,56. Disturbed sites, roadsides; < 2700 m. NCo, KR, NCoRI, CaRH, SNH (e slope), SCoRO, SCo, SnGb, SnBr, PR, GB; to AK, e N.Am; native to Eurasia. Cult widely for forage, revegetation after fire. Mar–Aug

B. japonicus Thunb. JAPANESE CHESS, JAPANESE BROME Ann 17–85 cm. **LF:** sheath hairy; ligule 1–2.2 mm; blade 1.5–6 mm wide, glabrous to hairy. **INFL:** 3–26 cm, ± open; branches spreading to ascending, lower nodding, ± wavy, branches mostly longer than spikelets. **SPIKELET:** 20–40 mm, not strongly flattened; glumes glabrous, lower 4–7 mm, 3–5-veined, upper 5–8 mm, 5–7-veined; lemma 7–10 mm, membranous margin 0.3–0.6 mm wide, back rounded, glabrous, veins not strongly raised, teeth < 1 mm, awn 5–11 mm, straight to slightly curved, from 2–5 mm below lemma tip; anthers 0.7–1 mm. 2*n*=14. Open, disturbed areas; < 2470 m. CA; to BC, e N.Am; native to Eurasia. May–Jul ❖

B. laevipes Shear CHINOOK BROME, WOODLAND BROME Per 50–160 cm. **LF:** glabrous; ligule (1.5)2–4 mm; blade 3–7 mm wide. **INFL:** 7–27 cm, open; branches nodding to spreading, upper ± ascending. **SPIKELET:** 23–35 mm, not strongly flattened; glumes glabrous, lower 6–9 mm, 3-veined, upper 7–11 mm, 5-veined; lemma 10–15 mm, back rounded, hairy, margin hairs sometimes denser, teeth < 1 mm, awn 3–6.5 mm; anthers 3.5–5 mm. 2*n*=14. Shrubland, conifer forest, shaded streambanks, roadsides; < 2500 m. NW, CaR, SN, ScV (Sutter Buttes), CW, SCo, ChI, WTR, PR; to WA, Baja CA. May–Jul

B. madritensis L. Ann 10–50 cm. **LF:** glabrous or hairy; ligule 1–3 mm; blade 1–4 mm wide. **INFL:** branches erect to ascending. **SPIKELET:** 20–50 mm, not strongly flattened; glumes glabrous to hairy, lower 3.5–13.5 mm, 1-veined, upper 6–20 mm, 3-veined; lemma 12–25 mm, back rounded, glabrous or hairy, teeth 1.5–3 mm, awn 10–25 mm, straight. 2*n*=14,28.

subsp. *madritensis* (p. 1435) FOXTAIL CHESS, MADRID BROME **INFL:** ± obovoid, ± dense, most branches visible, occ > spikelets, shortest branch on lowest node 6–24 mm, longest branch on lowest node branched 0–2 ×, infl internodes reduced upwards. **SPIKELET:** sterile florets ≤ 3; florets not overlapping at maturity. Disturbed areas, roadsides; < 2200 m. CA-FP, D (uncommon); s OR, Baja CA; native to Eur. Apr–Jun

subsp. *rubens* (L.) Husn. (p. 1435) RED BROME **INFL:** obovoid, dense; branches (exc lowest) obscure, < spikelets, shortest branch on lowest infl node ≤ 6 mm, longest branch on lowest node branched 2–5 ×, internodes much reduced upwards. **SPIKELET:** sterile florets 3+; florets overlapping at maturity. Disturbed areas, roadsides; < 3050 m. CA; to OR, se US, n Baja CA; native to Eur. [*B. r.* L.] Feb–Jun ❖

B. maritimus (Piper) Hitchc. (p. 1435) MARITIME BROME Per 20–80 cm. **LF:** sheath, blade glabrous, occ scabrous; ligule 1–6 mm; blade 3–12 mm wide. **INFL:** 9–20 cm, dense, spikelets overlapping; branches ascending to erect, < spikelets. **SPIKELET:** 20–40 mm, strongly flattened; glumes glabrous or hairy, lower 8–12 mm, 3–5-veined, upper 10–13 mm, 7–9-veined; lemma 12–14 mm, strongly keeled, hairy, awn 3.5–7 mm. 2*n*=84. Dunes, coastal meadows; < 200 m. NCo, CCo, SnFrB, n SCo, ChI; to OR. [*B. carinatus* var. *m.* (Piper) C.L. Hitchc.] Apr–Jul

B. orcuttianus Vasey ORCUTT'S BROME Per 75–150 cm; st nodes 2–3(4). **LF:** st sheath hairy, basal sheath glabrous; ligule 1–3 mm; blade 7–24 cm, 3–12 mm wide, glabrous, gen prow-tipped. **INFL:** 7–13.5 cm, open; branches gen ascending to spreading, ≤ 90° from st axis. **SPIKELET:** 20–40 mm, not strongly flattened; glumes glabrous, scabrous, or hairy, lower 5–9 mm, 1- or 3-veined, upper 7–11 mm, 3-veined; lemma 9–13 mm, back rounded, glabrous, scabrous, or hairy, awn (4)5–7(9) mm; anthers 3–5 mm. 2*n*=14. Dry places, meadows, scrub, open forest; 560–3500 m. NW, CaR, SN, SCoRO, TR, PR, MP; to WA, w NV, Baja CA. Jun–Jul

B. polyanthus Shear GREAT BASIN BROME, COLORADO BROME Per 60–120 cm. **LF:** sheath and throat glabrous; ligule 2–2.5 mm;

blade 6–12 mm wide, glabrous. **INFL:** 15–25 cm; branches erect, ascending or spreading. **SPIKELET:** 30–35 mm, strongly flattened; glumes glabrous or scabrous, upper (7.5)9–11(12.5) mm, 5–7-veined; lower (5.5)7–10(11.5) mm, 3-veined; lemma 12–15 mm, strongly keeled, 7–9-veined, awn 4–8 mm; anthers 1–5 mm; 2*n*=56. Open slopes, meadows; 1200–3100 m. c SNH (Mariposa, Tuolumne cos.); to OR, TX. [*B. polyanthus* var. *paniculatus* Shear] Aug

B. porteri (J.M. Coult.) Nash (p. 1435) NODDING BROME Per 32–100 cm. **LF:** ligule 1–2 mm; blade 2–5 mm wide, glabrous or slightly hairy. **INFL:** 6–20 cm, open; branches ascending to nodding. **SPIKELET:** 12–15 mm, not strongly flattened; glumes 3-veined, hairy, sometimes glabrous, lower 5–7 mm, upper 6–10 mm; lemma 7–13 mm, back ± rounded, hairy, awn 1.5–3(4) mm; anthers 1.5–3.5(4) mm. 2*n*=14. Exposed slopes, open woodland; 550–3500 m. NCoRI, s SN, SCoRI, SnBr, SNE; to BC, MB, TX. [*B. anomalus* E. Fourn., misappl.] Jul–Aug

B. pseudolaevipes Wagnon WOODLAND BROME Per 60–125 cm. **LF:** sheath glabrous or hairy; ligule 0.4–1(2) mm; blade 2–9 mm wide, glabrous, hairy on margin, or hairy throughout. **INFL:** 7.5–20 cm; branches erect to spreading. **SPIKELET:** 15–35 mm, not strongly flattened; glumes scabrous or hairy, lower 4–7 mm, 3-veined, upper 6–9 mm, 5-veined, lemma 10–12.5 mm, back rounded, hairy across back or only on margin, teeth < 1 mm, awn 2–5.5 mm; anthers 3.5–5.5 mm. 2*n*=14. Shaded or semi-shaded sites in chaparral, coastal-sage scrub, open woodland; 100–900 m. NCoRI, CaRH, SnFrB, SCoRO, SW. Apr–Jul

B. racemosus L. (p. 1435) SMOOTH BROME Ann 25–110 cm. **LF:** lower sheaths long-hairy; ligule 1–3 mm; blade 2–5 mm wide. **INFL:** 4–14 cm; narrow, branches < 4 cm, 1 spikelet per branch. **SPIKELET:** 11–18 mm, not strongly flattened, not widening substantially in fr, lemma obscuring most stalks in fr; glumes glabrous, lower 4–6 mm, 3–5-veined, upper 4–7 mm, 5–9-veined; lemma 7–9 mm, 4–5 mm wide, back rounded, glabrous, margin often smoothly curved, veins not strongly raised, teeth < 1 mm, awn 5–9 mm, ± equal, from < 1.5 mm below lemma tip; anthers 1.5–3 mm; fr flat or C-shaped in ×-section. 2*n*=28. Disturbed areas, roadsides; 60–1850 m. KR, NCoRO, NCoRI, SNF, GV, CCo, SnFrB, SnBr, MP; to BC, e US; native to Eur. May–Jul

B. richardsonii Link (p. 1435) RICHARDSON'S BROME Per 47–110 cm; upper st nodes gen glabrous. **LF:** basal sheath densely, short- to medium-hairy, upper sheath glabrous; ligule 1–2 mm; blade 3–9 mm wide. **INFL:** 9–22 cm, open; branches erect to nodding. **SPIKELET:** 24–35 mm, not strongly flattened; glumes glabrous or scabrous, lower 7.1–10 mm, 1(3)-veined, upper (7.8)8.9–11.3(13.2) mm, 3-veined; lemma 9–13.5 mm, margin hairy along lower 1/2 to 3/4, back rounded, sparsely to densely hairy with hairs > 0.1 mm, awn 3–6 mm; anthers (1.2)1.6–2.7(3.4) mm; fr (6.9)7.7–9.7(10.5) mm. 2*n*=28. Meadows, open woodland; 1200–3600 m. c&s SNH, SCoRO, SnBr, e PR, e DMoj; to AK, WY, w TX, Mex. [*B. ciliatus* var. *r.* (Link) B. Boivin] Jul–Sep

B. secalinus L. (p. 1435) RYE BROME Ann 45–100 cm. **LF:** sheath glabrous or sparsely short-hairy; ligule 1–3 mm; blade 4–12 mm wide, hairy. **INFL:** 8–17 cm, ± open; branches spreading to ascending, nodding in fr. **SPIKELET:** 12–24 mm, not strongly flattened, widening in fr, many stalks becoming visible as lemma wraps around fr; glumes glabrous, lower 5–7 mm, 3–5-veined, upper 6–9 mm, 5–9-veined; lemma 8–11 mm, 4–5 mm wide, back rounded, glabrous or hairy, veins not strongly raised, teeth < 1 mm, awn (0)3–9.5 mm, from < 1.5 mm below lemma tip; anthers 1.2–2.5 mm; fr U- and V-shaped in ×-section. 2*n*=14,28. Open, disturbed areas; < 1500 m. NW, CaR, n&c SN, n SnFrB, SnBr, MP; to BC, e US; native to Eur. May–Jul

B. sitchensis Trin. SITKA BROME, ALASKA BROME Per 40–145 cm. **LF:** sheath glabrous; ligule 3–4 mm; blade glabrous or hairy, 1–4 mm wide. **INFL:** 19–36 cm, open; branches ascending, spreading to nodding; lower branches 10–20 cm. **SPIKELET:** 25–41 mm, strongly flattened, keeled; glumes glabrous (hairy), lower 8–11 mm, 3–5-veined, upper 10–13 mm, 5–7-veined; lemma, 12–15 mm, back ± rounded, strongly keeled, glabrous, sometimes hairy, awn 5–10 mm; anthers to 6 mm. 2*n*=42,56. Rocky bluffs, cliffs, meadows, forest edges, disturbed areas; < 1670 m. SW; to AK. Mar–Jun

B. squarrosus L. (p. 1435) CORN BROME Ann 10–43 cm. **LF:** sheath hairy; ligule 0.5–2 mm; blade hairy or glabrous, 2.5–10 mm wide. **INFL:** 5–10.5 cm, open, often appearing 1-sided; spikelets 1(2) per branch; 1+ lower branches gen > spikelets, ± wavy. **SPIKELET:** 14–35 mm, not strongly flattened; glumes glabrous, lower 4.5–7 mm, 3–5(7)-veined, upper 6–9.5 mm, 7-veined; florets 7–18; lemma 8.2–11 mm, membranous margin 0.6–0.9 mm wide, back rounded, glabrous or minutely scabrous, teeth < 1 mm, awn 8–11.7 mm, from 1.5+ mm below tip; anthers 0.5–1.6 mm. 2*n*=14. Open, disturbed areas, roadsides; < 1494 m. MP; to BC, c N.Am; native to Eurasia. Jul

B. sterilis L. (p. 1435) POVERTY BROME Ann 25–85 cm. **LF:** sheath hairy; ligule 2–2.5 mm; blade 2–5 mm wide, hairy or glabrous. **INFL:** 10–25 cm, open; lower branches ascending to nodding, upper branches ascending, branches gen > spikelets (shorter when infl reduced to 1(3) spikelets), simple, lower branches sometimes branched 1 ×; spikelets 1(3) per branch. **SPIKELET:** 20–35 mm, not strongly flattened; glumes glabrous or scabrous, upper 7.5–21 mm, 3(5)-veined, lower 6–14 mm, 1(3)-veined; florets 6–11; lemma 13–20 mm, back rounded, glabrous to scabrous, teeth 0–22 mm, awn 15–30 mm, straight; anthers 0.5–2 mm. 2*n*=42,56. Open, disturbed areas; < 1100 m. NW, CaRF, SNF, n SNH, GV, SnFrB, SCoRO, s ChI, SCo, WTR (w Santa Susana foothills), PR (Otay Mtn); to BC, e N.Am; native to Eurasia. Mar–Jun

B. suksdorfii Vasey (p. 1435) SUKSDORF'S BROME Per 45–95 cm. **LF:** glabrous; ligule 0.5–2 mm; blade 4–11 mm wide. **INFL:** 6.5–13 cm, ≤ 2 cm wide, narrow; branches erect to ascending. **SPIKELET:** 15–30 mm, not strongly flattened; glumes glabrous or sparsely hairy, lower 7–10 mm, 1(3)-veined, upper 8–13 mm, 3-veined; lemma 10–14 mm, back rounded, glabrous to hairy, awn 2–5.5 mm; anthers 2.2–3.5 mm. 2*n*=14. Rocky slopes, meadows, conifer forest; 1250–3300 m. KR, NCoRH, CaR, SNH, SCoRO, MP; to WA. Jun–Aug

B. tectorum L. (p. 1435) CHEAT GRASS, DOWNY CHESS Ann 5–40 cm. **LF:** sheath hairy (sometimes glabrous); ligule 2–3 mm; blade glabrous to hairy, gen long-hairy near base, 1–5 mm wide. **INFL:** 6–22 cm; open; branches spreading to nodding; spikelets 1–14 per branch; branches < or > spikelets, 1–5 × branched. **SPIKELET:** 10–20 mm, not strongly flattened; glumes glabrous to hairy, lower 4–9 mm, 1(3)-veined, upper 7–13.5 mm, 3(5)-veined; florets 3–7; lemma 9–13 mm, back rounded, glabrous to hairy, teeth 1–3 mm, awn 8–18 mm, straight; anthers 0.5–1.3 mm. 2*n*=14. Open, disturbed areas; < 3400 m. CA; N.Am; native to Eurasia. [*B. t.* var. *glabratus* Spenn.] Invasive. May–Aug ❖

B. vulgaris (Hook.) Shear (p. 1435) COLUMBIA BROME Per 45–110 cm; nodes (3)4–6(7). **LF:** ligule (2)3–6(7) mm; blade 13–25(33) cm, 3–14 mm wide, gen hairy on upper surface. **INFL:** 8–22 cm, > 2 cm wide, open; branches ascending to nodding, glabrous or scabrous. **SPIKELET:** 15–30 mm, not strongly flattened; glumes glabrous or hairy, lower 4–9 mm, 1(3)-veined, upper 5–10 mm, 3-veined; lemma 10–16 mm, margin hairy, back rounded, glabrous to hairy, awn (4)6–11 mm; anthers 2–3.5(4) mm. 2*n*=14. Shady to open rocky woodland, ravines, meadows; < 1900 m. NW, CaR, n&c SN, CW (exc SCoRI); to BC, MT, WY. May–Aug

CALAMAGROSTIS REED GRASS

Paul M. Peterson, Jeffery M. Saarela & Craig W. Greene

Per, gen from rhizomes. **ST:** 1–15 dm, gen not branched, ± smooth; nodes (1)2–8. **LF:** gen basal and cauline; sheath smooth or scabrous; ligule membranous; blade flat to inrolled. **INFL:** panicle-like, open to dense; branches ± drooping to appressed; spikelets ascending to appressed. **SPIKELET:** glumes subequal, gen lanceolate, acute to acuminate, lower gen 1-veined, upper

3-veined; floret 1, breaking above glumes; axis prolonged beyond floret, hairy; callus hairy; lemma < glumes, awned from below middle to near base, tip gen 4-toothed, veins 3–5, awn straight to twisted, bent; palea ± = lemma, thin. ± 265 spp. (incl *Deyeuxia*): cool temp (esp moist montane); some forage value. (Greek: reed grass) [Marr et al. 2007 FNANM 24:706–732] Hybridization, polyploidy (diploids unknown), and asexual seed set contribute to taxonomic difficulty.

1. Awn exserted 1–10 mm beyond glume tips, twisted and bent
　2. Panicle open, branches spreading to ascending
　　3. Pl from rhizomes; st 5–15 dm; lf blade gen cauline, 3–10 mm wide, flat; coastal, < 500 m . . . . **C. bolanderi**
　　3′ Pl cespitose, rhizomes 0; st gen 1.2–5.4 dm; lf blade gen basal, 0.2–1.7 mm wide, inrolled to flat; > 1500 m
　　　4. Lf blade 7–11-veined, tip prow-shaped; infl (4)5.7–8.5 cm; callus hairs 0.3–1.2 mm . . . . ²**C. breweri**
　　　4′ Lf blade 3–5-veined, tip straight-sided; infl 1.9–5.7(7.5) cm; callus hairs 0.3–0.6 mm. . . . . ²**C. muiriana**
　2′ Panicle dense, narrow, branches ascending to appressed
　　5. Awn 12–15(17) mm, exserted 4–10 mm beyond the glume tips; glumes 8–10 mm. . . . . **C. foliosa**
　　5′ Awn (4.6)5–7.5(9) mm, exserted 1–3 mm beyond the glume tips; glumes 4.5–8 mm
　　　6. Lf blade with small white-opaque hooks between the veins visible only with magnification, upper
　　　　surface scabrous; gen on serpentine soils, < 1065 m . . . . ²**C. ophitidis**
　　　6′ Lf blade without small white-opaque hooks between the veins, upper surface soft-hairy; gen on rocky
　　　　slopes, sandy soils, 1300–4000 m. . . . . ²**C. purpurascens**
1′ Awn incl or occ prolonged < 1 mm beyond glume tips, straight or twisted and bent
　7. Lf blade 0.2–1.7 mm wide; panicle 1.9–8.5 cm; callus hairs sparse
　　8. Lf blade 7–11-veined, tip prow-shaped; panicle (4)5.7–8.5 cm; callus hairs 0.3–1.2 mm . . . . ²**C. breweri**
　　8′ Lf blade 3–5-veined, tip straight-sided; panicle 1.9–5.7(7.5) cm; callus hairs 0.3–0.6 mm . . . . ²**C. muiriana**
　7′ Lf blade (1.5)2–10(20) mm wide; panicle 4–30 cm; callus hairs abundant or sparse
　　9. Panicle ± open, lower branches spreading to ascending
　　　10. Lf collar gen puberulent or hairy-tufted; callus hairs < 2 mm . . . . ²**C. rubescens**
　　　10′ Lf collar smooth or scabrous, rarely hairy; callus hairs (1.5)2–3.5(4.5) mm
　　　　11. Glumes 5–7(7.5) mm; sts gen unbranched below with 1–2(3) nodes per st; ligules 1–4 mm, entire,
　　　　　truncate, gen hidden by expanded collar below. . . . . **C. nutkaensis**
　　　　11′ Glumes 3–4.5 mm; sts gen branched below with (2)3–8 nodes per st; ligules 3–8 mm, irregularly
　　　　　cut, obtuse to acute, not hidden by expanded collar below . . . . **C. canadensis**
　　　　　12. Glumes smooth or scabrous, projections along keel straight, tips acute (acuminate); spikelets
　　　　　　2.5–3.5(4) mm. . . . . var. **canadensis**
　　　　　12′ Glumes scabrous across entire surface, rarely smooth, projections along keel often bent, tips
　　　　　　acuminate; spikelets (3.5)4–4.5(5.2) mm. . . . . var. **langsdorffii**
　　9′ Panicle ± dense, narrow, lower branches appressed (occ ascending in *C. rubescens*, *C. stricta*)
　　　13. Glumes gen < than 3 × longer than wide. . . . . **C. stricta**
　　　　14. Glumes 3–6 mm, gen thick, margin opaque; callus hairs 2–4.5 mm; spikelet axis 1–1.5 mm; longest
　　　　　panicle branches 1.5–9.5 cm. . . . . subsp. **inexpansa**
　　　　14′ Glumes 2–3(3.5) mm, gen thin, margin ± translucent; callus hairs 1–3 mm; spikelet axis 0.5–1 mm;
　　　　　longest panicle branches 1.5–4 cm . . . . subsp. **stricta**
　　　13′ Glumes gen > 3 × longer than wide
　　　　15. Lf blade with small white-opaque hooks between the veins visible only with magnification. . . . . ²**C. ophitidis**
　　　　15′ Lf blade without small white-opaque hooks between the veins
　　　　　16. Lf blade gen soft-hairy on the upper surface. . . . . ²**C. purpurascens**
　　　　　16′ Lf blade smooth or scabrous but not soft-hairy on the upper surface
　　　　　　17. Awn 4–5.5 mm; lf collar scabrous or smooth; rhizomes 2–6 cm, 2–4 mm thick, stout . . . . **C. koelerioides**
　　　　　　17′ Awn 2–4 mm; lf collar gen puberulent or hairy-tufted; rhizomes 10–20 cm, 1.5–2 mm thick,
　　　　　　　slender. . . . . ²**C. rubescens**

C. bolanderi Thurb. (p. 1435) BOLANDER'S REED GRASS Rhizomes present. **ST**: 5–15 dm; nodes gen 4. **LF**: ligule 3–5 mm; blade 3–10 mm wide, flat. **INFL**: 10–25 cm, ± open; lower branches 6–8 cm, spreading. **SPIKELET**: glumes 3–4(5) mm, smooth exc keel, tip scabrous; axis ± 1 mm, hairs 1 mm; callus hairs ± 1 mm, tufted; lemma 2.5–3 mm, scabrous, awned near base; awn (4)5–6 mm, exserted 1–3 mm beyond glume tips, strongly twisted, bent. 2*n*=56. Peatland, marshes, wet meadows in forest, coastal scrub and prairie; < 500 m. NCo; OR. Jun–Aug ★

C. breweri Thurb. (p. 1435) SHORT HAIR REED GRASS Cespitose; rhizomes 0. **ST**: 2–5.4 dm; nodes 1–3. **LF**: gen basal; ligule 1.7–4.1(6) mm; blade 0.4–1.7 mm wide, flat or inrolled, 7–11-veined, tip prow-shaped. **INFL**: (4)5.7–8.5 cm, open; lower branches 0.4–3(3.5) cm, ± spreading. **SPIKELET**: glumes 3.1–5 mm, smooth or keel scabrous; axis 1–2 mm, hairs 1–2 mm; callus hairs 0.3–1.2 mm, sparse; lemma 3.4–5.5 mm, awned near base; awn 3.5–5.5 mm, exserted gen (0.5)1–3 mm beyond the glume tips, twisted, bent. 2*n*=42. Moist woodland, meadows, lake margins, streambanks; 1700–2600 m. KR, n SNH; OR. Jul–Sep

C. canadensis (Michx.) P. Beauv. Cespitose, rhizomes 2–15+ cm, 1–3 mm thick, stout. **ST**: 6–15 dm, gen branched below; nodes (2)3–8. **LF**: sheath glabrous to hairy; collar scabrous, rarely smooth or hairy; ligule 3–8 mm, obtuse to acute, irregularly cut, not hidden by expanded collar below; blade 3–8 mm wide, ± drooping, flat, surfaces scabrous. **INFL**: 10–25 cm, open to dense when young; lower branches 3–8 cm, ± spreading. **SPIKELET**: glumes 3–4.5 mm, smooth or scabrous, gen ± purple; axis 0.5–1 mm, hairs 1.5–3.2 mm; callus hairs 2–3.5(4.5) mm, dense; lemma 2.5–4.5 mm, thin, awned just below middle; awn 1–3 mm, not exserted, gen straight. 2*n*=42–66. Many forms, vars., subspp. have been attributed to this sp. Some pls set seed asexually.

var. **canadensis** (p. 1435) BLUEJOINT REED GRASS **SPIKELET**: spikelets 2.5–3.5(4) mm; glumes smooth or scabrous along keel, projections straight, tip acute (acuminate). Moist meadows, thickets, peatland, open woodland; 1500–3400 m. KR, CaRH, SNH; Widespread in US. Jul–Sep

var. **langsdorffii** (Link) Inman **SPIKELET**: spikelets (3.5)4–4.5(5.2) mm; glumes scabrous across entire surface (smooth), projec-

tions along keel often bent, tips acuminate. Moist meadows, thickets, peatland, open woodland; 1500–3400 m. KR, CaRH, SNH; Mostly w US, Can, Greenland. Jul–Sep

C. foliosa Kearney (p. 1435) LEAFY REED GRASS Cespitose. **ST**: 3–6(7) dm, tufted; nodes 2–3. **LF**: gen basal; ligule 4–6 mm; blade 1–2 mm wide, inrolled, upper surface scabrous. **INFL**: 5–12 cm, dense, narrow; lower branches < 4 cm, ascending to appressed. **SPIKELET**: glumes 8–10 mm, scabrous; axis 1.5–4 mm, hairs 2–3 mm; callus hairs 2–3 mm; lemma 5–7(8) mm, awned near base; awn 12–15(17) mm, exserted 4–10 mm beyond glume tips, bent, twisted. 2*n*=28. Coastal scrub, forest, rock outcrops, crevices, cliffs; < 1250 m. NCo, KR, NCoRO. May–Aug ★

C. koelerioides Vasey (p. 1435) DENSE-PINE REED GRASS Cespitose, rhizomes 2–6 cm, 2–4 mm thick, stout. **ST**: 6–10+ dm, nodes 3–5. **LF**: sheath ± scabrous; scabrous or smooth near collar; ligule 3–7 mm; blade 3–7 mm wide, flat or inrolled, scabrous or smooth. **INFL**: 5–16 cm, gen dense; branches gen < 3 cm, appressed. **SPIKE-LET**: glumes 4–6 mm, scabrous esp on keel; axis ± 1 mm, hairs 1–2 mm; callus hairs < 2 mm, sparse, tufted; lemma (3.5)4–5.5(6) mm, awned near base; awn 4–5.5 mm, < 1 mm exserted beyond the glume tips or ± = glume tips, stiff, twisted, bent. 2*n*=28. Meadows, slopes, dry hills, ridges; < 2300 m. NW, CW, PR; to WA, ID, MT, WY. Larger pls like *C. nutkaensis*. Jun–Aug

C. muiriana B.L. Wilson & Sami Gray MUIR'S REED GRASS Cespitose; rhizomes 0. **ST**: 1.2–3.4 dm; nodes 1–3. **LF**: gen basal; ligule 0.8–2.5 mm; blade 0.2–0.4 mm wide, inrolled, 3–5-veined, tip straight-sided. **INFL**: 1.9–5.7(7.5) cm, open; lower branches 0.4–2(3.5) cm, ± spreading. **SPIKELET**: glumes 3–4.5 mm, smooth or keel scabrous; axis ± 2 mm, hairs 0.5–1 mm; callus hairs 0.3–0.6 mm, sparse; lemma 2.5–4 mm, awned near base; awn 3.3–6 mm, exserted gen (0.5)1–3 mm beyond the glume tips, twisted, bent. 2*n*=28. Meadows, lake margins, streambanks; 2480–3900 m. c&s SNH. Jul–Sep

C. nutkaensis (J. Presl) Steud. (p. 1435) PACIFIC REED GRASS Cespitose, rhizomes 3–6 cm, 1.5–3 mm thick, stout. **ST**: 6–11(15) dm, gen unbranched below; nodes 1–2(3). **LF**: sheath loosely open at st base; collar smooth; ligule 1–4 mm, truncate, entire, gen hidden by expanded collar; blade 4–10(20) mm wide, flat, upper surface smooth. **INFL**: 12–30 cm, ± open below, narrow above; branches < 5–7+ cm, ascending. **SPIKELET**: glumes 5–7(7.5) mm, smooth to scabrous esp on keel; axis 0.5–1 mm, hairs 0.5–1.2 mm, sparse; callus hairs (1.5)2–2.5(3) mm; lemma 4–5 mm, awned ± near middle; awn 1–3 mm, not exserted, straight or slightly twisted, bent. 2*n*=28. Wet areas, beaches, dunes, coastal woodland, inland marshes; < 1070 m. NCo, CCo, SnFrB; to AK. May–Aug

C. ophitidis (J.T. Howell) Nygren (p. 1435) SERPENTINE REED GRASS Cespitose, often with rhizomes 2–15 cm. **ST**: 5.5–10 dm, clumped; nodes 3–5. **LF**: ligule 2–5.5(7) mm; blade (1.5)2–4 mm wide, gen inrolled, scabrous, with small white-opaque hooks between veins, visible only with magnification. **INFL**: 8–15 cm, ± dense, narrow; branches < 4 cm, gen appressed. **SPIKELET**: glumes (5)6.5–8 mm, scabrous esp on keel, gen pale; axis ± 2 mm, hairs 1–2 mm; cal-

lus hairs < 2 mm; lemma 4.5–6.5 mm, awned near base; awn 5–6.5(8) mm, incl to exserted ± 2 mm beyond the glume tips, twisted, bent. 2*n*=28. Meadows, seeps, grassland, chaparral, forest, gen on serpentine soils; < 1065 m. s NCoRO, n CCo, n SnFrB. May–Jun ★

C. purpurascens R. Br. (p. 1435) PURPLE REED GRASS Cespitose; rhizomes 1–4 cm. **ST**: 1–8 dm; nodes (1)2–3. **LF**: collar glabrous to short-hairy; ligule 2–6 mm; blade 2–5 mm wide, flat, lower surface smooth, upper gen soft-hairy. **INFL**: 4–15 cm, dense, narrow; branches < 3.5 cm. **SPIKELET**: glumes 4.5–8 mm, scabrous; axis 1–2 mm, hairs 1–2 mm; callus hairs 1–2.5 mm; lemma 3.5–5 mm, awned near base; awn (4.6)6–7.5(9) mm, exserted (0.5)1–3 mm beyond glume tips, twisted, bent. 2*n*=28,40–58,84. Rocky slopes, grassland, meadows, forest, gen on sandy soils; 1300–4000 m. CaRH, SN, n SNE (Sweetwater Mtns), W&I; to AK, Siberia, Greenland. Some pls set seed asexually. Jul–Sep

C. rubescens Buckley (p. 1435) PINE REED GRASS Loosely cespitose; rhizomes 10–20 cm, 1.5–2 mm thick, slender. **ST**: 6–10 dm; nodes 2–3. **LF**: sheath smooth, gen puberulent or hairy-tufted near collar; ligule 3–5 mm; blade 2–5 mm wide, flat, lower surface smooth or scabrous, upper surface scabrous, occ short-hairy. **INFL**: 6–15(25) cm, dense to ± open; branches < 2–4 cm. **SPIKELET**: glumes 4–5 mm, smooth to scabrous esp on keel; axis ± 1 mm, hairs < 2 mm; callus hairs < 2 mm; lemma 3–4 mm, awned near base; awn 2–4 mm, exserted < 1 mm beyond glume tips or ± = glume tips, strongly twisted, bent. 2*n*=28,42,56. Wooded slopes, montane forest, chaparral, meadows; < 900 m. NCo, NCoRO, CW, n ChI (Santa Cruz Island); to sw Can, Rocky Mtns. Jun–Sep

C. stricta (Timm) Koeler SLIPSTEM REED GRASS Loosely cespitose. **ST**: 2–12 dm. **LF**: sheath smooth; ligule 1–5.5 mm; blade 2–5 mm wide, gen inrolled, lower surface gen smooth, upper surface smooth to scabrous. **INFL**: 5–20 cm, dense, narrow; branches < 1.5–9.5 cm, ascending to appressed. **SPIKELET**: glumes 2–6 mm, smooth to scabrous; axis 0.5–1.5 mm, hairs 1.5–3 mm; callus hairs 1–4.5 mm; lemma 2–5 mm, finely scabrous, awned at or below middle; awn ± = glume tip, gen straight. Subspp. intergrade.

subsp. ***inexpansa*** (A. Gray) C.W. Greene (p. 1439) **ST**: 4–12 dm. **LF**: ligule 2–5.5 mm; blade flat, strongly scabrous, upper surface gen glaucous. **INFL**: 6–20 cm; longest branches 1.5–9.5 cm. **SPIKELET**: glumes 3–6 mm, gen thick, margin opaque; axis 1–1.5 mm; callus hairs 2–4.5 mm; lemma 2.5–5 mm; awn occ twisted, bent, stiff; anthers gen sterile. Slopes, meadows, coastal marshes; < 3400 m. NW, CaR, n&c SNH, CCo; to AK, ne N.Am, ne Asia. [*C. crassiglumis* Thurb.] Some pls set seed asexually. Jun–Aug ★

subsp. ***stricta*** (p. 1439) **ST**: 2–9 dm. **LF**: ligule 1–3.5(4) mm; blade gen inrolled, upper surface smooth to scabrous. **INFL**: 5–12 cm; longest branches 1.5–4 cm. **SPIKELET**: glumes 2–3(3.5) mm, gen thin, margin ± translucent; axis 0.5–1 mm; callus hairs 1–3 mm; lemma 2–3.5 mm; awn straight, slender; anthers gen fertile. 2*n*=28,42,56,±70. Conifer forest, meadows, slopes; 1500–3350 m. SNH, W&I; to AK, ne N.Am, Eurasia. Jul–Aug

CENCHRUS SANDBUR

Ann, per. **ST**: 5–100(200) cm, erect or decumbent, gen bent; nodes and internodes gen glabrous; internode solid to spongy. **LF**: basal and cauline; sheath gen smooth; ligule short-hairy or membranous, ciliate; blade flat or folded, margins gen cartilaginous. **INFL**: panicle-like, central axis wavy, with reduced branches bearing burs, each consisting of 1–4(8) ± sessile spikelets gen enclosed by an involucre of flattened or cylindric bristle- or spine-like, ± fused bracts, these gen forming an inner and outer set; involucre and enclosed spikelets falling as 1 unit. **SPIKELET**: ± dorsally compressed; glumes strongly unequal, ovate, lower 1-veined, upper ± = florets, 3–9-veined; florets 2, lower floret sterile or staminate, lemma gen 5-veined, palea gen present, upper floret fertile, lemma thick, ± hard, palea ± = lemma; anthers 3. ± 16 spp.: warm temp Am, Afr, s Asia. (Uncertain. Greek: for millet or Latin for a precious stone) [Stieber & Wipff 2003 FNANM 25:529–535] *C. ciliaris* L. is now treated as *Pennisetum ciliare* (L.) Link.

1. Bur with 1 whorl of flattened, fused inner bracts subtended by 1–several whorls of smaller, finer bracts . . . . ***C. echinatus***
1′ Bur with several whorls of flattened spines, these at irregular intervals through body of bur
  2. Sheaths strongly keeled; bracts 45–75, slender, < 1 mm wide; spikelets 5.8–7.8 mm . . . . ***C. longispinus***
  2′ Sheaths compressed, but not strongly keeled; bracts 8–40, broader at base, 1–3 mm; spikelets 3.5–5.9 mm
  . . . . ***C. incertus***

C. echinatus L. SOUTHERN SANDBUR Ann. **ST**: 1–5(10) dm. **LF**: sheath 3–7 cm; ligule ± 1–1.5 mm; blade (4)6–20(35) cm, 2–10 mm wide, upper surface glabrous or hairy. **INFL**: 2–12 cm; outer bracts 10–20, cylindric, gen < 1/2 inner bristles; inner bracts 2–5 mm, flattened, gen erect, pubescent, fused below to form a spheric cup. **SPIKELET**: 2–3(4) per bur, 5–7 mm, lanceolate to ovate, green; lower glume 1–3.5 mm, upper 4–6 mm; lower floret sterile; lower and upper lemmas equal. 2*n*=68. Disturbed places, fields; gen < 700 m. s ScV (Solano Co.), SCo (San Diego), PR, D; native to s US, Mex, C.Am, S.Am. Oct ◆

C. incertus M.A. Curtis. COASTAL SANDBUR, COMMON SANDBUR Ann or short-lived per. **ST**: decumbent, 1–5(10) dm. **LF**: sheath 2.5–7 cm; ligule 0.5–1.5 mm; blade 3–28 cm, (1)2.5–7 mm wide, upper surface glabrous. **INFL**: 2–5(8.5) cm; inner bracts 8–40, fused at least 1/2, forming cup. **SPIKELET**: 3.5–5.9 mm, ovate, glabrous;

lower glume 1–3.5 mm, upper glume 3–5 mm; lower floret staminate, lemma 1–3.5 mm, 5–7-veined; upper lemma 3.5–5 mm. 2*n*=(32)34. Disturbed areas; 10–1500 m. GV, SCo, PR, DMoj (near Daggett), DSon; WA, OR; native to s US, Mex, C.Am, S.Am. [*C. pauciflorus* Benth.; *C. spinifex* Cav.] Jul–Sep ◆

C. longispinus (Hack.) Fernald (p. 1439) MAT SANDBUR Ann. **ST**: 1–9 dm, gen decumbent. **LF**: sheath strongly keeled, 2.5–8 cm; ligule ± 1 mm; blade 4–28 cm, 1.5–5 mm wide, upper surface glabrous. **INFL**: 1.5–8 cm; bracts 45–75, outer < inner, gen cylindric, reflexed; inner 3.5–7 mm, fused > 1/2, forming cup. **SPIKELET**: 5.8–7.8 mm, lanceolate to ovate, green; lower glume 1–3 mm, upper 4.5–6 mm; lower floret gen staminate, lemma 4–6.5 mm, 3–7-veined; upper lemma 4–7.6 mm. 2*n*=34(38). Disturbed areas; < 1500 m. s ScV, SnJV, SCoRO, PR, MP (Lassen Co.), D; widespread in US, Can; native to c&e US. Jul–Sep ◆

CHLORIS

Rosa Cerros-Tlatilpa

Ann, per, cespitose, rhizomatous or stoloniferous. **ST**: decumbent to erect, 1–7 dm. **LF**: ligule membranous or hairy-tufted; blade gen 10–50 cm, 0.2–1.5 cm wide, flat. **INFL**: gen digitate; branches 2–30, sometimes in distinct whorls, each raceme- or spike-like branch with 2 rows of overlapping spikelets on 1 side of axis. **SPIKELET**: laterally compressed; glumes unequal, < florets, 1–3-veined; axis breaking above glumes; lower florets fertile, 1–2, upper florets sterile or staminate, 1–4, < 1/2 lower floret length; fertile floret lemma ovate to lanceolate, back glabrous, midvein hairy, 3-veined, awn 1; palea < lemma, translucent, obscure; anthers 3. **FR**: ± fusiform, 3-angled. ± 50 spp.: warm temp, trop worldwide, esp s hemisphere. (Greek: goddess of fls)

1. Infl branches in 2–4 separate whorls, spreading .... *C. verticillata*
1′ Infl branches digitate, branches 4–30, ± erect
　2. Ann .... *C. virgata*
　2′ Per, gen from stolons
　　3. Lf blades 25–50 cm, 3–9 mm wide; upper florets gen (1)2–4, sterile or staminate; spikelets light brown to brown-orange .... *C. gayana*
　　3′ Lf blades 5–20 cm, 2–3 mm wide; upper florets 1(2), staminate; spikelets dark brown to black .... *C. truncata*

C. gayana Kunth RHODES GRASS Per, gen from stolons. **ST**: 2–5 dm. **LF**: sheath glabrous to scabrous, hairy near collar; ligule hairy; blade 25–50 cm, 3–9 mm wide. **INFL**: digitate; branches 9–30, erect, 8–15 cm. **SPIKELET**: 3–5 mm, light brown to brown-orange at maturity; glumes acute to short-awned, lower 1–3 mm, upper 2–4.5 mm; fertile florets 1–2, 2–4 mm, < 1 mm wide, ovate, obovate, or elliptic, lemma margin gen hairy-tufted near tip, awn 1–7.5 mm; sterile florets (1)2–4, lowest 1.8–3.2 mm, awn 0.8–3 mm. **FR**: 1–1.6 mm. 2*n*=20,30,40. Disturbed areas; < 300 m. GV, SCo, expected elsewhere; AZ, TX, LA, IL, FL; native to Afr. Cult for forage. All yr

C. truncata R. Br. BLACK WINDMILL GRASS Per from stolons. **ST**: 3–5 dm. **LF**: sheath glabrous; ligule membranous, short-ciliate; blade 5–20 cm, 2–3 mm wide, glabrous to occ scabrous. **INFL**: digitate; branches 5–13, ± erect, 5–23 cm. **SPIKELET**: 2–4.5 mm, dark brown to black; lower glume 1.4–2.3 mm, upper 2.8–4.2 mm; lower floret bisexual, 2–4 mm, upper florets 1(2), staminate, lemma awns 3–12+ mm, lower gen > upper. **FR**: dorsally compressed. 2*n*=40. Disturbed areas; weed of cult fields; < 500 m. GV, PR, DMoj; SC; native to Australia. Pls may contain cyanogenic glycosides and be potentially TOXIC to livestock. Apr–Nov

C. verticillata Nutt. (p. 1439) WINDMILL GRASS Per. **ST**: gen 1–4 dm. **LF**: sheath glabrous; ligule hairy; blade < 15 cm, ± 0.3 cm wide. **INFL**: digitate; branches in 2–4 whorls, spreading, 5–15 cm. **SPIKELET**: 2–3.5 mm; glumes lanceolate, lower 2–3 mm, upper 2.8–3.5 mm; fertile floret 1, 2–3.5 mm, 1.5–2 mm wide, lanceolate to elliptic, glabrous to hairy, awn 4–9 mm; sterile floret 1, 1–2.5 mm, awn 3–7 mm. **FR**: ± 1.5 mm. 2*n*=28,40,63. Uncommon. Disturbed areas; < 300 m. n SNF, SnFrB, expected elsewhere; native to s Great Plains. Jul–Oct

C. virgata Sw. (p. 1439) FEATHER FINGER GRASS Ann. **ST**: gen 1–7 dm. **LF**: sheath glabrous to hairy near collar; ligule glabrous to hairy; blade < 30 cm, 1.5 cm wide. **INFL**: digitate; branches 4–20, ± erect, 5–10 cm. **SPIKELET**: 1.5–4.5 mm; glumes acute, lower 1.5–3 mm, 0.2–0.5 mm wide, upper 2.5–4.5 mm, 0.3–0.5 mm wide; fertile floret 1, 2.5–4 mm, 0.5–1.5 mm wide, ovate, obovate, or elliptic, keel hairy near tip, margin gen hairy-tufted, awn 2.5–15 mm; sterile floret 1, 1.5–3 mm, awn 3–10 mm. **FR**: 1.5–2 mm. 2*n*=20,26,30,40. Disturbed areas; < 200 m. GV, SCoRO, SCo, PR, D; to s Great Plains, se US, n Mex; native to warm temp regions worldwide. Apr–Sep

CINNA WOODREED

David M. Brandenburg

Per. **ST**: erect. **LF**: gen cauline; sheath glabrous; blade flat, margin scabrous. **INFL**: panicle-like; branches spreading to ascending. **SPIKELET**: ± sessile to stalked, breaking below glumes, falling as 1 unit; glumes ± equal or lower < upper, lower 1-veined, upper 1–3-veined; floret 1, slightly < or > glumes, bisexual; axis gen prolonged behind palea, short, bristle-like; lemma faintly 3–5-veined, short-awned just below acute tip, or awnless; palea ± < lemma; stamens 1–2. 4 spp.: temp N.Am, n S.Am, Eurasia. (Greek: a grass)

1. Spikelet 4–5.5 mm; stamens 2, anthers gen 1–2.5 mm; lemma faintly 5-veined .... *C. bolanderi*
1′ Spikelet 2.5–4 mm; stamen 1, anthers < 1 mm; lemma gen 3-veined .... *C. latifolia*

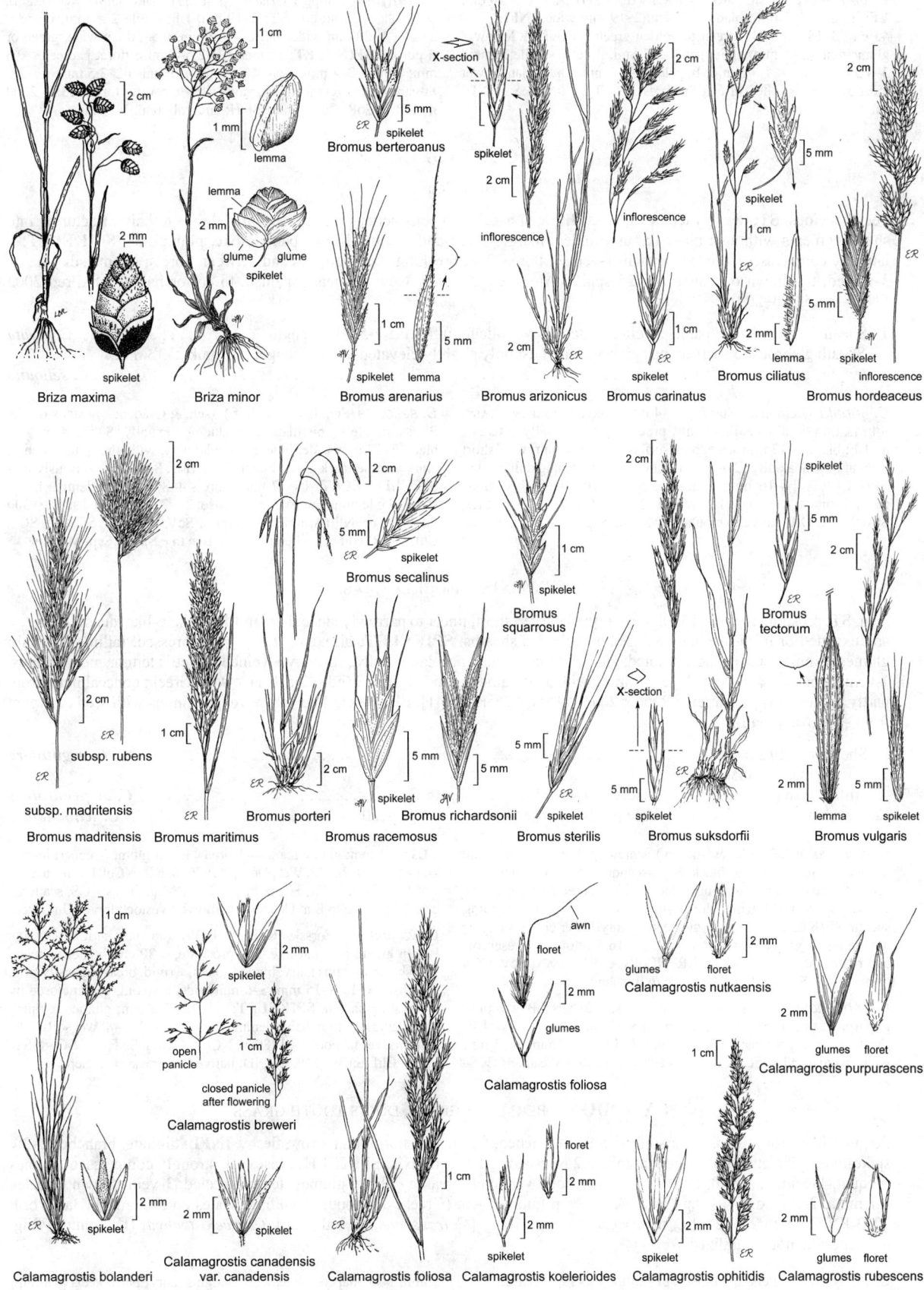

Briza maxima

Briza minor

Bromus berteroanus

Bromus arenarius

Bromus arizonicus

Bromus carinatus

Bromus ciliatus

Bromus hordeaceus

subsp. madritensis
subsp. rubens
Bromus madritensis

Bromus maritimus

Bromus secalinus

Bromus porteri

Bromus richardsonii

Bromus racemosus

Bromus squarrosus

Bromus sterilis

Bromus suksdorfii

Bromus tectorum

Bromus vulgaris

Calamagrostis bolanderi

Calamagrostis breweri

Calamagrostis canadensis var. canadensis

Calamagrostis foliosa

Calamagrostis foliosa

Calamagrostis koelerioides

Calamagrostis nutkaensis

Calamagrostis purpurascens

Calamagrostis ophitidis

Calamagrostis rubescens

C. bolanderi Scribn. BOLANDER'S WOODREED **ST**: 8.5–20 dm. **LF**: ligule 3.5–7 mm; blade < 40 cm, 2–19 mm wide. **INFL**: 7.5–43 cm, 3–18 cm wide, green to golden-green. **SPIKELET**: lower glume gen 3.5–5 mm, upper gen 4–5.5 mm; floret ± sessile; lemma 3–4.5 mm, awn < 1.5 mm or 0; palea 3–3.5 mm. Streambanks, wet meadows, moist sites in conifer forest; 1850–2400 m. c&s SNH. Jul–Sep ★

C. latifolia (Göpp.) Griseb. (p. 1439) DROOPING WOODREED, SLENDER WOODREED **ST**: 2–19 dm. **LF**: ligule 2–8 mm; blade < 28 cm, 1–20 mm wide. **INFL**: 3–46 cm, 0.5–20 cm wide, green or ± purple. **SPIKELET**: glumes gen 2.5–4 mm; floret pedicel < 0.5 mm; lemma 2–4 mm, awn < 2.5 mm or 0; palea 2–3.5 mm. 2*n*=28. Streambanks, wet meadows, moist sites in conifer forest; 1350–2800 m. KR, NCoR, CaRH, n&c SNH; circumboreal. Jul–Aug

CORTADERIA

H.E. Connor

Per; dioecious. **ST**: densely clumped, erect. **LF**: gen basal with conspicuous midrib; sheath glabrous to hairy, fracturing into short segments with age; blade flat and folded, margin sharp and cutting. **INFL**: panicle-like, plume-like. **SPIKELET**: ± laterally compressed; glumes unequal, 1-veined; florets 2–8, breaking above glumes and between florets; lemma silky-hairy, 3-veined, tip drawn out, awn-like. 25 spp.: S.Am, New Zealand, New Guinea. (Spanish, to cut or harvest) [Allred 2003 FNANM 25:298–299]

1. Sheath densely hairy; panicle well-elevated above foliage; infl stalk 4–5 × infl length . . . . **C. jubata**
1′ Sheath glabrous to sparsely hairy; panicle at most only slightly elevated above foliage; infl stalk 2–3 × infl length . . . . **C. selloana**

C. jubata (Lemoine) Stapf (p. 1439) PURPLE PAMPAS GRASS, JUBATA GRASS Pls pistillate only, producing fr asexually. **ST**: 2–7 m. **LF**: blade 2–12 mm wide, both surfaces green, with row of short hairs at base adaxially. **INFL**: 3–10 dm, violet becoming dull brown. **SPIKELET**: 14–16 mm; lemma hairy. 2*n*=108. Disturbed sites, many habitats, esp coastal; < 800 m. NCo, NCoRO, CCo, SnFrB, SCo, WTR; OR; native to montane w S.Am. Sep–Feb ◆

C. selloana (Schult. & Schult. f.) Asch. & Graebn. PAMPAS GRASS Pls staminate or pistillate, reproducing sexually. **ST**: 2–4 m. **LF**: blade 3–12 mm wide, blue-green adaxially, with small tuft of short hairs at base, dark green abaxially. **INFL**: 3–13 dm, variously colored. **SPIKELET**: 15–17 mm; florets 4–8; pistillate lemma hairy, staminate lemma glabrous exc at base. 2*n*=72. Disturbed sites; < 300 m. NCo, n SNF (American River), n ScV, CW (exc SCoRI), SCo, ChI, WTR, cult elsewhere; s US; native to e S.Am. Sep–Mar ❖

CRYPSIS PRICKLE GRASS

Ann. **ST**: prostrate to erect. **LF**: ligule hairy; blade gen short, linear to narrowly lanceolate. **INFL**: panicle-like, dense, cylindric and exserted or head-like and enclosed by enlarged sheaths. **SPIKELET**: bisexual, strongly compressed, falling as 1 unit; glumes, lemma acute or short-pointed; glumes ≤ floret, gen lanceolate, keeled, strongly 1-veined; floret 1; lemma membranous, 1-veined; palea gen 2-veined, gen splitting with age; stamens gen 3. 8 spp.: Medit Eur, Asia, c Afr. (Greek: concealment, from partly hidden infl) [Hammel & Reeder 2003 FNANM 25:139–141]. **FR**: utricle-like, when wet gelatinous with seed emergent from split ovary wall.

1. Sheath margins hairy . . . . **C. vaginiflora**
1′ Sheath margins glabrous
  2. Infl length gen > 5 × width, exserted from subtending sheath . . . . **C. alopecuroides**
  2′ Infl length gen < 5 × width, base partly enclosed by subtending sheath . . . . **C. schoenoides**

C. alopecuroides (Piller & Mitterp.) Schrad. (p. 1439) W PRICKLE GRASS Pl gen purple to black. **ST**: ascending, 5–75 cm; branches few. **LF**: sheath margin glabrous; blade 5–12 cm, green or ± glaucous. **INFL**: 15–65 mm, 4–6 mm wide, cylindric, exserted from subtending sheath. **SPIKELET**: 2–3 mm; glume keel hairy, lower glume < upper; lemma gen > glumes; palea 2-veined. 2*n*=16. Bottomland, reservoir and river margins; < 1700 m. KR, NCoRI, CaRH, n SN, ScV, CCo (Marin Co.), SNE; w US, PA; native to Eur. Jun–Sep

C. schoenoides (L.) Lam. SWAMP PRICKLE GRASS Pl gen pink to purple, gen mat-like. **ST**: decumbent, 5–75 cm; branches few. **LF**: sheath margin glabrous; blade 2–10 cm. **INFL**: 3–75 mm, 5–15 mm wide, ovoid to cylindric, partly enclosed by subtending sheath. **SPIKE-LET**: ± 3 mm; glume margin glabrous, lower glume < upper; lemma gen > glumes. 2*n*=32. Wet places; < 1970 m. KR, NCoRI, CaR, n SNF, GV, CW, SCo, WTR, SnGb, PR, MP, w DMoj; OR, sw US, scattered c&e US; native to Eur. Used for wildfowl, livestock forage. Jun–Oct

C. vaginiflora (Forssk.) Opiz (p. 1439) MODEST PRICKLE GRASS Pl gen green, gen mat-like. **ST**: prostrate, < 30 cm; branches many. **LF**: sheath margins hairy; blade 1–5 cm, ± rigid, breaking easily from sheaths. **INFL**: 3–15 mm, 3–6 mm wide, ± ovoid, gen enclosed by subtending sheath. **SPIKELET**: gen 2.5–3.2 mm; glumes ± equal, lower glume margin hairy; lemma ± = glumes. 2*n*=48. Wet soils, lake margins, vernal pools; < 1280 m. NCoR, CaR, n SNF, GV, SCoR, SW (exc n ChI, SnBr), MP; OR, ID; native to Eurasia. Jun–Sep

CYNODON BERMUDA GRASS, DOG'S-TOOTH GRASS

Per, mat-like, from rhizomes or stolons. **ST**: ± branched. **LF**: blade short, flat, narrow, fleshy. **INFL**: digitate; branches 2–20, spike-like, spikelets sessile, overlapping, in 2 rows along 1 side of axis. **SPIKELET**: bisexual, strongly compressed; glumes ± equal, 1-veined, awn 0; floret 1(2), upper floret vestigial, breaking above glumes; lemma keeled, 3-veined, awn 0; palea = lemma, 2-veined. 8–10 spp.: trop, warm temp Eurasia, Afr. (Greek: dog tooth, from hard scales on rhizomes) [Barkworth 2003 FNANM 25:235–241] *C. transvaalensis* Burtt Davy [*C. transvalensis*, orth. var.], *C. plectostachyus* (K. Schum.) Pilg. occ reported, not naturalized.

C. dactylon (L.) Pers. (p. 1439) BERMUDA GRASS Per from rhizomes or stolons. **ST**: gen erect, 1–4 dm. **LF**: ligule white-hairy; blade < 6 cm, glabrous or upper surface hairy. **INFL**: branches gen 4–7, 2.5–5 cm. **SPIKELET**: ± 2 mm; glumes ± 1.5 mm, gen ± purple; lemma ± 2 mm, boat-shaped, acute, keel, margins hairy; palea keels glabrous. 2*n*=36. Disturbed sites; < 1600 m. CA (exc MP); most of US, Can, warm temp, trop; native to Afr. TOXIC; important pollen source in hay fever; may produce contact dermatitis. Cult for lawns, forage. Jun–Aug ❖

CYNOSURUS DOGTAIL GRASS

Ann, per. **ST**: gen 1–8 dm, erect. **LF**: gen basal; ligule membranous, truncate, rough-edged, or ciliate; blade 3–15 cm, 2.5–14 mm wide, flat, glabrous to pubescent. **INFL**: terminal, spike- or panicle-like, cylindric, ovoid to head-like, dense; spikelets laterally compressed, gen paired, subsessile to short-stalked, of 2 kinds, 1 sterile, 1 fertile. **FERTILE SPIKELET**: bisexual; glumes ± equal, lanceolate, 1-veined, keeled, awn gen 0; florets 1–5; axis extended beyond uppermost floret, breaking above glumes and between florets; lemma back rounded, faintly 5-veined, awned [or not]; palea ± = lemma, 2-lobed. **STERILE SPIKELET**: sometimes obscuring fertile spikelet; glume narrow, linear; florets 6–18, axis not breaking apart; lemma narrow, lance-linear, rigid, 1-veined, awned or not. 2*n*=14. 8 spp.: Eur, w Asia, n Afr. (Greek: dog + tail, from shape of infl) [Long 2007 FNANM 24:685–687]

1. Per; infl cylindric; fertile lemma awn gen 0.5–1 mm . . . . *C. cristatus*
1′ Ann; infl ± head-like, occ interrupted; fertile lemma awn 3–20 mm . . . . *C. echinatus*

C. cristatus L. (p. 1439) CRESTED DOGTAIL GRASS Cespitose, ± puberulent. **ST**: 1.5–8 dm; base sometimes abruptly bent. **LF**: ligule 0.5–2.5 mm, truncate; blade 0.5–2 mm wide. **INFL**: 3–14 cm, < 1 cm wide; central axis ± zigzag, flat. **FERTILE SPIKELET**: glume 3–5 mm, acute; florets 2–5. **STERILE SPIKELET**: glume ± flat; florets gen 6–11; lemma awn < 1 mm. 2*n*=14. Fields, disturbed places; < 300 m. NCo, CCo, SCo; to WA, MT; also ne N.Am; native to Eur. Sometimes cult for forage. Jun–Aug

C. echinatus L. (p. 1439) BRISTLY DOGTAIL GRASS Tufted, glabrous. **ST**: 1–7 dm. **LF**: ligules 2.5–5 mm, obtuse, entire; blade 2–14 mm wide. **INFL**: gen 1–4 cm, ± 1-sided; central axis linear, not flat. **FERTILE SPIKELET**: glume 6–12 mm, awned; florets 1–5. **STERILE SPIKELET**: glume keeled, ciliate, long-pointed; florets 6–18; lemma awn < 8 mm. 2*n*=14,16. Open, disturbed sites; gen < 1000 m. NW, SNF, n&c SNH, ScV, CW, SW; to WA, e N.Am; native to s Eur. May–Jul ❖

DACTYLIS ORCHARD GRASS

Dieter H. Wilken

1 sp. (Greek: finger, from finger-like infl branches) [Allred 2007 FNANM 24:482–483]

D. glomerata L. (p. 1439) Per, cespitose, from short rhizomes. **ST**: (1)2–5, 3–20+ dm. **LF**: basal and cauline; sheath closed, keeled; auricles 0; ligule 4–9 mm, ± translucent, fringed to toothed at obtuse tip, glabrous to minutely soft-hairy; blade 3–6 mm wide, slightly scabrous. **INFL**: panicle-like, 4–20 cm; lower branches spike-like, spreading, upper stiffly erect to spreading. **SPIKELET**: crowded, gen on 1 side of branch tips, short-stalked to subsessile, laterally compressed; florets 2–6, gen bisexual (upper sometimes staminate), breaking above glumes and between florets; glumes, lemmas short-awned at tip, minutely ciliate to short-hairy on back; glumes 3–6 mm, subequal, 1–3-veined; lemma 4–7 mm, margin translucent, 5-veined; palea slightly < lemma, 2-forked at tip. 2*n*=14,21,27–31,42. Disturbed, often moist sites; < 2410 m. CA-FP, GB; to AK, Can, widespread in US; native to Eurasia. Cult for forage, hay. May–Aug ❖

DACTYLOCTENIUM

Ann, per; tufted, stoloniferous, or rhizomatous **ST**: erect or decumbent; 5–115 cm, often rooting at lower nodes. **LF**: ligule membranous; blade flat or inrolled, 4–25 cm, 2–8 mm wide. **INFL**: terminal, panicle-like, branches 2–11, finger-like, each with 2 rows of overlapping, sessile spikelets along lower side; axis extending beyond spikelets. **SPIKELET**: laterally compressed; glumes unequal, wide, 1-veined, keeled, lower persistent, acute, upper deciduous, short-awned, awn gen curved; axis breaking apart above glumes, florets falling as a unit; bisexual florets 3–7; lemma membranous, keeled, 3-veined, lateral veins gen faint, tip acuminate or short-awned, awn gen curved; palea ± = lemma. **FR**: utricle-like, seed falling free from outer wall. 10–13 spp.: Afr, Australia. (Greek: finger + a small comb, from spikelet arrangement on branches) [Hatch 2003 FNANM 25:112–114]

D. aegyptium (L.) Willd. (p. 1443) DURBAN CROWFOOT Ann, stoloniferous per. **ST**: 10–35(100) cm, rooting at lower nodes. **LF**: sheath keeled, with bulbous-based hairs; ligule 0.5–1 mm, membranous, ciliate; blade 5–20 cm, 2–8 mm wide, margin ciliate, hairs bulbous-based. **INFL**: branches gen 2–5, 1–6 cm. **SPIKELET**: ± 4 mm, ± 3 mm wide; glumes 1.5–2 mm, lower ovate, acute, upper elliptic, obtuse, awned, awn to 2.5 mm; florets 3; lemma ± 3 mm, ovate, midvein extended as curved awn to 1 mm; palea ± = lemma. 2*n*=20,36,40,45,48. Disturbed places; < 300 m. SnJV (Kern Co.), SCo, PR, DSon; s&e US, worldwide in warmer areas; native to Afr. Jul–Nov

DANTHONIA OAT GRASS

James P. Smith, Jr. & Kelly W. Allred

Per, cespitose. **ST**: erect. **LF**: gen basal and cauline; sheaths < internodes; ligule short, densely ciliate; blade narrow, flat to folded. **INFL**: gen raceme-like (occ panicle-like or spikelet 1). **SPIKELET**: ± laterally compressed; glumes ± equal, > florets, papery, 1–7-veined; florets 3–8, breaking above glumes and between florets; callus short-hairy; lemma rounded, (5)7–11-veined, tip 2-toothed, awn 0 or gen awned on back below teeth, awn gen bent, flat, coiled below bend, straight, ± cylindric above bend; palea = lemma; anthers 3. **FR**: elliptic. 20 spp.: warm temp, trop, Am, Eur, n Afr. (É. Danthoine, French botanist, agrostologist, 1739–1794) [Darbyshire 2003 FNANM 25:301–306] *D. pilosa* R. Br. now treated as *Rytidosperma*. Variation, esp in *D. californica*, and *D. unispicata*, needs study. *D. purpurea* L. f. [*Karroochloa purpurea* (L. f.) Conert & Türpe], grown at the Botanical Garden at Berkeley, but not naturalized.

1. Lemma awnless . . . . [*D. decumbens*]
1′ Lemma awned
  2. Upper lf blades abruptly spreading to reflexed; lower infl branches flexible, ± divergent to reflexed . . . . *D. californica*

2′ Upper lf blades ascending; lower infl branches or stalks stiff, erect

  3. Spikelets gen 5–10 per infl; lemma (exc awns) 3–6 mm; lvs gen glabrous, not papillate
    . . . . ***D. intermedia*** subsp. ***intermedia***

  3′ Spikelets 1(2–3) per infl; lemma (exc awns) 5.5–11 mm; lvs gen densely hairy, hairs papillate at base or
    lf surface papillate only. . . . . ***D. unispicata***

D. californica Bol. (p. 1443)   CALIFORNIA OAT GRASS   **ST:** (1)3–13 dm. **LF:** basal and cauline; sheath glabrous to densely hairy, base papillate; upper blades 8–30 cm, gen 2–5 mm wide, glabrous or hairy, flat, abruptly spreading to reflexed. **INFL:** gen raceme-like, 2–6 cm; lower branches flexible, divergent or reflexed at maturity. **SPIKELET:** gen 3–6, gen 14–25 mm; stalk gen spreading, puberulent; glumes 10–23 mm; florets 3–8; lemma 5–15 mm, base, lower margin hairy, back ± glabrous, teeth 2–5 mm, awn 4–12 mm. 2*n*=36. Gen moist meadows, open woodland; 45–2300 m. NW, CaR, SN, CW, SnBr, s PR, MP; to w US, w Can, S.Am (Chile). [*D. c.* var. *americana* (Scribn.) Hitchc.] Apr–Aug

D. intermedia Vasey subsp. ***intermedia*** (p. 1443)   INTERMEDIATE OAT GRASS   **ST:** 1–5 dm. **LF:** gen basal, gen glabrous exc near ligule; upper blades 5–10 cm, ± inrolled, ascending. **INFL:** 2–5 cm, narrow, compact; spikelets 4–10, lower branches or stalk erect to appressed, glabrous. **SPIKELET:** glumes 9–14 mm; florets 3–6; lemma 3–6 mm, margins, base hairy, teeth 1–2 mm, awn 5–9 mm. 2*n*=18,36. Meadows, bogs, damp banks, moist forest; 1460–3450 m. KR, NCoRH, CaRH, SNH, MP; to AK, e Can, w US. Another subsp. occurs in Russia. Jul–Aug

D. unispicata (Thurb.) Vasey   ONE-SPIKE OAT GRASS   **ST:** 1–3 dm. **LF:** basal and cauline; sheath densely hairy; blade sparsely to densely hairy, hairs papillate at base or lf surface papillate only; upper blades 3–8 cm, 1–3 mm wide, flat to ± inrolled, ascending. **SPIKELET:** gen 1(2–3); stalk erect, puberulent; glumes 9–25 mm; florets 3–6; lemma 5.5–11 mm, margin hairy, teeth 1–5 mm, awn 4–9 mm. 2*n*=36. Dry meadows, rocky slopes, open sites in conifer forest; 400–3200 m. KR, NCoR, CaRH, SNH, SnJV, WTR, SnBr, SnJt, MP; to w Can, w US. May–Aug

DASYOCHLOA   FLUFF GRASS

Jesús Valdés-Reyna

1 sp. (Greek: thick with hair and grass) [Valdés-Reyna 2003 FNANM 25:45–48]

D. pulchella (Kunth) Rydb. (p. 1443)   Per, stoloniferous or mat-forming. **ST:** gen 4–10 cm, scabrous or puberulent, initially erect, becoming bent, rooting at base of infl. **LF:** not basal; sheath with a tuft of hairs at throat, < 2 mm; ligule of hairs 3–5 mm; blade 2–6 cm, folded. **INFL:** panicle 1–2.5 cm, 1–1.5 cm wide, terminal, short, dense, branches spike-like, bearing 2–4 subsessile to short-pedicelled spikelets, subtended by lfy bracts, densely white-pubescent, light-green or purple-tinged. **SPIKELET:** gen 6–9 mm, laterally compressed, florets (4)6–10; axis breaking above glumes; glumes 6–9 mm, subequal to adjacent lemma, glabrous, 1-veined, short-awned to mucronate; florets bisexual; lemma 3–5.5 mm, densely long-hairy below and on margins, thinly membranous, 3-veined, 2-lobed to ± 1/2, obtuse, midvein extending into a straight awn; palea 2–3.5 mm; anthers 3. 2*n*=16. Sandy to rocky slopes, flats, desert shrubland, woodland; 300–1700 m. D; to CO, TX, c Mex. [*Erioneuron p.* (Kunth) Tateoka] Feb–May

DESCHAMPSIA   HAIR GRASS

Robert E. Preston & Dieter H. Wilken

Ann, per. **ST:** erect, 1 to densely clumped. **LF:** gen basal, tufted; ligule narrow, decurrent to sheath, glabrous to minutely hairy; blade flat to inrolled. **INFL:** panicle- to spike-like, open to narrow. **SPIKELET:** glumes, lemmas shiny; glumes ± equal, > lower floret; axis prolonged ≥ 0.75 mm beyond upper floret, conspicuously bristly (sometimes with vestigial floret at tip); florets 1–3, bisexual, breaking above glumes and between florets; callus soft-hairy; lemma rounded, 2–4-toothed at truncate tip, faintly 3–7-veined, awned at or below middle, awn straight to bent; palea ± = lemma. 30–40 spp.: temp Am, Eurasia, New Zealand, Antarctica. (Louis Auguste Deschamps, French naturalist, 1765–1842) *D. atropurpurea* is now treated as *Vahlodea atropurpurea*.

1. Ann; sts 1 or loosely clumped; basal lvs not tufted. . . . . ***D. danthonioides***
1′ Per; sts loosely to densely clumped; basal lvs tufted
  2. Infl < 1 cm wide; basal lf blades ± 1 mm wide . . . . ***D. elongata***
  2′ Infl > 1 cm wide; basal lf blades 1–4 mm wide. . . . . ***D. cespitosa***
    3. Infl compact, lower branches ascending to erect . . . . subsp. ***holciformis***
    3′ Infl open, lower branches spreading to drooping
      4. Pls gen glaucous; glumes 4.4–7.5 mm; awns gen exceeding lemmas . . . . subsp. ***beringensis***
      4′ Pls not glaucous; glumes 2–6 mm; awns gen exceeded by or exceeding lemmas. . . . . subsp. ***cespitosa***

D. cespitosa (L.) P. Beauv.   TUFTED HAIR GRASS   Per, densely to loosely clumped. **ST:** 2–10 dm. **LF:** glabrous to scabrous; ligule 3–8 mm, acute to obtuse, entire to toothed at tip; blade gen 8–20 cm, 1–4 mm wide, flat to inrolled. **INFL:** narrow to open; lower branches erect to drooping. **SPIKELET:** glumes, lemma tips ± purple; glumes ± equal, lanceolate, acute, lower 1-veined, upper 3-veined; florets gen 2; callus hairs gen < 1/3 lemma; lemma gen 4-toothed at tip, faintly 5-veined, awned at or below middle, awn straight to slightly bent. 2*n*=26–28. Subspp. intergrade, need study.

  subsp. ***beringensis*** (Hultén) W.E. Lawr.   Loosely clumped, gen glaucous. **INFL:** open; lower branches spreading to drooping. **SPIKELET:** 1 to clustered on exposed branchlets; glumes 4.4–7.5 mm; lemma 3–5(7) mm, awned below middle, awn 3.3–6.3 mm, gen exceeding lemma. Coastal marshes and meadows; < 800 m. NCo; AK; e Asia (Kamchatka Peninsula). Jul–Aug

  subsp. ***cespitosa*** (p. 1443)   Densely clumped, not glaucous. **INFL:** open (compact in some high elevation pls); lower branches spreading to drooping. **SPIKELET:** 1 to clustered on exposed branchlets; glumes 2–6 mm; lemma 2–4 mm, awn 1–8 mm, gen attached near base of lemma, gen not exceeding glumes (exceeding glumes in some high elevation pls). Meadows, streambanks, coastal marshes, forest, alpine; < 3820 m. NW, CaR, SN, CCo, SnFrB, TR, Wrn, n SNE, W&I; AK, Can, e US; Eurasia. Jul–Aug

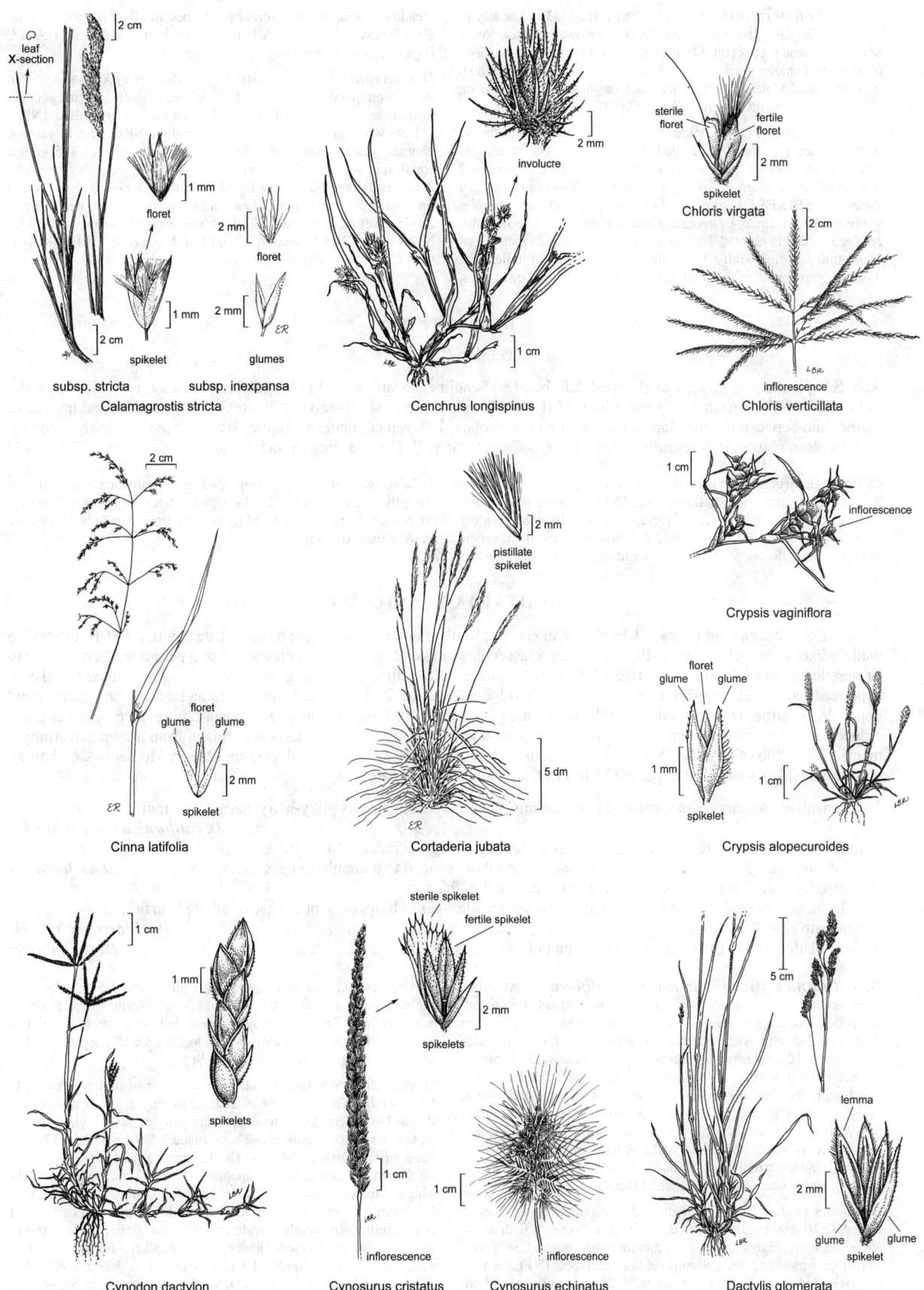

leaf
X-section

2 cm

floret
1 mm

floret
2 mm

spikelet
1 mm

glumes
2 mm

subsp. stricta   subsp. inexpansa
Calamagrostis stricta

involucre
2 mm

1 cm

Cenchrus longispinus

sterile
floret   fertile
floret

spikelet
2 mm

Chloris virgata

2 cm

inflorescence

Chloris verticillata

2 cm

floret
glume   glume

spikelet
2 mm

Cinna latifolia

pistillate
spikelet
2 mm

5 dm

Cortaderia jubata

1 cm

inflorescence

Crypsis vaginiflora

floret
glume   glume

1 mm

spikelet   1 cm

Crypsis alopecuroides

1 cm

1 mm

spikelets

Cynodon dactylon

sterile spikelet
fertile spikelet

spikelets
2 mm

1 cm

inflorescence

Cynosurus cristatus

1 cm

inflorescence

Cynosurus echinatus

5 cm

lemma

2 mm

glume   glume

spikelet

Dactylis glomerata

subsp. ***holciformis*** (J. Presl) W.E. Lawr. (p. 1443) Loosely to densely clumped, glaucous or not. **INFL**: narrow, compact; lower branches ascending to erect. **SPIKELET**: densely clustered on short, obscure branchlets; glumes 4.6–5.8 mm; lemma 3.8–4.5 mm, awn 2–3 mm, attached near middle of lemma. Coastal marshes, meadows; < 850 m. NCo, NCoRO, deltaic ScV, CCo; BC. May–Jul

D. danthonioides (Trin.) Munro (p. 1443) ANNUAL HAIR GRASS Ann. **ST**: gen 1 to loosely clumped, 1.2–6 dm. **LF**: gen basal, glabrous; ligule 2–4 mm, acute to acuminate, entire; blade 1–9 cm, 1–2 mm wide, gen inrolled. **INFL**: narrow to open; lower branches gen ascending. **SPIKELET**: glume and lemma tips sometimes ± purple; glumes 4–9 mm, equal, lanceolate, acute to acuminate, 3-veined; florets gen 2; callus hairs < 1/2 lemma length; lemma 2–3 mm, gen 2–3-toothed at tip, faintly 1–3-veined, awned below middle, awn 4–9 mm, gen bent at mid-length. 2*n*=26. Moist to drying, open sites,

meadows, streambanks, vernal pools, occ in alkali soil; < 2750 m. CA-FP (exc Teh, n ChI), MP, DMoj (uncommon); AK, ne US, AZ, Baja CA, S.Am. Mar–Aug

D. elongata (Hook.) Munro (p. 1443) SLENDER HAIR GRASS Per, densely clumped. **ST**: 1–7 dm. **LF**: glabrous; ligule 2–8 mm, acute to acuminate, entire; blade 4–8 cm, ± 1 mm wide, flat to inrolled. **INFL**: < 1 cm wide; branches spike-like, ascending. **SPIKELET**: glumes, lemmas green to tan, tips sometimes ± purple; glumes 3–5 mm, ± equal, narrow-lanceolate, acute to acuminate, 3-veined; florets gen 2; callus hairs ± 1/2 lemma length; lemma 2–3 mm, 2–4-toothed at tip, faintly 5-veined, awned near middle, awn 1–5 mm, gen straight. 2*n*=26. Wet sites, meadows, lakeshores, shaded slopes; < 3100 m. NW, CaRH, SN, CW (exc SCoRI), TR, PR, GB (exc W&I); AK, WY, Baja CA, S.Am. May–Sep

DESMAZERIA

Gordon C. Tucker

Ann. **ST**: prostrate to erect, 1 to clumped. **LF**: basal and cauline; sheath open; ligule membranous, toothed at tip; blade flat to inrolled. **INFL**: raceme- or panicle-like. **SPIKELET**: subsessile to short-stalked; florets 5–25, bisexual, breaking above glumes and between florets; glumes ± equal, lower lanceolate, 1–3-veined, upper ± elliptic, 3-veined; lemma firmly membranous, acute to obtuse, 1–5-veined; palea < lemma. 6–7 spp.: Eur. (J.B. Desmazières, French botanist, horticulturist, 1786–1862)

D. rigida (L.) Tutin **ST**: gen erect, 10–60 cm, glabrous. **LF**: blade 2–25 cm, 1–3 mm wide, flat, glabrous. **INFL**: panicle-like, ovoid, 4–8(12) cm; branches, pedicels rigid, 3-angled; branches widely spreading. **SPIKELET**: 4–10 mm, 2–5 borne on lower branches, lowest in axil; florets 5–12; glumes, lemmas gen glabrous; glumes 1–2.5 mm; lemmas spreading, 2–3 mm, translucent at acute to abruptly soft-pointed tip. 2*n*=14. Open, often sandy sites; < 700 m. s NCo, CCo, SnFrB, SCo, s ChI (Santa Catalina Island); N.Am; native to s&w Eur. Apr–Sep

DIGITARIA CRAB GRASS

Ann, per. **ST**: decumbent to erect. **LF**: basal and cauline; ligule membranous, ciliate or not; blade gen flat. **INFL**: digitate to panicle-like; 1° branches ± spike-like, spreading to ascending; spikelets gen many per branch, 2 or 3 per node, short-stalked to subsessile, on one side of axis. **SPIKELET**: dorsally compressed, falling as 1 unit; glumes unequal, upper glume ≤ spikelet, appressed-hairy, clearly 3–5-veined, veins minutely ridge-like; florets 2, lower floret sterile, lemma texture like upper glume, upper floret fertile, lemma ± thin, flexible, back facing away from infl axis, margin flat, tip gen obtuse, awn 0; lower palea reduced or 0, the upper ± = lemma; anthers 3. ± 200 spp.: warm temp, trop, worldwide. (Latin: finger, from infl branch arrangement) [Wipff 2003 FNANM 25:358–383] *D. bicornis* (Lam.) Roem. & Schult., collected in 1926 in Monterey Co., and *D. eriantha* Steud., collected in Imperial Co. in 1939, are presumed extirpated.

1. Infl panicle-like, branches appressed to ascending; spikelets conspicuously silky-hairy, hairs 1.5–5 mm
    . . . . ***D. californica*** var. ***californica***
1′ Infl branches ± digitate, spreading; spikelets glabrous or marginal hairs < 1 mm
  2. Spikelets gen 3 per node on a branch; lower glume 0 or reduced to a membranous scale or rim . . . . ***D. ischaemum***
  2′ Spikelets gen 2 per node on a branch; lower glume 0.2–0.8 mm
    3. Lf blade gen scabrous on both surfaces, exc few swollen-based hairs present at base of adaxial surface;
        sheath rounded . . . . ***D. ciliaris*** var. ***ciliaris***
    3′ Lf blade gen with swollen-based hairs on both surfaces; sheath keeled . . . . ***D. sanguinalis***

D. californica (Benth.) Henrard var. ***californica*** ARIZONA COTTONTOP Cespitose per. **ST**: gen erect, 40–100 cm. **LF**: sheath glabrous or long-hairy; ligule 1–6 mm, entire or ragged; blade gen 2–12 cm, 2–5 mm wide, glabrous to tomentose. **INFL**: panicle-like with 4–10 appressed to ascending 1° branches (2° branches occ present); spikelets paired, unequally stalked. **SPIKELET**: 3–4 mm (exc hairs), lanceolate; lower glume 0.4–0.6 mm, translucent, veinless; upper glume 2.5–5.1 mm, 3-veined; lemma 2.5–5 mm, 3–5(7)-veined; upper glume, lower lemma densely hairy, hairs 1.5–5 mm, white to purple. 2*n*=36,54,70,72. Rocky hillsides; < 1500 m. DMtns (San Bernardino Co.), DSon (San Diego Co.); sw US, Mex to S.Am. [*Trichachne c.* (Benth.) Hitchc.] Oct–Nov ★

D. ciliaris (Retz.) Koeler var. ***ciliaris*** SOUTHERN CRAB GRASS Ann. **ST**: 10–100 cm, erect and decumbent. **LF**: sheath with swollen-based hairs; ligule 2–3.5 mm, membranous; blade 1.5–25 cm, 3–10 mm wide, flat, gen scabrous on both surfaces. **INFL**: ± digitate (whorled) with 2–10 1° branches, 6–22 cm; spikelets paired, unequally stalked. **SPIKELET**: gen 2.5–4 mm, elliptical, acute;

lower glume 0.2–0.8 mm, ovate; upper glume gen 1.5–2.7 mm, lanceolate, 3-veined; lemma 2.5–4 mm, 3- or 7-veined, acute; upper palea leathery. 2*n*=54. Weedy in open, disturbed areas; gen < 150 m. SCo, PR (San Diego Co.); native to the c&e US, esp the se. [*D. sanguinalis* var. *c.* (Retz.) Parl.] Jun–Sep

D. ischaemum (Schreb.) Muhl. SMOOTH CRAB GRASS Ann. **ST**: 1.5–5(7) dm; decumbent or ascending, rooting at lower nodes. **LF**: sheath 1–12.5 cm, glabrous or sparsely hairy at base of blade or occ on lower surface; ligule ± 1–2 mm; blade 1.5–12 cm, 2.5–5(7) mm wide, upper surface glabrous. **INFL**: terminal and axillary (gen at least partially concealed), branches 2–8, 1.5–7 cm, axes narrowly winged; spikelets gen 3 per node, stalk 0.5–3 mm. **SPIKELET**: ± 2–2.5 mm, ± 1 mm wide, elliptic, purple; lower glume 0 or reduced to a membranous, veinless scale or rim, upper glume 3/4 spikelet, membranous, 3-veined; lower lemma = spikelet length, 7-veined, puberulent, acute. 2*n*=36. Disturbed areas; < 300 m. NCoRO, NCoRH, s SN, GV, CCo, SCo, D; to e US; native to Eur. Sep–Nov

D. sanguinalis (L.) Scop. (p. 1443) HAIRY CRAB GRASS Ann. **ST**: 2–7 dm; gen decumbent and rooting at lower nodes, often ± purple. **LF**: sheath 2.5–15 cm, keeled, hairy; ligule membranous, 1–3 mm; blade (2)3–10(14) cm, 3–8(14) mm wide, gen with swollen-based hairs on both surfaces. **INFL**: ± digitate, 1° branches 4–10(13), 3–9 cm, axes narrowly winged; spikelets gen 2 per node. **SPIKE-** LET: ± 2.5–3 mm, lanceolate to ovate, purple in fr; lower glume < 0.5 mm, veinless; upper glume ± 1/3–1/2 spikelet, 3-veined; lemma of lower floret 7-veined, acuminate to acute; upper lemma 2.5–3.3 mm, leathery, often brown at maturity; upper palea texture same as upper lemma. 2*n*=36,28,34,54. Disturbed areas; < 1250 m. CA (exc SNH, GB); to Can, widespread in US; native to Eur. Jun–Sep

DISSANTHELIUM

Nancy F. Refulio-Rodriguez & James P. Smith, Jr.

Ann, per, cespitose, sometimes from rhizomes. **ST**: erect to decumbent. **LF**: blade folded, inrolled, or flat. **INFL**: panicle-like, ± narrow; branches ± clustered, ascending to erect, ± dense. **SPIKELET**: laterally compressed; glumes ± equal, gen > lower floret, ± membranous, subacuminate, awn 0, lower glume ± 1-veined, upper 3-veined; axis breaking apart above glumes and between florets; florets 2(3), lower gen bisexual, upper pistillate, sometimes both bisexual; lemma membranous, keeled, 3-veined; palea < and enclosed by lemma. ± 20 spp.: sw CA (s ChI), Mex, w S.Am. (Greek: double small fl, from 2 small florets) Phylogenetically nested in, best treated in, *Poa* (Refulio-Rodriguez 2007 Ph.D. Dissertation Claremont Graduate Univ).

D. californicum (Nutt.) Benth. (p. 1443) CALIFORNIA DISSANTHELIUM Ann. **ST**: < ± 3 dm. **LF**: blade 1–1.5 dm, 2–4 mm wide. **INFL**: < 1.5 dm; branches ascending or curving. **SPIKELET**: 3–4 mm, equal; glumes > florets; lemma ± 1.5–2 mm, hairy, obtuse to acute. Coastal-sage scrub; < 500 m. s ChI (Santa Catalina, San Clemente islands); Baja CA (Guadalupe Island). Presumed extinct in TJM (1993); rediscovered on Santa Catalina Island in 2005, on San Clemente Island in 2010. Mar–May ★

DISTICHLIS

Hester L. Bell

Per from scaly rhizomes or glabrous stolons; dioecious (monoecious). **ST**: matted (< 5 cm) to ascending and erect (≤ 80 cm), stiff, glabrous, solid in ×-section. **LF**: conspicuously 2-ranked; ligule membranous or fringed; blade flat or ± rolled, if short awl-like, gen glabrous, occ with hairy tufts at collar. **INFL**: panicle- or raceme-like (a single spikelet); spikelets short-pedicelled (sessile), staminate infl sometimes above lvs. **SPIKELET**: unisexual; pistillate gen = staminate, gen laterally compressed; glumes unequal (0), firm, awns 0, lower 3(5)-veined, upper 5–7(9)-veined; axis breaking above glumes and between florets; florets 3–20; lemma wide narrowing to acuminate tip, 7–11-veined, awn 0; palea < or > lemma; keel minutely hairy or scabrous. 9 spp.: N & S.Am, 1 sp. in Australia. (Greek: in 2 rows, from lf arrangement) [Beetle 1943 Bull Torr Bot Club 70:638–650; Bell & Columbus 2008 Syst Bot 33:536–551]

1. Blade < 1.5 cm, awl-like; spikelet 1 per infl; glumes 0 . . . . ***D. littoralis***
1′ Blade > 1.5 cm, flat; spikelets 2–20 per infl; glumes 2 . . . . ***D. spicata***

D. littoralis (Engelm.) H.L. Bell & Columbus (p. 1443) SHORE GRASS Per, mat-like. **ST**: prostrate 3–8 dm; lateral 5–23 cm. **LF**: blade 4–12 mm, sharply folded. **INFL**: spikelet 1. **SPIKELET**: 8–13 mm, gen concealed by lvs; glumes 0; lemma veins 7–9; florets 3–5, 1–2 fertile, sterile above; lower lemma firmly enclosing fl and fr; palea ± = lemma Salt marshes; ± 0 m. SCo, ChI, DSon (Salton Sea); to TX, FL, Cuba, Mex. [*Monanthochloe l.* Engelm.] Apr–Aug

D. spicata (L.) Greene (p. 1443) SALT GRASS Per; rhizomes stout, scaly, ± yellow; stolons rarely present. **ST**: erect, 1–5(8) dm. **LF**: blade gen 2–10 cm, flat, stiff. **INFL**: panicle-like, 2–8 cm, narrowly ± cylindric to elliptic in outline, ± dense. **SPIKELET**: 2–20 per infl, 6–20(25) mm, straw-colored to ± purple; glumes 2, 1.5–4 mm; florets 5–15; lemma 3–6 mm; palea ± = lemma, keel ciliate in pistillate florets. 2*n*=40. Salt marshes, coastal dunes, moist, alkaline areas; < 1550 m. NCo, KR, NCoRI, NCoRO, CaR, SN, GV, CW, SW, GB (exc Wrn, W&I), D (exc DMtns); s Can, US, Mex, S.Am. Apr–Sep

ECHINOCHLOA BARNYARD GRASS

Scott Simono, adapted from Michael (2003)

Ann to per. **ST**: decumbent to erect; internode hollow or solid. **LF**: basal and cauline; sheath gen glabrous; ligule gen 0; blade gen flat, linear to linear lanceolate, midrib prominent, upper surface gen glabrous. **INFL**: panicle-like, of simple or compound branches; branches angular, gen ascending to appressed, axis gen glabrous; spikelets gen many, 1–2 per node, gen subsessile, densely packed on branches. **SPIKELET**: ovoid to compressed, falling as one unit, breaking free below glumes, or not at all; florets 2(3), lower floret sterile or staminate, upper florets bisexual, anthers 3; glumes membranous, unequal, lower < upper, short-bristly to hairy, gen green to ± purple, upper glume unawned or shortly awned; lower lemma similar to the upper glume in length and texture, unawned or awned, upper lemma leathery, dorsally rounded, mostly smooth, tip short or elongate, firm or membranous, unawned; upper palea free from lemma at tip, lower palea vestigial to well developed. 40–50 spp.: warm temp, subtrop, worldwide. (Greek: hedgehog grass, from bristly spikelet) [Webster 1993 TJM (1993):1252–1253; Michael 2003 FNANM 25:390–403]

1. Lower lemma gen unawned; spikelets, esp those near the base of infl, not breaking free at maturity; upper lemma > upper glume in length, width . . . . **[E. esculenta]**
1′ Lower lemma often awned; spikelets breaking free at maturity; upper lemma < to ± > upper glume in length, width

  2. St erect, densely tufted; spikelets 3.7–7 mm; pls resembling rice vegetatively, closely associated with rice cult. . . . . *E. oryzoides*
  2′ St sprawling, decumbent or erect; spikelets 2–5 mm; pl occ resembling rice vegetatively but, if so, the spikelets < 3 mm; pls gen not associated with rice cult, in wet, disturbed places and cult fields

    3. Infl branches 0.7–2(4) cm, without 2° branches; spikelet 2–3 mm, unawned . . . . *E. colona*
    3′ Infl branches 1–14 cm, gen rebranched, 2° branches often short and inconspicuous; spikelet 2.5–5 mm, awned or not

      4. Upper lemma broadly obovate, acute or acuminate, tip not sharply differentiated, not early withering; infl gen erect, branches gen spreading . . . . *E. muricata* var. *microstachya*
      4′ Upper lemma narrowly elliptic to broadly ovate, acute or obtuse, tip withering, differentiated from lemma body by a line of minute hairs or not; infl erect to strongly drooping

        5. Infl spreading to erect, occ nodding, not drooping 180°, 2° branches gen inconspicuous; awns straight or irregularly bent, ≤ 10 mm and only at branch tips or ≤ 50 mm and throughout infl . . . . *E. crus-galli*
        5′ Infl gen drooping 180°, 2° branches to 3 cm; awns curved, 3–15 mm, throughout infl
. . . . *E. crus-pavonis* var. *crus-pavonis*

E. colona (L.) Link   **ST**: cespitose or spreading, decumbent to erect, 1–7 dm, rooting from lower cauline nodes. **LF**: sheath 4–9 cm, glabrous; blade 8–22 cm, 3–6(10) mm wide. **INFL**: 2–13 cm; 1° branches 0.7–2(4) cm, axis glabrous to sparsely strigose. **SPIKE-LET**: 2–3 mm, 1–1.5 mm wide, pubescent to strigose; lower glume 1–1.5 mm; upper glume ± = spikelet; florets ± equal; lower floret sterile (staminate); lower lemma unawned; upper lemma with a strongly differentiated, early withering tip; palea ± = lemma. **FR**: ± white. Wet fields, disturbed areas; < 1400 m. e KR, e NCoRI, s SNF, n SNH, GV, CCo, s SnFrB, SW, n DMtns, DSon; s US; native to Old, New World trop. [*Echinochloa colonum*, orth. var.] Jun–Sep

E. crus-galli (L.) P. Beauv. (p. 1449)   **ST**: spreading, decumbent or stiffly erect, 3–20 dm. **LF**: sheath 3–7 cm, glabrous; blade to 65 cm, 5–30 mm wide. **INFL**: 5–25 cm, hairs few to many at or below nodes; 1° branches 1.5–10 cm, gen ascending to spreading, occ erect, scabrous to sparsely strigose; 2° branches gen inconspicuous. **SPIKE-LET**: 2.5–4 mm, 1–2.3 mm wide; lower glume ± 0.8–2 mm; upper glume ± = spikelet; florets subequal, lower sterile; lower lemma awn 0–50 mm; upper lemma broadly ovate to elliptic, leathery, ending in an early-withering, acuminate, membranous tip differentiated by a line of minute hairs; palea ± = lemma. **FR**: ± brown. Gen wet, disturbed sites, fields, roadsides; < 2000 m. CA (esp CA-FP); worldwide; native to Eurasia. [*E. c.* subsp. *spiralis* (Vasinger) Tzvelev; *E. c.* var. *zelayensis* (Kunth) Hitchc.] Jun–Oct

E. crus-pavonis (Kunth) Schult. var. *crus-pavonis*   **ST**: decumbent to erect, 3–15 dm. **LF**: sheath 7–20 cm, glabrous, often ± purple; blade 12–60 cm, 10–25 mm wide. **INFL**: gen drooping, 10–30 cm; 1° branches to 14 cm, 2° branches to 3 cm. **SPIKELET**: 2.5–3.4 mm, 1.2–1.4 mm wide; lower glume 1–1.75 mm; upper glume ± = spikelet; lower floret ± = upper, sterile; lower lemma awn (0)3–10(15) mm, curved; upper lemma narrowly elliptic, acute or obtuse, with a well-differentiated, early-withering tip; lower palea > 1/2 lower lemma length. **FR**: ± brown. Marshes, wet places, often in water; < 800 m. NCo, NCoRO, NCoRI, GV, SCo, PR, MP; BC, s US; native to Mex, S.Am. Pls with stiffly erect infls, palea of lower floret << 1/2 lemma, belong to *E. crus-pavonis* var. *macera* (Wiegand) Gould, reported but not documented in CA. Jun–Sep

E. muricata (P. Beauv.) Fernald var. *microstachya* Wiegand Ann, occ rooting at lower nodes, occ forming short 2° infl at upper nodes when mature. **ST**: erect, 8–16 dm. **LF**: sheath 6–20 cm, glabrous; blade 1–27 cm, 4–11 mm wide. **INFL**: gen erect, open, 7–35 cm; 1° branches 2–8 cm, gen spreading. **SPIKELET**: 2.5–3.8 mm, ± 1.5–2 mm wide; lower glume ± 1 mm; upper glume ± = spikelet; lower floret slightly < upper, sterile; lower lemma awned or not; upper lemma leathery, acute to acuminate, extending into an undifferentiated membranous tip; lower palea well developed. **FR**: ± yellow. Moist, often disturbed sites (rarely in rice fields); < 1800 m. NW (exc NCoRH), n SNF, GV, SCo, SnBr, PR, GB; w N.Am, Mex. Jul–Sept

E. oryzoides (Ard.) Fritsch   Ann. **ST**: erect, 4–12 dm, densely tufted. **LF**: sheath glabrous; blade gen drooping, gen glabrous, 7–20 cm, 4–12 mm wide. **INFL**: spreading to strongly drooping, 8–17(25) cm, nodes strigose, internodes glabrous; 1° branches to 5 cm, appressed, mostly simple. **SPIKELET**: 3.7–7 mm, 2–2.5 mm wide, broadly ovate to ovate; lower glume 0.9–2.8(4) mm, upper glume ± = spikelet; lower floret slightly < upper, sterile; lower lemma ± = spikelet, awn gen to 5 cm; upper lemma broadly elliptic with ± green tip; lower palea well developed. **FR**: ± brown. Wet places, esp rice fields, gen fl before rice; < 100 m. GV, DSon (Imperial Valley); LA; native to Eurasia. Jun–Oct

EHRHARTA   VELDT GRASS

Ann, per. **ST**: gen erect, 6–200 cm. **LF**: auricle gen ciliate; ligule gen membranous, truncate, irregularly torn; blade flat, linear to lanceolate. **INFL**: raceme- or panicle-like; branches spreading to erect; spikelets sessile to stalked. **SPIKELET**: 2–17 mm, cylindric or laterally compressed; glumes ± equal to unequal, < to > florets, gen ovate; axis breaking above glumes, falling as 1 unit; florets 3, lower 2 sterile; sterile lemma glabrous or pubescent, firmer than glumes, awned or not, lower gen auricled; palea 0; upper floret bisexual, lemma membranous, becoming hard at maturity, 5–7-veined, awn 0; palea 1–2(5)-veined; stamens 3 or 6 (in CA). ± 35 spp.: s Afr (25 spp.), New Zealand. (J.F. Ehrhart, German botanist, student of Linnaeus, 1742–1795) [Barkworth 2007 FNANM 24:33–36]

1. Sterile lemma awns 2–20 mm; stamens 3 . . . . *E. longiflora*
1′ Sterile lemma awns 0; stamens 6
  2. Glumes ± purple at maturity; sterile lemma soft-hairy, upper smooth . . . . *E. calycina*
  2′ Glumes ± green at maturity; sterile lemma ± glabrous, upper transversely wrinkled . . . . *E. erecta*

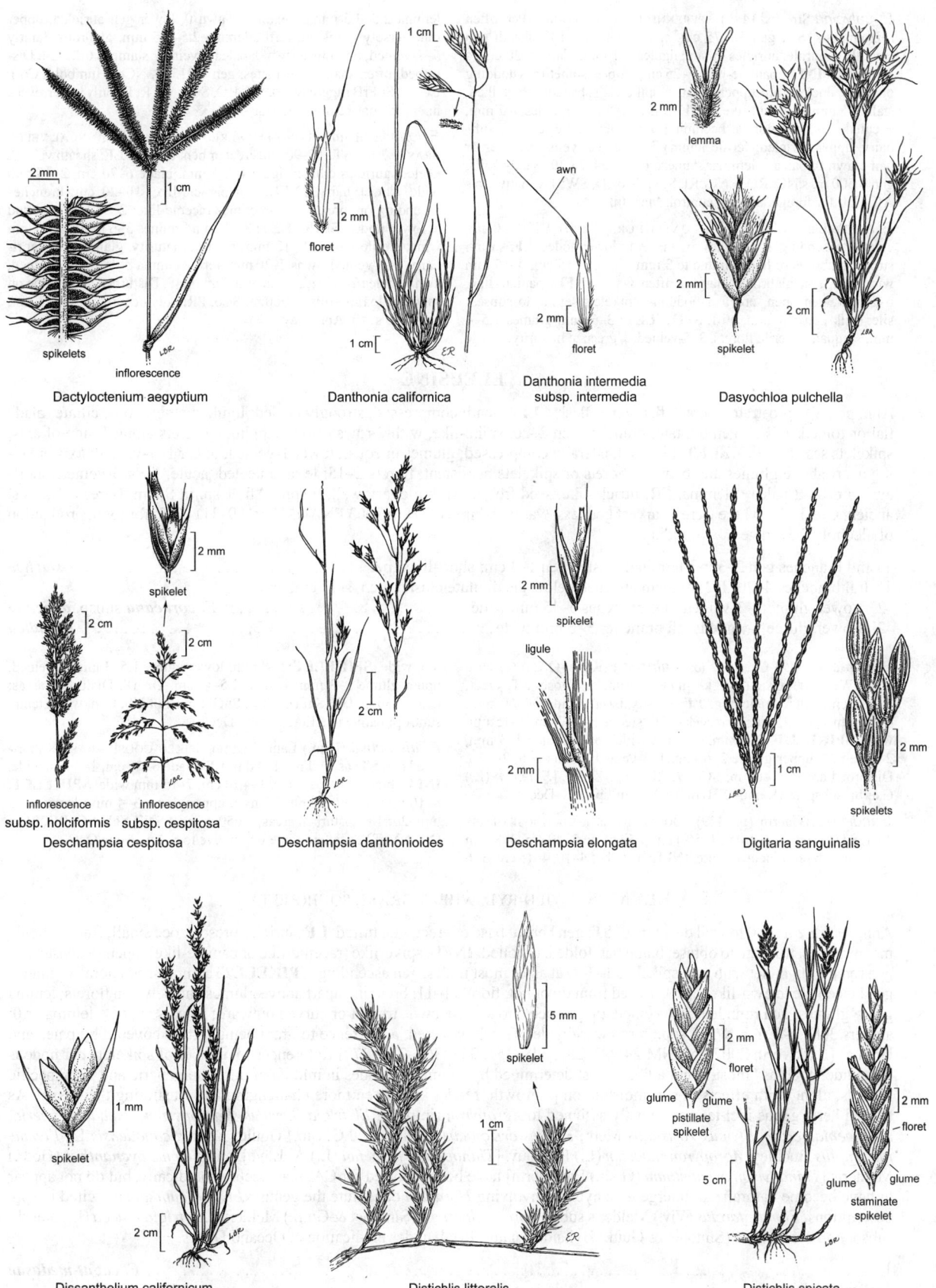

Dactyloctenium aegyptium

Danthonia californica

Danthonia intermedia
subsp. intermedia

Dasyochloa pulchella

Deschampsia cespitosa

Deschampsia danthonioides

Deschampsia elongata

Digitaria sanguinalis

Dissanthelium californicum

Distichlis littoralis

Distichlis spicata

E. calycina Sm. (p. 1449) PERENNIAL VELDT GRASS Per, often rhizomatous. **ST:** gen 30–75 cm, erect, glabrous. **LF:** sheath gen smooth, ± purple, auricles ciliate; ligules ± 1 mm; blade 5–20 cm, < 1 cm wide. **INFL:** panicle-like, 5–25 cm, ± open, sometimes nodding or partly enclosed in upper sheaths; spikelets subsessile to stalked, stalk < 5 mm, ± thread-like. **SPIKELET:** 4–8 mm; glumes 3–7 mm, ± equal, > sterile florets, becoming ± purple; sterile lemmas soft-hairy, upper auricled; fertile lemma 5–7-veined, veins glabrous or hairy, awn 0; palea < lemma; stamens 6. 2*n*=24–28,30. Sandy soils; gen < 500 m. s NCoRO, s NCoRI, ScV, SCoRO, SW; TX; native to s Afr. Cult for forage, erosion control. Mar–Jul ❖

E. erecta Lam. (p. 1449) PANIC VELDT GRASS Per. **ST:** 4–10 dm, erect or ascending, sometimes rooting from lower nodes. **LF:** sheath striate, glabrous or hairy; ligule to 3 mm; blade 5–15 cm, 4–15 mm wide, flat, gen glabrous, margins often wavy. **INFL:** panicle-like, 6–20 cm, gen open, erect or nodding; spikelets sessile to subsessile, stalk gen < 1 mm, stiff. **SPIKELET:** 3–6 mm; glumes 1.5–4 mm, ± equal, > sterile florets, 3–5-veined, ± green at maturity; sterile lemma 2.5–4.5 mm, ± glabrous, awn 0, lower gen auricled, upper transversely wrinkled; fertile lemma 2.5–3.5 mm, glabrous, faintly 5–7-veined, tip obtuse, awn 0; palea 2-veined; stamens 6 2*n*=24. Disturbed, often shady, moist sites; gen < 915 m. NCo (Humboldt Co.), CCo, e SnFrB (aggressive), SCoRO, SCo, WTR, D; only occurrences in N.Am; native to s Afr. Mar–Jun ❖

E. longiflora Sm. LONG-FLOWERED VELDT GRASS, ANNUAL VELDT GRASS Ann. **ST:** 15–90 cm, erect or bent basally. **LF:** sheath veined, keeled, auricles ciliate; ligule 1–2.5 mm, blade 6–20 cm, 2–15 mm wide, flat, ± hairy. **INFL:** gen panicle-like, 10–20 cm, branches ascending or spreading, sometimes raceme-like; 2° infl often found at lower nodes. **SPIKELET:** 8–18 mm; glumes 3–6 mm, to 1.5 mm wide; sterile lemma 9–12 mm, hard at maturity, glabrous or short-hairy, 3–7-veined, awns 2–20 mm; fertile lemma 8–10 mm, 7-veined, awn 0; palea 2-veined; stamens 3. 2*n*=24,48. Disturbed sites, often on sandy or loamy soils; < 900 m. SCo, PR; only occurrences in N.Am; native to s Afr. Apr–May ❖

ELEUSINE

Ann, per. **ST:** prostrate to erect, flat, gen ± fleshy. **LF:** sheath compressed, strongly keeled; ligule membranous, ciliate; blade flat or folded. **INFL:** gen digitate; branches gen 2–20, spike-like, with 2 rows of overlapping spikelets along 1 side of axis; spikelets sessile. **SPIKELET:** bisexual, laterally compressed; glumes unequal, lower 1-veined, upper 1–7-veined; axis breaking apart above glumes and between florets or spikelets persistent; florets 2–15; lemma keeled, acute, 1- or 3-veined, short-awned or not; palea < lemma. **FR:** utricle-like, seed free from thin ovary wall. 9 spp.: Afr, 1 sp. in S.Am. (Greek: Eleusis, ancient Greek city where Ceres, harvest goddess, was worshipped) [Hilu 2003 FNANM 25:109–111] Cult for food, production of alcoholic beverage in Afr, India.

1. Infl branches gen 2–3 in a terminal cluster, gen 1–4 cm; short-lived per . . . . *E. tristachya*
1′ Infl branches 4–20, 1(2) of them attached below the digitate cluster, gen 4–17 cm; ann
  2. Lower glume 2–3-veined; infl branches 5–15 mm wide . . . . *E. coracana* subsp. *africana*
  2′ Lower glume 1-veined; infl branches 3–6 mm wide. . . . . *E. indica*

E. coracana (L.) Gaertn. subsp. *africana* (Kenn.-O'Byrne) Hilu & de Wet AFRICAN FINGER MILLET Ann, cespitose. **ST:** erect, 20–60 cm, often branching. **LF:** mostly basal; blade 30–60 mm, 6–10 mm wide. **INFL:** branches 4–20, slender, 4–17 cm, 5–15 mm wide. **SPIKELET:** 5–8 mm, 3–4 mm wide; lower glume 1–3 mm, 2–3-veined; upper glume 2–6 mm, 1–3-veined; florets 2–6. 2*n*=36. Disturbed areas; < 400 m. SCo, WTR, PR; AZ, SC. [*E. indica* (L.) Gaertn. subsp. *a.* (Kenn.-O'Byrne) S.M. Phillips] Jul–Dec

E. indica (L.) Gaertn. (p. 1449) GOOSE GRASS, INDIA GOOSE GRASS Ann. **ST:** prostrate to erect, 1–5(9) dm. **LF:** blade 5–40 cm, 3–8 mm wide, midrib prominently white. **INFL:** branches 4–10, 4–15 cm, 3–6 mm wide. **SPIKELET:** 5–8 mm; lower glume 1.5–3 mm, 1-veined; upper glume 2–3 mm; lemma 2.5–4 mm. 2*n*=18. Disturbed areas; gen < 350 m. GV, SnFrB, SCo, SnGb; most of US, Can, warm temp, subtrop; native to s Eurasia. Jul–Dec

E. tristachya (Lam.) Lam. THREE-SPIKED GOOSE GRASS Short-lived per. **ST:** ascending, 1–4 dm. **LF:** blade 6–25 cm, 1–4 mm wide. **INFL:** branches (1)2–3(4), 1–4(6) cm, 7–15 mm wide. **SPIKELET:** 8–10 mm; lower glume 2–3 mm; upper glume 3–4 mm; lemma 3–5 mm. 2*n*=18. Disturbed areas; < 650 m. n&c SNF, ScV, SCoRO, SCo; OR, NV, TX, scattered in e US; native to S.Am. May–Oct

ELYMUS WILD-RYE, WHEAT GRASS, SQUIRRELTAIL

Ann, per, rhizomes 0 to well developed. **ST:** gen bent at base or erect, gen tufted. **LF:** auricles present, occ small, fragile; ligule membranous, truncate to obtuse; blade flat, folded, or rolled. **INFL:** spike-like (raceme-like or panicle-like), open to dense; axis gen remaining intact with age; spikelets 1–3(7) at all or most nodes, gen ascending. **SPIKELET:** compressed laterally, glumes gen lanceolate to awn-like, occ 0, awned from tip or not; florets 1–11; breaking apart above glumes and between florets; lemma gen > glumes, gen rounded, 5–7-veined, tip gen acute to awned, awn straight or curved outward; palea <, =, or > lemma or 0; anthers 3(1), 1–8 mm. ± 235 spp.: temp worldwide. (Greek: covered, a reference to grain being tightly covered by palea and lemma) [Barkworth 2007 FNANM 24:283–287, 348–351, 353–369, 373–378] References to number of spikelets per node is best understood as "most, if not all" and best determined by examining nodes in middle of infl. Intergeneric and interspecific hybrids, along with effects of soil moisture on pl growth, render keys even more challenging and frustrating than usual. As treated here, genus incl taxa previously assigned to *Agropyron* (in part), *Elytrigia*, *Leymus*, *Pascopyrum*, *Pseudoroegneria*, and *Taeniatherum*. *Elymus* ×*aristatus* Merr., *Elymus arizonicus* (Scribn. & J.G. Sm.) Gould, *Elymus canadensis* L., *Elymus interruptus* Buckley, *Agropyron junceum* (L.) P. Beauv. [*Thinopyrum junceum* (L.) Á. Löve], and *Elymus pycnanthus* (Godr.) Melderis [*Thinopyrum pycnanthum* (Godr.) Barkworth] have been reported for CA, may occur sporadically, but do not appear to have become naturalized. Intergeneric hybrids involving *Hordeum* constitute the genus ×*Elyhordeum* and are cited in spp. descriptions. *Elymus farctus* (Viv.) Melderis subsp. *boreo-atlanticus* (Simonet & Guin.) Melderis [*Elytrigia juncea* (L.) Nevski subsp. *boreo-atlantica* (Simonet & Guin.) Hylander] naturalized, under eradication at Oceano Dunes.

1. Ann. . . . . *E. caput-medusae*
1′ Per
  2. Infl branched, at least at base
    3. St 11–35 dm, 6–10 mm wide at base; blade 15–35 mm wide; spikelet 10–40 per node (inc branches) . . . *E. condensatus*
    3′ St 6.5–21 dm, 3.5–5 mm wide at base; blade 6–15 mm wide; spikelet 2–6 per node . . . . ²*E.* ×*gouldii*

2′ Infl spike-like (raceme-like), spikelets sessile or some slightly stalked (< 1 mm)
  4. Pls of coastal beaches, dunes, and bluffs
    5. Glumes awl-like . . . . *E. pacificus*
    5′ Glumes flat, lanceolate
      6. Glumes 1–2 mm wide at mid-length, keeled . . . . *E. ×vancouverensis*
      6′ Glumes 3–5 mm wide at mid-length, rounded . . . . *E. mollis* subsp. *mollis*
  4′ Pls of other habitats
    7. Glumes 0 or << lowest lemma . . . . *E. californicus*
    7′ Glumes 2 (on at least upper spikelets in *Elymus salina*), equal or unequal, but > 1/2 lowest lemma
      8. Infl axis breaking apart with age
        9. Spikelets gen 1 per node; glumes narrowly lanceolate . . . . *E. scribneri*
        9′ Spikelets 2(3) per node; glumes awn-like
          10. Glumes cleft into 3–9 awn-like divisions; auricles gen present . . . . *E. multisetus*
          10′ Glumes entire or 2(3)-cleft; auricles often lacking . . . . *E. elymoides*
            11. Glumes 2 on all spikelets; palea veins gen not extended as short bristles . . . . var. *brevifolius*
            11′ Glumes on spikelets at lower nodes appearing to be 3, 1 a glume-like sterile floret; palea veins
               gen extended as short bristles
              12. Glumes entire; lemma awns gen > glume awns . . . . var. *californicus*
              12′ Some glumes with 2–3 awn-like divisions; lemma awns gen < glume awns . . . . var. *elymoides*
      8′ Infl axis remaining intact with age
        13. Lemma awn gen 10–40 mm
          14. Lemma awn curving outward with age
            15. St erect at base, 30–100 cm; sagebrush steppe, woodland, < 1650 m . . . . *E. spicatus*
            15′ St decumbent at base, gen 20–50 cm; alpine, subalpine conifer forest, > 1800 m . . . . *E. sierrae*
          14′ Lemma awn straight or wavy, but not curving outward (sometimes slightly curved in *Elymus glaucus*)
            16. Spikelets gen 2 per node . . . . *E. glaucus* subsp. *glaucus*
            16′ Spikelets gen 1 per node
              17. Anthers 4–7 mm; infl internodes gen 9–27 mm . . . . [2]*E. stebbinsii*
              17′ Anthers 1.2–2.5 mm; infl internodes gen 7–9 mm . . . . [2]*E. trachycaulus* subsp. *trachycaulus*
        13′ Lemma awn 0–10(15) mm
          18. Glumes gen awl-like, or if lanceolate, then often inconspicuously 0–3(5)-veined, hard or tough
           textured, and awn-tipped or acute
           19. Spikelets 2–7 per node
            20. St node region gen densely pubescent; pls gen cespitose . . . . *E. cinereus*
            20′ St node region glabrous; pls rhizomatous
              21. Lf blades gen 3–6 mm wide; spikelets 2(3) per node, 3–7 florets. . . . . [2]*E. triticoides*
              21′ Lf blades gen 6–15 mm wide; spikelets 2–6 per node, with 6–9 florets . . . . [2]*E. ×gouldii*
          19′ Spikelets gen 1 per node (sometimes 2–3 in *Elymus salina*)
           22. Florets gen twisted so that back of lowermost lemma centered between glumes . . . . [2]*E. triticoides*
           22′ Florets not twisted, back of lowermost lemma not centered between glumes
            23. Pls gen cespitose (sometimes weakly rhizomed); st 30–90 cm; cauline lvs present; glumes not
              curved distally . . . . *E. salina*
            23′ Pls rhizomatous; st 20–100 cm; lvs mostly basal; glumes ± curved distally. . . . . *E. smithii*
        18′ Glumes flat, narrowly to broadly lanceolate, strongly 3–9-veined, thin (if thickened, tip obtuse)
          24. Glume tips rounded, obtuse to truncate
           25. Pls rhizomatous; midvein of glume more prominent than lateral veins . . . . *E. hispidus*
           25′ Pls cespitose; midvein of glume and lateral veins ± equally prominent . . . . *E. ponticus*
          24′ Glume tips acute to acuminate and/or awned
           26. Rhizomes present
            27. Lf blade 1–6 mm wide; lemma hairy . . . . *E. lanceolatus* subsp. *lanceolatus*
            27′ Lf blade 6–14 mm wide; lemma glabrous . . . . *E. repens*
          26′ Rhizomes gen 0
           28. Spikelets gen 2 per node . . . . *E. glaucus* subsp. *virescens*
           28′ Spikelets gen 1 per node
            29. Anthers 4–7 mm; infl internodes gen 9–27 mm . . . . [2]*E. stebbinsii*
            29′ Anthers 1.2–2.5 mm; infl internodes gen 7–9 mm . . . . [2]*E. trachycaulus* subsp. *trachycaulus*

E. californicus (Thurb.) Gould (p. 1449) CALIFORNIA BOTTLE-BRUSH GRASS Per. **ST**: 10–20 dm, rigid. **LF**: basal sheath hairs stiff, auricles 1–3 mm, slender; blade 10–28 mm wide, flat. **INFL**: 10–30 cm, erect; internodes 7–12 mm; spikelets 2–4(5) per node, appressed to spreading. **SPIKELET**: 12–17 mm; glumes 0 to < 1 mm; florets 2–5; lemma 10–15 mm, awn 16–33 mm, straight; anthers 6–8 mm. 2*n*=56. Conifer forest; < 500 m. NCo, NCoRO, n CCo, SnFrB (Santa Cruz Mtns). [*Leymus c.* (Thurb.) Barkworth] May–Aug ★

E. caput-medusae L. (p. 1449) MEDUSA HEAD Ann. **ST**: decumbent to ascending, 2–7 dm, slender. **LF**: sheath glabrous, auricles 0.1–0.5 mm, rarely 0; ligule membranous, 0.2–0.6 mm, truncate; blade 1–3 mm, ± inrolled, glabrous to puberulent, long-ciliate near collar. **INFL**: spike-like, dense, 1.5–6 cm (exc awns), spikelets 2-ranked, 2(3–4) per node. **SPIKELET**: glumes 2, equal, 5–80 mm, awl- to awn-like, erect to spreading or reflexed, stiff, fused at base, gen glabrous; axis breaking above glumes; florets (3), lower fertile, upper vestigial on prolonged axis; fertile lemma lanceolate

5–8 mm, 5-veined, awn (20)30–100+ mm, flat, straight to outwardly curved, palea = lemma; reduced florets with 3-veined lemmas, palea 0; anthers < 2 mm. 2*n*=14. Disturbed areas; < 2000 m. KR, NCoR, CaR, SNF, GV, SCoR, MP; to BC, Rocky Mtns, e US (rare); native to Eurasia. [*Taeniatherum c.* (L.) Nevski] Apr–Jul ◆

E. cinereus Scribn. & Merr. (p. 1449) GREAT BASIN WILD-RYE Pl gen cespitose; rhizomes 0 or short. **ST:** 7–27 dm, nodes, esp lower ones, gen densely hairy. **LF:** auricles to 1.5 mm, ligule 2.5–6.5 mm; blade 15–45 cm, 3–12 mm wide, strongly rolled to flat, upper surface scabrous. **INFL:** spike-like (rarely branched at lower nodes), 9–19 cm; spikelets 2–7 per node, sessile. **SPIKELET:** glumes 8–18 mm, awl-like, stiff, keeled; florets 3–7; lemma 6.5–12 mm, acute to awn-tipped, glabrous to short-hairy; anthers 4–7 mm. 2*n*=28,56. Streamsides, canyons, roadsides, sagebrush scrub, open woodland; < 3100 m. CaR, SN, ScV, TR, GB, DMtns; to Can, w US. [*Leymus c.* (Scribn. & Merr.) Á. Löve] Hybridizes with *E. triticoides.* Jun–Aug

E. condensatus J. Presl GIANT WILD-RYE Pl cespitose; rhizomes 0 or short. **ST:** 11–35 dm, glabrous, 6–10 mm wide at base. **LF:** auricles 0, ligule 0.7–6 mm, blade 15–35 mm wide. **INFL:** 17–44 cm, panicle-like, lower nodes with 2–6 short branches, spikelets 10–40 (incl branches) per node, reduced to 3–5 in upper infl, sessile or stalked. **SPIKELET:** 9–25 mm, glumes 6–16 mm, awl-like, keeled; florets 3–7; lemma 7–14 mm, glabrous to hairy, acute to awn-tipped, awn < 4 mm; anthers 3.5–7 mm. 2*n*=28,56. Dry slopes, open woodland; < 1830 m. CW, SW, DMoj; Mex. [*Leymus c.* (J. Presl) Á. Löve] Jun–Aug

E. elymoides (Raf.) Swezey (p. 1449) SQUIRRELTAIL Pl cespitose, often glaucous. **ST:** erect to decumbent, 1–6.5(8) dm. **LF:** sheath glabrous to densely long-white-hairy; auricles 0–1 mm; ligule < 1 mm; blade (1)2–4(6) mm wide, flat, folded, or rolled. **INFL:** 2.5–20 cm (exc awns), axis breaking apart with age; internodes 3–10 mm; spikelets 2(3) per node. **SPIKELET:** 10–20 mm; glumes 15–125 mm, awn-like, base narrow, thick, gen spreading, entire or split into 2–3 unequal divisions; florets 2–4(5); lemma 6–12 mm, awn 15–120 mm, spreading; palea 6–11 mm, veins sometimes extended as bristles; anthers ± 2 mm. 2*n*=28. Subsp. *hordeoides* (Suksd.) Barkworth doubtfully naturalized. If recognized taxonomically: sporadic hybrids with *E. trachycaulus* assignable to *E. ×saundersii* Vasey (Siskiyou, Modoc cos. to Monterey, Tulare cos.); widespread hybrids with *E. glaucus* assignable to *E. ×hansenii* Scribn.; hybrids with *E. spicatus* assignable to *E. ×saxicola* [×*Pseudelymus saxicola* (Scribn. & J.G. Sm.) Barkworth & D.R. Dewey], rarely collected.

var. ***brevifolius*** (J.G. Sm.) Dorn **SPIKELET:** glumes entire, awn 50–125 mm; lowest floret fertile, not glume-like; fertile florets 1+; lemma awn 50–105 mm. Dry, open areas; 450–3500 m. SnBr, PR, MP, W&I, DMoj; to Can, Great Plains, n Mex. [*E. e.* subsp. *b.* (J.G. Sm.) Barkworth] Jul–Aug

var. ***californicus*** (J.G. Sm.) J.P. Sm. **SPIKELET:** glumes entire, awn 15–40(70) mm; lowest floret gen sterile, glume-like, fertile florets 1+; lemma awn 25–70 mm. Dry, open areas; 275–4200 m. KR, CaR, SN, SnGb, SnBr, SnJt, SNE; w N.Am. [*E. e.* subsp. *c.* (J.G. Sm.) Barkworth] Jul–Aug

var. ***elymoides*** **SPIKELET:** some glumes split into 2–3 divisions, 35–85 mm; lowest floret gen sterile, awn-like, fertile florets 1+; lemma awn 25–75 mm. Desert shrubland, often in disturbed sites; 250–4300 m. SnFrB, TR, SnJt, GB, D; to WA, WY, CO. [*E. e.* subsp. *e.*] Jul–Aug

E. glaucus Buckley (p. 1449) BLUE OR WESTERN WILD-RYE Sometimes short-rhizomed. **ST:** 3–14 dm. **LF:** sheath glabrous or hairy, auricles ± 2 mm; ligule < 1 mm; blade 4–12(17) mm wide, gen flat. **INFL:** 5–21 cm (exc awns), erect or nodding, not breaking apart with age; internodes 4–8(12) mm; spikelets (1)2(3) per node. **SPIKELET:** 8–16 mm, appressed to divergent; glumes 6.5–19 mm, gen short-awned; florets (1)2–4(6), lower florets concealed; lemma 8.5–14 mm, awn 0–30 mm, gen straight; anthers 1.5–3.5 mm. 2*n*=28. Hybridizes with *E. elymoides, E. stebbinsii, E. trachycaulus* to produce highly variable populations.

subsp. ***glaucus*** **LF:** sheath, blade ± glabrous or scabrous. **SPIKELET:** lemma awn 20–30 mm. Open areas, chaparral, woodland, forest; < 2890 m. CA; w N.Am, n Mex. [*E. g.* subsp. *jepsonii* (Burtt Davy) Gould] Jun–Aug

subsp. ***virescens*** (Piper) Gould **LF:** sheath, blade ± glabrous or scabrous. **SPIKELET:** glume awns 0–2 mm; lemma awns (0)1–5(7) mm. Conifer forest, chaparral, rocky soils, riverbanks; < 2800 m. NCo, NCoRO, CW; to AK. Jun–Aug

E. ×gouldii J.P. Sm. & Columbus **ST:** 6.5–21 dm, 3.5–5 mm wide at base, gen glabrous. **LF:** ligule 0.5–2 mm; blade gen 6–15 mm wide, glabrous. **INFL:** 12–40 cm; spikelets 2–6 per node, sessile or short-stalked. **SPIKELET:** glumes > lower floret; lemma glabrous, awn to 2 mm. 2*n*=42. Open areas, often saline soils; < 1615 m. ScV, CW, SCo, SnBr, PR; Baja CA. [*Leymus ×multiflorus* (Gould) Barkworth & R.J. Atkins] A sterile hybrid between *E. triticoides* and *E. condensatus.* May–Jul

E. hispidus (Opiz) Melderis Rhizomatous. **ST:** 5–11.5 dm, glabrous or hairy. **LF:** sheath glabrous or ciliate; auricles 0.5–1.8 mm; ligule < 1 mm; blade 2–8 mm wide; veins many, weakly ribbed. **INFL:** 8–21 cm, erect or nodding; spikelets 1 per node. **SPIKELET:** 11–18 mm; glumes smooth or hairy, midvein more prominent than lateral veins; tips truncate to obtuse; florets 3–10; lemma 7.5–10 mm, gen glabrous, sometimes rough-hairy, awn gen 0; palea 7–9.5 mm; anthers 5–7 mm. 2*n*=42,43. Open areas, slopes; < 2500 m. KR, NCoRI (Yolo Co.), CaR, SN, WTR, SnBr, PR, MP, DMtns; to BC, Great Plains; scattered in e US; native to Eurasia. [*Elytrigia intermedia* (Host) Nevski subsp. *i.*; *Thinopyrum i.* (Host) Barkworth & D.R. Dewey; *T. i.* subsp. *barbulatum* (Schur) Barkworth & D.R. Dewey] Recognition of subspp. requires further study; introduced for erosion control. Jun–Aug

E. lanceolatus (Scribn. & J.G. Sm.) Gould subsp. ***lanceolatus*** THICK-SPIKE WHEAT GRASS From well developed rhizomes. **ST:** 2–13 dm. **LF:** sheath glabrous to hairy, auricles 0.5–2 mm; ligule 0.1–0.5 mm; blade 1–6 mm wide, rolled to flat. **INFL:** 10–22 cm, erect to slightly nodding; internodes 5–18 mm; spikelets gen 1 per node. **SPIKELET:** 10–28 mm; glumes 5–14 mm, awn 0, tips acute to acuminate; florets 3–11; lemma 7–11 mm, ± stiff-hairy, sometimes glabrous, acute to awn-tipped (to 2 mm); palea = lemma; anthers 3–5 mm. 2*n*=28. Open sites, woodland, conifer forest; 500–2250 m. KR, NCoRI, CaR, n SN, s SNH, SW, MP; to AK, Rocky Mtns. [*E. l.* subsp. *riparius* (Scribn. & J.G. Sm.) Barkworth] One other subsp. in se Can, ne US. May be confused with *E. smithii.* Jun–Sep

E. mollis Trin. subsp. ***mollis*** (p. 1449) AMERICAN DUNE GRASS, SEA LYME GRASS From well developed rhizomes. **ST:** 5–17 dm, densely hairy below infl. **LF:** sheath glabrous; auricles < 0.7 mm; ligule 0.2–2.5 mm; blade 10–95 cm, 3–15 mm wide, upper surface ± scabrous. **INFL:** 12–34 cm; spikelets gen 2 per node, sessile or stalked. **SPIKELET:** glumes 13–30 mm, lanceolate, flat, slightly to densely hairy; florets 3–6; lemma 12–20 mm, densely hairy, acute to awn-tipped; anthers 4–9 mm. 2*n*=28. Sandy beaches; gen < 10 m. NCo, CCo; to AK, Asia. [*Leymus m.* (Trin.) Pilg. subsp. *m.*] Other subsp. arctic. May–Jul

E. multisetus (J.G. Sm.) Burtt Davy BIG SQUIRRELTAIL Pl cespitose. **ST:** erect to ascending, 1.5–6 dm. **LF:** sheath glabrous or white-hairy, auricles gen 0.5–1.5 mm; ligule to 1 mm; blade 1.5–5 mm wide, flat or rolled. **INFL:** 3–20 cm (exc awns), axis breaking apart with age; internodes 4–8 mm; spikelets gen 2 per node. **SPIKELET:** 10–15 mm; glumes divided near base into 3–9 awn-like divisions, 25–200 mm, distal 1/2 curving outward with age; florets 2–4, lowest glume-like, vestigial; lemma 8–10 mm, tip gen 2-lobed, lobes awn-like, < 20 mm, awn between lobes 25–100 mm; palea 7–9 mm, veins gen extended into short bristles; anthers 1–2 mm. 2*n*=28. Open, sandy to rocky areas; < 3800 m. CA; to WA, Rocky Mtns. Hybridizes with *E. elymoides* var. *e.* in s CA. May–Jul

E. pacificus Gould PACIFIC WILD-RYE Pl strongly rhizomatous. **ST:** 1–3(6) dm, glabrous or sparsely hairy near nodes. **LF:** sheath glabrous; auricles to 1.4 mm; ligule 0.2–0.3; blade 10–30 cm, 2–4 mm wide, often > infl. **INFL:** 2–8 cm; spikelets 1 or 2 per node. **SPIKELET:** glumes (5)7–15 mm, awl-like; florets 4–6; lemma 7–11

mm, gen glabrous; tips acute to short-awned; anthers 3–4 mm. 2*n*=28. Coastal bluffs; < 230 m. s NCo, CCo, n ChI. [*Leymus p.* (Gould) D.R. Dewey] May–Jul

E. ponticus (Podp.) N. Snow (p. 1449) TALL WHEAT GRASS Cespitose, rhizomes 0. **ST:** 5–22 dm, glabrous. **LF:** sheath ciliate below middle; auricles 0.2–1.5 mm; ligule 0.3–1.5 mm; blade 2–6.5 mm wide, gen rolled; veins 1–8, strongly ribbed. **INFL:** 10–42 cm; spikelets 1 per node. **SPIKELET:** 13–30 mm; glumes 6.5–10 mm, midvein and laterals ± equally prominent, tips truncate; florets 6–12; lemma 9–12 mm, awn 0; palea 7.5–11 mm; anthers 2.5–6 mm. 2*n*=69,70. Disturbed, often alkaline areas; < 1600 m. SN, ScV, SW, MP; to BC, e US; native to se Eur, w Asia. [*Elytrigia elongata* (Host) Nevski; *E. p.* (Podp.) Holub; *Thinopyrum p.* (Podp.) Barkworth & D.R. Dewey] Planted along roadsides for soil stabilization. Jun–Jul

E. repens (L.) Gould (p. 1449) QUACK GRASS Pl strongly rhizomatous. **ST:** 5–10 dm. **LF:** sheath gen glabrous or lowermost soft-hairy; auricles 0.3–1 mm; ligule 0.25–1.5 mm; blade 6–14 mm wide, veins of 2 kinds, some faint, others strongly ribbed, widely spaced. **INFL:** 5–20 cm; 1(2) spikelets per node. **SPIKELET:** 10–27 mm; florets 4–7; glume tips acute, gen with awn 0.5–4 mm; lemma 8–12 mm, tapering to point or awn 0.5–10 mm; palea 7–9.5 mm; anthers 4–5.5 mm. 2*n*=22,42. Weed of disturbed areas, cult fields; < 2150 m. CA-FP, GB; to e US, Can; native to Eurasia. [*Elytrigia r.* (L.) Nevski] Apr–Aug ◆

E. salina M.E. Jones SALINA PASS WILD-RYE Pl gen cespitose. **ST:** 3–9 dm, gen glabrous. **LF:** ligule 0.5–1 mm; blade 1–5 mm wide, ± flat, adaxial surface evenly hairy; abaxial surface gen glabrous. **INFL:** 4–14 cm; spikelets gen 1 at lower, upper nodes, 2 at central nodes. **SPIKELET:** glumes 0 or < 13 mm, awl-like; florets 3–6; lemma 7–13 mm, gen glabrous, acute or awn-tipped, awn < 2.5 mm. N-facing slopes of pinyon/juniper woodland; 1350–2860 m. DMtns (Inyo, San Bernardino cos.); to ID, w CO, n AZ. [*Leymus s.* (M.E. Jones) Á. Löve subsp. *mojavensis* Barkworth & R.J. Atkins; *Leymus salinus* subsp. *mojavensis*, orth. var.] Study needed. Jun–Aug ★

E. scribneri (Vasey) M.E. Jones (p. 1449) SCRIBNER'S WHEAT GRASS Pl cespitose. **ST:** prostrate to decumbent, 1.5–5.5 dm. **LF:** sheath glabrous to short-hairy; auricles < 1 mm; ligule 0.2–0.7 mm; blade 1.5–4 mm wide, rolled or flat. **INFL:** 3.5–10 cm, breaking apart with age; internodes 3.5–8 mm; spikelet 1 per node, sometimes 2 at lower nodes. **SPIKELET:** 9–15 mm, appressed or ascending; glumes 4–9 mm, awn 10–30 mm, curving strongly outward; florets 2–12; lemma 7–10 mm, awn 15–30 mm, curving outward; pales > lemma; anthers 1–1.6 mm. 2*n*=28. Rocky areas, alpine; 2900–4200 m. SNH, n SNE, W&I; Can, Rocky Mtns. Hybridizes with *E. trachycaulus*, *E. elymoides*. Jul–Aug ★

E. sierrae Gould (p. 1449) Pl cespitose. **ST:** prostrate or decumbent, 2–5 dm. **LF:** sheath gen glabrous, auricles < 1 mm; ligule 0.2–0.5 mm; blade flat, 1–5 mm wide. **INFL:** 4–15 cm, erect to nodding; middle internodes 8–10 mm; spikelet 1 per node, sometimes 2 at lower nodes. **SPIKELET:** 15–20 mm; glumes 6–9 mm, awn 5 mm, straight; florets 3–7; lemma 12–16 mm, awn 15–30 mm, curving strongly outward with age; anthers 2–5 mm. 2*n*=28. Rocky slopes, ridges, conifer forest; 1800–3530 m. CaRH, SNH; to WA, NV. Jul–Aug

E. smithii (Rydb.) Gould (p. 1449) WESTERN WHEAT GRASS Per, from rhizomes. **ST:** 2–10 dm, erect, glaucous. **LF:** mostly basal, glaucous; auricles 1–2 mm; ligule 0.1 mm, membranous; clasping; blade 2–26 cm, 1–4.5 mm wide, flat, rolled when dry. **INFL:** 4–17 cm; spikelets 2-ranked, overlapping, 1 per node or sometimes 2 at lower nodes. **SPIKELET:** 5–15 mm; glumes narrowly lanceolate, subequal, ≤ lower floret, glabrous to rough, tapered from middle to acute tip, 3-veined in lower 1/2, midvein curving slightly to side; florets 4–11; lemma 8–14 mm, glabrous to hairy, awn < 5(15) mm;

anthers 2–4.5 mm. 2*n*=56. Uncommon. Dry, alkaline soils, flats; 1220–2200 m. n SNH, GB, DMtns (New York Mtns); widespread in US, to s Can. [*Pascopyrum s.* (Rydb.) Barkworth & D.R. Dewey] A polyploid hybrid, probably involving *E. lanceolatus* and *E. triticoides*. Jun–Aug

E. spicatus (Pursh) Gould (p. 1453) BLUE BUNCH WHEAT GRASS Pl cespitose, sometimes rhizomatous. **ST:** 3–10 dm, slender, green or glaucous. **LF:** auricles well developed; ligule 0.1–0.4 mm; blade 2–6 mm wide, flat to loosely rolled. **INFL:** 8–16 cm, narrow; spikelets 1 per node; middle internodes 0.8–2.5 cm. **SPIKELET:** 8–22(25) mm; glumes 6–13 mm, 0.9–2.2 mm wide, ± 1/2 spikelet length, veins evenly glabrous or scabrous; florets 4–9; lemma 9–14 mm, awn 0–25 mm, strongly divergent. 2*n*=14,28. Sagebrush steppe, open woodland; 670–2190 m. KR, NCoR, CaR, n&c SN, MP; to s Can, Rocky Mtns. [*Pseudoroegneria s.* (Pursh) Á. Löve] If recognized taxonomically, hybrids with *E. elymoides* assignable to *E. ×saxicola* Scribn. & J.G. Sm. Jun–Aug

E. stebbinsii Gould STEBBINS' WHEAT GRASS Pl cespitose, sometimes from short rhizomes. **ST:** 6–14 dm. **LF:** sheath glabrous or pubescent, auricles 0.5–2 mm, persistent; ligule 0.3–3.5 mm, membranous; blade 2–6 mm wide, flat or rolled. **INFL:** 15–30 cm; internodes (9)12–27 mm; spikelet gen 1 per node. **SPIKELET:** 12–29 mm; glumes 10–15 mm, awn 0; florets 5–7; lemma 8–12 mm, awn (0)1–28 mm, straight; anthers (3.5)4–7 mm. 2*n*=28. Dry slopes, chaparral, conifer forest; < 2230 m. NCoRI, SN, SCoR, TR, PR. Hybridizes with *E. glaucus* and *E. trachycaulus*. If recognized taxonomically, pls with ± longer awns, gen glabrous lf sheaths assignable to *E. stebbinsii* subsp. *septentrionalis* Barkworth. Jun–Jul

E. trachycaulus (Link) Shinners subsp. **trachycaulus** (p. 1453) SLENDER WHEAT GRASS Cespitose or weakly rhizomatous. **ST:** 3–15 dm. **LF:** sheath glabrous to hairy, auricles < 1 mm; ligule 0.2–0.8 mm; blade 2–5 (8) mm wide, flat to rolled. **INFL:** 4–25 cm; sometimes appearing 1-sided because of twisting of spikelets to 1 side; spikelet 1 per node; internodes gen 7–9 mm. **SPIKELET:** 10–20 mm; glumes 6–17 mm, lanceolate to narrowly ovate; florets 3–9; acute to short-awned; lemma 6–13 mm, glabrous to hairy, awn 1–40 mm; palea = lemma; anthers 1–2.5 mm. 2*n*=28. Dry to moist, open areas, forest, woodland; < 3400 m. CA (exc GV); to AK, e US, Greenland, Mex. [*E. t.* subsp. *subsecundus* (Link) Á. Löve & D. Löve] Another subsp. is found in Greenland; forms hybrids with *Agropyron cristatum*, *E. elymoides*, *E. glaucus*, *E. lanceolatus*, *E. scribneri*. Hybrids with *Hordeum jubatum* are called ×*Elyhordeum macounii* (Vasey) Barkworth & D.R. Dewey. Jun–Aug

E. triticoides Buckley (p. 1453) BEARDLESS WILD RYE Pl from rhizomes. **ST:** 4.5–12.5 dm, nodes glabrous. **LF:** sheath glabrous or hairy; auricles to 1 mm; ligule 0.2–1.3 mm; blade 10–35 cm, 3–6 mm wide, upper surface finely scabrous. **INFL:** 5–20 cm; spikelets (1)2(3) per node. **SPIKELET:** glumes 5–16 mm, awl-like; florets 3–7; lemma 5–12 mm, gen awn-tipped, awn to 3 mm. 2*n*=28. Dry to moist, often saline, meadows; < 2500 m. CA; to BC, TX, Baja CA. [*Leymus t.* (Buckley) Pilg.] Hybridizes with *E. condensatus*, *E. mollis*. Jun–Jul

E. ×vancouverensis Vasey Pl from rhizomes. **ST:** 6.5–13 dm, sparsely to densely hairy below infl. **LF:** sheath glabrous or hairy; auricles to 1 mm; ligule 0.4–1.2 mm; blade < 9 mm wide, prominently ribbed. **INFL:** 7–32 cm; spikelets 1–2 per node, sometimes short-stalked. **SPIKELET:** glumes 8–28 mm, lanceolate, often ± purple, sometimes glaucous; base flat, tapered, awn to 4 mm; florets 2–6; lemma base glabrous, scabrous to hairy near tip; anthers 3.3–7.5 mm. 2*n*=28,42. Sandy beaches; < 100 m. NCo, n CCo; to BC. [*Leymus ×v.* (Vasey) Pilg.] A sterile hybrid between *E. mollis* and *E. triticoides*. Jul

ENNEAPOGON PAPPUS GRASS

John R. Reeder & Dieter H. Wilken

Per, gen cespitose. **ST:** ascending to erect. **LF:** basal and cauline; ligule hairy; blade flat to inrolled. **INFL:** panicle- to spike-like, narrow, gen compact. **SPIKELET:** glumes ± equal, 3–9-veined; florets 3–6, breaking above glumes and weakly

between florets, lower 1–3 florets fertile, bisexual, upper gen sterile, gradually reduced; lemmas < glumes, elliptic to ovate, firmly membranous, rounded on back, gen 9-veined, awned at truncate tip, awns 9, plumose; fertile palea slightly > lemma. ± 30 spp.: warm temp N.Am, Afr, Asia, Australia. (Greek: nine beards, from 9 plumose lemma awns) [Reeder 2003 FNANM 25:286–287]

E. desvauxii P. Beauv. (p. 1453) NINE-AWNED PAPPUS GRASS **ST**: 1–4 dm; nodes dense, short-hairy; internodes 2–5 cm. **LF**: soft-hairy; sheath ciliate; ligule hairs < 1 mm; blade > 2 × internode, < 2 mm wide, ± inrolled. **INFL**: spike-like, 3–6 cm, ± gray. **SPIKELET**: glumes minutely soft-hairy on back, strongly veined, lower 3–5 mm, upper 4–6 mm; florets 3, lower 1 fertile, upper 2 sterile; lemma 1–3 mm, awns 2–5 mm, exserted. 2*n*=20. Rocky slopes, crevices, calcareous soils, desert woodland; 1275–1825 m. e DMoj; to CO, w TX, n Mex, S.Am. Lower sheaths sometimes enclose cleistogamous spikelets that disperse with st parts. Aug–Sep ★

ERAGROSTIS LOVE GRASS

John R. Reeder

Ann, per; often glandular, glands often wart-like, round, pitted. **LF**: sheath margin hairy on sides just below collar; ligules ciliate. **INFL**: gen panicle-like, open or dense, occ spike-like, often glandular. **SPIKELET**: laterally compressed; glumes ± unequal, acute or acuminate, 1(3)-veined; florets 3–many, axis breaking above glumes and between florets, or persistent, with glumes, lemmas deciduous, paleae remaining attached or not; lemma keeled or rounded, acute or obtuse, 3-veined, veins gen obvious; palea ± = lemma. **FR**: 0.4–2.4 mm, variously-shaped, occ longitudinally grooved, gen not noticeably compressed, gen red-brown. ± 300 spp.: trop, warm temp. (Greek: eros, love, agrostis, a kind of grass) [Peterson 2003 FNANM 25:65–105]

1. Per; spikelet axis breaking tardily above glumes and between florets
  2. Lf blade 12–50 (65) cm; spikelet 1.5–2 mm wide; lemma 1.8–3 mm; fr ± 1.5 mm . . . . **E. curvula**
  2′ Lf blade 2–12 cm; spikelet 0.8–1.2 mm wide; lemma 1.4–1.7 mm; fr 0.6–0.8 mm . . . . **E. lehmanniana**
1′ Ann; spikelet axis persistent, glumes and lemmas falling from axis, paleae gen persistent
  3. Pls mat-forming; basal portion of st prostrate, rooting at nodes . . . . **E. hypnoides**
  3′ Pls gen not mat-forming; basal portion of st erect, not rooting at nodes
    4. Fr with 1 or both ends truncate; surface checkered, with an evident groove on 1 side . . . . **E. mexicana**
    5. Spikelet ovate to oblong, 1.5+ mm wide; infl branches often with scattered glands . . . . subsp. **mexicana**
    5′ Spikelet linear to lance-linear, ± 1 mm wide; infl not glandular . . . . subsp. **virescens**
    4′ Fr variously-shaped, tip rounded; surface smooth, not grooved
      6. Pl without conspicuous glands or glandular areas (rarely so in *Eragrostis pilosa*)
      7. Palea deciduous; lower glume < 1/2 lowest floret; lowest infl branches whorled . . . . **E. pilosa** var. **pilosa**
      7′ Palea persistent; lower glume > 1/2 lowest floret; infl branches alternate or opposite . . . . **E. pectinacea**
        8. Spikelet stalks spreading . . . . var. **miserrima**
        8′ Spikelet stalks appressed to branches, diverging < 20° . . . . var. **pectinacea**
      6′ Pl with conspicuous glands or glandular areas on lf sheath, blade margin, infl axis and branches,
        spikelet stalk, or lemma keel
        9. Infl contracted, 0.5–2 cm wide, 1° branches gen ascending to appressed; spikelets light yellow . . . . . . **E. lutescens**
        9′ Infl open to contracted, 2–18 cm wide, 1° branches diverging; spikelets ± green, red-purple, or lead-colored
          10. Spikelets 2–4 mm wide; glume and lemma keels glandular; anthers yellow . . . . **E. cilianensis**
          10′ Spikelets 1.1–2.2 mm wide; glume and lemma keels not glandular; anthers red-brown
            11. Anthers 2; spikelet stalks gen with a distal ring of cup-shaped glands; lemma keels sometimes
              glandular . . . . **E. minor**
            11′ Anthers 3; spikelet stalks without a distal ring of cup-shaped glands; lemma keels not glandular
          . . . . **E. barrelieri**

E. barrelieri Daveau (p. 1453) MEDITERRANEAN LOVE GRASS Ann. **ST**: decumbent or erect, tufted, occ prostrate-spreading, branching at base, (1)2.5–3(6) dm, glandular below nodes. **LF**: sheath glabrous, soft-hairy near collar; ligule ± 0.5 mm; blade 2–10(15) cm, 2–5 mm wide, gen glabrous, flat to inrolled at tip, glandless. **INFL**: 3–15 cm, 2–6(8) cm wide, open; axis with glandular bands or patches below branches; branches, esp lowest, spreading to stiffly ascending. **SPIKELET**: ± 1 cm, 1–1.5 mm wide, slightly compressed, linear; lower glume 1–1.5 mm, slightly < upper; axis not breaking apart; florets 10–15(20); lemma ± 2 mm, gray-green or with ± red tinge, lateral veins conspicuous; palea persistent; anthers ± 0.2 mm. **FR**: ± 0.8 mm, elliptic-ovoid, not grooved. 2*n*=40,60. Disturbed soils; < 1200 m. GV, SCoRO, SCo, WTR, SnBr, PR; to KS, TX; native to s Eur. Jun–Jul

E. cilianensis (All.) Janch. (p. 1453) STINK GRASS Ann. **ST**: spreading or decumbent, occ abruptly bent, often branching, < 6 dm; glands gen present below nodes. **LF**: sheath glabrous, long-hairy below collar, keel glandular; blade 10–20 cm, 2–8 mm wide, flat or inrolled, margin with wart-like glands. **INFL**: < 20 cm, 5–6+ cm wide, ± compact, gen gray-green; spikelet stalk gen not glandular.

SPIKELET: 2.5–3 mm wide, linear to ovate; glumes 1.5–2 mm, ± equal, midvein often glandular; axis not breaking apart; florets (5)10–45; lemma 2–2.5 mm, lateral veins prominent, midvein gen glandular; palea persistent; anthers ± 0.2 mm. **FR**: ± 0.6 mm, ± spheric, not grooved. 2*n*=20. Disturbed soils; < 2000 m. CA; to e US; native to Eur. [*E. megastachya* (Koeler) Link] Jun–Oct

E. curvula (Schrad.) Nees (p. 1453) WEEPING LOVE GRASS Per. **ST**: erect, densely tufted, unbranched, 4–12 dm, glabrous. **LF**: sheaths < internodes, glabrous or hairy, gen with long hairs on collar and inside upper sheath margin; ligule ± 1 mm; blade (1)2–3(5) dm, inrolled, long-tapered, distally thread-like, scabrous. **INFL**: < 35 cm, 15 cm wide, open, gen nodding; branches flexible; lower axils long-hairy; spikelet short-stalked. **SPIKELET**: 5–8 mm, 1.5–2 mm wide, gen gray-green; glumes acute, 1-veined, lower 1.5–2 mm, upper 2.5–3 mm; axis breaking apart tardily; lemma obtuse to ± acute, 3-veined, lemma of lowest floret 1.8–3 mm; anthers 1–1.5 mm, purple. **FR**: ± 1.5 mm, ± ovoid or oblong, light brown. 2*n*=40,60,80. Roadsides; < 700 m. CaRF, GV, SCoRO, SnBr, PR, DMoj; to s US; native to s Afr. Orn, cult for erosion control. Aug–Oct

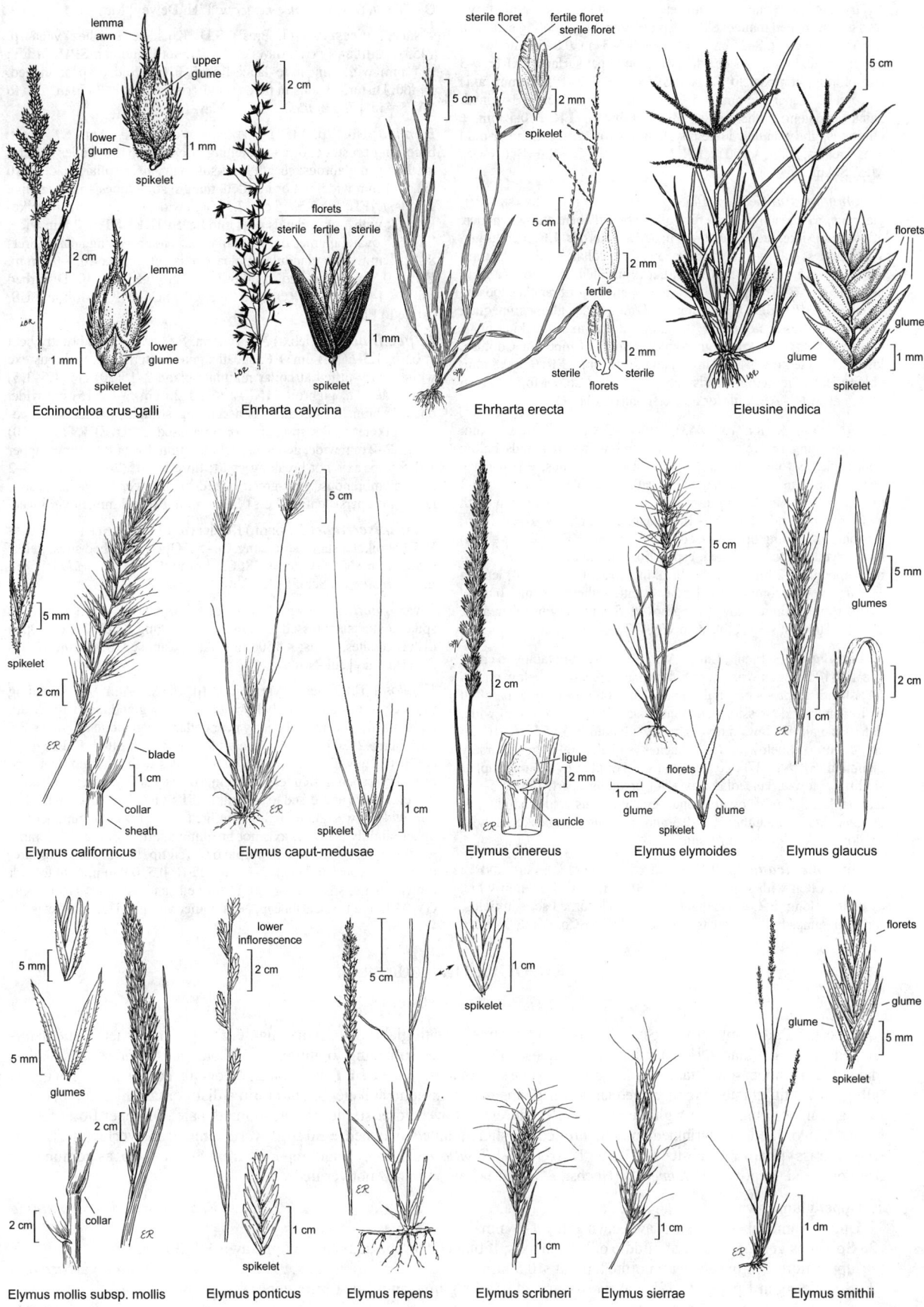

Echinochloa crus-galli

Ehrharta calycina

Ehrharta erecta

Eleusine indica

Elymus californicus

Elymus caput-medusae

Elymus cinereus

Elymus elymoides

Elymus glaucus

Elymus mollis subsp. mollis

Elymus ponticus

Elymus repens

Elymus scribneri

Elymus sierrae

Elymus smithii

E. hypnoides (Lam.) Britton et al. (p. 1453) CREEPING LOVE GRASS Ann, gen matted. **ST:** creeping, branching, rooting at nodes; fl branches ± erect, 2–25 cm. **LF:** blade 0.5–5 cm, 0.8–5.5 mm wide, flat or inrolled, glabrous or ± hairy esp on upper surface. **INFL:** 1–5 cm, elliptic, compact to ± open. **SPIKELET:** 5–10(20) mm; axis not breaking apart; lemma 1.5–2 mm, acute, translucent, 3-veined; palea persistent; anthers 0.2–0.3 mm, ± brown. **FR:** ± 0.5 mm, ± elliptic, slightly flattened laterally, light brown. 2*n*=20. Sand or mud near streams, lakes; < 500 m. KR, NCoR, c SNF, GV, SnFrB; to e US, Mex, S.Am. Jul–Sep

E. lehmanniana Nees (p. 1453) LEHMANN'S LOVE GRASS Per, glabrous. **ST:** decumbent to erect, often abruptly bent at lower nodes, often stolon-like, branched, 3–6 dm; nodes glabrous. **LF:** sheaths gen < internodes, glabrous or sparsely hairy near collar; ligule ± 0.5 mm; blade (5)8–15 cm, 1–3 mm wide, flat to inrolled, tapered to a rigid point. **INFL:** 6–20 cm, < 10 cm wide, open; lower branches loosely spreading. **SPIKELET:** 4–8 mm, ± 1 mm wide; glumes unequal, lower ± 1 mm, acute, upper 1–2 mm, obtuse; axis breaking apart tardily; florets 6–10, linear, gray-green; lemma membranous, obtuse, lowest floret lemma 1.4–1.7 mm; anthers ± 0.8 mm. **FR:** 0.6–0.8 mm, oblong, pale. 2*n*=40,60. Roadsides; < 1000 m. SCo, DMoj; to TX, n Mex; native to s Afr. Cult for erosion control. Mar–Oct

E. lutescens Scribn. (p. 1453) SIX-WEEKS LOVE GRASS Ann. **ST:** ascending or erect, 5–25(70) cm, with ring of glands below nodes; glands round, pitted. **LF:** sheath gen glabrous, ± soft-hairy near collar, veins ± glandular; ligule ± 0.5 mm; blade 2–10 cm, 1–3 mm wide, flat, folded, or inrolled, lower surface veins gen with many round glands. **INFL:** 2–10(25) cm, erect, narrow, occ ± open; branches ascending or appressed, glandular. **SPIKELET:** 2.5–7(10) mm, ± yellow, purple tinged, esp at tip; glumes acute, lower 1–1.5 mm, upper 1.5–2 mm; axis not breaking apart; florets 8–14; lemma ± 2 mm, veins prominent; palea persistent; anthers 0.2 mm, purple. **FR:** < 1 mm, elliptic, brown, not grooved. Sandy margins of streams, lakes; < 1050 m. s SNF, SnJV; to WA, CO, AZ. Jul–Oct

E. mexicana (Hornem.) Link Ann. **ST:** widely spreading to erect, gen 1.5–10 dm, gen with ring of glandular depressions below nodes. **LF:** sheath glabrous or papillate-soft-hairy on upper margins, often with glandular depressions on veins; blade 5–25 cm, 4–7 mm wide, flat, occ hairy below, midvein rarely glandular. **INFL:** (5)10–35 cm, open; axis below nodes, branches, spikelet stalks gen sparsely glandular. **SPIKELET:** gray-green to ± red; glumes ± 2 mm, upper slightly > lower, lanceolate; axis not breaking apart; florets 5–15; lemma 1.5–2.5 mm, ovate, acute, gen glabrous; palea slightly < lemma, persistent; anthers 0.2–0.5 mm. **FR:** surface checkered, with shallow to deep groove on 1 side.

subsp. ***mexicana*** (p. 1453) MEXICAN LOVE GRASS **INFL:** branches often with scattered glands. **SPIKELET:** 1.5+ mm wide, ovate to oblong. **FR:** ± rectangular. 2*n*=60. Disturbed soils in fields, forest margins, disturbed areas; < 1500 m. SnJV, SCo, WTR, SnBr; to

OK, TX, n S.Am. [*E. neomexicana* L.H. Dewey] May–Sep

subsp. ***virescens*** (J. Presl) S.D. Koch & Sánchez Vega (p. 1453) CHILEAN LOVE GRASS **INFL:** not glandular. **SPIKELET:** ± 1 mm wide, linear to lance-linear. **FR:** ovoid or pear-shaped. 2*n*=60. Disturbed soils in fields, sandy riverbanks; < 2470 m. CA; to NV, S.Am. [*E. orcuttiana* Vasey] May–Oct

E. minor Host (p. 1453) LITTLE LOVE GRASS Ann. **ST:** erect, becoming prostrate, often branching at base, < 6 dm, nodes glandular. **LF:** gen glabrous; sheath long-soft-hairy near collar; blade 2–10 cm, 1–4 mm wide, flat or inrolled, margin sometimes with wart-like glands. **INFL:** gen 3.5–15 cm, 2.5–6 cm wide, gen gray-green; spikelet stalk with 1 or 2 glands near middle. **SPIKELET:** ± 2 mm wide, linear to ovate; glumes not glandular; axis not breaking apart; florets 8–12; lemma not glandular; palea persistent; anthers 0.2–0.3 mm. **FR:** ± 0.5 mm, ± round to elliptic, not grooved. 2*n*=40. Disturbed soils; < 1400 m. NCoRO, n&c SNF, GV, SCo, MP, DMoj; to e US, native to Eur. Jun–Sep

E. pectinacea (Michx.) Nees Ann. **ST:** erect, occ abruptly bent at base, 1.5–6(7.5) dm. **LF:** sheaths gen < internodes, glabrous exc white- hairy-tufted at collar margins; blade 2–15(20) cm, 1.5–3(5) mm wide, flat, glabrous. **INFL:** (5)10–25 cm, 3–12(15) cm wide, open; 1° branches spreading or ascending, straight, alternate or opposite; spikelet stalks spreading or appressed. **SPIKELET:** 5–8(10) mm, 1.2–2 mm wide, gen linear; glumes thin, lower 0.5–1 mm, upper 1–1.5 mm; axis not breaking apart; florets 5–15(20); lemma 1.5–2 mm, membranous, gray-green or red-tinged near tip, veins prominent; palea persistent; anthers 0.2–0.7 mm. **FR:** ± 1 mm, not grooved.

var. ***miserrima*** (E. Fourn.) Reeder (p. 1453) GULF LOVE GRASS **INFL:** spikelet stalks spreading. 2*n*=60. Open disturbed sites, fields; < 500 m. n SNF, GV; to se US, Caribbean, C.Am. [*E. arida* Hitchc.; *E. tephrosanthos* Schult.] Jul–Nov

var. ***pectinacea*** (p. 1453) CAROLINA LOVE GRASS **INFL:** spikelet stalks appressed to branches, diverging < 20°. 2*n*=60. Open disturbed sites, fields; < 1500 m. CA; to Can, se US, C.Am. [*E. diffusa* Buckley] Jul–Nov

E. pilosa (L.) P. Beauv. var. ***pilosa*** (p. 1453) Ann. **ST:** ascending to erect, 1–6.5 dm; axis below nodes rarely glandular. **LF:** sheath glabrous, margin sparsely hairy near collar, not glandular; ligule < 0.5 mm; blade 2–20 cm, 1–3.5 mm wide, flat to ± inrolled. **INFL:** 4–20 cm, < 15 cm wide, ± open; lower 1–2 nodes with whorled branches; primary branches slender, spreading or ascending, rarely reflexed; spikelet stalks appressed to spreading. **SPIKELET:** 3.5–10 mm, 1–2 mm wide; lower glume 0.5–1 mm, gen < 1/2 lowest lemma length, upper slightly > lower; axis not breaking apart; florets 5–16; lemma ± 1.5 mm, gray-green with purple or ± red tip, lateral veins obscure; palea deciduous; anthers 0.2–0.3 mm. **FR:** 0.5–0.9 mm, light to dark brown, ovoid, smooth. 2*n*=40. Disturbed sandy soils; < 770 m. KR, GV, MP; to e US, Caribbean, Mex. Other var. in c US. Jul–Oct

ERIOCHLOA CUP GRASS

Robert Webster

Ann [per]. **ST:** decumbent to erect. **LF:** basal and cauline; sheath glabrous or hairy; ligule gen < 1 mm, membrane hairy-fringed; blade gen flat. **INFL:** panicle-like, ± dense; 1° branches spreading to appressed; 2° branches appressed; spikelets many, 1–2 per node, short-stalked to subsessile, on one side of axis. **SPIKELET:** lanceolate, ± dorsally compressed, gen green, falling as 1 unit; glumes strongly unequal, lower glume gen 0, fused to spikelet base to form a disk- or cup-like ring between stalk and upper glume, upper glume ± = spikelet; florets 2, lower floret sterile, gen acuminate, palea 0, upper floret fertile, lemma firm or hard, gen wrinkled, margin inrolled, tip short-pointed to awned. ± 30 spp.: warm temp, trop, worldwide. (Latin: woolly grass) [Shaw & Webster 1987 Sida 12:165–207] *E. villosa* (Thunb.) Kunth reported from Sonoma Co. as germinating from birdseed. Report of *E. fatmensis* (Hochst. & Steud.) Clayton in n D not verified.

1. Upper glume awnless. . . . . ***E. acuminata*** var. ***acuminata***
1′ Upper glume tapered to short awn, awn gen ± 1–3 mm
  2. Spikelets gen 2 per node at middle of branch axis; lf blade glabrous adaxially; glume awn ± 1–3 mm;
     upper floret lemma abruptly pointed, point < 0.5 mm. . . . . ***E. aristata*** var. ***aristata***
  2′ Spikelets gen 1 per node; lf blade short-hairy adaxially; glume awn 0.4–1 mm; upper floret lemma awn
     slightly < 1 mm. . . . . ***E. contracta***

E. acuminata (J. Presl) Kunth var. *acuminata* (p. 1453) SOUTHWESTERN CUP GRASS, SUMMER GRASS **ST**: 3–12 dm; nodes 2–5. **LF**: sheath 4–8 cm, glabrous or short-hairy; blade gen 5–12 cm, 5–12 mm wide, surface glabrous or short-hairy adaxially. **INFL**: main axis 7–16 cm; 1° branches 1–5 cm; spikelets gen 2 per node, 1 per node near the axis tip, stalk ≤ 1 mm. **SPIKELET**: 4–6 mm, ± 1–1.5 mm wide; upper glume awnless; lower floret lemma gen 5-veined; upper floret ± acuminate. Seasonal streams; < 200 m. GV, SW, DSon; to s US, Baja CA. [*E. gracilis* (E. Fourn.) Hitchc.] Other var. native to s AZ, NM, TX, n Mex. Aug–Sep

E. aristata Vasey var. *aristata* AWNED CUP GRASS **ST**: 4–10 dm; nodes 3–10. **LF**: sheath 4–13 cm, glabrous; blade gen 6–20 cm, 6–20 mm wide, surface gen glabrous adaxially. **INFL**: main axis 5–20 cm; 1° branches 2–3.5 cm; spikelets 1–2 per node; stalk 0.5–2 mm. **SPIKELET**: 4–7 mm (exc awn), ± 1–1.5 mm wide; glume awn ± 1–3 mm; lower floret lemma 3–7-veined; upper floret abruptly pointed, point < 0.5 mm. Seasonal streams, riverbanks; < 100 m. DSon (Imperial Co.); to s US, n Mex. Other var. in C.Am. Jun–Nov

E. contracta Hitchc. (p. 1453) PRAIRIE CUP GRASS **ST**: 2–10 dm; nodes 2–5. **LF**: sheath 4–8 cm, glabrous to short-hairy; blade gen 8–12 cm, 2–8 mm wide, gen short-hairy adaxially. **INFL**: main axis 6–20 cm; 1° branches 1.5–4.5(6) cm; spikelets gen 1 per node; stalk ≤ 1 mm. **SPIKELET**: 3.5–4.5 mm, ± 1–2 mm wide, becoming purple; lower glume ± = spikelet; glume awn 0.4–1 mm; lower floret lemma 3–7-veined; upper floret lemma awn slightly < 1 mm. Seasonal streams; < 100 m. GV, SW, DSon; native to c&s US. Aug–Oct

ERIONEURON

Jesús Valdés-Reyna

Per, tufted or mat-like, sometimes from short stolons. **ST**: spreading to erect. **LF**: gen basal; ligule short-hairy; blade ± stiff, pointed. **INFL**: raceme- or panicle-like, head-like, ± dense. **SPIKELET**: laterally compressed; glumes ± equal, membranous, gen acuminate, glabrous, 1-veined; axis breaking apart above glumes and between florets; florets 4–20, lower bisexual, upper staminate or sterile; lemma wide, back rounded, gen densely soft-hairy below middle, 3-veined, tip 2-lobed or not, short-awned from tip or not; palea densely hairy below middle; stamens 1 or 3. 5 spp.: sw US, Mex, S.Am. (Greek: woolly vein, from lemma, palea hairs) [Valdés-Reyna 2003 FNANM 25:44–46] *E. pulchellum* moved to *Dasyochloa*.

E. pilosum (Buckley) Nash (p. 1453) HAIRY ERIONEURON Pl cespitose. **ST**: erect, 1–3(4) dm. **LF**: sheath margin long-soft-hairy at collar; blade 3–6 cm, 1–1.5 mm wide, flat or folded, margin white. **INFL**: raceme- or panicle-like, 1.5–4 cm, 1–2 cm wide, stalked and elevated above terminal lf cluster. **SPIKELET**: 1–1.5 cm; glumes 3–7 mm, tan or ± purple, awn 0; lemma 4–7 mm, tip minutely 2-toothed, awn 0.5–2.5 mm. 2*n*=16. Rocky slopes, ridges, pinyon/juniper woodland; 1280–2000 m. SNE, e DMtns; to KS, TX, Mex. [*Tridens p.* (Buckley) Hitchc.] May–Jun ★

FESTUCA FESCUE, RYE GRASS

James P. Smith, Jr. & Susan G. Aiken

Ann, per, loosely to densely cespitose, gen ± glabrous, rhizomes present or 0; bisexual, dioecious in *Festuca kingii*. **ST**: decumbent, ascending to erect, to 2 m. **LF**: basal and cauline; sheath gen persisting; collar gen glabrous; ligule gen < 1 mm, membranous, truncate, minutely fringed; blade flat or rolled, claw-like basal auricles gen 0. **INFL**: panicle-like, branches dense and appressed to open and spreading or spike-like. **SPIKELET**: glumes 2, < lowest floret, unequal, lower sometimes minute, 1–3-veined, upper 3–5-veined; axis breaking above glumes and between florets, florets (1)2–10(17), gen bisexual; lemma 3–5(7)-veined, base gen glabrous, tip entire, occ minutely 2-toothed, awn 0 or terminal, gen straight, glabrous; palea ± = lemma; stamens 3 or 1. **FR**: obovoid to elongate, free from or ± adhering to palea and lemma. 500+ spp.: cosmopolitan. (Latin: straw, st or straw-like weed) [Darbyshire 2007 FNANM 24:443–448; Darbyshire & Pavlick 2007 FNANM 24:389–443; Lonard 2007 FNANM 24:448–454] As treated here, incl ×*Festulpia*, ×*Festulolium*, *Leucopoa*, *Lolium*, ×*Schedolium*, ×*Schedololium*, *Schedonorus*, and *Vulpia*. *Lolium rigidum* Gaudin (combination in *Festuca* evidently not available), native to Eur & Afr, has been found sporadically as an urban and agricultural weed, but has not been collected in recent yrs and is doubtfully naturalized.

1. Infl spike-like (sometimes sparingly branched); glume 1 exc 0 in uppermost spikelet (subg. *Lolium*)
  2. Glume < rest of spikelet; lower lemma firmly membranous, flat to rounded at base . . . . ***F. perennis***
  2′ Glume ≥ rest of spikelet; lower lemma becoming hard and thick at base . . . . ***F. temulenta***
1′ Infl panicle-like, branches dense and appressed to open and spreading, or raceme-like; glumes 2, lower sometimes minute
  3. Dioecious; spikelets staminate or pistillate; reduced, sterile pistils sometimes present (subg. *Leucopoa*) . . . . . ***F. kingii***
  3′ Bisexual; florets gen bisexual
    4. Lf blade with prominent claw-like or clasping basal auricles (subg. *Schedonorus*)
      5. Auricles hairy (sometimes reduced to only 1 or 2 hairs); lemma gen minutely hairy, awn 0.5–4 mm or 0 . . . . ***F. arundinacea***
      5′ Auricles glabrous; lemma gen smooth, awn << 0.5 mm or 0 . . . . ***F. pratensis***
    4′ Lf blade gen without prominent claw-like, clasping basal auricles (inconspicuous flap-like auricles sometimes present)
      6. Ann; stamen 1 (subg. *Vulpia*)
        7. Lower glume gen < 2 mm, < 1/2 upper glume, sometimes minute . . . . ***F. myuros***
        7′ Lower glume gen > 3 mm, ≥ 1/2 upper glume (2–3 mm, < 1/2 upper glume in some *Festuca microstachys*)
          8. Florets (5)7–12, closely overlapping, spikelet axis hidden; lemma awn 0.5–5 mm . . . . ***F. octoflora***
          8′ Florets gen 1–7, loosely overlapping, spikelet axis visible, each internode > 1 mm; lemma awn 3.5–12 mm
            9. Lowest infl branches appressed to erect at maturity; branches without basal swellings . . . . ***F. bromoides***
            9′ Lowest infl branches spreading or reflexed at maturity; branches with basal swellings . . . . ***F. microstachys***
      6′ Per; stamens gen 3 (subg. *Festuca*)
        10. Lf sheath closed, ± red, hairs ± downward-pointing . . . . ***F. rubra***

10′ Lf sheath open at least 1/2 its length, gen green, glabrous or hairy, hairs not downward-pointing
   11. Florets long-stalked, spikelet axis jointed below florets .... ***F. subuliflora***
   11′ Florets not long-stalked, spikelet axis not jointed
      12. Lf blade gen (1)2–10 mm wide, flat or rolled longitudinally
         13. Collar gen densely hairy (glabrous); lemma awn (0)1.5–2.5 mm, gen < 1/3 lemma; pls densely cespitose, with persistent dead lf sheaths at base .... [2]***F. californica***
         13′ Collar glabrous; lemma awn 2–20 mm, gen > 1/3 lemma; pls loosely cespitose, dead lf sheaths inconspicuous or 0
           14. Lemma distinctly 5-veined, awned from between 2 short teeth .... ***F. elmeri***
           14′ Lemma indistinctly veined, awn terminal .... ***F. subulata***
      12′ Lf blade < 2.5 mm wide, folded or rolled inward toward upper side
         15. Collar gen densely hairy, rarely glabrous; pls densely cespitose, with persistent dead lf sheaths at base .... [2]***F. californica***
         15′ Collar glabrous; pls loosely to densely cespitose, without persistent dead lf sheaths at base (exc *Festuca idahoensis*)
           16. Lemma awns 3–12 mm
              17. Ovary tip glabrous; awn < lemma body; lf blade often glaucous or ± blue .... ***F. idahoensis***
              17′ Ovary tip hairy; awn gen > lemma body; lf blade green .... ***F. occidentalis***
           16′ Lemma awns 0–2.5 mm
              18. Spikelets 9–12 mm; lower cauline lf blades much shorter and stiffer than the upper .... ***F. viridula***
              18′ Spikelets 2.5–9 mm; upper and lower cauline lf blades similar
                19. Anthers when dry 2–3.5 mm; pls of forested sites, gen 1100–1800 m .... ***F. trachyphylla***
                19′ Anthers when dry 0.5–2 mm; pls of subalpine or lower alpine habitats, gen > 3000 m
                  20. Anthers 0.9–1.2 mm when dry .... ***F. saximontana***
                  20′ Anthers 0.4–0.8(0.9) mm when dry
                    21. Infl branches at lowest node 0–1; ovary tip glabrous .... ***F. brachyphylla*** subsp. ***breviculmis***
                    21′ Infl branches at lowest node (1)2 or 3; ovary tip sparsely short-hairy .... ***F. minutiflora***

F. arundinacea Schreb. (p. 1457) TALL FESCUE Per, sometimes rhizomed. **ST**: 8–20 dm, robust; nodes visible. **LF**: sheath shredding with age; ligule 0.5–1 mm; blade 25–70 cm, 4–10 mm wide, flat or loosely rolled, hairy, ± rigid, prominently ribbed above, basal auricles ± clasping st, ± hairy. **INFL**: 15–35 cm; branches many, spreading. **SPIKELET**: 8–16 mm; lower glume 3–6 mm, upper 4–9 mm; florets 3–8; lemma 6–12 mm, gen minutely hairy, scabrous near tip, often tinged purple, awn 0.5–4 mm or 0; anthers 3–4 mm; ovary tip glabrous. 2n=28,42,56,63,70. Disturbed places; < 2700 (gen < 1000) m. CA-FP, W&I; to e N.Am; native to Eurasia. [*Schedonorus a.* (Schreb.) Dumort.] Used for forage and soil stabilization. Often infected with a fungus that is toxic to cattle. May–Jun ❖

F. brachyphylla Schult. & Schult. f. subsp. ***breviculmis*** Fred. (p. 1457) ALPINE FESCUE Per. **ST**: 0.4–2 dm, densely clumped; nodes ± concealed. **LF**: ligule < 0.5 mm; blade 1–6 cm, < 0.5 mm wide, folded. **INFL**: 0.8–2.5 cm, narrow, branches at lowest node 0–1. **SPIKELET**: 3.5–5.5 mm; lower glume 2–3 mm, upper 2.5–4.5 mm; florets 2–4; lemma 3–4.5 mm, slightly scabrous near tip, awn ± 1–1.5 mm; anthers ± 0.5–1.5 mm; ovary tip glabrous. 2n=28. Rocky places, subalpine or alpine; 2800–4300 m. c&s SNH, n SNE, W&I. *F. brachyphylla* subsp. *coloradensis* Fred. not in CA. Frequently incl in *F. ovina* L.; closely related to *F. minutiflora* Rydb. Jul–Sep

F. bromoides L. (p. 1457) BROME FESCUE Ann. **ST**: < 5 dm, glabrous or hairy. **INFL**: 1.5–15 cm, dense, clearly > uppermost lf; branches 1–3 per node; spikelet stalk flat or winged. **SPIKELET**: 5–10 mm; lower glume 3.5–5 mm, upper 4.5–8 mm; florets 4–7; lemma 5.5–8 mm, awn 2.5–12 mm, scabrous. **FR**: ± 3.5–5 mm. 2n=14. Uncommon. Dry, disturbed places, coastal-sage scrub, chaparral; gen < 1500 m. CA-FP, DMoj; to BC, e US, Baja CA; native to Eur. [*Vulpia b.* (L.) Gray] Apr–Jun

F. californica Vasey (p. 1457) CALIFORNIA FESCUE Per. **ST**: 4.5–12 dm, densely clumped; nodes visible. **LF**: sheath ± scabrous, conspicuously persisting; collar gen densely hairy, rarely glabrous; ligule 0.5–1.5(4) mm; blade 10–100 cm, 1.8–3.5 mm wide, flat or rolled, scabrous. **INFL**: 10–27 cm, open; spikelets borne near branch tips. **SPIKELET**: 13–18 mm; lower glume 4.5–7 mm, upper 6–8 mm; florets 4–6; lemma 7.5–11 mm, glabrous, awn 1.5–2.5 mm; anthers 4.5–5 mm; ovary tip hairy. 2n=56. Dry, open forest, moist streambanks, chaparral; < 1800 m. NW, CaR, n&c SN, CW, SnBr; OR. If recognized taxonomically: smaller pls with ± glabrous collars

in SnBr assignable to *F. californica* var. *parishii* (Piper) Hitchc. [*F. californica* subsp. *parishii* (Piper) Darbysh.]; pls in Santa Clara, San Luis Obispo cos. with longer ligules that gen lack cilia assignable to *F. californica* subsp. *hitchcockiana* (E.B. Alexeev) Darbysh. May–Jun

F. elmeri Scribn. & Merr. (p. 1457) Per. **ST**: 4–10 dm, loosely tufted; nodes visible. **LF**: ligule < 0.5 mm; blade 10–40 cm, ± 2–6 mm wide, flat or loosely rolled, scabrous or hairy above. **INFL**: 10–20 cm; branches slender, ± drooping. **SPIKELET**: 7–11 mm; lower glume 2–4 mm, upper 3–4.5 mm; florets 2–6; lemma 5.5–7 mm, very scabrous, awn 2–5 mm, subterminal from between 2 short teeth, scabrous; anthers 3.5–4 mm; ovary tip hairy. Moist, wooded slopes, under trees in rich soil; < 1830 m. NW, CW; OR. May–Jul

F. idahoensis Elmer (p. 1457) IDAHO FESCUE, BLUE BUNCHGRASS Per. **ST**: 3–10 dm, gen densely clumped; nodes visible. **LF**: sheath clearly persistent; ligule < 0.5 mm; blade 5–35 cm, < 2 mm wide, rolled, scabrous, ± stiff. **INFL**: 6–20 cm; branches ± appressed, scabrous. **SPIKELET**: 7–17 mm; lower glume ± 2.5–6 mm, upper 4–8 mm; axis gen visible, gen zigzag; florets 3–9; lemma ± 6–10 mm, scabrous near tip, awn 1–6 mm; anthers 3–4.5 mm; ovary tip glabrous. 2n=28. Dry, open or shady places; gen < 1800 m. NW, CaR, n&c SN, n&c CW, MP; to Can, CO. [*F. roemeri* (Pavlick) E.B. Alexeev] More studies are needed of the *F. idahoensis* complex. It is sometimes incl in *F. ovina* L. and often confused with it. Jul–Sep

F. kingii (S. Watson) Cassidy (p. 1457) Per, rhizomes short; dioecious. **ST**: 3–8 dm, tufted; nodes visible. **LF**: sheath glabrous to densely hairy, conspicuously persisting, gen red-brown with age; ligule 0.5–3.5 mm, back puberulent; blade 3–30 cm, 1.5–7 mm wide, ± flat, stiffly erect, glaucous. **INFL**: branches appressed, some with spikelets near base. **STAMINATE SPIKELET**: 5–10(12) mm; lower glume 3–5.5 mm, upper 4–6.5 mm; florets 2–5; lemma 5–8 mm, finely scabrous, awn 0; anthers 3.5–5 mm; pistil reduced or 0. **PISTILLATE SPIKELET**: 5–8 mm; lower glume 3–5 mm, upper 3.5–6 mm; florets 2–5; lemma 5–8 mm, finely scabrous, awn 0; stamens sterile or 0; ovary tip hairy. **FR**: plump, minutely beaked. 2n=56. Dry, sandy places, sagebrush plains to subalpine forest; > 1550 m. s SNH, SnBr, GB; to OR, NB. [*Leucopoa k.* (S. Watson) W.A. Weber] Jun–Aug

F. microstachys Nutt. Ann. **ST**: 1.5–7.5 dm, glabrous. **INFL**: 2–24 cm, ± open, at least lower branches spreading or reflexed;

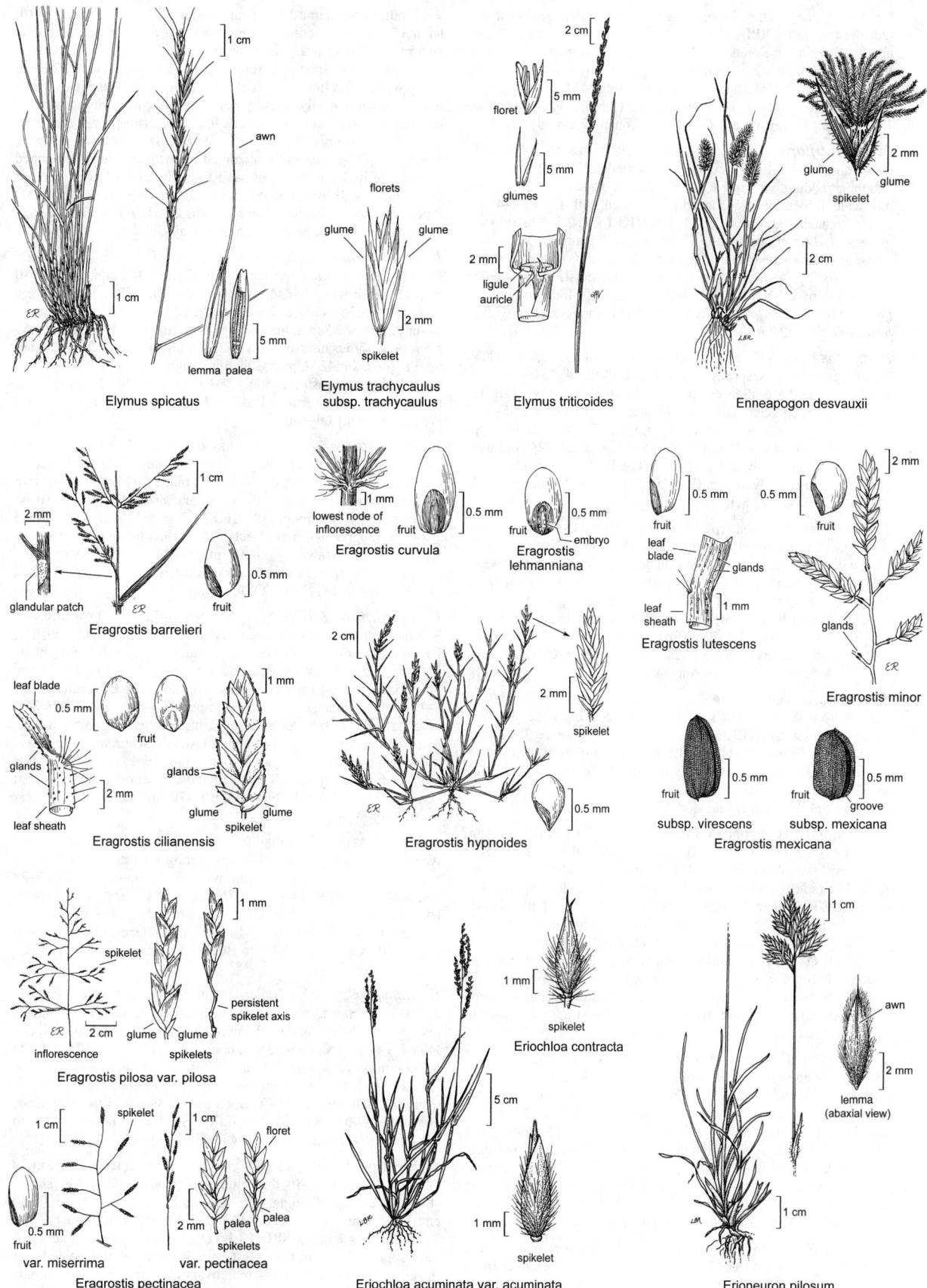

Elymus spicatus

Elymus trachycaulus
subsp. trachycaulus

Elymus triticoides

Enneapogon desvauxii

Eragrostis barrelieri

Eragrostis curvula

Eragrostis
lehmanniana

Eragrostis lutescens

Eragrostis minor

Eragrostis cilianensis

Eragrostis hypnoides

subsp. virescens

subsp. mexicana

Eragrostis mexicana

Eragrostis pilosa var. pilosa

Eriochloa contracta

var. miserrima

var. pectinacea

Eragrostis pectinacea

Eriochloa acuminata var. acuminata

Erioneuron pilosum

branches 1 per node, < 7 cm; spikelet stalk angular; gen most florets cleistogamous. **SPIKELET:** 5.5–10 mm; lower glume 2–3 mm, upper 3.5–7.5 mm; florets 2–4; lemma 4–9.5 mm, awn 3.5–12 mm. **FR:** ± 4–6 mm. $2n$=42. Disturbed, open, gen sandy soils; gen < 1500 m. CA; to BC, ID, NM, Baja CA. [*Vulpia m.* (Nutt.) Munro var. *ciliata* (Beal) Lonard & Gould; *V. m.* var. *confusa* (Piper) Lonard & Gould; *V. m.* var. *m.*; *V. m.* var. *pauciflora* (Beal) Lonard & Gould] Apr–Jun

F. minutiflora Rydb. SMALL-FLOWERED FESCUE Per. **ST:** 0.4–3 dm, loosely tufted; nodes ± concealed. **LF:** sheath open 1/2 length, shredding with age; ligule < 0.5 mm; blade 2–12 cm, << 0.5 mm wide, folded, often V-shaped in ×-section, soft. **INFL:** 1–4 cm, narrow, branches at lowest node 1–3. **SPIKELET:** 2.5–5 mm; lower glume ± 1–2.5 mm, upper 2–3.5 mm; florets 2; lemma ± 2–3.5 mm, sparsely scabrous near sharp tip, awn 0.5–1.5 mm; anthers ± 0.5–1 mm; ovary tip sparsely short-hairy. $2n$=28. Moist, shady banks; 2850–4050 m. c SNH (Mount Dana, Tuolumne, Mono cos.), W&I (White Mtns); to YT, w Can, WY, NM. Sometimes confused with *F. brachyphylla.* Jul ★

F. myuros L. (p. 1457) RATTAIL SIXWEEKS GRASS Ann. **ST:** < 7.5 dm, glabrous or scabrous only near infl. **INFL:** 4–25 cm, < 2 cm wide, ± dense; base often enclosed in sheath at maturity; branches 1–3 per node; spikelet stalk < 1 mm, slender. **SPIKELET:** 5–11.5 mm; lower glume gen < 2 mm, upper ± 2.5–5.5 mm; florets 3–6; lemma 4.5–6.5, awn 5–15 mm. **FR:** 3.5–4.5 mm. $2n$=42. Common. Gen open places, sandy soils; < 2000 m. CA-FP, D; worldwide; probably native to Eur. [*Vulpia m.* (L.) C.C. Gmel. var. *hirsuta* Hack.; *V. m.* f. *m.*; *V. m.* var. *m.*] Feb–May ❖

F. occidentalis Hook. (p. 1457) WESTERN FESCUE Per. **ST:** 4–11 dm, slender, loosely clumped; nodes visible. **LF:** sheath shredding with age; ligule << 0.5 mm; blade 5–25 cm, < 0.5 mm wide, folded, often V-shaped in ×-section, soft. **INFL:** 5–20 cm; branches 1–2, often drooping. **SPIKELET:** 6–12 mm; lower glume 2–5 mm, upper 3–6 mm; florets 2–6; lemma 4–8 mm, tip sparsely scabrous, awn 3–12 mm; anthers ± 2–3 mm; ovary tip hairy. $2n$=28. Open pine/oak woodland, redwood forest; gen < 1900 m. NW, CaR, SN, n&c CW, PR, MP; to BC, e N.Am. Apr–Jul

F. octoflora Walter (p. 1457) SIXWEEKS GRASS Ann. **ST:** < 6 dm, glabrous or hairy. **INFL:** 0.4–16 cm, 0.5–2 cm wide, dense; 1 branch per node. **SPIKELET:** 4.5–10 mm; lower glume ± 2–4.5 mm, upper ± 2.5–7 mm; florets (5)7–12; lemma ± 3–5 mm; awns 0.5–5 mm. **FR:** 2–3.5 mm. $2n$=14. Sandy to rocky soils, open sites; < 2000 m. CA; widespread Am, Eur. [*Vulpia o.* (Walter) Rydb.; *V. o.* var. *hirtella* (Piper) Henrard] Mar–Jun

F. perennis (L.) Columbus & J.P. Sm. (p. 1457) RYE GRASS Ann, per. **ST:** ascending to erect, (3)5–10 dm, gen glabrous. **LF:** ligule 1–3 mm; blade 4–30 cm, (1) 2–5 (6) mm wide. **INFL:** 3–30 cm, spike-like (sometimes panicle-like, branches spike-like, spreading). **SPIKELET:** 5–22 mm; glume < rest of spikelet, 3.5–15 mm, lance-linear; lemma 3.5–9 mm, ± lanceolate, firmly membranous, awn 0–8 mm. $2n$=14. Dry to moist disturbed sites, abandoned fields; < 1000 m. CA-FP, DMtns (Panamint Range); to AK, e N.Am; native to Eur. [*Lolium multiflorum* Lam.; *L. p.* L.; *L. p.* L. var. *multiflorum* (Lam.) Parnell] Because awned and awnless spikelets can occur on same pl, and variation in stature, spikelet size, duration appear related to environmental stress, these taxa are not recognized. May–Sep ❖

F. pratensis Huds. (p. 1457) MEADOW FESCUE Per. **ST:** 3–13 dm, loosely clumped; nodes visible. **LF:** sheath shredding with age; ligule < 0.5 mm; blade 10–30 cm, 2–7 mm wide, flat or loosely rolled, basal auricles ± clasping st, glabrous. **INFL:** 10–25 cm, narrow, branched only at lowest node. **SPIKELET:** 12–15.5 mm; lower glume 2.5–4 mm, upper 3.5–5 mm; florets 4–10; lemma 6–8 mm, awn << 0.5 mm or 0; anthers 2–4.5 mm; ovary tip glabrous. $2n$=14. Disturbed places; gen < 2000 m. CA-FP (less common SW), GB; to e N.Am; native to Eurasia. Grown for forage; potentially problematic weed. [*Schedonorus p.* (Huds.) P. Beauv.] May–Jul

F. rubra L. (p. 1457) RED FESCUE Per, rhizomes gen present, sometimes very short. **ST:** 3–8 dm, ± clumped, decumbent at base; nodes visible. **LF:** sheath closed, ± red, shredding with age, hairs ± downward-pointing; ligule < 0.5 mm; blade 5–30 cm, < 3 mm wide, ± folded. **INFL:** 5–20 cm, ± open; branches ± ascending. **SPIKELET:**

9–12 mm; lower glume 2.5–3.5 mm, upper 3.5–5.5 mm; florets 3–10; lemma 5–7 mm, sometimes scabrous near tip, awn < 4 mm, scabrous; anthers 2.5–4 mm; ovary tip glabrous. $2n$=14,28,42,56,70,128. Sand dunes, grassland, subalpine forest; gen < 2500 m. NW, CaR, n&c SN, CW, TR, W&I; worldwide. Distinct forms occur in NCo. If recognized taxonomically: loosely cespitose pls with awnless to acute lemmas on coastal sands assignable to *F. rubra* subsp. *arenicola* E.B. Alexeev [*F. ammobia* Pavlick; *F. rubra* subsp. *densiuscula* Piper]; loosely cespitose, obscurely rhizomed pls with acuminate or awned lemmas, in rocks, pebbles, or sand from the upper littoral zone to ocean bluffs and cliffs assignable to *F. rubra* subsp. *pruinosa* (Hack.) Piper. Pls assignable to *F. rubra* subsp. *fallax* (Thuill.) Nyman reported from CA but records lacking. May–Jul

F. saximontana Rydb. MOUNTAIN FESCUE Per. **ST:** 0.5–2.5 dm, clumped; nodes ± concealed. **LF:** ligule < 0.5 mm; blade 3–10 cm, < 0.5 mm wide, folded, scabrous near tip. **INFL:** 1–5 cm, very narrow, branched only at lowest node. **SPIKELET:** ± 5–7.5 mm; lower glume 2–3.5 mm, upper 3.5–5 mm; florets 2–5; lemma ± 3.5–5.5 mm, scabrous near awn, awn ± 1–2 mm; anthers 1–2 mm; ovary tip glabrous. $2n$=42. Alpine, subalpine summits, dry granitic gravel, talus fields, sagebrush scrub; gen > 3000 m. c&s SNH, SnBr, W&I; to AK; also reported from e Russia. [*F. s.* var. *purpusiana* (St.-Yves) Fred. & Pavlick] Jul–Aug

F. subulata Trin. (p. 1457) BEARDED FESCUE Per, rhizomes short. **ST:** 3.5–12 dm, loosely clumped; nodes visible. **LF:** sheath glabrous or ± hairy; ligule < 0.5 mm; blade 10–30 cm, 3–10 mm wide, flat, glabrous or upper surface minutely scabrous. **INFL:** 10–40 cm, open; branches drooping. **SPIKELET:** 8.5–12 mm; lower glume ± 2–3 mm, upper 3–6 mm; florets 3–5; lemma 6–9 mm, glabrous to sparsely scabrous, awn 5–20 mm; anthers 1.5–3 mm; ovary tip hairy. $2n$=14,28. Open places, moist banks, forest; gen < 2500 m. NW, CaR, n&c SN; to AK, MT, WY, CO. Jun–Aug

F. subuliflora Scribn. (p. 1457) CRINKLE-AWN FESCUE Per. **ST:** 6–10 dm, loosely clumped, lfy to near panicle; nodes visible. **LF:** sheath and collar glabrous or ± hairy; ligule < 0.5 mm; blade 15–30 cm, 2.5–8 mm wide, ± flat, ± hairy, soft. **INFL:** 10–20 cm, open; branches gen 1 per node, drooping. **SPIKELET:** panicle-like, 8–12.5 mm; lower glume 2–4 mm, upper 3.5–5.5 mm; florets 2–5, long-stalked, the spikelet axis appearing jointed just below florets; lemma 6–9 mm, ± strongly veined, sparsely scabrous, base hairy-tufted, tip slightly forked, awn 10–15 mm, crinkled; anthers 2.5–4 mm; ovary tip hairy. $2n$=28. Near streams, redwood, oak/pine forest; gen < 700 m. NCo, KR, NCoRO, n SNH, SnFrB; to BC, esp near coast. May–Aug

F. temulenta (L.) Columbus & J.P. Sm. (p. 1457) DARNEL, TARES Ann. **ST:** ascending to erect, 4–9 dm, glabrous to scabrous. **LF:** ligule < 2 mm; blade 5–20 cm, 2–7 mm wide, folded in bud. **INFL:** 7–28 cm, spike-like. **SPIKELET:** (15)20–30 mm; glume ≥ rest of spikelet exc awns, 10–17(20) mm, lanceolate to oblong; lemma 5–8 mm, awnless or awned at tip, awn 8–18 mm. $2n$=14. Open, disturbed sites; 150–1750 m. CA (exc SNE); to BC, e US; native to Medit. [*Lolium t.* L.] Apr–Jun

F. trachyphylla (Hack.) Krajina HARD OR SHEEP FESCUE Per. **ST:** 2–7.5 dm, clumped, sparsely hairy near infl; nodes visible. **LF:** sheath glabrous or ± hairy; ligule < 0.5 mm; blade 8–30 cm, < 1 mm wide, ± folded, ± hairy. **INFL:** 3–13 cm; branches ± appressed. **SPIKELET:** 5.5–9 mm; lower glume 2–4 mm, upper 3–5.5 mm; axis gen visible; florets 3–8; lemma 4–5 mm, awn 0.5–2.5 mm; anthers 2–3.5 mm; ovary tip glabrous. $2n$=28,42. Open places, slopes in forested areas; gen 1000–1800 m. KR, NCoRO, CaR, n SNH, expected elsewhere; to e N.Am; native to Eur. Frequently incl in *F. ovina* L. May–Jul

F. viridula Vasey (p. 1457) MOUNTAIN BUNCH GRASS, GREEN FESCUE Per. **ST:** 5–10 dm, clumped; nodes visible. **LF:** sheath strongly veined, shredding with age; ligule < 0.5 mm; blade 10–30 cm, 1–2.5 mm wide, ± rolled. **INFL:** 4–15 cm, open; branches gen 2 per node, ascending. **SPIKELET:** 9–12 mm; lower glume 3.5–5 mm, upper 4.5–7 mm; florets 3–6 mm; lemma 5–8 mm, folded near tip, awn 0–1 mm; anthers 3–4 mm; ovary tip hairy. $2n$=28. Subalpine meadows, open forest, rocky slopes; gen > 2000 m. KR, n SNH; to BC, ID. Important forage grass. Jun–Aug

GASTRIDIUM

James P. Smith, Jr. & Dieter H. Wilken

Ann. **ST**: ascending to erect. **LF**: basal and lower cauline; ligule ± translucent; blade flat. **INFL**: panicle-like, narrow, compact; branches gen appressed to ascending. **SPIKELET**: stalked, 3–5 mm, laterally compressed; glumes unequal, membranous at swollen base, 1-veined; floret 1, breaking above glumes; axis prolonged as a minute bristle; lemma << glumes, translucent, truncate to obtuse, awned or not, 5-veined; palea ± = lemma; anthers 3. 2 spp.: warm temp Eur, n Afr, w Asia. (Greek: small pouch, from swollen glume base) [Wipff 2007 FNANM 24:675–677]

G. phleoides (Nees & Meyen) C.E. Hubb. (p. 1457) NIT GRASS
Pl (1)2–4(7) dm, gen glabrous. **LF**: ligule 1–7 mm; blade (2)4–9(20) cm, (1)3–5 mm wide. **INFL**: 1.5–9 cm, 4–10 mm wide. **SPIKELET**: glumes translucent between veins, keel minutely scabrous, lower glume 6–7 mm, upper 3–5 mm; axis extended < 1 mm; lemma ± 1 mm, densely pubescent, awned below truncate, toothed tip, awn 3–6 mm, straight to curved. 2*n*=14. Open, gen dry, disturbed sites; gen < 1450 m. CA-FP, MP; to OR, Can, ne US; native to Eur. [*G. ventricosum* (Gouan) Schinz & Thell., misappl.] Apr–Nov

GLYCERIA MANNA GRASS

Gordon Leppig

Per from rhizomes (ann). **ST**: gen rhizomatous, base decumbent, erect above, gen rooting at lower nodes. **LF**: sheath closed to near top; ligule thin, membranous, acute; blade flat or folded, only midrib prominent. **INFL**: panicle-like, 15–50 cm, erect. **SPIKELET**: glumes < lowest floret, margin translucent, 1-veined; axis breaking above glumes and between florets; florets 8–16; lemma strongly 5–11-veined, veins not converging, ending short of tip, tip margin translucent; palea ± = lemma, keel prominent, ± curved; anthers 2(3). ± 50 spp.: temp worldwide, freshwater wetland and aquatic habitats. (Greek: sweet, from taste of grain) [Barkworth & Anderton 2007 FNANM 24:68–88]

1. Spikelet ovoid; infl open, branches spreading
    2. Lemma tip ± flat; anthers 3 . . . . *G. grandis*
    2′ Lemma tip boat-shaped; anthers 2
        3. Lf blade 6–15 mm wide; anther 0.5–0.8 mm; st 60–150 cm, 2.5–8 mm diam; lower glume 0.8–1.5 mm . . . . . *G. elata*
        3′ Lf blade 2–6 mm wide; anther 0.2–0.6 mm; st 20–130 cm, 1.5–5 mm diam; lower glume 0.4–1.2 mm . . . . . *G. striata*
1′ Spikelet cylindric; infl narrow, branches appressed
    4. Lemma glabrous between barely scabrous veins . . . . *G. borealis*
    4′ Lemma scabrous throughout
        5. Lemma 2.5–4(4.5) mm, tip truncate to broadly rounded, crenate to entire . . . . *G. leptostachya*
        5′ Lemma (4)4.5–8 mm, tip acute to acuminate, gen lobed or strongly dentate
            6. Spikelet (18)20–40 mm; lemma (5.5)6–8 mm, tip entire, acute to acuminate; anther 1.5–3 mm . . . . *G. fluitans*
            6′ Spikelet 10–23 mm; lemma 4–6 mm, tip gen lobed or irregularly dentate; anthers 0.3–1.5 mm
                7. Ann or per, tufted; rhizome gen absent; lf blade (2)3–12 cm, 1.5–6(8) mm wide; lemma tip gen
                   3-lobed; palea 0.2–1.5 mm ≥ lemma . . . . *G. declinata*
                7′ Per from creeping rhizome, often mat-forming; lf blade 20–30 cm, (3)4–13 mm wide; lemma tip
                   entire to variously dentate-crenate; palea ≤ lemma . . . . *G. ×occidentalis*

G. borealis (Nash) Batch. (p. 1457) BOREAL MANNA GRASS Per. **ST**: 8–15 dm, ± 2 mm diam. **LF**: ligule 4–12 mm; blade 2–7 mm wide. **INFL**: 18–50 cm, narrow; spikelets appressed. **SPIKELET**: 9–18 mm, cylindric; lower glume 1–2 mm, upper 2–3.5 mm; florets 6–11; lemma widest below middle, glabrous exc on veins. *n*=10. Shallow water, muddy shores, freshwater ponds, lakes in conifer forest; 800–2200 m. KR, NCoR, CaR, SN, Wrn; AK, e US, n Mex. Jun–Aug

G. declinata Bréb. LOW MANNA GRASS Ann, tufted per (rhizomatous). **ST**: (1)2–9 dm, 1.5–2.5 mm diam. **LF**: ligule 4–9 mm; blade (2)3–12 cm, 1.5–6 (8) mm wide. **INFL**: 6–30 cm, narrow; spikelets appressed. **SPIKELET**: 10–23 mm, cylindric; lower glume 1.4–3.5 mm, upper 2.5–4.9 mm; florets 5–14; lemma 4–5 mm; lemma tip gen 3-lobed; palea 0.2–1.5 mm ≥ lemma; anthers 0.3–1.5 mm. *n*=10. Common. Vernal pools, freshwater marshes, disturbed, seasonally wet areas; < 1300 m. n NCo, KR, NCoR, CaR, SNF, GV, CCo, SCo, expected elsewhere; BC, e US; native to Eur. May–Jun ❖

G. elata (Rydb.) M.E. Jones (p. 1457) FOWL MANNA GRASS Per. **ST**: 6–15 dm, 2.5–8 mm diam. **LF**: ligule 2–6 mm; blade 6–15 mm wide, thin. **INFL**: 15–25 cm, open; branches spreading. **SPIKELET**: 2.5–5 mm, ovoid; lower glume 0.8–1.5 mm, upper 1–1.5 mm; florets 4–7; lemma tip boat-shaped; anthers 2, 0.5–0.8 mm. *n*=10. Common. Wet places, conifer forest; < 2600 m. NW, SN, SnBr, SnJt, Wrn; BC, Rocky Mtns. Jul–Aug

G. fluitans (L.) R. Br. WATER MANNA GRASS Per, occ forming floating mats. **ST**: 2–15 dm, 2–4 mm diam. **LF**: ligule 5–15 mm;

blade 10–25 cm, 3–10 mm wide. **INFL**: 15–50 cm, narrow; spikelets, 1° branches appressed. **SPIKELET**: (18)20–40 mm, cylindric; lower glume 1.3–3.9 mm, upper 2.7–5 mm; florets 6–16; lemma (5.5)6–8 mm, tip acute to acuminate, entire; anther 1.5–3 mm. *n*=20. Uncommon. Wet places; < 500 m. NCo, NCoRI, CaRH, expected elsewhere; BC, e US; native to Eurasia. Jun–Jul

G. grandis S. Watson (p. 1457) AMERICAN MANNA GRASS Per. **ST**: 9–20 dm, 2–7 mm diam. **LF**: ligule 2–7 mm; blade 4–15 mm wide, ± thick. **INFL**: 16–40 cm; branches spreading. **SPIKELET**: 4–6.5 mm, ovoid; lower glume 1–1.8 mm, upper 1.5–2.5 mm; florets 4–7; lemma 2–2.8 mm, tip ± flat; anthers 3. *n*=10. Wet places, meadows, lake and stream margins; < 2100 m. NCo, NCoR, SN, SNE; AK, e US. Jun–Aug ★

G. leptostachya Buckley NARROW MANNA GRASS Per. **ST**: 10–15 dm, 2–4 mm diam. **LF**: ligule 4–9 mm; blade 4–10 mm wide, flat to rolled. **INFL**: 20–40 cm, narrow; spikelets appressed. **SPIKELET**: 10–20 mm, cylindric; florets 8–14; lower glume ± 1.5 mm, upper ± 3 mm; lemma 2.5–4(4.5) mm, tip truncate to broadly rounded, crenate to entire. *n*=20. Freshwater marshes, lakes; < 800 m. NW, CaRH, CCo, SnFrB; to s AK. [*G. davyi* (Merr.) Tzvelev] May–Jun

G. ×occidentalis (Piper) J.C. Nelson WESTERN MANNA GRASS Per, often forming floating mats. **ST**: 7–15 dm, 1.5–2 mm diam. **LF**: ligule 5–12 mm; blade 20–30 cm, (3)4–13 mm wide. **INFL**: 15–40 cm, narrow; spikelets appressed. **SPIKELET**: 15–20 mm,

cylindric; lower glume 1.5–3.5 mm, upper 2.5–5 mm; florets 6–13; lemma 4.5–6 mm, tip entire to dentate or crenate; palea ≤ lemma. *n*=20. Freshwater marshes, ponds and steams; < 1300 m. NW, n SNH, ScV, CCo, SnFrB; BC, ID. [*G. o.*] Taxonomic status and nativity are uncertain. Genetic evidence suggests this may be a hybrid of *G. leptostachya* and *G. fluitans,* see: Whipple et al. 2007 Amer J Bot 94:551–557. Jun–Aug

G. striata (Lam.) Hitchc. RIDGED MANNA GRASS Per from rhizomes. **ST:** 2–13 dm, 1.5–5 mm diam. **LF:** ligule 1–3.5 mm; blade 2–6 mm wide, firm. **INFL:** 6–25 cm, open; branches spreading. **SPIKELET:** 2.5–4.5 mm, ovoid; lower glume 0.4–1.2 mm, upper 0.7–1.2 mm; florets 3–6; lemma tip boat-shaped; anthers 2, 0.2–0.6 mm. *n*=20. Common. Wet meadows, streambanks, conifer forest; 1500–2500 m. CA-FP, MP; N.Am. Jul–Aug

HAINARDIA BARB GRASS

Thomas Worley

1 sp. (Pierre Hainard, Swiss geobotanist, b. 1936) [Smith 2007 FNANM 24:689–691] See *Parapholis.*

H. cylindrica (Willd.) Greuter (p. 1461) Ann. **ST:** ascending to erect, 2–5 dm, internodes solid, branched, glabrous. **LF:** sheath 1–6 cm; ligule 1 mm, membranous; blade 6–10 cm, 1–2 mm wide, gen flat, ribbed, upper surface scabrous. **INFL:** spike-like, 8–20 cm, cylindric, stiff, straight, breaking at nodes; spikelets alternate, 2-ranked, appressed, embedded in axis, falling with axis segment, lowest incl in sheath. **SPIKELET:** lower glume 0 on all but terminal spikelet, upper glume 1, gen 5–7 mm, thick, rigid, margin sometimes inrolled, 3–7(9)-veined; floret 1, bisexual, or 2, upper reduced and sterile; lemma 4–6 mm, translucent, back facing infl axis, 3-veined; palea ± = lemma, translucent; anther 2–3.5 mm. **FR:** 2.5–3.5 mm. 2*n*=14,26,52. Coastal salt marshes, alkaline soils; < 570 m. NCo, n SNF (Amador Co.), deltaic GV, CCo, SnFrB, SCo; Baja CA, TX, LA, SC; native to Eur. [*Rottboellia c.* Willd.; *Lepturus c.* (Willd.) Trin.; *Monerma c.* (Willd.) Coss. & Durieu] Mar–Jul

HETEROPOGON BARB GRASS

Elizabeth M. Skendzic

Ann or per, cespitose. **ST:** erect, densely clumped. **LF:** cauline; sheaths keeled; ligule membranous; blade flat or folded. **INFL:** raceme-like, numerous, long-stalked, with dissimilar spikelet pairs. **SPIKELET:** in pairs, on 1 side of axis; spikelets of lower 1–4 pairs ± equal, sterile or staminate; spikelets of upper pairs unequal, sessile spikelet bisexual, glumes tightly enclosing florets, florets 2, lower vestigial, upper fertile, lemma translucent, tip awned; stalked spikelet staminate or sterile, awn 0; pair breaking below sessile spikelet. **FR:** lanceolate, with a groove. 8–10 spp.: warm temp, trop. (Greek: different beard, from awned and awnless spikelets)

H. contortus (L.) Roem. & Schult. (p. 1461) TANGLEHEAD Per. **ST:** 2–10 dm. **LF:** blade 6–20 cm, 4–8 mm wide. **INFL:** 4–8 cm. **SPIKELET:** of lower pairs 6–10 mm; sessile spikelets of upper pairs 5–10 mm, glume dark brown, sharp-pointed, callus hairs red-brown, awn 6–10 cm, puberulent; stalked spikelets of upper pairs 6–10 mm, glumes glabrous or with hairs on papillae. 2*n*=40, 50, 60. Uncommon. Rocky slopes, washes, open areas; < 800 m. SCo (San Diego Co.), DSon; s US, Mex. Mar–Nov ◆

HILARIA

J. Travis Columbus

Per, cespitose, rhizomed. **ST:** ascending to erect, solid; nodes gen hairy. **LF:** ligule membranous, ciliate-fringed; blade firm, flat to inrolled, sharply acute. **INFL:** spike-like, gen cylindric; spikelets in clusters, 3 per node, ± equal; clusters wedge-shaped, overlapping, ascending to appressed, hairy-tufted at base, falling as 1 unit from zigzag axis; glumes of cluster together involucre-like. **CENTRAL SPIKELET:** subsessile, appressed to or nearest infl axis; glumes equal, < floret(s), oblanceolate, keeled, ciliate, tip deeply 2-lobed, lobes lanceolate, awns 3–9, 1 from ± mid-keel, others terminal; florets gen 1–2, lower floret bisexual, upper floret (if present) bisexual or staminate; lemma lanceolate, 3-veined, gen ciliate, tip gen 2-lobed, gen 1-awned ± from sinus; palea ± = lemma. **LATERAL SPIKELET:** sessile; glumes < to ± = florets, ciliate, lower glume asymmetric with 1 awn from ± middle near margin, gen 2-lobed, lobes unequal; florets 1–4, gen staminate; lemma 3-veined, tip gen ciliate; palea ± = lemma. 10 spp.: w US to Guatemala. (Auguste de Saint-Hilaire, French biologist, 1779–1853)

1. St internodes glabrous to puberulent . . . . ***H. jamesii***
1′ St internodes sparsely to densely tomentose . . . . ***H. rigida***

H. jamesii (Torr.) Benth. (p. 1461) GALLETA Pl 1.5–4(6.5) dm, unbranched above base. **ST:** ± 1 mm diam; node hairs ± straight. **LF:** gen basal; gen glabrous or scabrous exc long-ciliate near ligule; ligule membrane 1–3 mm, gen with lateral teeth; blade < 13(21) cm, 2–3 mm wide, adaxial surface sometimes with short, ± straight hairs. **INFL:** 3–7 cm; spikelet clusters 6–9 mm. **CENTRAL SPIKELET:** glume margin hairs < 0.5 mm; lower lemma awn < 2.5 mm. **LATERAL SPIKELET:** lower glume 1-awned; upper glume tip unlobed or 2-lobed, awn 0–0.4 mm, margin hairs < 0.5 mm; lemma tip unlobed or 2-lobed, awn 0–0.4 mm, margin hairs 0–0.2 mm. 2*n*=36,38,72. Dry, sandy to rocky slopes, flats, scrub, woodland; 1000–2700 m. SNE, DMoj (mostly DMtns); to WY, TX. [*Pleuraphis j.* Torr.] Apr–Nov

H. rigida (Thurb.) Scribn. (p. 1461) BIG GALLETA Pl 3.5–10 dm, branched above base, gen bush-like. **ST:** 1.5–3.5 mm diam at base; node hairs curly. **LF:** gen cauline, gen sparsely to densely tomentose, esp near and sometimes overlapping ligule; ligule membrane < 1 mm, lateral teeth 0; blade < 10 cm, 2–4 mm wide. **INFL:** 4–10 cm; spikelet clusters 7–11 mm. **CENTRAL SPIKELET:** glume margin hairs 0.5–3 mm; lower lemma tip sometimes 4-lobed, awns 3, ± from sinuses, central awn 2–5.5 mm. **LATERAL SPIKELET:** lower glume with 1+ subsidiary lobes, or larger lobe tip fringed, awns 2–4; upper glume tip with 2+ lobes or fringed, awns 1–3, 0.4–2.5 mm, margin hairs 0.5–2 mm; lemma tip 2-lobed, awn 1 ± from sinus, 0.4–2 mm, margin hairs 0.2–1 mm. 2*n*=18,36,±108. Common. Dry, open, sandy to rocky slopes, flats, and washes, sand dunes, scrub, woodland; < 1600 m. PR, e&s DMoj, DSon; to UT, nw Mex. [*Pleuraphis r.* Thurb.] Important forage; some pls from e DMoj and Ord Mtn (San Bernardino Co.), with ± straight internode hairs, intermediate to *H. jamesii.* All yr

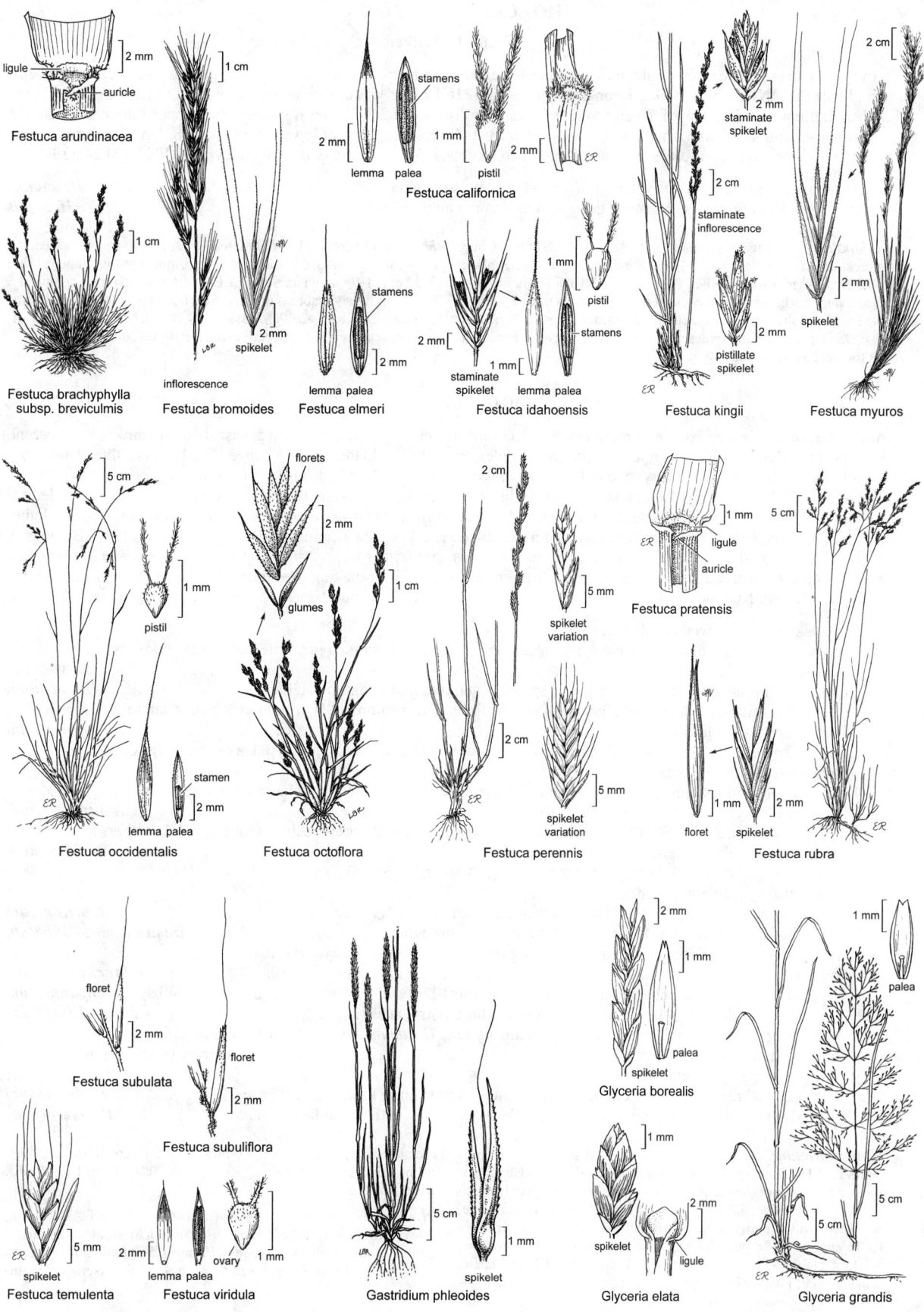

Festuca arundinacea

ligule
auricle
2 mm

Festuca brachyphylla
subsp. breviculmis

Festuca bromoides
inflorescence
1 cm
1 cm
spikelet
2 mm

Festuca californica
lemma palea
stamens
2 mm
pistil
1 mm
2 mm
ER

Festuca elmeri
lemma palea
stamens
2 mm

Festuca idahoensis
staminate
spikelet
2 mm
lemma palea
stamens
1 mm
pistil
1 mm

Festuca kingii
staminate
spikelet
2 mm
staminate
inflorescence
2 cm
pistillate
spikelet
2 mm
ER

Festuca myuros
2 cm
spikelet
2 mm

Festuca occidentalis
5 cm
pistil
1 mm
lemma palea
stamen
2 mm
ER

Festuca octoflora
florets
2 mm
glumes
1 cm

Festuca perennis
2 cm
spikelet
variation
5 mm
spikelet
variation
5 mm
2 cm
ER

Festuca pratensis
ligule
auricle
1 mm
ER

Festuca rubra
floret spikelet
1 mm   2 mm
5 cm
ER

Festuca subulata
floret
2 mm

Festuca subuliflora
floret
2 mm

Festuca temulenta
spikelet
5 mm
ER

Festuca viridula
lemma palea
2 mm
ovary
1 mm

Gastridium phleoides
5 cm
spikelet
1 mm

Glyceria borealis
spikelet
palea
2 mm
1 mm

Glyceria elata
spikelet
ligule
1 mm
2 mm

Glyceria grandis
palea
1 mm
5 cm
5 cm
ER

HOLCUS VELVET GRASS

Dieter H. Wilken

(Ann) per, gen cespitose or from rhizomes; glabrous to velvety soft-hairy. **LF**: gen basal; ligule membranous, truncate, puberulent; blade flat. **INFL**: panicle-like, ± congested to open. **SPIKELET**: laterally compressed, breaking below glumes, falling as 1 unit; glumes ± equal, lower 1-veined, upper 3-veined; florets gen 2(3), lower floret bisexual, upper staminate or sterile; callus hairy; lemma ± 2 mm, shiny, membranous, faintly 3–5-veined, lemma of lower floret awnless, lemma of upper floret awned near 2-lobed tip; palea ± = lemma. ± 8 spp.: temp Eurasia, Afr. (Latin: a grass) [Standley 2007 FNANM 24:739–741]

1. Sts tufted, internodes soft-hairy; lemma awn 1–2 mm, twisted to recurved . . . . ***H. lanatus***
1′ Sts 1–few, from rhizomes, internode gen glabrous; lemma awn 3–5 mm, bent to straight . . . . ***H. mollis***

H. lanatus L. (p. 1461) COMMON VELVET GRASS **ST**: ascending to erect, 2–10 dm. **LF**: ligule 1–4 mm, truncate, jagged-ciliate; blade 5–18 cm, 4–9 mm wide. **INFL**: 7–15 cm. **SPIKELET**: glumes 3–6 mm, ± purple, short-hairy on back, keel long-hairy; lemma 3–4 mm. 2*n*=14. Moist sites, roadbanks, cult fields, meadows; < 2400 m. CA (exc DSon); to AK, Can; widespread US, S.Am; native to Eur. Cult for forage, hay. Jun–Aug ❖

H. mollis L. (p. 1461) CREEPING VELVET GRASS **ST**: ascending to decumbent, 3–8 dm. **LF**: ligule 2–5 mm, obtuse, jagged; blade 3–15 cm, 4–9 mm wide. **SPIKELET**: glumes 4–5 mm, tan, tinged ± purple, minutely scabrous on back, keel and veins short-hairy; lemma 2–3 mm. 2*n*=28,35,42,49. Moist sites; < 700 m. NCo, n SNH, SnFrB; to Can; scattered in e US; native to temp n Eur. Jun–Aug

HORDEUM BARLEY

Ann to per, cespitose, occ from short rhizomes. **ST**: decumbent to erect, gen abruptly bent at base; nodes glabrous or pubescent. **LF**: sheath glabrous or hairy; ligules translucent; auricles present or 0; blade flat or ± rolled. **INFL**: spike-like, dense; axis breaking apart at nodes at maturity (exc in *Hordeum vulgare*); spikelets 2-ranked, strongly overlapping, 3 per node (central + 2 lateral). **CENTRAL SPIKELET**: bisexual, gen sessile; with 1 stalked or sessile floret; glumes awn-like, gen > floret; lemma awned. **LATERAL SPIKELET**: sterile or staminate (bisexual in *Hordeum vulgare*), gen short-stalked; floret 1, sessile; glumes awn-like, > floret, lemma gen awned. 32 spp.: temp worldwide exc Australia. (Latin: ancient name for barley) [von Bothmer et al. 2007 FNANM 24:241–252] As treated here, incl *Critesion*. Hybridizes with *Elymus* to produce ×*Elyhordeum* spp. Source of edible grains, sugar, orn, and weeds; enzymes from sprouts used in beer-making. Lemma or floret measurements do not incl awns. *H. pusillum* Nutt. has been reported for Ventura and San Diego cos., but it does not appear to be naturalized.

1. Auricles of upper lvs well developed, 1–4(8) mm
  2. Infl axis gen remaining intact at maturity; lateral spikelets sessile (stalked in sterile pls); glume margins not ciliate. . . . . ***[H. vulgare]***
  2′ Infl axis breaking apart at maturity; lateral spikelets stalked; glume margins ciliate . . . . ***H. murinum***
    3. Central spikelet stalk 0–0.5 mm; lemma of central floret gen = lemma of lateral floret; palea of lateral floret scabrous to glabrous. . . . . subsp. ***murinum***
    3′ Central spikelet stalk 1–2 mm; lemma of central floret < lemma of lateral floret; palea of lateral floret hairy or scabrous on lower 1/2
      4. Lemma of central floret ≤ those of lateral florets; palea of lateral florets long-hairy on lower 1/2; anthers of central florets < 1/2 those of lateral florets . . . . subsp. ***glaucum***
      4′ Lemma of central floret << those of lateral florets; palea of lateral florets scabrous on lower 1/2; anthers of central spikelets = those of lateral florets . . . . subsp. ***leporinum***
1′ Auricles of upper lvs gen 0 (vestigial rudiments occ present on lower lvs)
  5. Glumes strongly divergent at maturity
    6. Glumes of central spikelet 11–28 mm, flat near base . . . . ***H. arizonicum***
    6′ Glumes of central spikelet (15)35–85 mm, bristle-like throughout . . . . ***H. jubatum*** subsp. ***jubatum***
  5′ Glumes gen straight with age (slightly spreading in some *Hordeum brachyantherum*)
    7. Per; infl fully exserted at maturity. . . . . ***H. brachyantherum***
      8. Lf sheath gen glabrous, blade glabrous or sparsely short-hairy; st gen robust. . . . . subsp. ***brachyantherum***
      8′ Lf sheath gen densely hairy, blade gen short- and long-hairy; st slender. . . . . subsp. ***californicum***
    7′ Ann; infl gen ± enclosed in upper sheath at maturity (exc *Hordeum marinum* subsp. *gussoneanum*)
      9. Awn of lateral spikelet lemma 3–8 mm. . . . . ***H. marinum*** subsp. ***gussoneanum***
      9′ Awn of lateral spikelet lemma 0–1.2 mm
        10. Nodes glabrous; glumes of central spikelet rounded to slightly flattened near base . . . . ***H. depressum***
        10′ Nodes gen pubescent; glumes of central spikelet distinctly flattened at base . . . . ***H. intercedens***

H. arizonicum Covas ARIZONA BARLEY Ann, bien, sometimes per. **ST**: 2.1–7.5 dm, gen erect, tufted; nodes glabrous. **LF**: lower sheaths hairy, upper sheaths glabrous, auricles 0 or vestigial; blade to 13 cm, < 4 mm wide, flat, scabrous. **INFL**: 5–12 cm, 6–10 mm wide, pale green, often partly enclosed at maturity. **CENTRAL SPIKE-LET**: glumes 11–28 mm, flat near base, divergent with age; lemma 5–9 mm, glabrous, awn 10–22 mm. **LATERAL SPIKELET**: sterile, stalks curved; glumes 20–26 mm, base flat; lemma 2.5–5 mm, awn-

like, lemma tip tapered, awn 0–3 mm. 2*n*=42. Uncommon. Wet, often saline, places; 50–1400 m. CaRF, c SNH, ScV, DSon; to NV, NM, Mex. Apr–Jul

H. brachyantherum Nevski (p. 1461) Per. **ST**: 2–9.5 dm, loosely to densely tufted; nodes glabrous. **LF**: sheath glabrous to densely hairy; auricles 0; blade < 19 cm, 1.5–9 mm wide, glabrous or hairy. **INFL**: 3–10 cm, green to ± purple, fully exserted at matu-

rity. **CENTRAL SPIKELET**: glumes 7–19 mm, bristle-like, gen straight with age, base flat; lemma 5.5–10 mm, glabrous. **LATERAL SPIKELET**: staminate; glumes 6.5–19 mm, bristle-like, straight or spreading with age, lower glume occ flat at base; lemma < 7 mm or vestigial, awn < 7.5 mm or 0; anthers 0.9–5.5 mm.

subsp. ***brachyantherum*** NORTHERN BARLEY **ST**: 3–9.5 dm, gen robust. **LF**: sheath gen glabrous; blade < 19 cm, ≤ 9 mm wide, glabrous or sparsely short-hairy. **CENTRAL SPIKELET**: glumes 7–17 mm, gen straight with age; lemma awn < 6 mm; anthers 1–3.5 mm. 2*n*=28,42. Meadows, pastures, streambanks; < 3400 m. CA (exc D); to AK, w US, Baja CA. Hybridizes with *Elymus glaucus* to yield ×*Elyhordeum stebbinsianum* (Bowden) Bowden. May–Aug

subsp. ***californicum*** (Covas & Stebbins) Bothmer et al. CALIFORNIA BARLEY **ST**: 2–6.5 dm, slender. **LF**: sheath gen densely hairy; blade ≤ 11.5 cm, gen < 3.5 mm wide, gen short- and long-hairy, rarely glabrous. **CENTRAL SPIKELET**: glumes 9–19 mm, gen spreading with age; lemma awn to 7.5 mm; anthers 1.5–5.5 mm. 2*n*=14. Meadows, pastures, streambanks; < 3400 m. CA (exc D). [*H. c.* Covas & Stebbins] Mar–Jul

H. depressum (Scribn. & J.G. Sm.) Rydb. (p. 1461) LOW BARLEY, ALKALI BARLEY Ann. **ST**: 1–5.5 dm, erect, loosely tufted; nodes glabrous. **LF**: basal sheaths hairy, auricles 0; blade ≤ 7.5(13.5) cm, ≤ 4.5 mm wide, surfaces sparsely to densely hairy. **INFL**: 2.2–7 cm, 4–8 mm wide, pale green or ± red, gen ± enclosed at maturity. **CENTRAL SPIKELET**: glumes 5.5–20 mm, gen straight with age, rounded to slightly flattened near base; lemma 5–9 mm, glabrous, awned, lemma awn 3–12 mm. **LATERAL SPIKELET**: staminate or sterile, occ bisexual; glumes 5–20 mm, lower ± flattened at base; floret vestigial; lemma 1.8–8.5 mm, awn < 1 mm or 0; anthers 0.5–1.5 mm. 2*n*=28. Moist sites, vernal pools, gen alkaline soils; < 2200 m. CA (exc GB, DSon); to WA, ID. Apr–May

H. intercedens Nevski (p. 1461) BOBTAIL BARLEY Ann. **ST**: 0.5–4 dm, bent at base or erect, loosely tufted; nodes gen pubescent. **LF**: sheath hairs in vertical lines, auricles gen 0, occ < 2 mm; blade to 9 cm, to 4 mm wide, sparsely to densely long-spreading-hairy on both surfaces. **INFL**: 2.5–6.5 cm, pale green, gen ± enclosed at maturity. **CENTRAL SPIKELET**: glumes 9–17 mm, gen straight with age, base flat; floret 4.5–7.5 mm; lemma awn 5.5–10 mm, gen slightly spreading with age. **LATERAL SPIKELET**: gen sterile; glumes < 18 mm, base flat; lemma 1.7–4.4 mm, tip obtuse to acute, awn gen 0 or < 1.2 mm. 2*n*=14. Vernal pools, dry, saline streambeds, alkaline flats; < 500 m. SnJV, SCoRO, SCo, ChI, PR; nw Baja CA. Sometimes treated as synonym of *H. pusillum* Nutt., which has been reported from CA, but not naturalized; needs study. Mar–Jun ★

H. jubatum L. subsp. ***jubatum*** (p. 1461) FOXTAIL BARLEY, SQUIRRELTAIL BARLEY Ann, per. **ST**: 2–8 dm, bent at base or erect, densely tufted. **LF**: sheath glabrous to hairy, auricles 0; blade to 15 cm, < 5 mm wide, scabrous to short-hairy. **INFL**: 3–15 cm, white-green to light purple. **CENTRAL SPIKELET**: glumes (15)35–85 mm, bristle-like throughout, strongly divergent with age; lemma 4–8.5 mm, glabrous, awn 25–90 mm. **LATERAL SPIKELET**: sta-

minate or sterile; glumes 17–80 mm, base not flat; lemma 4–6.5 mm, awn 2–15 mm, divergent. 2*n*=28. Roadsides, disturbed areas, meadows, marshes; 20–3500 m. CA; to AK, e US, Mex, e Siberia. [*H. j.* var. *caespitosum* (Scribn.) Hitchc.] Hybridizes with *Elymus multisetus* to yield ×*Elyhordeum californicum* (Bowden) Barkworth and with *E. trachycaulus* to yield ×*E. macounii* (Vasey) Barkworth & D.R. Dewey. Awns can cause mechanical injury to animals. Another subsp. occurs in the n Rocky Mtns. May–Jul

H. marinum Huds. subsp. ***gussoneanum*** (Parl.) Thell. MEDITERRANEAN BARLEY Ann. **ST**: 1–5 dm, bent at base or erect; nodes glabrous. **LF**: basal sheaths ± hairy, auricles < 2 mm or 0; blade to 8 cm, 1–6 mm wide. **INFL**: 1.5–7 cm, 5–20 mm wide, green to purple, occ ± enclosed in upper sheaths at maturity, central and lateral spikelets falling together. **CENTRAL SPIKELET**: glumes 14–26 mm, gen straight with age, base not flat; lemma 5–8 mm, awn 6–18 mm. **LATERAL SPIKELET**: sterile; glumes 10–24 mm, bristle-like to ± flattened or winged; lemma gen 4–6 mm, awn 3–8 mm. 2*n*=28. Dry to moist, disturbed sites; < 1800 m. CA (exc SNE); to BC, MT, OK, AZ; scattered in e US; native to Medit. [*H. geniculatum* All.; *H. hystrix* Roth] Apr–Jun ❖

H. murinum L. WALL BARLEY Ann. **ST**: 1.5–11 dm, erect, sometimes ± prostrate; nodes glabrous. **LF**: basal sheaths glabrous to ± hairy, auricles 1–4(8) mm; blade ≤ 28 cm, 2–5 mm wide, glabrous, scabrous or sparsely long-hairy. **INFL**: 3–12 cm, 7–16 mm wide, green to glaucous, occ ± red or brown at maturity; axis breaking apart at maturity. **CENTRAL SPIKELET**: glumes 11–25 mm, margins gen ciliate, base flat; lemma 8–14 mm, awn 20–40 mm. **LATERAL SPIKELET**: staminate; glumes 11–35 cm, margins ciliate, base flat; lemma 8–15 mm, awn 20–50 mm. 2*n*=14,28,42. ❖

subsp. ***glaucum*** (Steud.) Tzvelev (p. 1461) SMOOTH BARLEY Summer ann. **ST**: 1.5–4 dm. **INFL**: green to glaucous, gen brown at maturity. **CENTRAL SPIKELET**: stalk 1–2 mm; lemma ≤ lateral floret lemma; lemma awn < lateral floret awn; anthers < 1/2 those of lateral florets. **LATERAL SPIKELET**: palea long-hairy on lower 1/2. 2*n*=14. Moist, gen disturbed sites; -60–1900 m. CA (exc KR, CaR, SNE); to BC, OK, AL, TX, Mex. [*H. g.* Steud.; *H. stebbinsii* Covas] Mar–Jul

subsp. ***leporinum*** (Link) Arcang. HARE BARLEY Winter ann. **ST**: 3–11 dm. **INFL**: gen green at fl, becoming ± purple. **CENTRAL SPIKELET**: stalk 1–2 mm; lemma << those of lateral florets; lemma awn slightly < lateral floret awn; anthers = those of lateral florets. **LATERAL SPIKELET**: palea scabrous on lower 1/2. 2*n*=28,42. Moist, gen disturbed sites; < 1900 m. CA (exc KR); to BC, OK, e US, Mex. [*H. l.* Link] Feb–May

subsp. ***murinum*** Winter ann. **ST**: 3–6 dm. **INFL**: green. **CENTRAL SPIKELET**: stalk 0–0.5 mm; lemma gen = those of lateral florets; lemma awn ± < awn of lateral floret. **LATERAL SPIKELET**: palea scabrous to glabrous. 2*n*=28. Moist, gen disturbed sites; 80–1000 m. CaRH, Teh, ScV, CW, DSon; to BC, CO, PA, MD. Feb–May

HYPARRHENIA

Elizabeth M. Skendzic

[Ann] per, gen cespitose. **ST**: erect, densely branched. **LF**: cauline; sheaths gen < internodes; ligule membranous; blade flat or folded. **INFL**: raceme-like, digitate, branches paired, long-stalked from upper axils, hairy. **SPIKELET**: in pairs; lower 1–2 pairs similar, staminate, awn 0; upper pairs dissimilar, sessile spikelet bisexual, glumes > lemmas, lanceolate, hairy, florets 2, lower vestigial, upper fertile, lemma translucent, awned; stalked spikelet staminate or sterile, awn 0; pairs breaking as a unit or spikelets breaking below glumes. 55 spp.: trop Afr, Asia. (Greek: below masculine, from basal staminate spikelets) [Barkworth 2003 FNANM 25:678–679]

H. hirta (L.) Stapf THATCHING GRASS Rhizomes short. **ST**: ± clumped, 3–10 dm. **LF**: blade 5–30 cm, < 3 mm wide, ± folded or inrolled. **INFL**: 2–4 cm, stalk long-soft-hairy, subtended by expanded sheath. **SPIKELET**: glumes densely hairy; sessile spikelet of lower pairs 5–5.6 mm; sessile spikelet of upper pairs 4–4.5 mm, lemma hairy in upper 1/2, awn 1–3.5 cm. 2*n*=30,44,45. Disturbed sites, canyons near urban areas; < 300 m. SCo (Los Angeles, San Diego cos.); native to warm temp Eurasia, Afr. Mar–Jul

IMPERATA

Elizabeth M. Skendzic

Per, rhizomed. **ST**: erect. **LF**: cauline; ligule membranous; blade flat, linear to lanceolate. **INFL**: panicle-like, ± cylindric; branches short, many, raceme-like, appressed, densely silky. **SPIKELET**: in pairs, similar, unequally stalked, breaking below glumes, glumes equal to subequal, membranous, florets 2, lower vestigial, upper bisexual, lemma reduced or 0, awn 0, palea vestigial, stamens 1–2, style exserted, stigmas plumose. **FR**: ovate. 9–10 spp.: warm temp, trop. (Ferrante Imperato, Italian naturalist, 1550–1625)

I. brevifolia Vasey (p. 1461)   CALIFORNIA SATINTAIL   Rhizomes hard, scaly. **ST**: 0.7–1.5 m. **LF**: ligule densely ciliate; blade 15–50 cm, 4–15 mm wide, narrow at collar. **INFL**: 1–3 dm, plume-like, densely white-silky-hairy, appearing speckled from adherent brown anthers, stigmas; hairs 8–15 mm. **SPIKELET**: 3–4 mm, silky hairs at base, lower lemmas > upper lemmas. 2*n*=20. Wet springs, meadows, streambanks, floodplains; < 500 m. NCoRO, CaRF, s SNF, SnJV, SCo, TR, D; to UT, TX, Mex. Sep–May ★

KOELERIA

Dieter H. Wilken

Ann, per. **ST**: erect. **LF**: basal to cauline; ligule membranous, glabrous to minutely ciliate, toothed at obtuse to truncate tip; blade narrow, flat to inrolled. **INFL**: panicle-like, gen compact, narrow, or wider in full fl. **SPIKELET**: laterally compressed; glumes unequal, upper > and wider than lower, keeled, acute, lower 1-veined, upper faintly 3–5-veined; axis prolonged beyond fertile floret, bristly (occ with vestigial floret at tip); florets 2–5, bisexual, breaking above glumes and between florets; lower lemmas gen > glumes, awned or not, 5-veined; palea ± < lemma, tip minutely 2-forked. ± 30 spp.: temp N.Am, Eurasia. (G.L. Koeler, German botanist, 1765–1807) [Standley 2007 FNANM 24:753–756]

1. Ann; infl axis, branches glabrous; glumes, lower lemmas papillate to coarse-hairy on back . . . . ***K. gerardi***
1′ Per; infl axis, branches puberulent; glumes, lower lemmas glabrous to minutely scabrous on back . . . . ***K. macrantha***

K. gerardi (Vill.) Shinners (p. 1461)   ANNUAL JUNE GRASS, BRISTLY KOELERIA   **ST**: 1–6(15), 5–40 cm, glabrous. **LF**: basal and cauline; basal loosely tufted, back and margins soft- to coarse-hairy; ligule < 1 mm; blade 1–5 cm, 1–2 mm wide, smooth to ridged. **INFL**: 8–30 cm, 4–10 mm wide, cylindric. **SPIKELET**: 3–4 mm, green to tan; lower glume ± 2 mm, upper ± 3 mm; lemma 2–3 mm, awned at tip, lower lemma awn 1–2 mm. 2*n*=26. Open, disturbed sites; < 600 m. NCoRI, CaRF, SNF, Teh, GV, s SCoRO, SCoRI, SCo, ChI, w TR; to WA, s US; native to Medit. [*K. phleoides* (Vill.) Pers., illeg.; *Lophochloa cristata* (L.) Hyl.; *Rostraria c.* (L.) Tzvelev] Apr–Jul

K. macrantha (Ledeb.) Schult. (p. 1461)   JUNE GRASS   Cespitose. **ST**: 2–8(10+) dm, glabrous to puberulent. **LF**: gen basal, tufted, glabrous to puberulent; ligule 1–2 mm; blade 2–20 cm, 1–2(3) mm wide, gen ridged. **INFL**: 2–15 cm, 1–2 cm wide, cylindric to narrowly conic, or wider in full fl, interrupted at base. **SPIKELET**: 4–6 mm, ± shiny, tan (occ ± purple); florets 2–4; lower glume ± 3 mm, upper ± 5 mm; lemma 3–5 mm, acute to small-pointed at tip. 2*n*=14. Dry, open sites, clay to rocky soils, shrubland, woodland, conifer forest; < 3840 m. NW, CaR, SN, ScV (Sutter Buttes), CW, TR, PR, GB, DMtns; to AK, e Can, c&e US, n Mex, Eurasia. [*K. cristata* Pers., in part; *K. c.* var. *longifolia* Burtt Davy; *K. gracilis* Pers., illeg.; *K. nitida* Nutt.; *K. pyramidata* (Lam.) P. Beauv., misappl.] May–Jul

LAGURUS   HARE'S TAIL GRASS

1 sp. (Greek: hare tail, from densely hairy infl) [Tucker 2007 FNANM 24:669–670]

L. ovatus L. (p. 1467)   Ann, tufted, soft-hairy. **ST**: ascending to erect, gen 10–50 cm, hairy. **LF**: basal and cauline; sheath inflated, loosely surrounding st; ligule membranous, truncate, densely hairy; blade 2–10 cm, 3–10 mm wide, flat. **INFL**: terminal, 1.5–3 cm, ovoid to ± cylindric, dense, hairy; spikelets gen subsessile. **SPIKELET**: 7–10 mm (exc awns), laterally compressed; glumes equal, > floret, narrowly lanceolate, soft-hairy to plumose, long-tapered; axis breaking apart above glumes, prolonged beyond floret; floret 1; lemma 3.5–4.5 mm, back glabrous, base puberulent, faintly 5-veined, awns 3, 2 from tip 1–2(6) mm, slender, 1 from lemma back gen 8–22 mm, ± stiff, bent; palea ± = lemma. 2*n*=14. Disturbed places; gen < 200 m. NCo (Humboldt Co.), ScV, CCo, SnFrB; OR, Can, e US, NM; native to Medit. Cult for orn, dried arrangements. May–Jul

LAMARCKIA   GOLDENTOP

Lynn G. Clark

1 sp.: native to Medit, naturalized in similar climates worldwide. (J.B. Lamarck, French botanist, 1744–1829)

L. aurea (L.) Moench (p. 1467)   Ann, cespitose, glabrous. **ST**: gen erect, 7–40 cm. **LF**: cauline, ± evenly distributed; ligule 3–7 mm, membranous, glabrous, tip ± irregularly cut; blade 2.5–9 cm, 2.5–7 mm wide, flat. **INFL**: panicle-like, terminal, 2–8 cm, dense, golden yellow to ± purple; axis short-white-hairy in branch axils; spikelets short-stalked, with 1 fertile and 1–3 sterile spikelets in spreading to drooping clusters, each cluster gen falling as 1 unit. **FERTILE SPIKELET**: glumes 2.5–4 mm, ± equal, gen = spikelet; florets 2; lower floret fertile, 2.5–3 mm, lemma awned from near tip, awn 6–7 mm, straight; upper floret sterile, ± 0.5 mm, awn 4–5 mm. **STERILE SPIKELET**: 6–9 mm, >> glumes, linear; glumes > lower floret; florets 5–8; lemmas ± overlapping, 1.5–2 mm, obtuse, tip ± fringed, awn 0 (stalk base sometimes with a reduced, sterile spikelet like fertile in size and shape). *n*=7. Open ground, moist seeps, rocky hillsides, sandy soil; < 1250 m. NCoRO, NCoRI, SNF, Teh, CW, SW; AZ; native to Medit. Somewhat weedy. Feb–May

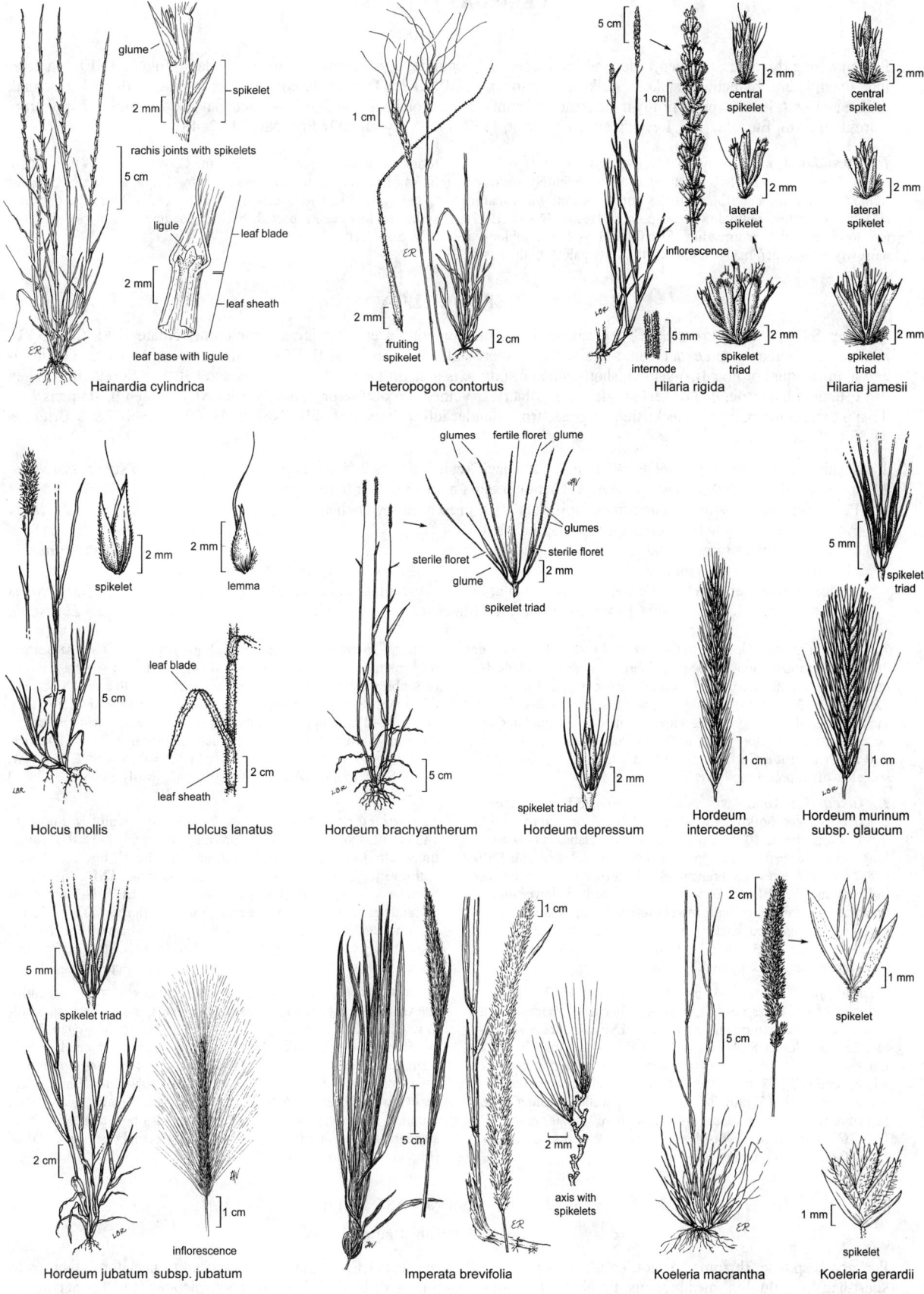

glume

spikelet

2 mm

rachis joints with spikelets

5 cm

ligule

leaf blade

2 mm

leaf sheath

leaf base with ligule

Hainardia cylindrica

1 cm

ER

2 mm

fruiting spikelet

2 cm

Heteropogon contortus

5 cm

1 cm

inflorescence

5 mm

internode

2 mm

central spikelet

2 mm

lateral spikelet

2 mm

spikelet triad

Hilaria rigida

2 mm

central spikelet

2 mm

lateral spikelet

2 mm

spikelet triad

Hilaria jamesii

2 mm

spikelet

2 mm

lemma

leaf blade

2 cm

leaf sheath

5 cm

Holcus mollis

Holcus lanatus

glumes   fertile floret   glume

sterile floret

glume

glumes

sterile floret

2 mm

spikelet triad

5 cm

Hordeum brachyantherum

2 mm

spikelet triad

Hordeum depressum

1 cm

Hordeum intercedens

5 mm

spikelet triad

1 cm

Hordeum murinum subsp. glaucum

5 mm

spikelet triad

2 cm

Hordeum jubatum subsp. jubatum

1 cm

inflorescence

5 cm

2 mm

axis with spikelets

1 cm

Imperata brevifolia

2 cm

5 cm

Koeleria macrantha

1 mm

spikelet

1 mm

spikelet

Koeleria gerardii

LEERSIA CUTGRASS

Dieter H. Wilken

Per from long rhizomes. **ST:** gen 1, decumbent to erect. **LF:** cauline; ligule membranous; blade flat to folded. **INFL:** panicle-like, open; lateral branchlets, spikelet stalks arched to wavy. **SPIKELET:** laterally compressed; glumes 0; floret 1, bisexual, falling as 1 unit; lemma, palea firmly membranous; lemma strigose on back, awnless, 5-veined; palea ± = lemma. 17 spp.: trop, warm temp Am, Eurasia. (J.D. Leers, German botanist, 1727–1774) [Pyrah 2007 FNANM 24:42–45]

L. oryzoides (L.) Sw. (p. 1467) RICE CUTGRASS **ST:** 1–1.5 m; nodes short, soft-hairy. **LF:** sheath glabrous to minutely scabrous; ligule ± 1 mm, truncate; blade 10–28 cm, 8–14 mm wide, margin strongly scabrous, with downward-pointing teeth. **INFL:** 12–20 cm; lower branches ± spreading. **SPIKELET:** 4–5 mm, oblong to narrowly elliptic; lemma 4–5 mm, width 3–5 × palea width. 2*n*=48.

Marshes, streams, ponds; < 1200 m. CA-FP (exc CW); to BC, e N.Am; Eurasia. Lateral infls enclosed by sheath, gen cleistogamous. *L. hexandra* Sw., with ascending to appressed infl branches, smaller spikelets, has been reported, but not documented, in rice fields in ScV. Aug–Oct

LEPTOCHLOA

Ann, per. **ST:** spreading to erect. **LF:** gen cauline; ligule membranous, ± entire to jagged, sometimes ciliate; blade flat. **INFL:** panicle-like; branches spike- or raceme-like; spikelets short-stalked or sessile. **SPIKELET:** laterally compressed or ± cylindrical; glumes equal or not, 1(3)-veined, short-awned or not; axis breaking apart above glumes and between florets; florets gen 2–12; lemma back rounded or keeled, glabrous or hairy, 3-veined, tip obtuse or minutely 2-lobed, awn gen 0; stamens 1–3. 32 spp.: warm temp, trop. (Greek: slender grass, from slender infl) [Snow 2003 FNANM 25:51–60; McNeill 1979 Brittonia 31:399–404]

1. Sheaths papillate-hairy; spikelets 1–3 mm; lemma back keeled; florets 2–4 . . . . **L. panicea** subsp. **brachiata**
1′ Sheaths glabrous or scabrous; spikelets 4–12 mm; lemma back rounded; florets gen 5–12
   2. Per; st internodes solid; cleistogamous spikelets often present in lower sheaths . . . . **L. dubia**
   2′ Ann; st internodes hollow; cleistogamous spikelets 0
      3. Lemma awnless or mucronate . . . . **L. fusca** subsp. **uninervia**
      3′ Lemma awn 0.5–3(5) mm
         4. Florets ± white; lower glume 2–3 mm, lanceolate; lemma lanceolate . . . . **L. fusca** subsp. **fascicularis**
         4′ Florets ± red; lower glume < 2 mm, triangular; lemma ovate . . . . **L. viscida**

L. dubia (Kunth) Nees GREEN SPRANGLETOP Per. **ST:** gen 30–100, internodes solid. **LF:** sheath glabrous; ligule 1–2; blade 8–35 cm, gen flat, ≤ 1 cm wide. **INFL:** racemes ascending, 2–15, 3–12 cm. **SPIKELET:** 5–12 mm, laterally compressed, glumes 2–6, lower < upper; florets (2)5–8, spreading widely at maturity; lemma midnerve occ extended as a short point, back rounded. Sandy or rocky soils; 300 m. ScV (Butte Co.); sw US, FL, Mex, S.Am. Cleistogamous spikelets often present in lower sheaths. Oct–Nov

L. fusca (L.) Kunth SPRANGLETOP Ann. **ST:** decumbent to erect; internodes hollow. **LF:** sheath glabrous or scabrous; ligule membranous; blade 10–55 cm, 1–7 mm wide, glabrous or scabrous, long-tapered to a fine tip, upper exceeding panicles or not. **INFL:** 0.7–7 dm, ≤ 20 cm wide, branches 3–35, lower sometimes enclosed in upper sheath. **SPIKELET:** ± round in ×-section, lower glume < upper; lemma midvein ciliate, margin soft-hairy, tip acute to truncate, mucronate or 1-awned, back rounded; palea back pubescent, hairy on margins; stamens 1–3. 2*n*=20. Subsp. *fusca* found once (1983, Butte Co.), but presumed extirpated.

   subsp. ***fascicularis*** (Lam.) N. Snow (p. 1467) BEARDED SPRANGLETOP **ST:** spreading to erect, 3–10 dm. **LF:** sheath gen glabrous; ligule (2)5–7 mm, jagged with age. **INFL:** branches spreading to ascending, lower 8–15 cm. **SPIKELET:** 6–12 mm; glumes 2–4 mm, lower lanceolate, upper > and wider than lower; florets 6–12, ± white; lemma 3.5–5 mm, lanceolate, often with dark spot on lower 1/2, awn at tip 1–3(5) mm. 2*n*=20. Marshes, wetlands, often wet disturbed areas, often associated with agriculture; < 1200 m. NCoRO, GV, CCo, GB, D; to e US, Mex, S.Am. [*L. fascicularis* (Lam.) A. Gray] Gen self-pollinating. Jun–Oct

subsp. ***uninervia*** (J. Presl) N. Snow MEXICAN SPRANGLETOP **ST:** erect, 3–10 dm, sometimes few-branched. **LF:** sheath glabrous or scabrous; ligule 2–6 mm, entire to jagged with age. **INFL:** 1–7 dm; branches gen ascending, lower 3–6 cm. **SPIKELET:** 5–7 mm; glumes 1–2 mm, upper > and wider than lower; florets 6–9; lemma 2–3 mm, tip obtuse to truncate, mucronate, awn 0. 2*n*=20. Ditches, drying ponds, disturbed wet areas; gen < 1000 m. s SNF, GV, CW, SW, GB, D; to s US, Mex, S.Am. [*L. u.* (J. Presl) Hitchc. & Chase] Mar–Dec

L. panicea (Retz.) Ohwi subsp. ***brachiata*** (Steud.) N. Snow (p. 1467) RED SPRANGLETOP Ann. **ST:** gen erect, 1–15 dm, sometimes ± red or purple. **LF:** sheath papillate-hairy; ligule 1–3 mm, entire to jagged; blade 5–30 cm, 3–20 mm wide. **INFL:** 3–20 cm; branches gen ascending, lower 5–10 cm. **SPIKELET:** 1–3 mm, compressed; glumes 1–2 mm, ± equal, ± purple; florets (2)3–4; lemma 1–2 mm, keeled, veins hairy, tip minutely 2-lobed, awn 0. 2*n*=20. Wet sites, drying ponds; < 100 m. D; to s US, Mex, S.Am. [*L. filiformis* (Pers.) P. Beauv.; *L. mucronata* (Michx.) Kunth] Sep–Dec

L. viscida (Scribn.) Beal STICKY SPRANGLETOP, SONORAN SPRANGLETOP Ann. **ST:** spreading to erect, 1–6 dm; internodes hollow. **LF:** sheath scabrous; ligule 1–3 mm, minutely jagged; blade 2–15 cm, 2–6 mm wide. **INFL:** 3–8 cm; branches ascending to ± appressed, lower 1–2 cm. **SPIKELET:** gen 4–6 mm, ± round in ×-section; glumes 1.6–2 mm, lower triangular, upper > and wider than lower; florets 4–7(8) ± red; lemma 1.5–3 mm, ovate, back rounded, ± sticky, veins hairy, tip minutely 2-lobed, awn 0.5–1 mm, slender. 2*n*=40. Wet places; < 100 m. SnJV (Kern, Fresno cos.), DSon (Imperial Co.); to TX, Mex. Jul–Sep

MELICA ONIONGRASS, MELIC

Susan J. Bainbridge

Per, gen cespitose, rhizomed or not, corms present or 0. **ST:** gen erect. **LF:** ± basal; sheath closed to near top, glabrous to short-hairy; ligule thin, membranous, tip obtuse to truncate, gen jagged; blade flat, veins inconspicuous. **INFL:** raceme- or

panicle-like, gen appressed to main axis. **SPIKELET**: glumes papery, translucent, back rounded, tip rounded, lower glume 3–5-veined, upper 1–3-veined; axis gen breaking above glumes; lower florets bisexual, 1–7, uppermost florets sterile, ± densely clustered at axis tip; lemma ± like glumes, prominently 5–7-veined, veins not converging, base occ red; palea < lemma. $n=9$ in all CA taxa. ± 80 spp.: gen temp, exc Australia. (Latin: honey, or old Italian name for pl with sweet sap) [Barkworth 2007 FNANM 27:88–102]

1. Glumes deciduous; spikelets falling as 1 unit, 1–2 per infl branch (sect. *Melica*) . . . . **M. stricta**
1′ Glumes persistent; spikelet axis breaking above glumes, spikelets > 2 per lower infl branch (sect. *Bromelica*)
  2. Lemma strongly acuminate or awned, glabrous or, if hairy, hairs longer on lower 2/3 of margins, marginal veins or basally
    3. Awn 5–12 mm; lemma surface glabrous, margin near base often hairy . . . . **M. aristata**
    3′ Awn 0–4 mm; lemma surface hairy near base
      4. Lemma obtuse to ± mucronate at tip, ciliate, with longer hairs on margins, awn 1–4 mm; corms 0 . . . . **M. harfordii**
      4′ Lemma strongly acuminate, hairy with longer hairs toward base, awn 0; corms present. . . . . **M. subulata**
  2′ Lemma rounded to acute, awn 0 (occ to 2 mm), glabrous to scabrous, longer hairs (if any) along entire margin or near tip
    5. Bisexual florets 1–2 in all spikelets
      6. Sterile cluster at axis tip widest at middle or below; axis < sterile cluster; lemma gen glabrous or minutely scabrous. . . . . **M. imperfecta**
      6′ Sterile cluster at axis tip truncate or widest above middle; axis > sterile cluster; lemma margin gen hairy. . . . . **M. torreyana**
    5′ Bisexual florets 3–7 in some or all spikelets
      7. Floret stalks swollen when fresh, ± wrinkled and brown when dry. . . . . **M. fugax**
      7′ Floret stalks not swollen when fresh or wrinkled when dry
        8. Palea 1/2–3/4 lemma; lvs 3–5 per st, blades 3–9 cm. . . . . **M. frutescens**
        8′ Palea ± 7/8 lemma; lvs 2–3 per st, blades gen >> 9 cm
          9. Sterile cluster at axis tip widest above middle, tip truncate (acute), not resembling floret. . . . . **M. californica**
          9′ Sterile cluster at axis tip widest below middle, tip acute to acuminate, resembling floret
            10. Both glumes ± 1/2 lowest floret; corms connected to rhizome by short stalk. . . . . **M. spectabilis**
            10′ 1 or both glumes ± 3/4 lowest floret; corms sessile on rhizome (or rhizome 0)
              11. Sheath of basal lf remaining intact; infl branches appressed to ascending . . . . **M. bulbosa**
              11′ Sheath of basal lf becoming fibrous; infl branches spreading to reflexed . . . . **M. geyeri**

M. aristata Bol. AWNED MELIC. **ST**: 4–12 dm; corm 0. **LF**: ligule 2.5–5 mm; blade 2–6 mm wide. **INFL**: 7–26 cm; branches appressed to narrowly spreading; spikelets 1–4 per branch. **SPIKELET**: 11–21 mm; glumes 7–11 mm, persistent, translucent margin 1 mm wide; bisexual florets 2–5; lemma 8–13 mm, glabrous or hairy, with longer hairs on marginal veins, awns 5–12 mm; palea ± 3/4 lemma; sterile cluster 2.5–6 mm, tapered, resembling bisexual floret. Dry open sites, conifer forest; 1000–3000 m. KR, NCoR, CaR, SNH, SnBr; s WA; possibly introduced to KY. Jun–Aug

M. bulbosa Porter & J.M. Coult. (p. 1467) ONIONGRASS Rhizomes short. **ST**: 2–10 dm; corms sessile. **LF**: ligule 2–6 mm, blade 1.5–5 mm wide; sheath of basal lf remaining intact. **INFL**: 5–30 cm; branches appressed to ascending; spikelets 1–5 per branch. **SPIKELET**: 6–24 mm; glumes persistent, lower 5–10.5 mm, upper 6–14 mm, ± 3/4 lowest floret; bisexual florets 2–7; lemma 6–12 mm, glabrous or scabrous, tip obtuse to acute; palea ± 7/8 lemma; sterile cluster 1.5–5 mm, truncate to tapered, resembling bisexual floret. Dry rocky slopes, conifer forest; < 3400 m. KR, NCoRH, NCoRI, CaRH, SN, GB, w edge DMoj; w Can, Rocky Mtns, TX. Jul–Aug

M. californica Scribn. CALIFORNIA MELIC **ST**: 4–14 dm; lowest internodes ± swollen; corm gen 0. **LF**: ligule 1.5–4 mm, blade 1.5–5 mm wide. **INFL**: 4–34 cm; branches appressed; spikelets 4–15 per branch. **SPIKELET**: 5–15 mm; glumes 3.5–13 mm, subequal, persistent; bisexual florets 2–5; lemma 5–9 mm, glabrous or scabrous, tip obtuse to weakly acute; palea ± 7/8 lemma; sterile cluster 1–3 mm, truncate (acute), not resembling floret. Open or rocky hillsides, oak woodland, conifer forest; < 2200 m. NW, CaR, SN, GV, CW, WTR, PR; OR. Apr–May

M. frutescens Scribn. WOODY MELIC **ST**: 4–20 dm, gen branching at basal nodes; corm 0. **LF**: 3–5 per st; ligule 2–8 mm; blade 3–9 cm, 1.5–5 mm wide. **INFL**: 12–40 cm; branches appressed; spikelets 4–15 per branch. **SPIKELET**: 9–18 mm; glumes persistent, lower 7–12 mm, 8–15 mm, translucent margin 2.5–3.5 mm wide; bisexual florets 3–5; lemma 8–11 mm, glabrous, tip round to acute; palea 1/2–

3/4 lemma; sterile cluster 2–5 mm, truncate, concealed by uppermost bisexual floret. Dry slopes, scrub, chaparral, woodland; 300–1500 m. s SNF, Teh, s SCoRO, SCo, WTR, PR, W&I, D; AZ, Baja CA. Mar–May

M. fugax Bol. LITTLE MELIC **ST**: 1–6 dm; corms clustered. **LF**: ligule 0.5–3 mm; blade 1.2–5 mm wide. **INFL**: 4.5–18 cm; branches appressed to spreading; spikelets 1–7 per branch. **SPIKELET**: 4–17 mm; glumes persistent, lower 3–5 mm, upper 3.5–7 mm; bisexual florets 2–5, floret stalk swollen when fresh, ± wrinkled, brown when dry; lemma 4–7 mm, glabrous or scabrous, tip obtuse; palea ± = lemma; sterile cluster 2–3.5 mm, tapered. Dry volcanic flats, hillsides, conifer forest; 1200–2200 m. KR, NCoRH, NCoRI, CaR, n&c SNH, GB; BC, ID, NV. May–Jul

M. geyeri Munro GEYER'S ONIONGRASS Rhizomes short to long. **ST**: 8–20 dm; corms sessile. **LF**: ligule 2–3.5 mm; blade 2–8 mm wide; sheath of basal lf becoming fibrous. **INFL**: 11–17 cm; branches spreading to reflexed; spikelets 1–6 per branch. **SPIKELET**: 8–24 mm; glumes persistent, lower 3.5–7 mm, upper 5.5–11 mm, tip extending to ± 3/4 lowest lemma; bisexual florets 2–6; lemma 8–11 mm, glabrous or scabrous, tip rounded to acute, sometimes notched, awn occ to 2 mm; palea ± 7/8 lemma; sterile cluster 3–7 mm, tapered, resembling bisexual floret. Dry open slopes, oak woodland, conifer forest; < 2200 m. KR, NCoR, CaRH, n&c SNH, SnFrB, n SCoRO; OR, BC. [*M. g.* var. *aristulata* J.T. Howell] Apr–Jul

M. harfordii Bol. **ST**: 3.5–12 dm; corm 0. **LF**: ligule 1–2 mm; blade 2–6 mm wide. **INFL**: 6–23 cm; branches appressed; spikelets 2–6 per branch. **SPIKELET**: 7–20 mm; glumes 5–11 mm, ± equal, obtuse to ± acute, persistent; bisexual florets 2–6; lemma 6–16 mm, ciliate, with longer hairs on lower 2/3 of margins, tip obtuse to ± mucronate, awn 1–4 mm; palea 3/4 lemma; sterile cluster 3–5 mm, lanceolate in outline, resembling bisexual floret. Dry slopes, conifer forest; < 2600 m. NW (exc NCo), CaRH, n&c SNH, SnJV, n SnFrB, n SCoR; BC. May–Jul

M. imperfecta Trin. (p. 1467) LITTLE CALIFORNIA MELICA **ST**: 3.5–12 dm; corm 0. **LF**: ligule 0.8–6.5 mm; blade 1–6 mm wide.

INFL: 5–36 cm; branches appressed to spreading; spikelets 3–30 per branch. **SPIKELET**: 3–7 mm; glumes 2–6 mm, ± equal, persistent; bisexual florets 1–2; lemma 3–7 mm, glabrous or scabrous, tip acute to obtuse, gen glabrous or minutely scabrous; palea ± = lemma; sterile cluster 0.5–4 mm, > axis, acute to obtuse, not resembling bisexual floret. Dry rocky hillsides, chaparral, woodland; < 1500 m. NCoRI, c&s SN, ScV (Sutter Buttes), CW, SW, DMoj; NV, AZ, Baja CA. Apr–May

M. spectabilis Scribn. (p. 1467) PURPLE ONIONGRASS Rhizomes short to long. **ST**: 3.5–10 dm; corms short-stalked, scattered. **LF**: ligule 0.1–3 mm; blade 2–5 mm wide. **INFL**: 5–26 cm; branches gen appressed; spikelets 1–4 per branch. **SPIKELET**: 7–19 mm; glumes 3.5–7 mm, ± equal, < 1/2 spikelet, obtuse, persistent; bisexual florets 3–7; lemma 6–9 mm, glabrous or minutely scabrous, gen widest above middle, tip acute to obtuse; palea ± 7/8 lemma; sterile cluster ± 2.5 mm, 1.5–3.5 mm, acute, tapered, gen concealed. Wet sites, meadows, conifer forest; 1200–2600 m. KR, n NCoR, n SNH, Wrn; BC, Rocky Mtns. May–Jul ★

M. stricta Bol. (p. 1467) ROCK MELIC Densely cespitose. **ST**: 1–9 dm; corms 0. **LF**: ligule 2.5–5 mm; blade 1.5–5 mm wide. **INFL**: 3–30 cm; branches very appressed; spikelets 1–2 per branch, pedicels sharply bent below spikelets. **SPIKELET**: 6–23 mm, open, appearing V-shaped; glumes 6–18 mm, ± equal, spreading, distal 1/2 translucent, deciduous; bisexual florets 2–4; lemma 8–16 mm, glabrous or

minutely scabrous, tip obtuse to acute, awn 0; palea 1/2–3/4 lemma; sterile cluster 2–7 mm, acute to acuminate, resembling lower florets. Open sites, conifer forest, rocky areas in alpine; 1200–3350 m. KR, NCoRH, CaR, SNH, Teh, s SCoR, TR, Wrn, SNE, DMtns; OR, UT. [*M. s.* var. *albicaulis* Boyle] Jun–Aug

M. subulata (Griseb.) Scribn. ALASKAN ONIONGRASS Rhizomes short. **ST**: 5–13 dm; corms sessile, clustered. **LF**: ligule 0.4–5 mm; blade 2–10 mm wide. **INFL**: 8–25 cm; branches appressed to spreading; spikelets 1–5 per branch. **SPIKELET**: 10–28 mm; glumes persistent, lower 4–8 mm, upper 6–12 mm, acute; bisexual florets 2–5; lemma 5–18 mm, gen hairy with longer hairs near base, tip strongly tapered, acuminate, awn 0; sterile cluster 4–9 mm, tapered, resembling bisexual florets. Moist sites, streambanks, conifer forest; < 2300 m. KR, NCoR, CaRH, n&c SNH, CCo, SnFrB, Wrn; AK, Rocky Mtns; also in S.Am. Apr–Jul

M. torreyana Scribn. (p. 1467) TORREY'S MELIC **ST**: decumbent to erect, 3–10 dm; corms 0. **LF**: ligule 1–5 mm; blade 1–2.5 mm wide. **INFL**: 6–25 cm; branches gen appressed; spikelets 4–38 per branch. **SPIKELET**: 3.5–7 mm; glumes 3–7 mm, ± equal, acute, persistent; bisexual florets 1–2; lemma 3.5–6 mm, back scabrous or occ hairy with longer hairs near tip, tip obtuse to weakly notched, awn 0 or to 2 mm; sterile cluster 0.5–4 mm, < axis, widest at distal end, truncate. Chaparral, conifer forest; < 1600 m. NW (exc NCoRH), CaRF, n&c SN, ScV (Sutter Buttes), CW. Mar–Jun

MELINIS

Ann, per. **ST**: erect to prostrate; internode solid to hollow. **LF**: gen cauline; ligule hairy. **INFL**: gen panicle-like, open; branches spreading to ascending; spikelets subsessile to stalked. **SPIKELET**: falling as 1 unit; glumes strongly unequal, lower << upper or 0, upper glume silky-hairy; florets 2, lower floret sterile or staminate, lemma ± = upper glume, upper floret fertile, lemma membranous to thick, firm, smooth, ± white in fr, margin flat, tip blunt, palea ± = lemma. 22 spp.: warm temp, subtrop, se Asia, Afr. (Greek: a cereal, probably millet) [Wipff 2007 FNANM 25:490–492; Zizka 1990 Mitt Inst Allg Bot Hamburg 23:563–572]

M. repens (Willd.) Zizka subsp. *repens* NATAL GRASS, RUBY GRASS Ann or short-lived per. **ST**: decumbent to erect, (2)4–15 dm; internodes glabrous or with swollen-based hairs; sheath 3–9 cm, glabrous; ligule hairs 0.5–1.5 mm; blade 3–27 cm, gen 2–9 mm wide, upper surface glabrous. **INFL**: (4)6–22 cm; 1° branches 2.5–6(11) cm, glabrous to puberulent; spikelet stalk 0.5–5 mm, ± wiry, hairy. **SPIKELET**: 2–5.5 mm, ± 1–2 mm wide, ovate to elliptic; lower glume < 1.5 mm, 0–1-veined, upper glume 2–5 mm, densely silky-

hairy, hairs ≤ 7 mm, rose to ± purple fading to pink or white in age; lower floret staminate or sterile, lemma ± like upper glume, 5-veined, tip minutely lobed, palea ± = lemma; upper floret ± 2/3 length lower floret, lemma firm, ± white, shiny; anthers 3. $2n=36$. Disturbed areas, slopes; < 850 m. NCoRO, SnJV, s CCo, SCoRO, SCo, WTR, PR; to s US; native to S.Afr, w Asia. [*Rhynchelytrum r.* (Willd.) C.E. Hubb.; *R. roseum* (Nees) Stapf & C.E. Hubb.; *Tricholaena r.* Nees] Used for soil stabilization. All yr

MUHLENBERGIA MUHLY

Paul M. Peterson

Ann, per, occ mat-like, often rhizomed. **ST**: decumbent to erect, ± clumped. **LF**: basal and cauline; sheath open; ligule membranous, entire to irregularly toothed, occ with 1 large tooth on each side; blade flat to rolled. **INFL**: panicle-like, narrow to open; branches spreading to appressed. **SPIKELET**: gen single (bisexual) occ paired (bisexual, staminate or sterile); glumes subequal, gen 1–3-veined, short-pointed to awned, upper glume occ 3-veined; florets 1, occ 2; axis breaking above glumes; lemma short-pointed to awned, glabrous to hairy, 3-veined; palea ≤ lemma. **FR**: ± fusiform, red-brown, gen falling with lemma and palea. 173 spp.: temp Am, s Asia. (G.H.E. Muhlenberg, PA botanist, 1753–1815) Reports of *M. dumosa* Vasey, *M. glauca* (Nees) B.D. Jacks., *M. thurberi* (Scribn.) Rydb. have proved to be erroneous.

1. Ann
  2. Lemma awned, awn 1–3 cm, lemma 2.5–6 mm
    3. Lemma gen 4.5–6 mm; infl gen < 1.5 cm wide, branches closely appressed; glumes gen 1–2 mm . . . . . . ***M. appressa***
    3′ Lemma gen 2.5–4.5 mm; infl 1–5 cm wide, branches spreading to ascending; glumes gen < 1 mm
    . . . . [2]***M. microsperma***
  2′ Lemma short-awned, awn < 1 mm, or awn 0; lemma < 2.5 mm
    4. Infl narrow, < 1 cm wide; branches closely appressed; sts often rooting at lower nodes . . . . [2]***M. filiformis***
    4′ Infl open, 1.5–8 cm wide; branches reflexed, ascending to spreading; sts not rooting at lower nodes
      5. Infl branches stiffly spreading to reflexed ± 90° from axis; glumes glabrous; ligule teeth 2, 1 on each
        side . . . . ***M. fragilis***
      5′ Infl branches ascending < 80° from axis; glume tip short-hairy; ligule truncate to obtuse, irregularly
        short-toothed . . . . ***M. minutissima***

1′ Per

  6. Lemma awn 0 or short-awned, awn < 1 mm

    7. Infl 5–14 cm wide, open, branches spreading . . . . ***M. asperifolia***

    7′ Infl ≤ 4 cm wide, narrow, branches ascending to appressed, occ loosely spreading

      8. Rhizomes creeping, ± scaly; sts ± decumbent to erect

        9. Glumes 2.5–4 mm, ± = lemma; lower 1/2 lemma short-soft-hairy. . . . . [2]***M. californica***

        9′ Glumes 0.5–1.8 mm, ± 1/2 lemma; lemma glabrous to ± scabrous

          10. Ligule gen 1–2.5 mm; infl axis gen obscured by branches, spikelets . . . . ***M. richardsonis***

          10′ Ligule gen < 1 mm; infl axis gen visible between branches . . . . ***M. utilis***

      8′ Rhizomes 0; sts clumped or decumbent and ± rooting at nodes

        11. Blade < 4 cm; lemma gen < 2 mm; st gen < 3 dm . . . . [2]***M. filiformis***

        11′ Blade ≥ 5 cm; lemma gen > 2.5 mm; st 2–15 dm

          12. Infl ≤ 15 cm, 1.5–4 cm wide; branches loosely fld . . . . ***M. jonesii***

          12′ Infl 15–60 cm, ≤ 1.2 cm wide; branches densely fld . . . . ***M. rigens***

6′ Lemma awn 1–30 mm

  13. Infl open, 6–15 cm wide, branches spreading. . . . . ***M. porteri***

  13′ Infl gen < 6 cm wide, branches ascending to appressed, occ spreading

    14. Glumes 1 or 2, gen < 0.5 mm, vein 0 . . . . [***M. schreberi***]

    14′ Glumes 2, gen > 0.5 mm, vein(s) evident

      15. Rhizome gen scaly, creeping

        16. Blade gen rolled, < 2 mm wide; anther purple, 1.5–3 mm

          17. Lemma, palea short-soft-hairy on lower 1/2; sts loosely clumped, decumbent. . . . . [2]***M. arsenei***

          17′ Lemma base sparsely short-hairy; palea glabrous to ± scabrous; sts erect, ± rooting at lower nodes

            . . . . [2]***M. pauciflora***

      16′ Blade gen flat, ≥ 2 mm wide; anther yellow, gen ≤ 1.5 mm

        18. Hairs at lemma base 2–3.5 mm. . . . . ***M. andina***

        18′ Hairs at lemma base < 1.5 mm

          19. Lemma awn gen < 2.2 mm; anther 1–1.5 mm . . . . [2]***M. californica***

          19′ Lemma awn 2.5–9 mm; anther < 0.5 mm . . . . ***M. mexicana***

  15′ Rhizomes 0

    20. Upper glume 3-veined, 3-toothed . . . . ***M. montana***

    20′ Upper glume gen 1-veined, obtuse, acute, or awned

      21. Lower glume 2-veined, 2-awned; infl 5–8 mm wide, spike-like, dense . . . . ***M. alopecuroides***

      21′ Lower glume 1-veined, with 1 awn or awnless; infl 5–30 mm wide, dense to loose

        22. Glumes 0.5–1 mm, obtuse; cleistogamous spikelets gen present in lowermost st axils . . . . . . [2]***M. microsperma***

        22′ Glumes gen 1.5–3.5 mm, acute, acuminate, or awned; cleistogamous spikelets 0

          23. Lemma, palea short-soft-hairy on lower 1/2; sts loosely clumped, decumbent. . . . . [2]***M. arsenei***

          23′ Lemma base sparsely short-hairy; palea glabrous to ± scabrous; sts erect, ± rooting at lower

            nodes . . . . [2]***M. pauciflora***

M. alopecuroides (Griseb.) P.M. Peterson & Columbus (p. 1467)
WOLFTAIL Per. **ST:** 3–6 dm. **LF:** ligule 2–12 mm, acuminate; blade 4–12 cm, 1–2 mm wide, flat; midribs, margins ± white; tip with awn-like extension of midrib, 3–10 mm. **INFL:** 4–10 cm, 5–8 mm wide, spike-like, densely fld. **SPIKELET:** paired, pair with subtending axis segment falling as 1 unit; lower spikelet staminate or sterile, upper spikelet bisexual; glumes 1–2 mm, unequal, awned, lower glume 2-veined, 2-awned, awns 1–3.5 mm, upper glume 1-veined, 1-awned, awns 2.5–5 mm; lemma 3–4 mm, short soft-hairy on lower 2/3, awn 1.5–3 mm; anthers 1.5–2 mm, yellow. 2*n*=40. Rocky slopes and mesas; ± 500 m. e DMtns (New York Mtns); sw US, Mex; also nw Argentina, Bolivia. [*Lycurus setosus* (Nutt.) C. Reeder; *L. phleoides* Kunth, misappl.] Jul–Oct ★

M. andina (Nutt.) Hitchc. FOXTAIL MUHLY Per; rhizome scaly, creeping. **ST:** 2.5–8.5 dm. **LF:** ligule 0.5–1.5 mm, truncate, ciliate; blade 4–16 cm, 2–4 mm wide, flat. **INFL:** 2–15 cm, 5–15 mm wide, narrow; branches appressed, loosely fld. **SPIKELET:** glumes 2–4 mm, acuminate or short-awned; lemma 2–3.5 mm, hairs at base = lemma, awn 1–7 mm; anthers 0.5–1.5 mm, yellow. 2*n*=20. Canyons, streambanks, wet meadows; < 3100 m. KR, NCoRI, SN, SCoRI, SnBr, SNE, DMtns; to Can, CO, w TX. Jul–Sep

M. appressa C.O. Goodd. APPRESSED MUHLY Ann. **ST:** 1–4 dm. **LF:** ligule 1.5–3 mm, truncate to obtuse, decurrent to sheath, toothed; blade 1–5 mm, 1–2 mm wide, flat or folded. **INFL:** 4–14 cm, 0.5–1.5 cm wide, narrow; branches appressed, loosely fld. **SPIKE-LET:** in lower branch axils cleistogamous, enclosed by tightly rolled sheath; glumes 1–2 mm, obtuse to acute; lemma 4.5–6 mm, hairs at base short-appressed between veins, awn 1–3 cm; anther 0.5–1 mm, purple. Open canyon bottoms, rocky slopes; 20–1600 m. s ChI (San Clemente Island), DMtns (Providence Mtns); s AZ, Baja CA. Apr–May ★

M. arsenei Hitchc. (p. 1467) TOUGH MUHLY Per; rhizomes occ ± short or appearing 0. **ST:** 1.5–4 dm, decumbent at base. **LF:** ligule 1–2 mm, acuminate, toothed, ± decurrent to sheath, with 1 large tooth on each side; blade 1–5 cm, < 2 mm wide, rolled. **INFL:** 4–12 cm, 1–3 cm wide, narrow; branches ascending to appressed, loosely fld. **SPIKELET:** glumes 2–3 mm, acute, ± short-awned, awn < 1 mm; lemma 3.5–5 mm, short-soft-hairy on lower 1/2, awn 4–12 mm; palea short-soft-hairy on lower 1/2; anther 1.6–3 mm, purple. Limestone rock outcrops, slopes; 1400–1860 m. DMtns (Clark Mtn Range); to se UT, n NM, n Baja CA. Aug–Sep ★

M. asperifolia (Trin.) Parodi (p. 1467) SCRATCH GRASS Per; rhizomes shiny, scaly; ± stoloned. **ST:** decumbent to erect, 1–6 dm. **LF:** ligule 0.2–1 mm, truncate, minutely ciliate; blade 2–6 cm, 1–2.8 mm wide, flat or folded. **INFL:** 6–17 cm, 5–14 cm wide, ovoid, open; branches 5–14 cm, spreading. **SPIKELET:** glumes 0.5–1.5 mm, acute; florets 1–2; lemma 1–2 mm, glabrous ± short-awned; anther 1–1.2 mm, purple. 2*n*=20,22,28. Moist, often alkaline meadows, seeps, hot springs; 120–2150 m. CA; w N.Am, s S.Am. Jul–Oct

M. californica Vasey (p. 1467) CALIFORNIA MUHLY Per; rhizomes short, scaly, creeping. **ST:** 3–7 dm. **LF:** ligule 0.8–2 mm, truncate, irregularly toothed, minutely ciliate; blade 4–16 cm, 2–6 mm wide, flat. **INFL:** 5–13 cm, < 2 cm wide, narrow; branches ascending

to erect, short, densely fld. **SPIKELET**: glumes 2.5–4 mm, acuminate, ± long-tapered to awned, awn < 1.2 mm; lemma 2.8–4 mm, short-soft-hairy on lower 1/2, hairs < 1.5 mm, awn < 2.2 mm; anther 1–1.5 mm, yellow. 2*n*=80. Streambanks, canyons; 100–2000 m. SCo, SnGb, SnBr, SnJt. Jun–Sep ★

M. filiformis (S. Watson) Rydb. PULL-UP MUHLY Ann. **ST**: decumbent, loosely clumped, rooting at lower nodes, 0.2–3 dm. **LF**: ligule 1–2.5 mm, obtuse to acute, margin serrate; blade 1–4 cm, 1–2 mm wide, flat or rolled. **INFL**: 1–6 cm, < 1 cm wide, cylindric, narrow; branches closely appressed. **SPIKELET**: glumes 0.5–1.2 mm, obtuse, ± toothed at 10×; lemma 1.5–2 mm, short-awned, awn < 1 mm; anther 0.5–1 mm, purple to yellow. 2*n*=18. Moist meadows, seeps, streambanks; 150–3350 m. NW, SN, WTR, SnBr, SnJt, GB; to BC, c US, NM, Mex. Jun–Aug

M. fragilis Swallen (p. 1467) DELICATE MUHLY Ann. **ST**: erect or spreading, 1–3.5 dm. **LF**: ligule 1–3 mm, decurrent to sheath, with 1 large tooth at each side; blade 1–6 cm, 1–2 mm wide, flat, margin, midvein strongly white-thickened. **INFL**: 10–30 cm, 3.5–8 cm wide, ovoid, open; branches thread-like, stiffly spreading to reflexed ± 90° from central axis. **SPIKELET**: glumes 0.5–1 mm, obtuse to acute, glabrous; lemma ± 1 mm, gen glabrous, margin, midvein occ short-hairy; anther < 0.5 mm, purple. 2*n*=20. Open, ± disturbed, limestone gravelly wash; ± 1600 m. e DMtns (Clark Mtn Range, New York Mtns); to w TX, Mex. Oct ★

M. jonesii (Vasey) Hitchc. JONES' MUHLY Per. **ST**: densely clumped, 2–5 dm. **LF**: basal gen tufted; ligule 2–4.5 mm, acute; blade 5–12 cm, 1–2.5 mm wide, flat to ± folded. **INFL**: 4–15 cm, 1.5–4 cm wide; branches ascending to loosely spreading, loosely fld. **SPIKELET**: glumes 0.6–1.8 mm, obtuse, upper irregularly toothed, ± 3-veined; lemma 2.8–3.5 mm, short-soft-hairy on lower 1/3, short-awned, awn < 1 mm; anther 1.4–2.2 mm, purple. 2*n*=20. Open slopes; 1130–2130 m. KR, CaRH, n SNH. Jun–Aug ★

M. mexicana (L.) Trin. Per; rhizomes scaly, creeping. **ST**: 3–7 dm. **LF**: ligule 0.4–1 mm, truncate, irregularly toothed; blade 4–12 cm, 2–5 mm wide, flat. **INFL**: 2–15 cm, 5–15 mm wide, narrow; branches stiffly ascending, densely fld. **SPIKELET**: glumes 1.5–3.5 mm, acuminate to short-awned, awn < 2 mm; lemma 1.8–3.4 mm, hairs at base < 1.5 mm, awn 2.5–9 mm; anther < 0.5 mm, yellow. 2*n*=40. Uncommon. Riverbanks, canyons; 60–1530 m. KR, NCoRO, n SN; to BC, e US. Jul–Aug

M. microsperma (DC.) Kunth (p. 1467) LITTLESEED MUHLY Ann, short-lived per. **ST**: 1–6 dm. **LF**: ligule 1–2 mm, decurrent to sheath, truncate to obtuse, toothed; blade 2–6 cm, 1–2.5 mm wide, flat or loosely rolled. **INFL**: 5–20 cm, 1–5 cm wide; branches spreading to ascending, loosely to densely fld. **SPIKELET**: in lower branch axils cleistogamous, enclosed by tightly rolled sheath; glumes 0.5–1 mm, obtuse; lemma gen 2.5–4.5 mm, short-soft-hairy at base, awn 1–3 cm; anther 0.5–1 mm, purple. 2*n*=20,40,60. Open, ± disturbed sites; < 1650 m. CCo, SCoRO, SW, D; to sw UT, AZ, Mex; also in S.Am. Mar–May

M. minutissima (Steud.) Swallen (p. 1467) Ann. **ST**: ascending to erect, 0.2–3 dm. **LF**: ligule 1–2 mm, truncate to obtuse, short-toothed; blade 0.5–4 cm, 1–2 mm wide, flat. **INFL**: 1–20 cm, narrowly ovoid, open; branches 1.5–5 cm, ascending < 80° from central axis. **SPIKELET**: glumes 0.5–1 mm, obtuse, tip short-hairy; lemma 1–1.5 mm, margins, midvein short-hairy; anther 0.5–1 mm, purple. 2*n*=60,80. Open, ± disturbed, sandy slopes, seeps; 400–2300 m. KR, n&c SNH, SnBr, SnJt, SNE; to WA, MT, w TX, Mex. Jul–Oct

M. montana (Nutt.) Hitchc. (p. 1467) MOUNTAIN MUHLY Per. **ST**: 1–4 dm, densely tufted. **LF**: ligule 4–10 mm, acute; blade 5–12 cm, 1–2.5 mm wide, flat, ± rolled. **INFL**: 4–15 cm, 2–6 cm wide, oblong; branches spreading to ascending, loosely fld. **SPIKELET**: glumes 1.5–3 mm, upper 3-toothed, 3-veined; lemma 3–4.2 mm, margin, midvein short-soft-hairy on lower 1/2, awn 6–18 mm; anther 1.5–2.2 mm, purple. 2*n*=20,40. Open slopes, granitic rock outcrops, dry meadows; 1640–3420 m. KR, SNH; w US to C.Am. Jun–Aug

M. pauciflora Buckley (p. 1473) FEW-FLOWERED MUHLY Per; rhizomes ± short, knot-like, or appearing 0. **ST**: erect, 3–5 dm, wiry, ± rooting at lower nodes; lower nodes knot-like. **LF**: ligule 1–2.5 mm, decurrent to sheath, with 1 large tooth on each side; blade 5–8 cm, 0.5–1.5 mm wide, flat to ± folded. **INFL**: 5–12 cm, 0.5–3 cm wide, narrow; branches ascending to appressed, loosely fld. **SPIKELET**: glumes 1.4–3.2 mm, acuminate to ± short-awned, awn < 1 mm; lemma 4–5 mm, base sparsely short-hairy, awn 5–20 mm; palea glabrous to ± scabrous; anther 1.8–2 mm, purple. Rocky slopes, ledges, canyons; 1755 m. e DMtns (New York Mtns); to s CO, w TX, Mex. Sep–Oct ★

M. porteri Beal (p. 1473) Per. **ST**: 2.5–8 dm, wiry; lower nodes knot-like. **LF**: ligule 1–2.5 mm, truncate, decurrent to sheath, toothed; blade 2–8 cm, 1–2 mm wide, flat to ± folded. **INFL**: 4–15 cm, 6–15 cm wide, ovoid, open; branches thread-like, spreading. **SPIKELET**: glumes 2–3 mm, acuminate, occ short-awned, awn < 1 mm; lemma 3–4.2 mm, hairy below middle, awn 2–10 mm; anther 1.5–2.3 mm, purple to yellow. 2*n*=20,23,24,40. Among boulders or shrubs, rocky slopes, cliffs; 610–1680 m. SnBr, PR, SNE, DMoj; to CO, w TX, Mex. Jun–Oct

M. richardsonis (Trin.) Rydb. (p. 1473) MAT MUHLY Per, matted; rhizome scaly. **ST**: decumbent to erect, 0.5–4 dm; lower nodes often swollen or knot-like. **LF**: ligule 1–2.5 mm, acute to truncate, decurrent to sheath; blade 1–5 cm, 1–2 mm wide, flat to ± rolled. **INFL**: 1–12 cm, 1–4 mm wide, cylindric, narrow; axis gen obscured by appressed branches. **SPIKELET**: glumes 0.8–1.8 mm, acute to ± short-awned; lemma 2–3 mm, glabrous, ± scabrous at tip, ± short-awned, awn < 0.5 mm; anther 1.2–1.5, yellow to purple. 2*n*=40. Open sites, ± moist meadows, talus slopes, along streams; 1220–3670 m. KR, CaRH, SNH, SCoRO, TR, SnJt, GB, DMtns; to Can, ne US, Mex. Jun–Aug

M. rigens (Benth.) Hitchc. (p. 1473) DEER GRASS Per. **ST**: densely clumped, 5–15 dm. **LF**: ligule 0.5–2 mm, truncate, ± ciliate; blade 10–50 cm, 1.5–6 mm wide, flat. **INFL**: 15–60 cm, 5–12 mm wide, cylindric, narrow; branches appressed, densely fld. **SPIKELET**: glumes 1.8–3 mm, acute or obtuse, ± scabrous; lemma 2.5–3.5 mm, base sparsely short-hairy, ± abruptly pointed; anther 1.3–1.7, yellow to purple. 2*n*=40. Sandy to gravelly places, canyons, stream bottoms; < 2150 m. CaRH, SN, GV, SCoRO, SCo, TR, SnJt, SNE, DMoj; to TX, Mex. Jun–Sep

M. utilis (Torr.) Hitchc. APAREJO GRASS Per; rhizome scaly. **ST**: decumbent, often creeping, 0.5–3 dm. **LF**: ligule 0.3–0.8 mm, truncate, decurrent to sheath; blade 1–3.5 cm, 0.6–1.2 mm wide, flat to ± rolled. **INFL**: 1–5 cm, 1–3 mm wide, narrow; branches short, appressed; axis gen visible between branches. **SPIKELET**: glumes 0.5–1.5 mm, acute; lemma 1.5–2.5 mm, acute, glabrous; anther 1–1.2 mm, yellow to purple. 2*n*=20. Wet sites along streams, ponds; 250–1000 m. SCoRO, SCo, WTR; to TX, C.Am. Oct–Mar

MUNROA

Jesús Valdés-Reyna

Ann, mat-forming. **ST**: stolon-like, 2–8 cm, terminating in clusters of lvs from which new sts arise; st to 15 cm. **LF**: mostly basal; sheath hairy-tufted at throat; ligule hairy; blade linear, gen inrolled, occ flat or folded, sharply pointed, margins white, thickened. **INFL**: terminal, head-like; spikelets 2–4, subsessile to pedicelled, clustered, subtended by lfy bract; axis breaking above glumes or below bract. **SPIKELET**: laterally compressed; florets 2–10, lower florets bisexual or pistillate; terminal florets sterile; glumes < spikelet, 1-veined, unawned; lower glumes gen present, upper absent or reduced on terminal spikelet;

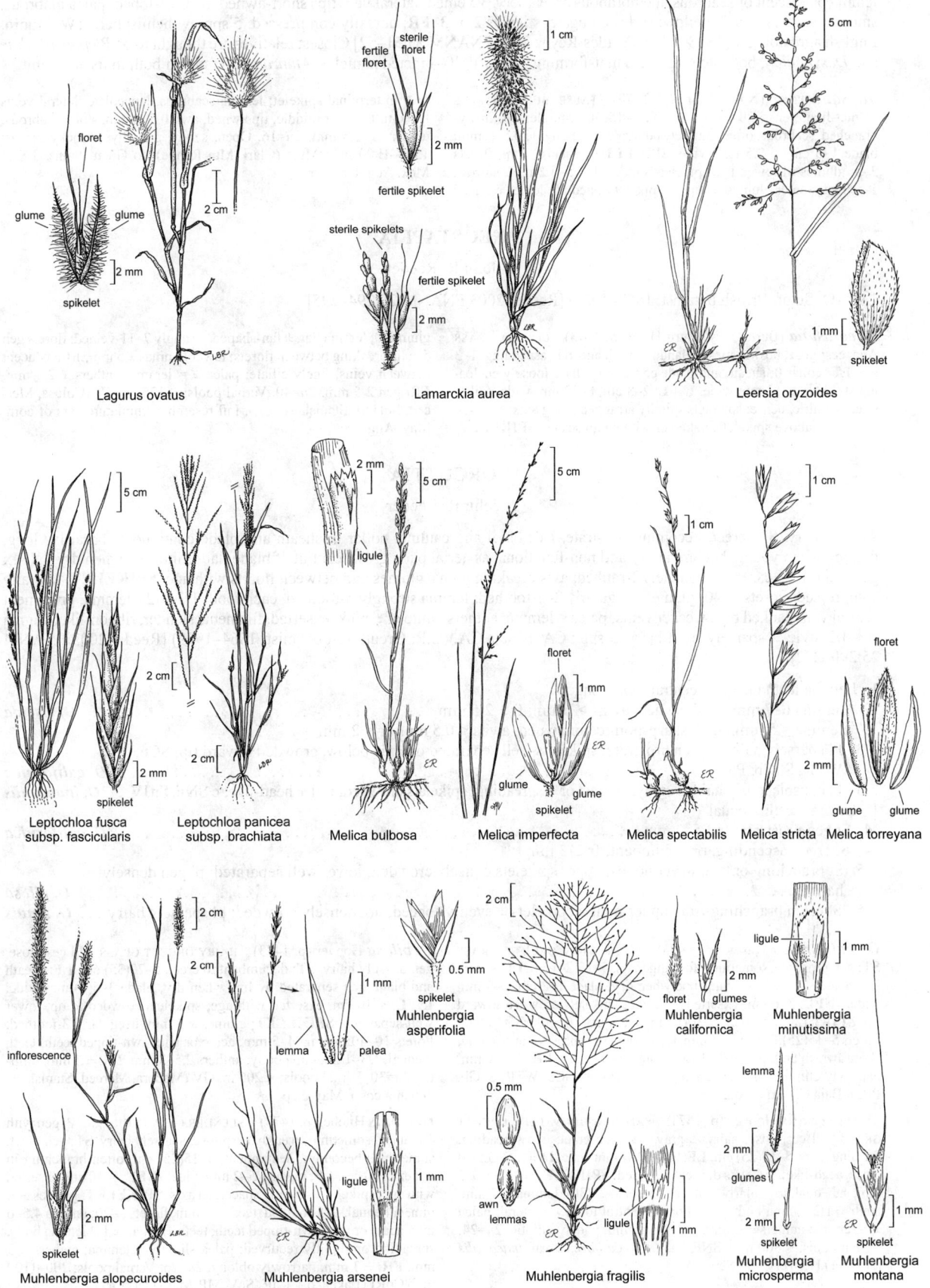

Lagurus ovatus

fertile floret
sterile floret
1 cm
2 mm
fertile spikelet
floret
glume
glume
2 mm
spikelet

Lamarckia aurea

sterile spikelets
fertile spikelet
2 mm

Leersia oryzoides

5 cm
1 mm
spikelet

Leptochloa fusca
subsp. fascicularis

5 cm
2 cm
2 mm
spikelet

Leptochloa panicea
subsp. brachiata

2 cm
2 cm

Melica bulbosa

2 mm
ligule
5 cm

Melica imperfecta

5 cm
floret
1 mm
glume
spikelet
glume

Melica spectabilis

1 cm

Melica stricta

1 cm

Melica torreyana

floret
2 mm
glume
glume

Muhlenbergia alopecuroides

2 cm
2 cm
inflorescence
spikelet
2 mm

Muhlenbergia arsenei

1 mm
lemma
palea
ligule
1 mm

Muhlenbergia asperifolia

0.5 mm
spikelet

Muhlenbergia fragilis

2 cm
0.5 mm
awn
lemmas
ligule
1 mm

Muhlenbergia californica

floret
glumes
2 mm

Muhlenbergia minutissima

ligule
1 mm

Muhlenbergia microsperma

lemma
2 mm
glumes
2 mm
spikelet

Muhlenbergia montana

1 mm
spikelet

lemma pubescent or scabrous, membranous or leathery, 3-veined, lateral vein tips short-awned, tips ± 2-lobed; palea glabrous, smooth; anthers 2 or 3, yellow; styles elongate, stigmas 2 or 3. **FR:** dorsally compressed. 5 spp.: w hemisphere. (W. Munro, English agrostologist, 1818–1880) [Valdés-Reyna 2003 FNANM 25:51–52] Closest relatives are thought to be *Blepharidachne* and *Dasyochloa*, both stoloniferous, mat-forming spp. with lfy-bracted panicles. *Munroa* differs from both in its ann habit.

M. squarrosa (Nutt.) Torr. (p. 1473) FALSE BUFFALOGRASS Pl mat-like, gen < 20 cm wide. **ST:** 3–15(30) cm, slender, many-branched, scabrous, often minutely puberulent. **LF:** ligule 0.5(1) mm; blade 1–5 cm, 1–2.5 mm wide. **SPIKELET:** 6–8(10) mm, florets 3–5; glumes of lower 1–2 spikelets subequal, 2.5–4.2 mm, narrow, 1-veined, acute; glumes of upper spikelets unequal, lower reduced or 0 in terminal spikelet; lemma scabrous, lanceolate, lateral veins hairy-tufted near middle, tip awned, awn 0.5–2 mm, stout, scabrous; anthers 1–1.5 mm. 2*n*=16. Open, sandy, gravelly or rocky places; 1500–1800 m. DMtns (Clark Mtn Range); to Great Plains, TX, n Mex. Aug–Oct ★

NEOSTAPFIA

John R. Reeder

1 sp. (O. Stapf, British botanist, 1857–1933) [Reeder 2003 FNANM 25:294–295]

N. colusana (Burtt Davy) Burtt Davy (p. 1473) COLUSA GRASS Ann, cespitose, brown-sticky-glandular with age. **ST:** ascending, 1–3 dm. **LF:** collar 0; sheath and blade continuous, base loosely enclosing st; blade 5–12 mm wide. **INFL:** 2–8 cm, 8–12 mm wide, spike-like, cylindric, dense; spikelets spirally arranged along axis; axis gen prolonged above spikelets, naked or with small scales. **SPIKELET:** glumes 0; lemma flat, ± fan-shaped, strongly 7–11-veined; florets gen 5, axis breaking between florets; lower lemmas ± 5 mm, ± translucent between veins, finely ciliate; palea ± = lemma; anthers ± 2.5 mm. **FR:** gen 2.5 mm. 2*n*=40. Vernal pools; < 125 m. GV (Colusa, Merced, Solano, Stanislaus cos.). Infl resembles miniature ears of corn. May–Aug ★

ORCUTTIA

John R. Reeder

Ann, ± hairy. **ST:** erect, becoming prostrate. **LF:** basal and cauline; collar 0; sheath and blade continuous; basal lvs long, floating when young, becoming dry and non-functional as aerial parts grow; cauline lf blade flat or inrolled when dry. **INFL:** gen spike-like, exserted; spikelets 2-ranked; axis breaking above glumes and between florets with age. **SPIKELET:** laterally compressed; florets 4–40; glumes irregularly 2–5-toothed; lemma strongly 5-toothed, each tooth 1/3–1/2+ lemma, prominent central vein flanked by 2 weaker veins; palea < lemma; anthers white or ± pink, exserted, filaments slender, ribbon-like; stigma 1/3–1/2 style, ± sparsely short-hairy. 5 spp.: CA, n Baja CA. (C.R. Orcutt, CA botanist, 1864–1929) [Reeder 2003 FNANM 25:290–292]

1. Lemma teeth unequal, central tooth longest
  2. Lemma 6–7 mm, teeth awned, awns > 1 mm; fr ± 2.5 mm . . . . ***O. viscida***
  2′ Lemma ≤ 5 mm, teeth sharp-pointed or awned, awns < 0.5 mm; fr ≤ 2 mm
    3. Pl sparsely hairy; st gen prostrate; spikelets well separated on axis below, crowded toward tip; SCo,
      WTR, SnGb, PR. . . . . ²***O. californica***
    3′ Pl conspicuously hairy, ± gray; st erect or ± spreading; spikelets crowded, infl ± head-like; c SNF, SnJV. . . ***O. inaequalis***
1′ Lemma teeth ± equal
  4. St gen prostrate; fr 1.5–2 mm. . . . . ²***O. californica***
  4′ St erect, ascending, or decumbent; fr 2–3 mm
    5. St branching only at lower nodes; upper spikelets densely crowded, lower well separated; pl gen densely
      hairy. . . . . ***O. pilosa***
    5′ St often branching from upper nodes; spikelets ± evenly spaced, not densely crowded; pl sparsely hairy . . . ***O. tenuis***

O. californica Vasey (p. 1473) CALIFORNIA ORCUTT GRASS **ST:** gen prostrate, sometimes forming mats, 5–15(20) cm. **LF:** sheath and blade separated by faint line when dry; blade 1–2 cm, 2–3 mm wide. **INFL:** 3–6 cm, exserted with age; spikelets crowded toward tip. **SPIKELET:** glumes 2–3(4) mm, subequal, irregularly toothed; florets 5–15(25); lemma ± 5 mm, teeth equal or central tooth > lateral, short-awn-tipped; palea ± = lemma; anthers ± 2 mm. **FR:** 1.5–2 mm, narrowly elliptic. 2*n*=32. Vernal pools; < 700 m. SCo, WTR, SnGb, PR; n Baja CA. Apr–Aug ★

O. inaequalis Hoover (p. 1473) SAN JOAQUIN VALLEY ORCUTT GRASS Pl cespitose, hairy, ± gray. **ST:** gen erect, occ spreading, forming mats, 5–15(25) cm. **LF:** blade 2–4 mm wide. **INFL:** 2–3.5(5) cm, ± head-like; spikelets densely crowded. **SPIKELET:** glumes ± 3 mm, subequal, irregularly toothed; florets 4–20(30); lemma 4–5 mm, teeth ± 1/2 lemma body, awn-tipped, central tooth >> lateral; palea ± = lemma; anthers ± 2 mm. **FR:** ± 1.5 mm, widely elliptic. 2*n*=24. Vernal pools; < 800 m. c SNF, SnJV. [*O. californica* var. *inaequalis* (Hoover) Hoover] Apr–Sep ★

O. pilosa Hoover (p. 1473) HAIRY ORCUTT GRASS Pl cespitose, gen densely hairy. **ST:** decumbent to erect, 5–20(35) cm. **LF:** sheath and blade gen separated by line when dry; blade 3–5(8) mm wide. **INFL:** < 10 cm, exserted with age; spikelets crowded at tip, lower well-separated. **SPIKELET:** glumes ± 3 mm, irregularly 3-toothed; florets 10–40; lemma 4–5 mm, acute or with awn-tipped teeth, teeth ± equal, 1/3–1/2 lemma body; anthers 2.5–3 mm. **FR:** ± 2 mm, elliptic. 2*n*=30. Vernal pools; < 200 m. GV (Madera, Merced, Stanislaus, Tehama cos.). May–Sep ★

O. tenuis Hitchc. (p. 1473) SLENDER ORCUTT GRASS Pl gen with 1 main st, sometimes weakly cespitose, sparsely hairy. **ST:** gen erect, sometimes becoming decumbent, 5–15(25) cm, often branching in upper 2–10 cm. **LF:** blade 1.5–2 mm wide. **INFL:** 5–10 cm, exserted with age; spikelets ± evenly spaced on axis. **SPIKELET:** glumes 3–6 mm, subequal, 3–5-toothed, teeth < 1 mm; florets 5–20; lemma 4.5–6 mm, acute or with awn-tipped teeth, teeth ± equal, ± 1/2 lemma body, spreading or slightly recurved; palea slightly < lemma; anthers ± 3 mm. **FR:** ± 3 mm, narrowly oblong. 2*n*=26. Vernal pools; 200–1100 m. NCoRI (Lake Co.), CaR, ScV, MP. May–Oct ★

O. viscida (Hoover) Reeder (p. 1473) SACRAMENTO ORCUTT GRASS Pl hairy, strongly aromatic, sticky. **ST**: erect to spreading with age, 3–10(15) cm. **LF**: blade 2–4 mm wide. **INFL**: 3–5 cm, exserted with age; spikelets ± crowded. **SPIKELET**: glumes 5–6 mm, subequal, unequally 3-toothed, teeth = glume body, awn-tipped; florets 6–20(30); lemma 6–7 mm, teeth = lemma body, awn-tipped, central tooth >> lateral; palea 3/4 to > lemma; anthers ± 2 mm. **FR**: ± 2.5 mm, widely elliptic. 2*n*=28. Vernal pools; < 100 m. ScV (Sacramento Co.). [*O. californica* var. *v.* Hoover] Apr–Jul ★

ORYZA RICE

Dieter H. Wilken

Ann, per. **ST**: gen 1, erect. **LF**: cauline; ligule membranous; blade flat to folded. **INFL**: panicle-like. **SPIKELET**: laterally compressed; glumes 2, reduced to 2-lobed cup at pedicel tip; florets 3, lower 2 reduced to glume-like sterile lemmas, uppermost bisexual; fertile lemma keeled, glabrous or stiff-hairy, tip short-beaked, awned or not; palea keeled, beaked, = lemma; stamens 6. ± 20 spp.: trop worldwide. (Greek: rice) There are historical collections of red rice, *O. rufipogon* Griff. ◆, from GV, but no evidence of its being naturalized.

O. sativa L. Ann (per). **ST**: 4–15(20) dm. **LF**: sheath gen glabrous; ligule 4–10 mm, acute; blade 15–35 cm, 3–11 mm wide, glabrous to minutely scabrous. **INFL**: 10–50 cm; branches capillary, drooping. **SPIKELET**: elliptic, 6–11 mm, 2–4 mm wide; glumes vestigial; sterile lemmas 1.5–3(10) mm, awns 0; fertile lemma 6–11 mm, 2–3 mm wide, 3-veined; fertile palea 1–2 mm wide. 2*n*=24. Wet sites; < 100 m. GV; TX to se US, Medit; native to se Asia. Perhaps our most important food pl, it is also used to feed cattle, as a starch source, and is fermented to make beer and sake. Sep–Nov

PANICUM PANIC GRASS

Robert W. Freckmann & Robert Webster

Ann, per. **ST**: gen erect or ascending; internode solid to hollow. **LF**: basal and cauline, similar or dissimilar, basal rosette well developed or not; blade gen flat or rolled under; sheath glabrous or hairy; ligule short-hairy or membranous, ciliate, hairs gen > membrane, with an adjacent false ligule of longer hairs in some spp. **INFL**: panicle-like, gen open; 1° branches spreading to ascending; 2° main branches simple or forked at base; spikelets many, 1–2 per node, gen stalked, on one side of axis or not, stalk tip expanded, one side concave. **SPIKELET**: falling as 1 unit, ± compressed, gen green to ± purple; glumes gen unequal, lower gen < upper, free, clasping, upper glume ± = spikelet, membranous, ± thin; florets 2, lower sterile or staminate, lemma texture like glumes, upper floret fertile, lemma leathery to hard, firm, gen shiny, smooth to rough, margin inrolled or partly flat, tip blunt, palea ± enclosed by lemma margin. ± 440 spp.: trop to warm temp, worldwide. (Latin: ancient name for millet) [Freckmann & Lelong 2003 FNANM 25:406–450, 450–488] Subg. *Dichanthelium* often recognized as distinct, monophyletic genus, as in FNANM by Freckmann; based on editorial decision, *Panicum* treated here to incl *Dichanthelium*, which is phylogenetically nested in *Panicum*.

1. Lvs in 2 forms, basal < and gen wider than cauline, rosette well developed, gen persistent; main panicle branches forked at base (subg. *Dichanthelium*)
   2. Spikelet 2.7–3.5 mm, 2–2.4 mm wide; upper glume 9-veined with a prominent orange spot at base, lower glume acuminate; ligule hairs 1–1.5 mm . . . . *P. oligosanthes* var. ***scribnerianum***
   2′ Spikelet 1–2 mm, ± < 1 mm wide; upper glume 7-veined, lacking orange spot at base, lower glume acute; ligule of hairs 2–4 mm . . . . *P. acuminatum*
      3. St, lower sheath gen glabrous . . . . var. ***lindheimeri***
      3′ St, sheath soft-hairy
         4. Lf blade puberulent abaxially; infl gen long-exserted . . . . var. ***fasciculatum***
         4′ Lf blade soft-hairy abaxially; infl gen scarcely exserted . . . . var. ***thermale***
1′ Lvs gradually reduced upward, basal rosette not well developed; main panicle branches simple at base (subg. *Panicum*)
   5. Sheath compressed-keeled, blade folded at base; ligule ragged . . . . *P. rigidulum* subsp. ***rigidulum***
   5′ Sheath not compressed-keeled, blade flat or rolled under; ligule ciliate
      6. Per, gen with rhizomes or stolons; lower floret staminate
         7. St hard, almost woody; upper lemma with flat margin . . . . *P. antidotale*
         7′ St not esp hard; upper lemma margin inrolled . . . . *P. urvilleanum*
      6′ Ann; lower floret sterile
         8. St, lvs ± glabrous; lower glume ± truncate, < 1/3 spikelet length . . . . . . *P. dichotomiflorum* subsp. ***dichotomiflorum***
         8′ St, lvs hairy; lower glume acute, > 1/3 spikelet length
            9. Spikelet > 4 mm; panicle dense, spikelets in overlapping clusters . . . . *P. miliaceum* subsp. ***miliaceum***
            9′ Spikelet < 4 mm; panicle open, most spikelets well separated
               10. Lower palea 0; base of upper floret without paired crescent-shaped scars . . . . *P. capillare*
               10′ Lower palea present; base of upper floret gen with paired crescent-shaped scars
                  11. Spikelet axis not elongated between glumes and florets; lower glume 1–1.5 mm, < 1/2 spikelet length; stalks gen spreading, gen > 2 mm . . . . *P. hillmanii*
                  11′ Spikelet axis ± elongated between glumes and florets, visible; lower glume 1.5–2.5 mm, > 1/2 spikelet length; stalks gen appressed, gen < 2 mm . . . . *P. hirticaule* subsp. ***hirticaule***

P. acuminatum Sw. **Per. LF:** sheath 2–7 cm; ligule of hairs 2–4 mm. **INFL:** 1° branches 2–4 cm, axis glabrous or short-hairy; spikelet 1 per node, stalk 0.5–3 mm. **SPIKELET:** ± < 1 mm wide; lower glume 1-veined, tip acute, upper glume 7-veined, lacking orange spot at base; lower floret sterile, lemma 7-veined, tip rounded, palea vestigial; upper floret ± < lower floret. 2*n*=18. [*Dichanthelium a.* (Sw.) Gould & C.A. Clark] Other vars., incl *P. acuminatum* var. *a.*, in Can, e US, C.Am, n S.Am.

var. ***fasciculatum*** (Torr.) Lelong PACIFIC PANIC GRASS **ST:** 1–6 dm, soft-hairy, hairs papillate-based. **LF:** sheath soft-hairy, hairs papillate-based; blade 5–10 cm, 5–12 mm wide, puberulent abaxially. **INFL:** 5–8 cm, gen long-exserted. **SPIKELET:** 1.5–2 mm, obovate to elliptic; lower glume 0.5–1 mm. Moist places, wet meadows, seeps, streambanks; < 2600 m. CA-FP, SNE, DMtns. [*Dichanthelium a.* subsp. *f.* (Torr.) Freckmann & Lelong; *P. huachucae* Ashe; *P. occidentale* Scribn.; *P. pacificum* Hitchc. & Chase] If recognized taxonomically, hybrids with *P. oligosanthes* var. *scribnerianum* assignable to *P.* ×*shastense* Scribn. & Merr. May–Nov

var. ***lindheimeri*** (Nash) Beetle **ST:** 3–7 dm, glabrous. **LF:** sheath glabrous; blade 3–15 cm, 3–11 mm wide, glabrous exc on margin near base. **INFL:** 4–12 cm. **SPIKELET:** 1.4–1.8 mm, gen elliptic; lower glume gen < 0.5 mm. Moist places, gen near springs or seeps; < 1000 m. NCo, CaRF, GV, SnGb; to e US, TX, n Mex. [*Dichanthelium a.* subsp. *l.* (Nash) Freckmann & Lelong; *P. l.* Nash] May–Oct

var. ***thermale*** (Bol.) Wipff (p. 1473) GEYSERS PANICUM **ST:** 1–8 dm. **LF:** sheath soft-hairy; blade 4–10 cm, 5–12 mm wide, soft-hairy abaxially. **INFL:** 5–8 cm, gen scarcely exserted. **SPIKELET:** 1.8–2 mm, obovate to elliptic; lower glume 0.5–1 mm. Peaty meadows and pockets, often at hot springs and fumaroles; 500–2700 m. NCoRI, CaRH. [*P. t.* Bol.; *Dichanthelium a.* subsp. *t.* (Bol.) Freckmann & Lelong; *D. lanuginosum* (Elliott) Gould var. *t.* (Bol.) Spellenb.; *P. lassenianum* Schmoll] Jun–Sep ★

P. antidotale Retz. BLUE PANIC GRASS **Per from rhizomes. ST:** erect, 5–25 dm, hard, almost woody; nodes 5–12. **LF:** sheath 4–8 cm, glabrous; ligule 1.5–2.5 mm, membranous, densely ciliate; blade 15–30 cm, 4–12 mm wide, upper surface glabrous. **INFL:** 13–28 cm; 1° branches 9–18 cm, glabrous; spikelets 1–2 per node, stalk < 2.5 mm. **SPIKELET:** 2.5–3 mm, ± 1 mm wide, ovate to elliptic, brown, green, or purple; lower glume 1.5–2.5 mm, 5-veined; lower floret staminate, lemma 7-veined, tip acute, palea ± = lemma; upper floret ± = lower floret, lemma margin flat. 2*n*=18. Open, gen disturbed areas, fields; 300–600 m. D; to UT, TX, n Mex; native to India. Apr–Oct ◆

P. capillare L. (p. 1473) WITCH GRASS **Ann. ST:** 2–10 dm. **LF:** sheath 4–8 cm, short-hairy; ligule membrane 0.5–1.5 mm, ciliate; blade 5–30 cm, 5–20 mm wide, upper surface short-soft- to rough-hairy. **INFL:** 15–40 cm, open; 1° branches 12–30 cm, axis sparsely short-hairy; spikelet 1 per node; stalk 2–6 mm. **SPIKELET:** 2–3.5 mm, ± 1 mm wide, lanceolate to elliptic; lower glume ± 1–1.5 mm, 3-veined, acute; lower floret sterile, lemma 7-veined, tapered to a tail-like tip, palea gen 0; upper floret ± 0.8 × lower floret. 2*n*=18. Open places, fields, roadsides; < 1500 m. CA; to Can, e US. Feb–Dec

P. dichotomiflorum Michx. subsp. ***dichotomiflorum*** FALL PANIC GRASS **Ann. ST:** decumbent to erect, 2–10 dm, ± succulent. **LF:** sheath 4–9 cm, ± glabrous, ± inflated, ± compressed; ligule membrane 1–2 mm, ciliate; blade 8–22 cm, 4–15 mm wide, upper surface glabrous. **INFL:** 7–20 cm; 1° branches 4–16 cm, axis glabrous; spikelets 1–2 per node, stalk 0.5–3 mm. **SPIKELET:** 2.2–3 mm, ± 1 mm wide, ovate; lower glume 0.5–1 mm, 1–3-veined, ± truncate; lower floret sterile, lemma 7-veined, acute, palea ± = lemma; upper floret ± 0.7–0.8 × lower floret. 2*n*=36, 54. Disturbed areas, wet ground, fields; < 1300 m. e KR, CaRF, w CaRH, SNF, SNH, GV, SnFrB, SCo, PR; native to e US. Jun–Nov

P. hillmanii Chase HILLMAN'S PANIC GRASS **Ann. ST:** 2–7 dm. **LF:** sheath 2.5–7 cm, rough-hairy; ligule membrane 1–2.5 mm, ciliate; blade 8–15 cm, 4–12 mm wide, upper surface short-soft-hairy.

INFL: 10–25 cm, open; 1° branches 5–13 cm, axis glabrous to short-hairy; spikelet 1 per node, stalk 1–6 mm. **SPIKELET:** 2.3–3 mm, ± 1 mm wide, elliptic; lower glume 1–1.5 mm, 3–5-veined, acute; lower floret sterile, lemma 7-veined, acute, palea gen < lemma; upper floret 0.8–0.9 × lower floret, with paired crescent-shaped scars at base. 2*n*=18. Disturbed areas, roadsides, fields; < 1200 m. NCoR, SNF, SNH, GV, SCo, PR, DSon; native to Great Plains. [*P. capillare* subsp. *h.* (Chase) Freckmann & Lelong] May–Oct

P. hirticaule J. Presl subsp. ***hirticaule*** ROUGHSTALK WITCH GRASS **Ann. ST:** 1–8 dm. **LF:** sheath 2–6 cm, axis glabrous to short-hairy; ligule membrane 0.5–2 mm, ciliate; blade 7–20 cm, 3–15 mm wide, upper surface gen sparsely short-hairy. **INFL:** 5–20 cm, open; 1° branches 3–8 cm, glabrous; spikelets 1–2 per node, stalk 0.5–3 mm, gen appressed. **SPIKELET:** ± 2.5–3 mm, ± 1 mm wide, lanceolate to ovate, green; axis between glumes and florets visible; lower glume ± 1.5–2.5 mm, gen 5-veined, acute; lower floret sterile, lemma 7-veined, acuminate to acute, palea gen < lemma; upper floret 0.7–0.8 × lower floret, stipitate, with paired crescent-shaped scars, often enlarged. Sandy soils, open sites, creosote-bush scrub; < 1400 m. D; to TX, S.Am. Aug–Dec

P. miliaceum L. subsp. ***miliaceum*** (p. 1473) BROOM CORN MILLET **Ann. ST:** 1–10 dm. **LF:** sheath 3.5–8 cm, rough-hairy; ligule membrane 1–3 mm, ciliate; blade 10–20 cm, 6–25 mm wide, upper surface velvety to rough-hairy. **INFL:** 10–40 cm, dense; 1° branches 4–18 cm, axis glabrous; spikelet 1 per node, stalk 2–10 mm. **SPIKELET:** 4.5–5.5 mm, ± 2–2.5 mm wide, elliptic, green to brown; lower glume 2.8–3.5 mm, 5–7-veined, acute; lower floret sterile, lemma 11-veined, acuminate to acute, palea < 2 mm; upper floret 0.8 × lower floret. 2*n*=36, 40, 42, 54, 72. Disturbed areas, fields, roadsides; < 1000 m. CA-FP (exc KR, CaR, Teh, ChI), D; to e US; native to Eurasia. Cult for seed, food. Feb–Nov

P. oligosanthes Schult. var. ***scribnerianum*** (Nash) Fernald SCRIBNER'S PANIC GRASS **Per. ST:** 3–6 dm. **LF:** sheaths 2–8 cm, glabrous or short-hairy; ligule hairs 1–1.5 mm; blade 3–14 cm, 3–15 mm, upper surface glabrous or short-hairy. **INFL:** 5–8 cm; 1° branches 2–4.5 cm, axis glabrous; spikelets 1 per node, stalk 2–5 mm. **SPIKELET:** 2.7–3.5 mm, 2–2.4 mm wide, elliptic, green; lower glume ± 1–1.5 mm, 1-veined, tip acuminate; upper glume 9-veined, with orange spot at base; lower floret sterile, lemma 9–11-veined, acute to rounded, palea ± = lemma to vestigial; upper floret ± < lower floret. 2*n*=18. Meadows, open sites in forest; < 1400 m. NW; to Can, e US. [*P. s.* Nash; *Dichanthelium o.* subsp. *s.* (Nash) Freckmann & Lelong] *P. oligosanthes* var. *o.* native to e N.Am; If recognized taxonomically, hybrids with *P. acuminatum* var. *fasciculatum* assignable to *P.* ×*shastense* Scribn. & Merr. May–Aug

P. rigidulum Nees subsp. ***rigidulum*** REDTOP PANICUM **Per,** cespitose. **ST:** 4–14 dm. **LF:** sheath 6–18 cm, glabrous or sparsely pubescent, compressed-keeled; ligule membrane 0.3–1 mm, ragged; blade 10–40 cm, 2–12 mm wide, folded at base, upper surface glabrous to sparsely long-hairy. **INFL:** 8–30 cm; 1° branches 6–12 cm, axis glabrous; spikelet 1–2 per node, crowded on adaxial side of branches; stalk 0.2–2 mm. **SPIKELET:** 1.6–2 mm, ± 0.6 mm wide, lanceolate; lower glume 0.7–1.5 mm, 3-veined; lower floret sterile, lemma 5-veined, tapered, palea ± 1 mm; upper floret 1.2–1.5 mm, with tuft of minute hairs at tip. 2*n*=18. Ditches, riverbanks; < 1600 m. SN, GV; e&se US, Mex. Jul–Oct

P. urvilleanum Kunth SILKY PANIC GRASS **Per from stolons or rhizomes. ST:** 4–10 dm. **LF:** sheath 12–35 cm, hairy; ligule membrane 1–2 mm, ciliate; blade 20–45 cm, 4–7 mm wide, upper surface short-hairy. **INFL:** 20–35 cm; 1° branches 8–13 cm, glabrous; spikelet 1 per node, stalk 1.5–10 mm. **SPIKELET:** 5–7.5 mm, 2–2.5 mm wide, elliptic, green; lower glume 4.5–6.5 mm, 5–7-veined; lower floret staminate, lemma 9-veined, acute, palea ± = lemma; upper floret 0.8–0.9 × lower floret, lower floret staminate; lemma margin inrolled. 2*n*=36. Sandy soils, dunes; < 1220 m. e SCo, D; AZ, n Mex, S.Am. Mar–May

PARAPHOLIS SICKLE GRASS

Thomas Worley

Ann. **ST**: decumbent to erect, 5–45 cm, branched, glabrous; nodes gen purple. **LF**: sheath 1–6 cm; ligule ± 1.5 mm, membranous; blade flat to ± inrolled, ribbed, upper surface scabrous. **INFL**: spike-like, 3–18 cm, rigid, cylindric, straight to strongly curved, breaking at nodes; spikelets alternate, 2-ranked, appressed, embedded in axis, falling with axis segment. **SPIKELET**: glumes 2, edge-to-edge or slightly overlapping, 4–7 mm, lanceolate, keel to 1 side near base or obscure, margin translucent, ± enclosing floret, 5-veined; florets 1–2, bisexual; lemma 3.5–5.5 mm, translucent, 3-veined; palea ± = lemma, translucent. **FR**: ± 3.5 mm. 5 spp.: native to Eur. (Greek: scales beside, from glume orientation) [Worley 2007 FNANM 24:687–688] See *Hainardia*.

1. Infl gen curved, gen twisted; upper lf sheath margins expanded, enclosing lower 1–4 spikelets; anther 0.5–1 mm . . . . *P. incurva*
1′ Infl gen straight; lf sheath margins all alike, all spikelets exserted from sheath; anther 2–4 mm . . . . *P. strigosa*

P. incurva (L.) C.E. Hubb. (p. 1473) **ST**: decumbent to erect, 3–35 cm, branched throughout or not. **LF**: sheath 1–4 cm; blade 1–10 cm, 1–3 mm wide, gen inrolled. **INFL**: 2–15 cm; spikelets 3–20. **SPIKELET**: glume keel ± 0.5 mm; floret occ cleistogamous. 2*n*=32,36,38,42. Disturbed, well drained soils of salt marshes, gen above highest tide level; < 100 m. NCo, e NCoRI, w ScV, n SnJV, CCo, SCo, ChI; to BC, Mex; native to Eur. Apr–Jun

P. strigosa (Dumort.) C.E. Hubb. **ST**: gen ascending to erect, 12–45 cm, branched at lower nodes. **LF**: sheath 2–6 cm; blade 1–10 cm, 1–3 mm wide, gen flat. **INFL**: 7–18 cm; spikelets 10–25. **SPIKELET**: glume keel obscure; floret never cleistogamous. 2*n*=14. Moist soils, gen salt marshes below highest tide level; < 5 m. NCo (esp n Humboldt Bay); BC; native to Eur. Apr–Jun

PASPALUM

Charles Allen

Per (in CA), tufted, rhizomatous, or stoloniferous. **ST**: decumbent to erect; internode solid to hollow. **LF**: basal and cauline; sheath glabrous or hairy; ligule membranous. **INFL**: racemes 1 to many, branches digitate or raceme-like, spreading to erect. **SPIKELET**: 2-fld (lower floret sterile, upper floret fertile), dorsally compressed, falling as one unit, fertile lemma convex, positioned adaxially, subsessile to short-pedicelled, 1 or 2, in 2 rows along 1 side of rachis; lower glume 0, rarely present and then not on every spikelet in a raceme; upper glume and sterile lemma membranous, subequal, tip rounded, sterile palea 0 or rudimentary; fertile lemma and palea hardened, smooth, straw-colored; fertile lemma margins scarious, inrolled, clasping palea. **FR**: white, yellow, or brown, round to elliptical, planoconvex or flattened. ± 300–400 spp.: trop, warm temp worldwide. (Greek: ancient name for a kind of millet) [Allen & Hall 2003 FNANM 25:566–599] *P. notatum* Flügge not naturalized in CA.

1. Margins of upper glume and sterile lemma glabrous; infl digitate; spikelets single
  2. Upper glume pubescent on back . . . . *P. distichum*
  2′ Upper glume glabrous on back. . . . . *P. vaginatum*
1′ Margins of upper glume and sterile lemma long-silky-hairy; infl a raceme; spikelets paired
  3. Racemes 2–7; spikelets gen ≥ 3 mm. . . . . *P. dilatatum*
  3′ Racemes gen 10–30; spikelets < 3 mm . . . . *P. urvillei*

P. dilatatum Poir. (p. 1473) DALLIS GRASS Cespitose, base knotty; rhizomes < 1 cm. **ST**: 50–175 cm, erect; nodes glabrous. **LF**: sheath glabrous to pubescent, esp lower; ligule 1.5–3.8 mm; blade to 35 cm, 2–16.5 mm wide, flat, gen glabrous, adaxial surface slightly long-hairy near base. **INFL**: racemes 2–7, 1.5–12 cm, terminal, ascending; axis 0.7–1.4 mm wide, winged, glabrous exc margins scabrous. **SPIKELET**: paired, (2.3)3–4 mm, 1.7–2.5 mm wide, appressed to rachis, ovate, straw-yellow (purple), tip tapered; lower glume 0; upper glume and sterile lemma 5–7-veined, long-silky-hairy on margins; fertile floret straw-colored. **FR**: 2–2.3 mm, white to brown. 2*n*=20,40,50. Disturbed areas; < 1160 m. CA-FP, DMoj; s US; native to S.Am. May–Nov

P. distichum L. (p. 1473) KNOT GRASS Creeping or cespitose. **ST**: 5–63 cm, erect; nodes glabrous. **LF**: sheath glabrous, sparsely long pubescent at tip; ligule 1–2 mm; blade < 14 cm, 1.8–11.5 mm wide, flat or folded, glabrous to pubescent, tip inrolled. **INFL**: racemes 2–3, 1.4–7 cm, terminal, digitate, ascending, often arched; axis 1.2–2.2 mm wide, winged, glabrous, margins scabrous. **SPIKELET**: 1 (rarely paired), appressed to axis, 2.4–3.2 mm, 1.1–1.6 mm wide, elliptic, straw-colored, sometimes partially purple, pubescent; lower glume 0, rarely present, < 1 mm, triangular; upper glume and sterile lemma 3-veined, margins glabrous; fertile floret straw-colored. **FR**: 1.9–2.1 mm, yellow. 2*n*=20,30,40,48,60. Edges of lakes, ponds, rice fields, wet roadside ditches; < 1770 m. CA-FP (exc mtns), GB, n DMoj; to WA, e US, S.Am; warm areas worldwide. [*P. d.* var. *indutum* Shinners; *P. paspalodes* (Michx.) Scribn.] Jun–Oct

P. urvillei Steud. Cespitose, base knotty, rhizomes < 1 cm. **ST**: 50–220 cm, erect; nodes glabrous or pubescent. **LF**: sheath glabrous to pubescent; ligule 1–4(7.7) mm; blade < 57 cm, 2–12 mm wide, flat, mostly glabrous but with few long hairs on adaxial surface near base. **INFL**: racemes (4)10–30, 1.2–11.5 cm, terminal, ascending; axis 0.5–1.1 mm wide, winged, glabrous, margins scabrous. **SPIKELET**: paired, appressed to axis, 2–2.8 mm, 1.1–1.5 mm wide, elliptic to slightly obovate, straw-yellow (purple); lower glume absent; upper glume, sterile lemma 3-veined, long-silky-hairy; fertile floret straw-colored. **FR**: 1.2–1.7 mm, white. 2*n*=40. Disturbed areas; < 100 m. ScV, SCo, expected elsewhere; se US; native to S.Am. Potentially problematic weed. Jun–Dec

P. vaginatum Sw. SEASHORE PASPALUM Creeping or cespitose. **ST**: 10–79 cm, erect; nodes glabrous. **LF**: sheath glabrous, sparsely long pubescent at tip; ligule 1–2 mm; blade to 19 cm, 1.4–8 mm wide, flat or folded, glabrous to pubescent, tip incurved. **INFL**: racemes 2–3, 1.1–7.9 cm, terminal, palmate, ascending to erect; axis 0.4–1.4 mm wide, winged, glabrous, margins scabrous. **SPIKELET**: 1, appressed to axis, 3–4.5 mm, 1.1–2 mm wide, elliptic to lanceolate, glabrous, light straw-colored; lower glume 0, rarely present; upper glume and sterile lemma 3-veined, margins glabrous; fertile floret white. **FR**: 2.8–3.1 mm, yellow. 2*n*=20,40. Brackish and salt marshes; < 100 m. SCo, DSon; s&e US, warm areas worldwide. Aggressively invasive in wetlands. Aug–Dec

PENNISETUM FOUNTAIN GRASS

Ann, per. **ST**: gen erect and tufted; internode solid or hollow. **LF**: basal and cauline; sheath gen glabrous; ligule short-hairy or membranous, ciliate. **INFL**: terminal and/or axillary, gen panicle-like, dense, ± cylindric (raceme-like in *Pennisetum clandestinum*); spikelets gen many, short-stalked to sessile, gen in clusters of 1–4 on highly reduced branches, subtended by 6–many inner and outer sets of bristles and a single terminal gen >> than the others; spikelet cluster and bristles gen falling as 1 unit. **SPIKELET**: dorsally compressed to ± round in ×-section; glumes 1–2, lower glume < upper when present; upper ± = spikelet; florets 2, lower floret sterile or staminate; upper floret fertile, lemma firm, ± thick or hard, smooth or scabrous, gen dull, margin flat to inrolled, tip blunt; palea < lemma, = in texture; anthers 3. ± 80 spp.: mostly warm temp, subtrop, trop. (Latin: feather or plume + bristle, conspicuous bristles subtending spikelets) [Wipff 2003 FNANM 25:517–529] Separation of *Pennisetum* from *Cenchrus* remains unsettled. *P. glaucum* (L.) R. Br., *P. latifolium* Spreng., *P. macrourum* Trin., *P. nervosum* (Nees) Trin., *P. purpureum* Schumach. have been collected in CA, but none has become naturalized. Some spp. cult for orn, food (pearl millet), fodder, fiber for paper-making; others are aggressive weeds.

1. Fl st decumbent; infl raceme-like, axis flattened, ± enclosed by sheath at maturity; subtending bristles
   < spikelets; spikelets in 1–6 clusters . . . . *P. clandestinum*
1′ Fl st erect; infl panicle-like, axis round in ×-section, clearly exserted from sheath at maturity; subtending
   bristles ≥ spikelet; spikelets in 10–many clusters
   2. Inner bristles fused to 1/4 length . . . . *P. ciliare*
   2′ Inner bristles free to base
      3. Pl cespitose; infl pink to dark red; spikelets 4.5–7 mm. . . . . *P. setaceum*
      3′ Pl rhizomatous; infl white to light brown; spikelets 9–12 mm. . . . . *P. villosum*

P. ciliare (L.) Link BUFFEL GRASS Per from knotty base, short rhizomes present or 0. **ST**: erect, 1–15 dm, glabrous (exc sometimes below infl). **LF**: sheath 2–7 cm, glabrous or hairy; ligule 0.5–3 mm; blade 3–25(50) cm, 2.5–13 mm wide, glabrous or pubescent on both sides, green or glaucous. **INFL**: panicle-like, 2–13 cm; main axis straight; outer subtending bristles 16–90, to 12 mm; inner bristles 7–20, 4–14 mm, fused to 1/4 length, terminal bristle 10–23 mm, long-ciliate. **SPIKELET**: 2.5–5.5 mm, lanceolate to ovate, gray to green; lower glume 1–3 mm, upper ± 1.5–3.5 mm; lower floret staminate or sterile, lemma 2.5–5.5 mm; palea present or 0. $2n$=45. Disturbed areas; < 110 m. SCo, PR, DSon; to TX, n Mex; native to Afr, w Asia, India. [*Cenchrus c.* L.] Apr–Oct

P. clandestinum Chiov. (p. 1473) KIKUYU GRASS Mat-forming per from stolons and rhizomes. **ST**: vegetative st spreading; fl st decumbent, 0.3–4.5 dm. **LF**: sheath 1–10 cm, glabrous or hairy, ± inflated; ligule ± 1–2 mm; blade 1.5–3 cm, 1–6 mm wide, upper surface glabrous to short-hairy, flat or folded, tip obtuse. **INFL**: raceme-like, 1–2.7 cm; 1° axis flattened, ± enclosed by sheath at maturity; branches < 0.5 cm, glabrous; inner and outer bristles 6–15, ≤ 11 mm; terminal bristle 10–14 mm. **SPIKELET**: 10–20 mm, ± 1 mm wide, lanceolate, gray-green; lower glume ± 1–2 mm, 0 or 1-veined; upper glume ± = spikelet length; lower floret lemma 9–13-veined, tip acuminate, palea 2–7-veined; upper floret slightly < lower floret. $2n$=36. Disturbed areas; < 100 m. NCo, CaRH, c SNF, ScV, CCo, SnFrB, SCo, n ChI (Santa Cruz, San Miguel islands), PR, D; AZ to S.Am; native to Afr. Often used for fodder or as lawn grass. Can sequester toxic levels of nitrates and soluble oxalates. All yr ◆

P. setaceum (Forssk.) Chiov. CRIMSON FOUNTAIN GRASS Ann or per, cespitose. **ST**: erect, 4–15 dm. **LF**: sheath 4–8 cm, glabrous; ligule 0.5–1 mm; blade 20–65 cm, 2–3.5 mm wide, upper surface glabrous. **INFL**: panicle-like, 8–30 cm, pink to dark red; outer bristles 28–65, 1–19 mm; inner bristles 8–16, 8–27 mm, free to base; terminal bristle 26–34 mm, ciliate. **SPIKELET**: 4.5–7 mm, lanceolate, lower glume 0–0.3 mm; upper glume 1.2–3.6 mm; lower floret lemma 4–6 mm, 3-veined, tip acuminate; upper lemma 4.5–6.5 mm, 5-veined, palea gen 0; $2n$=27. Disturbed areas; < 100 m. Deltaic GV, CCo, SnFrB, SCo, TR, PR, D; OR, NV to FL; Baja CA; native to Afr. Escaped orn and aggressive weed; closely related *P. advena* Wipff & Veldkamp occ escapes from cult in s CA (Sida 18:1035), where it is often sold as an orn under the name *P. setaceum* 'Rubrum'. Mar–Dec ❖

P. villosum Fresen. (p. 1473) FEATHERTOP Per, cespitose, from elongated rhizomes. **ST**: erect, 1.6–7.5 dm. **LF**: sheath 4–10 cm, glabrous; ligule ± 1–1.5 mm; blade 5–40 cm, 2–4.5 mm wide, upper surface glabrous to hairy, flat or folded. **INFL**: panicle-like, 2–10 cm, cylindric to head-like, white to light brown; outer bristles (0)1–8, 1–13.5 mm; inner bristles 20–41, 13–51 mm, free to base plumose, terminal bristle ciliate, not > inner bristles. **SPIKELET**: 9–12 mm, ± 1.5 mm wide, lanceolate, green to white; lower glume 0.5–1 mm, veins 0; upper glume 2.5–6 mm; lower floret sterile or staminate, upper lemma 9–11 mm, 5–7-veined, tip acuminate, palea ± = lemma or 0; upper floret ± = to slightly > lower floret. $2n$=45. Disturbed areas; < 200 m. SN, SnFrB, SCoR, SW, DSon; to e US; native to Afr. Grown as orn. Jun–Aug

PHALARIS CANARY GRASS

Riccardo M. Baldini

Ann, per, cespitose or from vigorous rhizomes. **ST**: gen erect, occ prostrate, 1–25 dm. **LF**: sheath open, glabrous, occ recurved; ligule membranous, glabrous, entire, acuminate or truncate; blade gen 2–5 dm, 1–2.5 cm wide, flat. **INFL**: panicle, gen ovoid to cylindric, dense, occ interrupted; branches ascending to appressed. **SPIKELET**: gen fertile, occ also sterile, laterally compressed, pedicelled; glumes equal, > and enclosing lemmas, glabrous to pubescent, awn 0, keeled, keel gen winged, > florets, 3–5-veined, wings gen entire; 3-fld, lower florets 1–2 vestigial or 0 reduced to rudimentary lemma, gen awl-shaped, to 1/2 fertile fl, occ scale-like, lower occ reduced to a vestigial knob, upper 1 fl fertile, bisexual, lemma gen lanceolate to ovate, glabrous or silky-hairy, faintly 5-veined, awn 0; palea ± = lemma, hairy, translucent; stamens 3; anthers 0.5–4 mm; stigmas 2, plumose. **FR**: plump, ± fusiform. ± 22 spp.: temp N.Am, Eurasia, C.Am, S.Am. (Greek: grass with shiny spikelets) [Barkworth 2007 FNANM 24:764–773]

1. Spikelets in clusters, lower 4–7 spikelets in each cluster with a staminate terminal floret
   2. Per; fertile spikelet surrounded by 6–7 staminate (or sterile) spikelets; panicle ovate; sterile lemma
      0.5–0.7 mm . . . . *P. coerulescens*
   2′ Ann; fertile spikelet surrounded by 5–6 abortive, club-shaped spikelets; panicle gen obovate; sterile
      lemma 0.2–0.4 mm. . . . . *P. paradoxa*

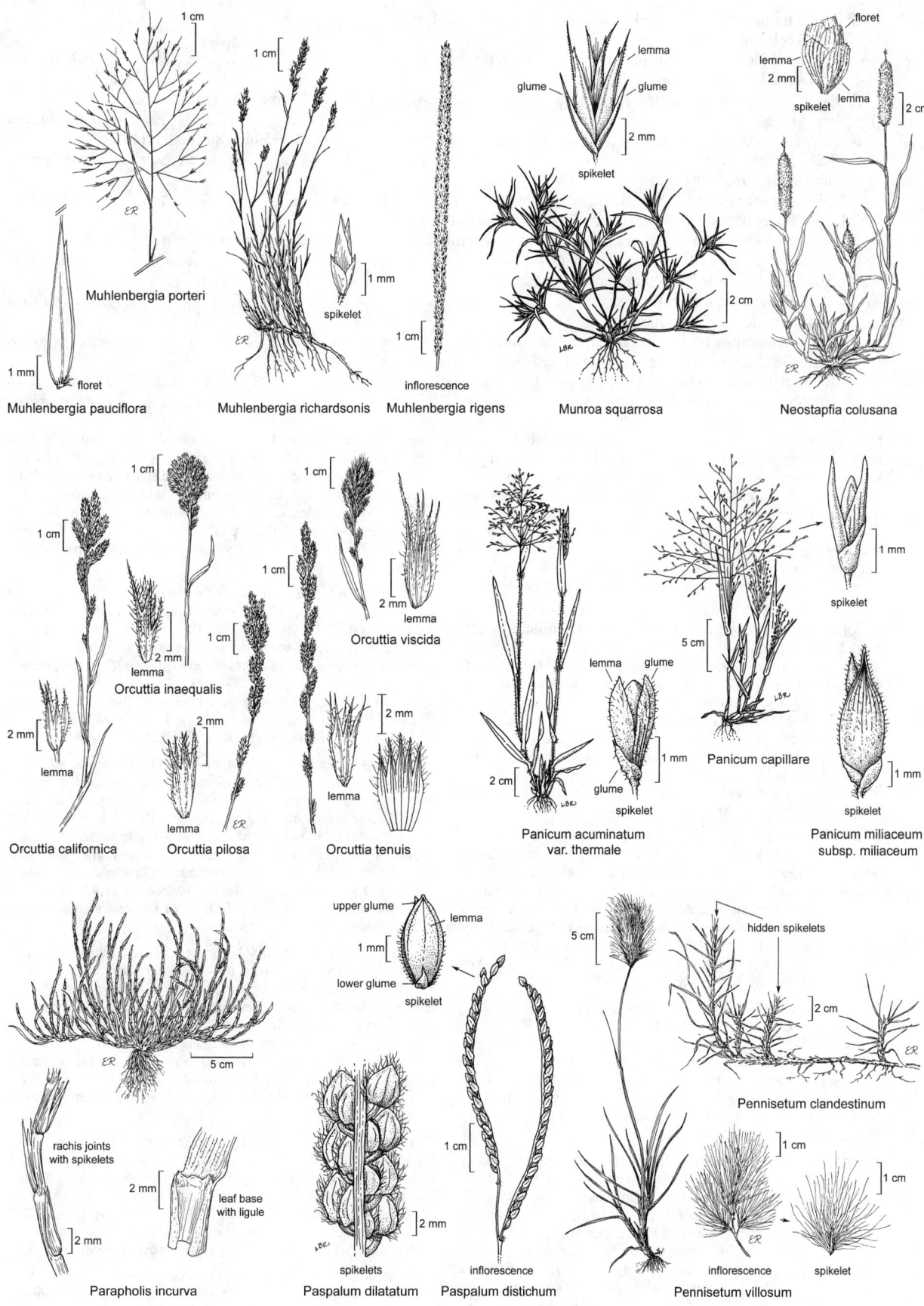

Muhlenbergia porteri

Muhlenbergia pauciflora

floret

Muhlenbergia richardsonis

spikelet

Muhlenbergia rigens

inflorescence

Munroa squarrosa

lemma

glume

glume

spikelet

Neostapfia colusana

floret

lemma

lemma

spikelet

Orcuttia inaequalis

lemma

Orcuttia viscida

lemma

Orcuttia californica

lemma

Orcuttia pilosa

lemma

Orcuttia tenuis

lemma

Panicum acuminatum
var. thermale

lemma

glume

glume

spikelet

spikelet

Panicum capillare

spikelet

Panicum miliaceum
subsp. miliaceum

Parapholis incurva

rachis joints
with spikelets

leaf base
with ligule

upper glume

lemma

lower glume

spikelet

spikelets

Paspalum dilatatum

inflorescence

Paspalum distichum

hidden spikelets

Pennisetum clandestinum

inflorescence

spikelet

Pennisetum villosum

1′ Spikelets borne singly, all spikelets with a bisexual terminal floret
  3. Glume keels not or narrowly winged
    4. Ann; fertile lemma lanceolate to ovoid, hairy exc tip, tip ± beaked. . . . . ***P. lemmonii***
    4′ Per; fertile lemma lanceolate, tip acute, ± sparsely hairy
      5. Panicle lanceolate, branched, lobed, occ ± purple; glumes 4–5 mm; fertile lemma 2.5–4 mm, sterile
        lemma 1.5–2 mm . . . . ***P. arundinacea***
      5′ Panicle gen ovoid, not lobed, ± purple; glumes gen 6–8 mm; fertile lemma 3.5–5 mm, sterile lemma
        2.5–3.5 mm. . . . . ***P. californica***
  3′ Glume keels broadly winged
    6. Keel wing entire to irregularly toothed, sterile lemma 1, 0.3–1.5 mm, awl-like . . . . ***P. minor***
    6′ Keel wing entire, sterile lemmas gen 2, up to 4 mm, not awl-like
      7. Glumes scabrous; panicle narrowly cylindric, base truncate, occ interrupted, fertile lemma gen
        2–3.5 mm . . . . ***P. angusta***
      7′ Glumes glabrous to hairy, not scabrous
        8. Per; glumes gen 4–5 mm . . . . ***P. aquatica***
        8′ Ann; glumes gen (4)6–10 mm
          9. Sterile lemmas 1/5 fertile lemma, ± fleshy to corky . . . . ***P. brachystachys***
          9′ Sterile lemma 1/2 fertile lemma, not fleshy
            10. Sterile lemma scale-like, 2–4 mm . . . . ***P. canariensis***
            10′ Sterile lemma feathery, 1.5–2 mm . . . . ***P. caroliniana***

P. angusta Trin. TIMOTHY CANARY GRASS Ann. **ST:** 1–17 dm. **INFL:** 2–20 cm, 0.5–2 cm wide, cylindric, narrow, occ interrupted along the rachis, base truncate. **SPIKELET:** glumes 2–5 mm, 0.5–2 mm wide, scabrous with scabrous prominent midvein, often ± purple, keel winged, mucronate; sterile lemmas 2, equal, 0.5–1.5 mm, feathery; fertile lemma 2–3.5(4) mm, 1–1.5(2) mm wide, lanceolate, sparsely hairy. **FR:** < 1.5 mm, 0.5–0.9 mm wide. 2*n*=14. Gen wet areas, marshes, sloughs, edge of chaparral; < 1400 m. NCoRO, NCoRI, s SNH, GV, SnFrB, SCoRO, SCo, PR, DMoj; to se US, S.Am. May–Jun

P. aquatica L. HARDING GRASS Per, cespitose, rhizomatous. **ST:** 4–20 dm, swollen at base. **INFL:** 1.5–15 cm, 1–2.5 cm wide, cylindric to ovate, often lobed below, occ interrupted. **SPIKELET:** glumes 4–5(7) mm, 1–2.5 mm wide, glabrous, keel winged, wing entire, not toothed; sterile lemmas 1–2, gen 0.2–2 mm, hairy; fertile lemma 3–4.5 mm, 1–1.5 mm wide, lanceolate to ovoid, hairy; occ bearing bulblets. **FR:** 2–2.5 mm, 1–1.5 mm wide. 2*n*=28. Disturbed areas, roadsides; < 1700 m. NW, CaR, n&s SNH, ScV, SnFrB, SCoRO, SCo, WTR, PR, DMoj; to OR, MT; native to Medit Eur; introduced Afr, Am, Australia. Cult for forage and erosion control along roads. Apr–Aug ❖

P. arundinacea L. (p. 1481) REED CANARY GRASS Per from distinct rhizomes. **ST:** 5–25 dm. **INFL:** 5–40 cm, 2–15 cm wide, lanceolate, interrupted near base, lobed and branched in age, often ± purple. **SPIKELET:** glumes 4–5 mm, midvein scabrous, wing 0, tip acute; sterile lemmas 2, equal, 1.5–2 mm, feathery; fertile lemma 2.5–4 mm, 1.5–1.8 mm wide, narrowly lanceolate, occ sparsely hairy. **FR:** 1.5–2 mm, < 1 mm wide. 2*n*=28. Moist habitats, woodland; < 1700 m. NW, CaRH, SN, GV, SCoRO, MP, W&I, DMoj; temp N.Am, Eurasia. Cult also for forage. May–Sep

P. brachystachys Link SHORT-SPIKED CANARY GRASS Ann. **ST:** 3–10 dm. **INFL:** 1.5–5 cm, ± 1.5–1.8 cm wide. **SPIKELET:** glumes 6–8 mm, 1.5–2.5 mm wide, glabrous to hairy, keel wing-like above lower 1/3 of glume, tip acute; sterile lemmas 2, 0.5–1.2 mm, ± 1 mm wide, gen 1/5 fertile lemma, ± fleshy to corky; fertile lemma 4.5–5 mm, 1.5–2 mm wide, ovoid, hairy. **FR:** 3–4 mm, ± 1.5 mm wide. 2*n*=12. Disturbed areas, cult fields, roadsides; < 500 m. ScV, SnFrB, SCoR; native to Medit Eur. May–Jul

P. californica Hook. & Arn. (p. 1481) CALIFORNIA CANARY GRASS Per, cespitose. **ST:** 5–16 dm, swollen at base. **INFL:** 1.5–5 cm, 1–3 cm wide, compact, ovoid to subcylindric, base ± truncate. **SPIKELET:** glumes (5)6–8 mm, 1–1.5 mm wide, glabrous, keel not winged or narrowly winged to 3/4 of glume, tip acute, gen ± purple; sterile lemmas 2, equal, 2–3.5 mm, arched, hairy; fertile lemma 3.5–5 mm, ± 1.5 mm wide, lanceolate, sparsely hairy. **FR:** ± 2.5 mm, ± 1 mm wide. 2*n*=28. Coastal open fields, disturbed areas, meadows, woodland; < 800 m. NW, CW (exc SCoRI); sw OR. Apr–Jun

P. canariensis L. (p. 1481) CANARY GRASS Ann. **ST:** 3–10 dm. **INFL:** 2–5 cm, 1.5–2 cm wide, ovoid to oblong. **SPIKELET:** glumes 6–10 mm, 1.5–2.5 mm wide, occ hairy, keel winged, tip acute; sterile lemmas 2, 2–4(4.5) mm, ± 1 mm wide, scale-like; fertile lemma 4–6(7) mm, 1.5–2.5 mm wide, ovoid, hairy. **FR:** 3–4 mm, ± 1.5 mm wide. 2*n*=12. Disturbed areas; < 400 m. NCo, KR, ScV, CW, SCo, WTR, PR, DMtns (Inyo Co.); to c N.Am; native to Medit Eur. Apr–Jul

P. caroliniana Walter CAROLINA CANARY GRASS Ann. **ST:** 5–15 dm. **INFL:** 0.5–8 cm, 1–2 cm wide, ovoid to subcylindric. **SPIKELET:** glumes 3.8–6(8) mm, 1–1.5(2) mm wide, keel narrowly winged, wing entire, scabrous on midvein, tip slightly scabrous, acute to mucronate; sterile lemmas 2, ± equal, 1.5–2(2.5) mm, arched, feathery; fertile lemma 2.8–3.5(4) mm, 1–2 mm wide, ± lanceolate, narrowed at tip, hairy. **FR:** ± 2 mm, ± 1 mm wide. 2*n*=14. Uncommon. Open fields, sandy soils, moist meadows; < 700 m. NCo, NCoRO, n SNF, GV, CCo, SCoRO, SW; native to e US; Mex. Apr–May

P. coerulescens Desf. SUNOL GRASS Per, cespitose. **ST:** 7–20 dm, swollen at base. **INFL:** 4–15 cm, 1.5–2.5 cm wide, cylindric, ovate to oblong; spikelets falling singly or in a cluster of 6–7, 1 or 2 fertile, the rest staminate, occ sterile. **FERTILE SPIKELET:** glumes 5–9 mm, ± 1–2 mm wide, glabrous, occ scabrous, wing ± dentate, tip acute; sterile lemma 0.5–0.7 mm, vestigial, knob-like; fertile lemma 2.5–5(7) mm, 0.5–1.5 mm wide, lanceolate to ovoid, tip occ hairy. **FR:** 2.5–3 mm, ± 1.5 mm wide. 2*n*=14,28,42. Cult fields, moist meadows; < 300 m. SCoRO, s ChI (Santa Catalina Island), DSon; native to Medit Eur. Apr–May

P. lemmonii Vasey (p. 1481) LEMMON'S CANARY GRASS Ann. **ST:** 5–15 dm. **INFL:** 4–20 cm, 0.7–1.5(2) cm wide, cylindric, interrupted at base; occ branched or lobed. **SPIKELET:** glumes 4.5–7 mm, 1–2 mm wide, keel not or weakly winged, veins scabrous, lanceolate, tip acute; sterile lemma (1)2, 1–1.5 mm, densely hairy; fertile lemma 4–5 mm, 1–1.5 mm wide, lanceolate to ovoid, tip ± beaked, hairy exc for the tip. **FR:** ± 2 mm, ± 1 mm wide. 2*n*=14. Gen moist areas with occ flooding, scrub, woodland; < 700 m. n SNF, n&s SNH, GV, SnFrB, SCoR, SCo, WTR, PR, DMoj; introduced in Australia. Apr–Jun

P. minor Retz. (p. 1481) LITTLE-SEEDED CANARY GRASS Ann. **ST:** 2–10 dm. **INFL:** 2–8 cm, 1–2 cm wide, oblong to ovoid. **SPIKELET:** glumes (4)5–6(6.5) mm, 1–1.5 mm wide, glabrous, keel winged, wing entire or toothed; sterile lemma 1, 0.3–1.5(2) mm, awl-like; fertile lemma 2.5–4 mm, 1–2 mm wide, ovate, hairy. **FR:** ± 2–2.5 mm, ± 1.5 mm wide. 2*n*=28. Disturbed areas, cult fields; < 1000 m. CA-FP (exc CaR), D; to e US; native to Medit. Apr–Jul

P. paradoxa L. (p. 1481) HOOD CANARY GRASS Ann. **ST:** 2–10 dm. **INFL:** 3–9 cm, 1–2 cm wide, obovate to oblong; base tapered; tip truncate to acuminate; spikelets in clusters of 6–7, cluster falling

as 1 unit; fertile spikelet gen surrounded by 5–6(7) sterile spikelets; sterile spikelet reduced, vestigial or = fertile. **FERTILE SPIKELET**: glumes 5–8 mm, ± 1 mm wide, glabrous, wing lobed to toothed near tip, conspicuously veined; sterile lemma 0.2–0.4 mm, vestigial, knob-like; fertile lemma 2.5–3.5 mm. 1–1.5 mm wide, ovoid, gen glabrous or sparsely hairy near the tip. **FR**: ± 2.5 mm, ± 1 mm wide. 2*n*=14. Disturbed areas, cult fields; < 300 m. NCoRO, NCoRI, SN, GV, CCo, SnFrB, SCo, DSon; native to Medit Eur. Small pls gen have many, vestigial sterile spikelets. May–Aug

PHLEUM TIMOTHY

Dieter H. Wilken

Ann, per, cespitose or rhizomatous. **ST**: ascending to erect. **LF**: basal and cauline; appendages 0 or small, acute to obtuse; ligule membranous to translucent, obtuse to truncate; blade gen flat, margin minutely scabrous. **INFL**: panicle-like, cylindric to ovoid, dense; branches spike-like, short. **SPIKELET**: ± sessile, strongly laterally compressed; glumes ± equal, membranous, keel gen stiff-ciliate (comb-like), pointed to awned at obtuse to truncate tip, 3-veined; floret 1, breaking above glumes, bisexual; lemma gen awnless at wide, truncate tip, 3–7-veined; palea ± = lemma. 15 spp.: temp Am, Eurasia. (Greek: a marsh reed) [Barkworth 2007 FNANM 24:670–675; Humphries 1978 Bot J Linn Soc 76:337–340]

1. Infl 1–6 cm, gen ovoid, often cylindric at low elevations; glume awn 2–3 mm . . . . *P. alpinum*
1′ Infl 4–18 cm, cylindric; glume awn < 2 mm . . . . *P. pratense*

P. alpinum L. (p. 1481) ALPINE TIMOTHY, MOUNTAIN TIMOTHY Per, cespitose, occ rhizomatous. **ST**: gen clumped, 2–6 dm. **LF**: basal loosely tufted; cauline blade 2–12 cm, 3–8 mm wide. **INFL**: 7–12 mm wide. **SPIKELET**: glumes 2–5 mm, scabrous on back; lemma 2–3 mm, puberulent on back. 2*n*=14,28. Wet meadows, streambanks, conifer forest, alpine; < 3700 m. NCo, KR, n&c NCoR, CaRH, SNH, n CCo (Marin Co.), SnBr, SnJt, Wrn, SNE; to AK, e Can, ne US, Mex; also in S.Am, Eur. Jul–Aug

P. pratense L. CULTIVATED TIMOTHY Per, loosely to densely cespitose. **ST**: 1 to loosely clumped, 5–10 dm; base gen swollen. **LF**: basal few, gen spreading; cauline blade 4–20 cm, 3–6 mm wide. **INFL**: 5–8 mm wide. **SPIKELET**: glumes 2–3 mm, lower 1/2 scabrous on back; lemma 1–2.5 mm, veins puberulent. 2*n*=14,28,42. Disturbed sites, roadsides, cult fields; < 2750 m. CA-FP, GB; N.Am, Mex; native to Eurasia. Widely cult for forage, hay; depauperate pls not easily distinguished from *P. alpinum*. May–Jun

PHRAGMITES COMMON REED, CARRIZO

James P. Smith, Jr. & Kelly W. Allred

Per with thick rhizomes or stolons, forming dense stands. **ST**: tall, erect. **LF**: cauline; sheaths open; ligule short, membranous, truncate or hairy; blade flat or folded, gen deciduous. **INFL**: panicle-like. **SPIKELET**: glumes unequal, lower < upper, 1–3-veined; axis long-soft-hairy; florets 1–10, breaking above glumes and between florets; lower florets sterile or staminate, upper bisexual; lemma lanceolate, glabrous, gen 3–5-veined; palea << lemma; stamens gen 2–3. 1 sp.: cosmopolitan. (Greek: fence, a reference to its growth form) [Allred 2003 FNANM 25:10,12]

P. australis (Cav.) Steud. (p. 1481) COMMON REED **ST**: 2–4 m. **LF**: blade gen 20–45 cm, 1–5 cm wide, margins scabrous, gen breaking at collar. **INFL**: 15–50 cm, plume-like, oblong to obovoid, ± purple to ± white. **SPIKELET**: 10–16 mm; lower glume 3–7 mm, upper glume 5–10 mm; florets 2–10. 2*n*=36,44,46,48,49–52, 54,72,84,96. Pond and lake margins, sloughs, marshes; gen < 1600 m. CA; worldwide. [*P. a.* var. *berlandieri* (E. Fourn.) C.F. Reed; *P. b.* E. Fourn.; *P. communis* Trin.; *P. c.* var. *b.* (E. Fourn.) Fernald] 3 different genetic lineages may occur in CA: *P. australis* subsp. *americanus* Saltonstall et al. (native), *P. australis* subsp. *berlandieri* (E. Fourn.) Saltonstall & Hauber (native), and a 3rd naturalized entity whose taxonomic identity has yet to be determined [Saltonstall & Hauber 2007 J Bot Res Inst Texas 1:385–388]. Jul–Nov

PLEUROPOGON SEMAPHORE GRASS

Susan J. Bainbridge

Ann, per. **ST**: gen erect. **LF**: gen cauline; sheath margins fused > 1/2; ligule membranous; blade flat or folded, drooping. **INFL**: raceme-like. **SPIKELET**: bisexual; 1–8 cm; laterally compressed; glumes << lowest floret, translucent, lower 1-veined, upper 1–3-veined; breaking apart above glumes and beneath florets; florets 5–20; lemma 7–9-veined, veins prominent, not converging, extending to tip, awn (0)0.2–22 mm; palea ± = lemma, veins with appendages; anthers 3. 5 spp.: temp N.Am, 1 sp. in e Asia. (Greek: side beard, from awn at palea base in some spp.) [But 2007 FNANM 24:103–109, Novon 4:16–17]

1. Lowest floret lemma 8–12 mm; st 7–16 dm; rhizome evident
  2. Awn 0.2–4 mm; spikelets erect or ascending at maturity . . . . *P. hooverianus*
  2′ Awn 5–22 mm; spikelets reflexed at maturity . . . . *P. refractus*
1′ Lowest floret lemma 4.5–7.5 mm; st (0.4)1.5–9.5 dm; rhizome 0 . . . . *P. californicus*
  3. Ann; lemma awn gen present, (0)5–11 mm; palea appendages 0.5–2.5 mm . . . . var. *californicus*
  3′ Per; lemma awn occ present, ≤ 1.5 mm; palea appendages 0.5–1 mm . . . . var. *davyi*

P. californicus (Nees) Vasey Occ weakly rooting at nodes. **ST**: decumbent to erect, (0.4)1.5–9.5 dm. **LF**: ligule 2–6 mm; blade 3–8 mm wide. **INFL**: 6–35 cm. **SPIKELET**: 10–65 mm, spreading to erect; lower glume 1–4.5 mm, upper 2–7 mm; lemma 4–7.5 mm; palea appendages 0.5–2.5 mm. 2*n*=16.

var. ***californicus*** Ann. **SPIKELET**: 10–30 mm; lemma awn gen present, 5–11 mm; palea appendages 0.5–2.5 mm. Wet places, vernal pools, wet grassland; < 650 m. NW, CaRF, n&c SNF, GV, CW (exc SCoRI). Mar–Jun

var. ***davyi*** (L.D. Benson) But DAVY'S SEMAPHORE GRASS Cespitose per. **SPIKELET**: 15–65 mm; lemma awn occ present, ≤ 1.5 mm, occ longer on florets at tip; palea appendages 0.5–1 mm. Vernal pools, sloughs, marshy grassland; < 650 m. NCoRO, NCoRI. Mar–Jun ★

P. hooverianus (L.D. Benson) J.T. Howell NORTH COAST SEMAPHORE GRASS Per from rhizomes. **ST**: erect, 8–16 dm. **LF**: ligule 3–7 mm; blade 3–10 mm wide. **INFL**: 15–35 cm. **SPIKELET**: 2–4.5 cm, erect or ascending at maturity; lower glume 3–6 mm, upper 4.5–7.5 mm; lemma 7–12 mm, awn 0.2–4 mm, occ longer on florets

at base; palea appendages 0.6–1.5 mm. 2*n*=16,36. Wet grassy areas; < 1300 m. s NCo, NCoRO, NCoRI, n SnFrB. Mar–Jun ★

P. refractus (A. Gray) Vasey (p. 1481) NODDING SEMAPHORE GRASS Per from rhizomes. **ST**: erect, 7–15 dm, often rooting at lower nodes. **LF**: ligule 2–9 mm; blade (3)5–14 mm wide. **INFL**: (10)20–35 cm. **SPIKELET**: (1.5)2.5–5 cm, reflexed at maturity; lower glume 3–7 mm, upper 4–9 mm; lemma 8–10.5 mm, awn 5–22 mm; palea appendages 0.2–0.6(1) mm. 2*n*=32,36. Wet meadows, shady banks; < 1600 m. NCo, KR, NCoRO; to BC. Apr–Jul ★

POA BLUE GRASS

Robert J. Soreng

Ann, per; some ± dioecious. **ST**: 0.3–12 dm. **LF**: sheath open most of length to closed (best observed on upper st lf); ligule thin, flexible, without a rim of hairs; blade grooved above on both sides of midvein, flat, folded, or inrolled, gen smooth or scabrous on veins, gen prow-tipped. **INFL**: panicle-like; branches appressed to drooping. **SPIKELET**: gen compressed, breaking between florets; glumes 2, ± equal, gen < lowest lemma, awnless; florets gen 2–6; callus indistinct, often with obvious tuft of long cobwebby hairs; lemma gen keeled to base, like glumes, awnless, veins gen 5, hairy or glabrous, upper margins thin-membranous; palea well developed, keel gen scabrous. **FL**: gen bisexual, sometimes pistillate and/or staminate; fertile anthers 0.2–4.5 mm; ovary glabrous; fr firm. ± 500 spp.: temp and cool regions. (Greek: ancient name applied to various grasses, fodder pls) [Soreng 2007 FNANM 24:486–601] CA is center of diversity in N.Am. Spikelet data recorded from lowest florets of spikelet at 25×.

1. Ann (bien); anthers gen 0.2–1(1.8) mm; fl gen bisexual
  2. Callus glabrous — upper fls of spikelet often pistillate
    3. Spikelet axis gen hidden, terminal internode < 1/2 terminal floret; anthers 0.6–1 mm; widespread . . . . ***P. annua***
    3′ Spikelet axis visible, terminal internode > 1/2 terminal floret; anthers 0.2–0.5 mm; esp s CA-FP . . . . ***P. infirma***
  2′ Callus with cobwebby hairs
    4. Infl branches appressed; lf blade abruptly prow-tipped; lemma veins hairy . . . . ***P. bigelovii***
    4′ Lower infl branches eventually spreading; lf blade abruptly or slightly prow-tipped; lemma short-hairy
      or surface glabrous or scabrous
      5. Lemma glabrous or scabrous; lf blade abruptly prow-tipped; some spikelet internodes > 1 mm . . . . ***P. bolanderi***
      5′ Lemma short-hairy; lf blade tapered, slightly prow-tipped; spikelet internodes gen < 1 mm. . . . . ***P. howellii***
1′ Per; anthers often > 1.2 mm; fls occ unisexual
  6. Rhizomes or stolons present; anthers 1–4 mm
    7. St and nodes distinctly compressed, keeled, wiry; lf sheath open nearly to base; infl branches scabrous . . . ***P. compressa***
    7′ St and nodes little or not compressed or wiry; lf sheath sometimes not open as far; infl branches
      sometimes smooth
      8. Lemma glabrous or minutely scabrous; callus long-cobwebby
        9. Sheaths of upper st lvs open most of their length; lf blade gen flat; upper surface glabrous or sparsely
          scabrous; fls bisexual . . . . ***P. kelloggii***
        9′ Sheaths of upper st lvs open 1/3–7/10 their length; lf blade flat, folded, or inrolled, upper surface
          sometimes finely hairy; fls gen unisexual
          10. Lf blades thin, soft, gen 0.8–1.5 mm wide, folded, or up to 2.5 mm wide and flat; st lvs gradually
            reduced in length upward; lemma (3.2)4.25–5 mm; pls with bisexual fls or some (rarely all)
            pistillate or staminate . . . . ***P. diaboli***
          10′ Lf blades ± thick, firm, gen 1.5–3 mm wide, never flat; st lvs strongly reduced in length upward;
            lemma gen 4–7 mm; pls nearly all pistillate- or staminate-fld . . . . ***P. piperi***
      8′ Lemma hairy at least on keel or callus glabrous
        11. Lemma gen < 3.5 mm, lemma and callus smooth, glabrous; st nodes gen hidden in basal tuft;
          infl < 8 cm, branches appressed, gen smooth; fls unisexual . . . . ***P. atropurpurea***
        11′ Lemma gen > 3.5 mm or ± scabrous or hairy (if < 4 mm and smooth and glabrous, then some st
          nodes well exposed and infl > 8 cm); infl 2–30 cm, branches sometimes ± spreading, sometimes
          scabrous; fls bisexual and/or unisexual
          12. Lf blade ± firm, inrolled, upper surface gen densely and finely hairy or scabrous; fls gen unisexual
            (exc *Poa confinis*); coastal dunes
            13. Lemma ≤ 4.5 mm, glabrous to sparsely hairy; pl ≤ 30 cm; infl ± open. . . . . ***P. confinis***
            13′ Lemma >> 4.5 mm, glabrous to densely hairy; pl sometimes > 30 cm; infl dense, branches appressed
              14. Upper glume < 7 mm; lowest lemma ≤ 7.5 mm; st very scabrous or coarsely hairy below infl;
                s NCo, CCo, n ChI . . . . ***P. douglasii***
              14′ Upper glume > 7 mm; lowest lemma > 7.5 mm; st smooth below infl; n&c NCo . . . . ***P. macrantha***
          12′ Lf blade gen not firm or inrolled, upper surface glabrous or sparsely soft-hairy; fls bisexual and/or
            unisexual; widespread, gen montane, rarely on coastal dunes
            15. Sheath of upper st-lf closed nearly to top, blade > sheath . . . . ***P. sierrae***

15′ Sheath of upper st-lf open at least 1/4 length, blade ≤ sheath

  16. Callus glabrous (rarely short-cobwebby); lemma glabrous or hairy; fls often pistillate

    17. Lemma gen < 3.5 mm, sometimes glabrous; infl linear in outline, short-branched, interrupted, many-fld; upper st-lf blade > 1/5 sheath length; fls appearing bisexual but anthers gen sterile; tall lfy-stemmed pls of moist, low-elevation meadows. . . . . **P. pratensis** × **P. secunda** (see *Poa secunda* subsp. *juncifolia*)

    17′ Lemma gen > 3.5 mm; infl lanceolate to ovate in outline (otherwise sometimes as above); upper st-lf blade sometimes < 1/6 sheath length; fls gen pistillate (anthers minute); cespitose or tall pls of well drained soils of slopes of forests

      18. St-lf blades strongly reduced upward, uppermost < 1/8 sheath length; lemma keel and marginal veins clearly hairy; pl densely cespitose (see 31. for subspp.) . . . . ²**P. fendleriana**

      18′ St-lf blades gen longest at middle of st, uppermost gen > 1/5 sheath length; lemma glabrous to sparsely scabrous or keel and marginal veins sparsely hairy; pl loosely tufted . . . . **P. wheeleri**

  16′ Callus long-cobwebby (hairs > 1/2 lemma length); lemma keel and marginal veins hairy, glabrous between veins; fls bisexual (exc *Poa rhizomata*)

    19. Sheath gen open > 3/4 length; stolons gen present; st sometimes branching above base; lemma sometimes hairy only on keel

      20. First glume gen 3-veined, not arched; lemma obviously hairy on keel and marginal veins . . . . . ²**P. palustris**

      20′ First glume 1-veined, ± sickle-shaped; lemma sparsely hairy on keel and rarely on marginal veins. . . . . ²**P. trivialis**

    19′ Sheath open ≤ 3/4 length; pl with obvious rhizomes; st not branching above base

      21. Pl often forming a dense sod; ligule truncate to rounded; infl densely fld, branches at lower nodes gen 4+ (some may be small); lemma gen < 4 mm; fls rarely pistillate. . . . **P. pratensis** subsp. **pratensis**

      21′ Pl not sod-forming; ligule acute; infl sparsely fld, branches at lower nodes gen 1–2; lemma gen > 4 mm; fls occ pistillate . . . . **P. rhizomata**

6′ Rhizomes and stolons 0; anthers 0.2–4.5 mm

  22. St base ± bulbous; most spikelets producing lfy bulblets or not. . . . . **P. bulbosa**

  23. Spikelets not producing lfy bulblets . . . . [subsp. **bulbosa**]

  23′ Spikelets gen producing lfy bulblets . . . . subsp. **vivipara**

  22′ St base not bulbous; florets not mostly replaced by bulblets

    24. Callus cobwebby; lemma glabrous . . . . **P. abbreviata** subsp. **marshii**

    24′ Callus glabrous, with a ring of short hairs, or cobbwebby; if callus cobbwebby then lemma also hairy to some degree

      25. Callus ± cobwebby; lemma hairy on keel and gen on marginal veins but not between veins

        26. Infl < 10 cm, sparse or dense; pl densely cespitose

          27. Pl << 20 cm; sheaths open > 3/4 length; fl gen bisexual, anthers gen 0.7–1.2 mm . . . . **P. abbreviata** subsp. **pattersonii**

          27′ Pl > 20 cm; sheaths open ≤ 1/2 length; fls pistillate, anthers vestigial . . . . ²**P. cusickii** subsp. **purpurascens**

        26′ Infl gen > 10 cm or open; pl loosely cespitose

          28. Lower glume gen 3-veined; sheaths open 3/4 length or more

            29. Ligule of upper st-lf truncate, gen < 0.5 mm . . . . **P. nemoralis**

            29′ Ligule of upper st-lf acute to rounded, gen > 1 mm. . . . . ²**P. palustris**

          28′ Lower glume gen 1-veined; sheaths open to 3/4 length

            30. Anthers ≤ 1 mm; infl branches with fls in top 1/3; high montane to alpine . . . . **P. leptocoma** subsp. **leptocoma**

            30′ Anthers > 1.3 mm; some infl branches with fls from mid-length; low elevation . . . . ²**P. trivialis**

      25′ Callus glabrous or not cobwebby (ring of short callus hairs present in some *Poa secunda*, *Poa unilateralis*); lemma often hairy between veins

        31. Lemma obviously hairy at least on keel and marginal veins; fls unisexual; sheath open ± 1/3 length; uppermost st-lf blade firm, often highly reduced; callus glabrous . . . . ²**P. fendleriana**

          32. Ligule 0.2–1.5 mm, margin scabrous or short-ciliate . . . . subsp. **fendleriana**

          32′ Ligule 1–18 mm, margins smooth, glabrous . . . . subsp. **longiligula**

        31′ Lemma often glabrous (if obviously hairy on keel and marginal veins or between, then fls bisexual, sheaths open > 3/4 or < 1/2 length; uppermost st-lf blade often thin and withering or not highly reduced); callus sometimes with a crown of short hairs

          33. Spikelet lanceolate to narrowly ovate in outline, little compressed; lemma weakly keeled; sheath open > 3/4 length; fls bisexual

            34. Infl branches widely spreading in fl and fr, gen densely scabrous; ligule 0.5–1.5(2.5) mm, truncate to obtuse (rarely acute), scabrous; pl delicate, spring-active; often on serpentine . . . . **P. tenerrima**

            34′ Infl branches gen ascending to appressed in fr (gen spreading only in fl), ± scabrous; pl gen more robust, or summer-active, of high elevation, or delicate with branches ± smooth and ligules longer and ± smooth; on various substrates . . . . **P. secunda**

              35. Lemma glabrous (sometimes scabrous, rarely sparsely short-hairy); ligule of sterile shoots and often st < 2 mm, ± truncate, scabrous; lf blade ± firm, retaining shape; infl branches appressed . . . . subsp. **juncifolia**

              35′ Lemma ± evenly short-hairy on keel and sides across base; ligule of sterile shoots and st gen > 2 mm, acute or acuminate, smooth or sparsely scabrous; lf blade gen ± soft, soon withering; infl branches appressed or spreading . . . . subsp. **secunda**

33′ Spikelet ± ovate in outline, obviously compressed; lemma obviously keeled to base; sheaths
   sometimes open < 3/4 length; fls sometimes unisexual
   36. Uppermost sheath open ≤ 1/2 length; lemma > 4.5 mm, sparsely hairy on keel, base, and
      sometimes marginal veins; fls pistillate . . . . [2]*P. cusickii* subsp. ***purpurascens***
   36′ Uppermost sheath gen open > 1/2 length; lemma gen < 4.5 mm or glabrous (if hairy, sheath open
      > 3/4 length, fls bisexual)
      37. Lemma obviously hairy, < 3.6 mm; subalpine or higher . . . . *P. glauca* subsp. ***rupicola***
      37′ Lemma often glabrous, length various (if lemma obviously hairy, pl coastal)
         38. Sheath open ± 9/10 length; fls bisexual; low salty places; coast or w CA
            39. Infl branches ascending, very scabrous, with spikelets only in top 1/2; near hot springs,
               NCoRO, Napa Co. . . . . *P. napensis*
            39′ Infl branches appressed, smooth to sparsely scabrous, with spikelets from near base; sea
               bluffs, NCo, n&c CCo . . . . *P. unilateralis* subsp. ***unilateralis***
         38′ Sheath open < 9/10 length; fls sometimes unisexual; gen ± salt-free places; montane or GB
            40. Pl gen < 1 dm; longest anthers 0.2–2 mm; upper glume gen ≥ first lemma; lemma sometimes
               sparsely short-hairy; fls bisexual
               41. Lf firm; lemma 3–5 mm; anthers 0.7–1.8 mm . . . . *P. keckii*
               41′ Lf soft; lemma ≤ 3 mm; anthers 0.2–0.7 mm . . . . *P. lettermanii*
            40′ Pl gen 1–7 dm; longest fertile anthers 2–4.5 mm; upper glume gen < first lemma; lemma
               glabrous; fls sometimes unisexual
               42. Ligule of uppermost lf of sterile sts 2.5–6 mm, acute, smooth; lemma ≤ 5.5 mm; lowest true
                  lvs lacking blades; fls bisexual or pistillate (rarely staminate); SNH . . . . *P. stebbinsii*
               42′ Ligule of uppermost lf of sterile sts gen < 2 mm (if 2–3 mm, lemma > 5.5 mm), truncate to
                  obtuse, scabrous; lowest true lvs sometimes with blades; all fls gen unisexual; more widespread
                  43. Pl gen < 30 cm, base often decumbent; lf blades of sterile sts gen > 1.5 mm wide, arched,
                     their ligules sometimes > 2 mm; upper glume sometimes = first lemma; KR, CaRH, n SNH
                     . . . . *P. pringlei*
                  43′ Pl 25–70 cm, base rarely decumbent; lf blades of sterile sts < 1.5 mm wide, erect, their
                     ligules < 2 mm; upper glume obviously < first lemma; widespread, esp GB . . . . *P. cusickii* (in part)
                     44. Basal tuft of lvs sparse; infl branches ± smooth; 1–2 st nodes exposed; fls pistillate . . . . . . subsp. ***epilis***
                     44′ Basal tuft of lvs dense; infl branches obviously scabrous; 0–1 st nodes exposed; fls
                        sometimes staminate (rarely bisexual)
                        45. Infl branches slender, longest gen > 17 mm; moist sagebrush to dry montane slopes . . . subsp. ***cusickii***
                        45′ Infl branches stout, longest gen < 15 mm; dry ± alpine ridges — uncommon in CA . . . . subsp. ***pallida***

P. abbreviata R. Br. DWARF BLUE GRASS Per, densely cespitose, gen 0.5–1.5 dm; sterile shoots many. **LF:** sheath open 3/4–9/10 length, basal sheaths persisting; ligule 1–3.5 mm, acute, smooth; blade 1.5–2 mm wide, folded, ± inrolled, closely ascending. **INFL:** 2–6 cm, lanceolate, dense; branches appressed, smooth or sparsely scabrous. **SPIKELET:** glumes ± equal, upper 3.5–4.2 mm, gen = first lemma; callus gen cobwebby; lemma 3.5–4 mm, keel, marginal veins gen hairy; palea keels sparsely scabrous above, some sparsely short-hairy below. **FL:** anthers gen 0.7–1.2 mm. 2*n*=42.

subsp. ***marshii*** Soreng MARSH'S BLUE GRASS **LF:** ligule 1–3 mm. **SPIKELET:** callus cobwebby; lemma glabrous. **FL:** anthers gen 0.6–1.2 mm. High alpine open ground; gen > 3300 m. W&I (White Mtns); to ID, MT, WY, NV. Jul ★

subsp. ***pattersonii*** (Vasey) Á. Löve et al. PATTERSON'S BLUE GRASS **LF:** ligule 1–3.5 mm. **SPIKELET:** callus gen cobwebby; lemma keel, marginal veins gen hairy. **FL:** anthers gen 0.7–1.2 mm. 2*n*=42. High alpine open ground; gen > 3300 m. SNH, W&I (White Mtns); AK to NM; Far East Russia. [*P. p.* Vasey] Jul ★

P. annua L. (p. 1481) ANNUAL BLUE GRASS Ann, bien, cespitose or with stolons, 0.2–2 dm. **LF:** sheath open ± 2/3 length; ligule 1–5 mm, rounded to obtuse; blade gen 1–3 mm wide, soft, gen flat, bright- or yellow-green. **INFL:** 1–10 cm, triangular, 1.2–1.6 × longer than wide, open in fr; branches spreading, smooth, with spikelets only in top 1/2. **SPIKELET:** axis ± hidden; callus glabrous; lemma 2.5–4 mm, smooth, veins soft-hairy or glabrous; palea keels hairy. **FL:** bisexual or upper 1–2 pistillate; anther 0.6–1 mm. 2*n*=28. Abundant. Disturbed moist ground; gen < 2000 m. CA (esp near coast); ± worldwide; native to Eur. Feb–Sep

P. atropurpurea Scribn. SAN BERNARDINO BLUE GRASS Per from rhizomes, tufted, 1–5.5 dm; dioecious. **LF:** sheath open ± 2/3 length; ligule 1–2 mm, truncate to rounded, smooth; blade 1.5–3 mm wide, ± firm, folded and inrolled. **INFL:** 3–7 cm, lanceolate; branches appressed, gen smooth. **SPIKELET:** callus glabrous; lemma 2.5–3.5 mm, glabrous, smooth. **FL:** unisexual; anthers 1.5–2 mm. 2*n*=28. Moist meadows; 1500–2200 m. SnBr, PR. Threatened by development, grazing, vehicles. PR pls pistillate. Apr–Jun ★

P. bigelovii Vasey & Scribn. Ann, cespitose, gen 1.5–4 dm. **LF:** sheath open 1/2–3/4 length; ligule 1.5–6 mm, truncate to obtuse, minutely scabrous; blade 1.5–5 mm wide, soft, gen flat, abruptly prow-tipped. **INFL:** 5–15 cm, ± linear; branches appressed, with spikelets from near base. **SPIKELET:** callus cobwebby; lemma 2.5–4 mm, veins (sometimes between) hairy; palea keels hairy. **FL:** anthers 0.2–1 mm. 2*n*=28. Uncommon. Shady places in desert scrub, yellow-pine forest; < 1500 m. SW, D; to CO, TX, nw Mex. Mar–May

P. bolanderi Vasey Ann, cespitose, 2–6 dm. **LF:** sheath open 1/4–1/2 length; ligule 2–7 mm, rounded to obtuse; blade 1.5–5 mm wide, short, soft, gen flat, abruptly prow-tipped. **INFL:** gen 10–15 cm; branches appressed in fl, spreading in fr, with spikelets in top 1/2. **SPIKELET:** internodes elongated, glabrous; callus cobwebby; lemma 2.5–3 mm, minutely scabrous to nearly smooth; palea keels scabrous. **FL:** anthers 0.5–1.8 mm. 2*n*=28. Mtns, esp in open pine forest; gen 1280–3275 m. NW, CaRH, SNH, SnJt (probably MP); to WA, ID, UT. Jun–Aug

P. bulbosa L. Per, densely cespitose, 1.5–6 dm. **ST:** base ± bulbous. **LF:** sheath open to near base; ligule 2–4 mm, obtuse; blade 1–2 mm wide, soft, flat or folded, soon withering. **INFL:** 3–10 cm, ovate to lanceolate; branches gen ascending, smooth. **FL:** anthers 1.2–1.5 mm in fertile fls. 2*n*=21–42.

subsp. ***vivipara*** (Koeler) Arcang. (p. 1481) **SPIKELET:** gen producing lfy bulblets; callus cobwebby or not, lower lemma keel and marginal veins hairy. Disturbed places; gen < 2000 m. CA; ± worldwide temp; native to Eur. Mar–Jul

P. compressa L. CANADIAN BLUE GRASS Per from long, stout rhizomes, 1.5–6 dm. **ST:** (incl nodes) flattened, keeled, wiry; nodes obviously exposed. **LF:** sheath open 3/4 length to near base; ligule 1–3 mm, rounded; blade 1.5–4 mm wide, soft to ± firm, flat or folded. **INFL:** 2–9 cm, lanceolate to ovate, dense (or sparse and interrupted); branches ± ascending, short, densely scabrous on angles. **SPIKE-LET:** callus glabrous or ± cobwebby; lemma 2.3–3.5 mm, keel, marginal veins hairy. **FL:** anthers 1.3–1.8 mm. 2*n*=42. Moist, often disturbed low ground; < 1800 m. CA; ± worldwide temp; native to Eur. Apr–Aug

P. confinis Vasey BEACH BLUE GRASS Per from slender rhizomes or stolons, delicate, tufted, 0.7–3 dm; ± dioecious. **LF:** sheath open 1/2–2/3 length; ligule < 2 mm, obtuse to acute, scabrous; blade 1–2 mm wide, ± firm, folded, inrolled, upper blade surface on sterile sts gen finely hairy. **INFL:** gen 1–5 cm, ovate, ± tawny; branches ± appressed, ± scabrous. **SPIKELET:** callus glabrous or diffusely short-cobwebby; lemma 2–4.5 mm, glabrous or sparsely hairy. **FL:** bisexual and pistillate, fertile anthers 1.5–2 mm. 2*n*=42. Ocean beaches, stabilized dunes; < 100 m. NCo; to BC. Apr–Jun

P. cusickii Vasey Per, ± densely cespitose, 1–6 dm; ± dioecious. **LF:** ligule gen 1–6 mm (on sterile sts < 2 mm, truncate, scabrous); blade longest at mid-st, on sterile sts gen 0.5–1 mm wide, ± firm, inrolled (sometimes also folded), upper surface finely hairy. **INFL:** lanceolate to ovate, gen dense; branches ascending to appressed, slender, smooth or scabrous. **SPIKELET:** callus gen glabrous; lemma keeled, gen glabrous (rarely keel sparsely hairy), smooth or scabrous; palea keels scabrous. **FL:** unisexual; fertile anthers 2–3.5 mm.

subsp. **cusickii** **LF:** basal tuft dense; sheath open 1/2–3/4 length; 0–1 nodes barely exposed. **INFL:** 3–12 cm; branches slender, obviously scabrous, longest gen > 17 mm. **SPIKELET:** lemma 3.5–7.5 mm, glabrous or scabrous. **FL:** gen staminate or pistillate. 2*n*=28. Moist meadows to dry slopes in sagebrush scrub or montane forest; 1500–2500 m. e KR, n SNH, GB; to WA, ID, n NV. [*P. han-senii* Scribn.] May–Jul

subsp. **epilis** (Scribn.) W.A. Weber (p. 1481) SKYLINE BLUE GRASS **LF:** basal tuft sparse; sheath open 1/3–2/3 length; 1–2 nodes well exposed. **INFL:** 3–6 cm, dense; branches appressed, ± smooth, longest gen 10–22 mm. **SPIKELET:** lemma 4–5.8 mm, glabrous or scabrous. **FL:** pistillate only; fr produced asexually. 2*n*=56. Moist subalpine, esp snowbeds; 2400–3600 m. SNH; to BC, MT, CO. [*P. e.* Scribn.] Jul–Aug

subsp. **pallida** Soreng **LF:** basal tuft dense; sheath open 1/2–4/5 length; 0–1 nodes barely exposed. **INFL:** 3–5 cm, ± dense; branches appressed, obviously scabrous, stout, longest to 17 mm. **SPIKELET:** lemma 5–6.5 mm, glabrous or scabrous. **FL:** pistillate (in CA); fr produced asexually. 2*n*=56,59. Uncommon. High montane to lower alpine dry meadows, ridges; 2000–3500 m. c&s SNH, W&I; w Can, ND, CO. [*P. subaristata* Beal, illeg.] Jul–Aug

subsp. **purpurascens** (Vasey) Soreng **LF:** basal tuft sparse; sheath open 1/4–1/2 length; 1–2 nodes exposed. **INFL:** 4–8 cm, sparse; branches ascending, slender, sparsely scabrous, 17–30 cm, few-fld. **SPIKELET:** callus sometimes sparsely short-cobwebby; lemma 4.5–7 mm, keel, marginal veins gen sparsely short-hairy near base. **FL:** pistillate only; fr produced asexually. 2*n*=28. Moist subalpine meadows, ledges; 2100–3500 m. KR, n SNH; to BC. [*P. alpina* L. var. *p.* Vasey; *P. p.* Vasey, illeg.] Jul–Sep

P. diaboli Soreng & D.J. Keil DIABLO CANYON BLUE GRASS Per from slender rhizomes or stolons; delicate, loosely tufted mounds to 30 cm across, 2.5–5 dm. **LF:** sheath open 1/5–7/10 length; ligule 1–3 mm (on st), obtuse to acute, scabrous; blade 0.8–2 (2.5) mm wide, ± soft, flat or folded, upper blade surface on sterile sts sometimes finely hairy. **INFL:** 4–8.5 (10.5) cm, pyramidal to ovate, ± green; branches ascending to spreading, ± scabrous. **SPIKELET:** callus ± diffusely short-cobwebby; lemma (3.2)4.25–5 mm. **FL:** bisexual or pistillate, rarely staminate; fertile anthers (1.4)1.7–2.6 mm. Thin soils on Edna shale slopes, upper coastal scrub, live-oak woodland, Bishop-pine forest, near coast; 120–400 m. s CCo (San Luis Obispo Co.). Mar–Apr ★

P. douglasii Nees (p. 1481) SAND DUNE BLUE GRASS Per from long rhizomes or stolons, tufted, 1–3 dm; dioecious. **LF:** sheath open ± 1/2 length; ligule 1–2 mm, finely scabrous; blade 1–2.5 mm wide, firm, slender, folded, inrolled, upper surface finely hairy. **INFL:** 2–6 cm, narrowly ovate, dense; peduncle, branches densely long scabrous to coarsely hairy. **SPIKELET:** callus glabrous or with a ring of hairs; lemma 5–7.5 mm, glabrous or veins hairy. **FL:** unisexual; fertile anthers 1.5–2 mm. 2*n*=28. Coastal dunes; gen < 100 m. s NCo, CCo, n ChI. Threatened by alien spp. Mar–Jul

P. fendleriana (Steud.) Vasey Per, densely cespitose, gen with short rhizomes, 1.5–7 dm; dioecious. **LF:** sheath open ± 1/3 length; ligule 0.2–18 mm, truncate to acuminate; blade 1–4 mm wide, firm, folded, inrolled, uppermost st-lf blade gen ± vestigial, sterile st blade upper surface gen finely hairy. **INFL:** gen 2–12 cm, lanceolate to ovate, dense; branches smooth or scabrous. **SPIKELET:** callus glabrous; lemma 3.5–6 mm, keel, marginal veins, sometimes intermediate veins hairy. **FL:** unisexual, gen pistillate; fertile anthers 2–4 mm, sterile anthers minute. [*Eragrostis f.* Steud.]

subsp. **fendleriana** MUTTON GRASS **LF:** ligule 0.2–1.2(1.5) mm, truncate, margin scabrous or short-ciliate; blade 1–3 mm wide. **SPIKELET:** lemma keel and marginal and sometimes intermediate veins hairy. Uncommon. Slopes, yellow-pine forest to subalpine; 3000–3200 m. SnBr, SnJt; to BC, SK, TX, nw Mex. Gen more e than *P. fendleriana* subsp. *longiligula*. May–Jul

subsp. **longiligula** (Scribn. & T.A. Williams) Soreng (p. 1481) LONG-TONGUE MUTTON GRASS **LF:** ligule 1–18 mm, truncate to acuminate, margins smooth, glabrous; blade 1.5–4 mm wide. **SPIKELET:** lemma keel and marginal veins (sometimes between) hairy. 2*n*=56. Slopes, sagebrush scrub to subalpine; 2000–3200 m. SNH, SnBr, PR, GB, DMtns; to BC, SD, NM, Baja CA. [*P. l.* Scribn. & T.A. Williams] Apr–Jul

P. glauca Vahl subsp. **rupicola** (Rydb.) W.A. Weber TIMBERLINE BLUE GRASS Per, densely cespitose, all current shoots fl, gen 0.5–1.5 dm. **LF:** sheath open > 4/5 length; ligule 0.5–3 mm, truncate and finely scabrous at margin to acute and smooth; blade 1–2 mm wide, soft, flat or folded, abruptly ascending or spreading. **INFL:** 1–5 cm, lanceolate to ovate; branches ascending to appressed, gen < 1.5 cm, gen scabrous on angles. **SPIKELET:** upper internodes < 1 mm, not elongated; glumes 3-veined, upper 2.5–3.5 mm (< 3/4 length of lower); callus glabrous; lemma 2.5–3.5 mm, veins, base hairy. **FL:** anthers 1.2–1.8 mm. 2*n*=42–±100. Dry alpine slopes, ridges; 3300–4100 m. SNH, W&I; to sw Can, NM. [*P. r.* Rydb.] *P. glauca* circumboreal, and S.Am. Jul–Sep

P. howellii Vasey & Scribn. Ann, cespitose, 3–8 dm. **LF:** sheath open 1/8–1/2 length; ligule 2.5–5 mm, truncate to acute; blade gen 1–6 mm wide, soft, gen flat, barely prow-tipped. **INFL:** gen 20–25 cm; branches ascending in fl, spreading to reflexed in fr, with spikelets only in top 1/2, densely scabrous. **SPIKELET:** internodes gen < 1 mm, some hairy; callus cobwebby; lemma ± 3 mm, short-hairy over body; palea keels scabrous or short-hairy. **FL:** anthers 0.2–1 mm. Rocky banks, shaded slopes of woodland, chaparral, disturbed places; gen < 1700 m. CA-FP (exc s SN, SnJV, SCo); to BC. [*P. bolanderi* var. *h.* (Vasey & Scribn.) M.E. Jones] Apr–Jun

P. infirma Kunth WEAK BLUE GRASS Ann, cespitose, 0.2–1.5 dm. **LF:** sheath open ± 2/3 length; ligule 1–5 mm; blade 1–4 mm wide, gen flat, yellow-green. **INFL:** 1–6 cm, 1.5–3 5 × longer than wide; branches ascending, smooth, spikelets crowded in top 1/2. **SPIKELET:** axis plainly visible, terminal internode 1/2–3/4 length of terminal lemma; callus glabrous; lemma 2–2.5 mm, veins curly-hairy; palea keels hairy, hooks absent. **FL:** upper 1–2 pistillate; fertile anthers 0.2–0.5 mm. 2*n*=14. Disturbed low ground, esp sandy places; < 500 m. s GV, s CW, SW; native to Medit. Mostly confined to s CA-FP in US. Like *P. annua*. Mar–May

P. keckii Soreng Per, densely cespitose, gen 0.3–1 dm. **LF:** sheath open 4/5–9/10 length; ligule of uppermost sterile st lf < 2.5 mm, smooth or sparsely scabrous; blade 1–2 mm wide, firm, folded, inrolled. **INFL:** 1.5–6 cm, lanceolate to narrowly ovate. **SPIKELET:** callus glabrous; lemma 3–5 mm, glabrous or sparsely short-hairy. **FL:** anthers 0.7–1.8 mm. High alpine, often on open ground; > 3000 m. SNH, SNE (Sweet-

water, White mtns). [*P. suksdorfii* (Beal) Piper, misappl.] Some alpine pls difficult to distinguish from *P. secunda.* Jul–Sep

P. kelloggii Vasey KELLOGG'S BLUE GRASS Per from long rhizomes, 2.5–8.5 dm. **LF:** sheath open to near base; ligule 1–4 mm, finely scabrous, truncate to obtuse; blade 2–5 mm wide, soft, gen flat. **INFL:** 10–20 cm, ovate to triangular, open, sparse; branches ascending in fl, ± drooping in fr, scabrous, with spikelets only in top 1/3. **SPIKELET:** callus cobwebby; lemma 4–5 mm, glabrous. **FL:** anthers ± 2 mm. 2*n*=56. Shady openings in mixed-conifer and redwood forest; < 500 m. NCo, n CCo. Apr–Jun

P. leptocoma Trin. subsp. **leptocoma** BOG BLUE GRASS Per, loosely cespitose, 1–7 dm. **LF:** sheath open 1/2–3/4 length; ligule 1.5–4 mm, truncate to obtuse, smooth; blade 1–4 mm wide, soft, gen flat. **INFL:** 4–15 cm, open; branches ascending in fl, ± drooping in fr, with spikelets in top 1/3. **SPIKELET:** lower glume 1-veined; callus cobwebby; lemma 3–4 mm, keel, marginal veins sparsely hairy. **FL:** anthers 0.2–1 mm. 2*n*=42. Moist subalpine, lower alpine meadows; 1800–3200 m. KR, CaRH, n&c SNH, GB; to AK, MT, NM. Jun–Sep

P. lettermanii Vasey LETTERMAN'S BLUE GRASS Per, cespitose, delicate, gen 0.2–0.9 dm. **LF:** sheath open 3/4–9/10 length; ligule 1–4 mm, truncate to acute, smooth; blade 0.5–1.5 mm wide, soft, flat or folded. **INFL:** gen 1–3 cm, narrowly lanceolate. **SPIKELET:** glumes (at least upper) > lowest lemma; callus glabrous; lemma 2.5–3 mm, glabrous or very sparsely short-hairy. **FL:** anthers 0.2–0.7 mm. 2*n*=14. High alpine, in sandy soil around boulders; > 3500 m. s SNH, W&I; to sw Can, CO. Jul–Aug ★

P. macrantha Vasey SEASHORE BLUE GRASS Per from very long, stout rhizomes or stolons, tufted, 1–6 dm; dioecious. **LF:** sheath open ± 1/2 length; ligule < 5 mm; blade 2–4 mm wide, firm, folded and inrolled, upper surface finely hairy. **INFL:** 3–15 cm, narrowly ovate, dense, ± interrupted; branches smooth to ± scabrous. **SPIKELET:** callus glabrous or with a ring of hairs; lemma 6–11 mm, glabrous or keel, marginal veins hairy. **FL:** unisexual; fertile anthers 2–5 mm. 2*n*=28. Coastal dunes; < 100 m. n&c NCo; to AK. [*P. douglasii* subsp. *m.* (Vasey) D.D. Keck] May–Jul

P. napensis Beetle (p. 1481) NAPA BLUE GRASS Per, densely cespitose, gen 3–10 dm. **LF:** sheath open to near base; ligule 4–6 mm, obtuse to acute, scabrous; blade 1–3 mm wide, ± firm, folded to inrolled, scabrous. **INFL:** gen 5–15 cm, ± open; branches ascending, spikelets dense, only in top 1/2, densely scabrous. **SPIKELET:** callus glabrous; lemma 3–4 mm, glabrous or sparsely short-hairy, scabrous. **FL:** anthers 1.2–1.8 mm. 2*n*=42. Low sterile ground near hot springs; 100–200 m. se NCoRO (near Calistoga, Napa Co.). May ★

P. nemoralis L. WOOD BLUE GRASS Per, cespitose, all current shoots fl, 3–7 dm. **LF:** sheath open 3/4 length to near base; ligule < 0.5 mm, minutely scabrous, truncate; blade 1.5–4 mm wide, soft, gen flat, abruptly spreading, barely prow-tipped. **INFL:** 5–15 cm, open, lanceolate to narrowly ovate; branches ascending, scabrous on angles. **SPIKELET:** lower glume gen 3-veined; callus sparsely cobwebby; lemma 3–3.5 mm, keel, marginal veins hairy. **FL:** anthers 0.3–1.4 mm. 2*n*=28,42. Disturbed moist places in forest, rocks near streams; gen < 2000 m. NW, SN; cool temp; native to Eurasia. May–Jun

P. palustris L. FOWL BLUE GRASS Per, cespitose or with stolons, gen 2.5–12 dm. **LF:** sheath open 3/4 length to near base; ligule 1–3 mm, acute to rounded; blade 1.5–6 mm wide, soft, gen flat, often > sheath, narrowly prow-tipped, base closely ascending. **INFL:** gen 10–30 cm, eventually open, lanceolate to narrowly triangular, many-fld; branches ascending to spreading in fr, scabrous on angles. **SPIKELET:** lower glume gen 3-veined; callus cobwebby; lemma gen 2–3 mm, keel, marginal veins hairy. **FL:** anthers 0.8–1.4 mm. 2*n*=28. Disturbed ground in moist forest or sagebrush scrub, meadows, along streams; 1500–2000 m. KR, CaRH, n&c SNH, SnGb, SnBr, GB; cool temp; native to N.Am boreal forest, Eurasia. May–Sep

P. piperi Hitchc. (p. 1487) PIPER'S BLUE GRASS Per from rhizomes, tufted, 2–5.5 dm; dioecious. **LF:** sheath open 1/3–2/3 length; ligule 1–2 mm, scabrous; blade 1.5–3 mm wide, firm, folded, inrolled, collar, upper surface of sterile st blades finely hairy. **INFL:** gen 4–8 cm, sparse; branches ascending. **SPIKELET:** few; callus diffusely cobwebby; lemma 4–7 mm, glabrous, smooth to scabrous. **FL:** unisexual; fertile anthers 2–3 mm. 2*n*=28. Serpentine, talus, chaparral,

forest openings; 100–500 m. nw KR (Del Norte Co.); sw OR. [*P. rhizomata,* misappl.] Apr–Jun ★

P. pratensis L. subsp. **pratensis** (p. 1487) KENTUCKY BLUE GRASS Per from multiple, long, stout rhizomes, tufted or loose, gen 2–7 dm. **LF:** sheath open 1/2–3/4 length; ligule 1–4 mm, truncate to rounded, smooth to minutely scabrous at margin; blade gen 2–4 mm wide, soft to ± firm, flat or folded. **INFL:** gen 6–15 cm, ovate to triangular; branches gen spreading, smooth or scabrous. **SPIKELET:** callus densely long-cobwebby; lemma 3–4 mm, keel, marginal veins hairy. **FL:** anthers 1.2–2 mm. 2*n*=21–144. Common. Disturbed habitats, incl saline or alkaline soils; < 3500 m. CA; n temp; native to Eurasia. Widely planted as lawn or pasture grass. *P. pratensis* subsp. *agassizensis* (B. Boivin & D. Löve) R.L. Taylor & MacBryde is widespread and hypothetically native in CA and N.Am, has dense infl < 6 cm with smooth branches, lf firm and folded, sterile shoot blade upper surface often sparsely soft-hairy; *P. pratensis* subsp. *angustifolia* (L.) Lej., native of Medit Eurasia, probably introduced in CA, has long, folded sterile ± firm shoots gen < 0.5 mm wide, sometimes hairy as in *P. pratensis* subsp. *agassizensis,* narrowly triangular open infl with smooth ascending branches. Mar–Jul ❖

P. pringlei Scribn. Per, densely cespitose, 0.5–3.5 dm; dioecious. **LF:** sheath open 2/3–6/7 length; ligule 1–6 mm, truncate to acute, scabrous or smooth; blade 1.5–3 mm wide, ± firm, folded, inrolled, upper surface of sterile st blades finely hairy. **INFL:** 1–6 cm, gen ± lanceolate, dense; branches steeply ascending or appressed. **SPIKELET:** glumes ± = 1st lemma, thin, shiny; callus glabrous; lemma 5–8 mm, glabrous. **FL:** pistillate in n SNH, staminate or pistillate in KR, CaRH; fertile anthers 2–4 mm. Open places, esp snowbeds; 2000–3000 m. KR, CaRH, n SNH; sw OR. Hybridizes with *P. cusickii* (esp Mount Shasta and Mount Lassen). Jun–Aug

P. rhizomata Hitchc. TIMBER BLUE GRASS Per from rhizomes, 2.5–6 dm; ± dioecious. **LF:** sheath open 1/2–2/3 length; ligule 2–8 mm, acute, smooth; blade 1.5–3.5 mm wide, soft, gen flat. **INFL:** 4–10 cm, ovate, sparse; branches gen 1–2 per node, with few spikelets; top ± nodding. **SPIKELET:** callus cobwebby; lemma 4–6.5 mm, keel, marginal veins hairy. **FL:** pistillate or bisexual (rarely staminate); fertile anthers 2.5–4 mm. 2*n*=28. Shady moist slopes in forest, in rich loose soils, on ultramafic substrates; 400–1000 m. KR (Siskiyou, n Trinity cos.); sw OR. Apr–Jul ★

P. secunda J. Presl (p. 1487) NEVADA BLUE GRASS Per, densely cespitose. **LF:** sheath open 3/4 length to near base; ligule 0.5–10 mm, truncate to acuminate, sometimes scabrous; blade gen 0.5–3 mm wide, soft to firm, flat to folded or inrolled. **INFL:** 2–25 cm, often ± 1-sided, gen linear to lanceolate, gen dense; branches gen appressed to ascending (gen spreading only in fl), ± scabrous. **SPIKELET:** ± cylindric or little compressed; upper internodes gen > 1.2 mm; callus glabrous or with a ring of short hairs; lemma 3.5–5 mm, weakly keeled to rounded across lower back, glabrous to ± evenly short-hairy across body (rarely soft-hairy only on veins), smooth to scabrous. **FL:** anthers 1.5–3 mm. Many ecological forms; subspp. tend to intergrade.

subsp. **juncifolia** (Scribn.) Soreng Pl 3–12 dm. **LF:** ligule 0.5–6 mm, gen truncate to acuminate, scabrous (those on lateral shoots 0.5–2 mm); blade gen < 1.5 mm wide, ± firm, tightly folded and inrolled, retaining shape, often glaucous. **INFL:** gen 6–25 cm. **SPIKELET:** lemma 3.5–6 mm, glabrous (rarely sparsely short-hairy on keel, marginal veins near base); palea keels scabrous. 2*n*=±63. Sagebrush scrub to lower montane forest, often in alkaline depressions; 900–3000 m. NW, CaRH, SNH (esp e slope), GB; to s Can, ND, NM. *Poa* ×*limosa* Scribn. & T.A. Williams (Lassen Co. blue grass) is a set of hybrids between *P. secunda* subsp. *juncifolia* and *P. pratensis;* 2*n*=63–64; low ground, often saline meadows; 800–2000 m; range of subsp. *juncifolia.* Apr–Jul

subsp. **secunda** ONE-SIDED BLUE GRASS Pl 1.5–10 dm. **LF:** ligule 2–10 mm, acute to acuminate, smooth or sparsely scabrous; blade 0.5–3 mm wide, soft, gen flat, soon withering, sometimes glaucous, basal often thread-like. **INFL:** gen 2–15 cm. **SPIKELET:** lemma 4–5 mm, base ± evenly short-hairy (sometimes nearly glabrous); palea gen hairy in lower 1/2. 2*n*=±84 most often. Common. Dry slopes to saline/alkaline meadows to alpine; < 3900 m. CA; to s Can, Baja CA; Patagonia, S.Am. [*P. canbyi* (Scribn.) Howell; *P. gracillima* Vasey; *P. incurva* Scribn. & T.A. Williams; *P. sandbergii*

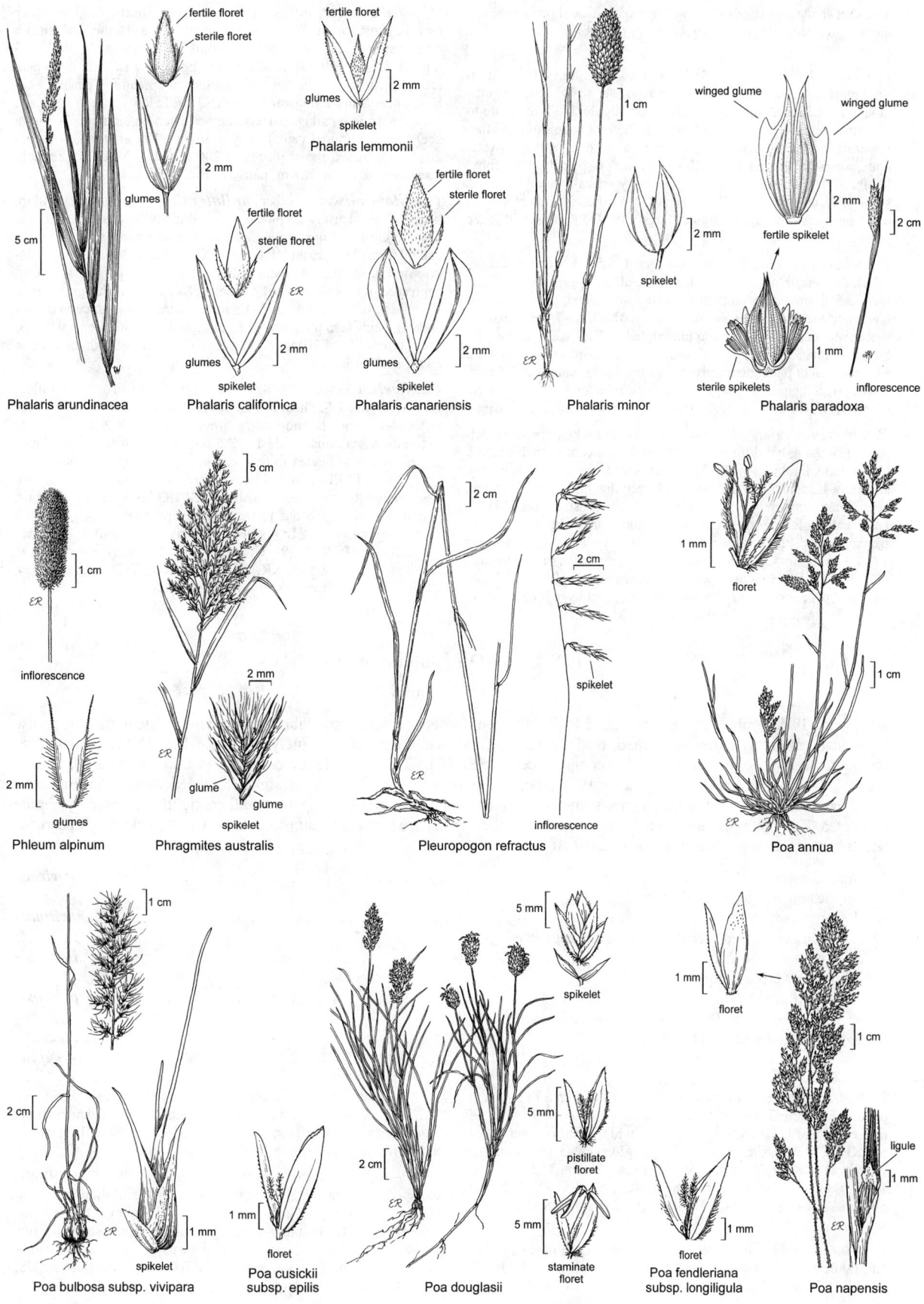

Phalaris arundinacea

Phalaris californica

Phalaris canariensis

Phalaris minor

Phalaris paradoxa

Phalaris lemmonii

Phleum alpinum

Phragmites australis

Pleuropogon refractus

Poa annua

Poa bulbosa subsp. vivipara

Poa cusickii subsp. epilis

Poa douglasii

Poa fendleriana subsp. longiligula

Poa napensis

Vasey; *P. scabrella* (Thurb.) Vasey] Many ecological forms have been named, all intergrade completely, probably do not warrant taxonomic recognition. Mar–Aug

P. sierrae J.T. Howell (p. 1487) SIERRA BLUE GRASS Per from rhizomes, 2–6 dm, slender, gen with ± purple, scaly, axillary buds; dioecious. **LF:** sheath open < 1/10 length; ligule 3–6 mm, acute to acuminate, minutely scabrous; blade 1.5–2.5 mm wide, soft, gen flat, uppermost > 1.5 × sheath length. **INFL:** 5–15 cm, narrowly ovate, open, sparse. **SPIKELET:** callus glabrous or sparsely cobwebby; lemma 4–7 mm, glabrous to sparsely hairy on base, keel, marginal veins. **FL:** unisexual; fertile anthers 2–4 mm. Shady moist slopes, often on mossy rocks, in canyons, forest; 350–1500 m. n SNF, n&c SNH. Apr–Jun ★

P. stebbinsii Soreng Per, cespitose, gen 1–3.5 dm, slender; ± dioecious. **LF:** sheath open 3/5–4/5 length; ligule of uppermost lf of sterile st 2.5–6 mm, clear, smooth; blade ± firm, folded, inrolled, upper blade surface on sterile sts ± finely hairy. **INFL:** 3–7 cm, narrowly lanceolate, dense to sparse and interrupted. **SPIKELET:** callus glabrous; lemma 3.7–5.5 mm, glabrous, smooth or barely scabrous, thin. **FL:** bisexual or pistillate (rarely staminate); fertile anthers 2–4.5 mm. 2*n*=42,81. Subalpine to lower alpine mesic meadows; 2700–3700 m. c&s SNH. Hybridizes with *P. keckii, P. secunda* subsp. *s.* Jun–Aug

P. tenerrima Scribn. DELICATE BLUE GRASS Per, cespitose, delicate, 1.5–5 dm; basal lf tuft gen < 8 cm. **LF:** sheath open 3/4 length to near base; ligule 0.5–1.5(2.5) mm, truncate to obtuse (rarely acute), scabrous; blade 0.5–1.5 mm wide, soft, gen flat, fine. **INFL:** 7–9 cm, open; branches widely spreading in fr, gen densely scabrous. **SPIKELET:** callus glabrous or with a ring of short hairs; lemma 3–4.2 mm, weakly keeled, short-hairy near base to sparsely bristly. **FL:** anthers 1.6–2.1 mm. 2*n*=42. Uncommon. Thin, drying soils, on serpentine; < 700 m. SNF. Previous reports for se SnFrB, SCoRO, WTR, are based on pls considered *P. secunda* subsp. *s.* that have longer, acute, smoother, ligules, and smoother branches. Mar–Jun

P. trivialis L. ROUGH BLUE GRASS Per, short-lived, cespitose or with short stolons from a dense basal tuft, 3–10 dm. **LF:** sheath open gen 3/4, scabrous; ligule 3–7 mm, acute, scabrous; blade 1–5 mm wide, soft, gen flat, barely prow-tipped. **INFL:** 8–25 cm, open, lanceolate to ovate; branches ascending to spreading, densely scabrous, many-fld from lower 1/2. **SPIKELET:** lower glume 1-veined, ± sickle-shaped; callus sparsely long-cobwebby; lemma ± 2–3 mm, keel (rarely marginal veins) hairy near base. **FL:** anthers 1.4–2 mm. 2*n*=14. Disturbed moist places; < 700 m. NCo, NCoRO, probably elsewhere; to AK, e N.Am; native to Eur. Apr–Jul

P. unilateralis Vasey subsp. ***unilateralis*** SAN FRANCISCO BLUE GRASS Per, densely cespitose, 0.5–4 dm. **LF:** sheath open to near base; ligule 1.5–5 mm, smooth; blade 1–5 mm wide, soft, gen moderately thick, broader and flat on st. **INFL:** gen 3–7 cm, dense, cylindric or ± lanceolate; branches short, smooth or sparsely scabrous, appressed, with spikelets from base. **SPIKELET:** callus glabrous or with a ring of hairs; lemma 3–4.5 mm, glabrous or with curved hairs on keel, veins (esp marginal). **FL:** anthers 1.5–3 mm. Coastal bluffs, in ± saline soils; < 410 m. NCo, n&c CCo; to n OR. Other subsp. in n OR, s WA. Mar–Jun

P. wheeleri Vasey (p. 1487) Per from short rhizomes, ± tufted, gen 3.5–8 dm. **LF:** sheath open 1/3–2/3 length, lower often finely reflexed-scabrous or short-hairy; lower ligules 0.5–2 mm, truncate to rounded, scabrous; blade 1.5–3.5 mm wide, soft to ± firm, flat or folded; sterile st blades folded and ± inrolled, upper surface gen ± finely hairy. **INFL:** gen 5–12 cm, ovate, sparse; branches gen 1–4, ± ascending to spreading, ± scabrous. **SPIKELET:** callus glabrous; lemma 3–6 mm, glabrous to sparsely scabrous or sparsely hairy on keel, marginal veins. **FL:** pistillate (anthers ± vestigial); fr produced asexually. 2*n*=56,63,70–91. Common (esp SN). Mtns, open forest in rich soil; 1300–3800 m. KR, NCoRH, CaRH, SNH, GB; to s Can, CO. [*P. nervosa* (Hook.) Vasey, misappl.] May–Aug

POLYPOGON BEARD GRASS

Steven A. Conley

Ann, per. **ST:** decumbent to erect, simple. **LF:** sheath open, loosely enclosing st, glabrous; ligule thinly membranous, obtuse to truncate, minutely ciliate to toothed; blades ± cauline, flat, scabrous, veins minutely prickly at 10×. **INFL:** panicle-like, oblong to narrowly ovoid, interrupted to compact, dense. **SPIKELET:** floret and glumes falling as 1 unit at maturity; glumes ± equal, 1-veined, entire or 2-lobed, awns 0, or arising at tip or between lobes, awn straight; floret 1; lemma ± 0.5 × glumes, translucent, 5-veined, tip toothed, awn < glume awn; palea slightly < lemma, transparent; anthers tightly enclosed by lemma and palea. **FR:** oblong, smooth, enclosed by lemma, palea. 18 spp.: warm temp Eurasia, Afr, S.Am. (Greek: much bearded) Some spp. orn. *P. fugax* Steud. an historical waif.

1. Glumes awnless . . . . ***P. viridis***
1′ Glumes awned, awns 1–12 mm
  2. Lemma awnless . . . . ***P. maritimus***
  2′ Lemma gen awned (occ 0 in *Polypogon imberbis*)
    3. Spikelet stalk > 2 mm . . . . ***P. imberbis***
    3′ Spikelet stalk < 2 mm
      4. Ann; glume lobed; infl dense . . . . ***P. monspeliensis***
      4′ Per; glume lobes 0; infl lobed or interrupted
        5. Glume awn (2)4–7(11) mm . . . . ***P. australis***
        5′ Glume awn 1.5–4.5 mm . . . . ***P. interruptus***

P. australis Brongn. CHILEAN BEARD GRASS Per. **ST:** 1–10 dm, erect. **LF:** ligule 2–4 mm, truncate, ± entire; blade 3–14 cm, 5–7 mm wide. **INFL:** 6–16.5 cm, 1–3.5 cm wide, lobed or interrupted. **SPIKELET:** ± sessile, stalk < 0.5 mm; glumes 2–3 mm, minutely bristly, awn ± purple in fl, (2)4–7(11) mm; lemma 1–2 mm, awn 1.2–5(7) mm. Streambanks; < 1000 m. NCo, NCoRO, CaRF, SNF, n SNH, GV, CCo, SnFrB, SCo, SnGb, SnBr; Baja CA; native to S.Am. Jun–Oct

P. imberbis (Phil.) Johow Per. **ST:** 6–10 dm, decumbent to erect. **LF:** ligule 6–8 mm, toothed; blade 5.5–11.5 cm, 1 cm wide. **INFL:** 19–30 cm, 2 cm wide; lower branches stiffly erect. **SPIKELET:** stalk

> 2 mm; glumes 2.5–4 mm, acute, sparsely bristly below middle, awn 1–2.5 mm; lemma 1.5–2.5 mm, awn 0–2 mm. 2*n*=28,56. Uncommon. Sand dunes, salt marshes; < 100 m. SnJV, CCo, SnFrB; native to S.Am. [*P. elongatus* Kunth, misappl.] Jun–Sep

P. interruptus Kunth DITCH BEARD GRASS Per. **ST:** 5–9 dm, clumped. **LF:** ligule 2–8(13) mm, obtuse to truncate, minutely hairy; blade 0.5–19.5 cm, 3–6 mm wide. **INFL:** 1.5–18 cm, lobed or interrupted. **SPIKELET:** stalk < 1 mm; glumes 1.5–3 mm, scabrous, awn 1.5–4.5 mm, ± purple; lemma 1–2 mm, awn 0.5–3 mm. 2*n*=42. Common. Streambanks; < 1300 m. CA (exc Teh, GB, DMtns); to s-c US; native to S.Am. May–Aug

P. maritimus Willd. (p. 1487) MEDITERRANEAN BEARD GRASS Ann. **ST**: 0.5–5 dm. **LF**: ligule < 6 mm, irregularly toothed, minutely hairy; blade 1–14 cm, 2–4 mm wide. **INFL**: 1–8.5(15) cm, plume-like, dense. **SPIKELET**: stalk < 1 mm; glumes 1–2 mm, minutely bristly below middle, ciliate-fringed, tip lobes 0.5–1 mm, awn 4.5–12 mm; lemma 0.5–1.5 mm, awn 0. $2n=14,42$. Common. Moist places; < 600 m. NW (exc KR), CaR, SN, GV, SnFrB, SCo, WTR, PR, DMoj; native to Medit Eur, Afr. Jun

P. monspeliensis (L.) Desf. (p. 1487) ANNUAL BEARD GRASS, RABBITFOOT GRASS Ann. **ST**: 2–10 dm. **LF**: ligule 2–12 mm, irregularly toothed, minutely hairy; blade 1–20.5 cm, 4–6 mm wide. **INFL**: 1–17 cm, plume-like, dense. **SPIKELET**: stalk << 0.5 mm; glumes 1–2.5 mm, 2-lobed, minutely bristly, awn 2–10 mm; lemma 0.5–1.5

mm, awn 0.5–4.5 mm. $2n=28,35$. Common. Moist places, along streams; < 2100 m. CA; N.Am; native to s&w Eur. Apr–Aug ❖

P. viridis (Gouan) Breistr. (p. 1487) WATER BEARD GRASS Per. **ST**: 10–100 cm, decumbent to long-trailing, rooting at nodes. **LF**: ligule 1–6 mm; lower blades 3–18 cm, 2–10 mm wide, flat. **INFL**: 2–15 cm, narrowly lanceolate to elliptic in outline, dense, often interrupted; 1° branches erect to ± appressed, lower 1–4 cm. **SPIKELET**: glumes 1.5–2.5 mm, back finely scabrous; lemma 1–1.5 mm, awn 0. $2n=28$. Common. Disturbed areas, wet areas, ponds, streambanks; < 3000 m. NCo, KR, NCoRO, SN (exc s SNH), GV, CW, SW, SNE, D; sw US; native to Eur, n Afr, w Asia. [*Agrostis v.* Gouan; *A. semiverticillata* (Forssk.) C. Chr.; *P. s.* (Forssk.) Hyl.] May–Jun

PUCCINELLIA ALKALI GRASS

Jerrold I. Davis

Ann, per; stolons and rhizomes gen 0. **ST**: decumbent to erect. **LF**: basal and cauline; sheath open ± to base; ligule thinly membranous, acute to truncate, sometimes toothed. **INFL**: panicle-like; lower branches reflexed to erect in fr; spikelets stalked. **SPIKELET**: bisexual; glumes < lowest floret, lower glume gen 1-veined, upper 3-veined; florets 2–9; lemma gen firm, back gen rounded, sometimes weakly keeled near tip, margin entire to scabrous-serrate near tip at 10×, glabrous to ± puberulent at base and along the veins in lower 1/2, gen glabrous elsewhere on surface, gen faintly 5-veined; awn 0; palea ± = lemma. ± 120 spp.: temp to arctic, N.Am, Eurasia, s hemisphere. (B. Puccinelli, botanist, Italy, 1808–1850) [Davis & Consaul 2007 FNANM 24:459–477] Gen on wet saline or alkaline soils in coastal and inland habitats; some spp. difficult to separate without hand lens. *P. maritima* (Huds.) Parl. not in CA.

1. Ann; remains of previous yr's growth not present
  2. Lowest lemma 1.8–2.2 mm; lemma tip obtuse to truncate . . . . **P. parishii**
  2′ Lowest lemma 2.5–4 mm; lemma tip acute. . . . . **P. simplex**
1′ Per; previous yr's lvs persistent
  3. Spikelet stalk glabrous to sparsely scabrous
    4. Lemma margin near tip entire or sparsely finely scabrous at 10×; anthers of lowest floret 0.5–1.2 mm;
      coastal habitats, NCo . . . . **P. pumila**
    4′ Lemma margin near tip scabrous-serrate at 10×; anthers of lowest floret 0.4–2 mm; coastal and inland habitats
      5. Anthers of lowest floret 1.5–2 mm; inland habitats, KR (Shasta Co.) . . . . **P. howellii**
      5′ Anthers of lowest floret 0.5–1.4 mm; coastal habitats, NCo. . . . . ²**P. nutkaensis**
  3′ Spikelet stalk uniformly scabrous
    6. Lowest lemma 1.5–2(2.2) mm; lemma tip obtuse to truncate; lower infl branches spreading to reflexed
      in fr; anthers of lowest floret 0.4–0.8 mm. . . . . **P. distans**
    6′ Lowest lemma 2–5 mm; lemma tip acute or acuminate to obtuse; lower infl branches erect to reflexed in
      fr; anthers of lowest floret 0.5–2 mm
      7. Cauline lf blade inrolled, 1–2 mm wide when flat; lemma tip acute, margin near tip gen entire,
        sometimes sparsely finely scabrous or scabrous-serrate at 10× anthers of lowest floret 1–2 mm . . . . **P. lemmonii**
      7′ Cauline lf blade flat to inrolled, 1–6 mm wide when flat; lemma tip acute to obtuse, margin near tip
        scabrous-serrate at 10× anthers of lowest floret 0.5–2 mm
        8. Cauline lf blade 1.5–6 mm when flat; lowest lemma (2.2)3–4.5(5) mm; coastal habitats, NCo . . . . . ²**P. nutkaensis**
        8′ Cauline lf blade 1–4 mm when flat; lowest lemma (2)2.2–3(3.5) mm; gen inland habitats, occ coastal
        . . . . **P. nuttalliana**

P. distans (Jacq.) Parl. EUROPEAN ALKALI GRASS Per. **LF**: cauline blade flat to inrolled, 1–7 mm wide when flat. **INFL**: 2.5–20 cm; lower branches spreading to reflexed in fr; spikelet stalk scabrous. **SPIKELET**: lemma tip widely obtuse to truncate, margin near tip scabrous-serrate, lowest lemma 1.5–2(2.2) mm; anthers of lowest floret 0.4–0.8 mm. $2n=14,28,42$. Saline meadows, flats; < 2700 m. CA-FP, GB; to AK, ne N.Am; native to Eurasia. Jun–Jul

P. howellii J.I. Davis (p. 1487) HOWELL'S ALKALI GRASS Per. **LF**: cauline blade inrolled, 1.4–2.2 mm wide when flat. **INFL**: 2–13 cm; lower branches erect to reflexed in fr; spikelet stalk glabrous to sparsely scabrous. **SPIKELET**: lemma tip acute to obtuse, margin near tip scabrous-serrate, lowest lemma 2.4–3.3 mm; anthers of lowest floret 1.5–2 mm. Mineral springs; ± 425 m. KR (Shasta Co.). Apr–Jun ★

P. lemmonii (Vasey) Scribn. (p. 1487) LEMMON'S ALKALI GRASS Per. **LF**: cauline blade inrolled, 1–2 mm wide when flat. **INFL**: 2–10(18) cm; lower branches ascending to reflexed in fr; spikelet stalk scabrous. **SPIKELET**: lemma tip acute, margin near tip entire

to scabrous-serrate, lowest lemma 2.4–4 mm; anthers of lowest floret 1–2 mm. $2n=14$. Saline meadows, flats; 700–2000 m. KR, CaR, GB; to OR, ID, WY. May–Aug

P. nutkaensis (J. Presl) Fernald & Weath. ALASKA ALKALI GRASS Per. **LF**: cauline blade flat to inrolled, 1.5–6 mm wide when flat. **INFL**: 5–30 cm; lower branches gen erect to ascending in fr, sometimes spreading to reflexed; spikelet stalk sparsely to densely scabrous. **SPIKELET**: lemma tip acute to obtuse, margin near tip scabrous-serrate, lowest lemma (2.2)3–4.5(5) mm; anthers of lowest floret 0.5–1.4 mm. $2n=42,56$. Marshes, wet sites; < 10 m. NCo; to AK. Jun

P. nuttalliana (Schult.) Hitchc. NUTTALL'S ALKALI GRASS Per. **LF**: cauline blade flat to inrolled, 1–4 mm wide when flat. **INFL**: 5–30 cm; lower branches erect to reflexed in fr; spikelet stalk scabrous. **SPIKELET**: lemma tip acute to obtuse, margin near tip scabrous-serrate, lowest lemma (2)2.2–3(3.5) mm; anthers of lowest floret 0.6–2 mm. $2n=28,42,56$. Saline meadows, flats; < 2640 m. CA-FP, GB; to AK, e N.Am. Valuable forage grass. Jun–Sep

P. parishii Hitchc. PARISH'S ALKALI GRASS Ann. **LF**: cauline blade gen inrolled, < 1.2 mm wide when flat. **INFL**: 1–8.5 cm; lower branches erect to reflexed in fr; spikelet stalk scabrous (occ sparsely scabrous). **SPIKELET**: lemma tip obtuse to truncate, margin near tip entire to sparsely finely scabrous, veins densely hairy in lower 1/2, lowest lemma ± 2 mm; anthers of lowest floret ± 0.5 mm. 2*n*=14. Mineral springs; 700–1000 m. w DMoj (sw San Bernardino Co.); to NM. Apr–May ★

P. pumila (Vasey) Hitchc. DWARF ALKALI GRASS Per. **LF**: cauline blade flat to inrolled, 1–3 mm wide when flat. **INFL**: 3–20 cm; lower branches reflexed to ascending in fr; spikelet stalk glabrous to sparsely scabrous. **SPIKELET**: lemma tip acute to obtuse, margin near tip entire to sparsely finely scabrous, lower lemma 2.5–4.6 mm; anthers of lowest floret 0.5–1.2 mm. 2*n*=14,42,56. Marshes, flats; < 10 m. NCo; to AK, ne N.Am, ne Asia. Jul ★

P. simplex Scribn. (p. 1487) Ann. **LF**: cauline blade gen inrolled, 0.7–2 mm wide when flat. **INFL**: 1–18 cm; lower branches erect in fr; spikelet stalk ± glabrous to scabrous, often also faintly short-soft-hairy. **SPIKELET**: lemma rounded to weakly keeled at tip, acute, margin near tip entire to sparsely finely scabrous, back gen sparsely short-hairy throughout (hairs 0.1 mm), gen also with longer sometimes curly hairs along veins near base, lowest lemma 2.5–4 mm; anthers of lowest floret < 0.5 mm. 2*n*=56. Saline flats, mineral springs; < 900 m. Teh, GV, SnFrB, w DMoj; to UT. Mar–May ★

RYTIDOSPERMA

H.E. Connor & S.J. Darbyshire

Per, cespitose, occ shortly rhizomed. **ST**: erect to nodding, up to 1 m. **LF**: sheath glabrous or hairy, with tuft of hairs at tip; ligule a rim of hairs; blades persistent or breaking apart at ligules, flat or inrolled, glabrous or hairy. **INFL**: gen raceme-like or few-branched, panicle-like. **SPIKELET**: laterally compressed; glumes subequal, veins 3–13; florets 3–10, bisexual, breaking apart above glumes and between florets; callus sharp or ± blunt with lateral tufts of stiff hairs; lemma ovate to lanceolate with 2 complete or incomplete transverse rows of tufts of hairs occ reduced to marginal tufts, veins 5–9, 2-lobed, lobes often bristle-tipped, central awn from between lobes gen with twisted column. **FR**: obovate to elliptic. 45–50 spp.: temperate S.Am, New Zealand, Australia, New Guinea. (Greek: wrinkled seed) [Darbyshire & Connor 2003 FNANM 25:309–312] Florets cross-pollinated and anthers long, or self-pollinated and anthers short. *R. biannulare* (Zotov) Connor & Edgar and *R. richardsonii* (Cashmore) Connor & Edgar occ escaped from research plots, not known to have persisted.

1. Upper and lower lemma hairs (exc callus) in 2 continuous rows of tufts . . . . [*R. caespitosum*]
1′ Upper lemma hairs (exc callus) in isolated tufts or in tufts at margins only, lower row of tufts gen continuous
  2. Callus hairs gen overlapping lower row of lemma hairs; lateral lemma lobes gradually narrowed to awn; central awn 5–17.5 mm . . . . *R. penicillatum*
  2′ Callus hairs rarely reaching lower row of lemma hairs; lateral lemma lobes abruptly narrowed to awn; central awn 11–14 mm . . . . *R. racemosum*

R. penicillatum (Labill.) Connor & Edgar (p. 1487) HAIRY OAT GRASS, POVERTY GRASS **ST**: 3–9 dm, short rhizomed. **LF**: gen basal, < st, < 30 cm, flat or inrolled; **INFL**: 4–10 cm. **SPIKELET**: 10–15 mm; glumes 7.5–11 mm; florets 6–7; callus 0.5–1.5 mm, marginal hair tufts thick, overlapping lower row of lemma hairs; lemma 2–4.2 mm, gen glabrous above, a continuous row of hairs below, lateral lobes 5–13 mm, gradually narrowed to awns; central awn 5–17.5 mm, column 1.5–4 mm, reflexed, revealing palea tip. 2*n*=24. Disturbed areas; < 800 m. NCo, NCoRO, CCo, WTR (Santa Ynez Mtns); to OR; HI; native to Australia; naturalized in New Zealand. [*Danthonia pilosa* R. Br., misappl.] Jun–Jul ❖

R. racemosum (R. Br.) Connor & Edgar (p. 1487) **ST**: 2–6 dm, short rhizomed. **LF**: gen basal, 15–75 cm, flat or inrolled. **INFL**: 5–15 cm. **SPIKELET**: 10–13 mm; glumes 8–13 mm; florets 6–7; callus 1–1.5 mm, marginal hair tufts not reaching lower lemma hairs; lemma 2.5–3.5 mm, isolated tufts of hairs above, continuous lower row of tufts of hairs, lateral lobes 5–10 mm abruptly narrowed to awns; central awn 11–14 mm, column 2–3 mm, reflexed, revealing palea tip. 2*n*=24. Disturbed areas; < 200 m. n CCo (Alameda Co.); native to Australia; naturalized in New Zealand. Jun

SACCHARUM

Elizabeth M. Skendzic

Per, cespitose, or rhizomed. **ST**: erect. **LF**: cauline; sheaths > internodes; ligule short, membranous or hairy; blade flat. **INFL**: panicle-like, branches numerous, raceme-like, densely silky-hairy. **SPIKELET**: in pairs, similar, sessile and stalked; pair breaking with subtending axis segment as a unit or at base of stalked spikelet; glumes > florets, florets 2, lower vestigial, upper fertile; lemma awned or awn 0; palea < lemma or 0. 35–40 spp.: trop, subtrop, se Asia. Some orn, sugarcane (*Saccharum officinarum* L.) widely cult for sugar. (Latin: sugar)

S. ravennae (L.) L. RAVENNA GRASS **ST**: densely tufted, gen 2–4 m. **LF**: sheath glabrous; ligule < 1 mm; blade 5–10 dm, < 12 mm wide, gen densely hairy near ligule, strongly serrate. **INFL**: plume-like, 2.5–6 dm. **SPIKELET**: 3.5–7 mm; glumes lanceolate, base densely silky-hairy; lemma awned, awn 3–5 mm; stalked spikelet deciduous. 2*n*=20. Ditch banks, marshes; < 300 m. NCoRI, GV, DSon (Imperial Co.); scattered across US; native to Eurasia. Jul–Sep ❖

SCHISMUS MEDITERRANEAN GRASS

Kelly W. Allred

Ann, bien, cespitose. **ST**: erect to prostrate. **LF**: gen basal, tufted, glabrous; ligule short-hairy; blade flat or inrolled. **INFL**: panicle-like, dense, 1–7 cm, 0.5–2(3) cm wide. **SPIKELET**: ± laterally compressed; axis breaking above glumes and between florets; glumes lanceolate, membranous, 5–7-veined; florets 3–8, bisexual; lemma 9-veined, ciliate proximally, toothed to notched, awn ± 0 (mucro occ ≤ 1.5 mm); palea < to > lemma. 5 spp.: warm temp, subtrop, Eurasia, Afr. (Greek: split, from notched lemma) [Kellogg 2003 FNANM 25:307–308]

1. Lower lemma gen > 2.5 mm, teeth narrowly triangular; palea of lowest floret acute, < lemma . . . . ***S. arabicus***
1′ Lower lemma gen < 2.5 mm, teeth obtuse to widely triangular; palea of lowest floret obtuse, > lemma . . . . . ***S. barbatus***

S. arabicus Nees (p. 1487) Ann. **ST**: gen 5–20 cm. **LF**: blade < 1 mm wide, thread-like. **SPIKELET**: glumes 4.5–6.5 mm; lemma 2.5–4 mm, teeth ± 0.3 × lemma; palea 2–3 mm. 2*n*=12. Dry, open, gen disturbed areas; < 1900 m. SnJV, CW, SCo, s ChI, TR, W&I, D; to TX, AZ; native to Eurasia. Mar–May ❖

S. barbatus (L.) Thell. (p. 1487) Ann. **ST**: gen 2–16 cm. **LF**: blade < 2 mm wide, thread-like. **SPIKELET**: glume 4–5 mm; lemma 2–2.5 mm, teeth < ± 0.2 × lemma; palea 1.5–2.5 mm. 2*n*=12. Dry, open, gen disturbed areas; < 1200 m. Teh, SnJV, SCoRO, SW, D; to TX, n Mex; native to s Eur, Afr. Mar–Apr ❖

SCLEROCHLOA

David M. Brandenburg

Ann. **ST**: gen prostrate to ascending, less often erect. **LF**: basal and cauline; sheath glabrous, upper often inflated; blade flat or folded, glabrous or minutely scabrous. **INFL**: raceme-like or panicle-like, contracted, 1-sided, often partially enclosed in upper lf sheath(s). **SPIKELET**: bisexual, sessile or short-stalked, axis breaking tardily and irregularly; glumes unequal, florets 2–7; lemma prominently 5–9-veined; palea < lemma. 2 spp.: Eurasia. (Greek: hard grass, from leathery glumes and lemmas) [Brandenburg 2007 FNANM 24:480–482]

S. dura (L.) P. Beauv. (p. 1495) HARDGRASS, FAIRGROUND GRASS **ST**: 2–15(30) cm. **LF**: blade 0.5–5(7) cm, 1–4 mm wide, tip prow-shaped. **INFL**: 1–4 cm, 0.5–2(4) cm wide, dense, spikelets overlapping. **SPIKELET**: (3.4)5–15 mm; florets (2)3–4(7); glumes << lower lemmas, rounded, lower ± 1–3 mm, upper 2.5–6 mm; lemma 3–7 mm, strongly keeled, obtuse, awn 0. 2*n*=14. Disturbed sites; 700–800 m. w CaRH (Shasta Valley); widespread in US; nearly cosmopolitan; native from s-c Eur to c Asia. May

SCLEROPOGON

John R. Reeder & Dieter H. Wilken

1 sp. (Greek: hard beard, from firm awns) [Reeder 2003 FNANM 25:42–43]

S. brevifolius Phil. (p. 1495) BURRO GRASS Per, from short stolons, mat-like; gen dioecious or monoecious, rarely bisexual. **ST**: erect, 1–2 dm. **LF**: gen basal, densely tufted; sheath smooth; ligule short-hairy; blade 2–8 cm, 1–3 mm wide, flat, firm, sharp-pointed. **STAMINATE INFL**: 3–7 cm, raceme- or panicle-like. **PISTILLATE INFL**: 10–20 cm, spike-like. **STAMINATE SPIKELET**: 1–2.5 cm, compressed; glumes 3–8 mm, ± equal, lanceolate, 1(3)-veined, awn 0; axis breaking above glumes and falling as 1 unit; florets 5–10(20); lemma ± = glumes, 3(5)-veined, short-awned. **PISTILLATE SPIKELET**: 8–15 cm incl awns; body exc awns ± cylindric, gen subtended by 1 bract; bract ± = lower glume; glumes 3-veined, acute, awn 0, lower glume 1–2 cm, upper 1.5–3 cm; axis breaking above glumes and ± between florets; florets 3–5, lower with sharp-pointed callus, upper reduced to awns; lemma 7–11 mm, narrow, awns 3, 6–12 cm, ± spreading to ascending. 2*n*=40. Open creosote-bush scrub; ± 1600 m. e DMtns (New York Mtns); to CO, TX, Mex, w S.Am. Sep–Oct ★

SCRIBNERIA SCRIBNER GRASS

Thomas Worley

1 sp. (Frank L. Scribner, Am agrostologist, 1851–1938) [Smith 2007 FNANM 24:689–690]

S. bolanderi (Thurb.) Hack. (p. 1495) Ann. **ST**: ascending to erect, 5–30 cm, branched, glabrous; nodes purple. **LF**: sheath 1–5 cm; ligule 2–4 mm, translucent; blade midrib, margins scabrous. **INFL**: 4–11 cm, straight, ± purple; spikelets 1–2(4) per node, appressed, lower ± embedded in axis, upper, if present, with stalk 1–3 mm, overlapping node above; cleistogamous spikelets occ present within lf sheaths. **SPIKELET**: glumes 2, 5–7 mm, edge-to-edge or overlapping, lower 2-veined, upper 4-veined; floret 1, breaking above glumes; callus short-bristly; lemma 4–6 mm, translucent, scabrous on upper 1/3, tip notched, awned, 5-veined; awn 2–4 mm, ± straight; palea gen < lemma, notched; stamen 1, anther gen > 1 mm. **FR**: ± 2.5 mm. 2*n*=26. Dry, disturbed areas; 500–2500 m. CA-FP, MP; to WA. Inconspicuous and easily overlooked. Mar–Jun

SECALE RYE

Ann, per. **ST**: gen erect. **LF**: sheath appendaged; ligule membranous; blade gen flat. **INFL**: spike-like, dense, ± flat; axis occ breaking at nodes in fr; spikelets 2-ranked, 1 per node, sessile, not sunken. **SPIKELET**: glumes narrow, rigid, keeled, vein gen 1; florets 2, fertile, sessile and side-to-side, occ with vestigial floret between; lemma with keel near margin, keel, margins ciliate, veins 5, tip tapered, awn straight, scabrous. 5 spp.: Eurasia. (Latin: ancient name for a cereal, probably for rye) [Barkworth 2007 FNANM 24:259–761]

S. cereale L. (p. 1495) Ann, occ bien. **ST**: 6–12.5 dm, glabrous exc below infl. **LF**: sheath glabrous, appendages ± 1 mm; blade 3–10 mm wide. **INFL**: 8–17 cm, nodding at maturity, gen not breaking apart. **SPIKELET**: glumes 6–17 mm, keeled; lemma 10–16 mm, awn 2–7 cm; anthers 7.5–8.5 mm. 2*n*=14. Disturbed slopes, roadsides; < 1800 m. KR, NCoRI, n SNH, Teh, sw SnFrB, TR, PR, s MP, W&I, DMoj, expected elsewhere; to AK, Can, US; native to sw Asia. Important cereal; planted to stabilize soil; fr milled for flour, fermented to produce whiskey. May–Aug

SETARIA BRISTLE GRASS

Robert E. Preston & Robert Webster

Ann, per. ST: gen erect; internode solid to hollow inside. **LF**: basal and cauline; sheath gen glabrous, margins occ scabrous or ciliate; ligule short-hairy or membranous, ciliate; blade glabrous, scabrous, or hairy on one or both surfaces. **INFL**: panicle-like, dense, gen cylindric; 1° branches spreading to appressed; spikelets many, gen clustered on one side of short 2° branches, short-stalked to subsessile, subtended by 1–15 bristles, bristles gen scabrous. **SPIKELET**: falling as 1 unit, gen elliptic; glumes unequal; florets gen 2, ± equal, lower floret sterile or staminate, palea gen < lemma, upper floret fertile, firm, gen hard, rough, margin inrolled, tip blunt. ± 140 spp.: warm temp, trop Eurasia, Afr. (Latin: bristly) [Rominger 2003 FNANM 25:539–558] Some spp. cult for food. *S. italica* (L.) P. Beauv., millet, common component of birdseed, occ collected but not persisting. *S. megaphylla* (Steud.) T. Durand & Schinz, *S. verticilliformis* Dumort. recorded in CA from a few sites, but do not appear to be naturalized.

1. Bristles at base of spikelets with short stiff hairs pointed toward base
  2. Lf sheath margins glabrous; lf blade sparsely hairy abaxially . . . . ***S. adhaerens***
  2′ Lf sheath margins ciliate; lf blade scabrous abaxially. . . . . ***S. verticillata***
1′ Bristles at base of spikelets with short stiff hairs pointed toward tip
  3. Upper glume ± = upper lemma; upper lemma ± smooth, dull or shiny
    4. Spikelet ± 2 mm; panicle 3–8 cm; lf blades 3–12 mm wide . . . . ***S. viridis***
    4′ Spikelet 2.5–3 mm; panicle 8–30 cm; lf blades 10–30 mm wide . . . . ***S. faberi***
  3′ Upper glume < upper lemma; upper lemma strongly ridged
    5. Rhizomes absent; upper glume ≤ 0.5 × upper lemma; spikelet gen > 2.8 mm, gen 1.5–2 mm wide
      . . . . ***S. pumila*** subsp. ***pumila***
    5′ Rhizomes present; upper glume > 0.5 × upper lemma; spikelet gen < 2.8 mm, gen ≤ 1.5 mm wide
      6. Panicle < 10 cm, 3–4 mm wide; bristles green or straw-colored; stigma dark brown or purple; base of fl sts with hard, knot-like swellings . . . . ***S. parviflora***
      6′ Panicle > 10 cm, 4–6 mm wide; bristles gen orange; stigma pale; base of fl sts not swollen . . . . ***S. sphacelata***

S. adhaerens (Forssk.) Chiov. BUR BRISTLE GRASS **Ann. ST**: 2.5–6 dm. **LF**: ligule 1–2 mm; blade ≤ 10 cm, 5–10 mm wide, sparsely hairy abaxially. **INFL**: 2–6 cm; bristle 1 per spikelet. **SPIKELET**: 1.5–2.2 mm; lower glume ± 1 mm, upper glume = spikelet; 2*n*=18. Disturbed areas; < 130 m. SnJV, DSon (Coachella Valley); s US, C.Am, widespread in subtropics. Has been confused with *S. verticillata*. Jul–Oct

S. faberi R.A.W. Herrm. CHINESE FOXTAIL **Ann. ST**: 1–20 dm. **LF**: sheath 6–11 cm; ligule ± 1.5–2.5 mm; blade 15–30 cm, 1–2 cm wide, adaxial surface short-soft-hairy. **INFL**: 7–18 cm; 1° branches 0.5–4 mm; bristles 2–3; spikelet stalk ± 0.5–1 mm. **SPIKELET**: 2.5–3 mm, 1.4–1.8 mm wide; lower glume 1–1.5 mm, 3–5-veined, upper glume 0.7–0.9 × spikelet; lower floret sterile; lemma 7-veined, tip acute; palea ± 2/3 lemma. 2*n*=36. Disturbed areas; < 100 m. n SNF, deltaic GV, CCo, SCo; to e US; native to Asia. Jun–Nov ◆

S. parviflora (Poir.) Kerguélen (p. 1495) KNOTROOT BRISTLE GRASS Per, cespitose, from short rhizomes. **ST**: 7–12 dm; base with hard, knot-like swellings. **LF**: sheath 4–9 cm; ligule < 1 mm; blade < 25 cm, 2–8 mm wide, adaxial surface glabrous. **INFL**: 3–8 cm; 1° branches 3–8 mm; axis glabrous; bristles 4–12; spikelet stalk << 0.5 mm. **SPIKELET**: 2–3 mm, ± 1–1.5 mm wide; lower glume 1–1.5 mm, 3-veined, upper glume 0.5–0.8 × spikelet length; lower floret sterile or staminate; lemma 5–7-veined, tip acute; palea ± = lemma. Moist, disturbed areas, roadsides, streambanks, canal banks; < 1470 m. n SNF, GV, CW (exc SCoRI), SW (exc ChI, SnJt), s DMoj (uncommon); to e US, C.Am, S.Am. [*S. gracilis* Kunth; *S. geniculata* (Willd.) P. Beauv., misappl.] May–Sep

S. pumila (Poir.) Roem. & Schult. subsp. *pumila* YELLOW BRISTLE GRASS **Ann. ST**: 2–13 dm. **LF**: sheath 4–9 cm; ligule ± 0.5–1 mm; blade 5–30 cm, 3–10 mm wide, adaxial surface glabrous. **INFL**: 2–6 cm; 1° branches 5–10 mm, axis short-hairy; bristles 4–12; spikelet stalk << 0.5 mm. **SPIKELET**: 3–3.5 mm, ± 1.5–2 mm wide; lower glume 1–1.5 mm, 3–5-veined, upper glume 0.5–0.7 × spikelet length; lower floret staminate; lemma 5-veined, tip acute; palea = lemma. 2*n*=36,72. Gen moist sites, disturbed areas, streambanks; < 1680 m.

w KR, NCoRO, CaRF, n&c SN, s SNF, GV, SnFrB, SCoRO, SCo, SnBr, PR, SNE, sw DMoj (exc DMtns); to s Can, e US, Mex; native to Eur. Other subsp. invasive in se US. Jun–Oct

S. sphacelata (Schumach.) Stapf & C.E. Hubb. AFRICAN BRISTLE GRASS Per, cespitose or not. **ST**: 6–30 dm. **LF**: sheath 4–10 cm, glabrous or hairy; ligule 1–2 mm; blade 10–50 cm, 3–15 mm wide, adaxial surface glabrous. **INFL**: 9–40 cm; 1° branches 5–10 mm, axis glabrous; bristles 5–15; spikelet stalk < 0.5 mm. **SPIKELET**: ± 2–3 mm, 1–1.5 mm wide; lower glume ± 1–1.5 mm, 2-veined, tip acute, upper glume 0.6–0.8 × spikelet; lower floret staminate; lemma 5-veined, tip acute; palea = lemma. 2*n*=18,36. Ditch banks; < 300 m. CaRF, GV; se US, native to Afr. Planted experimentally as a pasture grass and occ escaping. Jul–Oct

S. verticillata (L.) P. Beauv. HOOKED BRISTLE GRASS **Ann. ST**: decumbent to erect, 1.6–10 dm. **LF**: sheath 2–10 cm; ligule 1–2 mm; blade 5–25 cm, 3–12 mm wide, adaxial surface glabrous to hairy. **INFL**: 3–13 cm; 1° branches 1.1–2 cm, axis glabrous; bristles 1–2; spikelet stalk < 0.5 mm. **SPIKELET**: ± 1.7–2.5 mm, 1–1.5 mm wide; lower glume ± 0.5–1.5 mm, 1–3-veined, upper glume ± = spikelet length; lower floret sterile; lemma 5–7-veined, tip acute to rounded; palea vestigial. Disturbed places, fields, roadsides; < 550 m. SCo, WTR, PR, DSon (Coachella Valley); widespread in N.Am, native to Eur. [*S. carnei* Hitchc., misappl.] May–Nov

S. viridis (L.) P. Beauv. (p. 1495) GREEN BRISTLE GRASS **Ann. ST**: decumbent to erect, 2–10 dm. **LF**: sheath 5–15 cm, scabrous, upper margins ciliate; ligule 1–2 mm; blade 8–20 cm, 3–12 mm wide, upper surface glabrous. **INFL**: 2–15 cm; 1° branches 3–10 mm, glabrous or hairy; bristles 1–3; spikelet stalk < 0.5 mm. **SPIKELET**: ± 2 mm, ± 1 mm wide; lower glume ± 1 mm, 3-veined, upper glume = spikelet length; lower floret sterile; lemma 5-veined, tip acute; palea vestigial. 2*n*=18. Moist, disturbed areas, roadsides, streambanks; < 1525 m. NW (exc NCoRH), CaRF, SN (exc s SNH, Teh), GV, SnFrB, SCoRO, SW (exc ChI), SNE, sw DMoj, DSon (Coachella, Imperial, Palo Verde valleys); widespread N.Am; native to Eurasia. May–Oct

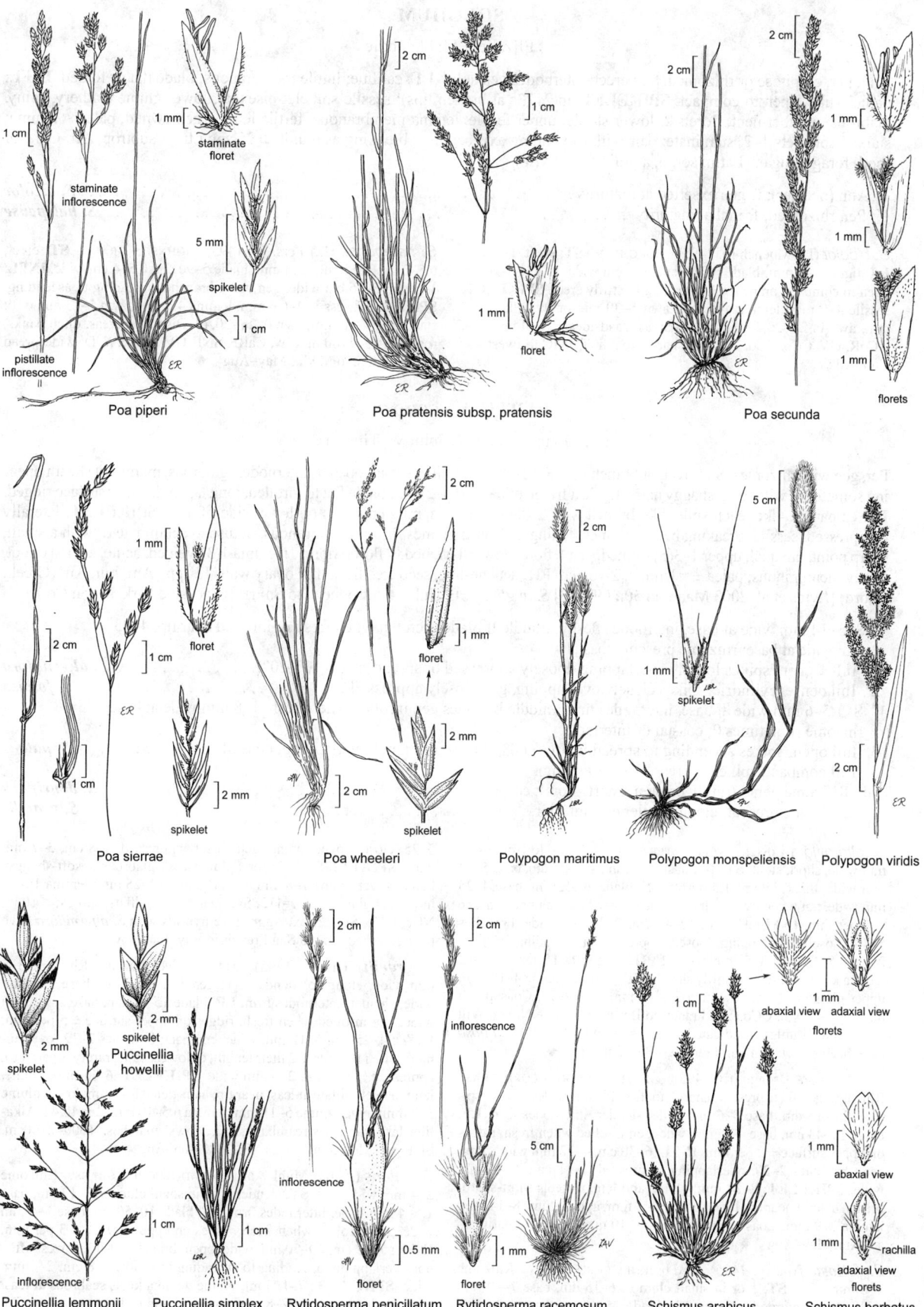

Poa piperi

Poa pratensis subsp. pratensis

Poa secunda

Poa sierrae

Poa wheeleri

Polypogon maritimus

Polypogon monspeliensis

Polypogon viridis

Puccinellia howellii

Puccinellia lemmonii

Puccinellia simplex

Rytidosperma penicillatum

Rytidosperma racemosum

Schismus arabicus

Schismus barbatus

SORGHUM

Elizabeth M. Skendzic

Ann, per, cespitose or rhizomed. **ST**: erect; internodes gen solid. **LF**: cauline; ligule membranous; blade flat or folded. **INFL**: panicle-like, open to compact. **SPIKELET**: in 2s (3s at branch tips); sessile spikelet bisexual, lower glume leathery, shiny, glabrous to puberulent, florets 2, lower sterile, upper fertile, lemma membranous, fertile lemma gen awned, palea < lemma; stalked spikelets 1–2, staminate; pair with subtending axis segment breaking as a unit. ± 25 spp.: trop, subtrop, Afr. Cult for food, forage, sugar. (Latin: sorgo, a grain from Syria)

1. Ann (occ per); fr gen exserted at maturity . . . . ***S. bicolor***
1′ Per, rhizomed; fr incl at maturity . . . . ***S. halepense***

S. bicolor (L.) Moench (p. 1495) SORGHUM **ST**: erect, 1–2.5 m. **LF**: ligule 1–4 mm, blade 5–100 cm, 3–5 cm wide. **INFL**: 1–6 dm, open to compact; branches ± spreading to stiffly erect. **SPIKELET**: sessile 4–9 mm; lemma 4–5 mm, awn 5–30 mm or 0; stalked 3–6 mm, awn 0. 2*n*=20,40. Disturbed areas, roadsides; < 600 m. NCo, NCoR, GV, CCo, SCo, WTR, D; native to Afr. TOXIC to livestock. Cult. All yr

S. halepense (L.) Pers. (p. 1495) JOHNSON GRASS **ST**: erect, 0.5–2 m. **LF**: ligule 2–6 mm, blade 5–90 cm, 0.5–4 cm wide. **INFL**: 1–5 dm, 5–25 cm wide, gen open; branches spreading to ascending. **SPIKELET**: sessile 4–6.5 mm; lemma 4–5 mm, awn 5–13 mm or 0; stalked 3.6–5.6 mm, awn 0. 2*n*=20,40. Disturbed areas, ditch banks, roadsides; < 1160 m. NW, CaRF, SNF, GV, CW, SW, D; widespread in US; native to Medit. May–Aug ◆

SPARTINA CORD GRASS

John R. Baird & John W. Thieret

Per, gen with rhizomes. **ST**: erect, unbranched. **LF**: basal and cauline; sheath open, > internodes, glabrous, margin of sheath opening sometimes with long, shaggy hairs; ligule a fringe of hairs, 0.5–2 mm; blade flat to inrolled, long tapered, upper surface ridged. **INFL**: panicle-like; each spike-like branch with 2 rows of overlapping spikelets on lower side of axis. **SPIKELET**: laterally compressed, sessile, breaking below glumes, falling as 1 unit; glumes firmly membranous, obtuse to acuminate or with a small, sharp point, unequal, upper 1–3(5)-veined, gen > floret, lower 1-veined, < floret; floret 1; lemma 1(3)-veined, acute, awned or not, firmly membranous; palea ± = lemma, 2-veined. **FR**: achene-like, seed free from thin ovary wall. 15 spp.: Am, Eur, Afr. (Greek: a cord) [Ayres et al. 2003 Madroño 50:209–213] *S. anglica* C.E. Hubb ◆ introduced to SnFrB (Creekside Park, Marin Co.).

1. St 5–14 mm wide at base, internodes fleshy; middle lf blades gen flat when fresh or inrolled near tip, 4–25 mm wide at base; rhizome present; coastal
    2. Infl ± open, spikes loosely overlapping, loosely appressed or spreading (to 10°–20°) . . . . ***S. alterniflora***
    2′ Infl dense, cylindric, spikes closely overlapping, gen closely appressed . . . . ***S. foliosa***
1′ St 1.5–6 mm wide at base, internodes firm; middle lf blades gen inrolled when fresh, 1–8 mm wide at base; rhizome sometimes 0; coastal or interior
    3. Infl open, spikes ascending to spreading (to ± 60°), lowest often not overlapping; coastal . . . . ***S. patens***
    3′ Infl compact, spikes appressed, overlapping
        4. Rhizome gen 0 (rarely present, short, stout); coastal . . . . ***S. densiflora***
        4′ Rhizome present, elongate, slender; interior . . . . ***S. gracilis***

S. alterniflora Loisel. SALT-WATER CORD GRASS Rhizome 4–7 mm wide, short, stout. **ST**: 1 or in small clumps, 6–25 dm, base 5–14 mm wide, thick, internodes fleshy. **LF**: blade 20–55 cm, base 4–25 mm wide, flat when fresh or inrolled near tip; ridges on upper surface ± 6 per mm. **INFL**: ± open, 10–40 cm, 7–22 mm wide; branches 5–30, loosely overlapping, loosely appressed or spreading (to 10°–20°), 4.5–13 cm, 3–5 mm wide. **SPIKELET**: 8–15 mm; glume, lemma keels glabrous to softly shaggy-hairy; lower glume 4–10 mm; upper glume 8–15 mm; lemma 7.5–11 mm. 2*n*=56,70. Coastal salt marshes; < 10 m. CCo (San Francisco Bay); native to e N.Am. Will likely spread unless eradicated; aggressive hybrids with *S. foliosa* have been reported in San Francisco Bay. Jul–Sep ◆

S. densiflora Brongn. (p. 1495) DENSE-FLOWERED CORD GRASS Rhizome gen 0 (if present, short, stout, < 10 mm wide). **ST**: cespitose, 2.7–15 dm, base 3–6 mm wide, internodes firm. **LF**: blade 12–43 cm, base 4–8 mm wide, gen inrolled when fresh, ridges on upper surface ± 2 per mm. **INFL**: 6–30 cm, 4–12 mm wide, compact; branches 2–20, overlapping, appressed, 1–11 cm, 2.5–6 mm wide. **SPIKELET**: 8–14 mm; glume and lemma keels short-sharp-bristly, at least near tip; lower glume 4–7 mm; upper glume 8–14 mm; lemma 7–9 mm. Coastal salt marshes; < 10 m. NCo, CCo; native to s S.Am. May–Aug ◆

S. foliosa Trin. (p. 1495) CALIFORNIA CORD GRASS Rhizome 2–6 mm wide. **ST**: 1 or in small clumps, 6–15 dm, base 7–12 mm wide, thick, internodes fleshy. **LF**: blade 15–45 cm, base 5–17 mm wide, flat when fresh or inrolled near tip, ridges on upper surfaces ± 5 per mm. **INFL**: 9–25 cm, 5–13 mm wide, dense, cylindric; branches 3–25, closely overlapping, gen closely appressed, 2–8 cm, 3–7 mm wide. **SPIKELET**: 10–25 mm; glume keels glabrous or soft-shaggy-hairy; lower glume 7–8 mm; upper glume 12–25 mm; lemma 10–12 mm, keel glabrous. 2*n*=112. Salt marshes, mudflats, shores; < 10 m. NCo, CCo, SCo; Mex. Aggressive hybrids with *S. alterniflora* have been reported around San Francisco Bay. Jun–Nov

S. gracilis Trin. (p. 1495) ALKALI CORD GRASS Rhizome 3–5 mm wide, elongate, slender. **ST**: gen 1, 1.8–10 dm, base 2–5 mm wide, slender, internodes firm. **LF**: blade 15–27 cm, base 2.5–6 mm wide, gen inrolled when fresh, ridges on upper surface ± 5 per mm. **INFL**: 4–25 cm, 5–12 mm wide, compact; branches 2–12, overlapping (often for only 1/2 their length, or lowest spike rarely separated), appressed, 1.5–8 cm, 2–6 mm wide. **SPIKELET**: 6–11 mm; glume, lemma keels ciliate at least near tip, hairs gen 0.3–1 mm; lower glume 3–7 mm; upper glume 5–11 mm; lemma 6.5–10 mm. 2*n*=40,42. Alkaline lake shores, streambanks, meadows, marshes; 1000–2100 m. SNE, n DMoj; to n&e Can, KS, NM. Jun–Aug ★

S. patens (Aiton) Muhl. SALT-MEADOW CORD GRASS Rhizome 2–4 mm wide or 0. **ST**: slender, 1 or in small clumps, 3–12 dm, base 1.5–4 mm wide, internodes firm. **LF**: blade 10–50 cm, base 1–4 mm wide, gen inrolled when fresh, ridges on upper surface ± 3 per mm. **INFL**: 5–22 cm, 10–90 mm wide, open; branches 2–13, lowest often not overlapping, ascending to spreading to ± 60°, 1–8 cm, 2–4 mm wide. **SPIKELET**: 7–12 mm; glume, lemma keels scabrous at least near tip; lower glume 2–8 mm; upper glume 6–12 mm; lemma 5–8 mm. 2*n*=28. Coastal salt marshes; < 10 m. CCo; native to se US. Nov ◆

SPHENOPHOLIS WEDGE GRASS

Thomas F. Daniel & Dieter H. Wilken

Ann, per. **ST**: 1 to clumped, erect. **LF**: cauline, glabrous to hairy; ligule ± membranous, obtuse to truncate; blade flat to inrolled. **INFL**: panicle-like, open to compact; lower branches spreading to erect. **SPIKELET**: subsessile to stalked, ± laterally compressed; glumes unequal, lower < upper, lower linear to lanceolate, 1-veined, upper obovate, 3–5-veined; florets 2–3, bisexual; lemma membranous, rounded on back, compressed at acute to obtuse tip, weakly 5-veined, awned or not; palea < lemma. 6 spp.: N.Am, Caribbean. (Greek: wedge scale, from upper glume shape) [Daniel 2007 FNANM 24:620–624]

S. obtusata (Michx.) Scribn. (p. 1495) PRAIRIE WEDGE GRASS
Per (occ fl 1st yr). **ST**: 2–8 dm. **LF**: sheath glabrous to scabrous; ligule 1–4 mm, tip jagged; blade 5–8 cm, flat. **INFL**: 4–12 cm, gen erect, compact; lower branches ± ascending. **SPIKELET**: 2–5 mm; lower glume ± 2 mm, ± linear, keel minutely scabrous, acute; upper glume 2–3 mm, widely obovate, obtuse; lemma 2–3 mm, lower > glumes, acute to obtuse. 2*n*=14. Wet meadows, streambanks, ponds; 240–2870 m. n SNF (Amador Co.), s SNH (Fresno Co.), ne SCo (Santa Ana River), SnBr, s-c PR (Cuyamaca Mtns), SNE; N.Am, Caribbean. Apr–Jun ★

SPOROBOLUS DROPSEED, SACATON

Michael Curto

Ann, per. **ST**: gen ascending to erect, 2–20 dm, gen tufted, ± solid in ×-section. **LF**: gen basal; cauline few, ascending or curving away; distal sheath margin and collar glabrous or hairy; ligule < 1 mm, hairy or membranous, fringed; blade flat to inrolled, gen glabrous or scabrous, occ short-soft-hairy. **INFL**: terminal, also occ axillary, panicle- or spike-like, gen partly enclosed by sheath; branches spreading or appressed. **SPIKELET**: < 6 mm, gen pale to gray-green or ± purple; glumes gen unequal, upper < or > lemma, membranous to translucent, 1-veined; floret bisexual, gen breaking above glumes; lemma texture gen like glumes, 1(3)-veined; palea < or > lemma. **FR**: utricle-like, when wet gelatinous with seed emergent from split ovary wall. ± 150 spp.: Am, Eurasia, Afr, Australia. (Greek: to throw seed, from deciduous seeds) [Peterson et al. 2003 FNANM 25:115–139]

1. Ann; infl 1–5 cm, terminal and axillary at ± most st nodes; upper glume 2.5–5 mm; lemma leathery, tip beaked . . . . ***S. vaginiflorus***
1′ Per; infl 8–80 cm, gen terminal, sometimes axillary at some st nodes; upper glume 0.5–3 mm; lemma membranous, tip obtuse to acute
  2. Sheath collar with conspicuous hairs to 4 mm; glume back keeled, upper glume midvein ± scabrous at 20×
    3. Lower 1° infl branches gen < 2 cm, bearing 2° infl branches and spikelets to near base . . . . ***S. contractus***
    3′ Lower 1° infl branches > 3 cm, bearing 2° infl branches and spikelets along upper 3/4, base naked
      4. 2°, 3° infl branches weakly spreading or appressed, ± obscure . . . . ***S. cryptandrus***
      4′ 2°, 3° infl branches spreading, evident, entangled . . . . ***S. flexuosus***
  2′ Sheath collar glabrous or with sparse long hairs to 6 mm; glume back rounded, upper glume midvein ± smooth at 20×
    5. Upper glume length < 0.6 × lemma, tip obtuse, minutely toothed; anthers 0.5–1.1 mm . . . . ***S. indicus***
    5′ Upper glume length > 0.6 × lemma, tip acute, entire; anthers 1.1–1.8 mm
      6. St gen 3–10 dm, to ± 3 mm thick at base; 2° infl branches spreading, bearing spikelets distally, base naked, 3° infl branches spreading . . . . ***S. airoides***
      6′ St gen 9–25 dm, to ± 9 mm thick at base; 2° infl branches weakly spreading or appressed, bearing spikelets to near base, 3° infl branches appressed . . . . ***S. wrightii***

S. airoides (Torr.) Torr. (p. 1495) ALKALI SACATON Per. **ST**: tufted, erect, 3–10(15) dm. **LF**: base shiny, straw-colored; distal sheath margin glabrous to short-hairy; collar glabrous or with sparse long hairs to 6 mm; ligule < 0.5 mm, fringe hairs to 2 mm; blade 10–40(60) cm, 2–4(6) mm wide. **INFL**: gen terminal, exserted or sheathed basally, pyramid-shaped, diffuse; 1° branches spreading widely, lowest gen ± 6 cm, > internodes, 2° spreading, bearing 3°, spikelets distally; base naked. **SPIKELET**: 1–3 mm, green to purple; glumes narrowly lanceolate, tip acute to obtuse, unequal, lower 0.5–2 mm, upper 1.1–2.5, > 0.6 × lemma; lemma 1.2–2.5 mm, glabrous, ovate to lanceolate, tip acute; anthers 1.1–1.8 mm. **FR**: 1–1.4 mm, ellipsoid, striate, red-brown. 2*n*=80,90,108,126. Seasonally moist, alkaline areas; < 2100 m. SNF, Teh, s ScV, SnJV, s SCoRO, SW, GB, D; to BC, c&s US, n&c Mex. Seeded for erosion control or wetland revegetation. Used occ as an orn. Apr–Oct

S. contractus Hitchc. (p. 1495) SPIKE DROPSEED Per. **ST**: tufted, erect, 4–12 dm. **LF**: base dull, green to brown, margin ciliate, collar with conspicuous hairs to 4 mm; ligule < 0.5–1 mm, fringe hairs to 4 mm; blade 5–35 cm, 3–8 mm wide. **INFL**: gen terminal, exserted or sheathed basally, spike-like, dense; branches, spikelets to near base; 1° branches appressed, lowest gen < 2 cm, > internodes, 2° appressed, bearing 3°. **SPIKELET**: 1.7–3 mm, green to lead-colored; glumes narrowly lanceolate, tip acute, unequal, lower 0.7–1.7 mm, upper 2–3 mm, > 0.6 × lemma; lemma 2–3 mm, glabrous, lanceolate, tip acute; anthers 0.2–0.5 mm. **FR**: 0.8–1.2 mm, ellipsoid, striate, light-brown. 2*n*=36. Rocky to sandy washes, slopes, scrub, woodland; 400–2300 m. W&I, D; to UT, CO, TX, n Mex. [*S. cryptandrus* var. *strictus* Scribn.] *S. cryptandrus* pls with contracted, sheathed panicles are often misidentified as *S. contractus*. Note length difference in lowest 1° infl branches. Aug–Oct

S. cryptandrus (Torr.) A. Gray (p. 1495) SAND DROPSEED Per. **ST**: tufted, erect, 3–10(12) dm. **LF**: base dull, green to brown, margin glabrous or ciliate, collar with conspicuous hairs to 4 mm; ligule < 0.5–1 mm, fringe hairs to 4 mm; blade 5–25 cm, 1–6 mm wide. **INFL**: gen terminal, exserted to sheathed completely, gen spike-like initially, gen pyramid-shaped at maturity, open; 1° branches ascending, lowest gen > 3 cm, > internodes, 2° weakly spreading or appressed, bearing 3°, spikelets to near obscure base. **SPIKELET**: 1.5–3 mm, green to purple or lead-colored; glumes narrowly lanceolate, tip acute, unequal, lower 0.5–1.1 mm, upper 1.5–2.7 mm, > 0.6 × lemma; lemma 1.5–2.5 mm, glabrous, ovate to lanceolate, tip acute; anthers 0.2–1 mm. **FR**: 0.7–1 mm, ellipsoid, striate, light- to red-brown. 2*n*=36,38,72. Rocky to sandy washes, slopes, scrub, woodland; 350–2800 m. SNH, SnGb, SnBr, PR, SNE, D; naturalized CaRF, ScV; waif in SnJV, SCoRO, SCo; widespread s Can, US (exc FL), n Mex, Argentina. [*S. nealleyi* Vasey misappl. from San Luis Obispo Co.] Apr–Oct

S. flexuosus (Vasey) Rydb. MESA DROPSEED Per. **ST**: tufted, erect or decumbent, 3–10(12) dm. **LF**: base dull, green to brown, margin glabrous or ciliate, collar with conspicuous hairs to 4 mm; ligule < 0.5–1 mm, fringe hairs to 4 mm; blade 5–25 cm, 1–6 mm wide. **INFL**: gen terminal, exserted or base sheathed, gen ovate at maturity, open; central axis nodding, 1° branches gen spreading widely, lowest gen > 3 cm, > internodes, 2°, 3° spreading, entangled, base naked, evident. **SPIKELET**: 1.5–2.5 mm, green to purple or lead-colored; glumes lanceolate, tip acute, unequal, lower 0.9–1.5 mm, upper 1.5–2.5, > 0.6 × lemma; lemma 1.5–2.5 mm, glabrous, ovate to lanceolate, tip acute; anthers 0.2–0.7 mm. **FR**: 0.6–1 mm, ellipsoid, striate, light- to red-brown. 2*n*=36,38. Rocky to sandy washes, slopes, scrub, woodland; 100–2100 m. SNE, D; to NV, CO, OK, w TX, n Mex. May–Oct

S. indicus (L.) R. Br. SMUT GRASS Per. **ST**: tufted, erect, 3–6(12) dm. **LF**: base dull, green to brown, margin glabrous or ciliate, collar glabrous; ligule < 0.5 mm, fringe hairs to 1 mm; blade 6–30 cm, 1–5 mm wide. **INFL**: gen terminal, exserted or base sheathed, contracted, dense, branches and spikelets to near base; 1° branches appressed, lowest gen < 2 cm, > internodes, 2° appressed, bearing 3°, spikelets to near base, appearing dense. **SPIKELET**: 1.5–2.5 mm, green or lead-colored; glumes ovate, tip obtuse, often jagged, unequal, lower 0.5–1 mm, upper 0.8–1.6, < 0.6 × lemma; lemma 1.8–2.5 mm, glabrous, ovate, tip obtuse; anthers 0.5–1 mm. **FR**: 1–1.2 mm, 4-angled, striate, red-brown. 2*n*=36. Open, disturbed areas, roadsides; < 1200 m. ScV, n SnJV, CW (exc SCoRI), SCo, PR; to se US, Caribbean, Mex; pantrop. [*S. creber* DeNardi] Apr–Nov

S. vaginiflorus (A. Gray) Alph. Wood POVERTY DROPSEED Ann. **ST**: 1–several, tufted, erect or decumbent, 2–8 dm. **LF**: base dull, purple to straw-colored, margin glabrous or ciliate, collar glabrous or with hairs to 2 mm; ligule < 0.3 mm, fringe hairs to 1 mm; blade 2–15 cm, 0.5–2 mm wide. **INFL**: gen terminal and axillary, exserted to mostly sheathed, contracted, dense; 1° branches appressed, lowest gen < 3 cm, > internodes, 2° appressed, bearing 3°, spikelets to near base. **SPIKELET**: 3–6 mm, green or lead-colored; glumes lanceolate, tip acute, entire, unequal, lower 2.5–4.7 mm, upper 2.5–5, > 0.6 × lemma; lemma 3–5.5 mm, strigose, lanceolate, hairs < 0.5 mm, tip acute to beaked; anthers 1.2–3.2 mm. **FR**: 1.8–2.7 mm, obovoid, striate, light-brown. 2*n*=54. Open, gen sandy, disturbed areas; < 600 m. e KR (nw Shasta Co.), n SNH (Grass Valley, Truckee Canyon); native to se Can, c&e US, ne Mex; naturalized in ID, AZ. Possibly only a waif in CA. Apr–Nov

S. wrightii Scribn. BIG SACATON Per. **ST**: tufted, erect, 9–25 dm. **LF**: base, shiny, straw-colored, distal sheath margin glabrous to short-hairy, collar glabrous or with sparse long hairs to 6 mm; ligule < 0.5 mm, fringe hairs to 2 mm; blade 20–70 cm, 3–10 mm wide. **INFL**: gen terminal, exserted, lanceolate, open; 1° branches ascending, lowest gen > 6 cm, > internodes, 2° weakly spreading or appressed, bearing 3°, spikelets to near base, appearing dense. **SPIKELET**: 1.5–3 mm, green to purple; glumes narrowly lanceolate, tip acute to obtuse, unequal, lower 0.5–1 mm, upper 0.8–2, > 0.6 × lemma; lemma 1.2–2.5 mm, glabrous, lanceolate to ovate, tip acute; anthers 1.1–1.3 mm. **FR**: 1–1.4 mm, ellipsoid, striate, red-brown. 2*n*=36. Seasonally moist, alkaline areas; < 1000 m. SW, DSon; UT; AZ to OK, TX, n&c Mex. Seeded for erosion control or wetland revegetation. Apr–Nov

STENOTAPHRUM

James P. Smith, Jr. & Robert Webster

Ann, per. **ST**: prostrate to ascending, gen compressed; internode solid to spongy inside. **LF**: cauline; ligule mostly hairy, membrane < 0.5 mm; blade folded. **INFL**: spike-like, infl branches, pedicels reduced, ± embedded, with spikelet bases, in flattened, straight or wavy, thickened axis. **SPIKELET**: compressed, gen falling as 1 unit, sometimes with infl axis; glumes equal to unequal; florets 2, lower floret sterile or staminate, lemma ± = upper glume, upper floret bisexual, lemma thick, firm, smooth or minutely rough, margin flat or inrolled, tip blunt. 7 spp.: trop worldwide. (Greek: narrow trench, from spikelet scars on infl axis)

S. secundatum (Walter) Kuntze SAINT AUGUSTINE GRASS Per from long stolons. **ST**: vegetative gen prostrate; fl st decumbent, 1–4 dm. **LF**: sheath 1.2–10 cm, glabrous; ligule ± 0.5 mm; blade 5–15 cm, 5–15 mm wide, upper surface glabrous. **INFL**: 5–10 cm; axis straight or wavy, thickened, flat; each spikelet subtended by 1 bristle; spikelet stalk < 0.5 mm. **SPIKELET**: 4–5 mm, ± 1.5–2 mm wide, lanceolate to elliptic; lower glume 1–2 mm; upper glume ± = spikelet; lower floret sterile, lemma 7–9-veined, tip acute to acuminate, palea ± = lemma; upper floret slightly < lower floret. Fields, roadsides; < 150 m. CCo, SCo; s US, Afr, Pacific islands; probably native to S.Am. Jul–Sep

STIPA NEEDLE GRASS

J. Travis Columbus, James P. Smith, Jr. & Douglas H. Goldman

Gen per, tufted or occ rhizomatous. **ST**: gen erect, gen unbranched at distal nodes, persistent or not. **LF**: basal or cauline, sheaths open, auricles 0, ligules scarious to membranous, gen ciliate, blade flat to inrolled. **INFL**: gen panicle-like, gen narrow. **SPIKELET**: gen 1-fld; glumes gen > floret exc awn, membranous, flexible, veins 1–10, awns 0; floret 1(6), gen round (compressed) in ×-section; axis breaking between glumes and floret; callus blunt to sharp, glabrous to hairy; lemma membranous to leathery or hardened, glabrous to hairy, obscurely 3–7-veined, margins overlapping or not, tip entire to 2-lobed or forked, occ fused into a crown, 1-awned at or near tip (mucronate), awn (0)3–225(300) mm, unbranched, straight to bent or curled, twisted or not, persistent or not; palea < to > lemma, gen flat, glabrous to hairy, veins (0)2(3). **FR**: ovoid to fusiform, beak 0. ± 600 spp.: temp, trop, gen ± arid, ± worldwide; many spp. valuable cattle forage, cordage, some weedy. (Greek: stupe or stuppeion, for fiber or cordage, referring to plumose awns of Eurasian spp. or fibers from esp *Stipa tenacissima* L.) [Barkworth 2007 FNANM 24:109–186; Columbus & Smith 2010 Aliso 28:65–67] Floret, lemma lengths and shapes in key and text exclude awn. Many related and some poorly defined genera placed here in a broadly treated genus. Hybrids not rare. 4 waifs in CA, *S. caudata* Trin. [*Amelichloa caudata* (Trin.) Arriaga & Barkworth], *S. clandestina* Hack. [*A. clandestina* (Hack.) Arriaga & Barkworth], *S. plumosa* Trin. [*Jarava plumosa* (Spreng.) S.W.L. Jacobs & J. Everett], *S. tenuissima* Trin. [*Nassella tenuissima* (Trin.) Barkworth] ◆. *S. viridula* Trin. [*N. viridula* (Trin.) Barkworth], possibly in CA but records unclear.

1. Florets 2–6 per spikelet . . . . ***S. mauritanica***
1′ Florets 1 per spikelet
  2. Palea grooved lengthwise
    3. Blade 0.8–1.5 mm wide; awns 10–16 mm . . . . ***S. purpurata***

3′ Blade 0.2–0.4 mm wide; awns 15–25 mm . . . . *S. chaetophora*
2′ Palea flat
  4. Lf tips sharp, stiff; lemma awn glabrous to scabrous; fr ribbed. . . . . *S. brachychaeta*
  4′ Lf tips acute to acuminate, not both sharp and stiff; lemma awn glabrous to scabrous or hairy; fr not ribbed
    5. Lemma margins in age strongly overlapping entire length, lemma body gen bumpy throughout, tip
      with crown; palea 1/4–1/2 lemma, glabrous, veins 0
      6. Lemma crown gen sharply constricted proximally, distinct from rest of lemma, ± 1/5 lemma, lemma
        body in age distally ± glabrous . . . . *S. manicata*
      6′ Lemma crown gradually expanded proximally, weakly to ± distinct from lemma, < 1/6 lemma,
        lemma body in age distally hairy to ± glabrous, or if crown constricted proximally then lemma body hairy in age
        7. Lemma crown hairs 0.3–0.6 mm, awn 12–55 mm, ± 0.1 mm diam, callus 0.4–1.6 mm, lemma body
          gen hairy in age; st 0.8–1.2 mm diam. . . . . *S. lepida*
        7′ Lemma crown hairs 0.8–1.1 mm, awn 38–110 mm, 0.2–0.4 mm diam, callus 1.4–3.6 mm, lemma
          body hairy to mostly glabrous in distal 3/4 in age; st 1–3.1 mm diam
          8. Lemma body in age glabrous in distal 3/4 exc on veins . . . . *S. cernua*
          8′ Lemma body in age hairy throughout . . . . *S. pulchra*
    5′ Lemma margins in age gen not or only ± overlapping at least part of length, or strongly overlapping in
      some spp. with smooth lemmas; lemma bodies gen smooth proximally, tip gen 1–2-lobed; palea 1/3 to
      ± > lemma, 2-veined at least proximally, gen hairy, or both lemma and palea glabrous
      9. Floret callus 1.7–6 mm, sharp-pointed; awn 35–225 mm; per or ann
        10. Lemma awn densely long-hairy and feather-like proximally; per . . . . ²*S. speciosa*
        10′ Lemma awn proximally scabrous to densely hairy but neither long-hairy nor feather-like; per or ann
          11. Ann, florets 4–7 mm; callus 1.7–2.3 mm; awns proximally hairy, hairs gen visible without a hand
            lens. . . . . *S. capensis*
          11′ Per; florets 7–13 mm; callus 2–4 mm; awn proximally hairy but hairs gen not easily visible without
            a hand lens. . . . . *S. comata*
            12. Distal awn segment 40–120 mm, wavy to curled in age; proximal st nodes gen concealed by lf
              sheaths; panicles in age gen partially enclosed in most distal sheath . . . . var. *comata*
            12′ Distal awn segment 30–80 mm, straight; proximal st nodes gen exposed; panicles in age gen
              fully exserted . . . . var. *intermedia*
      9′ Floret callus 0.1–2 mm, blunt- to sharp-pointed, awn 3–80 mm; per
        13. Florets ≤ 2.5 mm; lemma glabrous, occ sparsely hairy in *Stipa divaricata*
          14. Lf blade 0.5–2.5 mm wide, margins gen inrolled; panicle 5–20 cm, proximal infl nodes
            1–3-branched. . . . .*S. divaricata*
          14′ Lf blade 2–10 mm wide, flat; panicle 10–40 cm, proximal infl nodes gen 3–7 branched
             . . . . *S. miliacea* var. *miliacea*
        13′ Florets ≥ 2.8 mm; lemma hairy throughout or at least in proximal 3/4
          15. Florets ± obovoid, length gen ≤ 3 × width; lemma densely long-hairy at least in youth, awn
            early-deciduous. . . . . *S. hymenoides*
          15′ Florets elongate-ellipsoid to fusiform, length ≥ 4 × width; lemma hairy to densely hairy (partly
            glabrous), awn persistent or not, or if length < 4 × width then awn persistent and floret not densely long-hairy
            16. Awns persistent, proximally densely long-hairy, at least some hairs 0.5–8 mm
              17. Distal lvs on st with ligules 3–8 mm; lemma tip 1-lobed, lobe ≤ 0.1 mm, thick, leathery . . . . . *S. thurberiana*
              17′ Distal lvs on st with ligules 0.3–3 mm; lemma tip gen 2-lobed or lobes occ 0, lobes ≤ 1 mm,
                thin, membranous
                18. Lemma awn bent 1 ×, proximally densely long-hairy, feather-like, hairs 3–8 mm. . . . . ²*S. speciosa*
                18′ Lemma awn bent 2 ×, proximally hairy, ± feather-like or not, hairs ≤ 2 mm
                  19. Awns proximally with hairs gradually, evenly decreasing in length distally, or hairs of equal
                    length; hairs at lemma tip < or gen ± = longest proximal awn hairs
                    20. Floret 8–9 mm; lemma 1.2–1.7 × palea; glumes 1.3–1.9 mm wide from midvein to margin
                     . . . . ²*S. latiglumis*
                    20′ Floret 5.5–7.5 mm; lemma 1.7–2.5 × palea; glumes < 1 mm wide from midvein to margin
                     . . . . ²*S. occidentalis*
                      21. Distal awn segment gen densely hairy; st 0.3–1 mm diam, internodes glabrous; glumes gen
                        ± purple. . . . . ²var. *occidentalis*
                      21′ Distal awn segment gen scabrous or glabrous, occ densely hairy at segment base; st 0.5–2
                        mm diam, internodes glabrous to densely hairy; glumes gen green
                        22. Proximal 2 awn segments scabrous or densely hairy, hairs of mixed lengths; lemma tip
                          hairs > proximal awn hairs. . . . . ²var. *californica*
                        22′ Proximal 2 awn segments densely hairy, hairs gradually and evenly shorter towards 1st
                          bend; lemma tip hairs ± = proximal awn hairs . . . . ²var. *pubescens*
                  19′ Awn proximally with mixed-length hairs, longer hairs mixed with shorter; hairs at lemma tip
                    longer than most proximal awn hairs
                    23. Florets 8–9 mm; glumes 1.3–1.9 mm wide from midvein to margin. . . . . ²*S. latiglumis*

23′ Florets 5–7.5 mm; glumes 0.6–1 mm wide from midvein to margin

    24. Callus 0.5–0.7 mm; lemma 1.3–2 × palea; palea tip hairs gen ± 1 mm . . . . **S. nevadensis**

    24′ Callus 0.8–1.2 mm; lemma 1.7–2.5 × palea; palea tip hairs gen < 1 mm . . . . ²**S. occidentalis**

        25. Distal awn segment gen densely hairy; st 0.3–1 mm diam, internodes glabrous; glumes gen

            ± purple . . . . ²var. **occidentalis**

        25′ Distal awn segment gen scabrous or glabrous, occ densely hairy at segment base; st 0.5–2

            mm diam, internodes glabrous to densely hairy; glumes gen green

            26. Proximal 2 awn segments scabrous or densely hairy, hairs of mixed lengths; lemma tip

                hairs > proximal awn hairs . . . . ²var. **californica**

            26′ Proximal 2 awn segments densely hairy, hairs gradually and evenly shorter towards 1st

                bend; lemma tip hairs ± = proximal awn hairs . . . . ²var. **pubescens**

16′ Awns persistent or not, proximally scabrous or with hairs < 0.5 mm

  27. Glumes ovate to narrowly ovate, 3–6 mm; lemma awn 3.9–14 mm

    28. Lf blade ± 0.3 mm wide, thread-like, wavy or curled; infl branches appressed to main axis or

        not; glumes 3–4.5 mm; lemma awn 10–14 mm . . . . **S. kingii**

    28′ Lf blade 0.6–1.4 mm wide, straight, erect, gen folded; infl branches ± appressed to main axis;

        glumes 3.5–6 mm; lemma awn 3.9–7 mm . . . . **S. exigua**

  27′ Glumes narrowly lanceolate to lanceolate, ≥ 6 mm; lemma awn (4)10–80 mm

    29. Lemma evenly hairy, hairs 1.2–6 mm, hairs at lemma middle gen ± = hairs at tip of lemma body

      30. Awns early-deciduous . . . . **S. webberi**

      30′ Awns persistent

        31. Sheaths not flat and ribbonlike in age; blade margins gen inrolled, 0.2–0.4 mm thick, 0.5–1

            mm wide when flat; awn bent 2 × . . . . **S. pinetorum**

        31′ Sheaths flat and ribbonlike in age; blade 0.5–1.5 mm thick when margins inrolled, ≤ 7 mm

            wide when flat; awn bent 1–2 ×

            32. Awn bent 2 ×; st 3–6 mm diam . . . . ²**S. coronata**

            32′ Awn bent 1 ×; st 0.8–2 mm diam . . . . ²**S. parishii** var. **parishii**

    29′ Lemma glabrous or hairs 0.2–1.5(2) mm at lemma middle, distally glabrous or hairy, hairs at

        lemma middle gen < hairs at tip of lemma body

      33. Lemma tip hairs 2–7 mm, gen ≥ 1 mm > hairs at lemma middle

        34. Awn bent 2 ×; sts 3–6 mm diam . . . . ²**S. coronata**

        34′ Awn bent 1 ×; sts 0.8–2 mm diam . . . . ²**S. parishii** var. **parishii**

      33′ Lemma tip hairs ≤ 2.2 mm or 0, gen < 1 mm > hairs at lemma middle

        35. Distal awn segment ± wavy . . . . **S. arida**

        35′ Distal awn segment straight or ± arched

          36. Lemma tip 2-lobed, lobes 1–3 mm; palea ≥ lemma, tip 2-lobed, veins extending into lobes . . . . **S. stillmanii**

          36′ Lemma tip not lobed or lobes ≤ 1.2 mm; palea ≤ lemma, tip unlobed, palea veins ending

            before or at palea tip

            37. Lemma tip lobes thick, stiff, ± 0.1 mm; florets ± compressed . . . . **S. lemmonii**

            38. Proximal lf sheaths, st glabrous to hairy, not tomentose, hairs ≤ 0.2 mm . . . . var. **lemmonii**

            38′ Proximal lf sheaths, st tomentose, hairs 0.4–0.6 mm . . . . var. **pubescens**

          37′ Lemma tip lobes membranous, 0.1–1.2 mm; florets not compressed

            39. Proximal internodes densely hairy 3–9 mm below nodes, elsewhere less densely hairy and

               hairs shorter . . . . **S. diegoensis**

            39′ Proximal internodes glabrous or ± hairy ≤ 5 mm below nodes, elsewhere gen glabrous

             40. Lemma (1.1)1.2–1.3 × palea; palea tip hairs extending beyond tip; lemma awn 12–25

                mm; lf blades 0.5–2 mm wide . . . . **S. lettermanii**

             40′ Lemma 1.5–3 × palea; palea tip hairs gen not extending beyond tip; lemma awn 19–31

                mm; lf blades (0.5)1.2–5 mm wide . . . . **S. nelsonii** var. **dorei**

S. arida M.E. Jones MORMON NEEDLE GRASS **ST**: 3.5–8.5 dm.
LF: proximal sheaths glabrous; blade 0.9–3 mm wide. **INFL**: 5–17
cm, dense. **SPIKELET**: glumes 8–15 mm, lanceolate; floret 4–6.5
mm; callus 0.2–1 mm, sharp; lemma 1.3–2 × palea, hairy throughout
or occ glabrous near tip, awn 40–80 mm, obscurely bent, persistent,
minutely scabrous, distal segment ± wavy. Outcrops, shrub land,
pinyon/juniper woodland; 1100–1850 m. DMoj; to CO, NM. [*Achnatherum a.* (M.E. Jones) Barkworth] May–Aug ★

S. brachychaeta Godr. PUNA NEEDLE GRASS **ST**: 4–9 dm. **LF**:
proximal sheaths mostly glabrous; blade 8–35 cm, 2–3 mm wide
when flat, margins inrolled, tip sharp, stiff. **INFL**: 10–25 cm, ±
dense to spreading. **SPIKELET**: glumes 6–8 mm, lance-linear; floret 4–5.5 mm, cylindric; callus 0.4–0.5 mm, blunt; lemma 1.1–1.3 ×
palea, hairy proximally, glabrous distally, awn 10–18 mm, gen bent
1 ×, persistent, glabrous to scabrous. Grassland, riparian woodland,
open conifer woodland, disturbed areas, roadsides; 60–1300 m. SnJV,
CCo, SCoR, SCo; native to S.Am. [*Achnatherum b.* (Godr.) Barkworth; *Amelichloa b.* (Godr.) Arriaga & Barkworth] May–Oct ◆

S. capensis Thunb. CAPE RICE GRASS Ann. **ST**: 1–10 dm. **LF**:
sheath glabrous to densely hairy; blade ≤ 3 mm wide. **INFL**: 3–15
cm, dense. **SPIKELET**: glumes 12–20 mm, narrowly lanceolate to
linear; floret 4–7 mm; callus 1.7–2.3 mm, sharp; lemma constricted
below tip, awn 50–100 mm, bent 2 ×, persistent, proximally dense-
hairy, distally scabrous to glabrous. 2*n*=36. Desert scrub, roadsides,
disturbed areas; 80–400 m. DSon; native to s Eur, c&w Asia, Afr.
Mar–Apr ❖

S. cernua Stebbins & Love NODDING NEEDLE GRASS **ST**: 3–10
dm, 1–3.1 mm diam. **LF**: sheath gen glabrous; blade 3–26 cm, 0.4–
1.2 mm wide, flat to inrolled. **INFL**: 15–80 cm, open. **SPIKELET**:

1-fld; glumes 12–22 mm, narrowly lanceolate; floret 4–9 mm; callus 1.4–3.6 mm, sharp; lemma hairy in proximal 1/4, distally hairy only on veins, margin strongly overlapping entire length, crown 0.2–0.5 mm, ± distinct from lemma, hairs 0.8–1.1 mm, awn 50–100 mm, 2 × (occ obscurely) bent, persistent, distally wavy; palea 1/4–1/2 lemma, glabrous, veins 0. 2*n*=70. Grassland, chaparral, juniper woodland; 5–1700 m. NCoR, CaRF, n&s SNF, Teh, GV, SnFrB, SCoR, SW; nw Baja CA. [*Nassella c.* (Stebbins & Love) Barkworth] Feb–Jul

S. chaetophora (Griseb.) Columbus & J.P. Sm. STIPOID RICE GRASS **ST:** 2–6 dm. **LF:** sheath glabrous distally minutely bristly; blade (5)14–30 cm, 0.2–0.4 mm wide. **INFL:** 4–15 cm, ± dense. **SPIKELET:** glumes 4–8.5 mm, ± ovate; floret 3–4(5) mm, obovoid; callus 0.5–0.6 mm, blunt; lemma dark brown to black, shiny, smooth to bumpy, glabrous, awn 15–25 mm, bent 2 ×; palea grooved lengthwise, late-deciduous. Meadows, open woodland, disturbed areas; 100–120 m. CCo (Marin Co.); native to S.Am. [*Piptochaetium stipoides* (Trin. & Rupr.) Arechav.] In CA occ grows with *S. purpurata*. May–Jun

S. comata Trin. & Rupr. NEEDLE-AND-THREAD **ST:** 1.2–11 dm. **LF:** proximal sheaths glabrous or hairy; blade 10–30 cm, 0.5–4 mm wide, margins gen inrolled. **INFL:** 10–32 cm, dense. **SPIKELET:** glumes 16–35 mm, lance-linear; floret 7–13 mm, narrow; callus 2–4 mm, sharp; lemma white-hairy gen throughout, awn clearly to obscurely bent 2 ×, persistent, gen scabrous. [*Hesperostipa c.* (Trin. & Rupr.) Barkworth]

var. **comata** (p. 1495) **ST:** proximal nodes gen concealed by sheaths. **INFL:** panicles gen only partially exserted. **SPIKELET:** awn 75–225 mm, distal segment 40–120 mm, wavy to curled in age. 2*n*=38,44,46. Grassland, sagebrush scrub; 1500–3100 m. SN, PR, GB, DMtns; to YT, ON, MI, TX. [*Hesperostipa c.* (Trin. & Rupr.) Barkworth subsp. *c.*] May–Jul

var. **intermedia** Scribn. & Tweedy (p. 1495) **ST:** proximal nodes gen exposed. **INFL:** panicles gen fully exserted. **SPIKELET:** awn 55–130 mm, distal segment 30–80 mm, straight. 2*n*=44,46. Pinyon/juniper woodland; 1350–3100 m. n SNH; to sw Can, NM. [*Hesperostipa c.* subsp. *i.* (Scribn. & Tweedy) Barkworth] May–Aug

S. coronata Thurb. (p. 1499) CRESTED NEEDLE GRASS **ST:** 5.5–21 dm, 3–6 mm diam. **LF:** proximal sheaths mostly glabrous; blade 2.5–7 mm wide, gen flat. **INFL:** 15–60 cm, dense. **SPIKE-LET:** glumes 11–21 mm, lanceolate; floret 6.5–10 mm; callus 0.5–2 mm, blunt to acute; lemma 1.1–1.7 × palea, densely hairy, awn 25–45 mm, bent 2 ×, persistent, scabrous. 2*n*=40. Gen rocky or gravelly slopes, in chaparral, coastal-sage scrub, pinyon/juniper woodland; < 1500(2300) m. SCoR, SCo, ChI, WTR, PR; Baja CA. [*Achnatherum c.* (Thurb.) Barkworth] Apr–Jul

S. diegoensis Swallen SAN DIEGO NEEDLE GRASS **ST:** 11–14 dm; proximal internodes densely hairy 3–9 mm below nodes. **LF:** proximal sheaths gen puberulent; blade 15–40 cm, 1–3.5 mm wide. **INFL:** 21–25 cm, ± open to dense. **SPIKELET:** glumes 8–11.5 mm, narrowly lanceolate; floret 5.5–7.5 mm; callus 0.25–1.2 mm, acute; lemma 1.3–2 × palea, hairy throughout, awn 20–50 mm, bent 2 ×, persistent, scabrous. Chaparral, coastal-sage scrub, in rocky soil near streams or the coast; < 2280 m. n ChI, s ChI (San Nicolas Island), PR; Baja CA. [*Achnatherum d.* (Swallen) Barkworth] Mar–Jun ★

S. divaricata Columbus & J.P. Sm. (p. 1499) SMALL-FLOWERED RICE GRASS **ST:** 2–8.5 dm. **LF:** sheaths glabrous; blade 5–16 cm, 0.5–2.5 mm wide, margins gen inrolled. **INFL:** 5–20 cm, open, proximal nodes 1–3-branched. **SPIKELET:** glumes 2.5–3.5 mm, ± lanceolate; floret 1.5–2.5 mm, compressed; callus 0.1–0.2 mm, blunt; lemma 2–2.5 mm, gen glabrous, occ sparse-hairy, awn 4–8 mm, ± straight, deciduous. 2*n*=22. Gravel benches, rocky slopes, creek banks; 800–3100 m. W&I, DMtns; to BC, MB, ND, TX. [*Oryzopsis micrantha* (Trin. & Rupr.) Thurb.; *Piptatherum m.* (Trin. & Rupr.) Barkworth] Jun–Sep ★

S. exigua (Thurb.) Columbus & J.P. Sm. LITTLE RICE GRASS **ST:** 1.2–4 dm. **LF:** sheaths ± glabrous, occ scabrous distally; blade 9–30 cm, 0.6–1.4 mm wide, gen folded. **INFL:** 3.5–9 cm, dense, branches ± appressed to main axis. **SPIKELET:** glumes 3.5–6 mm, ovate; floret 3–6 mm; callus 0.2–0.5 mm, blunt; lemma ± = glume

length, hairy throughout, awn 3.9–7 mm, bent 1 ×, persistent. 2*n*=22. Rocky slopes in sagebrush scrub; 1800–2400 m. CaRH, MP; to BC, AB, MT, CO. [*Oryzopsis e.* Thurb.; *Piptatherum e.* (Thurb.) Dorn] Possibly more widespread in ne CA. Jun ★

S. hymenoides Roem. & Schult. (p. 1499) SAND RICE GRASS **ST:** 2.5–7 dm. **LF:** sheaths glabrous to minutely scabrous or distally puberulent; blade ± = sts, 0.1–1 mm wide, margins gen inrolled. **INFL:** 9–20 cm, open. **SPIKELET:** glumes 5–9 mm, ± lance-ovate, sac-like; floret 3–4.5 mm, 1–2 mm wide; callus 0.4–1 mm, sharp; lemma ± 3 mm, densely long-hairy throughout, hardened, dark brown to black, hairs easily removed, awn 3–6 mm, not bent, early-deciduous, scabrous. 2*n*=46,48. Desert or sagebrush scrub, pinyon/juniper woodland, in dry soil, gen sandy; 60–3500 m. CaR, SN, SnJV, SW, GB, D; to YT, MB, MN, AR, TX, n Mex. [*Achnatherum h.* (Roem. & Schult.) Barkworth] Hybrids with *S. occidentalis* var. *o.* called *S.* ×*bloomeri* Bol. [*Achnatherum* ×*bloomeri* (Bol.) Barkworth; *Oryzopsis bloomeri* (Bol.) Ricker ex Piper]; Barkworth 2007 FNANM 24:142. Apr–Jul

S. kingii Bol. (p. 1499) KING'S RICE GRASS **ST:** 1.5–3.8 dm. **LF:** blade ± 0.3 mm wide, thread-like, wavy or curled. **INFL:** 6–10 cm, ± dense, branches appressed to main axis or not. **SPIKELET:** glumes 3–4.5 mm, narrowly ovate; floret 2.8–4.2 mm; callus 0.3–0.7 mm, blunt; lemma ± hairy, awn 10–14 mm, obscurely bent, persistent. 2*n*=22. Damp streambanks and meadows; 2000–3650 m. c&s SNH. [*Oryzopsis k.* (Bol.) Beal; *Ptilagrostis k.* (Bol.) Barkworth] The hybrid *Stipa* ×*bloomeri* may key to this sp. See note under *S. hymenoides.* Jul–Sep

S. latiglumis Swallen WIDE-GLUMED NEEDLE GRASS **ST:** 5–11 dm. **LF:** proximal sheaths gen hairy; blade 0.7–3 mm wide. **INFL:** 15–30 cm, dense. **SPIKELET:** glumes 12–15 mm, lanceolate; floret 8–9 mm; callus 0.7–1 mm, blunt to sharp; lemma 1.2–1.7 × palea, hairy throughout, awn 33–45 mm, bent 2 ×, persistent, proximally densely hairy, distally ± scabrous. 2*n*=70. Dry slopes in conifer forest; 1100–2200 m. c&s SN, TR. [*Achnatherum l.* (Swallen) Barkworth] Jun–Aug

S. lemmonii (Vasey) Scribn. LEMMON'S NEEDLE GRASS **ST:** 1.5–9 dm. **LF:** proximal sheaths glabrous to tomentose; blade 0.5–2.5 mm wide, folded or margins inrolled. **INFL:** 7–21 cm, dense. **SPIKELET:** glumes 7–11.5 mm, lanceolate; floret 5.5–7 mm, ± compressed; callus 0.4–1.2 mm, blunt; lemma 1–1.3 × palea, hairy throughout, tip lobes thick, stiff, ± 0.1 mm, awn 16–30 mm, bent (1)2 ×, persistent, scabrous. 2*n*=34.

var. **lemmonii** (p. 1499) Proximal pl hairs ≤ 0.2 mm. **ST:** proximally glabrous to hairy. **LF:** proximal sheaths glabrous to hairy. Pine, mixed-evergreen forest, sagebrush scrub, chaparral, grassland, serpentine or not; 50–2300 m. KR, NCoR, CaR, SN, SnFrB, TR, PR, GB; to BC, ID, UT. [*Achnatherum l.* (Vasey) Barkworth subsp. *l.*] May–Jul

var. **pubescens** Crampton Proximal pl hairs 0.4–0.6 mm. **ST:** proximally tomentose. **LF:** proximal sheaths tomentose. Chaparral, open pine forest, on serpentine; 1000–1300 m. NCoRH, NCoRI (Lake, Tehama cos.). [*Achnatherum l.* subsp. *p.* (Crampton) Barkworth] Jun–Jul ★

S. lepida Hitchc. (p. 1499) FOOTHILL NEEDLE GRASS **ST:** 3.5–10 dm, 0.8–1.2 mm diam; proximal internodes glabrous or ± hairy ≤ 5 mm below nodes. **LF:** sheaths glabrous to coarsely hairy or minutely scabrous; blade 12–23 cm, 1–3.5 mm wide, flat to margins inrolled. **INFL:** 9–55 cm, open. **SPIKELET:** glumes 5.5–15 mm, narrowly lanceolate; floret 4–7 mm; callus 0.4–1.6 mm, sharp; lemma hairy throughout when young, hairy on veins in age, margin strongly overlapping entire length, crown weakly distinct from lemma, hairs 0.3–0.6 mm, awn 12–55 mm, ± 0.1 mm diam, bent ± 2 ×, persistent; palea 1/4–1/2 lemma, glabrous, veins 0. 2*n*=34. Dry slopes, chaparral, grassland, savanna, coastal scrub; < 1400 m. NW, CW, SW; Baja CA. [*Nassella l.* (Hitchc.) Barkworth] Mar–Jun

S. lettermanii Vasey LETTERMAN'S NEEDLE GRASS **ST:** 1.5–9 dm. **LF:** proximal sheaths glabrous; blade 0.5–2 mm wide. **INFL:** 7–19 cm, dense. **SPIKELET:** glumes 6.5–9 mm, lanceolate; floret

4.5–6 mm; callus 0.4–1 mm, blunt; lemma (1.1)1.2–1.3 × palea, hairy throughout, awn 12–25 mm, bent 2 ×, persistent, scabrous. 2*n*=32. Uncommon. Meadows, dry slopes, sagebrush scrub, conifer forest, subalpine; 880–3200 m. KR, CaR, n&c SN, SnBr, PR, GB, DMtns; to OR, MT, NM. [*Achnatherum l.* (Vasey) Barkworth] May–Aug

S. manicata E. Desv. ANDEAN TUSSOCK GRASS **ST:** 4–15 dm. **LF:** sheaths glabrous; blade 12–25 cm, 1.5–2.5 mm wide, flat to margins inrolled. **INFL:** 10–20 cm, open, occ forming side-panicles from old infl sts. **SPIKELET:** glumes 10–15 mm, narrowly lanceolate; floret 6–8 mm; callus 1.5–2.5 mm, sharp; lemma margins in age strongly overlapping entire length, crown gen sharply constricted proximally, distinct from rest of lemma, ± 1/5 lemma, lemma body in age distally ± glabrous, awn 30–50 mm, bent 2 ×, persistent; palea 1/4–1/2 lemma, glabrous, veins 0. Moist meadows, disturbed areas; < 400 m. NCo, n SNF, CCo; MS, native to S.Am. [*Nassella m.* (E. Desv.) Barkworth; *Nassella formicara* (Delile) Barkworth, orth. var., misappl.; *N. formicarum*, misappl.; *S. f.* Delile, misappl.] Jun ❖

S. mauritanica (Poir.) Columbus & J.P. Sm. MAURITANIAN GRASS **ST:** 6–35 dm. **LF:** sheaths glabrous, ± grooved; blade ≤ 100 cm, 3–9 mm wide, margins serrate. **INFL:** ≤ 50 cm, open. **SPIKELET:** florets 2–6; glumes 1–12 mm, lanceolate; floret 10–12 mm; callus 0.2–0.5 mm, rounded; lemma ± longer than palea, hairy proximally, glabrous distally, mucronate or awn 0–2 mm. 2*n*=48. Mixed-evergreen forest; < 1000 m. NCoRI; native to s Eur, n Afr, w Asia. [*Ampelodesmos mauritanica* (Poir.) T. Durand & Schinz, orth. var.; *A. m.*] May–Nov

S. miliacea (L.) Hoover var. *miliacea* SMILO GRASS **ST:** 4–15 dm. **LF:** sheath glabrous; blade 5–30 cm, 2–10 mm wide, flat. **INFL:** 10–40 cm, open, proximal nodes gen 3–7-branched. **SPIKELET:** glumes 2.5–3.5 mm, lanceolate to narrow-ovate; floret 1.5–2 mm, compressed; callus 0.3 mm, blunt; lemma ± 2 mm, glabrous, awn 3–4 mm, weakly bent, persistent or not. 2*n*=24. Salt marshes, dry washes, streambanks, chaparral, open woodland, disturbed areas; < 1550 m. NCo, GV, CW, SCo; native to Eur, n Afr, w Asia. [*Piptatherum m.* (L.) Coss.] Other var. in Eurasia. Mar–Sep ❖

S. nelsonii Scribn. var. *dorei* (Barkworth & J.R. Maze) Dorn MOUNTAIN NEEDLE GRASS **ST:** 4–17.5 dm; proximal internodes glabrous or ± hairy ≤ 5 mm below nodes. **LF:** proximal sheaths glabrous to densely hairy; blade (0.5)1.2–5 mm wide. **INFL:** 9–36 cm, dense. **SPIKELET:** glumes 6–12.5 mm, lanceolate; floret 4.5–7 mm; callus 0.2–1 mm, blunt; lemma 1.5–3 × palea, hairy throughout, awn 19–31 mm, bent 2 ×, persistent, proximally minutely hairy to scabrous. For the sp. 2*n*=36,44. Openings, sagebrush scrub, meadows; 450–3500 m. CaR, SN, MP; to YT, SK, SD, TX. [*Achnatherum n.* (Scribn.) Barkworth subsp. *d.* (Barkworth & J.R. Maze) Barkworth] Other var. in Rocky Mtns. Jun–Sep

S. nevadensis B.L. Johnson NEVADA NEEDLE GRASS **ST:** 2–8.5 dm. **LF:** proximal sheaths glabrous to hairy, occ minutely scabrous; blade gen 10–25 cm, 1–3 mm wide, margins gen inrolled. **INFL:** 6–25 cm, dense. **SPIKELET:** glumes 7–14 mm, narrowly lanceolate; floret 5–6.5 mm; callus 0.5–0.7 mm, sharp; lemma 1.3–2 × palea, hairy throughout, awn 20–35 mm, bent 2 ×, persistent, proximally densely hairy, distally minutely scabrous to glabrous. 2*n*=68. Sagebrush scrub, open woodland; 1000–3450 m. CaRH, SNH, GB; to WA, CO, UT. [*Achnatherum n.* (B.L. Johnson) Barkworth] Jun–Aug

S. occidentalis S. Watson **ST:** 1.4–12(18) dm. **LF:** proximal sheaths glabrous to densely hairy; blade 0.3–3 mm wide when flat, margins gen inrolled. **INFL:** 5–30 cm, dense. **SPIKELET:** glumes 9–15 mm, narrowly lanceolate; floret 5.5–7.5 mm; callus 0.8–1.2 mm, sharp; lemma 1.7–2.5 × palea, hairy throughout, awn 15–55 mm, bent 2 ×, persistent, hairy to distally glabrous to scabrous.

var. *californica* (Merr. & Burtt Davy) C.L. Hitchc. (p. 1499) CALIFORNIA NEEDLE GRASS **ST:** 3–10(18) dm, 0.5–2 mm diam, internodes glabrous to densely hairy. **LF:** sheaths glabrous to densely hairy; blade 0.8–2 mm wide. **INFL:** 8–30 cm. **SPIKELET:** glumes gen green; lemma tip hairs > proximal awn hairs, awn 18–55 mm, proximal 2 segments scabrous or densely hairy, hairs of mixed lengths. 2*n*=36. Sagebrush scrub, conifer forest; 150–3450 m. KR, CaR, SN, SnBr, PR, GB; to WA, MT, WY, UT. [*Achnatherum o.*

(S. Watson) Barkworth subsp. *c.* (Merr. & Burtt Davy) Barkworth] Hybridizes with *S. occidentalis* var. *pubescens*. May–Aug

var. *occidentalis* (p. 1499) WESTERN NEEDLE GRASS **ST:** 1.4–5 dm, 0.3–1 mm diam, internodes glabrous. **LF:** proximal sheaths gen glabrous; blade 0.3–2 mm wide. **INFL:** 5–20 cm. **SPIKELET:** glumes gen ± purple; lemma tip hair length ± = proximal awn hairs, awn 15–42 mm, proximal 2 segments densely hairy, distal segment gen densely hairy, occ scabrous or glabrous. Conifer forest, alpine areas; 1200–3450 m. SN, PR, SNE; to WA, ID, UT. [*Achnatherum o.* subsp. *o.*] Hybrids with *S. hymenoides* called *S.* ×*bloomeri* [*Achnatherum* ×*bloomeri*]. Jun–Aug

var. *pubescens* (Vasey) J.R. Maze et al. COMMON WESTERN NEEDLE GRASS **ST:** 3–12 dm, 0.8–1.3(2) mm diam, proximal internodes short-hairy to hairy. **LF:** proximal sheaths gen hairy; blade 1–3 mm wide. **INFL:** 10–30 cm. **SPIKELET:** glumes gen green; lemma tip hairs ± = proximal awn hairs, awn 24–50 mm, proximal 2 segments densely hairy, distal segment scabrous to glabrous. 2*n*=36. Sagebrush scrub, conifer forest; 1200–3500 m. KR, CaR, SN, SnBr; to sw Can, MT, CO, AZ. [*Achnatherum o.* subsp. *p.* (Vasey) Barkworth] Hybridizes with *S. occidentalis* var. *californica*. May–Aug

S. parishii Vasey var. *parishii* (p. 1499) **ST:** 2–8 dm, 0.8–2 mm diam. **LF:** proximal sheaths mostly glabrous; blade 11–30 cm, 2.5–4.2 mm wide. **INFL:** 11–15 cm, dense. **SPIKELET:** glumes 8–15 mm, narrowly lanceolate; floret 5.5–6.5 mm; callus 0.2–0.8 mm, acute; lemma 1.2–2 × palea, densely hairy throughout, awn 15–35 mm, bent 1 ×, persistent, gen scabrous proximally. Dry rocky slopes, scrub, pinyon/juniper woodland; 700–3050 m. s SN, TR, PR, W&I, DMtns; to UT, AZ, Baja CA. [*Achnatherum p.* (Vasey) Barkworth var. *p.*; *A. p.* subsp. *p.*] Other var. in c NV, w UT. Apr–Aug

S. pinetorum M.E. Jones PINEWOODS NEEDLE GRASS **ST:** 1.4–5(8) dm. **LF:** proximal sheaths gen glabrous; blade 0.5–1 mm wide when flat, margins gen inrolled. **INFL:** 4.5–20 cm, dense. **SPIKELET:** glumes 7–11 mm, lanceolate; floret 3.5–5.5 mm; callus 0.4–0.6 mm, sharp; lemma 1–1.5 × palea, densely hairy throughout, awn 13–25 mm, bent 2 ×, persistent, scabrous proximally. 2*n*=32. Rocky soil, pinyon/juniper woodland, conifer forest; 2000–3800 m. SN, SnBr, SNE, DMtns; to OR, MT, CO. [*Achnatherum p.* (M.E. Jones) Barkworth] Jun–Aug

S. pulchra Hitchc. (p. 1499) PURPLE NEEDLE GRASS **ST:** 3.5–10 dm, 1–3.1 mm diam. **LF:** sheaths glabrous to hairy; blade 10–20 cm, 0.8–4.9 mm wide, flat or margins inrolled. **INFL:** 18–60 cm, open. **SPIKELET:** glumes 12–20 mm, narrowly lanceolate; floret 7.5–11.5 mm; callus 1.8–3.5 mm, sharp; lemma hairy throughout, margin strongly overlapping entire length, crown 0.6–1.1 mm, weakly distinct from lemma, hairs 0.8–1.1 mm, awn 38–100 mm, 0.2–0.4 mm diam, bent 2 ×, persistent; palea 1/4–1/2 × lemma, glabrous, veins 0. 2*n*=64. Oak woodland, chaparral, grassland; < 1700 m. NW, n&c SNF, ScV, CW, SCo, ChI, WTR, PR; Baja CA. [*Nassella p.* (Hitchc.) Barkworth] Mar–Jun

S. purpurata Columbus & J.P. Sm. BRISTLY NEEDLE GRASS **ST:** 2–4 dm. **LF:** sheaths glabrous; blade (3)5–12.5 cm, 0.8–1.5 mm wide. **INFL:** 3–15 cm, dense. **SPIKELET:** glumes 5–7 mm, ± ovate; floret 2.5–3 mm, obovoid; callus 0.2–0.5 mm, blunt; lemma dark brown in age, glabrous, awn 10–16 mm, bent 1–2 ×, deciduous; palea grooved lengthwise. Meadows, grassland, open woodland, disturbed areas; 100–200 m. NCo, CCo; native to S.Am. [*Piptochaetium setosum* (Trin.) Arechav.] In CA occ grows with *S. chaetophora*. Apr–Jun

S. speciosa Trin. & Rupr. (p. 1499) DESERT NEEDLE GRASS **ST:** 3–6 dm. **LF:** sheaths mostly glabrous; blade 10–30 cm, 0.5–2 mm wide when flat, margins gen inrolled. **INFL:** 10–15 cm, dense. **SPIKELET:** glumes 13–24 mm, lance-linear; floret (6)8–10 mm; callus 0.8–1.6(3) mm, sharp; lemma (1.2)1.5–2.5 × palea, densely hairy throughout, awn 35–45(80) mm, bent 1 ×, persistent, proximally feather-like, densely long-hairy, distally glabrous. 2*n*=66,68,±74. Rocky slopes, canyons, washes; < 2500 m. s SN, Teh, SnJV, SCoR, SW, GB, D; to OR, CO, NM, n Mex, S.Am. [*Achnatherum s.* (Trin. & Rupr.) Barkworth; *Jarava s.* (Trin. & Rupr.) Peñailillo] Apr–Jul

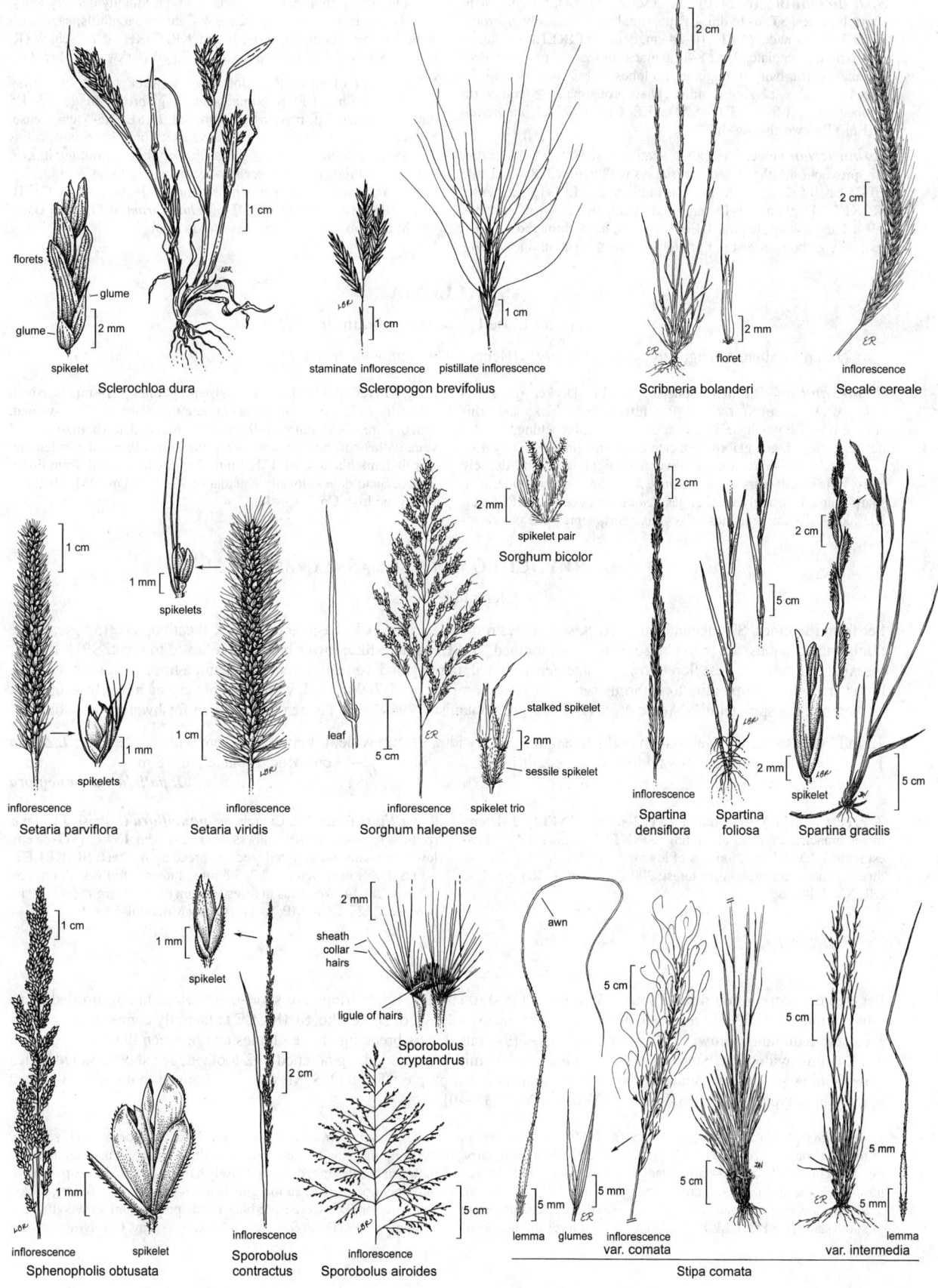

florets

glume

glume

spikelet

Sclerochloa dura

staminate inflorescence   pistillate inflorescence

Scleropogon brevifolius

floret

Scribneria bolanderi

inflorescence

Secale cereale

spikelets

spikelets

inflorescence

Setaria parviflora

inflorescence

Setaria viridis

spikelet pair

Sorghum bicolor

leaf

inflorescence   spikelet trio

Sorghum halepense

stalked spikelet

sessile spikelet

inflorescence

Spartina densiflora

Spartina foliosa

spikelet

Spartina gracilis

inflorescence   spikelet

Sphenopholis obtusata

spikelet

inflorescence

Sporobolus contractus

sheath collar hairs

ligule of hairs

Sporobolus cryptandrus

inflorescence

Sporobolus airoides

awn

lemma   glumes   inflorescence var. comata

var. intermedia

lemma

Stipa comata

S. stillmanii Bol. (p. 1499) STILLMAN'S NEEDLE GRASS With short rhizomes. **ST:** 6–15 dm. **LF:** proximal sheaths mostly glabrous; blade 3–7 mm wide. **INFL:** 10–24 cm, dense. **SPIKELET:** glumes 14–16 mm, lanceolate; floret 8–12 mm; callus 0.5–1.2 mm, rounded; lemma ≤ palea, hairy throughout, tip lobes 2, 1–3 mm, awn 18–25 mm, bent 1–2 ×, persistent, minutely scabrous; palea ≥ lemma, tip 2-lobed. Conifer forest; 350–1920 m. KR, CaR, n SN. [*Achnatherum s.* (Bol.) Barkworth] Jun–Jul

S. thurberiana Piper THURBER'S NEEDLE GRASS **ST:** 3–7.5 dm. **LF:** proximal sheaths glabrous; distal lvs with ligules 3–8 mm; blade 10–25 cm, 0.5–2 mm wide, margins inrolled. **INFL:** 7–15 cm, dense. **SPIKELET:** glumes 9–15 mm, lanceolate; floret 6–9 mm; callus 0.9–1.5 mm, sharp; lemma 1.1–1.3 × palea, hairy throughout or occ abaxially glabrous near tip, tip 1-lobed, lobe ≤ 0.1 mm, thick, leath-ery, awn 32–56 mm, bent 2 ×, persistent, proximally densely hairy, distally glabrous to scabrous. 2*n*=34. Canyons, foothills, sagebrush scrub, juniper woodland; 800–3100 m. KR, CaRH, SNH, Teh, WTR, GB; to WA, MT, UT. [*Achnatherum t.* (Piper) Barkworth] May–Jul

S. webberi (Thurb.) B.L. Johnson WEBBER'S NEEDLE GRASS **ST:** 1.2–3.5 dm. **LF:** proximal sheaths glabrous; blade 0.5–1.5 mm wide when flat, margins gen inrolled. **INFL:** 2.5–7 cm, dense. **SPIKELET:** glumes 6–10 mm, lanceolate; floret 4.5–6 mm; callus 0.3–0.8 mm, blunt; lemma ≤ palea, densely hairy throughout, awn 4–11 mm, straight to bent 1 ×, early-deciduous, scabrous. 2*n*=32. Dry, open flats, rocky slopes, gen with sagebrush; 1450–3500 m. CaRH, n&c SNH, MP, W&I; to OR, ID. [*Achnatherum w.* (Thurb.) Bark-worth; *Oryzopsis w.* (Thurb.) Vasey] May–Jul

SWALLENIA

Hester L. Bell & James P. Smith, Jr.

1 sp. (Jason Swallen, Am agrostologist, 1903–1991) [Henry 1979 Fremontia 7(2):3–6]

S. alexandrae (Swallen) Soderstr. & H.F. Decker (p. 1499) EUREKA VALLEY DUNE GRASS Per, tufted, from thick, scaly rhi-zomes; nodes ± woolly. **ST:** ascending to erect, branching, 1.5–3.5 dm, stiff, ridged, gen glabrous; sheath open, margin short-hairy. **LF:** gen cauline, awl-like, strongly divergent; ligule ± 1 mm, densely hairy, tufts of soft hairs at collar; blade 5–14 cm, 3–6 mm wide, stiff, sharp-pointed, strongly veined; gen deciduous at collar. **INFL:** pan-icle-like, 4–10 cm; branches spike-like, short, appressed; axis short-hairy. **SPIKELET:** < 1.5 cm, persistent; glumes 9–14 mm, ± equal, > florets, wide narrowing to acuminate tip, glabrous, 7–11-veined, awn 0; florets 3–7; lemma 7–9 mm, back rounded, lower margin, mid vein, callus soft-wavy-hairy, 5–7-veined, awn 0; palea ± = lemma, margin hairs like lemma. **FR:** 4 mm, 2 mm wide, falling from floret. 2*n*=20. Sand dunes (locally abundant); 900–1200 m. DMoj (Eureka Valley, ne Inyo Co.). Apr–Jun ★

TORREYOCHLOA FALSE MANNA GRASS

Jerrold I. Davis

Per from rhizomes. **ST:** decumbent to erect, sometimes rooting at nodes. **LF:** basal and cauline; sheath open ± to base; ligule thinly membranous, acute to obtuse, sometimes toothed. **INFL:** panicle-like; lower branches reflexed to erect. **SPIKELET:** bisexual; glumes < lowest floret, lower glume gen 1-veined, upper gen 3-veined; florets 2–8; lemma firm, back rounded, sca-brous, margin near tip entire to scabrous-serrate at 10×, prominently (5)7–9-veined, veins not converging at tip; awn 0; palea ± = lemma. ± 5 spp.: temp N.Am, e Asia. (J. Torrey, Am botanist, 1796–1873) Pls gen occurring in freshwater wet habitats.

1. Infl linear to narrowly elliptic in outline, length 5–19 × width, < 1 cm wide; lf blade 3.4–7 mm wide . . . . *T. erecta*
1′ Infl ovate to elliptic or obovate in outline, length 1–6(7.5) × width, (1)2–14 cm wide; lf blade 3.6–18 mm wide . . . . *T. pallida* var. *pauciflora*

T. erecta (Hitchc.) G.L. Church **ST:** 2–6 dm. **INFL:** 5.5–11 cm; lower branches erect to ascending. **SPIKELET:** florets 4–7; low-est lemma 2.3–3.1 mm; anthers of lowest floret 0.6–0.8 mm. 2*n*=14. Stream, lake margins, conifer forest; 2000–3500 m. CaRH, SNH; to OR, NV. Jul–Sep

T. pallida (Torr.) G.L. Church var. *pauciflora* (J. Presl) J.I. Davis (p. 1499) WEAK MANNA GRASS **ST:** 2–15 dm. **INFL:** (3)5–25 cm; lower branches weakly reflexed to spreading or erect. **SPIKELET:** florets 3–8; lowest lemma 2.2–3.3 mm; anthers of lowest floret 0.5–0.7 mm. 2*n*=14. Wet areas in forest, stream or lake margins; < 3500 m. NW, CaR, SN, CCo, MP; to AK, Rocky Mtns. Jul–Sep

TRIDENS

Per, gen cespitose, often rhizomatous. **ST:** gen erect, 5–180 cm. **LF:** ligule fringed or short-hairy; blade flat or inrolled, mar-gins not thickened. **INFL:** terminal, gen panicle-like, open to dense, or spike-like. **SPIKELET:** laterally compressed; glumes ± equal, membranous, lower 1(3)-veined, upper 1–3(9)-veined; axis breaking above glumes and between florets; florets gen 4–11; lemma wide, thin, back rounded, veins hairy below middle, 3-veined, tip notched or 2-toothed, gen short-pointed; palea gen < lemma, glabrous or minutely hairy below; anthers 3, red-purple. 15 spp.: US, Mex, Cuba. (Latin: 3 tooth, from 3-veined lemma tip in type sp.) [Valdés-Reyna 2003 FNANM 25:33–40]

T. muticus (Torr.) Nash var. *muticus* (p. 1499) SLIM TRIDENS Pl tufted, shortly rhizomatous. **ST:** gen 20–50 cm. **LF:** sheath hairy, esp near collar; ligule short-hairy; blade 3–25 cm, 1–4 mm wide, gen inrolled, ± fine scabrous, sometimes sparsely hairy. **INFL:** 4–20 cm, 3–8 mm wide, narrow; branches short, appressed; spikelets subsessile to short-stalked. **SPIKELET:** 8–13 mm, ± cylindric; glumes 3–10 mm, lower 1–3-veined, upper glume 1-veined; florets 5–11, strongly overlapping, pale to light purple; callus densely hairy; lemma 3–7 mm, gen purple-tinged, veins densely hairy below middle, tip entire to minutely notched; palea margins hairy, < lemma. 2*n*=40. Dry, rocky, gen limestone soils, creosote-bush scrub, pinyon/juniper woodland; < 2000 m. ne PR, D; sw&s-c US, n Mex. Apr–May, Oct–Nov

TRISETUM

Robert E. Preston & Dieter H. Wilken

Ann, per. **ST**: ascending to erect, gen clumped. **LF**: gen basal and cauline; ligule membranous, obtuse to truncate, toothed, tip ciliate or not; blade flat to inrolled. **INFL**: panicle- to spike-like, open to compact, cylindric to narrowly conic. **SPIKELET**: glumes ± unequal, gen ≤ lower floret, keeled, acute, lower 1-veined, upper 3-veined; axis stiff- to soft-hairy, gen prolonged behind upper floret, bristly or with vestigial floret; florets 2–3, bisexual, breaking above glumes and between florets (sometimes below glumes); callus short-hairy; lemma ± keeled, tip 2-bristled or not, awned on back near tip or not, awn straight or wavy to bent; palea = or < lemma; ovary glabrous or with short, stiff hairs at tip. 70–75 spp.: worldwide (exc Afr) temp, subarctic, alpine. (Latin: 3 bristles, from its 3-awned lemma) [Rumely 2007 FNANM 24:744–753] Some spp. intergrade; needs study in w N.Am. *T. flavescens* (L.) P. Beauv. extirpated, last collected in CA 1917.

1. Lemma awn 0(2) mm. . . . . ***T. wolfii***
1′ Lemma awn 4–12 mm
  2. Infl gen panicle-like, ± dense to open; awns gen > 7 mm; ovary hairy at tip
    3. Florets gen to base of panicle branches; lower infl branches ascending to erect; lower glume 3–5 mm . . . ***T. canescens***
    3′ Florets only at tips of panicle branches; lower infl branches gen spreading; lower glume < 3 mm . . . . ***T. cernuum***
  2′ Infl gen spike-like, dense; awns gen ≤ 7 mm; ovary glabrous at tip
    4. Panicle 8–23 cm, interrupted; lvs densely soft-hairy; upper glumes 5.5–7 mm . . . . ***T. projectum***
    4′ Panicle 2.5–7(10) mm, not interrupted; lvs glabrous or densely soft-hairy; upper glumes 4–5 mm . . . . ***T. spicatum***

T. canescens Buckley (p. 1499) TALL FALSE OAT Per. **ST**: 5–8 dm, clumped. **LF**: glabrous, scabrous, or sparsely hairy; ligule 1–4 mm; cauline blade 2–5(8) mm wide. **INFL**: gen panicle-like, 6–20 cm, narrow, compact to open; lower branches ascending to erect; central axis gen exposed, sparsely hairy. **SPIKELET**: gen on distal 2/3 of lower branches; glumes lanceolate, lower 3–5 mm, upper 6–7 mm, acute; lemma 4–6 mm, awn 6–11 mm; ovary with short, stiff hairs at tip. 2*n*=28,42. Open to shaded sites, meadows, chaparral, conifer forest; < 2830 m. NW, CaRH, SNH, CW, SnJt, MP; to AK, MT, UT. [*T. cernuum* var. *c.* (Buckley) Beal] Some coastal populations with spike-like panicles; apparently intergrades with *T. cernuum* outside of CA. May–Aug

T. cernuum Trin. NODDING FALSE OAT Per. **ST**: 4–10 dm, 1 to loosely clumped. **LF**: ligule 1–3 mm; cauline blade gen (4)6–12 mm wide. **INFL**: panicle-like, 10–30 cm, open, triangular in outline; lower branches ± wiry, gen spreading; axis mostly exposed, glabrous. **SPIKELET**: gen on distal 1/2 of lower branches; lower glume < 3 mm, lance-linear to ovate, upper 3–4 mm, narrowly elliptic, tip 1–2-toothed; axis conspicuous, segments between florets 1–2 mm; upper floret ± exserted; lemma 4–6 mm, awn 7–12 mm, bent ± midlength. 2*n*=42. Moist, shaded sites, redwood, conifer forest; < 1000 m. NCo, NCoRO; to AK, MT. May–Jul

T. projectum Louis-Marie INTERRUPTED FALSE OAT Per. **ST**: 3.5–9 dm, clumped. **LF**: densely soft-hairy; ligule 0.4–1.5 mm, tip ciliate; cauline blade 2–3 mm wide. **INFL**: spike-like, interrupted, 8–23 cm; branches 5–30 mm, appressed to ascending. **SPIKELET**: on branches from base to tip, green (incl awns); glumes unequal, lower 4–5.5 mm, upper 5.5–7 mm; lemma 5–6 mm, awn < 6.5 mm,

curved or wavy, gen not bent. Dry to moist sites, meadows, streambanks, open areas in conifer forest; 1220–2870 m. s CaRH, SNH; w NV. [*T. cernuum* var. *p.* (Louis-Marie) Beetle; *T. spicatum* var. *p.* (Louis-Marie) J.T. Howell] Intergrades somewhat with *T. canescens*, *T. spicatum*; needs critical study. Jul–Sep

T. spicatum (L.) K. Richt. (p. 1499) SPIKE FALSE OAT Per. **ST**: 0.5–4 dm, densely clumped. **LF**: mostly basal, tufted, glabrous or densely soft-hairy; ligule 1–3 mm; cauline blade gen 1–4 mm wide. **INFL**: spike-like, 2.5–7(10) cm, dense, cylindric to narrowly elliptic in outline; lower branches erect, ± appressed; axis hairy, hidden by spikelets. **SPIKELET**: on branches from base to tip, purple-tinged (incl awns); glumes lanceolate, acute, lower 2–4 mm, upper 4–5 mm; lemma 2–3 mm, awn 4–8 mm, often bent near base. 2*n*=14,28,42. Dry to moist sites, meadows, streambanks, rock outcrops, open areas in conifer forest, sagebrush scrub; 1370–3900 m. KR, CaRH, SNH, SnGb, SnBr, SnJt, Wrn, n SNE, W&I; worldwide (exc Afr), temp. [*T. montanum* Vasey] Highly variable, with much ecological variation; many infraspecific taxa described, not universally accepted. Jul–Aug

T. wolfii Vasey (p. 1499) WOLF'S FALSE OAT Per, sometimes from short rhizomes. **ST**: 3–8 dm, clumped. **LF**: ligule 1–5 mm; cauline blade 2–7 mm wide. **INFL**: gen panicle-like, 3–9 cm, compact, narrow; lower branches ascending to stiffly erect; axis exposed to ± hidden by spikelets. **SPIKELET**: gen on distal 4/5 of lower branches (sometimes at base); glumes lanceolate, acute, lower 4–6 mm, upper 5–7 mm; lemma 5–6 mm, short-awned or not, awn < 2 mm or 0. 2*n*=14. Open, gen dry sites, meadows, conifer forest; 1740–3300 m. CaRH, c&s SNH; to sw Can, w US. [*Graphephorum w.* (Vasey) J.M. Coult.] Jun–Sep

TRITICUM WHEAT

Ann. **ST**: 3–16 dm, gen erect to abruptly bent at base, nodes glabrous or pubescent. **LF**: sheath appendaged; ligule membranous, truncate; blade flat, glabrous or pubescent. **INFL**: spike-like, spikelets 2-ranked, sessile, 1 per node, lateral to infl axis; axis remaining intact or breaking apart at maturity. **SPIKELET**: 9–16 mm, laterally compressed, glumes 6–12 mm, ± equal, thick, stiff, keeled, 5–11-veined, loosely enclosing lower florets, mucronate, awned or awn 0; florets 3–9; lemmas like glumes, keeled or rounded, toothed, gen awned; palea 2-veined, 2-keeled. 10–20+ spp.: Medit, w Asia. (Latin: ancient common name for wheat) [Morrison 2007 FNANM 24:268–277]

T. aestivum L. Ann, cespitose. **ST**: stout, 6–15 dm. **LF**: appendages claw-like; blade 10–60 cm, 10–15 (20) mm wide. **INFL**: 6–18 cm, not breaking apart at maturity. **SPIKELET**: 10–15 mm, gen appressed to ascending; glumes 6–12 mm, prominently keeled toward base, 5–9-veined, toothed or awned to 4 cm; lemmas simi-

lar, 10–15 mm, gen awned to 12 cm, 5–9-veined. 2*n*=42. Escaped cereal crop along roadsides and in disturbed places; gen < 500 m. NCoR, ScV, CW, SW, D; N.Am; native to Near East. The world's most widely planted crop. Apr–Jul

TUCTORIA

John R. Reeder

Ann, ± hairy. **ST:** ascending to erect, unbranched, often fragile, easily breaking apart at nodes. **LF:** basal and cauline; collar 0; sheath and blade continuous; blade flat or becoming inrolled when dry. **INFL:** spike-like, partly enclosed by upper sheath or exserted with age; spikelets spirally arranged on axis. **SPIKELET:** laterally compressed; glumes entire or irregularly short-toothed; florets 5–40, tardily breaking above glumes and between florets; lemma entire or minutely toothed, gen with central sharp point, 11–17-veined; palea ± = lemma; anthers exserted, filaments slender, ribbon-like; stigmas 1/3–1/2 style length, sparsely short-hairy. **FR:** obovoid or oblong. 3 spp.: CA, s Baja CA. (Anagram of *Orcuttia*) [Reeder 2003 FNANM 25:293–294]

1. Infl exserted from upper lf sheath; lemma tip ± truncate; fr 2 mm, minutely wrinkled . . . . **T. greenei**
1' Infl partly enclosed by upper lf sheath; lemma tapered gradually, tip sharp-pointed; fr 3 mm, smooth . . . . . **T. mucronata**

T. greenei (Vasey) Reeder (p. 1505) GREENE'S TUCTORIA **ST:** erect, becoming decumbent, 5–15(30) cm; nodes often ± purple. **LF:** 2–3 cm, < 5 mm wide, curved outward. **INFL:** < 8 cm; upper spikelets crowded, lower ± separated. **SPIKELET:** glumes 3–5 mm, subequal, tip irregularly short-toothed, strongly veined; florets < 40; lemma gen 4–5(6) mm, 9–13-veined; anthers 3–3.5 mm, ± white. **FR:** slightly flattened laterally, oblong. 2*n*=24. Vernal pools; < 1050 m. GV, MP (Shasta Co.). [*Orcuttia g.* Vasey] May–Sep ★

T. mucronata (Crampton) Reeder (p. 1505) CRAMPTON'S TUCTORIA, SOLANO GRASS **ST:** ascending, becoming decumbent, < 12 cm. **LF:** 2–4 cm, curved outward, inrolled, tapered to fine point. **INFL:** 1.5–6 cm; spikelets crowded. **SPIKELET:** glumes 4–7 mm, subequal, short-pointed, sometimes with 1 or 2 short lateral teeth; florets 5–10; lemma 5–7 mm, 11–15-veined, central vein ending in a sharp point < 1 mm; anthers ± 3 mm, yellow, drying ± pink. **FR:** laterally flattened, widely oblong. 2*n*=40. Vernal pools, grassland; < 10 m. sw ScV (Solano, Yolo cos.). [*Orcuttia m.* Crampton] Apr–Aug ★

VAHLODEA MOUNTAIN HAIR GRASS

Robert E. Preston, Bruce G. Baldwin & Dieter H. Wilken

1 sp. (J.L.M. Vahl, Danish botanist, 1796–1854) [Chiapella 2007 Taxon 56:55–64]

V. atropurpurea (Wahlenb.) Hartm. (p. 1505) Per. **ST:** erect, 1 to densely clumped, 1.5–6 dm. **LF:** basal and cauline, basal tufted, gen glabrous; ligule 1–3 mm, obtuse to truncate, minutely ciliate at tip; blade 4–9 cm, 3–7 mm wide, flat. **INFL:** panicle-like, > 1 cm wide, open; lower branches spreading to drooping. **SPIKELET:** ± purple; glumes 4–6 mm, equal, > lower floret, elliptic, acute, glabrous or scabrous, lower 1-veined, upper 3-veined; axis inconspicuous beyond upper floret, ± 0.5 mm, flat, minutely ciliate (sometimes with vesti-gial floret at tip); florets gen 2, bisexual, breaking above glumes and between florets; callus soft-hairy, hairs ± 1/2 lemma length; lemma 2.5–4 mm, gen 2-toothed at tip, faintly 3-veined, awned near middle, awn 2–3 mm, straight to slightly bent; palea ± = lemma. 2*n*=14. Wet sites, meadows, streambanks, in conifer forest; 2000–2300 m. KR (Marble Mtns, Trinity Alps), CaRH (Mount Shasta); AK, e Can, ne US, n Eurasia. [*Deschampsia a.* (Wahlenb.) Scheele] Jul–Sep ★

VENTENATA

Ann, tufted. **ST:** ascending to erect, 10–75 cm. **LF:** sheath, blade glabrous to ± hairy; ligule membranous, gen acute. **INFL:** panicle-like, open or dense. **SPIKELET:** bisexual, compressed; glumes unequal, < lemma, lower glume 3–7-veined, upper 3–9-veined; axis breaking above lower floret and between distal florets; florets 2–10; lemma ± leathery, lemma of lower florets short-awned or not; uppermost awned from back, awn bent. 5 spp.: s Eur, n Afr, w Asia. (E.P. Ventenat, French clergyman, botanist, 1757–1808) [Crins 2007 FNANM 24:683–684]

V. dubia (Leers) Coss. NORTH AFRICA GRASS **ST:** nodes purple-black, puberulent below nodes. **LF:** ligule 1–8 mm; blade 2–7(12) cm, 1–3 mm wide, rolled under with age. **INFL:** gen 15–20 cm, open; branches spreading to drooping; spikelets near branch tips, stalked. **SPIKELET:** 10–15 mm; glumes lanceolate, acuminate, lower 4.5–6 mm, 3–7-veined, upper 6–8 mm, 3–9-veined; florets 2 or 3, lowest gen staminate, lemma awn straight when present; upper 1 or 2 florets bisexual, lemma awn 1–1.5 cm. 2*n*=14. Dry, open disturbed sites; 500–1500 m. KR, NCoRH (Lake Co.), CaRH, n SNH (Nevada Co.), MP; to Can, ne US; native to c&s Eur. Mature specimens gen retain 1 floret after upper florets have broken away, which can lead to mis-interpretation of the spikelet. Jun–Sep

PONTEDERIACEAE PICKEREL-WEED FAMILY

Charles N. Horn & Elizabeth McClintock

Ann, per, submersed, emergent, floating, or on wet ground. **LF:** simple, alternate or whorled, ± in basal rosette or not; blade linear (esp underwater) to round, sometimes sagittate or cordate, base gen sheathing, veins parallel; petiole inflated or not. **INFL:** raceme, spike, panicle, or 1-fld, terminal but often appearing ± axillary, subtended by sheathing bract. **FL:** bisexual, radial or ± bilateral; perianth lobes 6 (in 2 series) [3,4], petal-like; stamens 3, 6 (in 2 series) [1,4], from various levels on peri-anth tube, equal or not, sometimes 3 sterile, modified; ovary superior, 1 or 3-chambered. **FR:** many-seeded, loculicidal capsule [1-seeded utricle]. **SEED:** < 1 cm, longitudinally ribbed [or not]. ± 6 genera, 30 spp.: most pantrop, some temp. [Horn 2003 FNANM 26:22–45] Some cult as orns, some weeds, esp in rice fields. *Pontederia cordata* L. ◆ establishes from garden waste. Scientific Editor: Thomas J. Rosatti.

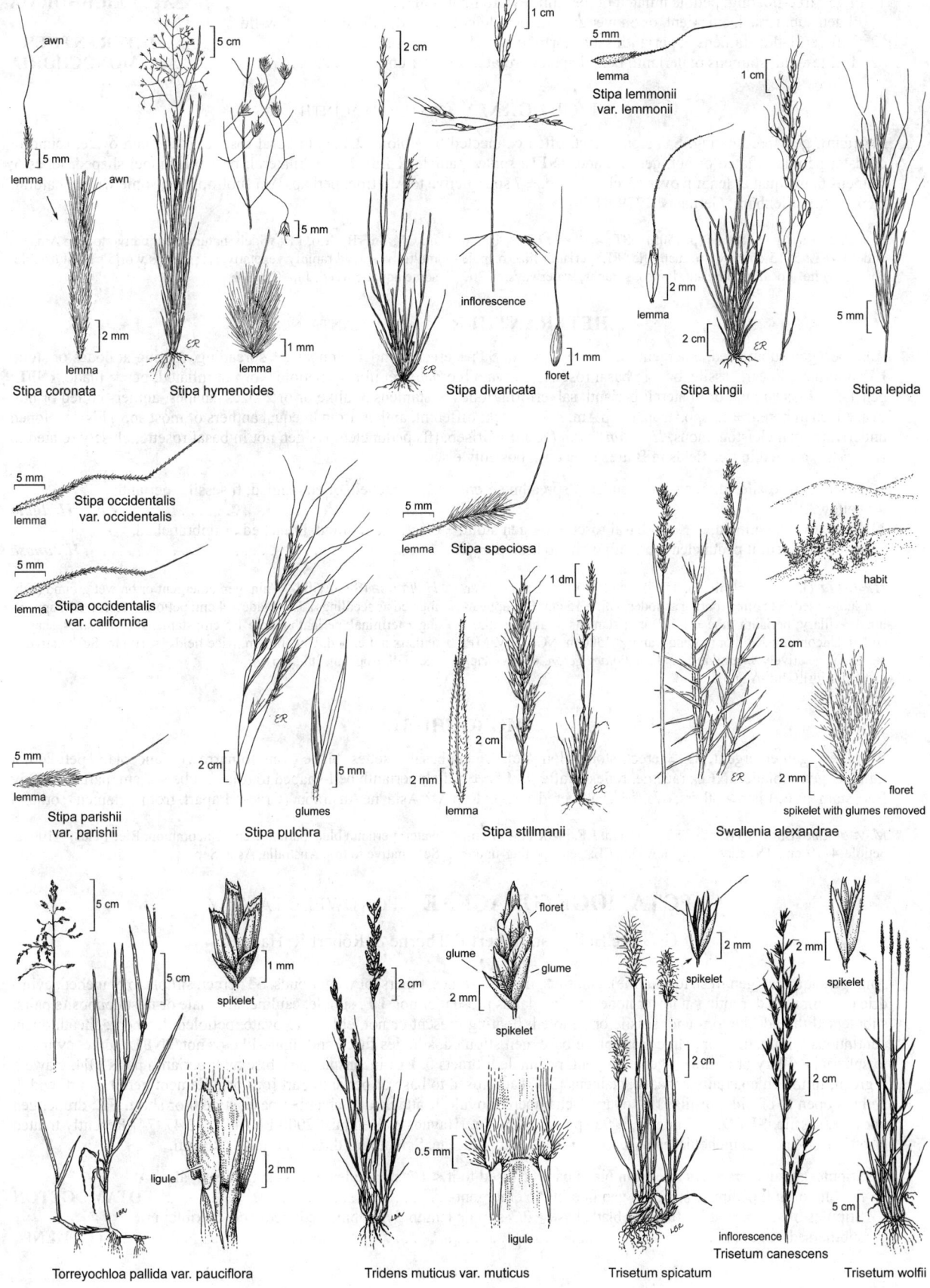

awn

5 cm

lemma
5 mm

awn

lemma
2 mm

Stipa coronata

ER

5 mm

lemma
1 mm

Stipa hymenoides

2 cm

inflorescence

floret
1 mm

ER

Stipa divaricata

1 cm

lemma
5 mm

**Stipa lemmonii
var. lemmonii**

1 cm

lemma
2 mm

2 cm

ER

Stipa kingii

5 mm

Stipa lepida

5 mm
lemma
**Stipa occidentalis
var. occidentalis**

5 mm
lemma
**Stipa occidentalis
var. californica**

5 mm
lemma
**Stipa parishii
var. parishii**

ER

2 cm

5 mm
glumes

Stipa pulchra

5 mm
lemma
Stipa speciosa

1 dm

2 cm

2 mm
lemma

2 mm

ER

Stipa stillmanii

2 cm

habit

ER

2 mm

floret

spikelet with glumes removed

Swallenia alexandrae

5 cm

5 cm

spikelet
1 mm

ligule
2 mm

Torreyochloa pallida var. pauciflora

2 cm

floret

glume

glume

2 mm
spikelet

spikelet

0.5 mm

ligule

Tridens muticus var. muticus

2 mm
spikelet

2 cm

2 cm

Trisetum spicatum

2 mm
spikelet

2 cm

inflorescence
Trisetum canescens

5 cm

Trisetum wolfii

1. Pl gen free-floating; petiole inflated or not; infl few- to many-fld. . . . . **EICHHORNIA**
1′ Pl gen submersed, emergent, or on wet ground; petiole not inflated; infl gen 1- or few-fld
  2. Infl gen 1-fld; stamens 3; perianth salverform, tube > 1 cm . . . . **HETERANTHERA**
  2′ Infl few-fld; stamens 6; perianth bell-shaped to rotate, tube < 1 cm. . . . . **MONOCHORIA**

EICHHORNIA   WATER HYACINTH

Per [ann], gen free-floating. **ST**: stout, erect, often connected by stolons. **LF**: ± in basal rosette; blade gen ovate, round or cordate; petiole inflated or not, gen > blade. **INFL**: spike [panicle, 1-fld]. **FL**: ± bilateral; perianth funnel-shaped, lobes 6; stamens 6, unequal in length; ovary 3-chambered. ± 7 spp.: native to Am trop, perhaps Am subtrop, Afr; some widely naturalized. (J.A.F. Eichhorn, Germany, 1779–1856)

E. crassipes (Mart.) Solms (p. 1505)   **ST**: > 30 cm. **LF**: < 10 cm wide. **INFL**: 5–15 cm, few- to many-fld. **FL**: perianth lilac or pale blue to white. Locally abundant. Ponds, sloughs, waterways; < 200 m. GV, SnFrB, SCo, PR; widely naturalized, native to trop Am. Pls multiply, spread rapidly vegetatively; perhaps world's most troublesome aquatic weed. Jun–Oct ◆

HETERANTHERA   MUD PLANTAIN

Ann, per, gen submersed, emergent, or on wet ground. **ST**: ± erect, slender, elongate, ± spreading, rooting at nodes or stout. **LF**: alternate, linear, sessile, or ± in basal rosettes, ± ovate [cordate, reniform]; petiole 0 or not inflated, gen > blade. **INFL**: gen 1-fld. **FL**: ± radial or bilateral; perianth salverform, lobes 6; stamens 3, alike or of 2 sizes, shapes, anthers coiled or not; ovary 1-chambered. ± 12 spp.: trop, temp Am, Afr. (Greek: different, anther, from unequal anthers of most spp.) Fls developed underwater gen cleistogamous; *H. rotundifolia* (Kunth) Griseb. (fls peduncled; lvs gen not in basal rosette; closely related to *H. limosa*) a weed in rice fields in Butte, Yuba cos., possibly elsewhere.

1. Lvs linear, sessile, alternate; sts slender, ± spreading, gen much-branched, often tangled; fl sessile; perianth
   yellow. . . . . *H. dubia*
1′ Lvs ± ovate, petioled, gen ± in basal rosette; sts gen stout, ± erect, often tufted, branched or unbranched,
   rarely tangled; fl peduncled; perianth white to blue-purple. . . . . *H. limosa*

H. dubia (Jacq.) MacMill. (p. 1505)   WATER STAR-GRASS   Per, gen submersed. **ST**: often rooting at nodes. **LF**: > 15 cm. **FL**: appearing ± axillary; perianth tube ± 1.5(7) cm; stamens 3, alike, anthers coiled. Uncommon. Still or moving water; < 1500 m. NCo, ScV, MP; N.Am. Vegetatively ± similar to spp. of *Potamogeton*, exc lvs lack distinct midrib. Jul–Aug

H. limosa (Sw.) Willd.   Ann, gen emergent or on wet ground (submersed as seedlings). **LF**: blade > 4 cm; petiole > 15 cm. **FL**: appearing ± terminal; perianth tube > 1.5 cm; stamens 3, of 2 sizes, shapes, anthers not coiled. Uncommon. Rice fields; < 100 m. ScV; native to c&e US, trop Am. Jul–Sep

MONOCHORIA

Ann [per], gen emergent. **ST**: ± erect, stout [elongate]. **LF**: in basal rosettes; blade ovate to narrowly lanceolate; petiole not inflated, gen > blade. **INFL**: raceme, reflexed after fl. **FL**: ± radial; perianth bell-shaped to rotate, tube < 1 cm, parts 6, nearly free; stamens 6, 1 gen > others; ovary 3-chambered. 6 spp.: trop Afr, Asia, ne Australia. (Greek: 1 apart, from 1 stamen > others)

M. vaginalis (Burm. f.) Kunth   **ST**: > 12 cm. **LF**: blade 2.5–11 cm; petiole 4–70 cm. **INFL**: > 5 cm, few-fld. **FL**: gen opening under water; perianth blue, parts < 2 cm. Uncommon. Rice fields; > 100 m. ScV; native to trop Australia, Asia. Sep

POTAMOGETONACEAE   PONDWEED FAMILY

C. Barre Hellquist, Robert F. Thorne & Robert R. Haynes

Ann, per, aquatic, (gen fresh to alkaline), glabrous, from rhizomes, tubers, or winter buds. **ST**: erect, simple to branched, cylindric to compressed, rooting at lower nodes; nodal glands present or not. **LF**: simple, cauline, alternate or in subopposite pairs; submersed thread-like to round, sessile or petioled; floating present or not, elliptic to ovate, petioled, leathery; sheath open, continuous with petiole or ± free from blade base, gen stipuled, stipules fused and ligule-like or not. **INFL**: spike, cylindric to spheric, axillary or terminal, gen emergent, peduncled; bracts 0. **FL**: inconspicuous, bisexual; perianth parts [0]4, clawed, ± green, limb gen adaxially concave; stamens [2]4, each fused to base of perianth part [or not], filament gen 0 [short, wide], anthers open to outside; pistils [1]4, ovary 1-chambered, ovule 1, attached to chamber base, style 0 or short. **FR**: drupe, gen obovate, sessile. **SEED**: 1. 3 genera, ± 95 spp.: worldwide. [Haynes & Hellquist 2000 FNANM 22:47–74] Recently treated elsewhere to incl Zannichelliaceae. *Ruppia* moved to Ruppiaceae. Scientific Editor: Thomas J. Rosatti.

1. Stipules of submersed lvs free from blade base or fused to it < 1/2 stipule length; lvs submersed, also
   floating or not; peduncle stiff; infl gen floating to emergent . . . . **POTAMOGETON**
1′ Stipules of submersed lvs fused to blade base ≥ 2/3 stipule length; lvs submersed; peduncle flexible; infl
   submersed . . . . **STUCKENIA**

POTAMOGETON PONDWEED

Ann, per, rhizomed or not, tubers 0, winter buds gen 0. **ST**: simple or branched, cylindric to compressed; nodal glands present or 0. **LF**: submersed, also floating or not, gen flat, gen green, entire to finely serrate; submersed sessile or not, linear to round, tip rounded to acuminate, veins 1–35; floating 0 or gen petioled, elliptic to ovate, leathery; stipules of submersed lvs free from blade base or fused to it < 1/2 stipule length. **INFL**: gen floating to emergent; peduncle stiff. **FR**: gen beaked, abaxially with 0 or 1 central keel, 0 or 2 lateral, sometimes lower keels, sides concave to convex. ± 88 spp.: gen temp n hemisphere. (Greek: river neighbor, from aquatic habitat) Other taxa in TJM (1993) moved to *Stuckenia*.

1. Lvs submersed and floating
  2. Submersed lvs linear, < 1 cm wide
    3. Stipules fused to blade base; infls of 2 kinds, lower submersed, spheric, upper emergent, cylindric; fr compressed . . . . [2]*P. diversifolius*
    3′ Stipules free from blade base; infls all emergent, ± cylindric; fr ± compressed or not
      4. Submersed lvs with a net-like band along midrib, ribbon-like, 2–10 mm wide . . . . [2]*P. epihydrus*
      4′ Submersed lvs without a net-like band along midrib, not ribbon-like, either cylindric, often petiole-like, < 2 mm wide, or linear to lance-elliptic, 10–40 mm wide . . . . *P. natans*
  2′ Submersed lvs linear or wider, gen 1–7 cm wide
    5. Submersed lvs curved backward, gen folded along midvein, gen lanceolate to ovate; floating lvs 27–50-veined . . . . [2]*P. amplifolius*
    5′ Submersed lvs not curved backward, not folded along midvein, lanceolate to elliptic to oblanceolate; floating lvs < 30-veined
      6. Submersed lvs petioled
        7. Submersed lvs microscopically serrate near tip, petioles 0.5–4 cm . . . . [4]*P. illinoensis*
        7′ Submersed lvs entire, petioles 2–13 cm. . . . . [2]*P. nodosus*
      6′ Submersed lvs sessile (petioles rarely ≤ 3 cm in *P. gramineus*)
        8. Submersed lvs 3–9-veined; fr 1.9–2.3 mm . . . . [2]*P. gramineus*
        8′ Submersed lvs 7–19-veined; fr 2.5–3.6 mm
          9. Floating lvs 7–15-veined; submersed lvs 7–9-veined, entire. . . . . [2]*P. alpinus*
          9′ Floating lvs 13–29-veined; submersed lvs 7–19-veined, microscopically serrate near tip . . . . [4]*P. illinoensis*
1′ Lvs submersed
  10. Lvs linear or ovate to lanceolate to elliptic to oblanceolate, sessile or petioled
    11. Lf blade serrate; fr beak 2–3 mm. . . . . *P. crispus*
    11′ Lf blade entire or microscopically serrate near tip; fr beak < 2 mm
      12. Lf base not clasping
        13. Petioles 0.5–26 cm
          14. Lf microscopically serrate near tip, tip acute-mucronate . . . . [4]*P. illinoensis*
          14′ Lf entire, tip acute
            15. Lvs 19–50-veined, curved backward, gen folded along midvein. [2]*P. amplifolius*
            15′ Lvs 7–15-veined, not curved backward, not folded along midvein. [2]*P. nodosus*
        13′ Petioles 0
          16. Lvs entire, 7–9-veined; fr pedicelled . . . . [2]*P. alpinus*
          16′ Lvs minutely to microscopically serrate near tip, 3–19-veined; fr sessile
            17. Lf blade 3–9-veined, 2–10 cm, (3)10–27 mm wide; fr 1.9–2.3 mm, beak erect, 0.3–0.5 mm. . . . . . [2]*P. gramineus*
            17′ Lf blade 7–19-veined, 6–20 cm, 15–50 mm wide; fr 2.5–3.6 mm, beak erect to ± recurved, 0.5–0.8 mm. . . . . [4]*P. illinoensis*
      12′ Lf base clasping
        18. Lf lance-linear, 5–20 cm, tip hood-like, splitting when dry; stipules 2–10 cm, persistent; sts often zigzag; fr 4–5 mm, keels 1(3). . . . . *P. praelongus*
        18′ Lf linear- to lance-ovate, 1–12 cm, tip flat, not hood-like, not splitting when dry; stipules < 2 cm, persistent as fibers; sts gen straight; fr 2–4 mm, keels 0(1) . . . . *P. richardsonii*
  10′ Lvs linear (linear to lanceolate in *P. robbinsii*), sessile
    19. Stipules fused to blade base for ± 1/4 to 1/2 stipule length, or 1–15 mm, sheath-like basally, liguled.
      20. Lvs thin, not conspicuously 2-ranked, vein 1, not fine . . . . [2]*P. diversifolius*
      20′ Lvs ± stiff, 2-ranked, 3–4(8) mm wide, veins 20–60, fine. . . . . *P. robbinsii*
    19′ Stipules free from or fused to blade base for ≤ 1/5 stipule length, or 1–2 mm, not sheath-like basally, not liguled
      21. Lf veins 15–35; sts compressed, winged . . . . *P. zosteriformis*
      21′ Lf veins 1–13; sts ± compressed or cylindric
        22. Rhizomes present, winter buds 0; fr 2.5–4.5 mm, 2–3.6 mm wide; sts compressed . . . . [2]*P. epihydrus*
        22′ Rhizomes 0, winter buds present; fr 1.4–2.7 mm, 1.1–2.2 mm wide; sts rounded to ± compressed
          23. Fr with wing-like keel; peduncles 0.3–1.1(3.7) cm; nodal glands 0 (faint) . . . . *P. foliosus*
            24. Stipule veins in age persisting as fibers; infl interrupted; fr pale green, 1.4–1.7 mm, 1.1–1.2 mm wide, keel < 0.2 mm high, beak ≤ 0.2 mm. . . . . subsp. *fibrillosus*
            24′ Stipule veins in age decaying, occ persistent as delicate fibers; infl gen not interrupted; fr olive to green-brown, 1.5–2.7 mm, 1.2–2.2 mm wide, keel 0.2–0.4 mm high, beak 0.2–0.6 mm . . . . . subsp. *foliosus*

23′ Fr without keel; peduncles 0.5–8 cm; nodal glands present
25. Fr widest at or below middle, sides convex (concave); peduncles per pl > 3; infl continuous;
stipules free . . . . *P. berchtoldii*
25′ Fr widest above middle, sides concave; peduncles per pl 1–3; infl interrupted; stipules fused below
middle . . . . *P. pusillus*

P. alpinus Balb. Per, slender, rhizomed. **ST:** < 100 cm, branches 0–few, cylindric. **LF:** ± red; submersed sessile, 2–8 cm, 1–2 cm wide, lanceolate to elliptic-oblong, tapered to both ends, entire, tip blunt, veins 7–9; floating 0 or short-petioled, lanceolate, base tapered, veins 7–15; stipules < 3 cm, free, membranous, early-deciduous. **INFL:** gen < 3 cm. **FR:** ± 2.5 mm, keels 1 or 3. *n*=26. Uncommon. Ponds, lakes, marshes; 1300–2350 m. KR, SNH, CCo, MP; to AK, Greenland, CO; e Asia. [*P. a.* subsp. *tenuifolius* (Raf.) Hultén] Hybridizes with *P. gramineus, P. nodosus.* Jul–Sep

P. amplifolius Tuck. (p. 1505) BROAD-LEAVED PONDWEED Per, rhizome stout. **ST:** < 90 cm, branches 0 or above, cylindric. **LF:** often red-brown; submersed 8–20 cm, 2–7 cm wide, gen lanceolate to ovate, gen folded along midvein, curved backward, acute at both ends, veins 19–50; floating 5–10 cm, 2.5–5 cm wide, elliptic to ovate, tapered or rounded at both ends, veins 27–50; stipules gen 3–10 cm, free, fibrous in age, stringy. **INFL:** gen < 5 cm. **FR:** 3–5 mm, keels 1 or 3, lateral 0 or lower, sides flat. *n*=26. Uncommon. Deep, clear-water lakes; 600–1850 m. NCoR, CaRH, SNH, Wrn; to BC, e N.Am. Hybridizes with *P. illinoensis, P. praelongus, P. richardsonii.* Jul–Aug

P. berchtoldii Fieber Ann, rhizomes 0. **ST:** < 80 cm, winter buds often terminal, branches 0–many; nodal glands present. **LF:** submersed, sessile, < 6 cm, gen < 1.5(2.5) mm wide, tip acute to obtuse, veins 1–3(5); stipules 3–9 mm, free, flat or margins inrolled. **INFL:** short-cylindric to spheric, whorls 1–2; peduncle 0.5–4 cm. **FR:** keels 0, sides convex (concave), tip asymmetric, beak to 1 side. *n*=13. Uncommon. Shallow water, mostly cold, acidic lakes, ponds; < 2100 m. NCo, KR, n SNF, SNH, SnJV, SCoRO, SnBr, GB; circumboreal. [*P. pusillus* var. *tenuissimus* Mert. & W.D.J. Koch; *P. p.* subsp. *t.* (Mert. & W.D.J. Koch) R.R. Haynes & Hellq.] Jul–Aug

P. crispus L. (p. 1505) CRISP-LEAVED PONDWEED Ann, from hard winter buds. **ST:** < 90 cm, branches above, ± compressed. **LF:** submersed, sessile, 4–8 cm, 0.5–0.8 cm wide, oblong, thick, ± wavy, serrate, tip rounded, veins 3–5; stipules gen < 1 cm, ± free, in age fibrous. **INFL:** < 2 cm. **FR:** 4–6 mm, keels 0, sides ± concave to convex, beak 2–3 mm. 2*n*=52. Uncommon. Shallow water, ponds, streams; < 2100 m. KR, NCoR, CaR, s SN, n SNH, GV, CCo, SnFrB, SCo, ChI, SnGb, SnBr, PR, D; ± worldwide; native to Eurasia, Australia. Jul–Sep ❖

P. diversifolius Raf. (p. 1505) DIVERSE-LEAVED PONDWEED Per, delicate. **ST:** < 50 cm, branches many, above, cylindric. **LF:** submersed sessile, ± 6 cm, 1–5 mm wide, linear, thin, tip acute, vein 1; stipules sheath-like basally, liguled, < 1 cm, fused to blade base for ± 1/4–1/2 stipule length, or 1–15 mm; floating petiole < 2 × blade, blade 0.5–3 cm, 0.3–0.8 mm wide, elliptic, tapered at base, tapered or rounded at tip. **INFL:** < 1.5 cm, of 2 kinds; lower submersed, spheric, subsessile, upper emergent, cylindric, long-peduncled. **FR:** 1.1–1.8 mm, ± spheric, compressed; keels 3, middle highest, sides flat to concave. Uncommon. Shallow water, ponds, lakes; < 2500 m. NCoRI, CaRF, n&c SNF, n SNH, GV, SCo, SnJt, MP; to BC, e US, AZ, Mex. Apr–Sep

P. epihydrus Raf. (p. 1505) NUTTALL'S RIBBON-LEAVED PONDWEED Per, rhizomes slender, matted, winter buds 0. **ST:** < 170 cm, branches 0–few, compressed. **LF:** submersed sessile, 5–25 cm, 2–10 mm wide, linear, ribbon-like, with a net-like band along midrib, tip acute, veins 3–13; floating gen opposite, petioled, blade 2–8 cm, 0.4–3.5 cm wide, gen oblong to elliptic, base tapered, tip rounded, veins 3–13; stipules gen < 4 cm, free from blade base, not sheathing, ligule-like. **INFL:** 1–4 cm. **FR:** 2.5–4.5 mm, 2–3.6 mm wide, spheric to obovate, keels 3, sides concave. *n*=13. Shallow water, ponds, lakes, streams; 400–1900 m. NCoRO, SNH, MP; to AK, e N.Am, CO. [*P. e.* subsp. *nuttallii* (Cham. & Schltdl.) Calder & Roy L. Taylor] Hybridizes with *P. nodosus, P. gramineus.* Jul–Aug ★

P. foliosus Raf. LEAFY PONDWEED Ann (per), rhizomes rare, slender, winter buds present. **ST:** branches many, ± compressed;

nodal glands 0 (faint). **LF:** submersed, sessile, 1–10 cm, linear, veins 1–3(5); stipules fused, sheath-like. **INFL:** 1.5–7 mm, short-cylindric to spheric; peduncle 0.3–1.1(3.7) cm, club-shaped. **FR:** olive to green-brown, keel 1, wavy, wing-like, sides convex to ± concave.

subsp. **fibrillosus** (Fernald) R.R. Haynes & Hellq. FIBROUS PONDWEED **ST:** < 60 cm; nodal glands present. **LF:** 2–4 cm, 1–2 mm wide, tip acute; stipules < 12 cm, veins in age persisting as fibers. **INFL:** interrupted. **FR:** 1.4–1.7 mm, 1.1–1.2 mm wide, pale green, keel < 0.2 mm high, beak ≤ 0.2 mm. Shallow water, small streams; < 1300 m. n NCo (Crescent City), w MP, s SNE; to WA, ID, WY, UT. [*P. foliosus* var. *fibrillosus* (Fernald) R.R. Haynes & Reveal] All pls in Yellowstone National Park sampled for DNA are hybrids between *P. foliosus* and *P. pusillus.* Jul–Oct ★

subsp. **foliosus** (p. 1505) Rhizomes slender, rare. **ST:** < 100 cm, compressed; nodal glands faint or 0. **LF:** 1–10 cm, 0.3–2.5 mm wide, tip acute to abruptly pointed; stipules < 2 cm, veins in age decaying, occ persistent as delicate fibers. **INFL:** ± spheric, gen not interrupted. **FR:** 1.5–2.7 mm, 1.2–2.2 mm wide, olive to green-brown, keel 0.2–0.4 mm high, beak 0.2–0.6 mm. 2*n*=28. Common. Ponds, lakes, streams; < 2300 m. NW, CaRH, SNF, n SNH, GV, CW, SW (exc ChI), GB, D; to AK, e Can, C.Am. [*P. f.* var. *f.*] Jul–Oct

P. gramineus L. (p. 1505) GRASS-LEAVED PONDWEED Per, rhizomes matted. **ST:** < 100 cm, branches gen many, short, cylindric. **LF:** submersed blade 2–10 cm, (3)10–27 mm wide, elliptic, minutely serrate, tip acute or long-tapered, veins 3–9, petiole 0 (≤ 3 cm); floating gen on short, axillary branches, blade < petiole, 1.5–7 cm, 1–3 cm wide, elliptic to ovate, tip ± obtuse, veins 11–13; stipules < 3 cm, free, persistent. **INFL:** 1–4 cm. **FR:** 1.9–2.3 mm, keels 3, beak erect, 0.3–0.5 mm. *n*=26. Uncommon. Shallow water, ponds, lakes, bogs; 900–2750 m. KR, NCoRI, CaRH, SNH, SnFrB, SnBr, GB; to AK, Greenland, AZ; Eurasia. Hybridizes with *P. alpinus, P. illinoensis, P. natans, P. nodosus, P. richardsonii.* Jul–Aug

P. illinoensis Morong (p. 1505) SHINING PONDWEED Per, rhizomed. **ST:** < 150 cm, branches gen many, slender, cylindric. **LF:** submersed blade 6–20 cm, 15–50 mm wide, elliptic to oblanceolate, microscopically serrate near tip, tip acute-mucronate, veins 7–19, petiole 0 or 0.5–4 cm; floating 0 or blade gen > petiole, 4–12 cm, 2–6 cm wide, widely elliptic to oblong-elliptic, tip rounded-mucronate, veins 13–29; stipules 2.5–7 cm, free, persistent. **INFL:** gen < 6 cm. **FR:** 2.5–3.6 mm, keels 3, lateral lower, beak erect to ± recurved, 0.5–0.8 mm. *n*=52. Lakes, ponds, streams; 400–2350 m. NCoR, CaR, SNH, GV, CW (exc SCoRI), SnGb, SnBr, PR, GB; to BC, e Can, TX, Baja CA; also C.Am, S.Am, Caribbean. Hybridizes with *P. amplifolius, P. gramineus, P. nodosus, P. richardsonii.* Jun–Aug

P. natans L. (p. 1505) FLOATING-LEAVED PONDWEED Per, rhizomed. **ST:** < 160 cm, gen simple, subcylindric. **LF:** submersed 10–30 cm, < 2 mm wide, linear, sessile; floating 6–11 cm, < 60 mm wide, oblong to widely elliptic, base ± lobed, tip rounded, veins 17–37, petiole lighter green near blade; stipules 6–8 cm, free, persistent. **INFL:** < 5 cm. **FR:** 3.5–5 mm, keels ± 0, sides concave. 2*n*=52. Shallow, fresh or brackish water, lakes, ponds, bogs, marshes, lagoons, streams; < 2700 m. NCo, NCoR, CaRH, SNH, SnJV, SnFrB, SnBr, PR, MP; to AK, e N.Am, Baja CA; Eurasia. Hybridizes with *P. gramineus, P. nodosus.* Jun–Aug

P. nodosus Poir. (p. 1505) LONG-LEAVED PONDWEED Per, rhizomed. **ST:** gen < 300 cm, branches 0 to above, subcylindric. **LF:** submersed 2–15 cm, 10–40 mm wide, linear to lance-elliptic, tapered at both ends, veins 7–15, petioles 2–13 cm; floating 5–10 cm, < 5 cm wide, elliptic to ovate, base tapered to rounded, tip rounded, veins 9–21, petiole 3.5–26 cm; stipules 3–9 cm, free, breaking apart early. **INFL:** < 5 cm. **FR:** 3–5 mm, keels 3, sides flat. *n*=26. Shallow water, lakes, ponds, streams; 100–2750 m. NCoR, CaRF, c&s SNF, n SNH, GV, CCo, SnFrB, SCo, SnBr, PR, GB, DMoj; worldwide (exc Aus-

tralia). Hybridizes with *P. epihydrus*, *P. gramineus*, *P. illinoensis*, *P. natans*, *P. richardsonii*. May–Aug

P. praelongus Wulfen (p. 1505) WHITE-STEMMED PONDWEED Per, rhizome stout. **ST**: gen < 30 cm, branches few, often zigzag, subcylindric, ± white. **LF**: submersed, sessile, 5–20 cm, 10–30 mm wide, lance-linear, often wavy, base gen lobed, clasping, tip hood-like, splitting when dry, veins 11–33; stipules 2–10 cm, free, ± white, persistent. **INFL**: 3–7.5 mm; peduncle 10–30 cm. **FR**: 4–5 mm, keels 1(3). 2n=52. Deep water, lakes; 1800–3000 m. CaRH, n&c SNH; to AK, Greenland, Mex; Eur, e Asia. Hybridizes with *P. amplifolius*, *P. richardsonii*. Jul–Aug ★

P. pusillus L. (p. 1505) SMALL PONDWEED Ann (per), rhizomes rare, slender, winter buds terminal or lateral. **ST**: < 100 cm, cylindric; nodal glands present. **LF**: submersed, sessile, thread-like to linear, veins 1–3(7); stipules fused below middle. **INFL**: ± interrupted, whorls 2–4; peduncles 0.5–8 cm. **FR**: < 2.5 mm, keels 0, sides concave, tip symmetric, beak at top. Shallow water, ponds, lakes, vernal pools, slow streams; < 2700 m. NCo, KR, n SNF, SNH, GV, SnFrB, SCoR, SW (exc ChI), GB, DMoj; circumboreal. May–Jun

P. richardsonii (A. Benn.) Rydb. (p. 1505) RICHARDSON'S PONDWEED Per, rhizome matted. **ST**: < 70 cm, branches few, gen straight, cylindric. **LF**: submersed, sessile, 1–12 cm, gen 10–20 mm wide, linear- to lance-ovate, margin crinkly near tip, base ± cordate, clasping, tip acute to rounded, flat, not hood-like, not splitting when dry, veins 3–35; stipules < 2 cm, free, persistent as fibers. **INFL**: < 4 cm, ± < peduncle. **FR**: 2–4 mm, keels 0(1), sides concave to convex. 2n=52. Uncommon. Shallow to deep water, ponds, lakes, lagoons, streams; < 2800 m. NCo, KR, CaRH, c SN, n SNH, MP; to AK, CO, e N.Am. Jul–Aug

P. robbinsii Oakes (p. 1505) ROBBINS' PONDWEED Per, rhizomed. **ST**: < 200 cm, branches gen many, cylindric. **LF**: submersed, ± stiff, 2-ranked, sessile, < 12 cm, 3–4(8) mm wide, linear to lanceolate, gen finely serrate, veins 20–60, fine; stipules 0.5–2 cm, fused to blade base, fibrous, sheath-like basally, free, liguled, shredding at tips, green-brown to white. **INFL**: < 2 cm, fls few, paired. **FR**: < 4 mm, keels 3, sides convex. Deep water, lakes; 1600–3300 m. KR, c&s SNH; to AK, e US. Fr rarely collected. Aug–Sep ★

P. zosteriformis Fernald (p. 1505) EEL-GRASS PONDWEED Ann, winter buds present. **ST**: < 60 cm, 0.7–4 mm wide, branches gen few, compressed, winged; nodal glands few. **LF**: submersed, sessile, 5–20 cm, 2–5 mm wide, linear, ± rigid, tip obtuse to abruptly pointed, veins 15–35; stipules 2–6 cm, free, breaking apart. **INFL**: < 3 cm; peduncle < 10 cm. **FR**: ± 5 mm, keel 1, sides convex. Ponds, lakes, streams; < 1300 m. s NCoRI (Clear Lake), GV, MP; to BC, e N.Am. Jun–Jul ★

STUCKENIA PONDWEED

Per, rhizomed, tubered or not, winter buds 0. **ST**: branched, cylindric. **LF**: submersed, sessile, channeled, entire, tip acute to abruptly pointed, notched, or rounded, veins 1–5; stipules fused to lf blade ≥ 2/3 stipule length. **INFL**: submersed; peduncle flexible. **FR**: beaked or not, keel 0. x=13. ±7 spp.: worldwide.

1. Lf tip notched, blunt, rounded, rarely with abrupt, short point; stipule sheaths on proximal sts often inflated, others not; fr beak inconspicuous . . . . *S. filiformis* subsp. *alpina*
1′ Lf tip acute or with abrupt, short to long point, rarely rounded; stipule sheaths not inflated; fr beak conspicuous
  2. Lvs 0.2–1 mm wide, tip acute or with abrupt, short to medium-long point; infl axillary or terminal; fr 2.5–5 mm . . . . *S. pectinata*
  2′ Lvs (0.4)1.5–5(8.5) mm wide, tip with abrupt, medium-long to long point, rarely rounded; infl axillary, rarely terminal; fr 3–3.9 mm . . . . *S. striata*

S. filiformis (Pers.) Börner subsp. *alpina* (Blytt) R.R. Haynes et al. (p. 1505) SLENDER-LEAVED PONDWEED Slender, rhizomed, tubered. **ST**: < 65 cm, branches below, subcylindric. **LF**: < 12 cm, < 3 mm wide, linear to bristle-like, tip notched, blunt, rounded (or with abrupt, short point); stipules < 3 cm, sheaths on proximal sts often inflated, others not. **INFL**: terminal, interrupted below, 5–55 mm, whorls 2–6(9); peduncle 2–10(15) cm. **FR**: 2–3 mm, beak inconspicuous. 2n=66. Shallow, clear water of lakes, drainage channels; 300–2150 m. KR, c SNH, GV, CCo, SnFrB, GB; to AK, AZ, Greenland, Eurasia. [*Coleogeton f.* (Pers.) Les & R.R. Haynes subsp. *a.* (Blytt) Les & R.R. Haynes; *Potamogeton f.* Pers. var. *a.* (Blytt) Asch. & Graebn.; *P. f.* var. *macounii* Morong] Unrecognized subsp. under *Potamogeton filiformis* in TJM (1993). May–Jul ★

S. pectinata (L.) Börner (p. 1505) FENNEL-LEAF PONDWEED Tubered, rhizomes matted. **ST**: < 80 cm, branches many, ± cylindric. **LF**: < 15(35) cm, 0.2–1 mm wide, gen thread-like, tip acute or with abrupt, short to medium-long point; stipules 2–5 cm, sheaths not inflated. **INFL**: axillary or terminal, interrupted in fr, 14–22 mm, whorls 2–6; peduncle 4.5–11.4 cm. **FR**: 2.5–5 mm, beak conspicuous. Common. Ponds, lakes, marshes, streams; < 2400 m. CA; worldwide. [*Potamogeton p.* L.; *Coleogeton p.* (L.) Les & R.R. Haynes] Often weedy in reservoirs, irrigation canals; important food for waterfowl. May–Jul

S. striata (Ruiz & Pav.) Holub (p. 1505) NEVADA PONDWEED Rhizomed, tubers 0. **ST**: < 60 cm, branches many, subcylindric, ± white. **LF**: 2–10 cm, (0.4)1.5–5(8.5) mm wide, linear, veins 3–5, tip with abrupt, medium-long to long point (rounded); stipules < 3 cm, sheaths not inflated. **INFL**: axillary (terminal), interrupted below, 13–45 mm, whorls 4–9; peduncle 1.2–5.2 cm. **FR**: 3–3.9 mm, beak conspicuous. Uncommon. Shallow, alkaline water, ponds, lakes; < 1450 m. SN, GV, SnFrB, GB, DMoj; to OR, CO, TX, Mex, C.Am, S.Am. [*Coleogeton s.* (Ruiz & Pav.) Les & R.R. Haynes; *Potamogeton latifolius* (J.W. Robbins) Morong] May be locally abundant. Jul–Aug

RUPPIACEAE DITCH-GRASS FAMILY

Robert F. Thorne, C. Barre Hellquist & Robert R. Haynes

Ann [(per)], submersed. **ST**: gen many-branched, thread-like, rooting at lower nodes. **LF**: cauline, alternate to ± opposite, sessile, linear, < 10 cm, < 1 mm wide; stipules < 15 mm, ± completely fused to blade, sheath-like, ± open. **INFL**: terminal, head-like spikes with subtending bract, < 20-fld; peduncle elongating after fl [or not], coiling or not, ovary base elongating after fertilization, becoming stalk-like. **FL**: bisexual, minute; perianth 0; stamens 2, anthers ± sessile; pistils (2)4(8), simple, ovule attached to chamber top, stigma sessile, peltate. **FR**: drupe, ± nut-like, stalk-like at base, beaked, ovoid-oblique. 1 genus, ± 10 spp.: ± worldwide. Scientific Editor: Thomas J. Rosatti.

RUPPIA DITCH-GRASS

(H.B. Ruppius, German botanist, 1688–1719) [Haynes 2000 FNANM 22:75–76] 2 N.Am spp. ± indistinct; study needed.

1. Peduncle in fr 30–300 mm, coils 5–30; lf tip acute . . . . **R. cirrhosa**
1′ Peduncle in fr 2–25 mm, coils 0–4; lf tip ± obtuse . . . . **R. maritima**

R. cirrhosa (Petagna) Grande (p. 1509) 2*n*=40. Freshwater marshes, ponds, sloughs; < 2045 m. NCo, NCoRI, Teh, SnJV, CW, SCo, SnBr, GB, D; to AK, e-c US; ± worldwide. Apr–Jul

R. maritima L. 2*n*=20. Brackish or saline waters, often coastal, marshes, ponds, sloughs; < 600 m. NCo, sw SnJV (Soda Lake, San Luis Obispo Co.), CW, SCo, ChI; ± worldwide. Mar–Aug

RUSCACEAE BUTCHER'S-BROOM FAMILY

Dale W. McNeal, except as noted

Per to shrub, rhizomed or not. **ST**: lfy or scapose. **LF**: 2–15, basal or cauline, cauline clasping or sheathing, petioled [or reduced to scales, with short branches flattened, green, appearing lf-like]. **INFL**: terminal or axillary, panicle or raceme. **FL**: unisexual or bisexual, gen white (± pink); perianth parts 4 or 6, fused or not, erect, nodding or pendent; stamens (4)6, ± fused to perianth, anthers attached near base; ovary superior, chambers 2–3, style 1, stigma ± 3-lobed. **FR**: berry, spheric, red, orange-red, or blue-black, or capsule, papery. **SEED**: 1–12. 26 genera, 475 spp.: n hemisphere, S.Afr, n Australia; spp. in several genera cult as orn, house pls. *Smilacina* moved to *Maianthemum*. Scientific Editor: Thomas J. Rosatti.

1. Per; lvs cauline; fls on 1 pl all bisexual . . . . **MAIANTHEMUM**
1′ Shrub; lvs in dense rosettes; fls on 1 pl all unisexual or unisexual and bisexual . . . . **NOLINA**

MAIANTHEMUM FALSE LILY-OF-THE-VALLEY, FALSE SOLOMON'S SEAL

Rhizome creeping. **ST**: erect, simple [0]. **LF**: alternate, gen ± clasping, lower reduced to sheathing scales. **INFL**: terminal, raceme to panicle. **FL**: perianth parts 4 or 6 in 2 petal-like whorls, white; stamens 4 or 6; ovary superior, chambers 2–3, style 1, stigmas 2–3. **FR**: berry. **SEED**: 1–3. 3 spp.: n temp. (Greek: May fl, from fl season) [LaFrankie 2002 FNANM 26:206–210] Historically divided into 2 genera, *Maianthemum*, *Smilacina* (1. vs 1′ in key); otherwise ± equal morphologically, unique chromosomally.

1. Perianth parts 4; ovary chambers 2 . . . . **M. dilatatum**
1′ Perianth parts 6; ovary chambers 3
  2. Infl a panicle; perianth 1–2 mm, << stamens . . . . **M. racemosum**
  2′ Infl a raceme (or few-branched near base); perianth 4–7 mm, > stamens . . . . **M. stellatum**

M. dilatatum (Alph. Wood) A. Nelson & J.F. Macbr. (p. 1509) Rhizome 1–1.5 mm diam. **ST**: 10–40 cm. **LF**: 2–3, 5–20 cm, cordate to sagittate, acute to acuminate, glabrous to sparsely hairy below, few; petioles 5–15 cm. **INFL**: raceme, 2–8 cm, 10–50-fld. **FL**: perianth parts 4, ± 2.5 mm, > stamens, reflexed to spreading, ovate to elliptic; ovary chambers 2. **FR**: ± 6 mm, red. **SEED**: ± 3 mm, brown. *n*=18. Moist, shady, conifer forest; < 500 m. w NW, CCo; to AK, ID. May–Jun

M. racemosum (L.) Link (p. 1509) Rhizome 5–10 mm diam. **ST**: 30–90 cm, glabrous to finely hairy above. **LF**: > 5, 7–20 cm, ovate to oblong-elliptic, acute to acuminate, glabrous to finely hairy abaxially; petioles ± 0. **INFL**: panicle, 5–12 cm; fls > 20. **FL**: perianth parts 6, 1–2 mm, << stamens, ± erect, narrowly oblong; ovary chambers 3.

FR: 5–7 mm, red, dotted purple. **SEED**: 2.5–6 mm, brown. *n*=±18. Moist, open woodland, streambanks; < 2000 m. NW, CaR, SN, CW, TR; to AK, e N.Am. [*Smilacina r.* (L.) Desf.; *M. r.* subsp. *amplexicaule* (Nutt.) LaFrankie] Mar–Jul

M. stellatum (L.) Link (p. 1509) Rhizome 3–5 mm diam. **ST**: 30–70 cm, straight or ± zigzag above, glabrous to puberulent. **LF**: > 5, 5–17 cm, (ob) lanceolate to elliptic, acuminate, abaxially puberulent; petioles 0. **INFL**: raceme (or few-branched near base), gen 2–8 cm; fls 5–15. **FL**: perianth parts 6, 4–7 mm, > stamens, spreading, oblong to lanceolate; ovary chambers 3. **FR**: 7–10 mm, red-purple to black. **SEED**: 2.5–3 mm, brown. *n*=18. Moist woodland, streambanks, open slopes; < 2400 m. NW, CaR, SN, CW, SW, Wrn, SNE; to BC, e N.Am. [*Smilacina s.* (L.) Desf.] Apr–Jun

NOLINA BEARGRASS

Dale W. McNeal & James C. Dice

ST: thick, woody or ± below ground. **LF**: rosettes 6–20 dm, stiff, sword-like, base much expanded, white, fleshy. **INFL**: panicle or raceme-like, scapose, bracted, < 4 m. **FL**: perianth parts 6 in 2 petal-like whorls, < 6 mm, ± white; stamens 6, filaments slender; pistil reduced in staminate fls, pistillate fls with staminodes; ovary superior, 3-chambered, style and 3 stigmas short, ovules 2 per chamber. **FR**: capsule, papery. **SEED**: 1–3 per fr, ovoid. ± 30 spp.: s US, Mex. (P.C. Nolin, French agriculturist, b. 1717) [Hess 2003 FNANM 26:415–421] Lf widths taken just above expanded lf base.

1. St below ground, horizontal; lf blade 12–35 mm wide . . . . **N. interrata**
1′ St gen above ground, erect; lf blade 12–48 mm wide
  2. Lf margin in age fibrous-shredding; seed gray-white; DMtns, DSon . . . . **N. bigelovii**
  2′ Lf margin in age not fibrous-shredding; seed red-brown; s SNH (Kern Plateau), SCo, WTR, e SnBr, PR
    3. Lf base 3–8.5 cm wide, infl 0.9–1.8 m; dry chaparral of coastal mtns; SCo, WTR, PR . . . . **N. cismontana**
    3′ Lf base 5–16.5 cm wide; infl 1.4–3.8 m; pinyon/juniper woodland, rocky desert slopes; s SNH (Kern Plateau), e SnBr, PR, DMtns, DSon . . . . **N. parryi**

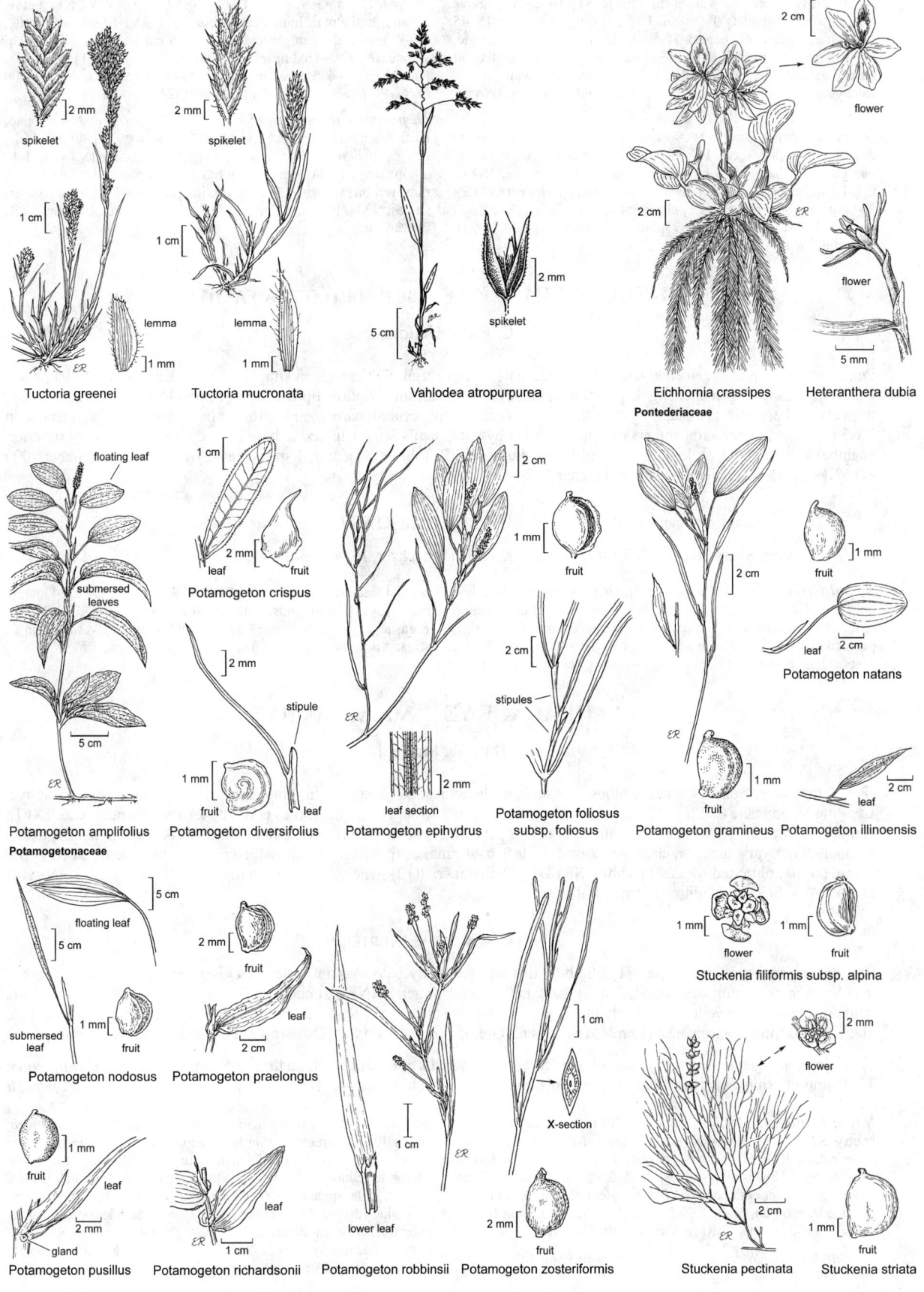

2 cm — flower

spikelet

2 mm

spikelet

2 mm

1 cm

lemma

lemma

1 mm

1 mm

Tuctoria greenei

Tuctoria mucronata

5 cm

2 mm

spikelet

Vahlodea atropurpurea

2 cm

2 cm

ER

flower

5 mm

Eichhornia crassipes
Pontederiaceae

Heteranthera dubia

floating leaf

submersed leaves

5 cm

ER

Potamogeton amplifolius
Potamogetonaceae

1 cm

2 mm

leaf

fruit

2 mm

1 mm

fruit

stipule

leaf

Potamogeton crispus

Potamogeton diversifolius

2 cm

1 mm

fruit

2 cm

stipules

2 mm

leaf section

Potamogeton epihydrus

Potamogeton foliosus
subsp. foliosus

2 cm

1 mm

fruit

2 cm

leaf

Potamogeton natans

ER

1 mm

fruit

leaf

2 cm

Potamogeton gramineus **Potamogeton illinoensis**

5 cm

floating leaf

5 cm

submersed leaf

1 mm

fruit

Potamogeton nodosus

2 mm

fruit

leaf

2 cm

Potamogeton praelongus

1 cm

1 cm

ER

lower leaf

1 cm

X-section

2 mm

fruit

Potamogeton robbinsii **Potamogeton zosteriformis**

1 mm

flower

1 mm

fruit

Stuckenia filiformis subsp. alpina

1 mm

fruit

leaf

2 mm

gland

Potamogeton pusillus

ER

leaf

1 cm

Potamogeton richardsonii

2 mm

flower

ER

2 cm

Stuckenia pectinata

1 mm

fruit

Stuckenia striata

N. bigelovii (Torr.) S. Watson (p. 1509) **ST**: 10–25 dm, above ground, erect, branches 0–several. **LF**: 34–160 per rosette, 15–45 mm wide, ± glaucous, base 5–11 cm wide, margin minute-serrate, in age fibrous-shredding. **INFL**: 13–37 dm; scape 15–45 cm diam at base; bracts early-deciduous. **SEED**: 2.5–3.5 mm, gray-white. *n*=19. Rocky slopes, ridges; 300–1500 m. se PR, s DMoj, DMtns, DSon; s NV, w AZ, nw Mex. May–Jun

N. cismontana Dice CHAPARRAL NOLINA **ST**: gen < 3(15) dm, above ground, erect. **LF**: 30–90 per rosette, 12–30 mm wide, occ glaucous, base 3–8.5 cm wide, margin minute-serrate. **INFL**: 0.9–1.8 m; scape 14–35 mm diam at base; bracts persistent. **SEED**: 3–4 mm, red-brown. *n*=19. Dry chaparral of coastal mtns; 200–1300 m. SCo, WTR, PR. Undescribed sp. under *N. parryi* in TJM (1993). May–Jul ★

N. interrata Gentry (p. 1509) DEHESA NOLINA **ST**: below ground, horizontal, branched, rosettes ≥ 1. **LF**: 10–45 per rosette, 12–35 mm wide, gen glaucous, base 2–6 cm wide, margin minute-serrate. **INFL**: 5–16 dm; scape 5–16 mm diam at base; bracts persistent. **SEED**: 4–6 mm, red-brown. Foothills on gabbro soils; 200–700 m. sw PR (sw San Diego Co.); nw Baja CA. Jun–Jul ★

N. parryi S. Watson (p. 1509) **ST**: 3–21 dm, gen aboveground, erect, branches 0–several. **LF**: 65–200 per rosette, 20–40 mm wide, gen green, base 5–16.5 cm wide, margin minute-serrate. **INFL**: 1.4–3.8 m; scape 26–90 mm diam at base; bracts persistent. **SEED**: 3–4 mm, red-brown. *n*=19,20. Pinyon/juniper woodland, rocky desert slopes; 900–2100 m. s SNF, s SNH (Kern Plateau), e SnBr, e PR, D. May–Jun

SCHEUCHZERIACEAE SCHEUCHZERIA FAMILY

C. Barre Hellquist

Per, rhizomed, aquatic, gen emergent, on floating mats, or terrestrial. **ST**: erect, gen simple, slender. **LF**: basal, cauline, alternate, simple; sheath gen open; ligule prominent; blade erect, linear, subcylindric, tip with a large pore. **INFL**: raceme, terminal, fl bracts lf-like, pedicels elongating after fl. **FL**: bisexual, radial; perianth parts 6, in 2 ± dissimilar whorls, free; stamens 6, in 2 whorls, fused to perianth, anthers opening outward by slits; pistils 3(6), ± united at base or not, free above, ovary superior, chamber 1, ovules 1–3, style stout [0], papillate or plumose. **FR**: follicles. 1 genus, 1 sp.: w&e N.Am, Eurasia. [Nienaber 2000 FNANM 22:41–42] Scientific Editor: Thomas J. Rosatti.

SCHEUCHZERIA

(J.J. Scheuchzer, Swiss botanist, 1672–1733, and brother, J. Scheuchzer, 1684–1738)

S. palustris L. (p. 1509) SCHEUCHZERIA Glabrous. **ST**: 1–4 dm, zigzag. **LF**: striate; ligule 2–12 mm; basal 1–2(3) dm; cauline reduced upward, blade 1–3 mm wide. **INFL**: peduncle in fr < 2.5 dm; pedicels spreading, 5–25 mm; fls 3–12. **FL**: perianth parts 2–3 mm, 1-veined, green-white. **FR**: gen 1–4, 4–10 mm, dehiscing on curved, adaxial side, beak 0.5–1 mm. **SEED**: 1–3, 4–5 mm. 2*n*=22. Floating mats, bogs, lake margins; 1400–2000 m. s CaRH; to AK, e N.Am; Eurasia. [*S. p.* subsp. *americana* (Fernald) Hultén] Variation in fr, stigmas does not support infraspecific taxa. Jul–Aug ★

SMILACACEAE SMILAX FAMILY

Dale W. McNeal

[Per, shrub] vine, from rhizomes, stolons, or caudices, dioecious. **ST**: erect or climbing, gen prickly, scaly below, lfy above. **LF**: simple, opposite or alternate, veins net-like between gen 3 stronger veins; petioles gen with 2+ tendrils near base. **INFL**: gen umbel [raceme, spike]. **FL**: perianth parts 6, in 2 petal-like whorls, free [rarely united], stamens 6, free (as staminodes in pistillate fls); ovary superior, chambers 2 or 3, style ± 0, stigmas 3, (pistils 0 or reduced, sterile in staminate fls). **FR**: berry, black, purple, blue, red [orange] (white). **SEED**: 1–2/chamber. 4(12) genera, 375 spp.: temp, trop. [Holmes 2002 FNANM 26:468–478] Scientific Editor: Thomas J. Rosatti.

SMILAX GREENBRIER

Per; caudex gen large, tuber-like. **ST**: climbing or trailing, woody below or not, often prickly ± throughout. **LF**: alternate, deciduous or persistent; base rounded to sagittate; petiole tendrils gen 2. **INFL**: umbels or clusters, axillary. **FL**: perianth parts white to ± green or ± yellow; ovary chambers gen 3, stigmas spreading. **FR**: black, purple, blue, or red (white). **SEED**: 1–6, black. ± 350 spp.: esp trop, also temp N.Am, e Asia. (Greek: origin uncertain) [Cameron & Fu 2006 Aliso 22:598–605]

1. St not shiny, gen armed with needle-like prickles; staminate fls 5–6 mm; pistillate fls 3–3.5 mm . . . . ***S. californica***
1′ St smooth, shiny, unarmed; staminate and pistillate fls ± 2.5 mm . . . . ***S. jamesii***

S. californica (A. DC.) A. Gray (p. 1509) Vine; caudex short, knotty. **ST**: 2–5 m, ± woody; central pith 0. **LF**: 5–10 cm, ovate, acuminate, dull green, abaxially not glaucous, base ± cordate. **INFL**: staminate gen 15–30-fld; pistillate gen 15–35-fld; peduncles 2–5 cm, drooping. **FL**: perianth parts ± 1.5 mm wide, reflexed from middle, linear. **FR**: ± 6 mm, black. **SEED**: 3. *n*=16. Streambanks in conifer forest; < 1600 m. NW, CaRF, n SNF; sw OR. May–Jun

S. jamesii G.A. Wallace (p. 1509) ENGLISH PEAK GREENBRIER Vine; rhizome long, ± zigzag. **ST**: 2–3 m, herbaceous; central pith present. **LF**: 5–8 cm, triangular to ± ovate, acute or rigid-tipped, adaxially dark green, abaxially glaucous, minutely papillate, base truncate to ± cordate. **INFL**: staminate gen < 20-fld; pistillate gen < 40-fld; peduncles 5–13.5 cm, stout, ± erect. **FL**: perianth parts ± 1.5 mm wide, spreading or reflexed, elliptic to oblong. **FR**: 6–8 mm, dark blue (drying dull maroon). **SEED**: 6. Lakesides, streambanks, alder thickets in montane conifer forest; gen 1500–2500 m. KR, CaR. Only herbaceous sp. in w N.Am; member of *S. herbacea* L. complex, otherwise of e N.Am. May–Jul ★

TECOPHILAEACEAE   TECOPHILAEA FAMILY

Dale W. McNeal

Per; corm deep, outer coat fibrous. **ST**: 1, erect, ± zigzag. **LF**: mostly basal, linear [to lanceolate], sheathing. **INFL**: panicle, in smaller pls raceme; bracts scarious. **FL**: perianth parts 6 in 2 petal-like whorls, free or fused, often reflexed; stamens 6, fused to perianth parts, alternate 6 staminodes, anthers attached at base; ovary superior, chambers 3, style 1, stigma ± 3-lobed. **FR**: capsule, loculicidal. **SEED**: 3, brown. 9 genera: 23 spp.: w US, Afr, Chile. Scientific Editor: Thomas J. Rosatti.

ODONTOSTOMUM

1 sp. (Greek: tooth mouth, from staminodes ± at top of perianth tube)

O. hartwegii Torr. (p. 1509)   Corm 2.5–3 cm, ± ovoid. **ST**: ± scapose, 12–50 cm, ± curved, gen branched. **LF**: 10–30 cm; upper reduced. **INFL**: bracts 3–10 mm, linear; pedicel 3–30 mm, with scarious bractlet above middle. **FL**: perianth parts fused below, petal-like, white to ± yellow, tube 4–5 mm, strongly veined, lobes ± = tube, spreading, reflexed in age, 3–7-veined; stamens 6, filaments 1–2 mm, staminodes 0.5–1 mm, anthers dehiscing by pores at tip; style 4–5 mm, thread-like. **SEED**: obovoid. $n=10$. Clay, often serpentine soils; < 600 m. KR, NCoRI, CaR, n&c SNF. Apr–May

THEMIDACEAE   BRODIAEA FAMILY

J. Chris Pires, except as noted

Per from corm, outer coat fibrous [membranous]. **LF**: basal, 1–10, linear to narrow-lanceolate. **INFL**: scapose, gen umbel-like; scape erect, gen 1(2), cylindric, gen rigid, occ wavy to twining; fl bracts 2–4[10], not enclosing fl buds. **FL**: perianth parts 6 in 2 petal-like whorls, free or ± fused below into tube; staminodes 0 or 3; stamens 3 or 6, free or fused to perianth, occ append-aged; ovary superior, chambers 3, ovules 2–several per chamber. **FR**: capsule, loculicidal. 10–11 genera, 60–70 spp.: w N.Am. [Pires & Sytsma 2002 Amer J Bot 89:1342–1359] Scientific Editors: Dale W. McNeal, Thomas J. Rosatti.

1. Perianth parts ± free at base, not forming obvious tube
  2. Filaments basally dilated, fused to form nectar cup, distally thread-like . . . . **BLOOMERIA**
  2′ Filaments thread-like to dilated, occ overlapped or fused but not forming cup . . . . **MUILLA**
1′ Perianth parts fused at base, forming obvious tube
  3. Stamens 3, equal (or 6, 3 smaller in *Dichelostemma capitatum*)
    4. Scape gen straight; staminodes (0)3, alternate 3 stamens; filaments gen not forming crown-like tube
      . . . . **BRODIAEA**
    4′ Scape gen curved, twisted, or twining; staminodes 0 (3, stamen-like in *Dichelostemma volubile*);
      filaments forming crown-like tube outside anthers . . . . **DICHELOSTEMMA**
  3′ Stamens 6, equal or not
    5. Filaments fused into tube with appendages forming crown . . . . **ANDROSTEPHIUM**
    5′ Filaments free, appendages forming crown or gen 0 . . . . **TRITELEIA**

ANDROSTEPHIUM

Corm spheric. **LF**: basal, linear, channeled. **INFL**: scape straight; bracts 3, lanceolate; pedicels not jointed, erect. **FL**: perianth parts 6 in 2 petal-like whorls, tube funnel-shaped, lobes narrow-oblong; stamens 6, filaments fused into nectar tube with erect toothed appendages forming crown between anthers, anthers attached near base; style persistent, stigma small, 3-lobed. **FR**: ± spheric, obtuse-3-angled. **SEED**: several per chamber, flat, black-crusted. 3 spp.: w US, n Mex. (Greek: stamen crown, for appendages of fused filaments)

A. breviflorum S. Watson (p. 1509) SMALL-FLOWERED ANDROSTEPHIUM   **LF**: 10–30 cm, 1.5–2 mm wide. **INFL**: scape 10–35 cm, scabrous near base; pedicels 15–30 mm; fls 3–12. **FL**: perianth white to light violet drying yellow-brown, tube 5–7 mm, lobes 10–14 mm; filaments 8–10 mm, tube ± funnel-shaped, appendages ± 2 mm, anthers 2.5–3 mm. **FR**: 10–15 mm, deeply 3-lobed. Open desert scrub, sandy to rocky soil; 100–1600 m. s DMoj, n DSon; to w CO. Mar–Jun   ★

BLOOMERIA   GOLDENSTAR

Corm ± spheric, daughter corms 1–3 on stalks. **LF**: 1–8, keeled, entire, withered in fl. **INFL**: scape stiff, straight, cylindric, minute-scabrous; bracts 2–4, not enclosing fl buds, scarious in fl; pedicels ray-like, ascending to erect, jointed; fls 10–35+. **FL**: perianth parts ± free at base, not forming obvious tube, golden-yellow, striped ± brown or green, lobes ascending to spreading; stamens 6, filaments ± 6 mm, parallel to style, dilated bases fused to form nectar cup with shallow to awned cusps, or leaning away from style, leaving dilated bases separated, not forming cup, distally thread-like, anthers attached near base; style ± 5 mm, thread- or club-like, persistent, stigma 3-lobed. **FR**: 5–6 mm, ± spheric, 3-angled, sessile. **SEED**: angled, wrinkled, black-crusted. 3 spp.: c&s CA, n Baja CA. (H.G. Bloomer, early San Francisco botanist)

1. Filaments leaning away from style, leaving dilated bases separated, not forming cup; style ≤ ovary; lvs 2–8,
1–3 mm wide . . . . ***B. clevelandii***
1′ Filaments parallel to style, dilated bases forming nectar cup; style > ovary; lvs 1–2, 3–15 mm wide
  2. Scape 5–10 cm; perianth lobes ascending at base, gradually spreading above; lvs gen 1–2 . . . . ***B. humilis***
  2′ Scape 15–70 cm; perianth lobes abruptly spreading at base; lvs gen 1 . . . . ***B. crocea***
    3. Nectar cup cusps < 1 mm . . . . var. ***crocea***
    3′ Nectar cup cusps > 1 mm
      4. Nectar cup cusps 1–2 mm, < 1/2 filaments, tips acute . . . . var. ***aurea***
      4′ Nectar cup cusps 3–3.5 mm, ± 1/2 filaments, tips tapered . . . . var. ***montana***

B. clevelandii S. Watson (p. 1509) SAN DIEGO GOLDENSTAR
LF: 2–8, 6–15 cm, 1–3 mm wide. **INFL**: scape 15–70 cm; pedicels
0.5–6.5 cm. **FL**: perianth lobes abruptly spreading at base, 5–10 mm,
yellow with green midvein; filaments 3–5 mm, leaning away from
style, cup 0, anthers 0.4–1 mm; ovary 2–3 mm, style 2–3 mm, ≤
ovary. *n*=7. Coastal scrub, mesa grassland; < 100 m. s SCo (sw San
Diego Co.); n Baja CA. [*Muilla c.* (S. Watson) Hoover] May ★

B. crocea (Torr.) Coville COMMON GOLDENSTAR **LF**: gen 1, <
30 cm, 3–15 mm wide. **INFL**: scape 15–70 cm; pedicels 0.5–6.5 cm.
FL: perianth 5–13 mm, lobes abruptly spreading at base, yellow with
brown midvein; filaments 3–10 mm, parallel to style, nectar cup shallow, cusps obtuse to acute, anthers 1.5–3 mm; ovary 2–3 mm, style
± 6 mm, > ovary. *n*=9.

  var. **aurea** (Kellogg) J.W. Ingram **FL**: perianth 11–12 mm, yellow; nectar cup cusps 1–2 mm, < 1/2 filaments, tips acute. Foothill
woodland; < 1700 m. CW. Apr–Jun

  var. **crocea** (p. 1509) **FL**: perianth 5–12 mm, yellow-orange,
with 2 dark parallel lines; nectar cup cusps < 1 mm. Dry flats, hillsides, chaparral, coastal-sage scrub, valley grassland, oak woodland;
< 1700 m. CW, SW; n Baja CA. Apr–Jun

  var. **montana** (Greene) J.W. Ingram MOUNTAIN GOLDENSTAR
FL: perianth 11–13 mm, yellow; nectar cup cusps 3–3.5 mm, ± 1/2
filaments, tips tapered. Chaparral, yellow-pine forest; 300–1700 m.
Teh, SCoR, WTR. Apr–Jun

B. humilis Hoover (p. 1509)   DWARF GOLDENSTAR **LF**: gen
1–2, 5–30 cm, 3–15 mm wide. **INFL**: scape 5–10 cm; pedicels 1–5
cm. **FL**: perianth lobes ascending at base, gradually spreading above,
7–11 mm, yellow with ± brown-purple midvein; filaments parallel
to style, nectar cup present, cusps 0, anthers 1.5–1.8 mm; ovary ± 2
mm, style ± 6 mm, > ovary. Grassland/chaparral edges, open mesas
on ocean bluffs; < 100 m. CCo, expected in adjacent c SCoR (s Monterey, n San Luis Obispo cos.). Jun ★

BRODIAEA

J. Chris Pires & Robert E. Preston

Daughter corms often present. **LF**: gen 1–6, linear, gen crescent-shaped in ×-section, glabrous, entire, often withered at fl.
INFL: open; scape 2–70 cm, gen slender, gen straight, cylindric; bracts scarious; pedicels gen > fls, gen < 13 cm. **FL**: perianth
tube bell- to funnel-shaped, green-white, lobes erect to spreading, violet to lavender, occ pink, midribs purple or green; staminodes (0)3, gen ± erect, opposite outer lobes, white to violet or lavender, lateral margins flat, incurved, or inrolled; stamens 3,
equal, fused to perianth, opposite inner lobes, filaments occ winged or appendaged, free or fused basally to staminodes, gen
not forming crown-like tube, anthers attached at base, appressed to style, abaxially papillate; style 1, stigma lobes 3, spreading
and recurved. **FR**: sessile, ovoid. **SEED**: oblong, black, lined, angles ridged. ± 18 spp.: mostly endemic to CA. (James Brodie,
Scottish botanist, 1744–1824) [Preston 2006 Novon 16:254–259] Several undescribed taxa, hybrids under study. *B. matsonii*
R.E. Preston described from Shasta Co., 2010.

1. Staminodes 0 . . . . ***B. orcuttii***
1′ Staminodes gen present (see *Brodiaea santarosae*)
  2. Staminodes thread-like or narrow, uniformly tapered base to tip
    3. Perianth 13–28 mm; filaments 0.5–1.5 mm; staminodes 1–4 mm, reflexed against perianth . . . . ***B. filifolia***
    3′ Perianth 24–36.5 mm; filaments 2.4–8.2 mm; staminodes 0 in ± 10% of fls, or ≤ 7 mm, erect to recurved
      or reflexed, the shorter against perianth . . . . ***B. santarosae***
  2′ Staminodes linear to oblong, uniformly wide or abruptly narrowed at tip
    4. Perianth tube narrowed above ovary
      5. Staminode tips spreading to erect, margins 3/4 inrolled; stamens narrow-notched between anther sacs;
        filaments T-shaped in ×-section . . . . ***B. minor***
      5′ Staminode tips erect, margins flat to 1/2 inrolled; stamens wide-notched between anther sacs; filaments
        V- or Y-shaped in ×-section
        6. Perianth lobes not paler toward base; perianth tube narrowed above ovary; staminodes in width <
          outer perianth lobes . . . . ***B. nana***
        6′ Perianth lobes paler toward base; perianth tube ± narrowed above ovary; staminodes in width ≥ outer
          perianth lobes . . . . ***B. pallida***
    4′ Perianth tube not narrowed above ovary
      7. Staminodes gen > 15 mm
        8. Staminode margins wavy or toothed, 1/4–1/2 inrolled, midrib not noticeably thickened . . . . ***B. californica***
        8′ Staminode margins gen flat, 1/4 inrolled, midrib noticeably thickened
          9. Perianth lobes violet with green midrib; staminodes 14.5–21.5 mm; style 15.5–23.5 mm . . . . ***B. leptandra***
          9′ Perianth lobes violet (pink); staminodes 20.5–30 mm; style 20–31.5 mm . . . . ***B. sierrae***
      7′ Staminodes < 15 mm
        10. Perianth ≤ 25(28) mm; filaments ≤ 3 mm, channeled abaxially, basal 1–2 mm fused with staminodes;
          style gen < ovary
          11. Staminodes held away from stamens, tip narrowed abruptly to sharp point . . . . ***B. kinkiensis***

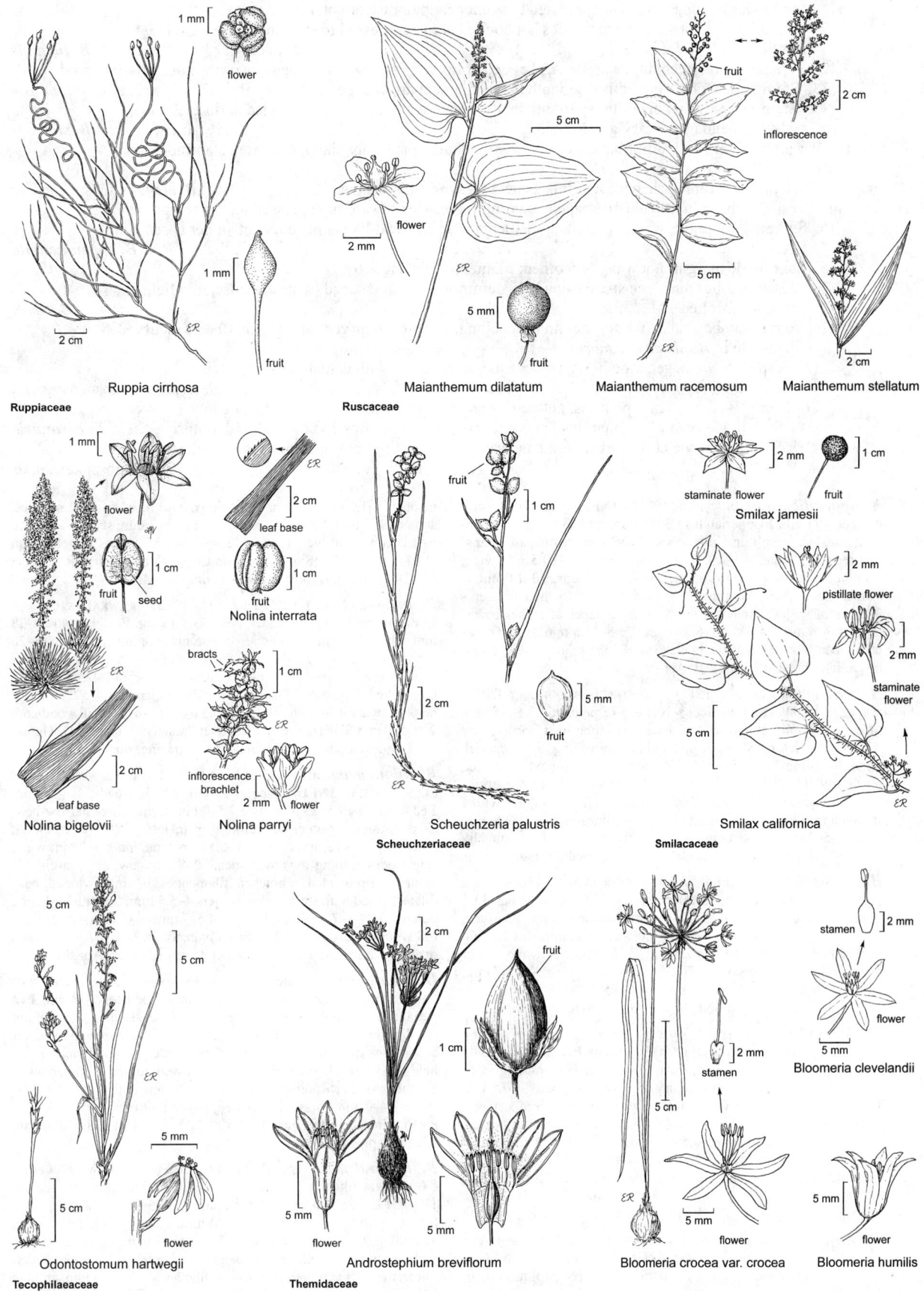

flower

fruit

Ruppia cirrhosa

Ruppiaceae

flower

fruit

Maianthemum dilatatum

fruit

Maianthemum racemosum

inflorescence

Maianthemum stellatum

Ruscaceae

flower

fruit   seed

Nolina bigelovii

leaf base

fruit

Nolina interrata

bracts

inflorescence
brachlet   flower

Nolina parryi

fruit

fruit

Scheuchzeria palustris

Scheuchzeriaceae

staminate flower   fruit

Smilax jamesii

pistillate flower

staminate
flower

Smilax californica

Smilacaceae

flower

Odontostomum hartwegii

Tecophilaeaceae

fruit

flower   flower

Androstephium breviflorum

Themidaceae

stamen

flower

Bloomeria crocea var. crocea

stamen

flower

Bloomeria clevelandii

flower

Bloomeria humilis

11′ Staminodes leaning toward or appressed to stamens, tip obtuse or notched
  12. Perianth lobes spreading; staminodes not hooded at tip, appressed to stamens, margins 1/4–3/4
    inrolled . . . . ***B. insignis***
  12′ Perianth lobes ascending; staminodes hooded at tip, leaning toward stamens, margins not to 1/4 incurved
    13. Staminodes violet, in width < perianth lobes; filament appendages 0; ovary purple . . . . ***B. jolonensis***
    13′ Staminodes white, in width = perianth lobes; filament abaxial wings extended at tip as 2 wide,
      white appendages; ovary green . . . . ***B. stellaris***
10′ Perianth (15)21.5–47 mm; filaments gen ≥ 3 mm, flat to convex abaxially, free from staminodes to
  base; style gen ≥ ovary
  14. Perianth tube funnel-shaped; staminodes held away from, < stamens . . . . ***B. elegans*** subsp. ***elegans***
  14′ Perianth tube cylindric to bell-shaped; staminodes leaning toward or appressed to, > stamens
    15. Staminode margins wavy or toothed; filaments with 2 thread-like appendages at anther bases
    . . . . ***B. appendiculata***
    15′ Staminode margins not wavy or toothed; filament appendages 0
      16. Perianth lobes pink, occ streaked purple; staminodes bottle-shaped (widest at base, middle),
        margins inrolled . . . . ***B. rosea***
      16′ Perianth lobes violet (white); staminodes oblong, margins incurved, often to tip (flat or incurved
        below tip in *Brodiaea coronaria*)
        17. Scape 0.5–7 cm, gen < pedicels; anther tips gen reflexed, with dentate lobe in notch
        . . . . ***B. terrestris*** subsp. ***terrestris***
        17′ Scape 2–25 cm, gen > pedicels; anther tips erect to reflexed, gen without dentate lobe in notch
          18. Staminodes gen white, margins flat or incurved below tip, tips erect to reflexed, obtuse . . . . ***B. coronaria***
          18′ Staminodes violet (white), margins incurved, often to tip, tips erect, notched or with small
            tooth . . . . ***B. terrestris*** subsp. ***kernensis***

B. appendiculata Hoover (p. 1515) **INFL**: scape 10–45 cm; pedicel ≤ 11 cm. **FL**: perianth 21.5–35 mm, tube 7–11.5 mm, cylindric, translucent, splitting in fr, lobes 13.5–24 mm, violet, ascending, tips recurved, outer 4.5–10.5 mm wide, inner 6.5–10.5 mm wide; staminodes leaning toward or appressed to stamens, 10–14 mm, > stamens, gen white, margins 1/2 inrolled, wavy or toothed; filaments 3–5.5 mm with 2 thread-like forked appendages at anther bases, anthers 3–6 mm; ovary 4.5–6.5 mm, style 8–11.5 mm. *n*=6. Grassland, open woodland, gravelly clay soil; 8–335 m. n&c SNF, e GV; historically in NCoRO, SnFrB. Apr–May

B. californica Lindl. (p. 1515) CALIFORNIA BRODIAEA **INFL**: scape 20–70 cm, stout; pedicel ≤ 10 cm. **FL**: perianth 29.5–44 mm, tube 8–12.5 mm, cylindric, translucent, splitting in fr, lobes 20.5–32.5 mm, violet, lavender, or pink (white), ascending, tips recurved, outer 4–7.5 mm wide, inner 6–10.5 mm wide; staminodes reflexed above middle, 14.5–22 mm, linear, white, narrowed at tip, margins 1/4–1/2 inrolled, wavy or toothed; filaments 6.5–11 mm, occ with a forked appendage at tip, anthers 9–13 mm, linear; ovary 6–9.5 mm, style 15.5–23 mm. *n*=6. Grassland, open woodland, rocky or gravelly soils; 50–315 m. KR, CaRF, n ScV; s OR (reported). Apr–Jun

B. coronaria (Salisb.) Engl. GARLAND BRODIAEA **INFL**: scape 5–25 cm; pedicel ≤ 7 cm. **FL**: perianth 23–37.5 mm, tube 8.5–13.5 mm, cylindric, lobes 14–25 mm, violet, ascending, tips recurved, outer 3.5–5.5 mm wide, inner 5–8 mm wide; staminodes leaning toward but not appressed to stamens, 7.5–12 mm, gen white, flat or incurved below tip, tips obtuse; filaments 3–6 mm, anthers 4.5–7 mm; ovary 6–10 mm, style 6.5–11 mm. *n*=12,24,42. Vernal pools, grassland; 30–760 m. CaRF, n SNF, ScV; to BC. Apr–Jun

B. elegans Hoover subsp. ***elegans*** (p. 1515) HARVEST BRODIAEA **INFL**: scape 10–50 cm, stout; pedicels ≤ 8 cm. **FL**: perianth 29–47 mm, tube 10–17.5 mm, funnel-shaped, opaque, not splitting in fr, lobes 17.5–30 mm, violet, ascending, tips recurved, outer 4–7.5 mm wide, inner 6–10.5 mm wide; staminodes held away from stamens, 6–10.5 mm, < stamens, flat, gen white, tip obtuse, occ acute or toothed, erect or ± recurved; filaments 4.5–8.5 mm, anthers 5.5–11 mm, linear; ovary 7.5–12.5 mm, style 8.5–18 mm. *n*=8,16,20,21,24. Grassland, meadows, open woodland, chaparral, occ serpentine; < 2430 m. KR, NCoR, CaRF, SN (exc Teh), GV, SnFrB, SCoRI, SnBr, PR. Subsp. *hooveri* T.F. Niehaus in OR. Hybrids reported with other sp. Apr–Aug

B. filifolia S. Watson (p. 1515) THREAD-LEAVED BRODIAEA **INFL**: scape 20–30 cm, slender; pedicels ≤ 6 cm. **FL**: perianth 13–28 mm, tube 4.5–9 mm, narrow-cylindric, transparent, splitting in fr, lobes 7–20 mm, violet, spreading, outer 3–5 mm wide, inner 5–7 mm wide; staminodes reflexed against perianth, 1–4 mm, awl-like; fila-

ments 0.5–1.5 mm, with narrow wings, base fused with staminodes, anthers 3–6 mm, tip wide-notched; ovary 3–7 mm, style 4.5–9 mm. *n*=12. Grassland, vernal pools; 25–860 m. SCo (Los Angeles, San Diego cos.), SnBr (San Bernardino Co.), w PR (Orange, Riverside, San Diego cos.). Hybridizes with *B. orcuttii*. Mar–Jun ★

B. insignis (Jeps.) T.F. Niehaus (p. 1515) KAWEAH BRODIAEA **INFL**: scape 5–15 cm, slender; pedicels ≤ 13 cm. **FL**: perianth 20–28 mm, tube 6.5–8.5 mm, cylindric, opaque, not splitting in fr, lobes 13–20 mm, violet, spreading, outer 5–7 mm wide, inner 6–8 mm wide; staminodes appressed to stamens, 6–10 mm, white, margin 1/4–3/4 inrolled, tip notched; filaments 1.5–3 mm, base dilated, anthers 4–5.5 mm, tips hooked; ovary 4.5–6 mm, style 3.5–5.5 mm. *n*=16. Foothill woodland; 200–500 m. s SNF (Kaweah, Tule River drainages, Tulare Co.). Threatened by development, road maintenance, grazing. May–Jun ★

B. jolonensis Eastw. (p. 1515) MESA BRODIAEA, CHAPARRAL CLUSTER-LILY **INFL**: scape 5–15 cm, slender; pedicels ≤ 6 cm. **FL**: perianth 19.5–25 mm, tube 7.5–9 mm, cylindric or narrow-bell-shaped, thick, opaque, not splitting in fr, lobes 12–16 mm, violet, ascending, tips recurved, outer 4–5.5 mm wide, inner 5–7 mm wide; staminodes leaning toward stamens, 7–8.5 mm, violet, margin 1/4 incurved, tip hooded, ± notched; filaments 1.5–3 mm, winged, base dilated, fused with staminodes, anthers 4–5.5 mm, tip with V-shaped notch; ovary 5–7 mm, purple, style 4.5–7 mm. *n*=6. Grassland, foothill woodland; < 300 m. SCoRO. Pls in SW, n Baja CA with green ovaries, white staminodes may be undescribed taxon. Apr–Jun

B. kinkiensis T.F. Niehaus (p. 1515) SAN CLEMENTE ISLAND BRODIAEA **INFL**: scapes 2–3 cm, slender; pedicels ≤ 8 cm. **FL**: perianth 14–24 mm, tube 9–12 mm, narrow-bell-shaped, opaque, not splitting in fr, lobes 12–18 mm, purple or violet, spreading, outer 5–7 mm wide, oblong, inner 8–11 mm wide, ± round; staminodes held away from stamens, 3–7 mm, narrowed abruptly, tip narrowed abruptly to sharp point; filaments 1.5–2 mm, with narrow wings, anthers 4–7 mm, tip wide-notched; ovary 6–9 mm, style 5–7 mm. *n*=16. Grassland on clay flats; 100–200 m. s ChI (San Clemente Island). Apr–May ★

B. leptandra (Greene) Baker (p. 1515) NARROW-FLOWERED CALIFORNIA BRODIAEA **INFL**: scape 20–70 cm, stout; pedicels ≤ 11 cm. **FL**: perianth 28.5–42 mm, tube 8–12.5 mm, cylindric, translucent, splitting in fr, lobes 19.5–30 mm, violet with green midrib, ascending, tips recurved, outer 3.5–5.5 mm wide, inner 5.5–8 mm wide; staminodes reflexed at tip, 14.5–21.5 mm, white to lavender, linear, margins gen flat, 1/4 inrolled; filaments 7.5–12.5 mm, anthers 8–11.5 mm, linear; ovary 5.5–8 mm, style 15.5–23.5 mm. *n*=12. Open mixed-evergreen forest, chaparral, gravelly soil; 40–1220 m.

NCoRI (Napa, Lake, Sonoma cos.). [*B. californica* var. *l.* (Greene) Hoover] May–Jul ★

B. minor (Benth.) S. Watson (p. 1515) SMALL BRODIAEA **INFL**: scape 10–25 cm, slender; pedicels ≤ 4 cm. **FL**: perianth 18.5–29 mm, tube 6.5–10.5 mm, urn-shaped, narrowed above ovary, tough, opaque, not splitting in fr, lobes 11–19 mm, violet, spreading, outer 3–5 mm wide, inner 3.5–6 mm wide; staminodes near stamens, 7–11.5 mm, white, tips spreading to erect, notched, margins 3/4 inrolled; filaments 1–3.5 mm, T-shaped in ×-section, anthers 4–6.5 mm; ovary 3.5–6.5 mm, style 6.5–11 mm. *n*=12,16,24. Vernal pools, grassland, meadows, open foothill woodland, chaparral, occ serpentine or gabbro; 55–1500 m. s CaRF, n SN, n ScV; reported from s OR. [*B. purdyi* Eastw.] Apr–Jul

B. nana Hoover DWARF BRODIAEA **INFL**: scape 2–10 cm, slender; pedicels ≤ 4 cm. **FL**: perianth 16.5–26 mm, tube 6–9 mm, urn-shaped, narrowed above ovary, tough, opaque, not splitting in fr, lobes 10.5–18 mm, violet, spreading, outer 3–5 mm wide, inner 4–7 mm wide; staminodes near stamens, 6–8.5 mm, white, margins 1/2 inrolled at middle, tip notched; filaments 1–2 mm, narrow-winged, V- or Y-shaped in ×-section, anthers 3–5 mm, abaxially with prominent, rounded papillae; ovary 2.5–4.5 mm, style 4–6.5 mm. *n*=6. Vernal swales, pools, shallow soil over bedrock; 10–410 m. s CaRF, n&c SNF, e ScV, ne SnJV. Apr–May

B. orcuttii (Greene) Baker (p. 1515) ORCUTT'S BRODIAEA **INFL**: scape 8–25 cm, slender; pedicels ≤ 6 cm. **FL**: perianth 17–32 mm, tube 4–10 mm, transparent, splitting in fr, lobes spreading, inner 3–5 mm wide, outer 5–7 mm wide; staminodes 0; filaments 4–8 mm, anthers 4–6.5 mm, linear, tip with V-shaped notch; ovary 3–7 mm, style 5.8–15.1 mm. *n*=12. Grassland near streams, vernal pools; < 1600 m. PR (s Riverside, San Diego cos.). Hybridizes with *B. filifolia*. Apr–Jul ★

B. pallida Hoover (p. 1515) CHINESE CAMP BRODIAEA **INFL**: scape 10–20 cm, slender; pedicels ≤ 3.5 cm. **FL**: perianth 19–28 mm, tube 7.5–12.5 mm, ± urn-shaped, ± narrowed above ovary, thin, opaque, not splitting in fr, lobes 10.5–16.5 mm, violet near tips, paler toward base, ascending to spreading, outer 3–4.5 mm wide, inner 4–5.5 mm wide; staminodes near stamens, 7–9.5 mm, white, in width ≥ outer perianth lobes, margins flat to 1/2 inrolled above middle, tip deep-notched; filaments 2–3 mm, winged, V- or Y-shaped in ×-section, anthers 4–5.5 mm, abaxially with prominent, finger-like papillae; ovary 3–5 mm, style 7–9.5 mm. *n*=6. Intermittent streambeds, vernal swales, serpentine or not; 160–390 m. n&c SNF (Calaveras, Tuolumne cos.). Hybrids with *B. elegans* reported. May–Jun ★

B. rosea (Greene) Baker (p. 1515) INDIAN VALLEY BRODIAEA Corm coat thin. **INFL**: scape 4–7 cm, stout; pedicels ≤ 4 cm. **FL**: perianth 21.5–30 mm, tube bell-shaped, opaque, not splitting in fr, 7.5–11 mm, lobes 13.5–19 mm, pink, occ streaked purple, ascending, tips recurved, outer 3.5–5 mm wide, inner 4–6 mm wide; staminodes appressed to stamens, 8–13 mm, bottle-shaped (widest at base, middle), tips erect to recurved, margins inrolled; filaments 3–4 mm, base dilated, anthers 4–5.5 mm; ovary 6–8 mm, style 5.5–7.5 mm. *n*=6. Openings, along drainages, chaparral, closed-cone forest, on serpentine; 450–600 m. KR, NCoRI. [*B. coronaria* subsp. *r.* (Greene) T.F. Niehaus] May–Jun ★

B. santarosae T.J. Chester et al. SANTA ROSA BASALT BRODIAEA **INFL**: scape 9–36 cm; pedicels 1.8–10.7 cm. **FL**: perianth 24–36.5 mm, tube 6–11.2 mm, lobes 15.4–29.5 mm, violet, ascending; staminodes 0 in ± 10% of fls, or ≤ 7 mm, gen variable in length within fl, tapered to tip, white to lavender; filaments 2.4–8.2 mm, anthers 5.4–8.9 mm; ovary 3.5–8.2 mm, style 10.5–17 mm. Grassland, on soils derived from Santa Rosa basalt; 580–1045 m. PR (Santa Ana Mtns). May–Jun ★

B. sierrae R.E. Preston SIERRA FOOTHILLS BRODIAEA **INFL**: scape 20–80 cm, stout; pedicels ≤ 10 cm. **FL**: perianth 35–59 mm, tube 8.5–14 mm, cylindric, translucent, splitting in fr, lobes 25–38 mm, violet (pink), ascending, tips recurved, outer 4.5–7.5 mm wide, inner 6.5–11 mm wide; staminodes reflexed at tip, 20.5–30 mm, linear, gen flat, white, margins 1/4 inrolled, entire, midrib thickened; filaments 9–14.5 mm, anthers 10.5–15 mm, linear; ovary 6–10 mm, style 20–31.5 mm. *n*=6. Open areas in chaparral, foothill woodland (dry meadows), gen on soils derived from basic and ultramafic intrusive rocks; 180–945 m. n SNF. Jun–Jul

B. stellaris S. Watson (p. 1515) STAR BRODIAEA **INFL**: scape 2–6 cm, slender; pedicels ≤ 11 cm. **FL**: perianth 17.5–25 mm, tube 6–9 mm, bell-shaped, transparent, not splitting in fr, lobes 10.5–17 mm, violet, ascending, tips recurved, outer 4–6.5 mm wide, inner 4.5–7,5 mm wide; staminodes leaning toward stamens, 4.5–7.5 mm, white, erect, margins 1/4 incurved, tip hooded, notched; filaments 1–3 mm, fused at base to staminodes, wings extended at tip as 2 wide, white appendages, anthers 4–5.5 mm, tip notched; ovary 4.5–8.5 mm, green, style 4–5.5 mm. *n*=6. Openings in chaparral, coastal forest, gen serpentine; < 900 m. c&s NCo, NCoRO. May–Jun

B. terrestris Kellogg **INFL**: scape 0.5–20 cm, slender; pedicels ≤ 13 cm. **FL**: perianth 15–39.5 mm, tube 7.5–13 mm, narrow-bell-shaped, opaque, not splitting in fr, lobes 10–24 mm, violet (white), ascending, tips recurved; staminodes leaning toward but not appressed to stamens, 5.5–12 mm, white to violet, margins 1/4 incurved, tip occ ± hooded; filaments 2–4.5 mm, base dilated, anthers 2.5–7 mm, tips gen reflexed, with dentate lobe in notch or not; ovary 4–10 mm, style 4–10 mm.

subsp. ***kernensis*** (Hoover) T.F. Niehaus (p. 1515) **INFL**: scape 2–20 cm; pedicels ≤ 7.5 cm. **FL**: perianth 26–36 mm, tube 10–15 mm, lobes 15.5–23 mm, outer 4.5–6.5 mm wide, outer 6–9 mm wide; staminodes 7–12 mm, violet (white), margins 1/4 incurved, often to tip; filaments 2–4 mm, anthers 4.5–7 mm, tips erect to reflexed, gen without dentate lobe in notch; ovary 7–10 mm, style 6.5–10 mm. *n*=24. Grassland, open foothill woodland; < 1500 m. s SNF, Teh, TR, PR. Apr–Jun

subsp. ***terrestris*** (p. 1515) **INFL**: scape 0.5–7 cm; pedicels ≤ 13 cm. **FL**: perianth 15–39.5 mm, tube 7.5–13 mm, lobes 10–24 mm, violet, outer 3.5–6.5 mm wide, inner 4.5–8.5 mm wide; staminodes 5.5–7 mm, white (pale violet), margins 1/4 incurved below tip; filaments 2–4.5 mm, anthers 2.5–6 mm, tips gen reflexed, with dentate lobe in notch; ovary 4–9 mm, style 4–9 mm. *n*=6,18. Coastal prairie, foothill woodland; < 450 m. NCo, w NCoRO, ScV, CW; sw OR. Apr–Jul

DICHELOSTEMMA

J. Chris Pires & Glenn Keator

Daughter corms gen sessile. **LF**: 2–5, gen keeled, entire, glabrous, occ withered by fl. **INFL**: raceme- or umbel-like, gen dense; scape gen curved, twisted, or twining, cylindric; bracts ± papery; pedicels gen < fls. **FL**: perianth tube cylindric to bell-shaped, appendages at junction of tube, lobes forming crown; staminodes 0 (3, stamen-like); stamens 3, equal, or 6, unequal, 3 smaller, filaments fused to perianth tube, forming crown-like tube outside anthers, anthers attached at base; style 1, stigma weakly 3-lobed. **FR**: gen sessile, gen ovoid, 3-angled. **SEED**: sharp-angled, black-crusted. 5 spp.: w US, esp n CA, n Mex. (Greek: toothed crown, from perianth appendages) [Keator 1992 Four Seasons 9:24–39]

1. Perianth tube cylindric, ≥ 20 mm; infl open; perianth tube red, lobes yellow-green . . . . ***D. ida-maia***
1′ Perianth tube cylindric to ± spheric, < 20 mm; infl dense; perianth tube pink or not but not red, lobes not yellow-green
  2. Scape twining; perianth pink (white); staminodes 3, stamen-like, alternate stamens . . . . ***D. volubile***
  2′ Scape straight or ± curved; perianth blue, purple, or white; staminodes 0
    3. Stamens 6, unequal; perianth tube not narrowed above ovary . . . . ***D. capitatum***
      4. Bracts gen dark purple (paler, striped dark purple); fls 6–15; pedicels 1–15 mm, gen < bracts . . . . subsp. ***capitatum***

4′ Bracts ± white or streaked purple; fls 2–5; pedicels 6–35 mm, > bracts . . . . subsp. ***pauciflorum***
3′ Stamens 3, equal; perianth tube narrowed above ovary or ± so
    5. Perianth crown deep-notched; infl raceme-like; perianth tube ± narrowed above ovary . . . . ***D. congestum***
    5′ Perianth crown truncate or rounded; infl umbel-like; perianth tube narrowed above ovary. . . . . ***D. multiflorum***

D. capitatum (Benth.) Alph. Wood BLUE DICKS **LF:** 2–3, 10–70 cm, ± keeled. **INFL:** umbel-like, dense; scape 5–70 cm; bracts wide-lanceolate. **FL:** perianth blue, blue-purple, pink-purple, or white, tube 3–12 mm, narrow-cylindric to short-bell-shaped, lobes gen ascending, 7–12 mm, crown erect, leaning toward anthers, outward at tip, 4–6 mm, lanceolate, deep-notched, white; stamens 6, smaller 3 on outer perianth parts, outer filaments wider than inner at base, outer anthers 2–3 mm, inner anthers 3–4 mm; ovary sessile 4–8 mm, style 4–8 mm.

    subsp. ***capitatum*** (p. 1515) **INFL:** bracts gen dark purple (paler, striped dark purple); pedicels 1–15 mm, gen < bracts; fls 6–15. **FL:** perianth lobes gen ascending. *n*=9,18,27,36,45. Open woodland, scrub, desert, grassland; < 2300 m. CA; to OR, n Mex. Mar–Jun

    subsp. ***pauciflorum*** (Torr.) Keator (p. 1515) **INFL:** bracts ± white or streaked purple; pedicels 6–35 mm, > bracts; fls 2–5. **FL:** perianth lobes spreading. *n*=9,18,27. Deserts, open scrub; 300–2100 m. SNE, D; to UT, NM, n Mex. Mar–Jun

D. congestum (Sm.) Kunth (p. 1515) FORK-TOOTHED OOKOW **LF:** 3–4, 4–35 cm, ± glaucous, keeled. **INFL:** raceme-like, dense; scape 30–90 cm, ± scabrous; bracts wide-lanceolate, pale purple to green; pedicels 1–6 mm; fls 6–15. **FL:** perianth blue-purple, tube 8–10 mm, narrow-ovoid, ± narrowed above ovary, lobes ascending, 8–10 mm, crown erect, leaning away from anthers, 5–6 mm, narrow-lanceolate, deep-notched, ± purple; stamens 3, anthers 4–5 mm; ovary sessile 4–6 mm, style 5–6 mm. *n*=9,18. Open woodland, grassland; < 2000 m. NW, CaR, n SN, SnFrB; to BC. Apr–Jun

D. ida-maia (Alph. Wood) Greene (p. 1515) FIRECRACKER FLOWER **LF:** 3–5, 30–50 cm, glaucous, keeled. **INFL:** umbel-like, open; scape 30–100 cm; bracts lanceolate, ± red; pedicels 1–5 cm; fls 6–20 (nodding in fl, erect in fr). **FL:** perianth tube 20–27 mm, cylindric, red, with 6 sac-like swellings on lower 1/3, lobes recurved, erect in fr, 7–9 mm, yellow-green, crown erect, 3–4 mm, entire, rounded, inrolled, ± white; stamens 3, anthers 6–8 mm; ovary 7–12 mm, narrow-ovoid, stalk 3–6 mm, style 7–20 mm. *n*=24. Forest edges, grassland; < 2000 m. NW; sw OR. May–Jul

D. multiflorum (Benth.) A. Heller (p. 1515) WILD HYACINTH **LF:** 3–4, 30–85 cm, glaucous, unkeeled. **INFL:** umbel-like, dense; scape 20–90 cm, scabrous; bracts 7–12 mm, ovate, acuminate, streaked purple; pedicels 3–15 mm; fls 6–35, erect. **FL:** perianth pink- to blue-purple, tube 8–10 mm, ± cylindric, narrowed above ovary, tough, persistent in fr, lobes spreading, 8–10 mm, crown erect, arching inward toward anthers, outward at tip, 5–6 mm, truncate to rounded, entire to minute-dentate, inrolled, white or pale purple; stamens 3, anthers 4–5 mm; ovary sessile, 4–6 mm, spheric, style 5–8 mm. *n*=9,18,27; 2*n*=45. Foothill grassland, open woodland, scrub; < 2000 m. NW, n&c SN, ScV, SnFrB (uncommon), MP; sw OR. May–Jun

D. volubile (Kellogg) A. Heller (p. 1515) TWINING BRODIAEA, SNAKE LILY **LF:** 3–4, 30–70 cm, keeled. **INFL:** umbel-like, dense; scape 40–150 cm, twining, scabrous; bracts 12–15 mm, wide-ovate, acute, ± pink; pedicels 10–40 mm; fls 6–30, occ horizontal in fl, erect in fr. **FL:** perianth pink (white), tube 5–7 mm, ± spheric, narrowed above ovary, angles 6, sac-like, lobes ± spreading (ascending in fr), 5–7 mm, crown folding inward toward, hiding anthers, 3–4 mm, narrow-lanceolate, white; staminodes 3, alternate stamens, ± 3 mm, linear-oblong, stamen-like, cream-white, ciliate-dentate, tip gen notched; stamens 3, anthers 3–4 mm; ovary 4–5 mm, ± spheric, stalk 0 or short, style 3–4 mm. *n*=9,18. Foothill woodland, chaparral scrub; 100–1600 m. NCoRI, CaRF, SNF. May–Jun

MUILLA

LF: 1–10, basal, linear. **INFL:** scape 3–50 cm; bracts several, papery, acuminate; pedicels slender, not jointed, erect. **FL:** perianth parts ± free at base, not forming obvious tube, lobes spreading, gen equal, lanceolate or oblong, white to green-white or ± blue; stamens 6, filaments thread-like to dilated, occ overlapped or fused but not forming cup, anthers attached at middle; style short, club-shaped, persistent, stigma 3-lobed. **FR:** ± spheric, 3-angled. **SEED:** irregularly angled, black. 3 spp.: sw US, n Mex. (Anagram of *Allium*, from superficial resemblance) [Shevock 1984 Aliso 10:621–627] *M. clevelandii* moved to *Bloomeria*.

1. Anthers green, blue, or purple; perianth lobes green-white with ± brown midvein; filaments thread-like or wider at base but not dilated . . . . ***M. maritima***
1′ Anthers yellow; perianth lobes white often tinged lilac or ± white or ± blue with green midvein abaxially; filaments wider than thread-like, dilated at base or throughout
  2. Scape 3–5 cm; perianth lobes 3–6 mm, ± white or ± blue with green midvein abaxially; lvs 1(2); filaments dilated throughout. . . . . ***M. coronata***
  2′ Scape 15–50 cm; perianth lobes 5–8 mm, white often tinged lilac; lvs 3–5; filaments dilated basally
  . . . . ***M. transmontana***

M. coronata Greene (p. 1515) CROWNED MUILLA **LF:** 1(2), 7–18 cm. **INFL:** scape 3–5 cm; pedicels 10–30 mm; fls 2–10. **FL:** perianth lobes 3–6 mm, ± white or ± blue, midvein green abaxially; stamens 2–4 mm, filaments dilated throughout, free but with wide overlapped margins forming nectar tube with cylindric crown, anthers yellow. **FR:** 3–7 mm. *n*=7. Open desert scrub, woodland; 1000–1600 m. s SNH, SNE, DMoj, expected Teh; w NV. Mar–Apr ★

M. maritima (Torr.) S. Watson (p. 1515) COMMON MUILLA **LF:** 3–10, 10–60 cm. **INFL:** scape 10–50 cm; pedicels 10–50 mm; fls 4–70. **FL:** perianth lobes 3–6 mm, green-white, midveins ± brown; stamens 2–5 mm, filaments thread-like or wider at base but not dilated, anthers green, blue, or purple. **FR:** 5–8 mm. *n*=7,8,10. Coastal scrub, grassland, woodland, valleys; < 2300 m. NCoR, GV, CW, SW (exc ChI), w D; Baja CA. Mar–Jun

M. transmontana Greene **LF:** 3–5, 5–30 cm. **INFL:** scape 15–50 cm; pedicels 10–30 mm; fls 12–30. **FL:** perianth lobes 5–8 mm, white often tinged lilac; stamens 4–6 mm, filaments erect, dilated, fused at base, forming nectar cup, anthers yellow. **FR:** 8–10 mm. *n*=10. High desert scrub, conifer woodland; 1400–2500 m. GB; w NV. May–Jul

TRITELEIA

J. Chris Pires & Glenn Keator

Corm spheric, tan; daughter corms sessile. **LF:** 1–3, gen narrow-lanceolate, keeled, glabrous, entire, often withered at fl. **INFL:** umbel-like, open; bracts ± lanceolate, ± scarious; pedicels ± erect, gen > perianth; fls gen many. **FL:** perianth tube gen funnel-shaped, lobes gen ascending to spreading; stamens 6, attached to perianth tube at 1 level or alternately at 2 levels, equal or short

alternating with long, filaments free, appendages forming crown or gen 0, anthers attached at middle, gen angled away from stigma; ovary stalked, style 1, stigma ± 3-lobed. **FR:** gen stalked, ovoid. **SEED:** ± spheric, black-crusted. 15 spp.: w N.Am, esp n&c CA. (Greek: 3 complete, for fl parts in 3s)

1. Stamens attached alternately at 2 levels
  2. Ovary 2 × stalk; perianth tube rounded at base . . . . *T. grandiflora*
  2′ Ovary ≤ stalk; perianth tube acute to long-tapered at base (rounded at base in *Triteleia hyacinthina*,
    Triteleia lilacina)
   3. Stamens unequal
    4. Pedicel 7–20 mm; perianth 12–19 mm, bright yellow or pale blue; ovary green . . . . *T. crocea*
    4′ Pedicel 20–180 mm; perianth 15–28 mm, white, often flushed violet abaxially; ovary bright yellow . . . *T. peduncularis*
   3′ Stamens ± equal
    5. Perianth 16–27 mm, lavender; anthers ± 1.5 mm, purple; s ChI (San Clemente Island) . . . . *T. clementina*
    5′ Perianth 18–47 mm, blue, blue-purple, or white; anthers 2–5 mm, white to ± blue; NW, CaR, SN, CW, TR . . . *T. laxa*
1′ Stamens attached at 1 level
  6. Stamens ± equal
   7. Perianth 27–45 mm, tube tapered at base, > lobes . . . . *T. bridgesii*
   7′ Perianth 7–26 mm, tube bowl-shaped to tapered at base, < lobes
    8. Perianth tube yellow or white, often tinged or fading blue, or yellow aging purple
     9. Perianth 18–26 mm, filaments < 1/2 perianth lobes . . . . *T. hendersonii*
     9′ Perianth 12–17 mm, filaments > 1/2 perianth lobes . . . . *T. montana*
    8′ Perianth tube white (occ flushed purple abaxially), blue, or lilac
     10. Filaments gen triangular; anthers ± white to yellow . . . . *T. hyacinthina*
     10′ Filaments linear; anthers lilac or lilac-dotted . . . . *T. lilacina*
  6′ Stamens unequal
   11. Tips of longer filaments rounded, without appendages . . . . *T. lugens*
   11′ Tips of longer filaments not rounded, gen with appendages
    12. Filament tip appendages 0 or short, blunt; perianth tube = lobes . . . . *T. dudleyi*
    12′ Filament tip appendages forked, pointed; perianth tube ≤ lobes . . . . *T. ixioides*
     13. Lf 1, 10–20 cm . . . . subsp. *unifolia*
     13′ Lvs 1–2, 10–50 cm
      14. Perianth tube ± = lobes, white to pale straw-colored — w-c SCoRO (s Santa Lucia Range) . . . . subsp. *cookii*
      14′ Perianth tube < lobes, yellow (drying ± blue) or gold-yellow or straw-colored to pale yellow
       15. Anthers gen blue or cream; filament tip appendages straight or incurved — KR, CaR, SN . . . . . . subsp. *anilina*
       15′ Anthers cream or yellow, occ blue; filament tip appendages ± straight to recurved
        16. Perianth tube 7–10 mm; c&s CCo, sw SnFrB, SCoRI . . . . subsp. *ixioides*
        16′ Perianth tube 3–7 mm; KR, CaR, SN. . . . . subsp. *scabra*

T. bridgesii (S. Watson) Greene (p. 1515) **LF:** 20–55 cm, 3–10 mm wide. **INFL:** scape 10–60 cm, smooth exc ± scabrous near base; pedicels 20–90 mm. **FL:** perianth 27–45 mm, lilac, blue-purple, pink, or red-purple, tube 17–25 mm, tapered at base, throat translucent, shiny, lobes 10–20 mm, < tube, abruptly spreading; stamens attached at 1 level, equal, filaments 3–4 mm, triangular, anthers 3.5–4.5 mm, ± blue; ovary 1/4–1/3 stalk. *n*=8. Foothills, forest edges, often on rocks, serpentine; < 100 m. KR, NCoR, CaRF, n&c SNF, n SNH; sw OR. Apr–Jun

T. clementina Hoover (p. 1515) SAN CLEMENTE ISLAND TRITELEIA **LF:** 30–100 cm, 4–30 mm wide. **INFL:** scape 30–90 cm, smooth; pedicels 30–80 mm. **FL:** perianth 16–27 mm, lavender, tube 7–12 mm, tapered at base, lobes 9–15 mm, erect; stamens attached alternately at 2 levels, equal, anthers ± 1.5 mm, purple; ovary = stalk, white. *n*=8. Damp clefts, rocky walls, coastal-sage scrub; < 460 m. s ChI (San Clemente Island). Mar–Apr ★

T. crocea (Alph. Wood) Greene (p. 1515) **LF:** 9–40 cm, 2–10 mm wide. **INFL:** scape 10–30 cm, smooth exc ± scabrous near base; pedicels 7–20 mm. **FL:** perianth 12–19 mm, bright yellow or pale blue, tube 5–10 mm, tapered at base, lobes 5–11 mm, spreading, striped ± green; stamens attached alternately at 2 levels, unequal, filaments ± 1 mm and ± 3 mm, linear or ± wider at base, anthers 1–2 mm; ovary ± = stalk, green. *n*=8. Open conifer forest, dry slopes; 650–2200 m. KR, CaRH; sw OR. [*T. c.* var. *modesta* (H.M. Hall) Hoover] May–Jul ★

T. dudleyi Hoover (p. 1515) **LF:** 10–30 cm, 3–11 mm wide. **INFL:** scape 10–35 cm, smooth; pedicels 15–40 mm, slender. **FL:** perianth 18–24 mm, pale yellow, drying ± purple, tube 8–12 mm, ± cylindric to narrowly funnel-shaped, lobes 8–12 mm, spreading; stamens attached at 1 level, unequal, filaments ± 2 mm and ± 3.5 mm, narrow-triangular, tip appendages 0 or short, blunt, anthers ± 1 mm, lavender; ovary = stalk. *n*=8. Subalpine forest; 1200–3500 m. c&s SNH, Teh, TR. Jul

T. grandiflora Lindl. (p. 1515) LARGE-FLOWERED TRITELEIA **LF:** 20–70 cm, 4–10 mm wide. **INFL:** scape 20–75 cm, smooth; pedicels 10–40 mm, ascending. **FL:** perianth 17–35 mm, white to blue-purple, tube 8–20 mm, rounded at base, lobes 9–13 mm, spreading; stamens attached alternately at 2 levels, unequal, filaments 1–2 mm and 3–4 mm, triangular, anthers 2–4 mm, yellow or purple; ovary 2 × stalk. *n*=8,12,16,20,24,28. Grassland, sagebrush, pine forests, hills; 100–3000 m. n KR?, n CaR; to BC, MT, UT. [*T. g.* subsp. *howellii* (S. Watson) Hoover, ined.] Possibly extirpated in CA. Apr–Jun ★

T. hendersonii Greene **LF:** 15–40 cm, 3–12 mm wide. **INFL:** scape 10–35 cm, smooth or ± scabrous near base; pedicels 15–40 mm. **FL:** perianth 18–26 mm, yellow or white, often tinged or fading blue, tube 6–10 mm, ± tapered at base, lobes 12–16 mm, spreading, with conspicuous dark purple midvein; stamens attached at 1 level, ± equal, filaments 3–4 mm, awl-like, anthers 1.5–2 mm, blue or occ white; ovary 1/2 stalk. *n*=16. Dry slopes; 100–3000 m. n KR; sw OR. May–Jul ★

T. hyacinthina (Lindl.) Greene (p. 1515) WHITE BRODIAEA, FOOL'S ONION **LF:** 10–40 cm, 4–22 mm wide. **INFL:** scape 30–60 cm, smooth to scabrous; pedicels 5–50 mm. **FL:** perianth 9–16 mm, white, occ flushed purple abaxially, tube 2–4 mm, shallowly bowl-shaped, lobes 7–12 mm, ascending to spreading, midvein green; stamens attached at 1 level, equal, filaments 2–4 mm, gen triangular; anthers 1–2 mm, ± white to yellow; ovary 2 × stalk. *n*=14,28,35,42. Grassland, vernally wet meadows, occ drier slopes; < 2000 m. NW, CaR, SN, GV, n&c CW, SnBr; to BC, ID. Mar–Jul

T. ixioides (W.T. Aiton) Greene **LF:** 1–2, 10–50 cm, 3–15 mm wide, linear. **INFL:** scape 10–80 cm, gen smooth (scabrous near base); pedicels 10–90(120) mm, upcurved. **FL:** perianth 12–27 mm, straw-colored or gold-yellow striped dark (white flushed purple), tube 3–10 mm, acute at base, lobes 6–20 mm, ascending to reflexed, midvein green, brown, or purple; stamens close to pistil, attached at 1 level, unequal, filaments flat, tip appendages forked, pointed, anthers 1–2 mm, white, cream, yellow, or blue; ovary > stalk.

subsp. ***anilina*** (Greene) L.W. Lenz (p. 1519) PRETTY FACE **LF:** 10–25 cm. **INFL:** scape 10–30 cm, smooth; pedicels 10–70 mm. **FL:** perianth yellow (drying ± blue), tube 4–7 mm, lobes spreading, > tube; filaments 1–2 mm and 3–3.5 mm, tip appendages gen short, straight or incurved, anthers gen blue or cream. *n*=5,7,20,21,25. Conifer forest edges, often in moist gravel or sand; 600–3000 m. KR, CaR, SN. May–Jul

subsp. ***cookii*** (Hoover) L.W. Lenz (p. 1519) COOK'S TRITELEIA **LF:** 10–20 cm. **INFL:** scape 20–25 cm, ± scabrous; pedicels 10–120 mm. **FL:** perianth white to pale straw-colored, tube 6–10 mm, ± = lobes, lobes reflexed; filaments ± 3 mm and ± 4 mm, tip appendages straight or ± recurved, anthers white. *n*=7,14. Streamsides, wet ravines on serpentine, often near cypresses; < 700 m. w-c SCoRO (s Santa Lucia Range). May–Jun ★

subsp. ***ixioides*** (p. 1519) GOLDEN BRODIAEA **LF:** 10–40 cm. **INFL:** scape 20–80 cm, smooth or scabrous; pedicels 10–70 mm. **FL:** perianth gen gold-yellow, tube 7–10 mm, lobes 10–15 mm, ascending to spreading; filaments ± 3 mm and 4–5.5 mm, tip appendages ± straight to recurved, anthers cream or yellow, occ blue. *n*=7. Closed-cone-pine forest, foothill woodland; < 1500 m. c&s CCo, sw SnFrB, SCoRI. May–Jul

subsp. ***scabra*** (Greene) L.W. Lenz (p. 1519) **LF:** 10–50 cm. **INFL:** scape 20–50 cm, scabrous; pedicels 10–90 mm. **FL:** perianth gen straw-colored to pale yellow, tube 3–7 mm, lobes 10–20 mm, ascending to reflexed; filaments 4–5 mm and 5–7 mm, tip appendages ± straight to recurved, anthers cream or yellow, occ blue. *n*=5,6,8,12,16. Scrub edges, mixed or conifer forest, foothill woodland and grassland, clay, granite soils; 150–2200 m. KR, CaR, SN. Mar–May

subsp. ***unifolia*** L.W. Lenz **LF:** 1, 10–20 cm. **INFL:** scape 10–30 cm, smooth or ± scabrous; pedicels 10–40 mm. **FL:** perianth pale yellow, tube 5–7 mm, lobes 9–12 mm, spreading; filaments 2–4 mm and 3–5 mm, tip appendages short, straight or incurved, anthers cream or yellow. *n*=7,8. Conifer forest, clay, granite soils; 600–2000 m. CaR, n&c SN; sw OR. May–Jul

T. laxa Benth. (p. 1519) ITHURIEL'S SPEAR, COMMON TRITELEIA **LF:** 20–40 cm, 4–25 mm wide. **INFL:** scape 10–70 cm, smooth or scabrous; pedicels 10–100 mm. **FL:** perianth 18–47 mm, blue, blue-purple, or white, tube 12–25 mm, tapered at base, lobes 8–20 mm, gradually spreading; stamens attached alternately at 2 levels, ± equal, filaments 3–6 mm, linear, anthers 2–5 mm, white to ± blue; ovary 1/3–1/2 stalk. *n*=8,9,14,15,16,21,24. Common. Open forest, conifer or foothill woodland, grassland on clay soil; < 1500 m. NW, CaR, SN, CW, TR; sw OR. Highly variable; more study needed. Apr–Jun

T. lilacina Greene **LF:** 10–40 cm, 4–22 mm wide. **INFL:** scape 30–60 cm, smooth to scabrous; pedicels 5–50 mm. **FL:** perianth 7–10 mm, white, unstriped, tube 2–4 mm, shallowly bowl-shaped, glassy-beaded (translucent-shiny when dry), lobes 7–12 mm; stamens attached at 1 level, equal, filaments 2–4 mm, linear, anthers 1–2 mm, lilac or lilac-dotted; ovary 2 × stalk. *n*=8. Dry rocky outcrops, volcanic hills, mesas; 70–200 m. s CaRF, n&c SNF. Mar–Jul

T. lugens Greene (p. 1519) DARK-MOUTHED TRITELEIA **LF:** 10–40 cm, 3–10 mm wide. **INFL:** scape 10–40 cm, smooth; pedicels 10–25 mm. **FL:** perianth 12–15 mm, pale yellow to gold, tube 4–6 mm, lobes 6–9 mm, spreading, with dark midvein; stamens attached at 1 level, unequal, filaments 1–2 mm and 2–3 mm, bases wide (bases of inner ± = perianth lobe width), tips rounded, without appendages on longer, anthers 1.5–2 mm, yellow or blue; ovary ± = stalk. *n*=8. Edges of chaparral, mixed forest, foothill woodland, streambanks; 100–1000 m. s NCoR, n&s SNH, n SCoR, SnGb. Related to *T. ixioides*; scattered occurrences need study. Apr–Jun ★

T. montana Hoover (p. 1519) MOUNTAIN TRITELEIA **LF:** 10–30 cm, 2–5 mm wide. **INFL:** scape 5–25 cm, ± scabrous; pedicels 5–30 mm. **FL:** perianth 12–17 mm, yellow aging purple, tube 4–7 mm, narrowly funnel-shaped, tapered at base, lobes 8–10 mm, ± spreading, midveins brown; stamens attached at 1 level, equal, filaments 5–6 mm, linear, anthers 1–1.5 mm, cream to blue; ovary = stalk. *n*=8. Open montane conifer forest, gravelly plains, granite ridges; 1200–3000 m. n&c SNH. Jun–Jul

T. peduncularis Lindl. (p. 1519) LONG-RAYED BRODIAEA **LF:** 20–40 cm, 5–15 mm wide. **INFL:** scape 10–80 cm, smooth; pedicels 20–180 mm, straight. **FL:** perianth 15–28 mm, white, often flushed violet abaxially, tube 7–11 mm, acute at base, lobes 10–16 mm; stamens attached alternately at 2 levels, unequal, filaments 1–1.5 mm and 2–3 mm, ± linear, anthers 2–4 mm, white; ovary = stalk, bright yellow. *n*=7,14. Wet grassland, vernal streams and pools, conifer and foothill woodland, often serpentine; < 800 m. NW, n&c CW. May–Jul

TOFIELDIACEAE   FALSE-ASPHODEL FAMILY

Dale W. McNeal

Per, rhizomes short, spreading. **ST:** erect, ± glabrous near base, dense-glandular-hairy above [or not]. **LF:** gen basal, ± 2-ranked, ± linear; cauline 0–3, on lower 1/3 of st, sheathing. **INFL:** terminal raceme, ± head- or spike-like, gen longer in fr. **FL:** perianth parts 6 in 2 petal-like whorls, white to ± green or ± yellow; stamens 6, filaments ± flat, glabrous, wider at base, anthers attached at base; ovary superior, stalked, 3-lobed, chambers 3, styles 3. **FR:** capsule [follicle], 3-beaked from persistent styles, septicidal below, loculicidal above. **SEED:** ± brown; coat white, loose, spongy, inflated; appendage gen at 1 end, rarely 0 or at both ends. 3 genera, 18 spp.: se US, nw S.Am, n temp. *Tofieldia occidentalis* moved to *Triantha*. Scientific Editor: Thomas J. Rosatti.

TRIANTHA

ST: scapose, 30–80 cm. **INFL:** axes ± glandular-hairy; fls in clusters of 2–7, each cluster subtended by cluster of ± few bracts; bractlets 3, fused, appearing calyx-like. **FL:** perianth parts in ± dissimilar whorls, free, persistent. **FR:** ovoid to wide-ellipsoid. 4 spp.: N.Am, Japan. (Greek: 3, fl, from fls in clusters of 3s) [Packer 2003 FNANM 26:61–64]

T. occidentalis (S. Watson) R.R. Gates subsp. ***occidentalis*** (p. 1519) **LF:** 5–20 cm, 3–8 mm wide. **INFL:** 1–3 cm (to 7 cm in fr), ± spheric to ovoid-cylindric; pedicels 1–6 mm. **FL:** perianth 3–6 mm, parts oblong-ovate, inner narrower, > outer; stamens ± = perianth; styles 1.3–3 mm. **FR:** 5–9 mm. **SEED:** gen 1, 0.5–1.5 mm, 1–3 × longer than wide, red-brown; appendages (0)1(2). 2*n*=30. Wet meadows, bogs; < 3100 m. NW, CaR, SNH; sw OR. [*Tofieldia o.* S. Watson subsp. *o.*] Jul–Sep

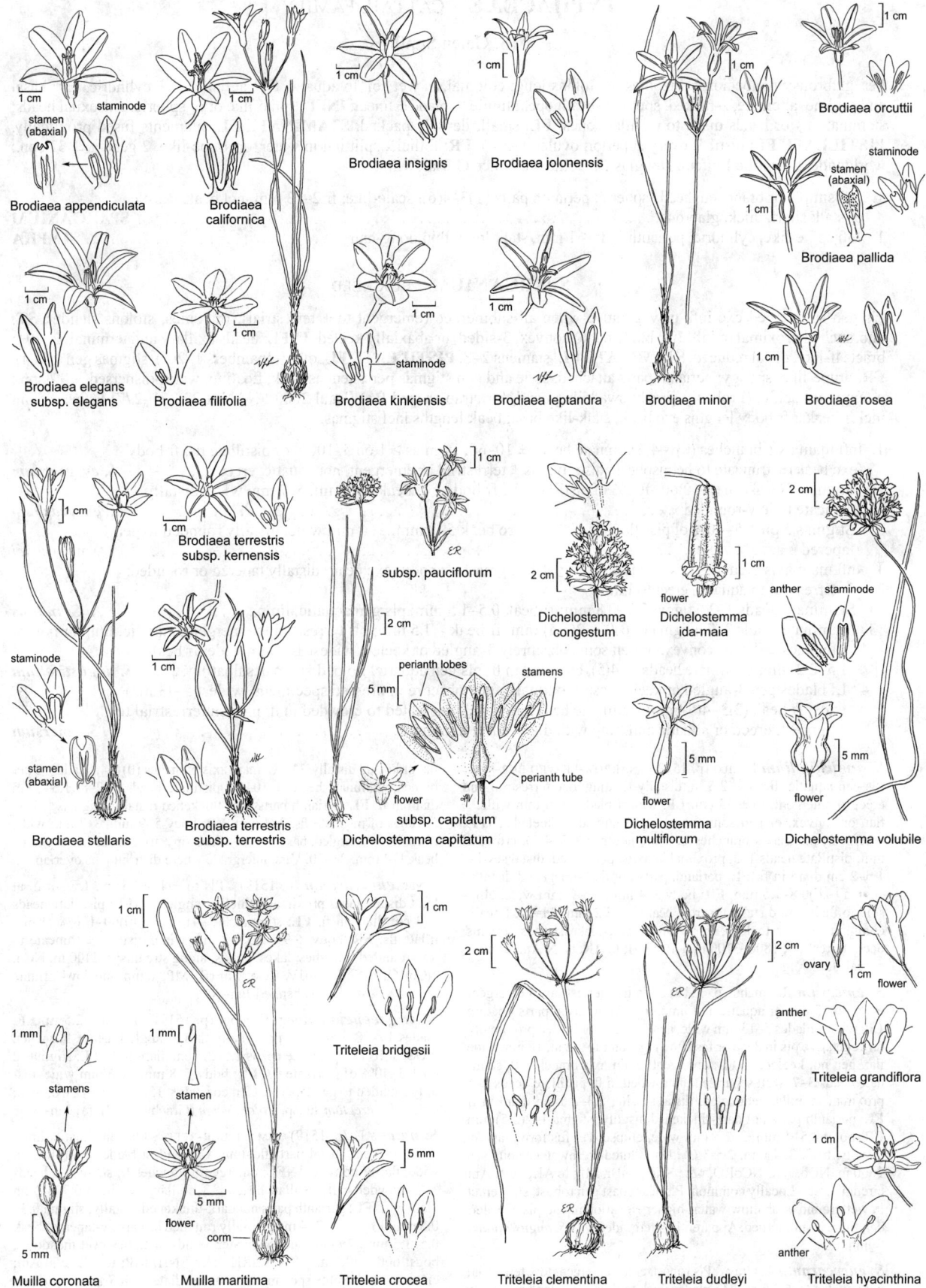

Brodiaea appendiculata

stamen (abaxial)
staminode

Brodiaea californica

Brodiaea insignis

Brodiaea jolonensis

Brodiaea orcuttii

staminode
stamen (abaxial)

Brodiaea pallida

Brodiaea elegans subsp. elegans

Brodiaea filifolia

staminode

Brodiaea kinkiensis

Brodiaea leptandra

Brodiaea minor

Brodiaea rosea

Brodiaea terrestris subsp. kernensis

subsp. pauciflorum

Dichelostemma congestum

Dichelostemma ida-maia

anther   staminode

flower

staminode

stamen (abaxial)

Brodiaea stellaris

Brodiaea terrestris subsp. terrestris

perianth lobes   stamens

perianth tube

flower

subsp. capitatum
Dichelostemma capitatum

flower

Dichelostemma multiflorum

flower

Dichelostemma volubile

stamens

Muilla coronata

stamen

flower

corm

Muilla maritima

Triteleia bridgesii

Triteleia crocea

Triteleia clementina

Triteleia dudleyi

ovary   flower

anther

Triteleia grandiflora

anther

Triteleia hyacinthina

TYPHACEAE CATTAIL FAMILY

S. Galen Smith

Per, glabrous; monoecious; rhizomes or stolons long; colonial, in wet soil to aquatic. **ST:** aerial st 1, ± cylindric. **LF:** basal and cauline, alternate, 2-ranked, spongy; sheath open; ligule 0; blade ± linear. **INFL:** spike-like or of spheric, unisexual heads; staminate fls or heads distal to pistillate ones. **FL:** small, densely-packed. **STAMINATE FL:** filaments fused proximally. **PISTILLATE FL:** pistil 1, ovary superior, ovules 1–2(4). **FR:** follicle, splitting in water, or drupe-like. 2 genera, ± 32 spp.: worldwide. Scientific Editors: Douglas H. Goldman, Bruce G. Baldwin.

1. Infl simple or branched, heads spheric; perianth parts (1)3–4(6), scale-like; fr 2–13 mm incl beak, sessile, or stalk short, thick, glabrous . . . . **SPARGANIUM**
1′ Infl spike-like, cylindric; perianth 0; fr ≤ 1 mm, stalk long, thin, long-hairy . . . . **TYPHA**

SPARGANIUM BUR-REED

Pls gen submerged exc infl, or vegetative when all-submerged (emergent to ± terrestrial); rhizomes, stolons slender. **ST:** unbranched proximal to infl. **LF:** blade flat to convex, 3-sided, or abaxially keeled. **INFL:** heads axillary and terminal, 2–60+; bracts lf-like, distal reduced. **STAMINATE FL:** stamens 2–8. **PISTILLATE FL:** ovary chambers 1–2(4), stigmas gen 1(2–4). **FR:** drupe-like, spongy; perianth parts attached, style and gen stigmas persistent as beak; floating, water-dispersed. ± 17 spp.: n temp, w Pacific. (Probably Greek: swaddling band, for strap-shaped lvs) [Kaul 2000 FNANM 22:271–277] Fr head diam incl fr beaks; fr body lengths exc beak, stalk-like base; beak lengths incl stigmas.

1. Infl main axis branches (0)1–4; staminate heads ± 10–60; stigmas > 1 on 5–100% of pistillate fls; fr body (exc beak) ± truncate to hemispheric distally; pls ± terrestrial or emergent, not aquatic, erect . . . . *S. eurycarpum*
    2. Stigmas 2(3–4) on ± (40)60–100% of pistillate fls; fr body exc beak 8–9 mm, 5–7 mm wide, distally ± truncate to low-rounded . . . . var. *eurycarpum*
    2′ Stigmas 2 on ± 5–40% of pistillate fls; fr body exc beak 5–8 mm, 3–6 mm wide, distally rounded to gen tapered . . . . var. *greenei*
1′ Infl main axis branches 0; staminate heads ≤ 10; stigma 1; fr body (exc beak) distally tapered or rounded; pls terrestrial to aquatic, erect to limp
    3. Staminate heads 1–2; stigma 0.3–0.8 mm; fr beak 0.5–1.5 mm; pls gen aquatic, limp . . . . *S. natans*
    3′ Staminate heads 3–7; stigma (0.6)0.8–2(2.5) mm; fr beak ≥ 1.5 mm; pls terrestrial to emergent, aquatic, limp or not
        4. Lf blades ± flat, or convex, or gen some obscurely 3-angled or keeled, widest 1–5 mm wide; fr beak 1.5–2.2 mm; staminate heads 3–4(5), crowded in fl; pls gen aquatic, limp, distal parts floating . . . . . . *S. angustifolium*
        4′ Lf blades gen 3-angled or keeled, esp proximally, gen obscure in pressed specimens, widest 5–13 mm wide; fr beak (2)3–4(5) mm; staminate heads 3–7, well-separated to crowded in fl; pls gen terrestrial to emergent, erect, or aquatic and limp with distal parts floating . . . . *S. emersum*

S. angustifolium Michx. (p. 1519) NARROW-LEAVED BUR-REED
Pls gen aquatic, limp, ≤ 2.5 m, distally floating, occ terrestrial and erect to emergent, to ± 35 cm. **LF:** widest blades 1–5 mm wide, ± flat, or convex, or gen some obscurely 3-angled or keeled. **INFL:** emergent; main axis branches 0; staminate heads 3–4(5), crowded in fl; pistillate heads 1–3, proximal in axils, peduncled, distal sessile, 1.5–2 cm diam in fr. **FL:** perianth parts not dark-thickened distally; stigma (0.6)0.8–1.5 mm. **FR:** body 3–4 mm, 1.5–2 mm wide, ellipsoid to fusiform, distally tapered, beak 1.5–2.2 mm. 2*n*=30. Locally common. Shallow to deep, nutrient-poor water, lakes, ponds, streams, occ ± terrestrial; 1800–3100 m. KR, SNH; to AK, ne N.Am; circumboreal. Jun–Aug

S. emersum Rehmann (p. 1519) Pls gen terrestrial to emergent, erect, to 75 cm, or aquatic and limp, to 1(2) m, distal parts floating. **LF:** widest blades 5–13 mm wide, gen 3-angled or keeled proximally; on vegetative pls in deep or fast-flowing water to 2.2 m, 18 mm wide, flat, gen not keeled. **INFL:** emergent; main axis branches 0; staminate heads 3–7, well separated to crowded in fl; pistillate heads 1–6, proximal in axils, peduncled, distal sessile, 1.3–2.5 cm diam in fr. **FL:** perianth parts not dark-thickened distally; stigma 1–2(2.5) mm. **FR:** body 3.5–4 mm, 2–2.5 mm wide, ellipsoid to fusiform, tapered distally, beak 3–4 mm. 2*n*=30. ± Unpolluted freshwater habitats; < 2600 m. NCo, KR, NCoRO, n&c SNH, SnBr, MP; to AK, ne N.Am; circumboreal. Locally common. Pls near coast gen robust, stiff, erect, in wet ground or shallow water; higher elevation inland pls gen slender, limp, ± submerged. Apparently intergrades with *S. angustifolium*. Summer

S. eurycarpum A. Gray Pls robust, erect, emergent to ± terrestrial, not aquatic, to ± 2.5 m. **LF:** blades thick to thin, 3-angled proximally, flat or keeled distally. **INFL:** main axis branches (0)1–4, heads ≥ 2 per branch; staminate heads ± 10–60; pistillate heads (1)2–8, 1.5–3 cm diam in fr. **FL:** perianth parts dark-thickened distally; stigmas > 1 on 5–100% of pistillate fls, 3–4 mm. **FR:** body 5–9 mm, 3–7 mm wide, obconic, 3–6-angled, base not stalk-like, tip ± truncate to hemispheric, beak 1–4 mm. 2*n*=30. Vars. intergrade where distributions overlap.

var. *eurycarpum* (p. 1519) Pls to ± 1.5(2.5) m. **LF:** firm; at least distal blades proximally thick, 3-angled. **INFL:** pistillate heads 2–3 cm diam in fr. **FL:** stigmas 2(3–4) on ± (40)60–100% of pistillate fls. **FR:** body 8–9 mm, 5–7 mm wide, distally ± truncate to low-rounded. Marshes, lakes, ponds, along streams; < 2400 m. NCo, NCoRO, n SNH, SnJV, CCo, SnFrB, MP; temp and low-latitude boreal N.Am. [*S. e.* subsp. *e.*] Jun–Aug

var. *greenei* (Morong) Graebn. (p. 1519) Pls to ± 2.5 m. **LF:** blades 10–28 mm wide, proximally thin, 3-angled, gen ± flat when pressed. **INFL:** pistillate heads 1.5–2.5 cm diam in fr. **FL:** stigmas 2 on ± 5–40% of pistillate fls. **FR:** body 5–8 mm, 3–6 mm wide, distally rounded to gen tapered. Gen coastal; < 150 m. NCo, CCo, SCo; to BC. [*S. erectum* subsp. *stoloniferum* (Graebn.) H. Hara] Jun–Aug

S. natans L. (p. 1519) SMALL BUR-REED Pls gen aquatic, limp, ≤ 20(100) cm, distal parts floating. **LF:** widest blades 1–4(10) mm wide, flat or convex. **INFL:** emergent; branches 0; staminate heads 1–2, crowded in fl; pistillate heads 1–3, axillary, sessile, ± 0.7–1.2 cm diam in fr. **FL:** perianth parts not dark-thickened distally; stigma 0.3–0.8 mm. **FR:** body 2–4 mm, distally rounded or narrow-tapered; beak 0.5–1.5 mm. 2*n*=30. Shallow water, ponds, marshes, wet meadows, bogs; 600–2400 m. e KR, CaRH, n&c SNH, MP; to AK, e N.Am; circumboreal. Some specimens ± intermediate with *S. angustifolium*. Jun–Aug ★

TYPHA CATTAIL

ST: erect, simple, cylindric, firm, air cavities 0. **LF:** ascending; blade C-shaped or planoconvex in ×-section proximally, flat distally, internal air cavities large; sheath-tip lobes present or not. **INFL:** terminal; fls 1000+; staminate fls distal, mixed with many papery scales; pistillate fls proximal, clustered on peg-like compound pedicels; bractlets many, thread-like with enlarged tips gen visible at spike surface, or 0. **STAMINATE FL:** stamens 2–7 on slender stalk; filaments slender, gen deciduous in fr. **PISTILLATE FL:** stalk long-hairy, persistent; ovary chambers 1, style 1, persistent, stigma 1; many modified pistils with enlarged sterile ovary, style deciduous. **FR:** fusiform, thin-walled, yellow-brown, wind-dispersed. ± 15 spp.: boreal to trop worldwide. (Greek: to smoke or emit smoke) [Smith 2000 FNANM 22:278–285] Dissecting microscope ideal for *Typha* identification (fl structures small), which is complicated by hybridization.

1. Pistillate bractlets 0; stigma ovate to lance-ovate; naked st between staminate, pistillate fls gen 0(8) cm; widest lf blades 10–29 mm wide (fresh), 5–20 mm (dry) . . . . ***T. latifolia***
1′ Pistillate bractlets present, tip width ≥ stigma width; stigma linear; naked st between staminate, pistillate fls (0)1–12 cm; widest lf blades 4–18 mm wide (fresh), 3–15 mm (dry)
  2. Lf blade glands 0; pistillate spike dark brown; pistillate bractlet tip dark brown, gen darker than stigma, blunt . . . . ***T. angustifolia***
  2′ Lf blade gland-dotted adaxially near base; pistillate spike cinnamon to medium brown; pistillate bractlet tip straw-colored (to light brown), gen paler than stigma, acute to acuminate. ***T. domingensis***

T. angustifolia L. (p. 1519) NARROW-LEAVED CATTAIL Pl 15–30 dm. **ST:** 2–3 mm diam at infl. **LF:** sheath-tip lobes ear-like, membranous, disintegrating with age; widest fresh blades 4–15 mm wide, dry 3–8 mm wide, glands 0. **INFL:** naked st between staminate, pistillate fls 1–12 cm; staminate scales hair- to strap-like, brown; pistillate spike dark brown; compound pedicels short, ± blunt, peg-like in fr, ± 0.5 mm; pistillate bractlets blunt, length = pistillate fl hairs, width ≥ stigma width, dark brown, gen darker than stigma. **FL:** pollen grains single; stigma linear, ± white in fl, light brown in age; sterile ovary green, drying brown, visible at spike surface, reaching pistil hair tips; pistil hair tips swollen, uniformly brown. 2n=30. Nutrient-rich freshwater to brackish marshes, wet disturbed places; < 2000 m. NW, SN, GV, CCo, SnFrB, SCo; to s BC, c&e temp N.Am, Eurasia. Possibly naturalized in CA. May–Aug

T. domingensis Pers. (p. 1519) SOUTHERN CATTAIL Pl 15–40 dm. **ST:** 3–4 mm diam near infl. **LF:** sheath-tip lobes ear-like, membranous or 0; widest fresh blades 6–18 mm wide, dry 5–15 mm wide, orange-brown gland-dotted adaxially on proximal 1–10 cm. **INFL:** naked st between staminate, pistillate fls (0)1–8 cm; staminate scales gen strap-like, tips widened, gen irregularly dissected, yellow-brown; pistillate spike cinnamon to medium brown; compound pedicels short, ± blunt, peg-like in fr, ± 0.7 mm; pistillate bractlets acute to acuminate, length > pistillate fl hairs, width ≥ stigma width, straw-colored to light brown, gen paler than stigma. **FL:** pollen grains single; stigma linear, ± white in fl, medium- to yellow-brown in fr; sterile ovary visible at spike surface, ± = pistil hair tips, straw-colored; pistil hair tips ± swollen, straw-colored with large orange-brown spot. 2n=30. Nutrient-rich freshwater to brackish marshes, wet disturbed places; < 1500 m. NCo, NCoRO, GV, CW, SW, GB, D; to AK, e N.Am, worldwide warm temp, trop. Very variable worldwide; needs study. Jun–Jul

T. latifolia L. (p. 1519) BROAD-LEAVED CATTAIL Pl 15–30 dm. **ST:** 3–7 mm diam near infl. **LF:** sheath tip lobes ear-like, papery, or 0; widest fresh blades 10–29 mm wide, dry 5–20 mm wide, glands 0. **INFL:** naked st between staminate, pistillate fls gen 0(8) cm; staminate scales hair- to strap-like, colorless; pistillate spikes medium- to black- or red-brown, gen white-mottled in age; compound pedicels elongate, bristle-like in fr, 1.5–3.5 mm; pistillate bractlets 0. **FL:** pollen grains in 4s; stigma ovate to lance-ovate, green in fl, medium- to red- or black-brown in age; sterile ovary straw-colored, not visible at spike surface, < pistil hair tips. 2n=30. Unpolluted to nutrient-rich freshwater (brackish) marshes; < 2300 m. CA; boreal N.Am to n S.Am, Eurasia, n Afr, Tasmania (introduced). Jun–Jul

ZANNICHELLIACEAE HORNED-PONDWEED FAMILY

Robert F. Thorne, Robert R. Haynes & C. Barre Hellquist

[Ann] per from slender, creeping rhizome, glabrous, aquatic, submersed; monoecious. **ST:** thread-like, weak. **LF:** alternate to ± whorled, linear; stipules fused to blade or free, sheath-like. **INFL:** axillary, gen ± cyme, sessile, bract 1, cup-like; staminate fls below pistillate. **FL:** unisexual [bisexual]. **STAMINATE FL:** perianth parts 0 (or 3, minute, scale-like); stamens 1 (or possibly 2–3 but then united, appearing as 1). **PISTILLATE FL:** perianth parts 0; pistils (1)4–5(9), ovaries superior, 1-chambered, 1-ovuled, 1-styled. **FR:** achene-like drupe. 4 genera, 10–12 spp.: ± worldwide. [Haynes & Hellquist 2000 FNANM 22:84–85] Scientific Editor: Thomas J. Rosatti.

ZANNICHELLIA HORNED-PONDWEED

Per. **ST:** gen few-branched. **LF:** entire. **INFL:** staminate fls gen 1, pistillate 2–4[6], pedicelled in fr or not. **STAMINATE FL:** stamen 1, filament slender, anther 2–4-chambered. **PISTILLATE FL:** style short, stigma peltate. **FR:** compressed, abaxially curved, ridged, toothed, adaxially ± straight, base stalk-like, tip beaked. ± 5 spp.: ± worldwide. (G.G. Zannichelli, Venetian botanist, 1662–1729)

Z. palustris L. (p. 1519) **ST:** 3–10 dm. **LF:** 2–10 cm; blade < 1 mm wide, tip acute. **FR:** 1.7–4 mm; beak 1–1.5 mm. 2n=12,24,28,32,36. Streams, ponds, lakes; < 2200 m. ± CA (exc possibly KR, SN); ± worldwide. Mar–Nov

ZOSTERACEAE EEL-GRASS FAMILY

Robert F. Thorne, Robert R. Haynes, C. Barre Hellquist & Gordon Leppig

Per (ann), submersed marine aquatic, rhizomed, glabrous; monoecious or dioecious. **ST**: flat, lfy, creeping or short with wide, thick base. **LF**: alternate, 2-ranked or tufted, < 2 m; sheath open or closed; blade linear, ribbon-like, tip entire to notched; stipules 0. **INFL**: axillary or terminal spike, of limited distribution on pl, ± flat, enclosed by membranous, lf-like bract in fl, also in fr or not; fls many, 2-rowed on 1 side of axis, each embedded in axis, enclosed by translucent, scale-like bract or not; peduncle not elongating. **STAMINATE FL**: perianth 0; stamen 1, anther gen sessile, 2-chambered; pollen thread-like, < 2 mm. **PISTILLATE FL**: perianth 0; ovary 1, chamber 1, ovule 1, style 1, stigmas 2. **FR**: ± achene. 3 genera, ± 18 spp.: sea-coasts, worldwide; some spp., esp *Zostera marina*, source of food for marine animals, some aboriginal humans. [Haynes 2000 FNANM 22:90–94] Scientific Editor: Thomas J. Rosatti.

1. Dioecious; attached by short, tuber-like rhizomes, to coastal, wave-swept rocks; lvs gen tufted on rhizomes
   at nodes; fr beaked, base with 2 horn-like lobes . . . . **PHYLLOSPADIX**
1′ Monoecious; attached by long, slender, creeping rhizomes to bottoms of bays, estuaries, gen deep waters
   below wave action; lvs not tufted on rhizomes; fr ± not beaked, base ± truncate, unlobed . . . . **ZOSTERA**

PHYLLOSPADIX SURF-GRASS

Rhizomes < 12 cm. **ST**: 10–60 cm, slender to thick, below flat, winged. **LF**: 0.2–1 m, 0.5–6 mm wide. **INFL**: enclosed by subtending, membranous, sheathing, lf-like bract in youth, exserted from it in age; fls each enclosed by translucent, scale-like bract; pistillate fls alternate fls with vestigial anthers. **PISTILLATE FL**: ovary tapered to short style. 5 spp.: n Pacific coasts. (Greek: lf-like infl) Island distributions of CA spp. need study.

1. Pistillate infls 1(2) per st, ± basal; peduncle 10–60 mm; lvs 1–4 mm wide, flat . . . . *P. scouleri*
1′ Pistillate infls (1)3–5(7) per st, cauline; peduncle 8–45 mm; lvs 0.5–2 mm wide, in age folded or cylindric . . . . *P. torreyi*

P. scouleri Hook. (p. 1519) **PISTILLATE FL**: bract not narrowed at base. **FR**: ± 3.5 mm. Surf zones, rocky shores, intertidal, subtidal; 0–2 m below mean low tide. NCo, CCo, SCo, ChI; to AK, Baja CA. Apr–Jul

P. torreyi S. Watson (p. 1519) **PISTILLATE FL**: bract narrowed at base. **FR**: ± 2–3 mm. Surf zones, rocky shores, intertidal, subtidal; 0–15 m below mean low tide. NCo, CCo, SCo, ChI; to BC, Baja CA. May–Nov

ZOSTERA EEL-GRASS

ST: < 3 m. **LF**: 0.5–1.5(2) m, (0.5)1–18.5 mm wide. **INFL**: enclosed by subtending, membranous, sheath-like lf base; staminate, pistillate fls alternate. **PISTILLATE FL**: ovary tapered to short style. 12 spp.: marine waters worldwide. (Greek: from ribbon-like lvs) [Coyer et al. 2008 Ann Bot 101:73–87; Dean et al. 2008 Madroño 55:93–112] Important food source for marine animals, some aboriginal humans; *Z. asiatica* Miki reported for CA (Coyer et al., Dean et al.), possibly naturalized, but may not be distinct from *Z. marina*.

1. Per (ann); roots 2 per node; lf sheath open, with 2 membranous flaps; lf blade to 20(25) cm, (0.5) 1–1.5 (2)
   mm wide; staminate fls each subtended by 1 bract . . . . *Z. japonica*
1′ Per; roots 5–20 per node; lf sheath closed, without membranous flaps; lf blade to 110 cm, 1.5–18.5 mm
   wide; staminate fls not subtended by bract (or lowermost subtended by 1 bract)
   2. Lvs 1.5–12(16) mm wide, tip gen obtuse; seed with longitudinal ridges; pls of shallow waters in bays,
      estuaries, rooted gen 1–4 m below mean low tide; fl Mar–Jun, fr May–Aug . . . . *Z. marina*
   2′ Lvs 12–18.5 mm wide, tip gen notched; seed smooth or with weak longitudinal ridges; pls of subtidal
      waters, rooted gen 4–17 m below mean low tide; fl (May) Aug, fr Sep–Oct . . . . *Z. pacifica*

Z. japonica Aschers. & Graebn. Per (ann); roots 2 per node. **LF**: sheath open, with 2 membranous flaps; blade to 20(25) cm, (0.5)1–1.5(2) mm wide. **INFL**: staminate fls each subtended by 1 bract. *n*=6. Occ. Shallow water, bays, estuaries; rooted gen 0–1 m below mean low tide. NCo, expected elsewhere; to BC; native to Pacific Asia. Apr–Aug ◆

Z. marina L. (p. 1519) Per; roots 5–20 per node. **LF**: sheath closed, without membranous flaps; blade to 150(200) cm, 1.5–12(16) mm wide, tip gen obtuse. **INFL**: staminate fls not subtended by bract (or lowermost subtended by 1 bract). *n*=6. Common. Shallow water,

bays, estuaries; rooted gen 1–4 m below mean low tide. NCo, CCo, SCo, ChI; n hemisphere. Rhizomes, lf bases edible. Mar–Jun

Z. pacifica S. Watson Per; roots 5–20 per node. **LF**: sheath closed, without membranous flaps; blade to 110 cm, 12–18.5 mm wide, tip gen notched. **INFL**: staminate fls not subtended by bract (or lowermost subtended by 1 bract). Common. Subtidal waters; rooted gen 4–17 m below mean low tide. NCo, CCo, SCo, ChI; to BC. Probably indistinct from *Z. marina*, yet recognition supported by microsatellite data (Coyer et al. 2008). Aug

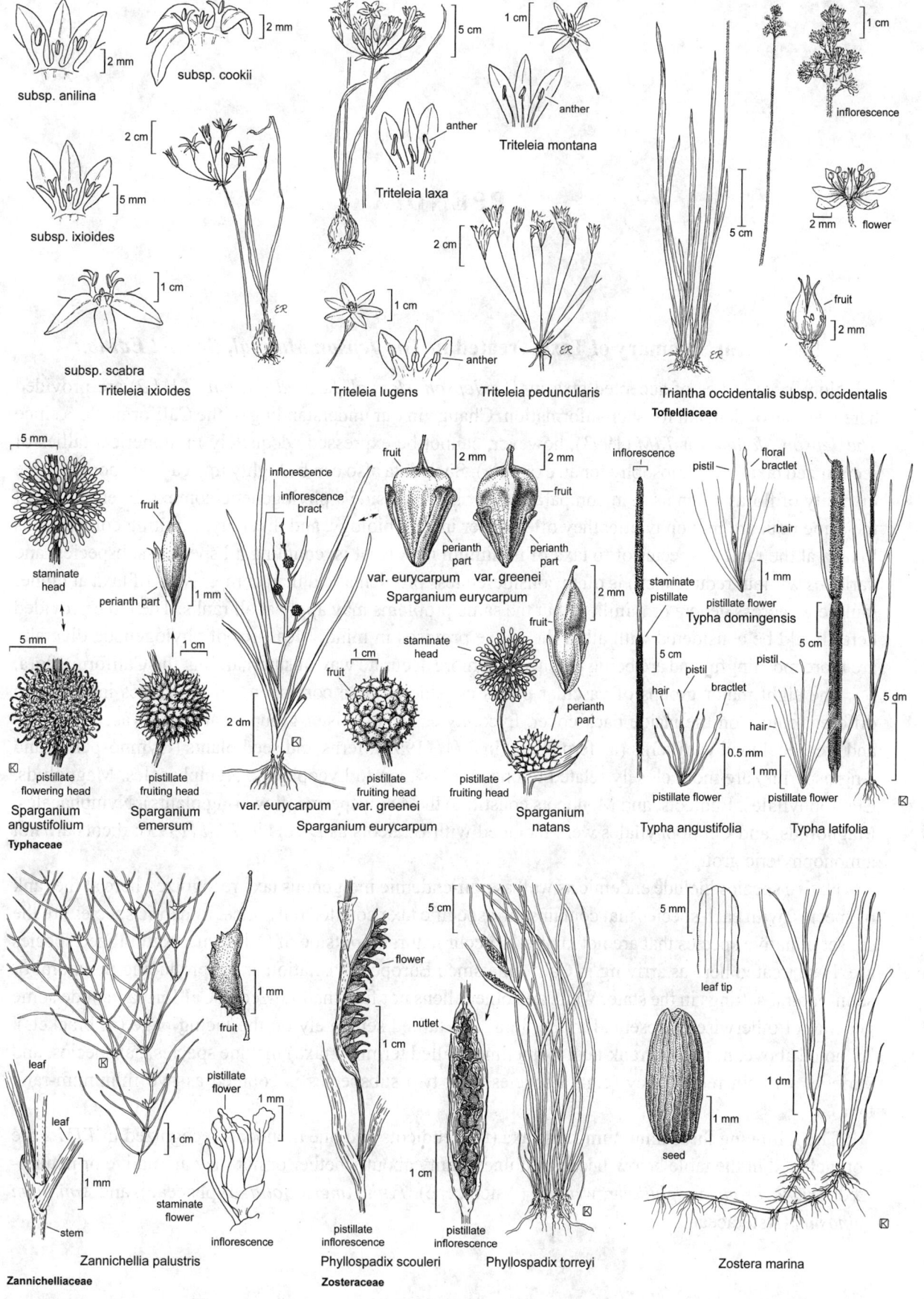

subsp. anilina

subsp. cookii

subsp. ixioides

subsp. scabra

Triteleia ixioides

anther

Triteleia laxa

anther

Triteleia montana

Triteleia lugens

anther

Triteleia peduncularis

inflorescence

flower

fruit

Triantha occidentalis subsp. occidentalis
Tofieldiaceae

staminate head

fruit

perianth part

pistillate flowering head

Sparganium angustifolium

pistillate fruiting head

Sparganium emersum

Typhaceae

inflorescence

inflorescence bract

fruit

var. eurycarpum

var. eurycarpum

var. greenei

Sparganium eurycarpum

fruit

staminate head

perianth part

pistillate fruiting head

var. greenei

Sparganium eurycarpum

staminate head

fruit

perianth part

pistillate fruiting head

Sparganium natans

inflorescence

pistil

floral bractlet

hair

staminate pistillate

pistillate flower

Typha domingensis

hair

pistil

bractlet

pistillate flower

Typha angustifolia

pistil

hair

pistillate flower

Typha latifolia

leaf

leaf

stem

fruit

pistillate flower

staminate flower

inflorescence

Zannichellia palustris
Zannichelliaceae

flower

nutlet

pistillate inflorescence

Phyllospadix scouleri

pistillate inflorescence

Phyllospadix torreyi
Zosteraceae

leaf tip

seed

Zostera marina

APPENDIX

Numerical Summary of Taxa Treated in *The Jepson Manual, Second Edition*

Numerical tallies of accepted taxa in *The Jepson Manual, Second Edition* (*TJM 2*) are provided here because of demand for such information. Change in our understanding of the California flora since *The Jepson Manual*, or *TJM* (1993), however, cannot be expressed adequately in numerical tallies of recognized taxa (see Philosophy for an example). Such data also can be highly misleading about relative diversity of floras when used in comparative floristics, outside a phylogenetic context. Even if taxa of the same rank are monophyletic, they often differ in age, biology, and diversity. Limiting comparisons to taxa at the rank of species or to taxa of minimum rank (that is, counting all species, subspecies, and varieties without redundancy) is often considered less problematical than comparisons of taxa at higher rank (e.g., across genera or families), but the same problems may apply at all ranks. The table provided here should be considered with all of the above provisos in mind. Measures of phylogenetic diversity are more meaningful and are being pursued as a more accurate way of summarizing the California flora.

The eight major groups of vascular plants considered here correspond to the clades indicated in different colors on the inside back cover. Lycophytes are the sister group to all other vascular plants and were included with ferns (as fern allies) in *TJM* (1993); ferns and seed plants (Gymnosperms and Angiosperms) are more closely related to one another than to Lycophytes. Nymphaeales, Magnoliids, Ceratophyllales, Eudicots, and Monocots constitute the Angiosperms (flowering plants). Nymphaeales, Magnoliids, and Ceratophyllales were included with Eudicots as Dicots in *TJM* (1993); dicots are not a monophyletic group.

Native species include endemic as well as non-endemic indigenous taxa recognized here at the rank of species, with each species that contains infraspecific taxa counted only once. Endemic species are the subset of native species that are not known to occur naturally outside of California. Naturalized species are those treated here as arriving in California since European visitation and reproducing in natural or semi-natural settings in the state. Waifs and other aliens not fully naturalized in California include some genera not otherwise represented in the flora, and tallied separately on the facing page (in brackets). As noted above, minimum-rank taxa (sometimes called terminal taxa) include species, subspecies, and varieties without redundancy (e.g., a species with two subspecies is counted as two minimum-rank taxa).

The following three minimum-rank taxa (two eudicots and one monocot) recognized in *TJM 2* are not included in the table below because of uncertainty about whether or not they are native or naturalized: *Trichocoronis wrightii* var. *wrightii* (Asteraceae), *Typha angustifolia* (Typhaceae), and *Xanthium spinosum* (Asteraceae).

	Lycophytes	Ferns	Gymnosperms	Nymphaeales	Magnoliids	Ceratophylales	Eudicots	Monocots	TOTAL
Families	3	13	4	2	4	1	125	33	185
Genera	4[0]	31[2]	16[0]	4[0]	6[2]	1[0]	1032[133]	220[36]	1314
Native species	18	79	59	2	8	1	3976	833	4976
Endemic species	1	4	10	0	3	0	1119	178	1315
Naturalized species	0	5	1	3	1	0	747	234	991
Native minimum-rank taxa	18	93	71	2	8	1	5313	996	6502
Naturalized minimum-rank taxa	0	5	1	3	1	0	826	263	1099
Not fully naturalized alien taxa (mostly waifs); in keys only (descriptions online)	0	2	0	0	2	0	196	38	239

INDEX

The following index includes all common names, scientific names of families and genera, and synonyms used in *The Jepson Manual, Second Edition*. It does not include scientific names of accepted species or infraspecific taxa, which are arranged alphabetically in the text (within the genus or species to which they belong), unless such names appear in places other than the main entry for the taxa involved. Italics are not used for scientific names of genera or species in the index. The index applies only to contents of the taxonomic treatments; introductory sections and the key to families are not referenced here.

Page numbers for all common names, synonyms, and miscellaneous entries are given in regular type. Accepted scientific names of families are followed by the page number (in **bold type**) of the beginning of the family description. Scientific names of described genera are often followed by three or more page numbers. The first, in regular type, refer(s) to the page(s) where the genus appears in a key to genera within a family treatment. Next, a single **bold type** entry indicates the page on which the generic description begins. Third, there may be one or more page numbers in *italics* that refer to illustration plate(s) for the genus.

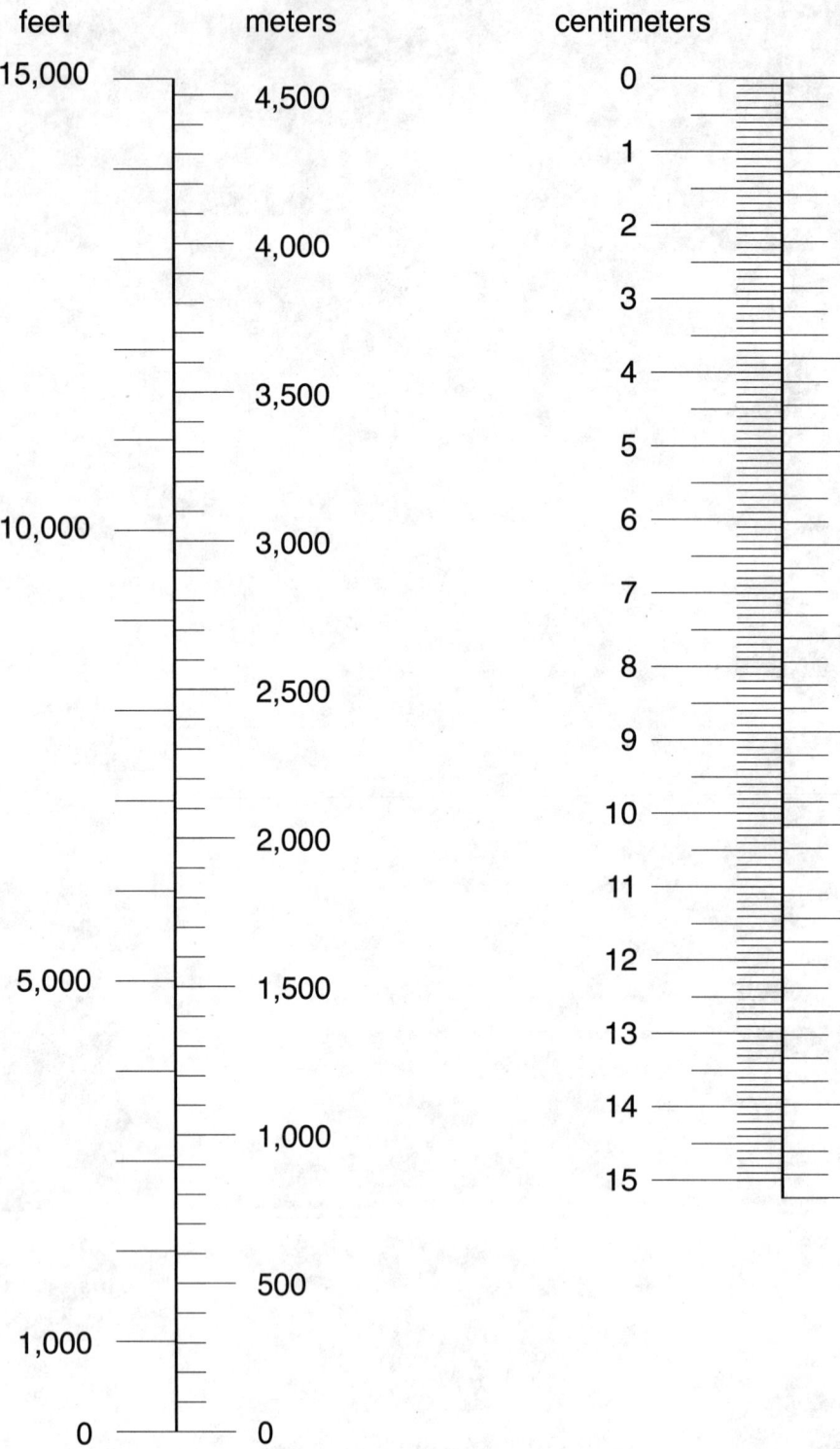